AF817452

The ARRL Handbook FOR RADIO COMMUNICATIONS

101

Editor
Gregory D. Lapin, N9GL, PhD, PE

Production Editors
Mark Derks, KC1RVQ
Makenzie Ozycz

Technical Consultants
George Spatta, W1GKS
Stephen Anderson, W1EMI
John McAuliffe, W1DRF

Cover Design
Lori Pedrick

Production
Michelle Bloom, WB1ENT
David F. Pingree, N1NAS

Editorial Assistant
Maty Weinberg, KB1EIB

Additional Contributors to Handbook 101

Jim Andrews, KH6HTV
Alan Applegate, KØBG
Jim Brown, K9YC
Orv Beach, W6BI
John Burningham, W2XAB
Bob Clarke, N1RC
Mike Crownover, AD5A
Brian Daly, WB7OML
John Davis, WB4QDX
Jean-Francois Debroux
Lynn Deffenbaugh, KJ4ERJ
Frank Donovan, W3LPL
Joe Eisenberg, KØNEB
Philip Erickson, W1PJE
Adam Farson, AB4OJ
Steven R. Ford, WB8IMY
Al Francisco, K7NHV
Steve Galchutt, WGØAT
Doug Grant, K1DG
Dale Grover, KD8KYZ
Matt Heere, N3NWV
Alan Higbie, KØAV
Ken Humbertson, WØKAH
Jim Idelson, K1IR
Jim Klitzing, W6PQL
John Langner, WB2OSZ
James Lux, W6RMK
Fred Maas, KT5X
Ethan Miller, K8GU
Doc Murphy, KØGRM
Wayne Mueller, W1QC
Ken Norris, KK9N
John Portune, W6NBC
Warren Pratt, NRØV
Jim Rautio, AJ3K
Rich Rosen, K1DS
Richard Russel, ACØUB
Hermann Schumacher, DF2DR
Cory Sickles, WA3UVV
Steve Stroh, N8GNJ
Jim Tonne, W4ENE
Mel Whitten, KØPFX
Ed Woodrick, WA4YIH

ARRL The National Association for Amateur Radio®
225 Main Street, Newington, CT 06111-1400 USA
www.arrl.org

Printed in the USA

ISBN: Softcover ISBN: 978-1-62595-201-1
ISBN: 6-volume set ISBN: 978-1-62595-207-3

One hundred and first edition

Kindle ebook editions are based on the 101st edition of
The ARRL Handbook for Radio Communications
ISBN: 978-1-62595-208-0 — Volume 1
ISBN: 978-1-62595-209-7 — Volume 2
ISBN: 978-1-62595-210-3 — Volume 3
ISBN: 978-1-62595-211-0 — Volume 4
ISBN: 978-1-62595-212-7 — Volume 5
ISBN: 978-1-62595-213-4 — Volume 6

Contents

A more detailed table of contents is included at the beginning of each chapter.

Radio Propagation and Antenna Systems

Safe Practices and Station Construction

Foreword

Welcome to *Handbook 101*. For 98 years *The ARRL Handbook for Radio Communications* has been the stalwart companion of the radio amateur, capturing the state of radio science and technology in one authoritative work. With this new edition, the book welcomes a new editor, Gregory D. Lapin, N9GL, PhD, PE. Greg brings a fresh perspective and wide expertise to the *Handbook*. Readers will undoubtedly notice the changes he has made, including the contributions of many subject matter experts in our community.

Information on electrical and radio fundamentals has been refreshed, as have the chapters on transmission lines, general safety, assembling a station, and more. Major updates include:

- Information on electromagnetic analysis and its use for modeling circuits, antennas, and propagation, with an emphasis on inexpensive tools that are readily accessible.
- New material on higher level modeling of transmitters and receivers.
- A new section on preparing your station for emergency operations by Brian Daly, WB7OML.
- Radio astronomy receiver and antenna design information by Dr. Richard Russel, ACØUB, and the Society of Amateur Radio Astronomers, SARA.
- A new section on batteries and battery safety.
- A new section on NEC4 and the antenna modeling software built on it.
- New material on SWR meters and related tests.
- The latest on RF safety and compliance with FCC exposure regulations.

This edition of the *Handbook* represents a step forward, not just with a new editor, new topics, and updated information, but with a new outlook on the future of amateur radio. We invite you to look hard at where amateur radio stands and to think equally hard about where it might go. SDRs and digital technology provide countless avenues for innovation. New networking protocols and modulation techniques utilizing LoRa and other technologies on the ISM (industrial, scientific, and medical) and similar bands offer new ways for those who are radio-interested to experiment and connect with others who share their curiosity. Technologies like these can serve as onramps to the hobby for new amateurs and offer fulfilling challenges to those even with decades of radio experience. Changes to FCC regulations make experimentation with high-speed digital modes and technologies on HF more practical and useful than ever before. The future of amateur radio is bright, and the possibilities are limitless. We expect you'll experience that excitement when you discover something new or build a project from the *Handbook,* or as you delve into its wealth of online content.

Again, welcome to *Handbook 101*. We're glad you've included *The ARRL Handbook for Radio Communications* in your pursuit of wireless technology experimentation, practice, and development.

David A. Minster, NA2AA
ARRL Chief Executive Officer
Newington, Connecticut
August 2024

The Amateur's Code

The Radio Amateur is:

Considerate...never knowingly operates in such a way as to lessen the pleasure of others.

Loyal...offers loyalty, encouragement and support to other amateurs, local clubs, and the American Radio Relay League, through which Amateur Radio in the United States is represented nationally and internationally.

Progressive...with knowledge abreast of science, a well-built and efficient station and operation above reproach.

Friendly...slow and patient operating when requested; friendly advice and counsel to the beginner; kindly assistance, cooperation and consideration for the interests of others. These are the hallmarks of the amateur spirit.

Balanced...radio is an avocation, never interfering with duties owed to family, job, school or community.

Patriotic...station and skill always ready for service to country and community.

—The original Amateur's Code was written by Paul M. Segal, W9EEA, in 1928.

Preface

Welcome to *Handbook 101*! I took over the job of editor this year, receiving the baton from Ward Silver, NØAX, who admirably performed this job for the past 14 editions. I had big shoes to fill! I owe huge thanks to Mark Derks, KC1RVQ, ARRL's acquisitions editor, who brought the new guy up to speed and helped with the scores of details that were needed to put a work of this magnitude together, to all the ARRL staff who helped pull this book together, and to the authors who contributed their expertise to improve the *Handbook*.

I received my first *ARRL Handbook*, the 51st edition, in 1974. What a great gift! I spent hours perusing the information that was jam-packed into that book. That edition was half the size of the current *Handbook* and I was never able to read it from cover-to-cover. I have had many editions since then and, to be frank, have never read every word in a *Handbook* until this year. Regardless, it has always been one of the great references on my bookshelf; I refer to it often. I never imagined that fifty years later I would be the editor of the 101st edition of the *ARRL Handbook*.

When people hear that I am a radio amateur, they frequently ask if ham radio still exists, or why people would want to be hams now that most people have cell phones and the internet. My initial reply is that when I was first licensed, the United States had 288,000 licensed radio amateurs. Today that number has blossomed to over 750,000. I explain that if you are satisfied buying a device and using the technologies of several huge companies to communicate with others, then that's fine. Radio amateurs are different. We want to understand how the devices work and how radio waves travel. We want to build new gadgets and methods for sending radio waves, and we want to develop our on-air skills so we can be as efficient and capable as possible.

I spend time working with a university amateur radio club, and am continually impressed by the brains, ingenuity, and excitement of the students who have come to love amateur radio as I do. They are our future, and it is clear that they are helping to insert new technology into our more-than 100-year-old hobby. It is with their encouragement that I made my first QSO via satellite, started working with WSPR, and set up a 100-watt HF station in my backpack to hike to POTA locations. I, in turn, showed them the wonder of making HF QSOs across the world. Clearly, our hobby is alive and thriving.

The *Handbook* mirrors our hobby's progress. While it contains topics that have always been of interest to hams, it has stayed current and includes information on SDR, digital modulation, the latest battery technologies, new and innovative antenna configurations, satellites, radio astronomy, propagation monitoring, 3D printing, computer modeling, emergency preparation, and many other innovations that were not even dreamed of when I was first licensed. Hams have so many interests that we have supplemented the more-than 1,200-page paper book, which is also available electronically, with a wealth of online information that delves deeper into the topics that grab your interest.

I hope you enjoy the 101st edition of the *ARRL Handbook* and gain lots of knowledge to help our hobby continue moving forward into the next century and beyond.

73, Greg Lapin, N9GL
Editor, *The ARRL Handbook for Radio Communications*

ARRL Handbook Online Content

(print edition only)

A wealth of additional material for this *Handbook* is available in the online content. As a purchaser of the print edition, you are entitled to download this material. If you purchased the book directly from ARRL through **arrl.org/shop** you should have received an email with directions for downloading the content. If you purchased the *Handbook* from another retailer, lost, or never received the email with the link to the online content, go to **arrl.org/handbookdownload** to complete a quick form. Proof of purchase is required before download instructions are provided.

Searchable Edition of the *Handbook*

The online content includes a PDF version of the one-hundredth edition of *The ARRL Handbook*, including text, drawings, tables, illustrations, and photographs. Using *Adobe Reader*, you can view, print, or search the entire book.

Supplemental Files for Each Chapter

The online content contains supplemental information for most chapters of this book. This includes articles from *QST*, *QEX*, and other sources; material from previous editions of *The ARRL Handbook*; tables and figures in support of the chapter material; and files that contain information to build and test the projects provided in the chapters. The online information is arranged in folders for each chapter.

Companion Software

The following software is included with the online content:

TubeCalculator — a Windows application by Bentley Chan and John Stanley, K4ERO, accompanies the tube type RF power amplifier discussion in the **RF Power Amplifiers** chapter.

The following *Windows* programs by Tonne Software (**www.tonnesoftware.com**) are provided by Jim Tonne, W4ENE.

ClassE — Designs single-ended Class E RF amplifiers.

Diplexer — Designs both high-pass/low-pass and band-pass/band-stop types of diplexer circuits.

Helical — Designs and analyzes helical-resonator bandpass filters for the VHF and UHF frequency ranges.

JJSmith — A graphics-intensive transmission-line calculator based on the Smith chart.

Elsie — The free student edition of Elsie, a lumped-element filter design and analysis program.

MeterBasic — Designs and prints professional-quality analog meter scales on your printer. The full featured version of *Meter* is available from Tonne Software.

OptLowpass — Designs and analyzes very efficient transmitter output low-pass filters.

PI-EL — Designs and analyzes pi-L networks for transmitter output.

Pizza — A mapping program that generates printable azimuth-equidistant or rectangular maps showing the great-circle path and the sunrise-sunset terminator between your location and selectable locations.

QuadNet — Designs and analyzes active quadrature ("90-degree") networks for use in SSB transmitters and receivers.

SVCfilter — Standard-value component routine to design low-pass and high-pass filters and delivers exact-values as well as nearest-5% values.

Tower — Analyzes vertical antennas. Plots resistance, reactance, and impedance at the base as a function of frequency.

Common Schematic Symbols Used in Circuit Diagrams

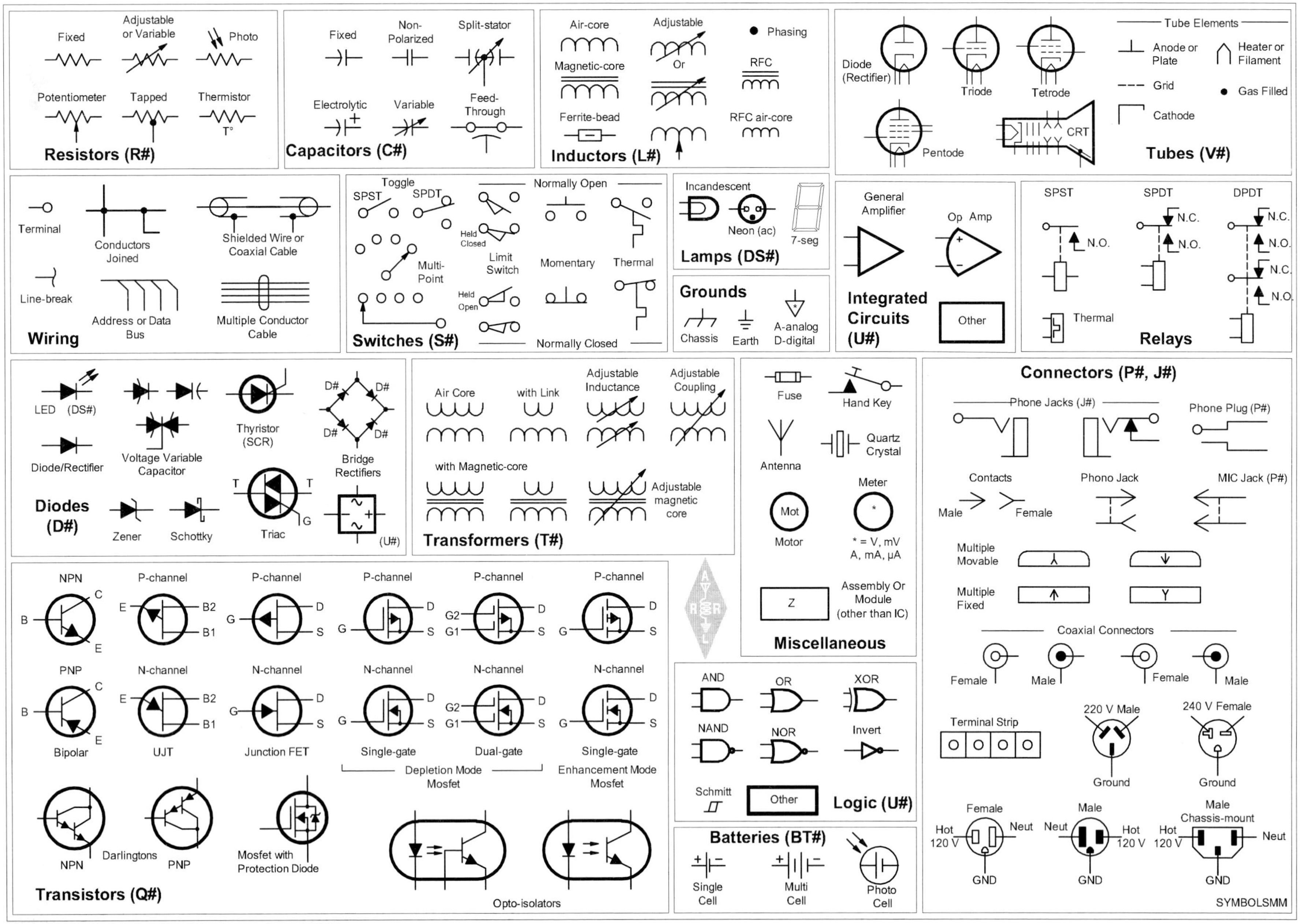

About ARRL

We're the American Radio Relay League, Inc. — better known as ARRL. We're the largest membership association for the amateur radio hobby and service in the US. For over 100 years, we have been the primary source of information about amateur radio, offering a variety of benefits and services to our members, as well as the larger amateur radio community. We publish books on amateur radio, as well as four magazines covering a variety of radio communication interests. In addition, we provide technical advice and assistance to amateur radio enthusiasts, support several education programs, and sponsor a variety of operating events.

One of the primary benefits we offer to the ham radio community is in representing the interests of amateur radio operators before federal regulatory bodies advocating for meaningful access to the radio spectrum. ARRL also serves as the international secretariat of the International Amateur Radio Union, which performs a similar role internationally, advocating for amateur radio interests before the International Telecommunication Union and the World Radiocommunication Conference.

Today, we proudly serve nearly 150,000 members, both in the US and internationally, through our national headquarters and flagship amateur radio station, W1AW, in Newington, Connecticut. Every year we welcome thousands of new licensees to our membership, and we hope you will join us. Let us be a part of your amateur radio journey. Visit www.arrl.org/join for more information.

ARRL The National Association for Amateur Radio®

225 Main Street
Newington, CT 06111-1400 USA
Tel: 860-594-0200
FAX: 860-594-0259
Email: membership@arrl.org
arrl.org

US Amateur Radio Bands

US AMATEUR POWER LIMITS — FCC 97.313 An amateur station must use the minimum transmitter power necessary to carry out the desired communications. (b) No station may transmit with a transmitter power exceeding 1.5 kW PEP.

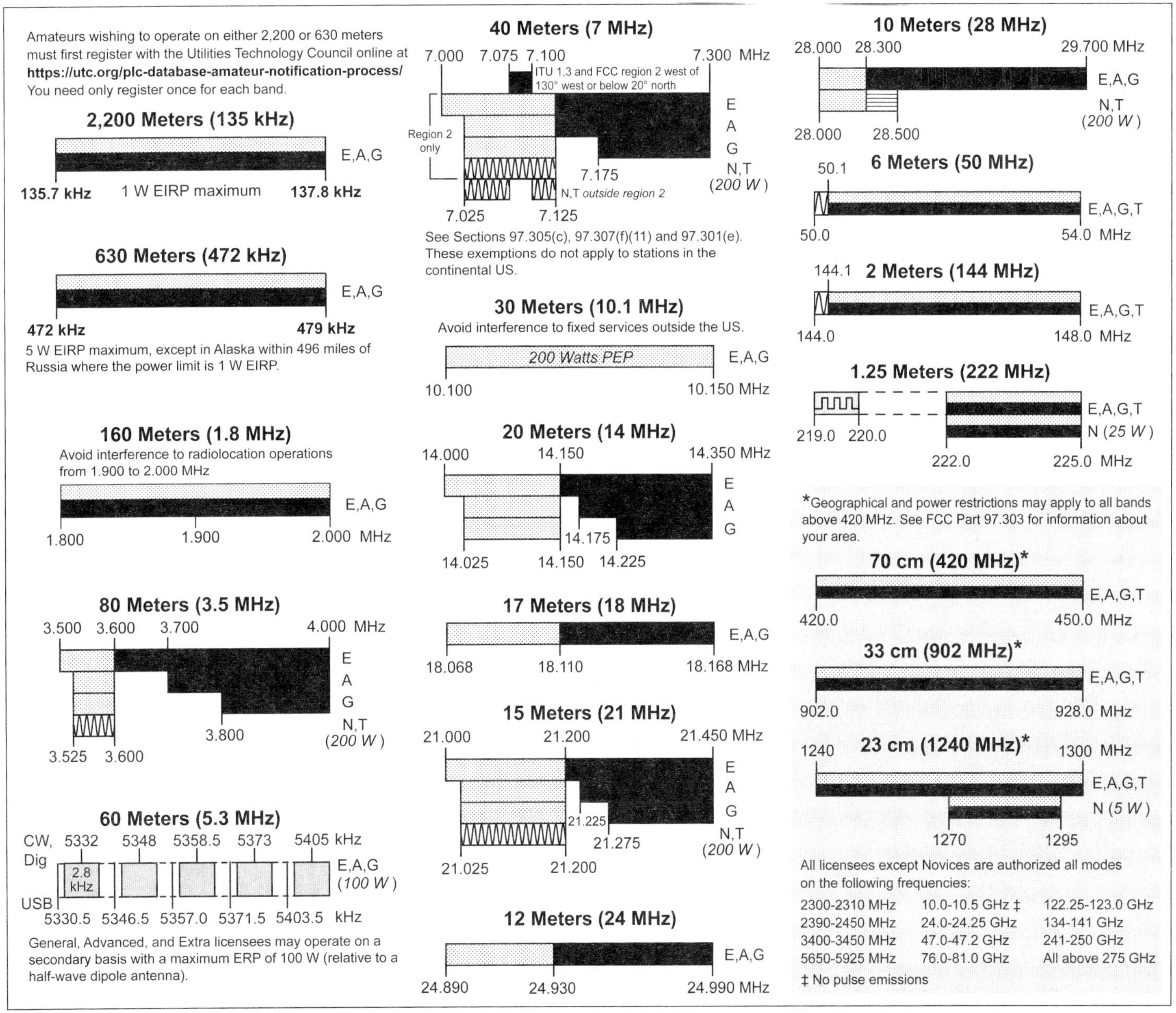

KEY

Note:
CW operation is permitted throughout all amateur bands.
MCW is authorized above 50.1 MHz, except for 144.0-144.1 and 219-220 MHz.
Test transmissions are authorized above 51 MHz, except for 219-220 MHz

- = RTTY and data
- = phone and image
- = CW *only*
- = SSB phone
- = USB phone, CW, RTTY, and data.
- = Fixed digital message forwarding systems *only*

E = Amateur Extra
A = Advanced
G = General
T = Technician
N = Novice

See *www.arrl.org/band-plan* for detailed band plans.

ARRL
We're At Your Service

ARRL Headquarters:
860-594-0200 (Fax 860-594-0259)
email: hq@arrl.org

Publication Orders:
www.arrl.org/shop
Toll-Free 1-888-277-5289 (860-594-0355)
email: orders@arrl.org

Membership/Circulation Desk:
www.arrl.org/membership
Toll-Free 1-888-277-5289 (860-594-0338)
email: membership@arrl.org

Getting Started in Amateur Radio:
Toll-Free 1-800-326-3942 (860-594-0355)
email: newham@arrl.org

Exams: 860-594-0300 email: vec@arrl.org

 rev. 07/25/2024

Contents

Chapter 1
What is Amateur (Ham) Radio?

Amateur radio enthusiasts — hams — often have been at the forefront of the sweeping blaze of progress over the past century in wireless and electronics, leading to technology that has broadened our horizons and touched virtually all of our lives. In the days before the telephone and household electricity were commonplace and the internet not yet conceived, hams pioneered personal communication. Amateur radio was the first wireless social medium.

From sophisticated smartphones, tablets, and devices worn on your wrist to diminutive laptop PCs that go anywhere, wireless technology is changing so rapidly that it can be difficult to keep current. People-to-people communication is the goal, whether by voice, text, or image, but amateur radio remains vital and active today. In this chapter, we present an overview of what to expect from *The ARRL Handbook*, as well as a brief look at amateur radio activities and licensing requirements.

Amateur radio, better known as "ham radio," is many things to many people — more than 750,000 of them in the US alone, and over 2.7 million worldwide. Communication with foreign countries, or DX, is one of the thrilling aspects of ham radio. You may want to practice your language skills over the radio, but for those languages that you have not mastered, fortunately, most hams across the world speak enough English to be understood.

Ham radio hobbyists have at their fingertips the ability to directly contact diverse and fascinating people they may never meet who live in distant places they'll never visit. They do this without any external infrastructure, such as a cellular network or the Internet, using their own equipment and antennas. They contact one another directly, and do it because it is fun and offers a sense of accomplishment — "*I* made this happen!" As a radio amateur, you can meet new friends, win awards, experiment with, and learn about radio science and technology, challenge yourself in on-the-air competitions (radiosport), contribute to your community, travel, promote international goodwill, and continue a more than century-old wireless communication tradition.

A NEW WINDOW ON THE WORLD AND BEYOND

Amateur radio is almost unique in allowing hams to experience the world in ways of which many people are unaware by using the atmosphere to send signals around the Earth. Reaching to the edge of outer space, the ionosphere bends and reflects radio signals. This upper region of the atmosphere changes with solar activity, terrestrial seasons, and the hours of light and darkness, so radio conditions are continually changing, too. In fact, hams can literally "hear the world turning" as different areas become illuminated by the Sun at sunrise, pass through the day, and then enter the nighttime hours after sunset.

Lower down where the air is denser, there are clouds and precipitation — this is the troposphere — and hams interact with that part of the atmosphere, too. By using their knowledge of how radio waves travel, hams can communicate far beyond the "line of sight," extending their radio horizon to hundreds of miles. Hams even bounce signals off mountains and buildings!

Space weather also gets into the act. If you are interested in what's happening on the Sun or in the magnetic fields around Earth, radio signals are a great way to measure or observe these effects in real time. A beautiful aurora can reflect radio signals. When a new sunspot, crackling with energy, makes its appearance on the Sun's surface, hams can sense its presence as it enables communication at higher and higher frequencies. The HamSCI community of researchers and amateurs (**hamsci.org**) constantly collaborates in group experiments and measurements.

Many radio amateurs have an interest in listening to radio signals from deep space, also known as *radio astronomy*. Hearing signals that originated many light years away from Earth can be more challenging than hearing signals from around the world. Sections in the **Antennas** and **Receiving** Chapters share some of the methods that amateur radio astrono-

Figure 1.1 — Tom Gaines, KB5FHK, and Sloan Davis, N3UPS (not shown) demonstrates how to make contacts through amateur radio satellites from the parking lot at the Orlando HamCation ham convention.

Figure 1.3 — Audrey Hance, KN4TMU, enjoys operating from a state park in Tennessee. Parks On the Air (POTA) is a program that encourages hams to enjoy the outdoors while using temporary ham radio setups.

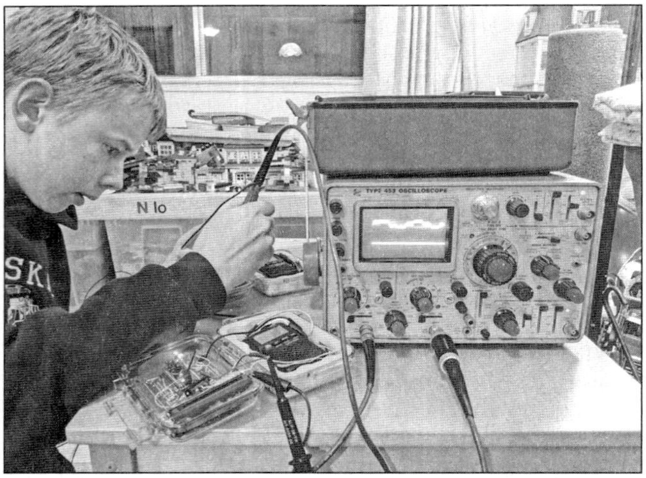

Figure 1.2 — Roy Gross, KM6EOO, got interested in ham radio in the second grade. He enjoys building and experimenting with antennas and radios. Here, he debugs a science fair project using an oscilloscope.

Figure 1.4 — Elixander Valladares, Jr., N4EVJ, takes part in the ARRL Rookie Roundup, an operating event for newcomers. [Elixander Valladares, Sr., W7HU, photo]

ARRL The National Association for Amateur Radio®

ARRL is the internationally recognized society representing amateur radio in the US. Since its founding in 1914, ARRL has grown and evolved along with amateur radio. ARRL Headquarters and the Maxim Memorial Station, W1AW, are in Newington, Connecticut, near Hartford. Through its dedicated volunteers and a professional staff, ARRL promotes the advancement of the amateur service in the US and around the world.

ARRL is a nonprofit educational and scientific organization dedicated to the promotion and protection of the many privileges that ham radio operators enjoy. Of, by, and for the radio amateur, ARRL numbers some 150,000 members — the vast majority of active amateurs in North America. Licensees can become Full Members, while unlicensed persons are eligible to become Associate Members with all membership privileges except for voting in ARRL elections. Anyone with a genuine interest in amateur radio belongs in the ARRL.

The ARRL volunteer corps is called the Field Organization. Working at the state and local level, these individuals tackle ARRL's goals to further amateur radio. They organize emergency communication in times of disaster and work with agencies such as the American Red Cross and Citizen Corps. Other volunteers keep government officials abreast of the good that hams do at the state and local level.

When you join ARRL, you add your voice to those who are most involved with ham radio. The most prominent membership benefits are its membership journal *QST*, published monthly in print and digital form, and *On the Air*, a bi-monthly magazine aimed at amateur radio newcomers and hams gaining experience.

QST includes stories you'll want to read; articles on projects to undertake; announcements of upcoming hamfests, conventions, contests, and other on-air activities; reviews of new equipment; reports on the role hams play in emergencies; and much more.

On the Air helps you make the most of your amateur radio license with helpful articles and infographics explaining the fundamentals of ham radio, including popular antenna setups and projects, best practices for operating your handheld transceiver, tips for participating in field day events, and many other topics — all written without a lot of math or jargon. The accompanying *On the Air* podcast dives deep into a single subject drawn from the magazine.

Members may elect to receive the print edition of either *QST* or *On the Air* for an additional fee. In addition, ARRL publishes two bimonthly specialty magazines aimed at particular segments of amateur radio: *QEX: A Forum for Communications Experimenters* and *NCJ*: *National Contest Journal*, which caters to those interested in amateur radio contesting, or "radiosport." These magazines are available online to all ARRL members as a member service.

Being an ARRL member is far more than a magazine subscription. ARRL represents your interests before the FCC and Congress, sponsors operating events throughout the year, and offers membership services at a personal level. These include:

- low-cost ham radio equipment insurance
- the Volunteer Examiner program
- the Technical Information Service, which answers your technical questions about amateur radio
- the ARRL Learning Center, with online courses for getting on the air, emergency communications, and electronics and technology
- Logbook of The World (LoTW), an online database for recording your contacts and cross-checking them with other amateurs around the world for award programs
- the QSL Service, which helps you exchange paper QSL cards in bulk with hams in other countries to confirm your contacts with them

For answers to any questions about amateur radio, contact ARRL Headquarters. See the Resources section at the end of this chapter for more information.

ARRL The National Association for Amateur Radio

Figure 1.5 — NASA Astronaut Tim Kopra, KE5UDN, makes the 1000th ARISS school contact from NA1SS aboard the International Space Station, where he spent one year in 2016. [NASA Photo]

mers use to detect those very weak signals. Listening for signals from the heavens does not require an amateur radio license, but the many parallels between the two hobbies have resulted in many amateur radio astronomers holding ham licenses. Certainly, the techniques used by each hobby can be helpful to the operation of the other.

THE ORIGINAL SOCIAL MEDIUM

In this age of multiple sophisticated communication platforms, it's not uncommon for people to ask, "Ham radio? Do they still do that?" Yes, "they" do. But, given the proliferation of communication alternatives, the larger question may be, *Why*?

Ham radio is a hands-on technological and social medium — personal communication with no bills, minutes, or data plans. It's communication that's "off the grid," and a wireless service you can rely on when other services become unavailable.

It doesn't cost a lot to get into amateur radio, and participation is open and accessible to everyone. Hams are mothers, fathers, and children of all ages, ethnic backgrounds, physical abilities, and walks of life who belong to a unique worldwide community of licensed radio hobbyists. Some are even well-known celebrities. All find joy and excitement in experiencing radio communication and electronics on a very personal level across a spectrum of activities.

HAMS ARE EVERYWHERE

The driver of that car sporting an odd-looking antenna may be a radio amateur equipped to enjoy his or her hobby while on the road — called mobile operation. Your neighbor on the next block with the wires strung between trees or, perhaps, a tower supporting what looks like a very large television antenna probably is one too.

Modern technology continues to make ham radio more accessible to all, including those living on tight budgets or facing physical challenges. People lacking mobility may find the world of amateur radio a rewarding place to find lasting friendships — on the next block, in the next state, or around the world.

Hams are ambassadors, contacting and making friends with other enthusiasts in other countries and sometimes traveling to far-flung locations to get on the air and meet the people who live there. Amateur radio recognizes no international or political boundaries, and it brings the world together in friendship.

The Second Century of Ham Radio

We may think of "wireless" as a relatively modern term that applies to a wide variety of electronic devices, but it's been around since the late 19th century. At some point along the way, "radio" (and later "television") took over as the preferred term. Being able to communicate from one place to another without connecting wires was a goal of late 19th and early 20th century experimenters. Equipment and methods for early wireless were rudimentary — a simple "crystal" radio to listen, and a "spark gap" transmitter (it *actually* generated sparks) to send Morse code, coupled with what was then called an "aerial" — we'd call it an antenna today. Little to no ready-made equipment was available, and parts for these early radio do-it-yourselfers were expensive and hard to obtain. On a good night, their transmissions might even span 50 miles. In the early 20th century, when few households had telephones and calling long-distance was pricey, ham radio was, in more contemporary terms, "really cool technology."

In 1914, just two years after the federal government required hams to hold licenses, inventor and industrialist Hiram Percy Maxim, 1AW, and radio enthusiast Clarence Tuska, 1WD, established the American Radio Relay League (ARRL) to bring US radio hobbyists under one tent to serve their common interests. These two founding fathers of ham radio and their peers would be awestruck to see how the world of amateur radio and wireless technology has expanded and evolved in the intervening years.

While Maxim and Tuska were not the first hams, the organization they founded, now known as ARRL® The National Association for Amateur Radio®, has championed and sustained these radio pioneers and their successors. Now more than 100 years down the road — light years in terms of radio science and technology — amateur radio continues to adapt to the times. While many traditions continue, today's ham radio is *not* the ham radio of yesteryear.

This iconic brick building houses W1AW, the station operated by the ARRL in Newington, Connecticut, and known around the world. W1AW memorializes Hiram Percy Maxim, one of the founders of the ARRL. Visitors are welcome and often operate the station. [Rick Lindquist, WW1ME, photo]

1.1 About this Book

For over a century, *The ARRL Handbook* has helped amateurs explore and learn about our great hobby. Of course, amateur radio covers a very wide range of interests, and the *Handbook* is a great resource for theoretical and practical information no matter what areas catch your interest. Here's a sampling:

• If you're putting together a home, portable, or mobile station, the **Power Sources**, **Transmission Lines**, **Antennas**, **Safe Practices**, **Assembling a Station**, and **RFI and EMC** chapters will be of special interest.

• Emergency communications (EmComm)/public service operators will find useful material in the **Power Sources**, **Repeater Systems**, **Antennas**, **Safe Practices**, and **Assembling a Station** chapters

• If your interests lie in equipment and circuit design and building, there is a wealth of material in the **Circuits and Components**, **RF Techniques**, **Computer Modeling and Simulation**, **DSP and SDR Fundamentals**, **Oscillators and Synthesizers**, **Analog and Digital Filtering**, **Modulation**, **Receiving**, **Transmitting**, **Transceiver Design Topics**, and **Power Amplifiers** chapters, among others.

• If you're more involved with repairing equipment or restoring vintage gear, the **Test Equipment and Measurements** and **Troubleshooting and Maintenance** chapters will be valuable.

• For antenna design and building, the **Antennas** chapter is filled with theoretical and practical information. Also check out the antenna-related sections of the **Test Equipment and Measurements** and **Troubleshooting and Maintenance** chapters.

• Amateur satellite theory, station design, and antennas are covered in detail in the online **Space Communications** chapter.

• Digital mode operators can take a deep dive into the theory behind many digital modes in the **Digital Protocols and Modes** chapter or turn to the online **Digital Communications** chapter for practical advice on putting these modes to use.

• For insight into amateur radio applications for balloons, buoys, autonomous craft, UAV (drones), radio controlled (R/C) craft, and robotics, try the **Amateur Radio Data Platforms** chapter. There's some useful material in the online **Image Communications** chapter as well.

• Those interested in STEM education topics and learning about how electronics and radio work will find a wealth of useful theoretical and practical material throughout this book. In particular, the **Electrical Fundamentals**, **Radio Fundamentals**, **Circuits and Components**, and **RF Techniques** chapters, along with the online **Digital Basics** chapter, offer a well-rounded look at fundamental electronic and radio theory. The **Propagation of Radio Signals** and **Amateur Radio Data Platforms** chapters offer background for those interested in scientific experimentation, research, and geophysics.

And the list goes on and on.

In addition to amateur radio operators, electronics hobbyists, engineers, and technicians turn to *The ARRL Handbook* for practical information about electronics and radio communication topics.

1.1.1 A Brief Handbook Tour

This *Handbook* presents a mix of theory and practical applications presented in the following chapters, which are divided into five broad sections:

- Fundamentals of Radio Electronics
- Principles of Radio Technology
- Radio Propagation and Antenna Systems
- Safe Practices and Station Construction
- Test Equipment, Troubleshooting, and RFI

In addition to the material in the printed book, companion online resources include a fully searchable version of the *Handbook* and a wealth of additional chapters, articles, tools, and other useful information.

The next sections are a brief overview of the *Handbook's* contents. For a more in-depth look, check out the detailed **Contents** page at the beginning of each chapter.

FUNDAMENTALS OF RADIO ELECTRONICS

The three chapters in the Fundamentals section cover basic electronic and radio concepts and components that are common throughout our station equipment and accessories.

Chapter 2 — Electrical Fundamentals is an overview of fundamental electrical and electronic concepts, including voltage and current, units of measurement, series and parallel circuits, direct and alternating current, and basic circuit components.

Chapter 3 — Radio Fundamentals moves into the realm of ac waveforms and signals and radio frequency (RF). The chapter discusses ac voltages in capacitors and inductors, along with reactance, impedance, Q, and resonant circuits. It wraps up with a discussion of analog signal processing and electromagnetic waves.

Chapter 4 — Circuits and Components gets into the practical devices used throughout electronic equipment. The chapter covers passive components such as resistors, capacitors, inductors, and transformers, as well as semiconductors and basic amplifier circuits.

PRINCIPLES OF RADIO TECHNOLOGY

The 14 chapters in this section discuss analog and digital techniques, circuits, building blocks, and systems essential to understanding how radios work.

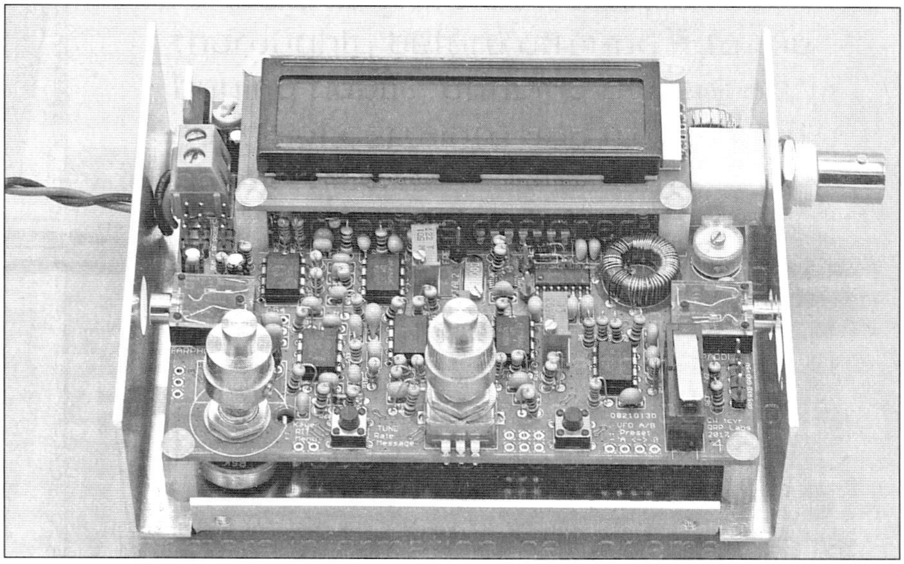

Figure 1.6 — Hams enjoy learning how their equipment works, as well as the satisfaction that comes from building their own equipment. This CW transceiver was assembled from a kit that included all of the parts, a circuit board, and detailed instructions.

Figure 1.7 — An amateur radio balloon project by Bill Brown, WB8ELK; Paul Verhage, KD4STH; and Ann Boes, KDØQCA; carries a 4-H Lab Revolution student experiment at 53,000 feet. [Jeff Ducklow, NØNQN, photo]

Figure 1.8 — Antennas are always a popular topic among hams. The Lake Area Radio Klub (LARK) in South Dakota embarked on a club project to design and build baluns for use with wire dipole antennas. It offered club members a chance to learn more about antennas and build a useful project for their station. Rich Grant, KEØEPY, helped club members with assembly.

Chapter 5 — RF Techniques covers a wide range of topics relating to how components behave when used in radio frequency (RF) circuits. Other popular topics include ferrite materials, impedance matching networks, and RF transformers.

Chapter 6 — Computer Modeling and Simulation is an overview of computer-aided tools and techniques for electronic design and circuit simulation. Later sections discuss special considerations for simulation of RF circuits, as well as electromagnetic analysis of RF circuits.

Chapter 7 — Power Sources covers the wide range of options for powering electronic circuits. The chapter covers analog power supply concepts as well as switch-mode dc-dc power conversion and high voltage techniques. Extensive coverage of batteries is essential for portable, mobile, and emcomm operators. Practical construction projects round out the chapter.

Chapter 8 — DSP and SDR Fundamentals explores the fundamentals of digital signal processing (DSP) and software defined radio (SDR), which are at the heart of new radio designs.

Chapter 9 — Oscillators and Synthesizers is an overview of the many ways RF signals can be generated. The chapter covers analog LC oscillators, crystal oscillators, and UHF techniques, as well as digital frequency synthesizers. The chapter wraps up with an in-depth look at phase noise and its effects on receiver and transmitter performance.

Chapter 10 — Analog and Digital Filtering starts with a discussion of basic filter concepts and characteristics, followed by sections on many types of analog and digital filters. The chapter includes several practical filter projects you can build.

Chapter 11 — Modulation covers the wide range of modulation types used in amateur radio. Topics include CW and analog voice modulation, as well as digital modulation techniques, image modulation, and spread spectrum.

Chapter 12 — Receiving discusses characteristics common to all receivers, such as sensitivity and dynamic range, and then explores heterodyne and SDR receiver architecture. The chapter also examines various building blocks such as mixers, demodulators, automatic gain control circuits, and noise reduction techniques.

Chapter 13 — Transmitting is the counterpart to Chapter 12. Topics include FCC rules for transmitter signal purity, performance measurements, heterodyne and SDR transmitter architecture, and basic building blocks.

Chapter 14 — Transceiver Design Topics includes a mix of some system level topics that deal with integrating receiving

Figure 1.9 — Hams operate using a wide variety of digital modes. Rosemarie Lones, K1AQT, operated radioteletype (RTTY) as part of the Nashua (New Hampshire) Area Radio Society team during an on-air radiosport competition. [Nashua Area Radio Society photo]

Figure 1.10 — Amateur radio frequency allocations extend well into the microwaves. Here, Chip Taylor, W1AIM, hunts contacts on the 10 GHz band from Mount Washington in New Hampshire, some 6,200 feet above sea level. [Chris Craig, K1MHZ, photo]

and transmitting functions into an effective transceiver. SDR technology is covered in more depth, along with user interface, panadapters, computer control interfaces, and transverters.

Chapter 15 — Digital Protocols and Modes takes a detailed look at the many different digital protocols and modes used by amateurs for transferring data. These are grouped into several categories: unstructured modes, fuzzy modes, structured modes, and networking modes.

Chapter 16 — Amateur Radio Data Platforms addresses the increasing use of amateur communication as a key element of scientific experimentation. Topics include different types of sensors, data, telemetry, and payloads. Then the chapter looks at vehicles to carry the payloads — high altitude balloons, UAVs, rockets, and robotics platforms.

Chapter 17 — RF Power Amplifiers covers RF power amplifiers beyond the typical 100 to 150 W transceiver and up to the legal power limit of 1.5 kW. Coverage includes the theory, characteristics, construction, and operation of both vacuum tube and solid-state amplifiers.

Chapter 18 — Repeater Systems is an overview of analog and digital VHF/UHF repeater technologies used by amateurs today. FM voice, D-STAR, DMR, and System Fusion systems are covered.

RADIO PROPAGATION AND ANTENNA SYSTEMS

The three chapters in this section discuss how signals generated by our transmitters leave our stations and travel to distant receivers.

Chapter 19 — Propagation of Radio Signals explores the principles of electromagnetic waves, the structure of the Earth's atmosphere, and solar-terrestrial interactions necessary for a working knowledge of radio propagation at MF, HF, VHF, UHF, and microwaves.

Chapter 20 — Transmission Lines starts with fundamental transmission line concepts and practical considerations for their selection, installation, and use. Related topics include impedance transformation, transmission line stub filters, antenna system impedance matching, antenna tuners, baluns, chokes, transmission line transformers, and waveguides.

Chapter 21 — Antennas is a perennial favorite. This chapter covers basic antenna theory, and then goes into a broad assortment of HF antennas, including dipoles, verticals, Yagis, and loops of various types. Antennas for HF mobile and portable operation are covered in detail, as are VHF/UHF home station and mobile antennas. A final section covers rotators. Throughout the chapter you'll find antenna projects that you can build or modify.

SAFE PRACTICES AND STATION CONSTRUCTION

The three chapters in this section are devoted to combining the concepts in the earlier sections into a working amateur station.

Chapter 22 — Safe Practices discusses electrical safety inside your station, along with best practices for grounding, bonding, and lightning protection. Antenna and tower safety includes some information on trees and masts but focuses primarily on safely installing and working on towers. In the RF Safety section, you'll find coverage of RF

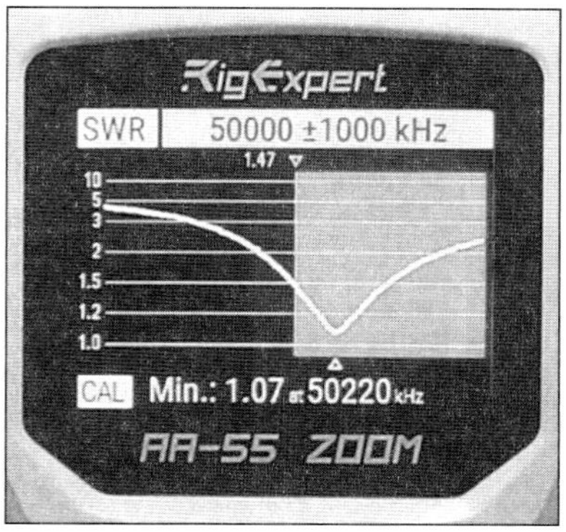

Figure 1.11 — Learning to use basic test equipment such as an antenna analyzer will help you keep your amateur station running at peak efficiency,

exposure concerns and regulations, along with guidelines for safe operation of your station.

Chapter 23 — Construction Techniques is a great resource for anyone who likes to build their own equipment. Topics include soldering tools and techniques, working with surface mount devices (SMD), and computer-aided design (CAD) using free or low-cost schematic capture and PC board design software. Microwave construction techniques and mechanical fabrication (chassis, panels) are covered, along with a comprehensive section on 3D printing.

Chapter 24 — Assembling a Station is all about station installation and integration for fixed, mobile, portable, emergency, and remote stations. The chapter covers power sources, antennas, and setup techniques for portable and temporary stations. The section on remote control covers network configuration and important performance issues like latency.

TEST EQUIPMENT, TROUBLESHOOTING, AND RFI

This section provides information on tools, equipment, and their uses to deal with the inevitable problems that crop up in any amateur radio station.

Chapter 25 — Test Equipment and Measurements explains the fundamentals of dc and ac measurements, including the use of multimeters, oscilloscopes, frequency counters, and function generators. RF measurements and equipment including spectrum analyzers, network analyzers, and signal generators are covered, as well as details of receiver and transmitter measurements in the ARRL Lab. Antenna system measurements and instruments are included, along with some practical projects.

Chapter 26 — Troubleshooting and Maintenance offers techniques for troubleshooting equipment problems at the component, circuit, and system levels. The chapter offers insight on typical problems with different types of equipment, troubleshooting antenna systems, and working on vintage gear.

Chapter 27 — RFI and EMC is an in-depth look at common types of interference both from and to amateur radio equipment. Basic concepts of electromagnetic compatibility and associated components are presented.

ONLINE CONTENT, TOOLS, AND RESOURCES

In addition to the printed *ARRL Handbook*, a number of additional chapters, articles, and tools are available online, along with a fully searchable copy of the book. See the page at the front of this book for details on how to access this information.

Additional chapters included with the fully searchable online version of the *ARRL Handbook* include:

• **Space Communications** with details of amateur satellite and EME (earth-moon-earth, or moonbounce) stations.

• **Digital Communications** discusses station setup, technical details, and operating tips for many popular digital data and voice modes.

• **Image Communications** is all about analog and digital flavors of fast scan amateur television (ATV) and slow scan television (SSTV).

• **Digital Basics** covers digital theory fundamentals and some applications of that theory in amateur radio.

• **Station Accessories and Projects** is a collection of useful projects that you can undertake.

• **Radio Mathematics** is a collection of tutorial information on a variety of mathematics used in amateur radio.

In addition to these chapters, there is a collection of useful design software and a wide variety of supplemental articles and files for each chapter. A list of relevant online information and software is included at the beginning of each *Handbook* chapter.

Figure 1.12 — Nathan Charles, N3QKA, installed and maintained the solar power system at the Signal Hill Amateur Radio Club's ARRL Field Day site atop a mountain ridge on the Virginia/West Virginia border. Batteries were sheltered in tents to keep them out of the rain. [Jonathan Charles, NB3I, photo]

1.2 Structure of Amateur Radio

International and national radio regulations govern the operational and technical standards of all radio stations. The International Telecommunication Union (ITU) governs telecommunication on the international level. In the US, the Federal Communication Commission (FCC) is the agency that administers and oversees the operation of nongovernmental and nonmilitary stations — including amateur radio. Title 47 of the *US Code of Federal Regulations* governs telecommunication, and Part 97 of the Code spells out the Amateur Service rules. (See **www.arrl.org/part-97-amateur-radio**.)

Figure 1.13 — Providing communication during emergencies is part of the Basis and Purpose of amateur radio as spelled out in the FCC rules. James Plumlee, KI5DAZ, serves as net control operator for the Hospital Net during the annual Simulated Emergency Test (SET) training exercise. [Paul Teel, WB5ANX, photo]

Part 97 includes a definition of the Amateur Service: "A radiocommunication service for the purpose of self-training, intercommunication, and technical investigations carried out by licensed individuals interested in radio technique solely with a personal aim and without pecuniary interest." (*Pecuniary* means payment of any type, whether money or goods.) Amateur radio is a recognized national asset, a noncommercial service providing trained operators, technical specialists, and disaster response communications in time of need. It was created for people who have an interest in radio communications (see the sidebar, "Basis and Purpose of Amateur Radio").

Amateur radio will surprise you with all its different activities. If you've encountered amateur radio in a public service role or if someone you know has a ham radio in their home or car, then you already have some ideas.

Some hams prefer to focus on the technology and science of radio. Competitive events and award programs hold the interest of others. Some train to use radio in support of emergency response efforts, to provide public service, or to keep in touch with family. There are many hams who simply like to talk with other hams. The ARRL website (**www.arrl.org**) has much more information on these activities in the On the Air, Public Service, Technology, and Get Involved sections.

1.2.1 Bands and Modes

Anyone can listen to amateur radio signals, but an FCC license is required to transmit. There are currently three levels of license available — Technician, General, and Amateur Extra — each requiring an exam and conveying a set of privileges. We'll explain more about licensing and testing later in this chapter. Licensed hams have access to small frequency bands (segments) throughout the radio spectrum (see **Figure 1.14**), with some limitations depending on license class.

Each band has different characteristics, opportunities, and challenges to learn about — part of what makes ham radio so interesting. Depending upon license class, hams have access to up to 10 distinct "high frequency" (HF or shortwave) bands in the range from 1.8 to 29.7 MHz, where most direct international communication happens. All licensees have all amateur privileges in the VHF-UHF and microwave spectrum (50 MHz and higher), which allow more local and regional operation on widely available FM voice repeaters and digital voice networks. Frequency and wavelength terms are explained in the **Electrical Fundamentals** chapter.

Amateurs also use a wide variety of "modes" for making contacts. Some have been around since amateur radio began, while others are very recent creations. Much more information about various modes may be found in the **Modulation**, **Digital Protocols and Modes**, and **Repeater Systems** chapters.

VOICE MODES

Hams use a number of analog and digital voice modes (methods of adding voice to a radio signal). Analog voice modes are amplitude modulation (AM), which includes the narrower-bandwidth single sideband (SSB), and frequency modulation (FM). For the most part, SSB is heard on HF, while FM is the typical voice mode employed on

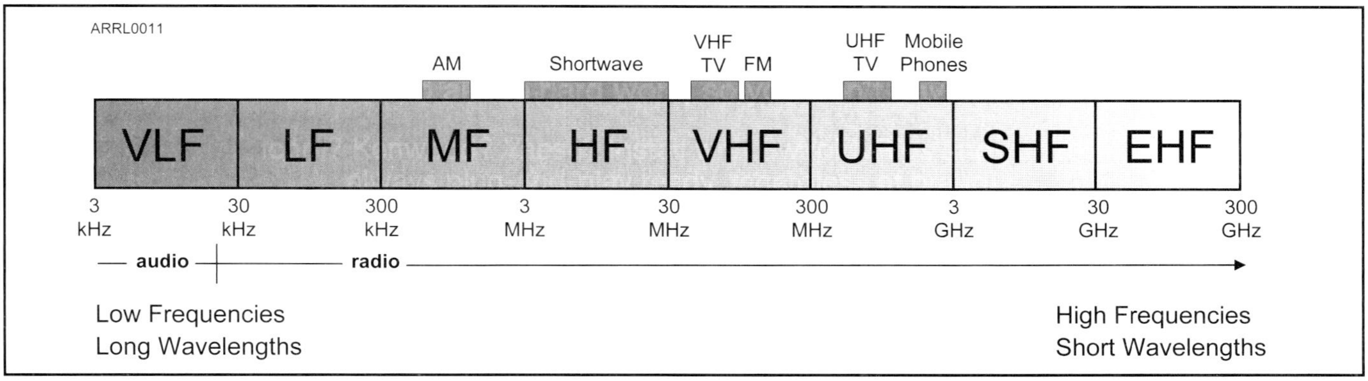

Figure 1.14 — The radio spectrum extends over a very wide range of frequencies. The drawing shows the frequency ranges used by broadcast stations and mobile phones. Amateurs can use small frequency bands in the LF and higher frequency regions of the spectrum.

VHF, UHF, and microwave bands.

Amateurs also use digital voice modes that convert speech into a bit stream for transmission. Systems such as D-STAR, System Fusion, and DMR are being rapidly adopted for repeater use. The FreeDV system is designed for use on the HF bands along with SSB.

The great majority of ham radio HF phone operators use SSB (subdivided further into upper sideband and lower sideband), but a few still enjoy and experiment with heritage "full-carrier AM" with its warm, rich audio quality. See **www.arrl.org/amplitude-modulation**.

REPEATERS

Hams often make their first contacts on local voice repeaters, using analog FM or one of the digital voice modes. Repeaters can greatly extend the useful range of a typical handheld or mobile transceiver much in the same way a cell tower retransmits voice or text messages. Repeaters carry the vast majority of VHF/UHF traffic, making local and even regional mobile communication possible for many hams. Located in high spots, repeaters strengthen signals and retransmit them to provide communication over much greater distances than would be possible when operating point-to-point or "simplex." The wider coverage can be especially important if the repeater is ever pressed into service during an emergency.

Most repeaters are maintained by clubs or groups of hams. If you use a particular repeater frequently, you should join and support the repeater organization. Some hams set up their own repeaters as a service to the community. The best way to learn the customs of a particular repeater is to listen for a while before transmitting.

Most repeaters are *open*, meaning that any amateur may use the repeater, although FM repeaters typically require users to transmit an access tone (which any modern FM transceiver can transmit). Some repeaters are *closed*, meaning that usage is restricted to members of the club or group that owns and operates the repeater. The *ARRL Repeater Directory* shows repeater locations, frequencies, and other details. See the **Repeater Systems** chapter for more information.

MORSE CODE

Morse code was the very first radio transmission mode, although it wasn't long before early experimenters figured out how to transmit the human voice and even music over the airwaves. Morse is also the original digital mode; the message is transmitted by turning a radio signal on and off ("1" and "0" in digital terms) in a prescribed pattern to represent individual letters, numerals, and characters. This pattern is the International Morse Code. Hams often refer to Morse transmissions as "CW," after an archaic definition for "continuous wave," which described the type of radio wave transmitted.

Although proficiency in Morse code is no longer required for any class of amateur radio license in the US, many hams still embrace CW as a favorite mode and use it routinely. Hams typically send Morse code signals by manipulating a manual telegraph key, a "semi-automatic" key (called a "bug"), or a CW "paddle" and an electronic keyer to form the dots and dashes. Experienced hams decipher Morse code "by ear," either writing down the letters, numerals, and characters as they come through the receiver's headphones or speaker or simply reading it in their heads. Some use one of the available computer-based or stand-alone accessories that can translate CW into plain text without the need to learn the code.

Hams who enjoy CW cite its narrow bandwidth — a CW signal takes up very little of the radio spectrum — simpler equipment, and the ability of a CW signal to "get through" noise and interference with minimal transmitting power. CW is a common low-power (QRP) mode.

Basis and Purpose of Amateur Radio

Experimentation has always been the backbone of amateur radio, and the Amateur Service rules provide a framework within which hams enjoy wide latitude to operate and experiment. The FCC's rules governing amateur radio recognize five aspects, paraphrased below, in the Basis and Purpose of the Amateur Service.

- Amateur radio's value to the public, particularly with respect to providing emergency communication support
- Amateur radio's proven ability to contribute to the advancement of the radio art
- Encouraging and improving the Amateur Service through rules that help advance communication and technical skills
- Maintaining and expanding the Amateur Service as a source of trained operators, technicians, and electronics experts
- Continuing and extending the radio amateur's unique ability to enhance international goodwill

The Amateur Service rules, Part 97, are in six sections: General Provisions, Station Operation Standards, Special Operations, Technical Standards, Providing Emergency Communication, and Qualifying Examination Systems. Part 97 is available in its entirety online (see the Resources section at the end of this chapter for further information).

Table 1.1
Maximum Symbol Rates and Bandwidth

Band (meters)	*Symbol Rate (baud)*	*Bandwidth (kHz)*
160 through 10 m	—	2.8
6 m, 2 m	19.6k	20
1.25 m, 70 cm	56k	100
33 cm and above	no limit	no limit

DIGITAL MODES

Digital modes are used to exchange information between computers as individual characters, either in complete files or one character at a time in "keyboard-to-keyboard" contacts. Amateurs use modes originally invented for commercial or military applications and have invented several of their own! Innovation in digital communications is one of amateur radio's most important and active contributions. Digital modes are described in detail in the **Digital Protocols and Modes** chapter and in the **Digital Communications** supplement available online. Note that the FCC rules spell out limitations to the bandwidth digital modes may occupy, as shown in **Table 1.1**.

Two popular digital modes today are FT8, one of several modes in the *WSJT-X* software package (**wsjt.sourceforge.io/wsjtx.html**), and either PACTOR or ARDOP and VARA, which are used to exchange email and other types of information as part of the Winlink system (**winlink.org**). FT8, PACTOR, and VARA use your PC with its sound card interfaced to a transceiver. The later versions of PACTOR that are most widely used require a special modem between the PC and radio.

The free package of *WSJT-X* software includes a number of specialized digital modes optimized for different uses: JT65 for EME, MSK144 for meteor scatter, FT8 for HF operation (a similar FT4 protocol is customized for radio contesting), WSPR for very low-power beacons, and more. FT8 is extremely popular due to its ability to exchange data with signals many times weaker than the noise. FT8 is a great mode for hams in neighborhoods with many radio noise sources or who can't put up big antennas. Hams have also developed a number of versions of FT8 that add features or variations of messages.

Hams interested in public service, espe-

Figure 1.15 — Hams use repeaters to relay signals from low-power radios over a wide area. Repeaters are a popular on-the-air meeting place for hams.

Figure 1.16 — ARRL Field Day is an annual event that combines emergency preparedness training and radiosport as individuals and clubs set up stations off the grid with temporary antennas and try to make as many contacts as possible. The Huntsville Amateur Radio Club's K4BFT setup in Alabama attracted a group of teachers from the nearby US Space and Rocket Center Space Camp. Todd Cline, K7KDT, demonstrates a digital "waterfall" as club members John Boyette, KK5KKT (leaning at left), and Geoff Suiter, KK4IV (at keyboard), operate the station. [William Martin, KK4FDF, photo]

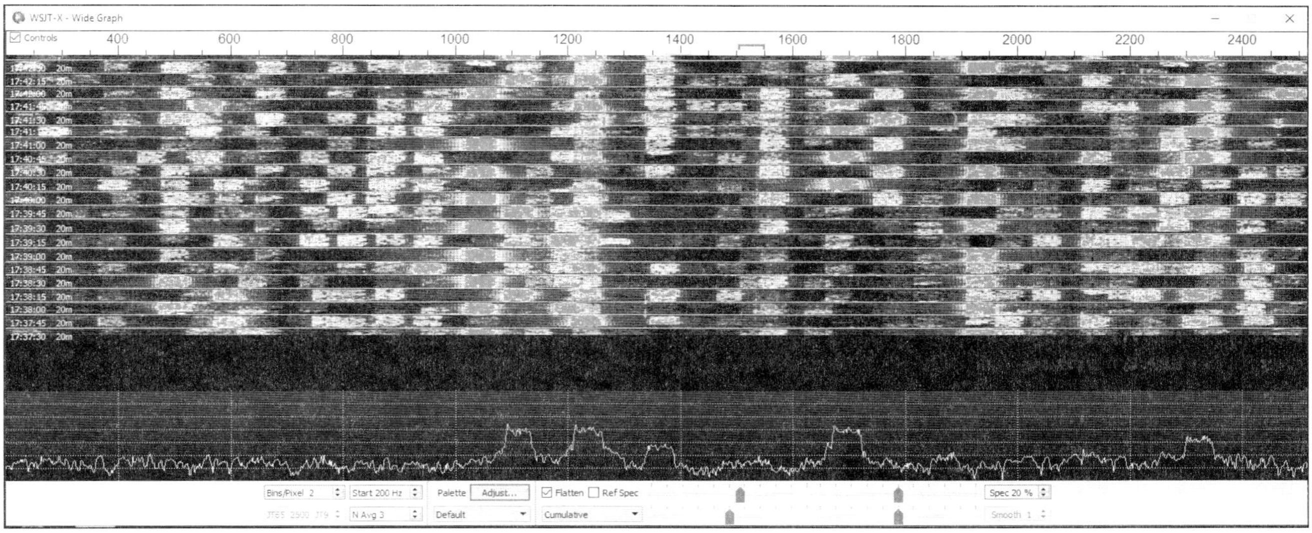

Figure 1.17 — Waterfall display of stations using the FT8 digital mode on a busy 20 meter band. Each small rectangle represents a transmission from a different station. Using state-of-the-art signal processing technology, it allows an astonishing number of stations to squeeze into a 2.5 kHz passband that would accommodate just one SSB voice signal. It also allows decoding of weak signals inaudible to the human ear, making it ideal for stations with low power and compromise antennas that would have difficulty making contacts using other modes.

cially emergency communications, make use of the Winlink system to maintain contact while camping, boating, or just traveling. Many public service teams include Winlink as part of their training and operation programs. The system acts as a worldwide gateway between amateur radio and the Internet, although the FCC restrictions on commercial content must be strictly followed.

Two other keyboard-to-keyboard modes used by hams to "talk by typing" include radioteletype (RTTY) and PSK31. Often pronounced "ritty," radioteletype was originally developed for point-to-point communications. Software such as *fldigi* supports many different modes including PSK31 and RTTY (see **www.w1hkj.com**).

Another amateur innovation is the Automatic Packet Reporting System or APRS, which combines position information (such as from GPS) with text messages and packet radio. This worldwide system (**www.aprs.org**) uses continuously monitoring "gateway" receivers to listen for the low-power APRS signals, relaying the information to Internet servers, where the information is viewed. Hams use APRS in their vehicles, while biking or boating, for weather reporting, and even to track high-altitude balloons during experiments.

The digital modes mentioned here are just a fraction of the many different digital modes used in amateur radio. New variations and innovations are being released all the time as experimenters try a new modulation or encoding scheme. This is a great example of the Amateur Service fulfilling its Basis and Purpose by "contribut(ing) to the advancement of the radio art."

Figure 1.18 — Ricardo "Rick" Navarrete Lopez, EA4M, used this nicely appointed station during the CW weekend of ARRL International DX Contest.

IMAGE COMMUNICATION

Users of current cell phone and internet technology often enjoy sharing photos or talking face-to-face. While not as sophisticated, several ham radio communication modes allow the exchange of still or moving images over the air. Amateur radio communication takes on an exciting, new dimension when you can actually see the person you're communicating with.

Amateur TV (ATV) is full-motion video over the air, sometimes called "fast-scan TV." A newer form is Digital Amateur TV (DATV), which is used to transmit distinctly sharper images and voice. Amateur groups in some areas have set up ATV repeaters, allowing lower-power stations to communicate over a fairly wide area. Since this is a wide-bandwidth mode, operation is limited to the UHF bands (70 centimeters and higher).

DATV folds nicely into an amateur radio technological initiative called high-speed multimedia (HSMM), which supports networks such as Broadband-Hamnet and Amateur Radio Emergency Data Network (AREDN). The ham bands above 50 MHz can support computer-to-computer communication at speeds high enough to sustain multimedia applications — voice, data, and image. One approach adapts IEEE 802 technologies, particularly 802.11b, to operating on specific amateur radio frequencies in the 2.4, 3.4, and 5.6 GHz microwave bands.

SSTV, or "slow-scan TV," is an older, narrow-bandwidth image mode similar to fax. Instead of full-motion video, SSTV enthusiasts exchange photographs and other static images. Individual SSTV pictures take anywhere from 8 seconds to about 2 minutes to send, depending on the transmission method. Most SSTV operation is done in color, using computers and sound cards in conjunction with custom software that converts images into a series of audio tones representing brightness level and colors that are "painted" on the receiver's screen line by line. Since SSTV is a narrow-band mode, it is popular on HF on the same frequencies used for voice operation.

For more information, see the online Image Communications chapter for details on these modes.

1.2.2 A Sampling of FCC On-Air Operating Rules

FCC amateur radio licenses convey privileges on their holders that do not exist for other radio services in the United States. If someone buys a radio in another service, the FCC requires that they use that radio as is, and they are not permitted to change it in any way. A licensed radio amateur, on the other hand, can make changes to a purchased radio to make it work better. If you buy a handheld radio and need to improve its signal to reach a far-away repeater, you are free to change the antenna to a more efficient one. ***No other radio service can do that.*** If you determine that your radio needs more power to communicate with others, you are free to add an amplifier to boost the power as high as necessary, up to the legal limit. Again, ***no other services can do this.*** If you have a software defined radio, and you determine that a change to the software would make it work better, you are free to make that change. ***No other service can do that.*** Other services must operate their radio equipment exactly as it was designed and certified by an FCC-approved testing lab.

However, with those privileges come responsibilities. The FCC expects hams to control their transmitted signals so that they do not interfere with other hams or services. While most commercial radio transmitters used by hams produce clean signals, it is the control operator's ultimate responsibility to make sure no interference occurs. Every radio amateur should read about the different ways to test their transmitted signals in the **Test Equipment and Measurement** chapter.

Figure 1.19 — Sean Kutzko, KX9X, uses a handheld antenna to make contacts through an amateur radio satellite.

There are quite a few FCC rules and regulations covering behavior on the air, but much of it boils down to polite behavior and common sense. In the FCC rules, you'll see references to the *control operator*. That's simply the licensed amateur designated to be responsible for making sure that transmissions comply with FCC rules. Any licensed amateur can be a control operator, and that doesn't have to be the same person as the station owner. The station licensee is responsible for designating the control operator.

Proper identification of your station while on the air is important so the FCC knows who is transmitting and for any other station that wants to contact you or to know where your signal is coming from. The identification rules are simple — give your call sign at least once every 10 minutes during a contact and when the communication is finished. Unidentified communications are not allowed.

As noted earlier, amateur radio is a noncommercial service. Transmissions related to conducting your own business or your employer's activities are not permitted, and you can't be paid for using your amateur station. There are plenty of other radio services available for commercial activities. (Note that your personal activities don't count as "business" communications. For example, it's perfectly okay for you to use ham radio to talk to your spouse about doing some shopping or to confer about what to pick up at the store.)

Prohibited transmissions include false distress or emergency signals, obscene or indecent speech, or encrypted transmissions intended to hide the meaning. You're also not allowed to engage in one-way broadcasting for the general public or transmit music. And it is never okay to intentionally interfere with other hams using the band.

1.2.3 Station Considerations

The FCC rules include some station requirements as well. For example, the maximum allowed transmitter power is 1500 W PEP output on most bands. There are lower power limits for the 2200-, 630-, 60-, and 30-meter bands, and for Novice and Technician licensees in their HF allocations. Full details are available from the US Amateur Bands chart at the front of this book or online from **www.arrl.org/band-plan**.

There are also some rules about signal quality because a poor quality signal may interfere with other hams or even other radio services. Signal quality for various modes is covered in the **Modulation**, **Transmitters**, and **Test Equipment and Measurements** chapters. Fortunately, most commercially produced transceivers generate good quality voice, CW, and digital signals when properly adjusted, and you should take the time to read the instructions for your gear. ARRL Lab tests and *QST* Product Reviews are also a good resource for learning about how well equipment complies with FCC rules and good engineering practice for signal quality. See **www.arrl.org/arrl-lab**.

Amateur radio is basically a safe activity. In recent years, however, there has been considerable discussion and concern about the possible hazards of electromagnetic radiation, including both RF energy and power-frequency (50 to 60 Hz) electromagnetic fields. To help allay such concerns, the FCC set limits on the amount of RF energy to which people can be exposed from any transmitter, including ham transmitters. Amateur radio stations need to be evaluated to see if they comply with the rules. Detailed information on RF safety may be found in the **Safe Practices** chapter of this book and online at **www.arrl.org/rf-exposure**.

Figure 1.20 — Johannes Hafkenscheid, PA5X, operated from Mauritania as 5T5PA, while working there for several months.

STATION EQUIPMENT AND ANTENNAS

Amateur radio costs as much or as little as your budget and enthusiasm dictate. Today's radio amateurs most often start out using off-the-shelf, commercially made transceivers purchased new or on the used market, perhaps at a ham radio flea market ("hamfest") or through an internet auction or classified ad site. An abundance of ham gear is readily available, and it's not hard to find something that meets your needs and suits your budget. Used higher-end VHF or VHF-UHF (or "dual-band") handheld transceivers often are available for $100 to $200 or so. The entrance of Chinese manufacturers into the ham radio market has dramatically driven down the cost of a basic VHF-UHF handheld to well under $100. To find hamfests in your area, try **www.arrl.org/hamfests-and-conventions-calendar**.

Those interested in HF work can get in on the ground floor with a used, but serviceable, transceiver in the $200 to $500 range, and an excellent selection of new transceivers is available in the $500 to $1,500 range. Low-power CW (Morse code) transceivers covering single bands sell new for less than $100 in kit form. An HF antenna such as a simple backyard dipole suspended from available trees is both inexpensive and effective. You'll find a wide variety of equipment advertised in ARRL's monthly journal, *QST*, which is available to ARRL members and includes comprehensive equipment reviews (**www.arrl.org/product-review**). Some helpful tips and resources may be found at **www.arrl.org/what-rig-should-i-buy**.

More elaborate antennas and various accessories can add appreciably to the cost of your station, but less-expensive alternatives are available, including building your own. The **Antennas** chapter features several simple and effective antennas that you can build from wire and readily available parts.

Early radio amateurs generally built their own gear, mainly out of necessity. Constructing ham radio equipment from kits became popular in the mid-20th century. Several manufacturers still provide parts kits and circuit boards to make it easier to build equipment yourself. Some hams still like to design and build their own equipment (called "homebrewing"), saving money in the bargain and enjoying the challenge. Check out the **Construction Techniques** chapter for a look at how you can get involved in building equipment or kits, even using 3D printing.

COMPUTERS AND HAM RADIO

Amateurs discovered decades ago that interconnecting their PCs with their ham stations not only makes operating more convenient but can open the door to additional activities on the ham bands. Most ham radio stations include a computer with software for many ham radio applications, from record keeping to antenna and circuit design.

Probably the most common use for a computer in the ham station is logging your contacts — keeping a record of the stations you have communicated with on the air. Some computer logging applications also let you control many of your radio's functions, such as frequency or band selection, without having to leave your logging program. It's also possible to control various accessories, such as antenna rotators or selection switches, by software.

You can enjoy digital modes with nothing more than a couple of simple connections, operating software (typically free), and an interface. Most current generation transceivers have a built-in sound card device, and all that's needed is a USB cable and appropriate software. Everything you need to know to get started with FT8 and other digital modes may be found in the online **Digital Communications** chapter.

Computers also can alert you to DX activity on the bands, help you practice taking amateur radio license examinations, or improve your Morse code skill. Many ham radio organizations and interest groups maintain online discussion groups with valuable information about a wide variety of topics, and, of course, equipment vendors maintain websites too.

STATION LOCATION

Many hams set up home-based stations. By tradition the room or place where a station is located is your "ham shack," but many lower-profile hams carry their stations with them in the form of relatively inexpensive handheld VHF/UHF transceivers. Without requiring any antenna beyond the one attached to the radio itself, such radios can accompany you when you're out and about or traveling.

On the other end of the scale, radio amateurs serious about radiosport competitions or DXing (contacting distant stations in other countries) often invest in top-tier equipment and extensive antenna systems.

Figure 1.21 — Dennis Lazar, W4DNN, using his vintage station on CW. Some hams enjoy restoring and using classic tube-type equipment, often called "boat anchors" because of their weight.

Figure 1.22 — Amateur Radio Direction Finding (ARDF) equipment need not be expensive. Yagis made from measuring tapes and PVC pipe are very popular for direction finding on 2 meters with a handheld transceiver. Here, Dan Slater, AG6HF (left), tests an antenna he built. [Joe Moell, KØOV, photo]

Most hams fall somewhere between these extremes. They have a modest equipment complement, simple wire antennas suspended from trees, and maybe a small "beam" (directional antenna) on a backyard tower or mast. Whatever your investment level, you'll be able to talk around the world.

ANTENNA RESTRICTIONS

In today's world, it's common for hams to face limitations on installing antennas if they live in an apartment or condo, or in a development governed by a homeowner's association (HOA) with restrictions on how they can use their property. Some areas have zoning requirements such as height limits or setback from adjacent properties. In most cases a building permit from the local town offices is needed for a larger installation such as a permanent mast or tower. This is something to explore when buying a property or planning to install an antenna. It's worth some time to review the ARRL web pages on this topic at **www.arrl.org/antenna-regulation-and-zoning**.

In some cases, hams are able to install "low profile" indoor or outdoor antennas. In recent years there has been increased interest in portable operation from outdoor locations such as local, state, and national parks. This allows temporary installation of a more effective antenna on a mast or in a tree (if allowed) but usually requires a portable power source such as a battery.

Other hams install VHF/UHF and/or HF equipment and antennas in their vehicles for mobile operation. Some installations are for a single band (say, 2 meters), while others use complex antenna systems to cover a number of HF bands. It's not unusual for hams to use their mobile stations to chat with friends during commuting hours, to stay in touch with friends while traveling, or to travel to rare locations to contact hams chasing various awards. An effective mobile station has a more serious side — use during disasters or other interruptions in normal communication channels.

Portable and mobile stations, antennas, and batteries are discussed in the **Power Sources**, **Antennas**, and **Assembling a Station** chapters. Another good resource is *Portable Operating for Amateur Radio* by Stuart Thomas, KB1HQS, available from the ARRL Store.

REMOTE AND AUTOMATIC OPERATION

It is becoming common to operate a remote HF station via a link over the Internet. You may have your own remote station, belong to a group with a shared club station with remote access, or you may use one of the stations available for rent. Availability of remote hardware and widespread access to reliable, high-speed Internet service has been a huge benefit to hams living in locations where outdoor antennas are limited or prohibited. Remote operation is discussed in detail in the **Assembling a Station** chapter.

Many stations, such as repeaters and beacons, operate without a human control operator present to perform control functions. This is called automatic operation.

Remote and automatic operation are defined in the FCC rules, but the requirement remains the same — the station must be operated in compliance with FCC rules, no matter where the control point is located. See the sidebar, "Local, Remote, and Automatic Control" for more information.

1.3 Amateur Radio Licensing in the US

As noted earlier, the Federal Communications Commission (FCC) is charged with administering all of the radio signals transmitted by US radio stations. The vast majority of radio users must have a license or be employed by a company that has a license. This section explains how licensing works for amateur radio.

1.3.1 Licensing Overview

The FCC has a different set of rules for each type of radio use. These uses are called *services*. Each service was created for a specific purpose — Land Mobile, Aviation, and Broadcasting, for example. Nearly all services require that a license be obtained before transmissions are made. These are called licensed services, and the Amateur service is one of them.

Most services do not require an examination to be licensed. This is because the FCC sets strict technical standards for the radio equipment used in these services and restricts how those radios may be used. This trade-off reduces the training required for those radio users. Licensing in these services is primarily a method to control access to the airwaves.

UNLICENSED PERSONAL RADIOS

The most popular personal radios are the FRS/GMRS handheld radios that are seemingly sold everywhere. FRS stands for Family Radio Service and GMRS stands for General Mobile Radio Service. These radios use a set of 22 channels in a narrow frequency band best suited for short-range, line-of-sight communications.

Citizens Band remains popular in the

Why Get A License?

Amateurs are free to choose from many types of radios and activities — that's what you get in return for passing the license exam. If you can learn the basics of radio and the rules of amateur radio, then the opportunities of ham radio are all yours! Just remember that the license is there to ensure that you understand the basics before transmitting. This helps keep amateur radio useful and enjoyable to everyone.

Why don't people just buy radios and transmit anyway? (This is called "bootlegging" or "pirating.") First of all, it's quite apparent to hams who has and who hasn't passed a license exam. You'll find yourself attracting the attention of the FCC, but more importantly, you won't fit in and you won't have fun.

One of the most important benefits to being licensed is that you have the right to be protected from interference by signals from unlicensed devices, such as consumer electronics. Your right to use the amateur bands is similarly protected. The protection doesn't work perfectly all the time, but nevertheless, as a licensed amateur operator your license is recognized by law. This is a big improvement over unlicensed radio users. It's definitely worth the effort to get that license!

Local, Remote, and Automatic Control

Local control — a control operator is physically present at the control point, the physical location from which the station controls are operated. This is the situation for nearly all amateur stations, including mobile operation. Any type of station can be locally controlled.

Remote operation — the control point is located away from the transmitter and the control operator adjusts or operates the transmitter indirectly via some kind of *control link*. The control operator must be present at the control point during all transmission. Many stations operate under remote control over an internet link. Any station can be remotely controlled.

Automatic operation — the station operates completely under the control of devices and procedures that ensure compliance with FCC rules. A control operator is still required, but need not be at the control point when the station is transmitting. Repeaters, beacons and space stations are allowed to be automatically controlled. Digipeaters that relay messages, such as for the APRS network, are also automatically controlled.

Table 1.2
Types of Personal Radio

Service	*Channels*	*Intended Use*	*Range*
Citizens Band (CB)	40	Private/Business	10 miles +
Marine VHF	50	Maritime	20 miles +
Family Radio Service (FRS)	22	Personal	2 miles
Multi-Use Radio Service (MURS)	5	Personal	5 miles +

Table 1.3

Amateur License Class Examinations

License Class	*Exam Element*	*Number of Questions*	*Privileges*
Technician	2	35 (passing grade is 26 correct)	All VHF and UHF privileges, with some HF privileges
General	3	35 (passing grade is 26 correct)	All VHF, UHF and most HF privileges
Amateur Extra	4	50 (passing grade is 37 correct)	All amateur privileges

Table 1.4

ARRL License Study Guides

Technician	*The ARRL Ham Radio License Manual ARRL's Tech Q&A*
General	*The ARRL General Class License Manual ARRL's General Q&A*
Amateur Extra	*The ARRL Extra Class License Manual ARRL's Extra Q&A*

Question pools are revised every four years. When new question pools are released, ARRL will produce new study materials before the effective date. Be sure to check the book cover or copyright page to ensure that you're using a study guide for the question pool currently in use.

Figure 1.23 — ARRL License Manuals are a great way to prepare for the Technician, General, and Extra class exams. Each book thoroughly covers the exam questions and topics.

The FCC Universal Licensing System (ULS)

The Universal Licensing System (ULS) and FCC registration and filing systems were deployed in the Amateur Service on August 16, 1999. To conduct business with the FCC, you must register through the FCC Commission Registration System (CORES) and be assigned an FCC Registration Number (FRN). This number will be used to uniquely identify you in all transactions with the FCC. You must do this before you can use ULS to get your first license, upgrade your license, renew your license, change your address, obtain a duplicate license, or use any of the other services that the FCC offers. Not sure if you have an FRN? Check your license to determine if it's been assigned an FRN.

The FCC has taken great strides to make the ULS and the CORES easy to use and has devoted a section of their website to the Amateur Service. Helpful web pages include:

- New Users Guide To Getting Started With Universal Licensing System (ULS)
- Common Amateur Filing Tasks
- Commission Registration System Video Tutorials

These web pages offer instruction on setting up a user account, registering for a FRN, and filing electronically via the ULS. For more information on the ULS and links to these FCC online resources, visit **www.arrl.org/universal-licensing-system**.

Figure 1.24 — Amateur radio licensing test sessions are administered by volunteer examiners — hams just like you. They grade the exams, help you fill out the necessary forms, and take care of all the paperwork for your ham radio license.

applications for which it was originally intended. Mobile radios in vehicles, boats, and farm equipment provide useful, medium-range radio communications to other vehicles or with radios at home or at work. Communication is fairly reliable over a range of several miles.

Boaters will be familiar with marine VHF radios used for boats to communicate with each other and with stations on shore. These radios can use up to 50 channels for communicating around harbors and for short-range needs during both fresh and salt-water travel.

All three of these services are designed to use a set of channels selected for a single type of communication as shown in **Table 1.2**. They do their designated job well. If you find your personal radio interesting, but limited, then amateur radio is definitely the place for you.

Amateurs have access to a much broader range of communications options and create new ways of communicating that are more powerful and flexible than those of the unlicensed radio services. We can build, modify, and repair our own radios. The procedures we use to communicate are completely up to us. We can operate however we want, with few restrictions. This flexibility, in order to not cause interference to other radio services, requires that amateurs be more knowledgeable than the typical user in other services. That is why amateurs have to pass a licensing examination.

AMATEUR LICENSES

Amateurs give and grade the exams themselves under the guidance of a Volunteer Examiner Coordinator (VEC). There are currently 14 different clubs or organizations recognized as VECs by the FCC. Each VEC also certifies Volunteer Examiners (VEs), who administer the exam sessions. The VEC then handles the paperwork for each license exam and application. The examination process is not as imposing as it may seem.

The result of passing the exam is an *operator license* (or "ticket") granted by the FCC after it receives the necessary paperwork from the VEC that administered your exam session. The license also specifies a call sign that becomes your radio identity.

There are three classes of license being granted today: the Technician, General, and Amateur Extra. The exam for each of the three license classes is called an *element*. Passing each of the elements grants the licensee more and more *privileges* allowed by the FCC's Amateur service rules. **Table 1.3** shows the elements and privileges for each of the Amateur license classes, and the US Amateur Bands chart at the front of this book shows the privileges by license class on the various frequency bands. (You may see references to Novice and Advanced amateur licenses. These have not been issued in many years, but some amateurs still hold them.)

1.3.2 Getting Your Ham Radio License

A comprehensive overview and many helpful resources for newcomers to ham radio may be found at **www.arrl.org/getting-licensed**. The Technician license is the first one for most people, and it requires passing a 35-question, multiple-choice exam on the rules of ham radio, simple operating procedures, and basic electronics.

The next step is the General license, which grants many more privileges, including access to large portions of the HF amateur bands and operation using all legal modes. The General exam is 35 questions about rules applicable to the expanded General privileges, more advanced electronics and radio theory, and operating procedures with a focus on modes and activities used on the HF bands.

The Amateur Extra license is the "all access pass" to amateur privileges. The exam is 50 questions, and it takes some dedica-

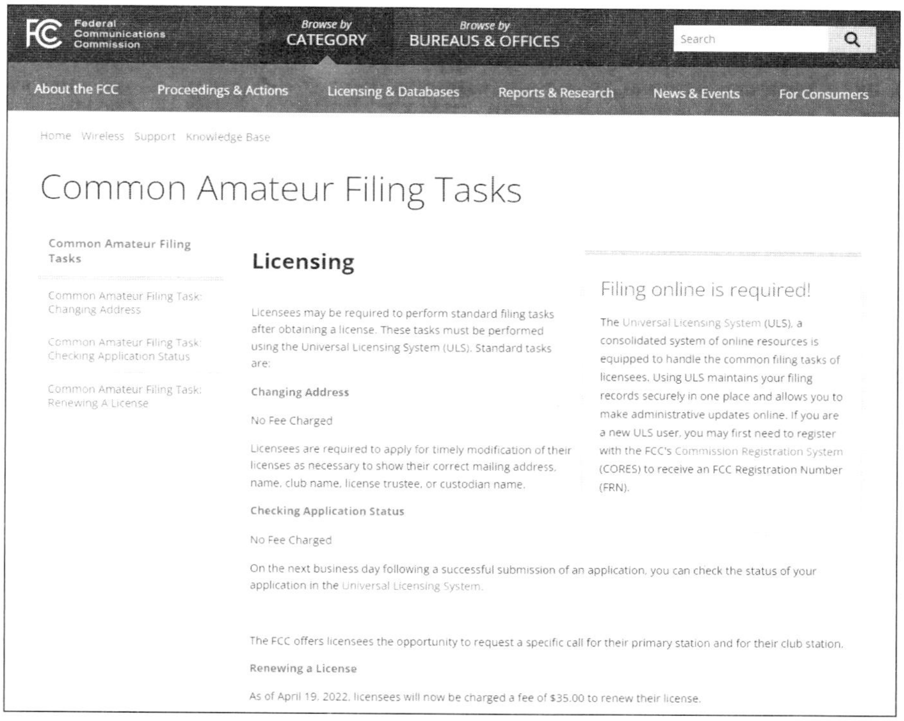

Figure 1.25 — The FCC website includes a Common Amateur Filing Tasks page to help licensees navigate the Universal Licensing System.

Operating from Other Countries

There are three basic types of agreements that allow amateurs licensed in one country to operate from the territory of another country.

- **European Conference of Postal and Telecommunications Administrations (CEPT) radio-amateur license** — allows US amateurs to travel to and operate from most European countries and their overseas territories without obtaining an additional license or permit. (This does not automatically confer permission to enter restricted areas. You must also carry with you a copy of FCC Public Notice DA 11-221.) Amateurs from countries that participate in CEPT may also operate in the US. The CEPT treaty does not automatically allow remote-control operation across international borders. Be sure you know the rules for transmitting in the host country!
- **International Amateur Radio Permit (IARP)** — For operation in certain countries of Central and South America, the IARP allows US amateurs to operate without seeking a special license or permit to enter and operate from that country.
- **ITU Reciprocal Permit** — a reciprocal agreement between the US and a country that does not participate in either CEPT or IARP agreements.

More information about obtaining permission to use your license elsewhere in the world is available online from **www.arrl.org/international-operating** or **www.arrl.org/us-amateurs-operating-overseas**.

Figure 1.26 — At Eisenhower Middle School in Lawton, Oklahoma, Jada, KF5TAT (left) and Kerson, KF5TAQ sit at the Viking Radio Club station in the classroom of teacher Clifton Harper, KE5YZB. The school received a ham radio equipment grant through the ARRL Foundation. Harper also attended an ARRL Education & Technology Program Teacher's Institute on Wireless Technology at ARRL Headquarters. [Pamely Harper, KF5JXO, photo]

Figure 1.27 — Then-13-year-old Aidan helped his grandfather, Chuck Schneebeli, KI9A, during the 2018 ARRL November Sweepstakes, which attracts thousands of hams to the airwaves each fall.

tion to learn about all of the advanced topics covered. Once you've passed the Extra exam, you're done with exams as long as you keep your license renewal up to date.

You can study on your own or you can enroll in a licensing class. Log on to the ARRL website where you can search for classes being held near you (**www.arrl.org/find-an-amateur-radio-license-class**). **Table 1.4** shows the appropriate ARRL study materials for the various exams. A note of caution: No matter what study guide you use, be sure that the questions used are current — the Technician, General, and Extra question pools are revised and updated every four years on a rotating schedule.

TESTING PROCESS

When you are comfortable with the study material, it's time to locate an exam session. If you're part of a study class, the instructor will make the necessary arrangements. For solo students, you can find a nearby exam session by visiting the ARRL website (**www.arrl.org/find-an-amateur-radio-license-exam-session**). Some organizations such as the Greater Los Angeles Area Repeater Group (GLAARG, **glaarg.org**) are offering online exam sessions to anyone.

Whether the exam is in-person or online, be sure to come prepared (see **www.arrl.org/what-to-bring-to-an-exam-session**) and contact the exam sponsors if you have any questions. Note that before the exam session, you *must* create an online account through the FCC Commission Registration System (CORES) and be assigned an FCC Registration Number (FRN). You will need to bring your FRN to the exam session, whether it's for your first license or an upgrade. See the sidebar, "The FCC Universal Licensing System (ULS)," for more information.

Once you're signed in at (or logged on to) the exam session, you pay the test fee (check with the test session administrator) and get ready. Amateur radio tests usually take less than an hour. You will be given a question sheet and an answer sheet or presented with an online screen. As you answer each question, mark a box on the answer sheet or select your desired answer. Once you've answered all of the questions, the volunteer examiners will grade and verify your test results.

Assuming you've passed (congratulations!) you'll fill out a Certificate of Successful Completion of Examination (CSCE) and a NCVEC Form 605. The exam organizers will submit your results to the FCC while you keep the CSCE as evidence that you've passed your latest test. Licensed hams who already have a call sign and are upgrading may use their new privileges right away. If this is your first amateur exam, in a week to 10 days your name and new call sign will appear in the FCC's database of licensees, and you can start transmitting. Although the FCC no longer routinely prints or mails license documents, you can print your own "official copy" from the FCC Universal Licensing System (ULS) site.

In 2022, the FCC introduced a $35 application fee that applies to new, renewal, rule waiver, and modification applications that request a new vanity call sign. The fee is per application and must be paid online directly to the FCC and not at your exam session. For details, see **www.arrl.org/fcc-application-fee**.

Once you have your license, you can operate from all 50 states, the District of Columbia, and all US territories. Your US amateur license is also recognized in many countries outside the US, although the specific requirements vary widely. See the sidebar "Operating from Other Countries."

1.3.3 Your Ham Radio Identity

A ham radio operator is known and recognized by a unique call sign (some hams shorten this to simply "call") that the FCC issues when granting your license. Your call sign not only identifies your station on the air, it's an individual ham radio identity, and many hams become better known by their call signs than by their names!

A call sign also identifies the issuing country. US call signs, for example, begin with W, K, N, or A. Canadian call signs usually start with VE or VA. See **www.arrl.org/**

international-call-sign-series for a list of worldwide call sign allocations.

US call signs start with one or two letters, a numeral from 0 to 9, and up to three more letters. The first part of a call sign is called the *prefix*. The part following the numeral is called the *suffix* and is unique to a specific licensee. For example, one well-known ham radio call sign is W1AW, assigned to the Hiram Percy Maxim Memorial Station at ARRL Headquarters in Newington, Connecticut. "W1" is the prefix and "AW" is the suffix. At one time, the numeral indicated a US station's geographical region — 9 or Ø (zero) for the Midwest, for example — but that's no longer the case. So, a call sign with "9" may belong to a ham located in Florida.

The shortest call signs — known as 1×2 or 2×1— are reserved for Extra class hams, such as NØAX, or NN1N. Technician call signs are the longest — 2×3, such as KA1JPA. Details of how amateur call signs are allocated for the various license classes may be found online at **www.fcc.gov/wireless/bureau-divisions/mobility-division/amateur-radio-service/amateur-call-sign-systems**.

You don't have to keep the call sign the FCC assigns. The FCC's *vanity call sign* program permits a ham to select a new call sign from among the database of certain unassigned call signs, based on the applicant's license class, and to file an application for it along with the application fee. See **www.arrl.org/vanity-call-signs** for details.

1.4 Resources

1.4.1 ARRL Resources

ARRL—the National Association for Amateur Radio
225 Main St.
Newington, CT 06111-1494
860-594-0200
Fax: 860-594-0259
email: **hq@arrl.org**
www.arrl.org

Prospective hams call
1-800-32 NEW HAM (1-800-326-3942)

ARRL is the membership organization for US ham radio operators and those interested in ham radio. ARRL publishes study guides for all amateur radio license classes, a monthly journal (*QST*), and many books on amateur radio and electronics. ARRL also sponsors a variety of activities to encourage amateurs to get on the air and use their radios.

Amateur Radio Service Rules & Regulations — FCC Part 97
Available on the ARRL website:
www.arrl.org/part-97-amateur-radio

ARRL Technical Information Service
Member resource for technical problems:
www.arrl.org/technical-information-service
Technical pages on various topics:
www.arrl.org/radio-technology-topics

Interference and RF Safety
All about interference to and from amateur stations: **www.arrl.org/radio-frequency-interference-rfi**
All about RF safety:
www.arrl.org/rf-exposure

Licensing and Training
For those new to ham radio:
www.arrl.org/new-ham-resources
All about training opportunities:
www.arrl.org/licensing-education-training
All about license exams:
www.arrl.org/getting-licensed
ARRL Learning Center for skill development: **learn.arrl.org**

Local Contacts, Clubs, and Events
ARRL Field Organization:
www.arrl.org/sections
Find a club nearby:
www.arrl.org/find-a-club
Find a hamfest or convention:
www.arrl.org/hamfests-and-conventions-calendar

Publications
ARRL offers a wide variety of books, periodicals, electronic publications, maps, kits, and other products of interest to radio amateurs at all levels.
www.arrl.org/shop

Station Construction
Overview:
www.arrl.org/setting-up-a-station
Grounding, bonding, lightning protection, electrical safety: **www.arrl.org/safety**
Buying equipment:
www.arrl.org/buying-your-first-radio
New equipment reviews, available to ARRL Members:
www.arrl.org/product-review
Antenna ideas:
www.arrl.org/building-simple-antennas

1.4.2 Other Organizations

AMSAT NA (The Radio Amateur Satellite Corporation)
Membership organization for those interested in amateur radio satellites.
www.amsat.org

Amateur Radio on the International Space Station, Inc (ARISS-USA)
Promotes amateur radio and science, technology, engineering, arts, and math within educational organizations; arranges ham radio contacts with ISS crew members for schools and educational organizations.
www.ariss-usa.org

Courage Kenny Handiham Program
Provides assistance to persons with disabilities who want to earn a ham radio license or set up a station.
handiham.org

Contents

Chapter 2
Electrical Fundamentals

Collecting material on fundamental concepts from previous editions, this chapter summarizes the basic ideas of electricity and electronics. It covers the physical quantities, elementary circuits, basic components, and the laws that govern their behavior. These are the foundations on which all of electronics is constructed. Glossaries are included for each group of topics, as well.

Since many of the basic ideas are expressed or defined in terms of mathematics, the collection of tutorials "Radio Mathematics" is available online in the Handbook's reference web page at **www.arrl.org/arrl-handbook-reference**. Tutorials on scientific notation and using a scientific calculator are also provided.

Chapter 2 — Online Content

Articles

- Hands On Radio: Kirchoff's Laws by Ward Silver, NØAX
- Hands On Radio: Laying Down the Laws by Ward Silver, NØAX
- Hands On Radio: Putting the Laws to Work by Ward Silver, NØAX
- Hands On Radio: Thevenin Equivalents by Ward Silver, NØAX

2.1 Introduction to Electricity

The *atom* is the primary building block of matter and is made up of a *nucleus* containing *protons* and *neutrons* surrounded by *electrons*. Protons have a positive electrical charge, electrons a negative charge, and neutrons have no electrical charge. An *element* (or *chemical element*) is a type of atom that has a specific number of protons, the element's *atomic number*. Each different element, such as iron, oxygen, silicon, or bromine, has a distinct chemical and physical identity determined primarily by the number of protons. A *molecule* is two or more atoms bonded together and acting as a single particle.

Unless modified by chemical, mechanical, or electrical processes, all atoms are electrically neutral because they have the same number of electrons as protons. If an atom loses electrons, it has more protons than electrons and thus has a net positive charge. If an atom gains electrons, it has more electrons than protons and a net negative charge. Atoms or molecules with a positive or negative charge are called *ions*. Electrons not bound to any atom, or *free electrons*, can also be considered as ions because they have a negative charge.

2.1.1 Electric Charge, Voltage, and Current

Any piece of matter that has a net positive or negative electrical charge is said to be *electrically charged*. An electrical force exists between electrically charged particles, pushing charges of the same type apart (like charges repel each other) and pulling opposite charges together (opposite charges attract). Moving charges in a magnetic field also generate an electrical force. This is the *electromotive force* (or EMF), the source of energy that causes charged particles to move. *Voltage* is the general term for the strength of the electromotive force or the difference in electrical potential between two points. Voltage and EMF are often used interchangeably in radio. A good diagram showing the relationship between EMF and voltage is available at **hyperphysics.phy-astr.gsu.edu/hbase/electric/elevol.html#c2**.

Under most conditions, the number of positive and negative charges in any volume of space is very close to balanced and so the region has no net charge. When there are extra positive ions in one region and extra negative ions (or electrons) in another region, the resulting EMF attracts the charges toward each other. The direction of the force, from the positive region to the negative region, is called its *polarity*. Because an imbalance of charge between two regions generates an EMF, its voltage is always measured between two points, with positive voltage defined as being in the direction from the positively charged to the negatively charged region.

If there is no path by which electric charge can move in response to an EMF (called a *conducting path*), the charges cannot move together and so remain separated. If a conducting path is available, then the electrons or ions will flow along the path, neutralizing the net imbalance of charge. The movement of electrical charge is called *electric current*. Materials through which current flows easily are called *conductors*. Most metals, such as copper or aluminum, are good conductors. Materials in which it is difficult for current to flow are *insulators*. *Semiconductors*, such as silicon or germanium, are materials with much poorer conductivity than metals. Semiconductors can be chemically altered to acquire properties that make them useful in solid-state devices such as diodes, transistors, and integrated circuits.

Voltage differences can be created in a variety of ways. For example, chemical ions can be physically separated to form a battery. The resulting charge imbalance creates a voltage difference at the battery terminals so that if a conductor is connected to both terminals at once, electrons flow between the terminals and gradually eliminate the charge imbalance, discharging the battery's stored energy. Mechanical means such as friction (static electricity, lightning) and moving conductors in a magnetic field (generators) can also produce voltages. Devices or systems that produce voltage are called *voltage sources*.

2.1.2 Electronic and Conventional Current

Electrons move in the direction of positive voltage — this is called *electronic current*. *Conventional current* takes the other point of view — of positive charges moving in the direction of negative voltage. Conventional current was the original model for electricity and results from an arbitrary decision made by Benjamin Franklin in the 18th century when the nature of electricity and atoms was still unknown. It can be imagined as electrons flowing "backward" and is completely equivalent to electronic current.

Conventional current is used in nearly all electronic literature and is the standard used in this book. The direction of conventional current establishes the polarity for most electronics calculations and circuit diagrams. The arrows in the drawing symbols for transistors point in the direction of conventional current, for example.

2.1.3 Units of Measurement

Measurement of electrical quantities is made in several standard units. Charge is measured in *coulombs* (C) and represented by q in equations. One coulomb is equal to the combined charge of 6.24×10^{18} electrons (or protons). Current, the flow of charge, is measured in *amperes* (A) and represented by i or I in equations. One ampere represents one coulomb of charge flowing past a point (or through a specific area) in one second so 1 A = 1 C/s. Electromotive force (EMF) is measured in *volts* (V) and represented by e, E, v, or V in equations. One volt is defined as the EMF required for one ampere of current to do one joule (J, a measure of energy) of work and 1 V = 1 J/C.

When Is E a V and V an E?

Beginners in electronics are often confused about the interchange of *V* and *E* to refer to voltage in a circuit. When should each be used? Unfortunately, there is no universal convention. *E* or *e* usually refers to force created by an electric or magnetic field. *E* is also commonly used in the equation for Ohm's Law: $I = E/R$. Capital V is used when describing a measured difference in voltage between two points in a circuit or the terminal voltage of a power supply or battery.

The Origin of Unit Names

Many units of measure carry names that honor scientists who made important discoveries in or advanced the state of scientific knowledge of electrical and radio phenomena. For example, Georg Ohm (1787–1854) discovered the relationship between current, voltage and resistance that now bears his name as Ohm's Law and as the unit of resistance, the ohm. The following table lists the most common electrical units, what they are used to measure, and the scientists for whom they are named. You can find more information on these and other notable scientists in encyclopedia entries on the units that bear their names.

Electrical Units and Their Namesakes

Unit	*Measures*	*Physical Quantities*	*Named for*
Ampere (A)	Current	Coulombs per second	Andree Ampere 1775 – 1836
Coulomb (C)	Charge	Quantity of charge	Charles Coulomb 1736 – 1806
Farad (F)	Capacitance	Coulombs per volt	Michael Faraday 1791 – 1867
Henry (H)	Inductance	Volts per amp per second	Joseph Henry 1797 – 1878
Hertz (Hz)	Frequency	Cycles per second	Heinrich Hertz 1857 – 1894
Joule (J)	Energy	Kg m^2 per s^2	James Joule 1818 – 1889
Ohm (Ω)	Resistance	Volts per amp	Georg Simon Ohm 1787 – 1854
Siemens (S)	Conductance	Ω^{-1}	Ernst von Siemens 1816 – 1892
Volt (V)	Voltage	Joules per coulomb	Alessandro Volta 1745 – 1827
Watt (W)	Power	Joules per second	James Watt 1736 – 1819

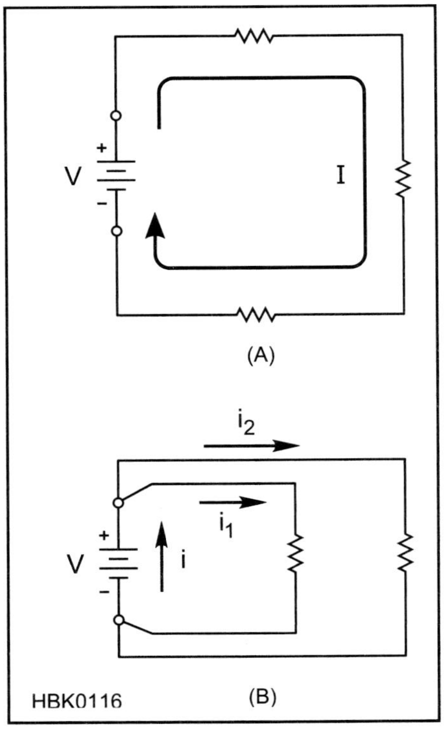

Figure 2.1 — A series circuit (A) has the same current through all components. Parallel circuits (B) apply the same voltage to all components.

2.1.4 Series and Parallel Circuits

A *circuit* is any conducting path through which current can flow between two points that have different voltages. An *open circuit* is a circuit in which a desired conducting path is interrupted, such as by a broken wire or a switch. A *short circuit* is a circuit in which a conducting path allows current to flow directly between the two points at different voltages.

The two fundamental types of circuits are shown in **Figure 2.1**. Part A shows a *series circuit* in which there is only one current path. The current in this circuit flows from the voltage source's positive terminal (the symbol for a battery is shown with its voltage polarity as + and –) in the direction shown by the arrow through three *resistors* (electronic components discussed later in this chapter) and back

Schematic Diagrams

The drawing in Figure 2.1 is a *schematic diagram*. Schematics are used to show the electrical connections in a circuit without requiring a drawing of the actual components or wires, called a *pictorial diagram*. Pictorials are fine for very simple circuits like these, but quickly become too detailed and complex for everyday circuits. Schematics use lines and dots to represent the conducting paths and connections between them. Individual electrical devices and electronic components are represented by *schematic symbols* such as the resistors shown here. A set of the most common schematic symbols is provided in the front matter at the beginning of this book. You will find additional information on reading and drawing schematic diagrams in the ARRL website Technology section at **www.arrl.org/circuit-construction**.

to the battery's negative terminal. Current is the same at every point in a series circuit.

Part B shows a *parallel circuit* in which there are multiple paths for the current to take. One end of each resistor is connected to the battery's positive terminal. The other end of each resistor is connected to the battery's negative terminal. Current flowing out of the battery's positive terminal divides into smaller currents that flow through the individual resistors and then recombine at the battery's negative terminal. All of the components in a parallel circuit experience the same voltage. All circuits are made up of series and parallel combinations of components and sources of voltage and current.

2.1.5 Direct and Alternating Current

A circuit is a complete conductive path for current to flow from a source, through a load and back to the source. *Direct current* or *dc* flows in only one direction. *Alternating current* or *ac* changes direction. **Figure 2.2** illustrates the two types of circuits. Circuit A shows the source as a battery, a typical dc source. Circuit B shows a voltage source symbol to indicate ac such as from a generator or household power outlet. In an ac circuit, both the current and the voltage reverse direction. For nearly all ac signals in electronics and radio, the reversal is *periodic*, meaning that the change in direction occurs on a regular basis. The rate of reversal may range from a few times per second to many billion times per second.

Graphs of current or voltage, such as Figure 2.2, begin with a horizontal axis that represents time. The vertical axis represents the amplitude of the current or the voltage, whichever is graphed. Distance above the zero line indicates larger positive amplitude; distance below the zero line means larger negative amplitude. Positive and negative only designate the opposing directions in which current may flow in an alternating current circuit or the opposing *polarities* of an ac voltage.

If the current and voltage never change direction, then we have a dc circuit, even if the level of dc constantly changes. **Figure 2.3A** shows a current that is always positive with respect to 0. It varies periodically in amplitude, however. Whatever the shape of the variations, the current can be called *pulsating dc*. If the current periodically reaches 0, it can be called *intermittent dc*.

We can also look at intermittent and pulsating dc as a combination of an ac and a dc current (Figures 2.3B and 2.3C). Special circuits can separate the two currents into ac and dc *components* for separate analysis or use. There are circuits that combine ac and dc currents and voltages, as well.

2.1.6 Glossary — Basic Electricity

Alternating current (ac) — A flow of charged particles through a conductor, first in one direction, then in the other direction.

Ampere — A measure of flow of charged particles per unit of time. One ampere (A) represents one coulomb of charge flowing past a point in one second.

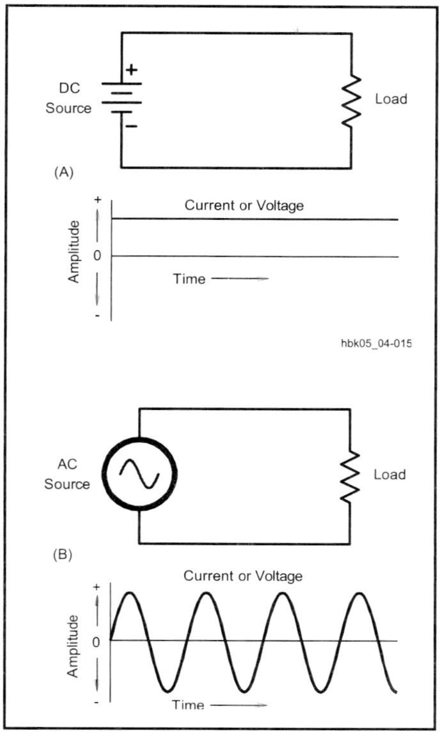

Figure 2.2 — Basic circuits for direct (A) and alternating (B) currents. With each circuit is a graph of the current, constant for the dc circuit, but periodically changing direction in the ac circuit.

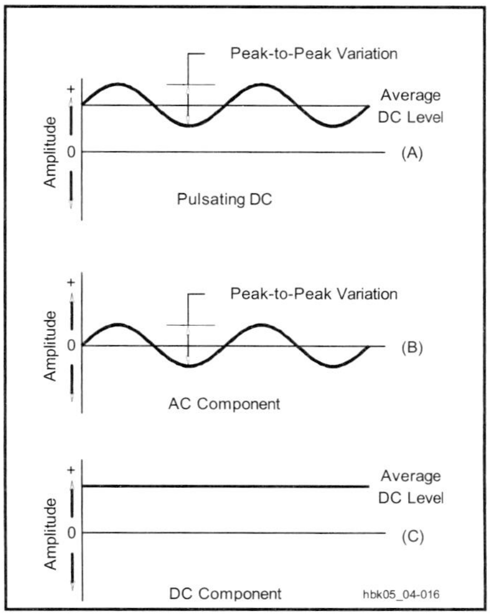

Figure 2.3 — A pulsating dc current (A) and its resolution into an ac component (B) and a dc component (C).

Atom — The smallest particle of matter that makes up a distinct chemical element. Atoms consist of protons and neutrons in the central region called the nucleus, and electrons orbiting the nucleus.

Circuit — Conducting path between two points of different voltage. In a *series circuit*, there is only one current path. In a *parallel circuit*, there are multiple current paths.

Conductor — Material in which electrons or ions can move easily.

Conventional current — Current defined as the flow of positive charges in the direction of positive to negative voltage. Conventional current flows in the opposite direction of electronic current, the flow of negative charges (electrons) from negative to positive voltage.

Coulomb — A unit of measure of a quantity of electrically charged particles. One coulomb (C) is equal to the combined charge of 6.24×10^{18} electrons (or protons).

Current (I) — The movement of electrical charge, measured in amperes and represented by i or I in equations.

Direct current (dc) — A flow of charged particles through a conductor in one direction only.

Electronic current — see ***Conventional Current***

Electromotive force (EMF) — The source of energy that creates a force between charged particles or regions. The term used to define the force of attraction or repulsion between electrically charged regions. Also see ***Voltage***.

Energy — Capability of doing work. It is usually measured in electrical terms as the number of watts of power consumed during a specific period of time, such as watt-seconds or kilowatt-hours.

Insulator — Material through which it is difficult for electrons or ions to move.

Ion — Atom or molecule with a positive or negative electrical charge.

Joule — Measure of a quantity of energy. One joule is defined as one newton (a measure of force) acting over a distance of one meter.

Polarity — The direction of EMF or voltage, from positive to negative.

Potential — See ***Voltage***

Power — Power is the rate at which work is done. One watt of power is equal to one volt of EMF causing a current of one ampere through a one ohm resistor.

Voltage — The general term for the difference in electrical potential energy between two points. Measured in volts or joules/coulomb.

Voltage source — Device or system that creates a voltage difference at its terminals.

2.2 Resistance and Conductance

2.2.1 Resistance and Resistors

Any conductor connected to points at different voltages will allow current to pass between the points. No conductor is perfect or lossless, at least not at normal temperatures. The moving electrons collide with the atoms making up the conductor and lose some of their energy by causing the atoms to vibrate, which is observed externally as heat. The property of energy loss due to interactions between moving charges and the atoms of the conductor is called *resistance*. The amount of resistance to current is measured in *ohms* (Ω) and is represented by *r* or *R* in equations.

Suppose we have two conductors of the same size and shape, but of different materials. Because all materials have different internal structures, the amount of energy lost by current flowing through the material is also different. The material's ability to impede current flow is its *resistivity*. Numerically, the resistivity of a material is given by the resistance, in ohms, of a cube of the material measuring one centimeter on each edge. The symbol for resistivity is the Greek letter rho, ρ.

The longer a conductor's physical path, the higher the resistance of that conductor. For direct current and low-frequency alternating currents (up to a few thousand hertz) the conductor's resistance is inversely proportional to the cross-sectional area of the conductor. Given two conductors of the same material and having the same length, but differing in cross-sectional area, the one with the larger area (for example, a thicker wire or sheet) will have the lower resistance.

One of the best conductors is copper, and it is frequently convenient to compare the resistance of a material under consideration with that of a copper conductor of the same size and shape. **Table 2.1** gives the ratio of the resistivity of various conductors to the resistivity of copper.

A package of material exhibiting a certain amount of resistance and made into a single unit or component is called a *resistor*. There are many types of resistors described in the **Circuits and Components** chapter, each suited to different applications and power levels. Next to the transistors built into microprocessors by the billion, resistors are the most commonly used electronic component.

2.2.2 Conductance

The reciprocal of resistance (1/R) is *conductance*. It is usually represented by the symbol G. A circuit having high conductance has low resistance, and vice versa. In radio work, the term is used chiefly in connection with electron-tube and field-effect transistor characteristics. The units of conductance are siemens (S). A resistance of 1 Ω has a conductance of 1 S, a resistance of 1000 Ω has a conductance of 0.001 S, and so on. A unit frequently used in regard to vacuum tubes and the field-effect transistor is the μS or one millionth of a siemens. It is the conductance of a 1M Ω resistance. Siemens have replaced the obsolete unit *mho* (abbreviated as an upside-down Ω symbol).

Table 2.1
Relative Resistivity of Metals

Material	*Resistivity Compared to Copper*
Aluminum (pure)	1.60
Brass	3.7-4.90
Cadmium	4.40
Chromium	8.10
Copper (hard-drawn)	1.03
Copper (annealed)	1.00
Gold	1.40
Iron (pure)	5.68
Lead	12.80
Nickel	5.10
Phosphor bronze	2.8-5.40
Silver	0.94
Steel	7.6-12.70
Tin	6.70
Zinc	3.40

Ohm's Law Timesaver

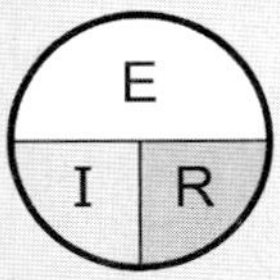

This simple diagram presents the mathematical equations relating voltage, current, and resistance. Cover the unknown quantity (E,I, or R) and the remaining symbols are shown as in the equation. For example, covering I shows E over R, as they would be written in the equation I=E/R.

When the current is small enough to be expressed in milliamperes, calculations are simplified if the resistance is expressed in kilohms rather than in ohms. With voltage in volts, if resistance in kilohms is substituted directly in Ohm's Law, the current will be milliamperes. Expressed as an equation: V = mA × kΩ.

2.2.3 Ohm's Law

The amount of current that will flow through a conductor when a given voltage is applied will vary with the resistance of the conductor. The lower the resistance, the greater the current for a given EMF. One ohm (Ω) is defined as the amount of resistance that allows one ampere of current to flow between two points that have a potential difference of one volt. This proportional relationship is known as *Ohm's Law*:

$$R = E / I$$

where

R = resistance in ohms,
E = voltage or EMF in volts, and
I = current in amperes.

Rearranging the equation gives the other common forms of Ohm's Law as:

$$E = I \times R$$

and

$$I = E / R$$

All three forms of the equation are used often in electronics and radio. You must remember that the quantities are in volts, ohms and amperes; other units cannot be used in the equations without first being converted. For example, if the current is in milliamperes you must first change it to the equivalent fraction of an ampere before substituting the value into the equations.

The following examples illustrate the use of Ohm's Law in the simple circuit of **Figure 2.4**. If 150 V is applied to a circuit and the current is measured as 2.5 A, what is the resistance of the circuit? In this case R is the unknown, so we will use:

$$R = \frac{E}{I} = \frac{150\ \text{V}}{2.5\ \text{A}} = 60\ \Omega$$

No conversion of units was necessary because the voltage and current were given in volts and amperes.

If the current through a 20 kΩ resistance is 150 mA, what is the voltage across that resistor? To find voltage, use E = I × R. Convert the current from milliamperes to amperes by dividing by 1,000 mA / A (or multiplying by 10^{-3} A / mA) so that 150 mA becomes 0.150 A. Likewise, convert the 20 kΩ resistance to 20,000 Ω, since kΩ represents 1,000 Ω. (Notice the conversion factor of 1,000 does not limit the number of significant figures in the calculated answer.)

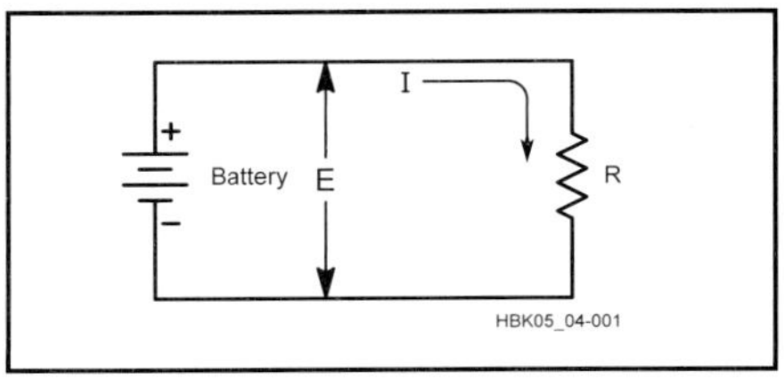

Figure 2.4 — A simple circuit consisting of a battery and a resistor.

$$I = \frac{150 \text{ mA}}{1000 \frac{\text{mA}}{\text{A}}} = 0.150 \text{ A}$$

Then:

$$E = 0.150 \text{ A} \times 20{,}000\ \Omega = 3{,}000 \text{ V}$$

In a final example, how much current will flow if 250 V is applied to a 5 kΩ resistor, which is equal to 5,000 Ω. Since I is unknown,

$$I = \frac{E}{R} = \frac{250 \text{ V}}{5{,}000\ \Omega} = 0.05 \text{ A}$$

This value of current is more conveniently stated in mA, and 0.05 A × 1,000 mA / A = 50 mA.

It is important to note that Ohm's Law applies in any portion of a circuit as well as to the circuit as a whole. No matter how many resistors are connected together or how they are connected together, the relationship between the resistor's value, the voltage across the resistor, and the current through the resistor still follows Ohm's Law.

2.2.4 Glossary — Conductance and Resistance

Conductance (G) — The reciprocal of resistance, measured in siemens (S).

Ohm (Ω) — Unit of resistance. One ohm is defined as the resistance that will allow one ampere of current when one volt of EMF is impressed across the resistance.

Ohm's law — The expression that describes resistance (R) as the proportional relationship between voltage (E) and current (I); R = E / I.

Resistance (R) — Opposition to current by conversion into other forms of energy, such as heat, measured in ohms (Ω).

2.3 Basic Circuit Principles

Circuits are composed of *nodes* and *branches*. A node is any point in the circuit at which current can divide between conducting paths. For example, in the parallel circuit of **Figure 2.5A**, the node is represented by the schematic dot. A branch is any unique conducting path between nodes. A series of branches that make a complete current path, such as the series circuit of Figure 2.5B, is called a *loop*.

Very few actual circuits are as simple as those shown in Figure 2.5. However, all circuits, no matter how complex, are constructed of combinations of these series and parallel circuits. We will now use these simple circuits of resistors and batteries to illustrate two fundamental rules for voltage and current, known as *Kirchoff's Laws*.

2.3.1 Kirchhoff's Current Law

Kirchhoff's Current Law (KCL) states, *"The sum of all currents flowing into a node and all currents flowing out of a node is equal to zero."* KCL is stated mathematically as:

$$(I_{in1} + I_{in2} + ...) - (I_{out1} + I_{out2} + ...) = 0$$

The dots (...) indicate that as many currents as necessary may be added.

An equivalent way of stating KCL is that the sum of all currents flowing into a node must balance the sum of all currents flowing out of the node:

$$(I_{in1} + I_{in2} + ...) = (I_{out1} + I_{out2} + ...)$$

KCL is illustrated by the following example. Suppose three resistors (R1 = 5.0 kΩ, R2 = 20.0 kΩ, and R3 = 8.0 kΩ) are connected in parallel as shown in Figure 2.5A. The same voltage, 250 V, is applied to all three resistors. The current through R1 is I1, I2 is the current through R2, and I3 is the current through R3.

The current in each can be found from Ohm's Law, as shown below. Note that when we use units of kΩ (1,000 Ω) we must balance that with units of mA (milliamperes, 1/1,000 A = 0.001 A).

$$I1 = \frac{E}{R1} = \frac{250 \text{ V}}{5.0 \text{ k}\Omega} = 50.0 \text{ mA}$$

$$I2 = \frac{E}{R2} = \frac{250 \text{ V}}{20.0 \text{ k}\Omega} = 12.5 \text{ mA}$$

$$I3 = \frac{E}{R3} = \frac{250 \text{ V}}{8.0 \text{ k}\Omega} = 31.2 \text{ mA}$$

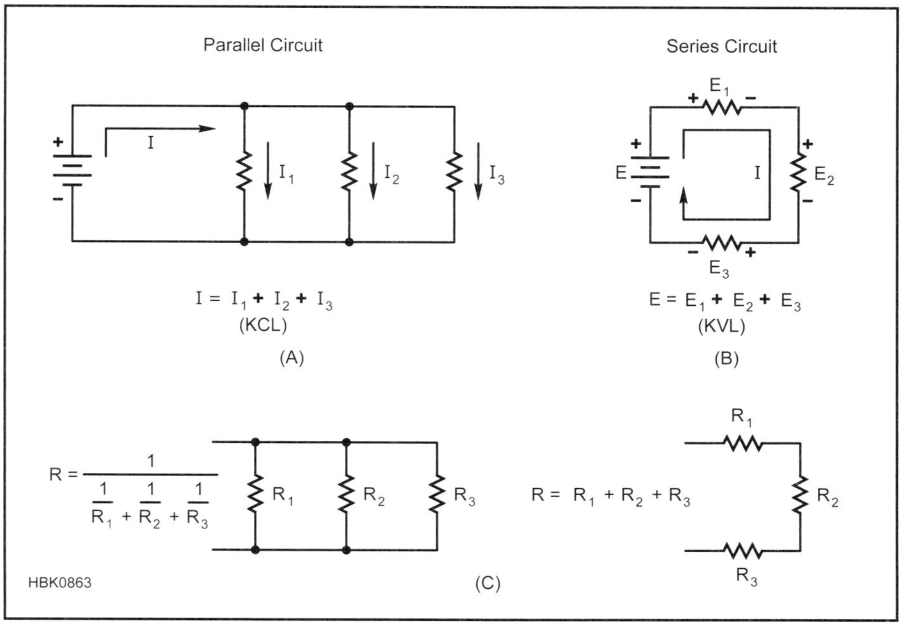

Figure 2.5 — An example of resistors in parallel (A) and series (B). In series circuits, the current is the same in all components, and voltages are summed. In parallel circuits, voltage across all components is the same and the sum of currents into and out of circuit junctions must be equal. (C) shows how to calculate equivalent values for series and parallel combinations.

Notice that the branch currents are inversely proportional to the resistances. The 20 kΩ resistor has a resistance four times larger than the 5 kΩ resistor, and has a current one-quarter as large. If a resistor has a value twice as large as another, it will have half as much current through it when they are connected in parallel.

Using the balancing form of KCL the current that must be supplied by the battery is:

$$I_{Batt} = I1 + I2 + I3$$

$$I_{Batt} = 50.0 \text{ mA} + 12.5 \text{ mA} + 31.2 \text{ mA}$$

$$I_{Batt} = 93.7 \text{ mA}$$

2.3.2 Resistors in Parallel

In a circuit made up of resistances in parallel, the resistors can be represented as a single *equivalent* resistance that has the same value as the parallel combination of resistors. In a parallel circuit, the equivalent resistance is less than that of the lowest resistance value present. This is because the total current is always greater than the current in any individual resistor. The formula for finding the equivalent resistance of resistances in parallel is:

$$R_{EQUIV} = \frac{1}{\frac{1}{R1} + \frac{1}{R2} + \frac{1}{R3} + \frac{1}{R4} \ldots}$$

where the dots (...) indicate that any number of parallel resistors can be combined by the same method. The equation is often referred to as the "reciprocal of reciprocals." The left side of Figure 2.5C shows the general rule on a schematic.

In the example of the previous section, the equivalent resistance is:

$$R = \frac{1}{\frac{1}{5\text{ k}\Omega} + \frac{1}{20\text{ k}\Omega} + \frac{1}{8\text{ k}\Omega}} = 2.67\text{ k}\Omega$$

The notation "//" (two slashes) is frequently used to indicate "in parallel with." Using that notation, the preceding example would be given as "5.0 kΩ // 20 kΩ // 8.0 kΩ."

If all the resistors in parallel have the same value, divide the resistor value by the number of resistors, N, to get the parallel resistance. For example, for five 50 Ω resistors in parallel, the equivalent resistance is R = 50 / N = 50 / 5 = 10 Ω.

For only two resistances in parallel (a very common case) the formula can be reduced to the much simpler (and easier to remember):

$$R_{EQUIV} = \frac{R1 \times R2}{R1 + R2}$$

Example: If a 500 Ω resistor is connected in parallel with a 1200 Ω resistor, what is the total resistance?

$$R = \frac{R1 \times R2}{R1 + R2} = \frac{500\ \Omega \times 1{,}200\ \Omega}{500\ \Omega + 1{,}200\ \Omega}$$

$$R = \frac{600{,}000\ \Omega^2}{1{,}700\ \Omega} = 353\ \Omega$$

Any number of parallel resistors can be combined two at a time by using this equation until all have been combined into a single equivalent.

Another useful equation finds the value of a parallel resistor, R_{PAR}, that when connected across a known resistor, R_1, creates a desired equivalent resistance, R_{DES}:

$$R_{PAR} = \frac{R_1 R_{DES}}{R_1 - R_{DES}}$$

For example, what value of resistor connected across a 10 kΩ resistor results in a 3 kΩ equivalent resistance?

$$R_{PAR} = \frac{10\text{ k}\Omega \times 3\text{ k}\Omega}{10\text{ k}\Omega - 3\text{ k}\Omega} = 4.28\text{ k}\Omega$$

If a calculated resistor value is not available, use the closest manufactured value, in this case, with 5% resistors, 4.3 kΩ is the closest available value.

2.3.3 Kirchhoff's Voltage Law

Kirchhoff's Voltage Law (KVL) states, *"The sum of the voltages around a closed current loop is zero."* Where KCL is somewhat intuitive, KVL is not as easy to visualize. In the circuit of Figure 2.5B, KVL requires that the battery's voltage must be balanced exactly by the voltages that appear across the three resistors in the circuit. If it were not, the "extra" voltage would create an infinite current with no limiting resistance, just as KCL prevents charge from "building up" at a circuit node. KVL is stated mathematically as:

$$E_1 + E_2 + E_3 + \ldots = 0$$

where each E represents a voltage encountered by current as it flows around the circuit loop.

This is best illustrated with an example. Although the current is the same in all three of the resistances in the previous example, the total voltage divides between them, just as current divides between resistors connected in parallel. The voltage appearing across each resistor (the *voltage drop*) can be found from Ohm's Law. (Voltage across a resistance is often referred to as a "drop" or "I-R drop" because the value of the voltage "drops" by the amount E = I × R.)

For the purpose of KVL, it is common to assume that if current flows *into* the more positive terminal of a component the voltage is treated as positive in the KVL equation. If the current flows *out* of a positive terminal, the voltage is treated as negative in the KVL. Positive voltages represent components that consume or "sink" power, such as resistors. Negative voltages represent components that produce or "source" power, such as batteries. This allows the KVL equation to be written in a balancing form, as well:

$$(E_{source1} + E_{source2} + \ldots) = (E_{sink1} + E_{sink2} + \ldots)$$

All of the voltages are treated as positive in this form, with the power sources (current flowing *out* of the more positive terminal) on one side and the power sinks (current flowing *into* the more positive terminal) on the other side.

Note that it doesn't matter what a component terminal's *absolute* voltage is with respect to ground, only which terminal of the component is more positive than the other. If one side of a resistor is at +1,000 V and the other at +998 V, current flowing into the first terminal and out of the second experiences a +2 V voltage drop. Similarly, current supplied by a 9 V battery with its positive terminal at –100 V and its negative terminal at –108.5 V still counts for KVL as an 8.5 V power source. Also note that current can flow *into* a battery's positive terminal, such as during recharging, making the battery a power sink, just like a resistor.

Here's an example showing how KVL works: In Figure 2.5B, if R1 is 5 kΩ, R2 is 20 kΩ, R3 is 8 kΩ, and the battery voltage is 250 V, the voltage across R1 is E_1, that across R2 is E_2 and that across R3 is E_3, then:

$$-250 + I \times R1 + I \times R2 + I \times R3 = 0$$

Since the same current flows through all three resistors, this equation can be simplified to:

$$-250 + I\,(5{,}000 + 20{,}000 + 8{,}000) =$$

$$-250 + I\,(33{,}000\ \Omega) = 0$$

Solving for I gives I = 250 / 33,000 = 0.00758 A = 7.58 mA. This allows us to calculate the value of the voltage across each resistor:

$$E_1 = I \times R1 = 0.00758\text{ A} \times 5{,}000\ \Omega = 37.9\text{ V}$$

$$E_2 = I \times R2 = 0.00758\text{ A} \times 20{,}000\ \Omega = 152\text{ V}$$

$$E_3 = I \times R3 = 0.00758\text{ A} \times 8{,}000\ \Omega = 60.6\text{ V}$$

Verifying that the sum of E_1, E_2, and E_3 does indeed equal the battery voltage of 250 V ignoring rounding errors:

$$E_{BATTERY} = E_1 + E_2 + E_3$$

$$E_{BATTERY} = 37.9\text{ V} + 152\text{ V} + 60.6\text{ V}$$

$$E_{BATTERY} = 250\text{ V}$$

2.3.4 Resistors in Series

The previous example illustrated that in a circuit with a number of resistances connected in series, the equivalent resistance of the circuit is the sum of the individual resistances. If these are numbered R1, R2, R3 and so on, then:

$$R_{EQUIV} = R1 + R2 + R3 + R4 \ldots$$

The right side of Figure 2.5C shows the general rule on a schematic.

Example: Suppose that three resistors are connected to a source of voltage as shown in Figure 2.5B. The voltage is 250 V, R1 is 5.0 kΩ, R2 is 20.0 kΩ and R3 is 8.0 kΩ. The total resistance is then

$$R_{EQUIV} = R1 + R2 + R3$$

$$R_{EQUIV} = 5.0\ k\Omega + 20.0\ k\Omega + 8.0\ k\Omega$$

$$R_{EQUIV} = 33.0\ k\Omega$$

The current in the circuit is then

$$I = \frac{V}{R} = \frac{250\ V}{33.0\ k\Omega} = 7.58\ mA$$

2.3.5 Conductances in Series and Parallel

Since conductance is the reciprocal of resistance, G = 1/R, the formulas for combining resistors in series and in parallel can be converted to use conductance by substituting 1/G for R. Conductances in series are thus combined similarly to resistors in parallel:

$$G = \frac{1}{\frac{1}{G1} + \frac{1}{G2} + \frac{1}{G3} + \frac{1}{G4} \ldots}$$

and two conductances in series may be combined in a manner similar to two parallel resistors:

$$G_{EQUIV} = \frac{G1 \times G2}{G1 + G2}$$

Conductances in parallel are combined similarly to resistances in series:

$$G_{TOTAL} = G1 + G2 + G3 + G4 \ldots$$

This also shows that when faced with a large number of parallel resistances, converting them to conductances may make the math a little easier to deal with.

2.3.6 Equivalent Circuits

A circuit may have resistances both in parallel and in series, as shown in **Figure 2.6A**. In

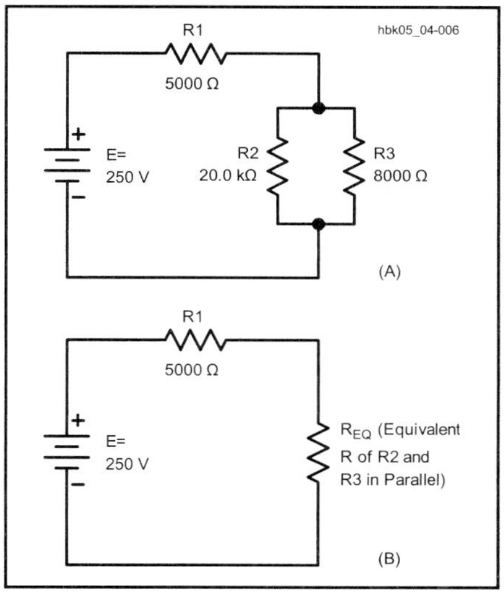

Figure 2.6 — At (A), an example of resistors in series-parallel. The equivalent circuit is shown at (B).

order to analyze the behavior of such a circuit, *equivalent circuits* are created and combined by using the equations for combining resistors in series and resistors in parallel. Each separate combination of resistors, series or parallel, can be reduced to a single equivalent resistor. The resulting combinations can be reduced still further until only a single resistor remains.

The simplest process begins with combining any two of the resistors into a single equivalent resistance using the formulas for series or parallel resistances. Then combine the resulting equivalent resistance with any single remaining resistor into a new equivalent resistance. Repeat the process of combining the equivalent resistance with a single resistor until all resistances have been combined into a single equivalent resistance. For example, to find the equivalent resistance for the circuit in Figure 2.5A: Combine R2 and R3 to create the equivalent single resistor, R_{EQ}, whose value is equal to R2 and R3 in parallel.

$$R_{EQ} = \frac{R2 \times R3}{R2 + R3} = \frac{20000\ \Omega \times 8000\ \Omega}{20000\ \Omega + 8000\ \Omega}$$

$$= \frac{1.60 \times 10^8\ \Omega^2}{28000\ \Omega} = 5710\ \Omega = 5.71\ k\Omega$$

This resistance in series with R1 then forms a simple series circuit, as shown in Figure 2.5B. These two resistances can then be combined into a single equivalent resistance, R_{TOTAL}, for the entire circuit:

$$R_{TOTAL} = R1 + R_{EQ} = 5.0\ k\Omega + 5.71\ k\Omega$$

$$R_{TOTAL} = 10.71\ k\Omega$$

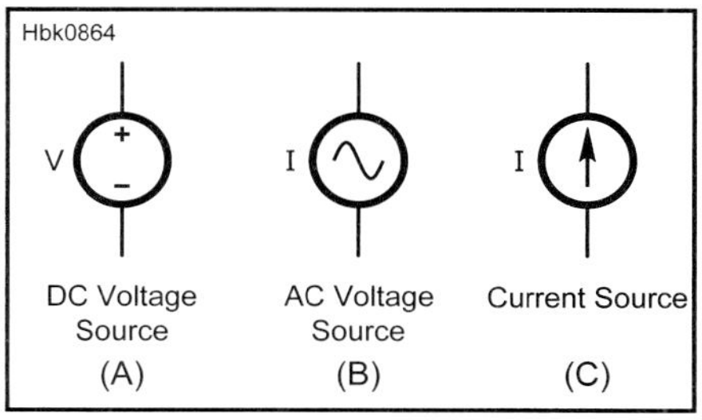

Figure 2.7— Voltage sources for dc (A) and ac (B) and current source (C) are examples of ideal energy sources.

The battery current is then:

$$I = \frac{E}{R} = \frac{250\ V}{10.71\ k\Omega} = 23.3\ mA$$

The voltage drops across R1 and R_{EQ} are:

$$E1 = I \times R1 = 23.3\ mA \times 5.0\ k\Omega = 117\ V$$

$$E2 = I \times R_{EQ} = 23.3\ mA \times 5.71\ k\Omega = 133\ V$$

These two voltage drops total 250 V, as described by Kirchhoff's Voltage Law. E2 appears across both R2 and R3 so,

$$I2 = \frac{E2}{R2} = \frac{133\ V}{20.0\ k\Omega} = 6.65\ mA$$

$$I3 = \frac{E3}{R3} = \frac{133\ V}{8.0\ k\Omega} = 16.6\ mA$$

where

I2 = current through R2, and
I3 = current through R3.

The sum of I2 and I3 is equal to 23.3 mA, conforming to Kirchhoff's Current Law.

2.3.7 Voltage and Current Sources

In designing circuits and describing the behavior of electronic components, it is often useful to use *ideal sources*. The two most common types of ideal sources are the *voltage source* and the *current source*, symbols for which are shown in **Figure 2.7**. These sources are considered ideal because no matter what circuit is connected to their terminals, they continue to supply the specified amount of voltage or current. Practical voltage and current sources can approximate the behavior of an ideal source over certain ranges, but are limited in the amount of power they can supply, and so under excessive load their output will drop.

Voltage sources are defined as having zero *internal impedance*, where impedance is a more general form of resistance as described in the sections of this chapter dealing with alternating current. A short circuit across an ideal voltage source would result in the source providing an infinite amount of current. Practical voltage sources have non-zero internal impedance, and this also limits the amount of power they can supply. For example, placing a short circuit across the terminals of a practical voltage source such as 1.5 V dry-cell battery may produce a current of several amperes, but the battery's internal impedance acts to limit the amount of current produced in accordance with Ohm's Law — as if resistor R1 in Figure 2.6 was inside of or internal to the battery.

Current sources are defined to have infinite internal impedance. This means that no matter what is connected to the terminals of an ideal current source, it will supply the same amount of current. An open circuit across the terminal of an ideal current source will result in the source generating an infinite voltage at its terminals. Practical current sources will raise their voltage until the internal power supply limits are reached and then reduce output current.

2.3.8 Thevenin's Theorem and Thevenin Equivalents

Thevenin's Voltage Theorem (usually just referred to as "Thevenin's Theorem") is a useful tool for simplifying electrical circuits or *networks* (the formal name for circuits) by allowing circuit designers to replace a circuit with a simpler equivalent circuit. Thevenin's Theorem states, "*Any two-terminal network made up of resistors and voltage or current sources can be replaced by an equivalent network made up of a single voltage source and a series resistor.*"

Thevenin's Theorem can be readily applied to the circuit of Figure 2.6A, to find the current through R3. In this example, illustrated in **Figure 2.8**, the circuit is redrawn to show R1 and R2 forming a voltage divider, with R3 as the load (Figure 2.8A). The current drawn by the load (R3) is simply the voltage across R3, divided by its resistance. Unfortunately, the value of R2 affects the voltage across R3, just as the presence of R3 affects the voltage appearing across R2. Some means of separating the two is needed; hence the *Thevenin-equivalent circuit* is constructed, replacing everything connected to terminals A and B with a single voltage source (the *Thevenin-equivalent voltage,* E_{THEV}) and series resistor (the *Thevenin-equivalent resistance,* R_{TH}).

The first step of creating the Thevenin-equivalent of the circuit is to determine its *open-circuit voltage,* measured when there is no load current drawn from either terminal A or B. Without a load connected between A and B, the total current through the circuit is (from Ohm's Law):

$$I = \frac{E}{R1 + R2}$$

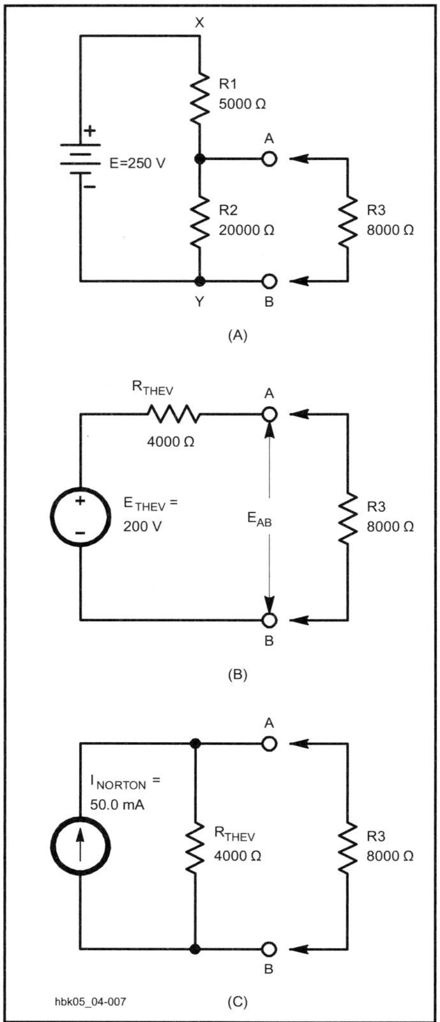

Figure 2.8 — Equivalent circuits for the circuit shown in Figure 2.6. (A) shows the circuit to be replaced by an equivalent circuit from the perspective of the resistor (R3 load). (B) shows the Thevenin-equivalent circuit, with a resistor and a voltage source in series. (C) shows the Norton-equivalent circuit, with a resistor and current source in parallel.

and the voltage between terminals A and B (E_{AB}) is:

$$E_{AB} = I \times R2$$

By substituting the equation for current for I, we have an expression for E_{AB} in which all values are known:

$$E_{AB} = \frac{R2}{R1 + R2} \times E$$

Using the values in our example, this becomes:

$$E_{AB} = \frac{20.0\ k\Omega}{25.0\ k\Omega} \times 250\ V = 200\ V$$

when nothing is connected to terminals A or B. E_{THEV} is equal to E_{AB} with no current drawn.

The equivalent resistance between terminals A and B is R_{THEV}. R_{THEV} is calculated as the equivalent circuit at terminals A and B with all sources, voltage or current, replaced by their internal impedances. The ideal voltage source, by definition, has zero internal resistance and is replaced by a short circuit. The ideal current source has infinite internal impedance and is replaced by an open circuit.

Assuming the battery to be a close approximation of an ideal source, replace it with a short circuit between points X and Y in the circuit of Figure 2.8A. R1 and R2 are then effectively placed in parallel, as viewed from terminals A and B. R_{THEV} is then:

$$R_{THEV} = \frac{R1 \times R2}{R1 + R2}$$

$$R_{THEV} = \frac{5{,}000\ \Omega \times 20{,}000\ \Omega}{5{,}000\ \Omega + 20{,}000\ \Omega}$$

$$R_{THEV} = \frac{1.0 \times 10^8\ \Omega^2}{25{,}000\ \Omega} = 4{,}000\ \Omega$$

This gives the Thevenin-equivalent circuit as shown in Figure 2.8B. The circuits of Figures 2.8A and 2.8B are completely equivalent from the perspective of R3, so the circuit becomes a simple series circuit.

Once R3 is connected to terminals A and B, there will be current through R_{THEV}, causing a voltage drop across R_{THEV} and reducing E_{AB}. The current through R3 is equal to

$$I3 = \frac{E_{THEV}}{R_{TOTAL}} = \frac{E_{THEV}}{R_{THEV} + R3}$$

Substituting the values from our example:

$$I3 = \frac{200\ V}{4{,}000\ \Omega + 8{,}000\ \Omega} = 16.7 mA$$

This agrees with the value calculated earlier.

The Thevenin-equivalent circuit of an ideal voltage source in series with a resistance is a good model for a real voltage source with non-zero internal resistance. Using this more realistic model, the maximum current that a real voltage source can deliver is seen to be

$$I_{sc} = \frac{E_{THEV}}{R_{THEV}}$$

and the maximum output voltage is $V_{oc} = E_{THEV}$.

Sinusoidal voltage or current sources can be modeled in much the same way, keeping in mind that the internal impedance, Z_{THEV}, for such a source may not be purely resistive, but may have a reactive component that varies with frequency.

2.3.9 Norton's Theorem and Norton Equivalents

Norton's Theorem is another method of creating an equivalent circuit. Norton's Theorem states, "*Any two-terminal network made up of resistors and current or voltage sources can be replaced by an equivalent network made up of a single current source and a parallel resistor.*" Norton's Theorem is to current sources what Thevenin's Theorem is to voltage sources. In fact, the Thevenin-resistance calculated previously is also the *Norton-equivalent resistance*.

The circuit just analyzed by means of Thevenin's Theorem can be analyzed just as easily by Norton's Theorem. The equivalent Norton circuit is shown in Figure 2.8C. The short circuit current of the equivalent circuit's current source, I_{NORTON}, is the current through terminals A and B with the load (R3) replaced by a short circuit. In the case of the voltage divider shown in Figure 2.8A, the short circuit completely bypasses R2 and the current is:

$$I_{AB} = \frac{E}{R1}$$

Substituting the values from our example, we have:

$$I_{AB} = \frac{E}{R1} = \frac{250\ V}{5{,}000\ \Omega} = 50.0 mA$$

The resulting Norton-equivalent circuit consists of a 50.0 mA current source placed in parallel with a 4000 Ω resistor. When R3 is connected to terminals A and B, one-third of the supply current flows through R3 and the remainder through R_{THEV}. This gives a current through R3 of 16.7 mA, again agreeing with previous conclusions.

A Norton-equivalent circuit can be transformed into a Thevenin-equivalent circuit and vice versa. The equivalent resistor, R_{THEV}, is the same in both cases; it is placed in series with the voltage source in the case of a Thevenin-equivalent circuit and in parallel with the current source in the case of a Norton-equivalent circuit. The voltage for the Thevenin-equivalent source is equal to the open-circuit voltage appearing across the resistor in the Norton-equivalent circuit. The current for a Norton-equivalent source is equal to the short circuit current provided by the Thevenin source. A Norton-equivalent circuit is a good model for a real current source that has a less than infinite internal impedance.

2.4 Power and Energy

Regardless of how voltage is generated, energy must be supplied if current is drawn from the voltage source. The energy supplied may be in the form of chemical energy or mechanical energy. This energy is measured in *joules* (J). One joule is defined from classical physics as the amount of energy or *work* done when a force of one newton (a measure of force) is applied to an object that is moved one meter in the direction of the force.

Power is another important concept and measures the rate at which energy is generated or used. One *watt* (W) of power is defined as the generation (or use) of one joule of energy (or work) per second.

One watt is also defined as one volt of EMF causing one ampere of current to flow through a resistance. Thus,

$$P = I \times E$$

where

P = power in watts,
I = current in amperes, and
E = EMF in volts.

(This discussion pertains only to direct current in resistive circuits. See the **Radio Fundamentals** chapter for a discussion about power in ac circuits, including reactive circuits.)

Common fractional and multiple units for power are the milliwatt (mW, one thousandth of a watt) and the kilowatt (kW, 1000 W).

Example: The plate voltage on a transmitting vacuum tube is 2000 V and the plate current is 350 mA. (The current must be changed to amperes before substitution in the formula, and so is 0.350 A.) Then:

$$P = I \times E = 2{,}000\ V \times 0.350\ A = 700\ W$$

Power may be expressed in *horsepower* (hp) instead of watts, using the following conversion factor:

1 horsepower = 746 W

This conversion factor is especially useful if you are working with a system that converts electrical energy into mechanical energy, and vice versa, since mechanical power is often expressed in horsepower in the U.S. In metric countries, mechanical power is usually expressed in watts. All countries use the metric power unit of watts in electrical systems, however. The value 746 W/hp assumes lossless conversion between mechanical and electrical power; practical efficiency is taken up shortly.

2.4.1 Energy

When you buy electricity from a power company, you pay for electrical energy, not power. What you pay for is the work that the electrical energy does for you, not the rate at which that work is done. Like energy, work is equal to power multiplied by time. The common unit for measuring electrical energy is the *watt-hour* (Wh), which means that a power of one watt has been used for one hour. That is:

$$Wh = P \times t$$

where

Wh = energy in watt-hours,
P = power in watts, and
t = time in hours.

Actually, the watt-hour is a fairly small energy unit, so the power company bills you for *kilowatt-hours* (kWh) of energy used. Another energy unit that is sometimes useful is the *watt-second* (Ws), which is equivalent to joules.

It is important to realize, both for calculation purposes and for efficient use of power resources, that a small amount of power used for a long time can eventually result in a power bill that is just as large as if a large amount of power had been used for a very short time.

A common use of energy units in radio is in specifying the energy content of a battery. Battery energy is rated in *ampere-hours* (Ah) or *milliampere-hours* (mAh). While the multiplication of amperes and hours does not result in units of energy, the calculation assumes the result is multiplied by a specified (and constant) battery voltage. For example, a rechargeable NiMH battery rated to store 2,500 mAh of energy is assumed to supply that energy at a terminal voltage of 1.2 V. Thus, after converting 2,500 mA to 2.5 A, the actual energy stored is:

$$\text{Energy} = 1.2 \text{ V} \times 2.5 \text{ A} \times 1 \text{ hour} = 3 \text{ Wh}$$

Another common energy unit associated with batteries is *energy density*, with units of Ah per unit of volume or weight.

One practical application of energy units is to estimate how long a radio (such as a hand-held unit) will operate from a certain battery. For example, suppose a fully charged battery stores 900 mAh of energy and that the radio draws 30 mA on receive. A simple calculation indicates that the radio will be able receive 900 mAh / 30 mA = 30 hours with this battery, assuming 100% efficiency. You shouldn't expect to get the full 900 mAh out of the battery because the battery's voltage will drop as it is discharged, usually causing the equipment it powers to shut down before the last fraction of charge is used. Any time spent transmitting will also reduce the time the battery will last. The **Power Sources** chapter includes additional information about batteries and their charge/discharge cycles.

2.4.2 Generalized Definition of Resistance

Electrical energy is not always turned into heat. The energy used in running a motor, for example, is converted to mechanical motion. The energy supplied to a radio transmitter is largely converted into radio waves. Energy applied to a loudspeaker is changed into sound waves. In each case, the energy is converted to other forms and can be completely accounted for. None of the energy just disappears! These are examples of the Law of Conservation of Energy. When a device converts energy from one form to another, we often say it *dissipates* the energy, or power. (Power is energy divided by time.) Of course the device doesn't really "use up" the energy, or make it disappear, it just converts it to another form. Proper operation of electrical devices often requires that the power be supplied at a specific ratio of voltage to current. These features are characteristics of resistance, so it can be said that any device that "dissipates power" has a definite value of resistance.

This concept of resistance as something that absorbs power at a definite voltage-to-current ratio is very useful; it permits substituting a simple resistance for the load or power-consuming part of the device receiving power, often with considerable simplification of calculations. Of course, every electrical device has some resistance of its own in the more narrow sense, so a part of the energy supplied to it is converted to heat in that resistance, even though the major part of the energy may be converted to another form.

2.4.3 Efficiency

In devices such as motors and transmitters, the objective is to convert the supplied energy (or power) into some form other than heat. In such cases, power converted to heat is considered to be a loss because it is not useful power. The efficiency of a device is the useful power output (in its converted form) divided by the power input to the device. In a transmitter, for example, the objective is to convert power from a dc source into ac power at some radio frequency. The ratio of the RF power output to the dc input is the *efficiency* (*Eff* or η) of the transmitter. That is:

$$\text{Eff} = \frac{P_O}{P_I}$$

where

Eff = efficiency (as a value or fraction between 0 and 1),

P_O = power output (W), and

P_I = power input (W).

Example: If the dc input to the transmitter is 100 W, and the RF power output is 60 W, the efficiency is:

$$\text{Eff} = \frac{P_O}{P_I} = \frac{60 \text{ W}}{100 \text{ W}} = 0.6$$

Efficiency is usually expressed as a percentage — that is, it expresses what percent of the input power will be available as useful output. To calculate percent efficiency, multiply the efficiency ratio by 100%. The efficiency in the example above is 60%.

Suppose a mobile transmitter has an RF power output of 100 W with 52% efficiency at 13.8 V. The vehicle's alternator system charges the battery at a rate of 5.0 A at this voltage. Assuming an alternator efficiency of 68%, how much horsepower must the engine produce to operate the transmitter and charge the battery? Solution: To charge the battery, the alternator must produce 13.8 V × 5.0 A = 69 W. The transmitter dc input power is 100 W / 0.52 = 190 W. Therefore, the total electrical power required from the alternator is 190 + 69 = 259 W. The engine load then is:

$$P_I = \frac{P_O}{\text{Eff}} = \frac{259 \text{ W}}{0.68} = 381 \text{ W}$$

We can convert this to horsepower using the conversion factor given earlier to convert between horsepower and watts:

$$\frac{381 \text{ W}}{746 \text{ W / hp}} = 0.51 \text{ horsepower (hp)}$$

Ohm's Law and Power Circle

During the first semester of my *Electrical Power Technology* program, one of the first challenges issued by our dedicated instructor — Roger Crerie — to his freshman students was to identify and develop 12 equations or formulas that could be used to determine voltage, current, resistance, and power. Ohm's Law is expressed as R = E / I and it provided three of these equation forms, while the basic equation relating power to current and voltage (P = I × E) accounted for another three. With six known equations, it was just a matter of applying mathematical substitution for his students to develop the remaining six. Together, these 12 equations compose the *circle* or *wheel* of voltage (E), current (I), resistance (R) and power (P) shown in **Figure 2.A1**. Just as Roger's previous students had learned at the Worcester Industrial Technical Institute (Worcester, Massachusetts), our Class of '82 now held the basic electrical formulas needed to proceed in our studies or professions. As can be seen in Figure 2.A1, we can determine any one of these four electrical quantities by knowing the value of any two others. You may want to keep this page bookmarked for your reference. You'll probably be using many of these formulas as the years go by — this has certainly been my experience. — *Dana G. Reed, W1LC*

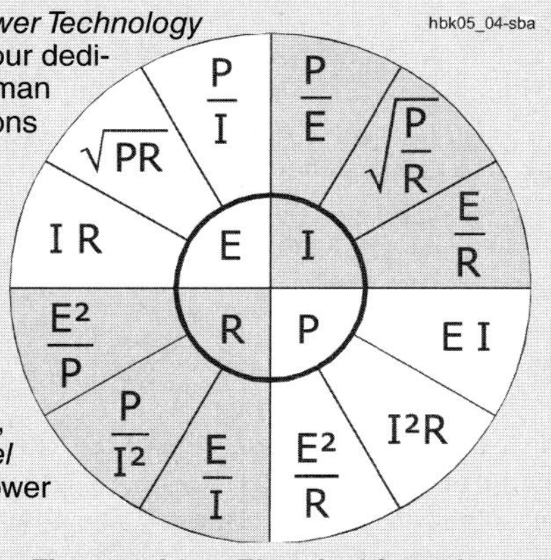

Figure 2.A1 — Electrical formulas

2.4.4 Ohm's Law and Power Formulas

Electrical power in a resistance is turned into heat. The greater the power, the more rapidly the heat is generated. By substituting the Ohm's Law equivalent for E and I, the following formulas are obtained for power:

$$P = E^2 / R$$

and

$$P = I^2 \times R$$

These formulas are useful in power calculations when the resistance and either the current or voltage (but not both) are known.

Example: How much power will be dissipated by (converted to heat in) a 4000-Ω resistor if the potential applied to it is 200 V?

$$P = \frac{E^2}{R} = \frac{40,000\ V^2}{4,000\ \Omega} = 10.0 W$$

As another example, suppose a current of 20 mA flows through a 300-Ω resistor. Then:

$$P = I^2 \times R = 0.020^2\ A^2 \times 300\ \Omega$$

$$P = 0.00040\ A^2 \times 300\ \Omega$$

$$P = 0.12\ W$$

Note that the current was changed from milliamperes to amperes before substitution in the formula.

Resistors for radio work are made in many sizes, the smallest being rated to safely operate at power levels of about 1/16 W. The largest resistors commonly used in amateur equipment are rated at about 100 W. Large resistors, such as those used in dummy-load antennas, are often cooled with oil to increase their power-handling capability.

2.5 Circuit Control Components

2.5.1 Switches

Switches are used to allow or interrupt a current flowing in a particular circuit. Most switches are mechanical devices, although the same effect can be achieved with solid-state devices.

Switches come in many different forms and a wide variety of ratings. The most important ratings are the *voltage-handling* and *current-handling* capabilities. The voltage rating usually includes both the *breakdown voltage rating* and the *interrupt voltage rating*. The breakdown rating is the maximum voltage that the switch can withstand when it is open before the voltage will arc between the switch's terminals. The interrupt voltage rating is the maximum amount of voltage that the switch can interrupt without arcing. Normally, the interrupt voltage rating is the lower value, and therefore the one given for (and printed on) the switch.

Switches typically found in the home are usually rated for 125 V ac and 15 to 20 A. Switches in cars are usually rated for 12 V dc and several amperes. The breakdown voltage rating of a switch primarily depends on the insulating material surrounding the contacts and the separation between the contacts. Plastic or phenolic material normally provides both structural support and insulation. Ceramic material may be used to provide better insulation, particularly in rotary (wafer) switches.

A switch's current rating includes both the *current-carrying capacity* and the *interrupt capability*. The current-carrying capacity of the switch depends on the contact material and size, and on the pressure exerted to keep the contacts closed. It is primarily determined from the allowable contact temperature rise. On larger ac switches and most dc switches, the interrupt capability is usually lower than the current carrying value.

Most power switches are rated for alternating current use. Because ac current goes through zero twice in each cycle, switches can successfully interrupt much higher alternating currents than direct currents without arcing. A switch that has a 10 A ac current rating may arc and damage the contacts if used to turn off more than an ampere or two of dc.

Switches are normally designated by the number of *poles* (circuits controlled) and *throws* or *positions* (circuit path choices). The simplest switch is the on-off switch, which is a single-pole, single-throw (SPST) switch as shown in **Figure 2.9A**. The off position does not direct the current to another circuit. The next step would be to change the current path to another path. This would be a single-pole, double-throw (SPDT) switch as shown in Figure 2.9B. Adding an off position would give a single-pole, double-throw, center-off (ON-OFF-ON) switch as shown in Figure 2.9C.

Several such switches can be "ganged" to or actuated by the same mechanical activator to provide double-pole, triple-pole or even more, separate control paths all activated at once. Switches can be activated in a variety of ways. The most common methods include lever or toggle, push-button, and rotary switches. Samples of these are shown in Figure 2.9D. Most switches stay in position once set, but some are spring-loaded so they

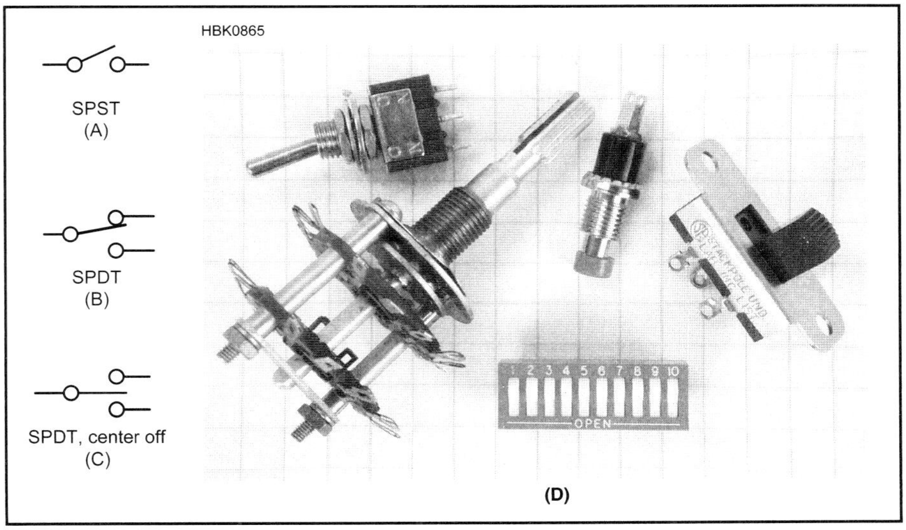

Figure 2.9 — Schematic diagrams of various types of switches. (A) is an SPST, (B) is an SPDT, and (C) is an SPDT switch with a center-off position. The photo (D) shows examples of various styles of switches. The ¼-inch-ruled graph paper background provides for size comparison.

only stay in the desired position while held there. These are called *momentary* switches.

Rotary/wafer switches can provide very complex switching patterns. Several poles (separate circuits) can be included on each wafer. Many wafers may be stacked on the same shaft. Not only may many different circuits be controlled at once, but by wiring different poles/positions on different wafers together, a high degree of circuit switching logic can be developed. Such switches can select different paths as they are turned and can also "short" together successive contacts to connect numbers of components or paths.

Rotary switches can also be designed to either break one contact before making another (*break-before-make*), or to short two contacts together before disconnecting the first one (*make-before-break*) to eliminate arcing or perform certain logic functions. The two types of switches are generally not interchangeable and may cause damage if inappropriately substituted for one another during circuit construction or repair. When buying rotary switches from a surplus or flea-market vendor, check to be sure the type of switch is correct.

Microswitches are designed to be actuated by the operation of machine components, opening or closing of a door, or some other mechanical movement. Instead of a handle or button-type actuator that would be used by a human, microswitches have levers or buttons more suitable for being actuated as part of an enclosure or machine.

When choosing a switch for a particular task, consideration should be given to function, voltage and current ratings, ease of use, availability, and cost. If a switch is to be operated frequently, a better-quality switch is usually less costly over the long run. If signal noise or contact corrosion is a potential problem (usually in low-current signal applications), it is best to get gold-plated contacts. Gold does not oxidize or corrode, thus providing surer contact, which can be particularly important at very low signal levels. Gold plating will not hold up under high-current-interrupt applications, however.

2.5.2 Fuses and Circuit Breakers

Fuses self-destruct to protect circuit wiring or equipment. The fuse *element* that melts or *blows* is a carefully shaped piece of soft metal, usually mounted in a cartridge of some kind. The element is designed to safely carry a given amount of current and to melt at a current value that is a certain percentage above the rated value.

The most important fuse rating is the *nominal current rating* that it will safely carry for an indefinite period without blowing. A fuse's melting current depends on the type of material, the shape of the element and the heat dissipation capability of the cartridge and holder, among other factors.

Next most important are the timing characteristics, or how quickly the fuse element blows under a given current overload. Slow-blow fuses are designed to carry an overload for a short period of time. They typically are used in motor-starting and power-supply circuits that have a large inrush current when first started. Other fuses are designed to blow more quickly to protect delicate instruments and solid-state circuits.

A fuse also has a voltage rating, both a value in volts and whether it is expected to be used in ac or dc circuits. The voltage rating is the amount of voltage an open fuse can withstand without arcing. While you should never substitute a fuse with a higher current rating than the one it replaces, you may use a fuse with a higher voltage rating.

Figure 2.10A shows typical cartridge-style cylindrical fuses likely to be encountered in ac-powered radio and test equipment. Automotive style fuses, shown in the lower half of Figure 2.10A, have become widely used in low-voltage dc power wiring of amateur stations. These are called "blade" fuses. Rated for vehicle-level voltages, automotive blade fuses should never be used in ac line-powered circuits.

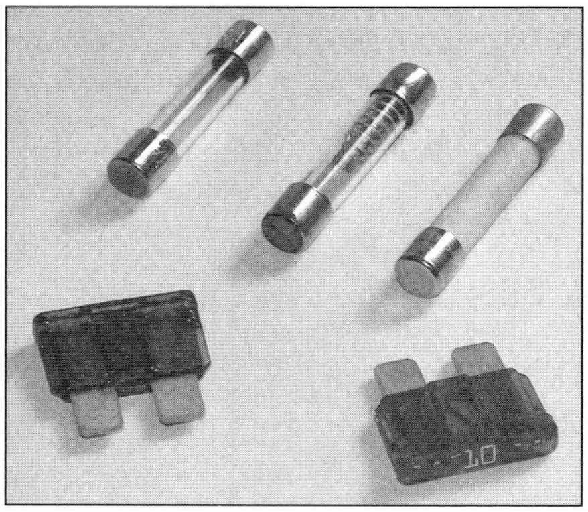

(A)

(B)

Figure 2.10 — These photos show examples of various styles of fuses. Cartridge-type fuses (A, top) can use glass or ceramic construction. The center fuse is a slow-blow type. Automotive blade-type fuses (A, bottom) are common for low-voltage dc use. A typical home circuit breaker for ac wiring is shown at (B).

Circuit breakers perform the same function as fuses — they open a circuit and interrupt current flow when an overload occurs. Instead of a melting element, circuit breakers use spring-loaded magnetic mechanisms to open a switch when excessive current is present. Once the overload has been corrected, the circuit-breaker can be reset. Circuit breakers are generally used by amateurs in home ac wiring (a typical ac circuit breaker is shown in Figure 2.10B) and in dc power supplies.

A replacement fuse or circuit breaker should have the same current rating and the same characteristics as the fuse it replaces. Never substitute a fuse with a larger current rating. You may cause permanent damage (maybe even a fire) to wiring or circuit elements by allowing larger currents to flow when there is an internal problem in equipment. (Additional discussion of fuses and circuit breakers is provided in the chapter on **Safety**.)

Fuses blow and circuit breakers open for several reasons. The most obvious reason is that a problem develops in the circuit, causing too much current to flow. In this case, the circuit problem needs to be fixed. A fuse can fail from being cycled on and off near its current rating. The repeated thermal stress causes metal fatigue and eventually the fuse blows. A fuse can also blow because of a momentary power surge, or even by rapidly turning equipment with a large inrush current on and off several times. In these cases it is only necessary to replace the fuse with the same type and value.

Panel-mount fuse holders should be wired with the hot lead of an ac power circuit (the black wire of an ac power cord) connected to the end terminal, and the ring terminal is connected to the power switch or circuit inside the chassis. This removes voltage from the fuse as it is removed from the fuse holder. This also locates the line connection at the far end of the fuse holder where it is not easily accessible.

2.5.3 Relays and Solenoids

Relays are switches controlled by an electrical signal. *Electromechanical relays* consist of a electromagnetic *coil* and a moving *armature* attracted by the coil's magnetic field when energized by current flowing in the coil. Movement of the armature pushes the switch contacts together or apart. Many sets of contacts can be connected to the same armature, allowing many circuits to be controlled by a single signal. In this manner, the signal voltage that energizes the coil can control circuits carrying large voltages and/or currents.

Relays have two positions or *states* — *energized* and *de-energized*. Sets of contacts called *normally closed* (NC) are closed when the relay is de-energized and open when it is energized. *Normally open* (NO) contact sets

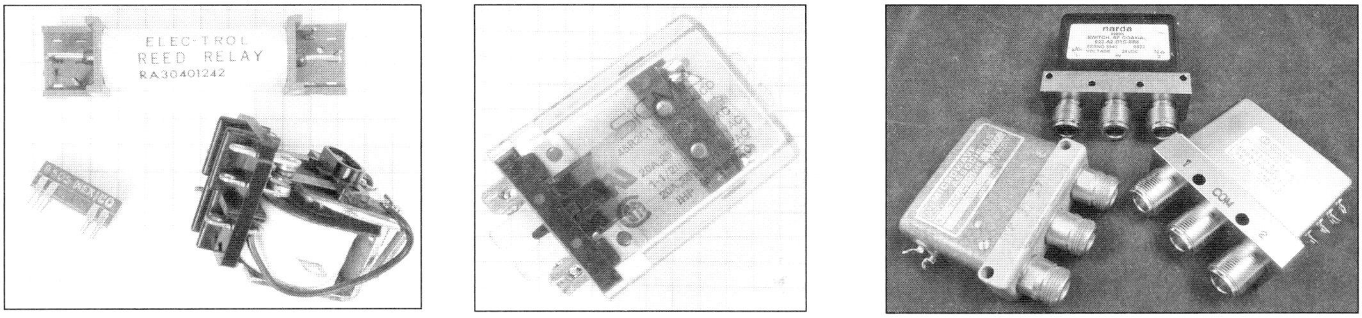

(A) (B) (C)

Figure 2.11 — These photos show examples of various styles and sizes of relays. Photo (A) shows a large reed relay, and a small reed relay in a package the size of an integrated circuit. The contacts and coil can clearly be seen in the open-frame relay. Photo (B) shows a relay inside a plastic case. The background grid for Photos (A) and (B) is ¼ inch to provide a size comparison. Photo (C) shows three SPDT coaxial relays with type N connectors used for RF switching at VHF and higher frequencies.

are closed when the relay is energized.

Like switches, relay contacts have breakdown voltage, interrupting, and current-carrying ratings. These are not the same as the voltage and current requirements for energizing the relay's coil. Relay contacts (and housings) may be designed for ac, dc or RF signals. The most common control voltages for relays used in amateur equipment are 12 V dc or 120 V ac. Relays with 6, 24, and 28 V dc, and 24 V ac coils are also common. **Figure 2.11** shows some typical relays found in amateur equipment.

Relays with dc coils store energy in the coil when energized. When voltage to the coil is removed, the stored energy creates a large voltage transient that can damage the relay driver circuit. Techniques for managing the coil's stored energy are presented in "Using the Transistor as a Switch" in the **Circuits and Components** chapter.

A relay's *pull-in voltage* is the minimum voltage at which the coil is guaranteed to cause the armature to move and change the relay's state. *Hold-in voltage* is the minimum voltage at which the relay is guaranteed to hold the armature in the energized position after the relay is activated. A relay's pull-in voltage is higher than its hold-in voltage due to magnetic hysteresis of the coil (see the section on magnetic materials later in this chapter). *Current-sensing relays* activate when the current through the coil exceeds a specific value, regardless of the voltage applied to the coil. They are used when the control signal is a current rather than a voltage.

Latching relays have two coils; each moves the armature to a different position where it remains until the other coil is energized. These relays are often used in portable and low-power equipment so that the contact configuration can be maintained without the need to supply power to the relay continuously.

Reed relays have no armature. The contacts are attached to magnetic strips or "reeds" in a glass or plastic tube, surrounded by a coil. The reeds move together or apart when current is applied to the coil, opening or closing contacts. Reed relays can open and close very quickly and are often used in transmit-receive switching circuits.

Solid-state relays (SSR) use transistors instead of mechanical contacts and electronic circuits instead of magnetic coils. They are designed as substitutes for electromechanical relays in power and control circuits and are not used in low-level ac or dc circuits.

Coaxial relays have an armature and contacts designed to handle RF signals. The signal path in coaxial relays maintains a specific characteristic impedance for use in RF systems. Coaxial connectors are used for the RF circuits. Coaxial relays are typically used to control antenna system configurations or to switch a transceiver between a linear amplifier and an antenna.

A *solenoid* is very similar to a relay, except that instead of the moving armature actuating switch contacts, the solenoid moves a lever or rod to actuate some mechanical device. Solenoids are not commonly used in radio equipment, but may be encountered in related systems or devices.

2.6 Capacitance and Capacitors

It is possible to build up and hold an electrical charge in an *electrostatic field*. This phenomenon is called *capacitance*, and the devices that exhibit capacitance are called *capacitors*. (Old articles and texts use the obsolete term *condenser*.) **Figure 2.12** shows schematic symbols for capacitors: a fixed capacitor with a single value of capacitance (Figure 2.12A) and variable capacitors adjustable over a range of values (Figure 2.12B). If the capacitor is of a type that is *polarized*, meaning that dc voltages must be applied with a specific polarity, the straight line in the symbol should be connected to the most positive voltage, while the curved line goes to the more negative voltage. For clarity, the positive terminal of a polarized capacitor symbol is usually marked with a + symbol, though not always. The symbol for *non-polarized* capacitors may be two straight lines or the + symbol may be omitted. When in doubt, consult the capacitor's specifications or the circuits parts list.

2.6.1 Electrostatic Fields and Energy

An *electrostatic field* is created wherever a voltage exists between two points, such as two opposite electric charges or regions that contain different amounts of charge. The field causes electric charges (such as electrons or ions) in the field to feel a force in the direction of the field. If the charges are not free to move, as in an insulator, they store the field's energy as *potential energy*, just as a weight held in place by a surface stores gravitational energy. If the charges are free to move, the field's stored energy is converted to *kinetic energy* of motion just as if the weight is released to fall in a gravitational field.

The field is represented by *lines of force* that show the direction of the force felt by the electric charge. Each electric charge is surrounded by an electric field. The lines of force of the field begin on the charge and extend away from charge into space. The lines of force can terminate on another charge (such as lines of force between a proton and an electron), or they can extend to infinity.

The strength of the electrostatic field is measured in *volts per meter* (V/m). Stronger fields cause the moving charges to accelerate more strongly and store more energy in fixed charges. The stronger the field in V/m, the more force an electric charge in the field will feel. The strength of the electric field diminishes with the square of the distance from its source, the electric charge.

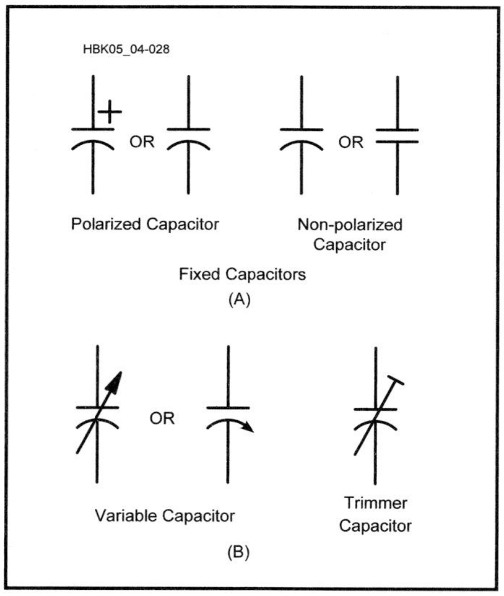

Figure 2.12 — Schematic symbol for a fixed capacitor is shown at (A). The symbols for a variable capacitor are shown at (B).

2.6.2 The Capacitor

Suppose two flat metal plates are placed close to each other (but not touching) and are connected to a battery through a switch, as illustrated in **Figure 2.13A**. At the instant the switch is closed, electrons are attracted from the upper plate to the positive terminal of the battery, while the same quantity is repelled from the negative battery terminal and pushed into the lower plate. This imbalance of charge creates a voltage between the plates. Eventually, enough electrons move into one plate and out of the other to make the voltage between the plates the same as the battery voltage. At this point, the voltage between the plates opposes further movement of electrons and no further current flow occurs.

If the switch is opened after the plates have been charged in this way, the top plate is left with a deficiency of electrons and the bottom plate with an excess. Since there is no current path between the two plates, they remain charged despite the fact that they are no longer connected to the battery, which is the source of the voltage. In Figure 2.13B, the separated charges create an electrostatic field between the plates. The electrostatic field contains the energy that was expended by the battery in causing the electrons to flow off of or onto the plates. These two plates create a *capacitor*, a device that has the property of storing electrical energy in an electric field, a property called *capacitance*.

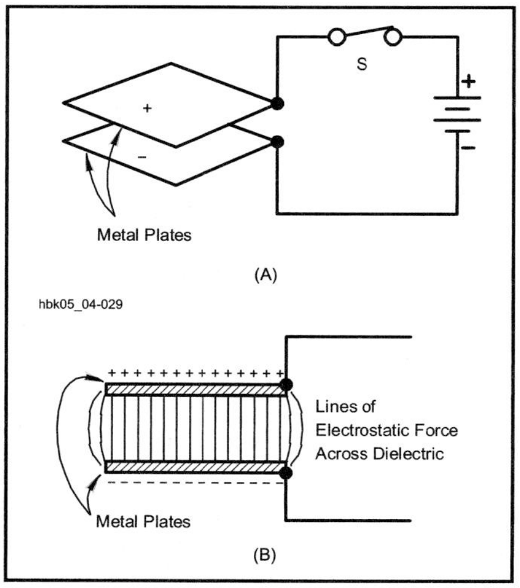

Figure 2.13 — A simple capacitor showing the basic charging arrangement at (A), and the retention of the charge due to the electrostatic field at (B).

The more general name for the capacitor's plates is *electrodes*. However, amateur literature generally refers to a capacitor's electrodes as *plates,* and that is the convention in this text.

The amount of electric charge that is held on the capacitor plates is proportional to the applied voltage and to the capacitance of the capacitor:

$$Q = CV$$

where

Q = charge in coulombs,

C = capacitance in farads (F), and

V = electrical potential in volts. (The symbol E is also commonly used instead of V in this and the following equation.)

The energy stored in a capacitor is also a function of voltage and capacitance:

$$U = \frac{V^2 C}{2}$$

where U = energy in joules (J) or watt-seconds.

If a wire is simultaneously touched to the two plates (short circuiting them), the voltage between the plates causes the excess electrons on the bottom plate to flow through the wire to the upper plate, restoring electrical neutrality. The plates are then *discharged*.

Figure 2.14 illustrates the voltage and current in the circuit, first, at the moment the switch is closed to charge the capacitor and, second, at the moment the shorting switch is closed to discharge the capacitor. Note that the periods of charge and discharge are very short, but that they are not zero. This finite charging and discharging time can be controlled, which will prove useful in the creation of timing circuits.

During the time the electrons are moving — that is, while the capacitor is being charged or discharged — a current flows in the circuit even though the circuit apparently is broken by the gap between the capacitor plates. The current flows only during the time of charge

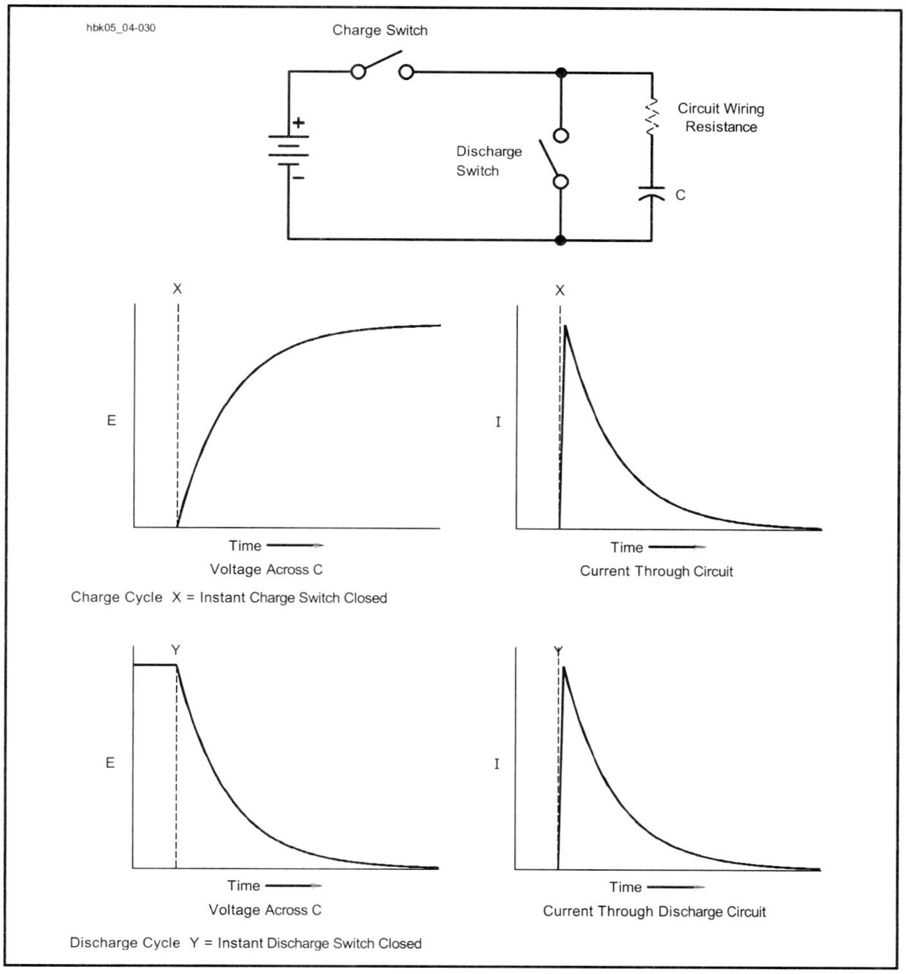

Figure 2.14 — The flow of current during the charge and discharge of a capacitor. The upper charging graphs assume that the charge switch is closed and the discharge switch is open. The lower discharging graphs assume just the opposite.

and discharge, however, and this time is usually very short. There is no continuous flow of direct current through a capacitor.

Although dc cannot pass through a capacitor, alternating current can. At the same time one plate is charged positively by the positive excursion of the alternating current, the other plate is being charged negatively at the same rate. (Remember that conventional current is shown as the flow of positive charge, equal to and opposite the actual flow of electrons.) The reverse process occurs during the second half of the cycle as the changing polarity of the applied voltage causes the flow of charge to change direction, as well. The continual flow into and out of the capacitor caused by ac voltage appears as an ac current, although with a phase difference between the voltage and current flow as described below.

UNITS OF CAPACITANCE

The basic unit of capacitance, the ability to store electrical energy in an electrostatic field, is the *farad*. This unit is generally too large for practical radio circuits, although capacitors of several farads in value are used in place of small batteries or as a power supply filter for automotive electronics. Capacitance encountered in radio and electronic circuits is usually measured in microfarads (abbreviated µF), nanofarads (abbreviated nF), or picofarads (pF). The microfarad is one millionth of a farad (10^{-6} F), the nanofarad is one thousandth of a microfarad (10^{-9} F), and the picofarad is one millionth of a microfarad (10^{-12} F). Old articles and texts use the obsolete term micromicrofarad (mmF or µµF) in place of picofarad.

The voltage rating of a capacitor indicates the maximum voltage that can be applied across the plates before the dielectric material between them begins to break down. That rating should never be exceeded, even for an instant. For nonpolar capacitors, the voltage rating applies for both positive and negative voltages. For polar capacitors, the rating applies to positive voltages and there should never be negative voltages applied.

CAPACITOR CONSTRUCTION

An idealized capacitor is a pair of parallel metal plates separated by an insulating or *dielectric* layer, ideally a vacuum. The capacitance of a vacuum-dielectric capacitor is given by

$$C = \frac{A\,\varepsilon_r\,\varepsilon_0}{d}$$

where

C = capacitance, in farads,
A = area of plates, in cm^2,
d = spacing of the plates in cm,
ε_r = dielectric constant of the insulating material, and
ε_0 = permittivity of free space, 8.85×10^{-14} F/cm.

The actual capacitance of such a parallel-plate capacitor is somewhat higher due to *end effect* caused by the electric field that exists just outside the edges of the plates.

The *larger* the plate area and the *smaller* the spacing between the plates, the *greater* the amount of energy that can be stored for a given voltage, and the *greater* the capacitance.

The amount of capacitance also depends on the material used as insulating material between the plates; capacitance is smallest with air or a vacuum as the insulator. Substituting other insulating materials for air may greatly increase the capacitance.

The ratio of the capacitance with a material other than a vacuum or air between the plates to the capacitance of the same capacitor with air insulation is called the *dielectric constant* (ε_r or K) of that particular insulating material. The dielectric constants of a number of materials commonly used as dielectrics in capacitors are given in **Table 2.2**. For example, if polystyrene is substituted for air in a capacitor, the capacitance will be 2.6 times greater.

In practice, capacitors often have more than two plates, with alternating plates being connected in parallel to form two sets, as shown in **Figure 2.15**. This practice makes it possible to obtain a fairly large capacitance in a

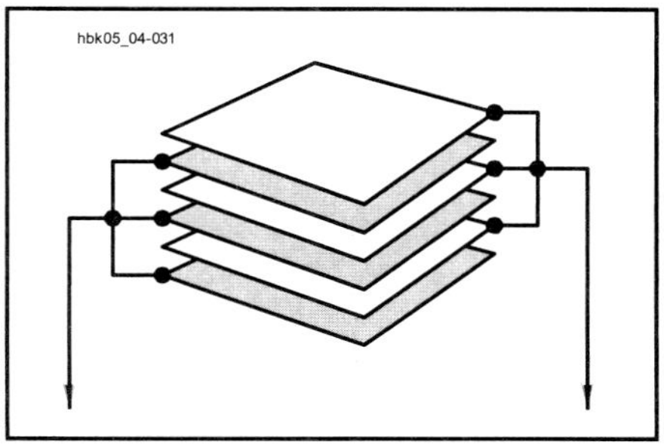

Figure 2.15 — A multiple-plate capacitor. Alternate plates are connected to each other, increasing the total area available for storing charge.

small space, since several plates of smaller individual area can be stacked to form the equivalent of a single large plate of the same total area. Also, all plates except the two on the ends of the stack are exposed to plates of the other group on both sides, and so are twice as effective in increasing the capacitance.

The formula for calculating capacitance from these physical properties is:

$$C = \frac{0.2248\ K\ A\ (n-1)}{d}$$

where

C = capacitance in pF,

K = dielectric constant of material between plates,

A = area of one side of one plate in square inches,

d = separation of plate surfaces in inches, and

n = number of plates.

If the area (A) is in square centimeters and the separation (d) is in centimeters, then the formula for capacitance becomes

$$C = \frac{0.0885\ K\ A\ (n-1)}{d}$$

If the plates in one group do not have the same area as the plates in the other, use the area of the smaller plates.

Example: What is the capacitance of two copper plates, each 1.50 square inches in area, separated by a distance of 0.00500 inch, if the dielectric is air?

$$C = \frac{0.2248\ K\ A\ (n-1)}{d}$$

$$C = \frac{0.2248 \times 1 \times 1.50\ (2-1)}{0.00500}$$

$$C = 67.4\ \text{pF}$$

What is the capacitance if the dielectric is mineral oil? (See Table 2.2 for the appropriate dielectric constant.)

$$C = \frac{0.2248 \times 2.23 \times 1.50\ (2-1)}{0.00500}$$

$$C = 150.3\ \text{pF}$$

2.6.3 Capacitors in Series and Parallel

When a number of capacitors are connected in parallel, as in the right side of **Figure 2.16**, the total capacitance of the group is equal to the sum of the individual capacitances:

$$C_{total} = C_1 + C_2 + C_3 + C_4 \ldots + C_n$$

Table 2.2

Relative Dielectric Constants of Common Capacitor Dielectric Materials

Material	*Dielectric Constant (k)*	*(O)rganic or (I)norganic*
Vacuum	1 (by definition)	I
Air	1.0006	I
Ruby mica	6.5 - 8.7	I
Glass (flint)	10	I
Barium titanate (class I)	5 - 450	I
Barium titanate (class II)	200 - 12000	I
Kraft paper	≈ 2.6	O
Mineral Oil	≈ 2.23	O
Castor Oil	≈ 4.7	O
Halowax	≈ 5.2	O
Chlorinated diphenyl	≈ 5.3	O
Polyisobutylene	≈ 2.2	O
Polytetrafluoroethylene	≈ 2.1	O
Polyethylene terephthalate	≈ 3	O
Polystyrene	≈ 2.6	O
Polycarbonate	≈ 3.1	O
Aluminum oxide	≈ 8.4	I
Tantalum pentoxide	≈ 28	I
Niobium oxide	≈ 40	I
Titanium dioxide	≈ 80	I

(Adapted from: Charles A. Harper, *Handbook of Components for Electronics*, p 8-7.)

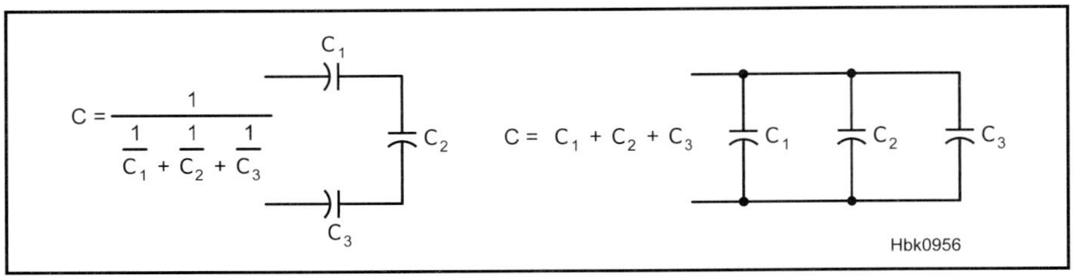

Figure 2.16 — Capacitors in series and parallel.

When two or more capacitors are connected in series, as in the left side of Figure 2.16, the total capacitance is less than that of the smallest capacitor in the group. The rule for finding the capacitance of a number of series-connected capacitors is the same as that for finding the resistance of a number of parallel-connected resistors.

$$C_{total} = \frac{1}{\frac{1}{C_1} + \frac{1}{C_2} + \frac{1}{C_3} + \ldots + \frac{1}{C_n}}$$

For only two capacitors in series, the formula becomes:

$$C_{total} = \frac{C_1 \times C_2}{C_1 + C_2}$$

The same units must be used throughout; that is, all capacitances must be expressed in µF, nF or pF, etc. Different units cannot be combined in the same equation.

Capacitors are often connected in parallel to obtain a larger total capacitance than is available in one unit. The voltage rating of capacitors connected in parallel is the lowest voltage rating of any of the capacitors.

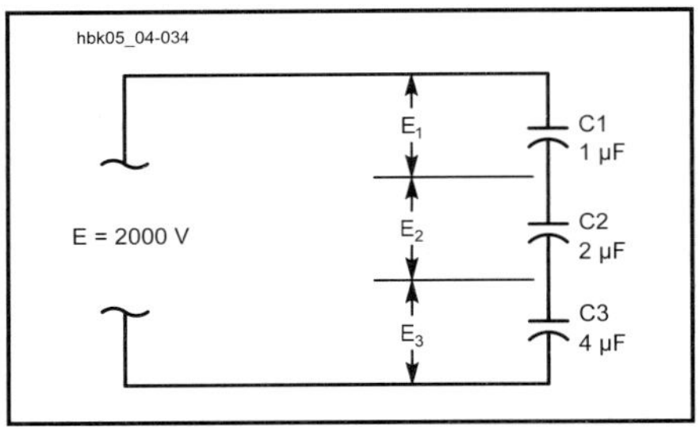

Figure 2.17 — An example of capacitors connected in series. The text shows how to find the voltage drops, E_1 through E_3.

When capacitors are connected in series, the applied voltage is divided between them according to Kirchhoff's Voltage Law. The situation is much the same as when resistors are in series and there is a voltage drop across each. The voltage that appears across each series-connected capacitor is inversely proportional to its capacitance, as compared with the capacitance of the whole group. (This assumes ideal capacitors.)

Example: Three capacitors having capacitances of 1, 2, and 4 µF, respectively, are connected in series as in **Figure 2.17**. The voltage across the entire series is 2000 V. What is the total capacitance? (Since this is a calculation using theoretical values to illustrate a technique, we will not follow the rules of significant figures for the calculations.)

$$C_{total} = \frac{1}{\dfrac{1}{C_1} + \dfrac{1}{C_2} + \dfrac{1}{C_3}}$$

$$= \frac{1}{\dfrac{1}{1\ \mu F} + \dfrac{1}{2\ \mu F} + \dfrac{1}{4\ \mu F}}$$

$$= \frac{1}{\dfrac{7}{4\ \mu F}} = \frac{4\ \mu F}{7} = 0.5714\ \mu F$$

The voltage across each capacitor is proportional to the total capacitance divided by the capacitance of the capacitor in question. So the voltage across C_1 is:

$$E_1 = \frac{0.5714\ \mu F}{1\ \mu F} \times 2{,}000\ V = 1{,}143\ V$$

Similarly, the voltages across C_2 and C_3 are:

$$E_2 = \frac{0.5714\ \mu F}{2\ \mu F} \times 2{,}000\ V = 571\ V$$

and

$$E_3 = \frac{0.5714\ \mu F}{4\ \mu F} \times 2{,}000\ V = 286\ V$$

The sum of these three voltages equals 2,000 V, the applied voltage. The voltage ratings of these three capacitors do not have to be able to withstand the full 2,000 V, but rather the voltages that will be across each one.

Capacitors may be connected in series to enable the group to withstand a larger voltage than any individual capacitor is rated to withstand. The trade-off is a decrease in the total capacitance. As shown by the previous example, the applied voltage does not divide equally between the capacitors except when all the capacitances are precisely the same. Use care to ensure that the voltage rating of any capacitor in the group is not exceeded. It is good practice to use capacitors with voltage ratings that are higher than the bare minimum, to allow for voltage fluctuations that may damage them. If you use capacitors in series to withstand a higher voltage, you should also connect an "equalizing resistor" across each capacitor as described in the **Power Sources** chapter.

2.6.4 RC Time Constant

An Excel spreadsheet by Lou Ernst, WA2GKH, illustrates charging and discharging time constants. The file name is Capacitor_TC_Calc_Version_2.1r.xls, and it is available in this book's online content, found at the website, **www.arrl.org/arrl-handbook-reference**.

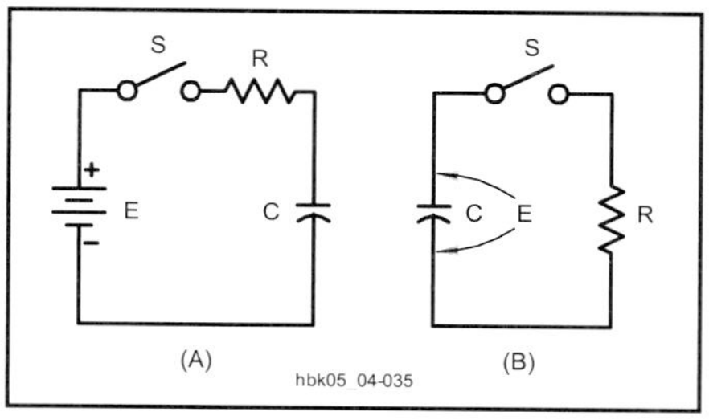

Figure 2.18 — An RC circuit. The series resistance delays the process of charging (A) and discharging (B) when the switch, S, is closed.

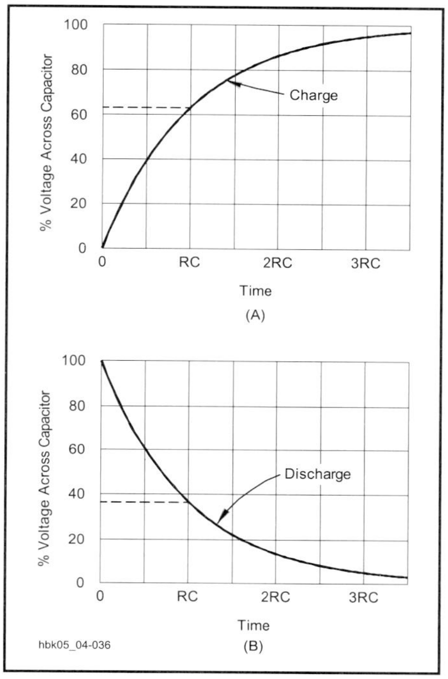

Figure 2.19 — At (A), the curve shows how the voltage across a capacitor rises, with time, when charged through a resistor. The curve at (B) shows the way in which the voltage decreases across a capacitor when discharging through the same resistance. For practical purposes, a capacitor may be considered charged or discharged after five RC periods.

Connecting a dc voltage source directly to the terminals of a capacitor charges the capacitor to the full source voltage almost instantaneously. Any resistance added to the circuit (as R in **Figure 2.18A**) limits the current, lengthening the time required for the voltage between the capacitor plates to build up to the source-voltage value. During this charging period, the current flowing from the source into the capacitor gradually decreases from its initial value. The increasing voltage stored in the capacitor's electric field offers increasing opposition to the steady source voltage.

While it is being charged, the voltage between the capacitor terminals is an exponential function of time, and is given by:

$$V(t) = E\left(1 - e^{-\frac{t}{RC}}\right)$$

where

$V(t)$ = capacitor voltage at time t,

E = power source potential in volts,

t = time in seconds after initiation of charging current,

e = natural logarithmic base = 2.718,

R = circuit resistance in ohms, and

C = capacitance in farads.

(References that explain exponential equations, e, and other mathematical topics are found in the "Radio Mathematics" article in this book's online content.)

By letting t = RC, the above equation becomes:

$$V(RC) = E\,(1 - e^{-1}) \cong 0.632\ E$$

The product of R in ohms times C in farads is called the *time constant* (also called the *RC time constant*) of the circuit and is the time in seconds required to charge the capacitor to 63.2% of the applied voltage. (The lower-case Greek letter tau, τ, is often used to represent the time constant in electronics circuits.) After two time constants ($t = 2\tau$) the capacitor charges another 63.2% of the difference between the capacitor voltage at one time constant and the applied voltage, for a total charge of 86.5%. After three time constants the capacitor reaches 95% of the applied voltage, and so on, as illustrated in the curve of **Figure 2.19A**. After five time constants, a capacitor is considered fully charged, having reached 99.24% of the applied voltage. Theoretically, the charging process is never really finished, but eventually the charging current drops to an immeasurably small value and the voltage is effectively constant.

If a charged capacitor is discharged through a resistor, as in Figure 2.18B, the same time constant applies to the decay of the capacitor voltage. A direct short circuit applied between the capacitor terminals would discharge the capacitor almost instantly. The resistor, R, limits the current, so a capacitor discharging through a resistance exhibits the same time-constant characteristics (calculated in the same way as above) as a charging capacitor. The voltage, as a function of time while the capacitor is being discharged, is given by:

$$V(t) = E\left(e^{-\frac{t}{RC}}\right)$$

RC Timesaver

When calculating time constants, it is handy to remember that if R is in units of MΩ and C is in units of μF, the result of R × C will be in seconds. Expressed as an equation: MΩ × μF = seconds.

where t = time in seconds after initiation of discharge and E is the fully charged capacitor voltage prior to beginning discharge.

Again, by letting t = RC, the time constant of a discharging capacitor represents a decrease in the voltage across the capacitor of about 63.2%. After five time-constants, the capacitor is considered fully discharged, since the voltage has dropped to less than 1% of the full-charge voltage. Figure 2.19B is a graph of the discharging capacitor voltage in terms of time constants..

Time constant calculations have many uses in radio work. The following examples are all derived from practical-circuit applications.

Example 1: A 100-μF capacitor in a high-voltage power supply is shunted by a 100-kΩ resistor. What is the minimum time before the capacitor may be considered fully discharged? Since full discharge is approximately five RC periods,

$$t = 5 \times RC = 5 \times 100 \times 10^3\ \Omega \times 100 \times 10^{-6}\ F = 50{,}000 \times 10^{-3} = 50\ s$$

Caution: Although waiting almost a minute for the capacitor to discharge seems safe in this high-voltage circuit, *never* rely solely on capacitor-discharging resistors (often called *bleeder resistors*). Be certain the power source is removed and the capacitors are totally discharged before touching any circuit components. (See the **Power Sources** chapter for more information on bleeder resistors.)

Example 2: Smooth CW keying without clicks requires both the rising and falling edges of the waveform to take approximately 5 ms (0.005 s). If an RC delay circuit in a keyed voltage line is used to set the rise and fall time, what values of R and C should be used? Since full charge and discharge require 5 RC periods,

$$RC = \frac{t}{5} = \frac{0.005\ s}{5} = 0.001\ s$$

Any combination of resistor and capacitor whose values, when multiplied together, equal 0.001 would do the job. A typical capacitor might be 0.05 μF. In that case, the necessary resistor would be:

$$R = \frac{0.001\ s}{0.05 \times 10^{-6}\ F}$$

$$= 0.02 \times 10^6\ \Omega = 20{,}000\ \Omega = 20\ k\Omega$$

In practice, a builder would use the calculated value as a starting point. The final value would be selected by monitoring the waveform on an oscilloscope.

Example 3: The popular 555 timer IC activates its output pin when the trigger input reaches 0.667 of the supply voltage. What value of capacitor and resistor would be required for a 4.5-second timing period?

First we will solve the charging equation for the time constant, RC. The threshold voltage is 0.667 times the supply voltage, so we use this value for V(t).

$$V(t) = E\left(1 - e^{-\frac{t}{RC}}\right)$$

$$0.667\ E = E\left(1 - e^{-\frac{t}{RC}}\right)$$

$$e^{-\frac{t}{RC}} = 1 - 0.667$$

$$\ln\left(e^{-\frac{t}{RC}}\right) = \ln(0.333)$$

$$-\frac{t}{RC} = -1.10$$

We want to find a capacitor and resistor combination that will produce a 4.5 s timing period, so we substitute that value for t.

$$RC = \frac{4.5\ s}{1.10} = 4.1\ s$$

If we select a value of 10 μF, we can solve for R.

$$R = \frac{4.1\ s}{10 \times 10^{-6}\ F} = 0.41 \times 10^6\ \Omega = 410\ k\Omega$$

A 1% tolerance resistor and capacitor will give good results. You could also use a variable resistor and an accurate method of measuring time to set the circuit to a 4.5 s period.

2.7 Inductance and Inductors

A complete treatment of magnetism and magnetic properties of materials can be found in *Electricity and Magnetism* by Edward Purcell (see **References and Bibliography**). This section is limited to a discussion of inductance, inductors, and related materials.

A second way to store electrical energy is in a *magnetic field*. This phenomenon is called *inductance*, and the devices that exhibit inductance are called *inductors*. Inductance is derived from some basic underlying magnetic properties.

2.7.1 Magnetic Fields and Magnetic Energy Storage

As an electric field surrounds an electric charge, magnetic fields surround *magnets*. You are probably familiar with metallic bar, disc, or horseshoe-shaped magnets. **Figure 2.20** shows a bar magnet, but particles of matter as small as an atom can also be magnets.

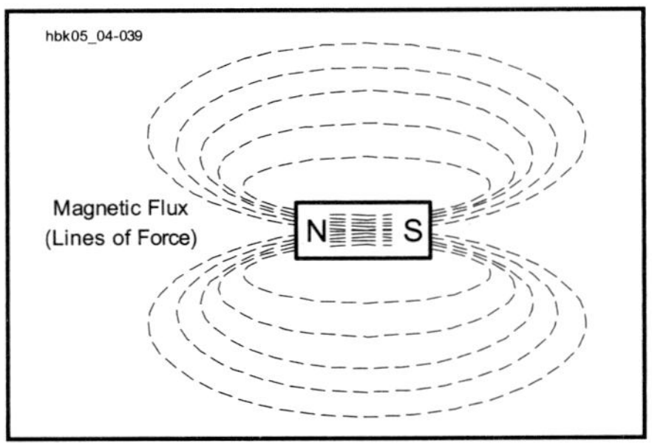

Figure 2.20 — The magnetic field and poles of a permanent magnet. The magnetic field direction is from the north to the south pole.

Figure 2.20 also shows the magnet surrounded by lines of force called *magnetic flux*, representing a *magnetic field*. (More accurately, a *magnetostatic field*, since the field is not changing.) Similar to those of an electric field, magnetic lines of force (or *flux lines*) show the direction in which a magnet would feel a force in the field.

There is no "magnetic charge" comparable to positive and negative electric charges. All magnets and magnetic fields have a polarity, represented as *poles*, and every magnet — from atoms to bar magnets — possesses both a *north* and *south pole*. The size of the source of the magnetism makes no difference. The north pole of a magnet is defined as the one attracted to the Earth's north magnetic pole. (Confusingly, this definition means the Earth's North Magnetic Pole is magnetically a south pole!) Like conventional current, the direction of magnetic lines of force was assigned arbitrarily by early scientists as pointing *from* the magnet's north pole *to* the south pole.

An electric field is *open* — that is, its lines of force have one end on an electric charge and can extend to infinity. A magnetic field is *closed* because all magnetic lines of force form a loop passing through a magnet's north and south poles.

The Right-Hand Rule

How do you remember which way the magnetic field around a current is pointing? Luckily, there is a simple method, called the *right-hand rule*. Make your right hand into a fist, then extend your thumb, as in the figure below. If your thumb is pointing in the direction of conventional current flow, then your fingers curl in the same direction as the magnetic field.

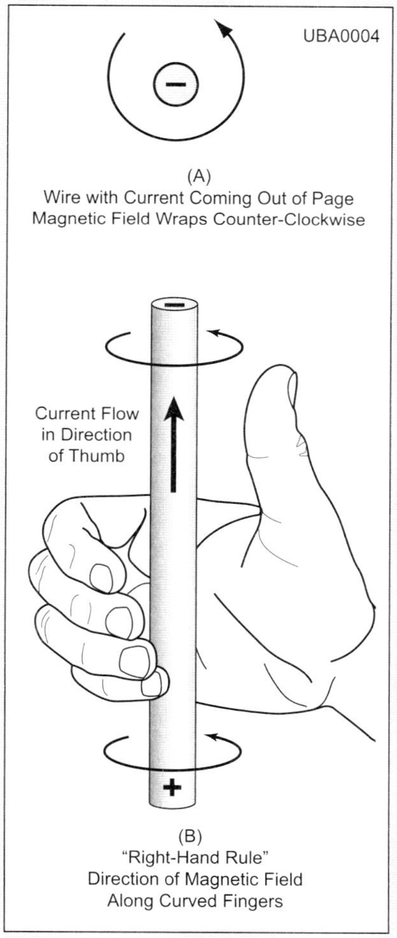

Figure 2.A2 — Use the right-hand rule to determine magnetic field direction from the direction of current flow.

Magnetic fields exist around two types of materials; *permanent magnets* and *electromagnets*. Permanent magnets consist of *ferromagnetic* and *ferrimagnetic* materials whose atoms are or can be aligned so as to produce a magnetic field. Ferro- or ferrimagnetic materials are strongly attracted to magnets. They can be *magnetized*, meaning to be made magnetic, by the application of a magnetic field. Lodestone, magnetite, and ferrites are examples of ferrimagnetic materials. Iron, nickel, cobalt, alnico alloys and other materials are ferromagnetic. Magnetic materials with high *retentivity* form permanent magnets because they retain their magnetic properties for long periods. Other materials, such as soft iron, yield temporary magnets that lose their magnetic properties rapidly.

Paramagnetic substances are very weakly attracted to a magnet and include materials such as platinum, aluminum, and oxygen. *Diamagnetic* substances, such as copper, carbon, and water, are weakly repelled by a magnet.

The second type of magnet is an electrical conductor with a current flowing through it. As shown in **Figure 2.21**, moving electrons are surrounded by a closed magnetic field, illustrated as the circular lines of force around the wire lying in planes perpendicular to the current's motion. The magnetic needle of a compass placed near a wire carrying direct current will be deflected as its poles respond to the forces created by the magnetic field around the wire.

If the wire is coiled into a *solenoid* as shown in **Figure 2.22**, the magnetic field greatly intensifies. This occurs as the magnetic fields from each successive turn in the coil add together because the current in each turn is flowing in the same direction.

Note that the resulting *electromagnet* has magnetic properties identical in principle to those of a permanent magnet, including poles and lines of force or flux. The strength of the magnetic field depends on several factors: the number and shape of turns of the coil, the magnetic properties of the materials surrounding the coil (both inside and out), the length of the coil, and the amplitude of the current.

Magnetic fields and electric current have a special two-way connection: voltage causing an electrical current (moving charges) in a

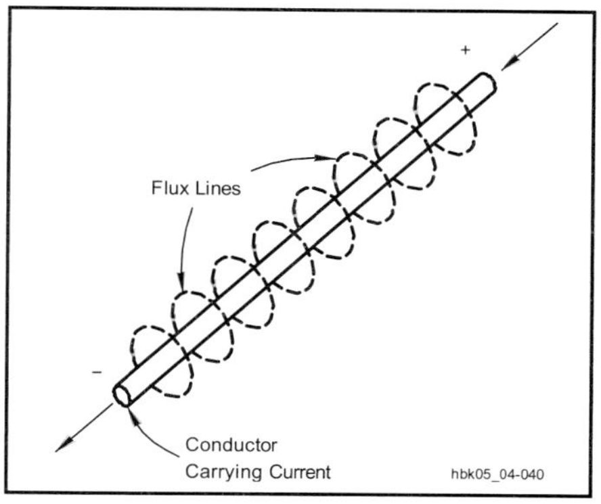

Figure 2.21 — The magnetic field around a conductor carrying an electrical current. If the thumb of your right hand points in the direction of the conventional current (plus to minus), your fingers curl in the direction of the magnetic field around the wire.

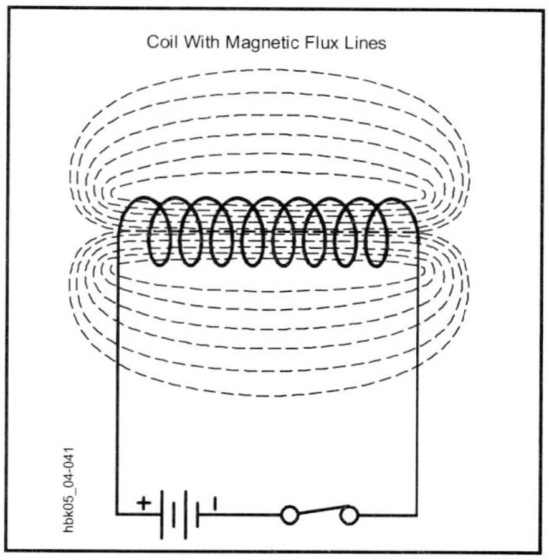

Figure 2.22 — Cross section of an inductor showing its flux lines and overall magnetic field.

Table 2.3
Magnetic Quantities

Value	*Symbol*	*MKS*	*cgs*
Magnetic Flux	lines	Weber, Wb = V-s	Maxwell, Mx = 10^{-8} Wb
Magnetic Flux Density	B	Tesla, T = Wb/m^2	Gauss, G = Mx/ cm^2 T = 10,000 G
Magnetomotive Force	[T]	Amp-turn = A	Gilbert, Gb = 0.79577 A
Magnetic Field Strength	H	A / meter	Oersted, Oe = Gb/cm = 79.58 A/m

Magnetic Circuit Analogies

Electric Circuit	*Magnetic Circuit*
Voltage drop V	Hl magnetovoltage drop
Voltage source V	nI magnetomotive force
Current I	psi = BA magnetic flux

Note – Magnetic circuit analogies as described by Shen and Kong, *Applied Electromagnetism* conductor will produce a magnetic field and a moving magnetic field will create an electrical field (voltage) that produces current in a conductor. This is the principle behind motors and generators, converting mechanical energy into electrical energy and vice-versa. **Table 2.3** shows the similarities: magnetic quantities and circuits have analogues in electrical quantities and circuits. This illustrates the deep relationship between electricity and magnetism, two sides of the same coin.

MAGNETIC FLUX

Magnetic flux is measured in the SI unit (International System of Units) of the weber, which is a volt-second (Wb = V-s). In the *centimeter-gram-second* (*cgs*) metric system units, magnetic flux is measured in maxwells (1 Mx = 10^{-8} Wb). The volt-second is used because of the relationship described in the previous paragraph: 1 volt of electromotive force will be created in a loop of wire in which magnetic flux through the loop changes at the rate of 1 weber per second. The relationship between current and magnetic fields is one of motion and change.

Magnetic field intensity, known as *flux density*, decreases with the square of the distance from the source, either a magnet or current. Flux density (*B*) is represented in gauss (G), where one gauss is equivalent to one line of force (1 Mx) per square centimeter of area measured perpendicularly to the direction of the field (G = Mx / cm^2). The Earth's magnetic field at the surface is approximately one gauss. The gauss is a *cgs* unit. In SI units, flux density is represented by the tesla (T), which is one weber per square meter (T = Wb/m^2 and 1T = 10,000 G).

Magnetomotive Force and Field Strength

The magnetizing or *magnetomotive force* ($\Im$) that produces a flux or total magnetic field is measured in gilberts (Gb). Magnetomotive force is analogous to electromotive force in that it produces the magnetic field. The SI unit of magnetomotive force is the ampere-turn, abbreviated A, just like the ampere. (1 Gb = 0.79577 A)

$$\Im = \frac{10\,N\,I}{4\pi}$$

where

$\Im$ = magnetomotive strength in gilberts,
N = number of turns in the coil creating the field,
I = dc current in amperes in the coil, and
π = 3.1416.

The magnetic field strength (*H*) measured in oersteds (Oe) produced by any particular magnetomotive force (measured in gilberts) is given by:

$$H = \frac{\Im}{\ell} = \frac{10\,N\,I}{4\,\pi\,\ell}$$

where

H = magnetic field strength in oersteds, and
ℓ = mean magnetic path length in centimeters.

The *mean magnetic path length* is the average length of the lines of magnetic flux. If the inductor is wound on a closed core as shown in the next section, *l* is approximately the average of the inner and outer circumferences of the core. The SI unit of magnetic field strength is the ampere-turn per meter. (1 Oe = 79.58 A/m)

2.7.2 Magnetic Core Properties

PERMEABILITY

The nature of the material within the coil of an electromagnet, where the lines of force are most concentrated, has the greatest effect upon the magnetic field established by the coil. All core materials are compared relatively to air. The ratio of flux density produced by a given material compared to the flux density produced by an air core is the *permeability* (μ) of the material. Air and non-magnetic materials have a permeability of one.

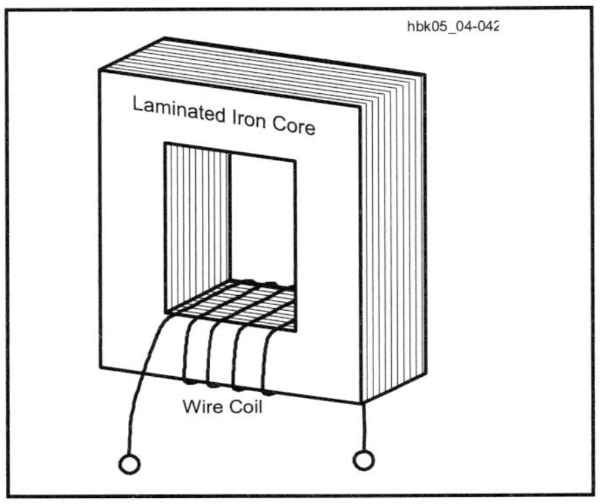

Figure 2.23 — A coil of wire wound around a laminated iron core.

Suppose the coil in **Figure 2.23** is wound on an iron core having a cross-sectional area of 2 square inches. When a certain current is sent through the coil, it is found that there are 80,000 lines of force in the core. Since the area is 2 square inches, the magnetic flux density is 40,000 lines per square inch. Now suppose that the iron core is removed and the same current is maintained in the coil. Also suppose the flux density without the iron core is found to be 50 lines per square inch. The ratio of these flux densities, iron core to air, is 40,000 / 50 or 800. This ratio is the core's permeability.

Permeabilities as high as 10^6 have been attained. The three most common types of materials used in magnetic cores are:

A) stacks of thin steel laminations (for power and audio applications, see the discussion on eddy currents below);

B) various ferrite compounds (for cores shaped as rods, toroids, beads, and numerous other forms); and

C) powdered iron (shaped as slugs, toroids and other forms for RF inductors).

The permeability of silicon-steel power-transformer cores approaches 5,000 in high-quality units. Powdered-iron cores used in RF tuned circuits range in permeability from 3 to about 35, while ferrites of nickel-zinc and manganese-zinc range from 20 to 15,000. Not all materials have permeabilities higher than

Table 2.4
Properties of Some High-Permeability Materials

Material	*Approximate Percent Composition*					*Maximum Permeability*
	Fe	Ni	Co	Mo	Other	
Iron	99.91	—	—	—	—	5000
Purified Iron	99.95	—	—	—	—	180,000
4% silicon-iron	96	—	—	—	4 Si	7000
45 Permalloy	54.7	45	—	—	0.3 Mn	25,000
Hipernik	50	50	—	—	—	70,000
78 Permalloy	21.2	78.5	—	—	0.3 Mn	100,000
4-79 Permalloy	16.7	79	—	—	0.3 Mn	100,000
Supermalloy	15.7	79	—	5	0.3 Mn	800,000
Permendur	49.7	—	50	—	0.3 Mn	5000
2V Permendur	49	—	49	—	2 V	4500
Hiperco	64	—	34	—	2 Cr	10,000
2-81 Permalloy*	17	81	—	2	—	130
Carbonyl iron*	99.9	—	—	—	—	132
Ferroxcube III**	($MnFe_2O_4$ + $ZnFe_2O_4$)			1500		

Note: all materials in sheet form except * (insulated powder) and ** (sintered powder).
(Reference: L. Ridenour, ed., *Modern Physics for the Engineer*, p 119.)

air. Brass has a permeability of less than one. A brass core inserted into a coil will decrease the magnetic field compared to an air core.

Table 2.4 lists some common magnetic materials, their composition and their permeabilities. Core materials are often frequency sensitive, exhibiting excessive losses outside the frequency band of intended use. (Ferrite materials are discussed separately in a later section of the chapter on **RF Techniques**.)

As a measure of the ease with which a magnetic field may be established in a material as compared with air, permeability (μ) corresponds roughly to electrical conductivity. Higher permeability means that it is easier to establish a magnetic field in the material. Permeability is given as:

$$\mu = \frac{B}{H}$$

where

B is the flux density in gauss, and

H is the magnetic field strength in oersteds.

RELUCTANCE

That a force (the magnetomotive force) is required to produce a given magnetic field strength implies that there is some opposition to be overcome. This opposition to the creation of a magnetic field is called *reluctance*. Reluctance ($\Re$) is inversely proportional to permeability and corresponds roughly to resistance in an electrical circuit. Carrying the electrical resistance analogy a bit further, the magnetic equivalent of Ohm's Law relates reluctance, magnetomotive force, and flux density: $\Re = \Im / B$. There are two related properties. *Reluctivity* is the reciprocal of permeability, and *permeance* is the reciprocal of reluctance.

HYSTERESIS

Retentivity in magnetic core materials is caused by atoms retaining their alignment from an applied magnetizing force. Retentivity is desirable if the goal is to create a permanent magnet. In an electronic circuit, however, the changes caused by retentivity cause the properties of the core material to depend on the history of how the magnetizing force was applied.

Figure 2.24 illustrates the change of flux density (*B*) with a changing magnetizing force (*H*). From starting point A, with no flux in the core, the flux reaches point B at the maximum magnetizing force. As the force decreases, so too does the flux, but it does not reach zero simultaneously with the force at point D. As the force continues in the opposite direction, it brings the flux density to point C. As the force decreases to zero, the flux once more lags behind. This occurs because some of the atoms in the core retain their alignment, even after the external magnetizing force is removed. This creates *residual flux* that is present even with no applied magnetizing force. This is the property of *hysteresis*.

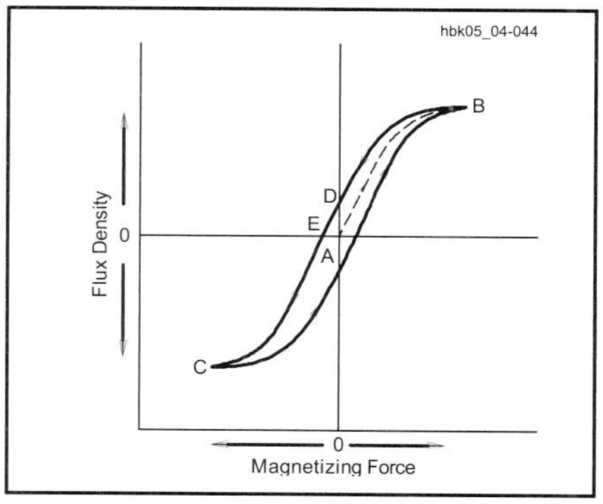

Figure 2.24 — A typical hysteresis curve for a magnetic core, showing the additional energy needed to overcome residual flux.

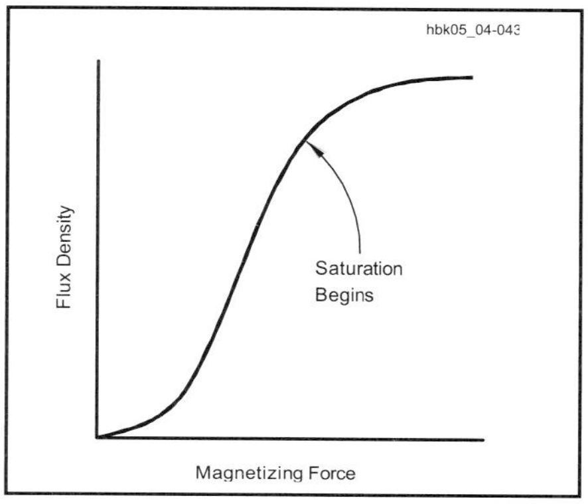

Figure 2.25 — A typical permeability curve for a magnetic core, showing the point where saturation begins.

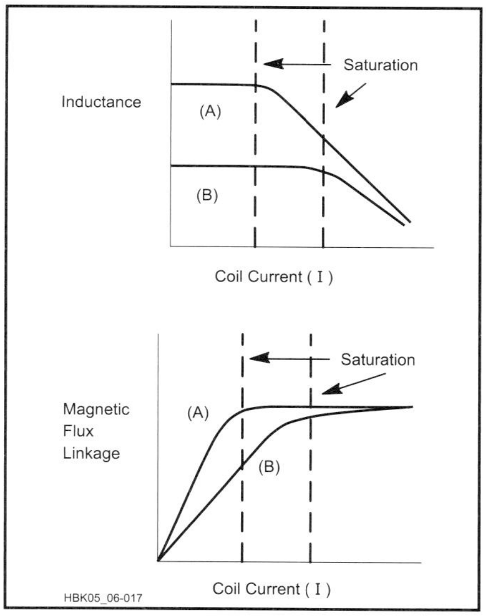

Figure 2.26 — Magnetic flux linkage and inductance plotted versus coil current for (A) a typical iron-core inductor. As the flux linkage Nϕ in the coil saturates, the inductance begins to decrease since inductance = flux linkage / current. The curves marked (B) show the effect of adding an air gap to the core. The current-handling capability has increased, but at the expense of reduced inductance.

In effect, a *coercive force* is necessary to reverse or overcome the residual magnetism retained by the core material. If a circuit carries a large ac current (that is, equal to or larger than saturation), the path shown in Figure 2.24 will be retraced with every cycle and the reversing force each time. The result is a power loss to the magnetic circuit, which appears as heat in the core material. Air cores are immune to hysteresis effects and losses.

SATURATION

Unlike electrical conductivity, which is independent of other electrical parameters, the permeability of a magnetic material varies with the flux density. At low flux densities (or with an air core), increasing the current through the coil will cause a proportionate increase in flux. This occurs because the current passing through the coil forces the atoms of the iron (or other material) to line up, just like many small compass needles. The magnetic field that results from the atomic alignment is *much* larger than that produced by the current with no core. As more and more atoms align, the magnetic flux density also increases.

At very high flux densities, increasing the current beyond a certain point may cause no appreciable change in the flux because all of the atoms are aligned. At this point, the core is said to be *saturated*. Saturation causes a rapid decrease in permeability, because it decreases the ratio of flux lines to those obtainable with the same current using an air core. **Figure 2.25** displays a typical permeability curve, showing the region of saturation. The saturation point varies with the makeup of different magnetic materials. Air and other nonmagnetic materials do not saturate.

EFFECTS OF SATURATION

An important concept for using inductors is that as long as the coil current remains below saturation, the inductance of the coil is essentially constant. **Figure 2.26** shows graphs of magnetic flux linkage (ψ) and inductance (L) vs. current (I) for a typical iron-core inductor both saturated and non-saturated. These quantities are related by the equation

$$\psi = N\phi = LI$$

where

ψ = the flux linkage,

N = number of turns,

ϕ = flux in webers,

L = inductance in henrys, and

I = current in amperes.

In the lower graph, a line drawn from any point on the curve to the (0,0) point will show

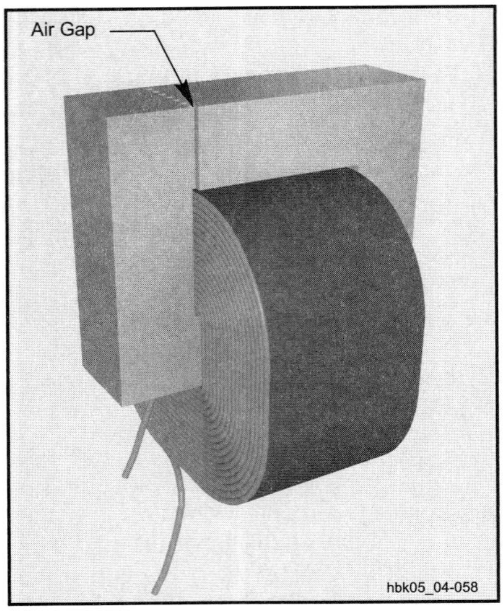

Figure 2.27 — Typical construction of a magnetic-core inductor. The air gap greatly reduces core saturation at the expense of reducing inductance. The insulating laminations between the core layers help to minimize eddy currents, as well.

the effective inductance, L = Nϕ / I, at that current. These results are plotted on the upper graph.

Note that below saturation, the inductance is constant because both ψ and I are increasing at a steady rate. Once the saturation current is reached, the inductance decreases because ψ does not increase anymore (except for the tiny additional magnetic field the current itself provides).

One common method of increasing the saturation current level is to cut a small air gap in the core (see **Figure 2.27**). This gap forces the flux lines to travel through air for a short distance, reducing the permeability of the core. Since the saturation flux linkage of the core is unchanged, this method works by requiring a higher current to achieve saturation. The price that is paid is a reduced inductance below saturation. The curves in Figure 2.26B show the result of an air gap added to that inductor.

Manufacturer's data sheets for magnetic cores usually specify the saturation flux density. Saturation flux density (ϕ) in gauss can be calculated for ac and dc currents from the following equations:

$$\varphi_{ac} = \frac{3.49\ V}{fNA}$$

$$\varphi_{dc} = \frac{NIA_L}{10A}$$

where

V = RMS ac voltage,
f = frequency, in MHz,
N = number of turns,
A = equivalent area of the magnetic path in square inches (from the data sheet)
I = dc current, in amperes, and
A_L = inductance index (also from the data sheet).

EDDY CURRENT

Since magnetic core material is usually conductive, the changing magnetic field produced by an ac current in an inductor also induces a voltage in the core. This voltage causes a current to flow in the core. This *eddy current* (so-named because it moves in a closed path, similarly to eddy currents in water) serves no useful purpose and results in energy being dissipated as heat from the core's resistance. Eddy currents are a particular problem in inductors with iron cores. Cores made of thin strips of magnetic material, called *laminations*, are used to reduce eddy currents. (See also the section on Practical Inductors in the **Circuits and Components** chapter.)

2.7.3 Inductance and Direct Current

In an electrical circuit, any element whose operation is based on the transfer of energy into and out of magnetic fields is called an *inductor* for reasons to be explained shortly. **Figure 2.28** shows schematic-diagram symbols and photographs of a few representative inductors. The photograph shows an air-core inductor, a slug-tuned (variable-core) inductor with a nonmagnetic core, and an inductor with a magnetic (iron) core. Inductors are often called *coils* because of their construction.

As explained above, when current flows through any conductor — even a straight wire — a magnetic field is created. The transfer of energy to the magnetic field represents work performed by the source of the voltage. Power is required for doing work, and since power is equal to current multiplied by voltage, there must be a voltage drop across the inductor while energy is being stored in the field. This voltage drop, exclusive of any voltage drop caused by resistance in the conductor, is the result of an opposing voltage created in the conductor while the magnetic field is building up to its final value. Once the field becomes constant, the *induced voltage* or *back-voltage* disappears, because no further energy is being stored. Back voltage is analogous to the opposition to current flow in a capacitor from the increasing capacitor voltage.

The induced voltage opposes the voltage of the source, preventing the current from rising rapidly when voltage is applied. **Figure 2.29A** illustrates the situation of energizing an inductor or magnetic circuit, showing the relative amplitudes of induced voltage and the delayed rise in current to its full value.

Rate of Change

The symbol Δ represents change in the following variable, so that ΔI represents "change in current" and Δt "change in time." A rate of change per unit of time is often expressed in this manner. When the amount of time over which the change is measured becomes very small, the letter *d* replaces Δ in both the numerator and denominator to indicate infinitesimal changes. This notation is used in the derivation and presentation of the functions that describe the behavior of electric circuits.

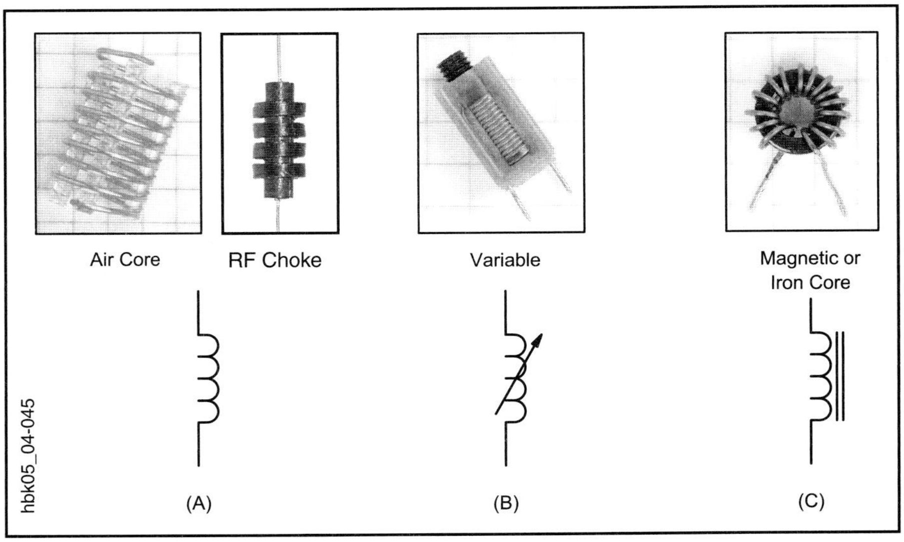

Figure 2.28 — Photos and schematic symbols for representative inductors. (A), an air core inductor and a pie-wound RF choke; (B), a variable inductor with a nonmagnetic slug, and (C), an inductor with a toroidal magnetic core. The ¼-inch-ruled graph paper background provides a size comparison.

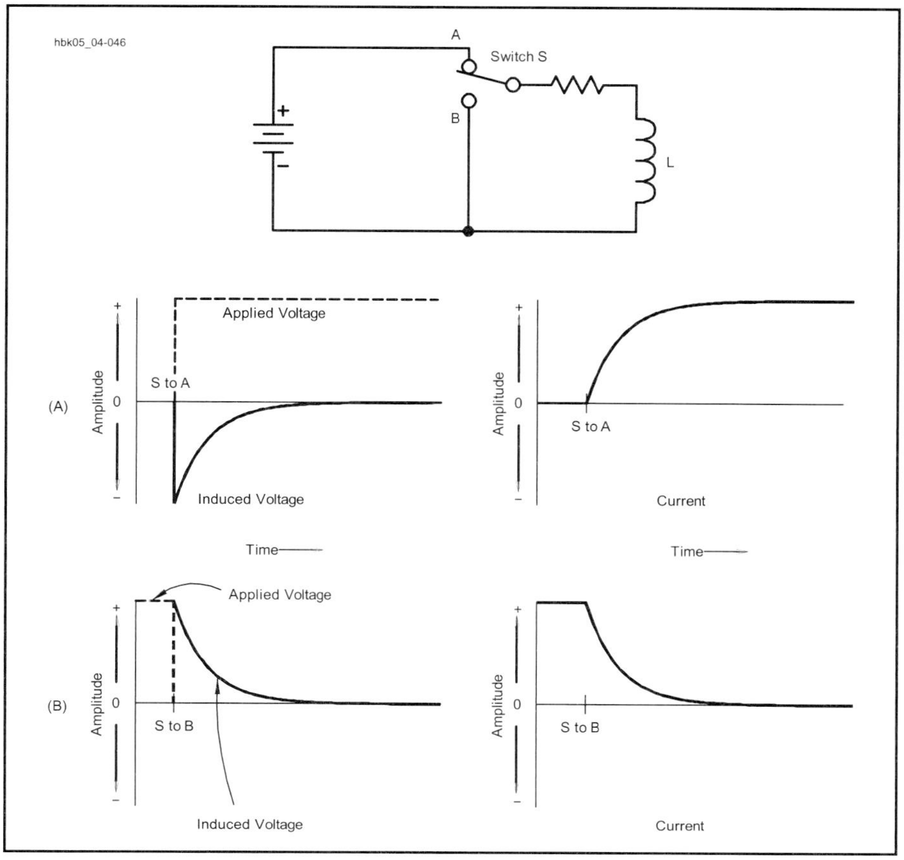

Figure 2.29 — Inductive circuit showing the generation of induced voltage and the rise of current when voltage is applied to an inductor at (A), and the decay of current as the coil shorted at (B).

The amplitude of the induced voltage is proportional to the rate at which the current changes (and consequently, the rate at which the magnetic field changes) and to a constant associated with the inductor itself, *inductance* (*L*). (*Self-inductance* is sometimes used, distinguishing it from *mutual inductance*, as described below.) The basic unit of inductance is the *henry* (abbreviated H).

$$V = -L\frac{\Delta I}{\Delta t}$$

where

V is the induced voltage in volts,

L is the inductance in henries, and

$\Delta I/\Delta t$ is the rate of change of the current in amperes per second.

An inductance of 1 H generates an induced voltage of one volt when the inducing current is varying at a rate of one ampere per second. The minus sign (–) indicates that the induced voltage has a polarity opposing the change in current.

The energy stored in the magnetic field of an inductor is given by the formula:

$$U = \frac{I^2 L}{2}$$

where

U = energy in joules,

I = current in amperes, and

L = inductance in henrys.

This formula corresponds to the energy-storage formula for capacitors: energy storage is a function of current squared. Inductance is proportional to the amount of energy stored in an inductor's magnetic field for a given amount of current. The magnetic field strength, H, is proportional to the number of turns in the inductor's winding, N, (see the equation for magnetic field strength given previously) and for a given amount of current, to the value of μ for the core. Thus, inductance is directly proportional to both N and μ.

The polarity of the induced voltage is always such as to oppose any change in the circuit current. (This is why the term "back" is used, as in back-voltage or *back-EMF* for this reason.) This means that when the current in the circuit is increasing, work is being done against the induced voltage by storing energy in the magnetic field. Likewise, if the current in the circuit tends to decrease, the stored energy of the field returns to the circuit, and adds to the energy being supplied by the voltage source. The net effect of storing and releasing energy is that inductors oppose changes in current just as capacitors oppose changes in voltage. This phenomenon tends to keep the current flowing even though the applied voltage may be decreasing or be removed entirely. Figure 2.29B illustrates the decreasing but continuing flow of current caused by the induced voltage after the source voltage is removed from the circuit.

Inductance depends on the physical configuration of the inductor. All conductors, even straight wires, have inductance. Coiling a conductor increases its inductance. In effect, the growing (or shrinking) magnetic field of each turn produces magnetic lines of force that — in their expansion (or contraction) — intercept the other turns of the coil, inducing a voltage in every other turn. (Recall the two-way relationship between a changing magnetic field and the voltage it creates in a conductor.) The mutuality of the effect, called *magnetic flux linkage* (ψ), multiplies the ability of the coiled conductor to store magnetic energy.

A coil of many turns will have more inductance than one of few turns, if both coils are otherwise physically similar. Furthermore, if an inductor is placed around a magnetic core, its inductance will increase in proportion to the permeability of that core, if the circuit current is below the point at which the core saturates.

In various aspects of radio work, inductors may take values ranging from a fraction of a nanohenry (nH) through millihenrys (mH) up to about 20 H.

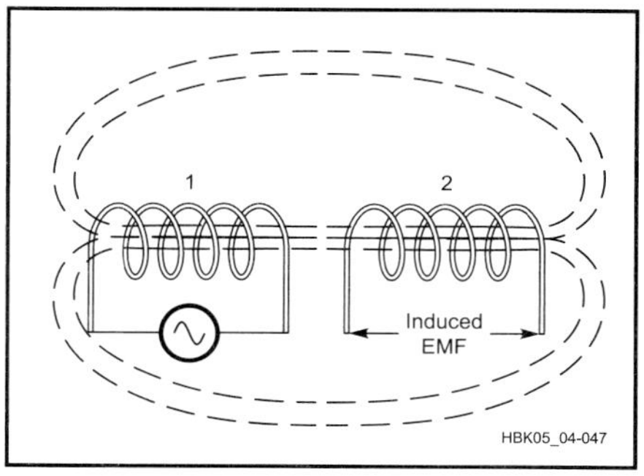

Figure 2.30 — Mutual inductance: When S is closed, current flows through coil number 1, setting up a magnetic field that induces a voltage in the turns of coil number 2.

2.7.4 Mutual Inductance and Magnetic Coupling

When two inductors are arranged with their axes aligned as shown in **Figure 2.30**, current flowing in through inductor 1 creates a magnetic field that intercepts inductor 2. Consequently, a voltage will be induced in inductor 2 whenever the field strength of inductor 1 is changing. This induced voltage is similar to the voltage of self-induction, but since it appears in the second inductor because of current flowing in the first, it is a mutual effect and results from the *mutual inductance* between the two inductors.

When all the flux set up by one coil intercepts all the turns of the other coil, the mutual inductance has its maximum possible value. If only a small part of the flux set up by one coil intercepts the turns of the other, the mutual inductance is relatively small. Two inductors having mutual inductance are said to be *coupled*.

The ratio of actual mutual inductance to the maximum possible value that could theoretically be obtained with two given inductors is called the *coefficient of coupling* between the inductors. It is expressed as a percentage or as a value between 0 and 1. Inductors that have nearly the maximum possible mutual inductance (coefficient = 1 or 100%) are said to be closely, or tightly, coupled. If the mutual inductance is relatively small the inductors are said to be loosely coupled. The degree of coupling depends upon the physical spacing between the inductors and how they are placed with respect to each other. Maximum coupling exists when they have a common or parallel axis and are as close together as possible (for example, one wound over the other). The coupling is least when the inductors are far apart or are placed so their axes are at right angles.

The maximum possible coefficient of coupling is closely approached when the two inductors are wound on a closed iron core. The coefficient with air-core inductors may run as high as 0.6 or 0.7 if one inductor is wound over the other, but will be much less if the two inductors are separated. Although unity coupling is suggested by Figure 2.30, such coupling is possible only when the inductors are wound on a closed magnetic core.

Coupling between inductors can be minimized by using separate closed magnetic cores for each. Since an inductor's magnetic field is contained almost entirely in a closed core, two inductors with separate closed cores, such as the toroidal inductor in Figure 2.28C, can be placed close together in almost any relative orientation without coupling.

UNWANTED COUPLING

The inductance of a short length of straight wire is small, but it may not be negligible. (In free-space, round wire has an inductance on the order of 1 μH/m, but this is affected by wire diameter and the total circuit's physical configuration.) Appreciable voltage may be induced in even a few inches of wire carrying ac by changing magnetic fields with a frequency on the order of 100 MHz or higher. At much lower frequencies or at dc, the inductance of the same wire might be ignored because the induced voltage would be very small.

There are many phenomena, both natural and man-made, that create sufficiently strong or rapidly changing magnetic fields to induce voltages in conductors. Many of them create brief but intense pulses of energy called *transients* or "spikes." The magnetic fields from these transients intercept wires leading into and out of — and wires wholly within — electronic equipment, inducing unwanted voltages by mutual coupling.

Lightning is a powerful natural source of magnetically coupled transients. Strong transients can also be generated by sudden changes in current in nearby circuits or wiring. High-speed digital signals and pulses can also induce voltages in adjacent conductors.

2.7.5 Inductances in Series and Parallel

When two or more inductors are connected in series (left side of **Figure 2.31**), the total inductance is equal to the sum of the individual inductances, provided that the inductors are sufficiently separated so that there is no coupling between them (see the preceding section):

$$L_{total} = L_1 + L_2 + L_3 \ldots + L_n$$

If inductors are connected in parallel (right side of Figure 2.31), again assuming no mutual coupling, the total inductance is given by:

$$L_{total} = \frac{1}{\frac{1}{L_1} + \frac{1}{L_2} + \frac{1}{L_3} + \ldots + \frac{1}{L_n}}$$

For only two inductors in parallel, the formula becomes:

$$L_{total} = \frac{L_1 \times L_2}{L_1 + L_2}$$

Thus, the rules for combining inductances in series and parallel are the same as those for resistances, assuming there is no coupling between the inductors. When there is coupling between the inductors, the formulas given above will not yield correct results.

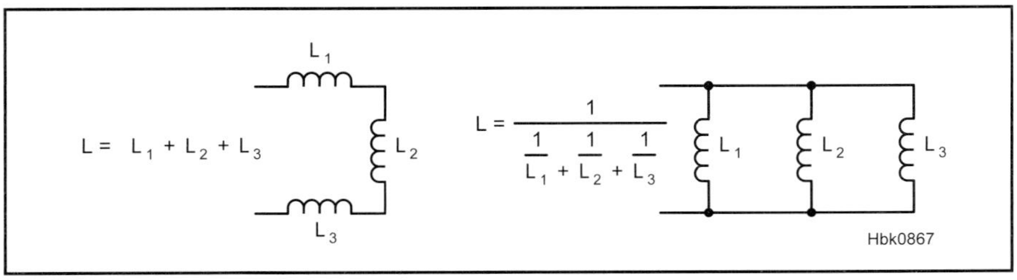

Figure 2.31 — Inductors in series and parallel.

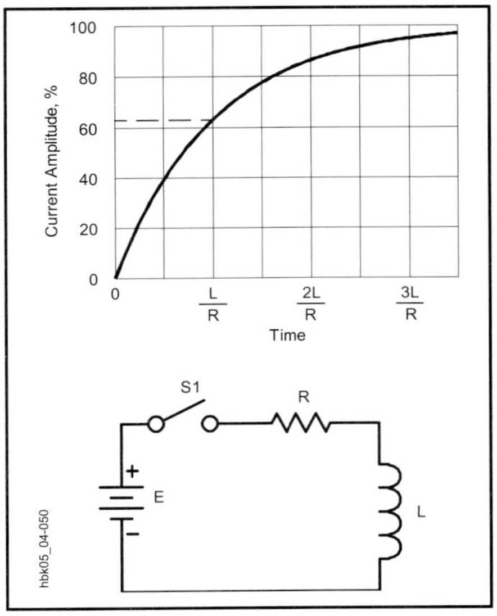

Figure 2.32 — Time constant of an RL circuit being energized.

2.7.6 RL Time Constant

As with capacitors, the time dependence of inductor current is a significant property. A comparable situation to an RC circuit exists when resistance and inductance are connected in series. In **Figure 2.32**, first consider the case in which R is zero. Closing S1 sends a current through the circuit. The instantaneous transition from no current to a finite value, however small, represents a rapid change in current, and an opposing voltage is induced in L. The value of the opposing voltage is almost equal to the applied voltage, so the resulting initial current is very small.

The opposing voltage is created by change in the inductor current and would cease to exist if the current did not continue to increase. With no resistance in the circuit, the current would increase forever, always growing just fast enough to keep the self-induced opposing voltage just below the applied voltage.

When resistance in the circuit limits the current, the opposing voltage induced in L must only equal the difference between E and the drop across R, because that is the voltage actually applied to L. This difference becomes smaller as the current approaches its final value, limited by Ohm's Law to I = E/R. Theoretically, the opposing voltage never quite disappears, and so the current never quite reaches the Ohm's Law limit. In practical terms, the difference eventually becomes insignificant, just as described above for capacitors charging to an applied voltage through a resistor.

The inductor current at any time after the switch in Figure 2.32 has been closed can be found from:

$$I(t) = \frac{E}{R}\left(1 - e^{-\frac{tR}{L}}\right)$$

where

I(t) = current in amperes at time t,
E = power source potential in volts,
t = time in seconds after application of voltage,
e = natural logarithmic base = 2.718,
R = circuit resistance in ohms, and
L = inductance in henrys.

(References that explain exponential equations, e, and other mathematical topics are found in the "Radio Mathematics" article in this book's online information.) The term E/R in this equation represents the dc value of I, or the value of I(t) when t becomes very large; this is the *steady-state value* of I. If t = L/R, the above equation becomes:

$$V(L/R) = \frac{E}{R}(1 - e^{-1}) \approx 0.632 \frac{E}{R}$$

The time in seconds required for the current to build up to 63.2% of the maximum value is called the *time constant* (also the *RL time constant*), and is equal to L/R, where L is in henrys and R is in ohms. (Time constants are also discussed in the section on RC circuits above.) After each time interval equal to this constant, the current increases by an additional 63.2% closer to the final value of E/R. This behavior is graphed in Figure 2.32. As is the case with capacitors, after five time constants the current is considered to have reached its maximum value. As with capacitors, we use the lower-case Greek tau (τ) to represent the time constant.

Example: If a circuit has an inductor of 5.0 mH in series with a resistor of 10 Ω, how long will it take for the current in the circuit to reach full value after power is applied? Since achieving maximum current takes approximately five time constants,

$$t = \frac{5\,L}{R} = \frac{5 \times 5.0 \times 10^{-3}\ H}{10\ \Omega}$$

$$= 2.5 \times 10^{-3}\ \text{seconds} = 2.5\ ms$$

Note that if the inductance is increased to 5.0 H, the required time increases by a factor of 1,000 to 2.5 seconds. Since the circuit resistance didn't change, the final current is the same for both cases in this example. Increasing inductance increases the time required to reach full current.

Zero resistance would prevent the circuit from ever achieving full current. All practical inductors have some resistance in the wire making up the inductor.

An inductor cannot be discharged in the simple circuit of Figure 2.32 because the magnetic field ceases to exist or "collapses" as soon as the current ceases. Opening S1 does not leave the inductor charged in the way that a capacitor would remain charged. Energy storage in a capacitor depends on the separated charges staying in place. Energy storage in an inductor depends on the charges continuing to move as current.

The energy stored in the inductor's magnetic field attempts to return instantly to the circuit when S1 is opened. The rapidly changing (collapsing) field in the inductor causes a very large voltage to be induced across the inductor. Because the change in current is now in the opposite direction, the induced voltage also reverses polarity. This induced voltage (called *inductive kick-back*) is usually many times larger than the originally applied voltage, because the induced voltage is proportional to the rate at which the field changes.

The common result of opening the switch in such a circuit is that a spark or arc forms at the switch contacts during the instant the switch opens. When the inductance is large and the current in the circuit is high, large amounts of energy are released in a very short time. It is not at all unusual for the switch contacts to burn or melt under such circumstances.

The spark or arc at the opened switch can be reduced or suppressed by connecting a suitable capacitor and resistor in series across the contacts to absorb the energy non-destructively. Such an RC combination is called a *snubber network*. The current rating for a switch may be significantly reduced if it is used in an inductive circuit.

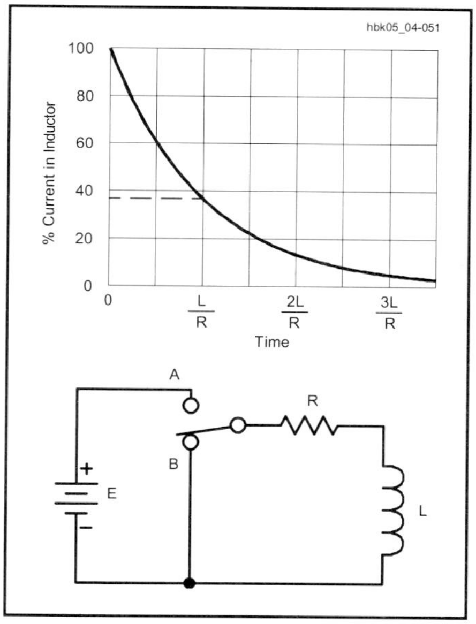

Figure 2.33 — Time constant of an RL circuit being de-energized. This is a theoretical model only, since a mechanical switch cannot change state instantaneously.

Transistor switches connected to and controlling inductors, such as relays and solenoids, also require protection from the high kick-back voltages. In most cases, a small power diode connected across the relay coil so that it does not conduct current when the inductor is energized (called a *kick-back diode*) will protect the transistor.

If the excitation is removed without breaking the circuit, as shown in **Figure 2.33**, the current will decay according to the formula:

$$I(t) = \frac{E}{R}\left(e^{-\frac{tR}{L}}\right)$$

where t = time in seconds after removal of the source voltage.

After one time constant the current will decay by 63.2% to be 36.8% of the steady-state value.The graph in Figure 2.33 shows the current-decay waveform to be identical to the voltage-discharge waveform of a capacitor. Be careful about applying the terms *charge* and *discharge* to an inductive circuit, however. These terms refer to energy storage in an electric field. An inductor stores energy in a magnetic field and the usual method of referring to the process is *energize* and *de-energize* (although it is not always followed).

2.8 Semiconductor Devices

2.8.1 Introduction to Semiconductors

In a conductor, such as a metal, some of the outer, or *valence*, electrons of each atom are free to move about between atoms. These *free electrons* are the constituents of electrical current. In a good conductor, the concentration of these free electrons is very high, on the order of 10^{22} electrons/cm^3. In an insulator, nearly all the electrons are tightly held by their atoms and the concentration of free electrons is very small — on the order of 10 electrons/cm^3.

Between the classes of materials considered to be conductors and insulators is a class of elements called *semiconductors*, materials with conductivity much poorer than metals and much better than insulators. (In electronics, "semiconductor" means a device made from semiconductor elements that have been chemically manipulated as described below, leading to interesting properties that create useful applications.)

Semiconductor atoms (silicon, Si, is the most widely used) share their valence electrons in a chemical bond that holds adjacent atoms together, forming a three-dimensional *lattice* that gives the material its physical characteristics. A lattice of pure semiconductor material (one type of atom or molecule) can form a crystal, in which the lattice structure and orientation is preserved throughout the material. *Monocrystalline* or "single-crystal" is the type of material used in electronic semiconductor devices. *Polycrystalline* material is made of up many smaller crystals with their own individual lattice orientations.

MAJORITY AND MINORITY CARRIERS

Crystals of pure semiconductor material are called *intrinsic* semiconductors. When energy, generally in the form of heat, is added to a semiconductor crystal lattice, some electrons are liberated from their bonds and move freely throughout the lattice. The bond that loses an electron is then unbalanced and the space that the electron came from is referred to as a *hole*. In these materials the number of free electrons is equal to the number of holes.

Electrons from adjacent bonds can leave their positions to fill the holes, thus leaving behind a hole in their old location. As a consequence of the electron moving, two opposite movements can be said to occur: negatively charged electrons move from bond to bond in one direction and positively charged holes move from bond to bond in the opposite direction. Both of these movements represent forms of electrical current, but this is very different from the current in a conductor. While a conductor has free electrons that flow independently from the bonds of the crystalline lattice, the current in a pure semiconductor is constrained to move from bond to bond.

MATERIAL DOPING

Impurities can be added to intrinsic semiconductors (by a process called *doping*) to enhance the formation of electrons or holes and thus improve conductivity. These materials are *extrinsic* semiconductors. Since the additional electrons and holes can move, their movement is current and they are called *carriers*. The type of carrier that predominates in the material is called the *majority carrier*. In N-type material the majority carriers are electrons and in P-type material, holes.

There are two types of impurities that can be added: a *donor impurity* with five valence electrons donates free electrons to the crystalline structure; this is called an *N-type* impurity, for the negative charge of the majority carriers. Some examples of donor impurities are antimony (Sb), phosphorus (P), and arsenic (As). N-type extrinsic semiconductors have more electrons and fewer holes than intrinsic semiconductors. *Acceptor impurities* with three valence electrons accept free electrons from the lattice, adding holes to the overall structure. These are called P-type impurities, for the positive charge of the majority carriers; some examples are boron (B), gallium (Ga), and indium (In).

It is important to note that even though N-type and P-type material have different numbers of holes and free electrons than intrinsic material, they are still electrically neutral. When an electron leaves an atom, the positively charged atom that remains in place in the crystal lattice electrically balances the roaming free electron. Similarly, an atom gaining an electron acquires a negative charge that balances the positively charged atom it left. At no time does the material acquire a net electrical charge, positive or negative.

Compound semiconductor material can be formed by combining equal amounts of N-type and P-type impurity materials. Some examples of this include gallium-arsenide (GaAs), gallium-phosphate (GaP), and indium-phosphide (InP). To make an N-type compound semiconductor, a slightly higher amount of N-type material is used in the mixture. A P-type compound semiconductor has a little more P-type material in the mixture.

Impurities are introduced into intrinsic semiconductors by diffusion, the same physical process that lets you smell cookies baking from several rooms away. (Molecules diffuse through air much faster than through solids.) Rates of diffusion are proportional to temperature, so semiconductors are doped with impurities at high temperature to save time. Once the doped semiconductor material is cooled, the rate of diffusion of the impurities is so low that they are essentially immobile for many years to come. If an electronic device made from a structure of N- and P-type materials is raised to a high temperature, such as by excessive current, the impurities can again migrate and the internal structure of the device may be destroyed. The maximum operating temperature for semiconductor devices is specified at a level low enough to limit additional impurity diffusion.

The conductivity of an extrinsic semiconductor depends on the charge density (in other words, the concentration of free electrons in N-type, and holes in P-type, semiconductor material). As the energy in the semiconductor increases, the charge density also increases. This is the basis of how all semiconductor devices operate: the major difference is the way in which the energy level is increased. Variations include: The *transistor*, where conductivity is altered by injecting current into the device via a wire; the *thermistor*, where the level of heat in the device is detected by its conductivity, and the *photoconductor*, where light energy that is absorbed by the semiconductor material increases the conductivity.

2.8.2 The PN Semiconductor Junction

If a piece of N-type semiconductor material is placed against a piece of P-type semiconductor material, the location at which they join is called a *PN junction*. The junction has characteristics that make it possible to develop diodes and transistors. The action of the junction is best described by a diode operating as a rectifier.

Initially, when the two types of semiconductor material are placed in contact, each type of material will have only its majority carriers: P-type will have only holes and N-type will have only free electrons. The presence of the positive charges (holes) in the P-type material attracts free electrons from the N-type material immediately across the junction. The opposite is true in the N-type material.

These attractions lead to diffusion of some of the majority carriers across the junction, which combine with and neutralize the majority carriers immediately on the other side (a process called *recombination*). As distance from the junction increases, the attraction quickly becomes too small to cause the carriers to move. The region close to the junction is then *depleted* of carriers, and so is named the *depletion region* (also the *space-charge region* or the *transition region*). The width of the depletion region is very small, on the order of 0.5 µm.

If the N-type material (the *cathode*) is placed at a more negative voltage than the P-type material (the *anode*), current will pass through the junction because electrons are attracted from the lower potential to the higher potential and holes are attracted in the opposite direction. This *forward bias* forces the majority carriers toward the junction where recombination occurs with the opposite type of majority carrier. The source of voltage supplies replacement electrons to the N-type material and removes electrons from the P-type material so that the majority carriers are continually replenished. Thus, the net effect is a *forward current* flowing through the semiconductor, across the PN junction. The *forward resistance* of a diode conducting current is typically very low and varies with the amount of forward current.

When the polarity is reversed, majority-carriers are attracted away from the junction, not toward it. Very little current flows across the PN junction — called *reverse leakage current* — in this case. Allowing only unidirectional current flow is what allows a semiconductor diode to act as rectifier.

2.8.3 Junction Semiconductors

Semiconductor devices that operate using the principles of a PN junction are called *junction semiconductors*. These devices can have one or several junctions. The properties of junction semiconductors can be tightly controlled by the characteristics of the materials used and the size and shape of the junctions.

SEMICONDUCTOR DIODES

Diodes are commonly made of silicon and occasionally germanium. Although they act similarly, they have slightly different characteristics. The *junction threshold voltage*, or *junction barrier voltage,* is the forward bias voltage (V_F) at which current begins to pass through the device. This voltage is different for the two kinds of diodes. In the diode response curve of **Figure 2.34**, V_F corresponds to the voltage at which the positive portion of the curve begins to rise sharply from the x-axis. Most silicon diodes have a junction threshold voltage of about 0.7 V, while the voltage for germanium diodes is typically 0.3 V. Reverse leakage current is much lower for silicon diodes than for germanium diodes.

The characteristic curve for a semiconductor diode junction is given by the following equation (slightly simplified) called the *Fundamental Diode Equation* because it describes the behavior of all semiconductor PN junctions.

$$I = I_S \left(e^{\frac{V}{\eta V_t}} - 1 \right)$$

where

I = diode current,

V = diode voltage,

I_s = reverse-bias saturation current,

V_t = kT/q, the thermal equivalent of voltage (about 25 mV at room temperature), and

η = emission coefficient

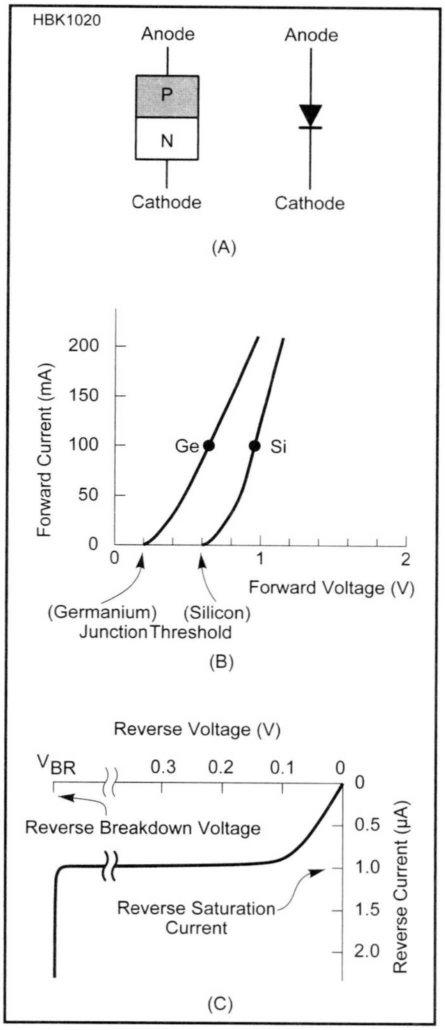

Figure 2.34 — Semiconductor diode (PN junction) structure (A). Characteristic curve (B) shows forward-biased (anode positive with respect to cathode) response for germanium, Ge, and silicon, Si, devices. Each curve breaks away from the X-axis at its junction threshold voltage. The slope of each curve is its forward resistance. (C) Reverse-biased response. Very small reverse current increases until it reaches the reverse saturation current I_0. The reverse current increases suddenly and drastically when the reverse voltage reaches the reverse breakdown voltage, V_{BR}.

The value of I_s varies with the type of semiconductor material, with the value of 10^{-12} used for silicon. η also varies from 1 to 2 with the type of material and method of fabrication. (η is close to 1 for silicon at normal current values, increasing to 2 at high currents.) This curve is shown in **Figure 2.35B**.

The obvious differences between Figure 2.35A and B are that the semiconductor diode has a finite *turn-on* voltage — it requires a small but nonzero forward bias voltage before it begins conducting. Furthermore, once conducting, the diode voltage continues to increase very slowly with increasing current, unlike a true short circuit. Finally, when the applied voltage is negative, the reverse current is not exactly zero but very small (microamperes). The reverse current flow rapidly reaches a level that varies little with the reverse bias voltage. This is the *reverse-bias saturation current*, I_s.

For bias (dc) circuit calculations, a useful model for the diode that takes these two effects into account is shown by the artificial I-V curve in Figure 2.35C. This model neglects the negligible reverse bias current I_s.

When converted into an equivalent circuit, the model in Figure 2.35C yields the circuit in Figure 2.35D. The ideal voltage source V_a represents the turn-on voltage and R_f represents the effective resistance caused by the small increase in diode voltage as the diode current increases. The turn-on voltage is material-dependent: approximately 0.3 V

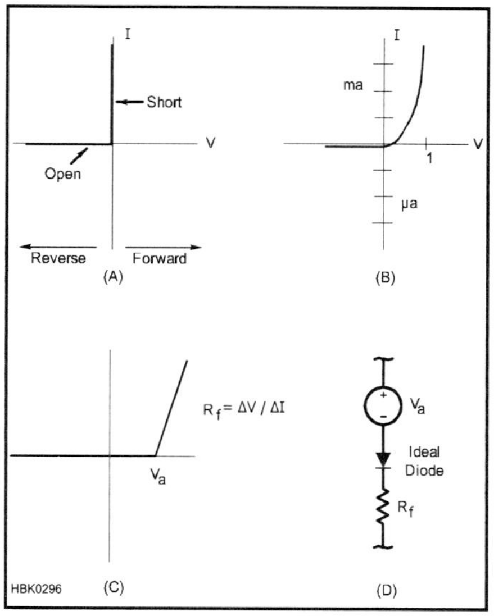

Figure 2.35 — Circuit models for rectifying switches (diodes). (A): I-V curve of the ideal rectifier. (B): I-V curve of a typical semiconductor diode showing the typical small leakage current in the reverse direction. Note the different scales for forward and reverse current. (C) shows a simplified diode I-V curve for dc-circuit calculations (at a much larger scale than (B)). (D) is an equivalent circuit for (C).

for germanium diodes and 0.7 for silicon. R_f is typically on the order of 10 Ω, but it can vary according to the specific component. R_f can often be completely neglected in comparison to the other resistances in the circuit. This very common simplification leaves only a pure *voltage drop* for the diode model.

THYRISTORS

Thyristors are semiconductors made with four or more alternating layers of P- and N-type semiconductor material. In a four-layer thyristor, when the anode is at a higher potential than the cathode, the first and third junctions are forward biased and the center junction reverse biased. In this state, there is little current, just as in the reverse-biased diode. The different types of thyristor have different ways in which they turn on to conduct current and in how they turn off to interrupt current flow.

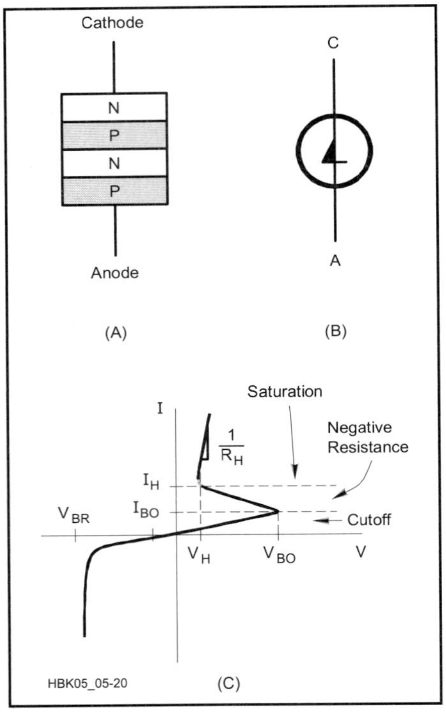

Figure 2.36 — PNPN diode. (A) Alternating layers of P-type and N-type semiconductor. (B) Schematic symbol with cathode, C, and anode, A, leads. (C) I-V curve. Reverse-biased response is the same as normal PN junction diodes. Forward biased response acts as a hysteresis switch. Resistance is very high until the bias voltage reaches V_{BO} (where the center junction breaks over) and exceeds the cutoff current, I_{BO}. The device exhibits a negative resistance when the current increases as the bias voltage decreases until a voltage of V_H and saturation current of I_H is reached. After this, the resistance is very low, with large increases in current for small voltage increases.

PNPN Diode

The simplest thyristor is a PNPN (usually pronounced like *pinpin*) diode with three junctions (see **Figure 2.36**). As the forward bias voltage is increased, the current through the device increases slowly until the *breakover* (or *firing)* voltage, V_{BO}, is reached and the flow of current abruptly increases. The PNPN diode is often considered to be a switch that is off below V_{BO} and on above it.

Bilateral Diode Switch (Diac)

A semiconductor device similar to two PNPN diodes facing in opposite directions and attached in parallel is the *bilateral diode switch* or *diac*. This device has the characteristic curve of the PNPN diode for both positive and negative bias voltages. Its construction, schematic symbol and characteristic curve are shown in **Figure 2.37**.

Silicon Controlled Rectifier (SCR)

Another device with four alternate layers of P-type and N-type semiconductor is the *silicon controlled rectifier (SCR)*. (Some sources refer to an SCR as a thyristor, as well.) In addition to the connections to the outer two layers, two other terminals can be brought out for the inner two layers. The connection to the P-type material near the cathode is called the *cathode gate* and the N-type material near the anode is called the *anode gate*. In nearly all commercially available SCRs, only the cathode gate is connected (**Figure 2.38**).

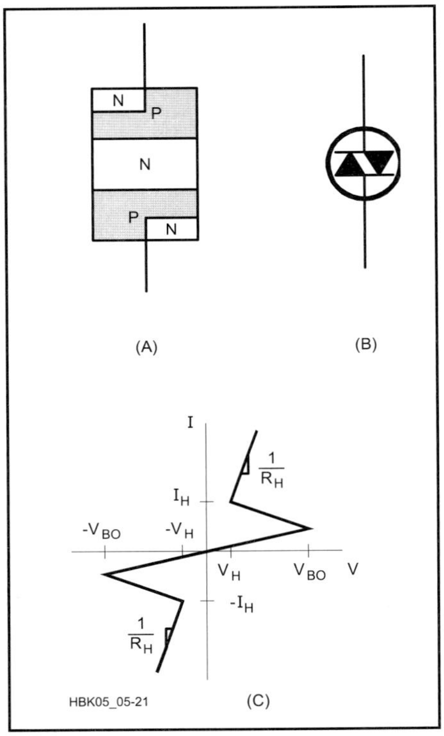

Figure 2.37 — Bilateral switch. (A) Alternating layers of P-type and N-type semiconductor. (B) Schematic symbol. (C) I-V curve. The right-hand side of the curve is identical to the PNPN diode response in Figure 2.36. The device responds identically for both forward and reverse bias so the left-hand side of the curve is symmetrical with the right-hand side.

Like the PNPN diode switch, the SCR is used to abruptly start conducting when the voltage exceeds a given level. By biasing the gate terminal appropriately, the breakover voltage can be adjusted.

Triac

A five-layered semiconductor whose operation is similar to a bidirectional SCR is the *triac* (**Figure 2.39**). This is also similar to a bidirectional diode switch with a bias control gate. The gate terminal of the triac can control both positive and negative breakover voltages, and the devices can pass both polarities of voltage.

Thyristor Applications

The SCR is highly efficient and is used in power control applications. SCRs are available that can handle currents of greater than 100 A and voltage differentials of greater

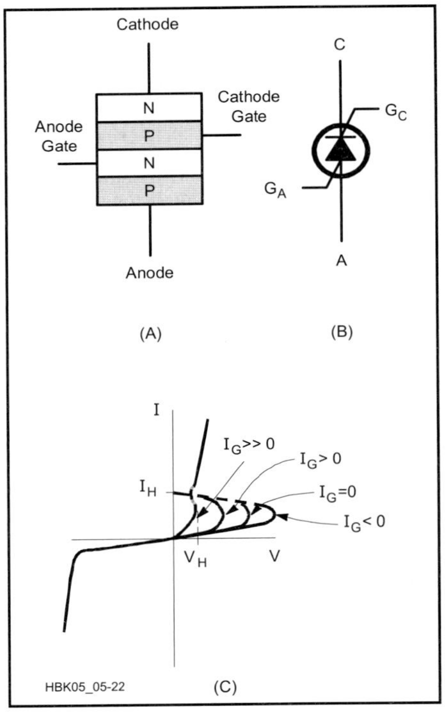

Figure 2.38 — SCR. (A) Alternating layers of P-type and N-type semiconductor. This is similar to a PNPN diode with gate terminals attached to the interior layers. (B) Schematic symbol with anode, A, cathode, C, anode gate, G_A, and cathode gate, G_C. Many devices are constructed without G_A. (C) I-V curve with different responses for various gate currents. $I_G = 0$ has a similar response to the PNPN diode.

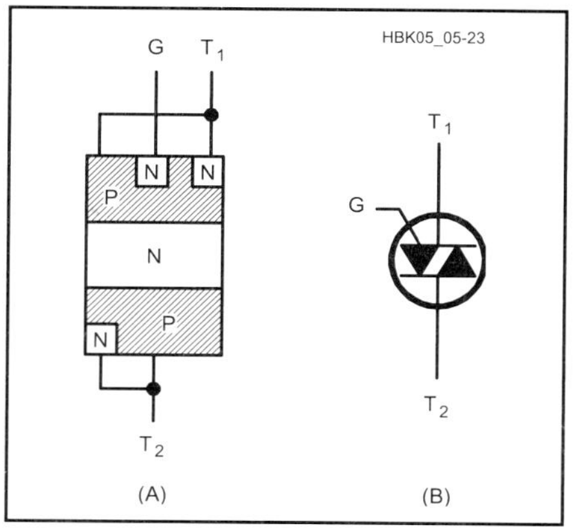

Figure 2.39 — Triac. (A) Alternating layers of P-type and N-type semiconductor. This behaves as two SCR devices facing in opposite directions with the anode of one connected to the cathode of the other and the cathode gates connected together. (B) Schematic symbol.

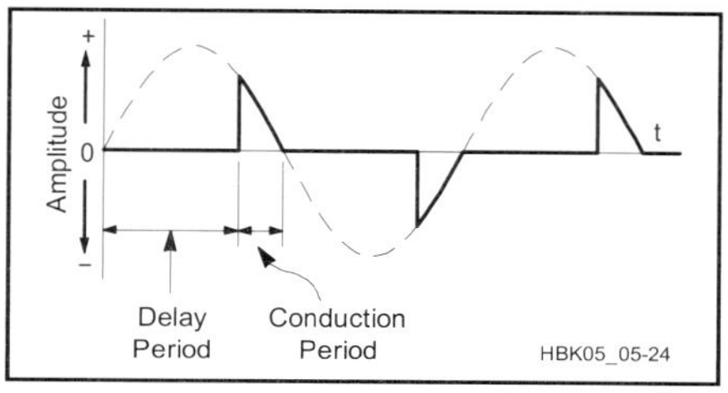

Figure 2.40 — Triac operation on sine wave. The dashed line is the original sine wave and the solid line is the portion that conducts through the triac. The relative delay and conduction period times are controlled by the amount or timing of gate current, I_G. The response of an SCR is the same as this for positive voltages (above the X-axis) and with no conduction for negative voltages.

than 1,000 V, yet can be switched with gate currents of less than 50 mA. Because of their high current-handling capability, SCRs are used as "crowbars" in power supply circuits, to short the output to ground and blow a fuse when an overvoltage condition exists.

SCRs and triacs are often used to control ac power sources. A sine wave with a given RMS value can be switched on and off at preset points during the cycle to decrease the RMS voltage. When conduction is delayed until after the peak (as **Figure 2.40** shows) the peak-to-peak voltage is reduced. If conduction starts before the peak, the RMS voltage is reduced, but the peak-to-peak value remains the same. This method is used to operate light dimmers and 240 V ac to 120 V ac converters. The sharp switching transients created when these devices turn on are common sources of RF interference. (See the chapter on **RF Interference** for information on dealing with interference from thyristors.)

BIPOLAR TRANSISTOR

A *bipolar junction transistor (BJT)* is created by three alternating layers or regions of N- and P-type material as shown in **Figure 2.41**. The designation NPN or PNP describes the order of the regions as shown in the figure.

Physically, we can think of the transistor as two PN junctions back-to-back, such as two diodes connected at their *anodes* (the positive terminal) for an NPN transistor or two diodes connected at their *cathodes* (the negative terminal) for a PNP transistor. The connection point is the base of the transistor. (You can't actually make a transistor this way — this is a representation for illustration only.)

A transistor conducts when the base-emitter junction is forward biased and the base-collector is reverse biased. Under these conditions, the emitter region emits majority carriers into the base region, where they become minority carriers because the materials of the emitter and base regions have opposite polarity. The excess minority carriers in the base are then attracted across the very thin base to the base-collector junction, where they are collected and are once again considered majority carriers before they can flow to the base terminal.

The flow of majority carriers from emitter to collector can be modified by the application of a bias current to the base terminal. If the bias current causes majority carriers to be injected into the base material (electrons flowing into an N-type base or out of a P-type base), the emitter-collector current increases. In this way, a transistor allows a small base current to control a much larger collector current.

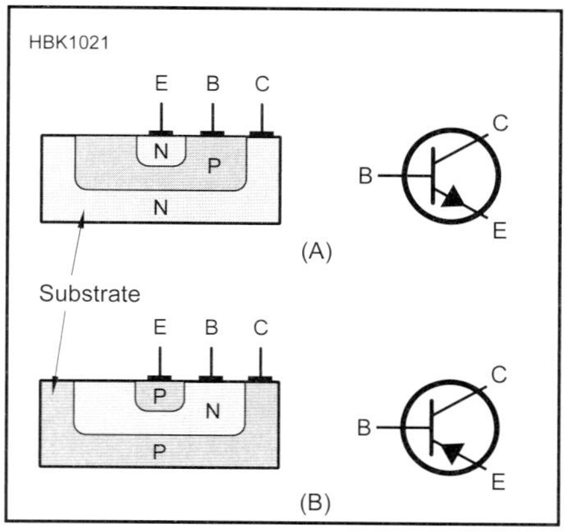

Figure 2.41 — Bipolar junction transistors (BJT) are usually constructed as three regions of N- or P-type material with a substrate at the bottom, the base region embedded into the substrate, and the emitter region embedded in the base. The order of the layers determines whether the device is NPN (A) or PNP (B). There are three electrodes: collector, C, base, B, and emitter, E. In the schematic symbol for an NPN device the arrow points away from the base; for a PNP device it points toward the base.

As in a semiconductor diode, the forward biased base-emitter junction has a threshold voltage (V_{BE}) that must be exceeded before the emitter current increases. As the base-emitter current continues to increase, the point is reached at which further increases in base-emitter current cause no additional change in collector current. This is the condition of *saturation*. Conversely, when base-emitter current is reduced to the point at which collector current ceases to flow, that is the situation of *cutoff*.

2.8.4 Field-Effect Transistors (FET)

The *field-effect transistor (FET)* controls the current between two points but does so differently than the bipolar transistor. The FET operates by the effects of an electric field on the flow of electrons through a single type of semiconductor material. This is why the FET is sometimes called a *unipolar* transistor. Unlike bipolar semiconductors that can be arranged in many configurations to provide diodes, transistors, photoelectric devices, temperature sensitive devices, and so on, the field effect technique is usually only used to make transistors, although FETs are also available as special-purpose diodes, for use as constant current sources.

FET devices are constructed on a *substrate* of doped semiconductor material. The channel is formed within the substrate and has the opposite polarity (a P-channel FET has N-type substrate). Most FETs are constructed with silicon.

Within the FET, current moves in a *channel* as shown in **Figure 2.42**. The channel is made

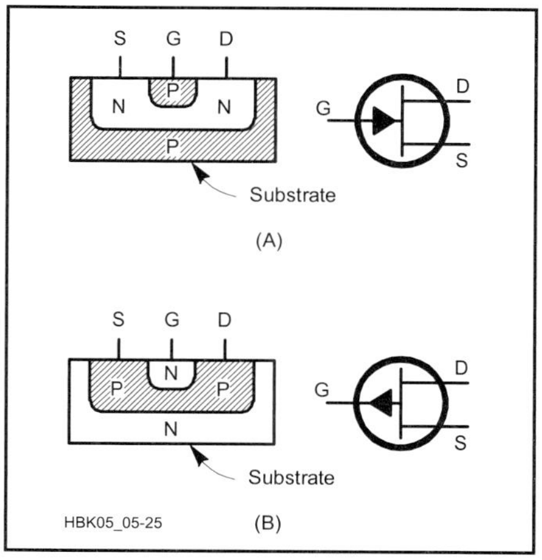

Figure 2.42 — JFET devices with terminals labeled: source, S, gate, G and drain, D. (A) Pictorial of N-type channel embedded in P-type substrate and schematic symbol. (B) P-channel embedded in N-type substrate and schematic symbol.

of either N-type or P-type semiconductor material; an FET is specified as either an N-channel or P-channel device. Current flows from the *source* terminal (where majority carriers are injected) to the *drain* terminal (where majority carriers are removed). A *gate* terminal generates an electric field that controls the current in the channel.

In N-channel devices, the drain potential must be higher than that of the source (V_{DS} > 0) for electrons (the majority carriers) to flow in the channel. In P-channel devices, the flow of holes requires that $V_{DS} < 0$. The polarity of the electric field that controls current in the channel is determined by the majority carriers of the channel, ordinarily positive for P-channel FETs and negative for N-channel FETs.

Variations of FET technology are based on different ways of generating the electric field. In all of these, however, electrons at the gate are used only for their charge in order to create an electric field around the channel. There is a minimal flow of electrons through the gate. This leads to a very high dc input resistance in devices that use FETs for their input circuitry. There may be quite a bit of capacitance between the gate and the other FET terminals, however, causing the input impedance to be quite low at high frequencies.

The current through an FET only has to pass through a single type of semiconductor material. Depending on the type of material and the construction of the FET, drain-source resistance when the FET is conducting ($r_{DS(ON)}$) may be anywhere from a few hundred ohms to much less than an ohm. The output impedance of devices made with FETs is generally quite low. If a gate bias voltage is added to operate the transistor near cutoff, the circuit output impedance may be much higher.

In order to achieve a higher gain-bandwidth product, other materials have been used. Gallium-arsenide (GaAs) has *electron mobility* and *drift velocity* (both are measures of how easily electrons are able to move through the crystal lattice) far higher than the standard doped silicon. Amplifiers designed with GaAsFET devices operate at much higher frequencies and with a lower noise factor at VHF and UHF than those made with silicon FETs (although silicon FETs have improved dramatically in recent years).

JFET

One of two basic types of FET, the *junction FET (JFET)* gate material is made of the opposite polarity semiconductor to the channel material (for a P-channel FET the gate is made of N-type semiconductor material). The gate-channel junction is similar to a diode's PN junction with the gate material in direct contact with the channel. JFETs are used with the junction reverse-biased, since any current in the gate is undesirable. The reverse bias of the junction creates an electric field that "pinches" the channel. Since the magnitude of the electric field is proportional to the reverse-bias voltage, the current in the channel is reduced for higher reverse gate bias voltages. When current in the channel is completely halted by the electric field, this is called *pinch-off,* which is analogous to cutoff in a bipolar transistor. The channel in a JFET is at its maximum conductivity when the gate and source voltages are equal ($V_{GS} = 0$).

Because the gate-channel junction in a JFET is similar to a bipolar junction diode, this junction must never be forward biased; otherwise large currents will pass through the gate and into the channel. For an N-channel JFET, the gate must always be at a lower potential than the source ($V_{GS} < 0$). The prohibited condition is for $V_{GS} > 0$. For P-channel JFETs these conditions are reversed (in normal operation $V_{GS} > 0$ and the prohibited condition is for $V_{GS} < 0$).

MOSFET

Placing an insulating layer between the gate and the channel allows for a wider range of control (gate) voltages and further decreases the gate current (and thus increases the device input resistance). The insulator is typically made of an oxide (such as silicon dioxide, SiO_2). This type of device is called a *metal-oxide-semiconductor FET (MOSFET)* or *insulated-gate FET (IGFET).*

The substrate is often connected to the source internally. The insulated gate is on the opposite side of the channel from the substrate (see **Figure 2.43**). The bias voltage on the gate terminal either attracts or repels the majority carriers of the substrate across its PN-junction with the channel. This narrows (*depletes*) or widens (*enhances*) the channel, respectively, as V_{GS} changes polarity. For example, in the N-channel enhancement-mode MOSFET, positive gate voltages with respect to the substrate and the source (V_{GS} > 0) repel holes from the channel into the substrate, thereby widening the channel and

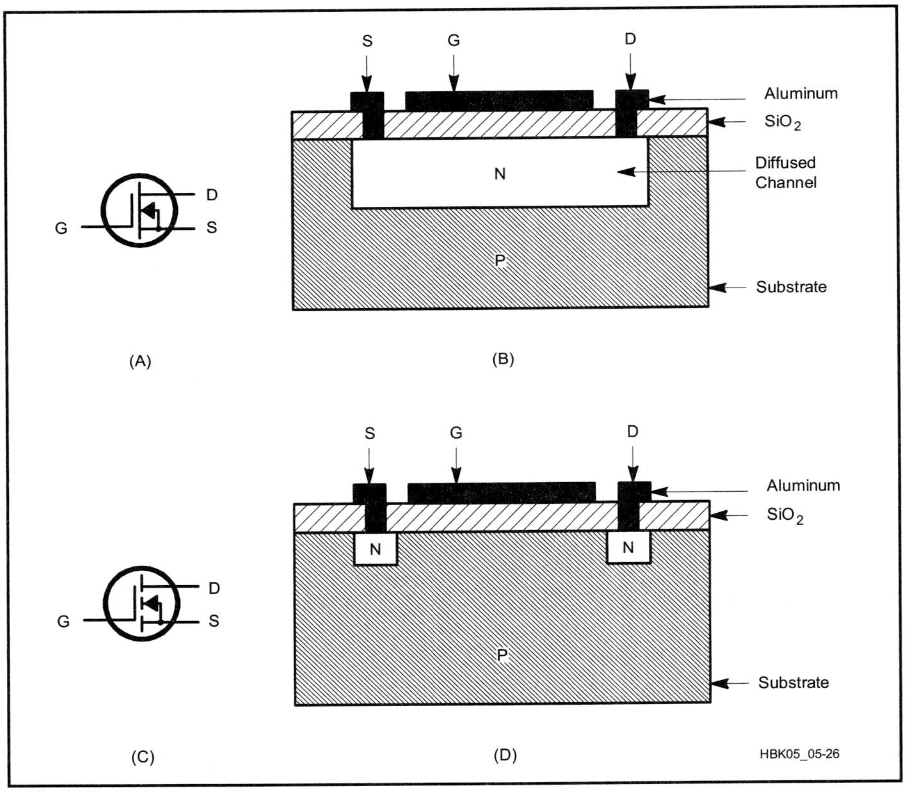

Figure 2.43 — MOSFET devices with terminals labeled: source, S, gate, G, and drain, D. N-channel devices are pictured. P-channel devices have the arrows reversed in the schematic symbols and the opposite type semiconductor material for each of the layers. (A) N-channel depletion mode device schematic symbol and (B) pictorial of P-type substrate, diffused N-type channel, SiO_2 insulating layer, and aluminum gate region and source and drain connections. The substrate is connected to the source internally. A negative gate potential narrows the channel. (C) N-channel enhancement mode device schematic and (D) pictorial of P-type substrate, N-type source and drain wells, SiO_2 insulating layer, and aluminum gate region and source and drain connections. Positive gate potential forms a channel between the two N-type wells by repelling the P-carriers away from the channel region in the substrate.

decreasing channel resistance. Conversely, $V_{GS} < 0$ causes holes to be attracted from the substrate, narrowing the channel and increasing the channel resistance. Once again, the polarities discussed in this example are reversed for P-channel devices. The common abbreviation for an N-channel MOSFET is *NMOS*, and for a P-channel MOSFET, *PMOS*.

Because of the insulating layer next to the gate, input resistance of a MOSFET is usually greater than 10^{12} Ω (a million megohms). Since MOSFETs can both deplete the channel, like the JFET, and also enhance it, the construction of MOSFET devices differs based on the channel size in the quiescent state, $V_{GS} = 0$.

A *depletion mode* device (also called a *normally-on MOSFET*) has a channel in the quiescent state that gets smaller as a reverse bias is applied; this device conducts current with no bias applied (see Figure 2.43A and B). An *enhancement mode* device (also called a *normally-off MOSFET*) is built without a channel and does not conduct current when $V_{GS} = 0$; increasing forward bias forms a temporary channel that conducts current (see Figure 2.43C and D).

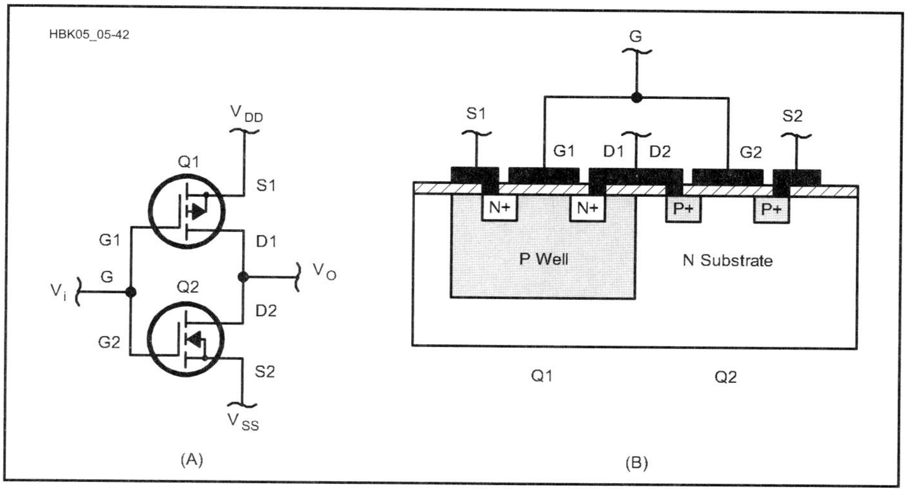

Figure 2.44 — Complementary metal oxide semiconductor (CMOS). (A) CMOS device is made from a pair of enhancement mode MOS transistors. The upper is a P-channel device, and the lower is an N-channel device. When one transistor is biased on, the other is biased off; therefore, there is minimal current from V_{DD} to ground. (B) Implementation of a CMOS pair as an integrated circuit.

Complementary Metal Oxide Semiconductors (CMOS)

Power dissipation in a circuit can be reduced to very small levels (on the order of a few nanowatts) by using MOSFET devices in complementary pairs (CMOS). Each amplifier is constructed of a series circuit of MOSFET devices, as in **Figure 2.44**. The gates are tied together for the input signal, as are the drains for the output signal. In saturation and cutoff, only one of the devices conducts. The current drawn by the circuit under no load is equal to the OFF leakage current of either device and the voltage drop across the pair is equal to V_{DD}, so the steady-state power used by the circuit is always equal to $V_{DD} \times I_{D(OFF)}$. Power is only consumed during the switching process, so for ac signals, power consumption is proportional to frequency.

CMOS circuitry could be built with discrete components, but the number of extra parts and the need for the complementary components to be matched has made that an unusual design technique. The low power consumption and ease of fabrication has made CMOS the most common of all IC technologies. Although CMOS is most commonly used in digital integrated circuitry, its low power consumption has also been put to work by manufacturers of analog as well as digital ICs.

2.9 References and Bibliography

BOOKS AND ARTICLES

Alexander and Sadiku, *Fundamentals of Electric Circuits* 7th ed. (McGraw-Hill, 2021).

Banzhaf, W., WB1ANE, *Understanding Basic Electronics*, 2nd ed. (ARRL, 2010).

Glover, T., *Pocket Ref* 4th ed. (Sequoia Publishing, 2010).

Horowitz and Hill, *The Art of Electronics* 3rd ed. (Cambridge University Press, 2015).

Kaplan, S., *Wiley Electrical and Electronics Dictionary* (Wiley Press, 2004).

Orr, W., *Radio Handbook* 23rd ed. (Newnes, 1997).

Purcell, E., *Electricity and Magnetism*, 3rd ed. (Cambridge University Press, 2013).

Silver, W., NØAX, "Experiment #117 — Laying Down the Laws," *QST*, Oct 2012, pp. 60-61.

Silver, W., NØAX, "Experiment #118 — The Laws at Work," *QST*, Nov 2012, pp. 70-71.

Silver, W., NØAX, "Experiment #138 — E vs V," *QST*, Jul 2014, pp. 59–60.

Terman, F., *Radio Engineers' Handbook* (McGraw-Hill, 1943).

Contents

Chapter 3 — Online Content

Articles

- Digital Electronic Basics by Dale Botkin, NØXAS
- Hands On Radio: Maxwell's Equations — Grad, Div, and Curl by Ward Silver, NØAX
- Hands On Radio: Maxwell's Equations — The Wave Emerges by Ward Silver, NØAX
- Hands On Radio: The Effects of Gain-Bandwidth Product by Ward Silver, NØAX
- Maxwell Without Tears by Paul Shuch, N6TX
- Untangling the Decibel Dilemma — Expanded by Ward Silver, NØAX

*Also see these articles with the **Electrical Fundamentals** chapter Online Content:*

- Radio Mathematics
- Radio Math Formulas and Notes

Tools and Data

- Frequency Response Spreadsheet

Chapter 3
Radio Fundamentals

Radio begins with an understanding of alternating current waveforms and how they are measured. This chapter also examines the relationship between ac voltage and current in energy-storing components like capacitors and inductors that defines reactance and impedance. We can then explore quality factor (Q) and the properties of resonant circuits.

Analog system concepts are introduced to explain the concepts and techniques of working with electronic circuits in radio. Finally, we discuss electromagnetic waves that carry information between stations.

As for the previous chapter, additional mathematics resources are available in the article "Radio Mathematics" in this book's online content.

3.1 AC Waveforms

A *waveform* is the pattern of *amplitudes* reached by voltage or current measured over time, including combinations of ac and dc voltage and current. For example, **Figure 3.1** shows two ac waveforms fairly close in frequency and their combination. **Figure 3.2** shows two ac waveforms dissimilar in both frequency and wavelength, along with the resultant combined waveform.

3.1.1 Sine Waves and Rotation

Not only is a sine wave the most fundamental ac waveform — energy at a single frequency — it also describes rotation. The cyclical nature of the sine wave is at the heart of much of radio technology, whether analog or digital. A good grasp of the sine wave and the closely related cosine wave are key to understanding the techniques that make up radio. (Sine and cosine functions as well as vectors and phasors are discussed in the "Radio Mathematics" article in this book's online content.)

Figure 3.3 illustrates the relationship. Imagine a wheel rotating counterclockwise with a visible dot at a point anywhere along the circumference (rim) of the wheel. If you spin the wheel at a constant rate and watch the wheel on-edge as in Figure 3.3A, the dot will just move

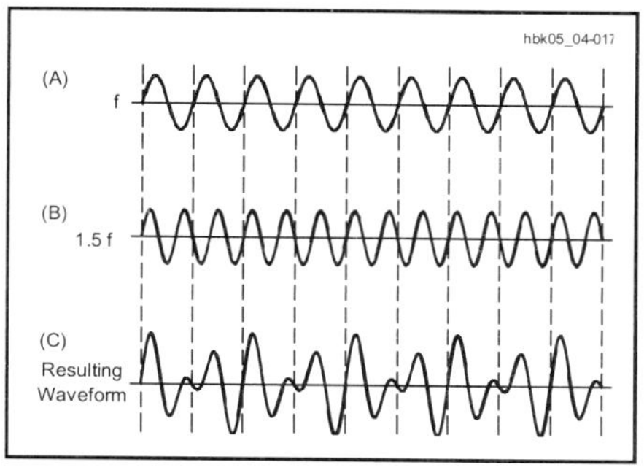

Figure 3.1 — Two ac waveforms of similar frequencies ($f_B = 1.5\ f_A$) with amplitudes added together to form a composite wave. Note the points where the positive peaks of the two waves combine to create high composite peaks at a frequency that is the difference between f_B and f_A. The beat note frequency is $1.5\ f_A - f_A = 0.5 f_A$ and is visible in the drawing.

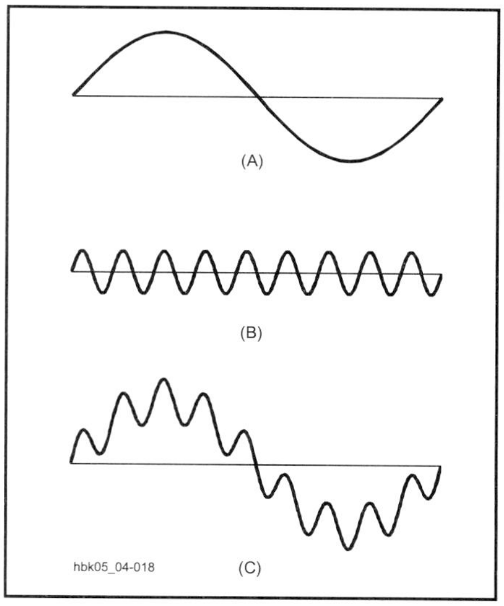

Figure 3.2 — Two ac waveforms of widely different frequencies and amplitudes form a composite wave in which one wave appears to ride upon the other.

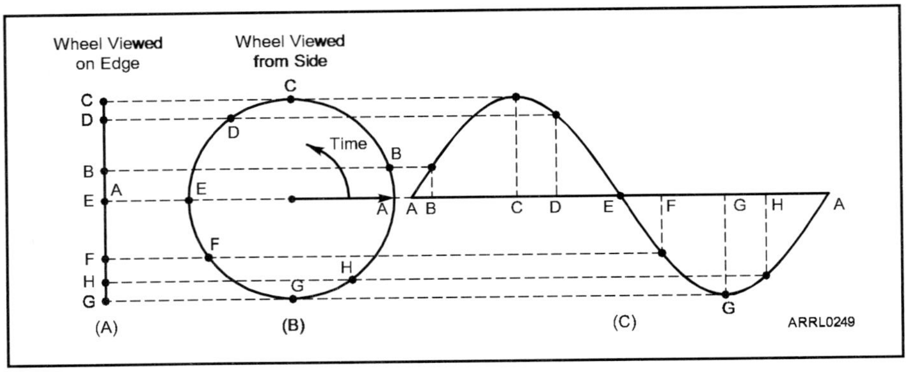

Figure 3.3 — This diagram illustrates the relationship between a sine wave and circular rotation. You can see how various points on the circle correspond to values on the sine wave.

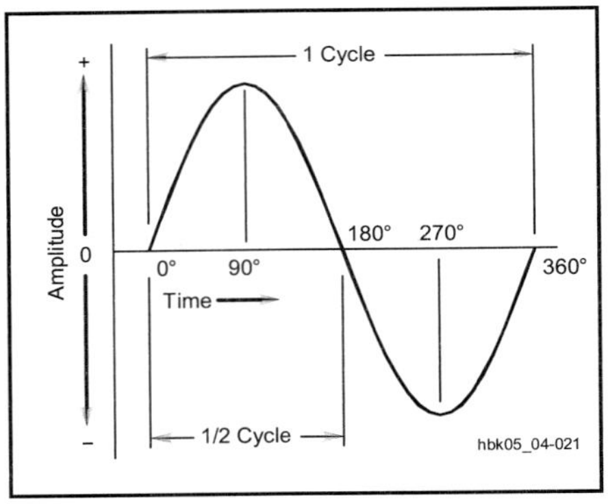

Figure 3.4 — An ac cycle is divided into 360 degrees that are used as a measure of time or phase.

up and down. If we designate the heights at points A and E as 0, at point C as +1 and at point G as –1, then make a table of values as the wheel rotates through 360°, the height values will correspond exactly to the values of the sine function (*sin*). When the wheel is at 0° in position A, sin (0°) = 0; when at 90° in position C, sin (90°) = 1, and so forth all the way around.

Now look at the wheel from the side as in Figure 3.3B. Plotting the dot's height around its circular path against degrees on the horizontal axis will then trace out a sine wave as in Figure 3.3C. Each rotation of the wheel corresponds to one *cycle* of the sine wave. The amplitude, A, (not to be confused with position A in Figure 3.3) of the sine wave is equal to the sine of the wheel's *angular position* in degrees (θ):

$$A = \sin(\theta)$$

Figure 3.4 shows how each cycle of a sine wave is divided into 360° to measure angular position.

3.1.2 Frequency, Period, and Harmonics

With a continuously rotating generator, alternating current or voltage will pass through many equal cycles over time. This is a *periodic waveform*, composed of repeated identical cycles. An arbitrary point on any one cycle can be used as a marker of position on a periodic waveform. For this discussion, the positive peak of the waveform will work as an unambiguous marker. The number of times per second that the current (or voltage) reaches this positive peak in any one second is called the *frequency* of the waveform. In other words, frequency expresses the *rate* at which current (or voltage) cycles occur. The unit of frequency is *cycles per second*, or *hertz* — abbreviated Hz (after Heinrich Hertz, the 19th century physicist who demonstrated the existence of radio waves).

If the sine wave has a constant frequency, every complete cycle takes the same amount of time, the *period*, T, as in **Figure 3.5**. The signal's period is the reciprocal of its frequency:

$$\text{Frequency (f) in Hz} = \frac{1}{\text{Period (T) in seconds}}$$

and

$$\text{Period (T) in seconds} = \frac{1}{\text{Frequency (f) in Hz}}$$

Example: What is the period of 60 Hz ac current?

$$T = \frac{1}{f} = \frac{1}{60\ \text{Hz}} = 0.01666\text{s} = 16.7$$

To calculate the amplitude, A, of the sine wave at any point in time, t, we need to be able to convert time to the angle (in degrees), θ:

$$\theta = 360\frac{t}{T}$$

The sine wave equation is now:

$$A = \sin\left(360\frac{t}{T}\right) = \sin\left(360 \times \frac{1}{T} \times t\right)$$

$$= \sin(360 \times f \times t)$$

HARMONICS

A *harmonic* is a signal with a frequency that is some integer multiple (2, 3, 4 and so on) of a signal at a *fundamental frequency* (which also can be called the first harmonic). Figure 3.5 shows a simple example of harmonics. The harmonic at twice the fundamental's frequency is called the *second harmonic*, at three times the fundamental frequency the *third harmonic*, and so forth. For example, if a complex waveform is made up of sine waves with frequencies of 10 kHz, 20 kHz, and 30 kHz, 10 kHz is the fundamental and the other two are harmonics. Signals at the frequency of harmonics are said to be *harmonically related* to the fundamental.

3.1.3 Phase

Now let's make the connection between angular position and time. Although time is measurable in parts of a second, it is more convenient to treat each cycle as a complete time unit divided into 360°. The conventional starting point for phase in a sinusoidal wave-

Degrees, Radians, and Angular Frequency

While most electronic and radio mathematics use degrees as a measure of phase, you will occasionally encounter *radians*. Radians are used because they are more convenient mathematically in certain types of equations and computations. There are 2π radians in a circle, just as there are 360°, so one radian = 360/2π ≈ 57.3°. In this book, unless it is specifically noted otherwise, the convention will be to use degrees in all calculations of phase or angle.

In engineering textbooks and other electronic references, you will often encounter the symbol ω used to represent *angular frequency*. The sine wave equation would then be written:

$$A = \sin(2\pi\,\text{ft}) = \sin(\omega\,\text{t})$$

where ω=2πf. The use of angular frequency is more straightforward in many types of engineering calculations. Radians are used for angular position when angular frequency (ω) is being used.

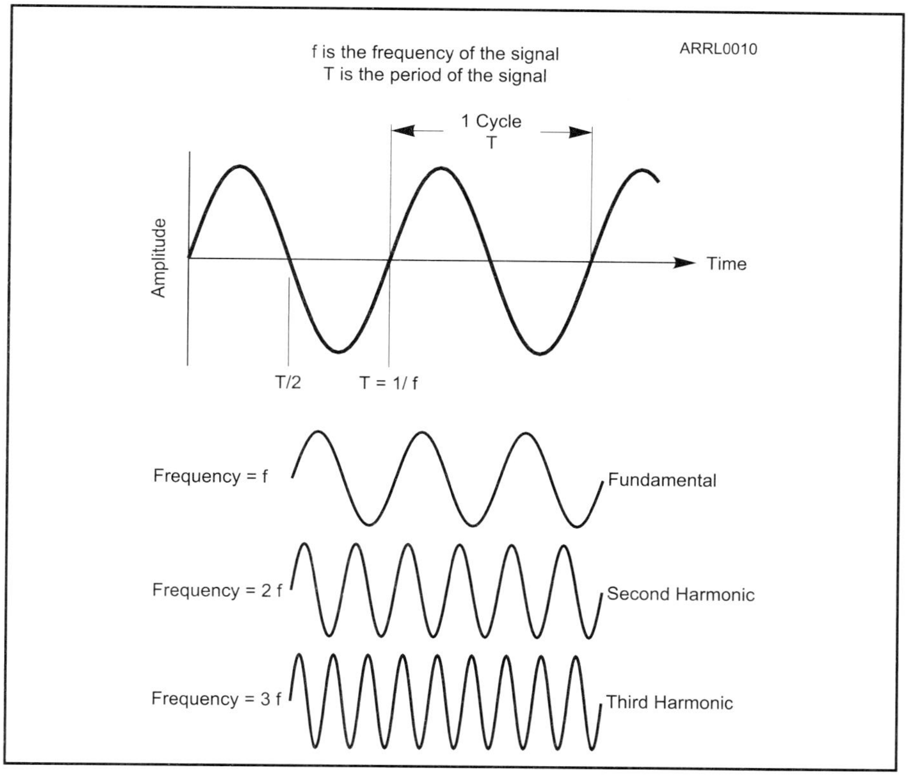

Figure 3.5 — The frequency of a signal and its period are reciprocals. A higher frequency means a shorter period and vice versa. Harmonics are signals with frequencies that are integer multiples of a fundamental frequency.

Phase versus Polarity

It is important to distinguish between polarity and phase. Polarity is the assigned conventions or directions for positive and negative voltage or current. Phase is a function of time or position in a waveform. It is quite possible for two signals to have opposite polarities, but still be in phase, for example. In a multi-phase ac power system, "phase" refers to one of the distinct voltage waveforms generated by the utility.

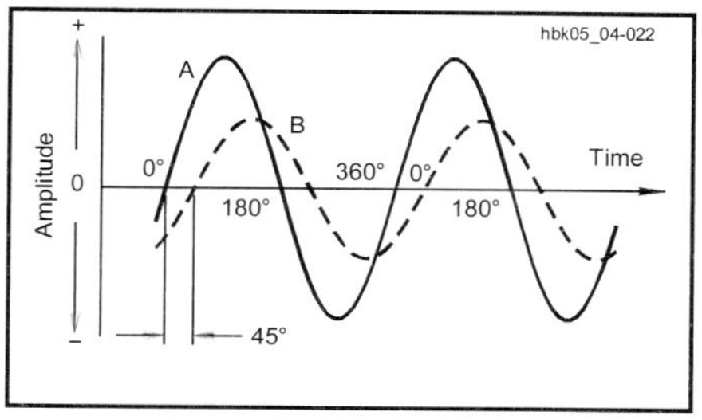

Figure 3.6 — When two waves of the same frequency start their cycles at slightly different times, the time difference, or phase difference, is measured in degrees. In this drawing, wave B starts 45° (one-eighth cycle) later than wave A, and so lags 45° behind A.

form is the *zero point* at which the positive *half cycle* begins as shown in Figure 3.4.

The advantage of treating the ac cycle in this way is that many calculations and measurements can be taken and recorded in a manner that is independent of frequency. The positive peak voltage or current occurs at 90° into the cycle. Relative to the starting point, 90° is the *phase* of the ac at that point. Phase is the position within an ac cycle expressed in degrees or *radians*. Thus, a complete description of an ac voltage or current involves reference to three properties: frequency, amplitude, and phase.

Each degree of phase represents the same amount of time. For example, a sine wave with a frequency of four cycles per second has a period T = 0.25 second and each degree of phase is equivalent to T/360 = 0.25 / 360 = 0.00069 second.

Phase relationships also permit the comparison of two ac voltages or currents at the same frequency. If the zero point of two signals with the frequency occur at the same time, there is zero phase difference between the signals and they are said to be *in phase.*

Figure 3.6 illustrates two equal frequency waveforms with a constant phase difference. Since B crosses the zero point in the positive direction after A has already done so, there is a *phase difference* between the two waves. In the example, B *lags* A by 45°, or A *leads* B by 45°. If A and B occur in the same circuit, their composite waveform will also be a sine wave at an intermediate phase angle relative to each. Adding any number of sine waves of the same frequency always results in a sine wave at that frequency. Adding sine waves of different frequencies, as in Figure 3.1, creates a complex waveform with a *beat frequency* that is the difference between the two sine waves.

Figure 3.6 might equally apply to a voltage and a current measured in the same ac circuit. Either A or B might represent the voltage; that is, in some instances voltage will lead the current and in others voltage will lag the current.

Two important special cases appear in **Figure 3.7**. In the upper drawing, line B lags 90° behind line A. Its cycle begins exactly one quarter cycle later than the A cycle. When one wave is passing through zero, the other just reaches its maximum value. In this example, the two sine waves are said to be *in quadrature*. If waveform B is a sine wave, waveform A is a *cosine wave*, leading waveform B by 90°. Quadrature signals form the basis of I/Q modulation, as is described in the **Modulation** chapter. (I and Q stand for In-Phase and Quadrature.)

In the lower drawing, lines A and B are 180° *out of phase*, sometimes called *anti-*

Vectors and Phasors

In Figure 3.3B, the arrow drawn from the center of the wheel to point A is a *vector* which has both an amplitude (its length) and direction (its angular position). The amplitude of the vector is equal to its length, which we arbitrarily decided would be 1 when constructing the table of sine values. Since the vector is pointing exactly along the horizontal axis, its direction is 0°. Thus, the vector is described as "1 at an angle of 0°" or "1 with a phase of 0°." In the *phasor notation* commonly used in radio, this is written $1\angle 0°$. The rotation of the wheel and the repeated cycles of the sine wave can also be described by a vector that is rotating (spinning around like the hand of a clock) at the frequency of the sine wave. You can find more information about vectors and phasors in the ARRL's "Radio Mathematics" article in the online content for this book.

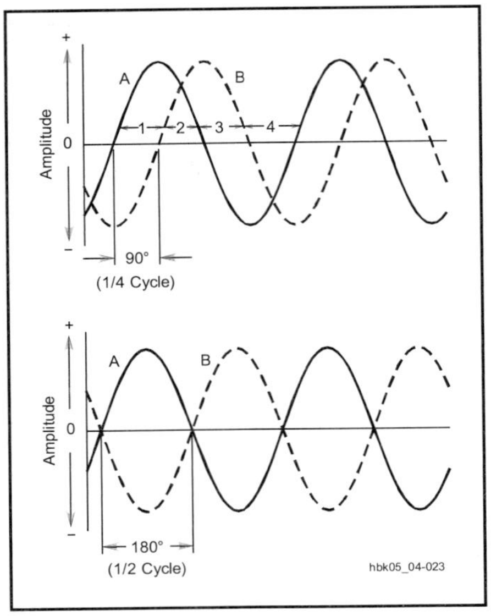

Figure 3.7 — Two important special cases of phase difference: In the upper drawing, the phase difference between A and B is 90°; in the lower drawing, the phase difference is 180°.

phase. In this case, it does not matter which one is considered to lead or lag. Line B is always positive while line A is negative, and vice versa. If the two waveforms are of two voltages or two currents in the same circuit and if they have the same amplitude, they will cancel each other completely.

3.1.4 Time and Frequency Domain

To this point in the chapter, our discussions and illustrations have been in the *time domain,* in which some characteristic of the signal (usually amplitude) is presented in relation to time. On a graph, this means the horizontal axis represents time. Events to the right take place later; events to the left occur earlier. For pure sinusoids (sine waves), this is enough to describe the signal.

For a complex signal with more than one sine wave, the time domain is insufficient to describe the necessary frequency and time information. The *frequency domain* is better for showing the characteristics of these signals as illustrated in **Figure 3.8**. A sine wave signal is represented as a vertical line, with the height of the line showing the signal's amplitude. Note that a sine wave signal occupies a single frequency.

To better understand the relationship between the time and frequency domains, refer to **Figure 3.9**. In Figure 3.9A, the three-dimensional coordinates show time (as the line sloping toward the bottom right); frequency (as the line sloping toward the top right); and amplitude (as the vertical axis).

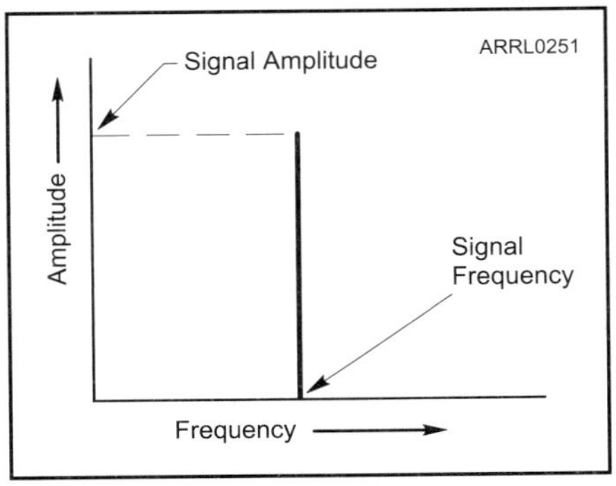

Figure 3.8 — A frequency domain or spectrum graph shows a sine wave as a single vertical line. The horizontal axis represents frequency and the vertical axis represents amplitude. The height of the line representing the sine wave shows its amplitude.

The two frequencies shown are harmonically related (f_1 and $2f_1$). The time domain is represented in Figure 3.9B, in which all frequency components are added together. If the two frequencies were applied to the input of an oscilloscope, we would only see the bold line that represents the amplitudes of the signals added together. The frequency domain contains information not found in the time domain, and vice versa.

The display shown in Figure 3.9C is typical of a spectrum analyzer's display of a complex waveform. (The spectrum analyzer is described in the **Test Equipment and Measurements** chapter.) In the figure the signal is separated into its individual frequency components, and a measurement made of the amplitude of each signal component. A signal's amplitude can be represented on the vertical scale as its voltage or as its power. You can see that using the frequency domain gives more information about the composition of the signal.

3.1.5 Complex Waveforms

All signals, whether audio or radio, are made up of the sum of sine waves. The signal that is composed of more than one sine wave is called a *complex waveform.* A common example of a complex waveform is the audio transmitted by an amateur radio handheld radio when the numbers on the keypad are pressed. This sound is made up of two tones at different frequencies, thus the name "dual-tone multi-frequency" or DTMF (DTMF is the main method of dialing numbers on the wired telephone system).

There are certain well-known and common complex waveforms that are made up of a sine wave and its harmonics. These are termed *regular* waveforms because the harmonic relationship of all the sine waves results in a waveform with a single overall frequency

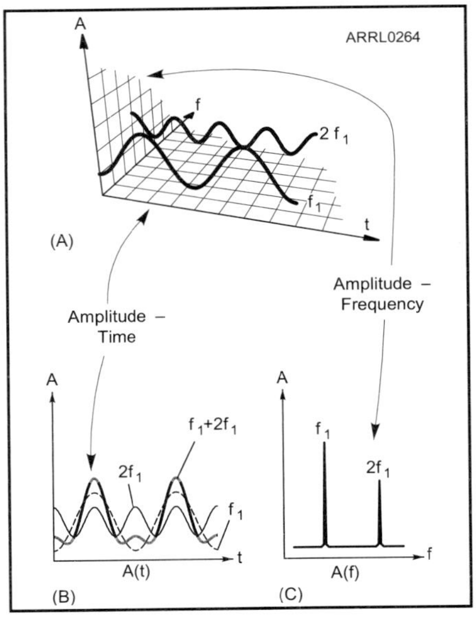

Figure 3.9 — This diagram shows how a complex waveform may be displayed in either the time domain or frequency domain. Part (A) is a three-dimensional display of amplitude, time, and frequency of the two component waveforms, f1 and 2f1. At (B), the sum of the component waveforms is shown in the time domain, as on an oscilloscope. At (C), the signal's frequency domain information is shown, as on a spectrum analyzer.

and period. A waveform that is made of sine waves that are not harmonically related, such as human speech, is an *irregular* waveform. Whether regular or irregular, the sine waves that make up a complex waveform are called its *components*. (Analysis of complex signals in terms of its individual components is addressed in the **DSP and SDR Fundamentals** chapter.) A spreadsheet tool by Lou Ernst, WA2GKH, for experimenting with and comparing waveforms is available on this book's reference website in the Supplemental Information and Files section.

It is common for complex ac signals to contain a fundamental signal and a series of harmonics. Which harmonics are combined with the fundamental and the relative amplitude of each determine the final shape of the waveform as you will see in the following two sections. The set of all components that make up a signal is called the signal's *spectrum*. (More than one spectrum is *spectra*.)

SAWTOOTH WAVES

A *sawtooth* waveform, as shown in **Figure 3.10**, has a significantly faster *rise time* (the time it takes for the wave to reach a maximum value) compared to its *fall time* (the time it takes for the wave to reach a minimum value). A sawtooth wave is made up of a sine wave at its fundamental frequency and all of its har-

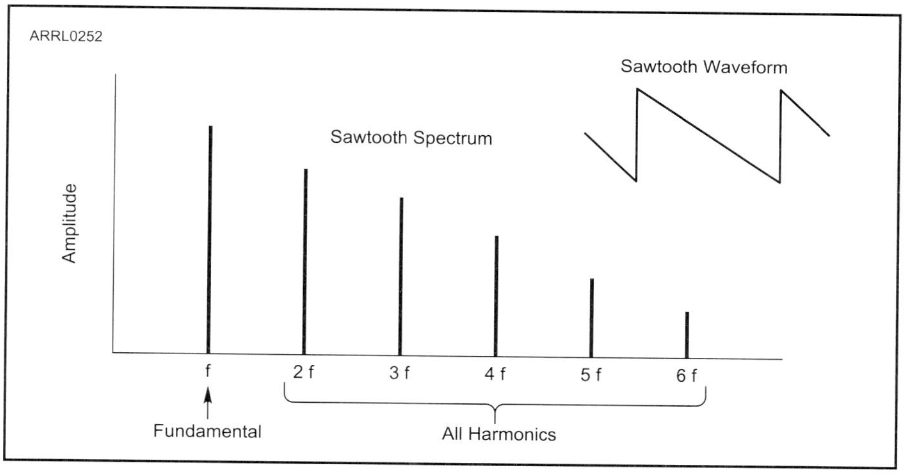

Figure 3.10 — The sawtooth waveform is made up of sine waves at the fundamental frequency and all of its harmonics. The amplitude of the harmonics decreases as their frequency increases.

monics. The sawtooth's spectrum is shown in Figure 3.10. The *ramp* waveform is similar to the sawtooth but slowly rises (the ramp) then has a fast fall, the opposite of the sawtooth. Both the sawtooth and ramp waveforms are useful in timing circuits.

SQUARE WAVES

A square wave is one that abruptly changes back and forth between two voltage levels and remains an equal time at each level as in **Figure 3.11**. (If the wave spends an unequal time at each level, it is known as a *rectangular wave*.) A square wave is made up of sine waves at the fundamental and all the *odd* harmonic frequencies as shown in Figure 3.11.

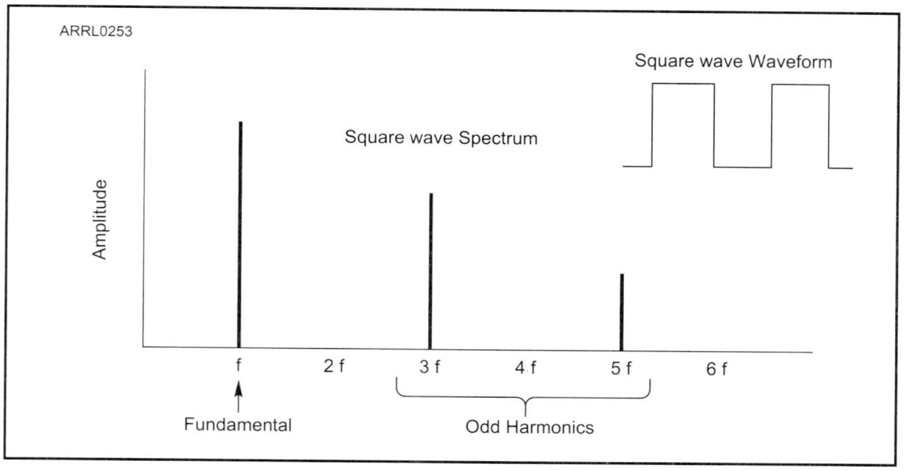

Figure 3.11 — The square wave is made up of sine waves at the fundamental frequency and only the odd harmonics. amplitude of the harmonics decreases as their frequency increases.

3.2 Measuring AC Voltage, Current, and Power

Measuring the voltage or current in a dc circuit is straightforward, as in the upper circuit in **Figure 3.12** demonstrates. Since the current flows in only one direction, the voltage and current have constant values until the resistor values are changed.

The lower circuit in Figure 3.12 illustrates a perplexing problem encountered when measuring voltages and currents in ac circuits — the current and voltage continuously change direction and value. Which values are meaningful? How are measurements performed? In fact, there are several methods of measuring sine-wave voltage and current in ac circuits with each method providing different information about the waveform. Note that the following sections assume the waveform is a sine wave unless otherwise noted.

3.2.1 Instantaneous Values

By far, the most common waveform associated with ac of any frequency is the sine wave.

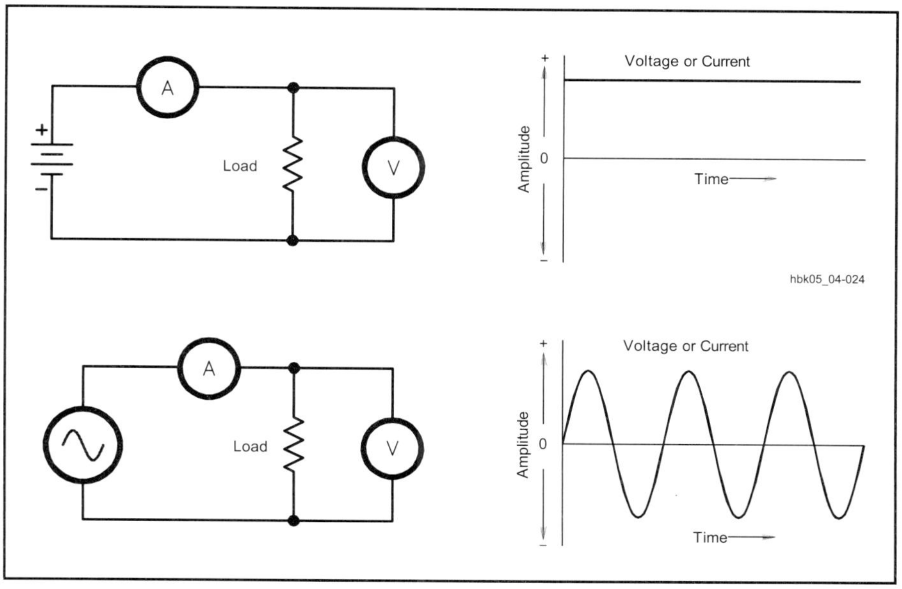

Figure 3.12 — Voltage and current measurements in dc (upper) and ac circuits (lower).

Unless otherwise noted, it is safe to assume that measurements of ac voltage or current are of a sinusoidal waveform. **Figure 3.13** shows a sine wave representing a voltage or current of some arbitrary frequency and amplitude. The *instantaneous* voltage (or current) is the value at one instant in time. If a series of instantaneous values are plotted against time, the resulting graph will show the waveform.

In the sine wave of Figure 3.13, the instantaneous value of the waveform at any point in time is a function of three factors: the maximum value of voltage (or current) above zero along the curve (point B, E_{PK}), the frequency of the wave, f, and the time elapsed from the preceding positive-going zero crossing, t, in seconds or fractions of a second. Thus,

$$E_{INST} = E_{PK} \sin(ft)$$

where the product ft represents the phase at time t, in degrees. (See the sidebar "Degrees, Radians and Angular Frequency." If the sine calculation is done in radians, substitute $2\pi ft$ for ft in the equation.)

If the point's phase is known — the position along the waveform — the instantaneous voltage at that point can be calculated directly as:

$$E_{INST} = E_{PK} \sin \theta$$

where θ is the number of degrees of phase difference from the beginning of the cycle.

Example: What is the instantaneous value of voltage at point D in Figure 3.13, if the maximum voltage value is 120 V and point D's phase is 60.0°?

$$E_{INST} = 120\text{ V} \times \sin 60° = 120 \times 0.866 = 104\text{ V}$$

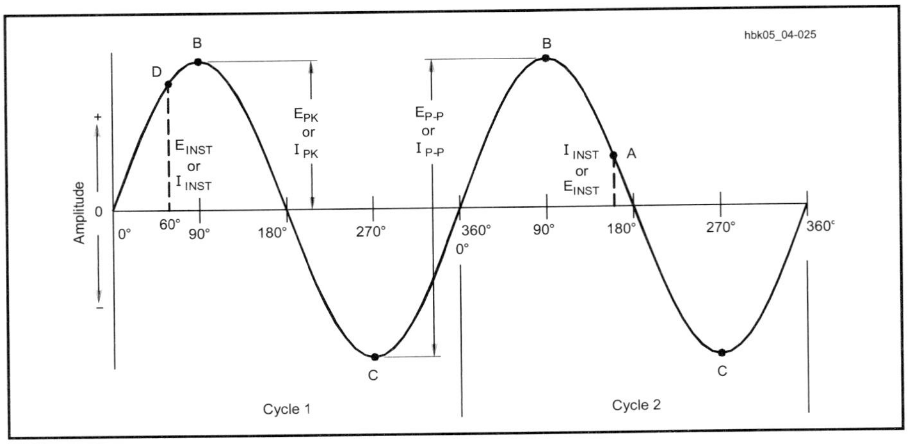

Figure 3.13 — Two cycles of a sine wave to illustrate instantaneous, peak, and peak-to-peak ac voltage and current values. The horizontal axis is calibrated in degrees of phase, which range from 0° to 360° for each sine wave cycle.

3.2.2 Peak and Peak-to-Peak Values

The most important of an ac waveform's instantaneous values are the maximum or *peak values* reached on each positive and negative half cycle. In Figure 3.13, points B and C represent the positive and negative peaks. Peak values (indicated by a "PK" or "P" subscript) are especially important with respect to component ratings, which the voltage or current in a circuit must not exceed without danger of component failure.

The *peak power* in an ac circuit is the product of the peak voltage and the peak current, or

$$P_{PK} = E_{PK} \times I_{PK}$$

The span from points B to C in Figure 3.13 represents the largest difference in value of the sine wave. Designated the *peak-to-peak* value (indicated by a "P-P" or "PK-PK" subscript), this span is equal to twice the peak value of the waveform. Thus, peak-to-peak voltage is:

$$E_{PP} = 2\,E_{PK}$$

3.2.3 RMS Values

The *root mean square* or *RMS* values of voltage and current are the most common values encountered in electronics. Sometimes called *effective* values, the RMS value of an ac voltage or current is the value of a dc voltage or current that would cause a resistor to dissipate the same average amount of power as the ac waveform. This measurement became widely used in the early days of electrification when both ac and dc power utility power were in use. Even today, the values of the ac line voltage available from an electrical power outlet are given as RMS values. Unless otherwise specified, unlabeled ac voltage and current values found in most electronics literature are normally RMS values.

The RMS values of voltage and current get their name from the mathematical method used to derive their value relative to peak voltage and current. This procedure provides the RMS value for any type of periodic waveform, sinusoidal or not. Start by *squaring* the individual values of all the instantaneous values of voltage or current during an entire single cycle of ac. Take the average (*mean*) of these squares (this is done by computing an integral of the waveform) and then find the square *root* of that average.

SINE WAVE RMS VALUES

This section applies only when the waveform in question is a sine wave. The simple formulas and conversion factors in this section are generally *not* true for non-sinusoidal waveforms such as square or triangle waves (see the sidebar "Measuring Non-Sinusoidal Waveforms"). The following formulas are true *only* if the waveform is a sine wave and the circuit is *linear* — that is, raising or lowering the voltage will raise or lower the current

Measuring Nonsinusoidal Waveforms

Making measurements of ac waveforms is covered in more detail in the **Test Equipment and Measurements** chapter. However, this is a good point in the discussion to reinforce the dependence of RMS and values on the nature of the waveform being measured.

Analog meters and other types of instrumentation that display RMS values may only be calibrated for sine waves, those being the most common type of ac waveform. Using that instrumentation to accurately measure waveforms other than sine waves — such as speech, intermittent sine waves (such as CW from a transmitter), square waves, triangle waves or noise — requires the use of *calibration factors* or the measurement may not be valid.

To make calibrated, reliable measurements of the RMS or value of these waveforms requires the use of *true-RMS* instruments. These devices may use a balancing approach to a known dc value or if they are microprocessor-based, may actually perform the full root-mean-square calculation on the waveform. Be sure you know the characteristics of your test instruments if an accurate RMS value is important.

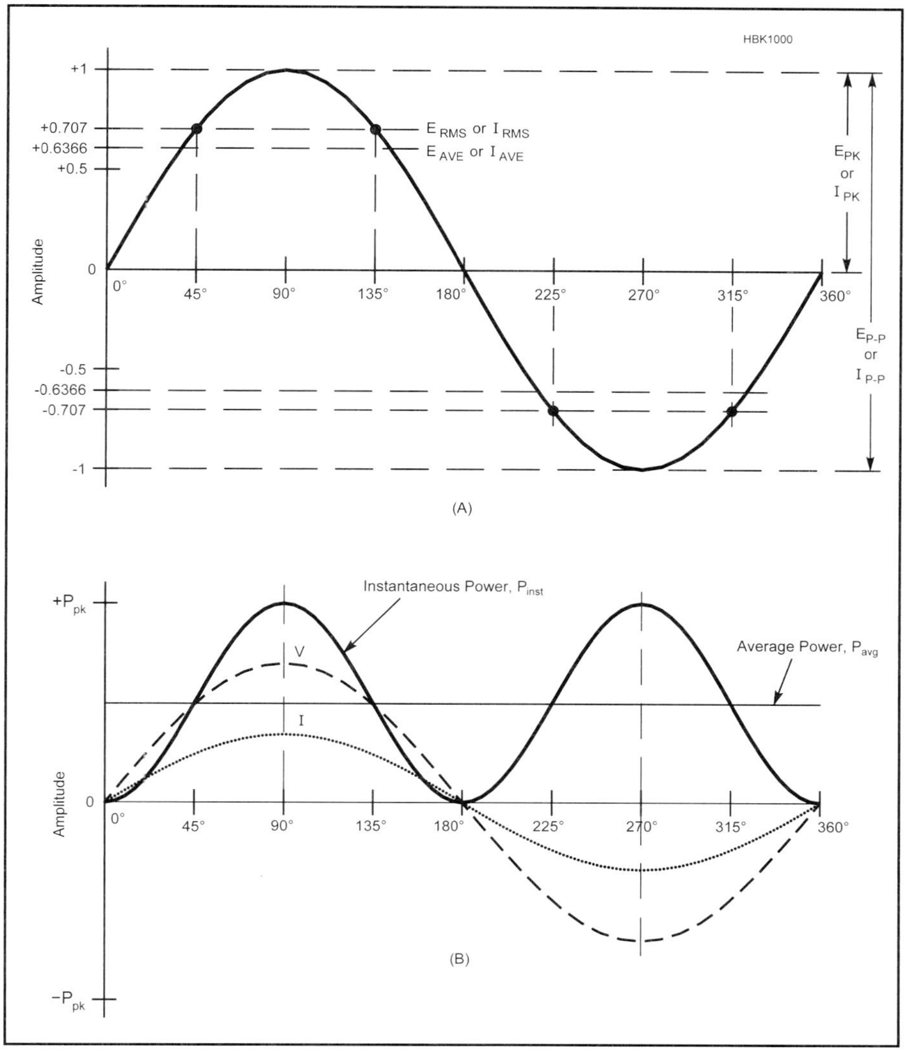

Figure 3.14 — (A) The relationships between RMS, average, peak, and peak-to-peak values of ac voltage and current for a sine wave normalized to a peak voltage or current of 1. The numeric constants are different for non-sinusoidal waveforms. (B) The value of instantaneous power is the product of instantaneous voltage and current. Average power is the product of the RMS values of voltage and current.

Table 3.1
Conversion Factors for Sinusoidal AC Voltage or Current

From	*To*	*Multiply By*
Peak	Peak-to-Peak	2
Peak-to-Peak	Peak	0.5
Peak	RMS	$1/\sqrt{2}$ or 0.707
RMS	Peak	$\sqrt{2}$ or 1.414
Peak-to-Peak	RMS	$1/(2\times\sqrt{2})$ or 0.35355
RMS	Peak-to-Peak	$2\times\sqrt{2}$ or 2.828
Peak	Average	$2/\pi$ or 0.6366
Average	Peak	$\pi/2$ or 1.5708
RMS	Average	$(2\times\sqrt{2})/\pi$ or 0.90
Average	RMS	$\pi/(2\times\sqrt{2})$ or 1.11

Note: These conversion factors apply only to continuous pure sine waves.

proportionally. If those conditions are true, the following conversion factors have been computed and can be used without any additional mathematics.

For a sine wave to produce heat equivalent to a dc waveform the peak ac power required is twice the dc power. Therefore, the average ac power equivalent to a corresponding dc power is half the peak ac power.

$$P_{avg} = \frac{P_{pk}}{2}$$

A sine wave's RMS voltage and current values needed to arrive at average ac power are related to their peak values by the conversion factors:

$$E_{RMS} = \frac{E_{PK}}{\sqrt{2}} = \frac{E_{PK}}{1.414} = E_{PK} \times 0.707$$

$$I_{RMS} = \frac{I_{PK}}{\sqrt{2}} = \frac{I_{PK}}{1.414} = I_{PK} \times 0.707$$

RMS voltages and currents are what is displayed by most volt and ammeters.

If the RMS voltage is the peak voltage divided by $\sqrt{2}$, then the peak voltage must be the RMS voltage multiplied by $\sqrt{2}$, or

$$E_{PK} = E_{RMS} \times 1.414$$

$$I_{PK} = I_{RMS} \times 1.414$$

Example: What is the peak voltage and the peak-to-peak voltage at the usual household ac outlet, if the RMS voltage is 120 V?

$$E_{PK} = 120 \text{ V} \times 1.414 = 170 \text{ V}$$

$$E_{P\text{-}P} = 2 \times 170 \text{ V} = 340 \text{ V}$$

In the time domain of a sine wave, the instantaneous values of voltage and current correspond to the RMS values at the 45°, 135°, 225° and 315° points along the cycle shown in **Figure 3.14A**. (The sine of 45° is approximately 0.707.) The instantaneous value of voltage or current is greater than the RMS value for half the cycle and less than the RMS value for half the cycle.

Since circuit specifications will most commonly list only RMS voltage and current values, these relationships are important in finding the peak voltages or currents that will

Sine and Square Wave Measurement Definitions

Since square waves are very common waveforms, the following table provides definitions for the two types of waveforms. These are *not* conversion factors between two measurements unless a true-RMS instrument is used.

AC Measurements for Sine and Square Waves

	Sine Wave	*Square Wave*
Peak-to-Peak	2 × Peak	2 × Peak
Peak	0.5 × Peak-to-Peak	0.5 × Peak-to-Peak
RMS	0.707 × Peak	Peak
Peak	1.414 × RMS	RMS
Average	0 (full cycle)	0 (full cycle)
	0.637 × Peak (half cycle)	0.5 × Peak (half cycle)

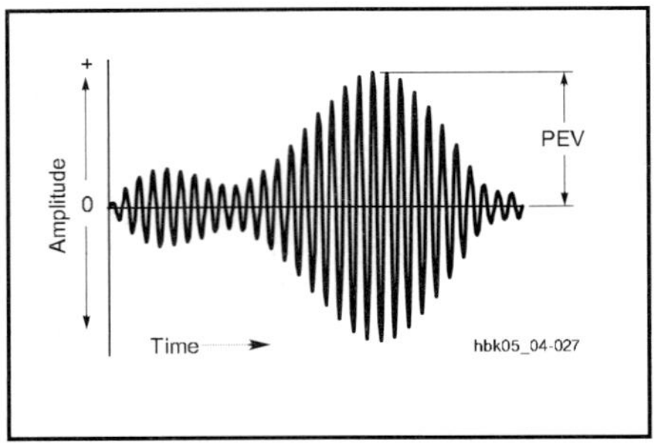

Figure 3.15 — The peak envelope voltage (PEV) for a composite waveform.

stress components.

Example: What is the peak voltage across a capacitor if the RMS voltage of a sinusoidal waveform signal across it is 300 V ac?

$$E_{PK} = 300 \text{ V} \times 1.414 = 424 \text{ V}$$

The capacitor must be able to withstand this higher voltage, plus a safety margin. (The capacitor must also be rated for ac use because of the continually reversing polarity and ac current flow.) In power supplies that convert ac to dc and use capacitive input filters, the output voltage will approach the peak value of the ac voltage rather than the RMS value. (See the **Power Sources** chapter for more information on specifying components in this application.)

3.2.4 Average Values of AC Waveforms

Certain kinds of circuits respond to the *average* voltage or current (not power) of an ac waveform. Among these circuits are analog electrodynamic meter movements and power supplies that convert ac to dc and use heavily inductive ("choke") input filters, both of which work with the pulsating dc output of a full-wave rectifier. The average value of each ac half cycle is the *mean* of all the instantaneous values in that half cycle. (The average value of a sine wave or any symmetric ac waveform over an entire cycle is zero!) Related to the peak values of voltage and current, average values for each half-cycle of a sine wave are 2/π (or 0.6366) times the peak value.

$$E_{AVE} = 0.6366\ E_{PK}$$

$$I_{AVE} = 0.6366\ I_{PK}$$

For convenience, **Table 3.1** summarizes the relationships between all of the common ac values. All of these relationships apply *only* to sine waves in linear circuits.

3.2.5 Average Power of AC Waveforms

The *instantaneous power*, P_{inst}, of an ac waveform as shown in **Figure 3.14B** is equal to the instantaneous values of voltage and current multiplied together:

$$P_{inst} = E_{inst} \times I_{inst}$$

Note that the frequency of the instantaneous power waveform is twice the frequency of voltage and current.

Average power, P_{avg}, is used for measuring the power of ac signals. It is determined by multiplying the RMS values of voltage and current. RMS values are calculated over an entire cycle, so average power applies only to an entire cycle:

$$P_{avg} = E_{RMS} \times I_{RMS}$$

This equation assumes the voltage and current are in-phase as shown. If reactance is present, there will be a phase shift between voltage and current — see the section "Reactive Power and Power Factor" later in this chapter. Although an RMS value can be calculated for the instantaneous power waveform, it is not equal to the average power and does not have any physical significance.

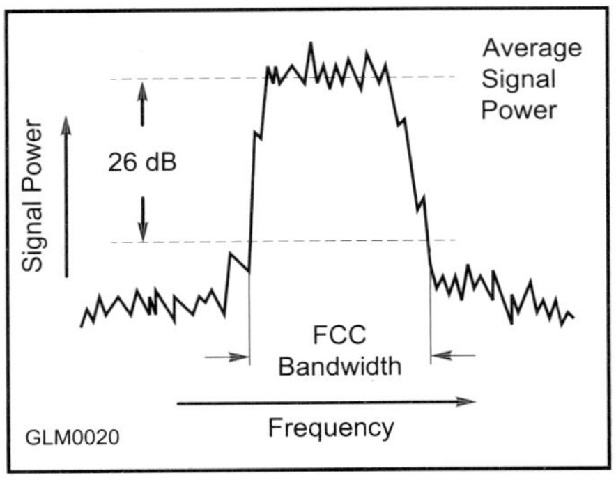

Figure 3.16 — The FCC defines bandwidth as "the width of a frequency band outside of which the mean (average) power of the transmitted signal is attenuated at least 26 dB below the mean power of the transmitted signal within the band" (FCC §97.3(a)(8))

3.2.6 Complex Waveforms and Peak Envelope Values

Complex waveforms, as shown earlier in Figures 3.10 and 3.11, differ from sine waves. For speech signals, the peak voltage may vary significantly from one cycle to the next, for example. Therefore, other amplitude measures are required, especially for accurate measurement of voltage and power with transmitted speech or data waveforms.

An SSB waveform (either speech or data) contains an RF ac waveform with a frequency many times that of the audio-frequency ac waveform with which it is combined. Therefore, the resultant *composite* waveform appears as an amplitude envelope superimposed upon the RF waveform as illustrated by **Figure 3.15**. For a complex waveform such as this, the *peak envelope voltage* (PEV) is the maximum or peak value of voltage anywhere in the waveform.

Peak envelope voltage is used in the calculation of *peak envelope power* (PEP). The Federal Communications Commission (FCC) sets the maximum power levels for amateur transmitters in terms of peak envelope power. PEP is the *average* power supplied to the antenna transmission line by the transmitter during one RF cycle at the crest of the modulation envelope, taken under normal operating conditions. That is, the average power for the RF cycle during which PEV occurs.

Since calculation of PEP requires the average power of the cycle, and the deviation of the modulated RF waveform from a sine wave is very small, the error incurred by using the conversion factors for sine waves is insignificant. Multiply PEV by 0.707 to obtain an RMS value. Then calculate PEP by using the square of the RMS voltage divided by the load resistance (R).

$$PEP = \frac{(PEV \times 0.707)^2}{R}$$

Example: What is the PEP of a transmitter's output with a PEV of 100 V into a 50-ohm load?

$$PEP = \frac{(100 \times 0.707)^2}{R} = \frac{(70.7)^2}{50} = 100 \text{ W}$$

COMPOSITE WAVEFORMS AND BANDWIDTH

Composite signals are groups of individual signals that combine to create a complex signal. Composite signals have *components* that generally cover a range of frequencies. The general definition of a signal's *bandwidth* is the difference in frequency between the two points at which the signal's amplitude falls to 3 dB below its peak value. For some kinds of very simple and very complex signals this may not be a useful definition.

The FCC has a more specific definition of bandwidth in section §97.3(a)(8): "*Bandwidth.* The width of a frequency band outside of which the mean power of the transmitted signal is attenuated at least 26 dB below the mean power of the transmitted signal within the band." In this context, the word, mean, is synonymous with the word, average, as it has been used in the preceding discussions. **Figure 3.16** illustrates how this measurement is made. This definition is used when evaluating a signal's occupied bandwidth to determine whether it satisfies FCC rules.

3.3 Effective Radiated Power

When evaluating total station performance, accounting for the effects of the entire system is important, including antenna gain. This allows you to evaluate the effects of changes to the station. Transmitting performance is usually computed as *effective radiated power* (*ERP*). ERP is calculated with respect to a reference antenna system — usually a dipole but occasionally an isotropic antenna — and answers the question, "How much power does my station radiate as compared to that if my antenna was a simple dipole?" Effective isotropic radiated power (EIRP) results when an isotropic antenna is used as the reference. If no antenna reference is specified, assume a dipole reference antenna. For directional antennas, the gain is defined for the main beam of the antenna, and the subsequent calculated ERP or EIRP is valid only in that direction.

ERP is especially useful in designing and coordinating repeater systems. The effective power radiated from the antenna helps establish the coverage area of the repeater. In addition, the height of the repeater antenna as compared to valleys and mountains in the surrounding area (*height above average terrain*, or *HAAT*) has a large effect on the repeater coverage. In general, for a given coverage area, with a greater antenna HAAT, less effective radiated power (ERP) is needed. A frequency coordinator may even specify a maximum ERP for a repeater, to help reduce interference between stations using the same frequencies.

ERP calculations begin with the *transmitter power output* (*TPO*). (This is assumed to be the output of the final power amplification stage if an external power amplifier is used.) Then the *system gain* of the entire antenna system including the antenna, the transmission line, and all transmission line components is applied to TPO to compute the entire station's output power.

System Gain = Transmission Line Loss – Transmission Components Loss + Antenna Gain

There is always some power lost in the feed line and often there are other devices inserted in the line, such as a filter or an impedance-matching network. In the case of a repeater system, there is usually a duplexer, so the transmitter and receiver can use the same antenna, and perhaps a circulator to reduce the possibility of intermodulation interference. These devices also introduce some loss to the system. The antenna system then usually returns some gain to the system. (See the **Antennas** chapter for information on antenna gain and the **Transmission Lines** chapter for information on feed line loss.)

ERP = TPO × System Gain

Since the system gains and losses are usually expressed in decibels, they can simply be added together, with losses written as negative values. System Gain must then be converted back to a linear value from dB to calculate ERP ($\log_{-1(x)}$ represents the inverse log of x, which is also written as $10_{(x)}$).

$$ERP = TPO \times \log^{-1}\left(\frac{\text{System Gain(dB)}}{10}\right)$$

It is also common to work entirely in dBm and dB until the final result for ERP is obtained and then converted back to watts. (dBm represents "decibels with respect to one mW" such that 0 dBm = 1 mW, 30 dBm = 1 W, and so forth.)

ERP (in dBm) = TPO (in dBm) + System Gain (in dB)

Suppose we have a repeater station that uses a 50 W transmitter and a feed line with 4 dB of loss. There is a duplexer in the line that exhibits 2 dB of loss and a circulator that adds another 1 dB of loss. This repeater uses an antenna that has a gain of 6 dBd. Our total System Gain looks like:

System Gain = –4 dB + –2 dB + –1 dB + 6 dBd = –1 dB

Note that this is a loss of 1 dB total for the system from TPO to radiated power. The effect on the 50 W of TPO results in:

$$ERP = 50 \text{ W} \times \log^{-1}\left(\frac{\text{System Gain (dB)}}{10}\right)$$

$$= 50 \times \log^{-1}(-0.1) = 50 \times 0.79 = 39.7 \text{ W}$$

This is consistent with the expectation that with a 1 dB system loss we would have somewhat less ERP than transmitter output power.

As another example, suppose we have a transmitter that feeds a 100 W output signal into a feed line that has 1 dB of loss. The feed line connects to an antenna that has a gain of 6 dBd. What is the effective radiated power from the antenna? To calculate the total system gain (or loss) we add the decibel values given:

System Gain = – 1 dB + 6 dBd = 5 dB

and

$$ERP = 100 \text{ W} \times \log^{-1}\left(\frac{\text{System Gain (dB)}}{10}\right)$$

$$= 100 \times \log^{-1}(0.5) = 100 \times 3.16 = 316 \text{ W}$$

The total system has positive gain, so we should expect a larger value for ERP than TPO. Keep in mind that the gain antenna concentrates more of the signal in a desired direction, with less signal in undesired directions. So the antenna doesn't really increase the total available power. If directional antennas are used, ERP will change with direction.

Example: What is the effective radiated power of a repeater station with 150 W trans-

The Decibel

The decibel (dB) is discussed in the section on Gain, later in this chapter. It is also covered in the "Untangling the Decibel Dilemma" item in the online content.

mitter power output, 2 dB feed line loss, 2.2 dB duplexer loss and 7 dBd antenna gain?

System Gain = –2 dB – 2.2 dB + 7 dBd = 2.8 dB

$$\text{ERP} = 150\ \text{W} \times \log^{-1}\left(\frac{\text{System Gain (dB)}}{10}\right)$$

$$= 150 \times \log^{-1}(0.28) = 150 \times 1.9 = 285\ \text{W}$$

Example: What is the effective radiated power of a repeater station with 200 W transmitter power output, 4 dB feed line loss, 3.2 dB duplexer loss, 0.8 dB circulator loss and 10 dBd antenna gain?

System Gain = –4 – 3.2 – 0.8 + 10 = 2 dB

$$\text{ERP} = 200\ \text{W} \times \log^{-1}\left(\frac{\text{System Gain (dB)}}{10}\right)$$

$$= 200 \times \log^{-1}(0.2) = 200 \times 1.58 = 317\ \text{W}$$

What is the effective isotropic radiated power of a repeater station with 200 W transmitter power output, 2 dB feed line loss, 2.8 dB duplexer loss, 1.2 dB circulator loss and 7 dBi antenna gain?

System Gain = –2 – 2.8 – 1.2 + 7 = 1 dB

$$\text{ERP} = 200\ \text{W} \times \log^{-1}\left(\frac{\text{System Gain (dB)}}{10}\right)$$

$$= 100 \times \log^{-1}(0.1) = 200 \times 1.26 = 252\ \text{W}$$

3.4 AC in Capacitors and Inductors

Capacitors store electrical energy and inductors store magnetic energy. When an ac signal is applied to either of them, the storing and releasing of energy results in a phase shift between the current and voltage waveforms. The phase shift varies with frequency and makes capacitors and inductors behave differently than resistors. They also interact to create frequency-sensitive tuned circuits, filters, impedance-matching circuits, and more.

3.4.1 Alternating Current in Capacitance

While a capacitor in a dc circuit will appear as an open circuit except for the brief charge and discharge periods, the same capacitor in an ac circuit will both pass and oppose current. A capacitor in an ac circuit does not handle electrical energy like a resistor, however. Instead of converting the energy to heat and dissipating it, capacitors store electrical energy when the applied voltage is greater than the electrical field already stored within them, and return it to the circuit when the opposite is true.

In **Figure 3.17** a sine-wave ac voltage having a maximum value of 100 V is applied to a capacitor. In the period OA, the applied voltage increases from 0 to 38 V, storing energy in the capacitor; at the end of this period the capacitor is charged to that voltage. In interval AB the voltage increases an additional 33 V to 71 V. During this interval a smaller quantity of charge has been added than in OA, because the voltage rise during interval AB is smaller. Consequently the average current during interval AB is smaller than during OA. In the third interval, BC, the voltage rises another 21 V to 92 V. This is less than the voltage increase during AB, so the quantity of charge added is less; in other words, the average current during interval BC is still smaller. In the fourth interval, CD, the voltage increases only 8 V; the charge added is smaller than in any preceding interval and therefore the current also is smaller.

By dividing the first quarter-cycle, in the time frame from 0 to D, into a very large number of such intervals, it can be shown that the current charging the capacitor has the shape of a sine wave, just as the applied voltage does. The current is largest at the beginning of the cycle and becomes zero at the maximum value of the voltage, so there is a phase difference of 90° between the voltage and the current. During the first quarter-cycle the current is flowing in the original (positive) direction through the circuit as indicated by the dashed line in Figure 3.17, since the capacitor is being charged. The increasing capacitor voltage indicates that energy is being stored in the capacitor.

Capacitive Reactance Timesaver

The fundamental units for frequency and capacitance (hertz and farads) are inconvenient for practical use in radio circuits. If the capacitance is specified in microfarads (µF) and the frequency is in megahertz (MHz), however, the reactance is calculated in ohms (Ω).

In the second quarter-cycle, in the time from D to H, the voltage applied to the capacitor decreases. During this time the capacitor loses charge, returning the stored energy to the circuit. Applying the same reasoning, it is evident that the current is small in interval DE and continues to increase during each succeeding interval. The current is flowing *against* the applied voltage, which is decreasing, causing the capacitor to return energy (discharge into) the circuit. The current thus flows in the *negative* direction during this quarter-cycle.

The third and fourth quarter-cycles repeat the events of the first and second, respectively, although the polarity of the applied voltage has reversed, and so the current correspondingly changes. In other words, an alternating current flows in the circuit because of the alternate charging and discharging of the capacitance. As shown in Figure 3.17, the

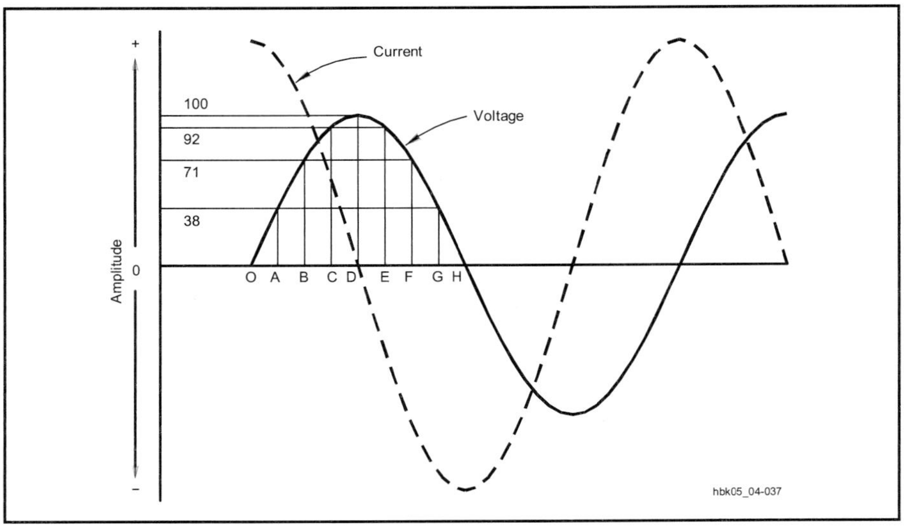

Figure 3.17 — Voltage and current phase relationships when an alternating current is applied to a capacitor.

current starts its cycle 90° before the voltage, so the current in a capacitor *leads* the applied voltage by 90°. You might find it helpful to remember the word "ICE" as a mnemonic because the current (I) in a capacitor (C) comes before voltage (E). (See the sidebar "ELI the ICE man" in the section on inductors.) We can also turn this statement around, to say the voltage in a capacitor *lags* the current by 90°.

3.4.2 Capacitive Reactance and Susceptance

The quantity of electric charge that can be placed on a capacitor is proportional to the applied voltage and the capacitance. If the applied voltage is ac, this amount of charge moves back and forth in the circuit once each cycle. Therefore, the rate of movement of charge (the current) is proportional to voltage, capacitance and frequency. Stated in another way, capacitor current is proportional to capacitance for a given applied voltage and frequency.

When the effects of capacitance and frequency are considered together, they form a quantity called *reactance* that relates voltage and current in a capacitor, similar to the role of resistance in Ohm's Law. Because the reactance is created by a capacitor, it is called *capacitive reactance*. The units for reactance are ohms, just as in the case of resistance. Although the units of reactance are ohms, there is no power dissipated in reactance. The energy stored in the capacitor during one portion of the cycle is simply returned to the circuit in the next.

The formula for calculating the magnitude of the capacitive reactance is:

$$X_C = \frac{1}{2\pi f C}$$

where:

X_C = magnitude of capacitive reactance in ohms,
f = frequency in hertz,
C = capacitance in farads, and
π = 3.14.

By convention, capacitive reactance is assigned a negative value whereas inductive reactance (discussed below) is assigned a positive value.

Note: In many references and texts, angular frequency (in radians) ω=2πf is used and the equation reads:

$$X_C = \frac{1}{\omega C}$$

Example: What is the reactance of a capacitor of 470 pF (0.000470 μF) at a frequency of 7.15 MHz?

$$X_C = \frac{1}{2\pi f C}$$

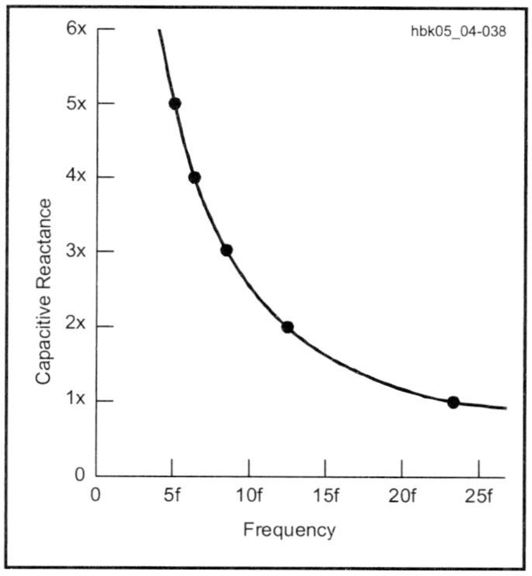

Figure 3.18 — A graph showing the general relationship of reactance to frequency for a fixed value of capacitance.

$$= \frac{1}{2\pi \times 7.15\ \text{MHz} \times 0.000470\ \mu\text{F}}$$

$$= \frac{1}{0.0211} = 47.4\ \Omega$$

Example: What is the reactance of the same capacitor, 470 pF (0.000470 μF), at a frequency of 14.3 MHz?

$$X_C = \frac{1}{2\pi f C}$$

$$= \frac{1}{2\pi \times 14.3\ \text{MHz} \times 0.000470\ \mu\text{F}}$$

$$= \frac{1}{0.0422} = 23.7\ \Omega$$

Current in a capacitor is directly related to the rate of change of the capacitor voltage. The maximum rate of change of voltage in a sine wave increases directly with the frequency, even if its peak voltage remains fixed. Therefore, the maximum current in the capacitor must also increase directly with frequency. Since, if voltage is fixed, an increase in current is equivalent to a decrease in reactance, the reactance of any capacitor decreases proportionally as the frequency increases. **Figure 3.18** illustrates the decrease in reactance of an arbitrary-value capacitor with respect to increasing frequency. The only limitation on the application of the graph is the physical construction of the capacitor, which may favor low-frequency uses or high-frequency applications.

CAPACITIVE SUSCEPTANCE

Just as conductance is sometimes the most useful way of expressing a resistance's ability to conduct current, the same is true for capacitors and ac current. This ability is called *susceptance* (abbreviated *B*). The units of susceptance are siemens (S), the same as that of conductance and admittance.

Susceptance in a capacitor is *capacitive susceptance*, abbreviated B_C. In an ideal capacitor with no resistive losses, susceptance is simply the negative reciprocal of reactance.

$$B_C = -\frac{1}{X_C}$$

where

X_C is the capacitive reactance, and
B_C is the capacitive susceptance.

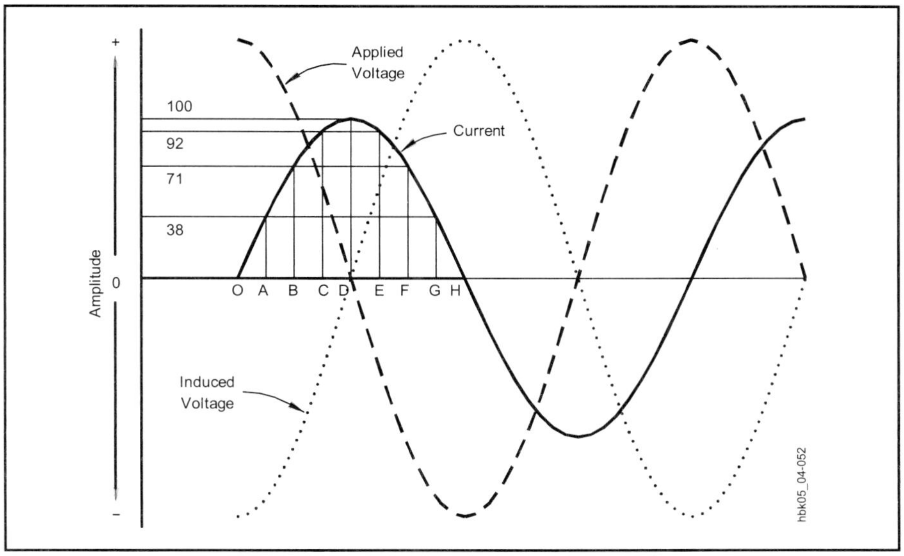

Figure 3.19 — Phase relationships between voltage and current when an alternating current is applied to an inductance.

3.4.3 Alternating Current in Inductors

For reasons similar to those that cause a phase difference between current and voltage in a capacitor, when an alternating voltage is applied to an ideal inductance with no resistance, the current is 90° out of phase with the applied voltage. In the case of an inductor, however, the current *lags* 90° behind the voltage as shown in **Figure 3.19**, the opposite of the capacitor current-voltage relationship. (Equivalently, the voltage across an inductor *leads* the current by 90°.) Interpreting Figure 3.19 begins with understanding that the cause for current lag in an inductor is the opposing voltage that is induced in the inductor and that the amplitude of the opposing voltage is proportional to the rate at which the inductor current changes.

In time segment OA, when the applied voltage is at its positive maximum, the rate at which the current is changing is also the highest, a 38% change. This means that the opposing voltage is also maximum, allowing the least current to flow. In the segment AB, as a result of the decrease in the applied voltage, current changes by only 33% inducing a smaller opposing voltage. The process continues in time segments BC and CD, the latter producing only an 8% rise in current as the applied and induced opposing voltage approach zero.

In segment DE, the applied voltage changes polarity, causing current to begin to decrease, returning stored energy to the circuit from the inductor's magnetic field. As the current rate of change is now negative (decreasing) the induced opposing voltage also changes polarity. Current flow is still in the original direction (positive), but is decreasing as less energy is stored in the inductor.

As the applied voltage continues to increase negatively, the current, although still positive, continues to decrease in value, reaching zero as the applied voltage reaches its negative maximum. The energy once stored in the inductor has now been completely returned to the circuit. The negative half-cycle then continues just as the positive half-cycle.

Similar to the capacitive circuit discussed earlier, by dividing the cycle into a large number of intervals, it can be shown that the current and voltage are both sine waves with a difference in phase.

Compare Figure 3.19 with Figure 3.17. Whereas in a pure capacitive circuit, the current *leads* the voltage by 90º, in a pure inductive circuit, the current *lags* the voltage by 90º. These phenomena are especially important in circuits that combine inductors and capacitors. Remember that the phase difference between voltage and current in both types of circuits is a result of energy being stored and released as voltage across a capacitor and as current in an inductor.

ELI the ICE Man

If you have difficulty remembering the phase relationships between voltage and current with inductors and capacitors, you may find it helpful to think of the phrase, "ELI the ICE man." This will remind you that voltage across an inductor leads the current through it, because the E comes before (leads) I, with an L between them, as you read from left to right. (The letter L represents inductance.) Similarly, I comes before (leads) E with a C between them.

3.4.4 Inductive Reactance and Susceptance

The amount of current that can be created in an inductor is proportional to the applied voltage but inversely proportional to the inductance because of the induced opposing voltage. If the applied voltage is ac, the rate of change of the current varies directly with the frequency and this rate of change also determines the amplitude of the induced or reverse voltage. Hence, the opposition to the flow of current increases proportionally to frequency. Stated in another way, inductor current is inversely proportional to inductance for a given applied voltage and frequency.

The combined effect of inductance and frequency is called *inductive reactance*, which, like capacitive reactance, is expressed in ohms. As with capacitive reactance, no power is dissipated in inductive reactance. The energy stored in the inductor during one portion of the cycle is returned to the circuit in the next portion.

The formula for calculating the magnitude of the inductive reactance is:

$$X_L = 2 \pi f L$$

where

X_L = magnitude of inductive reactance in ohms,
f = frequency in hertz,
L = inductance in henrys, and
$\pi = 3.14$
(If $\omega = 2 \pi f$, then $X_L = \omega L$.)

Example: What is the reactance of an inductor having an inductance of 8.00 H at a frequency of 120 Hz?

$$X_L = 2 \pi f L$$

$$= 6.28 \times 120 \text{ Hz} \times 8.00 \text{ H}$$

$$= 6{,}030 \ \Omega$$

Inductive Reactance Timesaver

Similarly to the calculation of capacitive reactance, if inductance is specified in microhenrys (μH) and the frequency is in megahertz (MHz), the reactance is calculated in ohms (Ω). The same is true for the combination of mH and kHz.

Example: What is the reactance of a 15.0-microhenry inductor at a frequency of 14.0 MHz?

$$X_L = 2 \pi f L$$

$$= 6.28 \times 14.0 \text{ MHz} \times 15.0 \ \mu\text{H}$$

$$= 1{,}320 \ \Omega$$

The resistance of the wire used to make the inductor has no effect on the reactance, but simply acts as a separate resistor connected in series with the inductor.

Example: What is the reactance of the same inductor at a frequency of 7.00 MHz?

$$X_L = 2 \pi f L$$

$$= 6.28 \times 7.00 \text{ MHz} \times 15.0 \ \mu\text{H}$$

$$= 659 \ \Omega$$

The direct relationship between frequency and reactance in inductors, combined with the inverse relationship between reactance and frequency in the case of capacitors, will be of fundamental importance in creating resonant circuits.

INDUCTIVE SUSCEPTANCE

As a measure of the ability of an inductor to limit the flow of ac in a circuit, inductive reactance is similar to capacitive reactance in having a corresponding susceptance, or ability to pass ac current in a circuit. In an ideal inductor with no resistive losses — that is, no energy lost as heat — susceptance is simply the negative reciprocal of reactance.

$$B_L = -\frac{1}{X_L}$$

where

X_L = the inductive reactance, and
B_L = the inductive susceptance.

The unit of susceptance for both inductors and capacitors is the *siemens*, abbreviated S.

3.5 Working with Reactance

3.5.1 Ohm's Law for Reactance

Only ac circuits containing capacitance or inductance (or both) have reactance. Despite the fact that the voltage in such circuits is 90° out of phase with the current, circuit reactance does oppose the flow of ac current in a manner that corresponds to resistance. That is, in a capacitor or inductor, reactance is equal to the ratio of ac voltage to ac current and the equations relating voltage, current and reactance take the familiar form of Ohm's Law:

$$E = I\,X$$

$$I = E / X$$

$$X = E / I$$

where

E = RMS ac voltage in volts,

I = RMS ac current in amperes, and

X = inductive or capacitive reactance in ohms.

Example: What is the RMS across a capacitor of 200 pF at 7.15 MHz, if the RMS current through the capacitor is 50 mA RMS?

Since the reactance of the capacitor is a function of both frequency and capacitance, first calculate the reactance:

$$X_C = \frac{1}{2\,\pi\, f C}$$

$$= \frac{1}{2 \times 3.14 \times 7.15 \times 10^6\,\text{Hz} \times 200 \times 10^{-12}\,\text{F}}$$

$$= \frac{1}{0.00898} = 111\ \Omega$$

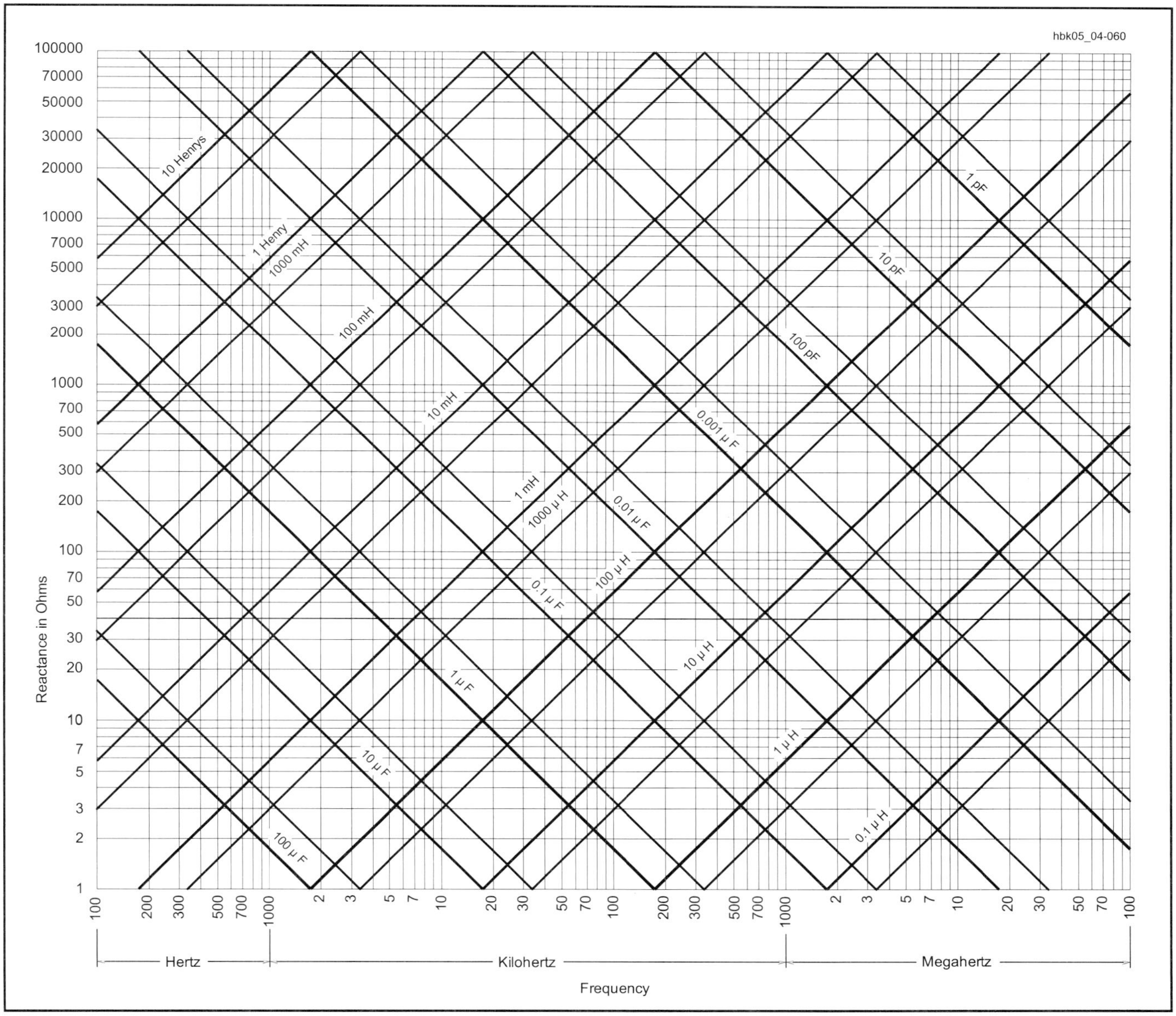

Figure 3.20 — Inductive and capacitive reactance vs frequency. Heavy lines represent multiples of 10, intermediate lines multiples of 5. For example, the light line between 10 µH and 100 µH represents 50 µH; the light line between 0.1 µF and 1 µF represents 0.5 µF, and so on. Other values can be extrapolated from the chart. For example, the reactance of 10 H at 60 Hz can be found by taking the reactance of 10 H at 600 Hz and dividing by 10 for the 10 times decrease in frequency. (Originally from Terman, *Radio Engineer's Handbook*. See references.)

Next, use:

$E = I \times X_C = 0.050 \text{ A RMS} \times 111\ \Omega$
$= 5.6 \text{ V RMS}$

Example: What is the RMS current through an 8.00 H inductor at 120 Hz, if 420 V RMS is applied?

$X_L = 2\ \pi\ f\ L$

$= 2 \times 3.14 \times 120 \text{ Hz} \times 8.00 \text{ H}$

$= 6{,}030\ \Omega$

$I = E / X_L = 420 / 6{,}030 = 69.6 \text{ mA RMS}$

Figure 3.20 charts the reactances of capacitors from 1 pF to 100 µF, and the reactances of inductors from 0.1 µH to 10 H, for frequencies between 100 Hz and 100 MHz. Approximate values of reactance can be read or interpolated from the chart. The formulas will produce more exact values, however. (The chart can also be used to find the frequency at which an inductor and capacitor have equal reactances, creating resonance as described in the section "Reactances At and Near Resonance" below.)

Although both inductive and capacitive reactance oppose the flow of ac current, the two types of reactance differ. With capacitive reactance, the current *leads* the voltage by 90°, whereas with inductive reactance, the current *lags* the voltage by 90°. The convention for charting the two types of reactance appears in **Figure 3.21**. On this graph, inductive reactance is plotted along the +90° vertical line, while capacitive reactance is plotted along the –90° vertical line. This convention of assigning a positive value to inductive reactance and a negative value to capacitive reactance results from the mathematics used for working with impedance as described elsewhere in this chapter.

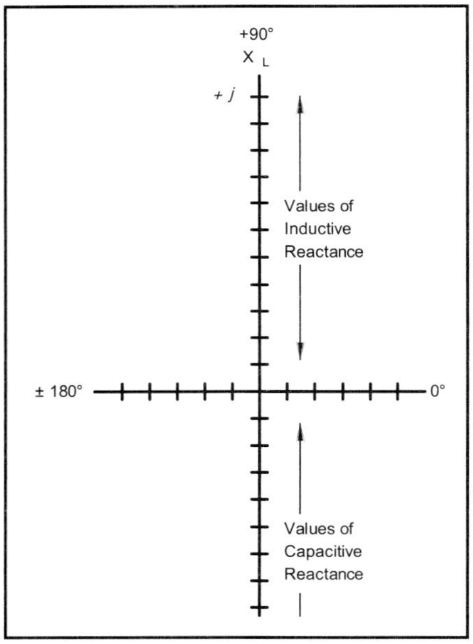

Figure 3.21 — The conventional method of plotting reactances on the vertical axis of a graph, using the upward or "plus" direction for inductive reactance and the downward or "minus" direction for capacitive reactance. The horizontal axis will be used for resistance in later examples.

3.5.2 Reactances in Series and Parallel

If a circuit contains two reactances of the same type, whether in series or in parallel, the resulting reactance can be determined by applying the same rules as for resistances in series and in parallel. Series reactance is given by the formula

$$X_{total} = X_1 + X_2 + X_3 ... + X_n$$

Example: Two noninteracting inductances are in series. Each has a value of 4.0 µH, and the operating frequency is 3.8 MHz. What is the resulting reactance?

The reactance of each inductor is:

$X_L = 2\ \pi\ f\ L$

$= 2 \times 3.14 \times 3.80 \times 10^6 \text{ Hz} \times 4.00 \times 10^{-6} \text{ H}$

$= 96.0\ \Omega$

$X_{total} = X_1 + X_2 = 96.0\ \Omega + 96.0\ \Omega = 192\ \Omega$

We might also calculate the total reactance by first adding the inductances:

$L_{total} = L_1 + L_2 = 4.00\ \mu H + 4.00\ \mu H = 8.0\ \mu H$

$X_{total} = 2\ \pi\ f\ L$

$= 2 \times 3.14 \times 3.80 \times 10^6 \text{ Hz} \times 8.00 \times 10^{-6} \text{ H}$

$= 191\ \Omega$

(The fact that the last digit differs by one illustrates variability caused by the limited precision of the measured values in the problem, and differences caused by rounding off the calculated values. This also shows why it is important to follow the rules for significant figures.)

Example: Two noninteracting capacitors are in series. One has a value of 10.0 pF, the other of 20.0 pF. What is the resulting reactance in a circuit operating at 28.0 MHz?

$$X_{C1} = \frac{1}{2\ \pi\ f\ C_1}$$

$$= \frac{1}{2 \times 3.14 \times 28.0 \times 10^6 \text{ Hz} \times 10.0 \times 10^{-12} \text{F}}$$

$$= \frac{1}{0.00176} = 568\ \Omega$$

$$X_{C2} = \frac{1}{2\ \pi\ f\ C_2}$$

$$= \frac{1}{2 \times 3.14 \times 28.0 \times 10^6 \text{ Hz} \times 20.0 \times 10^{-12} \text{F}}$$

$$= \frac{1}{0.00352} = 284\ \Omega$$

$$X_{total} = X_{C1} + X_{C2} = 568\ \Omega + 284\ \Omega = 852\ \Omega$$

Alternatively, combining the series capacitors first, the total capacitance is 6.67×10^{-12} F or 6.67 pF. Then:

$$X_{total} = \frac{1}{2\ \pi\ f\ C_{total}}$$

$$= \frac{1}{2 \times 3.14 \times 28.0 \times 10^6 \text{ Hz} \times 6.67 \times 10^{-12} \text{F}}$$

$$= \frac{1}{0.00117} = 855\ \Omega$$

(Within the uncertainty of the measured values and the rounding of values in the calculations, this is the same result as the 852 Ω we obtained with the first method.)

This example serves to remind us that *series capacitance* is not calculated in the manner used by other series resistance and inductance, but *series capacitive reactance* does follow the simple addition formula.

For reactances of the same type in parallel, the general formula is:

$$X_{total} = \frac{1}{\frac{1}{X_1} + \frac{1}{X_2} + \frac{1}{X_3} + ... + \frac{1}{X_n}}$$

or, for exactly two reactances in parallel

$$X_{total} = \frac{X_1 \times X_2}{X_1 + X_2}$$

Example: Place the capacitors in the last example (10.0 pF and 20.0 pF) in parallel in the 28.0 MHz circuit. What is the resultant reactance?

$$X_{total} = \frac{X_1 \times X_2}{X_1 + X_2}$$

$$= \frac{568\ \Omega \times 284\ \Omega}{568\ \Omega + 284\ \Omega} = 189\ \Omega$$

Alternatively, two capacitors in parallel can be combined by adding their capacitances.

$$C_{total} = C_1 + C_2 = 10.0 \text{ pF} + 20.0 \text{ pF} = 30 \text{ pF}$$

$$X_{C1} = \frac{1}{2 \pi f C_{total}}$$

$$= \frac{1}{2 \times 3.14 \times 28.0 \times 10^6 \text{ Hz} \times 30.0 \times 10^{-12} \text{ F}}$$

$$= \frac{1}{0.00528} = 189 \ \Omega$$

Example: Place the series inductors above (4.0 µH each) in parallel in a 3.8-MHz circuit. What is the resultant reactance?

$$X_{total} = \frac{X_{L1} \times X_{L2}}{X_{L1} + X_{L2}}$$

$$= \frac{96 \ \Omega \times 96 \ \Omega}{96 \ \Omega + 96 \ \Omega} = 48 \ \Omega$$

Of course, a number (N) of equal reactances (or resistances) in parallel yields a reactance that is the value of one of them divided by N, or:

$$X_{total} = \frac{X}{N} = \frac{96 \ \Omega}{2} = 48 \ \Omega$$

All of these calculations apply only to reactances of the same type; that is, all capacitive or all inductive. Mixing types of reactances requires a different approach.

UNLIKE REACTANCES IN SERIES

When combining unlike reactances — that is, combinations of inductive and capacitive reactance — in series, it is necessary to take into account that the voltage-to-current phase relationships differ for the different types of reactance. **Figure 3.22** shows a series circuit with both types of reactance. Since the reactances are in series, the current must be the same in both. The voltage across each circuit element differs in phase, however. The voltage across the inductor, E_L, *leads* the current by 90°, and voltage across the capacitor, E_C, *lags* the current by 90°. Therefore, E_L and E_C have opposite polarities and cancel each other in whole or in part. The line E in Figure 3.22 approximates the resulting voltage, which is the *difference* between E_L and E_C.

Since, for a constant current, the reactance is directly proportional to the voltage, the net reactance is still the sum of the individual reactances. Because inductive reactance is considered to be positive and capacitive reactance negative, the resulting reactance can be either positive (inductive) or negative (capacitive) or even zero (no reactance).

$$X_{total} = X_L - X_C$$

The convention of using absolute values for the reactances and building the sense of positive and negative into the formula is the preferred method used by hams and will be used in all of the remaining formulas in this chapter. Nevertheless, before using any formulas that include reactance, determine whether this convention is followed before assuming that the absolute values are to be used.

Example: Using Figure 3.22 as a visual aid, let $X_C = 20.0 \ \Omega$ and $X_L = 80.0 \ \Omega$. What is the resulting reactance?

$$X_{total} = X_L - X_C$$

$$= 80.0 \ \Omega - 20.0 \ \Omega = 60.0 \ \Omega$$

Since the result is a positive value, total reactance is inductive. Had the result been a negative number, the total reactance would have been capacitive.

When reactance types are mixed in a series circuit, the resulting reactance is always smaller than the larger of the two reactances. Likewise, the resulting voltage across the series combination of reactances is always smaller than the larger of the two voltages across individual reactances.

Every series circuit of mixed reactance types with more than two circuit elements can be reduced to this simple circuit by combining all the reactances into one inductive and one capacitive reactance. If the circuit has more than one capacitor or more than one inductor in the overall series string, first use the formulas given earlier to determine the total series inductance alone and the total series capacitance alone (or their respective reactances). Then combine the resulting single capacitive reactance and single inductive reactance as shown in this section.

UNLIKE REACTANCES IN PARALLEL

The situation of parallel reactances of mixed type appears in **Figure 3.23**. Since the elements are in parallel, the voltage is common to both reactive components. The current through the capacitor, I_C, *leads* the voltage by 90°, and the current through the inductor, I_L, *lags* the voltage by 90°. In this case, it is the currents that are 180° out of phase and thus cancel each other in whole or in part. The total current is the difference between the individual currents, as indicated by the line I in Figure 3.23.

Since reactance is the ratio of voltage to current, the total reactance in the circuit is:

$$X_{total} = \frac{E}{I_L - I_C}$$

In the drawing, I_C is larger than I_L, and the resulting differential current retains the phase of I_C. Therefore, the overall reactance, X_{total}, is capacitive in this case. The total reactance of the circuit will be smaller than the larger of the individual reactances, because the total current is smaller than the larger of the two individual currents.

In parallel circuits, reactance and current are inversely proportional to each other for a constant voltage and the equation for two

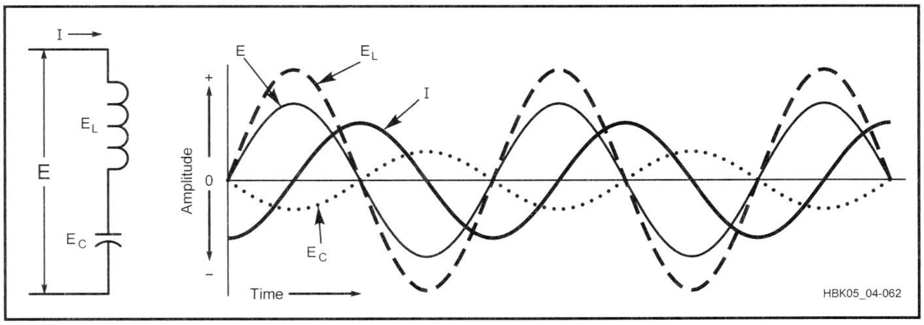

Figure 3.22 — A series circuit containing both inductive and capacitive components, together with representative voltage and current relationships.

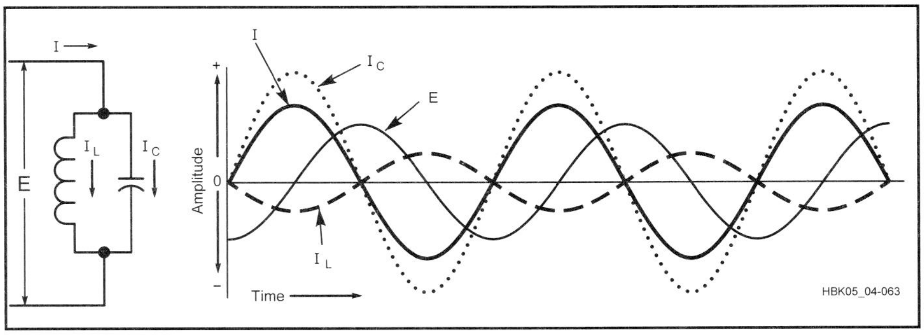

Figure 3.23 — A parallel circuit containing both inductive and capacitive components, together with representative voltage and current relationships.

reactances in parallel can be used, carrying the positive and negative signs:

$$X_{total} = \frac{X_L \times (-X_C)}{X_L - X_C} = \frac{-X_L \times X_C}{X_L - X_C}$$

As with the series formula for mixed reactances, follow the convention of using absolute values for the reactances, since the minus signs in the formula account for capacitive reactance being negative. If the solution yields a negative number, the resulting reactance is capacitive, and if the solution is positive, then the reactance is inductive.

Example: Using Figure 3.23 as a visual aid, place a capacitive reactance of 10.0 Ω in parallel with an inductive reactance of 40.0 Ω. What is the resulting reactance?

$$X_{total} = \frac{-X_L \times X_C}{X_L - X_C}$$

$$= \frac{-40.0\ \Omega \times 10.0\ \Omega}{40.0\ \Omega - 10.0\ \Omega}$$

$$= \frac{-400\ \Omega}{30.0\ \Omega} = -13.3\ \Omega$$

The reactance is capacitive, as indicated by the negative solution. Moreover, like resistances in parallel, the resultant reactance is always smaller than the larger of the two individual reactances.

As with the case of series reactances, if each leg of a parallel circuit contains more than one reactance, first simplify each leg to a single reactance. If the reactances are of the same type in each leg, the series reactance formulas for reactances of the same type will apply. If the reactances are of different types, then use the formulas shown above for mixed series reactances to simplify the leg to a single value and type of reactance.

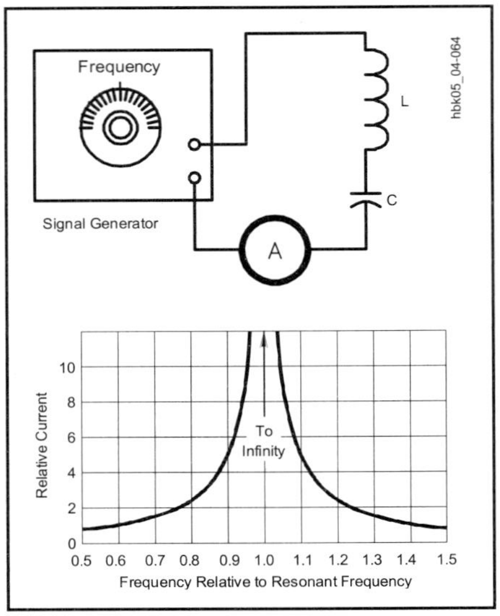

Figure 3.24 — The relative generator current with a fixed voltage in a series circuit containing inductive and capacitive reactances as the frequency approaches and departs from resonance.

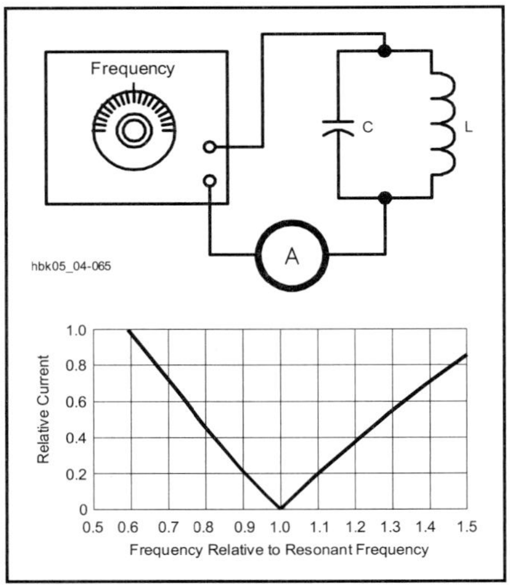

Figure 3.25 — The relative generator current with a fixed voltage in a parallel circuit containing inductive and capacitive reactances as the frequency approaches and departs from resonance. (The circulating current through the parallel inductor and capacitor is a maximum at resonance.)

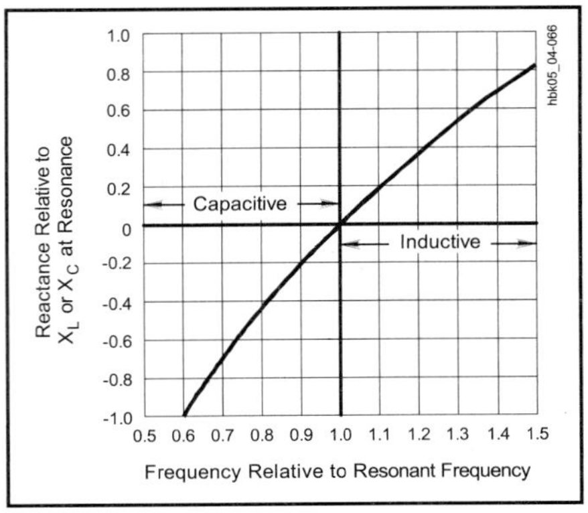

Figure 3.26 — The transition from capacitive to inductive reactance in a series-resonant circuit as the frequency passes resonance.

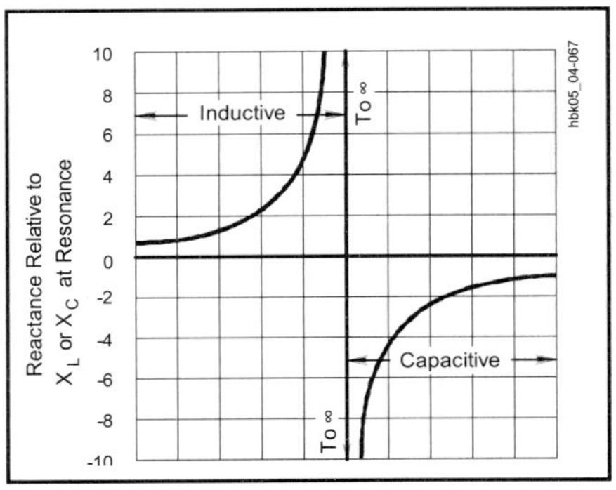

Figure 3.27 — The transition from inductive to capacitive reactance in a parallel-resonant circuit as the frequency passes resonance.

3.5.3 Reactances At and Near Resonance

Any series or parallel circuit in which the values of the two unlike reactances are equal is said to be *resonant*. For any given inductance or capacitance, it is theoretically possible to find a value of the opposite reactance type to produce a resonant circuit for any desired frequency. This is discussed below in the section Resonant Circuits.

When a series circuit like the one shown in Figure 3.22 is resonant, the voltages E_C and E_L are equal and cancel; their sum is zero. This is a *series-resonant* circuit. Since the reactance of the circuit is proportional to the sum of these voltages, the net reactance also goes to zero. Theoretically, the current, as shown in **Figure 3.24**, can become infinite. In fact, it is limited only by losses in the components and other resistances that would exist in a real circuit of this type. As the frequency of operation moves slightly off resonance and the reactances no longer cancel completely, the net reactance climbs as shown in the figure. Similarly, away from resonance the current drops to a level determined by the net reactance.

In a *parallel-resonant* circuit of the type in Figure 3.23, the current I_L and I_C are equal and cancel to zero. Since the reactance is inversely proportional to the current, as the current approaches zero, the reactance becomes infinite. As with series circuits, component losses and other resistances in the circuit prevent the current from reaching zero. **Figure 3.25** shows the theoretical current curve near and at resonance for a purely reactive parallel-resonant circuit. Note that in both Figure 3.24 and Figure 3.25, the departure of current from the resonance value is close to, but not quite, symmetrical above and below the resonant frequency. This is discussed below in the section Resonant Circuits.

Example: What is the reactance of a series L-C circuit consisting of a 56.04-pF capacitor and an 8.967-µH inductor at 7.000, 7.100 and 7.200 MHz? Using the formulas from earlier in this chapter, we calculate a table of values:

Freq (MHz)	X_L *(Ω)*	X_C *(Ω)*	X_{total} *(Ω)*
7.000	394.4	405.7	–11.30
7.100	400.0	400.0	0
7.200	405.7	394.4	11.30

The exercise shows the manner in which the reactance rises rapidly as the frequency moves above and below resonance. Note that in a series-resonant circuit, the reactance at frequencies below resonance is capacitive (negative), and above resonance, it is inductive (positive). **Figure 3.26** displays this fact graphically. In a parallel-resonant circuit, where the reactance becomes infinite at resonance, the opposite condition exists: above resonance, the reactance is capacitive (negative) and below resonance it is inductive (positive), as shown in **Figure 3.27**. Of course, all graphs and calculations in this section are theoretical and presume a purely reactive circuit. Real circuits are never purely reactive; they contain some resistance that modifies their performance considerably. Real resonant circuits will be discussed later in this chapter.

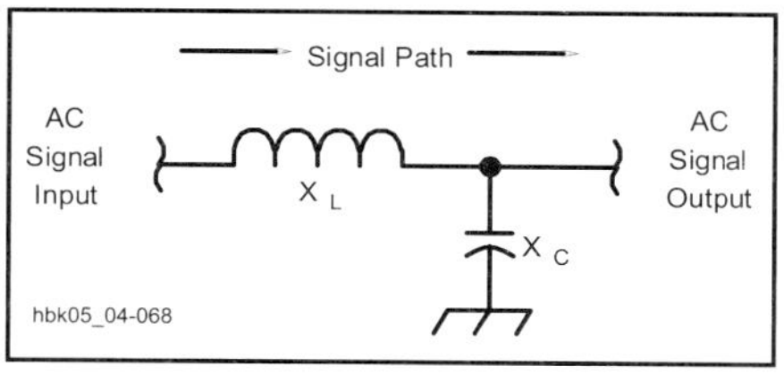

Figure 3.28 — A signal path with a series inductor and a shunt capacitor. The circuit presents different reactances to an ac signal and to its harmonics.

3.5.4 Reactance and Complex Waveforms

All of the formulas and relationships shown in this section apply to alternating current in the form of regular sine waves. Complex wave shapes complicate the reactive situation considerably. A complex or nonsinusoidal wave can be treated as a sine wave of some fundamental frequency and a series of harmonic frequencies whose amplitudes depend on the original wave shape. When such a complex wave, or collection of sine waves, is applied to a reactive circuit, the current through the circuit will not have the same wave shape as the applied voltage. The difference results because the reactance of inductors and capacitors vary with the applied frequency.

For the second-harmonic component of the complex wave, the reactance of the inductor is twice and the reactance of the capacitor is half their respective values at the fundamental frequency. A third-harmonic component produces inductive reactances that are triple and capacitive reactances that are one-third those at the fundamental frequency. Thus, the overall circuit reactance is different for each harmonic component.

The frequency sensitivity of a reactive circuit to various components of a complex wave shape creates both difficulties and opportunities. On the one hand, calculating the circuit reactance in the presence of highly variable as well as complex waveforms, such as speech, is difficult at best. On the other hand, the frequency sensitivity of reactive components and circuits lays the foundation for filtering, that is, for separating signals of different frequencies or acting upon them differently. For example, suppose a coil is in the series path of a signal and a capacitor is connected from the signal line to ground, as represented in **Figure 3.28**. The reactance of the coil to the second harmonic of the signal will be twice that at the fundamental frequency and oppose more effectively the flow of harmonic current. Likewise, the reactance of the capacitor to the harmonic will be half that to the fundamental, allowing the harmonic an easier current path toward ground and away from the signal line. The result is a low-pass filter that attenuates the harmonic more than the fundamental signal. (See the **Analog and Digital Filtering** chapter for detailed information on filter theory and construction.)

3.6 Impedance

When a circuit contains both resistance and reactance, the combined opposition to current is called *impedance*. Symbolized by the letter Z, impedance is a more general term than either resistance or reactance. Frequently, the term is used even for circuits containing only resistance or reactance. Qualifications such as "resistive impedance" are sometimes added to indicate that a circuit has only resistance, however.

The reactance and resistance comprising an impedance may be connected either in series or in parallel, as shown in **Figure 3.29**. In these circuits, the reactance is shown as a box to indicate that it may be either inductive or capacitive. In the series circuit at A, the current is the same in both elements, but there may be different voltages appearing across the resistance and reactance. In the parallel circuit at B, the same voltage is applied to both elements, but different currents may flow in the two branches.

In a resistance, the current is in phase with the applied voltage, while in a reactance it is 90° out of phase with the voltage. Thus, the phase relationship between current and voltage in the circuit as a whole may be anything between zero and 90°, depending on the relative amounts of resistance and reactance.

As shown in Figure 3.21 in the preceding section, reactance is graphed on the vertical axis to record the phase difference between the voltage and the current. **Figure 3.30** adds resistance to the graph. Since the voltage is in phase with the current, resistance is recorded on the horizontal axis, using the positive or right side of the scale.

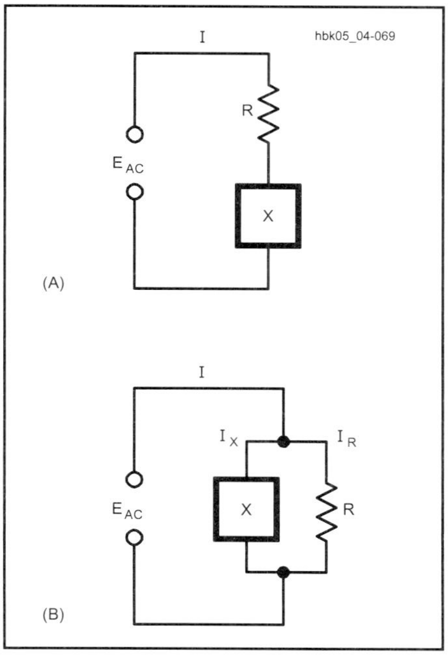

Figure 3.29 — Series and parallel circuits containing resistance and reactance.

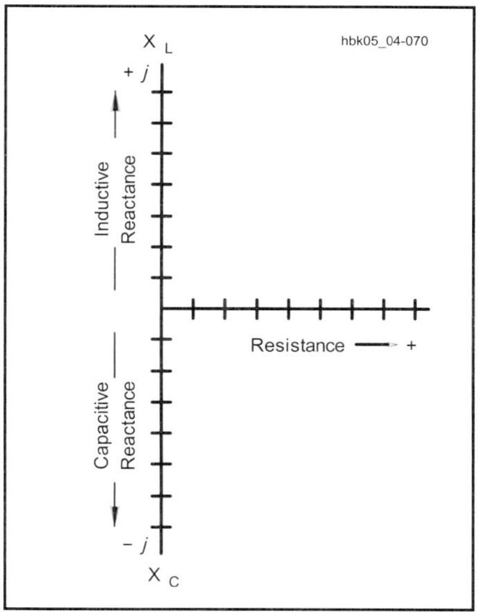

Figure 3.30 — The conventional method of charting impedances on a graph, using the vertical axis for reactance (the upward or "plus" direction for inductive reactance and the downward or "minus" direction for capacitive reactance), and using the horizontal axis for resistance.

3.6.1 Calculating Z from R and X in Series Circuits

Impedance is the complex combination of resistance and reactance. Since there is a 90° phase difference between resistance and reactance (whether inductive or capacitive), simply adding the two values does not correspond to what actually happens in a circuit and will not give the correct result. Therefore, expressions such as "Z = R + X" are incorrect because they show resistance and reactance being added directly. The correct expression is "Z = R + *j*X" showing that complex mathematics must be used. In pure mathematics, "*i*" indicates an imaginary number. Because *i* represents current in electronics, we use the letter "*j*" for the same mathematical operator, although there is nothing imaginary about what it represents in electronics. (References to explain imaginary numbers, rectangular coordinates, polar coordinates and how to work with them are provided in the "Radio Mathematics" article in this book's online content.) With respect to resistance and reactance, the letter *j* is normally assigned to those figures on the vertical axis, 90° out of phase with the horizontal axis. The presence of *j* means that impedance is a vector and calculating impedance from resistance and reactance involves *vector addition.*

As noted in the sidebar Vectors and Phasors at the start of this chapter, a vector is a value with both magnitude and direction, such as velocity; "10 meters/second to the north." Impedance also has a "direction" derived from the phase differences between voltage and current as described below. In vector addition, the result of combining two values with a phase difference is a quantity different from the simple *algebraic addition* of the two values. The result will have a phase difference intermediate between two vectors.

RECTANGULAR FORM OF IMPEDANCE

Because this form for impedances, Z = R ± *j*X, can be plotted on a graph using rectangular coordinates, this is called the *rectangular form* of impedance. The rectangular coordinate system in which one axis represents real number and the other axis imaginary numbers is called the *complex plane* and impedance with both real (R) and imaginary (X) components is called *complex impedance.* Unless specifically noted otherwise, assume that "impedance" means "complex impedance" and that both R and X may be present.

Consider **Figure 3.31**, a series circuit consisting of an inductive reactance and a resistance. As given, the inductive reactance is 100 Ω and the resistance is 50 Ω. Using *rectangular coordinates*, the impedance becomes

$$Z = R + jX$$

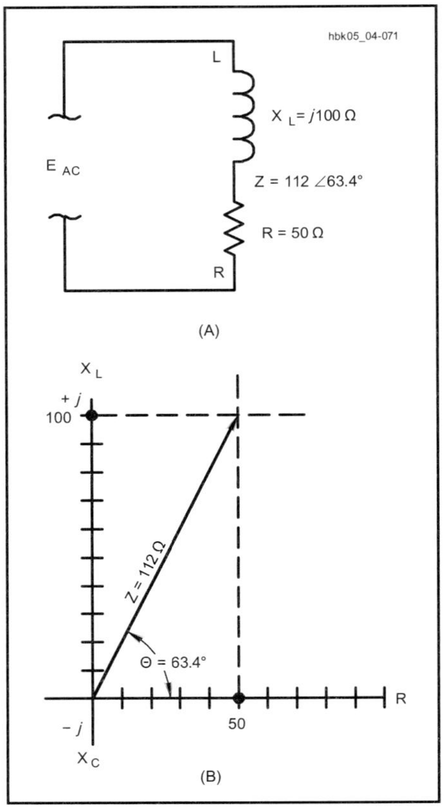

Figure 3.31 — A series circuit consisting of an inductive reactance of 100 Ω and a resistance of 50 Ω. At B, the graph plots the resistance, reactance, and impedance.

where
Z = the impedance in ohms,
R = the resistance in ohms, and
X = the reactance in ohms.

In the present example,

$$Z = 50 + j100\ \Omega$$

This point is located at the tip of the arrow drawn on the graph where the dashed lines cross.

POLAR FORM OF IMPEDANCE AND PHASE ANGLE

As the graph in Figure 3.31 shows, the impedance that results from combining R and X can also be represented by a line completing a right triangle whose sides are the resistance and reactance. The point at the end of the line, the complex impedance, can be described by how far it is from the origin of the graph where the axes cross (the *magnitude* of the impedance indicated by vertical bars around the variable, |Z|) and the angle made by the line with the horizontal axis representing 0° (the *phase angle* of the impedance, θ). This is the *polar form* of impedance and it is written in the form

$$Z = |Z| \angle\theta$$

Occasionally, θ may be given in radians. The convention in this handbook is to use degrees unless specifically noted otherwise.

The length of the hypotenuse of the right triangle represents the magnitude of the impedance and can be calculated using the formula for calculating the hypotenuse of a right triangle, in which the square of the hypotenuse equals the sum of the squares of the two sides:

$$|Z| = \sqrt{R^2 + X^2}$$

In this example:

$$|Z| = \sqrt{(50\ \Omega)^2 + (100\ \Omega)^2}$$

$$= \sqrt{2{,}500\ \Omega^2 + 10{,}000\ \Omega^2}$$

$$= \sqrt{12{,}500\ \Omega^2} = 112\ \Omega$$

The magnitude of the impedance that results from combining 50 Ω of resistance with 100 Ω of inductive reactance is 112 Ω. From trigonometry, the tangent of the phase angle is the side opposite the angle (X) divided by the side adjacent to the angle (R), or

$$\tan\theta = \frac{X}{R}$$

where
X = the reactance, and
R = the resistance.

Find the angle by taking the inverse tangent ($\tan^{-1}$), or arctan:

$$\theta = \arctan\frac{X}{R}$$

Remember to be sure your calculator is set to use the right angular units, either degrees or radians.

In the example shown in Figure 3.31,

$$\theta = \arctan\frac{100\ \Omega}{50\ \Omega} = \arctan 2.0 = 63.4°$$

Using the information just calculated, the complex impedance in polar form is:

$$Z = 112\ \angle 63.4°\ \Omega$$

This is stated verbally as "112 ohms at an angle of 63 point 4 degrees."

POLAR TO RECTANGULAR CONVERSION

The expressions R ± *j*X and |Z| ∠θ both provide the same information, but in two different forms. The procedure just given permits conversion from rectangular coordinates into polar coordinates. The reverse procedure is also important. **Figure 3.32** shows an impedance composed of a capacitive reactance and a resistance. Since capacitive reactance appears as a negative value, the impedance will be at a negative phase angle (remember, magnitude is always positive), in this case, 12.0 Ω at a phase angle of – 42.0° or Z = 12.0 ∠ – 42.0° Ω.

Since the impedance forms a triangle with the values of X and R from the rectangular coordinates, the reactance axis forms the side opposite the angle θ.

$$\sin\theta = \frac{\text{side opposite}}{\text{hypotenuse}} = \frac{X}{|Z|}$$

Solving this equation for reactance, we have:

$$X = |Z| \times \sin\theta \text{ (ohms)}$$

Likewise, the resistance forms the side adjacent to the angle.

$$\cos\theta = \frac{\text{side adjacent}}{\text{hypotenuse}} = \frac{R}{|Z|}$$

Solving for resistance, we have:

$$R = |Z| \times \cos\theta \text{ (ohms)}$$

Then from our example:

$$X = 12.0\ \Omega \times \sin(-42.0°)$$
$$= 12.0\ \Omega \times -0.669 = -8.03\ \Omega$$

$$R = 12.0\ \Omega \times \cos(-42.0°)$$
$$= 12.0\ \Omega \times 0.743 = 8.92\ \Omega$$

Since X is a negative value, it is plotted on the lower vertical axis, as shown in Figure 3.32, indicating capacitive reactance. In rectangular form, Z = 8.92 Ω – *j*8.03 Ω.

In performing impedance and related calculations with complex circuits, rectangular coordinates are most useful when formulas require the addition or subtraction of values. Polar notation is most useful for multiplying and dividing complex numbers. (See the article "Radio Mathematics" in this book's online content for references dealing with the mathematics of complex numbers.)

All of the examples shown so far in this section presume a value of reactance that contributes to the circuit impedance. Reactance is a function of frequency, however, and many impedance calculations may begin with a value of capacitance or inductance and an operating frequency. In terms of these values, |Z| can be calculated in either of two ways, depending on whether the reactance is inductive or capacitive:

$$|Z| = \sqrt{R^2 + (2\pi f L)^2}$$

$$|Z| = \sqrt{R^2 + \left(\frac{1}{2\pi f C}\right)^2}$$

Example: What is the impedance of a circuit like Figure 3.31 but with different component values: A resistance of 100 Ω and a 7.00 µH inductor operating at a frequency of 7.00 MHz? Using the equation appropriate for inductive reactance,

$$|Z| = \sqrt{R^2 + (2\pi f L)^2}$$

$$= \sqrt{(100\ \Omega)^2 + (2\pi \times 7.0 \times 10^{-6}\ H \times 7.0 \times 10^{6}\ Hz)^2}$$

$$= \sqrt{10{,}000\ \Omega^2 + (308\ \Omega)^2}$$

$$= \sqrt{10{,}000\ \Omega^2 + 94{,}900\ \Omega^2}$$

$$= \sqrt{104{,}900\ \Omega^2} = 323.9\ \Omega$$

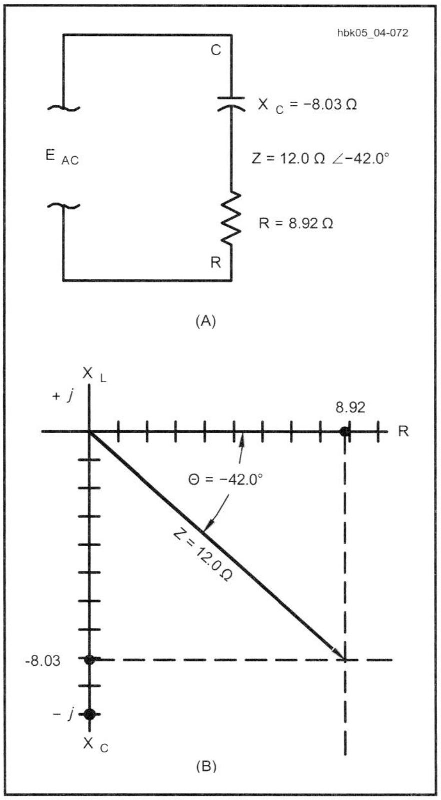

Figure 3.32 — A series circuit consisting of a capacitive reactance and a resistance: the impedance is given as 12.0 Ω at a phase angle θ of –42.0 degrees. At B, the graph plots the resistance, reactance, and impedance.

Since 308 Ω is the value of inductive reactance of the 7.00-µH coil at 7.00 MHz, the phase angle calculation proceeds as given in the earlier example:

$$\theta = \arctan\frac{X}{R} = \arctan\left(\frac{308.0\ \Omega}{100.0\ \Omega}\right)$$

$$= \arctan(3.08) = 72.0°$$

Since the reactance is inductive, the phase angle is positive.

3.6.2 Calculating Z from R and X in Parallel Circuits

In a parallel circuit containing reactance and resistance, such as shown in **Figure 3.33**, calculation of the resultant impedance from the values of R and X does not proceed by direct combination as for series circuits. The general formula for parallel circuits is:

$$|Z| = \frac{R|X|}{\sqrt{R^2 + |X|^2}}$$

where the formula uses the absolute (unsigned) reactance value. The phase angle for the parallel circuit is given by:

$$\theta = \arctan\left(\frac{R}{X}\right)$$

The sign of θ has the same meaning in both series and parallel circuits: if the parallel reactance is capacitive, then θ is a negative angle, and if the parallel reactance is inductive, then θ is a positive angle.

Example: A resistor of 30 Ω is in parallel with an inductor with a reactance of 40 Ω. What is the resulting impedance and phase angle?

$$|Z| = \frac{R|X|}{\sqrt{R^2 + |X|^2}} = \frac{30.0\ \Omega \times 40.0\ \Omega}{\sqrt{(30.0\ \Omega)^2 + (40.0\ \Omega)^2}}$$

$$= \frac{1{,}200\ \Omega^2}{\sqrt{900\ \Omega^2 + 1{,}600\ \Omega^2}} = \frac{1{,}200\ \Omega^2}{\sqrt{2{,}500\ \Omega^2}}$$

$$= \frac{1{,}200\ \Omega^2}{50.0\ \Omega} = 24.0\ \Omega$$

$$\theta = \arctan\left(\frac{R}{X}\right) = \arctan\left(\frac{40.0\ \Omega}{30.0\ \Omega}\right)$$

$$\theta = \arctan(1.33) = 53.1°$$

In polar form,

$$Z = 24.0\angle 53.1°\ \Omega.$$

Since the parallel reactance is inductive, the resultant angle is positive.

Example: A capacitor with a reactance of 16.0 Ω is in parallel with a resistor of 12.0 Ω.

What is the resulting impedance and phase angle? (Remember that capacitive reactance is negative when used in calculations.)

$$|Z| = \frac{R|X|}{\sqrt{R^2 + |X|^2}} = \frac{12.0\ \Omega \times 16.0\ \Omega}{\sqrt{(12.0\ \Omega)^2 + (16.0\ \Omega)^2}}$$

$$= \frac{192\ \Omega^2}{\sqrt{144\ \Omega^2 + 256\ \Omega^2}} = \frac{192\ \Omega^2}{\sqrt{400\ \Omega^2}}$$

$$= \frac{192\ \Omega^2}{20.0\ \Omega} = 9.60\ \Omega$$

$$\theta = \arctan\left(\frac{R}{X}\right) = \arctan\left(\frac{12.0\ \Omega}{-16.0\ \Omega}\right)$$

$$\theta = \arctan(-0.750) = -36.9°$$

Because the parallel reactance is capacitive and the reactance negative, the resultant phase angle is negative.

3.6.3 Admittance

Impedance also has an inverse: *admittance (Y)*, measured in siemens (symbol S, also Ω^{-1} and ℧ are used). Thus,

$$Y = 1 / Z$$

Since resistance, reactance, and impedance are inversely proportional to the current (Z = E / I), conductance, susceptance, and admittance are directly proportional to current. That is,

$$Y = I / E$$

Admittance can be expressed in rectangular and polar forms, just like impedance,

$$Y = G + jB = |Y|\angle\theta$$

The phase angle for admittance has the opposite sign convention as for impedance; if the susceptance component is inductive, the phase angle is negative, and if the susceptive component is capacitive, the phase angle is positive.

One handy use for admittance is in simplifying parallel circuit impedance calculations. Similar to series combinations, the admittance of a parallel combination of reactance and resistance is the vector addition of susceptance and conductance. In other words, for parallel circuits:

$$|Y| = \sqrt{G^2 + B^2}$$

where

|Y| = magnitude of the admittance in siemens,

G = conductance or 1 / R in siemens, and

B = susceptance or –1 / X in siemens.

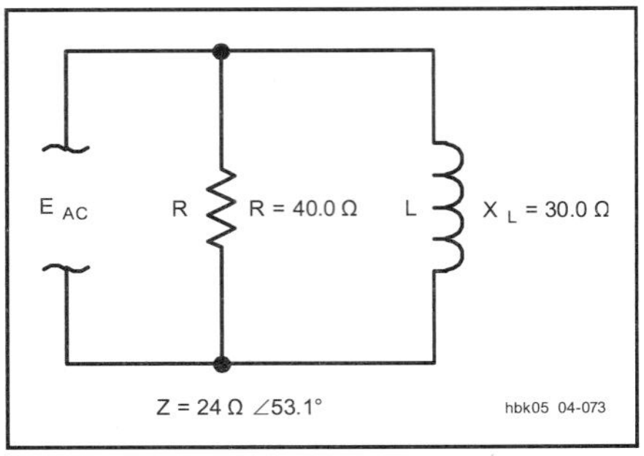

Figure 3.33 — A parallel circuit containing an inductive reactance of 30.0 Ω and a resistor of 40.0 Ω. No graph is given, since parallel impedances cannot be manipulated graphically in the simple way of series impedances.

Repeating the previous example from Figure 3.33 for an inductor with a reactance of 30.0 Ω in parallel with a resistor of 40.0 Ω. Calculate the resulting impedance and phase angle by using admittance.

The inductor's susceptance is –1 / 30.0 Ω = –0.0333 S and the resistor's conductance is 1 / 40.0 Ω = 0.0250 S.

$$|Y| = \sqrt{(0.0250\,S)^2 + (-0.0333\,S)^2}$$

$$= \sqrt{0.00173\,S^2} = 0.0417\,S$$

The admittance's phase angle in terms of conductance and susceptance is:

$$\theta = \arctan\left(\frac{B}{G}\right)$$

$$\theta = \arctan\left(\frac{-0.0333\,S}{0.0250\,S}\right) = \arctan(-1.33) = -53.1°$$

In polar form,

$$Y = 0.0417\angle -53.1°$$

Converting back to impedance:

$$|Z| = \frac{1}{|Y|} = \frac{1}{0.0417\,S} = 24.0\ \Omega$$

The impedance's phase angle is the negative of the admittance's phase angle, so:

$$\angle Z = -\angle Y = \angle 53.1°$$

and in polar form

$$Z = 24.0\angle 53.1°\ \Omega$$

Conversion between resistance, impedance and admittance is very useful in working with complex circuits and in impedance matching of antennas and transmission lines. There are many on-line calculators that can perform these operations and many programmable calculators and suites of mathematical computer software have these functions built-in. Knowing when and how to use them, however, demands some understanding of the fundamental strategies shown here.

3.6.4 More than Two Elements in Series or Parallel

When a circuit contains several resistances or several reactances in series, simplify the circuit before attempting to calculate the impedance. Resistances in series add, just as in a purely resistive circuit. Series reactances of the same kind, that is, all capacitive or all inductive, also add, just as in a purely reactive circuit. The goal is to produce a single value of resistance and a single value of reactance that can be used in the impedance calculation.

Figure 3.34A illustrates a case in which a circuit consists of two different types of reactance in series, along with series resistance.

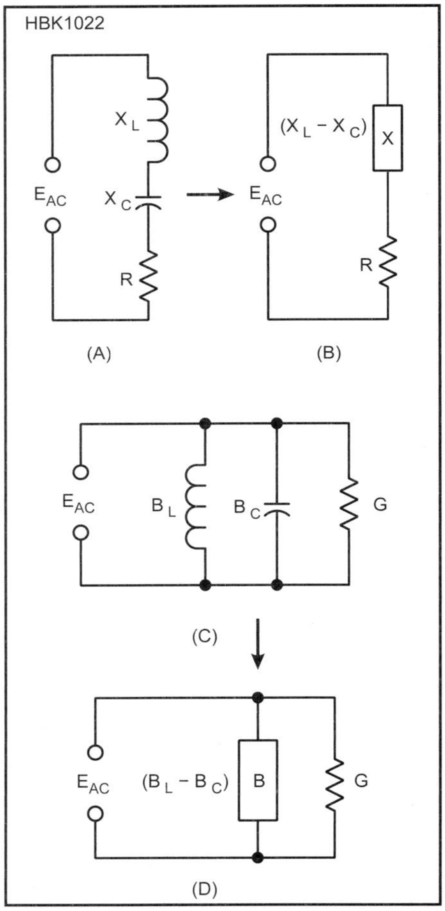

Figure 3.34 — A series circuit (A) containing both capacitive and inductive reactances can be simplified to a circuit with a single reactance (B) by adding the reactance values together. A parallel circuit (C) is simplified using the formulas in the previous section, converting the parallel reactances to susceptances, adding the values to get circuit (D) and then converting the resulting susceptance back to reactance.

The series combination of X_C and X_L reduces to a single value using the same rules of combination discussed in the section on purely reactive components. As Figure **3.34B** shows, the resulting reactance is the difference of the two reactances in series; $X_{TOTAL} = X_L - X_C$.

For parallel circuits with multiple resistances or multiple reactances of the same type, use the rules of parallel combination to reduce the resistive and reactive components to single elements. Where two or more reactive components of different types appear in the same circuit, they can be combined using formulas shown earlier for pure reactances. As **Figure 3.34C** suggests, however, they can also be combined as susceptances. Parallel susceptances of different types add, with attention to their differing signs. The resulting single susceptance can then be combined with the conductance to arrive at the overall circuit admittance whose inverse is the final circuit impedance.

3.6.5 Equivalent Series and Parallel Circuits

The two circuits shown in Figure 3.29 are equivalent if the same current flows when a given voltage of the same frequency is applied, and if the phase angle between voltage and current is the same in both cases. It is possible, in fact, to transform any given series circuit into an equivalent parallel circuit, and vice versa.

Using the conventions shown in **Figure 3.35**, a series RX circuit can be converted into its parallel equivalent by means of the formulas:

$$R_P = \frac{R_S^2 + X_S^2}{R_S}$$

$$X_P = \frac{R_S^2 + X_S^2}{X_S}$$

where the subscripts P and S represent the parallel- and series-equivalent values, respectively. If the parallel values are known, the equivalent series circuit can be found from:

$$R_S = \frac{R_P\, X_P^2}{R_P^2 + X_P^2}$$

and

$$X_S = \frac{R_P^2\, X_P}{R_P^2 + X_P^2}$$

Example: Let the series circuit in Figure 3.29(A) have a series reactance of –50.0 Ω (indicating a capacitive reactance) and a resistance of 50.0 Ω. What are the values of the equivalent parallel circuit?

$$R_P = \frac{R_S^2 + X_S^2}{R_S} = \frac{(50.0\ \Omega)^2 + (-50.0\ \Omega)^2}{50.0\ \Omega}$$

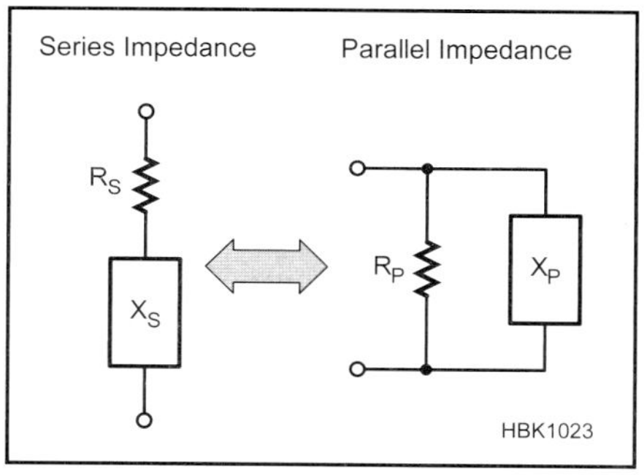

Figure 3.35 — Circuits containing reactance and resistance can be transformed into their series and parallel equivalents using the formulas in this section. Remember to use positive values for inductive reactance (X_L) and negative for capacitive reactance (X_C).

$$= \frac{2{,}500\ \Omega^2 + 2{,}500\ \Omega^2}{50.0\ \Omega} = \frac{5{,}000\ \Omega^2}{50.0\ \Omega} = 100\ \Omega$$

$$X_P = \frac{R_S^2 + X_S^2}{X_S} = \frac{(50.0\ \Omega)^2 + (-50.0\ \Omega)^2}{-50.0\ \Omega}$$

$$= \frac{2{,}500\ \Omega^2 + 2{,}500\ \Omega^2}{-50.0\ \Omega} = \frac{5{,}000\ \Omega^2}{-50.0\ \Omega} = 100\ \Omega$$

A capacitive reactance of 100 Ω in parallel with a resistance of 100 Ω is the equivalent circuit to the series circuit.

3.6.6 Ohm's Law for Impedance

Ohm's Law applies to circuits containing impedance just as readily as to circuits having resistance or reactance only. The formulas are:

$$E = I\,Z$$

$$I = E / Z$$

$$Z = E / I$$

where
- E = voltage in volts,
- I = current in amperes, and
- Z = impedance in ohms.

Z must now be understood to be a complex number, consisting of resistive and reactive components. If Z is complex, then so are E and I, with magnitudes and phase angles. The rules of complex mathematics are then applied and the variables are written in boldface type, or an arrow is added above them to indicate that they are complex, such as,

$$\mathbf{E} = \mathbf{IZ} \text{ or } \overleftarrow{E} = \overleftarrow{I}\overleftarrow{Z} \text{ or } \vec{E} = \vec{I}\vec{Z}$$

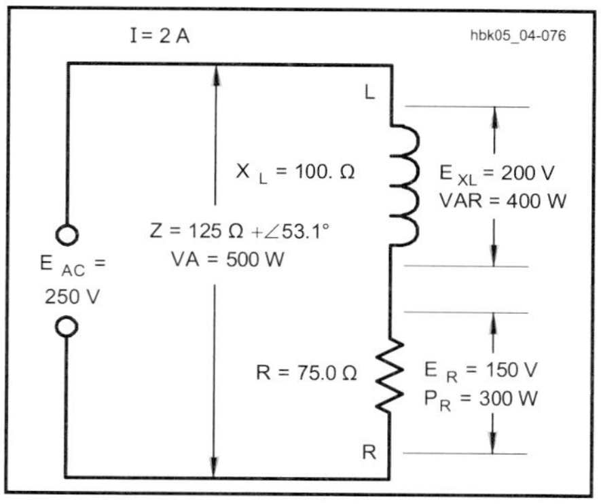

Figure 3.36 — A series circuit consisting of an inductive reactance of 100 Ω and a resistance of 75.0 Ω. Also shown is the applied voltage, voltage drops across the circuit elements, and the current.

If only the magnitude of impedance, voltage, and currents are important, however, then the magnitudes of the three variables can be combined in the familiar ways without regard to the phase angle. In this case E and I are assumed to be RMS values (or some other steady-state value such as peak, peak-to-peak, or average). **Figure 3.36** shows a simple circuit consisting of a resistance of 75.0 Ω and a reactance of 100 Ω in series. From the series-impedance formula previously given, the impedance is

$$Z = \sqrt{R^2 + X_L^2} = \sqrt{(75.0\ \Omega)^2 + (100\ \Omega)^2}$$

$$= \sqrt{5{,}630\ \Omega^2 + 10{,}000\ \Omega^2}$$

$$= \sqrt{15{,}600\ \Omega^2} = 125\ \Omega$$

If the applied voltage is 250 V, then

$$I = \frac{E}{Z} = \frac{250\ \text{V}}{125\ \Omega} = 2.00\ \text{A}$$

This current flows through both the resistance and reactance, so the voltage drops are:

$$E_R = I\,R = 2.00\ \text{A} \times 75.0\ \Omega = 150\ \text{V}$$

$$E_{XL} = I\,X_L = 2.00\ \text{A} \times 100\ \Omega = 200\ \text{V}$$

Illustrating one problem of working only with RMS values, the simple arithmetical sum of these two drops, 350 V, is greater than the applied voltage because the two voltages are 90° out of phase. When phase is taken into account,

$$E = \sqrt{(150\ \text{V})^2 + (200\ \text{V})^2}$$

$$= \sqrt{22{,}500\ \text{V}^2 + 40{,}000\ \text{V}^2}$$

$$= \sqrt{62{,}500\ \text{V}^2} = 250\ \text{V}$$

3.6.7 Reactive Power and Power Factor

The charge placed on a capacitor during part of an ac cycle is returned to the circuit during the next part of a cycle. Likewise, the energy stored in the magnetic field of an inductor returns to the circuit later in the ac cycle. A reactive circuit simply cycles and recycles energy into and out of the reactive components. If a purely reactive circuit were possible in reality, it would consume no energy at all.

In reactive circuits, circulation of energy accounts for seemingly odd phenomena. For example, in a series circuit with capacitance and inductance, the voltages across the components may exceed the supply voltage. That condition can exist because, while energy is being stored by the inductor, the capacitor is returning energy to the circuit from its previously charged state, and vice versa. In a parallel circuit with inductive and capacitive branches, the current circulating through the components may exceed the current drawn from the source. Again, the phenomenon occurs because the inductor's collapsing magnetic field supplies current to the capacitor, and the discharging capacitor provides current to the inductor.

Reactance creates a phase difference between voltage and current as shown in **Figure 3.37**, which shows current lagging behind voltage by about 60°. Instantaneous power is still calculated as the product of instantaneous voltage and current, but notice that it is negative for part of the cycle when voltage is positive and current negative, or vice versa. This causes average power dissipated in the resistance over the entire cycle to be lower than when voltage and current are in-phase. If current and voltage are 90° out of phase as in a purely reactive circuit, average power is reduced to zero. So, although reactive circuits show a measurable RMS value for ac voltage and current, we cannot simply multiply the values together as for the purely resistive circuit of Figure 3.14B.

To distinguish between the non-dissipated energy circulating in a purely reactive circuit and the dissipated or *real power* in a resistive circuit, the unit of *reactive power* is called the *volt-ampere reactive*, or *VAR*. The term watt is not used and sometimes reactive power is called "wattless" power. VAR has only limited use in radio circuits. Formulas similar to those for resistive power are used to calculate VAR:

$$VAR = I \times E$$

$$VAR = I^2 \times X$$

$$VAR = E^2 / X$$

where E and I are RMS values of voltage and current.

AC Component Summary

	Resistor	Capacitor	Inductor
Basic Unit	ohm (Ω)	farad (F)	henry (H)
Units Commonly Used		microfarads (μF) picofarads (pF)	millihenrys (mH) microhenrys (μH)
Time constant	(None)	RC	L/R
Voltage-Current Phase	In phase	Current leads voltage Voltage lags current	Voltage leads current Current lags voltage
Resistance or Reactance	R	$X_C = 1 / 2\pi fC$	$X_L = 2\pi fL$
Change with increasing frequency	No	X_C decreases	R_L increases
Q of circuit	Not defined	X_C / R	X_L / R

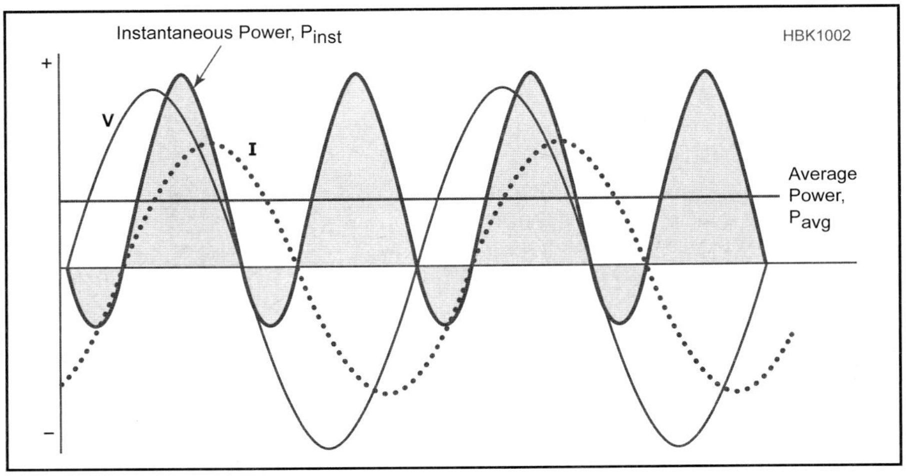

Figure 3.37 — Instantaneous power is the product of instantaneous voltage and current but may be positive or negative depending on the relative phase of voltage or current. This reduces the average power that is calculated over the entire cycle.

Real, or dissipated, power is measured in watts. *Apparent power* is the product of the voltage across and the current through an impedance. To distinguish apparent power from real power, apparent power is measured in *volt-amperes* (*VA*).

In the circuit of Figure 3.36, an applied voltage of 250 V results in a current of 2.00 A, giving an apparent power of 250 V × 2.00 A = 500 W. Only the resistance actually consumes power, however. The real power dissipated by the resistance is:

$$E = I^2 R = (2.00 \text{ A})^2 \times 75.0 \text{ V} = 300 \text{ W}$$

and the reactive power is:

$$VAR = I^2 \times X_L = (2.00 \text{ A})^2 \times 100\ \Omega = 400 \text{ VA}$$

The ratio of real power to the apparent power is called the circuit's *power factor* (PF).

$$PF = \frac{P_{consumed}}{P_{apparent}} = \frac{R}{Z}$$

Power factor is frequently expressed as a percentage. The power factor of a purely resistive circuit is 1.0 or 100% while the power factor of a pure reactance is zero. In the example of Figure 3.36 the power factor would be 300 W / 500 W = 0.600 or 60%.

Apparent power has no direct relationship to the power actually dissipated unless the power factor of the circuit is known.

$$P = \text{Apparent Power} \times \text{Power Factor}$$

An equivalent definition of power factor is:

$$PF = \cos \theta$$

where θ is the phase angle of the circuit impedance.

Since the phase angle in the example equals:

$$\theta = \arctan\left(\frac{X}{R}\right) = \arctan\left(\frac{100\ \Omega}{75.0\ \Omega}\right)$$

$$\theta = \arctan(1.33) = 53.1°$$

and the power factor is:

$$PF = \cos 53.1° = 0.600$$

as the earlier calculation confirms.

Since power factor is always rendered as a positive number, the value must be followed by the words "leading" or "lagging" to identify the phase of the voltage with respect to the current. Specifying the numerical power factor is not always sufficient. For example, many dc-to-ac power inverters can safely operate loads having a large net reactance of one sign but only a small reactance of the opposite sign. Hence, the final calculation of the power factor in this example would be reported as "0.600, leading."

3.7 Quality Factor (Q) of Components

Components that store energy, such as capacitors and inductors, may be compared in terms of *quality factor* or *Q factor*, abbreviated *Q*. The concept of *Q* originated in 1914 (then dubbed *K*) and first appeared in print in 1923 when Kenneth S. Johnson used it to represent the ratio of reactance to resistance as a "figure of merit" for inductors in US patent 1,628,983. For a series or parallel representation of a reactive circuit element:

$$Q = \frac{X_S}{R_S} = \frac{R_P}{X_P}$$

where for series-connected reactance and its series loss resistance (such as an inductor)

- Q = quality factor (no units),
- X_S = series reactance of the component (in ohms), and
- R_S = the sum of all series resistances associated with the energy losses in the component (in ohms).

For a parallel connected reactance and its parallel loss resistance (such as a capacitor)

- Q = quality factor (no units),
- X_P = parallel-connected reactance of the component (in ohms), and
- R_P = the total parallel resistance associated with the energy losses in the component (in ohms).

Table 3.2
Equivalent Formulas for Expressing Q and Their Uses

[a]	$Q = \frac{\text{Series Reactance}}{\text{Series Resistance}}$	Johnson's historical definition of Q for inductors, used for series circuits.
[b]	$Q = \frac{\text{Parallel Resistance}}{\text{Parallel Reactance}}$	Parallel equivalent circuit definition of Q, useful for capacitors.
[c]	$Q = \frac{2\pi \times \text{Stored energy}}{\text{Energy lost in one cycle}}$	Fundamental energy definition, useful with antennas, reactive components, and mechanical systems.
[d]	$Q = \frac{\sqrt{f_U f_L}}{f_U f_L} = \frac{f_r}{\text{FWMH}}$	Bandwidth formula for simple resonant circuits. Impedance Z = R + *j*X, and f_r is the resonant frequency, f_U is the upper frequency where R = X, and f_L is the lower frequency where R = –X, and $f_U - f_L$ represents the –3 dB bandwidth, also called Full-Width Half-Maximum (FWHM).

Several exactly equivalent formulas for Q may be seen in **Table 3.2**. In Table 3.2, where equation [a] most naturally represents the Q of an inductor, and equation [b] is useful for a capacitor. Both representations are equivalent to equation [c], which relates the energy storage to energy losses in inductors and capacitors. Note that in a series circuit representation, the series resistance is proportional to energy loss, and the series reactance is proportional to stored energy. In a parallel circuit, however, the reciprocal of the resistance is proportional to the lost energy and the reciprocal of the reactance is proportional to the stored energy. See the referenced paper by Ohira in the References section for a theoretical discussion of Q as applied to different circuit characteristics.

The Q of a tuned circuit may be found by measuring the upper and lower frequencies where the resistance equals the magnitude of the reactance, and applying equation [d]. The geometrical mean frequency is

$$f = \sqrt{f_U f_L}$$

and may be replaced by the center frequency for high-Q circuits. In circuits having several reactive components, such as the tuned circuits in Figure 3.42 later in this chapter, the circuit Q is the parallel combination of the individual Q factors. For example:

$$Q = 1 / \left(\frac{1}{Q_C} + \frac{1}{Q_L} \right)$$

where Q_C is the capacitor Q (sometimes specified by a manufacturer) and the inductor Q is Q_L.

The Q of capacitors is ordinarily high. Good quality ceramic capacitors and mica capacitors may have Q values of 1,200 or more. Microwave capacitors can have poor Q values — 10 or less at 10 GHz and higher frequencies because X_C will be low. Capacitors are predominantly subject to dielectric losses which are modeled as a parallel loss resistance across the capacitive reactance. Capacitors also have a series loss resistance associated with the conductor leads and capacitor plates, but this loss is often small enough to ignore.

Inductors are subject to several types of electrical energy losses such as wire resistance (including skin effect) and core losses. All electrical conductors have some resistance through which electrical energy is lost as heat. Wire conductors suffer additional ac losses because alternating current tends to flow on the conductor surface due to the skin effect discussed in the chapter on **RF Techniques**. If the inductor's core is iron, ferrite, or brass, the core will introduce additional losses of energy. Note that core losses for inductors are modeled as a resistor in parallel with the inductor (analogous to capacitor dielectric losses). The specific details of these losses are discussed in connection with each type of core material.

The sum of all core losses may be depicted by showing an equivalent series connected resistor with the inductor (see the section on Practical Inductors in the **Circuits and Components** chapter), although there is no separate component represented by the resistor symbol. As a result of inherent energy losses, inductor Q rarely approaches capacitor Q in a circuit where both components work together. Although many circuits call for the highest Q inductor obtainable, other circuits may call for a specific Q, even a very low one.

Q is also discussed in the following section on Resonant Circuits. For these circuits, Q is closely related to circuit bandwidth and selectivity. In addition, the circuit Q depends on whether or not a load is attached.

3.8 Resonant Circuits

A circuit containing both an inductor and a capacitor, and therefore, both inductive and capacitive reactance, is often called a *tuned circuit* or a *resonant circuit*. For any such circuit, there is a particular frequency at which the inductive and capacitive reactances are the same, that is, $X_L = X_C$. For most purposes, this is the *resonant frequency* of the circuit. At the resonant frequency — or at *resonance*, for short:

$$X_L = 2\,\pi\,f\,L = X_C = \frac{1}{2\,\pi\,f\,C}$$

By solving for f, we can find the resonant frequency of any combination of inductance and capacitance from the formula:

$$f = \frac{1}{2\,\pi\,\sqrt{L\,C}}$$

where

f = frequency in hertz (Hz),
L = inductance in henrys (H),
C = capacitance in farads (F)

For most high-frequency (HF) radio work, smaller units of inductance and capacitance and larger units of frequency are more convenient. The basic formula becomes:

$$f = \frac{10^3}{2\,\pi\,\sqrt{L\,C}}$$

where

f = frequency in megahertz (MHz),
L = inductance in microhenrys (μH),
C = capacitance in picofarads (pF)

Example: What is the resonant frequency of a circuit containing an inductor of 5.0 μH and a capacitor of 35 pF?

$$f = \frac{10^3}{2\pi\sqrt{L\ C}} = \frac{10^3}{6.28\sqrt{5.00\ \times\ 35.0}}$$

$$= \frac{10^3}{83.0} = 12\text{ MHz}$$

To find the matching component (inductor or capacitor) when the frequency and one component is known (capacitor or inductor) for general HF work, use the formula:

$$f^2 = \frac{10^6}{4\,\pi^2\,L\,C}$$

where f, L, and C are in Hz, μH, and pF. To find L or C, the formula is converted to:

$$L = \frac{25{,}330}{f^2\,C}$$

$$C = \frac{25{,}330}{f^2\,L}$$

For most radio work, these formulas will permit calculations of frequency and component values well within the limits of component tolerances.

Example: What value of capacitance is needed to create a resonant circuit at 21.1 MHz, if the inductor is 2.00 μH?

$$C = \frac{25{,}330}{f^2 L} = \frac{25{,}330}{\left(21.1^2 \times 2.00\right)}$$

$$= \frac{25{,}330}{890} = 28.5\text{ pF}$$

Figure 3.20 can also be used if an approximate answer is acceptable. From the horizontal axis, find the vertical line closest to the desired resonant frequency. Every pair of diagonals that cross on that vertical line represent a combination of inductance and capacitance that will resonate at that frequency. For example, if the desired frequency is 10 MHz, the pair of diagonals representing 5 μH and 50 pF cross quite close to that frequency. Interpolating between the given diagonals will provide more resolution. Remember that all three sets of lines are spaced logarithmically.

Resonant circuits have other properties of importance, in addition to the resonant frequency, however. These include impedance, voltage drop across components in series-resonant circuits, circulating current in parallel-resonant circuits, and bandwidth. These properties determine such factors as the selectivity of a tuned circuit and the component ratings for circuits handling significant amounts of power. Although the basic determination of the tuned-circuit resonant frequency ignores any resistance in the circuit, that resistance plays a vital role in the circuit's other characteristics.

3.8.1 Series-Resonant Circuits

Figure 3.38 presents a basic schematic diagram of a *series-resonant circuit*. Although most schematic diagrams of radio circuits show only the inductor and the capacitor, resistance is always present in such circuits. The most notable resistance is associated with the series resistance losses in the inductor at HF. The dominant losses in the capacitor may be modeled as a parallel resistance (not shown), but these losses are low enough at HF to be ignored. The current meter shown in the circuit is a reminder that in series circuits, the same current flows through all elements.

At resonance, the reactance of the capacitor cancels the reactance of the inductor. The voltage and current are in phase with each other, and the impedance of the circuit is determined solely by the resistance. The actual current through the circuit at resonance, and for frequencies near resonance, is determined by the formula:

$$I = \frac{E}{Z} = \frac{E}{\sqrt{R^2 + \left[2 \pi f L - \frac{1}{(2 \pi f C}\right]^2}}$$

where all values are in basic units (A, V, Ω, Hz, H, F).

At resonance, the reactive factor in the formula is zero (the bracketed expression under the square root symbol) and Z = R. As the frequency is shifted above or below the resonant frequency without altering component values, however, the reactive factor becomes significant, and the amplitude of the current becomes smaller than at resonance. At frequencies far from resonance, the reactive components become dominant, and the resistance no longer significantly affects the current amplitude.

The exact curve created by recording the current as the frequency changes depends on the ratio of reactance to resistance. When the reactance of either the coil or capacitor is of the same order of magnitude as the resistance, the current decreases rather slowly as the frequency is moved in either direction away from resonance. Such a curve is said to be *broad.* Conversely, when the reactance is considerably larger than the resistance, the current decreases rapidly as the frequency moves away from resonance, and the circuit is said to be *sharp.* A sharp circuit responds a great deal more readily to the resonant frequency than to frequencies quite close to resonance; a broad circuit responds almost equally well to a group or band of frequencies centered around the resonant frequency.

Both types of resonance curves are useful. A sharp circuit gives good selectivity — the ability to respond strongly (in terms of current amplitude) at one desired frequency and to discriminate against others. A broad circuit is used when the apparatus must give about the same response over a larger range of frequencies, rather than at only a single frequency.

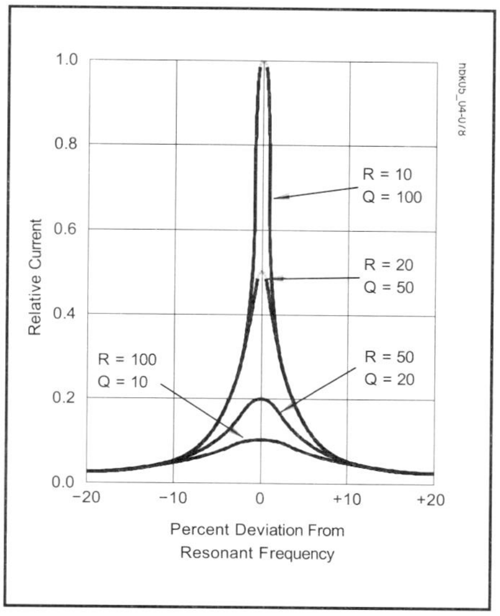

Figure 3.39A — Relative current in series-resonant circuits with various values of series resistance and Q. (An arbitrary maximum value of 1.0 represents current at resonance.) The reactance at resonance for all curves is 1000 Ω. Note that the current is hardly affected by the resistance in the circuit at frequencies more than 10% away from the resonant frequency.

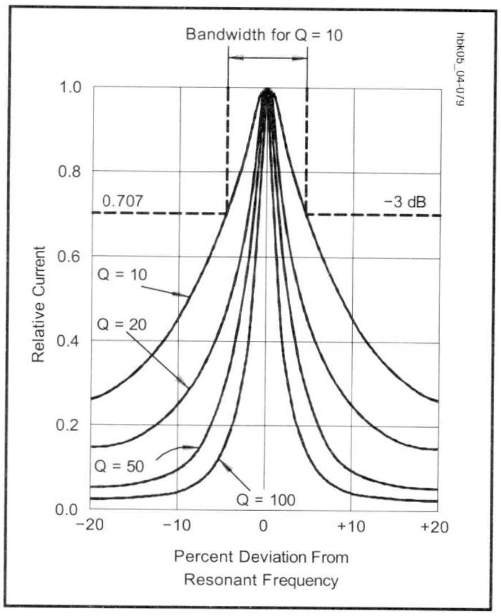

Figure 3.39B — Relative current in series-resonant circuits having different values of Q_U. The current at resonance is normalized to the same level for all curves in order to show the rate of change of decrease in current for each value of Q_U. The half-power points are shown to indicate relative bandwidth of the response for each curve. The bandwidth is indicated for a circuit with a Q_U of 10.

Figure 3.39A presents a family of curves, showing the decrease in current as the frequency deviates from resonance. In each case, the inductive and capacitive reactances are assumed to be 1,000 Ω. The maximum current, shown as a relative value on the graph, occurs with the lowest resistance, while the lowest peak current occurs with the highest resistance. Equally important, the rate at which the current decreases from its maximum value also changes with the ratio of reactance to resistance. It decreases most rapidly when the ratio is high and most slowly when the ratio is low.

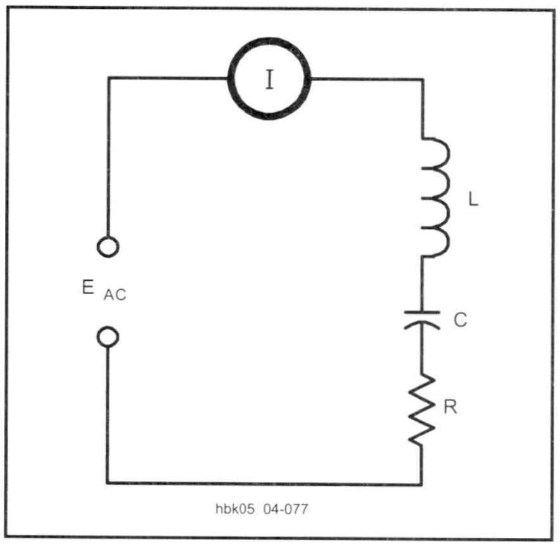

Figure 3.38 — A series circuit containing L, C, and R is resonant at the applied frequency when the reactance of C is equal to the reactance of L. The I in the circle is the schematic symbol for an ammeter.

UNLOADED Q

As noted in equation [a] of Table 3.2 above, Q is the ratio of series reactance representing 2π times the stored energy (equation [c] in Table 3.2) to series resistance or consumed energy. Since both terms of the ratio are measured in ohms, Q has no units and is known as the *quality factor* (and less frequently, the *figure of merit* or the *multiplying factor*). The series resistive losses of the coil often dominate the energy consumption in HF series-resonant circuits, so the inductor Q largely determines the resonant-circuit Q. Since this value of Q is independent of any external load to which the circuit might transfer power, it is called the *unloaded Q* or Q_U of the circuit.

Example: What is the unloaded Q of a series-resonant circuit with a series loss resistance of 5 Ω and inductive and capacitive components having (1) a reactance of 500 Ω each? (2) a reactance of 50 Ω each?

$$Q_{U1} = \frac{X_1}{R} = \frac{500\ \Omega}{5\ \Omega} = 100$$

$$Q_{U2} = \frac{X_2}{R} = \frac{50\ \Omega}{5\ \Omega} = 10$$

BANDWIDTH

Figure 3.39B is an alternative way of drawing the family of curves that relate current to frequency for a series-resonant circuit. By assuming that the peak current of each curve is the same, the rate of change of current for various values of Q_U and the associated ratios of reactance to resistance are more easily compared. From the curves, it is evident that the lower Q_U circuits pass current across a greater *bandwidth* of frequencies than the circuits with a higher Q_U. For the purpose of comparing tuned circuits, bandwidth is often defined as the frequency spread between the two frequencies at which the current amplitude decreases to 0.707 (or $\sqrt{2}/2$) times the maximum value. Since the power consumed by the resistance, R, is proportional to the square of the current, the power at these points is half the maximum power at resonance, assuming that R is constant for the calculations. The half-power, or –3 dB, points are marked on Figure 3.39B. The -3 dB bandwidth is also called the Full-Width Half-Maximum (FWHM) bandwidth.

For Q values of 10 or greater, the curves shown in Figure 3.39B are approximately symmetrical. On this assumption, bandwidth (BW) can be easily calculated by inverting equation [d] in Table 3.2, and approximating the geometrical mean –3 dB frequency by f:

$$BW = \frac{f}{Q_U}$$

where BW and f are in the same units, that is, in Hz, kHz, MHz, or GHz.

Example: What is the 3 dB bandwidth of a series-resonant circuit operating at 14 MHz with a Q_U of 100?

$$BW = \frac{f}{Q_U} = \frac{14\ \text{MHz}}{100} = 0.14\ \text{MHz} = 140\ \text{kHz}$$

The relationship between Q_U, f and BW provides a means of determining the value of circuit Q when inductor losses may be difficult to measure. By constructing the series-resonant circuit and measuring the current as the frequency varies above and below resonance, the half-power points can be determined. Then:

$$Q_U = \frac{f}{BW}$$

Example: What is the Q_U of a series-resonant circuit operating at 3.75 MHz, if the –3 dB bandwidth is 375 kHz?

$$Q_U = \frac{f}{BW} = \frac{3.75\ \text{MHz}}{0.375\ \text{MHz}} = 10.0$$

If the loss resistance of the inductor is much greater than of the capacitor (the usual case), BW is approximately R/L. The Q of a series resonant circuit can also be stated

$$Q_U = \frac{f_0}{BW} = \frac{1}{R}\sqrt{\frac{L}{C}}$$

The illustrations of the relationship between Q and bandwidth at **www.hyperphysics.phy-astr.gsu.edu/hbase/electric/serres.html** are helpful in understanding the concept. **Table 3.3** provides some simple formulas for estimating the maximum current and phase angle for various bandwidths, if both f and Q_U are known.

Table 3.3
The Selectivity of Resonant Circuits

Approximate percentage of current at resonance[1] or of impedance at resonance[2]	*Bandwidth (between half-power or –3 dB points on response curve)*	*Series circuit current phase angle (degrees)*
95.0	f / 3Q	18.5
90.0	f / 2Q	26.5
70.7	f / Q	45.0
44.7	2f / Q	63.5
24.2	4f / Q	76.0
12.4	8f / Q	83.0

[1]For a series resonant circuit
[2]For a parallel resonant circuit

VOLTAGE DROP ACROSS COMPONENTS

The voltage drop across the coil and across the capacitor in a series-resonant circuit are each proportional to the reactance of the component for a given current (since E = I X). These voltages may be many times the applied voltage for a high-Q circuit. In fact, at resonance, the voltage drop is:

$$E_X = Q_U\, E_{AC}$$

where

E_X = the voltage across the reactive components,
Q_U = the circuit unloaded Q, and
E_{AC} = the applied voltage in Figure 3.38.

(Note that the voltage drop across the inductor is the vector sum of the voltages across the resistance and the reactance; however, for Q greater than 10, the error created by using this is not ordinarily significant.) Since the calculated value of E_X is the RMS voltage, the peak voltage will be higher by a factor of 1.414. Antenna couplers and other high-Q circuits handling significant power may experience arcing from high values of E_X, even though the source voltage to the circuit is well within component ratings.

CAPACITOR LOSSES

Although capacitor energy losses tend to be insignificant compared to inductor losses up to about 30 MHz, the losses may affect circuit Q in the VHF range. Leakage resistance, principally in the solid dielectric that forms the insulating support for the capacitor plates, appears as a resistance in parallel with the capacitor plates. Instead of forming a series resistance, capacitor leakage usually forms a parallel resistance with the capacitive reactance. If the leakage resistance of a capacitor is significant enough to affect the Q of a series-resonant circuit, the parallel resistance (R_P) may be converted to an equivalent series resistance (R_S) before adding it to the inductor's resistance.

$$R_S = \frac{{X_C}^2}{R_P} = \frac{1}{R_P \times (2\,\pi\, f\, C)^2}$$

Example: A 10.0 pF capacitor has a leakage resistance of 10,000 Ω at 50.0 MHz. What is the equivalent series resistance?

$$R_S = \frac{1}{R_P \times (2\,\pi\, f\, C)^2}$$

$$= \frac{1}{1.00 \times 10^4 \times \left(6.28 \times 50.0 \times 10^6 \times 10.0 \times 10^{-12}\right)^2}$$

$$= \frac{1}{1.00 \times 10^4 \times 9.87 \times 10^{-6}}$$

$$= \frac{1}{0.0987} = 10.1\ \Omega$$

Equivalent series resistance (ESR) is a parameter that is commonly specified in the data sheets for capacitors. Many modern ceramic capacitors are advertised as having "ultra-low ESR," making them ideal for use in resonant circuits.

In calculating the impedance, current and bandwidth for a series-resonant circuit in which this capacitor might be used, the ESR of the unit is added to the loss resistance of the coil. Since inductor losses tend to increase with frequency because of skin effect in conductors, and capacitor dielectric losses also tend to increase with frequency, the combined losses in the capacitor and the inductor can seriously reduce circuit Q.

3.8.2 Parallel-Resonant Circuits

Although series-resonant circuits are common, the vast majority of resonant circuits used in radio work are *parallel-resonant circuits*. **Figure 3.40** represents a typical HF parallel-resonant circuit. As is the case for series-resonant circuits, the inductor is the chief source of resistive losses (that is, the parallel loss resistance across the capacitor is not shown), and these losses appear in series with the coil. Because current through parallel-resonant circuits is lowest at resonance, and impedance is highest, they are sometimes called *antiresonant* circuits. (You may encounter the old terms *acceptor* and *rejector* referring to series- and parallel-resonant circuits, respectively.)

Because the conditions in the two legs of the parallel circuit in Figure 3.40 are not the same (the resistance is shown in only one of the legs) all of the conditions by which

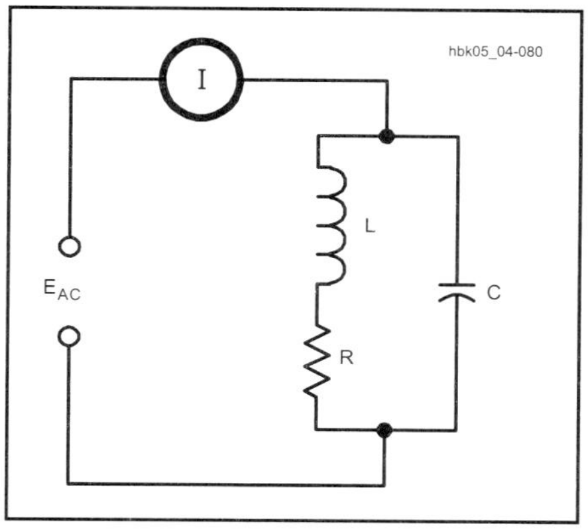

Figure 3.40 — A typical parallel-resonant circuit, with the resistance shown in series with the inductive leg of the circuit. Below a Q_U of 10, resonance definitions may lead to three separate frequencies which converge at higher Q_U levels. See text.

series resonance is determined do not occur simultaneously in a parallel-resonant circuit. **Figure 3.41** graphically illustrates the situation by showing the currents through the two components. (Currents are drawn in the manner of complex impedances shown previously to show the phase angle for each current.) When the inductive and capacitive reactances are identical, the condition defined for series resonance is met as shown at point (a). The impedance of the inductive leg is composed of both X_L and R, which yields an impedance greater than X_C and that is not 180° out of phase with X_C. The resultant current is greater than the minimum possible value and is not in phase with the voltage.

By altering the value of the inductor slightly (and holding the Q constant), a new frequency can be obtained at which the current reaches its minimum. When parallel circuits are tuned using a current meter as an indicator, this point (b) is ordinarily used as an indication of resonance. The current "dip" indicates a condition of maximum impedance, and thus minimum current, and is sometimes called the *antiresonant* point or *maximum impedance resonance* to distinguish it from the condition at which $X_C = X_L$. Maximum impedance is achieved at this point by vector addition of X_C, X_L and R, however, and the result is a current somewhat out of phase with the voltage.

Point (c) in the figure represents the *unity-power-factor* resonant point. Adjusting the inductor value and hence its reactance (while holding Q constant) produces a new resonant frequency at which the resultant current is in phase with the voltage. The new value of inductive reactance is the value required for a parallel-equivalent inductor and its parallel-equivalent resistor (calculated according to the formulas in the last section) to just cancel the capacitive reactance. The value of the parallel-equivalent inductor is always smaller than the actual inductor in series with the resistor and has a proportionally smaller reactance. (The parallel-equivalent resistor, conversely, will always be larger than the coil-loss resistor shown in series with the inductor.) The result is a resonant frequency slightly different from the one for minimum current and the one for $X_L = X_C$.

The points shown in the graph in Figure 3.41 represent only one of many possible situations, and the relative positions of the three resonant points do not hold for all possible cases. Moreover, specific circuit designs can draw some of the resonant points together, for example, compensating for the resistance of the coil by retuning the capacitor. The differences among these resonances are significant for circuit Q below 10, where the inductor's series resistance is a significant percentage of the reactance. Above a Q of 10, the three points converge to within a percent of the frequency and the differences between them can be ignored for practical calculations. Tuning for minimum current will not introduce a sufficiently large phase angle between voltage and current to create circuit difficulties.

PARALLEL CIRCUITS OF MODERATE TO HIGH Q

The resonant frequencies defined above converge in parallel-resonant circuits with Q higher than about 10. Therefore, a single set of formulas will sufficiently approximate circuit performance for accurate predictions. Indeed, above a Q of 10, the performance of a parallel circuit appears in many ways to be simply the inverse of the performance of a series-resonant circuit using the same components.

Accurate analysis of a parallel-resonant circuit requires the substitution of a parallel-equivalent resistor for the actual inductor-loss series resistor, as shown in **Figure 3.42**. Sometimes called the *dynamic resistance* of the parallel-resonant circuit, the parallel-equivalent resistor value will increase with circuit Q, that is, as the series resistance value decreases. To calculate the approximate parallel-equivalent resistance, use the formula:

$$R_P = \frac{{X_L}^2}{R_S} = \frac{(2\,\pi\,f\,L)^2}{R_S} = Q_U X_L$$

for $R_S << X_C << R_P$ and $X_P \approx X_S$ in the equations for series-parallel conversion in the section on Impedance.

Example: What is the parallel-equivalent resistance for a coil with an inductive reactance of 350 Ω and a series resistance of 5.00 Ω at resonance?

$$R_P = \frac{{X_L}^2}{R_S} = \frac{(350\ \Omega)^2}{5.00\ \Omega}$$

$$= \frac{122,500\ \Omega^2}{5.00\ \Omega} = 24,500\ \Omega$$

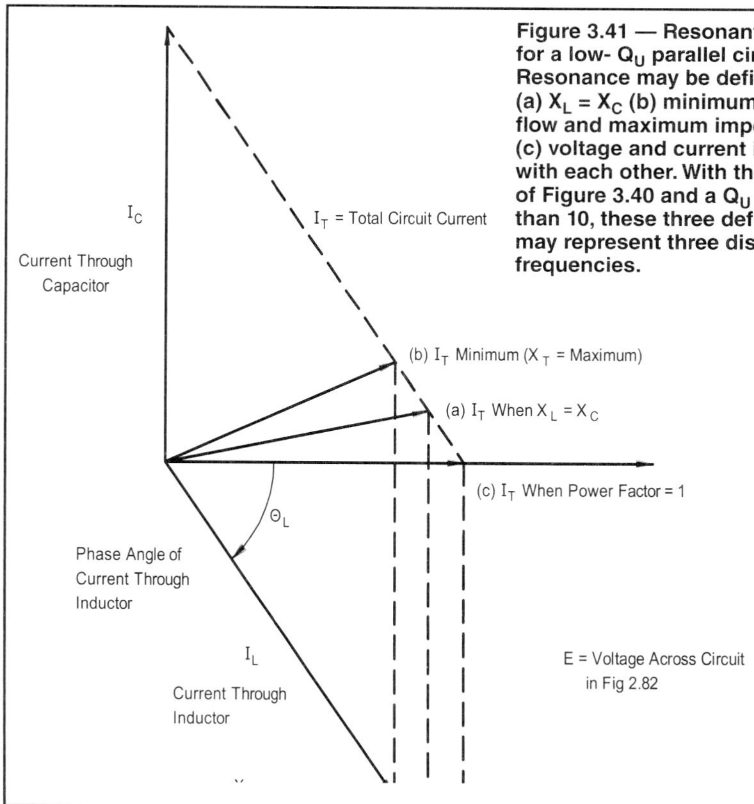

Figure 3.41 — Resonant conditions for a low- Q_U parallel circuit. Resonance may be defined as (a) $X_L = X_C$ (b) minimum current flow and maximum impedance or (c) voltage and current in phase with each other. With the circuit of Figure 3.40 and a Q_U of less than 10, these three definitions may represent three distinct frequencies.

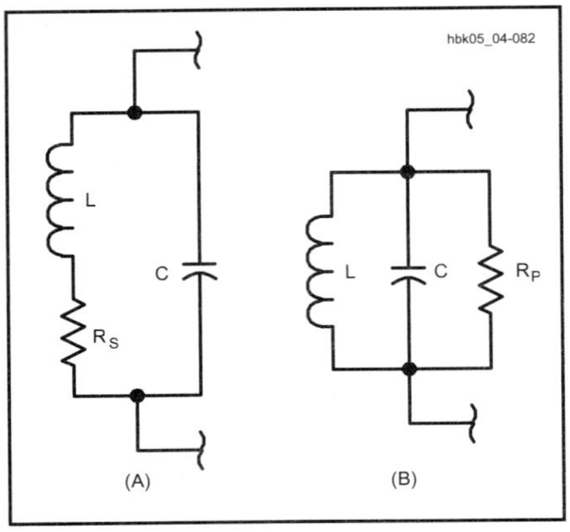

Figure 3.42 — Series and parallel equivalents when both circuits are resonant. The series resistance, R_S in A, is replaced by the parallel resistance, R_p in B, and vice versa. $R_P = X_L^2 / R_S$.

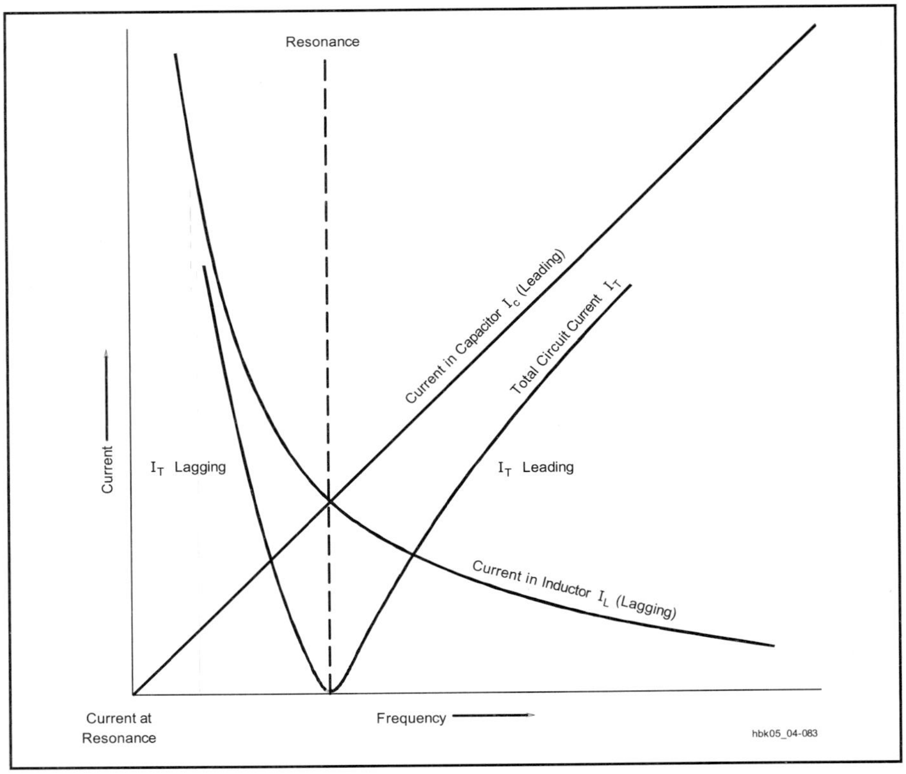

Figure 3.43 — The currents in a parallel-resonant circuit as the frequency moves through resonance. Below resonance, the current lags the voltage; above resonance the current leads the voltage. The base line represents the current level at resonance, which depends on the impedance of the circuit at that frequency.

Since the coil Q_U remains the inductor's reactance divided by its series resistance, the coil Q_U is 70. Multiplying Q_U by the reactance also provides the approximate parallel-equivalent resistance of the coil series resistance.

At resonance, where $X_L = X_C$, R_P defines the impedance of the parallel-resonant circuit. The reactances equal each other, leaving the voltage and current in phase with each other. In other words, the circuit shows only the parallel resistance. Therefore, the equation for R_P can be rewritten as:

$$Z = \frac{X_L^2}{R_S} = \frac{(2 \pi f L)^2}{R_S} = Q_U X_L$$

In the previous example, the circuit impedance, Z, at resonance is 24,500 Ω.

At frequencies below resonance the current through the inductor is larger than that through the capacitor, because the reactance of the coil is smaller and that of the capacitor is larger than at resonance. There is only partial cancellation of the two reactive currents, and the total current therefore is larger than the current taken by the resistance alone. At frequencies above resonance the situation is reversed and more current flows through the capacitor than through the inductor, so the total current again increases.

The current at resonance, being determined wholly by R_P, will be small if R_P is large, and large if R_P is small. **Figure 3.43** illustrates the relative current flows through a parallel-tuned circuit as the frequency is moved from below resonance to above resonance. The base line represents the minimum current level for the particular circuit. The actual current at any frequency off resonance is simply the vector sum of the currents through the parallel equivalent resistance and through the reactive components.

To obtain the impedance of a parallel-tuned circuit either at or off the resonant frequency, apply the general formula:

$$Z = \frac{Z_C \, Z_L}{Z_S}$$

where

Z = overall circuit impedance

Z_C = impedance of the capacitive leg (usually, the reactance of the capacitor),

Z_L = impedance of the inductive leg (the vector sum of the coil's reactance and resistance), and

Z_S = series impedance of the capacitor-inductor combination as derived from the equation for current in a series-resonant circuit.

After using vector calculations to obtain Z_L and Z_S, converting all the values to polar form, as described earlier in this chapter, will ease the final calculation. Of course, each impedance may be derived from the resistance and the application of the basic reactance formulas on the values of the inductor and capacitor at the frequency of interest.

Since the current rises away from resonance, the parallel-resonant circuit impedance must fall. It also becomes complex, resulting in an ever-greater phase difference between the voltage and the current. The rate at which the impedance falls is a function of Q_U. **Figure 3.44** presents a family of curves showing the impedance drop from resonance for circuit Q ranging from 10 to 100. The curve family for parallel-circuit impedance is essentially the same as the curve family for series-circuit current.

As with series-resonant circuits, the higher the Q of a parallel-tuned circuit, the sharper will be the response peak. Likewise, the lower the Q, the wider the band of frequencies to which the circuit responds. Using the half-power (–3 dB) points as a comparative measure of circuit performance as in series-resonant circuits, $BW = f / Q_U$ and $Q_U = f / BW$, where the resonant frequency and the bandwidth are in the same units. Also similarly to the series-resonant circuit:

$$Q_U = R\sqrt{\frac{C}{L}}$$

As a handy reminder, **Table 3.4** summarizes the performance of parallel-resonant circuits at high and low Q and above and below resonant frequency.

It is possible to use either series- or parallel-resonant circuits to do the same work in many circuits, thus giving the designer considerable flexibility. **Figure 3.45** illustrates this general principle by showing a series-resonant circuit

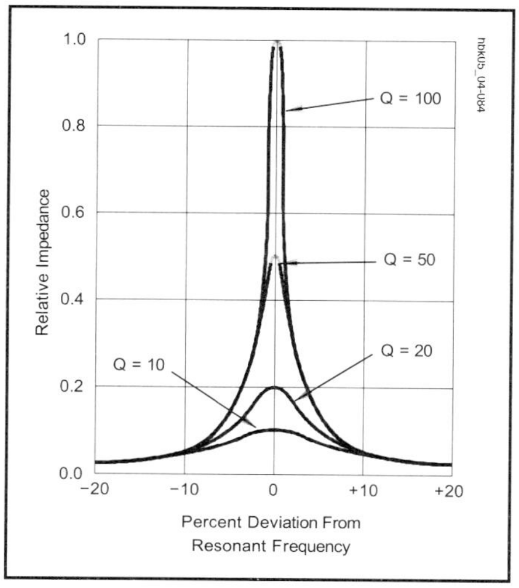

Figure 3.44 — Relative impedance of parallel-resonant circuits with different values of Q_U. The curves are similar to the series-resonant circuit current level curves of Figure 3.39A. The effect of Q_U on impedance is most pronounced within 10% of the resonance frequency.

Table 3.4

The Performance of Parallel-Resonant Circuits

A. High- and Low-Q Circuits (in relative terms)

Characteristic	*High-Q Circuit*	*Low-Q Circuit*
Selectivity	high	low
Bandwidth	narrow	wide
Impedance	high	low
Total current	low	high
Circulating current	high	low

B. Off-Resonance Performance for Constant Values of Inductance and Capacitance

Characteristic	*Above Resonance*	*Below Resonance*
Inductive reactance	increases	decreases
Capacitive reactance	decreases	increases
Circuit resistance	unchanged*	unchanged*
Relative impedance	decreases	decreases
Total current	increases	increases
Circulating current	decreases	decreases
Circuit impedence	capacitive	inductive

*This is true for frequencies near resonance. At distant frequencies, skin effect may alter the resistive losses of the inductor.

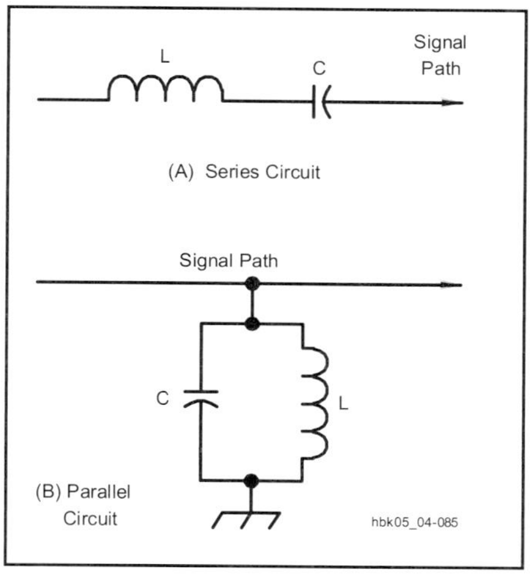

Figure 3.45 — Series- and parallel-resonant circuits configured to perform the same theoretical task: passing signals in a narrow band of frequencies along the signal path. A real design example would consider many other factors.

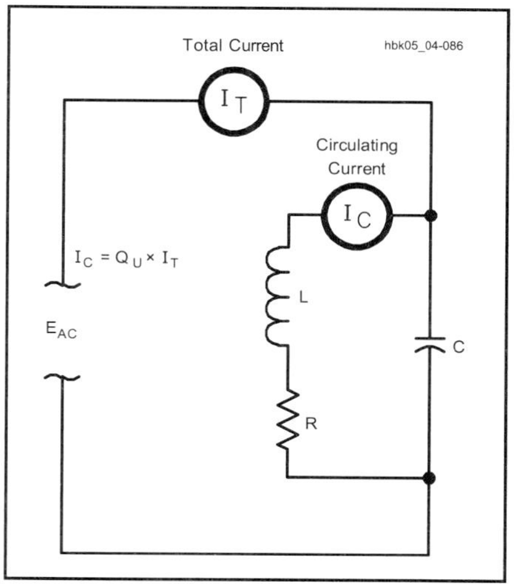

Figure 3.46 — A parallel-resonant circuit redrawn to illustrate both the total current and the circulating current.

in the signal path and a parallel-resonant circuit shunted from the signal path to ground. Assume both circuits are resonant at the same frequency, f, and have the same Q. The series-resonant circuit at A has its lowest impedance at f, permitting the maximum possible current to flow along the signal path. At all other frequencies, the impedance is greater and the current at those frequencies is less. The circuit passes the desired signal and tends to impede signals at undesired frequencies. (This is also called a band-pass filter). The parallel circuit at B provides the highest impedance at resonance, f, making the signal path the lowest impedance path for the signal. At frequencies off resonance, the parallel-resonant circuit presents a lower impedance, thus presenting signals with a path to ground and away from the signal path. (This is also called a band-stop or band-reject filter. For more on filters see the following section and the **Analog and Digital Filtering** chapter).

CIRCULATING CURRENT

In a parallel-resonant circuit, the source voltage is the same for all the circuit elements. The current in each element, however, is a function of the element's reactance. **Figure 3.46** redraws the parallel-resonant circuit to indicate the total current and the current circulating between the coil and the capacitor. The current drawn from the source may be low, because the overall circuit impedance is high. The current through the individual elements may be high, however, because there is little resistive loss as the current circulates through the inductor and capacitor. For parallel-resonant circuits with an unloaded Q of 10 or greater, this *circulating current* is approximately:

$$I_C = Q_U I_T$$

where

I_C = circulating current in A, mA or μA,
Q_U = unloaded circuit Q, and
I_T = total current in the same units as I_C.

Example: A parallel-resonant circuit permits an ac or RF total current of 30 mA and has a Q of 100. What is the circulating current through the elements?

$$I_X = Q_U I = 100 \times 30 \text{ mA} = 3{,}000 \text{ mA} = 3 \text{ A}$$

Circulating currents in high-Q parallel-tuned circuits can reach a level that causes component heating and power loss. Therefore, components should be rated for the anticipated circulating currents, and not just the total current.

LOADED Q

In many resonant-circuit applications, the only power lost is that dissipated in the resistance of the circuit itself. At frequencies below 30 MHz, most of this resistance is in the coil. Within limits, increasing the number of turns in the coil increases the reactance faster than it raises the resistance, so coils for circuits in which the Q must be high are made with relatively large inductances for the frequency.

When the circuit delivers energy to a load (as in the case of the resonant circuits used in transmitters), the energy consumed in the circuit itself is usually negligible compared with that consumed by the load. The equivalent of such a circuit is shown in **Figure 3.47**, where the parallel resistor, R_L, represents the load to which power is delivered. If the power dissipated in the load is at least 10 times as great as the power lost in the inductor and capacitor, the parallel impedance of the resonant circuit itself will be so high compared with the resistance of the load that for all practical purposes the impedance of the combined circuit is equal to the load impedance. Under these conditions, the load resistance replaces the circuit impedance in calculating Q. The Q of a parallel-resonant circuit loaded by a resistive impedance is:

$$Q_L = \frac{R_L}{X}$$

where

Q_L = circuit loaded Q,
R_L = parallel load resistance in ohms, and
X = reactance in ohms of the inductor and capacitor at resonance

Example: A resistive load of 3000 Ω is connected across a resonant circuit in which the inductive and capacitive reactances are each 250 Ω. What is the circuit Q?

$$Q_L = \frac{R_L}{X} = \frac{3{,}000\ \Omega}{250\ \Omega} = 12$$

The effective Q of a circuit loaded by a parallel resistance increases when the reactances are decreased. A circuit loaded with a relatively low resistance (a few thousand ohms) must have low-reactance elements (large capacitance and small inductance) to have reasonably high Q. Many power-handling circuits, such as the output networks of transmitters, are designed by first choosing a loaded Q for the circuit and then determining component values. See the chapter on **RF Power Amplifiers** for more details.

Parallel load resistors are sometimes added to parallel-resonant circuits to lower the circuit Q and increase the circuit bandwidth. By using a high-Q circuit and adding a parallel resistor, designers can tailor the circuit response to their needs. Since the parallel resistor consumes power, such techniques ordinarily apply to receiver and similar low-power circuits, however.

Example: Specifications call for a parallel-resonant circuit with a bandwidth of 400 kHz at 14.0 MHz. The circuit at hand has a Q_U of 70.0 and its components have reactances of 350 Ω each. What is the parallel load resistor that will increase the bandwidth to the specified value? The bandwidth of the existing circuit is:

$$BW = \frac{f}{Q_U} = \frac{14.0\ \text{MHz}}{70.0} = 0.200\ \text{MHz}$$

$$= 200\ \text{kHz}$$

The desired bandwidth, 400 kHz, requires a circuit with a Q of:

$$Q = \frac{f}{BW} = \frac{14.0\ \text{MHz}}{0.400\ \text{MHz}} = 35.0$$

Since the desired Q is half the original value, halving the resonant impedance or parallel-resistance value of the circuit is in order. The present impedance of the circuit is:

$$Z = Q_U X_L = 70.0 \times 350\ \Omega = 24{,}500\ \Omega$$

The desired impedance is:

$$Z = Q_U X_L = 35.0 \times 350\ \Omega$$

$$= 12{,}250\ \Omega = 12.25\ \text{k}\Omega$$

or half the present impedance.

A parallel resistor of 24,500 Ω, or the nearest lower value (to guarantee sufficient bandwidth), will produce the required reduction in Q and bandwidth increase. Although this example simplifies the situation encountered in real design cases by ignoring such factors as the shape of the band-pass curve, it illustrates the interaction of the ingredients that determine the performance of parallel-resonant circuits.

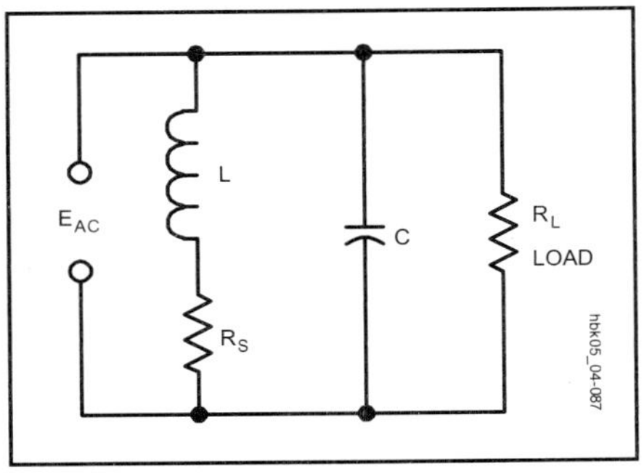

Figure 3.47 — A loaded parallel-resonant circuit, showing both the inductor-loss resistance and the load, R_L. If smaller than the inductor resistance, R_L will control the loaded Q of the circuit (Q_L).

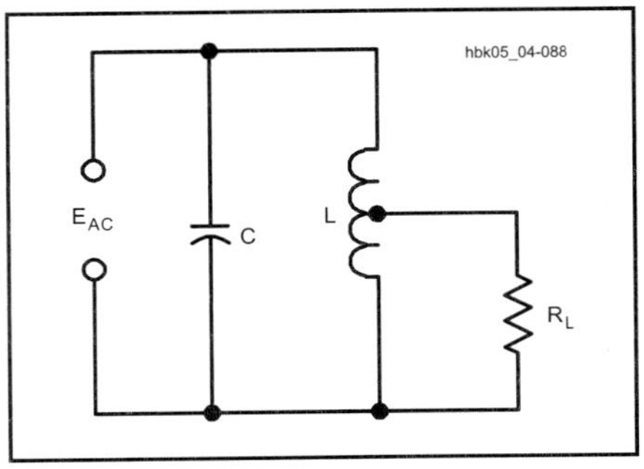

Figure 3.48 — A parallel-resonant circuit with a tapped inductor to effect an impedance match. Although the impedance presented to E_{AC} is very high, the impedance at the connection of the load, R_L, is lower.

IMPEDANCE TRANSFORMATION

An important application of the parallel-resonant circuit is as an impedance matching device. Circuits and antennas often need to be connected to other circuits or feed lines that do not have the same impedance. Transferring power effectively requires a circuit that will convert or "transform" the impedances so that each connected device or system can operate properly.

Figure 3.48 shows such a situation where the source, E_{AC}, operates at a high impedance, but the load, R_L, operates at a low impedance. The technique of impedance transformation shown in the figure is to connect the parallel-resonant circuit, which has a high impedance, across the source, but connect the load across only a portion of the coil. (This is called *tapping the coil* and the connection point is a *tap*.) The coil acts as an *autotransformer*, with the magnetic field of the coil shared between what are effectively two coils in series, the upper coil having many turns and the lower coil fewer turns. Energy stored in the field induces larger voltages in the many-turn coil than it does in the fewer-turn coil, "stepping down" the input voltage so that energy can be extracted by the load at the required lower voltage-to-current ratio (i.e., impedance). The correct tap point on the coil usually has to be experimentally determined, but the technique is very effective.

When the load resistance has a very low value (say below 100 Ω) it may be connected in series in the resonant circuit (such as R_S in Figure 3.42A, for example), in which case the series L-R circuit can be transformed to an equivalent parallel L-R circuit as previously described. If the Q is at least 10, the equivalent parallel impedance is:

$$Z_R = \frac{X^2}{R_L}$$

where

Z_R = resistive parallel impedance at resonance,
X = reactance (in ohms) of the coil and the capacitor at the resonant frequency, and
R_L = load resistance inserted in series.

If the Q is lower than 10, the reactance will have to be adjusted somewhat, for the reasons given in the discussion of low-Q parallel resonant circuits, to obtain a resistive impedance of the desired value.

These same techniques work in either "direction," with a high-impedance source and low-impedance load or vice versa. Using a parallel-resonant circuit for this application does have some disadvantages. For instance, the common connection between the input and the output provides no dc isolation. Also, the common ground is sometimes troublesome with regard to ground-loop currents. Consequently, a circuit with only mutual magnetic coupling is often preferable. With the use of ferrite cores, constructing impedance transformers that are both broadband and permit operation well up into the VHF portion of the spectrum has become relatively easy. The basic principles of broadband impedance transformers appear in the **RF Techniques** chapter.

3.9 Analog Signal Processing

The term *analog signal* refers to voltages, currents and waves that make up ac radio and audio signals, dc measurements, even power. The essential characteristic of an analog signal is that the information or energy it carries is continuously variable. Even small variations of an analog signal affect its value or the information it carries. This stands in contrast to *digital signals* that have values only within well-defined and separate ranges called *states*. To be sure, at the fundamental level all circuits and signals are analog: Digital signals are created by designing circuits that restrict the values of analog signals to those discrete states.

Analog signal processing involves various electronic stages to perform functions on analog signals such as amplifying, filtering, modulation and demodulation. A piece of electronic equipment, such as a radio, is constructed by combining a number of these circuits. How these stages interact with each other and how they affect the signal individually and in tandem is the subject of sections later in the chapter.

3.9.1 Terminology

A similar terminology is used when describing active electronic devices. The letter V or v stands for voltages and I or i for currents. Capital letters are often used to denote dc or bias values (bias is discussed later in the **Circuits and Components** chapter). Lower-case often denotes instantaneous or ac values.

Voltages generally have two subscripts indicating the terminals between which the voltage is measured (V_{BE} is the dc voltage between the base and the emitter of a bipolar transistor). Currents have a single subscript indicating the terminal into which the current flows (I_C is the dc current into the collector of a bipolar transistor). If the current flows out of the device, it is generally treated as a negative value.

Resistance is designated with the letter R or r, and impedance with the letter Z or z. For example, r_{DS} is resistance between drain and source of an FET and Z_i is input impedance. Qualifiers are sometimes added to the subscripts to indicate certain operating modes of the device. SS for saturation, BR for breakdown, ON and OFF are all commonly used.

Power supply voltages have two subscripts that are the same, indicating the terminal to which the voltage is applied. V_{DD} would represent the power supply voltage applied to the drain of a field-effect transistor.

Since integrated circuits are collections of semiconductor components, the abbreviations for the type of semiconductor used also apply to the integrated circuit. For example, V_{CC} is a power supply voltage for an integrated circuit made with bipolar transistor technology in which voltage is applied to transistor collectors.

3.9.2 Linearity

The premier properties of analog signals are *superposition* and *scaling*. At the beginning of this chapter we discussed how sine waves of different frequencies can add together to create a composite waveform. This is an excellent example of superposition, which is the property by which signals are combined, whether in a circuit, in a piece of wire, or even in air, as the sum of the individual signals. This is to say that at any one point in time, the voltage of the combined signal is the sum of the voltages of the original signals at the same time. In a *linear system* any number of signals will add in this way to give a single combined signal. (Mathematically, this is a *linear combination*.) For this reason, analog signals and components are often referred to as *linear signals* or *linear components*. A

The Decibel

The decibel (dB) is the standard unit for comparing two quantities, such as power or voltage, as a ratio. It is logarithmic so very large and very small ratios are easy to work with. The formula for calculating decibels is:

$$dB = 10\log\left(\frac{P_2}{P_1}\right) = 20\log\left(\frac{V_2}{V_1}\right)$$

where P represents power and V represents voltage (power is proportional to voltage squared, which is why the factor 10 is used with powers and the factor 20 is used with voltages.) Decibels for a circuit will the same whether calculated from the power or voltage measurements.

When calculating decibels using voltage (or current), remember that both voltages must be measured at the same value of circuit impedance. For example, the 50 Ω input and output impedance of a preamplifier. If this is not the case, convert each voltage to a power level (V^2/Z) and calculate decibels using a power ratio.

Another caution is rounding decibel values. Because decibels are a logarithmic value, a change of 1 dB, whether from 10 dB or 10,000 dB, represents a change of nearly 26% so that ±1 dB is also ±26%. Be careful when rounding or specifying tolerances in dB.

For more on working with decibels, read the articles "Untangling the Decibel Dilemma" found in this book's online content. Rohde & Schwarz has published Application Note IMA98 "dB or not dB? Everything you ever wanted to know about decibels but were afraid to ask…" at the website listed in the References and Bibliography section of this chapter.

linear system whose characteristics do not change, such as a resistive voltage divider, is called *time-invariant*. If the system changes with time, it is *time-varying*. The variations may be random, intermittent (such as being adjusted by an operator) or periodic.

One of the more important features of superposition, for the purposes of signal processing, is that signals that have been combined by superposition can be separated back into the original signals. This is what allows multiple signals that have been received by an antenna to be separated back into individual signals by a receiver.

3.9.3 Linear Operations

Any operation that modifies a signal and obeys the rules of superposition and scaling is a *linear operation*. The following sections explain the basic linear operations from which linear systems are made.

AMPLIFICATION AND ATTENUATION

Amplification and *attenuation* scale signals to be larger and smaller, respectively. The operation of *scaling* is the same as multiplying the signal at each point in time by a constant value; if the constant is greater than one then the signal is amplified, if less than one then the signal is attenuated.

An *amplifier* is a circuit that increases the amplitude of a signal. Schematically, a generic amplifier is signified by a triangular symbol, its input along the left face and its output at the point on the right (see **Figure 3.49**). The linear amplifier multiplies every value of a signal by a constant. Amplifier gain is often expressed as a multiplication factor (× 5, for example).

$$\text{Gain} = V_o/V_i$$

where V_o is the output voltage from an amplifier when an input voltage, V_i, is applied.

An *attenuator* is a circuit that reduces the amplitude of a signal. Attenuators can be constructed from passive components, using resistors, as described in the chapter on **Test Equipment and Measurements**. Active attenuator circuits include amplifiers whose gain is less than one or circuits with adjustable resistance in the signal path, such as a PIN diode attenuator.

GAIN AND TRANSCONDUCTANCE

The operation of an amplifier is specified by its *gain*. Gain in this sense is defined as the change (Δ) in the output parameter divided by the corresponding change in the input parameter. If a particular device measures its input and output as currents, the amplifier produces *current gain*. If the input and output are voltages, the amplifier is defined by its *voltage gain*. *Power gain* is often used, as well. Gain is technically unitless, but is often given in V/V. Decibels (dB) are often used to specify gain, particularly power gain. (See the sidebar "The Decibel".)

If an amplifier's input is a voltage and the output is a current, the ratio of the change in output current (I_o) to the change in input voltage (V_i) is called *transconductance*, g_m.

$$g_m = \frac{\Delta I_o}{\Delta V_i}$$

Transconductance has the same units as conductance and admittance, siemens (S), but is only used to describe the operation of active devices, such as transistors or vacuum tubes.

Ideal linear amplifiers have the same gain for all parts of a signal. Thus, a gain of 10 changes 10 V to 100 V, 1 V to 10 V and –1 V to –10 V.

The ability of an amplifier to change a signal's level is limited by the amplifier's *dynamic range*, however. An amplifier's dynamic range is the range of signal levels over which the amplifier produces the required gain linearly (without distortion). The lower limit of dynamic range is caused by noise, distortion and other nonlinearities. Dynamic range is limited for large signals because an amplifier can only produce output voltages (and currents) that are within the range of its power supply. (The upper and lower power supply voltages, one of which may be ground, are also called the *rails* of a circuit.) As the amplified output approaches one of the rails, the output cannot exceed a given voltage near the rail and the operation of the amplifier becomes nonlinear as described below in the section on Clipping and Rectification.

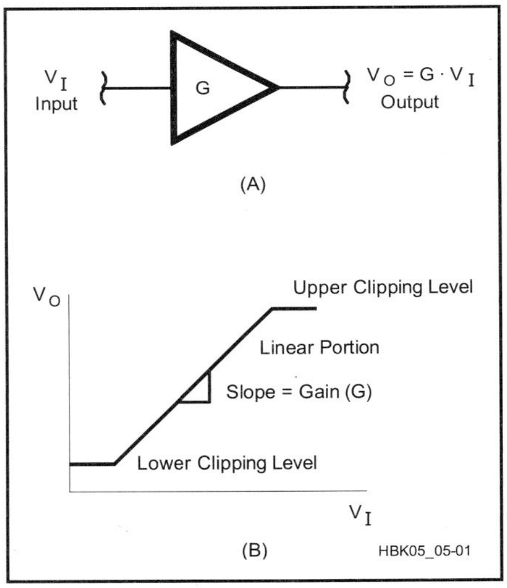

Figure 3.49 — Generic amplifier. (A) Symbol. For the linear amplifier, gain is the constant value, G, and the output voltage is equal to the input voltage times G; (B) Transfer function, input voltage along the x-axis is converted to the output voltage along the y-axis. The linear portion of the response is where the plot is diagonal; its slope is equal to the gain, G. Above and below this range are the clipping limits, where the response is not linear and the output signal is clipped.

Another similar limitation on amplifier linearity is called *slew rate*. Applied to an amplifier, this term describes the maximum rate at which a signal can change levels and still be accurately amplified in a particular device. *Input slew rate* is the maximum rate of change to which the amplifier can react linearly. *Output slew rate* refers to the maximum rate at which the amplifier's output circuit can change. Slew rate is an important concept, because there is a direct correlation between a signal level's rate of change and the frequency content of that signal. The amplifier's ability to react to or reproduce that rate of change affects its frequency response.

FREQUENCY RESPONSE AND BODE PLOTS

Another important characteristic of a circuit is its *frequency response*, a description of how it modifies a signal of any frequency. Frequency response can be stated in the form of a mathematical equation called a *transfer function*, but it is more conveniently presented as a graph of gain vs frequency. The absolute value of output amplitude to input amplitude is often called the circuit's *magnitude response*. A plot of the circuit's magnitude response in dB versus frequency on a logarithmic scale, such as in **Figure 3.50A**, is called a *Bode plot* (after Hendrik Wade Bode). The combination of decibel and log-frequency scales is used because the behavior of most circuits depends on ratios of magnitude and frequency and thus appears linear on a graph in which both the vertical and horizontal scales are logarithmic.

Most circuits also affect a signal's phase along with its magnitude. This is called *phase shift*. A plot of phase shift from the circuit's input to its output is called the *phase response*, seen in Figure 3.50B. Positive phase greater than 0° indicates that the output signal *leads* the input signal, while *lagging* phase shift has a negative phase. The combination of a magnitude and phase response plot gives a good picture of what effect the circuit has on signals passing through it.

TRANSFER CHARACTERISTICS

Transfer characteristics are the ratio of an output parameter to an input parameter, such as output current divided by input current, h_{FE} of a bipolar transistor). There are different families of transfer characteristics, designated by letters such as h, s, y or z. Each family compares parameters in specific ways that are useful in certain design or analysis methods. The most common transfer characteristics used in radio are the h-parameter family (used in transistor models) and the s-parameter family (used in RF design, particularly at VHF and above). See the **RF Techniques** chapter for more discussion of transfer characteristics.

COMPLEX FREQUENCY

We are accustomed to thinking of frequency as a real number, like cycles per second, but frequency can also be a complex number, s, with both a real part, designated by σ, and an imaginary part, designated by *j*ω. (ω is equal to 2πf.) The resulting complex frequency is written as s = σ + *j*ω. At the lower left of **Figure 3.51** a pair of real and imaginary axes is used to plot values of s. This is called the s-plane. Complex frequency is used in Laplace transforms, a mathematical technique used for circuit and signal analysis. (Thorough treatments of the application of complex frequency can be found in college-level textbooks on circuit and signal analysis.)

When complex frequency is used, a sinusoidal signal is described by Ae^{st}, where A is the amplitude of the signal and t is time. Because s is complex, $Ae^{st} = Ae^{(\sigma + j\omega)t} = A(e^{\sigma t})(e^{j\omega t})$. The two exponential terms describe independent characteristics of the signal. The second term, $e^{j\omega t}$, is the sine wave with which we are familiar and that has frequency f, where f = ω/2π. The first term, $e^{\sigma t}$, represents the rate at which the signal increases or decreases. If σ is negative, the exponential term decreases with time and the signal gets smaller and smaller. If σ is positive, the signal gets larger and larger. If σ = 0, the exponential term equals 1, a constant, and the signal amplitude stays the same.

Complex frequency is very useful in describing a circuit's stability. If the response to an input signal is at a frequency on the right-hand side of the s-plane, for which σ > 0, the system is *unstable* and the output signal will get larger until it is limited by the circuit's power supply or some other mechanism. If the response is on the left-hand side of the s-plane, the system is *stable* and the response to the input signal will eventually die out. The larger the absolute value of σ, the faster the response changes. If the response is precisely on the *j*ω axis, where σ = 0, the response will persist indefinitely.

In Figure 3.51 the equation for the simple RC-circuit's transfer function is shown at the left of the figure. It describes the circuit's behavior at real-world frequencies as well as imaginary frequencies whose values contain *j*. Because complex numbers are used for f, the transfer function describes the circuit's phase response, as well as amplitude. At one such frequency, f = –*j*/2πRC, the denominator of the transfer function is zero, and the gain is infinite! Infinite gain is a pretty amazing thing to achieve with a passive circuit — but because this can only happen at an imaginary frequency, it does not happen in the real world.

The practical effects of complex frequency can be experienced in a narrow CW crystal or LC filter. Such a filter has complex frequency with σ_0, so the input signal causes the filters to "ring", or output a damped sine wave along with the desired signal. Similarly, the complex frequency of an oscillator's output at power-up must have σ_0 or the oscillation would never start! The output amplitude continues to grow until limiting takes place, reducing gain until σ = 1 for a steady output.

POLES AND ZEROES

Frequencies that cause the transfer function to become infinite are called *poles*. This is shown at the bottom right of Figure 3.51 in the graph of the circuit's magnitude response for imaginary frequencies shown on the horizontal axis. (The pole causes the graph to extend up "as a pole under a tent," thus the name.) Similarly, circuits can have *zeroes,* which occur at imaginary frequencies that cause the transfer function to be zero, a less

Obtaining a Frequency Response

With the computer tools such as spreadsheets, it's easy to do the calculations and make a graph of frequency response. If you don't have a spreadsheet program, then use semi-log graph paper with the linear axis used for dB or phase and the logarithmic axis for frequency. An Excel spreadsheet set up to calculate and display frequency response is available in this book's online content. You can modify it to meet your specific needs.

Follow these rules whether using a spreadsheet or graph paper:

- Measure input and output in the same units, such as volts, and use the same conventions, such as RMS or peak-to-peak.
- Measure phase shift from the input to the output. (The **Test Equipment and Measurements** chapter discusses how to make measurements of amplitude and phase.)
- Use 10 log (P_O/P_I) for power ratios and 20 log (V_O/V_I) for voltage or current.

To make measurements that are roughly equally spaced along the logarithmic frequency axis, follow the "1-2-5 rule." Dividing a range this way, for example 1-2-5-10-20-50-100-200-500 Hz, creates steps in approximately equal ratios that then appear equally spaced on a logarithmic axis.

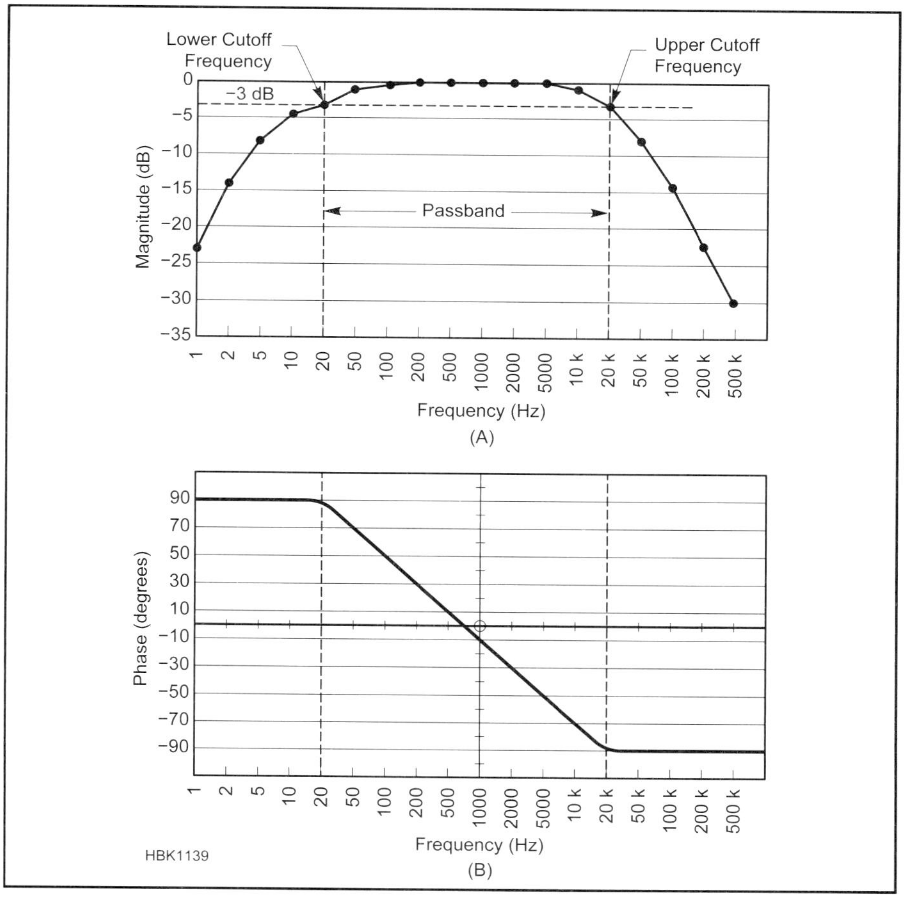

Figure 3.50 — Bode plot of (A) band-pass filter magnitude response and (B) band-pass filter phase response.

imaginative name, but quite descriptive.

A circuit can also have poles and zeroes at frequencies of zero and infinity. For example, the circuit in Figure 3.51 has a zero at infinite frequency because the capacitor's reactance is zero and the transfer function is zero, as well. If the resistor and capacitor are exchanged, so that the capacitor is in series with the output, then at zero frequency (dc), the output becomes zero because the capacitor's reactance is infinite, creating a zero.

Complex circuits can have multiple poles or multiple zeroes at the same frequency. Poles and zeroes can also occur at complex-number frequencies. The poles and zeroes of a circuit form a pattern in the complex plane that corresponds to certain types of circuit behavior. (The relationships between the pole-zero pattern and circuit behavior is beyond the scope of this book, but are covered in textbooks on circuit theory.)

What is a Pole?

Poles cause a specific change in the circuit's magnitude and phase response for real-world frequencies. A pole is associated with a bend in a magnitude response plot that changes the slope of the response downward with increasing frequency by 6 dB per octave or 20 dB per decade. (An octave increase is double the frequency and an octave decrease is half the frequency; a decade increase is 10 times the frequency and a decade decrease is 1/10th the frequency.)

There are four ways to identify the existence and frequency of a pole, as shown in Figure 3.51:

1) For a downward bend in the magnitude versus frequency plot, a single pole is at the –3 dB frequency. If the bend causes a change in slope of more than 6 dB/octave, there must be multiple poles overlaid at this frequency.

2) A 90° lagging change on a phase versus frequency plot, where the lag increases with frequency. The pole is at the point of 45° added lag on the S-shaped transition. Multiple poles will add their phase lags, as with magnitude.

3) On a circuit diagram, a single pole looks like a simple RC low-pass filter. The pole is at the –3 dB frequency (f = 1/2πRC Hz). Any other type of circuit that has the same response also has a pole at the same frequency.

4) In an equation for the transfer function of a circuit, a pole is a theoretical value of frequency that would result in infinite gain. This is clearly impossible, but as the value of frequency will either be absolute zero, or will have an imaginary component, it is impossible to make an actual real world signal at a pole frequency.

For example, comparing the magnitude responses at top and bottom of Figure 3.51 shows that the frequency of the pole is equal to the circuit's –3 dB cutoff frequency (1/2πRC) multiplied by *j*, which is also the frequency at which the circuit causes a –45° (lagging) phase shift from input to output.

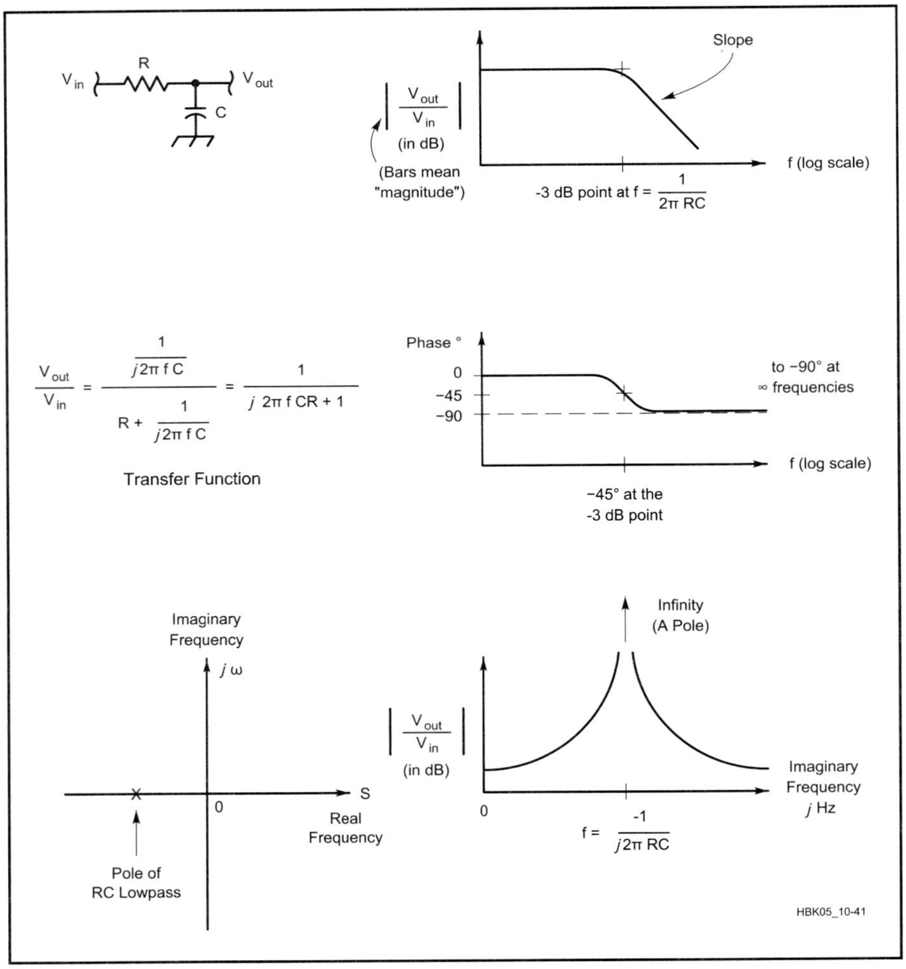

Figure 3.51 — The transfer function for a circuit describes both the magnitude and phase response of a circuit. The RC circuit shown at the upper left has a pole at f = 2πRC, the filter's –3 dB or cutoff frequency, at which the phase response is a 45° lagging phase shift. Poles cause an infinite response on the imaginary frequency axis.

What Is a Zero?

A zero is the complement of a pole. In math, it is a frequency at which the transfer function equation of a circuit is zero. This is not impossible in the real world (unlike the pole), so zeroes can be found at real-number frequencies as well as complex-number frequencies.

Each zero is associated with an *upward* bend of 6 dB per octave in a magnitude response. Similar to a pole, the frequency of the zero is at the +3 dB point. Each zero is associated with a transition on a phase-versus-frequency plot that reduces the lag by 90°. The zero is at the 45° leading phase point. Multiple zeroes add their phase shifts just as poles do.

In a circuit, a zero creates gain that increases with frequency forever above the zero frequency. This requires active circuitry that would inevitably run out of gain at some frequency, which implies one or more poles up there. In real-world circuits, zeroes are usually not found by themselves, making the magnitude response go up, but rather paired with a pole of a different frequency, resulting in the magnitude response having a slope between two frequencies but flat above and below them.

Real-world circuit zeroes are only found accompanied by an equal or greater number of poles. Consider a classic RC high-pass filter, such as if the resistor and capacitor in Figure 3.51 were exchanged. The response of such a circuit increases at 6 dB per octave from 0 Hz (so there must be a zero at 0 Hz) and then levels off at 1/2πRC Hz. This leveling off is due to the presence of a pole adding its 6 dB-per-octave roll-off to cancel the 6 dB-per-octave roll-up of the zero. The transfer function for such as circuit would equal zero at zero frequency and infinity at the imaginary pole frequency.

FEEDBACK AND OSCILLATION

The *stability* of an amplifier refers to its ability to provide gain to a signal without tending to oscillate. For example, an amplifier just on the verge of oscillating is not generally considered to be "stable." If the output of an amplifier is fed back to the input, the feedback can affect the amplifier stability. If the amplified output is added to the input, the output of the sum will be larger. This larger output, in turn, is also fed back. As this process continues, the amplifier output will continue to rise until the amplifier cannot go any higher (clamps). Such *positive feedback* increases the amplifier gain, and is called *regeneration*. (The chapter on **Oscillators and Synthesizers** includes a discussion of positive feedback.)

Most practical amplifiers have some intrinsic and unavoidable feedback either as part of the circuit or in the amplifying device(s) itself. To improve the stability of an amplifier, *negative feedback* can be added to counteract any unwanted positive feedback. Negative feedback is often combined with a phase-shift *compensation* network to improve the amplifier stability.

Although negative feedback reduces amplifier or stage gain, the advantages of *stable* gain, freedom from unwanted oscillations and the reduction of distortion are often key design objectives and advantages of using negative feedback.

The design of feedback networks depends on the desired result. For amplifiers, which should not oscillate, the feedback network is customized to give the desired frequency response without loss of stability. For oscillators, the feedback network is designed to create a steady oscillation at the desired frequency.

SUMMING

In a linear system, nature does most of the work for us when it comes to adding signals; placing two signals together naturally causes them to add according to the principle of superposition. When processing signals, we would like to control the summing operation so the signals do not distort or combine in a nonlinear way. If two signals come from separate stages and they are connected together directly, the circuitry of the stages may interact, causing distortion of either or both signals.

Summing amplifiers generally use a resistor in series with each stage, so the resistors connect to the common input of the following stage. This provides some *isolation* between the output circuits of each stage. **Figure 3.52** illustrates the resistors connecting to a summing amplifier. Ideally, any time we wanted to combine signals (for example, combining an audio signal with a sub-audible tone in a 2 meter FM transmitter prior to modulating the RF signal) we could use a summing amplifier.

FILTERING

A *filter* is a common linear stage in radio equipment. Filters are characterized by their ability to selectively attenuate certain frequencies in the filter's *stop band*, while passing or amplifying other frequencies in the *passband*. If the filter's passband extends to or near dc, it is a *low-pass* filter, and if it extends toward infinity (or at least very high frequencies for the circuitry involved), it is a *high-pass* filter. Filters that pass a range of frequencies are *band-pass* filters. *All-pass* filters are designed to affect only the phase of a signal without changing the signal magnitude. The range of frequencies between a band-pass circuit's low-pass and high-pass regions is its *passband*.

Figure 3.50A is the magnitude response for a typical band-pass audio filter. It shows that the input signal is passed to the output with no loss (0 dB) between 200 Hz and 5 kHz. This is the filter's *mid-band response*. Above and below those frequencies the response of the filter begins to drop. By 20 Hz and 20 kHz, the magnitude response has been reduced to one-half of (–3 dB) the mid-band response. These points are called the circuit's *cutoff* or *corner* or *half-power frequencies*. The range between the cutoff frequencies is the filter's passband, or full-width half maximum (FWHM). Outside the filter's passband, the magnitude response drops at the lower cutoff frequency from –3 dB at 20 Hz to about –13 dB at 2 Hz, which is 10 dB/decade and, thus is due to 1 pole in this part of the circuit. At the upper cutoff frequency, the response drops from –3 dB at 20 kHz to –23 dB at 200 kHz, which is 20 dB/decade, indicating that this part of the circuit has 2 poles.

Figure 3.50B represents the phase response of the same filter. At low frequencies, the capacitor has little effect on phase shift, which is unchanging with frequency, leading at 90°. As the signal frequency increases to the filter's lower –3 dB inflection point, the phase shift starts decreasing with frequency. It switches from a leading to a lagging phase shift at the center of the passband. The lagging phase shift continues to increase (i.e., becomes more negative) until the frequency reaches the upper –3 dB inflection point, where it levels off, lagging at 90° (i.e., –90°).

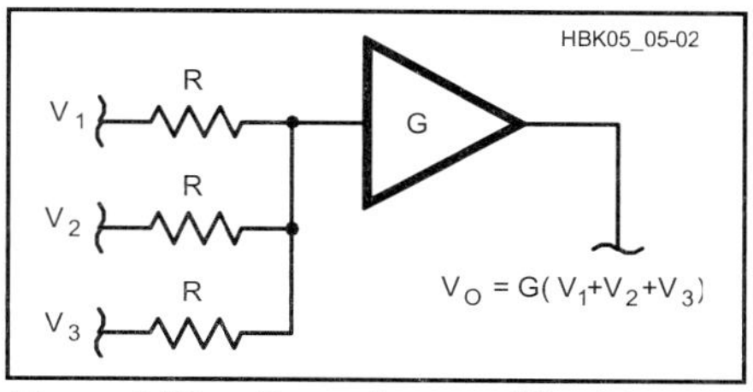

Figure 3.52 — Summing amplifier. The output voltage is equal to the sum of the input voltages times the amplifier gain, G. As long as the resistance values, R, are equal and the amplifier input impedance is much higher, the actual value of R does not affect the output signal.

Practical analog (both passive and active) and digital filters are discussed in the chapter **Analog and Digital Filtering**. Filters at RF may also be created by using transmission lines as described in the **Transmission Lines** chapter.

3.9.4 Nonlinear Operations

All signal processing doesn't have to be linear. Any time that we treat various signal levels differently, the operation is called *nonlinear*. This is not to say that all signals must be treated the same for a circuit to be linear. High-frequency signals are attenuated in a low-pass filter while low-frequency signals are not, yet the filter can be linear. The distinction is that the amount of attenuation at different frequencies is always the same, regardless of the amplitude of the signals passing through the filter.

What if we do not want to treat all voltage levels the same way? This is commonly desired in analog signal processing for clipping, rectification, compression, modulation and switching.

CLIPPING AND RECTIFICATION

Clipping is the process of limiting the range of signal voltages passing through a circuit (in other words, *clipping* those voltages outside the desired range of the signals). There are a number of reasons why we would like to do this. As shown in Figure 3.49, clipping is the process of limiting the positive and negative peaks of a signal. (Clipping is also called *clamping*.)

Clipping might be used to prevent a large audio signal from causing excessive deviation in an FM transmitter that would interfere with communications on adjacent channels. Clipping circuits are also used to protect sensitive circuitry from excessive voltages. Clipping distorts the signal, changing it so that the original signal waveform is lost.

Another kind of clipping results in *rectification*. A *rectifier* circuit clips off all voltages of one polarity (positive or negative) and passes only voltages of the other polarity, thus changing ac to pulsating dc (see the **Power Sources** chapter). Another use of rectification is in a *peak detection* circuit that measures the peak value of a waveform. Only one polarity of the ac voltage needs to be measured and so a rectifier clips the unwanted polarity.

LIMITING

Another type of clipping occurs when an amplifier is intentionally operated with so much gain that the input signals result in an output that is clipped at the limits of its power supply voltages (or some other designated voltages). The amplifier is said to be driven into *limiting* and an amplifier designed for this behavior is called a *limiter*. Limiters are used in FM receivers to amplify the signal until all amplitude variations in the signal are removed and the only characteristic of the original signal that remains is the frequency.

LOGARITHMIC AMPLIFICATION

It is sometimes desirable to amplify a signal logarithmically, which means amplifying low levels more than high levels. This type of amplification is often called *signal compression*. Speech compression is sometimes used in audio amplifiers that feed modulators. The voice signal is compressed into a small range of amplitudes, allowing more voice energy to be transmitted without overmodulation (see the **Modulation** and the **Transmitting** chapters).

3.9.5 System Design Functions

Many kinds of electronic equipment are developed by combining basic analog signal processing circuits, often treating them as independent functional blocks. This section describes several topics associated with building systems from multiple blocks. Because analog circuits often interface with digital circuits or include digital elements in a hybrid circuit, some topics associated with digital systems are included. Many similar functions are implemented as part of Digital Signal Processing (DSP) systems and have similar behaviors and concerns. Although not all basic electronic functions are discussed here, the concepts associated with combining them can be applied generally.

Since our main concern is the effect that circuitry has on a signal, we often describe the circuit by its actions rather than by its specific components. A *black box* is a circuit that can be described entirely by the behavior of its interfaces with other blocks and circuitry. When circuits are combined in such as way as to perform sequential operations on a signal, the individual circuits are called *stages*.

The most general way of referring to an analog circuit is as a *network*. Two basic properties of analog networks are of principal concern: the effect that the network has on a signal and the interaction that the network has with the circuitry surrounding it. Interfaces between the network and the rest of the network are called *ports*.

Many analog circuits are analyzed as *two-port networks* with an input and an output port. The signal is fed into the input port, is modified inside the network and then exits from the output port. (See the chapter on **RF Techniques** for more information on two-port networks.)

TRANSFER FUNCTIONS

The specific way in which the analog circuit modifies the signal can be described mathematically as a transfer function. The mathematical operation that combines a signal with a transfer function is pictured symbolically in **Figure 3.53**. The transfer function, h(t) or h(f), describes the circuit's modification of the input signal in the time domain where all values are functions of time, such as a(t) or b(t), or in the frequency domain where all values are functions of frequency, such as a(f) or b(f). The mathematical operation by which h(t) operates on a(t) is called *convolution* and is represented as a dot, as in a(t) • h(t) = b(t). In the frequency domain, the transfer function multiplies the input, as in a(f) × h(f) = b(f).

While it is not necessary to understand transfer functions mathematically to work with analog circuits, it is useful to realize that they describe how a signal interacts with other signals in an electronic system. In general, the output signal of an analog system depends not only on the input signal at the same time, but also on past values of the input signal. This is a very important concept and is the basis of such essential functions as analog filtering.

CASCADING STAGES

If an analog circuit can be described with a transfer function, a combination of analog circuits can also be described similarly. This description of the combined circuits depends upon the relationship between the transfer functions of the parts and that of the combined circuits. In many cases this relationship allows us to predict the behavior of large and complex circuits from what we know about the parts of which they are made. This aids in the design and analysis of analog circuits.

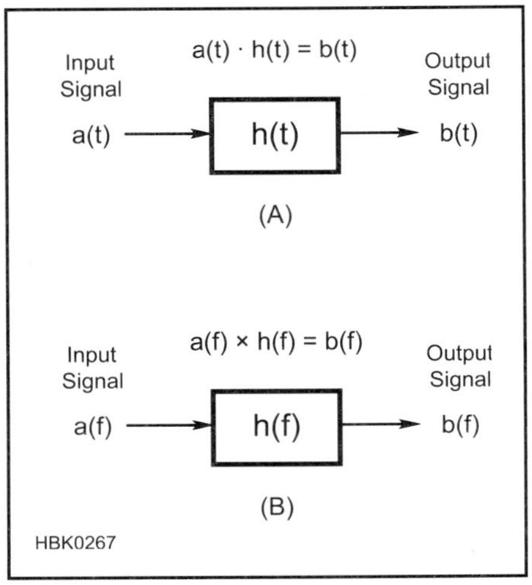

Figure 3.53 — Linear function blocks and transfer functions. The transfer function can be expressed in the time domain (A) or in the frequency domain (B). The transfer function describes how the input signal a(t) or a(f) is transformed into the output signal b(t) or b(f).

When two analog circuits are cascaded (the output signal of one stage becomes the input signal to the next stage) their transfer functions are combined. The mechanism of the combination depends on the interaction between the stages. The ideal case is the functions of the stages are completely independent. In other words, when the action of a stage is unchanged, regardless of the characteristics of any stages connected to its input or output.

Just as the signal entering the first stage is modified by the action of the first transfer function, the ideal cascading of analog circuits results in changes produced only by the individual transfer functions. For any number of stages that are cascaded, the combination of their transfer functions results in a new transfer function. The signal that enters the circuit is changed by the composite transfer function to produce the signal that exits from the cascaded circuits.

While each stage in a series may use feedback within itself, feedback around more than one stage may create a function, and resultant performance, different from any of the included stages. Examples include oscillation or negative feedback.

AMPLIFIER FREQUENCY RESPONSE

At higher frequencies a typical amplifier acts as a low-pass filter, decreasing amplification with increasing frequency. Signals within a range of frequencies are amplified consistently but outside that range the amplification changes. At high gains many amplifiers work properly only over a small range of frequencies. The combination of gain and frequency response is often expressed as a *gain-bandwidth product*. For many amplifiers, gain times bandwidth is approximately constant. As gain increases, bandwidth decreases, and vice versa.

Performance at lower frequencies depends on whether the amplifier is *dc- or ac-coupled*. Coupling refers to the transfer of signals between circuits. A dc-coupled amplifier amplifies signals at all frequencies down to dc. An ac-coupled amplifier acts as a high-pass filter, decreasing amplification as the frequency decreases toward dc. Ac-coupled circuits usually use capacitors to allow ac signals to flow between stages while blocking the dc bias voltages of the circuit.

INTERSTAGE LOADING AND IMPEDANCE MATCHING

Every two-port network can be further defined by its input and output impedance. The input impedance is the opposition to current, as a function of frequency, seen when looking into the input port of the network.

Likewise, the output impedance is similarly defined when looking back into a network through its output port.

If the transfer function of a stage changes when it is cascaded with another stage, we say that the second stage has *loaded* the first stage. This often occurs when an appreciable amount of current passes from one stage to the next. Interstage loading is related to the relative output impedance of a stage and the input impedance of the stage that is cascaded after it.

In some applications, the goal is to transfer a maximum amount of power from the output of the stage to a load connected to the output. In this case, the output impedance of the stage is *matched* or transformed to that of the load (or vice versa). This allows the stage to operate at its optimum voltage and current levels. In an RF amplifier, the impedance at the input of the transmission line feeding an antenna is transformed by means of a matching network to produce the resistance the amplifier needs in order to efficiently produce RF power.

In contrast, it is the goal of most analog signal processing circuitry to modify a signal rather than to deliver large amounts of power. Thus, an impedance-matched condition may not be required. Instead, current between stages can be minimized by using mismatched impedances. Ideally, if the output impedance of a network is very low and the input impedance of the following stage is very high, very little current will pass between the stages, and interstage loading will be negligible.

NOISE

Generally we are only interested in specific man-made signals. Nature allows many signals to combine, however, so the desired signal becomes combined with many other unwanted signals, both man-made and naturally occurring. The broadest definition of noise is any signal that is not the one in which we are interested. One of the goals of signal processing is to separate desired signals from noise. (See the **RF Techniques** chapter for a more complete discussion on noise, including calculation and use of noise factor and noise figure.)

One form of noise that occurs naturally and must be dealt with in low-level processing circuits is called *thermal noise*, or *Johnson noise*. Thermal noise is produced by random motion of free electrons in conductors and semiconductors. This motion increases as temperature increases, hence the name. This kind of noise is present at all frequencies and is proportional to temperature. Naturally occurring noise can be reduced either by decreasing the circuit's bandwidth or by reducing the temperature in the system. Thermal noise voltage and current vary with the circuit impedance and follow Ohm's Law. Low noise amplifier design techniques are based on these relationships.

Analog signal processing stages are characterized in part by the noise they add to a signal. A distinction is made between enhancing existing noise (such as amplifying it) and adding new noise. The noise added by analog signal processing is commonly quantified by the *noise factor, f*. Noise factor is the ratio of the total output noise power (thermal noise plus noise added by the stage) to the amplifier input noise power when the termination is at the standard temperature of 290 K (17 °C). When the noise factor is expressed in dB, we often call it *noise figure, NF*.

In a system of many cascaded signal processing stages, such as a communications receiver, each stage contributes to the total noise of the system. The noise factor of the first stage dominates the noise factor of the entire system because noise added at the first stage is then multiplied by each following stage. Noise added by later stages is not multiplied to the same degree and so is a smaller contribution to the overall noise at the output. Designers try to optimize system noise factor by using a first stage with a minimum possible noise factor and maximum possible gain. However, if the gain it set too high and the circuit overloads, noise is not the biggest problem.

BUFFERING

It is often necessary to isolate the stages of an analog circuit. This isolation reduces the loading, coupling and feedback between stages. When connecting circuits that operate at different impedance levels between stages, an intervening stage, a type of amplifier called a *buffer*, is often used for this purpose. If signal level is sufficient, an attenuator can alsoserve as a buffer at the expense of some signal loss.

Buffers can have high values of amplification but this is unusual. A buffer used for impedance transformation generally has a low or unity gain. In some circuits, notably power amplifiers, the desired goal is to deliver a maximum amount of power to the output device (such as a speaker or an antenna). Matching the amplifier output impedance to the output-device impedance provides maximum power transfer. A buffer amplifier may be just the circuit for this type of application. Such amplifier circuits must be carefully designed to avoid distortion. Combinations of buffer stages can also be effective at isolating the stages from each other as well as making impedance transformations.

TRANSITION TIME

The transition between the binary 0 and binary 1 states of a digital signal or circuit does not occur instantly. There is a *transition time* between states. This transition time is a result of the time it takes to charge or discharge the stray capacitance in wires and other components because voltage cannot change instantaneously across a capacitor. Stray inductance in the wires also has an effect because the current through an inductor can't change instantaneously. The rate at which the digital circuit's output transistors can change state may also be a factor.

Distributed inductances and capacitances in wires or PC-board traces may cause rise and fall times to increase as the pulse moves away from the source. Ringing and reflections may occur due to transmission line effects as discussed in the **Transmission Lines** chapter.

The transition from a 0 to a 1 state is called the *rise time*, and is usually specified as the time for the pulse to rise from 10% of its final value to 90% of its final value. Similarly, the transition from a 1 to a 0 state is called the *fall time*, with a similar 10% to 90% definition. Note that these times need not be the same. **Figure 3.54A** shows an ideal signal, or *pulse*, with zero-time switching. Figure 3.54B shows a typical pulse, as it changes between states in a smooth curve.

The faster the rise or fall time (RT), the wider the bandwidth (BW) of signals associated with the transition. Such high frequency signals can be radiated, causing noise and interfering signals in receivers or sensitive circuits. The general rule of thumb for the bandwidth of digital signals is:

$$\mathrm{BW(GHz)} = \frac{0.35}{\mathrm{RT(ns)}} \text{ and}$$

$$\mathrm{RT(ns)} = \frac{0.35}{\mathrm{BW(GHz)}}$$

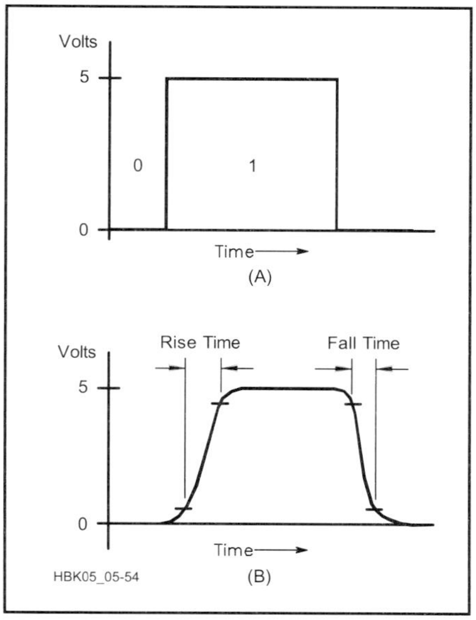

Figure 3.54 — (A) An ideal digital pulse and (B) a typical actual pulse, showing the gradual transition between states.

Rise and fall times for digital integrated circuits vary with the logic family used and the location in a circuit. Typical values of transition time are in the range of microseconds (4000-series CMOS with high-impedance loads) to the sub-nanosecond range for current digital logic technology.

PROPAGATION DELAY

Rise and fall times only describe a relationship within a pulse. A pulse input into the circuit must propagate through each component in the circuit until eventually it arrives at the output. The time delay between providing an input to a circuit and seeing a response at the output is the *propagation delay* and is illustrated by **Figure 3.55**.

For modern switching logic, typical propagation delay values are in the sub-nanosecond range. (It is useful to remember that the propagation delay along a wire or printed-circuit-board trace adds about 1.0 to 1.5 ns per inch.) Propagation delay is the result of cumulative transition times as well as transistor switching delays, reactive element charging times and the time for signals to travel through wires. In complex circuits, varying propagation delays through different paths can cause problems when pulses must arrive somewhere at exactly the same time.

The effect of these delays on digital devices can be seen by looking at the speed of the digital pulses. Most digital devices use *clock pulses*. If two pulses are supposed to arrive at a logic circuit at, or very close to, the same time, the path length for the two signals cannot differ by more than two to three inches. This can be a very significant design problem for high-speed logic designs.

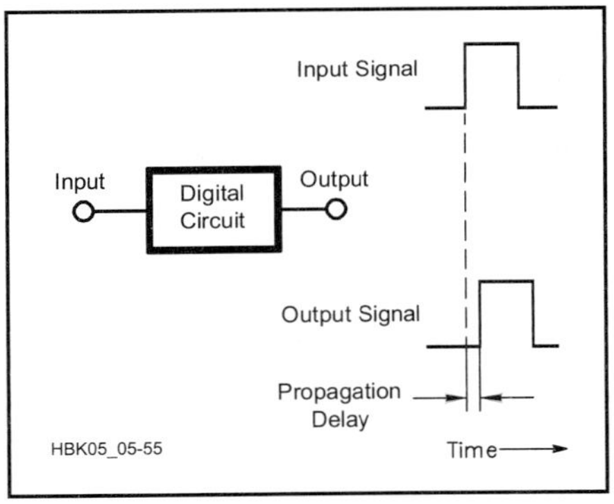

Figure 3.55 — Propagation delay in a digital circuit.

3.10 Electromagnetic Waves

Audio or *sonic* energy is imparted by the mechanical movement of a medium, which can be air, metal, water or even the human body. Sound that humans can hear normally requires the movement of air between 20 Hz and 20 kHz, although the human ear loses its ability to detect the extremes of this range as we age. Some animals, such as elephants, can detect air vibrations well below 20 Hz, while others, such as dogs and cats, can detect air vibrations well above 20 kHz.

Electrical circuits do not directly produce air vibrations. Sound production requires a *transducer*, a device that transforms one form of energy into another form of energy; in this case electrical energy into sonic energy. The speaker and the microphone are the most common audio transducers. There are numerous ultrasonic transducers for various applications.

Radio frequency energy exists at frequencies for which it is practical to generate and detect waves that exist independently of the movement of electrical charge, such as a radio signal. As with sonic energy, a transducer is required to convert electrical energy in a circuit into electromagnetic waves. We call that type of transducer an antenna. In a physical circuit, such as a wire, electromagnetic energy exists as both electromagnetic waves and the physical movement of electrical charge.

Electromagnetic waves have been generated and detected in many forms, with frequencies from below 1 Hz to above 10^{21} Hz, including at the higher frequencies of infrared, visible, and ultraviolet light, and a number of energy forms of greatest interest to physicists and astronomers. **Table 3.5** provides a brief glimpse at the total spectrum of electromagnetic energy. The *radio spectrum* is generally

Table 3.5
Key Regions of the Electromagnetic Energy Spectrum

Region Name	*Frequency Range*		
Radio frequencies*	3.0×10^3 Hz	to	3.0×10^{11} Hz
Infrared	3.0×10^{11} Hz	to	4.3×10^{14} Hz
Visible light	4.3×10^{14} Hz	to	7.5×10^{14} Hz
Ultraviolet	7.5×10^{14} Hz	to	6.0×10^{16} Hz
X-rays	6.0×10^{16} Hz	to	3.0×10^{19} Hz
Gamma rays	3.0×10^{19} Hz	to	5.0×10^{20} Hz
Cosmic rays	5.0×10^{20} Hz	to	8.0×10^{21} Hz

Range Name	*Abbreviation*	*Frequency Range*
Very Low Frequency	VLF	3 kHz - 30 kHz
Low Frequency	LF	30 kHz - 300 kHz
Medium Frequency	MF	300 kHz - 3 MHz
High Frequency	HF	3 MHz - 30 MHz
Very High Frequency	VHF	30 MHz - 300 MHz
Ultra High Frequency	UHF	300 MHz - 3 GHz
Super High Frequency	SHF	3 GHz - 30 GHz
Extremely High Frequency	EHF	30 GHz - 300 GHz

*Note: Radio frequencies are commonly considered to cover 3 kHz to 300 GHz, VLF to EHF

considered to begin around 3 kHz and end below infrared light.

Within the part of the electromagnetic energy spectrum of most interest to radio amateurs, frequencies have been classified into groups and given names. Table 3.5 provides a reference list of these classifications. To a significant degree, the frequencies within each group exhibit similar properties, both in circuits and as RF waves. For example, HF or high frequency waves, with frequencies from 3 to 30 MHz, all exhibit ionospheric refraction that permits regular long-range radio communications. This property also applies occasionally both to MF (medium frequency) and to VHF (very high frequency) waves, as well.

Despite the close relationship between electromagnetic energy and waves, it remains important to distinguish the two. To a circuit producing or amplifying a 15-kHz alternating current, the ultimate transformation and use of the electrical energy may make no difference to the circuit's operation. By choosing the right transducer, one can produce either a sonic wave or an electromagnetic wave — or both. Such is a common problem of video monitors and switching power supplies; forces created by the ac currents cause electronic parts both to vibrate audibly *and* to radiate electromagnetic energy.

3.10.1 Electric and Magnetic Fields

Electrical and magnetic energy are invisible. You can't detect them with any of your senses. All you can do is observe their effects such as when a resistor gets hot, a motor spins, or an electromagnet picks up iron or steel. The energy exists as a *field* — a region of space in which energy is stored and through which electrical and magnetic forces act. (For serious inquiries as to the nature of fields, see **www.en.wikipedia.org/wiki/Electric_field** and **www.en.wikipedia.org/wiki/Magnetic_field**.)

You are already quite familiar with fields in the form of gravity. You are being pulled down toward the Earth as you read this because you are in the Earth's *gravitational field*. Because your body has mass it interacts with the gravitational field in such a way that the Earth attracts you. Think of a bathroom scale as a "gravitational voltmeter" that instead of reading "volts," reads "pounds." The heavier something is, the stronger the Earth is attracting it. Weight is the same as force.

This field makes you do work, such as when you climb stairs. Work has a precise definition when it comes to fields: *Work* equals force times distance moved in the direction of the field's force. For example, let's say you pick up a stone that weighs 1 pound and lift it to a shelf 10 feet above where it previously lay. How much work did you do? You moved a weight of 1 pound a distance of 10 feet against the attraction of the field, so you have done 10 foot-pounds of work. (It doesn't count if you move the stone sideways instead of vertically.)

What did that work accomplish? You stored gravitational energy in the stone equal to the amount of work that you performed. This stored energy is called *potential energy*, whether gravitational, electrical or magnetic. You could store the same amount of gravitational energy by lifting a 10-pound stone 1 foot or by lifting a stone that weighs 1/10th of a pound 100 feet. If you drop the stone (or it falls off the shelf), the same amount of potential energy is converted back to *kinetic energy* as the stone moves toward the Earth in the gravitational field.

In electronics we are interested in two types of fields: *electric fields* and *magnetic fields*. Electric fields can be detected as voltage differences between two points. The electric field's analog to gravitational mass is electric charge. Every electric charge has its own electric field, just as every mass has its own gravitational field. The more charged a body is, the "heavier" it is in terms of an electric field. Just as a body with mass feels a force to move in a gravitational field, so does an electric charge in an electric field. Electrical energy is stored by moving electrical charges apart so that there is a voltage between them. If the field does not change with time, it is called an *electrostatic field*.

Magnetic energy is detected by its effects on moving electrical charges or current. Magnetic energy is stored through the motion of electric charge (current) creating a magnetic field. Magnetic fields that don't change with time, such as from a stationary permanent magnet, are called *magnetostatic fields*.

The potential energy is released by allowing the charges to move in the field. For example, electric energy is released when a current flows from a charged-up capacitor. Magnetic energy stored by current flowing in an inductor is released when the current is allowed to change, such as a relay's armature does when the coil is de-energized.

3.10.2 Electromagnetic Fields and Waves

An *electromagnetic field* is created when the potential energy stored in an electric field or magnetic field changes. The changing electric and magnetic fields create *electromagnetic waves* (what we call *radio waves*) that propagate through space carrying both electric and magnetic energy. The electric and magnetic fields in the wave vary with time in a sinusoidal pattern. The potential energy is shared between the electric and magnetic fields making up the electromagnetic field.

The *field strength* of an electromagnetic wave can be measured either by the electric field (volts/meter) or the magnetic field (amps/meter). Usually the wave's field strength is stated only in volts/meter since that is easier to measure than the magnetic field. If we multiply the electric and magnetic field strengths, we have power per unit area:

$$\frac{E}{m} \times \frac{H}{m} = \frac{\text{watts}}{m^2}$$

MAXWELL'S EQUATIONS

The basic theory of electromagnetic fields was established by James Maxwell in 1860-1864. The behavior of the fields are described by the four equations known today as Maxwell's equations. (This form of the equations was actually produced by Oliver Heaviside in his work prior to 1890.) The existence of the electromagnetic waves predicted by Maxwell in 1864 was demonstrated by Heinrich Hertz in 1886. For more information about the experiment, see the References and Bibliography listing at the end of this chapter for G.S. Smith's article analyzing Hertz's experiments.

A discussion of Maxwell's equations and their application to electromagnetic simulation is presented in the **Modeling and Simulation** chapter. A more complete treatment is provided by Bob Zavrel, W7SX, in his book *Antenna Physics: An Introduction* (see the References and Bibliography section of this chapter). A summary treatment of the equations in a pair of Hands-On Radio columns by Ward Silver, NØAX, is included in the online content. These include an explanation of the vector calculus concepts of gradient, divergence, and curl, as well as illustrating how waves are created by moving electric charge.

3.10.3 Electromagnetic Wave Propagation

All electromagnetic energy has one thing in common: it travels, or *propagates,* at the speed of light, abbreviated *c*. This speed is approximately 300,000,000 (or 3×10^8) meters per second in a vacuum, termed *free space*.

In general, the speed at which electromagnetic waves propagate depends on the permittivity ε, (ε_0 is the value for free space) and permeability μ, (μ_0 is the value for free space) of the medium through which they travel.

$$c = \frac{1}{\sqrt{\varepsilon_0 \mu_0}}$$

The speed of light is highest in the vacuum of free space and only slightly lower in air. In materials such as glass or plastic, however, velocity can be quite a bit lower. For example, in polyethylene (commonly used as a center insulator in coaxial cable), the *velocity of propagation* is about two-thirds (67%) of that in free space.

Electromagnetic waves have a wavelength uniquely associated with each possible frequency. (See **Figure 3.56**) The *wavelength* (λ) is the speed of propagation, c, divided by the frequency (f) in hertz.

$$\lambda\,(m) = \frac{3.0\times10^{8}\left(\frac{m}{s}\right)}{f\,(Hz)}$$

and

$$f\,(Hz) = \frac{3.0\times10^{8}\left(\frac{m}{s}\right)}{\lambda\,(m)}$$

Example: What is the frequency of an RF wave with wavelength of 80 meters?

$$f\,(Hz) = \frac{3.0\times10^{8}\left(\frac{m}{s}\right)}{\lambda\,(m)}$$

$$= \frac{3.0\times10^{8}\left(\frac{m}{s}\right)}{80.0\ m}$$

$$= 3.75\times10^{6}\ Hz$$

This is 3.750 MHz or 3750 kHz, a frequency in the middle of the ham band known as "80 meters."

A similar equation is used to calculate the wavelength of a sound wave in air, substituting the speed of sound for the speed of light in the numerator. The speed of propagation of the mechanical movement of air that we call sound varies considerably with air temperature and altitude. The speed of sound at sea level is about 331 m/s at 0 °C and 344 m/s at 20 °C.

To calculate the frequency of an electromagnetic wave directly in kilohertz, change the speed constant to 300,000 (3×10^{5}) km/s.

$$f\,(kHz) = \frac{3.0\times10^{5}\left(\frac{km}{s}\right)}{\lambda\,(m)}$$

and

$$\lambda\,(m) = \frac{3.0\times10^{5}\left(\frac{km}{s}\right)}{f\,(kHz)}$$

For frequencies in megahertz, change the speed constant to 300 (3×10^{2}) Mm/s (million meters per second).

$$f\,(MHz) = \frac{300\left(\frac{Mm}{s}\right)}{\lambda\,(m)}$$

and

$$\lambda\,(m) = \frac{300\left(\frac{Mm}{s}\right)}{f\,(MHz)}$$

Stated as it is usually remembered and used, "wavelength in meters equals 300 divided by frequency in megahertz." Assuming the proper units for the speed of light constant simplify the equation.

$$\lambda\,(in\,m) = \frac{300}{f\,(in\,MHz)}$$

$$\text{and } f\,(in\,MHz) = \frac{300}{\lambda\,(in\,m)}$$

Example: What is the wavelength of an RF wave whose frequency is 4.0 MHz?

$$\lambda(m) = \frac{300}{f\,(MHz)} = \frac{300}{4.0} = 75\ m$$

At higher frequencies, circuit elements with lengths that are a significant fraction of a wavelength can act like transducers. This property can be useful, but it can also cause problems for circuit operations. Therefore, wavelength calculations are of some importance in designing ac circuits for those frequencies.

3.10.4 Electromagnetic Wave Structure

Electromagnetic waves move through space independently of any component or conductor. The electric and magnetic fields of the wave are oriented at right angles to each other as shown by **Figures 3.57 – 3.59**. The direction of the right angle between the electric and magnetic fields determines the direction the wave travels, as illustrated by Figure 3.59. The term "lines of force" in Figure 3.57 means the direction in which a force would be felt by an electron (from the electric field) or by a magnet (from the magnetic field).

An important note about electromagnetic waves: The electric and magnetic fields making up the wave are not just perpendicular electric and magnetic fields that simply happen to be in the same place at the same time! The fields are *coupled*; that is they are both aspects of the same entity — the electromagnetic wave. The fields cannot be separated although the energy in the wave can be detected as either electric or magnetic force. The electromagnetic wave is created as a single entity by the motion of electrons, such as in a transmitting antenna.

WAVEFRONTS

To an observer staying in one place, such as a fixed station's receiving antenna, the electric and magnetic fields of the wave appear to oscillate as the wave passes. That is, the fields create forces on electrons in the antenna that increase and decrease in the sine wave pattern. Some of the energy in the propagating wave is transferred to the electrons as the forces from the changing fields cause them to move. This creates a sine wave current in the antenna with a frequency determined by the rate at which the field strength changes in the passing wave.

If the observer is moving along with the wave at the same speed, however, the strength of the fields will not change. To that observer, the electric and magnetic field strengths are fixed, as in a photograph. This is a *wavefront* of the electromagnetic wave — a flat surface

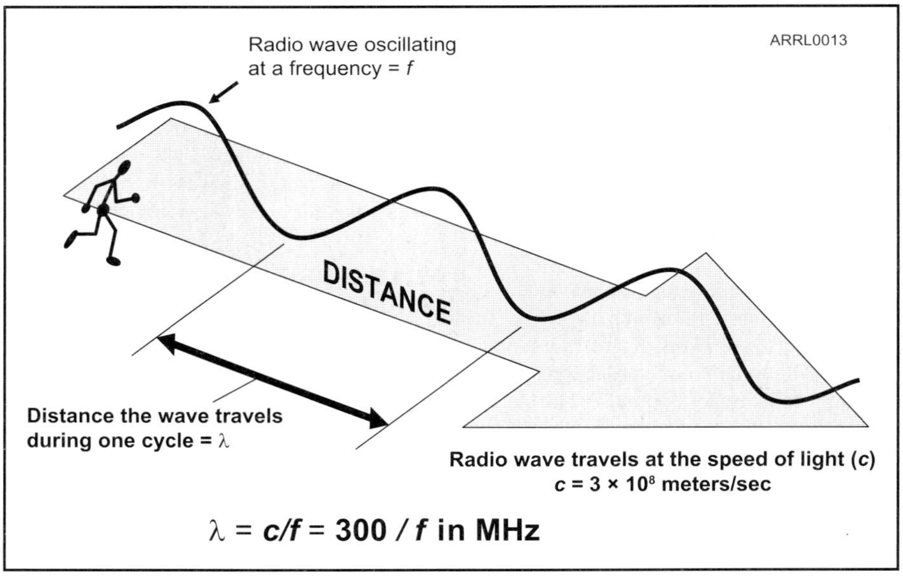

Figure 3.56 — As a radio wave travels, it oscillates at the frequency of the signal. The distance covered by the wave during the time it takes for one complete cycle is its wavelength.

or plane moving through space on which the electric and magnetic fields have a constant value as illustrated in Figure 3.57.

Just as an ac voltage is made up of an infinite sequence of instantaneous voltages, each slightly larger or smaller than the next, an infinite number of wavefronts make up an electromagnetic wave, one behind another like a deck of cards. The direction of the wave is the direction in which the wavefronts move. The fields on each successive wavefront have a slightly different strength, so as they pass a fixed location the detected field strength changes as well. The result is that the fixed observer "sees" fields with strengths varying as a sine wave.

Figure 3.58 is a drawing of what would happen if we could suddenly freeze all of the wavefronts in the wave and take measurements of the electric and magnetic field strengths in each. In this diagram, the electric field is oriented vertically and the magnetic field horizontally. (Each of the vertical lines in the electric field can be thought of as representing an individual wavefront.) **Figure 3.59** illustrates the right-angle relationship of the *E* and *H* fields, and the direction of their motion.

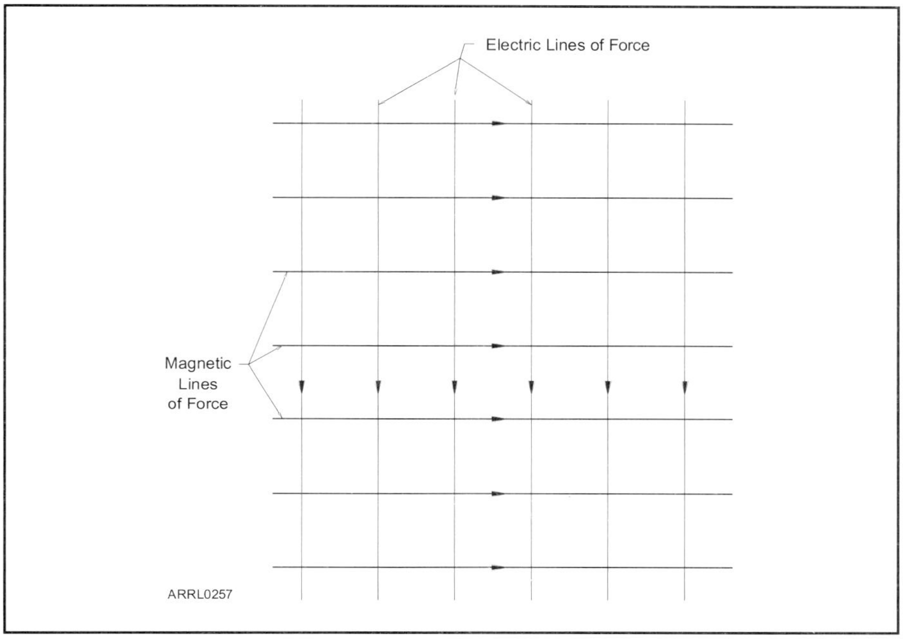

Figure 3.57 — Representation of electric and magnetic lines of force in an electromagnetic wavefront. Arrows indicate the instantaneous directions of the fields for a wavefront in a wave traveling toward you, out of the page. Reversing the direction of either of the fields would also reverse the direction of the wave.

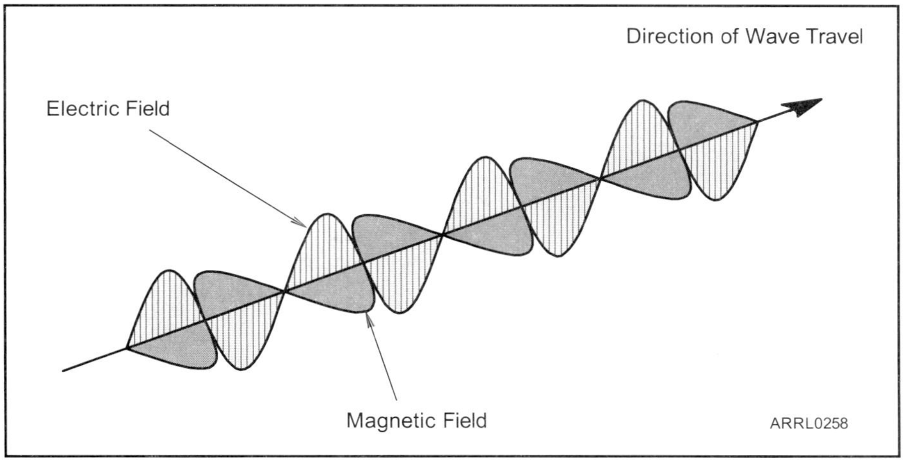

Figure 3.58 — Representation of the magnetic and electric field strengths of a vertically polarized electromagnetic wave. In the diagram, the electric field is oriented vertically and the magnetic field horizontally.

All of the wavefronts are moving in the direction indicated. The whole set of them moves together at the same speed. As the wave moves past the receiving antenna, the varying field strengths of the different wavefronts are perceived as a continuously changing wave. What we call a "wave" is really this entire group of wavefronts moving through space.

POLARIZATION

The orientation of the pair of fields in an electromagnetic wave can face any direction with respect to the surface of the Earth, but the electric and magnetic fields will always be at right angles to each other. The orientation of the wave's electric field determines the *polarization* of the wave. If the electric field's lines of force are parallel to the surface of the Earth (meaning those of the magnetic field are perpendicular to the Earth), the wave is *horizontally polarized.* Conversely, if the electric field's lines of force are perpendicular to the surface of the Earth (meaning those of the magnetic field are parallel to the Earth), the wave is *vertically polarized.* Knowing the polarization of the wave allows the receiving antenna to be oriented so that the passing wave will exert the maximum force on the electrons in the antenna, maximizing received signal strength.

For the most part, the wave's polarization is determined by the type of transmitting antenna and its orientation. For example, a Yagi antenna with its elements parallel to the Earth's surface transmits a horizontally polarized wave. On the other hand, an amateur mobile whip antenna, mounted vertically on an automobile, radiates a vertically polarized wave. If a vertically polarized antenna is used

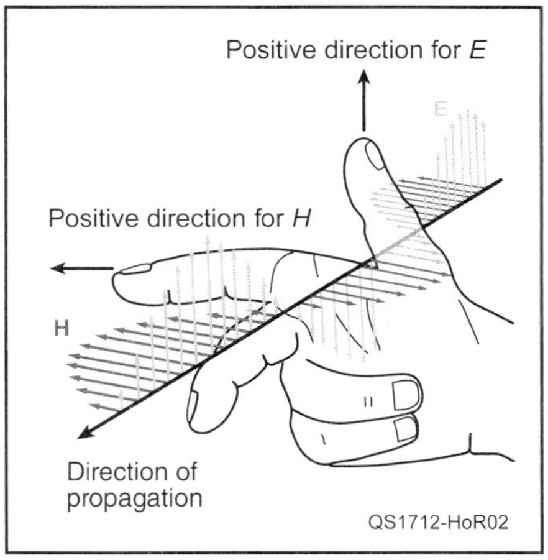

Figure 3.59 — The right-hand rule shows how to determine direction of propagation of an electromagnetic wave. Point your thumb in the positive direction for the *E* field, your Index finger in the positive direction for *H*, and your middle finger will point in the direction the wave is traveling.

to receive a horizontally polarized radio wave (or vice versa), received signal strength can be reduced by more than 20 dB as compared to using an antenna with the same polarization as the wave. This is called *cross-polarization*.

It is also possible to generate electromagnetic waves in which the orientation of successive wavefronts rotates around the direction of travel — both the electric and magnetic fields. This is called *circular polarization*. Imagine the wave of Figure 3.58 being twisted so at one point the direction of the electric field is horizontal and a bit further along the wave it is vertical. As the twisted, circularly polarized wave passes the receiving antenna, the polarization of its fields will appear to rotate. The rate at which the polarization changes and the direction of the rotation — *right-handed* or *left-handed* — is determined by the construction of the transmitting antenna. Note that the electric and magnetic fields rotate together so the right-angle between them remains fixed. Polarization that does not rotate is called *linear polarization* or *plane polarization*. Horizontal and vertical polarization are examples of linear polarization.

To best receive a circularly polarized wave, the structure of the receiving antenna should match that of the transmitting antenna. It is particularly helpful to use circular polarization in satellite communication, where polarization tends to shift with the orientation of the satellite and the path of its signal through the atmosphere. Circular polarization is usable with linearly polarized antennas at one end of the signal's path. There will be some small loss in this case, however.

IMPEDANCE OF FREE SPACE

Maxwell's equations provide the relationships (direction and ratio) between the magnetic and electric fields associated with an electromagnetic wave. Far from an antenna or other distorting surfaces, the wave is treated as a *plane* wave in which the wavefronts are infinite, flat planes. In a plane wave, the ratio of the two fields' amplitudes remains constant and the two fields are always at right angles. (There are exceptions but they are not discussed here.)

Since the ratio of the two fields is constant in free space, that gives rise to the idea of an intrinsic impedance. If the impedance is zero or infinite, the magnetic and electric fields would have to be infinite or zero and there could be no electromagnetic radiation.

The impedance can be derived from the permittivity (ε_0) and permeability (μ_0) of free space:

$$\varepsilon_0 = \frac{\text{F(farads)}}{\text{meter}} \text{ and } \mu_0 = \frac{\text{H(henries)}}{\text{meter}}$$

Taking the ratio of the two, noting that farads have units of joules/volt^2 and henries have units of joules/ampere^2:

$$\sqrt{\frac{\mu_0}{\varepsilon_0}} = \sqrt{\frac{\text{H}}{\text{F}}} = \sqrt{\frac{\text{J}/\text{I}^2}{\text{J}/\text{V}^2}} = \sqrt{\frac{\text{V}^2}{\text{I}^2}} = \frac{\text{V}}{\text{I}} \approx 377\Omega$$

It is interesting to remember that the speed of light is:

$$c = \frac{1}{\sqrt{\varepsilon_0 \mu_0}}$$

This links the fundamental electrical and magnetic constants to both velocity of electromagnetic waves and an impedance describing how energy is distributed between the electric and magnetic fields.

3.11 References and Bibliography

BOOKS AND ARTICLES

Alexander and Sadiku, *Fundamentals of Electric Circuits* 7th ed (McGraw-Hill, 2021).

Banzhaf, W., WB1ANE, *Understanding Basic Electronics*, 2nd ed (ARRL, 2010).

Horowitz and Hill, *The Art of Electronics* 3rd ed (Cambridge University Press, 2015).

Kaplan, S., *Wiley Electrical and Electronics Dictionary* (Wiley Press, 2004).

Millman and Grabel, *Microelectronics: Digital and Analog Circuits and Systems* (McGraw-Hill, 1988).

Ohira, T. "What in the World is Q?," *IEEE Microwave Magazine*, Jun 2016, pp 42–49.

Orr, W., *Radio Handbook* 23rd ed (Newnes, 1997).

Rautio, J., AJ3K, "The Long Road to Maxwell's Equations," *IEEE Spectrum*, Dec 2014, **www.spectrum.ieee.org/the-long-road-to-maxwells-equations**

Silver, W., NØAX, "Experiment #74 — Resonant Circuits," *QST*, Mar 2009, pp 72-73.

Silver, W., NØAX, "Experiment #178 — Maxwell's Equations — Grad, Div, and Curl," *QST*, Nov 2017, pp 77-78.

Silver, W., NØAX, "Experiment #179 — Maxwell's Equations — The Wave Emerges," *QST*, Dec 2017, pp 59-60.

Smith, GS, "Analysis of Hertz's Experimentum Crucis on Electromagnetic Waves," IEEE Antennas and Propagation Magazine, Oct 2016, pp. 96-108.

Terman, F., *Radio Engineers' Handbook* (McGraw-Hill, 1943).

Zavrel, R., W7SX, *Antenna Physics: An Introduction*, 2nd ed (ARRL, 2020).

WEBSITES

Oliver Heaviside — **en.wikipedia.org/wiki/Oliver_Heaviside**

History of Maxwell's Equations — **en.wikipedia.org/wiki/History_of_Maxwell's_equation**

Rohde & Schwarz, "dB or not dB? Everything you ever wanted to know about decibels but were afraid to ask…," Application Note 1MA98, **scdn.rohde-schwarz.com/ur/pws/dl_downloads/dl_application/application_notes/1ma98/1MA98_13e_dB_or_not_dB.pdf**

Contents

Chapter 4

Circuits and Components

This chapter begins by covering the various aspects of dealing with real components — resistors, capacitors, inductors, and transformers. (A sheet of standard schematic symbols is available in the online content.) All of these have special features and behaviors you must take into account, especially at RF. We then cover common semiconductors you'll encounter in radio and their important characteristics.

The discussion then proceeds to building-block circuits, beginning with various types of amplifiers constructed with bipolar and field-effect transistors. Building-block circuits including op amps and miscellaneous analog ICs are also covered. Finally, a section on Heat Management discusses how to deal with heat in electronic devices.

Chapter 4 — Online Content

Articles

- Common Schematic Symbols
- Copper Wire Tables
- Digital Electronics Tutorial
- Hands On Radio: Basic Operational Amplifiers by Ward Silver, NØAX
- Hands On Radio: Field Effect Transistors by Ward Silver, NØAX
- Hands On Radio: Load Lines by Ward Silver, NØAX
- Hands On Radio: The Common Base Amplifier by Ward Silver, NØAX
- Hands On Radio: The Common Emitter Amplifier by Ward Silver, NØAX
- Hands On Radio: The Emitter-Follower Amplifier by Ward Silver, NØAX
- Large Signal Transistor Operation

Tools and Data

- *LTSpice* Simulation Files
- Frequency Response Spreadsheet

4.1 EIA and Industry Standards

The American National Standards Institute (ANSI), the Electronic Industries Alliance (EIA), and the Electronic Components Association (ECA) establish the US standards for most electronic components, connectors, wire and cables. These standards establish component sizes, wattages, "standard values," tolerances and other performance characteristics. A branch of the EIA sets the standards for Mil-spec (standard military specification) and special electronic components used by defense and government agencies. The Joint Electron Devices Engineering Council (JEDEC), another branch of the EIA, develops the standards for the semiconductor industry. The EIA cooperates with other standards agencies such as the International Electrotechnical Commission (IEC), a worldwide standards agency. You can often find published EIA standards in the engineering library of a college or university.

And finally, the International Organization of Standardization (ISO), headquartered in Geneva, Switzerland, sets the global standards for nearly everything from paper sizes to photographic film speeds. ANSI is the US representative to the ISO.

These organizations, or their acronyms, are familiar to most of us. They are much more than a label on a component. EIA and other industry standards are what mark components for identification, establish the "preferred standard values" and ensure their reliable performance from one unit to the next, regardless of their source. Standards require that a 1.2 kΩ 5% resistor from Ohmite Corp. has the same performance as a 1.2 kΩ 5% resistor from Vishay, or a 2N3904 has the same performance characteristics and physical packaging whether from ON Semi or NTE.

By selecting components manufactured under these industry standards, building a project from the *Handbook* or other source will ensure nearly identical performance to the original design.

There are many sources you can consult for detailed component data but the best source of component information and data sheets is the internet. Most manufacturers maintain extensive websites with information and data on their products. Often, the quickest route to detailed product information is to search for "data sheet" and the part number. Distributors such as Digi-Key and Mouser include links to useful information in their online catalogs as well. Some manufacturers publish parts catalogs which also good sources of component data and application notes and bulletins.

The ARRL Technical Information Service on the ARRL website (**www.arrl.org/technical-information-service**) provides technical assistance to members and nonmembers, including information about components and useful references. The TIS includes links to detailed, commonly needed information in many technical areas. Questions may also be submitted via email (**tis@arrl.org**); fax (860-594-0259); or mail (TIS, ARRL, 225 Main St, Newington, CT 06111).

4.1.1 Component Packaging

Through-hole or *leaded* components have wire leads intended to be inserted into holes in printed-circuit boards or used in point-to-point wiring. Passive leaded components come in two styles: *axial* and *radial*. Axial leads are aligned in opposite directions along a common axis, usually the largest dimension of the component. Radial leads leave the component body in the same direction and are usually arranged radially about the center of the component body. Many types of leaded passive components (R, L, and C) are available in both styles. As an example, see the Capacitors section for drawings of typical axial and radial lead packages for electrolytic capacitors.

"SMT" is used throughout this book to refer to components, printed-circuit boards, or assembly techniques that involve surface-mount technology. SMT components are often referred to by the abbreviations "SMD" and "SMC," but all three abbreviations are considered to be effectively equivalent.

Many different types of electronic components, both active and passive, are now available in surface-mount packages. Each package is identified by a code, such as 1802 or SOT. Resistors in SMT packages are referred to by package code and not by power dissipation, as through-hole resistors are. SMT package outlines and PCB pad placement for a variety of standard package sizes are provided in the **Construction Techniques** chapter's section on PCB Layout.

The very small size of these components leaves little space for marking with conventional codes, so brief alphanumeric codes are used to convey the most information in the smallest possible space. You will need a magnifying glass to read the markings on the bodies of SMT components.

In many cases, vendors will deliver SMT components packaged in tape from master reels and the components will not be marked. This is often the case with SMT resistors and small capacitors. However, the tape will be marked or the components are delivered in a plastic bag with a label. Take care to keep the components separated and labeled or you'll have to measure their values one by one!

4.2 Practical Resistors

4.2.1 Resistance of Wires

The problem of determining the resistance of a round wire of given diameter and length — or the converse, finding a suitable size and length of wire to provide a desired amount of resistance — can easily be solved with the help of the copper wire tables in the online content which give the resistance, in ohms per 1,000 feet, of each standard wire size. For example, suppose you need a resistance of 3.5 Ω, and some #28 AWG wire is on hand. The wire table shows that #28 AWG wire has a resistance of 63.31 Ω / 1,000 ft. Since the desired resistance is 3.5 Ω, the required length of wire is:

$$\text{Length} = \frac{R_{DESIRED}}{\frac{R_{WIRE}}{1000\ \text{ft}}} = \frac{3.5\ \Omega}{\frac{63.31\ \Omega}{1000\ \text{ft}}}$$

$$= \frac{3.5\ \Omega \times 1000\ \text{ft}}{63.31\ \Omega} = 53.6\ \text{ft}$$

As another example, suppose that the resistance of wire in a radio's power cable must not exceed 0.05 Ω and that the length of wire required for making the connections totals 14 ft. Then:

$$\frac{R_{WIRE}}{1000\ \text{ft}} < \frac{R_{MAXIMUM}}{\text{Length}} = \frac{0.05\ \Omega}{14.0\ \text{ft}}$$

$$= 3.57 \times 10^{-3} \frac{\Omega}{\text{ft}} \times \frac{1000\ \text{ft}}{1000\ \text{ft}}$$

$$\frac{R_{WIRE}}{1000\ \text{ft}} < \frac{3.57\ \Omega}{1000\ \text{ft}}$$

Find the value of R_{WIRE} / 1,000 ft that is less than the calculated value. The wire table shows that #15 AWG is the smallest size having a resistance less than this value. (The resistance of #15 AWG wire is given as 3.1810 Ω / 1,000 feet.) Select any wire size larger than this for the connections in your circuit to ensure that the total wire resistance will be less than 0.05 Ω.

When the wire in question is not made of copper, the resistance values in the wire table should be multiplied by the ratios shown in Table 2.1 (Relative Resistivity of Metals) to obtain the resulting resistance. If the wire in the first example were made from nickel instead of copper, the length required for 3.5 Ω would be:

$$\text{Length} = \frac{R_{DESIRED}}{\frac{R_{WIRE}}{1000\ \text{ft}}}$$

$$= \frac{3.5\ \Omega}{\frac{66.17\ \Omega}{1000\ \text{ft}} \times 5.1} = \frac{3.5\ \Omega \times 1000\ \text{ft}}{66.17\ \Omega \times 5.1}$$

$$\text{Length} = \frac{3500\ \text{ft}}{337.5} = 10.5\ \text{ft}$$

Component Tolerance and Temperature Coefficient

Electronic components such as resistors, capacitors, and inductors are manufactured having a *nominal* value — the value with which they are labeled. The component's *actual* value is what is measured with a suitable measuring instrument. The actual value of resistance varies from the nominal value because of random variations in the manufacturing process. If the nominal value is given as text characters, an "R" in the value (for example "4R7") stands for radix and is read as a decimal point, thus "4.7."

The maximum allowable amount of variation is the component's *tolerance*, and it is expressed in percent. For example, a 1,000-Ω resistor with a tolerance of 5% could have any value of resistance between 95% and 105% of 1,000 Ω; 950 to 1,050 Ω. In most circuits, this small variation doesn't have much effect, but it is important to be aware of tolerance and choose the correct value (10%, 5%, 1%, or even tighter tolerance values are available for *precision components*) of tolerance for the circuit to operate properly, no matter what the actual value of resistance. All components have this same nominal-to-actual value relationship.

The *temperature coefficient* or *tempco* of a component describes its change in value with temperature. Tempco may be expressed as a change in unit value per degree (ohms per degree Celsius) or as a relative change per degree (parts per million per degree). Except for temperature sensing components that may use Fahrenheit or Kelvin, Celsius is almost always used for the temperature scale. Temperature coefficients may not be linear, such as those for capacitors, thermistors, or quartz crystals. In such cases, tempco is specified by an identifier such as Z5U or C0G and an equation or graph of the change with temperature is provided by the manufacturer.

TEMPERATURE EFFECTS ON RESISTORS

Current through a resistance causes the conductor to become heated; the higher the resistance and the larger the current, the greater the amount of heat developed. The power to be dissipated is $P = I^2R$ or $P = V^2/R$ using RMS values for the current through the resistor, I, or voltage across the resistor, V.

Resistors intended for carrying large currents must be physically large so the heat can be radiated quickly to the surrounding air or some type of heat sinking material. If the resistor does not dissipate the heat quickly, it may get hot enough to melt or burn.

The amount of heat a resistor can safely dissipate depends on the material, surface area and design. Typical resistors used in amateur electronics (⅛ to 2-W resistors) dissipate heat primarily through the surface area of the case, with some heat also being carried away through the connecting leads. Wirewound, metal oxide, and thick-film resistors are usually used for higher power levels. Some have finned cases for better convection cooling and/or special metal cases for better conductive cooling.

4.2.2 Component Resistors

TYPES OF RESISTORS

The size and construction of resistors having the same value of resistance in ohms may vary considerably based on how much power they are intended to dissipate, how much voltage is expected to be applied to them, and so forth (see **Figure 4.1** and **Table 4.1**).

Resistors are made in several different ways: carbon composition, metal oxide, carbon film, metal film, and wirewound. In some circuits, the resistor value may be critical. In this case, precision resistors are used. These are typically wirewound or carbon-film devices whose values are carefully controlled during manufacture. In addition, special material or construction techniques may be used to provide temperature compensation, so the value does not change (or changes in a precise manner) as the resistor temperature changes.

Carbon composition resistors are simply small cylinders of carbon mixed with various binding agents to produce any desired resistance. The most common sizes of "carbon comp" resistors are ½ and ¼ W resistors. They are moderately stable from 0 to 60 °C (their resistance increases above and below this temperature range). They can absorb short overloads better than film-type resistors, but they are relatively noisy and have relatively wide tolerances. Because carbon composition resistors tend to be affected by humidity and other environmental factors and because they are difficult to manufacture in surface-mount packages, they have largely been replaced by film-type resistors.

Metal-film resistors are made by depositing a thin film of aluminum, tungsten, or other metal on an insulating substrate. Their resistances are controlled by careful adjustments of the width, length and depth of the film. As a result, they have very tight tolerances. They are used extensively in surface-mount technology. As might be expected, their power handling capability is somewhat limited. They also produce very little electrical noise. For new designs metal-film and carbon-film (see next entry) resistors should be used for their improved characteristics and lower cost compared to the older carbon composition resistors. Metal-film resistors have lower residual inductance and are often preferred at VHF.

Carbon-film resistors use a film of carbon mixed with other materials instead of metal. They are not quite as stable as other film resistors and have wider tolerances than metal-film resistors, but they are still as good as (or better than) carbon composition resistors.

Metal-oxide resistors are similar to carbon composition resistors in that the resistance is supplied by a cylinder of metal oxide. Metal-oxide resistors have replaced carbon composition resistors in higher power applications because they are more stable and can operate at higher temperatures. These resistors are available in leaded styles and in transistor-style TO-220 packages for use at high power and RF.

Wirewound resistors are made from wire that is wound on a coil form (usually ceramic). They are capable of handling high power, their values are very stable, and they are manufactured to close tolerances. The wound-wire construction creates inductance so these

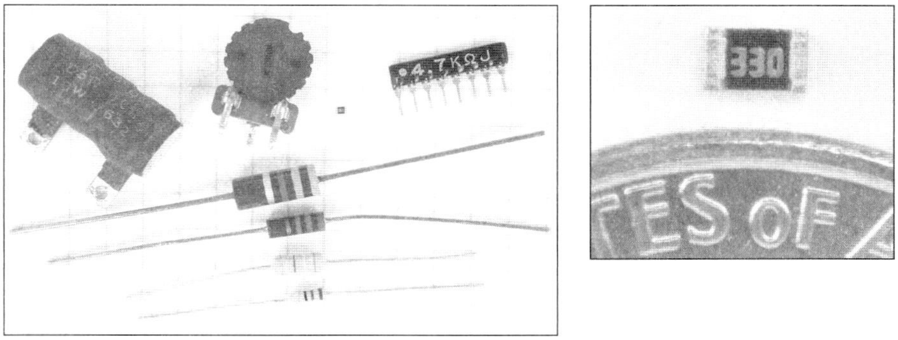

Figure 4.1 — Examples of various resistor types. At the top left is a small 10-W wirewound resistor. A single in-line package (SIP) of resistors is at the top right. At the top center is a small PC-board-mount variable resistor. A tiny surface-mount (chip) resistor is also shown at the top. Below the variable resistor is a 1-W carbon composition resistor and then a ½-W composition unit. The dog-bone-shaped resistors at the bottom are ½-W and ¼-W film resistors. The ¼-inch-ruled graph paper background provides a size comparison. The photo on the right shows the chip resistor with a penny for size comparison. The drawing in Table 4.1 shows several different types of wirewound power resistors.

Table 4.1
Power Resistor Types and Wattage Ranges

HBK0463

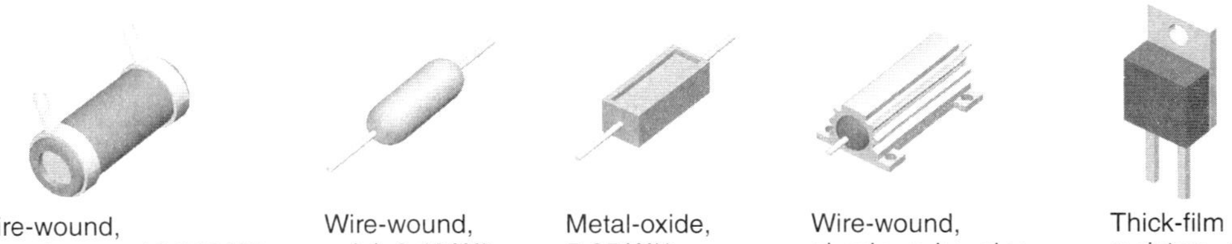

Wire-wound, ceramic core, 10-300 W*	Wire-wound, axial, 3-10 W*	Metal-oxide, 5-25 W**	Wire-wound, aluminum housing, 3-50 W*	Thick-film resistors, 15-100 W**

*Wire-wound resistors are inductive, though seldom noted as such on the data sheets, and are not recommended for RF.

**Thick-film and metal-oxide power resistors are low inductance or noninductive.

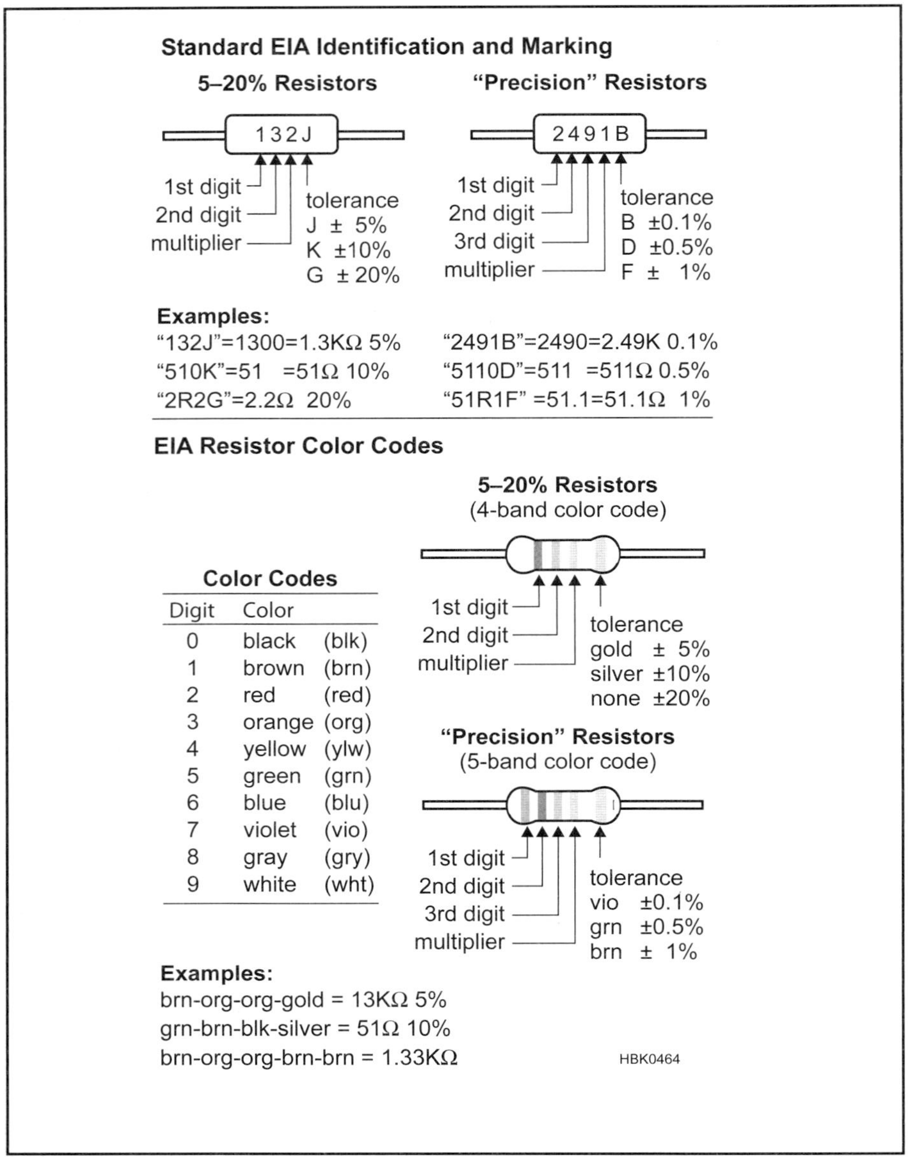

Figure 4.2 — Resistor value identification.

resistors are not suitable for ac circuits above a few kHz. They have power ratings from 3 to 300 W.

Thick-film resistors are also available at high power ratings (up to 100 W) in a transistor-like TO-220 package. Because of their packaging, they can be mounted to heat sinks and printed-circuit boards. These resistors also have low inductance for use in switch-mode circuits and high-frequency applications.

RESISTOR IDENTIFICATION

Resistors are identified by the IEC 60062:2016 standard as shown in **Figure 4.2**. The IEC numerical code is used worldwide in industry. The nominal resistance, expressed in ohms, is identified by three digits for 2% (and greater) tolerance devices. The first two digits represent the significant figures; the last digit specifies the multiplier as the exponent of 10. (The multiplier is simply the number of zeros following the significant numerals.) For values less than 100 Ω, the letter R is substituted for one of the significant digits and represents a decimal point. An alphabetic character indicates the tolerance as follows:

D	±0.5%
F	±1.0%
G	±2.0%
J	±5.0%

For example, a resistor marked with "122J" would be a 1,200 Ω, or a 1.2 kΩ 5% resistor. If the tolerance of the unit is narrower than ±2%, the code used is a four-digit code where the first three digits are the significant figures and the last is the multiplier. A resistor containing four digits, such as "1211F," would be a 1,210 Ω, or a 1.21 kΩ 1% precision resistor.

The letter R is used in the same way to represent a decimal point. For example, 22R0 indicates a 22-Ω unit.

Here are some additional examples of resistor value markings:

Code	*Value*
101	10 and 1 zero = 100 Ω
224	22 and 4 zeros = 220,000 Ω
1R0	1 point zero = 1 Ω
22R	22.0 and no zeros = 22 Ω
R10	0 point 1 zero = 0.1 Ω

The resistor color code, used only with through-hole components, assigns colors to the numerals one through nine and zero, as shown in **Table 4.2**, to represent the significant numerals, the multiplier and the tolerance. The color code is often memorized with a mnemonic such as "Big boys race our young girls, but Violet generally wins" to represent the colors black (0), brown (1), red (2), orange (3), yellow (4), green (5), blue (6), violet (7), gray (8) and white (9). You will no doubt discover other versions of this memory aid made popular over the years.

For example, a resistor with color bands black (1), red (2), red (2) and gold would be a 1,200 Ω, or 1.2 kΩ 5% resistor, with the gold band signifying 5% tolerance.

The resistor color code should be memorized as it is also used for identifying capacitors and inductors. It is also handy to use when connecting multiconductor or ribbon cables.

Resistors are also identified by an "E" series classification, such as E12 or E48. The number following the letter E signifies the number of logarithmic steps per decade. The more steps per decade, the more the choices of resistor values and the tighter the tolerances can be. For example, in the E12 series, there are twelve resistor values between 1 kΩ and 10 kΩ with 10% tolerance; E48 provides 48 values between 1 kΩ and 10 kΩ at 1% tolerance. This system is often used with online circuit calculators to indicate the resistor accuracy and tolerance desired. The standard resistor values of the E12 (±10%), E24 (±5%), E48 (±2%) and E96 (±1%) series are listed in **Table 4.3**.

Resistors used in military electronics (Mil-spec) use the type identifiers listed in **Table 4.4**. In addition, Mil-spec resistors with paint-stripe value bands have an extra band indicating the reliability level to which they are certified.

Table 4.2
Resistor Color Codes

Color	*Significant Figure*	*Decimal Multiplier*	*Tolerance (%)*
Black	0	1	
Brown	1	10	1
Red	2	100	2
Orange	3	1,000	
Yellow	4	10,000	
Green	5	100,000	0.5
Blue	6	1,000,000	0.25
Violet	7	10,000,000	0.1
Gray	8	100,000,000	0.05
White	9	1,000,000,000	
Gold		0.1	5
Silver		0.01	10
No color			20

Table 4.3
EIA Standard Resistor Values

±10% (E12)	*±5% (E24)*	*±2% (E48)*		*±1% (E96)*			
100	100	100	316	100	178	316	562
120	110	105	332	102	182	323	576
150	120	110	348	105	187	332	590
180	130	115	365	107	191	340	604
220	150	121	383	110	196	348	619
270	160	127	402	113	200	357	634
330	180	133	422	115	205	365	649
390	200	140	442	118	210	374	665
470	220	147	464	121	215	383	681
560	240	154	487	124	221	392	698
680	270	162	511	127	226	402	715
820	300	169	536	130	232	412	732
	330	178	562	133	237	422	750
	360	187	590	137	243	432	768
	390	196	619	140	249	442	787
	430	205	649	143	255	453	806
	470	215	681	147	261	464	825
	510	226	715	150	267	475	845
	560	237	750	154	274	487	866
	620	249	787	158	280	499	887
	680	261	825	162	287	511	909
	750	274	866	165	294	523	931
	820	287	909	169	301	536	953
	910	301	953	174	309	549	976

Use Table 4.3 values for each decade.
Example: 133 = 13.3 Ω, 133 Ω, 1.33 kΩ, 13.3 Ω, 133 kΩ, 1.33 MΩ

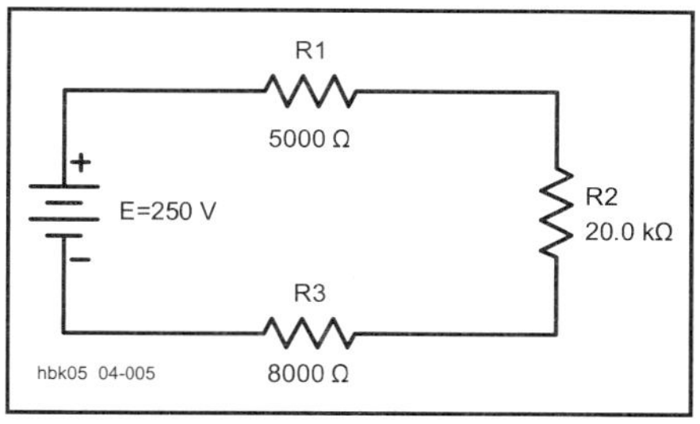

Figure 4.3 — An example of resistors in series.

Surface-mount (SMT) resistors are labeled with an alphanumeric code. There are several identification conventions, including the three-digit and four-digit value-and-exponent and an EIA-96 labeling standard described at **www.hobby-hour.com/electronics/smdcalc.php**.

4.2.3 Effect of Temperature on Resistors

The major departure of resistors from ideal behavior at low frequencies is their *temperature coefficient* (TC or TCR for resistors). (See the chapter on **RF Techniques** for a discussion of the behavior of resistors at high frequencies.) The resistivity of most materials changes with temperature, and typical TC values for resistor materials are given in **Table 4.5**. TC values are usually expressed in parts-per-million (ppm) for each degree (centigrade) change from some nominal temperature, usually room temperature (77 °F or 27 °C). A positive TC indicates an increase in resistance with increasing temperature, while a negative TC indicates a decreasing resistance. For example, if a 1,000-Ω resistor with a TC of +300 ppm/°C is heated to 50 °C,

Table 4.4
Mil-Spec Resistors

Wattage	*Metal Film Types*	*Fixed Film Types*	*Composition Types*	
1/10 W	RN50			
1/8 W	RN55	RL05	RLR05	RCR05
1/4 W	RN60	RL07	RLR07	RCR07
1/2 W	RN65	RL20	RLR20	RCR20
1 W	RN75	RL32	RLR32	RCR32
2 W	RN80	RL42	RLR62	RCR42

Tolerance Codes

B	±0.1%
C	±0.25%
D	±0.5%
F	±1%
G	±2%
J	±5%
K	±10%

Examples:
RN60D-2202F = 22 kΩ 1%
RL07S-471J = 470 Ω ±5%
RLR07C-471J = 470 Ω ±5%

Note: The RN Mil-Spec was discontinued in 1996. It is still used by some manufacturers such as Vishay-Dale.

Table 4.5

Temperature Coefficients for Various Resistor Compositions

1 PPM = 1 part per million = 0.0001%

Type	*TC (PPM/°C)*
Wire wound	±(30 - 50)
Metal film	±(100 - 200)
Carbon film	+350 to –800
Carbon composition	±800

the *change* in resistance is 300 × (50 – 27) = 6,900 ppm, yielding a new resistance of

$$1000\left(1+\frac{6900}{1000000}\right)=1006.9\ \Omega$$

Carbon-film and carbon-composition resistors are unique among the major resistor families because they alone have a negative temperature coefficient. They are often used to "offset" the thermal effects of the other components.

If the temperature increase is small (less than 30 – 40 °C), the resistance change with temperature is nondestructive — the resistor will return to normal when the temperature returns to its nominal value. Resistors that get too hot to touch, however, may be permanently damaged even if they appear normal. For this reason, be conservative when specifying power ratings for resistors. It's common to specify a resistor rated at 200% to 400% of the expected dissipation.

4.2.4 Voltage Dividers

According to Kirchoff's Voltage Law (KVL), the voltage drop across each resistor in a series circuit is directly proportional to the resistance. When connected in series, a resistor that has a value twice as large as another will have twice the voltage drop across it.

Resistors in series without any other connections form a *resistive voltage divider*. (Other types of components can form voltage dividers, too.) The voltage across any specific resistor in the divider, R_n, is equal to the voltage across the entire string of resistors multiplied by the ratio of R_n to the sum of all resistors in the string.

For example, in the circuit of **Figure 4.3**, the voltage across the 5,000 Ω resistor is:

$$E1=250\frac{5000}{5000+20000+8000}=37.9\text{ V}$$

This is a more convenient method than calculating the current through the resistor and using Ohm's law.

Voltage dividers can be used as a source of voltage. As long as the device connected to the output of the divider has a much higher

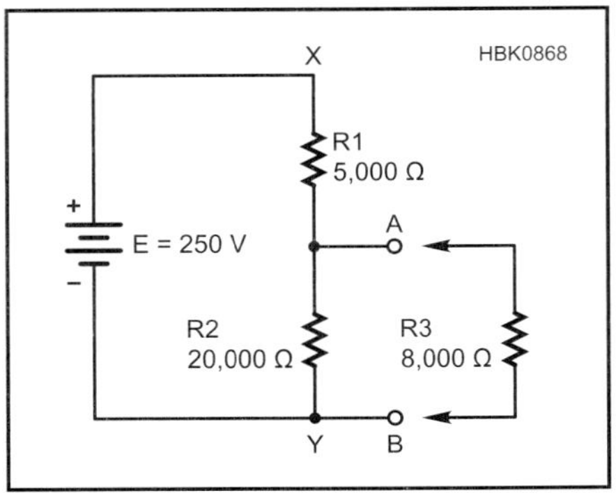

Figure 4.4 — A voltage divider showing both unloaded (R3 not connected) and loaded (R3 connected) conditions.

resistance than the resistors in the divider, there will be little effect on the divider output voltage. For example, for a voltage divider with a voltage of E = 15 V and two resistors of R1 = 5 kΩ and R2 = 10 kΩ, the voltage across R2 will be 10 V measured on a high-impedance voltmeter because the measurement draws very little current from the divider. However, if the measuring device or load across R2 draws significant current, it will increase the amount of current drawn through the divider and change the output voltage.

The following equations show how to calculate the voltage produced by a voltage divider. Using the circuit in **Figure 4.4**, the *unloaded* output voltage (with R3 not connected) is:

$$V_{OUT}=V_{IN}\left(\frac{R2}{R1+R2}\right)$$

If R3 is connected, the *loaded* output voltage is:

$$V_{OUT}=V_{IN}\left(\frac{R2//R3}{R1+R2//R3}\right)$$

where // indicates "in parallel with."

A good rule of thumb to keep the loaded output voltage within about 10% of the unloaded voltage is for the load resistance to be at least 10 times higher than the output resistor of the divider. As the load resistance approaches the value of the output resistor, the additional current through the load causes additional voltage drop across the divider's input resistor.

Potentiometers (variable resistors described in the next section) are often used as adjustable voltage dividers, which is how they got their name. Potential is an older name for voltage and a "potential-meter" is a device that can "meter" or adjust potential, thus potentiometer.

ATTENUATORS

A special type of voltage divider is used to reduce signal power levels while maintaining a constant resistance through the divider. The two primary types of divider circuits are the Pi-network and T-network shown at the bottom of **Table 4.6** and **Table 4.7**. These circuits are symmetric: the input and output resistor values are equal. The values for R1 and R2 in the tables are chosen to present a 50-Ω resistance at both the input and output. See the **Test Equipment and Measurements** chapter for a discussion of switched-step attenuators and how to use them.

4.2.5 Current Dividers

Resistors connected in parallel form a circuit called a *resistive current divider*. For any number of resistors connected in parallel (R1, R2, R3, ... R4), the current through one of the resistors, R_n, is equal to the sum of all resistor currents multiplied by the ratio of the equivalent of all parallel resistors *except* R_n to the sum of R_n and the equivalent value.

$$I_n = I_{TOT}\, R_{EQ}/(R_n + R_{EQ})$$

For example, in a circuit with three parallel resistors R1, R2, and R3, the current through

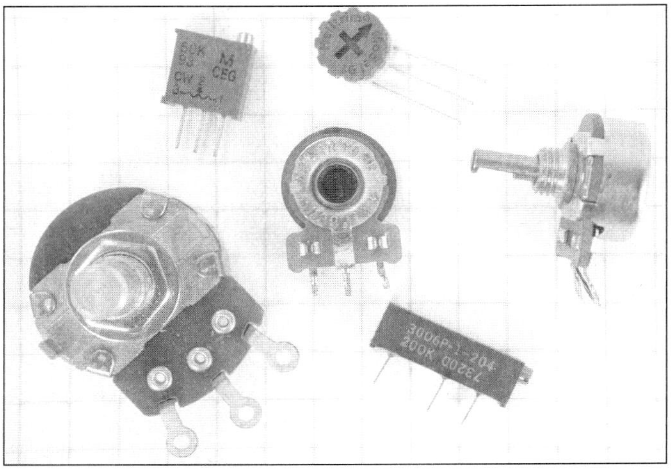

Figure 4.5 — This photo shows examples of different styles of potentiometers. The ¼-inch-ruled graph paper background provides a size comparison.

Table 4.6
Pi-Network Resistive Attenuators (50 Ω)

dB Atten.	*R1 (Ohms)*	*R2 (Ohms)*
1.0	870	5.77
2.0	436	11.6
3.0	292	17.6
4.0	221	23.8
5.0	178	30.4
6.0	150	37.4
7.0	131	44.8
8.0	116	52.8
9.0	105	61.6
10.0	96.2	71.2
11.0	89.2	81.7
12.0	83.5	93.2
13.0	78.8	106
14.0	74.9	120
15.0	71.6	136
16.0	68.8	154
17.0	66.4	173
18.0	64.4	195
19.0	62.6	220
20.0	61.1	248
21.0	59.8	278
22.0	58.6	313
23.0	57.6	352
24.0	56.7	395
25.0	56.0	443
30.0	53.2	790
35.0	51.8	1405
40.0	51.0	2500
45.0	50.5	4446
50.0	50.3	7906
55.0	50.2	14,058
60.0	50.1	25,000

An RF Step Attenuator project is shown in the **Test Equipment and Measurements** chapter of this *Handbook*, and a Low Power Step Attenuator PC board is available from FAR Circuits at **www.farcircuits.net/test2.htm**.

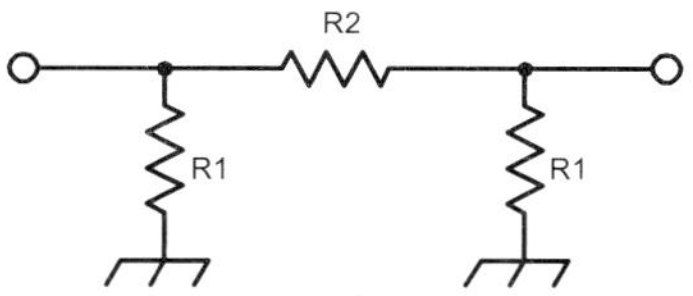

Table 4.7
T-Network Resistive Attenuators (50 Ω)

dB Atten.	*R1 (Ohms)*	*R2 (Ohms)*
1.0	2.88	433
2.0	5.73	215
3.0	8.55	142
4.0	11.3	105
5.0	14.0	82.2
6.0	16.6	66.9
7.0	19.1	55.8
8.0	21.5	47.3
9.0	23.8	40.6
10.0	26.0	35.1
11.0	28.0	30.6
12.0	30.0	26.8
13.0	31.7	23.5
14.0	33.3	20.8
15.0	35.0	18.4
16.0	36.3	16.2
17.0	37.6	14.4
18.0	38.8	12.8
19.0	40.0	11.4
20.0	41.0	10.0
21.0	41.8	9.0
22.0	42.6	8.0
23.0	43.4	7.1
24.0	44.0	6.3
25.0	44.7	5.6
30.0	47.0	3.2
35.0	48.2	1.8
40.0	49.0	1.0
45.0	49.4	0.56
50.0	49.7	0.32
55.0	49.8	0.18
60.0	49.9	0.10

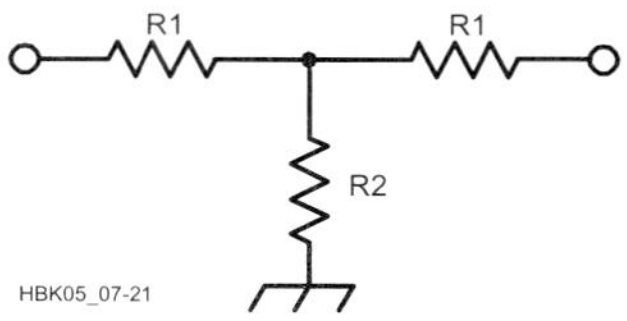

R2 is equal to:

$$I_2 = I\frac{R1 + R3}{R1 + R2 + R3}$$

where I is the total current through all the resistors. If I = 100 mA, R1 = 100 Ω, R2 = 50 Ω, and R3 = 200 Ω:

$$100\text{ mA}\frac{100 + 200}{100 + 50 + 200} = 85.7\text{ mA}$$

4.2.6 Potentiometers

Potentiometer (pronounced po-ten-tchee-AH-meh-tur) is a formal name for a variable resistor and the common name for these components is "pots." A typical potentiometer consists of a circular *element* of resistive material which can be a carbon compound similar to that used in carbon composition resistors or a conductive plastic. One contact is made to each end of the element. A variable-position *wiper* makes contact with the element at different positions. As the wiper moves along the material, more resistance is introduced between the wiper and one of the element's contacts. The wiper is turned by a shaft to move across the material.

For higher power applications, the element may be wire wound around a core, like a wirewound resistor. Like the wirewound resistor, this type of pot is not suitable for high-frequency applications due to inductance.

A potentiometer may be used to control current, voltage, or resistance in a circuit. **Figure 4.5** shows several different types of potentiometers. **Figure 4.6** shows the schematic symbol for a potentiometer and how changing the position of the shaft changes the resistance between its three terminals. The figure shows a *panel pot*, designed to be mounted on an equipment panel and adjusted by an operator. The small rectangular *trimmer* potentiometers in Figure 4.6 are adjusted with a screwdriver and have wire terminals.

Typical specifications for a potentiometer include element resistance, power dissipation, voltage and current ratings, number of turns (or degrees) the shaft can rotate, type and size of shaft, mounting arrangements, and resistance *taper*. Taper describes how the resistance of the element changes with position along it.

Pots with a *linear taper* have equal change in resistance with position along the element. That is, the change in resistance is the same for a given number of degrees of shaft rotation

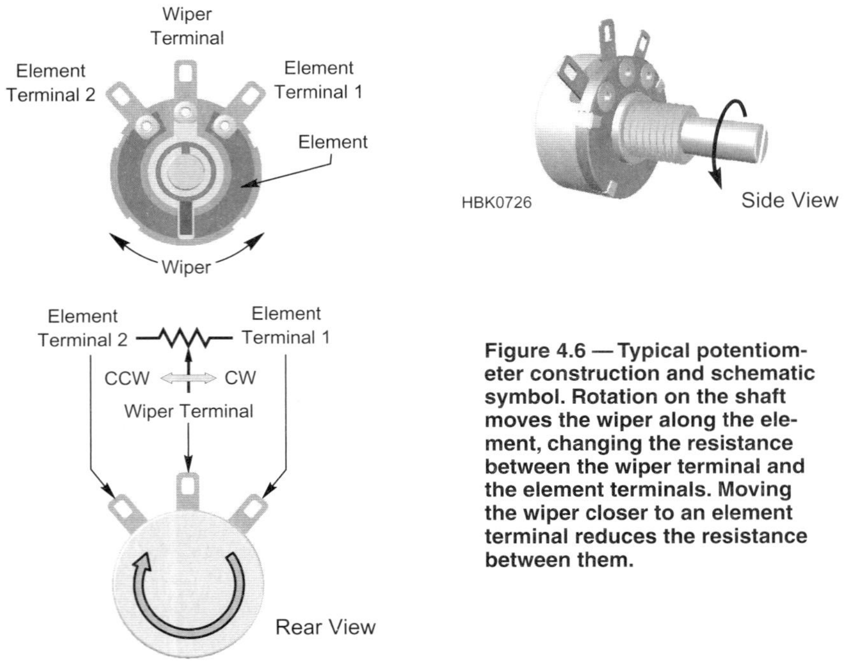

Figure 4.6 — Typical potentiometer construction and schematic symbol. Rotation on the shaft moves the wiper along the element, changing the resistance between the wiper terminal and the element terminals. Moving the wiper closer to an element terminal reduces the resistance between them.

anywhere along different portions of the resistive material.

Some pots have *non-linear tapers*. A typical use for a nonlinear taper is as a volume control in an audio amplifier. Since the human ear has a logarithmic response to sound, a volume control must change the amplifier output much more near one end of the element than the other (for a given amount of rotation) so that the "perceived" change in volume is about the same for a similar change in the control's position. This is commonly called an *audio taper* or *log taper* as the change in resistance per degree of rotation attempts to match the response of the human ear. Tapers can be designed to match almost any desired control function for a given application. Linear and audio tapers are the most common tapers.

4.3 Practical Capacitors

Like resistors, capacitors are available in EIA standard series of values, E6 and E12, shown in **Table 4.8**. Most capacitors have a tolerance of 5% or greater. High-value capacitors used for filtering may have asymmetric tolerances, such as –5% and +10%, since the primary concern is for a guaranteed minimum value of capacitance.

SMT capacitors are generally film, ceramic, or tantalum electrolytics. Although the IEC scheme is the standard method of labeling capacitor value, you may encounter a two-character alphanumeric code (see **Table 4.9**) consisting of a letter indicating the significant digits and a number indicating the multiplier. The code represents the capacitance in picofarads. For example, a chip capacitor marked "A4" would have a capacitance of 10,000 pF, or 0.01 µF. A unit marked "N1" would be a 33-pF capacitor. If there is sufficient space on the device package, a tolerance code may be included.

Table 4.8
EIA Standard Capacitor Values

±20% Capacitors (E6)

pF	*pF*	*pF*	*µF*	*µF*	*µF*	*µF*
1.0	10	100	0.001	0.01	0.1	1
1.5	15	150	0.0015	0.015	0.15	1.5
2.2	22	220	0.0022	0.022	0.22	2.2
3.3	33	330	0.0033	0.033	0.33	3.3
4.7	47	470	0.0047	0.047	0.47	4.7
6.8	68	680	0.0068	0.068	0.68	6.8

±10%, ±5% Capacitors (E12)

pF	*pF*	*pF*	*µF*	*µF*	*µF*	*µF*
1.0	10	100	0.001	0.01	0.1	1
1.2	12	120	0.0012	0.012	0.12	
1.5	15	150	0.0015	0.015	0.15	
1.8	18	180	0.0018	0.018	0.18	
2.2	22	220	0.0022	0.022	0.22	2.2
2.7	27	270	0.0027	0.027	0.27	
3.3	33	330	0.0033	0.033	0.33	3.3
3.9	39	390	0.0039	0.039	0.39	
4.7	47	470	0.0047	0.047	0.47	4.7
5.6	56	560	0.0056	0.056	0.56	
6.8	68	680	0.0068	0.068	0.68	
8.2	82	820	0.0082	0.082	0.82	

4.3.1 Capacitor Types and Characteristics

Capacitors exhibit the largest variety of electronic components. So many varieties and types are available that selecting the proper capacitor for a particular application can be overwhelming. Capacitors are generally grouped by the type of dielectric; ceramic, mica, film, polymer, and electrolytic are the most common. Some of the different types are shown in **Figure 4.7** and many other types and packages are available. (See the **RF Techniques** chapter for information about capacitor characteristics at RF.) The type of capacitor can often be determined by its appearance, as shown in **Figures 4.8** and **4.9**.

The type of dielectric determines many properties of the capacitor, although the construction of the electrodes or plates strongly affects the capacitor's ac performance and

Table 4.9

SMT Capacitor Two-Character Labeling

Significant Figure Codes

Character	*Significant Figures*	*Character*	*Significant Figures*
A	1.0	T	5.1
B	1.1	U	5.6
C	1.2	V	6.2
D	1.3	W	6.8
E	1.5	X	7.5
F	1.6	Y	8.2
G	1.8	Z	9.1
H	2.0	a	2.5
J	2.2	b	3.5
K	2.4	d	4.0
L	2.7	e	4.5
M	3.0	f	5.0
N	3.3	m	6.0
P	3.6	n	7.0
Q	3.9	t	8.0
R	4.3	y	9.0
S	4.7		

Multiplier Codes

Numeric Character	*Decimal Multiplier*
0	1
1	10
2	100
3	1,000
4	10,000
5	100,000
6	1,000,000
7	10,000,000
8	100,000,000
9	0.1

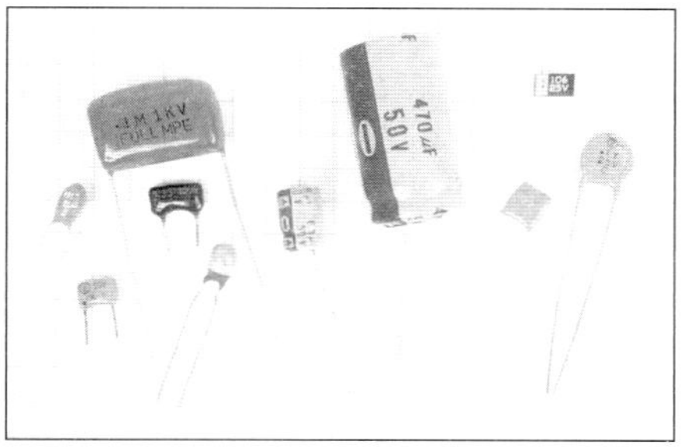

(A)

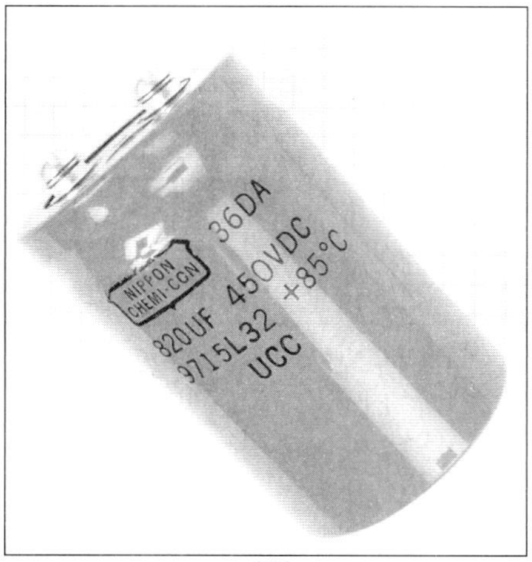

(B)

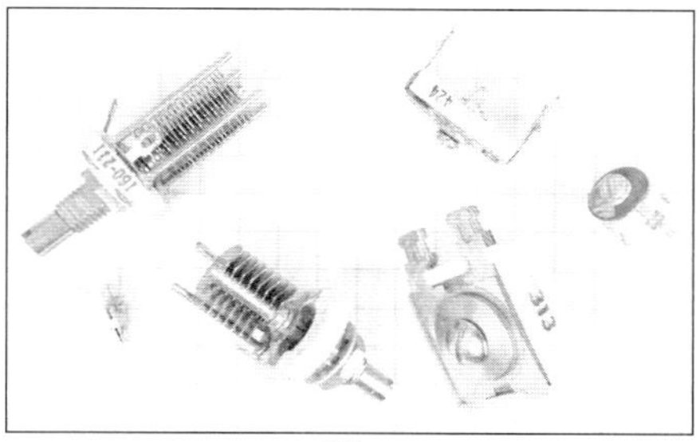

(C)

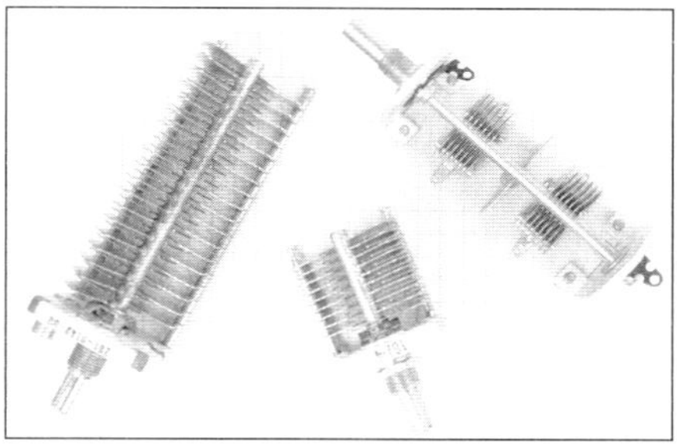

(D)

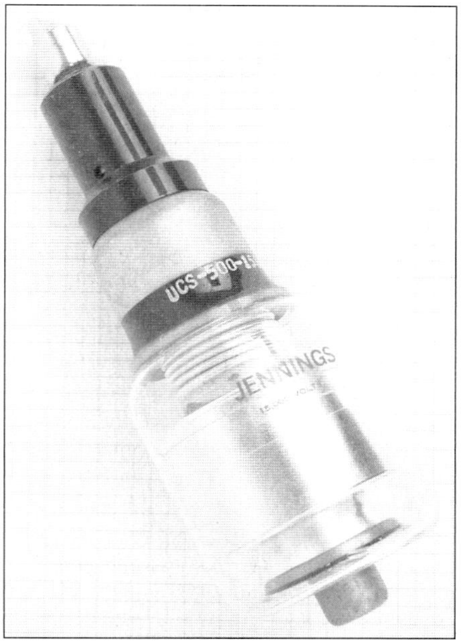

(E)

Figure 4.7 — Fixed-value capacitors are shown in parts A and B. Aluminum electrolytic capacitors are pictured near the center of photo A. The small tear-drop units to the left of center are tantalum electrolytic capacitors. The rectangular units are silvered-mica, polystyrene film and monolithic ceramic. At the right edge is a disc-ceramic capacitor and near the top right corner is a surface-mount capacitor. B shows a large "computer-grade" electrolytic. These have very low equivalent series resistance (ESR) and are often used as filter capacitors in switch-mode power supplies, and in series-strings for high-voltage supplies of RF power amplifiers. Parts C and D show a variety of variable capacitors, including air variable capacitors and mica compression units. Part E shows a vacuum variable capacitor such as is sometimes used in high-power amplifier circuits. The ¼-inch-ruled graph paper backgrounds provide size comparisons.

some dc parameters. Various materials are used for different reasons such as working voltage and current, availability, cost, and desired capacitance range.

Disc* (or *disk*) *ceramic capacitors consist of two metal plates separated by a ceramic dielectric that establishes the desired capacitance. Due to their low cost, they are the most common capacitor type. The main disadvantage is their sensitivity to temperature changes (that is, a high temperature coefficient).

Monolithic ceramic capacitors are made by sandwiching layers of metal electrodes and ceramic layers to form the desired capacitance as shown in **Figure 4.10**. "Monolithics" are physically smaller than disc ceramics for the same value of capacitance and cost, but exhibit the same high temperature coefficients. They are made from a stack of metal plating on ceramic layers that are pressed together and coated with an insulating jacket.

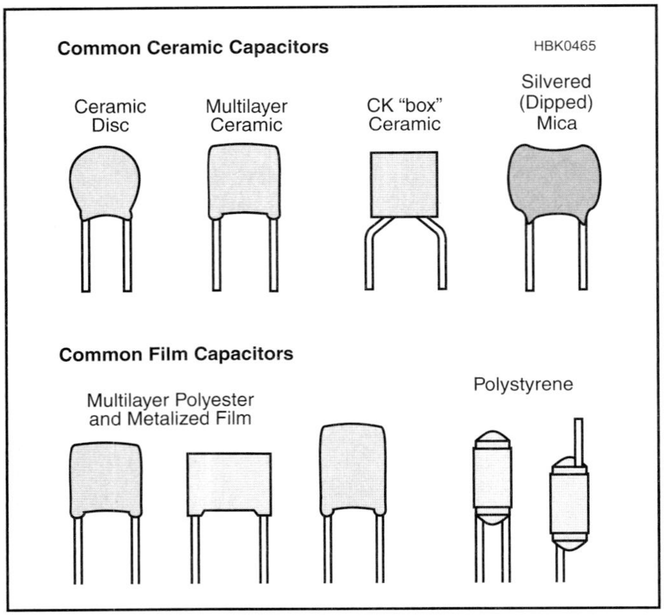

Figure 4.8 — Common capacitor types and package styles.

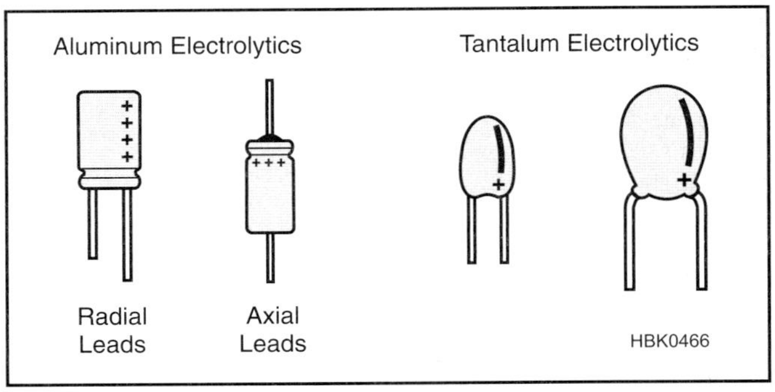

Figure 4.9 — Aluminum and tantalum electrolytic capacitors.

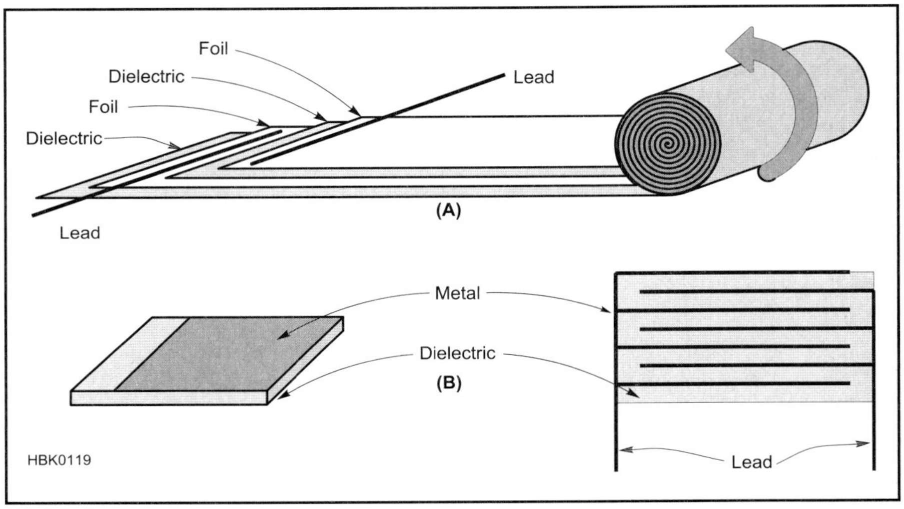

Figure 4.10 — Two common types of capacitor construction. A shows the roll method for film capacitors with axial leads. B shows the alternating layer method for ceramic capacitors with radial leads.

Plastic film capacitors use layers of metal and a plastic such as polystyrene, polypropylene, or polyester (Mylar) to make a wide range of capacitances. Values range from tens of pF to 1 µF. They have high leakage resistances (even at high temperatures) and low TCs. Smaller sizes can be used up to several megahertz. Film capacitors are often made using roll construction as in Figure 4.10. This construction has relatively high inductance and the capacitor may not be suitable for RF applications. Plastic-film variable capacitors are also available.

Most film capacitors are not polarized; however, the body of the capacitor is usually marked with a color band at one end. The band indicates the terminal that is connected to the outermost plate of the capacitor. This terminal should be connected to the side of the circuit at the lower potential as a safety precaution.

Paper capacitors are generally not used in new designs and are largely encountered in older equipment; capacitances from 500 pF to 50 µF are available. High working voltages are possible, but paper-dielectric capacitors have low leakage resistances and tolerances are no better than 10 to 20%. Paper capacitors are not available as new stock (except possibly for specialty restoration applications) and should not be used in new equipment.

Oil-filled capacitors use special high-strength dielectric oils to achieve voltage ratings of several kV. Values of up to 100 µF are commonly used in high-voltage applications such as high-voltage power supplies and energy storage. (See the chapter on **Power Sources** for additional information about the use of oil-filled and electrolytic capacitors.)

Mica capacitors are very stable with respect to time, temperature, and electrical stress. Leakage and losses are very low and they are often used in transmitting equipment. Values range from 1 pF to 0.1 µF. High working voltages are possible, but they must be derated severely as operating frequency increases. Note that many WWII-era mica capacitors are still available on the used and surplus market. Given the age of these components (now approaching 80 years) either use newer components or test them before using them in a high-power or high-voltage circuit.

Silver-mica capacitors are made by depositing a thin layer of silver on the mica dielectric. This makes the value even more stable, but it presents the possibility of silver migration through the dielectric. The migration problem worsens with increased dc voltage, temperature and humidity. Avoid using silver-mica capacitors under such conditions. Silver-mica capacitors are often used in RF circuits requiring stable capacitor values, such as oscillators and filters.

Transmitting ceramic capacitors are made, like transmitting air-variables, with heavy electrodes and high voltage ratings. They are relatively large (often called "doorknobs"), but very stable and have nearly as low losses as mica capacitors at HF.

Vacuum capacitors, both fixed and variable, are available. They are rated by their maximum working voltages (3 to 60 kV) and currents. Losses are specified as negligible for most applications. The high working voltage and low losses make vacuum capacitors widely used in transmitting applications. Vacuum capacitors are also unaffected by humidity, moisture, contamination, or dust, unlike air-dielectric capacitors discussed next. This allows them to be used in environments for which air-dielectric capacitors would be unsuitable.

Air variable capacitors. Since K ≈ 1 for air, air-dielectric capacitors are large when compared to those of the same value using other dielectrics. Their capacitance is very stable over a wide temperature range, leakage losses are low, and therefore a high Q can be obtained. They also can withstand high voltages. Values range from a few tens to hundreds of pF.

Air-variable capacitors have one set of plates that is movable with respect to the other set to vary the area of overlap and thus the capacitance. A *transmitting-variable* capacitor has heavy plates far enough apart to withstand the high voltages and currents encountered in a transmitter. (Air variable capacitors with more closely spaced plates are often referred to as *receiving-variables.*)

Trimming capacitors. Small-value variable capacitors are often referred to as *trimmers* because they are used for fine-tuning or frequency adjustments, called *trimming*. Trimmers have dielectrics of Teflon, plastic film, air, or ceramic and generally have values of less than 100 pF. *Compression trimmers* constructed with mica dielectrics have higher values of up to 1,000 pF and are no longer manufactured but widely available as surplus or NOS (new old stock).

TEMPERATURE COEFFICIENT

Capacitors are particularly sensitive to temperature changes because the physical dimensions of the capacitor determine its value. Standard temperature coefficient codes are shown in **Table 4.10**. Each code is made up of one character from each column in the table. For example, a capacitor marked Z5U is suitable for use between +10 and +85 °C, with a maximum change in capacitance of –56% or +22%.

Capacitors with highly predictable temperature coefficients of capacitance are sometimes used in circuits whose performance must remain stable with temperature. If an application called for a temperature coefficient of –750 ppm/°C (N750), a capacitor marked U2J would be suitable. The older industry code for these ratings is being replaced with the EIA code shown in Table 4.10. NP0 (that is, N-P-zero) means "negative, positive, zero." It is a characteristic often specified for RF circuits requiring temperature stability, such as VFOs. A capacitor of the proper value marked C0G is a suitable replacement for an NP0 unit.

Table 4.10
Ceramic Temperature Characteristics

Common EIA Types:

EIA Class	*EIA Code*	*Characteristics*	*Temp. Range**
1	C0G	0 ± 30 ppm/°C	–55 °C to +125 °C
2	Y5P	±10%	–30 °C to + 85 °C
2	X7R	±15%	–55 °C to +125 °C
2	Y5U	±20%	–10 °C to + 85 °C
2	Z5U	±20%	+10 °C to + 85 °C
2	Z5V	+80%, –20%	–30 °C to + 85 °C
3	Y5V	+80%, –20%	–10 °C to + 85 °C

Common Industry Types:

EIA Class	*EIA Code*	*Characteristics*	*Temp. Range**
1	NP0	0 ± 30 ppm/°C	–55 °C to +125 °C
2	CK05	±10%	–55 °C to +125 °C

*Temp. range for which characteristics are specified and may vary slightly between different manufacturers

Temperature Coefficient Codes

Minimum Temperature	*Maximum Temperature*	*Maximum capacitance change over temp range*
X –55 °C	2 +45°C	A ±1.0%
Y –30 °C	4 +65°C	B ±1.5%
Z +10 °C	5 +85°C	C ±2.2%
6 +105 °C		D ±3.3%
7 +125 °C		E ±4.7%
		F ±7.5%
		P ±10%
		R ±15%
		S ±22%
		T –33%, +22%
		U –56%, +22%
		V –82%, +22%

LEAKAGE RESISTANCE

If we use anything other than a vacuum for the insulating layer, even air, two imperfections are created. Because there are atoms between the plates, some electrons will be available to create a current between the plates when a dc voltage is applied. The magnitude of this *leakage current* will depend on the insulator quality, and the current is usually very small. Leakage current can be modeled by a resistance R_L in parallel with the capacitance (in an ideal capacitor, R_L is infinite). **Table 4.11** shows typical dc leakage resis-

Table 4.11
Typical Temperature Coefficients and Leakage Resistances for Various Capacitor Constructions

Type	*TC @ 20°C (PPM/°C)*	*DC Leakage Resistance (Ω)*
Ceramic Disc	±300(NP0)	> 10 M
	+150/–1500(GP)	> 10 M
Mica	–20 to +100	> 100,000 M
Polyester	±500	> 10 M
Tantalum Electrolytic	±1500	> 10 MΩ
Small Alum Electrolytic(≈ 100 µF)	–20,000	500 k - 1 M
Large Alum Electrolytic(≈ 10 mF)	–100,000	10 k
Vacuum (glass)	+100	≈ ∞
Vacuum (ceramic)	+50	≈ ∞

tances for different dielectric materials. Leakage also generally increases with increasing temperature.

CAPACITOR LOSSES

When an ac current flows through the capacitor (even at low frequencies), capacitors dissipate some of the energy stored in the dielectric due to the electromagnetic properties of dielectric materials. This loss can be thought of as a resistance in series with the capacitor and it is often specified in the manufacturer's data for the capacitor as *effective (or equivalent) series resistance (ESR)* which is specified in ohms. (Capacitors that are constructed using roll construction as in Figure 4.10 also have significant inductance specified as *equivalent series inductance (ESL)* which is specified in micro- or nano-henries.)

Loss can also be specified as the capacitor's *loss angle,* θ. (Some literature uses δ for loss angle.) Loss angle is the angle between X_C (the reactance of the capacitor without any loss) and the impedance of the capacitor made up of the combination of ESR and X_C. Increasing loss increases loss angle. The loss angle is usually quite small and is zero for an ideal capacitor.

Dissipation Factor (DF) or *loss tangent* is the ratio of loss resistance to reactance.

$$DF = \tan \theta = ESR / X_C$$

The loss angle of a given capacitor is relatively constant over frequency, meaning that $ESR = (\tan \theta) / 2\pi fC$ goes down as frequency goes up. For this reason, ESR must be specified at a given frequency.

4.3.2 Electrolytic Capacitors

Aluminum electrolytic capacitors use aluminum foil "wetted" with a chemical agent and formed into layers to increase the effective area, and therefore the capacitance. Aluminum electrolytics provide high capacitance in small packages at low cost. Most varieties are polarized, that is, voltage should only be applied in one "direction." Polarized capacitors have a negative (–) and positive (+) lead. EIA standard values for aluminum electrolytics are given in **Table 4.12**.

Table 4.12
Aluminum Electrolytic Capacitors
EIA ±20%Standard Values

µF	*µF*	*µF*	*µF*	*µF*
0.1	1.0	10	100	1,000
0.22	2.2	22	220	2200
0.33	3.3	33	330	3300
0.47	4.7	47	470	4700
0.68	6.8	68	680	6800
0.82	8.2	82	820	8200

Very old electrolytic capacitors should be used with care or, preferably, replaced. The wet dielectric agent can dry out during prolonged periods of non-use, causing the internal capacitor plates to form a short circuit when energized. Applying low voltage and gradually increasing it over a period of time may restore the capacitor to operation, but if the dielectric agent has dried out, the capacitor will have lost some or most of its value and will likely be lossy and prone to failure.

Tantalum electrolytic capacitors consist of an extremely porous tantalum (a rare-earth metallic element) pentoxide powder mixed with a wet or dry electrolyte, then formed into a pellet or slug for a large effective area. Tantalums also provide high capacitance values in very small packages. Tantalums tend to be more expensive than aluminum electrolytic capacitors. Like the aluminum electrolytic capacitor, tantalum capacitors are also polarized for which care should be exercised. Some varieties of tantalums can literally explode or burst open if voltage is applied with reverse polarity or the voltage rating is exceeded. Tantalum electrolytics are used almost exclusively as high-value SMT components due to their small sizes. Capacitance values up to 1,000 µF at 4 V are available with body sizes about a quarter-inch square.

Identifying the polarity markings of aluminum and tantalum electrolytics (shown in Figure 4.9) can be confusing. Most tantalum electrolytics are marked with a solid band indicating the positive lead. Aluminum electrolytics are available with bands or symbols marking *either* the negative or positive lead. The positive lead of axial-lead electrolytic capacitors is usually manufactured to be longer than the negative lead and often enters the capacitor through fiber or plastic insulating material while the negative lead is connected directly to the metallic case of the capacitor. Misidentifying the polarity of capacitors is a common error during assembly or repair.

Supercapacitors are a special type of electrolytic capacitor with very high capacitance (greater than 1 F) and a low working voltage (a few volts, typically). Supercapacitors have one electrode that is a porous material with an extremely high ratio of surface area to volume and is immersed in electrolyte fluid or gel. The porous electrode is contained in a metal can that forms the other electrode. The dielectric forms on the porous material's surface, similarly to foil-type electrolytics. "Supercaps" are used as short-term power sources and as filter capacitors for power connections with large, sudden, short-term current demands.

4.3.3 Capacitor Voltage Ratings

Capacitors are also rated by their maximum operating voltage. The importance of selecting a capacitor with the proper voltage rating is often overlooked. Exceeding the voltage rating, even momentarily, can cause excessive heating, a permanent shift of the capacitance value, a short circuit, or outright destruction. As a result, the voltage rating should be at least 25% higher than the working voltage across the capacitor; many designers use 50 – 100%.

Following the 25% guideline, filter capacitors for a 12-V system should have at least a 15-V rating (12 V × 1.25). However, 12-V systems such as 12-V power supplies and automotive 12-V electrical systems actually operate near 13.8 V and in the case of automotive systems, as high as 15 V. In such cases, capacitors rated for 15 V would be an insufficient margin of safety; 20 to 25-V capacitors should be used in such cases.

In large signal ac circuits, the maximum voltage rating of the capacitor should be based on the peak-to-peak voltages present. For example, the output of a 5-W QRP transmitter is 16 V_{RMS}, or about 45 V_{P-P}. Capacitors exposed to the 5 W RF power, such as in the output low-pass filter, should be rated well above 50 V for the 25% rule. A 100 W transmitter produces RF voltages of about 200 V_{P-P}. Remember, too, that ac line voltage is given as RMS, with peak-to-peak voltage 2.83 times higher: 120 V_{RMS} = 339 V_{P-P}.

Capacitors that are to be connected to primary ac circuits (directly to the ac line) for filtering or coupling *must* be rated for ac line use. These capacitors are listed as Class X and Class Y with several levels of line voltage ratings. (See **www.allaboutcircuits.com/technical-articles/safety-capacitor-class-x-and-class-y-capacitors**). They are designed to have failure modes that minimize fire and other hazards in case of failure.

Applying peak-to-peak voltages approaching the maximum voltage rating will cause excessive heating of the capacitor. This, in turn, will cause a permanent shift in the capacitance value. This could be undesirable in the output low pass filter example cited above in trying to maintain the proper impedance match between transmitter and antenna.

Exceeding the maximum voltage rating can also cause a breakdown of the dielectric material in the capacitor. The voltage can jump between the plates causing momentary or permanent electrical shorts between the capacitor plates.

In electrolytic and tantalum capacitors, exceeding the voltage rating can produce extreme heating of the oil or wetting agent used as the dielectric material. The expanding gases can cause the capacitor to burst or explode.

These over-voltage problems are easily avoided by selecting a capacitor with a voltage rating 25 – 50% above the normal peak-to-peak operating voltage. **Table 4.13** lists standard working voltages for common capacitor types.

DIELECTRIC BREAKDOWN

When voltage is applied to the plates of a capacitor, force is exerted on the atoms and molecules of the dielectric by the electrostatic field between the plates. If the voltage is high enough, the atoms of the dielectric will ionize (one or more of the electrons will be pulled away from the atom), causing a large dc current to flow discharging the capacitor. This is *dielectric breakdown,* and it is generally destructive to the capacitor because it creates punctures or defects in solid dielectrics that provide permanent low-resistance current paths between the plates. (*Self-healing* dielectrics have the ability to seal off this type of damage.) With most gas dielectrics such as air, once the voltage is removed, the arc ceases and the capacitor is ready for use again.

The *breakdown voltage* of a dielectric depends on the chemical composition and thickness of the dielectric. Breakdown voltage is not directly proportional to the thickness; doubling the thickness does not quite double the breakdown voltage. A thick dielectric must be used to withstand high voltages. Since capacitance is inversely proportional to dielectric thickness (plate spacing) for a given plate area, a high-voltage capacitor must have more plate area than a low-voltage one of the same capacitance. High-voltage, high-capacitance capacitors are therefore physically large. *Dielectric strength* is specified in terms of a *dielectric withstanding voltage* (DWV), given in volts per mil (0.001 inch) at a specified temperature.

Dielectric breakdown in a gas or air dielectric capacitor occurs as a spark or arc between the plates. Spark voltages are generally given with units of *kilovolts per centimeter*. For air, the spark voltage or V_s may range from more than 120 kV/cm for gaps as narrow as 0.006 cm down to 28 kV/cm for gaps as wide as 10 cm. In addition, a large number of variables enter into the actual breakdown voltage in a real situation. Among the variables are the plate shape, the gap distance, the air pressure or density, the voltage, impurities in the air (or any other dielectric material) and the nature of the external circuit (with air, for instance, the humidity affects conduction on the surface of the capacitor plate).

Dielectric breakdown occurs at a lower voltage between pointed or sharp-edged surfaces than between rounded and polished surfaces. This can be a problem when constructing or repairing high-voltage power supplies such as are used in RF power amplifiers. Consequently, the breakdown voltage between metal plates of any given spacing in air can be increased by buffing the edges of the plates. If the plates are damaged so they are no longer smooth, they may have to be polished or the capacitor replaced.

Table 4.13
Capacitor Standard Working Voltages

Ceramic	*Polyester*	*Electrolytic*	*Tantalum*
		6.3 V	6.3 V
		10 V	10 V
16 V		16 V	16 V
			20 V
25 V		25 V	25 V
		35 V	35 V
50 V	50 V	50 V	50 V
		63 V	63 V
100 V	100 V	100 V	
	150 V	150 V	
200 V	200 V		
	250 V	250 V	

4.3.4 Capacitor Identification

Capacitors are identified by the IEC numerical code as shown in **Figure 4.11**. Color coding schemes used prior to 2000 are becoming rare, used only by a few non-US manufacturers. Some thru-hole "gum drop" tantalum capacitors also still use the color codes of **Figure 4.12**. Electrolytic and tantalum capacitors are often labeled with capacitance and working voltage in μF and V as in Figure 4.12C.

The IEC capacitor identification scheme is shown in **Figure 4.13**. Similar to the resistor IEC code, numerals are used to indicate the significant numerals and the multiplier, followed by an alphabetic character to indicate the tolerance. The multiplier is simply the number of zeros following the significant numerals. For example, a capacitor marked with "122K" would be a 1,200 pF 10% capacitor. The use of R to denote a decimal point in a value can be confusing if pF or μF are not specified. Generally, an inspection of the capacitor will determine which is correct, but a capacitance meter may be required. Additional digits and codes may be encountered

Table 4.14
European Marking Standards for Capacitors

Marking	*Value*
1p	1 pF
2p2	2.2 pF
10p	10 pF
100p	100 pF
1n	1 nF (= 0.001 μF)
2n2	2.2 nF (= 0.0022 μF)
10n	10 nF (= 0.01 μF)
100n	100 nF (= 0.1 μF)
1u	1 μF
5u6	5.6 μF
10u	10 μF
100u	100 μF

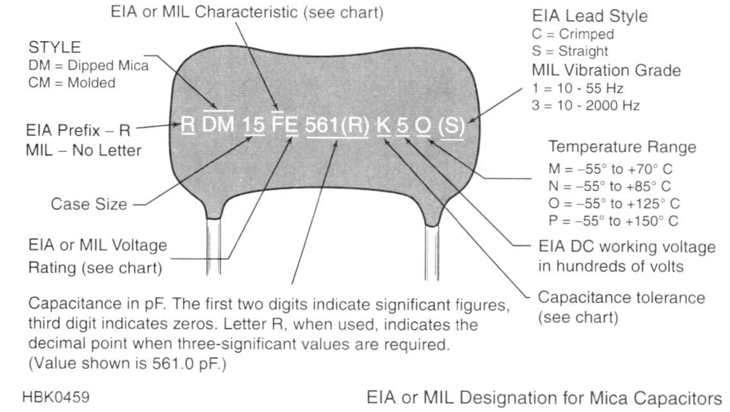

Letter Designator	*"Characteristic" Max Capacitance Drift*	*"Characteristic" Max Range of Temp Coeff (ppm / deg. C)*	*MIL Voltage Rating (V)*	*Capactitance Tolerance (Percent)*
A	-	-	100	-
B	Not Specified	Not Specified	250	-
C	±(0.5% + 0.1 pF)	±200	300	-
D	±(0.3% + 0.1 pF)	±100	500	-
E	±(0.1% + 0.1 pF)	–20 to +100	600	-
F	±(0.05% + 0.1 pF)	0 to +70	1000	±1
G	-	-	1200	±2
H	-	-	1500	-
J	-	-	2000	±5
K	-	-	2500	±10
L	-	-	3000	-
M	-	-	4000	±20

MIL voltage ratings for other letter designators: N=5000 V, P=6000 V, Q=8000 V, R=10,000 V, S-12,000 V, T=15,000 V, U=20,000 V, V=25,000 V, W=30,000 V, X=35,000 V.

Figure 4.11 — Complete capacitor labeling scheme.

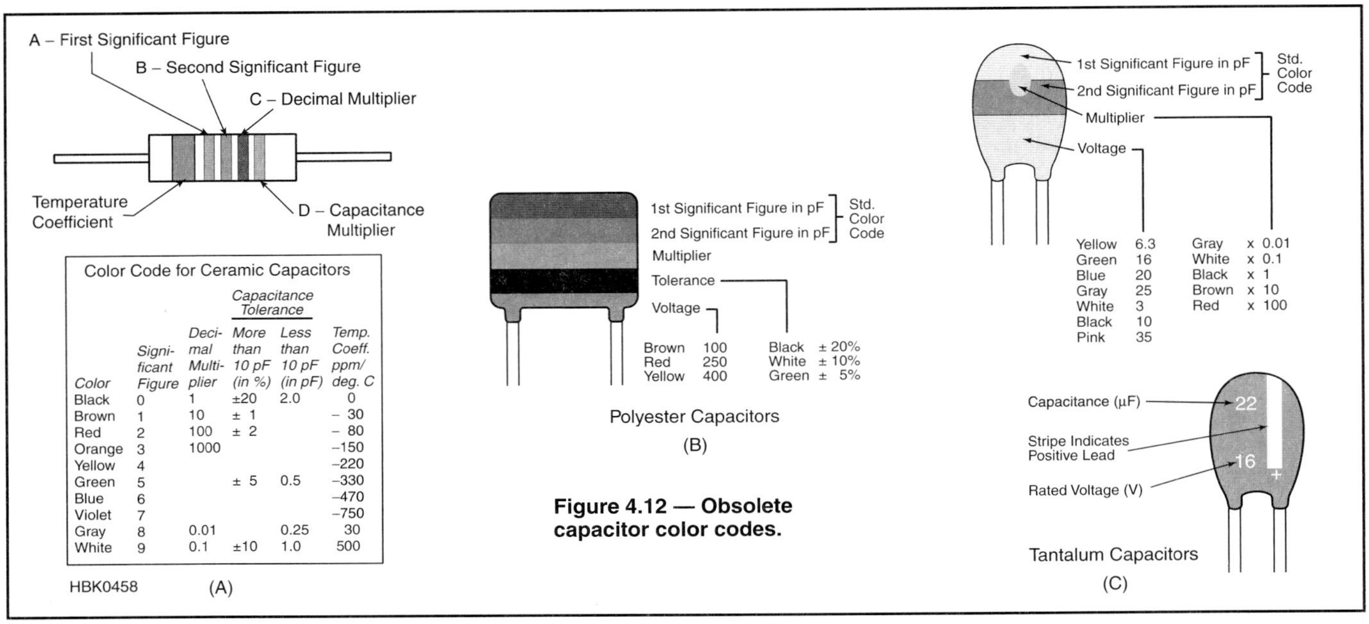

Color Code for Ceramic Capacitors

Color	Significant Figure	Decimal Multiplier	Capacitance Tolerance More than 10 pF (in %)	Capacitance Tolerance Less than 10 pF (in pF)	Temp. Coeff. ppm/deg. C
Black	0	1	±20	2.0	0
Brown	1	10	± 1		– 30
Red	2	100	± 2		– 80
Orange	3	1000			–150
Yellow	4				–220
Green	5		± 5	0.5	–330
Blue	6				–470
Violet	7				–750
Gray	8	0.01		0.25	30
White	9	0.1	±10	1.0	500

Figure 4.12 — Obsolete capacitor color codes.

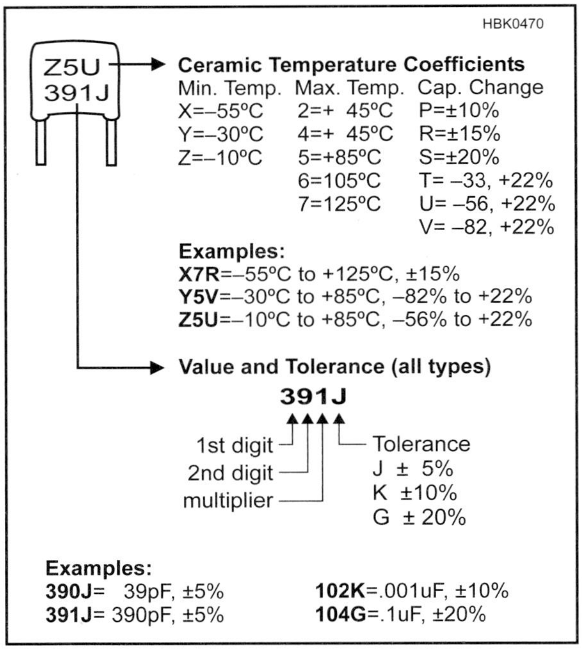

Figure 4.13 — Abbreviated IEC capacitor identification. This method is used on SMT capacitors. An "R" in the numeric field stands for "radix" and represents a decimal point so the "4R7" indicates "4.7" for example.

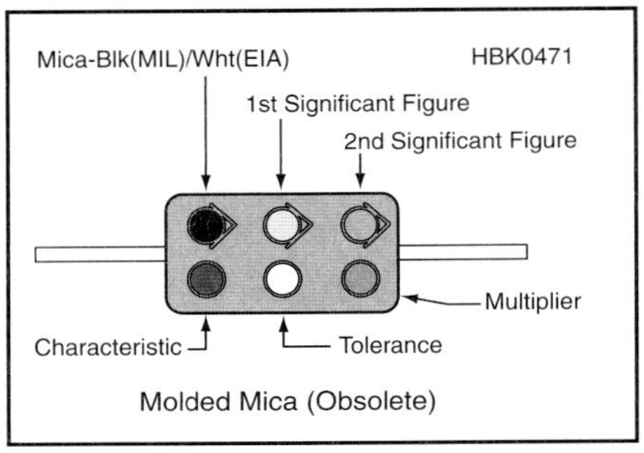

Figure 4.14 — Obsolete JAN "postage stamp" capacitor labeling.

as shown in Figure 4.11.

Military-surplus equipment using the obsolete "postage stamp" capacitors is still encountered in amateur radio. These capacitors used the colored dot method of value identification shown in **Figure 4.14**.

European manufacturers often use nanofarads or nF, such that 10 nF, or simply 10N, indicates 10 nanofarads. This is equivalent to 10,000 pF or 0.01 µF. This notational scheme, shown in **Table 4.14**, is more commonly found on schematic diagrams than actual part markings.

4.4 Practical Inductors

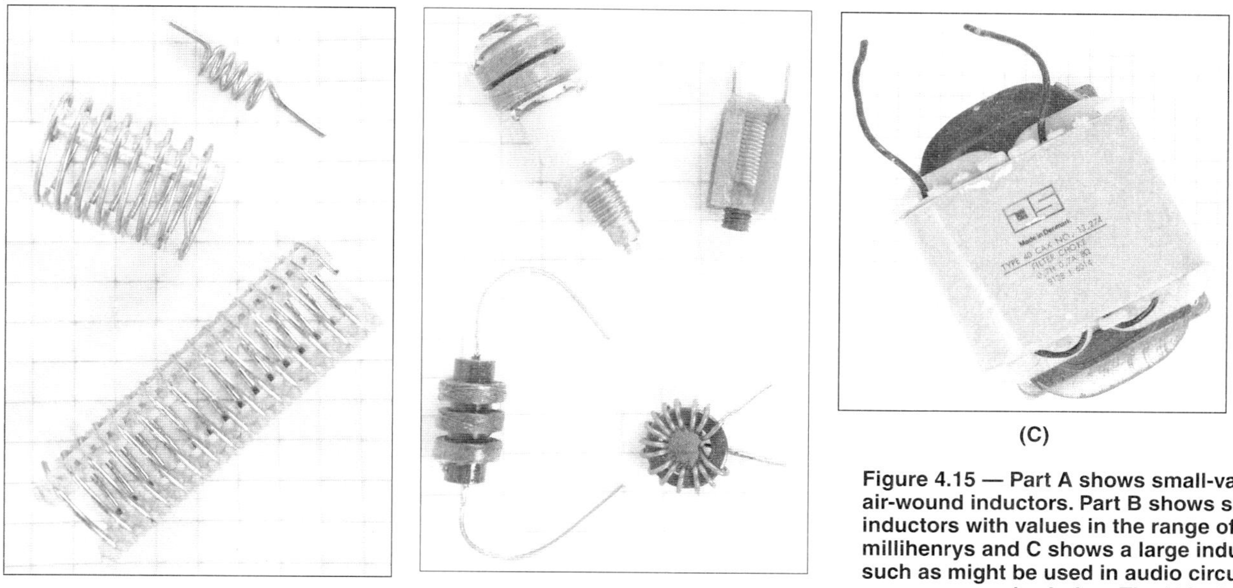

(A) (B) (C)

Figure 4.15 — Part A shows small-value air-wound inductors. Part B shows some inductors with values in the range of a few millihenrys and C shows a large inductor such as might be used in audio circuits or as power-supply chokes. The ¼-inch-ruled graph paper background provides a size comparison.

4.4.1 Component Inductors

Various facets of radio circuits make use of inductors ranging from the tiny up to the massive. Small values of inductance, such as those inductors in **Figure 4.15A**, serve mostly in RF circuits. They may be self-supporting, air-core or air-wound inductors, or the winding may be supported by nonmagnetic strips or a form. Phenolic, certain plastics and ceramics are the most common *coil forms* for air-core inductors. These inductors range in value from a few hundred µH for medium- and high-frequency circuits down to tenths of a µH at VHF and UHF. **Table 4.15** is a list of the EIA standard inductor values.

The most common inductor in small-signal RF circuits is the *encapsulated inductor* or *RF choke*. These components look a lot like carbon-composition or film resistors and are often marked with colored paint stripes to indicate value. (See **Figure 4.16**) The body of these components is often green to identify them as inductors and not resistors. Measure the resistance of the component with an ohmmeter if there is any doubt as to the identity of the component. Pay attention to the limit current-carrying abilities of these inductors. These inductors have values from less than 1 µH to a few mH.

It is possible to make solenoid inductors variable by inserting a moveable *slug* in the center of the inductor. (Slug-tuned inductors normally have a ceramic, plastic, or phenolic insulating form between the conductive slug and the inductor winding.) If the slug material is magnetic, such as powdered iron, the induc-

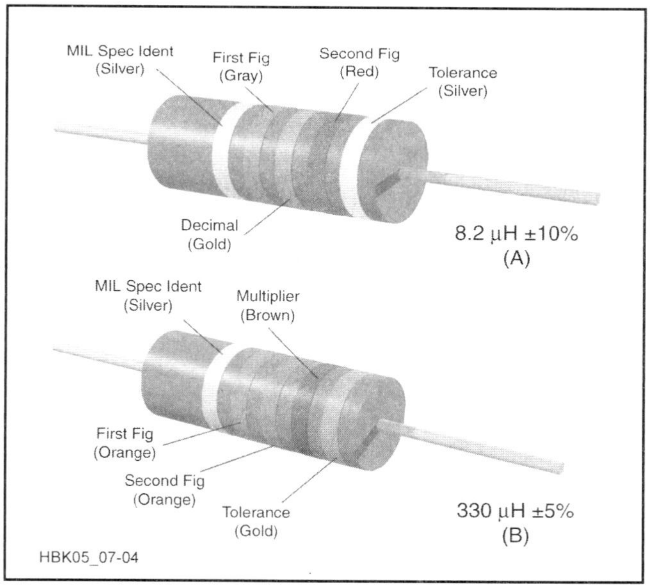

Figure 4.16 — Color coding for cylindrical encapsulated RF chokes. At A, an example of the coding for an 8.2-µH choke is given. At B, the color bands for a 330-µH inductor are illustrated. The color code is given in Table 4.2.

Table 4.15
EIA Standard Inductor Values

µH	*µH*	*µH*	*mH*	*mH*	*mH*
1.0	10	100	1.0	10	100
1.2	12	120	1.2	12	120
1.5	15	150	1.5	15	150
2.2	22	220	2.2	22	220
2.7	27	270	2.7	27	270
3.3	33	330	3.3	33	330
3.9	39	390	3.9	39	390
4.7	47	470	4.7	47	470
5.6	56	560	5.6	56	560
6.8	68	680	6.8	68	680
8.2	82	820	8.2	82	820

tance increases as the slug is moved into the center of the inductor. If the slug is brass or some other nonmagnetic material, inserting the slug will reduce the inductor's inductance.

Alternatives to air-core inductors for RF work are *toroidal* inductors (or *toroids*) wound on powdered-iron or ferrite cores. The availability of many types and sizes of powdered-iron cores has made these inductors popular for low-power fixed-value service. The toroidal shape concentrates the inductor's field nearly completely inside the inductor, eliminating the need in many cases for other forms of shielding to limit the interaction of the inductor's magnetic field with the fields of other inductors. (Ferrite core materials are discussed here and in the chapter on **RF Techniques**.)

Figure 4.15B shows samples of inductors in the millihenry (mH) range. Among these inductors are multi-section RF chokes designed to block RF currents from passing beyond them to other parts of circuits. Low-frequency radio work may also use inductors in this range of values, sometimes wound with *Litz wire*. Litz wire is a special version of stranded wire, with each strand insulated from the others, and is used to minimize losses associated with skin effect.

For audio filters, toroidal inductors with values up to 100 mH are useful. Resembling powdered-iron-core RF toroids, these inductors are wound on ferrite or molybdenum-permalloy cores having much higher permeabilities.

Audio and power-supply inductors appear in Figure 4.15C. Lower values of these iron-core inductors, in the range of a few henrys, are useful as audio-frequency chokes. Larger values up to about 20 H may be found in power supplies, as choke filters, to suppress 120-Hz ripple. Although some of these inductors are open frame, most have iron covers to confine the powerful magnetic fields they produce.

Although builders and experimenters rarely construct their own capacitors, inductor fabrication for low- and high-power RF circuits is common. In fact, it is often necessary, since commercially available units may be unavailable or expensive. Even if available, they may consist of inductor stock to be trimmed to the required value. Core materials and wire for winding both solenoid and toroidal inductors are readily available. The following information includes fundamental formulas and design examples for calculating practical inductors, along with additional data on the theoretical limits in the use of some materials.

4.4.2 Air-Core Inductors

Many circuits require air-core inductors using just one layer of wire. The approximate inductance of a single-layer air-core inductor may be calculated from the simplified formula:

$$L\,(\mu H) = \frac{d^2 n^2}{18d + 40\ell}$$

where

L = inductance in microhenrys,

d = inductor diameter in inches (from wire center to wire center),

ℓ = inductor length in inches, and

n = number of turns.

If dimensions are given in cm, the equation is:

$$L(\mu H) = \frac{d^2 n^2}{6.4\left(7.2d + 15.8\ell\right)}$$

The notation is illustrated in **Figure 4.17**. This formula is a close approximation for inductors having a length equal to or greater than 0.4d. (Note: Inductance varies as the square of the turns. If the number of turns is doubled, the inductance is quadrupled. This relationship is inherent in the equation but is often overlooked. For example, to double the inductance, add additional turns equal to 1.4 times the original number of turns, or 40% more turns.)

Example: What is the inductance of an inductor if the inductor has 48 turns wound at 32 turns per inch and a diameter of ¾ inch? In this case, d = 0.75, ℓ = 48/32 = 1.5 and n = 48.

$$L\,(\mu H) = \frac{0.75^2 \times 48^2}{(18 \times 0.75) + (40 \times 1.5)}$$

$$= \frac{1300}{74} = 18\ \mu H$$

To calculate the number of turns of a single-layer inductor for a required value of inductance, the formula becomes:

$$n = \frac{\sqrt{L(18d + 40\ell)}}{d}\ \text{(inches)}$$

$$n = \frac{\sqrt{L(7.2d + 15.8\ell)}}{d}\ \text{(cm)}$$

Example: Suppose an inductance of 10.0 µH is required. The form on which the inductor is to be wound has a diameter of one inch and is long enough to accommodate an inductor of 1¼ inches. Then d = 1.00 inch, ℓ= 1.25 inches and L = 10.0. Substituting:

$$n = \frac{\sqrt{10.0\,[\,(18 \times 1.0) + (40 \times 1.25)]}}{1}$$

$$= \sqrt{680} = 26.1 \text{ turns}$$

A 26-turn inductor would be close enough in practical work. Since the inductor will be 1.25 inches long, the number of turns per inch will be 26.1 / 1.25 = 20.9. Consulting the wire table, we find that #17 AWG enameled wire

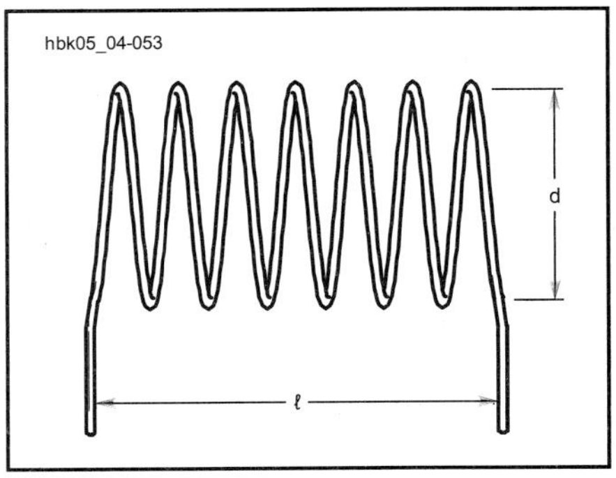

Figure 4.17 — Coil dimensions used in the inductance formula for air-core inductors.

(or anything smaller) can be used. The proper inductance is obtained by winding the required number of turns on the form and then adjusting the spacing between the turns to make a uniformly spaced inductor 1.25 inches long.

Most inductance formulas lose accuracy when applied to small inductors (such as those used in VHF work and in low-pass filters built for reducing harmonic interference to televisions) because the conductor thickness is no longer negligible in comparison with the size of the inductor. **Figure 4.18** shows the measured inductance of VHF inductors and may be used as a basis for circuit design. Two curves are given: curve A is for inductors wound to an inside diameter of ½ inch; curve B is for inductors of ¾ -inch inside diameter. In both curves, the wire size is #12 AWG and the winding pitch is eight turns to the inch (⅛-inch turn spacing). The inductance values include leads ½-inch long.

Machine-wound inductors with the preset diameters and turns per inch were sold as Miniductor and Airdux coils. These are no longer manufactured but are commonly available as surplus and NOS (new old stock). Charts and tables of the inductance of these coils is provided in the online content.

Forming a wire into a solenoid increases its inductance but also introduces distributed capacitance. Since each turn is at a slightly different ac potential, each pair of turns effectively forms a capacitor in parallel with part of the inductor. (See the chapter on **RF Techniques** for information on the effects of these and other factors that affect the behavior of the "ideal" inductors discussed in this chapter.)

Moreover, the Q of air-core inductors is, in part, a function of the inductor shape, specifically its ratio of length to diameter. Q tends to be highest when these dimensions are nearly equal. With wire properly sized to the current carried by the inductor, and with high-caliber construction, air-core inductors can achieve Q above 200.

For a large collection of formulas useful in

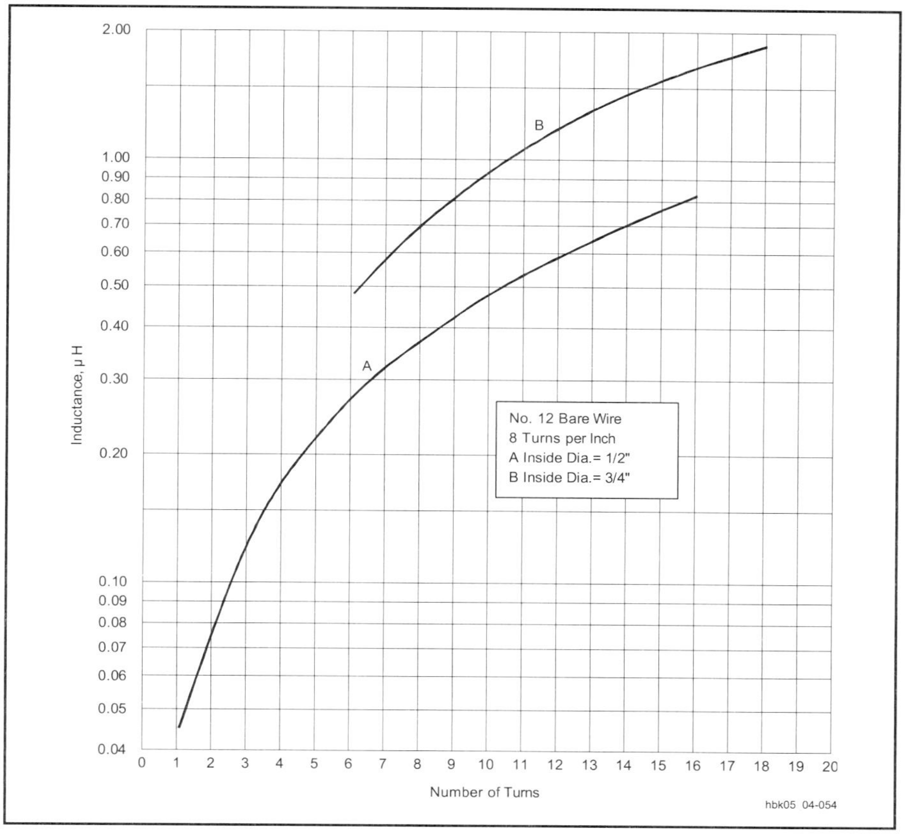

Figure 4.18 — Measured inductance of coils wound with #12 bare wire, eight turns to the inch. The values include half-inch leads.

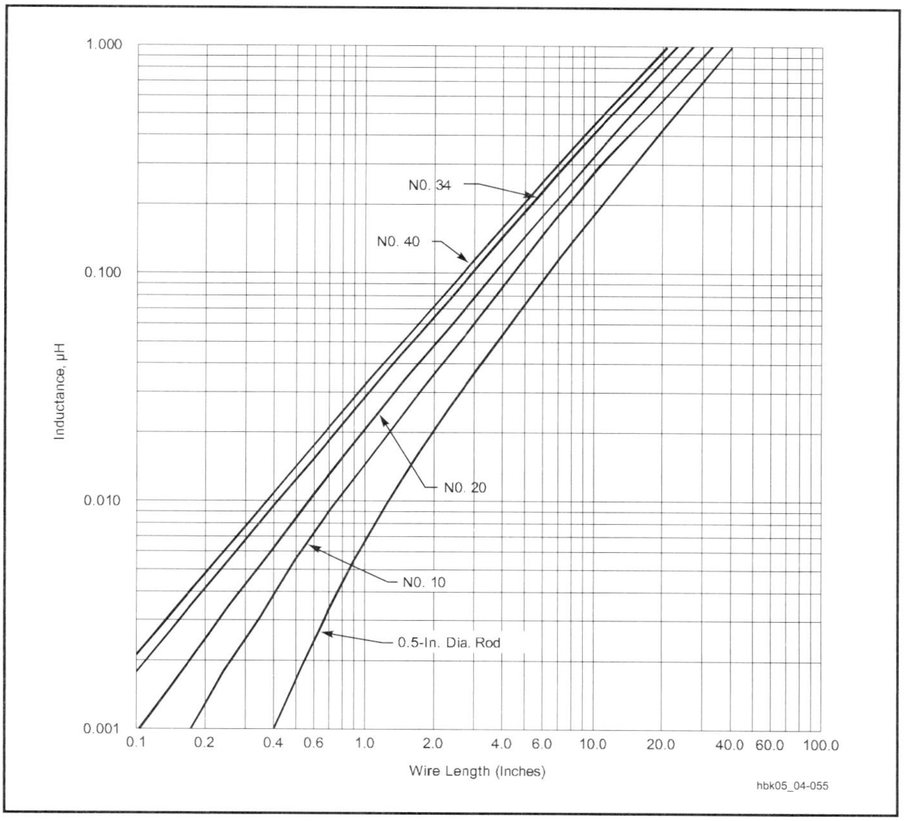

Figure 4.19 — Inductance of various conductor sizes as straight wires.

constructing air-core inductors of many configurations, see the "Circuit Elements" section in Terman's *Radio Engineers' Handbook* (listed in the References section).

4.4.3 Straight-Wire Inductance

At low frequencies the inductance of a straight, round, nonmagnetic wire in free space is given by:

$$L = 0.00508\, b \left\{ \left[\ln\left(\frac{2\,b}{a}\right) \right] - 0.75 \right\}$$

where

L = inductance in µH,

a = wire radius in inches,

b = wire length in inches, and

ln = natural logarithm = 2.303 × common logarithm (base 10).

If the dimensions are expressed in millimeters instead of inches, the equation may still be used, except replace the 0.00508 value with 0.0002.

Skin effect reduces the inductance at VHF and above. As the frequency approaches infinity, the 0.75 constant within the brackets approaches unity. As a practical matter, skin effect will not reduce the inductance by more than a few percent.

Example: What is the inductance of a wire that is 0.1575 inch in diameter and 3.9370 inches long? For the calculations, a = 0.0787 inch (radius) and b = 3.9370 inch.

$$L = 0.00508\, b \left\{ \left[\ln\left(\frac{2\,b}{a}\right) \right] - 0.75 \right\}$$

$$= 0.00508\,(3.9370) \times \left\{ \left[\ln\left(\frac{2 \times 3.9370}{0.0787}\right) \right] - 0.75 \right\}$$

$$= 0.20 \times [\ln(100) - 0.75)]$$

$$= 0.20 \times (4.60 - 0.75)$$

$$= 0.20 \times 3.85 = 0.077\ \mu H$$

Figure 4.19 is a graph of the inductance for wires of various radii as a function of length.

A VHF or UHF tank circuit can be fabricated from a wire parallel to a ground plane, with one end grounded. A formula for the inductance of such an arrangement is given in **Figure 4.20**.

Example: What is the inductance of a wire 3.9370 inches long and 0.0787 inch in radius, suspended 1.5748 inch above a ground plane? (The inductance is measured between the free end and the ground plane, and the formula includes the inductance of the 1.5748-inch

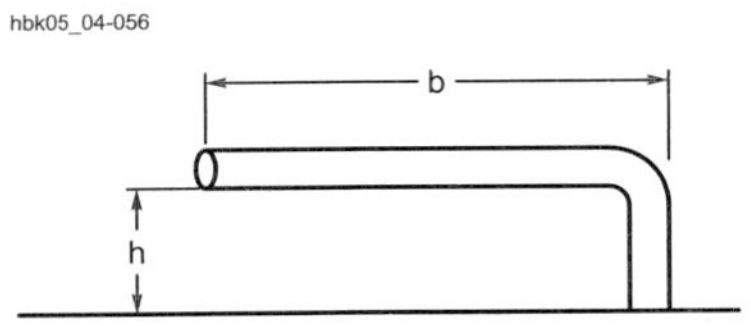

$$L = 0.0117b\left\{\text{Log}_{10}\left[\frac{2h}{a}\left(\frac{b+\sqrt{b^2+a^2}}{b+\sqrt{b^2+4h^2}}\right)\right]\right\} + 0.00508\left(\sqrt{b^2+4h^2} - \sqrt{b^2+a^2} + \frac{b}{4} - 2h + a\right)$$

where
L = inductance in µH
a = wire radius in inches
b = wire length parallel to ground plane in inches
h = wire height above ground plane in inches

Figure 4.20 — Equation for determining the inductance of a wire parallel to a ground plane, with one end grounded. If the dimensions are in millimeters, the numerical coefficients become 0.0004605 for the first term and 0.0002 for the second term.

grounding link.) To demonstrate the use of the formula in Figure 4.20, begin by evaluating these quantities:

$$b + \sqrt{b^2 + a^2}$$

$$= 3.9370 + \sqrt{3.9370^2 + 0.0787^2}$$

$$= 3.9370 + 3.94 = 7.88$$

$$b + \sqrt{b^2 + 4\,(h^2)}$$

$$= 3.9370 + \sqrt{3.9370^2 + 4\,(1.5748^2)}$$

$$= 3.9370 + \sqrt{15.50 + 4\,(2.480)}$$

$$= 3.9370 + \sqrt{15.50 + 9.920}$$

$$= 3.9370 + 5.0418 = 8.9788$$

$$\frac{2\,h}{a} = \frac{2 \times 1.5748}{0.0787} = 40.0$$

$$\frac{b}{a} = \frac{3.9370}{4} = 0.098425$$

Substituting these values into the formula yields:

$$L = 0.0117 \times 3.9370\left\{\log_{10}\left[40.0 \times \left(\frac{7.88}{8.9788}\right)\right]\right\}$$

$$+ 0.00508 \times (5.0418 - 3.94 + 0.98425 - 3.1496 + 0.0787)$$

$$L = 0.0662\ \mu\text{H}$$

Another conductor configuration that is frequently used is a flat strip over a ground plane. This arrangement has lower skin-effect loss at high frequencies than round wire because it has a higher surface-area to volume ratio. The inductance of such a strip can be found from the formula in **Figure 4.21**.

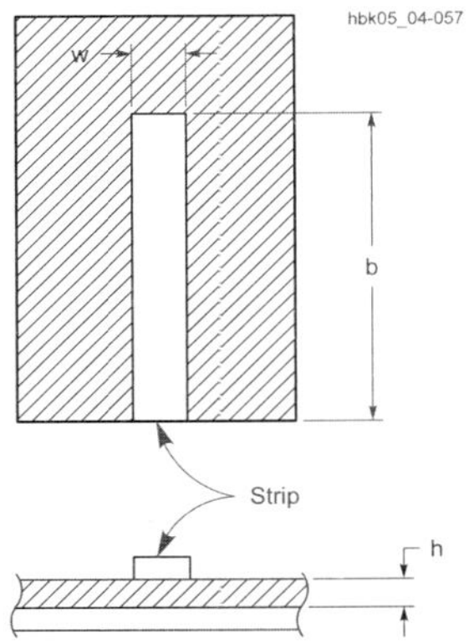

$$L = 0.00508\,b\left(\ln\frac{2b}{w+h} + 0.5 + 0.2235\,\frac{w+h}{b}\right)$$

where
L = inductance in µH
b = length in inches
w = width in inches
h = thickness in inches

Figure 4.21 — Equation for determining the inductance of a flat strip inductor.

4.4.4 Iron-Core Inductors

If the permeability of an iron core in an inductor is 800, then the inductance of any given air-wound inductor is increased 800 times by inserting the iron core. The inductance will be proportional to the magnetic flux through the inductor, other things being equal. The inductance of an iron-core inductor is highly dependent on the current flowing in the inductor, in contrast to an air-core inductor, where the inductance is independent of current because air does not saturate.

Iron-core inductors are used chiefly in power-supply equipment. They usually have direct current flowing through the winding, and any variation in inductance with current is usually undesirable. Inductance variations may be overcome by keeping the flux density below the saturation point of the iron. Opening the core so there is a small air gap will achieve this goal, as discussed in the earlier section on inductors. The reluctance or magnetic resistance introduced by such a gap is very large compared with that of the iron, even though the gap is only a small fraction of an inch. Therefore, the gap — rather than the iron — controls the flux density. Air gaps in iron cores reduce the inductance, but they hold the value practically constant regardless of the current magnitude.

When alternating current flows through an inductor wound on an iron core, a voltage is induced. Since iron is a conductor, eddy currents also flow in the core as discussed earlier. Eddy currents represent lost power because they flow through the resistance of the iron and generate heat. Losses caused by eddy currents can be reduced by laminating the core (cutting the core into thin strips). These strips or laminations are then insulated from each other by painting them with some insulating material such as varnish or shellac. Eddy current losses add to hysteresis losses, which are also significant in iron-core inductors.

Eddy-current and hysteresis losses in iron increase rapidly as the frequency of the alternating current increases. For this reason, ordinary iron cores can be used only at power-line and audio frequencies — up to approximately 15,000 Hz. Even then, a very good grade of iron or steel is necessary for the core to perform well at the higher audio frequencies. Laminated iron cores become completely useless at radio frequencies because of eddy current and hysteresis losses.

4.4.5 Slug-Tuned Inductors

For RF work, the losses in iron cores can be reduced to a more useful level by grinding the iron into a powder and then mixing it with a binder of insulating material in such a way that the individual iron particles are insulated from each other. Using this approach, cores can be made that function satisfactorily even into the VHF range. Because a large part of the magnetic path is through a nonmagnetic material (the binder), the permeability of the powdered iron core is low compared with the values for solid iron cores used at power-line frequencies.

The slug is usually shaped in the form of a cylinder that fits inside the insulating form on which the inductor is wound. Despite the fact that the major portion of the magnetic path for the flux is in air, the slug is quite effective in increasing the inductor inductance. By pushing (or screwing) the slug in and out of the inductor, the inductance can be varied over a considerable range.

4.4.6 Powdered-Iron Toroidal Inductors

For fixed-value inductors intended for use at HF and VHF, the powdered-iron toroidal core has become the standard in low- and medium-power circuits. **Figure 4.22** shows the general outlines of a toroidal inductor on a magnetic core.

Manufacturers offer a wide variety of core materials, or *mixes*, to create conductor cores that will perform over a desired frequency range with a reasonable permeability. Permeabilities for powdered-iron cores fall in the range of 3 to 35 for various mixes. In addition, core sizes are available in the range of 0.125-inch outside diameter (OD) up to 1.06-inch OD, with larger sizes to 5-inch OD available in certain mixes. The range of sizes permits the builder to construct single-layer inductors for almost any value using wire sized to meet the circuit current demands.

The use of powdered iron in a binder reduces core losses usually associated with iron, while the permeability of the core permits a reduction in the wire length and associated resistance in forming an inductor of a given inductance. Therefore, powdered-iron-core toroidal inductors can achieve Q well above 100, often approaching or exceeding 200 within the frequency range specified for a given core. Moreover, these inductors are considered *self-shielding* since most of the magnetic flux is within the core, a fact that simplifies circuit design and construction.

Table 4.16 lists the properties of common powdered-iron cores. Formulas are given for calculating the number of required turns based on a given inductance and for calculating the inductance given a specific number of turns. Most powdered-iron toroid cores that amateurs use are manufactured by Micrometals (**www.micrometals.com**). Paint is used to identify the material used in the core. The Micrometals color code is part of Table 4.16. **Table 4.17** gives the physical dimensions of powdered-iron toroids.

Each powdered-iron core has an *inductance factor* or *index* A_L determined by the manufacturer. Amidon specifies A_L in μH per 100 turns-squared and other manufacturers specify A_L in μH or nH per turns-squared. Check the manufacturer's website or product information for the correct units for A_L and method of calculating the inductance or desired number of turns.

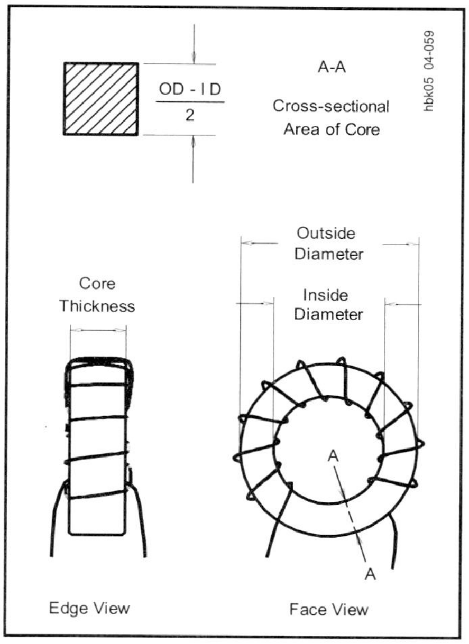

Figure 4.22 — A typical toroidal inductor wound on a powdered-iron or ferrite core. Some key physical dimensions are noted. Equally important are the core material, its permeability, its intended range of operating frequencies, and its A_L value. This is an 11-turn toroid.

$$L = \frac{A_L \times N^2}{10000} \text{ (Amidon) or}$$

$$L = A_L \times N^2 \text{ (Other manufacturers)}$$

where

L = the inductance in μH,

A_L = the inductance index (see text above)

N = the number of turns.

The builder must then ensure that the core is capable of holding the calculated number of turns of wire of the required wire size.

Example: What is the inductance of a 60-turn inductor on a core with an A_L of 55 μH/100-turns²?

$$L = \frac{A_L \times N^2}{10000} = \frac{55 \times 60^2}{10000}$$

$$= \frac{198000}{10000} = 19.8 \text{ μH}$$

Example: What is the inductance of a 20-turn inductor on a core with an A_L of 0.3 μH/turns²?

$$L = A_L \times N^2 = 0.3 \times 20^2$$
$$= 0.3 \times 400 = 120 \text{ μH}$$

To calculate the number of turns needed for a particular inductance, use the formula:

$$N = 100\sqrt{\frac{L}{A_L}} \text{ (Amidon) or}$$

$$N = \sqrt{\frac{L}{A_L}} \text{ (Other manufacturers)}$$

Example: How many turns are needed for a 12.0-μH inductor if the A_L for the selected core is 49 μH/100-turns²?

$$N = 100\sqrt{\frac{L}{A_L}} = 100\sqrt{\frac{12.0}{49}}$$

$$= 100\sqrt{0.245} = 100 \times 0.495 = 49.5 \text{ turns}$$

Example: How many turns are needed for a 300-nH inductor if the A_L for the selected core is 12 nH/turns²?

$$N = \sqrt{\frac{L}{A_L}} = \sqrt{\frac{300}{12}} = \sqrt{25} = 5 \text{ turns}$$

Count turns by each pass of the wire through the center of the core. (*A straight wire through a toroidal core counts as a one-turn inductor.*) Fine adjustment of the inductance may be possible by spreading or compressing inductor turns.

If the value is critical, experiment by starting with an extra turn or two, then measure the inductance or test the circuit. Core characteristics may vary slightly from batch to batch and winding style has a small effect on inductance, as well.

The power-handling ability of toroidal cores depends on many variables, which include the cross-sectional area through the core, the core material, the numbers of turns in the inductor, the applied voltage and the operating frequency. Although powdered-iron cores can withstand dc flux densities up to 5,000 gauss without saturating, ac flux densities from sine waves above certain limits can overheat cores. Manufacturers provide guideline limits for ac flux densities to avoid overheating. The limits range from 150 gauss at 1 MHz to 30 gauss at 28 MHz, although the curve is not linear. To calculate the maximum anticipated flux density for a particular inductor, use the formula:

$$B_{max} = \frac{E_{RMS} \times 10^8}{4.44 \times A_e \times N \times f}$$

where

B_{max} = the maximum flux density in gauss,

E_{RMS} = the voltage across the inductor,

A_e = the cross-sectional area of the core in square centimeters,

N = the number of turns in the inductor, and

f = the operating frequency in Hz.

Example: What is the maximum ac flux density for an inductor of 15 turns if the frequency is 7.0 MHz, the RMS voltage is 25 V

Table 4.16
Powdered-Iron Toroidal Cores: Magnetic Properties

There are differing conventions for referring to the type of core material: #, mix and type are all used. For example, all of the following designate the same material: #12, Mix 12, 12-Mix, Type 12 and 12-Type.

Inductance and Turns Formula

The turns required for a given inductance or inductance for a given number of turns can be calculated from:

$$N = 100\sqrt{\frac{L}{A_L}} \qquad L = A_L\left(\frac{N^2}{10{,}000}\right) \qquad \text{(Amidon cores)}$$

$$N = \sqrt{\frac{L}{A_L}} \qquad L = A_L \times N^2 \qquad \text{(Non-Amidon cores)}$$

where N = number of turns; L = desired inductance; A_L = inductance index (µH per 100 turns-squared for Amidon cores; nH or µH per turns-squared for non-Amidon cores; see core data sheet)

Amidon Associates literature gives the value of A_L as inductance per 100 turns but the correct units are inductance per 100 turns-squared. The units of inductance are generally in nH but may also be mH. Make sure you understand which units apply and use the A_L value and formula provided by the manufacturer of the core to calculate number of turns or inductance.

Toroid diameter is indicated by the number following "T" — T-200 is 2.00 in. dia; T-68 is 0.68 in. diameter, etc.

A_L Values

	Mix										
Size	*26**	*3*	*15*	*1*	*2*	*7*	*6*	*10*	*12*	*17*	*0*
T-12	na	60	50	48	20	18	17	12	7.5	7.5	3.0
T-16	145	61	55	44	22	na	19	13	8.0	8.0	3.0
T-20	180	76	65	52	27	24	22	16	10.0	10.0	3.5
T-25	235	100	85	70	34	29	27	19	12.0	12.0	4.5
T-30	325	140	93	85	43	37	36	25	16.0	16.0	6.0
T-37	275	120	90	80	40	32	30	25	15.0	15.0	4.9
T-44	360	180	160	105	52	46	42	33	18.5	18.5	6.5
T-50	320	175	135	100	49	43	40	31	18.0	18.0	6.4
T-68	420	195	180	115	57	52	47	32	21.0	21.0	7.5
T-80	450	180	170	115	55	50	45	32	22.0	22.0	8.5
T-94	590	248	200	160	84	na	70	58	32.0	na	10.6
T-106	900	450	345	325	135	133	116	na	na	na	19.0
T-130	785	350	250	200	110	103	96	na	na	na	15.0
T-157	870	420	360	320	140	na	115	na	na	na	na
T-184	1640	720	na	500	240	na	195	na	na	na	na
T-200	895	425	na	250	120	105	100	na	na	na	na

*Mix-26 is similar to the older Mix-41, but can provide an extended frequency range.

Magnetic Properties Iron Powder Cores

Mix	*Color*	*Material*	µ	*Temp stability (ppm/°C)*	*f (MHz)*	*Notes*
26	Yellow/white	Hydrogen reduced	75	825	dc - 1	Used for EMI filters and dc chokes
3	Gray	Carbonyl HP	35	370	0.05 - 0.50	Excellent stability, good Q for lower frequencies
15	Red/white	Carbonyl GS6	25	190	0.10 - 2	Excellent stability, good Q
1	Blue	Carbonyl C	20	280	0.50 - 5	Similar to Mix-3, but better stability
2	Red	Carbonyl E	10	95	2 - 30	High Q material
7	White	Carbonyl TH	9	30	3 - 35	Similar to Mix-2 and Mix-6, but better temperature stability
6	Yellow	Carbonyl SF	8	35	10 - 50	Very good Q and temperature stability for 20-50 MHz
10	Black	Powdered iron W	6	150	30 - 100	Good Q and stability for 40 - 100 MHz
12	Green/white	Synthetic oxide	4	170	50 - 200	Good Q, moderate temperature stability
17	Blue/yellow	Carbonyl	4	50	40 - 180	Similar to Mix-12, better temperature stability, Q drops about 10% above 50 MHz, 20% above 100 MHz
0	Tan	phenolic	1	0	100 - 300	Inductance may vary greatly with winding technique

Courtesy of Amidon Assoc and Micrometals
Note: Color codes hold only for cores manufactured by Micrometals, which makes the cores sold by most amateur radio distributors.

Table 4.17

Powdered-Iron Toroidal Cores: Dimensions

Toroid diameter is indicated by the number following "T" — T-200 is 2.00 in. dia; T-68 is 0.68 in. diameter, etc.
See Table 4.16 for a core sizing guide.

Red E Cores—500 kHz to 30 MHz (μ = 10)

No.	*OD (in)*	*ID (in)*	*H (in)*
T-200-2	2.00	1.25	0.55
T-94-2	0.94	0.56	0.31
T-80-2	0.80	0.50	0.25
T-68-2	0.68	0.37	0.19
T-50-2	0.50	0.30	0.19
T-37-2	0.37	0.21	0.12
T-25-2	0.25	0.12	0.09
T-12-2	0.125	0.06	0.05

Black W Cores—30 MHz to 200 MHz (μ=6)

No.	*OD (In)*	*ID (In)*	*H (In)*
T-50-10	0.50	0.30	0.19
T-37-10	0.37	0.21	0.12
T-25-10	0.25	0.12	0.09
T-12-10	0.125	0.06	0.05

Yellow SF Cores—10 MHz to 90 MHz (μ=8)

No.	*OD (In)*	*ID (In)*	*H (In)*
T-94-6	0.94	0.56	0.31
T-80-6	0.80	0.50	0.25
T-68-6	0.68	0.37	0.19
T-50-6	0.50	0.30	0.19
T-26-6	0.25	0.12	0.09
T-12-6	0.125	0.06	0.05

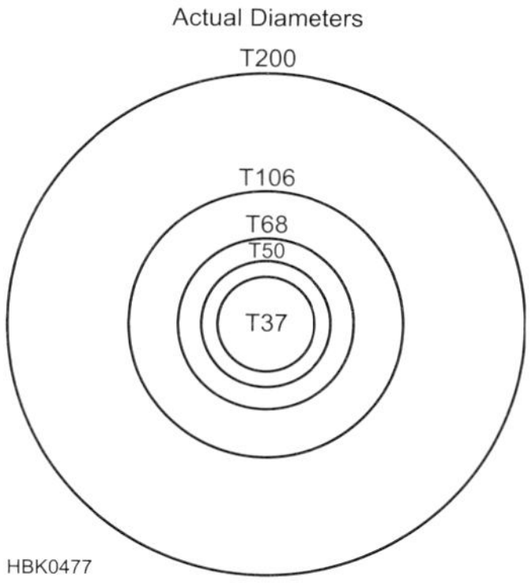

Number of Turns vs Wire Size and Core Size

Approximate maximum number of turns—single layer wound—enameled wire.

Wire Size	*T-200*	*T-130*	*T-106*	*T-94*	*T-80*	*T-68*	*T-50*	*T-37*	*T-25*	*T-12*
10	33	20	12	12	10	6	4	1		
12	43	25	16	16	14	9	6	3		
14	54	32	21	21	18	13	8	5	1	
16	69	41	28	28	24	17	13	7	2	
18	88	53	37	37	32	23	18	10	4	1
20	111	67	47	47	41	29	23	14	6	1
22	140	86	60	60	53	38	30	19	9	2
24	177	109	77	77	67	49	39	25	13	4
26	223	137	97	97	85	63	50	33	17	7
28	281	173	123	123	108	80	64	42	23	9
30	355	217	154	154	136	101	81	54	29	13
32	439	272	194	194	171	127	103	68	38	17
34	557	346	247	247	218	162	132	88	49	23
36	683	424	304	304	268	199	162	108	62	30
38	875	544	389	389	344	256	209	140	80	39
40	1103	687	492	492	434	324	264	178	102	51

Actual number of turns may differ from above figures according to winding techniques, especially when using the larger size wires. Chart prepared by Michel J. Gordon, Jr, WB9FHC. Courtesy of Amidon Assoc.

and the cross-sectional area of the core is 0.133 cm^2?

$$B_{max} = \frac{E_{RMS} \times 10^8}{4.44 \times A_e \times N \times f}$$

$$= \frac{25 \times 10^8}{4.44 \times 0.133 \times 15 \times 7.0 \times 10^6}$$

$$= \frac{25 \times 10^8}{62 \times 10^6} = 40 \text{ gauss}$$

Since the recommended limit for cores operated at 7 MHz is 57 gauss, this inductor is well within guidelines.

4.4.7 Ferrite Toroidal Inductors

Although nearly identical in general appearance to powdered-iron cores, ferrite cores differ in a number of important characteristics. Composed of nickel-zinc ferrites for lower permeability ranges and of manganese-zinc ferrites for higher permeabilities, these cores span a permeability range from 20 to above 1,0000. Nickel-zinc cores with permeabilities from 20 to 800 are useful in high-Q applications, but function more commonly in amateur applications as RF chokes. They are also useful in wide-band transformers, discussed in the **RF Techniques** chapter. An excellent design resource for ferrite-based components is the Fair-Rite Materials Corp on-line catalog at **www.fair-rite.com**. The Fair-Rite website's "Technical Resources" section also has free papers on the use of ferrites for EMI suppression and broadband transformers. The following list presents the general characteristics (material and composition and intended application) of Fair-Rite's ferrite materials:

- Type 31 (MnZn) — EMI suppression applications from 1 MHz up to 500 MHz.
- Type 43 (NiZn) — Suppression of conducted EMI, inductors and HF common-mode chokes from 20 MHz to 250 MHz.
- Type 44 (NiZn) — EMI suppression from 30 MHz to 500 MHz.
- Type 61 (NiZn) — Inductors up to 25 MHz and EMI suppression above 200 MHz.
- Type 67 (NiZn) — Broadband transformers, antennas and high-Q inductors up to 50 MHz.
- Type 73 (MnZn) — Suppression of conducted EMI below 50 MHz.
- Type 75 (MnZn) — Broadband and pulse transformers.

See **Table 4.18** for information about the dimensions and magnetic properties of ferrite cores. Identification can often be difficult because ferrite cores are often unpainted, unlike powdered-iron toroids. Ferrite toroids and rods often have sharp edges, while powdered-iron toroids usually have rounded edges.

Table 4.18
Ferrite Toroids: Magnetic Properties and Dimensions

There are differing conventions for referring to the type of ferrite material: #, mix and type are all used. For example, all of the following designate the same ferrite material: #43, Mix 43, 43-Mix, Type 43, and 43-Type.

Fair-Rite Corporation (**www.fair-rite.com**) and Amidon (**www.amidoncorp.com**) ferrite toroids can be cross-referenced as follows:

For Amidon toroids, "FT-XXX-YY" indicates a ferrite toroid, with XXX as the OD in hundredths of an inch and YY the mix. For example, an FT-23-43 core has an OD of 0.23 inch and is made of type 43 material. Additional letters (usually "C") are added to indicate special coatings or different thicknesses.

For Fair-Rite toroids, digits 1 and 2 of the part number indicate product type (59 indicates a part for inductive uses), digits 3 and 4 indicate the material type, digits 5 through 9 indicate core size, and the final digit indicates coating (1 for Paralene and 2 for thermo-set). For example, Fair-Rite part number 5943,000101 is equivalent to the Amidon FT-23-43 core.

Ferrite Toroids: A_L Chart (mH per 1,000 turns-squared)

Toroid diameter is specified as the outside diameter of the core. See Table 4.16 for a core sizing guide.

Core Size (in)	*63/67-Mix* $\mu = 40$	*61-Mix* $\mu = 125$	*43-Mix* $\mu = 850$	*77 (72)-Mix* $\mu = 2{,}000$	*J (75)-Mix* $\mu = 5{,}000$
0.23	7.9	24.8	188	396	980
0.37	19.7	55.3	420	884	2196
0.50	22.0	68.0	523	1100	2715
0.82	22.4	73.3	557	1170	NA
1.14	25.4	79.3	603	1270	3170
1.40	45	140	885	2400	5500
2.40	55	170	1075	2950	6850

31-Mix is an EMI suppression material and not recommended for inductive use.

Inductance and Turns Formula

The turns required for a given inductance or inductance for a given number of turns can be calculated from:

$$N = 100\sqrt{\frac{L}{A_L}} \qquad L = A_L\left(\frac{N^2}{10{,}000}\right) \quad \text{(Amidon cores)}$$

$$N = \sqrt{\frac{L}{A_L}} \qquad L = A_L \times N^2 \quad \text{(Non-Amidon cores)}$$

where N = number turns; L = desired inductance; A_L = inductance index. Amidon specifies A_L as mH per 1,000 turns-squared. See non-Amidon manufacturer's core data sheet to determine appropriate units for L and A_L, usually nH and nH/turns-squared. Make sure you understand which units apply and use the A_L value appropriate for your core.

Ferrite Magnetic Properties

Property	*Unit*	*63/67-Mix*	*61-Mix*	*43-Mix*	*77 (72)-Mix*	*J (75)-Mix*	*31-Mix*
Initial perm.	(μ_i)	40	125	850	2,000	5,000	1500
Max. perm.		125	450	3,000	6000	8000	Not spec.
Saturation flux density @ 10 oe	gauss	1850	2350	2750	4600	3900	3400
Residual flux density	gauss	750	1,200	1,200	1150	1250	2500
Curie temp.	°C	450	350	130	200	140	>130
Vol. resistivity	ohm/cm	1×10^8	1×10^8	1×10^5	1×10^2	5×10^2	3×10^3
Resonant circuit frequency	MHz	15-25	0.2-10	0.01-1	0.001-1	0.001-1	*
Specific gravity		4.7	4.7	4.5	4.8	4.8	4.7
Loss factor	$\frac{1}{\mu_i Q}$	110×10^{-6} @25 MHz	32×10^{-6} @2.5 MHz	120×10^{-6} @1 MHz	4.5×10^{-6} @0.1 MHz	15×10^{-6} @0.1 MHz	20×10^{-6} @0.1 MHz
Coercive force	Oe	2.40	1.60	0.30	0.22	0.16	0.35
Temp. Coef. of initial perm.	%/°C (20°-70°)	0.10	0.15	1.0	0.60	0.90	1.6

*31-Mix is an EMI suppression material and not recommended for inductive uses.

Ferrite Toroids—Physical Properties

All physical dimensions in inches.

OD (in)	*ID (in)*	*Height (in)*	A_e	ℓ_e	V_e
0.230	0.120	0.060	0.00330	0.529	0.00174
0.375	0.187	0.125	0.01175	0.846	0.00994
0.500	0.281	0.188	0.02060	1.190	0.02450
0.825	0.520	0.250	0.03810	2.070	0.07890
1.142	0.750	0.295	0.05810	2.920	0.16950
1.400	0.900	0.500	0.12245	3.504	0.42700
2.400	1.400	0.500	0.24490	5.709	1.39080

Different height cores may be available for each core size.
A_e — Effective magnetic cross-sectional area (in)^2
ℓ_e — Effective magnetic path length (inches)
V_e — Effective magnetic volume (in)^3
To convert from (in)^2 to (cm)^2, divide by 0.155
To convert from (in)^3 to (cm)^3, divide by 0.0610
Courtesy of Amidon Assoc. and Fair-Rite Corp.

Because of their higher permeabilities, the A_L values for ferrite cores are higher than for powdered-iron cores. Amidon Corp. is the most common supplier of cores for amateurs and specifies A_L in mH per 1,000 turns-squared. Other manufacturers specify A_L in nH per turns-squared.

To calculate the inductance of a ferrite toroidal inductor when the number of turns and the core material are known:

$$L = \frac{A_L \times N^2}{1000000} \text{ (Amidon)}$$

where
L = the inductance in mH,
A_L = the inductance index in mH per 1,000 turns-squared, and
N = the number of turns.

$$L = A_L \times N^2 \text{ (Other manufacturers)}$$

where
L = the inductance in nH,
A_L = the inductance index in nH per turns-squared, and
N = the number of turns.

Calculations are performed similarly to the examples given for powdered-iron cores in the previous section. The builder must then ensure that the core is capable of holding the calculated number of turns of wire of the required wire size.

For inductors carrying both dc and ac currents, the upper saturation limit for most ferrites is a flux density of 2,000 gauss, with power calculations identical to those used for powdered-iron cores.

4.5 Transformers

When the ac source current flows through every turn of an inductor, the generation of a counter-voltage and the storage of energy during each half cycle is said to be by virtue of self-inductance. If another inductor — not connected to the source of the original current — is positioned so the magnetic field of the first inductor intercepts the turns of the second inductor, coupling the two inductors and creating mutual inductance as described earlier, a voltage will be induced and current will flow in the second inductor. A load such as a resistor may be connected across the second inductor to consume the energy transferred magnetically from the first inductor.

Figure 4.23 illustrates a pair of coupled inductors, showing an ac energy source connected to one, called the *primary inductor*, and a load connected to the other, called the *secondary inductor*. If the inductors are wound tightly on a magnetic core so that nearly all magnetic flux from the first inductor intersects with the turns of the second inductor, the pair is said to be *tightly coupled*. Inductors not sharing a common core and separated by a distance would be *loosely coupled*.

The signal source for the primary inductor may be household ac power lines, audio or other waveforms at low frequencies, or RF currents. The load may be a device needing power, a speaker converting electrical energy into sonic energy, an antenna using RF energy for communications or a particular circuit set up to process a signal from a preceding circuit. The uses of magnetically coupled energy in electronics are innumerable.

Mutual inductance (M) between inductors is measured in henrys. Two inductors have a mutual inductance of 1 H under the following conditions: as the primary inductor current changes at a rate of 1 A/s, the voltage across the secondary inductor is 1 V. The level of mutual inductance varies with many factors: the size and shape of the inductors, their relative positions and distance from each other, and the permeability of the inductor core material and of the space between them.

If the self-inductance values of two inductors are known (self-inductance is used in this section to distinguish it from the mutual inductance), it is possible to derive the mutual inductance by way of a simple experiment schematically represented in **Figure 4.24**. Without altering the physical setting or position of two inductors, measure the total coupled inductance, L_C, of the series-connected inductors with their windings complementing each other and again with their windings opposing each other. Since, for the two inductors, $L_C = L1 + L2 + 2M$, in the complementary case, and $L_O = L1 + L2 - 2M$ for the opposing case,

$$M = \frac{L_C - L_O}{4}$$

The ratio of magnetic flux set up by the secondary inductor to the flux set up by the primary inductor is a measure of the extent to which two inductors are coupled, compared to the maximum possible coupling between them. This ratio is the *coefficient of coupling*

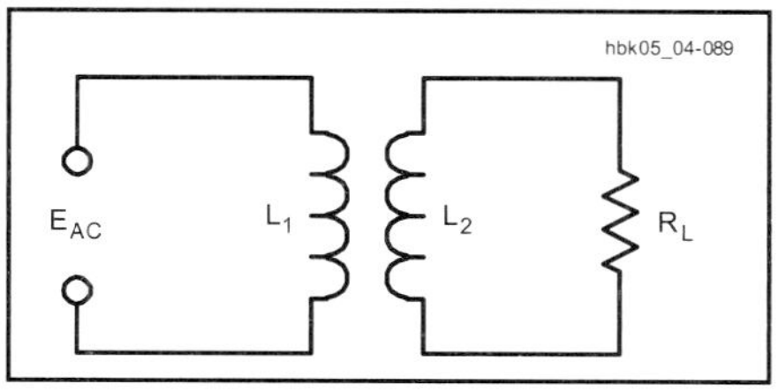

Figure 4.23 — A basic transformer: two inductors — one connected to an ac energy source, the other to a load — with coupled magnetic fields.

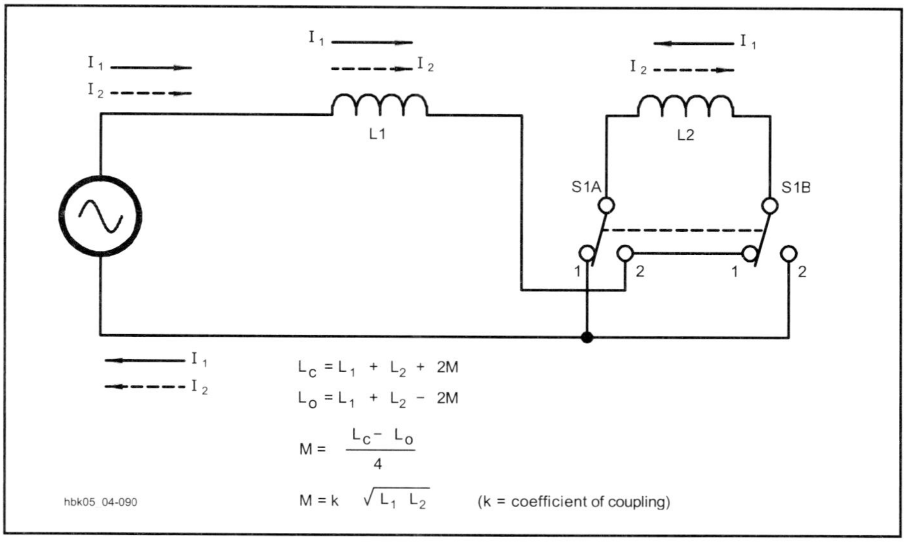

Figure 4.24 — An experimental setup for determining mutual inductance. Measure the inductance with the switch in each position and use the formula in the text to determine the mutual inductance.

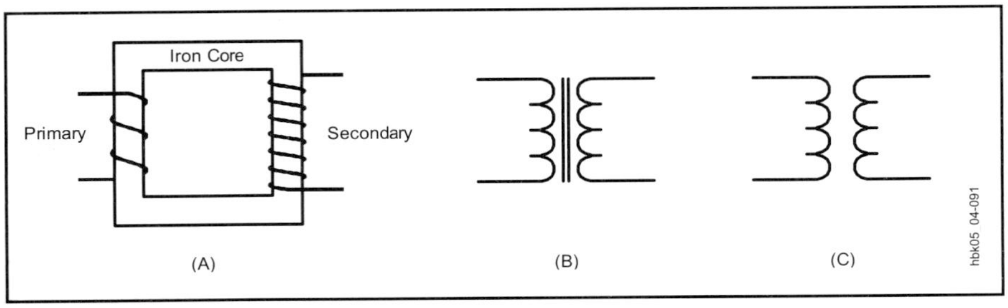

Figure 4.25 — A transformer. A is a pictorial diagram. Power is transferred from the primary coil to the secondary by means of the magnetic field. B is a schematic diagram of an iron-core transformer, and C is an air-core transformer.

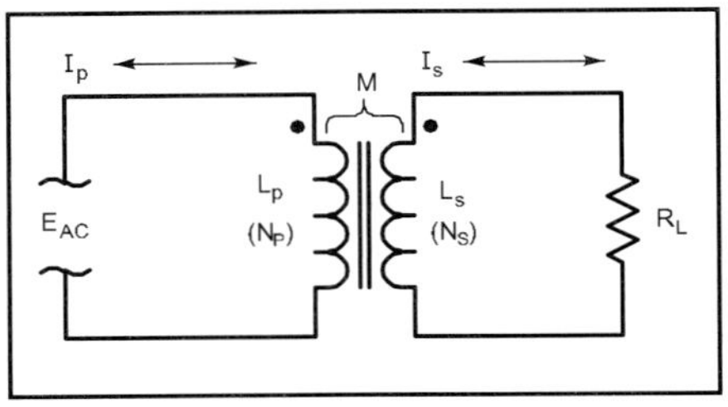

Figure 4.26 — The conditions for transformer action: two coils that exhibit mutual inductance, an ac power source, and a load. The magnetic field set up by the energy in the primary circuit transfers energy to the secondary for use by the load, resulting in a secondary voltage and current. The phasing dots at the top of each winding indicate the points at which ac voltages are in-phase.

(k) and is always less than 1. If k were to equal 1, the two inductors would have the maximum possible mutual coupling. Thus:

$$M = k\sqrt{L1\,L2}$$

where
M = mutual inductance in henrys,
L1 and L2 = individual coupled inductors, each in henrys, and
k = the coefficient of coupling.

Using the experiment above, it is possible to solve for k with reasonable accuracy.

Any two inductors having mutual inductance comprise a *transformer* having a *primary winding* or inductor and a *secondary winding* or inductor. The word "winding" is generally dropped so that transformers are said to have "primaries" and "secondaries." **Figure 4.25** provides a pictorial representation of a typical iron-core transformer, along with the schematic symbols for both iron-core and air-core transformers. Conventionally, the term *transformer* is most commonly applied to coupled inductors having a magnetic core material, while coupled air-wound inductors are not called by that name. They are still transformers, however.

We normally think of transformers as ac devices, since mutual inductance only occurs when magnetic fields are changing. A transformer connected to a dc source will exhibit mutual inductance only at the instants of closing and opening the primary circuit, or on the rising and falling edges of dc pulses, because only then does the primary winding have a changing field. There are three principle uses of transformers: to physically isolate the primary circuit from the secondary circuit, to transform voltages and currents from one level to another, and to transform circuit impedances from one level to another. These functions are not mutually exclusive and have many variations.

4.5.1 Basic Transformer Principles

The primary and secondary windings of a transformer may be wound on a core of magnetic material. The permeability of the magnetic material increases the inductance of the windings so a relatively small number of turns may be used to induce a given voltage value with a small current. A closed core having a continuous magnetic path, such as that shown in Figure 4.25, also tends to ensure that practically all of the field set up by the current in the primary winding will intercept or "cut" the turns of the secondary winding.

For power transformers and impedance-matching transformers used at audio frequencies, cores made of soft iron strips or sheets called *laminations* are most common and generally very efficient. At higher frequencies, ferrite or powdered-iron cores are more frequently used. This section deals with basic transformer operation at audio and power frequencies. RF transformer operation is discussed in the chapter on **RF Techniques**.

The following principles presume a coefficient of coupling (k) of 1, that is, a perfect transformer. The value k = 1 indicates that all the turns of both windings link with all the magnetic flux lines, so that the voltage induced per turn is the same with both windings. This condition makes the induced voltage independent of the inductance of the primary and secondary inductors. Iron-core transformers for low frequencies closely approach this ideal condition. **Figure 4.26** illustrates the conditions for transformer action.

The *phasing dots* at the top of each winding in Figure 4.26 are polarity markings that indicate phase relationships in transformer schematic diagrams. If the dots are placed at the same ends of the primary and secondary windings as in the figure, the polarity of the instantaneous voltage across the primary winding will be the same as that across the secondary winding. That means the phase shift between the primary and secondary winding will be zero and the directions of the secondary current (I_S) and primary current (I_P) will be the same. If the dots are at the opposite ends of the windings, the primary and secondary voltages will be out of phase and the primary and secondary currents will be in the opposite directions.

Winding polarity is important because it is determined by the direction in which the windings are wrapped on the transformer core. This also determines the direction of magnetic flux in the core and the polarity of the resulting induced voltage and current. This relationship is illustrated with diagrams at **circuitdigest.com/article/understanding-dot-convention-in-transformers**. Dot convention is also used in switchmode power conversion circuits (see the **Power Sources** chapter) to show how inductors and transformers store and release magnetic energy.

VOLTAGE RATIO

For a given varying magnetic field, the voltage induced in an inductor within the field is proportional to the number of turns in the inductor. When the two windings of a transformer are in the same field (which is the case when both are wound on the same closed core), it follows that the induced voltages will be proportional to the number of turns in each winding. In the primary, the induced voltage practically equals, and opposes, the applied voltage, as described earlier. Hence:

$$E_S = E_P\left(\frac{N_S}{N_P}\right)$$

where
E_S = secondary voltage,
E_P = primary applied voltage,
N_S = number of turns on secondary, and
N_P = number of turns on primary.

Example: A transformer has a primary with 400 turns and a secondary with 2,800 turns, and a voltage of 120 V is applied to the primary. What voltage appears across the secondary winding?

$$E_S = 120\text{ V}\left(\frac{2800}{400}\right) = 120\text{ V} \times 7 = 840\text{ V}$$

(Notice that the number of turns is taken as a known value rather than a measured quantity, so they do not limit the significant figures in the calculation.) Also, if 840 V is applied to the 2,800-turn winding (which then becomes the primary), the output voltage from the 400-turn winding will be 120 V.

Either winding of a transformer can be used as the primary, provided the winding has enough turns (enough inductance) to induce a voltage equal to the applied voltage without requiring an excessive current. The windings must also have insulation with a voltage rating sufficient for the voltages applied or created. Transformers are called *step-up* or *step-down* transformers depending on whether the secondary voltage is higher or lower than the primary voltage, respectively.

CURRENT OR AMPERE-TURNS RATIO

The current in the primary when no current is taken from the secondary is called the *magnetizing current* of the transformer. An ideal transformer, with no internal losses, would consume no power, since the current through the primary inductor would be 90° out of phase with the voltage. In any properly designed transformer, the power consumed by the transformer when the secondary is open (not delivering power) is only the amount necessary to overcome the losses in the iron core and in the resistance of the wire with which the primary is wound.

When power is transferred from the secondary winding to a load, the secondary current creates a magnetic field that opposes the field established by the primary current. For the induced voltage in the primary to equal the applied voltage, the original magnetizing field must be maintained. Therefore, enough additional current must flow in the primary to create a field exactly equal and opposite to the field set up by the secondary current, leaving the original magnetizing field.

In practical transformer calculations it may be assumed that the entire primary current is caused by the secondary load. This is justifiable because the magnetizing current should be very small in comparison with the primary load current at rated power output.

If the magnetic fields set up by the primary and secondary currents are to be equal, the number of ampere-turns must be equal in each winding. (See the previous discussion of magnetic fields and magnetic flux density.) Thus, primary current multiplied by the primary turns must equal the secondary current multiplied by the secondary turns.

$$I_P N_P = I_S N_S$$

and

$$I_P = I_S \left(\frac{N_S}{N_P} \right)$$

where

I_P = primary current,
I_S = secondary current,
N_P = number of turns in the primary winding, and
N_S = number of turns in the secondary winding.

Example: Suppose the secondary of the transformer in the previous example is delivering a current of 0.20 A to a load. What will be the primary current?

$$I_P = 0.20 \text{ A} \left(\frac{2800}{400} \right) = 0.20 \text{ A} \times 7 = 1.4 \text{ A}$$

Although the secondary voltage is higher than the primary voltage, the secondary current is lower than the primary current, and by the same ratio. The secondary current in an ideal transformer is 180° out of phase with the primary current, since the field in the secondary just offsets the field in the primary. The phase relationship between the currents in the windings holds true no matter what the phase difference between the current and the voltage of the secondary. In fact, the phase difference, if any, between voltage and current in the secondary winding will be *reflected* back to the primary as an identical phase difference.

POWER RATIO

A transformer cannot create power; it can only transfer it and change the voltage and current ratios. Hence, the power taken from the secondary cannot exceed that taken by the primary from the applied voltage source. There is always some power loss in the resistance of the windings and in the iron core, so in all practical cases the power taken from the source will exceed that taken from the secondary.

$$P_O = \eta \, P_I$$

where

P_O = power output from secondary,
P_I = power input to primary, and
η = efficiency.

The efficiency, η, is always less than 1 and is commonly expressed as a percentage: if η is 0.65, for instance, the efficiency is 65%.

Example: A transformer has an efficiency of 85.0% at its full-load output of 150 W. What is the power input to the primary at full secondary load?

$$P_I = \frac{P_O}{\eta} = \frac{150 \text{ W}}{0.850} = 176 \text{ W}$$

A transformer is usually designed to have the highest efficiency at the power output for which it is rated. The efficiency decreases with either lower or higher outputs. On the other hand, the losses in the transformer are relatively small at low output but increase as more power is taken. The amount of power that the transformer can handle is determined by its own losses, because these losses heat the wire and core. There is a limit to the temperature rise that can be tolerated, because too high a temperature can either melt the wire or cause the insulation to break down. A transformer can be operated at reduced output, even though the efficiency is low, because the actual loss will be low under such conditions. The full-load efficiency of small power transformers such as are used in radio receivers and transmitters usually lies between about 60 and 90%, depending on the size and design.

IMPEDANCE RATIO

In an ideal transformer — one without losses or leakage inductance (see Transformer Losses) — the primary power, $P_P = E_P I_P$, and secondary power, $P_S = E_S I_S$, are equal. The relationships between primary and secondary voltage and current are also known. Since impedance is the ratio of voltage to current, $Z = E/I$, the impedances represented in each winding are related as follows:

$$Z_P = Z_S \left(\frac{N_P}{N_S} \right)^2$$

where

Z_P = impedance at the primary terminals from the power source,
Z_S = impedance of load connected to secondary, and
N_P/N_S = turns ratio, primary to secondary.

The transformer is converting input power at one ratio of voltage to current (i.e., impedance) to output power at a different ratio of voltage to current (i.e., a different impedance).

A load of any given impedance connected to the transformer secondary will thus be transformed to a different value at the primary terminals. The impedance transformation is proportional to the square of the primary-to-secondary turns ratio. (Take care to use the primary-to-secondary turns ratio, since the secondary-to-primary ratio is more commonly used to determine the voltage transformation ratio.)

The term *looking into* is often used to mean the conditions observed from an external perspective at the terminals specified. For example, "impedance looking into" the transformer primary means the impedance measured externally to the transformer at the terminals of the primary winding.

Example: A transformer has a primary-to-secondary turns ratio of 0.6 (the primary has six-tenths as many turns as the secondary) and a load of 3,000 Ω is connected to the secondary. What is the impedance looking into the primary of the transformer?

$Z_P = 3{,}000\ \Omega \times (0.6)^2 = 3{,}000\ \Omega \times 0.36$

$Z_P = 1{,}080\ W$

By choosing the proper turns ratio, the impedance of a fixed load can be transformed to any desired value, within practical limits. If transformer losses can be neglected, the transformed (reflected) impedance has the same phase angle as the actual load impedance. Thus, if the load is a pure resistance, the load presented by the primary to the power source will also be a pure resistance. If the load impedance is complex, that is, if the load current and voltage are out of phase with each other, then the primary voltage and current will have the same phase angle.

Many devices or circuits require a specific value of load resistance (or impedance) for optimum operation. The impedance of the actual load that is to dissipate the power may be quite different from the impedance of the source device or circuit, so an *impedance-matching transformer* is used to change the actual load into an impedance of the desired value. The turns ratio required is:

$$\frac{N_P}{N_S} = \sqrt{\frac{Z_P}{Z_S}}$$

where

N_P / N_S = required turns ratio, primary to secondary,

Z_P = primary impedance required, and

Z_S = impedance of load connected to secondary.

Example: A transistor audio amplifier requires a load of 150 Ω for optimum performance, and is to be connected to a loudspeaker having an impedance of 4.0 Ω. What primary-to-secondary turns ratio is required in the coupling transformer?

$$\frac{N_P}{N_S} = \sqrt{\frac{Z_P}{Z_S}} = \sqrt{\frac{150\ \Omega}{4.0\ \Omega}} = \sqrt{38} = 6.2$$

The primary therefore must have 6.2 times as many turns as the secondary.

These relationships may be used in practical circuits even though they are based on an ideal transformer. Aside from the normal design requirements of reasonably low internal losses and low leakage reactance, the only other requirement is that the primary has enough inductance to operate with low magnetizing current at the voltage applied to the primary.

The primary terminal impedance of an iron-core transformer is determined wholly by the load connected to the secondary and by the turns ratio. If the characteristics of the transformer have an appreciable effect on the impedance presented to the power source, the transformer is either poorly designed or is not suited to the voltage and frequency at which it is being used. Most transformers will operate quite well at voltages from slightly above to well below the design figure.

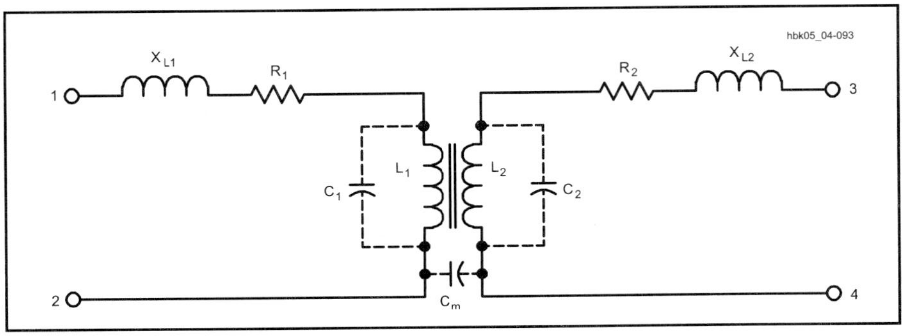

Figure 4.27 — A transformer as a network of resistances, inductances and capacitances. Only L1 and L2 contribute to the transfer of energy.

TRANSFORMER LOSSES

In practice, none of the formulas given so far provides truly exact results, although they afford reasonable approximations. Transformers in reality are not simply two coupled inductors, but a network of resistances and reactances, most of which appear in **Figure 4.27**. Since only the terminals numbered 1 through 4 are accessible to the user, transformer ratings and specifications take into account the additional losses created by these complexities.

In a practical transformer not all of the magnetic flux is common to both windings, although in well-designed transformers the amount of flux that cuts one winding and not the other is only a small percentage of the total flux. This *leakage flux* causes a voltage by self-induction in the winding creating the flux. The effect is the same as if a small *leakage inductance* existed independently of the main windings. Leakage inductance acts in exactly the same way as inductance inserted in series with the winding. It has, therefore, a certain reactance, depending on the amount of leakage inductance and the frequency. This reactance is called *leakage reactance* and is shown as X_{L1} and X_{L2} in Figure 4.27.

Current flowing through the leakage reactance causes a voltage drop. This voltage drop increases with increasing current (or frequency); hence, it increases as more power is taken from the secondary. The resistances of the transformer windings, R1 and R2, also cause voltage drops when there is current. Although these voltage drops are not in phase with those caused by leakage reactance, together they result in a lower secondary voltage under load than is indicated by the transformer turns ratio. Thus, in a practical transformer, the greater the

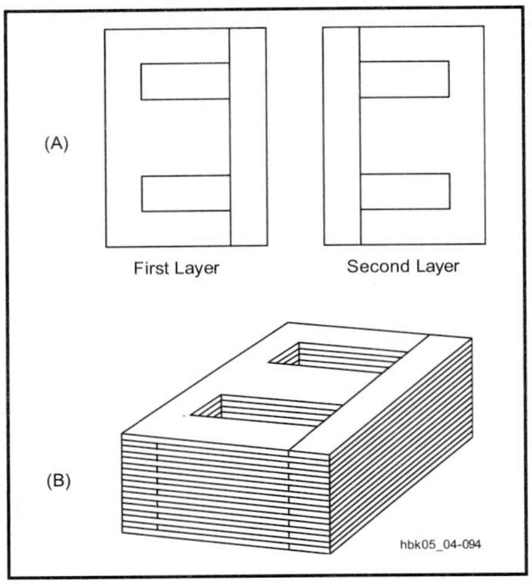

Figure 4.28 — A typical transformer iron core. The E and I pieces alternate direction in successive layers to improve the magnetic path while attenuating eddy currents in the core.

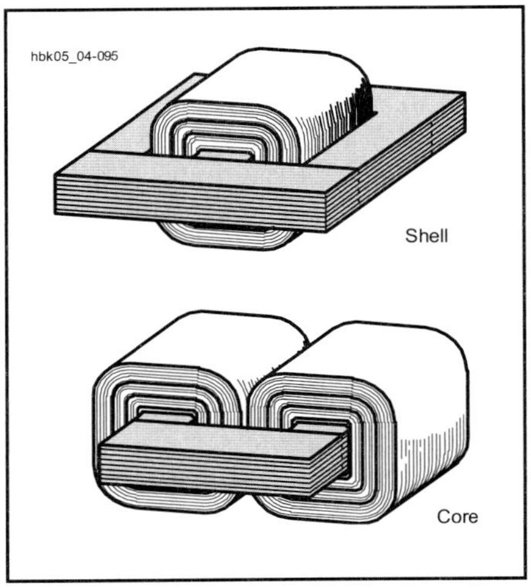

Figure 4.29 — Two common transformer constructions: shell and core.

secondary current, the smaller the secondary terminal voltage becomes.

At ac line frequencies (50 or 60 Hz), the voltage at the secondary, with a reasonably well-designed iron-core transformer, should not drop more than about 10% from open-circuit conditions to full load. The voltage drop may be considerably more than this in a transformer operating at audio frequencies, because the leakage reactance increases with frequency.

In addition to wire resistances and leakage reactances, certain unwanted or "stray" capacitances occur in transformers. The wire forming the separate turns of the windings acts as the plates of a small capacitor, creating a capacitance between turns and between the windings. This *distributed capacitance* appears in Figure 4.27 as C1, C2, and C_M. Moreover, transformer windings can exhibit capacitance relative to nearby metal, for example, the chassis, the shield and even the core. When current flows through a winding, each turn has a slightly different voltage than its adjacent turns. This voltage causes a small current to flow in these *interwinding* and *winding-to-winding capacitances.*

Although stray capacitances are of little concern with power and audio transformers, they become important as the frequency increases. In transformers for RF use, the stray capacitance can resonate with either the leakage reactance or, at lower frequencies, with the winding reactances, L1 or L2, especially under very light or zero loads. In the frequency region around resonance, transformers no longer exhibit the properties formulated above or the impedance properties to be described below.

Iron-core transformers also experience losses within the core itself. *Hysteresis losses* include the energy required to overcome the retentivity of the core's magnetic material. Circulating currents through the core's resistance are *eddy currents*, which form part of the total core losses. These losses, which add to the required magnetizing current, are equivalent to adding a resistance in parallel with L1 in Figure 4.27.

CORE CONSTRUCTION

Audio and power transformers usually employ silicon steel as the core material. With permeabilities of 5,000 or greater, these cores saturate at flux densities approaching 10^5 (Mx) per square inch of cross section. The cores consist of thin insulated laminations to break up potential eddy current paths.

Each core layer consists of an "E" and an "I" piece butted together, as represented in **Figure 4.28**. The butt point leaves a small gap. Each layer is reversed from the adjacent layers so that each gap is next to a continuous magnetic path, minimizing the effect of the gaps. This is different from an air-gapped inductor in which the air gap is maintained for all layers of laminations.

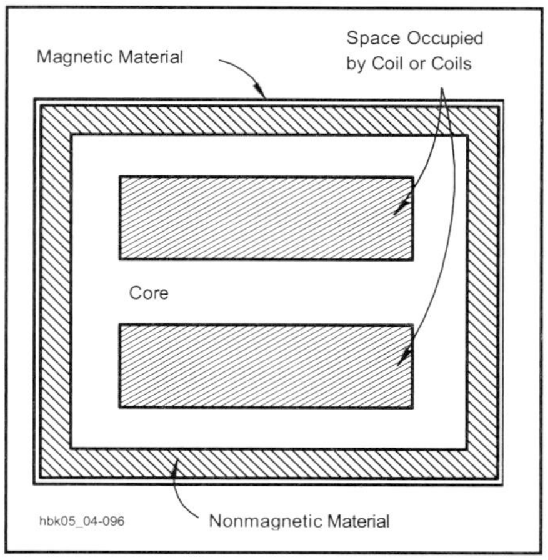

Figure 4.30 — A shielded transformer cross-section: the core plus an outer shield of magnetic material contain nearly all of the magnetic field.

Two core shapes are in common use, as shown in **Figure 4.29**. In the shell type, both windings are placed on the inner leg, while in the core type the primary and secondary windings may be placed on separate legs, if desired. This is sometimes done when it is necessary to minimize capacitance between the primary and secondary, or when one of the windings must operate at very high voltage.

The number of turns required in the primary for a given applied voltage is determined by the size, shape, and type of core material used, as well as the frequency. The number of turns required is inversely proportional to the cross-sectional area of the core. As a rough indication, windings of small power transformers frequently have about six to eight turns per volt on a core of 1-square-inch cross-section and have a magnetic path 10 or 12 inches in length. A longer path or smaller cross-section requires more turns per volt, and vice versa.

In most transformers the windings are wound in layers, with a thin sheet of treated-paper insulation between each layer. Thicker insulation is used between adjacent windings and between the first winding and the core.

SHIELDING

Because magnetic lines of force are continuous loops, shielding requires a complete path for the lines of force of the leakage flux. The high permeability of iron cores tends to concentrate the field, but additional shielding is often needed. As depicted in **Figure 4.30**, enclosing the transformer in a good magnetic material can restrict virtually all of the magnetic field in the outer case. The nonmagnetic material between the case and the core creates a region of high reluctance, attenuating the field before it reaches the case.

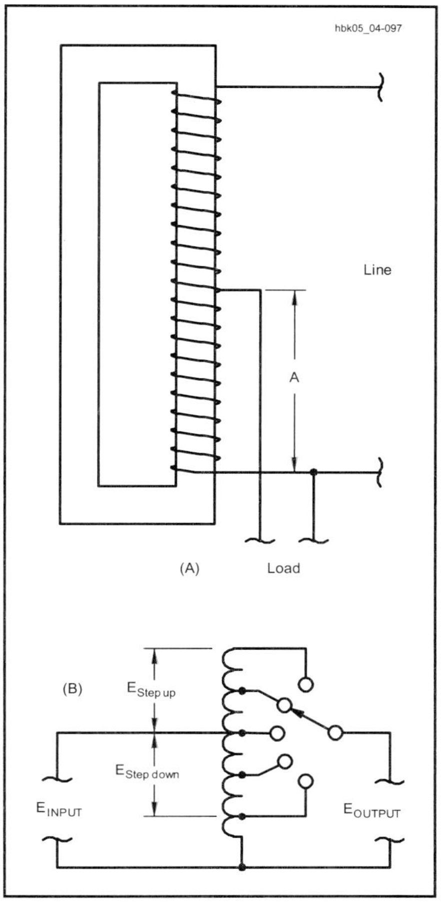

Figure 4.31 — The autotransformer is based on the transformer but uses only one winding. The pictorial diagram at A shows the typical construction of an autotransformer. The schematic diagram at B demonstrates the use of an autotransformer to step up or step down ac voltage, usually to compensate for excessive or deficient line voltage.

4.5.2 Transformer Identification

Many transformers, including power transformers, IF transformers, and audio transformers, are made to be installed on PC boards, and have terminals designed for that purpose. Some transformers are manufactured with wire leads that are color-coded to identify each connection. When colored wire leads are present, the color codes in **Tables 4.19**, **4.20**, and **4.22** usually apply. In addition, many miniature IF transformers are tuned with slugs, color-coded to signify their application. **Table 4.21** lists application versus slug color.

4.5.3 Autotransformers

The transformer principle can be used with only one winding instead of two, as shown in **Figure 4.31A**. The principles that relate volt-

Table 4.19
Power-Transformer Wiring Color Codes

Non-tapped primary leads:	Black
Tapped primary leads:	Common: Black Tap: Black/yellow striped Finish: Black/red striped
High-voltage plate winding:	Red
Center tap:	Red/yellow striped
Rectifier filament winding:	Yellow
Center tap:	Yellow/blue striped
Filament winding 1:	Green
Center tap:	Green/yellow striped
Filament winding 2:	Brown
Center tap:	Brown/yellow striped
Filament winding 3:	Slate
Center tap:	Slate/yellow striped

Table 4.22
Audio Transformer Wiring Color Codes

Plate lead of primary	Blue
B+ lead (plain or center-tapped)	Red
Plate (start) lead on center-tapped primaries	Brown (or blue if polarity is not important)
Grid (finish) lead to secondary	Green
Grid return (plain or center tapped)	Black
Grid (start) lead on center tapped secondaries	Yellow (or green if polarity not important)

Note: These markings also apply to line-to-grid and tube-to-line transformers.

Table 4.20
IF Transformer Wiring Color Codes

Plate lead:	Blue
B+ lead:	Red
Grid (or diode) lead:	Green
Grid (or diode) return:	Black

Note: If the secondary of the IF transformer is center-tapped, the second diode plate lead is green-and-black striped, and black is used for the center-tap lead.

Table 4.21
IF Transformer Slug Color Codes

Frequency	*Application*	*Slug color*
455 kHz	1st IF	Yellow
	2nd IF	White
	3rd IF	Black
	Osc tuning	Red
10.7 MHz	1st IF	Green
	2nd or 3rd IF	Orange, Brown or Black

age, current and impedance to the turns ratio also apply equally well. A one-winding transformer is called an *autotransformer*. The current in the common section (A) of the winding is the difference between the line (primary) and the load (secondary) currents, since these currents are out of phase. Hence, if the line and load currents are nearly equal, the common section of the winding may be wound with comparatively small wire. The line and load currents will be equal only when the primary (line) and secondary (load) voltages are not very different.

Autotransformers are used chiefly for boosting or reducing the power-line voltage by relatively small amounts. Figure 4.31B illustrates the principle schematically with a switched, stepped autotransformer. Continuously variable autotransformers are commercially available under a variety of trade names; Variac and Powerstat are typical examples.

Technically, tapped air-core inductors, such as the one in the network in Figure 3.48 at the close of the discussion of resonant circuits in the chapter on **Radio Fundamentals**, are also autotransformers. The voltage from the tap to the bottom of the winding is less than the voltage across the entire winding. Likewise, the impedance of the tapped part of the winding is less than the impedance of the entire winding. Because in this case, leakage reactances are great and the coefficient of coupling is quite low, the relationships that are true in a perfect transformer grow quite unreliable in predicting the exact values. For this reason, tapped inductors are rarely referred to as transformers. The stepped-down situation in Figure 3.48 is better approximated — at or close to resonance — by the formula

$$R_P = \frac{R_L \, X_{COM}^2}{X_L}$$

where

R_P = tuned-circuit parallel-resonant impedance,
R_L = load resistance tapped across part of the winding,
X_{COM} = reactance of the portion of the winding common to both the resonant circuit and the load tap, and
X_L = reactance of the entire winding.

The result is approximate and applies only to circuits with a Q of 10 or greater.

4.6 Practical Semiconductors

There are several different kinds of semiconductors used to build analog electronic circuits, including bipolar diodes, thyristors, transistors, field-effect transistors, and integrated circuits. (Vacuum tubes are discussed in the chapter on **RF Power Amplifiers**, their primary application in amateur radio.) Manufacturer websites are often a rich source of information on applying semiconductors, both in general and about the specific devices they offer.

Most semiconductors are labeled with industry standard part numbers, such as 1N4148 or 2N3904, and possibly a date or batch code. You will also encounter numerous manufacturer-specific part numbers and the so-called "house numbers" (marked with codes used by an equipment manufacturer instead of the standard part numbers). In such cases, it is often possible to find the standard equivalent or a suitable replacement by using one of the semiconductor cross-reference directories available from various replacement parts distributors. If you look up the house number and find the recommended replacement part, you can often find other standard parts that are replaced by that same part.

4.6.1 Device Characteristics

CHARACTERISTIC CURVES

Analog devices are described most completely with their *characteristic curves*. The characteristic curve is a plot of the interrelationships between two or three variables. The vertical (y) axis parameter is the output, or result of the device being operated with an input parameter on the horizontal (x) axis. Often the output is the result of two input values. The first input parameter is represented along the X-axis and the second input parameter by several curves, each for a different value. Characteristic curves are used to graphically describe a device's operation in both its linear and nonlinear regions.

Figure 4.32A shows the characteristic curve for a semiconductor diode with the Y-axis showing the forward current, I_F, flowing through the diode and the X-axis showing forward voltage, V_F, across the diode. This curve shows the relationship between current and voltage in the diode when it is conducting current. Characteristic curves showing voltage and current in two-terminal devices such as diodes are often called *I-V curves*. Characteristic curves may include all four quadrants of operation in which both axes include positive and negative values. It is also common for different scales to be used in the different quadrants, so inspect the legend for the curves carefully.

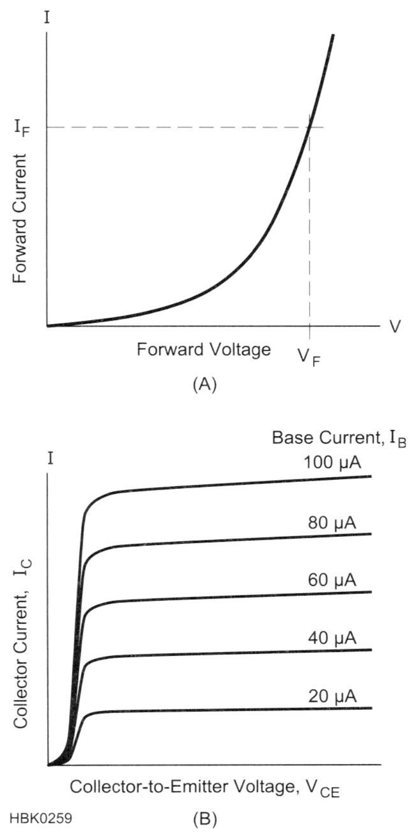

Figure 4.32 — Characteristic curves. A forward voltage vs forward current characteristic curve for a semiconductor diode is shown at (A). (B) shows a set of characteristic curves for a bipolar transistor in which the collector current vs collector-to-emitter voltage curve is plotted for five different values of base current.

The parameters plotted in a characteristic curve depend on how the device will be used so that the applicable design values can be obtained from the characteristic curve. The slope of the curve is often important because it relates changes in output to changes in input. To determine the slope of the curve, two closely spaced points along that portion of the curve are selected, each defined by its location along the x and y axes. If the two points are defined by (x_1,y_1) and (x_2,y_2), the slope, m, of the curve (which can be a gain, a resistance or a conductance, for example) is calculated as:

$$m = \frac{\Delta y}{\Delta x} = \frac{y_1 - y_2}{x_1 - x_2}$$

It is important to pick points that are close together or the slope will not reflect the actual behavior of the device. A device whose characteristic curve is not a straight line will not have a linear response to inputs because the slope changes with the value of the input parameter.

For a device in which three parameters interact, such as a transistor, sets of characteristic curves can be drawn. Figure 4.32B shows a set of characteristic curves for a bipolar transistor where collector current, I_C, is shown on the y axis and collector-to-emitter voltage, V_{CE}, is shown on the x axis. Because the amount of collector current also depends on base current, I_B, the curve is repeated several times for different values of I_B. From this set of curves, an amplifier circuit using this transistor can be designed to have specific values of gain.

BIASING

The operation of an analog signal-processing device is greatly affected by which portion of the characteristic curve is used to do the processing. The device's *bias point* is its set of operating parameters when no input signal is applied. The bias point is also known as the *quiescent point* or *Q-point*. By changing the bias point, the circuit designer can affect the relationship between the input and output signal. The bias point can also be considered as a dc offset of the input signal. Devices that perform analog signal processing require appropriate input signal biasing.

As an example, consider the characteristic curve shown in **Figure 4.33**. (The exact types of device and circuit are unimportant.) The characteristic curve shows the relationship between an input voltage and an output current. Increasing input voltage results in an increase in output current so that an input signal is reproduced at the output. The characteristic curve is linear in the middle but is quite nonlinear in its upper and lower regions.

In the circuit described by the figure, bias points are established by adding one of the three voltages, V_1, V_2 or V_3 to the input signal. Bias voltage V_1 results in an output current of I_1 when no input signal is present. This is shown as Bias Point 1 on the characteristic curve. When an input signal is applied, the input voltage varies around V_1 and the output current varies around I_1 as shown. If the dc value of the output current is subtracted, a reproduction of the input signal is the result.

If Bias Point 2 is chosen, we can see that the input voltage is reproduced as a changing output current with the same shape. In this case, the device is operating linearly. If either Bias Point 1 or Bias Point 3 is chosen, however, the shape of the output signal is distorted because the characteristic curve of the device is nonlinear in this region. Either the increasing portion of the input signal results in more variation than the decreasing portion (Bias

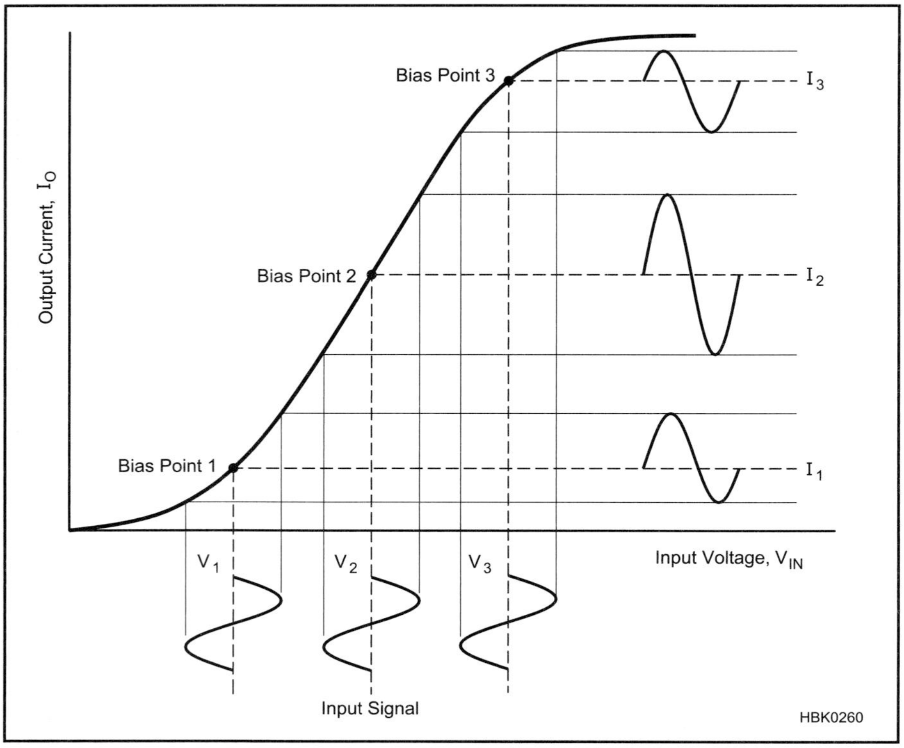

Figure 4.33 — Effect of biasing. An input signal may be reproduced linearly or nonlinearly depending on the choice of bias points.

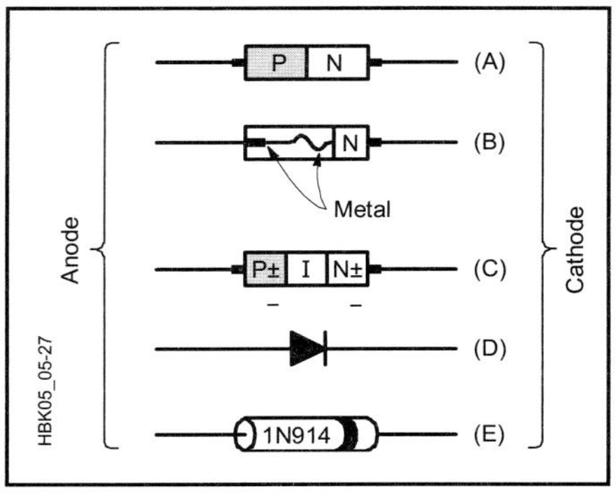

Figure 4.34 — Practical semiconductor diodes. All devices are aligned with anode on the left and cathode on the right. (A) Standard PN junction diode. (B) Point-contact or "cat's whisker" diode. (C) PIN diode formed with heavily doped P-type (P+), undoped (intrinsic) and heavily doped N-type (N+) semiconductor material. (D) Diode schematic symbol. (E) Diode package with marking stripe on the cathode end.

Point 1) or vice versa (Bias Point 3). Proper biasing is crucial to ensure that a device operates linearly.

MANUFACTURER'S DATA SHEETS

Manufacturer's data sheets list device characteristics, along with the specifics of the part type (polarity, semiconductor type), identity of the pins and leads (*pinouts*), and the typical use (such as small signal, RF, switching or power amplifier). The pin identification is important because, although common package pinouts are normally used, there are exceptions. Manufacturers may differ slightly in the values reported, but certain basic parameters are listed. Different batches of the same devices are rarely identical, so manufacturers specify the guaranteed limits for the parameters of their device. There are usually three values listed in the data sheet for each parameter: guaranteed minimum value, the guaranteed maximum value, and/or the typical value.

Another section of the data sheet lists ABSOLUTE MAXIMUM RATINGS, beyond which device damage may result. For example, the parameters listed in the ABSOLUTE MAXIMUM RATINGS section for a solid-state device are typically voltages, continuous currents, total device power dissipation (P_D) and operating- and storage-temperature ranges.

Rather than plotting the characteristic curves for each device, the manufacturer often selects key operating parameters that describe the device operation for the configurations and parameter ranges that are most commonly used. For example, a bipolar transistor data sheet might include an OPERATING PARAMETERS section. Parameters are listed in an OFF CHARACTERISTICS subsection and an ON CHARACTERISTICS subsection that describe the conduction properties of the device for dc voltages. The SMALL-SIGNAL CHARACTERISTICS section might contain a minimum Gain-Bandwidth Product (f_T or GBW), maximum output capacitance, maximum input capacitance, and the range of the transfer parameters applicable to a given device. Finally, the SWITCHING CHARACTERISTICS section might list absolute maximum ratings for Delay Time (t_d), Rise Time (t_r), Storage Time (t_s), and Fall Time (t_f). Other types of devices list characteristics important to operation of that specific device.

When selecting equivalent parts for replacement of specified devices, the data sheet provides the necessary information to tell if a given part will perform the functions of another. Lists of *cross-references* and *substitution guides* generally only specify devices that have nearly identical parameters. There are usually a large number of additional devices that can be chosen as replacements. Knowledge of the circuit requirements adds even more to the list of possible replacements. The device parameters should be compared individually to make sure that the replacement part meets or exceeds the parameter values of the original part required by the circuit. Be aware that in some applications a far superior part may fail as a replacement, however. A transistor with too much gain could easily oscillate if there were insufficient negative feedback to ensure stability.

4.6.2 Diodes

Although many types of semiconductor diodes are available, they share many common characteristics. The different types of diodes have been developed to optimize particular characteristics for one type of application. You will find many examples of diode applications throughout this book.

The diode symbol is shown in **Figure 4.34**. Forward current flows in the direction from anode to cathode, in the direction of the arrow. Reverse current flows from cathode to anode. (Current is considered to be conventional current as described in the **Electrical Fundamentals** chapter.) The anode of a semiconductor junction diode is made of P-type material and the cathode is made of N-type material, as indicated in Figure 4.34.

Most diodes are marked with a part number and some means of identifying the anode or cathode. A thick band or stripe is commonly used to identify the cathode lead or terminal. Stud-mount diodes are usually labeled with a small diode symbol to indicate anode and cathode. Diodes in axial lead packages are sometimes identified with a color scheme as shown in **Figure 4.35**. (Diode packaging dimensions and marking are given in the **Construction Techniques** chapter's section on PCB layout.) Many surface mount diodes are packaged in the same SMT packages as resistors.

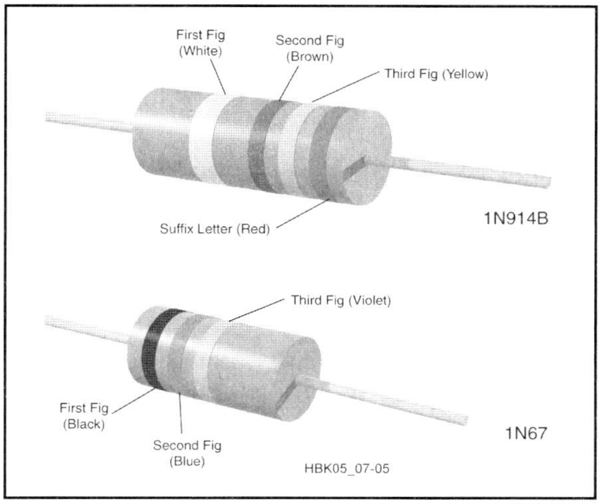

Figure 4.35 — Color-coding for semiconductor diodes. At A, the cathode is identified by the double-width first band. At B, the bands are grouped toward the cathode. Two-Figure designations are signified by a black first band. The color code is given in Table 4.2. The suffix-letter code is A-Brown, B-red, C-orange, D-yellow, E-green, F-blue. The 1N prefix is assumed.

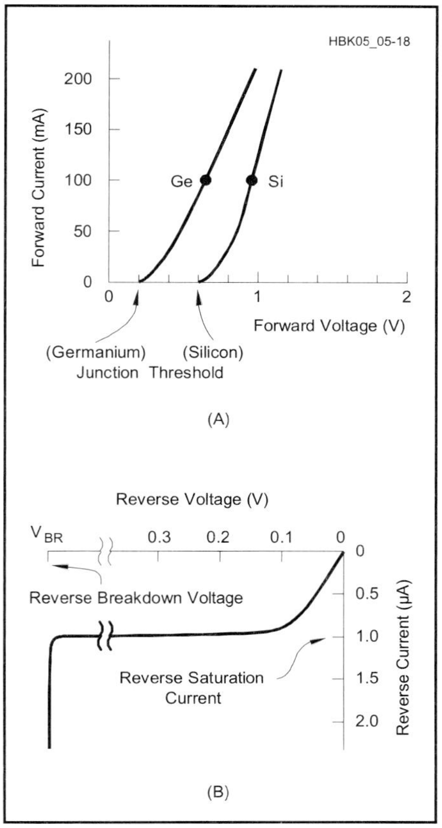

Figure 4.36 — Semiconductor diode (PN junction) characteristic curve. (A) Forward-biased (anode voltage higher than cathode) response for Germanium (Ge) and Silicon (Si) devices. Each curve breaks away from the X-axis at its junction threshold voltage. The slope of each curve is its forward resistance. (B) Reverse-biased response. Very small reverse current increases until it reaches the reverse saturation current (I_0). The reverse current increases suddenly and drastically when the reverse voltage reaches the reverse breakdown voltage, V_{BR}.

Packages containing multiple diodes and rectifier bridge configurations are also commonly available. Full-wave bridge packages are labeled with tildes (~) for the ac inputs and + and – symbols for the rectifier outputs. High-power diodes are often packaged in TO-220 packages with two leads. The package may be labeled with a diode symbol but if not, you will have to obtain the manufacturer's data sheet to identify the anode and cathode leads.

DIODE RATINGS

Five major characteristics distinguish standard junction diodes from one another: current handling capacity, maximum voltage rating, response speed, reverse leakage current and junction forward voltage. Each of these characteristics can be manipulated during manufacture to produce special purpose diodes.

Current Capacity

The ideal diode would have zero resistance in the forward direction and infinite resistance in the reverse direction. This is not the case for actual devices, which behave as shown in the plot of a diode response in **Figure 4.36A**. Note that the scales of the two graphs are drastically different. The inverse of the slope of the line (the change in voltage between two points on a straight portion of the line divided by the corresponding change in current) on the upper right is the resistance of the diode in the forward direction, R_F.

The range of voltages is small and the range of currents is large since the forward resistance is very small (in this example, about 2 Ω). Nevertheless, this resistance causes heat dissipation according to $P = I_F^2 \times R_F$.

In addition, there is a forward voltage, V_F, whenever the forward current is flowing. This also results in heat dissipation as $P = I \times V_F$. In power applications where the average forward current is high, heating from forward resistance and the forward voltage drop can be significant. Since forward current determines the amount of heat dissipation, the diode's power rating is stated as a *maximum average current*. Exceeding the current rating in a diode will cause excessive heating that leads to PN junction failure as described earlier.

Although fixed voltages are often used for diodes in small-signal applications (0.6 V for silicon PN-junction diodes and 0.3 V for germanium, for example), the actual forward voltage at higher currents can be significantly higher and must be taken into account in high-current applications, such as power supplies.

Peak Inverse Voltage (PIV)

In Figure 4.36B, the lower left portion of the curve illustrates a much higher resistance that increases from tens of kilohms to thousands of megohms as the reverse voltage gets larger, and then decreases to near zero (a nearly vertical line) very suddenly. This sudden change occurs because the diode enters *reverse breakdown* or when the reverse voltage becomes high enough to push current across the junction. The voltage at which this occurs is the *reverse breakdown voltage*. Unless the current is so large that the diode fails from overheating, breakdown is not destructive and the diode will again behave normally when the bias is removed. The maximum reverse voltage that the diode can withstand under normal use is the *peak inverse voltage (PIV)* rating. A related effect is *avalanche breakdown* in which the voltage across a device is greater than its ability to control or block current flow.

Response Speed

The speed of a diode's response to a change in voltage polarity limits the frequency of ac current that the diode can rectify. The diode response in Figure 4.36 shows how that diode will act at dc. As the frequency increases, the diode may not be able to turn current on and off as fast as the changing polarity of the signal.

Diode response speed mainly depends on *charge storage* in the depletion region. When forward current is flowing, electrons and holes fill the region near the junction to recombine. When the applied voltage reverses, these excess charges move away from the junction so that no recombination can take place. As reverse bias empties the depletion region of excess charge, it begins to act like a small capacitor formed by the regions containing majority carriers on either side of the junction and the depletion region acting as the dielectric. This *junction capacitance* is inversely proportional to the width of the depletion region and directly proportional to the cross-sectional surface area of the junction.

The effect of junction capacitance is to allow current to flow for a short period after the applied voltage changes from positive to negative. To halt current flow requires that the junction capacitance be charged. Charging this capacitance takes some time; a few μs for regular rectifier diodes and a few hundred nanoseconds for *fast-recovery* diodes. This is the diode's *charge-storage time*. The amount of time required for current flow to cease is the diode's *recovery time*.

Reverse Leakage Current

Because the depletion region is very thin, reverse bias causes a small amount of reverse leakage or reverse saturation current to flow from cathode to anode. This is typically 1 μA or less until reverse breakdown voltage is reached. Silicon diodes have lower reverse leakage currents than diodes made from other materials with higher carrier mobility, such as germanium.

The reverse saturation current I_S is not constant but is affected by temperature, with higher temperatures increasing the mobility of the majority carriers so that more of them cross the depletion region for a given amount of reverse bias. For silicon diodes (and transistors) near room temperature, I_S increases by a factor of 2 every 4.8 °C. This means that for every 4.8 °C rise in temperature, either the current doubles (if the voltage across it is constant), or if the current is held constant by other resistances in the circuit, the diode voltage will *decrease* by $V_T \times \ln 2 = 18$ mV. For germanium, the current doubles every 8 °C and for gallium-arsenide (GaAs), 3.7 °C. This dependence is highly reproducible and may actually be exploited to produce temperature-measuring circuits.

While the change resulting from a rise of several degrees may be tolerable in a circuit design, that from 20 or 30 degrees may not. Therefore, it's a good idea with diodes, just as with other components, to specify power ratings conservatively (2 to 4 times margin) to prevent self-heating.

While component derating does reduce self-heating effects, circuits must be designed for the expected operating environment. For example, mobile radios may face temperatures from –20 to +140 °F (–29 to 60 °C).

Forward Voltage

The amount of voltage required to cause majority carriers to enter the depletion region and recombine, creating full current flow, is called a diode's *forward voltage*, V_F. It depends on the type of material used to create the junction and the amount of current. For silicon junction diodes at normal currents, $V_F = 0.7$ V, and for germanium diodes, $V_F = 0.3$ V. As stated earlier, V_F also affects power dissipation in the diode. Schottky diodes are used when low forward voltages are required, particularly at high-currents, and exhibit forward voltages of 0.1 – 0.2 V at low currents.

POINT-CONTACT DIODES

One way to decrease charge storage time in the depletion region is to form a metal-semiconductor junction for which the depletion is very thin. This can be accomplished with a *point-contact diode*, where a thin piece of aluminum wire, often called a *whisker*, is placed in contact with one face of a piece of lightly doped N-type material. In fact, the original diodes used for detecting radio signals ("cat's whisker diodes") were made with a steel wire in contact with a crystal of impure lead (galena). Point-contact diodes have high response speed but poor PIV and current-handling ratings. The 1N34 germanium point-contact diode is the best-known example of a point-contact diode still in common use.

SCHOTTKY DIODES

An improvement to point-contact diodes, the *hot-carrier diode* is similar to a point-contact diode, but with more ideal characteristics attained by using more efficient metals, such as platinum and gold, that act to lower forward resistance and increase PIV. This type of contact is known as a *Schottky barrier*, and diodes made this way are called *Schottky diodes*. The junctions of Schottky diodes, being smaller, store less charge and as a result have shorter switching times and junction capacitances than standard PN-junction diodes. Their forward voltage is also lower, typically less than 0.4 V at low currents (see the device data sheet). In most other respects they behave similarly to PN diodes.

PIN DIODES

The PIN diode, shown in Figure 4.34C, is a *slow response* diode that is capable of passing RF and microwave signals when it is forward biased. This device is constructed with a layer of *intrinsic* (undoped) semiconductor placed between very highly doped P-type and N-type material (called P^+-type and N^+-type material to indicate the extra amount of doping), creating a *PIN junction*. These devices provide very effective switches for RF signals and are often used in transmit-receive switches in transceivers and amplifiers. The majority carriers in PIN diodes have longer than normal lifetimes before recombination, resulting in a slow switching process that causes them to act more like resistors than diodes at high radio frequencies. The amount of resistance can be controlled by the amount of forward bias applied to the PIN diode, which allows them to act as current-controlled attenuators. (For additional discussion of PIN diodes and projects in which they are used, see the chapters on **Transmitting**, **RF Power Amplifiers**, and **Test Equipment and Measurements**.)

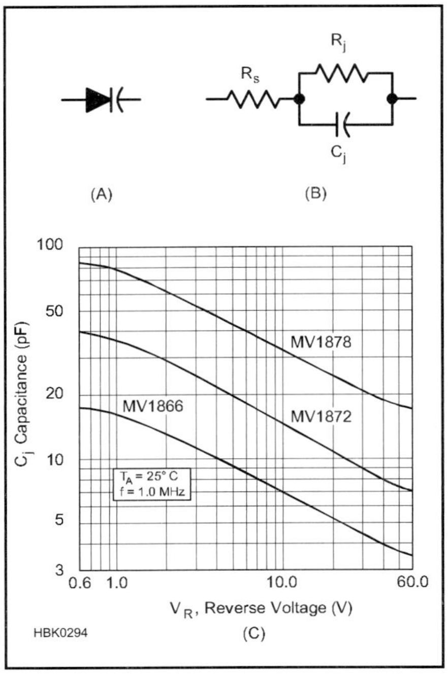

Figure 4.37 — Varactor diode. (A) Schematic symbol. (B) Equivalent circuit of the reverse biased varactor diode. R_S is the junction resistance, R_J is the leakage resistance and C_J is the junction capacitance, which is a function of the magnitude of the reverse bias voltage. (C) Plot of junction capacitance, C_J, as a function of reverse voltage, V_R, for three different varactor devices. Both axes are plotted on a logarithmic scale.

VARACTOR DIODES

Junction capacitance can be used as a circuit element by controlling the reverse bias voltage across the junction, creating a small variable capacitor. A *varactor* (also known by the trade name Varicap) is a diode with a junction specially formulated to have a relatively large range of capacitance values for a modest range of reverse bias voltages (**Figure 4.37**). They are used in oscillator and tuned circuits where a variable capacitor is needed. Operated with reverse bias, the depletion region forms a capacitor of variable width with a well-controlled voltage vs. capacitance function. Standard tuning diodes produce capacitances in the range of 5 to 40 pF. Hyper-abrupt tuning diodes produce variable capacitances to 100 pF or more for low frequency or wide-tuning-range applications.

The diode junction capacitance (C_j) under a reverse bias of V volts is given by

$$C_j = \frac{C_{j0}}{\sqrt{V_{on} - V}}$$

where C_{j0} = measured capacitance with zero applied voltage. As the reverse bias voltage on a diode increases, the width of the depletion region increases, decreasing its capacitance. Maximum capacitance occurs with minimum reverse voltage.

Note that the quantity under the radical is a large *positive* quantity for reverse bias. As seen from the equation, for large reverse biases C_j is inversely proportional to the square root of the voltage. Varactors are usually operated over a smaller range of bias so that the change in capacitance versus voltage is fairly linear.

Tuning diodes are specified by the capacitance produced at two reverse voltages, usually 2 and 30 V. This is called the *capacitance ratio* and is specified in units of pF per volt. Beyond this range, capacitance change with voltage can become non-linear and may cause signal distortion.

All diodes exhibit some capacitance when reverse biased. Amateurs have learned to use reverse biased Zener and rectifier diodes to form tuning diodes with 20 – 30 pF maximum capacitance. These "poor man's" tuning diodes are widely used in homebrew projects. However, because the capacitance ratio varies widely from one diode to the next, requiring experimentation to find a suitable diode, they are seldom used in published construction articles. When designing with varactor diodes, the reverse bias voltage must be absolutely free of noise since any variations in the bias voltage will cause changes in capacitance. For example, if the varactor is used to tune an oscillator, unwanted frequency shifts or instability will result if the reverse bias voltage is noisy. It is possible to frequency modulate a signal by adding the audio signal to the reverse bias on a varactor diode used in the carrier oscillator. (For examples of the use of varactors in oscillators and modulators, see the chapters on **Transmitting** and **Oscillators and Synthesizers**.)

ZENER DIODES

When the PIV of a reverse-biased diode is exceeded, the diode begins to conduct current as it does when it is forward biased. This current will not destroy the diode if it is limited to less than the device's maximum allowable value. By using heavy levels of doping during manufacture, a diode's PIV can be precisely controlled to be at a specific level, called the *Zener voltage*, creating a type of voltage reference. These diodes are called Zener diodes after their inventor, American physicist Clarence Zener.

When the Zener voltage is reached, the reverse voltage across the Zener diode remains constant even as the current through it changes. With an appropriate series current-limiting resistor, the Zener diode provides an accurate voltage reference (see **Figure 4.38**).

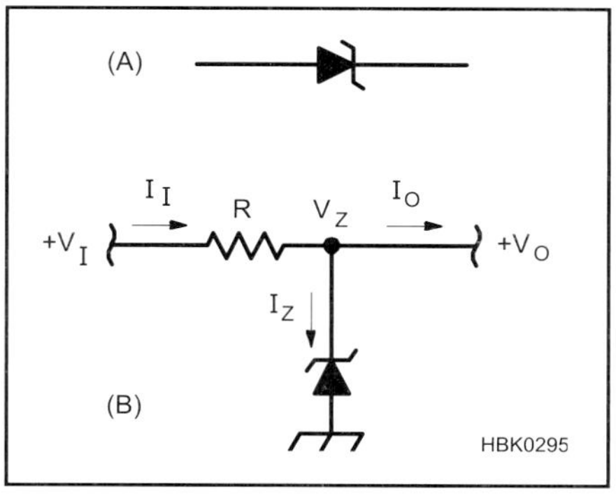

Figure 4.38 — Zener diode. (A) Schematic symbol. (B) Basic voltage regulating circuit. V_Z is the Zener reverse breakdown voltage. Above V_Z, the diode draws current until $V_I - I_I R = V_Z$. The circuit design should select R so that when the maximum current is drawn, $R < (V_I - V_Z) / I_O$. The diode should be capable of passing the same current when there is no output current drawn.

Zener diodes are rated by their reverse-breakdown voltage and their power-handling capacity, where $P = V_Z \times I_Z$. Since the same current must always pass though the resistor to drop the source voltage down to the reference voltage, with that current divided between the Zener diode and the load, this type of power source is very wasteful of current.

The Zener diode does make an excellent and efficient voltage reference in a larger voltage regulating circuit where the load current is provided from another device whose voltage is set by the reference. (See the **Power Sources** chapter for more information about using Zener diodes as voltage regulators.) When operating in the breakdown region, Zener diodes can be modeled as a simple voltage source.

Zener diodes are manufactured in a wide range of voltages and power handling capacities. Power dissipation in a Zener diode is equal to the Zener voltage multiplied by the average reverse current. The Zener voltage has a significant temperature coefficient and also varies with reverse current. To avoid excessive variations in Zener voltage, limit the diode's power dissipation to no more than 1/2 of the rated value; for precision uses, 1/5 to 1/10 of the rated power dissipation is recommended. Temperature-compensated Zener diodes are available with temperature coefficients as low as 5 parts per million per °C. If this is unacceptable, voltage reference integrated circuits based on Zener diodes have been developed that include additional circuitry to counteract temperature effects.

A variation of Zener diodes, *transient voltage suppressor* (*TVS*) diodes are designed to dissipate the energy in short-duration, high-voltage transients that would otherwise damage equipment or circuits. TVS diodes have large junction cross-sections so that they can handle large currents without damage. These diodes are also known by the trade name TransZorbs. Since the polarity of the transient can be positive, negative, or both, transient protection circuits can be designed with two devices connected with opposite polarities.

RECTIFIERS

The most common application of a diode is to perform rectification; that is, permitting current flow in only one direction. Power rectification converts ac current into pulsating dc current. There are three basic forms of power rectification using semiconductor diodes: half wave (1 diode), full-wave center-tapped (2 diodes) and full-wave bridge (4 diodes). These applications are shown in **Figure 4.39A**, **B** and **C** and are more fully described in the **Power Sources** chapter.

The most important diode parameters to consider for power rectification are the PIV and current ratings. The peak negative voltages that are blocked by the diode must be smaller in magnitude than the PIV and the peak current through the diode when it is forward biased must be less than the maximum average forward current.

Rectification is also used at much lower current levels in modulation and demodulation and other types of analog signal processing circuits. For these applications, the diode's response speed and junction forward voltage are the most important ratings.

4.6.3 Bipolar Junction Transistors (BJT)

The bipolar junction transistor is a *current-controlled device* with three basic terminals; *emitter*, *collector* and *base*. The current between the emitter and the collector is controlled by the current between the base and emitter. The convention when discussing transistor operation is that the three currents into the device are positive (I_c into the collector, I_b into the base and I_e into the emitter). Kirchhoff's Current Law applies to transistors just as it does to passive electrical networks: the total current entering the device must be zero. Thus, the relationship between the currents into a transistor can be generalized as

$$I_c + I_b + I_e = 0$$

which can be rearranged as necessary. For example, if we are interested in the emitter current,

$$I_e = -(I_c + I_b)$$

The back-to-back diode model shown in Figure 2.36 is appropriate for visualization of transistor construction. In actual transistors, however, the relative sizes of the collector, base and emitter regions differ. A common transistor configuration that spans a distance of 3 mm between the collector and emitter contacts typically has a base region that is only 25 μm across.

The operation of the bipolar transistor is described graphically by characteristic curves as shown in **Figure 4.40**. These are similar to the I-V characteristic curves for the two-terminal devices described in the preceding sections. The parameters shown by the curves depend on the type of circuit in which they are measured, such as common emitter or common collector. The output characteristic shows a set of curves for either collector or emitter current versus collector-emitter voltage at various values of input current (either base or emitter). The input characteristic shows the voltage between the input and common terminals (such as base-emitter) versus the input current for different values of output voltage.

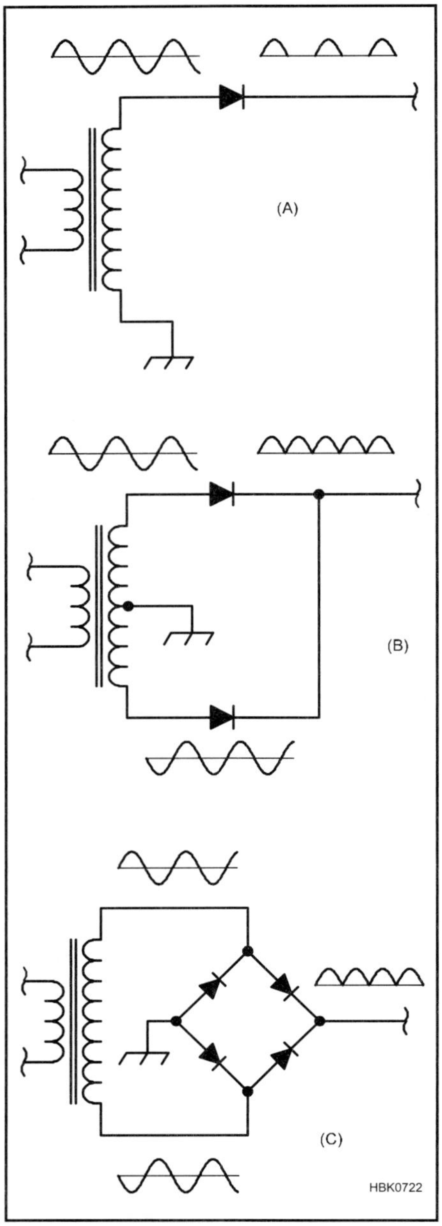

Figure 4.39 — Diode rectifier circuits. (A) Half wave rectifier circuit. Only when the ac voltage is positive does current pass through the diode. Current flows only during half of the cycle. (B) Full-wave center-tapped rectifier circuit. Center-tap on the transformer secondary is grounded and the two ends of the secondary are 180° out of phase. (C) Full-wave bridge rectifier circuit. In each half of the cycle two diodes conduct.

CURRENT GAIN

Two parameters describe the relationships between the three transistor currents at low frequencies:

$$\alpha = -\frac{\Delta I_C}{\Delta I_E} = 1$$

$$\beta = \frac{\Delta I_C}{\Delta I_B}$$

The relationship between α and β is defined as

$$\alpha = -\frac{\beta}{1+\beta}$$

Another designation for β is often used: h_{FE}, the *forward dc current gain.* (The "h" refers to "h parameters," a set of transfer parameters for describing a two-port network and described in more detail in the **RF Techniques** chapter.) The symbol h_{fe}, in which the subscript is in lower case, is used for the forward current gain of ac signals.

OPERATING REGIONS

Current conduction between collector and emitter is described by *regions* of the transistor's characteristic curves in Figure 4.40. (References such as *common-emitter* or *common-base* refer to the configuration of the circuit in which the parameter is measured.) The transistor is in its *active* or *linear region* when the base-collector junction is reverse biased and the base-emitter junction is forward biased. The slope of the output current, I_O, versus the output voltage, V_O, is virtually flat, indicating that the output current is nearly independent of the output voltage. In this region, the output circuit of the transistor can be modeled as a constant-current source controlled by the input current. The slight slope that does exist is due to base-width modulation (known as the *Early effect*).

When both the junctions in the transistor are forward biased, the transistor is said to be in its *saturation region*. In this region, V_O is nearly zero and large changes in I_O occur for very small changes in V_O. Both junctions in the transistor are reverse-biased in the *cutoff region*. Under this condition, there is very little current in the output, only the nanoamperes or microamperes that result from the very small leakage across the input-to-output junction. Finally, if V_O is increased to very high values, avalanche breakdown begins as in a PN-junction diode and output current increases rapidly. This is the *breakdown region*, not shown in Figure 4.40.

These descriptions of junction conditions are the basis for the use of transistors. Various configurations of the transistor in circuitry make use of the properties of the junctions to

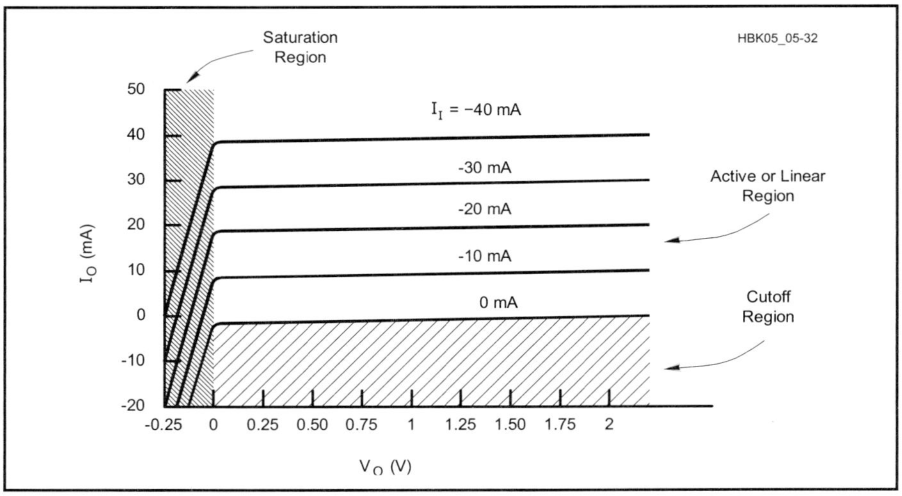

Figure 4.40 — Transistor response curve output characteristics. The X-axis is the output voltage, and the Y-axis is the output current. Different curves are plotted for various values of input current. The three regions of the transistor are its cutoff region, where no current flows in any terminal, its active region, where the output current is nearly independent of the output voltage and there is a linear relationship between the input current and the output current, and the saturation region, where the output current has large changes for small changes in output voltage.

serve different purposes in analog signal processing.

OPERATING PARAMETERS

A typical general-purpose bipolar-transistor data sheet lists important device specifications. Parameters listed in the ABSOLUTE MAXIMUM RATINGS section are the three junction voltages (V_{CEO}, V_{CBO} and V_{EBO}), the continuous collector current (I_C), the total device power dissipation (P_D) and the operating and storage temperature range. Exceeding any of these parameters is likely to cause the transistor to be destroyed. (The "O" in the suffixes of the junction voltages indicates that the remaining terminal is not connected, or open.)

In the OPERATING PARAMETERS section, three guaranteed minimum junction breakdown voltages are listed $V_{(BR)CEO}$, $V_{(BR)CBO}$ and $V_{(BR)EBO}$. Exceeding these voltages is likely to cause the transistor to enter avalanche breakdown, but if current is limited, permanent damage may not result.

Under ON CHARACTERISTICS are the guaranteed minimum dc current gain (β or h_{FE}), guaranteed maximum collector-emitter saturation voltage, $V_{CE(SAT)}$, and the guaranteed maximum base-emitter on voltage, $V_{BE(ON)}$. Two guaranteed maximum collector cutoff currents, I_{CEO} and I_{CBO}, are listed under OFF CHARACTERISTICS.

The next section is SMALL-SIGNAL CHARACTERISTICS, where the guaranteed minimum current gain-bandwidth product, BW or f_T, the guaranteed maximum output capacitance, C_{obo}, the guaranteed maximum input capacitance, C_{ibo}, the guaranteed range of input impedance, h_{ie}, the small-signal current gain, h_{fe}, the guaranteed maximum voltage feedback ratio, h_{re} and output admittance, h_{oe} are listed.

Finally, the SWITCHING CHARACTERISTICS section lists absolute maximum ratings for delay time, t_d; rise time, t_r; storage time, t_s; and fall time, t_f.

4.6.4 Field-Effect Transistors (FET)

FET devices are controlled by the voltage level of the input rather than the input current, as in the bipolar transistor. FETs have three basic terminals, the *gate*, the *source* and the *drain*. They are analogous to bipolar transistor terminals: the gate to the base, the source to the emitter, and the drain to the collector. Symbols for the various forms of FET devices are pictured in **Figure 4.41**.

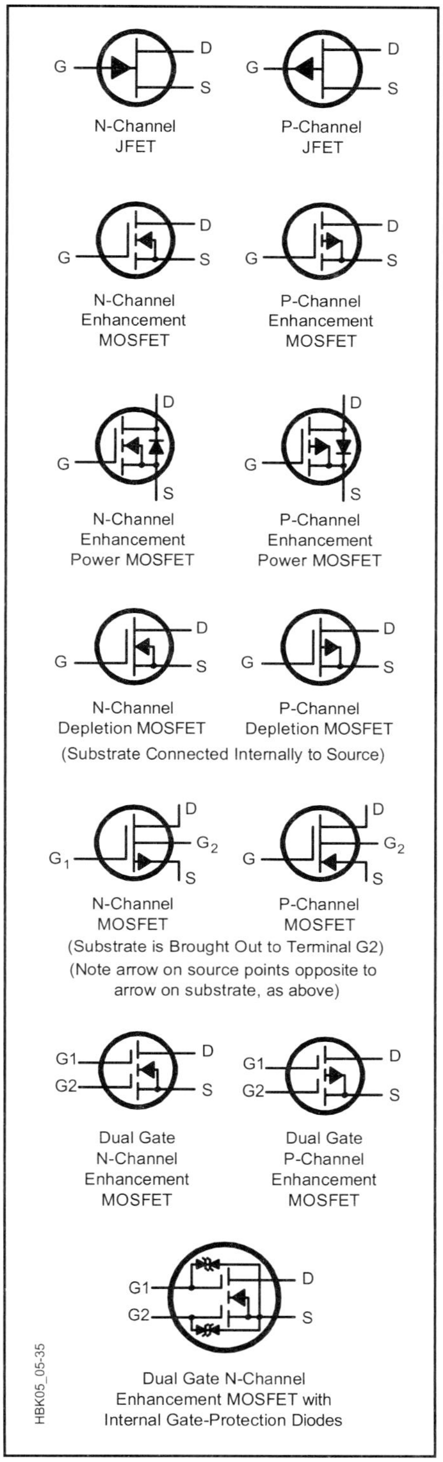

Figure 4.41 —FET schematic symbols.

The FET gate has a very high impedance, so the input can be modeled as an open circuit. The voltage between gate and source, V_{GS}, controls the resistance of the drain-source channel, r_{DS}, and so the output of the FET is modeled as a current source, whose output current is controlled by the input voltage.

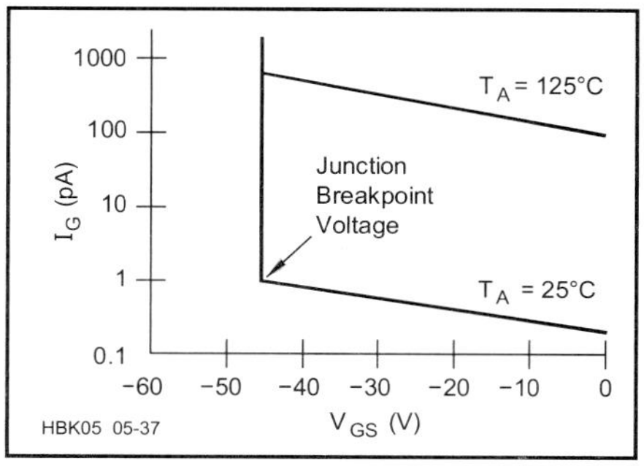

Figure 4.42 — JFET input leakage curves for common source amplifier configuration. Input voltage (V_{GS}) on the X-axis versus input current (I_G) on the Y-axis, with two curves plotted for different operating temperatures, 25 °C and 125 °C. Input current increases greatly when the gate voltage exceeds the junction breakpoint voltage.

The action of the FET channel is so nearly ideal that, as long as the JFET gate does not become forward biased and inject current from the base into the channel, the drain and source currents are virtually identical. For JFETs the *gate leakage current*, I_G, is a function of V_{GS} and this is often expressed with an *input curve* (see **Figure 4.42**). The point at which there is a significant increase in I_G is called the *junction breakpoint voltage*. Because the gate of MOSFETs is insulated from the channel, gate leakage current is insignificant in these devices.

The dc channel resistance, r_{DS}, is specified in data sheets to be less than a maximum value when the device is biased on ($r_{DS(on)}$). When the gate voltage is maximum ($V_{GS} = 0$ for a JFET), $r_{DS(on)}$ is minimum. This describes the effectiveness of the device as an analog switch. Channel resistance is approximately the same for ac and dc signals until at high frequencies the capacitive reactances inherent in the FET structure become significant.

FETs also have strong similarities to vacuum tubes in that input voltage between the grid and cathode controls an output current between the plate and cathode. (See the chapter on **RF Power Amplifiers** for more information on vacuum tubes.)

Both JFET and MOSFET devices can be used in a wide variety of voltage-controlled circuits from variable attenuators to variable-gain amplifiers to servo circuits. A complete tutorial on these circuits is beyond the scope of this section but the subject is well-covered by the five-part series of articles by Ron Quan of Linear Integrated Circuits. It was published by *EDN* at **www.edn.com/a-guide-to-using-fets-for-voltage-controlled-circuits-part-1**. (Replace -1 with the part you wish to read.)

FORWARD TRANSCONDUCTANCE

The change in FET drain current caused by a change in gate-to-source voltage is called *forward transconductance, g_m*.

$$g_m = \frac{\Delta I_{DS}}{\Delta V_{GS}}$$

or

$$\Delta I_{DS} = g_m \, DV_{GS}$$

The input voltage, $V_{GS,}$ is measured between the FET gate and source and drain current, I_{DS}, flows from drain to source. Analogous to a bipolar transistor's current gain, the units of transconductance are siemens (S) because it is the ratio of current to voltage. (Both g_m and g_{fs} are used interchangeably to indicate transconductance. Some sources specify g_{fs} as the *common-source forward transconductance*. This chapter uses g_m, the most common convention in the reference literature.)

OPERATING REGIONS

The most useful relationships for FETs are the output and transconductance response characteristic curves in **Figure 4.43**. (References such as *common-source* or *common-gate* refer to the configuration of the circuit in which the parameter is measured.) Transconductance curves relate the drain current, I_D, to gate-to-source voltage, V_{GS}, at various drain-source voltages, V_{DS}. The FET's forward transconductance, g_m, is the slope of the lines in the forward transconductance curve. The same parameters are interrelated in a different way in the output characteristic, in which I_D is shown versus V_{DS} for different values of V_{GS}.

Like the bipolar transistor, FET operation can be characterized by regions. The *ohmic region* is shown at the left of the FET output characteristic curve in **Figure 4.44** where I_D is increasing nearly linearly with V_{DS} and the FET is acting like a resistance controlled by V_{GS}. As V_{DS} continues to increase, I_D saturates and becomes nearly constant. This is the FET's *saturation region* in which the channel of the FET can be modeled as a constant-current source. V_{DS} can become so large that V_{GS} no longer controls the conduction of the device and avalanche breakdown occurs as in bipolar transistors and PN-junction diodes. This is the *breakdown region*, shown in Figure 4.44 where the curves for I_D break sharply upward. If V_{GS} is less than V_P, so that transconductance is zero, the FET is in the *cutoff region*.

OPERATING PARAMETERS

A typical FET data sheet gives ABSOLUTE MAXIMUM RATINGS for V_{DS}, V_{DG}, V_{GS} and I_D, along with the usual device dissipation (P_D) and storage temperature range. Exceeding these limits usually results in destruction of the FET.

Under OPERATING PARAMTERS the OFF CHARACTERISTICS list the gate-source breakdown voltage, $V_{GS(BR)}$, the reverse gate current, I_{GSS} and the gate-source cutoff voltage, $V_{GS(OFF)}$. Exceeding $V_{GS(BR)}$ will not permanently damage the device if current is limited. The primary ON CHARACTERISTIC parameters are the channel resistance, r_{DS}, and the zero-gate-voltage drain current (I_{DSS}). An FET's dc channel resistance, r_{DS}, is specified in data sheets to be less than a maximum value when the device is biased on ($r_{DS(on)}$). For ac signals, $r_{ds(on)}$ is not necessarily the same as $r_{DS(on)}$, but it is not very different as long as the frequency is not so high that capacitive reactance in the FET becomes significant.

The SMALL SIGNAL CHARACTERISTICS include the forward transfer admittance, y_{fs}, the output admittance, y_{os}, the static drain-source on resistance, $r_{ds(on)}$ and various capacitances such as input capacitance, C_{iss}, reverse transfer capacitance, C_{rss}, and the drain-substrate capacitance, $C_{d(sub)}$. FUNCTIONAL CHARACTERISTICS include the noise figure, NF, and the common source power gain, G_{ps}.

MOSFETS

As described earlier, the MOSFET's gate is insulated from the channel by a thin layer of nonconductive oxide, doing away with any appreciable gate leakage current. Because of this isolation of the gate, MOSFETs do not need input and reverse transconductance curves. Their output curves (**Figure 4.45**) are similar to those of the JFET. The gate acts as a small capacitance between the gate and both the source and drain.

The output and transconductance curves in Figure 4.45A and 4.45B show that the depletion-mode N-channel MOSFET's transconductance is positive at $V_{GS} = 0$, like that of the N-channel JFET. Unlike the JFET, however, increasing V_{GS} does not forward-bias the gate-source junction and so the device can be operated with $V_{GS} > 0$.

In the enhancement-mode MOSFET, transconductance is zero at $V_{GS} = 0$. As V_{GS} is increased, the MOSFET enters the ohmic region. If V_{GS} increases further, the saturation region is reached and the MOSFET is said to be *fully-on*, with r_{DS} at its minimum value. The behavior of the enhancement-mode MOSFET is similar to that of the bipolar transistor in this regard.

The relatively flat regions in the MOSFET output curves are often used to provide a con-

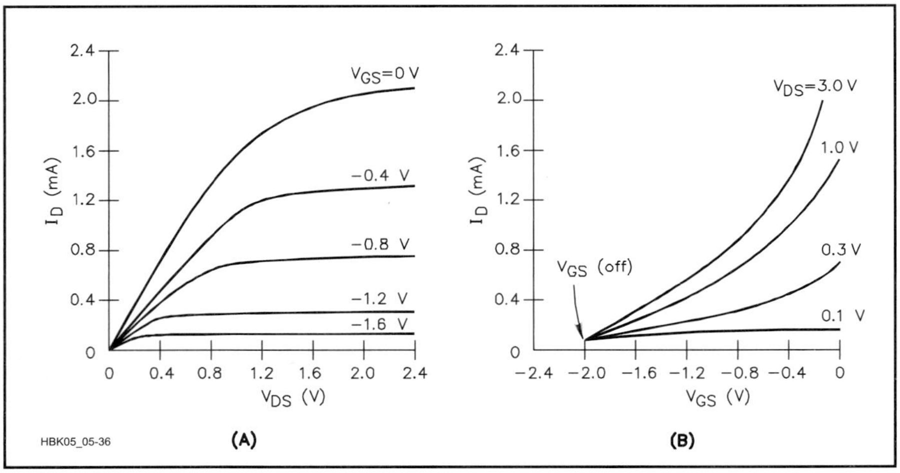

Figure 4.43 — JFET output and transconductance response curves for common source amplifier configuration. (A) Output voltage (V_{DS}) on the X-axis versus output current (I_D) on the Y-axis, with different curves plotted for various values of input voltage (V_{GS}). (B) Transconductance curve with the same three variables rearranged: V_{GS} on the X-axis, I_D on the Y-axis and curves plotted for different values of V_{DS}.

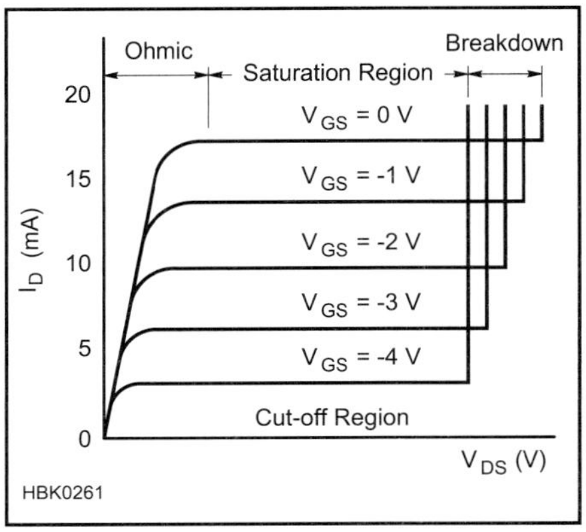

Figure 4.44 — JFET operating regions. At the left, I_D is increasing rapidly with V_{GS} and the JFET can be treated as resistance (R_{DS}) controlled by V_{GS}. In the saturation region, drain current, I_D, is relatively independent of V_{GS}. As V_{DS} increases further, avalanche breakdown begins and I_D increases rapidly.

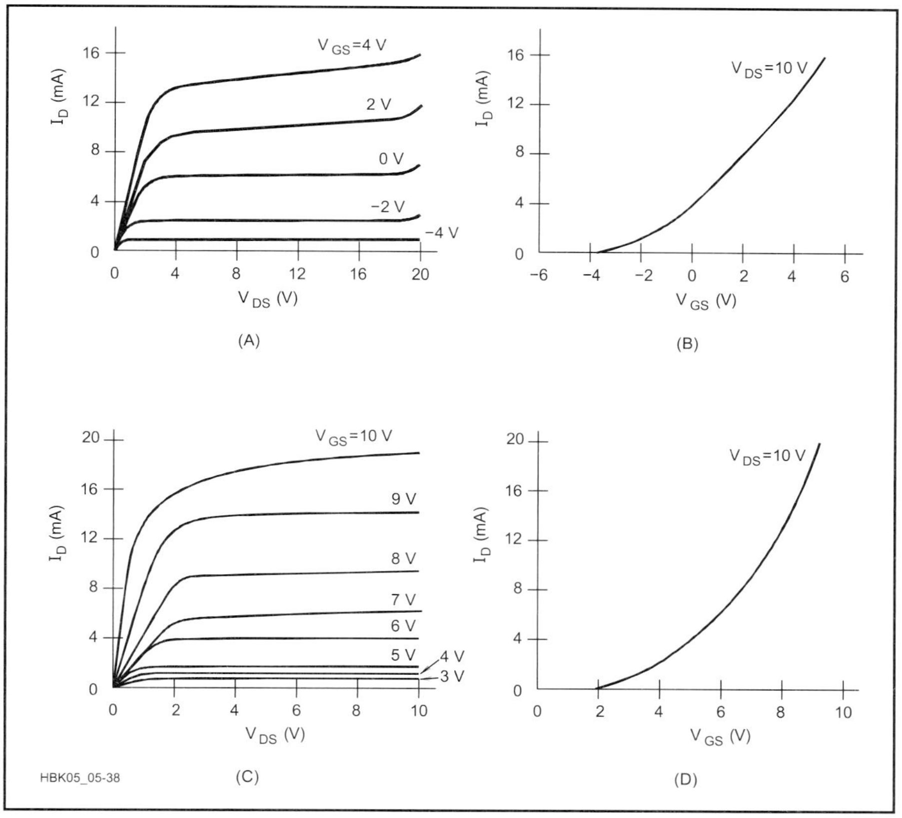

Figure 4.45 — MOSFET output [(A) and (C)] and transconductance [(B) and (D)] response curves. Plots (A) and (B) are for an N-channel depletion mode device. Note that V_{GS} varies from negative to positive values. Plots (C) and (D) are for an N-channel enhancement mode device. V_{GS} has only positive values.

stant current source. As is plotted in these curves, the drain current, I_D, changes very little as the drain-source voltage, V_{DS}, varies in this portion of the curve. Thus, for a fixed gate-source voltage, V_{GS}, the drain current can be considered to be constant over a wide range of drain-source voltages.

Multiple-gate MOSFETs are also available. Due to the insulating layer, the two gates are isolated from each other and allow two signals to control the channel simultaneously with virtually no loading of one signal by the other. A common application of this type of device is an automatic gain control (AGC) amplifier. The signal is applied to one gate and a rectified, low-pass filtered form of the output (the AGC voltage) is fed back to the other gate. Another common application is as a mixer in which the two input signals are applied to the pair of gates.

These two application notes are excellent sources of information for using these devices in radio applications. They can be found online by searching for the full name. *AN1020 — Active Mixer Design Using the NE25139 Dual Gate MESFET* from California Eastern Laboratories and *AN423 — Field Effect Transistor RF Amplifier Design Techniques* from Freescale Semiconductor.

MOSFET Gate Protection

The MOSFET is constructed with a very thin layer of SiO_2 for the gate insulator. This layer is extremely thin in order to improve the transconductance of the device, but this makes it susceptible to damage from high voltage levels, such as *electrostatic discharge* (ESD) from static electricity. If enough charge accumulates on the gate terminal, it can *punch through* the gate insulator and destroy it. The insulation of the gate terminal is so good that virtually none of this potential is eased by leakage of the charge into the device. While this condition makes for nearly ideal input impedance (approaching infinity), it puts the device at risk of destruction from even such seemingly innocuous electrical sources as static electrical discharges from handling.

Some MOSFET devices contain an internal Zener diode with its cathode connected to the gate and its anode to the substrate. If the voltage at the gate rises to a damaging level the Zener diode junction conducts, bleeding excess charges off to the substrate. When voltages are within normal operating limits the Zener has little effect on the signal at the gate, although it may decrease the input impedance of the MOSFET.

This solution will not work for all MOSFETs. The Zener diode must always be reverse biased to be effective. In the enhancement-mode MOSFET, $V_{GS} > 0$ for all valid uses of the part, keeping the Zener reverse biased. In depletion mode devices however, V_{GS} can be both positive and negative; when negative, a gate-protection Zener diode would be forward biased and the MOSFET gate would not be driven properly. In some depletion mode MOSFETs, back-to-back Zener diodes are used to protect the gate.

MOSFET devices are at greatest risk of damage from static electricity when they are out of circuit. Even though an electrostatic discharge is capable of delivering little energy, it can generate thousands of volts and high peak currents. When storing MOSFETs, the leads should be placed into conductive foam. When working with MOSFETs, it is a good idea to minimize static by wearing a grounded wrist strap and working on a grounded workbench or mat. A humidifier may help to decrease the static electricity in the air. Before inserting a MOSFET into a circuit board it helps to first touch the device leads with your hand and then touch the circuit board. This serves to equalize the excess charge so that little excess charge flows when the device is inserted into the circuit board.

Power MOSFETs

Power MOSFETs are designed for use as switches, with extremely low values of $r_{DS(on)}$; values of 50 milliohms (mΩ) are common. The largest devices of this type can switch tens of amps of current with V_{DS} voltage ratings of hundreds of volts. The schematic symbol for power MOSFETs (see Figure 4.30) includes a *body diode* that allows the FET to conduct in the reverse direction, regardless of V_{GS}. This is useful in many high-power switching applications. Power MOSFETs used for RF amplifiers are discussed in more detail in the **RF Power Amplifiers** chapter.

While the maximum ratings for current and voltage are high, the devices cannot withstand both high drain current and high drain-to-source voltage at the same time because of the power dissipated; $P = V_{DS} \times I_D$. It is important to drive the gate of a power MOSFET such that the device is fully on or fully off so that either V_{DS} or I_D is at or close to zero. When switching, the device should spend as little time as possible in the linear region where both current and voltage are nonzero because their product (P) can be substantial. This is not a big problem if switching only takes place occasionally, but if the switching is repetitive (such as in a switching power supply) care should be taken to drive the gate properly and remove excess heat from the device.

Because the gate of a power MOSFET is capacitive (up to several hundred pF for large devices), charging and discharging the gate quickly results in short current peaks of more than 100 mA. Whatever circuit is used to drive the gate of a power MOSFET must be able to handle that current level, such as an integrated circuit designed for driving the capacitive load an FET gate presents.

The gate of a power MOSFET should not be left open or connected to a high-impedance circuit. Use a pull-down or pull-up resistor connected between the gate and the appropriate power supply to ensure that the gate is placed at the right voltage when not being driven by the gate drive circuit.

GaAsFETs

FETs made from gallium-arsenide (GaAs) material are used at UHF and microwave frequencies because they have gain at these frequencies and add little noise to the signal. The reason GaAsFETs have gain at these frequencies is the high mobility of the electrons in GaAs material. Because the electrons are more mobile than in silicon, they respond to the gate-source input signal more quickly and strongly than silicon FETs, providing gain at higher frequencies (f_T is directly proportional to electron mobility). The higher electron mobility also reduces thermally generated noise in the FET, making the GaAsFET especially suitable for weak-signal preamps.

Because electron mobility is always higher than hole mobility, N-type material is used in GaAsFETs to maximize high-frequency gain. Since P-type material is not used to make a gate-channel junction, a metal Schottky junction is formed by depositing metal directly on the surface of the channel. This type of device is also called a *MESFET* (metal-semiconductor field-effect transistor).

4.6.5 Comparison of BJT and FET Devices

Analog signal processing deals with changing a signal to a desired form. The three primary types of devices — bipolar transistors, field-effect transistors and integrated circuits — perform similar functions, each with specific advantages and disadvantages. The vacuum tube, once the dominant signal processing component, is relegated to high-power amplifier and display applications and is found only in the **RF Power Amplifiers** chapter of this *Handbook*. *Cathode-ray tubes* (CRTs) are covered in a separate article in this book's online material.

Bipolar transistors, when treated properly, can have virtually unlimited life spans. They are relatively small and, if they do not handle high currents, do not generate much heat. They make excellent high-frequency amplifiers. Compared to MOSFET devices they are less susceptible to damage from electrostatic discharge. Bipolar transistors and ICs, like all semiconductors, are susceptible to damage from power and lightning transients.

There are many performance advantages to FET devices, particularly MOSFETs. The extremely low gate currents allow the design of analog stages with nearly infinite input resistance. Signal distortion due to loading is minimized in this way. FETs are less expensive to fabricate in ICs and so are gradually replacing bipolar transistors in many IC applications.

RF amplifiers are now designed almost exclusively using some variety of MOSFET in their final amplifiers. The transistors are often integrated into modules (a.k.a. "pallets") that include circuitry to protect the transistors from the high voltages generated by reflections under high SWR conditions. See the **RF Power Amplifier** chapter for more information on advances in RF amplifier technology.

An important consideration in the use of analog components is the future availability of parts. At an ever-increasing rate, as new components are developed to replace older technology, the older components are discontinued by the manufacturers and become unavailable for future use. ASIC and PGA technology, discussed along with integrated circuits, brings the power of custom electronics to the radio, but can make it nearly impossible to repair by replacing an IC, even if the problem is known. If field repair and service at the component level are to be performed, it is important to use standard ICs wherever possible. Even so, when demand for a particular component drops, a manufacturer will discontinue its production. This happens on an ever-decreasing timeline.

A further consideration is the trend toward digital signal processing and software-defined radio systems. (See the chapter on **DSP and SDR Fundamentals**.) More and more analog functions are being performed by microprocessors and the analog signals converted to digital at higher and higher frequencies. It is now common practice to digitize the incoming RF signal directly at the antenna system interface.

There will always be a need for analog circuits, but the balance point between analog and digital is accelerating towards the latter. In future years, radio and test equipment will consist of a powerful, general-purpose digital signal processor, surrounded by the necessary analog circuitry to convert the signals to digital form and supply the processor with power.

4.6.6 Optical Semiconductors

In addition to electrical energy and heat energy, light energy also affects the behavior of semiconductor materials. If a device is made to allow photons of light to strike the surface of the semiconductor material, the energy absorbed by electrons disrupts the bonds between atoms, creating free electrons and holes. This increases the conductivity of the material (*photoconductivity*). The photon can also transfer enough energy to an electron

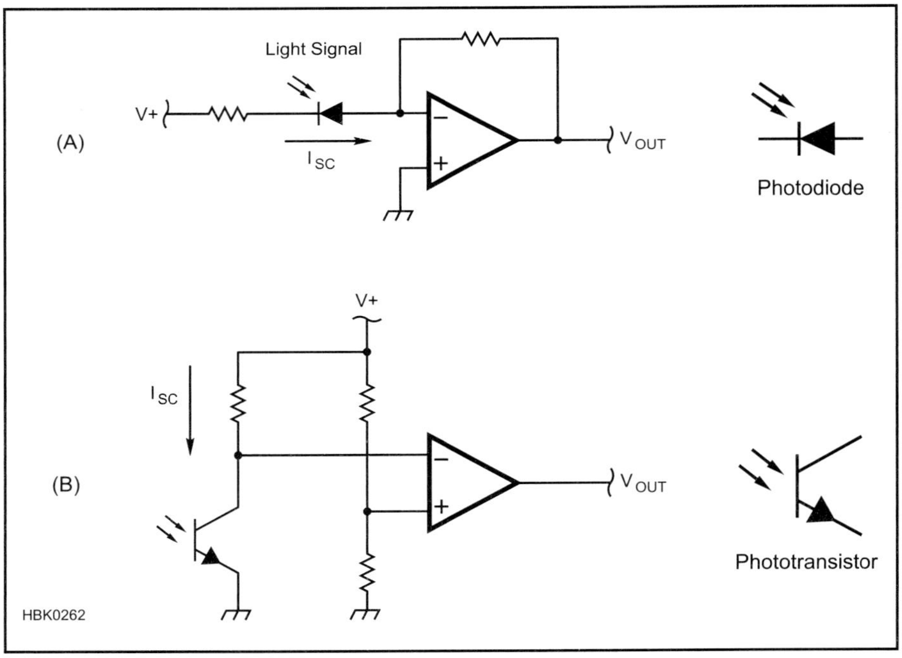

Figure 4.46 — The photodiode (A) is used to detect light. An amplifier circuit changes the variations in photodiode current to a change in output voltage. At (B), a photo-transistor conducts current when its base is illuminated. This causes the voltage at the collector to change causing the amplifier's output to switch between ON and OFF.

to allow it to cross a PN junction's depletion region as current flow through the semiconductor (*photoelectricity*).

PHOTOCONDUCTORS

In commercial *photoconductors* (also called *photoresistors*) the resistance can change by as much as several kilohms for a light intensity change of 100 ft-candles. The most common material used in photoconductors is cadmium sulfide (CdS), with a resistance range of more than 2 MΩ in total darkness to less than 10 Ω in bright light. Other materials used in photoconductors respond best at specific colors. Lead sulfide (PbS) is most sensitive to infrared light and selenium (Se) works best in the blue end of the visible spectrum.

PHOTODIODES

A similar effect is used in some diodes and transistors so that their operation can be controlled by light instead of electrical current biasing. These devices, shown in **Figure 4.46**, are called *photodiodes* and *phototransistors*. The flow of minority carriers across the reverse biased PN junction is increased by light falling on the doped semiconductor material. In the dark, the junction acts the same as any reverse biased PN junction, with a very low current, I_{SC}, (on the order of 10 μA) that is nearly independent of reverse voltage. The presence of light not only increases the current but also provides a resistance-like relationship (reverse current increases as reverse voltage increases). See **Figure 4.47** for the characteristic response of a photodiode. Even with no reverse voltage applied, the presence of light causes a small reverse current, as indicated by the points at which the lines in Figure 4.47 intersect the left side of the graph.

Photoconductors and photodiodes are generally used to produce light-related analog signals that require further processing. For example, a photodiode is used to detect infrared light signals from remote control devices as in Figure 4.46A. The light falling on the reverse-biased photodiode causes a change in I_{SC} that is detected as a change in output voltage.

Light falling on the phototransistor acts as base current to control a larger current between the collector and emitter. Thus, the phototransistor acts as an amplifier whose input signal is light and whose output is current. Phototransistors are more sensitive to light than the other devices. Phototransistors have lots of light-to-current gain, but photodiodes normally have less noise, so they make more sensitive detectors. The phototransistor in Figure 4.46B is being used as a detector. Light falling on the phototransistor causes collector current to flow, dropping the collector voltage below the voltage at the amplifier's + input and causing a change in V_{OUT}.

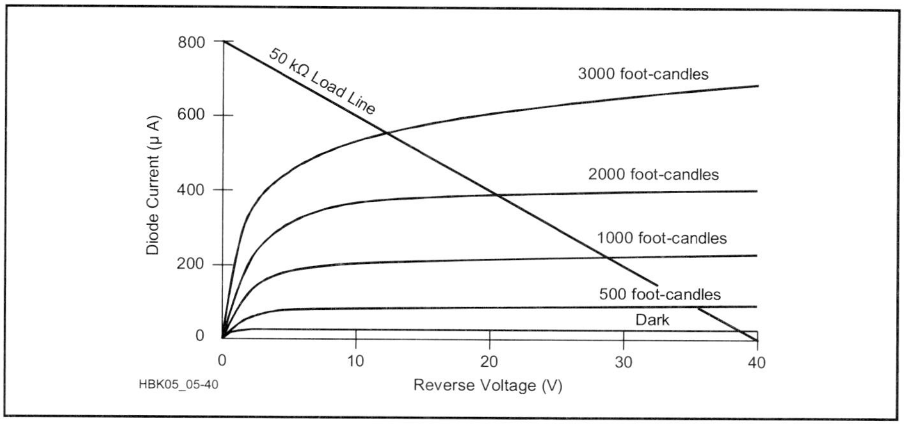

Figure 4.47 — Photodiode I-V curve. Reverse voltage is plotted on the X-axis and current through diode is plotted on the Y-axis. Various response lines are plotted for different illumination. Except for the zero illumination line, the response does not pass through the origin since there is current generated at the PN junction by the light energy. A load line is shown for a 50-kΩ resistor in series with the photodiode.

PHOTOVOLTAIC CELLS

When illuminated, the reverse-biased photodiode has a reverse current caused by excess minority carriers. As the reverse voltage is reduced, the potential barrier to the forward flow of majority carriers is also reduced. Since light energy leads to the generation of both majority and minority carriers, when the resistance to the flow of majority carriers is decreased these carriers form a forward current. The voltage at which the forward current equals the reverse current is called the *photovoltaic potential* of the junction. If the illuminated PN junction is not connected to a load, a voltage equal to the photovoltaic potential can be measured across it as the *terminal voltage,* V_T, or *open-circuit voltage,* V_{OC}.

Devices that use light from the sun to produce electricity in this way are called *photovoltaic (PV)* or *solar cells* or *solar batteries*. The symbol for a photovoltaic cell is shown in **Figure 4.48A**. The electrical equivalent circuit of the cell is shown in Figure 4.48B. The cell is basically a large, flat diode junction exposed to light. Metal electrodes on each side of the junction collect the current generated.

When illuminated, the cell acts like a current source, with some of the current flowing through a diode (made of the same material as the cell), a shunt resistance for leakage current and a series resistor that represents the resistance of the cell. Two quantities define the electrical characteristics of common silicon photovoltaic cells. These are an open-

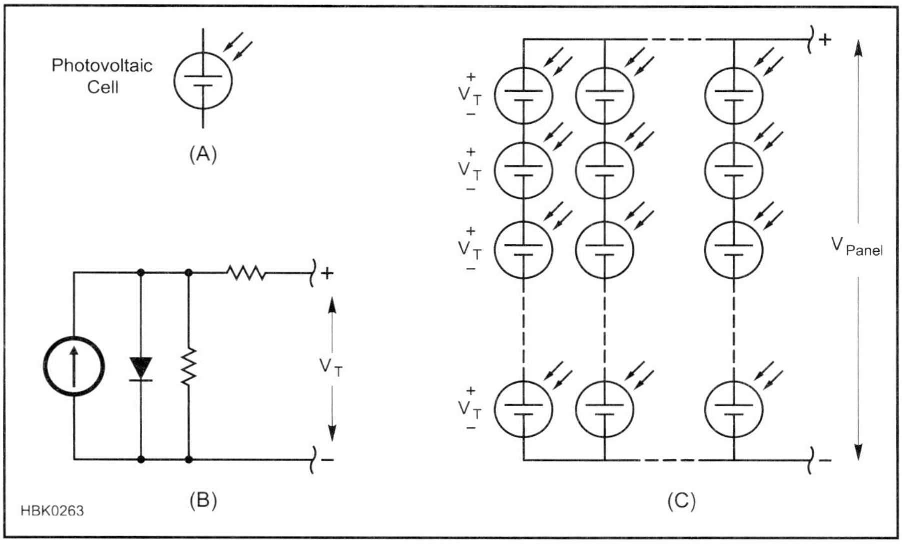

Figure 4.48 — A photovoltaic cell's symbol (A) is similar to a battery. Electrically, the cell can be modeled as the equivalent circuit at (B). Solar panels (C) consist of arrays of cells connected to supply power at a convenient voltage.

circuit voltage, V_{OC} of 0.5 to 0.6 V and the output *short-circuit current*, I_{SC} as above, that depends on the area of the cell exposed to light and the degree of illumination. A measure of the cell's effectiveness at converting light into current is the *conversion efficiency*. Typical silicon solar cells have a conversion efficiency of 10 to 15% although special cells with stacked junctions or using special light-absorbing materials have shown efficiencies as high as 40%.

Solar cells are primarily made from single-crystal slices of silicon, similar to diodes and transistors, but with a much greater area. *Polycrystalline silicon* and *thin-film* cells are less expensive but have lower conversion efficiency. Technology is advancing rapidly in the field of photovoltaic energy and there are a number of different types of materials and fabrication techniques that have promise in surpassing the effectiveness of the single-junction silicon cells.

Solar cells are assembled into arrays called *solar panels*, shown in Figure 4.48C. Cells are connected in series so that the combined output voltage is a more useful voltage, such as 12 V. Several strings of cells are then connected in parallel to increase the available output current. Solar panels are available with output powers from a few watts to hundreds of watts. Note that unlike batteries, strings of solar cells can be connected directly in parallel because they act as sources of constant current instead of voltage. (More information on the use of solar panels for powering radio equipment can be found in the chapter on **Power Sources**.)

LIGHT EMITTING DIODES (LEDs) AND LASER DIODES

In the photodiode, energy from light falling on the semiconductor material is absorbed to create additional electron-hole pairs. When the electrons and holes recombine, the same amount of energy is given off. In normal diodes, the energy from recombination of carriers is given off as heat. In certain forms of semiconductor material, the recombination energy is given off as light with a mechanism called *electroluminescence*. Devices made for this purpose are called *light emitting diodes (LEDs)*.

The LED emits light when it is forward biased and excess carriers are present. As the carriers recombine, light is produced with a color that depends on the properties of the semiconductor material used. Gallium-arsenide (GaAs) generates light in the infrared region, gallium-phosphide (GaP) gives off red light when doped with oxygen or green light when doped with nitrogen. Orange light is attained with a mixture of GaAs and GaP (GaAsP). Silicon-carbide (SiC) creates a blue LED. White LEDs are made by coating the inside of the LED lens with a white-light emitting phosphor and illuminating the phosphor with light from a single-color LED.

LEDs have the advantages of low power requirements, fast switching times (on the order of 10 ns), and narrow spectra (relatively pure color). The primary application of low-power LEDs is as an illuminated visual indicator, replacing miniature incandescent bulbs for indicators and illumination. (High-power LEDs used for lighting and large displays are not discussed here.) LEDs are specified primarily by their color, size, shape, and output light intensity.

The "standard" size LED used in electronics is the T-13/4, which is 5 mm or 0.20 inches in diameter. The "miniature" size is the T-1, which is 3 mm or 0.125 inches in diameter. Today, the "standard" and "miniature" size is a bit of a misnomer due to the wide variety of LED sizes and shapes, including SMT varieties. However, the T-1 and T-13/4 remain the most common for homebrew projects due to their inexpensive availability and ease of mounting with a simple panel hole. Their long leads are ideal for prototyping. The LED, shown in **Figure 4.49**, is very simple to use. It is connected to a voltage source (V) with a series resistor (R) that limits the current to the desired level (I_F) for the amount of light to be generated.

$$R = \frac{V - V_F}{I_F}$$

where V_F is the forward voltage of the LED.

The cathode lead is connected to the lower potential and is specially marked as shown in the manufacturer's data sheet. LEDs may be connected in series for additional light, with the same current flowing in all of the diodes. Diodes connected in parallel without current-limiting resistors for each diode are likely to share the current unequally, thus the series connection is preferred.

The laser diode operates by similar principles to the LED except that all of the light produced is *monochromatic* (of the same color and wavelength) and it is *coherent*, meaning that all of the light waves emitted by the device are in phase. Laser diodes generally require higher current than an LED and will not emit light until the *lasing current* level is reached. Because the light is monochromatic and coherent, laser diodes can be used for applications requiring precise illumination and modulation, such as high-speed data links, and in data storage media such as CD-ROM and DVD. LEDs are not used for high-speed or high-frequency analog modulation because of recovery time limitations, just as in regular rectifiers.

OPTOISOLATORS

An interesting combination of optoelectronic components proves very useful in many

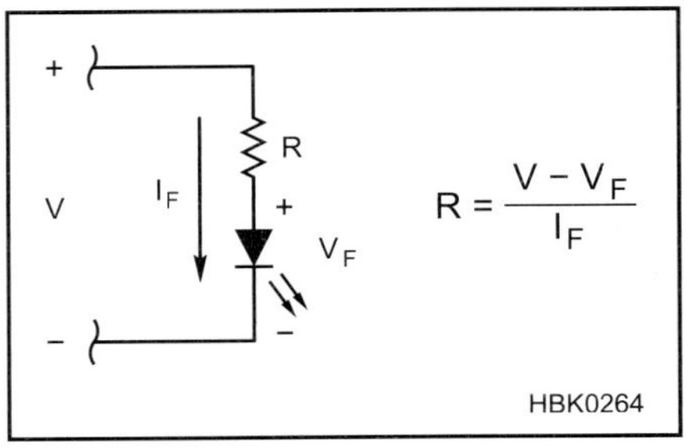

Figure 4.49 — A light-emitting diode (LED) emits light when conducting forward current. A series current-limiting resistor is used to set the current through the LED according to the equation.

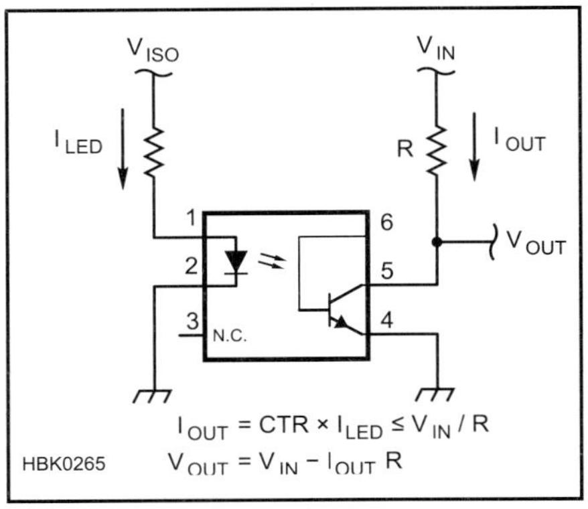

Figure 4.50 — The optoisolator consists of an LED (input) that illuminates the base of a phototransistor (output). The phototransistor then conducts current in the output circuit. CTR is the optoisolator's current transfer ratio.

analog signal processing applications. An *optoisolator* consists of an LED optically coupled to a phototransistor, usually in an enclosed package (see **Figure 4.50**). The optoisolator, as its name suggests, isolates different circuits from each other. Typically, isolation resistance is on the order of $10^{11}\ \Omega$ and isolation capacitance is less than 1 pF. Maximum voltage isolation varies from 1,000 to 10,000 V ac. The most common optoisolators are available in six-pin DIP packages.

Optoisolators are primarily used for voltage level shifting and signal isolation. Voltage level shifting allows signals (usually digital signals) to pass between circuits operating at greatly different voltages. The isolation has two purposes: to protect circuitry (and operators) from excessive voltages and to isolate noisy circuitry from noise-sensitive circuitry.

Optoisolators also cannot transfer signals with high power levels. The power rating of the LED in a 4N25 device is 120 mW. Optoisolators have a limited frequency response due to the high capacitance of the LED. A typical bandwidth for the 4N25 series is 300 kHz. Optoisolators with bandwidths of several MHz are available, but are somewhat expensive.

As an example of voltage level shifting, an optoisolator can be used to allow a low-voltage, solid-state electronic Morse code keyer to activate a vacuum-tube grid-block keying circuit that operates at a high negative voltage (typically about –100 V) but low current. No common ground is required between the two pieces of equipment.

Optoisolators can act as input protection for circuits that are exposed to high voltages or transients. For example, a short 1,000-V transient that can destroy a semiconductor circuit will only saturate the LED in the optoisolator, preventing damage to the circuit. The worst that will happen is the LED in the optoisolator will be destroyed, but that is usually quite a bit less expensive than the circuit it is protecting.

Optoisolators are also useful for isolating different ground systems. The input and output signals are totally isolated from each other, even with respect to the references for each signal. A common application for optoisolators is when a computer is used to control radio equipment. The computer signal, and even its ground reference, typically contains considerable wide-band noise caused by the digital circuitry. The best way to keep this noise out of the radio is to isolate both the signal and its reference; this is easily done with an optoisolator.

The design of circuits with optoisolators is not greatly different from the design of circuits with LEDs and with transistors. On the input side, the LED is forward-biased and driven with a series current-limiting resistor whose value limits current to less than the maximum value for the device (for example, 60 mA is the maximum LED current for a 4N25). This is identical to designing with standalone LEDs.

On the output side, instead of current gain for a transistor, the optoisolator's *current transfer ratio* (*CTR*) is used. CTR is a ratio given in percent between the amount of current through the LED to the output transistor's maximum available collector current. For example, if an optoisolator's CTR = 25%, then an LED current of 20 mA results in the output transistor being able to conduct up to $20 \times 0.25 = 5$ mA of current in its collector circuit.

If the optoisolator is to be used for an analog signal, the input signal must be appropriately dc shifted so that the LED is always forward biased. A phototransistor with all three leads available for connection (as in Figure 4.50) is required. The base lead is used for biasing, allowing the optical signal to create variations above and below the transistor's operating point. The collector and emitter leads are used as they would be in any transistor amplifier circuit. (There are also linear optoisolators that include built-in linearizing circuitry.) The use of linear optoisolators is not common.

FIBER OPTICS

An interesting variation on the optoisolator is the *fiber-optic* connection. Like the optoisolator, the input signal is used to drive an LED or laser diode that produces modulated light (usually light pulses). The light is transmitted in a fiber optic cable, an extruded glass fiber that efficiently carries light over long distances and around fairly sharp bends. The signal is recovered by a photo detector (photoresistor, photodiode or phototransistor). Because the fiber optic cable is nonconductive, the transmitting and receiving systems are electrically isolated.

Fiber optic cables generally have far less loss than coaxial cable transmission lines. They do not leak RF energy, nor do they pick up electrical noise. Fiber optic cables are virtually immune to electromagnetic interference! Special forms of LEDs and phototransistors are available with the appropriate optical couplers for connecting to fiber optic cables. These devices are typically designed for higher frequency operation with gigahertz bandwidth.

4.6.7 Integrated Circuits (ICs)

If you look inside a transistor, the actual size of the semiconductor is quite small compared to the size of the packaging. For most semiconductors, the packaging takes considerably more space than the actual semiconductor device. Thus, an obvious way to reduce the physical size of circuitry is to combine more of the circuit inside a single package.

HYBRID INTEGRATED CIRCUITS

It is easy to imagine placing several small semiconductor chips in the same package. This is known as *hybrid circuitry*, a technology in which several semiconductor chips are placed in the same package and miniature wires are connected between them to make complete circuits.

Hybrid circuits miniaturize analog or analog/digital electronic circuits by eliminating much of the packaging and interconnections inherent in discrete electronics. The term *discrete* refers to the use of individual components to make a circuit, each in its own package. The individual components are attached together on a small circuit board or with bonding wires. Manufacturers often use hybrids when small size and specialized techniques are needed, but there is insufficient volume to justify the expense of a custom IC.

A current application for hybrid circuitry is UHF and microwave amplifiers — they are in wide use by the mobile phone industry. For example, the NXP MW7IC930N Wideband Integrated Power Amplifier is a complete transmitting module used in cellular base stations. Its TO-272 package in **Figure 4.51** is only about 1 inch long by 3/8-inch wide. This particular device is designed for use between 728 and 960 MHz and can be adapted for use on the amateur 902 MHz band. Other devices available as hybrid circuits include oscillators, signal processors, preamplifiers, and so forth. Surplus hybrids can be hard to adapt to amateur use unless they are clearly identified with manufacturing marks that will help in obtaining a data sheet.

MONOLITHIC INTEGRATED CIRCUITS

In order to build entire circuits on a single piece of semiconductor, it must be possible to fabricate resistors and capacitors, as well as transistors and diodes. Only then can the entire circuit be created on one piece of silicon called a *monolithic integrated circuit*. The following description is a greatly simplified illustration of monolithic integrated circuit structure and fabrication techniques.

An integrated circuit (IC) or "chip" is fabricated in layers. An example of a semiconductor circuit schematic and its implementation in an IC is pictured in **Figure 4.52**. The base layer of the circuit, the *substrate*, is made of P-type semiconductor material. Although less common, the polarity of the substrate can also be N-type material. Since the mobility of electrons is about three times higher than that of holes, bipolar transistors made with N-type collectors and FETs made with N-type channels are capable of higher speeds and power handling. Thus,

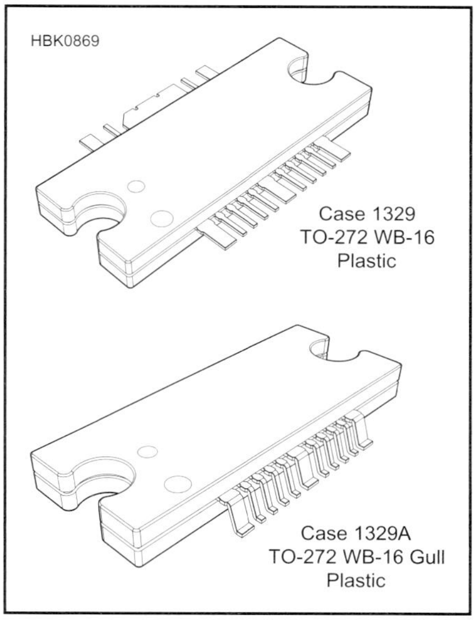

Figure 4.51 — The TO-272 is typical of packages used for hybrid IC RF amplifier modules at UHF and microwave frequencies.

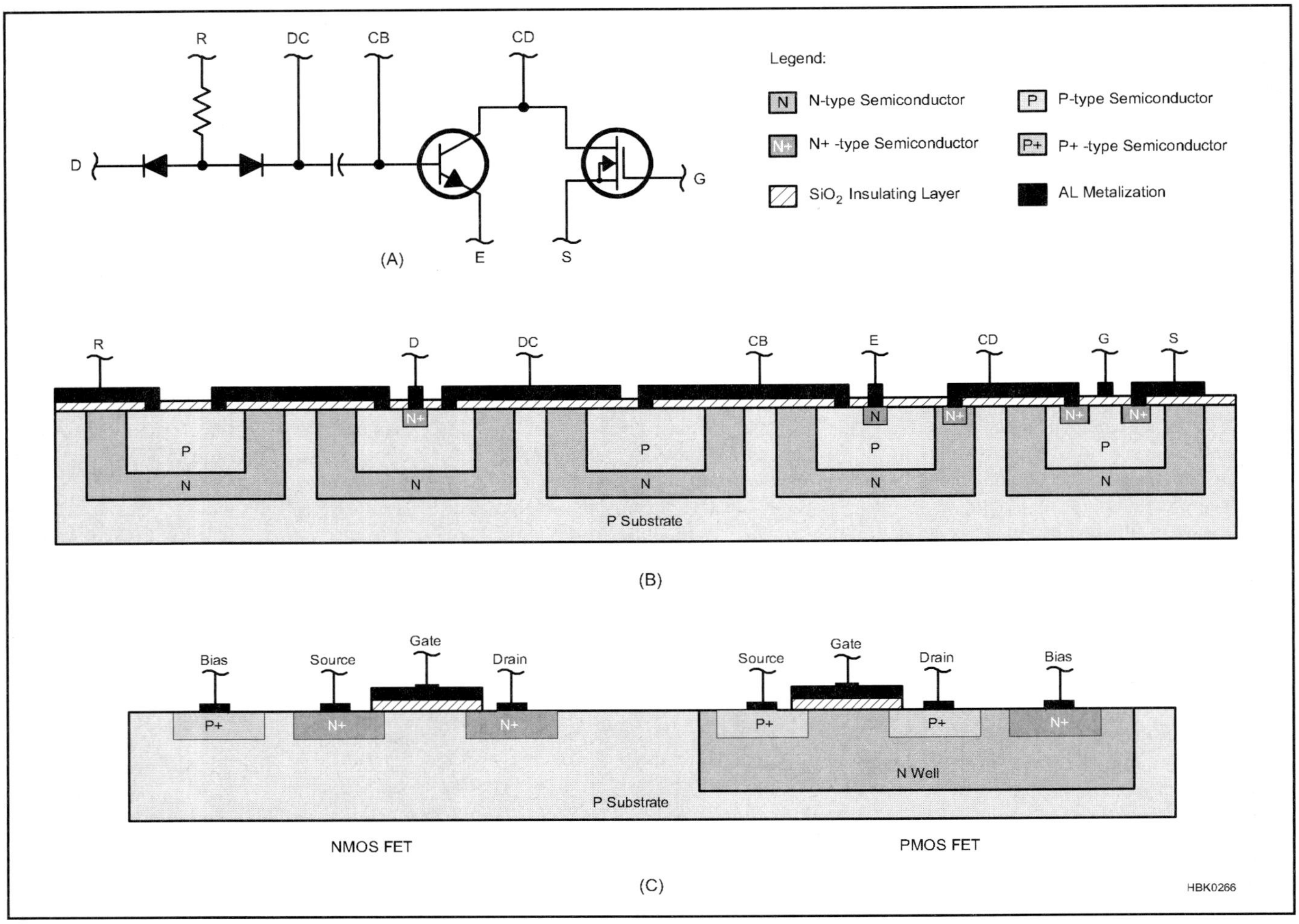

Figure 4.52— Integrated circuit construction. (A) Circuit containing two diodes, a resistor, a capacitor, an NPN transistor and an N-channel MOSFET. Labeled leads are D for diode, R for resistor, DC for diode-capacitor, E for emitter, S for source, CD for collector-drain and G for gate. (B) Integrated circuit that is identical to circuit in (A). Same leads are labeled for comparison. Circuit is built on a P-type semiconductor substrate with N-type wells diffused into it. An insulating layer of SiO_2 is above the semiconductor and is etched away where aluminum metal contacts are made with the semiconductor. Most metal-to-semiconductor contacts are made with heavily doped N-type material (N^+-type semiconductor).

P-type substrates are far more common. For devices with N-type substrates, all polarities in the ensuing discussion would be reversed. Many other types of substrates are employed in various special applications.

On top of the P-type substrate is a thin layer of N-type material in which the active and passive components are built. Impurities are diffused into this layer to form the appropriate component at each location. To prevent random diffusion of impurities into the N-layer, its upper surface must be protected. This is done by covering the N-layer with a layer of silicon dioxide (SiO_2). Wherever diffusion of impurities is desired, the SiO_2 is etched away. The precision of placing the components on the semiconductor material depends mainly on the fineness of the etching. The fourth layer of an IC is made of aluminum or copper and is used to make the interconnections between the components.

Different components are made in a single piece of semiconductor material by first diffusing a high concentration of acceptor impurities into the layer of N-type material. This process creates a P-type semiconductor — often referred to as a P^+-type semiconductor because of its high concentration of acceptor atoms — that isolates regions of N-type material. Each of these regions is then further processed to form single components.

A component is produced by the diffusion of a lesser concentration of acceptor atoms into the middle of each isolation region. This results in an N-type *isolation well* that contains P-type material, is surrounded on its sides by P^+-type material and has P-type material (substrate) below it. The cross-sectional view in Figure 4.52B illustrates the various layers. Connections to the metal layer are often made by diffusing high concentrations of donor atoms into small regions of the N-type well and the P-type material in the well. The material in these small regions is N^+-type and facilitates electron flow between the metal contact and the semiconductor. In some configurations, it is necessary to connect the metal directly to the P-type material in the well.

Fabricating Resistors and Capacitors

An isolation well can be made into a resistor by making two contacts into the P-type semiconductor in the well. Resistance is inversely proportional to the cross-sectional area of the well. An alternate type of resistor that can be integrated in a semiconductor circuit is a *thin-film resistor*, where a metallic film is deposited on the SiO_2 layer, masked on its upper surface by more SiO_2 and then etched to make the desired geometry, thus adjusting the resistance.

There are two ways to form capacitors in a semiconductor. One is to make use of the PN junction between the N-type well and the

P-type material that fills it. Much like a varactor diode, when this junction is reverse biased a capacitance is created. Since a bias voltage is required, this type of capacitor is polarized, like an electrolytic capacitor. Nonpolarized capacitors can also be formed in an integrated circuit by using thin film technology. In this case, a very high concentration of donor ions is diffused into the well, creating an N^+-type region. A thin metallic film is deposited over the SiO_2 layer covering the well and the capacitance is created between the metallic film and the well. The value of the capacitance is adjusted by varying the thickness of the SiO_2 layer and the cross-sectional size of the well. This type of thin film capacitor is also known as a metal oxide semiconductor (MOS) capacitor.

Unlike resistors and capacitors, it is very difficult to create inductors in integrated circuits. Generally, RF circuits that need inductance require external inductors to be connected to the IC. In some cases, particularly at lower frequencies, the behavior of an inductor can be mimicked by an amplifier circuit. In many cases the appropriate design of IC amplifiers can reduce or eliminate the need for external inductors. As the frequency of operation increases, however, the amount of inductance needed falls for an equivalent amount of reactance. This had led to several innovative techniques for creating small inductors using the metallization layers of the IC.

Fabricating Diodes and Transistors

The simplest form of diode is generated by connecting to an N^+-type connection point in the well for the cathode and to the P-type well material for the anode. Diodes are often converted from NPN transistor configurations. Integrated circuit diodes made this way can either short the collector to the base or leave the collector unconnected. The base contact is the anode and the emitter contact is the cathode.

Transistors are created in integrated circuitry in much the same way that they are fabricated in their discrete forms. The NPN transistor is the easiest to make since the wall of the well, made of N-type semiconductor, forms the collector, the P-type material in the well forms the base and a small region of N^+-type material formed in the center of the well becomes the emitter. A PNP transistor is made by diffusing donor ions into the P-type semiconductor in the well to make a pattern with P-type material in the center (emitter) surrounded by a ring of N-type material that connects all the way down to the well material (base), which is surrounded by another ring of P-type material (collector). This configuration results in a large base width separating the emitter and collector, causing these devices to have much lower current gain than the NPN form. This is one reason why integrated circuitry is designed to use many more NPN transistors than PNP transistors.

FETs can also be fabricated in IC form as shown in Figure 4.52C. Due to its many functional advantages, the MOSFET is the most common form used for digital ICs. MOSFETs are made in a semiconductor chip much the same way as MOS capacitors, described earlier. In addition to the signal processing advantages offered by MOSFETs over other transistors, the MOSFET device can be fabricated in 5% of the physical space required for bipolar transistors. CMOS ICs can contain 20 times more circuitry than bipolar ICs with the same chip size, making the devices more powerful and less expensive than those based on bipolar technology. CMOS is the most popular form of integrated circuit.

The final configuration of the switching circuit is CMOS as described in a previous section of this chapter. CMOS gates require two FETs, one of each form (NMOS and PMOS as shown in the figure). NMOS requires fewer processing steps, and the individual FETs have lower on-resistance than PMOS. The fabrication of NMOS FETs is the same as for individual semiconductors; P+ wells form the source and drain in a P-type substrate. A metal gate electrode is formed on top of an insulating SiO_2 layer so that the channel forms in the P-type substrate between the source and drain. For the PMOS FET, the process is similar but begins with an N-type well in the P-type substrate.

MOSFETs fabricated in this manner also have bias (B) terminals connected to the positive power supply to prevent destructive *latch-up*. This can occur in CMOS gates because the two MOSFETs form a *parasitic SCR*. If the SCR mode is triggered and both transistors conduct at the same time, large currents can flow through the FET and destroy the IC unless power is removed. Just as discrete MOSFETs are at risk of gate destruction, IC chips made with MOSFET devices have a similar risk. They should be treated with the same care to protect them from static electricity as discrete MOSFETs.

While CMOS is the most widely used technology, integrated circuits need not be made exclusively with MOSFETs or bipolar transistors. It is common to find IC chips designed with both technologies, taking advantage of the strengths of each.

4.6.8 Integrated Circuit Packaging

ICs come in a variety of packages, primarily dual and single in-line packages (DIPs and SIPs) and surface-mount packages. Most are marked with a part number and a four-digit manufacturer's date code indicating the year (first two digits) and week (last two digits) that the component was made. As mentioned in the introduction to this chapter, ICs are frequently house-marked and cross-reference directories can be helpful in identification and replacement.

IC part numbers provide a complete description of the device's function and ratings. For example, a 4066 IC contains four independent CMOS SPST switches. The 4066 is a CMOS device available from a number of different manufacturers in different package styles and ratings. The two- or three-letter prefix of the part number is generally associated with the part manufacturer. Next, the part type (4066 in this case) shows the function and pin assignments or "pin outs." Following the part type is an alphabetic suffix that describes the version of the part, package code, temperature range, reliability rating and possibly other information. For complete information on the part — any or all of which may be significant to circuit function — use the websites of the various manufacturers or do an online search for "data sheet" and the part number.

When choosing ICs that are not exact replacements, be wary of substituting "similar" devices, particularly in demanding applications, such as high-speed logic, sensitive receivers, precision instrumentation, and similar uses. In particular, substitution of one type of logic family for another — even if the device functions and *pin outs* (connections between the internal circuitry and external pins) are the same — can cause a circuit to not function or function erratically, particularly at temperature extremes. For example, substituting LS TTL devices for HCMOS devices will result in mismatches between logic level thresholds. Substituting a lower-power IC may result in problems supplying enough output current. Even using a faster or higher clock-speed part can cause problems if signals change faster or propagate more quickly than the circuit was designed for. Problems of this sort can be extremely difficult to troubleshoot unless you are skilled in circuit design. When necessary, you can add interface circuits or buffer amplifiers that improve the input and output capabilities of replacement ICs, but auxiliary circuits cannot improve basic device ratings, such as speed or bandwidth. Whenever possible, substitute ICs that are guaranteed or "direct" replacements and that are listed as such by the manufacturer.

4.6.9 Integrated Circuit Temperature Ranges

ICs are available in different operating temperature ranges. Four standard ranges are common:

- Commercial: 0 °C to 70 °C
- Industrial: –25 °C to 85 °C
- Automotive: –40 °C to 85 °C
- Military: –55 °C to 125 °C

In some cases, part numbers reflect the

temperature ratings. For example, an LM301A op amp is rated for the commercial temperature range, an LM201A op amp for the industrial range and an LM101A for the military range. It is usually acceptable, all other things being equal, to substitute ICs rated for a wider temperature range, but there are often other performance differences associated with the devices meeting wider temperature specifications that should be evaluated before making the substitution.

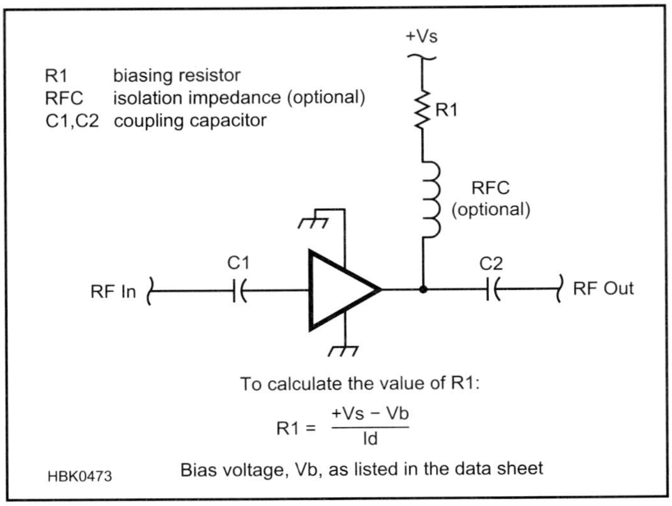

Figure 4.53 — MMIC application.

4.6.10 MMIC Amplifiers

Monolithic microwave integrated circuit (MMIC) amplifiers are single-supply 50-Ω wideband gain blocks offering high dynamic range for output powers to about +15 dBm. MMIC amplifiers are becoming increasingly popular in homebrew communications circuits. With bandwidths over 1 GHz, they are well suited for HF, VHF, UHF, and lower microwave frequencies.

MMIC amplifiers produce power gains from 10 dB to 30 dB. They also have a high third-order intercept point (IP3), usually in the +20 to +30 dBm range, easing the concerns about amplifier compression for most applications. They are used for RF and IF amplifiers, local oscillator amplifiers, transmitter drivers, and other medium power applications in 50-Ω systems. MMICs are especially well suited for driving 50-Ω double-balanced mixers (DBM). **Figure 4.53** shows the typical circuit arrangement for most MMIC amplifiers.

MMICs are available in a variety of surface mount packages requiring very few external components. (Package dimensions and outlines are provided in the **Construction Techniques** chapter on PCB layout.) Vendor data sheets and application notes, found on the manufacturer's websites, should be used for the proper selection of the biasing resistor, coupling capacitors, and other design criteria.

The main disadvantage of MMIC amplifiers is their relatively high current demands, usually in the 30 mA to 80 mA range per device, making them unsuitable for battery-powered portable equipment. On the other hand, the high current demand is what establishes their high gain and high IP3 characteristics with 50-Ω loads..

Their wide gain-bandwidth should be controlled by input and output tuned circuits or filters to reduce the gain outside the desired ranges. For example, for an HF amplifier, 30 MHz low-pass filters can be used to reduce the gain outside the HF spectrum, or a band-pass filter used for the frequency band of interest.

Selecting the proper MMIC amplifier is fairly straightforward. First, select a device for the desired frequency bandwidth, gain, and output power. Ensure device current is compatible with the design application. Calculate the value for the bias resistor (R1 in Figure 4.53) based on the device's biasing voltage (V_b) and whatever value of supply voltage (V_s) is available.

4.7 Amplifiers

By far, the most common type of analog circuit is the amplifier. The basic component of most electronics — the transistor — is an amplifier in which a small input signal controls a larger signal. Most of modern electronics, both analog and digital, are based on the transistor amplifier or switch, regardless of whether the input signal is amplified at the output.

4.7.1 Amplifier Configurations

Amplifier configurations are described by the *common* part of the device. The word "common" is used to describe the connection of a lead directly to a reference that is used by both the input and output ports of the circuit. The most common reference is ground, but positive and negative power sources are also valid references.

The type of circuit reference used depends on the type of device (transistor [NPN or PNP] or FET [P-channel or N-channel]), which lead is chosen as common, and the range of signal levels. Once a common lead is chosen, the other two leads are used for signal input and output. Based on the biasing conditions, there is only one way to select these leads. Thus, there are three possible amplifier configurations for each type of three-lead device. (Vacuum tube amplifiers are discussed in the chapter on **RF Power Amplifiers**.)

DC power sources are usually constructed so that ac signals at the output terminals are bypassed to ground through a very low impedance. This allows the power source to be treated as an *ac ground*, even though it may be supplying dc voltages to the circuit. When a circuit is being analyzed for its ac behavior, ac grounds are usually treated as ground, since dc bias is ignored in the ac analysis. Thus, a transistor's collector can be considered the "common" part of the circuit, even though in actual operation, a dc voltage is applied to it.

Figure 4.54 shows the three basic types of bipolar transistor amplifiers: the common-base, common-emitter, and common-collector. The common terminal is shown connected to ground, although as mentioned earlier, a dc bias voltage may be present. Each type of amplifier is described in the following sections. Following the description of the amplifier, additional discussion of biasing transistors and their operation at high frequencies and for large signals is presented.

4.7.2 Transistor Amplifiers

Creating a useful transistor amplifier depends on using an appropriate model for the transistor itself, choosing the right configuration of the amplifier, using the design equations for that configuration, and ensuring that the amplifier operates properly at different temperatures. This section follows that sequence, first introducing simple transistor

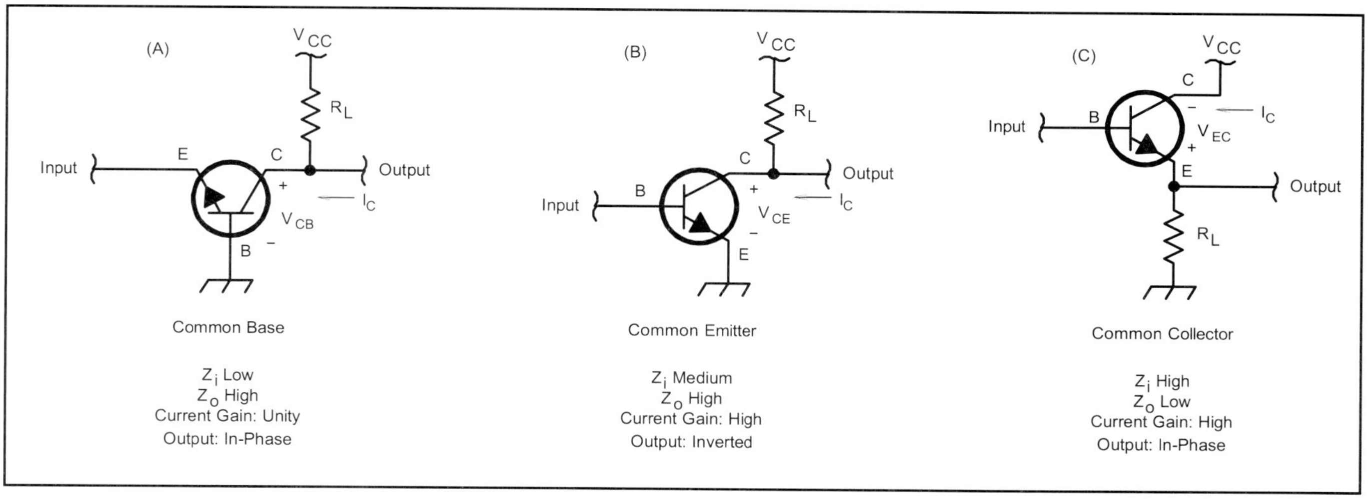

Figure 4.54 — The three configurations of bipolar transistor amplifiers with the basic characteristics of each circuit.

models and then extending that knowledge to the point of design guidelines for common circuits that use bipolar and FETs.

DEVICE MODELS AND CLASSES

Semiconductor circuit design is based on equivalent circuits that describe the physics of the devices. These circuits, made up of voltage and current sources and passive components such as resistors, capacitors and inductors, are called models. A complete model that describes a transistor exactly over a wide frequency range is a fairly complex circuit. As a result, simpler models are used in specific circumstances. For example, the *small-signal model* works well when the device is operated close to some nominal set of characteristics such that current and voltage interact fairly linearly. The *large-signal model* is used when the device is operated so that it enters its saturation or cut-off regions, for example.

Different frequency ranges also require different models. The *low-frequency models* used in this chapter can be used to develop circuits for dc, audio and very low RF applications. At higher frequencies, small capacitances and inductances that can be ignored at low frequencies begin to have significant effects on device behavior, such as gain or impedance. In addition, the physical structure of the device also becomes significant as gain begins to drop or phase shifts between input and output signals start to grow. In this region, *high-frequency models* are used.

Amplifiers are also grouped by their *operating class*, which describes the way in which the input signal is amplified. There are several classes of analog amplifiers; A, B, AB, AB1, AB2 and C.

The analog class designators specify over how much of the input cycle the active device is conducting current. A class-A amplifier's active device conducts current for 100 percent of the input signal cycle, such as shown in Figure 4.33. A class-B amplifier conducts during one-half of the input cycle, class-AB, AB1, and AB2 some fraction between 50 and 100 percent of the input cycle, and class-C for less than 50 percent of the input signal cycle.

Digital amplifiers, in which the active device is operated as a switch that is either fully-on or fully-off, similarly to switchmode power supplies, are also grouped by classes beginning with the letter D and beyond. Each different class uses a different method of converting the switch's output waveform to the desired RF waveform.

Amplifier classes, models, and their use at high-frequencies are discussed in more detail in the chapter on **RF Techniques**. In addition, the use of models for circuit simulation is discussed at length in the **Computer Modeling and Simulation** chapter.

4.7.3 Bipolar Transistor Amplifiers

In this discussion, we will focus on simple models for bipolar transistors (BJTs). This discussion is centered on NPN BJTs but applies equally well to PNP BJTs if the bias voltage and current polarities are reversed. This section assumes the small-signal, low-frequency models for the transistors.

SMALL-SIGNAL BJT MODEL

The transistor is usually considered as a *current-controlled* device in which the base current controls the collector current:

$$I_c = \beta I_b$$

where
I_c = collector current,
I_b = base current, and
β = common-emitter current gain, beta.

(The term "common-emitter" refers to the type of transistor circuit described below, in which the transistor operates with base current as its input and collector circuit as its output.) Current is positive if it flows *into* a device terminal.

The transistor can also be treated as a voltage-controlled device in which the transistor's emitter current, I_e, is controlled by the base-emitter voltage, V_{be}:

$$I_c = I_{es}\,[e^{(qV_{be}/kT)} - 1] \approx I_{es}e^{(qV_{be}/kT)}$$

where
q = electronic charge,
k = Boltzmann's constant,
T = temperature in degrees Kelvin (K), and
I_{es} = emitter saturation current, typically 1×10^{-13} A.

The subscripts for voltages indicate the direction of positive voltage, so that V_{be} indicates positive is from the base to the emitter. It is simpler to design circuits using the current-controlled device, but accounting for the transistor's behavior with temperature requires an understanding of the voltage-controlled model.

Transistors are usually driven by both biasing and signal voltages. Both equations above for collector current apply to both transistor dc biasing and signal design. Both of these equations are approximations of the more complex behavior exhibited by actual transistors. The second equation applies to a simplification of the first *Ebers-Moll model* (see references). More sophisticated models for BJTs are described by Getreu (see references). Small-signal models treat only the signal components. We will consider bias later.

The next step is to use these basic equations to design circuits. We will begin with small-signal amplifier design and the limits of where the techniques can be applied. Later, we'll discuss large-signal amplifier design and the distortion that arises from operating the transistor in regions where the relationship between the input and output signals is nonlinear.

Common-Emitter Model

Figure 4.55 shows a BJT amplifier connected in the common-emitter configuration. (The emitter, shown connected to ground, is common to both the input circuit with the voltage source and the output circuit with the transistor's collector.) The performance of this circuit is adequately described by the simple equation $I_c = \beta I_b$. **Figure 4.56** shows the most common of all transistor small-signal models, a controlled current source with emitter resistance.

There are two variations of the model shown in the figure. Figure 4.56B shows the base as a direct connection to the junction of a current-controlled current source ($I_c = \beta I_b$) and a resistance, r_e, the *dynamic emitter resistance* representing the change in V_{be} with I_e. This resistance also changes with emitter current:

$$r_e = \frac{kT}{qI_c} \approx \frac{26}{I_e}$$

where I_e is the dc bias current in milliamperes.

The simplified approximation only applies at a typical ambient temperature of 300 K because r_e increases with temperature. In Figure 4.56A, the emitter resistance has been moved to the base connection, where it has the value $(\beta+1)r_e$. These models are electrically equivalent.

The transistor's output resistance (the Thevenin or Norton equivalent resistance between the collector and the grounded emitter) is infinite because of the current source. This is a good approximation for most silicon transistors at low frequencies (well below the transistor's gain-bandwidth product, F_T) and will be used for the design examples that follow.

As frequency increases, the capacitance inherent in BJT construction becomes significant and the *hybrid-pi model* shown in **Figure 4.57** is used, adding C_π in parallel with the input resistance. In this model the transfer parameter h_{ie} often represents the input impedance, shown here as a resistance at low frequencies.

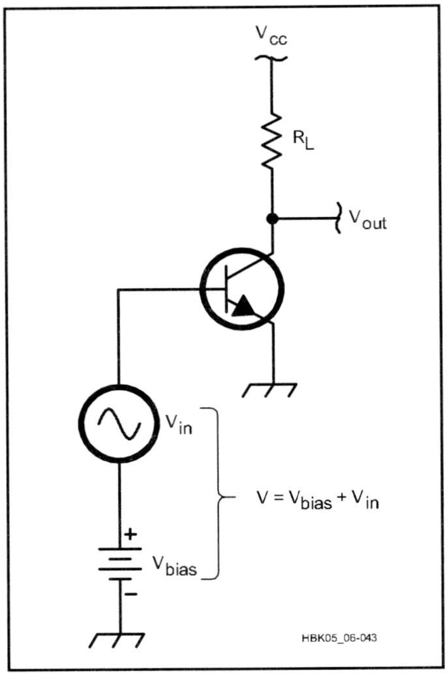

Figure 4.55 — Bipolar transistor with voltage bias and input signal.

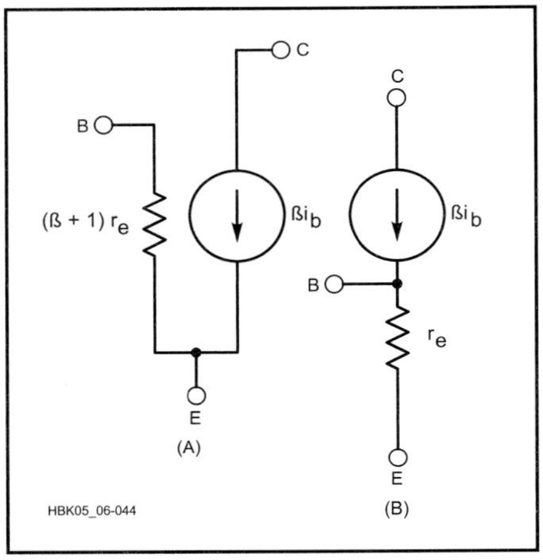

Figure 4.56 — Simplified low-frequency model for the bipolar transistor, a "beta generator with emitter resistance." $r_e = 26\ /\ I_e$ (mA dc).

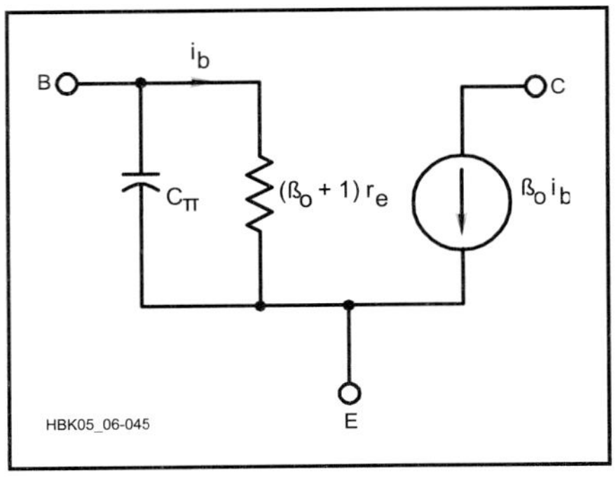

Figure 4.57 — The hybrid-pi model for the bipolar transistor.

THREE BASIC BJT AMPLIFIERS

Figure 4.58 shows a small-signal model applied to the three basic bipolar junction transistor (BJT) amplifier circuits: *common-emitter* (CE), *common-base* (CB) and *common-collector* (CC), more commonly known as the *emitter-follower* (EF). As defined earlier, the word "common" indicates that the referenced terminal is part of both the input and output circuits. The three transistor amplifier configurations are shown as simple circuits in Figure 4.44. The different transistor amplifier configurations have different gains, input and output impedances and phase relationships between the input and output signals.

In these simple models, transistors in both the CE and CB configurations have infinite output resistance because the collector current source is in series with the output current.

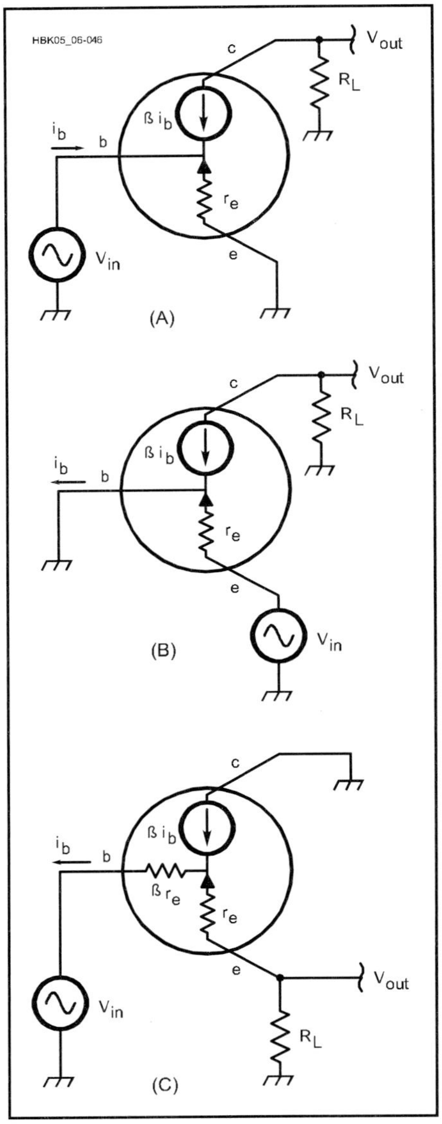

Figure 4.58 — Application of small-signal models for analysis of (A) the CE amplifier, (B) the CB and (C) the EF (CC) bipolar junction transistor amplifiers.

(The amplifier circuit's output impedance must include the effects of R_L.) The transistor connected in the EF configuration, on the other hand, has a finite output resistance because the current source is connected in parallel with the base circuit's equivalent resistance. Calculating the EF amplifier's output resistance requires including the input voltage source, V_s, and its impedance.

Examining the performance needs of the amplifier (engineers refer to these as the circuit's *performance requirements*) determines which of the three circuits is appropriate. Then, once the amplifier configuration is chosen, the equations that describe the circuit's behavior are used to turn the performance requirements into actual circuit component values.

This text presents design information for the CE amplifier in some detail, then summarizes designs for the CC and CB amplifiers. Detailed design analysis for all three amplifiers is described in the texts listed in the reference section for this chapter. All of the analysis in the following sections assume the small-signal, low-frequency model and ignore the effects of the coupling capacitors. High-frequency considerations are discussed in the **RF Techniques** chapter and some advanced discussion of biasing and large signal behavior of BJT amplifiers is available in the online content.

LOAD LINES AND Q-POINT

Transistors can operate with an infinite number of combinations of current (collector, emitter, and base) and voltage (collector-emitter, collector-base, or emitter-collector). Characteristic curves (see Figure 4.32 and the sections on Bipolar Junction and Field-Effect Transistors) show these combinations as a family of graphs in which one of the currents or voltages is varied in steps. A different set of curves can be drawn for each type of transistor and each type of amplifier circuit.

The particular combination at which the amplifier is operating is its *operating point*. The operating point is controlled by the selection of component values that make up the amplifier circuit so that it has the proper combination of gain, linearity and so forth. The result is that the operating point is restricted to a set of points that fall along a *load line*. The operating point with no input signal applied is the circuit's *quiescent point* or *Q-point*. As the input signal varies, the operating point moves along the load line, but returns to the Q-point when the input signal is removed.

More complete information on design using characteristic curves and load lines is available in texts: see the references for Banzhaf or Millman or the online example for BJT circuits at **www.zen22142.zen.co.uk/Design/bjtbias.htm**.

Figure 4.59 shows the load line and Q-point for an amplifier drawn on a transistor's set of characteristic curves for the CE amplifier circuit. The two end-points of the load line correspond to transistor saturation (I_{Csat} on the I_C current axis) and cutoff (V_{CC} on the V_{CE} voltage axis).

When a transistor is in saturation, further increases in base current do not cause a further increase in collector current. In the CE amplifier, this means that V_{CE} is very close to zero and I_C is at a maximum. In the circuit of Figure 4.44B, imagine a short circuit across the collector-to-emitter so that all of V_{CC} appears across R_L. Increasing base current will not result in any additional collector current. At cutoff, base current is so small that V_{CE} is at a maximum because no collector current is flowing and further reductions in base current cause no additional increase in V_{CE}.

In this simple circuit, $V_{CE} = V_{CC} - I_C R_L$ and the relationship between I_C and V_{CE} is a straight line between saturation and cutoff. This circuit's load line has a slope of $R_L = (V_{CC} - V_{CE}) / I_C$. No matter what value of base current is flowing in the transistor, the resulting combination of I_C and V_{CE} will be somewhere on the load line.

With no input signal to this simple circuit, the transistor is at cutoff where $I_C = 0$ and $V_{CE} = V_{CC}$. As the input signal increases so that base current gets larger, the operating point begins to move along the load line to the left, so that I_C increases and the voltage drop across the load, $I_C R_L$, increases, reducing V_{CE}. Eventually, the input signal will cause enough base current to flow that saturation is reached, where $V_{CE} \approx 0$ (typically 0.1 to 0.3 V for silicon transistors) and $I_C \approx V_{CC} / R_L$. If R_L is made smaller, the load line will become steeper and if R_L increases, the load line's slope is reduced.

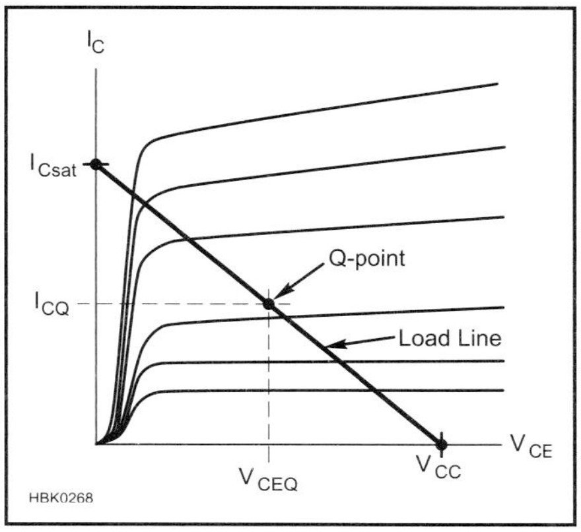

Figure 4.59 — A load line drawn on a transistor's characteristic curves. A circuit's load line shows all of the possible operating points with the specific component values chosen. If there is no input signal, the operating point is the quiescent or Q-point.

This simple circuit cannot reproduce negative input signals because the transistor is already in cutoff with no input signal. In addition, the shape and spacing of the characteristic curves show that the transistor responds nonlinearly when close to saturation and cutoff (the nonlinear regions) than it does in the middle of the curves (the linear or active region). Biasing is required so that the circuit does not operate in nonlinear regions, distorting the signal as shown in Figure 4.33.

If the circuit behaves differently for ac signals than for dc signals, a separate *ac load line* can be drawn as discussed below in the section "AC Performance" for the common-emitter amplifier. For example, in the preceding circuit, if R_L is replaced by a circuit that includes inductive or capacitive reactance, ac collector current will result in a different voltage drop across the circuit than will dc collector current. This causes the slope of the ac load line to be different than that of the dc load line.

The ac load line's slope will also vary with frequency, although it is generally treated as constant over the range of frequencies for which the circuit is designed to operate. The ac and dc load lines intersect at the circuit's Q-point because the circuit's ac and dc operation is the same if the ac input signal is zero.

COMMON-EMITTER AMPLIFIER

The *common-emitter amplifier* (*CE*) is the most common amplifier configuration of all — found in analog and digital circuits, from dc through microwaves, made of discrete components and fabricated in ICs. If you understand the CE amplifier, you've made a good start in electronics.

The CE amplifier is used when modest voltage gain is required along with an *input impedance* (the load presented to the circuit supplying the signal to be amplified) of a few hundred to a few kΩ. The current gain of the CE amplifier is the transistor's current gain, β.

The simplest practical CE amplifier circuit is shown in **Figure 4.60**. This circuit includes both coupling and biasing components. The

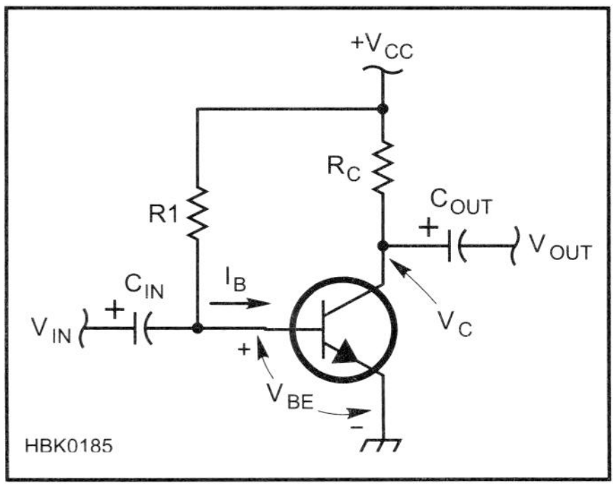

Figure 4.60 — Fixed-bias is the simplest common-emitter (CE) amplifier circuit.

capacitors at the input (C_{IN}) and output (C_{OUT}) block the flow of dc current to the load or to the circuit driving the amplifier. This is an ac-coupled design. These capacitors also cause the gain at very low frequencies to be reduced — gain at dc is zero, for example, because dc input current is blocked by C_{IN}. Resistor R_1 provides a path for bias current to flow into the base, offsetting the collector current from zero and establishing the Q-point for the circuit.

As the input signal swings positive, more current flows into the transistor's base through C_{IN}, causing more current to flow from the collector to emitter. This causes more voltage drop across R_L and so the voltage at the collector also drops. The reverse is true when the input signal swings negative. Thus, the output from the CE amplifier is inverted from its input.

Kirchoff's Voltage Law is used to analyze the circuit. We'll start with the collector circuit and treat the power supply as a voltage source.

$$V_{cc} = I_c R_c + V_{ce}$$

We can determine the circuit's voltage gain, A_V, from the variation in output voltage caused by variations in input voltage. The output voltage from the circuit at the transistor collector is

$$V_c = V_{CC} - I_c R_c = V_{CC} - \beta I_B R_C$$

It is also necessary to determine how base current varies with input voltage. Using the transistor's equivalent circuit of Figure 4.56A,

$$I_B = \frac{V_B}{(\beta + 1)\, r_e}$$

so that

$$V_c = V_{CC} - V_B \frac{\beta}{\beta + 1} \times \frac{R_C}{r_e}$$

We can now determine the circuit's *voltage gain*, the variation in output voltage, DV_C, due to variations in input voltage, DV_B. Since V_{CC} is constant and β is much greater than 1 in our model:

$$A_V \approx -\frac{R_C}{r_e}$$

Because r_e is quite small (typically a few ohms, see the equation for r_e in the section on the Common-Emitter Model), A_V for this circuit can be quite high.

The circuit load line's end-points are $V_{CE} = V_{CC}$ and $I_C = V_{CC} / R_C$. The circuit's Q-point is determined by the collector resistor, R_C, and resistor R_1 that causes bias current to flow into the base. To determine the Q-point, again use KVL starting at the power source and assuming that $V_{BE} = 0.7$ V for a silicon transistor's PN junction when forward-biased.

$$V_{CC} - I_B R_1 = V_B = V_{BE} = 0.7 \text{ V}$$

so

$$I_B = \frac{V_{CC} - 0.7\text{V}}{R_1}$$

And the Q-point is therefore

$$V_{CEQ} = V_{CC} - \beta I_B R_C$$

and

$$I_{CQ} = \beta I_B$$

The actual V_{BE} of silicon transistors will vary from 0.6-0.75 V, depending on the level of base current, but 0.7 V is a good compromise value and widely used in small-signal, low-frequency design. Use 0.6 V for very low-power amplifiers and 0.75 V (or more) for high-current switch circuits.

This simple *fixed-bias* circuit is a good introduction to basic amplifiers, but is not entirely practical because the bias current will change due to the change of V_{BE} with temperature, leading to thermal instability. In addition, the high voltage gain can lead to instability due to positive feedback at high frequencies.

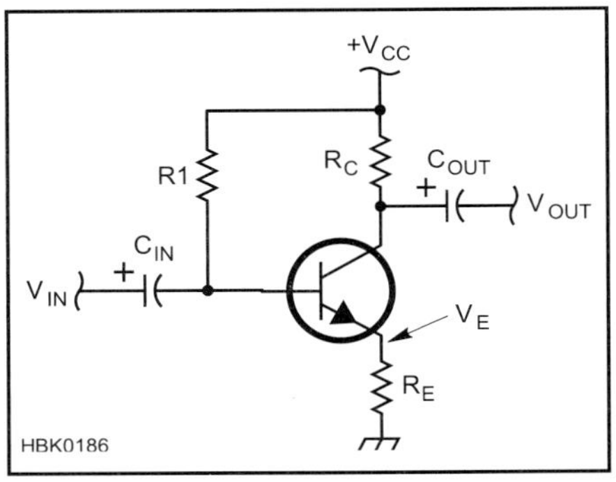

Figure 4.61 — Emitter degeneration. Adding R_E produces negative feedback to stabilize the bias point against changes due to temperature. As the bias current increases, the voltage drop across R_E also increases and causes a decrease in V_{BE}. This reduces bias current and stabilizes the operating point.

To stabilize the dc bias, **Figure 4.61** adds R_E, a technique called *emitter degeneration* because the extra emitter resistance creates negative feedback: as base current rises, so does V_E, the voltage drop across R_E. This reduces the base-emitter voltage and lowers base current. The benefit of emitter degeneration comes from stabilizing the circuit's dc behavior with temperature, but there is a reduction in gain because of the increased resistance in the emitter circuit. Ignoring the effect of R_L for the moment,

$$A_V \approx -\frac{R_C}{R_E}$$

In effect, the load resistor is now split between R_C and R_E, with part of the output voltage appearing across each because the changing current flows through both resistors. While somewhat lower than with the emitter connected directly to ground, voltage gain becomes easy to control because it is the ratio of two resistances.

Biasing the CE Amplifier

Figure 4.62 adds R_1 and R_2 from a voltage divider that controls bias current by fixing the base voltage at:

$$V_B = V_{CC} \frac{R_2}{R_1 + R_2}$$

Since

$$V_B = V_{BE} + (I_B + I_C)\, R_E = 0.7 \text{ V} + (\beta + 1)\, I_B R_E$$

base current is

$$I_B = \frac{V_B - 0.7 \text{ V}}{(\beta + 1)\, R_E}$$

and Q-point collector current becomes for high values of β

$$I_{CQ} = \beta I_B \approx \frac{V_{CC} \frac{R_2}{R_1 + R_2} - 0.7}{R_E}$$

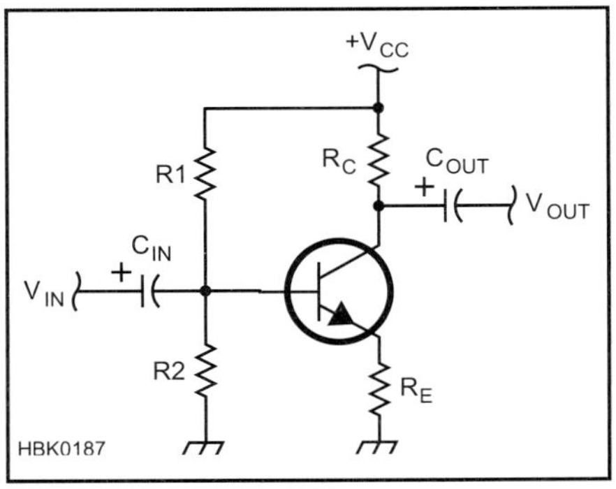

Figure 4.62 — Self-bias or self-emitter bias. R1 and R2 form a voltage divider to stabilize V_B and bias current. A good rule of thumb is for current flow through R1 and R2 to be 10 times the desired bias current. This stabilizes bias against changes in transistor parameters and component values.

This is referred to as *self-bias* or *self-emitter bias* in which the Q-point is much less sensitive to variations in temperature that affect β and V_{BE}.

A good rule-of-thumb for determining the sum of R_1 and R_2 is that the current flowing through the voltage divider, $V_{CC}/(R_1+R_2)$, should be at least 10 times the bias current, I_B. This keeps V_B relatively constant even with small changes in transistor parameters and temperature.

Q-point V_{CEQ} must now also account for the voltage drop across both R_C and R_E.

$$V_{CEQ} \approx V_{CC} - \beta I_B (R_C + R_E)$$

More sophisticated techniques for designing the bias networks of bipolar transistor circuits are described in reference texts listed at the end of this chapter.

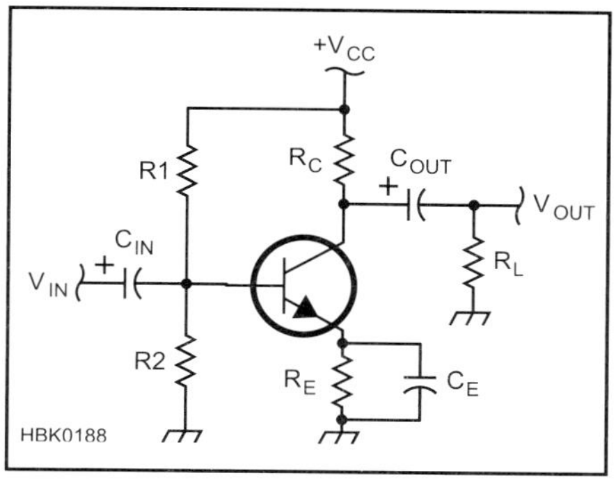

Figure 4.63 — Emitter bypass. Adding C_E allows ac currents to flow "around" R_E, returning ac gain to the value for the fixed-bias circuit while allowing R_E to stabilize the dc operating point.

Input and Output Impedance

With R_E in the circuit, the small changes in input current, I_B, when multiplied by the transistor's current gain, β, cause a large voltage change across R_E equal to $\beta I_B R_E$. This is the same voltage drop as if I_B were flowing through a resistance equal to bR_E. Thus, the effect of β on impedance at the base is to multiply the emitter resistance, R_E by β, as well. At the transistor's base,

$$Z_B \approx (\beta + 1) R_E$$

The input source doesn't just drive the base, of course, it also has to drive the combination of R1 and R2, the biasing resistors. From an ac point of view, both R1 and R2 can be considered as connected to ac ground, and they can be treated as if they were connected in parallel. When R1//R2 are considered along with the transistor base impedance, Z_B, the impedance presented to the input signal source is:

$$Z_{IN} = R1 // R2 // (\beta + 1) R_E$$

where // designates "in parallel with."

For both versions of the CE amplifier, the collector output impedance is high enough that

$$Z_{OUT} \approx R_C$$

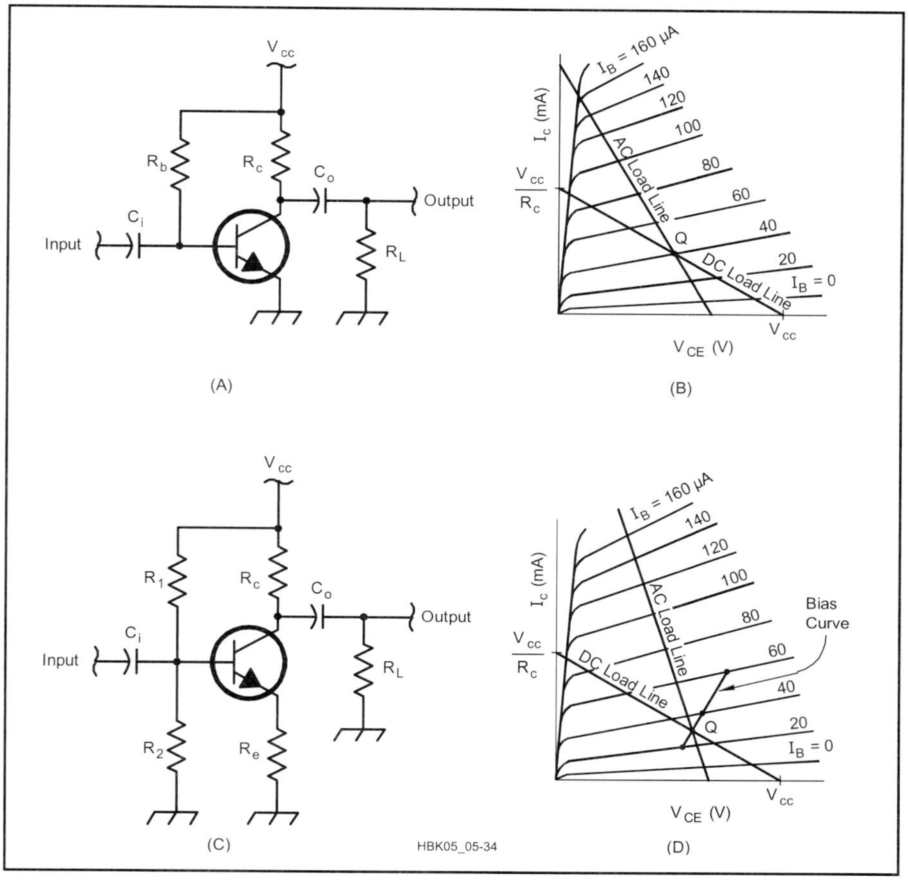

Figure 4.64 — Amplifier biasing and ac and dc load lines. (A) Fixed bias. Input signal is ac coupled through C_i. The output has a voltage that is equal to $V_{CC} - I_C \times R_C$. This signal is ac coupled to the load, R_L, through C_O. For dc signals, the entire output voltage is based on the value of R_C. For ac signals, the output voltage is based on the value of R_C in parallel with R_L. (B) Characteristic curve for the transistor amplifier pictured in (A). The slope of the dc load line is equal to $-1 / R_C$. For ac signals, the slope of the ac load line is equal to $-1 / (R_C // R_L)$. The quiescent-point, Q, is based on the base bias current with no input signal applied and the point where this characteristic line crosses the dc load line. The ac load line must also pass through point Q. (C) Self-bias. Similar to fixed bias circuit with the base bias resistor split into two: R1 connected to V_{CC} and R2 connected to ground. Also an emitter bias resistor, R_E, is included to compensate for changing device characteristics. (D) This is similar to the characteristic curve plotted in (B) but with an additional "bias curve" that shows how the base bias current varies as the device characteristics change with temperature. The operating point, Q, moves along this line and the load lines continue to intersect it as it changes. If C_E was added as in Figure 4.63, the slope of the ac load line would increase further.

CE Amplifier Design Example

The general process depends on the circuit's primary performance requirements, including voltage gain, impedances, power consumption, and so on. The most common situation in which a specific voltage gain is required and the circuit's Q-point has been selected based on the transistor to be used, and using the circuit of Figure 4.62, is as follows:

1) Start by determining the circuit's design constraints and assumptions: power supply $V_{CC} = 12$ V, transistor $\beta = 150$ and $V_{BE} = 0.7$ V. State the circuit's design requirements: $|A_V| = 5$, Q-point of $I_{CQ} = 4$ mA and $V_{CEQ} = 5$ V. (A $V_{CEQ} \approx 1/2\ V_{CC}$ allows a wide swing in output voltage with the least distortion.)

2) Determine the values of R_C and R_E: $R_C + R_E = (V_{CC} - V_{CEQ})/I_{CQ} = 1.75\ k\Omega$

3) $A_V = -5$, so $R_C = 5\ R_E$, thus $6R_E = 1.75$ kΩ and $R_E = 270\ \Omega$

4) Determine the base bias current, $I_B = I_{CQ}/\beta = 27\ \mu A$. By the rule of thumb, current through R_1 and $R_2 = 10\ I_B = 270\ \mu A$

5) Find the voltage across $R_2 = V_B = V_{BE} + I_C R_E = 0.7 + 4$ mA (0.27 kΩ) = 1.8 V. Thus, $R_2 = 1.8$ V / 270 μA = 6.7 kΩ

6) The voltage across $R_1 = V_{CC} - V_{R2} = 12 - 1.8 = 10.2$ V and $R_1 = 10.2$ V / 270 μA = 37.8 kΩ

Use the nearest standard values (R_E = 270 Ω, R_1 = 39 kΩ, R_2 = 6.8 kΩ) and circuit behavior will be close to predicted performance.

AC Performance

To achieve high gains for ac signals while maintaining dc bias stability, the *emitter-bypass capacitor*, C_E, is added in **Figure 4.63** to provide a low impedance path for ac signals around R_E. In addition, a more accurate formula for ac gain includes the effect of adding R_L through the dc blocking capacitor at the collector. In this circuit, the ac voltage gain is

$$A_V \approx -\frac{R_C\ //\ R_L}{r_e}$$

Because of the different signal paths for ac and dc signals, the ac performance of the circuit is different than its dc performance. This is illustrated in **Figure 4.64** by the intersecting load lines labeled "AC Load Line" and "DC Load Line." The load lines intersect at the Q-point because at that point dc performance is the same as ac performance if no ac signal is present.

The equation for ac voltage gain assumes that the reactances of C_{IN}, C_{OUT}, and C_E are small enough to be neglected (less than one-tenth that of the components to which they are connected at the frequency of interest). At low frequencies, where the capacitor reactances become increasingly large, voltage gain is reduced. Neglecting C_{IN} and C_{OUT}, the low-frequency 3 dB point of the amplifier, f_L, occurs where $X_{CE} = 0.414\ r_e$,

$$f_L = \frac{2.42}{2\pi r_e C_E}$$

This increases the emitter circuit impedance such that A_V is lowered to 0.707 of its mid-band value, lowering gain by 3 dB. (This ignores the effects of C_{IN} and C_{OUT}, which will also affect the low-frequency performance of the circuit.)

The ac input impedance of this version of the CE amplifier is lower because the effect of R_E on ac signals is removed by the bypass capacitor. This leaves only the internal emitter resistance, r_e, to be multiplied by the current gain,

$$Z_{IN} \approx R1\ //\ R2\ //\ \beta r_e$$

and

$$Z_{OUT} \approx R_C$$

again neglecting the reactance of the three capacitors.

The power gain, A_P, for the emitter-bypassed CE amplifier is the ratio of output power, V_O^2/Z_{OUT}, to input power, V_I^2/Z_{IN}. Since $V_O = V_I A_V$,

$$A_P = A_V^2\ \frac{R1\ //\ R2\ //\ \beta r_e}{R_C}$$

COMMON-COLLECTOR (EMITTER-FOLLOWER) AMPLIFIER

The common-collector (CC) amplifier in **Figure 4.65** is also known as the *emitter-follower* (EF) because the emitter voltage "follows" the input voltage. In fact, the amplifier has no voltage gain (voltage gain ≈ 1) but is used as a buffer amplifier to isolate sensitive circuits such as oscillators or to drive low-impedance loads, such as coaxial cables. As in the CE amplifier, the current gain of the emitter-follower is the transistor's current gain, β. It has relatively high input impedance with low output impedance and good power gain.

The collector of the transistor is connected directly to the power supply without a resistor and the output signal is created by the voltage drop across the emitter resistor. There is no 180° phase shift as seen in the CE amplifier; the output voltage follows the input signal with 0° phase shift because increases in the input signal cause increases in emitter current and the voltage drop across the emitter resistor.

The EF amplifier has high input impedance: following the same reasoning as for the CE amplifier with an unbypassed emitter resistor,

$$Z_{IN} = R1\ //\ R2\ //\ (\beta + 1)\ R_E$$

The impedance at the EF amplifier's output consists of the emitter resistance, R_E, in parallel with the series combination of the internal emitter resistance, r_e, the parallel combination of biasing resistors R1 and R2, and the internal impedance of the source providing the input signal. In this case, current gain acts to *reduce* the effect of the input circuit's impedance on output impedance:

$$Z_{OUT} = \left[\frac{R_S\ //\ R1\ //\ R2}{(\beta + 1)}\right]//\ R_E$$

In practice, with transistor β of 100 or more, $Z_{OUT} \approx R_S/\beta$. However, if a very high impedance source is used, such as a crystal microphone element or photodetector, the effects of the biasing and emitter resistors must be considered.

Because the voltage gain of the EF amplifier is unity, the power gain is simply the ratio of input impedance to output impedance,

$$A_P \approx \frac{R1\ //\ R2\ //\ (\beta + 1)\ R_E}{R_E}$$

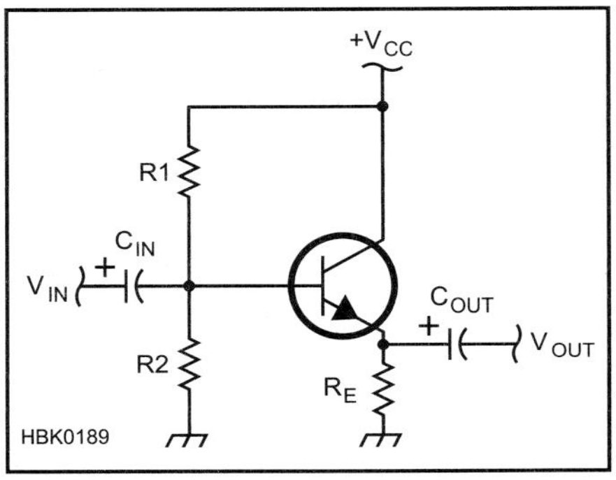

Figure 4.65 — Emitter follower (EF) amplifier. The voltage gain of the EF amplifier is unity. The amplifier has high input impedance and low output impedance, making it a good choice for use as a buffer amplifier.

EF Amplifier Design Example

The following procedure is similar to the design procedure in the preceding section for the CE amplifier, except $A_V = 1$.

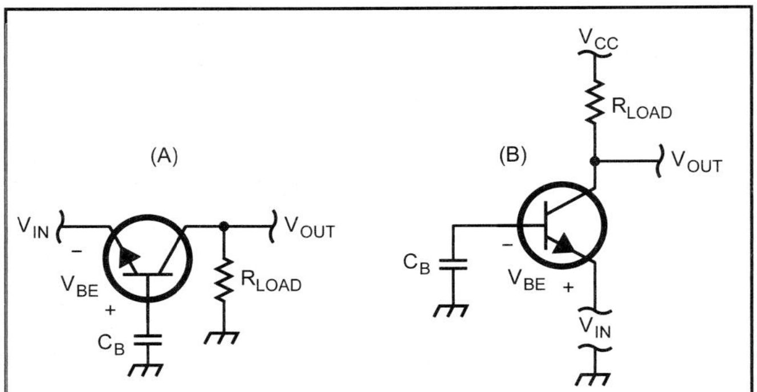

Figure 4.66 — The common-base (CB) amplifier is often drawn in an unfamiliar style (A), but is more easily understood when drawn similarly to the CE and EF amplifiers (B). The input signal to the CB amplifier is applied to the emitter instead of the base.

1) Start by determining the circuit's design constraints and assumptions: V_{cc} = 12 V (the power supply voltage), a transistor's β of 150 and V_{BE} = 0.7 V. State the circuit's design requirements: Q-point of I_{CQ} = 5 mA and V_{CEQ} = 6 V.

2) $R_E = (V_{CC} - V_{CEQ})/I_{CQ}$ = 1.2 kΩ

3) Base current, $I_B = I_{CQ}/\beta$ = 33 μA

4) Current through R1 and R2 = 10 I_B = 330 μA (10 I_B rule of thumb as with the CE amplifier)

5) Voltage across R2 = $V_{BE} + I_C R_E$ = 0.7 + 5 mA (1.2 kΩ) = 6.7 V and R2 = 6.7 V / 330 μA = 20.3 kΩ (use the standard value 22 kΩ)

6) Voltage across R1 = V_{CC} – 6.7 V = 5.3 V

7) R1 = 5.3 V / 330 μA = 16.1 kΩ (use 16 kΩ)

8) $Z_{IN} = R1 \,//\, R2 \,//\, R_E(\beta + 1) \approx 8.5$ kΩ

COMMON-BASE AMPLIFIER

The common-base (CB) amplifier of **Figure 4.66** is used where low input impedance is needed, such as for a receiver preamp with a coaxial feed line as the input signal source. Complementary to the EF amplifier, the CB amplifier has unity current gain and high output impedance.

Figure 4.66A shows the CB circuit as it is usually drawn, without the bias circuit resistors connected and with the transistor symbol turned on its side from the usual orientation so that the emitter faces the input. In order to better understand the amplifier's function, Figure 4.66B reorients the circuit in a more familiar style. We can now clearly see that the input has just moved from the base circuit to the emitter circuit.

Placing the input in the emitter circuit allows it to cause changes in the base-emitter current as for the CE and EF amplifiers, except that for the CB amplifier a positive change in input amplitude reduces base current by lowering V_{BE} and raising V_C. As a result, the CB amplifier is noninverting, just like the EF, with output and input signals in-phase.

A practical circuit for the CB amplifier is shown in **Figure 4.67**. From a dc point of view (replace the capacitors with open circuits), all of the same resistors are there as in the CE amplifier. The input capacitor, C_{IN}, allows the dc emitter current to bypass the ac input signal source and C_B places the base at ac ground while allowing a dc voltage for biasing. (All voltages and currents are labeled to aid in understanding the different orientation of the circuit.)

The CB amplifier's current gain,

$$A_I = \frac{i_C}{i_E} = \frac{\beta}{\beta + 1}$$

is relatively independent of input and output impedance, providing excellent isolation between the input and output circuits. Output impedance does not affect input impedance, allowing the CB amplifier to maintain stable input impedance, even with a changing load.

Following reasoning similar to that for the CE and EF amplifiers for the effect of current gain on R_E, we find that input impedance for the CB amp is

$$Z_{IN} = R_E \,//\, (\beta + 1)\, r_e$$

With high-gain transistors having a β > 100, for typical values of r_e (about 1 kΩ) the input impedance is approximately R_E. If R_E is chosen to be 50 Ω, the result will be a good input impedance match to 50 Ω feed lines and signal sources.

The output impedance for the CB amplifier is approximately

$$Z_{OUT} = R_C \,//\, \frac{1}{h_{oe}} \approx R_C$$

where h_{oe} is the transistor's collector output admittance. The reciprocal of h_{oe} is in the range of 100 kΩ at low frequencies.

Voltage gain for the CB amplifier is

$$A_V \approx \frac{R_C \,//\, R_L}{r_e}$$

As a result, the usual function of the CB amplifier is to convert input current from a low-impedance source into output voltage at a higher impedance.

Power gain for the CB amplifier is approximately the ratio of output to input impedance,

$$A_P \approx \frac{R_C}{R_E \,//\, (\beta + 1)\, r_e}$$

CB Amplifier Design Example

Because of its usual function as a current-to-voltage converter, the design process for the CB amplifier begins with selecting R_E and A_V, assuming that R_L is known.

1) Start by determining the circuit's design constraints and assumptions: V_{cc} = 12 V (the power supply voltage), a transistor's β of 150 and V_{BE} = 0.7 V. State the circuit's design requirements: R_E = 50 Ω, R_L = 1 kΩ, I_{CQ} = 5 mA, V_{CEQ} = 6 V.

2) Base current, $I_B = I_{CQ}/\beta$ = 33 μA

3) Current through R1 and R2 = 10 I_B = 330 μA (10 × I_B rule of thumb as with the CE amplifier)

4) Voltage across R2 = $V_{BE} + I_C R_E$ = 0.7 + 5 mA (50 Ω) = 0.95 V and R2 = 0.95 V / 330 μA = 2.87 kΩ (use the standard value 2.7 kΩ)

5) Voltage across R1 = V_{CC} – 0.95 V = 11.05 V

6) R1 = 11.05 V / 330 μA = 33.5 kΩ (use 33 kΩ)

7) $R_C = (V_{CC} - I_{CQ} R_E - V_{CEQ}) / I_{CQ}$ = (12 – 0.25 – 6) / 5 mA = 1.15 kΩ (use 1.2 kΩ)

8) A_V = (1.2 kΩ // 1 kΩ) / (26 mV / I_E) = 105

4.7.4 FET Amplifiers

The field-effect transistor (FET) is widely used in radio and RF applications. There are many types of FETs, with JFETs (junction FET) and MOSFETs (metal-oxide-semiconductor FET) being the most common types. In this section we will discuss JFETs, with the understanding that the use of MOSFETs is similar. (This discussion is based on N-channel JFETs, but the same discussion applies to P-channel devices if the bias voltages and currents are reversed.)

SMALL-SIGNAL FET MODEL

While bipolar transistors are most commonly viewed as current-controlled devices, the JFET, however, is purely a voltage-controlled device — at least at low frequencies. The input gate is treated as a reverse-biased diode junction with virtually no current flow. As with the bipolar transistor amplifier circuits, the circuits in this section are very basic and more thorough treatments of FET amplifier design can be found in the references at

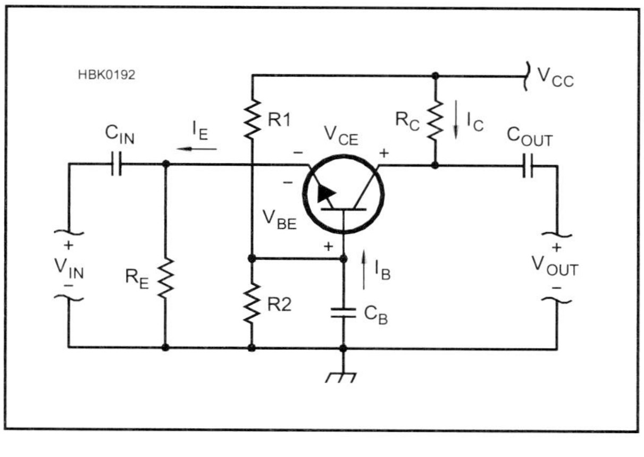

Figure 4.67 — A practical common-base (CB) amplifier. The current gain of the CB amplifier is unity. It has low input impedance and high output impedance, resulting in high voltage gain. The CB amplifier is used to amplify signals from low-impedance sources, such as coaxial cables.

the end of the chapter.

The operation of an N-channel JFET for both biasing and signal amplification can be characterized by the following equation:

$$I_D = I_{DSS}\left(\frac{1 - V_{SG}}{V_P}\right)^2$$

where

I_{DSS} = drain saturation current
V_{GS} = the gate-source voltage
V_P = the pinch-off voltage.

I_{DSS} is the maximum current that will flow between the drain and source for a given value of drain-to-source voltage, V_{DS}. Note that the FET is a *square-law* device in which output current is proportional to the square of an input voltage. (The bipolar transistor's output current is an exponential function of input current.)

Also note that V_{GS} in this equation has the opposite sense of the bipolar transistor's V_{BE}. For this device, as V_{GS} increases (making the source more positive than the gate), drain current decreases until at V_P the channel is completely "pinched-off" and no drain current flows at all. This equation applies only if V_{GS} is between 0 and V_P. JFETs are seldom used with the gate-to-channel diode forward-biased ($V_{GS} < 0$).

None of the terms in this equation depend explicitly on temperature. Thus, the FET is relatively free of the thermal instability exhibited by the bipolar transistor. As temperature increases, the overall effect on the JFET is to reduce drain current and to stabilize the operation.

The small-signal model used for the JFET is shown in **Figure 4.68**. The drain-source channel is treated as a current source whose output is controlled by the gate-to-source voltage so that $I_D = g_m V_{GS}$. The high input impedance allows the input to be modeled as an open circuit (at low frequencies). This simplifies circuit modeling considerably as biasing of the FET gate can be done by a simple voltage divider without having to consider the effects of bias current flowing in the JFET itself.

The FET has characteristic curves as shown in Figure 4.43 that are similar to those of a bipolar transistor. The output characteristic curves are similar to those of the bipolar transistor, with the horizontal axis showing V_{DS} instead of V_{CE} and the vertical axis showing I_D instead of I_C. Load lines, both ac and dc, can be developed and drawn on the output characteristic curves in the same way as for bipolar transistors.

The set of characteristic curves in Figure 4.43 are called *transconductance response curves* and they show the relationship between input voltage (V_{GS}), output current (I_D) and output voltage (V_{DS}). The output characteristic curves show I_D and V_{DS} for different

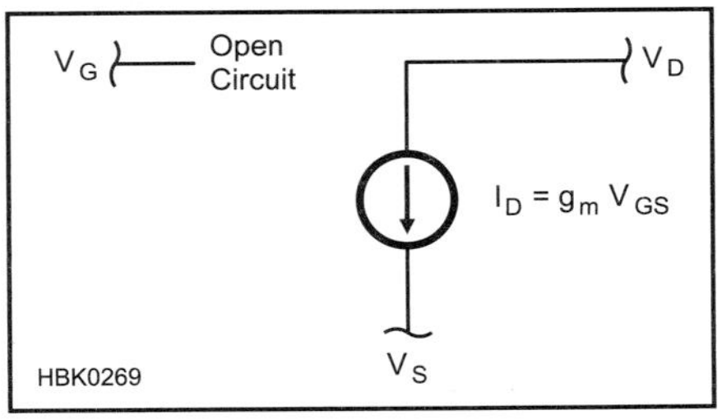

Figure 4.68 — Small-signal FET model. The FET can be modeled as a voltage-controlled current source in its saturation region. The gate is treated as an open-circuit due to the reverse-biased gate-channel junction.

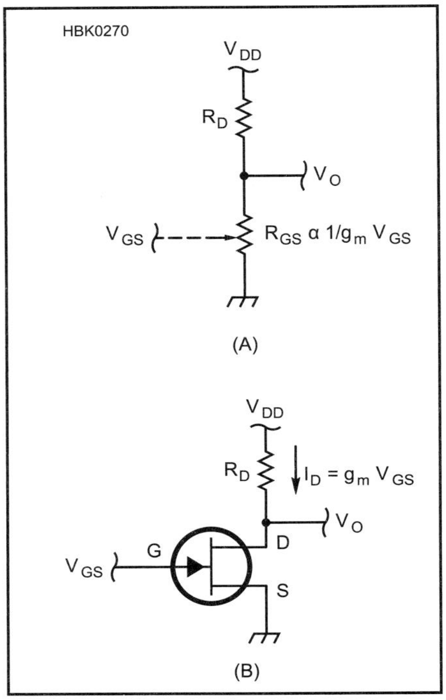

Figure 4.69 — In the ohmic region (A), the FET acts like a variable resistance, R_{DS}, with a value controlled by V_{GS}. The alpha symbol (α) means "is proportional to." In the saturation region (B), the drain-source channel of the FET can be treated like a current source with $I_D = g_m V_{GS}$.

values of V_{GS} and are similar to the BJT output characteristic curve. The input characteristic curves show I_D versus V_{GS} for different values of V_{DS}.

MOSFETs act in much the same way as JFETs when used in an amplifier. They have a higher input impedance, due to the insulation between the gate and channel. The insulated gate also means that they can be operated with the polarity of V_{GS} such that a JFET's gate-channel junction would be forward biased, beyond V_P. Refer to the discussion of depletion- and enhancement-mode MOSFETs in the previous section on Practical Semiconductors.

THREE BASIC FET AMPLIFIERS

Just as for bipolar transistor amplifiers, there are three basic configurations of amplifiers using FETs: the *common-source* (CS) (corresponding to the common-emitter), *common-drain* (CD) or *source-follower* (corresponding to the emitter-follower) and the *common-gate* (CG) (corresponding to the common base). Simple circuits and design methods are presented here for each, assuming low-frequency operation and a simple, voltage-controlled current-source model for the FET. Discussion of the FET amplifier at high frequencies is available in the **RF Techniques** chapter and an advanced discussion of biasing FET amplifiers and their large-signal behavior is contained in this book's online information.

COMMON-SOURCE AMPLIFIER

The basic circuit for a common-source FET amplifier is shown in **Figure 4.69**. In the ohmic region (see the previous discussion on FET characteristics), the FET can be treated as a variable resistance as shown in Figure 4.69A where V_{GS} effectively varies the resistance between drain and source. However, most FET amplifiers are designed to operate in the saturation region and the model of Figure 4.68 is used in the circuit of Figure 4.69B in which,

$$I_D = g_m V_{GS}$$

where g_m is the FET's forward transconductance.

If V_O is measured at the drain terminal (just as the common-emitter output voltage is measured at the collector), then

$$\Delta V_O = -g_m \Delta V_{GS} R_D$$

The minus sign results from the output voltage decreasing as drain current and the voltage drop across R_D increases, just as in the CE amplifier. Like the CE amplifier, the input and output voltages are thus 180° out of phase. Voltage gain of the CS amplifier in terms of transconductance and the drain resistance is:

$$A_V = -g_m R_D$$

As long as $V_{GS} < 0$, this simple CS amplifier's input impedance at low frequencies is that of a reverse-biased diode — nearly infinite with a very small leakage current. Output impedance of the CS amplifier is approximately R_D because the FET drain-to-source channel acts like a current source with very high impedance.

$$Z_{IN} = \infty \text{ and } Z_{OUT} \approx R_D$$

As with the BJT, biasing is required to

create a Q-point for the amplifier that allows reproduction of ac signals. The practical circuit of Figure 4.69B is used to allow control of V_{GS} bias. A load line is drawn on the JFET output characteristic curves, just as for a bipolar transistor circuit. One end point of the load line is at $V_{DS} = V_{DD}$ and the other at $I_{DS} = V_{DD} / R_D$. The Q-point for the CS amplifier at I_{DQ} and V_{DSQ} is thus determined by the dc value of V_{GS}.

The practical JFET CS amplifier shown in **Figure 4.70** uses self-biasing in which the voltage developed across the source resistor, R_S, raises V_S above ground by $I_D R_S$ volts and $V_{GS} = -I_D R_S$ since there is no dc drop across R_G. This is also called *source degeneration*. The presence of R_S changes the equation of voltage gain to

$$A_V = -\frac{g_m R_D}{1 + g_m R_S} \approx -\frac{R_D}{R_S}$$

The value of R_S is obtained by substituting $V_{GS} = I_D R_S$ into the small-signal model's equation for I_D and solving for R_S as follows:

$$R_S = \frac{-V_P}{I_{DQ}} \left(1 - \sqrt{\frac{I_{DQ}}{I_{DSS}}}\right)$$

Once R_S is known, the equation for voltage gain can be used to find R_D.

The input impedance for the circuit of Figure 4.60 is essentially R_G. Since the gate of the JFET is often ac coupled to the input source through a dc blocking capacitor, C_{IN}, a value of 100 kΩ to 1 MΩ is often used for R_G to provide a path to ground for gate leakage current. If R_G is omitted in an ac-coupled JFET amplifier, a dc voltage can build up on the gate from leakage current or static electricity, affecting the channel conductivity.

$$Z_{IN} = R_G$$

Because of the high impedance of the drain-source channel in the saturation region, the output impedance of the circuit is:

$$Z_{OUT} \approx R_D$$

Designing the Common-Source Amplifier

The design of the CS amplifier begins with selection of a Q-point $I_{DQ} < I_{DSS}$. Because of variations in V_P and I_{DSS} from JFET to JFET, it may be necessary to select devices individually to obtain the desired performance.

1) Start by determining the circuit's design constraints and assumptions: V_{DD} = 12 V (the power supply voltage) and the JFET has an I_{DSS} of 35 mA and a V_P of –3.0V, typical of small-signal JFETs. State the circuit's design requirements: $|A_V| = 10$ and I_{DQ} = 10 mA.

2) Use the equation above to find R_S = 139 Ω.

3) Since $|A_V| = 10$, $R_D = 10\ R_S$ = 1390 Ω.

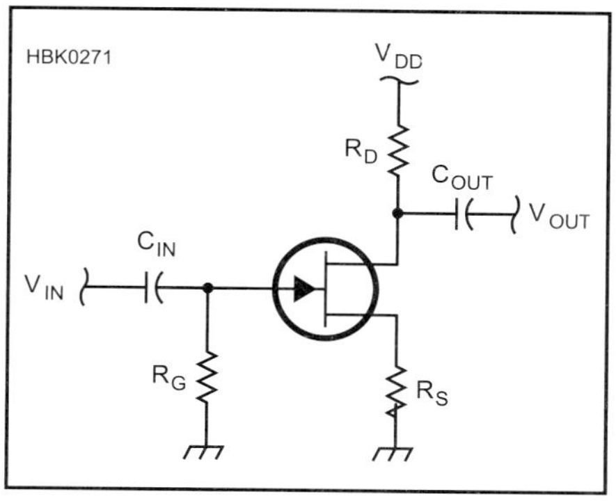

Figure 4.70 — Common-source (CS) amplifier with self-bias.

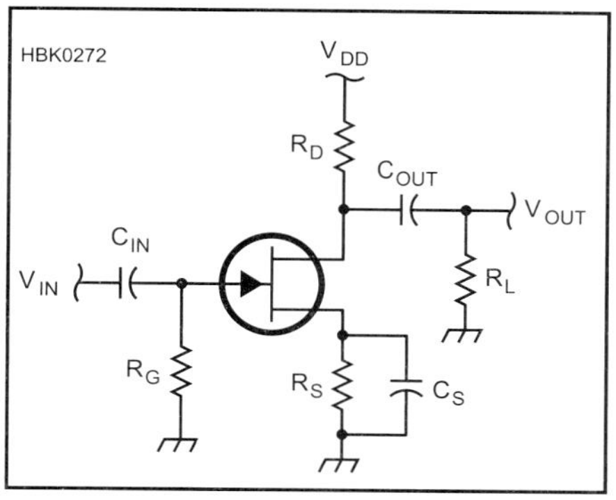

Figure 4.71 — Common-source amplifier with source bypass capacitor, C_S, to increase voltage gain without affecting the circuit's dc performance.

Use standard values for R_S = 150 Ω and R_D = 1.5 kΩ.

AC Performance

As with the CE bipolar transistor amplifier, a bypass capacitor can be used to increase ac gain while leaving dc bias conditions unchanged as shown in **Figure 4.71**. In the case of the CS amplifier, a *source bypass* capacitor is placed across R_S and the load, R_L, connected through a dc blocking capacitor. In this circuit voltage gain becomes:

$$A_V = -g_m\ (R_D\ //\ R_L)$$

Assuming C_{IN} and C_{OUT} are large enough to ignore their effects, the low-frequency cutoff frequency of the amplifier, f_L, is approximately where $X_{CS} = 0.707\ (R_D // R_L)$,

$$f_L = \frac{1.414}{2\pi\,(R_D\ //\ R_L)\,C_S}$$

as this reduces A_V to 0.707 of its mid-band value, resulting in a 3 dB drop in output amplitude.

The low-frequency ac input and output impedances of the CS amplifier remain

$$Z_{IN} = R_G \text{ and } Z_{OUT} \approx R_D$$

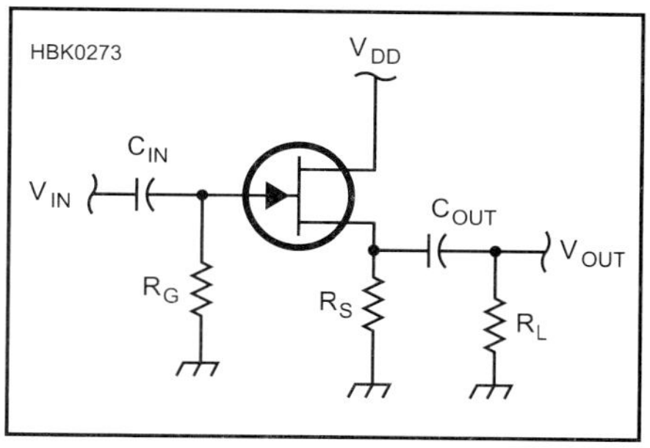

Figure 4.72 — Similar to the EF amplifier, the common-drain (CD) amplifier has a voltage gain of unity, but makes a good buffer with high input and low output impedances.

COMMON-DRAIN (SOURCE-FOLLOWER) AMPLIFIER

The *common-drain* amplifier in **Figure 4.72** is also known as a *source-follower* (SF) because the voltage gain of the amplifier is unity, similar to the emitter follower (EF) bipolar transistor amplifier. The SF amplifier is used primarily as a buffer stage and to drive low-impedance loads.

At low frequencies, the input impedance of the SF amplifier remains nearly infinite. The SF amplifier's output impedance is the source resistance, R_S, in parallel with the impedance of the controlled current source, $1/g_m$.

$$Z_{OUT} = R_S\ //\ \frac{1}{g_m}$$

$$= \frac{R_S}{g_m R_S + 1} \approx \frac{1}{g_m} \text{ for } g_m R_S \gg 1$$

Design of the SF amplifier follows essentially the same process as the CS amplifier, with $R_D = 0$.

THE COMMON-GATE AMPLIFIER

The *common-gate* amplifier in **Figure 4.73** has similar properties to the bipolar transistor common-base (CB) amplifier: unity current gain, high voltage gain, low input impedance and high output impedance. (Refer to the discussion of the CB amplifier regarding placement of the input and how the circuit schematic is drawn.) It is used as a voltage amplifier, particularly for low-impedance sources, such as coaxial cable inputs.

The CG amplifier's voltage gain is

$$A_V = g_m (R_D\ \ //\ R_L)$$

Since the output impedance of the CG amplifier is very high, we must take into account the output resistance of the controlled current source, r_o. This is analogous to the appearance of h_{oe} in the equation for output impedance of the bipolar transistor CG amplifier.

$$Z_O \approx r_o\ (g_m R_S + 1)\ //\ R_D$$

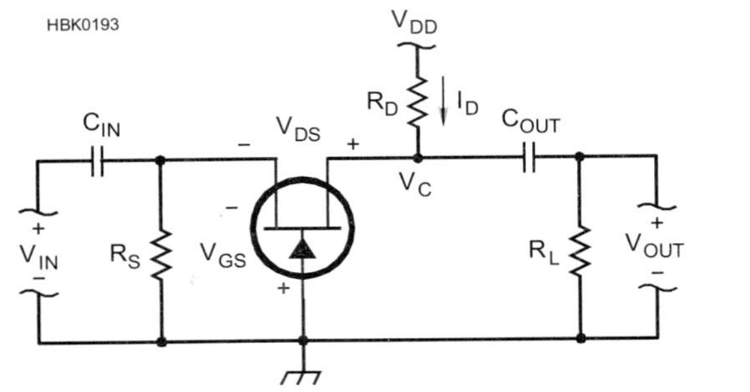

Figure 4.73 — FET common-gate (CG) amplifiers are often used as preamplifiers because of their high voltage gain and low input impedance. With the proper choice of transistor and quiescent-point current, the input impedance can match coaxial cable impedances directly.

The CG amplifier input impedance is approximately:

$$Z_I = R_S\ //\ \frac{1}{g_m}$$

Because the input impedance is quite low, the cascode circuit described later in the section on buffers is often used to present a higher-impedance input to the signal source.

Occasionally, the value of R_S must be fixed in order to provide a specific value of input impedance. Solving the small-signal FET model equation for I_{DQ} results in the following equation:

$$I_{DQ} = \frac{V_P}{2R_S{}^2 I_{DSS}}\left(V_P + \sqrt{V_P{}^2 - 4R_S I_{DSS} V_P}\right) - \frac{V_P}{R_S}$$

Designing the Common-Gate Amplifier

Follow the procedure for designing a CS amplifier, except determine the value of R_D as shown in the equation above for voltage gain, A_V.

1) Start by determining the circuit's design constraints and assumptions: V_{DD} = 12 V (the power supply voltage) and the JFET has a g_m of 15 mA/V, an I_{DSS} of 60 mA and V_P = –6 V. State the circuit's design requirements: A_V = 10, R_L = 1 kΩ and R_S = 50 Ω.

2) Solve the A_V equation for R_D: 10 = 0.015 × $R_D//R_L$, so $R_D//R_L$ = 667 Ω. R_D = 667 R_L / (R_L – 667) = 2 kΩ.

3) Use the equation above to find I_{DQ} = 10 mA. If I_{DQ} places the Q-point in the ohmic region, reduce A_V and repeat the calculations.

4.7.5 Buffer Amplifiers

Figure 4.74 shows common forms of buffers with low-impedance outputs: the *emitter follower* using a bipolar transistor, the *source follower* using a field-effect transistor and the *voltage follower* using an operational amplifier. (The operational amplifier is discussed later in this chapter.) These circuits are called "followers" because the output "follows" the input very closely with approximately the same voltage and little phase shift between the input and output signals.

4.7.6 Cascaded Buffers

THE DARLINGTON PAIR

Buffer stages that are made with single active devices can be more effective if cascaded. Two types of such buffers are in common use. The *Darlington pair* is a cascade of two transistors connected as emitter followers as shown in **Figure 4.75.** The current gain of the Darlington pair is the product of the current gains for the two transistors, $b_1 \times b_2$.

What makes the Darlington pair so useful is that its input impedance is equal to the load impedance times the current gain, effectively multiplying the load impedance:

$$Z_I = Z_{LOAD} \times \beta_1 \times \beta_2$$

For example, if a typical bipolar transistor has β = 100 and Z_{LOAD} = 15 kΩ, a pair of these transistors in the Darlington-pair configuration would have:

$$Z_I = 15\ \text{k}\Omega \times 100 \times 100 = 150\ \text{M}\Omega$$

This impedance places almost no load on the circuit connected to the Darlington pair's input. The shunt capacitance at the input of real transistors can lower the actual impedance as the frequency increases.

Drawbacks of the Darlington pair include lower bandwidth and switching speed. The extremely high dc gain makes biasing very sensitive to temperature and component tolerances. For these reasons, the circuit is usually used as a switch and not as a linear amplifier.

CASCODE AMPLIFIERS

A common-emitter amplifier followed by a common-base amplifier is called a *cascode buffer*, shown in its simplest form in

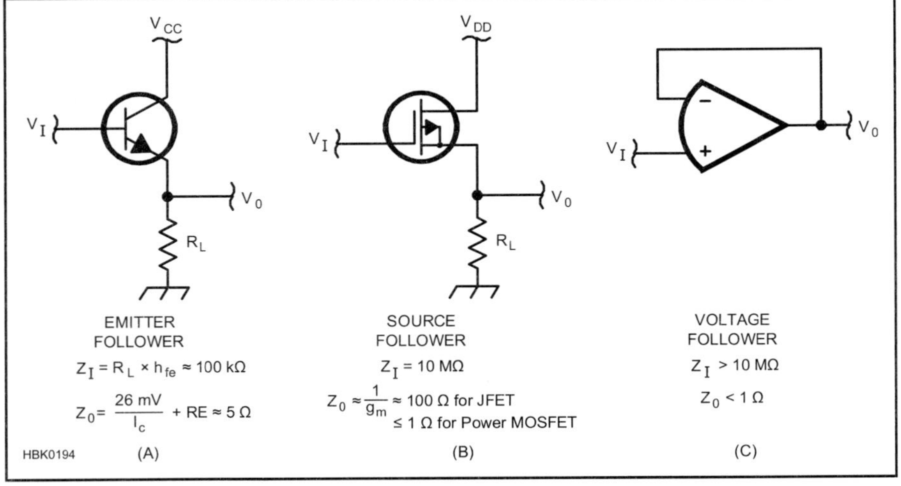

Figure 4.74 — Common buffer stages and some typical input (Z_I) and output (Z_O) impedances. (A) Emitter follower, made with an NPN bipolar transistor; (B) Source follower, made with an FET; and (C) Voltage follower, made with an operational amplifier. All of these buffers are terminated with a load resistance, R_L, and have an output voltage that is approximately equal to the input voltage (gain ≈ 1).

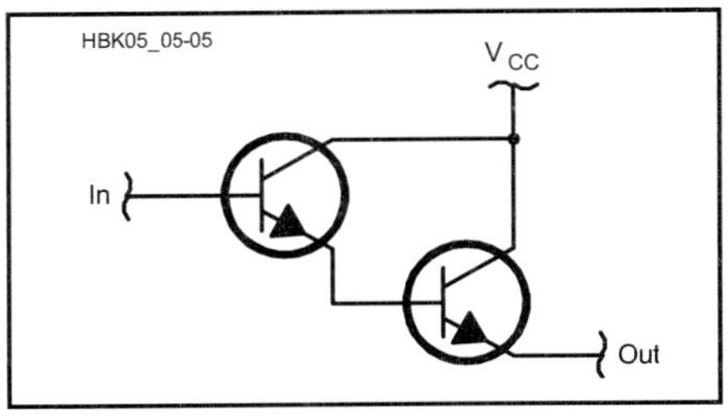

Figure 4.75 — Darlington pair made with two emitter followers. Input impedance, Z_I, is far higher than for a single transistor and output impedance, Z_O, is nearly the same as for a single transistor. DC biasing has been omitted for simplicity.

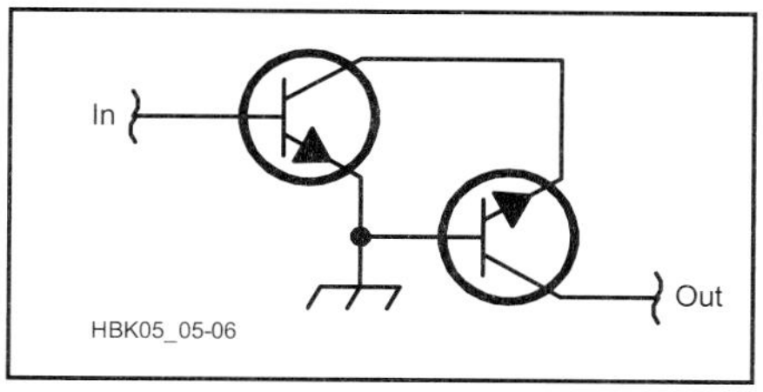

Figure 4.76 — Cascode buffer made with two NPN bipolar transistors has a medium input impedance and high output impedance. DC biasing has been omitted for simplicity.

Figure 4.76. (Biasing and dc blocking components are omitted for simplicity — replace the transistors with the practical circuits described earlier.) Cascode stages using FETs follow a common-source amplifier with a common-gate configuration. The input impedance and current gain of the cascode amplifier are approximately the same as those of the first stage. The output impedance of the common-base or -gate stage is much higher than that of the common-emitter or common-source amplifier. The power gain of the cascode amplifier is the product of the input stage current gain and the output stage voltage gain.

As an example, a typical cascode buffer made with BJTs has moderate input impedance (Z_{IN} = 1 kΩ), high current gain (h_{fe} = 50), and high output impedance (Z_{OUT} = 1 MΩ). Cascode amplifiers have excellent input/output isolation (very low unwanted internal feedback), resulting in high gain with good stability. Because of its excellent isolation, the cascode amplifier has little effect on external tuning components. Cascode circuits are often used in tuned amplifier designs for these reasons.

More on Transistor Amplifiers at RF

See the **RF Techniques** chapter for more information on transistor amplifier design at RF, including a list of RF Amplifier Design resources in the References section.

4.7.7 Using the Transistor as a Switch

When designing amplifiers, the goal was to make the transistor's output a replica of its input, requiring that the transistor stay within its linear region, conducting some current at all times. A switch circuit has completely different properties — its output current is either zero or some maximum value. **Figure 4.77** shows both a bipolar and metal-oxide semiconductor field-effect transistor (or MOSFET) switch circuit. Unlike the linear amplifier circuits, there are no bias resistors in either circuit. When using the bipolar transistor as a switch, it should operate in saturation or in cutoff. Similarly, an FET switch should be either fully-on or fully-off. The figure shows the waveforms associated with both types of switch circuits.

This discussion is written with power control in mind, such as to drive a relay or motor or lamp. The concepts, however, are equally applicable to the much lower-power circuits that control logic-level signals. The switch should behave just the same — switch between on and off quickly and completely — whether large or small.

DESIGNING SWITCHING CIRCUITS

First, select a transistor that can handle the load current and dissipate whatever power is dissipated as heat. Second, be sure that the input signal source can supply an adequate input signal to drive the transistor to the required states, both on and off. Both of these conditions must be met to insure reliable driver operation.

To choose the proper transistor, the load current and supply voltage must both be known. Supply voltage may be steady, but sometimes varies widely. For example, a car's 12 V power bus may vary from 9 to 18 V, depending on battery condition. The transistor must withstand the maximum supply voltage, V_{MAX}, when off. The load resistance, R_L, must also be known. The maximum steady-state current the switch must handle is:

$$I_{MAX} = \frac{V_{MAX}}{R_L}$$

If you are using a bipolar transistor, calculate how much base current is required to drive the transistor at this level of collector current. You'll need to inspect the transistor's data sheet because β decreases as collector current increases, so use a value for β specified at a collector current at or above I_{MAX}.

$$I_B = \frac{I_{MAX}}{\beta}$$

Now inspect the transistor's data sheet values for V_{CEsat} and make sure that this value of I_B is sufficient to drive the transistor fully into saturation at a collector current of I_{MAX}. Increase I_B if necessary — this is I_{Bsat}. The transistor must be fully saturated to minimize heating when conducting load current.

Using the *minimum* value for the input voltage, calculate the value of R_B:

$$R_B = \frac{V_{IN(min)} - V_{BE}}{I_{Bsat}}$$

The minimum value of input voltage must be used to accommodate the *worst-case* combination of circuit voltages and currents.

Designing with a MOSFET is a little easier because the manufacturer usually specifies the value V_{GS} must have for the transistor to be fully on, $V_{GS(on)}$. The MOSFET's gate, being insulated from the conducting channel, acts like a small capacitor of a few hundred pF and draws very little dc current. R_G in Figure 4.67 is required if the input voltage source does not actively drive its output to

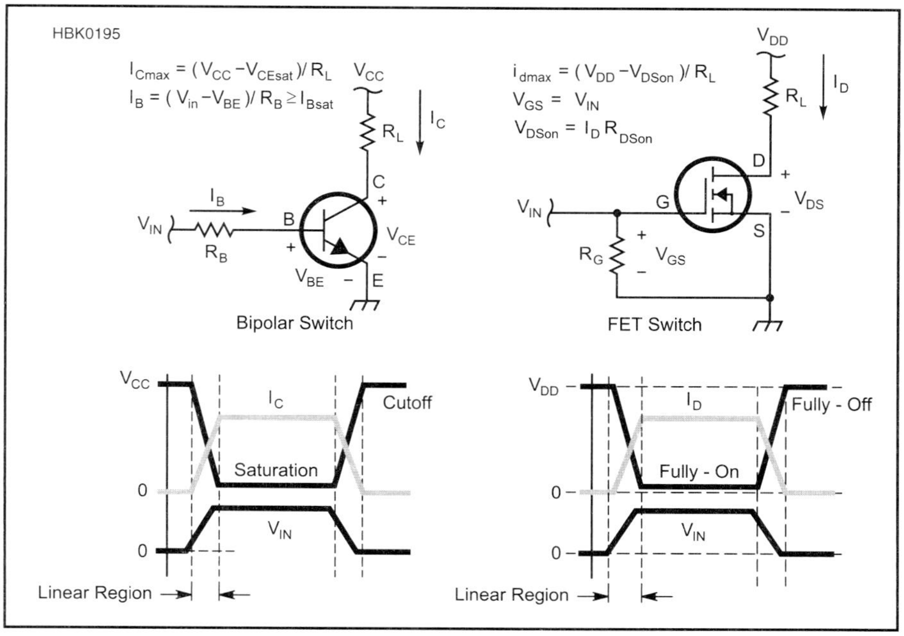

Figure 4.77 — A pair of transistor driver circuits using a bipolar transistor and a MOSFET. The input and output signals show the linear, cutoff and saturation regions.

zero volts when off, such as a switch connected to a positive voltage. The MOSFET won't turn off reliably if its gate is allowed to "float." R_G pulls the gate voltage to zero when the input is open-circuited.

Power dissipation is the next design hurdle. Even if the transistors are turned completely on, they will still dissipate some heat. Just as for a resistor, for a bipolar transistor switch the power dissipation is:

$$P_D = V_{CE} I_C = V_{CE(sat)}\, I_{MAX}$$

where $V_{CE(sat)}$ is the collector-to-emitter voltage when the transistor is saturated.

Power dissipation in a MOSFET switch is:

$$P_D = V_{DS} I_D = R_{DS(on)}\, I_{MAX}^2$$

$R_{DS(on)}$ is the resistance of the channel from drain to source when the MOSFET is on. MOSFETs are available with very low on-resistance, but still dissipate a fair amount of power when driving a heavy load. The transistor's data sheet will contain $R_{DS(on)}$ specification and the V_{GS} required for it to be reached.

Power dissipation is why a switching transistor needs to be kept out of its linear region. When turned completely off or on, either current through the transistor or voltage across it are low, also keeping the product of voltage and current (the power to be dissipated) low. As the waveform diagrams in Figure 4.67 show, while in the linear region, both voltage and current have significant values and so the transistor is generating heat when changing from off to on and vice versa. It's important to make the transition through the linear region quickly to keep the transistor cool.

The worst-case amount of power dissipated during each on-off transition is approximately

$$P_{transition} = \frac{1}{4} V_{MAX}\, I_{MAX}$$

assuming that the voltage and current increase and decrease linearly. If the circuit turns on and off at a rate of f, the total average power dissipation due to switching states is:

$$P_D = \frac{f}{2} V_{MAX}\, I_{MAX}$$

since there are two on-off transitions per switching cycle. This power must be added to the power dissipated when the switch is conducting current.

Once you have calculated the power the switch must dissipate, you must check to see whether the transistor can withstand it. The manufacturer of the transistor will specify a *free-air dissipation* that assumes no heat-sink and room temperature air circulating freely around the transistor. This rating should be at least 50% higher than your calculated power dissipation. If not, you must either use a larger transistor or provide some means of getting rid of the heat, such as heat sink. Methods of dissipating heat are discussed in this chapter's section on Heat Management.

INDUCTIVE AND CAPACITIVE LOADS

Voltage transients for inductive loads, such as solenoids or relays, can easily reach dozens of times the power supply voltage when load current is suddenly interrupted. To protect the transistor, the voltage transient must be clamped or its energy dissipated.

Where switching is frequent, such as for a variable frequency drive or switchmode power control, a series-RC *snubber* circuit (see **Figure 4.78**) is connected across the load. The RC circuit dissipates the transient's energy as heat. The following procedure will suffice for most amateur designs, and a more comprehensive analysis by Rudy Severns, N6LF, is listed in the References section.

1) Determine the transistor's output capacitance plus the device mounting capacitance. Multiply that value by 2 and select the nearest standard value for C_S. For example, if the transistor's output capacitance is 150 pF and the mounting capacitance is 70 pF, the total of 220 pF × 2 = 440 pF, and the nearest standard value is 470 pF. Use a low-loss film or ceramic component.

2) The required resistance for $R_S = V_{CC} / I_L$ where I_L is the load current. Power dissipation, $P_{DISS} = C_S V_{CC}^2 f_S$ where f_S is the switching frequency. Use a noninductive component such as carbon composition or metal oxide.

The most common method of dealing with the transient if switching is not frequent or continuous is to employ a "kickback" diode that is reverse-biased when the load is energized. (See Figure 4.68B.) When the load current is interrupted, the diode routes the transient's energy back to the power supply. See this chapter's section on Analog-Digital Interfacing for more information about kickback diode use. If the solenoid or relay is going to be used in an amateur station, add a small bypass capacitor (0.001 - 0.01 µF, value is not critical) across the diode to prevent it from generating harmonics or mixing products from strong RF.

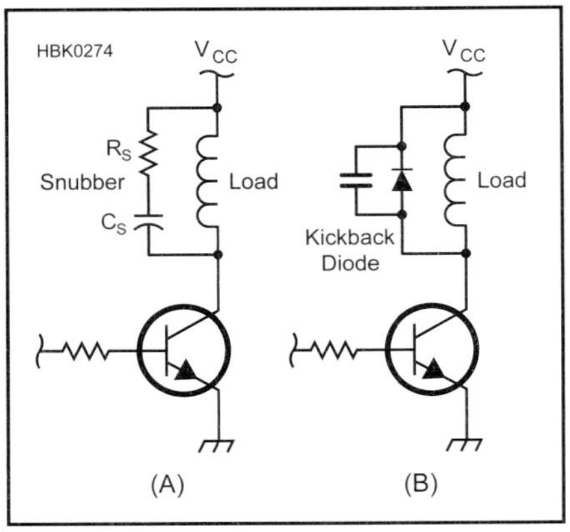

Figure 4.78 — The snubber RC circuit at (A) absorbs energy from transients with fast rise- and fall-times. At (B) a kickback diode protects the switching device when current is interrupted in the inductive load, causing a voltage transient, by conducting the energy back to the power source.

Capacitive loads such as heavily filtered power inputs may temporarily act like short circuits when the load is energized or de-energized. The surge current is only limited by the internal resistance of the load capacitance. The transistor will have to handle the temporary overloads without being damaged or overheating. The usual solution is to select a transistor with an I_{MAX} rating greater than the surge current. Sometimes a small cur-rent-limiting resistor can be placed in series with the load to reduce the peak surge current at the expense of dissipating power continuously when the load is drawing current.

HIGH-SIDE AND LOW-SIDE SWITCHING

The switching circuits shown in Figure 4.78 are *low-side switches*. This means the switch is connected between the load and ground. A *high-side switch* is connected between the power source and the load. The same concerns for power dissipation apply, but the methods of driving the switch change because the voltage of the emitter or source of the switching device will be at or near the power supply voltage when the switch is on.

To drive an NPN bipolar transistor or an N-channel MOSFET in a high-side circuit requires the switch input signal to be at least $V_{BE(sat)}$ or $V_{GS(on)}$ *above* the voltage supplied to the load. If the load expects to see the full power supply voltage, the switch input signal will have to be *greater* than the power supply voltage. A small step-up or boost dc-to-dc converter is often used to supply the extra voltage needed for the driver circuit.

One alternate method of high-side switching is to use a PNP bipolar transistor as the switching transistor. A small input transistor turns the main PNP transistor on by controlling the larger transistor's base current. Similarly, a P-channel MOSFET could also be employed with a bipolar transistor or FET acting as its driver. P-type material generally does not have the same high conductivity as N-type material, so these devices dissipate somewhat more power than N-type devices under the same load conditions.

4.7.8 Choosing a Transistor

With all the choices for transistors — websites and catalogs can list hundreds — selecting a suitable transistor can be intimidating. Start by determining the maximum voltage (V_{CEO} or $V_{DS(MAX)}$), current (I_{MAX}) and power dissipation ($P_{D(MAX)}$) the transistor must handle. Determine what dc current gain, β, or transconductance, g_m, is required. Then determine the highest frequency at which full gain is required and multiply it by either the voltage or current gain to obtain f_T or h_{fe}. This will reduce the number of choices dramatically.

4.8 Operational Amplifiers

An *operational amplifier*, or *op amp*, is one of the most useful linear devices. While it is possible to build an op amp with discrete components, and early versions were, the symmetry of the circuit demanded for high performance requires a close match of many components. It is more effective, and much easier, to implement as an integrated circuit. (The term "operational" comes from the op amp's origin in analog computers, where it was used to implement mathematical operations.)

The op amp's performance approaches that of an ideal analog circuit building block: an infinite input impedance (Z_i), a zero-output impedance (Z_o) and an open-loop voltage gain (A_v) of infinity. Obviously, practical op amps do not meet these specifications, but they do come closer than most other types of amplifiers. These attributes allow the circuit designer to implement many different functions with an op amp and only a few external components.

4.8.1 Characteristics of Practical Op-Amps

An op amp has three signal terminals (see **Figure 4.79**). There are two input terminals, the *noninverting input* marked with a + sign and the *inverting input* marked with a – sign. Voltages applied to the noninverting input cause the op amp output voltage to change with the same polarity.

The output of the amplifier is a single terminal with the output voltage referenced to the external circuit's reference voltage. Usually, that reference is ground, but the op amp's internal circuitry allows all voltages to *float*, that is, to be referenced to any arbitrary voltage between the op amp's power supply voltages. The reference can be negative, ground or positive. For example, an op amp powered from a single power supply voltage amplifies just as well if the circuit reference voltage is halfway between ground and the supply voltage.

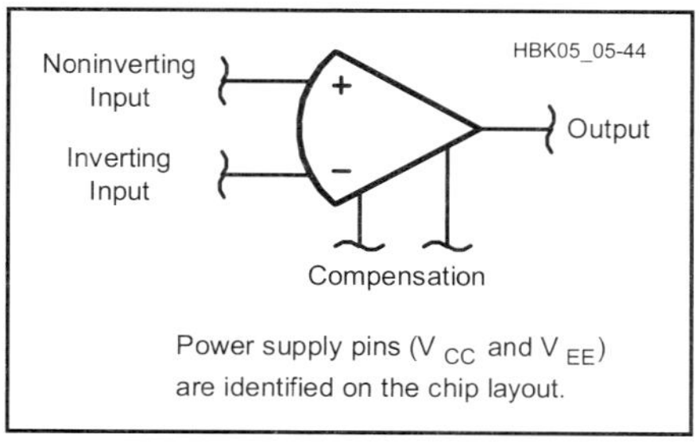

Figure 4.79 — Operational amplifier schematic symbol. The terminal marked with a + sign is the noninverting input. The terminal marked with a – sign is the inverting input. The output is to the right. On some op amps, external compensation is needed and leads are provided, pictured here below the device. Usually, the power supply leads are not shown on the op amp itself but are specified in the data sheet.

GAIN-BANDWIDTH PRODUCT AND COMPENSATION

An ideal op amp would have infinite frequency response, but just as transistors have an f_T that marks their upper frequency limit, the op amp has a *gain-bandwidth product* (GBW or BW). GBW represents the maximum product of gain and frequency available to any signal or circuit: voltage gain × frequency = GBW. If an op-amp with a GBW of 10 MHz is connected as a ×50 voltage amplifier, the maximum frequency at which that gain could be guaranteed is GBW / gain = 10 MHz / 50 = 200 kHz. GBW is an important consideration in high-performance filters and signal processing circuits whose design equations require high-gain at the frequencies over which they operate.

Older operational amplifiers, such as the LM301, have an additional two connections for *compensation*. To keep the amplifier from oscillating at very high gains it is often necessary to place a capacitor across the compensation terminals. This also decreases the frequency response of the op amp but increases its stability by making sure that the output signal cannot have the right phase to create positive feedback at its inputs. Most modern op amps are internally compensated and do not have separate pins to add compensation capacitance. Additional compensation can be created by connecting a capacitor between the op amp output and the inverting input.

CMRR AND PSRR

One of the major advantages of using an op amp is its very high *common mode rejection ratio* (CMRR). *Common mode* signals are those that appear equally at all terminals. For example, if both conductors of an audio cable pick up a few tenths of a volt of 60 Hz signal from a nearby power transformer, that 60 Hz signal is a common-mode signal to whatever device the cable is connected. Since the op amp only responds to *differences* in voltage at its inputs, it can ignore or reject common mode signals. CMRR is a measure of how well the op amp rejects the common mode signal. High CMRR results from the symmetry between the circuit halves. CMRR is important when designing circuits that process low-level signals, such as microphone audio or the mV-level dc signals from sensors or thermocouples.

The rejection of power-supply imbalance is also an important op amp parameter. Shifts in power supply voltage and noise or ripple on the power supply voltages are coupled directly to the op amp's internal circuitry. The op amp's ability to ignore those disturbances is expressed by the *power supply rejection ratio* (PSRR). A high PSRR means that the op amp circuit will continue to perform well even if the power supply is imbalanced or noisy.

INPUT AND OUTPUT VOLTAGE LIMITS

The op amp is capable of accepting and amplifying signals at levels limited by the power supply voltages, also called *rails*. The difference in voltages between the two rails limits the range of signal voltages that can be processed. The voltages can be symmetrical positive and negative voltages (±12 V), a positive voltage and ground, ground and a negative voltage or any two different, stable voltages.

In most op amps the signal levels that can be handled are one or two diode forward voltage drops (0.7 V to 1.4 V) away from each rail. Thus, if an op amp has 15 V connected as its upper rail (usually denoted V^+) and ground connected as its lower rail (V^-), input signals can be amplified to be as high as 13.6 V and as low as 1.4 V in most amplifiers. Any values that would be amplified beyond those limits are clamped (output voltages that should be 1.4 V or less appear as 1.4 V and those that should be 13.6 V or more appear as 13.6 V). This clamping action is illustrated in Figure 3.49 in the **Radio Fundamentals** chapter.

"Rail-to-rail" op amps have been developed to handle signal levels within a few tens of mV of rails (for example, the MAX406, from Maxim Integrated Products processes signals to within 10 mV of the power supply voltages). Rail-to-rail op-amps are often used in battery-powered products to allow the circuits to operate from low battery voltages for as long as possible.

INPUT BIAS AND OFFSET

The inputs of an op amp, while very high impedance, still allow some input current to flow. This is the *input bias current*, and it is in the range of nA in modern op amps. Slight asymmetries in the op amp's internal circuitry result in a slight offset in the op amp's out-put voltage, even with the input terminals shorted together. The amount of voltage difference between the op amp's inputs required to cause the output voltage to be exactly zero is the *input offset voltage*, generally a few mV or less. Some op amps, such as the LM741, have special terminals to which a potentiometer can be connected to *null* the offset by correcting the internal imbalance. Introduction of a small dc correction voltage to the noninverting terminal is sometimes used to apply an offset voltage that counteracts the internal mismatch and centers the signal in the rail-to-rail range.

DC offset is an important consideration in op amps for two reasons. Actual op amps have a slight mismatch between the inverting and noninverting terminals that can become a substantial dc offset in the output, depending on the amplifier gain. The op amp output voltage must not be too close to the clamping limits or distortion will occur.

A TYPICAL OP AMP

As an example of typical values for these parameters, one of today's garden-variety op amps, the TL084, which contains both JFET and bipolar transistors, has a guaranteed minimum CMRR of 80 dB, an input bias current guaranteed to be below 200 pA (1 pA = 1 millionth of a µA), and a gain-bandwidth product of 3 MHz. Its input offset voltage is 3 mV. CMRR and PSRR are 86 dB, meaning that an unwanted signal or power supply imbalance of 1 V will only result in a 2.5 nV change at the op amp's output! All this for less than 50 cents even purchased in single quantities, and there are four op-amps per package — that's a lot of performance.

4.8.2 Basic Op-Amp Circuits

If a signal is connected to the input terminals of an op amp without any other circuitry attached, it will be amplified at the device's *open-loop gain* (typically 200,000 for the TL084 at dc and low frequencies, or 106 dB). This will quickly saturate the output at the power supply rails. Such large gains are rarely used. In most applications, negative feedback is used to limit the circuit gain by providing a feedback path from the output terminal to the inverting input terminal. The resulting *closed-loop gain* of the circuit depends solely on the values of the passive components used to form the loop (usually resistors and, for frequency-selective circuits, capacitors). The higher the op-amp's open-loop gain, the closer the circuit's actual gain will approach that predicted from the component values. Note that the gain of the op amp itself has not changed — it is the configuration of the external components that determines the overall gain of the circuit. Some examples of different circuit configurations that manipulate the closed-loop gain follow.

INVERTING AND NONINVERTING AMPLIFIERS

The op amp is often used in either an *inverting* or a *noninverting* amplifier circuit as shown in **Figure 4.80**. (Inversion means that the output signal is inverted from the input signal about the circuit's voltage reference as described below.) The amount of amplification is determined by the two resistors: the feedback resistor, R_f, and the input resistor, R_i.

In the noninverting configuration shown in Figure 4.80A, the input signal is connected to the op-amp's noninverting input. The feedback resistor is connected between the output and the inverting input terminal. The inverting input terminal is connected to R_i, which is connected to ground (or the circuit reference voltage).

This circuit illustrates how op amp circuits use negative feedback, the high open-loop gain of the op amp itself, and the high input impedance of the op amp inputs to create a stable circuit with a fixed gain. The signal applied to the noninverting input causes the output voltage of the op-amp to change with the same polarity. That is, a positive input signal causes a positive change in the op amp's output voltage. This voltage causes current to flow in the voltage divider formed by R_f and R_i. Because the current into the inverting input is so low, the current through R_f is the same as R_i.

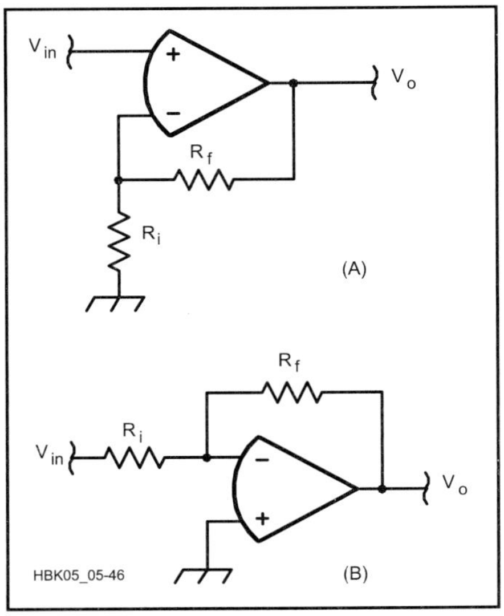

Figure 4.80 — Operational amplifier circuits. (A) Noninverting configuration. (B) Inverting configuration.

The voltage at the *summing junction*, the connection point for the two resistors and the inverting terminal, V_{INV}, is:

$$V_{INV} = V_O\left(\frac{R_i}{R_i + R_f}\right)$$

The op amp's output voltage will continue to rise until the *loop error signal*, the difference in voltage between the inverting and noninverting inputs, is close to zero. At this point, the voltage at the inverting terminal is approximately equal to the voltage at the noninverting terminal, V_{in}, so that $V_{INV} = V_{in}$. Substituting in the equation for V_{INV}, the gain of this circuit is:

$$\frac{V_O}{V_{in}} = \left(1 + \frac{R_f}{R_i}\right)$$

where

V_o = the output voltage, and

V_{in} = the input voltage.

The higher the op amp's open-loop gain, the closer will be the voltages at the inverting and noninverting terminals when the circuit is balanced and the more closely the circuit's closed-loop gain will equal that in the equation. So the negative feedback creates an electronic balancing act with the op amp increasing its output voltage so that the input error signal is as small as possible.

In the inverting configuration of Figure 4.80B, the input signal (V_{in}) is connected through R_i to the inverting terminal. The feedback resistor is again connected between the inverting terminal and the output. The noninverting terminal is connected to ground (or the circuit reference voltage). In this configuration the feedback action results in the output voltage changing to whatever value is needed such that the current through R_i is balanced by an equal and opposite current through R_f. The gain of this circuit is:

$$\frac{V_O}{V_{in}} = -\frac{R_f}{R_{in}}$$

where V_{in} represents the voltage input to R_{in}.

For the remainder of this section, "ground" or "zero voltage" should be understood to be the circuit reference voltage. That voltage may not be "earth ground potential." For example, if a single positive supply of 12 V is used, 6 V may be used as the circuit reference voltage. The circuit reference voltage is a fixed dc voltage that can be considered an ac ground because of the reference source's extremely low ac impedance.

The negative sign in the voltage gain equation indicates that the signal is inverted. For ac signals, inversion represents a 180° phase shift. The gain of the noninverting configuration can vary from a minimum of one to the maximum of which the op amp is capable, as indicated by A_v for dc signals, or the gain-bandwidth product for ac signals. The gain of the inverting configuration can vary from a minimum of zero (gains from zero to one attenuate the signal while gains of 1 and higher amplify the signal) to the maximum of which the device is capable.

The inverting amplifier configuration results in a special condition at the op amp's inverting input called *virtual ground*. Because the op amp's high open-loop gain drives the two inputs to be very close together, if the noninverting input is at ground potential, the inverting input will be very close to ground as well and the op amp's output will change with the input signal to maintain the inverting input at ground. Measured with a voltmeter, the input appears to be grounded, but it is merely maintained at ground potential by the action of the op amp and the feedback loop. This point in the circuit may not be connected to any other ground connection or circuit point because the resulting additional current flow will upset the balance of the circuit.

The *voltage follower* or *unity-gain buffer* circuit of **Figure 4.81** is commonly used as a buffer stage. The voltage follower has the input connected directly to the noninverting terminal and the output connected directly to the inverting terminal. This configuration has unity gain because the circuit is balanced when the output and input voltages are the same (error voltage equals zero). It also provides the maximum possible input impedance and the minimum possible output impedance of which the device is capable.

Differential and Difference Amplifier

A *differential amplifier* is a special application of an operational amplifier (see **Figure 4.82**). It amplifies the difference between two analog signals and is very useful to cancel noise under certain conditions. For instance, if an analog signal and a reference signal travel over the same cable they may pick up noise, and it is likely that both signals will have the same amount of noise. When the differential amplifier subtracts them, the signal will be unchanged but the noise will be completely removed, within the limits of the CMRR. The equation for differential amplifier operation is:

$$V_O = \frac{R_f}{R_i}\left[\frac{1}{\frac{R_n}{R_g}+1}\left(\frac{R_i}{R_f}+1\right)V_n - V_i\right]$$

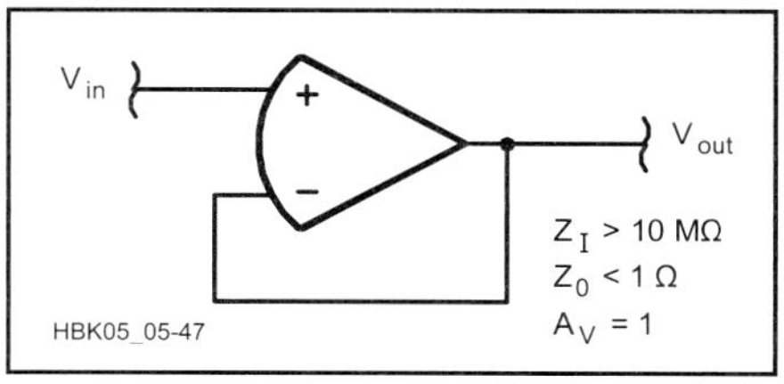

Figure 4.81 — Voltage follower. This operational amplifier circuit makes a nearly ideal buffer with a voltage gain of about one, and with extremely high input impedance and extremely low output impedance.

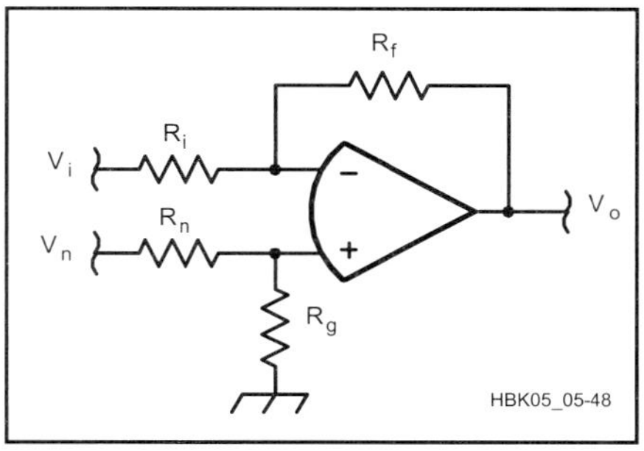

Figure 4.82 — Difference amplifier. This operational amplifier circuit amplifies the difference between the two input signals.

which, if the ratios R_i/R_f and R_n/R_g are equal, simplifies to:

$$V_O = \frac{R_f}{R_i}(V_n - V_i)$$

Note that the differential amplifier gain is identical to the inverting amplifier gain if the voltage applied to the noninverting terminal is equal to zero. If the voltage applied to the inverting terminal (V_i) is zero, the analysis is a little more complicated, but it is possible to derive the noninverting amplifier gain from the differential amplifier gain by taking into account the influence of R_n and R_g. If all four resistors have the *same* value the *difference amplifier* is created and V_O is just the difference of the two voltages:

$$V_O = V_n - V_l$$

Instrumentation Amplifier

Just as the symmetry of the transistors making up an op amp leads to a device with high values of Z_i, A_v and CMRR and a low value of Z_o, a symmetric combination of op amps is used to further improve these parameters. This circuit, shown in **Figure 4.83** is called an *instrumentation amplifier*. It has three parts; each of the two inputs is connected to a noninverting buffer amplifier with a gain of 1 + R2/R1. The outputs of these buffer amplifiers are then connected to a differential amplifier with a gain of R4/R3. V2 is the circuit's inverting input and V1 the noninverting input.

The three amplifier modules are usually all part of the same integrated circuit. This means that they have essentially the same temperature and the internal transistors and resistors are very well matched. This causes the subtle gain and tracking errors caused by temperature differences and mismatched components between individual op amps to be cancelled out or dramatically reduced. In addition, the external resistors using the same designators (R2, R3, R4) are carefully matched as well, sometimes being part of a single integrated *resistor pack*. The result is a circuit with better performance than any single-amplifier circuit over a wider temperature range.

Summing Amplifier

The high input impedance of an op amp makes it ideal for use as a *summing amplifier*. In either the inverting or noninverting configuration, the single input signal can be

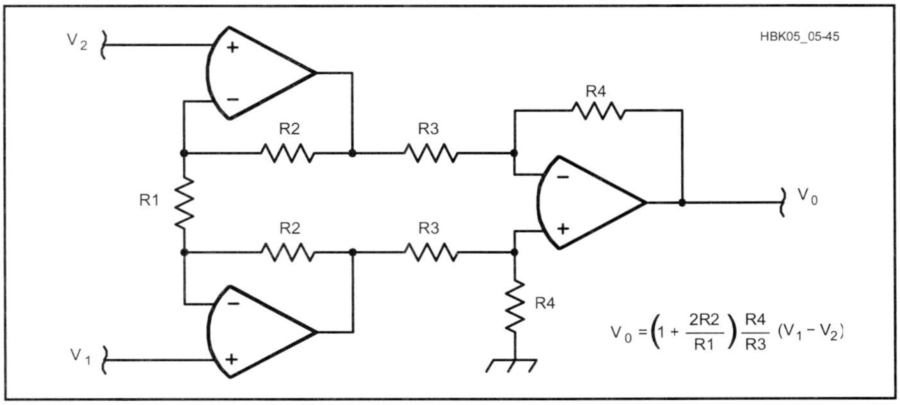

Figure 4.83 — Operational amplifiers arranged as an instrumentation amplifier. The balanced and cascaded series of op amps work together to perform differential amplification with good common-mode rejection and very high input impedance (no load resistor required) on both the inverting (V_1) and noninverting (V_2) inputs.

replaced by multiple input signals that are connected together through series resistors, as shown in **Figure 4.84**. For the inverting summing amplifier, the gain of each input signal can be calculated individually using the equation for inverting amplifier gain and, because of the superposition property of linear circuits, the output is the sum of each input signal multiplied by its gain. In the noninverting configuration, the output is the gain times the weighted sum of the m different input signals:

$$V_n = V_{n1}\frac{R_{p1}}{R_1 + R_{p1}} + V_{n2}\frac{R_{p2}}{R_2 + R_{p2}} + \ldots + V_{nm}\frac{R_{pm}}{R_m + R_{pm}}$$

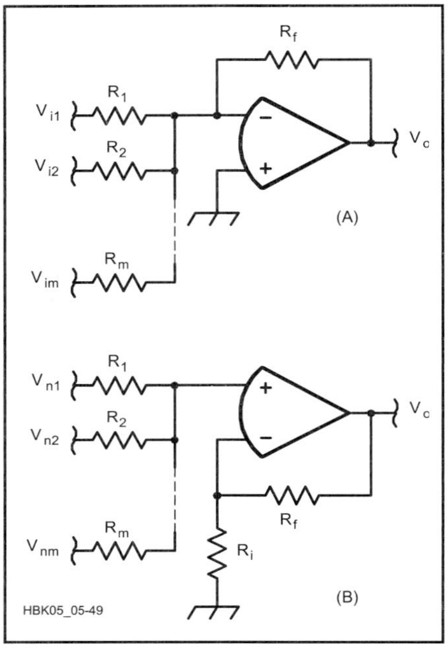

Figure 4.84 — Summing operational amplifier circuits. (A) Inverting configuration. (B) Noninverting configuration.

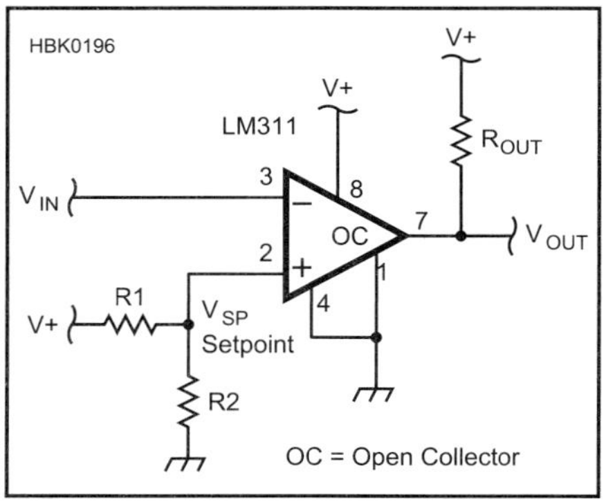

Figure 4.85 — A comparator circuit in which the output voltage is low when voltage at the inverting input is higher than the setpoint voltage at the noninverting input.

where R_{pm} is the parallel resistance of all m resistors excluding R_m. For example, with three signals being summed, R_{p1} is the parallel combination of R_2 and R_3.

Comparators

A *voltage comparator* is another special form of op amp circuit, shown in **Figure 4.85**. It has two analog signals as its inputs, and its output is either TRUE or FALSE depending on whether the noninverting or inverting signal voltage is higher, respectively. Thus, it "compares" the input voltages. TRUE generally corresponds to a positive output voltage and FALSE to a negative or zero voltage. The circuit in Figure 4.85 uses external resistors to generate a reference voltage, called the *setpoint*, to which the input signal is compared. A comparator can also compare two variable voltages.

A standard operational amplifier can be made to act as a comparator by connecting the two input voltages to the noninverting and inverting inputs with no input or feedback resistors. If the voltage of the noninverting input is higher than that of the inverting input, the output voltage will be driven to the positive clamping limit. If the inverting input is at a higher potential than the noninverting input, the output voltage will be driven to the negative clamping limit. If the comparator is comparing an unknown voltage to a known voltage, the known voltage is called the *setpoint* and the comparator output indicates whether the unknown voltage is above or below the setpoint.

An op amp that has been intended for use as a comparator, such as the LM311, is optimized to respond quickly to the input signals. In addition, comparators often have *open-collector outputs* that use an external *pull-up* resistor, R_{OUT}, connected to a positive power supply voltage. When the comparator output is TRUE, the output transistor is turned off and the pull-up resistor "pulls up" the output volt-

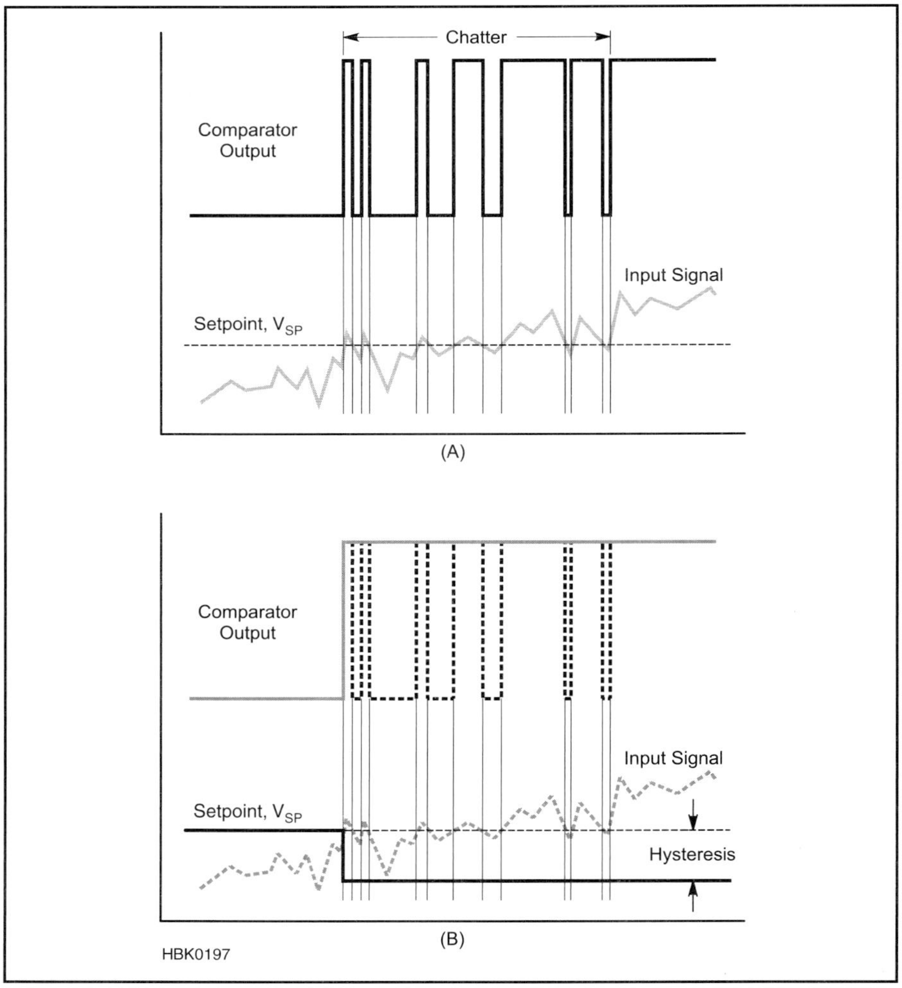

Figure 4.86 — Chatter (A) is caused by noise when the input signal is close to the setpoint. Chatter can also be caused by voltage shifts that occur when a heavy load is turned on and off. Hysteresis (B) shifts the setpoint a small amount by using positive feedback in which the output pulls the setpoint farther away from the input signal after switching.

age to the positive power supply voltage. When the comparator output is FALSE, the transistor is driven into the saturation and the output voltage is the transistor's $V_{CE(sat)}$.

Hysteresis

Comparator circuits also use *hysteresis* to prevent "chatter" — the output of the comparator switching rapidly back and forth when the input voltage is at or close to the setpoint voltage. There may be noise on the input signal, as shown in **Figure 4.86A**, that causes the input voltage to cross the setpoint threshold repeatedly. The rapid switching of the output can be confusing to the circuits monitoring the comparator output.

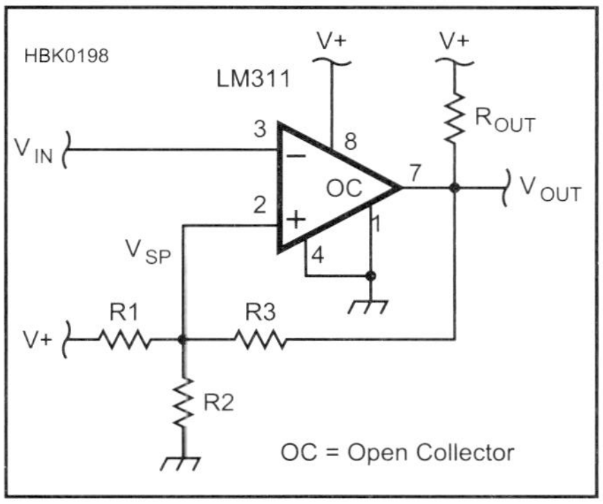

Figure 4.87 — Comparator circuit with hysteresis. R3 causes a shift in the comparator setpoint by allowing more current to flow through R1 when the comparator output is low.

Hysteresis is a form of positive feedback that "moves" the setpoint by a few mV in the direction *opposite* to that in which the input signal crossed the setpoint threshold. As shown in Figure 4.86B, the slight shift in the setpoint tends to hold the comparator output in the new state and prevents switching back to the old state. **Figure 4.87** shows how the output of the comparator is fed back to the positive input through resistor R3, adding or subtracting a small amount of current from the divider and shifting the setpoint. If V_{HYS} is the amount of hysteresis desired (the shift in the setpoint voltage):

$$V_{HYS} \approx (V_{OH})\ (R1\ //\ R2)\ /\ [R3 + (R1\ //\ R2)]$$

where V_{OH} is the high-level output voltage with the comparator's output is off. Solving for R3 if the other values are known:

$$R3 \approx [(V_{OH})(R1\ //\ R2)\ /\ V_{HYS}] - (R1\ //\ R2)$$

Some applications of a voltage comparator are a zero crossing detector, a signal squarer (which turns other cyclical wave forms into square waves) and a peak detector. An amateur station application: Circuits that monitor the CI-V band data output voltage from ICOM HF radios use a series of comparators to sense the level of the voltage and indicate on which band the radio is operating.

FILTERS

One of the most important types of op amp circuits is the *active filter*. Two examples of op amp filter circuits are shown in **Figure 4.88.** The simple noninverting low-pass filter in Figure 4.88A has the same response as a passive single-pole RC low-pass filter, but unlike the passive filter, the op amp filter circuit has a very high input impedance and a very low output impedance, so that the filter's frequency and voltage response are relatively unaffected by the circuits connected to the filter input and output. This circuit is a low-pass filter because the reactance of the feedback capacitor decreases with frequency, requiring less output voltage to balance the voltages of the inverting and noninverting inputs.

The *multiple-feedback* circuit in Figure 4.88B results in a band-pass response while using only resistors and capacitors. This circuit is just one of many different types of active filters. Active filters are discussed in the **Analog and Digital Filtering** chapter.

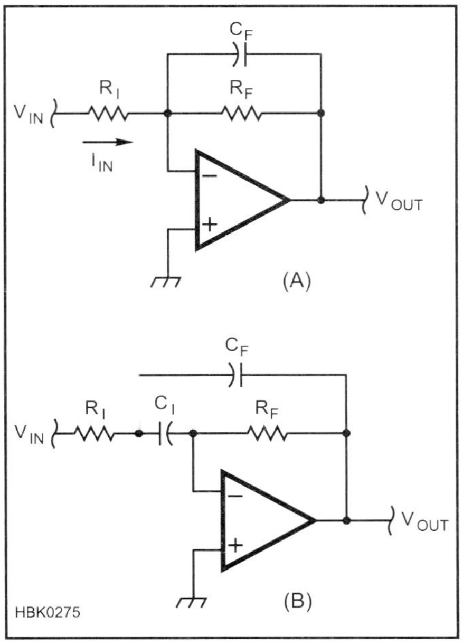

Figure 4.88 — Op amp active filters. The circuit at (A) has a low-pass response identical to an RC filter. The –3 dB frequency occurs when the reactance of C_F equals R_F. The band-pass filter at (B) is a multiple-feedback filter.

RECTIFIERS AND PEAK DETECTORS

The high open-loop gain of the op amp can also be used to simulate the I-V characteristics of an ideal diode. A *precision rectifier* circuit is shown in **Figure 4.89** along with the I-V characteristics of a real (dashed lines) and ideal (solid line) diode. The high gain of the op amp compensates for the V_F forward voltage drop of the real diode in its feedback loop with an output voltage equal to the input voltage plus V_F. Remember that the op amp's output increases until its input voltages are balanced. When the input voltage is negative, which would reverse-bias the diode, the op amp's output can't balance the input because the diode blocks any current flow through the feedback loop. The resistor at the output holds the voltage at zero until the input voltage is positive once again. Precision half-wave and full-wave rectifier circuits are shown in **Figure 4.90**, and their operation is described in many reference texts.

One application of the precision rectifier circuit useful in radio is the *peak detector*, shown in **Figure 4.91**. A precision rectifier is used to charge the output capacitor which holds the peak voltage. The output resistor sets the time constant at which the capacitor discharges. The resistor can also be replaced by a transistor acting as a switch to reset the detector. This circuit is used in AGC loops,

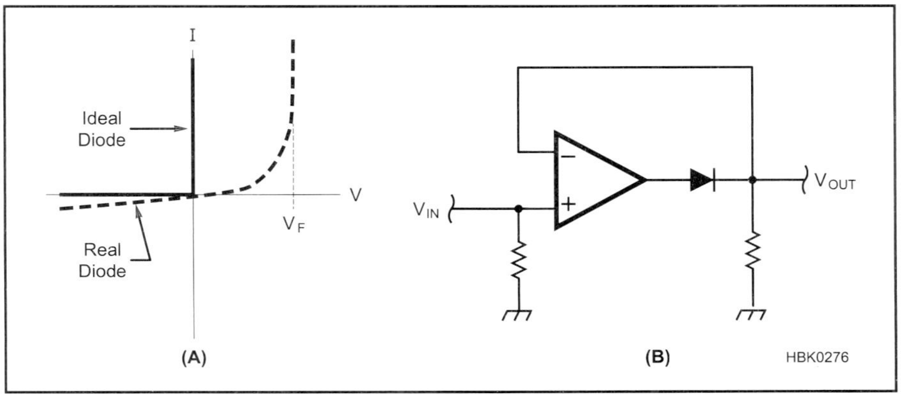

Figure 4.89 — Ideal and real diode I-V characteristics are shown at (A). The op amp precision rectifier circuit is shown at (B).

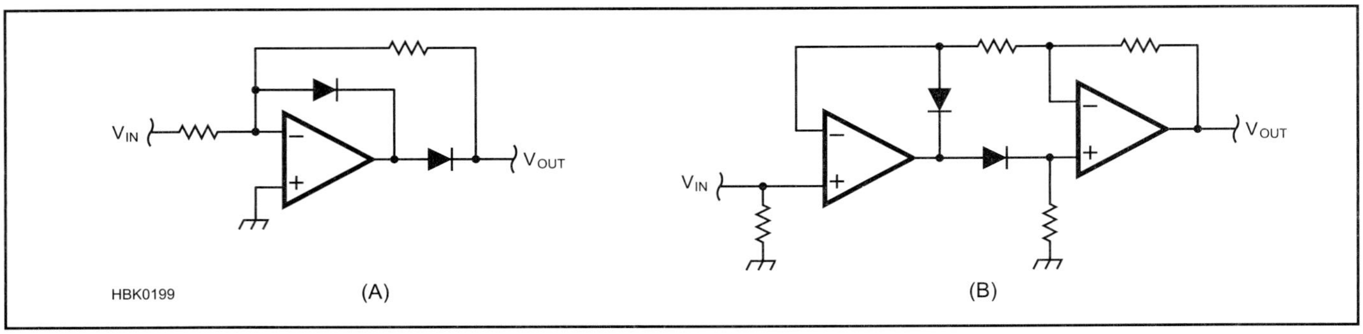

Figure 4.90 — Half-wave precision rectifier (A). The extra diode at the output of the op amp prevents the op amp from saturating on negative half-cycles and improves response time. The precision full-wave rectifier circuit at (B) reproduces both halves of the input waveform.

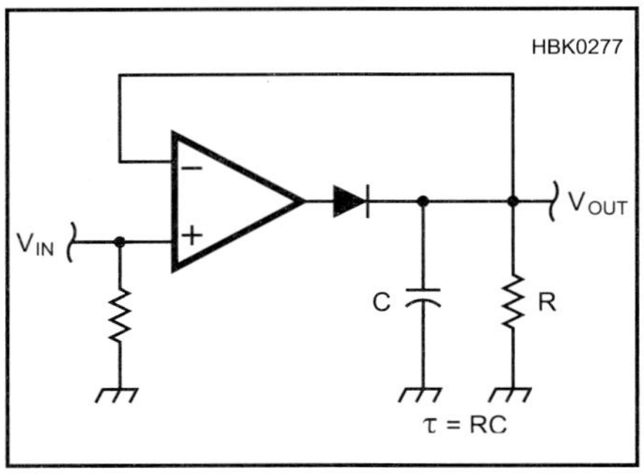

Figure 4.91 — Peak detector. Coupling a precision diode with a capacitor to store charge creates a peak detector. The capacitor will charge to the peak value of the input voltage. R discharges the capacitor with a time constant of τ = RC and can be omitted if it is desired for the output voltage to remain nearly constant.

spectrum analyzers, and other instruments that measure the peak value of ac waveforms.

LOG AMPLIFIER

There are a number of applications in radio in which it is useful for the gain of an amplifier to be higher for small input signals than for large input signals. For example, an audio compressor circuit is used to reduce the variations in a speech signal's amplitude so that the average power output of an AM or SSB transmitter is increased. A *log amplifier* circuit whose gain for large signals is proportional to the logarithm of the input signal's amplitude is shown in **Figure 4.92**. The log amp circuit is used in compressors and limiter circuits.

At signal levels that are too small to cause significant current flow through the diodes, the gain is set as in a regular inverting amplifier, $A_V = -R_f / R_i$. As the signal level increases, however, more current flows through the diode according to the Fundamental Diode Equation (see the **Electronic Fundamentals** chapter). That means the op amp output voltage has to increase less (lower gain) to cause enough current to flow through R_i such that the input voltages balance. The larger the input voltage, the more the diode conducts and the lower the gain of the circuit. Since the diode's current is exponential in response to voltage, the gain of the circuit for large input signals is logarithmic.

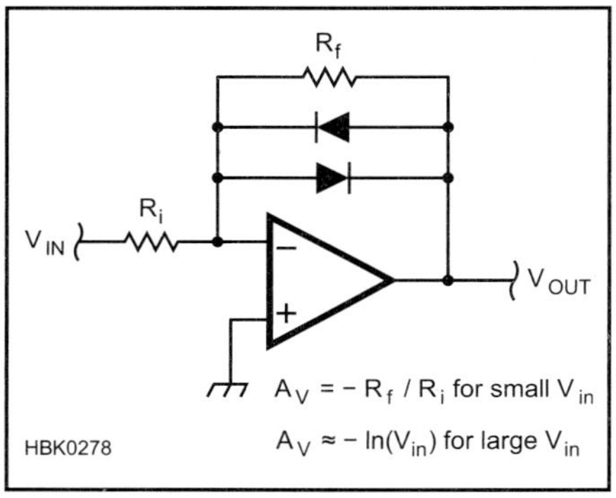

Figure 4.92 — Log amplifier. At low voltages, the gain of the circuit is $-R_f/R_i$, but as the diodes begin to conduct for higher-voltage signals, the gain changes to –ln (V_{in}) in because of the diode's exponential current as described in the Fundamental Diode Equation.

Voltage-Current Converters

Another pair of useful op amp circuits convert voltage into current and current into voltage. These are frequently used to convert currents from sensors and detectors into voltages that are easier to measure. **Figure 4.93A** shows a voltage-to-current converter in which the output current is actually the current in the feedback loop. Because the op amp's high open-loop gain insures that its input voltages are equal, the current $I_{R1} = V_{IN} / R1$. Certainly, this could also be achieved with a resistor and Ohm's law, but the op amp circuit's high input impedance means there is little interaction between the input voltage source and the output current.

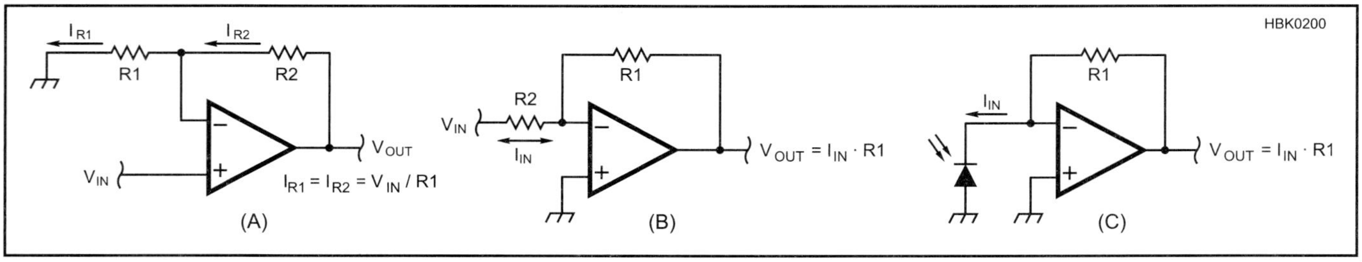

Figure 4.93 — Voltage-current converters. The current through R1 in (A) equals V_{in}/R1 because the op amp keeps both input terminals at approximately the same voltage. At (B), input current is balanced by the op amp, resulting in $V_{OUT} = I_{IN}R1$. Current through a photodiode (C) can be converted into a voltage in this way.

Going the other way, Figure 4.93B is a current-to-voltage converter. The op amp's output will change so that the current through the feedback resistor, R1, exactly equals the input current, keeping the inverting terminal at ground potential. The output voltage, $V_O = I_{IN}$ R1. Again, this could be done with just a resistor, but the op amp provides isolation between the source of input current and the output voltage. Figure 4.93C shows an application of a current-to-voltage converter in which the small currents from a photodiode are turned into voltage. This circuit can be used as a detector for amplitude modulated light pulses or waveforms.

4.9 Miscellaneous Analog ICs

The three main advantages of designing a circuit into an IC are to take advantage of the matched characteristics of its components, to make highly complex circuitry more economical, and to miniaturize the circuit and reduce power consumption. As circuits standardize and become widely used, they are often converted from discrete components to integrated circuits. Along with the op amp described earlier, there are many such classes of linear ICs.

4.9.1 Transistor and Driver Arrays

The most basic form of linear integrated circuit and one of the first to be implemented is the component array. The most common of these are the resistor, diode and transistor arrays. Though capacitor arrays are also possible, they are used less often. Component arrays usually provide space saving, but this is not the major advantage of these devices. They are the least densely packed of the integrated circuits because each device requires a separate off-chip connection. While it may be possible to place over a million transistors on a single semiconductor chip, individual access to these would require a total of three million pins, which is beyond the limits of practicability. More commonly, resistor and diode arrays contain from 5 to 16 individual devices and transistor arrays contain from three to six individual transistors. The advantage of these arrays is the very close matching of component values within the array. In a circuit that needs matched components, the component array is often a good method of obtaining this feature. The components within an array can be internally combined for special functions, such as termination resistors, diode bridges and Darlington pair transistors. A nearly infinite number of possibilities exists for these combinations of components, many of which are available in arrays.

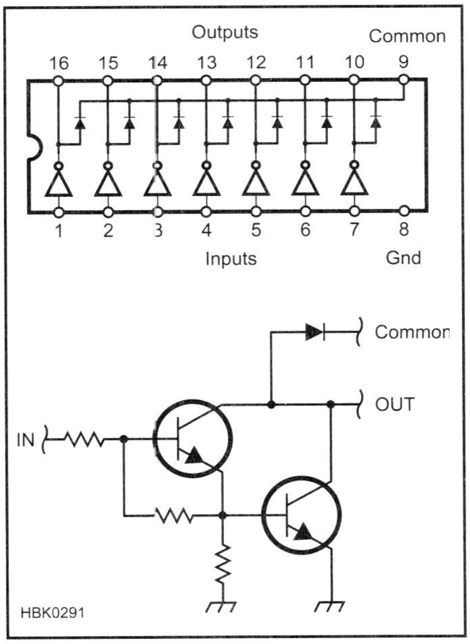

Figure 4.94 — Typical ULN2,000-series driver array configuration and internal circuit. The use of driver array ICs is very popular as an interface between microprocessor or other low-power digital circuits and loads such as relays, solenoids or lamps.

Driver arrays, such as the ULN2,000-series devices shown in **Figure 4.94** are very useful in creating an interface between low-power circuits such as microprocessors and higher-power loads and indicators. Each driver consists of a Darlington pair switching circuit as described earlier in this chapter. There are different versions with different types and arrangements of resistors and diodes.

Many manufacturers offer driver arrays. They are available with built-in kickback diodes to allow them to drive inductive loads, such as relays, and are heavy enough to source or sink current levels up to 1 A. (All of the drivers in the array cannot operate at full load at the same time, however. Read the data sheet carefully to determine what limitations on current and power dissipation may exist.)

4.9.2 Voltage Regulators and References

One of the most popular linear ICs is the voltage regulator. There are two basic types, the three-terminal regulator and the regulator controller. Examples of both are described in the **Power Sources** chapter.

The three-terminal regulator (input, ground, output) is a single package designed to perform all of the voltage regulation functions. The output voltage can be fixed, as in the 7800-series of regulators, or variable, as in the LM317 regulator. It contains a voltage reference, comparator circuits, current and temperature sensing protective circuits, and the main pass element. These ICs are usually contained in the same packages as power transistors and the same techniques of thermal management are used to remove excess heat.

Regulator controllers, such as the popular 723 device, contain all of the control and voltage reference circuitry but require external components for the main pass element, current sensing, and to configure some of their control functions.

Voltage references such as the Linear Technology LT1635 are special semiconductor diodes that have a precisely controlled I-V characteristic. A buffer amplifier isolates the sensitive diode and provides a low output impedance for the voltage signal. Voltage references are used as part of power regulators and by analog-digital converter circuits.

4.9.3 Timers (Multivibrators)

A *multivibrator* is a circuit that oscillates between two states, usually with a square wave or pulse train output. The frequency of oscillation is accurately controlled with the addition of appropriate values of external resistance and capacitance. The most common multivibrator in use today is the 555 timer IC (NE555 by Signetics [now Philips] or LM555 by National Semiconductor). This very simple eight-pin device has a frequency range from less than one hertz to several hundred kilohertz. Such a device can also be used in *monostable* operation, where an input pulse generates an output pulse of a different duration, or in *a stable* or *free-running* operation, where the device oscillates continuously. Other applications of a multivibrator include a frequency divider, a delay line, a pulse width modulator and a pulse position modulator. (These can be found in the IC's data sheet or in the reference listed at the end of this chapter.)

Figure 4.95 shows the basic components of a 555. Connected between power input (V_{cc}) and ground, the three resistors labeled "R" at the top left of the figure form a *voltage divider* that divides V_{CC} into two equal steps—one at 2/3 V_{CC} and one at 1/3 V_{CC}.

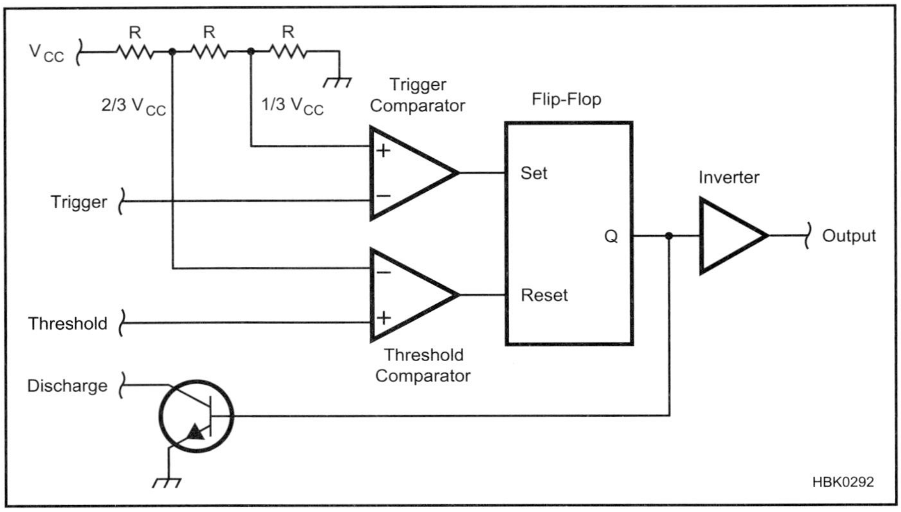

Figure 4.95 — Internal NE555 timer components. This simple array of components combines to make one of the most popular analog ICs. The 555 timer IC uses ratios of internal resistors to generate a precise voltage reference for generating time intervals based on charging and discharging a capacitor.

These serve as reference voltages for the rest of the circuit.

Connected to the reference voltages are blocks labeled *trigger comparator* and *threshold comparator*. (Comparators were discussed in a preceding section.) The trigger comparator in the 555 is wired so that its output is high whenever the trigger input is *less* than ⅓ V_{CC} and vice versa. Similarly, the threshold comparator output is high whenever the threshold input is *greater* than ⅔ V_{CC}.

These two outputs control a digital *flip-flop* circuit with an output, *Q*, that changes to high or low when the state of its *set* and *reset* input changes. The Q output stays high or low (it *latches* or *toggles*) until the opposite input changes. When the set input changes from low to high, Q goes low. When reset changes from low to high, Q goes high. The flip-flop ignores any other changes. An inverter makes the 555 output high when Q is low and vice versa — this makes the timer circuit easier to interface with external circuits.

The transistor connected to Q acts as a switch. When Q is high, the transistor is on and acts as a closed switch connected to ground. When Q is low, the transistor is off and the switch is open. These simple building blocks — voltage divider, comparator, flip-flop and switch — build a surprising number of useful circuits.

THE MONOSTABLE OR "ONE-SHOT" TIMER

The simplest 555 circuit is the monostable circuit. This configuration will output one fixed-length pulse when triggered by an input pulse. **Figure 4.96** shows the connections for this circuit.

Starting with capacitor C discharged, the flip-flop output, Q, is high, which keeps the discharge transistor turned on and the voltage on C below ⅔ V_{CC}. The circuit is in its stable state, waiting for a trigger pulse.

When the voltage at the trigger input drops below ⅓ V_{CC}, the trigger comparator output changes from low to high, which causes Q to toggle to the low state. This turns off the transistor (opens the switch) and allows C to begin charging toward V_{CC}.

When C reaches ⅔ V_{CC}, this causes the threshold comparator to switch its output from low to high, which resets the flip-flop. Q returns high, turning on the transistor and discharging C. The circuit has returned to its stable state. The output pulse length for the monostable configuration is:

$$T = 1.1R\,C_1$$

Notice that the timing is independent of the absolute value of V_{CC} — the output pulse width is the same with a 5 V supply as it is with a 15 V supply. This is because the 555 design is based on ratios and not absolute voltage levels.

THE ASTABLE MULTIVIBRATOR

The complement to the monostable circuit is the astable circuit in **Figure 4.97**. Pins 2, 6 and 7 are configured differently and the timing resistor is now split into two resistors, R1 and R2.

Start from the same state as the monostable circuit, with C completely discharged. The monostable circuit requires a trigger pulse to initiate the timing cycle. In the astable circuit, the trigger input is connected directly to the capacitor, so if the capacitor is discharged, then the trigger comparator output must be high. Q is low, turning off the discharge transistor, which allows C to immediately begin charging.

C charges toward V_{CC}, but now through the combination of R1 and R2. As the capacitor voltage passes ⅔ V_{CC}, the threshold comparator output changes from low to high, resetting Q to high. This turns on the discharge transistor and the capacitor starts to discharge through R2. When the capacitor is discharged below ⅓ V_{CC}, the trigger comparator changes from high to low and the cycle begins again, automatically. This happens over and over, causing a train of pulses at the output while C charges and discharges between ⅓ and ⅔ V_{CC} as seen in the figure.

The total time it takes for one complete cycle is the charge time, T_c, plus the discharge time, T_d:

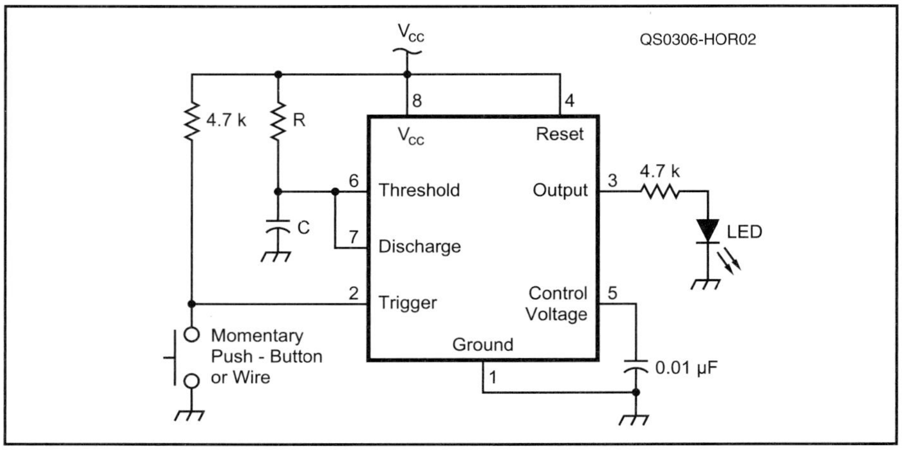

Figure 4.96 — Monostable timer. The timing capacitor is discharged until a trigger pulse initiates the charging process and turns the output on. When the capacitor has charged to ⅔ V_{CC}, the output is turned off, the capacitor is discharged and the timer awaits the next trigger pulse.

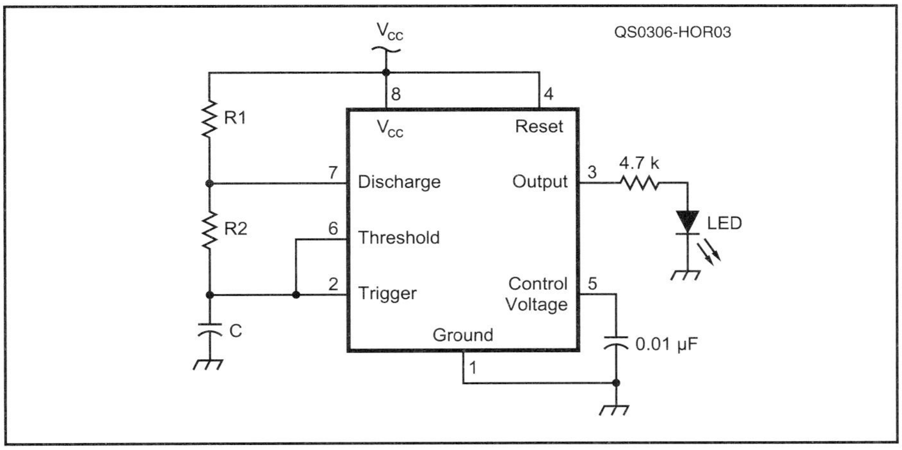

Figure 4.97 — Astable timer. If the capacitor discharge process initiates the next charge cycle, the timer will output a pulse train continuously.

$T = T_c + T_d = 0.693(R_1 + R_2)C + 0.693 R_2C = 0.693\ (R_1 + 2R_2)C$

and the output frequency is:

$$f = \frac{1}{T} = \frac{1.443}{(R_1 + 2\,R_2)\,C}$$

When using the 555 in an application in or around radios, it is important to block any RF signals from the IC power supply or timing control inputs. Any unwanted signal present on these inputs, especially the Control Voltage input, will upset the timer's operation and cause it to operate improperly. The usual practice is to use a 0.01 µF bypass capacitor (shown on pin 5 in both Figure 4.96 and 4.97) to bypass ac signals such as noise or RF to ground. Abrupt changes in V_{CC} will also cause changes in timing, and these may be prevented by connecting filter capacitors at the V_{CC} input to ground.

4.9.4 Analog Switches and Multiplexers

Arrays of analog switches, such as the Maxim MAX312-series, allow routing of audio through lower frequency RF signals without mechanical switches. There are several types of switch arrays. Independent switches have isolated inputs and outputs and are turned on and off independently. Both SPST and SPDT configurations are available. Multiple switches can be wired with common control signals to implement multiple-pole configurations.

Use of analog switches at RF through microwave frequencies requires devices specifically designed for those frequencies. The Analog Devices ADG901 is a switch usable to 2.5 GHz. It absorbs the signal when off, acting as a terminating load. The ADG902 instead reflects the signal as an open circuit when off. Arrays of three switches called "T-switches" are used when very high isolation between the input and output is required.

Multiplexers or "muxes" are arrays of SPST switches configured to act as a multiposition switch that connects one of four to sixteen input signals to a single output. Demultiplexers ("demuxes") have a single input and multiple outputs. Multiplexer ICs are available as single N-to-1 switches (the MAX4617 is an 8-to-1 mux) or as groups of N-to-1 switches (the MAX4618 is a dual 4-to-1 mux).

Crosspoint switch arrays are arranged so that any of four to sixteen signal inputs can be connected to any of four to sixteen output signal lines. The Analog Devices AD8108 is an 8-by-8 crosspoint switch with eight inputs and eight outputs. These arrays are used when it is necessary to switch multiple signal sources among multiple signal receivers. They are most commonly used in telecommunications.

All analog switches use FET technology as the switching element. To switch ac signals, most analog switches require both positive and negative voltage power supplies. An alternative is to use a single power supply voltage and ground, but bias all inputs and output at one-half the power supply voltage. This requires dc blocking capacitors in all signal paths, both input and output, and loading resistors may be required at the device outputs. The blocking capacitors can also introduce low-frequency roll-off.

The impedance of the switching element varies from a few ohms to more than 100 ohms. Check the switch data sheet to determine the limits for how much power and current the switches can handle. Switch arrays, because of the physical size of the array, can have significant coupling or *crosstalk* between signal paths. Use caution when using analog switches for high-frequency signals as coupling generally increases with frequency and may compromise the isolation required for high selectivity in receivers and other RF signal processing equipment.

4.9.5 Audio Output Amplifiers

While it is possible to use op amps as low power audio output drivers for headphones, they generally have output impedances that are too high for most audio transducers such as speakers and headphones. The LM380 series of audio driver ICs have been used in radio circuits for many years, and a simple schematic for a speaker driver is shown in **Figure 4.98**.

The popularity of personal music players has resulted in the creation of many new and inexpensive audio driver ICs, such as the National Semiconductor LM4800- and LM4900-series. Drivers that operate from voltages as low as 1.5 V for battery-powered devices and up to 18 V for use in vehicles are now available.

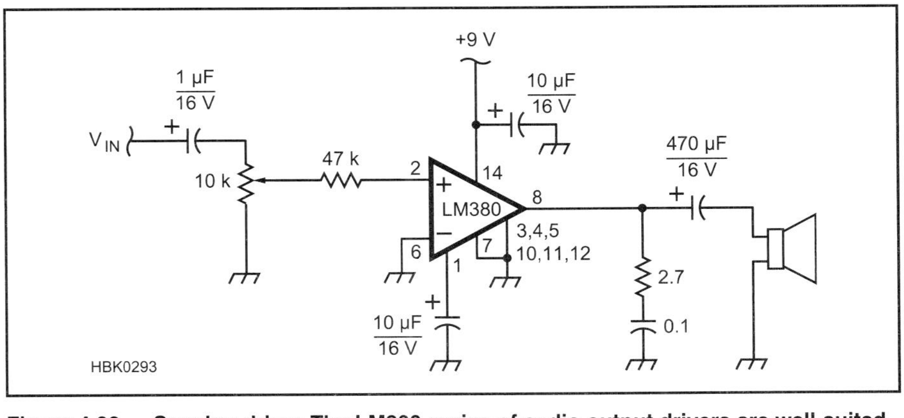

Figure 4.98 — Speaker driver. The LM380-series of audio output drivers are well-suited for low-power audio outputs, such as for headphones and small speakers. When using IC audio output drivers, be sure to refer to the manufacturer's data sheet for layout and power supply guidelines.

When choosing an audio driver IC for communications audio, the most important parameters to evaluate are its power requirements and power output capabilities. An overloaded or underpowered driver will result in distortion. Driver ICs intended for music players have frequency responses well in excess of the 3,000 Hz required for communications. This can lead to annoying and fatiguing hiss unless steps are taken to reduce the circuit's frequency response.

Audio power amplifiers should also be carefully decoupled from the power supply and the manufacturer may recommend specific circuit layouts to prevent oscillation or feedback. Check the device's data sheet for this information.

4.9.6 Temperature Sensors

Active temperature sensors use the temperature-dependent properties of semiconductor devices to create voltages that correspond to absolute temperature in degrees Fahrenheit (LM34) or degrees Celsius (LM35). These sensors (of which many more are available than the two examples given here) are available in small plastic packages, both leaded and surface-mount, that respond quickly to temperature changes. They are available with 1% and better accuracy, requiring only a source of voltage at very low current and ground. Complete application information is available in the manufacturer data sheets. Thermistors, a type of passive temperature sensor, are discussed in the Thermal Management section of this chapter. Temperature sensors are used in radio mostly in cooling and temperature control circuits.

4.10 Analog-Digital Interfacing

Quite often, logic circuits must either drive or be driven from non-logic sources. A very common requirement is sensing the presence or absence of a high (as compared to +5 V) voltage or perhaps turning on or off a 120 V ac device or moving the motor in an antenna rotator. A similar problem occurs when two different units in the shack must be interfaced because induced ac voltages or ground loops can cause problems with the desired signals.

A slow speed but safe way to interface such circuits is to use a relay. This provides absolute isolation between the logic circuits and the load. **Figure 4.99A** shows the correct way to provide this connection. The relay coil is selected to draw less than the available current from the driving logic circuit.

When current through the coil (or any inductive load) is suddenly interrupted by the switching transistor, a voltage transient called *back EMF* is generated at the switching device. The transient can easily reach dozens of times the power supply voltage, damaging or destroying a switching transistor. To clamp the voltage and return the stored energy to the power supply, a *kickback* or *flyback diode* is used. The diode limits the voltage at the switching transistor to the power supply voltage plus the diode's forward voltage drop.

Use a diode with a PIV rating of at least twice the operating voltage (5, 12, or 24 V in Figure 4.99). The diode's peak current rating should be several times the coil's steady-state current. In almost all solid-state circuits, an inexpensive 1N4001 (100 V, 1 A) diode will suffice.

The use of a kickback diode can increase the turn-off time of the relay. If the turn-off time is an important part of the design, a Zener diode in series with the kickback diode can be used. The Zener voltage should be approximately twice the supply voltage (a 1N5252 is a 24V Zener). This will still result in some turn-off delay. Experimentation may be necessary to find the best trade-off between Zener voltage and turn-off time for critical circuits.

Tranzorbs and MOVs can also be used.

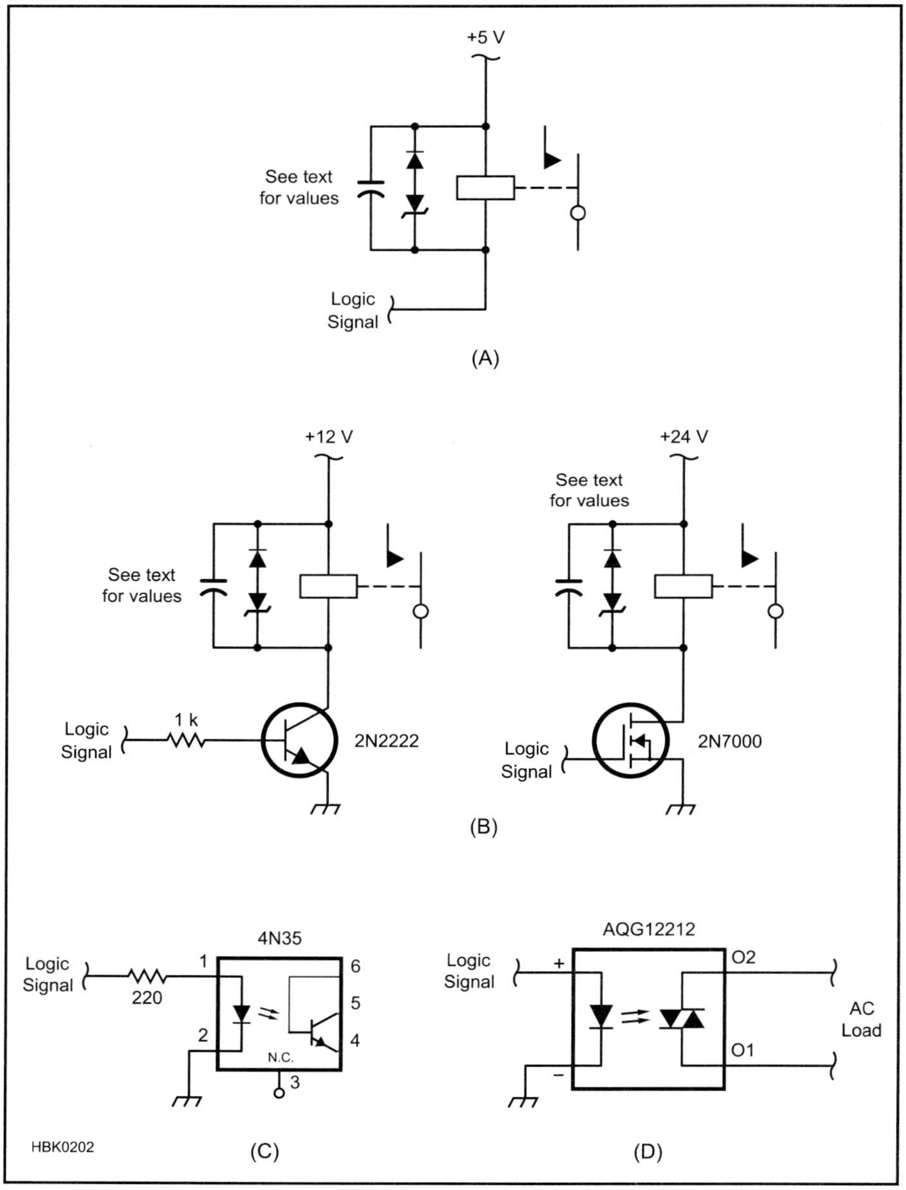

Figure 4.99 — Interface circuits for logic driving real-world loads. (A) driving a relay from a logic output; (B) using a bipolar transistor or MOSFET to boost current capacity; (C) using an optoisolator for electrical isolation; (D) using a solid-state relay for switching ac loads.

MOVs may change clamping voltages after extended use. A thorough discussion of relay drive protection can be found in STMicroelectronics Application Note AN-319, "Relay Drive Protection," which is available online from several sources.

For applications in which the relay is connected to circuits where strong RF might be present, such as antenna switches, rotator control cables, or tuning circuits, add a 0.01 to 0.001 µF ceramic capacitor (rated at 50 V or more) in parallel with the coil to prevent the diode(s) or clamping devices from generating mixing products or harmonics.

It is often not possible to find a relay that meets the load requirements *and* has a coil that can be driven directly from the logic output. Figure 4.99B shows two methods of using transistors to allow the use of higher power relays with logic gates.

Electro-optical couplers such as optoisolators and solid-state relays can also be used for this circuit interfacing. Figure 4.99C uses an optoisolator to interface two sets of logic circuits that must be kept electrically isolated, and Figure 4.99D uses a solid-state relay to control an ac line supply to a high current load. Note that this example uses a solid-state relay with internal current limiting on the input side; the LED input has an impedance of approximately 300 Ω. Some devices may need a series resistor to set the LED current; always consult the device data sheet to avoid exceeding device limits of the relay or the processor's I/O pin.

For safely using signals with voltages higher than logic levels as inputs, the same simple resistor and Zener diode circuit similar to that shown in Figure 4.38 can be used to clamp the input voltage to an acceptable level. Care must be used to choose a resistor value that will not load the input signal unacceptably.

4.11 Heat Management

While not strictly an electrical fundamental, managing the heat generated by electronic circuits is important in nearly all types of radio equipment. Thus, the topic is included in this chapter. Information on the devices and circuits discussed in this section may be found in other chapters.

Any actual energized circuit consumes electric power because any such circuit contains components that convert electricity into other forms of energy. This dissipated power appears in many forms. For example, a loudspeaker converts electrical energy into sound, the motion of air molecules. An antenna (or a light bulb) converts electricity into electromagnetic radiation. Charging a battery converts electrical energy into chemical energy (which is then converted back to electrical energy upon discharge). But the most common transformation by far is the conversion, through some form of resistance, of electricity into heat.

Sometimes the power lost to heat serves a useful purpose — toasters and hair dryers come to mind. But most of the time, this heat represents a power loss that is to be minimized wherever possible or at least taken into account. Since all real circuits contain resistance, even those circuits (such as a loudspeaker) whose primary purpose is to convert electricity to some *other* form of energy also convert some part of their input power to heat. Often, such losses are negligible, but sometimes they are not.

If unintended heat generation becomes significant, the involved components will get warm. Problems arise when the temperature increase affects circuit operation by either

- causing the component to fail, by explosion, melting, or other catastrophic event, or, more subtly,
- causing a slight change in the properties of the component, such as through a temperature coefficient (TC).

In the first case, we can design conservatively, ensuring that components are rated to safely handle two, three or more times the maximum power we expect them to dissipate. In the second case, we can specify components with low TCs, or we can design the circuit to minimize the effect of any one component. Occasionally we even exploit temperature effects (for example, using a resistor, capacitor or diode as a temperature sensor). Let's look more closely at the two main categories of thermal effects.

Not surprisingly, heat dissipation (more correctly, the efficient removal of generated heat) becomes important in medium- to high-power circuits: power supplies, transmitting circuits, and so on. While these are not the only examples where elevated temperatures and related failures are of concern, the techniques we will discuss here are applicable to all circuits.

4.11.1 Thermal Resistance

The transfer of heat energy, and thus the change in temperature, between two ends of a block of material is governed by the following heat flow equation and illustrated in **Figure 4.100**):

$$P = \frac{kA}{L}\Delta T = \frac{\Delta T}{\theta}$$

where

P = power (in the form of heat) conducted between the two points,

k = *thermal conductivity*, measured in W/(m °C), of the material between the two points, which may be steel, silicon, copper, PC board material, and so on,

L = length of the block,

A = area of the block, and

DT = *difference* in temperature between the two points.

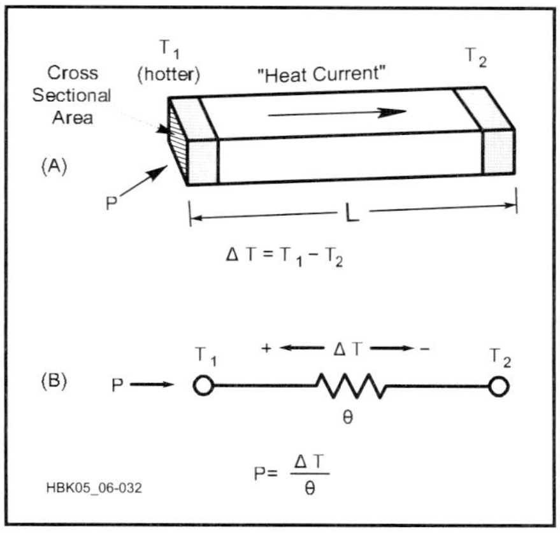

Figure 4.100 — Physical and "circuit" models for the heat-flow equation.

Thermal conductivities of various common materials at room temperature are given in **Table 4.23**.

The heat flow equation has the same form as the variation of Ohm's law relating current flow to the ratio of a difference in potential to resistance; I = E/R. In this case, what's flowing is heat (P), the difference in potential

Table 4.23

Thermal Conductivities of Various Materials

Gases at 0 °C, Others at 25 °C; from *Physics,* by Halliday and Resnick, 3rd Ed.

Material	*k in units of W/m°C*
Aluminum	200
Brass	110
Copper	390
Lead	35
Silver	410
Steel	46
Silicon	150
Air	0.024
Glass	0.8

is a temperature difference (DT), and what's resisting the flow of heat is the *thermal resistance*:

$$\theta = \frac{L}{kA}$$

with units of °C/W. (The units of resistance are equivalent to V/A.) The analogy is so apt that the same principles and methods apply to heat flow problems as circuit problems. The following correspondences hold:

- Thermal conductivity W/(m °C) ↔ Electrical conductivity (S/m).
- Thermal resistance (°C/W) ↔ Electrical resistance (Ω).
- Thermal current (heat flow) (W) ↔ Electrical current (A).
- Thermal potential (T) ↔ Electrical potential (V).
- Heat source ↔ Power source.

For example, calculate the temperature of a 2-inch (0.05 m) long piece of #12 copper wire at the end that is being heated by a 25 W (input power) soldering iron, and whose other end is clamped to a large metal vise (assumed to be an infinite heat sink), if the ambient temperature is 25 °C (77 °F).

First, calculate the thermal resistance of the copper wire (diameter of #12 wire is 2.052 mm, cross-sectional area is 3.31 × 10^{-6} m^2)

$$\theta = \frac{L}{k\,A} = \frac{(0.05\ \text{m})}{(390\ \text{W}/(\text{m}\ °\text{C})) \times (3.31 \times 10^{-6}\ \text{m}^2)}$$

= 38.7 °C/W

Then, rearranging the heat flow equation above yields (after assuming the heat energy actually transferred to the wire is around 10 W)

$$DT = P\,\theta = (10\ \text{W}) \times (38.7\ °\text{C/W}) = 387\ °\text{C}$$

So the wire temperature at the hot end is 25 C + DT = 412 °C (or 774 °F). If this sounds a little high, remember that this is for the steady state condition, where you've been holding the iron to the wire for a long time.

From this example, you can see that things can get very hot even with the application of moderate power levels. For this reason, circuits that generate sufficient heat to alter, not necessarily damage, the components must employ some method of cooling, either active or passive. Passive methods include heat sinks or careful component layout for good ventilation. Active methods include forced air (fans) or some sort of liquid cooling (in some high-power transmitters).

4.11.2 Heat Sink Selection and Use

The purpose of a heat sink is to provide a high-power component with a large surface area through which to dissipate heat. To use the models above, it provides a low thermal-resistance path to a cooler temperature, thus allowing the hot component to conduct a large "thermal current" away from itself.

Power supplies probably represent one of the most common high-power circuits amateurs are likely to encounter. Everyone has certainly noticed that power supplies get warm or even hot if not ventilated properly. Performing the thermal design for a properly cooled power supply is a very well-defined process and a good illustration of heat-flow concepts.

This material was originally prepared by ARRL Technical Advisor Dick Jansson, KD1K, during the design of a 28-V, 10-A power supply. (**Power Sources** chapter has more information on power supply design.) An outline of the design procedure shows the logic applied:

1. Determine the expected power dissipation (P_{in}).
2. Identify the requirements for the dissipating elements (maximum component temperature).
3. Estimate heat-sink requirements.
4. Rework the electronic device (if necessary) to meet the thermal requirements.
5. Select the heat exchanger (from heat sink data sheets).

The first step is to estimate the filtered, unregulated supply voltage under full load. Since the transformer secondary output is 32 V ac (RMS) and feeds a full-wave bridge rectifier, let's estimate 40 V as the filtered dc output at a 10-A load.

The next step is to determine the critical components and estimate their power dissipations. In a regulated power supply, the pass transistors are responsible for nearly all the power lost to heat. Under full load and allowing for some small voltage drops in the power-transistor emitter circuitry, the output of the series pass transistors is about 29 V for a delivered 28 V under a 10-A load. With an unregulated input voltage of 40 V, the total energy heat dissipated in the pass transistors is (40 V – 29 V) × 10 A = 110 W. The heat sink for this power supply must be able to handle that amount of dissipation and still keep the transistor junctions below the specified safe operating temperature limits. It is a good rule of thumb to select a transistor that has a maximum power dissipation of twice the desired output power.

Now, consider the ratings of the pass transistors to be used. This supply calls for 2N3055s as pass transistors. The data sheet shows that a 2N3055 is rated for 15-A service and 115-W dissipation. But the design uses *four* in parallel. Why? Here we must look past the big, bold type at the top of the data sheet to such subtle characteristics as the junction-to-case thermal resistance, q_{jc}, and the maximum allowable junction temperature, T_j.

The 2N3055 data sheet shows q_{jc} = 1.52 °C/W, and a maximum allowable case (and junction) temperature of 220 °C. While it seems that one 2N3055 could barely, on paper at least, handle the electrical requirements — at what temperature would it operate?

To answer that, we must model the entire "thermal circuit" of operation, starting with the transistor junction on one end and ending at some point with the ambient air. A reasonable model is shown in **Figure 4.101**. The ambient air is considered here as an infinite heat sink; that is, its temperature is assumed to be a constant 25 °C (77 °F). q_{jc} is the thermal resistance from the transistor junction to its case. q_{cs} is the resistance of the mounting interface between the transistor case and the heat sink. q_{sa} is the thermal resistance between the heat sink and the ambient air. In this "circuit," the generation of heat (the "thermal current source") occurs in the transistor at P_{in}.

Proper mounting of most TO-3 package power transistors such as the 2N3055 requires that they have an electrical insulator between the transistor case and the heat sink. However, this electrical insulator must at the same time

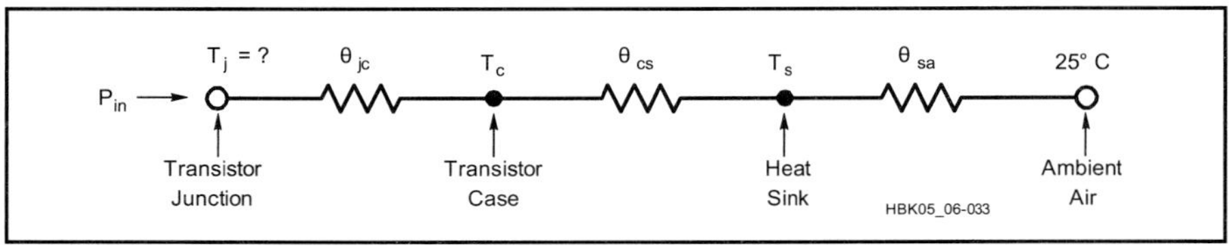

Figure 4.101 — Resistive model of thermal conduction in a power transistor and associated heat sink. See text for calculations.

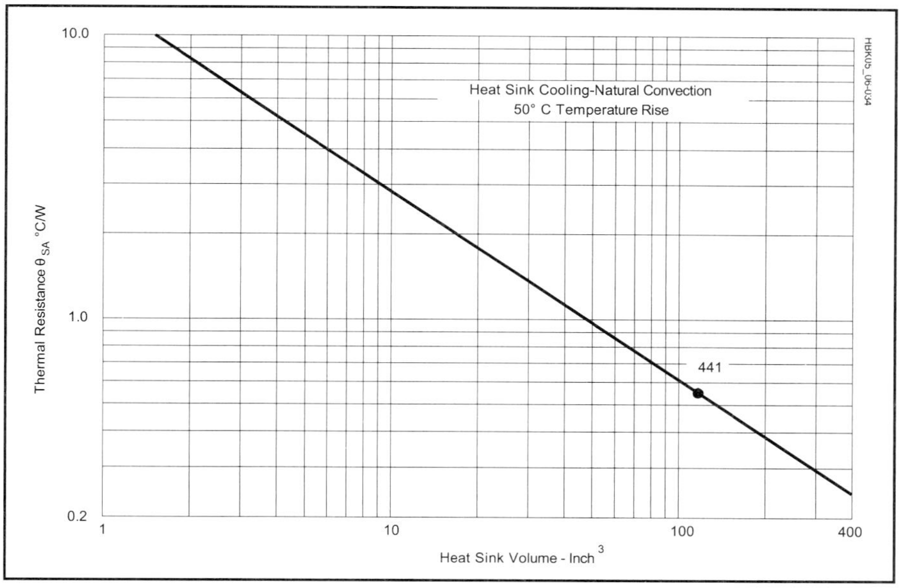

Figure 4.102 — Thermal resistance vs. heat-sink volume for natural convection cooling and 50 °C temperature rise. The graph is based on engineering data from Wakefield Thermal Solutions, Inc.

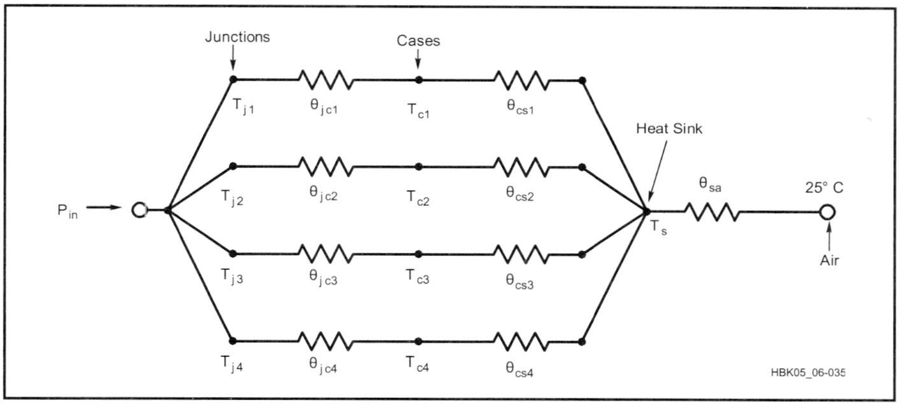

Figure 4.103 — Thermal model for multiple power transistors mounted on a common heat sink.

exhibit a low thermal resistance. To achieve a quality mounting, use thin polyimide or mica formed washers and a suitable thermal compound to exclude air from the interstitial space. "Thermal greases" are commonly available for this function. Any silicone grease may be used, but filled silicone oils made specifically for this purpose are better.

Using such techniques, a conservatively high value for q_{cs} is 0.50 °C/W. Lower values are possible, but the techniques needed to achieve them are expensive and not generally available to the average amateur. Furthermore, this value of q_{cs} is already much lower than q_{jc}, which cannot be lowered without going to a somewhat more exotic pass transistor.

Finally, we need an estimate of q_{sa}. **Figure 4.102** shows the relationship of heat-sink volume to thermal resistance for natural-convection cooling. This relationship presumes the use of suitably spaced fins (0.35 inch or greater) and provides a "rough order-of-magnitude" value for sizing a heat sink. For a first calculation, let's assume a heat sink of roughly 6 × 4 × 2 inch (48 cubic inches). From Figure 4.102, this yields a q_{sa} of about 1 °C/W.

Returning to Figure 4.101, we can now calculate the approximate temperature increase of a single 2N3055:

$$dT = P\, q_{total}$$

$$= 110 \text{ W} \times (1.52 \text{ °C/W} + 0.5 \text{ °C/W} + 1.0 \text{ °C/W})$$

$$= 332 \text{ °C}$$

Given the ambient temperature of 25 °C, this puts the junction temperature T_j of the 2N3055 at 25 + 332 = 357 °C! This is clearly too high, so let's work backward from the air end and calculate just how many transistors we need to handle the heat.

First, putting more 2N3055s in parallel means that we will have the thermal model illustrated in **Figure 4.103**, with several identical q_{jc} and q_{cs} in parallel, all funneled through the same q_{as} (we have one heat sink).

Keeping in mind the physical size of the project, we could comfortably fit a heat sink of approximately 120 cubic inches (6 × 5 × 4 inches), well within the range of commercially available heat sinks. Furthermore, this application can use a heat sink where only "wire access" to the transistor connections is required. This allows the selection of a more efficient design. In contrast, RF designs require the transistor mounting surface to be completely exposed so that the PC board can be mounted close to the transistors to minimize parasitics. Looking at Figure 4.102, we see that a 120-cubic-inch heat sink yields a q_{sa} of 0.55 °C/W. This means that the temperature of the heat sink when dissipating 110 W will be 25 °C + (110 W × 0.55 °C/W) = 85.5 °C.

Industrial experience has shown that silicon transistors suffer substantial failure when junctions are operated at highly elevated temperatures. Most commercial and military specifications will usually not permit design junction temperatures to exceed 125 °C. To arrive at a safe figure for our maximum-allowed T_j, we must consider the intended use of the power supply. If we are using it in a 100% duty-cycle transmitting application such as RTTY or FM, the circuit will be dissipating 110 W continuously. For a lighter duty-cycle load such as CW or SSB, the "key-down" temperature can be slightly higher as long as the average is less than 125 °C. In this intermittent type of service, a good conservative figure to use is T_j = 150 °C.

Given this scenario, the temperature rise across each transistor can be 150 – 85.5 = 64.5 °C. Now, referencing Figure 2.4.93, remembering the total θ for each 2N3055 is 1.52 + 0.5 = 2.02 °C/W, we can calculate the maximum power each 2N3055 can safely dissipate:

$$P = \frac{\delta T}{\theta} = \frac{64.5 \text{ °C}}{2.02 \text{ °C / W}} = 31.9 \text{ W}$$

Thus, for 110 W full load, we need four 2N3055s to meet the thermal requirements of the design. Now comes the big question: What is the "right" heat sink to use? We have already established its requirements: it must be capable of dissipating 110 W and have a q_{sa} of 0.55 °C/W (see above).

A quick consultation with several manufacturer's catalogs reveals that Wakefield Thermal Solutions, Inc. model nos. 441 and

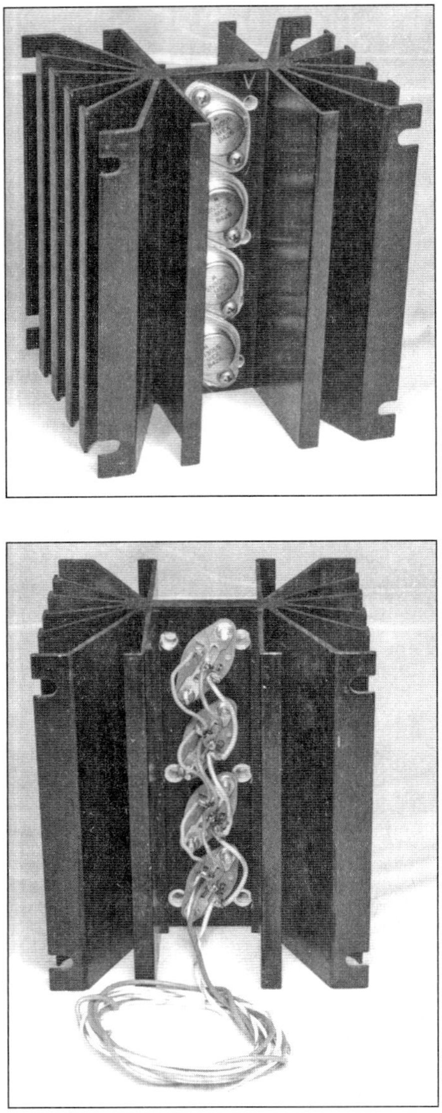

Figure 4.104 — A Wakefield 441 heat sink with four 2N3055 transistors mounted.

435 heat sinks meet the needs of this application. A Thermalloy model no. 6441 is suitable as well. Data published in the catalogs of these manufacturers show that in natural-convection service, the expected temperature rise for 100 W dissipation would be just under 60 °C, an almost perfect fit for this application. Moreover, the No. 441 heat sink can easily mount four TO-3-style 2N3055 transistors as shown in **Figure 4.104**. Remember: heat sinks should be mounted with the fins and transistor mounting area vertical to promote convection cooling.

The design procedure just described is applicable to any circuit where heat buildup is a potential problem. By using the thermal-resistance model, we can easily calculate whether or not an external means of cooling is necessary, and if so, how to choose it. Aside from heat sinks, forced air cooling (fans) is another common method. In commercial transceivers, heat sinks with forced-air cooling are common.

4.11.3 Semiconductor Temperature Effects

The number of excess holes and electrons in semiconductor material is increased as the temperature of a semiconductor increases. Since the conductivity of a semiconductor is related to the number of excess carriers, this also increases with temperature. With respect to resistance, semiconductors have a negative temperature coefficient. The resistance of silicon *decreases* by about 8% per °C and by about 6% per °C for germanium. Semiconductor temperature properties are the opposite of most metals, which *increase* their resistance by about 0.4% per °C. These opposing temperature characteristics permit the design of circuits with opposite temperature coefficients that cancel each other out, making a temperature insensitive circuit.

Semiconductor devices can experience an effect called *thermal runaway* as the current causes an increase in temperature. (This is primarily an issue with bipolar transistors.) The increased temperature decreases resistance and may lead to a further increase in current (depending on the circuit) that leads to an additional temperature increase. This sequence of events can continue until the semiconductor destroys itself, so circuit design must include measures that compensate for the effects of temperature.

Semiconductor Failure Caused by Heat

There are several common failure modes for semiconductors that are related to heat. The semiconductor material is connected to the outside world through metallic *bonding leads*. The point at which the lead and the semiconductor are connected is a common place for the semiconductor device to fail. As the device heats up and cools down, the materials expand and contract. The rate of expansion and contraction of semiconductor material is different from that of metal. Over many cycles of heating and cooling the bond between the semiconductor and the metal can break. Some experts have suggested that the lifetime of semiconductor equipment can be extended by leaving the devices powered on all the time, but this requires removal of the heat generated during normal operation.

A common failure mode of semiconductors is caused by the heat generated during semiconductor use. If the temperatures of the PN junctions remain at high enough levels for long enough periods of time, the impurities resume their diffusion across the PN junctions. When enough of the impurity atoms cross the depletion region, majority carrier recombination stops functioning properly and the semiconductor device fails permanently.

Excessive temperature can also cause failure anywhere in the semiconductor from heat generation within any current-carrying conductor, such as an FET channel or the bonding leads. Integrated circuits with more than one output may have power dissipation limits that depend on how many of the outputs are active at one time. The high temperature can cause localized melting or cracking of the semiconductor material, causing a permanent failure.

Another heat-driven failure mode, usually not fatal to the semiconductor, is excessive leakage current or a shift in operating point that causes the circuit to operate improperly. This is a particular problem in complex integrated circuits — analog and digital — dissipating significant amounts of heat under normal operating conditions. Computer microprocessors are a good example, often requiring their own cooling systems. Once the device cools, normal operation is usually restored.

To reduce the risk of thermal failures, the designer must comply with the limits stated in the manufacturer's data sheet, devising an adequate heat removal system. (Thermal issues are discussed in the **Electrical Fundamentals** chapter.)

4.11.4 Safe Operating Area (SOA)

Devices intended for use in circuits handling high currents or voltages are specified to have a *safe operating area* (*SOA*). This refers to the area drawn on the device's characteristic curve containing combinations of voltage and current that the device can be expected to control without damage under specific conditions. The SOA combines a number of limits — voltage, current, power, temperature and various breakdown mechanisms — in order to simplify the design of protective circuitry. The SOA is also specified to apply to specific durations of use — steady-state, long pulses, short pulses and so forth. The device may have separate SOAs for resistive and inductive loads.

You may also encounter two specialized types of SOA for turning the device on and off. *Reverse bias safe operating area* (*RBSOA*) applies when the device is turning off. *Forward bias safe operating* area (*FBSOA*) applies when turning the device on. These SOAs are used because the high rate-of-change of current and voltage places additional stresses on the semiconductor.

4.11.5 Semiconductor Derating

Maximum ratings for power transistors are usually based on a case temperature of 25 °C. These ratings will decrease with increasing operating temperature. Manufacturer's data sheets usually specify a *derating* figure or curve that indicates how the maximum ratings change per degree rise in temperature. If such information is not available (or even if it is!), it is a good rule of thumb to select a power transistor with a maximum power dissipation of at least twice the desired output power.

RECTIFIERS

Diodes are physically quite small, and they operate at high current densities. As a result their heat-handling capabilities are somewhat limited. Normally, this is not a problem in high-voltage, low-current supplies in which rectifiers in axial-lead DO-type packages are used. The use of high-current (2 A or greater) rectifiers at or near their maximum ratings, however, requires some form of heat sinking. The average power dissipated by a rectifier is

$$P = I_{AVG} \times V_F$$

where

I_{AVG} is the average current, and
V_F is the forward voltage drop.

Average current must account for the conduction duty cycle and the forward voltage drop must be determined at the average current level.

Rectifiers intended for such high-current applications are available in a variety of packages suitable for mounting to flat surfaces. Frequently, mounting the rectifier on the main chassis (directly, or with thin mica insulating washers) will suffice. If the diode is insulated from the chassis, thin layers of thermal compound or thermal insulating washers should be used to ensure good heat conduction. Large, high-current rectifiers often require special heat sinks to maintain a safe operating temperature. Forced-air cooling is sometimes used as a further aid.

4.11.6 RF Heating

RF current often causes component heating problems where the same level of dc current may not. An example is the tank circuit of an RF oscillator. If several small capacitors are connected in parallel to achieve a desired capacitance, skin effect will be reduced and the total surface area available for heat dissipation will be increased, thus significantly reducing the RF heating effects as compared to a single large capacitor. This technique can be applied to any similar situation; the general idea is to divide the heating among as many components as possible.

4.11.7 Forced-Air and Water Cooling

In amateur radio today, forced-air cooling is most commonly found in vacuum-tube circuits or in power supplies built in small enclosures, such as those in solid-state transceivers or computers. Fans or blowers are commonly specified in cubic feet per minute (CFM). While the nomenclature and specifications differ from those used for heat sinks, the idea remains the same: to offer a low thermal resistance between the inside of the enclosure and the (ambient) exterior.

For forced air cooling, we basically use the "one resistor" thermal model of Figure 4.100. The important quantity to be determined is heat generation, P_{in}. For a power supply, this can be easily estimated as the difference between the input power, measured at the transformer primary, and the output power at full load. For variable-voltage supplies, the worst-case output condition is minimum voltage with maximum current. A discussion of forced-air cooling for vacuum tube equipment appears in the **RF Power Amplifiers** chapter. For an in-depth discussion of fan cooling, see the References entry for Rabassa's February 2019 *QST* article.

Dust build-up is a common problem for forced-air cooling systems, even with powerful blowers and fans. If air intake grills and vents are not kept clean and free of lint and debris, air flow can be significantly reduced, leading to excessive equipment temperature and premature failure. Cleaning of air passageways should be included in regular equipment maintenance for good performance and maximum equipment life.

Water cooling systems are much less common in amateur equipment, used primarily for high duty cycle operating, such as RTTY, and at frequencies where the efficiency of the amplifier is relatively low, such as UHF and microwaves. In these situations, water cooling is used because water can absorb and transfer more than 3,000 times as much heat as the same volume of air!

The main disadvantage of water cooling is that it requires pumps, hoses, and reservoirs whereas a fan or blower is all that is required for forced-air cooling. For high-voltage circuits, using water cooling also requires special insulation techniques and materials to allow water to circulate in close contact with the heat source while remaining electrically isolated.

Nevertheless, the technique can be effective. The increased availability of inexpensive materials designed for home sprinkler and other low-pressure water distribution systems make water-cooling less difficult to implement. It is recommended that the interested reader review articles and projects in the amateur literature to observe the successful implementation of water cooling systems.

4.11.8 Heat Pipe Cooling

A heat pipe is a device containing a *working fluid* in a reservoir where heat is absorbed and a channel to a second reservoir where heat is dissipated. Heat pipes work by *evaporative cooling*. The heat-absorbing reservoir is placed in thermal contact with the heat source which transfers heat to the working fluid, usually a liquid substance with a boiling point just above room temperature. The working fluid vaporizes and the resulting vapor pressure pushes the hot vapor through the channel to the cooling reservoir.

In the cooling reservoir, the working fluid gives up its heat of vaporization, returning to the fluid state. The cooled fluid then flows back through the channel to the heat-absorbing reservoir where the process is repeated.

Heat pipes require no fans or pumps — movement of the working fluid is driven entirely from the temperature difference between the two reservoirs. The higher the temperature difference between the absorbing and dissipating reservoirs, the more effective the heat pump becomes, up to the limit of the dissipating reservoir to dissipate heat.

The primary application for heat pipes is for space operations. Heat pipe applications in terrestrial applications are limited due to the gravity gradient sensitivity of heat pipes. At present, computers and certain amplifier modules are the only amateur equipment making use of heat pipes. Nevertheless, as more general-purpose products become available, this technique will become more common.

4.11.9 Thermoelectric Cooling

Thermoelectric cooling makes use of the *Peltier effect* to create heat flow across the junction of two different types of materials. This process is related to the *thermoelectric effect* by which thermocouples generate voltages based on the temperature of a similar junction. A *thermoelectric cooler* or *TEC* (also known as a *Peltier cooler*) requires only a source of dc power to cause one side of the device to cool and the other side to warm. TECs are available with different sizes and power ratings for different applications.

TECs are not available with sufficient heat transfer capabilities that they can be used in high-power applications, such as RF amplifiers. However, they can be useful in lowering the temperature of sensitive receiver circuits, such as preamplifiers used at UHF and microwave frequencies, or imaging devices, such as charge-coupled devices (CCDs). Satellites use TECs as *radiative coolers* that dissipate heat directly as thermal or infrared radiation. TECs are also found in some computing equipment where they are used to remove heat from microprocessors and other large integrated circuits.

4.11.10 Temperature Compensation

Aside from catastrophic failure, temperature changes may also adversely affect circuits if the temperature coefficient (TC) of one or more components is too large. If the resultant change is not too critical, adequate temperature stability can often be achieved simply by using higher-precision components with low TCs (such as NP0/C0G capacitors or metal-film resistors). For applications where this is impractical or impossible (such as many solid-state circuits), we can minimize temperature sensitivity by *compensation* or *matching* — using temperature coefficients to our advantage.

Compensation is accomplished in one of two ways. If we wish to keep a certain circuit quantity constant, we can interconnect pairs of components that have equal but opposite TCs. For example, a resistor with a negative TC can be placed in series with a positive TC resistor to keep the total resistance constant. Conversely, if the important point is to keep the *difference* between two quantities constant, we can use components with the *same* TC so that the pair "tracks." That is, they both change by the same amount with temperature.

An example of this is a Zener reference circuit. Since a diode is strongly affected by operating temperature, circuits that use diodes or transistors to generate stable reference voltages must use some form of temperature compensation. Since, for a constant current, a reverse-biased PN junction has a negative voltage TC while a forward-biased junction has a positive voltage TC, a good way to temperature-compensate a Zener reference diode is to place one or more forward-biased diodes in series with it.

4.11.11 Thermistors

Thermistors can be used to control temperature and improve circuit behavior or protect against excessive temperatures, hot or cold. Circuit temperature variations can affect gain, distortion or control functions like receiver AGC or transmitter ALC. Thermistors can be used in circuits that compensate for temperature changes.

A *thermistor* is a small bit of intrinsic (no N or P doping) metal-oxide semiconductor compound material between two wire leads. As temperature increases, the number of liberated hole/electron pairs increases exponentially, causing the resistance to decrease exponentially. You can see this in the resistance equation:

$$R(T) = R(T0)e^{-\beta(1/T0-1/T)}$$

where T is some temperature in Kelvins and T0 is a reference temperature, usually 298 K (25°C), at which the manufacturer specifies R(T0).

The constant β is experimentally determined by measuring resistance at various temperatures and finding the value of β that best agrees with the measurements. A simple way to get an approximate value of β is to make two measurements, one at room temperature, say 25 °C (298 K) and one at 100 °C (373 K) in boiling water. Suppose the resistances are 10 kΩ and 938 Ω.

$$\beta = \frac{\ln\left(\frac{R(T)}{R(T0)}\right)}{\frac{1}{T} - \frac{1}{T0}} = \frac{\ln\left(\frac{938}{1000}\right)}{\frac{1}{373} - \frac{1}{298}} = 3507$$

With the behavior of the thermistor known — either by equation or calibration table — its change in resistance can be used to create an electronic circuit whose behavior is controlled by temperature in a known fashion. Such a circuit can be used for controlling temperature or detecting specific temperatures.

See "Termistors in Homebrew Projects" and "Thermistor Based Temperature Controller" both by Bill Sabin, WØIYH in this book's online information for practical projects using thermistors.

4.12 References and Bibliography

BOOKS AND PERIODICALS

Alexander, C., and Sadiku, M., *Fundamentals of Electric Circuits* 7th Edition (McGraw-Hill, 2021).

ARRL Lab Staff, "Lab Notes — Capacitor Basics," *QST*, Jan. 1997, pp. 85 – 86.

Banzhaf, W., WB1ANE, *Understanding Basic Electronics*, 2nd Edition (ARRL, 2010).

Bergeron, B., NU1N, "Under the Hood II: Resistors," *QST,* Nov. 1993, pp. 41 – 44.

Bergeron, B., NU1N, "Under the Hood III: Capacitors," *QST,* Jan. 1994, pp. 45 – 48.

Bergeron, B., NU1N, "Under the Hood IV: Inductors," *QST,* Mar. 1994, pp 37 – 40.

Bergeron, B., NU1N, "Under the Hood: Lamps, Indicators, and Displays," *QST,* Sep. 1994, pp. 34 – 37.

Ebers, J., and Moll, J., "Large-Signal Behavior of Junction Transistors," *Proceedings of the IRE*, Vol. 42, No. 12, Dec. 1954, pp. 1761 – 1772.

Getreu, I., *Modeling the Bipolar Transistor* (Elsevier, 1979). Also available from Tektronix, Inc, Beaverton, Oregon, in paperback form. Must be ordered as Part Number 062-2841-00.

Glover, T., *Pocket Ref*, 4th Edition (Sequoia Publishing, 2010).

Grover, F., *Inductance Calculations*, Reprint (Dover Publications, 2009).

Hayward, W., W7ZOI, *Introduction to Radio Frequency Design* (ARRL, 2004).

Hayward, W., Campbell, R., and Larkin, B., *Experimental Methods in RF Design* (ARRL, 2009).

Horowitz, P., and Hill, W., *The Art of Electronics*, 3rd Edition (Cambridge University Press, 2015).

Jung, W., *IC Op Amp Cookbook* (Prentice Hall, 1986).

Kaiser, C., *The Capacitor Handbook*, 2nd Edition (CJ Publishing, 1995).

Kaiser, C., *The Diode Handbook* (CJ Publishing, 1999).

Kaiser, C., *The Inductor Handbook* (CJ Publishing, 1996).

Kaiser, C., *The Resistor Handbook*, 2nd Edition (CJ Publishing, 1998).

Kaiser, C., *The Transistor Handbook* (CJ Publishing, 1999).

Kaplan, S., *Wiley Electrical and Electronics Dictionary* (Wiley-IEEE Press, 2004).

McClanahan, J., W4JBM, "Understanding and Testing Capacitor ESR," *QST*, Sep. 2003, pp. 30 – 32.

Millman, J., and Grabel, A., *Microelectronics: Digital and Analog Circuits and Systems* (McGraw-Hill, 1988).

Mims, F., *Timer, Op Amp & Optoelectronic Circuits & Projects* (Master Publishing, 2004).

Orr, W., *Radio Handbook* 23rd Edition (Newnes 1997).

Rabassa, A., NW2M, "The Basics of Fan Cooling," *QST*, Feb. 2019, pp. 34 – 35.

Severns, R., N6LF, "Design of Snubbers for Power Circuits," Cornell-Dublier (CDE). **www.cde.com/resources/technical-papers/design.pdf**.

Silver, W., NØAX, "Experiment #24 — Heat Management," *QST*, Jan. 2005, pp. 64 –65.

Silver, W., NØAX, "Experiment #33 — The Transformer," *QST*, Oct. 2005, pp. 62 –63.

Silver, W., NØAX, "Experiment #62 — About Resistors," *QST*, Mar. 2008, pp. 66 –67.

Silver, W., NØAX, "Experiment #63 — About Capacitors," *QST*, Apr. 2008, pp. 70 – 71.

Silver, W., NØAX, "Experiment #132 — Resistor Networks," *QST*, Jan. 2014, pp. 63 – 64.

Smith, J., K8ZOA, "Carbon Composition, Carbon Film and Metal Oxide Film Resistors," *QEX*, Mar./Apr. 2008, pp. 46-57.

Terman, F., *Radio Engineers' Handbook* (McGraw-Hill, 1943).

White, J., "Transistor Amplifier Design" *Microwaves & RF*, Parts 1 – 8, Jul. 2004 to Jan. 2005.

WEBSITES

"dB or not dB? Everything you ever wanted to know about decibels but were afraid to ask…," Rohde & Schwarz, Application Note 1MA98, **www.rohde-schwarz.us/en/applications/db-or-not-db-application-note_56280-15534.html**.

Hyperphysics Op-Amp Circuit Tutorials, **hyperphysics.phy-astr.gsu.edu/Hbase/Electronic/opampvar.html#c2**.

Munir, U., "Selecting the Right CMOS Analog Switch," Maxim Semiconductor, Application Note 5299, **www.maximintegrated.com/en/design/technical-documents/app-notes/5/5299.html**.

Schultz, W., "Power Transistor Safe Operating Area: Special Considerations for Switching Power Supplies," onsemi, AN-875-D, **www.onsemi.cn/pub/Collateral/AN875-D.PDF**.

Contents

Chapter 5
RF Techniques

This chapter is a compendium of material from ARRL publications and other sources. It assumes the reader is familiar with the concepts introduced in the **Electrical Fundamentals**, **Radio Fundamentals**, and **Circuits and Components** chapters. The topics and techniques discussed here are associated with the special demands of circuit design in the HF and VHF ranges. The material is collected from previous editions of this book written by Leonard Kay, K1NU; *Introduction to Radio Frequency Design* by Wes Hayward, W7ZOI; and *Experimental Methods in RF Design* by Wes Hayward, W7ZOI, Rick Campbell, KK7B, and Bob Larkin, W7PUA. Material on ferrites is drawn from publications by Jim Brown, K9YC. The section on Noise was written by Paul Wade, W1GHZ, with contributions from Joe Taylor, K1JT.

5.1 Introduction

When is an inductor not an inductor? When it's a capacitor! This statement may seem odd, but it suggests the main message of this chapter. In the earlier chapter, **Electrical Fundamentals**, the basic components of electronic circuits were introduced. As you may know from experience, those simple component pictures are ideal. That is, an ideal component (or element) by definition behaves exactly like the mathematical equations that describe it, and only in that fashion. For example, the current through an ideal capacitor is equal to the capacitance times the rate of change of the voltage across it without consideration of the materials or techniques by which a real capacitor is manufactured.

It is often said that, "Parasitics are anything you don't want," meaning that the component is exhibiting some behavior that detracts from or compromises its intended use. Real components only approximate ideal components, although sometimes quite closely. Any deviation from ideal behavior a component exhibits is called *non-ideal, parasitic,* or *stray.* One way of thinking about parasitic and stray effects — although this is by no means universal — is that parasitic effects are intrinsic to the component and stray effects include both parasitics and environmental effects such as coupling to nearby materials. Since most of what we'll discuss in the following sections deals with the characteristics of actual components, we'll use the term parasitic. Remember that stray and parasitic are often treated as interchangeable terms.

The important thing to realize is that *every* component has parasitic aspects that become significant when it is used in certain ways. This chapter deals with parasitic effects that are commonly encountered at radio frequencies. Knowing to what extent and under what conditions real components cease to behave like their ideal counterparts, and what can be done to account for these behaviors, allows the circuit designer or technician to work with circuits at radio frequencies. We will explore how and why the real components behave differently from ideal components, how we can account for those differences when analyzing circuits and how to select components to minimize, or exploit, nonideal behaviors.

Chapter 5 — Online Content

Articles

- Designing Wide-band Transformers for HF and VHF Power Amplifiers by Chris Trask, N7ZWY
- Reflections on the Smith Chart by Wes Hayward, W7ZOI
- Simplified Design of Impedance-Matching Networks, Parts I through III by George Grammer, W1DF
- The Galactic Background in the Upper HF Band by Dave Typinski, AJ4CO

Tools and Data

- *LTSpice* simulation files for Effects of Parasitic Characteristics (Section 5.3)
- *MATCH.EXE* software (for use with Tuned (Resonant) Networks discussion)

5.2 Lumped-Element versus Distributed Characteristics

Most electronic circuits that we use every day are inherently and mathematically considered to be composed of *lumped elements*. That is, we assume each component acts at a single point in space, and the wires that connect these lumped elements are assumed to be perfect conductors (with zero resistance and insignificant length). This concept is illustrated in **Figure 5.1**. These assumptions are perfectly reasonable for many applications, but they have limits. Lumped element models break down when:

- Circuit impedance is so low that the small, but non-zero, resistance in the wires is important. (A significant portion of the circuit power may be lost to heat in the conductors.)
- Lead and interconnection inductance is high enough (or the frequency is high enough) that the additional reactance affects circuit behavior.
- Operating frequency is high enough that the length of the connecting wires is a significant fraction (> 0.1) of the wavelength causing the propagation delay along the conductor

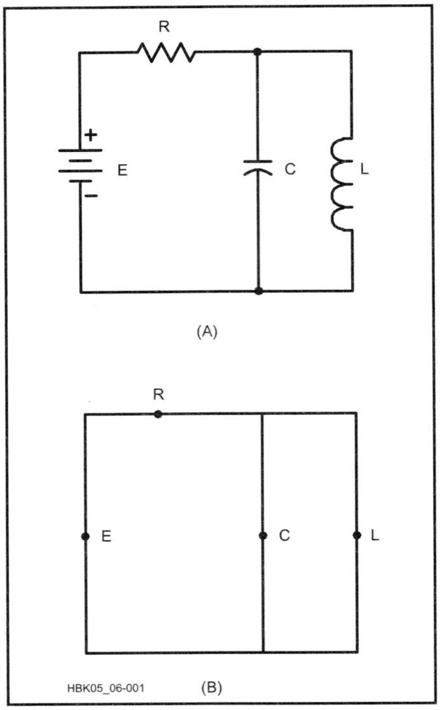

Figure 5.1 — The lumped element concept. Ideally, the circuit at A is assumed to be as shown at B, where the components are isolated points connected by perfect conductors. Many components exhibit nonideal behavior when these assumptions no longer hold.

or radiation from it to affect the circuit in which it is used.

- Transmission lines are used as conductors. (Their characteristic impedance is usually significant, and impedances connected to them are transformed as a function of the line length. See the **Transmission Lines** chapter for more information.)

Effects such as these are called *distributed*, and we talk of *distributed elements* or effects to contrast them to lumped elements.

To illustrate the differences between lumped and distributed elements, consider the two resistors in **Figure 5.2**, which are both 12 inches long. The resistor at A is a uniform rod of carbon. The second "resistor" B is made of two 6-inch pieces of silver rod (or other highly conductive material), with a small resistor soldered between them. Now imagine connecting the two probes of an ohmmeter to each of the two resistors, as in the figure. Starting with the probes at the far ends, as we slide the probes toward the center, the carbon rod will display a constantly decreasing resistance on the ohmmeter. This represents a distributed resistance. On the other hand, the ohmmeter connected to the other 12-inch "resistor" will

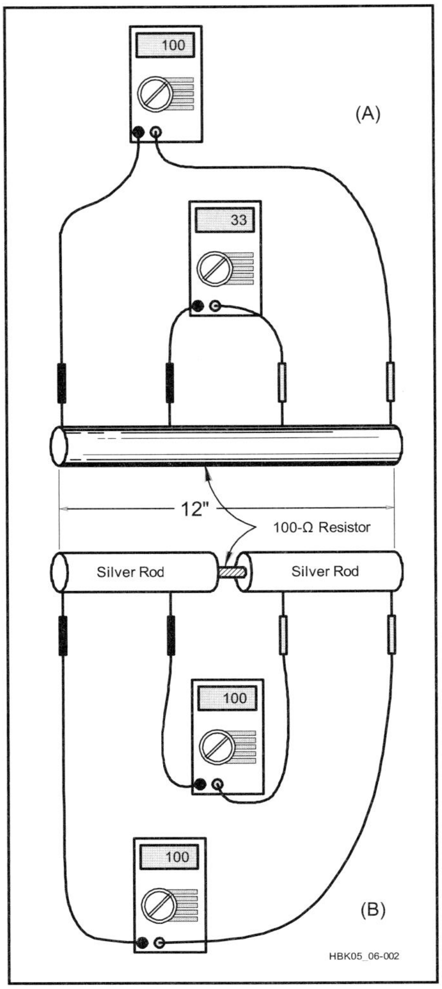

Figure 5.2 — Distributed (A) and lumped (B) resistances. See text for discussion.

display a constant resistance as long as one probe remains on each side of the small resistance and as long as we neglect the resistance of the silver rods! This represents a lumped resistance connected by perfect conductors.

Lumped elements also have the very desirable property that they introduce no phase shift resulting from propagation delay through the element. (Although combinations of lumped elements can produce phase shifts by virtue of their R, L, and C properties.) Consider a lumped element that carries a sinusoidal current, as in **Figure 5.3A**. Since the element has negligible length, there is no phase difference in the current between the two sides of the element — *no matter how high the frequency* — precisely *because* the element length is negligible. If the physical length of the element were long, say 0.25 wavelength (0.25 λ) as shown in Figure 5.3B, the current

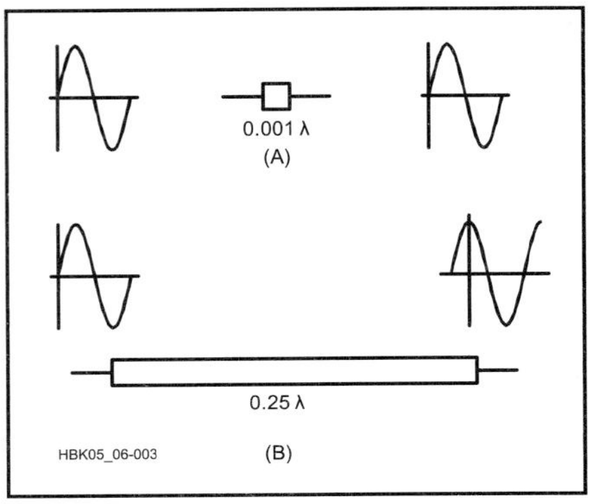

Figure 5.3 — The effects of distributed resistance on the phase of a sinusoidal current. There is no phase delay between ends of a lumped element.

phase would *not* be the same from end to end. In this instance, the current is delayed by 90 electrical degrees as it moves along the element. The amount of phase difference depends on the circuit's electrical length.

Because the relationship between the physical size of a circuit and the wavelength of an ac current present in the circuit will vary as the frequency of the ac signal varies, the ideas of lumped and distributed effects actually occupy two ends of a spectrum. At HF (30 MHz and below), where $\lambda \geq 10$ m, the lumped element concept is almost always valid. In the UHF and microwave region (300 MHz and above), where $\lambda \leq 1$ m and physical component size can represent a significant fraction of a wavelength, nearly all components and wiring exhibits distributed effects to one degree or another. From roughly 30 to 300 MHz, whether the distributed effects are significant must be considered on a case-by-case basis.

Of course, if we could make resistors, capacitors, inductors and so on, very small, we could treat them as lumped elements at much higher frequencies. For example, surface-mount components, which are manufactured in very small, leadless packages, can be used at much higher frequencies and with fewer nonideal effects than can components with leads.

It is for these reasons that circuits and equipment are often specified to work within specific frequency ranges. Outside of these ranges the designer's assumptions about the physical characteristics of the components and the methods and materials of the circuit's assembly become increasingly invalid. At frequencies sufficiently removed from the design range, circuit behavior often changes in unpredictable ways.

5.3 Effects of Parasitic (Stray) Characteristics

Parasitic effects can be important at almost any frequency where performance is held to tight specifications. Stray reactances can have a big effect even at audio frequencies, for example. Lightning protection is very sensitive to grounding conductor inductance. Power connections to high-current solid-state amplifiers must have very low resistance, and so on. At HF and above (where we do much of our circuit design) these considerations become very important, in some cases dominant, in the models we use to describe our components. To understand what happens to circuits at RF we turn to a brief discussion of some electromagnetic and microwave theory concepts.

Parasitic effects due to component leads, packaging, leakage, and so on are relatively common to all components. When working at frequencies where many or all of the parasitics become important, a complex but completely general model such as that in **Figure 5.4** can be used for just about any component, with the actual component placed in the box marked *. Parasitic capacitance, C_p, and leakage conductance, G_L, appear in parallel across the device, while series resistance, R_s, and parasitic inductance, L_s, appear in series with it. Package capacitance, C_{pkg}, appears as an additional capacitance in parallel across the whole device.

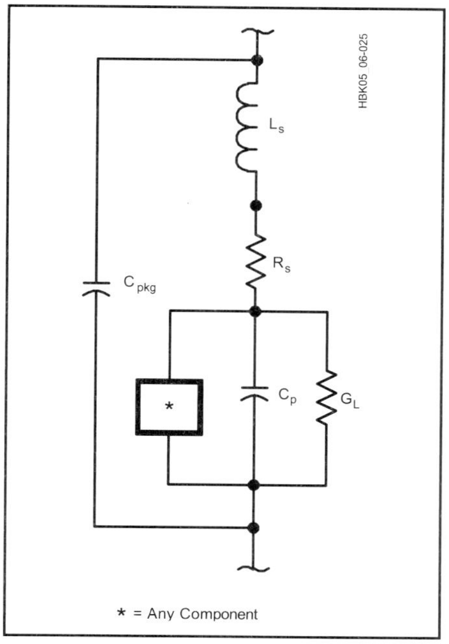

Figure 5.4 — A general model for electrical components at VHF frequencies and above. The box marked * represents the component. See text for discussion.

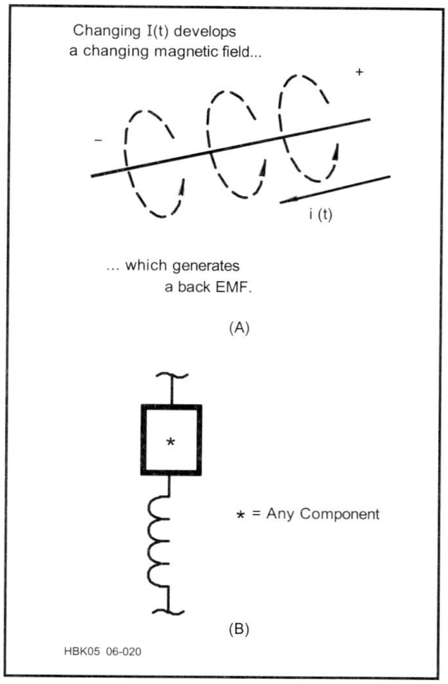

Figure 5.5 — Inductive consequences of Maxwell's equations. At A, any wire carrying a changing current develops a voltage difference along it. This can be mathematically described as an effective inductance. B adds parasitic inductance to a generic component model.

These small parasitics can significantly affect frequency responses of RF circuits. Either take steps to minimize or eliminate them, or use simple circuit theory to predict and anticipate changes. This maze of effects may seem overwhelming, but remember that it is very seldom necessary to consider all parasitics at all frequencies and for all applications. The **Computer Modeling and Simulation** chapter shows how to incorporate the effect of multiple parasitics into circuit design and performance modeling. Files for the *LTSpice* simulation package that include parasitic characteristics for a resistor, capacitor, and inductor are provided in the online content for this *Handbook*.

5.3.1 Parasitic Inductance

Maxwell's equations — the basic laws of electromagnetism that govern the propagation of electromagnetic waves and the operation of all electronic components — tell us that any wire carrying a current that changes with time (one example is a sine wave) develops a changing magnetic field around it. This changing magnetic field in turn induces an opposing voltage, or *back EMF*, on the wire. The back EMF is proportional to how fast the current changes (see **Figure 5.5**).

We exploit this phenomenon when we make an inductor. The reason we typically form inductors in the shape of a coil is to concentrate the magnetic field and thereby maxi-

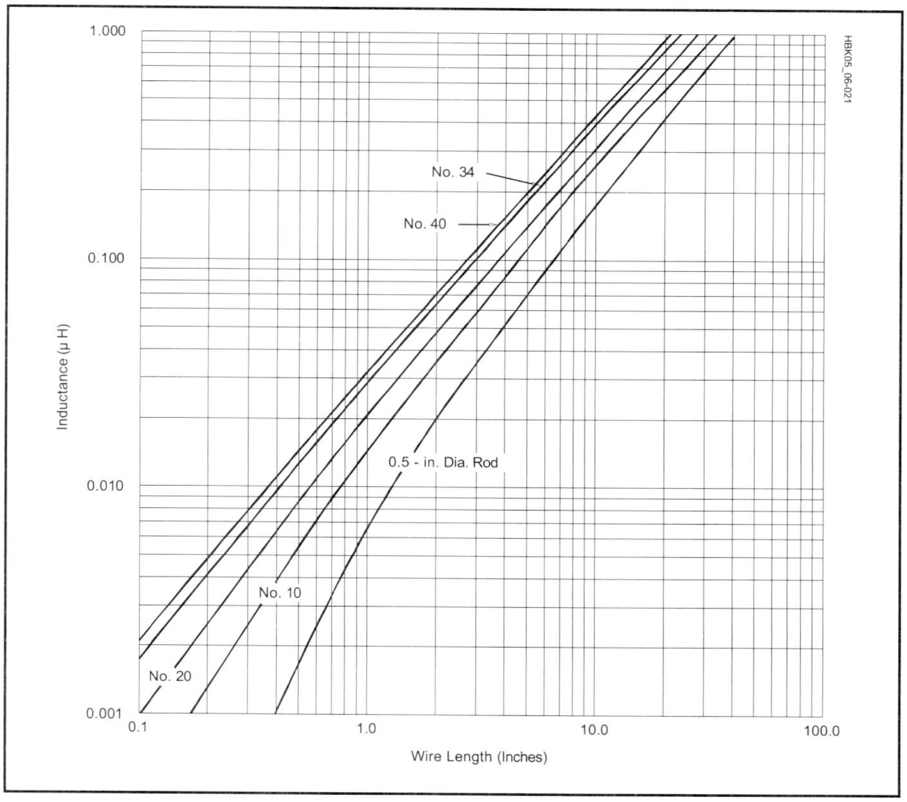

Figure 5.6 — A plot of inductance vs. length for straight conductors in several wire sizes.

mize the inductance for a given physical size. However, *all* wires carrying varying currents have these inductive properties. This includes the wires we use to connect our circuits, and even the *leads* of capacitors, resistors and so on. The inductance of a straight, round, nonmagnetic wire in free space is given by:

$$L = 0.00508\, b \left[\ln\left(\frac{2b}{a}\right) - 0.75 \right] \qquad (1)$$

where

L = inductance, in µH,
a = wire radius, in inches,
b = wire length, in inches, and
ln = natural logarithm (2.303 × $\log_{10}$).

Skin effect (discussed below) changes this formula slightly at VHF and above. As the frequency approaches infinity, the value 0.75 in the above equation increases to approach 1. This effect usually causes a change of no more than a few percent.

As an example, let's find the inductance of a typical #18 wire (diameter = 0.0403 inch and a = 0.0201) that is 4 inches long (b = 4):

$$L = 0.00508\,(4) \left[\ln\left(\frac{8}{0.0201}\right) - 0.75 \right]$$

$$= 0.0203\,[5.98 - 0.75] = 0.106\ \mu H$$

Wire of this diameter has an inductance of about 25 nH per inch of length. In circuits operating at VHF and higher frequencies, including high-speed digital circuits, the inductance of component leads can become significant. (The #24 AWG wire typically used for component leads has an inductance on the order of 20 nH per inch.) At these frequencies, lead inductance can affect circuit behavior, making the circuit hard to reproduce or repair. Good design and construction practice is to minimize the effects of lead inductance by using surface-mount components or trimming the leads to be as short as possible.

The impact of reactance due to parasitic inductance is usually very small; at audio frequencies (AF) or LF, parasitic inductive reactance of most components is practically zero. To use this example, the reactance of a 0.106 µH inductor even at 10 MHz is only 6.6 Ω. **Figure 5.6** shows a graph of the inductance for wires of various gauges (radii) as a function of length. Whether the reactance is significant or not depends on the application and the frequency of use.

We can represent parasitic inductance in component models by adding an inductor of appropriate value in series with the component since the wire leads are in series with the element. This (among other reasons) is why minimizing lead lengths and interconnecting wires becomes very important when designing circuits for VHF and above.

PARASITIC INDUCTANCE IN RESISTORS

The basic construction of common resistor types is shown in **Figure 5.7**. The primary parasitic effect associated with resistors is parasitic inductance. (Some parasitic capacitance exists between the leads or electrodes due to packaging.) **Figure 5.8** shows some more accurate circuit models for resistors at low to medium frequencies. The type of resistor with the most parasitic inductance are wire-wound resistors, essentially inductors used as resistors. Their use is therefore limited to dc or low-frequency ac applications where their reactance is negligible. Remember that this inductance will also affect switching transient waveforms, because the component will act as an RL circuit. The inductive effects of wire-wound resistors begin to become significant in the audio range above a few kHz.

As an example, consider a 1 Ω wire-wound resistor formed from 300 turns of #24 wire closely wound in a single layer 6.3 inches long on a 0.5-inch diameter form. What is its approximate inductance? From the inductance formula for air-wound coils in the **Electrical Fundamentals** chapter:

$$L = \frac{d^2 n^2}{18d + 40l} = \frac{0.5^2 \times 300^2}{(18 \times 0.5) + (40 \times 6.3)} = 86\ \mu H$$

If we want the inductive reactance to be less than 10% of the resistor value, then this resistor cannot be used above f = 0.1 / (2π × 86 µH) = 185 Hz! Real wire-wound resistors have multiple windings layered over each other to minimize both size and parasitic inductance (by winding each layer in opposite directions, much of the inductance is canceled). If we assume a five-layer winding, the length is reduced to 1.8 inches and the inductance to approximately 17 µH, so the resistor can then be used below 937 Hz. (This has the effect of increasing the resistor's parasitic capacitance, however.)

The resistance of certain types of tubular film resistors is controlled by inscribing a spiral path through the film on the inside of the tube. This creates a small inductance that may be significant at and above the higher audio frequencies.

NON-INDUCTIVE RESISTORS

The resistors with the least amount of para-

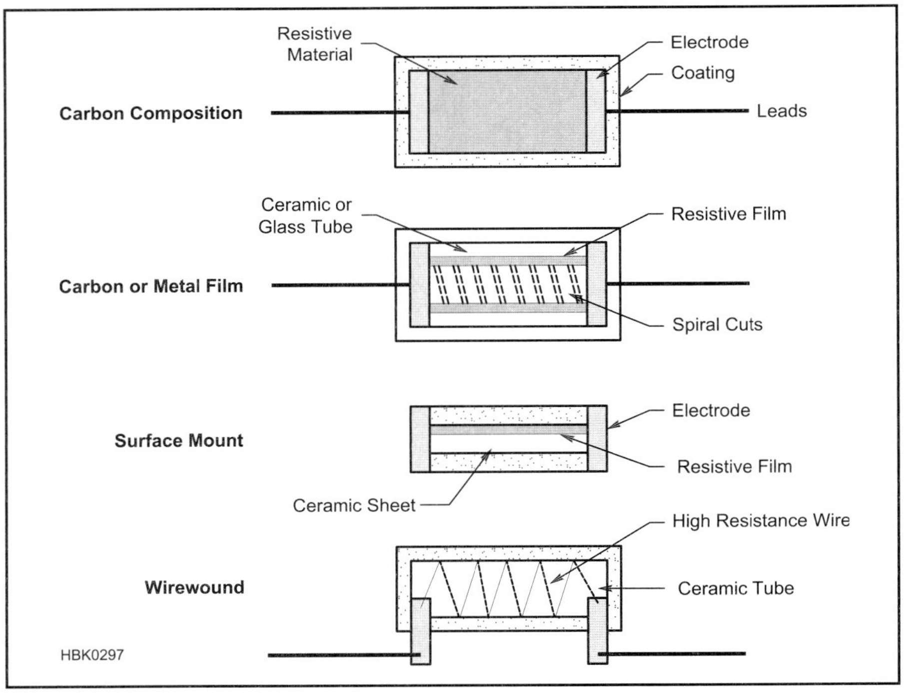

Figure 5.7 — The electrical characteristics of different resistor types are strongly affected by their construction. Reactance from parasitic inductance and capacitance strongly impacts the resistor's behavior at RF.

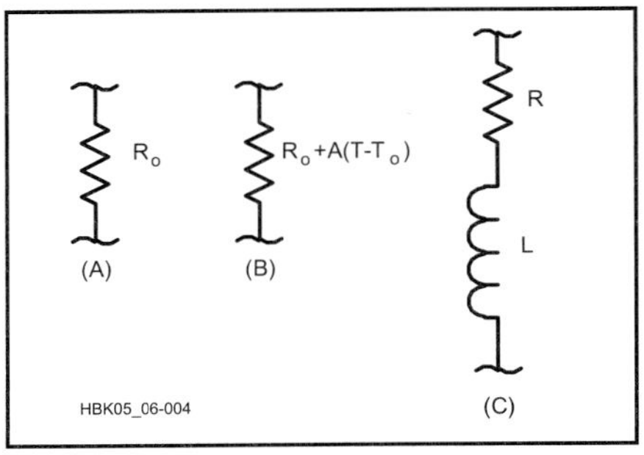

Figure 5.8 — Circuit models for resistors. The wire-wound model with associated inductance is shown at C. B includes the effect of temperature (T). For designs at VHF and higher frequencies, the model at C could be used with L representing lead inductance.

sitic inductance are the bulk resistors, such as carbon-composition, metal-oxide, and ceramic resistors. These resistors are made from a single linear cylinder, tube or block of resistive material so that inductance is minimized. Each type of resistor has a maximum usable frequency, above which parasitic capacitance and inductance begin to become significant. Review the manufacturer's data sheet for the component to learn about its performance at high frequencies.

Some resistors advertised as "noninductive" are actually wire-wound resistors with a special winding technique that minimizes inductance. These resistors are intended for use at audio frequencies and are not suitable for use at RF. If you are not sure, ask the vendor if the resistors are suitable for use in RF circuits.

Because resistors are manufactured with an insulating coating, it can be difficult to determine their internal structure and thus estimate their parasitic inductance. In cases where a surplus or used component is to be included, it is recommended that you test the component with an impedance meter or make some other type of reactance measurement if you are unable to access the manufacturer's specifications for the resistor.

PARASITIC INDUCTANCE IN CAPACITORS

The size and shape of a capacitor's plates and the leads used to connect them to circuits create parasitic inductance, often referred to as *equivalent series inductance* (ESL) by capacitor manufacturers. **Figures 5.9** and **5.10** show reasonable models for capacitors that are good up to VHF.

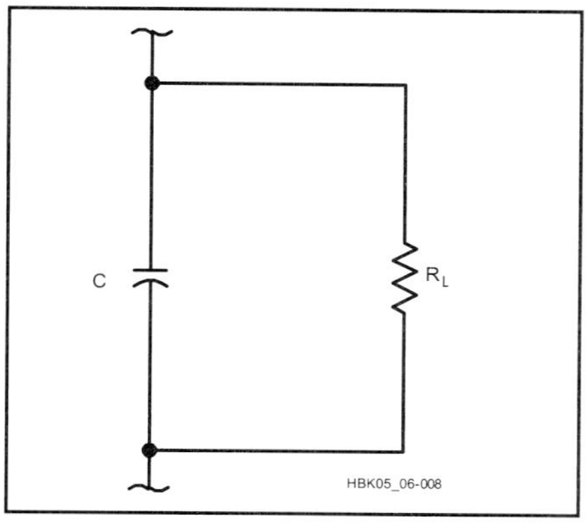

Figure 5.9 — A simple capacitor model for frequencies well below self-resonance.

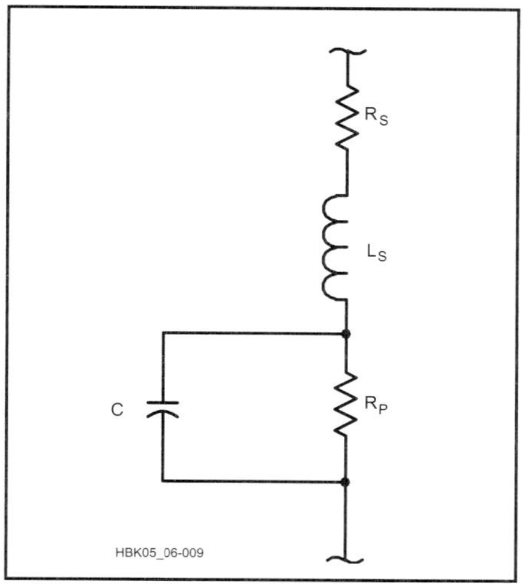

Figure 5.10 — A capacitor model for VHF and above including series resistance and distributed inductance.

Figure 5.11 shows a roll-type capacitor made of two strips of very thin metal foil and separated by a dielectric. After leads are attached to the foil strips, the sandwich is rolled up and either placed in a metal can or coated with plastic. *Radial leads* both stick out of one end of the roll and *axial leads* from both ends along the roll's axis. Because of the rolled strips, the ESL is high. Electrolytic and many types of film capacitors are made with roll construction. As a result, they are generally not useful in RF circuits.

In the stack capacitor, thin sheets of dielectric are coated on one side with a thin metal layer. A stack of the sheets is placed under pressure and heated to make a single solid unit. Metal side caps with leads attached contact the metal layers. The ESL of stack capacitors is very low and so they are useful at high frequencies. Ceramic and mica capacitors are the most common stack-style capacitor.

Parallel-plate air and vacuum capacitors used at RF have relatively low parasitic inductance, but transmitting capacitors made to withstand high voltages and current are large enough that parasitic inductance becomes significant, limiting their use to low-VHF and lower frequencies. Adjustable capacitors (air variables and compression or piston trimmer capacitors) for tuning low-power circuits are much smaller and so have correspondingly lower parasitic inductance.

It is difficult for a single capacitor to work well over a very wide frequency range, so capacitors are often placed in parallel as discussed in the section below on Bypassing and Decoupling. It is often suggested that different types of capacitors be connected in parallel to avoid the effects of parasitic inductance at different frequencies, but without testing and careful modeling the results are often unpredictable or even counterproductive as the referenced discussion shows.

5.3.2 Parasitic Capacitance

Maxwell's equations also tell us that if the voltage between any two points changes with time, a displacement current is generated between these points as illustrated in **Figure 5.12.** This *displacement current* results from the propagation of the electromagnetic field between the two points and is not to be confused with *conduction current*, which is the movement of electrons. Displacement current is directly proportional to the rate at which the voltage is changing.

When a capacitor is connected to an ac voltage source, a steady ac current can flow because taken together, conduction current and displacement current "complete the loop" from the positive source terminal, across the plates of the capacitor, and back to the negative terminal.

In general, parasitic capacitance shows up *wherever* the voltage between two points is changing with time, because the laws of

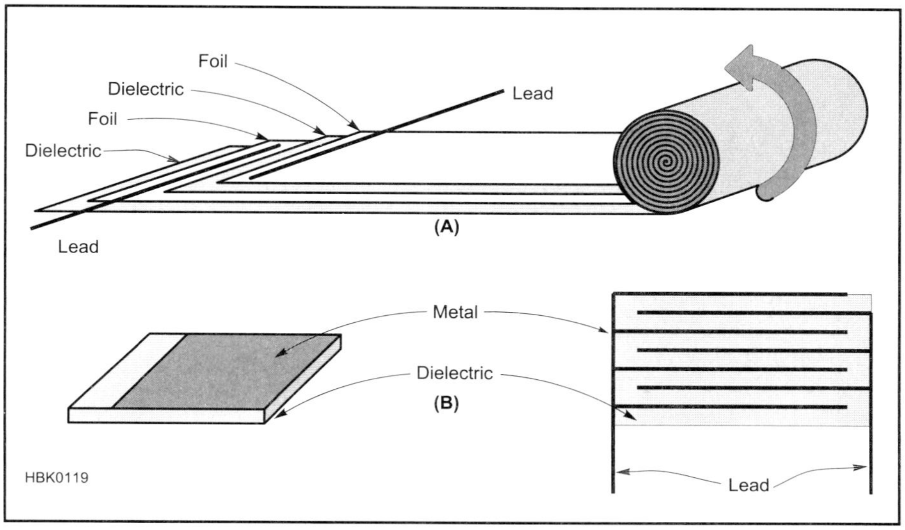

Figure 5.11 — Two common types of capacitor construction. (A) Roll construction uses two strips of foil separated by a strip of dielectric. (B) Stack construction layers dielectric material (such as ceramic or film), one side coated with metal. Leads are attached and the assembly coated with epoxy resin.

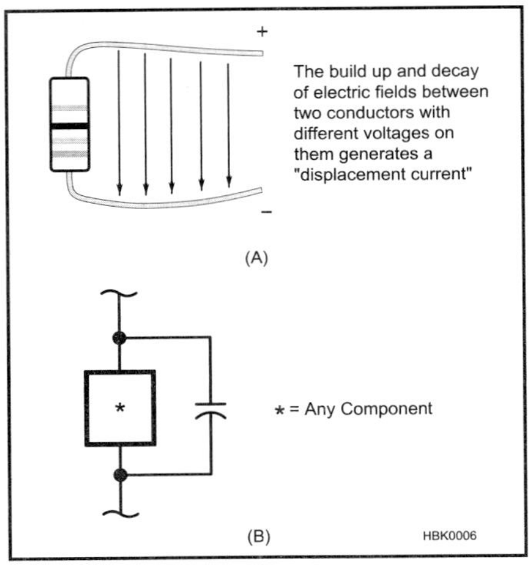

Figure 5.12 — Capacitive consequences of Maxwell's equations. A: Any changing voltage between two points, for example along a bent wire, generates a displacement current running between them. This can be treated mathematically as a capacitance. B adds parasitic capacitance to a generic component model.

electromagnetics require a displacement current to flow. Since this phenomenon represents an *additional* current path from one point in space to another, we can add this parasitic capacitance to our component models by adding a capacitor of appropriate value in *parallel* with the component. These parasitic capacitances are typically less than 1 pF, so that below VHF they can be treated as open circuits (infinite reactances) and thus neglected.

PACKAGE CAPACITANCE

Another source of capacitance, also in the 1 pF range and therefore important only at VHF and above, is the packaging of the component itself. For example, a power transistor packaged in a TO-220 case (see **Figure 5.13**), often has either the emitter or collector connected to the metal tab itself. This introduces an extra *inter-electrode capacitance* across the junctions.

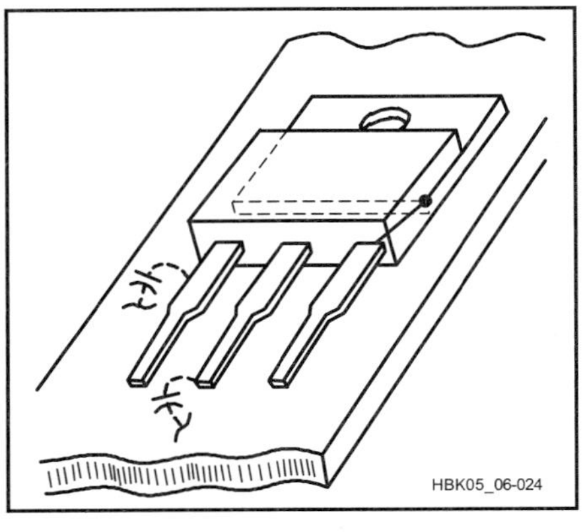

Figure 5.13 — Unexpected stray capacitance. The mounting tab of TO-220 transistors is often connected to one of the device leads. Because one lead is connected to the chassis, small capacitances from the other lead to the chassis appear as additional package capacitance at the device. Similar capacitance can appear at any device with a conductive package.

The copper traces on a PC board also create capacitance with the circuit components, to other traces, and to power and ground planes. Double-sided PC boards have a certain capacitance per square inch between the layers of copper on each side of the board. (Multilayer PC boards have higher values of capacitance due to the smaller separation between layers.) It is possible to create capacitors by leaving unetched areas of copper on both sides of the board. The dielectric constant of inexpensive PC board materials intended for use at low-frequencies is not well-controlled, however, leading to significant variations in capacitance. For this reason, the copper on one side of a double-sided board should be completely removed under frequency-determining circuits such as VFOs.

Stray capacitance (a general term used for any "extra" capacitance that exists due to physical construction) appears in any circuit where two metal surfaces exist at different voltages. Such effects can be modeled as an extra capacitor in parallel with the given points in the circuit. A rough value can be obtained with the parallel-plate formula given in the chapter on **Electrical Fundamentals**. Similar to parasitic inductance, *any* circuit component that has wires attached to it, or is fabricated from wire, or is near or attached to metal, will have a parasitic capacitance associated with it, which again, becomes important only at RF.

Stray capacitance can be difficult to account for in circuit design because it exists *between* components and other circuit structures, depending on the physical orientation of the component. Its presence may allow signals to flow in ways that disrupt the normal operation of a circuit and may have a greater affect in a high-impedance circuit because the capacitive reactance may be a greater percentage of the circuit impedance. Also, because stray capacitance often appears in parallel with the circuit, the stray capacitor may bypass more of the desired signal at higher frequencies. Careful physical design of an RF circuit and selection of components can minimize the effects of stray capacitance.

5.3.3 Self Resonance

Because of parasitic effects, a capacitor or inductor — all by itself — exhibits the properties of a resonant RLC circuit at frequencies for which the parasitic effects are significant. Figures 5.10 and 5.14 illustrate RF models for the capacitor and inductor, which are based on the general model in Figure 5.4, leaving out the packaging capacitance. Note the slight difference in configuration; the pairs C_p-R_p and L_s-R_s are in series in the capacitor but in parallel in the inductor. This is because of the different physical structure of the components.

At some sufficiently high frequency, both inductors and capacitors become *self-resonant* when the parasitic reactance cancels or equals the intended reactance, creating a series-resonant circuit. Similar to a series-resonant circuit made of discrete components, above the self-resonant frequency a capacitor will appear inductive, and an inductor will appear capacitive.

For an example, let's calculate the approximate self-resonant frequency of a 470 pF capacitor whose leads are made from #20 AWG wire (0.032-inch diameter), with a total length of 1 inch. Using equation 1, we calculate the approximate parasitic inductance

$$L\,(\mu H) = 0.00508\,(1)\left[\ln\left(\frac{2(1)}{(0.032/2)}\right) - 0.75\right]$$

$$= 0.021\ \mu H$$

and the self-resonant frequency is roughly

$$f = \frac{1}{2\pi\sqrt{LC}} = 50.6\ \text{MHz}$$

Similarly, an inductor can also have a parallel self resonance.

The purpose of making these calculations is to provide a feel for actual component values. They could be used as a rough design guideline, but should not be used quantitatively. Other factors such as lead orientation, shielding and so on, can alter the parasitic effects to a large extent. Large-value capacitors tend to have higher parasitic inductances (and therefore a lower self-resonant frequency) than small-value capacitors.

Self resonance becomes critically important at VHF and UHF because the self-resonant frequency of many common components is at or below the frequency where the component will be used. In this case, either special techniques can be used to construct components to operate at these frequencies, by reducing the parasitic effects, or else the idea of lumped elements must be abandoned altogether in favor of microwave techniques such as striplines and waveguides.

5.3.4 Inductors and RF Chokes at Radio Frequencies

Inductors are perhaps the component with the most significant parasitic effects. Where there are many different varieties of form for capacitors and resistors, most inductors are fundamentally similar: a coil of wire on a tubular or toroidal form. As such, they are affected by both parasitic resistance and capacitance as shown in the simple inductor model of **Figure 5.14**.

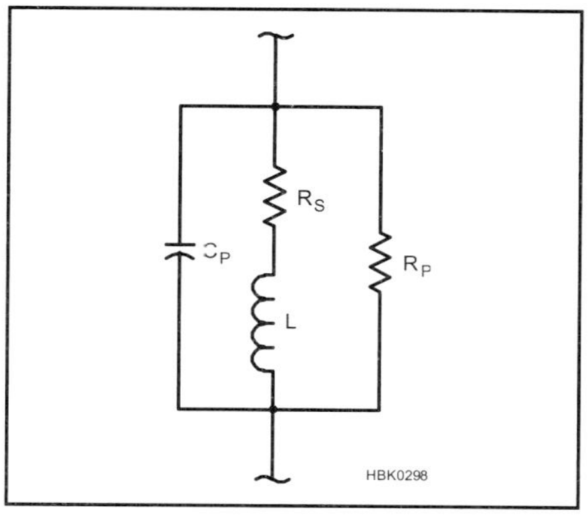

Figure 5.14 — General model for inductor with parasitic capacitance and resistance. The parasitic capacitance represents the cumulative effect of the capacitance between the different turns of wire. Parasitic resistance, R_S, depends on frequency due to the skin effect. R_P represents leakage resistance. At low frequencies, parasitic capacitance, C_P, can be neglected.

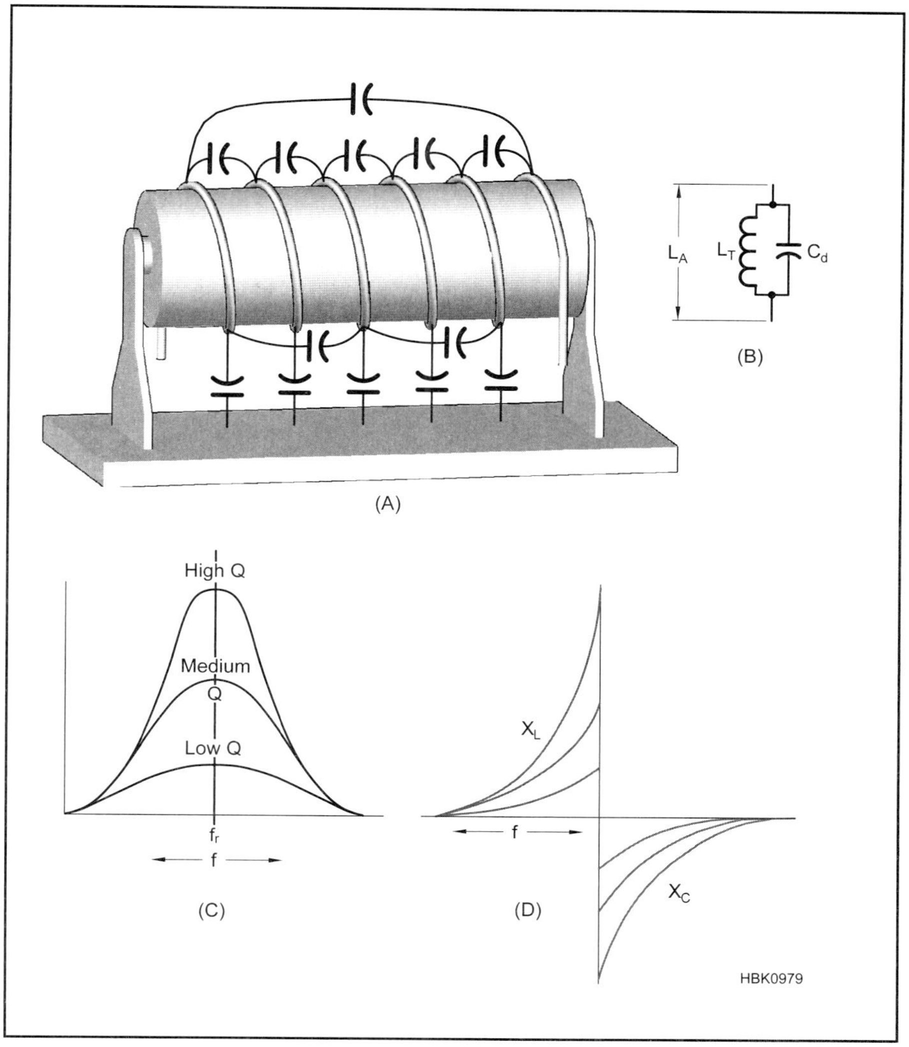

Figure 5.15 — Coils have distributed capacitance (A) between windings and from the coil to adjacent structures. This forms a parallel tuned circuit as shown in B with a resonant frequency, f_r. The coil's impedance and reactance vary with frequency as at C and D. Below resonance, the reactance is inductive and above resonance, the reactance is capacitive.

While the leakage conductance of a capacitor is usually negligible, the series resistance of an inductor often is not. This is caused by the long lengths of thin wire needed to create typical inductance values used in RF circuits and the skin effect (discussed below). Consider a typical air-core inductor, with L = 33 µH and a minimum Q of 30 measured at 2.5 MHz. This would indicate a series resistance of $R_s = 2\pi f L / Q = 17\ \Omega$ that could significantly alter the bandwidth of a circuit. The skin effect also results in parasitic resistance in the upper HF ranges and above. To minimize losses due to the skin effect, inductors at these frequencies, particularly those intended for use in transmitters and amplifiers and that are expected to carry significant currents, are often made from large-diameter wire or tubing or even flat strap to maximize the surface area for current flow.

For coils with many turns for which large conductors are impractical, *Litz wire* is sometimes used. Litz wire is made of many fine insulated strands woven together, each with a diameter smaller than the skin depth at the expected frequency of use, thus presenting a larger surface area than a solid wire or normal stranded wire. This creates multiple resistive inductors in parallel, which reduces the total impedance. The reduction is not proportional to the number of parallel paths, however, because they are inductively coupled.

If an inductor is wound on a magnetic core, the core itself can have losses that are treated as parasitic resistance. Each type of core — iron, powdered iron, or ferrite — has a frequency range over which it is designed for maximum efficiency. Outside of that range, the core may have significant losses, raising the parasitic resistance of the inductor.

Parasitic capacitance is a particular concern for inductors because of their construction. Consider the inductor in **Figure 5.15**. If this coil has n turns, then the ac voltage between identical points of two neighboring turns is 1/n times the ac voltage across the entire coil. When this voltage changes due to an ac current passing through the coil, the effect is that of many small capacitors acting in parallel with the inductance of the coil. Thus, in addition to the capacitance resulting from the leads, inductors have higher parasitic capacitance due to their physical shape.

These various effects illustrate why inductor construction has such a large effect on performance. As an example, assume you're working on a project that requires you to wind a 5 µH inductor. Looking at the coil inductance formula in the **Electrical Fundamentals** chapter, it comes to mind that many combinations of length and diameter could yield the desired inductance. If you happen to have both 0.5 and 1-inch coil forms, why should you select one over the other? To eliminate some other variables, let's make both coils 1 inch long, close-wound, and give them 1-inch leads on each end.

Let's calculate the number of turns required for each. On a 0.5-inch-diameter form:

$$n = \frac{\sqrt{L\,(18d + 40l)}}{d} =$$

$$\frac{\sqrt{5\,[(18 \times 0.5) + (40 \times 1)]}}{0.5} = 31.3 \text{ turns}$$

This means coil 1 will be made from #20 AWG wire (29.9 turns per inch). Coil 2, on the 1-inch form, yields

$$n = \frac{\sqrt{5\,[(18 \times 1) + (40 \times 1)]}}{1} = 17 \text{ turns}$$

which requires #15 AWG wire in order to be close-wound.

What are the series resistances associated

with each? For coil 1, the total wire length is 2 inches + (31.3 × π × 0.5) = 51 inches, which at 10.1 Ω/1,000 ft gives R_S = 0.043 Ω at dc. Coil 2 has a total wire length of 2 inches + (17.0 × π × 1) = 55 inches, which at 3.18 Ω/1,000 ft gives a dc resistance of R_S = 0.015 Ω, or about ⅓ that of coil 1. Furthermore, at RF, coil 1 will begin to suffer from skin effect at a frequency about 3 times lower than coil 2 because of its smaller conductor diameter. Therefore, if Q were the sole consideration, it would be better to use the larger diameter coil.

Q is not the only concern, however. Such coils are often placed in shielded enclosures. A common rule of thumb says that to prevent the enclosure from affecting the inductor, the enclosure should be at least one coil diameter from the coil on all sides. That is, 3×3×2 inches for the large coil and 1.5×1.5×1.5 for the small coil, a volume difference of over 500%.

RF CHOKES

An RF choke is an inductor intended to pass low-frequency ac and dc power while presenting a high impedance to RF. They are not intended for use in tuned circuits, filters, or other high-Q applications. (A detailed online article "How to Select R.F. Chokes" is available on the RF Café website, **www.rfcafe.com/references/electronics-world/select-rf-chokes-may-1966-electronics-world.htm**.)

Miniature and subminiature chokes are similar in size to resistors, both through-hole and SMT. They use a small ferrite core and very fine wire. Chokes with a metal shield around them to reduce coupling (see next section) are available. DC current ratings for these chokes are fairly low — typically tens of mA at most. Inductance values range from nanohenries to millihenries for use in low-power audio and RF circuits or for managing EMI.

Power supplies often use high-current chokes wound on ferrite rods to block common-mode RF from getting into or out of equipment. Inductance of 10 µH to several hundred µH and current ratings up to 10 A are typical, with wire windings using heavy solid enameled wires.

Medium-duty chokes of several hundred µH to several mH are often wound on ceramic forms in several sections to minimize distributed capacitance. Each section is made of multiple turns formed into a small disk. The disks look a bit like a pie so these chokes are called "pie-wound" (also "pi-wound"). These are common in older tube equipment and high-power RF circuits. A photo of this type of RF choke is shown in the **Electrical Fundamentals** chapter's section on inductors.

Heavy-duty transmitting chokes used in tube equipment are usually employed in the plate circuit to block RF from getting back into the high-voltage supply. Values of 100 µH to several hundred µH are typical with ratings of several amps for plate chokes used at HF. Filament chokes have to handle many amps of low-voltage ac with the same range of inductance values.

An RF choke performs best well below its self-resonant frequency (SRF) but it can be used throughout the range for which it has an acceptably high impedance. The choke's impedance will be inductive below the SRF and capacitive above the SRF (see Figure 5.15). If the choke is going to be used in a high-power transmitting circuit, resonances may produce very high voltages at one or more points along the coil. This can cause arcing and damage to the choke. These chokes are often specially wound to minimize the effect of resonances but this must be done to suit the exact application since distributed capacitance to the enclosure and other components affects the SRF.

Tom Rauch, W8JI, has written an informative paper about SRF and plate chokes which can be viewed at **www.w8ji.com/rf_plate_choke.htm**. His companion web page on inductors used in high-power RF circuits (**www. w8ji.com/loading_inductors.htm**) is also very informative.

Selecting the right type of choke is important to the success of the circuit. Like capacitors, there are many types of chokes with different characteristics. A reliable guide is to inspect well-engineered commercial equipment of the same type and frequency range to see what type of inductors are used. Handbooks and articles on equipment and choke design are also helpful, particularly for high-power applications. Manufacturer application notes and selection guides, such as the Coilcraft "RF Inductor Design Center" (**www.coilcraft.com/rfdc.cfm**) are also helpful for lower-power circuits.

When selecting the choke, your choke will need to be rated adequately for the dc current load with a low dc voltage drop. The SRF should be higher than the frequency of use or at least have a high impedance. Be sure the Q is high enough to avoid excessive RF power dissipation. Finally, any magnetic core (powdered iron or ferrite) should be intended for the frequency of use.

INDUCTOR COUPLING

Mutual inductance (see the **Electrical Fundamentals** chapter) will also have an effect on the resonant frequency and Q of RF circuits. For this reason, inductors in frequency-critical circuits should always be oriented and with sufficient spacing to minimize coupling. For example, mount coils near each other with their axes perpendicular as in **Figure 5.16**. The use of ferrous cores also tends to keep magnetic fields within the

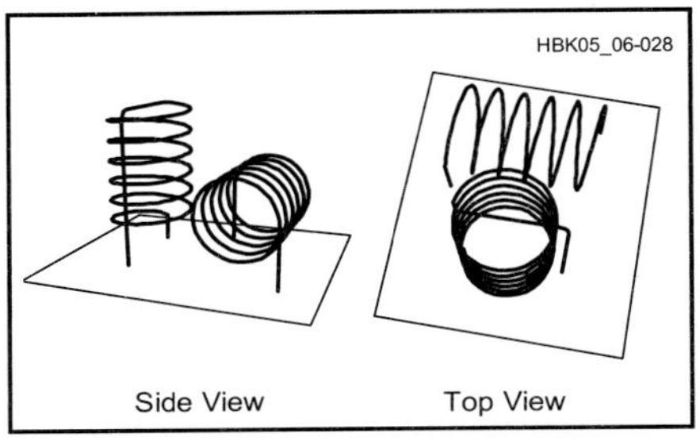

Figure 5.16 — Unshielded coils in close proximity should be mounted perpendicular to each other to minimize coupling.

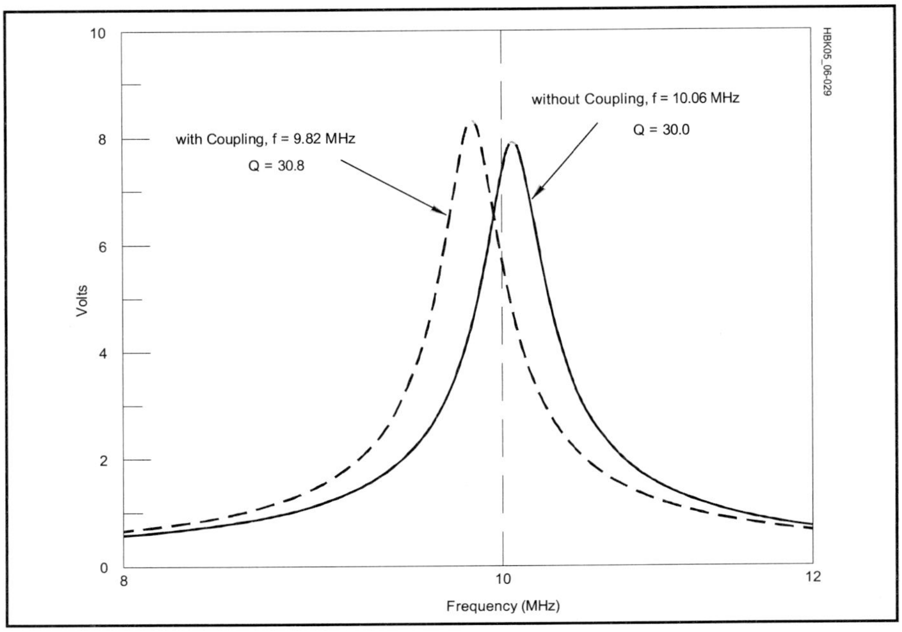

Figure 5.17 — Result of light coupling (k = 0.05) between two identical circuits of Figure 5.19A on their frequency responses.

core, reducing unwanted coupling.

As an example, assume we build an oscillator circuit that has both input and output resonant circuits. If we are careful to keep the two coils in these circuits uncoupled, the frequency response of either of the two circuits is shown by the solid line in **Figure 5.17**.

If the two coils are coupled either through careless placement or improper shielding, the resonant frequency and Q will be affected. The dashed line in Figure 5.17 shows the frequency response that results from a coupling coefficient of k = 0.05, a reasonable value for air-wound inductors mounted perpendicularly in close proximity on a circuit chassis. Note the resonant frequency shifted from 10.06 to 9.82 MHz, or 2.4%. The Q has gone up slightly from 30.0 to 30.8 as a result of the slightly higher inductive reactance at the resonant frequency.

Capacitance between coils can also affect a circuit's frequency response. Using shielded inductors or can-mounted coils reduces the effects of stray capacitance between components.

5.3.5 Skin Effect

The resistance of a conductor to ac is different than its value for dc because of the way ac fields interact with the conductor. As a result, thick, near-perfect conductors (such as metals) conduct ac only to a certain *skin depth*, δ, inversely proportional to the square root of the frequency of the current. Called the *skin effect*, this decreases the effective cross-section of the conductor at high frequencies and thus increases its resistance.

$$\delta = \frac{1}{\sqrt{\pi f \mu \sigma}} \qquad (2)$$

where

μ is the conductor's permeability, and

σ is the conducting material's conductance.

The increase in resistance caused by the skin effect is insignificant at and below audio frequencies, but beginning around 1 MHz (depending on the size of the conductor) it is so pronounced that practically all the current flows in a very thin layer near the conductor's surface. For example, at 10 MHz, the skin depth of a copper conductor is about 0.02 mm (0.00079 inch). For this reason, at RF a hollow tube and a solid rod of the same diameter and made of the same metal will have the same resistance and the RF resistance of a conductor is often much higher than its dc resistance. A rough estimate of the frequency above which a nonmagnetic wire (one made of nonferrous metal) will begin to show appreciable skin effect can be calculated from

$$f = \frac{124}{d^2} \qquad (3)$$

where

f = frequency, in MHz, and

d = diameter in mils (a mil is 0.001 inch).

Above this frequency, increase the resistance of the wire by 10× for every 2 decades of frequency (roughly 3.2× for every decade). For example, say we wish to find the RF resistance of a 2-inch length of #18 AWG copper wire at 100 MHz. From the wire tables in this book's online content, we see that this wire has a dc resistance of 2 inches × 6.386 Ω / 1,000 ft = 1.06 mΩ. From the above formula, the frequency is found to be 124 / 40.3^2 = 76 kHz. Since 100 MHz is roughly three decades above this (100 kHz to 100 MHz), the RF resistance will be approximately 1.06 mΩ × 3.2^3 = 1.06 mΩ × 32.8 = 34.8 mΩ. Again, values calculated in this manner are approximate and should be used qualitatively — like when you want an answer to a question such as, "Can I neglect the RF resistance of this length of connecting wire at 100 MHz?" Several useful charts regarding skin effect are available in *Reference Data for Engineers*, listed in the References section of this chapter.

Losses associated with skin effect can be reduced by increasing the surface area of the conductor carrying the RF current. Flat, solid strap and tubing are often used for that reason. In addition, because the current-carrying layer is so thin at UHF and microwave frequencies, a thin highly conductive layer at the surface of the conductor, such as silver plating, can lower resistance. Silver plating is too thin to improve HF conductivity significantly, however.

5.3.6 RF Heating

RF current often causes component heating problems where the same level of dc or low frequency ac current may not. These losses result from both the skin effect and from dielectric losses in insulating material, such as a capacitor dielectric.

An example is the tank circuit of an RF oscillator. If several small capacitors are connected in parallel to achieve a desired capacitance, skin effect will be reduced and the total surface area available for heat dissipation will be increased, thus significantly reducing the RF heating effects as compared to a single large capacitor. This technique can be applied to any similar situation; the general idea is to divide the heating among as many components as possible.

An example is shown in the circuit block in **Figure 5.18**, which is representative of the input tank circuit used in many HF VFOs. Along with L, C_{main}, C_{trim}, C1 and C3 set the oscillator frequency. Therefore, temperature effects are critical in these components. By using several capacitors in parallel, the RF current (and resultant heating, which is proportional to the square of the current) is reduced in each component. Parallel combinations are used for the feedback capacitors for the same reason.

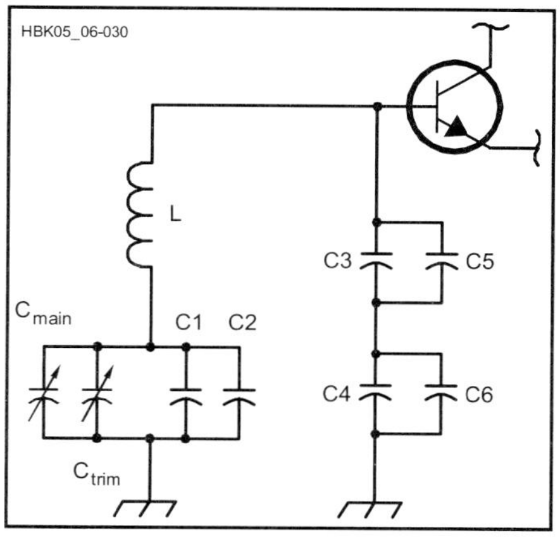

Figure 5.18 — A tank circuit of the type commonly used in VFOs. Several capacitors are used in parallel to distribute the RF current, which reduces temperature effects.

At high power levels, losses due to RF heating — either from resistive or dielectric losses — can be significant. Insulating materials may exhibit dielectric losses even though they are excellent insulators. For example, nylon plastic insulators may work very well at low frequencies, but being quite lossy at and above VHF, they are unsuitable for use in RF circuits at those frequencies. To determine whether material is suitable for use as an insulator at RF, a quick test can be made by heating the material in a microwave oven for a few seconds and measuring its temperature rise. (Take care to avoid placing parts or materials containing any metal in the microwave!) Insulating materials that heat up are lossy and thus unsuitable for use in RF circuits.

5.3.7 Effect on Q

Recall from the **Electrical Fundamentals** chapter that circuit Q, a useful figure of merit for tuned RLC circuits, can be defined in several ways:

$$Q = \frac{X_L \text{ or } X_C}{R}$$

$$= \frac{\text{energy stored per cycle}}{\text{energy dissipated per cycle}} \qquad (4)$$

Q is also related to the bandwidth of a tuned circuit's response by

$$Q = \frac{f_0}{BW_{3\,dB}} \qquad (5)$$

Parasitic inductance, capacitance and resistance can significantly alter the performance and characteristics of a tuned circuit if the design frequency is close to the self-resonant frequencies of the components.

As an example, consider the resonant circuit of **Figure 5.19A**, which could represent the frequency-determining resonant circuit of an oscillator. Neglecting any parasitics,

$$f_0 = \frac{1}{2\pi\sqrt{LC}} = 10.06 \text{ MHz}$$

As in many practical cases, assume the resistance arises entirely from the inductor series resistance. The data sheet for the inductor specified a minimum Q of 30, so assuming Q = 30 yields an R value of

$$\frac{X_L}{Q} = \frac{2\pi\,(10.06 \text{ MHz})\,(5\ \mu\text{H})}{30} = 10.5\ \Omega$$

Next, let's include the parasitic inductance of the capacitor (Figure 5.19B). A reasonable assumption is that this capacitor has the same physical size as the example from the section on Parasitic Inductance for which we calculated L_s = 0.106 µH. This would give the capacitor a self-resonant frequency of 434 MHz — well above our frequency of interest. However, the added parasitic inductance does account for an extra 0.106/5.0 = 2% inductance. Since this circuit is no longer strictly series or parallel, we must convert it to an equivalent form before calculating the new f_0.

An easier and faster way is to *simulate* the altered circuit by computer. This analysis was performed on a desktop computer using *SPICE*, a standard circuit simulation program described in the **Modeling and Design** chapter. The voltage response of the circuit (given an input current of 1 mA) was calculated as a function of frequency for both cases, with and without parasitics. The results are shown in the plot in Figure 5.19C, where we can see that the parasitic circuit has an f_0 of 9.96 MHz (a shift of 1%) and a Q (measured from the –3 dB points) of 31.5. For comparison, the simulation of the unaltered circuit does in fact show f_0 = 10.06 MHz and Q = 30.

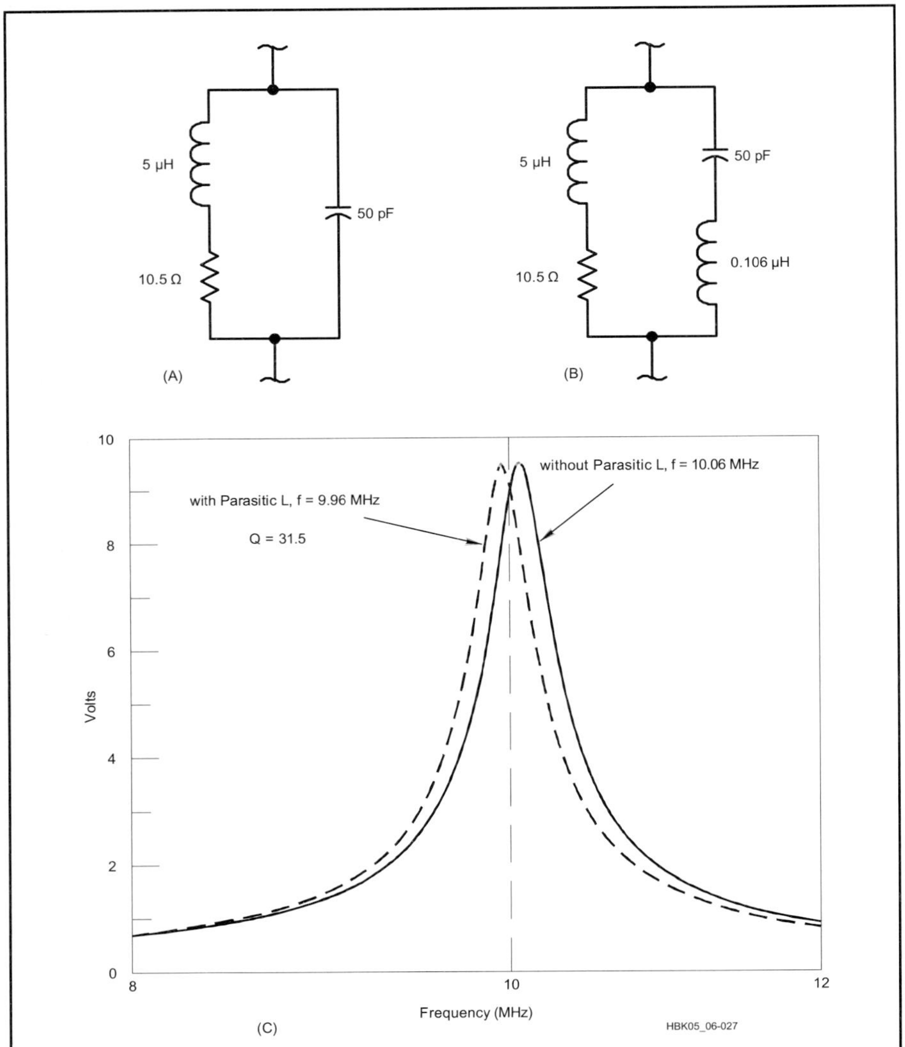

Figure 5.19 — A is a tank circuit, neglecting parasitics. B is the same circuit including L_p on capacitor. C is the frequency response curves for A and B. The solid line represents the unaltered circuit (also see Figure 5.17) while the dashed line shows the effects of adding parasitic inductance.

5.3.8 Dielectric Breakdown and Arcing

Anyone who has ever seen an arc form across a transmitting capacitor's plates, seen static discharges jump across an antenna insulator or touched a metal doorknob on a dry day has experienced the effects of dielectric breakdown.

In the ideal world, we could take any two conductors and put as large a voltage as we want across them, no matter how close together they are. In the real world, there is a voltage limit (*dielectric strength*, measured in kV/cm and determined by the insulating material between the two conductors) above which the insulator will break down.

Because they are charged particles, the electrons in the atoms of a dielectric material feel an attractive force when placed in an electric field. If the field is sufficiently strong, the force will strip the electron from the atom. This electron is available to conduct current, and furthermore, it is traveling at an extremely high velocity. It is very likely that this electron will hit another atom, and free another electron. Before long, there are many stripped electrons producing a large current, forming an *arc*. When this happens, we say the dielectric has suffered *breakdown*. Arcing in RF circuits is most common in transmitters and transmission line components where high voltages are common, but it is possible anywhere two components at significantly different voltage levels are closely spaced.

If the dielectric is liquid or gas, it will repair itself when the applied voltage is removed and the molecules in the dielectric return to their normal state. A solid dielectric, however, cannot repair itself because its molecules are fixed in place and the low-resistance path created by the arc is permanent. A good example of this is a CMOS integrated circuit. When exposed to the very high voltages associated with static electricity, the electric field across the very thin gate oxide layer exceeds the dielectric strength of silicon dioxide, and the device is permanently damaged by the resulting hole created in the oxide layer.

The breakdown voltage of a dielectric layer depends on its composition and thickness (see **Table 5.1**). The variation with thickness is not linear; doubling the thickness does not quite double the breakdown voltage. Breakdown voltage is also a function of geometry: Because of electromagnetic considerations, the breakdown voltage between two conduc-

Table 5.1

Dielectric Constants and Breakdown Voltages

Material	*Dielectric Constant**	*Puncture Voltage***
Aisimag 196	5.7	240
Bakelite	4.4-5.4	240
Bakelite, mica filled	4.7	325-375
Cellulose acetate	3.3-3.9	250-600
Fiber	5-7.5	150-180
Formica	4.6-4.9	450
Glass, window	7.6-8	200-250
Glass, Pyrex	4.8	335
Mica, ruby	5.4	3800-5600
Mycalex	7.4	250
Paper, Royalgrey	3.0	200
Plexiglas	2.8	990
Polyethylene	2.3	1200
Polystyrene	2.6	500-700
Porcelain	5.1-5.9	40-100
Quartz, fused	3.8	1000
Steatite, low loss	5.8	150-315
Teflon	2.1	1000-2000

*At 1 MHz
**In volts per mil (0.001 in.)

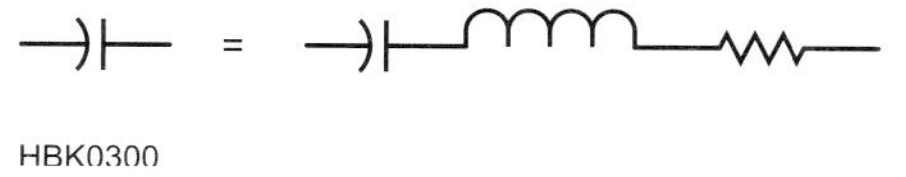

Figure 5.20 — Model for a bypass capacitor.

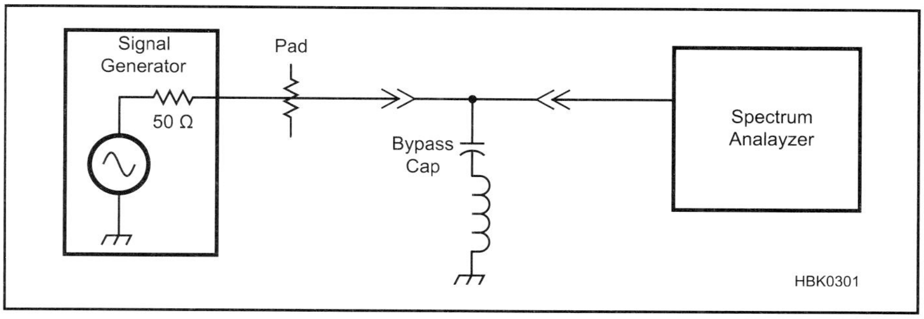

Figure 5.21 — Test set for home lab measurement of a bypass capacitor.

tors separated by a fixed distance is less if the surfaces are pointed or sharp-edged than if they are smooth or rounded. Therefore, a simple way to help prevent breakdown in many projects is to file and smooth the edges of conductors. (See the **Power Sources** and **RF Power Amplifier** chapters for additional information on high voltage applications.)

Capacitors are, by nature, the component most often associated with dielectric failure. To prevent damage, the working voltage of a capacitor — and there are separate dc and ac ratings — should ideally be two or three times the expected maximum voltage in the circuit. Capacitors that are not air-insulated or have a *self-healing dielectric* should be replaced if a dielectric breakdown occurs.

Resistors and inductors also have voltage ratings associated with breakdown of their insulating coating. High-value resistors, in particular, can be bypassed by leakage current flowing along the surface of the resistor. High-voltage resistors often have elongated bodies to create a long *leakage path* to present a high resistance to leakage current. Cleaning the bodies of resistors and inductors in high-voltage circuits helps prevent arcs from forming and minimizes leakage current.

5.3.9 Radiative Losses

Any conductor placed in an electromagnetic field will have a current induced in it. We put this principle to good use when we make an antenna. The unwelcome side of this law of nature is the phrase "any conductor;" even conductors we don't intend to act as antennas will respond this way.

Fortunately, the efficiency of such "antennas" varies with conductor length. They will be of importance only if their length is a significant fraction of a wavelength. When we make an antenna, we usually choose a length on the order of λ/2. Therefore, when we *don't* want an antenna, we should be sure that the conductor length is *much less* than λ/2, no more than 0.1 λ. This will ensure a very low-efficiency antenna. This is why even unpaired 60-Hz power lines do not lose a significant fraction of the power they carry — at 60 Hz, 0.1 λ is about 300 miles!

In addition, we can use shielded cables. Such cables do allow some penetration of EM fields if the shield is not solid, but even 95% coverage is usually sufficient, especially if some sort of RF choke is used to reduce shield current.

Radiative losses and coupling can also be reduced by using twisted or parallel pairs of conductors — the fields tend to cancel. In some applications, such as audio cables, this may work better than shielding. Critical stages such as tuned circuits should be placed in shielded compartments where possible.

This argument also applies to large components — remember that a component or long wire can both radiate and receive RF energy. Measures that reduce radiative losses will also reduce unwanted RF pickup. See the **RFI and EMC** chapter for more information.

5.3.10 Bypassing and Decoupling

BYPASSING

Circuit models showing ac behavior often show ground connections that are not at dc ground. Bias voltages and currents are neglected in order to show how the circuit responds to ac signals. Rather, those points are "signal grounded" through bypass capacitors. Obtaining an effective bypass can be difficult and is often the route to design difficulty. The problem is parasitic inductance. Although we label and model parts as "capacitors," a more complete model is needed. The better model is a series RLC circuit, shown in **Figure 5.20**. Capacitance is close to the marked value while inductance is a small value that grows with component lead length. Resistance is a loss term, usually controlled by the Q of the parasitic inductor. Even a leadless SMT (surface-mount technology) component will display inductance commensurate with its dimensions. As shown in the Parasitic Inductance section, wire component leads have an inductance of about 1 nH per mm of length (20 – 25 nH per inch).

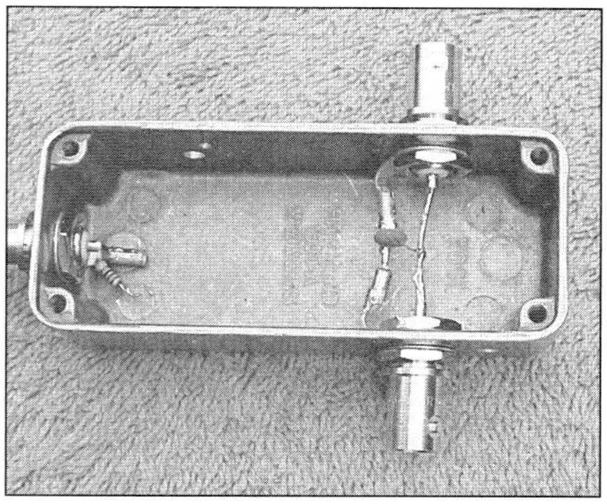
Figure 5.22 — Test fixture for measuring self-resonant frequency of capacitors.

Bypass capacitor characteristics can be measured in the home lab with the test setup of **Figure 5.21**. **Figure 5.22** shows a test fixture with an installed 470 pF leaded capacitor. The fixture is used with a signal generator and spectrum analyzer to evaluate capacitors. Relatively long capacitor leads were required to interface to the BNC connectors, even though the capacitor itself was small. The signal generator was tuned over its range while examin-

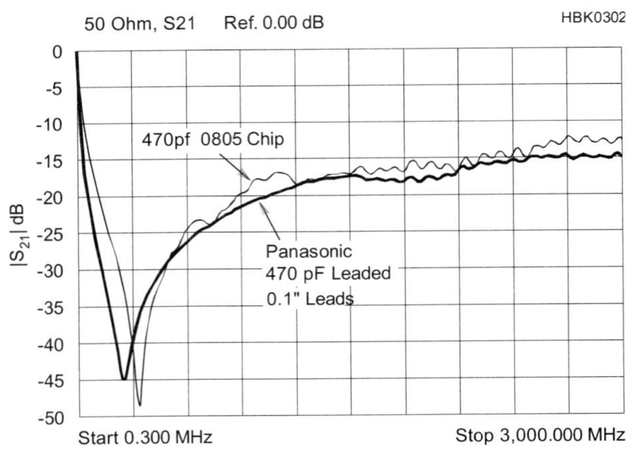

Figure 5.23 — Network analyzer measurement of 470-pF shunt capacitors. Both SMT and leaded parts are studied.

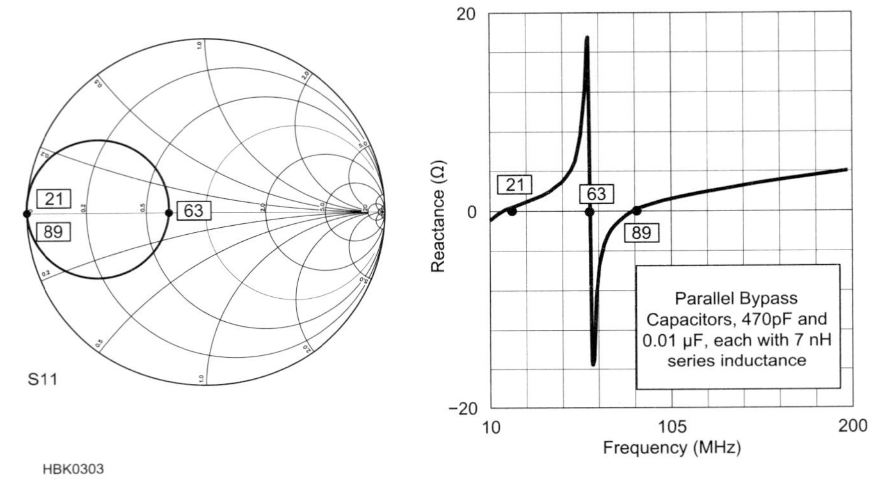

Figure 5.24 — The classic technique of paralleling bypass capacitors of two values, here 470 pF and 0.01 μF. This is a *terrible* bypass! See text.

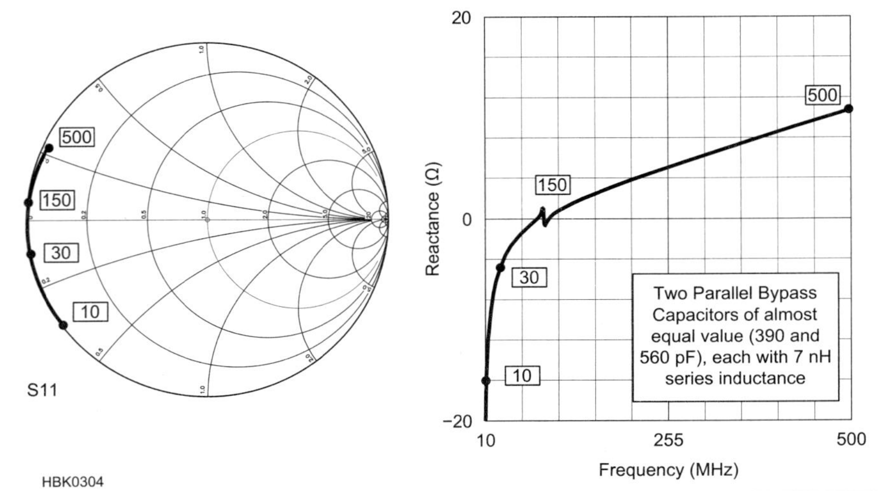

Figure 5.25 — Paralleling bypass capacitors of nearly the same value. This results in improved bypassing without complicating resonances.

ing the spectrum analyzer response, showing a minimum at the series resonant frequency. Parasitic inductance is calculated from this frequency. The C value was measured with a low-frequency LC meter. Additional test instruments and techniques are discussed in the **Test Equipment and Measurement** chapter.

The measured 470 pF capacitor is modeled as 485 pF in series with an inductance of 7.7 nH. The L is larger than we would see with shorter leads. A 0.25-inch 470 pF ceramic disk capacitor with zero lead length will show a typical inductance closer to 3 nH. The measured capacitor Q was 28 at its self-resonance of 82 MHz but is higher at lower frequency. Data from a similar measurement, but with a network analyzer is shown in **Figure 5.23**. Two 470 pF capacitors are measured, one surface mounted and the other a leaded part with 0.1-inch leads.

It is often suggested that the bandwidth for bypassing can be extended by paralleling a capacitor that works well at one frequency with another to accommodate a different part of the spectrum. Hence, paralleling the 470 pF with a 0.01 μF capacitor should extend the bypassing to lower frequencies. The calculations are shown in the plots of **Figure 5.24**. The results are terrible! While the low frequency bypassing is indeed improved, a high impedance response is created at 63 MHz. This complicated behavior is again the result of inductance. Each capacitor was assumed to have a series inductance of 7 nH. A parallel resonance is approximately formed between the L of the larger capacitor and the C of the smaller. The Smith Chart plot shows us that the impedance is nearly 50 Ω at 63 MHz. Impedance would be even higher with greater capacitor Q.

Bypassing can be improved by paralleling capacitors. However, the capacitors should be approximately identical. **Figure 5.25** shows the result of paralleling two capacitors of about the same value. They differ slightly at 390 and 560 pF, creating a hint of resonance. This appears as a small perturbation in the reactance plot and a tiny loop on the Smith Chart. These anomalies disappear as the C values become equal. Generally, paralleling is the scheme that produces the best bypassing. The ideal solution is to place an SMT capacitor on each side of a printed circuit trace or wire at a point that is to be bypassed.

Matched capacitor pairs form an effective bypass over a reasonable frequency range. Two 0.01 μF disk capacitors have a reactance magnitude less than 5 Ω from 2 to 265 MHz. A pair of the 0.1 μF capacitors was even better, producing the same bypassing impedance from 0.2 to 318 MHz. The 0.1 μF capacitors are chip-style components with attached wire leads. Even better results can be obtained with multilayer ceramic chip capacitors. Construction with multiple layers creates an integrated paralleling.

Some applications (for example, IF amplifiers) require effective bypassing at even lower frequencies. Modern tantalum electrolytic capacitors are surprisingly effective through the RF spectrum while offering high enough C to be useful at audio. In critical applications, however, the parts should be tested to be sure of their effectiveness.

DECOUPLING

The bypass capacitor usually serves a dual role, first creating the low impedance needed to generate a "signal" ground. It also becomes part of a decoupling low-pass filter that passes dc while attenuating signals. The attenuation must function in both directions, suppressing noise in the power supply that might reach an amplifier while keeping amplifier signals from reaching the power supply. A low-pass filter is formed with alternating series and parallel component connections. A parallel bypass is followed by a series impedance, ideally a resistor.

Additional shunt elements can then be added, although this must be done with care. An inductor between shunt capacitors should have high inductance. It will resonate with the shunt capacitors to create high impedances just like those that came from parasitic inductance in the bypasses. This makes it desirable to have an inductance that is high enough that any resonance is below the band of interest. But series inductors have their own problems; they have parasitic capacitance, creating their own self-resonance. As an example, a pair of typical RF chokes (RFCs) were measured (now as series elements) as described earlier. A 2.7 µH molded choke was parallel resonant at 200 MHz, indicating a parallel capacitance of 0.24 pF. The Q at 20 MHz was 52. A 15 µH molded choke was parallel resonant at 47 MHz, yielding a parallel C of 0.79 pF. This part had a Q of 44 at 8 MHz.

Large inductors can be fabricated from series connections of smaller ones. The best wideband performance will result only when all inductors in a chain have about the same value. The reasons for this (and the mathematics that describe the behavior) are identical with those for paralleling identical capacitors.

Low inductor Q is often useful in decoupling applications, encouraging us to use inductors with ferrite cores. Inductors using the Fair-Rite Type #43 material have low Q in the 4 to 10 range over the HF spectrum. One can also create low-Q circuits by paralleling a series inductor of modest Q with a resistor.

Figure 5.26 shows a decoupling network and the resulting impedance when viewed from the "bypass" end. The 15 µH RFC resonates with a 0.1 µF capacitor to destroy the bypass effect just above 0.1 MHz. A low-value parallel resistor fixes the problem.

A major reason for careful wideband bypassing and decoupling is the potential for amplifier oscillation. Instability that allows oscillations is usually suppressed by low impedance terminations. The base and collector (or gate and drain) should both "see" low impedances to ensure stability. But that must be true at all frequencies where the device can produce gain. It is never enough to merely consider the operating frequency for the amplifier. A parallel resonance in the base or collector circuits can be a disaster. When wideband bypassing is not possible, negative feedback that enhances wideband stability is often used.

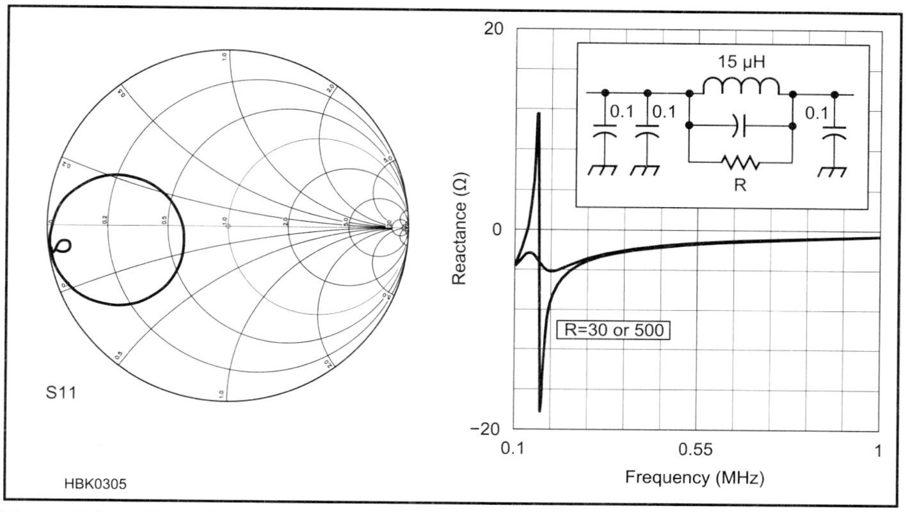

Figure 5.26 — Two different resistor values parallel a decoupling choke. The lower, 30 Ω value is more effective. See text.

Emitter bypassing is often a critical application. As demonstrated in the **Circuits and Components** chapter, a few extra ohms of impedance in the emitter circuit can drastically alter amplifier performance. A parallel-resonant emitter bypass could be a profound difficulty while a series-resonant one can be especially effective.

Capacitors also appear in circuits as blocking elements. A blocking capacitor, for example, appears between stages, creating a near short circuit for ac signals while accommodating different dc voltages on the two sides. A blocking capacitor is not as critical as a bypass, for the impedances on either side will usually be higher than that of the block.

With parasitic effects having the potential to strongly affect circuit performance, the circuit designer must account for them wherever they are significant. The additional complexity of including parasitic elements can result in a design too complex to be analyzed manually. Clearly, detailed modeling is the answer to component selection and the control of parasitic effects.

5.3.11 Effects on Filter Performance

LC bandpass filters perform a critical function in determining the performance of a typical RF system such as a receiver. An input filter, usually a bandpass, restricts the

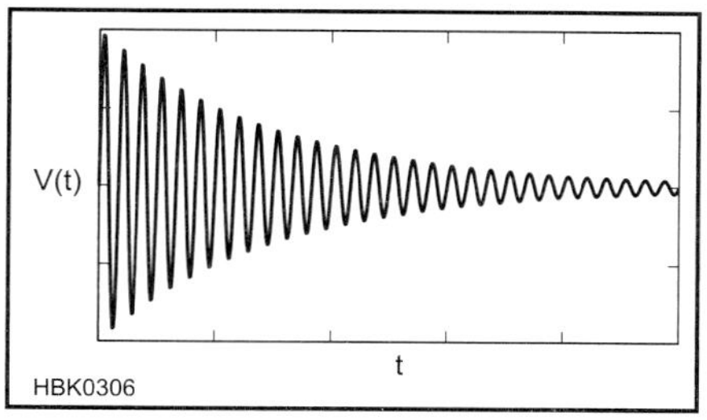

Figure 5.27 — The amplitude of a chime's ring after being struck by a hammer. Units are arbitrary.

frequency range that the receiver must process. Transmitters use LC filters to reduce harmonic output. Audio and LF filters may also use LC elements. The LC filters we refer to in this section are narrow with a bandwidth from 1 to 20% of the center frequency. Even narrower filters are built with resonators having higher Q; the quartz crystal filter discussed in the **Analog and Digital Filtering** chapter is used where bandwidths of less than a part per thousand are possible. The basic concepts that we examine with LC circuits will transfer to the crystal filter.

LOSSES IN FILTERS AND Q

The key elements in narrow filters are tuned circuits made from inductor-capacitor pairs, quartz crystals, or transmission line sections. These resonators share the property that they store energy, but they have losses. A chime is an example. Striking the chime with a hammer produces the waveform of **Figure 5.27**. The rate at which the amplitude decreases with time after the hammer strike is determined by the filter's Q, which is discussed in the chapter on **Circuits and Components**. The higher the Q, the longer it takes for the sound to disappear. The oscillator amplitude would not decrease if it were not for the losses that expend energy stored in the resonator. The mere act of observing the oscillation will cause some energy to be dissipated.

A chime is an acoustic resonator, but the same behavior occurs in electric resonators. A pulse input to an LC circuit causes it to ring; losses cause the amplitude to diminish. The most obvious loss in an LC circuit is conductor resistance, including that in the inductor wire. This resistance is higher than the dc value owing to the skin effect, which forces high frequency current toward the conductor surface. Other losses might result from hysteresis losses in an inductor core or dielectric losses in a capacitor.

An inductor is modeled as an ideal part with a series or a parallel resistance. The resistance will depend on the Q if the inductor was part of a resonator with that quality. The two resistances are shown in **Figure 5.28**.

$$R_{Series} = \frac{2\pi f L}{Q} \quad \text{and}$$

$$R_{Parallel} = Q \times 2\pi f L \qquad (6)$$

The higher the inductor Q, the smaller the series resistance, or the larger the parallel resistance is needed to model that Q. It does not matter which configuration is used. The Q of a resonator is related to the bandwidth of the tuned circuit by

$$BW = \frac{f_C}{Q} \qquad (7)$$

where f_C is the tuned circuit's center frequency. This is also the Q of an inductor in a tuned circuit if the capacitor is lossless.

The single-tuned circuit is presented in two different forms in **Figure 5.29**. In the top, a parallel tuned circuit consisting of L and C has loss modeled by three resistors. The resistor R_p is the parallel loss resistance representing the non-ideal nature of the inductor. (Another might be included to represent capacitor losses.) But the LC is here paralleled by three resistors: the source, the load, and the loss element. R_p would disappear if the tuned circuit was built from perfect components. The source and load remain; they represent the source resistance that must be present

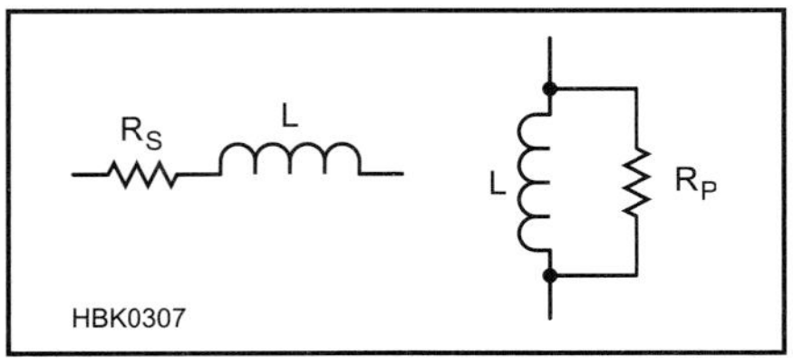

Figure 5.28 — Inductor Q may be modeled with either a series or a parallel resistance.

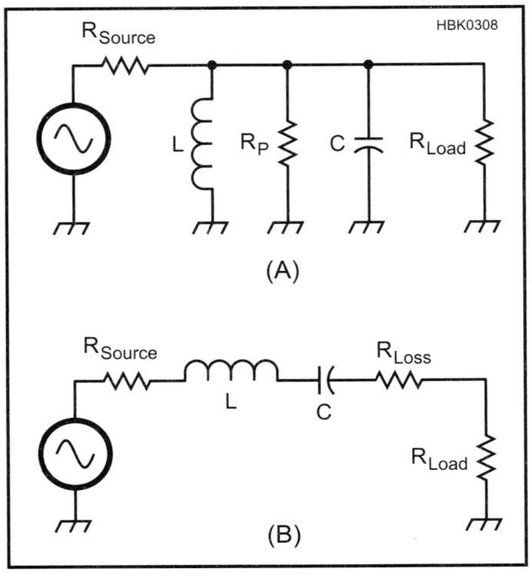

Figure 5.29 — Two simple forms of the single-tuned circuit.

if power is available and a load resistance that must be included if power is to be extracted.

Equation 6 and 7 can be applied in several ways. If the resonator is evaluated with only its intrinsic loss resistance (in either series or parallel form) the resulting Q is called the *unloaded Q*, or Q_u. If, however, the net resistance is used, which is the parallel combination of the load, the source, and the loss in the parallel tuned circuit, the resulting Q is called the *loaded Q*, Q_L. If we were working with the series tuned circuit form, the loaded Q would be related to the total series R.

Consider an example, a parallel tuned circuit (Figure 5.29A) with a 2 µH inductor tuned to 5 MHz with a 507-pF capacitor. Assume the parallel loss resistor was 12.57 kΩ. The unloaded Q calculated from equation 6 is 200. The unloaded bandwidth would be 5 MHz / 200 = 25 kHz. Assume that the source and load resistors were equal, each 2 kΩ. The net resistance paralleling the LC would then be the combination of the three resistors, 926 Ω. The loaded Q becomes 14.7 with a loaded bandwidth of 339 kHz. The loaded Q is also the filter Q, for it describes the bandwidth of the single-tuned circuit, the simplest of band-pass filters.

This filter also has *insertion loss* (IL). This is illustrated in **Figure 5.30** which shows the parallel-resonant LC combination removed (at resonance, the LC combination has infinite impedance), leaving only the loss resistance of 12.57 kΩ. We use an arbitrary open circuit source voltage of 2. The available power to a load is then 1 V across a resistance equaling the 2 kΩ source. If the resonator had no internal losses, this available power would be delivered to the 2 kΩ load. However, the loss R parallels the load, causing the output voltage to be 0.926 V, a bit less than the ideal 1 V. Calculation of the output power into the 2 kΩ load resistance and the available power shows that the insertion loss is 0.67 dB.

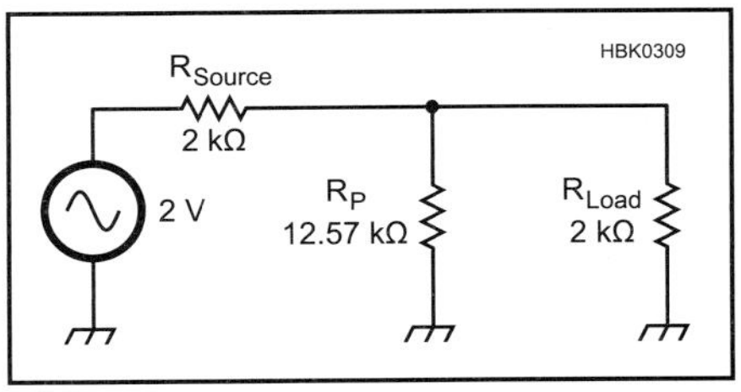

Figure 5.30 — Simplified parallel tuned circuit at resonance. The effect of loss is illustrated by removing the parallel-LC combination which at resonance has infinite impedance.

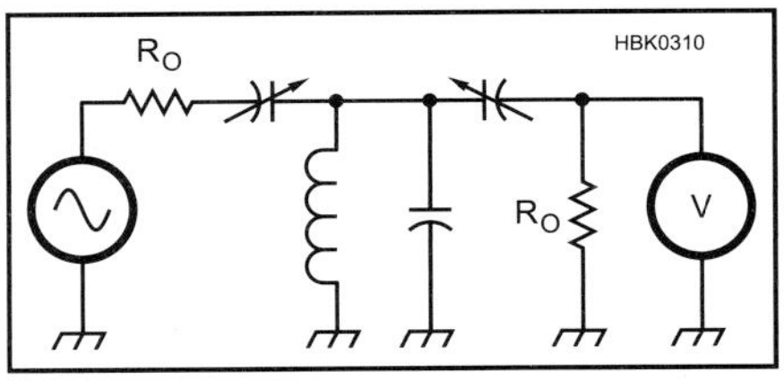

Figure 5.31 — Test setup for measuring the Q of a resonator. The source and load impedances, R_O, are assumed to be identical. The two coupling capacitors are adjusted to be equal to each other. The output signal is measured with an appropriate ac voltmeter, an oscilloscope or a spectrum analyzer.

This exercise illustrates two vital points that are general for all bandpass filters. First, the bandwidth of any filter must always be larger than the unloaded bandwidth of the resonators used to build the filter. Second, any filter built from real-world components will have an insertion loss. The closer the Q of the filter approaches the unloaded resonator Q, the greater the insertion loss becomes. A parallel tuned circuit illustrated these ideas; the series tuned filter would have produced identical results. Generally, the insertion loss of a single-tuned circuit relates to loaded and unloaded Q by

$$IL(dB) = -20 \log \left(1 - \frac{Q_L}{Q_U} \right) \qquad (8)$$

The Q of a tuned LC circuit is easily measured with a signal generator of known output impedance (usually 50 Ω) and a sensitive detector, again with a known impedance level, often equal to that of the generator. The test configuration is shown in **Figure 5.31**. It uses equal loads and equal capacitors to couple from the terminations to the resonator. Equal capacitors, C1 and C2 guarantees that each termination contributes equally to the resonator parallel load resistance. The voltmeter across the load is calibrated in dB.

To begin measurement we remove the tuned circuit and replace it with a direct connection from generator to load. The available power delivered to the load is calculated after the voltage is measured. The resonator is then inserted between the generator and load, and the generator is tuned for a peak. The measured power is less than that available from the source, with the difference being the insertion loss for the simple filter. Capacitors C1 and C2 are adjusted until the loss is 30 dB or more. With loss this high, the intrinsic loss resistance of the resonator will dominate the loss. The generator is now tuned first to one side of the peak, and then to the other, noting the frequencies where the response is lower than the peak by 3 dB. The unloaded bandwidth, ΔF, is the difference between the two 3 dB frequencies. The unloaded Q is calculated as

$$Q_U = \frac{F}{\Delta F} \tag{9}$$

This method for Q measurement is quite universal, being effective for audio tuned circuits, simple LC RF circuits, VHF helical resonators, or microwave resonators. The form of the variable capacitors, C1 and C2, may be different for the various parts of the spectrum, but the concepts are general. Indeed, it is not even important how the coupling occurs. The Q measurement normally determines an unloaded value, but loaded values are also of interest when testing filters.

5.4 Semiconductor Circuits at RF

The models used in the **Circuits and Components** chapter are reasonably accurate at low frequencies, and they are of some use at RF, but more sophisticated models are required for consistent results at higher frequencies. This section notes several areas in which the simple models must be enhanced. Circuit design using models accurate at RF, particularly for large signals, is performed today using design software as described in the **Computer Modeling and Simulation** chapter of this book. In-depth discussions of the elements of RF circuit design appear in Hayward's *Introduction to Radio Frequency Design*, an excellent text for the beginning RF designer (see the References section). See also the section of RF Amplifier Design papers and application notes in the References section of this chapter.

5.4.1 The Diode at High Frequencies

A DIODE AC MODEL

At high frequencies, the diode's behavior becomes less like a switch, especially for small signals that do not cause the diode's operating point to move by large amounts. In this case, the diode's small-signal dynamic behavior becomes important.

Figure 5.32 shows a simple resistor-diode circuit to which is applied a dc bias voltage plus an ac signal. Assuming that the voltage drop across the diode is 0.6 V and R_f is negligible, we can calculate the bias current to be I = (5 – 0.6) / 1 kΩ = 4.4 mA. This point is marked on the diodes I-V curve in Figure 5.32. If we draw a line tangent to this point, as shown, the slope of this line represents the *dynamic resistance* of the diode, R_d, experienced by a small ac signal. At room temperature, dynamic resistance is approximately

$$R_d = \frac{25}{I}\Omega \tag{10}$$

where I is the diode current in mA. Note that this resistance changes with bias current and should not be confused with the dc forward resistance discussed in the **Circuits and Components** chapter, which has a similar value but represents a different concept. **Figure 5.33** shows a low-frequency ac model for the diode, including the dynamic resistance and junction capacitance. This model should only be used when the diode's dc operating parameters can be neglected.

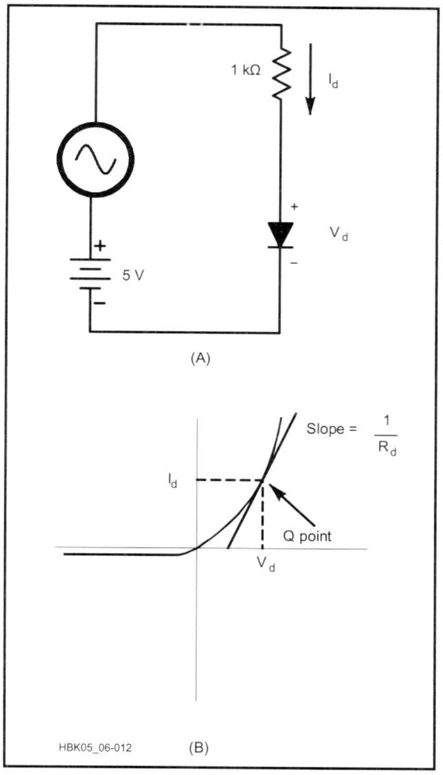

Figure 5.32 — A simple resistor-diode circuit used to illustrate dynamic resistance. The ac input voltage "sees" a diode resistance whose value is the slope of the line at the Q-point, shown in B.

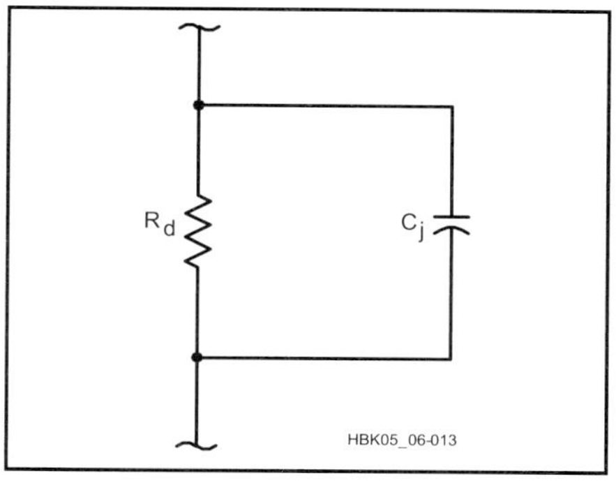

Figure 5.33 — An ac model for diodes. R_d is the dynamic resistance and C_j is the junction capacitance.

SWITCHING TIME

If you change the polarity of a signal applied to the ideal diode, current flow stops or starts instantaneously. Current in a real diode cannot do this, as a finite amount of time is required to move electrons and holes in or out of the diode as it changes states. Effectively, the diode junction capacitance, C_J, in Figure 5.33 must be charged or discharged. As a result, diodes have a maximum useful frequency when used in switching applications.

The operation of diode switching circuits can often be modeled by the circuit in **Figure 5.34**. The approximate switching time (in seconds) for this circuit is given by

$$t_s = \tau_p \frac{\left(\frac{V_1}{R_1}\right)}{\left(\frac{V_2}{R_2}\right)} = \tau_p \frac{I_1}{I_2} \tag{11}$$

where τ_p is the minority carrier lifetime of the diode, a material constant determined during manufacture, on the order of 1 ms. I_1 and I_2 are currents that flow during the switching process. The minimum time in which a diode can switch from one state to the other and back again is therefore $2t_s$, and thus the maximum usable switching frequency is f_{sw} (Hz) = 1 / ($2t_s$). It is usually a good idea to stay below this by a factor of two. Diode data sheets usually give typical switching times and show the circuit used to measure them.

Note that f_{sw} depends on the forward and reverse currents, determined by I_1 and I_2 (or equivalently V_1, V_2, R_1, and R_2). Within a reasonable range, the switching time can be reduced by manipulating these currents. Of course, the maximum power that other circuit elements can handle places an upper limit on switching currents.

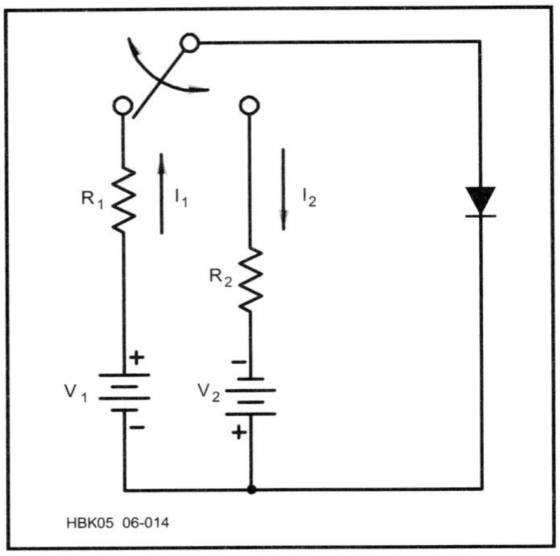

Figure 5.34 — Circuit used for computation of diode switching time.

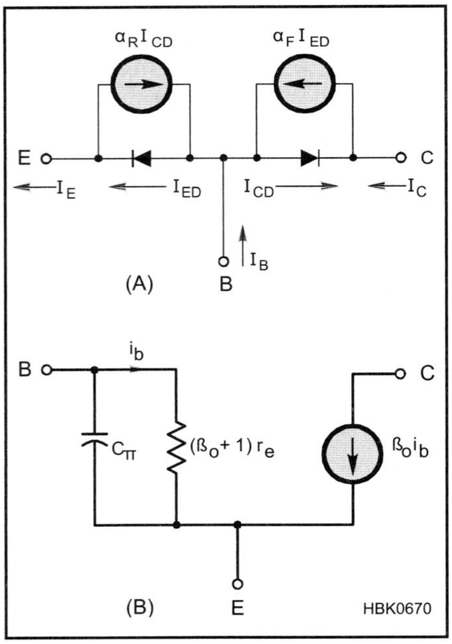

Figure 5.35 — (A) The Ebers-Moll model of the bipolar junction transistor (BJT) is used at dc and low frequencies when the transistor is in its active region. (B) The hybrid-pi model includes frequency dependence.

5.4.2 The Transistor at High Frequencies

The development and selection of suitable models for active devices at RF is an involved and nuanced process. This section presents a few of the models used for bipolar junction transistors (BJT) and field-effect transistors (FET) as examples of the issues involved. For more information on model details, start with the *SPICE* home page at **www.web.archive.org/web/20231208045915/http://bwrcs.eecs.berkeley.edu/Classes/IcBook/SPICE/** and the user's manual for the simulation software you intend to use.

GAIN VERSUS FREQUENCY

The circuit design equations in **Circuits and Components** generally assumed that current gain in the bipolar transistor was independent of frequency. As signal frequency increases, however, current gain decreases. The low-frequency current gain, β_0, is constant through the audio spectrum, but it eventually decreases, and at some high frequency it will drop by a factor of 2 for each doubling of signal frequency. (The h-parameter [h_{FE}] is often substituted for β at dc and h_{FE} for ac current gain. H-parameters are discussed later in this chapter.) A transistor's frequency vs. current gain relationship is specified by its *gain-bandwidth product* (GBW), or F_T, the frequency at which the current gain is 1. Common transistors for lower RF applications might have $\beta_0 = 100$ and $F_T = 500$ MHz. The frequency at which current gain equals β_0 is called F_b and is related to F_T by $F_b = F_T / \beta_0$.

The *Ebers-Moll model* for the bipolar transistor, shown in **Figure 5.35A**, is a "large signal model." It is used when the transistor is in its active mode and gives good results for dc collector and emitter currents. This would be a good model to use when designing a pass-transistor voltage regulator circuit, for example.

The frequency dependence of current gain is modeled by adding a capacitor across the base resistor of Figure 5.35B, the *hybrid-pi* model results. The capacitor's reactance should equal the low-frequency input resistance, $(\beta + 1)\ r_e$, at F_b. This simulates a

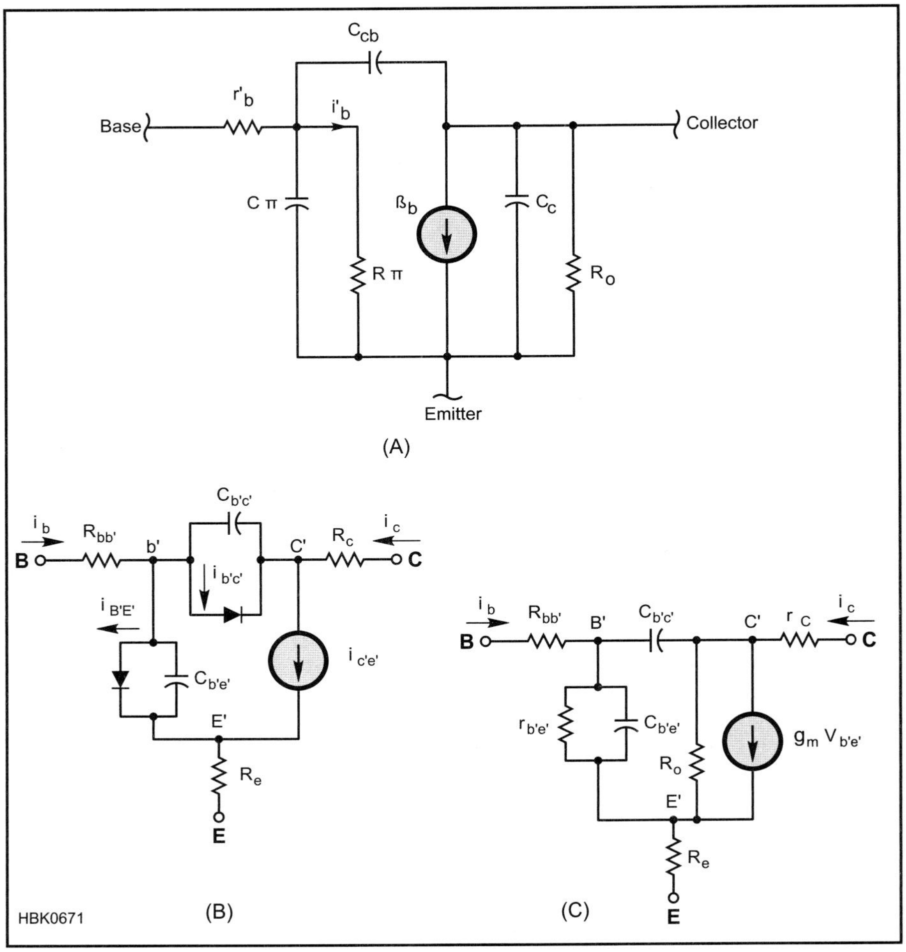

Figure 5.36 — High-frequency models for the bipolar junction transistor. (A) An improved version of the hybrid-pi model including additional device capacitances. (B) is the large-signal Gummel-Poon model used at dc and low-frequencies — the standard used for bipolar transistors by simulators based on *SPICE3* and (C) is the small-signal ac version.

frequency-dependent current gain.

Similar considerations affect the transconductance of FETs as the gate-to-source and gate-to-drain capacitances act to reduce high-frequency gain. Instead of F_T, most FETs designed for amplifier use specify input and output transconductance, susceptance, and power gain at different frequencies.

Recognizing that transistor gain varies with frequency, it is important to know the conditions under which the simpler model is useful. Calculations show that the simple model is valid, with $\beta = F_T / F$, for frequencies well above F_b. The approximation worsens, however, as the operating frequency (f_0) approaches F_T.

SMALL-SIGNAL DESIGN AT RF

The simplified hybrid-pi model of Figure 5.35B is often suitable for noncritical designs, but **Figure 5.36A** shows a better small-signal model for RF design that expands on the hybrid-pi. Consider the physical aspects of a real transistor: There is some capacitance across each of the PN junctions (C_{cb} and C_p) and capacitance from collector to emitter (C_c). There are also capacitances between the device leads (C_e, C_b and C_c). There is a resistance in each current path, emitter to base and collector. From emitter to base, there is r_π from the hybrid-pi model and r´b, the "base spreading" resistance. From emitter to collector is R_o, the output resistance. The leads that attach the silicon die to the external circuit present three inductances.

Manual circuit analysis with this model is best tackled with the aid of a computer and specialized software. Other methods are presented in Hayward's *Introduction to Radio Frequency Design*. Surprisingly accurate results may be obtained, even at RF, from the simple models. Simple models also give a better "feel" for device characteristics that might be obscured by the mathematics of a more rigorous treatment. Use the simplest model that describes the important features of the device and circuit at hand.

The *Gummel-Poon model* shown in Figures 5.36B and 5.36C is the standard used by simulation software based on *SPICE3*. By adding additional parasitic elements, such as lead inductance and lead-to-package capacitances, the small-signal ac model can be used accurately at high frequencies. Note how the ac model becomes progressively more sophisticated from Figure 5.35A through 5.36C.

HIGH FREQUENCY FET MODELS

At low frequencies the FET can be treated as a simple current source controlled by the gate-to-source voltage as in Figure 3.49 in the **Circuits and Components** chapter. As frequency increases, the interelectrode capacitances become significant and must be included, such as in the model in **Figure 5.37A**. As with the BJT, separate models are used for small-signal and large-signal applications. Figure 5.37B shows a small-signal *SPICE3* simulation model for the JFET. All of the model's capacitances vary with temperature and device operating characteristics. Figure 5.37C shows a typical high-frequency MOSFET model used by *SPICE3* (there are several) that includes the effect of the body or substrate elements.

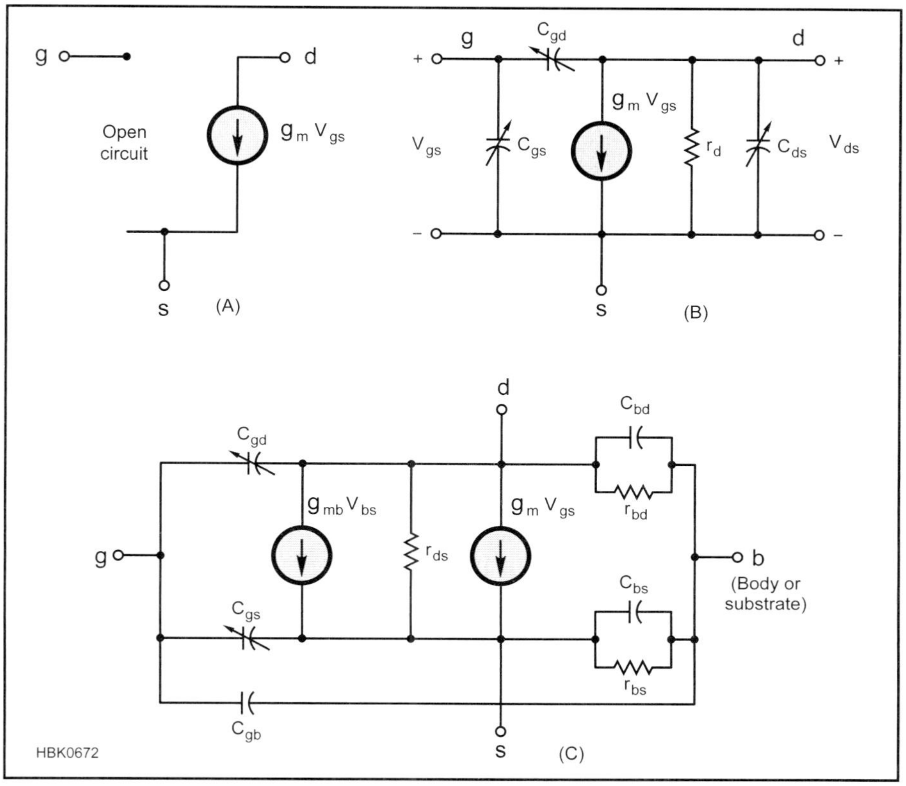

Figure 5.37 — Small-signal models for the field-effect transistor (FET). (A) is a simple low-frequency model. (B) is a common *SPICE3* large-signal model and (C) is a typical high-frequency model for the JFET.

The advances in CMOS integrated circuit technology have resulted in transistors (and capacitors and inductors) with sufficient performance for use in RF design work. In fact, most radio transceivers for cellular telephones, wireless LANs and similar high-volume applications are implemented almost exclusively in CMOS technology due to the high integration available. MOSFET RF design principles are similar to BJT and JFET designs, and most circuit simulators provide good models for MOSFET devices. These additional techniques are enabled by the excellent device matching available in IC technology and the ability to integrate additional circuitry at low incremental cost.

PARASITIC ELEMENTS AND NOISE

For accurate RF simulation the effects of additional parasitic elements must be included. The most significant is lead inductance — generally inserted in series with the device's external connections. Lead-to-package and lead-to-lead capacitances are also included in most high-frequency simulations. Depending on the simulation package used, the user may be required to install these parasitic elements as separate circuit elements or the simulator or manufacturer may provide special models for high-frequency simulation.

Simulation of device noise is not covered here, but many simulator packages include separate noise models for various active devices. Review the simulator's documentation for information about how to include noise in your simulation.

5.4.3 Amplifier Classes

The class of operation of an amplifier is determined by the fraction of a drive cycle during which conduction occurs in the amplifying device or for switchmode devices. (Switchmode amplifiers use different criteria.) The Class A amplifier conducts for 100% of the cycle. It is characterized by constant flow of supply current, regardless of the strength of the driving signal. Most of the amplifiers we use for RF applications and many audio circuits in receivers operate in Class A.

A Class B amplifier conducts for 50% of the cycle, which is 180 degrees if we examine the circuit with regard to a driving sine wave.

A Class B amplifier draws no dc current when no input signal is applied, but current begins to flow with any input, growing with the input strength. A Class B amplifier can display good *envelope linearity*, meaning that the output amplitude at the drive frequency changes linearly with the input signal. The total absence of current flow for half of the drive cycle will create harmonics of the signal drive.

A Class C amplifier is one that conducts for less than half of a cycle. No current flows without drive. Application of a small drive produces no output and no current flow. Only after a threshold is reached does the device begin to conduct and provide output. A bipolar transistor with no source of bias for the base typically operates in Class C.

The large-signal models discussed earlier are suitable for the analysis of all amplifier classes. Small-signal models are generally reserved for Class A amplifiers. The most common power amplifier class is a cross between Class A and B — the Class AB amplifier that conducts for more than half of each cycle. A Class AB amplifier at low drive levels is indistinguishable from a Class A design. However, increasing drive produces greater collector (or drain) current and greater output.

Amplifier class letter designators for vacuum tube amplifiers were augmented with a numeric subscript. A Class AB_1 amplifier operates in AB, but with no grid current flowing. A Class AB_2 amplifier's grid is driven positive with respect to the cathode and so some grid current flows. In solid-state amplifiers, which have no grids, no numeric subscripts are used.

While wide-bandwidth Class A and Class B amplifiers are common, most circuits operating in Class C and higher are tuned at the output. The tuning accomplishes two things. First, it allows different terminations to exist for different frequencies. For example, a resistive load could be presented at the drive frequency while presenting a short circuit at some or all harmonics. The second consequence of tuning is that reactive loads can be created and presented to the amplifier collector or drain. This then provides independent control of current and voltage waveforms.

While not as common as A, B, and C, Class D and E amplifiers are of increasing interest. The Class D circuit is a balanced (two transistor) switching format where the input is driven hard enough to produce square wave collector waveforms. Class E amplifiers usually use a single switching device with output tuning that allows high current to flow in the device only when the output voltage is low. Other "switching amplifier" class designators refer to the various techniques of controlling the switched currents and voltages.

Class A and AB amplifiers are capable of good envelope linearity, so they are the most common formats used in the output of SSB amplifiers. Class B and, predominantly, Class C amplifiers are used for CW and FM applications, but lack the envelope linearity needed for SSB.

Efficiency varies considerably between amplifier class. The Class A amplifier can reach a collector efficiency of 50%, but no higher, with much lower values being typical. Class AB amplifiers are capable of higher efficiency, although the wideband circuits popular in HF transceivers typically offer only 30% at full power. A Class C amplifier is capable of efficiencies approaching 100% as the conduction cycle becomes small, with common values of 50 to 75%. Both Class D and E are capable of 90% and higher efficiency. An engineering text treating power amplifier details is Krauss, Bostian, and Raab's *Solid State Radio Engineering*. A landmark paper by a Cal Tech group led by David Rutledge, KN6EK, targeted to the home experimenter, *High-Efficiency Class-E Power Amplifiers*, was presented in *QST* for May and June, 1997.

5.4.4 RF Amplifiers with Feedback

The feedback amplifier appears frequently in amateur RF circuits. This is a circuit with two forms of negative feedback with (usually) a single transistor to obtain wide bandwidth, well-controlled gain and well-controlled, stable input and output resistances. Several of these amplifiers can be cascaded to form a high gain circuit that is both stable and predictable.

The small-signal schematic for the feedback amplifier is shown in **Figure 5.38** without bias components or power supply details. The design begins with an NPN transistor biased for a stable dc current. Gain is reduced with emitter degeneration, increasing input resistance while decreasing gain. Additional feedback is then added with a parallel feedback resistor, R_f, between the collector and base. This is much like the resistor between an op-amp output and the inverting input which reduces gain and *decreases* input resistance.

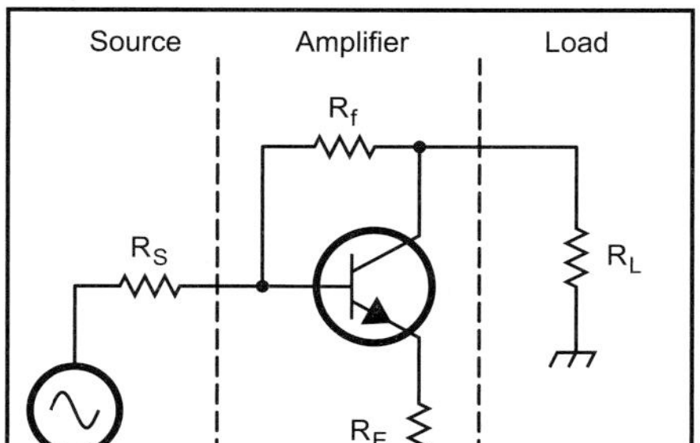

Figure 5.38 — Small-signal circuit for a feedback amplifier.

Several additional circuits are presented showing practical forms of the feedback amplifier. **Figure 5.39** shows a complete circuit. The base is biased with a resistive divider from the collector. However, much of the resistor is bypassed, leaving only R_f active for actual signal feedback. Emitter degeneration is ac coupled to the emitter. The resistor R_E dominates the degeneration since R_E is normally much smaller than the emitter bias resistor. Components that are predominantly used for biasing are marked with "B." This amplifier would normally be terminated in 50 Ω at both the input and output. The transformer has the effect of making the 50 Ω load "look like" a

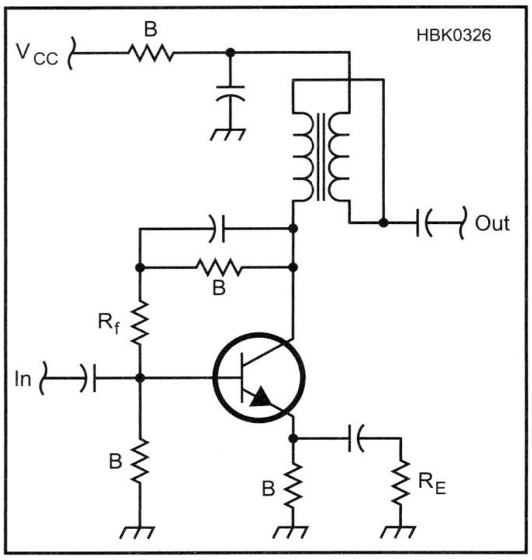

Figure 5.39 — A practical feedback amplifier. Components marked with "B" are predominantly for biasing. The 50 Ω output termination is transformed to 200 Ω at the collector. A typical RF transformer is 10 bifilar turns of #28 on a FT-37-43 ferrite toroid. The inductance of one of the two windings should have a reactance of around 250 Ω at the lowest frequency of operation.

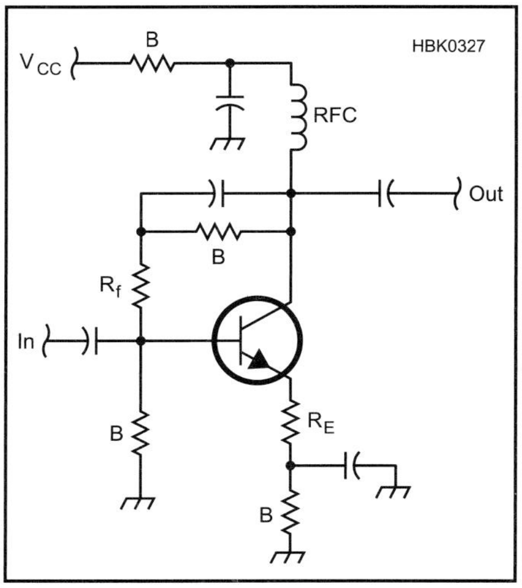

Figure 5.40 — A variation of the feedback amplifier with a 50 Ω output termination at the collector.

larger load value, $R_L = 200\ \Omega$ to the collector. This is a common and useful value for many HF applications.

Figure 5.40 differs from Figure 5.39 in two places. First, the collector is biased through an RF choke instead of a transformer. The collector circuit then "sees" 50 Ω when that load is connected. Second, the emitter degeneration is in series with the bias, instead of the earlier parallel connection. Either scheme works well, although the parallel configuration affords experimental flexibility with isolation between setting degeneration and biasing. Amplifiers without an output transformer are not constrained by degraded transformer performance and often offer constant or "flat" gain to several GHz.

The variation of **Figure 5.41** may well be the most general. It uses an arbitrary transformer to match the collector. Biasing is traditional and does not interact with the feedback.

Feedback is obtained directly from the output tap in the circuit of **Figure 5.42**. While this scheme is common, it is less desirable than the others, for the transformer is part of the feedback loop. This could lead to instabilities. Normally, the parallel feedback tends to stabilize the amplifiers.

The circuit of **Figure 5.43** has several features. Two transistors are used, each with a separate emitter biasing resistor. However, ac coupling causes the pair to operate as a single device with degeneration set by R_E. The parallel feedback resistor, R_f, is both a signal feedback element and part of the bias divider. This constrains the values slightly. Finally, an arbitrary output load can be presented to the composite collector through a π-type matching network. This provides some low pass filtering, but constrains the amplifier bandwidth.

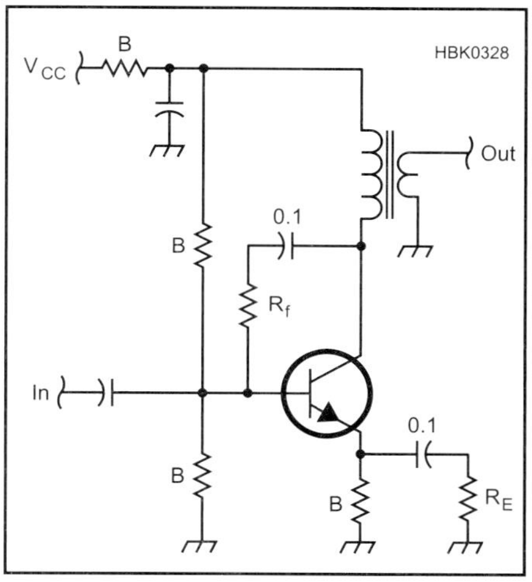

Figure 5.41 — This form of the feedback amplifier uses an arbitrary transformer. Feedback is isolated from bias components.

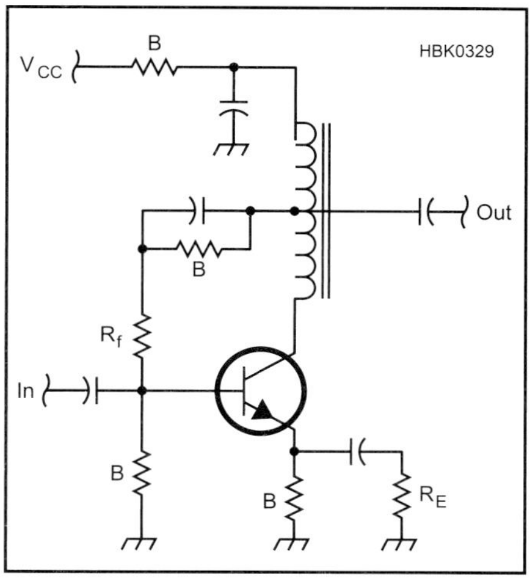

Figure 5.42 — A feedback amplifier with feedback from the output transformer tap. This is common, but can produce unstable results.

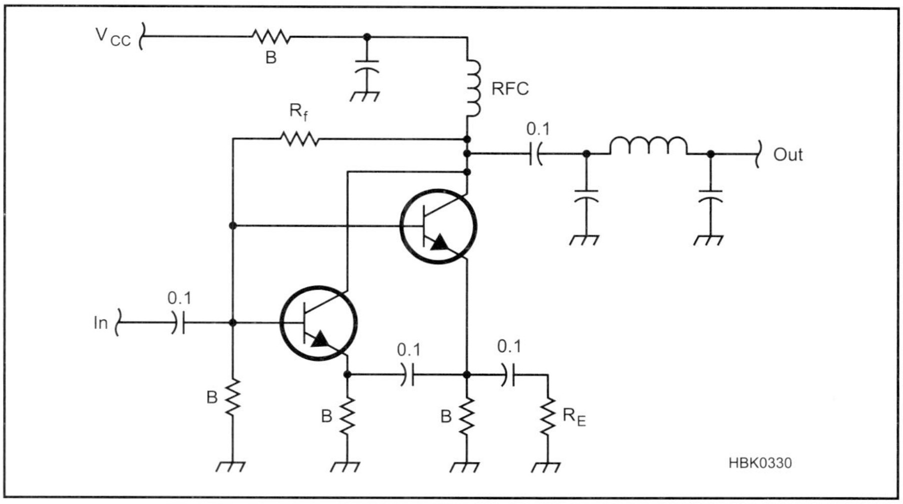

Figure 5.43 — Feedback amplifier with two parallel transistors.

5.5 Ferrite Materials

Ferrites are ceramics consisting of various metal oxides formulated to have very high permeability. Iron, manganese, manganese zinc (MnZn), and nickel zinc (NiZn) are the most commonly used oxides. Ferrite cores and beads are available in many styles, as shown in **Figure 5.44**. Wires and cables are then passed through them or wound on them.

When ferrite surrounds a conductor, the high permeability of the material provides a much easier path for magnetic flux set up by current flow in the conductor than if the wire were surrounded only by air. The short length of wire passing through the ferrite will thus see its self-inductance "magnified" by the relative permeability of the ferrite. The ferrites used for suppression are soft ferrites — that is, they are not permanent magnets.

Recalling the definition in the **Electrical Fundamentals** chapter, *permeability* (μ) is the characteristic of a material that quantifies the ease with which it supports a magnetic field. Relative permeability is the ratio of the permeability of the material to the permeability of free space. The relative permeability of nonmagnetic materials like air, copper, and aluminum is 1, while magnetic materials have a relative permeability much greater than 1. Typical values (measured at power frequencies) for stainless steel, steel and mu-metal are on the order of 500, 1,000 and 20,000 respectively. Various ferrites have values from the low tens to several thousand. The permeability of these materials changes with frequency and this affects their suitability for use as an inductor's magnetic core or as a means of providing EMI suppression.

Manufacturers vary the chemical composition (the *mix* or *material type*) and the dimensions of ferrites to achieve the desired electrical performance characteristics. It is important to select a mix with permeability and loss characteristics that are appropriate for the intended frequency range and application. (The **Construction Techniques** chapter includes tables showing the data for different types of ferrite and powdered-iron cores and appropriate frequency ranges for each.)

Fair-Rite Products Corp. supplies most of the ferrite materials used by amateurs, and their website (**www.fair-rite.com**) includes extensive technical data on both the materials and the many parts made from those materials. Much can be learned from the study of this data, the most extensive of any manufacturer. The website includes two detailed application notes on the use of ferrite materials, *How to Choose Ferrite Components for EMI Suppression* and *Use of Ferrites in Broadband Transformers*.

5.5.1 Ferrite Permeability and Frequency

Product data sheets characterize ferrite materials used as chokes by graphing their series equivalent impedance, and chokes are usually analyzed as if their equivalent circuit had only a series resistance and inductance, as shown in **Figures 5.45A** and 5.45B. (Figure 5.45 presents small-signal equivalent circuits.

Figure 5.44 — Ferrites are made in many forms. Beads are cylinders with small center holes so that they can be slid over wires or cables. Toroidal cores are more ring-like so that the wire or cable can be passed through the center hole multiple times. Snap-on or split cores are made to be clamped over cables too large to be wound around the core or for a bead to be installed.

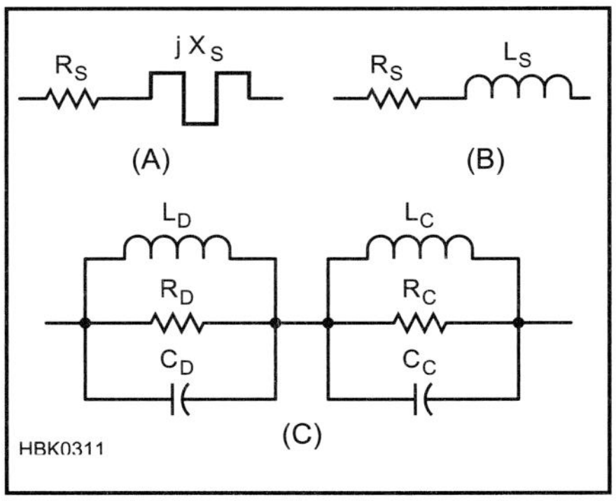

Figure 5.45 — Equivalent circuits for ferrite material. (A) shows the equivalent series circuit specified by the data sheet. (B) shows an over-simplified equivalent circuit for a ferrite choke. (C) shows a better equivalent circuit for a ferrite choke.

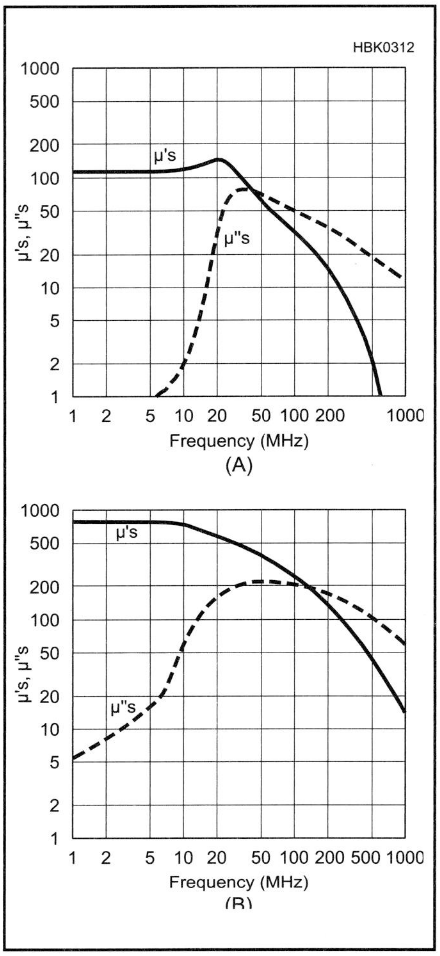

Figure 5.46 — Permeability of a typical ferrite material, Fair-Rite Type #61 (A) and of Type #43 material (B). (Based on product data published by Fair-Rite Products Corp.)

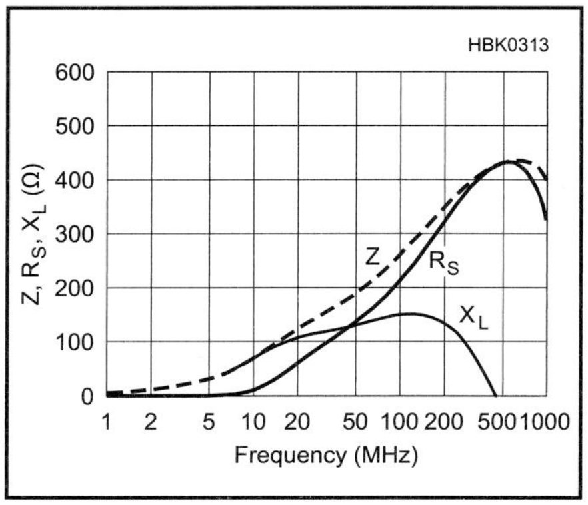

Figure 5.47 — Impedance of a bead for EMI suppression at UHF, Fair-Rite Type #61. (Based on product data published by Fair-Rite Products Corp.)

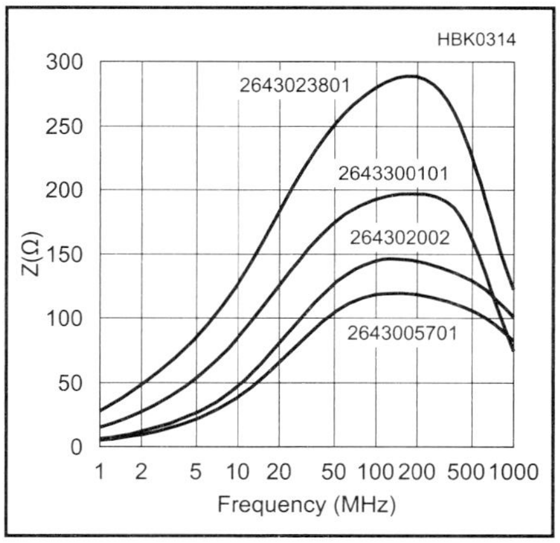

Figure 5.48 — Impedance of different length beads of Fair-Rite Type #43 material. The longer the bead, the higher the impedance. (Based on product data published by Fair-Rite Products Corp.)

Issues of temperature and saturation at high power levels are not addressed in this section.) The actual equivalent circuit is closer to Figure 5.45C. The presence of both inductance and capacitance creates resonances visible in the graphs of impedance versus frequency as discussed below. One resonance is created by the pair L_D C_D and the other by L_C C_C.

Figure 5.46A graphs the permeabilities μ'_S and μ''_S for Fair-Rite Type #61 ferrite material (Fair-Rite products are identified by a material "Type" and a number), one of the many mix choices available. This material is recommended for use in inductive applications below 25 MHz and for EMI suppression at frequencies above 200 MHz.

For the simple series equivalent circuit of Figure 5.45A and B, the permeability constant for ferrite is actually complex; $\mu = \mu'_S + j\,\mu''_S$. In this equation, μ'_S represents the component of permeability defining ordinary inductance, and μ''_S describes the component that affects losses in the material. You can see that up to approximately 20 MHz, μ'_S is nearly constant, meaning that an inductor wound on a core of this material will have a stable inductance. Below about 15 MHz, the chart shows that μ''_S is much smaller than μ'_S, so that losses are small, making this material good for high power inductors and transformers in this frequency range.

Above 15 MHz, μ''_S increases rapidly and so do the material's associated losses, peaking between 300 and 400 MHz. That makes the inductor very lossy at those frequencies and good for suppressing EMI by absorbing energy in the unwanted signal. **Figure 5.47** shows the manufacturer's impedance data for one "turn" of wire through a cylindrical ferrite bead (a straight wire passing through the bead) made of mix #61 with the expected peak in impedance above 200 MHz. Interestingly,

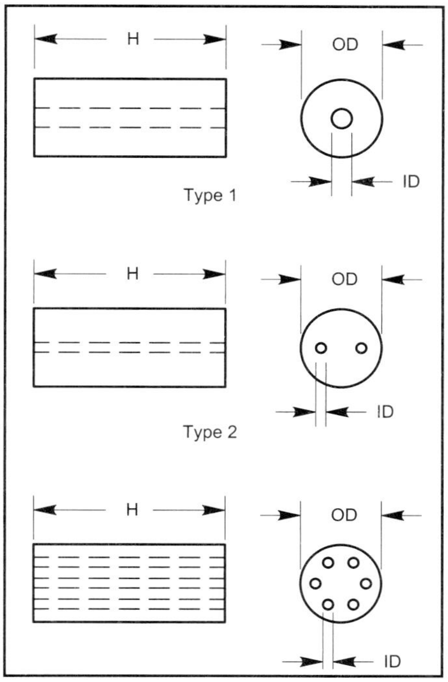

Figure 5.49 — Typical one-piece ferrite bead configurations. Component leads or cables can be inserted through beads one or more times to create inductance and loss as described in the text.

X_L goes below the graph above resonance, but it isn't zero. The negative reactance is created by the capacitors in Figure 5.45C.

The Type #43 mix (see Figure 5.46B) used for the beads of **Figure 5.48** is optimized for suppression at VHF (30 - 300 MHz). Type #43 material's μ'_S is much higher than type #61, meaning that fewer turns are required to achieve the needed inductance. But μ''_S for #43 remains significant even at low frequencies, limiting its usefulness for inductor and transformer cores that must handle high power. The figure shows the impedance data for several beads of different lengths. The longer the bead, the higher the impedance. The same effect can be obtained by stringing multiple beads together on the same wire or cable.

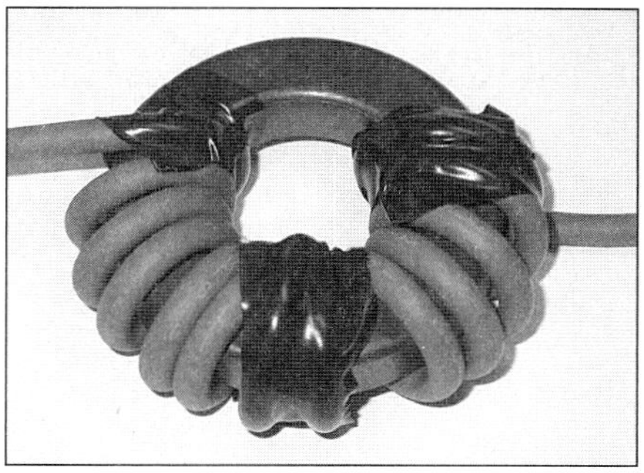

Figure 5.50 — A ferrite choke consisting of multiple turns of cable wound on a 2.4-in. OD × 1.4-in. ID × 0.5 in. toroid core.

5.5.2 Resonances of Ferrite Cores

Below resonance, the impedance of a wire passing through a ferrite cylinder is propor-

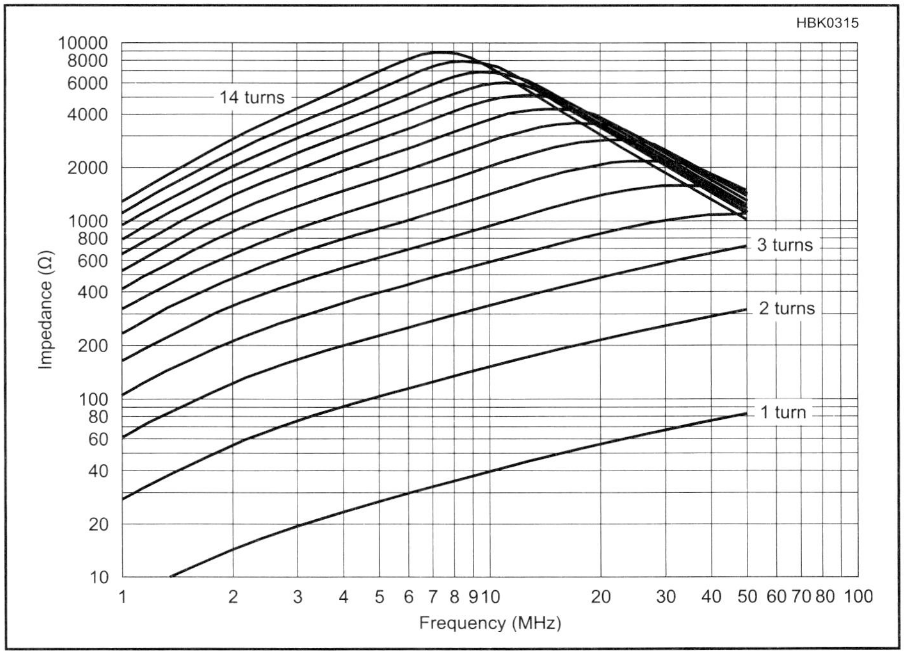

Figure 5.51 — Impedance of multi-turn choke wound on a Fair-Rite Type #43 core as shown in Figure 5.50. Type #43 material is optimized for use as a choke for the VHF range. (Measured data)

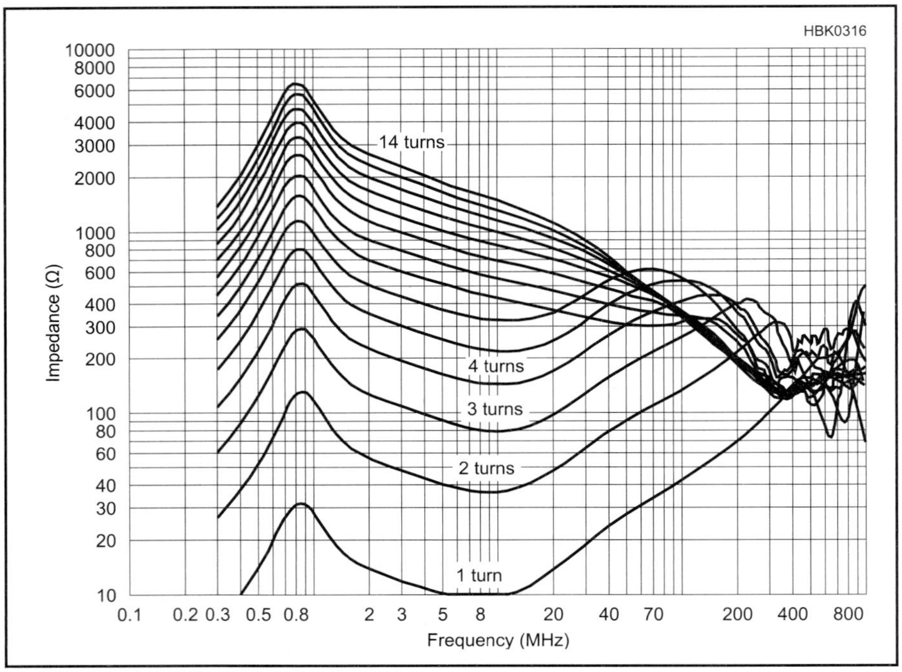

Figure 5.52 — Impedance of multi-turn choke wound on a Fair-Rite Type #78 core as shown in Figure 5.50. Type #78 material is optimized for use as a choke below 2 MHz. (Measured data)

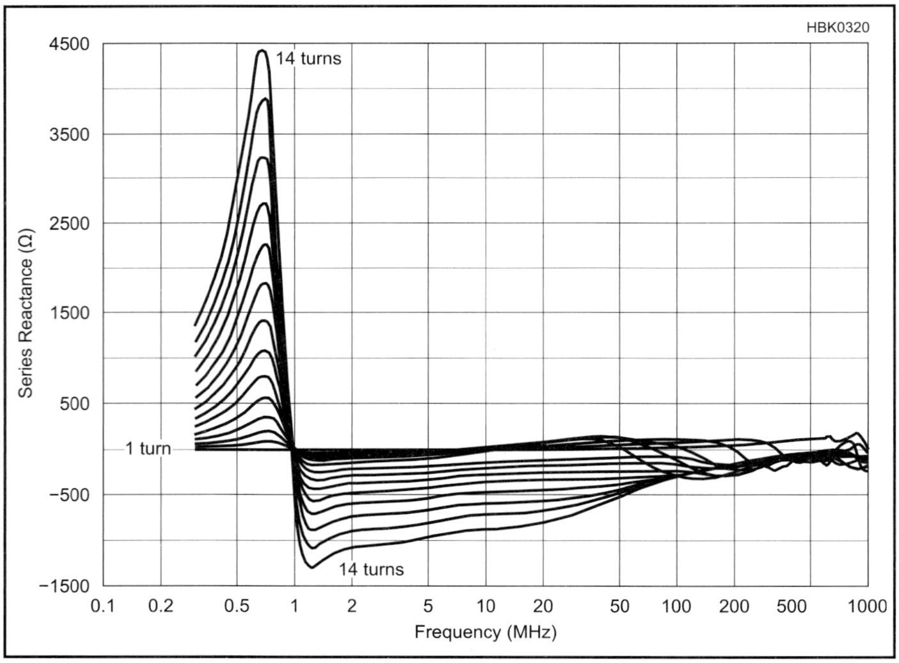

Figure 5.53 — Series reactive component of the chokes of Figure 5.52. (Measured data)

tional to the length of the cylinder. Figure 5.48 shows the impedance of a family of beads that differ primarily in their length. There are also small differences in their cross section, which is why the resonant frequency shifts slightly as discussed below. **Figure 5.49** shows some typical ferrite bead configurations.

Like all inductors, the impedance of a ferrite choke below resonance is approximately proportional to the square of the number of turns passing through the core. **Figure 5.50** shows a multi-turn choke wound on a ferrite toroid core. **Figure 5.51** is measured data for the choke in Figure 5.50 made of ferrite optimized for the VHF range (30 - 300 MHz). The data of **Figure 5.52** are for toroids of the same size, but wound on a material optimized for use below 2 MHz.

The measured data for Figures 5.51, 5.52, and 5.53 are for chokes wound with small diameter wire. The choke of Figure 5.50 will exhibit a somewhat lower resonant frequency for the same number of turns because the larger diameter cable has more capacitance between turns. The cable in the photo is a high quality braid-shielded twisted pair with an outside diameter comparable to RG-59.

We'll study the L_D C_D resonance first. A classic text (*Soft Ferrites, Properties and Applications*, by E. C. Snelling, published in 1969), shows that there is a dimensional resonance within the ferrite related to the velocity of propagation (V_P) within the ferrite and standing waves that are set up in the cross-sectional dimensions of the core. In general, for any given material, the smaller the core, the higher will be the frequency of this resonance, and to a first approximation, the resonant frequency will double if the core dimension is halved. In Figure 5.45C, L_D and C_D account for this dimensional resonance, and R_D for losses within the ferrite. R_D is mostly due to eddy currents (and some hysteresis) in the core.

Now it's time to account for R_C, L_C, and C_C. Note that there are two sets of resonances for the chokes wound around the Type #78 material (Figure 5.52), but only one set for the choke of Figure 5.51. For both materials, the upper resonance starts just below 1 GHz for a single turn and moves down in frequency as the number of turns is increased. **Figure 5.53**, the reactance for the chokes of Figure 5.52, also shows both sets of resonances. That's why the equivalent circuit must include two parallel resonances!

The difference between these materials that accounts for this behavior is their chemical composition (the mix). Type #78 is a MnZn ferrite, while Type #43 is a NiZn ferrite. The velocity of propagation (V_P) in NiZn ferrites is roughly two orders of magnitude higher than for MnZn, and, at those higher frequencies, there is too much loss to allow the standing waves that establish dimensional resonance to exist.

To understand what's happening, we'll return to our first order equivalent circuit of a ferrite choke (Figure 5.45C). L_C, and R_C, and C_C are the inductance, resistance, (including the effect of the μ of the ferrite), and stray capacitance associated with the wire that passes through the ferrite. This resonance moves down in frequency with more turns because both L and C increase with more turns. The dimensional resonance does not move, since it depends only on the dimensions V_P of the ferrite.

What is the source of C_C if there's no "coil," only a single wire passing through a cylinder? It's the parasitic capacitance from the wire at

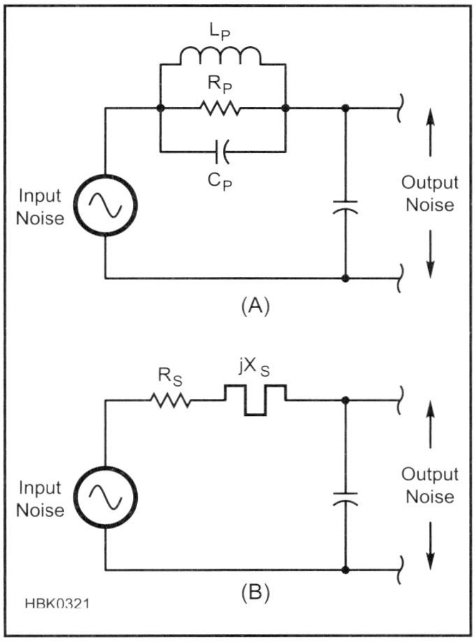

Figure 5.54 — At A, the series element of the divider is a parallel-resonant circuit. At B, the series element of the divider is the series-equivalent circuit used by ferrite data sheets.

one end of the cylinder to the wire at the other end, with the ferrite acting as the dielectric. Yes, it's a very small capacitance, but you can see the resonance it causes in the measured data and on the data sheet.

5.5.3 Ferrite Series and Parallel Equivalent Circuits

Let's talk briefly about series and parallel equivalent circuits. Many impedance analyzers express the impedance between their terminals as Z with a phase angle, and the series equivalent R_S, and X_S. They could just have easily expressed that same impedance using the parallel equivalent R_P and X_P but R_P and X_P will have values that are numerically different from R_S and X_S. (See the section on series-parallel impedance transformation in the **Electrical Fundamentals** chapter.)

It is important to remember that in a series circuit, the larger value of R_S and X_S has the greatest influence, while in a parallel circuit, the smaller value of R_P and X_P is dominant. In other words, for R_P to dominate, R_P must be small.

Both expressions of the impedance (series or parallel) are correct at any given frequency, but whether the series or parallel representation is most useful will depend on the physics of the device being measured and how that device is used in a circuit. We've just seen, for example, that the parallel equivalent circuit is a more realistic representation of a ferrite choke — the values of R_P, L_P, and C_P will come much closer to remaining constant as frequency changes than if we use the series equivalent. (R_P, L_P, and C_P won't be precisely constant though, because the physical properties of all ferrites — permeability, resistivity, and permittivity — all vary with frequency.)

But virtually all product data for ferrite chokes is presented as the series equivalent R_S and X_S. Why? First, because it's easy to measure and understand, second, because we tend to forget there is stray capacitance, and third because ferrite beads are most often used as chokes to reduce current in a series circuit! **Figure 5.54A** and Figure 5.54B are both useful representations of the voltage divider formed by a ferrite choke and a small bypass capacitor across the device input. Which we use will depend on what we know about our ferrite.

If we know R_P, L_P, and C_P and they are constant over the frequency range of interest, Figure 5.54A may be more useful, because we can insert values in a circuit model and perhaps tweak the circuit. But if we have a graph of R_S and X_S vs. frequency, Figure 5.54B will give us a good answer faster. Because we will most often be dealing with R_S and X_S data, the series circuit equivalents are used most often. Another reason for using R_S and X_S is that the impedance of two or more ferrite

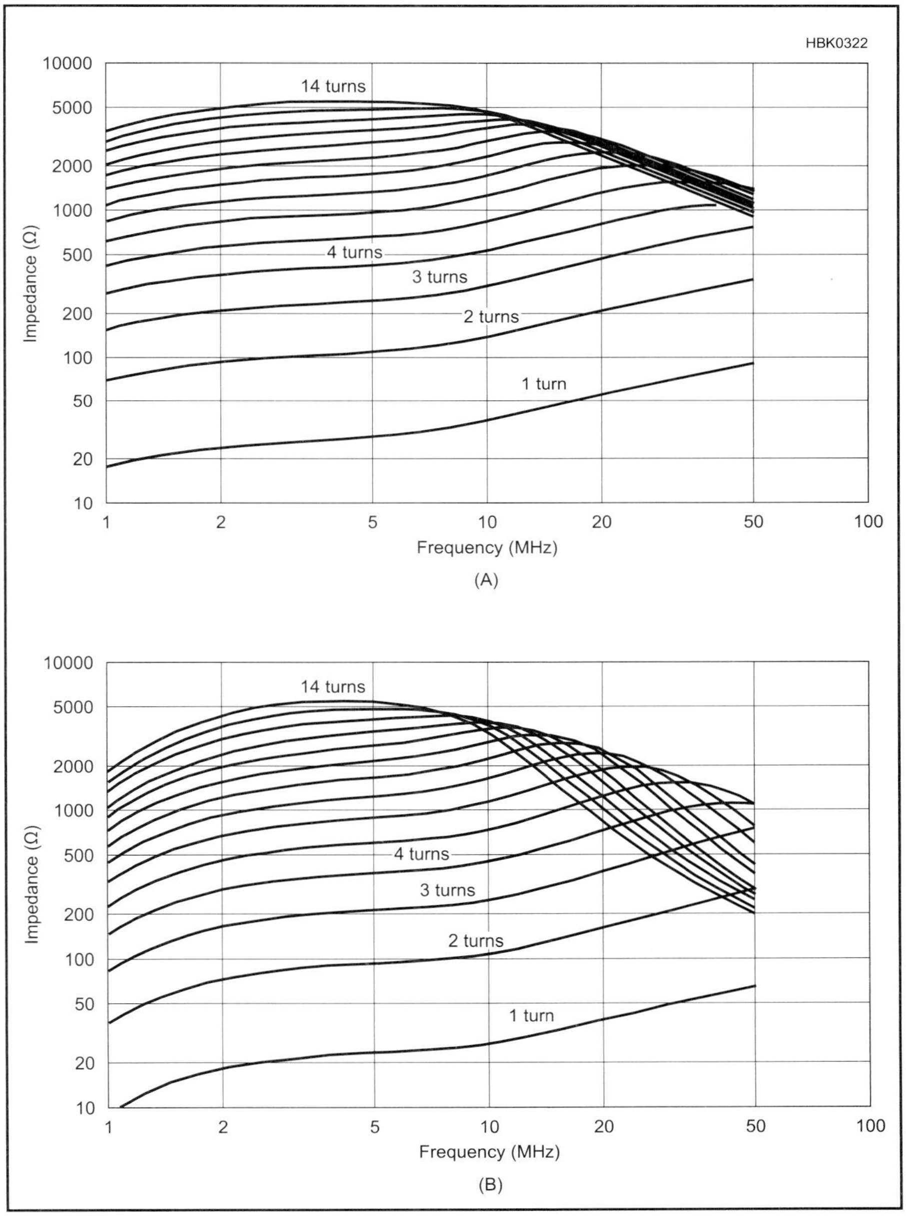

Figure 5.55 — At A, the impedance of multi-turn chokes on a Type #31 2.4-in. OD toroidal core. At B, the equivalent series resistance of the chokes of (A). (Measured data)

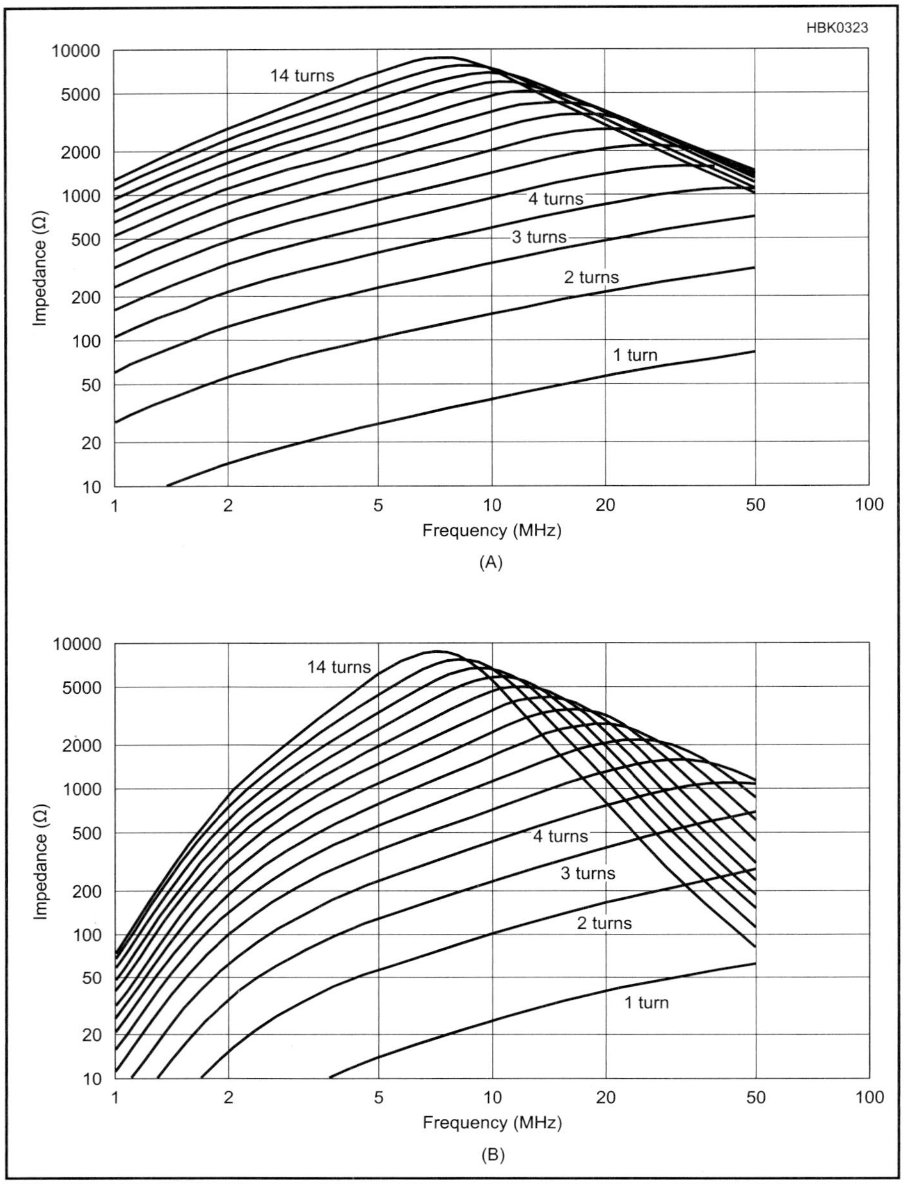

Figure 5.56 — At A, the impedance of multi-turn chokes wound with small-diameter wire on a Type #43 2.4-in. OD toroidal core. At B, the equivalent series resistance of the chokes of A. (Measured data).

chokes in series can be computed simply by adding their R_S and X_S components, just as with any other series impedances! When you look at the data sheet plots of R_S, X_L, and Z for a standard ferrite part, you are looking at the series equivalent parameters of their dominant resonance. For most MnZn materials, it is dimensional resonance, while, for most NiZn materials, it is the circuit resonance.

Values for R_P, L_P, and C_P for nearly any NiZn ferrite choke can be obtained by curve-fitting the data for its parallel-resonance curve. This is much more difficult for MnZn ferrite chokes, especially type #31. (Self resonance is discussed earlier in this chapter.) Because the C_P of many practical chokes is quite small and their impedance at resonance is often rather high, they are quite difficult to measure accurately, especially with reflection-based measurement systems. See the **Test Equipment and Measurements** chapter for a simple measurement method thatcan provide good results. More information on the use of ferrite cores and beads for EMI suppression is available in the chapter on **RF Interference** and in *The ARRL RFI Book*. A thorough treatment of the use of these ferrite materials for EMI suppression is continued in the online publication *A Ham's Guide to RFI, Ferrites, Baluns, and Audio Interfacing* at **k9yc.com/publish.htm**. The use of ferrite beads and cores for transmitting chokes is presented in the **Transmission Lines** chapter.

5.5.4 Type 31 Material

Type #31 material made by Fair-Rite Products is extremely useful as a choke core, especially if some component of your problem occurs below 5 MHz. Measured data for the new material is displayed in **Figures 5.55A** and 5.55B. Compare it with **Figures 5.56A** and 5.56B, which are corresponding plots for the older Type #43 material. By comparison, Type #31 provides nearly 7 dB greater choking impedance at 2 MHz, the top of the 160-meter band, and at least 3 dB more between 3.5 and 4 MHz, the 80-meter band. At 10 MHz and above, the two materials are nearly equivalent, with Type #43 being about 1 dB better. If your goal is EMI suppression or a feed line choke (a so-called "current balun"), the Type #31 material is the best all round performer to cover all HF bands, and is clearly the material of choice at 5 MHz and below. Between 5 MHz and 20 MHz, Type #43 has a slight edge (about 1 dB), and above 20 MHz they're equivalent. (Baluns are discussed more in the section below on Transmission Line Transformers and in the chapters on **Transmission Lines** and **Antennas**.)

Type #31 material is useful because it exhibits both of the resonances in our equivalent circuit — that is, the dimensional resonance of the core and the resonance of the choke with the lossy permeability of the core material. Below 10 MHz, these two resonances combine (in much the manner of a stagger-tuned IF) to provide significantly greater suppression bandwidth (roughly one octave, or one additional harmonically related ham band). The result is that a single choke on a Type #31 core can be made to provide very good suppression over about 8:1 frequency span, as compared to 4:1 for Type #43. Type #31 also has somewhat better temperature characteristics at HF.

5.5.5 Determining Ferrite Type

The type of mix for ferrite cores and beads without identification is not easily determined, unfortunately. This makes ferrite components of unknown type hard to use properly. Unlike powdered-iron cores, most ferrite components are not manufactured with a colored coating that identifies the type of material. It is a good idea to label these components as soon as they are purchased. Colored tape using standard color codes or labels work well. (As of early 2022, DX Engineering adds a paint stripe to the ferrite cores they distribute.) A light-color paint or silver Sharpie-type marker will write indelibly on the charcoal-color cores.

Three electrical tests described below can

identify an unknown mix with reasonable confidence:

1. Begin with an ohmmeter to determine the general ceramic compound: MnZn or NiZn.

2. Measure the material's relative permeability, μ_I, using an inductance meter or impedance analyzer and compare it to published values for different mixes.

3. Finally make swept-frequency measurements using a vector impedance analyzer (VIA) or vector network analyzer (VNA) to confirm the most likely mix based on the behavior of the material's reactance and impedance magnitude with frequency.

OBTAINING CORE DATA

The vast majority of ferrite cores encountered by hams are manufactured by Fair-Rite. Data for their cores can be accessed from the PRODUCTS tab at **fair-rite.com**, then selecting SUPPRESSION COMPONENTS or POWER & INDUCTIVE COMPONENTS. For POWER AND INDUCTIVE COMPONENTS, select CLOSED MAGNETIC CIRCUIT, then TOROIDS (assuming that is the type of component being tested).

For SUPPRESSION COMPONENTS, select CABLE COMPONENTS, then either ROUND CABLE EMI SUPPRESSION CORES (toroids and beads) or ROUND CABLE SNAP-ITS (also known as "snap-on cores"). Selecting the type of core will display a table listing components of the type selected.

For both types of materials, the dropdown showing ALL MATERIALS / FREQS can limit the listing to only one suppression mix. From that table, clicking on the part number of a core will display its data sheet.

IDENTIFYING FERRITE WITH AN OHMMETER

Ferrites are semiconductors with resistivity varying by seven orders of magnitude from 100 Ω cm to 10^9 Ω cm. In general, MnZn ferrite materials have lower values of resistivity; in the range of 100 – 300 Ω cm for Fair-Rite type #73, #75, #77, and #78, and 3,000 Ω cm for type #31. NiZn ferrites have much higher bulk resistivities: 10^5 Ω cm for type #43, 10^7 Ω cm for type #67, 10^8 Ω cm for type #61, and 10^9 Ω cm for types #44 and #52.

First suggested by W1HIS, these bulk resistivity differences allow us to partially identify the mix of MnZn cores with an ohmmeter. Of the MnZn cores tested by K9YC, types #75, #77, and #78 measure in the range of 1 – 10 kΩ while type #31 measured in the range of 100 kΩ – 700 kΩ.

Ferrite parts are likely to have a protective coating, so test probes must scratch aggressively through that coating to make contact with the core. K9YC used three multimeters; a Simpson 260 analog volt/ohm meter (VOM), a Fluke 8060A digital VOM, and an inexpensive digital VOM. All three meters made the same identifications with good contact between the probe and the ferrite material.

Note that these are not quantitative measurements: They don't show how good or bad the part is, they simply allow us to differentiate between MnZn and NiZn mixes, and for the MnZn materials to tentatively identify the difference between type #31 and types #75, #77, and #78. To get a better idea of the exact mix, we must make additional measurements.

IDENTIFYING FERRITE BY INDUCTANCE

In this method a coil is wound on a ferrite core and the resulting inductance is measured at low frequencies, well below any dimensional resonance. An LCR meter can be used with a test frequency at 50 kHz or lower. Some analyzers and/or their software can compute and display inductance vs. frequency from a swept measurement. A sweep starting at 50 kHz or an impedance bridge or meter that can measure in this range is recommended. This method may not yield definitive results for cores of very low permeability like Fair-Rite type #61. For very low permeability cores, the swept impedance method described below should be used.

The resonance of a multiturn winding and the dimensional resonance in MnZn materials reduce the measured inductance as frequency increases approaching either resonance. Use a value for inductance measured at lower frequencies where it is approximately constant. An instrument that can plot inductance vs. frequency from a swept measurement should display a straight line in the frequency range where an inductance measurement is valid for this method. (See the multiturn plots below for examples.)

From the inductance and number of turns, relative permeability μ_i, is calculated from the equation $\mu_i = (79.55\ N^2\ L\ l_e)/A_e$, where N is the number of turns, L is in μH, l_e is in cm, and A_e is in cm^2. A_e is the cross-sectional area of the core and l_e is the length of the path for magnetic flux around the core. (These mechanical dimensions can usually be found in catalog data for cores with the same dimensions, regardless of the type of material.) Remember that the number of turns is the number of times the winding passes through the core. The calculated value of μ_i is then compared to values for cores of the same dimension from manufacturer data sheets. (See the **Circuits and Components** chapter for tables of relative permeability for some ferrite materials.)

From the Fair-Rite data listing the type of component (see above), clicking on the selector at the top of the page showing ALL MATERIALS / FREQS will list those materials and their μ_i values. A_L and l_e values are published in the component data sheet. POWER & INDUCTIVE COMPONENTS and EMI SUPPRESSION parts having the same dimensions will have the same l_e and A_e values.

If no catalog data is available, A_e and l_e can often be computed from dimensions of the core. A_e is the cross-sectional area of the core and l_e is the length of the path for magnetic flux around the core. For a cylindrical core (including toroids), A_e and l_e are easy to calculate from their dimensions: $A_e = (OD - ID) \times$ (core length) / 2, and to a first approximation, $l_e = \pi\ (OD + ID) / 2$, the average of the inner and outer diameters of the core.

For cores of other shapes it's not so simple. For cores with a circular internal diameter and a square, rectangular, hexagonal, or octagonal outer shape (typical of snap-on cores), A_e is the area of that outer shape minus the area of the internal circular cross section. But a lot more math is needed to determine l_e for these cores with sufficient precision to be meaningful.

IDENTIFYING FERRITE WITH SWEPT IMPEDANCE

In this test, one or more turns of wire are close-wound on the core or bead and a vector analyzer measures the resulting impedance over a wide frequency range. The component's impedance and reactance are then plotted against frequency. The shape of the plots is characteristic of the mix type. It is important that the turns be close-wound. Wide turn spacing will move measured resonances higher in frequency, particularly for the NiZn and type #31 mixes.

The analyzer measures S parameter S11, the reflection coefficient of a single-port network or device, from which its complex impedance can be computed at each frequency. (See the Two-Port Networks section later in this chapter for information on S parameters.) The type of ferrite can then be determined from the shape of the reactance and impedance plots. The technique of using a vector

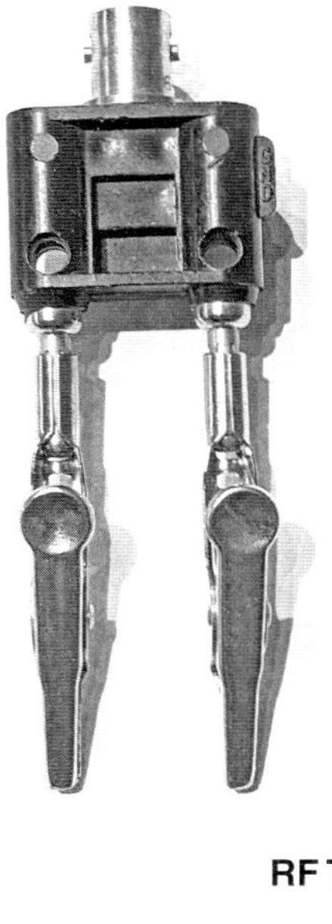

Figure 5.57 — A simple test fixture for measuring component impedance with a vector analyzer. The adaptor is a BNC-to-Banana Plug adapter. Large alligator clips will usually slide on to the banana plugs. The cable used to connect the adaptor to the analyzer must be included in the fixture calibration. (Photo by Jim Brown, K9YC)

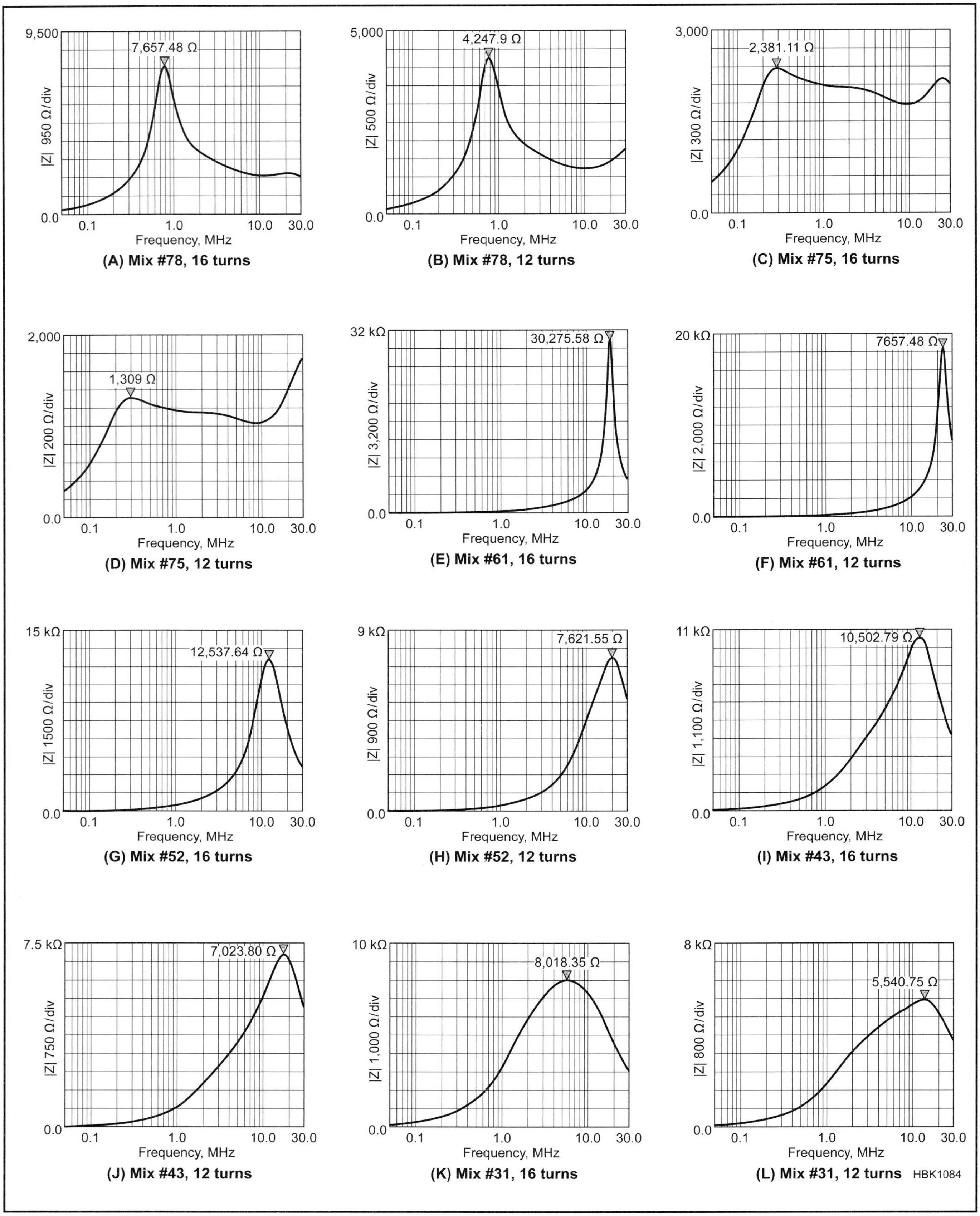

Figure 5.58 — Measured impedance magnitude (Zmag) for Fair-Rite Type #78, #75, #61, #52, #43, and #31 toroidal cores with 16 and 12 turns of #18 AWG insulated wire.

analyzer to make swept-frequency impedance measurements is explained in more detail in the **Test Equipment and Measurements** chapter.

For the purpose of measuring windings on a ferrite core, it is most convenient to make a simple test fixture as in **Figure 5.57** from a BNC-to-Banana Plug adapter with alligator clips mounted on the banana plugs. This test fixture is then attached to the analyzer with a convenient length of high-quality cable such as RG-400. The measurement plane for this fixture is at the tips of the alligator clip jaws. This fixture can be used to measure any component that can be attached to the clips, provided the leads are short enough to not add enough inductance or stray capacitance to create a resonance in or near the measured frequency range.

Particularly for the higher-frequency data, be sure the core and test fixture are not near conducting objects or surfaces. Letting the fixture hang over the edge of a work surface or chair with the clips facing down and the core suspended in the air from the clips is a good arrangement.

When evaluating measurements, remember that ferrite parts have wide manufacturing tolerances, so don't expect to duplicate these plots, but do expect parts of the same type of material to show similar shapes. Fair-Rite's inductive parts are specified with +/- 25% tolerances for inductance; their suppression parts are specified only for minimum impedance values at selected frequencies.

Materials from other manufacturers are likely to behave very differently from Fair-Rite materials having similar values for μ_i, but will likely have similar manufacturing tolerances. Both identification methods should be used to make sure it is or is not a Fair-Rite part, and if not, to learn more about its characteristics.

In general, manufacturers do not provide impedance versus frequency data, only permeability, for parts sold as Power and Inductive components under the assumption they will be used at low enough frequencies that their loss component is insignificant. Fair-Rite types #43, #61, and #75 are sold both as Power & Inductive Components at low frequencies and as EMI Suppression Components at higher frequencies but carry different part numbers for the two applications. The difference is testing for production tolerances — EMI Suppression Components are controlled only for minimum impedance at specific high frequencies, Power and Inductive Components are specified by their initial relative permeability, μ_i, which is measured at very low frequencies.

DATA FOR COMMON FERRITE MIXES

Figures 5.58 parts A – L shows data from testing randomly selected 2.4-inch O.D. cores of different ferrite types from Fair-Rite. To compare a core to this data, close-wind 16 turns of #18 AWG insulated wire on the core, trim leads to minimum length, make a measurement, and save the data. Remove four turns, trimming leads to minimum length, and repeat the measurement. Smaller cores may need to be swept over a higher frequency range to see comparable plot shapes.

In these data plots the curve is the magnitude of the impedance (Z_{MAG}) with zero at bottom of the vertical scale. The frequency scale is logarithmic and the same for all cores. (Note that the plot shapes will be very different on a linear frequency scale.) The amplitude scale is varied to set the resonant peak value near the top of the plot so that curve shapes can be best compared. Scales are shown at the left edge of each plot.

The general shape of the Z_{MAG} curves and the frequency range in which the impedance peak occurs are what distinguishes the different mixes. MnZn cores of types #75, #77, and #78 clearly show dimensional resonance, even for only a single turn. These occur at frequencies that are *not* dependent on the number of turns but *do* depend on the size and/or shape of the core. For these materials, the resonances don't move (much) in frequency with the number of turns. The peak simply gets stronger with more turns. Type #31 will show a broad resonance that moves and changes shape with the number of turns as the dimensional resonance combines with the resonance of the winding. NiZn mixes will show a single resonance, usually in the upper

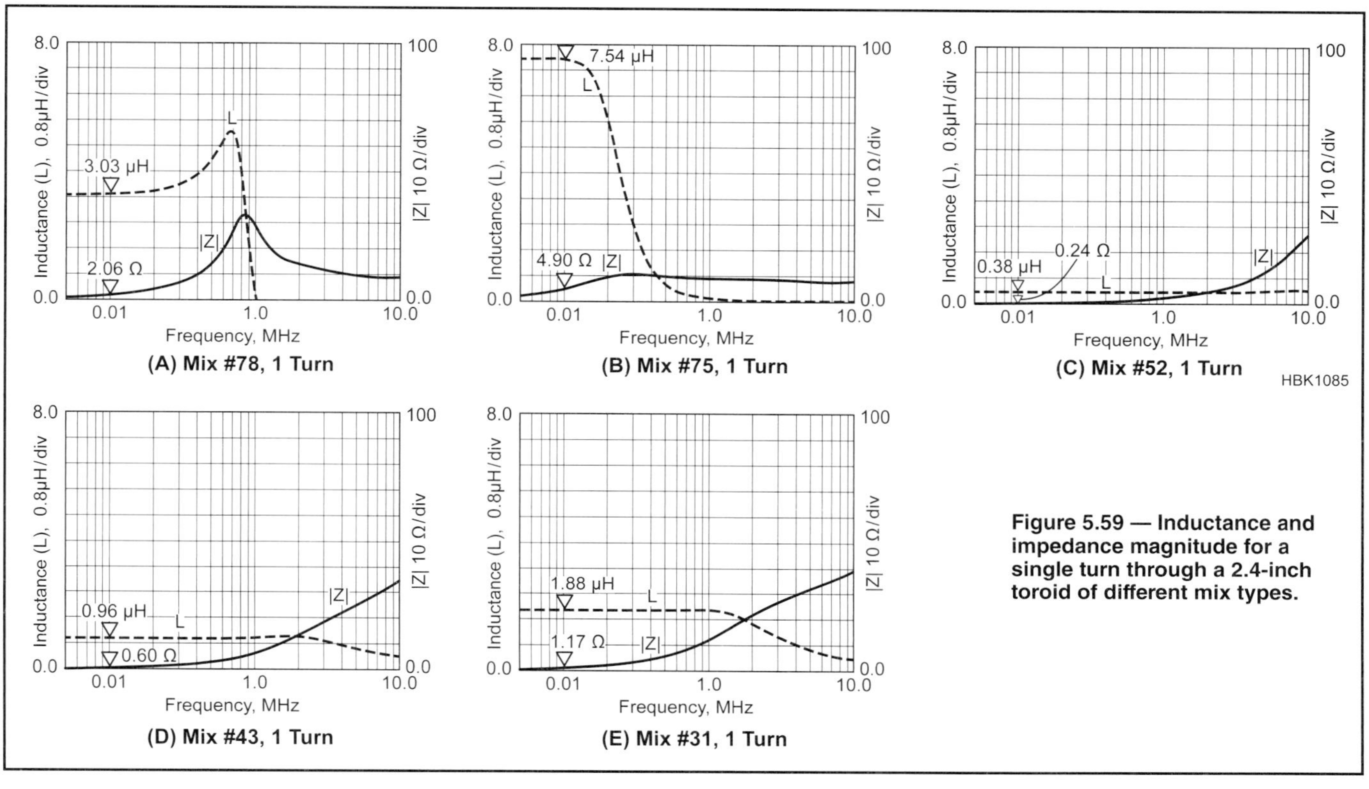

Figure 5.59 — Inductance and impedance magnitude for a single turn through a 2.4-inch toroid of different mix types.

HF range, that moves higher in frequency with fewer turns. It's formed by the inductance of the winding and the stray (parasitic) capacitance between turns, as shown in the plots for types #43, #52, and #61 material.

Note that these data for #18 insulated wire differ from comparable measurements for transmitting chokes because the transmission lines used for the chokes have larger conductors, different insulation, and greater interwinding capacitance. (See the **Transmission Lines** chapter's section Transmitting Ferrite Chokes and Baluns.)

Figures 5.59 parts A – E are for a single turn through a 2.4-inch O.D. toroid core. The dashed curve is inductance and the solid curve is the magnitude of the impedance, Z_{MAG}, both with zero at the bottom of the vertical scale.

• Type #78: resonances should appear in the range of 700 – 1000 kHz, they should not move much in frequency, and the peak's amplitude should get lower as turns are removed.

• Type #77: should measure about the same as type #78, except that the resonance will be in the range of 1300 – 1500 kHz.

• Type #75: resonances will be below 700 kHz.

• Type #61: Swept-frequency measurements of Type #61 were unreliable due to the material's very low resistive component and are not shown here as a graph. If the permeability falls below 200 and the multiturn measurements show a strong, high-Q resonance, the material is probably Type #61 or a similar material from another manufacturer.

• Type #43 and #52: multiturn tests can be ambiguous, but measuring A_L will be clear for distinguishing between these types. Both of these types are not recommended for chokes at HF in favor of Type #31.

• Type #31: the resonance will be much broader, will move up in frequency as turns are removed, and the impedance at resonance will be lower with fewer turns.

For use in RF chokes, only Type #31 is recommended for chokes at HF and 160 meters, Type #75 is useful only on 630 meters, and mixes like Type #43, #52, and #61 should only be used for chokes at VHF and above.

For use in RF transformers and inductors, Type #43, #52, and #61 are useful at HF.

Surplus round cores are most likely to be either intended for EMC compliance above 30 MHz, which are likely to be Type #43 or something similar; or those used for power magnetics at low frequencies, which are likely to be MnZn mixes; or those used for transformers, which are most likely to be MnZn if manufactured for use at low frequencies or NiZn if manufactured for use at at high frequencies.

5.6 Impedance Matching Networks

An impedance transforming or matching network is one that accepts power from a generator with one characteristic impedance, the source, and delivers virtually all of that power to a load at a different impedance. The simple designs in this section provide matching at only one frequency. More refined methods discussed in the reference texts can encompass a wide band of frequencies.

Both source and load impedances are likely to be complex with both real and imaginary (reactive) parts. The procedures given in this section allow the reader to design the common impedance-matching networks based on the simplification that both the source and load impedance are resistive (with no reactive component). To use these procedures with a complex load or source impedance, the usual method is to place a reactance in series with the impedance that has an equal and opposite reactance to cancel the reactive component of the impedance to be matched. Then treat the resulting purely resistive impedance as the resistance to be transformed. For example, if a load impedance to be matched is $120 + j\,40\ \Omega$, add capacitive reactance of $-j\,40\ \Omega$ in series with the load, resulting in a load impedance of $120 + j\,0\ \Omega$ so that the following procedures can be used. The series reactance-canceling component may then be combined with the matching network's output component in some configurations.

A set of 14 simple resonant networks, and their equations, is presented in **Figure 5.60**. Note that in these diagrams R_S is the low impedance side and R_L is the high impedance side and that the X values are calculated in the top-down order given.

The formulas for the various networks use and produce values of reactance. To convert the reactances to L and C values, use the formulas

$$C = \frac{1}{2\pi f X_c} \text{ and } L = \frac{X_L}{2\pi f}$$

The program *MATCH.EXE* in the online content can perform the calculations.

You may wish to review the section on series-parallel impedance transformations in the **Electrical Fundamentals** chapter as those techniques form the basis of impedance-matching network design. There is additional discussion of L networks in the chapters on **Transmission Lines** and of Pi networks in the **RF Power Amplifiers** chapter. These discussions apply to more specific applications of the networks.

5.6.1 L Networks

Perhaps the most common LC impedance transforming network is the L network, so named because it uses two elements — one series element and one parallel — resembling the capital L on its side. There are eight different types of L networks as shown in **Figure 5.61**. (Types G and H are not as widely used and not covered here.) The simplified Smith Chart sketches show what range of impedance values can be transformed to the required impedance Z_0 and the path on the chart by which the transformation is achieved by the two reactances. (An introductory discussion of the Smith Chart is provided with the online content, and a detailed treatment is available in *The ARRL Antenna Book*.) Z_{DEV} represents the output impedance of a device, such as a transistor amplifier or any piece of equipment.

The process of transformation works the same in both "directions." That is, a network designed to transform Z_{DEV} to Z_0 will also transform Z_0 to Z_{DEV} if reversed. The L network (and the Pi and T networks discussed below) is *bilateral,* as are all lossless networks.

Purely from the standpoint of impedance matching, the L network can be constructed with inductive or capacitive reactance in the series arm and the performance will be exactly the same in either case. However, the usual case in amateur circuits is to place the inductive reactance in the series arm to act as a low pass filter. Your particular circumstances may dictate otherwise, however. For example, you may find it useful that the series-C/parallel-L network places a dc short circuit across one side of the network while blocking dc current through it.

The design procedure for L network configurations A through F in Figure 5.60 is as follows:

Given the two resistance values to be matched, connect the series arm of the circuit to the smaller of the two (R_S) and the parallel arm to the larger (R_P).

Find the ratio R_P / R_S and the L network's Q:

$$Q = \sqrt{\frac{R_P}{R_S} - 1}$$

Calculate the series reactance $X_S = QR_S$
Calculate the parallel reactance $X_P = R_P/Q$.

5.6.2 Pi Networks

Another popular impedance matching network is the Pi network shown in circuits 1 and 2 of Figure 5.60. The Pi network

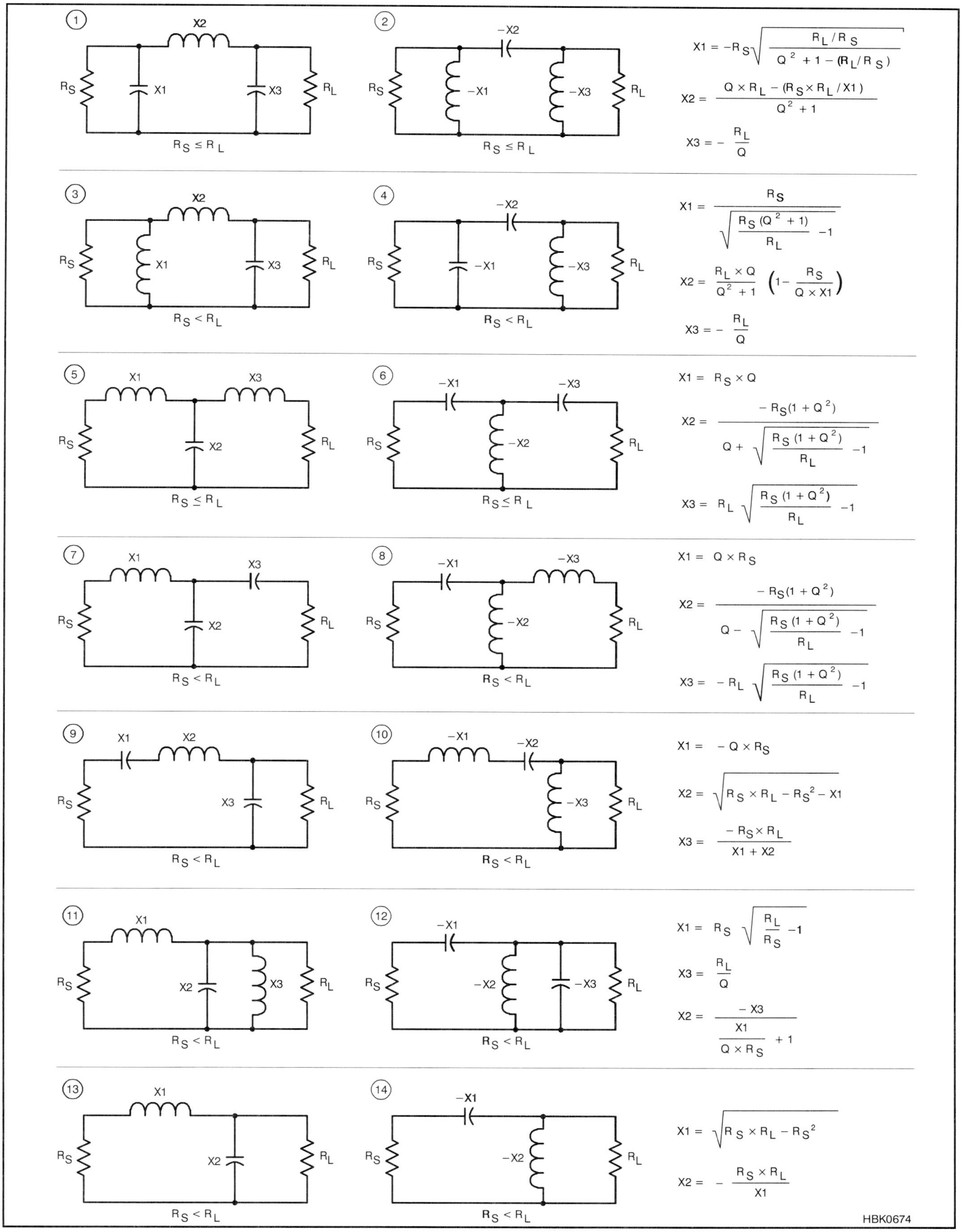

Figure 5.60 — Fourteen impedance transforming networks with their design equations (for lossless components).

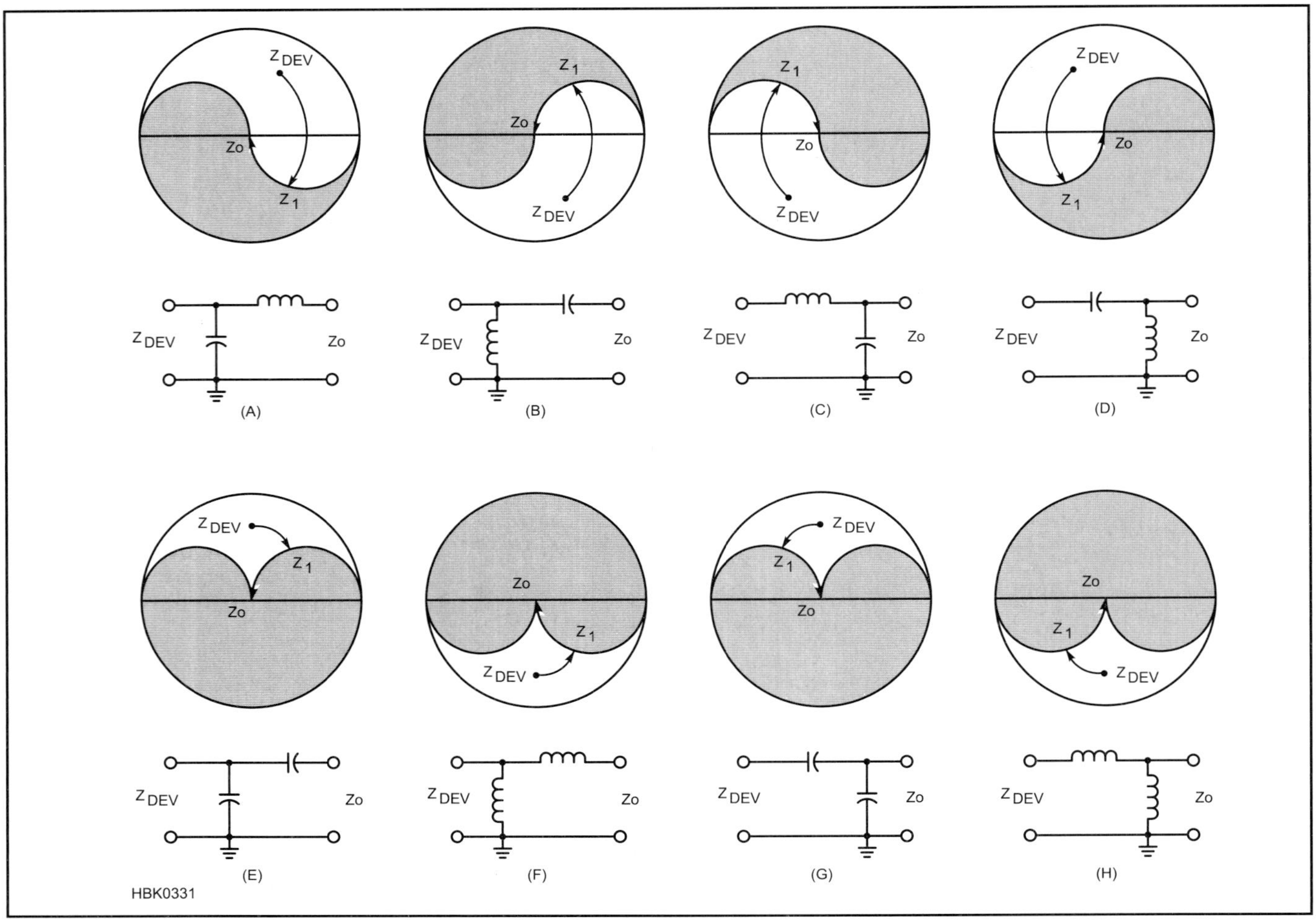

Figure 5.61 — L networks which will match a complex impedance (shown here as Z_{DEV}, the output impedance of a device) to Z_0, a resistive source or load. Impedances within the shaded portion of the simplified Smith Chart cannot be matched by the network. Z_1 represents the impedance that is transformed from Z_{DEV} by the series element.

can be thought of as two L networks "back to back." For example, the Pi network in circuit 1 can be split into the L network of circuit 13 on the right and its mirror image on the left. The two inductors in series are combined into the single inductance of the Pi network. There are other forms of the Pi network with different configurations of inductance and capacitance, but the version shown is by far the most common in amateur circuits.

The use of two transformations allows the designer to choose Q for the Pi network, unlike the L network for which Q is determined by the ratio of the impedances to be matched. This is particularly useful for matching amplifier outputs as discussed in the **RF Power Amplifier** chapter because it allows more control of the network's frequency response and of component values. Q must be high enough that $(Q^2 + 1) > (R_1 / R_2)$. If these two quantities are equal, X_{C2} becomes infinite, meaning zero capacitance, and the Pi network reduces to the L network in Figure 5.60A. The design procedure for the Pi network in Figure 5.61 is as follows:

Determine the two resistance values to be matched, $R_1 > R_2$, and select a value for Q. Follow the calculation sequence for circuit 1 or 2 in Figure 5.61.

Calculate the value of the parallel reactance $X_{C1} = R_1 / Q$.

Calculate the value of the parallel reactance X_{C2}

$$X_{C2} = R_2 \sqrt{\frac{R_1 / R_2}{Q^2 + 1 - R_1 / R_2}}$$

Calculate the value of the series reactance X_L

$$X_L = \frac{QR_1 + R_1 R_2 / X_{C2}}{Q^2 + 1}$$

5.6.3 T Networks

Many amateur transmission line impedance matching units ("antenna tuners") use the version of the T network in circuit 6 of Figure 5.60. The circuits are constructed using variable capacitors and a tapped inductor. This is easier and less expensive to construct than a fully-adjustable version of circuit 5 in which two variable inductors are required. Circuit 5 is especially useful in matching relatively low input impedances from solid-state amplifier outputs to 50 Ω loads with low Q and good harmonic suppression due to the series inductances. Circuit 7 is also useful in solid-state amplifier design as described in the reference texts listed at the end of this chapter.

The T network shown in circuit 5 of Figure 5.60 is especially useful for matching to relatively low impedance from 50 Ω sources with practical components and low Q. Like the Pi network, Q must be high enough that $(Q^2 + 1) > (R_1 / R_2)$. Designing the component values for this network requires the calculation of a pair of intermediate values, A and B, to make the equations more manageable.

Determine the two resistance values to be matched, $R_1 > R_2$, and select a value for Q.

Calculate the intermediate variables A and B

$$A = R_1(Q^2 + 1) \text{ and } B = \sqrt{\left(\frac{A}{R_2} - 1\right)}$$

Calculate the value of the input series reactance $X_{L1} = R_1 Q$.

Calculate the value of the output series reactance $X_{L2} = R_2 B$.

Calculate the value of the parallel reactance $X_C = A / (Q + B)$.

Convert the reactances to component values:

$$C = \frac{1}{2\pi f X_c} \text{ and } L = \frac{X_L}{2\pi f}$$

5.6.4 Impedance Inversion

Symmetrical Pi and T networks have the useful property of *impedance inversion* when the reactances of all elements are the same at the design frequency: $X_C = X_L = |X|$, resulting in a Q of 1. For either type of network, the impedance looking into the network, Z_{IN}, will be the load impedance, Z_{OUT}, inverted about X:

$$Z_{IN} = \frac{X^2}{Z_{OUT}}$$

This is the same effect as if the network were replaced with a ¼-wavelength transmission line with $Z_0 = |X|$. Since the network is symmetrical, the inversion occurs in either direction through the network. The result is true only at the frequency for which all reactances are equal.

For example, to invert all impedances about 50 Ω, set X = 50 Ω at the design frequency. The input impedance will then be $50^2/Z_{OUT}$. If $Z = 10 + j10$ Ω is connected to one end of the network, the impedance looking into the other end of the network will be $2{,}500 / (10 + j10) = 125 - j125$ Ω.

5.7 RF Transformers

5.7.1 Air-Core Nonresonant RF Transformers

Air-core transformers often function as mutually coupled inductors for RF applications. They consist of a primary winding and a secondary winding in close proximity. Leakage reactances are ordinarily high, however, and the coefficient of coupling between the primary and secondary windings is low. Consequently, unlike transformers having a magnetic core, the turns ratio does not have as much significance. Instead, the voltage induced in the secondary depends on the mutual inductance.

In a very basic transformer circuit operating at radio frequencies, such as in **Figure 5.62**, the source voltage is applied to L1. R_S is the series resistance inherent in the source. By virtue of the mutual inductance, M, a voltage is induced in L2. A current flows in the secondary circuit through the reactance of L2 and the load resistance of R_L. Let X_{L2} be the reactance of L2 independent of L1, that is, independent of the effects of mutual inductance. The impedance of the secondary circuit is then:

$$Z_S = \sqrt{R_L^2 + X_{L2}^2} \qquad (12)$$

where

Z_S = the impedance of the secondary circuit in ohms,

R_L = the load resistance in ohms, and

X_{L2} = the reactance of the secondary inductance in ohms.

The effect of Z_S upon the primary circuit is the same as a coupled impedance in series with L1. Figure 5.63 displays the coupled impedance (Z_P) in a dashed enclosure to indicate that it is not a new physical component. It has the same absolute value of phase angle as in the secondary impedance, but the sign of the reactance is reversed; it appears as a capacitive reactance. The value of Z_P is:

$$Z_P = \frac{(2\pi f M)^2}{Z_S} \qquad (13)$$

where

Z_P = the impedance introduced into the primary,

Z_S = the impedance of the secondary circuit in ohms, and

$2\pi f M$ = the mutual reactance between the reactances of the primary and secondary coils (also designated as X_M).

5.7.2 Air-Core Resonant RF Transformers

The use of at least one resonant circuit in place of a pair of simple reactances eliminates the reactance from the transformed impedance in the primary. For loaded or operating Q of at least 10, the resistances of individual components is negligible. **Figure 5.63** represents just one of many configurations in which at least one of the inductors is in a resonant circuit. The reactance coupled into the primary circuit is cancelled if the circuit is tuned to resonance while the load is connected. If the reactance of

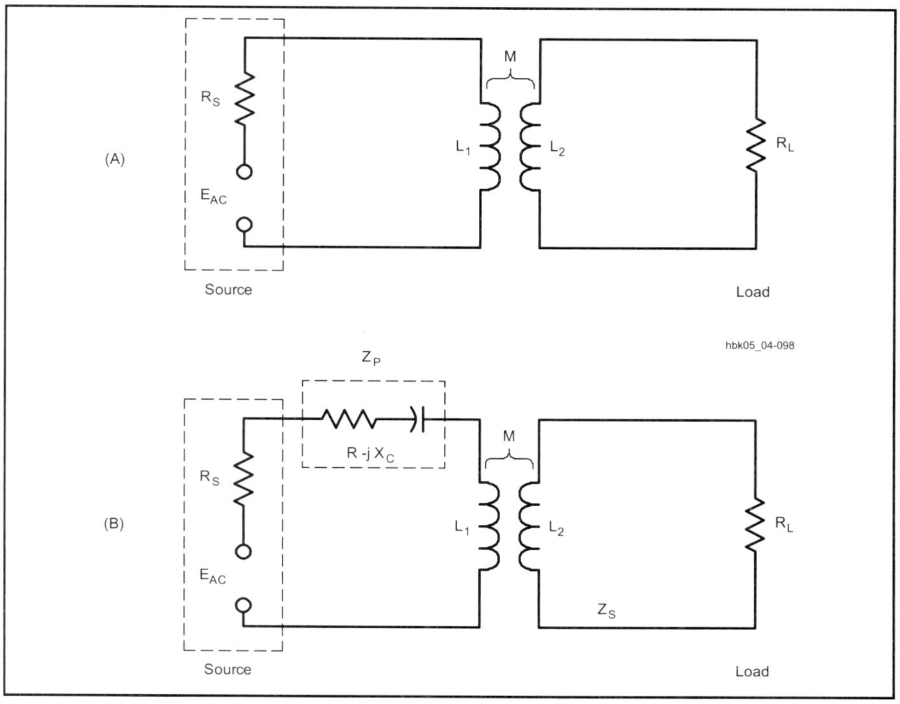

Figure 5.62 — The coupling of a complex impedance back into the primary circuit of a transformer composed of nonresonant air-core inductors.

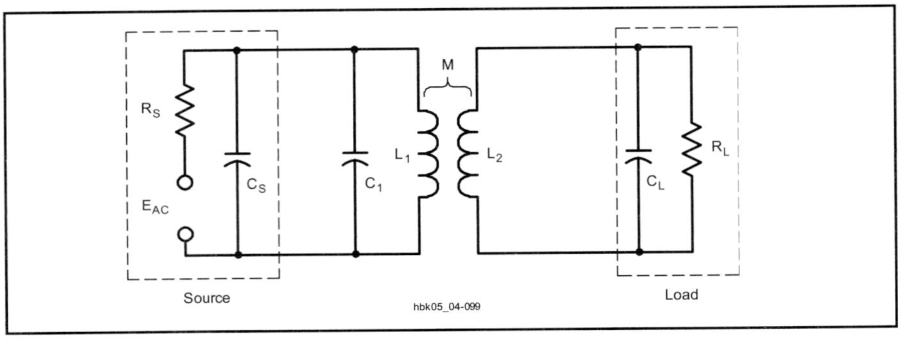

Figure 5.63 — An air-core transformer circuit consisting of a resonant primary circuit and an untuned secondary. R_S and C_S are functions of the source, while R_L and C_L are functions of the load circuit.

the load capacitance, C_L is at least 10 times any stray capacitance in the circuit, as is the case for low impedance loads, the value of resistance coupled to the primary is

$$R1 = \frac{{X_M}^2 \, R_L}{{X_2}^2 + {R_L}^2}$$

where:

R1 = series resistance coupled into the primary circuit,
X_M = mutual reactance,
R_L = load resistance, and
X_2 = reactance of the secondary inductance.

The parallel impedance of the resonant circuit is just R1 transformed from a series to a parallel value by the usual formula, $R_P = X_2 / R1$.

The higher the loaded or operating Q of the circuit, the smaller the mutual inductance required for the same power transfer. If both the primary and secondary circuits consist of resonant circuits, they can be more loosely coupled than with a single tuned circuit for the same power transfer. At the usual loaded Q of 10 or greater, these circuits are quite selective, and consequently narrowband.

Although coupling networks have to a large measure replaced RF transformer coupling that uses air-core transformers, these circuits are still useful in antenna tuning units and other circuits. For RF work, powdered-iron toroidal cores have generally replaced air-core inductors for almost all applications except where the circuit handles very high power or the coil must be very temperature stable. Slug-tuned solenoid coils for low-power circuits offer the ability to tune the circuit precisely to resonance. For either type of core, reasonably accurate calculation of impedance transformation is possible. It is often easier to experiment to find the correct values for maximum power transfer, however.

5.7.3 Broadband Ferrite RF Transformers

The design concepts and general theory of ideal transformers presented in the **Electrical Fundamentals** chapter apply also to transformers wound on ferromagnetic-core materials (ferrite and powdered iron). As is the case with stacked cores made of laminations in the classic I and E shapes, the core material has a specific permeability factor that determines the inductance of the windings versus the number of wire turns used. (See the earlier discussion on Ferrite Materials in this chapter.)

One of the most common ferromagnetic transformers used in amateur circuits is the conventional broadband transformer. Broadband transformers with losses of less than 1 dB are employed in circuits that must have a uniform response over a substantial frequency range, such as a 2 – 30-MHz broadband amplifier. In applications of this sort, the reactance of the windings should be at least four times the impedance that the winding is designed to look into at the lowest design frequency.

Example: What should be the winding reactances of a transformer that has a 300 Ω primary and a 50 Ω secondary load? Relative to the 50 Ω secondary load:

$$X_S = 4\ Z_S = 4 \times 50\ \Omega = 200\ \Omega$$

and the primary winding reactance (X_P) is:

$$X_P = 4\ Z_P = 4 \times 300\ \Omega = 1200\ \Omega$$

The core-material permeability plays a vital role in designing a good broadband transformer. The effective permeability of the core must be high enough to provide ample winding reactance at the low end of the operating range. As the operating frequency is increased, the effects of the core tend to disappear until there are scarcely any core effects at the upper limit of the operating range. The limiting factors for high frequency response are distributed capacity and leakage inductance due to uncoupled flux. A high-permeability core minimizes the number of turns needed for a given reactance and therefore also minimizes the distributed capacitance at high frequencies.

Ferrite cores with a permeability of 850 are common choices for transformers used between 2 and 30 MHz. Lower frequency ranges, for example, 1 kHz to 1 MHz, may require cores with permeabilities up to 2,000. Permeabilities from 40 to 125 are useful for VHF transformers. Conventional broadband transformers require resistive loads. Loads with reactive components should use appropriate networks to cancel the reactance. See **www.w8ji.com/core_selection.htm** for more guidelines on core material selection. Also see the article by Trask, which is included in the online content and listed in the References section.

The equivalent circuit in Figure 5.45 applies to any coil wound on a ferrite core, including transformer windings. (See the section on Ferrite Materials.) However, in the series-equivalent circuit, μ_S is not constant with frequency as shown in Figure 5.46A and 5.46B. Using the low-frequency value of μ_S is a useful approximation, but the effects of the parallel R and C should be included. In high-power transmitting and amplifier applications, the resistance R may dissipate some heat, leading to temperature rise in the core. The parasitic capacitances of each winding are shown as C_{PP} and C_{SP} in parallel with the primary and secondary circuits, respectively. These capacitances act to reduce high-frequency response.

Regarding C, there are at least two forms of stray capacitance between windings of a transformer as shown in **Figure 5.64A**; from wire-to-wire through air and from wire-to-wire through the ferrite, which acts as a dielectric material. These capacitances are combined as C_W from the primary to secondary circuits. (Ferrites with low iron content have a relative dielectric constant of approximately 10 to 12.)

For further information on conventional transformer matching using ferromagnetic materials, see the **RF Power Amplifiers** chapter. Refer to the **Circuits and Components** chapter for more detailed information on available ferrite cores. A standard reference on conventional broadband transformers using ferromagnetic materials is *Ferromagnetic Core Design and Applications Handbook* by Doug DeMaw, W1FB, published by MFJ Enterprises.

TOROIDAL CORE TRANSFORMERS

Toroidal cores are useful from a few hundred hertz well into the UHF spectrum. The principal advantage of this type of core is the self-shielding characteristic. Another

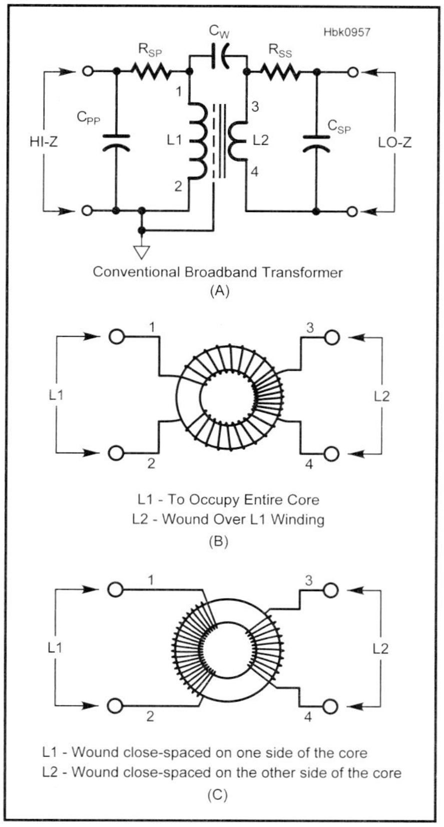

Figure 5.64 — (A) Schematic representation of a conventional broadband transformer wound on a toroid core. A Faraday shield (see text) can also be used to reduce capacitive coupling between the primary and secondary circuits. At (B) a pictorial showing the secondary winding (L2) is wound over the primary winding (L1) which provides very good coupling but low isolation. For designs emphasizing isolation over coupling, wind the transformer as in C with the primary and secondary windings separated on the core.

feature is the compactness of a transformer or inductor. Therefore, toroidal-core transformers are excellent for use not only in dc-to-dc converters, where tape-wound steel cores are employed, but at frequencies up to at least 1000 MHz with the selection of the proper core material for the range of operating frequencies. Toroidal cores are available from micro-miniature sizes up to several inches in diameter that can handle multi-kW military and commercial powers.

Figure 5.64B illustrates one method of transformer construction using a single toroid as the core. The primary of a step-down impedance transformer is wound to occupy the entire core, with the secondary wound over the primary. Conventional broadband transformers provide dc isolation between the primary and secondary circuits.

Winding-to-winding capacitance C_W reduces isolation between primary and secondary circuits. If isolation is an important characteristic of the transformer, then the windings should *not* be layered but separated on the core as in Figure 5.64C. The effect of C_W will increase with increasing frequency. This is a likely path for coupling of HF noise onto Ethernet cables if the network interface uses transformers.

A Faraday shield can be used to minimize C_W. It is made of conductive material, such as aluminum foil, that is connected to the circuit reference as in Figure 5.64A. (Faraday shields can also be connected to the secondary circuit reference.) Faraday shields are also used between windings in transformers for use in power, data, and audio circuits to prevent capacitive coupling of noise between windings.

The high voltages encountered in high-impedance-ratio step-up transformers may require that the core be wrapped with glass electrical tape before adding the windings (as an additional protection from arcing and voltage breakdown), especially with ferrite cores that tend to have rougher edges. In addition, high voltage applications should also use wire with high-voltage insulation and a high temperature rating.

The first step in designing the transformer is to select a core of the desired permeability. Convert the required reactances determined earlier into inductance values for the lowest frequency of use. To find the number of turns for each winding, use the A_L value for the selected core and the equation for determining the number of turns:

$$L = \frac{A_L \times N^2}{1{,}000{,}000} \qquad (15)$$

where

L = the inductance in mH,

A_L = the inductance index in mH per 1,000 turns, and

N = the number of turns.

Be certain the core can handle the power by calculating the maximum flux and comparing the result with the manufacturer's guidelines.

$$B_{max} = \frac{E_{RMS} \times 10^8}{4.44 \times A_e \times N \times f} \qquad (16)$$

where

B_{max} = the maximum flux density in gauss,

E_{RMS} = the voltage across the inductor,

A_e = the cross-sectional area of the core in square centimeters,

N = the number of turns in the inductor, and

f = the operating frequency in Hz.

(Both equations are from the section on ferrite toroidal inductors in the **Electrical Fundamentals** chapter and are repeated here for convenience.)

Example: Design a small broadband transformer having an impedance ratio of 16:1 for a frequency range of 2.0 to 20.0 MHz to match the output of a small-signal stage (impedance ≈ 500 Ω) to the input (impedance ≈ 32 Ω) of an amplifier.

Since the impedance of the smaller winding should be at least 4 times the lower impedance to be matched at the lowest frequency,

$$X_S = 4 \times 32\ \Omega = 128\ \Omega$$

The inductance of the secondary winding should be

$$L_S = \frac{X_S}{2\pi f} = \frac{128}{6.2832 \times 2.0 \times 10^6\ \text{Hz}}$$

$$= 0.0101\ \text{mH}$$

Select a core with a suitable size for the power level and ferrite mix for the frequency range. For this low-power application, a ⅜ inch. ferrite core with permeability of 850 is suitable. The core has an A_L value of 420. Calculate the number of turns for the secondary.

$$N_S = 1000\sqrt{\frac{L}{A_L}} = 1000\sqrt{\frac{0.010}{420}}$$

$$= 4.88\ \text{turns}$$

A 5-turn secondary winding should suffice. The primary winding derives from the impedance ratio:

$$N_P = N_S\sqrt{\frac{Z_P}{Z_S}} = 5\sqrt{\frac{16}{1}}$$

$$= 5 \times 4 = 20\ \text{turns}$$

This low-power application will not approach the maximum flux density limits for the core, and #28 AWG enamel wire should both fit the core and handle the currents involved.

NOTES ON TOROID WINDINGS

For a toroidal (cylindrical) core, the number of turns is the number of times the conductor passes through the core. A wire passing once through a cylindrical core constitutes one turn. Likewise, a split or "clamp-on" core that is simply clamped onto a conductor forms a single-turn choke. A wire passing twice through the core is a two-turn choke, even though there is only one pass external to the core.

The inductance of a toroid can be adjusted. If the turns can be pressed closer together or separated a little, inductance variations of a few percent are possible.

In general, all of the flux associated with ferrite inductors (and chokes) is confined to the core material — for all practical purposes, there is no inductive coupling between inductors (or chokes) that are physically adjacent but wound on different cores. Coupling between adjacent coils can be eliminated by placing a Faraday shield between them as discussed earlier in this section.

Toroidal windings do exhibit a small amount of leakage flux. Toroid coils are wound in the form of a helix (screw thread) around the circular length of the core. This means that there is a small component of the flux from each turn that is perpendicular to the circle of the toroid (parallel to the axis through the hole) and is therefore not adequately linked to all the other turns. This effect is responsible for a small leakage flux and the effect is called the "one-turn" effect.

BINOCULAR CORE TRANSFORMERS

Broadband transformers can also be made using a binocular core — a large bead-style core with two parallel holes as shown in **Figure 5.65B**. The windings on a binocular core have tighter coupling between the primary and secondary than separate windings on a toroid core. (Both the binocular and toroid transformers can use multifilar windings for even tighter coupling.) Binocular cores are used in broadband impedance transformers with ratios of 9:1 or 16:1 for receiving antennas, such as Beverages and small loops.

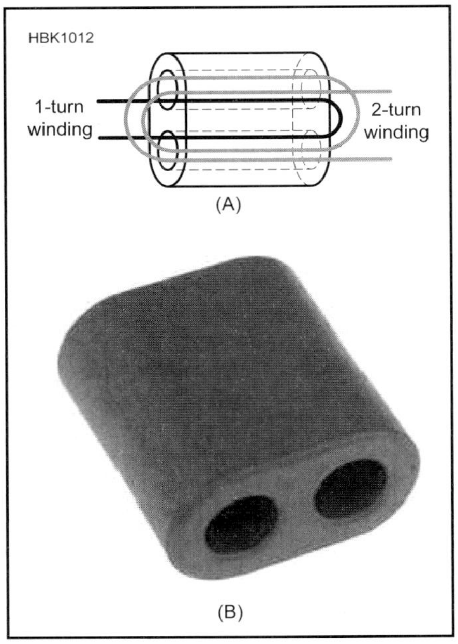

Figure 5.65 — Binocular ferrite cores. When used as a transformer (A) each full turn of a winding must pass through both holes. Cores (B) are available as small beads or large cores for high-power applications.

To make one complete turn in a binocular core, the wire must go through one of the holes and back through the other hole. This is different than a toroid where the wire passing through the core just once counts as a complete turn. Like toroids, binocular transformers must be wound with insulated or enameled wire so there is no direct contact between the metal and the core material.

A high-power version of a binocular core transformer can be made from two stacks of ferrite cores or a single large core. A single-turn winding can be made with copper or brass tubing through the stack that is soldered to strips of PC-board at each end. See the **RF Power Amplifiers** chapter and the previously referenced article by Trask for examples of this type of transformer.

5.8 Noise

The following material was contributed by Paul Wade, W1GHZ. The section on background noise by Joe Taylor, K1JT, is reproduced from his discussion of Earth-Moon-Earth (EME) communications in this books's online content. Additional discussion of noise measurement is available in the **Test Instruments and Measurements** *chapter and in the Noise Instrumentation document provided with the online content.*

As anyone who has listened to a receiver suspects, everything in the universe generates noise. In communications, the goal is to maximize the desired signal in relation to the undesired noise we hear. In order to accomplish this goal, it would be helpful to understand where noise originates, how much our own receiver adds to the noise we hear, and how to minimize it.

It is difficult to improve something unless we are able to measure it. Measurement of noise in receivers does not seem to be clearly understood by many amateurs, so this section attempts to explain the concepts and clarify the techniques, and to describe the standard "measure of merit" for receiver noise performance: "noise figure." In addition, the Noise Instrumentation document with the online content describes how to build your own noise generator for noise figure measurements.

A number of equations are included, but only a few need be used to perform noise figure measurements. The rest are included to as an aid to understanding supported by explanatory text.

5.8.1. Noise Power

The most pervasive source of noise is *thermal noise* (also called *Johnson* or *Johnson-Nyquist noise*), due to the motion of thermally agitated free electrons in a conductor. Since everything in the universe is at some temperature above absolute zero, every conductor must generate noise.

Every resistor (and all conductors have resistance) generates an RMS noise voltage:

$$e = \sqrt{4kTRB} \qquad (17)$$

where R is the resistance, T is the absolute temperature in kelvins (K), B is the bandwidth in hertz, and k is Boltzmann's constant, 1.38×10^{-23} joules /K (or J K^{-1}).

Converting to power, e^2/R, and adjusting for the Gaussian distribution of noise voltage, the noise power generated by the resistor is:

$$P_n = kTB \text{ (watts)} \qquad (18)$$

which is independent of the resistance. Thus, all resistors at the same temperature generate the same noise power.

Thermal noise is *white noise*, meaning that the power density does not vary with frequency, but always has a *power density* or *spectral density* of kT watts/Hz. (The corresponding noise voltage distribution is a *spectral voltage density*, measured in volts / $\sqrt{Hz}$, spoken as "volts per root hertz.") More important is that the noise power is directly proportional to absolute temperature T, since k is a constant. At the nominal ambient temperature of 290 K, we can calculate this power; converted to dBm, we get the familiar –174 dBm/Hz. Multiply by the bandwidth in hertz to get the available noise power at ambient temperature. The choice of 290 K for ambient might seem a bit cool, since the equivalent 17° C or 62° F would be a rather cool room temperature, but the value 290 makes for an easier-to-remember numeric calculation of $P_n = (1.38 \times 10^{-23} \times 290)\, B = 400 \times 10^{-23}\, B$.

The *instantaneous* noise voltage has a *Gaussian distribution* around the RMS value. The Gaussian distribution has no limit on the peak amplitude so at any instant the noise voltage may have any value from –infinity to +infinity. For design purposes we can use a value that will not be exceeded more than 0.01% of the time. This voltage is 4 times the RMS value, or 12 dB higher, so our system must be able to handle peak powers 12 dB higher than the average noise power if we are to measure noise without errors. (See Pettai in the Reference section.)

5.8.2. Signal to Noise Ratio

Now that we know the noise power in a given bandwidth, we can easily calculate how much signal is required to achieve a desired *signal to noise ratio, S/N* or *SNR*. For SSB, perhaps 10 dB SNR is required for good communications; since ambient thermal noise in a 2.5 kHz bandwidth is –140 dBm, calculated as follows:

$P_n = kTB = 400 \times 10^{-23} \times 2500$
$= 1.0 \times 10^{-17}$ W

dBm = 10 log ($P_n \times 1000$) [multiplying watts by 1000 converts to milliwatts]

The signal power must be 10 dB larger, so minimum signal level of –130 dBm is required for a 10 dB S/N. This represents the noise and signal power levels at the antenna. We are then faced with the task of amplifying the signal without degrading SNR.

5.8.3. Noise Temperature

There are many types of noise, but most have similar characteristics to thermal noise and are often added together, creating a single equivalent noise source whose output power per unit of bandwidth is P_N. The *noise temperature* of the source is defined as the temperature $T = P_N / k$ at which a resistor would generate the same noise power per unit of bandwidth as the source. This is a useful way to characterize the various sources of noise in a communications system.

All amplifiers add additional noise to the noise present at their input. The input noise per unit of bandwidth is $N_i = kT_g$, where T_g is the noise temperature at the amplifier's input. Amplified by power gain G, the output noise is kT_gG. The additional noise contributed by the amplifier can also be represented as a noise temperature, T_n. The noise power added by the amplifier, kT_n, is then added to the amplified input noise to produce a total output noise:

$N_o = kT_gG + kT_n$

We can treat the amplifier as ideal and noise-free but with an additional noise-generating resistor of temperature $T_e = T_n / G$ at the input so that all sources of noise can be treated as inputs to the amplifier as illustrated by **Figure 5.66**. The output noise is then:

$N_o = kG\ (T_g + T_e)$

The noise added by an amplifier can then be represented as kGT_e, which is amplifier's noise temperature amplified by the amplifier gain. T_e is sometimes referred to as *excess temperature*.

Note that while the noise temperature of a resistor is the same as its physical temperature, the noise temperature of a device such as a diode or transistor can be many times the physical temperature.

SINAD

Signal-to-noise and distortion ratio (SINAD) is often used to measure of the quality of a demodulated signal.

$$\text{SINAD} = \frac{P_{signal} + P_{noise} + P_{distortion}}{P_{noise} + P_{distortion}} \qquad (19)$$

where P is an average power. SINAD is usually expressed in dB and is often used as a condition at which a receiver's RF sensitivity is measured. For example: a sensitivity of 0.1 μV for 12 dB SINAD. (A thorough explanation of SINAD and several related terms such as THD is provided by the Analog Devices tutorial MT-003 by Kester listed in the References section of this chapter.)

5.8.4. Noise Factor and Noise Figure

The *noise factor,* F, of an amplifier is the ratio of the total noise output of an amplifier with an input T_g of 290 K to the noise output of an equivalent noise-free amplifier. A more useful definition is to calculate it from the excess temperature T_e:

$$F = 1 + T_e / T_g, \text{ where } T_g = 290 \text{ K} \qquad (20)$$

It is often more convenient to work with *noise figure, NF*, the logarithm of noise factor expressed in dB:

$$NF = 10 \log (1 + T_e / T_g) = 10 \log F \qquad (21)$$

$F = \log^{-1} (NF/10)$

Expressed in terms of signal, S, and noise power, N, at the input and output of a device:

$F = (S_{in}/N_{in})/(S_{out}/N_{out})$ and

$F = G_{noise}/G_{signal}$

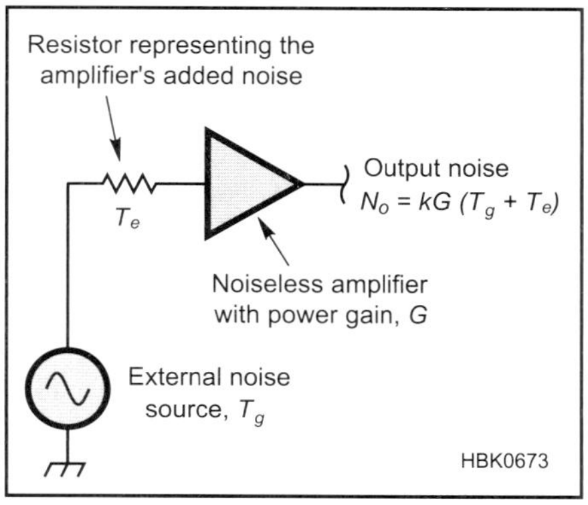

Figure 5.66 — The noise generated by an amplifier can be represented as an external resistor with a noise temperature of T_e connected at the input of a noiseless amplifier.

where G_{signal} is the device's power gain and G_{noise} is the device's *noise gain*. If SNR in dB is known at the input and output:

$NF = SNR_{in} - SNR_{out}$

If NF or F is known, then T_e may be calculated as:

$T_e = (F - 1)\ T_g$

Typically, T_e is specified for very low noise amplifiers where the NF would be a fraction of a dB. NF is used when it seems a more manageable number than thousands of K.

Noise figure is sometimes stated as *input noise figure* to emphasize that all noise sources and noise contributions are converted to an equivalent set of noise sources at the input of a noiseless device. In this way, noise performance can be compared on equal terms across a wide variety of devices.

Noise figure is particularly important at VHF and UHF where atmospheric and other artificial noise is quite low. Typical noise figures of amateur amplifiers range from 1 to 10 dB. Mixers are generally toward the high end of that range. Modern GaAsFET and HEMT preamplifiers are capable of attaining an NF of 0.1 to 0.2 dB at UHF with NF under 1 dB even at 10 GHz.

5.8.5. Losses

We know that any loss or attenuation in a system reduces the signal level. If attenuation also reduced the noise level then we could suppress thermal noise by adding attenuation. We know intuitively that this can't be true — the attenuator or any lossy element has a noise temperature, T_x, which contributes noise to the system while the input noise is being attenuated.

The output noise after a loss L (expressed as ratio) expressed as an equivalent input noise temperature is:

$T_g' = T_g / L + [(L - 1)/L]\ T_x$

If the original source temperature, T_g, is higher than the attenuator temperature, T_x, then the noise contribution is found by adding the loss in dB to the NF. However, for low source temperatures the degradation can be much more dramatic. If we do a calculation for the effect of 1 dB of loss (L = 1.26) on a T_g of 25 K:

$T_g' = 25/1.26 + (0.26/1.26) \times 290 = 80$ K

The resulting T_g' is 80 K, a 5 dB increase in noise power (or 5 dB degradation of signal to noise ratio). Since noise power = kT and k is a constant, the increase is the ratio of the two temperatures, 80/25, or in dB, 10 log (80/25) = 5 dB.

It is also useful to note that for linear, passive devices, such as resistors or resistive attenuators, noise figure is the same as loss in dB. A resistive attenuator with 6 dB loss has a noise figure of 6 dB which is equal to a noise factor of 4.

5.8.6. Cascaded Amplifiers

If several amplifiers are cascaded, the output noise N_o of each becomes the input noise Tg to the next stage. We can create a single equation for the total system of amplifiers. After removing the original input noise term, we are left with the added noise:

$$N_{added} = (k\,T_{e1}G_1G_2 \ldots G_N) + (kT_{e2}G_2 \ldots G_N) + \ldots + (kT_{eN}G_N)$$

where N is the number of stages cascaded. Substituting in the total gain $G_T = (G_1G_2 \ldots G_N)$ results in the total excess noise:

$$T_{eT} = T_{e1} + \frac{T_{e2}}{G_1} + \frac{T_{e2}}{G_1G_2} + \ldots + \frac{T_{eN}}{G_1G_2 \ldots G_{N-1}}$$

with the relative noise contribution of each succeeding stage reduced by the gain of all preceding stages.

The *Friis formula for noise (a.k.a. the Friis equation)* expresses this in terms of noise factor:

$$F = F_1 + \frac{F_2 - 1}{G_1} + \frac{F_3 - 1}{G_1G_2} + \ldots + \frac{F_N - 1}{G_1G_2 \ldots G_{N-1}} \quad (22)$$

Clearly, if the gain of the first stage, G_1, is large, then the noise contributions of the succeeding stages become too small to be significant. In addition, the noise temperature of the first stage is the largest contributor to the overall system noise because it is amplified by all remaining stages. The effect on overall noise figure of adding a low-noise preamplifier ahead of a noisy receiver are illustrated in **Figure 5.67**, in which the system's noise figure changes from 20 dB for the receiver alone to 7.1 dB with the preamplifier added.

Any lossy component of an antenna system, such as the feed line, increases the noise figure at its input by an amount equal to the loss. As a result, it is important to concentrate noise-reduction efforts on the first amplifier or preamplifier in a system. Because noise performance is so important in early stages of cascaded systems such as receivers, low-noise VHF+ preamplifiers are usually mounted at the antenna so that their gain occurs ahead of the feed line loss. **Figure 5.68** compares the results of adding a preamplifier before and after 1.5 dB of feed line loss. Moving the preamplifier to the antenna improves the system's noise figure from 2.57 to 1.13 dB.

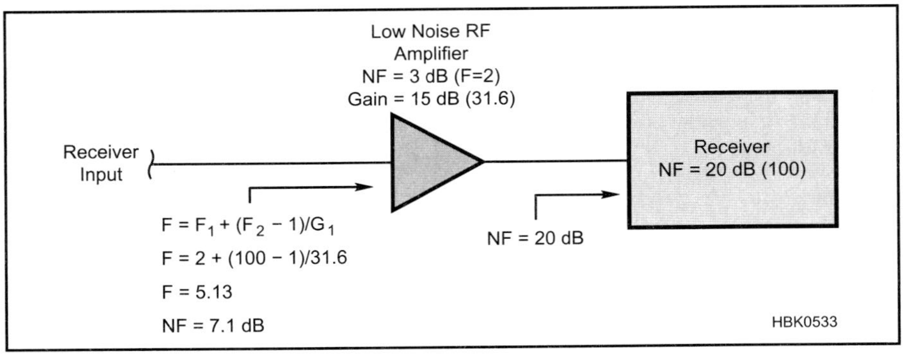

Figure 5.67 — The effect of adding a low-noise preamplifier in front of a noisy receiver system.

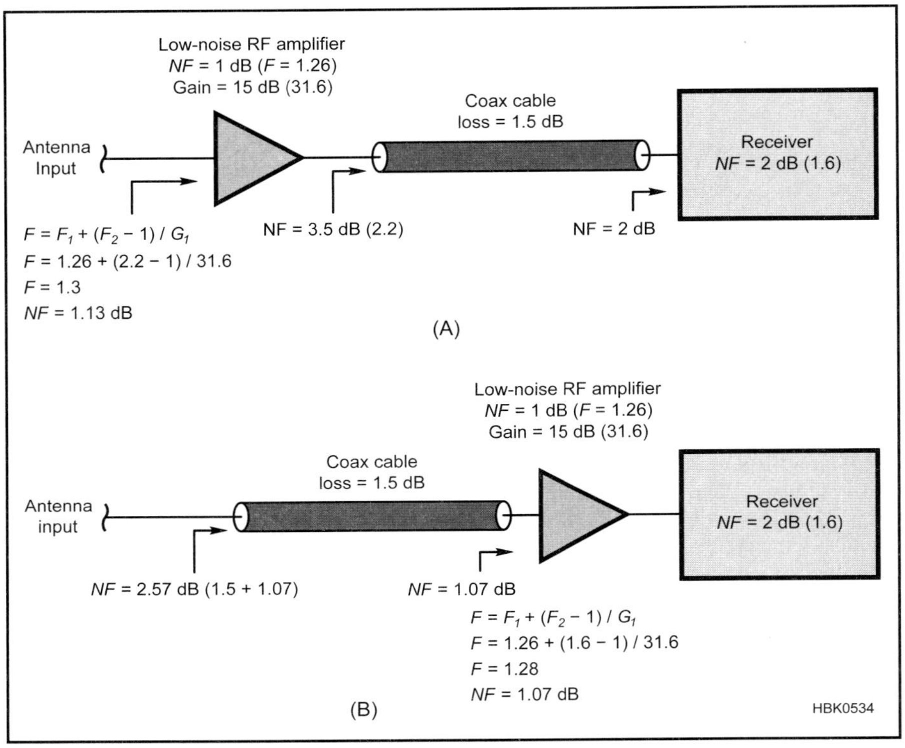

Figure 5.68 — The effect of adding a low-noise preamplifier at the antenna (A) compared to adding it at the receiver input (B).

5.8.7. Antenna Temperature

Antenna temperature, T_A, is a way of describing how much noise an antenna produces. It is not the physical temperature of the antenna because the antenna gathers noise from the environment according to its radiation pattern. If the antenna is directional and looks at a warm environment, T_A will be higher than if the antenna is looking at something cooler.

For example, if a lossless dish antenna is receiving signals from space rather than the warm Earth then the background noise is much lower than the warmer ambient temperature of 290 K or so. The background temperature of the universe has been measured as about 3.2 K. An empirical temperature for a 10 GHz antenna pointing into clear sky is about 6 K, since the antenna must always look through attenuation and temperature of the atmosphere. (See Graves in the Reference section.)

If the antenna's radiation pattern has any sidelobes that must be accounted for in the total noise received by the antenna. This raises the noise temperature. If a warm body, such as the Sun, moves into the antenna's view, the additional *sun noise* will raise T_A as well. If the antenna is looking directly at the Earth, T_A will be close to ambient temperature. As an example, T_A will vary with frequency, but a good EME antenna might have a T_A of around 20 K at UHF and higher frequencies.

5.8.8. Image Response

Most receiving systems use at least one frequency converting mixer which has two responses: the desired frequency and an image frequency above and below the frequency of the local oscillator. If the image response is not filtered out, it will add additional noise to the mixer output. Since most preamps are sufficiently broadband to have significant gain (and thus, noise output) at both the desired frequency, $G_{desired}$, and at the image frequency, G_{image}, an image filter must be placed between the preamplifier and the mixer. The total NF including image response is:

$$NF = 10\log\left[\left(\frac{1+T_e}{T_0}\right)\left(1+\frac{G_{image}}{G_{desired}}\right)\right] \quad (23)$$

assuming equal noise bandwidth for the desired and image responses.

Without any filtering, $G_{image} = G_{desired}$ so $G_{image}/G_{desired} = 1$, doubling the noise figure, which is the same as adding 3 dB. Thus, without any image rejection, the overall noise figure is at least 3 dB regardless of the NF of the preamplifier. For the image to add less than 0.1 dB to the overall NF, gain at the image frequency must be at least 16 dB lower than at the operating frequency.

As the state of the art improves beyond the typical numbers in this and previous sections, system performance also improves. The very best low-noise preamplifiers today have noise figures as low as 0.2 dB, or a T_r of about 14 K, at UHF and 1296 MHz. The best EME dishes can have a T_a in the neighborhood of 20 K at 1296 MHz when pointed at clear sky. Thus the potential T_{sys} is perhaps 40 K. At these low noise temperatures any small loss or stray noise is significant — a loss of just 0.2 dB will reduce the signal to noise ratio by 1 dB, and low SNR communications such as EME rarely have many dB to spare. The preamp must be right at the antenna for optimum performance, and have sufficient gain so that subsequent stages have little effect.

5.8.9. Background Noise

A received signal at VHF and higher frequencies necessarily competes with noise generated in the receiver as well as that picked up by the antenna, including contributions from the warm Earth, the atmosphere, the lunar surface, the diffuse galactic and cosmic background and possibly the Sun and other sources, filling the whole sky. If P_n is the total noise power collected from all such noise sources expressed in dBW, we can write the expected signal-to-noise ratio of a communications link as

$$SNR = P_r - P_n = P_t + G_t + L + G_r - P_n \quad (24)$$

where P_r is received power, P_n is noise power, P_t is transmitted power, G_t of the transmitting antennas, L is *isotropic path loss*, and G_r is the gain of the receiving antennas. All powers are expressed in dBW and all gains in dBi. (Isotropic path loss is explored further in the material on Earth-Moon-Earth (EME) communications in this book's online content on **Space Communications**.)

Since isotropic path loss L is essentially fixed by choice of a frequency band, optimizing the signal-to-noise ratio generally involves trade-offs designed to maximize P_r and minimize P_n — subject, of course, to such practical considerations as cost, size, maintainability, and licensing constraints.

It is convenient to express P_n (in dBW) in terms of an equivalent system noise temperature T_s in kelvin (K), the receiver bandwidth B in Hz, and Boltzmann's constant $k = 1.38 \times 10^{-23}$ J K^{-1}:

$$P_n = 10\log(kT_sB)$$

The system noise temperature may in turn be written as

Table 5.2

Typical Contributions to System Noise Temperature

Freq (MHz)	*CMB (K)*	*Atm (K)*	*Moon (K)*	*Gal (K)*	*Side (K)*	*T_a (K)*	*T_r (K)*	*T_s (K)*
50	3	0	0	2400	1100	3500	50	3500
144	3	0	0	160	100	260	50	310
222	3	0	0	50	50	100	50	150
432	3	0	0	9	33	45	40	85
902	3	0	1	1	30	35	35	70
1296	3	0	2	0	30	35	35	70
2304	3	0	4	0	30	37	40	77
3456	3	1	5	0	30	40	50	90
5760	3	3	13	0	30	50	60	110
10368	3	10	42	0	30	85	75	160
24048	3	70	170	0	36	260	100	360

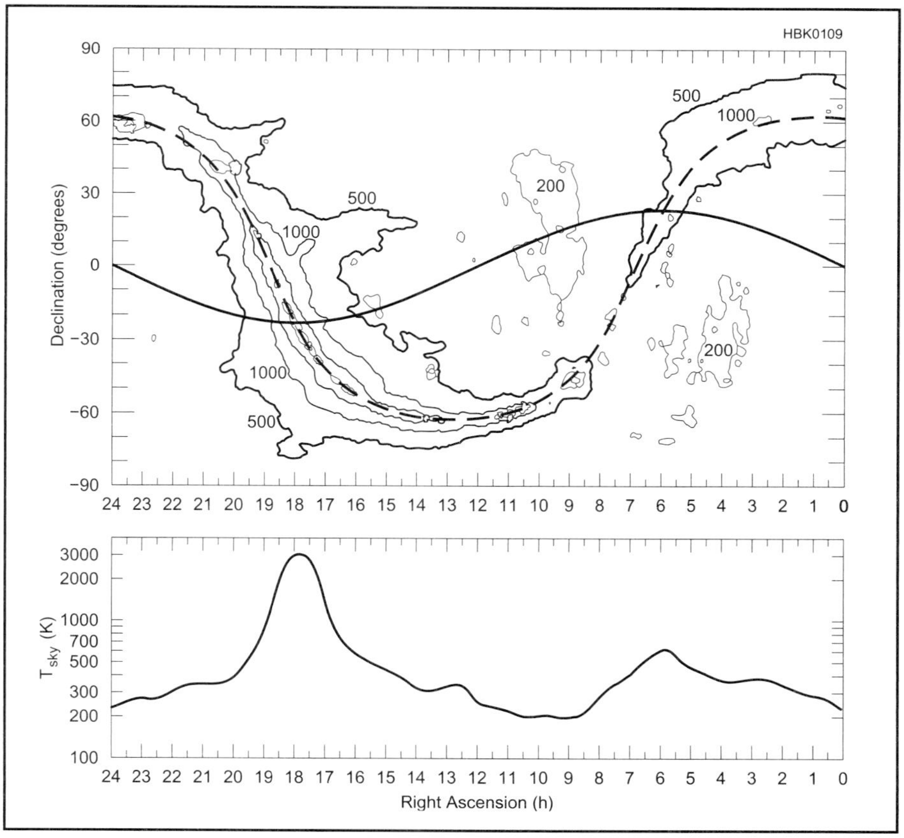

Figure 5.69 —Top: All-sky contour map of sky background temperature at 144 MHz. The dashed curve indicates the plane of our galaxy, the Milky Way; the solid sinusoidal curve is the plane of the ecliptic. The sun follows a path along the ecliptic in one year; the Moon moves approximately along the ecliptic (± 5°) each month. Map contours are at noise temperatures 200, 500, 1,000, 2,000 and 5,000 K. Bottom: One-dimensional plot of sky background temperature at 144 MHz along the ecliptic, smoothed to an effective beamwidth of 15°.

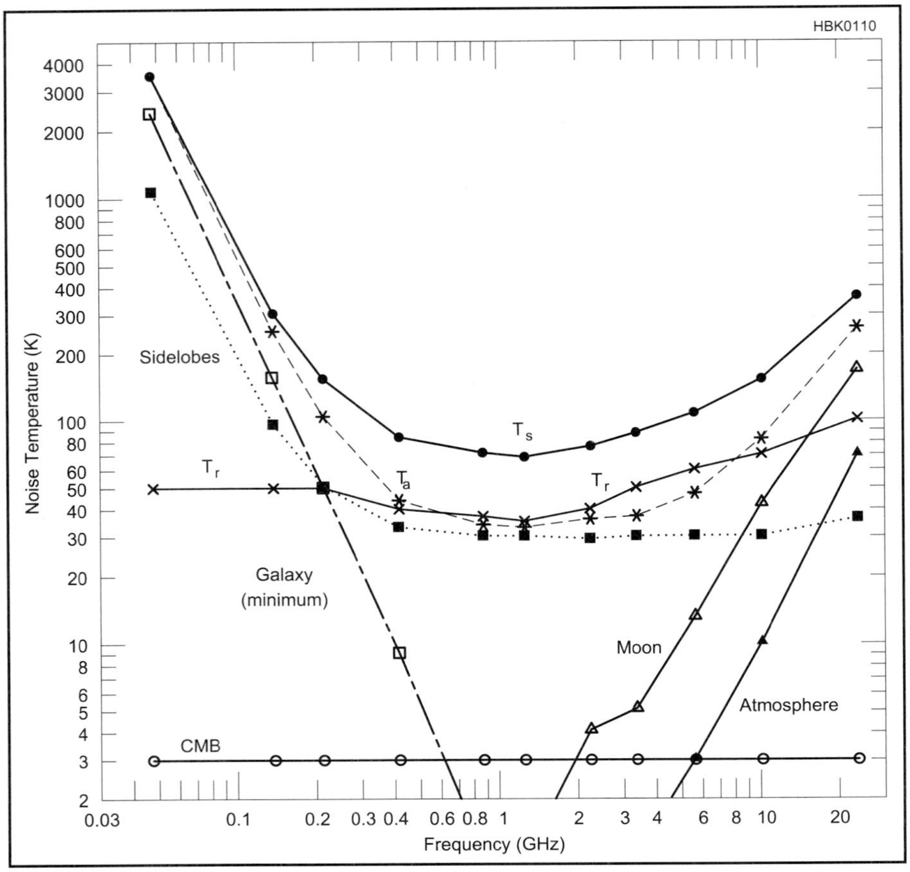

Figure 5.70 — Typical contributions to system noise temperature T_s as function of frequency. See text for definitions and descriptions of the various sources of noise.

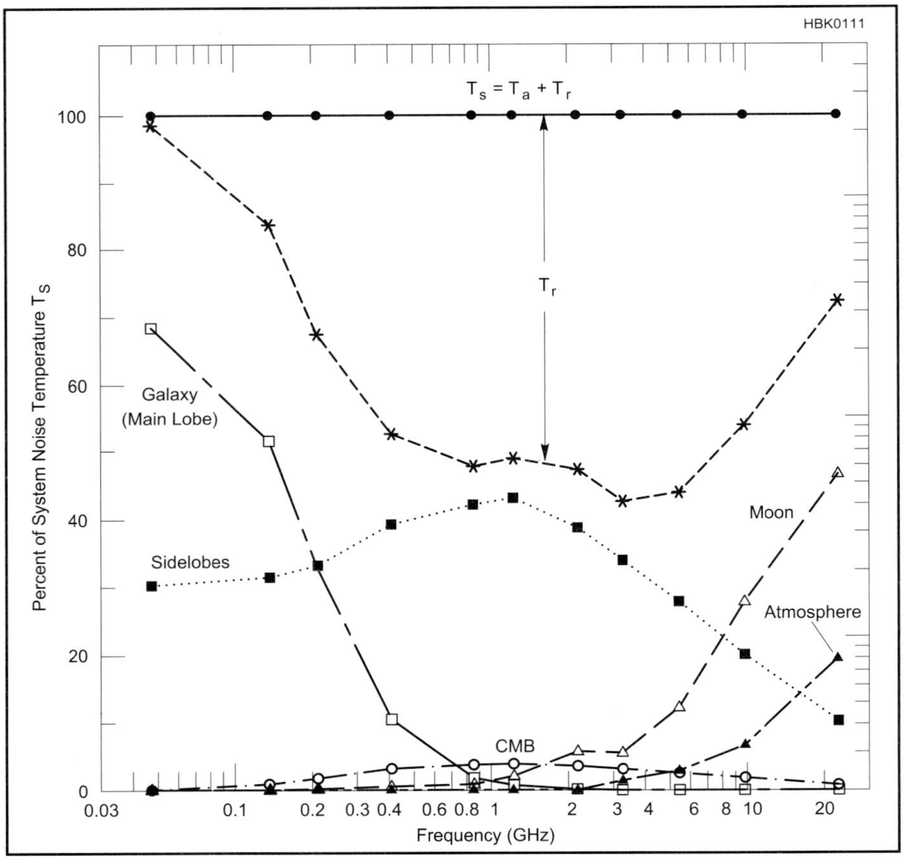

Figure 5.71 — Percentage contributions to system noise temperature as a function of frequency.

$T_s = T_r + T_a$

Here T_r is receiver noise temperature, related to the commonly quoted noise figure (NF) in dB by

$T_r = 290\,(10^{0.1NF} - 1)$

Antenna temperature T_a includes contributions from all noise sources in the field of view, weighted by the antenna pattern. Sidelobes are important, even if many dB down from the main beam, because their total solid angle is large and therefore they are capable of collecting significant unwanted noise power.

At VHF the most important noise source is diffuse background radiation from our galaxy, the Milky Way. An all-sky map of noise temperature at 144 MHz is presented in the top panel of **Figure 5.69**. This noise is strongest along the plane of the galaxy and toward the galactic center. Galactic noise scales as frequency to the –2.6 power, so at 50 MHz the temperatures in Figure 5.69 should be multiplied by about 15, and at 432 divided by 17. At 1296 MHz and above galactic noise is negligible in most directions. (See the previously mentioned online content on EME for the effects of lunar noise.)

The galactic background (GB) is a factor for HF reception as well. For daytime communications, it is less obvious due to the contributions of daytime band noise. Somewhere above 10 MHz, however, what today's quiet HF receivers hear at night becomes dominated by the GB. The frequency at which GB noise overtakes band noise depends on sunspot activity, the strength of atmospheric noise sources such as storms, geomagnetic conditions, the position of the galaxy in the sky, and the antenna's radiation pattern.

The GB has a negative spectral index, meaning it gets weaker with increasing frequency, but is still strong in the 15, 17, and 20 meter bands. Below 10 MHz the GB continues to increase, peaking around 3 MHz, but ionospheric attenuation increases with decreasing frequency, making it less of a factor than atmospheric noise. (The article "The Galactic Background in the Upper HF Band" by Dave Typinksi, AJ4CO discusses the GB at HF and is available with this book's online content.)

By definition the Sun also appears to an observer on Earth to move along the ecliptic, and during the day solar noise can add significantly to P_n if the antenna has pronounced sidelobes. At frequencies greater than about 5 GHz the Earth's atmosphere also contributes significantly. An ultimate noise floor of 3 K, independent of frequency, is set by cosmic background radiation that fills all space. A practical summary of significant contributions to system noise temperature for the amateur bands 50 MHz through 24 GHz is presented in Table 5.2, and **Figures 5.70** and **5.71**.

5.9 Two-Port Networks

A *two-port network* is one with four terminals. The terminals are arranged into pairs, each being called a *port*. The general network schematic is shown in **Figure 5.72**. The input port is characterized by input voltage and current, V_1 and I_1, and the output is described by V_2 and I_2. By convention, currents into the network are usually considered positive.

Many devices of interest have three terminals rather than four. Two-port methods are used with these by choosing one terminal to be common to both input and output ports. The two-port representations of the common emitter, common base and common collector connections of the bipolar transistor are shown in **Figure 5.73**. Similar configurations may be used with FETs, vacuum tubes, ICs or passive networks.

The general concepts of two-port theory are applicable to devices with a larger number of terminals. The theory is expandable to any number of ports. Alternatively, the bias on some terminals can be established with attention fixed only upon two ports of a multielement device. An example would be a dual-gate MOSFET in a common-source configuration as shown in **Figure 5.74**. The input port contains the source and gate 1 while the output port contains the source and drain leads. The fourth device terminal, gate 2, has a fixed bias potential and is treated as an ac ground. Signal currents at this terminal are ignored in the analysis.

5.9.1 Two-Port Parameters

There are four variables associated with any two-port network; two voltages and two currents. These are signal components. Any two variables may be picked as independent. The remaining variables are then dependent variables. These are expressed as an algebraic linear combination of the two independent quantities. The following overview is intended to define the parameters. A complete discussion of the use of two-port parameters can be found in the reference texts at the end of this chapter and examples of their use in RF circuit design in Hayward's *Introduction to Radio Frequency Design*.

Y AND Z PARAMETERS

Assume that the two voltages are chosen as independent variables. The two currents are then expressed as linear combinations of the voltages, $I_1 = K_a V_1 + K_b V_2$ and $I_2 = K_c V_1 + K_d V_2$. The constants of proportionality, K_a through K_d, have the dimensions of admittance. The usual representation is

$$I_1 = y_{11}V_1 + y_{12}V_2$$
$$I_2 = y_{21}V_1 + y_{22}V_2$$

The independent and dependent variable sets are column vectors, leading to the equivalent matrix representation

$$\begin{pmatrix} I_1 \\ I_2 \end{pmatrix} = \begin{pmatrix} y_{11} & y_{12} \\ y_{21} & y_{22} \end{pmatrix} \begin{pmatrix} V_1 \\ V_2 \end{pmatrix} \qquad (25)$$

The y matrix for a two-port network uniquely describes that network. The set of y_{11} through y_{22} are called the two-port network's *Y parameters* or *admittance parameters*. Consider the y parameters from an experimental viewpoint. The first y parameter, y_{11}, is the input admittance of the network with V_2 set to zero. Hence, it is termed the *short-circuit input admittance*. y_{21} is the *short-circuit forward transadmittance*, the reciprocal of transconductance. Similarly, if V_1 is set to zero, by short circuiting the input, y_{22} is the *short-circuit output admittance* and y_{12} is the *short-circuit reverse transadmittance*.

The matrix subscripts are sometimes replaced by letters. The set of y parameters can be replaced by y_i, y_r, y_f, and y_o where the subscripts indicate respectively input, reverse, forward, and output. The subscripts are sometimes modified further to indicate the connection of the device. For example, the short circuit forward transfer admittance of a common emitter amplifier would be y_{21e} or y_{fe}.

The y parameters are only one set of two-port parameters. The open-circuited *Z parameters* or *impedance parameters* result if the two currents are treated as independent variables

$$\begin{pmatrix} V_1 \\ V_2 \end{pmatrix} = \begin{pmatrix} z_{11} & z_{12} \\ z_{21} & z_{22} \end{pmatrix} \begin{pmatrix} I_1 \\ I_2 \end{pmatrix} \qquad (26)$$

The parameter sets describe the same device; hence, they are related to each other. If

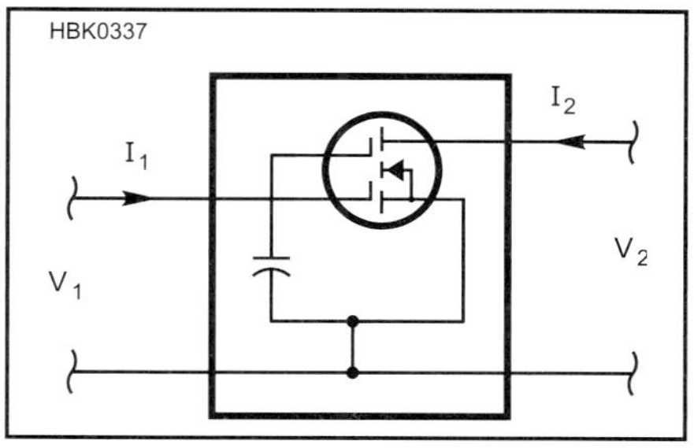

Figure 5.74 — A dual-gate MOSFET treated as a three-terminal device in a two-port network.

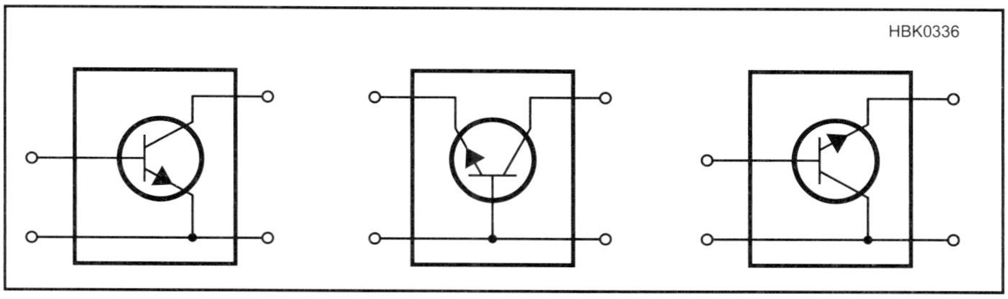

Figure 5.73 — Two-port representations of the common emitter, common base, and common collector amplifiers.

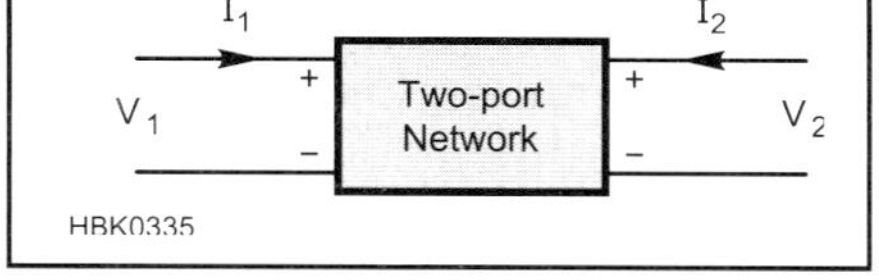

Figure 5.72 — General configuration of a two-port network. Note the voltage polarities and direction of currents.

the KVL equations are multiplied by y_{12} and the resulting equations subtracted, the result is the input voltage as a function of the currents

$$V_1 = \frac{I_1 y_{22} - y_{12} I_2}{y_{11} y_{22} - y_{12} y_{21}}$$

A similar procedure is used to find the output voltage as a function of the currents, leading to the general relationships

$$z_{11} = \frac{y_{22}}{\Delta y} \quad z_{12} = \frac{-y_{12}}{\Delta y}$$

$$z_{21} = \frac{-y_{21}}{\Delta y} \quad z_{22} = \frac{y_{11}}{\Delta y}$$

where Δy is the determinant of the y matrix, $y_{11}y_{22}$-$y_{12}y_{21}$. The inverse transformations, yielding the y parameters when z parameters are known, are exactly the same except that the y_{jk} and z_{jk} values are interchanged. The similarity is useful when writing transformation programs for a programmable calculator or computer.

H PARAMETERS

The *H parameters* or *hybrid parameters* are defined if the input current and output voltage are selected as independent variables

$$\begin{pmatrix} V_1 \\ I_2 \end{pmatrix} = \begin{pmatrix} h_{11} & h_{12} \\ h_{21} & h_{22} \end{pmatrix} \begin{pmatrix} I_1 \\ V_2 \end{pmatrix} \quad (27)$$

The input term, h_{11}, is an impedance, while h_{22} represents an output admittance. The forward term, h_{21}, is the ratio of the output to the input current, beta for a bipolar transistor. The reverse parameter, h_{12}, is a voltage ratio. The mixture of dimensions accounts for the "hybrid" name of the set.

SCATTERING (S) PARAMETERS

The two-port parameters presented above deal with four simple variables; input and output voltage and current at the ports. The variables are interrelated by appropriate matrices. The choice of which matrix is used depends upon which of the four variables are chosen to be independent.

There is no reason to limit the variables to simple ones. Linear combinations of the simple variables are just as valid. The more complicated variables chosen should be linearly independent and, ideally, should have some physical significance.

A transformation to other variables is certainly not new. For example, logarithmic transformations such as the dB or dBm are so common that we used them interchangeably with the fundamental quantities without even mentioning that a transformation has occurred. Such a new viewpoint can be of great utility in working with transmission line when an impedance is replaced by a reflection coefficient, $\Gamma = (Z - Z_0) / (Z + Z_0)$.

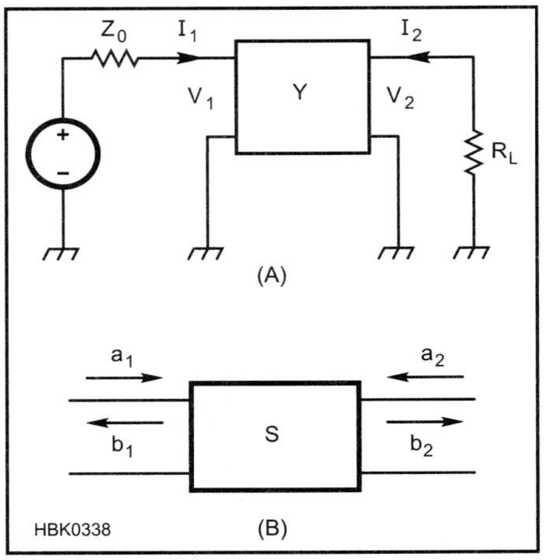

Figure 5.75 — A two-port network viewed as being driven by voltages and currents (A) or voltage waves (B). The voltages and currents are related by Y parameters while the voltage waves are related by scattering or S parameters.

Scattering parameters or *S parameters* are nothing more than a repeat of this viewpoint. Instead of considering voltages and currents to be the fundamental variables, we use four "voltage waves." They are interrelated through an appropriate matrix of s parameters.

Figure 5.75 shows the traditional two-port network and an alternate one with voltage waves incident on and reflected from the ports. The voltage waves are defined with the letters a_1, b_1, a_2, and b_2. The a waves are considered to be incident waves on the parts and are the independent variables. The b waves are the result of reflection or "scattering" and are the dependent variables. The waves are related to voltages and currents and defined with respect to a characteristic impedance, Z_0.

The scattered waves are related to the incident ones with a set of linear equations just as the port currents were related to the port voltages with y parameters. The relating equations are

$$b_1 = S_{11}a_1 + S_{12}a_2$$

$$b_2 = S_{21}a_1 + S_{22}a_2$$

or, in matrix form

$$\begin{pmatrix} b_1 \\ b_2 \end{pmatrix} = \begin{pmatrix} S_{11} & S_{12} \\ S_{21} & S_{22} \end{pmatrix} \begin{pmatrix} a_1 \\ a_2 \end{pmatrix} \quad (28)$$

Consider the meaning of S_{11}. If the incident wave at the output, a_2, is set to zero, the set of equations 27 reduce to $b_1 = S_{11}a_1$ and $b_2 = S_{21}a_1$. S_{11} is the ratio of the input port reflected wave to the incident one. This reduces, using the defining equations for a_1 and b_1 to

$$S_{11} = \frac{Z - Z_0}{Z + Z_0} \quad (29)$$

This is the *input port reflection coefficient*. Similarly, S_{21} is the voltage wave emanating from the output as the result of an incident wave at the network input. In other words, S_{21} represents a forward gain. The other two S parameters have similar significance. S_{22} is the *output reflection coefficient* when looking back into the output port of the network with the input terminated in Z_0. S_{12} is the reverse gain if the output is driven and the signal at the input port detected.

The reflection coefficient nature of S parameters makes them especially convenient for use in design and specification and even more so when displayed on a Smith Chart.

5.9.2 Return Loss

Although SWR as described in the **Transmission Lines** chapter is usually used by amateurs to describe the relationship between a transmission line's characteristic impedance and a terminating impedance, the engineering community generally finds it more convenient to use *return loss, RL,* instead.

Return loss and SWR measure the same thing — how much of the incident power, P_{INC}, in the transmission line is transferred to the load and how much is reflected by it, P_{REFL} — but state the result differently.

$$\text{Return Loss(dB)} = -10\log\left(\frac{P_{REFL}}{P_{INC}}\right) \quad (30)$$

Because P_{REFL} is never greater than P_{FWD}, RL is always positive. The more positive RL, the less the amount of power reflected from the load compared to forward power. If all the power is transferred to the load because $Z_L = Z_0$, RL = ∞ dB. If none of the power is transferred to the load, such as at an open- or short-circuit, RL = 0 dB. (You may encounter negative values for RL in literature or data sheets. Use the absolute magnitude of these values — the negative value does not indicate power gain.)

RL can also be calculated directly from power ratios, such as dBm (decibels with respect to 1 mW) or dBW (decibels with respect to 1 watt). In this case, RL = $P_{INC} - P_{REFL}$ because the logarithm has already been taken in the conversion to dBm or dBW. (Ratios in dB are computed by subtraction, not division.) For example, if P_{INC} = 10 dBm and P_{REFL} = 0.5 dBm, RL = 10 - 0.5 = 9.5 dB. Both power measurements must have the same units (dBm, dBW, and so on) for the subtraction to yield the correct results — for example, dBW can't be subtracted from dBm directly.

Since SWR and RL measure the same thing — reflected power as a fraction of forward power — they can be converted from one to the other. Start by converting RL back to a power ratio:

$$\frac{P_{REFL}}{P_{INC}} = \log^{-1}(-0.1 \times RL) \quad (31)$$

Now use the equation for computing SWR from forward and reflected power (see the **Transmission Lines** chapter):

$$SWR = \frac{\left[1+\sqrt{\frac{P_{REFL}}{P_{INC}}}\right]}{\left[1-\sqrt{\frac{P_{REFL}}{P_{INC}}}\right]} \quad (32)$$

SWR can also be converted to RL by using the equation for power ratio in terms of SWR:

$$\frac{P_{REFL}}{P_{INC}} = \left[\frac{SWR-1}{SWR+1}\right]^2 \quad (33)$$

Then convert to RL using equation 30.

5.10 References and Bibliography

Alley, C., and Atwood, K., *Electronic Engineering* (John Wiley & Sons, 1973)

Brown, J., K9YC, "Measured Data For HF Ferrite Chokes," **www.audiosystemsgroup.com/publish.htm**.

Brown, J., K9YC, "A Ham's Guide to RFI, Ferrites, Baluns, and Audio Interfacing," **www.audiosystemsgroup.com/publish.htm**.

Bruene, W., W5OLY, "Introducing the Series-Parallel Network," *QST*, June 1986, pp. 21 – 23.

Counselman, C., W1HIS, "Common-Mode Chokes," **www.yccc.org/Articles/W1HIS/mmonModeChokesW1HIS2006Apr06.pdfb**.

DeMaw, D., W1FB, "Under Construction: Understanding and Constructing RF Chokes," *QST*, Feb. 1987, pp. 19 – 21, 22.

Dorf, R., Ed., *The Electrical Engineering Handbook* (CRC Press, 2006).

Grammer, G., W1DF, "Simplified Design of Impedance-Matching Networks," *QST*, Part 1, Mar. 1957, pp. 38 – 42; Part 2, Apr. 1957, pp. 32 – 35; and Part 3, May 1957, pp. 32 – 35.

Hall, C. "RF chokes — Their Performance Above and Below Resonance," *ham radio*, June 1978, pp. 40 – 42.

Kaiser, C., *The Resistor Handbook* (CJ Publishing, 1994).

Kaiser, C., *The Capacitor Handbook* (CJ Publishing, 1995).

Kaiser, C., *The Inductor Handbook* (CJ Publishing, 1996).

Kester, W., "Understand SINAD, ENOB, SNR, THD, THD + N, and SFDR so You Don't Get Lost in the Noise Floor," Analog Devices, MT-003 Tutorial, **www.analog.com/media/en/training-seminars/tutorials/MT-003.pdf**.

Maxwell, W., W2DU, "Reflections III," *CQ Communications*, 2010.

Pettai, R., *Noise in Receiving Systems* (Wiley, 1984).

Silver, W., NØAX "About Impedance Matching Circuits," *QST*, Oct. 2019, pp. 43 – 46.

Trask, C., "Designing Wide-band Transformers for HF and VHF Power Amplifiers," *QEX*, Mar./Apr. 2005, pp. 3 – 15.

Van Valkenburg, M., *Reference Data for Engineers* (Newnes, 2001).

Zavrel, R., W7SX, *Antenna Physics: An Introduction* (ARRL, 2016).

RF AMPLIFIER DESIGN

Carr, J., *Secrets of RF Circuit Design* (McGraw-Hill/TAB Electronics, 2000).

DeMaw, D., W1FB, *Practical RF Design Manual* (MFJ Enterprises, 1997).

"FET RF Amplifier Design Techniques," Motorola Application Note AN423.

Hayward, W., W7ZOI, *Introduction to RF Design* (ARRL, 1994, out of print)

Hayward, W., W7ZOI, "A SMT Dual Gate MOSFET Preamplifier for 50 MHz," **www.w7zoi.net/lna50.pdf**.

Hayward, W., Campbell, R., and Larkin, B., *Experimental Methods in Radio Frequency Design* (ARRL, 2009).

Hayward, W., W7ZOI, and DeMaw, D., W1FB, *Solid-State Design for the Radio Amateur* (ARRL, 1994).

Pozar, D., *Microwave Engineering*, 4th Edition (John Wiley & Sons, 2012).

Raab, F., et al., "RF and Microwave Power Amplifier and Transmitter Technologies — Parts 1– 5," *High Frequency Electronics*, May, July, Sep., Nov., 2003, and Jan. 2004 issues.

"RF Small Signal Design Using Two-Port Parameters," Freescale Application Note AN215A.

"Small-Signal RF Design with Dual-Gate MOSFETS," Motorola Application Note AN478A, **www.archive.org/details/AN478A**.

Terman, F., *Electronic and Radio Engineering* (McGraw-Hill, 1955).

Contents

Chapter 6 — Online Content

Articles
- Mathematical Stability Problems in Modern Non-Linear Simulation Programs by Dr. Ulrich Rohde, N1UL and Rucha Lakhe
- The Dangers of Simple Usage of Microwave Software by Dr. Ulrich Rohde, N1UL and Hans Hartnagel
- Harmonic Balance by Dr. Ulrich Rohde, N1UL, Ajay Poddar, and Matthias Rudolph
- Using Simulation at RF by Dr. Ulrich Rohde, N1UL
- Harmonic Balance by Dr. Ulrich Rohde, N1UL, Ajay Poddar, and Matthias Rudolph
- A *Sonnet* Tutorial using Port Tuning

Tools and Data
- SON and data files for the Electromagnetic Analysis of RF Circuits section

Note that the color image files, SON/data files, and the paper by Rohde are in the Chapter 6 Working Files folder “Chapter 6 — Online Content”

Chapter 6

Computer Modeling and Simulation

This chapter, which in the past has been entitled Computer-Aided Design (CAD) and Electronic Design Automation (EDA), surveys modeling techniques that can be used to simulate the operation of electronic circuits and electromagnetic waves to aid in the design and analysis of RF circuits and antennas. Sophisticated modeling software is available to any amateur with a personal computer. It predicts how circuit designs will behave and how antennas will radiate before those objects are built, saving considerable time and expense that in the past was spent getting things to work the way they should.

Jean-Francois Debroux addresses circuit simulation tools. Dr. Ulrich Rohde, N1UL, surveys issues associated with linear and nonlinear RF simulation and contributes three extensive papers to the online content. Dr. Wolfgang Griebel, DL1DWG, provides SPICE examples. Jim Rautio, AJ3K, presents an overview of electromagnetic (EM) simulation.

The purpose of this chapter is not to provide detailed instructions for using any particular software package, but to explain the basic operations, limitations and vocabulary for circuit and RF design software. Software to aid design and analysis in specialized areas, for example filter design, switch mode power supplies, transmission lines, and RF power amplifiers, is covered in other chapters.

6.1 Circuit Simulation Overview

Circuit simulation is a branch of a more general computer-based technique: simulation. If the laws of physics and a technical device or system can be described appropriately for a computer program (a simulator) to process them, the result mimics the behavior of the actual system without the need to build it. Because computers are more and more powerful and simulation programs are more and more efficient, simulation is now widely used in the development of almost every product. There are simulators in fields like mechanics, electronics, and many others. Here, the focus is placed on circuit simulation.

Circuit simulators have been around for decades. Significant progress has been made. One of the milestones was *SPICE,* developed at the University of California at Berkeley in the 1970s. *SPICE* is an acronym standing for **S**imulation **P**rogram with **I**ntegrated **C**ircuit **E**mphasis. Most circuit simulators today still use the *SPICE* syntax and philosophy even though their capabilities and performance have improved by orders of magnitude. Over time, versions of SPICE, such as PSPICE, were adapted for use on personal computers rather than mainframe computers. However, as the power of a personal computer has exceeded that of the mainframe computers of the 1970s, the difference is no longer meaningful.

Many circuit simulators today are associated with *schematic capture* programs that allow drawing schematics the way it used to be done by hand. (See the **Construction Techniques** chapter's section on PCB CAD and Fabrication for more information on schematic capture.) The simulation *results viewer* graphically displays voltages and currents like an oscilloscope or a spectrum analyzer does.

6.1.1 The Design Cycle with Simulation

Using simulation has a significant impact on the development cycle. The classic cycle is shown in **Figure 6.1**. The process starts with a specification that defines the goal in terms of functionalities and performances. Functionalities define architecture (the way parts are assembled and connected) while performance parameters (power, speed, etc.) define part selection and size. After the project sketch is defined, it is time to build, which requires buying the parts and assembling everything.

Next comes the test phase, where you compare the prototype to the specification. To determine whether the prototype matches the specifications, you must have the right test equipment. If the prototype performs well, the project is done, but if not, a potentially painful and costly phase begins to identify what's wrong, fix the issues, and eventually iterate the process until everything is okay.

If you're using a simulation, then you modify the cycle as in **Figure 6.2**. This cycle is slightly more complex, but it saves time and money. You don't need to buy parts or test equipment during the first phase — the simulator does it all. After you've run the simulation successfully, having addressed any issues flagged during earlier attempts, you are ready to begin the implementation phase. The simulation should normally reduce the risk of damaged

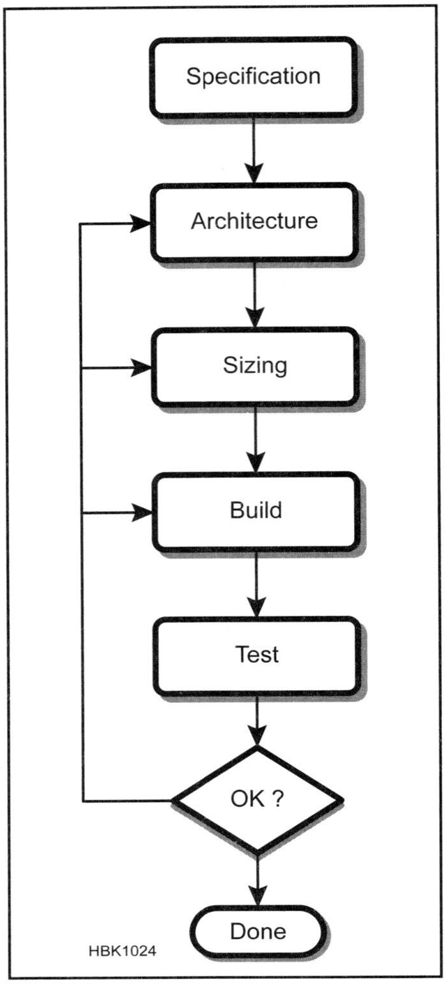

Figure 6.1 — Classic design cycle.

parts and limit the number of iterations you must go through during the implementation.

For interested readers / To go further

https://en.wikipedia.org/wiki/Computer_simulation

https://en.wikipedia.org/wiki/SPICE

www.analog.com/en/design-center/design-tools-and-calculators/ltspice-simulator

6.1.2 Hobby vs. Professional Circuit Simulation Tools

The market for simulation tools has evolved in recent years because systems are getting more and more complex with more and more interactions requiring very complex and expensive software packages that target professionals.

On the other hand, open-source software is now very common and addresses almost every domain. Several open-source circuit simulators exist with active communities improving them. *Ngspice* is an example. These tools are available under a variety of licensing models (gpl, lgpl, etc...) The same is true for schematic capture, printed circuit board (PCB) design, and IC layout. *Kicad, KLayout* and *gEDA* are some examples. These open-source tools may not offer the same performance as professional tools, but they are a good choice for hobbyists.

Another family of circuit simulators, that bridges professional and hobbyists, has appeared with tools like *LTSPICE*. It was developed by the IC manufacturer Linear Technology (now part of Analog Devices) for users to validate an application using the company's products before buying the components and building a board. These circuit simulators are free but not open source. They come with a comprehensive library of components, although much of the library is dedicated to the components made by that company, and are associated with schematic capture and a waveform viewer for ease of use. Additionally, previously professional circuit simulators, that once sold for thousands of dollars, have been discontinued and made available to everyone without cost. *MICRO-CAP* from Spectrum-Soft is an example of such a simulator.

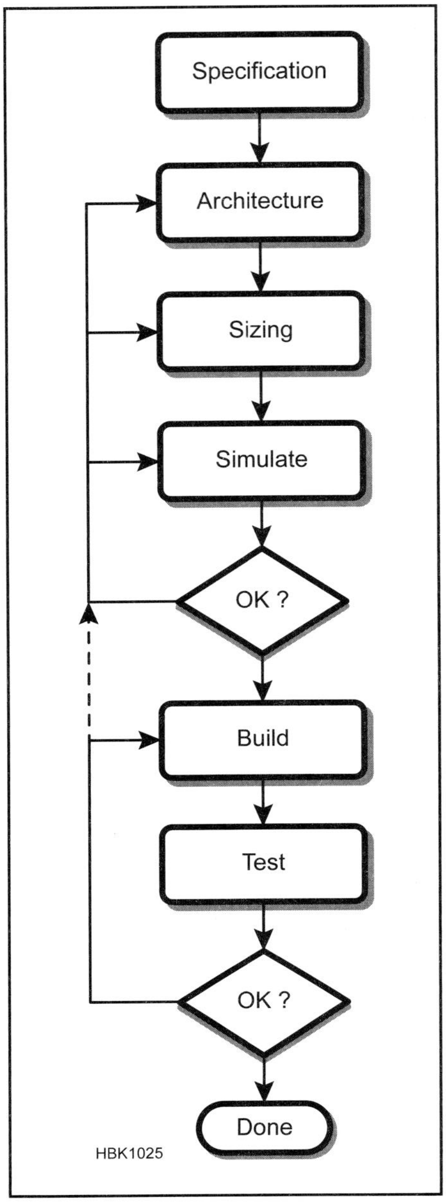

Figure 6.2 — Design cycle with simulation.

6.2 Interests and Limitations of Circuit Simulation

There are four major benefits to simulating a circuit rather than building it:

1. It is fast
2. It is cost effective
3. It gives deep insight on circuit behavior
4. It is safer

But there are limitations:

1. Only what is modeled can be simulated
2. Only what is measured can be monitored
3. Only analysis types supported by a given simulator can be performed

It is important to understand these points well so we will focus on them.

Simulation is fast and cost effective: With today's free circuit simulators and the power of available computers, it takes only minutes to draw a simple schematic and simulate it. There is no need to buy components, no delay in getting them and building the circuits, no need for expensive signal generators, oscilloscopes and spectrum analyzers — the circuit simulator has all that inside!

Simulation gives deep insight: Once the simulation is done, every voltage and every current in any circuit device can be displayed. This is not so easy to do in a physical circuit. These "measurements" do not perturb the circuit. Simulation can be done at different temperatures and consider component tolerances, giving some information on circuit robustness.

Simulation is safer: There is no risk in simulating a high voltage or a high-current circuit. No smoke is produced in case a component is not used properly!

Simulation is model limited: A circuit simulator uses models to describe components. A model is a set of equations with parameters that describes the relationship between voltages and currents.

For instance, a resistor model is defined by Ohm's law: $V = I \cdot R$. This model has only one parameter: R, the resistor value. Such a resistor is stable over temperature since it has no temperature coefficient.

A more realistic resistor would be described by the same Ohm's law, but in addition, R would depend on temperature. For example, $R = R_0 (1 + TC (Temp - T_0))$, where R_0 is nominal resistance at a nominal temperature, T_0, and TC is the temperature coefficient of the resistor that describes how much its resistance changes as the temperature changes.

Simulation can only show what is described and what is watched: In a circuit simulator, components are connected to *nets*, the connections between components. The list of all connections is called a *netlist*. All components sharing a net are connected by a zero-ohm resistance, which does not exist. If the actual connections exhibit some resistance and inductance, these must be added explicitly to the model or as separate components. In the same way, if two nets exhibit some parasitic capacitance between each other, that must be added explicitly.

Accounting for parasitic elements must be done wisely. (See the **RF Techniques** chapter for more about parasitic components.) There is no need to consider connection resistance in a circuit that draws micro-amps. Similarly, there is no need to consider parasitic inductance and capacitance at low frequencies.

Model limitations and parasitic elements are the major sources of difference between simulation and experiment. For example, if a 0.25 W resistor dissipates 2 W in a breadboard circuit, it will probably be destroyed. But in a simulation, it is not. The designer must check dissipated power and eventually use that information to choose the proper resistor size. Some simulators have a feature called "Smoke alarm" that can highlight overloaded components provided that the model includes some voltage/current/power limits.

Simulation is limited to available analysis: Many simulators inherit the analysis types that were defined by *SPICE*. But these do not cover all the fields of circuit analysis. For instance, noise analysis in *SPICE* is limited to small signal ac simulation. So, *SPICE* is not suited to perform a mixer noise analysis since such a circuit's operation is fundamentally non-linear.

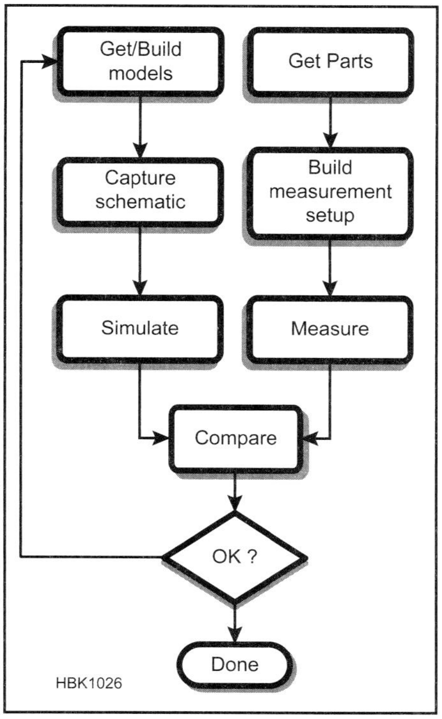

Figure 6.3 — Model validation process.

6.2.1 Circuit Simulation Process

To simulate a circuit, you need six things:

1. A method
2. A circuit description (so that the simulator knows what to simulate)
3. Models for the components used in the circuit
4. One or more simulation commands defining the simulation to be performed
5. A circuit simulator
6. A computer

SIMULATION METHOD

Most circuit simulators are embedded in *frameworks* that manage schematic, layout, parasitic extraction, etc. In a professional setting, models are usually validated by dedicated support teams. This organization and these tools are expensive but, as a result, simulation results are reliable, prototypes operate as expected in most cases, and production yield is high and consistent.

For the hobbyist, the situation is very different. Models are supplied without a real description of validity, and parasitic elements are hard to predict, so the prototype may behave somewhat differently from the simulation. A model validation phase shown in **Figure 6.3** is usually required, especially in high frequency applications.

The principle of accurate simulation is to compare modeling results to measurement results and trim the model until the difference is acceptable. Trimming differs according to model origin. If it is a manufacturer model, trimming consists mainly of adding environmental parasitics, such as PCB trace resistance, inductance, and capacitance.

If it is a homemade model, trimming implies changing model parameters and adding environmental parasitic elements. It is a much more complex process that requires a good knowledge of physics, electronics, and the underlying math.

The key here is not to blame the simulator or the model or the test setup if differences exist. The important thing is to understand why there are differences and what can influence them to bring the simulation schematic and model closer and closer to the test setup until results match. It is also important to keep in mind that the errors can be located on the test side and that the setup might require improvement.

CIRCUIT DESCRIPTION

Most modern *SPICE*-type simulators have graphical input interfaces, such as *LTSPICE* and *MICRO-CAP*, and automate the generation of the *SPICE* command language. The casual user will not need to write command lines in most cases when using programs like these. However, it may be helpful to be aware of the *SPICE* command language in order to customize a design in which the graphical interface does not exactly model the desired circuit behavior. See **Figure 6.4** for a simple schematic with its command language description (called a "netlist") and the sidebar, "*SPICE* Command Structure," for more detail.

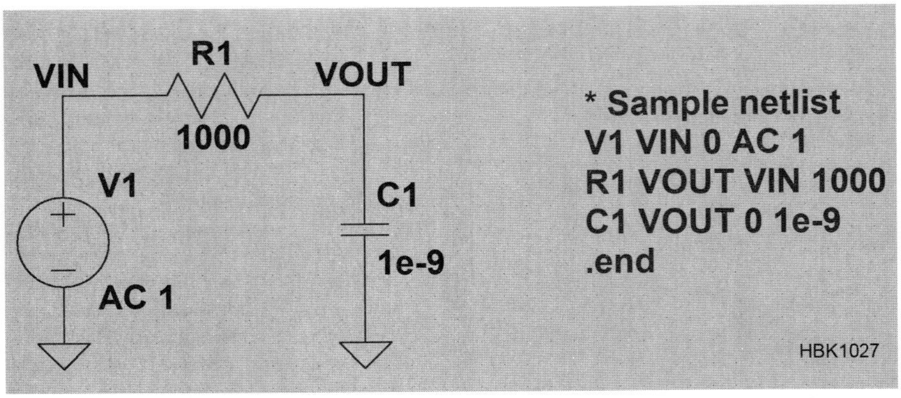

Figure 6.4 — An example of how a schematic of an ac voltage source driving a first order low-pass filter is described by a netlist. Components and nets are labeled. Here, there are three components and two nodes, plus ground. On the right is the circuit netlist. It is a text form description of a schematic that can be parsed by the simulator to build an internal description that can be simulated.

6.2.2 Models

Some components, like resistors, can be modeled with a single value, but other components, like diodes, must be defined by multiple values. All diodes share some common characteristics; they all behave in a similar way and can be described by the same set of equations. However, a small signal diode and a power rectifier do not have the same characteristics. To make that distinction, circuit simulators use parameterized models. For instance, a diode instance can be:

D1 Node1 Node2 D1N4148

This defines that diode D1 has its anode connected to Node1, its cathode connected to Node2, and it behaves according to model D1N4148. There must be a model definition in the netlist:

.MODELD1N4148D(is=2.52E-9rs=.568 n=1.752 cjo=4E-12 m=.4 tt=20E-9)

In contrast, a fast rectifier, like MURS320, has the following model:

.MODEL MURS320 D (is=1.06E-9 rs=.0111 n=1.367 cjo=135E-12 m=.45 tt=45E-9)

These two diodes use different parameter values that are passed to the equations so that they behave differently. A quick look at the parameters shows that the biggest difference between these two diodes lies in series resistance, rs, and junction capacitance at 0 V, cjo. A detailed description of all *SPICE* model equations can be found online, but in most cases modern modeling tools convert schematic diagrams into a series of *SPICE* model equations, and there is little need to directly manipulate those.

From the beginning, *SPICE* has been IC-oriented. Differences between a discrete transistor and one that is embedded in an IC include:

- A discrete transistor is usually packaged separately. The package exhibits significant parasitic capacitance between pins and significant series inductance on each pin.
- A transistor inside an IC has no package, only the entire IC is packaged. So, parasitic capacitance for individual components in the IC is about two orders of magnitude smaller. The same applies for inductance since connections are much shorter inside ICs. But in an IC, the transistor is built on the same substrate as other components and has an isolation diode between collector and substrate.

For a discrete transistor, especially if it is to be used at high frequencies, the model describing the transistor should be embedded in a package description that includes parasitic inductances and capacitances. Not doing so is a major source of differences between simulation and measurements. **Figure 6.5** shows what a high frequency device model should look like.

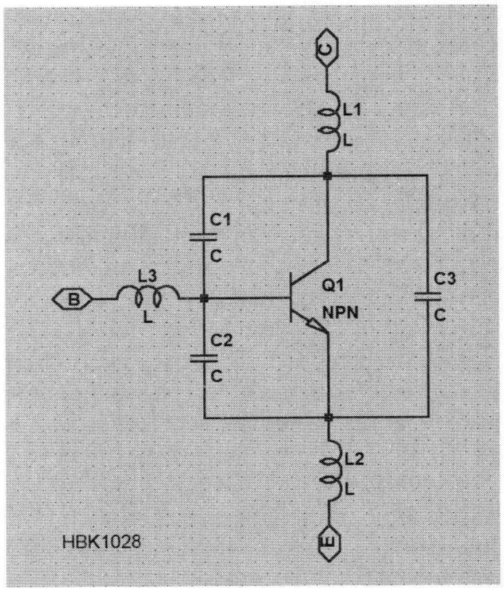

Figure 6.5 — Transistor model showing parasitic inductance and capacitance.

For bipolar transistors, the *SPICE* model includes series resistance on collector, base, and emitter. These are named RC, RB, and RE in the model parameter list. But the model does not include series inductance. If the package has significant leads length and if operating frequency is high, these inductances should be added. Fortunately, high frequency transistors usually come in packages with short leads or no leads.

The model includes both B-C and B-E junction capacitances, but these are the internal junction capacitance. The reverse-biased junction (usually B-C) capacitance results from the depleted region and depends on C-B voltage. For the forward-biased junction, capacitance (usually B-E) results from the depleted region, which is very narrow, and from the minority carrier transit time. At low collector currents, depleted region capacitance dominates and is stable, while at high currents, transit time capacitance dominates and is proportional to current.

The model does not include the fixed package parasitic capacitances. Connecting these fixed parasitic capacitances in parallel with the transistor's varying capacitances results in larger values that vary less. Unfortunately, especially for low frequency devices, manufacturers provide tweaked C_{BE} and C_{BC} values to include package capacitance values. The result is a poorly matched description of how these values vary with a signal. A more accurate model consists of putting together a transistor model as a chip without additional parasitic elements, and a separate package model extracted by measuring an empty package.

Nothing prevents users from building their own models. But this requires adding package parasitic elements through the difficult work of measuring them and building the transistor chip model through the even

SPICE Command Structure

Most circuit simulators accept a netlist input. The netlist format was popularized by *SPICE* in the 1970s. Even though the "*SPICE* netlist" is in fact a components and commands list, here is a description of *SPICE* netlists syntax for those who need to write netlists manually or need to read and understand netlists.

- *SPICE* netlists are not case sensitive.
- A *SPICE* netlist file is basically organized as separate lines.
- Each line ends with a return or newline character.
- Each line contains different fields. Fields are separated by a space character.
- The line's first character defines the line's function.

SPICE uses SI units (volts, amperes, ohms, farads, henrys, hertz, seconds, etc.). It accepts letters for exponents (k for kilo, p for pico, etc.). Confusion is possible with "m" that stands for milli while "meg" stands for mega, and "mil" stands for 0.001 inch. A good practice to avoid this confusion is to use a floating-point format (1.5E3 for 1.5 k). The default temperature for *SPICE* is 27 °C (300 K, 80.6 °F). It can be changed and even swept by dedicated commands.

Comment Lines

The netlist file's first line is always considered to be a title line and is not interpreted. It is treated like a comment line but appears in detailed result files. Any line beginning with an asterisk (*) character is a comment line. It is a good practice to begin the title line with an asterisk so that an error is not caused if a blank line is unintentionally added at the beginning of the file.

Continuation Lines

A line beginning with a plus (+) character is a continuation of previous line. There can be more than one continuation line.

Component Lines

A line beginning with a letter (A to Z or a to z) is an instance of a component. The line's first letter defines the component type. For instance, C stands for a capacitor and R stands for a resistor. The line's first field (beginning with the identifier letter) defines the component name or designator, such as R1, L4, D2, etc. The name can contain letters and numbers and must be unique. Other fields define the net names the component is connected to. The number of nets depends on the component type and model. Most recent simulators accept alphanumerical net names, but the old *SPICE* syntax allowed only numeric net names. Additional fields depending on the component nature define model name, sizing, etc.

The following is a reduced components list from strict *SPICE* syntax. Many more recent simulators have additional letters for additional component types.

C: *Capacitors*. Next two fields are net names. Then, either a value or a model name followed by a value.

D: *Diodes*. Next two fields define Anode and Cathode net names in that sequence, then a model name and optionally a multiplier.

E: *Voltage-controlled voltage sources*. Next two fields are output nodes followed by two fields defining controlling nodes and finally a field defining gain value.

F: *Current-controlled current sources*. Next two fields are output nodes followed by one field defining controlling source name and finally a field defining gain value.

G: *Voltage-controlled current sources*. Next two fields are output nodes followed by two fields defining controlling nodes and finally a field defining gain value.

H: *Current-controlled voltage sources*. Next two fields are output nodes followed by one field defining controlling source name and finally a field defining gain value.

I: *Independent current sources*. Next two fields are output nodes followed by current type and parameter(s).

J: *JFETs*.

K: *Inductor coupling factor*.

L: *Inductors*.

M: *MOSFETs*.

Q: *Bipolar transistors*. Next four fields define Collector, Base, Emitter, and Substrate node names in that sequence. Then a field defines model name and optionally a multiplier.

R: *Resistors*.

T: *Transmission lines*.

V: *Voltage sources*.

X: *Subcircuit calls*. Next fields are subcircuit pins. Last field is subcircuit name.

For a detailed description of these components, the simulator reference manual is the most valuable source of information. A description of model syntax is available in the *LTSPICE* guide:

www.analog.com/media/en/simulation-models/spice-models/LTspiceGettingStarted Guide.pdf?modelType=spice-models.

Command Lines

A line beginning with a dot (.) is a simulation command. Most commands define simulation to be performed or contain simulation directives. These commands are associated with fields defining simulation parameters.

A special command, .SUBCKT is used to define a subcircuit, allowing a hierarchical description of a complex circuit.

A reduced commands list from strict *SPICE* syntax follows. Many more recent simulators have additional commands or, confusingly, have modified the significance of some commands. Be sure you understand a command's function and syntax before you use it.

Simulation Setup Commands

These commands are interpreted prior to running the analysis. They define global variables, parameters, or libraries for models.

.LIB: *Defines a model library*. Only the required devices are loaded.

.STEP: *Defines a parameter sweep*. Analysis is performed once for each defined parameter value.

.TEMP: *Defines temperature*.

.IC: *Defines initial conditions that exist on some nets*. This command is particularly helpful to get an oscillator to start.

Analysis Commands

.AC: *Defines small signal analysis frequency sweep*. Parameters are sweep type, start frequency, stop frequency, and step. Operating point is computed first, the small signal equivalent circuit for non-linear devices is extracted, and ac analysis is performed on the circuit's equivalent small signal model.

.DC: *Defines dc sweep analysis*. A voltage, a current, or other parameter can be swept to plot any output vs. the swept value. It is a static sweep with no timing involved.

.NOISE: *Defines small signal noise analysis*. Parameters are output, name of input source, and a step parameter. Frequency range is defined by .AC command. Summary is plotted every "step" value of AC point.

.OP: *Defines operating point analysis*.

.TRAN: *Defines transient analysis*. Parameters are start time, stop time, and time step.

Post-Processing Commands

.FOUR: Defines a Fourier series analysis on TRAN results.

.SENS: Defines a sensitivity analysis on OP results.

more difficult task of getting bare transistors and measuring them.

Many models, and parameters for models, can be found on manufacturer websites. These models should be used carefully as they are usually supplied "as is," without documentation or validity information.

Do not be discouraged if your simulation does not exactly describe the nuances of a circuit, such as the inclusion of parasitic elements for many components. Although the model may not produce an exact simulation of the final circuit, it will provide valuable insight into the behavior of a design and allow the designer to provide circuit modifications to make it behave as desired. Once the circuit is built, small deviations from the simulation can generally be corrected with the minor modification of some component values.

PASSIVE COMPONENT PHYSICAL MODELS

Several simulators have introduced "physical models" as opposed to *SPICE* ideal models. *SPICE* is somewhat of a hybrid with respect to this issue. It does include, for instance, series resistance on active device pins (with values defaulting to 0), but not for passive devices.

Passive physical models include parasitic elements such as series resistance or inductance to provide a more realistic modeling. But there is no standard definition, and such models provide more confusion than accuracy!

The key is to be aware of what the simulator is modeling and what it is not, and to add elements missing from the simulation to make it more realistic. The **RF Techniques** chapter shows component construction and explains the effect of parasitic elements on component behavior. Fortunately, in most modern designs, the size reduction provided by surface mount components has significantly reduced parasitic elements.

6.2.3 Example: Series Voltage Regulator

This and the following sections show different examples of simulation methodology to design and evaluate most aspects of circuit operation. These examples are based on *LTSPICE* for its ease of use and power.

The target specification is:

- Input voltage: 15 V to 25 V
- Output voltage: 12 V +/- 5%
- Output current: 0 to 1 A

The first step is choosing an architecture. For simplicity, and since the specification is not stringent, the design can be a Zener reference, open loop regulator. The initial schematic is shown in **Figure 6.6**.

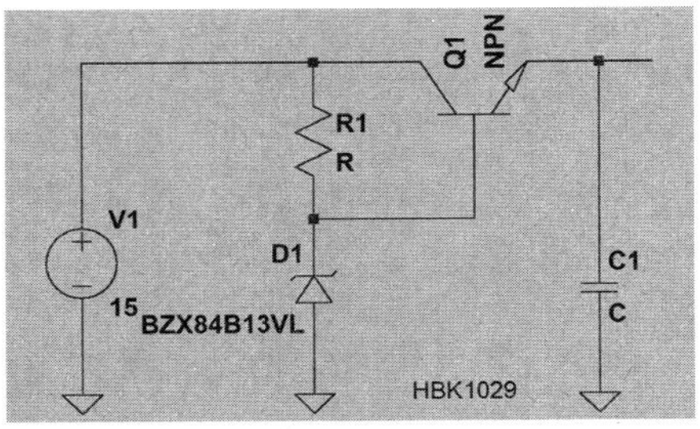

Figure 6.6 — Initial schematic for a series voltage regulator circuit.

The specification states that the pass transistor should be able to handle at least 1 A and 25 V. Simple calculations show that at maximum input voltage (25 V) and maximum current, the pass transistor should be able to dissipate P = 25 – 12 × 1 = 13 W.

R1 should bias the Zener diode and supply pass transistor base current. The Zener should be biased between 5 and 20 mA to benefit from a low dynamic impedance. Base current would be load current divided by current gain, β (beta).

$$R1 = \frac{V_{in} - V_{ref}}{I_{bias} + \frac{I_{load}}{\beta}}$$

The next step is choosing a pass transistor that can handle 1 A and dissipate at least 13 W. A safety factor of 3X changes the requirement to 3 A and 40 W. Devices like BD241C or 2N3055 fit that requirement. The transistor datasheet minimum β is listed as 25, so the base current could be as large as 40 mA. This is more than most Zener diodes can handle. There are two design options to increase pass transistor β:

- Using a Darlington configuration at the expense of an increased voltage drop
- Using a composite NPN-PNP configuration.

The second option is preferable but requires a power PNP together with a small signal NPN. A BD242C or 2N2955 that are complementary to the BD241C or 2N3055 are suitable. A 2N2222 is fully capable of delivering the PNP base current. The 2N2222 minimum β at 40 mA is 50, so the base current should not increase more than 1 mA. This value is much lower than the maximum Zener current.

Output voltage will be the Zener voltage minus 2N2222 V_{BE} (about 0.7 V, temperature dependent). An elegant solution uses a 12 V Zener diode in series with a diode-connected 2N2222 that approximately compensates for the pass driver. The tuned-up schematic is shown in **Figure 6.7**.

Models for the BD242C and 2N2955 can be found at **www.onsemi.com** and copied into *LTSPICE*. 2N2222 and 12 V Zener models are already available in the *LTSPICE* library. The following simulations have been done with a BD242C power transistor, but results are very similar with the 2N2955. **Table 6.1** shows an example netlist.

It is now time to simulate this circuit to check the assumptions used to build it. In *LTSPICE*, the simulation command lines are placed on the screen with the schematic diagram. Only one active simulation is allowed at a time, which can be a cumbersome process when several simulations are desired.

Table 6.1

***LTSPICE* Netlist**

```
* C:\Users\JF\Desktop\Travail\LTC\DiscreteSeriesRegulator.asc
V1 N001 0 15
R1 N001 N004 500
I1 N002 0 0.5 AC 1
Q2 N003 N004 N002 0 2N2222
Q3 N002 N003 N001 0 Qbd242c
R2 N001 N003 100
Q4 N004 N004 N005 0 2N2222
D2 0 N005 BZX84C12VL
.lib C:\Users\JF\Documents\LTspiceXVII\lib\cmp\standard.dio
.lib C:\Users\JF\Documents\LTspiceXVII\lib\cmp\standard.bjt
.MODEL Qbd242c pnp
+IS=1e-09 BF=144.298 NF=0.85 VAF=15.6106
+IKF=1.39466 ISE=8.61573e-12 NE=3.32689 BR=0.1
+NR=0.75 VAR=19.1801 IKR=10 ISC=3.98813e-12
+NC=3.60783 RB=0.1 IRB=0.1 RBM=0.1
+RE=0.0001 RC=0.0601441 XTB=0.828305 XTI=1
+EG=1.05 CJE=3.18622e-10 VJE=0.49128 MJE=0.481942
+TF=1e-08 XTF=4.23532 VTF=7.00536 ITF=0.001
+CJC=3.12657e-10 VJC=0.630571 MJC=0.450516 XCJC=0.799996
+FC=0.8 CJS=0 VJS=0.75 MJS=0.5
+TR=9.57121e-06 PTF=0 KF=0 AF=1
.backanno
.end
```

To streamline changing commands, here is a useful tip: turn commands into comments so that the simulation ignores them, but you have a full list of the commands you want to run available on the schematic diagram. To comment a command:

1. Right click on the command.
2. Choose "Cancel" in the dialog box that pops up.
3. Choose "Comment" in the second dialog box.

This way you can enter all the commands at once while turning on only the commands you want to run in each simulation.

The commands that are embedded in the schematic to provide the simulation results can be selected from drop-down menus in other simulation programs, such as *MICRO-CAP*.

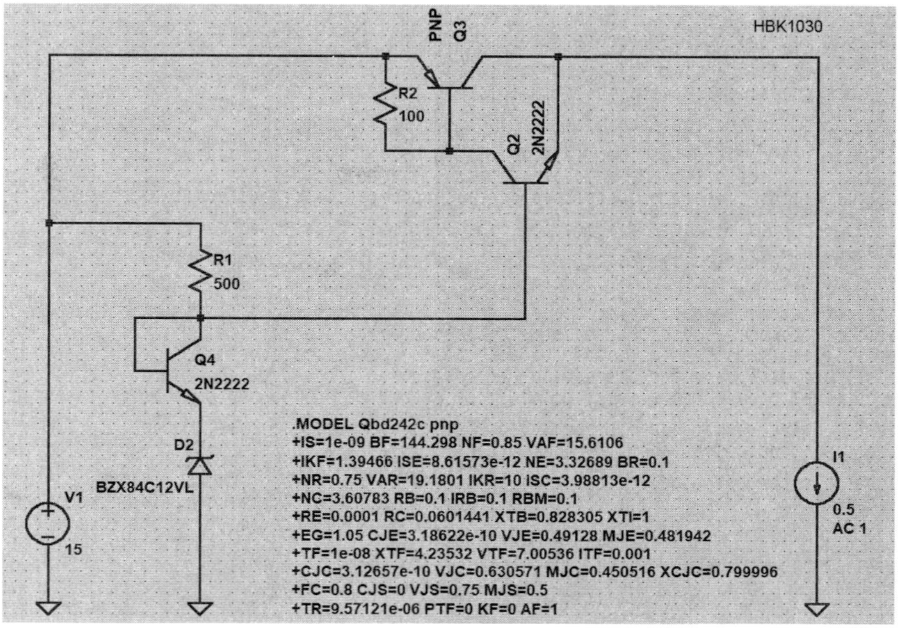

Figure 6.7 — Regulator circuit with improved pass transistor drive capability.

SWEEPING OUTPUT CURRENTS

The command to sweep the load current from 0 to 1 A in 1 mA steps is:

.DC I1 0 1 0.001

Figure 6.8 shows the output voltage vs. load current.

The output voltage is stable, but at very small loads, the output voltage rises close to the input voltage. This is a design flaw: At no load, both transistors' V_{BE} are zero, so an output that is V_{IN} minus the sum of the two V_{BE} is equal to V_{IN}. To avoid this, a small output current, in the mA range, should be added, for instance by connecting a 12 kΩ resistor on the output. **Figure 6.9** shows that with the addition of this small extra load, the output voltage is acceptable.

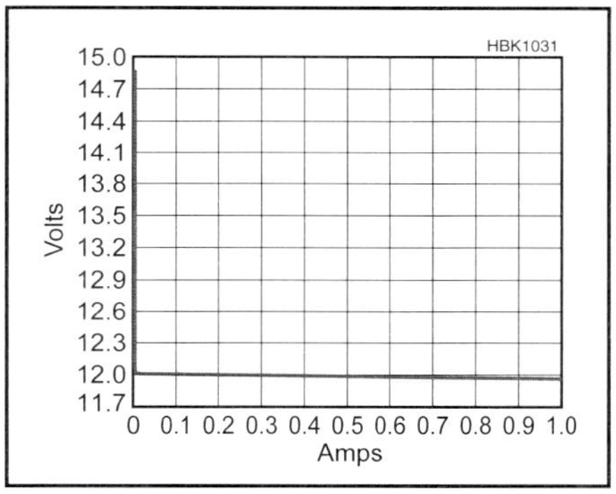

Figure 6.8 — Output voltage versus load current of the circuit in Figure 6.7.

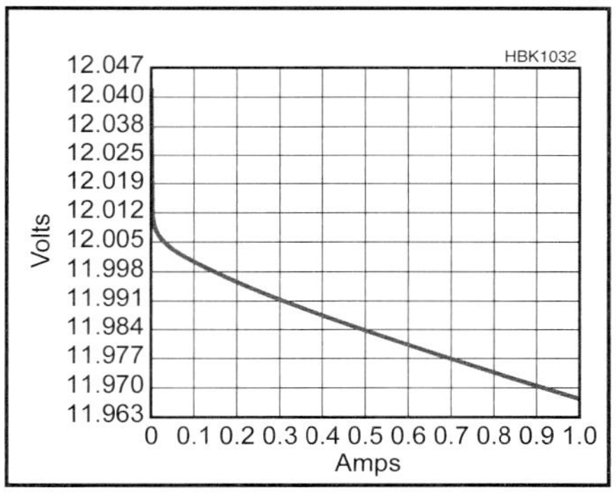

Figure 6.9 — Output voltage versus load current with a 12 kΩ load.

SWEEPING INPUT VOLTAGE FOR VARIOUS LOAD CURRENTS

The commands to sweep the load current and output voltage are:

.STEP I1 0 1 0.5

Load current swept from 0 to 1 A in 0.5 A steps.

.DC V1 15 25 0.01

Input voltage swept from 15 to 25 V in 10 mV steps.

Figure 6.10 shows that, as expected, output voltage increases with input voltage and decreases with load current.

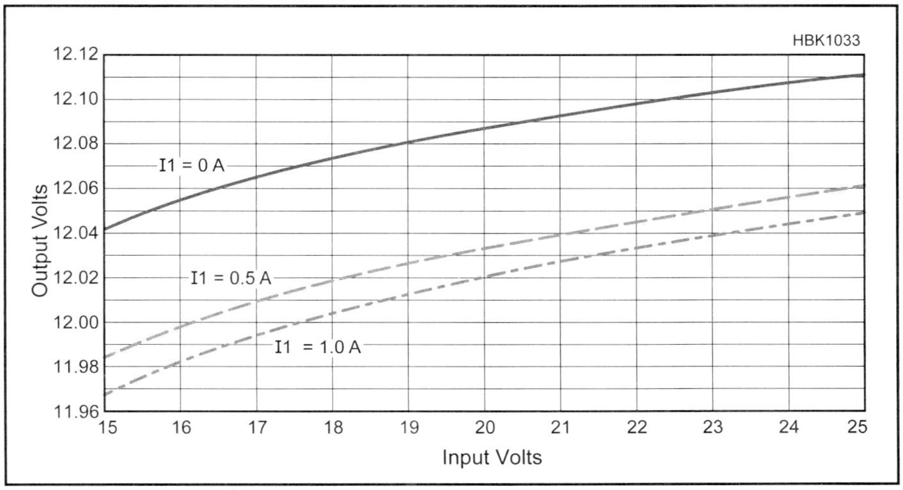

Figure 6.10 — A swept simulation of output voltage for a range of input voltages.

OUTPUT IMPEDANCE VS. FREQUENCY

To measure output impedance, a 1 A ac current is superimposed on the ac load current. Then ac output voltage gives a direct reading of output impedance. One should not be confused that ac current can be larger than dc current. It does not mean current changes direction. It just means that a small-signal model is computed from the operating point and then the linear model is used to compute ac response. That is another benefit of simulation, since directly superimposing ac and dc currents would be hard to do with real components!

The commands to sweep load current and frequency are:

.STEP I1 0 1 0.5

Load current swept from 0 to 1 A in 0.5 A steps.

.AC DEC 20 1 1E8

Frequency is swept from 1 Hz to 100 MHz with 20 steps per decade.

Figure 6.11 shows that low frequency output impedance depends on current. At high current, impedance is low but increases with frequency. The usual fix is adding a capacitor across the output. Pay attention to the Equivalent Series Resistance (ESR) of that capacitor and, for large values of capacitance, the Equivalent Series Inductance (ESL).

The simulation shows output impedance up to 100 MHz. Keep in mind that if wiring or PCB traces have significant length, parasitic inductance and resistance will affect this result. Add them to the schematic to get more realistic results.

Figure 6.12 shows the effect of a 470 µF capacitor with 90 mΩ ESR connected across the output.

The improvement is significant above 100 kHz. This example shows that simulation can reveal many things, but requires including realistic components, especially for parasitic elements. For instance, make sure that wires and traces exhibit some resistance (and inductance) while wiring this supply or laying out a PCB.

INPUT REJECTION VS. FREQUENCY

When a 1 V ac source is superimposed on the input dc voltage, output ac voltage (in dBV by default) gives a direct reading of input rejection.

The commands to measure input rejection by sweeping the load current and input frequency are:

.STEP I1 0 1 0.5

Load current is swept from 0 to 1 A in 0.5 A steps.

.AC DEC 20 1 1E8

Frequency is swept from 1 Hz to 100 MHz with 20 steps per decade.

Figure 6.13 shows that input rejection is better than –35 dB.

LOAD TRANSIENT

The pulse is defined as follows:

I1 N002 0 PULSE(0 1 1e-3 1e-6 1e-6 999e-6 2e-3)

And the command to simulate the circuit with the pulsed load is:

.TRAN 5e-3

Figure 6.14 shows that the output voltage exhibits some transient changes on load current transitions but remains quite stable.

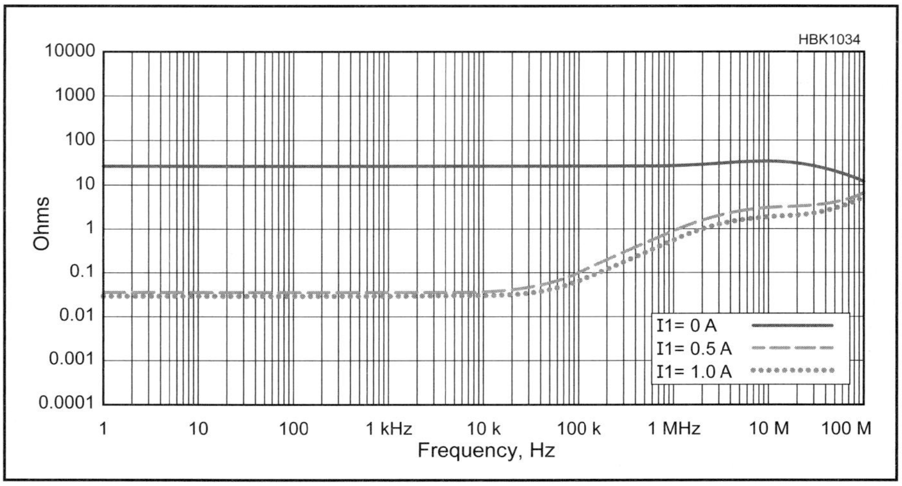

Figure 6.11 — Output impedance versus frequency from 1 Hz to 100 MHz.

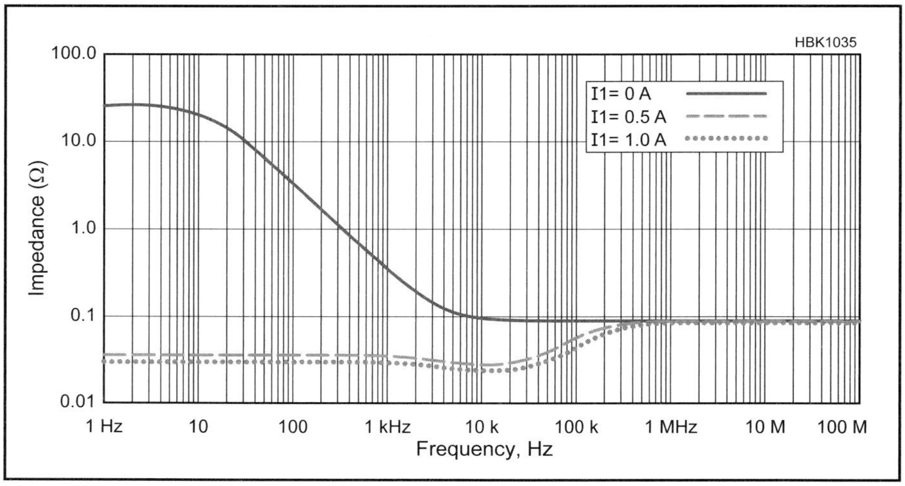

Figure 6.12 — The effect on output impedance of adding a 470 µF capacitor at the output.

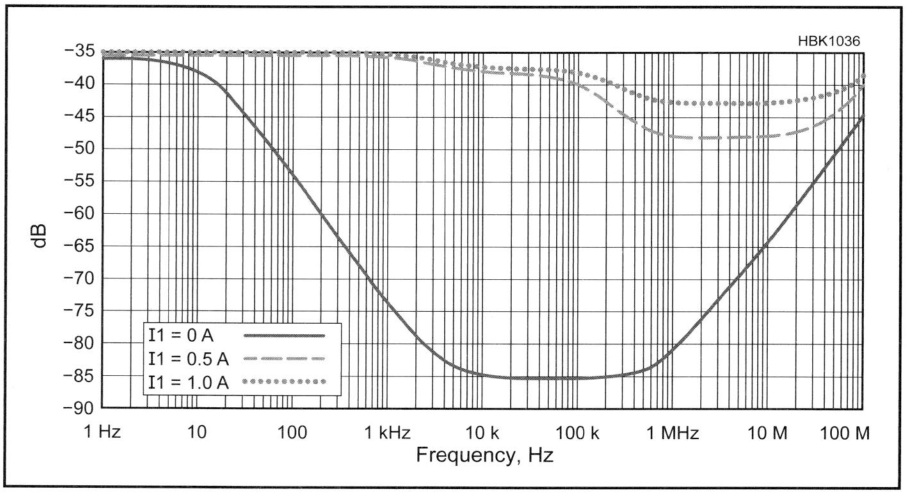

Figure 6.13 — Rejection of ac voltage on the input versus frequency from 1 Hz to 100 MHz.

FINAL SCHEMATIC

The final schematic after optimization is given in **Figure 6.15**.

EXAMPLE CONCLUSION

Using *LTSPICE*, it is possible to check a circuit's operation and to optimize it with respect to its specifications. As you should have noted, it is easy to characterize a circuit using a simulator, since it does not require complex (and expensive) instruments and allows measurements that are quite difficult to do on a bench.

However, you must remember that a simulator models only what it is given as input. Non-ideal components should be modeled properly, and the effects of connections should be considered, if relevant.

6.2.4 Example: Fourth Order Band-Pass Filter

The example circuit in **Figure 6.16** is taken from the **Analog and Digital Filtering** chapter's section on Passive LC Filters. This simulation is shown using *LTSPICE*.

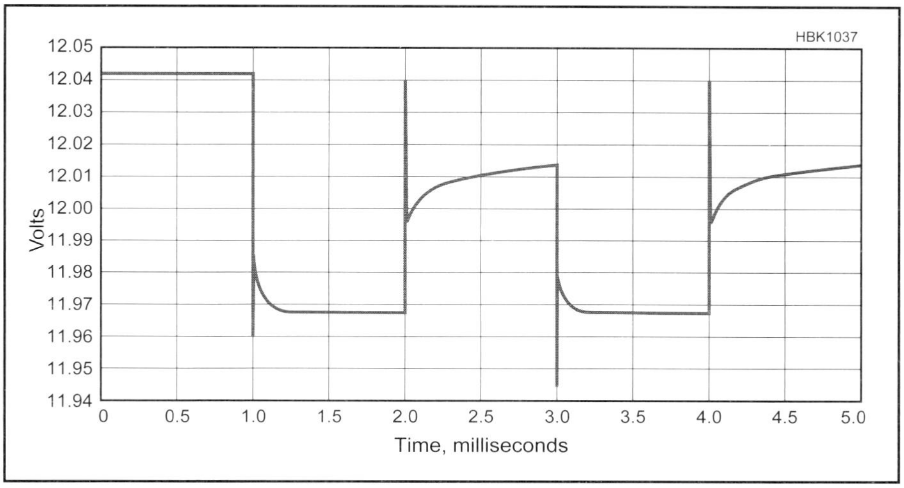

Figure 6.14 — Output voltage transients resulting from load current changes.

FREQUENCY RESPONSE

Figure 6.17 shows the simulated frequency response. This represents the S21 response of this filter, which is also known as the filter transfer function, and is calculated in *LTSPICE* as Vout/Vin (which gives a valid result if the input and output impedances are the same). The center frequency is 2.818 MHz and gain at that frequency is 0 dB (i.e., no insertion loss). The –3 dB (with respect to center gain) bandwidth extends from 1.463 MHz to 5.459 MHz. Depending on the actual value of the components and circuit wiring, parasitic elements may affect this result on the high frequency side.

INPUT IMPEDANCE

Figure 6.18 shows the simulated input impedance. Input impedance is calculated by the model by displaying input voltage [v(V1)] divided by input current [i(V1)]. It is close to 50 Ω at the center frequency and fluctuates over frequency.

CHOOSING STANDARD COMPONENT VALUES

The component values on the schematic result from computations. It is not likely that you will be able to find components that were manufactured to the exact values that you calculate. All components are manufactured with standard nominal values and are not guaranteed to exactly match those values. Rather, a tolerance range is specified,

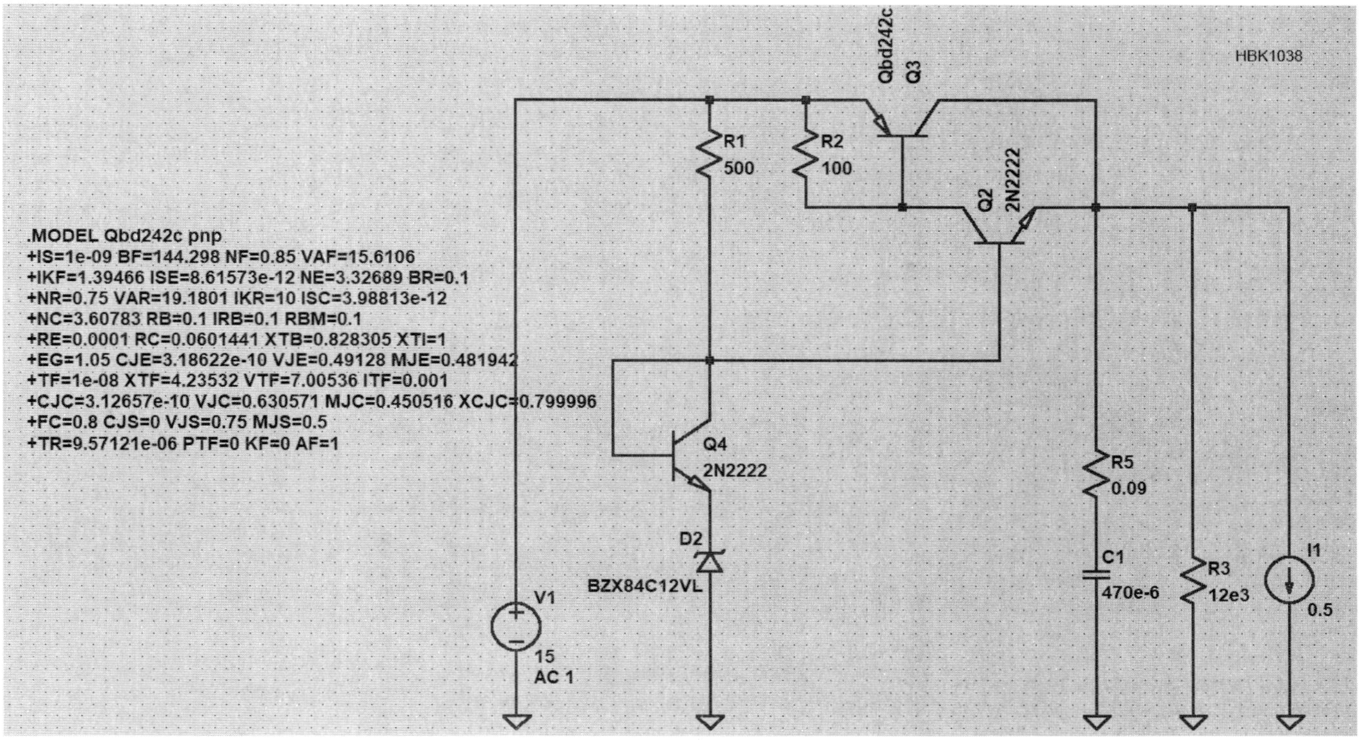

Figure 6.15 — The final schematic.

and all components are guaranteed to have values that are within the range of the nominal ± tolerance percentage value.

With respect to your ideally calculated circuit, two questions arise:

- What if values are rounded to standard values for availability?
- What if components values are affected by tolerances?

Modeling provides easy answers to both questions. You can find standard component values by searching a component distributor's website, such as **digikey.com**. By searching for a type of component (such as ceramic capacitor) you will be given choices for the various component parameters. Select the necessary parameters, such as voltage and package type for the capacitor. Next select the desired component tolerance(s) and then peruse the list of component values for the one closest to the theoretical value in your *SPICE* model. Substitute the standard value for the theoretical value in the model and then run the *SPICE* simulation again to see if the results are close enough to your design requirements. You can do this for all components in the design at once or try one at a time to see if the circuit works with that component replaced by its standard value. If any of the component substitutions cause an unacceptable change in the modeled circuit results, try changing that component's tolerance to a tighter version. For instance, if you originally specified a 5% tolerance component, fewer standard nominal values would be available. By specifying a 1% tolerance component, you see more standard nominal values and one should be closer in value to your theoretical value.

Substituting standard nominal values of 2 µH for the three inductors and 1,600 pF for the three capacitors in the *SPICE* model, the central frequency shifts to 2.816 MHz (–2 kHz from theoretical component values) and gain at that frequency is unaffected. The –3 dB bandwidth extends from 1.453 MHz to 5.432 MHz (shifts of –10 to –27 kHz).

EFFECT OF COMPONENT TOLERANCES

For the effect of tolerance, we can use the Monte-Carlo feature of *LTSPICE*. Monte-Carlo analysis is a broad class of algorithms used to study problems involving random variables. In circuit simulation, this technique uses random numbers to mimic what happens when combining components affected by tolerances. Depending on the actual value of each component that is randomly defined by tolerance, the circuit may behave differently.

To achieve this analysis, component values are defined by a parameter involving a nominal value, a tolerance, and a random variable that is used to generate an actual value. If a circuit exhibits a significant sensitivity to tolerances, it may behave improperly in some cases. That is an issue for production as it affects yield, and it is also an issue for hobbyists as the item they build may not work properly.

It would be possible for a hobbyist to sit

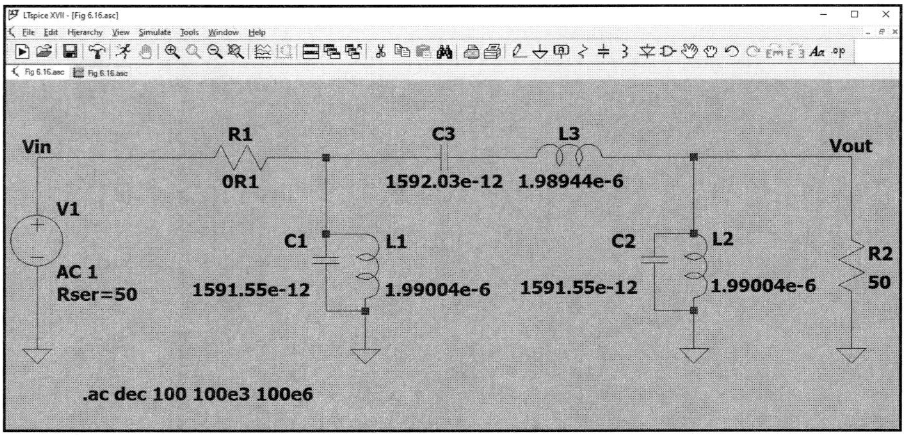

Figure 6.16 — A fourth-order band-pass LC filter. Note that the input series resistor, R2, has been changed to a very low resistance, since its actual resistance of 50Ω is included in the definition of the voltage source, V1.

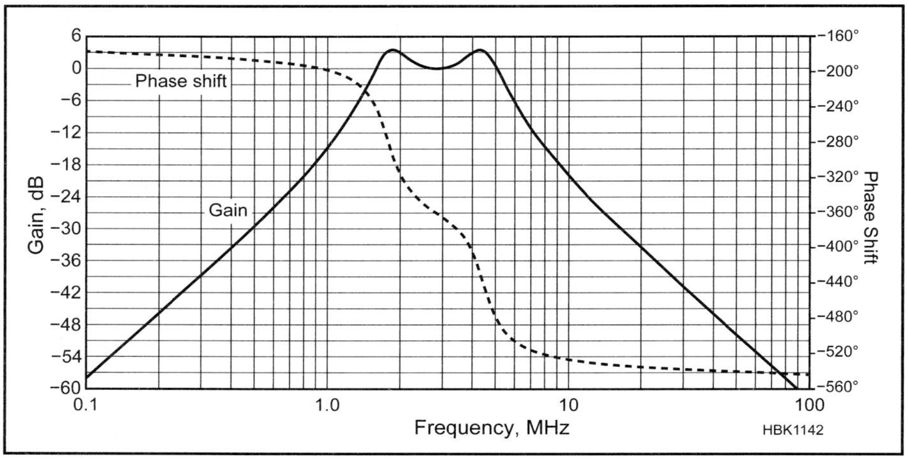

Figure 6.17 — Frequency amplitude and phase response of the filter in Figure 6.16. The amplitude is calculated as S21 (transfer function) by displaying Vout/Vin.

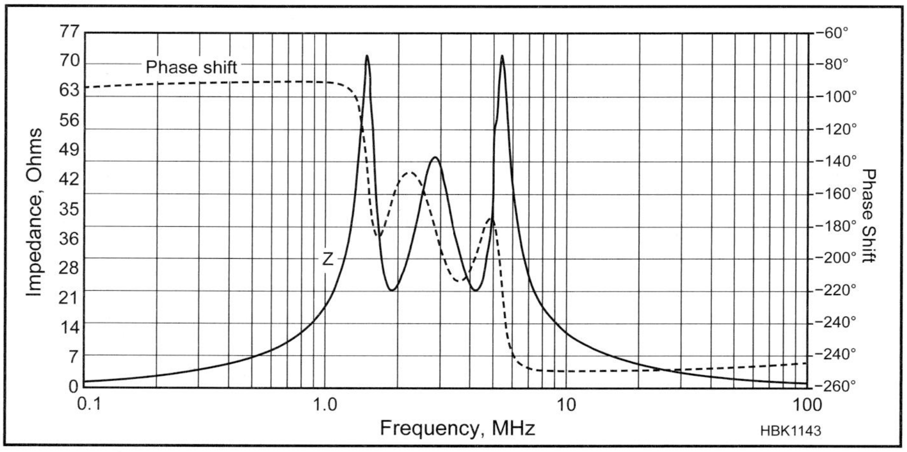

Figure 6.18 — Input impedance vs. frequency of the filter in Figure 6.16.

at the workbench and measure a pile of 5% tolerance resistors with an ohmmeter until values that are very close to those in the design are found. However, it is much easier to model the effects of value deviations from the nominal and then determine how important exact values are to obtaining the desired results. The model is able to confirm that the circuit will work properly with all possible values of components within their range of tolerances. If not, it will show which components must be purchased with tighter tolerances (or manually measured). Running sensitivity analysis indicates which components can have high tolerances without affecting the desired operation of the circuit, and conversely, which components need to have tight tolerances for the circuit to perform as desired.

Figure 6.19 shows the commands to perform the analysis. With all components and their tolerances specified, Monte Carlo analysis is performed for variations of the values to determine which combinations deviate from the nominal results by the largest margins.

In this example, tolerance is defined to be 5% on all components. The AC analysis is then performed to run many simulations. For each simulation, a new set of random variables is computed but nothing else changes.

For each component, its value is computed by multiplying the nominal value by 1 plus a Gaussian random number that is within the specified tolerance. Since 99% of values fit within ±3 times the standard deviation, defining that standard deviation as tolerance is a good approximation. The "fuzziness" of the amplitude and phase curves represents the range of variations of those parameters for randomly selected sets of components with the given tolerances.

Figure 6.20 is the simulated frequency response for 101 runs. It shows some variations in the band pass filter response. Should these variations be unacceptable for the application, components with tighter tolerance should be used.

Monte Carlo Analysis can also be used to determine circuit operation under varying conditions, such as temperature or power supply variation. The analyzer only needs to know how much each component varies as temperatures change or the possible range of power supply voltage variations.

EXAMPLE CONCLUSION

This example shows an additional benefit of simulators: predicting real-world behavior of a circuit before building the first prototype.

6.2.5 Example: Crystal Oscillator

When it comes to simulating a crystal oscillator, several issues arise, and some designers may just give up the task. But if addressed properly, a crystal oscillator can be simulated, providing valuable insight on its operation. Some classic issues of crystal oscillator simulation follow.

```
.PARAM TOL=0.05
.param valC1={1591.55e-12*(1+gauss(tol/3))}
.param valL1={1.99004e-6*(1+gauss(tol/3))}
.param valC2={1591.55e-12*(1+gauss(tol/3))}
.param valL2={1.99004e-6*(1+gauss(tol/3))}
.param valC4={1592.03e-12*(1+gauss(tol/3))}
.param valL4={1.98944e-6*(1+gauss(tol/3))}

.step param X 0 100 1 ; a dummy parameter to cycle Monte Carlo runs
```
HBK1042

Figure 6.19 — Commands to perform Monte-Carlo analysis of the filter by randomly changing component values.

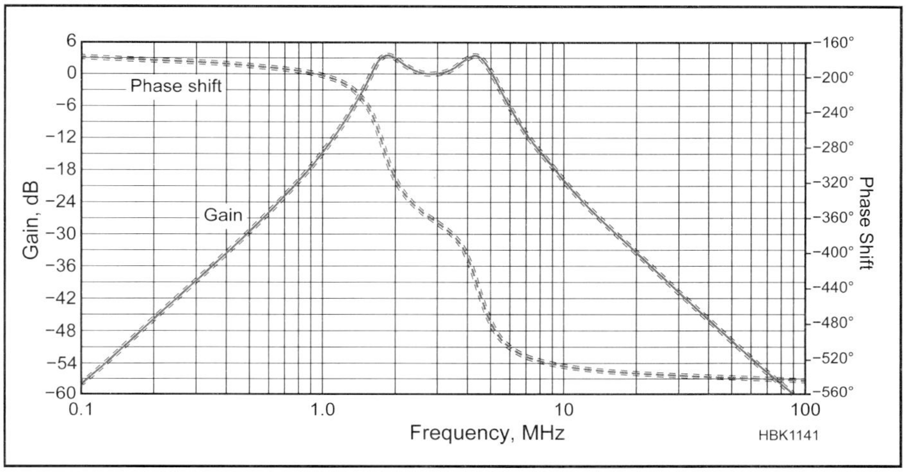

Figure 6.20 — The results of the Monte-Carlo analysis of the filter.

CRYSTAL MODEL AND PARAMETERS

A crystal equivalent schematic is shown in **Figure 6.21**. The schematic shows a series RLC branch in parallel with a capacitor. The series branch is the electrical representation of the crystal mechanical resonator, as transformed by the piezoelectric effect. (See the **Oscillators and Synthesizers** chapter for more information about crystal parameters.)

Intuitively, one can feel that at low frequencies, when inductor impedance approaches zero and capacitor impedance is large with respect to resistance, the equivalent impedance is that of the two capacitors in parallel. Similarly, at high frequencies, when inductor impedance is large, equivalent impedance is that of the parallel capacitor only. In between, things are more difficult to guess and calculation is a bit complex, but fortunately simulation can help. **Figure 6.22** shows the impedance vs. frequency of the crystal's equivalent schematic.

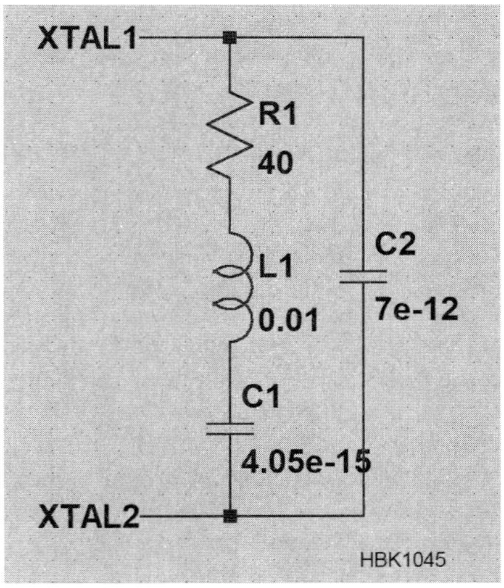

Figure 6.21 — A crystal's equivalent model circuit.

Overall, impedance follows a 1/f law,

confirming the capacitive behavior at high and low frequencies. Around 25 MHz, two resonances occur. A series resonance where impedance reaches a minimum and a parallel resonance where impedance reaches a maximum. The series resonance frequency is lower than the parallel resonance frequency. This is because series resonance involves only L1 and C1 while the parallel resonance involves L1 and C1 in series with C2, resulting in a slightly lower capacitance value (since C2 is much larger than C1). Both resonances are very narrow because the crystal's quality factor, Q, is very high.

DEFINING CRYSTAL MODEL PARAMETERS

A concern here is to find the component values to use in the crystal equivalent schematic. Crystal manufacturers rarely supply these values. Fortunately, there is a relatively simple method to find them. In most cases, C2 and R1 are specified in the crystal data sheet. R1 is named series resistance and C2 is named parallel or shunt capacitance, not to be confused with the load capacitance.

Now what about L1 and C1 values? L1 can be estimated from R1 and Q values. R1 is generally available. Q is sometimes specified but it stands in the 10,000 to 100,000 region. We'll see later that this value is not critical.

From the equation for series resonator Q, L1 can be found:

$$Q = \frac{2\pi f L1}{R1} \quad \text{so } L1 = \frac{Q R1}{2\pi f}$$

C1 and L1 define the series resonance that is either defined as, or very close to, the crystal parallel frequency.

$$F0 = \frac{1}{2\pi\sqrt{L1C1}} \quad \text{so } C1 = \frac{1}{(2\pi F0)^2 L1}$$

Finally, knowing F0, Q, R1, and C2 is sufficient to calculate the missing values L1 and C1.

The equivalent schematic used so far was extracted from a 25 MHz miniature SMD crystal with a specified series resistance R1 = 40 Ω and shunt capacitance C2 = 7 pF. Assuming a Q of 40,000 leads to L1 = 0.0101 H (rounded to 0.01 H). And then C1 = 4.05 fF. The model is complete! (Note the very large value of L1 and very small value of C1 as a result of the high Q!)

THE OSCILLATOR

There are many options for a crystal oscillator. A Colpitts type using a follower NPN is used in **Figure 6.23**. A simple bias is used to fix dc base voltage around 3.3 V and dc collector current around 5 mA. Loop gain is defined by a capacitive divider that

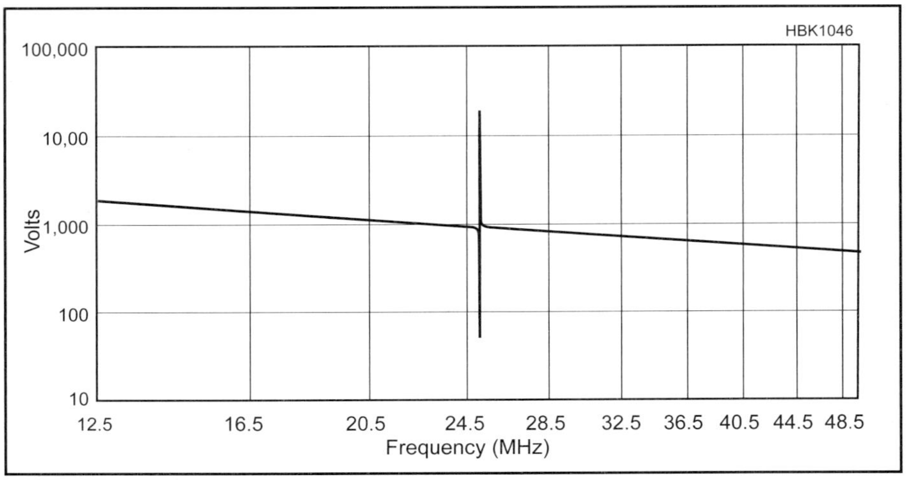

Figure 6.22 — Impedance vs. frequency of the crystal in Figure 6.21.

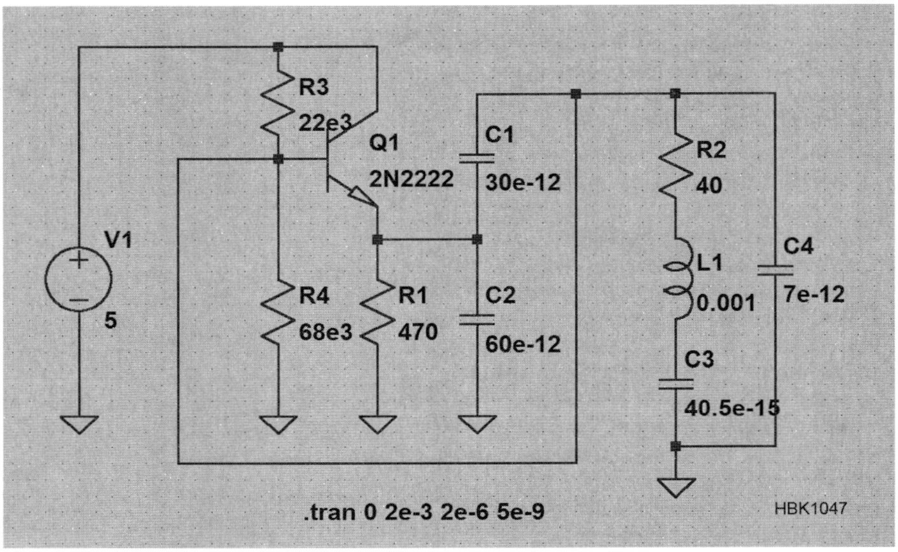

Figure 6.23 — The schematic for a Colpitts crystal oscillator.

also defines crystal load capacitance.

OSCILLATOR LOOP GAIN

First, let's look at loop gain by connecting a series ac source. **Figure 6.24** shows the gain vs. frequency (zoomed into the region of interest). This shows that gain is larger than 1 at a frequency where phase is 0° so the conditions for oscillation are met.

OSCILLATION STARTUP

Now, let's look at base and emitter signals in the transient simulation of **Figure 6.25**. These two signals are perfectly stable, but the oscillator seems not to work! What happened? The oscillation condition is met but the oscillator does not start.

In fact, a real circuit always has some noise. This noise power spreads over frequency and then has a non-zero value at the frequency where the oscillation condition is met. So, oscillation may start from noise and increase in amplitude over time until it reaches some nonlinearity that limits amplitude.

In a transient simulation, there is no explicit noise and computation noise is lower than default accuracy so oscillation cannot start. Could this be related to simulator settings?

SIMULATOR SETTINGS

Forcing accuracy to 1E-9 instead of default 1E-3 and limiting time steps to 5 ns so that there are at least several points per period does not improve the situation.

"SNAP KICKING"

A classic method of simulating an oscillator starting up is "putting a snap" in the crystal to cause it to start. For some oscillator circuits, it is sufficient to set an initial

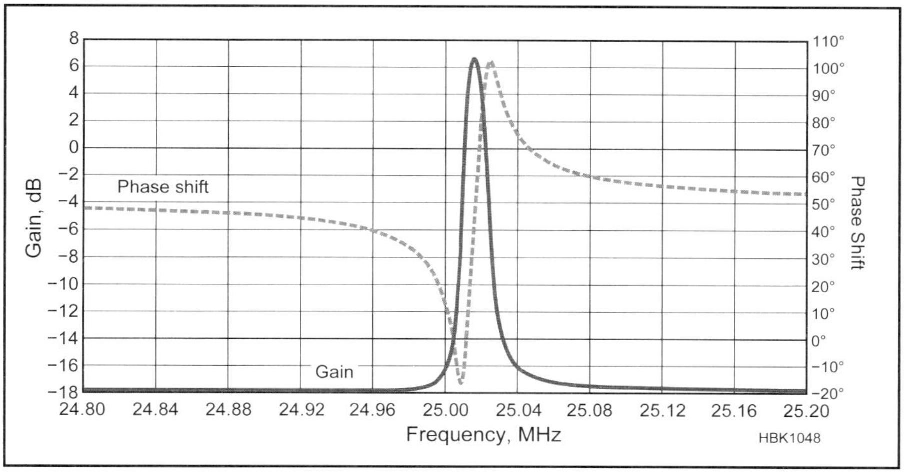

Figure 6.24 — Loop gain of the oscillator circuit in Figure 6.23.

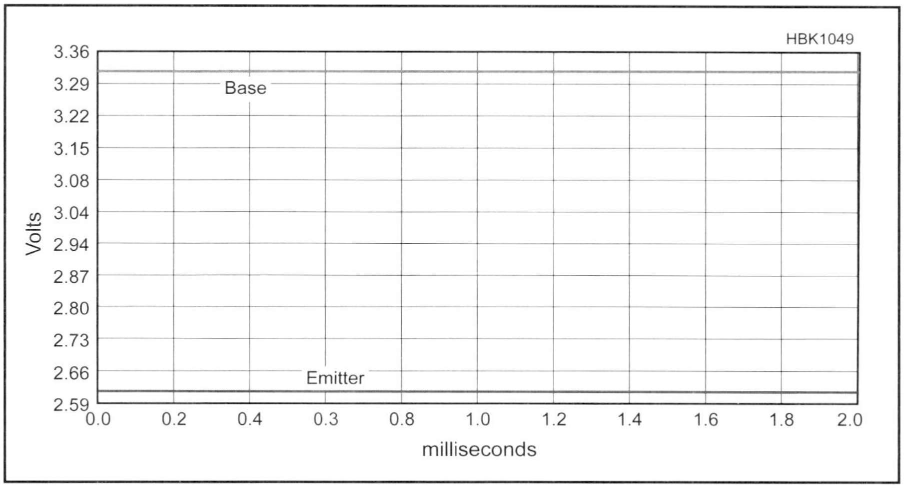

Figure 6.25 — Base and emitter signals for the oscillator circuit in Figure 6.23.

condition to start oscillations. In this circuit, an appropriate initial condition would be to place a voltage on the base of the transistor. When the simulation starts, the initial condition disappears and, if its influence is strong enough before disappearing, the oscillator starts. In *SPICE*, the command that has to be added to the simulation is: ".IC V(3)=4.0V," which puts a voltage of 4.0 V on Net 3 (the transistor base) at the beginning of the simulation. For this oscillator, the initial condition on the base of the transistor was not sufficient to start it oscillating.

Figure 6.26 shows another way of stimulating the oscillator to start by adding one period of an ac source in parallel with the crystal. After that transient, current is zero and has no more effect. Even with that trick, after a large initial transient and despite an oscillation at the right frequency, the oscillation is damped and would stop after some time as shown in **Figure 6.27**. What happened? In fact, Q is so large that it would require a very high dynamic range to handle the actual amplitude increase from one period to the next. So, what if Q is reduced by a factor of 10?

REDUCING Q FACTOR

This is easily done by dividing the L1 value by 10 and multiplying C1 by 10 to keep the oscillation frequency the same value. Now, with a lower quality factor and nothing else changed, the oscillator does start as **Figure 6.28** shows. Oscillation amplitude increases exponentially with time until it reaches a limitation by circuit non-linearity.

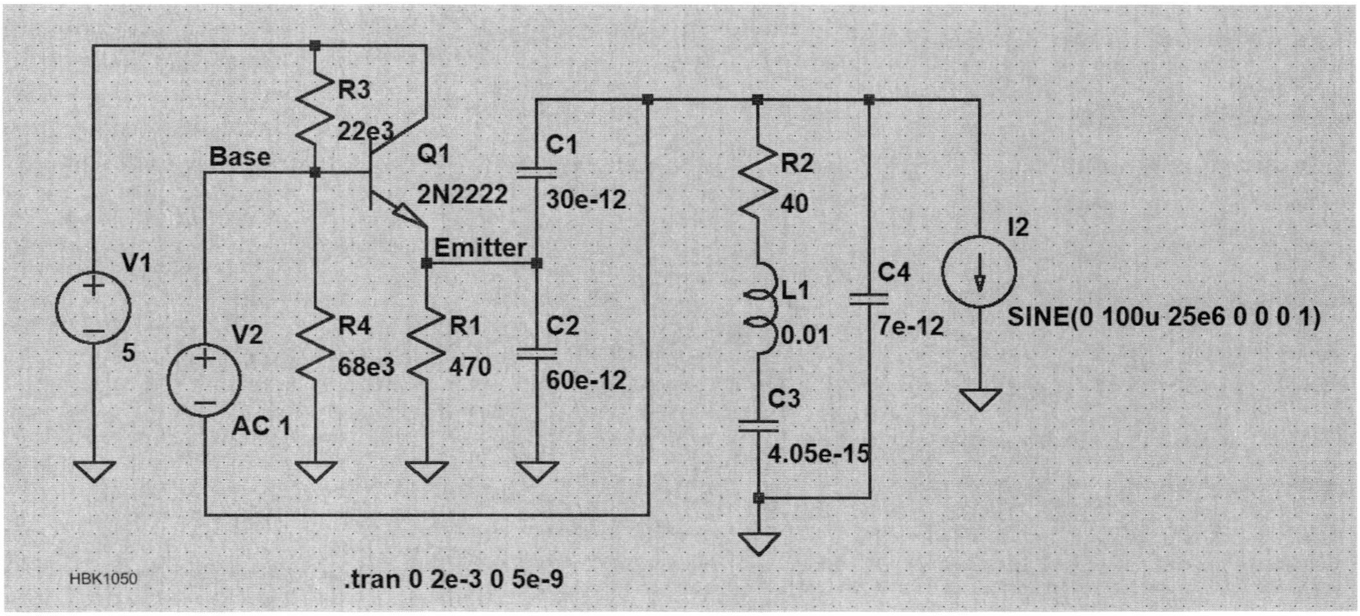

Figure 6.26 — Adding a "snap" transient to the crystal as a current source that is active for only one simulation period.

EXAMPLE CONCLUSION

When it comes to simulation of harmonic oscillators that operate through loop small-signal instability, adding a short transient is often necessary to cause the oscillation to actually start in transient simulation. In cases where the Q factor is very large, artificially reducing it helps and speeds up transient simulation. Oscillation conditions are correctly predicted by ac small-signal analysis. In the case of relaxation oscillators, the type using hysteresis in the loop, they always start properly in transient simulations but cannot be analyzed with ac simulation as they have no stable operating point.

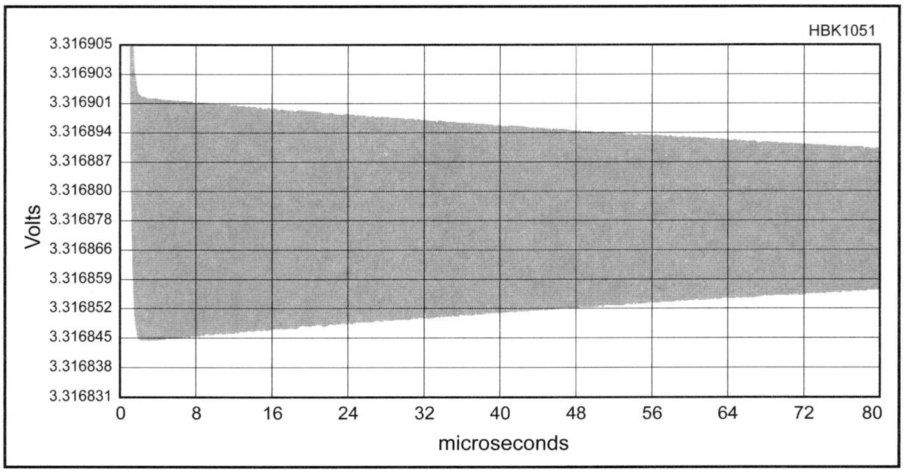

Figure 6.27 — Output waveform for the oscillator with the snap transient.

6.2.6 Non-*SPICE* Simulators

SPICE set a standard for circuit simulation and many simulators evolved from *SPICE*. The breakthrough was such that most simulators that did exist before *SPICE* simply disappeared. For quite some time, electronic circuit simulation was completely ruled by *SPICE*, *SPICE* compatibility, and extensions to *SPICE*. But as already stated, *SPICE* types of analysis suffer from some intrinsic limitations. As a result, some recognition of other needs slowly emerged and non-*SPICE* circuit simulators appeared.

Probably the most significant concept having been developed is *Harmonic Balance*. This approach is so successful at fixing some *SPICE*-type analysis issues that some major commercial *SPICE* products have included Harmonic Balance features and have even significantly extended and improved it. Harmonic Balance is discussed in more detail in this chapter's section on the "Limitations of Simulation at RF."

Another field is *Behavioral Simulation*. This approach has been integrated in most *SPICE*-derived simulators through syntax extensions without changing simulation types. This is also true for mixed signal, analog-digital simulation.

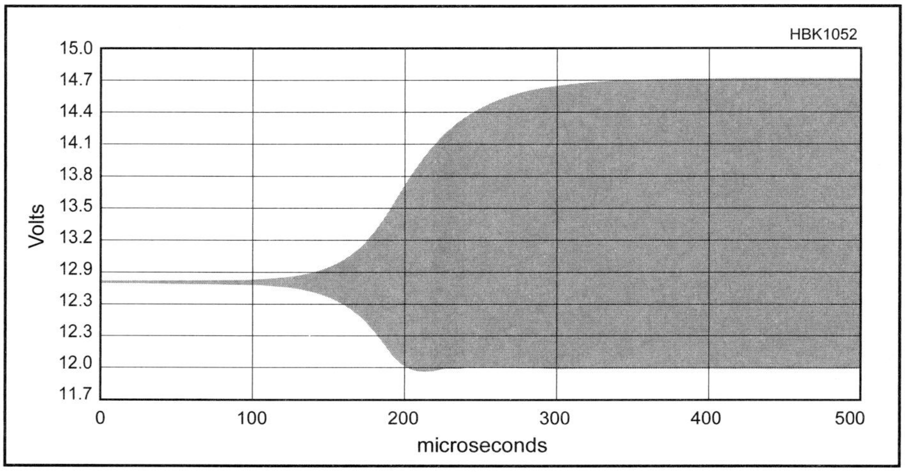

Figure 6.28 — Final output waveform for the oscillator after crystal Q was reduced.

ELECTROMAGNETIC SIMULATORS

Another type of simulator does not target circuit simulation but extracts equivalent schematics or S-parameters from a physical structure description. These are *electromagnetic simulators*. Confusion exists because some of these simulators allow mixing physical description and active devices described by appropriate models. At higher frequencies, for which propagation effects cannot be neglected, electromagnetic simulators are mandatory. For professional development there are several (expensive) tools. Electromagnetic simulators are discussed later in this chapter.

6.3 Limitations of Simulation at RF

Experienced users of circuit simulation software are wary of using any software beyond the boundaries of circuits and parameters for which it was intended and tested. RF simulation can present just such situations, leading to software failure and unrealistic results. Introduced and summarized in this section, several detailed papers by Dr. Ulrich Rohde, N1UL, exploring simulation at RF are provided with the online content. The papers are:

• "Using Simulation at RF" by Rohde, a survey of the issues of RF simulation and the techniques used in current modeling programs.

• "The Dangers of Simple Usage of Microwave Software" by Rohde and Hartnagel, a discussion of inaccuracies introduced by device parameter measurement and model characteristics.

• "Mathematical Stability Problems in Modern Non-Linear Simulations Programs" by Rohde and Lakhe, presenting various approaches to dealing with nonlinear circuit simulation.

• "Transistor Applications from RF to Microwave Frequencies" by Rohde, Poddar, and Rudolph, a comprehensive paper showing examples of circuit simulation techniques through microwave frequencies.

• "Harmonic Balance" by Rohde, Poddar, Rudolph, a discussion of how harmonic balance modeling techniques work.

In addition, there are many online resources to help you obtain trustworthy simulation results with a simulator designed for RF. For the interested reader with some technical background, the online paper "Introduction to RF Simulation and its Application" by Ken Kundert (**kenkundert.com/docs/bctm98-RFsim.pdf**) introduces RF simulation methods and how they account for the characteristics of RF circuits when generating common RF measurements. The website The Designer's Guide (**www.designers-guide.org)** also provides many tutorials, technical guides, models, and other resources for analog and RF simulation users.

While the lower bound of "RF" is not precisely defined, RF effects start at about 100 kHz. This was first noticed as self-resonance of high-Q inductors for receivers. In response, Litz wire was invented, in which braided copper wires were covered with cotton and then braided again to reduce self-resonance effects.

As frequencies get higher, passive elements show the effects of parasitic elements such as lead inductance and stray capacitance. At very high frequencies, the physical dimensions of components and their

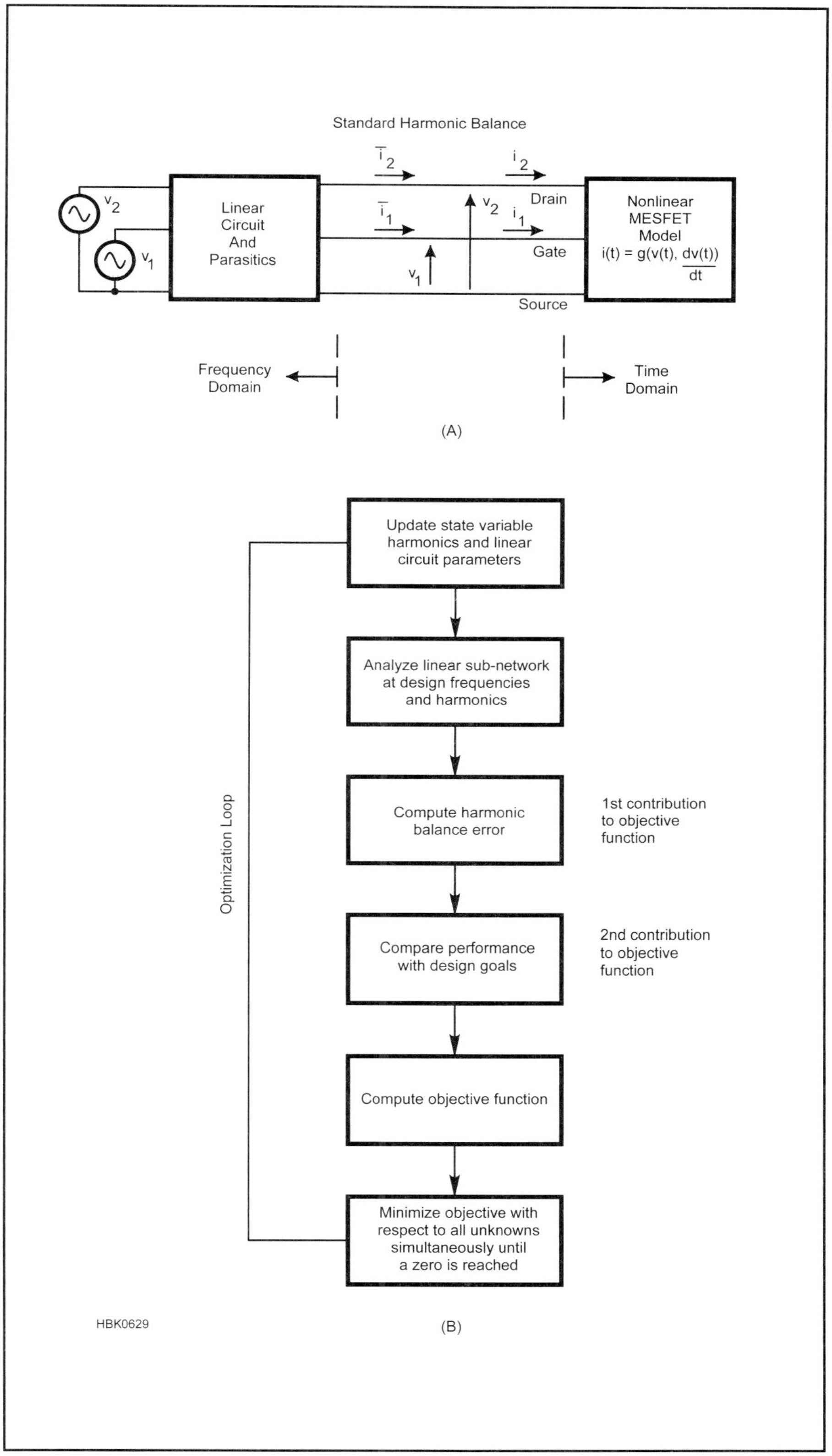

Figure 6.29 — (A) MESFET circuit partitioned into linear and nonlinear subcircuits for harmonic balance analysis. Applied gate and drain voltages, and relevant terminal voltages and currents, are indicated. (B) Flowchart of a general-purpose harmonic balance design algorithm that includes optimization.

interconnections reach an appreciable fraction of the signal wavelength and their RF performance can change drastically.

RF simulators fall in the categories of *SPICE*, harmonic balance (HB) programs, and EM (electromagnetic) programs. The EM simulators are more exotic programs. Two types are common, the 2-dimensional and the full 3-dimensional versions. They are used to analyze planar circuits, including vias (connections between layers) and wraparounds (top-to-ground plane connections), and solid-shapes at RF. They go far beyond the *SPICE* concept.

6.3.1 *SPICE*-Based Simulators

SPICE was originally developed for low frequency and dc analysis. While dc, frequency, and time-domain simulations work very well, *SPICE*-based simulation has limitations. The time-domain calculation uses the very complex mathematics of the Newton-Raphson solution to nonlinear equations. These methods are not always stable. Many adjustments to the program settings may be necessary for the calculations to converge properly. Knowledge of the specifics of different types of electronic circuits can assist the user in finding an accurate solution by specifying appropriate analysis modes, options, tolerances, and suitable model parameters. For example, oscillators require certain initializations not necessary for amplifiers and bipolar transistors may need different convergence tolerances than do MOS circuits. Generally, *SPICE* finds a solution to most circuit problems. However, because of the nonlinearity of the circuit equations and a few imperfections in the analytical device models, a solution is not always guaranteed when the circuit and its specification are otherwise correct.

The next problem at RF is that the basic *SPICE* simulator uses ideal elements and some transmission line models. As we approach higher frequencies, where the lumped elements turn into distributed elements and special connecting elements become necessary, the use of the standard elements ends. To complicate matters, active elements, such as diodes and transistors, force the designer to develop more complex simulations. Adding the missing component elements leads to highly complex models and problems of convergence in which the simulator gives an error, suggesting a numerical problem, or, more likely, by failing to generate a solution.

SPICE also has problems with very high-Q circuits and noise analysis. Questions of the noise figure of amplifiers or phase noise of an oscillator cannot be answered accurately by a *SPICE*-based program. Noise analysis, if not based on the noise correlation matrix approach, will not be correct if the feedback capacitance [Im(Y12), the imaginary component of Y-parameter Y12] is significant at the frequencies involved. Analysis of oscillators in *SPICE* does not give a reliable output frequency and some of the latest *SPICE* programs resort to approximation calculations.

6.3.2 Harmonic Balance (HB) Simulators

There are now nonlinear microwave simulators using the *harmonic balance* (HB) technique. Harmonic balance (HB) analysis is performed using a spectrum of harmonically related frequencies or tones, like what you would see by measuring signals on a spectrum analyzer. The fundamental frequencies are those whose integral combinations form the spectrum of harmonic frequency components used in the analysis. On a spectrum analyzer you may see many signals, even if the input to your circuit is only one or two tones. The harmonic balance analysis must truncate the number of harmonically related signals so it can be

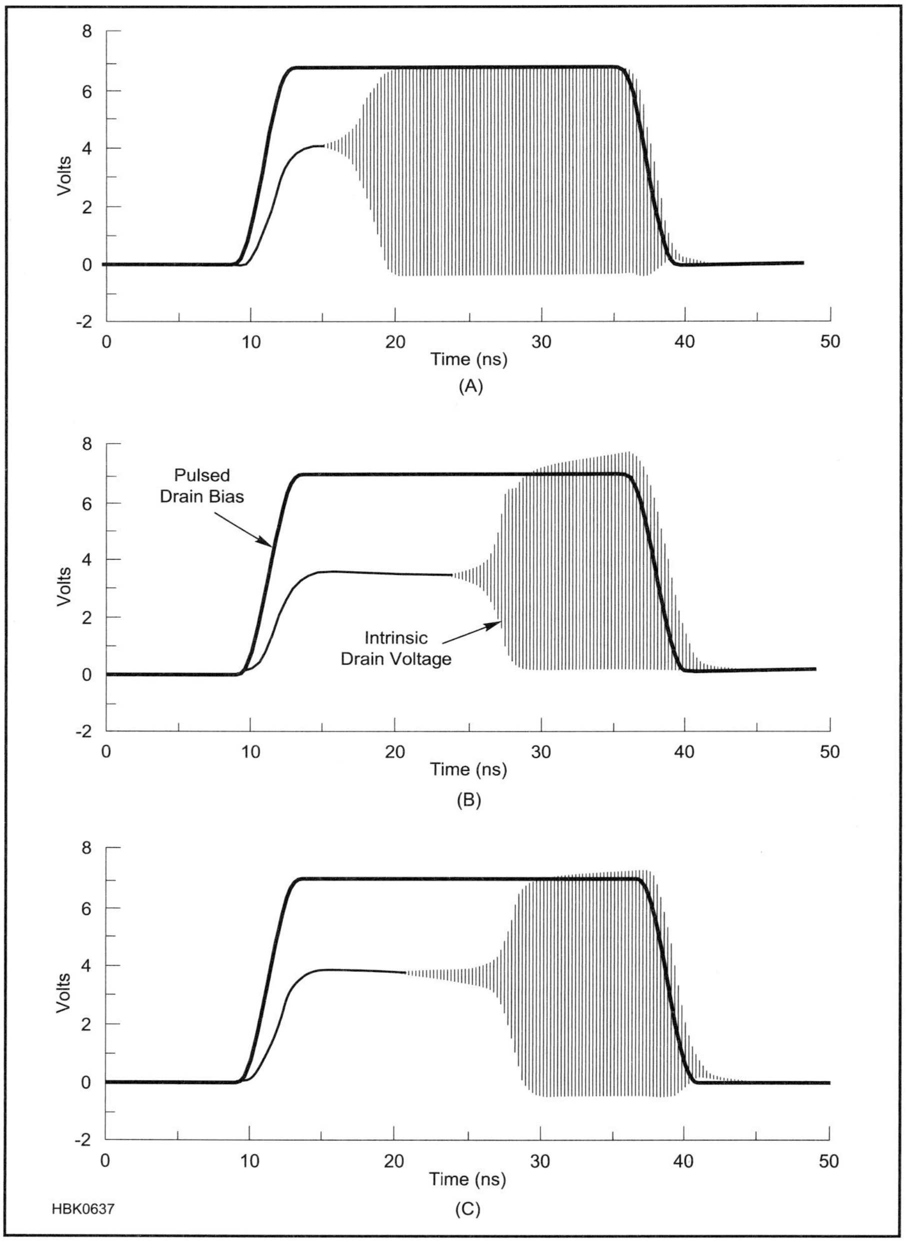

Figure 6.30 — (A) is the initial simulation of a *SPICE*-based simulator. (B) is the correct response of a pulsed microwave oscillator obtained by harmonic balance simulation using the Krylov-subspace solution. (C) is the *SPICE*-based simulation after 80 pulses of the drain voltage.

analyzed on a computer.

HB performs steady-state analysis of periodically excited circuits. The circuit to be analyzed is split into linear and nonlinear sub-circuits. The linear sub-circuit is analyzed in the frequency domain by using distributed models. This enables straightforward intermodulation calculations and mixer analysis. The nonlinear sub-circuit is calculated in the time domain by using nonlinear models derived directly from device physics. This allows a more intuitive and logical circuit representation.

Because of its nonlinear optimization capabilities, HB is uniquely suited to the task of oscillator design, mixers, and similar circuits that are difficult or impossible to simulate with linear techniques. The harmonic balance method seeks a solution to a steady-state nonlinear design problem by iteratively solving for a set of variables, referred to as *state variables*. The state variables can typically be chosen as the voltages at the linear–nonlinear interface in a circuit partitioned into linear and nonlinear segments. They are expressed as the phasor components of a sinusoidal excitation frequency and their harmonics. The state variables are usually found iteratively by a gradient-based technique that seeks a simultaneous solution for Kirchhoff's equations applied to the linear and nonlinear sides of the network separately.

Analysis parameters such as Number of Harmonics specify the truncation and the set of fundamental frequencies used in the analysis. The fundamental frequencies are typically not the lowest frequencies (except in the single-tone case) nor must they be the frequencies of the excitation sources. They define the base frequencies upon which the complete analysis spectrum is built.

Figure 6.29A diagrams the harmonic balance approach for a MESFET amplifier. (The online content paper by Rohde, Poddar, and Rudolph presents an in-depth example of HB analysis and simulation of a microwave oscillator for the interested reader.) Figure 6.29B charts a general-purpose nonlinear design algorithm that includes optimization. Modern analysis tools that must provide accurate phase noise calculation should be based on the principle of harmonic balance.

Modern HB programs have developed solutions for handling very large numbers of transistors (>1 million) through efficient math solutions. Memory management using matrix formulations reduces the number of internal nodes and solving nonlinear equations for transient analysis are some of the key factors to their success.

6.3.3 Contrasts in Results

The following time domain analysis is a good example of differences between *SPICE* and harmonic balance simulation. A microwave oscillator is keyed on and off and a transient analysis is performed. When using the standard *SPICE*, the initial calculation shows an incorrect response after one iteration, as seen in **Figure 6.30A**. It takes about 80 pulses (80th period of the pulsed drain voltage) until the simulation attains the correct answer (Figure 6.30C) of the Krylov-subspace-based harmonic balance in Figure 6.30B.

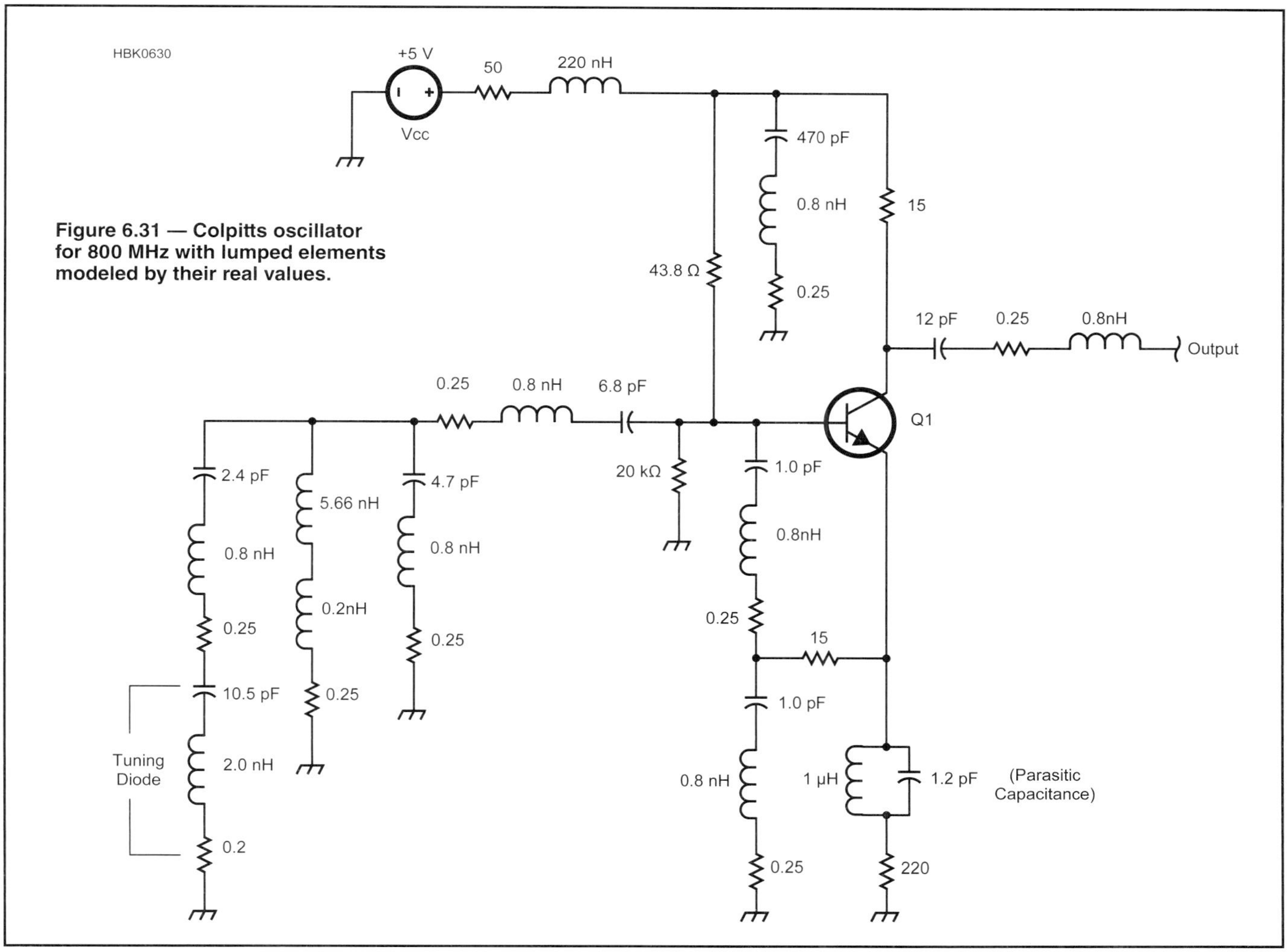

Figure 6.31 — Colpitts oscillator for 800 MHz with lumped elements modeled by their real values.

The frequencies involved need not be in the GHz range. Oscillators can be very difficult to analyze at any frequency as shown by simulations of a low MHz phase shift oscillator and a 10 MHz Colpitts oscillator in the referenced papers.

Validating the harmonic balance approach, **Figure 6.31** shows a BJT microwave oscillator entered with the schematic capture module of a commercially available HB simulator (Ansoft *Serenade 8.0*); **Figure 6.32** compares this oscillator's simulated phase noise to measured data. HB analysis gives similarly accurate results for mixers.

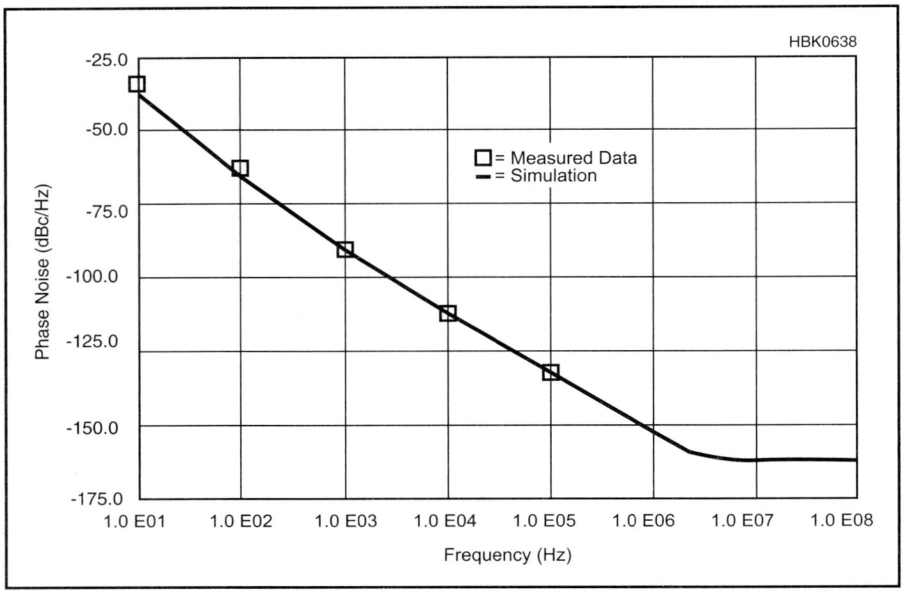

Figure 6.32 — Comparison between predicted and measured phase noise for the oscillator shown in Figure 6.31.

6.3.4 Block-Level Simulation

Rather than modeling every discrete component or integrated circuit, simulation tools exist to model systems made of functional blocks. This is particularly useful in the design of receivers and transmitters, where each functional block is defined by predefined characteristics of its design: input impedance (Z_{in}), output impedance (Z_{out}), power gain (G), second order input intercept point (IIP2), third order input intercept point (IIP3), input power 1 dB compression point (IP1dB), and Noise Figure (NF). The model calculates interactions between the stages and determines most of the system functional parameters that would be of interest.

For a given input signal to a receiver, the model determines the output signal power and voltage, the input and output power 1 dB compression points, the overall power and voltage gains, the overall noise figure, the output noise spectral density, the output noise floor, the SNR, the input sensitivity, the overall output third order intercept point, the overall input third order intercept point, the third order intermodulation distortion, the overall output second order intercept point, the overall input second order intercept point, the second order intermodulation distortion, the spurious free dynamic range, and an estimate of the adjacent channel leakage ratio. Similar calculated parameters are determined for transmitter systems. Analog Device's ADISimRF is a block-level simulation program that can be used to predict the behavior of receivers and transmitters based on the transfer characteristics of the stages used in the design. The program can be obtained without cost from Analog Devices (**www.analog.com/en/lp/resources/adisimrf.html**). **Figure 6.33** shows a block level model of a typical superheterodyne receiver simulated by ADISimRF.

6.3.5 Simulation Tools

PSPICE: PSPICE is a popular version available from Cadence PCB Solutions (**www.cadence.com**) and a "Lite" version is available for free download. (The Lite version is limited in the size and complexity of designs that can be created). *AIM-spice*

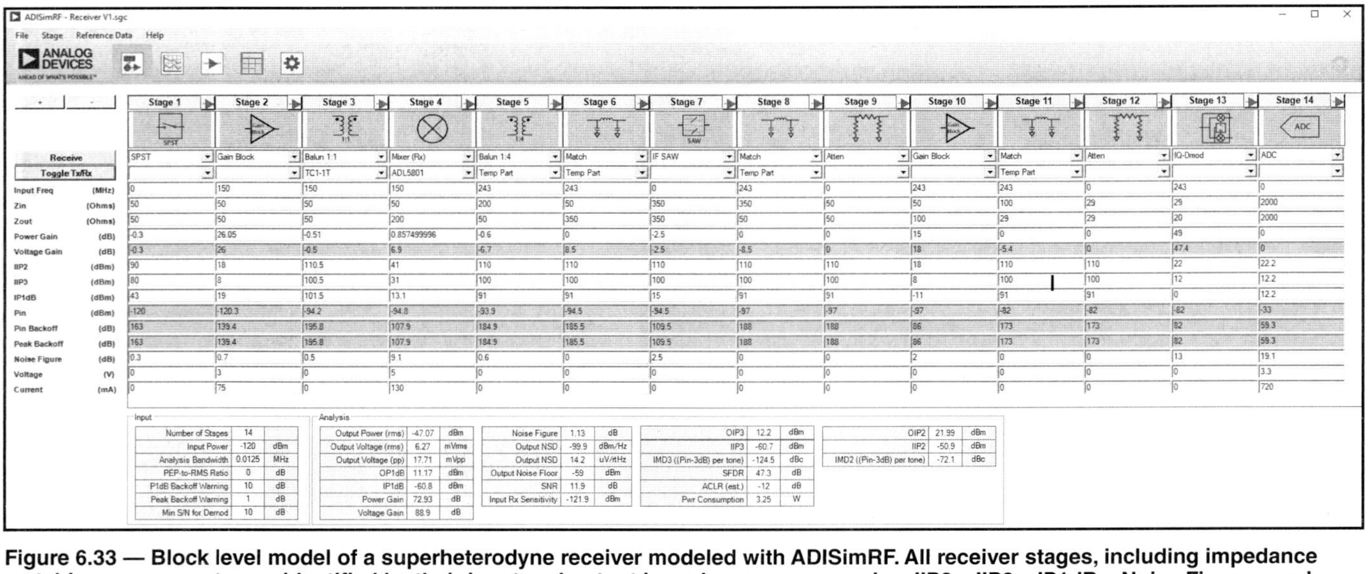

Figure 6.33 — Block level model of a superheterodyne receiver modeled with ADISimRF. All receiver stages, including impedance matching components, are identified by their input and output impedances, power gains, IIP2s, IIP3s, IP1dBs, Noise Figures, and source voltages and currents. The model calculates the interrelationships of the adjacent stages and determines overall system parameters, shown along the bottom.

(**www.aimspice.com**) is a PC version of *SPICE* with a revised user interface, simulation control, and with extra models. A student version can be downloaded.

There are several PC-based *SPICE* programs in the $1,000 range, but they are designed more for switching power supplies and logic circuits optimization than RF. *ICAP4* (**www.intusoft.com/demos.htm**) has an evaluation version available for download.

Agilent, AWR, Ansys, and Synopsis offer very modern mixed-mode CAD tools and they combine the concept of *SPICE* with the advanced technologies. These are professional quality tools, but if one can arrange to make use of them through a friend or associate, the results are worth investing the time to learn their use.

Linear Technologies, a division of Analog Devices, also offers a free version, *LTSPICE*, with a graphical user interface that simplifies schematic capture. *LTSPICE* supports switch mode products but handles all types of circuits and has many users. (For a *QST* tutorial and a set of references and downloadable component libraries, see the Reference entries for Silver and Würth Elektronik*)*. The program can be downloaded without cost from Analog Devices (**www.analog.com/en/resources/design-tools-and-calculators/ltspice-simulator.html**).

Spectrum Software's *MICRO-CAP* is another popular SPICE-based program with a similar graphical user interface and a large component library. When Spectrum Software closed in 2019, it released *MICRO-CAP* into the public domain and its final released version (12.2.0.5) has been made available online without cost (**archive.org/details/mc12cd_202110**).

6.4 Electromagnetic Analysis

Computers can be used to solve the equations that govern electromagnetic fields. This first became technically feasible in the late 1970's and is now a well-developed technology. A wide variety of software is available, all developed over decades of intense research. Today, you can take advantage of the fruits of this research without learning the equations and without writing any software. In this section we provide some historical background, discuss the different types of EM analysis techniques, and outline how electromagnetic software is used today. Then we show how you can use some of these tools for your RF designs.

A Biography of Maxwell

The best-known biography of James Clerk Maxwell is the first one that was written, *The Life of James Clerk Maxwell*. The authors, Lewis Campbell and William Garnett, knew Maxwell personally and they concentrated on his youth. We relive young Maxwell's wonder at multicolored light reflecting from an oil slick on a puddle of water; his experiments with Iceland Spar and thus learning about the polarization of light; and other personal experiences that lead to his becoming, along with Einstein and Newton, one of the three greatest physicists of all time. Unique material from the second edition is also presented along with a few pages of modern material on Maxwell.

6.4.1 Historical Background

In 1865 James Clerk Maxwell, **Figure 6.34**, published the equations that govern electric and magnetic fields. Maxwell's equations, as they are now called, put in solid mathematical terms the following principles:

1. An electric field that changes with time creates a magnetic field that changes with distance and is exactly at right angles to the electric field.
2. A magnetic field that changes with time creates an electric field that changes with distance and is exactly at right angles to the magnetic field.

Maxwell recognized an important implication of his equations. If both an electric and magnetic field could somehow exist at the same time, in the same place, and be exactly at right angles to each other, both changing with time and with distance in just the right way, each would continually create the other and a propagating wave would result. Using his equations, he calculated the speed of that hypothetical "electromagnetic" wave. Amazingly, he saw that it was virtually identical to the mechanically measured speed of light. His conclusion was earth shaking, "that light is an electromagnetic disturbance propagated through the field according to electromagnetic laws." However, Maxwell felt that the chances of using electricity to actually create such a wave would be difficult, if not impossible. For more information about Maxwell, go to the James Clerk Maxwell Foundation website (**www.clerkmaxwellfoundation.org**).

After Maxwell's death, a German professor, Heinrich Hertz, was experimenting with a spark gap. Almost by accident, he realized he had succeeded in creating and detecting electromagnetic waves. He then refined how he created and detected the waves and conducted experiments reflecting, diffracting, focusing, and polarizing electromagnetic waves. This provided definitive proof that Maxwell's equations were correct. However, Hertz felt there was little, if any, practical use for such waves. An Italian entrepreneur, Guglielmo Marconi, thought otherwise

Figure 6.34 — James Clerk Maxwell (1831 – 1879) gave us the equations that govern the interaction between changing electric and magnetic fields. These equations can be solved on a computer, revealing how our RF and microwave circuits will perform.

and was spectacularly successful in commercializing use of electromagnetic waves for communications. Today, every time we key our radio, work a DX station, or even just talk on our cell phone, we honor these amazing electromagnetic pioneers.

Figure 6.35 shows how an electromagnetic wave propagates. We key our handheld transceiver (HT) and it produces an RF (i.e., changing with time) voltage across the antenna terminals. For example, if it produces 10 volts, and the terminals are 1 cm apart, we have an electric field of 10 volts/cm. This is indicated by the long vertical arrow coming out of the HT. (The arrow indicates the strength and direction of the electric field right at the antenna terminals. The electric field does not physically extend out of the HT!) This electric field is changing with time, so it creates a magnetic field that changes with distance and is at right angles to the electric field. Since the electric field at the antenna (that creates the magnetic field) is changing with time, the resulting magnetic field also changes with time. That means the magnetic field then creates an electric field that changes with distance. The electric field and magnetic field keep creating each other, resulting in an electromagnetic wave. The wave travels into space until it runs into something.

You may be familiar with the concept of a *photon*, the particle of light. A photon is indivisible. We cannot have half of a photon. The light we see can only be composed of an integer number of photons. A sensitive photo-multiplier tube can receive one photon, or two, or three, but never 2.853 photons.

Maxwell's equations, summarized in the sidebar, form a *field theory*; there are no photons. Mathematically, we can write equations for an electromagnetic wave with the energy of partial photons, but we know that such a wave cannot exist. It is here that something called *quantum electrodynamics* (QED) takes over. This is part of quantum theory. See the sidebar presenting a very readable explanation of QED by its co-inventor, Richard Feynman. Fortunately, for virtually everything we do, it is not needed. Because the number of photons is so incredibly huge, and the errors due to calculating with fractions of photons are vanishingly small, we can just pretend that we are dealing with continuous electric and magnetic fields and use Maxwell's equations for our work with radio waves. Maxwell's equations were created before quantum theory was discovered, yet they still faithfully represent the observed behavior of electromagnetic waves.

We are especially fortunate to have Maxwell's equations. In nearly all other fields of

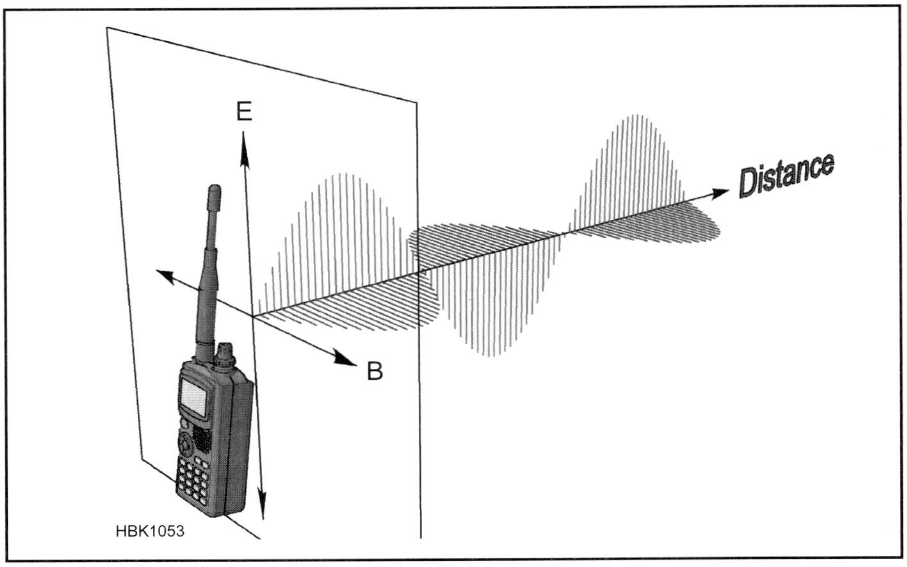

Figure 6.35 — We create one of Maxwell's electromagnetic waves when we key our radio. That generates a voltage across the antenna terminals that creates a time-varying electric field, which creates a magnetic field that varies with both time and distance that creates more electric field, which creates more magnetic field, and on and on. One and one-half cycles of an electromagnetic wave are shown.

The Basic Elements of Maxwell's Equations

The complete set of Maxwell's equations can be described as, "A set of four simultaneous partial differential equations operating on four vector fields in four dimensions and driven by two scalar fields." The four dimensions are the three spatial dimensions (*x*, *y*, *z*) and time. The four vector fields are electric, **E** and **D**, and magnetic, **H** and **B**. The two scalar fields are electric and magnetic charge. (Magnetic charge has yet to be discovered, but that does not stop the mathematics.)

You may also know these equations as laws named for scientists who originally discovered the relationships. It was Maxwell who brought them all together into one set of equations that define electromagnetics. (For an explanation of vectors, scalars, and what the various symbols mean, see "Radio Mathematics" in the online content.)

Gauss's Law for Electric Fields:

$$\nabla \bullet \mathbf{D} = \rho$$

Gauss's Law for Magnetic Fields:

$$\nabla \bullet \mathbf{B} = 0$$

Faraday's Law: $\nabla \times \mathrm{E} = \frac{\partial \mathrm{B}}{\partial t}$

Ampere – Maxwell Law:

$$\nabla \times \mathrm{H} = \mathbf{J} + \frac{\partial \mathrm{D}}{\partial t}$$

Where ρ represents the density of electric charge and **J** represents a distribution of electric current (moving charge). There are also forms of Maxwell's Laws that are written as integrals, that incorporate relativity, that use mathematical objects known as tensors, and various other ways. The four equations above, though, are the most widely known.

J is a vector because current has a direction. A distribution of *electric* charge changing with time is exactly what makes electric current. So, for RF, we can use either current or charge (both of which can vary with time) to drive the equations. All sorts of other abstract mathematical quantities are also helpful, including some strange beasts known as vector and scalar potentials. Fortunately, we can often simplify things by assuming that **D** is proportional to **E** (**D** = ε**E**), and **B** is proportional to **H** (**B** = μ**H**). Then, we have only two vector fields to deal with, but even that is still a big problem.

By doing this extra math, Maxwell's equations and all their vector and scalar fields can be turned into a big pile of discrete numbers that can be stored on and crunched in a computer. It has only taken a couple decades with hundreds — if not thousands — of researchers all over the world, but it seems like today we are pretty much there.

For a detailed introduction to Maxwell's equations, if you have some calculus background see *A Student's Guide to Maxwell's Equations* by Daniel Fleisch from Cambridge Press. It is an excellent tutorial on this important topic.

science and engineering, exact solutions to practical problems are often impossible. In contrast, given a precisely known passive RF circuit, we can use Maxwell's equations to get precise results for how it will operate. (Active RF circuits, typically made with transistors, are another matter.) Appropriate circuits include filters, matching networks, couplers, power dividers, connectors, transformers, and any passive structure involved in handling RF.

6.4.2 Types of Electromagnetic Analysis

Maxwell liked to think "analogically," reasoning that completely different areas of physics and nature somehow use similar equations. For example, he suspected that incompressible fluid flow (think of a submarine powering through the ocean depths) and electrostatics (think about rubbing a balloon on your hair) use the same equations. It sounds ridiculous, but he gave it a try anyway and he derived the equations for incompressible fluid flow and found that they magically worked just fine for electrostatics. That was his first step in coming up with the full set of Maxwell's equations.

We can use analogical thinking in solving Maxwell's equations on a computer too. The problem is that we have continuous electric and magnetic fields everywhere. In contrast, a computer can only store numbers, and we must be careful that we do not overrun the computer's memory capacity.

First, the fields and currents are *discretized*, that is, they are divided into small subsections, just like an image being converted to pixels. However, an image is usually two-dimensional while EM fields are three-dimensional in space, and they vary with time. Thus, we must deal with pixels in four dimensions. To make the problem worse, we must discretize the electric field, the magnetic field, the charge (or, equivalently, the current), and even the circuit itself.

Fortunately, today's computers have gotten good at storing lots of numbers. Once we have the numbers stored, we can use software to do what we want with them. This is where the last few decades of EM research has concentrated.

SURFACE MESHING — METHOD OF MOMENTS (MOM)

Method of Moments (MoM) was developed in the 1960s. It was fully formalized and popularized in the west by Prof. Roger Harrington of Syracuse University. He named it after similar work that had previously been performed in Russia. The equations needed to understand MoM can be difficult. Fortunately, the concept is easy to understand.

The Basic Elements of Maxwell's Equations Quantum Electrodynamics (QED)

Maxwell's electromagnetic theory does not consider photons, i.e., "particles" of light. QED predicts everything Maxwell's theory does, and it does so assuming that electromagnetic waves are composed of photons. *QED — The Strange Theory of Light and Matter* by Richard Feynman, the world-famous Nobel laureate and popularizer of science, explains QED in simple terms that anyone can understand. When you read this book, which is available from many sources, be prepared to enter a world that is unlike anything we have ever known in our day-to-day lives.

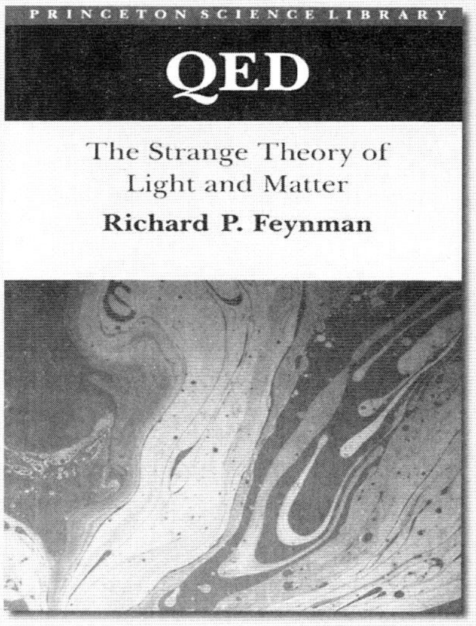

Imagine we have a planar circuit, say a filter on a PCB. In the first step, we discretize (i.e., divide into small subsections) only the metal of the circuit. Next, we put current (at a given RF frequency) on one subsection of that circuit. We calculate how much voltage that bit of RF current induces on all other subsections. We can view this as one tiny transmitting antenna (the subsection with current) and all the other subsections are tiny receiving antennas. Typically, high voltage is induced on nearby subsections, and lower voltage on more distant subsections. The effect of surrounding dielectric material is included in this calculation.

This calculation (i.e., current on one subsection induces voltage on all other subsections) is repeated for all subsections in the circuit, one at a time. All this information is stored on the computer in the form of a matrix (i.e., like a spreadsheet with rows and columns). Next, the computer puts current on all subsections at the same time. This, as previously calculated, induces voltage in all the subsections in the circuit. Every subsection has voltage induced in it from all the other subsections in the circuit. In fact, each subsection also has voltage induced in it due to itself. All the voltages induced in each subsection add together.

But a perfect conductor (which is our assumption for this description) cannot have any voltage across it! So, the computer adjusts all the currents in all the subsections so that there is zero total voltage across all the subsections. In practice, this is done by a process called matrix inversion. Of course, the subsection that we are using for the input to the circuit must have some voltage across it. Otherwise, we will have zero current (and zero voltage) everywhere.

The currents in all the subsections that give us zero total voltage everywhere that the conductor exists (given an input voltage in the input subsection) represent the current distribution in the conductor. That is our solution. Next, we take the now known current on the input subsection and, given the voltage we assumed on that subsection, we can calculate the input impedance. Repeat this process for each frequency of interest, and we have a plot of the input impedance as a function of frequency. We now have all this information without ever having built anything.

When modeling circuit boards, this implementation of the MoM is restricted to planar circuits only. Arbitrary 3D structures are not appropriate for this approach. The reason planar MoM has seen wide-spread use is that it can handle much larger planar circuits at high levels of accuracy. Appropriate planar circuits can include many layers. PCBs in a computer typically use 4 to 8 layers but can have as many as 100 layers. This is no problem for the MoM. Sometimes planar problems are referred to as having 2.5 dimensions, rather than full 3D. However, full 3D fields and full 3D current are calculated (conducting connections between layers in a circuit board are called vias and they allow current to flow vertically), so describing this kind of problem as 3D planar is more accurate. Fields and currents are 3D, while the circuit is formed on layered (i.e., planar) dielectric. That said, enhancing MoM to include 3D arbitrary volume meshing problems has seen substantial progress recently. See, for example, the MoM book described in the sidebar.

MoM has been developed for a wide variety of problems. For example, there are a few ways to implement MoM that make it amazingly efficient for doing extremely large problems, like a radar signal illuminating an airplane, or even an entire aircraft carrier.

Antenna Modeling with MoM

The same concept of modeling electromagnetics in circuit boards or along the metal surfaces that form an airplane can also be applied to the metal that forms an antenna. This method is commonly used by radio amateurs to determine radiation patterns. A tool developed in the 1960s, called Numerical Electromagnetics Code (NEC) at Lawrence Livermore National Laboratories is still in common use today, after several revisions. NEC is a 3D MoM tool that calculates both the E- and H-fields as they radiate out from an antenna. It accounts for interactions with the ground and gives field strengths and power densities in all directions, in both the near and far fields.

NEC Version 2 was successfully applied to antenna analysis software with a much improved user interface by Roy Lewallen, W7EL, who sold his product as EZNEC. After retiring, W7EL made his software available at no cost with the proviso that it was available "as-is" and would not be supported further. See the section below, **Commercial Electromagnetic Analysis Tools**, for information on obtaining this software.

Further development of NEC at Lawrence Livermore has taken place over the years, culminating in the most recent version, NEC Version 5, which requires a license to use. See the section below for more information about obtaining it. The user interface for NEC-5 is far superior to that of earlier NEC versions. An option in EZNEC allows it to use any other version of NEC for its calculations so that its user interface can benefit from the improvements in the NEC modeling software.

NEC-2 does not permit analysis of wires that are too close to the ground, making it unable to properly calculate the effects of buried ground radials such as are often found in a vertical ground plane antenna. Subsequent versions of NEC, including NEC-5, have utilized Sommerfeld equations to overcome that limitation, allowed more exact definitions of ground type, allowed the use of either voltage or current sources and added features to work with very small antennas and conducting surfaces.

Planar Circuit Modeling with MoM

There are two widely used approaches that are preferred for MoM analysis of planar circuits. Both approaches can handle large planar circuits, even when they have multiple layers. One approach performs the required current-induces-voltage calculation assuming the planar circuit is unshielded. The other assumes the circuit is shielded. Both approaches are useful. The unshielded approach is a natural for unshielded circuits, like antennas. The shielded approach is a natural for shielded circuits, like a filter inside a shielding box. However, each approach can flex into the other's domain. The shielded approach benefits from a more accurate induced voltage calculation but it requires an underlying, fine pixel-like grid for meshing. The unshielded approach benefits from allowing arbitrary subsection size, which can be handy, provided the required less accurate induced-voltage calculation is not of concern.

Figure 6.36 shows a typical circuit analyzed using shielded 3D planar MoM. This is a spiral inductor on silicon, a critical com-

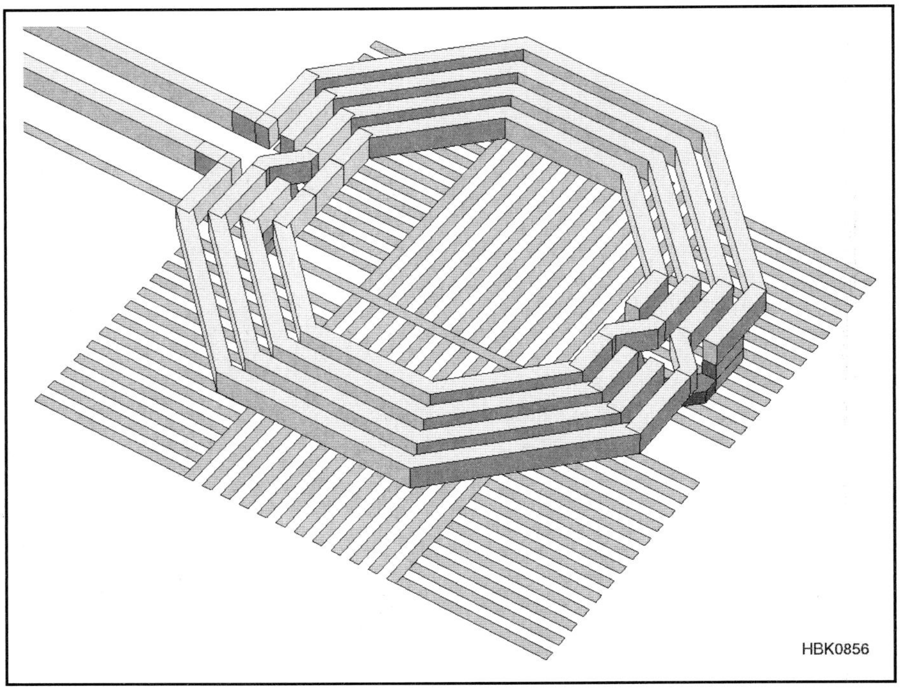

Figure 6.36 — Method of Moments (MoM) can be used to analyze planar 3D circuits like this spiral inductor over silicon. Many such inductors are present in the RF chip set for every cell phone. Silicon dioxide insulator (not shown) gives some separation between the bottom of the inductor and the top of the semi-conducting silicon. To reduce inductor losses, a grid-like ground shield is placed directly on the silicon to try to keep the inductor fields out of the silicon. (Courtesy Sonnet, with permission.)

Method of Moments (MoM)

The math involved in the MoM uses vectors, matrices, and calculus no more complicated than various sine and cosine integrals. However, some derivations and results are lengthy and complicated. If you can handle this level of mathematics and write computer programs, then you can use *Method of Moments – Analysis of Structures in Shielded Layered Media* by James Rautio, AJ3K, as a recipe to write your own MoM program. MoM programs typically deal only with surface current (infinitely thin sheets of current). A recent development, fully detailed in this book, is how to use MoM for volume current, i.e., full 3D. Ultimately, this could supplement, or possibly supplant, the common volume meshing techniques like Finite Elements as MoM meshes only the structure of interest while other volume meshing techniques must also mesh all dielectric, including the air surrounding the structure of interest.

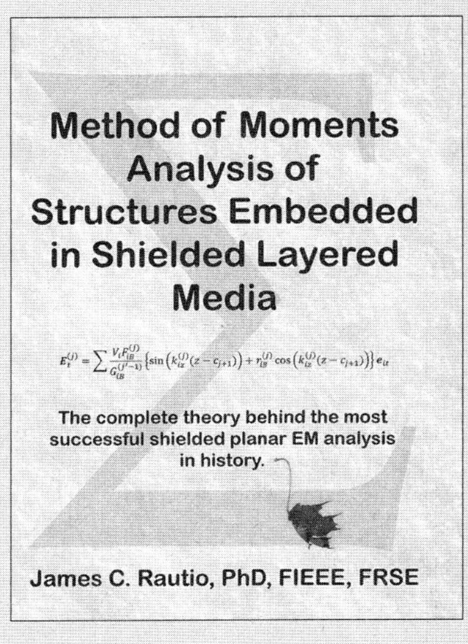

ponent used in all cell phone RF chip sets. For example, when you make a call, the cell phone computer moves the transmitter and receiver to the channel assigned by the cell site for your call. To do this, it programs a VCO (voltage-controlled oscillator) that is inside a PLL (phase locked loop) to move to the required frequency. The VCO has an inductor as part of its tuned circuit. Compared to silicon chips, inductors are gigantic. They take up a lot of space, and space (or "real estate" as the RFIC designers call it) is expensive. The solution is to make the spiral turns even smaller and use more of them. However, that increases the loss. If there is too much loss, the VCO noise goes up and tuning range goes down. In the worst case, loss might be so high that the VCO does not even start oscillating.

A typical silicon wafer fabrication ("wafer fab" for short) can easily cost $1 million. Fabricators do not have the time or the money to repeatedly build and tweak a circuit. It is critical that the VCO, and everything else on that wafer, work the first time. This is where EM analysis saves the day. The designer does all the trial and error on the computer, and then builds the circuit once.

Silicon is a lossy conductor, i.e., a semiconductor. Any RF designer knows that when we put an inductor close to a lossy conductor, the inductor loss increases. To reduce the loss, we try to get the inductor as far as possible from the silicon. A given process might have a few microns of silicon dioxide, or other insulators, deposited on top of the silicon. So, the designer puts the inductor on top of all available silicon dioxide. In Figure 6.36, these insulating layers are not shown, so it looks like the inductor is floating in air.

Another trick to reduce loss is to add a ground shield. This has been done for the inductor in Figure 6.36. This is the grid-like conductor pattern under the inductor. It is important that the ground shield conductors are, more or less, at right angles to the spiral inductor turns above them. If the ground shield conductors are parallel, they start to act like a shorted turn in the secondary winding of a transformer — not good!

Figure 6.37 shows the current in the surface of the silicon directly under the spiral inductor when there is no ground shield. The figure's caption explains current levels. At first, it is a surprise that the direction of the current (not shown) is radial, flowing toward or away from the inductor. This means the current is driven by the electric field around the inductor, not by the magnetic field. The designer wants this current to be as low as possible to reduce loss.

Figure 6.38 shows the substrate current in the silicon under a small portion of the inductor with a ground shield. This was analyzed using *Sonnet*, a shielded 3D planar EM tool. (For a Sonnet tutorial, see the PDF "A Sonnet Tutorial Using Port Tuning" in this book's online content.) As mentioned above, the induced voltage calculation is very accurate for shielded tools, but a fine, underlying rectangular mesh is required. If you look closely on this zoomed image, you can see the *pixelation* caused by the fine mesh.

Normally, we view only the current in the conductors of our circuit with planar EM tools. A description of how to view tangential (i.e., parallel to the surface of the substrate) electric field using something called "sense metal" in planar MoM is given the Sonnet User's manual (**www.sonnetsoftware.com**), search for "sense metal." By Ohm's law, the tangential electric field on the surface of the silicon substrate is proportional to the current flowing in the silicon. If we make the sense metal surface resistance, say, 1 million ohms/square, then it will have little effect on the overall fields. The (very tiny) currents in the sense metal (due to the current flowing in the lossy silicon) are proportional to the tangential electric field required by Ohm's law.

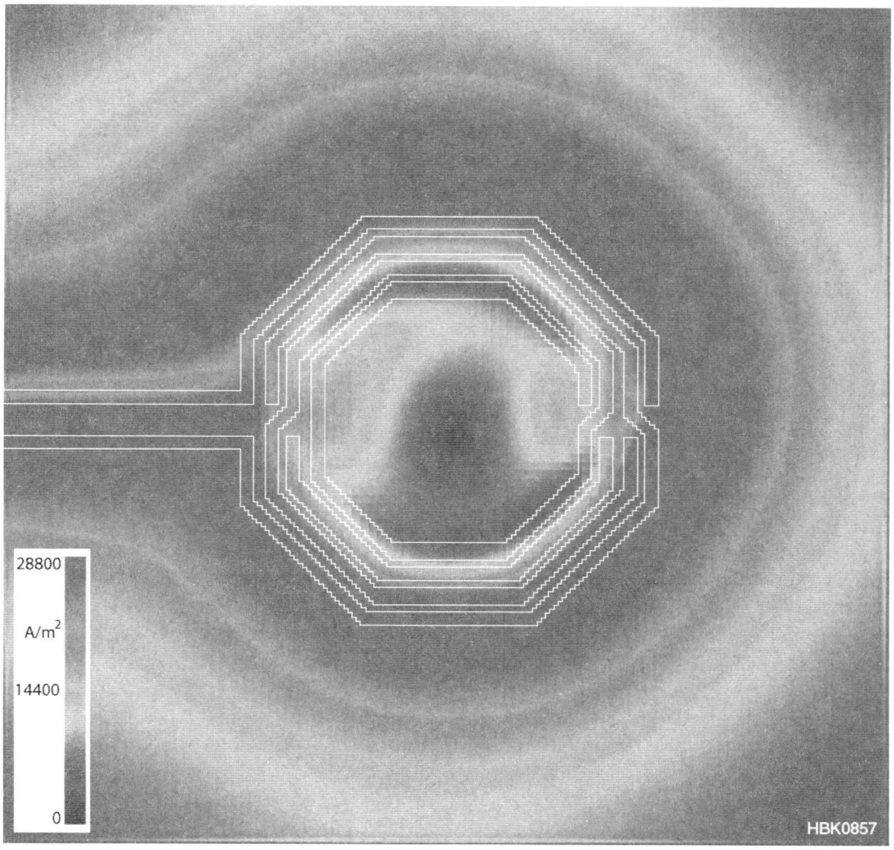

Figure 6.37 — The spiral inductor, here without the ground shield, induces current in the silicon. This current increases loss and must be minimized. Brighter shades indicate mid-level current (see the scale at the lower left). Highest current is in the dark area directly under the outer turns of the inductor. (Courtesy Sonnet, with permission.)

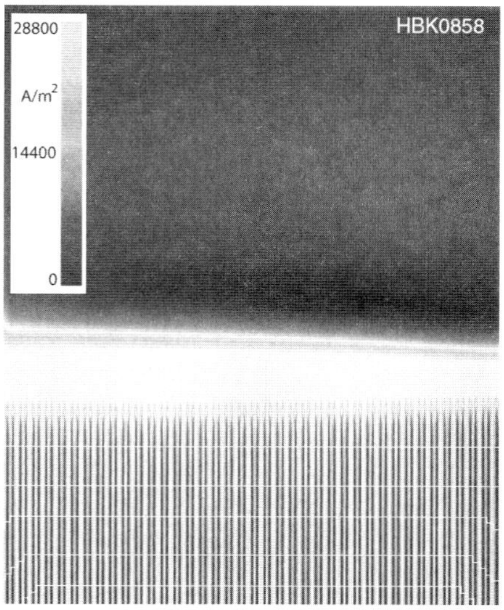

Figure 6.38— A small section of the current in the silicon under the inductor with ground shield in place shows a strongly modified current in the silicon. Whether or not inductor loss is actually decreased, or increased, strongly depends on the ground shield design. (Courtesy Sonnet, with permission.)

VOLUME MESHING — FDTD

MoM does not consider all of the electromagnetic effects of the materials surrounding conductors. When those materials cause changes in the electromagnetic properties predicted by MoM, another method is needed to add precision to the calculations. The easiest to understand of these is *finite difference time domain* (FDTD) analysis. Practical computer-based FDTD analysis was pioneered by Allen Taflove, WA9JLV, at Northwestern University in the 1970s.

In its simplest form, we divide the entire volume of the RF structure into tiny rectangular hexahedral cells (six sides, e.g., cubes, rectangular cuboids, etc.). This is called *meshing*. Note that we must also mesh all of the space that might have any EM fields. This includes any air (e.g., free space) surrounding the structure. Next, we assume an initial condition. This is the RF input into the structure, like when you first key your transmitter. Let's say we have 1 volt/meter of electric field at the input right at the start (time: $t = 0$). This input is a single impulse, we put it there for only a single instant (i.e., time step). This is like hitting a bell to start it ringing.

And indeed it does ring. Changing the input voltage, which we placed on an edge of one of the hexahedrals, from 0 volts to 1 volt right at the start gives us a changing electric field. In addition to meshing the structure, we also have to discretize time, so we pick a tiny time-step. The software uses Maxwell's equations to calculate exactly how much of a magnetic field is created by the changing electric field at the input. In addition, the software knows, because of Maxwell's equations, that the changing magnetic field must be at right angles to the changing electric field and it must also vary with distance. The changing magnetic field is now created on the edges of every hexahedral surrounding the hexahedral edge that has the changing electric field input.

In the next time-step, we have both changing magnetic fields and changing electric fields near the original input. Each hexahedral edge with a changing magnetic field creates a changing electric field in all surrounding hexahedral edges. Each changing electric field creates a changing magnetic field in all surrounding edges. Using Maxwell's equations, the software calculates how much electric and magnetic field is created and the process is repeated. Soon electric and magnetic fields propagate throughout the entire structure at the speed of light. For every hexahedral in the structure, the computer now stores a list of numbers that tell us how the electric and magnetic fields change as a function of time. If there is a little bit of loss (and in real life, there is always at least a little bit of either conductor, dielectric, or radiation loss), the ringing electric and magnetic fields gradually die out, and the solution is complete.

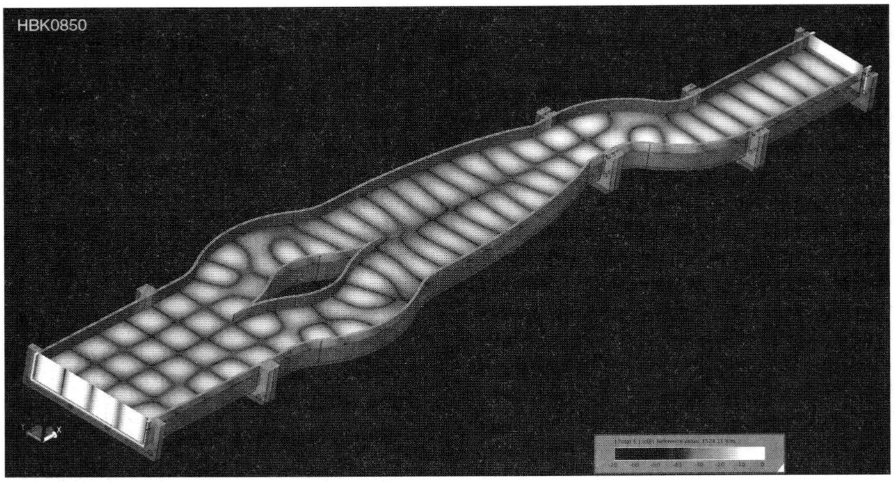

Figure 6.39 — A complex rectangular waveguide mode converter as analyzed by FDTD. Electric field is shown with bright shades indicating high field strength. Input is on the right. (Courtesy Remcom, with permission.)

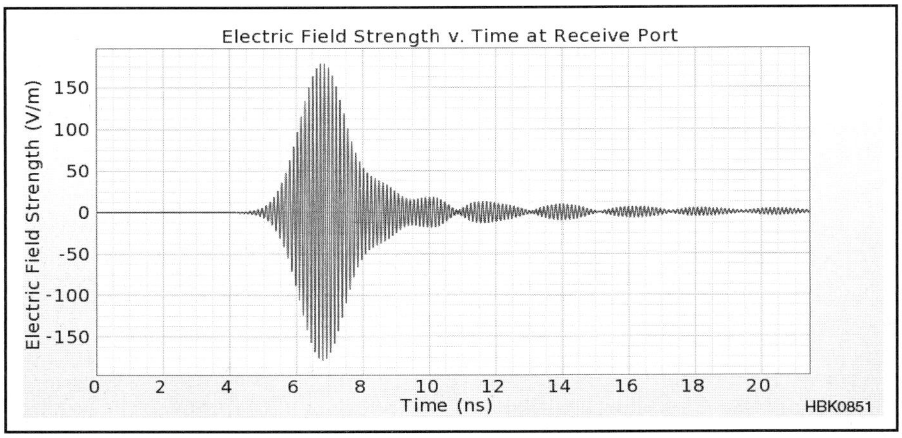

Figure 6.40 — The FDTD calculated time-domain response to a brief RF pulse on the input of the mode converter of Figure 6.36 resembles the ringing of a bell. This is called the *impulse response*. (Courtesy Remcom, with permission.)

Often, we are not interested in the electric and magnetic fields in the entire volume. For example, we might be interested only in an input and an output. We still need to calculate all the field information everywhere to get to the final desired result. But at any given time-step during analysis, only the last several time-steps of complete field information are needed. In this case, we discard all internal field information after it is no longer needed for calculating the next time-step. In other cases, we want to see the complete internal fields as they develop during the analysis and we keep that information available, usually by storing it conveniently on the computer's hard drive. **Figure 6.39** shows a typical FDTD result. Bright shades indicate high electric fields as a continuous RF wave at a single frequency propagates along a complex rectangular waveguide circuit. (Color versions of the figures are available as online content.) The field strength is shown across the entire cross-section of the guide at the ports, and on a horizontal plane in the middle of the waveguide along its length. The hexahedral meshing is so fine that it is not visible in this image.

A little more detail is important. Due to problems calculating rate-of-change with time (known in calculus as a *time derivative*) and rate-of-change across space (a *gradient*), there are actually two full and nearly identical discretizations of the structure being analyzed. Both discretizations cover the same volume. The only difference is one discretization is offset from the other by one-half of a hexahedral. The changing electric field is calculated on one discretization, and the changing magnetic field is calculated on the other. This means that every sample of magnetic field is midway between all adjacent samples of the electric field. Likewise, the electric field calculation is always for points that are midway between all adjacent samples of the magnetic field. Without this special modification, there can be stability problems.

In practice, the input is typically not a simple impulse. A pure impulse, an infinitely short pulse, can cause numerical instability. So, we typically use a "gentler" input, like a Gaussian pulse. This kind of pulse is the well-known "bell curve" used to represent the probability of many random processes. If we are especially interested in results over a specific band of frequencies, we can use the Gaussian pulse to modulate an RF carrier centered on the band of interest.

For most circuits, the electric and magnetic fields at a given point, plotted as a function of time, look just like what we would expect for a ringing bell, or a ringing resonant LC circuit, **Figure 6.40**. For this case, the input is a Gaussian pulse modulating an RF carrier. The rapid oscillation is from the RF carrier, and the envelope is the result of the "bell" ringing.

The input pulse that started everything ringing has a wide band of frequencies present in it. If we think of our circuit as being a filter, the output is just a filtered version of that broad band impulse we put on the input. The spectrum of that output is the filter response that we want to see. The computer can take the numbers that were calculated for the impulse response and simulate a spectrum analyzer by using a mathematical algorithm, i.e., a mathematical recipe, called a Fourier transform. The fastest way of doing a Fourier transform on a computer is creatively called the fast Fourier transform, or FFT.

The software then performs an FFT on the impulse response, and we have the current and voltage output of our circuit. This output is then converted to S-parameters and plotted, as in **Figure 6.41**. The indicated S21 is the ratio of the amplitude of the output voltage wave (i.e., coming out of port 2) to the amplitude of the input voltage wave (i.e., going into port 1). S11 is the reflection coefficient, i.e., the ratio of the reflected wave to the input wave. (See the **RF Techniques** chapter for more on S-parameters).

There are other time domain approaches like FDTD. Another approach is called transmission line method (TLM), where Maxwell's equations are modeled as a fine mesh of short transmission lines. However, the basic idea of all time domain volume meshing EM tools is similar to what is described above. The entire volume is meshed, an input pulse is applied, the time-domain response is calculated one time-step at a time. The final result is usually converted from time-domain to frequency-domain.

VOLUME MESHING — FINITE ELEMENT METHOD (FEM)

FDTD is a *time domain* analysis where the EM fields are evaluated as they evolve in time. We obtain the frequency response by performing Fourier analysis on the calculated impulse response.

Finite Element Method (FEM) tools perform their calculations in the *frequency domain*. We start by specifying a frequency, and the FEM software analyzes the response at that frequency. Then we specify a second frequency and rerun the analysis for it. This process repeats until we have results for all frequencies of interest.

The advantage of frequency domain analysis is that we analyze only the frequencies that we want. In addition, we now need to solve only for the three spatial dimensions.

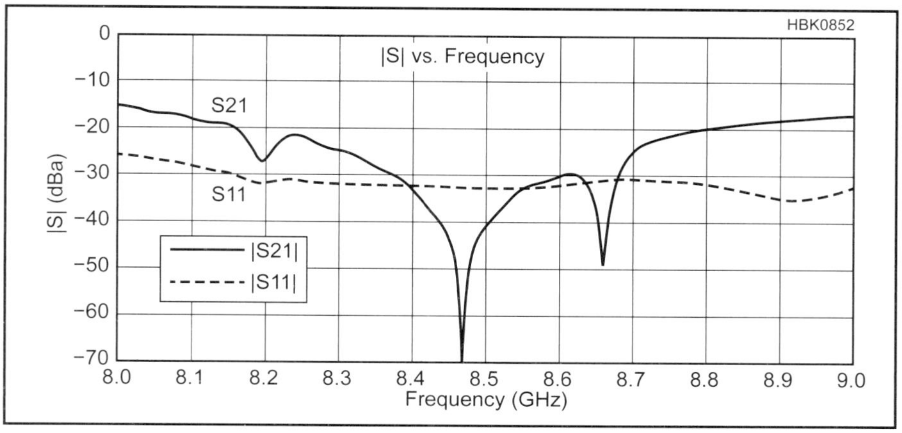

Figure 6.41 — Performing a Fourier transform on the circuit response is equivalent to using a spectrum analyzer to see the frequency domain response of the mode converter given the broad band impulse used to excite the filter. (Courtesy Remcom, with permission.)

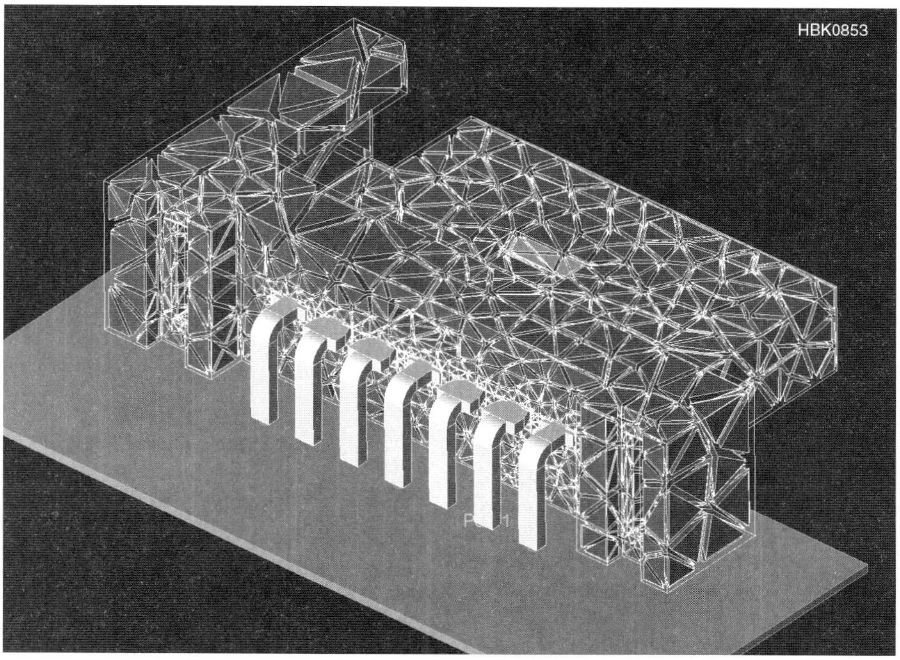

Figure 6.42 — Extremely arbitrary structures can be analyzed with FEM. Here we see the meshing for a PCB multi-pin connector. High accuracy analysis at high frequency is critical when high performance is needed. The entire volume is meshed with tetrahedra. For clarity, the tetrahedra used for meshing are outlined only where they intersect with a surface of the connector. (Courtesy Keysight, with permission.)

We already know that the fields all vary in time with a sine wave at a given frequency. We are however, restricted to working with linear circuits. Given input at a single frequency, linear circuits have only that frequency present everywhere in the circuit. An amplifier or a mixer could have several frequencies bouncing around even if only one frequency is provided at the input. Given that only one frequency is present everywhere, our EM analysis now needs to solve only for the magnitude and phase of that sine wave. In contrast, the advantage of time domain analysis is that we get all the frequencies with a single analysis.

FEM starts just like FDTD, by discretizing the structure we are analyzing. As with FDTD, FEM must mesh all space that contains electromagnetic fields, including air. For RF work, FEM tools mesh the structure into tetrahedra. **Figure 6.42** shows an example of a multi-pin connector meshing. The rectangular hexahedral mesh used for structures in FDTD often leaves a "staircase" appearance. The staircase can be reduced by making the mesh very fine, but this increases memory requirements and analysis time.

FEM can use tetrahedra of any size; curving surfaces appear to be meshed into triangles (the exposed surface of each tetrahedron). This is often a better representation of the surface than is possible with the hexahedral cells of FDTD. In addition, since different size tetrahedra can be used as desired, the accuracy of the mesh can be varied. Very tiny tetrahedra can be used where the highest accuracy is needed, and large tetrahedra can be used elsewhere. An initial solution can be performed, and the error estimated everywhere. Adaptive meshing then refines the mesh (i.e., makes more and smaller tetrahedra) wherever the error is too high. The fine mesh, which can consume a lot of analysis time and computer memory, is used only where it is needed. Care should be taken because the location of the fine mesh might require modification depending on the frequency of analysis and on how the structure is excited.

The mathematics of FEM can get very involved, and there are many ways to implement it. At a high level, we must cast the problem into what is called a *variational form*. For example, an equation for the energy in, say, a microwave filter, given some boundary conditions (like an RF voltage source on the input), is a variational form. FEM then finds the EM fields consistent with Maxwell's equations and the given boundary conditions that give the minimum total energy. That field is the solution.

We can intuitively see why the minimum energy solution is the correct solution. Imagine a length of thin, moderately wide conductor on a printed circuit board. Next, we place static electric charge on the conductor. Electrons are all negative, and repel each other, trying to push each other to opposite edges of the conductor as best they can. The highest charge is on the edges of the conductor and lower charge are towards the center of the conductor. This configuration of charge gives the minimum potential energy for the charge. This is the solution that FEM finds. It makes sense: highest charge on the edges means we have the highest voltage there too. If we add too much charge, the sharp edges are where the arcs and sparks will come from.

Mathematically, the situation for RF is nearly identical. It is just that instead of charge, we have RF current flowing. The RF current flows most strongly on the edges of a conductor. In fact, if we run too much RF current, the edges literally start evaporating. This can happen even with low power if the conductor dimensions are very small, for example, on a silicon RF integrated circuit (SiRFIC) that is widely used in cell phones.

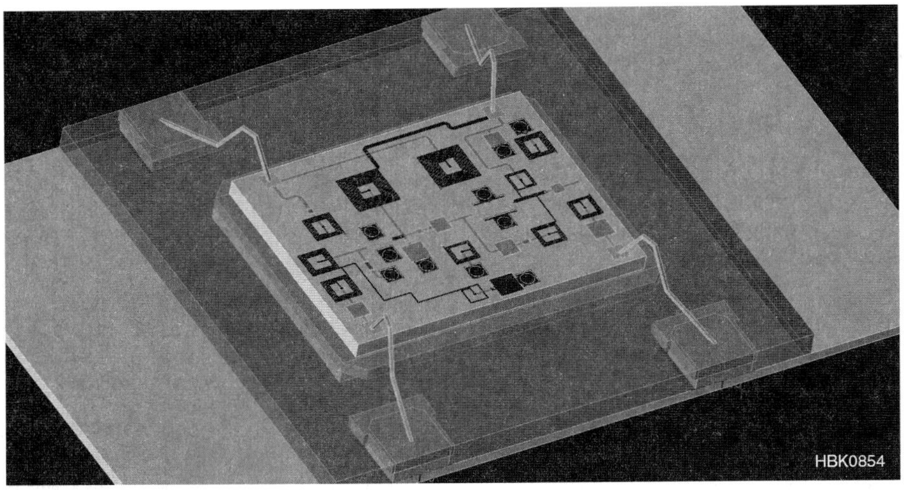

Figure 6.43 — This RF integrated circuit is mounted on a carrier with wire bonds making connections to the outside world. This is a view of the circuit prior to meshing. FEM can analyze the entire structure including all electromagnetic interactions. (Courtesy Keysight, with permission.)

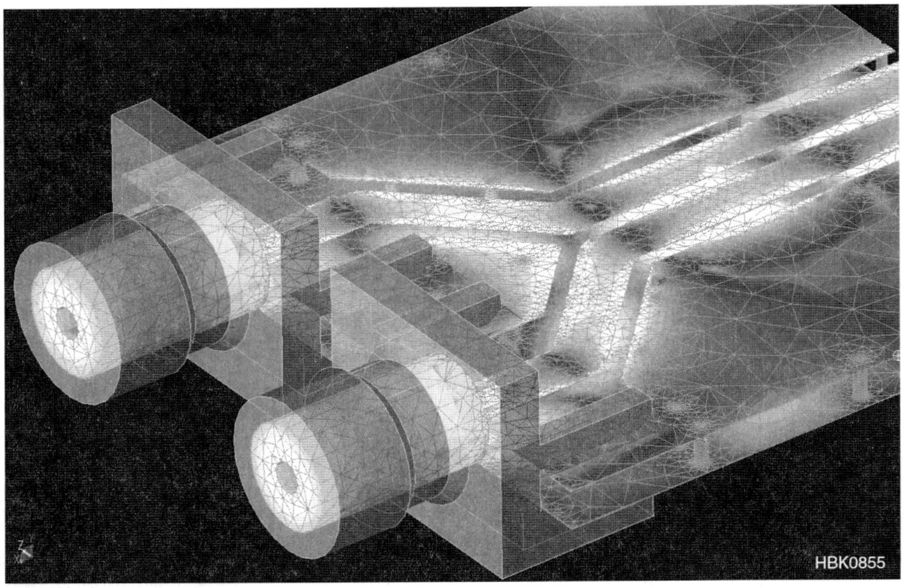

Figure 6.44 — FEM has meshed a circuit with two coaxial connectors transitioning to two PCB conductors. The PCB conductors then come close to each other to form the start of a coupled line. The shading shows current density on the conductors with bright shades indicating high current. This includes a ground plane on the underside, which we can see through the narrow slots between the PCB conductors and the ground plane that surrounds them on the top side of the substrate. This is known as *coplanar waveguide,* or CPW. Note that the current is smooth with high current on the edges, indicating a high-quality result. (Courtesy Keysight, with permission.)

Figure 6.43 shows a drawing of a typical RFIC made using GaAs (instead of silicon) mounted on a carrier. This was prepared using CAD (*computer aided design*) software and is ready for FEM meshing.

FEM is an optimization problem, and as such it is often iterative. It will try an initial (usually fast, coarse mesh, with high error) analysis and evaluate the error by finding non-minimum energy everywhere in that solution. Then the FEM software refines the meshing and refines the solution. This process repeats until the error is below a user-specified threshold. If you feel you need a better solution, just make the error threshold smaller. For high-accuracy work, it is best to refine the solution until all FEM calculated fields and currents appear smooth (error appears as noise in the fields). Sharp edges (as in PCB and SiRFIC) should have smooth high current. However, results can be useful even if error is still apparent in the field display. **Figure 6.44** shows a high-quality result for a complex coaxial connector-to-PCB transition.

As we can see above, a major strength of the volume meshing techniques, like FEM and FDTD, is that they can easily analyze just about anything, no matter what shape it is. These techniques excel by being able to handle extremely arbitrary structures, provided (as with all EM analyses) the problem does not get too big for our computer memory or our willingness to wait for an answer.

6.4.3 Commercial Electromagnetic Analysis Tools

Below are descriptions of the most common RF and microwave EM software modeling packages.

Ansoft (a division of Ansys): Widely used finite element code specialized for microwave work, *HFSS* (High Frequency Structure Simulator). Ansoft provides a variety of other EM tools, including planar unshielded MoM. All tools are integrated into a comprehensive microwave design framework with a wide variety of circuit theory and systems simulation tools: **www.ansys.com**.

AWR (a division of Cadence): Their tool, *Axiem*, uses unshielded planar MoM integrated into a comprehensive widely used microwave design framework (*Microwave Office*) incorporating a wide variety of circuit theory and systems simulation tools: **www.awrcorp.com**.

CST (a division of Dassault Systèmes): A wide range of nearly every kind of EM solver available today. Their primary tool is based on a technique closely related to FDTD providing time-domain full 3D volume meshing EM analysis. Their wide range of tools are organized into various suites. For example, CST *Microwave Studio* includes transient, frequency domain, eigenmode, resonant, integral equation, asymptotic, and TLM solvers: **www.cst.com**.

EM Software and Systems (a division of Altair): Their primary product is *FEKO*, based on a fast implementation of MoM intended for very large problems. It is hybridized with several other EM techniques including finite elements, physical optics, geometric optics, and uniform theory of diffraction. Suitable for EMC, cable harnesses, antenna placement, bio-medical, microwave circuits, etc.:**www.altair.com/feko-applications/**.

EZNEC: Based on an older version (2.0) of the popular NEC (see below) MoM code, an easy-to-use interface was added by Roy Lewallen, W7EL, and has been a popular modeling product for over 30 years. After retirement, W7EL made his latest version: EZNEC Pro+ v.7.0 available at no cost: **eznec.com**.

gprMax: Open-source software that simulates electromagnetic wave propagation. It solves Maxwell's equations in 3D using the Finite-Difference Time-Domain (FDTD) method. gprMax was designed for modeling Ground Penetrating Radar (GPR) but can also be used to model electromagnetic wave propagation for many other applications. Free download from **www.gprmax.com**.

IMST: Their main software product, *Empire XPU*, provides time domain 3D volume meshing FDTD analysis of microwave structures. They also offer a tool, *MultiLib*, that provides models for planar multi-layer microwave circuits that can be used directly in the Keysight Technologies microwave design framework ADS: **www.imst.com**.

Keysight Technologies: Two main EM tools, unshielded 3D planar MoM EM analysis Momentum, and 3D volume meshing finite elements EMPro. Formerly Agilent Technologies Electronic Measurement Group, they offer the largest and most widely used range of integrated microwave tools including their microwave design framework, ADS, Advanced Design System: **www.keysight.com/us/en/products/software/pathwave-design-software.html**.

Mician: Mode matching tool *μWave Wizard*, uses the fact that every finite volume structure has a set of characteristic modes, sometimes also called resonances. Any structure that can be decomposed into a set of simpler building blocks that have known modes solved for by any of the many different EM tools included can be analyzed by adding all modes together and weighted so that all fields match at the boundaries between the blocks. Very fast and accurate analysis of waveguide filters, for example, is possible. A free version is available at **www.mician.com**.

Microwave Innovation Group: Mode matching tool *WASP-NET*, uses the fact that every finite volume structure has a set of characteristic modes, sometimes also called resonances. Any structure that can be decomposed into a set of simpler building blocks that have known modes solved for by any of the many different EM tools included can be analyzed by adding all modes together and weighted so that all fields match at the boundaries between the blocks. Very fast and accurate analysis of waveguide filters, for example, is possible: **www.wasp-net.com**.

Numerical Electromagnetic Code (NEC): The most popular implementation of MoM was originally produced in the 1960s under U.S. government contract by Lawrence Livermore National Laboratories. NEC has been refined and improved to the current version 5, though older versions are commonly used and produce acceptable results (see EZNEC above). A license to use the latest version of NEC can be purchased at **softwarelicensing.llnl.gov/product/nec-v50**.

Remcom: Provider of *XFdtd*, a full wave, 3D time domain, EM solver for antenna design and placement, mobile device design, biomedical, microwave, automotive radar, etc. Other products include *Wireless InSite*, site-specific radio propagation software for analysis of wireless communication systems, and *XGtd*, a ray-based EM analysis tool for antenna placement on large structures such as ships and aircraft: **www.remcom.com**.

Sonnet Software: Main tool, Sonnet Suites, integrates a shielded, 3D planar, MoM, EM analysis with plotting, modeling, etc., tools. Sonnet also integrates well with Keysight Technologies ADS, AWR Microwave Office, and Cadence Virtuoso microwave design frameworks. A free version (Sonnet Lite), with limited capabilities, is available from **www.sonnetsoftware.com**.

WIPL-D: A range of products all based on unshielded MoM. Emphasis is on fast solution of very large problems. Primary applications include large antennas, including dish antennas, antenna placement, EM compatibility analysis (EMC), and microwave circuits. A free version with a limited model size that is active for 30 days is available from **www.wipl-d.com**.

Contents

Chapter 7

Power Sources

Our transceivers, amplifiers, accessories, computers and test equipment all require power to operate. This chapter illustrates the various techniques, components and systems used to provide power at the voltage and current levels our equipment needs. Topics range from basic transformers, rectifiers and filters to linear voltage regulation, switchmode power conversion, high voltage techniques and batteries. Material on switchmode conversion was contributed by Rudy Severns, N6LF and Chuck Mullett, KR6R. The section on batteries was contributed by Isidor Buchmann from his book *Batteries in a Portable World*. The sections on selecting batteries were contributed by, Alan Applegate, KØBG, on mobile use and James Duffey, KK6MC, on portable use.

Chapter 7 — Online Content

Articles

- Testing and Monitoring Batteries — Excerpts from *Batteries in a Portable World* by Isidor Buchmann
- Vacuum Tube and Obsolete Rectifiers

Projects

- 12 V 15 A Power Supply — Article and PCB Template
- 13.8 V 5 A Power Supply — PCB Template
- A Deluxe High Voltage Supply by James Garland, W8ZR
- A Small, Lightweight High-Voltage Switch-Mode Power Supply by Ralph Crumrine, NØKC
- Build an Inverting DC-DC Converter by Jim Stewart
- Diode Voltage Drops Raise Battery Power Drain by Ward Silver, NØAX
- Four Output Bench Supply by Larry Cicchinelli, K3PTO — Article and support files
- Overvoltage Crowbar Circuit
- Power Gadgets from USB Sources with the Simple Switching Boost Supply by Tom Wheeler, NØGSG
- Revisiting the 12 V Power Supply
- Simple Adjustable Tracking Power Supply by Bryant Julstrom, KCØZNG

The title of this chapter reflects the broad assortment of methods for powering amateur equipment. More mobile and portable operation relies on power from batteries, for example. Hybrids of ac and dc power sources are becoming more common, blurring what has traditionally been known as a "power supply." (Generators are covered in the Portable Installations section of the chapter on **Assembling a Station**.)

7.1 Power Processing

Figure 7.1 illustrates the concept of a power processing unit inserted between the energy source and the electronic equipment or load. The *power processor* is often referred to as the *power supply*. That's a bit misleading in that the energy "supply" actually comes from some external source (battery, utility power and so forth), which is then converted to useful forms by the power processor. Be that as it may, in practice the terms "power supply" and "power processor" are used interchangeably.

The real world is even more arbitrary. Power processors are frequently referred to as *power converters* or simply as *converters*, and we will see other terms used later in this chapter. It is usually obvious from the context of the discussion what is meant and the glossary at the end of this chapter gives some additional information.

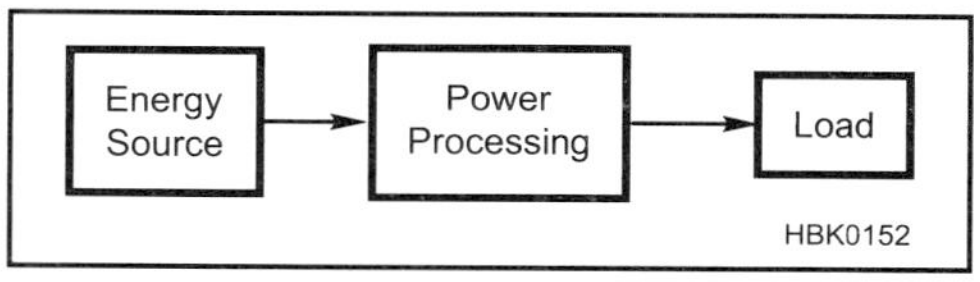

Figure 7.1 — Basic concept of power processing.

Power conversion schemes can take the form of: ac-to-ac (usually written

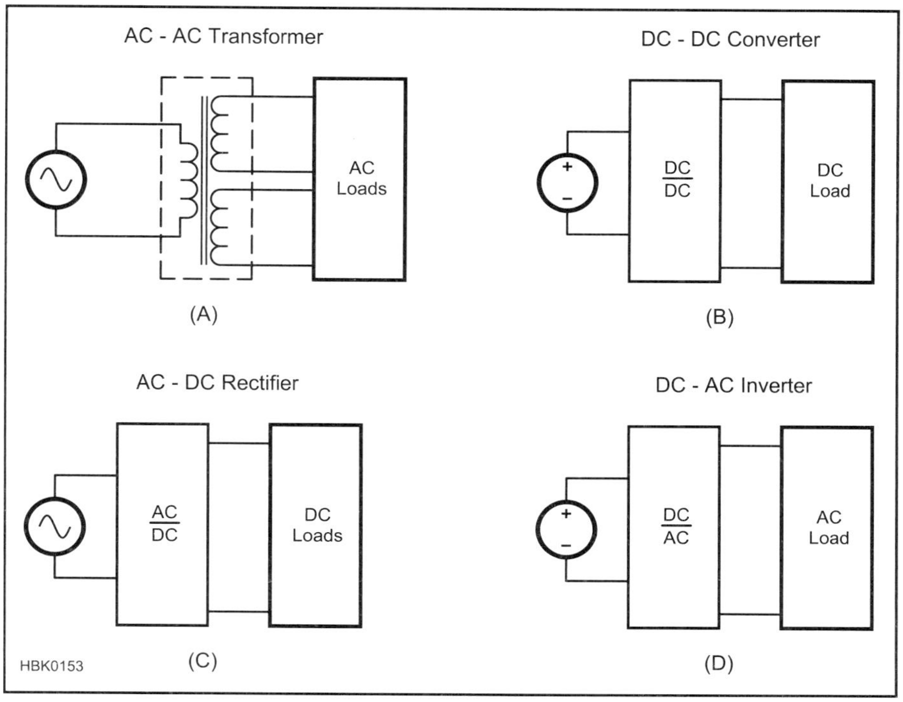

Figure 7.2 — Four power processing schemes: ac-ac, dc-dc, ac-dc and dc-ac.

ac-ac), ac-dc, dc-ac and dc-dc. Examples of these schemes are given in **Figure 7.2**. Specific names may be given to each scheme: ac-dc => rectifier, dc-dc => converter and dc-ac => inverter. These are the generally recognized terms but you will see exceptions.

Power conversion normally includes voltage and current regulation functions. For example, the voltage of a vehicle battery may vary from more than 14 V when being charged down to 10 V or less when discharged. A converter and regulator are required to maintain adequate voltage to mobile equipment at both over- and under-voltage conditions. Commercial utility power may vary from 90 to 270 V ac depending on where you are in the world. AC power converters are frequently required to handle that entire voltage range while still providing tightly regulated dc power.

7.2 AC-AC Power Conversion

In most US residences, three wires are brought in from the outside electrical-service mains to the house distribution panel. In this three-wire system, one wire is neutral and should be connected to a ground electrode. (See the **Safety** chapter for information on electrical safety.) The neutral connection to a ground rod or electrode is usually made at the distribution panel. The voltage between the other two wires is 60-Hz ac with a potential difference of approximately 240 V RMS. Half of this voltage appears between each of these wires and the neutral, as indicated in **Figure 7.3A**. In systems of this type, the 120 V household loads are divided at the breaker panel as evenly as possible between the two sides of the power mains. Heavy appliances such as electric stoves, water heaters, central air conditioners and so forth, are designed for 240 V operation and are connected across the two ungrounded wires.

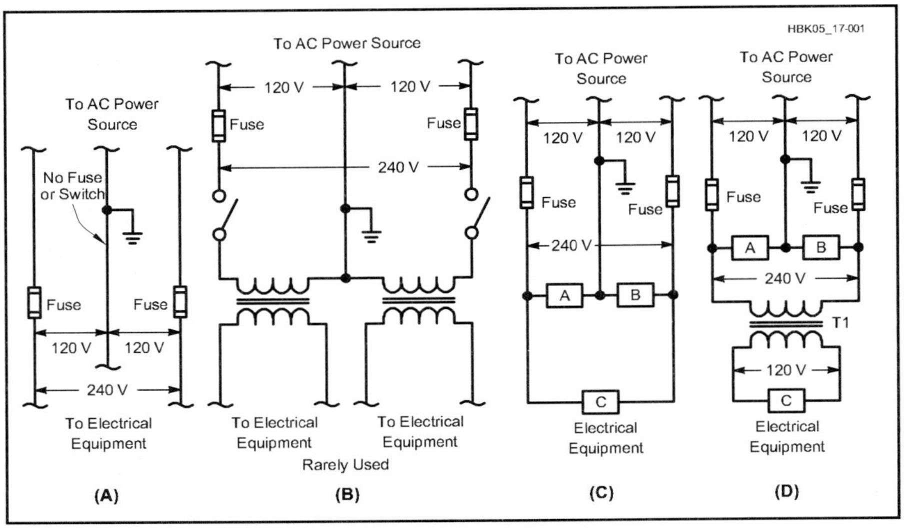

Figure 7.3 — Three-wire power-line circuits. At A, normal three-wire-line termination. No fuse should be used in the grounded (neutral) line. The ground symbol is the power company's ground, not yours! Do not connect anything other than power return wiring, including the equipment chassis, to the power neutral wire. At B, the "hot" lines each have a switch, but a switch in the neutral line would not remove voltage from either side of the line and should never be used. At C, connections for both 120 and 240 V loads. At D, operating a 120 V load through a 240-to-120 V step-down transformer (T1) to reduce the current drawn from the power source.

Both hot wires for 240 V circuits and the single hot wire for 120 V circuits should be protected by either a fuse or breaker. A fuse or breaker or any kind of switch should *never* be used in the neutral wire. Opening the neutral wire does not disconnect the equipment from an active or "hot" line, possibly creating a potential shock hazard between that line and earth ground.

Another word of caution should be given at this point. Since one side of the ac line is grounded (through the green or bare wire — the standard household wiring color code) to earth, all communications equipment should be reliably connected to the ac-line ground through a heavy ground conductor made of strap, heavy flat-weave braid or #14 AWG or heavier wire. This safety conductor must be separate from the power wiring neutral conductor. (A properly-wired 120 V outlet with a ground terminal uses one wire for the ac hot connection, one wire for the ac neutral connection and a third wire for the safety ground connection.) This provides a measure of safety for the operator in the event of accidental short or leakage of one side of the ac line to the chassis.

Remember that the antenna system is frequently bypassed to the chassis via an RF choke or tuned circuit, which could make the antenna electrically "live" with respect to the earth ground and create a potentially lethal shock hazard. A *ground fault circuit interrupter* (GFCI or GFI) is also desirable for safety reasons, and should be a part of the shack's electrical power wiring. (See the reference item for Silver for more information on grounding and bonding.)

7.2.1 Fuses and Circuit Breakers

All transformer primary circuits should be fused properly and multiple secondary outputs should also be individually fused. To determine the approximate current rating of the fuse or circuit breaker on the line side of a power supply it is necessary to determine the total load power. This can be done by multiplying each current (in amperes) being drawn by the load or appliance, by the voltage at which the current is being drawn. In the case of linear regulated power supplies, this voltage has to be the voltage appearing at the output of the rectifiers before being applied to the regulator stage. Include the current drawn by bleeder resistors and voltage dividers. Also include filament power if the transformer is supplying vacuum tube filaments. The National Electrical Code (NEC) also specifies maximum fuse ratings based on the wire sizes used in the transformer and connections.

After multiplying the various voltages and currents, add the individual products. This is the total power drawn from the line by the supply. Then divide this power by the line voltage and add 10 to 30% to account for the inefficiency of the power supply itself. Use a fuse or circuit breaker with the nearest larger current rating. Remember that the charging of filter capacitors can create large surges of current when the supply is turned on. If fuse blowing or breaker tripping at turn on is a problem, use slow-blow fuses, which allow for high initial surge currents.

For low-power semiconductor circuits, use fast-blow fuses. As the name implies, such fuses open very quickly once the current exceeds the fuse rating by more than 10%.

7.3 Power Transformers

Numerous factors are considered to match a transformer to its intended use. Some of these parameters are:

1. Output voltage and current (volt-ampere rating).
2. Power source voltage and frequency.
3. Ambient temperature.
4. Duty cycle and temperature rise of the transformer at rated load.
5. Mechanical considerations like weight, shape and mounting.

7.3.1 Volt-Ampere Rating

In alternating-current equipment, the term *volt-ampere* (*VA*) is often used rather than the term watt. This is because ac components must handle reactive power as well as real power. If this is confusing, consider a capacitor connected directly across the secondary of a transformer. The capacitor appears as a reactance that permits current to flow, just as if the load were a resistor. The current is at a 90° phase angle, however. If we assume a perfect capacitor, there will be no heating of the capacitor, so no real power (watts) will be delivered by the transformer. The transformer must still be capable of supplying the voltage, and be able to handle the current required by the reactive load. The current in the transformer windings will heat the windings as a result of the I^2R losses in the winding resistances. The product of the voltage and current in the winding is referred to as "volt-amperes," since "watts" is reserved for the real, or dissipated, power in the load. The volt-ampere rating will always be equal to, or greater than, the power actually being drawn by the load.

The number of volt-amperes delivered by a transformer depends not only upon the dc load requirements, but also upon the type of dc output filter used (capacitor or choke input), and the type of rectifier used (full-wave center tap or full-wave bridge). With a capacitive-input filter, the heating effect in the secondary is higher because of the high peak-to-average current ratio. The volt-amperes handled by the transformer may be several times the power delivered to the load. The primary winding volt-amperes will be somewhat higher because of transformer losses. This point is treated in more detail in the section on ac-dc conversion. (See the **Electrical Fundamentals** chapter for more information on transformers and reactive power.)

7.3.2 Source Voltage and Frequency

A transformer operates by producing a magnetic field in its core and windings. The intensity of this field varies directly with the instantaneous voltage applied to the transformer primary winding. These variations, coupled to the secondary windings, produce the desired output voltage. Since the transformer appears to the source as an inductance in parallel with the (equivalent) load, the primary will appear as a short circuit if dc is applied to it. The unloaded inductance of the primary (also known as the *magnetizing inductance*) must be high enough so as not to draw an excess amount of input current at the design line frequency (normally 60 Hz in the US). This is achieved by providing a combination of sufficient turns on the primary and enough magnetic core material so that the core does not saturate during each half-cycle.

The voltage across a winding is directly related to the time rate of change of magnetic flux in the core. This relationship is expressed mathematically by $V = N\, d\Phi/dt$ as described in the section on Inductance in the **Electrical Fundamentals** chapter. The total flux in turn is expressed by $\Phi = A_e B$, where A_e is the cross-sectional area of the core and B is the flux density.

The maximum value for *flux density* (the magnetic field strength produced in the core) is limited to some percentage (< 80% for example) of the maximum flux density that the core material can stand without saturating, since in saturation the core becomes ineffective and causes the inductance of the primary to plummet to a very low level and input current to rise rapidly. Saturation causes high primary currents and extreme heating in the primary windings.

At a given voltage, 50 Hz ac creates more flux in an inductor or transformer core because the longer time period per half-cycle results in more flux and higher magnetizing current than the same transformer when excited by same 60-Hz voltage. For this reason, transformers and other electromagnetic equipment designed for 60-Hz systems must not be used on 50-Hz power systems unless specifically designed to handle the lower line frequency.

7.3.3 How to Evaluate an Unmarked Power Transformer

Hams who regularly visit hamfests frequently develop a junk box filled with used and unmarked transformers. Over time, transformer labels or markings on the coil wrappings may come off or be obscured. There is a good possibility that the transformer is still useable, but the problem is to determine what voltages and currents the transformer can supply. First consider the possibility that you may have an audio transformer or other impedance-matching device rather than a power transformer. If you aren't sure, don't connect it to ac power!

If the transformer has color-coded leads, you are in luck. There is a standard for transformer lead color-coding, as is given in the Transformers section of the **Circuits and Components** chapter. Where two colors are listed, the first one is the main color of the insulation; the second is the color of the stripe.

Check the transformer windings with an

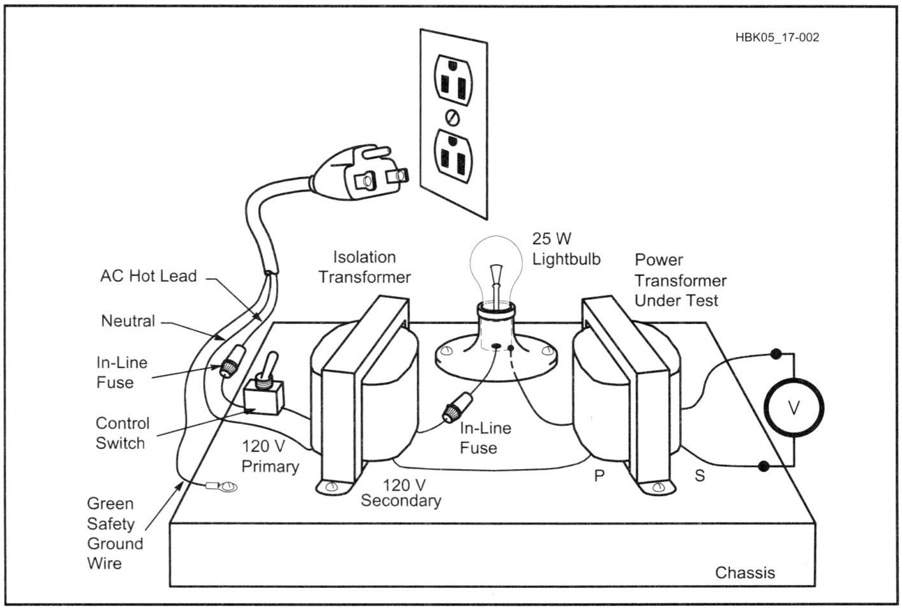

Figure 7.4 — Use a test fixture like this to test unknown transformers. Don't omit the isolation transformer, and be sure to insulate all connections before you plug into the ac mains.

ohmmeter to determine that there are no shorted (or open) windings. In particular, check for continuity between any winding and the core. If you find that a winding has been shorted to the core, do not use the transformer! The primary winding usually has a resistance higher than a filament winding and lower than a high-voltage winding.

Figure 7.4 shows that a convenient way to test the transformer is to rig a pair of test leads to an electrical plug with a 25 W household light bulb in series to limit current to safe (for the transformer) levels. For safety reasons use an isolation transformer and be sure to insulate all connections before you plug into the ac mains. Switch off the power while making or changing any connections. You can be electrocuted if the voltmeter leads or meter insulation are not rated for the transformer output voltage! If in doubt, connect the meter with the circuit turned off, then apply power while you are not in contact with the circuit. *Be careful! You are dealing with hazardous voltages!*

Connect the test leads to each winding separately. The filament/heater windings will cause the bulb to light to full brilliance because a filament winding has a very low impedance and almost all the input voltage will be across the series bulb. The high-voltage winding will cause the bulb to be extremely dim or to show no light at all because it will have a very high impedance, and the primary winding will probably cause a small glow. The bulb glows even with the secondary windings open-circuited because of the small magnetizing current in the transformer primary.

When the isolation transformer output is connected to what you think is the primary winding, measure the voltages at the low-voltage windings with an ac voltmeter. If you find voltages close to 6 V ac or 5 V ac, you know that you have identified the primary and the filament windings. Label the primary and low voltage windings.

Even with the light bulb, a transformer can be damaged by connecting ac mains power to a low-voltage or filament winding. In such a case the insulation could break down in a primary or high-voltage winding because of the high turns ratio stepping up the voltage well beyond the transformer ratings.

Connect the voltmeter to the high-voltage windings. Remember that transformers from vacuum tube equipment may supply as much as 800 V_{pk} or so across the winding, so make sure that your meter can withstand these potentials without damage and that you use the voltmeter safely.

Divide 6.3 (or 5) by the voltage you measured across the 6.3 V (or 5 V) winding in this test setup. This gives a multiplier that you can use to determine the actual no-load voltage rating of the high-voltage secondary. Simply multiply the ac voltage measured across the high-voltage winding by the multiplier.

You may find that with vintage equipment, the power transformers were designed for an ac line voltage of 110 V ac. The higher secondary voltages from today's 120 V ac lines can damage older equipment, particularly the tube filaments. Primary voltage can be reduced without a variable input transformer as described in the Repair and Restoration of Vintage Equipment section of this book's **Troubleshooting and Maintenance** chapter.

The current rating of the windings can be determined by loading each winding with the primary connected directly (no bulb) to the ac line. Using power resistors, increase loading on each winding until its voltage drops by about 10% from the no-load figure. The current drawn by the resistors is the approximate winding load-current rating.

7.4 AC-DC Power Conversion

One of the most common power supply functions is the conversion of ac power to dc, or *rectification*. The output from the rectifier will be a combination of dc, which is the desired component, and ac *ripple* superimposed on the dc. This is an undesired but inescapable component. Since most loads cannot tolerate more than a small amount of ripple on the dc voltage, some form of filter is required. The result is that ac-dc power conversion is performed with a rectifier-filter combination as shown in **Figure 7.5**.

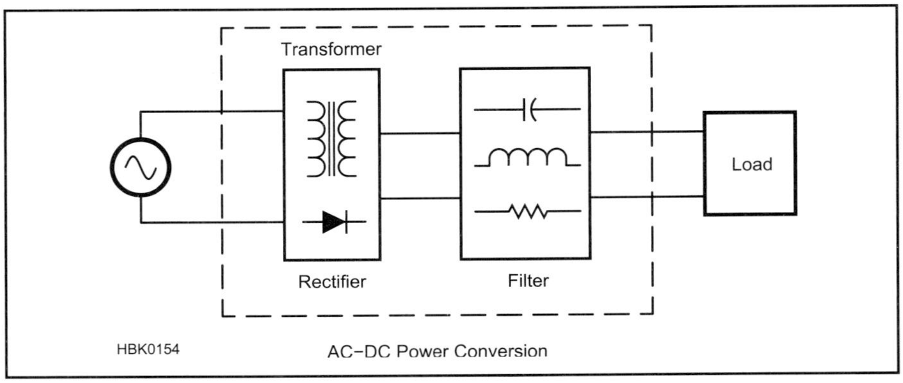

Figure 7.5 — Ac-dc power conversion with a rectifier and a filter.

As we will see in the rectifier circuit examples given in the next sections, sometimes the rectifier and filter functions will be separated into two distinct parts but very often the two will be integrated. This is particularly true for voltage and current multipliers as described in the sections on multipliers later in the chapter. Even when it appears that the rectifier and filter are separate elements, there will still be a strong interaction where the design and behavior of each part depends heavily on the other. For example the current waveforms in the rectifiers and the input source are functions of the load and filter characteristics. In turn the voltage waveform applied to the filter depends on the rectifier circuit and the input source voltages. To simplify the discussion we will treat the rectifier connections and the filters separately but always keeping in mind their interdependence.

The following rectifier-filter examples assume a conventional 60 Hz ac sine wave source, but these circuits are frequently used in switching converters at much higher frequencies and with square wave or quasi-square wave voltage and current waveforms. The component values may be different but the basic behavior will be very similar.

There are many different rectifier circuits or "connections" that may be used depending on the application. The following discussion provides an overview of some of the more common ones. The circuit diagrams use the symbol for a semiconductor diode, but the same circuits can be used with the older types of rectifiers that may be encountered in older equipment.

For each circuit we will show the voltage and current waveforms in the circuit for resistive, capacitive and inductive loads. The inductive and capacitive loads represent commonly used filters. We will be interested in the peak and average voltages as well as the RMS currents.

7.4.1 Half-Wave Rectifier

Figure 7.6 shows several examples of the half-wave rectifier circuit. It begins with a simple transformer with a resistive load (Figure 7.6A) and goes on to show how the output voltage and transformer current varies when a diode and filter elements are added.

Without the diode (Figure 7.6A) the output voltage (V_R) and current are just sine waves, and the RMS current in the transformer windings will be the same as the load (R) current.

Next, add a rectifier diode in series with the load (Figure 7.6B). During one half of the ac cycle, the rectifier conducts and there is current through the rectifier to the load. During the other half cycle, the rectifier is reverse-biased and there is no current (indicated by the broken line in Figure 7.6B) in R. The output voltage is pulsating dc, which is a combination of two components: an average dc value of 0.45 E_{RMS} (the voltage read by a dc voltmeter) and line-frequency ac ripple. The transformer secondary winding current is also pulsating dc. The power delivered to R is now ½ that for Figure 7.6A but the secondary RMS winding current in Figure 7.6B is still 0.707 times what it was in Figure 7.6A.

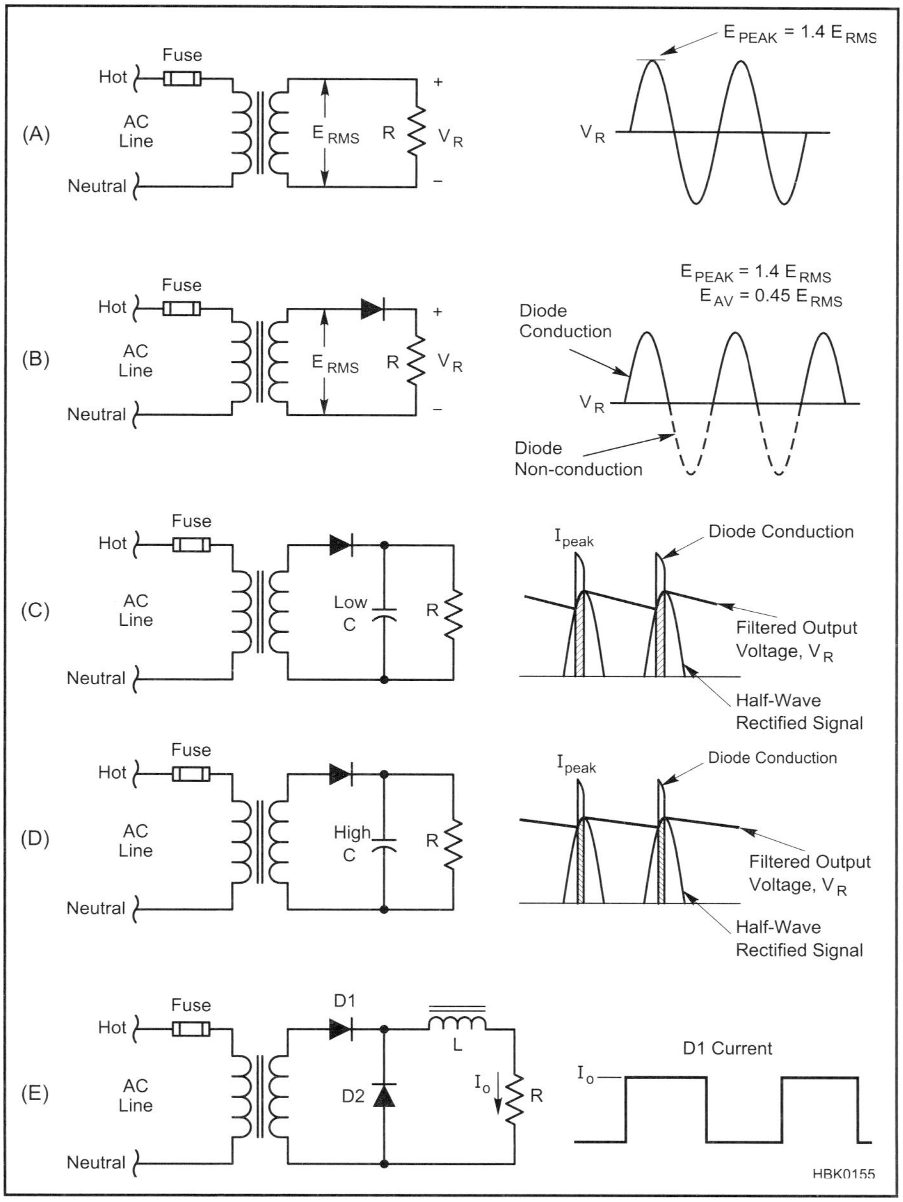

Figure 7.6 — Half-wave rectifier circuits. A illustrates the voltage waveform at the output without a rectifier. B represents the basic half-wave rectifier and the output waveform. C and D illustrate the impact of small and large filter capacitors on the output voltage and input current waveforms. E shows the effect of using an inductor filter with the half-wave rectifier. Note the addition of the shunt diode (D2) when using inductive filters with this rectifier connection.

For the same winding resistance, the winding loss, in proportion to the output power, is twice what it was in Figure 7.6A. This is an intrinsic limitation of the half-wave rectifier circuit — the RMS winding current is larger in proportion to the load power. In addition, the dc component of the secondary winding current may bias the transformer core toward saturation and increased core loss.

A filter can be used to smooth out these variations and provide a higher average dc voltage from the circuit. Because the frequency of the pulses (the ripple frequency) is low (one pulse per cycle), considerable filtering is required to provide adequately smooth dc output. For this reason the circuit is usually limited to applications where the required current is small. Parts C, D and E in Figure 7.6 show some possible capacitive and inductive filters.

As shown in Figure 7.6C and D, when a capacitor is used for filtering the output dc voltage will approach

$$V_{pk} = \sqrt{2} \times E_{RMS} = 1.4 \times E_{RMS} \qquad (1)$$

and the larger we make the filter capacitance, the smaller the ripple will be.

Unfortunately, as we make the filter capacitance larger, the diode, capacitor and transformer winding currents all become high-amplitude narrow pulses which will have a very high RMS value in proportion to the power level. These current pulses are also transmitted to the input line and inject currents at harmonics of the line frequency into the power source, which may result in interference to other equipment. Narrow high-amplitude current pulses are characteristic of capacitive-input filters in all rectifier connections when driven from voltage sources.

As shown in Figure 7.6E, it is possible to use an inductive filter instead, but a second diode (D2, sometimes called a *free-wheeling diode*) should be used. Without D2 the output voltage will get smaller as we increase the size of L to get better filtering, and the output voltage will vary greatly with load. By adding D2 we are free to make L large for small output ripple but still have reasonable voltage regulation. Currents in D1 and the winding will be approximately square waves, as indicated. This will reduce the line harmonic currents injected into the source but there will still be some.

Peak inverse voltage (PIV) is the maximum voltage the rectifier must withstand when it isn't conducting. This varies with the load and rectifier connection. In the half-wave rectifier, with a resistive load the PIV is the peak ac voltage ($1.4 \times E_{RMS}$). With a capacitor-input filter (Figure 7.6C and 7.6D), the capacitor is assumed to stay charged between the charging half-cycles. This means the rectifier output

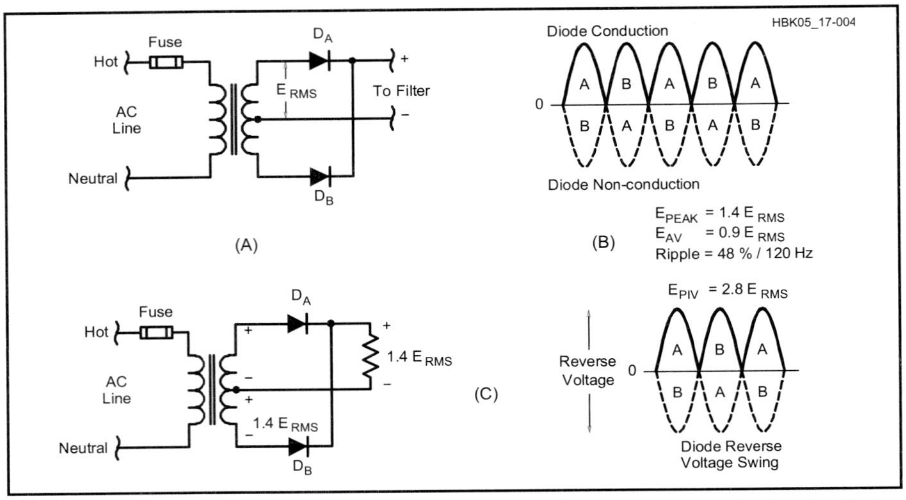

Figure 7.7 — Full-wave center-tap rectifier circuits. A illustrates the basic circuit. Diode conduction is shown at B with diodes A and B alternately conducting. The peak inverse voltage for each diode is 2.8 E_{RMS} as depicted at C.

stays near 1.4 × E_{RMS} while its input reaches the full opposing peak voltage. Thus, the rectifier experiences a PIV of 2 (1.4 × E_{RMS}) = 2.8 x E_{RMS}.

7.4.2 Full-Wave Center-Tapped Rectifier

The full-wave center-tapped rectifier circuit is shown in **Figure 7.7**. The operation of this circuit can be imagined as the secondary of the transformer producing two waveforms (A and B) that are 180° out of phase that are each connected to a half-wave rectifier. The half-wave rectifier outputs are then combined so that both halves of the ac cycle are used to deliver power to the output. A transformer with a center-tapped secondary is required.

The average output voltage of this circuit is 0.9 × E_{RMS} of half the transformer secondary (the center-tap to one side); this is the maximum that can be obtained with a suitable choke-input filter. The peak output voltage is 1.4 × E_{RMS} of half the transformer secondary; this is the maximum voltage that can be obtained from a capacitor-input filter.

As can be seen in Figure 7.7C, the PIV impressed on each diode is independent of the type of load at the output. This is because the peak inverse voltage condition occurs when diode D_A conducts and diode D_B is not conducting. The positive and negative voltage peaks occur at precisely the same time, a condition different from that in the half-wave circuit. As the cathodes of diodes D_A and D_B reach a positive peak (1.4 E_{RMS}), the anode of diode D_B is at a negative peak, also 1.4 E_{RMS}, but in the opposite direction. The total peak inverse voltage is therefore 2.8 E_{RMS}. If a capacitor-input filter is used, the capacitor stays charged to nearly 1.4 × E_{RMS} during each diode's non-conducting half cycles and so the PIV is 2.8 × E_{RMS}.

Figure 7.7C shows that the ripple frequency is twice that of the half-wave rectifier (two times the line frequency). Substantially less filtering is required because of the higher ripple frequency. Since the rectifiers work alternately, each handles half of the load current. The current rating of each rectifier need be only half the total current drawn from the supply.

The problem with dc bias in the transformer core associated with the half-wave connection is largely eliminated with this circuit and the RMS current in the primary winding will also be reduced.

7.4.3 Full-Wave Bridge Rectifier

Another commonly used rectifier circuit that does not require a center-tapped transformer is illustrated in **Figure 7.8**. In this arrangement, two rectifiers operate in series on each half of the cycle, one rectifier being in the lead supplying current to the load, the other being the current return lead. As shown in Figures 7.8A and B, when the top lead of the transformer secondary is positive with respect to the bottom lead, diodes D_A and D_C will conduct while diodes D_B and D_D are reverse-biased. On the next half cycle, when the top lead of the transformer is negative with respect to the bottom, diodes D_B and D_D will conduct while diodes D_A and D_C are reverse-biased.

The output voltage wave shape and ripple frequency are the same as for the full-wave center-tapped circuit. The average dc output voltage into a resistive load or choke-input filter is 0.9 times E_{RMS} delivered by the transformer secondary; with a capacitor filter and a light load, the maximum output voltage is 1.4 times the secondary E_{RMS} voltage.

Figure 7.8C shows the PIV to be 1.4 E_{RMS} for each diode which is half that of the full-wave center-tapped circuit for the same output voltage. When an alternate pair of diodes (such as D_A and D_C) is conducting, the other diodes are essentially connected in parallel (the conducting diodes are essentially short circuits) in a reverse-biased direction. This limits the diode PIV to 1.4 × E_{RMS} even if a capacitor-input filter is used. Each pair of

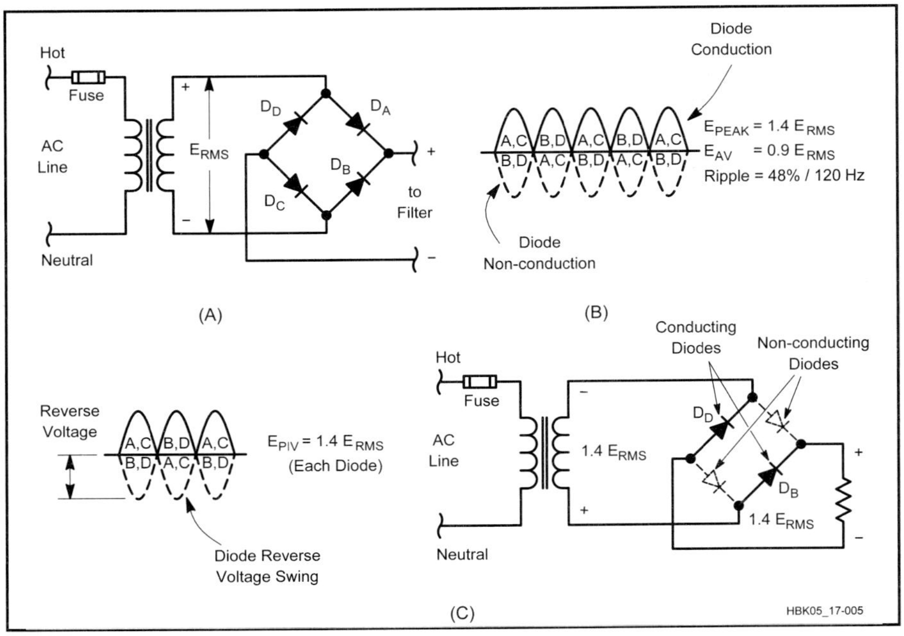

Figure 7.8 — Full-wave bridge rectifier circuits. The basic circuit is illustrated at A. Diode conduction and nonconduction times are shown at B. Diodes A and C conduct on one half of the input cycle, while diodes B and D conduct on the other. C displays the peak inverse voltage for one half cycle. Since this circuit reverse-biases two diodes essentially in parallel, 1.4 E_{RMS} is applied across each diode.

diodes conducts on alternate half cycles, with the full load current through each diode during its conducting half cycle. Since each diode is not conducting during the other half cycle the average diode current is one-half the total load current drawn from the supply.

Compared to the half-wave and full-wave center-tapped circuit, the full-wave bridge circuit further reduces the transformer RMS winding currents. In the case of a resistive load the winding currents are the same as when the resistive load is connected directly across the secondary. The RMS winding currents will still be higher when inductive and especially capacitive filters are used because of the pulsating nature of the diode and winding currents.

7.4.4 Comparison of Rectifier Circuits

Comparing the full-wave center-tapped and the full-wave bridge circuits, we can see that the center-tapped circuit has half the number of rectifiers as the bridge but these rectifiers have twice the PIV rating requirement of the bridge diodes. The diode current ratings are identical for the two circuits. The bridge makes better use of the transformer's secondary than the center-tapped rectifier, since the transformer's full winding supplies power during both half cycles, while each half of the center-tapped circuit's secondary provides power only during its positive half-cycle.

The full-wave center-tapped rectifier is typically used in high-current, low-voltage applications because only one diode conducts at a time. This reduces the loss associated with diode conduction. In the full-wave bridge circuit there are two diodes in series in conduction simultaneously, which leads to higher loss. The full-wave bridge circuit is typically used for higher output voltages where this is not a serious concern. The lower diode PIV and better utilization of the transformer windings makes this circuit very attractive for higher output voltages and higher powers typical of high voltage amplifier supplies.

Because of the disadvantages pointed out earlier, the half-wave circuit is rarely used in 60-Hz rectification except for bias supplies or other small loads. It does see considerable use, however, in high-frequency switchmode power supplies.

7.5 Voltage Multipliers

Other rectification circuits are sometimes useful, including *voltage multipliers*. These circuits function by the process of charging one or more capacitors in parallel on one half cycle of the ac waveform, and then connecting that capacitor or capacitors in series with the opposite polarity of the ac waveform on the alternate half cycle. In full-wave multipliers, this charging occurs during both half-cycles.

Voltage multipliers, particularly *voltage doublers*, find considerable use in high-voltage supplies. When a doubler is employed, the secondary winding of the power transformer need have only half the voltage that would be required for a bridge rectifier. This reduces voltage stress in the windings and decreases the transformer insulation requirements. This is not without cost, however, because the transformer-secondary *current* rating has to be correspondingly doubled for a given load current and charging of the capacitors leads to narrow high-RMS current waveforms in the transformer windings and the capacitors.

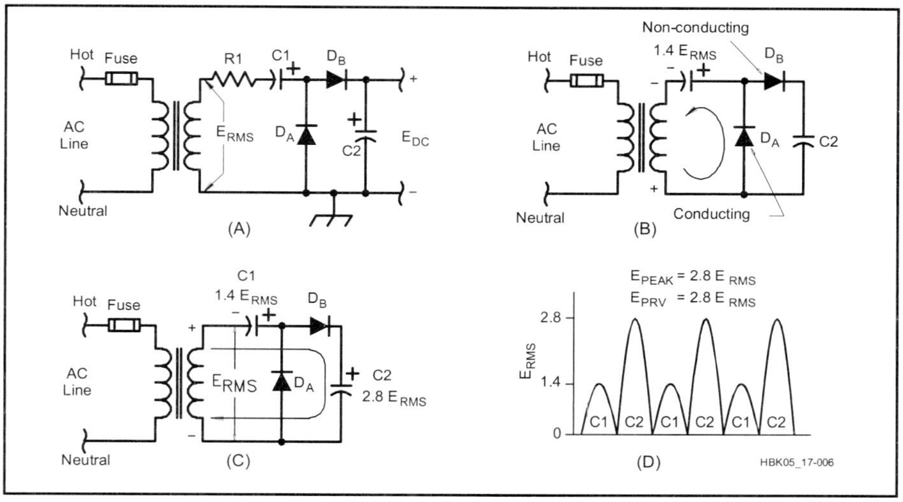

Figure 7.9 — Part A shows a half-wave voltage-doubler circuit. B displays how the first half cycle of input voltage charges C1. During the next half cycle (shown at C), capacitor C2 charges with the transformer secondary voltage plus that voltage stored in C1 from the previous half cycle. The arrows in parts B and C indicate the conventional current. D illustrates the levels to which each capacitor charges over several cycles.

7.5.1 Half-Wave Voltage Doubler

Figure 7.9 shows the circuit of a half-wave voltage doubler and illustrates the circuit operation. For clarity, assume the transformer voltage polarity at the moment the circuit is activated is that shown at Figure 7.9B. During the first negative half cycle, D_A conducts (D_B is in a nonconductive state), charging C1 to the peak rectified voltage (1.4 E_{RMS}). C1 is charged with the polarity shown in Figure 7.9B. During the positive half cycle of the secondary voltage, D_A is cut off and D_B conducts, charging capacitor C2. The amount of voltage delivered to C2 is the sum of the transformer peak secondary voltage plus the voltage stored in C1 (1.4 E_{RMS}). On the next negative half cycle, D_B is non-conducting and C2 will discharge into the load. If no load is connected across C2, the capacitors will remain charged — C1 to 1.4 E_{RMS} and C2 to 2.8 E_{RMS}. When a load is connected to the circuit output, the voltage across C2 drops during the negative half cycle and is recharged up to 2.8 E_{RMS} during the positive half cycle.

The output waveform across C2 resembles that of a half-wave rectifier circuit because C2 is pulsed once every cycle. Figure 7.9D illustrates the levels to which the two capacitors are charged throughout the cycle. In actual operation the capacitors will usually be large enough that they will discharge only partially, not all the way to zero as shown.

7.5.2 Full-Wave Voltage Doubler

Figure 7.10 shows the circuit of a full-wave voltage doubler and illustrates the circuit operation. During the positive half cycle of the transformer secondary voltage, as shown in Figure 7.10B, D_A conducts charging capacitor C1 to 1.4 E_{RMS}. D_B is not conducting at this time.

During the negative half cycle, as shown in Figure 7.10C, D_B conducts, charging capacitor C2 to 1.4 E_{RMS}, while D_A is nonconducting. The output voltage is the sum of the two capacitor voltages, which will be 2.8 E_{RMS} under no-load conditions. Figure 7.10D illustrates that each capacitor alternately receives a charge once per cycle. The effective filter capacitance is that of C1 and C2 in series, which is less than the capacitance of either C1 or C2 alone.

Resistors R1 and R2 in Figure 7.10A are used to limit the surge current through the rectifiers. Their values are based on the transformer voltage and the rectifier surge-current rating, since at the instant the power supply is turned on, the filter capacitors look like a short-circuited load. Provided the limiting resistors can withstand the surge current, their current-handling capacity is based on the maximum load current from the supply. Output voltages approaching twice the peak voltage of the transformer can be obtained with the voltage doubling circuit shown in Figure 7.10.

Figure 7.11 shows how the voltage depends upon the ratio of the series resistance to the load resistance, and the load resistance times the filter capacitance. The peak inverse voltage across each diode is 2.8 E_{RMS}. As indicated by the curves in Figure 7.11, the output voltage regulation of this doubler connection is not very good and it is not attractive for providing high voltages at high power levels.

There are better doubler connections for higher power applications, and two possibilities are shown in **Figure 7.12**. The connection in Figure 7.12A uses two bridge rectifiers in series with capacitive coupling between the ac terminals of the bridges. At the expense of more diodes, this connection will have much better output voltage regulation at higher power levels. Even better regulation can be achieved by using the connection shown in Figure 7.12B. In this example, two windings on the transformer are used. It is not essential that both windings have the same voltage, but both must be capable of providing the desired output current. In addition, the insulation of the upper winding must be adequate to accommodate the additional dc bias applied to it from the lower winding.

7.5.3 Voltage Tripler and Quadrupler

Figure 7.13A shows a voltage-tripling circuit. On one half of the ac cycle, C1 and C3 are charged to the source voltage through D1, D2 and D3. R1 represents the resistance of the transformer secondary winding, which limits the amount of current available for

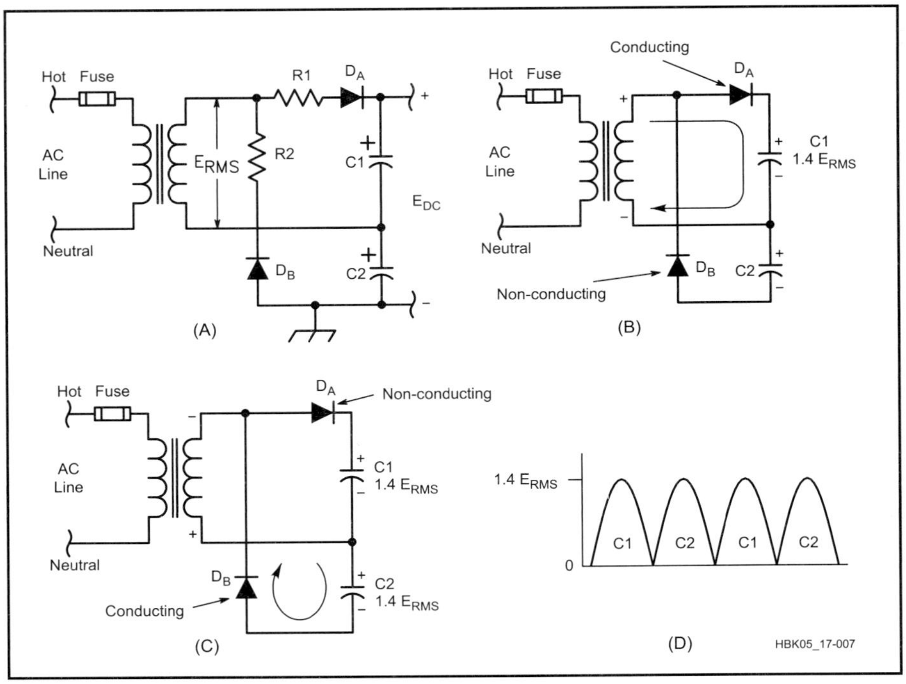

Figure 7.10 — Part A shows a full-wave voltage-doubler circuit. One-half cycle is shown at B and the next half cycle is shown at C. Each capacitor receives a charge during every input-voltage cycle. D illustrates how each capacitor is charged alternately.

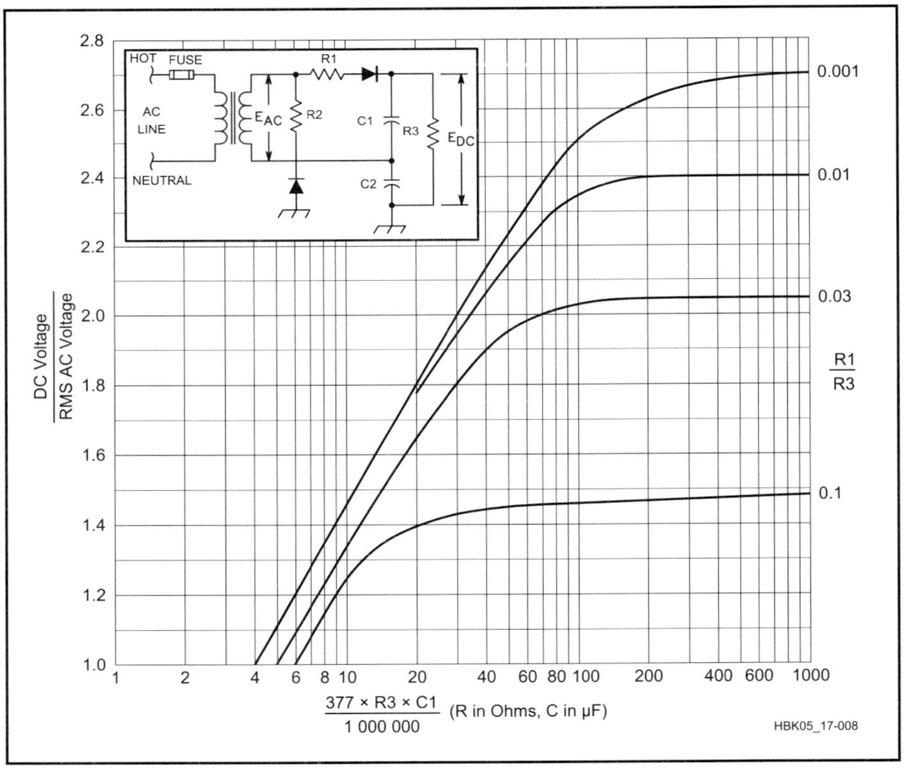

Figure 7.11 — DC output voltages from a full-wave voltage-doubler circuit as a function of the filter capacitances and load resistance. For the ratio R1 / R3 and for the R3 × C1 product, resistance is in ohms and capacitance is in microfarads. Equal resistance values for R1 and R2, and equal capacitance values for C1 and C2 are assumed. (From Schade, see References)

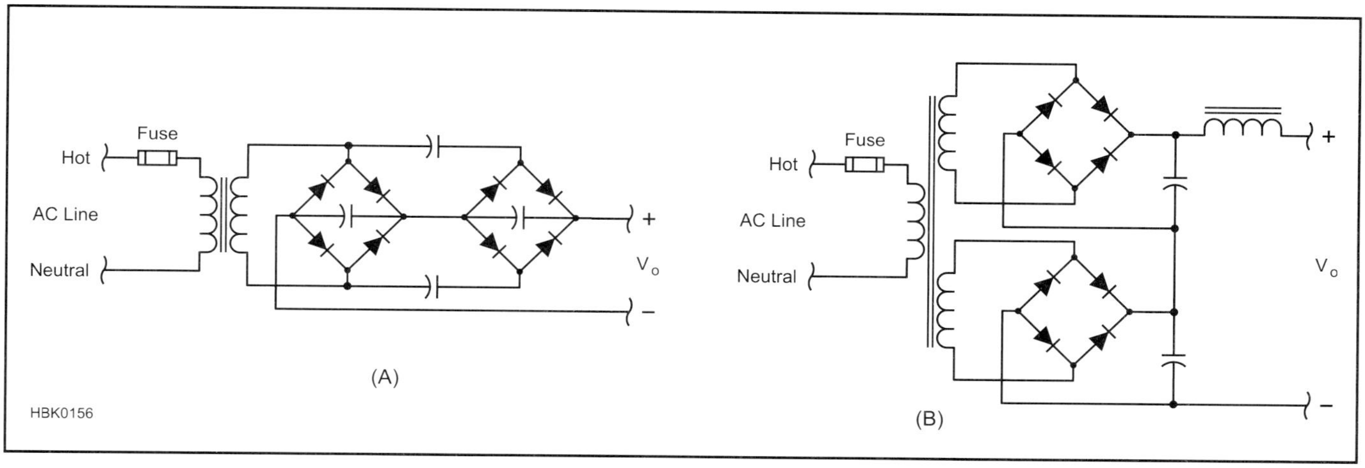

Figure 7.12 — Voltage-doubler rectifier connections for higher power levels. A is a capacitor-coupled doubler that can be extended to more sections for a higher multiplying factor. The circuit in B uses multiple transformer windings to boost the output voltage.

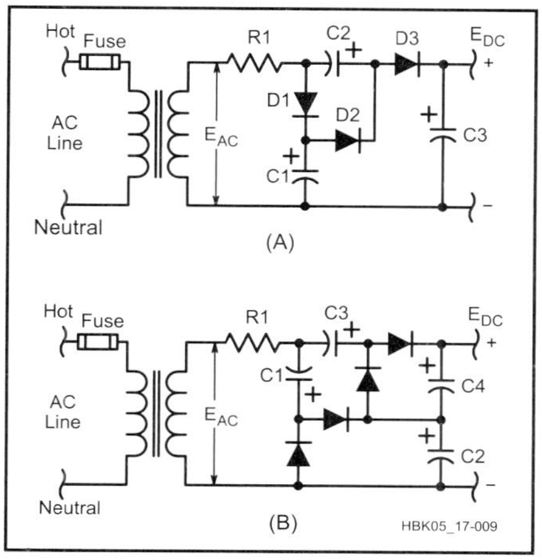

Figure 7.13 — Voltage-multiplying circuits with one side of the transformer secondary used as a common connection. A shows a voltage tripler and B shows a voltage quadrupler. Capacitances are typically 20 to 50 µF, depending on the output current demand. Capacitor dc ratings are related to E_{PEAK} (1.4 E_{RMS}):
C1 — Greater than E_{PEAK}
C2 — Greater than 2 E_{PEAK}
C3 — Greater than 3 E_{PEAK}
C4 — Greater than 2 E_{PEAK}
R1 — Resistance of the transformer secondary winding

charging the capacitors. On the opposite half of the cycle, D2 conducts and C2 is charged to twice the source voltage, because it sees the transformer plus the charge in C1 as its source (D1 is cut off during this half cycle). At the same time, D3 conducts, and with the transformer and the charge in C2 as the source, C3 is charged to three times the transformer voltage.

The voltage-quadrupling circuit of Figure 7.13B works in similar fashion. In either of the circuits of Figure 7.13, the output voltage will approach an exact multiple of the peak ac voltage when the output current drain is low and the capacitance values are large.

The secondary resistance, R1, limits the charging current for the capacitors, and thus the maximum output voltage. This is illustrated in a complementary way by the flattening of the output voltage curve in Figure 7.11 as filter capacitance increases. Increasing R1 has the same effect. For high-voltage supplies, a low-resistance secondary is often required to reach a full multiple of the available ac voltage.

7.6 Current Multipliers

Just as there are voltage multiplier connections for high-voltage, low-current loads, there are current multiplier connections for low-voltage, high-current loads. An example of a current-doubler is given in **Figure 7.14A**.

To make the circuit operation easier to visualize, we can represent L1 and L2 as current sources (Figure 7.14B) which is a good approximation for steady-state operation. When terminal 1 of the secondary winding is positive with respect to terminal 2, diode D_A will be reverse-biased and therefore non-conducting. The current flows within the circuit are shown in Figure 7.14B. Note that all of the output current (I_o) flows through D_B but only half of I_o flows through the winding. At the cathode of D_B the current divides with half going to L2 and the other half to the transformer secondary. The output voltage will be one-half the voltage of the average winding voltage (0.45 E_{RMS}). This rectifier connection divides the voltage and multiplies the current! Because of the need for two inductors, this circuit is seldom used in line-frequency applications but it is very useful in high-frequency switchmode regulators with very low output voltages (<10 V) because it makes the secondary winding design easier and can improve circuit efficiency. At high frequencies, the inductors can be quite small.

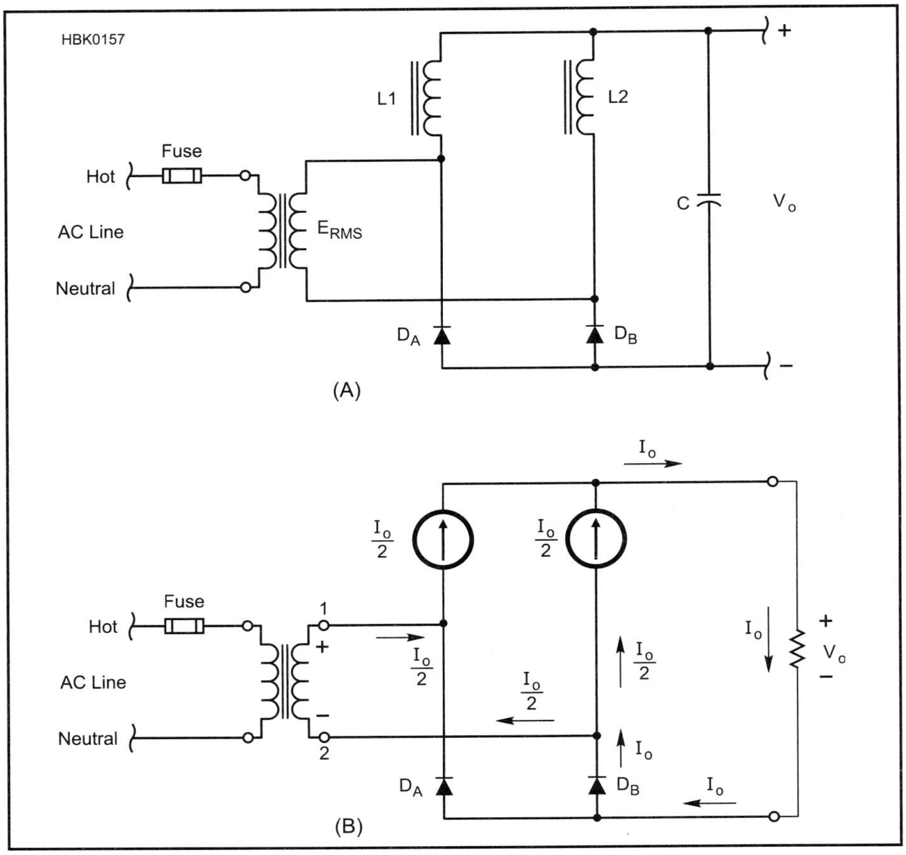

Figure 7.14 — A current doubler rectifier connection. A is the basic circuit. B illustrates current flow within the circuit.

7.7 Rectifier Types

Rectifiers have a long history beginning with mechanical rectifiers in the 1800s to today's abundant variety of semiconductor devices. While many different devices have been created for this purpose, they all have the characteristic that they block current flow in the reverse direction, withstanding substantial reverse voltage and allowing current flow in the forward direction with minimum voltage drop. The simplest rectifiers are diodes, but it is also possible to have three-terminal devices (such as a thyristor) that can be controlled to regulate the output dc in addition to providing rectification. It is also possible to use devices like MOSFETs as synchronous rectifiers with very low forward drop during conduction. This is typically done to improve efficiency for very low voltage outputs. The following is a brief description of several of the more common examples. The sections on vacuum tube and other obsolete types of rectifiers from previous editions are available as a PDF article with this book's online content.

7.7.1 Semiconductor Diodes

Rectifier diodes can be made from a number of different semiconductor materials such as germanium, silicon, silicon-carbide or gallium-arsenide, and no doubt other materials will appear in the future. The choice will depend on the application and as always cost is a factor.

Germanium diodes were the first of the solid-state semiconductor rectifiers. They have an extremely low forward voltage drop but are relatively temperature sensitive, having high reverse leakage currents at higher temperatures. They can be easily destroyed by overheating during soldering as well. Germanium diodes are no longer used as power rectifiers.

Today, silicon diodes are the primary choice for virtually all power rectifier applications. They are characterized by extremely high reverse resistance (low reverse leakage), forward drops of a volt or less and operation at junction temperatures up to 125 °C. Some multi-junction HV diodes will have forward drops of several volts, but that is still low compared to the voltage at which they are being used.

Many different types of silicon diodes are available for different applications. Silicon rectifiers fall into to two general categories: PN-junction diodes and Schottky barrier diodes (see the **Circuits and Components** chapter). Schottky diodes are the usual choice for low output voltages (<20 V) where their low forward conduction drop is critical for efficiency. For higher voltages however, the high reverse leakage of Schottky diodes is not acceptable and PN-junction diodes are normally chosen.

For 50/60 Hz applications, diodes with reverse recovery times of a microsecond or even more are suitable and very economical. For switchmode converters and inverters that regularly operate at 25 kHz and higher frequencies, fast-recovery diodes are needed. These converters typically have waveform transitions of less than 1 μs within the circuit. MOSFET power transistors often have transitions of less than 100 ns.

During the switching transitions, previously conducting diodes see a reversal of current direction. This change tends to reverse-bias those diodes, and thereby put them into an open-circuit condition. Unfortunately, as explained in the **Circuits and Components** chapter, solid-state rectifiers cannot be made to cease conduction instantaneously. As a result, when the opposing diodes in a bridge rectifier or full-wave rectifier become conductive at the time the converter switches states, the diodes being turned off will actually conduct in the reverse direction for a brief time. That effectively short circuits the converter for a period of time depending on the reverse recovery characteristics of the rectifiers. This characteristic can create high current transients that stress the switching transistors and lead to increased loss and electromagnetic interference. As the switching frequency increases, more of these transitions happen each second, and more power is lost because of diode cross-conduction.

These current transients and associated losses are reduced by using *fast recovery* diodes, which are specially doped diodes designed to minimize storage time. Diodes with recovery times of 50 ns or less are available.

7.7.2 Rectifier Strings or Stacks

DIODES IN SERIES

When the PIV rating of a single diode is not sufficient for the application, similar diodes may be used in series. (Two 500 PIV diodes in series will withstand 1,000 PIV and so on.) There used to be a general recommendation to place a resistor across each diode in the string to equalize the PIV drops. With modern diodes, this practice is no longer necessary.

Modern silicon rectifier diodes are constructed to have an avalanche characteristic. Simply put, this means that the diffusion process is controlled so the diode will exhibit a Zener characteristic in the reverse-biased direction before destructive breakdown of the junction can occur. This provides a measure of safety for diodes in series. A diode will go into Zener conduction before it self-destructs. If other diodes in the chain have not reached their avalanche voltages, the current through the avalanched diode will be limited to the leakage current in the other diodes. This should normally be very low. For this reason, shunting resistors are generally not needed across diodes in series rectifier strings. In fact, shunt resistors can actually create problems because they can produce a low-impedance source of damaging current to any diode that may have reached avalanche potential.

DIODES IN PARALLEL

Diodes can be placed in parallel to increase current-handling capability. Equalizing resistors should be added as shown in **Figure 7.15**. Without the resistors, one diode may take most of the current. The resistors should be selected to have a drop of several tenths of a volt at the expected peak current. A disadvantage of this form of forced current sharing will be the increase in power loss because of the added resistors.

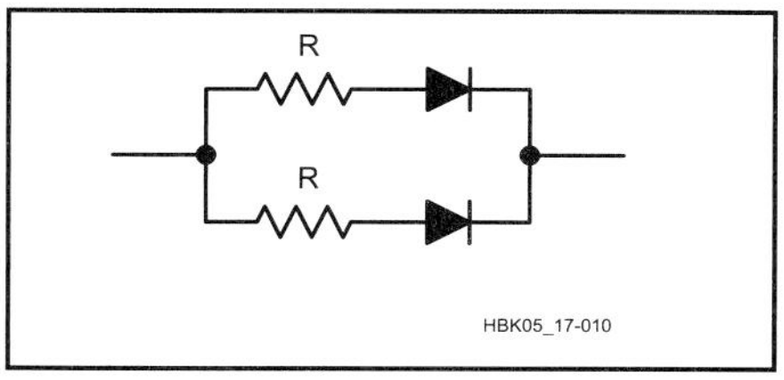

Figure 7.15 — Diodes can be connected in parallel to increase the current-handling capability of the circuit. Each diode should have a series current-equalizing resistor, with a value selected to have a drop of several tenths of a volt at the expected current.

7.7.3 Rectifier Ratings versus Operating Stress

Power supplies designed for amateur equipment use silicon rectifiers almost exclusively. These rectifiers are available in a wide range of voltage and current ratings: PIV ratings of 600 V or more and current ratings as high as 400 A are available. At 1,000 PIV, the current ratings may be several amperes. It is possible to stack several units in series for higher voltages. Stacks are available commercially that will handle peak inverse voltages up to 10 kV at a load current of 1 A or more.

7.7.4 Rectifier Protection

The discussion of rectifier circuits included the peak reverse voltage seen by the rectifiers in each circuit. You will need this information to select the voltage rating of the diodes in given application. It is normal good practice to not expose the diodes to more than 75% of their rated voltage for the worst case reverse voltage. This will probably be when operating at the highest input voltage but should also take into account transients that may occur.

The important specifications of a silicon diode are:

1. PIV — the peak inverse voltage.
2. I_0 — the average dc current rating.
3. I_{REP} — the peak repetitive forward current.
4. I_{SURGE} — a non-repetitive peak half-sine wave of 8.3 ms duration (one-half cycle of 60-Hz line frequency).
5. Switching speed or reverse recovery time.
6. Power dissipation and thermal resistance.

The first two specifications appear in most catalogs. I_{REP} and I_{SURGE} are not often specified in catalogs, but they are very important. Except in some switching regulator and capacitive filter circuits, rectifier current typically flows half the time — when it does conduct, the rectifier has to pass at least twice the average direct current. With a capacitor-input filter, the rectifier conducts much less than half the time. In this case, when it does conduct, it may pass as much as 10 to 20 times the average dc current, under certain conditions.

CURRENT INRUSH

When the supply is first turned on, the filter capacitors are discharged and act like a dead short. The result can be a very heavy current surge through the diode for at least one half-cycle and sometimes more. This current transient is called I_{SURGE}. The maximum surge current rating for a diode is usually specified for a duration of one-half cycle (at 60 Hz), or about 8.3 ms. Some form of surge protection is usually necessary to protect the diodes until the filter capacitors are nearly charged, unless the diodes used have a very high surge-current rating (several hundred amperes). If a manufacturer's data sheet is not available, an educated guess about a diode's capability can be made by using these rules of thumb for silicon diodes commonly used in Amateur Radio power supplies:

Rule 1. The maximum I_{REP} rating can be assumed to be approximately four times the maximum I_0 rating, where I_0 is the average dc current rating.

Rule 2. The maximum I_{SURGE} rating can be assumed to be approximately 12 times the maximum I_0 rating. This figure should provide a reasonable safety factor. Silicon rectifiers with 750 mA dc ratings, for example, seldom have 1-cycle surge ratings of less than 15 A; some are rated up to 35 A or more. From this you can see that the rectifier should be selected on the basis of I_{SURGE} and not on I_0 ratings.

Although you can sometimes rely on the resistance of the transformer windings to provide surge-current limiting, this is seldom adequate in high-voltage power supplies. Series resistors are often installed between the secondary and the rectifier strings or in the transformer's primary circuit, but these can be a deterrent to good voltage regulation.

One way to have good surge current limiting at turn-on without affecting voltage regulation during normal operation is to have a resistor in series with the input, along with a relay across the resistor that shorts it out after 50 ms or so. This kind of arrangement is particularly important in HV supplies.

VOLTAGE TRANSIENTS

Vacuum-tube rectifiers had little problem with voltage transients on the incoming power lines. The possibility of an internal arc was of little consequence, since the heat produced was of very short duration and had little effect on the massive plate and cathode structures.

Unfortunately, such is not the case with silicon diodes. Because of their low forward voltage drop, silicon diodes create very little heat with high forward current and therefore have tiny junction areas. However, conduction in the reverse direction beyond the normal reverse recovery time (reverse avalanching) can cause junction temperatures to rise extremely rapidly with the resulting destruction of the semiconductor junction.

To protect semiconductor rectifiers from voltage transients, special surge-absorption devices are available for connection across the incoming ac bus or transformer secondary. (See the **RF Interference and Electromagnetic Compatibility** chapter for more information.) These devices operate in a fashion similar to a Zener diode; they conduct heavily when a specific voltage level is reached. Unlike Zener diodes, however, they have the ability to absorb very high transient energy levels without damage. With the clamping level set well above the normal operating voltage range for the rectifiers, these devices normally appear as open circuits and have no effect on the power-supply circuits. When a voltage transient occurs, however, these protection devices clamp the transient and thereby prevent destruction of the rectifiers.

Transient protectors are available in three basic varieties:

1. *Silicon Zener diodes* — large junction Zeners specifically made for this purpose and available as single junction for dc (unipolar) and back-to-back junctions for ac (bipolar). These silicon protectors are available under the trade name of TransZorb from General Semiconductor Corporation and are also made by other manufacturers. They have the best transient-suppressing characteristics of the three varieties mentioned here, but are expensive and have the least energy absorbing capability per dollar of the group.

2. *Varistors* — made of a composition metal-oxide material that breaks down at a certain voltage. Metal-oxide varistors, also known as MOVs, are cheap and easily obtained, but have a higher internal resistance, which allows a greater increase in clamped voltage than the Zener variety. Varistors can also degrade with successive transients within their rated power handling limits (this is not usually a problem in the ham shack where transients are few and replacement of the varistor is easily accomplished).

Varistors usually become short-circuited when they fail. Large energy dissipation can result in device explosion. Therefore, it is a good idea to include a fuse that limits the short-circuit current through the varistor, and to protect people and circuitry from debris.

3. *Gas tube* — similar in construction to the familiar neon bulb, but designed to limit conducting voltage rise under high transient currents. Gas tubes can usually withstand the highest transient energy levels of the group. Gas tubes suffer from an ionization time problem, however. A high voltage across the tube will not immediately cause conduction. The time required for the gas to ionize and clamp the transient is inversely proportional to the level of applied voltage in excess of the device ionization voltage. As a result, the gas tube will let a little of the transient through to the equipment before it activates.

In installations where reliable equipment operation is critical, the local power is poor and transients are a major problem, the usual practice is to use a combination of protectors. Such systems consist of a varistor or Zener protector, combined with a gas-tube device. Often there is an indicator light to warn when a surge has blown out the varistor. Operationally, the solid-state device clamps the surge immediately, with the beefy gas tube firing shortly thereafter to take most of the surge from the solid-state device.

HEAT

The junction of a diode is quite small, so it must operate at a high current density. The heat-handling capability is, therefore, quite small. Normally, this is not a prime consideration in high-voltage, low-current supplies. Use of high-current rectifiers at or near their maximum ratings (usually 2 A or larger, stud-mount rectifiers) requires some form of heat sinking. Frequently, mounting the rectifier on the main chassis — directly or with thin thermal insulating washers — will suffice.

When a rectifier is directly mounted on the heatsink it is good practice to use a thin layer of thermal grease between the diode and the heat sink to assure good heat conduction. Most modern insulating thermal washers do not require the use of grease, but the older mica and other washers may benefit from a *very thin layer* of grease. Thermal grease and heat conducting insulating washers and pads are standard products available from mail-order component sellers.

Large, high-current rectifiers often require special heat sinks to maintain a safe operating temperature. Forced-air cooling from a fan is sometimes used as a further aid. Safe case temperatures are usually given in the manufacturer's data sheets and should be observed if the maximum capabilities of the diode are to be realized. See the thermal design section in the chapter on **Circuits and Components** for more information.

7.8 Power Filtering

Most loads will not tolerate the ripple (an ac component) of the pulsating dc from the rectifiers. Filters are required between the rectifier and the load to reduce the ripple to a low level. As pointed out earlier, some capacitances or inductances may be inherent in the rectifier connection, reducing the ripple amplitude. In most cases, however, additional filtering is required. The design of the filter depends to a large extent on the dc voltage output, the desired voltage regulation of the power supply and the maximum load current. Power-supply filters are low-pass devices using series inductors and/or shunt capacitors.

7.8.1 Load Resistance

In discussing the performance of power-supply filters, it is sometimes convenient to characterize the load connected to the output as a resistance. This *load resistance* is equal to the output voltage divided by the total load current, including the current drawn by the bleeder resistor.

7.8.2 Voltage Regulation

In an unregulated supply, the output voltage usually decreases as more current is drawn. This happens not only because of increased voltage drops in the transformer and filter chokes, but also because the output voltage at light loads tends to soar to the peak value of the transformer voltage as a result of charging the first capacitor. Proper filter design can reduce this effect. The change in output voltage with load is called the *voltage regulation* and is expressed as a percentage.

$$\text{Percent Regulation} = \frac{(E1 - E2)}{E2} \times 100\% \qquad (2)$$

where

E1 = the no-load voltage, and
E2 = the full-load voltage.

A steady load, such as that represented

by a receiver, speech amplifier or unkeyed stages of a transmitter, does not require good (low) regulation as long as the proper voltage is obtained under load conditions. The filter capacitors must have a voltage rating safe for the highest value to which the voltage will rise when the external load is removed.

Typically the output voltage will display a larger change with long-duration changes in load resistance than with short transient changes. The reason for this is that transient load currents are supplied from energy stored in the output capacitance. The regulation with long-term changes is often called the *static regulation*, to distinguish it from the *dynamic regulation* (transient load changes). A load that varies at a syllabic or keyed rate, as represented by some audio and RF amplifiers, usually requires good dynamic regulation (<15%) if distortion products are to be held to a low level. The dynamic regulation of a power supply can be improved by increasing the output capacitance.

When essentially constant voltage is required, regardless of current variation (for stabilizing an oscillator, for example), special voltage regulating circuits described later in this chapter are used.

7.8.3 Bleeder Resistors

A *bleeder resistor* is a resistance (R) connected across the output terminals of the power supply as shown in **Figure 7.16A**. Its functions are to discharge the filter capacitors as a safety measure when the power is turned off and to improve voltage regulation by providing a minimum load resistance. When voltage regulation is not of importance, the resistance may be as high as 100 Ω per volt of output voltage. The resistance value to be used for voltage-regulating purposes is discussed in later sections. From the consideration of safety, the power rating of bleeder resistors should be as conservative as possible — having a burned-out bleeder resistor is dangerous!

7.8.4 Ripple Frequency and Voltage

The ripple at the output of the rectifier is an alternating current superimposed on a steady direct current. From this viewpoint, the filter may be considered to consist of: 1) shunt capacitors that short circuit the ac component while not interfering with the flow of the dc component; and/or 2) series chokes that readily pass dc but will impede the ac component.

The effectiveness of the filter can be expressed in terms of percent ripple, which is the ratio of the RMS value of the ripple to the dc value in terms of percentage.

$$\text{Percent Ripple (RMS)} = \frac{E1}{E2} \times 100\% \quad (3)$$

where

E1 = the RMS value of ripple voltage, and
E2 = the steady dc voltage.

Any frequency multiplier or amplifier supply in a CW transmitter should have less than 5% ripple. A linear amplifier can tolerate about 3% ripple on the plate voltage. Bias supplies for linear amplifiers should have less than 1% ripple. VFOs, speech amplifiers and receivers may require no greater than 0.01% ripple.

Ripple frequency refers to the frequency of the pulsations in the rectifier output waveform — the number of pulsations per second. The ripple frequency of half-wave rectifiers is the same as the line-supply frequency — 60 Hz with a 60-Hz supply. Since the output pulses are doubled with a full-wave rectifier, the ripple frequency is doubled — to 120 Hz with a 60-Hz supply.

The amount of filtering (values of inductance and capacitance) required to give adequate smoothing depends on the ripple frequency. More filtering is required as the ripple frequency is reduced. This is why the filters used for line-frequency rectification are much larger than those used in switchmode converters where the ripple frequency is often in the hundreds of kHz.

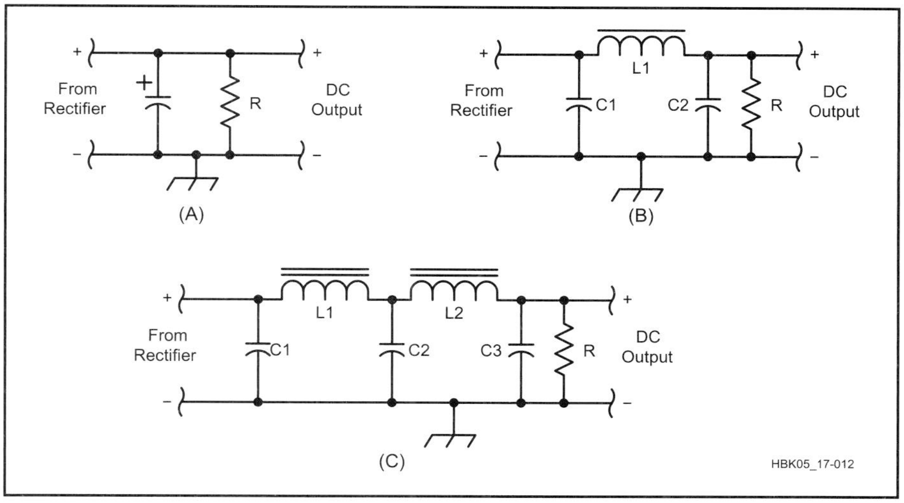

Figure 7.16 — Capacitor-input filter circuits. At A is a simple capacitor filter. B and C are single- and double-section filters, respectively.

7.8.5 Capacitor-Input Filters

Typical capacitor-input filter systems are shown in Figure 7.16. The ripple can be reduced by making C1 larger, but that can lead to very large capacitances and high inrush currents at turn-on. Better ripple reduction will be obtained when moderate values for C1 are employed and LC sections are added as shown in Figure 7.16B and C.

INPUT VERSUS OUTPUT VOLTAGE

The average output voltage of a capacitor-input filter is generally poorly regulated with load-current variations. As shown earlier (Figure 7.6) the rectifier diodes conduct for only a small portion of the ac cycle to charge the filter capacitor to the peak value of the ac waveform. When the instantaneous voltage of the ac passes its peak, the diode ceases to conduct. This forces the capacitor to support the load current until the ac voltage on the opposing diode in the bridge or full wave rectifier is high enough to pick up the load and recharge the capacitor. For this reason, the peak diode currents are usually quite high.

Since the cyclic peak voltage of the capacitor-filter output is determined by the peak of the input ac waveform, the minimum voltage and, therefore, the ripple amplitude, is determined by the amount of voltage discharge, or "droop," occurring in the capacitor while it is discharging and supporting the load. Obviously, the higher the load current, the proportionately greater the discharge, and therefore the lower the average output.

There is an easy way to approximate the

peak-to-peak ripple for a certain capacitor and load by assuming a constant load current. We can calculate the droop in the capacitor by using the relationship:

$$C \times E = I \times t \qquad (4)$$

where

C = the capacitance in microfarads,

E = the voltage droop, or peak-to-peak ripple voltage,

I = the load current in milliamperes, and

t = the length of time in ms per cycle during which the rectifiers are not conducting, during which the filter capacitor must support the load current. For 60-Hz, full-wave rectifiers, t is about 7.5 ms.

As an example, let's assume that we need to determine the peak-to-peak ripple voltage at the dc output of a full-wave rectifier/ filter combination that produces 13.8 V dc and supplies a transceiver drawing 2.0 A. The filter capacitor in the power supply is 5,000 µF. Using the above relationship:

$$C \times E = I \times t \qquad (5)$$

$$5{,}000\ \mu F \times E = 2{,}000\ mA \times 7.5\ ms$$

$$E = \frac{2{,}000\ mA \times 7.5\ ms}{5{,}000\ \mu F} = 3\ V\ P\text{-}P$$

Obviously, this is too much ripple. A capacitor value of about 20,000 µF would be better suited for this application. If a linear regulator is used after this rectifier/filter combination, then it is possible to trade off higher ripple voltage against high power dissipation in the regulator. A properly designed linear regulator can reduce the ripple amplitude to a very small value.

7.8.6 Choke-Input Filters

Choke-input filters provide the benefits of greatly improved output voltage stability over varying loads and low peak-current surges in

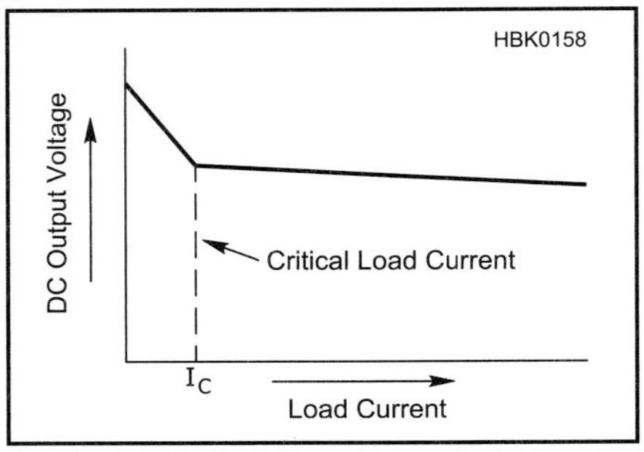

Figure 7.17 — Inductive filter output voltage regulation as a function of load current. The transition from capacitive peak charging to inductive averaging occurs at the critical load current, I_C.

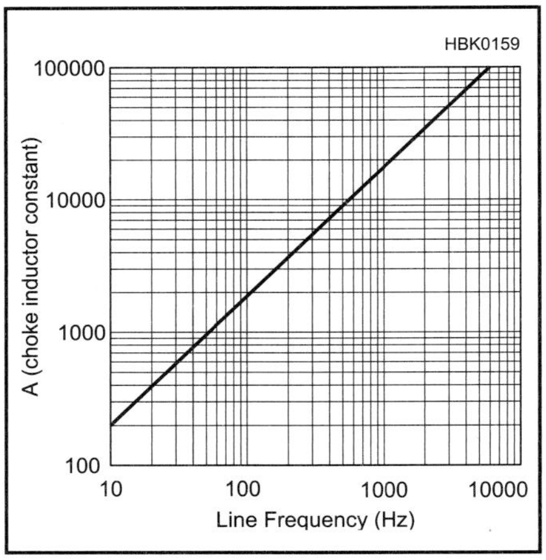

Figure 7.18 — The choke inductor constant, A, is used to solve equation 6.

the rectifiers. On the negative side, the output voltage will be lower than that for a capacitor-input filter.

In line-frequency power supplies, choke-input filters are less popular than they once were. This change came about in part because of the high surge current capability of silicon rectifiers, but more importantly because size, weight and cost are reduced when large filter chokes are eliminated. However, choke input filters are frequently used in high-frequency switchmode converters where the chokes will be much smaller.

As long as the inductance of the choke is large enough to maintain a continuous current over the complete cycle of the input ac waveform, the filter output voltage will be the average value of the rectified output. The average dc value of a full-wave rectified sine wave is 0.637 times its peak voltage. Since the RMS value is 0.707 times the peak, the output of the choke input filter will be (0.637 / 0.707), or 0.9 times the RMS ac voltage.

As shown in **Figure 7.17**, there is a minimum or "critical" load current below which the choke does not provide the necessary filtering. For light loads, there may not be enough energy stored in the choke during the input waveform crest to allow continuous current over the full cycle. When this happens, the filter output voltage will rise as the filter assumes more and more of the characteristics of a capacitor-input filter. One purpose of the bleeder resistor is to keep the minimum load current above the critical value.

The value for the critical (or minimum) inductance for a given maximum value of load resistance in a single phase, full-wave rectifier with a sine wave source voltage can be approximated from:

$$L_c = R/A \qquad (6)$$

where

R = the maximum load resistance, and

A = a constant obtained from **Figure 7.18**, derived from the frequency of the input current (see Reference 1).

Low values for minimum load current (high minimum load R) can lead to large values for L_C which may not be practical. Standard filter inductors typically have a relatively constant value for L as the dc current is varied but it is possible to use a swinging choke instead. This is an inductor which has a high inductance at low currents and much lower inductance at high currents. Using a swinging choke will usually result in a much smaller filter choke and/or better output regulation.

7.9 Power Supply Regulation

The output of a rectifier/filter system may be usable for some electronic equipment, but for today's transceivers and accessories, further measures may be necessary to provide power sufficiently clean and stable for their needs. Voltage regulators are often used to provide this additional level of conditioning.

Rectifier/filter circuits by themselves are unable to protect the equipment from the problems associated with input-power-line fluctuations, load-current variations and residual ripple voltages. Regulators can eliminate these problems, but not without costs in circuit complexity and power-conversion efficiency.

There are many types of regulator circuits, both linear and switchmode. This section addresses the most common circuits used by amateurs. A broader introduction is available in the Linear Technology application note AN140, "Basic Concepts of Linear Regulator and Switching Mode Power Supplies" which is available online at **www.analog.com/media/en/technical-documentation/app-notes/an140.pdf**.

7.9.1 Zener Diodes

A Zener diode (developed by Dr Clarence Zener) can be used to maintain the voltage applied to a circuit at a practically constant value, regardless of the voltage regulation of the power supply or variations in load current. The typical circuit is shown in **Figure 7.19**. Note that the cathode side of the diode is connected to the positive side of the supply.

Zener diodes are available in a wide variety of voltages and power ratings. The voltages range from less than 2 V to a few hundred volts, while the power ratings (power the diode can dissipate) run from less than 0.25 W to 50 W. The ability of the Zener diode to stabilize a voltage depends on the diode's conducting impedance. This can be as low

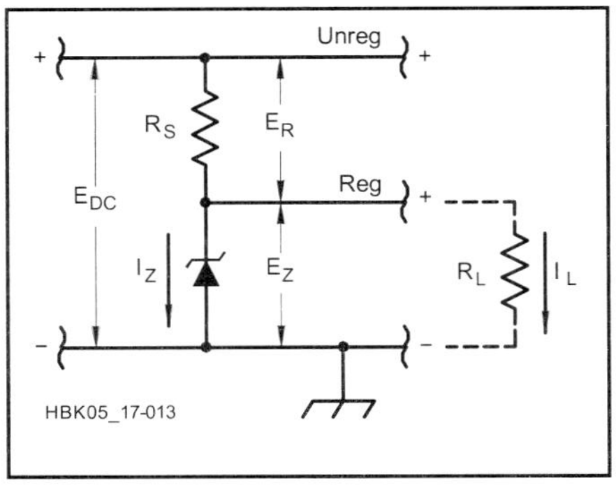

Figure 7.19 — Zener-diode voltage regulation. The voltage from a negative supply may be regulated by reversing the power-supply connections and the diode polarity.

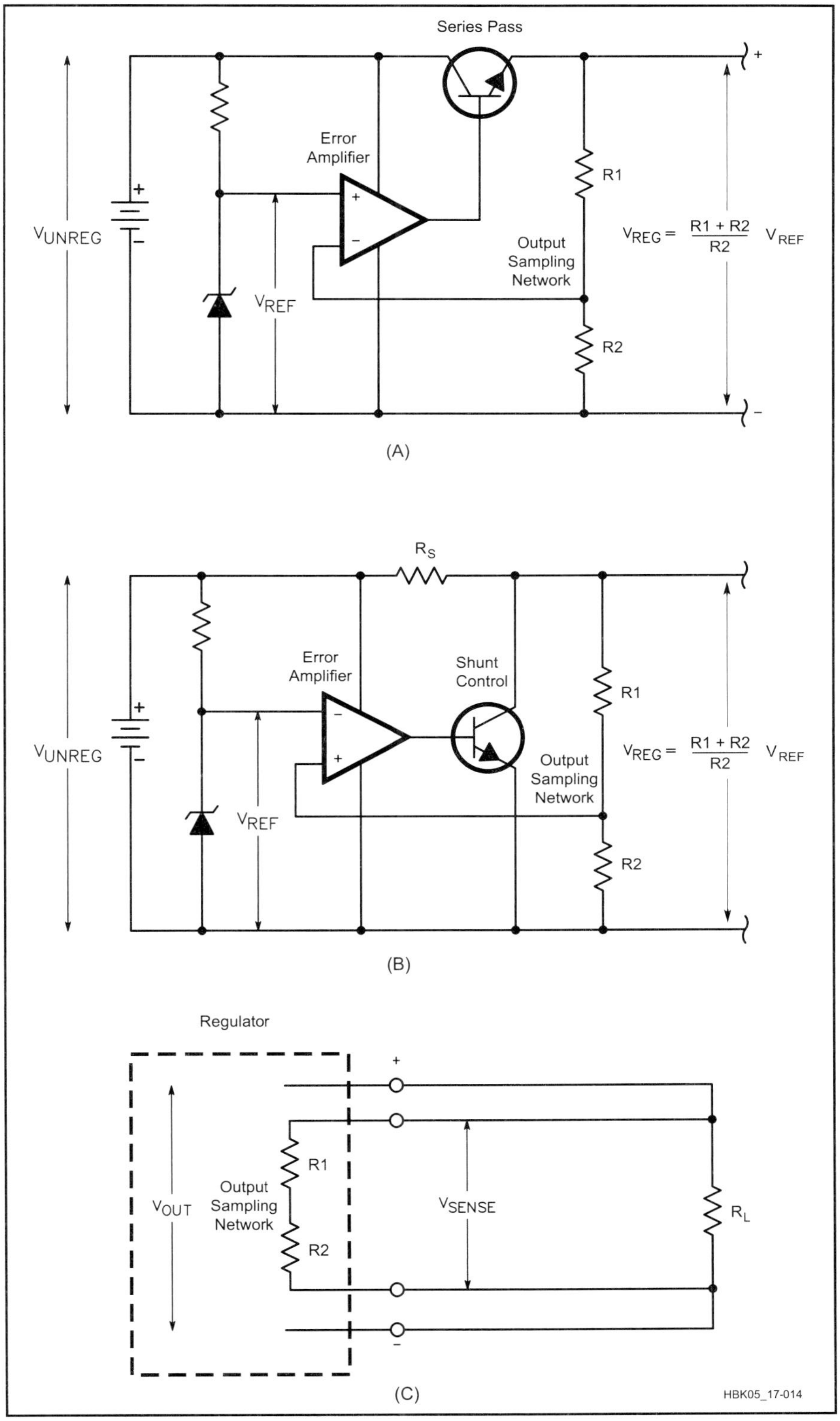

Figure 7.20 — Linear electronic voltage regulator circuits. In these diagrams, batteries represent the unregulated input-voltage source. A transformer, rectifier and filter would serve this function in most applications. Part A shows a series regulator and Part B shows a shunt regulator. Part C shows how remote sensing overcomes poor load regulation caused by the IR drop in the connecting wires by bringing them inside the feedback loop. The use of extra connections to sense voltage is called a "four-wire Kelvin connection."

as 1 Ω or less in a low-voltage, high-power diode or as high as 1,000 Ω in a high-voltage, low-power diode.

The circuit in Figure 7.19 is a *shunt* regulator in that it "shunts" current through a controlling device (the Zener diode) to maintain a constant output voltage. To design a Zener shunt regulator, you must know the minimum and maximum input voltage (E_{DC}); the output voltage, which is equal to the Zener diode voltage (E_Z); and the minimum and maximum load current (I_L) through R_L. If the input voltage is variable, you must specify the maximum and minimum values for E_{DC}. As a rule of thumb, the current through the Zener should be 10% of the maximum load current for good regulation and must be greater than the $I_{Z(min)}$ at which the Zener diode maintains its constant voltage drop. Once these quantities are known the series resistance, R_S, can be determined:

$$R_S = \frac{E_{DC(min)} - E_Z}{1.1\, I_{L(max)}} \quad (7)$$

The power dissipation of the Zener diode, P_{DZ}, is

$$P_{DZ} = \left[\frac{E_{DC(max)} - E_Z}{R_S} - I_{L(min)}\right] E_Z \quad (8)$$

and of the series resistor, R_S,

$$P_{DR} = \frac{\left(E_{DC(max)} - E_Z\right)^2}{R_S} \quad (9)$$

It is good practice to provide a five times rated power dissipation safety margin for both the series resistor and the Zener diode. This avoids heating in the Zener and the resulting drift in voltage. High-power Zener diodes (10 W dissipation or more) will require heat-sinking as discussed in the section on Heat Management in the **Circuits and Components** chapter.

7.9.2 Linear Regulators

Linear regulators come in two varieties, *series* and *shunt*, as shown in **Figure 7.20**. The shunt regulator is simply an electronic (also called "active") version of the Zener diode. For the most part, the active shunt regulator (Figure 7.20B) is rarely used since the series regulator is a superior choice for most applications.

The series regulator (Figure 7.20A) consists of a stable voltage reference, which is usually established by a Zener diode, a transistor in series between the power source and the load (called a *series pass transistor*), and an error amplifier. In critical applications a temperature-compensated reference diode would be used instead of the Zener diode.

The output voltage is sampled by the error amplifier, which compares the output (usually scaled down by a voltage divider) to the reference. If the scaled-down output voltage becomes higher than the reference voltage, the error amplifier reduces the drive current to the pass transistor, thereby allowing the output voltage to drop slightly. Conversely, if the load pulls the output voltage below the desired value, the amplifier drives the pass transistor into increased conduction.

The "stiffness" or tightness of regulation of a linear regulator depends on the gain of the error amplifier and the ratio of the output scaling resistors. In any regulator, the output is cleanest and regulation stiffest at the point where the sampling network or error amplifier is connected. If heavy load current is drawn through long leads, the voltage drop can degrade the regulation at the load. To combat this effect, the feedback connection to the error amplifier can be made directly to the load. This technique, called *remote sensing*, or a *four-wire Kelvin connection*, moves the point of best regulation to the load by bringing the connecting loads inside the feedback loop. This is shown in Figure 7.20C.

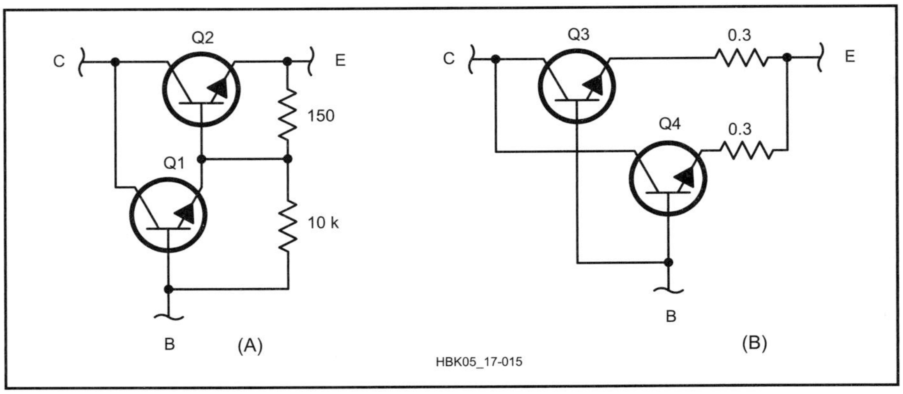

Figure 7.21 — At A, a Darlington-connected transistor pair for use as the pass element in a series-regulating circuit. At B, the method of connecting two or more transistors in parallel for high-current output. Resistances are in ohms. The circuit at A may be used for load currents from 100 mA to 5 A, and the one at B may be used for currents from 6 A to 10 A.

Q1 — NPN transistor, MJE340 or equivalent

Q2-Q4 — Power transistor such as 2N3055 or 2N3772

INPUT VERSUS OUTPUT VOLTAGE

In a series regulator, the pass-transistor power dissipation is directly proportional to the load current and input/output voltage differential. The series pass element can be located in either leg of the supply. Either NPN or PNP devices can be used, depending on the ground polarity of the unregulated input.

The differential between the input and output voltages is a design tradeoff. If the input voltage from the rectifiers and filter is only slightly higher than the required output voltage, there will be minimal voltage drop across the series pass transistor. A small drop results in minimal thermal dissipation and high power-supply efficiency. The supply will have less capability to provide regulated power in the event of power line brownout and other reduced line voltage conditions, however. Conversely, a higher input voltage will provide operation over a wider range of input voltage, but at the expense of increased heat dissipation.

7.9.3 Linear Regulator Pass Transistors

DARLINGTON PAIRS

A simple Zener-diode reference or IC op-amp error amplifier may not be able to source enough current to a pass transistor that must conduct heavy load current. The Darlington configuration of **Figure 7.21A** multiplies the pass transistor beta, thereby extending the control range of the error amplifier. If the Darlington arrangement is implemented with discrete transistors, resistors across the base-emitter junctions may be necessary to prevent collector-to-base leakage currents in Q1 from being amplified and turning on the transistor pair. These resistors are contained in the envelope of a monolithic Darlington device.

When a single pass transistor is not available to handle the current required from a regulator, the current-handling capability may be increased by connecting two or more pass transistors in parallel. The circuit of Figure 7.21B shows the method of connecting these pass transistors. The resistances in the emitter leads of each transistor are necessary to equalize the currents.

TRANSISTOR RATINGS

When bipolar (NPN, PNP) power transistors are used in applications in which they are called upon to handle power on a continuous basis, rather than switching, there are four parameters that must be examined to see if any maximum limits are being exceeded.

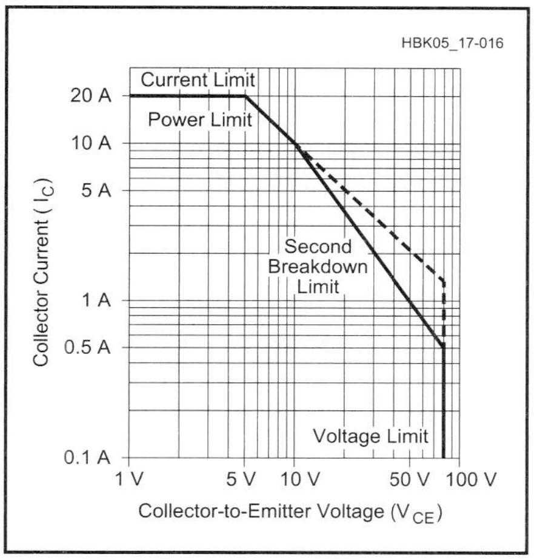

Figure 7.22 — Typical graph of the safe operating area (SOA) of a transistor. See text for details. Safe operating conditions for specific devices may be quite different from those shown here.

Operation of the transistor outside these limits can easily result in device failure, and these parameters must be considered during the design process.

The four limits are maximum collector current (I_C), maximum collector-emitter voltage (V_{CEO}), maximum power and *second breakdown* (I_{SB}). All four of these parameters are graphically shown on the transistor's data sheet on what is known as a *safe operating area (SOA)* graph. (see **Figure 7.22**) The first three of these limits are usually also listed prominently with the other device information, but it is often the fourth parameter — secondary breakdown — that is responsible for the "sudden death" of the power transistor after an extended operating period.

The maximum current limit of the transistor ($I_{C\ MAX}$) is usually the current limit for fusing of the bond wire connected to the emitter, rather than anything pertaining to the transistor chip itself. When this limit is exceeded, the bond wire can melt and open circuit the emitter. On the operating curve, this limit is shown as a horizontal line extending out from the Y-axis and ending at the voltage point where the constant power limit begins.

The maximum collector-emitter voltage limit of the transistor ($V_{CE\ MAX}$) is the point at which the transistor junctions can no longer withstand the voltage between collector and emitter.

With increasing collector-emitter voltage drop at maximum collector current, a point is reached where the power in the transistor will cause the junction temperature to rise to a level where the device leakage current rapidly increases and begins to dominate. In this region, the product of the voltage drop and the current would be constant and represent the maximum power (P_t) rating for the transistor; that is, as the voltage drop continues to increase, the collector current must decrease to maintain the power dissipation at a constant value.

With most transistors rated for higher voltages, a point is reached on the constant power portion of the curve whereby, with further increased voltage drop, the maximum power rating is *not* constant, but decreases as the collector to emitter voltage increases. This decrease in power handling capability continues until the maximum voltage limit is reached.

This special region is known as the *forward bias second breakdown* (FBSB) area. Reduction in the transistor's power handling capability is caused by localized heating in certain small areas of the transistor junction ("hot spots"), rather than a uniform distribution of power dissipation over the entire surface of the device.

The region of operating conditions contained within these curves is called the safe operating area, or SOA. If the transistor is always operated within these limits, it should provide reliable and continuous service for a long time.

MOSFET TRANSISTORS

The bipolar junction transistor (BJT) is rapidly being replaced by the MOSFET in new power supply designs because MOSFETs are easier to drive. The N-channel MOSFET (equivalent to the NPN bipolar) is more popular than the P-channel for pass transistor applications.

There are some considerations that should be observed when using a MOSFET as a linear regulator series pass transistor. Several volts of gate drive are needed to start conduction of the device, as opposed to less than 1 V for the BJT. MOSFETs are inherently very-high-frequency devices and will readily oscillate with stray-circuit capacitances. To prevent oscillation in the transistor and surrounding circuits, it is common practice to insert a small resistor of about 100 Ω directly in series with the gate of the series-pass transistor to reduce the gate circuit Q.

OVERCURRENT PROTECTION

Damage to a pass transistor can occur when the load current exceeds the safe amount. **Figure 7.23A** illustrates a simple current-limiter circuit that will protect Q1. All of the load current is routed through R1. A voltage difference will exist across R1; the value will depend on the exact load current at a given time. When the load current exceeds a predetermined safe value, the voltage drop across R1 will forward-bias Q2 and cause it to

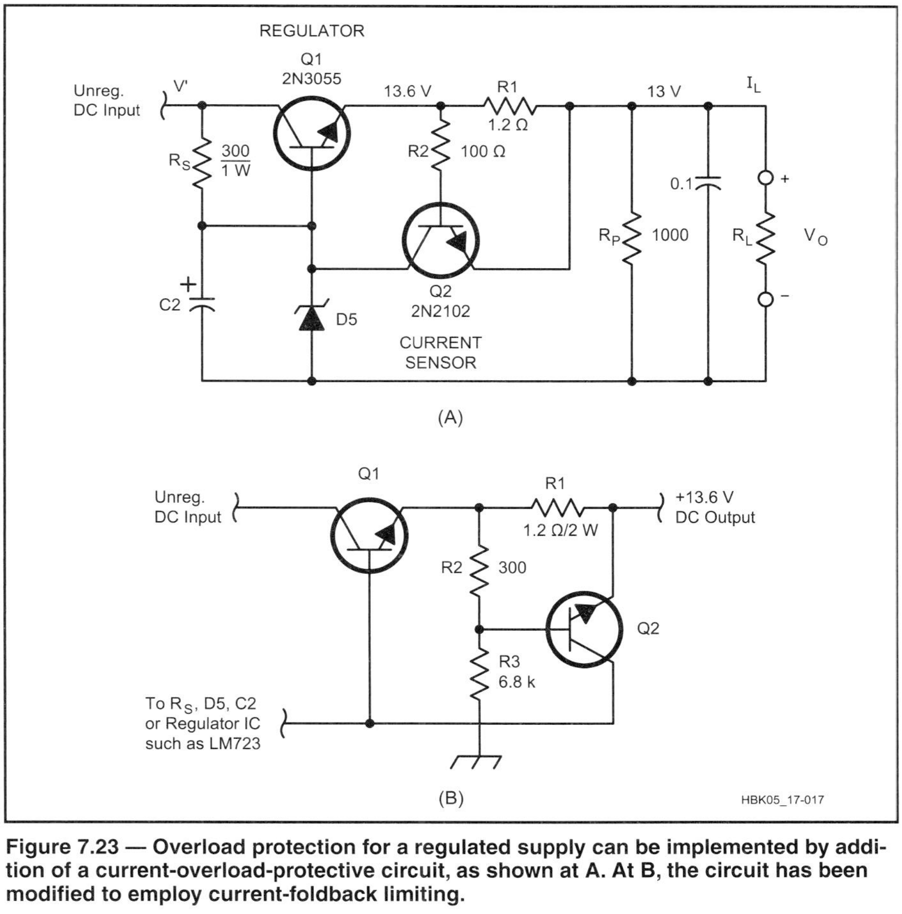

Figure 7.23 — Overload protection for a regulated supply can be implemented by addition of a current-overload-protective circuit, as shown at A. At B, the circuit has been modified to employ current-foldback limiting.

conduct. Because Q2 is a silicon transistor, the voltage drop across R1 must exceed 0.6 V to turn Q2 on. This being the case, R1 is chosen for a value that provides a drop of 0.6 V when the maximum safe load current is drawn. In this instance, the drop will be 0.6 V when I_L reaches 0.5 A. R2 protects the base-emitter junction of Q2 from current transients, or from destruction in the event Q1 fails under short-circuit conditions.

When Q2 turns on, some of the current through R_S flows through Q2, thereby depriving Q1 of some of its base current. This action, depending upon the amount of Q1 base current at a precise moment, cuts off Q1 conduction to some degree, thus limiting the current through it.

FOLDBACK CURRENT LIMITING

Under short-circuit conditions, a constant-current type current limiter must still withstand the full source voltage and limited short-circuit current simultaneously, which can impose a very high power dissipation or second breakdown stress on the series pass transistor. For example, a 12 V regulator with current limiting set for 10 A and having a source of 16 V will have a dissipation of 40 W [(16 V – 12 V) × 10 A] at the point of current limiting (knee). But its dissipation will rise to 160 W under short-circuit conditions (16 V × 10 A).

A modification of the limiter circuit can cause the regulated output current to decrease with decreasing load resistance beyond the over-current knee. With the output shorted, the output current is only a fraction of the knee current value, which protects the series-pass transistor from excessive dissipation and possible failure. Using the previous example of the 12 V, 10 A regulator, if the short-circuit current is designed to be 3 A (the knee is still 10 A), the transistor dissipation with a short circuit will be only 16 V × 3 A = 48 W.

Figure 7.23B shows how the current-limiter example given in the previous section would be modified to incorporate foldback limiting. The divider string formed by R2 and R3 provides a negative bias to the base of Q2, which prevents Q2 from turning on until this bias is overcome by the drop in R1 caused by load current. Since this hold-off bias decreases as the output voltage drops, Q2 becomes more sensitive to current through R1 with decreasing output voltage. See **Figure 7.24**.

The circuit is designed by first calculating the value of R1 for short-circuit current. For example, if 0.5 A is chosen, the value for R1 is simply 0.6 V / 0.5 A = 1.2 Ω (with the output shorted, the amount of hold-off bias supplied by R2 and R3 is very small and can be neglected). The knee current is then chosen. For this example, the selected value will be 1.0 A. The divider string is then proportioned to provide a base voltage at the knee that is just sufficient to turn on Q2 (a value of 13.6 V for 13.0 V output). With 1.0 A flowing through R1, the voltage across the divider will be 14.2 V. The voltage dropped by R2 must then be 14.2 V –13.6 V, or 0.6 V. Choosing a divider current of 2 mA, the value of R2 is then 0.6 V / 0.002 A = 300 Ω. R3 is calculated to be 13.6 V / 0.002 A = 6800 Ω.

7.9.4 Three-Terminal Voltage Regulators

The modern trend in regulators is toward the use of three-terminal devices commonly referred to as *three-terminal regulators.* Inside each regulator is a voltage reference, a high-gain error amplifier, temperature-compensated voltage sensing resistors and a pass element. Many currently available units have thermal shut-down, overvoltage protection and current foldback, making them virtually destruction-proof. It is easy to see why regulators of this sort are so popular when you consider the low price and the number of individual components they can replace.

Three-terminal regulators have connections for unregulated dc input, regulated dc output and ground, and they are available in a wide range of voltage and current ratings. Fixed-voltage regulators are available with output ratings in most common values between 5 and 28 V. Other families include devices that can be adjusted from 1.25 to 50 V.

The regulators are available in several different package styles, depending on current ratings. Low-current (100 mA) devices frequently use the plastic TO-92 and DIP-style cases. TO-220 packages are popular in the 1.5 A range, and TO-3 cases house the larger 3 A and 5 A devices. They are available in surface mount packages too.

Three-terminal regulators are available as positive or negative types. In most cases, a positive regulator is used to regulate a positive voltage and a negative regulator a negative voltage. Depending on the system ground requirements, however, each regulator type may be used to regulate the "opposite" voltage.

Figure 7.25A and B illustrate how the regulators are used in the conventional mode. Several regulators can be used with a common-input supply to deliver several voltages with a common ground. Negative regulators may be used in the same manner. If no other common supplies operate from the input supply to the regulator, the circuits of Figure 7.25C and D may be used to regulate positive voltages with a negative regulator and vice versa. In these configurations the input supply is floated; neither side of the input is tied to the system ground.

Manufacturers have adopted a system of family numbers to classify three-terminal

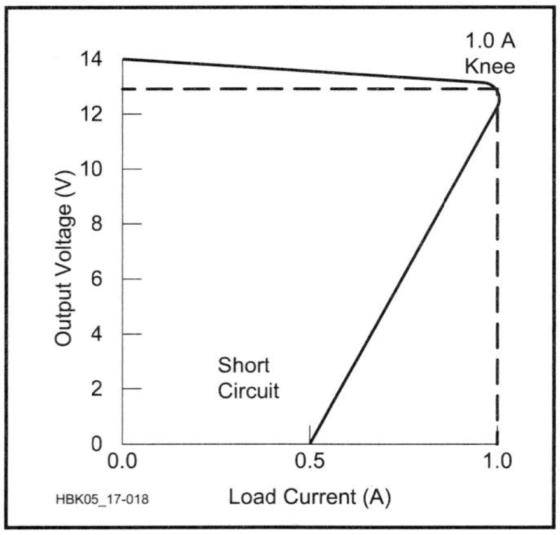

Figure 7.24 — The 1 A regulator shown in Figure 7.23B will fold back to 0.5 A under short-circuit conditions. See text.

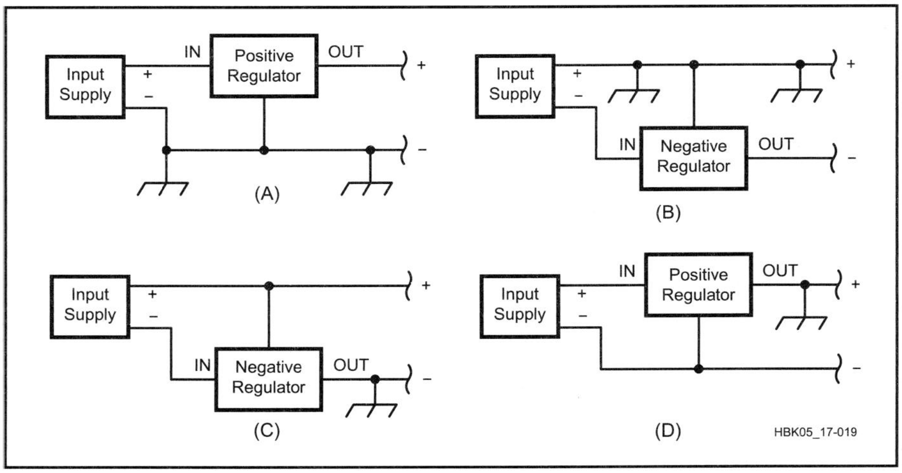

Figure 7.25 — Parts A and B illustrate the conventional manner in which three-terminal regulators are used. Parts C and D show how one polarity regulator can be used to regulate the opposite-polarity voltage.

regulators in terms of supply polarity, output current and regulated voltage. For example, 7805 describes a positive 5 V, 1.5 A regulator and 7905 a negative 5 V, 1.5 A unit. Depending on the manufacturer, the full part number might have a prefix such as LM, UA or MC, along with various suffixes (for example, LM7805CT or MC7805CTG). There are many such families with widely varied ratings available from manufacturers.

Make sure to check the pin assignments (pinouts) for all voltage regulators. While the fixed-voltage positive regulators generally share a common pattern of input, output, and ground, negative-voltage and adjustable regulators do not. Installing a regulator with the wrong connections will usually destroy it and may allow excessive voltage to be applied to the circuit it supplies.

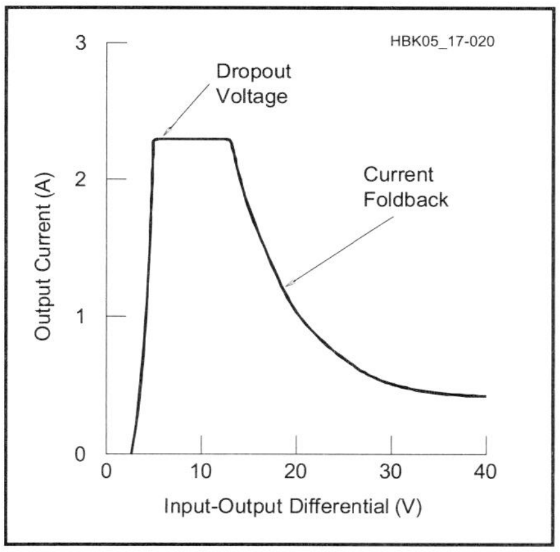

Figure 7.26 — Effects of input-output differential voltage on three-terminal regulator current.

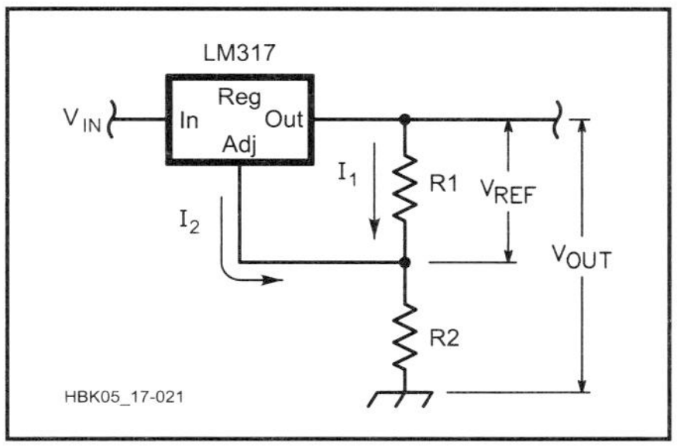

Figure 7.27 — By varying the ratio of R2 to R1 in this simple LM317 schematic diagram, a wide range of output voltages is possible. See text for details.

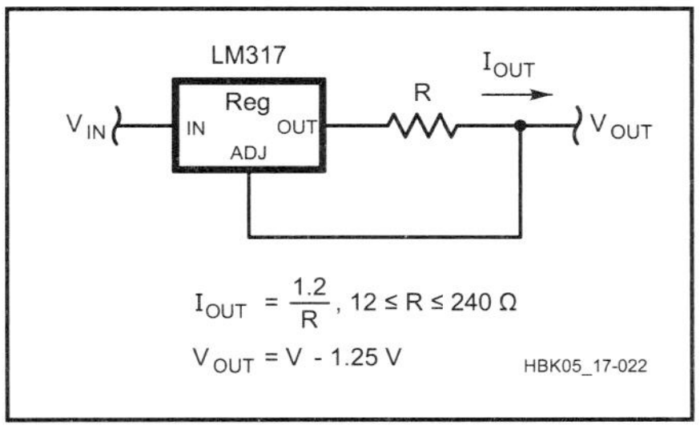

Figure 7.28 — The basic LM317 voltage regulator is converted into a constant-current source by adding only one resistor.

SPECIFYING A REGULATOR

When choosing a three-terminal regulator for a given application, the most important specifications to consider are device output voltage, output current, minimum and maximum input-to-output differential voltages, line regulation, load regulation and power dissipation. Output voltage and current requirements are determined by the load with which the supply will ultimately be used.

Input-to-output differential voltage is one of the most important three-terminal regulator specifications to consider when designing a supply. The differential value (the difference between the voltage applied to the input terminal and the voltage on the output terminal) must be within a specified range. The minimum differential value, usually about 2.5 V, is called the *dropout voltage*. If the differential value is less than the dropout voltage, no regulation will take place. Special *low dropout regulators* with lower minimum differential values are available as well. At the other end of the scale, maximum input-output differential voltage is generally about 40 V. If this differential value is exceeded, device failure may occur.

Increases in either output current or differential voltage produce proportional increases in device power consumption. By employing current foldback, as described above, some manufacturers ensure that maximum dissipation will never be exceeded in normal operation. **Figure 7.26** shows the relationship between output current, input-output differential and current limiting for a three-terminal regulator nominally rated for 1.5 A output current. Maximum output current is available with differential voltages ranging from about 2.5 V (dropout voltage) to 12 V. Above 12 V, the output current decreases, limiting the device dissipation to a safe value. If the output terminals are accidentally short circuited, the input-output differential will rise, causing current foldback, and thus preventing the power-supply components from being over stressed. This protective feature makes three-terminal regulators particularly attractive in simple power supplies.

When designing a power supply around a particular three-terminal regulator, input-output voltage characteristics of the regulator should play a major role in selecting the transformer-secondary and filter-capacitor component values. The unregulated voltage applied to the input of the three-terminal device should be higher than the dropout voltage, yet low enough that the regulator does not go into current limiting caused by an excessive differential voltage. If, for example, the regulated output voltage of the device shown in Figure 7.26 were 12 V, then unregulated input voltages of between 14.5 and 24 V would be acceptable if maximum output current is desired.

In use, all but the lowest-current regulators generally require an adequate external heat sink because they may be called on to dissipate a fair amount of power. Also, because the regulator chip contains a high-gain error amplifier, bypassing of the input and output leads is essential for stable operation.

Most manufacturers recommend bypassing the input and output directly at the leads where they protrude through the heat sink. Solid tantalum capacitors are usually recommended because of their good high-frequency capabilities.

External capacitors used with IC regulators may discharge through the IC junctions under certain circuit conditions, and high-current discharges can harm ICs. Look at the regulator data sheet to see whether protection diodes are needed, what diodes to use and how to place them in any particular application.

Adjustable Regulators

In addition to fixed-output-voltage ICs, high-current, adjustable voltage regulators are available. These ICs require little more than an external potentiometer for an adjustable output range from 5 to 24 V at up to 5 A. The unit price on these items is only a few dollars, making them ideal for test-bench power supplies. A very popular low current, adjustable output voltage three terminal regulator, the LM317, is shown in **Figure 7.27**. It develops a steady 1.25 V reference, V_{REF}, between the output and adjustment terminals. By installing R1 between these terminals, a constant current, I1, is developed, governed by the equation:

$$I1 = \frac{V_{REF}}{R1} \qquad (10)$$

Both I1 and a 100 µA error current, I2, flow through R2, resulting in output voltage V_O. V_O can be calculated using the equation:

$$V_O = V_{REF}\left(1 + \frac{R2}{R1}\right) + I2 \times R2 \qquad (11)$$

Any voltage between 1.2 and 37 V may be obtained with a 40 V input by changing the ratio of R2 to R1. At lower output voltages, however, the available current will be limited by the power dissipation of the regulator.

Figure 7.28 shows one of many flexible applications for the LM317. By adding only one resistor with the regulator, the voltage regulator can be changed into a constant-current source capable of charging NiCd batteries, for example. Design equations are given in the figure. The same precautions should be taken with adjustable regulators as with the fixed-voltage units. Proper heat sinking and lead bypassing are essential for proper circuit operation.

INCREASING REGULATOR OUTPUT CURRENT

When the maximum output current from an IC voltage regulator is insufficient to operate the load, discrete power transistors may be connected to increase the current capability. **Figure 7.29** shows two methods for boosting the output current of a positive regulator, although the same techniques can be applied to negative regulators.

In A, an NPN transistor is connected as an emitter follower, multiplying the output current capacity by the transistor beta. The shortcoming of this approach is that the base-emitter junction is not inside the feedback loop. The result is that the output voltage is reduced by the base-emitter drop, and the load regulation is degraded by variations in this drop.

The circuit at B has a PNP transistor "wrapped around" the regulator. The regulator draws current through the base-emitter junction, causing the transistor to conduct. R1 provides bias voltage for turning on Q1 so that U1 doesn't see the excess current. For example, a 6 Ω resistor will limit the current U1 sees to 100 mA. The IC output voltage is unchanged by the transistor because the collector is connected directly to the IC output (sense point). Any increase in output voltage is detected by the IC regulator, which shuts off its internal-pass transistor, and this stops the boost-transistor base current.

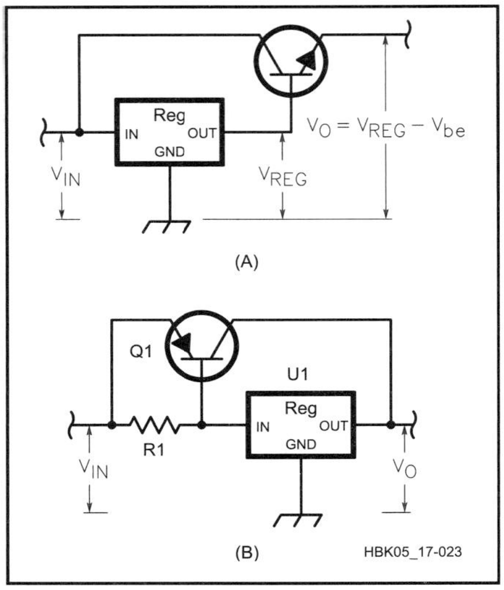

Figure 7.29 — Two methods for boosting the output-current capacity of an IC voltage regulator. Part A shows an NPN emitter follower and B shows a PNP "wraparound" configuration. Operation of these circuits is explained in the text.

7.10 "Crowbar" Protective Circuits

Electronic components *do* fail from time to time. In a regulated power supply, the only component standing between an elevated dc source voltage and your transceiver is one transistor, or a group of transistors wired in parallel. If the transistor, or one of the transistors in the group, happens to short internally, your equipment could suffer lots of damage.

To safeguard the load equipment against possible overvoltage, some power-supply manufacturers include a circuit known as a *crowbar*. This circuit usually consists of a silicon-controlled rectifier (SCR) or thyristor connected directly across the output of the power supply, with an over-voltage-sensing trigger circuit tied to its gate. The SCR is large enough to take the full short-circuit output current of the supply, as if a crowbar were placed across the output terminals, thus the name.

In the event the output voltage exceeds the trigger set point, the SCR will fire, and the output is short circuited. The resulting high current in the power supply (shorted output in series with a series pass transistor failed short) will blow the power supply's line fuses. This is a protection for the supply as well as an indicator that something has malfunctioned internally. For these reasons, never replace blown fuses with ones that have a higher current rating.

An example of a crowbar overvoltage protection circuit can be found as a project at the end of this chapter. It provides basic design equations that can be adapted to a wide range of power supply applications.

7.11 DC-DC Switchmode Power Conversion

Very often the power source is dc, such as a battery or solar cell, or the output of an unregulated rectifier connected to an ac source. In most applications high conversion efficiency is desired, both to conserve energy from the source and to reduce heat dissipation in the converter. When high efficiency is needed, some form of switching circuit will be employed for dc-dc power conversion. Besides being more efficient, switching circuits are usually much smaller and lighter than conventional 60 Hz, transformer-rectifier circuits because they operate at much higher frequencies — from 25 to 400 kHz or even higher. Switching circuits go by many names; switching regulators and switchmode converters are just two of the more common names.

The possibility of achieving high conversion efficiency stems directly from the use of switches for the power conversion process, along with low-loss inductive and capacitive elements. An active switch is a device that is either ON or OFF and the state of the switch (ON or OFF) can be controlled with an external signal. The loss in the switch is always the product of the voltage across the switch and the current flowing through it ($P = E \times I$).

In the ON (conducting) state the voltage drop across the switch is small, and in the OFF state the current through the switch is small. In both cases the losses can be small relative to the power level of the converter. During transitions between ON and OFF states, however, there will simultaneously be both substantial voltage across the switch and significant current flowing through it. This results in power dissipation in the switch, called switching loss. This loss is minimized by making the switching transitions as short as possible. In this way even though the instantaneous power dissipation may be high, the average loss is low because of the small duty cycle of the transitions. Of course, the more frequently the switch operates (higher switching frequency), the higher the average loss will be, and this eventually limits the maximum operating frequency. The need to be above audible frequencies (>25 kHz) puts a limit on the lower end of the switching frequency range.

This is quite different from the linear regulators discussed earlier in which there is always some voltage drop across the pass transistor (which is acting as a controlled variable resistor) while current is flowing through it. As a result, the efficiency of a linear regulator can be very low, often 65% or less. Switchmode converters, on the other hand, will typically have efficiencies in the range of 85 to 95%.

For converters in the power range typical of amateur applications (a few watts through 2 – 3 kW) the most common choices for switches are bipolar junction transistors (BJTs) and power MOSFETs. The circuit diagrams in this section will show MOSFETs or generic switch symbols for the switches.

Switchmode circuits can also generate radio-frequency interference (RFI) through VHF because of switching frequency harmonics and ringing induced by the rapid rise and fall times of voltage and current. In attempting to minimize the ON/OFF transition time, significant amounts of RF energy can be generated. To prevent RFI to sensitive receivers, careful bypassing, shielding, and filtering of both input and output circuits is required. (RFI from switchmode or "switching" supplies is also discussed in the chapter on **RF Interference and Electromagnetic Compatibility**.)

There are literally hundreds of different switchmode circuits or "topologies" (see Reference 2), but we will only look at a few of those most commonly used by amateurs. Fortunately, an understanding of the basic circuits provides an entry point to many other circuits.

The following discussion is only an introduction to the basics of switchmode converters. To really get a handle on designing these circuits, the reader will have to do some additional reading. Fortunately, a very large amount of useful information is freely available online. Many useful application notes are available from semiconductor manufacturers, and additional information can be found on the websites of filter capacitor and ferrite core manufacturers (see References 3 and 4). There are also numerous books on the subject, often available in libraries and used bookstores.

7.11.1 The Buck Converter

A schematic for a *buck converter* is shown in **Figure 7.30A**. This circuit is called a "buck" converter because the output voltage is always less than or equal to the input voltage. Power is supplied from the dc source (V_S) through the input filter (L1, C1) to the drain of the switch (Q1). The load (R) is connected across the output filter (L, C2).

The equivalent circuit when Q1 is ON is shown in Figure 7.30B (where the input filter components are assumed to be part of V_S). V_S is connected to one end of the output inductor (L) and, because $V_O < V_S$, current flows from the source to the output delivering energy from the source to the output and also storing energy in L. At some point Q1 is turned OFF and the current flowing in L commutates (switches) to D1, as shown in Figure 7.30C. (The current in an inductor cannot change instantaneously, so when Q1 turns OFF, the collapsing magnetic field in L pulls current through shunt diode D1, called

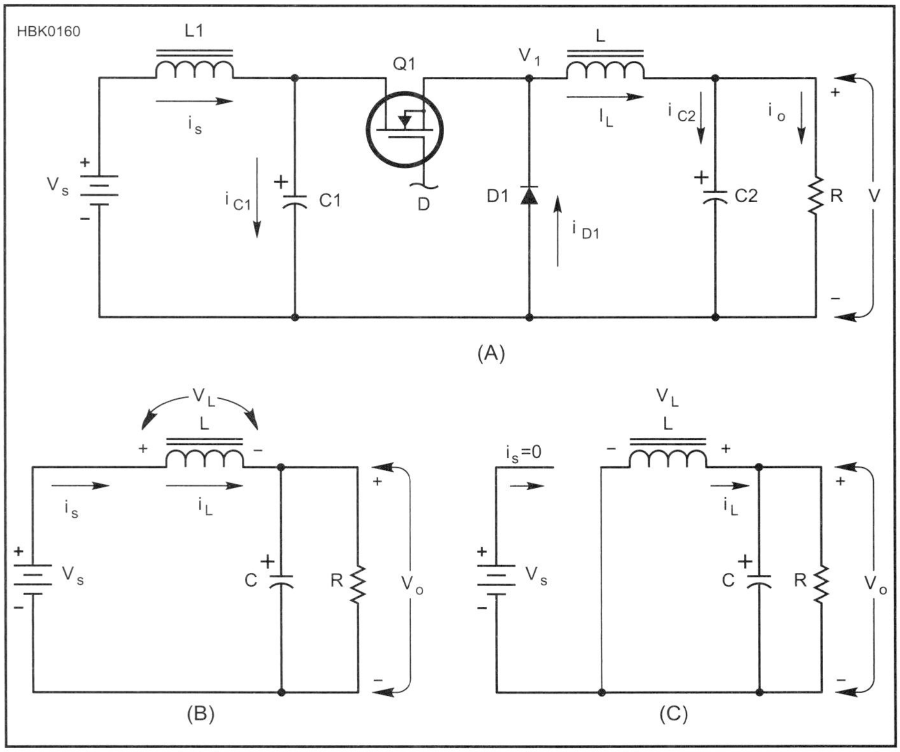

Figure 7.30 — Typical buck converter.

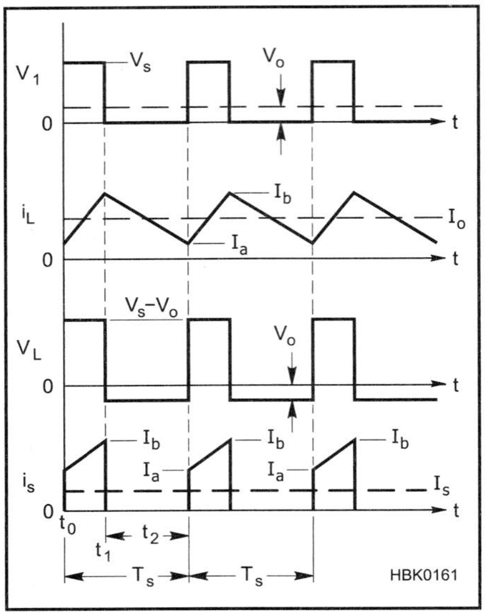

Figure 7.31 — Waveforms in a buck converter.

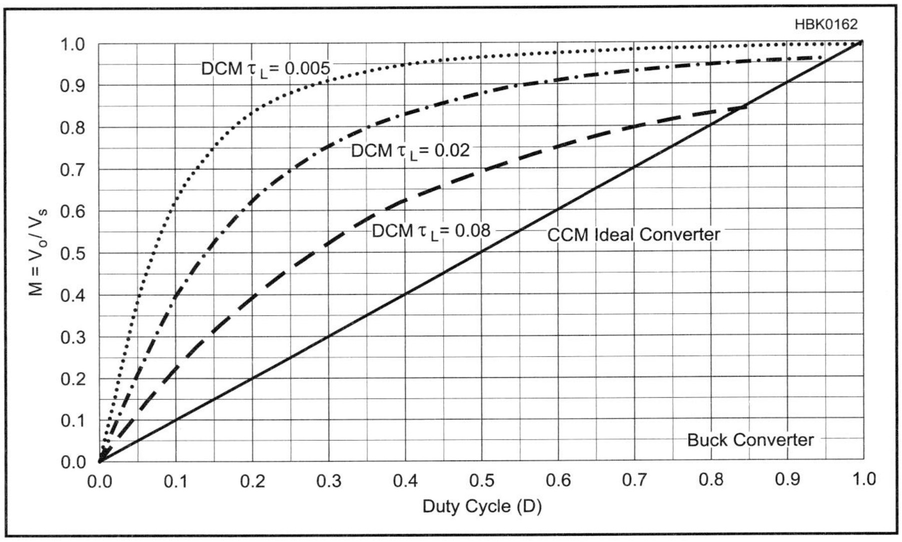

Figure 7.32 — Change in output voltage to input voltage ratio (M=VO/VS) as a function of switch duty cycle (D).

a *free-wheeling diode*.) The energy in L is now being discharged into the output. This cycle is repeated at the switching frequency (f_s). The ratio of the switch ON-time to the total switching period ($T_s = t_{on} + t_{off} = 1/f_s$) is called the *duty cycle (D)*.

Typical voltage and current waveforms for a buck converter are shown in **Figure 7.31** where V_1 is the voltage at the junction of Q1, D1 and L (see Figure 7.30A).

The interval 0-t_1 corresponds to the ON time of Q1 and $T_s - t_1$ corresponds to the OFF time ($T_s = t_1 + t_2$). From the waveforms it can be seen that the current in the inductor rises while Q1 is ON and falls while Q1 is OFF. The energy in the inductor is proportional to $LI^2/2$.

The input current (I_s) is pulsating at the switching frequency. This pulsation in the input current is the reason for the input filter (L1 and C1). We need to keep this high frequency noise (the ac component of the pulsation) out of the source. All switchmode converters require some form of filter on the input to keep switching noise out of the source. Some form of filter is also required on the output to keep the switching noise out of the load. In Figure 7.30, L and C2 form the output filter. For applications requiring low ripple it would not be unusual to see an additional stage of L-C filtering added to the output.

The voltage waveform (V_1) at the input to the output filter (L and C2) is pulse-width modulated (PWM) and V_O (the dotted line in the waveform for V_1 in Figure 7.31A) is the time average of V_1. V_O is controlled or regulated by adjusting the duty cycle of Q1 in response to input voltage or output load changes. If we increase D we increase V_O up to the point where D = 1 (the switch is ON all the time) and $V_O \approx V_S$. If we never turn Q1 on (D = 0) then $V_O = 0$. Normal operation will be somewhere between these extremes.

The inductor current waveform in Figure 7.31 does not go to zero while Q1 is OFF. In other words, not all of the energy stored in L is discharged into the output by the end of each switching cycle. This is a common mode of operation for heavier loads. It is referred to as the *continuous conduction mode* or CCM, where the conduction referred to is the current in L. There is another possibility: during the time Q1 is off all of the energy in L may be discharged and for some period of time the inductor current is zero until Q1 is turned ON again. This is referred to as the *discontinuous conduction mode* or DCM. A typical converter will operate in CCM for heavy loads but as the load is reduced, at some point the circuit operation will change to DCM. CCM is often referred to as the "heavy load" condition and DCM as the "light load" condition.

This distinction turns out to be very important because the behavior of the circuit, for both small signal and large signal, is radically different between the two modes. **Figure 7.32** shows the relationship between the duty cycle (D) of Q1 and the ratio of the output voltage to the input voltage ($M = V_O/V_S$) as a function of τ_L in Figure 7.32:

$$\tau_L = \left(\frac{L}{R}\right) f_S \qquad (12)$$

τ_L is just a convenient way to make Figure 7.32 more general by tying together the variables L, R and f_s. Smaller values for τ_L correspond to lighter loads (larger values for R). As can been seen in Figure 7.32, for very light loads the higher values for D have little effect on the output/input voltage ratio. This is basically charging of the output capacitor (C2) to the peak value of $V_1 \approx V_S$.

For CCM operation, M is a linear function of D. As we vary the load, V_O will remain relatively constant. But when we go into DCM, the relation between D and M is no longer linear and in addition is heavily dependent on the load: V_O will vary as the load is changed unless the duty cycle is varied to compensate. This kind of behavior is typical of all switching regulators, even those not directly related to the buck converter. In fact, we have seen this behavior already in the section on choke input filters in Figure 7.17 where the output voltage is close to the peak input voltage for light loads and decreases as the load increases until a point is reached (I_C) where V_O stabilizes.

In a buck converter the value for the critical inductance (L_C) can be found from:

$$L_c = R\left(\frac{1-M}{2\,f_S}\right) \qquad (13)$$

This looks just like equation 6 (in the earlier section on choke-input filters) with $A = 2fs / (1 - M)$! The two phenomena are the same: peak charging versus averaging of V1.

7.11.2 The Boost Converter

A *boost converter* circuit is shown in **Figure 7.33A**. This circuit is called a "boost" converter because the output voltage is always greater than or equal to the input voltage.

When Q1 is ON (Figure 7.33B), L is connected in parallel with the source (V_S) and energy is stored in it. During this time interval load energy is supplied from C. When Q1 turns OFF (Figure 7.33C), L is connected via D1 to the output and the energy in L, plus additional energy from the source, is discharged into the output. The value for L, V_S and length of time it is charged determines

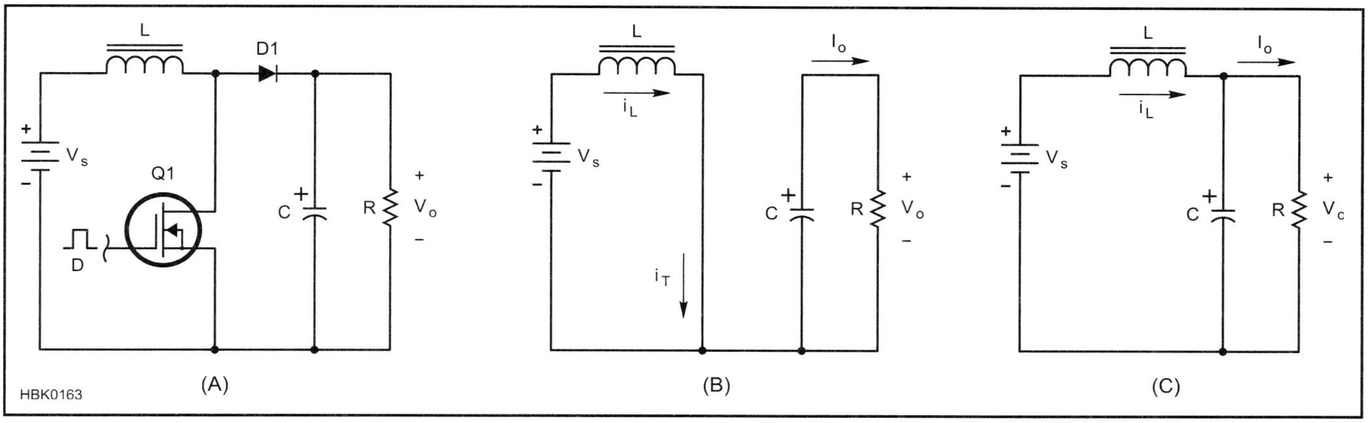

Figure 7.33 — Typical boost converter.

the energy stored in L. Again we have the possibilities that either some (CCM) or all (DCM) the energy in L is discharged during the OFF-time of Q1. The variation in the ratio of the output-to-input voltage (M) with duty cycle is shown in **Figure 7.34**.

As we saw in the buck converter, the conduction mode of L is important. In an ideal converter operating in CCM, the output voltage is substantially independent of load and M ≈ 1 / (1 – D). In realistic boost converters there is an important limit on the CCM value for M. In an ideal boost converter you could make the boost ratio (M) as large as you wish but in real converters the parasitic resistances associated with the components will limit the maximum value of M as indicated by the dashed line for CCM operation. The exact shape of this part of the control function will depend on the ratio of the parasitic resistance within the converter to the load resistance (see Reference 2).

There is also another very important practical effect of this limitation on the peak value for M. When the duty cycle is increased beyond the point of maximum M (this occurs at D = 0.9 in Figure 7.34), the sign of the slope of the control function changes so that the control loop will change from negative feedback to positive feedback. This can cause the converter to latch up under some load conditions if the control circuit allows D to exceed the value at maximum M. In boost converters the design of the control circuit must limit the maximum value for D so that latch-up is not possible although this may be difficult for overload conditions.

As in the case of the buck converter, in DCM the control function for M is very different from what it is in CCM and is strongly dependent on the load R. One advantage of the DCM operation is that the limitation on the maximum value for M because of parasitic circuit resistances is not nearly so pronounced. By operating in DCM it is possible to have M > 5.

An important limitation of the boost converter is that with Q1 turned OFF, you have no control over V_O. V_O will simply equal V_S. In addition, when V_S is turned on there will be an inrush current through D1, into C which cannot be controlled by Q1. In the case of the buck converter, if Q1 is kept OFF when V_S is turned on, there will be no inrush current charging the output filter capacitor (C2). In the buck converter C2 can be charged slowly by ramping up the duty cycle of Q1 during turn-on but this is not possible in the boost converter.

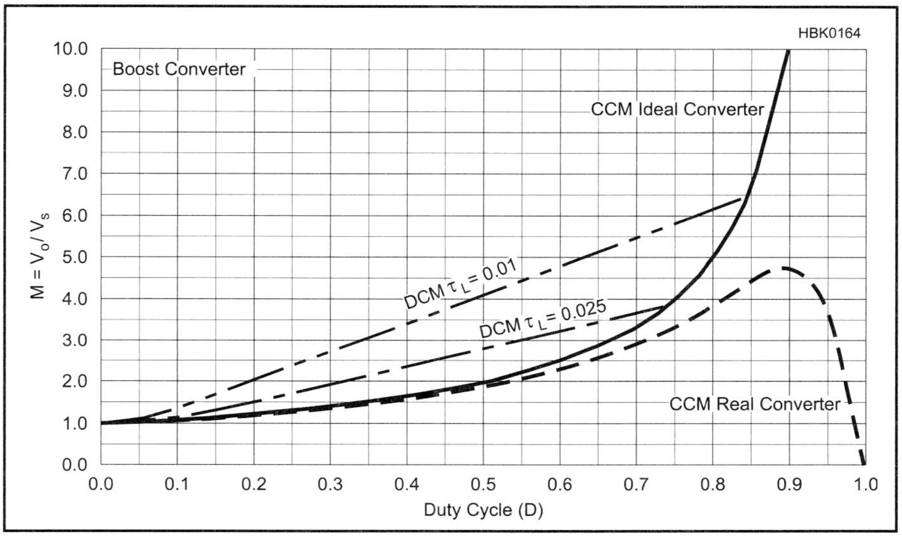

Figure 7.34 — DC control characteristic for a boost converter. M = V_O/V_S, D = duty cycle of Q1 and $\tau_L = f_s (L/R)$.

7.11.3 Buck-Boost and Flyback Converters

Figure 7.35 shows an example of a *buck-boost converter*. The name "buck-boost" comes from the fact that the magnitude of the output voltage can be either greater or less than the input voltage.

When S1 is ON, V_S is applied across L and energy is stored in it. During the ON time of Q1, D1 is reverse-biased and non-conducting. The output voltage (V_O) across the load (R) is supported solely from the energy stored in C. When S1 is OFF, the energy in L is discharged into C through D1. Note that the polarity of V_O is inverted from V_S: *positive V_S means negative V_O*. The relationship between V_O and V_S as a function of duty cycle is shown in **Figure 7.36**.

This graph closely resembles that for the boost converter (Figure 7.34) with one important exception: |M| begins at zero. This allows the magnitude of the output voltage to be either below or above the source voltage, hence the name "buck-boost." (A buck-boost converter construction project, "An Easy to Build Buck-Boost Converter" by N9DK was published in the October 2019 issue of *QRP Quarterly*.)

FLYBACK CONVERTERS

Simple buck-boost converters are occasionally used, but much more often it is the

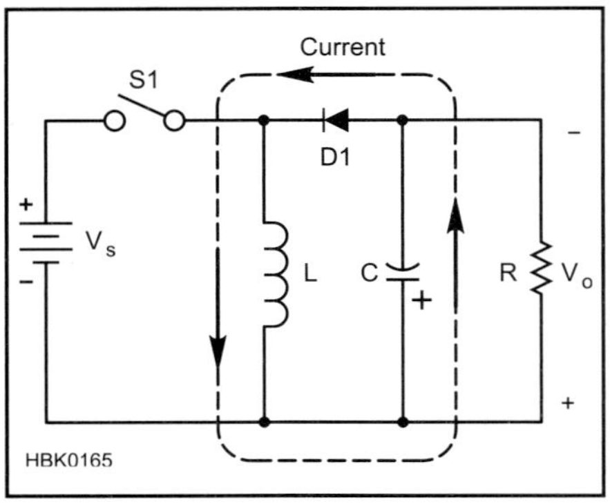

Figure 7.35 — An example of a buck-boost converter.

transformer-coupled version of this converter, referred to as a *flyback converter*, that is employed. The relationship between the flyback and buck-boost converters is illustrated in **Figure 7.37**. In Figure 7.37A we have a standard buck-boost converter, the only change from Figure 7.35 being two parallel and equal windings on the inductor. In Figure 7.37B we remove the links at the top and bottom of the two parallel windings, converting the inductor into a transformer with primary and secondary windings. The only change is that when S1 is ON, current flows in the primary of the transformer and when S1 is OFF, the current flows in the secondary winding delivering the stored energy to the output. At this point the circuit operation is the same except that we have introduced primary-to-secondary galvanic isolation.

We are now free to change the turns ratio from 1:1 to anything we wish. We can also change the polarity of the output voltages and/or add more windings with other voltages and additional loads as shown in Figure 7.37C, a typical example of a flyback converter. These are most often used in the power range of a few watts to perhaps 200 W. For higher power levels other circuits are generally more useful.

The advantages of the flyback converter lie in its simplicity. It requires only one power switch and one diode on each of the output windings. The inductor is also the isolation transformer so you have only one magnetic component. The disadvantage of this circuit is that both in the input and output current waveforms are pulsating. The result is that more filtering is required and the filter capacitors are exposed to high RMS currents relative to the power level.

The dots near the end of the inductors are *phasing dots* that indicate where instantaneous ac waveforms are in-phase. In these switchmode transformers, current flowing into a primary winding at the end with the dot will cause current to flow out of the secondary winding at the end with the dot. Phasing dots are discussed in the **Circuits and Components** chapter section on Transformers.

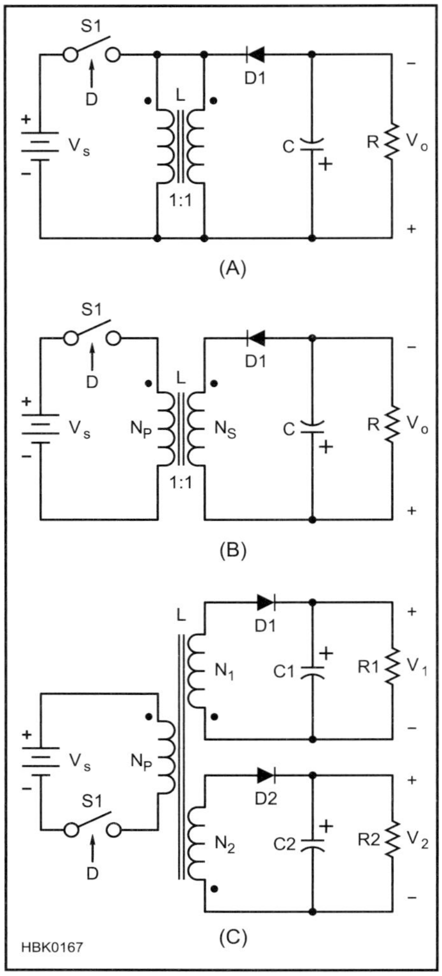

Figure 7.37 — The relationship between buck-boost and flyback converters.

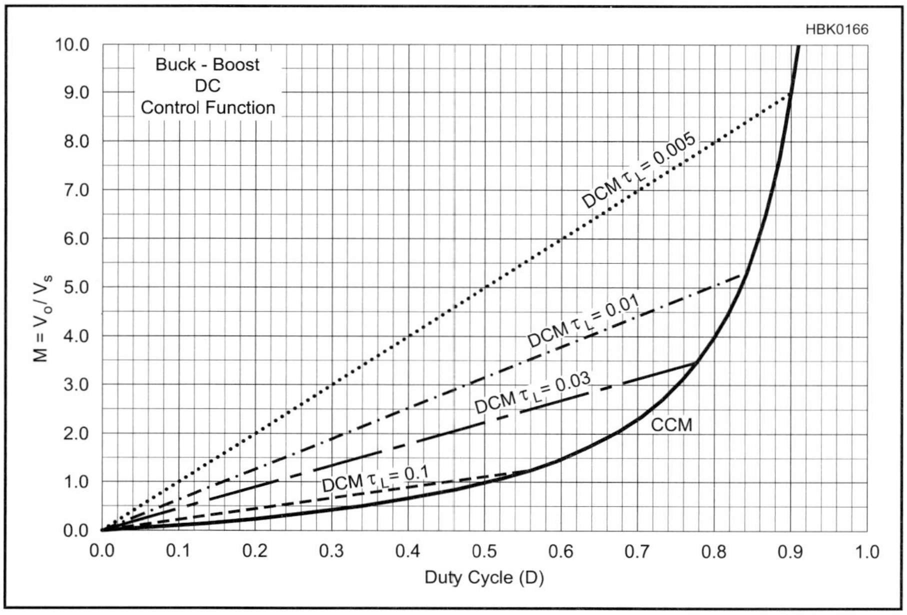

Figure 7.36 — DC control function for an ideal buck-boost converter. $M = V_O/V_S$, D = duty cycle of Q1 and $\tau_L = f_s (L/R)$.

7.11.4 The Forward Converter

The buck converter has many useful properties but it lacks input-to-output galvanic isolation, the ability to produce output voltages higher than the input voltage, and/or multiple isolated output voltages for multiple loads. We can overcome these drawbacks by inserting a transformer between the switch (S1) and the shunt diode (D1). To make this simple idea work however, we also have to add a diode in series with the output of the transformer (D2), and a third winding (N_R) with another diode (D3) to the transformer. This is done to provide a means for resetting the core (returning the magnetic flux to zero) by the end of each switching cycle. The result is the *forward converter* shown in **Figure 7.38**.

The circuit in A is the one just described. The variation in B uses two switches (S1 and S2, which switch ON and OFF *simultaneously*) instead of one but eliminates the need for a reset winding on the transformer. For the circuit in A and a given input voltage, the voltage across the switch (in the OFF state) will be equal to V_S plus the reset voltage during the OFF-time. Typically the peak switch voltage will be about $2V_S$. In the circuit shown in B the switch voltage is limited to V_S which is very helpful at higher line voltages. These circuits behave just like a buck converter except you can now have multiple isolated outputs with arbitrary voltages and polarities. These circuits are typically used in converters with power levels of 100 to 500 W.

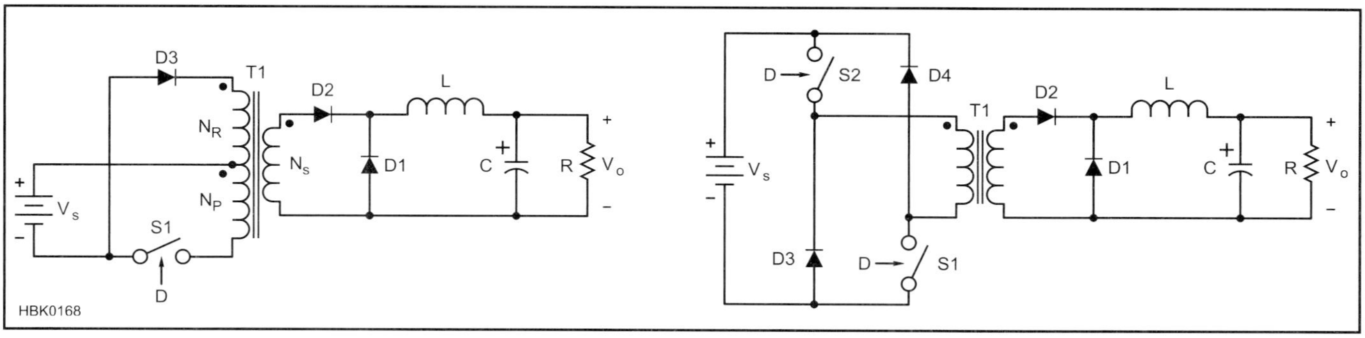

Figure 7.38 — Example of a forward converter.

Figure 7.39 — Example of a parallel quasi-square wave dc-dc

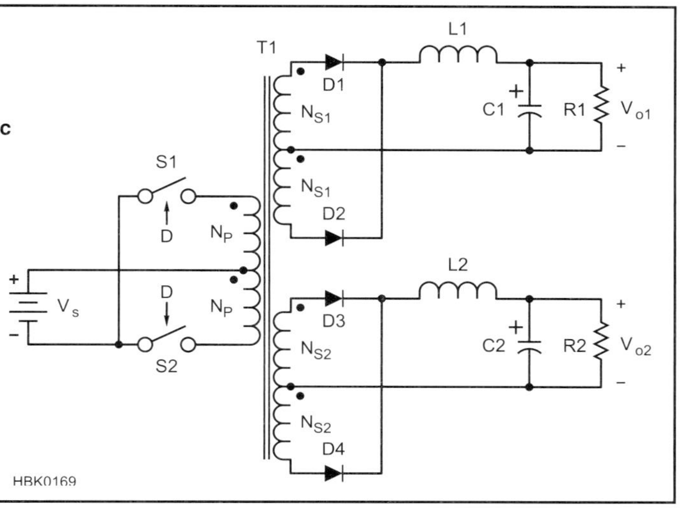

7.11.5 Parallel, Half and Full-Bridge Converters

As the power level increases it becomes advantageous to use more switches in a somewhat more complex circuit. For applications where the input voltage is low (<100V) and the current high, the push-pull *quasi-square wave* circuit shown in **Figure 7.39** is often used.

S1 and S2 are switched on alternately with duty cycles <0.5. The output voltages are controlled by the duty cycle, D. The peak switch currents are equal to the input current but the peak switch voltages will be $2V_S$.

This circuit is still just a buck converter, but with an isolation transformer added that allows multiple outputs with different voltages above or below the input voltage. It would be very common to have a 5 V, high-current output and ±12 V, lower-current outputs, for example. This converter is typically used for operation from vehicular power with loads up to several hundred watts.

Operating directly from rectified ac utility power usually means that V_S will be 200 V or more. For these applications, **Figure 7.40** shows how the switches are configured in either a half- (Figure 7.40A) or full-bridge (Figure 7.40B) circuit.

In A, S1 and S2 switch alternately and are pulse-width modulated (PWM) to control the output. CA and CB are large capacitors that form a voltage divider with $V_S/2$ across each capacitor. The peak switch voltages will be equal to V_S but the switch currents will be $2I_S$. The peak voltage across the primary winding will be $V_S/2$. This circuit would typically be used for off-line applications with output powers of 500 W or so.

In B, S1 and S4 switch simultaneously alternating with S2 and S3 which also switch simultaneously. The output is controlled by PWM. The peak switch voltage will be equal to V_S and the peak switch current close to I_S (the peak value is a little higher due to ripple on the inductor current which is reflected back into the primary winding). The full-bridge circuit is typically used for power levels of 500 W to several kW. C_S is present in the full-bridge circuit to prevent core saturation due to any asymmetry in the primary PWM voltage waveforms. Cs should be a non-polarized capacitor since differences in duty cycle between S1-4 can create dc voltages of either polarity across it.

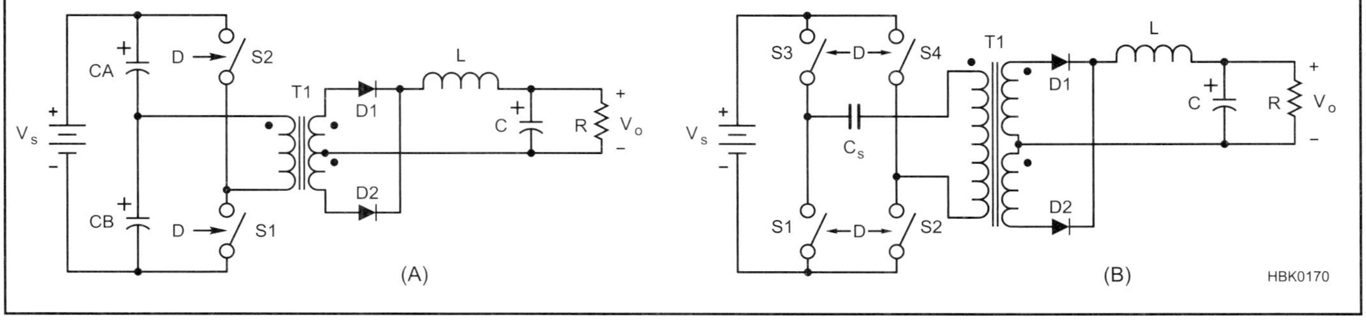

Figure 7.40 — Examples of half and full-bridge quasi-square wave dc-dc converters.

7.11.6 Switchmode Construction

Once a topology has been selected, the next step is to design the circuit. For nearly all applications, the necessary voltage reference, current switch or switch driver, output voltage control, and operating control circuitry are available on a single IC with many products available. Performing the calculations and selecting the energy storage inductor and capacitor can be complex, so the vendors also provide online tools to assist the switchmode circuit designer — see the sidebar "Switchmode Converter Design Aids" for a list of some of the better-known websites. The design tools will help you select the right IC for your application and generate the entire circuit, including component part numbers.

For a low-power converter, the current switch is included on the IC. Only the inductor, filter capacitor, and a few resistors and bypass capacitors are typically added to the converter IC. **Figure 7.41** shows an example of the LM20343 switchmode regulator IC configured as a buck converter with the associated components. For higher output power, ICs are available to drive an external transistor that can handle higher currents. **Figure 7.42** shows the LTC3872-1 IC configured as a boost converter with an external MOSFET switch.

LINE-OPERATED CONVERTERS

Some commercial non-isolated switchmode converters operate directly from the ac line. This type of circuit presents significant safety hazards that must be addressed carefully. Line operation requires special line-rated parts and mitigations against various shock and failure hazards. Commercial units are designed and tested to be completely safe by using a sophisticated process that is not recommended for the casual designer. With so many inexpensive power supplies available, it is easier and safer to buy rather than build, in this case.

PCB LAYOUT OF SWITCHMODE CONVERTERS

An important concern for switchmode converter PCB design is RFI created by switching current on and off. Switching rates from tens of kHz to more than 1 MHz are typical. To minimize switching losses, the rise and fall times of signals driving the current switch and the switched current itself are very short. The resulting switched current waveform includes components into the VHF and UHF range. The RF can be radiated directly from the PCB trace or conducted onto input and output connections. Careful attention to PCB layout minimizes the amount of time and expense spent on filtering, shielding, testing, and even re-laying out the PCB.

To minimize RFI from the converter, special attention must be paid to how the circuit traces and ground planes are laid out to minimize the area of *hot loops* that carry significant current and will radiate signals. Each hot loop acts as a small loop antenna, so minimizing loop area will also minimize radiation. Do not use point-to-point or breadboard construction for switchmode circuits if minimizing RFI is important to the application.

Figure 7.43 shows a simple converter cir-

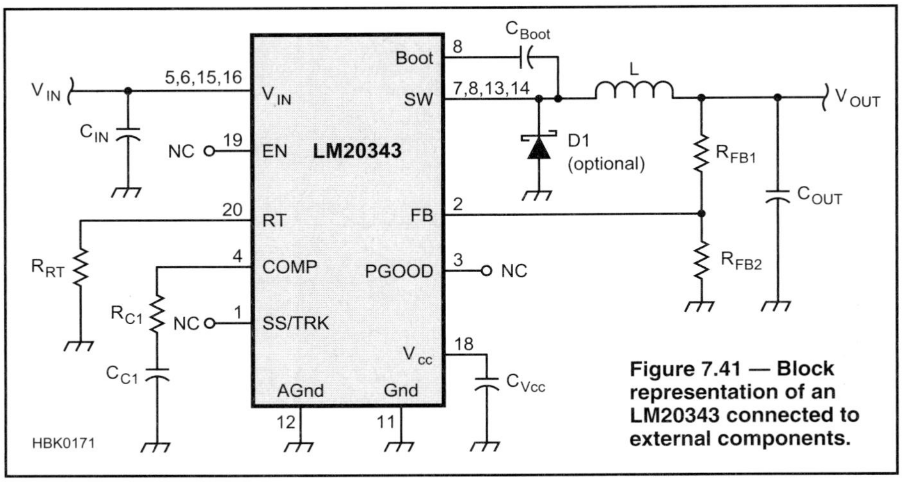

Figure 7.41 — Block representation of an LM20343 connected to external components.

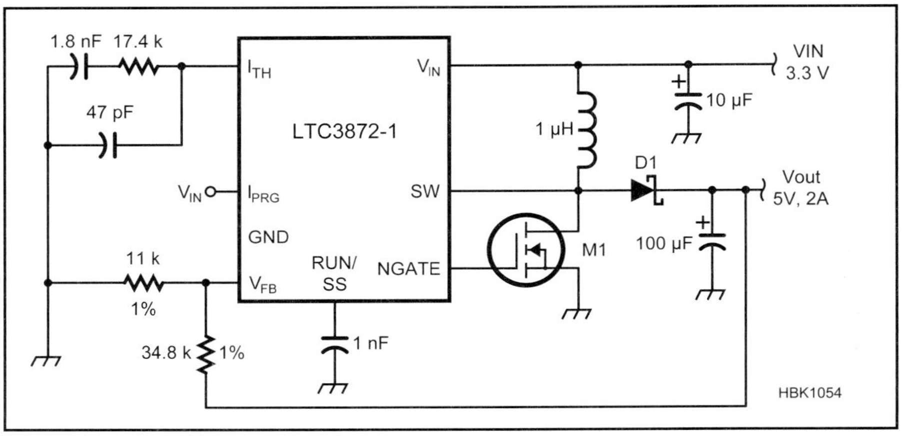

Figure 7.42 — The LTC3872-1 switchmode boost converter circuit showing the external current switch MOSFET (M1) and associated control and filter components. (After Linear Technology LTC3872-1 data sheet)

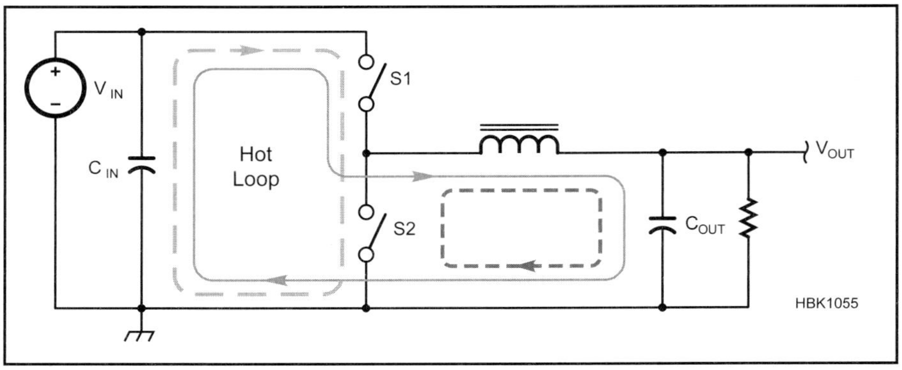

Figure 7.43 — Hot loops in a switchmode converter act as small loop antennas, radiating RF from the rapidly changing currents that becomes RFI. (After Linear Technology AN139)

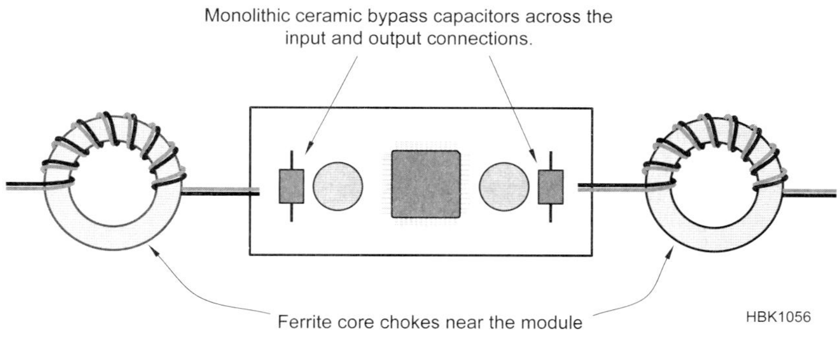

Figure 7.44 — Ferrite core chokes and bypass capacitors at the input and output of switchmode power supply PCBs are used to block common-mode RF on the wiring where it will radiate noise over a wide spectrum.

cuit that includes three hot loops: (1) through the switches and the input capacitor, CIN; (2) through S1–inductor–COUT–CIN; and (3) through S2–inductor–COUT. RF components are present in all three hot loops. By using wide traces, large ground planes, and keeping signal and current paths as short as possible, the production and radiation of RF is minimized. Techniques for laying out PCB switchmode converters are discussed in the Linear Technology application note, AN139, "Power Supply Layout and EMI" which is available online at **www.analog.com/media/en/technical-documentation/application-notes/an139f.pdf**.

RFI FROM SWITCHMODE CONVERTERS

Along with minimizing hot loop area, switchmode converter builders must also minimize the amount of RF that is conducted and radiated by the input and output wiring. RF into and out of the PCB can be blocked with a series inductor (10 to 100 μH values are typical) and filtered out with 0.1 to 1 nF bypass capacitors. If the supply is floating or not shielded by a metal enclosure, each lead into and out of the supply must be filtered. If the supply is in a metal enclosure, the input and output leads can be bypassed to the enclosure at the entry point. Due to the long circuit path through the enclosure, it is not recommended that the enclosure be used as circuit ground or circuit common. Use a direct ground-lead connection to the PCB from the enclosure entry point.

Another technique is to wind the input and output leads on a ferrite core to create an RF choke. Use a core that is designed for the frequency range involved — Type 31 for HF and upper MF; Type 43 for upper HF and VHF/UHF. The core should be as close to the converter PCB as possible as shown in **Figure 7.44**. If only one lead is wound on the choke, be sure the core will not saturate at maximum current. Saturation concerns are minimized if both the output and return wires are wound on the same core. Combining ferrite chokes and bypass capacitors is a good combination.

If filtering at the input and output is unsuccessful at reducing RFI, the design of the converter circuit and PCB should be inspected. Using a small RF "sniffer" loop (see the **RF Interference and Electromagnetic Compatibility** chapter) and an oscilloscope may allow you to isolate the source of the RFI to one or more of the hot loops. Correcting a PCB layout is rarely feasible, so you can either re-layout the board if it is your design or adjust the circuit design. Increasing the switching time (if possible) will reduce the amount of RF energy generated but will also increase power dissipation in the switch.

SWITCHMODE CONVERTER MODULES

The least expensive way of utilizing switchmode converters may be to purchase individual modules which are available at very low cost. Search online for "dc-dc converter module" and you will find a wide variety of options. Be sure you review the specifications carefully, including isolation (if needed), input voltage, whether the output voltage is adjustable, mounting options, etc. DC-DC converter modules are available from most component distributors and hobby electronics companies.

The more inexpensive converters will probably not have a lot of filtering for RFI, and you will have to provide that yourself with additional components. Converters that meet EU standards should be fairly clean, but it is prudent to assume they will produce some RF and plan to add filtering. Modules operated near their maximum power limits may also require heat sinking or additional air flow to keep from overheating.

7.11.7 Switchmode Converter Design Aids

Today, there are hundreds of ICs on the market, from dozens of manufacturers, aimed at controlling the power conversion process. These devices have grown from their simplest forms when introduced over 20 years ago to include a long list of ancillary functions aimed at reducing the supporting circuitry and enhancing the quality and efficiency of the power conversion process. As a result, the task of designing these ICs into the final power conversion circuit has actually become more complex. To the circuit designer who is below the expert level, the task can seem daunting, indeed.

In answer to this problem, many IC manufacturers provide computer-aided design tools that greatly simplify the task of using their products. These tools take several forms, from the most rudimentary cookbook-style guides, to full-fledged circuit simulation tools like *SPICE*. The purpose of these tools is to help the designer pick appropriate resistors, capacitors and magnetic components for the design, and also to estimate the stresses on the power-handling components.

These design tools fall generally into two types: equation-based design tools and iterative circuit simulators. The differences are that the equation-based tools simply automate the basic design equations of the circuit, provid-

Table 7.1
Switchmode Converter Design Aids

Vendor	*Main Web Site*	*Design Tools*
Analog Devices	**www.analog.com**	**www.analog.com/en/design-center.html**
Infineon	**www.infineon.com**	Look under the Tools menu
ON Semiconductor	**www.onsemi.com**	**www.onsemi.com/support/design-resources/tools**
Renesas	**www.renesas.com**	Look under the Design & Support menu
Texas Instruments	**www.ti.com**	**www.ti.com/power-management/overview.html**

ing recommended choices for the components and then solving equations to compute the voltage, current and power stresses on the components. In some cases, thermal analysis is also included.

The simulators, on the other hand, use detailed mathematical models for the components and provide both dc and ac simulation of the circuits. This allows the user to see the dynamic behavior of the circuit. The simulation includes details during the switching intervals and shows rise times, parasitic effects caused by component capacitance, internal resistance and other characteristics.

In modern power converter design, the magnetic components — transformers, power inductors and filter inductors — can be a challenge to the designer not versed in magnetic design. As a result, the vendors of these parts usually furnish design tools created specifically to the design of these components and/or the proper choice of off-the-shelf versions.

The reader is encouraged to explore design aids available from the websites of manufacturers (**Table 7.1**). The following is far from complete, but should give the reader valuable insight into the vast array of available tools. Remember also that manufacturers of passive components such as capacitors, resistors, heat sinks and thermal hardware may also have helpful and informative design aids on their websites. — *Chuck Mullett, KR6R*

7.12 High-Voltage Techniques

The construction of high-voltage supplies requires special considerations in addition to the normal design and construction practices used for lower-voltage supplies. In general, the builder needs to remember that physical spacing between leads, connections, parts and the chassis must be sufficient to prevent arcing. Also, the series connection of components such as capacitor and resistor strings needs to be done with consideration for the distribution of voltage stresses across the components. High-voltages can constitute a safety hazard and great care must be taken to limit physical access to components and wiring while high potentials are present.

7.12.1 High-Voltage Capacitors

For reasons of economy and availability, electrolytic capacitors are frequently used for output filter capacitors. Because these capacitors have relatively low voltage ratings (<600 V), in HV applications it will usually be necessary to connect them in series strings to form an equivalent capacitor with the capability to withstand the higher applied voltage. Electrolytic capacitors have relatively high leakage currents (low leakage resistance) especially at higher temperatures.

To keep the voltages across the capacitors in the series string relatively constant, equal-value bypassing resistors are connected across each capacitor. These *equalizing resistors* should have a value low enough to equalize differences in capacitor leakage resistance between the capacitors, while high enough not to dissipate excessive power. Also, capacitor bodies need to be insulated from the chassis and from each other by mounting them on insulating panels, thereby preventing arcing to the chassis or other capacitors in the string. The insulated mounting for the capacitors is often plastic or other insulating material in board form. A typical example from a commercial amplifier that implements some the guidelines given in this section is shown in **Figure 7.45**.

Equalizing resistors are needed because of differences in dc leakage current between different capacitors in the series string. The data sheet for an electrolytic capacitor will usually give the dc leakage current at 20 °C in the form of an equation. For example:

$$I_{leakage} = 3\sqrt{CV} \tag{14}$$

where
C = the value in μF,
V = the working voltage, and
$I_{leakage}$ = leakage current in μA.

Keep in mind that the leakage current will increase as the capacitor temperature rises above 20 °C. A value of 3 to 4 times the 20 °C value would not be unusual because of normal heating.

We can use an approximation which includes allowance for heating to determine the maximum value of the divider resistors:

$$R \le \frac{V_r - V_m}{I_{leakage}} \tag{15}$$

where
V_r = the voltage rating of the capacitor, and
V_m = the voltage across the capacitor during normal operation.

A typical 3,000 V power supply might use eight 330 μF, 450 V capacitors in series. In that case, V_r = 450 V, V_m = 375 V and $I_{leakage}$ = 1.06 mA. This would make R = 68 kΩ (standard value) or a total of 544 kΩ for the resistor string. With 3,000 V across the resistor string, the total power dissipation would be 16.5 W or about 2.06 W per resistor. To be conservative you should use resistors rated for at least 4 W each.

The equalizing resistors may dissipate significant power and become quite warm. It is important that these resistors do not heat the capacitors they are associated with, as this will increase the capacitor leakage current. You also have to be careful that the heat from the resistors is not trapped under the plastic support panels. The best practice is to place the resistors above the capacitors and their

Figure 7.45 — Example of the filter capacitor/bleeder resistor installation in a high voltage power supply.

mounting structure allowing the heat to rise unobstructed as shown in Figure 7.45.

OIL-FILLED CAPACITORS

For high voltages, oil-filled paper-dielectric capacitors are superior to electrolytics because they have lower internal impedance at high frequencies, higher leakage resistance and are available with much higher working voltages. These capacitors are available with values of several microfarads and have working voltage ratings of thousands of volts. On the other hand, they can be expensive, heavy and bulky.

Oil-filled capacitors are frequently offered for sale at flea markets at attractive prices. One caution: It is best to avoid older oil-filled capacitors because they may contain polychlorinated biphenyls (PCBs), a known cancer-causing agent. Newer capacitors have eliminated PCBs and have a notice on the case to that effect. Older oil-filled capacitors should be examined carefully for any signs of leakage. Contact with leaking oil should be avoided, with careful washing of the hands after handling. Do not dispose of any oil-filled capacitors with household trash, particularly older units. Contact your local recycling agency for information about how to dispose of them properly.

High Voltage Safety Considerations

We are all so besieged these days with verbose safety warnings on mostly harmless consumer goods that it is easy to forget that some things really are dangerous. High voltage power supplies definitely fall into this category, especially since many amateurs are accustomed to solid-state circuits and seldom encounter any dc voltage higher than 12 V. ***This power supply produces voltages that are highly lethal.*** So please take to heart the following 10 precautions. Furthermore, don't expect to learn from your mistakes, because if you don't exercise proper precautions the first time, you're unlikely ever to have a second chance.

1. Don't let your reach exceed your grasp. High voltage power supplies are not a project for beginners. You should not attempt to build this type of power supply unless you're a seasoned builder who has experience with high voltage circuitry.
2. Young amateurs should not attempt this type of project alone. Working with high voltages requires the maturity and patience that comes with age and experience.
3. Never work around high voltage when you are tired, stressed, or in a hurry.
4. Never work around high voltage after drinking alcohol. Even one beer or glass of wine can impair your judgment and make you careless.
5. Before working on a high voltage power supply, always follow these three steps: Unplug (the ac power cord), discharge (the filter capacitors) and verify (that the output voltage is truly zero). Time-honored practice is to use a "chicken stick" (a wooden dowel or PVC tube, with one end attached to a grounded wire) to make sure filter capacitors are completely discharged. See Figure 7.47 for an example.
6. When working on a high voltage power supply, remember that a dangerous time is after the power supply has just been turned off, but before the filter capacitors have fully discharged. A 50 µF capacitor charged to 4,000 V holds a potentially deadly 400 joules of energy. Even with bleeder resistors, it can take a minute or more to discharge fully.
7. When removing a recently discharged filter capacitor from a power supply, tie the two terminals together with wire. Large high voltage capacitors can self-charge to dangerous levels if the terminals are left floating.
8. Don't stake your life on the expectation that bleeder resistors, fuses, circuit breakers, relays, and switches are always going to do their job. Even though modern components are very reliable, it is safe practice always to assume the worst.
9. Don't build a high voltage power supply if you don't understand how the circuit works. High power amplifiers and power supplies are not "plug-and-play" projects with step-by-step instructions. Builders must be knowledgeable enough to improvise, make component substitutions, and implement design changes.
10. With high voltage projects, it doesn't pay to be "penny wise and pound foolish." Use high quality components throughout and save your forty-year-old junk box parts for projects where safety and reliability are not paramount requirements.

(The preceding guidelines are taken from the *QEX* article "A Deluxe High Voltage Supply" by Jim Garland, W8ZR, which is included with this book's online content.)

7.12.2 High-Voltage Bleeder Resistors

Bleeder resistors across the output are used to discharge the stored energy in the filter capacitors when the power supply is turned off and should be given careful consideration. These resistors provide protection against shock when the power supply is turned off and dangerous wiring is exposed. A general rule is that the bleeder should be designed to reduce the output voltage to 30 V or less within 30 seconds of turning off the power supply.

Take care to ensure that the maximum voltage rating of the resistor is not exceeded. In a typical divider string, the resistor values are high enough that the voltage across the resistor is not dissipation limited. The voltage limit is typically related to the insulation intrinsic to the resistor. Resistor maximum voltage ratings are usually given in the manufacturer's data sheet and can be found online. Two major resistor manufacturers are Ohmite Electronics (**www.ohmite.com**) and Stackpole Electronics (**www.seielect.com**).

A 2 W carbon composition resistor will have a maximum voltage rating of 500 V. As a rough estimate, larger wire-wound power resistors are typically rated at 500 V_{RMS} per inch of length — but check with the manufacturer to be sure.

The bleeder will consist of several resistors in series. Typically wire-wound power resistors are used for this application. One additional recommendation is that two separate (redundant) bleeder strings be used, to provide safety in the event one of the strings fails. When electrolytic capacitors are used, the equalizing resistors can also serve as the bleeder resistor but they should be redundant. Again, give careful attention to keeping the heat from the resistors away from the capacitors as shown in Figure 7.45.

In the example given above for calculating the equalizing resistor value, eight 330 µF capacitors in series created the equivalent of a 41 µF capacitor with a 3,000 V rating. The total resistance across the capacitors was 8 × 63 kΩ = 504 kΩ. This gives us a time constant (R × C) of about R × C ≈ 21 seconds. To discharge the capacitors to 30 V from 3,000 V would take about four time constants (about 84 seconds) which is well over a minute. To get the discharge time down to 30 seconds would require reducing the equalizing resistors to 25 kΩ each. The bleeder power dissipation would then be:

$$P = \frac{V^2}{R} = \frac{3{,}000^2}{200{,}000} = 45\,W \tag{16}$$

For a 2 or 3 kV power supply, this is a reasonable value but it is still a significant amount of power and you need to make sure the resulting heat is properly managed.

7.12.3 High-Voltage Metering Techniques

Special considerations should be observed for metering of high-voltage supplies, such as the plate supplies for linear amplifiers. This is to provide safety to both personnel and to the meters themselves.

To monitor the current, it is customary to place the ammeter in the supply return (ground) line. This ensures that both meter terminals are close to ground potential. Placing the meter in the positive output line creates a hazard because the voltage on each meter terminal would be near the full high-voltage potential. Also, there is the strong possibility that an arc could occur between the wiring and coils inside the meter and the chassis of the amplifier or power supply itself. This hazardous potential cannot exist with the meter in the negative leg.

Another good safety practice is to place a low-voltage Zener diode across the terminals of the ammeter. This will bypass the meter in the event of an internal open circuit in the meter. A 1 W Zener diode will suffice since the current in the metering circuit is low.

The chapter on **RF Power Amplifiers** contains examples of how to perform current and voltage metering in high voltage supplies and amplifiers. In the past, amplifier articles have shown the meters mounted on plastic boards with stand-off insulators behind a plastic window in the amplifier or power supply front panel. While this can work it is not considered good practice today. It is usually possible to arrange the metering so that the meters are close to ground potential and may be safely mounted on the front panel of either the amplifier or the power supply.

For metering of high voltage, the builder should remember that resistors to be used in multiplier strings will often have voltage-breakdown ratings well below the total voltage being sampled. Usually, several identical series resistors will be used to reduce voltage stress across individual resistors. A basic rule of thumb is that these resistors should be limited to a maximum of 200 V, unless rated otherwise. For example, in a 2,000 V power supply, the voltmeter multiplier should have a string of 10 resistors connected in series to distribute the voltage equally. The comments on resistor voltage rating in the sections on capacitor equalization and bleeder resistors apply to this case, as well.

7.12.4 High-Voltage Transformers and Inductors

Usable transformers and filter inductors can often be found at amateur flea markets but frequently these are very old units, often dating from WWII. The hermetically-sealed military components are likely to still be good, especially if they have not been used. Open-frame units, with insulation that has been directly exposed to the atmosphere and moisture for long periods of time, should be considered suspect. These units can be checked by running what is called a "hipot test" (high potential test). This involves a low current, variable output voltage power supply with a high-value resistor in series with the output. Unfortunately this equipment is seldom available to amateurs. Some motor repair shops will have an insulation tester called a "Megger" and may be willing to perform an insulation test on a transformer or inductor for you. This is not a completely definitive test but will certainly detect any gross problems with the insulation.

An alternative would be to perform the transformer tests discussed earlier with a variable autotransformer on the ac input. Slowly increase the input voltage until full line voltage is reached and let the transformer run for an hour or two while watching to see if there are any signs of failure. Doing this test in a dark room makes it easier to see any visible corona discharge, another sign of insulation problems. Because the transformer terminals will be at full voltage great care should be taken to avoid contact with the HV terminals. Some form of transparent insulating shield for the test setup is necessary.

7.12.5 Construction Techniques for High-Voltage Supplies

Layout and component arrangement in HV supplies requires some additional care beyond those for lower voltage projects. The photographs in the **RF Power Amplifiers** chapter are a good start, but there are some points which may not be obvious from the pictures.

SHARP EDGES

Sharp points, board edges and/or hardware with ragged edges can lead to localized intensification of the electric field's strength, resulting in a possible breakdown. One common offender is the component leads on the soldered side of a printed-circuit board. These are usually soldered and then cut off leaving a small but often very sharp point. The best way to handle this is to cut the lead as close to the board as practical and form a small solder ball or mound around the cutoff lead end. An example of these suggestions is given in **Figure 7.46**. The protruding component wire near the top would have a high field gradient. Below that we see a wire cut short with a rounded mound of solder covering the end of the wire.

The ends of bolts with sharp threads often protrude beyond the associated nut. One way to eliminate this is to use the dome-style nuts (cap nuts) that capture the end of the bolt or screw, forming a nicely rounded surface. An example of a cap nut is shown in Figure 7.46.

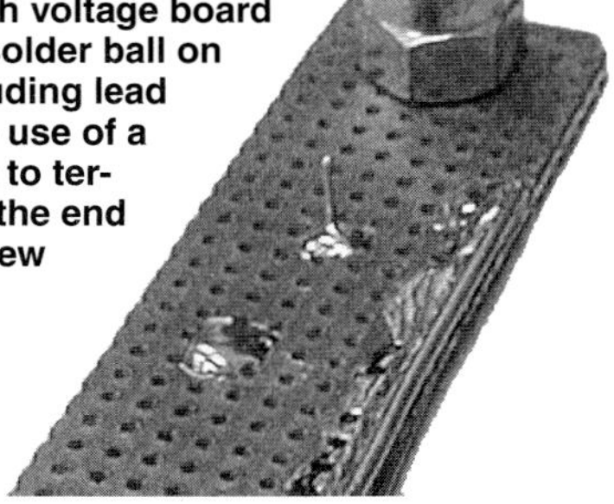

Figure 7.46 — Example of a high voltage board with a solder ball on a protruding lead and the use of a cap nut to terminate the end of a screw thread.

Sheet metal screws, with their needle-sharp ends, should be avoided if possible. Sheet metal screws used to close metal housings at high potential should have the screw tips inside the enclosure, which would be normal in most cases. You must also be careful of the tips of sheet metal screws that protrude on the inside of an outer enclosure. If these are in proximity to circuits at high potential, they can also lead to arcing. Keeping sheet metal screw tips well away from high voltage circuitry is the best defense.

Small pieces of sheet metal that may be part of a structure at high potential should have their edges rounded off with a file. Copper traces near the edges of circuit boards do not benefit from rounding, however. In fact, filing may make the edge sharper, ragged and more prone to breakdown. In critical areas, a small solid round wire can be soldered to the edges of a copper trace to form a rounded edge as illustrated at the right side of Figure 7.46.

INSULATORS

Some portions of the circuit may be mounted on insulators or plastic sheets. A new, clean insulator should easily withstand 10 kV per inch without creepage or breakdown across the insulating surface, but over time that surface will accumulate dirt and dust. Reducing the high-voltage stress across insulators to 5 kV per inch would be more conservative.

In theory the spacing between two smooth surfaces across an air gap can be much smaller, perhaps 20 to 30 kV per inch or even higher. But given the uncertainties of layout and voltage gradients around hardware, wiring, components and board edges, sticking with 5 kV per inch is good idea even for air gaps. This separation will seldom cause construction layout problems unless you are trying to build a very compact unit.

WIRE

Be sure to use wire that is rated for high voltage and for high temperatures. Pomona Electronics type 6733 test lead wire is stranded #18 AWG, rated for 10 kV and 150 °C with silicone insulation, available in both red and black. Flexible and durable, it is easy to route around a chassis or to fit inside sleeving with other wires. The outer diameter (0.144 inch) is thinner and more flexible than spark plug wire or most HV cable.

FUSES

Sometimes a fuse will be placed in series with a high voltage output to provide protection in case of load arcing. These fuses pose special problems. When a fuse blows, the fuse element will at least melt and perhaps even vaporize. There may be an interval when most, if not all, the output voltage appears across the fuse but the fuse has not stopped

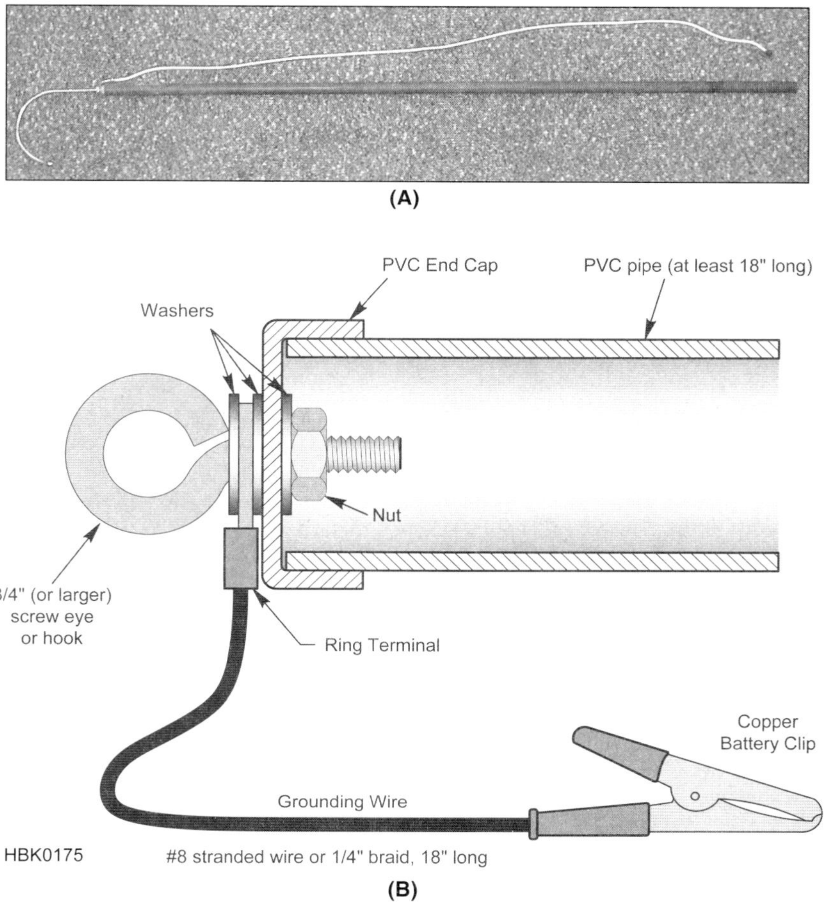

Figure 7.47 — Example of a grounding stick or hook to discharge capacitor energy safely. The end of the wire is connected to ground and the hook is touched to the capacitor terminals to discharge them. A diagram showing how to construct a grounding stick is shown at B.

conducting. That is, there can be a sustained arc in the fuse.

Fuses have voltage ratings that are the maximum voltages across the fuse for which the arc can be expected to quench quickly. The standard 0.25 × 1.25 inch fuses used in the input ac line are typically rated for 250 V. While no doubt these ratings are conservative, this type of fuse cannot be expected to reliably clear an arc with 2 to 3 kV across the fuse and should not be used in series with a high-voltage output.

Your first choice for HV fuses should be the recommended part specified by the power supply manufacturer. If those are unavailable or you are building the supply yourself, fuses for microwave oven transformer secondary use are often suitable. Do not substitute lower-voltage fuses or fuse holders not rated for HV service. When working on a HV power supply with the fuse visible and uncovered, be sure to wear safety glasses as the fuse may explode during a high-current overload, scattering shards of glass or ceramic at high speed.

7.12.6 Construction Practices for High-Voltage Safety

The voltages present in HV power supplies are lethal. Every effort must be made to restrict physical access to any high voltages. A number of steps can be taken:

1) Build the power supply within a closed box, preferably a metal one.

2) Install interlocks on removable panels used for access to the interior. An interlock is usually a normally-open microswitch in series with the input power line. The microswitch is positioned so that it can be closed only when the overlying panel has been secured in place. When the panel is removed the switch moves to the open position, removing power from the supply.

3) As noted previously, the use of a bleeder resistor to rapidly discharge the capacitors is mandatory.

4) To further protect the operator when accessing the inside of a high-voltage power supply, a grounding hook like that shown in **Figure 7.47A** should be used to positively discharge each capacitor by touching the hook to each capacitor terminal. The hook is connected to the chassis via the wire shown and held by the insulated handle. It is normal practice to leave the hook in place across the capacitors while the supply is being worked on, just in case primary power is inadvertently applied. When the work is finished, remove the hook and replace the covers on the power supply.

Figure 7.47B shows how to construct a grounding stick of your own. A large screw eye or hook is substituted for the hook. It is important that the grounding wire be left *outside* the handle so that any hazardous voltages or currents will not be present near your hand or body.

7.13 Batteries

A battery is a group of individual *cells*, usually series-connected to give some desired multiple of the *cell voltage*. The cell voltage is usually in the range of 1 to 4 V. Each chemical system used in the cell gives a particular nominal voltage that will vary with temperature and state of charge. This must be taken into account to make up a particular battery voltage. For example, four 1.5 V carbon-zinc cells make a 6 V battery and six 2 V lead-acid cells make a 12 V battery. **Table 7.2** lists the dimensions of commonly used batteries.

The following sections on battery types and usage consist primarily of excerpts from *Batteries in a Portable World* by Isidor Buchmann, CEO of Cadex Electronics (**www.cadex.com**), a leading manufacturer of battery charging and related equipment. The ARRL appreciates having been given permission to use this material. The book discusses the issues summarized here (and many others) in detail. The reader is directed to the original text for more complete information. This summary is intended to present and compare the various options commonly used by amateurs. Additional information and an extensive battery glossary is provided online at **www.batteryuniversity.com**.

Batteries are divided into two categories: *primary* (non-rechargeable) and *secondary* (rechargeable). Secondary batteries are expected to account for more than 80% of global battery use by 2015. The most common types of battery chemistry are lithium, lead and nickel-based systems.

Batteries store energy well and for a considerable length of time. Primary batteries hold more energy than secondary, and their *self-discharge* is lower. Alkaline cells are good for 10 years with minimal losses. Lead, nickel and lithium-based batteries need periodic recharges to compensate for lost power. **Figure 7.48** illustrates the energy and power densities of lead-acid, nickel-cadmium (NiCd), nickel-metal-hydride (NiMH), and the lithium-ion family (Li-ion).

Rather than giving batteries unique names by type, they are broadly distinguished by the following characteristics:

Chemistry — the families of chemicals used to make the battery. The most common chemistries are lead, nickel and lithium.

Voltage — the nominal open circuit voltage (OCV), which varies with chemistry and the number of cells connected in series.

Size — standard sizes of batteries or cells, such as AAA, AA, C and D.

Capacity (C) — the specific energy in ampere-hours (Ah).

Cold Cranking Amps (CCA) — a starter battery's ability to supply high load current at –18 °C (0 °F) as specified by the vehicle standard SAE J357.

Specific energy — battery capacity in watt-hours per kilogram (Wh/kg)

Specific power — indicates the loading capability or the amount of current the battery can provide in watts/kilogram (W/kg)

C-rate — charge and discharge rate specified in units of C (capacity). At 1 C, the battery charges or discharges at a current numerically equal to its Ah rating. For example, a 2,000 mAh battery discharging at 1 C is supplying a current of 2 A. At 0.5 C, the current would be 1 A, and so forth.

Load — whatever draws energy from the battery. Internal battery resistance and depleting the battery's state of charge cause the voltage to drop.

Primary batteries have one of the highest energy densities. One of the most common primary batteries is the *alkaline-manganese*, or alkaline battery. The *carbon-zinc* or *Leclanché battery* is less expensive but holds less energy than the alkaline battery. Although secondary batteries have improved, a regular household alkaline cell provides 50% more

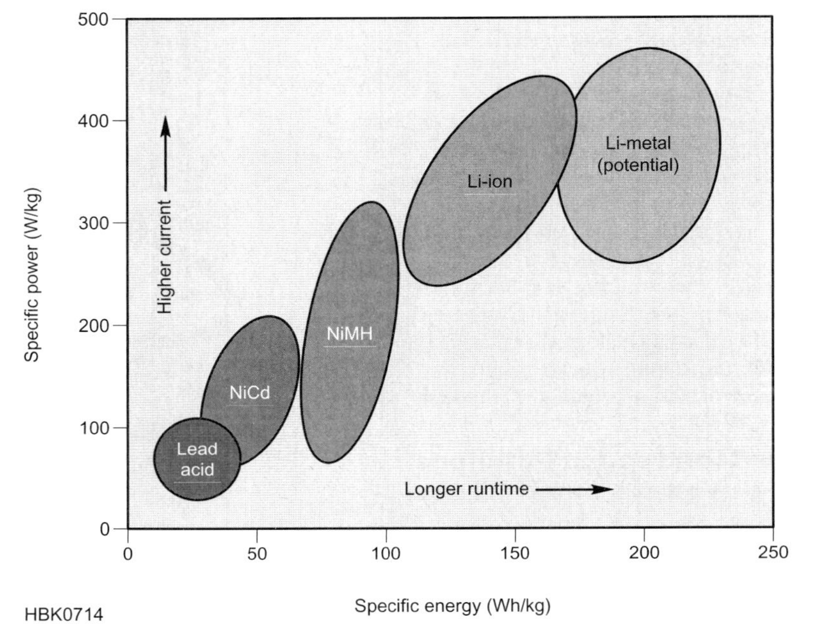

Figure 7.48 — Specific energy and specific power of rechargeable batteries. (Courtesy of Cadex)

Table 7.2
Dimensions of Common Standard Cells

Size	*Dimensions*	*Notes*
D	34 × 61 mm	
C	25.5 × 50 mm	
A	17 × 50 mm	Only available for NiCd; also available in half-length size
AA	14.5 × 50 mm	
AAA	10.5 × 44.5 mm	
AAAA	8.3 × 42.5 mm	Typical size of cell making up 9 V batteries
N	12 × 32 mm	
9 V	48.5 × 26.5 × 17.5 mm	Contains six AAAA cells in series, snap terminals
18650	18 × 65 mm	Commonly used in lithium-ion battery packs
Lantern	115 × 68.2 × 68.2	Spring terminals
CR2016	20 × 1.6 mm	Coin cell
CR2025	20 × 2.5 mm	Coin cell
CR2032	20 × 3.2 mm	Coin cell

Information courtesy of Cadex and from **en.wikipedia.org/wiki/List_of_battery_sizes**

Battery Safety

In addition to the precautions given for each type of battery, the following precautions apply to all battery types. Always follow the manufacturer's advice! Extensive application information can found on manufacturer's websites.

Hydrogen gas escaping from storage batteries can be explosive. Keep flames or any kind of burning material away, including cigarettes, cigars, pipes and so on. Use and charge batteries in well-ventilated areas to prevent hydrogen gas from building up.

No battery should be subjected to unnecessary heat, vibration or physical shock. The battery should be kept clean. Frequent inspection for leaks is a good idea. Electrolyte that has leaked or sprayed from the battery should be cleaned from all surfaces. The electrolyte is chemically active and electrically conductive, and may ruin electrical equipment. Acid may be neutralized with sodium bicarbonate (baking soda), and alkalis (found in NiCd batteries) may be neutralized with a weak acid such as vinegar. Both neutralizers will dissolve in water and should be quickly washed off. Do not let any of the neutralizer enter the battery.

Keep a record of the battery use, and include the last output voltage and, for lead-acid storage batteries, the hydrometer reading. This allows prediction of useful charge remaining, and the recharging or procuring of extra batteries, thus minimizing failure of battery power during an excursion or emergency.

Batteries can contain a number of hazardous materials such as lead, cadmium, mercury or acid, and some thought is needed for their disposal at the end of useful life. Municipal and county waste disposal sites and recycling centers will usually accept lead-acid batteries because they can readily be recycled. Other types of batteries are typically not recycled and should be treated as hazardous waste. Most disposal sites and recycling centers will have occasional special programs for accepting household hazardous waste. Hardware and electronic stores may have battery recycling programs, as well. Take advantage of them.

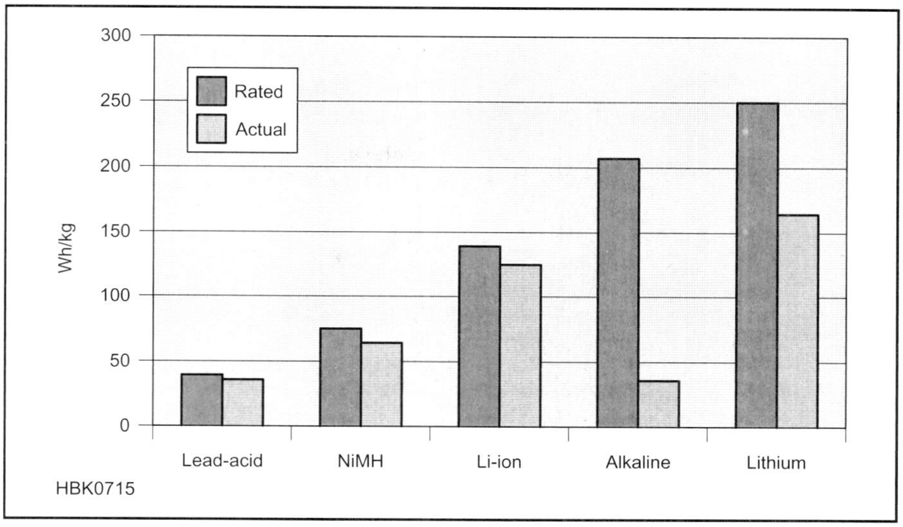

Figure 7.49 — Energy comparison of different battery types at a load of 1 C. (Courtesy of Cadex)

Table 7.3
Alkaline Specifications of Standard Batteries at Low Load

Battery Type	*Nominal Voltage (V)*	*Rated Capacity (mAh)*	*Voltage Cutoff (V)*	*Rated Load (Ω)*	*Discharge C-rate*
9 V	9	570	4.8	620	0.025
AAA	1.5	1150	0.8	75	0.017
AA	1.5	2870	0.8	75	0.007
C	1.5	7800	0.8	39	0.005
D	1.5	17,000	0.8	39	0.0022

(Table courtesy of Cadex)

energy than lithium-ion.

Primary batteries tend to have high internal resistance, which limits the discharge to light loads. This reduces the battery's specific power, even though its specific energy may be quite high. Non-rechargeable lithium metal and alkaline batteries are commonly used in low power applications such as clocks, meter LCDs, keyers and so forth.

Manufacturers of primary batteries specify specific energy at a small fraction of C and the batteries are allowed to discharge to a very low voltage of 0.8 volts per cell. **Figure 7.49** compares primary and secondary batteries discharged at a rate of 1 C. The primary (alkaline and lithium) batteries are unable to deliver their rated specific energy at heavy loads. **Table 7.3** gives typical specifications for alkaline batteries at light loads, such as operating a handheld radio during receive.

If recharging is available, using primary batteries can be expensive — about thirty-fold higher than for secondary batteries. Amateurs using batteries in the field, particularly for emergency communications, may want to maintain the ability to use primary batteries for times when recharging for secondary batteries is not available.

7.13.1 Choices of Secondary Batteries

The following discussion examines today's most popular secondary battery systems according to specific energy, years of service life, load characteristics, safety, price, self-discharge, environmental issues, maintenance and disposal. **Table 7.4** compares the characteristics of four commonly used rechargeable battery systems showing average performance rating at the time of publication in 2013.

The lithium-ion family is divided into three major battery types, so named by their cathode oxides, which are cobalt, manganese and phosphate. The characteristics of these Li-ion systems are as follows:

Lithium-ion-cobalt or **lithium-cobalt** ($LiCoO_2$) — Has high specific energy with moderate load capabilities and modest service life.

Lithium-ion-manganese or **lithium-manganese** ($LiMn_2O_4$) — Is capable of high

Table 7.4
Characteristics of Commonly Used Rechargeable Batteries

Specification	*Lead-Acid*	*NiCd*	*NiMH*	*Li-ion Cobalt*	*Li-ion Manganese*	*Li-ion Phosphate*
Specific energy (Wh/kg)	30-50	45-80	60-120	150-190	100-135	90-120
Internal resistance[1] (mΩ)	<100 12 V pack	100-200 6 V pack	200-300 6 V pack	150-300 7.2 V	25-75[2] per cell	25-50[2] per cell
Cycle life[4] (80% DoD)	200-300	1000[3]	300-500[3]	500-1,000	500-1,000	1,000-2,000
Fast-charge time	8-16 h	1 h typical	2-4 h	2-4 h	1 h or less	1 h or less
Overcharge tolerance	High	Moderate	Low	Low. Cannot tolerate trickle charge	Low. Cannot tolerate trickle charge.	Low. Cannot tolerate trickle charge.
Self-discharge/month (room temp)	5%	20%[5]	30%[5]	<10%[6]	<10%[6]	<10%[6]
Cell voltage (nominal)	2 V	1.2 V[7]	1.2 V[7]	3.6 V[8]	3.8 V[8]	3.3 V
Peak load current (best result)	5C[9] (0.2 C)	20 C (1 C)	5 C (0.5 C)	>3 C (<1 C)	>30 C (<10 C)	>30 C (<10 C)
Operating temp.[10] (discharge only)	–20 to 60 °C	–40 to 60 °C	–20 to 60 °C	–20 to 60 °C	–20 to 60 °C	–20 to 60 °C
Maintenance requirement	3-6 months[11]	30-60 days	60-90 days	Not required	Not required	Not required
Safety requirements	Thermally stable	Thermally stable, fuse protection common	Thermally stable, fuse protection common	Protection circuit mandatory	Protection circuit mandatory	Protection circuit mandatory
In use since	Late 1800s	1950	1990	1991	1996	2006
Toxicity	Very high	Very high	Low	Low	Low	Low

Table Notes:

[1]Internal resistance of a battery pack varies with milliampere-hour (mAh) rating, wiring and number of cells. Protection circuit of lithium-ion adds about 100 mΩ.

[2]Based on 18650 cell size. Cell size and design determines internal resistance.

[3]Cycle life is based on battery receiving regular maintenance.

[4]Cycle life is based on the depth of discharge (DoD). Shallow DoD improves cycle life.

[5] Self-discharge is highest immediately after charge. NiCd loses 10% in the first 24 hours, then declines to 10% every 30 days. High temperature increases self-discharge.

[6]Internal protection circuits typically consume 3% of the stored energy per month

[7]The traditional voltage is 1.25 V; 1.2 V is more commonly used.

[8]Low internal resistance reduces the voltage drop under load and Li-ion is often rated higher than 3.6 V/cell. Cells marked 3.7 V and 3.8 V are fully compatible with 3.6 V.

[9]Capable of high current pulses; needs time to recuperate.

[10]Applies to discharge only; charge temperature is more confined.

[11]Maintenance may be in the form of equalizing or topping charge to prevent sulfation.

(Table courtesy of Cadex)

charge and discharge currents but has low specific energy and modest service life.

Lithium-ion-phosphate or **lithium-phosphate** ($LiFePO_4$) — often abbreviated LiPo.Is similar to lithium-manganese; nominal voltage is 3.3 V/cell; offers long cycle life, has a good safety record but exhibits higher self-discharge than other Li-ion systems.

Another type of lithium-ion cell is the popular *lithium-ion-polymer* or *Li-polymer*. While Li-ion systems get their name from their unique cathode materials, Li-polymer differs by having a distinct architecture in which a gelled electrolyte replaces the usual porous separator between the anode and cathode. The gelled electrolyte acts as a catalyst to enhance the electrical activity of the other battery materials.

7.13.2 Lead-acid Batteries

Most lead-acid batteries used today are *maintenance-free* types in which the liquid electrolyte is sealed inside the battery in liquid or gel form. Two similar types are used: the *sealed lead-acid* (SLA) and the *valve-regulated lead-acid* (VRLA). SLA batteries have capacities up to about 30 Ah. VRLA batteries are larger and have capacities from 30 Ah to several thousand Ah. **Table 7.5** summarizes the advantages and limitations of lead-acid battery systems.

Applying the right voltage limit when charging lead-acid systems is critical and will be a compromise between capacity when recharged and maintaining the battery's internal materials. A low charge limit voltage may shelter the battery but this causes poor performance and a buildup of *sulfation* on the negative plate. A high voltage limit improves performance but it promotes irreversible grid corrosion on the positive plate. Temperature also changes the voltage threshold.

Lead-acid does not lend itself to fast charging and a fully saturated charge requires 14 to 16 hours. The battery must always be stored at full state-of-charge to avoid sulfation. Lead-acid is not subject to memory, but correct charge and float voltages are important to get a long life (see the section on charging below). While NiCd loses approximately 40% of its stored energy in three months, lead-acid self-discharges the same amount in one year.

Lead-acid batteries are inexpensive on cost-per-watt basis but are less suitable for repeated deep cycling. A full discharge causes strain and each discharge/charge cycle permanently robs the battery of a small amount of capacity. The fading becomes more acute as the battery falls below 80% of its nominal capacity. Depending on the depth of discharge and operating temperature, lead-acid for deep-cycle applications provides 200 to 300 discharge/charge cycles.

High temperature reduces the number of available cycles. As a guideline, every 8 °C (13 °F) rise above the optimum temperature of 25 °C (77 °F) cuts battery life in half.

Lead-acid batteries are rated at a 5-hour (0.2 C) and 20-hour (0.05 C) discharge, and the battery performs best when discharged slowly. Lead-acid can deliver high pulse currents of several C if done for only a few seconds, making lead-acid well suited as a *starter battery*.

Table 7.5

Advantages and Limitations of Lead-Acid Batteries

Advantages

Inexpensive and simple to manufacture; lowest cost per watt-hour
Mature and well-understood technology; provides dependable service
Low self-discharge; lowest among rechargeable batteries
High specific power, capable of high discharge currents

Limitations

Low specific energy; poor weight-to-energy ratio
Slow charge; fully saturated charge takes 14 hours
Must always be stored in charged condition
Limited cycle life; repeated deep-cycling reduces battery life
Flooded version requires watering
Not environmentally friendly
Transportation restrictions on the flooded type

(Table courtesy of Cadex)

Table 7.6

Cycle Performance of Starter and Deep-Cycle Batteries

Depth of Discharge	*Starter Battery*	*Deep-cycle Battery*
100%	12-15 cycles	150-200 cycles
50%	100-120 cycles	400-500 cycles
30%	130-150 cycles	1,000 and more cycles

(Table courtesy of Cadex)

Table 7.7

Advantages and Limitations of NiCd Batteries

Advantages	Fast and simple charging even after prolonged storage
	High number of charge/discharge cycles; provides over 1000 charge/discharge cycles with proper maintenance
	Good load performance; rugged and forgiving if abused
	Long shelf life; can be stored in a discharged state
	Simple storage and transportation; not subject to regulatory control
	Good low-temperature performance
	Economically priced; NiCd is the lowest in terms of cost per cycle
	Available in a wide range of sizes and performance options
Limitations	Relatively low specific energy compared with newer systems
	Memory effect; needs periodic full discharges
	Environmentally unfriendly; cadmium is a toxic metal and cannot be disposed of in landfills
	High self-discharge; needs recharging after storage

(Table courtesy of Cadex)

STARTER AND DEEP-CYCLE BATTERIES

The *starter* battery is designed to crank an engine with a momentary high power burst; the *deep-cycle* battery, on the other hand, is built to provide continuous power. The starter battery is made for high peak power and does not tolerate deep cycling well. The deep-cycle battery has a moderate power output but permits cycling.

Starter batteries have a CCA rating in amperes and a very low internal resistance. Deep-cycle batteries are rated in Ah or minutes of runtime. A starter battery cannot be swapped with a deep-cycle battery and vice versa because of their different internal construction. **Table 7.6** compares the typical life of starter and deep-cycle batteries when deep-cycled.

ABSORBENT GLASS MAT (AGM)

AGM is an improved lead-acid battery in which the electrolyte is absorbed in a mat of fine glass fibers. This makes the battery spill-proof. AGM has very low internal resistance, is capable of delivering high currents and offers long service even if *occasionally* deep-cycled. It also stands up well to high and low temperatures and has a low self-discharge. AGM has a higher specific power rating for high load currents and allows faster charge times (up to five times faster) than conventional lead-acid. AGM has a slightly lower specific energy and higher manufacturing cost.

AGM batteries are sensitive to overcharging. They can be charged to 2.40 V/cell without problems but the float charge should be reduced to 2.25 to 2.30 V/cell and summer temperatures may require lower voltages. Automotive charging systems designed for flooded lead-acid often have a fixed float voltage setting of 14.40 V (2.40 V/cell) and a direct replacement with an AGM battery could result in overcharge on a long drive.

Heat can be a problem for AGM and other

gelled electrolyte batteries. Manufacturers recommend halting charge if the battery core reaches 49 °C (120 °F). However, AGM batteries can sit in storage for long periods before a recharge to prevent sulfation becomes necessary.

7.13.3 Nickel-based Batteries

NICKEL-CADMIUM (NiCd)

The standard NiCd remains one of the most rugged and forgiving batteries but needs proper care to attain longevity. Nickel-based batteries also have a flat discharge curve that ranges from 1.25 to 1.0 V/cell. **Table 7.7** summarizes the advantages and limitations of NiCd battery systems.

NICKEL-METAL-HYDRIDE (NiMH)

NiMH provides 40% higher specific energy than standard NiCd, but the decisive advantage is the absence of toxic metals. NiMH also has two major advantages over Li-ion: price and safety. NiMH is offered in AA and AAA sizes.

NiMH is not without drawbacks. It has a lower specific energy than Li-ion and also has high self-discharge, losing about 20% of its capacity within the first 24 hours, and 10% per month thereafter. New types of NiMH such as the Eneloop from Sanyo, ReCyko by GP and others have reduced the self-discharge by a factor of six, increasing storage life by the same amount at the sacrifice of some capacity. **Table 7.8** summarizes the advantages and limitations of NiMH battery systems.

7.13.4 Lithium-based Batteries

The specific energy of Li-ion is twice that of NiCd and the high nominal cell voltage of 3.60 V as compared to 1.20 V for nickel systems contributes to this gain. The load characteristics are good, and the flat discharge curve offers effective utilization of the stored energy in a desirable voltage spectrum of 3.70 to 2.80 V/cell. Li-ion batteries vary in performance according to the choice of cathode materials. **Table 7.9** presents the characteristics of common Li-ion battery chemistries and **Table 7.10** summarizes the advantages and limitations of lithium-ion battery systems.

Lithium-polymer employs a gelled electrolyte in a micro-porous separator between the anode and cathode. Li-polymer is a construction technique and not a type of battery chemistry, so it can be applied to any of the lithium battery chemistries. Most Li-polymer battery packs are based on Li-cobalt. Charge and discharge characteristics of Li-polymer are identical to other Li-ion systems and this chemistry does not require a special charger.

Figure 7.50 compares the specific energy of lead, nickel and lithium-based systems. While Li-cobalt is the clear winner in terms of higher capacity than other systems, this only applies to specific energy. In terms of specific power and thermal stability, Li-manganese and Li-phosphate are superior.

7.13.5 Charging Methods

The performance and longevity of rechargeable batteries are to a large extent governed by the quality of the charger. Choosing a quality charger is important considering the costs of battery replacement and poor performance.

CHARGERS

This discussion focuses on third-party chargers designed to charge and maintain individual cells or batteries. *Fleet chargers* designed for maintaining many batteries of a common type in a company or agency environment are often provided by the equipment

Table 7.8
Advantages and Limitation of NiMH Batteries

Advantages	30-40% higher capacity than a standard NiCd
	Less prone to memory than NiCd
	Simple storage and transportation; not subject to regulatory control
	Environmentally friendly; contains only mild toxins
	Nickel content makes recycling profitable
Limitations	Limited service life; deep discharge reduces service life
	Requires complex charge algorithm
	Does no absorb overcharge well; trickle charge must be kept low
	Generates heat during fast-charge and high-load discharge
	High self-discharge; chemical additives reduce self-discharge at the expense of capacity
	Performance degrades if stored at elevated temperatures; should be stored in a cool place at about 40% state-of-charge

(Table courtesy of Cadex)

Table 7.9
Characteristics of the Four Most Commonly Used Lithium-ion Batteries

Specifications	*Li-cobalt* $LiCoO_2$ *(LCO)*	*Li-manganese* $LiMn_2O_4$ *(LMO)*	*Li-phosphate* $LiFePO_4$ *(LFP)*	*NMC*[1] $LiNiMnCoO_2$
Voltage (V)	3.60	3.80	3.30	3.60/3.70
Charge limit (V)	4.20	4.20	3.60	4.20
Cycle life[2]	500-1,000	500-1,000	1,000-2,000	1,000-2,000
Operating temperature	Average	Average	Good	Good
Specific energy (Wh/kg)	150-190	100-135	90-120	140-180
Specific Power (C)	1	10, 40 pulse	35 continuous	10
Safety	Average[3]	Average[3]	Very safe[4]	Safer than Li-cobalt[4]
Thermal runaway[5] (°C/°F)	150/302	250/482	270/518	210/410
Cost	Raw material high	30% less than cobalt	High	High
In use since	1994	1996	1999	2003

Table Notes
[1] NMC (nickel-manganese-cobalt), NCM, CMN, CNM, MNC and MCN are basically the same. The order of Ni, Mn and Co does not matter much
[2] Application and environment govern cycle life; the numbers do not always apply correctly
[3] Requires protection circuit and cell balancing of multi cell pack. Requirements for small formats with 1 or 2 cells can be relaxed
[4] Needs cell balancing and voltage protection
[5] A fully charged battery raises the thermal runaway temperature, a partial charge lowers it

(Table courtesy of Cadex)

Table 7.10

Advantages and Limitations of Li-ion Batteries

Advantages	High specific energy Relatively low self-discharge; less than half that of NiCd and NiMH Low maintenance. No periodic discharge is needed; no memory
Limitations	Requires protection circuit to limit voltage and current Subject to aging, even if not in use (life span is similar to other chemistries) Transportation regulations apply when shipping in larger quantities

(Table courtesy of Cadex)

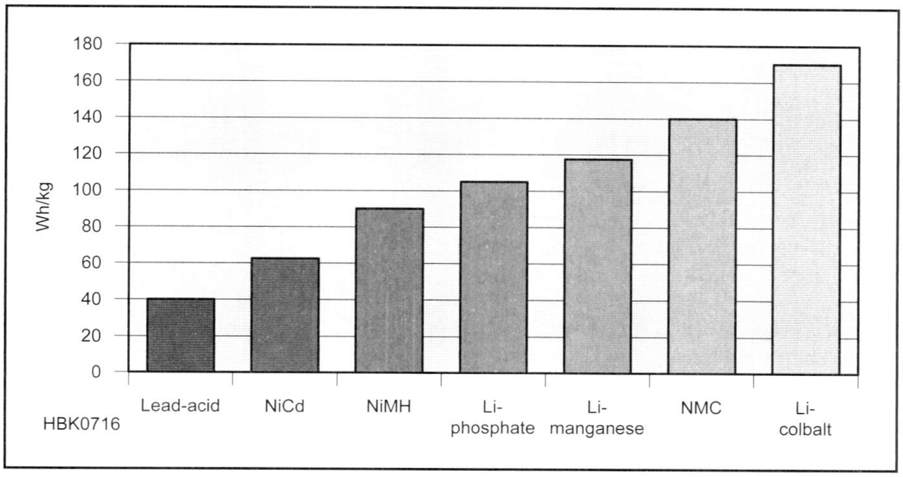

Figure 7.50 — Typical energy densities of lead, nickel- and lithium-based batteries.

OEM and usually have special features for battery conditioning.

"Smart" chargers include valuable additional features beyond simply applying current to a battery. Temperature protection is particularly important to slow or prevent charging below freezing or above recommended thresholds when the battery is hot. Advanced lead-acid chargers adjust the float and trickle charge thresholds based on temperature and battery age. **Table 7.11** summarizes the types of third-party chargers.

There are two common charge methods: voltage limiting (VL) and current limited (CL). Lead and lithium-based chargers cap the voltage at a fixed threshold. When reaching the cutoff voltage, the battery begins to saturate and the current drops while receiving the remaining charge on its own timetable. Full charge detection occurs when the current drops to a designated level.

Nickel-based batteries charge with a controlled current and the voltage is allowed to fluctuate freely. A slight voltage drop after a steady rise indicates a fully charged battery. The voltage drop method works well in terminating a fast charge, but the charger should detect and protect against shorted or mismatched cells. Most chargers include temperature sensors to end the charge if the temperature exceeds a safe level. (Some batteries also have internal temperature sensors.)

A temperature rise is normal as nickel-based batteries approach full charge. When in "ready" mode, the battery must cool down to room temperature. Heat causes stress and prolonged exposure to elevated temperature shortens battery life. Extended trickle charge also inflicts damage on nickel-based batteries, which should not be left in the charger continuously or beyond a few days.

A lithium-based battery should not get warm in a charger, and if this happens either the battery or charger may be faulty. Li-ion chargers do not apply a trickle charge and disconnect the battery electrically when fully charged. If the battery is left in the charger, a recharge may occur when its open circuit voltage drops below a set threshold. It is not necessary to remove Li-ion batteries from a charger.

SIMPLE GUIDELINES WHEN BUYING A CHARGER

- Use the correct charger for battery chemistry. Most chargers service one chemistry only.
- The battery voltage must agree with the charger. Do not charge if different.
- The Ah rating of a battery can be somewhat higher or lower than specified on the charger. A larger battery will take longer to charge than a smaller one and vice versa.
- The higher the amperage of the charger, the shorter the charge time will be, subject to limits on how fast the battery can be charged.
- Accurate charge termination and correct trickle charge prolong battery life.
- When fully saturated, a lead-acid charger should switch to a lower voltage; a nickel-based charger should have a trickle charge; a Li-ion charger provides no trickle charge.
- Chargers should have a temperature override to end charge on a malfunctioning battery.
- Observe the temperature of the charger and battery. Lead-acid batteries stay cool during charge; nickel-based batteries elevate the temperature toward the end of charge and should cool down after charge; Li-ion batteries should stay cool throughout charge.

SLOW CHARGERS

This type of charger applies a fixed current of about 0.1 C (one-tenth of the rated capacity) as long as the battery is connected. Slow chargers have no full-charge detection, charge current is always applied, and the charge time for a fully discharged battery is 14 to 16 hours.

Table 7.11

Charger Characteristics

Type	*Chemistry*	*C-rate (C)*	*Time*	*Charge Termination*
Slow	NiCd, Lead-acid	0.1	14 h	Continuous low charge or fixed timer. Subject to overcharge. Remove battery when charged.
Rapid	NiCd, NiMH, Li-ion	0.3-0.5	3-6 h	Senses battery by voltage, current, temperature, and time-out timer
Fast	NiCd, NiMH, Li-ion	1	1+ h	Same as rapid charger with faster service
Ultra-Fast	Li-ion, NiCd, NiMH	1-10	10-60 min	Applies ultra-fast charge to 70% SoC; limited to specialty batteries

All values specified over 0°C to 45°F (32°F to 113°F) range

(Table courtesy of Cadex)

Most slow chargers have no "ready" indicator.

When fully charged, a slow charger keeps NiCd batteries lukewarm to the touch. Some overcharge is acceptable and the battery does not need to be removed immediately when ready. Leaving the battery in the charger can cause internal crystal growth that leads to "memory effects" in NiCd batteries.

Charging a battery with a lower Ah rating than specified for the charger will cause the battery to heat up as it approaches full charge due to the higher charging rate. Because slow chargers have no provision to lower the current or terminate the charge, the excessive heat will shorten the life of the battery.

The opposite can occur when a slow charger is charging a larger battery than it is rated for. In this case, the battery may never reach full charge and remains cold. Battery performance will be poor because of insufficient charge. Repeated partial charging can also cause crystal growth and memory effects.

RAPID CHARGERS

Falling between a slow and fast charger, the rapid charger is designed for nickel and lithium-based batteries. Unless specially designed, the rapid charger cannot service both types of batteries.

Rapid chargers are designed to charge fully discharged batteries and battery packs in 3 to 6 hours. When full charge is reached, the charger switches to a "ready" state. Most rapid chargers include temperature protection.

FAST CHARGERS

The fast charger typically applies charge at a 1 C rate so that a fully discharged battery is recharged in a little over 1 hour. As the battery approaches full charge, the charger may reduce the charge current (particularly for NiCd), and when the battery is fully charged, the charger switches to a trickle or maintenance charge mode.

Most nickel-based fast chargers accommodate NiCd and NiMH batteries and apply the same charging algorithm, but cannot charge Li-ion batteries. To service a Li-ion pack, specialty dual-mode chargers can read a security code on the battery to switch to the right charger setting.

Lead-acid batteries cannot be fast-charged and the term "fast charge" is a misnomer for lead-acid chargers. Most lead-acid chargers charge the battery in 14 hours; anything slower may be a compromise. Lead-acid can be charged relatively quickly to 70% of full charge with the important saturation charge consuming the remaining time. A partial charge at high rate is acceptable provided the battery receives a fully saturated charge once every few weeks to prevent sulfation.

ULTRA-FAST CHARGERS

Large NiCd and Li-ion batteries can be charged at a very high rate (10 C is typical) up to 70% of full charge. Ultra-fast charging stresses batteries. If possible, charge the battery at a more moderate current. An ultra-fast charger should offer user-selectable rates to optimize the charging requirements.

At a rate of 10 C, a battery can be charged in a few minutes but several conditions must be observed:

• The battery must be designed to accept an ultra-fast charge.

• Ultra-fast charging only applies during the first charge phase and charge current must be lowered once the 70% state-of-charge (SoC) threshold is reached.

• All cells in a pack must be balanced and in good condition. Older batteries with high internal resistance will heat up and are no longer suitable for ultra-fast charging.

• Ultra-fast charging can only be done at moderate temperatures. Low temperatures slow the chemical reaction and the unabsorbed energy results in gassing and heat buildup.

• The charge must include temperature compensation and other safety provisions to end the charge if the battery is overly stressed.

CHARGING FROM A USB PORT

The universal serial bus (USB) interface has become ubiquitous on computers and consumer electronics. It is increasingly used on radio equipment.

USB hubs are specified to provide 5 V and 500 mA of current. (High-current charging ports can supply up to 2.0 A.) While this would be enough to charge a small Li-ion battery, it could overload the hub if other devices are attached. Many hubs limit the current and will shut down if overloaded, so charging capacity is quite limited.

The most common USB chargers are designed for single-cell Li-ion batteries. The charge begins with a constant current charge to 4.20 V/cell, at which point the voltage levels off and current begins to decrease. Due to voltage drops in the USB cable and charger circuit, the hub may not be able to fully charge the battery. This will not damage a Li-ion battery but it will deliver shorter than expected runtimes.

CHARGING LEAD-ACID

Lead-acid batteries should be charged in three stages as shown in **Figure 7.51**:

1 — constant-current charge

2 — topping charge

3 — float charge

The constant-current charge applies the bulk of the charge and takes up roughly half of

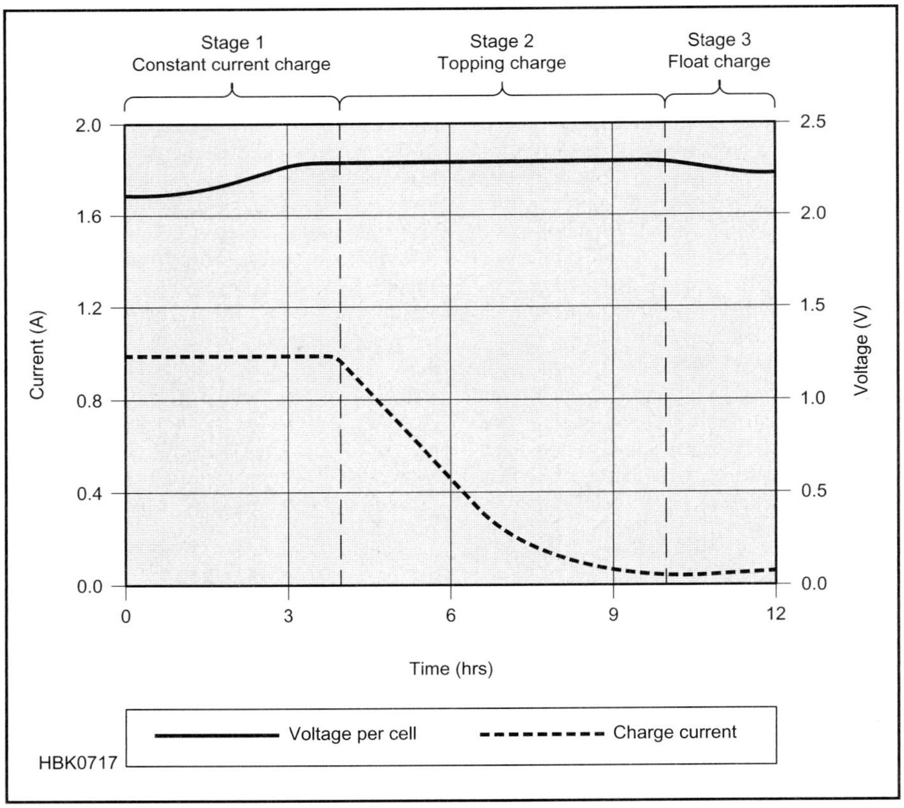

Figure 7.51 — Charge stages of a lead-acid battery. The battery is fully charged when the current drops to a pre-determined level or levels out in stage 2. The float voltage must be reduced at full charge. (Courtesy of Cadex)

the required charge time. The topping charge continues at a lower charge current and provides saturation. The float charge compensates for the loss caused by self-discharge.

During the constant-current charge the battery charges to 70% SoC in 5 to 8 hours and the remaining 30% is supplied by the slower topping charge that lasts another 7 to 10 hours. The topping charge is essential for the well-being of the battery. If topping charge is not performed, the battery will eventually lose the ability to accept a full charge and performance will decrease because of sulfation. The float charge maintains the battery at full charge.

The switch to topping charge happens when the battery reaches the set voltage limit. Current begins to drop as the battery starts to saturate and full charge is reached when the current decreases to 3% of the rated current. A battery with high leakage may never reach this level and a timer is required to start charging termination.

The correct setting of the charge voltage is critical, ranging from 2.3 to 2.45 V per cell. The threshold is selected as a compromise between charging to maximum capacity and creating internal corrosion and gassing. The battery voltage also shifts with temperature, with warmer temperatures requiring slightly lower voltage thresholds. If variable voltage thresholds are not available in the charger, it is better to use a lower voltage threshold for safety.

Once fully charged through saturation, the battery should not dwell at the *topping voltage* for more than 48 hours and must be reduced to the *float voltage* level. This is especially critical for sealed systems because they are less able to tolerate overcharge than the flooded type due to heating and gas (hydrogen) generation. The recommended float voltage of most lead-acid batteries is 2.25 to 2.27 V/cell. Manufacturers recommend lowering the float charge at ambient temperatures above 29 °C (85 °F). Not all chargers feature float charge. If the charger remains at the topping voltage and does not drop below 2.30 V/cell, remove the charger after a maximum of 48 hours of charge.

Aging batteries develop imbalances between cells that can result in overcharge and gassing from weak cells. This can also cause a strong cell to be undercharged and develop sulfation. Some battery manufacturers have developed cell-balancing devices that compensate for cell imbalance.

Lead-acid batteries must always be stored in a charged state. A topping charge should be applied every six months to prevent the voltage from dropping below 2.10 V/cell. With AGM, these requirements can be somewhat relaxed.

Measuring the open circuit voltage (OCV) while in storage provides a reliable indication of the battery's state-of-charge (SoC). A voltage of 2.10 V at room temperature indicates a charge of about 90%. (That is equivalent to 12.6 V for a typical six-cell "12 V" lead-acid battery.) Such a battery is in good condition and needs only a brief full charge prior to use. If the OCV is lower, the battery must be charged to prevent sulfation. Cool temperatures increase OCV slightly and warm temperatures lower it. Use OCV as an SoC indicator after the battery has rested for a few hours to allow the effects of charging to dissipate.

SIMPLE GUIDELINES FOR CHARGING LEAD-ACID BATTERIES

- Charge in a well-ventilated area. Hydrogen gas generated during charging is explosive.
- Choose the appropriate charge program for flooded, gel and AGM batteries. Follow the manufacturer's specifications of voltage thresholds.
- Charge lead-acid batteries after each use to prevent sulfation. Do not store in a low-charge condition.
- The plates of flooded batteries must always be fully submerged in electrolyte. Fill battery with distilled or de-ionized water to cover the plates if low. Never add electrolyte.
- Fill water level to designed level *after* charging. Overfilling a discharged battery can result in overflow and acid spillage.
- Formation of gas bubbles in a flooded lead-acid battery is an indication of approaching full charge.
- Reduce float charge if the ambient temperature is higher than 29 °C (85 °F).
- Do not allow a lead-acid battery to freeze and never charge a frozen battery.
- Do not charge at temperatures above 49 °C (120 °F).

CHARGING NICKEL-CADMIUM

Battery manufacturers recommend that new NiCd batteries be slow-charged for 16 to 24 hours before use. A slow charge brings all cells in a battery pack to an equal charge level. This is important because each cell within the NiCd battery may have self-discharged at its own rate. Furthermore, during long storage the electrolyte tends to gravitate to the bottom of the cell and the initial trickle charge helps redistribute the electrolyte to eliminate dry spots on the separator. The cells will reach optimal performance after several charge/discharge cycles.

Full-charge Detection by Temperature

Full-charge detection of sealed nickel-based batteries is more complex than for lead-acid and lithium-ion systems. Low-cost chargers often use temperature sensing to end the fast-charge, but this can be inaccurate due to internal and external temperature differences. Charger manufacturers use 50 °C (122 °F) as the temperature cutoff. Although any prolonged temperature above 45 °C (113 °F) is harmful, brief overshoot is acceptable if temperature will drop quickly when the charger changes to the "ready" state.

Some microprocessor-controlled chargers sense the rate of temperature increase with time, using the rapid temperature rise toward the end of charge to trigger the "ready" state. This is referred to as *delta temperature over delta time* or dV/dt. A rate of 1 °C (1.8 °F) per minute terminates charging. This keeps the battery cooler, but the cells need to charge reasonably fast for temperature to rise at the required rate. An absolute temperature of 60 °C (140 °F) terminates charging under any circumstances.

Chargers relying on temperature inflict harmful overcharges when fully charged batteries are inserted into the charger, such as if a hand-held radio is left in the charger between each use. This is not the case with Li-ion batteries where the charger uses voltage as the SoC indicator.

Full-charge Detection by Voltage Signature

Advanced chargers terminate charging when a defined voltage signature or profile with time occurs, referred to as *negative delta V* or NDV. This provides more precise full-charge detection for nickel-based batteries than temperature-based methods. Charging is terminated when the battery voltage drops as full charge is reached. NDV is the recommended method for NiCd cells that do not include an internal thermistor for temperature control and avoids overcharging of fully-charged batteries. NDV requires a charge rate of at least 0.5 C to generate a reliably measurable change in voltage and works best with fast charging. At a charge rate of 1 C, a fully discharged battery is recharged in about an hour.

Figure 7.52 illustrates the relationship of cell voltage, pressure, and temperature of a charging NiCd battery. Up to about 70% SoC, the battery accepts almost all of the energy supplied (called *charge efficiency*). Above 70%, the battery loses ability to accept charge, begins to generate gases so that pressure rises, and temperature increases rapidly.

Ultra-high-capacity NiCd batteries tend to heat up more than standard batteries when charging at 1 C and higher rates due to their higher internal resistance. Applying a high current during initial charge and tapering to a lower rate achieves good results with all nickel-based batteries and moderates temperature rise.

Some chargers can "burp" a charging battery by applying a load to generate a discharge pulse to cause gases to recombine and lower

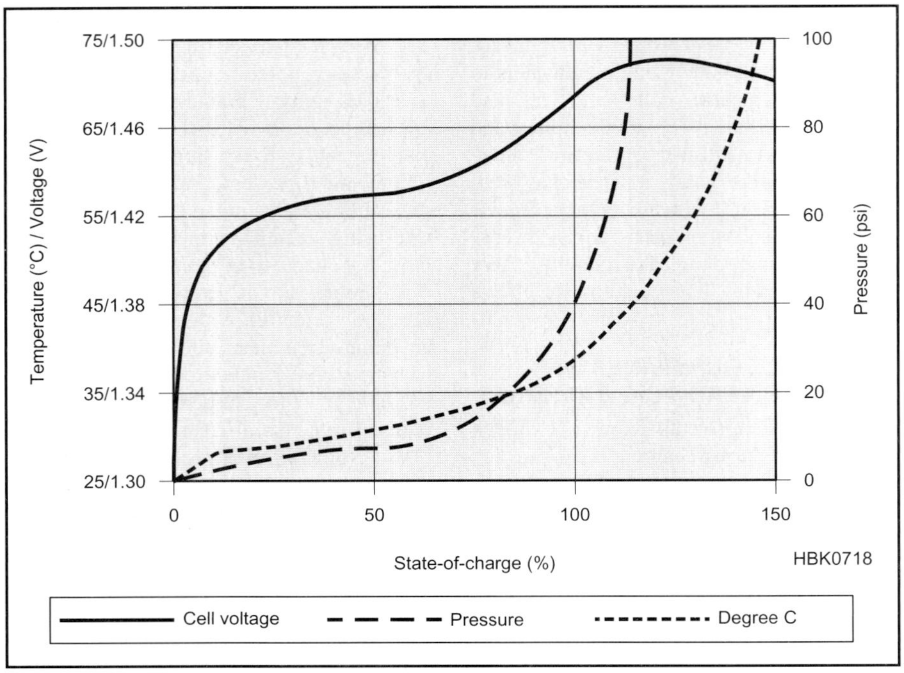

Figure 7.52 — Charge characteristics of a NiCd cell. Above 70% state-of-charge, temperature and cell pressure rise quickly. NiMH has similar charge characteristics. (Courtesy of Cadex)

internal pressure. The result is a cooler and more effective charge than with conventional dc charging. Pulse charging does not apply to lead and lithium-based systems.

After full charge, the NiCd battery receives a trickle charge of between 0.05 C and 0.1 C to compensate for self-discharge. To avoid possible overcharge, trickle charging should be done at the lowest possible rate and the batteries should be removed from the charger after more than a few days.

CHARGING NICKEL-METAL-HYDRIDE

The charging algorithm for NiMH is similar to NiCd with the exception that NiMH is more complex. The NDV method of measuring full charge has difficulty because the voltage drop at full charge is very small — about 5 mV/cell. As a result, modern chargers combine the various methods of measuring voltage and temperature into a composite algorithm that reacts depending on battery condition.

Some advanced chargers apply an initial fast charge at 1 C. After reaching a certain voltage threshold, a few minutes rest is taken to allow the battery to cool. Charging then resumes at lower currents as the charge progresses to full charge. This is known as the "step-differential charge" and works well for all nickel-based batteries, achieving an extra capacity of about 6% above basic chargers. A drawback of this method is that the fast-charge stress on the battery will shorten overall battery life by 10 to 20%.

NiMH cannot absorb overcharge well and the trickle charge current must be limited to around 0.05 C. In comparison, a basic NiCd charger trickle charges at 0.1 C. This higher trickle charge and the need for sensitive full-charge detection render the basic NiCd chargers unsuitable for NiMH batteries. On the other hand, NiCd cells can be charged in a NiMH charger at the lower trickle charge rate.

Slow charging should not be used for NiMH batteries. At the charging rate of 0.1 to 0.3 C, the voltage and temperature profiles make it very difficult to measure full charge accurately so the charger must depend on a timer. Harmful overcharge will occur if a fixed timer is used, particularly when charging partially or fully charged batteries. The same is true for charging old batteries with reduced capacity.

Inexpensive chargers are prone to incorrect charging because of the difficulty in correctly sensing full charge. Remove the batteries from the charger when you think they are fully charged. For high charge rates, remove the batteries when they are warm to the touch. It is better to remove the batteries and recharge them before use than to leave them in the charger where they might be overcharged and damaged.

SIMPLE GUIDELINES ON CHARGING NICKEL-BASED BATTERIES

- Do not charge at high or freezing temperatures; room temperature is best.
- Do not use chargers that allow the batteries to heat; remove the batteries when warm to the touch.
- Nickel-based batteries are best fast charged.
- NiMH chargers can charge NiCd batteries but not vice versa.
- High charge current or overcharging on an aging battery may cause heat build-up.
- Do not leave a nickel-based battery in the charger for long periods, even with correct trickle charge. Remove and apply a brief charge before use.
- Nickel- and lithium-based batteries require different charge algorithms and cannot share the same charger unless it can switch between the different chemistries.

CHARGING LITHIUM-ION

The Li-ion charger is a voltage-limiting device that is similar to the lead-acid system. The difference lies in a higher voltage per cell, tighter voltage tolerance and the absence of trickle or float charge at full charge. Li-ion cannot accept overcharge — any extra charging causes stress.

Most Li-ion cells charge to 4.20 V/cell with a tolerance of ±50 mV/cell. **Figure 7.53** shows the voltage and current signature as lithium-ion passes through the stages for constant current and topping charge. The charge rate of a typical consumer Li-ion battery is between 0.5 and 1 C in Stage 1 and the charge time is about three hours. Manufacturers recommend charging the 18650 cell at 0.8 C or less. The cell should remain cool during the charging process although there may be a slight temperature rise of a few degrees when reaching full charge. Full charge occurs when the battery reaches the voltage threshold and the current drops to 3% of the rated current, or if the charging current reaches a constant value and does not decrease further. The latter may be due to elevated self-discharge.

Li-ion does not need to be fully charged, as is the case with lead-acid, nor is it desirable to do so. In fact, it is better not to fully charge so as not to stress the battery. Choosing a lower voltage threshold or eliminating saturation charge prolongs battery life at the cost of reduced runtime. Without a saturation stage, the battery is usually charged to around 85% of capacity.

Once charging is terminated, the battery voltage begins to drop and this eases the voltage stress. Over time, the open circuit voltage (OCV) will settle to between 3.60 and 3.90 V/cell. A battery receiving a fully saturated battery families, only lead, nickel and lithium systems are covered. Along with these guidelines, you can prolong battery life by following three simple rules: keep the battery at moderate temperatures, control the level and rate of discharge, and avoid abusing the battery.

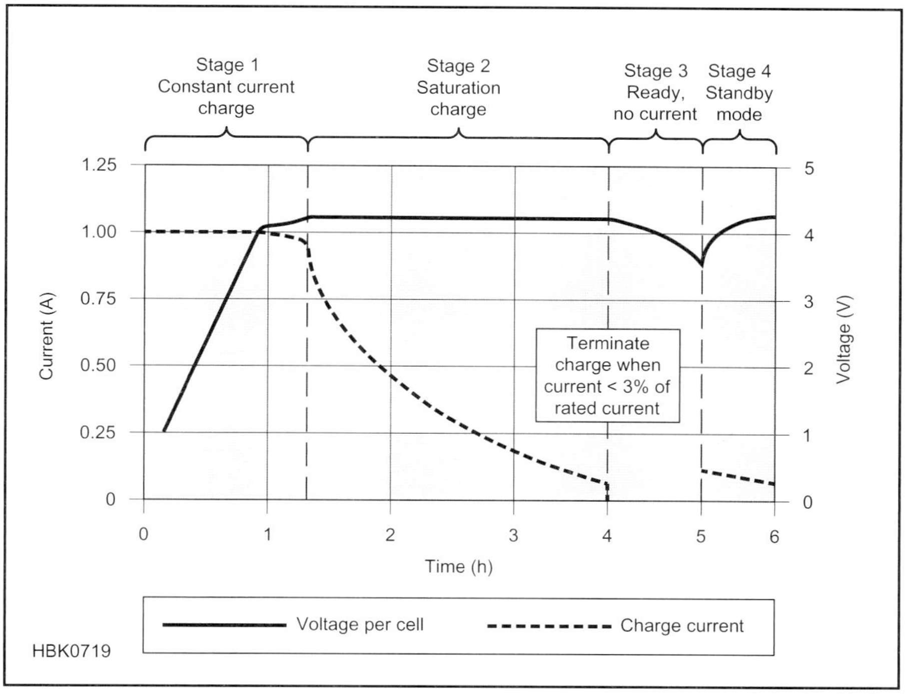

Figure 7.53 — Charge stages of lithium ion. Li-ion is fully charged when the current drops to a predetermined level or levels out at the end of Stage 2. In lieu of trickle charge, some chargers apply a topping charge when the voltage drops to 4.05 V/cell (Stage 4). (Courtesy of Cadex)

charge will keep the higher voltage longer than a battery that was fast charged and terminated without a saturation charge.

If a lithium-ion battery must be left in the charger for operational readiness, some chargers apply a brief topping charge to compensate for the small self-discharge of the battery and its protective circuit. It is common to let the battery voltage drop to 4.00 V/cell and then recharge to 4.05 V/cell to reduce voltage-related stress and prolong battery life. Battery-powered devices should be turned off when charging their battery. Otherwise, the *parasitic load* of the device can confuse the charger, distorting the charge cycle and stressing the battery.

Overcharging Lithium-ion

Lithium-ion systems operate safely within the designated operating voltages; however, the battery becomes unstable if inadvertently charged to a voltage higher than specified. Prolonged charging above 4.30 V/cell forms plating of metallic lithium on the anode, while the cathode material becomes an oxidizing agent, loses stability and produces CO_2. Cell pressure rises until the internal *current interrupt device* (CID) disconnects the current at 1380 kPa (200 psi).

Should the pressure rise further, a safety membrane bursts open at 3450 kPa (500 psi) and the cell might eventually vent with flame. The thermal runaway moves lower when the battery is fully charged; for Li-cobalt this threshold is between 130-150 °C (266-302 °F), for nickel-manganese-cobalt (NMC) between 170-180 °C (338-356 °F), and manganese is 250 °C (482 °F). Li-phosphate enjoys similar and better temperature stabilities than manganese.

Lithium-ion is not the only battery that is a safety hazard if overcharged. Lead and nickel-based batteries are also known to melt down and cause fires if improperly handled. Properly designed charging equipment is paramount for all battery systems.

SIMPLE GUIDELINES FOR CHARGING LITHIUM-BASED BATTERIES

• A battery-powered device should be turned off while charging.

• Charge at a moderate temperature. Do not charge below freezing.

• Lithium-ion does not need to be fully charged; a partial charge is better.

• Chargers use different methods for "ready" indication and may not always indicate a full charge.

• Discontinue using a charger and/or battery if the battery gets excessively warm.

• Before prolonger storage, apply some charge to bring a pack to about half charge.

SUMMARY OF CHARGING

Batteries have unique needs and **Table 7.12** explains how to satisfy these needs with correct handling. Because of similarities within the

7.13.6 Discharge Methods

C-RATE

According to the definition of *coulomb,* a current of 1 ampere is a flow of 1 coulomb (C) of charge per second. Today, the battery industry uses *C-rate* to scale the charge and discharge current of a battery.

Most portable batteries are rated at 1 C, meaning that a 1,000 mAh battery that is discharged at 1 C should under ideal conditions provide a current of 1,000 mA for 1 hr. The same battery discharging at 0.5 C would provide 500 mA for 2 hours, and at 2 C, the 1,000 mAh battery would deliver 2,000 mA for 30 minutes. 1 C is also known as a one-hour discharge; a 0.5 C is a two-hour discharge, a 2 C is a half-hour discharge, and so on.

A battery's capacity — the amount of energy a battery can hold — can be measured with a *battery analyzer*. The analyzer discharges the battery at a calibrated current while measuring the time it takes to reach its specified end-of-discharge voltage. If a 1,000 mAh battery could provide 1,000 mA for 1 hour, 100% of the battery's nominal energy rating would be reached. If the discharge only lasted 30 minutes before reaching the specified voltage, the battery has a capacity of 50% of its nominal rating.

When discharging a battery at different rates a higher C-rate will produce a lower capacity reading due to the internal resistance turning some of the energy into heat instead of delivering it as current to a load. Lower C-rate discharges will produce a higher capacity.

For example, to obtain a reasonably good capacity rating, manufacturers commonly rate lead-acid batteries at 0.05 C, or a 20-hour discharge. **Figure 7.54** illustrates the

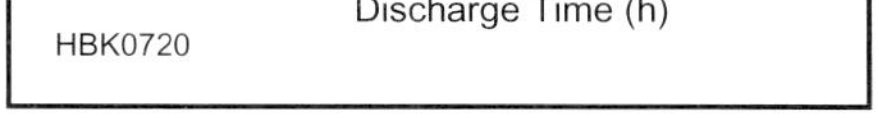

Figure 7.54 — Typical discharge curves of lead-acid as a function of C-rate. Smaller batteries are rated at a 1 C discharge rate. Due to sluggish behavior, lead-acid is rated at 0.2 C (5 hours) and 0.05 C (20 hours). (Courtesy of Cadex)

Table 7.12

Best Charging Methods

Frequently Asked Question	*Lead-acid (Sealed, flooded)*	*Nickel-Based (NiCd and NiMH)*	*Lithium-ion (Li-ion, Polymer)*
How should I prepare a new battery?	Battery comes fully charged. Apply topping charge	Charge 14-16 h. Priming may be needed.	Apply a topping charge before use. No priming needed
Can I damage a battery with incorrect use?	Yes, do not store partially charged, keep fully charged	Battery is robust and the performance will improve with use.	Keep some charge. Low charge can turn off protection circuit.
Do I need to apply a full charge?	Yes, partial charge causes sulfation.	Partial charge is fine.	Partial charge better than a full charge.
Can I disrupt a charge cycle?	Yes, partial charge causes no harm.	Interruptions can cause heat buildup.	Yes, partial charge causes no harm.
Should I use up all battery energy before charging?	No, deep discharge wears the battery down. Charge more often.	Apply scheduled discharges only to prevent memory.	No, deep discharge wears the battery down.
Do I have to worry about "memory"?	No memory	Discharge NiCd every 1-3 months.	No memory
How do I calibrate a "smart" battery?	Not applicable	Apply discharge/charge when the fuel gauge gets inaccurate. Repeat every 1-3 months.	Apply discharge/charge when the fuel gauge gets inaccurate. Repeat every 1-3 months.
Must I remove the battery when fully charged?	Depends on charger; needs correct float voltage	Remove after a few days in charger.	Not necessary; charger turns off
How do I store my battery?	Keep cells above 2.10 V, charge every 6 months	Store in cool place; a total discharge causes no harm.	Store in cool place partially charged, do not fully drain
Is the battery allowed to heat up during charge?	Battery may get lukewarm toward the end of charge.	Battery gets warm but must cool down on ready.	Battery may get lukewarm toward the end of charge.
How do I charge when cold?	Slow charge (0.1C): 0°-45 °C (32°-113 °F) Fast charge (0.5-1C): 5°-45 °C (41°-113 °F)	Slow charge (0.1C): 0°-45 °C (32°-113 °F) Fast charge (0.5-1C): 5°-45 °C (41°-113 °F)	Do not charge above 50 °C (122 °F) Do not charge above 50 °C (122 °F)
Can I charge at hot temperatures?	Above 25 °C, lower threshold by 3 mV/°C.	Battery will not fully charge when hot	Do not charge above 50 °C (122 °F)
What should I know about chargers?	Charger should float at 2.25-2.30 V/cell when ready.	Battery should not get too hot; should include temp sensor.	Battery must stay cool; no trickle charge when ready.

(Table courtesy of Cadex)

discharge times of a lead-acid battery at various loads expressed in C-rate.

DEPTH OF DISCHARGE

The end-of-discharge voltage for lead-acid is 1.75 V/cell; for nickel-based systems it is 1.00 V/cell; and for most Li-ion it is 3.00 V/cell. At this level, roughly 95% of the battery's stored energy has been spent and voltage would drop rapidly if discharge were to continue. Most devices prevent operation beyond the specified end-of-discharge voltage. When removing the load after discharge, the voltage of a healthy battery gradually recovers toward the nominal voltage.

Because of internal resistance, wiring, protection circuits and contact resistance, a high load current lowers the battery voltage and the end-of-discharge voltage threshold should be lowered accordingly. The cutoff voltage should also be lowered when discharging at very cold temperatures. **Table 7.13** shows typical end-of-discharge voltages of various battery chemistries. The lower end-of-discharge voltage for higher loads compensates for the losses from internal battery resistance.

Table 7.13

Recommended End-of-Discharge Voltages in V/Cell

End-of-discharge	*Li-manganese*	*Li-phosphate*	*Lead-acid*	*NiCd/NiMH*
Normal load	3.00	2.70	1.75	1.00
Heavy load	2.70	2.45	1.40	0.90

(Table courtesy of Cadex)

Since the cells in a battery pack can never be perfectly matched, a negative voltage potential can occur across a weaker cell on a multi-cell pack if the discharge is allowed to continue beyond a safe cutoff point. Known as *cell reversal*, the weak cell suffers damage to the point of developing a permanent electrical short circuit. The larger the number of cells in the pack, the greater the likelihood that a cell might reverse under heavy load. Over-discharge, particularly at low temperatures, is a large contributor to battery failure of cordless power tools, especially for nickel-based packs. Li-ion packs have protection circuits and the failure rate is lower.

DISCHARGING AT HIGH AND LOW TEMPERATURES

Batteries achieve optimum service life if used at 20 °C (68 °F) or slightly below, and nickel-based chemistries degrade rapidly when cycled at high ambient temperatures. Higher temperature operation lowers internal resistance and speeds up the chemical reactions but shortens service life if prolonged.

The performance of all battery chemistries drops drastically at low temperatures. At –20 °C (–4 °F) most nickel, lead, and lithium-based batteries stop functioning. Although NiCd can be used down to –40 °C (–40 °F), the permissible discharge is only 0.2 C (5-hour rate). Lead-acid also has the problem of the electrolyte freezing which can crack the enclosure. Lead-acid electrolyte also freezes more easily at a low charge.

SIMPLE GUIDELINES FOR DISCHARGING BATTERIES

• Battery performance decreases with cold temperature and increases with heat.

• Heat increases battery performance but shortens cycle life by a factor of two for every 10 °C (18 °F) above 25-30 °C (18 °F above 77-86 °F).

• Only charge at moderate temperatures. Check the manufacturer's specifications for charging below freezing.

• Use heating blankets if batteries need rapid charging at cold temperatures.

• Prevent over-discharging. Cell reversal can cause an electrical short circuit.

• Use a larger battery if repetitive deep discharge cycles cause stress.

• A moderate dc discharge is better for a battery than pulsed loads.

• Lead-acid systems are sluggish and require a few seconds of recovery between heavy loads.

7.13.7 Battery Handling

This section touches on the most important aspects of handling batteries when they are new, during their service life, and how to store and dispose of them.

FORMATTING AND PRIMING BATTERIES

Rechargeable batteries may not deliver their full rated capacity when new and will require *formatting* — a process that essentially completes the manufacturing process. Li-ion systems require less care in this regard, but cycling these batteries after long storage has been reported to improve performance. *Priming* is a conditioning cycle that is applied to improve battery performance during usage or after prolonged storage. Priming applies mainly to nickel-based batteries.

Formatting of lead-acid batteries occurs by applying a charge, followed by a discharge and recharge as part of regular use. Gradually increase the load on a new battery, allowing it to reach full capacity after 50 to 100 cycles.

Manufacturers advise to trickle charge a nickel-based battery pack for 16 to 24 hours when new and after a long storage. This allows the individual cells to reach an equal charge level. A slow charge also helps to redistribute the electrolyte to eliminate dry spots on the separator that may have formed due to gravity. Applying several charge/discharge cycles through normal use or with a battery analyzer completes the formatting process. This can require from five to seven cycles or as many as 50 cycles depending on battery quality.

Cycling also restores lost capacity when a nickel-based battery has been stored for six months or longer. Storage time, state-of-charge, and storage temperature all affect battery recovery. The longer the storage and the higher the temperature, the more cycles are required to regain full capacity.

Lithium-ion does not need formatting when new, nor does it require the level of maintenance that nickel-based batteries do. Maximum capacity is available immediately. A discharge/charge cycle may be beneficial for calibrating a "smart" battery but this does not improve the internal chemistry.

STORING BATTERIES

The recommended storage temperature for most batteries is 15 °C (59 °F) and the extreme allowable temperature is –40 °C to 50 °C (–40 °F to 122 °F) for most chemistries. While lead-acid must be kept at full charge during storage, nickel and lithium-based chemistries should be stored at around 40% state-of-charge.

Storage will always cause batteries to age. **Table 7.14** illustrates the *recoverable capacity* of lithium and nickel-based batteries at various temperatures and charge levels over one year. Recoverable capacity is the available battery capacity after storage with a full charge.

A sealed lead-acid battery can be stored up to two years. It is important to apply a charge when the battery falls to 70% SoC, typically 2.07 V/cell or 12.42 V for a 12 V pack.

Nickel-metal-hydride can be stored for about three years. The capacity drop that occurs during storage can partially be reversed with priming.

Primary alkaline and lithium batteries can be stored for up to 10 years with minimum capacity loss.

SIMPLE GUIDELINES FOR STORING BATTERIES

• Remove batteries from equipment and store in a dry and cool place.

• Avoid freezing. Batteries freeze more easily if in a discharged state.

• Charge lead-acid before storing and monitor the voltage frequently; apply a charge if below 2.10 V/cell.

• Nickel-based batteries can be stored for five years and longer, prime before use.

• Lithium-ion must be stored in a charged state, ideally 40%.

• Discard Li-ion if the voltage has stayed below 2.00 V/cell for more than a week.

RECYCLING BATTERIES

The main objective for recycling batteries is to prevent hazardous materials from entering landfills. Lead-acid and NiCd batteries are of special concern.

Testing and Monitoring Batteries

Several sections of the "Testing and Monitoring" chapter from *Batteries in a Portable World* are provided with this book's online content. The information covers measuring internal resistance and state-of-charge, measuring capacity, and special techniques for measuring nickel- and lithium-based batteries.

Table 7.14
Estimated Recoverable Capacity After Storage For 1 Year

Temp (°C)	*Lead-acid at full charge*	*Nickel-based at any charge*	*Lithium-ion (Li-cobalt) 40% charge*	*Lithium-ion (Li-cobalt) 100% charge*
0	97%	99%	98%	94%
25	90%	97%	96%	80%
40	62%	95%	85%	65%
60	38% (after 6 months)	70%	75%	60% (after 3 months)

(Table courtesy of Cadex)

Under no circumstances should batteries be incinerated, as fire can cause an explosion. Wear approved gloves when touching electrolyte. On exposure to skin, flush with water immediately. If eye exposure occurs, flush with water for 15 minutes and consult a physician immediately.

Automotive and larger lead-acid batteries can be recycled through auto parts stores and battery dealers. A recycling fee is usually charged when the battery is purchased.

Smaller batteries, including smaller SLA, can be recycled at many electronics and hardware stores, or your local municipal recycling center. Perform an Internet search for battery recyclers in your area.

It is helpful to create a specific location or designate a container for spent batteries at your home, office or workbench. This makes it easy to recycle the batteries by keeping them together in one spot.

7.13.8 DC-AC Inverters

For battery-powered operation of ac-powered equipment, dc-ac inverters are used. An inverter is a dc-to-ac converter that provides 120 V ac. Inverters come with varying degrees of sophistication. The simplest type of inverter switches directly at 60 Hz to produce a square-wave output. This is no problem for lighting and other loads that don't care about the input waveform. However, some equipment will work poorly or not at all when supplied with square wave power because of the high harmonic content of the waveform.

The harmonic content of the inverter output waveform can be reduced by the simple expedient of reducing the waveform duty cycle from 50% (for the square wave) to about 40%. For many loads, such as computers and other electronic devices, this may still not be adequate, and so many inverters use waveform shaping to approximate a sine-wave output. The simplest of these methods is a resonant inductor-capacitor filter. This adds significant weight and size to the inverter. Most modern inverters use high-frequency pulse-width modulation (PWM) techniques to synthesize the 60 Hz sinusoidal output waveform, much like a switching power supply. See **Figure 7.55**.

Inverters are usually rated in terms of their VA or "volt-ampere product" capability although sometimes they will be rated in watts. Care is required in interpreting inverter ratings. A purely resistive load operating from a sinusoidal voltage source will have a sinusoidal current flowing in phase with the voltage. In this case, VA, the product of the voltage (V) and the current (A), will equal the actual power in watts delivered to the load so that the VA and the watt ratings are the same. Some loads, such as motors and many rectifier power supplies, will shift the phase of load current away from the source voltage; or the load current will flow in short pulses as shown earlier for capacitor input filters. In these cases (which are very common), the VA product for that type of load can be much larger than the delivered power in watts. In the absence of detailed knowledge of the load characteristics it is prudent to select an inverter with a VA or wattage capability of 25% or more above the expected load.

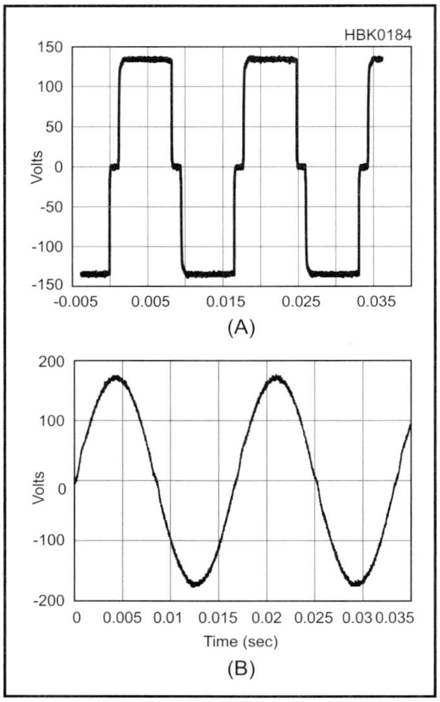

Figure 7.55 — Output waveforms of typical dc-ac inverters. At A, the output of a modified sine wave inverter. Note the stepped square waves. At B, the output of a "pure sine wave" inverter. Note the close approximation of a commercial ac sine wave.

7.13.9 Selecting a Battery for Mobile Operation

There are two basic modes of mobile operation: *in-motion*, and *stationary*. Each mode has unique power requirements and thus different battery requirements. To satisfy these needs, there are three basic battery types. It is important to understand the differences between the types and to apply them properly.

Standard vehicle batteries are referred to as *SLI* (starting, lights, ignition) or just *starter* batteries. Their primary function is to start the engine and then act as a power filter for the alternator which is the actual long-term power source. The most important rating of an SLI battery is the *cold cranking amps* (CCA) rating — the number of amps that the battery can produce at 32 °F (0 °C) for 30 seconds.

Batteries designed for repeated cycles of charging and discharging use are often called *deep-cycle* although the term is widely overused. A true deep-cycle battery is designed to be repeatedly discharged to 20% remaining capacity. The term "deep cycle" is a misnomer, as all lead-acid batteries are considered discharged when their output voltage drops below 10.5 V at some specified current draw as outlined by the Battery Council Institute (BCI — **www.batterycouncil.org**). At 10.5 V, a six-cell lead-acid battery is considered discharged.

Marine batteries are designed to be stored without charging for up to two years, yet still maintain enough power to start a marine engine. Contrary to common practice, they're not really designed for extended low-current power delivery. Marine batteries often have hybrid characteristics between SLI and deep-cycle batteries.

To differentiate true deep-cycle batteries from SLI and marine batteries, examine a battery's *reserve capacity* (RC). A battery's RC rating is the number of minutes that the battery can deliver 25 A while maintaining an output voltage above 10.5 V. Deep cycle batteries typically have RC ratings 20% or more higher than SLI batteries and perhaps 50% higher under ICAS (intermittent, commercial, and amateur service) conditions. A deep-cycle battery has a lower CCA rating than an SLI battery due to its internal construction that favors long-term power delivery over high-current starting loads.

With these facts in mind, we can now select the correct battery for our style of mobile operating. For in-motion operation with power outputs up to 200 W, a second trunk-mounted battery is seldom needed if the power cabling wire size is chosen correctly. (See the **Assembling a Station** chapter for information on wire sizes in mobile applications.)

When an amplifier is added for higher output power levels, it is often less expensive to add a second trunk-mounted battery than to install larger cables to the main battery. In these cases, the second battery can be of almost any type, as long as it is lead-acid. The second battery should be connected in parallel to the vehicle's main SLI battery. The battery's ampere-hour rating should be close to that of the vehicle's main SLI battery.

All secondary wiring should be properly fused, as outlined in the **Assembling a Station** chapter's section on mobile installations. The use of relays and circuit breakers should be avoided. Remember, should a short circuit occur, good-quality lead-acid batteries can deliver upwards of 3,000 A which exceeds the break circuit ratings of most relays and circuit breakers. A better solution is a FET switch such as those made by Perfect Switch (**www.perfectswitch.com**).

Assuming the second battery is mounted

inside the vehicle's passenger compartment or in the trunk, it should be an AGM type. AGM (Absorption Glass Mat) batteries do not outgas explosive hydrogen gas under normal operating conditions. Flooded (liquid electrolyte) batteries should *never* be used in an enclosed environment.

For stationary operation, select a battery with a large RC rating because it will not be continuously charged. There are two main considerations; the ampere-hour rating (Ah) and the reserve capacity rating, typically listed as C/8, C/10, or C/20, with units of hours. (C is the battery's capacity in Ah.) Dividing the Ah rating by the load amperage (8, 10 or 20 A) will give you the reserve capacity in hours, but the actual ampere-hours any given battery can deliver before the voltage reaches 10.5 V (nominal discharge level) will vary with the load, both average and peak. Heavier loads will reduce the actual ampere-hours available.

Automotive batteries are arranged in BCI group sizes (**www.batterystuff.com/kb/tools/bci-battery-group-sizes.html**), from 21 through 98. Generally speaking, the larger the group size the larger the battery and the higher the Ah rating. For example, size 24 (small car) has an average rating of 40 Ah, and size 34 (large car) has an average rating of 55 Ah. Exact ratings, including their reserve capacity, are available from the manufacturers' websites listed below. A good rule of thumb is to select a battery as physically large as you have room for, consistent with the highest RC rating for any given Ah rating.

Batteries are heavy, and need to be properly secured inside a battery box or by using factory-supplied brackets. For example, a BIC group 34 (average SLI size) battery weighs about 55 pounds. Some battery models (such as the Optima) come supplied with mounting brackets and terminal protection covers. Even though battery boxes aren't always needed, they should be used as a safety precaution to prevent accidental contact with the terminals and can protect the battery from external items. Battery restraints should be adequate to provide 6 Gs of lateral and 4 Gs of vertical retention, ruling out sheet metal screws and most webbing material. Use the proper brackets!

There are three other considerations: isolating the battery electrically, recharging the battery, and output voltage regulation. Diode-based battery isolators are not all equal. Models with FET bypass switches are the preferred type because of the low voltage-drop across the FET.

If you have wired the battery in parallel with the vehicle's main SLI battery, recharging is taken care of whenever the vehicle is running. If you plan on operating in stationary mode, you'll need a separate recharging system. Most vehicle factory-installed trailer wiring systems also include a circuit for charging RV or boat "house" batteries. Check with your dealer's service personnel about these options.

Voltage regulators, commonly called "battery boosters," are almost a necessity for stationary operation. A model with a low-voltage cutoff should be used to avoid discharging the battery below 10.5 V, as discharging a lead-acid battery beyond this point drastically reduces its charge-cycle life — the number of full-charge/full-discharge cycles. (See the November 2008 *QST* Product Review column.)

For additional information on battery ratings, sizes, and configuration, visit these websites:

optimabatteries.com, **www.exide.com**
www.interstatebatteries.com
www.lifelinebatteries.com

7.13.10 Selecting a Battery for Portable Operation

When choosing batteries for portable operation, consider the requirements that your type of operation will present and how your radio behaves when operated on batteries. The requirements are very different for casual operation while car camping compared to a Summits on the Air (SOTA) activation. If your radio will operate over a wide voltage range, the requirements for batteries are very different than if voltage must be in a critical range. If you are in a position to carry spares, then that will affect the requirements for battery capacity, too.

Developing a Battery Strategy

Powering portable equipment is much easier with a plan and a strategy. If possible, use the same battery type for all equipment or minimize the number of different types. This is not always possible, particularly with commercial equipment, but it is a good goal. Internal batteries are convenient, but they need to be checked, recharged, or replaced before an outing. It is good practice to be able to power devices with an external power source.

Physical size and weight are a consideration, particularly if you are backpacking or carrying your equipment a significant distance. The various lithium chemistry batteries, both primary and secondary, are the lightweight champs. Lead-acid chemistry is the heaviest with NiMH and NiCd chemistries in between. Pilots of RC (radio-controlled) model airplanes and drones are very sensitive to weight and size, so there are a lot of battery products from companies specializing in RC accessories, mostly lithium-based, that are of interest for amateur radio operations, particularly for backpacking.

AA batteries are a good standard choice. Alkaline AA cells are ubiquitous and commonly available from nearly any store or gas station. If you are camping, you may have other equipment powered by AA batteries which you can share for replacements. Rechargeable NiMH AA batteries are also available.

For higher power requirements, such as a QRP transmitter or even a modest power level transmitter, the 12 V, 7 Ah battery is common and available in both sealed lead-acid (SLA) and lithium iron phosphate ($LiFePO_4$) chemistries. See **www.powerstream.com/Size_SLA.htm** for a table of SLA battery dimensions.

One must have a way to recharge secondary (rechargeable) batteries, so standardizing on a battery chemistry and type is useful. Batteries should be charged before going to the field, but charging with a solar panel, even supplying just a fraction of the total power consumption, will extend operating time. Some operators also find it useful to be able to recharge batteries from a vehicle, but types other than lead-acid may require a special charger. Some RC suppliers can provide battery chargers that operate from a car battery.

Don't forget batteries for peripheral de-

Power Connections for Portable Operating

Try to use just one or two types of power connector. Anderson Powerpoles are a good choice for moderate to heavy-duty use and are the standard for many emergency communications and public service groups. When installed correctly, they are impossible to connect with reversed polarity.

The 5.5 mm OD × 2.1 mm ID coaxial connector is another common standard size. Some caution is necessary in using these connectors: While it is common for the inner conductor to carry positive voltage, it is not universal. If you have a piece of equipment with the opposite connector polarity, label the connector in large print, preferably on a brightly colored background as a warning.

It is easy to forget power cords when going to the field, so being able to interchange them is good. Another useful tip is to leave the power cord connected to the equipment when stored. That way it is always attached and if you select a piece of equipment without its accompanying cord, you immediately know it. You may wish to make up some cords with different power connectors on each end to do conversions from one connector to another. It is also helpful to have a cord or two with a car power outlet plug (formerly known as a cigarette lighter) on one end and your favorite power connector on the other end.

vices. When shopping for new peripherals, battery size should be a consideration. If you use standalone peripherals, a keyer, automatic antenna tuner, clock, flashlight, or GPS all use batteries. If these all use one battery type, such as AA cells, so much the better, but carry spares for any odd size batteries. Try not to duplicate peripherals, particularly when packing light. Using a radio that has peripherals built in, such as a keyer, automatic tuner, or clock, eliminates additional batteries and cables. A clock may be superfluous if you have a watch, GPS, or cell phone, for example.

When purchasing new equipment or building equipment for portable use, do so with an eye toward battery consumption. While some things like current draw, are obvious, other parameters, such as lowest stable operating voltage are not. Some "12 Volt" equipment is really designed for 13.8 V and does not operate properly much below 12 V. On the other hand, some equipment designed for portable operation will operate down to 9 V or less. This information is not always apparent when reviewing the specifications, but may be noted in the manual. Other users of the equipment can also be polled for their experiences, along with online reviews.

CALCULATING REQUIRED BATTERY CAPACITY

It is straightforward to calculate how much battery you need with some basic parameters. You should know the current draw of all the devices you are using, both on transmit and receive. These can usually be found in the equipment manual or are easily measured. Don't neglect the current drawn from a light if you are operating after dark, or a computer or laptop if you are using one for logging. You should also know the useful voltage end point of the battery. This can be found from manufacturer's data sheets.

The first step is to measure the current draw for of all devices you will be using. You should do this for both receive (I_{rec}) and full-power transmit (I_{tran}), as the current draw is usually significantly different in the two cases. $I_{peripherals}$ is the total current consumption of all peripherals using the same battery as the transceiver.

Next decide on what mode or modes you will be using and find the duty factor for each. (Duty factor is a value between 0 and 1. Duty cycle = duty factor × 100%.)

Next, determine your *average transmit time*, t_{xmt}, which is the ratio of transmitting time to total operating time. (*Average receive time,* t_{rcv}, equals $1 - t_{xmt}$.) For casual operating, assume an average transmit time of 0.25 (transmitting ¼ of the time). For contest or special event-style operating such as POTA or SOTA activation, assume an average transmit time of 0.5 (transmitting and receiving in equal amounts).

Operating duty factor is a measure of the time the transmitter is at full output when transmitting. Here are some operating duty factors for common modes taken from the RF Exposure safety calculation guidelines at **hintlink.com/power_density.htm**:

- 0.4 for CW
- 0.2 for SSB (0.4 if speech processing or if compression is used)
- 1.0 for FM
- 1.0 for continuous transmission digital modes such as RTTY and PSK
- 0.5 for interleaved transmission digital modes such as FT8 or FT4

The calculation for overall average current draw is then:

I_{avg} = Peripheral current + average receive current + average transmit current

$$I_{avg} = I_{peripherals} + (t_{rcv})(I_{rec}) + (t_{xmt})(\text{operating duty factor})(I_{tran})$$

Finally, determine the overall length of time, t, you will operate. The required battery capacity, C, is given by:

$$C = I_{avg} \times t$$

Add whatever margin you require (10% is typical) and look for batteries with that capacity at the acceptable discharge level and end voltage.

Here are two example calculations:

Example 1: For four hours of contest-style CW operating ($t_{rcv} = t_{xmt} = 0.5$) with $I_{peripherals}$ = 50 mA, I_{rec} = 100 mA, and I_{tran} = 1200 mA:

$$I_{avg} = 50 + (0.5)(100) + (0.5)(0.4)(1200) = 340 \text{ mA}$$

$$C = 340 \text{ mA} \times 4 \text{ hours} = 1.36 \text{ Ah}$$

Example 2: For ten hours of casual SSB operating ($t_{xnt} = 0.25$ and $t_{rcv} = 0.75$) without compression and using the same equipment:

$$I_{avg} = 50 + (0.75)(100) + (0.25)(0.2)(1200) = 185 \text{ mA}$$

$$C = 185 \text{ mA} \times 10 \text{ hours} = 1.85 \text{ Ah}$$

There are many batteries that will meet these requirements. A pack of 2.87 Ah alkaline AA batteries or a small 3 Ah SLA battery would provide plenty of margin for both types of operating. There are lithium chemistry batteries available with the same size and slightly higher capacity, so that would be a good choice, as well.

This calculation is a bit simplistic in that it assumes the voltage is more or less constant throughout the discharge of the battery, whereas there is some additional voltage drop under load, called *droop*, in nearly all battery chemistries. It is greatest (worse) for lead-acid and least (better) for lithium chemistries. As voltage droops, the current drawn by the radio will need to be increased to maintain the same power, so the transmitting current will increase. If one chooses a battery with little voltage droop during discharge or with a higher capacity, then this is not a significant problem.

Tables 7.3, 7.4, and 7.9 can help you choose. If you are buying a battery specifically for portable operation, Li-phosphate batteries are good choices because they are reliable, safe, can withstand deep discharge, and have much lower weight than SLA batteries with the same capacity. They cost more, but if properly maintained, they will outlast lead-acid batteries of similar size.

BATTERY SELECTION TIPS

If you only operate portable on occasion, or want to try portable operation at minimal expense, alkaline cells are probably the best choice. If weight is not a factor, you can operate a typical QRP radio from alkaline D-cells for months of casual operating and from a pair of alkaline lantern cells for almost a year. But their voltage drops as current is drawn from them, so, it is wise to carry spares.

For portable operations where weight is not a concern, sealed lead-acids were once the best bet. They are relatively inexpensive, particularly if you can get a surplus "pull" battery, and are easy to charge. Do some homework to size the battery properly.

For backpacking with efficient low power radios, where low weight is paramount, use a battery pack made from NiMH AA cells. If the capacity of an AA cell is not sufficient, use the larger sizes available in NiCd cells. NiMH cells must be charged with care to avoid damage.

Lithium chemistries have overtaken lead-acid and nickel chemistries, albeit at a higher cost. They offer a lot of advantages: lighter weight, flatter discharge curve, and higher usable endpoint voltage. Very lightweight and efficient battery packs are available from RC model airplane and drone suppliers. These are particularly attractive to backpackers, where every ounce counts.

Li-phosphate batteries are available in the same sizes for SLA batteries from 2 Ah to as large as vehicle starting batteries. They are a good initial battery choice for portable operation as they will last a very long time, even if abused.

If you want to try solar power, SLA batteries and NiCd batteries are the easiest to use and charge with solar panels. If the solar panel output is no larger than 0.1 C, the battery can be charged without a controller. If radio equipment has a similar current draw, there is little risk of overcharging and there will be sufficient charge to operate after dark.

Cold weather operation favors lithium, NiCd, and NiMH cells. SLA batteries and alkaline cells are poor choices for cold weather operation.

In order to make your rechargeable batteries last the longest, do not overcharge them, do not discharge them fully, store them fully charged, and top them off as required. Charging a stored battery every few months to its full capacity is better than leaving it on a trickle charger or battery maintainer.

7.14 References and Bibliography

REFERENCES

1) Landee, Davis, and Albrecht, *Electronic Designer Handbook* 2nd Edition, Giacoletto, L., Ed., (McGraw-Hill, 1977), p. 12-9. This book can frequently be found in technical libraries and used bookstores. The power supply section is well worth reading.
2) Severns, R., and Bloom, G., *Modern DC-to-DC Switchmode Power Converter Circuits* (Van Nostrand Reinhold, 1984). A reprint of this book is currently available at the Power Sources Manufacturers Association website, **www.psma.com**.
3) Fair-Rite website, **www.fair-rite.com**.
4) Magnetics website, **www.mag-inc.com**.

OTHER RESOURCES

Brown, M., *Power Supply Cookbook* (Newnes, 2001).

Buchmann, I., *Batteries in a Portable World* (Cadex Electronics, 2011).

Fielding, J., ZS5JF, *Power Supply Handbook* (RSGB, 2006).

Jeffrey, K., *Independent Energy Guide* (Orwell Cove Press, 1995). Includes information on batteries and inverters.

Kueck, C., "Power Supply Layout and EMI," Linear Technology AN139, 2012, **www.analog.com/media/en/technical-documentation/application-notes/an139f.pdf**.

Rosales, J., "High Voltage Boost and Inverting Converters for Communications," *Analog Dialogue*, Vol. 53, July 2019, pp. 20 – 22.

Schade, O., "Analysis of Rectifier Operation," *Proceedings of the I.R.E.*, Vol. 31, Is. 7, July 1943, pp. 341 – 361.

Silver, W., NØAX, *Grounding and Bonding for the Radio Amateur* (ARRL, 2017).

Zhang, H., "Basic Concepts of Linear Regulator and Switching Mode Power Supplies," Linear Technology AN140, 2013, **www.analog.com/media/en/technical-documentation/app-notes/an140.pdf**.

7.15 Power Source Projects

Construction of a power supply or accessory — basic to all of the radio equipment we operate and enjoy — can be one of the most rewarding projects undertaken by a radio amateur. Final testing and adjustment of most power-supply projects requires only a voltmeter, and perhaps an oscilloscope — tools commonly available to most amateurs.

General construction techniques that may be helpful in building the projects in this chapter are outlined in the **Construction Techniques** chapter. Other chapters in the *Handbook* contain basic information about the components that make up power supplies.

Safety must always be carefully considered during design and construction of any power supply. Power supplies contain potentially lethal voltages, and care must be taken to guard against accidental exposure. For example, electrical tape, insulated tubing ("spaghetti") or heat-shrink tubing is recommended for covering exposed wires, components leads, component solder terminals and tie-down points. Whenever possible, connectors used to mate the power supply to the outside world should be of an insulated type designed to prevent accidental contact.

Connectors and wire should be checked for voltage and current ratings. Always use wire with an insulation rating higher than the working voltages in the power supply. For supply voltages above 300 V, use wire with insulation rated accordingly. The **Component Data and References** chapter contains a table showing the current-carrying capability of various wire sizes. Scrimping on wire and connectors to save money could result in flashover, meltdown or fire.

All fuses and switches should be placed in the hot circuit(s) only. The neutral circuit should not be interrupted. Use of a three-wire (grounded) power connection will greatly reduce the chance of accidental shock. The proper wiring color code for 120 V circuits is: black — hot; white — neutral; and green — ground. For 240 V circuits, the second hot circuit generally uses a red wire.

POWER SUPPLY PRIMARY-CIRCUIT CONNECTOR STANDARD

The International Commission on Rules for the Approval of Electrical Equipment (CEE) standard for power-supply primary-circuit connectors for use with detachable cable assemblies is the CEE-22. The CEE-22 has been recognized by the ARRL and standards agencies of many countries. Rated for up to 250 V, 6 A at 65 °C, the CEE-22 is the most commonly used three-wire (grounded), chassis-mount primary circuit connector for electronic equipment in North America and Europe. It is often used in Japan and Australia as well.

When building a power supply requiring 6 A or less for the primary supply, a builder would do well to consider using a CEE-22 connector and an appropriate cable assembly, rather than a permanently installed line cord. Use of a detachable line cord makes replacement easy in case of damage. CEE-22 compatible cable assemblies are available with a wide variety of power plugs including most types used overseas.

Some manufacturers even supply the CEE-22 connector with a built-in line filter. These connector/filter combinations are especially useful in supplies that are operated in RF fields. They are also useful in digital equipment to minimize conducted interference to the power lines.

CEE-22 connectors are available in many styles for chassis or PC-board mounting. Some have screw terminals; others have solder terminals. Some styles even contain built-in fuse holders.

7.15.1 12 V, 15 A Linear Power Supply

This power supply is a linear 12 V, 15 A design by Ed Oscarson, WA1TWX. It is suitable for typical mobile radios and offers adjustable output voltage and current limiting. Supply regulation is excellent, typically exhibiting a change of less than 20 mV from no load to 15 A. This basic design, with heftier components and additional pass transistors, can deliver over 30 A — enough to supply a 100 W class transceiver. (All numbered notes, additional circuit design information, a discussion of how to change the supply voltage and/or current ratings, construction and testing notes, a PCB template and a complete parts list are included with this book's online content.)

CIRCUIT DESCRIPTION

Figure 7.56 is the supply's schematic. The ac line input is fused by F1, switched on and off by S1 and filtered by FL1. F1 and S1 are rated at about ¼ of the output current requirement (for 15 A output, use a 4 or 5 A slow-blow fuse or a similarly rated circuit breaker). FL1 prevents any RF from the secondary or load from coupling into the power line and prevents RF on the power line from disturbing supply operation. If your ac power line is clean, and you experience no RF problems, you can eliminate FL1, but it's inexpensive insurance.

When discharged, filter capacitor C1 looks like a short circuit across the output of rectifier U2 when ac power is applied. That usually subjects the rectifier and capacitor to a large inrush current, which can damage them. Fortunately a simple and inexpensive means of inrush-current limiting is available. Keystone Carbon Company (and others) produce a line of inrush-current limiters (thermistors) for this purpose. The device (RT1) is placed in series with one of the transformer primary leads. RT1 has a current rating of 6 A[1], and a cold resistance of 5 Ω. When it's hot, RT1's resistance drops to 0.11 Ω. Such a low resistance has a negligible effect on supply operation. Thermistors run *hot* so they must be mounted in free air, and away from anything that can be damaged by heat.[2,3]

The largest and most important part in the power supply is the transformer (T1). If purchased new, it can also be the most costly. Fortunately, a number of surplus dealers offer power transformers that can be used in this supply.

T1 produces 17 V ac RMS at 20 A; the center tap is not used. Bridge rectifier U2 provides full-wave rectification. Full-wave rectification reduces the ripple component of current that flows in the filter capacitor, resulting in less power dissipation in the capacitor's internal resistance. U2's voltage rating should be at least 50 V, and its current rating about 25% higher than the normal load requirement; a 2 A bridge rectifier will do. U2 is secured to the chassis or a heat sink because it dissipates heat.

C1 is a computer-grade electrolytic. Any capacitor value from 15,000 to 30,000 µF will suffice. This version uses a 19,000 µF, 40 V capacitor. The capacitor's voltage rating should be at least 50% higher than the expected no-load rectified de voltage. In this supply, that voltage is 25 V, and a 40 V capacitor provides enough margin,

R5, a 75 Ω, 20 W bleeder resistor, is connected across C1's terminals to discharge the supply when no load is attached or one is removed, Any resistance value from 50 Ω to 200 Ω is fine; adjust the resistor's wattage rating appropriately.

At the terminals of C1, we have a dc voltage, but it varies widely with the load applied. When keying a CW transmitter or switching a rig from receive to full output, 5 V swings can result. The dc voltage also has an ac ripple component of up to 1.5 V under full load. Adding a solid-state regulator (U1) provides a stable output voltage even with a varying input and load.

VOLTAGE REGULATOR IC AND PASS TRANSISTORS

The LM723 used at U1 has a built-in voltage reference and sense amplifier, and a 150 mA drive output for a pass-transistor array. U1's voltage reference provides a stable point of comparison for the internal regulator circuitry. In this supply, it's connected to the non-inverting input of the voltage-sense op amp. The reference is set internally to 7.15 V, but the absolute value is not critical because an output-voltage adjustment (R12) is provided. What is important is that the voltage is stable, with a specified variation of 0.05% per 1,000 hours of operation. This is more than adequate for the supply.

For the regulator to work properly, its ground reference must be at the same point as the output ground terminal. The best way to ensure this is to use the output GROUND terminal (J4) as a single-point ground for all of the supply grounds. Run wires to J4 from each component requiring a ground connection. Figure 7.56 attempts to show this graphically through the use of parallel connections to a single circuit node.

The output pass transistor array consists of a TIP112 Darlington-pair transistor (Q5) driving three 2N3055 power transistors (Q1-Q3). This two-stage design is less efficient than connecting the power transistors directly to the LM723, but Q5 can provide considerably more base current to the 2N3055s than the 150 mA maximum rating of the LM723. You can place additional 2N3055s in parallel to increase the output current capacity of the supply.

This design is not fussy about the pass transistors or the Darlington transistor used. Just ensure all of these devices have voltage ratings of at least 40 V. Q5 must have a 5 A (or greater) collector-current rating and a beta of over 100. The pass transistors should be rated for collector currents of 10 A or more, and have a beta of at least 10.[5]

Resistors R17, R18 and R19 prevent leakage current through the collector-base junction from turning on the transistor by diverting it around the base-emitter junction. When the pass transistors are hot, at the V_{CE} encountered in this design, the leakage current can be as high as 3 mA. The resulting drop across the 33 Ω resistors is 0.1 V — safely below the turn-on value for V_{BE}.

When unmatched transistors are simply connected in parallel they usually don't equally share the current.[6] By placing a low-value resistor in each transistor's emitter lead (emitter-ballasting resistors, R1-R3), equal current sharing is ensured. When a transistor with a lower voltage drop tries to pass more current, the emitter resistor's voltage drop increases, allowing the other transistors to provide more current. Because the voltage-sense point is on the load side of the resistors, the transistors are forced to dynamically share the load current.

With a 5 A emitter current, 0.25 V develops across each 0.05 Ω resistor, producing 1.25 W of heat. Ideally, a resistor's power rating should be at least twice the power it's called upon to dissipate. To help the resistors dissipate the heat, mount them on a heat sink, or secure them to a metal chassis. You can use any resistor with a value between 0.065 and 0.1 Ω, but remember that the power dissipated is higher with higher-value resistors (10 W resistors are used here).

At the high output currents provided by this supply, the pass transistors dissipate considerable power. With a current of 5 A through each transistor-and assuming a 9 V drop across the transistor — each device dissipates 45 W. Because the 2N3055's rating is 115 W when used with a properly sized heat sink, this dissipation level shouldn't present a problem. If the supply is to be used for continuous-duty operation, increase the size of the heat sink and mount it with the fins oriented vertically to assist in air circulation.

> **Switchmode DC-DC Converters**
>
> Inexpensive dc-dc converter modules available online can reduce or increase dc voltage from common power sources, avoiding the need for a full power supply. For example, a 5 V dc USB charging port can be converted to 9 V, as described by Tom Wheeler, NØGSG, in the online project article.

[1]See the full article included with this book's online content for a list of numbered notes.

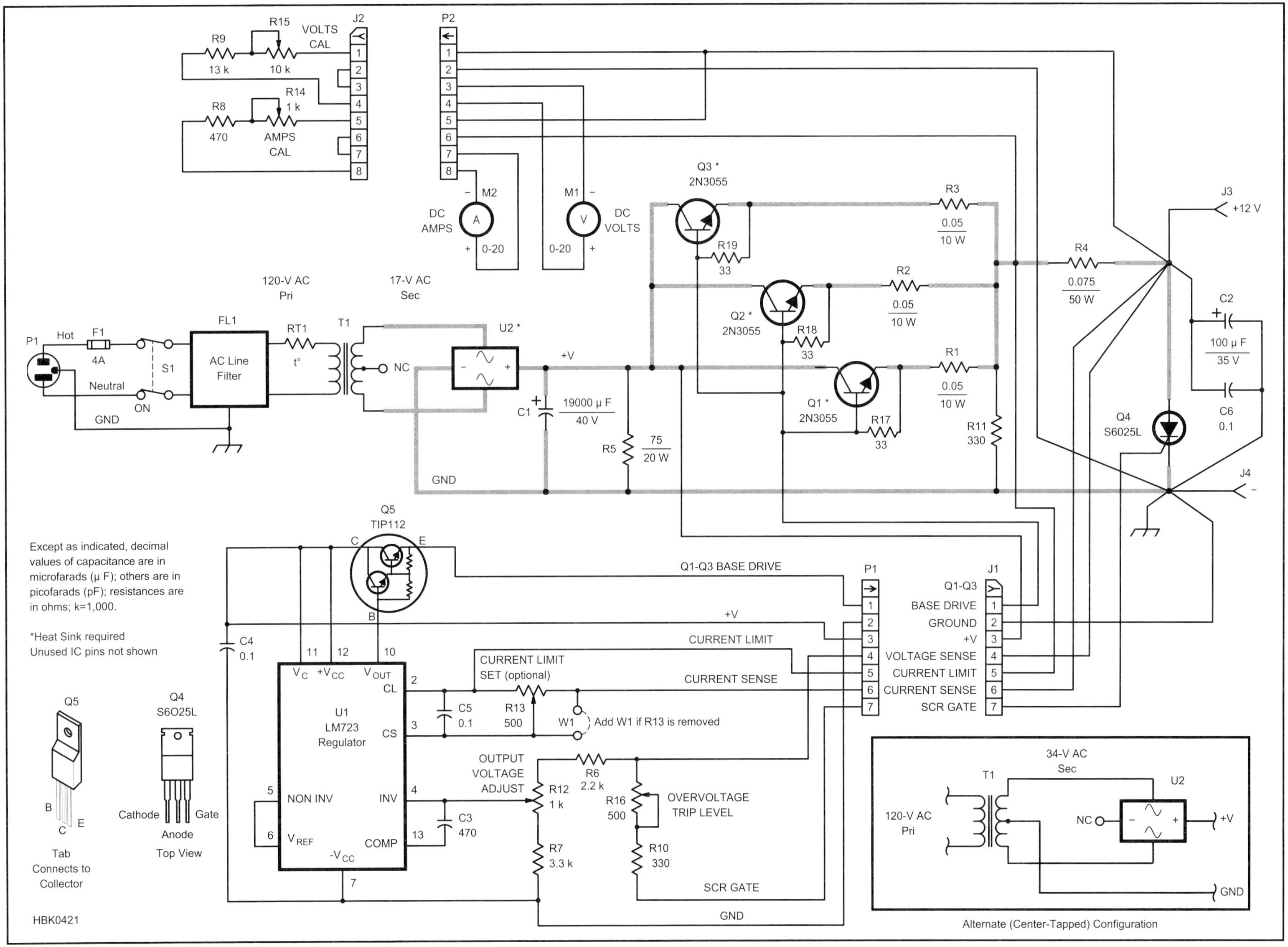

Figure 7.56 — Schematic of the 12 V, 15 A power supply. A parts list is included with this book's online content. Equivalent parts can be substituted. The bold lines indicate high-current paths that should use heavy-gauge (#10 or #12 AWG) wire. This schematic graphically shows wiring to a single-point ground; see text. The majority of the parts used in this supply are available as surplus components.

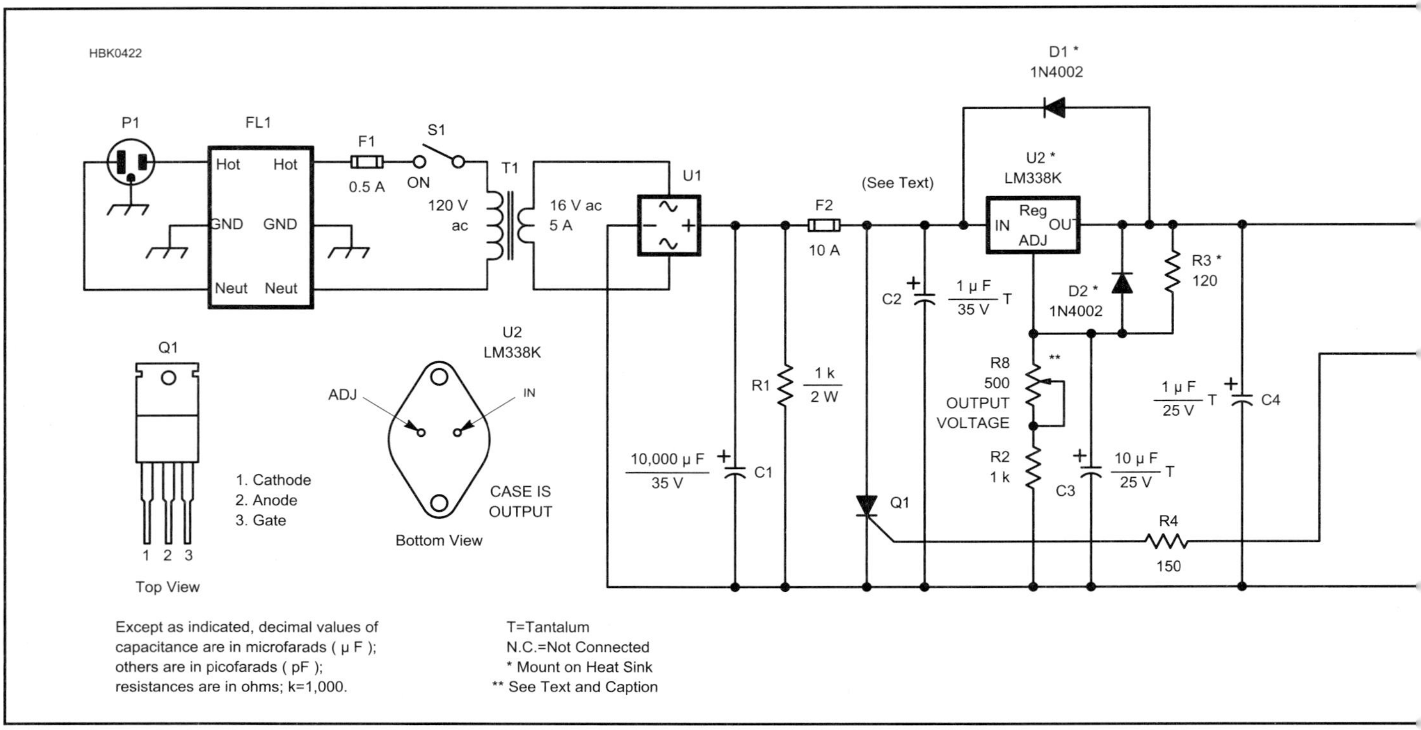

Figure 7.57 — Schematic for the 13.8 V, 5 A power supply. Unless otherwise specified, resistors are ¼ W, 5% tolerance. A PC board and U3 are available from FAR Circuits (www.farcircuits.net). Early versions of the circuit board from FAR mislabel U1 vs Q1. Check the board traces to be sure the parts are installed correctly. The PC board has mounting holes and pads to allow for handling different trimmer-potentiometer footprints. A PC board template is included with this book's online content. The author may be contacted at scskits@charter.net for assistance in obtaining parts and printed-circuit boards.

C1 — 10,000 μF, 35 V electrolytic.
C2, C4 — 1 μF, 35 V tantalum.
C3 — 10 μF, 35 V tantalum.
C5 — 0.1 μF, 25 V ceramic disc.
D1-D3 — 1N4002.
DS1 — Red LED.
F1 — Slow-blow 0.5 A fuse.
F2 — Fast-acting 10 A fuses; three required (see text).
FL1— Ac-line filter.
Q1 — BT152 400 V, 25 A SCR in TO-220A package (NTE5554)
R8 — 500 Ω, single-turn trimmer potentiometer.
R9 — 500 Ω or 1 kΩ, single-turn trimmer potentiometer.
S1 — SPST panel-mount switch.
T1 — 120 V primary, 16- to 20 V, 5 A secondary.
U1 — 100-PIV, 6 A bridge rectifier.
U2 — LM338K 5 A adjustable power regulator in a TO-3 package.
U3 — MC3423P1 overvoltage protection IC.
Misc: two panel-mount fuse holders; line cord; heat sinks for TO-3 case transistors; TO-3 mounting kit and heat-sink grease; black and red binding posts; chassis or cabinet; PC board; hardware, rubber hoods, heat-shrink tubing or electrical tape for F1 and FL1, hook-up wire.

The output-voltage sense is connected through a resistive divider to the negative input of U1. U1 uses the difference between its negative and positive inputs to control the pass transistors that in turn provide the output current. C3, a compensation capacitor, is connected between this input and a dedicated compensation pin to prevent oscillation. The output voltage is adjusted by potentiometer R12 and two fixed-value resistors, R6 and R7.[7] The voltage-sense input is connected to the supply's positive output terminal, J3.

Current sensing is done through R4, a 0.075 Ω, 50 W resistor connected between the emitter-ballasting resistors and J3. R4's power dissipation is much higher than that of R1, R2 or R3 because it sees the total output current. At 15 A, R4 dissipates 17 W. At 20 A, the dissipated power increases to 30 W.

U1 provides current limiting via two sense inputs connected across R4. Limiting takes place when the voltage across the sense inputs is greater than 0.65 V.[8] For a 15 A maximum output-current limit, this requires a 0.043 Ω resistor. By using a larger-value sense resistor and a potentiometer, you can vary the current limit. Connecting potentiometer R13 across R4 provides a current-limiting range from full limit voltage (8.7 A limit) to no limit voltage. This allows the current limit to be fine-tuned, if needed, and also permits readily available resistor values (such as the author's 0.075 Ω resistors) to be used. A current limit of 20 A is at the top end of the ammeter scale.

R20 maintains a small load of 35 – 40 mA depending on power supply output voltage. This reduces the effect of leakage current in the pass transistors and keeps the regulator's feedback action active, even if no external load is connected.

METERING

Voltmeter M1 is a surplus meter. R8 and potentiometer R15 provide for voltmeter calibration. If the correct fixed-value resistor is available, R15 can be omitted. The combined value of the resistor and potentiometer is determined by the full-scale current requirement of the meter used.[9]

Ammeter M2 is actually a voltmeter (also surplus) that measures the potential across R4. The positive side of M2 connects to the high side of R4. R8 and potentiometer R14 connect between the positive output terminal (J3) and the negative side of M2 to provide calibration adjustment. The values of R8 and R14 are determined by the coil-current requirements of the meter used. (Digital panel meters or a dedicated DMM can be used instead of separate analog meters.)

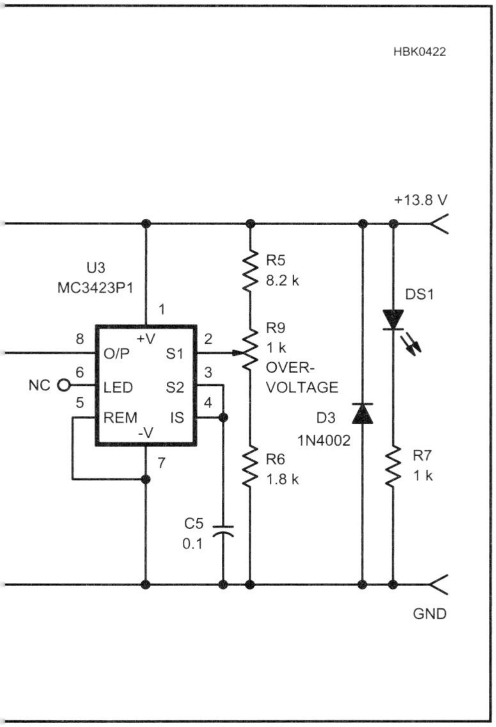

OUTPUT WIRING AND CROWBAR CIRCUIT

The supply output is connected to the outside world by two heavy-duty banana jacks, J3 and J4. C2, a 100 µF capacitor, is soldered directly across the terminals to prevent low-frequency oscillation. C6, a 0.1 µF capacitor, is included to shunt RF energy to ground. Heavy-gauge wire must be used for the connections between the pass transistors and J3 and between chassis ground and J4. The voltage-sense wire must connect directly to J3 and U2's ground pin must connect directly to J4 (see Figure 7.56). This provides the best output voltage regulation.

An over-voltage crowbar circuit prevents the output voltage from exceeding a preset limit. If that limit is exceeded, the output is shunted to ground until power is removed. If the current-limiting circuitry in the supply is working properly, the supply current-limits to the preset value. If the current limiting is not functioning, the crowbar causes the ac-line fuse to blow. Therefore, it's important to use the correct fuse size: 4 to 5 A for a 15 A supply.

The crowbar circuit is a simple design based on an SCR's ability to latch and conduct until the voltage source is removed. The SCR (Q4) is connected across output terminals J3 and J4. (The SCR can also be connected directly across the filter capacitor, C1, for additional protection.) R10 and potentiometer R16 in series with the Q4's gate provide a means of adjusting the trip voltage. The prototype crowbar is set to conduct at 15 V. The S6025L SCR is rated at 25 A and should be mounted on a metal chassis or heat sink. (Note: Some SCRs are isolated from their mountingtabs, others are not. The S6025L and the 65-ampere S4065J are isolated types. If the SCR you use is not isolated, use a mica washer or thermal pad to insulate it from the chassis or heat sink.)

The bold lines in Figure 7.56 indicate high-current paths that should use heavy-gauge (#10 or #12 AWG) wire. Traces that are connected to the output terminals in the schematic by individual lines should be connected directly to the terminals by individual wires. This establishes a 4-wire measurement, where the heavy wires carry the current (and have voltage drops) and the sense wires carry almost no current and therefore voltage errors are not caused by voltage drops in the wiring. If desired, the sense wire can be carried out to the load, but that may introduce noise into the sense feedback circuit, so use caution if that is done.

7.15.2 13.8 V, 5 A Linear Power Supply

This power supply was designed by Ben Spencer, G4YNM, provides 13.8 V dc at 5 A, suitable for many low-power transceivers and accessories. It features time-dependent current limiting and short-circuit protection, thermal overload protection within the safe operating area of the regulator IC, and overvoltage protection for the equipment it powers. The prototype supply powers a 25 W transmitter that continually draws 4.5 A.

Construction, testing and calibration are straightforward, requiring no special skills or equipment. Many of the components can be found in junk boxes, or purchased at hamfests or from mail-order suppliers.

CIRCUIT DESCRIPTION

Figure 7.57 is the power supply schematic. Incoming ac line current is filtered by a chassis-mounted line filter (FL1) and, after passing through the fuse (F1), is routed S1 to T1.

U1 rectifies, and C1 filters, the ac output of T1. U2 is an LM338K voltage regulator. This IC features a current-limited continuous output of 5 A, with a guaranteed peak output of 7 A. It also has on-chip thermal and safe-operating-area protection for itself. U2's output voltage is set by two resistors (R2 and R3) and a trimmer potentiometer (R8), which allows for adjustment over a small range. U2's input and output are bypassed by C2, C3, and C4. D1 and D2 protect U2 against these capacitors discharging through it.

Overvoltage protection is provided by an SCR, Q1, across the regulator input. Normally, Q1 presents an open circuit, but under fault conditions, it's triggered and short-circuits the unregulated dc input to ground. This discharges C1 through F2 which is rated at 10 A in order for C1 to quickly discharge below 12 V, avoiding damage to connected equipment.

U3, an overvoltage-protection IC, continuously monitors the output voltage. When the output voltage rises above a predetermined level, U3 starts charging C5. If the overvoltage duration is sufficiently long, U3 triggers Q1. This built-in delay (about 1 ms) allows short transient noise spikes on the output voltage to be safely ignored while still triggering the SCR if a true fault occurs. The monitored voltage is set by R5 and R6 and trimmer potentiometer R9, which allows for adjustment over a limited range.

D3 protects the supply from reverse-polarity discharge from connected equipment. The presence of output voltage is indicated by an LED, DS1. R7 is a current limiting resistor for DS1.

CONSTRUCTION

How you construct your supply depends on the size of the components and enclosure you use. General physical layout is not important, although there are a couple of areas that require some attention. In the unit shown in **Figure 7.58**, FL1, the fuse holders, S1, the heat sink. DS1 and the binding posts are mounted on the front and rear enclosure panels. T1 and the PC board are secured to the enclosure's bottom plate. C1's mounting clamp is attached to the rear panel. Bleeder resistor R1 is connected directly across C1's terminals. D3 is soldered directly across the output binding posts.

U2, D1, D2 and R3 are all mounted directly on the heat sink with the transistor pins and solder lugs acting as a terminal strip. It's important to keep R3 attached as closely as possible to U2's terminals to prevent instability. Use a TO-3 mounting kit and heat-conductive grease or thermal pad to electrically isolate U2 from the heat sink.

Mount U2, C1, and the PC board close to each other and keep the wire runs between these components as short as possible. Excessively long wire runs may lead to unpredictable behavior.

Cover all ac-input wiring (use insulated wire and heat-shrink tubing) to prevent electrical shock and route the ac wiring away from the dc wiring. Mount the heat sink on the enclosure with fins oriented vertically. Louvers or ventilation holes in the cabinet will help cool internal components.

TEST AND CALIBRATION

An accurate multimeter covering ranges of 30 V dc and 10 A dc is required. A variable resistive load with a power rating of 100 W is

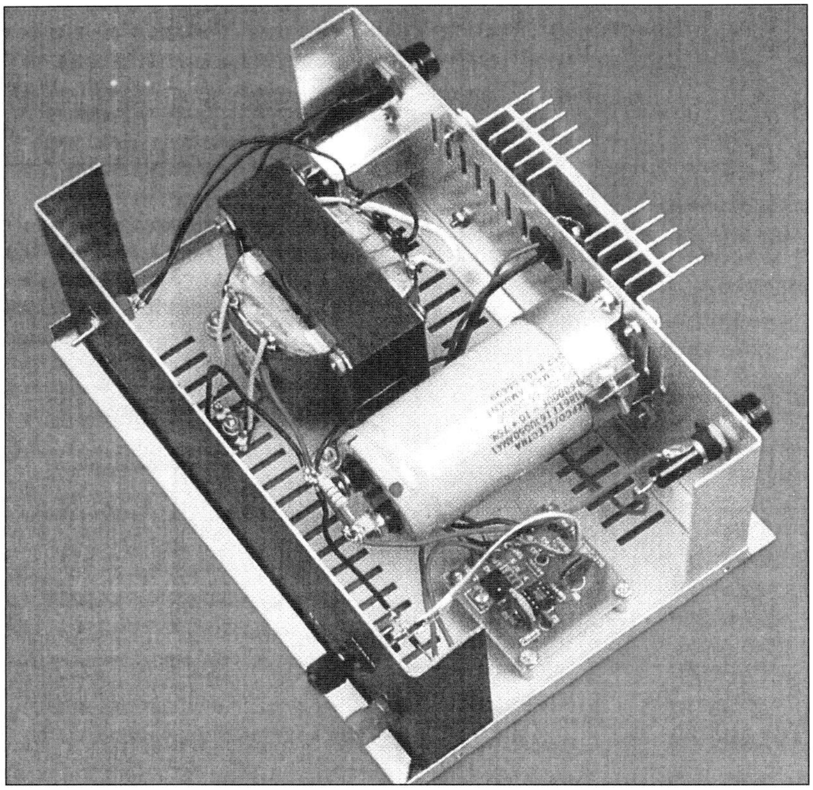

Figure 7.58 — Physical layout of the 13.8 V, 5 A power supply. On the rear panel, left, are the ac-line filter and F1. The regulator's heat sink is at the middle of the panel and F2 is to the right. At the bottom of the enclosure, in front of T1, is the diode bridge rectifier. Because C1 is too tall to mount vertically within the Hammond #1426O cabinet, its mounting clamp is secured to the inside rear panel. Immediately to the right of C1 is the PC board. On the front panel are the on/off switch, LED power-on indicator and output-voltage binding posts.

also needed; this can he made using a heat-sink-mounted 2N3055 power transistor and a couple of components as shown in the next project.

First, set R8 (OUTPUT VOLTAGE) fully clockwise and R9 (OVERVOLTAGE) fully counterclockwise. Insert a fuse in the dc line at F2. Connect the ac line, turn on S1 and check that DS1 lights. Measure the output voltage: it should be about 12 V. Adjust R8 counterclockwise until you obtain 14.2 V output; this sets the trip voltage. (Note that CCW adjustment *increases* voltage.)

While monitoring the output voltage, gradually adjust R9 clockwise until the voltage suddenly falls to zero. This indicates that the SCR has triggered and blown F2. Disconnect the ac line cord from the wall socket. *Don't make any adjustment to R9!* Instead, adjust R8 fully clockwise.

With the ac line cord removed, check that F2 is open. Replace F2 with a new fuse (now you know why two of the three fuses are called for). Reconnect the ac line cord, and while continually monitoring the output voltage, gradually adjust R8 until Q1 again triggers at 14.2 V, blowing F2. If you find the adjustment of R9 to be too sensitive, use a 500 Ω potentiometer in its place and reduce the value of R5 (if necessary) to provide the required adjustment range.

Again disconnect the line cord from the wall socket, set R8 fully clockwise, replace F2 (there's the third fuse!) and reset R8 for 13.8 V. This completes the voltage calibration and overvoltage protection tests. The power supply is now set to 13.8 V output, with the overvoltage protection set for 14.2 V.

Adjust R2 of the variable resistive load in Figure 7.59 to maximum resistance (minimum load current) and connect it to the power supply output in series with the ammeter. Turn on the power supply and gradually adjust R2 until a current of 5 A flows. Decrease the resistance further and check that the current limits between 5.5 A and 0.5 A.

Finally, be thoroughly unpleasant and apply a short circuit via the ammeter. Check that the current-limiting feature operates correctly. The prototype limited at approximately 3.5 A. Disconnect the ac line cord and test equipment, dose up the case and your power supply is ready for service.

7.15.3 Adjustable Resistive Load

Figure 7.59 shows the schematic of an adjustable resistive load that can be used to test and adjust power supplies at currents up to 10 A if the 2N3055 transistor is mounted on an adequate heat sink. R1 and R2 vary the base bias to control the collector current of Q1. For extended use, be sure to use a large heat sink with adequate ventilation. Use heavy wire through the current meter to the collector and from the emitter of Q1.

7.15.4 Inverting DC-DC Converter

It's often the case that you need +V and –V when all you have is +V. For example, you need +12 V and –12 V, but all that's available is +12 V. It would be really handy to have a "black box" that would give you –V out when you put +V in, and work over a range of voltages without adjustment. This project by Jim Stewart, which originally appeared in the January 2013 issue of *Nuts and Volts Magazine* (**www.nutsvolts.com**), fills that need by using a switchmode voltage mirror to supply more than 100 mA without a significant drop in voltage.

The following text summarizes how the circuit works. A PDF version of the complete article is included with this book's online content. It contains more details about the circuit's design, plus construction and testing information. More information and support files are available

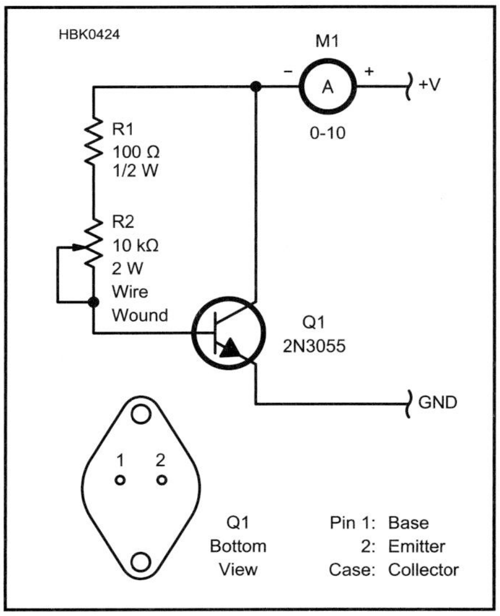

Figure 7.59 — An active resistive load to use in testing power supplies up to 10 A with adequate heat sinking. Adjustment of R2 is sensitive.
M1 — Multimeter or ammeter capable of measuring 10 A.
Q1 — 2N3055; mount on heat sink.

from **www.nutsvolts.com/magazine/article/build_an_inverting_dc_dc_converter**.

CIRCUIT DESIGN

The circuit in **Figure 7.60** is a buck-boost dc-dc converter as described earlier in this chapter. This particular design has five important parameters:

- Input voltage: V_{IN}
- Output voltage: V_{OUT}
- Load resistance: R_{LOAD}
- Oscillation frequency: f
- Inductance value: L

The parameters are related to each other by $f \times L < (V_{IN} / V_{OUT})^2 \times (R_{LOAD} / 8)$. For $V_{IN} = V_{OUT}$, the equation simplifies to $f \times L < (R_{LOAD} / 8)$. (See the complete article included with this book's online content for the derivation of these equations.)

Figure 7.60 shows the schematic of the circuit. A square wave oscillator (U2) drives a MOSFET (Q2) that, in turn, drives a PNP switching transistor (Q3). The oscillator is enabled/disabled by the output of the comparator (U1) with inputs that are a set point voltage (V_S on the comparator's + input) and a feedback voltage (V_F at the comparator's – input). (Comparators are described in the **Analog Fundamentals** chapter.) The frequency of the oscillator is $f = 1 / 2.2 (R7 \times C3)$.

When the output voltage, V_{OUT}, is the correct value, V_S exceeds V_F by a few µV and the oscillator is disabled by Q1 so that the PNP switch is off. When V_S is less than V_F, the oscillator is enabled and the PNP transistor switches on and off. V_S is half the input voltage $V_S = V_{IN}/2$ as set by the R3-R4 voltage divider.

V_F is set by the R1-R2 divider to equal $V_{IN}/2$ when the output voltage $V_{OUT} = -V_{IN}$. D2 is a commutating diode that blocks +V from the output while charging the inductor. It then provides a current path for the discharging inductor to transfer charge to the output capacitor.

The switching action continues until the voltage on the capacitor equals $-V_{IN}$. At that point, the comparator disables the oscillator and Q3 stays OFF. When voltage across the capacitor drops, the comparator enables the oscillator and the inductor is pumped up again.

Since there is a single input voltage, each op-amp uses a resistor divider to "split the rail" to create a signal ground. That allows the negative input to sense positive and negative voltages with respect to the positive input. C2 and C4 bypass the signal grounds to the supply voltage ground.

C1 and C5 are low-ESR (equivalent series resistance — see the **Electrical Fundamentals** chapter) aluminum electrolytic capacitors. Tantalum capacitors might be a bit better, but are more expensive. ESR determines how much power the capacitor can safely dissipate. $P_{DISS} = I^2_{RC} \times ESR$, where I_{RC} is the ac ripple current in the capacitor. The capacitors chosen are rated for a maximum ripple current of 840 mA.

D1 and D2 are Schottky diodes that can go from conducting to non-conducting very quickly, allowing a high switching frequency. They also have a low voltage drop when conducting to reduce power loss. D1 limits the output voltage in case the feedback fails.

Q3 is a ZTX550 PNP transistor, chosen because its specifications suit this application well:

- Maximum power dissipation, P_{MAX} = 1 W @ 25 °C
- Maximum continuous collector current, I_C = 1 A
- Maximum collector-base voltage, V_{CBO} = 60 V
- Maximum saturation voltage, V_{CE} = 0.25 V @ I_C = 150 mA
- Transition frequency, f_T = 150 MHz minimum
- Package: E-Line (slightly smaller than TO-92)

For this design, the chosen values are L = 220 µH, f = 45 kHz, and R_{LOAD} = 100 Ω. Verifying that f × L is less than R_{LOAD} / 8, $f \times L = (45 \times 10^3) \times (220 \times 10^{-6}) = 9{,}900 \times 10^{-3} = 9.9\ \Omega$ and $R_{LOAD} / 8 = 100/8 = 12.5\ \Omega$ which is less than R_{LOAD} / 8 = 12.5. Higher values of R_{LOAD} (lower output current) also satisfy the equation. Increasing the value of L or f will require that the circuit be tested to verify that it works.

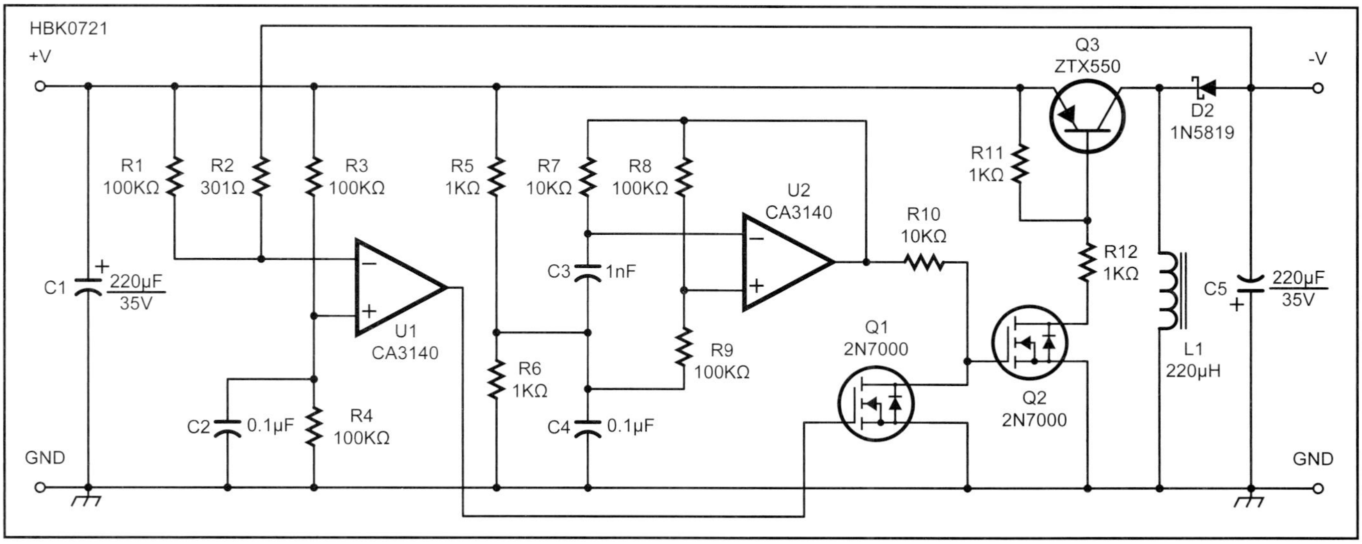

Figure 7.60 — A voltage mirror dc-dc power converter based on a buck-boost converter. The negative output –V is equal in magnitude to the positive input voltage +V at an output current up to approximately 100 mA

C1 — 220 µF, 35 V (Digi-Key part #493-1578-ND or equiv).
C2 — 0.1 µF, 50 V, ceramic.
C3 — 1 nF film, 5% (Digi-Key part #399-5871-ND or equiv).
C4 — 0.1 µF, 50 V, ceramic.
C5 — 220 µF, 35 V (Digi-Key part #493-1578-ND or equiv).
D1 — Not required, do not install.
D2 — Schottky, 1 A, 1N5819 or equiv.
L1 — 220 µH (Digi-Key part #811-1316-ND or equiv).
Q1, Q2 — 2N7000.
Q3 — ZTX550 PNP.
R1, R3, R4, R8, R9 — 100 kΩ, ¼ W, 1%.
R2 — 301 kΩ, ¼ W, 1%.
R5, R6, R11, R12 — 1 kΩ, ¼ W, 1%.
R7, R10 — 10 kΩ, ¼ W, 1%.
U1, U2 — CA3140.
Terminal Blocks (optional) — Two-position, 5 mm spacing (Jameco part #2094485 or equiv).

7.15.5 High-Voltage Power Supply

The online content includes "A Deluxe High Voltage Supply" by Jim Garland, W8ZR, which is discussed here, and "A Small, Lightweight High-Voltage Switch-Mode Power Supply" by Ralph Crumrine, NØKC.

This two-level, high-voltage power supply was designed and built by Dana G. Reed, W1LC. It was designed primarily for use with an RF power amplifier. The supply is rated at a continuous output current of 1.5 A, and will easily handle intermittent peak currents of 2 A. The 12 V control circuitry and the low-tap setting of the plate transformer secondary make it straightforward to adapt the design to homemade tube amplifiers.

The step-start circuit is straightforward and ensures that the rectifier diodes are current-limited when the power supply is first turned on. A 6 kV meter is used to monitor high-voltage output.

Figure 7.61 is a schematic diagram of the bi-level supply. An ideal power supply for a high-power linear amplifier should operate from a 240 V circuit, for best line regulation. A special, hydraulic/magnetic circuit breaker also serves as a disconnect for the plate transformer primary. Don't substitute a standard circuit breaker, switch or fuses for this breaker; fuses won't operate quickly enough to protect the amplifier or power supply in case of an operating abnormality. The 100 kΩ, 3 W bleeder resistors are of stable metal-oxide film design. These resistors are wired across each of the 14 capacitors to equalize voltage drops in the series-connected bank. This choice of bleeder resistor value provides a lighter load (less than 25 W total under high-tap output) and benefits mainly the capacitor-bank filter by yielding much less heat as a result. A reasonable, but longer bleed-down time to fully discharge the capacitors results — about nine minutes after power is removed. A small fan is included to remove any excess heat from the power supply cabinet during operation.

POWER SUPPLY CONSTRUCTION

The power supply can be built in a 23½ × 10¾ × 16-inch cabinet. The plate transformer is quite heavy, so use ⅛-inch aluminum for the cabinet bottom and reinforce it with aluminum angle for extra strength and stability. The capacitor bank will be sized for the specific capacitors used. This project employed ⅜-inch thick polycarbonate for reasonable mechanical stability and excellent high-voltage isolation. The full-wave bridge consists of four commercial diode block assemblies.

POWER SUPPLY OPERATION

When the front-panel breaker is turned on, a single 50 Ω, 100 W power resistor limits its primary inrush current to a conservative value as the capacitor bank charges. After approximately two seconds, step-start relay K1 actuates, shorting the 50 Ω resistor and allowing full line voltage to be applied to the plate transformer. No-load output voltages under low- and high-tap settings as configured and shown in Figure 7.61 are 3050 V and 5400 V, respectively. Full-load levels are somewhat lower, approximately 2800 V and 4900 V. If a tap-select switch is used as

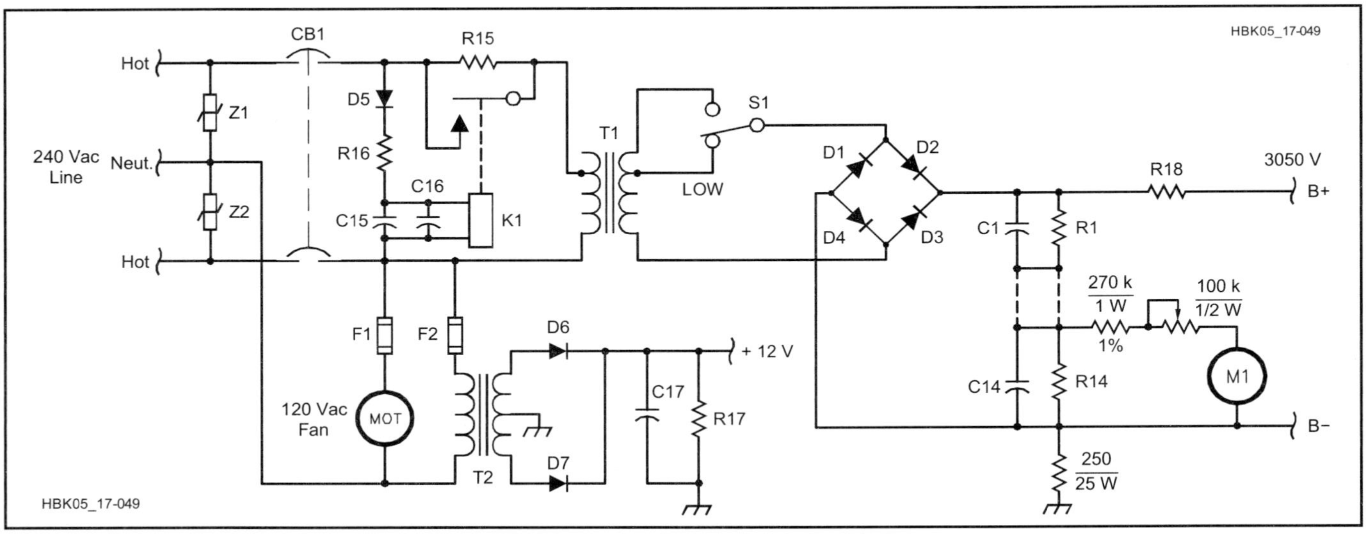

Figure 7.61 — Schematic diagram of the 3050 V/5400 V high-voltage power supply.

C1-C14 — 800 µF, 450 V electrolytic.
C15, C16 — 4700 µF, 50 V electrolytic.
C17 — 1000 µF, 50 V electrolytic.
CB1 — 20 A hydraulic/magnetic circuit breaker (TE Connectivity/Potter & Brumfield W68-X2Q12-20 or equiv). 40 A version required for commercial applications/ service (TE Connectivity/Potter & Brumfield W92-X112-40).
D1-D4 — String of 1000 PIV, 6 A diodes (6A10 or equiv).
D5 — 1000 PIV, 3 A, 1N5408 or equiv.
D6, D7 — 200 PIV, 3 A, 1N5402 or equiv.
F1, F2 — 0.5 A, 250 V (Littelfuse 313 Series, 3AG glass body or equiv).
K1 — DPDT power relay, 24 V dc coil; both poles of 240 V ac/25 A contacts in parallel (TE Connectivity/Potter & Brumfield PRD-11DYO-24 or equiv).
M1 — High-voltage meter, 6 kV dc full scale. (Important: Use a 1 mA or smaller meter movement to minimize parallel-resistive loading at R14. Also, select series meter-resistor and adjustment-potentiometer values to calibrate your specific meter. Values shown are for a 1 mA meter movement.)
MOT1 — Cooling fan, 119 mm, 120 V ac, 30 – 60 CFM, (EBM Pabst 4800Z or equiv).
R1-R14 — Bleeder resistor, 100 kΩ, 3 W, metal oxide film.
R15 — 50 Ω, 100 W.
R16 — 3.9 kΩ, 25 W.
R17 — 30 Ω, 25 W.
R18 — 20 Ω, 50 W.
S1 — Ceramic rotary, 2 position tap-select switch (optional). Voltage rating between tap positions should be at least 2.5 kV. Mount switch on insulated or ungrounded material such as a metal plate on standoff insulators, or an insulating plate, and use only a *nonconductive* or otherwise *electrically-isolated* shaft through the front panel for safety.
T1 — High-voltage plate transformer, 220/240 V primary, 2000/3500 V, 1.5 A CCS JK secondary, Hypersil C-core (Hammond Engineering, www.hammfg.com). Primary 220 V tap fed with nominal 240 V ac line voltage to obtain modest increase in specified secondary voltage levels.
T2 — 120 V primary, 18 V CT, 2 A secondary (Mouser 41FJ020).
Z1-Z2 — 130 V MOV.

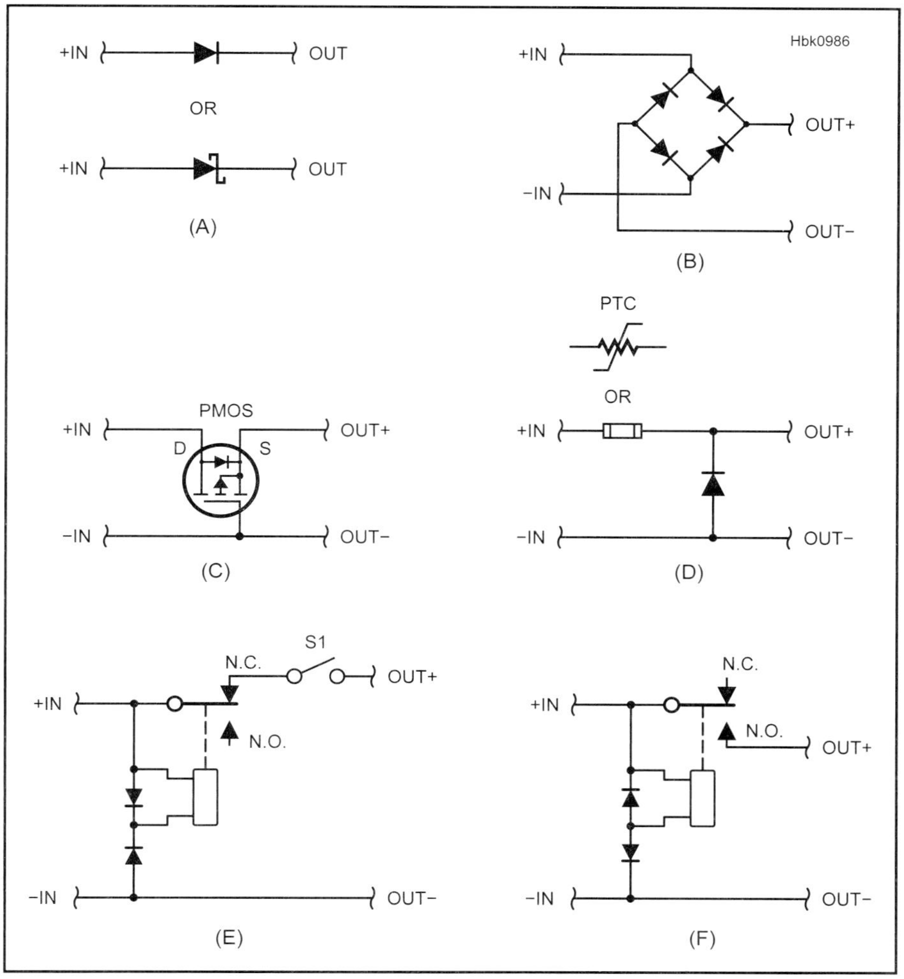

Figure 7.62 — Passive circuits for reverse-polarity protection. Series diode (A). Full-wave rectifier (B). PMOS MOSFET with integral body diode shown as separate component (C). Shunt diode (D). Schottky barrier diodes may be used in all circuits as a substitute for silicon junction rectifiers. See text for circuit comparison. Relay-based circuits can be based on normally-closed contacts (E) or normally-open contacts (F) as compared in the text. It is recommended that for use in amateur stations that a 0.001-0.01 uF capacitor be added across each diode at a power input. This will eliminate harmonics or mixing products being generated by any RF current on the power connection leads.

described in the schematic parts list, it should only be switched when the supply is off.

7.15.6 Reverse-Polarity Protection Circuits

*The following material was collected from various public-domain sources by Terry Fletcher, WAØITP(**www.wa0itp.com/revpro.html**) and published in the* QRP Quarterly, *Spring 2012 issue. (**www.qrparci.org**)*

DC power is the standard for most amateur radios and accessories, usually a nominal 12 V (10.5 to 13.8 V). There are many different types of connectors used for dc power — from screw terminals to custom-molded multi-pin designs. This makes it easy to accidentally apply power with reversed polarity and damage equipment. Even a few milliseconds of reversed power can be sufficient to destroy a semiconductor or burn out a narrow PCB trace. This can be a particular problem when using 9 V batteries as it is easy to reverse the snap-on connector when changing or installing a battery.

The following collection of circuits illustrates ways to protect equipment from reverse-polarity dc power. The suitability of the circuits depends on the equipment and power source. All dc power sources, particularly batteries, should be fused or current-limited to mitigate fire hazards and other damage from overheating wires and other conductors.

Figure 7.62A-D shows several passive circuits that dissipate some power due to the series forward voltage drop of the diodes. At currents above 1 A, the power dissipated can easily exceed 1 W and the maximum junction temperature of the diode can be exceeded without some sort of thermal protection or heat sinking. (The shunt diode circuit in Figure 7.66D does not dissipate power.) **Figure 7.62E-F** shows two methods of using an electromechanical relay that avoid the forward voltage drop of diode-based protection circuits. Which circuit you choose depends on the type of equipment and constraints on power dissipation and voltage drop.

PASSIVE CIRCUITS

Blocking diode (Figure 7.62A) — A series diode is very simple and inexpensive. Its PIV rating should be at least twice the expected applied voltage — 50 V PIV is a good minimum value for automotive and 12 V dc use. Its maximum average forward current rating should be several times the expected maximum steady-state current draw.

Remember that a silicon junction diode's forward voltage drop, V_f, is at least 0.6 V and can approach 1.0 V at high forward current. This can result in significant power dissipation ($P = V_f \times I_f$) and cause the diode's maximum rated junction temperature to be exceeded unless some means of cooling the diode is provided. See the article "Diode Voltage Drops Raise Battery Power Drain" in the online information for more details.

The forward voltage drop of the diode will also reduce the voltage available to the equipment being powered. This will raise the minimum allowable power supply voltage for the equipment to operate properly. For example, if a piece of equipment is rated to operate properly at or above 11 V, a series diode with V_f = 0.6 V raises the minimum allowable power supply voltage to 11 + 0.6 = 11.6 V. This may be significant in battery-powered installations.

The Schottky barrier diode shown as an alternate may be a better choice due to its forward voltage drop being lower by several tenths of a volt, reducing power dissipation. A Schottky diode may be used in place of any of the diodes in Figure 7.62. Be sure the reverse current leakage of the Schottky diode is acceptable.

Full-wave bridge rectifier (Figure 7.62B) — The full-wave circuit has the advantage of always supplying voltage with the proper polarity to the equipment being powered. Full-wave rectifiers are also available as integrated packages, making them easy to install. Remember that the forward voltage drop and power dissipation of this circuit will be twice that of the single series diode because two diodes are always in series with supply current.

PMOS P-channel MOSFET (Figure 7.62C) — This circuit uses a P-channel enhancement-mode (PMOS) MOSFET that conducts current with the gate connected as shown. PMOS devices have low on-resistance ($R_{ds(on)}$) and high maximum current ratings. Devices with on-resistance of 0.050 Ω and lower are commonly available. For more information about using PMOS and NMOS devices for polarity protection see Maxim Electronics Application Note

636, "Reverse-Current Circuitry Protection" (**www.maximintegrated.com/en/design/technical-documents/app-notes/6/636.html**). N-channel devices may also be used in the current return or ground lead, but opening the return connection can create other problems inside the equipment and with other devices on the same power circuits.

Shunt diode with fuse (Figure 7.62D) — These configurations use a single diode that acts to blow a fuse (either a fusible-link or positive temperature coefficient PTC resettable device) if reverse polarity voltage is applied. The advantage of this circuit is that no power is dissipated by the diode during normal operation. The diode must be sufficiently rated to handle the high surge current from shorting the power source and have current ratings significantly higher than the fuse current rating.

If the diode fails shorted or in a low-resistance state, it will continue to blow the fuse until replaced. If the diode fails open or high-resistance, it will no longer protect the circuit. If shunt diode protection is used and the fuse opens, check the diode to be sure it has not failed as well.

RELAY-BASED CIRCUITS

Relay-based circuits have the advantage of little to no voltage drop, even at high currents as long as the contact ratings are sufficient. The circuits are more complex than the diode-based circuits in the preceding section but can generally handle more current and are not damaged by reverse polarity voltages. The circuits reset themselves automatically.

Relay with normally-closed contacts (Figure 7.62E) — There is no current drain through the relay coil until reverse-polarity is applied. However, there will be a few milliseconds during which reverse polarity volt-age is applied if no power switch (S1) is used or the power switch is closed. This is generally enough time for damage to occur so this circuit is only recommended if a power switch is used to turn the equipment ON and OFF.

Relay with normally-open contacts (Figure 7.62F) — The relay contacts close and supply power to the equipment only when applied voltage has the proper polarity. The relay coil draws current continuously during normal operation. This may be unacceptable for low-power and battery-powered equipment.

ACTIVE CIRCUIT

The circuit in **Figure 7.63** uses a pair of P-channel MOSFETs to turn power on and off, while blocking reverse current flow. The circuit can be controlled manually or with a control circuit, such as a microprocessor. Using low ON-resistance MOSFETs avoids the voltage drop of a series diode. The circuit is part of the QSX transceiver designed by Hans Summers, GØUPL, of QRP Labs (**qrp-labs.com**) and is presented here with his permission.

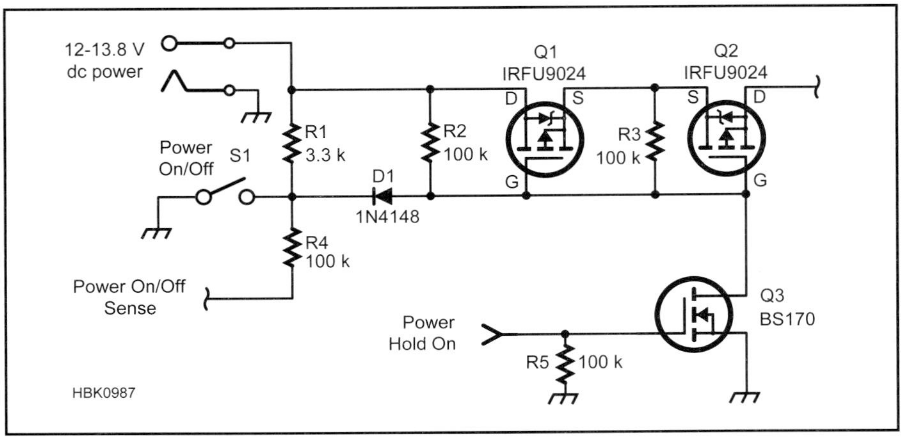

Figure 7.63 — Active switching circuit. See text for details.

Two P-channel MOSFETs in series provide the reverse-polarity protection. Note the orientation of drain and source and body diodes in Q1 and Q2 are reversed. If only Q2 is present, its body diode would allow reverse current to flow. Q1's body diode blocks reverse current. Both Q1 and Q2 must be on to allow forward current to flow. (If reverse polarity protection is not needed, Q1 and R3 can be removed.) To turn on a P-channel MOSFET, V_{gs}, the gate voltage must be negative with respect to the source by at several volts, typically 3V.

When external positive dc power is applied, the body diode of Q1 applies voltage to R3. This pulls the gates of both Q1 and Q2 to the external power voltage and turns the channels of both Q1 and Q2 off. The body diode of Q2 is reverse biased and non-conducting.

When S1 is closed, both gates are pulled to ground or common, switching on both transistors and allowing current to flow through the switch. If S1 is a toggle switch, both R5 and Q3 are not needed. When S1 is re-opened, the gates of Q1 and Q2 are pulled up again and the switch turns off.

If S1 is a momentary switch such as a push-button, R4 can supply a signal to a control circuit indicating S1 is closed. The control circuit can then generate the Power Hold On signal to Q3 (typically +3V or more). This turns Q3 on, pulling the gates of Q1 and Q2 to ground and holding the switch on. When S1 is closed again, the control circuit can remove the Power Hold On signal, allowing the switch to turn off as soon as S1 is opened.

7.15.7 Simple Sealed Lead-Acid Battery Float Charger and Switch

This charger was designed by John Boal, K9JEB, to keep a sealed lead-acid (SLA) battery charged from an external power source such as a vehicle, solar panel, or ac power supply. The battery will remain connected to the load if external power is removed or the power source voltage drops to below the

Figure 7.64 — The charging circuit shown mounted directly on a 9 Ah SLA battery. Note the USB charging module mounted next to J1 at the left-hand side of the board.

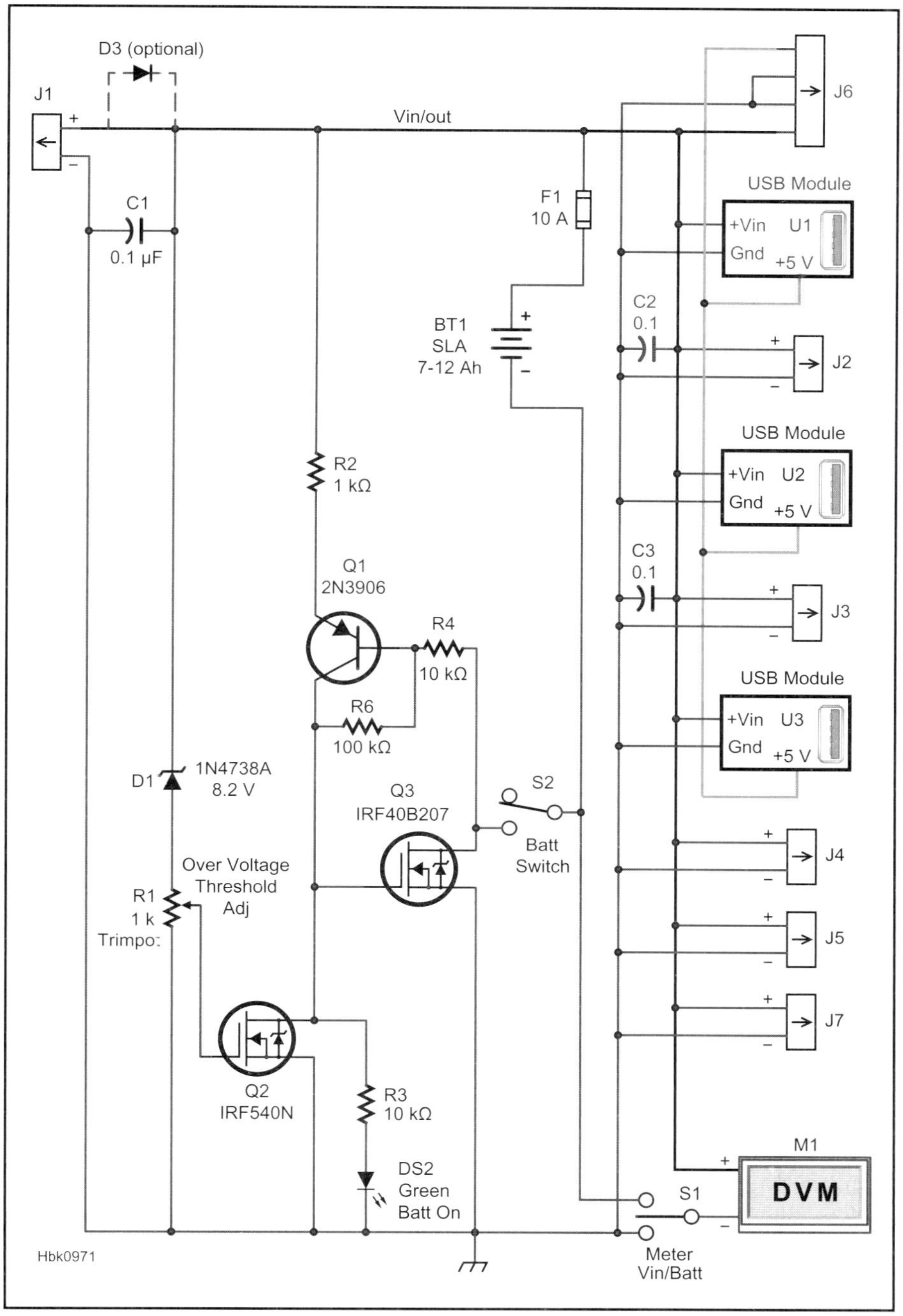

Figure 7.65 — Schematic of the battery charger circuit. Resistors are ⅛ W, 5% tolerance. Equivalent parts may be substituted. A complete kit of parts with PC board as well as separate PC boards are available from the author at his website, k9jeb.com. Enclosure design is the responsibility of the builder.

BT1 — SLA battery: 7-12 Ah (not supplied with parts kit)
C1, C2, C3 — 0.1 µF, 50 V ceramic
D1 — 1N4738A, 8.2 V Zener diode
D2 — LED, green
D3 — 10A10 10A Schottky diode (optional, install in series with J1)
F1 — 15A automotive blade fuse, standard
J1, J2, J3, J4, J5, J7 — Anderson Powerpole connector pair, including housing and contact
J6 — 0.1-inch DIP header, 2×2 (optional)
M1 — DVM module
Q1 — 2N3906 transistor
Q2 — RF540N MOSFET
Q3 — IRF40B207 MOSFET
R1 — 1 kΩ trimpot
R2 — 1 kΩ resistor
R3, R4 — 10 kΩ resistor
R6 — 100 kΩ resistor
S1 — Sub-micro toggle switch, On-Off-On (center off)
S2 — Mini toggle switch, On-On
U1, U2, U3 — USB charging module
Wire ties for securing the Powerpole connectors
PC board (available from K9JEB at k9jeb.com)

battery voltage. The circuit float-charges the battery from the external power source and includes an overvoltage detection circuit to prevent overcharging the battery.

The circuit has a low-profile (½-inch height) battery-top size that matches 7 to 12 Ah SLA form factor batteries with F1 or F2 connectors. The assembled unit is shown mounted on such a battery in **Figure 7.64**. The schematic and parts list for the circuit are provided by **Figure 7.65**. The circuit is intended for use with batteries from 1 to 35 Ah. Wet-cell lead-acid batteries may be used if the cells are kept topped off with electrolyte. Do not use this charger with lithium, nickel, or other battery chemistries.

POWER CONTROL

The input power source and the battery share the output load. The input source recharges the battery directly, as well. If the input source is disconnected, the battery takes over supplying the load. Low-power input sources can be used to charge the battery as long as their terminal voltage remains higher than the battery voltage.

D3 is an optional diode to prevent power from back-feeding current into the power source when the source voltage is lower than the battery voltage. D3 should be installed if the external power voltage can fall below the battery voltage, such as for automotive or solar power applications.

Q3 acts as the master battery ON/OFF switch. It is an ultra-low R_{ds}-ON 40B207 MOSFET rated to carry dozens of amps continuously. It is connected as a *low-side switch* from the battery negative terminal to the circuit common and input negative or return terminal (ground).

Q1, a PNP transistor, is biased on by R4 and R6. It supplies enough current through R3 and D2 to raise the gate voltage of Q3 so that it is fully on. This lowers the resistance between the drain and source of Q3 to about 4 mΩ. R2 limits the total current through Q1 to about 13 mA. D2 is the Battery ON indicator.

Q3 dissipates heat while conducting according to $P = I_{ds}^2 R_{ds}$. The design of the PC board includes bare heat-sinking traces on top and bottom of the board, as well as the FET mounting screw and nut. Installed directly on the board, Q3 can dissipate up to 2 W without damaging the device: if P = 2 W and R = 4 mΩ, then I_{dsMAX} = 22 A. The 15 A battery fuse in the parts list should open at about 18 A continuous or 20 A surge. Even a 20 A fuse can probably be used safely if required.

S2 is a Battery OFF switch that physically disconnects the battery's negative terminal. Since S2 carries the full battery current, be sure it is rated adequately for your battery and peak load current. (An alternative is to use a switch to ground the gate of Q3, turning it off along with D2. The battery remains connected to Q3 in this case.) The specified mini-toggle switch is rated at 5 A which is adequate for most 7 to 9 Ah battery applications. To carry higher currents, replace S2 with a heavier switch. If the Battery OFF function is not required, S2 can be replaced with a heavy jumper.

POWER DISTRIBUTION

Anderson Powerpole connectors are used for both input and output power. Up to six output power connections are available. Any of the connectors can be used as the input power source connection. All of the power connectors are connected in parallel on a large-trace bus on the bottom of the board which should carry 25 to30 A safely. If higher currents are required, #14 AWG wire can be soldered to the traces which are kept bare for this purpose. Power output cables may be hard-wired to the traces if required. It is recommended that the pairs of connectors be secured with wire-ties as shown in Figure 7.64.

While the circuit board and transistor can handle a higher load, the battery's terminals and internal plate construction should limit input or output current to 15 to 20 A maximum continuous load. Even loads of 25 A or more are okay for a few seconds but should not be continuous. Radios in the 100 W class can be used at full power on SSB or CW. Digital modes should be used at reduced power to prevent battery or circuit overload. (S2 will have to be replaced by a heavier switch, however. Battery voltage below 12 V may cause erratic or improper operation of transceivers designed for 13.8 V supply voltages.)

CHARGING CIRCUIT

The chemistry of the SLA battery allows it to be kept charged at a constant voltage between 13 and 13.8 V, with 13.8 V being the optimum voltage. The circuit keeps the battery connected to the input power source through Q3, so it can be kept fully charged. For a 7 to 12 Ah SLA battery discharged to about 11 V, inrush current to the battery through Q3 is about 4 A, decreasing to just a few mA within a few seconds as the battery voltage comes up to the input voltage. The battery is then float-charged by the external source. (A trickle-charger supplies a small amount of current to the battery at all times.)

OVER-VOLTAGE CUTOFF CIRCUIT

The battery is disconnected by turning off Q3 if the input charging voltage is too high, which can damage the battery by overcharging it. Voltages above 13.8 V are common in automotive electrical systems and solar panels. Zener diode D1 drops the input voltage across trimpot R1 to V_{in} – 8.2 V. As input voltage rises, the voltage on the wiper of R1 eventually turns on Q2, shorting the gate of Q3 to ground and turning it OFF. This also turns off the Battery ON indicator, D2.

The user should set the desired threshold voltage for the overvoltage protection circuit at about 13.7 or 13.8 V. Set the threshold by applying 13.8 V dc to the input, adjusting R1 until D2 goes out, then reverse the adjustment until D2 just turns back on. At this setting, the circuit will allow 13.8 V to charge the battery but disconnect the battery at any higher voltage. If the input voltage drops below 13.8 V, the battery will be reconnected and D2 will turn on again.

USB CHARGING MODULES

Up to three +5 V, 2.5 A USB charger modules can be added in three places on the board, each replacing an Anderson Powerpole connector pair. One can be seen at the left-hand side of the PC board in Figure 7.64. These modules are widely available as commodity-type parts from a variety of sources.

Modules can be added along any of three edges of the board for flexibility in possible mounting configurations. All of the module +5 V outputs are connected in parallel and to J6 for external use if needed. As with all dc-dc converters, these modules are switched (at about 250 kHz) and may occasionally cause RF interference (RFI). If RFI from the USB charging modules is objectionable, a linear regulator such as the 2 A, +5V L78S05CV (with heatsink) can be used as a noise-free 5 V power supply, but it will generate about 9 W of heat for every 5 W delivered to the load. It also requires filter capacitors of 0.01 µF and 100 µF.

MISCELLANEOUS

C1, C2, and C3 are 0.1 µF capacitors to reduce HF and VHF noise on the dc power busses. Different values may be substituted if noise at higher or lower frequencies is encountered.

The DVM is a typical miniature 3-digit meter/display module available from many vendors. S1 switches the DVM to either display input voltage, turns it off, or to display the actual battery voltage.

J6 is an optional 2×2 pin header that provides a connection to both input and output voltages, +5 V dc from the USB charging modules if installed, and common (ground). Note that the input source connection to J6 is not fused.

7.15.8 Simple Adjustable Tracking Power Supply

This project by Bryant Julstrom, KCØZNG, was originally published in the Spring 2014 issue of the QRP ARCI *QRP Quarterly*. The author also published "Another Low Voltage Power Supply" beginning on page 9 of the January 2019 issue of *QRP Quarterly*. A PDF version of the original article, including all figures, is available with this book's online content.

The typical experimenter's bench power supply provides an adjustable positive voltage of up to 25 V at up to 1 A of current. This suffices for many circuits and projects, but others require positive and negative voltages of equal magnitude. Op-amp circuits, for example, often require ±15 V.

THE CIRCUIT

The circuit is implemented using two garden-variety ICs, an LM317 positive regulator and an LM337 negative regulator, both in TO-220 packages. (Note that the orientation of the input-common-output connections are different between the two regulators!) The input voltages of the circuit are limited only by the maximum inputs of the two regulators, ±40 V. (An alternate implementation based on the LT1033 tracking regulator is shown in an article included with this book's online content.) A voltmeter is switched between either of the supply's outputs and common.

Input power to the tracking regulator is supplied by a power transformer with a center-tapped secondary. The peak secondary output voltage on each side of the center-tap should be several volts higher than the maximum supply output at full load. See this chapter's section on power supply ripple to determine the minimum required filter capacitance.

In the author's case, the heaviest transformer available provided only 26 V CT (center-tapped) and the input to the regulator of ±13 V was too low. One possible solution would be to use two identical transformers with their primary windings in parallel and their secondaries in series with the connection between secondaries serving as the center-tap.

Since 24 to 28 V CT is a common secondary voltage range, however, the author chose to use the available transformer and back-to-back half-wave voltage doublers, one on either side of the secondary winding's center tap, to provide about ±26 V to the tracking regulator. This halves the current that can be drawn from the transformer, but if the transformer secondary is rated at 2 A, the resulting maximum output current of 1 A in each leg of the supply is sufficient for a wide range of projects.

Figure 7.66 shows the circuit of the tracking power supply. The author used 1N5402 rectifiers in the voltage doublers because they were available, but any 100 V diode rated at 3 A or more of average forward current will work. The two 2.2 µF capacitors should be solid tantalum types.

R4, a front-panel potentiometer, is used

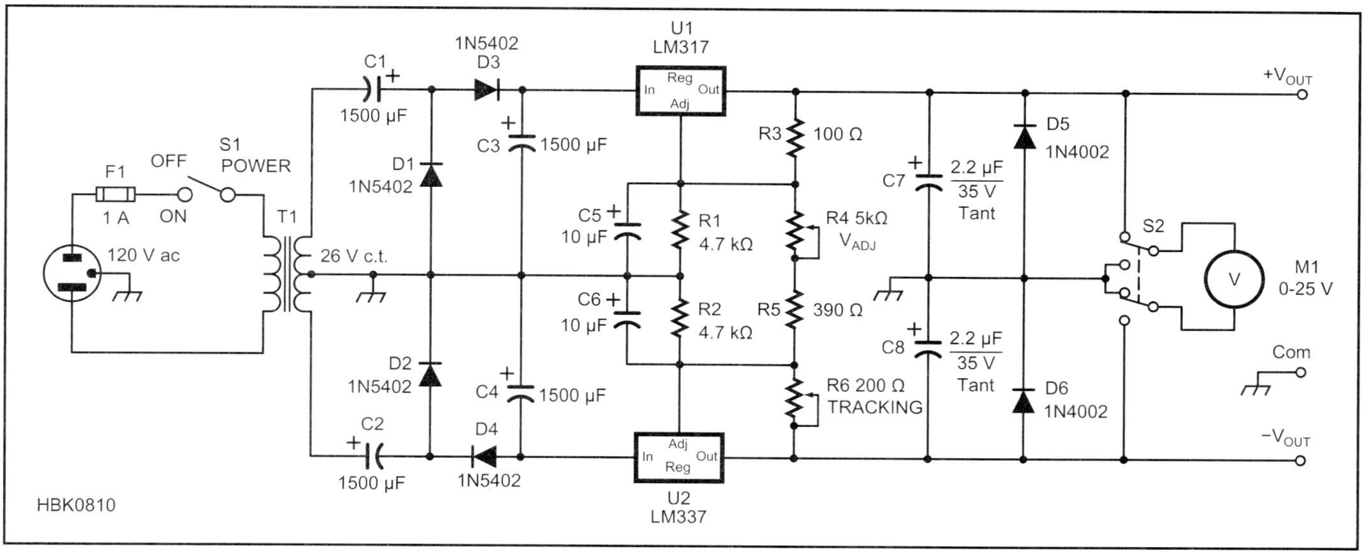

Figure 7.66 — Schematic of the adjustable tracking power supply. Regulators are mounted to the enclosure, which serves as a heat sink. Resistors are ¼ W film, 5% tolerance unless noted otherwise.

C1-C4 — 1500 µF, 50 V electrolytic
C5, C6 — 10 µF, 50 V electrolytic
C7, C8 — 2.2 µF, 35 V, solid tantalum
D1-D4 — 1N5402 or equivalent
D5, D6 — 1N4002 or equivalent
F1 — 250 V, 1 A ac fuse and fuse holder
M1 — 0-25 V voltmeter or equivalent
R1, R2 — 4.7 kΩ
R3 — 100 Ω
R4 — 5 kΩ, panel-mount potentiometer
R5 — 390 Ω
R6 — 200 Ω trimmer potentiometer
S1 — SPST toggle switch
S2 — DPDT miniature toggle switch
T1 — 26 V CT at 2 A secondary or equivalent (see text)
U1 — LM317adjustable positive voltage regulator
U2 — LM337adjustable negative voltage regulator

for voltage adjustment. R6, a 200 Ω trimmer potentiometer, provides for fine tracking adjustment. The 390 Ω resistor (R5) sets the regulator's minimum outputs at about ±3.6 V for reasons described below.

CONSTRUCTION

As constructed, the front panel holds three binding posts (positive output, negative output, and common), two miniature toggle switches (on/off and voltmeter switching), the voltage-adjustment potentiometer and its knob, and a voltmeter. Note that the common connection of the power supply is usually left floating from the enclosure ground which must be connected to the "third-wire" ac safety ground. If desired, a binding post connected to the enclosure can be added so that a jumper can be used to connect the power supply common to the ac safety ground.

The voltmeter used by the author is a three-digit LED unit from Marlin P. Jones (**www.mpja.com**; part number 30217 ME with blue digits, also available in green and red). The meter is powered by the voltage it measures, as long as the voltage is at least 3.6 V as determined by R5 as explained previously. An analog meter or a digital meter with a separate power supply would allow lower minimum output voltages. One could augment the supply's measurements with current metering or with simultaneous measurements of both outputs.

The enclosure holds the power transformer and two circuit boards. The two voltage doublers occupy one circuit board and the tracking regulator the other. The circuit is simple enough that perforated board ("Perf-board") was used with point-to-point wiring.

The LM317 and LM337 are mounted at the edge of the regulator board and attached to the back panel, which serves as a heat sink, using mica insulators, nylon bolts, and a thin coating of heat sink compound. The rear panel also holds a snap-in IEC three-wire ac line connector. (Photographs of the finished power supply are provided in an article included with this book's online content.)

ADJUSTMENT AND PERFORMANCE

The only internal adjustment in the supply is the tracking potentiometer (R6), which must be set so that the two voltages track each other accurately. This is touchy, but the two voltages can be made to match within 0.1 V throughout the supply's range. The completed supply provides closely matched positive and negative voltages from ±3.6 V to about ±23 V, and a maximum current of 1 A in each leg.

The tracking adjustment and the front-panel voltage adjustment (R4) are delicate. In both cases, multi-turn potentiometers would be much easier to set. The tracking control could also be replaced with a series resistor and a smaller-value potentiometer to make adjustment less sensitive.

7.15.9 Overvoltage Protection for AC Generators

When using portable generators, there is always a possibility of damage to expensive equipment as a result of generator failure, especially from overvoltage. If the generator supplying power to this equipment puts out too much voltage, you run the risk of burning up power supplies or other electronic components. This project, by Jerry Paquette, WB8IOW, addresses the problem of increased voltage (not lower voltage) or surges and spikes lasting for a few microseconds.

Using a portable generator overvoltage protection circuit and ground-fault circuit interrupter (GFCI) as shown in **Figure 7.67** is good insurance. This overvoltage protection device must be used in conjunction with a GFCI at each station! (More information on GFCIs may be found in the **Safety** chapter.)

CIRCUIT DESCRIPTION

Refer to Figure 7.67 for this description. R1 places an intentional fault on the load side of the GFCI. With the value resistor used, the fault is limited to 10 mA. (The normal tripping threshold of a GFCI is 5 mA. This current forces the GFCI to trip in just a few milliseconds. This circuit will not function at all without the use of a GFCI. A GFCI must be used at each station. If a single GFCI were used at the generator, rather than one at

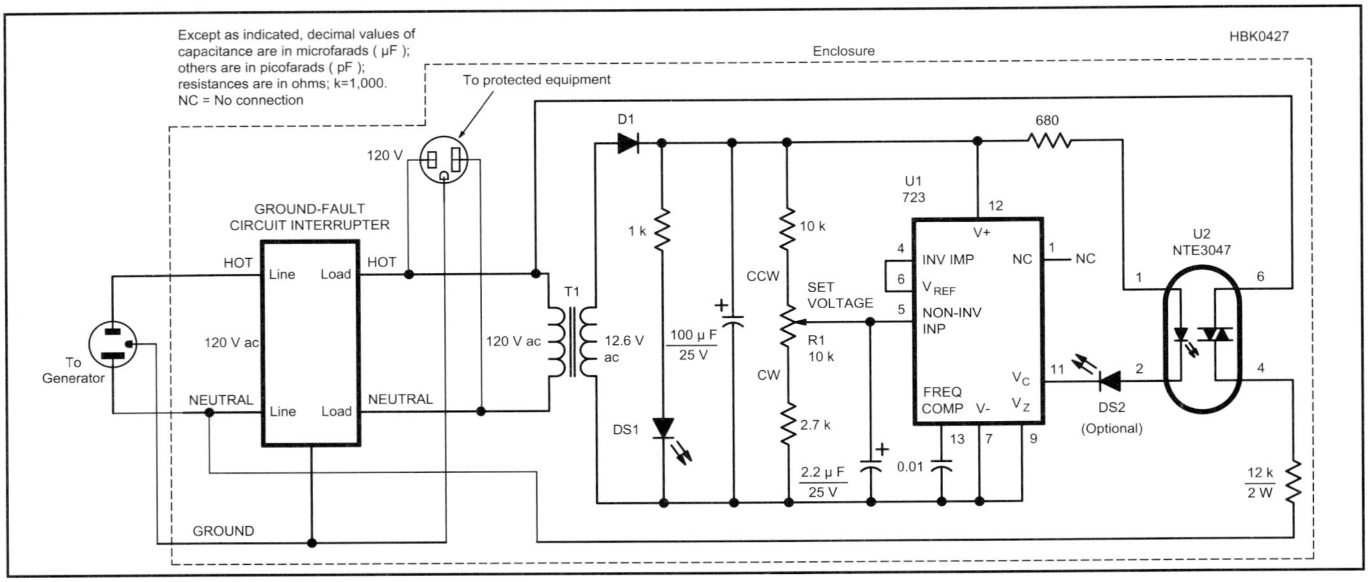

Figure 7.67 — Schematic of the Field Day equipment overvoltage-protection circuit. This circuit must be used in conjunction with a ground fault circuit interrupter (GFCI). A separate GFCI must be installed at each station. Unless otherwise specified, resistors are ¼ W, 5% tolerance. A PC board is available from FAR Circuits (www.farcircuits.net).

D1 — 200 PIV, 1 A diode; 1N4003 or equiv.
DS1, DS2 — Small LEDs
R1 — 10 kΩ board-mounted, multi-turn potentiometer

each location, premature tripping could occur. Several hundred feet of extension cords could have enough leakage to trip the GFCI.

You can see that the GFCI has separate lines (inputs) and loads (outputs). GFCI input terminals must be connected to the generator output. The GFCI ground must be tied to the ground of the generator. The load (computers, radios, etc) will plug into the GFCI or are wired to the load side of the GFCI. The primary of T1 is wired to the load side of the GFCI. The 12 kΩ/2 W resistor, however, must be wired to neutral on the *line* side of the GFCI in order for it to trip when used with generators that have windings isolated from ground. For safety, construct the entire unit in a single enclosure including the GFCI and its wiring. The generator connection can be made through a wired plug or using a male receptacle mounted on the enclosure.

T1 can be any 120 V to 12.6 V transformer capable of delivering 100 mA or more. Mounting of this transformer varies depending on the type used. All remaining components mount on a circuit board. D1 rectifies the ac from T1 and the 100 µF capacitor filters the dc. This voltage provides the power to the 723 voltage regulator.

Two fixed resistors and a potentiometer form the voltage-divider network supplying voltage to the LM723 input, pin 5. R1, the board-mounted potentiometer, has only three leads, but there are four pads on the circuit board, to accommodate different styles of pots. The 2.2 µF capacitor provides a slight delay, to prevent false tripping when the circuit is powered up. The 0.01 µF capacitor from pin 13 of the 723 to the negative supply bus should always be used. When the voltage at pin 5 goes higher than the reference voltage at pins 4 and 6, pin 11 goes low, turning on the trip indicator LED DS2 and the optical coupler LED. LED current is limited by the 1 kΩ resistor. The optical coupler turns on the TRIAC, which creates a 10 mA fault current between the hot wire and ground of the GFCI. DS2 will remain lit as an indicator until the 100 µF capacitor is discharged.

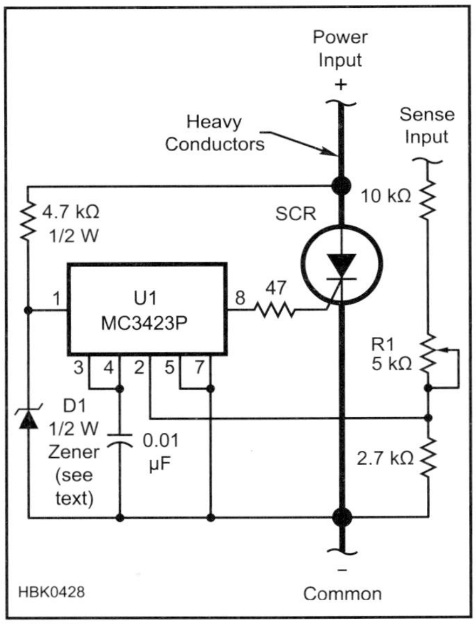

Figure 7.68 — Schematic of the overvoltage crowbar circuit. Unless otherwise specified, resistors are ¼ W, 5% tolerance. A PC board is available from FAR Circuits (www.farcircuits.net).
SCR — C38M stud-mount (TO-65 package)
U1 — MC3423P or NTE7172 voltage protection circuit, 8-pin DIP
D1 —Zener diode, ½ W (see text)
R1 — 5 kΩ, PC board mount trimmer potentiometer

ADJUSTMENT

Adjustment is simple. You'll need a variable ac transformer (Powerstat or Variac). Turn R1 fully clockwise and use the variable transformer to adjust input to 130 V ac. Turn the pot counterclockwise until the GFCI trips.

7.15.10 Overvoltage Crowbar Circuit

The overvoltage "crowbar" circuit in **Figure 7.68** is intended as a last line of defense against power supply output overvoltage failures. When an overvoltage condition is detected, the heavy-duty SCR is turned ON, presenting a short-circuit across the power supply output. The intent is to cause protective components in the supply, such as a fuse or circuit-breaker, to remove power. This protects equipment connected to the supply. A complete description of the circuit is provided in the online information for this chapter.

Additional Projects and Information

Additional power source projects and supporting files are included with this book's online content.

Contents

Chapter 8 — Online Content

- SDR Simplified — Fourier Transforms by Ray Mack, W5IFS
- SDR Simplified — Fundamentals of Sampling by Ray Mack, W5IFS
- SDR Simplified — More on Sampling by Ray Mack, W5IFS
- "Radio Mathematics" — see online content for the Electrical Fundamentals chapter

Chapter 8

DSP and SDR Fundamentals

This chapter was updated by Doug Grant, K1DG, from material originally created by Alan Bloom, N1AL. It explores the fundamentals of digital signal processing (DSP) and software defined radio (SDR). Material is also taken from *QEX* "SDR: Simplified" articles by Ray Mack, W5IFS. Key to DSP and SDR, analog-digital conversion and types of converters are covered, as well.

DSP technology has progressed to the point where it is the dominant technology in our radio equipment. DSP has largely replaced analog hardware circuits with digital processors and software, offering amateurs flexibility and features only dreamed of in the past.

Software defined radio has displaced the superheterodyne architecture that was dominant since its introduction in the 1920s. SDR is only possible through DSP techniques implemented on the advanced platforms that have become available.

This chapter begins with the fundamentals of DSP and extends them to SDR design. Other chapters will cover DSP methods in more detail for implementing oscillators, modulation, filters, and the functions associated with receiving and transmitting.

For the fullest understanding of this chapter, the reader should have a basic familiarity of the topics covered in the **Radio Fundamentals** chapter as well as some high school trigonometry.

References to math tutorials are provided in **Radio Mathematics** which is part of this book's online content, along with additional background and support materials, including a glossary.

8.1 Introduction to DSP

Digital signal processing (DSP) has been around a long time. The essential theory was developed by mathematicians such as Newton, Gauss, and Fourier in the 17th, 18th, and 19th centuries. It was not until the latter half of the 20th century, however, that digital computers became available that could do the calculations fast enough to process signals in real time. Today DSP is important in many fields, such as seismology, acoustics, radar, medical imaging, nuclear engineering, audio and video processing, as well as voice and data communications.

In all those systems, the idea is to process a digitized signal so as to extract information from it or to control its characteristics in some way. For example, an EKG monitor in a hospital extracts the essential characteristics of the signal from the patient's heart for display on a screen. A digital communications receiver uses DSP to filter and demodulate the received RF signal before sending it to the speaker, headphones, or waterfall/band-scope display. In some systems, the signal to be processed may have more than one dimension. An example is image data, which requires two-dimensional processing. Similarly, the controller for an electrically-steerable antenna array uses multi-dimensional DSP techniques to determine the amplitude and phase of the RF signal in each of the antenna elements.

A typical DSP system is conceptually very simple. It consists of only three sections, as illustrated in **Figure 8.1**. An *analog-to-digital converter* (ADC) at the input converts an analog signal into a series of digital numbers that represent snapshots of the signal at a series of equally spaced sample times. The digital signal processor does some kind of calculations on that digital signal to generate a new stream of numbers at its output. A *digital-to-analog converter* (DAC) then converts those numbers back into analog form.

Some DSP systems may not have all three components. For example, a DSP-based audio-frequency generator does not need an ADC. Similarly, there is no need for a DAC in a measurement system that monitors some sensor output, processes the signal, and stores the result in a computer file or displays it on a digital readout.

The term "DSP" is normally understood to imply processing that occurs in real time, at least in some sense. For example, an RF or microwave signal analyzer might include a DSP coprocessor that processes chunks of sampled data in batch mode for display a fraction of a second later.

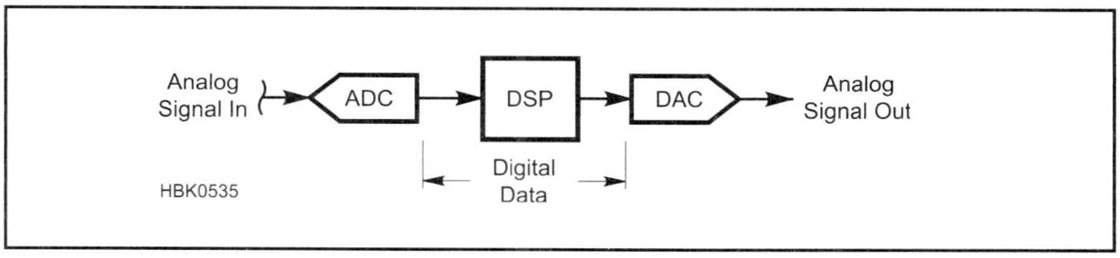

Figure 8.1 — A generic DSP system.

8.2 Introduction to SDR

The concept of a *software-defined radio* (SDR) has been around for a long time, but became popular in the 1990s. By then, DSP and data converter technology had developed to the point that it was possible to implement almost all the signal-processing functions of a transceiver using inexpensive programmable digital hardware. The frequency, bandwidth, modulation, filtering, and other characteristics can be changed under software control, rather than being fixed by the hardware design as in a conventional analog radio. Adding a new modulation type or a new improved filter design is a simple matter of downloading new software.

Compared to analog radios, SDR has some major advantages. In analog radios, passive components such as capacitors, resistors, and even inductors are subject to changes in value due to temperature drift and aging. They also have loose initial value tolerances and often require adjustments or calibration. Likewise, active components such as amplifiers and mixers also are subject to variations in performance. If the function of these components can be replaced with software, most of these problems disappear.

SDR is appealing to regulatory bodies such as the FCC because it makes a communications system called *cognitive radio* possible, in which multiple radio services can share the same frequency spectrum.[1] Each node in a wireless network is programmed to dynamically change its transmission or reception characteristics to avoid interference to or from other users. In this way, services that in the past enjoyed fixed frequency allocations but that only use their channels a small percentage of the time can share their spectrum with other wireless users with minimal interference.

There has been much, sometimes heated, discussion about the precise definition of an SDR. Most feel that, at minimum, an SDR must implement in software at least some of the functions that have traditionally been done in hardware. Others feel that a radio doesn't count as an SDR unless nearly all the signal-processing functions, from the RF input to the audio output (for the receiver), and from the microphone AD to the power amplifier input (for the transmitter), are done in software. Others add the requirement that the software must be re-configurable by downloading new code, preferably open-source. For our purposes, we will use a rather loose definition and consider any signal-processing function done in software to fall under the general category of SDR.

Some SDRs use a personal computer to do the computational work and external hardware to convert the transmitted and received RF signals to lower-frequency signals that the computer's audio interface can handle. Some SDRs avoid the use of the PC's sound card by including their own audio *codec* (short for "coder-decoder," a chip that includes both ADC and DAC converter functions) and using analog audio for the user interface. Some SDRs transfer the downconverted (and possibly filtered) data to the PC via a USB port. Modern PCs provide a lot of computational power for the money and are getting cheaper and more powerful all the time. They also come with a large color display, a keyboard and mouse for easy data entry and navigation, a large memory and hard disk, which allows running logging programs, and other software, while simultaneously doing the signal processing required by the SDR.

Smaller, even less expensive, computing platforms such as Arduino, Raspberry Pi, Beaglebone, Red Pitaya, and many others have become available. While they often lack the peripherals and user interfaces of a complete PC, some of them have sufficient computational power to be useful for SDR experimentation.

Some SDRs have almost no knobs or buttons on the box and none of the traditional features, such as frequency display, meters, and so on. In such radios, all control functions are done on an external PC with control software provided by the manufacturer. The user interface is the PC keyboard, monitor, and mouse. Other SDRs look more like conventional analog radios and don't need an external PC for control. While the signal processing is done with one or more embedded DSPs, the user interface consists of knobs and push buttons, which many users prefer to a mouse, keyboard, and pull-down menus. Some newer SDRs use a touchscreen for the user interface to emulate the buttons.

Either method offers all the important advantages of applying DSP techniques to signal processing. The channel filter can have a much better *shape factor* (the ratio between the width of the passband and the frequency difference of the stopband edges). FIR filters are linear phase and have less ringing than analog filters of the same bandwidth and shape factor. Once the signal is in the digital domain, all the fancy digital signal processing algorithms can be applied, such as automatic notch filters, adaptive channel equalization, noise reduction, noise blanking, and feed-forward automatic gain control. Correcting bugs, improving performance, or adding new features is as simple as downloading new software.

Following an overview of SDR systems, the details of the various blocks used in such systems, including analog-digital conversion, are then explored. The chapter concludes with a discussion of the basic theory of discrete-time and digital signals, with emphasis on topics relevant to radio communications.

8.2.1 SDR Architecture Options

The transition between analog and digital signals can occur at any of several places in the signal chain between the antenna and the user interface. This choice is an important factor in determining the overall architecture of the SDR. This section presents several block-diagram-level concepts for software-defined radio and compares architecture options.

DSP AT AUDIO FREQUENCIES

The initial use of DSP by amateur radio operators was to implement audio-frequency filtering. By the early 1990s, converter costs were reasonable and the processing power required was available in relatively inexpensive microprocessors with enhanced math function capability. The flexibility of digital filters (see the **Analog and Digital Filtering** chapter) and the ability to alter bandwidth and perform noise reduction resulted in quick adoption of DSP techniques in amateur equipment.

In 1992, Dave Hershberger, W9GR, designed an audio-frequency DSP filter based on the TMS320C10, one of the earliest practical DSP chips available.[2] This filter was an external standalone unit that plugged into the headphone jack of a receiver and included filters with various bandwidths, an automatic multi-frequency notch filter, and an adaptive noise filter.

The advantage of DSP at audio frequencies is that it can be easily added to an unmodified analog radio, as in **Figure 8.2**. Many amateurs use a similar approach to implement digital modulation modes, using a PC and software as the outboard DSP processor. The software produces the required transmission waveforms (and demodulates the received signal),

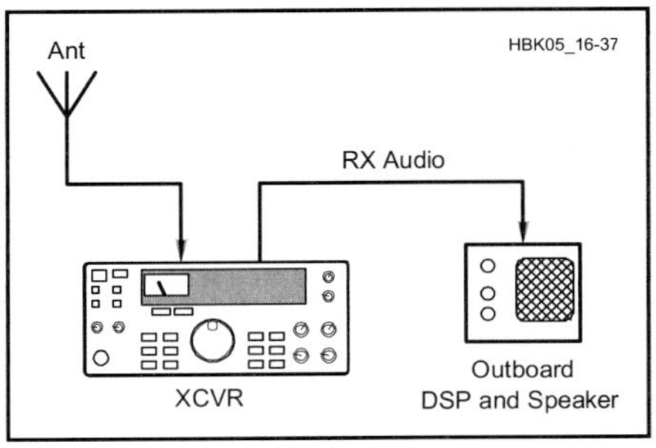

Figure 8.2 — An outboard DSP audio processor.

using the PC's audio input and output connected to the audio input and output of a conventional SSB transceiver.

DOWNCONVERSION TO ANALOG BASEBAND

A related technique is to *downconvert* a slice of the radio spectrum to baseband audio using a technique similar to the direct-conversion receivers popular with simple low-power CW transceivers. (See the **Receiving** chapter.) This idea was pioneered by Gerald Youngblood, AC5OG (now K5SDR), with the SDR-1000 transceiver, which he described in a series of *QEX* articles in 2002 and 2003.[3] The receiver block diagram is shown in **Figure 8.3**. It uses a unique I/Q demodulator designed by Dan Tayloe, N7VE, to convert the RF frequency directly to baseband I ("In-phase") and Q ("Quadrature") signals.[4] (I/Q modulation and quadrature signals are discussed in the **Modulation** chapter.)

The baseband signals are fed to the stereo input of a PC's sound card, represented by the low-pass filters and ADCs in the figure. Software in the PC does all the signal processing and demodulation, ultimately producing audio for the operator to hear or a display of decoded bits in the form of text on the screen. The transmitter is the same block diagram in reverse, with an I/Q modulator converting the I/Q signal from the sound card up to the RF frequency where it is filtered and amplified to the final power level.

The sound card method manages to achieve reasonable performance with simple, inexpensive hardware and resources already available in the PC. Once the ADCs in the sound card have digitized the signal, the DSP capability of the PC can do amazing things with it. In addition to implementing conventional transceiver functions such as several types of detector, variable-bandwidth filters, software AGC, an S-meter, and speech compression, the software can include some extra features such as an automatic notch filter, noise reduction, panadapter and waterfall spectrum displays, and decoding of signals such as RTTY, PSK, and WSJT modes.

The simple hardware of the SDR-1000 does impose some performance limitations. Because of imperfections in the analog downconverter, unwanted-sideband rejection is not perfect. This is called "image rejection" in the SDR-1000 literature. On the panadapter display, strong signals show up weakly on the opposite side of the display, equally-spaced from the center. DC offset in the analog circuitry causes a spurious signal to appear at the center of the bandwidth. To prevent an unwanted tone from appearing in the audio output, the software demodulator is tuned slightly off frequency, but that means interference at the image frequency can cause problems because of the imperfect image rejection. The dynamic range depends on the sound card performance as well as the RF hardware. Some newer SDRs include an integrated audio codec or other type of ADCs and DACs optimized for the application, so that the PC's sound card is not needed.

DIGITIZING AT IF

Another option for implementing software-defined radios involves performing the analog-digital conversion at an intermediate frequency. **Figure 8.4** shows such a design. In the receiver, placing the ADC after a crystal IF filter improves the *blocking dynamic range* (BDR) for interfering signals that fall outside the crystal filter bandwidth. BDR is the ratio, expressed in dB, between the noise level (normally assuming a 500 Hz bandwidth) and an interfering signal strong enough to cause 1 dB gain reduction of the desired signal. (See the **Receiving** chapter.) As shown, the downconversion to I/Q format still uses lower-speed ADCs but often the signal is actually at a low IF, say, 15 kHz or so. This allows an SSB-bandwidth signal to be con-

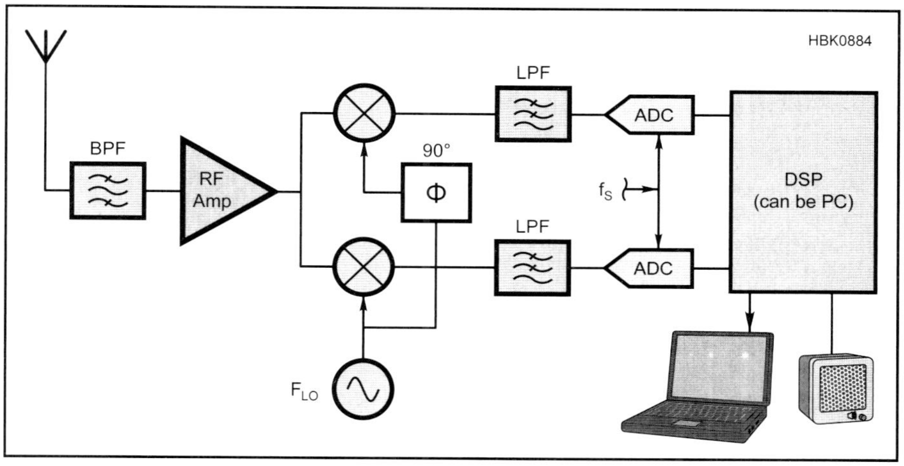

Figure 8.3 — Direct-to-baseband SDR receiver architecture.

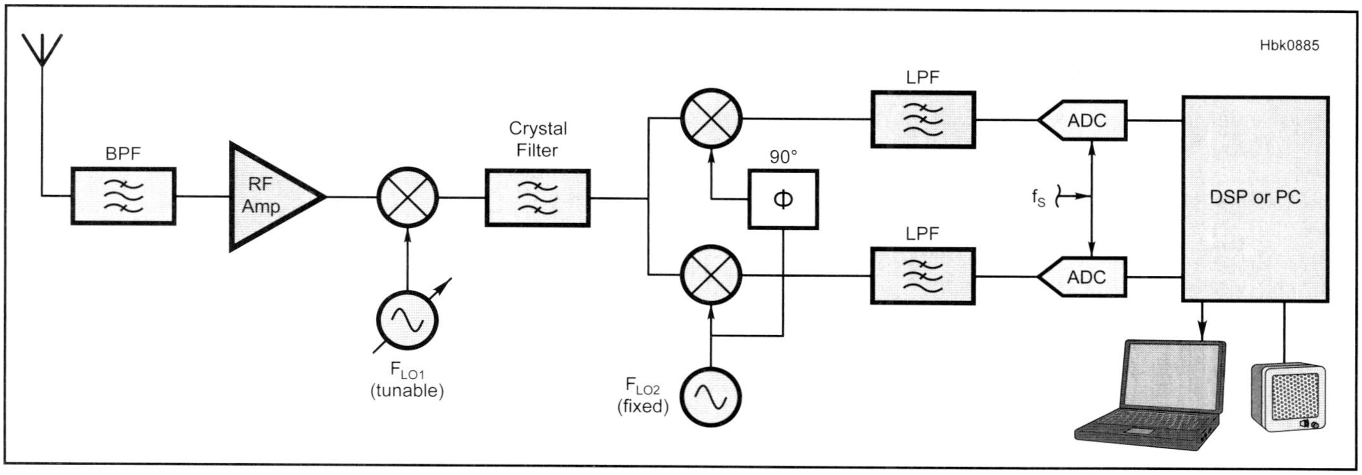

Figure 8.4 — Hybrid superhet/DSP SDR receiver architecture.

tained within the 20 kHz bandwidth of a typical audio codec and avoids errors due to dc offsets in the signal path. With careful design, a receiver with such an architecture can achieve 140 dB or more of BDR (if there are no other limiting factors such as LO phase noise). The third-order dynamic range is similar to that achieved with a conventional analog architecture since the circuitry up to the crystal filter, including amplifiers and mixer(s), is the same. The Elecraft K3 and Icom IC-7851 transceivers used this architecture and achieved very high receiver performance.

An advantage of the IF-based approach compared to directly sampling the RF frequency is that the ADC does not have to run at such a high sample rate. In fact, because the crystal filter acts as a high performance, narrow bandwidth, anti-aliasing filter, *undersampling* is possible if the ADC has sufficient sampling bandwidth (ADCs intended for audio applications generally do not). With bandwidths of a few kHz or less, sample rates in the tens of kHz can be used even though the center frequency of the IF signal is much higher, so long as the ADC's sample-and-hold circuit has sufficient bandwidth.

Another example of an IF-sampling SDR is the family of "SDR dongles" based on chips designed for TV tuner applications such as the RTL2832U manufactured by Realtek. Several different manufacturers produce similar devices. The RTL2832U IC includes a pair of 8-bit ADCs operating at a 28.8 Msps sample rate and a programmable DSP system including digital low-pass filters, demodulation, decoding, and other functions. The outputs of the ADCs and the outputs of the digital processor are accessible via the chip's integrated USB port. The RTL2832U device is usually preceded by either a superheterodyne (superhet) or quadrature zero-IF tuner. With a superhet receiver, only one of the on-chip ADCs is used, and the RTL2832U does the quadrature down-conversion digitally. Bandwidth-limited IFs up to 30+ MHz can be undersampled. With a quadrature zero-IF downconverting tuner, both ADCs are used. You can find a more detailed description of implementation options using this IC at **www.pa3fwm.nl/technotes/tn20.html**. A typical design (of many available), developed by **www.RTL-SDR.com** is shown in **Figure 8.5C**.

Some systems, especially in VHF/UHF/microwave applications, such as cellular base stations that deal with signals tens of MHz wide and in the GHz frequency range, often use the approach shown in **Figure 8.6**. Here, a high speed ADC digitizes an IF typically in the hundreds of MHz. The output data rate of

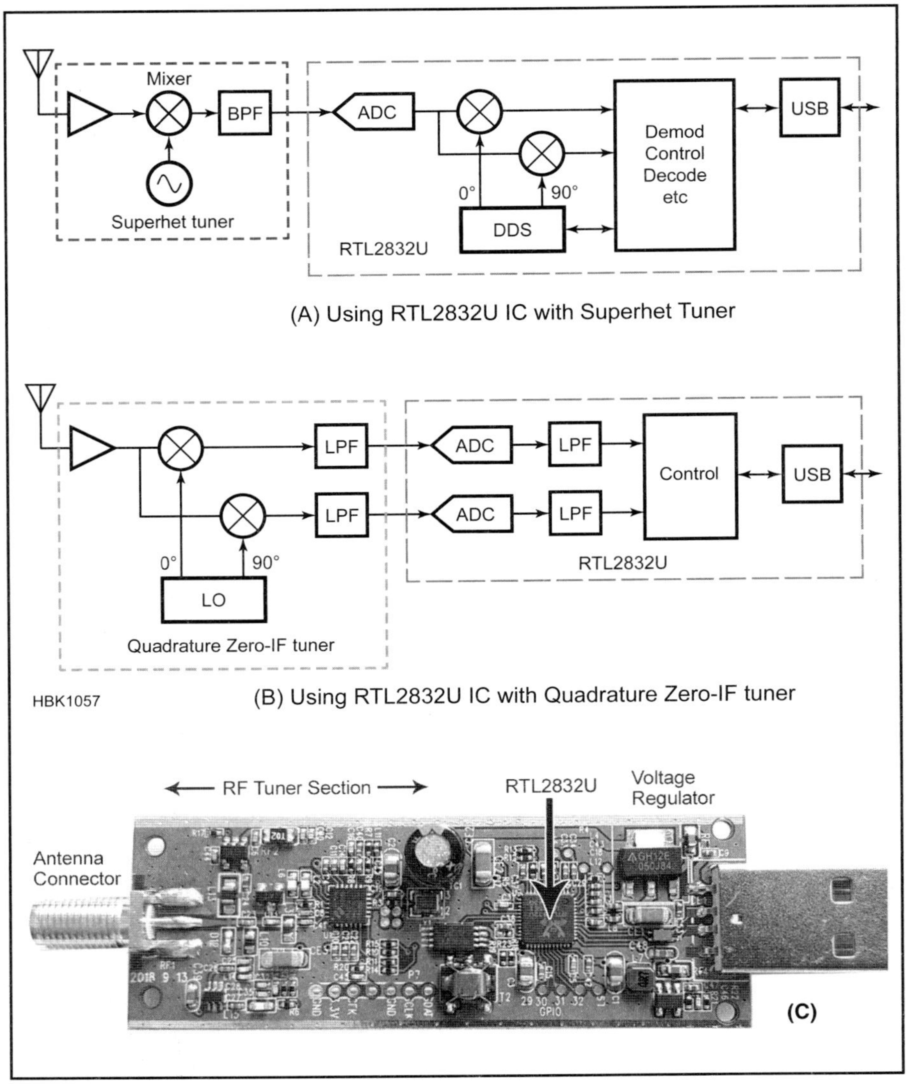

Figure 8.5 — Low cost SDR receiver using RTL2832U IC with a superhet tuner (A) and quadrature zero-IF tuner (B). An RTL-SDR dongle (C) plugs into a PC USB port. (photo courtesy of RTL-SDR.com)

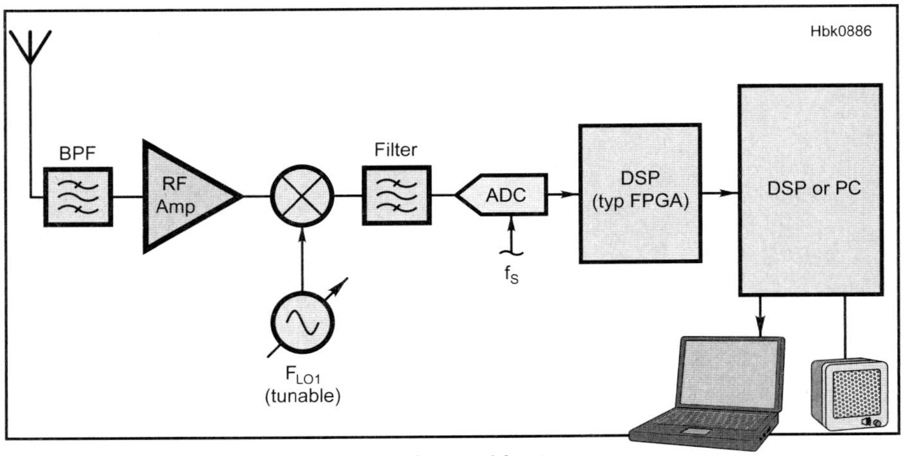

Figure 8.6 — High-IF sampling SDR receiver architecture.

such a converter is too high to be handled by a PC or even a low-cost programmable DSP microprocessor. The conversion from the original signal to baseband I/Q components, and then the demodulation to data streams for the network, is done in dedicated digital hardware blocks (such as FPGAs).

DIRECT RF DIGITIZING

The ultimate SDR architecture converts between the analog and digital domains right at the frequency to be transmitted or received, or converts a wide range of frequencies and does all filtering in the digital domain. The receive path of such a design is shown in **Figure 8.7**. In this receiver, the only remaining analog components in the signal chain are a wide-band anti-aliasing filter, similar to a preselector, and an amplifier to improve the noise figure of the ADC, if necessary. The FlexRadio Flex-6700, Elecraft K4, and Icom IC-7610 are examples of transceivers that use this basic architecture. The local oscillator, mixer, IF filters, AGC, demodulators, and other circuitry are all replaced by digital hardware and software. The implementations of these functions are perfectly stable with time and temperature, and need no adjustments. The K4 DSP FPGA actually delivers three outputs: a wideband path for the panadapter display, and two 48-kHz IF outputs (for "dual-watch" capability), on which the final high-performance filtering and demodulation occur.

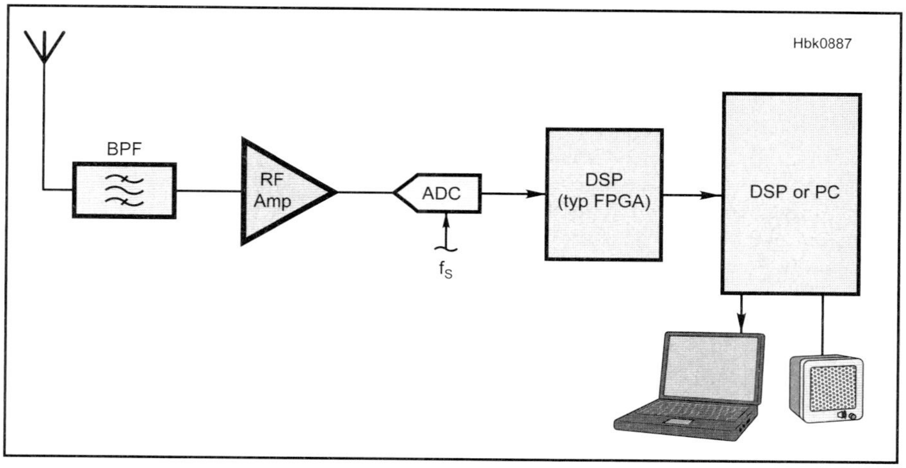

Figure 8.7 — Direct RF-sampling DSP SDR receiver architecture.

HYBRID SDR/ANALOG RECEIVERS

It has only been recently that low-cost, high-speed ADCs have become available with specifications good enough to allow reasonable performance in an RF-sampling communications receiver. Today it is possible to achieve blocking dynamic range (BDR) of 130 dB. Hybrid SDR/analog receivers can provide even higher performance.

The highest performance receivers currently available for HF applications use both analog and digital processing. The analog front end is amplified, filtered and downconverted (sometimes twice) to an intermediate frequency in order to enhance selectivity before the signal reaches the ADC. This approach is used in the Elecraft K3 and Icom IC-7851 transceivers.

Figure 8.8 shows two examples of hybrid receiver chains. The Yaesu FTDX101 series transceivers use two receive signal chains. One is a direct RF sampler, which is used only to provide a fast responding digital band scope display. The other path, which ultimately delivers audio to the user, is a hybrid design, with a 9 MHz IF and optional narrow crystal filters, offering very high receiver performance.

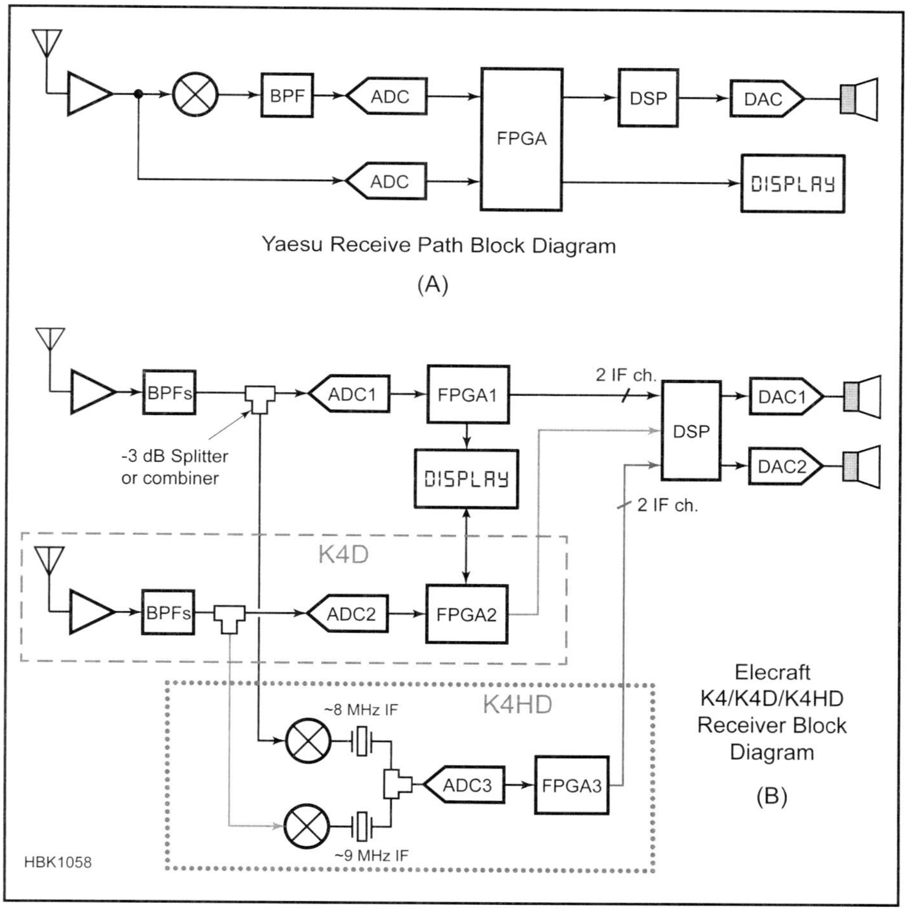

Figure 8.8 — Receiver architectures of current high-performance transceivers — the Yaesu FTDX101 series (A) and Elecraft K4 (B). K4D and K4HD refer to model numbers, see text. (Figure 8.8B courtesy of Elecraft)

The Elecraft K4 platform has three versions. The K4D adds a second direct-sampling receiver to the basic K4 to provide diversity or dual receive capability. The K4HD, intended for use in the most demanding environments, adds a two-channel *hybrid* signal chain to the K4D. Each channel includes an analog downconversion, one channel to an 8 MHz IF and one to a 9 MHz IF. These two analog IF signals are filtered by a bank of narrowband crystal filters, then combined and digitized by one ADC. The FPGA following this ADC delivers two 48 kHz digital IF signals to the DSP, which does the final filtering and demodulation as in the base model.

It is worth noting here that while huge BDR numbers can be measured in the laboratory, the performance achieved in a real-world environment with a receiver connected to an antenna is quite different. Often local noise sources raise the noise floor such that the receiver's full BDR cannot actually be utilized and other specifications become more important. In the real world, RF-sampling SDRs can often provide performance indistinguishable from conventional all-analog and hybrid receivers.

Third-order dynamic range (3IMD_DR or IP_3) is not a meaningful specification for this type of radio because it is based on the behavior of analog circuits that degrade gradually as the signal level increases. Calculation of 3IMD_DR assumes that distortion products increase 3 dB for each 1 dB increase in signal level, which is not always true for an ADC. The level of the distortion products in an ADC tends to be more or less independent of signal level until the signal peak exceeds the ADC's full-scale input, at which point the distortion increases dramatically. It is important to read the data sheet carefully and note the test conditions for the distortion measurements.

There are definite advantages to sampling at RF. For one thing, it saves a lot of analog circuitry. Even though a high-speed ADC is more expensive than an audio converter, the radio may end up being cheaper to build because of the reduced component count and fewer adjustments. Performance is improved in some areas. For example, image rejection is no longer a worry, as long as the anti-aliasing filter is doing its job. (See the **Receiving** chapter.) The dynamic range of an SDR theoretically does not depend on signal spacing. Close-in dynamic range is often better than with a conventional architecture that uses a wide IF filter. With no crystal filters in the signal chain, the entire system has a completely linear phase response, which can improve the quality of both analog and digital signals after demodulation.

The biggest challenge with RF sampling is what to do with the torrent of high-speed data coming out of the receiver's ADC and how to generate transmit data fast enough to keep up with the DAC sample rate. To cover 0 to 54 MHz without aliasing requires a sample rate of at least 130 MHz, and commercial products typically operate the ADC at sample rates well over 200 MHz. That is much faster than a typical microprocessor or programmable DSP can handle. The local oscillator, mixer, decimator, or interpolator must be implemented in digital hardware so that the DSP can send and receive data at a more reasonable sample rate. *Digital downconverters* (DDC) perform those functions and output a downsampled digital I/Q signal to the DSP. Stand-alone DDC ICs were available in the past, but the function is now usually integrated with the ADC. It is also possible to implement a DDC in a *field-programmable gate array* or FPGA. (See the **Transceiver Design Topics** chapter.) *Digital upconverters* (DUC) do the same conversion in reverse for the transmitter and are available integrated with the DAC, or can be implemented in an FPGA. Some commercial integrated DDC/DAC products even include the capability to encode several digital modulation formats such as GMSK, QPSK, and $\pi/4$ DQPSK. In an attempt to simplify the interface to the digital domain, many high-speed converters now use a standardized serial interface specification called JESD204B, capable of handling up to 12Gb/s. Code to implement this interface on the digital FPGA is readily available.

Some designers have been successful in repurposing a graphics processor (GPU) for this application, and some GPU manufacturers now offer FFT libraries to assist in the design process.

8.2.2 Advantages and Limitations of DSP and SDR

Digital signal processing has the reputation of being more complicated than the analog circuitry that it replaces. In reality, once the analog signal has been converted into the digital domain, complicated functions can be implemented in software much more simply than would be possible with analog components. For example, the traditional "phasing" method of generating an SSB signal without an expensive crystal filter requires various mixers, oscillators, filters, and a wideband audio frequency phase shift network built with a network of high precision resistors and capacitors. To implement the same function in a DSP system requires adding one additional subroutine to the software program, and no additional hardware.

Many features that are straightforward with DSP techniques are difficult or impractical to implement with analog circuitry. A few examples drawn just from the communications field are imageless mixing, noise reduction, OFDM (orthogonal frequency division multiplexing) modulation, and adaptive channel equalization. Digital signals can have much more dynamic range than analog signals, limited only by the number of bits used to represent the signal. For example, it is easy to add an extra 20 or 30 dB of headroom to the intermediate signal processing stages to ensure that there is no measurable degradation of the signal in a filter, for example. It would be difficult or impossible to add that much dynamic range with analog circuitry. Replacing analog circuitry with software algorithms eliminates the problems of nonlinearity and drift of component values with time and temperature. The programmable nature of most DSP systems means you can make the equivalent of circuit modifications without having to unsolder any components.

Despite its many advantages, we don't mean to imply that DSP is best in all situations. High-power and very high-frequency signals are still the domain of analog circuitry. Where simplicity and low power consumption are primary goals, a DSP solution may not be the best choice. For example, a simple CW receiver that draws a few milliamps from the power supply can be built with two or three analog ICs and a handful of discrete components. We are still a long way from that kind of low-power radio using SDR.

In many high performance systems, the performance of the analog-to-digital converter and digital-to-analog converter are the limiting factors. That is why, even with the latest generation of affordable ADC technology, it is still possible to obtain better blocking dynamic range in an HF receiver using a hybrid analog-digital system rather than going all digital by routing the RF input directly to an ADC. This may change as ADC technology continues to evolve and performance rises.

8.3 Analog-Digital Conversion

Analog-to-digital (and digital-to-analog) conversion consists of taking data in one form, such as digital binary data or an analog RF waveform, and creating an equivalent representation of it in the opposite domain. Converters that create a digital representation of analog voltages or currents are called *analog-to-digital converters* (ADC), *analog/digital converters, A/D converters,* or *A-to-D converters.* Similarly, converters that create analog voltages or currents from digital quantities are called *digital-to-analog converters* (DAC), *digital/analog converters, D/A converters,* or *D-to-A converters.* The word "conversion" in this first section on the properties of converting information between the analog and digital domains applies equally to analog-to-digital or digital-to-analog conversion.

Converters are typically implemented as integrated circuits that include all of the necessary interfaces and sub-systems to perform the entire conversion process. In some cases, the converter may be integrated on the same chip as other analog functions (such as buffer amplifiers, reference voltages, etc.) or digital functions (such as digital filters). Schematic symbols for ADCs and DACs are shown in **Figure 8.9**.

This section defines the basic elements of analog-digital conversion along with an overview of different types of converters and their key specifications and behaviors. The following section discusses the use of converters for DSP and SDR functions, which are the most demanding application in radio.

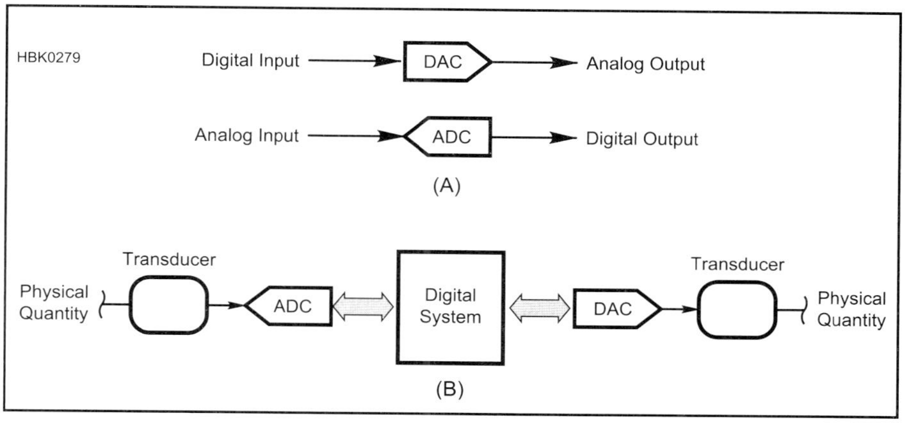

Figure 8.9 — Schematic symbols (A) for digital-to-analog converters (DAC) and analog-to-digital converters (ADC). The general block diagram of a system (B) that digitizes an analog signal, operates on it as digital data, then converts it back to analog form.

8.3.1 Basic Conversion Metrics

Figure 8.10 shows two different representations of the same physical phenomenon; an analog voltage changing from 0 to 1 V. In the analog world, the voltage is continuous and can be represented by any real number between 0 and 1. In the digital world, the number of possible values that can represent any phenomenon is limited by the number of bits contained in each value.

In Figure 8.10, there are only four two-bit digital values 00, 01, 10, and 11, each corresponding to the analog voltage being within a specific range of voltages. If the analog voltage is anywhere in the range 0 to 0.25 V, the digital value representing the analog voltage will be 00, no matter whether the voltage is 0.0001 or 0.24999 V. The range 0.25 to 0.5 V is represented by the digital value 01, and so forth.

The process of converting a continuous range of possible values to a limited number of discrete values is called *digitization* and each discrete value is called a *code* or a *quantization code*. If the code is a binary number, the number of possible codes that can represent an analog quantity is 2^N, where N is the number of bits in the code. A two-bit number can have four codes, as shown in Figure 8.10, a four-bit number can have sixteen codes, an eight-bit number 256 codes, and so forth. Assuming that the smallest change in code values is one bit, that value is called the *least significant bit* (LSB) regardless of its position in the format used to represent digital numbers.

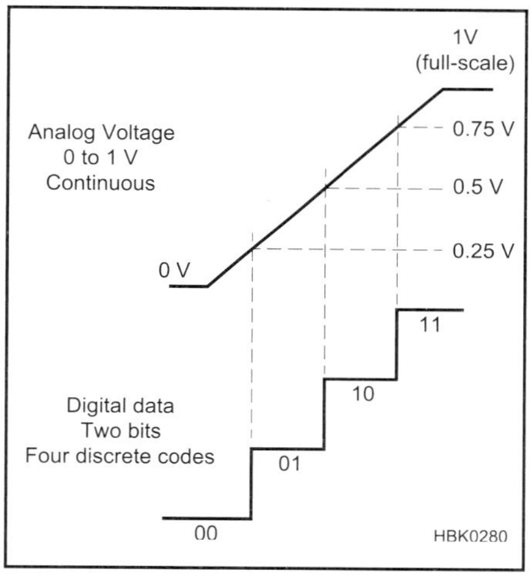

Figure 8.10 — The analog voltage varies continuously between 0 and 1 V, but the two-bit digital system only has four values to represent the analog voltage, so representation of the analog voltage is coarse.

Binary-coded-decimal (BCD) is a code in which groups of four bits represent individual decimal values of 0 to 9. In the *hexadecimal* code, groups of four bits represent decimal values of 0 to 15 and are represented by the digits 0 to 9 followed by the letters A through F. Other types of codes that may be encountered include *Gray code* and *octal.*

RESOLUTION AND RANGE

The *resolution* or *step size* of the conversion is the smallest change in the analog value that the conversion can represent. The *range* of the conversion is the total span of analog values that the conversion can process. The maximum value in the range is called the *full-scale* (F.S.) value. In Figure 8.10, the conversion range is 1 V. The resolution of the conversion is

$$resolution = \frac{range}{2^N}$$

In Figure 8.10, the conversion resolution is $1V/2^2 = 1V/4 = 0.25$ V in the figure. If each code had four bits instead, it would have a resolution of $1V/2^4 = 1V/16 = 0.0625$ V. Conversion range does not necessarily have zero as one end point. For example, a conversion range of 5 V may span 0 to 5 V, –5 to 0 V, –2.5 to +2.5 V, and so on.

Analog-digital conversion can have a range that is *unipolar* or *bipolar*. Unipolar means a conversion range that is entirely positive (or negative), usually referring to voltage. Bipolar means the range can take on both positive and negative values.

Because each code represents a range of possible analog values, the limited number of available codes creates *quantization error*. This is the maximum variation in analog values that can be represented by the same code. In Figure 8.10, any value from 0.25 through 0.50 V could be represented by the same code: 01. The quantization error in this case is 0.25 V.

Resolution can also be defined by the number of bits in the conversion. The higher the number of bits, the smaller the resolution as demonstrated above. Since many converters have variable ranges set by external components or voltages, referring to percent resolution or as a number of bits is preferred. The conversions between percent resolution and number of bits are as follows:

$$\%\text{resolution} = \frac{1}{2^N} \times 100\%$$

and

$$N = \frac{\log\left(\dfrac{100\%}{\%\,resolution}\right)}{\log 2}$$

Quantization error can also be specified as a number of least significant bits (LSB), where each bit is equivalent to the conversion's resolution.

ACCURACY

The number of bits of an ADC's resolution does not equate to the accuracy of the converter. A companion to resolution, *accuracy,* refers to the ability of the converter to either assign the correct code to an analog value or create the true analog value from a specific code. As with resolution, it is most convenient to refer to accuracy as either a percentage of full scale or in bits. *Full-scale error* is the maximum deviation of the code's value or the analog quantity's value as a percentage of the full scale value. If a converter's accuracy is given as 0.02% F.S., and the conversion range is 5 V, the conversion can be in error by as much as 0.02% × 5 V = 1 mV from the correct or expected value. Most ADCs include provision for user adjustment to calibrate the converter's full-scale range. Some ADCs include a reference voltage source internal to the device, while others require an external reference. External references can often provide higher accuracy and stability over temperature than internal references. *Offset* has the same meaning in conversion as it does in analog electronics: A consistent shift in the value of the conversion from the ideal value. It can also be adjusted if necessary by the user.

Linearity error represents the maximum deviation of the code transition points from the ideal code transition points after adjusting for the full scale and errors. This is also called *integral nonlinearity* (INL). In the converter of Figure 8.10, ideal step transitions occur at 0.25 V intervals. If the linearity error for the conversion was given as 0.05% F.S., any actual step size could be in error by as much as 0.05% × 5 V = 2.5 mV due to the transition to the next step being at the wrong point. *Differential nonlinearity* is a measure of how much any two adjacent step sizes deviate from the ideal step size. Errors can be represented as a number of bits, usually assumed to be least significant bits, or LSB, with one bit representing the same range as the conversion resolution. A typical ADC may specify its INL or DNL error as ± 0.5 LSB (least-significant bits).

CONVERSION RATE AND BANDWIDTH

Another important parameter of the conversion is the *conversion rate* or its reciprocal, *conversion speed.* A digital code that represents an analog value at a specific time is called a *sample,* so conversion rate, f_S, is specified in *samples per second* (sps) and conversion speed as some period of time per sample, such as 1 msec. Because of the mechanics by which conversion is performed, conversion speed can also be specified as a number of cycles of clock signal used by the digital system performing the conversion. Conversion rate then depends on the frequency of the clock.

According to the *Nyquist Sampling Theorem,* in order to accurately represent the input signal, a conversion must occur at a rate at least twice the bandwidth of the analog signal. This minimum rate is the *Nyquist rate* and the maximum frequency allowed in the analog signal is the *Nyquist frequency*. In this way, the converter *bandwidth* is limited to one-half the conversion rate.

Referring to the process of converting analog signals to digital samples, if a lower rate is used, called *undersampling,* false signals called *aliases* will be created in the digital representation of the input signal at frequencies related to the difference between the Nyquist sampling rate and f_S. This is called *aliasing*. Sampling faster than the Nyquist rate is called *oversampling* and can be used to advantage by following the converter with a digital filter.

Because conversions occur at some maximum rate, there is always the possibility of signals greater than the Nyquist frequency being present in an analog signal undergoing conversion, or that is being created from digital values. These signals would result in aliases and must be removed by *band-limiting filters,* also called *anti-aliasing filters,* that remove them prior to conversion. The mechanics of the sampling process are discussed later in this and following chapters as they apply to specific functions.

DISTORTION AND NOISE

Distortion and noise in a conversion are characterized by several parameters related to linearity and accuracy. THD+N (Total Harmonic Distortion + Noise) is a measure of how much distortion and noise is introduced by the conversion. THD+N can be specified in percent or in dB. Smaller values are better. SINAD (Signal to Noise and Distortion Ratio) is related to THD+N, generally specified along with a desired signal level to show what signal level is required to achieve a certain level of SINAD or the highest signal level at which a certain level of SINAD can be maintained. (See the Analog Devices application note MT-003 by Kester in the Bibliography for further information about SINAD and other noise metrics.)

ADC OVERLOAD

When *overloaded*, ADCs behave similarly to amplifiers driven into saturation or clipping. Consider the output of an ADC for various inputs as shown in **Table 8.1**. (This example assumes a 16-bit resolution and offset binary code.) "–FS" and "+FS" are the negative and positive Full-Scale input voltage levels of the ADC.

This table shows the ADC cannot distinguish between an input signal exactly at the top (or bottom) of its range and a signal that exceeds the input range. In fact, the ADC cannot tell if the input range has been slightly exceeded or grossly exceeded. This is an example of *hard clipping*.

Fortunately, most ADCs have an output signal to indicate the input range has been exceeded and the converter is in *overload* or *overflow/underflow* mode. This allows the system designer to take appropriate action — usually reducing the gain of an amplifier or increasing the attenuation of any stages ahead of the ADC.

In designing an SDR system, careful attention must be given to the peak signal reaching the ADC. Long periods of overload can create artifacts in the DSP after the converter. RF

Table 8.1
Output of an ADC for Various Inputs

Input voltage	*Output code (binary)*	*Output code (decimal)*
At or below -FS	0000 0000 0000 0000	0
Just above -FS	0000 0000 0000 0001	1
Middle of range	1000 0000 0000 0000	32768
Just below +FS	1111 1111 1111 1110	65534
At +FS-1 LSB	1111 1111 1111 1111	65535
At or above +FS	1111 1111 1111 1111	65535

signal levels are usually expressed in terms of power, rather than voltage, and often a signal is assumed to be a single sine wave. In a wideband receiver, the input signal is really the sum of many uncorrelated signals in the passband of the system, most of which are not even close to sinusoids.

In the **Radio Fundamentals** chapter, it is explained that ac waveforms are best characterized by their RMS (root-mean-square) values, and that a sine wave has an RMS value of 0.707 times the peak value. That means the peak value of the ac signal is 1.414 times the RMS value. This ratio is the *crest factor*. In terms of voltage, the 1.414 ratio is equal to 20 log (1.414), which is 3.01 dB. In terms of power, the 1.414 voltage ratio is a power ratio of $(1.414)^2$, which is 10 log (2), which is also 3.01 dB.

Other waveforms have different crest factors. A triangle wave has a crest factor of 1.732, or 4.77 dB. Signals that include multiple individual signals within the passband of the ADC tend to look almost like noise, which has a high crest factor. (True Gaussian noise has an infinite crest factor!). In addition, some environments have locally-generated noise from sources such as power-line leakage or arcing, electric fences, or nearby switch-mode power supplies, not to mention atmospheric noise. Any of these can cause an ADC to overload.

Typical high-speed ADCs are high-input-impedance devices and require a dc bias or offset on the input waveform when operated from a single-polarity power source. The input signal from preceding stages is usually transformer-coupled to the ADC input, biased from a signal generated by the ADC. Converter manufacturers usually give excellent guidance on how to couple the signal into the converter for optimal performance. Most ADCs utilize an input voltage range on the order of 2 V_{P-P}. It is good practice to plan for several dB of overhead in the signal level reaching the ADC. Allowing for 10 dB of headroom means that the expected signal levels should only use about ⅓ of the ADC's input voltage range.

8.3.2 Analog-to-Digital Converters

There are a number of methods by which the conversion from an analog quantity to a set of digital samples can be performed. Each method or *architecture* has its strong points — simplicity, speed, resolution, accuracy — and each affect the decision of which method to use for a particular application. In order to pick the right type of ADC, it is important to decide which of these criteria most strongly affect the performance of your application.

The following sections present ADC architectures used in amateur radio from the fastest to progressively slower conversion. For a more complete review of ADC architectures, see Kester's article "Which ADC Architecture Is Right for Your Application?" listed in the Reference section.

FLASH CONVERTER

The simplest type of ADC is the *flash converter*, shown in **Figure 8.11**. It continually generates a digital representation of the analog signal at its input. The flash converter uses an array of comparators that compare the amplitude of the input signal to a set of reference voltages. There is one reference voltage for each step.

The outputs of the comparator array represent a digital value in which each bit indicates whether the input signal is greater (1) or less than (0) the reference voltage for that comparator. A digital logic *priority encoder* then converts the array of bits into a digital output code. Each successive conversion is available as quickly as the comparators can respond and the priority encoder can create the output code. Flash converters are generally used for applications in which high speed is more important than bits of resolution.

Flash converters are the fastest of all ADCs (conversion speeds can be in the nanosecond

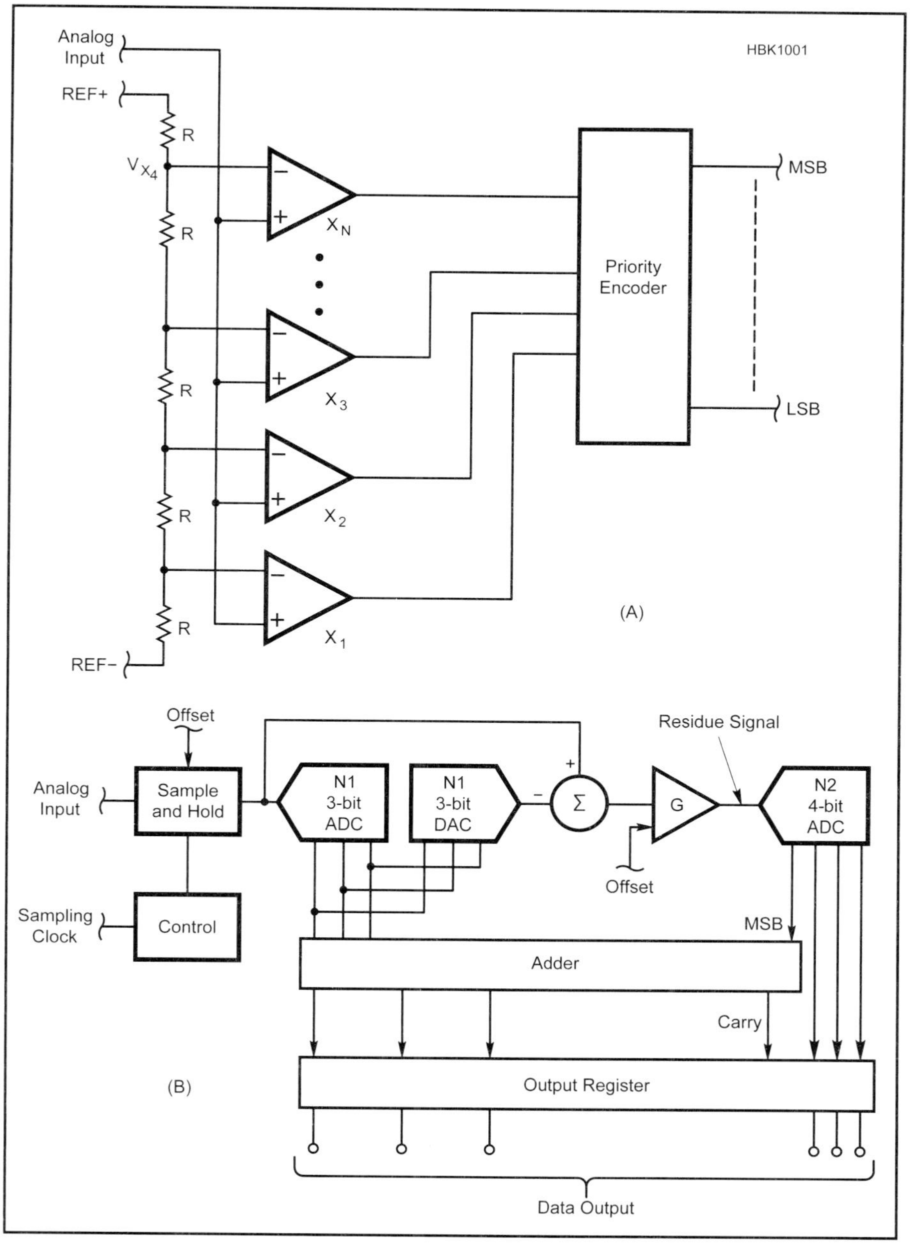

Figure 8.11 — The comparators in a flash converter (A) switch continuously to follow the input signal. The encoder converts the array of comparator outputs to a single digital word. Flash converters (N1 and N2) form the basis of the pipelined converter (B) that is used at the very high sample rates needed for SDR applications.

range, equivalent to sampling rates in the gigasamples/second) but do not have high resolution because of the number of comparators and reference voltages required. For example, an 8-bit flash converter requires 255 comparators, while a 12-bit version would require 4095 comparators.

One specialized variant of the flash converter is the *bar-graph display driver*. Such devices accept an analog input and deliver output signals capable of directly driving an LED display rather than a binary digital output word. These devices are commonly used to replace meters as front panel displays for signals that represent relatively slow-varying parameters such as transmitter output power. The LM3914 from Texas Instruments is one such device.

PIPELINED CONVERTER

A *pipelined converter* uses two or more stages of flash converters as shown in Figure 8.11B. The output of the first stage flash converter (N1) is changed back to an analog signal by a DAC and the difference between the DAC output and input signal is called the *residue signal*. The residue signal is amplified then digitized by a second flash converter (N2). The digital outputs of the two flash converters are combined by an adder and a register to create the overall converter output word. The digital outputs of multiple stages can be combined to yield more bits than a single-stage flash converter would produce. Pipelined converters are discussed in additional detail in the section Data Converters for SDR and DSP later in this chapter.

SUCCESSIVE-APPROXIMATION CONVERTER

The *successive-approximation* ADC is one of the most widely-used types of converters. As shown in **Figure 8.12**, it uses a single comparator and DAC to arrive at the value of the input voltage by comparing it to successive analog values generated by the DAC. This type of converter offers a good compromise of conversion speed and resolution.

The control logic begins a conversion by setting the output of the DAC to ½ of the conversion range. If the DAC output is greater than the analog input value, the output of the comparator is 0 and the most significant bit of the digital value is set to 0. If the DAC output is less than the analog input, the bit is set to 1. The DAC output then either increases or decreases by ¼ of the range, depending on whether the value of the first comparison was 1 or 0. At this point the input voltage is being compared to either ¼ of the scale or ¾ of the range. One test is made for each bit in digital output code and the result accumulated in a storage register, called the Successive-Approximation Register, which is why such converters are often called "SAR" converters.

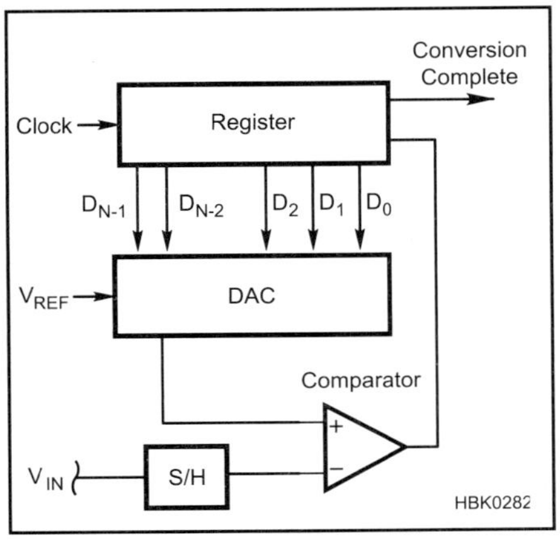

Figure 8.12 — The successive-approximation converter creates a digital word as it varies the DAC signal in order to keep the comparator's noninverting terminal close to the input voltage. A sample-and-hold circuit (S/H) holds the input signal steady while the measurement is being made.

The process is then repeated, forming a series of approximations, until a test has been made for all bits in the code.

While the digital circuitry to implement the SAR converter is more complex, it is less expensive to build and calibrate than the array of comparators and precision resistors of the flash converter, especially for higher resolutions. Each conversion also takes a known and fixed number of clock cycles. SAR ADCs are used for speeds up to a few Msps. They are often used with an analog multiplexer in front so that multiple signals in a system can be measured relatively quickly, rather than using a separate converter for every signal.

DUAL-SLOPE INTEGRATING CONVERTERS

The *dual-slope integrating ADC* is shown in **Figure 8.13**. It makes a conversion by integrating the input signal by charging a capacitor for a fixed period of time then measuring

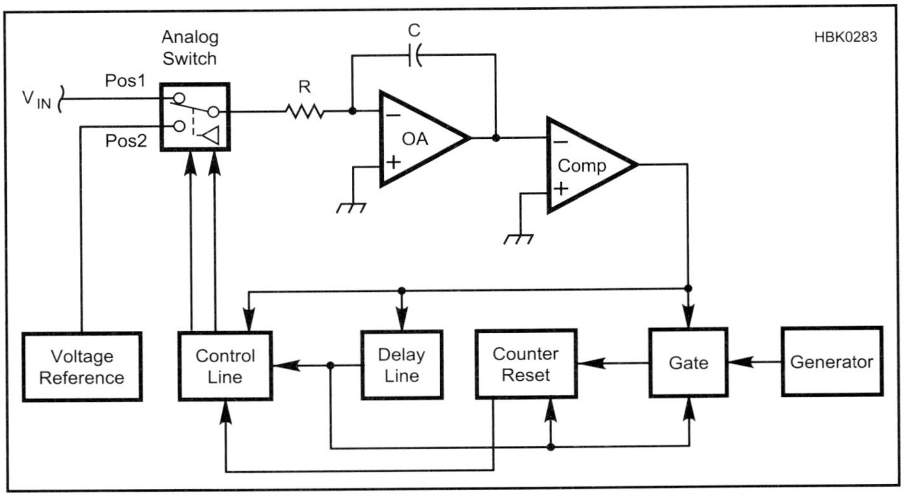

Figure 8.13 — Dual-slope integrating converter. By using a constant-current source to continually charge a capacitor to a known reference voltage and then discharge it, the resulting frequency is directly proportional to the resistor value.

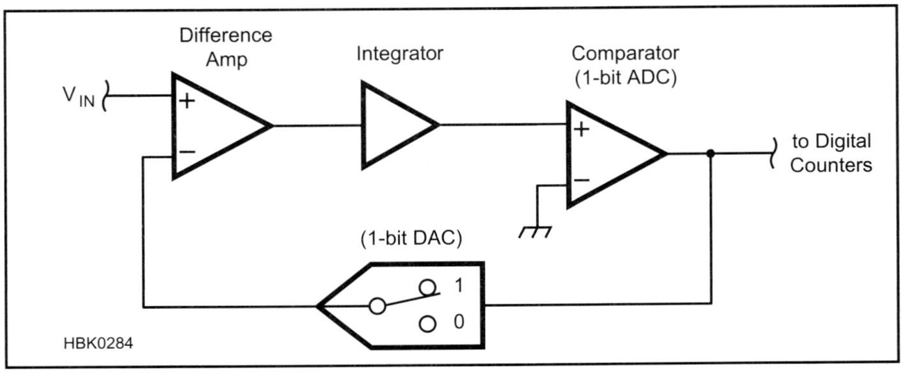

Figure 8.14 — Delta-encoded converter. The 1-bit DAC is operated in such a way that the bit stream out of the comparator represents the value of the input voltage.

the time it takes for the capacitor to discharge back to its starting value. The integration period is often set to reject an interfering signal by integrating it over an exact number of cycles. For example, setting the integration period to 100 milliseconds results in six full cycles of a 60 Hz sinusoidal interferer and five full cycles of a 50 Hz interferer.

Dual-slope ADCs are low-cost and relatively immune to temperature variations. Due to the slow speed of the conversion these converters are generally only used in test instruments such as multimeters or for measurement of slowly varying dc signals.

DELTA-ENCODED CONVERTERS

Instead of charging and discharging a capacitor from 0 V to the level of the input signal, and then back to 0 V, the *delta-encoded ADC* in **Figure 8.14** continually compares the output of a DAC to the input signal using a comparator. Whenever the signal changes, the DAC is adjusted until its output is equal to the input signal. Digital counter circuits keep track of the DAC value and generate the digital output code.

SIGMA-DELTA CONVERTERS

The *sigma-delta converter* also uses a DAC and a comparator in a feedback loop to generate a digital signal as shown in **Figure 8.15**. An integrator stores the sum of the input signal and the DAC output. (This is "sigma" or sum in the converter's name.) The comparator output indicates whether the integrator output is above or below the reference voltage and that signal is used to adjust the DAC's output so that the integrator output stays close to the reference voltage. (This is the "delta" in the name.) The stream of 0s and 1s from the comparator forms a high-speed digital bit stream that is digitally filtered to form the output code. Sigma-delta converters are used where high resolution (16 to 24 bits) is required at sampling rates in the ksps range, including very slowly-varying signals and even audio signals.

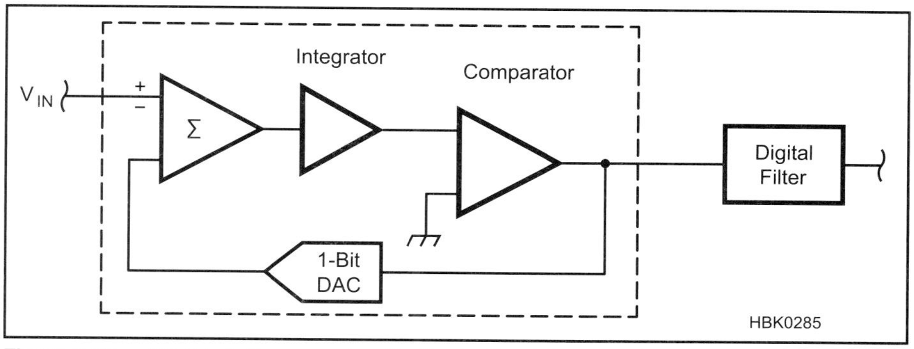

Figure 8.15 — Sigma-delta converter. Similar to the delta-encoded converter (Figure 8.14), the converter runs much faster than the output samples and uses a digital filter to derive the actual output value.

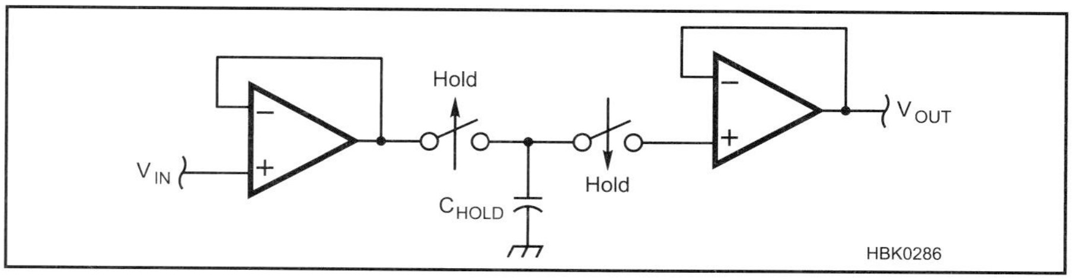

Figure 8.16 — Sample-and-hold (S/H). An input buffer isolates the sampled voltage from the input signal by charging the capacitor C_{HOLD} to that voltage with the input switch closed and the output switch open. When a measurement is being taken, the input switch is open to prevent the input signal from changing the capacitor voltage, and the output switch is closed so that the output buffer can generate a steady voltage at its output.

8.3.3 Analog-to-Digital Converter Subsystems

SAMPLE-AND-HOLD

ADCs that use a sequence of operations to create the digital output code must have a means of holding the input signal steady while the measurements are being made. This function is performed by the circuit of **Figure 8.16**. A high input-impedance buffer drives the external storage capacitor, C_{HOLD}, so that its voltage is the same as the input signal. Another high input-impedance buffer is used to provide a replica of the voltage on C_{HOLD} to the conversion circuitry.

When a conversion is started, a digital control signal opens the input switch, closes the output switch, and the capacitor's voltage is measured by the converter. It is important that the capacitor used for C_{HOLD} has low *leakage* and the output op amp has very high impedance so that while the measurement is being made, the voltage stays constant for the length of time required to complete the conversion process. This is of particular importance in high-precision conversion.

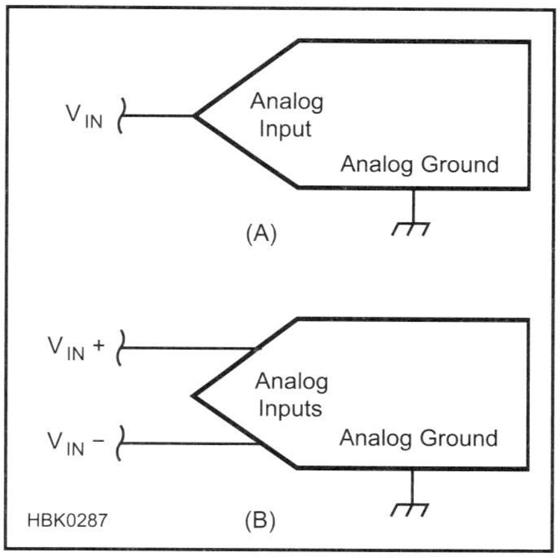

Figure 8.17 — Single-ended ADC inputs have a single active line and a ground or return line (A). Single-ended ADC input are often susceptible to noise and common mode signals or any kind of disturbance on their ground rails. In (B), the differential inputs used help the circuit "ignore" offsets and shifts in the input signal.

SINGLE-ENDED AND DIFFERENTIAL INPUTS

The input of most ADCs is *single-ended*, where the input signal is measured between the input pin and a common ground. Shown in **Figure 8.17A**, this is acceptable for most applications. If the voltage to be measured is small or is the difference between two nonzero voltages, an ADC with *differential inputs* should be used as in Figure 8.17B. Differential inputs are also useful when measuring current as the voltage across a small resistor in series with the current. In that case, neither side of the resistor is likely to be at ground, so a differential input is very useful. Differential inputs also help avoid the issue of noise contamination as discussed below.

INPUT BUFFERING AND FILTERING

The input impedance of most ADCs is high enough that the source of the input signal is unaffected. However, to protect the ADC input and reduce loading on the input source, an external buffer stage can be used. **Figure 8.18** shows a typical buffer arrangement with clamping diodes to protect against electrostatic discharge (ESD) and an RC-filter to

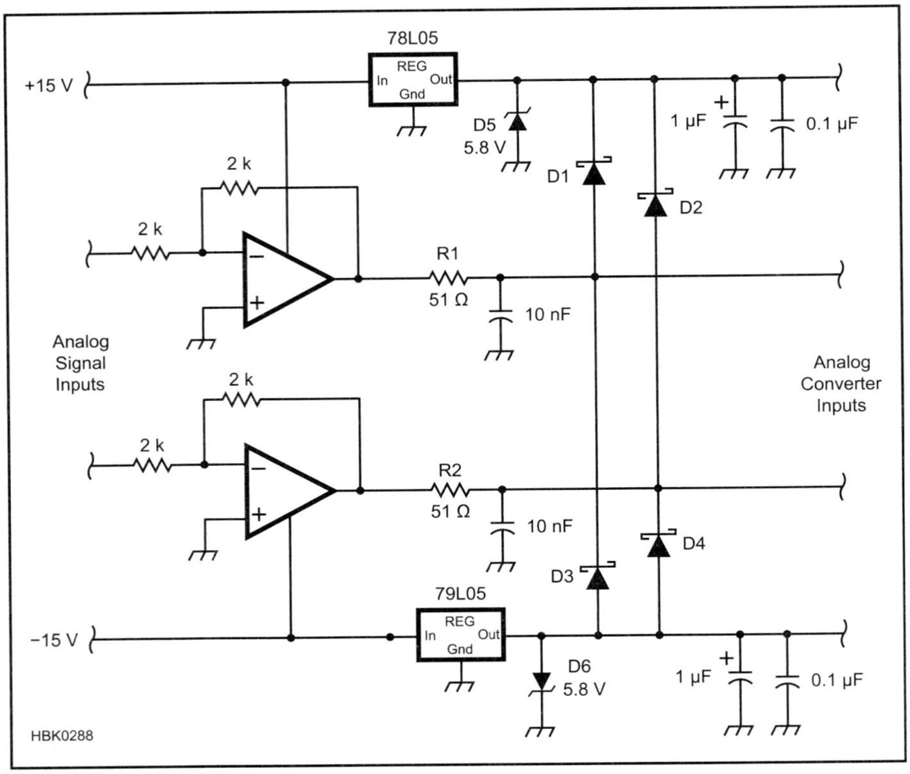

Figure 8.18 — Typical ADC input buffer-filter circuit. Unity-gain voltage followers help isolate the ADC from the input source. RC filters following the buffers act as band-limiting filters to prevent aliasing. Zener diodes are used to clamp the transient voltage and route the energy of transient into the power supply system.

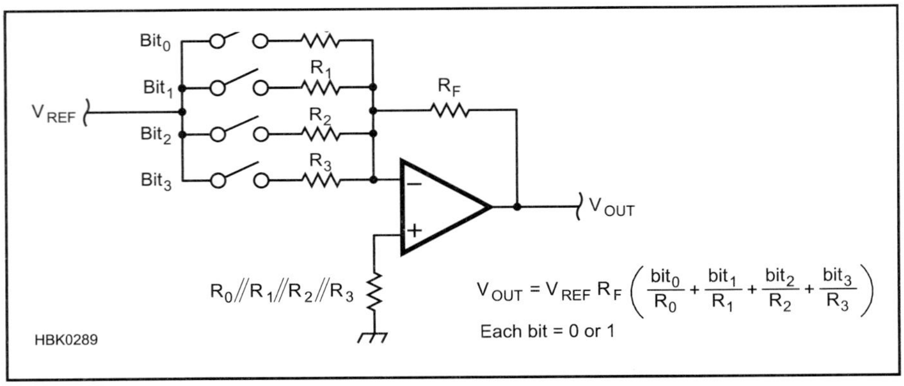

Figure 8.19 — Summing DAC. The output voltage is the inverted, weighted sum of the inputs to each summing resistor at the input. Digital data at the input controls the current into the summing resistors and thus, the output voltage.

prevent RF signals or noise from affecting the input signal. In addition, to attenuate higher frequency signals that might cause aliases, the input filters can also act as band-limiting filters.

ANALOG AND DIGITAL "GROUND"

By definition, ADCs straddle the analog and digital domain. In principle, the signals remain separate and isolated from each other. In practice, however, voltages and currents from the analog and digital circuitry can be mixed together. This can result in the contamination of an analog signal with components of digital signals and, rarely, vice versa. Mostly, this is a problem when trying to measure small voltages in the presence of large power or RF signals.

The usual problem is that currents from high-speed digital circuitry find their way into analog signal paths and create transients and other artifacts that affect the measurement of the analog signal. Thus, it is important to have separate current paths for the two types of signals. The manufacturer of the converter will provide guidance for the proper use of the converter either in the device's data sheet or as application notes. Look for separate pins on the converter, such as "AGND" (Analog Ground) or "DGND" (Digital Ground), that indicate how the two types of signal return paths should be connected.

8.3.4 Digital-to-Analog Converters (DAC)

Converting a digital value to an analog quantity is considerably simpler than the reverse, but there are several issues primarily associated with DACs that affect the selection of a particular converter.

As each new digital value is converted to analog, the output of the DAC makes an abrupt *step change*. Even if very small, the response of the DAC output does not respond perfectly or instantaneously. *Settling time* is the amount of time required for the DAC's output to stabilize within a certain amount of the final value. It is specified by the manufacturer and can be degraded if the load connected to the DAC is too heavy or if it is highly reactive. Settling time is critical in control applications and also sets the limit on update sampling rate.

Monotonicity is another aspect of characterizing the DAC's accuracy. A DAC is *monotonic* if increasing the digital input value causes the output of the DAC to increase with every step. Because of errors in the internal conversion circuitry, it is possible for there to be some steps that are too small or too large, leading to output values that seem "out of order." These are usually quite small, but if used in a precision control application where the DAC is part of a feedback loop, monotonicity is important because the control system can go unstable.

SUMMING DAC

A *summing DAC*, shown in **Figure 8.19**, is a summing amplifier with all of the inputs connected to a single reference voltage through switches. The digital value to be converted controls which switches are closed. The larger the digital value, the more switches are closed. Higher current causes the summing amplifier's output voltage to be higher, as well.

The input resistors are *binary-weighted* so that the summing network resistors representing the more significant bit values inject more current into the op amp's summing junction. Each resistor differs from its neighboring resistors in the amount of current it injects into the summing node by a factor of two, recreating the effect of each digital bit in the output voltage. At high resolutions, this becomes a problem because of the wide spread in resistor values. A 12-bit DAC requires a spread of 2048 between the largest and smallest resistor values. Summing DACs are generally only available with low resolution for that reason.

The *current output DAC* functions identically to the summing DAC, but does not have an op amp to convert current in the digitally-controlled resistor network to voltage. It consists only of the resistor network, so an external current-to-voltage circuit (discussed in the **Circuits and Components** chapter) is required to change the current to a voltage. In some applications, the conversion to voltage is not required, or it is already provided by some other circuit.

R-2R LADDER DAC

The summing and current output DACs both used binary weighted resistors to convert the binary digital value into the analog output. The practical limitation of this design is the large difference in value between the smallest and largest resistor. For example, in a 12-bit DAC, the smallest and largest resistors differ by a factor of 2048 (which is 2^{12-1}). This can be difficult to fabricate in an IC since it is difficult to match a wide range of resistors values closely enough to maintain the desired binary-weighted relationship with sufficient accuracy. For DACs with resolutions of 8 bits or more, the R-2R ladder DAC of **Figure 8.20** is a better design. R-2R DACs can operate in either voltage mode or current mode.

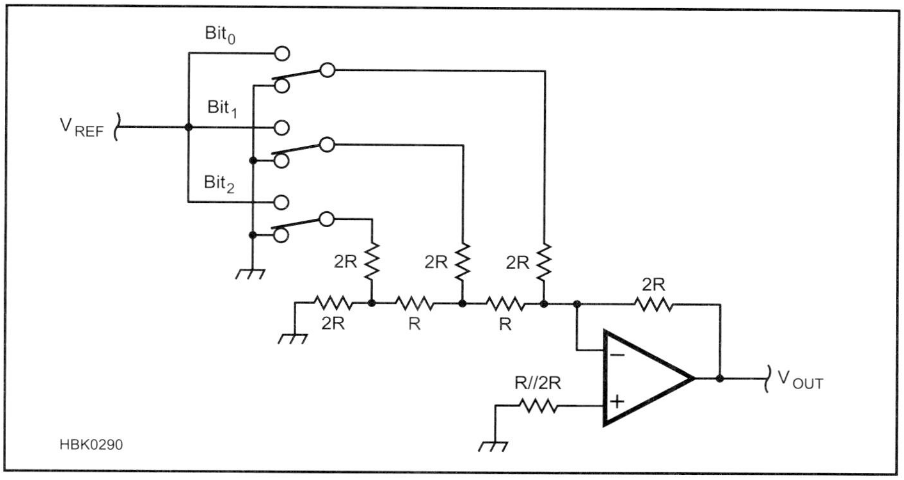

Figure 8.20 — R-2R Ladder DAC. This is the most common form of DAC because all of the resistor values are similar, making it easier to manufacture. The similarity in resistor values also means that there will be less variation of the comparator with temperature and other effects that affect all resistors similarly.

The problems of manufacturing are greatly reduced when resistances are fairly close in value. By using the *R-2R ladder* shown in the figure, the same method of varying current injected into an op amp circuit's summing junction can be accomplished with resistors of only two values, R and 2R. In fact, since the op amp feedback resistor is also one of the IC resistors, the absolute value of the resistance R is unimportant, as long as the ratio of R:2R is maintained. This simplifies manufacturing greatly and is an example of IC design being based on ratios instead of absolute values. For this reason, most DACs use the R-2R ladder design and the performance differences lie mostly in their speed and accuracy.

Similar ladder networks can be constructed using capacitors. Such networks are used on many SAR-type ADCs because in some fabrication processes, capacitors are easier to make than resistors.

DIGITAL POTENTIOMETERS

Since DACs are commonly used to provide digitally-controlled adjustments, a special class of DAC has evolved, specifically intended for use as calibration/adjustment devices. These are known as "digital potentiometers." They differ from traditional DACs in that they appear as three-terminal variable resistors of known value, and often include non-volatile control registers. They are specified in terms of resistance value, number of steps, and number of devices per package.

8.3.5 Choosing a Converter

From the point of view of performance, choosing a converter, either an ADC or a DAC, comes down to resolution, accuracy, and speed. Begin by determining the percent resolution or the dynamic range of the converter. Use the equations in the preceding section to determine the number of bits the converter must have. Select from converters with the next highest number of bits. For example, if you determine that you need 7 bits of resolution, use an 8-bit converter.

Next, consider accuracy. If the converter is needed for test instrumentation, you'll need to perform an *error budget* on the instrument's conversion processes, include errors in the analog circuitry. Once you have calculated percent errors, you can determine the requirements for FS error, offset error, and nonlinearities. If an ADC is going to be used for receiving applications, the spurious-free dynamic range may be more important than high dc precision.

The remaining performance criterion is the speed and rate at which the converter can operate. Conversions should be able to be made at a minimum of twice the bandwidth of the signal you wish to reproduce. If the converter will be running near its maximum rate, be sure that the associated digital interface, supporting circuitry, and software can support the required data rates, too!

Having established the conversion performance requirements, the next step is to consider cost, amount of associated circuitry, power requirements, and so forth. For example, a self-contained ADC is easier to use and takes up less PC board space, but may not be as accurate as one that allows the designer to use an external voltage reference to set the conversion range. Other considerations, such as the nature of the required digital interface, as discussed in the next section, can also affect the selection of the converter.

8.4 Data Converters for SDR and DSP

In this section we will focus more closely on converters for use with DSP and SDR. The first requirement when selecting a DAC or ADC is that it be able to handle the required *sample rate*. For communications-quality voice (300 – 3,000 Hz), a sample rate on the order of 8,000 samples per second (8 ksps) is adequate and has been used in the public switched telephone network for decades. For CD-quality music (20 – 20,000 Hz), the standard sample rate is 44.1 ksps, while 48 ksps, and even 96 or 192 ksps, are used for newer audio formats and applications such as professional studio equipment.

In an SDR system intended to use microphone inputs and speaker or headphone outputs, there is a need to convert between analog and digital at audio rates. Fortunately, combination ADC and DAC chips and software packages called *codecs* (short for coder-decoder) are readily available to do this job. Both "audio codecs," with sample rates consistent with high-fidelity audio up to 20 kHz or so, and "voice codecs," intended for voice-quality audio up to 4 or 8 kHz, are available. The switched telephone network originally used 8-bit logarithmic codecs to digitize and reconstruct voiceband audio signals. Logarithmic coding was used to compress the dynamic range of a voice signal to reduce the number of bits to be transported.

Today, linear-coded devices have become preferable since DSP algorithms work on linear-coded data and often perform more efficient compression algorithms to reduce the number of bits needed to transport digitized voice. Since audio codecs are used in PC and other consumer systems they are very low cost (or are free if the PC is part of the SDR system). They have more than adequate performance for SDR applications dealing with relatively narrowband signals, such as CW, SSB, AM, and digital signals such as RTTY, PACTOR, and WSJT modes that use bandwidths comparable to SSB.

Processing wideband RF signals requires data converters with sample rates in the megasamples per second (Msps) range. Fortunately, manufacturers of digital communications infrastructure equipment for the cellular industry have similar needs, and manufacturers of data converter ICs have developed products that are produced in reasonable volume (which lowers the cost) and can be repurposed for amateur SDRs.

The resolution of a data converter expressed as the number of bits in the data words gives an approximation of the converter's SNR. For example, an 8-bit ADC can only represent the sampled analog signal as one of $2^8 = 256$ possible numbers. The smallest signal that it can resolve is therefore 1/256 of full scale. In terms of signal-to-noise ratio (SNR), the rule of thumb is 6 dB per bit, so an ideal 8-bit converter should have an SNR of about 48 dB.

The actual formula is slightly different. In an ideal, error-free ADC, the *quantization error* is up to ±1/2 of one least-significant bit (LSB) of the digital word, or ±1/512 of full scale with 8-bit resolution. Similarly, a DAC can only generate the analog signal to within ±1/2 LSB of the desired value. These unavoidable errors are indistinguishable from noise in a signal-processing system.

It can be shown mathematically that a series of uniformly distributed random numbers between +0.5 LSB and –0.5 LSB has an RMS value of

$$\frac{LSB}{\sqrt{12}}$$

This represents the quantization noise of an ADC.

The **Radio Fundamentals** chapter shows that a sinusoidal signal of 2 V peak-to-peak (1 V peak) has an RMS voltage equal to $0.707 = \sqrt{2}\,/\,2$ V. In the case of a full-scale sinusoidal signal applied to an ADC of N bits (2^N LSBs), the RMS signal voltage can be written as $\left(2^{N-1} LSB\right) / \sqrt{2}$.

Combining that information results in the following equation for the SNR ratio in decibels for an ideal data converter word of width N bits:

$$SNR = 20\log\left(\frac{\dfrac{2^{N-1}LSB}{\sqrt{2}}}{\dfrac{LSB}{\sqrt{12}}}\right)$$

which simplifies to

SNR = (6.02N + 1.76) dB

The extra 1.76 dB arises from the fact that noise has a lower RMS value than a sine wave for the same peak-to-peak amplitude (see Analog Devices application note MT001 by Kester in the References and Bibliography section.) Therefore, with an ideal 8-bit converter, SNR = 49.9 dB. A perfect 16-bit ADC would achieve a 98.1 dB SNR. Of course, real-world devices are never perfect, so actual performance is always somewhat less. Internally-generated noise degrades the SNR, and errors in the linearity of the converter's transfer function give rise to distortion in the form of spurious spectral components and harmonics. (See also "Clocking the RF ADC: Should you worry about jitter or phase noise?" by Neu in the Reference and Bibliography section.)

8.4.1 Using Audio ADCs for SDR

ADCs and DACs used in audio systems and PC "sound cards" are often repurposed for use in hybrid SDRs. While they may offer as many as 24 bits of resolution, in reality, audio ADCs only provide 110 to 120 dB of SNR and "dynamic range," and the best "24-bit" audio DACs deliver 128 dB. According to the formula above, a 24-bit converter should have 1.76 + (6.02 × 24), or 146.2 dB. What happened to the remaining 20 to 40 dB?

Part of the reason that audio converters use 24-bit data words is that the recording and playback formats for many systems (such as Blu-Ray Disc and studio equipment) are designed to accommodate 24-bit data words, even if the lower-order bits are essentially meaningless. For reference an LSB for a typical 24-bit audio converter with a 5 V p-p signal range is about 100 nanovolts (0.1 microvolts) and it is difficult to keep noise as low as that in the real world.

Furthermore, the testing of ADCs intended for use in audio systems is different from how we might specify them for radio applications. For example, the "dynamic range" specification in digital audio systems is tested by measuring the THD+N (Total Harmonic Distortion plus Noise) for a 1 kHz input signal at a level 60 dB below full-scale, converting the number to a positive value, and then adding 60. For example, an audio DAC that delivers THD+N of –51 dB under this test condition can be specified as having a dynamic range of 111 dB. While this is not the same way that communications systems define dynamic range, the goal of both audio and radio systems is to evolve towards higher dynamic range, so the

Distortion and Noise

Distortion and noise in a conversion are characterized by several parameters all related to linearity and accuracy. THD+N (Total Harmonic Distortion + Noise) is a measure of how much distortion and noise is introduced by the conversion. THD+N can be specified in percent or in dB. Smaller values are better. SINAD (Signal to Noise and Distortion Ratio) is related to THD+N, generally specified along with a desired signal level to show what signal level is required to achieve a certain level of SINAD or the highest signal level at which a certain level of SINAD can be maintained. (See the Analog Devices application note MT-003 by Kester in the Bibliography for further information about SINAD and other noise metrics.)

performance generally heads in the right direction in both fields.

SIGMA-DELTA CONVERTERS

The noise spectrum of an ADC occupies the bandwidth from dc to one-half the sampling frequency, and is more or less uniformly spread out over that range. In any smaller slice of the output spectrum, the noise is much lower compared to the full-scale output of the converter. Thus, if we have an ADC sampling at a very high rate relative to the bandwidth of interest, and then add a digital filter after the ADC that eliminates the noise components outside the spectrum of interest. We can, in principle, increase the SNR in the desired bandwidth. An ADC system that operates this way is said to be *oversampling*.

Oversampling by a factor of N improves the SNR in a given bandwidth by the square root of N since the number of samples of the desired signal is multiplied by N, and the noise, which is uncorrelated, increases as the square root of N. This can be simplified to 3 dB SNR improvement for every doubling of the sample rate. Thus, a 4× oversampling rate is equal to adding another bit of resolution to a converter.

Audio converters are an example of oversampled converters, and usually use a technique known as *sigma-delta modulation*. The sigma-delta converter introduced earlier uses a DAC and a comparator in a feedback loop to generate a digital signal. An integrator stores the sum of the input signal and the DAC output. This is "sigma" or sum in the converter's name. The comparator is in essence a one-bit ADC, and its output indicates whether the integrator output is above or below the reference voltage. That signal is used to adjust the DAC's output so that the integrator output stays close to the reference voltage. This is the "delta" in the name. The stream of 0s and 1s from the comparator forms a high-speed digital bit stream that is digitally-filtered to form the output code, at a much lower output sample rate.

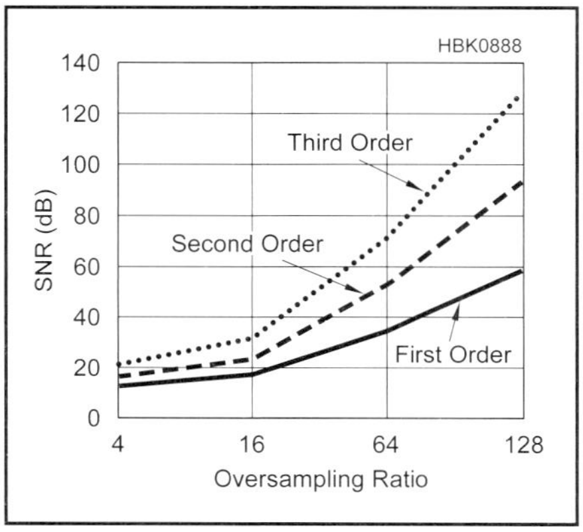

Figure 8.21 — The benefits of noise-shaping as the order of the sigma-delta converter feedback loop is increased.

Most ADCs and DACs in audio systems use an extreme form of oversampling, where the internal converter may oversample by a rate of 128 or 256 times the desired output rate. For example, the comparator of an audio ADC intended for a 48 ksps output rate may be clocked at 256 × 48 ksps, or 12.288 MHz. In addition, sigma-delta audio converters use a technique called noise shaping to push most of the quantization noise to frequencies above the audio band, thereby reducing it in the audio spectrum. Noise shaping is accomplished by adding more integrators in the feedback loop of the simple sigma-delta modulator to form a "higher-order loop" with a low-pass response for the signal, but a high-pass response for the noise. The digital low-pass filter then removes the noise and produces output data words of higher width at the desired sample rate. The high input sampling rate of the converter also relaxes the requirements for external band-limiting (anti-aliasing) filters on the input signal. **Figure 8.21** illustrates the relationship between loop order, sample rate, and improvements in SNR.

Since audio ADCs are the audio input device on many PC-based systems, they are quite inexpensive. The Left- and Right-channel stereo inputs can be repurposed and renamed "I" and "Q" for use in SDR applications. Even if the converter specifications are not exactly ideal they provide adequate performance for a hybrid SDR.

8.4.2 High-Speed ADCs for SDR

Commercial applications such as cellular base stations have driven the need for high-speed converters capable of sampling RF or high-IF signals directly. These have enabled the development of amateur radio HF transceivers using direct RF sampling. In addition to the usual specifications of linearity, etc., most manufacturers of high-speed data converters intended for use in wideband SDR systems now include a specification for *spurious-free dynamic range* (SFDR). This is the ratio, normally expressed in dB, between a (usually) full-scale sine wave and the worst-case spurious signal. It is a useful figure of merit for comparing such converters. (High-speed SDR hardware from several manufacturers and vendors is addressed in the column "SDR: Simplified" by Ray Mack, W5IFS, in the January/February 2013 issue of *QEX*.)

PIPELINED ADC ARCHITECTURE

Most high-speed converters intended for SDR applications use a *pipelined* architecture. This type of ADC consists of multiple stages, each of which includes a sample-and-hold function, low-resolution ADC and DAC, and a gain stage. The first stage performs a coarse quantization of the input signal, of say, "B" bits, where B is typically 1 to 4 in commercial products. The resultant data word is applied to a DAC, and the output analog voltage is

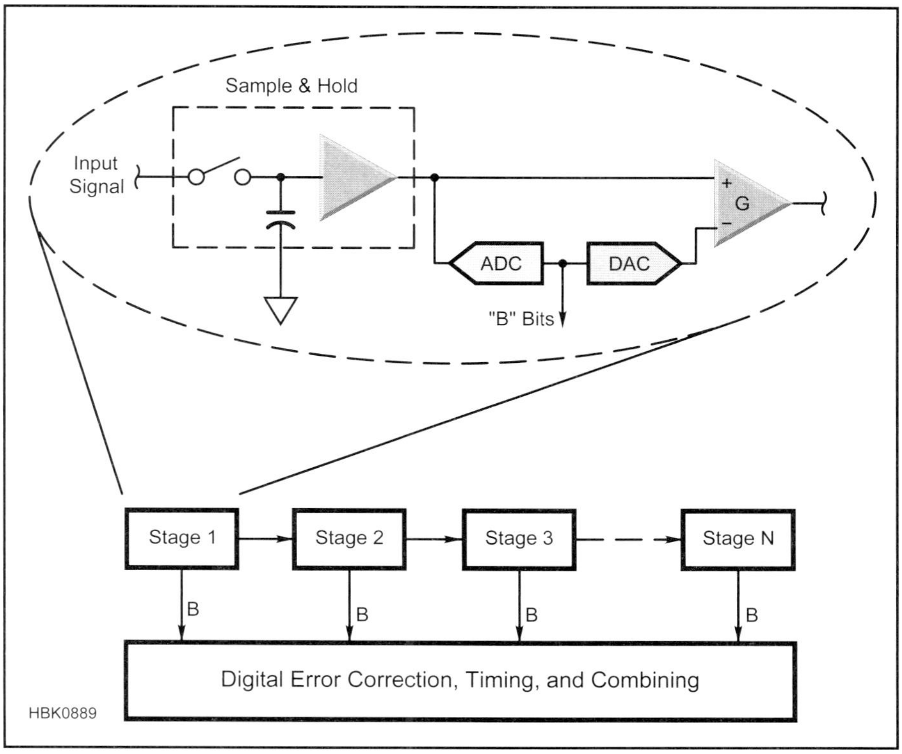

Figure 8.22 — Block diagram of pipelined high-speed ADC.

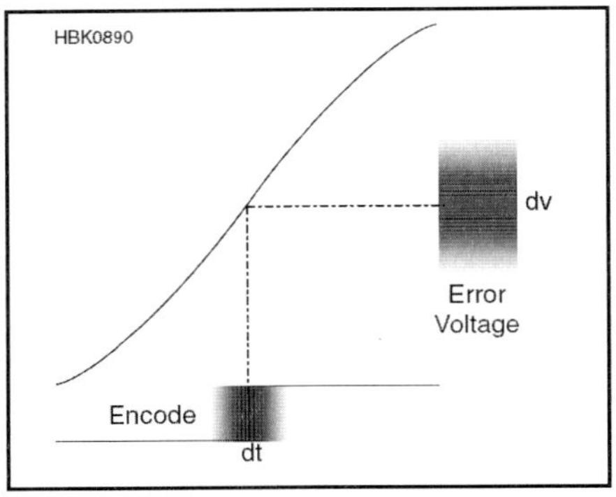

Figure 8.23 — RMS jitter vs. RMS noise effect of sampling uncertainty on SNR. [From Figure 1 of Analog Devices Application Note AN-501, available at www.analog.com, used with permission.]

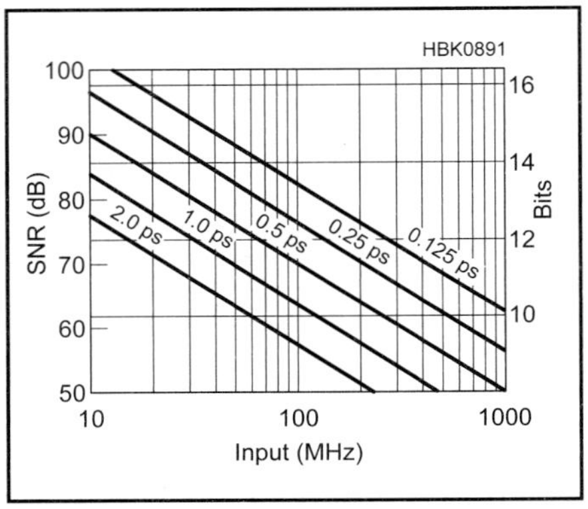

Figure 8.24 — Signal-to-noise ratio due to aperture jitter. [From Figure 1 of Analog Devices Application Note AN-501, available at www.analog.com, used with permission.]

subtracted from the original input signal, producing an error signal. That error signal is amplified and passed to another, usually identical, converter stage with enough overlap to compensate for any errors in the previous conversion. Many stages can be connected together, with the last stage only including an ADC. The digital outputs from each stage are combined and corrected, as well as time-aligned, to produce a higher resolution output. **Figure 8.22** illustrates the basic pipelined structure.

Since each stage is typically a low-resolution flash converter, pipelined converters are capable of quite high resolution — up to 16 bits — and sample rates in the hundreds of Msps, even into the Gsps (gigasamples per second) range. There is a delay before the first sample emerges, of course, since an N-stage pipeline converter must perform N conversions before the first valid output is available. However, each successive sample is produced at the full sample rate.

Some pipelined converters also include basic digital down conversion on chip, with a degree of programmability.

The cost of an ADC or DAC converter is generally proportional to both speed and resolution, so it is important to avoid over specifying these components. Later in this chapter we will discuss how to determine the required sample rate and resolution for a given application.

While sample rate, resolution, and SFDR are the principal selection criteria for data converters in a DSP system, other parameters such as SNR, harmonic and intermodulation distortion, full-power bandwidth, and aperture delay jitter can also affect system performance. Of course, basic specifications such as power requirements, interface type (serial or parallel), and cost also determine a device's suitability for a particular application. As with any electronic component, it is very important to read and fully understand the data sheet.

Most manufacturers of high-performance converters also include recommendations on support circuitry to be used with their devices to extract the best possible performance. For example, some ADCs require buffer amplifiers to drive the input, or external voltage references to set the full scale range. Many converters, especially high-speed and high-resolution devices, are very sensitive to circuit layout, power supply decoupling, and grounding. Most manufacturers provide evaluation boards to allow testing the converter IC on the bench, and the layout files can be copied into the final system design.

CLOCKING HIGH-SPEED CONVERTERS

ADCs generally require some kind of clock function to set the sample rate. Any cycle-to-cycle variation in the sample timing ("jitter") causes an error. It can be shown that in RF-sampling converters, sample-clock jitter is analogous to phase noise in the local oscillators of analog radios, and can degrade the performance of the converter substantially. **Figure 8.23** shows how this comes about.

As ADCs and DACs achieve higher and higher sample rates and signal bandwidths, the clocking becomes critical. Most logic families produce output waveforms that have more timing jitter than can be tolerated.

The degradation of a high-frequency sampled signal due to sample jitter can be calculated and plotted. In **Figure 8.24** you can see that a 100 MHz signal sampled with a 1ps jitter sample clock (typical of ACT-series logic gates) will have just over 60 dB of SNR, about equal to 10-bit performance, even if the ADC is capable of 16 bits! Improving the sample clock jitter to 0.125 ps using a specialized clock-distribution IC improves the system performance to nearly 14 bits.

8.5 Digital Signal Processors

The term *digital signal processor* (DSP) is commonly understood to mean a special-purpose microprocessor with an architecture that has been optimized for signal processing rather than data processing. And indeed, in many systems the box labeled "DSP" in Figure 8.1 is such a device. A microprocessor has the advantage of flexibility because it can easily be re-programmed. Even with a single program, it can perform many completely different tasks at different points in the code. On-chip hardware resources such as multipliers and other computational units are used efficiently because they are shared among various processes.

That is also the Achilles' heel of programmable DSPs. Any hardware resource that is shared among various processes can be used by only one process at a time. That can create bottlenecks that limit the maximum computation speed. Some DSP chips include multiple computational units or multiple *cores* (basically multiple copies of the entire processor) that can be used in parallel to speed up processing.

8.5.1 Microprocessor-type DSP ICs

Programming a DSP IC is relatively easy. C compilers are available for most devices, so you don't have to learn assembly language. Typically you include a connector on your circuit board into which is plugged an *in-circuit programmer* (ICP), which is connected to a PC via a serial or USB cable. The software is written and compiled on the PC and then downloaded to the DSP. The same hardware often also includes an *in-circuit debugger* (ICD) so that the program can be debugged on the actual circuitry used in the design. The combination of the editor, compiler, programmer, debugger, simulator, and related software is called an *integrated development environment* (IDE).

Until recently, you had to use an *in-circuit emulator* (ICE), which is a device that plugs into the circuit board in place of the microprocessor. The ICE provides sophisticated debugging tools that function while the emulator runs the user's software on the target device at full speed. Nowadays, however, it is more common to use the ICD function that is built into many DSP chips and which provides most of the functions of a full-fledged ICE. It is much cheaper and does not require using a socket for the microprocessor chip.

The architecture of a digital signal processor shares some similarities to that of a general-purpose microprocessor but also differs in important respects. For example, DSPs generally don't spend much of their lives handling large computer files, so they tend to have a smaller memory address space than processors intended to be used in computers. On the other hand, the memory they do have is often built into the DSP chip itself to improve speed and to reduce pin count by eliminating the external address and data bus.

Most microprocessors use the traditional *Von Neumann architecture* in which the program and data are stored in the same memory space. However, most DSPs use a *Harvard architecture*, which means that data and program are stored in separate memories. That speeds up the processor because it can be reading the next program instruction at the same time as it is reading or writing data in response to the previous instruction. Some DSPs have two data memories so they can read and/or write two data words at the same time. Most devices actually use a modified Harvard architecture by providing some (typically slower and less convenient) method for the processor to read and write data to program memory.

Probably the key difference between general-purpose and digital-signal processors is in the computational core, often called the *arithmetic logic unit* (ALU). The ALU in a traditional microprocessor only performs integer addition, subtraction, and bitwise logic operations such as AND, OR, one-bit shifting, and so on. More complicated calculations, such as multiplication, division, and operations with floating-point numbers, are done in software routines that exercise the simple resources of the ALU multiple times to generate the more complicated results.

In contrast, a DSP has special hardware to perform many of these operations much faster. For example, the *multiplier-accumulator* (MAC) multiplies two numbers and adds (accumulates) the product with the previous results in a single step. Many common DSP algorithms involve the sum of a large number of products, so nearly all DSPs include this function. **Figure 8.25** is a simplified block diagram of the dsPIC series from Microchip. Its architecture is basically that of a general purpose microcontroller to which a DSP engine has been added that includes a MAC, a barrel shifter, and other DSP features. It uses a modified Harvard architecture with two data memories that can be simultaneously accessed.

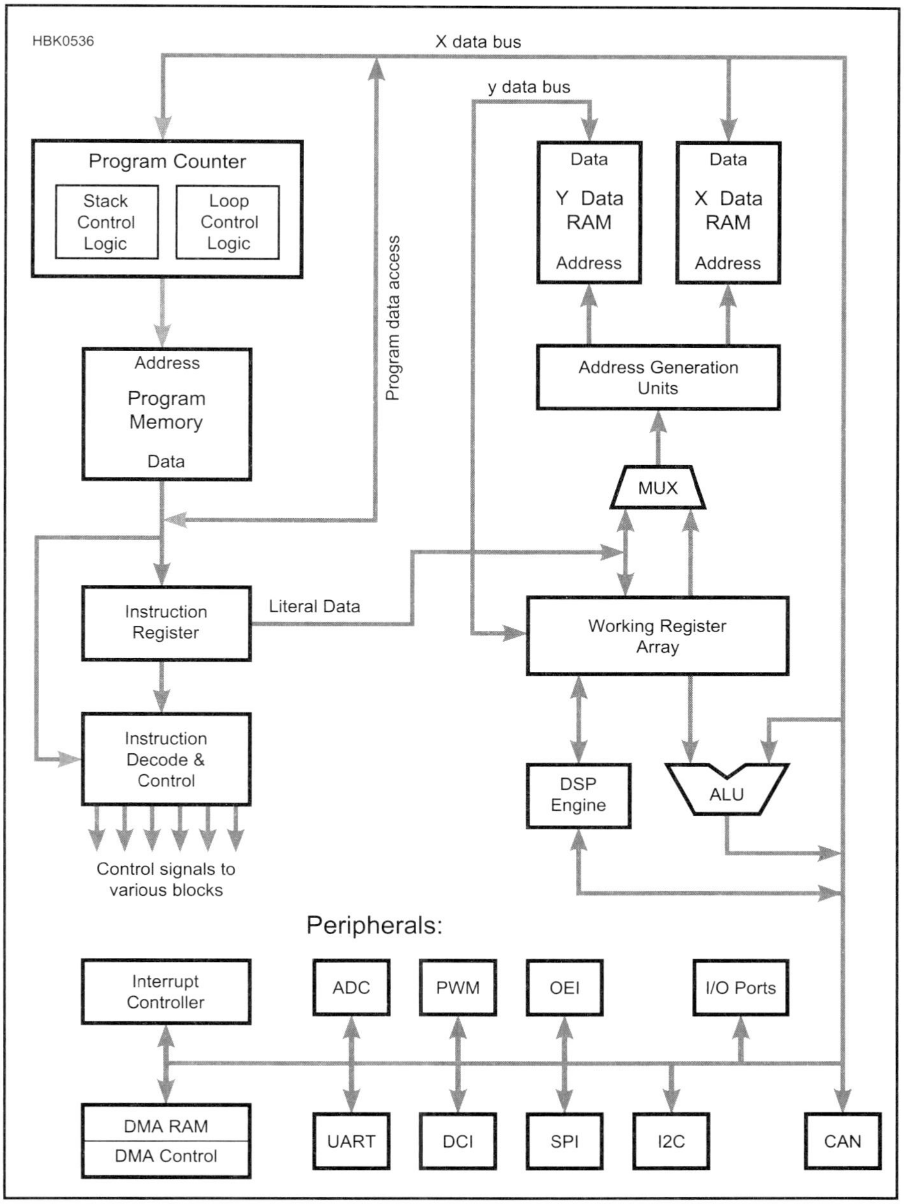

Figure 8.25 — Simplified block diagram of a dsPIC processor.

8.5.2 Fixed-Point vs. Floating-Point

DSP microprocessors are sometimes differentiated on their ability to handle fixed-point (integer) or *floating-point* numbers.

A floating-point number is the binary equivalent of scientific notation. Recall that the decimal integer 123,000 is expressed as 1.23×10^5 in scientific notation. It is common practice to place the decimal point after the first non-zero digit and indicate how many digits the decimal point must be moved by the *exponent* of 10, 5 in this case. The 1.23

part is called the *mantissa*. In a computer, base-2 binary numbers are used in place of the base-10 decimal numbers used in scientific notation. The *binary point* (equivalent to the decimal point in a decimal number) is assumed to be to the left of the first non-zero bit. For example the binary number 00110100, when converted to a 16-bit floating point number, would have an 11-bit mantissa of 11010000000 (with the binary point assumed to be to the left of the first "1") and a 5-bit exponent of 00110 (decimal +6).

A floating point number can represent a signal with much more dynamic range than an integer number with the same number of bits. For example, a 16-bit fixed-point signed integer can vary from –32768 to +32767. The difference between the smallest (1) and largest signal that can be represented is 20 log(65535) = 96 dB. If the 16 bits are divided into an 11-bit mantissa and 5-bit exponent, the available range is 20 log(2048) = 66 dB from the mantissa and 20 log(2^{32}) = 193 dB from the exponent for a total of 259 dB. Floating point formats greatly increase dynamic range at the expense of decreasing resolution. Less resolution can cause increased noise and distortion. Normally floating-point numbers are at least 32 bits wide to benefit from increased dynamic range and still have sufficient resolution.

Some DSPs can process floating-point numbers directly in hardware. Fixed-point DSPs can also handle floating-point numbers, but it must be done in software and that slows the computations down. The additional dynamic range afforded by floating-point processing is normally not needed for radio communications signals. The dynamic range of radio signals can usually be handled by the 16-bit data words used by most fixed-point DSPs. Using integer arithmetic and a fixed-point processor saves the additional cost of a floating-point processor, or the additional computational overhead of floating-point calculations on a fixed-point device. However, it requires careful attention to detail on the part of the programmer to make sure the signal can never exceed the maximum integer value or get so weak that the SNR is degraded. If cost or computation time is not an issue, it is much easier to program in floating point since dynamic range issues can be ignored for most computations.

The term *pipeline* refers to the ability of a microprocessor to perform portions of several instructions at the same time. The sequence of operations required to perform an instruction is broken down into steps. Since each step is performed by a different chunk of hardware, different chunks can be working on different instructions at the same time. Most DSPs have at least a simple form of pipelining in which the next instruction is being fetched while the previous instruction is being executed. Some DSPs can do a multiply-accumulate while the next two multiplicands are being read from memory and the previous accumulated result is being stored so that the entire operation can occur in a single clock cycle. MAC operations per second is a common figure of merit for measuring DSP speed. For conventional microprocessors, a more common figure of merit is millions of instructions per second (MIPS) or floating-point operations per second (FLOPS).

Many DSPs have a sophisticated address generation unit that can automatically increment one or more data memory pointers so that repetitive calculations can step through memory without the processor having to calculate the addresses. *Zero-overhead looping* is the ability to automatically jump the address pointer back to the beginning of the array when it reaches the end. That saves several microprocessor instructions per loop that normally would be required to check the current address and jump when it reaches a predetermined value.

While most DSPs do not include a full hardware divider, some do include special instructions and hardware to speed up division calculations. A *barrel shifter* is another common DSP feature. It allows shifting a data word a specified number of bits in a single clock cycle. *Direct memory access* (DMA) refers to special hardware that can automatically transfer data between memory and various peripheral devices or ports without processor overhead.

8.5.3 DSP in Embedded Systems

An *embedded system* is a device that is not a "computer" but nevertheless has a microprocessor or DSP chip embedded somewhere in its circuitry. Examples are microwave ovens, automobiles, mobile telephones, and software-defined radios. DSPs intended for embedded systems often include a wide array of on-chip peripherals such as various kinds of timers, multiple hardware interrupts, serial ports of various types, a real-time clock, pulse-width modulators, optical encoder interfaces, ADC and DACs, and lots of general-purpose digital I/O pins. Some DSPs have architectures that are well-suited for general-purpose control applications as well as digital signal processing.

Table 8.2 lists some manufacturers of DSP chips targeted to embedded systems. It should be mentioned that microprocessors made by Intel and AMD, and that are intended for use as CPUs in personal computers, also include extensive DSP capability. However, they are optimized for data-processing applications under an operating system such as Windows or Linux, and are not very efficient at signal-processing applications.

Table 8.2

Manufacturers of DSP Microprocessors for Communications Applications

Company	*URL*
Analog Devices	**www.analog.com**
Cirrus Logic	**www.cirrus.com**
Microchip	**www.microchip.com**
NXP	**www.nxp.com**
Texas Instruments	**www.ti.com**

8.5.4 Typical DSP Processors

When selecting a DSP device for a new design, often the available development environment is more important than the characteristics of the device itself. Microchip's dsPIC family of DSPs was chosen for the examples in this chapter because their integrated development environment is extensive and easy to use, and the IDE software is available for free download from their website.[5] The processor instruction set is a superset of the PIC24 family of general-purpose microcontrollers, with which many hams are already familiar. The company offers a line of low-cost evaluation boards and starter kits as well as an inexpensive in-circuit debugger, the ICD 3. The free IDE software includes a simulator that can run dsPIC software on a PC (at a much slower rate, of course) so that you can experiment with DSP algorithms before buying any hardware.

The Microchip DSP family is limited to 100 million instructions per second. In a system with, say, a 100 kHz sample rate, 1,000 instructions per sample are available, which should be plenty if the calculations are not too complex. However, if the sample rate is 1 MHz, then you get only 100 instructions per sample, which is unlikely to be sufficient.

If more horsepower is required, you'll need to select a processor from a different manufacturer. Look for one with a well-integrated suite of development software that is powerful and easy to use. Also check out the cost and availability of development hardware, such as evaluation kits, programmers, and debuggers. Once those requirements are met, you can move on to select a specific device with the performance and features required for your application. It can be helpful at the beginning of a project to first write some of the key software routines and test them on a simulator to estimate execution times, in order to determine how powerful a processor is needed.

When estimating execution time, don't forget to include the effect of interrupts. Most DSP systems require real-time response and make extensive use of interrupts to ensure that certain events happen at the correct times.

Although this is hidden from the programmer's view when programming in C, the interrupt service routines contain quite a bit of overhead each time they are called (to save the processor state when responding and to recall the state just before returning from the interrupt). Sometimes an interrupt may be called more often than you expect, which can eat up processor cycles, thus increasing the execution time of other unrelated routines.

In the past, many embedded systems were written in assembly language to save memory and increase processing speed. Many early microprocessors and DSPs did not have enough memory to support a high-level language. Today, most processors have sufficient memory and processing speed to support a C kernel and library without difficulty. For anything but the simplest of programs, it is not only faster and easier to develop software in C or C++ but it is easier to support and maintain as well, especially if people other than the original programmer might become involved. Far more people know the C programming language than any particular processor's assembly language. It is true that the version of C used on a DSP chip is usually modified from standard ANSI C to support specific hardware features, but it would still be far easier to learn for a programmer familiar with writing C code on a PC or on a different DSP.

A common technique is first to write the entire application in C. Then, if execution time is not acceptable, analyze the system to determine in which software routines the bottlenecks are occurring. You can then rewrite those routines in assembly language. Having an already working version written in C (even if too slow) can be helpful in testing and troubleshooting the equivalent assembly language.

8.5.5 DSP Without a Dedicated Processor

One way to speed up processing is to move all or part of the computations from the programmable DSP to an *application-specific integrated circuit* (ASIC), which has an architecture that has been optimized to perform some specific DSP function.

You could also design your own application-specific circuitry using a PC board full of discrete logic devices. Nowadays, however, it is more common to do that with a *programmable-logic device* (PLD). This is an IC that includes many general-purpose logic elements, but the connections between the elements are undefined when the device is manufactured. The user defines those connections by programming the device to perform whatever function is required. PLDs come in a wide variety of types, described by an alphabet soup of acronyms.

Programmable-array logic (PAL), *programmable logic array* (PLA), and *generic array logic* (GAL) devices are relatively simple arrays of AND gates, OR gates, inverters, and latches. They are often used as "glue logic" to replace the miscellaneous discrete logic ICs that would otherwise be used to interface various larger digital devices on a circuit board. They are sometimes grouped under the general category of *small PLD* (SPLD). A *complex PLD* (CPLD) is similar but bigger, often consisting of an array of PALs with programmable interconnections between them.

A *field-programmable gate array* (FPGA) is bigger yet, with up to millions of gates per device. An FPGA includes an array of *complex logic blocks* (CLB), each of which includes some programmable logic, often implemented with a RAM *look-up table* (LUT), and output registers. *Input/output blocks* (IOB) also contain registers and can be configured as input, output, or bi-directional interfaces to the IC pins. The interconnections between blocks are much more flexible and complicated than in CPLDs. Some FPGAs also include higher level circuit blocks such as general-purpose RAM, dozens or hundreds of hardware multipliers, and even entire on-chip microprocessors.

Some of the more inexpensive PLDs are *one-time programmable* (OTP), meaning you must throw the old device away if you want to change the programming. Other devices are re-programmable or even *in-circuit programmable* (ICP) which allows changing the internal circuit configuration after the device has been soldered onto the PC board, typically under the control of an on-board microprocessor. That offers the best of both worlds, with speed nearly as fast as an ASIC but retaining many of the benefits of the reprogrammability of a microprocessor-type DSP. Most large FPGAs store their programming in *volatile memory*, which is RAM that must be reloaded every time power is applied, typically by a ROM located on the same circuit board. Some FPGAs have programmable ROM on-chip.

Programming a PLD is quite different from programming a microprocessor. A microprocessor performs its operations sequentially — only one operation can be performed at a time. Writing a PLD program is more like designing a circuit. Different parts of the circuit can be doing different things at the same time. Special *hardware-description languages* (HDL) have been devised for programming the more complicated parts such as ASICs and FPGAs. The two most common industry-standard HDLs are Verilog and VHDL. (The arguments about which is "best" approach the fervor of the Windows vs Linux wars!) There is also a version of the C++ programming language called SystemC that includes a series of libraries that extend the language to include HDL functions. It is popular with some designers because it allows simulation and hardware description using the same software tool.

Despite the speed advantage of FPGAs, most amateurs use microprocessor-type devices for their DSP designs, supplemented with off-the-shelf ASICs where necessary. The primary reason is that the design process for an FPGA is quite complicated, involving obtaining and learning to use several sophisticated software tools. The steps involved in programming an FPGA are:

1. Simulate the design at a high abstraction level to prove the algorithms.
2. Generate the HDL code, either manually or using some tool.
3. Simulate and test the HDL program.
4. Synthesize the gate-level netlist.
5. Verify the netlist.
6. Perform a timing analysis.
7. Modify the design if necessary to meet timing constraints.
8. "Place and route" the chip design.
9. Program and test the part.

Many of the software tools needed to perform those steps are quite expensive, although some manufacturers do offer free proprietary software for their own devices. Some principal manufacturers are listed in **Table 8.3**.

8.5.6 Using Graphics Processors for DSP

As PCs have become used for graphics-intensive applications such as games, developers quickly found out that the microprocessor chip in the PC was unable to provide enough computing horsepower to do complex

Table 8.3
PLD Manufacturers

Company	*Devices*	*URL*
Achronix	FPGA	**www.achronix.com**
AMD	CPLD, FPGA	**www.amd.com**
Atmel	SPLD, CPLD, PGA, ASIC	**www.microchip.com**
Intel	CPLD, FPGA, ASIC	**www.intel.com**
Lattice Semiconductor	SPLD, CPLD, FPGA	**www.latticesemi.com**
Microsemi	FPGA	**www.microsemi.com**

graphics functions like 3D rendering, shading, etc. This gave rise to a new breed of specialized processors called Graphics Processing Units (GPU). These processors consist of simple computational blocks that are capable of only a few functions, like multiplication and addition. However, unlike a typical PC CPU with a handful of cores to speed up processing, GPUs have hundreds, or even thousands, of cores! In addition, they are optimized for handling large arrays of data. This makes them well suited for DSP functions such as Fast Fourier Transforms.

Fortunately, some manufacturers of GPUs have recognized that their products are useful for applications other than graphics, and are now providing libraries and programming support for using their products in communications systems such as SDR. GPU manufacturer Nvidia, for example, offers the CUDA parallel computing platform and programming model with the cuFFT library.

The inexpensive Linux-based Raspberry Pi computer board has become very popular among amateurs and experimenters. The BCM283x main processor is a system-on-chip device manufactured by Broadcom (now owned by Avago) that contains both a CPU and a GPU. At 700 MHz to 1.2 GHz, the CPU is too underpowered for many of the DSP algorithms used in SDR. However, the graphics processor included in the device can be programmed to do some of the FFT algorithms, and offers a 10x speed improvement.[6]

8.6 Digital (Discrete-time) Signals

Digital signals differ from analog signals in two ways. One is that they are digitized in time, a process called *sampling*. The other is that they are digitized in amplitude, a process called *quantization*. Sampling and quantization affect the digitized signal in different ways, so the following sections will consider their effects separately.

8.6.1 Sampling — Digitization in Time

Sampling is the process of measuring a signal at discrete points in time and recording the measured values. An example from history is recording the number of sunspots. If an observer goes out at noon every day and writes down the number of observed sunspots, then that data can be used to plot sunspot number vs. time. In this case, we say the *sample rate* is one sample per day. The data can then be analyzed in various ways to determine short and long-term trends. After recording only a few months of data it will quickly become apparent that sunspot number has a marked periodicity — the numbers tend to repeat every 27 days (which happens to be the rotation rate of the sun as seen from earth).

What if, instead of taking a reading once a day, the readings were taken only once per month? With a 30-day sample period, the 27-day periodicity would likely be impossible to see. Clearly, the sample rate must be at least some minimum value to accurately represent the measured signal. Based on earlier work by Harry Nyquist, Claude Shannon proved in 1948 that in order to sample a signal without loss of information, the sample rate must be greater than the *Nyquist rate*, which is two times the bandwidth of the signal. In other words, the bandwidth must be less than the *Nyquist frequency*, which is one-half the sample rate. This is known as the *Nyquist sampling criterion*.

That simple rule has some profound implications. If all the frequency components of a signal are contained within a bandwidth of B Hz, then sampling at a rate greater than 2B samples per second is sufficient to represent the signal with 100% accuracy and no loss of information. It is theoretically possible to convert the samples back to an analog signal that is exactly identical to the original.

Sampling theory assumes that there is absolutely no signal energy outside the specified bandwidth. In other words, the stopband attenuation is infinitely high. Any residual signal in the stop-band shows up as distortion or noise in the sampled signal.

To simplify the discussion, let's think about sampling a signal of a single frequency (a sine wave). **Figure 8.26** illustrates what happens if the sample rate is too low. As shown, the sample rate is approximately ⅞ the sine-wave frequency. You can see that the sampled signal has a period about eight times greater than the period of the sine wave, or ⅛ the frequency. The samples are the same as if the analog signal had been a sine wave of ⅛ the actual frequency.

That is an example of a general principle. Sampling a signal produces an output with components at frequencies equal to the difference between the actual frequency of the analog signal and the sample rate. If the sampling rate is too low (or the input signal too high in frequency) the signal at $(f_s - f_{sig})$ is indistinguishable from an in-band signal, which is why the new signal is called an *alias*. In the above example, the alias frequency f_o

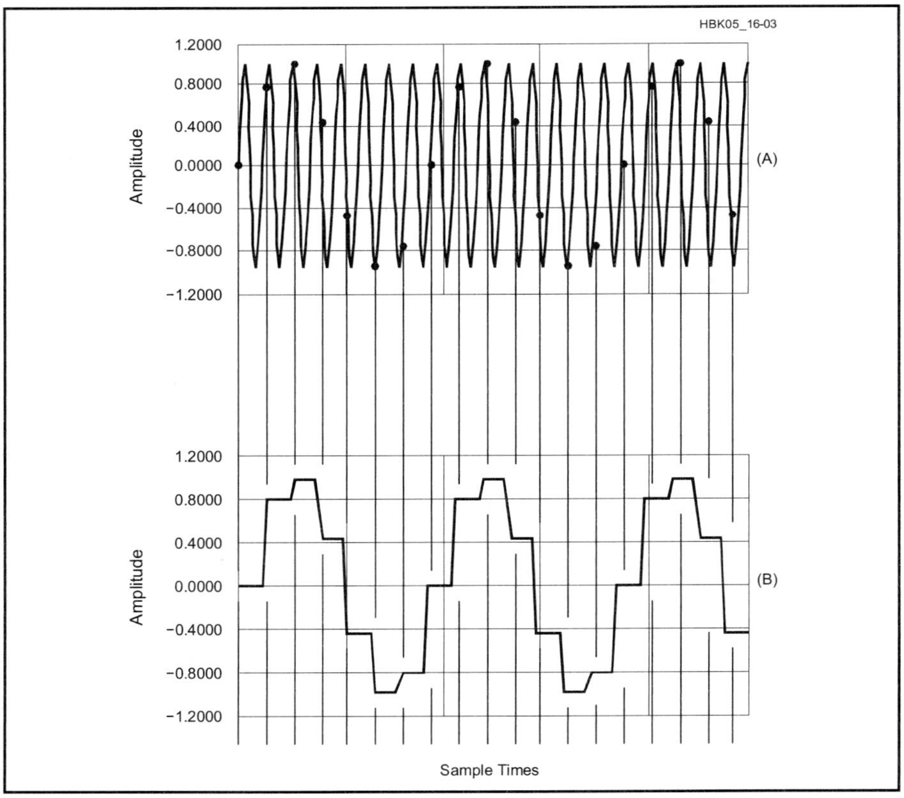

Figure 8.26 — Undersampled sine wave (A). Samples aliased to a lower frequency (B).

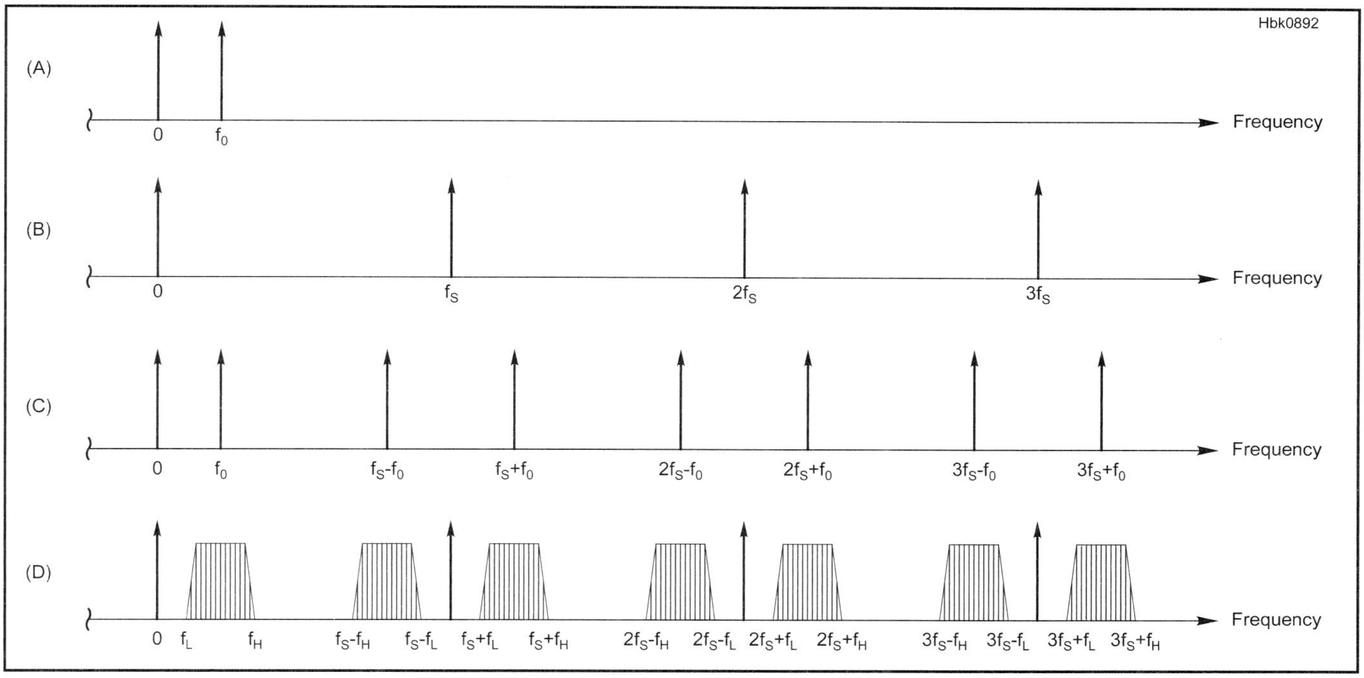

Figure 8.27 — Frequency relationships in a sampled-data system; spectrum of an analog sine wave (A). The spectrum of the sampling function, including all harmonics (B). The spectrum of the sampled sine wave (C). The spectrum of a sampled band of frequencies between f_L and f_H (D).

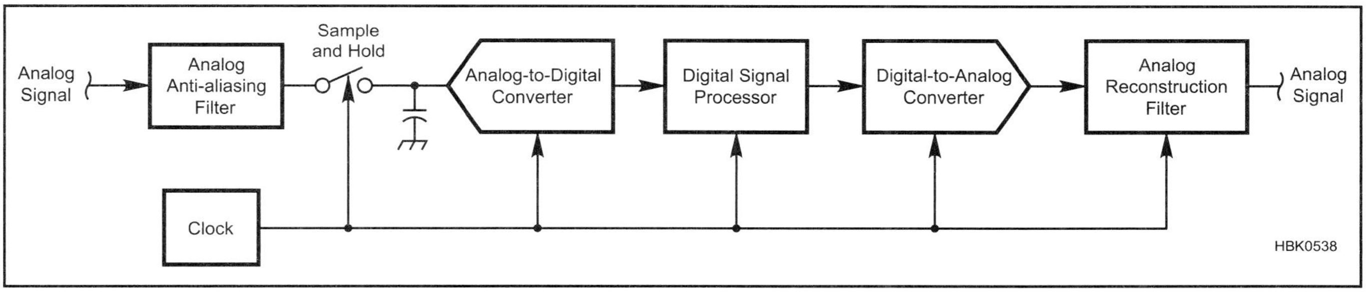

Figure 8.28 — A more complete block diagram of a DSP system.

is

$$f_o = f_{sig} - f_s = (1 - 7/8) f_{sig} = (1/8) f_{sig}$$

where f_{sig} is the frequency of the signal before sampling and f_s is the sample rate.

The sampling process also creates aliases from any input signals around harmonics of the sampling frequency. **Figure 8.27** shows the frequency-domain representation of the signal relationships in a sampled-data system. Figure 8.27A shows the spectrum of a single sine wave input signal, and Figure 8.27B shows the spectrum of the sampling clock and its harmonics. You can look at Figure 8.27C two ways. First, you can consider all the individual frequencies as the output signals generated by sampling the input signal at f_o. On the other hand, those are also the input signal frequencies that alias to f_o, calculated from the equation

$$f_{sig} = |f_o - Nf_s|$$

where N is the harmonic number. Experienced RF engineers will recognize this phenomenon as being very similar to image generation in the mixing process of a heterodyne receiver. The sampled signal (equivalent to the mixer output) contains the sum and difference frequencies of the input signal and all the harmonics of the sample frequency. Figure 8.27D shows the more general case of sampling a band of signals between f_L and f_H. Note the inversion of the aliased spectrum that extends below f_s. If f_H is higher than $f_S/2$, the higher frequencies in the band will overlap the original band in the output spectrum and create artifacts.

An *anti-aliasing filter* before the sampler prevents unwanted signals from creating artifacts in the sampled data output, as shown in **Figure 8.28**. For a baseband signal (one that extends to zero Hz), the anti-aliasing filter is a low-pass filter whose stopband extends from the Nyquist frequency to infinity. Of course, practical filters do not transition instantaneously from the passband to the stopband, so the bandwidth of the passband must be somewhat less than half the sample rate.

Figure 8.26 showed each sample being held at a constant value for the duration of one sample period. However, sampling theory actually assumes that the sample is only valid at the instant the signal is sampled; it is zero or undefined at all other times. A series of such infinitely narrow impulses has harmonics all the way to infinite frequency. Each harmonic has the same amplitude and is modulated by the signal being sampled. See Figure 8.27D.

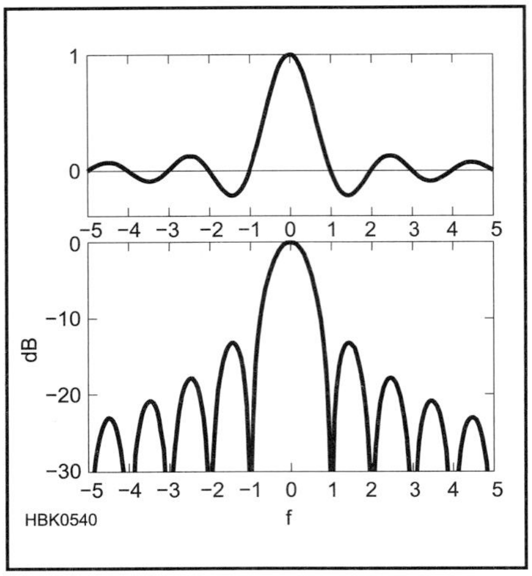

Figure 8.29 — The sinc function, where the horizontal axis is frequency normalized to the sample rate. At the bottom is the same function in decibels.

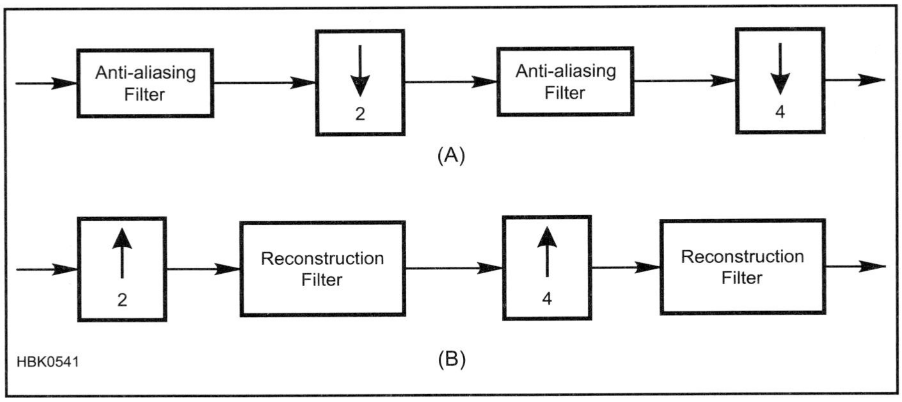

Figure 8.30 — Decimation (A) and interpolation (B). The arrow's direction indicates decimation (down) or interpolation (up) and the number is the factor.

When a digitized signal is converted back to analog form, unwanted harmonics must be filtered out by a *reconstruction filter* as shown in Figure 8.28. This is similar to the anti-aliasing filter used at the input in that its bandwidth should be no greater than one-half the sample rate. It is a low-pass filter for a baseband signal and a band-pass filter for an under-sampled signal.

Most DACs hold each sample value for the entire sample period. This is called *zero-order hold* and results in a frequency response in the shape of a sinc function

$$\text{sinc}(f) = \frac{\sin(\pi f)}{\pi f}$$

where f is normalized to the sample rate, f = frequency / sample rate.

The graph of the sinc function in **Figure 8.29** shows both positive and negative frequencies. Note that the logarithmic frequency response has notches at the sample rate and all of its harmonics. If the signal bandwidth is much less than the Nyquist frequency, then most of the signal at the harmonics falls near the notch frequencies, easing the task of the reconstruction filter. If the signal bandwidth is small enough (sample rate is high enough), the harmonics are almost completely notched out and a reconstruction filter may not even be required.

The sin(πf)/πf frequency response also affects the passband. For example, if the passband extends to f_S / 4 (f = 0.25), then the response is

$$20 \log \frac{\sin(\pi \cdot 0.25)}{\pi \cdot 0.25} = -0.9 \text{ dB}$$

at the top edge of the passband. At the Nyquist frequency, (f = 0.5), the error is 3.9 dB. If the signal bandwidth is a large proportion of the Nyquist frequency, then some kind of digital or analog compensation filter may be required to correct for the high-frequency rolloff. The interpolation filters used in digital audio systems have this correction included in the frequency response.

8.6.2 Decimation and Interpolation

The term *decimation* simply means reducing the sample rate. For example, to decimate by two, simply eliminate every second sample. That works fine as long as the signal bandwidth satisfies the Nyquist criterion at the lower, output sample rate. If the analog anti-aliasing filter is not narrow enough, then a digital anti-aliasing filter in the DSP can be used to reduce the bandwidth to the necessary value. This must be done *before* decimation to satisfy the Nyquist criterion.

If you need to decimate by a large amount, then the digital anti-aliasing filter must have a very small bandwidth compared to the sample rate. As we will see later, a digital filter with a small bandwidth is computationally intensive. For this reason, large decimation factors are normally accomplished in multiple steps, as shown in **Figure 8.30A**. The first decimation is by a small factor, typically 2, so that the first anti-alias filter can be as simple as possible. The second decimation stage then does not have to decimate by such a large factor, simplifying its task. In addition, since it is running at only half the input sample rate, it has more time to do its calculations. Generally it is most efficient to decimate by the smallest factor in the first stage, a larger factor in the second, and the largest factors in the third and any subsequent stages. The larger the total decimation factor, the greater the number of stages is appropriate, but more than three stages is uncommon.

Interpolation means increasing the sample rate. One way to do that is simply to insert additional zero-value samples, a process called *zero-stuffing*. For example, to interpolate by a factor of three, insert two zero-value samples after each input sample. That works, but may not give the results you expect. Recall that a sampled signal has additional copies of the baseband signal at all harmonics of the sample rate. All of those harmonics remain in the resampled signal, even though the sample rate is now higher. To eliminate them, the signal must be filtered after interpolation. After filtering, there is signal only at baseband and around the harmonics of the interpolated (higher-frequency) sample rate. It's as if the analog signal had been sampled at the higher rate to begin with, which relaxes the requirements on the reconstruction filter.

Just as with decimation, interpolation by a large factor is best done in stages, as shown in Figure 8.30B. In this case, the stage running at the lowest sample rate (again the first stage) is the one with the lowest interpolation factor.

Zero-stuffing followed by filtering is not the only way to interpolate. Really what you are trying to do is to fill in between the lower-rate samples with additional samples that "connect the dots" in as smooth a manner as possible. It can be shown that it is mathematically equivalent to zero-stuffing and filtering. For example, if instead of inserting zero-value samples you instead simply repeat the last input sample, you have a situation similar to the zero-order hold of a DAC output. It is equivalent to zero-stuffing followed by a low-pass filter with a frequency response of sin(πf)/πf. If you do a straight-line interpolation between input samples (a "first-order" interpolation), it turns out that it is equivalent to a low pass filter with a frequency response of $[\sin(\pi f)/\pi f]^2$, which has a sharper cutoff and better stop-band rejection than a zero-order interpolation. Higher-order interpolations have smoother responses in the time domain which translate to better filter responses in the frequency domain.

So far we have only covered decimation and interpolation by integer factors. It is also possible to change the sample rate by a non-integer factor, which is called *resampling* or *multi-rate* conversion. For example, if you want to increase the sample rate by a factor of 4/3, simply interpolate by 4 and then decimate by 3. That method can become impractical for some resample ratios. For example, to convert an audio file recorded from a computer sound card at 48 kHz to the 44.1 kHz required by a compact disc, the resample ratio is 44,100 / 48,000 = 147 / 160. After interpolation by 147, the 44.1 kHz input file is sampled at 6.4827 MHz, which would result in excessive processing overhead.

In addition, the interpolation/decimation method only works for resample ratios that are rational numbers (the ratio of two integers). To resample by an irrational number, a different method is required. The technique is as follows. For each output sample, first determine the two nearest input samples. Calculate the coefficients of the Nth-order equation that describes the trajectory between the two input samples. Knowing the trajectory between the input samples and the output sample's relative position between them, the value of the output sample can be calculated from the equation.

8.6.3 Quantization — Digitization in Amplitude

While sampling (digitization in time) theoretically causes no loss of signal information, quantization (digitization in amplitude) always does. For example, an 8-bit signed number can represent a signal as a value from –128 to +127. For each sample, the ADC assigns whichever number in that range is closest to the analog signal at that instant. If a particular sample has a value of 10, there is no way to tell if the original signal was 9.5, 10.5, or somewhere in between. That information has been lost forever.

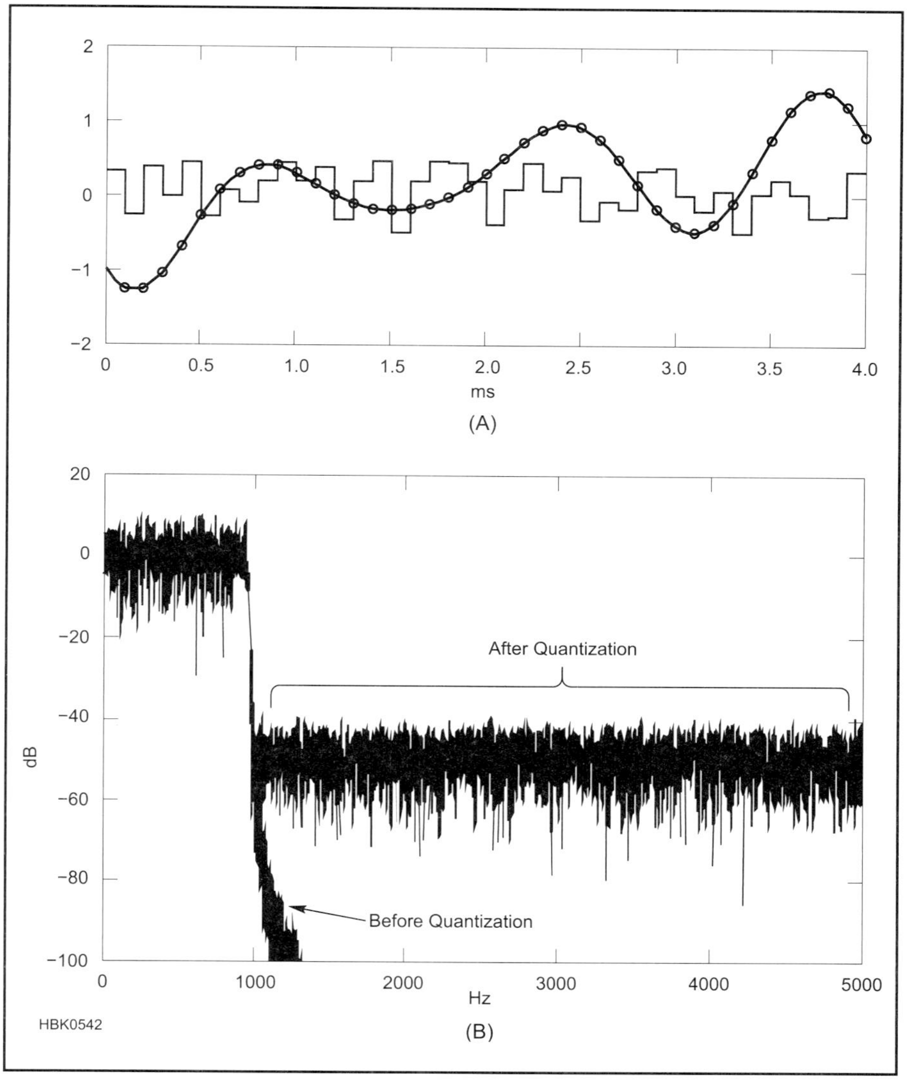

Figure 8.31 — Quantization error of a random noise signal that has been band-limited to 1 kHz to simulate an audio signal. (A) The sampler resolution is 8 bits and the sample rate is 10 kHz. Sample values are indicated by circles. Also shown is the quantization error, in units of LSB. (B) is the frequency spectrum of the signal before and after quantization.

QUANTIZATION NOISE

When quantizing a complex signal such as speech, this error shows up as noise, called *quantization noise* as discussed in section 8.2.2. The error is random. It is equally likely to be anywhere in the range of –½ to +½ of a single step of the ADC. In **Figure 8.31** the smooth waveform is a band-limited analog noise signal plotted over a 4 millisecond period. The vertical axis of Figure 8.31A indicates the quantization error for each sample in LSB, represented by the staircase-like waveform. The spectrum of the signal is shown in Figure 8.31B, both before quantization and after quantization. While the original signal has the frequency components above 1 kHz reduced by analog filtering, the digitized version has considerable noise in the 1 to 5 kHz band.

One critical point that is sometimes overlooked is that quantization noise is spread over the entire bandwidth from zero Hz to the sample rate. If you are digitizing a 3 kHz audio channel with a 48 ksps sampler, only a fraction of the noise power is within the channel. For that reason, the effective signal-to-noise ratio, SNR_{eff}, depends not only on the number of bits but also the sample rate, f_s, and the signal bandwidth, B:

$$SNR_{eff} = SNR + 10\log\left(\frac{f_s}{2B}\right) \text{ dB}$$

The reason for the factor of two in the denominator is that the bandwidth of a positive-frequency scalar signal should be compared to the Nyquist bandwidth, $f_s/2$. When filtering a complex signal (one that contains I and Q parts), the 2B in the denominator should be replaced by B.

When choosing an ADC, don't forget that SNR_{eff} depends on the sample rate. As an example, let's compare an Analog Devices AD9235 12-bit, 65 Msps ADC to an AD7653, which is a 16-bit 1 Msps ADC from the same manufacturer. Assume a 10 kHz signal bandwidth.

An ideal 12-bit ADC has a SNR of 1.76 + 6.02 × 12 = 74.0 dB. The AD9235's performance is not far from the ideal; its SNR is specified at 70.5 dB at its 65 Msps maximum sample rate. It costs about $12.50. In a 10 kHz bandwidth, the effective SNR is 70.5 + 10 log (65,000/20) = 105.6 dB.

An ideal 16-bit ADC has an SNR of

98.1 dB. The AD7653 is specified at 86 dB and costs about $15. The effective SNR is 86 + 10log(1000/20) = 93 dB.

So the 12-bit ADC with 70.5 dB SNR is actually 12.6 dB better than the 16-bit device with 86 dB SNR in this application! Even an ideal 16-bit, 1 Msps ADC would only have an effective SNR of 98.1 + 10log(1000/20) = 103.1 dB, still worse than the actual performance of the AD9235 when measured with the same bandwidth. Note that to actually realize 105.6 dB of dynamic range the signal from the ADC would need to be digitally filtered to a 10 kHz bandwidth while increasing the bits of data resolution from 12 to at least 18.

Oversampling is the name given to the technique of using a higher-than-necessary sample rate in order to achieve an improved SNR. Don't forget that when the high-sample-rate signal is decimated the data words must have enough bits to support the higher dynamic range at the lower sample rate. As a rule of thumb, the quantization noise should be at least 10 dB less than the signal noise in order not to significantly degrade the SNR. In the AD9235 example, assuming a 100 kHz output sample rate, about 18 bits would be required: 1.76 + 6.02 × 18 + 10log(100/20) = 117.1 dB, which is 11.5 dB better than the 105.6 dB dynamic range of the ADC in a 10 kHz bandwidth.

Most ADCs and DACs used in high-fidelity audio systems use an extreme form of oversampling, where the internal converter may oversample by a rate of 128 or 256 times, but with very low resolution. (In some cases, just a 1-bit ADC!). In addition, such converters use a technique called noise shaping to push most of the quantization noise to frequencies near the sample rate, and reduce it in the audio spectrum. The noise is then removed in the decimation filter.

Although quantization error manifests itself as noise when digitizing a complex non-periodic signal, it can show up as discrete spurious frequencies when digitizing a periodic signal. **Figure 8.32** illustrates a 1 kHz sine wave sampled with 8-bit resolution at a 9.5 kHz rate. On average there are 9.5 samples per cycle of the sine wave so that the sampling error repeats every second cycle. That 500 Hz periodicity in the error signal causes a spurious signal at 500 Hz and its harmonics. As the signal frequency is changed, the spurs move around in a complicated manner that depends on the ratio of sample rate to signal frequency. In real world ADCs, nonlinearities in the transfer function can also create spurious signals that vary unpredictably as a function of the signal amplitude, especially at low signal levels.

In many applications, broadband noise is preferable to spurious signals on discrete frequencies. The solution is to add *dithering*. This essentially involves adding a small amount of noise, typically on the order of an LSB or two, in order to randomize the quantization error. Some ADCs have dithering capability built in to improve the SFDR, even though it does degrade the SNR slightly. Dithering is also useful in cases where the input signal to an ADC is smaller than one LSB. Even though the signal would be well above the noise level after narrow-band filtering, it cannot be detected if the ADC input is always below one LSB. There is already sufficient noise in many systems from input amplifiers, other signals in the passband, and the ADC itself, to cause natural dithering.

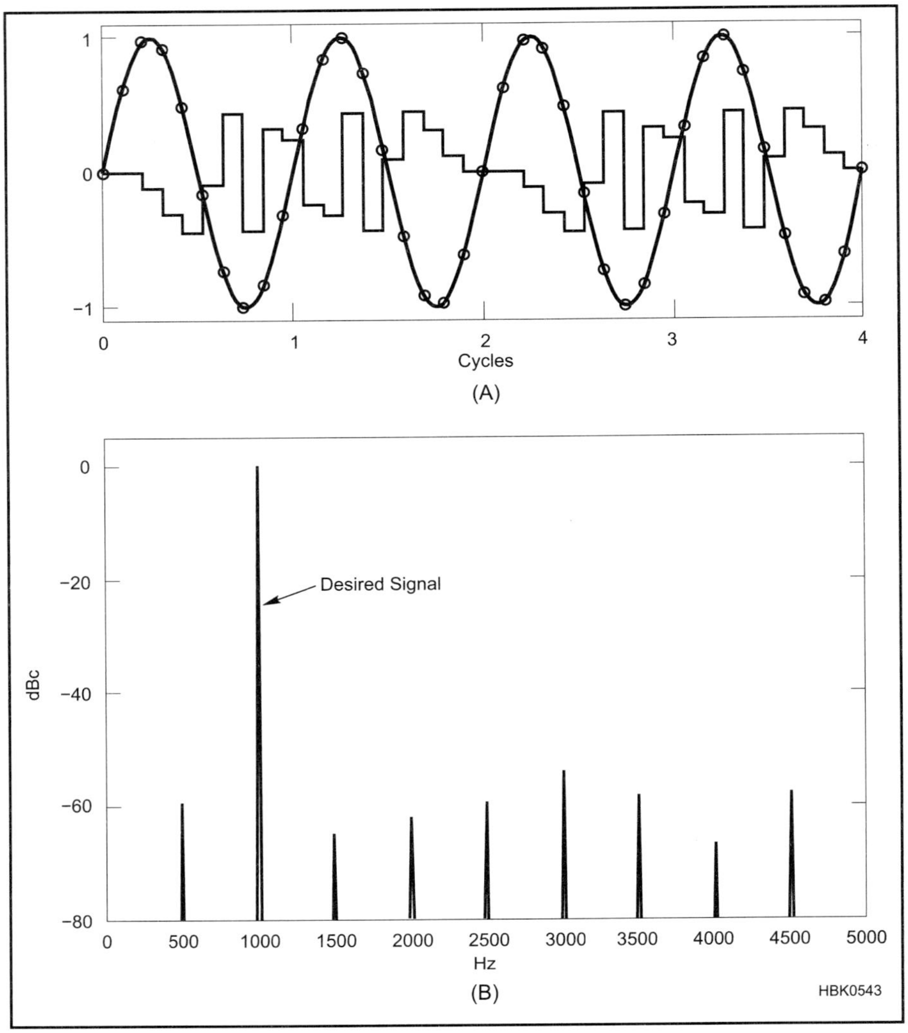

Figure 8.32 — Quantization error of a 1 kHz sine wave sampled at 9.5 kHz with 8-bit resolution (A). Sample values are indicated by circles. Also shown is the quantization error, in units of LSB. Below is the frequency spectrum, showing the spurious frequencies caused by the quantization (B).

8.7 The Fourier Transform

The *Fourier transform* is the software equivalent of a hardware spectrum analyzer. It takes a signal in the time domain and produces a set of data, that can also be thought of as a transformed signal, in the frequency domain that shows the spectral content of the time-domain signal. The Fourier transform works on both periodic and non-periodic signals, but since the periodic case is easier to explain, we will start with that. (Joseph Fourier was a French mathematician and physicist who introduced the transformation technique bearing his name in 1822. His contributions to physics, mathematics, and engineering were wide ranging and at the heart of many technologies used today.)

A *periodic* signal is one that repeats every τ seconds, where τ is the period. That means that the signal can consist only of frequencies whose sinusoidal waveforms have an integer number of cycles in τ seconds. In other words, the signal is made up of sinusoids that are at the frequency $1/\tau$ and its harmonics. Fourier's insight was that you can determine if a frequency is present by multiplying the waveform by a sinusoid of that frequency and integrating the result. The result of the integration yields the amplitude of that harmonic. If the integration yields zero, then that frequency is not present.

To see how that works, look at **Figure 8.33**. For the purpose of discussion, assume the signal to be tested consists of a single tone at the second harmonic as shown at (A). The first test frequency is the fundamental, shown at (B). When you multiply the two together you get the waveform at (C). Integrating that signal gives its average value, which is zero. However if you multiply the test signal by a sine wave at the second harmonic (D), the resulting waveform (E) has a large dc offset so the integration yields a large non-zero value. It turns out that all harmonics other than the second yield a zero result. That is, the second harmonic is *orthogonal* to all the others.

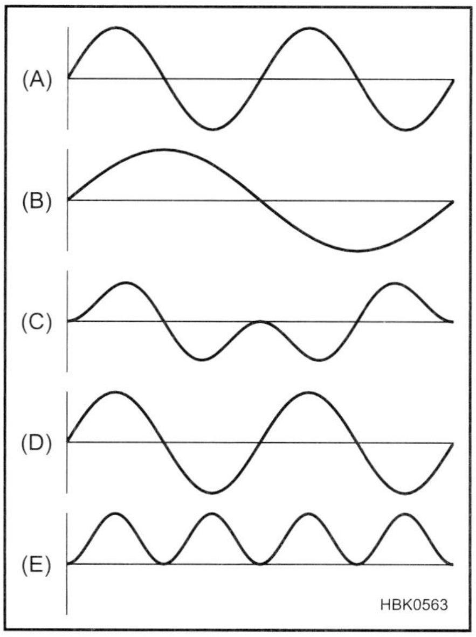

Figure 8.33 — Signal to be tested for frequency content (A). Fundamental test frequency (B). Product of signal and fundamental (C). Second harmonic test frequency (D). Product of signal and second harmonic (E).

If the test waveform included more than one frequency, each of those frequencies would yield a non-zero result when tested with the equivalent-frequency sine wave. The presence of additional frequencies does not disturb the tests for other frequencies since they are all orthogonal with each other.

You may have noticed that this method only works if the test sine wave is in phase with the one in the signal. If they are 90° out of phase, the integration yields zero. The Fourier transform therefore multiplies the signal by both a sine wave and a cosine wave at each frequency. The results of the two tests then yield both the amplitude and phase of that frequency component of the signal using the equations

$$A = \sqrt{a^2 + b^2}$$

and

$$\varphi = \arctan\left(\frac{b}{a}\right)$$

where A is the amplitude, φ is the phase, a is the cosine amplitude and b is the sine amplitude.

If one period of the signal contains, say, 256 samples, then testing a single frequency requires multiplying the signal by the sine wave and by the cosine wave 256 times, then adding the results 256 times for a total of 512 multiplications and additions. (This is referred to as a *multiply-accumulate* for the processing instructions that are required.) There are 128 frequencies that must be tested, since the 128th harmonic is at the Nyquist frequency. The total number of calculations is therefore 512 x 128 = 256^2 multiplications and additions. That is a general result. For any sample size, n, calculating the digital Fourier transform requires n^2 multiply-accumulates.

An excellent visual representation of Fourier's discovery is available online at **blog.matthen.com/post/42112703604/the-smooth-motion-of-rotating-circles-can-be-used**. The fundamental and harmonics are shown as a series of rotating vectors. Each vector traces out a circle, rotating at its assigned frequency and phase. When vectors are added together the result is the desired waveform!

8.7.1 The Fast Fourier Transform (FFT)

The number of calculations grows rapidly with sample size. Calculating the Fourier transform on 1,024 samples requires over a million multiply-accumulate cycles. However, you may notice that there is some redundancy in the calculations. When testing the second harmonic, for example, each of the two cycles of the test sine wave are identical. It is possible to pre-add signal data from the first and second halves of the sequence and then just multiply once by a single cycle of the test sine wave. Also, the first quarter cycle of a sine wave is just a mirror-image of the second quarter cycle, and the first half is just the negative of the second half.

In 1965, J. W. Cooley and John W. Tukey published an algorithm that takes advantage of all the symmetries inherent in the Fourier transform to speed up the calculations. The *Cooley-Tukey algorithm*, also called the *fast Fourier transform* (FFT), makes the number of calculations proportional to $n\log_2(n)$ instead of n^2. For a 1,024-point FFT, the calculation time is proportional to $1024\log_2(1024)$ = 10,240 instead of 1024^2 = 1,048,576, more than a 100-times improvement.

You'll notice that sample sizes are usually a power of two, such as $2^7 = 128$, $2^8 = 256$ and $2^9 = 512$. That is because the FFT algorithm is most efficient with sequences of such sizes. The algorithm uses a process called *radix-2 decimation in time*, that is, it first breaks the data into two equal size chunks, then breaks each of those chunks into two still-smaller chunks of equal size, and so on. It is possible to squeeze even a little more efficiency out of the algorithm with a *radix-4* FFT, which is based on decimation by four instead of by two. That is why you often see sample sizes that are powers of four, such as $4^3 = 64$, $4^4 = 256$, and $4^5 = 1{,}024$. Other variations on the algorithm include decimation in frequency rather than time, mixed-radix FFTs that use different decimation factors at different stages in the calculation, and in-place calculation that puts the results into the same storage buffer as the input data, saving memory. The latter method causes the order of the output data to be scrambled by bit-reversing the address words. For example, address 01010000 becomes 00001010.

The heart of most FFT algorithms is a data flow pattern called a *butterfly*. **Figure 8.34** shows a two element butterfly. The input contains two elements x(0) and x(1), and the output has two elements, y(0) and y(1). The equations for the butterfly are:

$y(0) = x(0) + x(1)\ W^k$

$y(1) = x(1) – x(1)\ W^k$

where W^k is the sum of the sine and cosine harmonics $\cos(2k) + j\sin(2k)$.

Figure 8.35 shows the data flow diagram of the scrambled in/natural out (the output is ordered from lowest to highest element). The number in the circle at each stage indicates the value of k for W^k. Notice that the first stage of calculation performs four 2 × 2 butterfly operations, the second is two 4 × 4 butterfly operations, and the final is one 8 × 8 butterfly. This is called a *decimation* in time algorithm because each stage has fewer butterfly operations.

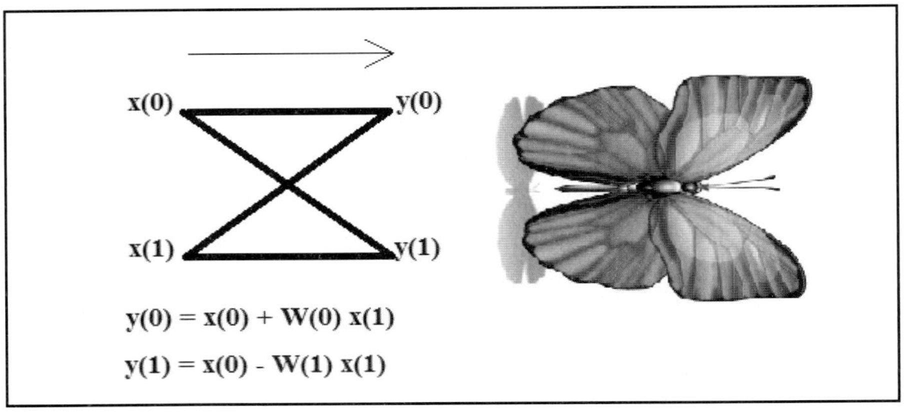

Figure 8.34 — A two-element "butterfly" signal flow diagram. The crossing of the lines makes the diagram look like stylized butterfly wings.

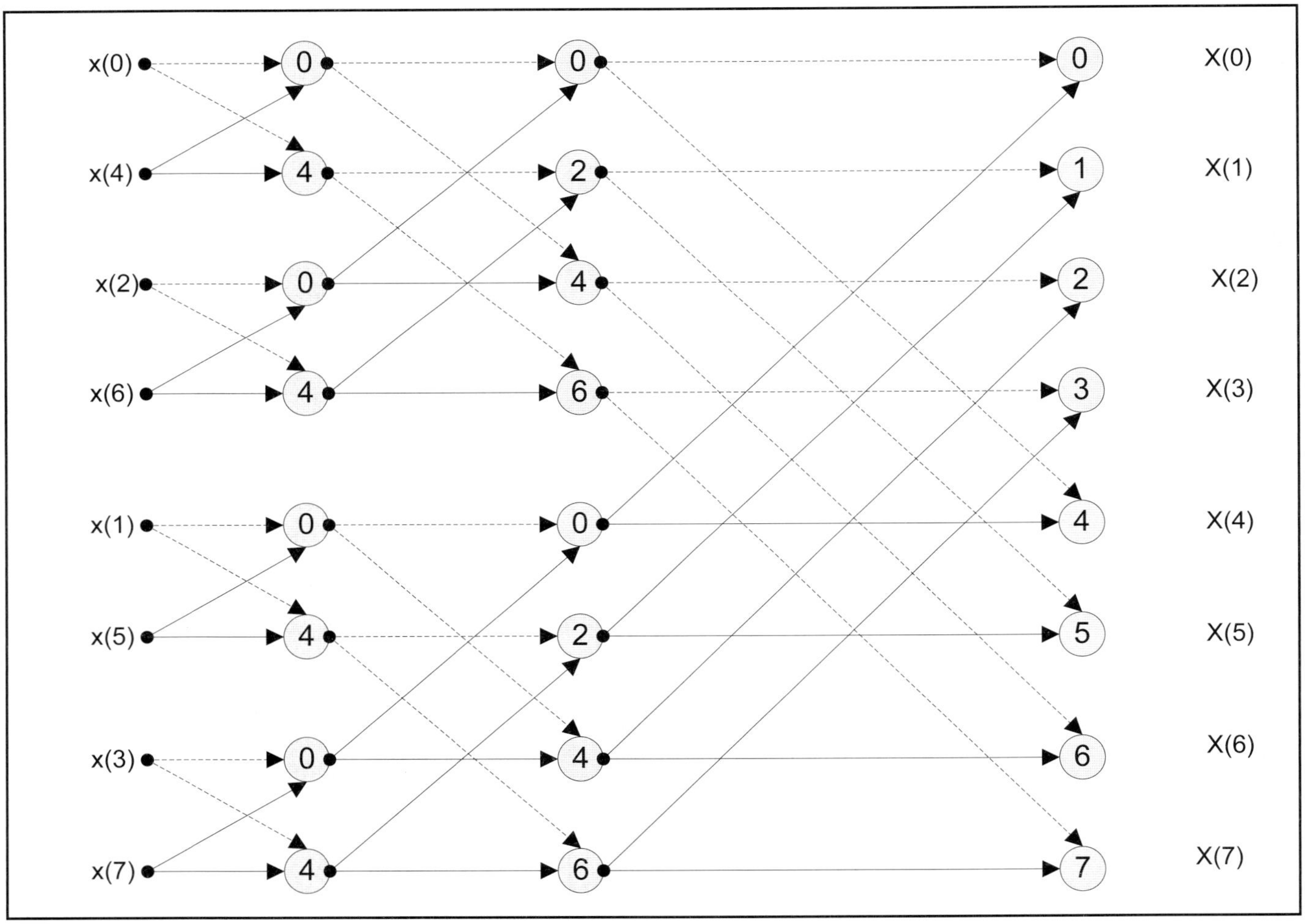

Figure 8.35 — A complete flow diagram for an 8-element FFT calculation. Each solid line represents a direct contribution of the element to the left. Each dashed line represents a contribution that is multiplied by the complex operator W^k. Each number in a circle represents the k value used for the multiplication. This example requires just 24 complex multiplication operations instead of a full 256 calculations.

8.7.2 Non-periodic Signals

So far we have assumed that the signal to be transformed is periodic so that there is an integer number of cycles of each sine wave harmonic in the sequence. With a non-periodic signal, that is not necessarily so. The various frequencies in the signal are not exact harmonics of $1/\tau$ and are no longer orthogonal to the test frequencies. The result is *spectral leakage*; a single frequency in the signal may give a non-zero result when tested at a number of different harmonic frequencies. **Figure 8.36B**, illustrates the FFT of a single sine wave at a non-harmonic frequency. You can see that the spurious response extends quite far from the actual frequency.

Those far-out spurious responses are primarily caused by the abrupt termination of the signal at the edges of the sequence. The spectrum can be cleaned up considerably by tapering the edges with a window, in the manner of digital filter coefficients as described in the **Analog and Digital Filtering** chapter. In fact, the same window functions are used for both. Figure 8.36C illustrates the result of applying a Hamming window to the signal in Figure 8.36A and the resulting improved spectrum is shown at Figure 8.36D. Just as with FIR filters, different windows excel in different areas. Windows with a gradual transition to zero at the edges do a better job of suppressing spurious responses, but smear adjacent spectral lines, analogous to using a wider resolution bandwidth in an analog spectrum analyzer. Windows with a wider center section and a more abrupt transition to zero at the edges have less smearing but more spurious responses.

8.7.3 The Inverse Fourier Transform (IFT)

It is also possible to run the Fourier transform "in reverse," changing frequency-domain data back into time-domain data. The IFT takes a collection of sine wave amplitude and phases and adds them together to create a waveform just as the "forward" Fourier transform changes the waveform into the collection of sine waves.

Just as clever ways of performing the Fourier transform led to the FFT, similarities in the calculations for the FFT and IFT allow the same calculations that perform the "forward" FFT to be used to perform the IFT. **Figure 8.37** shows two methods of implementing an IFT using the FFT algorithm on the I and Q channel data. Both methods have the frequency domain data as their input, I(m) and Q(m), which represent the amplitude and phase of sinusoidal components (harmonics). The output is a series of time domain samples, I(n) and Q(n), which represent the waveform. The divide-by-N functions are scaling factors required by the calculations. Advanced methods are even able to perform the FFT and IFT simultaneously on different streams of data, allowing one specialized set of hardware to do double duty.

Transceivers don't generate transmittable signals using the IFT which is included in this discussion for completeness. Transmit chains usually perform I/Q modulation with digital data on the I and Q channel carriers, filter the output (either with analog or digital filters), and amplify it conventionally to useful levels. Nevertheless, the elegance of using the very same algorithm to convert between the time and frequency domains illustrates the deep mathematical connection between time and frequency.

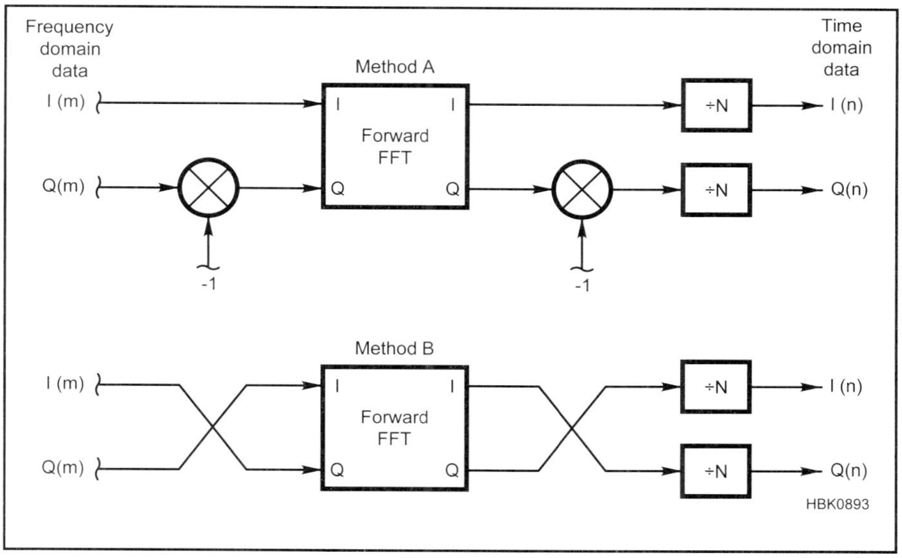

▲Figure 8.37 — Two methods of performing the Inverse Fourier Transform (IFT) using the "forward" FFT algorithm.

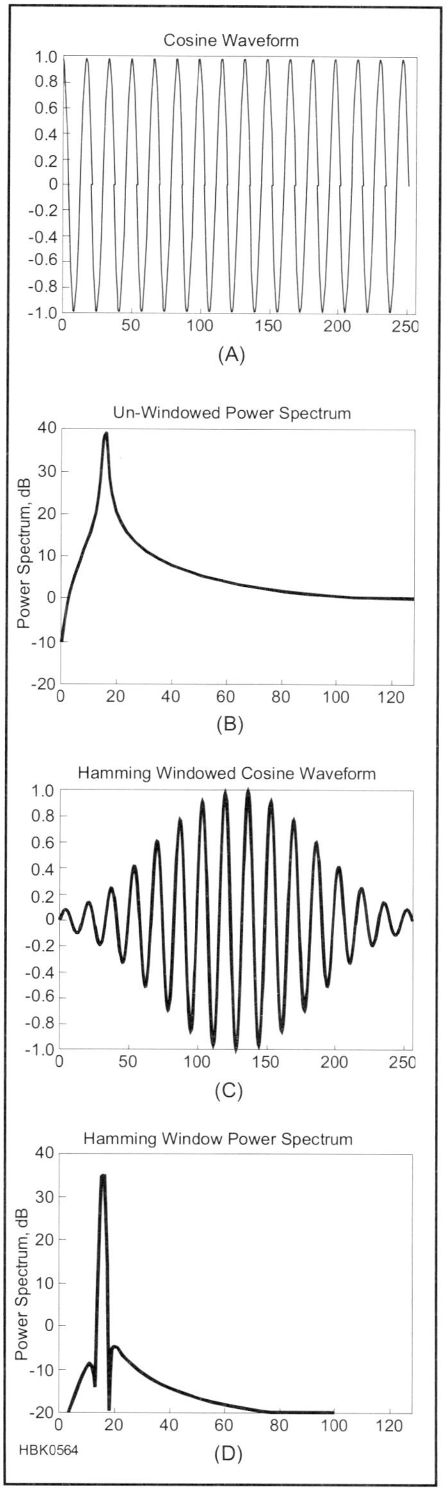

◄Figure 8.36 — Illustrating the use of windowing to minimize spectral leakage, the figures show (A) a cosine waveform not at a harmonic frequency, (B) the resulting unwindowed power spectrum, (C) the same cosine waveform with a Hamming window, and (D) the much narrower power spectrum of the windowed waveform.

8.8 References and Bibliography

References

1. FCC Report and Order FCC 01-264 authorized software-defined radios in the commercial service in 2001. **docs.fcc.gov/public/attachments/FCC-01-264A1.pdf**.
2. Hershberger, D., W9GR, "Low-Cost Digital Signal Processing for the Radio Amateur," *QST*, Sep. 1992, pp. 43 – 51.
3. Youngblood, G., AC5OG, "A Software-Defined Radio for the Masses," *QEX*; Part 1, July/Aug. 2002; Part 2, Sep./Oct. 2002; Part 3, Nov./Dec. 2002; Part 4, Mar./Apr. 2003.
4. Tayloe, D., N7VE, "Notes on 'Ideal' Commutating Mixers (Nov./Dec. 1999)" Letters to the Editor, *QEX*, Mar. 2001, p. 61.
5. The Microchip IDE software can be downloaded free at **www.microchip.com/en-us/tools-resources/develop/mplab-x-ide**.
6. For more details on using the Raspberry Pi GPU for SDR applications, see: **www.raspberrypi.org/blog/accelerating-fourier-transforms-using-the-gpu**
7. Blanchard, R., W6UYG, "Sugar-Coated Single Sideband," *QST*, Oct. 1952, pp. 38 – 39, 128.
8. Weaver, Jr, D. K., "A Third Method of Generation and Detection of Single-Sideband Signals" *Proc. IRE*, Dec. 1956.
9. Wright, Jr., H., W1PNB, "The Third Method of S.S.B.," *QST*, Sep. 1957, pp. 11 – 15.

Bibliography

DSP SOFTWARE TOOLS

The following books are referenced in the DSP files in the online content.

Alkin, O., *PC-DSP* (Prentice Hall, 1990).

Kamas, A. and Lee, E., *Digital Signal Processing Experiments* (Prentice Hall, 1989).

Momentum Data Systems, Inc., *QEDesign*, Costa Mesa, CA, 1990.

Stearns, S. D. and David, R. A., *Signal Processing Algorithms in FORTRAN and C* (Prentice Hall, 1993).

TEXTBOOKS

Hayward, W., Campbell, R., and Larkin, B., *Experimental Methods in RF Design* (ARRL, 2009). See Chapter 10 "DSP Components" and Chapter 11 "DSP Applications in Communications."

Lathi, B.P., *Signal Processing and Linear Systems* (Oxford University Press, 1998).

Lyons, R., *Understanding Digital Signal Processing* 3rd Edition (Pearson, 2012).

Madisetti, V. K. and Williams, D. B., Ed., *The Digital Signal Processing Handbook* (CRC Press, 1998).

Parks, T. W. and Burrus, C.S., *Digital Filter Design* (John Wiley and Sons, 1987).

Sabin, W. E. and Schoenike, E. O., Ed., *HF Radio Systems and Circuits* 2nd Edition (Noble Publishing Corp, 1998).

Widrow, B. and Stearns, S. D., *Adaptive Signal Processing* (Prentice Hall, 1985).

Zverev, A. I., *Handbook of Filter Synthesis* (John Wiley and Sons, 1967).

ARTICLES AND PAPERS

Ahlstrom, J., N2ADR, "An All-Digital SSB Exciter for HF," *QEX*, May/Jun 2008, pp. 3 – 10.

Ahlstrom, J., N2ADR, "An All-Digital Transceiver for HF," *QEX*, Jan./Feb. 2011, pp. 3 – 8.

Åsbrink, L., SM5BSZ, "Linrad: New Possibilities for the Communications Experimenter," *QEX*, Part 1, Nov./Dec. 2002; Part 2, Jan./Feb. 2003; Part 3 May/June 2003.

Bloom, J., KE3Z, "Measuring SINAD Using DSP," *QEX*, June 1993, pp. 9 – 18.

Bloom, J., KE3Z, "Negative Frequencies and Complex Signals," *QEX*, Sep. 1994, pp. 22 – 27.

Bloom, J., KE3Z, "Correlation of Sampled Signals," *QEX*, Feb. 1996, pp. 24 – 28.

Brannon, B., "Basics of Digital Receiver Design," *QEX*, Sep./Oct. 1999, pp. 36 – 44.

Cahn, H., WB2CPU, "Direct Digital Synthesis — An Intuitive Introduction," *QST*, Aug. 1994, pp. 30 – 32.

Cowling, S., WA2DFI, "Hands-On SDR," *QEX*, columns beginning Sep. 2014.

Danzer, P., N1II, "A Practical Demonstration of Fourier Series," *QST*, Apr. 2007, p. 37.

de Boer, P. T., PA3FWN, "How RTL-SDR dongles work,"**www.pa3fwm.nl/technotes/tn20.html**

Dobranski, L., VA3LGD, "The Need for Applications Programming Interfaces (APIs) in Amateur Radio," *QEX*, Jan./Feb. 1999, pp. 19 – 21.

Gradijan, S., WB5KIA, "Build a Super Transceiver — Software for Software Controllable Radios," *QEX*, Sep./Oct. 2004, pp. 30 – 34.

Hightower, M., KF6SJ, "Simple SDR Receiver," *QEX*, Mar./Apr. 2012, pp. 3 – 8.

Hill, C., W5BAA, "SDR2GO: A DSP Odyssey," *QEX*, Mar./Apr. 2011, pp. 36 – 43.

Kester, W., "Taking the Mystery out of the Infamous Formula, "SNR = 6.02N + 1.76dB," and Why You Should Care," Analog Devices, MT-001, **www.analog.com/media/en/training-seminars/tutorials/MT-001.pdf**.

Kester, W., "Understand SINAD, ENOB, SNR, THD, THD + N, and SFDR so You Don't Get Lost in the Noise Floor," Analog Devices, MT-003, **www.analog.com/media/en/training-seminars/tutorials/MT-003.pdf**.

Kester, W., "Which ADC Architecture Is Right for Your Application?," Analog Devices, **www.analog.com/media/en/analog-dialogue/volume-39/number-2/articles/the-right-adc-architecture.pdf**.

Mack, R., W5IFS, "SDR: Simplified," *QEX*, columns beginning Nov. 2009.

Martin, K., "Complex Signal Processing is NOT — Complex," **www.mikrocontroller.net/attachment/151612/complex_signals.pdf**.

Neu, T., "Clocking the RF ADC: Should you worry about jitter or phase noise?," **www.ti.com/lit/an/slyt705/slyt705.pdf**.

Nickels, R., W9RAN, "Cheap and Easy SDR," *QST*, Jan. 2013, pp. 30 – 35.

Scarlett, J., KD7O, "A High-Performance Digital Transceiver Design," *QEX*, Part 1, July/Aug. 2002; Part 2, Mar./Apr. 2003.

Smith, D., KF6DX, "Signals, Samples and Stuff: A DSP Tutorial, Parts 1 – 4," *QEX*, Mar./Apr. 1998 to Sep./Oct. 1998.

Stephensen, J., KD6OZH, "Software Defined Radios for Digital Communications," *QEX*, Nov./Dec. 2004, pp. 23 – 35. [Open-source platform]

Veatch, J., WA2EUJ, "The DSP-610 Transceiver," *QST*, Aug. 2012, pp. 30 – 32.

Ward, R., WA4ZZU, "Basic Digital Filters," *QEX*, Aug. 1993, pp. 7 – 8.

Youngblood, G., AC5OG, "A Software-Defined Radio for the Masses," *QEX*; Part 1, July/Aug. 2002; Part 2, Sep./Oct. 2002; Part 3, Nov./Dec. 2002; Part 4, Mar./Apr. 2003.

SDR AND DSP RECEIVER TESTING

Hakanson, E., "Understanding SDRs and their RF Test Requirements," Anritsu Application Note.

MacLeod, J.R. et al., "Software Defined Radio Receiver Test-Bed," IEEE Vehicular Tech Conf, Fall 2001, VTS 54th, Vol. 3, pp. 565 – 1569.

Sierra, R, Rhode & Schwarz, "Challenges In Testing Software Defined Radios," SDR Forum Sandiego Workshop, 2007.

Sirmans, D. and Urell, B., "Digital Receiver Test Results," Next Generation Weather Radar Program.

"Testing and Troubleshooting Digital RF Communications Receiver Designs," Agilent Technologies Application Note 1314. **www.keysight.com/us/en/assets/7018-06706/application-notes/5968-3579.pdf**

Contents

Chapter 9 — Online Content

Articles

- An Optimized Grounded Base Oscillator Design for VHF/UHF by Dr. Ulrich Rohde, N1UL, and Ajay Poddar, AC2KG
- Automatic Tracking Filter for DDS Generator by Riccardo Gionetti, IØFDH
- Calculation of FM and AM Noise Signals of Colpitts Oscillators in the Time Domain by Dr. Ulrich Rohde, N1UL
- Crystal Test Oscillators by Fred Brown, W6HPH
- Frequency Synthesis: Current and Future Projections by Alexander Chenakin
- Novel Grounded Base Oscillator Design for VHF-UHF by Dr. Ulrich Rohde, N1UL
- Optimized Oscillator Design by Dr. Ulrich Rohde, N1UL
- Oscillator Design Using LTSpice by David Stockton, GM4ZNX (also see LTSpice files in SwissRoll folder)
- Oscillator Phase Noise by Dr. Ulrich Rohde, N1UL
- Programming the DDS AD9850 Signal Generator Module by James Kretzschmar, AE7AX
- Simulation of the Low Noise Oscillator from Solid State Design for the Radio Amateur by Linley Gumm, K7HFD
- Some Thoughts on Crystal Oscillator Design by Dr. Ulrich Rohde, N1UL
- Some Thoughts on Designing Very High Performance VHF Oscillators by Dr. Ulrich Rohde, N1UL
- What You Always Wanted to Know About Colpitts Oscillators by Dr. Ulrich Rohde, N1UL, and Anisha M. Apte

Tools and Data

- Simulation files for Oscillator Design Using LTSpice (SwissRoll folder)

Chapter 9

Oscillators and Synthesizers

RF signals all begin with oscillators — circuits that generate a periodic output signal without any input signal. In general, we use sinusoidal waveforms for communication signals, and square-wave or pulse outputs for digital signals (clocks). Amateurs care greatly about oscillator design. In a crowded band, a stable and clean oscillator is necessary for the receiver to resolve each of the signals present. The same is required of oscillators in transmitters so that our signals use only the amount of spectrum necessary for communication. While this chapter includes a section on phase noise, the **RF Techniques** chapter discusses noise and noise metrics more generally.

This chapter has been updated by Bob Clarke, N1RC, including material from Earl McCune, WA-6SUH; Ulrich Rohde, N1UL; and David Stockton, GM4ZNX. These authors and others have contributed articles on phase noise and oscillator design that are included in the online content for the *Handbook*.

The sheer number of different oscillator circuits seen in the literature can be intimidating, but a manageably small number of oscillator types will cover all but very special requirements. Look at an oscillator circuit and "read" it: What form of filter — resonator — does it use? What form of amplifier? How have the amplifier's input and output been coupled into the filter? How is the filter tuned? These questions are easily answered, putting oscillator types into appropriate categories to make them understandable. The questions themselves make more sense when we understand the mechanics of oscillation, in which resonance plays a major role.

9.1 How Oscillators Work

9.1.1 Resonance

Figure 9.1 shows a simple LC circuit, also called a *tank circuit* or resonator. ("Tank" is used because the L and C components store energy.) The circuit is resonant (oscillates) at the frequency where the inductive reactance and capacitive reactance are equal, that is:

$Z = X_L + X_C = 0$. For this to occur, the magnitudes of the two reactances must be equal, or

$|X_L| = |X_C|$. Since $|X_L| = 2\pi fL$ and $|X_C| = 1/2\pi fC$, we can solve for the frequency, f, which yields

$$f^2 = \frac{1}{4\pi^2 LC}$$

Taking the square root of both sides give the familiar formula for the resonant frequency of an LC circuit,

$$f = \frac{1}{2\pi\sqrt{LC}}$$

Figure 9.2 illustrates the oscillating voltage and current in the circuit created by the regular exchange of energy between the components in Figure 9.1.

9.1.2 Maintained Resonance

The LC values of a tank circuit determine the resonant frequency but like all real-world circuits, the inductor and capacitor have real-world losses; the energy lost per cycle must be replaced for oscillation to occur. What we need is a way to make up for the resonator losses, but only just enough to maintain the oscillation indefinitely. By adding an amplifier to the tank circuit or *resonator*, we can take a small amount of the energy from the resonator, amplify it, and inject a part of the amplified tank signal back into the resonator. The resonator acts as a filter and the energy fed back to the amplifier input is *feedback*. The amount of feedback is represented as β. This process is shown in **Figure 9.3A**.

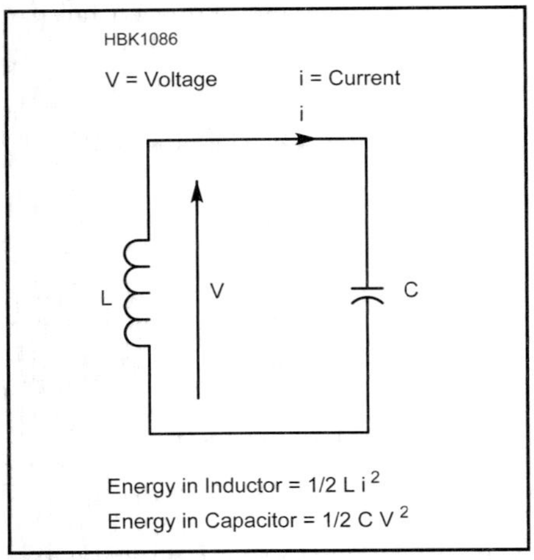

Figure 9.1 — The LC circuit stores energy in the inductor's magnetic field and capacitor's electric field.

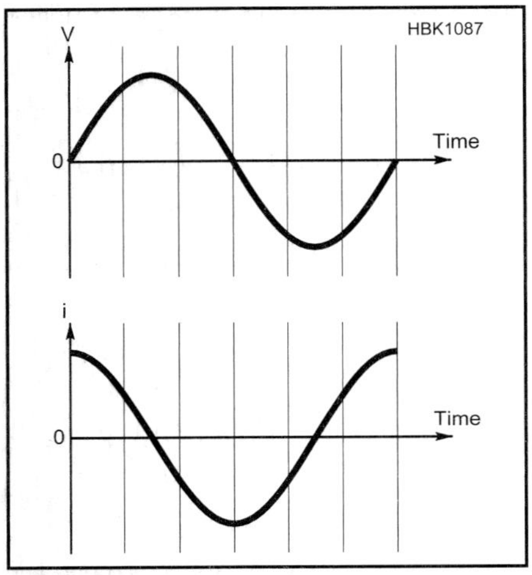

Figure 9.2 — The regular exchange of energy between the components in Figure 9.1 creates the oscillating voltage and current in the circuit.

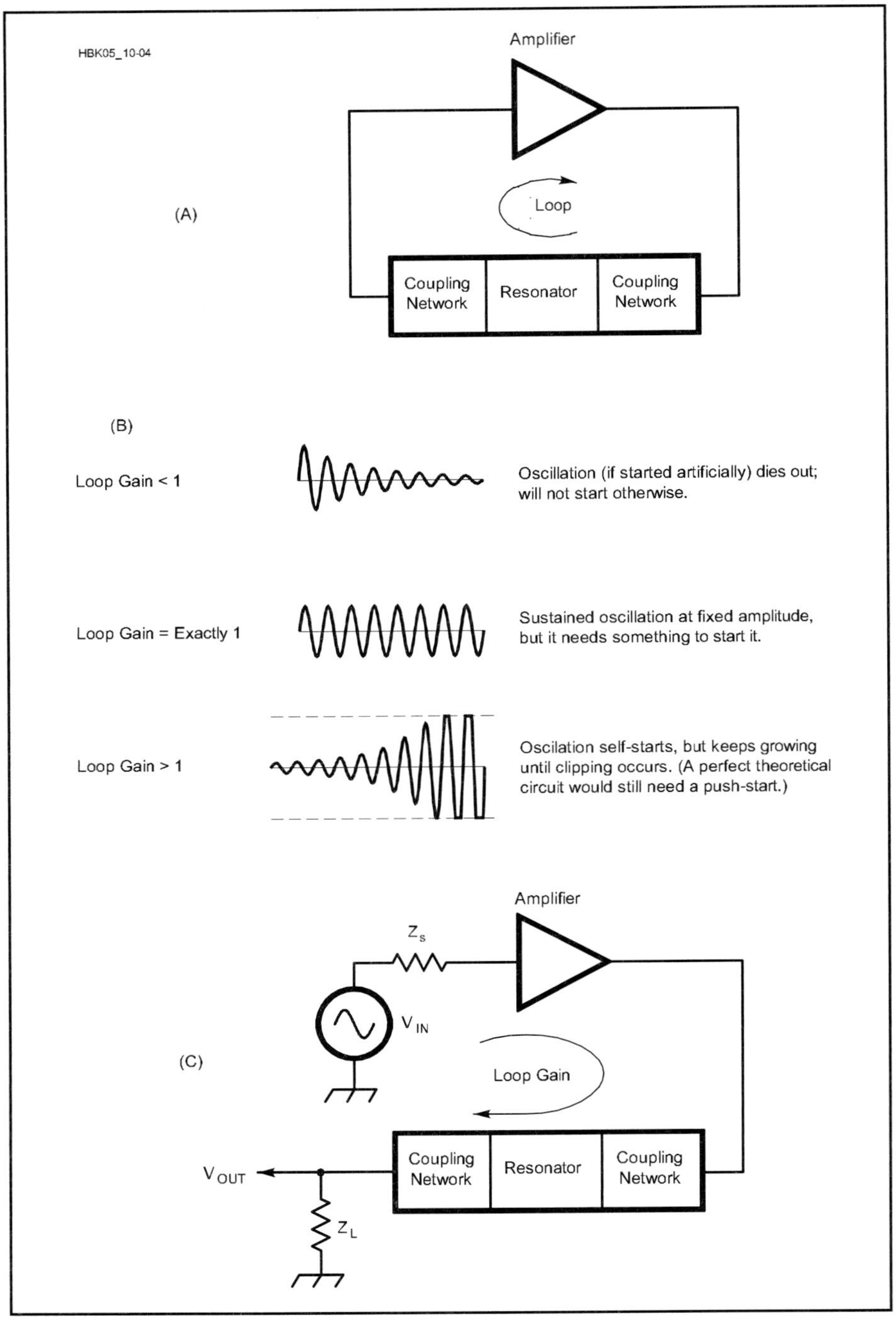

Figure 9.3 — A practical oscillator requires networks to couple power in and out of the resonator (A). At (B), we see the amount of loop gain determines whether the oscillations will die out (loop gain < 1), grow until the waveform is no longer a sine wave (>1), or sustain at a steady amplitude (=1). Breaking the loop, inserting a test signal and measuring the loop's overall gain allows us to determine whether the system can oscillate, sustain oscillation or clip (C).

Unlike in general amplifier design, here we are not interested in maximizing power transfer. Instead, we want to couple just enough energy into the resonator to overcome its losses, and to take only the minimum amount of energy out of the resonator needed to generate the restoring energy. Thus, Figure 9.3A shows coupling networks instead of matching networks. We end up with a loop: The amplifier input comes from the output of the resonator, and a portion of the amplifier output goes to the input of the resonator as feedback.

The trick to oscillator design is to get the coupling networks and the amplifier gain, *A*, working together just right. What this means is shown in Figure 9.3B. If we want to get a sine wave output, we need to get the loop gain — gain computed as the signal passes through the combination of amplifier, resonator, and both coupling networks — exactly equal to one, also called *unity*. If the loop gain is even slightly less than one, then not enough energy is supplied to overcome resonator losses and the damping still happens. If the loop gain is greater than one, then the magnitude of the sine wave will grow until the waveform is no longer sinusoidal.

There is more to making an oscillator work than getting the loop gain right. We also need to be sure that the energy from the amplifier output is applied to the resonator having the correct phase alignment with the oscillating signal in the resonator. Taken together these requirements are known as the *Barkhausen criteria*: to achieve oscillation, 1) the magni-

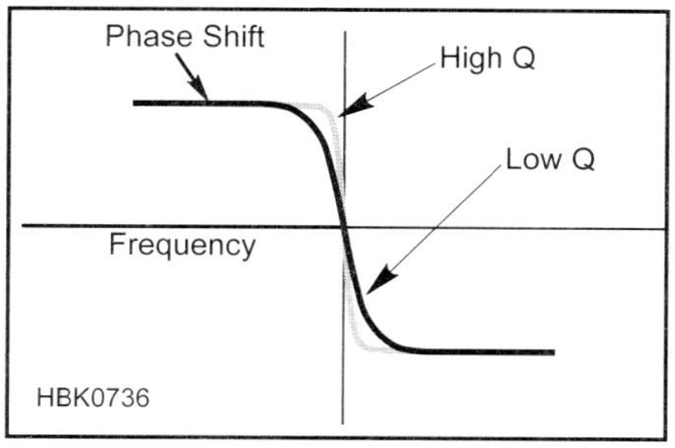

Figure 9.4 — Phase shift through the resonator depends on resonator Q. The higher resonator Q, the more abruptly phase changes through the resonator as frequency changes.

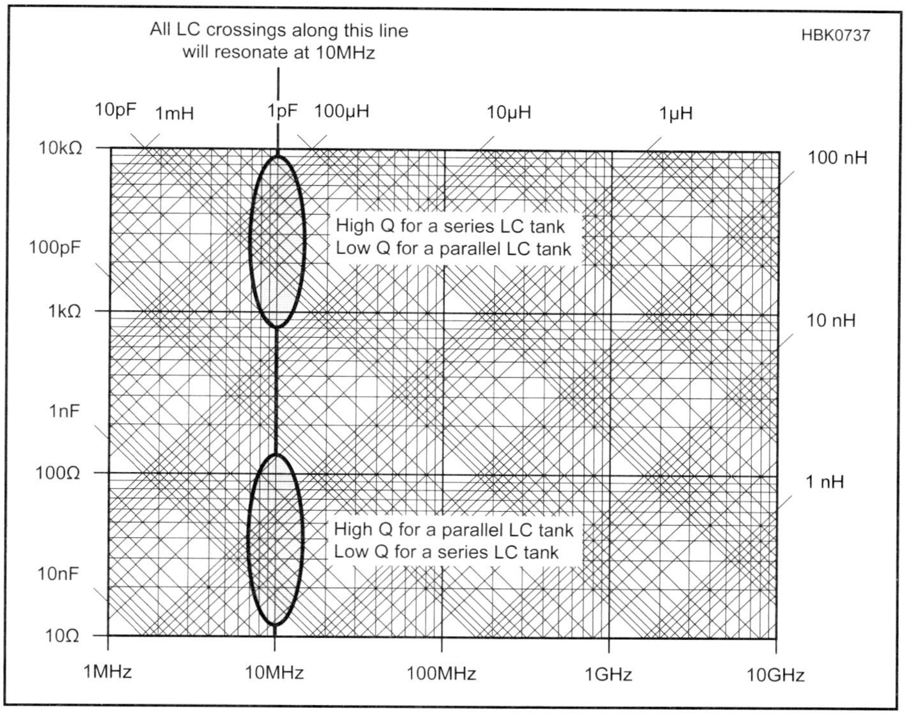

Figure 9.5 — There are an infinite number of combinations of L and C for a given resonant frequency. For a series-LC tank circuit, larger ratios of L to C result in higher values of Q and vice versa for parallel-LC tank circuits.

tude of the loop gain must be exactly equal to unity, $|\beta A| = 1$; and 2) the phase shift through the loop, $\angle\beta A$, must be zero or an integer multiple of 360°. Both of these are critically important. If either of these criteria is not met, the circuit will not oscillate. (See the following section on Oscillator Start-Up regarding loop gain greater than unity.)

How do we find out if we are meeting the Barkhausen criteria? We need to break the loop, usually at the input of the amplifier as shown in Figure 9.3C. We apply a signal slightly below the desired oscillation frequency, then sweep the frequency through and slightly past the desired oscillation frequency. During the sweep we monitor both the magnitude and phase response at the output of the resonator output coupling network. If everything is right, then at the desired frequency of oscillation the input and output signals will look exactly the same on an oscilloscope. If that doesn't happen, we have work to do.

Getting the phase response of the loop correct is usually the harder problem. Much of this difficulty comes from the many types of resonators that can be used, including:

- LC tank circuit (parallel or series)
- Quartz crystal (and other piezoelectric materials)
- Transmission line (stripline, microstrip, open-wire, coax, and so on)
- Microwave cavities, YIG spheres, dielectric resonators
- Surface-acoustic-wave (SAW) devices
- Ceramic Resonators

Each of these also can be called a filter. It is true that any filter can also be used as a resonator in an oscillator. The more complicated phase responses of filters make meeting both of the Barkhausen criteria more challenging, but certainly possible.

The phase response of the resonator and the energy lost per cycle are related to its Q (quality factor — see the **Radio Fundamentals** chapter). If we have a parallel LC tank circuit, then at frequencies well below resonance the tank circuit acts inductive ($X_L \ll X_C$) and the current lags the voltage. At frequencies well above resonance the tank acts capacitive ($X_C \ll X_L$) so the current leads the voltage. In between, the phase shift of the current relative to the voltage depends on the actual frequency. The rapidity of the phase shift with changing frequency is governed by resonator Q: the phase shift happens very rapidly when Q is high. This is shown in **Figure 9.4**. When the resonator Q is low then the phase changes much more slowly at different frequencies.

If we have a series resonator then the phase change profile is reversed (capacitive to inductive) but otherwise the shape is the same. There are an infinite number of combinations of L and C values that provide the same resonant frequency. If we are interested in high Q, for a series tank circuit we want to select large inductor values and small capacitor values. For a high-Q parallel tank circuit we want the opposite — large capacitor values and small inductor values. **Figure 9.5**, a modified version of the reactance vs frequency chart in the **Radio Fundamentals** chapter, shows these regions for a resonant frequency of 10 MHz.

9.1.3 Oscillator Start-Up

Looking at Figure 9.3B we see that the oscillation will build up in magnitude only when the loop gain is greater than unity. Thus, this is an important additional criterion for oscillator design. The loop gain must be slightly greater than unity for it to start oscillating. Otherwise, we have the undesired condition in which there is no signal in the resonator, so we sample nothing from it and put nothing back into it.

Still, we need one more thing — noise. Some kind of signal needs to be injected into the resonator for the oscillation to start building up. We are fortunate that all amplifiers have output noise when power is applied, even if there is no input signal. It is this noise that allows any oscillator to start. Amplifier noise initiates a resonant signal in the tank circuit, and if loop gain is slightly greater than unity, more signal is fed back into the resonator than the resonator loses. The output signal builds up until something stops the process.

Here we take advantage of an amplifier characteristic where gain is reduced as the output reaches some predefined value. This is referred to as *compression* or *limiting*: When the output signal gets large enough, the amplifier gain is reduced a little bit. When compression balances steady output amplitude with a loop gain of exactly unity, the oscillator has reached steady state operation.

When the loop gain and phase shift have stabilized at the correct values, the oscillation will be sustained at a steady value. This is called a *continuous wave* (CW). Amateurs use the term to refer to Morse transmissions (see the section on On-Off Keying (OOK) in the **Modulation** chapter) but it really means a steady, unchanging signal with a constant amplitude and frequency.

Why don't we have noise at the oscillator output instead of a sine wave? Because of the filtering action of the resonator, the filtered noise waveform appears mostly sinusoidal. The narrower the bandwidth of the resonator (the higher its Q), the more sinusoidal the waveform, but it actually consists of very well-filtered noise. It is impossible to get only a pure sine wave. Noise is inevitable, and tremendous efforts are spent in reducing this noise. This is discussed in this chapter's section on Phase Noise. A demonstration by David Stockton, GM4ZNX, of oscillator design is based on a simulation of how oscillators start up by the free *LTSpice* simulation program, which is available in the book's online material.

9.2 LC Variable Frequency Oscillator (VFO) Circuits

In this section we introduce some important oscillator circuits and provide well-tested guidelines on how to successfully build one. The following section presents a design example.

There are thousands of oscillator circuits, but just a few principal designs. One of the principal oscillator circuits is shown in **Figure 9.6**. An LC tank circuit is shown in Figure 9.6A. The usual single capacitance is replaced with two series capacitors having the same equivalent capacitance, C_{EQUIV}. The two capacitors act as an ac voltage divider, so that the voltage V1 at the midpoint is less than the total ac voltage of the tank, V_{TANK}.

What happens when we try to force V1 to be the same as V_{TANK} by adding an amplifier with unity voltage gain as shown in Figure 9.6B? The voltage division action of the capacitive divider does not change. If V1 is forced to be equal to V_{TANK} by the amplifier, then V_{TANK} will become greater than before. But then V1 will take on the new value of V_{TANK} and so forth. This is positive feedback and we have created an oscillator. Connecting an amplifier with low voltage gain to a split tank capacitance leads to the *Colpitts* group of oscillator designs.

It is certainly possible to "split" the inductor instead of the capacitor in a tank circuit and apply the same amplifier trick as before. This is shown in Figure 9.6C and describes the *Hartley* group of oscillators.

9.2.1 Basic Oscillator Circuits

(See also the papers "What You Always Wanted to Know About Colpitts Oscillators" by Ulrich Rohde, N1UL, and Anisha M. Apte, and "Calculation of FM and AM Noise Signals of Colpitts Oscillators in the Time Domain" by Ulrich Rohde, N1UL, included with the online material accompanying this book.)

The LC oscillators used in radio equipment are usually designed to be variable frequency oscillators (VFOs). Tuning is achieved by either varying part of the capacitance of the resonator or, less commonly, by using a movable magnetic core to vary the inductance. Since the early days of radio, there has been a huge quest for the ideal, low-drift VFO. Amateurs and professionals have made immense efforts in this pursuit. A brief search of the literature reveals a large number of VFO designs, many accompanied by claims of high stability. The quest for stability has been solved by the development of low-cost frequency synthesizers, which give crystal-controlled stability. Synthesizers are generally noisier than a good VFO, so the VFO still has much to offer in terms of signal cleanliness, cost, and power consumption, making it attractive for homebrew construction. No single VFO circuit has any overwhelming advantage over any other — component quality, mechanical design and the care taken in assembly are much more important.

Let's take a look at three popular oscillator circuits stripped of any non-essential components so they can be easily compared. The original Colpitts circuit (**Figure 9.7**) is now often referred to as the parallel-tuned Colpitts because its series-tuned derivative (**Figure 9.8**) has become the most common. The Hartley version of the oscillator is shown in **Figure 9.9**. All three of these circuits use an amplifier with a voltage gain less than unity, but large current gain. The N-channel JFET source follower shown is the most popular current choice for the transistor amplifier circuit.

A rule of thumb is that C3 and C4 in Figures 9.7 and 9.8 should be equal and valued such that their $X_C = 45\ \Omega$ at the operating frequency; likewise, for C2 in Figure 9.7, $X_C = 100\ \Omega$. For best stability, use C0G or NP0 units for all capacitors associated with the FET gates and sources. Depending on the FET chosen, the 1-kΩ source-bias-resistor value shown may require adjustment for reliable starting.

PARALLEL-TUNED COLPITTS VFO

In the parallel-tuned Colpitts, C3 and C4 are large values, perhaps 10 times larger than typical values chosen for C1 and C2 to resonate L at the desired frequency. This means only a small fraction of the total tank voltage is applied to the FET gate, and the FET can be considered to be only lightly coupled into the tank circuit. Equally important, the values of C3 and C4 must be much larger than the FET device internal capacitances to provide good stability for the output frequency. This

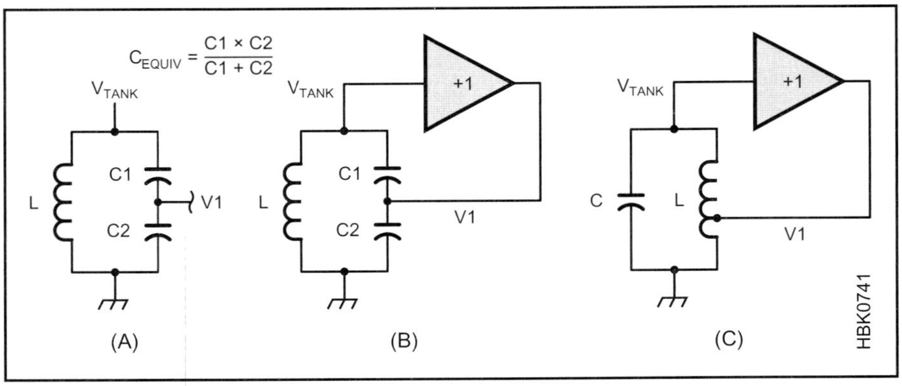

Figure 9.6 — The capacitor in an LC tank circuit (A) can be split into two capacitors with the same equivalent capacitance, C_{EQUIV}, so that the resonant frequency of the tank circuit is unchanged. (B) shows a unity-gain amplifier connected so that it forces V1 = V_{TANK}, creating positive feedback and a Colpitts oscillator results. (C) shows the same technique applied to the tank circuit inductor, creating a Hartley oscillator.

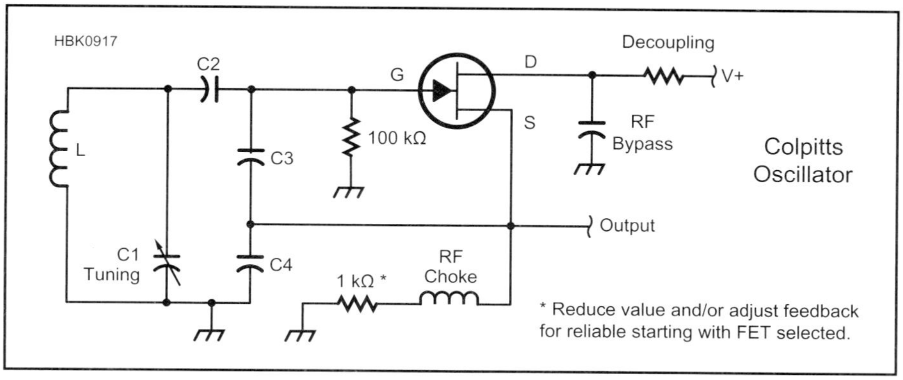

Figure 9.7 — The Colpitts oscillator circuit.

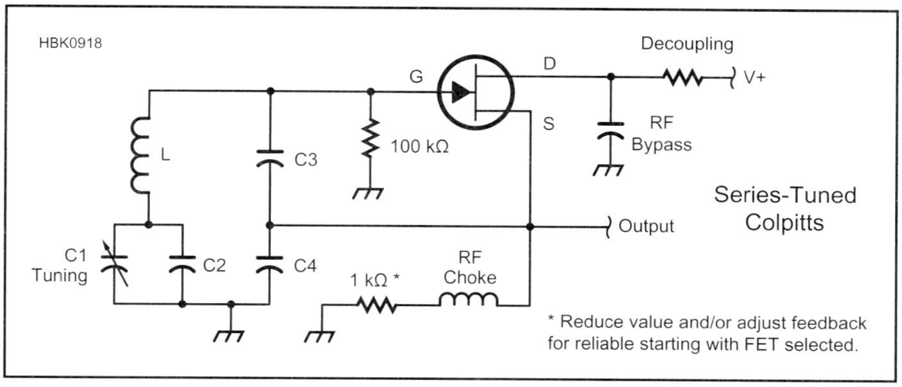

Figure 9.8 — The series-tuned Colpitts oscillator circuit.

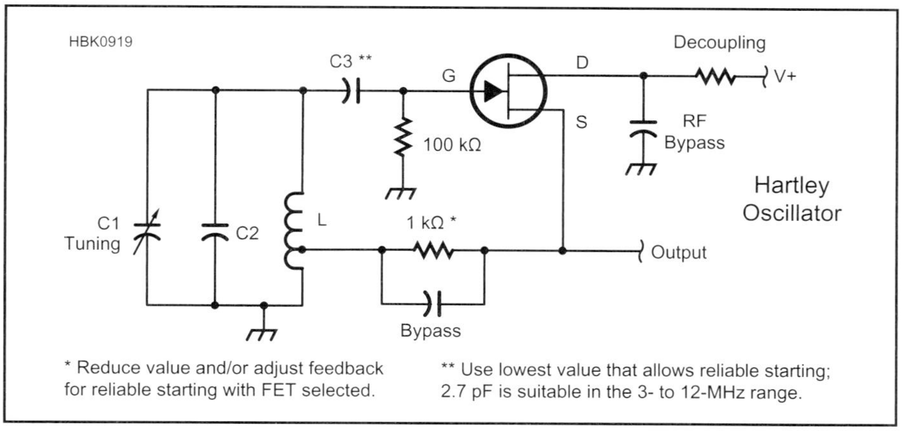

Figure 9.9 — The Hartley oscillator circuit.

keeps small variations in the FET capacitances from having a significant effect on oscillator operation.

The FET is driven by the sum of the voltages across C3 and C4, while it drives the voltage across C4 alone. This means that the tank circuit operates as a resonant, voltage-step-up transformer compensating for the less-than-unity-voltage-gain amplifier. The resonant circuit consists of L, C1, C2, C3, and C4. The resonant frequency can be calculated by using the formula derived at the beginning of this section.

The equation holds for all cases. Its accuracy is dependent on the designer's ability to account for all of the contributions to tank inductance and capacitance. There are several rules of thumb that help a VFO designer identify the various "stray" contributions to tank inductance and capacitance:

1. Wherever there is voltage, stray capacitance must be considered.

2. Wherever there is current, stray mutual inductance must be considered.

3. All currents form loops, creating inductance.

These "rules" will help point out where to look for contributions to inductance and capacitance. At frequencies below 10 MHz the actual components you see and touch tend to be all you need to consider. As the oscillator frequency increases, consideration of stray capacitance and stray inductance — reactance effects that are not associated directly with a visible component — becomes increasingly important. If the oscillator frequency is noticeably lower than what you predict, it is almost certain that there are important stray reactances that you need to account for.

Getting back to the circuit of Figure 9.7, for a wide tuning range C2 must be kept small to reduce the swamping effect of C3 and C4 on the variable capacitor C1. (For more information on component selection, the chapters on oscillators in *Experimental Methods for RF Design* and *Introduction to Radio Frequency Design* listed in the Reference section provide excellent material.) "Swamping" refers to a much larger value component reducing the effect of a small component connected to it.

A parallel-tuned Colpitts oscillator is the subject of the detailed paper "A Design Example for an Oscillator for Best Phase Noise and Good Output Power" by Ulrich Rohde, N1UL, available in the online information for this book. The paper shows the design process by which both noise and power can be optimized in a simple oscillator.

SERIES-TUNED COLPITTS VFO

The series-tuned Colpitts circuit (Figure 9.8) works in much the same way. The difference is that the variable capacitor, C1, is positioned so that it is well-protected from being swamped by the large values of C3 and C4. In fact, small values of C3 and C4 would act to limit the tuning range. Fixed capacitance, C2, is often added across C1 to allow the tuning range to be reduced to that required without interfering with C3 and C4, which set the amplifier coupling. The series-tuned Colpitts has a reputation for better stability than the parallel-tuned original. Note how C3 and C4 swamp the much smaller capacitances of the JFET amplifier in both versions.

HARTLEY VFO

The Hartley oscillator of Figure 9.9 is similar to the parallel-tuned Colpitts, but the JFET amplifier source is connected to a tap on the tank inductance instead of the tank capacitance. A typical tap placement is 10% to 20% of the total turns up from the cold end of the inductor. (It's common to refer to the lowest-signal-voltage end of an inductor as "cold" and the other, with the highest signal voltage as "hot.") C2 limits the tuning range as required.

C3 is reduced to the minimum value that allows reliable starting. This is necessary because the Hartley's lack of the Colpitts' capacitive divider would otherwise couple the FET's internal capacitances to the tank more strongly than in the Colpitts, potentially affecting the circuit's frequency stability.

9.2.2 VFO Design Topics

FREQUENCY SCALING

Generally, VFOs can be adapted to work at other frequencies (within the limits of the active device). To do so, compute an adjustment factor: f_{old}/f_{new}. Multiply the value of each frequency determining, feedback L, or C by that factor. As frequency increases and amplifier gain drops, it may be required to increase feedback more than indicated by the factor.

BIAS STABILIZATION

In all three circuits, there is a 1 kΩ resistor in series with the source bias circuit. This resistor does a number of desirable things. It reduces the Q of the inevitable low-frequency resonance of the choke with the tank tap circuit. It decreases tuning drift due to choke impedance and winding capacitance variations. It also protects against spurious oscillation at undesired frequencies due to internal choke resonances. Less obviously, it acts to stabilize the loop gain of the built-in AGC action of this oscillator. Stable operating conditions act to reduce frequency drift.

Some variations of these circuits may be found with added resistors providing a dc bias to stabilize the system quiescent current. More elaborate still are variations characterized by a constant-current source providing bias. This can be driven from a separate AGC detector system to give very tight level control.

GATE CLAMPING DIODE

It is important to note that the gate-to-ground clamping diode (1N914 or similar) long used by radio amateurs as a means of avoiding gate-source conduction has been shown by Ulrich Rohde, N1UL, to degrade oscillator phase-noise performance and its use is virtually unknown in professional circles.

OTHER AMPLIFYING DEVICES

Figure 9.10 shows two more VFOs to illustrate the use of a vacuum tube triode and a BJT as amplifying devices. Compared to the more frequently used JFET, bipolar transistors like the one used in Figure 9.10B are relatively uncommon in oscillators because their relatively low input and output impedances are more difficult to couple into a high-Q tank without loading it excessively. Bipolar devices do tend to give better sample-to-sample amplitude uniformity for a given oscillator circuit, because many of the bipolar transistor's characteristics are due more to device physics and are less influenced by manufacturing variations. This is not true for JFETs of any given type; JFETs tend to vary more in their characteristics because manufacturing variations directly affect their threshold or pinch-off voltage.

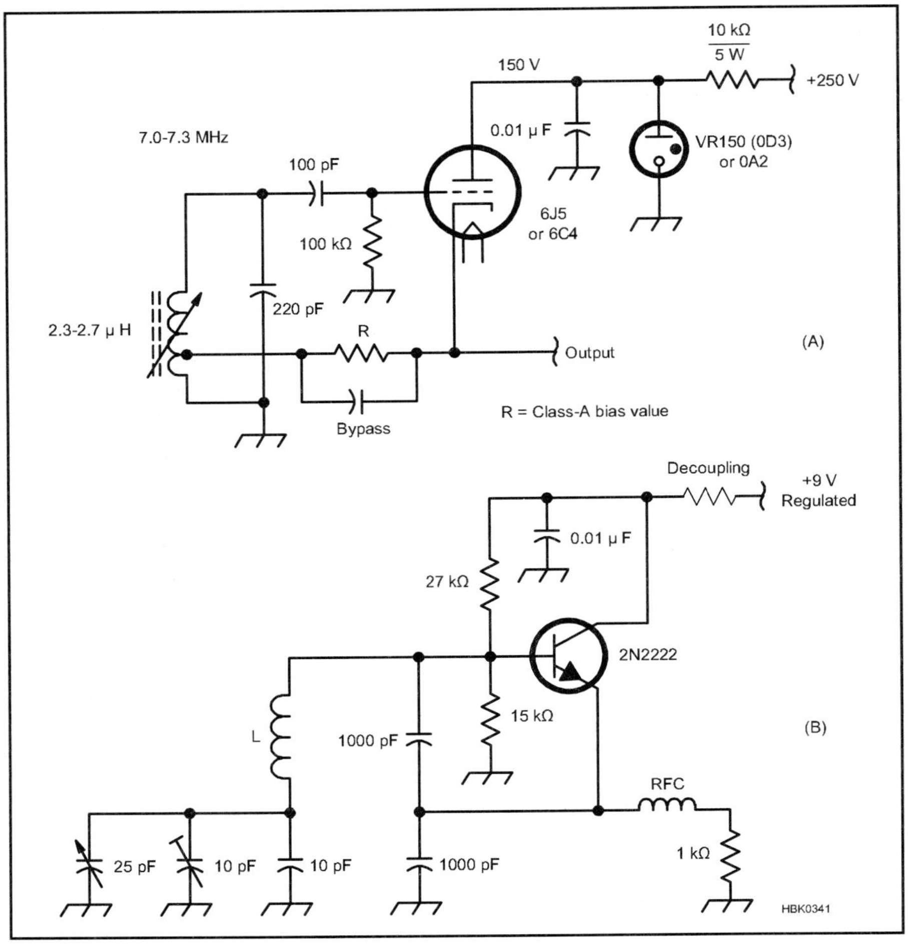

Figure 9.10 — Two additional oscillator examples: at A, a triode-tube Hartley; at B, a bipolar junction transistor in a series-tuned Colpitts.

SQUEGGING

The effect called *squegging* (or *squeeging*) can be loosely defined as oscillation on more than one frequency at a time, but may also manifest itself as RF oscillation interrupted at an audio frequency rate, as in a super-regenerative detector. One form of squegging occurs when an oscillator is fed from a power supply with high source impedance. The power supply charges up the decoupling capacitor until oscillation starts. The oscillator draws current and pulls down the capacitor voltage until it has starved itself of power and oscillation stops. The oscillator stops drawing current and the decoupling capacitor then recharges until oscillation restarts. The process, the low-frequency cycling of which is a form of relaxation oscillation, repeats indefinitely. Examination with an oscilloscope reveals the oscillator output is pulse modulated.

Squegging can be a consequence of poor design in battery-powered radios. As the battery becomes discharged, its internal resistance often rises and circuits that it powers can begin to misbehave. In audio stages, such misbehavior may manifest itself in the "putt-putt" sound of the slow relaxation oscillation called "motorboating."

9.2.3 Two Low-Noise VHF Oscillators

The following section is excerpted from the May/June 2018 QEX article "An Optimized Grounded Base Oscillator Design for VHF/UHF," by Ulrich Rohde, N1UL, and Ajay Poddar, AC2KG. The full article is available as a PDF file in this book's online content. Interested readers are encouraged to read the original article, which includes a comprehensive set of references and all design details.

The design of VHF/UHF oscillators has been described in many books and journals with most of the emphasis on frequency stability and, to a lesser extent, on output/efficiency.

Vacuum Tube Oscillators

Vacuum-tube radios are still maintained in operation and occasionally constructed for enjoyment but the semiconductor long ago achieved dominance in VFOs. Vacuum tubes cannot run as cool as competitive semiconductor circuits, so care is needed to keep the tank circuit away from the tube heat. LC VFOs are also no longer used in commercial transceivers today with the advent of DDS technology that is described elsewhere in this chapter.

In many amateur and commercial vacuum-tube oscillators, oscillation drives the tube into grid current at the positive peaks, causing rectification and producing a negative grid bias. The oscillator thus runs in Class C, in which the conduction angle reduces as the signal amplitude increases until the amplitude stabilizes. As in the JFET circuits of Figures 9.7 through 9.9, better stability and phase-noise performance can be achieved in a vacuum-tube oscillator by moving its operating point out of true Class C — that is, by including a bypassed cathode-bias resistor (the resistance appropriate for Class A operation is a good starting value).

The vacuum-tube triode Hartley shown in Figure 9.10A features permeability tuning, which has no sliding contact like that of a capacitor's rotor and can be made reasonably linear by artfully spacing the coil turns. The slow-motion drive can be done with a screw thread. The disadvantage is that special care is needed to avoid backlash and eccentric rotation of the core. If a non-rotating core is used, the slides have to be carefully designed to prevent motion in unwanted directions. The Collins Radio Company made extensive use of vacuum tube-based permeability tuners, and a semiconductor version was used in a number of Ten-Tec radios.

Since the introduction of reliable phase noise measurements and the ability to predict/simulate the phase noise, the improvement of this important parameter with the help of CAD and analytic equations has gained more attention. The novel design concept described in the article with the online content uses a time domain approach to provide both the best output power and the best phase noise performance. This very simple but powerfully scalable set of formulas for the oscillator synthesis provides extremely good results. In addition, the design principle for fixed or narrowband oscillators discussed here also applies to the broadband VCO design. This design methodology works over a multi-octave tuning range. The circuits here are shown as fixed-frequency designs. To use them as a VCO, variable capacitance diodes (varicaps) should be used in parallel with the output capacitors at P1 with very light coupling to the frequency-controlling LC circuit.

CIRCUIT DESIGN GUIDELINES

The results we have obtained so far were based on mathematical calculations. Some of these calculations are difficult to obtain (see the original *QEX* article). However, there are certain relationships between the values of the capacitance of the tuned circuit and the two feedback capacitors, the collector-emitter capacitor and the emitter-to-ground capacitor. The following shows the set of recommended steps for easy design of such oscillators.

A very popular circuit for VHF/UHF oscillators is the grounded-base configuration, which is shown in **Figure 9.11**. Its phase noise can be made very good, since the RF voltage swing at the collector can be close to the supply voltage. The circuit is simple and has been used for decades. This oscillator type works well from RF frequencies such as 10 MHz to above 1000 MHz.

The oscillator function is based on the principle that power from the output is taken and fed to the emitter. This feedback arrangement generates a negative resistance at the output, thus compensating the losses of the output-tuned circuit, starts oscillating, and then stabilizes the oscillation amplitude. The circuit has simple design rules where C_E and C_F are the feedback capacitors that generate the negative resistance to compensate for the loss resistance in the resonator network comprised of L_E and C^*_L.

Theoretically, the grounded-base configuration can be rotated to be the Colpitts circuit. If we look at the performance, in the case of the Colpitts oscillator the RF voltage swing is limited by the base-emitter and emitter-to-ground voltage and as a result there is less energy stored in the circuit and because of loading the operational Q can be less than in the grounded-base configuration. The Colpitts oscillator is popular because of its simplicity, and its perceived high isolation as the output power is taken at the collector. However, due to the strong Miller effect at very high frequencies, this is not a true statement.

SIMPLE DESIGN RULES

By setting the L/C ratio to a fixed value of 1,200 (for optimum energy storage, group delay, and energy transfer for a given cycle in the resonator network), the following should be used.

$$\frac{L}{C} = \left[\frac{L_E}{C^*_L}\right]_{\text{Grounded-Base}} = 1{,}200$$

$$\Rightarrow Z_0 = \sqrt{1{,}200} \cong 34.6\ \Omega$$

$$L = 1{,}200 \times C$$

$$\Rightarrow L_E = 1{,}200 \times C^*_L$$

$$f = \frac{1}{2\pi\sqrt{LC}} = \frac{1}{2\pi C\sqrt{1{,}200}}$$

$$C^*_L = C_L + \frac{C_E C_F}{C_E + C_F}$$

where C_E and C_F are feedback capacitors.

$$C_L = \frac{C_A C_B}{C_A + C_B}$$

where C_A and C_F are feedback capacitors.

$C_B = 10\ C_A$

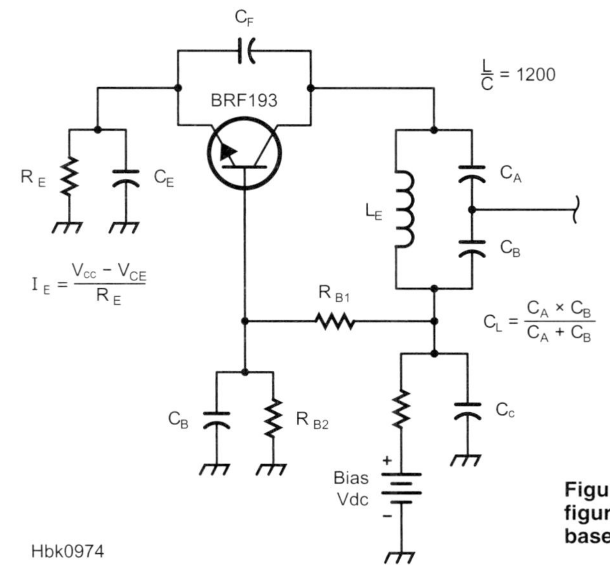

Figure 9.11 — Typical configuration of grounded-base oscillator circuit.

The reader is reminded that this "recipe" was developed based on the Y-parameters, transconductance, and package parasitics for the BRF193 transistor. Using a transistor greatly different than the BRF193 requires the full design process to be performed from the beginning using the new transistor as described in the original article.

144 MHz GROUNDED-BASE OSCILLATOR

The oscillator circuit in **Figure 9.12** is designed using the procedure of the preceding section.

$$C_L^* = \frac{1}{2\pi f\sqrt{1,200}} \Rightarrow C \cong 31 \text{ pF}$$

$$L_L = \frac{1}{(2\pi f)^2 C} \Rightarrow L \cong 39 \text{ nH}$$

$$C_F = 0.3 \times C_L^* \cong 11 \text{ pF}$$

$$C_L = 2 \times C_F \cong 22 \text{ pF}$$

$$C_E = 4 \times C_F \cong 44 \text{ pF}$$

$$C_A = 22 \text{ pF}$$

$$C_B = 220 \text{ pF}$$

These results are frequency scalable with minor corrections possibly necessary.

Figure 9.12A shows the schematic of the 144 MHz oscillator at $I_C = 10$ mA. The oscillator uses a lumped inductor of 39 nH and an unloaded Q of 200 at the operating frequency. Even at these frequencies the layout is quite critical. The example layout in Figure 9.12B shows an assembly of components where the lead inductances have been kept small. The inductor is a standard off the shelf component. (Note that the sample layout is given only as a guideline and not as a template.)

Figure 9.12C shows the CAD simulated phase noise plot. The output power is 11.5 dBm and the second and third harmonics are about –28 dBm and –34 dBm. Using phase noise simulation, the result is –134 dBc/Hz and –94 dBc/Hz at 10 kHz and 100 Hz offset. Calculated, simulated, and measured results closely agree within 1 dB.

If the same transistor is operated at 30 mA, the phase noise at 10 kHz offset will improve to be –144 dBc/Hz and the output power is increased to 20 dBm. This shows that for low phase noise design a more powerful transistor is a good choice. It is important to keep the dc dissipation of the device in mind, as the CAD process will probably not flag a misuse of the device.

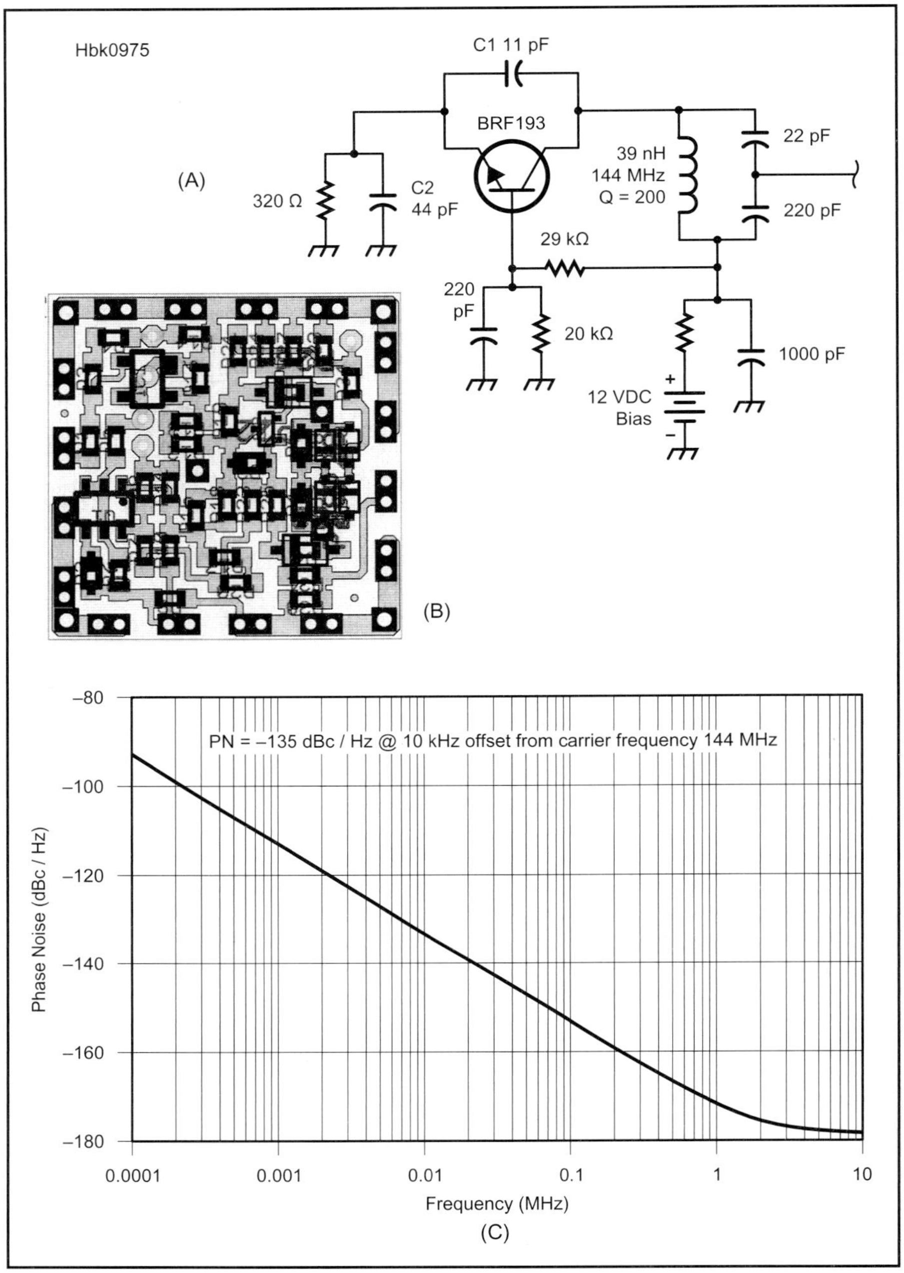

Figure 9.12 — 144 MHz oscillator circuit for I_E = 10 mA (A), an example of the circuit layout using LC lumped inductor-capacitor resonator network (B), and simulated phase noise plot (C).

MODIFIED CIRCUIT FOR UHF (432 MHz) AND HIGHER CURRENT

At 432 MHz and at $I_E = 30$ mA, the loading of the tank circuit decreases the operating Q significantly. The way around this to use a center-tapped inductor. As the coupling at these frequencies from winding to winding is not extremely high, two separate identical inductors can be used successfully.

Figure 9.13A shows the schematic of 432 MHz grounded base oscillator using the tapped inductor. This is a modification of the circuit we have used previously. In the case of a VCO, it would be advantageous to use a different output coupling scheme because in this configuration, the loading would vary with frequency. This can easily be achieved by adding some inductive coupling to the circuit. In case of a printed resonator this can be accomplished quite simply.

Figure 9.13B shows a sample guideline layout of the 432 MHz oscillator circuit using

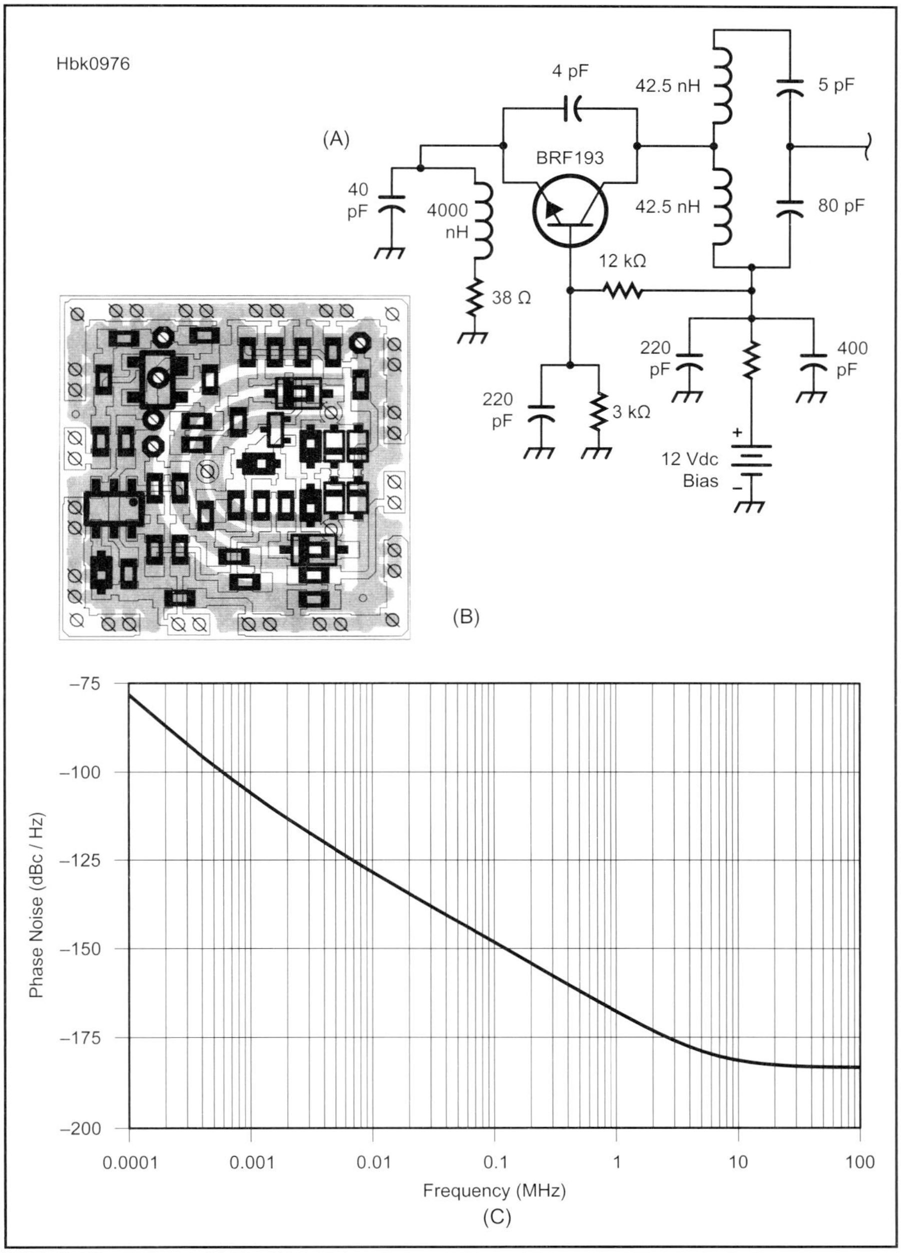

Figure 9.13 — 432 MHz oscillator circuit using tapped inductor and I_E = 30 mA (A), an example of the circuit layout using buried printed coupled line resonator (stripline resonator) in the middle layer of the board (B), and simulated phase noise plot (C).

The Vackar Oscillator

The original Vackar oscillator is named for Jirí Vackár, who invented the circuit in the late 1940s. The circuit's description — a refinement of the Clapp oscillator — can be found in older editions of the Radio Society of Great Britain's Radio Communication Handbook, with some further comments on the oscillator in RSGB's Amateur Radio Techniques. The circuit is also described in the Nov 1955 *QST* "Technical Correspondence" column by W9IK. The Vackar circuit optimized the Clapp oscillator for frequency stability: the oscillator transistor is isolated from the resonator, tuning does not affect the feedback coupling, and the transistor's collector output impedance is kept low so that gain is the minimum necessary to sustain oscillation.

a buried printed coupled line resonator network (stripline resonator in the middle layer). The actual resonator would not be visible if the oscillator is visually inspected.

Figure 9.13C shows the simulated phase noise plot with the expected noise degradation of 9 dB, as the frequency is approximately three times higher. The resulting simulated output power at 432 MHz is 16 dBm, compared to 18 dBm at 144 MHz. This is due to internal package parasitics, which could not be compensated externally. The second harmonic is suppressed by 38 dB; this is due to the higher operating Q.

9.2.4 Three High-Performance HF VFOs

THE N1UL MODIFIED VACKAR VFO

The oscillator circuit of **Figure 9.14A** is contributed by Ulrich Rohde, N1UL. It is a modified Vackar design (see the sidebar) in which a small coupling capacitor (8 pF) and voltage divider capacitor (18 pF) isolate the resonator circuit (10 µH and 50 pF tuning capacitor) from the oscillator transistor.

The oscillator transistor is followed by a buffer stage to isolate the oscillator from the load. Because the coupling between the transistor base and the resonator is fixed and light, stability of the oscillator is high across a wide tuning range from 5.5 to 6.6 MHz. Either the inductor or capacitor may be varied to tune the oscillator, but a variable capacitor is recommended as more practical and gives better performance.

Because of the oscillator transistor's large capacitors from base to ground (220 pF) and collector to ground (680 pF), the various parameters of the oscillator transistor have little practical influence on circuit performance. The widely available 2N3904 performs well for both the oscillator and buffer transistors. The BC546 is a suitable substitute.

Practical resonator coil and the tuning capacitors will have a positive temperature coefficient. The 8 pF and the 18 pF capacitors should have an N150 temperature coefficient to partially compensate for their drift. After 1 hour, the observed frequency drift for this circuit was less than 10 Hz/hour.

THE K7HFD LOW-NOISE-DIFFERENTIAL OSCILLATOR

The other high performance oscillator example, shown in **Figure 9.15**, is designed for low-noise performance by Linley Gumm, K7HFD, and appears on page 126 of the ARRL's *Solid State Design for the Radio Amateur* (out of print, but available used and through libraries). This circuit uses no unusual components and looks simple, yet it is a subtle and sophisticated circuit. (An analysis and simulation of this circuit by its designer is included in the online material for this book.)

The effects of limiting in reducing AM oscillator noise were covered previously. However, because AM noise sidebands can get translated into PM noise sidebands by imperfect limiting, there is an advantage to stripping off the AM as early as possible, in

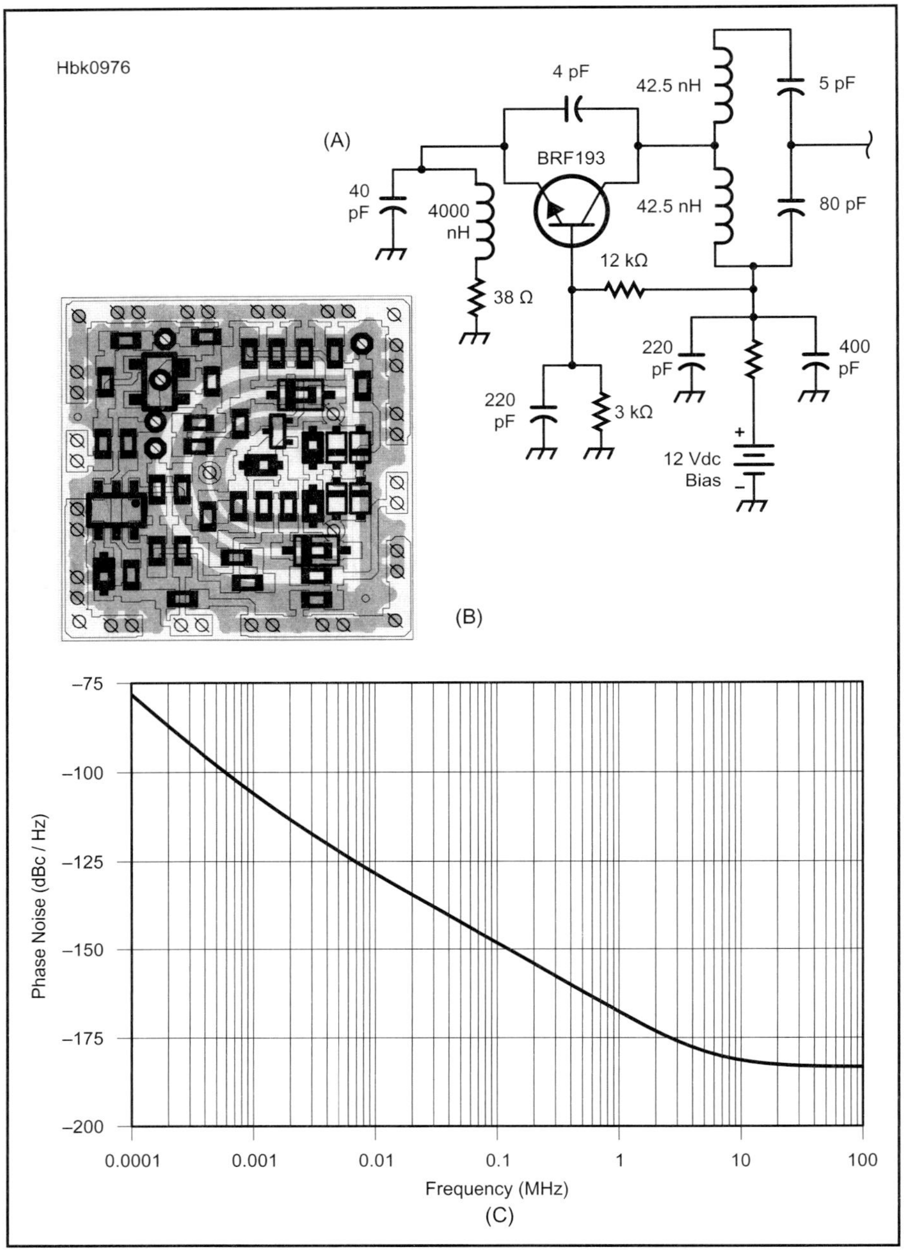

Figure 9.13 — 432 MHz oscillator circuit using tapped inductor and I_E = 30 mA (A), an example of the circuit layout using buried printed coupled line resonator (stripline resonator) in the middle layer of the board (B), and simulated phase noise plot (C).

the oscillator itself. An ALC system in the oscillator will counteract and cancel only the AM components within its bandwidth, but an oscillator based on a limiter will do this over a broad bandwidth. K7HFD's oscillator uses a differential pair of bipolar transistors as a limiting amplifier. The dc bias voltage at the bases and the resistor in the common emitter path to ground establishes a controlled dc bias current, here 25 to 27 mA. The ac voltage between the bases switches this current between the two collectors. This applies a rectangular pulse of current into link winding L2, which drives the series-resonant tank L1-C1.

The output impedance of the collector is high in both the current-on and current-off states. Along with the small number of turns of the link winding, this presents very high impedance to the tank circuit, which minimizes degradation of the tank Q. The input impedance of this limiter is also quite high and is applied across only a one-turn tap of L1, which similarly minimizes any impact on the tank Q. The input transistor base is driven into conduction only on one peak of the tank waveform. The output transformer has the inverse of the current pulse applied to it, so the output is not a low distortion sine wave, although the output harmonics will not be as extensive as simple theory would suggest because the circuit's high output impedance allows stray capacitances to attenuate high-frequency components. The low-frequency transistors used here also act to reduce the harmonic power.

With an output of +17 dBm, this is a power oscillator, running with nearly 300 mW of dc input power, so appreciable heating is present that can cause temperature-induced drift. The circuit's high-power operation is a deliberate ploy to create a high signal-to-noise ratio by having as high a signal power as possible. This also reduces the problem of the oscillator's broadband noise output. The limitation on the signal level in the tank is the transistors' base-emitter-junction breakdown voltage. The circuit runs with a few volts peak-to-peak across the one-turn tap, so the full tank is running at over 50 $V_{P\text{-}P}$.

Excessive voltage levels for the transistors can easily be generated by this circuit. The single easiest way to damage a bipolar transistor is to reverse bias the base-emitter junction until it avalanches. Most devices are rated to withstand only 5 V applied this way, the current needed to do damage is small, and very little power is needed. If the avalanche current is limited to less than that needed to perform immediate destruction of the transistor, it is likely that there will be some degradation of the device, a reduction in its bandwidth and gain along with an increase in its noise. These changes are irreversible and cumulative. Small, fast signal diodes have breakdown voltages of over 30 V and less capacitance than the transistor bases, so one possible experiment would be to try the effect of adding a fast signal diode in series with the base of each transistor and running the circuit at even higher levels.

The oscillation amplitude is controlled by the drive current limit. The voltage on L2 must never allow the collector of the transistor driving it to go into saturation, for if this happens the transistor presents very low impedance to L2 and badly loads the tank, wrecking the Q and the noise performance. The circuit can be checked to verify the margin from saturation by probing the hot end of L2 and the emitter with an oscilloscope. Another, less obvious, test is to vary the power-supply voltage and monitor the output power. While the circuit is under current control, there is very little change in output power, but if the supply is low enough to allow saturation, the output power will change significantly with varying supply voltage.

The use of the inexpensive, general-purpose 2N3904 is interesting, as it is not nor-

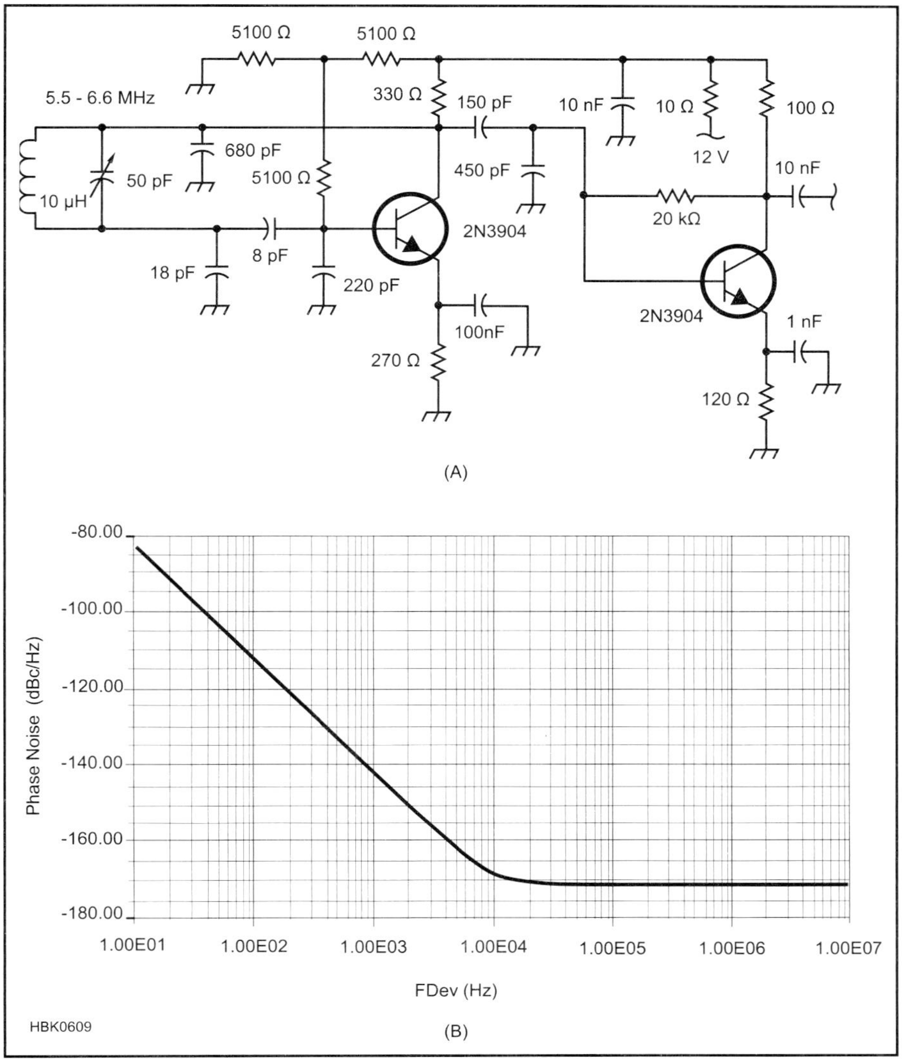

Figure 9.14 — At A, N1UL's Modified Vackar VFO is tuned from 5.5 to 6.6 MHz using the 50-pF capacitor. Tuning may be restricted to narrower ranges by placing a fixed capacitor in parallel with a smaller variable capacitor. The resonant frequency of the oscillator is determined by the 10 μH inductor and 50 pF tuning capacitor. B shows the excellent phase noise performance of the modified Vackar VFO in this *Harmonica* simulation. At 1 kHz from the carrier, noise is –144 dBc.

mally associated with RF oscillators, more often used at dc or audio frequencies. There is evidence that suggests some transistors that have good noise performance at RF have worse noise performance at low frequencies, and that the low-frequency noise they create can modulate an oscillator, creating noise sidebands. Experiments with low-noise audio transistors may be worthwhile, but many such devices have high junction capacitances.

In the description of this circuit in *Solid State Design for the Radio Amateur*, the results of a phase-noise test made using a spectrum analyzer with a crystal filter as a preselector are given. Ten kilohertz away from the carrier, in a 3 kHz measurement bandwidth, the noise was more than 120 dB below the carrier level. This translates into better than –120 – 10 log (3000), which equals –154.8 dBc/Hz, SSB, consistent with the modeled phase noise performance shown in Figure 9.15B. At this offset, –140 dBc is usually considered to be excellent. This VFO provides state-of-the art performance by today's standards — in a 1977 publication.

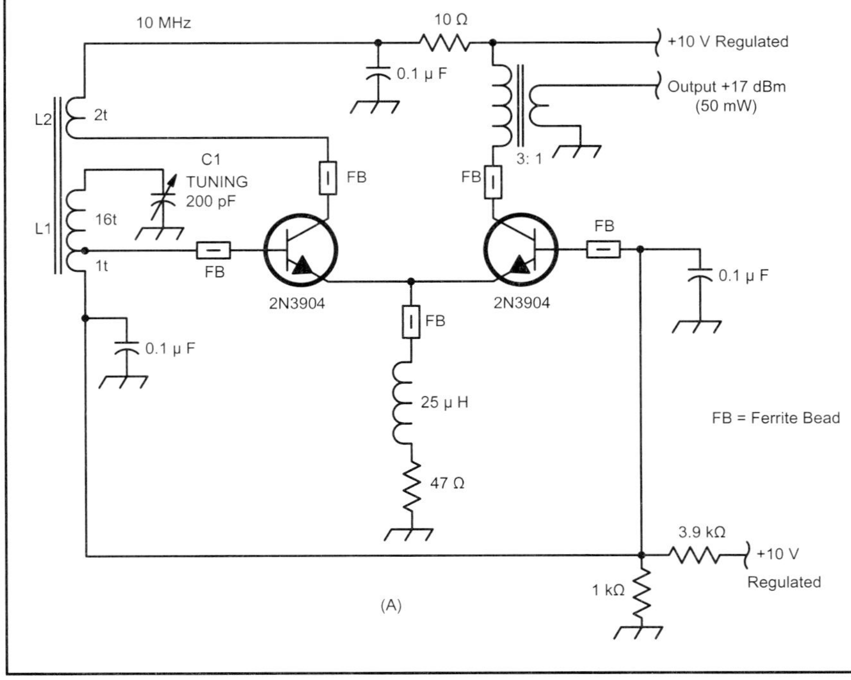

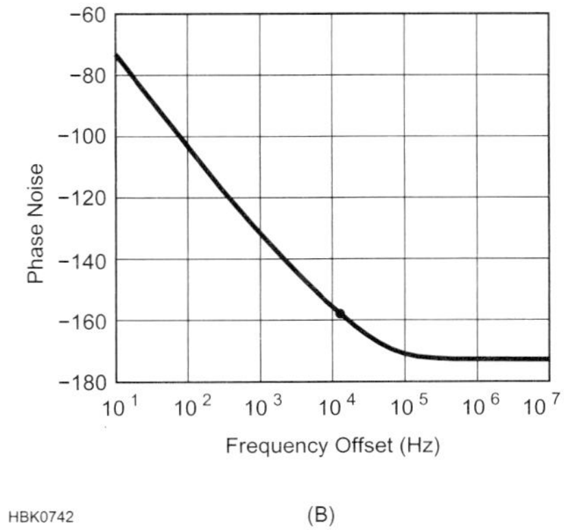

Figure 9.15 — At A, this low-noise oscillator design by K7HFD operates at an unusually high power level to achieve a high C/N (carrier-to-noise) ratio. L1 is 1.2 μH and uses 17 turns of wire on a T-68-6 toroid core. The tap is at 1 turn. Q at 10 MHz is 160. L2 is a 2-turn link over L1. At B, modeling of the differential oscillator by Ulrich Rohde, N1UL, shows its excellent phase-noise performance.

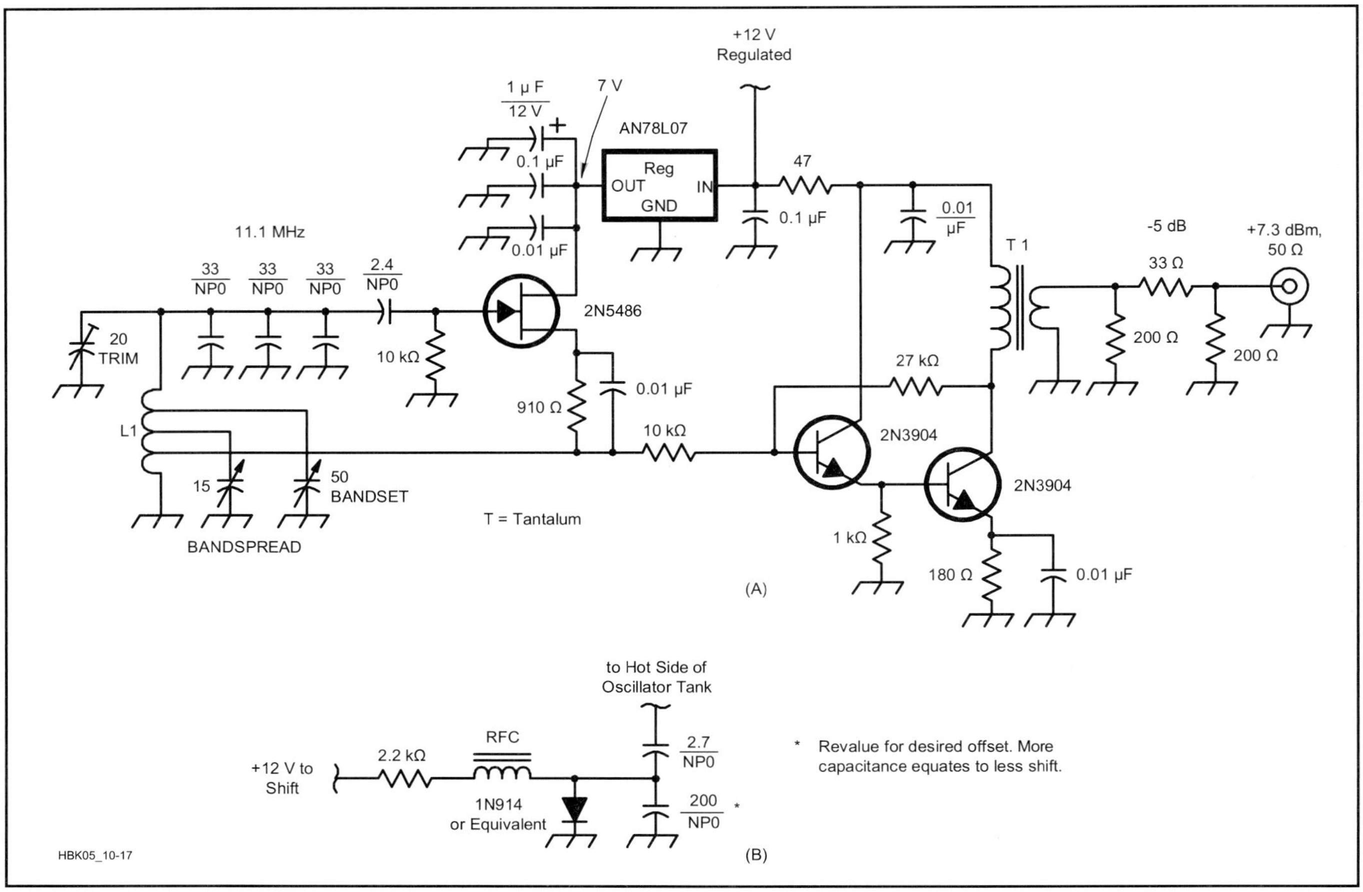

Figure 9.16 — Incorporating ideas from N1UL, KA7EXM, W7ZOI and W7EL, the oscillator at A achieves excellent stability and output at 11.1 MHz without the use of a gate-clamping diode, as well as end-running the shrinking availability of reduction drives through the use of bandset and bandspread capacitors. L1 consists of 10 turns of B & W #3041 Miniductor (#22 tinned wire, 5/8 inch in diameter, 24 turns per inch). The source tap is 2-1/2 turns above ground; the tuning-capacitor taps are positioned as necessary for bandset and bandspread ranges required. T1's primary consists of 15 turns of #28 enameled wire on an FT-37-72 ferrite core; its secondary, 3 turns over the primary. B shows a system for adding fixed TR offset that can be applied to any LC oscillator. The RF choke consists of 20 turns of #26 enameled wire on an FT-37-43 core.

A JFET HARTLEY VFO

Figure 9.16 shows an 11.1 MHz version of a VFO and buffer closely patterned after that used in 7 MHz transceiver designs published by Roger Hayward, KA7EXM, and Wes Hayward, W7ZOI ("The Ugly Weekender") and Roy Lewallen, W7EL ("The Optimized QRP Transceiver"). In it, a Hartley oscillator using a 2N5486 JFET drives the two-2N3904 buffer attributed to Lewallen. This version diverges from the originals in that its JFET uses source bias (the bypassed 910 Ω resistor) instead of a gate-clamping diode and is powered from a low-current 7 V regulator IC instead of a Zener diode and dropping resistor. The 5 dB pad sets the buffer's output to a level appropriate for "Level 7" (+7 dBm LO) diode ring mixers.

The circuit shown was originally built with a gate-clamping diode, no source bias and a 3 dB output pad. Adjusting the oscillator bias as shown increased its output by 2 dB without degrading its frequency stability (200 to 300 Hz drift at power up, stability within ±20 Hz thereafter at a constant room temperature).

In recognition that precision mechanical tuning components are hard to obtain, the resonator uses "Bandset" and "Bandspread" variable capacitors. These terms are from the early days of radio: *bandset* is for coarse-tuning and *bandspread* is for fine-tuning.

9.3 Building an Oscillator

We've covered a lot of ground about how oscillators work, their limitations and a number of interesting circuits, so the inevitable question arises of how to design one. Let's make an embarrassing confession right here: Very few oscillators you see in published circuits or commercial equipment were designed by the equipment's designer. Almost all have been adopted from other sources. While recycling in general is important for the environment, it means in this case that very few professional or amateur designers have ever designed an oscillator from scratch. We all have collections of circuits we've "harvested," and we adjust a few values or change a device type to produce something to suit a new project.

Oscillators aren't designed, they evolve. They seem to have a life of their own. The Clarke & Hess book listed in the references contains one of the few published classical design processes. The ARRL book *Experimental Methods in RF Design* contains extensive material on oscillator circuits that is well worth reading.

9.3.1 VFO Components and Construction

TUNING CAPACITORS AND REDUCTION DRIVES

As most commercially made radios now use DSP frequency synthesis, it has become increasingly difficult to find certain key components needed to construct a good VFO. Slow-motion drives and variable capacitors are available from *QST* advertiser National RF (**www.nationalrf.com**), Dan's Small Parts and Kits (**www.danssmallpartsandkits.net**), and Antique Electronic Supply (**www.tubesandmore.com**). Variable capacitors should be a high-quality component with double ball-bearings and silver-plated surfaces and contacts.

An alternate approach is also available: Scavenge suitable parts from old equipment; use tuning diodes instead of variable capacitors — an approach that, if uncorrected through phase locking, generally degrades stability and phase-noise performance; or use two tuning capacitors, one with a capacitance range ⅕ to ⅒ that of the other, in a bandset/bandspread approach.

Assembling a variable capacitor to a chassis and its reduction drive to a front panel can result in *backlash* — an annoying tuning effect in which rotating the capacitor shaft deforms the chassis and/or panel rather than tuning the capacitor. One way of minimizing this is to use the reduction drive to support the capacitor, and use the capacitor to support the oscillator circuit board, which is then attached to the chassis.

FIXED CAPACITORS

Use silvered-mica or other highly-stable capacitors for all fixed-value capacitors in the oscillator circuit. Power-supply decoupling capacitors may be any convenient type, such as ceramic or film.

Traditionally, silver-mica fixed capacitors have been used extensively in oscillators, but their temperature coefficient is not as low as can be achieved with other types, and some silver micas have been known to behave erratically. Polystyrene film has become a proven alternative. One warning is worth noting: polystyrene capacitors exhibit a permanent change in value should they ever be exposed to temperatures much above 70 °C; they do not return to their old value on cooling.

Particularly suitable for oscillator construction are the low-temperature-coefficient ceramic capacitors, often described as NP0 or C0G types. (NP0 and C0G are equivalent) These abbreviations are actually temperature-coefficient codes. **Figure 9.17** contains graphs showing the behavior of three common temperature coefficients. Some ceramic capacitors are available with deliberate, controlled temperature coefficients so that they can be used to compensate for other causes of frequency drift with temperature. For example, the code N750 denotes a part with a temperature coefficient of –750 parts per million per degree Celsius. These parts are now somewhat difficult to obtain, so other methods are needed. (Values for temperature coefficients and other attributes of capacitors are presented in the Component Information section of the **Construction Techniques** chapter.)

In a Colpitts circuit, the two large-value capacitors that form the voltage divider for the active device still need careful selection. It is tempting to use any available capacitor of the right value, because the effect of these components on the tank frequency is reduced by the proportions of the capacitance values in the circuit. This reduction is not as great as the difference between the temperature stability of an NP0 ceramic part and some of the low-cost, decoupling-quality X7R-dielectric ceramic capacitors. It's worth using low-temperature coefficient parts even in the seemingly less-critical parts of a VFO circuit — even for the bypass capacitors. Chasing the cause of temperature drift is more challenging than fun. Buy critical components like high-stability capacitors from trustworthy sources.

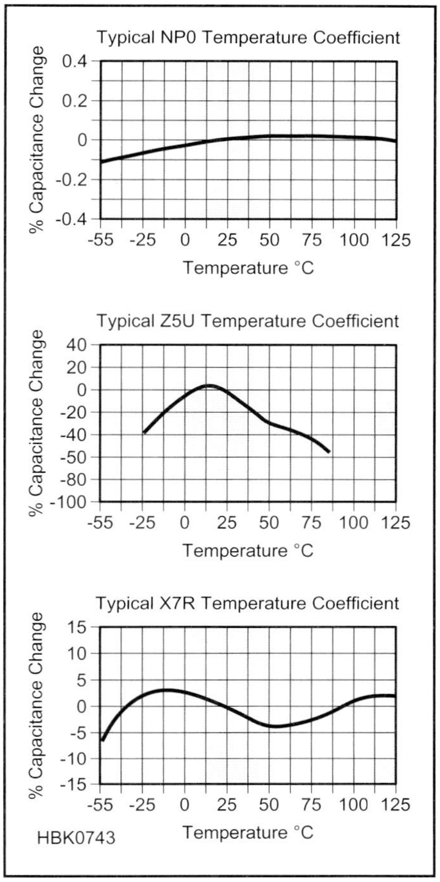

Figure 9.17 — EIA capacitor temperature coefficients specify change in capacitance with temperature. See the Component Data and References chapter for a complete table of temperature coefficient identifiers and characteristics.

TUNING CAPACITOR NETWORKS

Often an available variable capacitor has greater capacitance than required for a desired frequency range. While plates can sometimes be removed, a better solution embeds the variable capacitor in a network of fixed capacitors. The evolution of this network is shown in the middle section of **Figure 9.18**. The variable capacitor, C_V and C_2 are paralleled to form the equivalent C_{2V}. This is then placed in series with C_1 for the equivalent C_{12V}. This is, in turn, paralleled by C_3 to form the total capacitor, C_{NET}. The overall frequency is calculated from the usual resonance relationship. The equations are shown, with capacitance in farads, inductance in henrys, and frequency in Hz.

There is considerable flexibility available to the designer, afforded by picking C_1 and C_2 values. Some combinations with C_1 much smaller than the variable capacitor can produce highly nonlinear tuning.

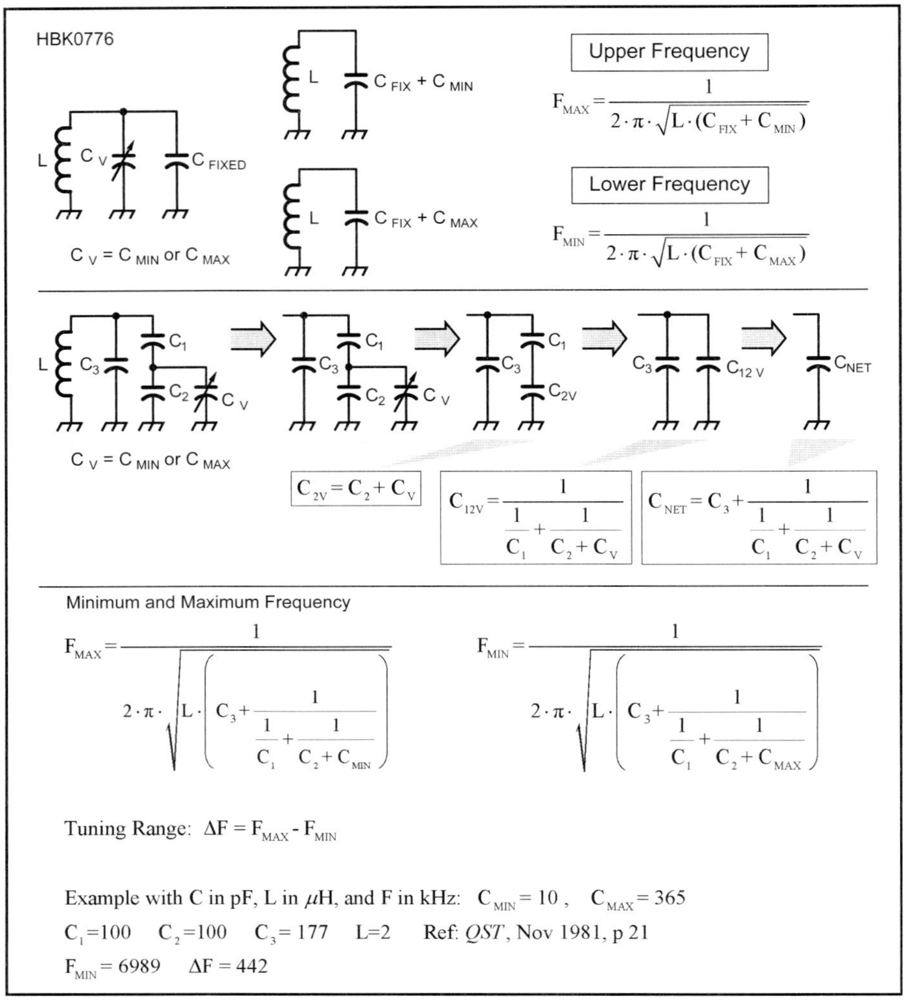

Figure 9.18 — A simple resonant circuit is tuned with parallel capacitors as shown in the top section. The tuning range is controlled by the ratio of the variable capacitance to the fixed capacitance.

INDUCTORS

Ceramic coil forms can give excellent results. If the required inductance is small enough, wind the coil on a ceramic form that is securely mounted. Self-supporting air-wound coils can also give good results if securely mounted and braced against vibration. B&W Miniductor (**www.bwantennas.com/coilcat.html** is the best-known commercially available product. It is not available to individuals from the manufacturer but there is a good selection available through **eBay.com**.

If you use a magnetic core, use powdered iron and support it securely. Do not use ferrite because of its temperature instability. Toroidal cores are preferred due to their self-shielding nature. Micrometals mix #6 has a low temperature coefficient and works well in conjunction with NP0 ceramic capacitors. Other materials have to be assessed on an individual basis.

A material's temperature stability will not be apparent until you try it in an oscillator, but you can apply a quick test to identify those nonmetallic materials that are lossy enough to spoil a coil's Q. Put a sample of the coil-form material into a microwave oven next to a glass of water and cook it about 10 seconds on low power. Do not include any metal fittings or ferromagnetic cores. Good materials will be completely unaffected; poor ones will heat and may even melt, smoke, or burst into flame. (This operation is a fire hazard if you try more than a tiny sample of an unknown material. Observe your experiment continuously and do not leave it unattended!)

W7ZOI suggests annealing toroidal VFO coils after winding. W7EL reports achieving success with this method by boiling his coils in water and letting them cool in air.

VOLTAGE REGULATORS

VFO circuits must be powered by well-regulated, low-noise supplies. Three-terminal regulators are inexpensive and have low output noise. It is easy to include them locally with the oscillator circuit. If the regulator is not part of the oscillator circuit, include a decoupling R-C combination to act as a noise filter.

Many oscillator circuits include Zener diode regulators. (See the **Power Sources** chapter.) Zener diodes have some idiosyncrasies that could spoil the oscillator. They are noisy, so decoupling is needed down to audio frequencies to filter this out. Zener diodes are often run at much less than their specified optimum bias current. Although this saves power, it results in a lower output voltage than intended and the diode's impedance is much greater, increasing its sensitivity to variations in input voltage, output current and temperature. Some common Zener types may be designed to run at as much as 20 mA; check the data sheet for your diode family to find the optimum current.

True Zener diodes are low-voltage devices; above a couple of volts, so-called Zener diodes are actually avalanche types. The temperature coefficient of these diodes depends on their breakdown voltage and crosses through zero for diodes rated at about 5 V. If you intend to use nothing fancier than a common-variety Zener, designing the oscillator to run from 5 V and using a 5.1 V Zener will give you a free advantage in voltage-versus-temperature stability.

There are some diodes available with especially low temperature coefficients, usually referred to as reference or temperature-compensated diodes. These usually consist of a series pair of diodes designed to cancel each other's temperature drift. The 1N821A diode has a temperature coefficient of ±100 ppm/°C. Running at 7.5 mA, the 1N829A provides 6.2 V ±5% and a temperature coefficient of just ±5 parts per million (ppm) maximum per degree Celsius. A change in bias current of 10% will shift the voltage less than 7.5 mV, but this increases rapidly for greater current variation. The curves in **Figure 9.19** show how the temperature coefficients of these diodes are dependent on bias current.

The LM399 is a complex IC that behaves like a superb Zener at 6.95 V, ±0.3 ppm/°C. Precision, low-power, three-terminal regulators are also available that are designed to be used as voltage references, some of which can provide enough current to run a VFO. There are comprehensive tables of all these devices on pages 614 and 615 of Horowitz and Hill, *The Art of Electronics, Third Edition*.

Over the last decade, specialized, low-noise, low dropout (LDO) regulators have been developed for oscillator circuits and are available from multiple manufacturers, including Analog Devices (including LTC and Maxim), Texas Instruments, and others. The noise performance of these parts is com-

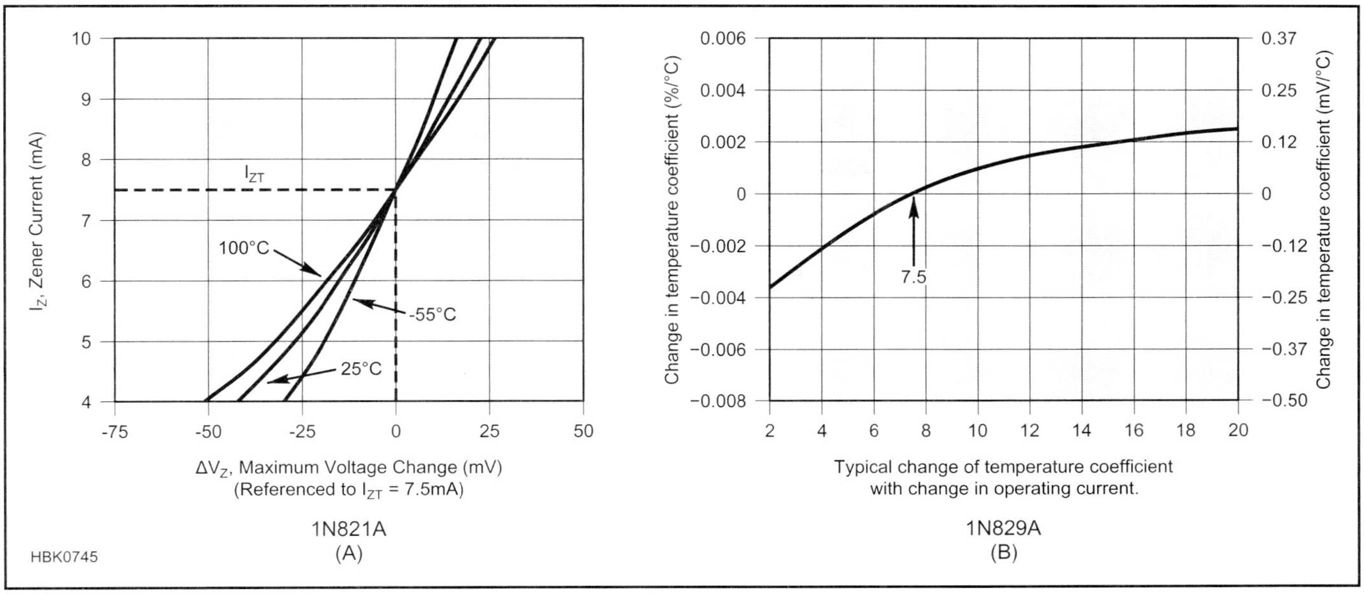

Figure 9.19 — The temperature coefficient of temperature-compensated diodes varies with bias current. To obtain the best temperature performance, use the specific bias current for the diode.

parable to that of batteries in bench tests comparing performance of a battery supply versus a supply using the low noise LDO. Texas Instruments' tutorial "LDO Noise Demystified – SLAA412B" (see **ti.com**) by Pithadia and Verma provides a good technical introduction to this product area. Note that you need a plot of noise versus bandwidth to make a useful decision on which part to use. Noise close to dc will modulate the VCO's output and be within the loop filter's bandwidth while noise at 100 kHz may be attenuated by the loop filter.

OSCILLATOR DEVICES

The 2N3819 FET, a classic from the 1960s, has proven to work well in VFOs but, like the MPF102, which is also long-popular with ham builders, it is manufactured to wide tolerances. Considering an oscillator's importance in receiver stability, you should not hesitate to spend a bit more on a better device. The 2N5484, 2N5485 and 2N5486 are worth considering; together, their transconductance ranges span that of the MPF102, but each is a better-controlled subset of that range. The 2N5245 is a more recent device with better-than-average noise performance that runs at low currents like the 2N3819. The 2N4416/A, also available as the plastic-cased PN4416, is a low-noise device, designed for VHF/UHF amplifier use, which has been featured in a number of good oscillators up to the VHF region. Its low internal capacitance contributes to low frequency drift. The J310 (plastic; the metal-cased U310 is similar) is another popular JFET for use in oscillators.

The 2N5179 (plastic, PN5179 or MPS5179) is a bipolar transistor capable of good performance in oscillators up to the top of the VHF region. Care is needed because its absolute-maximum collector-emitter voltage is only 12 V, and its collector current must not exceed 50 mA. Although these characteristics may seem to convey fragility, the 2N5179 is sufficient for circuits powered by stabilized 6 V power supplies.

Although discrete VCOs have been the mainstay of radio amateurs from the beginning of amateur radio; both the tubes and JFETs that were used to design oscillators are discontinued, or "no longer available" in industry language. Fortunately, Linear Systems (**www.linearsystems.com**) continues to make replacement JFETs for such VCO stalwarts as the MPF102 and U310 for fans of discrete oscillator designs from previous articles and books. Another supplier of JFETs, Interfet (**www.interfet.com**), still supplies the 2N3819.

VHF-UHF capable transistors are not really necessary in LC VFOs because such circuits are rarely used above 10 MHz. (High-bandwidth transistors also increase high-frequency harmonic content in the output signal.) Absolute frequency stability is progressively harder to achieve with increasing frequency, so free-running oscillators are used only rarely to generate VHF-UHF signals for radio communication. Instead, VHF-UHF radios usually use voltage-tuned, phase-locked oscillators in some form of synthesizer. Bipolar devices like the BFR90 and MRF901, with f_T in the 5 GHz region and mounted in stripline packages, are needed for successful oscillator design at UHF.

The popular SA/NE602 mixer IC has a built-in oscillator and can be found in many published circuits. This device has separate input and output pins to the tank and has proved to be quite tame. It may not have been "improved" yet (so far, it has progressed from the SA/NE602 to the SA/NE602A, the A version affording somewhat higher dynamic range than the original SA/NE602). It might be a good idea for anyone laying out a board using one to take a little extra care to keep PCB traces short in the oscillator section to build in some safety margin so that the board can be used reliably in the future. Professional designers know that their designs are going to be built for possibly more than 10 years and have learned to make allowances for the progressive improvement of semiconductor manufacture.

The number of sources for ICs has also shrunk due to consolidation in the semiconductor industry: As of early 2022, Analog Devices (**www.analog.com**) acquired Hittite, Linear Technology, and Maxim; Skyworks (**www.skyworksinc.com**) acquired Silicon Labs' clock products; and Texas Instruments (**www.ti.com**) acquired Burr-Brown and National Semiconductor; Hewlett Packard begat Agilent, which begat Avago (originally HP's semiconductor operation) and Keysight, and, finally, Avago acquired Broadcom and operates under that name, just to name a few. Fortunately, integrated solutions are mainstream products and new products are released on a regular basis.

MECHANICAL CONSTRUCTION

- All oscillator components should be clean and attached to a solid support to minimize thermal changes and mechanical vibration.
- The enclosure should be solid and isolated

from mechanical vibration.

• Keep component leads short and if point-to-point wiring is employed, use heavy wire (#16 to #18 AWG).

• Single-point grounding of the oscillator components is recommended to avoid stray inductance and to minimize noise introduced from other sources. If a PCB is used, include a ground plane.

It is often instructive to look at commercial or military equipment to see what techniques and materials are used for those demanding applications. The mechanical assemblies and parts can be removed from surplus equipment for home-built VFOs, as well.

9.3.2 Temperature Compensation

The general principle for creating a high-stability VFO is to use components with minimal temperature coefficients in circuits that are as insensitive as possible to changes in components' secondary characteristics. Even after careful minimization of the causes of temperature sensitivity, further improvement can still be desirable. The traditional method was to split one of the capacitors in the tank so that it could be apportioned between low-temperature-coefficient parts and parts with deliberate temperature dependency. Only a limited number of different, controlled temperature coefficients are available, so the proportioning between low coefficient and controlled coefficient parts was varied to "dilute" the temperature sensitivity of a part more sensitive than desired. This is a tedious process, involving much trial and error, an undertaking made more complicated by the difficulty of arranging means of heating and cooling the unit being compensated. (Hayward described such a means in December 1993 *QST*.) As commercial and military equipment have been based on frequency synthesizers for some time, supplies of capacitors with controlled temperature sensitivity are drying up. An alternative approach is needed.

A temperature-compensated crystal oscillator (TCXO) is an improved-stability version of a crystal oscillator that is used widely in industry. Instead of using controlled-temperature coefficient capacitors, most TCXOs use a network of thermistors and normal resistors to control the bias of a tuning diode. Manufacturers measure the temperature vs. frequency characteristic of sample oscillators, and then use a computer program to calculate the optimum normal resistor values for production. This can reliably achieve at least a tenfold improvement in stability. We here are not interested in mass manufacture, but the idea of a thermistor tuning a varactor is worth adopting. The parts involved are likely to be available for a long time.

Browsing through component suppliers' catalogs shows ready availability of 4.5 to 5 kΩ bead thermistors intended for temperature-compensation purposes, at less than a dollar each. **Figure 9.20** shows a circuit based on this form of temperature compensation. Commonly available thermistors have negative temperature coefficients, so as temperature rises, the voltage at the counterclockwise (CCW) end of R8 increases, while that at the clockwise (CW) end drops. Somewhere near the center there is no change. Increasing the voltage on the tuning diode decreases its capacitance, so settings toward R8's CCW end simulate a negative-temperature-coefficient capacitor; toward its clockwise end, a positive-temperature-coefficient part. Choose R1 to pass 8.5 mA from whatever supply voltage is available to the 6.2 V reference diode, D1. The 1N821A/1N829A-family diode used has a very low temperature coefficient and needs 7.5 mA bias for best performance; the bridge takes the other 1 mA. R7 and R8 should be good-quality multi-turn trimmers. D2 and C1 need to be chosen to suit the oscillator circuit. Choose the least capacitance that provides enough compensation range. This reduces the noise added to the oscillator. (It is possible, though tedious, to solve for the differential varactor voltage with respect to R2 and R5, via differential calculus and circuit theory. The equations in Hayward's 1993 article can then be modified to accommodate the additional capacitors formed by D2 and C1.) Use a single ground point near D2 to reduce the influence of ground currents from other circuits. Use good-quality metal-film components for the circuit's fixed resistors.

The novelty of this circuit is that it is designed to have an easy and direct adjustment process. The circuit requires two adjustments, one at each of two different temperatures, and achieving them requires a stable frequency counter that can be kept far enough from the radio so that the radio, not the counter, is subjected to the temperature extremes. (Using a receiver to listen to the oscillator under test can speed the adjustments.) After connecting the counter to the oscillator to be corrected, run the radio containing the oscillator and compensator in a room-temperature, draft-free environment until the oscillator's frequency reaches its stable operating temperature (rise over the ambient temperature). Lock its tuning, if possible. Adjust R7 to balance the bridge. This causes a drop of 0 V across R8, a condition you can reach by winding R8 back and forth across its range while slowly adjusting R7. When the bridge is balanced and 0 V appears across R8, adjusting R8 causes no frequency shift. When you've found this R7 setting, leave it there, set R8 to the exact center of its range and record the oscillator frequency.

Run the radio in a hot environment and allow its frequency to stabilize. Adjust R8 to restore the frequency to the recorded value. The sensitivity of the oscillator to temperature should now be significantly reduced between the temperatures at which you performed the adjustments. You will also have somewhat improved the oscillator's stability outside this range.

For best results with any temperature-compensation scheme, it's important to group all the oscillator and compensator components in the same enclosure, avoiding differ-

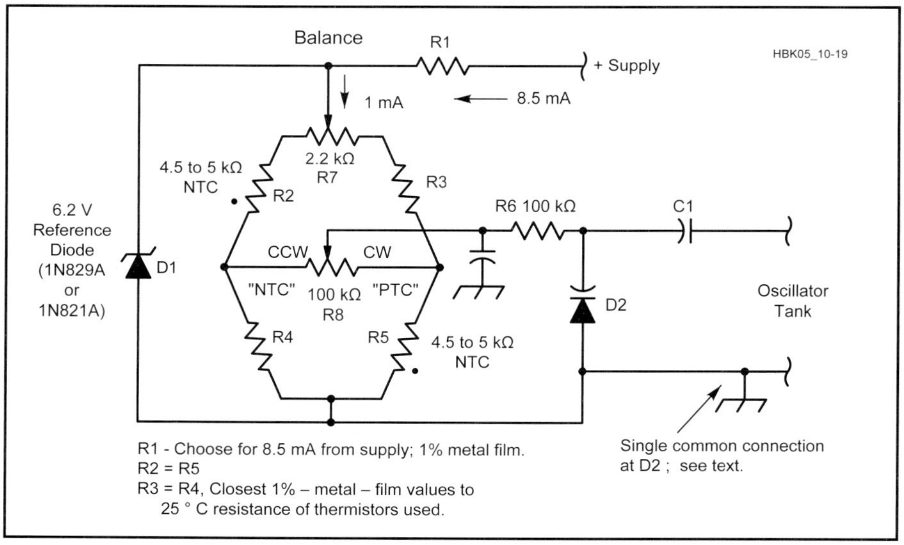

Figure 9.20 — Oscillator temperature compensation is difficult because of the scarcity of negative-temperature-coefficient capacitors. This circuit by GM4ZNX uses a bridge containing two identical thermistors to steer a tuning diode for drift correction. The 6.2 V Zener diode used (a 1N821A or 1N829A) must be a temperature-compensated part; just any 6.2 V Zener will not do.

ences in airflow over components. A good oscillator should not dissipate much power, so it's feasible, even advisable, to mount all of the oscillator components in an unventilated box. In the real world, temperatures change and if the components being compensated and the components doing the compensating have different thermal time constants, a change in temperature can cause a temporary change in frequency until the slower components have caught up. One cure for this is to build the oscillator in a thick-walled metal box that's slow to heat or cool, and so dominates and reduces the possible rate of change of temperature of the circuits inside. This is sometimes called a *cold oven*.

9.3.3 Shielding and Isolation

It is important to remember that any inductor acts as half of a transformer. Oscillators contain inductors running at moderate power levels and so can radiate strong enough signals to cause interference with other parts of a radio or with other radios. This is the tank (or other) inductor behaving as a transformer primary. Oscillators are also sensitive to radiated signals or other nearby varying magnetic fields. This is the tank (or other) inductor also behaving as a transformer secondary. Effective shielding is therefore vital.

Any oscillator is particularly sensitive to interference on the same or very nearby frequency. If this interference is strong enough an undesirable effect called *injection locking* will occur. The oscillator effectively stops oscillating and instead directly follows the interfering signal. For example, a VFO used to directly drive a power amplifier and antenna (to form a simple CW transmitter) can prove surprisingly difficult to shield well enough because of injection locking from any leakage of the power amplifier's high-level signal back into the oscillator. Even if injection locking does not fully kick in, the wrestling inside the VFO between its own oscillation and the interference can affect its frequency, resulting in an unstable transmitted signal. If the radio gear is in the antenna's near field, there are also strong fields that are coherent with the VFO oscillation, making sufficient shielding even more difficult.

The following rules of thumb continue to serve ham builders well:

• Use a complete metal box, with as few holes drilled in it as possible, with good contact around surface(s) where its lid(s) fit(s) on.

• Use feedthrough capacitors on power and control lines that pass in and out of the VFO enclosure and on the transmitter or transceiver enclosure as well.

• Use buffer amplifier circuitry that amplifies the signal by the desired amount and provides sufficient attenuation of signal energy flowing in the reverse direction. This is known as *reverse isolation* and is a frequently overlooked loophole in shielding. Figures 9.14 and 9.16 include buffer circuitry of proven performance. Another (and higher-cost) option is to consider using a high-speed buffer-amplifier IC (such as the LM6321N from Texas Instruments, a part that combines the high input impedance of an op amp with the ability to drive 50-Ω loads directly up into the VHF range).

• Use a mixing-based frequency-generation scheme instead of one that operates straight through or by means of multiplication. Such a system's oscillator stages can operate on frequencies with no direct frequency relationship to its output frequency. This essentially eliminates the possibility of injection locking the VFO.

• Use the time-tested technique of running your VFO at a sub-harmonic of the output signal desired — say, 3.5 MHz in a 7 MHz transmitter — and multiply its output frequency in a suitably nonlinear stage for further amplification at the desired frequency. This does reduce the tendency to injection lock.

9.4 Crystal Oscillators

Because crystals afford Q values and frequency stabilities that are orders of magnitude better than those achievable with LC circuits, fixed-frequency oscillators usually use quartz crystal resonators. Master references for frequency counters and synthesizers are always based on crystal oscillators.

So glowing is the crystal's reputation for stability that newcomers to radio experimentation naturally believe that the presence of a crystal in an oscillator will force oscillation at the frequency stamped on the can. This impression is usually revised after the first few experiences to the contrary! There is no sure-fire crystal oscillator circuit (although some are better than others); reading and experience soon provide a learner with plenty of anecdotes to the effect that:

• Some circuits have a reputation of being temperamental, even to the point of not always starting.

• Crystals sometimes mysteriously oscillate on unexpected frequencies.

Even crystal manufacturers have these problems, so don't be discouraged from building crystal oscillators. The occasional uncooperative oscillator is a nuisance, not a disaster, and it just needs a little individual attention. Knowing how a crystal behaves is the key to a cure.

Ulrich Rohde, N1UL, has generously contributed a pair of detailed papers that discuss the crystal oscillator along with several HF and VHF designs. Both papers, "Quartz Crystal Oscillator Design" and "A Novel Grounded Base Oscillator Design for VHF/UHF Frequencies" are included with the supplemental material accompanying this book.

9.4.1 Quartz and the Piezoelectric Effect

Quartz is a crystalline material with a regular atomic structure that can be distorted by the simple application of force. Remove the force, and the distorted structure springs back to its original form with very little energy loss. This property allows *acoustic waves* — sound — to propagate rapidly through quartz with very little attenuation, because the velocity of an acoustic wave depends on the elasticity and density (mass/volume) of the medium through which the wave travels.

If you heat a material, it expands. Heating may cause other characteristics of a material to change — such as elasticity, which affects the speed of sound in the material. In quartz, however, expansion and change in the speed of sound are very small and tend to cancel, which means that the transit time for sound to pass through a piece of quartz is very stable.

The third property of this wonder material is that it is *piezoelectric*. Apply an electric field to a piece of quartz and the crystal lattice distorts just as if a force had been applied. The electric field applies a force to electrical charges locked in the lattice structure. These charges are captive and cannot move around in the lattice as they can in a semiconductor, for quartz is an insulator. A capacitor's dielectric stores energy by creating physical distortion on an atomic or molecular scale. In a piezoelectric crystal's lattice, the distortion

An Oscillator for Crystal Testing

A test oscillator for determining the condition of a crystal and its actual frequency is quite useful. The circuit described in "Crystal Test Oscillators" by Fred Brown, W6HPH, will work with crystals from below 25 kHz to above 25 MHz in their fundamental mode. This article is available in the online material.

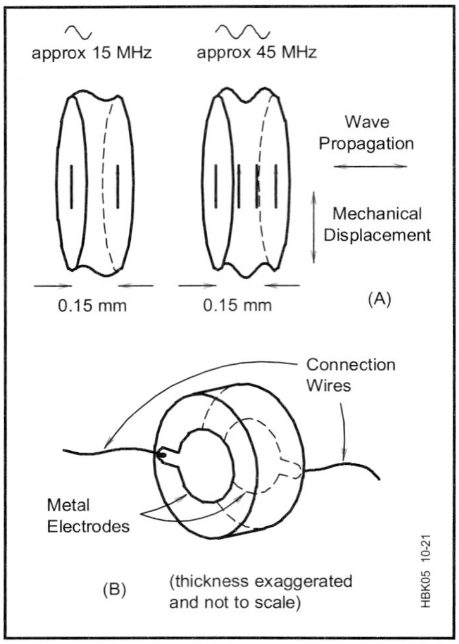

Figure 9.21 — Thickness-shear vibration at a crystal's fundamental and third overtone (A); B shows how the modern crystals commonly used by radio amateurs consist of etched quartz discs with electrodes deposited directly on the crystal surface.

affects the entire structure. In some piezoelectric materials, this effect is sufficiently pronounced that special shapes can be made that bend *visibly* when a field is applied.

Consider a rod made of quartz. Any sound wave propagating along it eventually hits an end, where there is a large and abrupt change in acoustic impedance. Just as when an RF wave hits the end of an unterminated transmission line, a strong reflection occurs. The rod's other end similarly reflects the wave. At some frequency, the phase shift of a round trip will be such that waves from successive round trips exactly coincide in phase and reinforce each other, dramatically increasing the wave's amplitude. This is *resonance*.

The passage of waves in opposite directions forms a standing wave with antinodes at the rod ends. Here we encounter a seeming ambiguity: not just one, but a family of different frequencies, causes standing waves — a family fitting the pattern of ½, ⅔, 5⁄2, 7⁄2 and so on, wavelengths into the length of the rod. And this is the case: A quartz rod can resonate at any and all of these frequencies.

The lowest of these frequencies, where the crystal is ½ wavelength long, is called the *fundamental* mode. The others are named the third, fifth, seventh and so on, *overtones*. There is a small phase-shift error during reflection at the ends, which causes the frequencies of the overtone modes to differ slightly from odd integer multiples of the fundamental. Thus, a crystal's third overtone is very close to, but not exactly, three times its fundamental frequency. Many people are confused by overtones and harmonics. Harmonics are additional signals at exact integer multiples of the fundamental frequency. Overtones are not signals at all; they are additional resonances that can be exploited if a circuit is configured to excite them.

The crystals we use most often resonate in the 1 to 30 MHz region and are of the *AT-cut, thickness shear* type, although these last two characteristics are rarely mentioned. A 15 MHz fundamental crystal of this type is about 0.15 mm thick. Because of the widespread use of pressure-mounted FT-243 crystals, you may think of crystals as small rectangles on the order of a half-inch in size. The crystals we commonly use today are discs, etched and/or doped to their final dimensions, with metal electrodes deposited directly on the quartz. A crystal's diameter does not directly affect its frequency; diameters of 8 to 15 mm are typical. (Quartz crystals are also discussed in the **Analog and Digital Filtering** chapter.)

AT-cut is one of a number of possible standard designations for the orientation at which a crystal disc is sawed from the original quartz crystal. The crystal lattice atomic structure is asymmetric, and the orientation of this with respect to the faces of the disc influences the crystal's performance. *Thickness shear* is one of a number of possible orientations of the crystal's mechanical vibration with respect to the disc. In this case, the crystal vibrates perpendicularly to its thickness. This is not easy to visualize, and diagrams don't help much, but **Figure 9.21** is an attempt at illustrating this. Place a moist bathroom sponge between the palms of your hands, move one hand up and down, and you'll see thickness shear in action.

There is a limit to how thin a disc can be made, given requirements of accuracy and price. Traditionally, fundamental-mode crystals have been made up to 20 MHz, although 30 MHz is now common at a moderately raised price. Using techniques pioneered in the semiconductor industry, crystals have been made with a central region etched down to a thin membrane, surrounded by a thick ring for robustness. This approach can push fundamental resonances to over 100 MHz, but these are more lab curiosities than parts for everyday use. The easy solution for higher frequencies is to use a manufacturably-thick crystal on an overtone mode. All crystals have multiple modes, so if you order a 28.060 MHz, third-overtone unit for a little QRP transmitter, you'll get a crystal with a fundamental resonance somewhere near 9.353333 MHz, but its manufacturer will have adjusted the thickness to plant the third overtone exactly on the ordered frequency. An accomplished manufacturer can do tricks with the flatness of the disc faces to make the wanted overtone mode a little more active and the other modes a little less active. (As some builders discover, however, this does not *guarantee* that the wanted mode is the most active!)

Quartz's piezoelectric property provides a simple way of driving the crystal electrically. Early crystals were placed between a pair of electrodes in a case. This gave amateurs the opportunity to buy surplus crystals, open them and grind them a little to reduce their thickness, thus moving them to higher frequencies. The frequency could be reduced very slightly by loading the face with extra mass, such as by blackening it with a soft pencil. Modern crystals have metal electrodes deposited directly onto their surfaces (Figure 9.21B), and such tricks no longer work.

The piezoelectric effect works both ways. Deformation of the crystal produces voltage across its electrodes, so the mechanical energy in the resonating crystal can also be extracted electrically by the same electrodes. Seen electrically, at the electrodes, the mechanical resonances look like electrical resonances. Their Q is very high. A Q of 10,000 would characterize a poor crystal today; 100,000 is often reached by high-quality parts. For comparison, a Q of over 200 for an LC tank is considered good.

9.4.2 Frequency Accuracy

A crystal's frequency accuracy is as outstanding as its Q. Several factors determine a crystal's frequency accuracy. First, the manufacturer makes parts with certain tolerances: ±200 ppm for a low-quality crystal for use as in a microprocessor clock oscillator, ±10 ppm for a good-quality part for professional radio use. Anything much better than this starts to get expensive! A crystal's resonant frequency is influenced by the impedance presented to its terminals, and manufacturers assume that once a crystal is brought within several parts per million of the nominal frequency, its user will perform fine adjustments electrically.

Second, a crystal ages after manufacture. Aging could give increasing or decreasing frequency; whichever, a given crystal usually keeps aging in the same direction. Aging is rapid at first and then slows down. Aging is influenced by the care in polishing the surface of the crystal (time and money) and by its holder style. The cheapest holder is a soldered-together, two-part metal can with glass bead insulation for the connection pins. Soldering debris lands on the crystal and affects its frequency. Alternatively, a two-part metal case can be made with flanges that are pressed together until they fuse, a process called *cold-welding*. This is much cleaner and improves aging rates roughly fivefold compared to soldered cans. An all-glass case can be made in two parts and fused together by heating in a

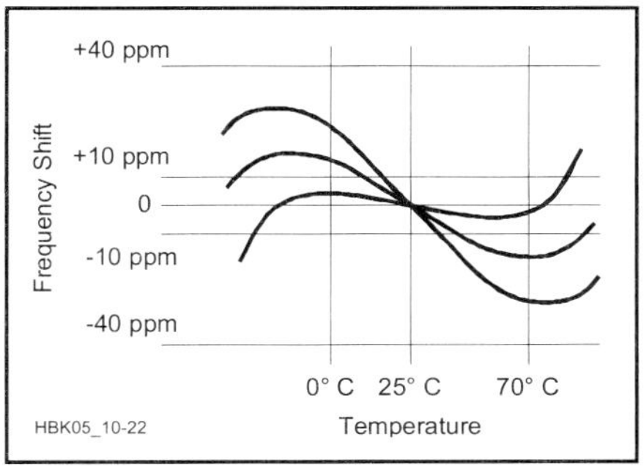

Figure 9.22 — Slight changes in a crystal cut's orientation shift its frequency-versus-temperature curve.

vacuum. The vacuum raises the Q, and the cleanliness results in aging that's roughly 10 times slower than that achievable with a soldered can. The best crystal holders borrow from vacuum-tube assembly processes and have a *getter*, a highly reactive chemical substance that traps remaining gas molecules, but such crystals are used only for special purposes.

Third, temperature influences a crystal. A reasonable, professional quality part might be specified to shift not more than ±10 ppm over 0 to 70 °C. An AT-cut crystal has an *S*-shaped frequency-versus-temperature characteristic, which can be varied by slightly changing the crystal cut's orientation. **Figure 9.22** shows the general shape and the effect of changing the cut angle by only a few seconds of arc. Notice how all the curves converge at 25 °C. This is because this temperature is normally chosen as the reference for specifying a crystal. The temperature stability specification sets how accurate the manufacturer must make the cut. Better stability may be needed for a crystal used as a receiver frequency standard, frequency counter clock and so on. A crystal's temperature characteristic shows a little hysteresis. In other words, there's a bit of offset to the curve depending on whether temperature is increasing or decreasing. This is usually of no consequence except in the highest-precision circuits.

It is the temperature of the quartz that is important, and as the usual holders for crystals all give effective thermal insulation, only a couple of milliwatts dissipation by the crystal itself can be tolerated before self-heating becomes troublesome. Because such heating occurs in the quartz itself and does not come from the surrounding environment, it defeats the effects of temperature compensators and ovens.

The techniques shown earlier for VFO temperature compensation can also be applied to crystal oscillators. An after-compensation drift of 1 ppm is routine and 0.5 ppm is good. The result is a *temperature-compensated crystal oscillator* (*TCXO*). Recently, oscillators have appeared with built-in digital thermometers, microprocessors and ROM look-up tables customized on a unit-by-unit basis to control a tuning diode via a digital-to-analog converter (DAC) for temperature compensation. These *digitally temperature-compensated oscillators* (*DTCXO*s) can reach 0.1 ppm over the temperature range. With automated production and adjustment, they promise to become the cheapest way to achieve this level of stability.

Oscillators have long been placed in temperature-controlled *ovens*, which are typically held at 80 °C. Stability of several parts per billion can be achieved over temperature, but this is a limited benefit as aging can easily dominate the accuracy. These are usually called *oven-controlled crystal oscillators* (*OCXOs*).

Fourth, the crystal is influenced by the impedance presented to it by the circuit in which it is used. This means that care is needed to make the rest of an oscillator circuit stable, in terms of impedance and phase shift.

Gravity can slightly affect crystal resonance. Turning an oscillator upside down usually produces a small frequency shift, usually much less than 1 ppm; turning the oscillator back over reverses this. This effect is quantified for the highest-quality reference oscillators.

9.4.3 The Equivalent Circuit of a Crystal

Because a crystal is a passive, two-terminal device, its electrical appearance is that of an impedance that varies with frequency. **Figure 9.23A** shows a very simplified sketch of the magnitude (phase is ignored) of the impedance of a quartz crystal. The general trend of dropping impedance with increasing frequency implies capacitance across the crystal. The sharp fall to a low value resembles a series-tuned tank, and the sharp peak resembles a parallel-tuned tank. These are referred to as series and parallel resonances. Figure 9.23B shows a simple circuit that will produce this impedance characteristic. The impedance looks purely resistive at the exact centers of both resonances, and the region between them has impedance increasing with frequency, which looks inductive.

C1 (sometimes called *motional capacitance*, C_m, to distinguish it from the lumped

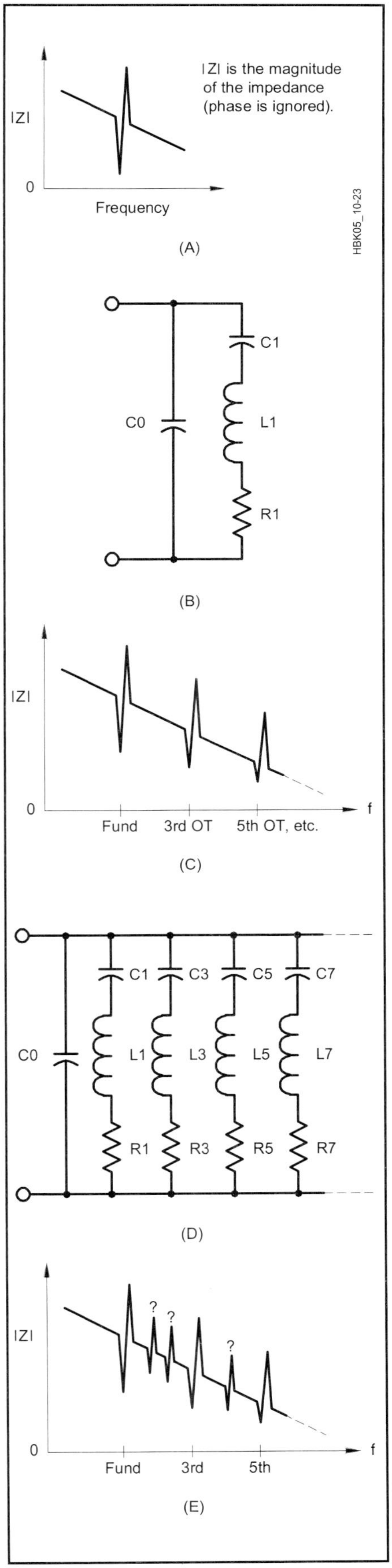

Figure 9.23 — Exploring a crystal's impedance (A) and equivalent circuit (B) through simplified diagrams. C and D extend the investigation to include overtones; E, to spurious responses not easily predictable by theory or controllable through manufacture. A crystal may oscillate on any of its resonances under the right conditions.

capacitance it approximates) and L1 (*motional inductance*, L_m) create the series resonance, and as C0 and R1 are both fairly small, the impedance at the bottom of the dip is very close to R1. At parallel resonance, L1 is resonating with C1 and C0 in series, hence the higher frequency. The impedance of the parallel tank is extremely high; the terminals are connected to a capacitive tap, which causes them to see only a small fraction of what is still a very large impedance. The overtones should not be neglected, so Figures 9.23C and 9.23D include them. Each overtone has series and parallel resonances and so appears as a series tank in the equivalent circuit. C0 again provides the shifted parallel resonance.

This is still simplified, because real-life crystals have a number of spurious, unwanted modes that add yet more resonances, as shown in Figure 9.23E. These are not well controlled and may vary a lot even between crystals made to the same specification. Crystal manufacturers work hard to suppress these spurs and have evolved a number of recipes for shaping crystals to minimize them. Just where they switch from one design to another varies from manufacturer to manufacturer.

Always remember that the equivalent circuit is just a representation of crystal behavior and does not represent circuit components actually present. Its only use is as an aid in designing and analyzing circuits using crystals. **Table 9.1** lists typical equivalent-circuit values for a variety of crystals. It is impossible to build a circuit with 0.026 to 0.0006 pF capacitors; such values would simply be swamped by strays. Similarly, the inductor must have a Q that is orders of magnitude better than is practically achievable, and impossibly low stray C in its winding.

The values given in Table 9.1 are nothing more than rough guides. A crystal's frequency is tightly specified, but this still allows inductance to be traded for capacitance. A good manufacturer could hold these characteristics within a ±25% band or could vary them over a 5:1 range by special design. Similarly marked parts from different sources vary widely in motional inductance and capacitance.

Quartz is not the only material that behaves in this way, but it is the best. Resonators can be made out of lithium tantalate and a group of similar materials that have lower Q, allowing them to be *pulled* over a larger frequency range in VXOs. Much more common, however, are ceramic resonators based on the technology of the well-known ceramic IF filters. These have much lower Q than quartz and much poorer frequency precision. They serve mainly as clock resonators for cheap microprocessor systems in which every last cent must be saved. A ceramic resonator could be used as the basis of a wide range, cheap VXO, but its frequency stability would not be as good as a good LC VFO.

Table 9.1
Typical Equivalent Circuit Values for a Variety of Crystals

Crystal Type	*Series L*	*Series C (pF)*	*Series R (Ω)*	*Shunt C (pF)*
1 MHz fundamental	3.5 H	0.007	340	3.0
10 MHz fundamental	9.8 mH	0.026	7	6.3
30 MHz third overtone	14.9 mH	0.0018	27	6.2
100 MHz fifth overtone	4.28 mH	0.0006	45	7.0

9.4.4 Crystal Oscillator Circuits

(See also the papers "What You Always Wanted to Know About Colpitts Oscillators," "Quartz Crystal Oscillator Design," "A Novel Grounded Base Oscillator Design for VHF/UHF Frequencies," and "Some Thoughts on Designing Very High Performance VHF Oscillators" by Ulrich Rohde, N1UL, included with the online material accompanying this book.)

Crystal oscillator circuits are usually categorized as series- or parallel-mode types, depending on whether the crystal's low- or high-impedance resonance comes into play at the operating frequency. The series mode is now the most common; parallel-mode operation was more often used with vacuum tubes. **Figure 9.24** shows a basic series-mode oscillator. Some people would say that it is an overtone circuit, used to run a crystal on one of its overtones, but this is not necessarily true. The tank (L-C1-C2) tunes the collector of the common-base amplifier. C1 is larger than C2, so the tank is tapped in a way that transforms to lower impedance, decreasing signal voltage, but increasing current. The current is fed back into the emitter via the crystal. The common-base stage provides a current gain of less than unity, so the transformer in the form of the tapped tank is essential to give loop gain. There are *two* tuned circuits, the obvious collector tank and the series-mode one "in" the crystal. The tank kills the amplifier's gain away from its tuned frequency, and the crystal will only pass current at the series resonant frequencies of its many modes. The tank resonance is much broader than any of the crystal's modes, so it can be thought of as the crystal setting the frequency, but the tank selecting which of the crystal's modes is active. The tank could be tuned to the crystal's fundamental, or one of its overtones.

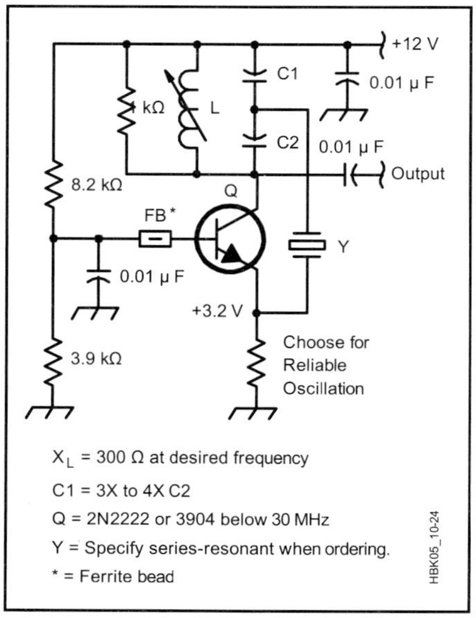

Figure 9.24 — A basic series-mode crystal oscillator. A 2N5179 can be used in this circuit if a lower supply voltage is used; see text.

Fundamental oscillators can be built without a tank quite successfully, but there is always the occasional one that starts up on an overtone or spurious mode. Some simple oscillators have been known to change modes while running (an effect triggered by changes in temperature or loading) or to not always start in the same mode! A series-mode oscillator should present low impedance to the crystal at the operating frequency. In Figure 9.24, the tapped collector tank presents a transformed fraction of the 1-kΩ collector load resistor to one end of the crystal, and the emitter presents a low impedance to the other. To build a practical oscillator from this circuit, choose an inductor with a reactance of about 300 Ω at the wanted frequency and calculate C1 in series with C2 to resonate with it. Choose C1 to be 3 to 4 times larger than C2. The amplifier's quiescent ("idling") current sets the gain and hence the operating level. This is not easily calculable, but can be found by experiment. Too little quiescent current and the oscillator will not start reliably; too much and the transistor can drive itself into saturation. If an oscilloscope is available, it can be used to check the collector waveform; otherwise, some form of RF voltmeter can be used to allow the collector voltage to be set to 2 to 3 V RMS. 3.3 kΩ would be a suitable starting point for the emitter bias resistor. The transistor type is not critical; 2N2222A or 2N3904 would be fine up to 30 MHz; a 2N5179 would allow operation as an overtone oscillator to over 100 MHz (because of the low collector voltage rating of the 2N5179, a supply voltage lower than 12 V is required). The ferrite bead on the base gives some protection against parasitic oscillation at UHF.

If the crystal is shorted, this circuit should still oscillate. This gives an easy way of adjusting the tank; it is even better to temporarily replace the crystal with a small-value (tens of ohms) resistor to simulate its *equivalent series resistance* (ESR), and adjust L until the circuit oscillates close to the wanted frequency. Then restore the crystal and set the quiescent current. If a lot of these oscillators were built, it would sometimes be necessary to adjust the current individually due to the different equivalent series resistance of individual crystals. One variant of this circuit has the emitter connected directly to the C1/C2 junction, while the crystal is a decoupler for the transistor base (the existing capacitor and ferrite bead not being used). This works, but with a greater risk of parasitic oscillation.

We commonly want to trim a crystal oscillator's frequency. While off-tuning the tank a little will pull the frequency slightly, too much detuning spoils the mode control and can stop oscillation (or worse, make the circuit unreliable). The answer to this is to add a trimmer capacitor, which will act as part of the equivalent series tuned circuit, in series with the crystal. This will shift the frequency in one way only, so the crystal frequency must be re-specified to allow the frequency to be varying around the required value. It is common to specify a crystal's frequency with a standard load (30 pF is commonly specified), so that the manufacturer grinds the crystal such that the series resonance of the specified mode is accurate when measured with a capacitor of this value in series. A 15 to 50 pF trimmer can be used in series with the crystal to give fine frequency adjustment. Too little capacitance can stop oscillation or prevent reliable starting. The Q of crystals is so high that marginal oscillators can take several seconds to start!

This circuit can be improved by driving the crystal's lower series-resonant impedance with an emitter follower as in **Figure 9.25**. This is the *Butler* oscillator. Again, the tank controls the mode to either force the wanted overtone or protect the fundamental mode. The tank need not be tapped because Q2 provides current gain, although the circuit is sometimes seen with C split, driving Q2 from a tap. The position between the emitters offers a good, low-impedance environment to keep the crystal's in-circuit Q high. R, in the emitter of Q1, is again selected to give reliable oscillation. The circuit has been shown with a capacitive load for the crystal, to suit a unit specified for a 30 pF load. An alternative circuit to give electrical fine tuning is also shown. The diodes across the tank act as limiters to stabilize the operating amplitude and limit the power dissipated in the crystal by clipping the drive voltage to Q2. The tank should be adjusted to peak at the operating frequency, not used to trim the frequency. The capacitance in series with the crystal is the proper frequency trimmer.

The Butler circuit works well, and has been used in critical applications to 140 MHz (seventh-overtone crystal, 2N5179 transistor). Although the component count is high, the extra parts are cheap ones. Increasing the capacitance in series with the crystal reduces the oscillation frequency but has a progressively diminishing effect. Decreasing the capacitance pulls the frequency higher, to a point at which oscillation stops; before this point is reached, start-up will become unreliable. The possible amount of adjustment, called *pulling range*, depends on the crystal; it can range from less than 10 to several hundred parts per million. Overtone crystals have much less pulling range than fundamental crystals on the same frequency; the reduction in pulling is roughly proportional to the square of the overtone number.

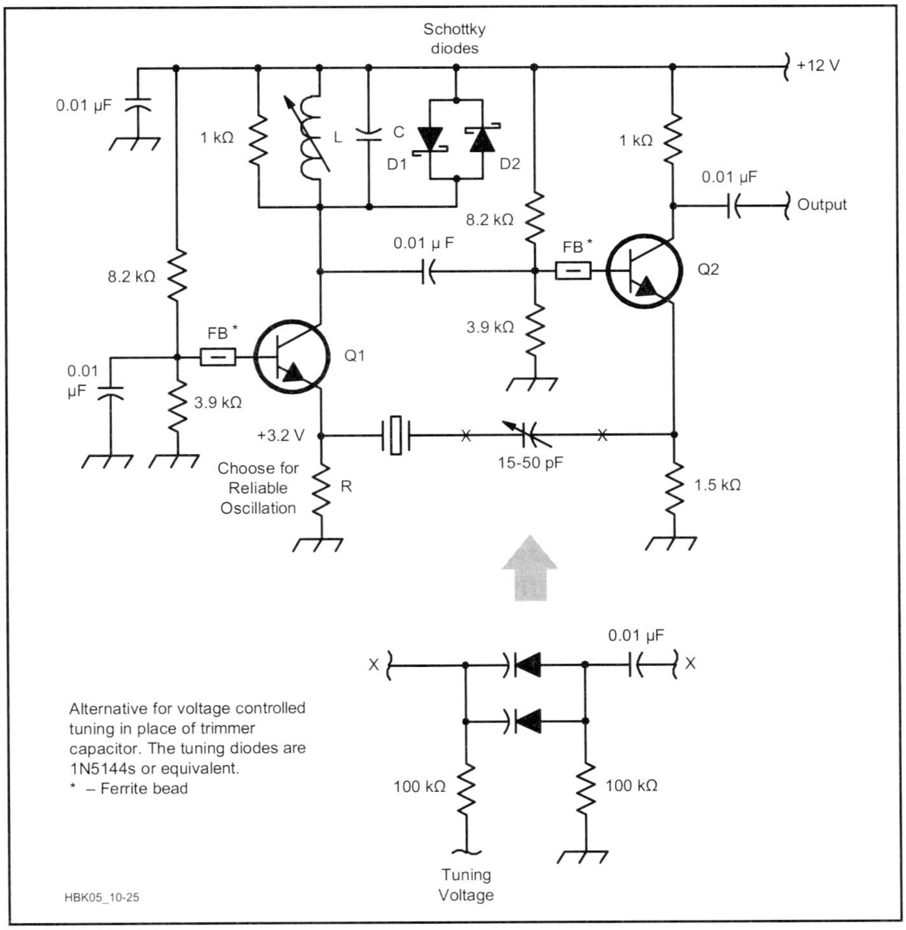

Figure 9.25 — A Butler crystal oscillator with Q2 connected as an emitter follower to drive the crystal's low series-resonant impedance.

LOW-NOISE CRYSTAL OSCILLATORS

Figure 9.26A shows a crystal operating in its series mode in a series-tuned Colpitts circuit. Because it does not include an LC tank to prevent operation on unwanted modes, this circuit is intended for fundamental mode operation only and relies on that mode being the most active. If the crystal is ordered for 30 pF loading, the frequency trimming capacitor can be adjusted to compensate for the loading of the capacitive divider of the Colpitts circuit. An unloaded crystal without a trimmer would operate slightly off the exact series resonant frequency in order to create an inductive impedance to resonate with the divider capacitors. Ulrich Rohde, N1UL, in Figure 4-47 of his book *Digital PLL Frequency Synthesizers — Theory and Design*, published an elegant alternative method of extracting an output signal from this type of circuit, shown in Figure 9.26B. This taps off a signal from the current in the crystal itself. This can be thought of as using the crystal as a band-pass filter for the oscillator output. The RF choke in the emitter keeps the emitter bias resistor from loading the tank and degrading the Q. In this case (3 MHz operation), it has

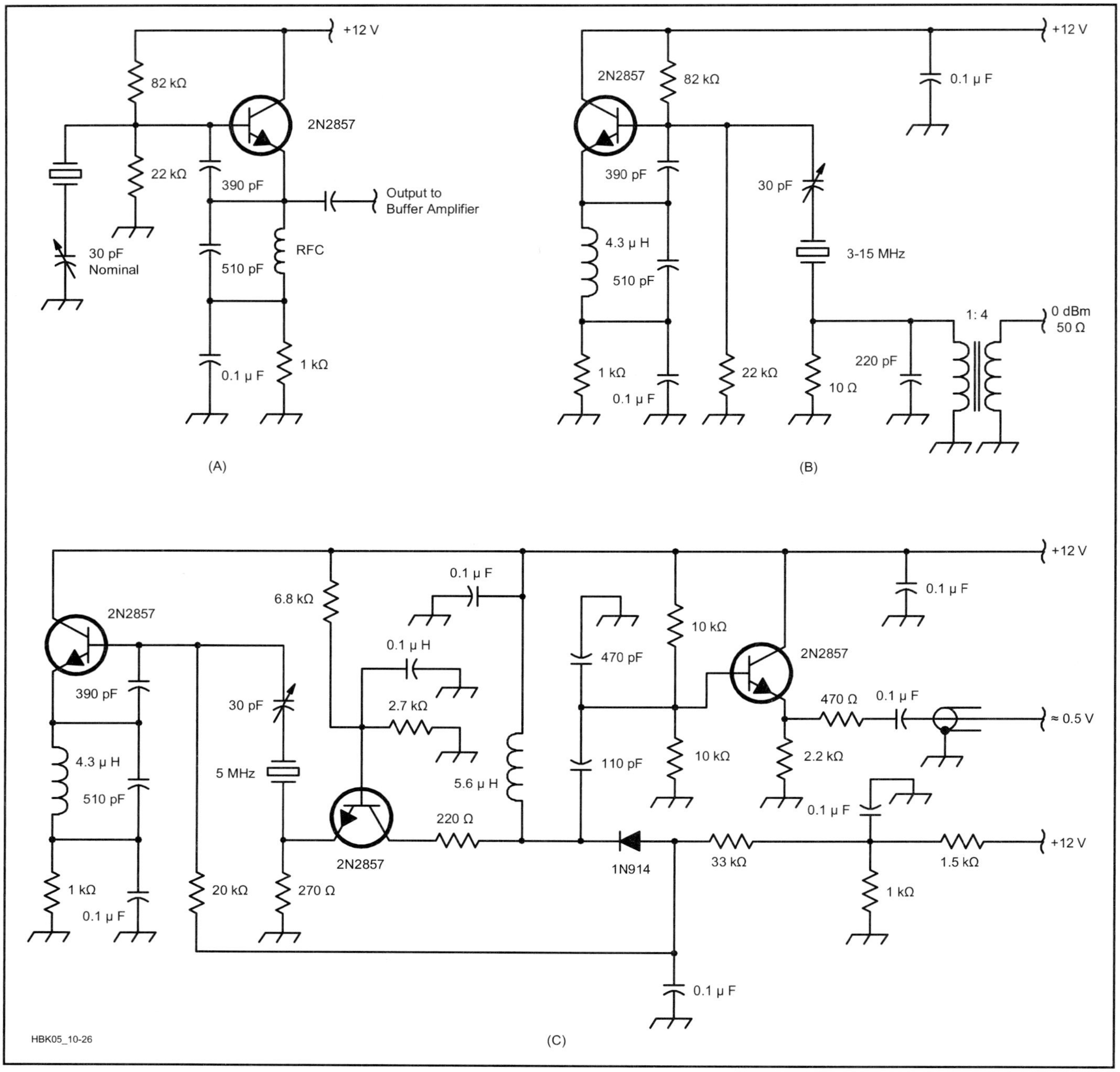

Figure 9.26 — The crystal in the series-tuned Colpitts oscillator at A operates in its series-resonant mode. B shows N1UL's low-noise version, which uses the crystal as a filter and features high harmonic suppression (from Rohde, *Microwave and Wireless Synthesizers Theory and Design*; see references). The circuit at C builds on the B version by adding a common-base output amplifier and ALC loop.

been chosen to resonate close to 3 MHz with the parallel capacitor (510 pF) as a means of forcing operation on the wanted mode. The 10-Ω resistor and the transformed load impedance will reduce the in-circuit Q of the crystal, so a further development substituted a common base amplifier for the resistor and transformer. This is shown in Figure 9.26C. The common-base amplifier is run at a large quiescent current to give a very low input impedance. Its collector is tuned to give an output with low harmonic content and an emitter follower is used to buffer this from the load. This oscillator sports a simple ALC system, in which the amplified and rectified signal is used to reduce the bias voltage on the oscillator transistor's base. This circuit is described as achieving a phase noise level of –168 dBc/Hz a few kilohertz out from the carrier. This may seem far beyond what may ever be needed, but frequency multiplication to high frequencies, whether by classic multipliers or by frequency synthesizers, multiplies the deviation of any FM/PM sidebands as well as the carrier frequency. This means that phase noise worsens by 20 dB for each tenfold multiplication of frequency. A clean crystal oscillator and a multiplier chain is still the best way of generating clean microwave signals for use with narrow-band modulation schemes.

It has already been mentioned that overtone crystals are much harder to pull than fundamental ones. This is another way of saying that overtone crystals are less influenced by their surrounding circuit, which is helpful in

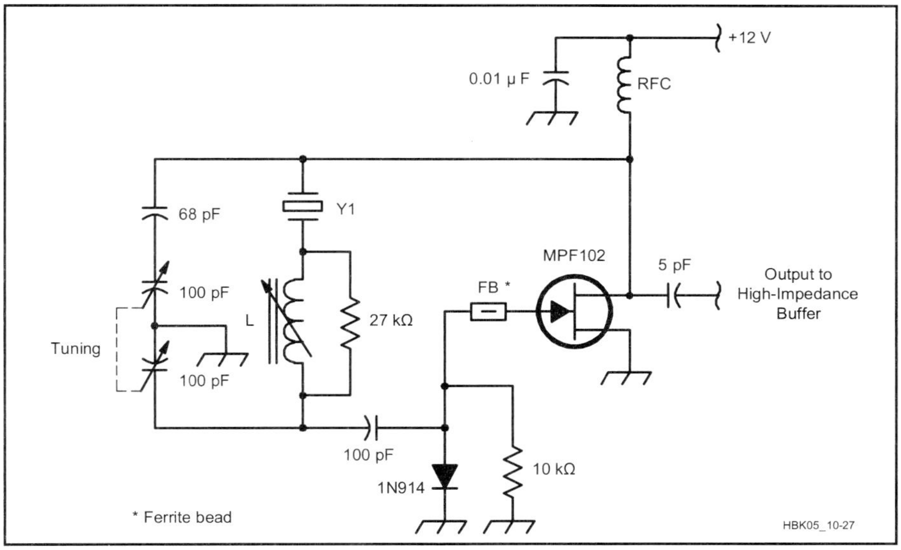

Figure 9.27 — A wide-range variable-crystal oscillator (VXO) by W7ZOI and W1FB. It was originally designed for use in low-power radios without the usual wide-range VFO.

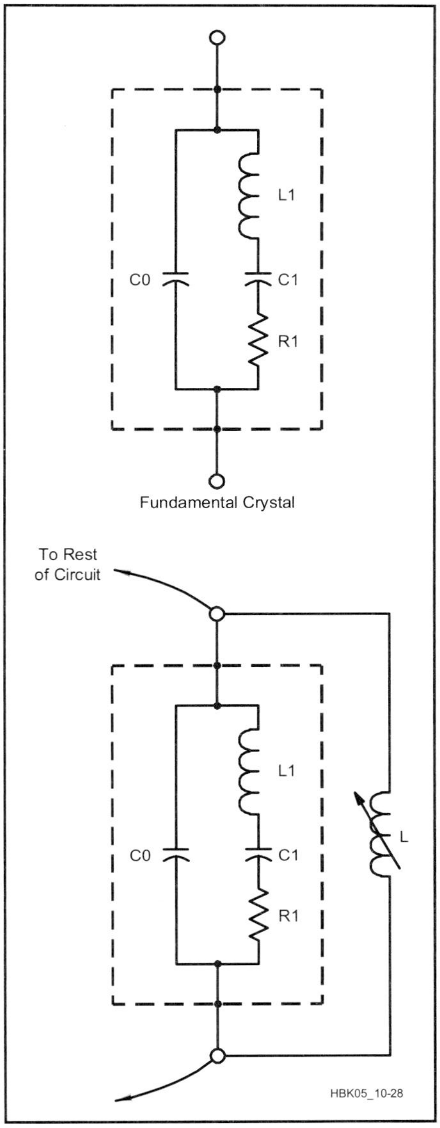

Figure 9.28 — Using an inductor to "tune out" C0 can increase a crystal oscillator's pulling range.

a frequency-standard oscillator like this one. Even though 5 MHz is in the main range of fundamental-mode crystals and this circuit will work well with them, an overtone crystal has been used. To further help stability, the power dissipated in the crystal is kept to about 50 μW. The common-base stage is effectively driven from a higher impedance than its own input impedance, under which conditions it gives a very low noise figure.

9.4.5 Variable-frequency Crystal Oscillators (VXOs)

Some crystal oscillators have frequency trimmers. If the trimmer is replaced by a variable capacitor as a front-panel control, we have a *variable crystal oscillator* (*VXO*): a crystal-based VFO with a narrow tuning range, but good stability and noise performance. VXOs are often used in small, simple QRP transmitters to tune a few kilohertz around common calling frequencies. Artful constructors, using optimized circuits and components, have achieved 1000-ppm tuning ranges. Poor-quality "soft" crystals are more pull-able than high-Q ones. Overtone crystals are not suited to VXOs. For frequencies beyond the usual limit for fundamental mode crystals, use a fundamental unit and frequency multipliers.

ICOM and Mizuho made some 2 meter SSB transceivers based on multiplied VXO local oscillators. This system is simple and can yield better performance than many expensive synthesized radios. SSB filters are available at 9 or 10.7 MHz, and yield sufficient image rejection with a single conversion. The choice of VXO frequency depends on whether the LO is to be above or below signal frequency and how much multiplication can be tolerated. Below 8 MHz, multiplier filtering is difficult. Above 15 MHz, the tuning range per crystal narrows. A 50-200 kHz range per crystal should work with a modern front-end design feeding a good 9 MHz IF, for a contest quality 2-meter SSB receiver.

The circuit in **Figure 9.27** is a JFET VXO from Wes Hayward, W7ZOI, and Doug DeMaw, W1FB, optimized for wide-range pulling. Published in *Solid State Design for the Radio Amateur*, many have been built and its ability to pull crystals as far as possible has been proven. Ulrich Rohde, N1UL, has shown that the diode arrangement as used here to make signal-dependent negative bias for the gate confers a phase-noise disadvantage, but oscillators like this that pull crystals as far as possible need any available means to stabilize their amplitude and aid start-up. In this case, the noise penalty is worth paying. This circuit can achieve a 2000-ppm tuning range with amenable crystals. If you have some overtone crystals in your junk box whose fundamental frequency is close to the wanted value, they are worth trying.

This sort of circuit doesn't necessarily stop pulling at the extremes of the possible tuning range; sometimes the range is set by the onset of such undesirable behavior as jumping mode or simply stopping oscillating. L was a 16 μH slug-tuned inductor for 10 MHz operation. It is important to minimize the stray and interwinding capacitance of L since this dilutes the range of impedance presented to the crystal.

One trick that can be used to aid the pulling range of oscillators is to tune out the C0 of the equivalent circuit with an added inductor. **Figure 9.28** shows how. L is chosen to resonate with C0 for the individual crystal, turning it into a high-impedance parallel-tuned circuit. The Q of this circuit is orders of magnitude less than the Q of the true series resonance of the crystal, so its tuning is much broader. The value of C0 is usually just a few picofarads, so L has to be a fairly large value considering the frequency at which it is resonated. This means that L has to have low stray capacitance or else it will self-resonate at a lower frequency. The tolerance on C0 and the variations of the stray C of the inductor means that individual adjustment is needed. This technique can also work wonders in crystal ladder filters.

9.4.6 Logic-Gate Crystal Oscillators

The frequency-determining network of an RC oscillator has a Q of less than one, which means that phase shift changes very slowly with different frequencies (see the prior section on RC Oscillators). The Pierce crystal oscillator discussed previously is a converted phase-shift oscillator, with the crystal taking the place of one series resistor. The crystal provides a much higher phase shift than an RC stage, so the crystal "controls" the frequency of oscillation.

The actual frequency of oscillation is the frequency at which the Barkhausen criteria are met. The crystal must operate near its series resonance in order for the loop gain to be unity. Oscillation then settles at the frequency where the crystal phase shift, added to the RC phase shifts, equals 180 degrees. The crystal usually provides between 45 and 60 degrees of phase shift, so the oscillation frequency is above the series resonance and below the parallel resonance of the crystal where the crystal behaves as a large inductor. (See the figure showing crystal response in the section Quartz Crystal Filters in the **Analog and Digital Filtering** chapter.)

The Pierce circuit is rarely seen in this full form. Instead, a cut-down version is the most common circuit in many microprocessors and other digital ICs that need a crystal-controlled clock. **Figure 9.29** shows this minimalist Pierce, using a CMOS logic inverter as the amplifier. R_{bias} provides dc negative feedback to bias the gate into its linear region. This value is not critical, anything between 100 kΩ and 10 MΩ works fine. The RC phase shifts needed to make this work come from R_{out}-C1 and $R_{crystal}$-C2. This circuit has a reputation of being temperamental, mainly because neither R_{out} nor $R_{crystal}$ are well documented or controlled in manufacturing. There is also a general belief that this circuit requires C1 = C2, which is not true.

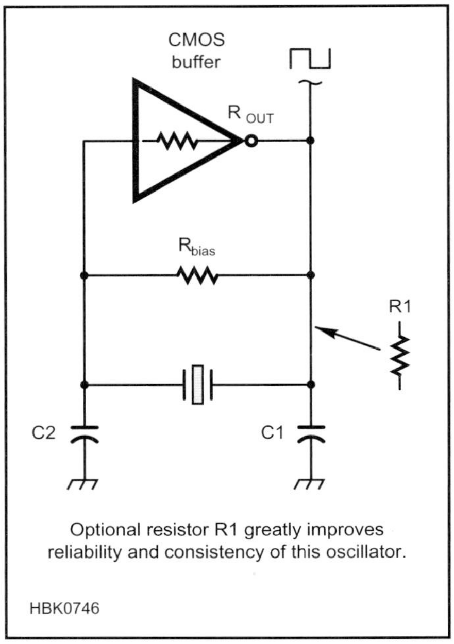

Figure 9.29 — The simplified Pierce oscillator using a logic-gate for the inverting amplifier. Adding R1 improves oscillator design and reliability.

A great improvement in performance is achieved by adding one resistor, R1, as shown in Figure 9.29. R_{bias} remains connected directly across the CMOS inverter. R1 is inserted in series from the gate output to the feedback network. The benefits are multiple:

- The edge speed of modern CMOS gates is extremely fast, so R1C1 eliminates the possibility of this fast edge exciting overtone operation.
- Phase shift into the crystal can be intentionally designed by R1 and C1 value selection.
- Drive from the inverter output is reduced into the crystal, possibly saving it from being damaged.
- Loop gain can now be controlled for best waveform and startup characteristics.

Design values are strongly dependent on the actual crystal frequency needed. C2 is selected first, to provide around 60 degrees of phase shift working against the crystal equivalent series resistance (usually a few tens of ohms, but the value needs to be verified!). The time constant R1 × C1 is usefully chosen to be the reciprocal of the crystal radian frequency ($\frac{1}{2}pf_{XTAL}$). Higher values of R1 reduce the loop gain and provide better protection for the crystal, until the loop gain gets too small and oscillation stops.

9.5 Oscillators at UHF and Above

The traditional way to make signals at higher frequencies is to make a signal at a lower frequency (where oscillators are easier) and multiply it up to the wanted range. Frequency multiplication is still one of the easiest and best ways of making a clean UHF/microwave signal. The design of a multiplier depends on whether the multiplication factor is an odd or even number. For odd multiplication, a Class-C biased amplifier can be used to create a series of harmonics; a filter selects the one wanted. For even multiplication factors, a full-wave-rectifier arrangement of diodes or other non-linear devices can be used to create a series of harmonics with strong even-order components, with a filter selecting the wanted component. At higher frequencies, diode-based passive circuits are commonly used. Oscillators using some of the LC circuits already described can, with care in construction, be used in the VHF range. At UHF, different approaches become necessary.

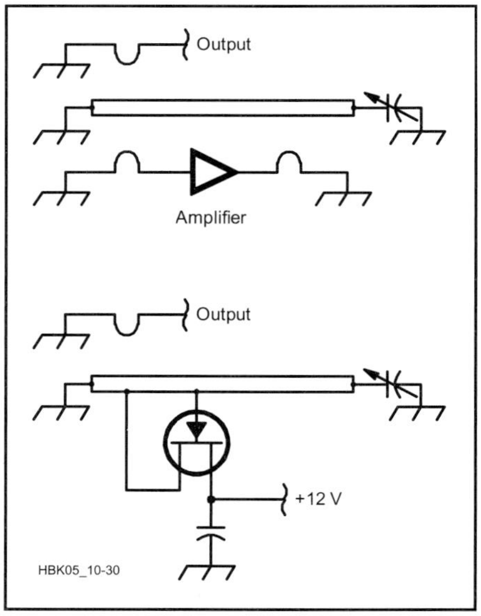

Figure 9.30 — Oscillators that use transmission-line segments as resonators. Such oscillators are more common than many of us may think, as Figure 9.31 reveals.

9.5.1 UHF Oscillators: Intentional and Accidental

The biggest change when working at UHF and microwave frequencies is that stray reactances and resistance dominate everything, making lumped circuit elements impractical. Success at these high frequencies requires making peace with the situation and developing a very good sense of where stray circuit elements reside in your layout and their approximate magnitude.

Figure 9.30 shows a pair of oscillators based on a resonant length of line, which is a distributed circuit element. The first one is a return to basics: a resonator, an amplifier and a pair of coupling loops. The amplifier can be a single bipolar or FET device or a monolithic microwave integrated circuit (MMIC) amplifier. The second circuit is really a Hartley oscillator; this one is a test oscillator for the 70 cm band made from a 10 cm length of wire suspended 10 mm over an unetched PC board as a ground plane, bent down, and soldered at one end, with a trimmer at the other end. The FET is a BF981 dual-gate device used as a source follower.

No free-running oscillator is really stable enough to be practical on these bands. Practical oscillators in this range are almost invariably tuned with tuning diodes controlled by phase-locked-loop synthesizers, which are themselves controlled by a crystal oscillator. This transfers the same stability of the crystal oscillator to the UHF oscillator.

There is one extremely common UHF oscillator that is almost always an undesired accident, being a very common form of spurious VHF/UHF oscillation in circuitry intended to process lower-frequency signals. **Figure 9.31A** shows the circuit in its simplest form. Analyzing this circuit using a comprehensive model of the UHF transistor reveals that the emitter presents an impedance that is small, resistive, and negative to the outside world. When this negative resistance is large enough to cancel the effective series resistance of a tank placed on the emitter, the circuit will oscillate.

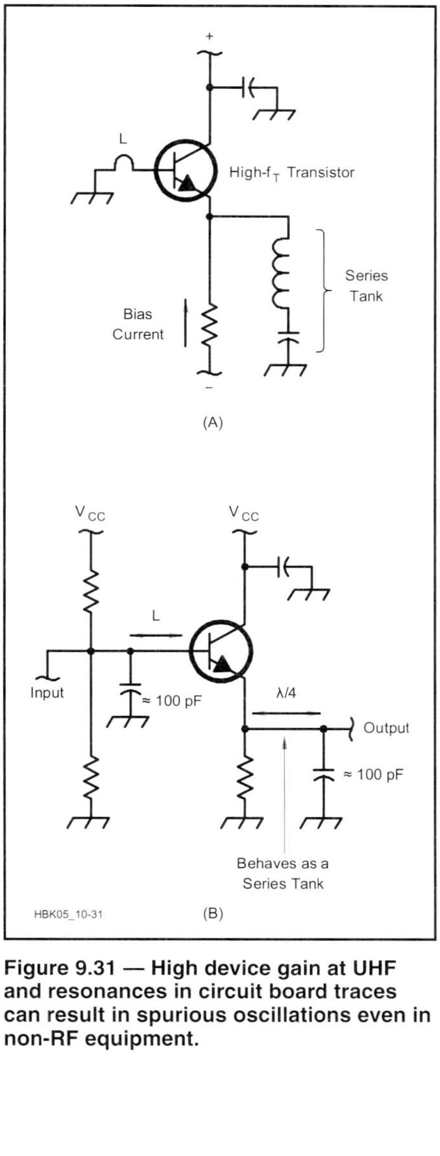

Figure 9.31 — High device gain at UHF and resonances in circuit board traces can result in spurious oscillations even in non-RF equipment.

At UHF, your schematic diagram will probably not show a tank circuit. But the high frequency transistor will know it is there, since it "sees" all of the stray reactances present in the layout. Figure 9.31B shows a very basic emitter-follower circuit with some capacitance to ground on both the input and output. If the capacitor shunting the input is a distance away from the transistor, the trace to the transistor's base looks like an inductor. The trace at the emitter of the transistor also looks like an inductor, and any nearby conductor will look like a capacitor to that trace (two conductors separated by a dielectric, which here is air and the PC board material). Any intentional capacitors add to this. If the transistor has gain at any frequency where the phasing from all of these stray reactances is right for oscillation (see the Barkhausen criteria in previous sections) then the circuit oscillates. This circuit is effectively the same as that in Figure 9.31A. This is a good reason to use the lowest-frequency transistor that you can for any application.

Stray reactances do not always have to cause headaches. Indeed, you can use them, knowing that they are there. The circuit of Figure 9.31A can be deliberately built as a useful wide-tuning oscillator covering say, 500 MHz to 1 GHz. This circuit is well-suited to construction with printed-circuit inductors. Common FR4 glass-epoxy board is lossy at these frequencies; better performance is achieved by using (the much more expensive) glass-Teflon board. If you can get surplus pieces of this type of material, it has many uses at UHF and microwave, but it is difficult to use as the adhesion between the copper and the substrate can be weak. A high-UHF transistor with a 5 GHz f_T such as the BFR90 is suitable; the base inductor can be 30 mm of 1 mm trace folded into a hairpin shape (inductance: less than 10 nH).

The upshot of this is that there is no longer any branch of electronics where RF design and layout techniques can be ignored safely. A circuit must not just be designed to do what it should do; it must also be designed so that it cannot do what it should not do.

If you have an accidental oscillator, there are three ways of taming such a circuit: adding a small resistor of perhaps 50 to 100 Ω in the collector lead close to the transistor, adding a similar resistor in the base lead, or by fitting a ferrite bead over the base lead under the transistor. Extra resistors can disturb dc conditions, depending on the circuit and its operating currents. Ferrite beads have the advantage that they can be easily added to existing equipment and have no effect at dc and low frequencies. Beware, however, that there are some electrically-conductive ferrite materials that can short transistor leads. If an HF oscillator uses beads to prevent any risk of spurs (such as shown in Figure 9.15), the beads should be anchored with a spot of adhesive to prevent movement which can cause small frequency shifts. Ferrite beads of Fair-Rite #43 material are especially suitable for this purpose; they are specified in terms of impedance, not inductance. Ferrites at frequencies above their inductive range become very lossy and can make a lead present a few tens of ohms of resistance.

9.5.2 Microwave Oscillators

Using conventional PC-board techniques with surface-mount components and extraordinarily careful layout allows the construction of circuits up to 4 GHz or so. Above this, commercial techniques and fancier materials become necessary unless we take the step to the ultimate distributed circuit — the cavity resonator.

Older than stripline techniques and far more amenable to home construction, cavity-based oscillators can give the highest possible performance at microwave frequencies. Air is a very low-loss dielectric with a dielectric constant of 1, so it gives high Q and does not force excessive miniaturization. **Figure 9.32** shows a series of structures used by G. R. Jessop, G6JP, to illustrate the evolution of a cavity from a tank made of lumped components. All cavities have a number of different modes of resonance, the orientation of the currents and fields are shown in **Figure 9.33**. The cavity can take different shapes, but the one shown here has proven to suppress

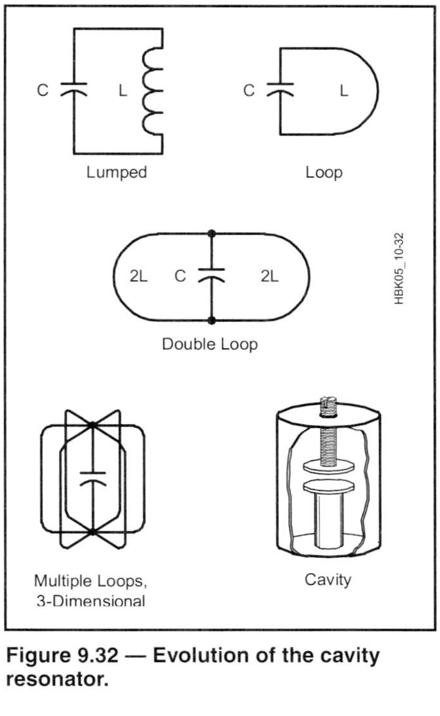

Figure 9.32 — Evolution of the cavity resonator.

unwanted modes well. The gap need not be central and is often right at the top. A screw can be fitted through the top, protruding into the gap, to adjust the frequency.

SEMICONDUCTOR CAVITY OSCILLATORS

To make an oscillator using a cavity, an amplifier is needed. Gunn and tunnel diodes have regions in their characteristics where their current falls with increasing bias voltage. This is negative resistance. If such a device is mounted in a loop in a cavity and bias is applied, the negative resistance can more than cancel the effective loss resistance of the cavity, causing oscillation. These diodes are capable of operating at extremely high frequencies and were discovered long before transistors were developed that had any gain at microwave frequencies.

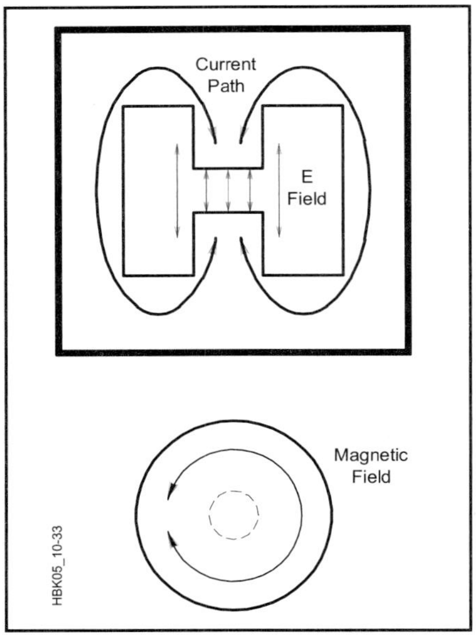

Figure 9.33 — Currents and fields in a cavity.

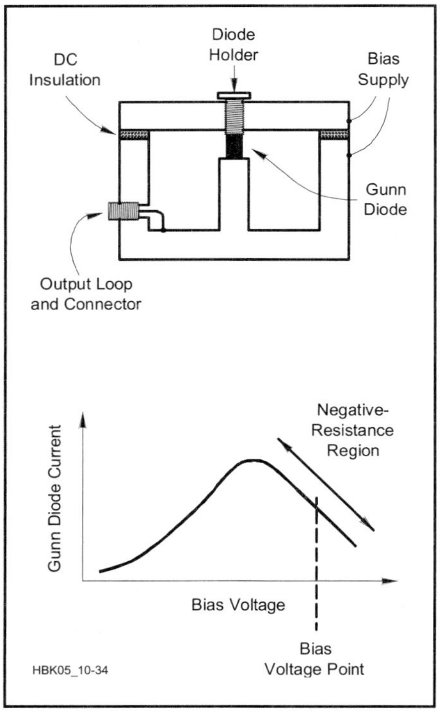

Figure 9.34 — A Gunn diode oscillator uses negative resistance and a cavity resonator to produce radio energy.

A Gunn-diode cavity oscillator is the basis of many of the Doppler radar modules used to detect traffic or intruders and of the Gunnplexer 10 GHz transceiver modules used by amateurs. **Figure 9.34** shows a common configuration. The coupling loop and coax output connector could be replaced with a simple aperture to couple into waveguide or a mixer cavity. **Figure 9.35** shows a transistor cavity oscillator version using a modern microwave transistor, which can be either a FET or bipolar device. The two coupling loops are electrically completed by the capacitance of the feedthrough capacitors.

DIELECTRIC-RESONATOR OSCILLATORS (DRO)

The dielectric-resonator oscillator (DRO) is a very common microwave oscillator, as it is used in the downconverter of satellite TV receivers. The dielectric resonator itself is a ceramic cylinder, like a miniature hockey puck, several millimeters in diameter. The ceramic has a very high dielectric constant, so the surface (where ceramic meets air in an abrupt dielectric mismatch) reflects electromagnetic waves and makes the ceramic body act as a resonant cavity. It is mounted on a substrate and coupled to the active device of the oscillator by a stripline that runs past it. At 10 GHz, a FET made of gallium arsenide (GaAsFET), rather than silicon, is normally used. The dielectric resonator elements are made at frequencies appropriate to mass applications like satellite TV. The setup charge to manufacture small quantities at special frequencies is likely to be prohibitive for the foreseeable future.

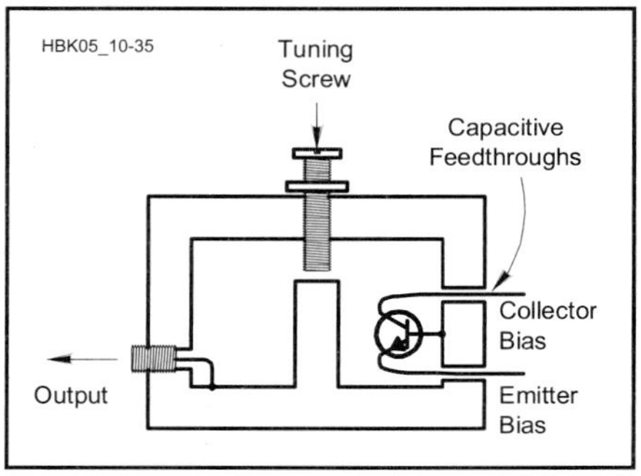

Figure 9.35 — A transistor can also directly excite a cavity resonator.

The challenge with these devices is to devise new ways of using oscillators on industry standard frequencies. Their chief attraction is their low cost in large quantities and compatibility with microwave stripline (microstrip) techniques. Frequency stability and Q are competitive with good cavities, but are inferior to that achievable with a crystal oscillator and chain of frequency multipliers.

YIG TUNED OSCILLATORS (YTO)

The yttrium-iron garnet (YIG) oscillator is a fundamental microwave source with a very wide tuning range and a linear tuning characteristic. Many YIG oscillators can be tuned over more than an octave and some tune more than five octaves. They are complete units that appear as heavy blocks of metal with low-frequency connections for power supplies and tuning, and an SMA connector for the RF output. The manufacturer's label usually states the tuning range and often the power supply voltages. This is very helpful because, with new units being very expensive, it is important to be able to identify the characteristics of surplus units. The majority of YTOs operate within the 2 to 18 GHz region, although units down to 500 MHz and up to 40 GHz are occasionally found. At microwave frequencies there is no octave-tunable device that can equal their signal cleanliness and stability. A typical stability for a 3 GHz YTO is less than 1 kHz per second drift. This may seem poor — until we realize that this is 0.33 ppm per second. Still, any YIG application involving narrow-band modulation will require some form of frequency stabilizer.

Nearly all surplus RF spectrum analyzers use YTOs as their first LO. For example, a 0 to 1500 MHz analyzer usually uses a 2 to 3.5 GHz YTO with a 2.0 GHz first IF. To understand the YTO tuning circuits, should there be need for troubleshooting and repair, a basic understanding of YIG oscillators definitely helps.

Figure 9.36 shows the construction of a YIG oscillator. A YTO is based on a YIG sphere that is carefully oriented within a coupling loop. This resonator is connected to a negative-resistance device and the whole assembly is placed between the poles of an electromagnet. Negative-resistance (Gunn) diodes were originally used in these oscillators, but transistor circuits are now essentially universal and use much less power. The support for the YIG sphere often contains a controlled heater to reduce temperature variation. YIG spheres are resonant at a frequency controlled not only by their physical dimensions, but also by any magnetic field around them. Hence the electromagnet: by varying the current through the electromagnet's windings, a controlled variable magnetic field is applied across the YIG sphere. This tunes the oscil-

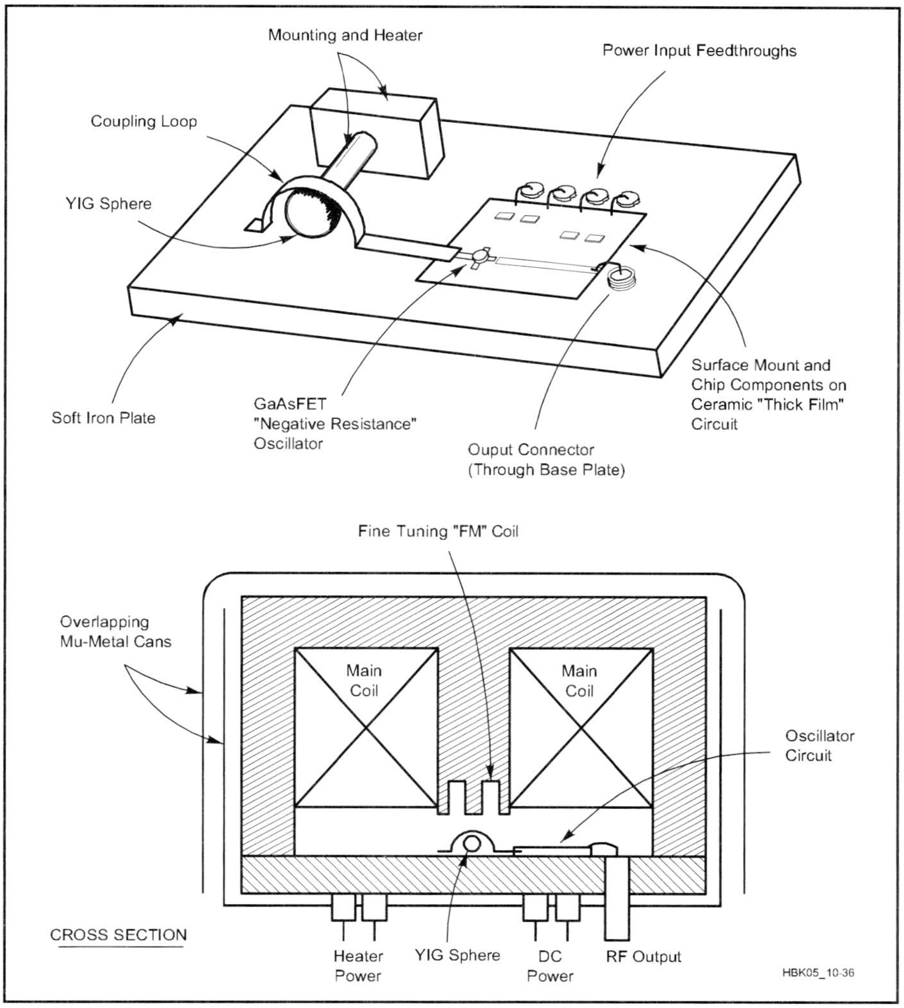

Figure 9.36 — A yttrium-iron-garnet (YIG) sphere serves as the resonator in the sweep oscillators used in many spectrum analyzers.

lator across a very wide range. The frequency/current relationship can have excellent linearity, typically around 20 MHz/mA. A YIG oscillator can be controlled by a PLL with the appropriate interface circuitry.

Magnetically-tuned oscillators bring some unique problems with them. The first problem is that magnetic fields, especially at low frequencies, are extremely difficult to shield. Therefore, YTO tuning is influenced by any local magnetic fields, causing frequency modulation. The YTO's magnetic core must be carefully designed to be all-enclosing in an attempt at self-shielding, and then one or more nested mu-metal cans are fitted around everything. It is still important to place the YTO away from obvious sources of magnetic fields, like power transformers. Cooling fans are also sources of fluctuating magnetic fields, with some fans generating fields 20 dB higher than from a well-designed 200 W 50/60-Hz transformer.

The second new problem is that the YTO's internal tuning coil needs significant current from the power supply to create the strong tuning field. This can be eased by adding a permanent magnet as a fixed "bias" field, but the bias shifts as the magnet ages. The main tuning coil still has many turns, and therefore high inductance (often more than 0.1 henrys). Large inductances require a high supply voltage for rapid tuning, with correspondingly high power consumption. The usual compromise is to have dual coils: One with many turns for slow tuning over a wide range, and a second coil with far fewer turns for fast tuning or FM over a limited range. This "FM coil" has a sensitivity around 1% to 2% of the main coil, perhaps 500 kHz/mA.

9.5.3 Klystrons, Magnetrons, and Traveling Wave Tubes

There are a number of thermionic (vacuum-tube) devices that are widely used as amplifiers or fundamental oscillators at microwave frequencies. Standard vacuum tubes (see the **RF Power Amplifiers** chapter for an introduction to vacuum tubes) work well for frequencies up to hundreds of megahertz. At frequencies higher than this, the amount of time that the electrons take to move between the cathode and the plate becomes a limiting factor. There are several special tubes designed to work at microwave frequencies, usually providing more power than can be obtained from solid-state devices.

Two of the following tubes (klystrons and traveling wave tubes) use the principle of *velocity-modulation* to extract RF energy from an electron beam. The general principles of velocity modulation and basic properties of devices using it are presented in the on-line tutorial **www.radartutorial.eu/druck/Book5.pdf**. Additional resources are described below.

THE KLYSTRON

The klystron tube uses the principle of velocity modulation of the electrons to avoid transit time limitations. The beam of electrons travels down a metal drift tube that has interaction gaps along its sides. RF voltages are applied to the gaps and the electric fields that they generate accelerate or decelerate the passing electrons. The relative positions of the electrons shift due to their changing velocities, causing the electron density of the beam to vary. This variation of electron beam density is used to perform amplification or oscillation.

Klystron tubes can be relatively large, and they can easily provide hundreds of watts to hundreds of kilowatts of microwave power. These power levels are useful for UHF broadcasting and particle accelerators, for example. Unfortunately, klystrons have relatively narrow bandwidths, and may not be re-tunable for operation on amateur frequencies. A video titled "How a Klystron Tube Works" can be found online on YouTube.com and a detailed history and tutorial is available at **www.slac.stanford.edu/cgi-bin/getdoc/slac-pub-7731.pdf**.

THE MAGNETRON

The magnetron tube is an efficient oscillator for microwave frequencies. Magnetrons are most commonly found in microwave ovens and high-powered radar equipment. The anode of a magnetron is made up of a number of coupled resonant cavities that surround the cathode. The applied magnetic field causes the electrons to rotate around the cathode and the energy that they give off as they approach the anode adds to the RF electric field. The RF power is obtained from the anode through a vacuum window.

Magnetrons are self-oscillating with the frequency determined by the construction of their anodes; however, they can be tuned by coupling either inductance or capacitance to the resonant anode. The range of frequencies depends on how fast the tuning must be accomplished. The tube may be tuned slowly over a range of approximately 10% of the center frequency. If faster tuning is necessary, such as is required for frequency modulation, the range decreases to about 5%.

Excellent drawings showing how magnetrons work are available at **hyperphysics.phy-astr.gsu.edu/hbase/waves/magnetron.html** and a thorough introduction for the interested reader is can be downloaded from **www.cpii.com/docs/related/2/Mag%20tech%20art.pdf**.

THE TRAVELING WAVE TUBE

A third type of tube operating in the microwave range is the traveling wave tube (TWT). For wide-band amplifiers in the microwave range this is the tube of choice. Either permanent magnets or electromagnets are used to focus the beam of electrons that travels through the TWT. This electron beam passes through a helical slow-wave structure, in which electrons are accelerated or decelerated, providing density modulation due to the applied RF signal, similar to that in the klystron. The modulated electron beam induces voltages in the helix that provides an amplified output signal whose gain is proportional to the length of the slow-wave structure. After the RF energy is extracted from the electron beam by the helix, the electrons are collected and recycled to the cathode. Traveling wave tubes can often be operated outside their designed frequencies by carefully optimizing the beam voltage.

9.6 Frequency Synthesizers

Like many of our modern technologies, the origins of frequency synthesis can be traced back to WWII. The driving force was the desire for stable, rapidly switchable, and accurate frequency control technology to meet the demands of narrow-band, frequency-agile HF communications systems without resorting to large banks of switched crystals. Early synthesizers were cumbersome and expensive, and therefore their use was limited to only the most sophisticated communications systems. With the help of the same technologies that have taken computers from room-sized to now fitting into the palms of our hands, frequency synthesis techniques have become one of the most enabling technologies in modern communications equipment.

Every communications device manufactured today, be it a handheld transceiver, cell phone, pager, AM/FM entertainment radio, scanner, television, HF communications equipment, or test equipment, contains a frequency synthesizer. Synthesis is the technology that allows an easy interface with both computers and microprocessor controllers. It provides amateurs with many desirable features, such as the feel of an analog knob with precision 10-Hz frequency increments, wideband accuracy and stability determined by a single precision crystal oscillator, frequency memories, and continuously variable precision frequency splits. Now reduced in size to small integrated circuits, frequency synthesizers have long replaced the cumbersome chains of frequency multipliers and filters in VHF, UHF, and microwave equipment, giving rise to the highly portable communications devices we use today. Frequency synthesis has also had a major impact in lowering the cost of modern equipment, particularly by reducing manufacturing complexity.

Frequency synthesizers are categorized in two general types: *direct synthesizers*, where the output signal is the result of operations directly performed on the input signal; and *indirect synthesizers*, where selected characteristics of the input signal are transferred onto a separately generated output signal. One defining feature of any direct synthesizer is that no feedback is used, so there is never any dynamic stability problem. Indirect synthesizers always include feedback control loops, so dynamic stability is a major design concern.

Direct synthesizers include the major techniques of *direct analog synthesis* (DAS) and *direct digital synthesis* (DDS). Direct analog synthesizers consist primarily of frequency multipliers, frequency dividers, mixers, and filters. DAS is very useful for generating small numbers of signals, with widely spaced frequencies, from a reference oscillator. DAS is particularly useful when more than one output signal is required at the same time.

Direct digital synthesizers are essentially dedicated microprocessors that have one function, to create an output signal waveform given a desired frequency (in digital form) using the applied reference signal. Unlike DAS, the output frequencies from a DDS can easily be separated by millihertz (0.001 Hz) while keeping all of the stability of the reference oscillator. Both being direct techniques, neither DAS nor DDS use feedback control loops, so switching from one frequency to another happens in nanoseconds.

Indirect synthesizers include the major techniques of frequency-locked loops (FLL) and phase-locked loops (PLL). The whole idea behind any indirect synthesizer is to transfer characteristics of one signal onto another separate signal. An FLL transfers only frequency characteristics of the input signal onto the output signal. Major FLL applications used by radio amateurs include *automatic frequency control* (AFC) loops and tone decoders. PLLs are more precise because not only frequency characteristics of the reference oscillator, but also its phase characteristics, are transferred to the oscillator generating the output signal.

There are two main application classes in which PLLs find wide use. The first is as a *frequency generator*, where we call it a *frequency synthesizer*. In transmitters this synthesizer may also include modulation, particularly for FM and FSK signals. The second PLL application class is as an angle (frequency or phase) demodulator, where we call it a *tracking demodulator* or *synchronizer*. Tracking demodulators used in deep space communication and clock recovery used in digital communication are major applications today.

It is curious to note that using modern digital circuitry, it actually is easier to make a PLL than an FLL. This turns out to be fortuitous. This section will focus on the PLL used as a frequency synthesizer. For readers interested in PLL use as a demodulator or as a synchronizer, there are many textbooks available that discuss these applications in great detail.

9.6.1 Digital Frequency Synthesis

A digital process is defined to be quantized in both value and time. (See the **DSP and SDR Fundamentals** chapter for background on digital signals and systems.) This is important to remember because there are sampled signal processes that are not digital. One is the sample-and-hold, which is discrete in time but analog in value. The opposite is the phase-frequency detector used in phase-locked loops, which is discrete in value but analog in time. Digital frequency synthesis is a true digital process.

The origins of digital frequency synthesis go back to 1972, even before the widespread use of microprocessors. In essence, these techniques directly calculate the desired output waveform, including any modulations. When no feedback is used these are called Direct Digital Synthesis (DDS), and sometimes you also see the name Direct Digital Frequency Synthesis (DDFS) used. The DDS name is older, though DDFS is more specific. Both names survive, but here we will use the original, DDS.

DDS has many attractive properties and two major drawbacks. Its attractions include very fine frequency resolution (microhertz is practical), essentially instant switching speed, no frequency drift, and absolute frequency set-ability. If modulation is included, this also happens with digital precision. Direct control of frequency, phase shift, and amplitude are available and independent of each other. For frequency and phase shift, this is new. Analog oscillators do not have the ability to independently address frequency and phase.

DDS drawbacks are also significant. Being a digital process, DDS is a sampled data system and so must adhere to the Nyquist criterion for waveform reconstruction as explained in the **DSP and SDR Fundamentals** chapter. This effectively means that the output frequency from a DDS is restricted to being less than half of the DDS clock frequency. The practical upper limit is around 35 – 40% of the clock frequency. Interestingly, there is no effective lower limit. Any DDS can tune all the way down to zero hertz (dc) since 0 is a legitimate tuning command.

The second drawback is more operationally serious. We demand a high-quality sine wave from our oscillators and DDS is no exception. But as a sampled data system, the DDS output is a sampled sine wave. Sampling a sine wave without error requires infinite resolution due to the properties of the sinusoid. We do not have infinite resolution available to our designs, so some errors creep in as the signal is constructed. These errors manifest themselves as non-harmonic spurious output signals, a.k.a. "spurs," which can be a huge problem. Much DDS design spends a lot of time dealing with spurs.

DDS PRINCIPLES

DDS is based on the simple counter structure of an accumulator, as shown in **Figure 9.37**. We choose to interpret the digital outputs from this counter as signal phase, and commonly represent the states as points around a circle, which means that one cycle through the accumulator states represents one output cycle from this synthesizer. The figure shows an example of the 16 digital states from a 4-bit accumulator. Action of this accumulator when its input number M is set to 3 is shown in (C), showing how the "wraparound" from one cycle to the next continues when the adder overflows. It is common to have the state progression from one output cycle to the next be a different state sequence. What is vitally important is that the jump size from one state to the next be exactly the same. This sets the phase change to be:

$$\frac{\Delta\theta}{\Delta t} = \frac{\Delta \text{ PHASE}}{1 \text{ CLOCK CYCLE}} = M$$

Frequency is simply how fast phase changes with time, so this accumulator input M is our direct tuning control. For an N-bit accumulator then the DDS frequency control is given by:

$$f_{DDS} = \frac{f_{CLOCK}}{2^N} \times M$$

Note that this relationship is perfectly linear. Also note that f_{DDS} is quantized, because the tuning number M must be an integer. There are some frequencies that a DDS cannot generate. But 0 Hz at M = 0 is legitimate.

One important property of DDS, which makes it particularly useful, is that N can be a reasonably large number. As examples, for N = 24 the denominator 2^N = 16,777,216. N = 32 is common, and then 2^N = 4,294,967,296. This means that the DDS frequency resolution, if a 100 MHz clock frequency is used, is 5.96 Hz when N = 24, and 0.023 Hz when N = 32. This is extremely fine frequency resolution and it is as accurate as the clock frequency. The only way that the DDS output frequency can drift is if the clock frequency drifts. Because everything else is digital, that cannot drift.

However, we need to have a sine wave output from our synthesizer. The full DDS block diagram is shown in **Figure 9.38**, where a new block called the waveform map is included. This is a digital process, usually implemented as a non-volatile memory, that converts the phase information from the accumulator into sinusoid waveform values which are then converted to a sampled waveform in a DAC. The output low-pass filter is critical to getting the desired sine wave.

DDS NOISE

DDS is a frequency divider and by the rules of frequency division the output phase noise is lower than the input phase noise, which in

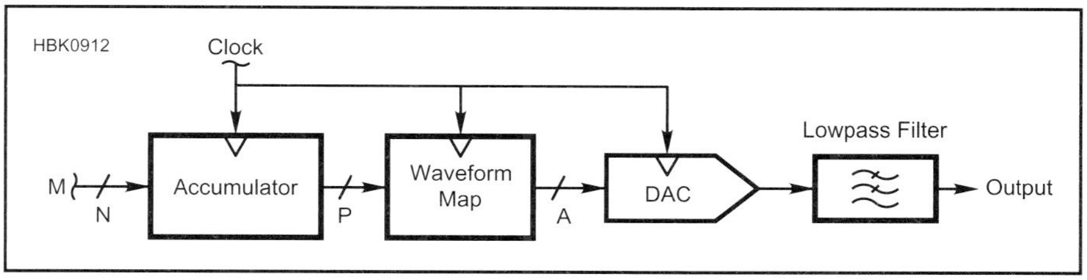

Figure 9.38 — Complete block diagram of a DDS.

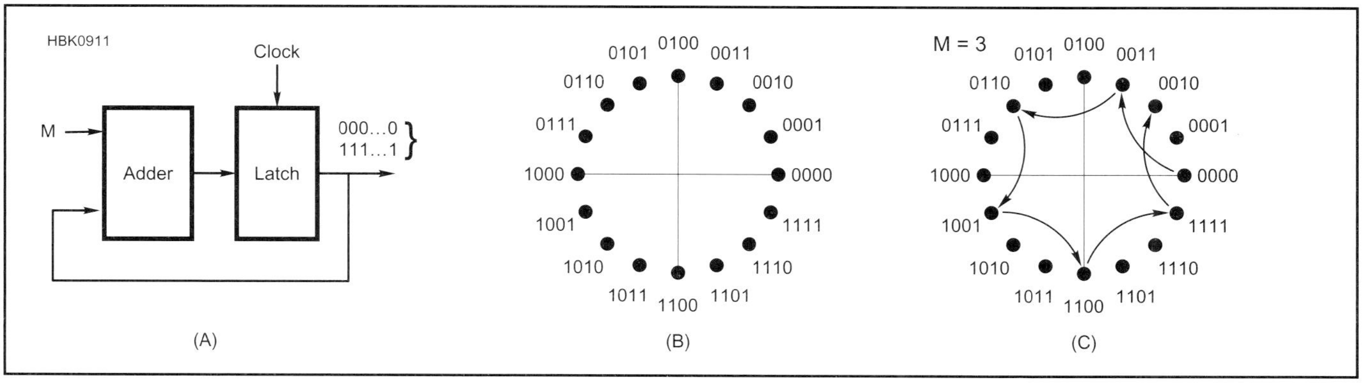

Figure 9.37 — Direct digital synthesis is based on the accumulator circuit (A). For our purposes, we assign all of the possible states that the accumulator can have around a circle and call them "phase" (B). The accumulator continuously adds its input number, once each clock cycle, to the present count. When the adder overflows, the overflow is dropped and counting continues around the circle as in (C).

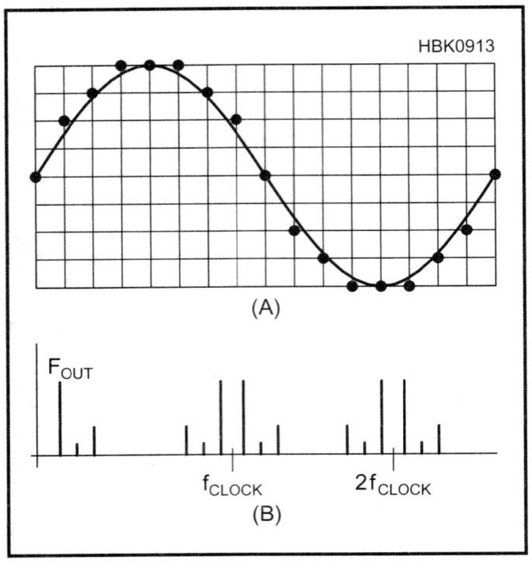

Figure 9.39 — Waveform map quantization forces output waveform errors because the output signal values must remain at the intersection of the grid lines (A). This results in output non-harmonic spurious signals from the DDS (B).

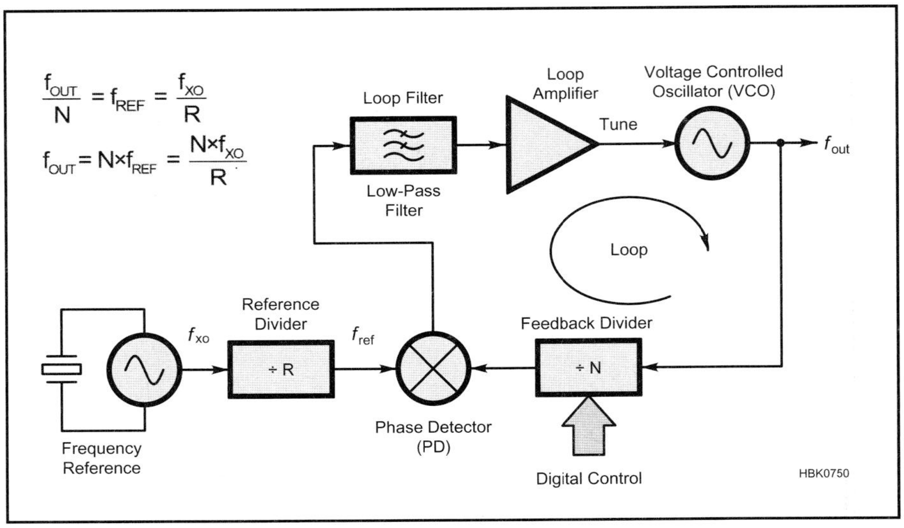

Figure 9.40 — A basic phase-locked-loop (PLL) synthesizer acts to keep the divided-down signal from its voltage-controlled oscillator (VCO) phase-locked to the divided-down signal from its reference oscillator. Fine tuning steps are therefore possible without the complication of direct synthesis.

any DDS comes from the clock. There is no VCO to add noise to the output signal, like we have in PLLs.

JITTER AND WANDER

Clock signals also exhibit *jitter* which is an irregular deviation from the exact clock period that creates noise in the DDS output. *Jitter period* is the interval between the maximum and minimum deviation from the nominal clock period. *Jitter frequency* is the inverse of jitter period. Jitter frequencies below 10 Hz are classified as *wander* and frequencies at or above 10 Hz as *jitter*.

ADDITIONAL DDS PRINCIPLES

In addition to very fine frequency resolution, extremely good tuning linearity, very low output phase noise, and tuning bandwidth that goes all the way down to dc, an additional important benefit of DDS is that it tunes very quickly. Indeed, as soon as the tuning number M is changed, the DDS immediately shifts to calculating the new output waveform. The tuning dynamics of PLLs are not present here. Likewise, the stability issues of a PLL do not happen in DDS designs.

The presence of non-harmonic spurious signals remains the biggest issue in successful application of DDS to amateur equipment. In principle, a 32-bit DDS can cover all HF amateur bands with 0.23 Hz resolution when clocked at 100 MHz. No VCOs, no band-switching, no tuning — it sounds like a dream! — but we need to apply our RF design skills to get rid of those pesky spurs. The quantization level of modern waveform maps is sufficient to keep the sinusoid quantization noise spurs below –100 dBc. The largest remaining problem is the DAC.

The signal sampling theorem does not predict spurs other than those generated by quantization. That means any additional spurs are a result of the actual output wave form not being exactly what the signal digital quantization requires. (See **Figure 9.39**.) Normal DAC specifications of *integral nonlinearity* (INL) and *differential nonlinearity* (DNL) turn out to not be very important in the DDS application. More important are the symmetry of output transition times for up and down steps, and having no overshoot on either transition direction. DACs with the latter two properties generally work very well for precision DDS use.

9.6.2 Phase-Locked Loops (PLL)

As mentioned in the introduction, the PLL is a key component of any indirect synthesizer. This section presents a detailed discussion of design and performance topics of PLL synthesizers and the circuits used to construct them.

ARCHITECTURE

The principle of the PLL synthesizer is straightforward. A tunable oscillator is first built to cover the required output frequency range; then the PLL system is constructed around this oscillator to keep it tuned to the desired frequency. This is done by continuously comparing the phase of the oscillator to a stable reference, such as a crystal oscillator, and using the result to steer tuning of the oscillator. If the oscillator frequency is too high, the phase of its output will start to lead that of the reference by an increasing amount. This is detected and is used to decrease the frequency of the oscillator. Too low a frequency causes an increasing phase lag, which is detected and used to increase the frequency of the oscillator. In this way, the oscillator is locked into a fixed-phase relationship with the reference, which means that their frequencies are held exactly equal.

A representative block diagram of a PLL is shown in **Figure 9.40**. Measurement of any error in the output signal phase is made by the *phase detector* (PD). The measured error is supplied to the *loop filter* and *loop amplifier* which work together to tune the *voltage-controlled oscillator* (VCO) just enough to make the error go to zero. The phase detector determines the actual frequency and phase of the VCO through the *feedback divider* (N). The phase detector operates at a lower frequency which is a sub-multiple of both the crystal oscillator frequency reference (f_{XO}) and the output frequency (f_{OUT}). This lower frequency is correctly called the PLL *reference frequency* (f_{REF}) because it sets the operating conditions of the PLL.

There is unfortunate history about the term reference frequency. It is often used to refer to both the operating frequency of the phase detector and to the frequency reference applied to the PLL. If the *reference divider* (R) is not present then of course these two frequencies are the same. But in general, the reference divider is present and these frequencies are different. Ambiguity here is widespread and a big problem, even today, many decades into the widespread use of PLLs. We must be very careful in the words we choose to use.

FREQUENCY RESOLUTION (STEP SIZE)

Frequency of the output signal can be easily changed by changing the divide ratio of the feedback divider. For example, if the reference frequency (at the phase detector) is 100 kHz and the output frequency is 147.5 MHz, the feedback divider number must be N = 1475. If we changed the value of N to 1474, the frequency at the output of the feedback divider becomes 100.068 kHz. The phase detector notices that the frequency of the VCO is too high, and "tells" the loop filter to reduce the frequency of the VCO until the frequency out of the feedback divider becomes exactly 100.000 kHz. This will happen when the VCO is retuned to a frequency of 147.4 MHz.

By changing the value of N by 1, we have just tuned the frequency synthesizer by its smallest available step. This is called the *frequency resolution* of the synthesizer, or equivalently the synthesizer *step size*. Here we note that this step size is (147.5 – 147.4) = 0.1 MHz. It is no accident that this frequency is exactly equal to the PLL reference frequency f_{REF}. This result is an important reason to avoid all ambiguity in the term "reference frequency." f_{REF} is equal to the synthesizer output frequency step size.

There certainly are applications where we want the step size to be a small number, say 10 Hz. You might guess — correctly — that there are problems in building a PLL synthesizer for a very low f_{REF}. The usual solution is to combine multiple PLLs with some DAS techniques to get around these problems.

DYNAMIC STABILITY AND LOOP BANDWIDTH

Because feedback control is required to transfer the frequency and phase characteristics from the crystal oscillator onto the VCO, all of the stability problems inherent in feedback control apply to PLL design. If the gain and phase responses are not well designed, the PLL can oscillate. In this case, instead of getting the stability transfer we want, the output is essentially a frequency modulated signal spread across the entire tuning bandwidth of the synthesizer — clearly a very bad thing. How to perform this design is discussed below.

For the moment let us assume that our PLL dynamics are properly designed. One other characteristic of feedback control is now important: how fast can the output frequency be changed from one value to another? This is controlled by the *loop bandwidth* of the PLL. For practical reasons the loop bandwidth should be less than 5% of f_{REF}. Using the example above in which f_{REF} = 100 kHz, the loop bandwidth we design for cannot exceed 5 kHz. To answer our question, we can use the rule of thumb that in a well-designed loop, the settling time should be between 1 and 2 times the reciprocal of the loop bandwidth. For a 5 kHz loop bandwidth the settling time should not exceed 2/5,000 = 400 µs. We can begin to see one problem with PLL design if f_{REF} is very small: To get a 1 kHz step size from a PLL synthesizer the maximum loop bandwidth available is 50 Hz. This is impractically small.

This loop bandwidth also influences how we can modulate our PLL synthesizer in an FM transmitter. Imagine now that we apply a very small amount of FM to the reference oscillator f_{XO}. The amount of deviation will be amplified by N/R — but this is true only for low modulating frequencies. If the modulating frequency is increased, the VCO has to change frequency at a higher rate. As the modulating frequency continues to increase, eventually the VCO has to move faster than the available *settling time*. The PLL now acts as a low-pass filter, reducing the available deviation from higher modulating frequencies. The cut-off frequency of this PLL low-pass action is equal to this same loop bandwidth.

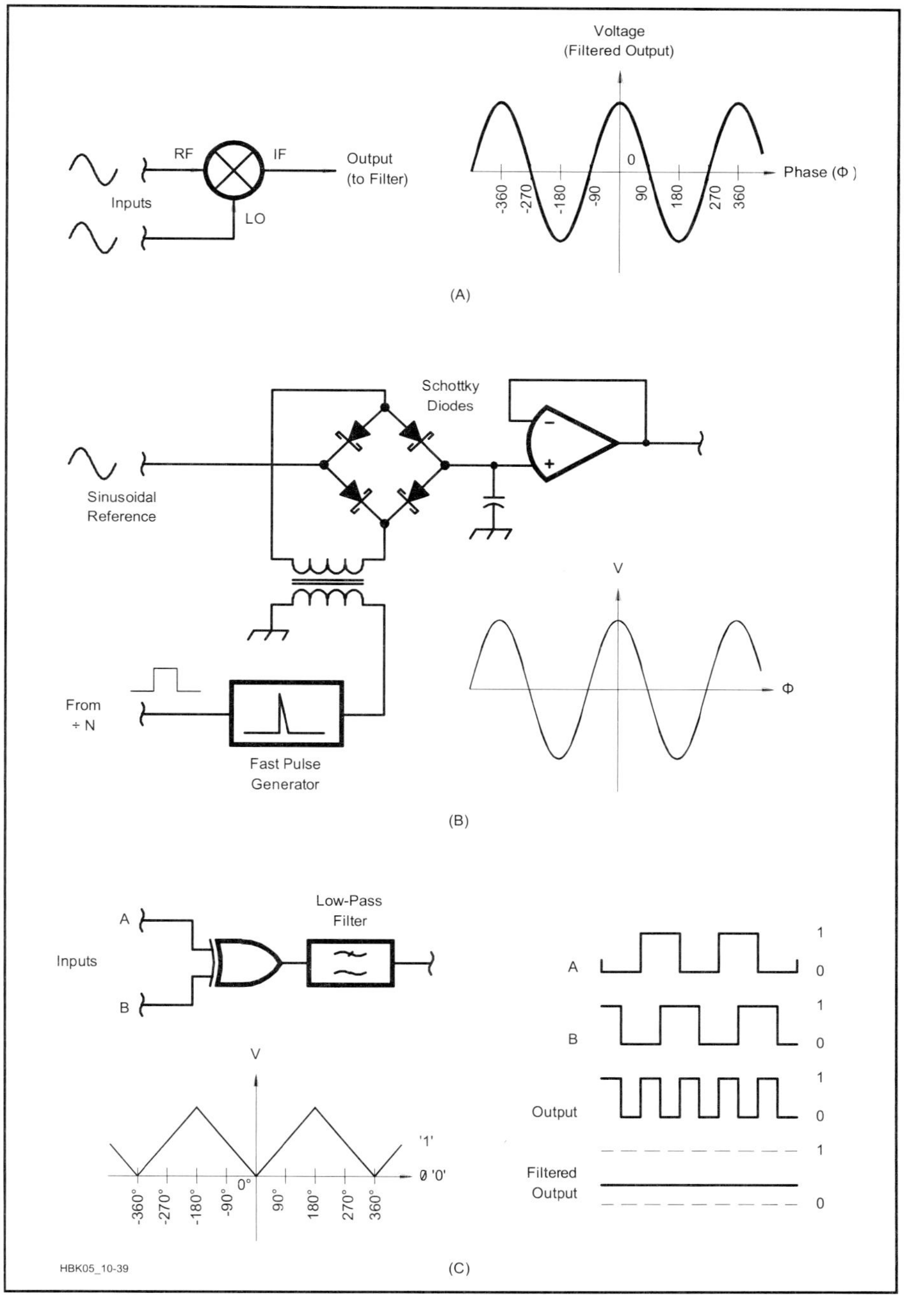

Figure 9.41 — Simple phase detectors: a mixer (A), a sampler (B) and an exclusive-OR gate (C).

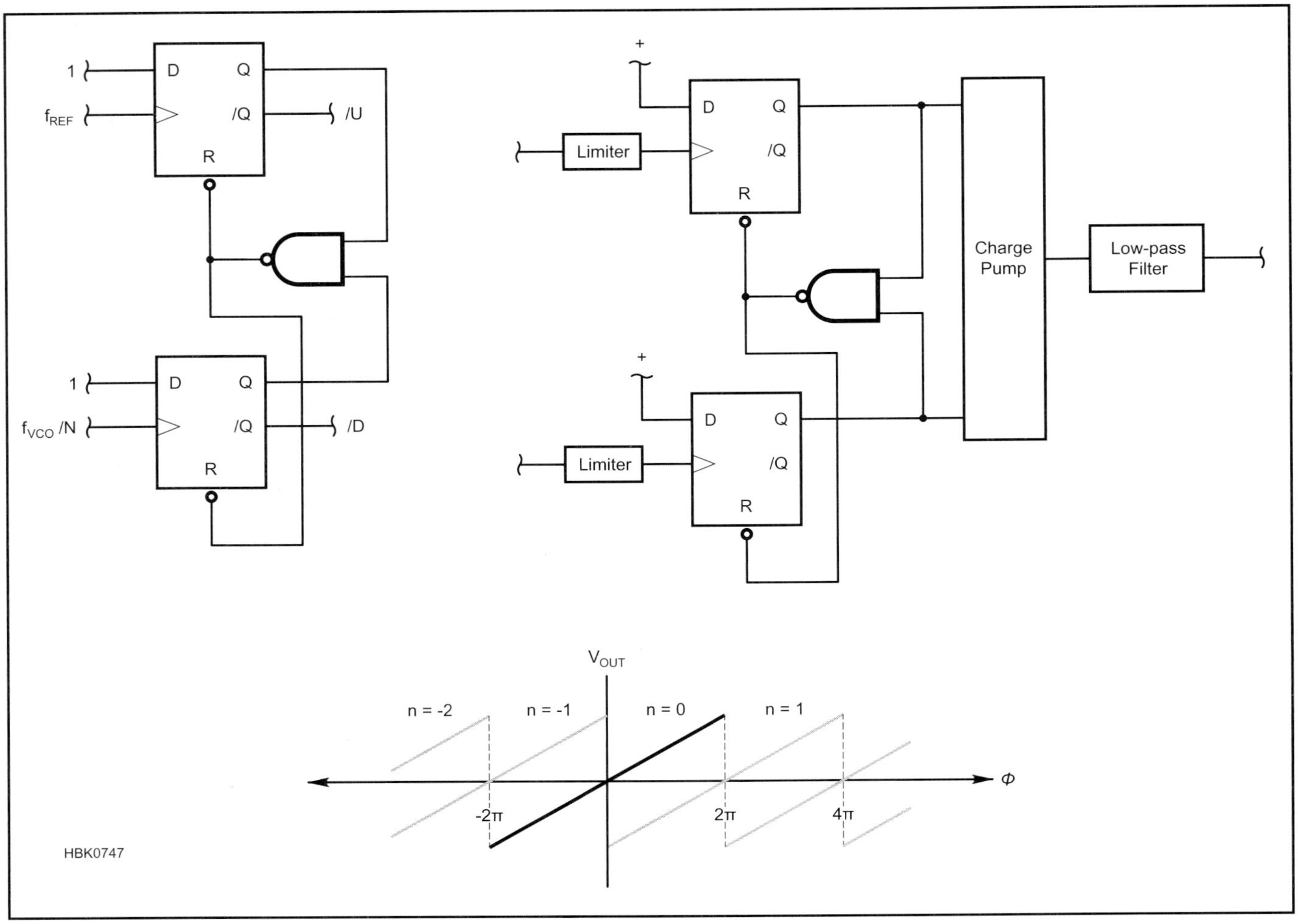

Figure 9.42 — The phase frequency detector (PFD) makes a measurement of the time difference between two rising edges of the input signals.

PLL COMPONENTS

Let us continue our discussion of phase-locked loop synthesizers by examining the role of each of the component pieces of the system. They are the phase detector, the VCO, the dividers (with possible prescalers), the loop filter, and of course the reference oscillator.

Phase Detectors

Phase detection is the key operation in any PLL. Remembering that we are making a phase locked loop, what we really need is not *absolute* phase measurement but a *relative* phase measurement. We need to know only how the phase of the VCO output signal f_{OUT} is changing with respect to the phase of the crystal reference f_{XO}.

As usual, there are both analog and digital circuit techniques available to perform phase detection. For frequency synthesizer design the analog techniques are seldom used. Analog phase detectors are used only in receiver and demodulator applications. There are two that sometimes still appear in ham equipment, the multiplier/mixer of **Figure 9.41A** and the sampling phase detector shown in Figure 9.41B.

The mixer is a very low noise device when used as a phase detector, which explains why it is not yet completely gone from synthesizer design. It remains only in PLL designs where f_{REF} is very high, greater than 10 – 100 MHz. The main reason for this is that a PLL using a mixer for the PD is subject to a phenomenon called *false lock*, where the PLL may act like it is locked even though it really isn't. Detecting and eliminating a false lock condition is much easier to do at higher frequencies. The sampling phase detector is now used only in synthesizers with output frequencies in the high microwave region, typically well above 10 GHz. The gain of this sampling circuit is very low, so it is used only if there is no viable alternative.

The most widely used digital phase detector is the exclusive-OR (XOR) gate shown in Figure 9.41C. If inputs A and B are almost in phase, the output will be low most of the time, and its average filtered value will be close to the logic 0 level. If inputs A and B are almost in phase opposition, the output will be high most of the time, and its average voltage will be close to logic 1. The average voltage is near its mid-value when the inputs are shifted in phase by 90 degrees. In this respect the XOR gate is much like the analog mixer. To achieve this circuit's full output-voltage range, it's important that the reference and VCO signals at its input have a 50% duty cycle.

Phase-Frequency Detector

One problem common to *all* phase detectors is the possibility of false lock. This happens because the output of any phase detector can have an average value of zero even when the frequencies at its inputs are not equal. This is a very serious problem that requires any PLL using them to have additional circuitry to serve as an acquisition aid. If the PLL tries to settle when the PD inputs are not at the same frequency, this acquisition aid must prod the PLL to keep moving toward the true lock frequency. Such acquisition aids are not

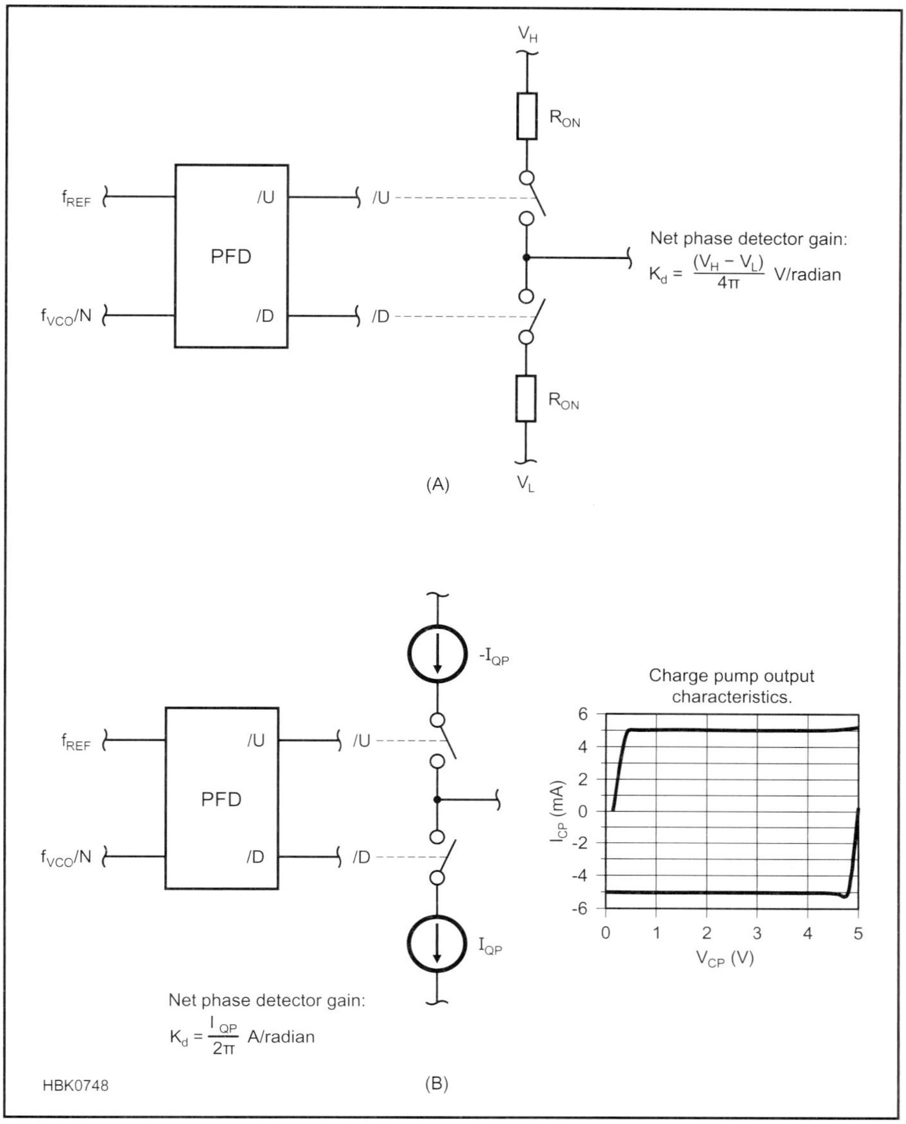

Figure 9.43 — The charge pump delivers short bursts of current to the loop filter under the control of the phase frequency detector (PFD) outputs. A voltage mode charge pump is shown at A and a current mode charge pump at B.

needed when a *phase-frequency detector* (PFD) is used.

The benefit of the PFD is that if the input signals are at different frequencies, even if they are at very different frequencies, the output is never zero. The PFD output always produces a dc shift in the direction the PLL needs to move until the lock condition is achieved. No acquisition aids are ever needed. This alone is enough to explain its nearly universal adoption today.

As seen in **Figure 9.42A**, the PFD is a simple digital circuit. This also helps make it attractive to use in this era of CMOS integrated circuits. Other designs for the PFD exist, but they all follow the logic shown in Figure 9.42A. Of particular importance is that the PFD has two outputs.

The PFD operates by measuring the time difference between the rising edges of the reference signal and the divided VCO. The first signal to arrive sets its flip-flop. The second edge to arrive also sets its flip-flop, which immediately causes both flip-flops to be reset, ready for the next set of signal edges. By operating at high speed, the PFD is very sensitive to phase differences between the input signals. The PFD does require an equal number of edges for each signal, so it is not useful when the signals have varying numbers of edges, such in a receiver's demodulator or a synchronizer.

If the VCO frequency is too low only the /U output is active, causing the loop filter to raise the VCO frequency. On the other hand, when the VCO frequency is high only the /D output is active, causing the loop filter to lower the VCO frequency. It is interesting to note that unlike the mixer and XOR phase detectors, which require the inputs to be in phase quadrature, the PFD locks when the input signals are in exact phase alignment.

Being an edge triggered circuit, PFD operation is essentially independent of input signal duty cycle. This eliminates the need for input waveform processing. If the input frequencies are different, the time differences measured by the PFD will be either constantly increasing or decreasing. As a result, only one of the PFD output signals will be active, informing the PLL unambiguously which way to correct the controlled source to achieve phase lock.

Charge Pumps

Because the PFD has two outputs and most loop filters have only one input, something is needed to bridge the PFD and the loop filter. The usual technique is the *charge pump*, shown in Figure 9.42B. The charge pump gets its name by allowing current to flow only for the brief interval, here when the /D or /U outputs are active. Current flow in short bursts (pulses) is equivalent to a finite charge transfer.

Charge pumps come in two configurations, voltage mode and current mode, as shown in **Figure 9.43**. Voltage mode (Figure 9.43A) was the original version, though now current mode (Figure 9.43B) is most widely used with integrated circuit PLLs.

The voltage mode charge pump directly connects voltage sources to the loop filter when the PFD output signals are active. When the PFD output signals are at rest (both high) both voltage sources are disconnected from the loop filter. These voltage spikes transfer charge to the integrator in accordance with the PLL loop filter time constant. One voltage source inputs current to the loop filter and the other removes it.

The current mode charge pump is architecturally similar to the voltage mode charge pump, with voltage sources replaced by current sources. Like the voltage mode charge pump, this circuit disconnects both sources when the PFD output signals are at rest. In this case, current spikes transfer charge to and from the loop filter integrator. This transfer is independent of the loop filter series resistances.

The current mode charge pump has some advantages to the frequency synthesizer designer. Three are particularly significant:

• Pump current flows at a constant value independent of the loop filter voltage. Unless V_H and V_L are both very large compared with any possible loop filter voltage, the amount of charge transfer will vary with different loop filter voltage values.

• During the PFD reset time, both output sources are active. The voltage mode charge pump will attempt to connect both voltage sources to the loop filter during this brief period, which can generate some undesirably large power supply current spikes. During PFD reset, the current mode charge pump loads the power supply with a single current value of I_{QP}. If both current sources are well matched, their currents completely cancel at the loop filter input, effectively disconnecting them that much earlier.

• At low offset frequencies, the loop gain flattens out with a voltage mode charge pump. Loop gain continues to increase with a current mode charge pump, effectively matching active filter performance.

VCO

The design and characteristics of oscillators, including tunable oscillators, is the major topic of this chapter so it will not be repeated here. Of particular importance to PLL design are the tuning characteristic of the VCO and the VCO tuning bandwidth.

The tuning characteristic of an oscillator shows the output frequency versus the tuning voltage, such as the example of **Figure 9.44A**. The slope of this curve is called the VCO's *modulation gain* (K_0) and it is this modulation gain that is most important to PLL design. For wideband VCOs the variation of K_0 can exceed 5:1 from minimum to maximum frequency. The design method below shows how a wide gain variation such as this is handled with a single loop filter design.

It is important to realize that the VCO has no idea that it is controlled by a PLL. While it is oscillating it continues to drift and jitter with time and temperature like any other oscillator. You must characterize these variations to be assured that the tuning capability has sufficient range for the PLL to "find" a tuning voltage value that will retune the VCO to the required frequency, under all circumstances.

The primary VCO characteristic of interest to the synthesizer designer is the tuning gain "constant," K_0. Almost never a constant, K_0 is a local measure of how much the VCO output frequency changes with a change in tuning voltage.

VCO designers go to great lengths to make K_0 an actual constant. This additional effort often increases the VCO design complexity and usually results in a more expensive design. If no linearization effort is undertaken, K_0 can easily vary from 3:1 to 10:1 over the tuning band. This also complicates the frequency synthesizer design. Specific applications will dictate where effort needs to be expended: in the VCO design or the PLL design.

All PLL design algorithms assume that there is no delay between when the loop filter presents a tuning change command to the VCO and when the VCO frequency actually changes. This situation is true when the VCO tuning bandwidth is much greater than the PLL loop bandwidth. If the VCO tuning bandwidth is not much greater than the PLL loop bandwidth, there are additional phase shifts that will cause problems with PLL stability. It is usually much easier to narrow the loop bandwidth than to redesign the VCO. If redesigning the VCO is a possibility, reduce the capacitance seen at the tuning input to increase the tuning bandwidth.

The VCO tuning bandwidth is important mainly for PLL designs that have wide loop bandwidths. Measuring the frequency deviation as the modulating frequency varies is best done by looking at the FM sidebands on a spectrum analyzer, as shown in Figure 9.46B. For constant deviation, FM sidebands follow the top profile. When the modulating frequency doubles, the sidebands drop below the profile by 6 dB, and if it triples, by 10 dB. The first FM sideband must exceed all other sidebands by 10 dB for proper VCO *characterization* — the process of determining the VCO's characteristic parameters such as the tuning sensitivity, noise spectrum, dynamic response to changes in the tuning input signal, and so on. (See the Keysight application note "Boosting PLL Design Efficiency" in the References section.) VCO tuning bandwidth is determined by the point at which increasing the modulating frequency deviates from

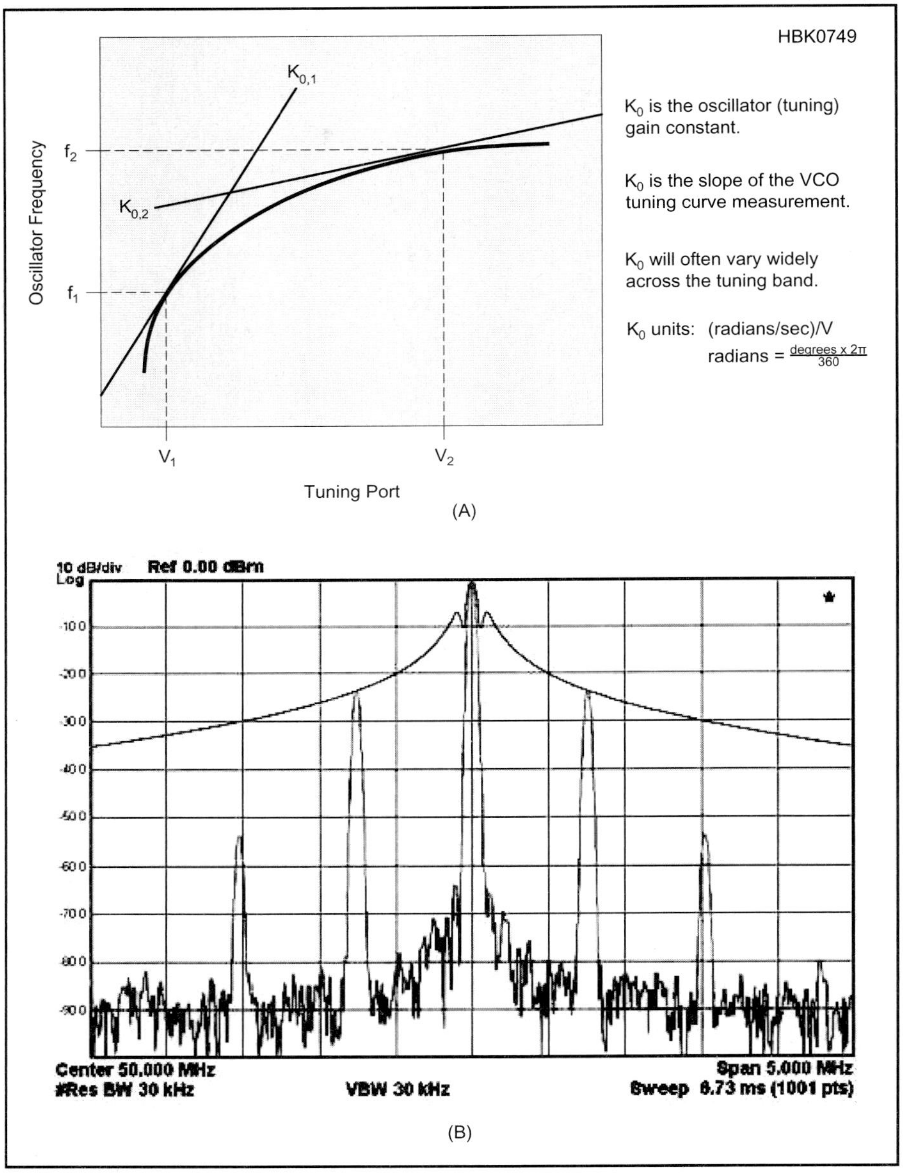

Figure 9.44 — (A) Measurement of the VCO tuning characteristic provides a curve describing frequency versus tuning voltage. The VCO tuning gain at a particular frequency (or tuning voltage) is the slope of this curve at that point. (B) A spectrum analyzer is used to measure frequency deviation of the VCO and the modulating frequency varies. FM sideband peak amplitudes follow the top profile when deviation is constant.

(drops below) the profile shown in Figure 9.44B.

Reference Divider

The reference divider is one of the easiest parts of a PLL to design, though technically it is outside the PLL. The input frequency is well known — it is the crystal reference. The output frequency is also well known, being the phase detector frequency f_{REF}. The ratio of these two frequencies is the required counter divide-by ratio or *modulus*.

When designing PLL synthesizers, f_{REF} is usually specified but f_{XO} is not. This provides some flexibility in design to use easy counter implementations if the resulting f_{XO} is also easy to get. The easiest digital counters to use for any divider are those operating with binary numbers (divide by 2, 4, 8, 16, …, 128, …). For example, if f_{REF} = 20 kHz and you choose R = 512 then f_{XO} = 10.24 MHz, which is a readily available crystal frequency.

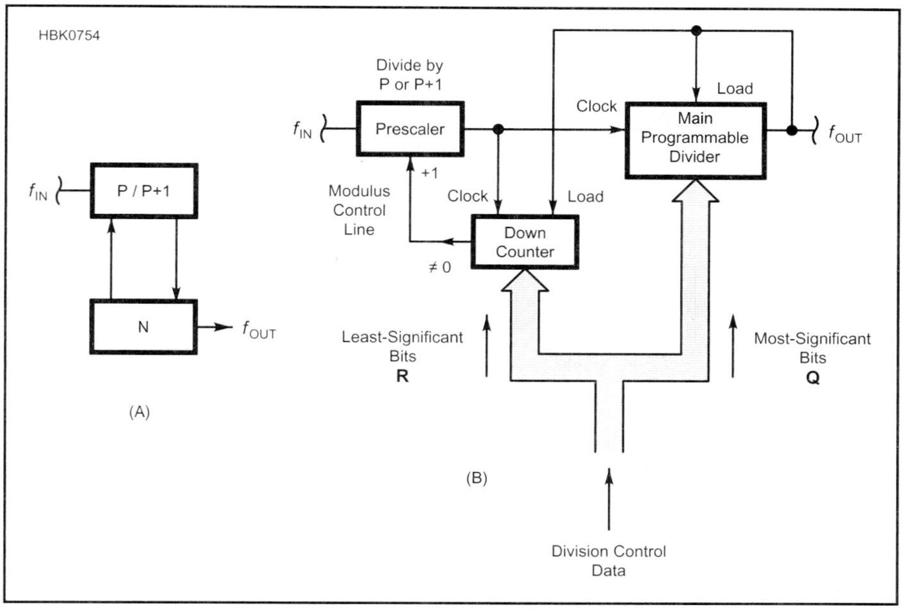

Figure 9.45 — A typical implementation of a dual-modulus prescaler.

Feedback Divider and Prescalers

The feedback divider is more of a challenge. First and foremost, this divider must work properly with whatever possible frequency it may ever see at its input. Whatever the highest and lowest frequencies the VCO may be, the divider must handle them all.

The feedback divider is almost always programmable so that the output frequency can be changed. Programmable counters are always slower than non-programmable designs, so some speed limitation occurs. Fortunately, in this era of BiCMOS integrated circuits the problems of past years in getting programmable counters to operate at hundreds of MHz are over. Older equipment still has feedback divider designs that are carefully crafted to handle the frequencies at which they must operate.

Fixed Prescaler

When the programmable counter does not have sufficient speed to handle the frequencies required, then the VCO frequency must be divided down ahead of the programmable counter to assure that everything works reliably. This additional divider or prescaler usually has a fixed binary value so it will be fast enough without using too much power. As long as the output frequency from the prescaler is always within the operating range of the programmable divider the PLL will be reliable.

Of course, there are consequences: when a prescaler is used, the PLL step size is multiplied by the prescaler divider value. For our f_{REF} = 100 kHz, if we adopt a divide-by-16 prescaler then the step size increases by a factor of 16 to 1.6 MHz. The direct way to correct for this is to reduce f_{REF} by the same factor of 16, which unfortunately means that the PLL loop bandwidth is also reduced by a factor of 16. Is there a solution with fewer compromises?

Dual-Modulus Prescalers

Yes, there is such a solution. If the prescaler is designed to divide by two values separated by 1, say 16 and 17, or 8 and 9, then we call this a *dual-modulus P/P+1 prescaler*. Using a dual-modulus prescaler, we can leave f_{REF} unchanged, which not only keeps our step size and loop bandwidth but also gains a much higher operating frequency for the feedback divider.

A dual-modulus prescaler is a cooperative effort between a high speed prescaler and a much lower speed programmable counter. The result of this cooperation is a programmable counter that operates at a very high input frequency while maintaining unit division resolution. A dual-modulus prescaler works by allocating quotient and remainder values from the division N/P as shown in **Figure 9.45A**.

The improvement comes from viewing the operation of division in a slightly different way. The division ratio (N) of any two integers will always provide a quotient (Q) and a remainder (R). If Q counts of f_{in}/P are followed by R counts of f_{in} directly, then N = QP + R. One way to build this counter would be to first count a quotient's worth of prescaled input, and then count a remainder's amount of the input frequency directly. This last step is hard because of the high frequency.

A second approach is to add the remainder counts to the prescaled output as follows: If Q – R counts of prescaled f_{in}/P are followed by R counts of prescaled $f_{in}/(P + 1)$, then N = (Q – R) P + R (P + 1). This is the same as N = QP – RP + RP + R = QP + R.

If the remainder counts are distributed among the quotient counts by increasing the prescaler modulus by one from P to P + 1 for the remainder's number of cycles, then returning the prescaler to its nominal division ratio while the rest of the quotient counts are made, the same result is achieved. This design always operates the programmable counter at a much lower frequency than the input frequency, which is a much more robust design.

The catch, however, is that there have to be enough quotient counts over which to distribute any amount of remainder counts. If you run out of quotient counts before remainder counts the technique falls apart and that particular loop divisor will not be *realizable*. For real dividers, all of the terms — Q, P, and R — must be positive integers, so Q – R must be greater than zero for N = QP+R to be realizable.

The maximum remainder, R_{max} = P – 1. The minimum quotient is equal to R_{max}. So the minimum remainder, R_{min} = 0. That means:

$$N_{min} = Q_{min} P + R_{min} = (P - 1) P + 0 = P2 - P$$

If the PLL design always requires N to be greater than P2 – P then there are no problems. If N does need to go below P2 – P, then the only solution is to choose a dual modulus prescaler with a smaller value of P. Various values of P and N_{min} are:

P	N_{min}
8	56
16	240
64	4032
128	16256

Dual-modulus prescalers therefore implies a minimum divisor value before continuous divider value coverage is realized. The value of this minimum divisor depends on the base modulus of the prescaler, P, and increases quadratically. Proper choice of prescaler modulus is one of the important decisions the frequency synthesizer designer has to make.

Figure 9.45B shows how a typical dual-modulus prescaler is implemented. Each cycle of the system begins with the last output pulse having loaded the frequency control word, shown as "Division Control Data," into both the main divider and the prescaler controller. If the division control data's least significant bits (which make up R) loaded into the prescaler controller are not zero, the prescaler is set to divide by P + 1, 1 greater than its normal ratio, P. The main divider counts Q pulses from the prescaler.

Each cycle out of the prescaler then clocks the down counter. Eventually, the down counter reaches zero, and two things happen: The counter is designed to hold (stop counting) at zero (and it will remain held until it is next reloaded) and the prescaler is switched back to its normal ratio, P, until the next reload.

Because the technique is widely used, dual-modulus prescaler ICs are widely used and widely available. Devices for use to a few hundred megahertz are cheap, and devices in the 2.5 GHz region are commonly available. Common prescaler IC division ratio pairs are: 8-9, 10-11, 16-17, 32-33, 64-65 and so on. Many ICs containing programmable dividers are available in versions with and without built-in prescaler controllers.

Loop Filter

When designing PLL synthesizers we usually just have to measure and accept the VCO characteristics of K_0 and the PFD output characteristics. Output frequency range and step size are also not flexible. We pull all of these together into a working and stable synthesizer by proper design of the loop filter.

Loop filter design is usually shrouded in mystery, which has given PLL synthesizer design the aura of a "black art." This is not at all warranted because very reliable design algorithms have been around for decades. Unfortunately, these have usually been published in obscure places so they are not well known. The best algorithms are presented below. These methods have served extremely well for over 30 years and should provide you with fast design times and very stable PLL synthesizers. No iteration should be needed.

Loop filters come in active and passive structures. In general, the easiest designs use passive (RC) loop filters if the output voltage range from the phase detector (including the PD or PFD and its charge pump) is sufficient to tune the VCO over its required frequency range. If the VCO needs a larger tuning voltage range then an active loop filter structure is necessary.

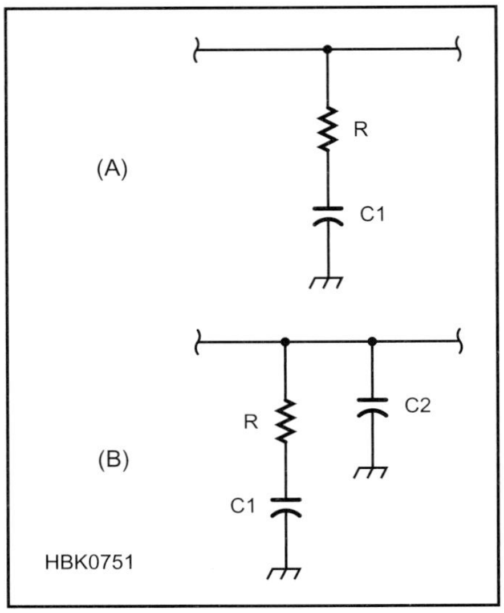

Figure 9.46 — Passive loop filters for current mode charge pump PFD outputs.

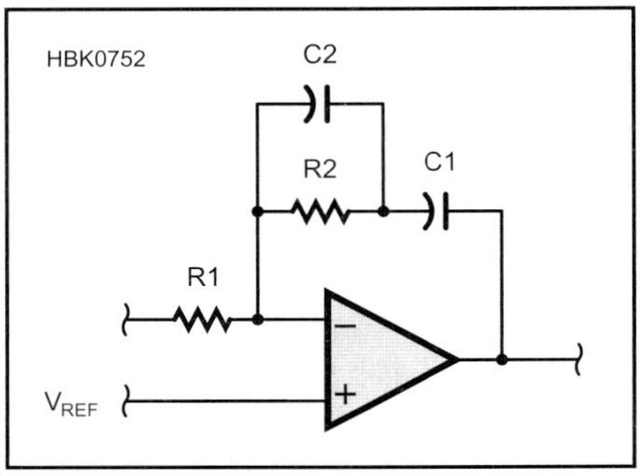

Figure 9.47 — Third-order active loop filter isolates the loop filter output voltage from the charge pump outputs.

Passive loop filter circuits for current mode charge pump outputs are shown in **Figure 9.46**. The loop filter time constant is set by the charge pump current and C1, independent of R and of the loop filter output voltage. R provides a phase leading zero for PLL stability. The first-order filter in Figure 9.46A has a minimum parts count and results in loop dynamics that are consistent with the output frequency.

A second-order filter has the undesirable characteristic of allowing step and impulse changes in the input to appear at the output. This results in significant reference sidebands — a bad thing. Figure 9.46B shows a third-order filter in which C2 changes step at the input to slower ramps by rolling off the high frequency response of the loop filter. This results in a significant reduction in reference sidebands.

Active loop filters offer more structures that work well and are easier to design, though more complicated to build. When a voltage mode charge pump is being used, an active loop filter is strongly desired since it removes the PLL response variations with different loop filter voltage values.

In **Figure 9.47**, the gain of the op-amp isolates the actual loop filter output voltage from the charge pump output voltage, controlling the loop so that the charge pump output voltage is V_{REF}. Each passive component has the same basic function here as in the passive filter.

The third-order active filter in Figure 9.47 adds an additional pole within the feedback loop and reference frequency sidebands are significantly reduced. Charge pump current impulses now cause ramps instead of steps in the output voltage, but they still cause problems at the amplifier inputs. Good design algorithms include this extra pole as part of the fundamental dynamics — not as a disturbance of conventional second-order dynamics.

Though the design in **Figure 9.48** adds another capacitor by splitting the input resistor, the added complexity is usually well worth it. Charge pumps put out very narrow, spiky signals that are usually of sufficient amplitude and duration to temporarily drive the op-amp input into saturation. If this happens the PLL response becomes non-linear and problems often arise.

The improvement of this design is to low-pass filter the charge pump pulses before the output is applied to the op amp input. The op amp far prefers the resulting ramps, remaining linear and therefore predictable in behavior. This results in further reduction in reference frequency sidebands and is easily adapted to difference-output digital PDs, including the PFD.

Design equations for these loop filters will be presented after the remaining PLL component blocks are discussed.

Reference Oscillator

Any PLL is simply a stability transfer mechanism, so the behavior of the reference oscillator will not be improved on. The fractional frequency error of the overall frequency synthesizer will match that of the reference oscillator. For example, if the output of the PLL synthesizer changes 1 kHz at an output frequency of 1 GHz [1000/1,000,000,000 = 1:1 million, or 1 part per million (ppm)] then the reference has changed that same fractional amount. If the crystal reference is 10 MHz (a very common frequency for crystal references) its frequency drift was (1 ppm) × (10 MHz) = 10 Hz. This is not much frequency drift, but it is significant to the ultimate output frequency.

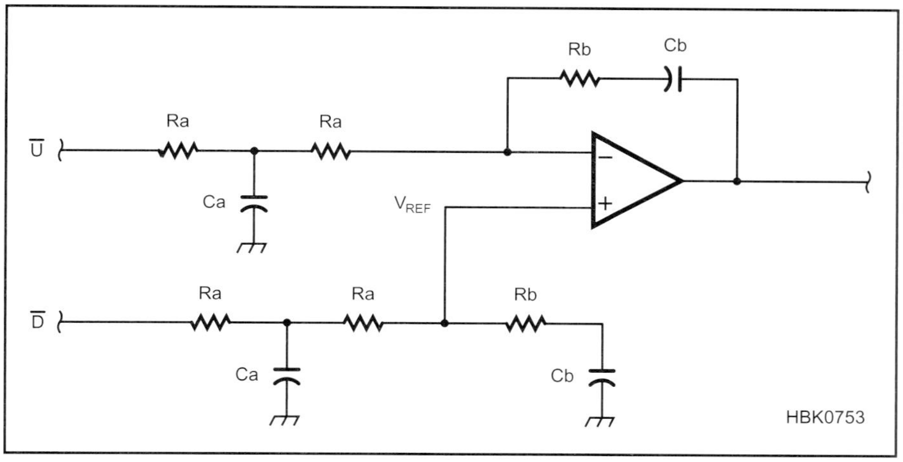

Figure 9.48 — Adding additional low-pass filtering by splitting the input resistor and adding C_a improves loop stability and reduces reference frequency sidebands. This design is well-suited for differential-output PFDs.

9.6.3 PLL Loop Filter Design

When it is time to design a PLL synthesizer, it is best to do it in steps. Begin with knowing how the hardware you are using behaves:

• VCO tuning characteristics, K_0

• PFD characteristic (with charge pump), K_s

• Crystal reference frequency, f_{XO}

Next, list the requirements of your application

• Output frequency range, f_{OUT}

• Output frequency step size, f_{STEP}

Now you are ready to choose the two main design parameters:

• Loop bandwidth — this should not exceed 5% of $f_{REF} = f_{STEP}$.

• Phase Margin — this sets the overall stability of the PLL and can be any value between 45 and 60 degrees. A phase margin of 45 degrees optimizes settling time for fast channel changes, as might be required in a frequency-hopping application. This level of speed is typically NOT needed in amateur radio applications. A phase margin of 60 degrees removes peaking in the transfer function and provides a small reduction in output phase noise, which is useful in amateur radio applications.

With all of this information in place it is time to calculate the values for the two frequency dividers and the loop filter components. Begin with the divider values:

$$R = f_{XO}/f_{REF}$$

$$N = f_{OUT}/f_{REF}$$

$$N_{MIN} = f_{OUT,MIN}/f_{REF}$$

$$N_{MAX} = f_{OUT,MAX}/f_{REF}$$

The best design algorithm for designing a stable PLL calls for using the geometric mean of the feedback divider value range in the loop filter design. This same idea is used to manage the range of VCO gain values for K_0.

$$N_{DESIGN} = \sqrt{N_{MAX} \times N_{MIN}}$$

$$K_0 \text{ design target value} = \sqrt{K_{0,MAX} \times K_{0,MIN}}$$

PFD gain depends on whether the charge pump is voltage mode or current mode:

(voltage mode) $K_d = (V_H - V_L)/4p$

(current mode) $K_d = I_{QP}/2\pi$

For all of the active loop filter structures, use the voltage mode value.

Everything is now ready. Place these design values into the appropriate design equations for the selected loop filter structure. For a passive loop filter use **Figure 9.49**. If you are using an active loop filter the appropriate equations are in **Figure 9.50** or **Figure 9.51**. The resistor and capacitor values from these calculations are not critical. Choose values within 30% of what you calculate, and the design will be fine.

Following any loop filter design, it is always a good idea to check the results by directly calculating the filter time constants from the components chosen. Compare these time constants with those derived theoretically. If they match within 10 – 20%, there should be no problem with loop stability.

The passive third order loop filter design equations of Figure 9.49 look very similar to those of its active filter counterpart. The major difference is that this procedure yields two time-constants instead of three, and a term equaling the sum of both filter capacitances.

Unlike the active filter form, there are no arbitrary component choices with the passive filter. From the time constants and the total capacitance, all three loop filter components are uniquely determined.

Since the concepts of natural frequency and damping no longer apply (by definition) in a third-order filter, the more general stability concepts of gain crossover and phase margin are used. A value of 50 – 60 degrees is usually used for the design phase margin.

The equations for Figure 9.50 and Figure 9.51 include the rolloff pole within the loop dynamics, rather than treating it as a loop perturbation. This allows a greater amount of high frequency rolloff to be achieved as it "kicks in" at a much lower frequency than perturbation techniques would allow.

$$\tau_2 = \frac{\sec\phi_0 - \tan\phi_0}{\omega_0} = \frac{1 - \sin\phi_0}{\omega_0 \cdot \cos\phi_0}$$

$$\tau_1 = \frac{1}{\omega_0^2 \cdot \tau_2}$$

$$C_T = \frac{K_0 \cdot K_d}{N \cdot \omega_0^2} \cdot \sqrt{\frac{1+(\omega_0 \cdot \tau_1)^2}{1+(\omega_0 \cdot \tau_2)^2}} = \frac{K_0 \cdot K_d}{N} \cdot \frac{\tau_1}{\omega_0}$$

$$\tau_1 = RC_1 \qquad C_2 = \frac{\tau_2}{\tau_1} \times C_T$$

$$\tau_2 = R\left(\frac{C_1 \times C_2}{C_1 + C_2}\right) \qquad C_1 = C_T - C_2$$

$$C_T = C_1 + C_2 \qquad R = \frac{\tau_1}{C_1}$$

ϕ_0 = phase margin (°)
ω_0 = open loop gain crossover frequency (radians/sec)

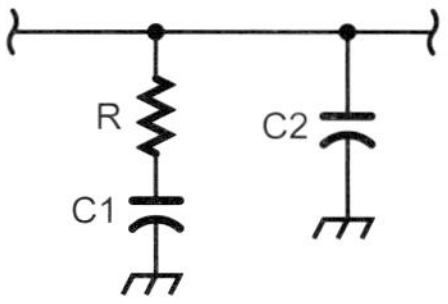

HBK0755

Figure 9.49 — Design equations for passive loop filters.

$$\tau_3 = \frac{\sec\phi_0 - \tan\phi_0}{\omega_0} = \frac{1-\sin\phi_0}{\omega_0 \cdot \cos\phi_0}$$

$$\tau_2 = \frac{1}{\omega_0^{\,2} \cdot \tau_3}$$

$$\tau_1 = \frac{K_0 \cdot K_d}{N \cdot \omega_0} \cdot \left| \frac{-j\omega_0\tau_2 - 1}{j\omega_0\tau_3 + 1} \right| = \frac{K_0 \cdot K_d}{N} \cdot \frac{\tau_2}{\omega_0}$$

$$\tau_1 = R_1 C_1$$

$$\tau_2 = R_2(C_1 + C_2) \qquad \frac{R_2}{R_1} = \frac{\tau_2 - \tau_3}{\tau_1}$$

$$\tau_3 = R_2 C_2$$

ϕ_0 = phase margin (°)
ω_0 = open loop gain crossover frequency (radians/sec)

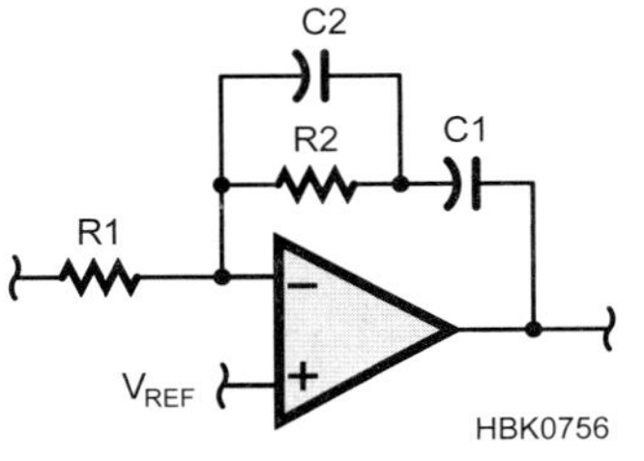

Figure 9.50 — Design equations for third-order active loop filters.

$$\tau_c = \frac{\sec\phi_0 + \tan\phi_0}{\omega_0} = \frac{1+\sin\phi_0}{\omega_0 \cdot \cos\phi_0}$$

$$\tau_a = \frac{2}{\omega_0^{\,2} \cdot \tau_c}$$

$$\tau_b = \frac{K_0 \cdot K_d}{N \cdot \omega_0} \cdot \frac{\tau_c}{2}$$

$$\tau_a = R_a \cdot C_a \qquad \frac{R_b}{R_a} = \frac{\tau_c}{\tau_b}$$

$$\tau_b = R_a \cdot C_b$$

$$\tau_c = R_b \cdot C_b$$

ϕ_0 = phase margin (°)
ω_0 = open loop gain crossover frequency (radians/sec)

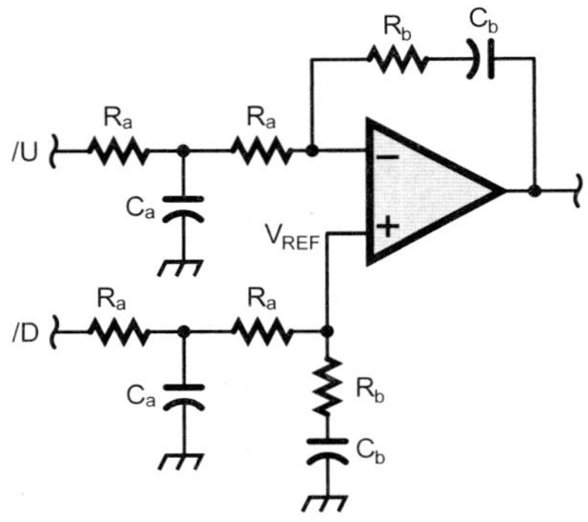

HBK0757

Figure 9.51 — Design equations for differential-input active loop filters.

A usual design strategy is to choose the resistors first to achieve the design ratio. This may provide several possible resistor pairs. Calculate the resulting capacitor values for each resistor pair possibility, and choose the set that provides the closest realizable values.

9.6.4 Fractional-N Synthesizers

The simplest frequency synthesizer is the single PLL with a programmable counter as a frequency divider in the feedback path. It can now all be integrated onto a single semiconductor device, though better performance may be available with some sections done discretely. These synthesizers are well suited to channelized radios where the channel spacing is fairly wide. They are common in VHF/UHF radios, cell phones, TV tuners and so on. However, apart from VHF/UHF FM transceivers, most amateur radio operation requires fine-resolution synthesizers to simulate the look and feel of a free-tuning VFO.

To this point in the chapter, the PLL synthesizer designs have assumed that the feedback divider operates as a single programmed integer value. This design is referred to as an *integer-N PLL*. These PLLs have an unavoidable trade-off of the step size versus every other performance parameter. Making a fine resolution loop by using a very high value of N forces a very low phase detector frequency and low open loop gain. This leads to extremely slow settling, bad phase noise, and poor suppression of spurs.

Multiple PLL loops can be used to give fine resolution without invoking other performance limitations other than cost, size, and power consumption. PLL/DDS hybrid designs can do the same job and save some cost, size, and power consumption. The holy grail is a fine resolution synthesizer with a clean output, small size, low cost and low power consumption but that has not yet been attained. There is another thread of development particularly suited to high levels of integration, although it, too, has limitations. It is called *fractional-N* or *frac-N synthesis*.

THE ORIGIN

A single PLL would have fine resolution, if the divide-by-N stage in its feedback path weren't constrained to divide only by integers. The dividers are implemented as digital counters, counting cycles of the VCO frequency. Counting in anything smaller than integers would need something running faster than the VCO to act as an interpolator. To get steps much finer than the increments given by integers would need an interpolator running at a very large multiple of the VCO frequency, which is impractical.

One solution is to vary the value of N so that the average value of N, taken over a long enough time period, gives the required resolution. This can be done, and it doesn't need impossibly fast logic. For example, consider the 2-meter synthesizer discussed in the previous section on Frequency Resolution: if N is varied in a repeating pattern of 8 divisions by 1474 followed by 2 divisions by 1475, the average is $(8 \times 1474 + 2 \times 1475)/10 = 1474.2$. A PLL operating with this type of variable division in the feedback loop is called a *fractional-N PLL*.

The resulting problem with such a PLL is that it is never exactly on the desired frequency. It switches between too-high and too-low, even though its average might be just right. We could view this as wide deviation FM, where the modulation is a rectangular pulse of controlled mark-space ratio. Such a spectrum will have high noise levels and huge sidebands which are quite undesirable.

Nevertheless, designers of such systems have tried to slow the loops so that they could not follow the switching and sat on the average frequency. The problem with obtaining sufficient filtering of the sidebands was that the loop became too slow and all the problems mentioned above appeared, making the solution worse than single-N synthesis.

ANALOG PHASE INTERPOLATION

Variable-N synthesizers were not used in amateur radio transceivers, but it was common in test equipment from the 1980s and 1990s without the tuning and noise requirements of radio equipment. This type of synthesizer is described in the context of using or repairing test equipment. (For example, see the HP 3335A Synthesizer service manual's Theory of Operation section.)

PLLs can be phase modulated, quite easily, within their bandwidths by summing in a modulating voltage after the phase detector. The phase effect of the switching of the N number can easily be calculated. A digital system can be made to do this, but an extra analog system must be added to interpolate the results of the digital system. This hybrid circuit can compute a phase modulation waveform which cancels all the effects of the frequency switching, but leaves the average intact. Cancellation is never perfect but manufactured synthesizers were developed that suppressed the unwanted sidebands to around –80 dBc (decibels with respect to the carrier level).

This is a complex system and can be hard to understand. Sometimes a second way of viewing it may be easier: If a ramp waveform is added after a phase detector in a PLL, the loop will be phase modulated with a ramping phase, or in other words, a frequency offset. If the ramp slope is controlled, any amount of resolution is possible. The problem is that ramps cannot go on forever, and the phase detector will soon hit the end of its range. The solution is to increase the N number of the divider just once at the same time that the ramp is reset by a cycle's worth of phase. A sampling system is used to disguise the disturbance of the transient phase shift and ramping resumes.

This is a large system needing careful, individual, trimming to get the analog circuitry to mesh seamlessly with the digital circuitry. The digital parts are integrated into a custom IC, but the discrete analog parts require a lot of board space. The performance was acceptable for mid-range equipment at the time. State of the art equipment used it as part of a hybrid structure as the least significant of three PLLs. The resulting fractional-N loop gave the system much finer resolution than the previous generation of 5-loop hybrids.

NOISE SHAPING

These early schemes switched the PLL divider between N and N+1 in simple, repetitive patterns, so the inevitable sideband energy was concentrated into large, obvious components. One variant of the filtering scheme used "rate multiplier" logic devices to control N to N+1 switching and reduced the amplitude of the unwanted components by spreading them out. While still not clean enough to be generally useful, other technologies were maturing: oversampling 1-bit DACs for audio, pseudo-random binary sequences for simulating telephone traffic for error rate testing, and the addition of dither to ADCs.

An integer-N PLL is a sort of DAC. You put data in as modulation, and the result is a change of frequency rather than voltage. The principles of an extreme oversampling DAC can be applied to it just as much as to the output buffer of a 1-bit DAC in a CD player. These systems scramble the sideband energy and make it much less conspicuous — a good thing. The sideband energy is also spread over a wider frequency range than the bandwidth of the loop filter. This is even better because it means more energy can be filtered out. Further, the scrambling process can be engineered to control the spectrum of the noise-like sidebands, called *noise shaping*, pushing most of the noise energy high enough in frequency that filtering can do a good job. The logic systems which shape the noise are higher-order sigma-delta modulators (See the **DSP and SDR Fundamentals** chapter). They take up a lot of logic elements on an IC, but can be highly integrated and are therefore small, cheap, and have low power consumption. As a result, the noise-shaped fractional-N synthesizer looks like a simple PLL with a large amount of logic processing the data fed to its programmable divider.

Delta-modulators above the second order are inherently unstable. Audio converters have various proprietary schemes to stabilize them, and these little subtleties are carefully guarded secrets. The MASH (Multi-stAge noise SHaping) DAC arrangement adds together a series of low-order modulators, rather than trying to make one very high-order modulator. Fixing the stability problem requires that the output doesn't just jump between two states, N and N+1 in PLL terms, but it dances over a limited range.

LIMITATIONS

With the division factor varying in a pseudo-random way, the operating point of the phase detector is also varying over a wide range and the phase detector needs to have a similarly wide operating range. Since filtering only happens *after* the phase detector and large amounts of noise at high offset frequencies are present, the phase detector must be very linear.

The noise-shaping scrambler keeps the close-in frequency range clean, but any nonlinearity in the phase detector allows the strong high frequency noise components to intermodulate. Intermodulation creates products at close-in frequencies, spoiling the noise performance of the close-in range.

This limits how clean the noise-shaping synthesizer can be made. The best examples are useful for many purposes but still aren't good enough for a high-grade HF receiver or any system requiring superior adjacent channel rejection. For such uses, they are combined with a high-performance integer-N PLL in a hybrid structure similar to how the original fractional-N loop was used.

AVAILABILITY

Fractional-N synthesizers have been available in chip form for several years. Analog Devices, Texas Instruments, and other manufacturers have offered families of PLL ICs for a long time, and their ranges include devices with all the noise shaping logic on-board. Many even have GHz VCOs on-board as well, with dividers down to more mundane frequency ranges.

These synthesizers have become an RF design building block. The RFMD RFFC2071, for example has a pair of medium-level active mixers with an entire synthesizer all on one die. Handheld transceivers that also receive 0.5 to 999.999999 MHz are likely to be made with these parts. If you are choosing one of these devices, look carefully at the noise sideband performance: the higher frequency version of the RFFC2071 IC has noise sidebands a few dB lower than the lower frequency version.

WHAT NEXT?

The RF semiconductor industry is currently focused on mass-market applications such as mobile phones, WiFi and so on. In particular, the requirements for high-speed data links (e.g., 5G) and the RF Instrumentation needed to test them continue to drive advances in integrated synthesizer, VCO, and DDS development with increases in operating frequency as well as decreasing phase noise in PLL/VCOs and the release of integrated vector signal generators such as the AD9164 with a 48-bit DDS operating as high as 12 GHz.

Development of fractional-N systems is within the range of the interested amateur, either by using the general-purpose parts on the market, or by developing their own using programmable logic array devices. A moderately-sized FPGA contains enough logic for a number of fractional-N synthesizers. For example, see the July/August 1998 *QEX* article by Ulrich L. Rohde, N1UL "A High-Performance Fractional-N Synthesizer" which describes a design patented in 2001.

A figure of merit for a fractional-N system is the maximum frequency at which the phase detector can operate. Higher frequencies allow the scrambled noise to be spread over a greater frequency range, making it easier to filter. The obvious approach is to develop faster and faster logic, but there is an alternative: Several fractional-N dividers with several separate phase detectors may have their outputs combined. The phases of the outputs from the separate dividers may be offset from each other by programming registers differently in each scrambler, and the reference signals to the phase detectors may be offset to match.

Several phase detectors can thus act sequentially over each detector's cycling period, making a system with an effective phase detector frequency several times that of a single-phase detector. The effective phase detector frequency can even be higher than the output frequency. A further elaboration would be to seed the pseudo-random noise generators in each divider differently. In this way the noise components from each divider will not correlate, though the tuning components correcting the VCO frequency will correlate. This means that the tuning components add more strongly, as voltages, while the noise components add more weakly as power. Each doubling of the number of dividers has the potential for a 3 dB improvement in noise performance of the system. This multiple divider arrangement is currently covered by a patent, but individuals are free to read them, to experiment and to perhaps find a better way. Patents eventually expire and their technology enters the public domain. Each increase in the density of programmable logic makes it easier.

9.6.5 A PLL Design Example

As our design example, let us consider a synthesized local oscillator chain for a 10 GHz transverter. **Figure 9.52** is a simplified block diagram of this 10 GHz converter. This example is chosen because it is a departure from the traditional multi-stage multiply-and-filter approach. It permits realization of the oscillator system with two simple loops and minimal RF hardware. It is also representative of what is achievable with current hardware, and can fit in a space of 2 to 3 square inches. This example is intended to be a vehicle to explore the loop design aspects and is not offered as a "construction project." The multiple design details required are beyond the scope of this chapter.

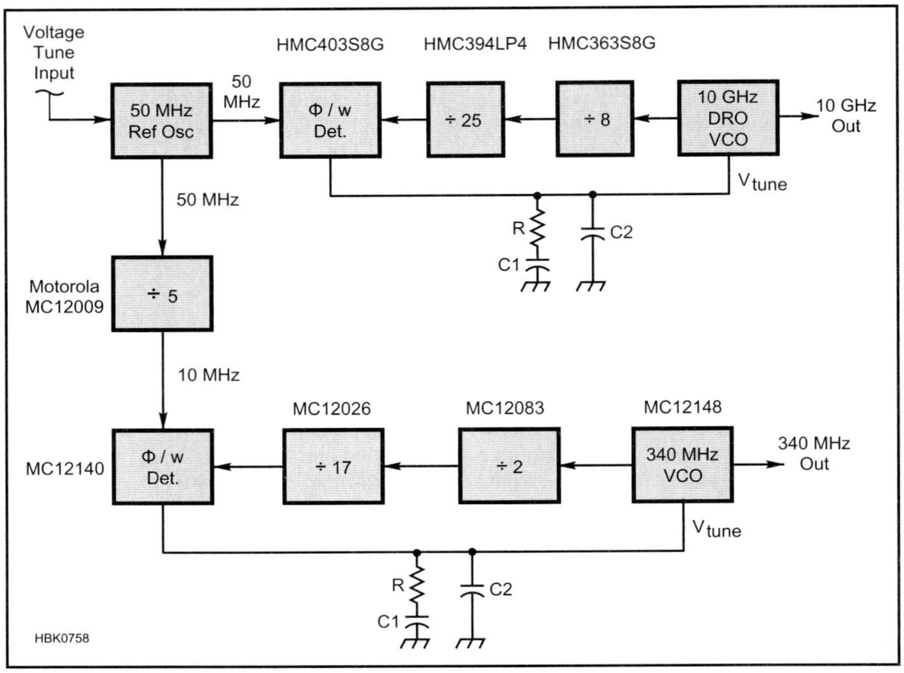

Figure 9.52 — A simplified block diagram of a PLL local oscillator for a 10 GHz converter.

Two synthesized frequencies, 10 GHz and 340 MHz, are required. Since 10.368 GHz is one of the popular X-band traffic frequencies, we initially mix this with the 10 GHz LO to produce an IF of 368 MHz. The 368 MHz IF signal is subsequently mixed with the 340 MHz LO to produce a 28 MHz final IF, which can be fed into the 10-meter input of any amateur transceiver. We focus our attention on the design of the 10 GHz synthesizer only. Once this is done, the same principles are applied to the 340 MHz section. Our goal is to design a low-noise LO system (by adopting minimum division ratios) with a loop reference oscillator that is an integer multiple of 10 MHz. Using this technique allows the entire system to be locked at a later time to a 10 MHz standard for precise frequency control.

For the microwave synthesizer we consider using a line of microwave integrated circuits made by Analog Devices in gallium-arsenide (GaAs) material. These devices include a selection of prescalers operating to 12 GHz, a 5-bit counter that operates to 2.2 GHz and a phase/frequency detector that operates up to 1.3 GHz. If we use a crystal reference frequency f_{XO} of 100 MHz, and also apply that as f_{REF} into the PFD (therefore R = 1), then the feedback divider number needs to be N = 10,000/100 = 100. One way to realize this in hardware is to represent 100 = 4 × 25, starting with a divide-by-4 prescaler and finishing with the 5-bit counter programmed to divide by 25. The divide-by-4 prescaler output at lock will be 2500 MHz, which is too high for the 2200 MHz limited programmable counter. We need a design change.

The obvious solution is to select a divide-by-8 prescaler, providing a 1250 MHz output from the 10 GHz input. We cannot program the counter to half of 25, so we need to leave it programmed at 25. This makes the output from the feedback divider at 10,000/(8×25) = 50 MHz. We need to set R = 2.

Before even thinking about designing the loop filter, we need to know the VCO gain, VCO noise performance, divider noise performance, phase detector gain, phase detector noise performance and finally reference noise performance. For the 10 GHz VCO, we are always looking for parts that are easily available, useable, and economical, so salvaging a dielectric resonator from a Ku band LNB is promising (see "SHF Super Regenerative Reception" by Andre Jamet, F9HX, in Jan-Feb 2002 *QEX*). These high-Q oscillators can be fitted with a varactor and tuned over a limited range with good results. The tuning sensitivity of our dielectric resonator VCO is about 10 MHz per volt and the phase noise at 10 kHz offset is –87 dBc/Hz, and –107 dBc/Hz at 100 kHz.

The phase detector and divider information is available from the device data sheets. The HMC363 divide-by-8 operates up to 12 GHz and has a programmable charge pump that sets the phase detector gain K_d. It is usually good to keep this gain high, so we choose to use a 2-mA charge pump current, giving K_d = 0.32 mA/radian. The data sheet also says that this PFD has an output noise floor of –153 dBc/Hz measured at 100 kHz offset. The HMC394 programmable divider operated up to 2.2 GHz with the same output noise floor. The HMC984 PFD operated up to 350 MHz with a noise figure of merit (FOM) of –231 dBc/Hz/Hz-f_{REF}, so with f_{REF} = 50 MHz the PFD output noise floor is –231 + 10 log (50,000,000) = –154 dBc/Hz. Assuming the R divider is implemented in CMOS, the output noise floor from it is –163 + 10 log (50,000,000/125,000) = – 137 dBc/Hz. This noise is much higher than the noise from the GaAs dividers, so it is actually better here to use a GaAs divide-by-2 for the R counter instead of using CMOS.

The logic noise floor for the entire PLL is the sum of the individual noises from each divider and the PFD. The result is –147 dBc/Hz. This is determined at the PFD. To calculate how this noise will measure at the PLL output we need to add 20 log N = 46 dB. Thus, the output noise floor due to PLL logic devices is –101 dBc/Hz. This is close to the VCO phase noise at an offset of 100 kHz, so minimum noise design suggests that we select 100 kHz as our loop bandwidth.

An alternative is to eliminate the R counter and directly use a 50 MHz reference. An excellent choice for a low noise reference is the one described by John Stephensen, KD6OZH in Nov/Dec 1999 *QEX*. The noise

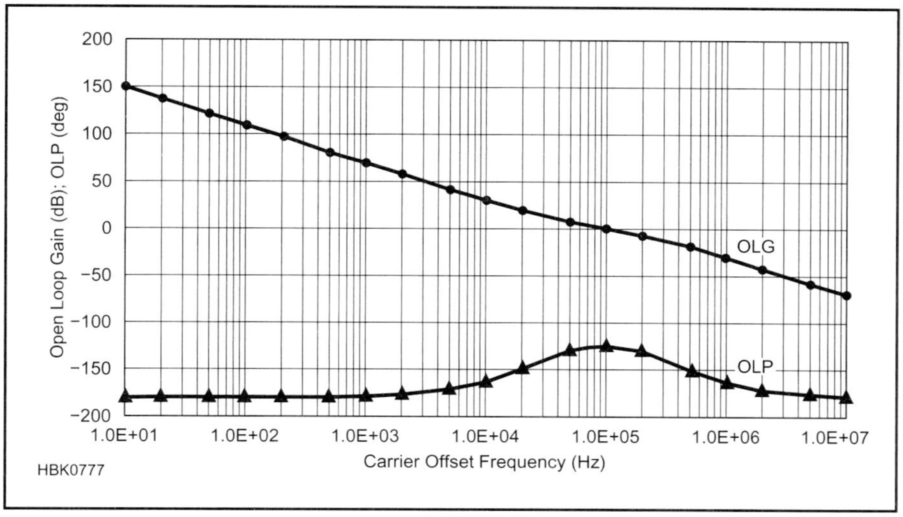

Figure 9.53 — Comparisons of the theoretical and actual PLL dynamics show extremely good correspondence.

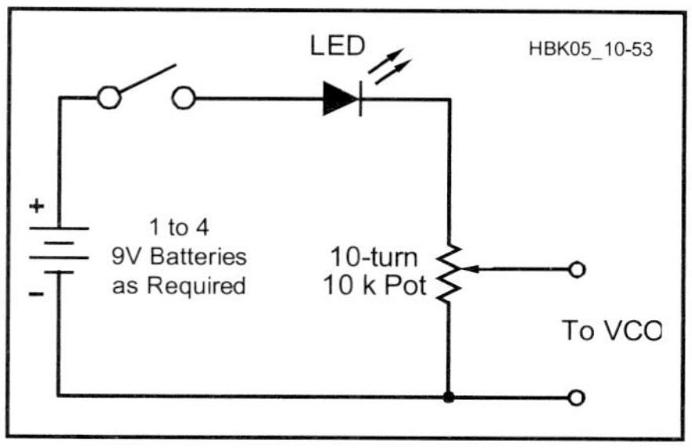

Figure 9.54 — Clean, variable dc voltage source used to measure VCO gain in a PLL design.

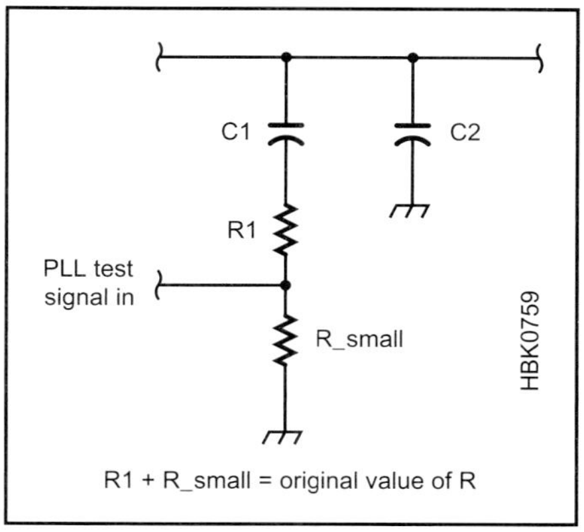

Figure 9.55 — A simple testing structure for PLL dynamics.

performance of this VCXO is in the order of –160 dBc/Hz at 10 kHz offset at the fundamental frequency. Translating this to the output frequency means adding 20 log N, for a result of –114 dBc/Hz. Eliminating the R counter also lowers the logic noise floor to –104 dBc/Hz at the output. The translated reference oscillator noise is 10 dB below the noise floor from the logic devices, so the logic noise floor dominates.

Using this information, we can now design the loop filter. Assuming that the VCO tuning voltage is between 0.5 and 4.5V we can directly use the charge pump included in this PFD and therefore select the passive third order loop filter from Figure 9.46. We use the equations in Figure 9.44 and calculate loop filter component values of R = 6,977 ohms, C1 = 724 pF, and C2 = 80 pF. We therefore select 6.8k, 820 pF, and 82 pF for these component values respectively. The Bode plot of the theoretical and actual PLL dynamics is shown in **Figure 9.53**. It is nearly impossible to tell them apart.

If the VCO requires higher tuning voltages, you can insert a non-inverting op-amp gain stage between the output of this loop filter and the VCO. Choose a low input current op-amp to minimize leakage from the loop filter capacitors. Be careful — the gain of this amplifier stage changes the effective VCO gain that the loop experiences. When designing the loop filter components, the "design" VCO gain must be replaced by the product of the actual VCO gain and the amplifier gain. It is also essential to assure that the 1 dB bandwidth of the voltage amplifier greatly exceeds the loop bandwidth so that it does not degrade the PLL phase response and hence loop stability.

9.6.6 PLL Measurements and Troubleshooting

VCO GAIN

One of the first things we need to measure when designing a PLL is the VCO gain. The tools needed include a voltmeter, some kind of frequency measuring device like a receiver or frequency counter and a clean source of variable dc voltage. The circuit in **Figure 9.54** containing one or more 9 V batteries and a 10-turn, 10 kΩ pot does nicely. One simply varies the voltage some amount and then records the associated frequency of the VCO. The gain of the VCO is then the change in f divided by the change in voltage (Hz/V).

LOOP BANDWIDTH

Measuring the actual bandwidth of the PLL usually means measuring the entire gain and phase responses of the PLL. There is a much easier way that needs only a square-wave generator and an oscilloscope. This method uses the PLL testing structure of **Figure 9.55** developed by Glenn Ewart that injects a low-value square wave into the loop at a low impedance point so that impulse response can be observed directly. The test signal is also ground-referenced and is independent of the loop locking voltage.

Start by swapping the series RC components in the loop filter so the resistor is connected to ground and the capacitor attaches to the VCO tuning line. Then split the resistor into two resistances that together add up to the total resistance needed by the filter. The bottom resistor connected to ground is very small — 50 Ω is a nice choice when the total resistor value is much greater than 50 Ω. The top resistor is temporarily replaced by a potentiometer that can make up the desired remaining resistance. Across the bottom resistor we connect the square-wave generator and set it to a small amplitude (100 or 200 mV is common) at a frequency a few percent of the loop bandwidth, and operate the PLL. Looking at the VCO tuning line with an oscilloscope we see the impulse response of the PLL.

Now adjust the potentiometer way off-value to the low side. The impulse response will now show ringing. The frequency of this ringing is a good approximation of the PLL loop bandwidth.

SETTLING TIME

Having measured the loop bandwidth, readjust the potentiometer back to its nominal value. The impulse response should look *much* cleaner! It is tempting to look at the settling time of the impulse response and say that this is the settling time of the PLL. This unfortunately is not quite true.

The PLL is not fully settled until capacitor C1 is fully charged. This usually takes slightly longer than what the impulse response itself shows. Move the scope probe to the "ground" side of C1 and measure when this voltage reaches zero. This means there is no more current flowing through the filter resistors and the capacitor is fully charged.

TROUBLESHOOTING PLLS

Here are some frequently encountered problems in PLL designs:

• The outputs of the phase detector are inverted. This results in the loop slewing to one or the other power supply rails. The loop cannot possibly lock in this condition. Solution: Swap the phase detector outputs.

• The loop cannot comply with the tuning voltage requirements of the VCO. If the loop runs out of tuning voltage before the required voltage for a lock is reached, the locked condition is not possible. Solution: Re-center the VCO at a lower tuning voltage or increase the rail voltages on the op amp.

The loop is very noisy and the tuning voltage is very low. The tuning voltage on the varactor diodes in the VCO should not drop below the RF voltage swing in the oscillator tank circuit. Solution: Adjust the VCO so that the loop locks with a higher tuning voltage.

9.6.7 Commercial Synthesizer ICs

In this section, we explore using commercially available synthesizer chips and the role of DDS in more recent hybrid architectures. The following is not intended to be project oriented, but rather is designed to expose the reader to additional concepts that cannot be fully explored here. The reader, being made aware of these ideas, may wish to examine them in detail using references at the end of this chapter.

Many synthesizer chips have been introduced in recent years for cellular and Wi-Fi applications and have some interesting properties and capabilities that can be exploited with additional design. Some properties that can be exploited are programmable charge pump current, a rich set of division options for creating multiple-loop synthesizers and typically low power consumption as will be shown in the next section.

For instance, an integrated PLL/VCO could be used as a 704 to 640 MHz synthesizer and its external divider set to 128. This division would reduce the phase noise profile by 40 dB, making a reasonably quiet synthesizer for the 5 to 5.5 MHz range. The step size would also be reduced by a factor of 100, making the spacing required in the 500 to 550 MHz range equal to 1 kHz for a step size of 10 Hz at 5 MHz. This would make a good local oscillator for a simple traditional radio that covers 80 and 20 meters using both mixer products against a 9 MHz SSB generator.

There are also other techniques that can prove helpful. Decoupling the VCO from the chip will permit one to avoid much of the "on chip" noise that VCOs are susceptible to at the expense of some more complexity. If the VCO is implemented externally, there is an opportunity to design it with increased operating Q, thereby improving the overall phase noise performance. An external VCO also opens other opportunities.

As mentioned earlier, these chips are designed for operation at a V_{CC} of 3 to 5 V, which limits the output tuning voltage swing to V_{CC}. One must be able to fit the entire tuning range voltage and the ac voltage excursion in the VCO tank circuit within the V_{CC} range. As mentioned at the end of the earlier section Improving VCO Noise Performance, this problem can be overcome with the addition of voltage gain in an external operational amplifier, running at a higher voltage. This allows the signal-to-noise ratio of the oscillator to be improved by increasing the tank voltage swing and still having adequate voltage range to perform the tuning function. There are certainly more examples of how the performance of these devices could be improved for amateur applications.

9.6.8 Synthesizer Architectures Using Commercial ICs

INTEGER-N VERSUS FRACTIONAL-N PLLS VERSUS DDS

One of the disadvantages of a single loop PLL frequency synthesizer is that the PFD frequency determines the step size of the output frequency, that is,

$$f_{OUT} = N \times f_{PFD}$$

where

$$f_{PFD} = \frac{f_{REF}}{R}$$

For two-way radio and other systems with fixed channel spacing this is generally not a problem. However, for CW and SSB operation, tuning steps of 10 Hz or smaller are almost a necessity. Both integer and fractional-N synthesizers can tune 10 Hz steps, but in a fractional-N PLL the cutoff frequency of the loop filter is still set by the channel spacing even if the PFD frequency is much higher. In the days of analog FM systems that used 25 kHz channel spacing, this wasn't an issue but a 10 Hz step is the same as 10 Hz channel spacing, which mandates an extremely long settling time in the loop filter. In an integer-N PLL, the PFD frequency is the same as the step size, so in this case both choices have the same settling time, set by the step size.

On the other hand, a DDS can easily tune in 10 Hz steps as fast as the steps can be programmed. However, it can have both phase truncation spurs (typically not an issue but worth mentioning) and (worth mentioning) spurs from the nonlinearity of its output DAC. Also, the resolution of the output DAC sets the noise floor, being approximately 6 dB per bit. Thus, the theoretical noise floors for 12-bit and 14-bit DACs are roughly 72 and 84 dB, respectively.

But the drawback of a DDS chip is its frequency coverage; the DDS chips used in amateur radio designs typically accept a 400 MHz clock input and provide a maximum output frequency that is 40% of that, or 160 MHz, using the industry rule-of-thumb of 2½ points per cycle. With none of these solutions having an obvious advantage, this brings us back to the original question of Integer-N versus fractional-N PLLs versus DDS: Which should we use?

Let's look at some modern synthesizer topologies, starting with the simple PLL/VCO combination shown in **Figure 9.56**, which is typically used for fixed-frequency clock or LO generation. Here, the VCO can be separate or integrated with the PLL and the PLL can be either an integer-N or fractional-N. Figure 9.56's output frequency is given by

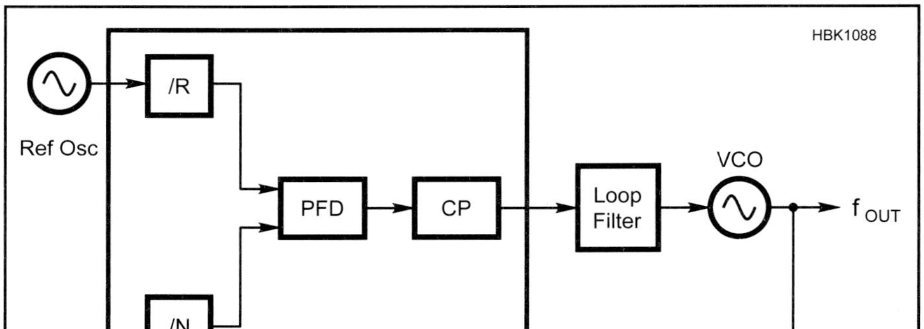

Figure 9.56 — An Integer-N PLL and VCO are the basic building blocks of frequency synthesizers.

$$f_{OUT} = \frac{N}{R} f_{REF}$$

where

f_{out} is the output frequency of the VCO,

N and R are the values of the N and R dividers, respectively, and

f_{ref} is the comparison frequency of the phase-frequency detector (PFD).

For the sake of example, let's say we are designing a 2 meter (144-148 MHz) all-mode receiver with a first IF of 45 MHz and high-side injection ($f_{LO} > f_{RF}$), giving us an LO tuning range of

144+45 = 189 MHz as the lowest LO frequency to

148+45=193 MHz as the highest LO frequency.

We will choose 25 MHz as f_{REF} and R=1 for several reasons. First, 25 MHz crystals are common. Second, no harmonics of 25 MHz fall either on the first IF (45 MHz), in the 2 meter band (144-148 MHz), or in the LO tuning range (189-193 MHz). And third, 25 MHz may also be used to clock the system's microprocessor, which minimizes both the number of oscillators to reduce cost and the number of "birdies" (low-level, in-band mixing products) heard in the receiver.

With R=1 and using the middle of the VCO's tuning range, 191 MHz, as f_{OUT},

$$N = \frac{f_{OUT}}{f_{REF}} = \frac{191}{24} = 7.64$$

where N_{INT}= 7 and N_{FRAC} = 0.64. Here, because N turns out to be a fractional value,

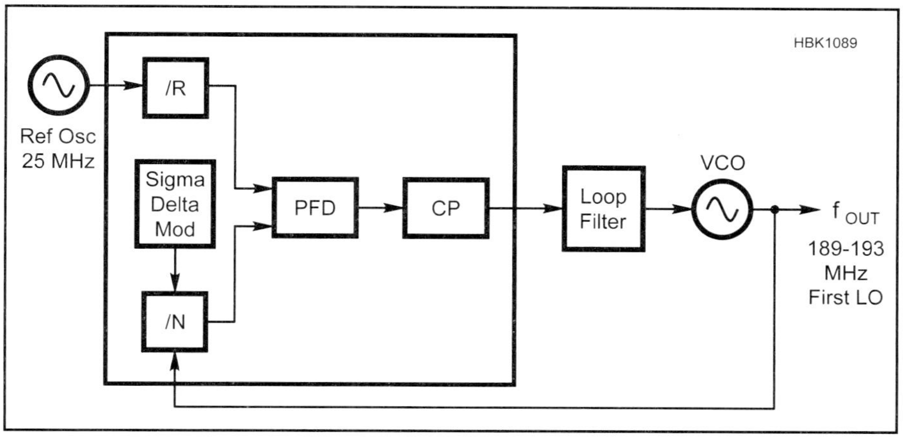

Figure 9.57 — A proposed fractional-N Synthesizer for the 2-meter band.

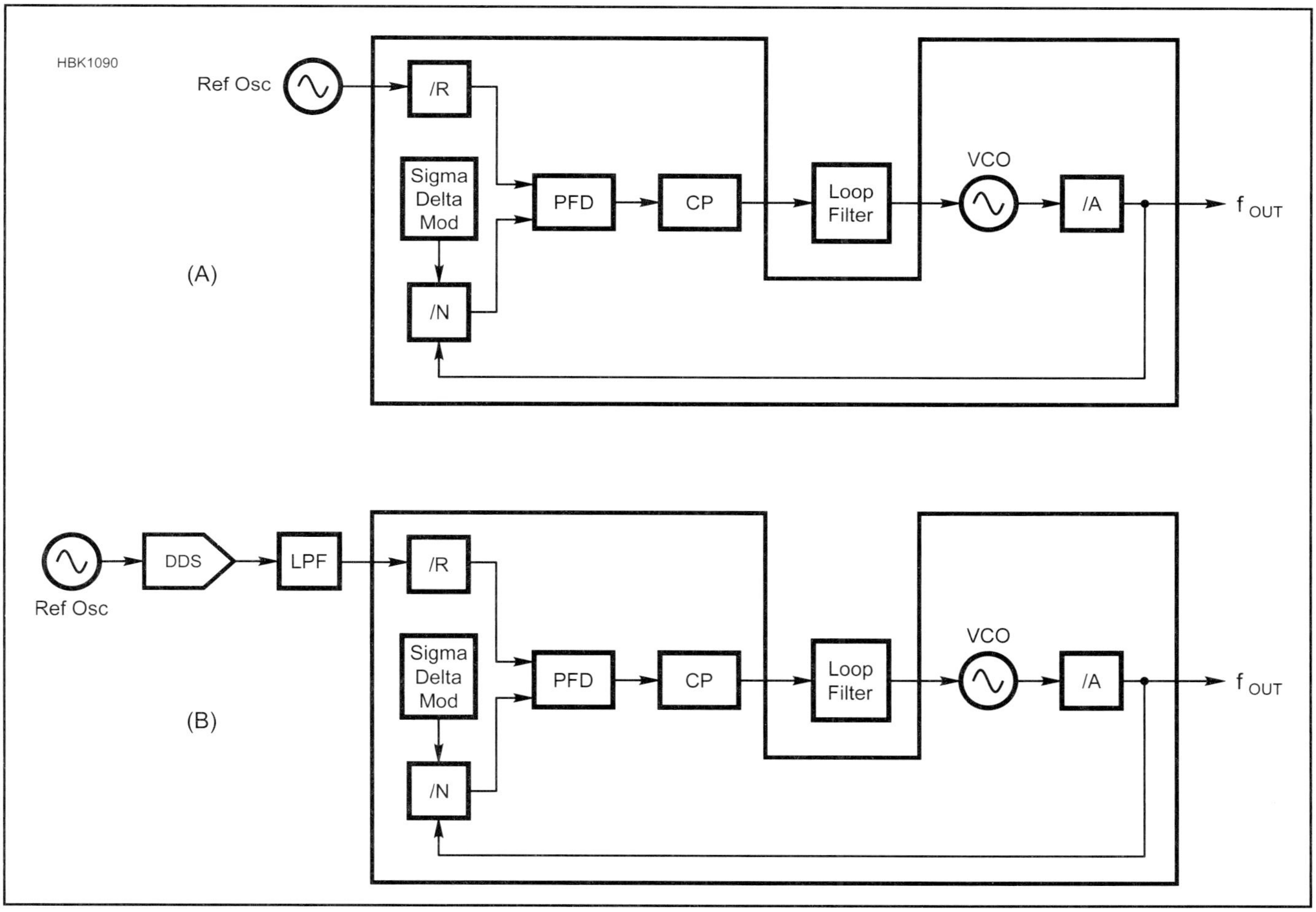

Figure 9.58 — A proposed synthesizer for the 2-meter band using a Fractional-N PLL with an integrated VCO and an auxiliary output divider. Suitable candidates include the Analog Devices ADF4351, which covers 35 – 4400 MHz; the STmicroelectronics STW81200, which covers 46.857 to 6000 MHz; and the Texas Instruments LMX2581, which covers 50 to 3760 MHz. Using a DDS as the reference allows 1 Hz resolution or better, depending on the resolution of the DDS.

we'll use a fractional-N PLL, which has a sigma-delta modulator.

A possible difficulty here is finding a fractional-N PLL with an N value as low as 7; many PLLs have prescalers that require a minimum N value in the 100s or even higher. In this case we would choose a PLL without a prescaler. **Figure 9.57** shows our completed design.

INTEGRATED PLL/VCOs

Fortunately, our baseline design and can also be implemented by using a fractional-N PLL with an integrated VCO and auxiliary divider A, as shown in **Figure 9.58**. The purpose of the auxiliary divider is twofold: it both improves the phase noise by 3 dB for each divide-by-two in the A divider and extends the frequency coverage of the PLL/VCO combination; it does this by dividing the integrated VCO's output frequency by 1, 2, 4, t etc. to turn a narrowband synthesizer into a wideband one. Figure 2's output frequency *at the output of the auxiliary divider A* is given by

$$f_{OUT} = \frac{f_{VCO}}{A} = \frac{N}{AR} f_{REF}$$

where

f_{OUT} is the output frequency of the PLL/VCO,

N and R are the values of the N and R dividers, respectively,

A is the value of the auxiliary divider, and

f_{ref} is the comparison frequency of the phase-frequency detector (PFD).

With this level of integration, integrated PLL/VCOs are specified by not only their available R, N, and A values but also their output range from the minimum possible output frequency to the maximum available output frequency.

Continuing our 2-meter design example, we need an integrated PLL/VCO with an output range that spans 189 to 193 MHz to provide the LO signal and a 25 MHz PFD frequency. Suitable candidates include the Analog Devices ADF4351, which covers 35-4400 MHz; the STmicroelectronics STW81200, which covers 46.857 to 6000 MHz; and the Texas Instruments LMX2581, which covers 50 to 3760 MHz. All three suppliers provide free PLL design software to use with these products.

The data sheet for the PLL/VCO may only have phase noise information (sometimes a single frequency measurement but more often a graph) with A=1 for the auxiliary output divider. In cases like this, you can use the following approximation:

$$PN_{dBc/Hz} = PN_{PLL/VCO\,(A=1)} - 3\frac{\log A}{\log 2}$$

where

$PN_{dBc/Hz}$ is the total output phase noise at the desired offset from the VCO frequency,

PN_{PLL} is the phase noise of the PLL/VCO at the same offset from the output frequency, usually listed as a typical specification on the PLL/VCO's data sheet, and

A is the value of the output divider.

The last piece of the equation (log A/log2) may need further explanation: it calculates the power of two used by the output divider. For example, if A=16 (2^4), then log A/log2 = log 16/log 2 = 4. Given that each divide-by-2 *decreases* the phase noise by 3 dB, the reduction in phase noise here is 4 × 3 dB = 12 dB.

So far, so good. But what if we want really small frequency steps, say 100 Hz or less. How do we get there? Let's use the circuit in Figure 9.58A and make tunable by using a DDS IC as the reference. Figure 9.58B's output frequency *at the output of the auxiliary divider* is now given by

$$f_{OUT} = \frac{N}{AR} f_{DDS}$$

where f_{DDS} is the output frequency of the DDS IC.

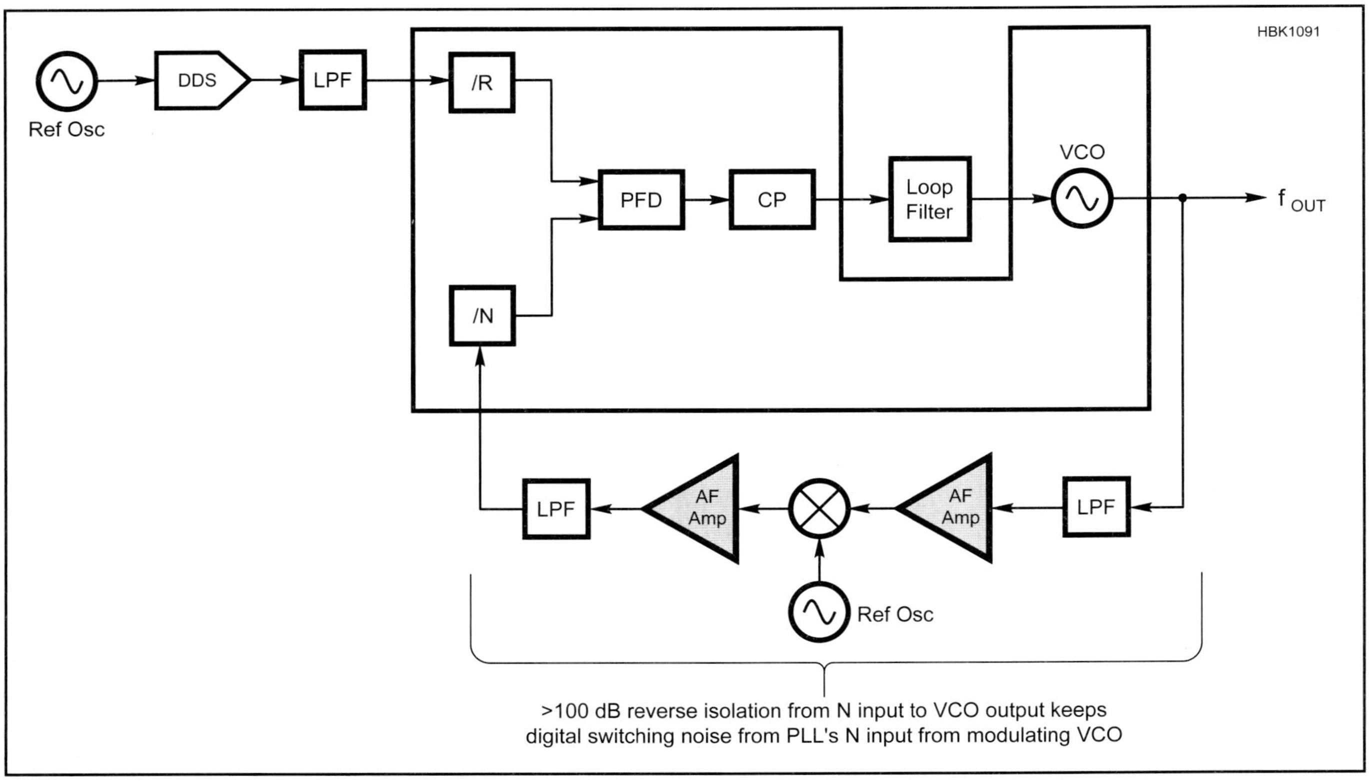

Figure 9.59 — A Frequency Translation Loop eliminates the noise introduced by the N divider in a frequency synthesizer at the expense of higher complexity.

For best performance, the DDS must be clocked by a low-phase-noise source such as a low-phase-noise crystal oscillator or a temperature-compensated crystal oscillator (TCXO) with a PLL/VCO and *not* a low-cost CMOS clock module used for microcontrollers.

Finally, note that the N divider multiplies the PFD frequency, which in turn comes from output of the R divider and before that the reference oscillator; any phase noise or drift from on the PFD's reference input signal is multiplied by N as part of the normal phase-locking process. Is there a way to reduce or eliminate this? Yes, by making N =1.

Figure 9.59 shows such a circuit, called a *frequency translation loop*. In this circuit, the N feedback divider is replaced by a mixer and the N divider is set to N=1. PLLs with built-in prescalers have minimum N values much greater than 1. If this is the case, then the R input can be used instead of the N input and the N input can be driven by the reference.

Assuming a PLL with N=R=1, Figure 9.59's output frequency can be shown to be

$$f_{OUT} = f_{PFD} + f_{EXT}$$

where

f_{OUT} = the synthesizer's output frequency in Hz,

f_{PFD} = the phase-frequency detector's comparison frequency, and

f_{EXT} = the external frequency driving the mixer in the feedback circuit

When f_{REF} is supplied by a DDS as shown here, then the output resolution of the synthesizer is that of the DDS for R =1 ($f_{PFD} = f_{DDS}$) and even smaller for R>1 (that is, $f_{PFD} = f_{DDS}/R$). Because the DDS determines the frequency resolution of this synthesizer, an integer-N PLL rated to only a few hundred MHz may be used with a high loop bandwidth for lower cost and faster frequency settling time. This architecture is used in such RF instruments as spectrum and network analyzers.

Typically, such circuits use either a PLL that allows N = R = 1 or a standalone PFD and charge pump. But what if N has a minimum value of, say, 100, due to a built-in prescaler? In this case, you could swap the N and R inputs: the R input is now the feedback input and the N input is now driven by the reference oscillator. Next use a reference oscillator based on a 2 GHz source, and set N at the lowest value that makes sense for the PFD frequency.

In a practical implementation of a translation loop synthesizer, the feedback path must have sufficient (e.g., 100 dB or more) isolation between the feedback input and the VCO to prevent digital switching noise from the PLL's digital dividers from feeding back to the VCO.

The choice of active or passive mixer can make a difference to the overall isolation in the feedback path. Let's look at the port-to-port isolation of a solid-state mixer compared to that of a diode ring mixer. To simplify the comparison, assume both mixers have identical 30 dB port-to-port isolation between all ports.

At first, all things seem equal with the same port-to port-isolation. However, the difference comes out when the LO drive levels are compared: an active mixer typically requires –10 to 0 dBm LO drive; a passive mixer typically requires +7 dBm or higher LO drive (**Figure 9.60**). Assuming identical port-to-port isolation specifications, the passive mixer can easily have 17 dB or more worse leakage between ports in this application. In practice, an active mixer may have 5 to 10 dB better port-to-port isolation than a passive mixer. Read the data sheets carefully and include the LO drive levels in your comparison.

A different approach is taken in **Figure 9.61**, showing a synthesizer with fractional division in the feedback loop but no fractional spurs. How does this work? The circuit uses a DDS, which functionally acts as a fractional divider. The key is that a DDS IC uses a DAC

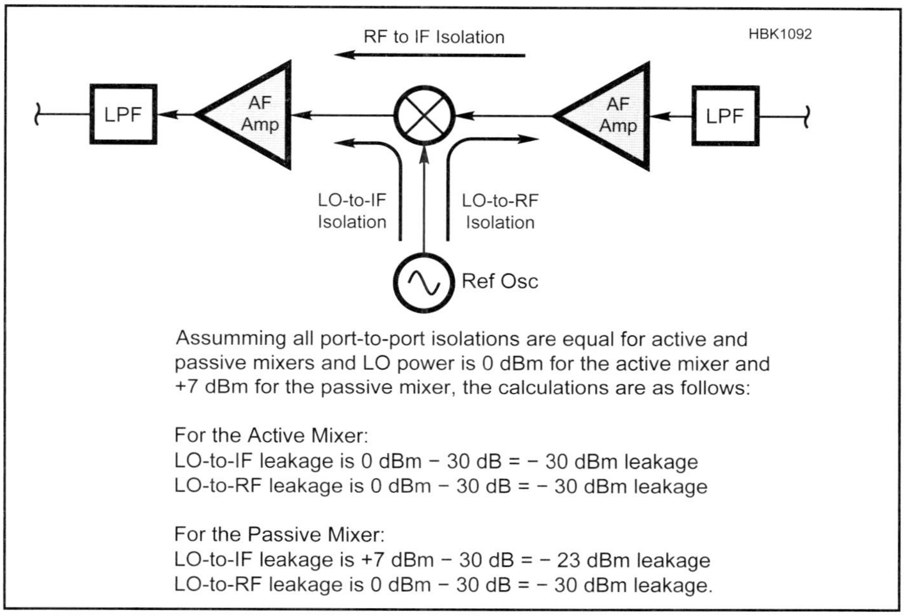

Figure 9.60 — Comparing the port-to-port isolation of active and passive mixers requires calculating the effect of the LO drive level on port-to-port leakage paths in a mixer.

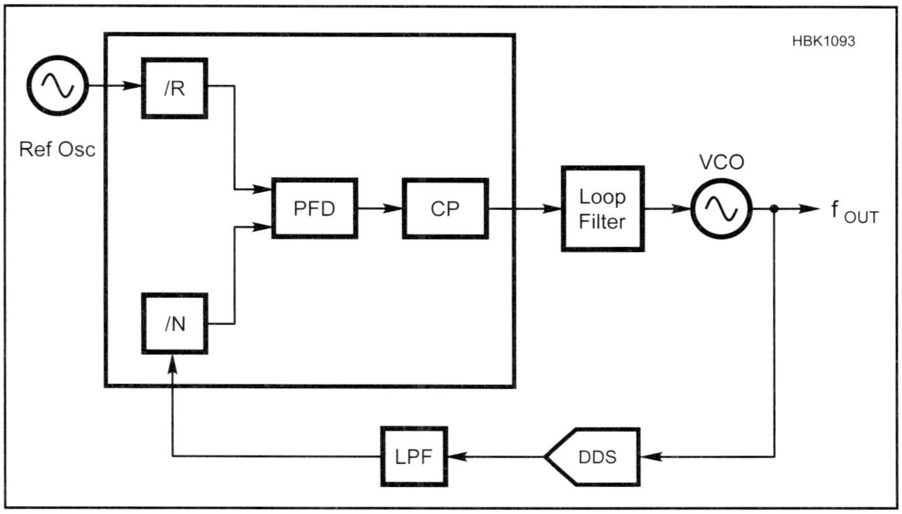

Figure 9.61 — Using a DDS as the "fractional divider" in the feedback path provides the functionality of a fractional-N PLL without the fractional spurs, resulting in a cleaner output spectrum: the DDS's output into the PFD is a single frequency synthesized by a DAC rather than the multiple frequencies created by a fractional-N divider.

and clock to generate a sinewave point-by-point; at all times the output is a single frequency. In contrast, a fractional-N PLL creates numerous output frequencies whose average is the desired frequency. Even with the averaging effect of the charge pump and loop filter, some vestige of the fractional spurs can modulate the VCO and show up in the output. Using the DDS allows a higher cutoff frequency in the loop filter and faster settling time.

A fractional-N divider, in contrast, uses the average of multiple divide-by-N cycles to create its fractional output frequency. The loop filter has to smooth out the variations in the VCO's tuning voltage caused by the multiple N divider values. What doesn't get filtered out becomes a fractional spur, hence the use of a DDS.

The VCO's output frequency also serves as the clock input, f_{clk}, to the DDS, that is, $f_{out} = f_{clk}$. The output of the DDS IC (f_{DDS}), and is one of the inputs to the PFD. Assuming R=1 so $f_{REF} = f_{PFD}$ and working our way back from the VCO to the PFD, these relationships give us

$$f_{REF} = f_{DDS} = \left\{\frac{FTW}{2^M}\right\} f_{CLK} = \left\{\frac{FTW}{2^M}\right\} f_{OUT}$$

Substituting for f_{DDS} and solving for f_{OUT} yields

$$f_{OUT} = \left\{\frac{2^M}{FTW}\right\} f_{REF}$$

Finally, solving for the *frequency tuning word* (FTW),

$$FTW = 2^M \frac{f_{REF}}{f_{OUT}}$$

where

f_{DDS} is the DDS's output frequency, in Hz,
f_{OUT} is the VCO's output frequency, in Hz,
f_{CLK} is the DDS's clock input frequency, in Hz, and
2^M is the fixed resolution of the DDS, here, m=32.

Using the same frequency plan as in our previous examples, that is, a 2 meter receiver covering 144-148 MHz with a 45 MHz first IF, high-side injection from a 189-193 MHz LO and a 25 MHz reference frequency, let's calculate the FTW. We'll use the midpoint of the VCO's 189-193 MHz tuning range in our calculations:

$$FTW = 2^M \frac{f_{REF}}{f_{OUT}} = 2^{32} \frac{25\ MHz}{191 MHz} =$$

$$4,294,967,296 \times \frac{25}{191} = 562,168,494.241$$

Rounding off the FTW to the nearest integer (562,168,494)and checking our work we get

$$f_{OUT} = \left\{\frac{2^M}{FTW}\right\} f_{REF} =$$

$$\left\{\frac{2^{32}}{562,168,494}\right\} 25\ MHz = 191.000000082\ MHz$$

Note that the frequency error is 0.082 Hz, which, besides being difficult to measure, is more than adequate for this application!

It's worth noting that a standalone DDS IC can synthesize frequencies from VLF to HF and higher. Let's look at an example. To generate a unique frequency, the DDS must generate at least two points per cycle. Industry practice is to generate at least 2½ points per cycle, or 40% of the clock frequency. In the case of DDS used above, we had to generate a 25 MHz clock frequency, which required a minimum of 2.5 points/Hz × 25 MHz, which yields 62.5 MHz as the minimum clock frequency. A clock input of 189–191 MHz easily meets this requirement. What frequency ranges can a DDS cover? **Table 9.2** shows the required minimum DDS clock frequencies for a few amateur frequency allocations.

Using the information in Table 9.2, we see that a DDS clocked at 125 MHz or higher will cover VLF through the 6 meter band. Once again, we'll use a 45 MHz IF and high-side injection. Searching through online DDS selection guides, we settle on the AD9951, a 14-bit DDS that accepts a 400 MHz input

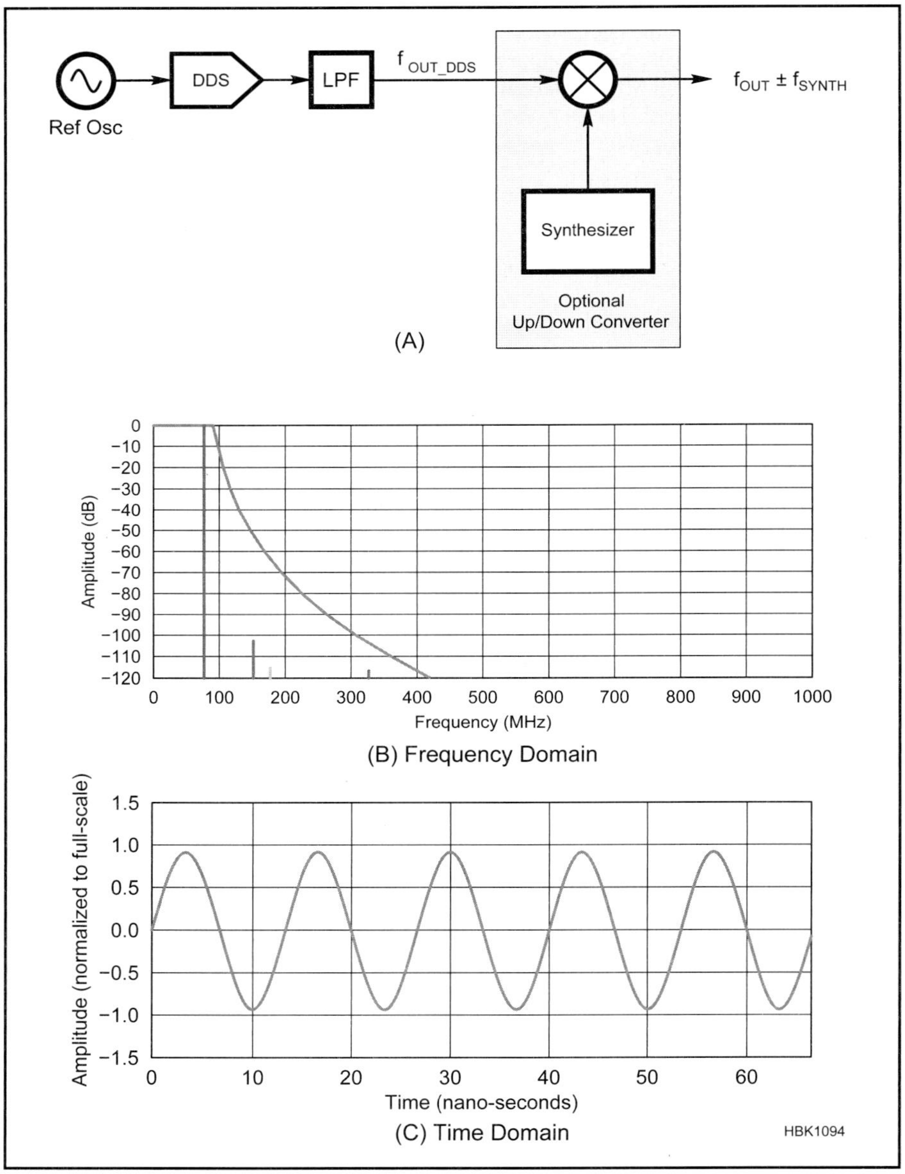

Figure 9.62 — Block diagram of a stand-alone DDS design using the AD9951 (A) with simulation results (B) from *ADIsimDDS* in the frequency and time domain. The filter is a 7th-order Chebyshev low-pass response.

Table 9.2

Frequency Band(s) and Minimum Required Clock for DDS using 2½ points per cycle.

Frequency Band(s)	*Minimum DDS Clock (MHz)*
VLF-30 MHz	75
50 MHz	125
70 MHz (UK)	175
144 MHz	360
225 MHz	562.5
432 MHz	1080

Phase Noise versus Jitter

The Si5351A was designed to replace the crystals used to generate the many clock frequencies required in digital systems: microprocessors, USB, video, data communications, etc. These clocks are specified by time jitter requirements rather than phase noise; that is, the system requirement is that the output signal has to fit within a time-based window for its leading and trailing edges. The phase noise in such a clock is higher than in its RF counterpart because there is no filtering of the output frequency in its fractional divider: the average output frequency is correct but all frequencies that were averaged will appear in the output waveform unless the output divider is programmed for integer mode. In addition, the integrated VCO is typically based on low-Q digital delay techniques (delay-locked loops, ring oscillators, etc.) rather an LC tank circuits, which again contributes to higher phase noise.

clock. We choose a DDS with a 14-bit DAC for better S/N ratio and lower spurious output than one with a 10- or 12-bit DAC. It also has an internal clock multiplier so we have the option of once again using a 25 MHz crystal (or slightly higher to keep its harmonics out of the 6 meter band). However, we'll assume a separate 400 MHz reference oscillator is available, which will have a lower noise floor than the crystal multiplier circuit. **Figure 9.62A** shows the block diagram for this synthesizer with an optional frequency converter.

The DDS manufacturer, Analog Devices, provides a web-based tool, *ADIsimDDS* for this and similar products. (**tools.analog.com/en/simdds**) Using this tool, we can generate an output spectrum and simulate several loop-pass filter topologies to choose the most effective. The tool also calculates the FTW value for a given combination of clock and output frequency in binary, decimal, and hex.

Figure 9.62A shows the simulator's output for the synthesizer in the frequency and time domain. In the frequency domain at the top, there are several spurious outputs shown as individual signals along with the low-pass filter's response. The output sine wave is also shown.

Figure 9.62's synthesizer architecture can also serve as a replacement VCO for vintage transceivers (for example, the Drake C-Line) that had optional external VCOs. Many of these transceivers typically used a tunable LO of 5-5.5 MHz based on a permeability-tuned oscillator (PTO, described previously) to tune a 10% frequency range with a linear frequency readout.

AFFORDABLE DESIGN SOLUTIONS

So far, we've looked at performance-driven synthesizer architectures. However, there are many designs where size and cost are the overriding concerns. For example, the Skyworks Si5153A is a low-cost, multiple-output digital clock synthesizer for crystal replacement in computers. It has three programable outputs that span 9 kHz to 200 MHz, which covers the VLF through 2m amateur bands.

The tradeoff in using a digital clock synthesizer is higher phase noise than a traditional PLL/VCO combination but satisfactory jitter for numerous digital applications. **Figure 9.63** shows a detailed block diagram based on the data sheet and includes references on using this particular IC. Note that the programmable output dividers are fractional-N dividers; this means that the average output frequency will have zero error but the instantaneous output frequency will vary. This and the noise from the integrated VCO accounts for the Si5153A's phase noise's being higher than that of a traditional PLL/VCO that uses an LC-based oscillator.

Not surprisingly, this combination of small size, high functionality, and low cost is used in numerous QRP designs to generate the LO

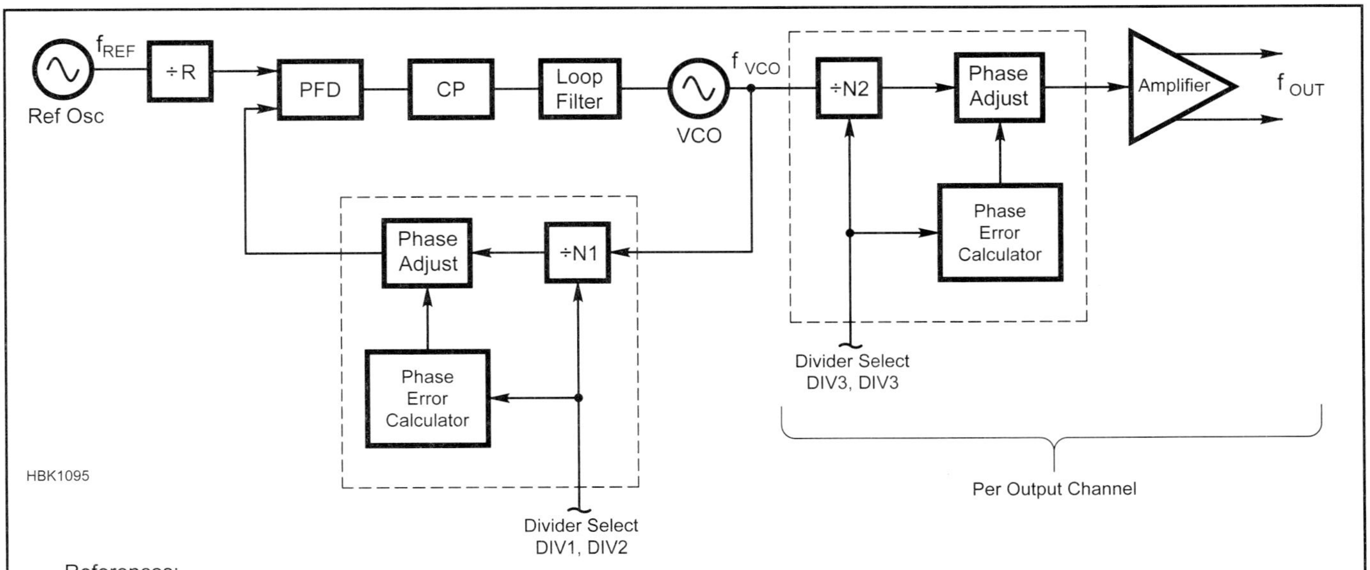

References:

1. Simplifying Timing Architectures with Flexible Clocks, Silicon Laboratories, Inc., Austin TX, now available from Skyworks, Inc.
2. Si5351A/B/C-BI2C-ProgrammableAny Frequency CMOS Clock Generator + VXCO Data Sheet.

Figure 9.63 — A detailed block diagram of the SI5351A's internal architecture.

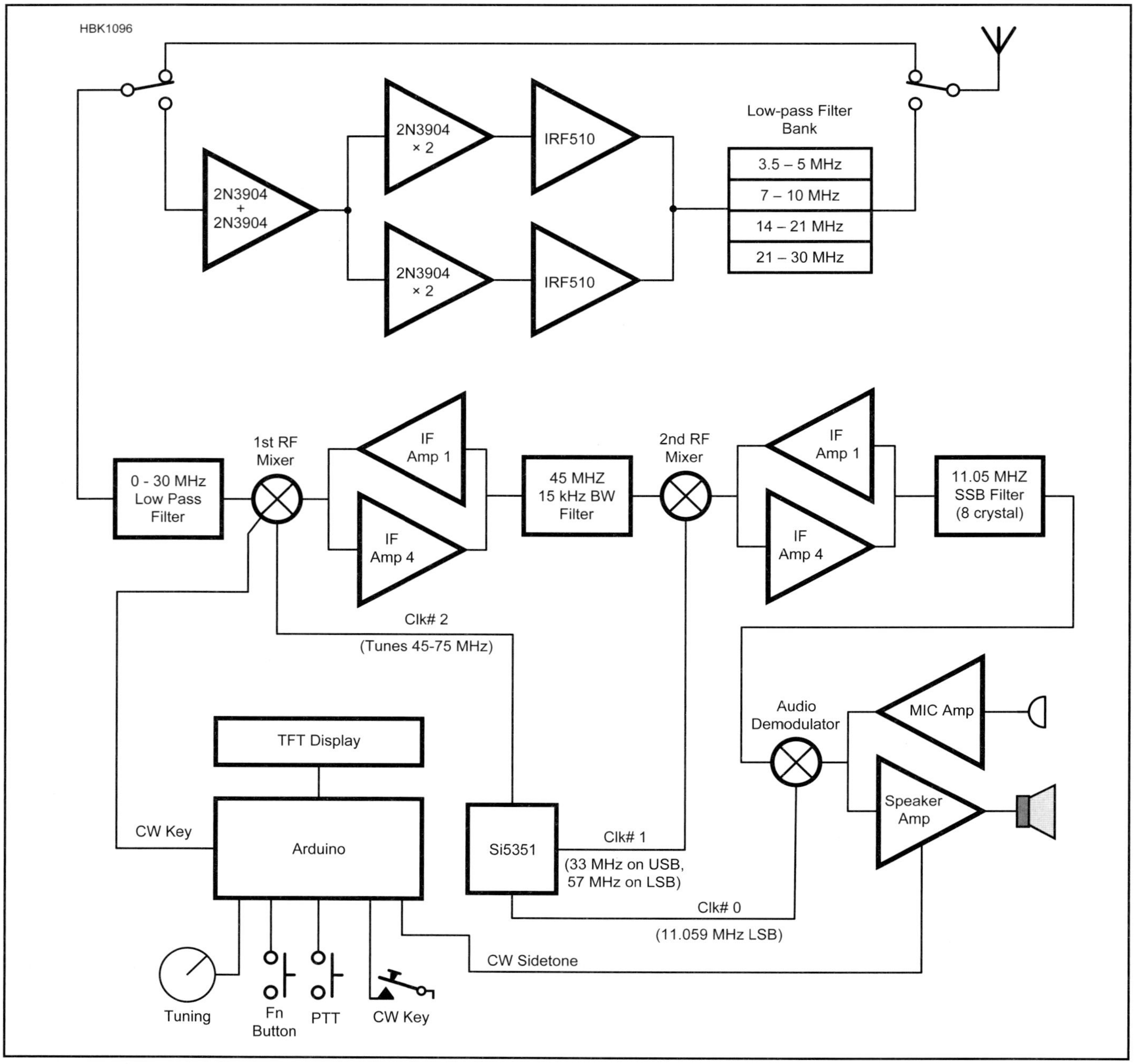

Figure 9.64 — A simplified block diagram of the uBitX6 and the Si5351a-based synthesizer, which is used in numerous QRP designs.

and BFO signals in single and dual conversion transceivers. One such example is the open-source uBitx family of SSB QRP transceivers designed by Ashhar Farhan, VU2ESE, and described in detail at **www.hfsignals.com**. In the synthesizer implementation used in the uBitx6 Five-Band Transceiver (**Figure 9.64**), the Si5351 provides three output frequencies: a 45-to-75 MHz first LO, a 57 MHz/33MHz LSB/USB second LO, and an 11.059 MHz BFO. Figure 9.64 shows a simplified version of the implementation of the receiver, including the synthesizer.

Another interesting synthesizer design is that of the AR7070 LF/HF Receiver, described in July 2013 *QEX* by Colin Horrabin, G3SBI. **Figure 9.65** shows the functional block diagram of the receiver section, including the synthesizer. Like the previous examples, it also uses a 45 MHz first IF. This synthesizer uses a mix of DDS, VCO, discrete PLL, frequency doublers to achieve its high performance.

The VCO is also worth mentioning; it uses a double-tuned tank circuit to achieve better performance than typical single-tuned tanks. Quoting from the article:

"The local oscillator in the HF7070 is a DDS/PLL design using a double-tank VCO. This was designed by John Thorpe and it is similar to the one used in the AR7030 receiver, but with improved performance. The basic double-tank oscillator is shown in **Figure 9.66**. The principle involved is that this circuit can only oscillate if the cold end of the active tank (the one with the J310) is a low impedance to ground. This can only occur if the dummy tank is series resonant at the same frequency as the active tank. This means that as you move away from the carrier two resonators are active, which increases the rate of phase noise fall-off with offset frequency. This circuit has never been analyzed from a phase noise point of view by a mathematician, but measurements show phase noise falls off at

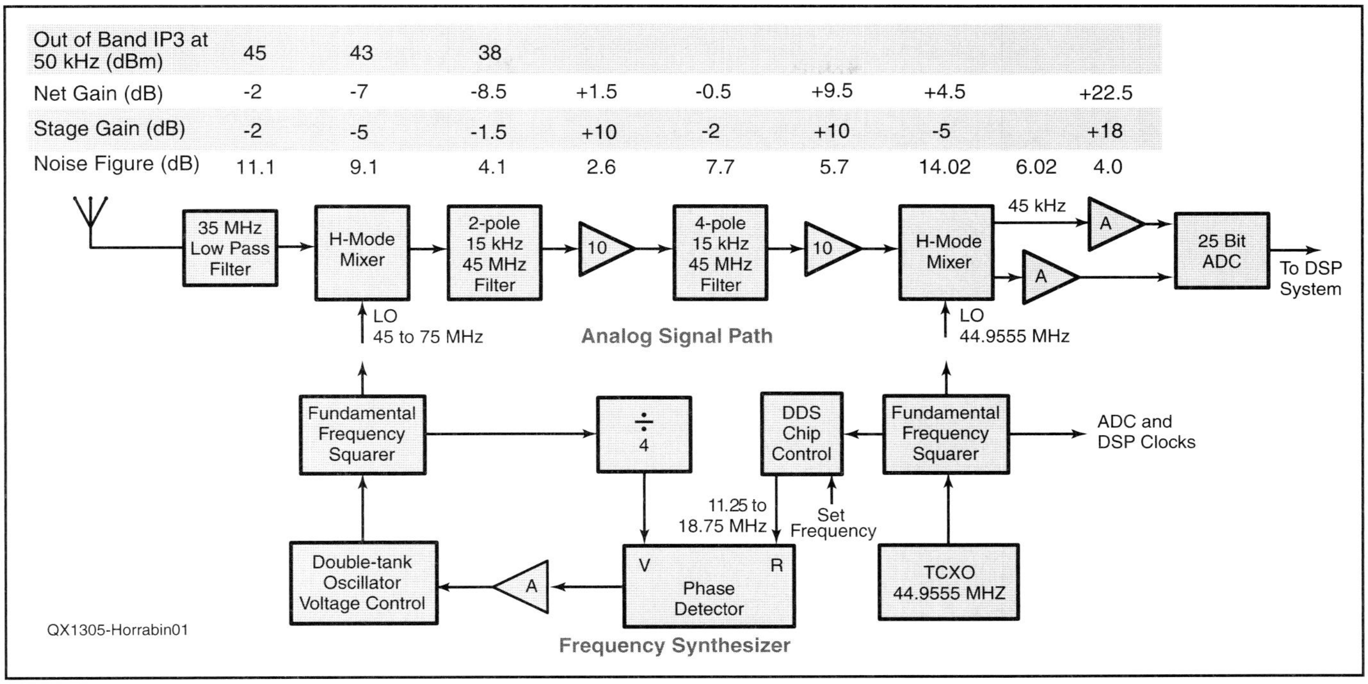

Figure 9.65 — Functional Block Diagram of the HF7070 LF/HF Receiver from *QEX* July 2013.

30 dB/decade compared to 20 dB/ decade in a single resonator oscillator."

Finally, let's look at a non-synthesizer design using a PLL as a programmable prescaler. Standalone prescalers are uncommon these days, yet a PLL can have a built-in programmable prescaler with several GHz of bandwidth. Fortunately, many PLLs have a MUXOUT pin, which lets the user access several internal signals during debug, including the output of the R and N programmable dividers. In a frequency counter, the R and N inputs can be used as user-programmable prescalers and the MUXOUT pin used to drive a microcontroller with a counter input to create a low-cost, wide bandwidth frequency counter. **Figure 9.67** shows an example.

INTEGRATING THE SYNTHESIZER COMPONENTS

Notwithstanding the previous block diagrams and text, there are still a few things missing from a typical design. These are the ubiquitous building blocks that are necessary to connect the elements in a synthesizer.

A PLL's feedback input (usually but not always driving the N counter input) requires both a sample of the VCO's output signal and the correct drive level at the PLL's input. A minimum-loss splitter (**Figure 9.68**) addresses the first issue. Many PLLs allow you to program the output power level to address the second issue.

There are other issues. For example, some PLLs require a minimum slew rate or a rail-to-rail input for proper operation of the R or N dividers when used at lower frequency inputs. For example, this type of requirement is stated as "For RFIN < 5 MHz, ensure slew rate (SR) > 4 V/us" and "For REFIN < 20 MHz, ensure slew rate (SR) > 50V/us". (These are minimum slew rate requirements from the ADF4002 data sheet.) In this case a high-speed Rail-to-Rail Input/Output comparator used to drive the input does the trick. The Texas Instruments TLV3601/TLV3603 or the Analog Devices LTC6752 are suitable choices for this application.

Many PLL/VCOs and DDS ICs have differential outputs; the output of the PLL is typically the differential output of a pair of NPN transistors and requires either inductors or resistors to the positive rail. Depending on the application, the output may connect directly to a mixer with differential inputs, in which case only a differential lowpass filter is needed (**Figure 9.69**).

In many cases, however, the differential output has to be converted to single-ended. Even though connecting only one output cheaply solves the problem, it would be a shame to lose 3 dB of the S/N ratio by discarding half the output signal. There are several ways to do this. **Figures 9.70** and **9.71** show how transformers and LC circuits may be used at higher frequencies.

For continuous coverage of the VLF through VHF bands (depending on the circuit), active circuits may be used, including a differential pair, active feedback circuits using op amps or ADC drivers, or even high-speed comparators when a square wave output is acceptable.

Figure 9.72A – D shows two circuit topologies for differential-to single ended conversion. A word of warning, a simulation is only as good as its designer and the models for the components; it can identify obvious design and wiring errors; and, finally, it lets you know that the circuit *may* work. There is no substitute for a physical prototype, such as breadboard version. Having said that, if you can simulate a circuit before construction, do so. It will give you an idea of how the circuit *should* behave (gain and bandwidth) and estimates for the voltages and currents you may end up measuring during debugging.

Figure 9.72A shows a simple, 4-transistor circuit comprising a differential pair (known as a long-tailed pair in the UK) of 2N3904s with an emitter follower output. Negative feedback (R15 and R16) reduces harmonics (simulation, Figure 9.72B) to about 30 dBc and distortion even more, to more than –45 dBc at frequencies at frequencies above 100 MHz. The circuit should be followed with a low pass filter. Figure 9.72C shows output (top) and input signals (bottom) simulated at 10 MHz. Bandwidth for the simulated circuit is about 30 MHz.

Figure 9.72D shows an op-amp-based circuit using ideal op amps for the simulation. The input here is really a discrete version of a differential ADC driver for high-speed

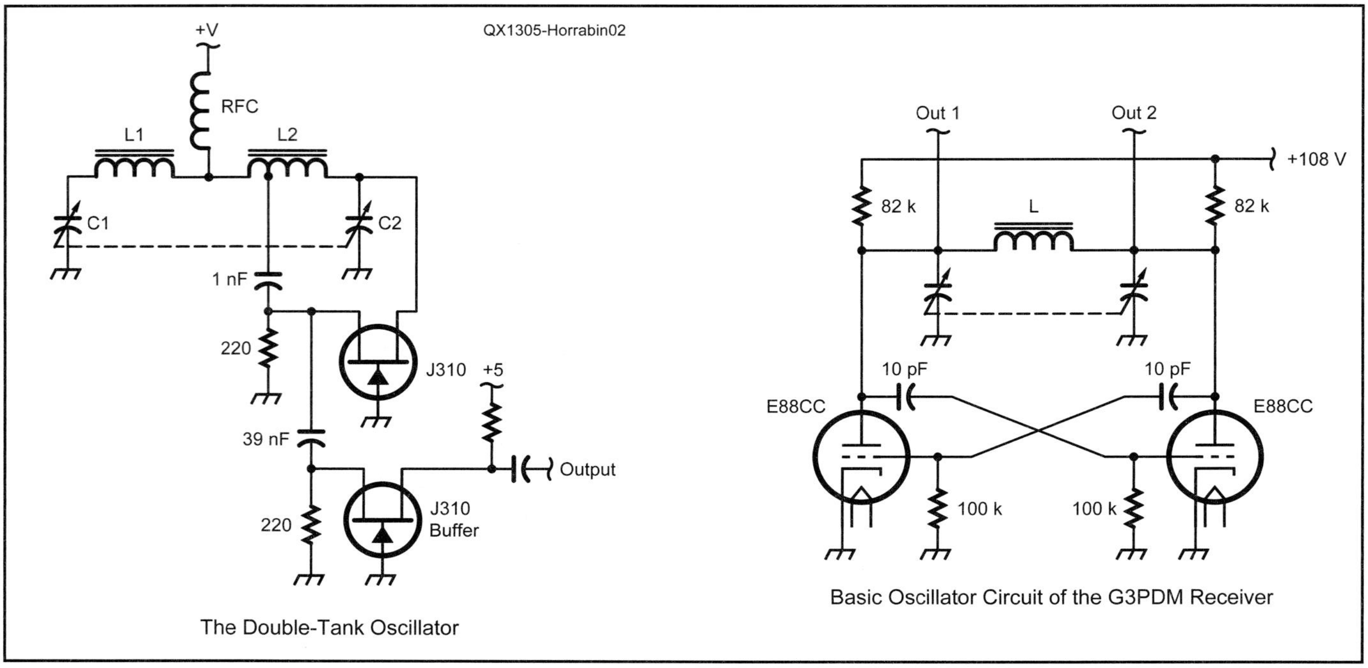

Figure 9.66 — The dual-tank circuit used in the HF7070's VCO achieves better phase noise than typical single-tank VCO designs.

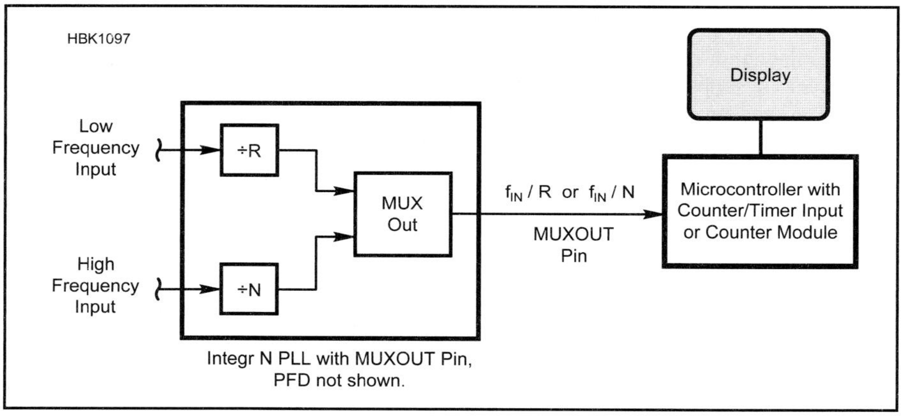

Figure 9.67 — A simple frequency counter uses a PLL as a programmable prescaler. Note that the MUXOUT pin has a limited frequency range (e.g., 10 MHz maximum). Check the data sheet for details.

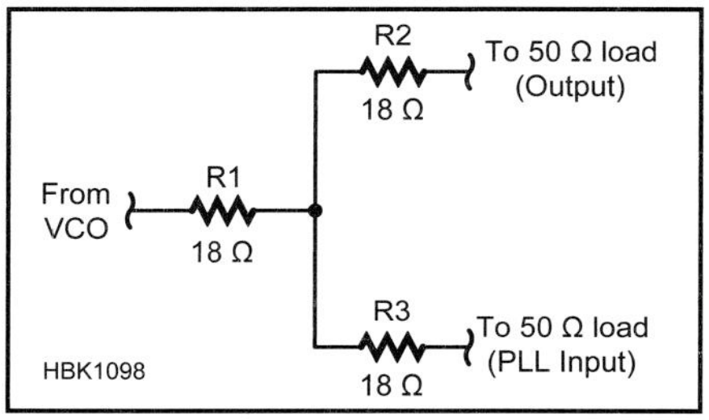

Figure 9.68 — A minimum-loss 3 dB signal splitter provides single-ended feedback from the PLL's output to its input and uses just three resistors.

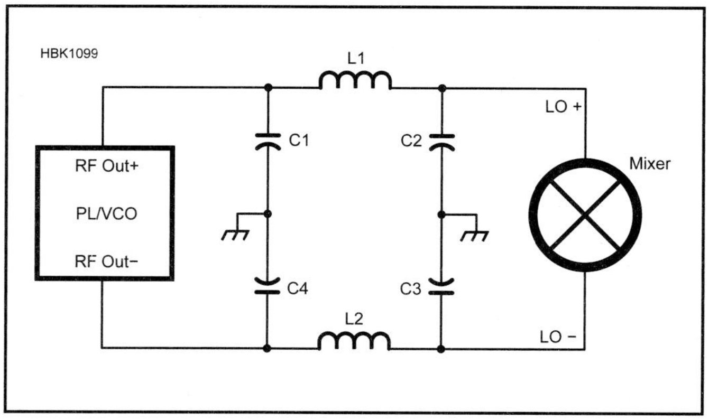

Figure 9.69 — The differential outputs of a PLL/VCO may connect directly to a mixer with a differential LO input. Pull-up inductors/resistors at the RFOUT pins are omitted for clarity.

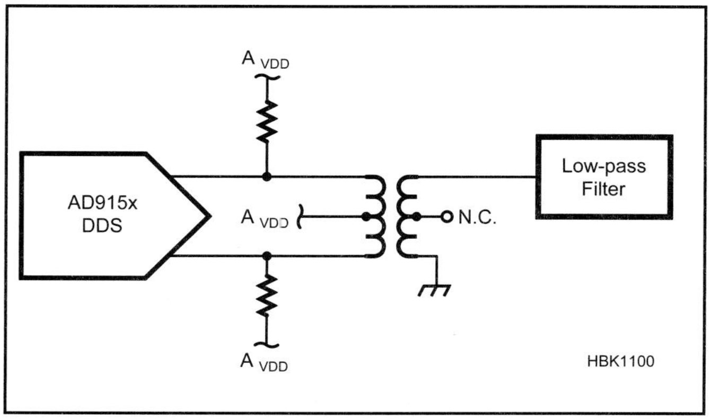

Figure 9.70 — A typical transformer circuit that converts the differential DAC outputs from a DDS IC to single-ended.

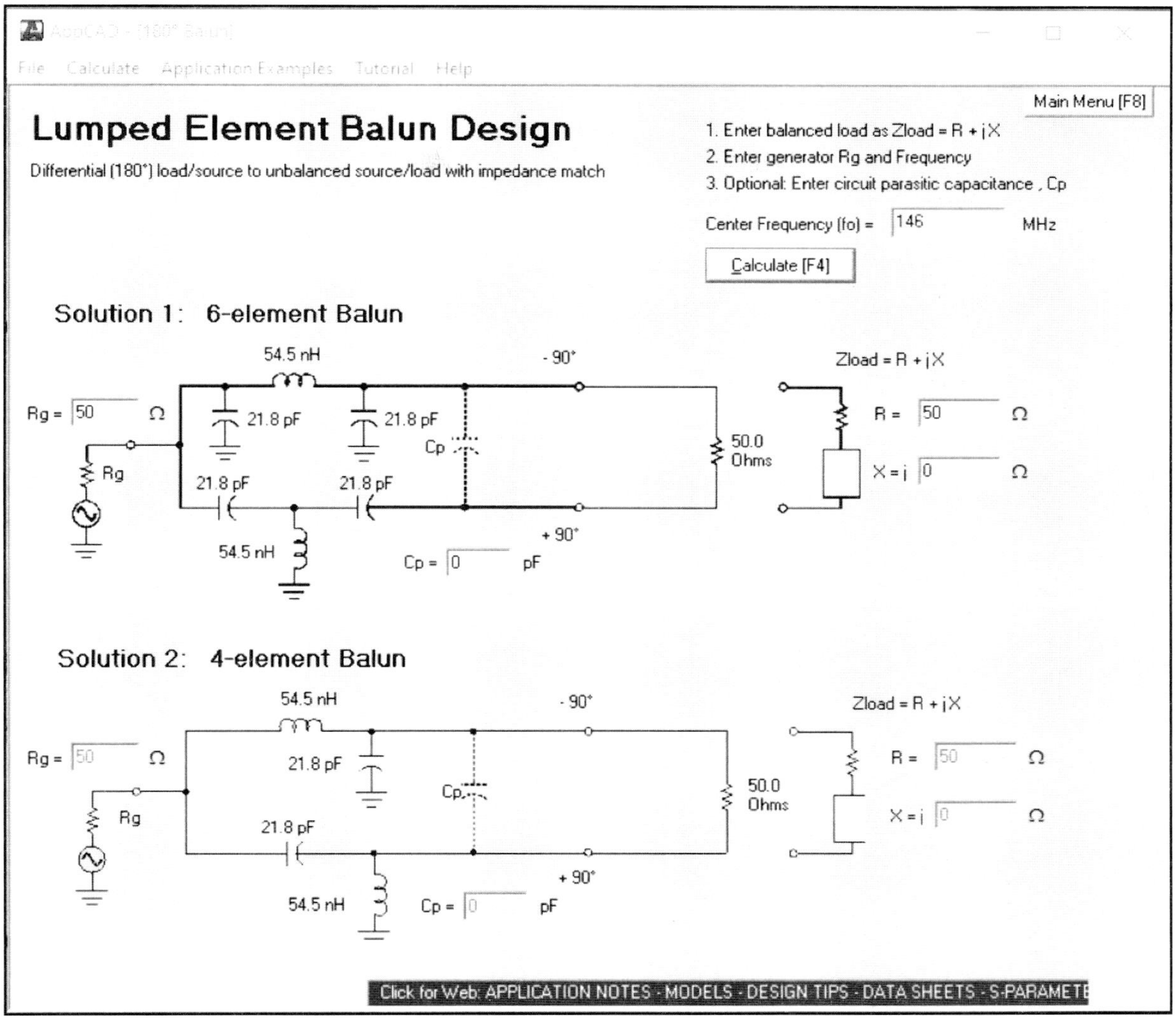

Figure 9.71 — Lumped-element baluns designed using *AppCAD* (now available free from Broadcom) convert differential outputs to single-ended signals. These circuits have about 5% bandwidth. Here the input for a differential output is to the right and the single-ended output is to the left. Suitable inductors are available from such vendors as Coilcraft, Mini-Circuits, Murata, and Toko.

Table 9.3
Popular Capacitor Types from Analog Devices' Tutorial MT-101

TECHNOLOGY	ADVANTAGES	DISADVANTAGES	APPLICATIONS
Aluminum Electrolytic, Switching Type. Avoid general purpose types	•High CV product/cost •Large energy storage •Best for 100V - 400V	•Temperature related wearout •High ESR/size •High ESR @ low temp	•Consumer products •Large bulk storage
Solid Tantalum	•High CV product/size •Stable @ cold temp •No wearout	•Fire hazard with reverse voltage •Expensive •Only rated up to 50V	•Popular in military •Concern for tantalum raw material supply
Aluminum-Polymer, Special-Polymer, Poscap, Os-Con	•Low ESR •Z stable over temp •Relatively small case	•Rapid degradation above 105°C •Relatively high cost	•Newest technology •CPU core regulators
Ceramic	•Lowest ESR, ESL •High ripple current •X7R good over wide temp	•CV product limited •Microphonics •C decreases with increasing voltage	•Excellent for HF decoupling •Good to 1GHz
Film (Polyester, Teflon, polypropylene, polystyrene, etc.	•Hi Q in large sizes •No wearout •High voltage	•CV product limited •Not popular in SMT •High cost	•High voltage, current •AC •Audio

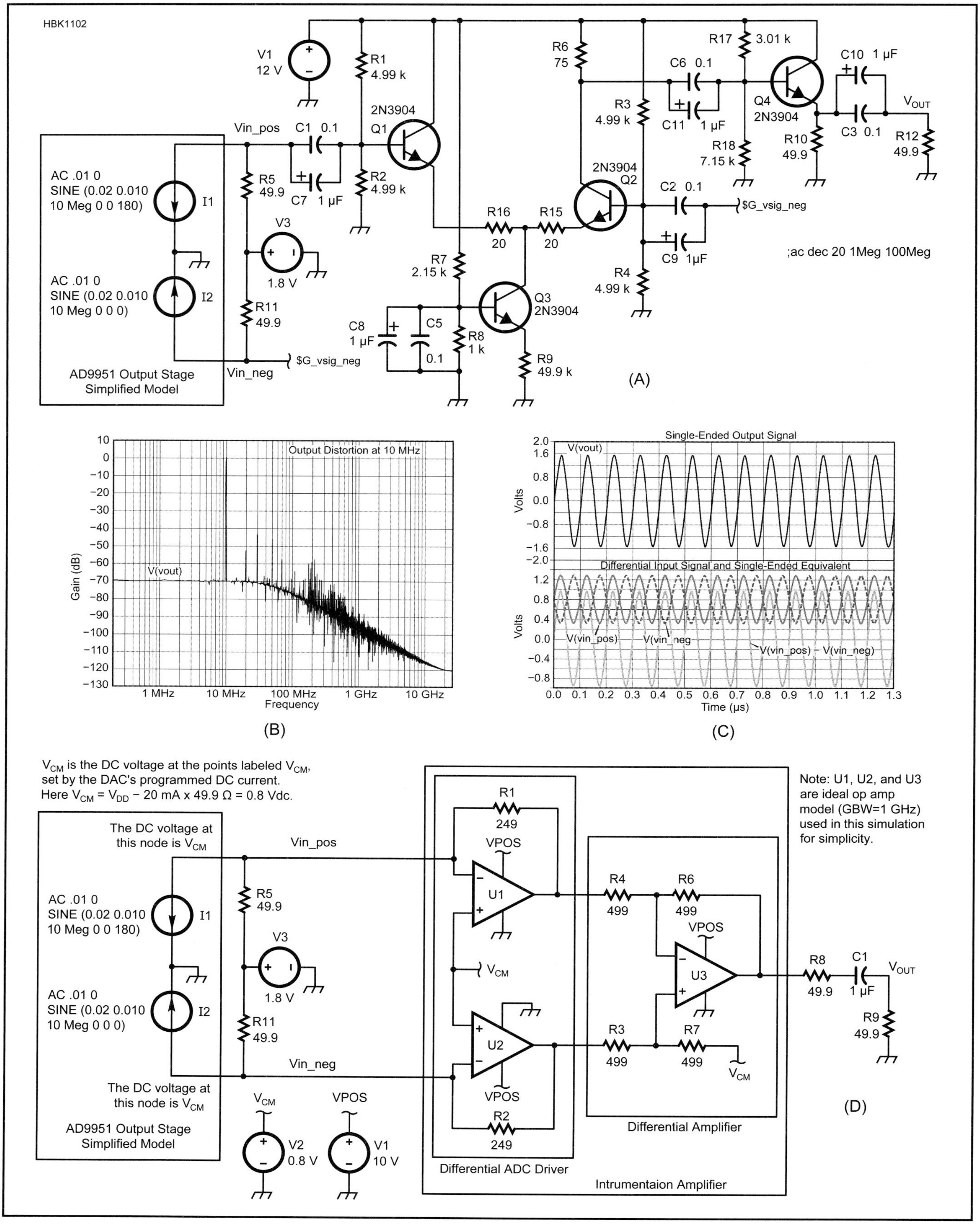

Figure 9.72 — Active differential-to-single-ended output topologies for audio through VHF operation of a DDS. See text for discussion of individual figures.

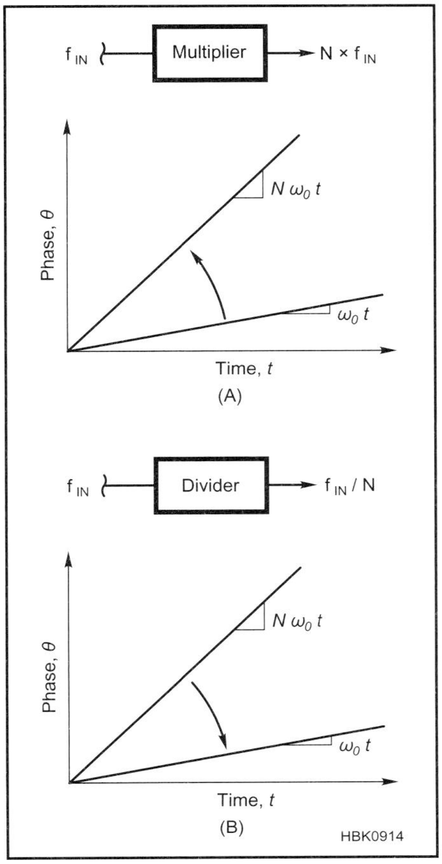

Figure 9.73 — Frequency multiplication and division: A — multiplication increases the rate that signal phase changes with time. This is often an integer (harmonic generation), but it can be non-integer in some applications (PLL); B — frequency division decreases the rate that signal phase changes with time. This is nearly always an integer.

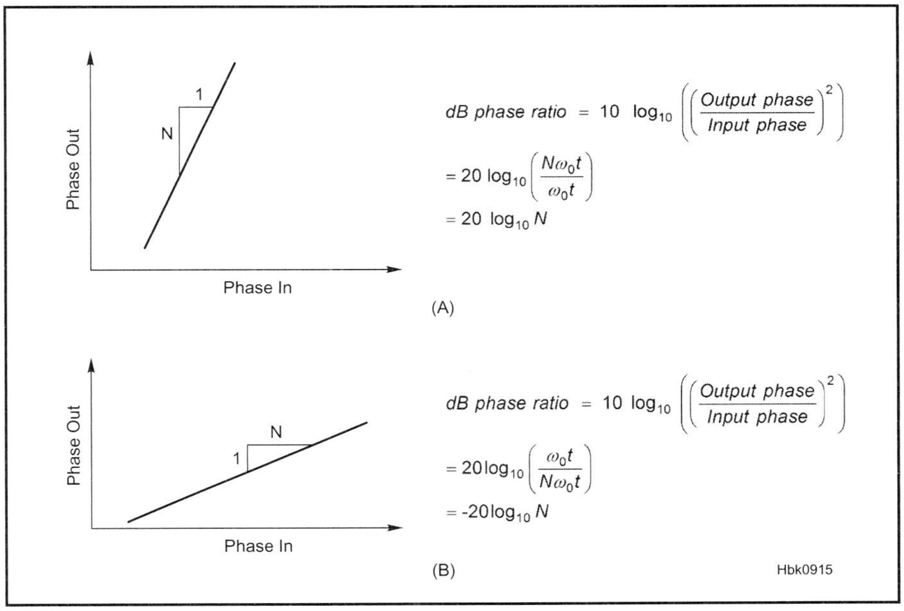

Figure 9.74 — Noise behavior of frequency multipliers and dividers: A — frequency multiplication increases output phase noise; B — frequency multiplication decreases output phase noise.

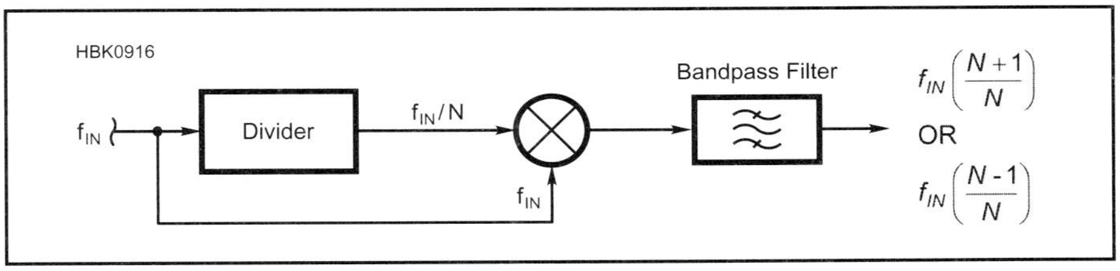

Figure 9.75 — One example of frequency mixing, where a signal is mixed with a divided version of itself to get a frequency ratio.

ADCs. Amplifiers A1 and A2 work as a pair of I/V converters, that is, current in, voltage out. The feedback resistors hold the amplifier's input voltages almost constant, which in turn stabilizes the voltage output of the DAC. A constant output voltage (see Figure 9.72E) minimizes the effect of voltage modulation on the DAC's output devices, which reduces distortion. It also maximizes output bandwidth: no change in voltage means any output capacitors have to be charged and discharged, minimizing their effect on distortion.

Figure 9.72D's circuit actually uses two standard circuits: a differential I/O ADC driver and a single op amp difference amplifier. Together, the two circuits form a third standard circuit: the classic three-op-amp instrumentation amplifier. This instrumentation amplifier, however, has a slightly unusual input configuration: the usual gain resistor is missing and in its place the instrumentation amplifier is connected to a pair of current inputs. In a real version of this circuit, op amps A1 and A2 would be current-feedback (transimpedance) op amps or a wideband-ADC driver and A3 would be a voltage-feedback op amp.

Although passive components have been covered elsewhere in this book, it doesn't hurt to list a summary of the advantages and disadvantages of different capacitor types (**Table 9.3**) from Analog Device's tutorial MT-101. Both ceramic (type X7R) and polystyrene capacitors have been popular in VFO tank circuits for HF. Many capacitor vendors (Murata and Toko, for example) have a line of ceramic capacitors designed specifically for RF applications. Murata, Toko, and Coilcraft, among others, also have inductors specifically for RF applications.

Ceramic capacitors are also used for grounding and bypassing. When drawing up schematics, consult the pin list on the data sheet, and determine a pin's function (input, output, analog or digital, power or ground) and its impedance (high impedance logic, or low impedance analog, etc.) and for each pin determine what should be connected to it and if bypassing is required.

At this point, review the schematic and layout of the evaluation board and see what the manufacturer does. Copy the relevant portions of the circuit, such as grounding, bypassing, routing the RF circuits, etc. Use the layout of the known-good evaluation board as a starting point in your board design. Do not be tempted to "improve" the layout by changing the routing other than removing what are obviously extra parts used only for evaluation.

The last component in a frequency synthesizer is the VCO. Although the text thus far has addressed both discrete (DIY) and integrated VCOs, many commercial products use VCO ICs or modules. **Table 9.4** lists some vendors.

Table 9.4

Manufacturers of VCO ICs and Modules

Manufacturer	*Website*
Analog Devices, Inc.	**www.analog.com**
Mini Circuits	**www.minicircuits.com**
Crystek Microwave, Inc.	**www.crystek.com**
Macom Technology Solutions	**www.macom.com**
Raltron	**www.raltron.com**
Synergy Microwave	**www.synergymicrowave.com**

Last but not least, there are many, many free tools for the designer and hobbyist provided by individuals, manufacturers, and universities. **Table 9.5** lists several, including some that have been mentioned previously.

Many of the preceding circuits can be used to generate a clock for an ADC. To conclude this section, it's worth looking at how the clock requirements are calculated. Oscillator signals for RF circuits are typically specified by their phase noise, shown as a curve in dBc/Hz below the LO across the spectrum of interest in the frequency domain. Jitter is how the total phase noise affects the same oscillator in the time domain and is measured in seconds.

For example, given an RMS jitter t_j in seconds and a sampling rate f_s in Hz, the SNR of an ADC is given by

$$SNR = 20log_{10}\left(\frac{1}{2\pi f_s t_j}\right)$$

Here, the time jitter has been calculated for the Nyquist frequency, or half the sampling rate (that is, $f_s/2$).

With a little bit of manipulation of the above equation, the jitter required for a given SNR and sampling frequency can be shown to be

$$t_j = \frac{1}{2\pi f_s 10^{\frac{SNR}{20}}}$$

A typical design starts by choosing an ADC that meets the system's requirements. Using the ADC's SNR specification, the designer next calculates the RMS jitter required to achieve that performance. For instance, when using an audio ADC specified for a 111 dB SNR at a sampling rate (f_s) of 192 kilosamples per second (ksps), the required RMS jitter,

$$t_j = \frac{1}{2\pi f_s 10^{\frac{SNR}{20}}} =$$

$$\frac{1}{2\pi \times 192,000 \times 10^{\frac{111}{20}}} = 2.34 \ \mu s$$

These numbers would be typical of a SDR using an audio ADC for sampling signals at baseband.

For GHz sampling rates for a direct-sampling radio, the jitter requirements are in femtoseconds, depending on the number of bits in the ADC and the sampling rate. In this case, we'll use a hypothetical ADC with an 80 dB SNR at a sampling rate of 1 Gsps. Using these numbers, the required clock jitter is

$$t_j = \frac{1}{2\pi f_s 10^{\frac{SNR}{20}}} =$$

$$\frac{1}{2\pi \times 1,000,000,000 \times 10^{\frac{80}{20}}} =$$

$$1.6 \times 10^{-14} s = 16 \ fs$$

The process of calculating time jitter from a phase noise curve is described in Analog Devices tutorials MT-007, "Aperture Time, Aperture Jitter, Aperture Delay Time—Removing the Confusion" and MT-008, "Converting Oscillator Phase Noise to Time Jitter" by Walt Kester, (**www.analog.com** — search for the tutorial IDs).

9.6.9 Analog Frequency Synthesis

Even though the term 'analog' seems old-fashioned in the present age, these techniques still have wide use in modern radio designs. They are briefly covered here.

FREQUENCY MULTIPLICATION

It is useful to view frequency multiplication as an increased slope of phase with time, as shown in **Figure 9.73A**. For harmonic generation, this multiplication factor is an integer. More complicated frequency multipliers, such as phase-locked loops, can readily provide frequency multiplication with non-integer values.

Table 9.5

Free RF Design Tools

Software	*Description*	*Download or Web Page*
ADIsimPLL	PLL design software	**www.analog.com**
ADIsimDAC	DDS design software	Web-based at preceding URL
AppCAD	Many useful RF design tools and calculators: lumped-element baluns, mixer spurs, etc.	**www.broadcom.com**
CPPSIM	System-level simulator for advanced users. Includes NGSpice analog simulator	**www.cppsim.com**
ELSIE (32-bit Windows)	Design and analyze lumped-element filters in the audio through microwave range. The free Student Version allows 7 components	**tonnesoftware.com/elsie.html**
Filter Design Tool	Audio frequency analog LP, BP, HP, Band Stop, and All-Pass filters using op amps	**www.TI.com**
GNU OCTAVE	Solves linear and nonlinear problems numerically using a language that is mostly compatible with Matlab An RF Toolbox is available.	**www.GNU.org**
KiCAD	Open-source schematic capture and layout software	**www.kicad.org**
LTspice	Analog Circuit Simulator	**www.analog.com**
PLLATINUMSIM-SW	PLL design software; TI PLLs only	**www.TI.com**
PSPICE for TI	Analog Circuit Simulator	**www.TI.com**
QUCS (Linux)	Quite Universal Circuit Simulator (QUCS) aims to support all kinds of circuit simulation types, e.g. DC, AC, S-parameter, Harmonic Balance analysis, noise analysis, etc.	**qucs.sourceforge.net**
RF Toolbox	Numerous RF Design Tools and Calculators	**RFtoolbox.dtu.dk** (Danish Technical University)
TinyCAD	Open-source schematic capture and drawing software	**www.tinycad.net**

FREQUENCY DIVISION

In the same way, frequency division is usefully viewed as a reduction in the slope of output signal phase with time. This is shown in Figure 9.73B. Frequency multiplication and division are inverse process of each other.

Noise behavior of frequency multipliers and dividers also has an inverse relationship. Phase noise encounters the same processes that the signal phase encounters. **Figure 9.74** shows that the phase noise out of any frequency multiplier is at least $+20\log_{10}(N)$ higher than the phase noise on the input signal. This is unavoidable. It is also a big problem when the multiplication factor N is large. We are motivated to have input signals into a phase multiplier with very low phase noise.

Fortunately, the output phase noise of a frequency divider is lower than that on its input signal. This effect is also symmetrical. This effect is often used in very low noise synthesizer design.

FREQUENCY MIXING

Mixers implement frequency addition and subtraction. The problem is that mixers do both at the same time, which means that some filtering is required to intentionally select whether the sum frequency, or the difference frequency, proceeds along to later stages of the radio. In **Figure 9.75** a particular example of frequency mixing is presented. Here the output is a close ratio of the input frequency, such a 2/3 or 3/2 when N = 2, or when N = 8 the outputs can be 7/8 or 9/8.

9.7 Phase Noise

(This section deals specifically with phase noise generated by oscillators. A more general discussion of noise can be found in the **RF Techniques** chapter.) No oscillator output signal is perfect. Viewing an oscillator as a filtered-noise generator is relatively modern. The older approach is to think of an oscillator making a pure sine wave with an added, unwanted noise signal. These are just different ways of visualizing the same thing. They are equally valid views which are used interchangeably, depending on which best makes some point clear.

For example, it is instructive to use the pure-sine-wave-plus-noise view to see relationships between AM (amplitude) noise and PM (phase) noise processes, shown in **Figure 9.76**. Adding Gaussian noise to a pure sine wave is usefully modeled using phasors as shown in Figure 9.76B. This generates both AM due to noise, and PM due to noise as shown in Figure 9.76C.

Passing the oscillator output signal through a limiter process, such as an amplifier in compression or through a switching mixer, leaves only the phase noise depicted in Figure 9.76D. Because it is so easy to remove AM noise but not phase noise, there is essentially no discussion about oscillator AM noise. Phase noise is the most important problem.

Phase modulation (PM) and frequency modulation (FM) are closely related. (See the **Modulation** chapter for a detailed description of each.) Phase is the integral of frequency, so phase modulation resembles frequency modulation, where the frequency deviation increases with increasing modulating frequency. Thus, there is no need to talk of "frequency noise" because phase noise already covers it.

A thorough analysis of oscillator phase noise is beyond the scope of this section. However, a detailed, state-of-the-art treatment by Ulrich Rohde, N1UL, of free-running oscillators using nonlinear harmonic-balance techniques and the sources of noise in the Colpitts oscillator circuit is presented in the online information accompanying this book. Also see the Reference entry for Dacus regarding low-noise synthesizer design.

Because of the dynamic range required to measure phase noise, it is one of the most difficult measurements. The section below on ARRL Lab Measurement of Transmitter Phase Noise illustrates the lengths to which one must go to obtain repeatable, reliable measurements. An additional article on measuring receiver phase noise is included with this book's online material.

9.7.1 Effects of Phase Noise

Phase noise becomes a problem when it is more noticeable than other limitations. It degrades all signals, but whether it is important or not depends on the application. For voice signals it sounds like background "hiss" in headphones or speakers. It also limits the dynamic range of receivers with closely separated signals, or receiving signals with widely different input powers.

Phase noise became a significant problem for amateurs when the use of frequency synthesizers supplanted conventional LC VFOs in amateur equipment. For reasons discussed in the Frequency Synthesizers section of this chapter, it is a major task to develop a synthesizer that tunes in steps fine enough for use with SSB and CW operation while competing with the phase-noise performance of a reasonable-quality LC VFO. Many synthesizers fall far short of this target. Along with the problems with frequency synthesizers, phase noise always gets worse at higher frequencies.

General-coverage, up-converting receiver architectures require local oscillators to operate at higher and higher frequencies, aggravating phase noise performance. As SDR techniques expand, however, phase noise performance is improving on both receive and transmit.

9.7.2 Reciprocal Mixing

All mixers are symmetrical, meaning that the output IF signal depends on the characteristics of both input signals: the local oscillator (LO) and the desired signal. A change to either signal shows up at the IF, where there is no way of knowing if the signal characteristics seen are from the input signal itself, or from the LO. We usually assume that the LO is very pure so only input signal modulations show up at the IF.

When there is phase noise on the LO signal, the situation changes. Noise on the LO transfers to *all* input signals at the mixer output as if it was originally present on each input signal and the LO was perfectly clean — the IF circuits can't tell the difference. This process is called *reciprocal mixing*, where noise on the LO appears as noise on the desired output signal. This is a serious limitation on a receiver's weak-signal ability.

How reciprocal mixing of LO phase noise can limit receiver dynamic range is shown in **Figure 9.77**. One possible scenario of band activity is shown as the set of input signals in Figure 9.77A. Figure 9.77B shows the phase noise profile for the LO of this receiver. Reciprocal mixing is illustrated in Figure 9.77C, showing that the LO phase noise is added to each of the signals in the band in proportion to its input power. The total noise seen by the receiver demodulator is the sum of all transferred phase noise profiles at the received frequency. This raises the apparent noise floor of the receiver.

Another receiver problem due to LO phase noise occurs if a very large signal appears in the receiver passband but at a frequency well removed from the desired signal. One example of this is shown in **Figure 9.78**. If the LO phase noise is excessive, then reciprocal mixing with this large signal can completely block the ability to demodulate the desired signals. Indeed, it is possible to make a wide swath of spectrum relatively useless.

Reciprocal mixing in a receiver does not affect the operating ability of other stations. The solution is to reduce the phase noise on the receiver's LO.

9.7.3 A Phase Noise Demonstration

Healthy curiosity demands some form of demonstration so the scale of a problem can

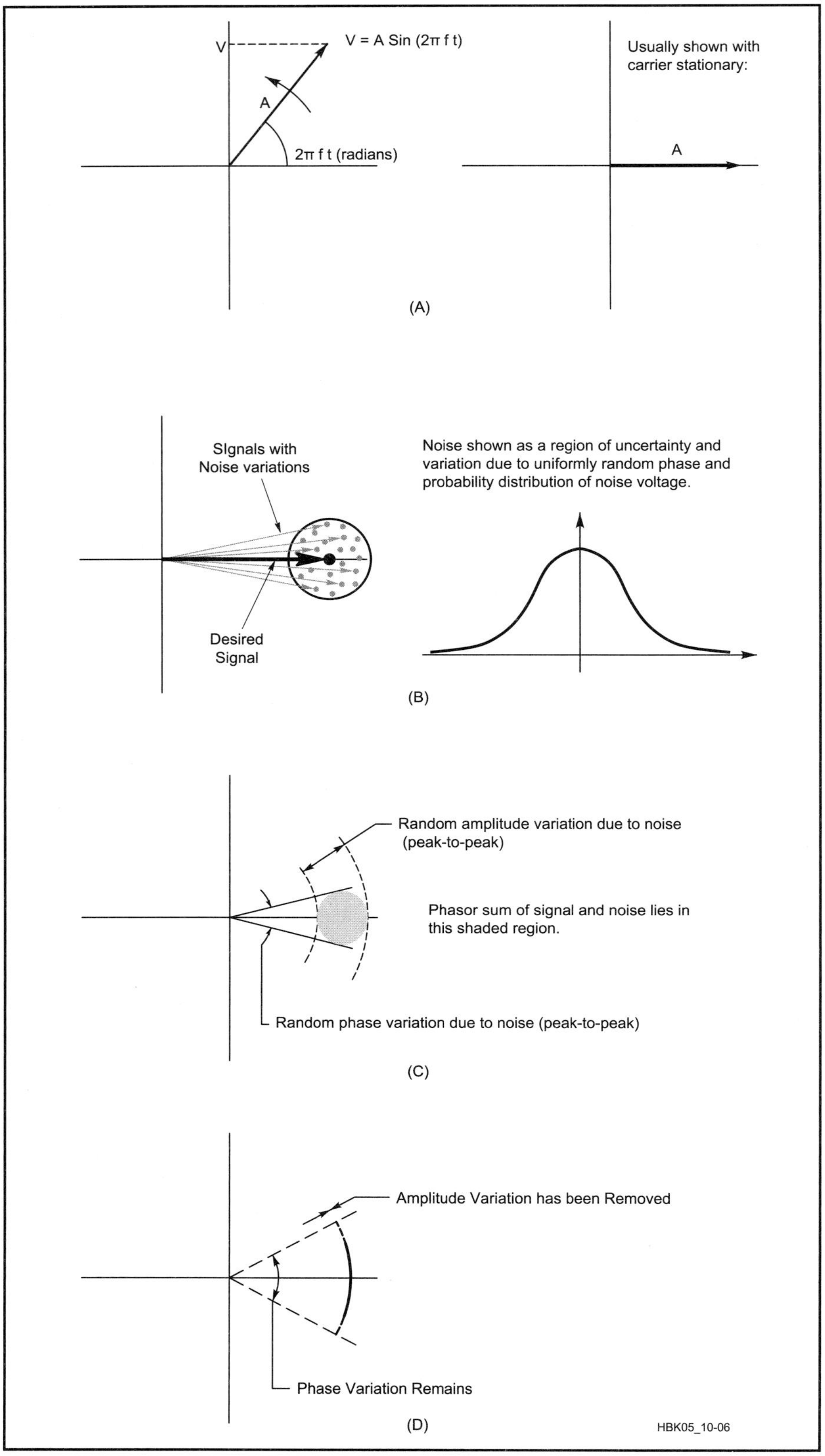

Figure 9.76 — At A, a vector (left) and phasor (right) diagram of an ideal oscillator with no noise. Added noise creates a region of uncertainty in the phasor's length and position (B). AM noise varies the phasor's length; PM noise varies the phasor's relative angular position (C). Limiting a signal that contains both AM and PM noise strips off the AM and leaves the PM (D).

be judged "by ear" before measurements are attempted. We need to be able to measure the noise of an oscillator alone (to aid in the development of quieter ones) and we also need to be able to measure the phase noise of the oscillators in a receiver (a transmitter can be treated as an oscillator). Conveniently, a receiver contains most of the functions needed to demonstrate its own phase noise.

Because reciprocal mixing adds the LO's sidebands to clean incoming signals, in the same proportion to the incoming carrier as they exist with respect to the LO carrier, all we need do is to apply a strong, clean signal wherever we want within the receiver's tuning range. This signal's generator must have lower phase noise than the radio being evaluated. A general-purpose signal generator is unlikely to be good enough; a crystal oscillator is needed.

It's appropriate to set the oscillator's signal level into the receiver to about that of a strong broadcast carrier, say S9 + 40 dB. Set the receiver's mode to SSB or CW and tune around the test signal, looking for an increasing noise floor (higher hiss level) as you tune closer toward the signal, as shown in Figure 9.77D. Switching in a narrow CW filter allows you to hear noise closer to the carrier than is possible with an SSB filter. This is also the technique used to measure a receiver's effective selectivity, and some equipment reviewers kindly publish their plots in this format. *QST* Product Reviews, done by the ARRL Lab, often include the results of specific phase-noise measurements.

9.7.4 Transmitted Phase Noise

Phase noise on an LO used to generate a transmitted signal will also be amplified and transmitted along with the desired output signal. This obscures reception by raising the noise floor even for receivers with low phase noise because they also receive the noise from the transmitter. In this case, the phase noise is generated externally to the receiver and must be removed at the transmitter.

If the transmitter is operating linearly, the strength of the transmitted noise is of the same proportion to transmitter output power as the phase noise is to the oscillator signal power. The noise may even extend well beyond the band in which the desired signal is transmitted unless the signal passes through narrow-band filtering that limits its bandwidth.

This transmitted noise is wasted power that is not useful for communication and unfortunately makes for a noisier band. If you are working a weak station and a nearby transmitter with a noisy oscillator comes on the air with a high-power signal on a different frequency, it is possible that the output noise from this off-frequency transmitter may completely block your ability to continue working that weak station.

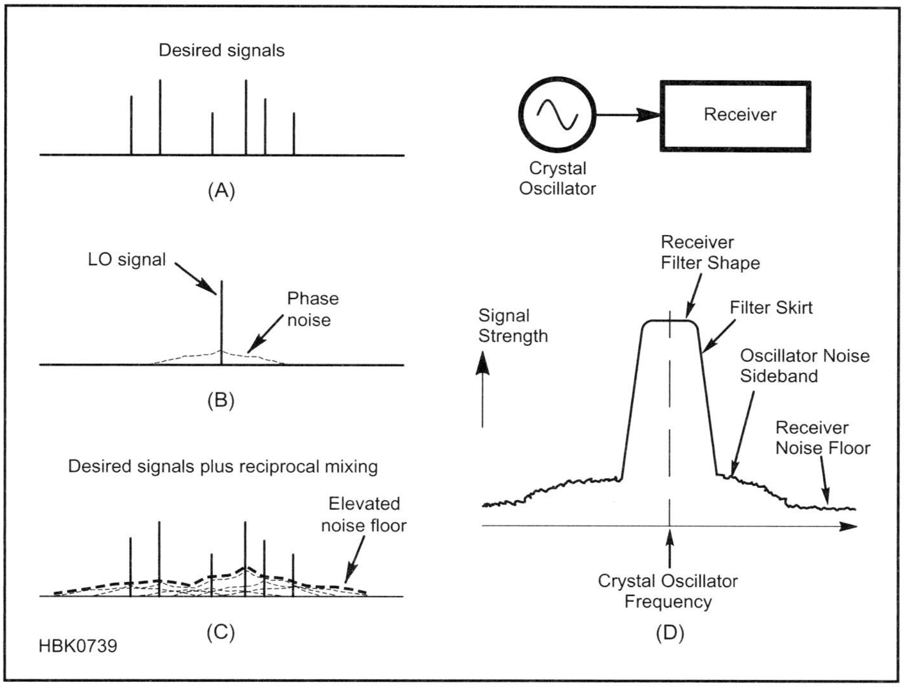

Figure 9.77 — A typical set of input signals is shown at A and an LO signal with phase noise at B. When the LO signal at B is mixed with the input signals at A, the result is a set of mixing products each having phase noise added, raising the noise floor across the band as shown at C. Phase noise in your receiver can be heard by tuning to a strong, clean crystal-oscillator signal as shown in D.

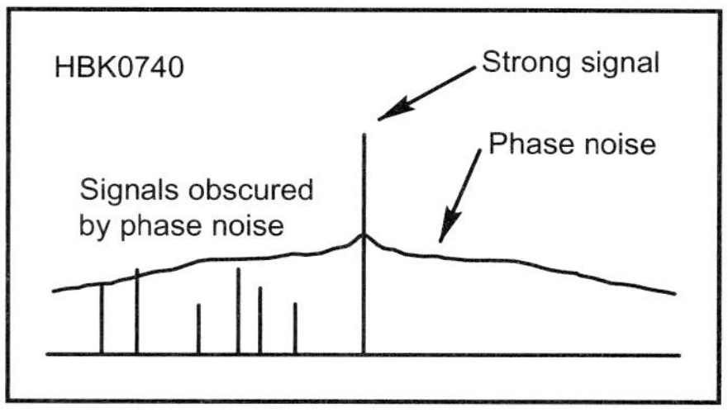

Figure 9.78 — A strong received signal can also cause such severe reciprocal mixing that desired signals are completely obscured by noise. If the signal is transmitted with phase noise from the transmitter LO, the effect is the same as noise covers the desired signals, sometimes across a very wide range of frequencies.

In bad cases, reception of nearby stations can be blocked over many tens of kilohertz above and below the frequency of the offending station. This is seen in Figure 9.80, where the offending signal and its associated noise are from a nearby transmitter. Frequencies close to that of the nearby transmitter are suddenly useless.

Transmitted phase noise can present serious problems for multi-station operation such as at Field Day or during emergency communications where several transmitters and receivers are in close proximity. This is a particular problem with transceivers that use early PLL-synthesized VFOs. If you are planning such an operation, be sure the level of transmitted phase noise is acceptable for the transceivers you plan on using.

At your receiver there is nothing you can do about transmitted phase noise. When there is noise power present at the same frequency as a weak (far) signal you are receiving, your receiver cannot separate them. The only solution is for the problem station to use a transmitter with a "cleaner" LO (less phase noise). This is a more serious problem than reciprocal mixing since this transmitted noise affects the ability of many other stations to use that band.

9.7.5 PLL Synthesizer Phase Noise

Differences in resonator Q usually make the phase-noise sidebands of a loop's reference oscillator much smaller than those of the VCO. Within its loop bandwidth, a PLL acts toward matching the phase-noise components of its VCO to those of the reference. There are several processes which get in the way of this actually happening, which are outlined here.

Dividing the reference oscillator signal f_{XO} to produce the signal f_{REF} which is applied to the phase detector also divides the deviation of the reference oscillator's phase-noise sidebands. (See the previous section on PLL.) This division results in a 20 dB reduction in phase noise per decade of division. This models as a factor of –20 log (R) dB, where R is the reference divisor value, since f_{REF} is less than f_{XO}. We also know that within its loop bandwidth the PLL acts as a frequency multiplier and this multiplies the deviation of the phase noise sidebands present at the PFD, again by 20 dB per decade or equivalently by a factor of 20 log (N) dB, where N is the loop divider's value. Between the reference oscillator and the PLL output, the sidebands on the reference signal are increased by 20 log (N/R) dB.

COUNTER AND PFD NOISE

Logic circuits have a noise characteristic that is extremely important to PLL design. Any circuit, when you really get down to it, is always an analog circuit. In this case the switching threshold of the logic circuits has a tiny amount of noise on it. This small amount of noise manifests itself in a very slight variation in when the logic circuit actually switches, even if the input clock is perfectly clean. We can think of this small timing jitter as a small shift in the output signal "zero crossing," which is directly equivalent to a phase modulation.

The amount of this timing jitter is constant, no matter the operating frequency. Thus, the amount of phase shift this represents depends on the operating frequency: 1 nanosecond of jitter at 1 kHz is very small, but represents 36 degrees at 100 MHz. As a result, we get different measures of the noise floor from digital circuits that depend on the frequency at which they are operating. This digital noise floor interferes with the PLL action to match the VCO noise to the divided reference oscillator noise as measured at the PFD inputs.

VCO NOISE

The PLL can track only slowly moving noise, well within its bandwidth. Therefore, VCO noise components that change slowly enough for the PLL to measure and track will be increasingly canceled successfully. Any VCO noise component that changes at a rate faster than the PLL bandwidth will not be changed at all. At offset frequencies beyond the PLL bandwidth the VCO noise is still present as if the PLL were not even there.

Does this imply that to get a low-noise synthesizer we are encouraged to make the PLL bandwidth very wide? Well, yes, to a certain degree. When we build PLL synthesizers we soon find that the logic noise from the digital section begins to get in the way if the loop bandwidth gets too wide. When this happens the total PLL output noise actually begins to increase as the bandwidth gets wider. There is a range of loop bandwidth values where the total PLL output noise has a minimum. This range is centered at the offset frequency where the logic noise floor, multiplied by 20 log (f_{OUT}/f_{REF}), has the same value as the VCO intrinsic phase noise.

Clearly, having a VCO design with minimum phase noise is very important to a good, low-noise PLL synthesizer. Much effort has been expended on this area in industry. A summary of important results from these noise reduction efforts is presented in the section Improving VCO Noise Performance below.

FRACTIONAL-DIVISION NOISE

From the brief discussion on fractional-N principles, we note that this whole idea is based on changing the feedback divider value in a semi-randomized way. This inherently jitters the output from the feedback divider, which is a source of phase noise into the PFD. This noise source is algorithmic instead of physical, but the noise is a big problem nonetheless.

OTHER SOURCES OF NOISE

Phase noise can be introduced into a PLL by other means. Any amplifier stages between the VCO and the circuits that follow it (such as the loop divider) will contribute some noise, as will microphonic effects in loop- and reference-filter components (such as those due to the piezoelectric properties of ceramic capacitors and the crystal filters sometimes used for reference-oscillator filtering). Noise on the power supply to the system's active components can modulate the loop. The fundamental and harmonics of the system's ac line supply can be coupled into the VCO directly or by means of ground loops.

9.7.6 Improving VCO Noise Performance

It is tempting to regard VCO design as a matter of coming up with a suitable oscillator topology with a variable capacitor, simply replacing the variable capacitor with a suitable varactor diode and applying a tuning voltage to the diode. Unfortunately, things are not this simple (as if even this is simple!). There is the matter of applying the tuning voltage to the diode without significantly disturbing the oscillator performance. There is also an issue of not introducing parasitic oscillation with the varactor circuit.

Next, as mentioned earlier, one would not like the tuning voltage to drop below the voltage swing in the oscillator tank. If this is allowed to occur, the tuning diodes go into conduction and the oscillator noise gets worse. A good first approach is to use varactor diodes back to back as shown in **Figure 9.79**. This allows the tank voltage swing to be developed across two diodes instead of one, as well as allow for a more balanced loading of the oscillator's tank. The semiconductor industry has realized this and there are a number of varactors available prepackaged in this configuration today.

IMPROVING INTEGRATED VCO OPERATING Q

The IC designer can improve operating Q with a variety of techniques. First, the VCO's tuning range can be divided into multiple, overlapping frequency ranges, with an optimized VCO per range and automatic switching between ranges. For example, improving the IC's metalization by lowering its resistance yields lower-loss inductors with improved Q. Moving the inductor away from lossy materials in the IC by placing the inductor on top of the glass passivation that coats the die also increases the inductor's Q.

Using Metal-Intrinsic-Metal capacitors improves the capacitor's Q. Finally, by multiplying the number of VCO cores and dividing the output frequency, the designer can trade die area and power for an improvement in phase noise. For example, by injection-locking two adjacent VCO cores, they will both oscillate in phase, which gives a 3 dB improvement in their phase noise. Four injection-locked VCOs gives a 6 dB improvement, 8 cores gives a 9 dB improvement, and so forth.

Each division by 2 of the output frequency also improves the phase noise by 3 dB per division. Thus, a division by 2^4 provides a 12 dB improvement. The combination of 8 VCO cores and dividing the output frequency by 8 yields a 24 dB improvement in phase noise at the expense of complexity, die area, and power consumption.

IMPROVING DISCRETE VCO OPERATING Q

Often the Q of passive capacitors available for use in an oscillator tank circuit exceeds the Q of available varactors. One way of reducing the influence of the varactor is to use only the amount of varactor capacitance required to tune the oscillator over its desired range. The balance of the capacitance is supplied to the tank in the form of a higher-Q fixed capacitor. This has the advantage of not requiring the varactor to supply all the capacitance needed to make the circuit function, and often allows for the use of a lower capacitance varactor.

Lower capacitance varactors typically exhibit higher Q values than their larger capacitance counterparts. This concept can be extended by splitting the oscillator tuning range, say 70 MHz to 98 MHz (for a typical lower-side up-converter receiver design popular in recent analog architectures), into multiple bands. For this example, let us consider four oscillators, each with 7 MHz of tuning range. We desire the varactor Q effects to be swamped by high-Q fixed capacitors. This can be further improved through the use of a "segment-tuned VCO" discussed below. A secondary but important benefit of this is to reduce the effective tuning gain (MHz/volt, K_0) of the oscillators, making them less susceptible to other noise voltage sources in the synthesizer loop. These noise sources can come from a variety of places including, but not limited to, varactor leakage current, varactor tuning drive impedance and output noise of the driving operational amplifier or charge pump.

Segment-tuned VCOs provide the designer with additional benefits, but also with additional challenges. By segmented we mean that circuit elements, which could be both inductance and capacitance, are selected for each range that the VCO is expected to tune. **Figure 9.80** shows the frequency range-switching section of a typical VCO in which several diode switches are used to alter the total tank circuit inductance in small steps. These segments create a type of "coarse tuning" for the VCO, and the output of the PLL loop filter performs "fine tuning" with the varactor diode. Usually, the component values are arranged in some sort of binary tree. In some integrated designs, 64 or even 128 sub-bands are available from the VCO.

Many times, the segmented-VCO concept is applied to an oscillator design that is required to cover several octaves in frequency. For example, this would be the case with a single-IF radio with the IF at about 8 MHz. The VCO

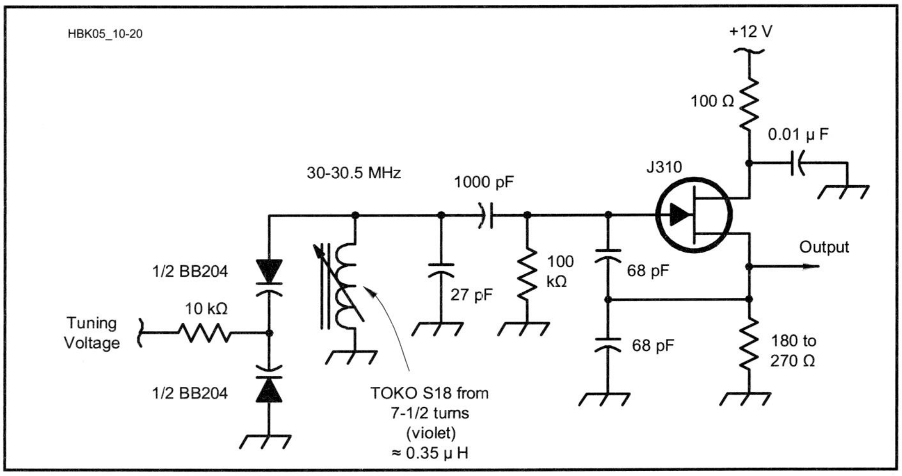

Figure 9.79 — A practical VCO. The tuning diodes are halves of a BB204 dual, common-cathode tuning diode (capacitance per section at 3 V, 39 pF) or equivalent. The ECG617, NTE617 and MV104 are suitable dual-diode substitutes, or use pairs of 1N5451s (39 pF at 4 V) or MV2109s (33 pF at 4 V).

would be required to cover 8 MHz to 48 MHz for 100 kHz to 50 MHz coverage of a typical transceiver.

While segmentation allows one to build a multi-octave tunable oscillator, segmentation also imposes some additional constraints. Recalling the earlier section in this chapter on oscillators, the condition for oscillation is an oscillator loop gain of 1 at a phase angle of 0 degrees. Over several octaves, the gain of the oscillating device (typically a transistor) varies, decreasing as operating frequency increases. While conventional limiting in the oscillator circuit will deal with some of this, it is not good practice to let the normal limiting process handle the entire gain variation. This becomes the role of automatic gain control (AGC) in multi-octave oscillator designs. Application of AGC allows for the maintenance of the oscillation criteria, a uniform output and good starting characteristics, and lower noise across the operating bandwidth.

Finally, a properly designed segmented VCO can improve synthesizer switching speed. The segmentation allows the designer to "pre-steer" the PLL system and reduce the time required for the loop filter to slew the VCO to the desired frequency for lock. Frequency-locked loops (FLL) are often employed for this steering operation to be sure that it happens properly.

9.7.7 ARRL Lab Transmitter Phase Noise Measurement

(This section was written by the ARRL Lab Staff.) Receiver performance has improved dramatically over the past generation or two of transceivers. With the widespread use of better filtering and improvements in digital signal processing, the effects of intermodulation distortion (IMD) at close signal spacing have been reduced greatly. Many software defined receivers (SDRs) now experience little to no reciprocal mixing or reduction of audio level on the received frequency from strong adjacent signals. Signals generated within the receiver during reception of one or more strong nearby signals are not the problem they once were, and it's possible to operate effectively in a crowded band.

Although today's receivers can hear weak signals very close to adjacent strong signals, even the best receivers cannot eliminate the effects of a wide signal from an adjacent transmitter. Transmitted signal issues include excessive keying bandwidth on CW, poor suppression of transmitted IMD products (which causes splatter on SSB), or transmitted phase noise (which raises a receiver's noise floor) in the speaker. ARRL Lab Product Review test reports include transmitter IMD products (both typical and worst case), a plot showing keying sidebands, and a plot showing transmitted phase noise.

PHASE NOISE

Phase noise is essentially the noise generated above and below an oscillator's frequency, also called sideband noise. All oscillators generate some level of phase noise. This is most evident when a close adjacent signal is very strong and the receiver's background noise increases. This is due to the mixing of the first local oscillator's phase noise with the incoming signal at the first IF (reciprocal mixing). The Lab reports the effects of reciprocal mixing in the receiver as "reciprocal mixing dynamic range" or RMDR. In many receivers we've tested, RMDR is the limiting dynamic range. In other words, third order IMD dynamic range and blocking dynamic range are better than RMDR. (See the **Receiving** chapter.)

A transmit oscillator, the heart of the transmitter, has phase noise too. A transmitter's phase noise is a fixed characteristic and, at times, can be a nuisance to other operators. A good example of observed phase noise can happen at Field Day or other environments where several transmitters are operating in close proximity. When two stations are operating on one band, the CW station blasts the phone operator's ears with bursts of noise, and vice versa. Using a transmitter exhibiting high phase noise with an RF amplifier magnifies the problem.

The solution for the reduction of both transmitted phase noise and receiver reciprocal mixing is the employment of high-quality oscillators by the manufacturer. Generally, the better the oscillator (the lower the phase noise), the better the RMDR and the lower the transmitted phase noise.

ARRL LAB TESTING

In the ARRL Laboratory, we use a Rohde & Schwarz FSUP 26 Signal Source Analyzer (**Figure 9.81**), which allows us to measure phase noise on any frequency up to 26.5 GHz. A crystal oscillator with very low phase noise is used to calibrate the system. We use an attenuator after the transmitter to bring the signal down to a suitable level for the phase noise test set.

A sample phase-noise plot for an amateur transceiver is shown in **Figure 9.82**. It was produced with the test setup shown in Figure 9.83. In this case, measurements for 14 MHz and 50 MHz are shown. These plots do not necessarily reflect the phase-noise characteristics of all units of a particular model.

The reference level (the top horizontal line on the scale in the plot) represents 0 dBc/Hz. Because each vertical division represents 20 dB, the plot shows the noise level between 0 dBc/Hz (the top horizontal line) and –180 dBc/Hz (the bottom horizontal line). The horizontal scale is logarithmic, with one decade per division (the first division shows noise from 100 Hz to 1 kHz Hz offset, whereas the last division shows noise from 100 kHz through 1 MHz offset).

WHAT DO THE PHASE-NOISE PLOTS MEAN?

Although they are useful for comparing different radios, plots can also be used to calculate the amount of interference you may receive from a nearby transmitter with known phase-noise characteristics. An approximation is given by

$$A_{QRM} = NL + 10 \times \log(BW)$$

where

A_{QRM} = Interfering signal level, dBc
NL = noise level on the receive frequency, dBc
BW = receiver IF bandwidth, in Hz

For instance, if the noise level is –90 dBc/Hz and you are using a 2.5 kHz SSB filter, the approximate interfering signal will be –56 dBc. In other words, if the transmitted signal is 20 dB over S-9, and each S unit is 6 dB, the interfering signal will be as strong as an S-3 signal.

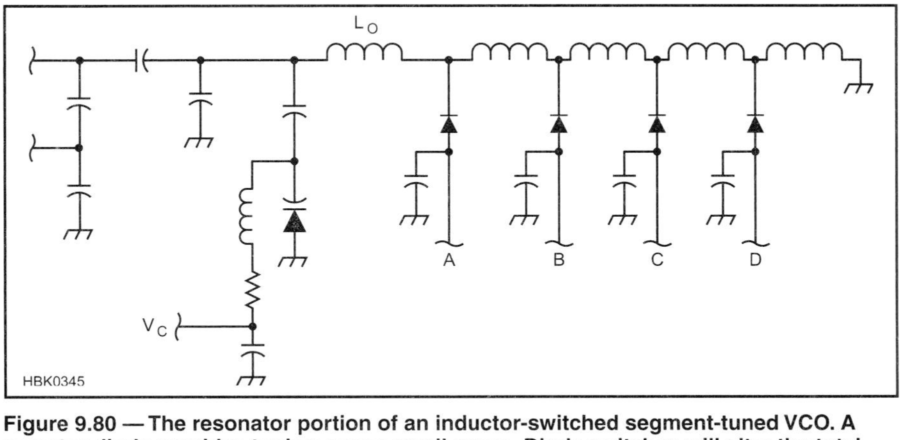

Figure 9.80 — The resonator portion of an inductor-switched segment-tuned VCO. A varactor diode provides tuning over a small range. Diode switches will alter the total inductance, changing the frequency in larger steps. Inductor-switched, capacitor-switched, and combinations are all used in VCO designs. (From Hayward, *Introduction to Radio Frequency Design*, Chapter 7. See references.)

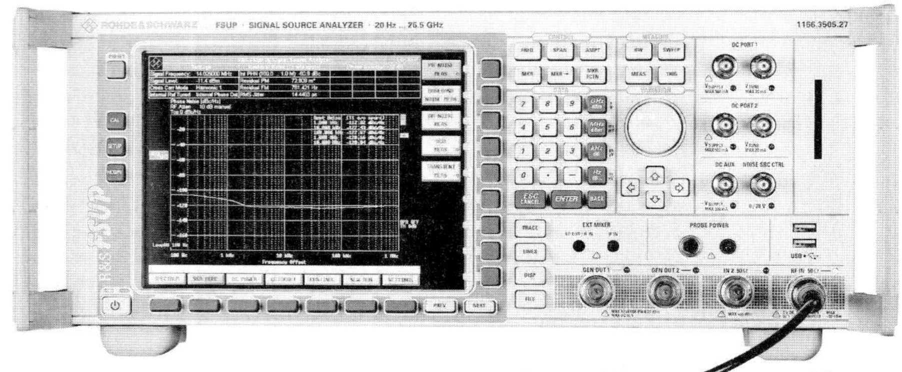

Figure 9.81 — The ARRL Lab's Rohde & Schwarz FSUP 26 Signal Analyzer, used for transmitter phase noise testing.

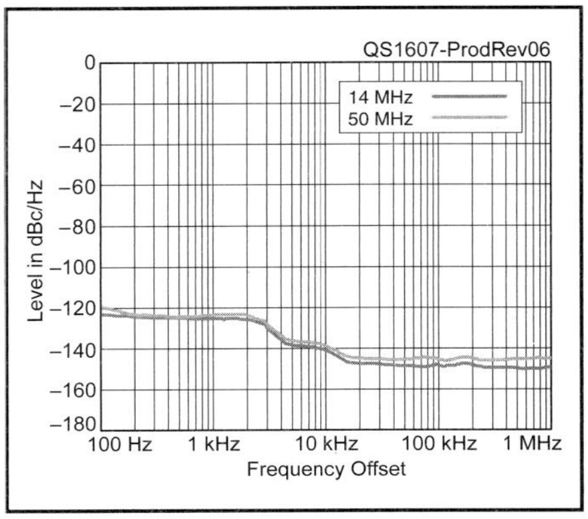

Figure 9.82 — Sample phase noise plot for an amateur HF transceiver as published in Product Review in *QST*. This is the Icom IC-7851.

The measurements made in the ARRL Lab apply only to transmitted signals. It is reasonable to assume that the phase-noise characteristics of most transceivers are similar on transmit and receive because the same oscillators are generally used in the local-oscillator (LO) chain or for generating DDS signals.

In some cases, the receiver may have better phase noise characteristics than the transmitter. Why the possible difference? The most obvious reason is that circuits often perform less than optimally in strong RF fields, as anyone who has experienced RFI problems can tell you. A less obvious reason results from the way that many high-dynamic-range superheterodyne receivers work.

To get good dynamic range in a superhet or in a hybrid receiver with mixers ahead of the DSP stages, a sharp crystal filter called a *roofing filter* is often placed immediately after the first mixer in the receive line. This filter removes all but a small slice of spectrum for further signal processing. If the desired filtered signal is a product of mixing an incoming signal with a noisy oscillator, signals far away from the desired one can end up in this slice. Once this slice of spectrum is obtained, however, unwanted signals cannot be reintroduced, no matter how noisy the oscillators used in further signal processing. As a result, some oscillators in receivers don't affect phase noise.

The difference between this situation and that in transmitters is that crystal filters are seldom used for reduction of phase noise in transmitting because of the high cost involved. Equipment designers have enough trouble getting smooth, click-free break-in operation in transceivers without having to worry about switching crystal filters in and out of circuits at 40 WPM keying speeds!

9.8 References and Bibliography

Analog Devices (**www.analog.com**), "ADI Converter Seminar."

Banerjee, D., *PLL Performance, Simulation, and Design, 6th Edition*, **deanbanerjeepll.com**

Brown, F., W6HPH, "Crystal Test Oscillators," *QEX*, Sep./Oct. 2016, pp. 11 – 13.

Clarke and Hess, *Communications Circuits, Analysis and Design* (Addison-Wesley, 1971). Wide coverage of transistor circuit design, including techniques suited to the design of integrated circuits. Its age shows, but it is especially valuable for its good mathematical treatment of oscillator circuits, covering both frequency- and amplitude-determining mechanisms. Look for a copy at a university library or initiate an interlibrary loan.

Gardner, F., *Phaselock Techniques*, (John Wiley and Sons, 1966).

Grebenkemper, J., KI6WX, "Phase Noise and Its Effects on Amateur Communications," *QST*, Part 1, Mar. 1988, pp. 14 – 20; *Part 2,* Apr. 1988 pp. 22 – 25. Also see Feedback, *QST,* May 1988, p. 44.

Hayward, W., W7ZOI, and DeMaw, D., W1CER, *Solid State Design for the Radio Amateur* (ARRL, 1986). Out of print, this is dated but a good source of RF design ideas.

Hayward, W., W7ZOI, Campbell, R., KK7B, and Larkin, B., W7PUA, *Experimental Methods in RF Design* (ARRL, 2003). A good source of RF design ideas, with good explanation of the reasoning behind design decisions.

Hayward, W., W7ZOI, *Introduction to Radio Frequency Design* (ARRL, 1994). Out of print, good in-depth treatments of circuits and techniques used at RF.

Hewlett-Packard Application Note 150-4, "Spectrum Analysis…Random Noise Measurements." **literature.cdn.keysight.com/litweb/pdf/5952-1147.pdf?id=800596**. Source of correction factors for noise measurements using spectrum analyzers.

Keyser, I., "An Easy to Set Up Amateur Band Synthesizer," *RADCOM* (RSGB), Dec. 1993, pp. 33 – 36.

"Boosting PLL Design Efficiency," Keysight application note 5989-9848EN **www.keysight.com/us/en/assets/7018-01964/application-notes/5989-9848.pdf**. Describes the characterization of VCOs using the E5202B Signal Source Analyzer.

Manassewitsch, V., *Frequency Synthesizers Theory and Design* (Wiley-Interscience, 1987).

Matthys, R., *Crystal Oscillator Circuits* (Wiley, 1983). Mini-Circuits collection of Application Notes on oscillators, synthesizers, and associated components and techniques. **blog.minicircuits.com/engineering-resources/**

Motorola Application Note AN-551, "Tuning Diode Design Techniques" No longer supplied by Freescale but available through various on-line sources. A concise explanation of varactor diodes, their characteristics and use.

Pappenfus. E., Bruene, W., Schoenike, E., *Single Sideband Circuits and Systems*, (McGraw-Hill, 1964). A book on HF SSB transmitters, receivers and accoutrements by Rockwell-Collins staff. Contains chapters on synthesizers and frequency standards. The frequency synthesizer chapter predates the rise of the DDS and other recent techniques, but good information about the effects of synthesizer performance on communi-cations is spread throughout the book.

Parzen, B., and Ballato. A., *Design of Crystal and Other Harmonic Oscillators* (Wiley-

Interscience, 1983). This book shows a thorough treatment of modern crystal oscillators including the optimum design. It is also the only book that looks at the base and collector limiting effect. The second author is affiliated with the US Army Electronics Technology & Devices Laboratory, Fort Monmouth, New Jersey.

Pontius, B., "Measurement of Signal Source Phase Noise with Low-Cost Equipment," *QEX*, May/Jun. 1998, pp. 38 – 49.

Rohde, U., N1UL, *Microwave and Wireless Synthesizers Theory and Design* (John Wiley and Sons, 1997). This book contains the textbook-standard mathematical analyses of frequency synthesizers combined with unusually good insight into what makes a better synthesizer and *a lot* of practical circuits to entertain serious constructors. A good place to look for low-noise circuits and techniques.

Rohde, U., N1UL, Newkirk, D., *RF/Microwave Circuit Design for Wireless Applications* (John Wiley and Sons, 2000). While giving a deep insight into circuit design for wireless applications, Chapter 5 (RF/Wireless Oscillators) shows a complete treatment of both discrete and integrated circuit-based oscillators. Chapter 6 (Wireless Synthesizers) gives more insight in synthesizers and large list of useful references.

Rohde, U., N1UL, Whitaker, J., *Communications Receivers DSP, Software Radios, and Design*, 3rd Edition (McGraw-Hill, 2001). While mostly dedicated to communication receivers, this book has a useful chapter on Frequency Control and Local Oscillators, specifically on Fractional Division Synthesizers.

Rohde, U., N1UL, Poddar, A., Boeck, G., *The Design of Modern Microwave Oscillators for Wireless Applications Theory and Optimization* (Wiley-Interscience, 2005). This is the latest and most advanced textbook on oscillator design for high frequency/microwave application. It shows in detail how to design and optimize high frequency and microwave VCOs. It also contains a thorough analysis of the phase noise as generated in oscillators. While highly mathematical, the appendix gives numerical solutions that can be easily applied to various designs. It covers both bipolar transistors and FETs.

Rohde, U., N1UL, "All About Phase Noise in Oscillators," *QEX*, Part 1, Dec. 1993; Part 2, Jan. 1994; Part 3, Feb. 1994.

Rohde, U., N1UL, "Key Components of Modern Receiver Design," *QST*, Part 1, May 1994; Part 2, Jun. 1994; Part 3 Jul. 1994. Includes discussion of phase-noise reduction techniques in synthesizers and oscillators.

Rohde, U., N1UL, "A High-Performance Hybrid Frequency Synthesizer," *QST*, Mar. 1995, pp. 30 – 38.

Stockton, D., "Polyphase noise-shaping fractional-N frequency synthesizer", US Patent 6,509,800.

Telewski, F., and Drucker, E., "Noise Reduction Method and Apparatus for Phase-locked Loops," US Patent 5,216,387.

Vendelin, G., Pavio, A., Rohde, U., *Microwave Circuit Design Using Linear and Nonlinear Techniques*, 2nd Edition, (Wiley-Interscience, 2005). This is a general handbook on microwave circuit design. It covers (in 200 pages) a large variety of oscillator design problems including microwave applications. It provides a step-by-step design guide for both linear and nonlinear oscillator designs.

Williams, J., *The Art and Science of Analog Circuit Design* (Butterworth-Heinemann, 1995).

Williams, J., *Analog Circuit Design: Art, Science and Personalities* (Butterworth-Heinemann, 1991).

INTERESTING MODERN JOURNAL PUBLICATIONS

Oscillator References

Analog Devices, "Fundamentals of Direct Digital Synthesis," Analog Devices (**www.analog.com**), MT-085.

Berny, A., Niknejad, A., and Meyer, R., "A 1.8 GHz LC VCO with 1.3 GHz Tuning Range and Digital Amplitude Calibration," *IEEE Journal of SSC*, vol. 40.
no. 4, 2005, pp. 909 – 917.

Chenakin, A., "Phase Noise Reduction in Microwave Oscillators," *Microwave Journal*, Oct. 2009, pp. 124 – 140.

Kester, W., "Aperture Time, Aperture Jitter, Aperture Delay Time—Removing the Confusion," Analog Devices, MT-007.

Khanna, A., "Microwave Oscillators: The State of The Technology," *Microwave Journal*, Apr. 2006, pp. 22 – 42.

Nomura, N., Itagaki, M., and Aoyagi, Y., "Small packaged VCSO for 10 Gbit Ethernet Application," IEEE IUFFCS, 2004, pp. 418 – 421.

Poddar, A., "A Novel Approach for Designing Integrated Ultra Low Noise Microwave Wideband Voltage-Controlled Oscillators," Dr.-Ing. Dissertation, TU-Berlin, Germany, 14 Dec. 2004.

Ravi, A., Carlton, B., Banerjee, G., Soumyanath, K., "A 1.4V, 2.4/5.2 GHz, 90 nm CMOS system in a Package Transreceiver for Next Generation WLAN," *IEEE Symp. on VLSI Tech.*, 2005.

Rohde, U., N1UL, "A New Efficient Method of Designing Low Noise Microwave Oscillators," Dr.-Ing. Dissertation, TU-Berlin, Germany, 12 Feb 2004.

Rohde, U., N1UL, "Calculation of FM and AM Noise Signals of Colpitts Oscillators in the Time Domain," *QEX*, Mar./Apr. 2016, pp. 22 – 39.

Rohde, U., N1UL, "Some Thoughts on Designing Very High Performance VHF Oscillators," *QEX*, Nov./Dec. 2015, pp. 32 – 40.

Sheng Sun and Lei Zhu, "Guided-Wave Characteristics of Periodically Nonuniform Coupled Microstrip Lines-Even and Odd Modes," *IEEE Transaction on MTT*, Vol. 53, No. 4, Apr. 2005, pp. 1221 – 1227.

Wei, C., Chiu, H., and Feng, W., "An Ultra-Wideband CMOS VCO with 3-5 GHz Tuning Range," IEEE, *International Workshop on Radio-Frequency Integration Technology*, Nov. 2005, pp. 87 – 90.

Yim, M., "Switched Resonators and Their Applications in a Dual Band Monolithic CMOS LC Tuned VCO," *IEEE MTT*, Vol. 54, Jan. 2006, pp. 74 – 81.

Synthesizer References

H. Arora, N. Klemmer, J. C. Morizio, and P. D. Wolf, "Enhanced Phase Noise Modeling of Fractional-N Frequency Synthesizer," *IEEE Trans. Circuits and Systems-I: Regular Papers*, Vol. 52, Feb. 2005, pp. 379 – 395.

van de Beek, R., Leenaerts, D., van der Weide, G., "A Fast-Hopping Single-PLL 3-band UWB Synthesizer in 0.25um SiGe BiCMOS," *Proc. European Solid-State Circuit Conf.* 2005, pp. 173 – 176.

Best, R., *Phase-Locked Loops: Design, Simulation, and Applications*, 5th Edition, McGraw-Hill, 2003.

Dacus, F., "Design Methods of Modern Ultra-Low-Noise Synthesizers," *Microwaves & RF*, Dec. 2018, pp. 36 – 42 (multi-part series).

Lam, C., Razavi, B., "A 2.6/5.2 GHz Frequency Synthesizer in 0.4um CMOS Technology," *IEEE JSSC*, Vol. 35, May 2000, pp. 788 – 79.

Lee, J., "A 3-to-8 GHz Fast-Hopping Frequency Synthesizer in 0.18-um CMOS Technology," *IEEE Journal of SSC*, Vol. 41, No. 3, Mar. 2006, pp. 566 – 573.

Natrajan, A., Komijani, A., and Hajimiri, A., "A Fully 24-GHz Phased-Array Transmitter in CMOS," *IEEE Journal of SSC*, Vol. 40, No. 12, Dec. 2005, pp. 2502 – 2514.

Rahajandraibe, W., Zaid, L., de Beaupre, V., and Bas, G., "Frequency Synthesizer and FSK Modulator For IEEE 802.15.4 Based Applications," *2007 RFIC Symp. Digest*, pp. 229 – 232.

Sandner, C., Wiesbauer, A., "A 3 GHz to 7 GHz Fast-Hopping Frequency Synthesizer For UWB," *Proc. Int. Workshop on Ultra Wideband Systems* 2004, pp. 405 – 409.

Staszewski R., et. al., "All Digital PLL and Transmitter for Mobile Phones," *IEEE J. Solid-State Circuits*, Vol. 40, No. 12, Dec. 2005, pp. 2469 – 2482.

Contents

Chapter 10 — Online Content

Articles

- 6-Meter Filter with Harmonic Suppression by Paul Wade, W1GHZ
- A High Performance, Low Cost 1.8 to 54 MHz Low Pass Filter by Bill Jones, K8CU
- Altoids Tin Filters by Paul Wade, W1GHZ
- Clean Up Your Signals with Band-Pass Filters — Parts 1 and 2 by Ed Wetherhold, W3NQN
- Combline Filters for VHF and UHF by Paul Wade, W1GHZ
- Crystal Filter Design and Crystal Characterization
- Crystal Parameter Measurements Simplified by Chuck Adams, K7QO
- Hands On Radio: *ELSIE* Filter Design — Parts 1 and 2 by Ward Silver, NØAX
- HF Yagi Triplexer Especially for Field Day by Gary Gordon, K6KV
- High-Power Harmonic Filters by George Cutsogeorge, W2VJN
- High-Power HF Band-Pass Filter Design by Jeff Crawford, KØZR
- Improved Audio-Frequency Bandpass Filter for Morse Code Reception by Jim Tonne, W4ENE
- Manual Filter Design Examples by Jim Tonne, W4ENE
- SDR Simplified — Cascaded Integrator Comb Filters by Ray Mack, W5IFS
- SDR Simplified — Filter Design Program by Ray Mack, W5IFS
- SDR Simplified — Fourier Transforms by Ray Mack, W5IFS
- SDR Simplified — More Filter Activities by Ray Mack, W5IFS
- Using Active Filter Design Tools by Dan Tayloe, N7VE

Design Software

The following Windows software by Jim Tonne, W4ENE, is available with the online content.

- *Diplexer* for design and analysis of diplexer filters
- *Elsie* for design and analysis of lumped-element LC filters
- *Helical* for design and analysis of helical-resonator bandpass filters
- *SVC Filter Designer* for design and analysis of lumped element high-pass and low-pass filters
- *QuadNet* for design and analysis of active all-pass networks for SSB operation

Chapter 10

Analog and Digital Filtering

This chapter discusses the most common types of filters used by radio amateurs, both analog and digital. Design information is supplied where appropriate or references to software or detailed design procedures are supplied.

The sections describing basic concepts, lumped element filters and some design examples were initially prepared by Jim Tonne, W4ENE, and updated by Ward Silver, NØAX. Digital filter material was updated by Doug Grant, K1DG, based on material originally developed by Alan Bloom, N1AL. Additional digital filter material was based on the SDR: Simplified column in *QEX* by Ray Mack, W5IFS. Material on cavity filters was provided by John Portune, W6NBC.

The online information's design example for active filters was provided by Dan Tayloe, N7VE. The online information's design example for crystal filters was developed by Dave Gordon-Smith, G3UUR.

10.1 Introduction

Electrical filters are circuits used to process signals based on their frequency. For example, most filters are used to pass signals of certain frequencies and reject others. The electronics industry has advanced to its current level in large part because of the successful use of filters. Filters are used in receivers so that the listener can hear only the desired signal; other signals are rejected. Filters are used in transmitters to pass only one signal and reject those that might interfere with other spectrum users. **Table 10.1** shows the usual signal bandwidths for several signal types.

The simplified receivers in **Figure 10.1** show filter use in both an analog, superheterodyne receiver (Figure 10.1A) and a simplified SDR receiver (Figure 10.1B). A *preselector* filter is placed between the antenna and the receiver's front end to pass all frequencies within a given amateur band with low loss. Strong out-of-band signals from broadcast, commercial, or military stations are rejected to prevent them from overloading the receiver input. The preselector filter is almost always built with *lumped-element* or "LC" technology.

In an analog receiver, there are one or more intermediate frequency (IF) filters to select only the desired signal. The IF filter closest to the antenna may be a relatively wide "roofing" filter with a bandwidth of several kHz to reject strong in-band signals not close to the desired signal frequency. Following detection or demodulation, an audio filter is placed somewhere ahead of the AF amplifier and speaker to rejects unwanted products from the noise. The audio filter is often implemented with *active filter* technology. For an SDR receiver, similar preselector and audio filters may be used but the filters which select a single signal are implemented as *digital filters* in the I/Q Processing block.

The complementary transmitter block diagrams are shown in **Figure 10.2** in which a similar array of filters appears in reverse order. In the analog transmitter, an audio filter between the microphone and the balanced mixer rejects noise and unwanted speech components. Since the balanced mixer generates both lower and upper sidebands, an IF filter is placed at the mixer output to pass only the desired lower (or upper) sideband. In the SDR version, the input signals are digitized and processed before the I/Q signals are filtered for transmission.

Finally, a filter at the output of the transmit mixer passes only signals within the amateur band in use rejects unwanted frequencies generated by the mixer to prevent them from being amplified and transmitted. Filters at the transmitter output attenuate the harmonics of the transmitted signals.

Table 10.1
Typical Filter Bandwidths for Typical Signals

Source	*Required Bandwidth*
Fast-scan analog television (ATV)	4.5 MHz
Broadcast-quality speech and music	15 kHz (from 20 Hz to 15 kHz)
Communications-quality speech	3 kHz (from 300 Hz to 3 kHz)
Slow-scan television (SSTV)	3 kHz (from 300 Hz to 3 kHz)
HF Digital (general)	100 to 1500 Hz (depends on modulation and bit rate)
HF RTTY (standard shift)	250 to 500 Hz
Radiotelegraphy (Morse code, CW)	200 to 500 Hz
PSK31 digital modulation	100 Hz

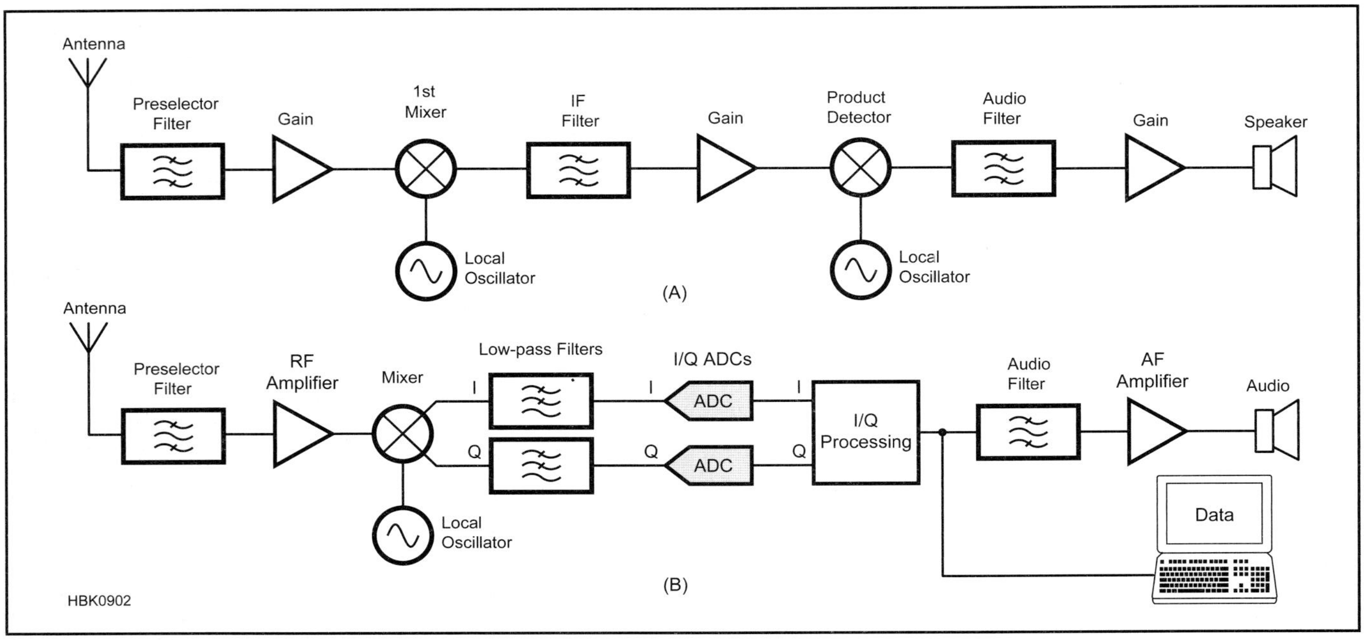

Figure 10.1 — Single-band SSB superheterodyne receiver (A) and SDR SSB/Data receiver (B) showing the filters typically used by each.

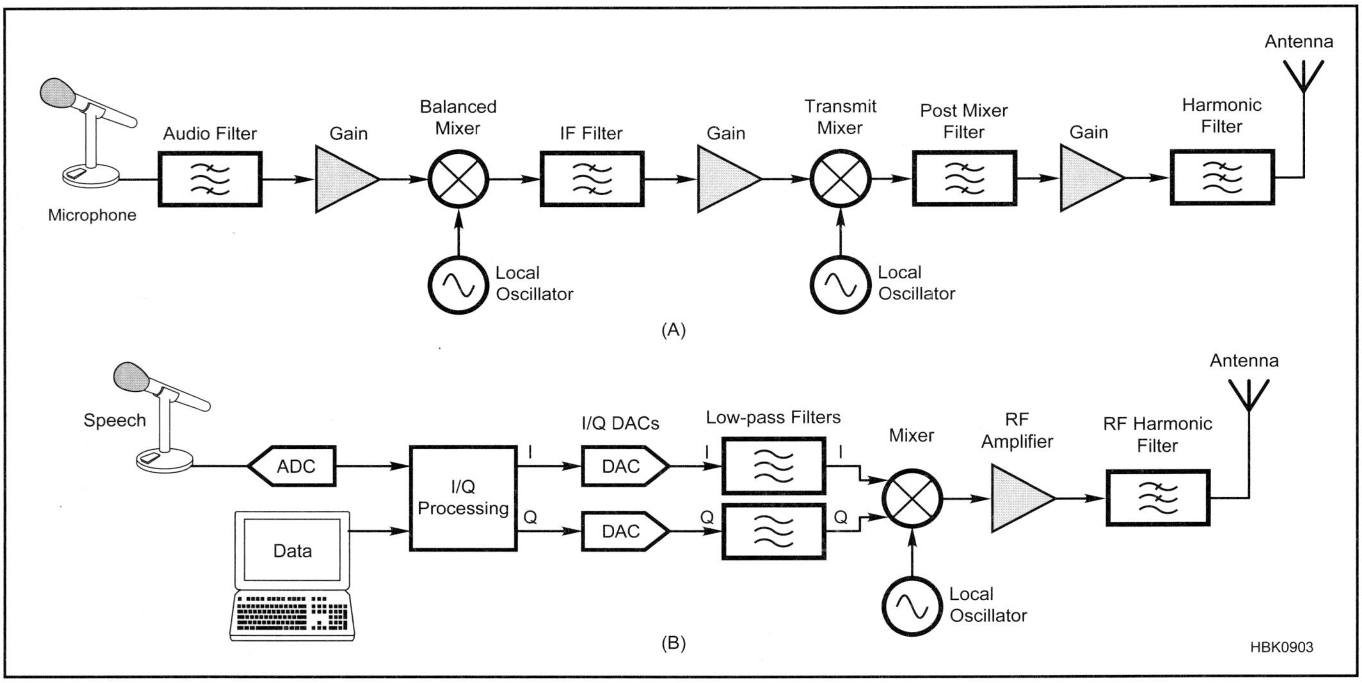

Figure 10.2 — Single-band SSB superheterodyne transmitter (A) and SDR SSB/Data transmitter (B) showing the filters typically used by each.

10.2 Filter Basics

10.2.1 Filter Magnitude Responses

A common type of filter, the *band-pass*, passes signals in a range of frequencies — the *passband* — while rejecting signals outside that range — the *stopband.* To pass signals from dc up to some *cutoff frequency* at which output power is halved (reduced by 3 dB) we would use a *low-pass* filter. To pass signals above a cutoff frequency (also called the "3-dB frequency") we would use a *high-pass* filter. Similarly, to pass signals within a range of frequencies we would use a *band-pass* filter. To pass signals at all frequencies *except* those within a specified range requires a *band-stop* or *notch filter*.

Figure 10.3A illustrates the *magnitude response* of a low-pass filter. Signals lower than the cutoff frequency (3 MHz in this case) are passed with some small amount of attenuation while signals higher than that frequency are attenuated. The degree of attenuation is dependent on several variables, filter complexity being a major factor.

Of the graphs in this chapter that show a filter's magnitude response, the vertical axes are labeled "Transmission (dB)" with 0 dB at the top of the axis and negative values increasing toward the X-axis. Increasingly negative values of transmission in dB are the same as increasingly positive values of attenuation in dB. For example, –40 dB transmission is the same as 40 dB of attenuation.

Figure 10.3B illustrates the magnitude response of a high-pass filter. Signals above the 3 MHz cutoff frequency are passed with minimum attenuation while signals below that frequency are attenuated. Again, the degree of attenuation is dependent on several variables.

Figure 10.3C illustrates the magnitude response of a band-pass filter. Signals within the band-pass range (between the lower and upper cutoff frequencies) are passed with minimum attenuation while signals outside that range are attenuated. In this example the filter was designed with cutoff frequencies of 2 MHz and 4 MHz, for a passband width of 2 MHz.

Figure 10.3D illustrates the magnitude response of a band-stop filter. Signals within the band-stop range are attenuated while all other signals are passed with minimum attenuation. A notch filter is a type of band-stop filter with a narrow stop-band in which the attenuation is a maximum at a single frequency.

An *ideal filter* — a low-pass filter, for example — would pass all frequencies up to some point with no attenuation at all and totally reject everything beyond that point. This is known as a *brick wall* response because the filter's passband and stopband are *flat*, meaning no attenuation, and the rolloff in the transition region is infinitely steep. The magnitude response of such a filter would be drawn as a flat line representing 0 dB attenuation up to the cutoff frequency that then abruptly changes at the cutoff frequency to a flat line at infinite attenuation throughout the stopband to infinite frequency.

Figure 10.4 shows all of the basic para-

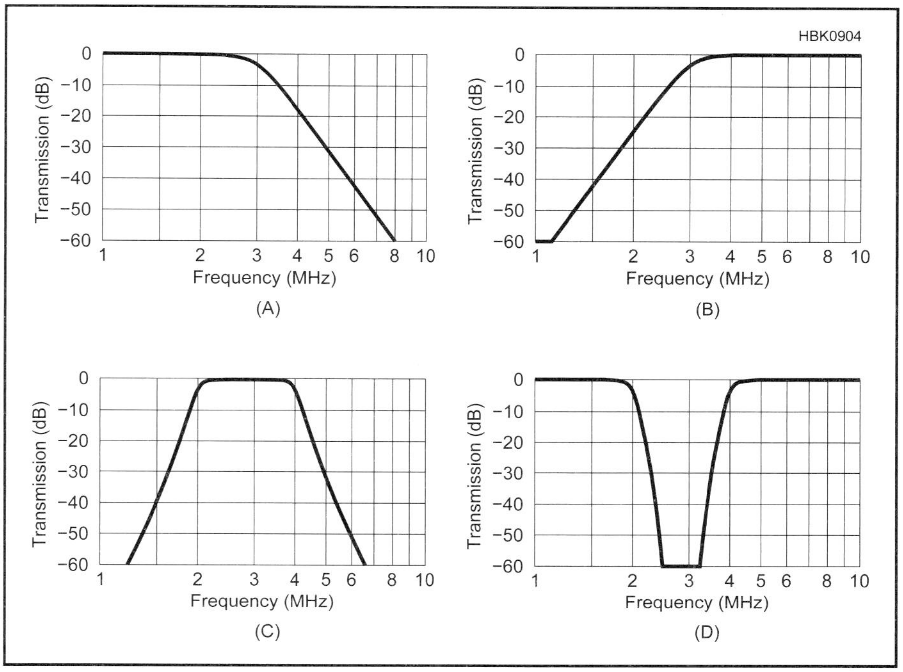

Figure 10.3 — Examples of a low-pass magnitude response (A), high-pass magnitude response (B), band-pass magnitude response (C), and band-stop magnitude response (D).

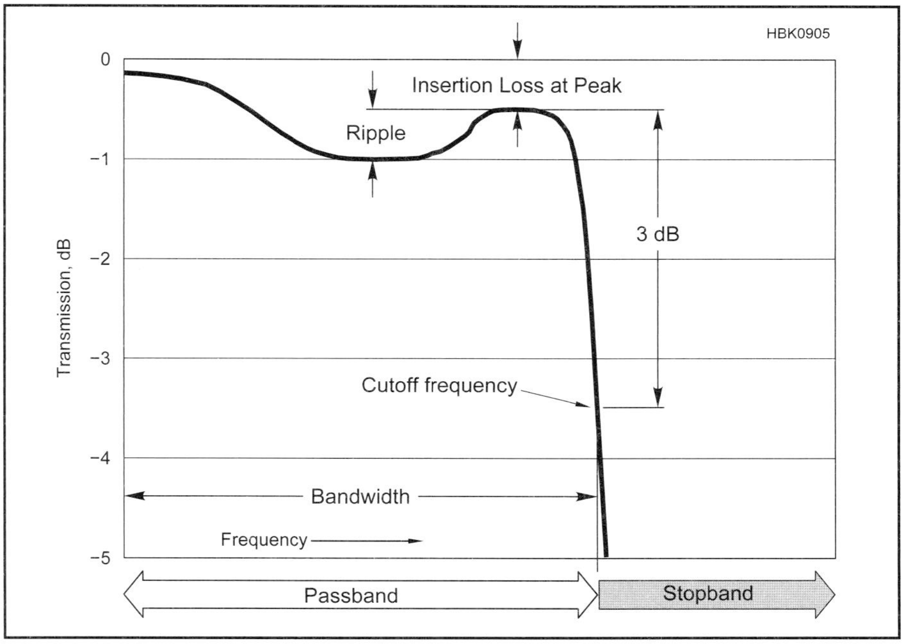

Figure 10.4 — Low-pass filter characteristics showing the passband and stopband, bandwidth, 3 dB cutoff, passband ripple, and insertion loss (IL). This filter has approximately 0.5 dB IL at the frequency of peak response while passband ripple is also 0.5 dB. The vertical axis shows gain (or loss) through the filter, assuming both the input and output are properly terminated. The horizontal axis represents frequency.

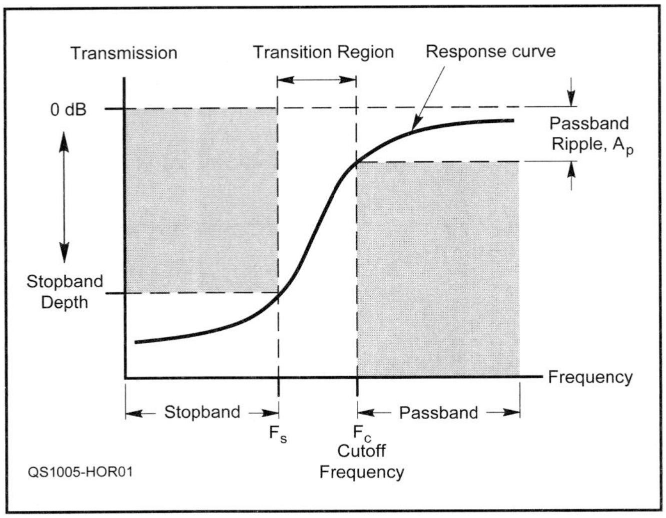

Figure 10.5 — The key specifications for a filter's amplitude response. As long as the filter response curve stays between the light gray boxes, the filter meets the design specification.

meters for a low-pass filter. Variations in amplitude of the response curve in the passband are called *ripple*. (Filters can also have ripple in the stopband for some design families as discussed later.) If the filter is made from passive components, there will be some losses in those components — the amount of loss is called *insertion loss* (IL). The attenuation for signal frequencies in the stopband far from the cutoff frequency is the filter's *ultimate attenuation*.

SPECIFYING A FILTER RESPONSE

Figure 10.5 shows a high-pass filter's *response curve* and some of the terms used to specify the filter's design. On the frequency axis, F_C is the cutoff frequency and F_S defines the *stopband width*. The passband for a high-pass filter consists of all frequencies above F_C. The stopband for a high-pass filter consists of all frequencies below F_S. The *transition region* is the range of frequencies between the stopband and passband. (For a low-pass filter, F_S and F_C are reversed along with the gray boxes they bound.)

The gray boxes leave a space between them through which the filter's response curve must pass. The borders of the gray boxes establish the required performance for the filter. The gray box defining the stopband is bounded on the bottom by the stopband depth. The filter's transmission in the stopband must be equal to or below the stopband depth. The gray box defining the passband is bounded on the top by the passband ripple. The filter's transmission in the passband must be equal to or greater than the passband ripple. Between F_S and F_C is the transition region in which the response curve passes between the stopband and passband. Any response curve that passes through the space between the gray boxes meets the performance requirements for the filter.

As frequency increases through the transition region from the passband into the stopband, the attenuation increases. *Rolloff*, the rate of change of the attenuation for frequencies above cutoff, is expressed in terms of dB of change per octave of frequency. A filter's response with high values of rolloff is called "sharp" or "steep" and "soft" or "shallow" if rolloff is low.

10.2.2 Filter Order

The steepness of the descent from the passband to the region of attenuation — the stopband — is dependent on the complexity of the filter, called the *order*. In a lumped-element filter made from inductors and capacitors, the order is determined by the number of separate energy-storing elements (either L or C) in the filter. For example, an RC filter with one resistor and one capacitor has an order of 1. A notch filter made of a single series-LC circuit has an order of 2. The definition of order for active and digital signal processing filters can be more complex, but the general understanding remains valid that the order of a filter determines how rapidly its response can change with frequency.

For example, **Figures 10.6** and **10.7** shows the magnitude response of one type of low-pass filter (Butterworth family) with orders varying from very simple ("N=2") to the more complex ("N=20"). For each order, frequency has been *normalized* to the ratio of frequency, f, to cutoff frequency, f_C. (i.e. Normalized cutoff frequency is always 1.0.) As the order increases, the rolloff also increases. Figure 10.6 has an expanded vertical scale to show the filter's behavior in the passband more clearly.

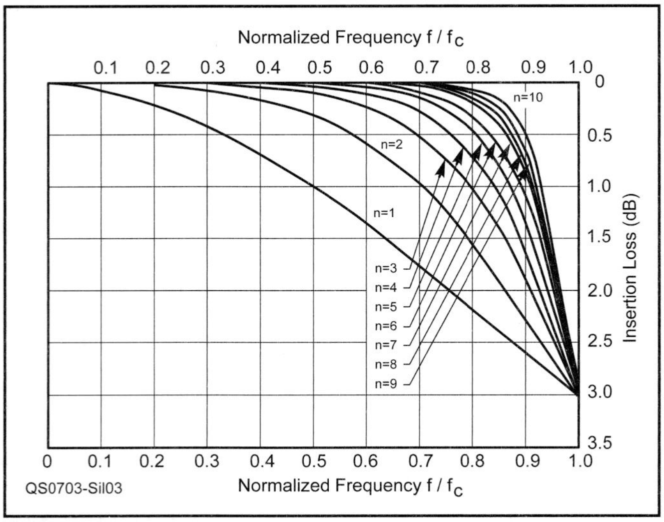

Figure 10.6 — Response of Butterworth filters in the passband below f_C. Frequency is shown as the normalized frequency, f/f_C.

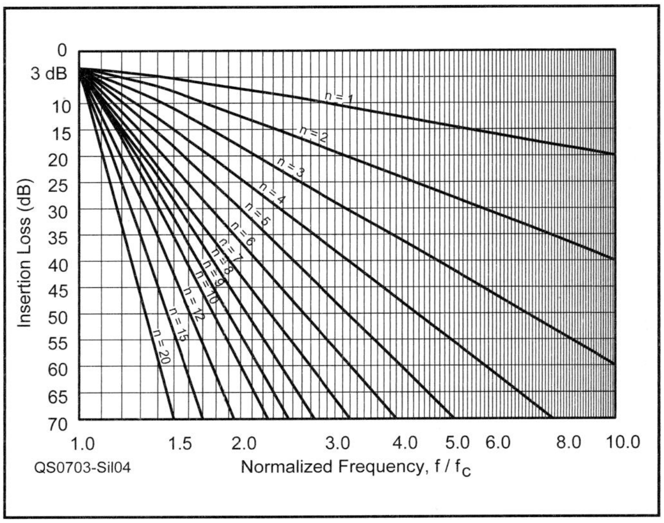

Figure 10.7 — Response of Butterworth filters in the transition and stop-bands above f_C. Frequency is shown as the normalized frequency, f/f_C.

10.2.3 Filter Families

There is no single "best" way to design a filter. Instead we have to decide on some traits and then choose the most appropriate *family* for our design. Different families have different traits, and filter families are commonly named after the mathematician or engineer responsible for defining their behavior mathematically. Each filter family is represented by a specific type of equation that describes the filter's behavior.

Two primary traits of the most importance to amateurs are used to describe the behavior of filter families: ripple (variations in the magnitude response within the passband and stopband) and rolloff. Different families of filters have different degrees of ripple and rolloff (and other characteristics, such as phase response).

With a flat magnitude response (and so no ripples in the passband) we have what is known as a *Butterworth* family design. This family also goes by the name of *Maximally Flat Magnitude* or *Maximally Flat Gain*. An example of the magnitude response of a filter from the Butterworth family is shown in **Figure 10.8**. (Most filter discussions are based on low-pass filters because the same concepts are easily extended to other types of filters and much of the mathematics behind filters is equivalent for the various types of frequency responses.)

By allowing magnitude response ripples in the passband, we can get a somewhat steeper rolloff from the passband into the stopband, particularly just beyond the cutoff frequency. A family that does this is the *Chebyshev* family. **Figure 10.9** illustrates how allowing magnitude ripple in the passband provides a sharper filter rolloff. This plot compares the 1 dB ripple Chebyshev with the no-ripple Butterworth filter down to 12 dB of attenuation.

Figure 10.9 also illustrates the usual definition of bandwidth for a low-pass Butterworth filter — the filter's 3-dB or cutoff frequency. For a low-pass Chebyshev filter, *ripple bandwidth* is used — the frequency range over which the filter's passband ripple is no greater than the specified limit. For example, the ripple bandwidth of a low-pass Chebyshev filter designed to have 1 dB of passband ripple is the highest frequency at which attenuation is 1 dB or less. The 3 dB bandwidth of the frequency of the filter will be somewhat greater.

A Butterworth filter is defined by specifying the order and bandwidth. The Chebyshev filter is defined by specifying the order, the ripple bandwidth, and the amount of passband ripple. In Figure 10.9, the Chebyshev filter has 1 dB of ripple; its ripple bandwidth is 1000 Hz. The Butterworth filter has a 3 dB bandwidth also of 1000 Hz. Some filter textbooks use the 3-dB point to define Chebyshev filters; most use the ripple bandwidth as illustrated here. The schematics (if you ignore parts values) of those two families are identical.

Even small amounts of ripple can be beneficial in terms of increasing a filter's rolloff. **Figure 10.10** compares a Butterworth filter (with the narrow line plot, no ripple in the passband) with a Chebyshev filter (wide line plot, 0.2 dB of ripple in the passband) down to 60 dB of attenuation. (For this comparison the cutoff frequencies at 3 dB of attenuation are the same for each filter.) Even that small amount of ripple in the Chebyshev filter passband allows a noticeably steeper rolloff between the passband and the stopband. As the passband ripple specification is increased, the steepness of the transition from passband to stopband increases, compared to the Butterworth family although the rolloff of the two families eventually becomes equal.

For Chebyshev filters, when the value of the passband ripple is changed, the magnitude response in the stopband region also changes. **Figure 10.11** compares the stopband response of Chebyshev filters with passband ripple ranging from 0.01 to 1 dB.

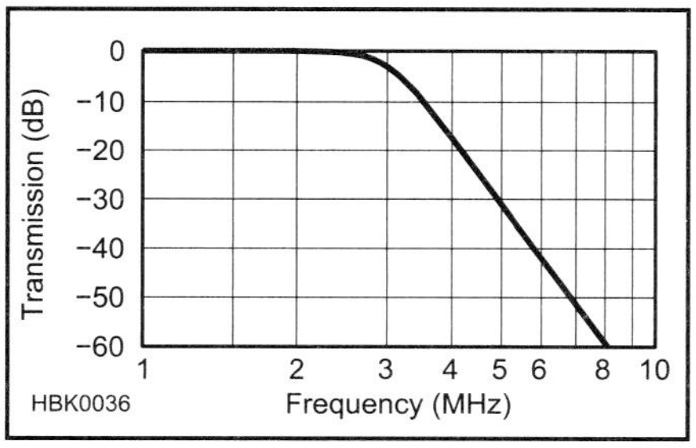

Figure 10.8 — Filters from the Butterworth family exhibit flat magnitude response in the passband.

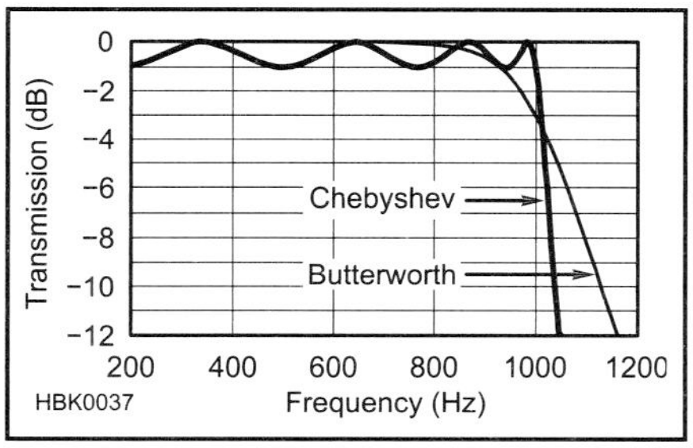

Figure 10.9 — This plot compares the response of a Chebyshev filter with a 1-dB ripple bandwidth of 1000 Hz and a Butterworth filter 3-dB bandwidth of 1000 Hz.

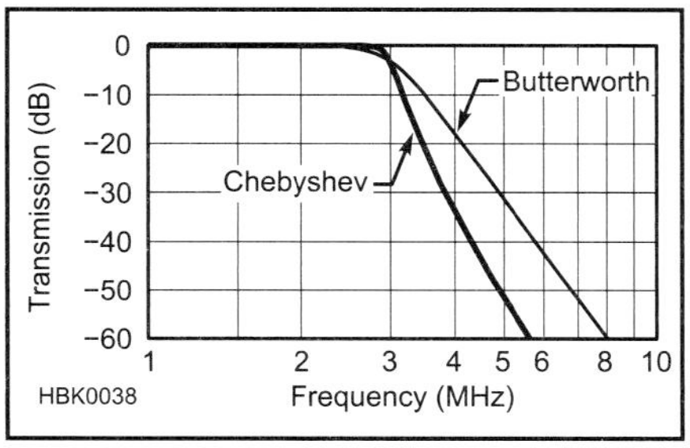

Figure 10.10 — A Chebyshev filter (0.2 dB passband ripple) allows a sharper cutoff than a Butterworth design with no passband ripple.

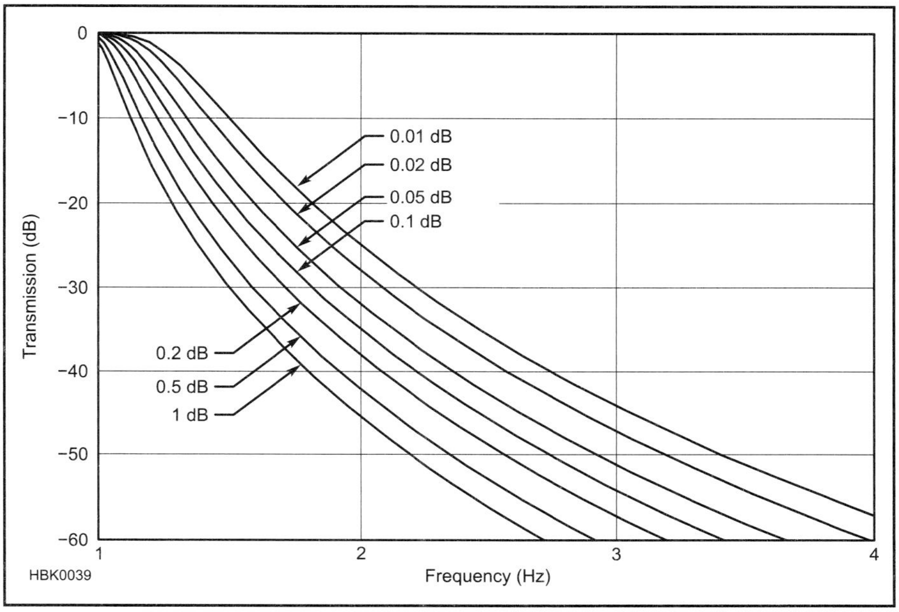

Figure 10.11 — These stop band response plots illustrate the Chebyshev family with various values of passband ripple. These plots are for a seventh-order low-pass design with ripple values from 0.01 to 1 dB. Ignoring the effects in the passband of high ripple values, increasing the ripple will allow somewhat steeper rolloff into the stop band area, and better ultimate attenuation in the stop band.

It is possible to obtain even steeper rolloff into the stopband by adding "traps" whose frequencies are carefully calculated. The resonant frequencies of those traps are in the stopband region and are set to yield best performance. When this is done, we have a *Cauer* family design (also called the *elliptic-function* design). The rolloff of the Cauer filter from the passband into the stopband is the steepest of all analog filter types provided that the behavior in the passband is uniform (either no ripple or a uniform amount of ripple). **Figure 10.12** shows the response of the Chebyshev and Cauer designs for comparison.

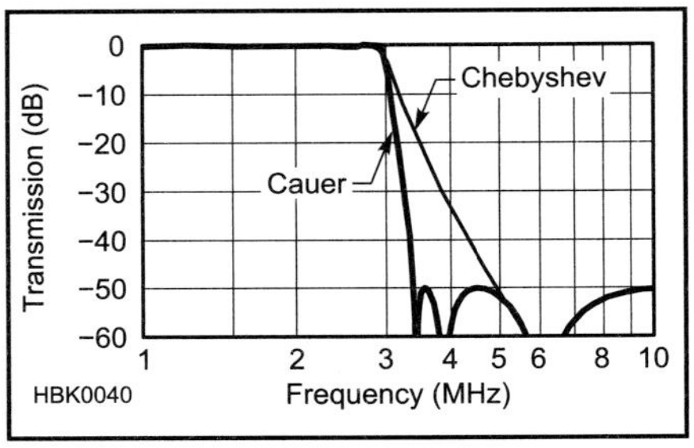

Figure 10.12 — The Cauer family has an even steeper rolloff from the passband into the stop band than the Chebyshev family. Note that the ultimate attenuation in the stop band is a design parameter rather than ever-increasing as is the case with other families.

The Cauer filter is defined by specifying the order, the ripple bandwidth, and the passband ripple, just as for the Chebyshev. Again, an alternative bandwidth definition is to use the 3 dB frequency instead of the ripple bandwidth. The Cauer family requires one more specification: the *stopband frequency* and/or the *stopband depth*. The stopband frequency is the lowest frequency of a null or notch in the stopband. The stopband depth is the minimum amount of attenuation allowed in the stopband. In Figure 10.12, the stopband frequency is about 3.5 MHz and the stopband depth is 50 dB. For this comparison both designs have the same passband ripple value of 0.2 dB.

A downside to the Cauer filter is that the ultimate attenuation is some chosen value rather than ever increasing, as is the case with the other families. In the far stopband region (where frequency is much greater than the cutoff frequency) the rolloff of Cauer filters ultimately reaches 6 or 12 dB per octave, depending on the order. Odd-ordered Cauer filters have an ultimate rolloff rate of about 6 dB per octave while the even-ordered versions have an ultimate rolloff rate of about 12 dB per octave.

10.2.4 Group Delay

Another trait that can influence the choice of which filter family to use is the way the transit time of signals through the filter varies with frequency. This is known as *group delay*. ("Group" refers to a group of waves of similar frequency and phase moving through a media, in this case, the filter.) The wide line (lower plot) in **Figure 10.13** illustrates the group delay characteristics of a Chebyshev low-pass filter, while the upper plot (narrow line) shows the magnitude response. As shown in Figure 10.13, the group delay of components near the cutoff frequency becomes quite large when compared to that of components at lower and higher frequencies. This is a result of the phase shift of the filter's transmission being nonlinear with frequency; it is usually greater near the cutoff frequency. By delaying signals at different frequencies different amounts, the signal components are "smeared" in time. This causes distortion of the signal and can seriously disrupt high-speed data signals.

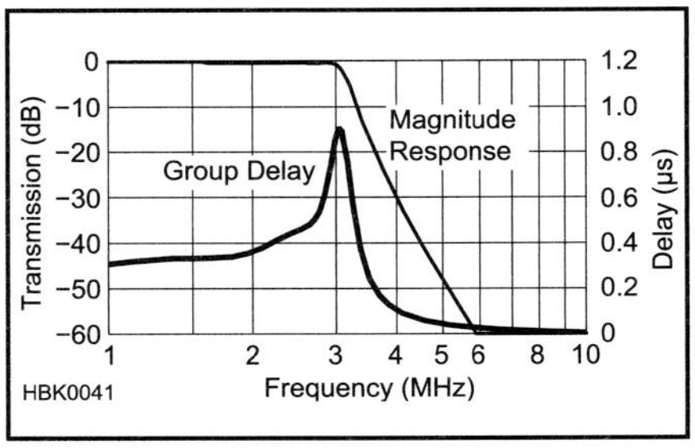

Figure 10.13 — Magnitude response and group delay of a Chebyshev low-pass filter.

If a uniform group delay for signals throughout the passband is needed, then the *Bessel* filter family should be selected. The Bessel filter can be used as a delay-line or time-delay element although the gentle rolloff of the magnitude response may need to be taken into account. The magnitude and delay characteristics for the Bessel family are shown in **Figure 10.14**. Comparing the Bessel filter's response to that of the Butterworth in Figure 10.8 shows the difference in roll off.

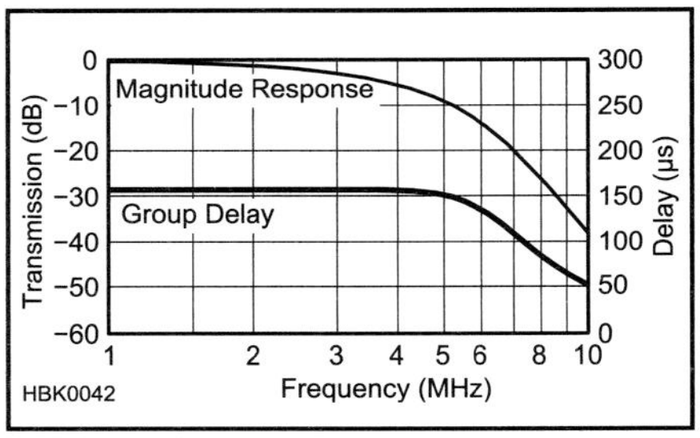

Figure 10.14 — Magnitude response and group delay of a Bessel low-pass filter.

A downside of Bessel family filters is that the rolloff characteristic is quite poor; they are not very good as a magnitude-response-shaping filter. The Bessel family is characterized largely by its constant group delay in the passband (for a low- or high-pass filter) shown as the bottom plot in Figure 10.14. The *constant-delay* characteristic of the Bessel extends into the stopband. The Bessel filter bandwidth is commonly defined by its 3-dB point (as with the Butterworth).

Because the Bessel filter is used when phase response is important, it is often characterized by the frequency at which a specific amount of phase shift occurs, usually one radian. (One radian is equal to 360 / 2π = 57.3°.) Since the delay causes the output signal to lag behind the input signal, the filter is specified by its *one-radian lag frequency*.

10.2.5 Transient Response

Some applications require that a signal with sharp rising and falling edges (such as a digital data waveform) applied to the input of a low-pass design have a minimum *overshoot* or *ringing* as seen at the filter's output. Overshoot (and undershoot) occurs when a signal exceeds (falls below) the final amplitude temporarily before settling at its final value. Ringing is a repeated sequence of overshoot and undershoot.

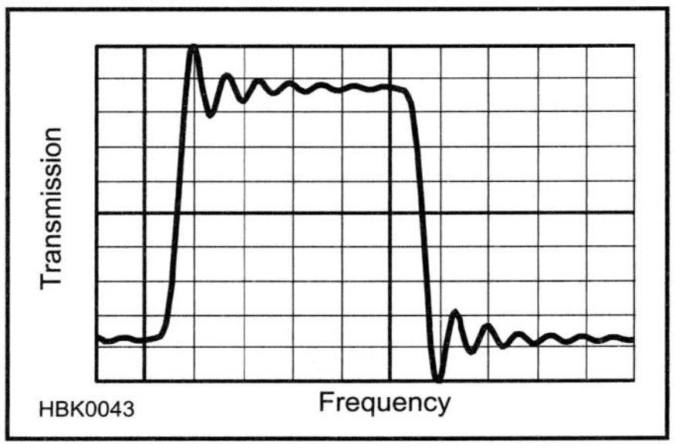

Figure 10.15 — Transient response of a Chebyshev low-pass filter.

If a sharp-cutoff analog filter is used in such an application, overshoot or ringing will occur. The appearance of a signal such as a square wave with sharp rising and falling edges as it exits from the filter may be as shown in **Figure 10.15**. The scales for both the X- and the Y-axes would depend on the frequency and magnitude of the waveform. (As described later in this chapter, digital filters can be designed not to exhibit this ringing, although they are much more complex than simple analog filters and have other tradeoffs.)

The sharp edge of the square wave is a type of *transient*, an abrupt change in a signal. When a signal is at a constant level and changes to another level very rapidly where it remains constant, that type of transient is a *step*. If the signal abruptly changes levels and immediately changes back again, that type of transient is a *pulse*. If the pulse is infinitely narrow, it is called an *impulse* and has interesting mathematical characteristics as discussed below. The output of a circuit when a transient occurs at its input is the circuit's *transient response*.

The square wave used in this discussion as a test waveform is composed of a fundamental and an endless series of odd harmonics. If harmonics only up to a certain order are used to create the square wave — that is, if the square wave is passed through a sharp-cutoff low-pass filter that attenuates higher frequencies — then that waveform will have the overshoot or ringing as shown as it exits from the filter. This is the filter's *step response*.

Transients can be repetitive, such as the edges of the square wave, but even a non-periodic waveform can be decomposed into sine waves, although in this case they are not harmonically related. For example a single pulse of width τ seconds has a frequency spectrum proportional to $\text{sinc}(f\tau) = \sin(\pi f\tau) / (\pi f\tau)$. You can think of this as an infinite number of sine waves spaced infinitely closely together with amplitudes that trace out that spectral shape. It is interesting to note that if τ is decreased, the value of f must increase by the same factor for any given value of $\sin(\pi f\tau) / (\pi f\tau)$. In other words, the narrower the pulse the wider the spectrum. Of course that applies to sine waves and other periodic waveforms as well — the smaller the wavelength the higher the frequency. In general, anything that makes the signal "skinnier" in the time domain makes it "fatter" in the frequency domain and vice versa.

As the pulse becomes narrower and narrower, the frequency spectrum spreads out more and more. In the limit, if the pulse is made infinitely narrow (an impulse), the spectrum becomes flat from zero hertz to infinity. The impulse is a very useful concept because of its flat frequency spectrum. The filter's *impulse response* has a frequency spectrum equal to the frequency response of the filter.

10.2.6 Filter Family Selection

Selecting a filter family is one of the first steps in filter design. To make that choice easier, the following list of filter family attributes is provided:

• *Butterworth* — No ripple in passband, smooth transition region, shallow rolloff for a given filter order, high ultimate attenuation, smooth group delay change across transition region. The smoothness of the response is particularly apparent near dc for the low-pass response, at the center frequency for a band-pass response, and at infinity for the high-pass response. The resulting magnitude response will also have a relatively gentle transition from the passband into the stop band.

• *Chebyshev* — Some passband ripple, abrupt transition region, steep rolloff for a given filter order, peak in group delay near cutoff frequency, high ultimate attenuation. The Chebyshev family is used when a sharper cutoff is desired for a given number of components and where at least a small amount of ripple is allowable in the passband.

• *Cauer* (or *Elliptical-function*) — Some passband ripple, abrupt transition region, steepest roll off, ripple in stopband due to traps, ultimate attenuation smaller than Butterworth and Chebyshev, group delay peaks near cutoff frequency and in stopband. When steepness of rolloff from passband into stop band is the item of greatest importance, then the Cauer filter family is used. Cauer filters involve a more complicated set of choices. In addition to selecting a passband ripple, the designer must also assign a stop band depth (or stop band frequency). Some of the items interact; they can't all be selected arbitrarily.

• *Bessel* — No ripple in passband, smooth transition region, constant group delay in passband, shallow rolloff for a given filter order, smooth changes in group delay, high ultimate attenuation

All characterizations such as "steep" and "abrupt" are relative with respect to filter designs from other families with similar orders. Other factors, such as number of components, sensitivity to component value and so on may need to be considered when selecting a filter family for a specific application.

All of the traits mentioned so far in this chapter apply to a filter regardless of how it is implemented, whether it is fabricated using *passive* lumped-element inductors and capacitors or using op amps with resistors and capacitors (an *active* filter) or in software as a digital filter. The traits are general descriptions of filter behavior and can be applied to any type of filter technology. Each type of technology (passive, active, digital) has strong points and tradeoffs.

These four families are the most common, but other families are used as well: for example, Gaussian, Constant-k, and M-derived are all supported by the *ELSIE* filter design package provided with this book's online information

Digital filters also implement these filter families and others, as well. They also have other classes, such as FIR and IIR, as described in this chapter's section on digital filters. For more information about analog and digital filter design, a list of references and articles is provided at the end of this chapter.

10.3 Passive LC Filters

This part of the chapter deals with passive LC filters fabricated using discrete inductors and capacitors (which gives rise to their name, *lumped-element*). We will begin with a discussion of basic low-pass filters and then generalize to other types of filters.

10.3.1 Low-Pass Filters

A very basic LC filter built using inductors and capacitors is shown in **Figure 10.16**. In the first case (Figure 10.16A), less power is delivered to the load at higher frequencies because the reactance of the inductor in series with the load increases as the test frequency increases. The voltage appearing at the load goes down as the frequency increases. This configuration would pass direct current (dc) and reject higher frequencies, and so it would be a low-pass filter.

With the second case (Figure 10.16B), less power is delivered to the load at higher frequencies because the reactance of the capacitor in parallel with the load decreases as the test frequency increases. Again, the voltage appearing at the load goes down as the frequency increases and so this, too, would be called a low-pass filter. In the real world, combinations of both series and parallel components are used to form a low-pass filter.

A high-pass filter can be made using the opposite configuration — series capacitors and shunt inductors. And a band-pass (or band-stop) filter can be made using pairs of series and parallel tuned circuits. These filters, made from alternating LC elements or LC tuned circuits, are called *ladder filters*.

We spoke of filter order, or complexity, earlier in this chapter. **Figure 10.17** illustrates *capacitor-input* low-pass filters with orders of 3, 4, and 5. Remember that the order corresponds to the number of energy-storing elements. For example, the third-order filter in Figure 10.17A has three energy-storing elements (two capacitors and one inductor), while the fifth-order design in Figure 10.17C has five elements total (three capacitors, two inductors). For comparison, a third-order filter is illustrated in **Figure 10.18**. The filters in Figure 10.17 are *capacitor-input* filters because a capacitor is connected directly across the input source. The filter in Figure 10.18 is an *inductor-input* filter.

As mentioned previously, the Cauer family has traps (series or parallel tuned circuits)

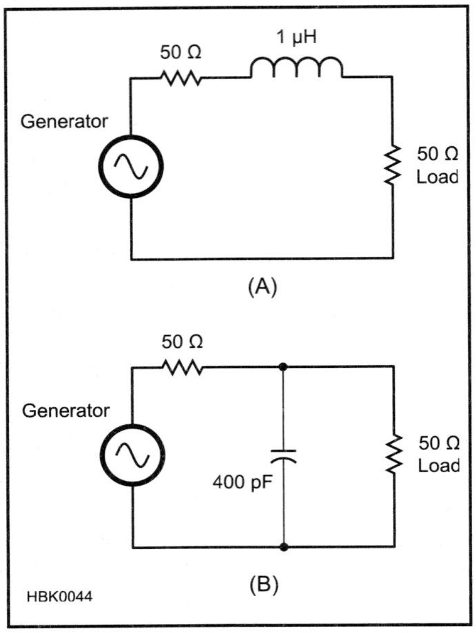

Figure 10.16 — A basic low-pass filter can be formed using a series inductor (A) or a shunt capacitor (B).

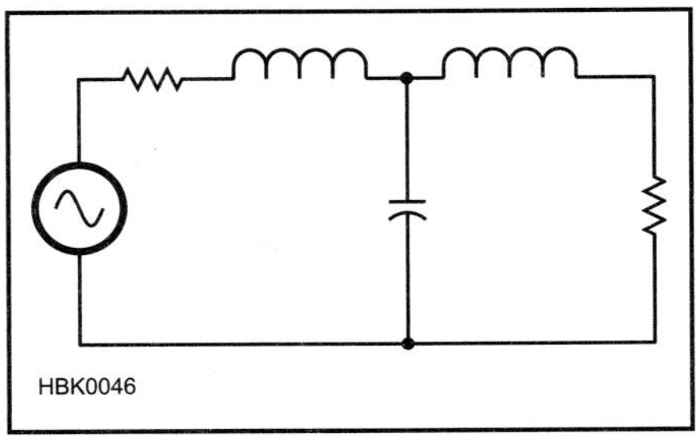

Figure 10.18 — A third-order inductor-input low-pass filter.

carefully added to produce dips or notches (properly called *zeros*) in the stop band. Schematics for the Cauer versions of capacitor-input low-pass filters with orders of 3, 4, and 5 are shown in **Figure 10.19**. The capacitors in parallel with the series inductors create the notches at calculated frequencies to allow the Cauer filter to be implemented.

The capacitor-input and the inductor-input versions of a given low-pass design have identical characteristics for their magnitude, phase and time responses, but they differ in the impedance seen looking into the filter. The capacitor-input filter has low impedance in the stop band while the inductor-input filter has high impedance in the stop band.

10.3.2 Low-Pass to Band-Pass Transformation

A band-pass filter is defined in part by a *bandwidth* and a *center frequency*. (An alternative method is to specify a lower and an upper cutoff frequency.) A low-pass design such as the one shown in Figure 10.17A can be converted to a band-pass filter by resonating each of the elements at the center frequency. **Figure 10.20** shows a third-order low-pass filter with a design bandwidth of 2 MHz for use in a 50 Ω system. If the shunt elements are now resonated with a parallel component, and if the series elements are resonated with a series component, the result is a band-pass filter as shown in **Figure 10.21**.

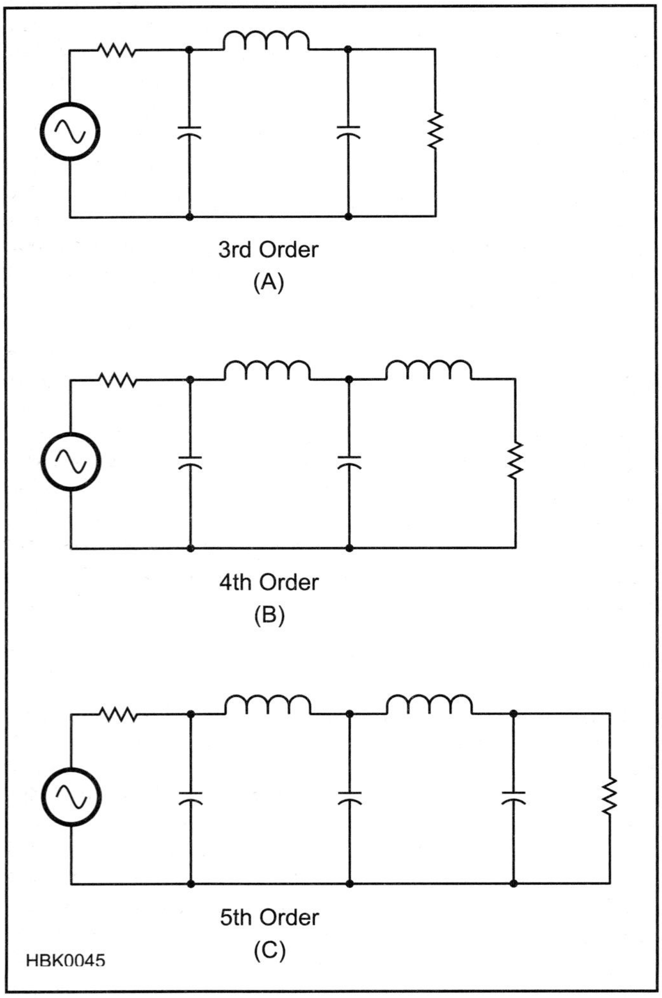

Figure 10.17 — Low-pass, capacitor-input filters for the Butterworth and Chebyshev families with orders 3, 4, and 5.

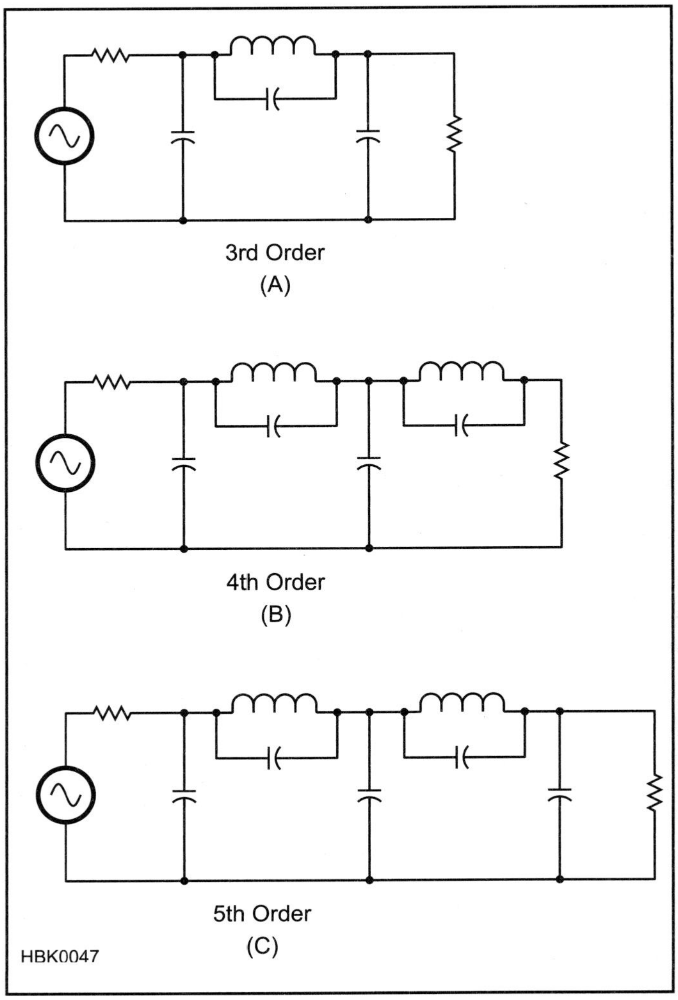

Figure 10.19 — Low-pass topologies for the Cauer family with orders 3, 4, and 5.

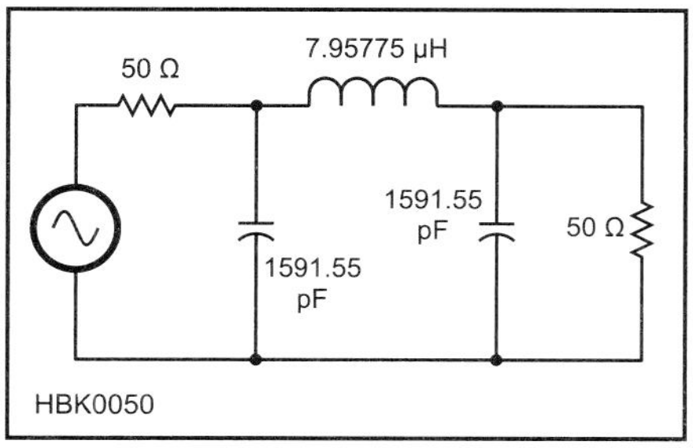

Figure 10.20 — Third-order low-pass filter with a bandwidth of 2 MHz in a 50 Ω system.

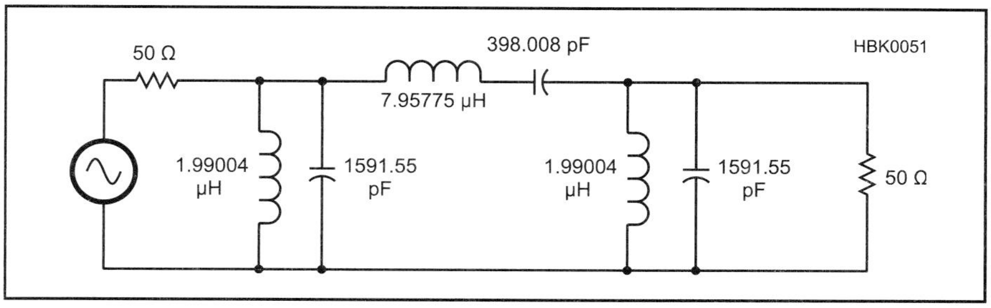

Figure 10.21 — The low-pass filter of Figure 10.20 can be transformed to a band-pass filter by resonating the shunt capacitors with a parallel inductor and resonating the series inductor with a series capacitor. Bandwidth is 2 MHz and the center frequency is 2.828 MHz.

The series inductor value and the shunt capacitor values are the same as those for the original low-pass design.

10.3.3 High-Pass Filters

A high-pass filter passes signals above its cutoff frequency and attenuates those below. Simple high-pass equivalents of the filters in Figure 10.16 are shown in **Figure 10.22.**

The reactance of the series capacitor in Figure 10.22A increases as the test frequency is lowered and so at lower frequencies there will be less power delivered to the load. Similarly, the reactance of the shunt inductor in Figure 10.22B decreases at lower frequency, with the same effect. Similarly to the low-pass filter designs presented in Figure 10.17, a high-pass filter in a real world design would typically use both series and shunt components but with the positions of inductors and capacitors exchanged.

An example of a high-pass filter application would be a broadcast-reject filter designed to pass amateur-band signals in the range of 3.5 MHz and above while rejecting broadcast signals at 1.7 MHz and below. A high-pass filter with a design cutoff of 2 MHz is illustrated in **Figure 10.23**. It can be implemented as a capacitor-input (Figure 10.23A) or inductor-input (Figure 10.23B) design. In each case, signals above the cutoff are passed with minimum attenuation while signals below the cutoff are attenuated, in a manner similar to the action of a low-pass filter.

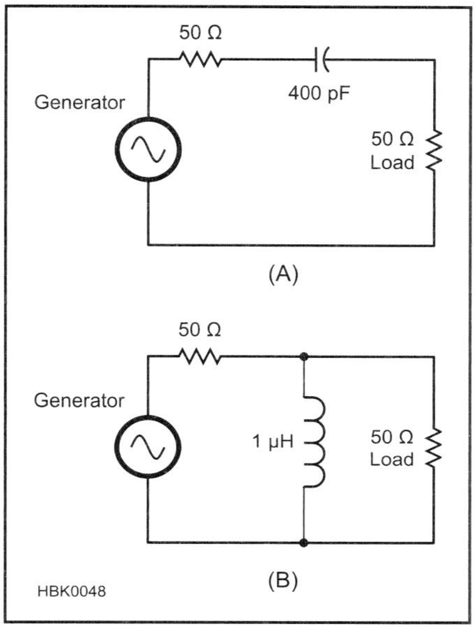

Figure 10.22 — A basic high-pass filter can be formed using a series capacitor (A) or a shunt inductor (B).

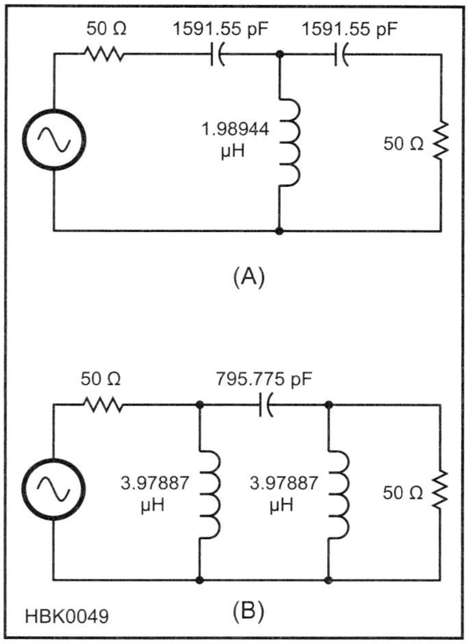

Figure 10.23 — Capacitor-input (A) and inductor-input (B) high-pass filters. Both designs have a 2 MHz cutoff.

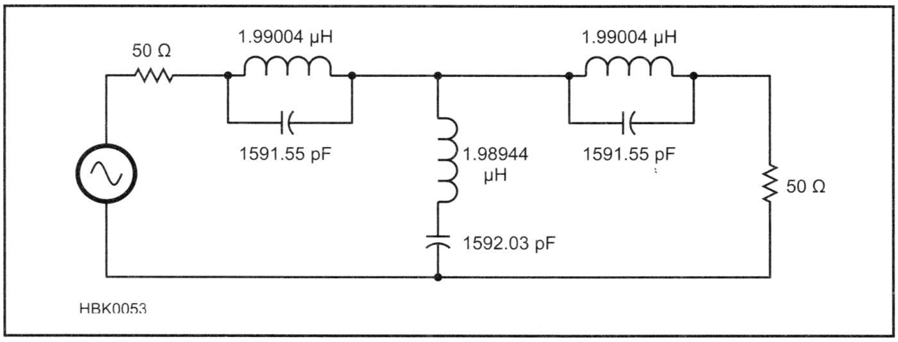

Figure 10.24 — The high-pass filter of Figure 10.23A can be converted to a band-stop filter with a bandwidth of 2 MHz, centered at 2.828 MHz.

10.3.4 High-Pass to Band-Stop Transformation

Just as a low-pass filter can be transformed to a band-pass type, a high-pass filter can be transformed into a band-stop (also called a *band-reject*) filter. The procedures for doing this are similar in nature to those of the transformation from low-pass to band-pass. As with the band-pass filter example, to transform a high-pass to a band-stop we need to specify a center frequency. The bandwidth of a band-stop filter is measured between the frequencies at which the magnitude response drops 3 dB in the transition region into the stop band.

Figure 10.24 shows how to convert the 2 MHz capacitor-input high-pass filter of Figure 10.20A to a band-stop filter centered at 2.828 MHz. The original high-pass components are resonated at the chosen center frequency to form a band-stop filter. Either a capacitor-input high-pass or an inductor-input high-pass may be transformed in this manner. In this case the series capacitor values and the shunt inductor values for the band-stop are the same as those for the high-pass. The series elements are resonated with an element in parallel with them. Similarly, the shunt elements are also resonated with an element in series with them. In each case the pair resonate at the center frequency of the band-stop.

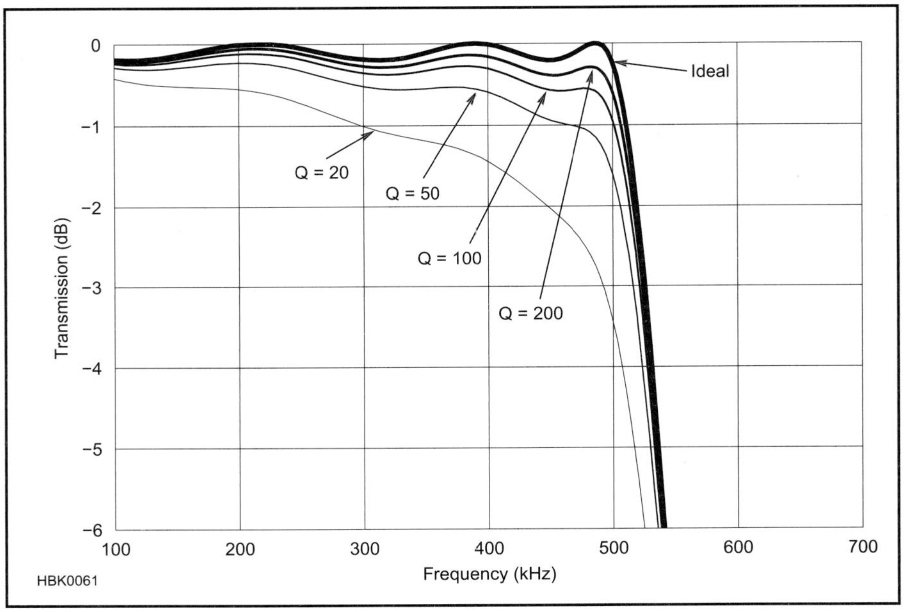

Figure 10.25 — Effect of inductor Q values on a low-pass filter.

10.3.5 Effect of Component Q

When components with less-than-ideal characteristics are used to fabricate a filter, the performance will also be less than ideal. One such item to be concerned about is component "Q." As described in the **Radio Fundamentals** chapter, Q is a measure of the loss in an inductor or capacitor, as determined by its resistive component. Q is the ratio of component's reactance to the loss resistance and is specified at a given test frequency. The loss resistance referred to here includes not only the value as measured by an ohmmeter but includes all sources of loss, such as skin effect, dielectric heating and so on.

The Q values for capacitors are usually greater than 500 and may reach a few thousand. Q values for inductors seldom reach 500 and may be as low as 20 or even worse for miniaturized parts. A good toroidal inductor can have a Q value in the vicinity of 250 to 400.

Q values can affect both the *insertion loss* of signals passing through the filter and the steepness of the filter's rolloff. Band-pass filters (and especially narrowband band-pass filters) are more vulnerable to this problem than low-pass and high-pass filters. **Figure 10.25** shows the effect of finite values of inductor Q values on the response of a low-pass filter. The Q values for each plot are as shown.

Inadequate component Q values introduce loss and more importantly they compromise the filter's response at cutoff, especially problematic in the case of narrowband band-pass filters. **Figure 10.26** illustrates the effect of finite inductor Q values on a narrowband band-pass filter. In the case of a band-pass filter, the Q values required to support a given response shape are much higher than those required for the low-pass or high-pass filter (by the ratio of center to width). Capacitor Q values are generally much high than inductor Q values and so contribute far less to this effect.

In general, component-value adjustment will not be able to fully compensate for inadequate component Q values. However, if the filter is deliberately mismatched (by changing the input and/or output terminations) then a limited amount of *response-shape* correction can sometimes be achieved by network component value optimization ("tweaking"). The loss caused by Q problems (at dc in the case of a low-pass or at the center frequency in the case of a band-pass) may increase if such correction is attempted.

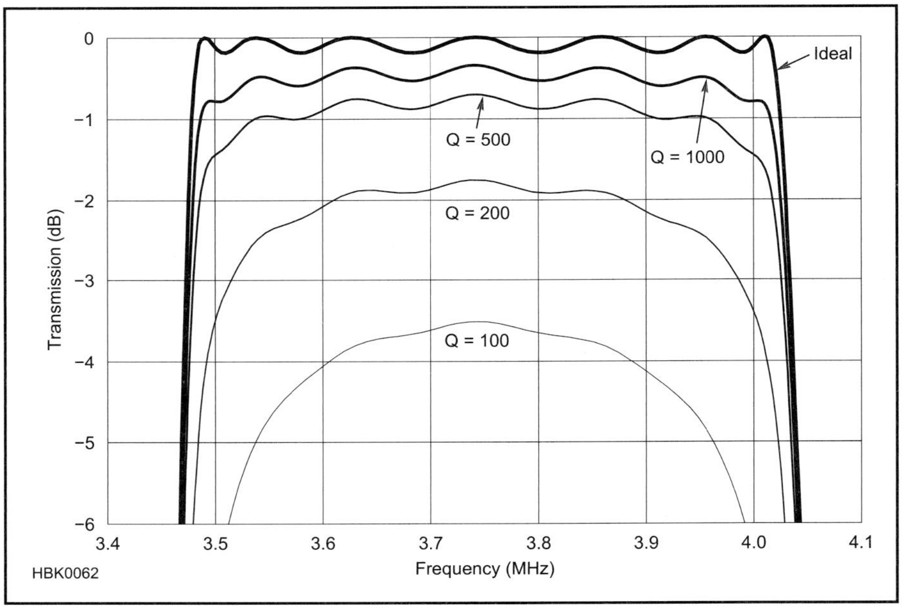

Figure 10.26 — Effect of inductor Q values on a narrowband band-pass filter.

10.3.6 Side Effects of Passband Ripple

Especially in RF applications it is desirable to design a filter such that the impedance seen looking into the input side remains fairly constant over the passband. Increasing the value of passband ripple increases the rate of descent from the passband into the stop band, giving a sharper cutoff. But it also degrades the uniformity of the impedance across the passband as seen looking into the input of the filter. This may be shown in terms of VSWR or return loss; those are simply different ways of stating the same effect. (*Return loss* is explained in the **RF Techniques** chapter.)

Designs using a low value of passband ripple are preferred for RF work. Audio-frequency applications are generally not as critical, and so higher ripple values (up to about 0.2 dB) may be used in audio work.

Table 10.2 shows the maximum value of VSWR and minimum value of return loss for various values of passband ripple. Note that

Table 10.2
Passband Ripple, VSWR and Return Loss

Passband Ripple (dB)	*VSWR*	*Return Loss (dB)*
0.0005	1.022	39.38
0.001	1.031	36.37
0.002	1.044	33.36
0.005	1.07	29.39
0.01	1.101	26.38
0.02	1.145	23.37
0.05	1.24	19.41
0.1	1.355	16.42
0.2	1.539	13.46
0.5	1.984	9.636
1	2.66	6.868

Note: As the passband ripple specification is changed so do the other items. Conversely, to limit VSWR to a maximum value or return loss to at least some minimum value, use this table to find the passband ripple that should be used to design the filter.

lower values for passband ripple yield better values for VSWR and return loss. Specifying a filter to have a passband ripple value of 0.01 dB will result in that filter's VSWR figure to be about 1.1:1. Or restated, the return loss will be about 26 dB. These values will be a function of frequency and at some test frequencies may be much better.

10.3.7 Use of Filters at VHF and UHF

Even when filters are designed and built properly, they may be rendered totally ineffective if not installed properly. Leakage around a filter can be quite high at VHF and UHF, where wavelengths are short. Proper attention to shielding and good grounding is mandatory for minimum leakage. Poor coaxial cable shield connection into and out of the filter is one of the greatest offenders with regard to filter leakage. Proper dc-lead bypassing throughout the receiving system is good practice, especially at VHF and above. Ferrite beads placed over the dc leads may help to reduce leakage. Proper filter termination is required to minimize loss.

Most VHF RF amplifiers optimized for noise figure do not have a 50 Ω input impedance. As a result, any filter attached to the input of an RF amplifier optimized for noise figure will not be properly terminated and filter loss may rise substantially. As this loss is directly added to the RF amplifier noise figure, carefully choose and place filters in the receiver.

10.3.8 Design Software for LC Filters

Previous editions of this book included an extensive set of design tables and formulas for manual calculation of filter component values. While this method was certainly instructive, it was tedious and error-prone, particularly for higher-order designs. The sections that described this method in previous editions have been extracted as a PDF document that is provided with the online information for this book.

Filter design is almost universally performed with software today and the *ARRL Handbook* is fortunate to include such a package. Jim Tonne, W4ENE, has made available a version of *ELSIE*, a filter design program, to amateurs at no charge. The latest version of *ELSIE* and related programs are available for downloading with the online content. The list of software is presented at the beginning of this chapter and the program capabilities are listed below.

Users unfamiliar with *ELSIE* will benefit from following the "walkthrough" accessible from the program's ABOUT menu tab. Numerous design examples and tutorials are available as online videos hosted on **YouTube.com** — search for "elsie filter design tutorial" to locate a good starting set. A two-part step-by-step tutorial for using *ELSIE* from the Hands-On Radio series of columns is available in the online material for this book as well.

• *ELSIE* — design and analysis of lumped-element LC filters. In addition to providing parts values for filters with various topologies from various families, tools are included to assist with practical construction.

• *SVC Filter Designer* — design of lumped-element high-pass and low-pass filters. The software shows ideal values and also the nearest 5% values for capacitors and inductors. It also analyzes those filters and shows the deviation of key responses from ideal when those 5% values are used.

• *QuadNet* — design and analysis of active quadrature ("90-degree") networks for use in SSB transmitters and receivers. It handles networks with orders from 2 to 10, odd and even, with tuning modes and analysis.

• *Helical* — design and analysis of helical-resonator band-pass filters usually used in the VHF and UHF frequency ranges.

• *Diplexer* — design and analysis of diplexer filters.

10.4 Active Audio Filters

Below RF, in what is broadly referred to as the "audio" range between a few Hz and a few hundred kHz, designers have several choices of filter technology.

• Passive LC
• DSP Digital Filters
• Switched-Capacitor Audio Filter (SCAF)
• Active RC

LC audio filters are not used much in current designs except in high-power audio applications, such as speaker crossover networks, and are not covered here. LC filters were once popular as external audio filters for CW reception. These designs tended to be large and bulky, often using large surplus 44 or 88 mH core inductors. LC filters used in very low-level receiver applications can be very compact, but these tend to suffer from relatively high insertion loss, which reduces the receiver sensitivity (if no preamp is used ahead of the filter) or its large-signal dynamic range (if a loss-compensating preamp is used ahead of the filter). In addition, LC filters used in low-level receiver applications tend to pick up 60/120 Hz hum from stray magnetic fields from ac power transformers. Even "self shielded" inductors can produce noticeable ac hum pick up when followed by 90 to 120 dB of audio gain!

Digital filtering (see the following section of this chapter) can yield filters that are superior to analog filters. High signal-level DSP-based external audio filters such as the popular MFJ Enterprises MFJ-784B (**www.mfjenterprises.com**) have been available for some time. In addition, if the filter is coupled to a high-performance audio A/D converter, digital filtering can provide excellent filtering for even very low-level audio signals.

The main drawback to standalone digital filters is that of expense. A low-end DSP external audio filter costs at least $100, while a DSP with a high-end A/D converter (such as an audio sound card front-end to a PC) can run in the $150 to $400+ price range, which does not include the cost of the host PC.

10.4.1 SCAF Filters

Simple SCAF filter designs can produce extremely effective low-power audio filters. The implementation of these filters in practical IC form usually involves small-value capacitors and high-value resistors. The use of high-value resistors tends to generate enough noise to make these filters unsuitable for very low-level audio processing such as the front end of a receiver audio chain. Thus, SCAF filters tend to be limited to filtering at the output end of the audio chain — headphone or speaker level audio applications, an area in which they very much excel. The cutoff frequency of a SCAF filter is set by an external clock signal. Thus, this class of filter naturally lends itself to use in variable frequency filters. For simplicity, excellent frequency selectivity and relatively low cost, SCAF filters are highly recommended when additional audio filtering is desired on the output of an existing receiver.

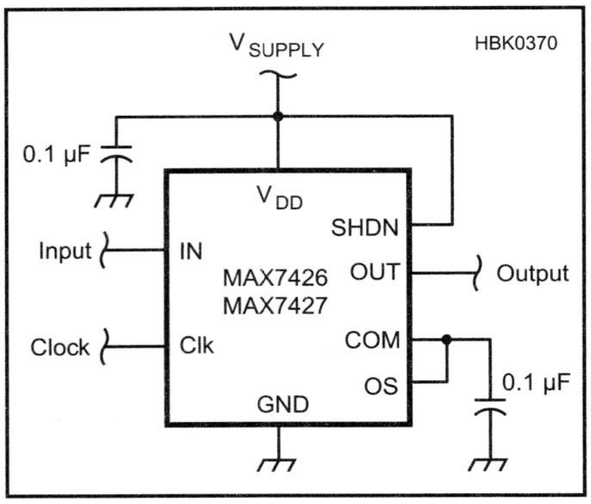

Figure 10.27 — Simple SCAF low pass filter, taken from the MAX7426 data sheet.

An example of a very simple, but highly effective SCAF low-pass filter IC is the Maxim MAX7426 (5 V supply) or MAX7427 (3 V supply) in **Figure 10.27**. The filter's cutoff frequency can be set by placing an appropriately sized capacitor across the clock oscillator inputs because the part can generate its own internal clock signal. For example, connecting a 180 pF capacitor across the MAX7426's clock inputs will produce a 1 kHz low-pass filter. The MAX7427 (3 V version) was used in the NC2030 QRP transceiver along with a MVAM108 varactor diode to create a low-pass filter that was tunable from 300 Hz to 1 kHz.

A portion of that schematic is shown in **Figure 10.28**. The SCAF low-pass filter is followed by a 3 V unity-gain headphone amplifier (U14C and U14D) as the filter IC itself cannot directly drive headphones. The low-pass cutoff frequency is tuned by using R83 to vary the voltage applied across the MVAM108 varactor diode D10, thus changing the capacitance across the clock input (pin 8) of the chip. C116 was used to isolate the dc voltage across the varactor diode from the bias voltage on the clock input line. The frequency response of such a filter is shown in **Figure 10.29**.

As can be seen, this is an extremely sharp filter, producing almost 40 dB of attenuation very close to the cut off frequency — not too bad for an inexpensive part. The slightly more expensive MAX7403 gives an even steeper 80 dB cutoff. Both parts are specified for an 80 dB signal-to-noise ratio based on an assumed 4 V_{P-P} signal. Sensitive modern in-ear type headphones require only 20 mV_{P-P} to produce a fairly loud signal. A 20 mV signal is 46 dB below 4 V_{P-P}, so at such headphone levels, the noise floor is actually only 34 dB below the 4 V_{P-P} signal — fairly quiet,

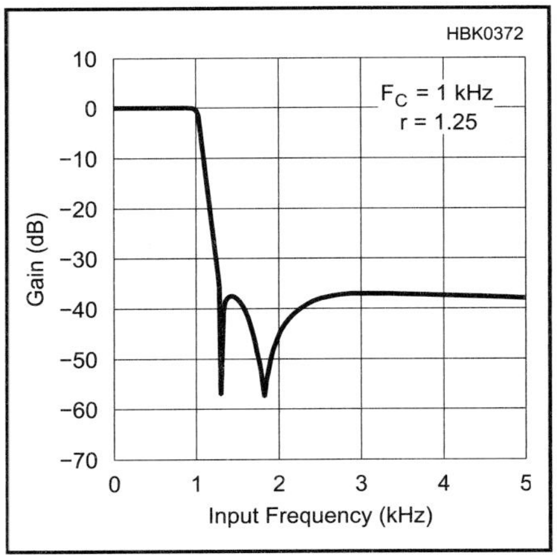

Figure 10.29 — Frequency response of a low-pass filter using the MAX7426 set to a 1 kHz cutoff frequency.

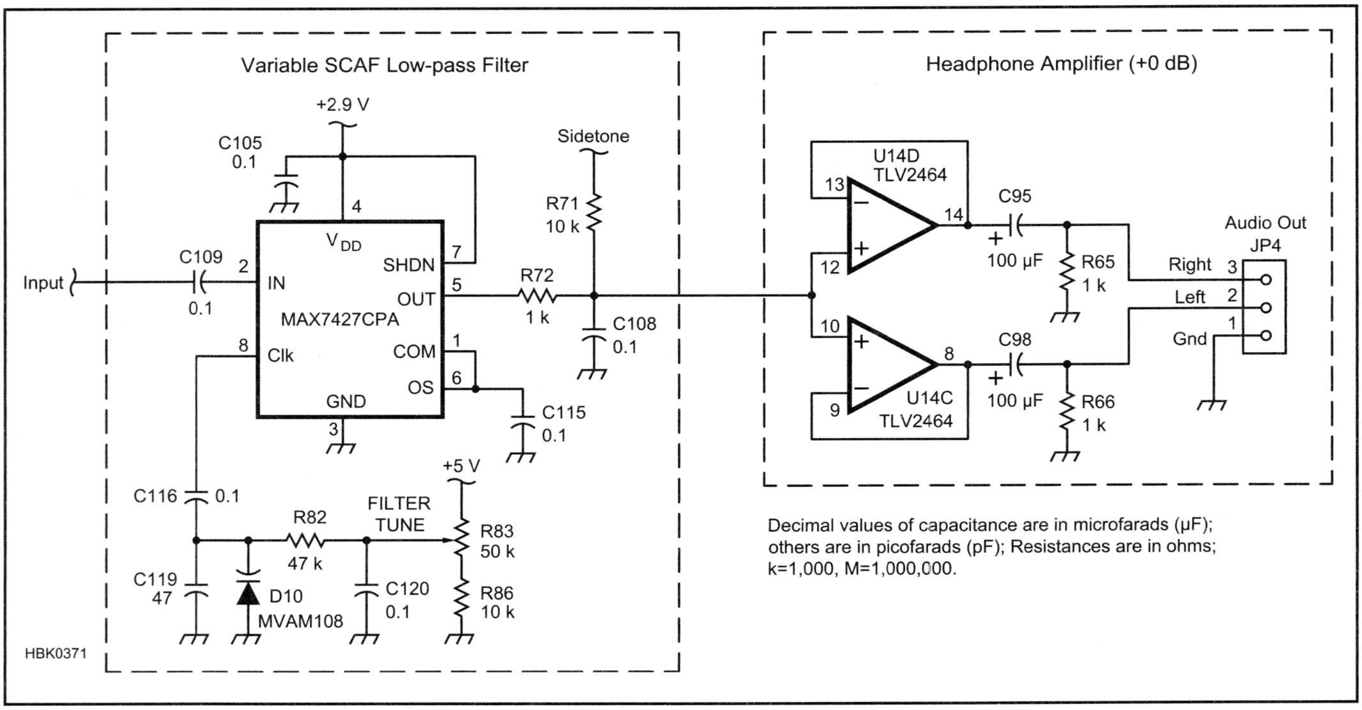

Figure 10.28 — SCAF with variable 300 – 1000 Hz low-pass cutoff and unity-gain headphone amplifier.

but is a lot less noise margin than one would tend to think given the 80 dB specification.

Again, with a noise floor only 34 dB below typical headphone levels, SCAF filters such as these are fine at the end of the receiver chain, but are not useful as an audio filter at very low signal levels early on in a receiver.

Figure 10.30 is an example of a slightly more complex SCAF band-pass filter. This filter features both a variable center frequency (450 Hz to 1000 Hz) and a variable bandwidth (90 Hz to 1500 Hz). This very popular filter, the NESCAF, was designed by the New England QRP Club (**www.newenglandqrp.org**) and is currently offered in kit form in batches. Contact the club if you wish to purchase the kit. The design is included here and is well within the range of a home-builder with intermediate construction skills. There are a number of online websites that discuss the filter and how to build it.

The NESCAF filter features both sections of the SCAF10 IC configured as identical band-pass filters. The bandwidth (Q) is adjustable via R7A and R7B, while the center frequency is adjustable by R10 and the trimmer R9 as these resistors set the frequency of the LM555 clock generator.

As in the previous SCAF filter example, this filter also includes an audio amplifier (LM386) for driving either an external speaker or headphones as none of the SCAF filter ICs are capable of driving headphones directly.

10.4.2 Active RC Filters

Active RC filters based on op amp circuits can be used in either high-level audio output or very low-level direct-conversion receiver front-end filtering applications and thus are extremely flexible. Unlike LC filters, active RC filters can provide both filtering and gain at the same time, eliminating the relatively high insertion loss of a physically small, sharp LC filter. In addition, an active RC filter is not susceptible to the same ac hum pickup in low signal level applications as LC filters.

Active filters can be designed for gain and they offer excellent stage-to-stage isolation. The circuits require only resistors and capacitors, avoiding the limitations associated with inductors. By using gain and feedback, filter Q is controllable to a degree unavailable to passive LC filters. Despite the advantages, there are also some limitations. They require power, and performance may be limited by the op amp's finite input and output levels, gain, and bandwidth. Active filters that drive speakers or other heavy loads usually employ an audio output amplifier circuit to boost the output power after a low-power filter stage.

A particular advantage of active RC filters for use in receivers is that they can be capable of extremely low noise operation, allowing the filtering of extremely small signals. At the same time they are capable of handling extremely large signals. Op amps are often capable of using ±18 V dual supply voltages or 36 V single supply voltages allowing the construction of a very high performance audio filter/amplifier chain that can handle signals up to 33 V_{P-P}. The ability to provide gain while also providing filtering provides a lot of flexibility in managing the sensitivity of a receiver audio chain.

The main disadvantage of active RC filters is their relatively high parts count compared to other filter types. In addition, active RC filters tend to be fixed-frequency designs unlike SCAF filters whose frequency can be moved simply by changing the clock frequency that drives the SCAF IC.

10.4.3 Active Filter Responses

Active filters can implement any of the

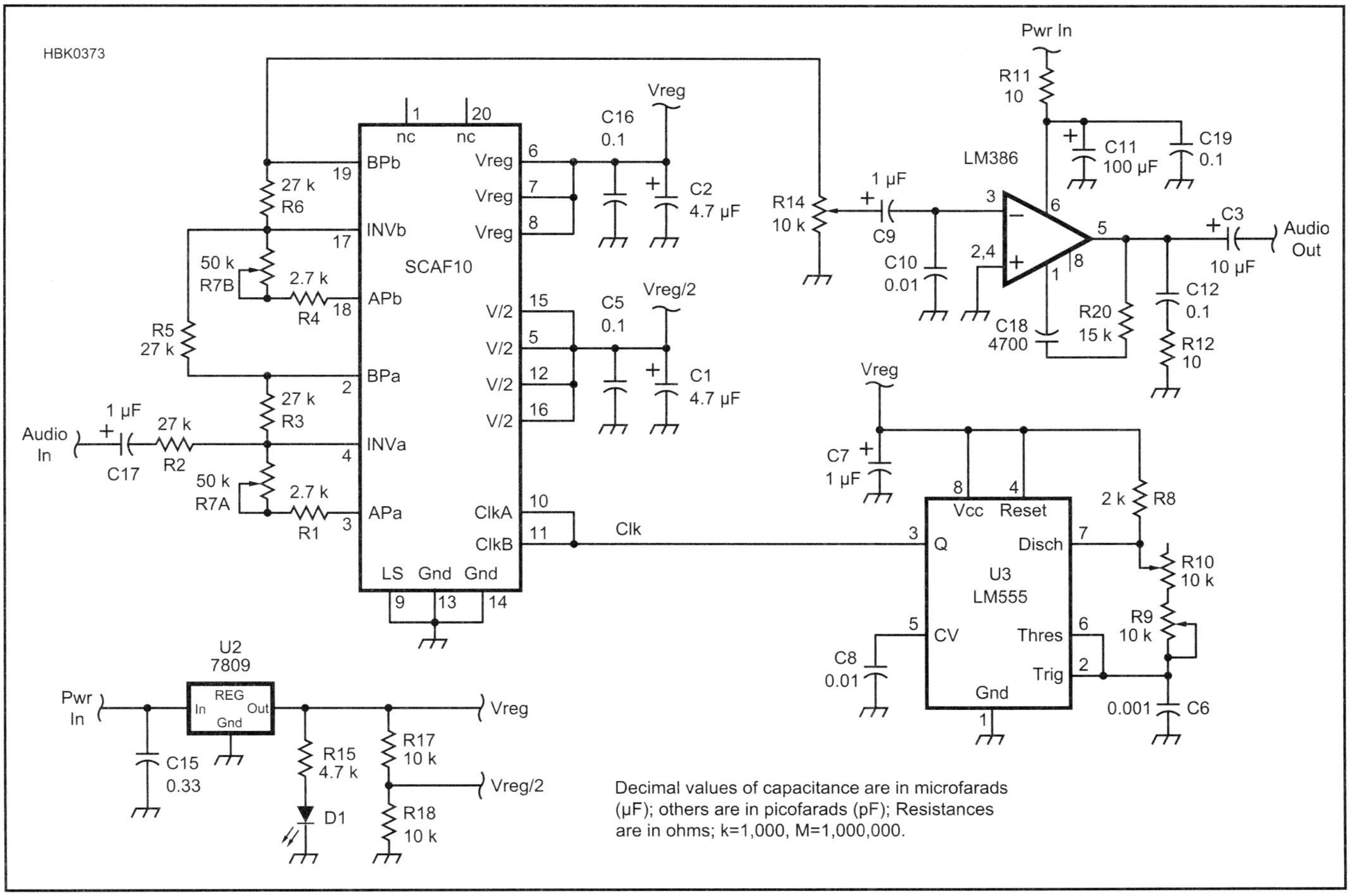

Figure 10.30 — Example of a simple SCAF band-pass filter from a New England QRP Club kit.

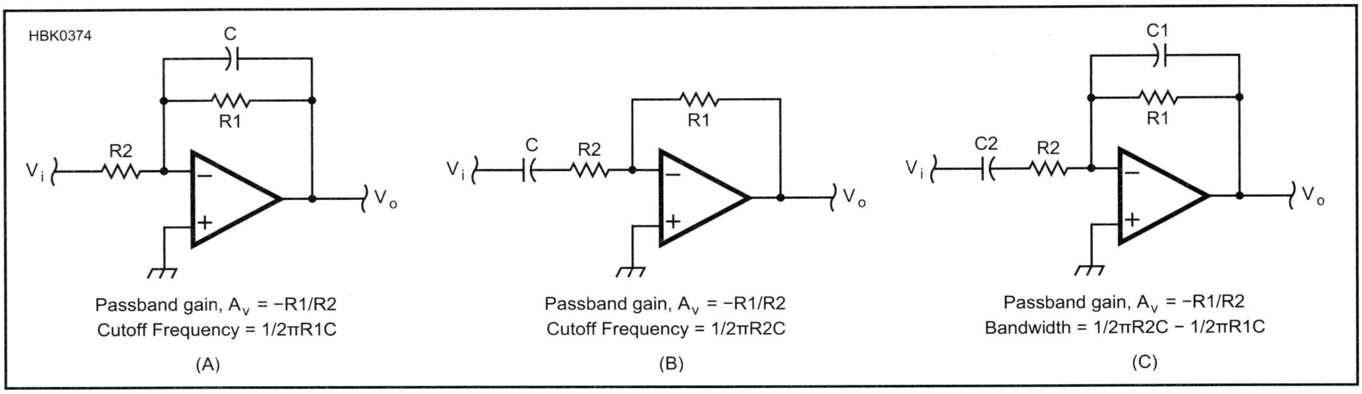

Figure 10.31 — Simple active filters. A low-pass filter is shown at A; B is a high-pass filter. The circuit at C combines the low- and high-pass filters into a wide band-pass filter.

Unless otherwise specified, values of R are in ohms, C is in farads, F in hertz, and ω in radians per second. Calculations shown here were performed on a scientific calculator.

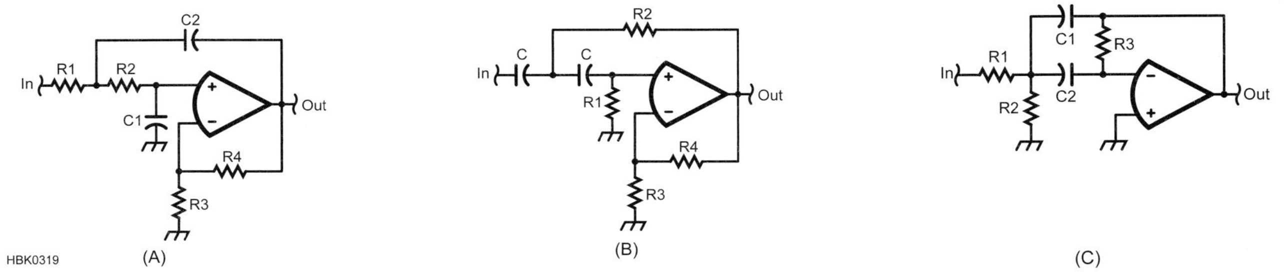

Low-Pass Filter

$$C_1 \le \frac{\left[a^2 + 4(K-1)\right]C_2}{4}$$

$$R_1 = \frac{2}{\left[aC_2 + \sqrt{\left[a^2 + 4(K-1)\right]C_2^{\,2} - 4C_1C_2}\right]\omega_C}$$

$$R_2 = \frac{1}{C_1C_2R_1\omega_C^{\,2}}$$

$$R_3 = \frac{K(R_1 + R_2)}{K-1} \quad (K > 1)$$

$$R_4 = K(R_1 + R_2)$$

where
- K = gain
- f_c = −3 dB cutoff frequency
- $\omega_c = 2\pi f_c$
- C_2 = a standard value near $10/f_c$ (in μF)

Note: For unity gain, short R4 and omit R3.

Example:
- a = 1.414 (see table, one stage)
- K = 2
- f = 2700 Hz
- ω_c = 16,964.6 rad/sec
- C_2 = 0.0033 μF
- C1 ≤ 0.00495 μF (use 0.0050 μF)
- R1 ≤ 25,265.2 Ω (use 24 kΩ)
- R2 = 8,420.1 Ω (use 8.2 kΩ)
- R3 = 67,370.6 Ω (use 68 kΩ)
- R4 = 67,370.6 Ω (use 68 kΩ)

High-Pass Filter

$$R_1 = \frac{4}{\left[a + \sqrt{a^2 + 8(K-1)}\right]\omega_C\, C}$$

$$R_2 = \frac{1}{\omega_C^{\,2}C^2R_1}$$

$$R_3 = \frac{KR_1}{K-1} \quad (K > 1)$$

$$R_4 = KR_1$$

where
- K = gain
- f_c = −3 dB cutoff frequency
- $\omega_c = 2\pi f_c$
- C = a standard value near $10/f_c$ (in μF)

Note: For unity gain, short R4 and omit R3.

Example:
- a = 0.765 (see table, first of two stages)
- K = 4
- f = 250 Hz
- ω_c = 1570.8 rad/sec
- C = 0.04 μF (use 0.039 μF)
- R1 = 11,123.2 Ω (use 11 kΩ)
- R2 = 22,722 Ω (use 22 kΩ)
- R3 = 14,830.9 Ω (use 15 kΩ)
- R4 = 44,492.8 Ω (use 47 kΩ)

Band-Pass Filter

Pick K, Q, $\omega_o = 2\pi f_c$ where f_c = center freq.
Choose C
Then

$$R1 = \frac{Q}{K_0\omega_0 C}$$

$$R2 = \frac{Q}{\left(2Q^2 - K_0\right)\omega_0 C}$$

$$R3 = \frac{2Q}{\omega_0 C}$$

Example:
- K = 2, f_o = 800 Hz, Q = 5 and C = 0.022 μF
- R1 = 22.6 kΩ (use 22 kΩ)
- R2 = 942 Ω (use 910 Ω)
- R3 = 90.4 kΩ (use 91 kΩ)

Figure 10.32 — Equations for designing a low-pass RC active audio filter are given at A. B, C, and D show design information for high-pass, band-pass and band-reject filters, respectively. All of these filters will exhibit a Butterworth response. Values of K and Q should be less than 10. See Table 10.3 for values of "a."

passive LC filter responses described in the preceding section: low-pass, high-pass, band-pass, band-stop and all-pass. Filter family responses such as Butterworth, Chebyshev, Bessel, and Cauer (elliptic) can be realized. All of the same family characteristics apply equally to passive and active filters and will not be repeated here. (Op amps are discussed in the **Circuits and Components** chapter.)

Figure 10.31 presents circuits for first-order low-pass (Figure 10.31A) and high-pass (Figure 10.31B) active filters. The frequency response of these two circuits is the same as a parallel and series RC circuit, respectively, except that these two circuits can have a pass-band gain greater than unity. Rolloff is shallow at 6 dB/octave. The two responses can be combined to form a simple band-pass filter (Figure 10.31C). This combination cannot produce sharp band-pass filters because of the shallow rolloff.

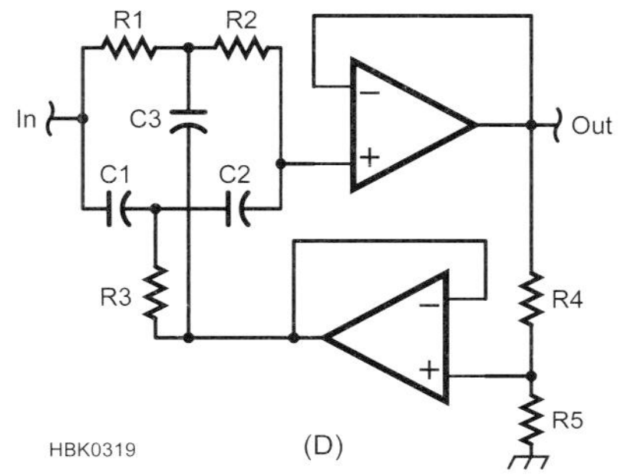

Band-Reject Filter

$$F_0 = \frac{1}{2\pi R1C1}$$

$$K = 1 - \frac{1}{4Q}$$

$$R >> (1-K)R1$$

where

$$C1 = C2 = \frac{C3}{2} = \frac{10\,\mu F}{f_0}$$

R1 = R2 = 2R3
R4 = (1 – K)R
R5 = K × R

Example:
f_0 = 500 Hz, Q=10
K = 0.975
C1 = C2 = 0.02 µF (or use 0.022 µF)
C3 = 0.04 µF (or use 0.044 µF)
R1 = R2 = 15.92 kΩ (use 15 kΩ)
R3 = 7.96 kΩ (use 8.2 kΩ)
R >> 1 kΩ
R4 = 25 Ω (use 24 Ω)
R5 = 975 Ω (use 910 Ω)

To achieve high-order responses with steeper rolloff and narrower bandwidths, more complex circuits are required in which combinations of capacitors and resistors create *poles* and *zeroes* in the frequency response. (Poles and zeros are described in the **Radio Fundamentals** chapter.) The various filter response families are created by different combinations of additional poles and zeroes. There are a variety of circuits that can be configured to implement the equations that describe the various families of filter responses. The most common circuits are *Sallen-Key* and *multiple-feedback*, but there are numerous other choices.

There are many types of active filters — this section presents some commonly used circuits as examples. A book on filter design (see the References section) will present more choices and how to develop designs based on the different circuit and filter family types. In addition, op amp manufacturer's publish numerous application notes and tutorials on active filter design. Several are listed in the References section of this chapter. The set of tutorials published by Analog Devices is particularly good.

SECOND-ORDER ACTIVE FILTERS

Figure 10.32 shows circuits for four second-order filters: low-pass (Figure 10.32A), high-pass (Figure 10.32B), band-pass (Figure 10.32C) and band-reject or notch (Figure 10.32D). Sequences of these filters are used to create higher-order circuits by connecting them in series. Two second-order filter stages create a fourth-order filter, and so forth.

The low-pass and high-pass filters use the Sallen-Key circuit. Note that the high-pass circuit is just the low-pass circuit with the positions of R1-R2 and C1-C2 exchanged. R3 and R4 are used to control gain in the low- and high-pass configurations. The band-pass filter is a multiple-feedback design. The notch filter is based on the twin-T circuit. All of the filter design equations and tables will result in a Butterworth family response. (For the Chebyshev and other filter responses, consult the references listed at the end of the chapter.)

The circuits in Figure 10.32 assume the op amp is operating from a balanced, bipolar power supply, such as ±12 V. If a single supply is used (such as +12 V and ground), the circuit must have a dc offset added and blocking capacitors between filter sections to prevent the dc offset from causing the op amp to saturate. The references listed at the end of the chapter provide more detailed information on single-supply circuit design.

Avoid electrolytic or tantalum capacitors as frequency-determining components in active filter design. These capacitors are best used for bypassing and power filtering as their tolerance is generally quite low, they have significant parasitic effects, and are usually polarized. Very small values of capacitance (less than 100 pF) can be affected by stray capacitance to other circuit components and wiring. High-order and high-Q filters require close attention to component tolerance and temperature coefficients, as well.

SECOND-ORDER ACTIVE FILTER DESIGN PROCEDURES

The following simple procedures are used to design filters based on the schematics in Figure 10.32. Equations and a design example are provided in the figure.

Low- and High-Pass Filter Design

To design a low- or high-pass filter using the circuits in Figure 10.32, start by determining your performance requirements for filter order (2, 4, 6, or 8), gain (K) and cutoff frequency (f_C). Calculate $\omega_C = 2\pi f_C$ and C_2 or C as required.

Table 10.3 provides design coefficients to create the Butterworth response from successive Sallen-Key low- and high-pass stages. A different coefficient is used for each stage. Obtain design coefficient "a" from Table 10.3. Calculate the remaining component values from the equations provided.

Band-Pass Filter Design

To design a band-pass filter as shown in Figure 10.32C, begin by determining the filter's required Q, gain, and center frequency, f_0. Choose a value for C and solve for the resistor values. Very high or very low values

Table 10.3
Factor "a" for Low- and High-Pass Filters in Figure 10.32

No. of Stages	*Stage 1*	*Stage 2*	*Stage 3*	*Stage 4*
1	1.414	–	–	–
2	0.765	1.848	–	–
3	0.518	1.414	1.932	–
4	0.390	1.111	1.663	1.962

These values are truncated from those of Appendix C of Williams, *Electronic Filter Design Handbook*, for even-order Butterworth filters of resistance (above 1 MΩ or lower than 10 Ω) should be avoided. Change the value of C until suitable values are obtained. High gain and Q may be difficult to obtain in the same stage with reasonable component values. Consider a separate stage for additional gain or to narrow the filter bandwidth.

Band-Reject (Notch) Filter Design

Band-reject filter design begins with the selection of center frequency, f_0, and Q. Calculate the value of K. Choose a value for C1 that is approximately 10 µF / f_0. This determines the values of C2 and C3 as shown. Calculate the value for R1 that results in the desired value for f_0. This determines the values of R2 and R3 as shown. Select a convenient value for R4 + R5 that does not load the output amplifier. Calculate R4 and R5 from the value of K.

The depth of the notch depends on how closely the values of the components match the design values. If possible, use 1% tolerance resistors and matched values of capacitance. If all identical components are used, two capacitors can be paralleled to create C3 and two resistors in parallel create R3. This helps to minimize thermal drift.

10.4.4 Active Filter Design Tools

While the simple active filter examples presented above can be designed manually, more sophisticated circuits are more easily designed using filter-design software. Follow the same general approach to determining the filter's performance requirements and then the filter family as was presented in the section on LC filters. You can then enter the values or make the necessary selections for the design software. Once a basic design has been calculated, you can then "tweak" the design performance, use standard value components and make other adjustments. The design example presented below shows how a real analog design is assembled by understanding the performance requirements and then using design software to experiment for a "best" configuration.

Op amp manufacturers such as Texas Instruments and National Semiconductor (originally separate companies but now merged) have made available sophisticated "freeware" filter design software. These packages are extremely useful in designing active RC filters. They begin by collecting specifications from the user and then creating a basic circuit. Once the initial circuit has been designed, the user can adjust specifications, component values, and op amp types until satisfied with the final design.

Free online filter design software is available from the following sources:

• Analog Devices — *Analog Filter Wizard 2.0* (**tools.analog.com/en/filterwizard/**)

• Texas Instruments — *FilterPro 3.1* (**www.ti.com**, search for "FilterPro")

Even if a manual design process is followed, using a software tool to double-check the results is a good way to verify the design before building the circuit. A circuit simulator such as *LTSpice* (see the **Computer Modeling and Simulation** chapter) can also verify a design before building the actual circuit.

An extended design example using the Texas Instruments *FilterPro* software is included in this book's online information. In the example, Dan Tayloe, N7VE explains the process of designing a high-performance 750 Hz low-pass filter, illustrating the power of using sophisticated interactive tools that enable design changes on-the-fly. The reader is encouraged to follow along and experiment with *FilterPro* as a means of becoming familiar with the software so that it can be used for other filter design tasks. Similar processes apply to other filter design software tools.

10.5 Digital Filters

Where an analog filter operates on a continuous signal in the time domain, digital filters operate on signals that have been converted to a digital stream of data. (See the discussion of digital signals in the **DSP and SDR Fundamentals** chapter.) The usual method of constructing a digital filters is as a series of registers with the stream of samples moving through them in a series of steps called *delays*. At each register (called a *tap*) the amplitude of the signal is modified (by an amount referred to as a *coefficient*) and the result added to the results from other taps. The signal stream is shifted to the next register and the process is repeated. The sum of the results from the taps constitutes the output of the filter.

The value of each coefficient and the configuration of how the filter's taps are added together determine the filter's response to the input signal. This section will discuss two basic filter types, the finite impulse response or FIR, the infinite impulse response or IIR.

The purpose of this section is not to give a tutorial in how each type of filter is designed but to explain the characteristics of each type of filter. As with analog filters, the design of digital filter is mostly done by software using validated algorithms that accept performance requirements as inputs and generate design information, in many cases as a data file that can be loaded directly into a DSP processor or subsystem. (The GNU Radio (**gnuradio.org**) toolkit, for example, treats digital filters as one of many functional blocks. The user configures the filter parameters and the software manages the organization of the actual computing elements in the DSP or SDR system.) For the reader interested in a deeper discussion of digital filter theory and design, several excellent references are listed at the end of the chapter.

10.5.1 FIR Filters

The earlier discussion of transient response introduced the impulse, an infinitely narrow pulse. The spectrum of an impulse is infinitely wide, containing all frequencies. If you feed an impulse into the input of a filter (analog or digital), the output of the filter is the filter's impulse response. Because the impulse is made up of all frequencies, it excites the filter at all frequencies. As a result, the impulse response has a frequency spectrum equal to the frequency response of the filter.

One way to design a filter is to determine the impulse response that corresponds to the desired frequency spectrum and then design the filter to have that impulse response. That method is the basis for designing *finite impulse response* (FIR) filters. (Finite refers to time, not amplitude.)

An FIR filter is a filter whose impulse response is finite, ending in some fixed time. Note that analog filters have an infinite impulse response — the output theoretically rings forever. Even a simple R-C low-pass filter's output decays exponentially toward zero but theoretically never quite reaches it. In contrast, an FIR filter's impulse response becomes exactly zero at some time after receiving the impulse and stays zero forever (or at least until another impulse comes along).

Given that you have somehow figured out the desired impulse response, how would you design a digital filter to have that response? The obvious method would be to pre-calculate a table of impulse response values, sampled at the sample rate. These are called the filter *coefficients*. When an impulse of a certain amplitude is received, you multiply that amplitude by the first entry in the coefficient table and send the result to the output. At the next sample time, multiply the impulse by the second entry, and so on until you have used up all the entries in the table.

A circuit to do that is shown in **Figure 10.33**. The input signal is stored in a shift register. Each block labeled "Delay" represents a delay of one sample time. At each sample time, the signal is shifted one register to the right. Each register feeds a multiplier and the other input to the multiplier comes from one of the coefficient table entries. All

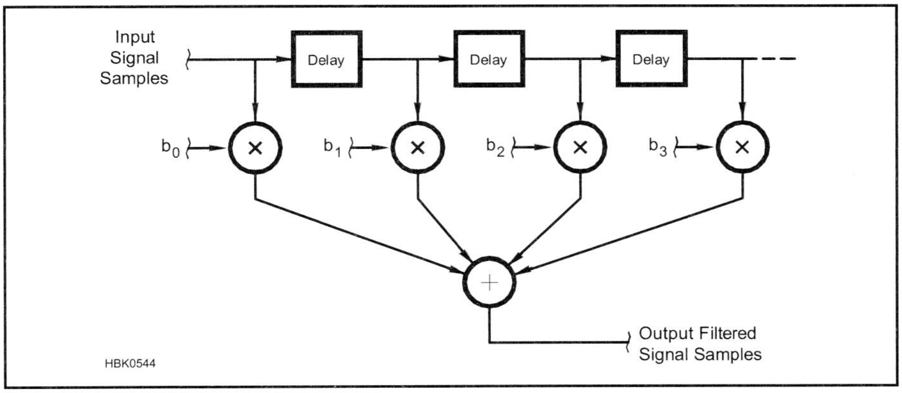

Figure 10.33 — A 4-tap FIR filter. The bn values are the filter coefficients.

the multiplier outputs are added together. Since the input is assumed to be a single impulse, at any given time all the shift registers contain zero except one, which is multiplied by the appropriate table entry and sent to the output.

We've just designed an FIR filter! By using a shift register with a separate multiplier for each tap, the filter works for continuous signals as well as for impulses. Since this is a linear system, the continuing signal is affected by the filter the same as an individual impulse.

It should be obvious from the diagram how to implement an FIR filter in software. You set up two buffers in memory, one for the filter coefficients and one for the data. The length of each buffer is the number of filter taps. (A *tap* is the combination of one filter coefficient, one shift register and one multiplier/accumulator.) Each time a new data value is received, it is stored in the next available position in the data buffer and the accumulator is set to zero. Next, a software loop is executed a number of times equal to the number of taps. During each loop, pointers to the two buffers are incremented, the next coefficient is multiplied by the next data value and the result is added to the current accumulator value. After the last loop, the accumulator contents are the output value. Normally the buffers are implemented as *circular buffers* — when the address pointer gets to the end it is reset back to the beginning.

Now you can see why a hardware multiplier-accumulator (MAC) is such an important feature of a DSP chip. Each tap of the FIR filter involves one multiplication and one addition. With a 1,000-tap FIR filter, 1,000 multiplications and 1,000 additions must be performed during each sample time. The ability to do each MAC operation in a single clock cycle saves a lot of processing time compared to a CPU type processor.

An FIR filter is a hardware or software implementation of the mathematical operation called *convolution*. We say that the filter *convolves* the input signal with the impulse response of the filter. It turns out that convolution in the time domain is mathematically equivalent to multiplication in the frequency domain. That means that the frequency spectrum of the output equals the frequency spectrum of the input times the frequency spectrum of the filter. Expressed in decibels, the output spectrum equals the input spectrum plus the filter frequency response, all in dB. If at some frequency the input signal is +3 dB and the filter is –10 dB compared to some reference, then the output signal will be 3 – 10 = –7 dB at that frequency.

An FIR filter whose bandwidth is very small compared to the sample rate requires a long impulse response with lot of taps. This is another consequence of the "skinny" versus "fat" relationship between the frequency and time domains. If the filter is narrow in the frequency domain, then its impulse response is wide. Actually, if you want the frequency response to go all the way to zero (minus infinity dB) throughout the stopband, then the impulse response theoretically becomes infinitely wide. Since we're designing a *finite* impulse response filter we have to truncate the impulse response at some point to get it to fit in the coefficient table. When you do that, however, you no longer have infinite attenuation in the stopband. The more heavily you truncate (the narrower the impulse response) the worse the stopband attenuation and the more ripple you get in the passband. Assuming optimum design techniques for selecting coefficients, you can estimate the minimum length L of the impulse response from the following equation:

$$L = 1 - \frac{10\log(\delta_1\delta_2) - 15}{14\left(\frac{f_T}{f_s}\right)} \quad \text{taps}$$

where

δ_1 and δ_2 = the passband and stopband ripple expressed as a fraction,

f_T = the transition bandwidth (frequency difference between passband and stopband edges), and

f_s = the sample rate.

For example, for a low-pass filter with a passband that extends up to 3 kHz, a stopband that starts at 4 kHz ($f_T = 4 - 3 = 1$ kHz), f_s = 10 kHz sample rate, ±0.1 dB passband ripple ($d_1 = 10^{0.1/20} - 1 = 0.0116$), and 60 dB stopband rejection ($d_2 = 10^{-60/20} = 0.001$), we get

$$L = 1 - \frac{10\log(0.0116 \times 0.001) - 15}{14\left(\frac{1}{10}\right)}$$

$$= 1 - \frac{-49.4 - 15}{1.4} = 47 \quad \text{taps}$$

Overflow is a potential problem when doing the calculations for an FIR filter. Multiplying two N-bit numbers results in a product with 2N bits, so space must be provided in the accumulator to accommodate that. Although the final result normally will be scaled and truncated back to N bits, it is best to carry through all the intermediate results with full resolution in order not to lose any dynamic range. In addition, the sum of all the taps can be a number with more than 2N bits. For example, if the filter width is 256 taps, then if all coefficients and data are at full scale, the final result could theoretically be 256 times larger, requiring an extra 8 bits in the accumulator. We say "theoretically" because normally most of the filter coefficients are much less than full scale and it is highly unlikely that all 256 data values would ever simultaneously be full-scale values of the correct polarity to cause overflow. The dsPIC processors use 16-bit multipliers with 32-bit results and a 40-bit accumulator, which should handle any reasonable circumstances.

After all taps have been calculated, the final result must be retrieved from the accumulator. Since the accumulator has much more resolution than the processor's data words, normally the result is truncated and scaled to fit. It is up to the circuit designer or programmer to scale by the correct value to avoid overflow. The worst case is when each data value in the shift register is full-scale — positive when it is multiplying a positive coefficient and negative for negative coefficients. That way, all taps add to the maximum value. To calculate the worst-case accumulator amplitude, simply add the absolute values of all the coefficients. However, that normally gives an unrealistically pessimistic value because statistically it is extremely unlikely that such a high peak will ever be reached. For a low-pass filter, a better estimate is to calculate the gain for a dc signal and add a few percent safety margin. The dc gain is just the sum of all the coefficients (not the absolute values). For a band-pass filter, add the sum of all the coefficients multiplied by a sine wave at the center frequency.

CALCULATING FIR FILTER COEFFICIENTS

So far we have ignored the question of how to determine the filter coefficients. For an ideal brick wall low-pass filter, the answer turns out to be pretty simple. A brick wall low-pass filter is one that has a constant response from zero hertz up to the cutoff frequency and zero response above. Its impulse response is proportional to the sinc function:

$$C(n) = C_0 \ \text{sinc}(2Bn) = \frac{\sin(2\pi Bn)}{2\pi Bn}$$

where C(n) are the filter coefficients, n is the sample number with n = 0 at the center of the impulse response, C_0 is a constant, and B is the single-sided bandwidth normalized to the sample rate, B = bandwidth / sample rate.

It is interesting that this has the same form as the frequency response of a pulse, as was shown in Figure 10.33. That is because a brick wall response in the frequency domain has the same shape as a pulse in the time domain. A pulse in one domain transforms to a sinc function in the other. This is an example of the general principle that the transformation between time and frequency domains is symmetrical. This is discussed in more detail in the section on Fourier transforms in the **DSP and SDR Fundamentals** chapter.

Normally, the filter coefficients are set up with the peak of the sinc function, sinc(0), at the center of the coefficient table so that there is an equal amount of "tail" on both sides. That points up the principle problem with this method of determining filter coefficients. Theoretically, the sinc function extends from minus infinity to plus infinity. Abruptly terminating the tails causes the frequency response to differ from an ideal brick wall filter. There is ripple in the passband and non-zero response in the stopband, as shown in the graph in the upper right of **Figure 10.34**. This is mainly caused by the abruptness of the truncation. In effect, all the coefficients outside the limits of the coefficient table have been set to zero. The passband and stopband response can be improved by tapering the edges of the impulse response instead of abruptly transitioning to zero.

The process of tapering the edges of the impulse response is called *windowing*. The impulse response is multiplied by a window, a series of coefficients that smoothly taper to zero at the edges. For example, a rectangular window is equivalent to no window at all. Many different window shapes have been developed over the years — at one time it seemed that every doctoral candidate in the field of signal processing did their dissertation on some new window. Each window has its advantages and disadvantages. A window that transitions slowly and smoothly to zero has excellent passband and stopband response but a wide transition band. A window that has a wider center portion and then transitions more abruptly to zero at the edges has a narrower transition band but poorer passband and stopband response. The equations for the windows in Figure 10.34 are included in a sidebar.

The routine shown in **Table 10.4** is written for a dsPIC processor so it can be used to calculate filter coefficients "on the fly" as the operator adjusts a bandwidth control. The same code should also work using a generic C compiler on a PC so the coefficients could be downloaded into an FIR filter implemented in hardware.

The windowed-sinc method works pretty well for a simple low-pass filter, but what if some more-complicated spectral shape is

Table 10.4
Routine for dsPIC Processor to Calculate Filter Coefficients

```
// Calculate FIR filter coefficients
// using the windowed-sinc method
void set_coef (
 double sample_rate;
 double bandwidth;)
{
extern int c[FIR_LEN]; // Coefficient array
int i; // Coefficient index
double ph; // Phase in radians
double coef; // Filter coefficient
int coef_int; // Digitized coefficient
double bw_ratio; // Normalized bandwidth

bw_ratio = 2 * bandwidth / sample_rate;
for (i = 0; i < (FIR_LEN/2); i++) {
 // Brick-wall filter:
 ph = PI * (i + 0.5) * bw_ratio;
 coef = sin(ph) / ph;
 // Hann window:
 ph = PI * (i + 0.5) / (FIR_LEN/2);
 coef *= (1 + cos(ph)) / 2;
 // Convert from floating point to int:
 coef *= 1 << (COEF_WIDTH - 1);
 coef_int = (int)coef;
 // Symmetrical impulse response:
 c[i + FIR_LEN/2] = coef_int;
 c[FIR_LEN/2 - 1 - i] = coef_int;
 }
}
```

Equations for Window Functions

For each window function, the center of the response is considered to be at time t = 0 and the width of the impulse is L. Each window is 1.0 when t = 0 and 0.0 when the | t | > L/2.

Rectangular:

$$w(t) = 1.0$$

Triangular (Bartlett):

$$w(t) = 2\left(\frac{L/2 - |t|}{L}\right)$$

Blackman:

$$w(t) = 0.42 + 0.5\cos\left(\frac{2\pi t}{L}\right) + 0.08\cos\left(\frac{4\pi t}{L}\right)$$

Hamming:

$$w(t) = 0.54 + 0.46\cos\left(\frac{2\pi t}{L}\right)$$

Hanning (Hann):

$$w(t) = 0.5 + 0.5\cos\left(\frac{2\pi t}{L}\right)$$

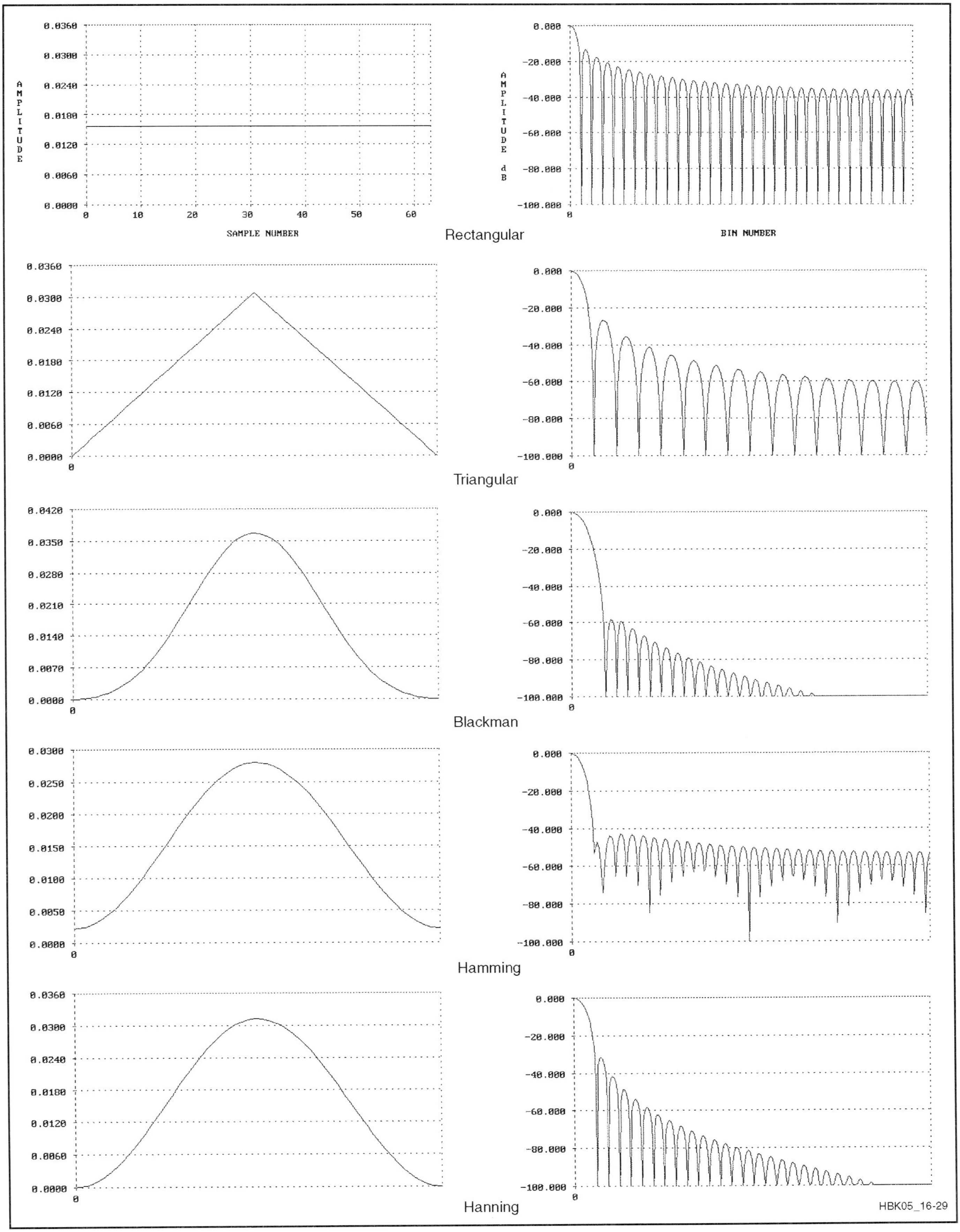

Figure 10.34 — Various window functions and their Fourier transforms.

desired? The same general design approach still applies. You determine the desired spectral shape, transform it to the time domain using an inverse Fourier transform, and apply a window. There is lots of (often free) software available that can calculate the fast Fourier transform (FFT) and inverse fast Fourier transform (IFFT). So the technique is to generate the desired spectral shape, transform to the time domain with the IFFT, and multiply the resulting impulse response by the desired window. Then you can transform back to the frequency domain with the FFT to see how the window affected the result. If the result is not satisfactory, you can either choose a different window or modify the original spectral shape and go through the process again.

Windowing methods are useful because they are simple to program and the resulting software routines execute quickly. For example, you can include a bandwidth knob on your DSP filter and calculate filter coefficients "on the fly" as the user turns the knob. However, while the filter performance that results is pretty good, it is not "optimum" in the sense that it does not have minimum passband ripple and maximum stopband attenuation for a given number of filter coefficients. For that, you need what is known as an *equal-ripple*, or *Chebyshev* filter. The calculations to determine Chebyshev filter coefficients are more complicated and time-consuming. For that reason, the coefficients are normally calculated in advance on a PC and stored in DSP program memory for retrieval as needed.

Engineers have not had much success in devising a mathematical algorithm to calculate the Chebyshev coefficients directly, but in 1972 Thomas Parks and James McClellan figured out a method to do it iteratively. The *Parks-McClellan algorithm* is supported in most modern filter-design software, including a number of programs available for free download on the Web. Typically you enter the sample rate, the passband and stopband frequency ranges, the passband ripple and the stopband attenuation. The software then determines the required number of filter coefficients, calculates them and displays a plot of the resulting filter frequency response.

Filter design software typically presents the filter coefficients as floating-point numbers to the full accuracy of the computer. You will need to scale the values and truncate the resolution to the word size of your filter implementation. Truncation of filter coefficients affects the frequency response of the filter but does not add noise in the same manner as truncating the signal data.

As you look at impulse responses for various FIR filters calculated by various methods you soon realize that most of them are symmetrical. If the center of the impulse response is considered to be at time zero, then the value at time t equals the value at time –t for all t.

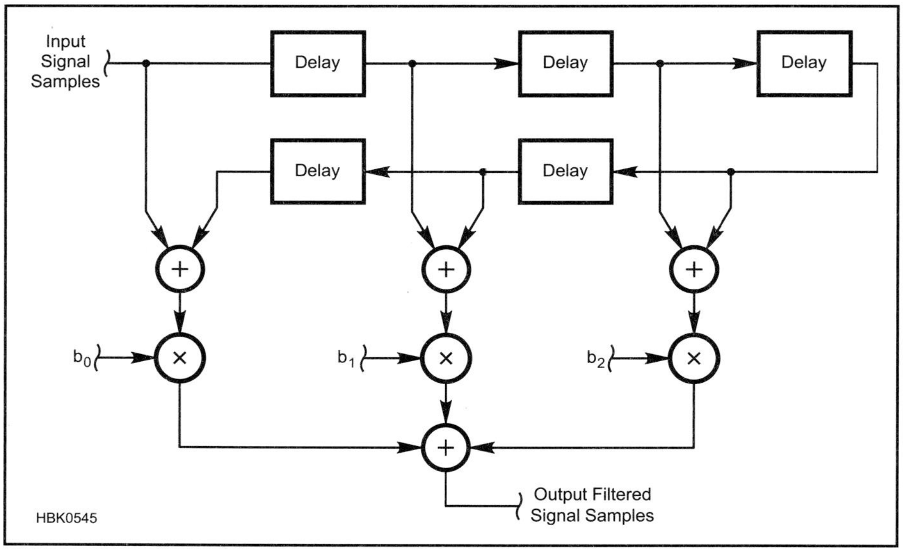

Figure 10.35 — A 6-tap FIR filter. Because the coefficients are symmetrical, the symmetrical taps may be combined before multiplication.

If you know in advance that the filter coefficients are symmetrical, you can take advantage of that in the filter design. By rearranging the adders and multipliers, the number of multipliers can be reduced by a factor of two, as shown in **Figure 10.35**. This trick is less useful in a software implementation of an FIR filter because the number of additions is the same and many DSPs take the same amount of time to do an addition as a multiply-accumulate.

In addition to the computational benefit, a symmetrical impulse response also has the advantage that it is *linear phase*. The time delay through such a filter is one-half the length of the filter for all frequencies. For example, for a 1,000-tap filter running at 10 kHz the delay is 500/10,000 = 0.05 second. Since the time delay is constant for all frequencies, the phase delay is directly proportional to the frequency. For example, if the phase delay at 20 Hz is one cycle (0.05 second) it is ten cycles at 200 Hz (still 0.05 second). Linear phase delay is important with digital modulation signals to avoid distortion and inter-symbol interference. It is also desirable with analog modulation where it can result in more natural-sounding audio. All analog filters are non-linear-phase; the phase distortion tends to be worse the more abrupt the transition between passband and stopband. That is why an SSB signal sounds unnatural after being filtered by a crystal filter with a small shape factor even though the passband ripple may be small and distortion minimal.

A band-pass filter can be constructed from a low-pass filter simply by multiplying the impulse response by a sine wave at the desired center frequency. This can be done before or after windowing. The linear-phase property is retained but with reference to the center frequency of the filter, that is, the phase shift is proportional to the difference in frequency from the center frequency. The frequency response is a double-sided version of the low-pass response with the zero-hertz point of the low-pass filter shifted to the frequency of the sine wave.

10.5.2 IIR Filters

An *infinite impulse response* (IIR) filter is a filter whose impulse response is infinite. After an impulse is applied to the input, theoretically the output never goes to zero and stays there. In practice, of course, the signal eventually does decay until it is below the noise level (analog filter) or less than one LSB (digital filter).

Unlike a symmetrical FIR filter, an IIR filter is not generally linear-phase. The delay through the filter is not the same for all frequencies. Also, IIR filters tend to be harder to design than FIR filters. On the other hand, many fewer adders and multipliers are typically required to achieve the same passband and stopband ripple in a given filter, so IIR filters are often used where computations must be minimized.

All analog filters have an infinite impulse response. For a digital filter to be IIR it must have feedback. That means a delayed copy of some internal computation is applied to an earlier stage in the computation. A simple but useful example of an IIR filter is the exponential decay circuit in **Figure 10.36**. In the absence of a signal at the input, the output on the next clock cycle is always (1–δ) times the current output. The time constant (the time for the output to die to 1/e = 36.8% of the initial value) is very nearly

$$\tau = f_s\left(\frac{1}{\delta} - \frac{1}{2}\right)$$

where f_s is the sample rate. The circuit is the digital equivalent of a capacitor with a resistor in parallel and might be useful for example in a digital automatic gain control circuit.

One issue with IIR filters is resolution. Because of the feedback, the number of bits of resolution required for intermediate computations can be much greater than at the input or output. In the previous example, δ is very small for very long time constants. When the value in the register falls below a certain level the multiplication by (1–δ) will no longer be accurate unless the bit width is increased. In practice, the increased resolution required with IIR filters often cancels out part of the savings in the number of circuit elements.

Another issue with IIR filters is stability. Because of the feedback it is possible for the filter to oscillate if care is not taken in the design. Stability can also be affected by nonlinearity at low signal levels. A circuit that is stable with large signals may oscillate with small signals due to the round-off error in certain calculations, which causes faint tones to appear when strong signals are not present. This is known as an unstable *limit cycle*. These issues are part of the reason that IIR filters have a reputation for being hard to design.

Design techniques for IIR filters mostly involve first designing an analog filter using any of the standard techniques and then transforming the design from the analog to the digital domain. The *impulse-invariant* method attempts to duplicate the filter response directly by making the digital impulse response equal the impulse response of the equivalent analog filter. It works fairly well for low-pass filters with bandwidths much less than the sample rate. Its problem is that it tries to duplicate the frequency response all the way to infinity hertz, but that violates the Nyquist criterion resulting in a folding back of the high-frequency response down into low frequencies. It is similar to the aliasing that occurs in a DSP system when the input signal to be sampled is not band-limited below the Nyquist frequency.

The *bilinear transform* method gets around that problem by distorting the frequency axis such that infinity hertz in the analog domain becomes sample rate / 2 in the digital domain. Low frequencies are fairly accurate, but high frequencies are squeezed together more and more the closer you get to the Nyquist frequency. It avoids the aliasing problem at the expense of a change in the spectrum shape, especially at the high-frequency end. For example, when designing a low-pass filter it may be necessary to change the cutoff frequency to compensate. Again, the method works best for filters with passband frequencies much less than the sample rate.

In general, the output of an IIR filter is a combination of the current and previous input values (feed-forward) and previous output values (feed-back). **Figure 10.37A** shows the so-called *direct form I* of an IIR filter. The b_i coefficients represent feed-forward and the a_i coefficients feed-back. For example the previous value of the y output is multiplied by a_1, the second previous value is multiplied by a_2, and so on. Because the filter is linear, it doesn't matter whether the feed forward or feed back stage is performed first. By reversing the order, the number of shift registers is reduced as in Figure 10.37B. There are other equivalent topologies as well. The mathematics for generating the a_i and b_i coefficients for both the impulse-invariant and bilinear transform methods is fairly involved, but fortunately some filter design programs can handle IIR as well as FIR filters.

10.5.3 CIC Filters

(This overview of CIC filters is taken from the 2011 Sep/Oct *QEX* column SDR: Simplified by Ray Mack, W5IFS. Additional background on CIC filters is available in the referenced articles by Donadio and by Lyons.)

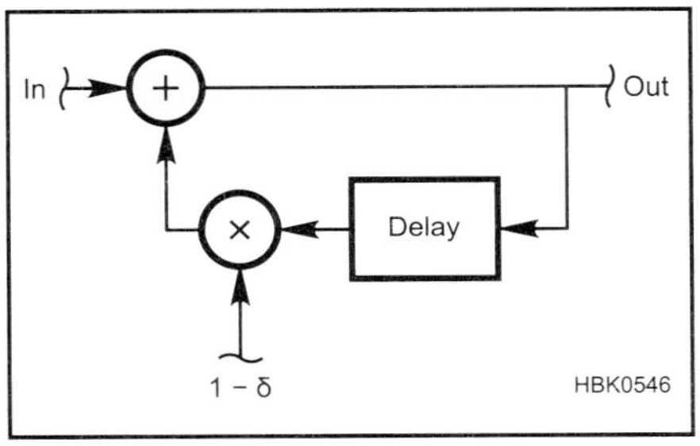

Figure 10.36 — An exponential decay implementation.

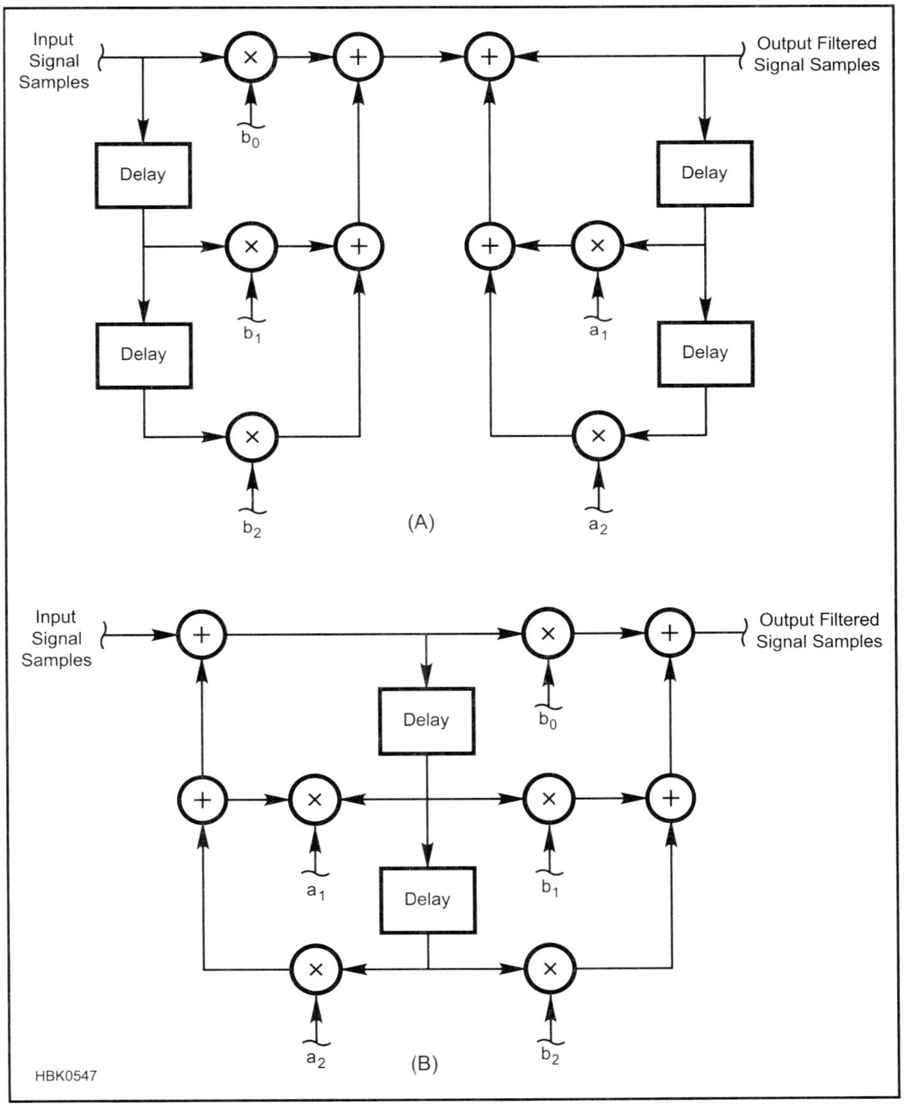

Figure 10.37 — An IIR filter with three feed-forward taps and two feed-back taps. Direct form I (A) and the equivalent direct form II (B).

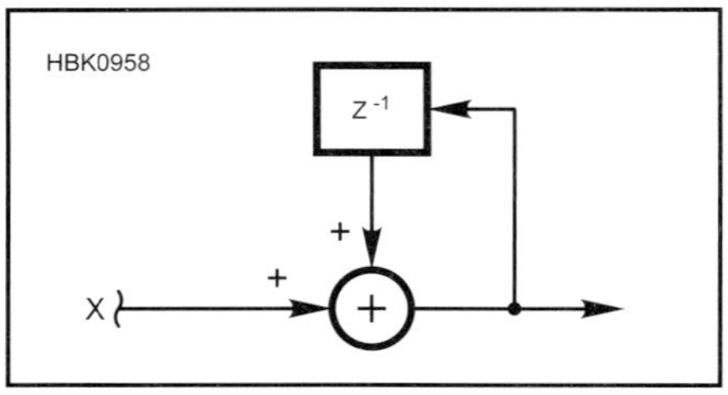

Figure 10.38 — Z transform diagram of an integrator. The new output is the sum of the previous output (represented by Z^{-1}) and the new sample. (The Z transform is explained at en.wikipedia.org/wiki/Z-transform.)

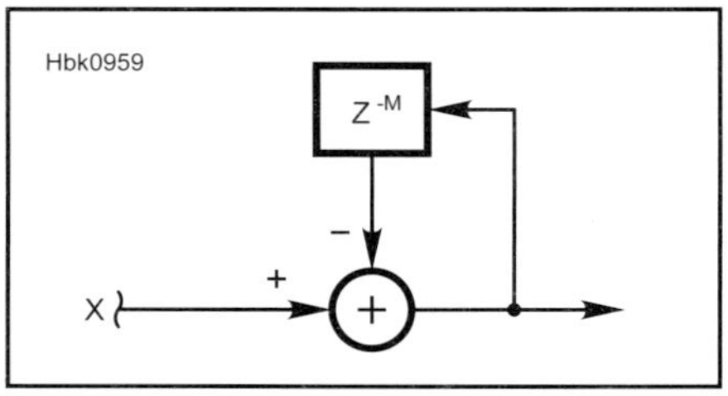

Figure 10.39 — Z transform diagram of a comb. The new output is the difference of the present sample and a delayed sample. The number M designates how many steps happened during the delay.

The problem with both FIR and IIR filters is that they require a large number of multiply operations, which consume a large number of DSP processor cycles. Eugene Hogenauer (see references) developed a very useful simplification of the sample conversion/filter configuration called a *cascaded integrator comb* (CIC) filter. The important aspect of CIC filters is that only addition, subtraction, and delay operations are required for implementation.

As with most things in life, improving one aspect of a system requires compromise in other aspects. This is also true of CIC filters where we trade the simplification of eliminating multipliers for restricting the filter response: A CIC filter can only have a low-pass response. Additionally, there is a limited subset of possible low-pass responses constrained by the sample rate change and number of stages in the comb and integrator stages. The most important property of a CIC filter is that it can be very easily implemented in hardware either in an FPGA or as part of the dedicated logic of an IC.

The integrator is an infinite impulse response filter. **Figure 10.38** shows how it works and how simple it is. The integrator holds a running total of all previous samples. The integrator adds the last output value (z–1) to the current input value (x). Ordinarily, we would worry about overflow in an integrator because a dc component in the signal will cause the integrator to overflow.

The combination of the comb and the integrator, however, cancels any problems with overflow. The integrator is a single pole low pass filter with infinite gain at dc. (The possibility of overflow doesn't matter as long as the integrators are implemented with adders using two's complement addition (**en.wikipedia.org/wiki/Two's_complement**) that allow wrap around when overflow occurs, and that the number of bits in the word is as big as the expected output word.)

The comb is a finite impulse response stage that subtracts a previous sample from the present sample. The amount of delay between the present sample and the delayed sample is called the differential delay and is denoted as M by most authors. **Figure 10.39** shows the operation of the comb.

An actual CIC filter is composed of multiple integrator-comb sections that are cascaded. A CIC filter has exactly the same number of integrators as combs. According to the associative property you can rearrange the order of the additions in a sequence and the result of the sequence does not change (a + (–b) + c + d + (–e) + (–f) is identical to a + c + d – b – e – f). A CIC filter with a sample rate change uses that property to group all of the integrators together and to group all of the combs together. We place either a downsample or upsample rate changer between the combs and integrators. (See the **DSP and SDR Fundamentals** chapter for a discussion of sample rate changing.)

Figure 10.40 shows that a decimator is an integrator section followed by a down rate change, which is then followed by a comb section. An interpolator turns the system around and puts the comb section first, followed by an up rate changer, which is followed by an integrator section. It is very useful for a hardware implementation that the number of integrators and combs is independent (within reason) from the rate change and that, in general, you can rearrange the inputs, outputs, and rate change to create a decimator and interpolator with the same blocks.

There are three parameters that affect the implementation of a CIC filter. "R" is the up sample rate or down sample rate. "M" is the delay in the comb section and is almost always either one or two. "N" is the number of stages in the comb section (which is required to be the same as the number of integrator sections). The simplification of separating the combs into one section and the integrators into a second section is advantageous for the speed required of the storage elements for the combs. The combs always operate at the low frequency end of the system and the integrators work at the high speed side of the system.

Notice that the associative property would also allow a decimator with the comb first and the integrator after the down converter. Either configuration will give the same results. The reason we always put the comb on the low sample rate side of the system is pragmatic. The comb requires one additional storage element (for the usual M = 1 situation) over what is required for an integrator. Each storage element consumes power when it is clocked. A faster clock and more storage registers translate directly into additional power dissipation. The additional power is an issue when implementing a CIC filter in hardware such as an FPGA.

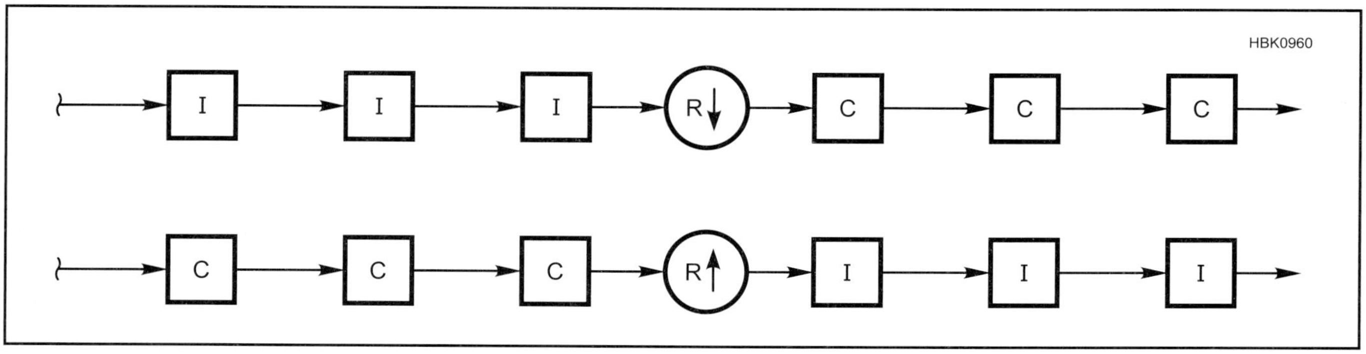

Figure 10.40 — The top CIC filter shows a decimator, where the sample rate is reduced by a value "R." The bottom CIC filter shows an interpolator where the sample rate is increased by a value "R." Note that the difference between the two is the order of the combs and integrators as well as the direction of the rate change. The combs are on the low sample rate side of both systems.

10.5.4 Adaptive Filters

An *adaptive filter* is one that automatically adjusts its filter coefficients under the control of some algorithm. This is often done in situations where the filter characteristics are not known in advance. For example, an adaptive channel equalizer corrects for the non-flatness in the amplitude and phase spectrum of a communications channel due to multipath propagation. Typically, the transmitting station periodically sends a known sequence of data, known as a *training sequence*, which is used by the receiver to determine the channel characteristics and adjust its filter coefficients accordingly.

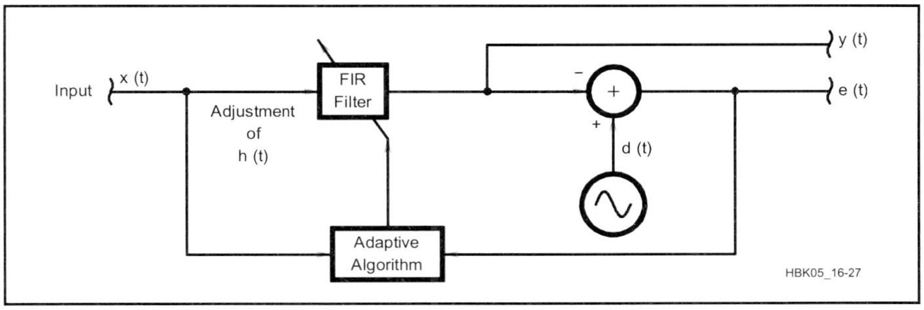

Figure 10.41 — An adaptive filter configuration.

Another example is an automatic notch filter. An algorithm determines the frequency of an interfering tone and automatically adjusts the notch frequency to remove the tone. Noise cancellation is another application. It can be thought of as the opposite of a notch filter. In this case, all the sine-wave tones in the input signal are considered to be desired and the filter coefficients are configured to enhance them. That method works not only for CW signals but for voice as well since the human voice consists largely of discrete frequencies.

A generic block diagram of an adaptive filter is shown in **Figure 10.41**. The variable filter is typically an FIR type with coefficients calculated by the update algorithm. By some means, an estimate of the desired, unimpaired signal, d, is generated and compared to the filter output y. The difference between y and d is the error, e, which is used by the update algorithm to modify the filter coefficients to improve the accuracy of y. The algorithm is capable of acting as a noise-reduction filter and a notch filter simultaneously. Assuming d is in the form of a pure tone (sine wave), the tone is simultaneously optimized in the y output and minimized in the e output.

A common algorithm for minimizing the error signal is called *least mean squares* (LMS). The LMS algorithm includes a performance parameter, μ, which can be adjusted between 0 and 1 to control the tradeoff between adjustment speed and accuracy. A value near 1 results in fast convergence but the convergence is not very accurate. For better accuracy at the cost of slower adjustment, lower the value of μ. Some implementations adjust μ on the fly, using a large value at first to get faster lock-in when the error is large then a smaller value after convergence to reduce the error. That works as long as the signal characteristics are not changing too rapidly.

10.6 Quartz Crystal Filters

(The sections on designing and building crystal filters from previous editions are available as a PDF document in the this book's online material.)

Inductor Q values effectively limit the minimum bandwidth that can be achieved with LC band-pass filters. Higher-Q circuit elements, such as quartz crystal, PLZT ceramic and constant-modulus metal alloy resonators, are required to extend these limits. Quartz crystals offer the highest Q and best stability with time and temperature of all available resonators. They are manufactured for a wide range of frequencies from audio to VHF, using cuts (crystal orientations) that suit the frequency and application of the resonator. The AT cut is favored for HF fundamental and VHF overtone use, whereas other cuts (DT, SL, and E) are more convenient for use at lower frequencies.

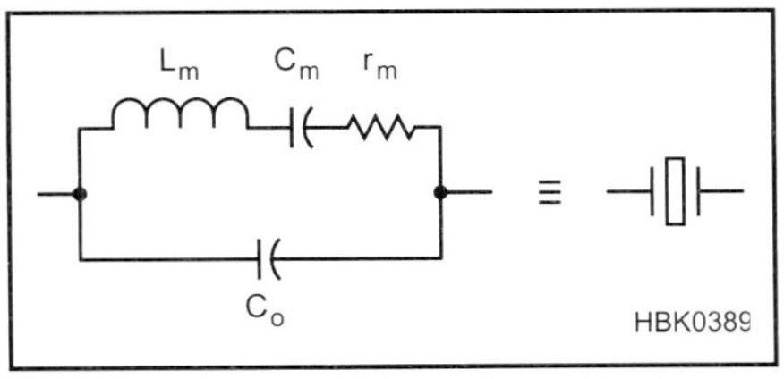

Figure 10.42 — Equivalent circuit of a quartz crystal and its circuit symbol.

Each crystal plate has several modes of mechanical vibration. These can be excited electrically thru the piezoelectric effect, but generally resonators are designed so as to maximize their response on a particular operating frequency using a crystal cut that provides low loss and a favorable temperature coefficient. Consequently, for filter design, quartz crystal resonators are modeled using the simplified equivalent circuit shown in **Figure 10.42**. Here L_m, C_m, and r_m represent the *motional* parameters of the resonator at the main operating frequency — r_m being the loss resistance, which is also known as the *equivalent series resistance*, or ESR. C_o is a combination of the static capacitance formed between the two metal electrodes with the quartz as dielectric ($e_r = 4.54$ for AT-cut crystals) and some additional capacitance introduced by the metal case, base, and mount. There is a physical relationship between C_m and the static capacitance formed by the resonator electrodes, but, unfortunately, the added holder capacitance masks this direct relationship causing C_o/C_m to vary from 200 to over 500. However, for modern fundamental AT-cut crystals between 1 and 30 MHz, their values of C_m are typically between 0.003 and 0.03 pF. Theoretically, the motional inductance of a quartz crystal should the same whether it is operated on the fundamental or one of its overtones, making the motional capacitance at the nth overtone $1/n^2$ of the value at the fundamental. However, this relationship is modified by the effect of the metal electrodes deposited on either side of the crystal plate and in practice the motional inductance increases with the frequency of the overtone. This makes the motional capacitance at the overtone substantially less than C_m/n^2.

An important parameter for crystal filter design is the unloaded Q at f_s, the resonant frequency of the series arm. This is usually denoted by Q_U.

$$Q_u = 2 \pi f_s L_m/r_m$$

Q_U is very high, often exceeding 100,000 in the lower HF region. Even VHF overtone crystals can have Q_U values over 20,000, making it possible to design quartz crystal band-pass filters with a tremendously wide range of bandwidths and center frequencies.

The basic filtering action of a crystal can be seen from **Figure 10.43**, which shows a plot of attenuation vs frequency for the test circuit shown in the inset. The series arm of the crystal equivalent circuit forms a series-tuned circuit, which passes signals with little attenuation at its resonant frequency, f_s, but appears inductive above this frequency and

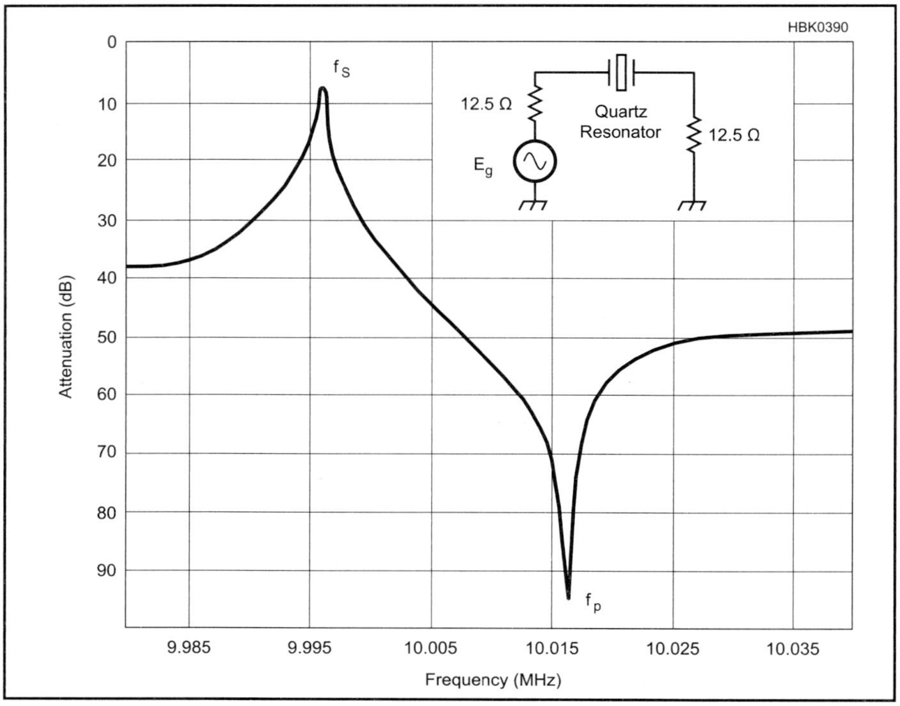

Figure 10.43 — Response of 10 MHz crystal (Cm = 0.0134 pF, Lm = 18.92 mH, ESR = 34 Ω) in a series test circuit (see inset) showing peak of transmission (lowest attenuation) at the series resonant frequency, f_s, and a null (maximum attenuation) caused by the parallel resonance due to C_o at f_p.

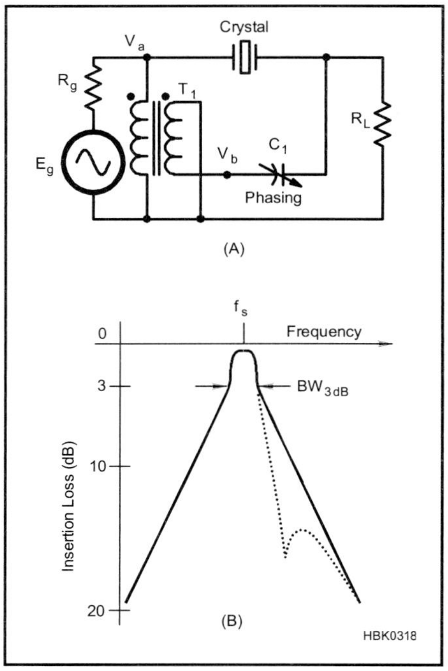

Figure 10.44 — Classic single-crystal filter in A has the response shown in B. The phasing capacitor can be adjusted to balance out C_o (solid line), or set to a lesser or greater value to create a movable null to one side, or other, of the passband (dotted line).

parallel resonates with C_o at f_p to produce a deep notch in transmission. The difference between f_s and f_p is known as the pole-zero separation, or PZ spacing, and is dependent on C_m/C_o as well as f_s. Further information on quartz crystal theory and operation can be found in the **Oscillators and Synthesizers** chapter and in the reference for Bottom.

A simple crystal filter developed in the 1930s is shown in **Figure 10.44**. The voltage-reversing transformer T1 was usually an IF transformer, but nowadays could be a bifilar winding on a ferrite core. Voltages V_a and V_b have equal magnitude but 180° phase difference. When $C_1 = C_o$, the effect of C_o will disappear and a well-behaved single resonance will occur as indicated by the solid line in Figure 10.44B. However, if C_1 is adjusted to unbalance the circuit, a transmission zero (notch) is produced well away from the pass band and by increasing the amount of imbalance this can be brought back toward the edge of the pass band to attenuate close-by interfering CW signals. If C_1 is reduced in value from the balanced setting, the notch comes in from the high side and if C_1 is increased, it comes in from the low side. The dotted curve in Figure 10.44B illustrates how the notch can be set with C_1 less than C_o to suppress adjacent signals just above the pass band. In practice a notch depth of up to 60 dB can be achieved. This form of "crystal gate" filter, operating at 455 kHz, was present in many high-quality amateur communications receivers from the 1930s through the 1960s. When the filter was switched into the receiver IF amplifier the bandwidth was reduced to a few hundred Hz for CW reception. The close-in range of the notch was sometimes improved by making C_1 part of a differential capacitor that could add extra capacitance to either the C_1 or C_o side of the IF transformer. This design could also be used to good effect at frequencies up to 1.7 MHz with an increased minimum bandwidth. However, any crystal gate requires considerable additional IF filtering to achieve a reasonable ultimate attenuation figure, so it should not be the only form of selectivity used in an IF amplifier.

The half-lattice filter shown in **Figure 10.45** offers an improvement in performance over a single-crystal filter. The quartz crystal static capacitors, C_o, cancel each other. The remaining series-resonant arms, if offset in frequency and terminated properly, will produce an approximate two-pole Butterworth or Chebyshev response. The crystal spacing for simple Chebyshev designs is usually around two-thirds of the bandwidth. Half-lattice filter sections can be cascaded to produce composite filters with multiple poles. Many of the older commercial filters are coupled half-lattice types using 4, 6, or 8

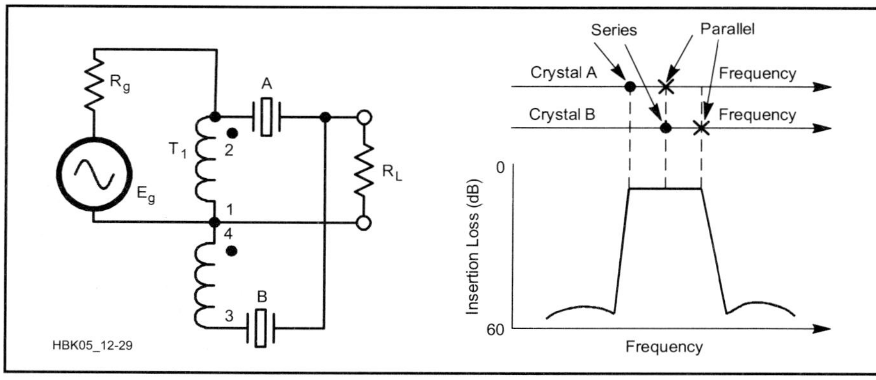

Figure 10.45 — A half-lattice crystal filter. The C_o of one crystal can be made to balance the C_o of the other, or C_o across the higher crystal can be deliberately increased to create nulls on either side of the passband.

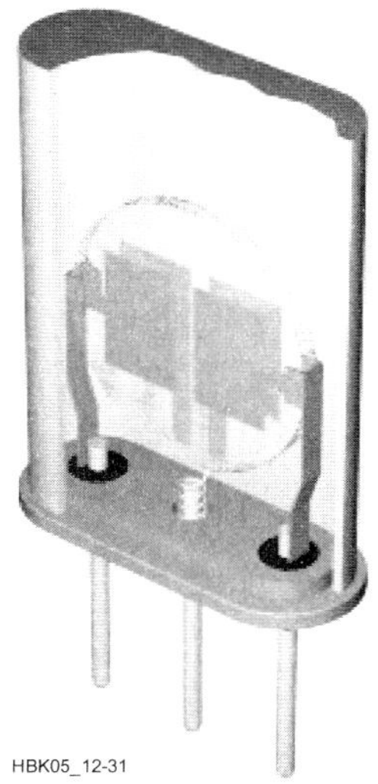

Figure 10.46 — A dual monolithic crystal filter has two sets of electrodes acoustically coupled to provide a two-pole response in a single crystal unit. The center lead is connected to the case and grounded in normal operation.

crystals, and this is still the favored technology for some crystal filters at lower frequencies. Very often extra capacitance was added across one crystal in each half lattice to unbalance C_0 and provide deep transmission zeroes on either side of the pass band to sharpen up the close-in response at the expense of the attenuation further out. The reference for Steder and Hardcastle discusses the computer design of half-lattice filters.

Most commercial HF and VHF crystal filters produced today use dual monolithic structures as their resonant elements. These are a single quartz plate onto which two sets of metal electrodes have been deposited, physically separated to control the acoustic (mechanical) coupling between them. An example of a dual monolithic filter (two-pole) is shown in **Figure 10.46**. These are available with center frequencies from 9 MHz to well over 120 MHz. In effect, the dual monolithic structure behaves just like a pair of coupled crystal resonators. There is a subtle difference, however, because the static capacitance in duals appears across the input and output terminations and doesn't produce a null above the pass band, as it would for two electrically coupled crystal resonators. Multipole filters can be built by coupling duals together using external capacitors, the C_0 from each dual input and output being absorbed in the coupling capacitors or terminations, or by including more acoustically coupled resonators in the electrode structure on a single quartz plate. Though not common, three-pole and four-pole monolithic filters housed in standard, single crystal holders are available from some manufacturers.

10.6.1 Filter Parameters

An ideal band-pass filter would pass the desired signal with no loss and completely attenuate everything else. Practically, it is not possible to achieve such a response with a finite number of resonators, and approximations to this ideal have to be accepted. Greater stop-band attenuation and steeper sides can be achieved if more and more crystals are used and the response nears the ideal brick wall. This feature of a filter is expressed as a *shape factor*, which specifies the ratio of the bandwidth at an attenuation of 60 dB to the bandwidth at 6 dB — both these levels being taken relative to the actual pass band peak to eliminate insertion loss from the calculation. An ideal brick wall filter would have a shape factor of 1, and practical filters have shape factors that depend on the number of crystals used in the design and the type of response chosen for the pass band. A 1 dB Chebyshev design, for example, can typically produce shape factors that vary from about 4.1 for a 4-pole to 1.5 for a 10-pole, but the actual figures obtained in practice are very much dependent on Q_u and how much greater it is than the filter Q, defined by f_0/BW, where f_0 is the center frequency and BW the 3 dB bandwidth. The ratio of Q_u to f_0/BW is often quoted as Q_0, or q_0, and this, along with the order and type of response, determines the insertion loss of a crystal filter. Q_0 also determines how closely the pass band follows the design response, and how much the passband ripple is smoothed out and the edges of the response rounded off by crystal loss.

Commercial filter manufacturers usually choose the Chebyshev equiripple design for SSB, AM, and FM bandwidths because it gives the best compromise between passband response and the steepness of the sides, and 1 dB ripple Chebyshev designs are pretty standard for speech bandwidths. Tolerances in component values and crystal frequencies can cause the ripple in the pass band to exceed 1 dB, so often the maximum ripple is specified as 2 dB even though the target ripple is lower. Insertion loss, the signal loss going through the filter, also varies with the type of response and increases as the order of the filter increases. The insertion loss for a given order and bandwidth is higher for high-ripple Chebyshev designs than it is for low-ripple ones, and Butterworth designs have lower insertion loss than any Chebyshev type for a given bandwidth. Pass-band amplitude response and shape factor are important parameters for assessing the performance of filters used for speech communications.

However, group delay is also important for data and narrow bandwidth CW reception. Differential group delay can cause signal distortion on data signals if the variations are greater than the automatic equalizer can handle. Ringing can be an annoying problem when using very narrow CW filters, and the group delay differential across the pass band must be minimized to reduce this effect. When narrow bandwidth filters are being considered, shape factor has to be sacrificed to reduce differential group delay and its associated ringing problems. It's no good having a narrow Chebyshev design with a shape factor of 2 if the filter produces unacceptable ringing and is intolerable to use in practice.

Both Bessel and linear phase (equiripple 0.05°) responses have practically constant, low group delay across the entire pass band and well beyond on either side, making either a good choice for narrow CW or specialized data use. They also have the great advantage of offering the lowest possible insertion loss of all the types of response currently in use, which is important when Q_0 is low, as it often is for very narrow bandwidth filters. The insertion loss of the Bessel design is marginally lower than that of the linear phase, but the latter has a superior shape factor giving it the best balance of low group delay and good selectivity. A six-pole linear phase (equiripple 0.05°) design has a shape factor of 3.39, whereas a six-pole Bessel has 3.96. The Gaussian to 12 dB response has a better shape factor than that of either the linear phase or Bessel designs but the group delay across the pass band is not as flat and pronounced peaks (ears) are beginning to appear at the band edges with a six-pole design. The Gaussian to six dB group delay is reasonably flat for three- and four-pole filters, but significant ears appear toward the passband edges in designs with five poles or more.

10.6.2 Crystal Filter Evaluation

The simplest means of assessing the performance of a crystal filter is to temporarily install it in a finished transceiver, or receiver, and use a strong on-air signal, or locally generated carrier, to run through the filter pass band and down either side to see if there are any anomalies. Provided that the filter crystals have been carefully characterized in the first instance, and computer modeling has shown that the design is close to what's required, this may be all that's required to confirm a successful project. However, more elaborate checks on both the pass band and stop band can be made if a DDS signal generator, or vector network analyzer (VNA), is available. A test setup for evaluating the response of a filter using a DDS generator requires an oscilloscope to display the response, whereas a VNA controlled via the USB port of a PC can display the response on the PC screen with suitable software — see the **Test Equipment and Measurements** chapter. In addition, it can measure the phase and work out the group delay of the filter. VNAs can also be used to characterize the crystals prior to making the filter as well as evaluating filter performance after completion.

10.7 SAW Filters

The resonators in a monolithic crystal filter are coupled together by bulk acoustic waves. These acoustic waves are generated and propagated in the interior of a quartz plate. It is also possible to launch, by an appropriate transducer, acoustic waves that propagate only along the surface of the quartz plate. These are called "surface-acoustic-waves" because they do not appreciably penetrate the interior of the plate.

A *surface acoustic wave* (SAW) filter consists of thin aluminum electrodes, or fingers, deposited on the surface of a piezoelectric substrate as shown in **Figure 10.47**. Lithium niobate ($LiNbO_3$) is usually favored over quartz because it yields less insertion loss. The electrodes make up the filter's transducers. RF voltage is applied to the input transducer and generates electric fields between the fingers. The piezoelectric material vibrates in response, launching an acoustic wave along the surface. When the wave reaches the output transducer it produces an electric field between the fingers. This field generates a voltage across the load resistor.

Since both input and output transducers are not entirely unidirectional, some acoustic power is lost in the acoustic absorbers located behind each transducer. This lost acoustic power produces a mid-band electrical insertion loss typically greater than 10 dB. The SAW filter frequency response is determined by the choice of substrate material and finger pattern. The finger spacing, (usually one-quarter wavelength) determines the filter center frequency. Center frequencies are available from 20 to 1000 MHz. The number and length of fingers determines the filter loaded Q and shape factor.

Loaded Qs are available from 2 to 100, with a shape factor of 1.5 (equivalent to a dozen poles). Thus the SAW filter can be made broadband much like the LC filters that it replaces. The advantage is substantially reduced volume and possibly lower cost. SAW filter research was driven by military needs for exotic amplitude-response and time-delay requirements. Before the switch to digital television, low-cost SAW filters were frequently found in television IF amplifiers where high midband loss could be tolerated.

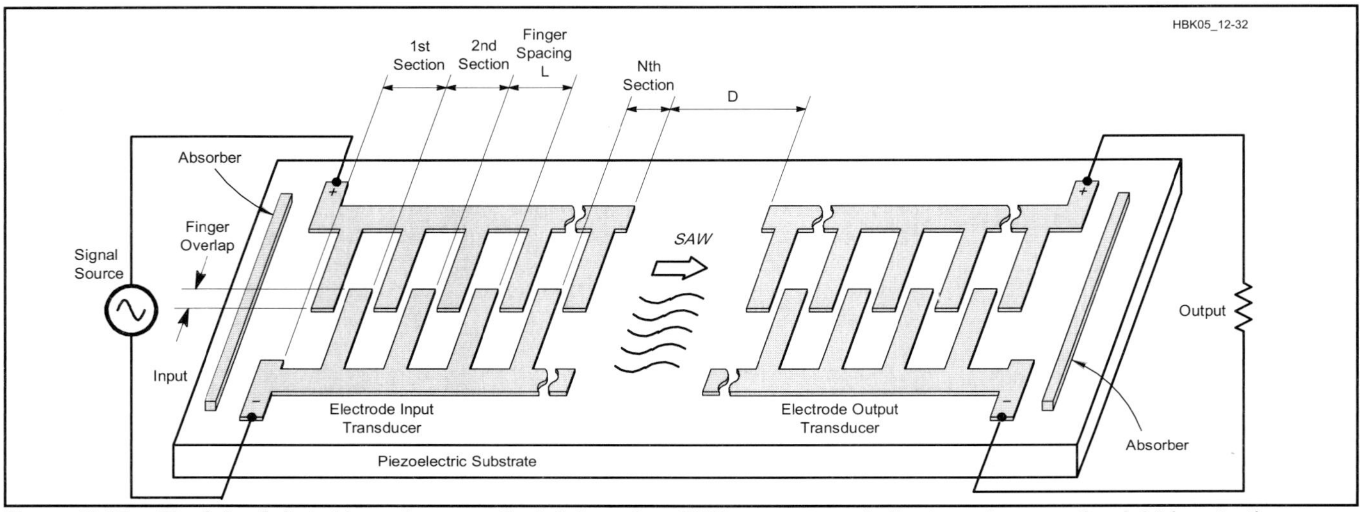

Figure 10.47 — The *interdigitated* transducer, on the left, launches SAW energy to a similar transducer on the right (see text).

10.8 Transmission Line VHF/UHF/Microwave Filters

LC filter calculations are based on the assumption that the reactances are *lumped*—that the physical dimensions of the components are considerably less than the operating wavelength. In such cases the unavoidable interturn parasitic capacitance associated with inductors and the unavoidable series parasitic inductance associated with capacitors are neglected as being secondary effects. If careful attention is paid to circuit layout and miniature components are used, lumped LC filter technology can be used up to perhaps 1 GHz.

Replacing lumped reactances with selected short sections of *Transverse Electromagnetic Mode* (TEM) transmission lines results in transmission line filters. (In TEM the electric and magnetic fields associated with a transmission line are at right angles (transverse) to the direction of wave propagation.) Coaxial cable, stripline and microstrip are examples of TEM components. Waveguides and waveguide resonators are not TEM components.

Coaxial cable transmission line filters are often used at HF and VHF frequencies. Stripline and microstrip transmission line filters predominate from 500 MHz to 10 GHz. In addition they are often used down to 50 MHz when narrowband ($Q_L > 10$) band-pass filtering is required. In this application they exhibit considerably lower loss than their LC counterparts and are useful at frequencies where coaxial transmission lines are too lossy. A detailed treatment of the use of coaxial cable to form transmission line filters is presented in the **Transmission Lines** chapter. This section focuses on stripline and microstrip filters used at VHF and above.

10.8.1 Stripline and Microstrip Filters

Figure 10.48 shows three popular transmission lines used in transmission line filters. The circular coaxial transmission line (*coax*) shown in Figure 10.48A consists of two concentric metal cylinders separated by dielectric (insulating) material. The first transmission-line filters were built from sections of coaxial line. Their mechanical fabrication is expensive and it is difficult to provide electrical coupling between line sections.

Fabrication difficulties are reduced by the use of shielded strip transmission line (*stripline*) shown in Figure 10.48B. The outer conductor of stripline consists of two flat parallel metal plates (ground planes) and the inner conductor is a thin metal strip. Sometimes the inner conductor is a round metal rod. The dielectric between ground planes and strip can be air or a low-loss plastic such as polyethylene. The outer conductors (ground planes or shields) are separated from each other by distance *b*.

Striplines can be easily coupled together by locating the strips near each other as shown in Figure 10.48B. Stripline Z_0 vs width (w)

is plotted in **Figure 10.49**. Air-dielectric stripline technology is best for low bandwidth (Q_L > 20) band-pass filters.

The most popular transmission line at UHF and microwave is *microstrip* (unshielded stripline), shown in Figure 10.48C. It can be fabricated with standard printed-circuit processes and is the least expensive configuration. In microstrip the outer conductor is a single flat metal ground-plane. The inner conductor is a thin metal strip separated from the ground-plane by a solid dielectric substrate. Typical substrates are 0.062 inch G-10 fiberglass (ε = 4.5) for the 50 MHz to 1 GHz frequency range and 0.031 inch Teflon (ε = 2.3) for frequencies above 1 GHz. Unfortunately, microstrip has the most loss of the three types of transmission line; therefore it is not suitable for narrow, high-Q, band-pass filters.

Conductor separation must be minimized or radiation from the line and unwanted coupling to adjacent circuits may become problems. Microstrip characteristic impedance and the effective dielectric constant (ε) are shown in **Figure 10.50**. Unlike coax and stripline, the effective dielectric constant is less than that of the substrate since a portion of the electromagnetic wave propagating along the microstrip travels in the air above the substrate.

The characteristic impedance for stripline and microstrip-lines that results in the lowest loss is not 75 Ω as it is for coax. Loss decreases as line width increases, which leads to clumsy, large structures. Therefore, to conserve space, filter sections are often constructed from 50 Ω stripline or microstrip stubs even though the loss at that characteristic impedance is not a minimum for that type of transmission line.

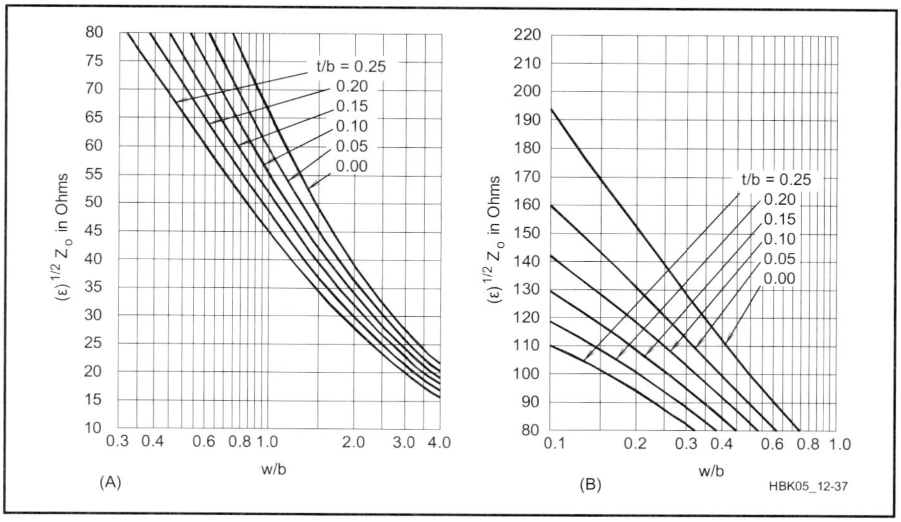

Figure 10.49 — The Z_0 of stripline varies with *w*, *b* and *t* (conductor thickness). See Figure 10.48B. The conductor thickness is *t* and the plots are normalized in terms of *t/b*.

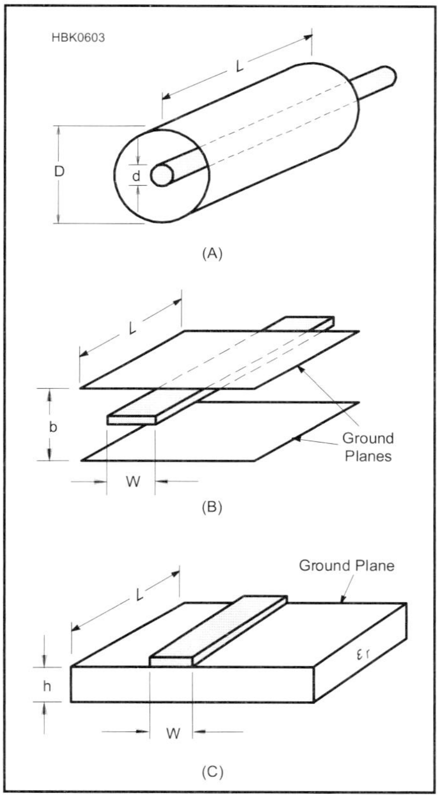

Figure 10.48 — Transmission lines. A: Coaxial line. B: Coupled stripline, which has two ground planes. C: Microstrip has only one ground plane.

10.8.2 Transmission Line Band-Pass Filters

Band-pass filters can also be constructed from transmission line stubs. (See the **Transmission Lines** chapter for information on stub behavior and their use as filters at HF and VHF.) At VHF the stubs can be considerably shorter than a quarter-wavelength (¼ λ), yielding a compact filter structure with less mid-band loss than its LC counterpart. The single-stage 146 MHz stripline band-pass filter shown in **Figure 10.51** is an example.

Z_0 Ω	ε = 1 (AIR) W/h	ε = 2.3 (RT/Duroid) W/h	$\sqrt{\varepsilon_e}$	ε = 4.5 (G-10) W/h	$\sqrt{\varepsilon_e}$
25	12.5	7.6	1.4	4.9	2.0
50	5.0	3.1	1.36	1.8	1.85
75	2.7	1.6	1.35	0.78	1.8
100	1.7	0.84	1.35	0.39	1.75
	$\sqrt{\varepsilon}$ = 1				

HBK05_12-38

Figure 10.50 — Microstrip parameters (after H. Wheeler, *IEEE Transactions on MTT*, March 1965, p 132). e_e is the effective ε.

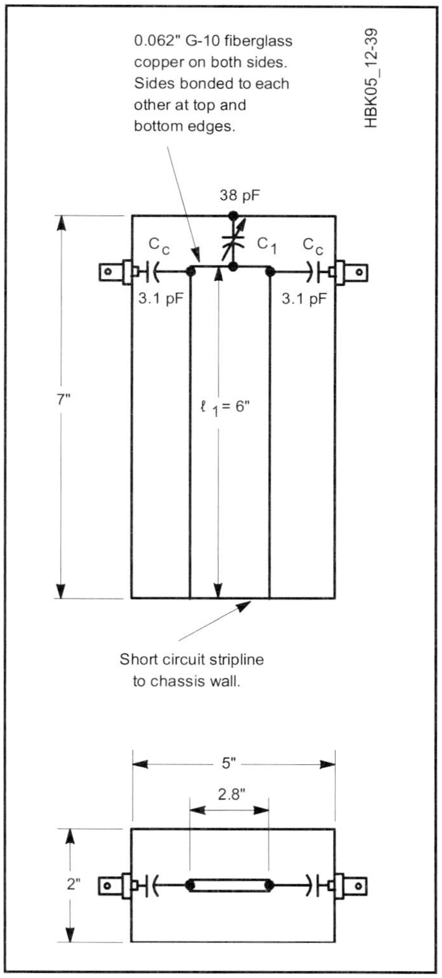

Figure 10.51 — This 146 MHz stripline band-pass filter has been measured to have a Q_L of 63 and a loss of approximately 1 dB.

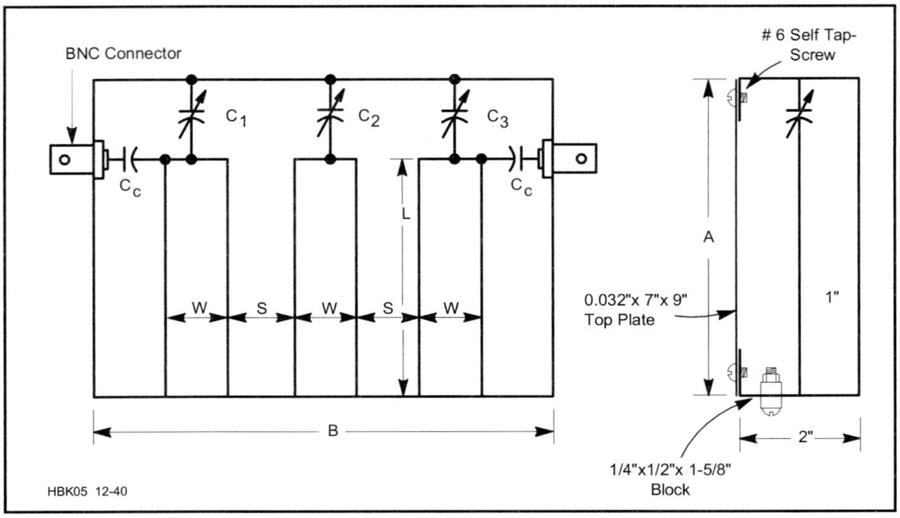

Figure 10.52 — This Butterworth filter is constructed in combline. It was originally discussed by R. Fisher in December 1968 *QST*.

This filter consists of a single inductive 50 Ω strip-line stub mounted into a 2 × 5 × 7 inch aluminum box. The stub is resonated at 146 MHz with the "APC" variable capacitor, C1. Coupling to the 50 Ω generator and load is provided by the coupling capacitors C_c. The measured performance of this filter is: f_o = 146 MHz, BW = 2.3 MHz (Q_L = 63) and midband loss = 1 dB.

Single-stage stripline filters can be coupled together to yield multistage filters. One method uses the capacitor coupled band-pass filter synthesis technique to design a three-pole filter. Another method allows closely spaced inductive stubs to magnetically couple to each other. When the coupled stubs are grounded on the same side of the filter housing, the structure is called a "combline filter." Three examples of combline band-pass filters are shown in **Figure 10.52** and a set of VHF/UHF filter designs by W1GHZ is included in the projects section of this chapter. These filters are constructed in 2 × 7 × 9 inch aluminum boxes. The article describing these filters by Reed Fisher, W2CQH, is available in this book's online information and is listed in the references, as well.

10.8.3 Quarter-Wave Transmission Line Filters

The reactance of a ¼ λ shorted-stub is infinite, as discussed in the **Transmission Lines** chapter. Thus, a ¼ λ shorted stub behaves like a parallel-resonant LC circuit. Proper input and output coupling to a ¼ λ resonator yields a practical band-pass filter. Closely spaced ¼ λ resonators will couple together to form a multistage band-pass filter. When the resonators are grounded on opposite walls of the filter housing, the structure is called an *interdigital filter* because the resonators look like interlaced fingers. Two examples of three-pole UHF interdigital filters are shown in **Figure 10.53**. Design graphs for round-rod interdigital filters are given in the reference for Metcalf. The ¼ λ resonators may be tuned by physically changing their lengths or by tuning the screw opposite each rod.

If the short-circuited ends of two ¼ λ resonators are connected to each other, the resulting ½ λ stub will remain in resonance, even when the connection to the ground-plane is removed. Such a floating ½ λ microstrip line, when bent into a U-shape, is called a *hairpin* resonator. Closely coupled hairpin resonators can be arranged to form multistage band-pass filters. Microstrip hairpin band-pass filters are popular above 1 GHz because they can be easily fabricated using photo-etching techniques. No connection to the ground-plane is required.

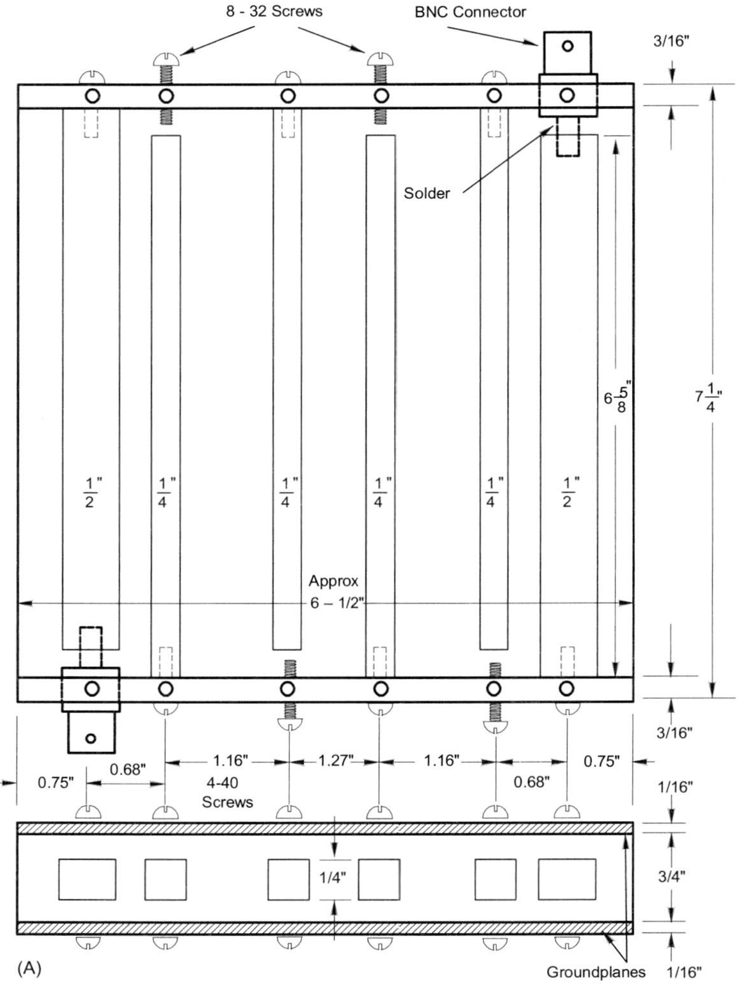

Figure 10.53 — These 3-pole Butterworth filters for 432 MHz (shown at A, 8.6 MHz bandwidth, 1.4 dB passband loss) and 1296 MHz (shown at B, 110 MHz bandwidth, 0.4 dB passband loss) are constructed as interdigitated filters. The material is from R. E. Fisher, March 1968 *QST*.

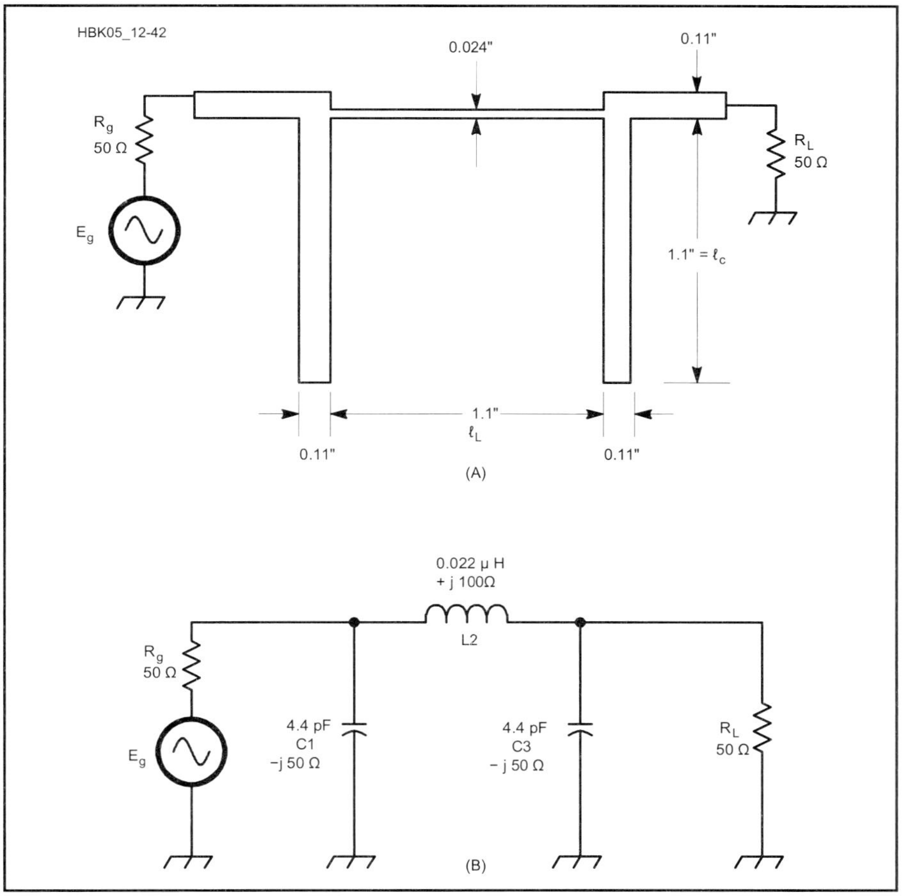

Figure 10.54 — A microstrip three-pole emulated-Butterworth low-pass filter with a cutoff frequency of 720 MHz. A: Microstrip version built with G-10 fiber-glass board (ε = 4.5, h = 0.062 inches). B: Lumped LC version of the same filter. To construct this filter with lumped elements very small values of L and C must be used and stray capacitance and inductance would have to be reduced to a tiny fraction of the component values.

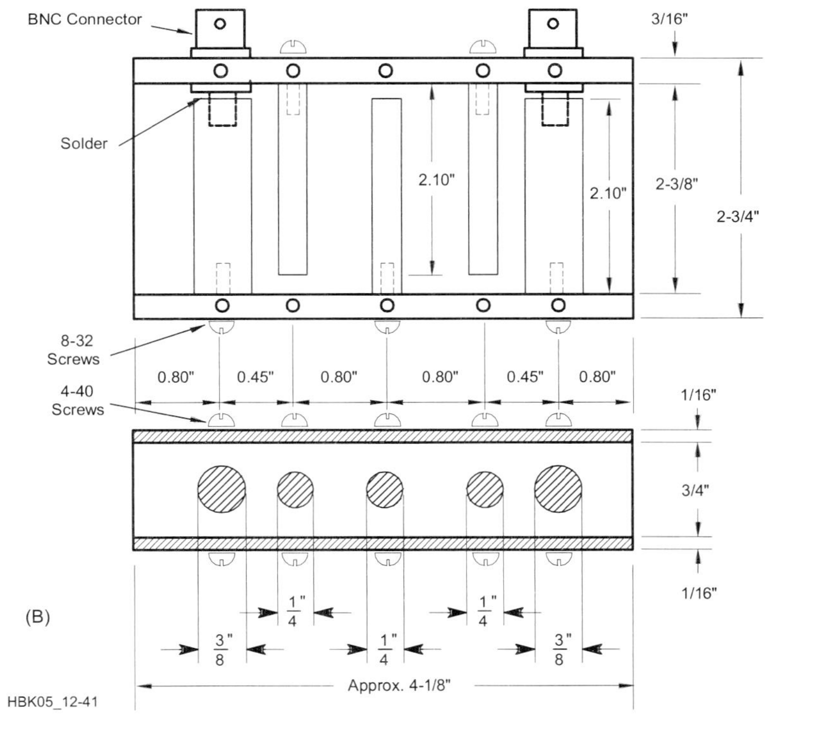

10.8.4 Emulating LC Filters with Transmission Line Filters

Low-pass and high-pass transmission-line filters are usually built from short sections of transmission lines (stubs) that emulate lumped LC reactances. Sometimes low-loss lumped capacitors are mixed with transmission line inductors to form a hybrid component filter. For example, consider the 720 MHz, three-pole microstrip low-pass filter shown in **Figure 10.54A** that emulates the LC filter shown in Figure 10.54B. C1 and C3 are replaced with 50 Ω open-circuit shunt stubs ℓ_C long. L2 is replaced with a short section of 100 Ω line ℓ_L long. The LC filter, Figure 10.54B, was designed for f_c = 720 MHz. Such a filter could be connected between a 432 MHz transmitter and antenna to reduce harmonic and spurious emissions. A reactance chart shows that X_C is 50 Ω, and the inductor reactance is 100 Ω at f_c. The microstrip version is constructed on G-10 fiberglass 0.062 inch thick, with ε = 4.5. Then, from Figure 10.50, *w* is 0.11 inch and ℓ_C = 0.125 l_g for the 50 Ω capacitive stubs. Also, from Figure 10.50, *w* is 0.024 inch and ℓ_L is 0.125 l_g for the 100 Ω inductive line. The inductive line length is approximate because the far end is not a short circuit. l_g is 300/(720 × 1.75) = 0.238 m, or 9.37 inches Thus ℓ_C is 1.1 inch and ℓ_L is 1.1 inches.

This microstrip filter exhibits about 20 dB of attenuation at 1296 MHz. Its response rises again, however, around 3 GHz. This is because the fixed-length transmission line stubs change in terms of wavelength as the frequency rises. This particular filter was designed to eliminate third-harmonic energy near 1296 MHz from a 432 MHz transmitter and does a better job in this application than the Butterworth filter in Figure 10.53 which has spurious responses in the 1296 MHz band.

10.9 Cavity and Helical Filters

10.9.1 Cavity Filters

The following material is based on the article "Understanding the Cavity Duplexer" by John Portune, W6NBC, available at **w6nbc.com/articles/duplexer.pdf**. A detailed analysis of cavities and other transmission line resonators is contained in *Microwave Engineering* by David Pozar. The website Repeater-Builders (**www.repeater-builder.com**) also contains additional information about the use of cavities as duplexers for repeater systems. (Also see this book's **Repeater Systems** chapter.)

This section discusses amateur radio's most common type of cavity filter, the ¼ λ coaxial cavity used in repeater duplexers. Overviews of commercial cavity filters used at VHF and UHF from Telewave can be found online by searching for "A Brief Guide to Telewave's Cavity Filters" and "Tips on Navigating Cavity Filter Types, Performance, and Features."

CAVITY BASICS

The ¼ λ cavity has three important properties: (1) the ability to handle power, (2) high Q, and (3) low loss. It is this unique combination of all three in that has long made the resonant cavity the dominant choice for use as duplexer filters.

A *coaxial cavity resonator* is an open cylinder with highly conductive walls. The cavity's size is such that it resonates at RF. A cavity resonator is analogous to an acoustic organ pipe, a penny whistle, or flute. The mathematical descriptions of electromagnetic and acoustic cavities are nearly identical.

The equivalent circuit of the ¼ λ coaxial cavity is a parallel LC "tank" circuit as shown in **Figure 10.55**. This makes the basic concept of the resonant cavity a little easier to visualize. Inside the outer cylinder is a smaller metal cylinder which forms the coaxial center conductor. See **Figure 10.56**. The center conductor is connected to the outer cylinder at one end of the cavity but not at the other. Figure 10.56 also shows the coupling loops and tuning adjustment which are discussed below.

This configuration is a short length of large-diameter air-insulated coaxial transmission line, shorted at one end. Thus, the common duplexer cavity is a simply a ¼ λ shorted coaxial stub. (See the **Transmission Lines** chapter.) The open end is covered with a conductive cap placed just a small distance away from the open end of the center conductor. The cover has little effect on the action of the filter and keeps the cavity free of dust and other contaminants.

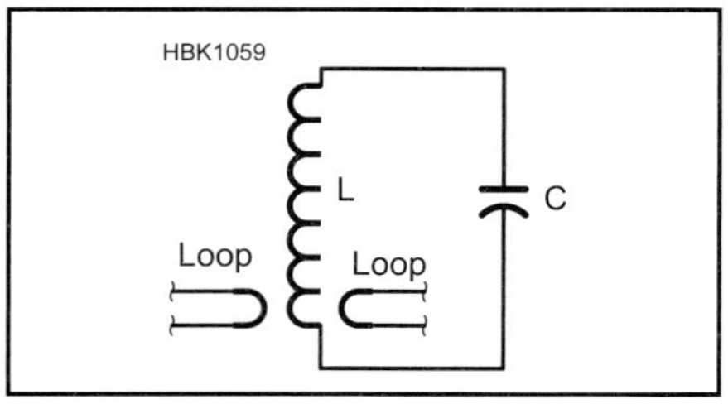

Figure 10.55 — The parallel LC tank circuit is an equivalent for the cavity filter.

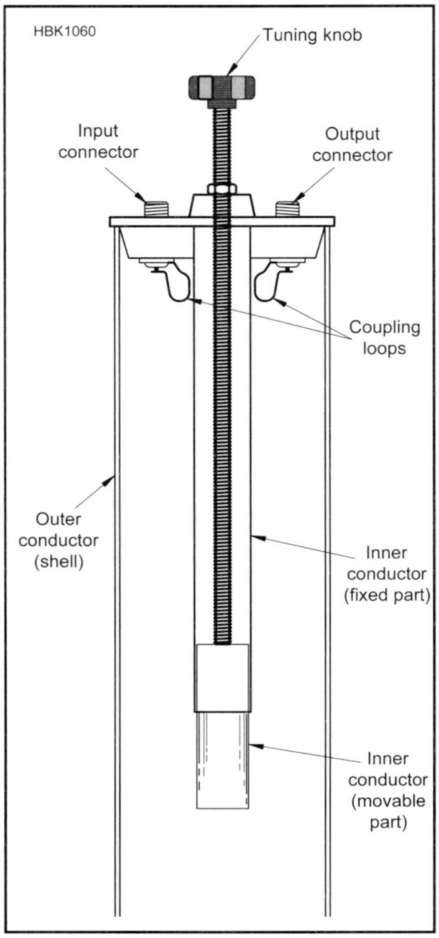

Figure 10.56 — Components of a ¼ λ coaxial cavity. The shorted end is at the top where connectors are located and the open end is effectively at the end of the inner conductor. The bottom of the cavity is covered by a conductive cap.

FIELD STRENGTH AND ORIENTATION

The energy in the cavity is in the form of an electromagnetic (EM) field. The electric component of the field (E) is parallel to the length of the cavity, as shown in **Figure 10.57**. The magnetic component of the field (H) is at right angles to the electric component, in concentric circles around the center conductor. The H component and the associated currents in the cavity walls are strongest at the shorted end of the cavity and weaker at the open end. The E component and its associated voltages across the cavity wall and center conductor are strongest at the open end and weakest at the shorted end. Both, however, are stronger near the center conductor and weaker near the outer conductor. The orientation of the field components is important when using loops and probes for coupling signals into and out of the cavity. It is also possible to create *overtone* or *harmonic* modes in which the E and H components take on a more complex structure in the cavity.

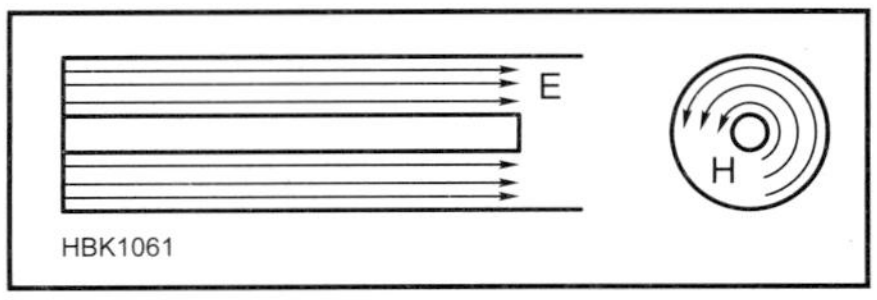

Figure 10.57 — Electric (E) and magnetic (H) field orientation.

CAVITY DIMENSIONS

The outer cylinder's diameter should not be larger than roughly ⅓ λ to avoid the overtone modes which result in higher losses. This means that the approximate diameter limit for 450 MHz cavities is roughly 8 inches and 25 inches for 2-meter cavities.

Optimum Q (for the sharpest filter response) is obtained for a ratio of 3.6 between the outer cylinder's inside diameter and the inner cylinder's outside diameter. This geometry results in the cavity having a characteristic impedance of roughly 77 Ω. The inner surfaces of the cavity are often silver plated for lower loss which also increases Q.

The resonant frequency is determined almost exclusively by the length of the center conductor which is approximately ¼ λ long. For the 70 cm band that is about 6 inches and for 2 meters roughly 19 – 20 inches. On 6 meters and 10 meters, these cavities become very large, making helical resonators (discussed in the following sections) a more practical choice. As shown in Figure 10.56, the center conductor is usually made so that its length is adjustable for tuning. (Note that the sliding contact surface must be carefully designed to avoid resistive losses.)

The length of the outer tube of a ¼ λ cavity has little effect on the tuning, so it is usually made a little longer than the center conductor for mechanical convenience in attaching a cap and in allowing room for adjusting the length of the inner (center) conductor.

10.9.2 Coupling to the Cavity

There are four practical ways to couple to a cavity: loops, probes, ports, and taps. All four coupling methods ultimately perform more or less the same after critical adjustments. However, loops are the most common choice since they are much easier to build and adjust.

LOOP COUPLING

A coupling loop acts as a single-turn coil excited from a connector mounted through the cavity wall. It is most often placed at the shorted end but may also be placed along the side of the cavity. The far end of the loop is connected to the cavity as in Figure 10.56. Loops are analogous to the link windings shown in Figure 10.55.

A loop couples to the H component of the field. Coupling is strongest when the loop is perpendicular to the H component. Since the H component, as we learned, lies in concentric circles around the center conductor, the loop is normally placed parallel to the length of the cavity and along the cavity's radius. It also couples best where the H component is strongest. This is near the shorted end and near the center conductor. Figure 10.56 shows loops in the maximum coupling position.

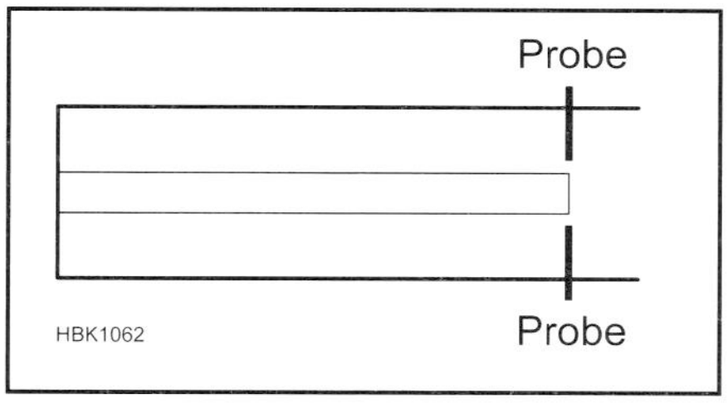

Figure 10.58 — Probe position for maximum coupling.

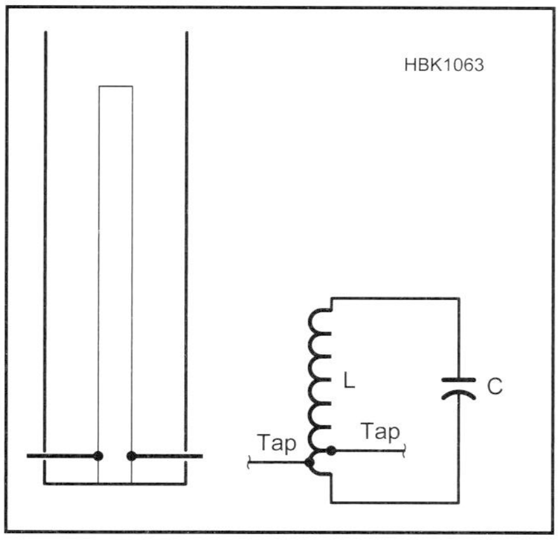

Figure 10.59 — Tap coupling and equivalent circuit.

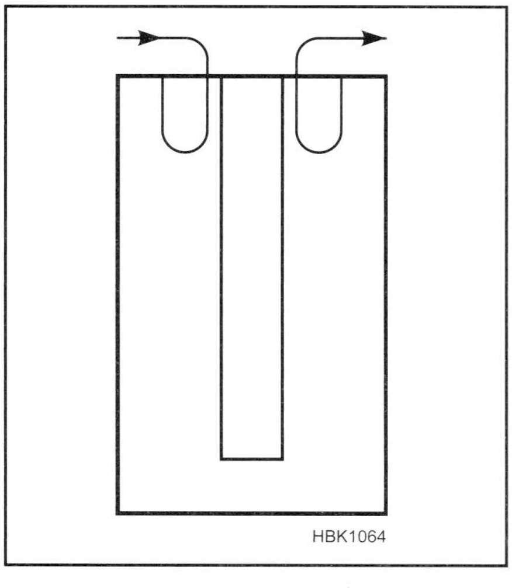

Figure 10.60 — Cavity in series with transmission line. This is a band-pass (BP) configuration.

PROBE COUPLING

A less frequently used coupling method is a probe as shown in **Figure 10.58**. The probe acts as one electrode of a capacitor that couples to the E field. The center conductor of the cavity acts as other half of the coupling capacitor. As opposed to a grounded loop, a probe is open at the end. A probe couples best when it is perpendicular to the E field and is placed where the E field is strongest. This is at the open end of the cavity, near the center conductor.

One disadvantage of probes is that they must be placed at the high voltage (open) end of the cavity. With probes, arcing is a potential problem, even at moderate power levels. In a good quality cavity, Q can easily reach 1,000. If 100 W in a 50 Ω feed line produces 71 V, a cavity with a Q of 1,000 connected to that feed line will experience 71 × 1,000 or 71 kV. Very high voltages can exist at the open end of a cavity where a probe is likely to be installed.

PORT COUPLING

A third way to couple energy, typically between adjacent cavities, is to cut a hole in the outer walls of both cavities to allow some of the field to leak through into the adjacent cavity. This is called *port coupling*. A number of duplexer designs successfully implement this method. Helical resonators, often found in receiver front ends, commonly use port coupling. It is economical and a space saver compared to loops or probes. Small mobile duplexers of rectangular cross section often use port coupling.

The main difficulty with port coupling is mechanical. Varying the position and coupling of loops and probes is easy. To change the amount of port coupling one must physically change the size of the port. This precludes easy experimentation. Since duplexer filters are often made of cylindrical tubing, a port between cavities is also not easy to fabricate and may not be practical for the home builder.

TAP COUPLING

The final method, illustrated in **Figure 10.59**, is *tap coupling*. On the left is an actual cavity with taps and on the right is the equivalent LC circuit with taps. By adjusting the position of the taps, one can achieve a good impedance match as well as efficient coupling. In small cavities, where loops could be too large for the space available, a tap is an easy way to obtain good coupling.

The disadvantage of tap coupling, however, is isolation. If the type of cavity you wish to use requires two ports, an input and an output tap, isolation between ports is difficult to achieve. Tap coupling finds its best application in single-port cavities, such as notch cavities. It is seldom used in band-pass cavities.

10.9.3 Cavity Filter Types

By selecting the coupling method and connection to the transmission line, as ¼ λ cavity can be configured to perform any of three basic filter types: band pass (BP), band-reject or notch (BR), and band-pass/band-reject (BP-BR). A BP cavity *passes* a small band of frequencies and all others are rejected. A BR cavity only rejects a small band of frequencies. All other frequencies are unaffected.

In the series configuration all of the energy

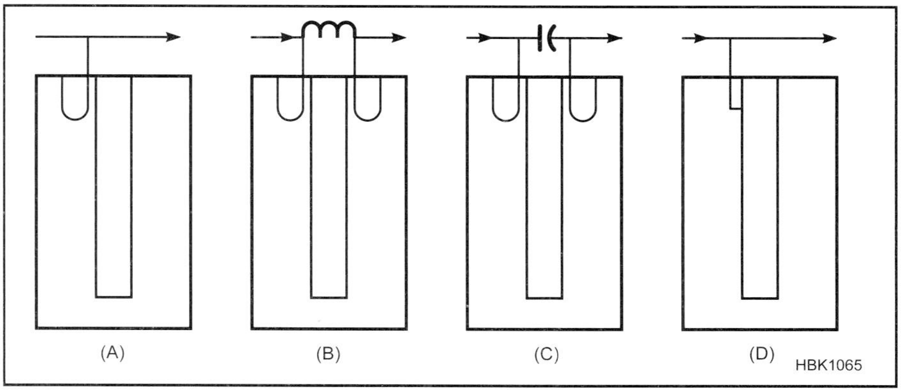

Figure 10.61 — Cavities in parallel with transmission line. These are band-reject or notch (BR) configurations. See text for explanation of different configuration responses.

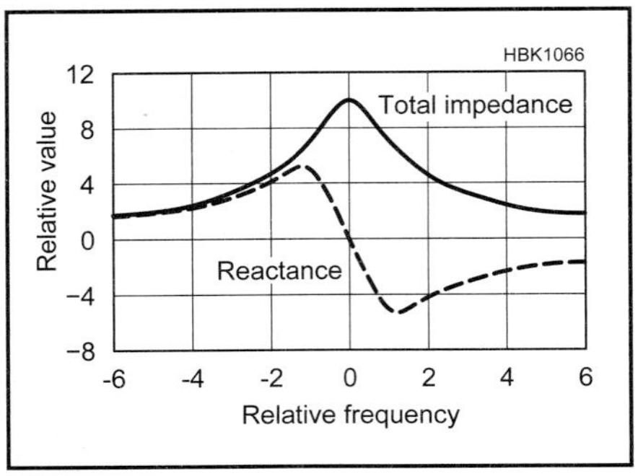

Figure 10.62 — Impedance and reactance of a parallel LC circuit.

passes through the cavity. It is coupled into the cavity by one loop and out by the other as in **Figure 10.60**. (Probes, ports, or taps could also be used.) In a parallel or shunt configuration as in **Figure 10.61**, energy is coupled into and out of the cavity so as to reinforce or cancel the energy in the transmission line.

To understand how each type of filtering is achieved, examine **Figure 10.62** which shows the impedance curve for a parallel LC circuit, the cavity's equivalent circuit. At resonance, impedance reaches a maximum. The absolute value depends on the Q of the cavity. Off resonance, like a parallel LC circuit, the cavity's total impedance is low.

BAND-PASS (BP) CAVITIES

To obtain a BP response, the cavity must be in series with the transmission line as in Figure 10.60. In this series configuration, all of the energy passes through the cavity. It is coupled into the cavity by one loop and out by the other loop. At the resonant frequency, the high-Q cavity readily accepts the energy supplied to it by the input loop. At the output loop the H field couples energy back into the transmission line. At resonance very little signal is lost. Essentially, to the energy in the transmission line, the cavity is invisible. Away from resonance, at a higher or lower frequency, the cavity couples less energy back into the transmission line. This suppresses off-frequency signals. A typical BP cavity frequency response is shown in **Figure 10.63A**. This cavity has a 2.5 MHz bandwidth at 50 dB below the peak response.

BAND-REJECT (BR) OR NOTCH CAVITIES

By placing the cavity in parallel with the line, as in Figure 10.61, we create a band-reject (BR) or notch cavity or shunt configuration. Parallel or shunt-connected cavities are sometimes called *suck out cavities*. A BR cavity is capable of many dB of signal rejection, far more than a BP cavity. The deep narrow notch shown in Figure 10.63B is why BR cavities are so useful in cavity duplexers.

In Figure 10.61A a single loop and in Figure 10.61D a single tap connects the cavity across or shunts the transmission line. At resonance, the open ¼ λ cavity returns energy to the transmission line through the loop, canceling energy in the line.

BP-BR CAVITIES

A third class of cavity filter has both BP and BR characteristics. In Figures 10.61B and C the cavity is connected across a coil or capacitor. All BR cavities using a shunt inductor or capacitor also have a band-pass response "bump" as shown in Figure 10.63B. The combination of the cavity's reactance (shown in Figure 10.62) in parallel with the reactance inserted into the transmission line by the inductor or capacitor create a high-Q resonance that blocks the energy in the line at the notch frequency which is offset from the cavity's natural resonance.

If a shunt inductor is used, as in Figure 10.61B, the notch will be above the cavity's natural resonance where its capacitive reactance resonates with the shunt inductive reactance. If a shunt capacitor is used, as in Figure 10.61C, the notch will be on the low frequency side. The amount of inductance or capacitance determines the separation of the notch and bump in the cavity's response. (If the band-pass bump is intentionally minimized, the cavity is called a *pure notch cavity*.)

To be used in a repeater duplexer, it is possible to configure the bump frequency to be above or below that of the notch. As in Figure 10.63B, if the transmitter frequency (the signal to be "notched out") is higher than the receiver frequency, such as for a 2 meter repeater with a transmit frequency of 146.97 MHz and a receive frequency of 146.37 MHz, the bump (the band-pass range) should be below the notch and vice versa. The bump is typically adjusted to be centered on the receive frequency.

For more detailed discussions of cavity operation and photos of a disassembled commercial cavity showing several of the techniques discussed here, see **www.repeater-builder.com/antenna/pdf/ve2azx-duplexer-info.pdf** or the online article by ZL1BTB at **www.amalgamate2000.com/radio-hobbies/radio/coaxial_resonators.htm**. The Repeater-Builders web page **www.repeater-builder.com/antenna/ant-sys-index.html#duplexers** offers a number of excellent articles, as well.

Building high-quality microwave cavities is illustrated in the *QEX* articles "Easy Microwave Filters Using Waveguides and Cavities" and "Very High Q Microwave Cavities and Filters" by Paolo Antoniazzi, IW2ACD, which are included with the online content.

10.9.4 Helical Resonators

Ever-increasing occupancy of the radio spectrum brings with it a parade of receiver overload and spurious responses. Overload problems can be minimized by using high-

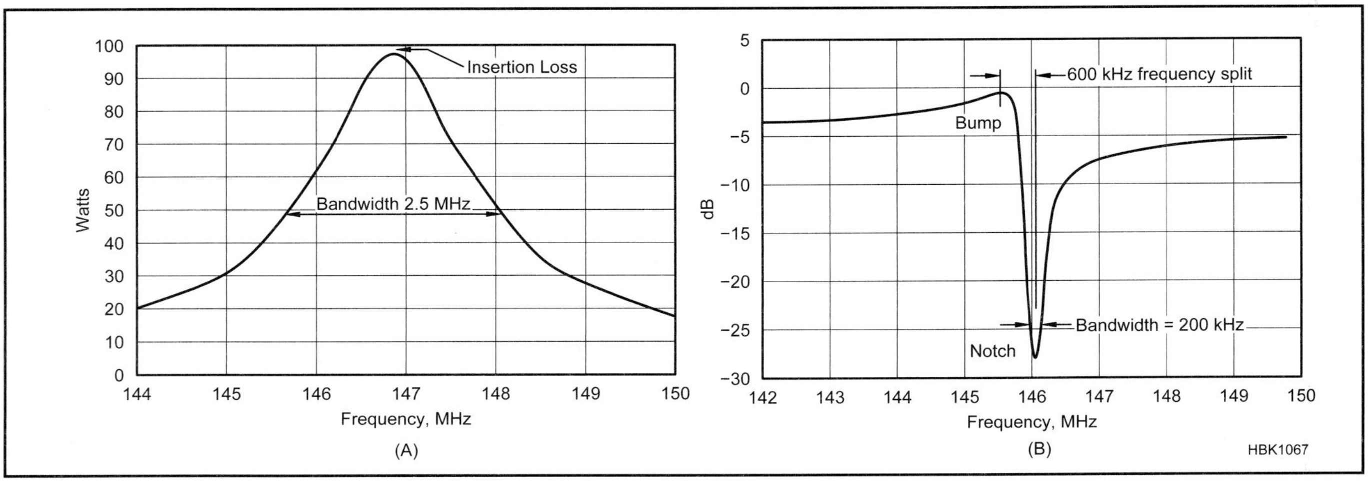

Figure 10.63 — Typical band-pass (BP) and notch (BR) response for 2 meter cavity filters.

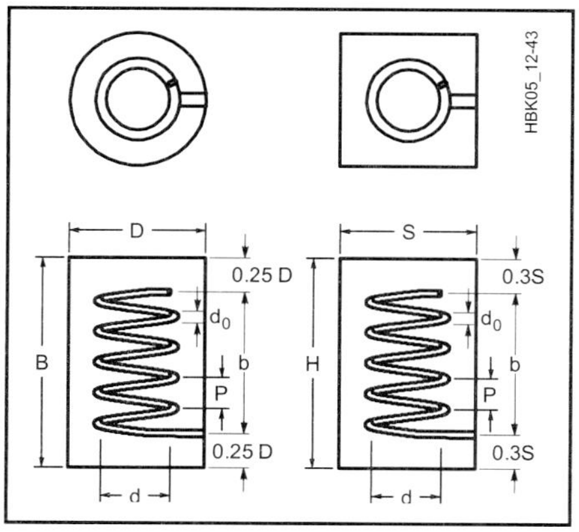

Figure 10.64 — Dimensions of round and square helical resonators. The diameter, D (or side, S) is determined by the desired unloaded Q. Other dimensions are expressed in terms of D or S (see text).

dynamic-range receiving techniques, but spurious responses (such as the image frequency) must be filtered out before mixing occurs. Conventional tuned circuits cannot provide the selectivity necessary to eliminate the plethora of signals found in most urban and many suburban environments. Other filtering techniques must be used.

Helical resonators are usually a better choice than ¼ λ cavities on 50, 144, and 222 MHz to eliminate these unwanted inputs because they are smaller and easier to build than coaxial cavity resonators, although their Q is not as high as that of cavities. In the frequency range from 30 to 100 MHz it is difficult to build high-Q inductors, and coaxial cavities are very large. In this frequency range the helical resonator is an excellent choice. At 50 MHz for example, a capacitively tuned, ¼ λ coaxial cavity with an unloaded Q of 3,000 would be about 4 inches in diameter and nearly 5 ft long. On the other hand, a helical resonator with the same unloaded Q is about 8.5 inches in diameter and 11.3 inches long. Even at 432 MHz, where coaxial cavities are common, the use of helical resonators results in substantial size reductions.

The helical resonator was described by the late Jim Fisk, W1HR, in a June 1976 *QST* article (see the Reference section). The resonator is described as a coil surrounded by a shield, but it is actually a shielded, resonant section of helically wound transmission line with relatively high characteristic impedance and low propagation velocity along the axis of the helix. The electrical length is about 94% of an axial ¼ λ or 84.6°. One lead of the helical winding is connected directly to the shield and the other end is open circuited as shown in **Figure 10.64**. Although the shield may be any shape, only round and square shields will be considered here.

10.9.5 Helical Resonator Design

The unloaded Q of a helical resonator is determined primarily by the size of the shield. For a round resonator with a copper coil on a low-loss form, mounted in a copper shield, the unloaded Q is given by

$$Q_U = 50D\sqrt{f_0}$$

where

D = inside diameter of the shield, in inches, and

f_0 = frequency, in MHz.

D is assumed to be 1.2 times the width of one side for square shield cans. This formula includes the effects of losses and imperfections in practical materials. It yields values of unloaded Q that are easily attained in practice. Silver plating the shield and coil increases the unloaded Q by about 3% over that predicted

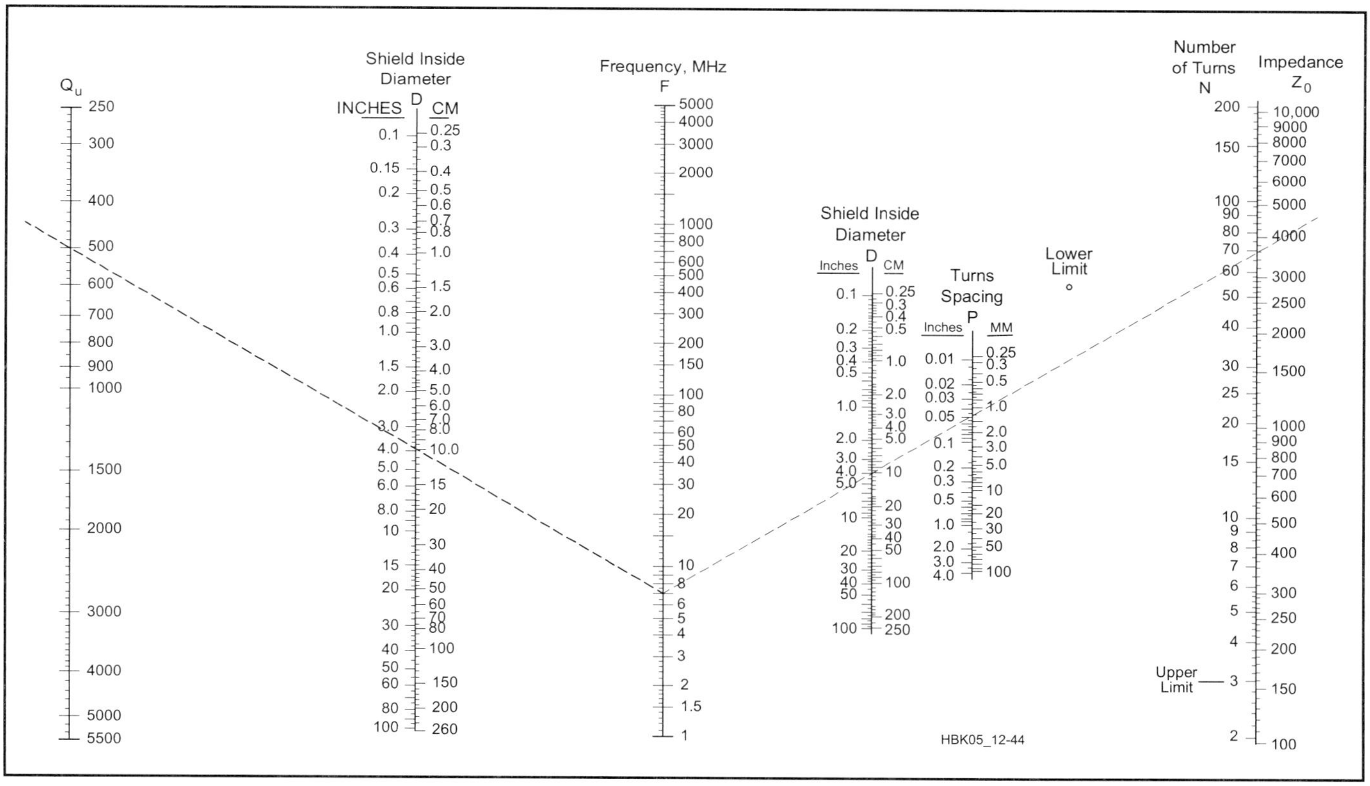

Figure 10.65 — The design nomograph for round helical resonators starts by selecting Q_U and the required shield diameter. A line is drawn connecting these two values and extended to the frequency scale (example here is for a shield of about 3.8 inches and Q_U of 500 at 7 MHz). Finally the number of turns, N, winding pitch, P, and characteristic impedance, Z_0, are determined by drawing a line from the frequency scale through selected shield diameter (but this time to the scale on the right-hand side. For the example shown, the dashed line shows P ≈ 0.047 inch, N = 70 turns, and Z_n = 3600 Ω).

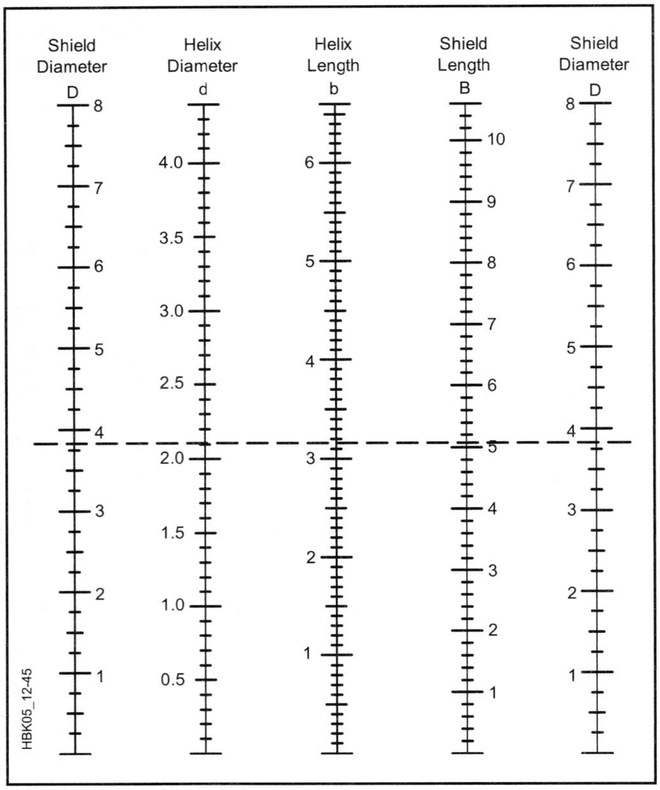

Figure 10.66 — The helical resonator is scaled from this design nomograph. Starting with the shield diameter, the helix diameter, d, helix length, b, and shield length, B, can be determined with this graph. The example shown has a shield diameter of 3.8 inches. This requires a helix mean diameter of 2.1 inches, helix length of 3.1 inches, and shield length of 5 inches.

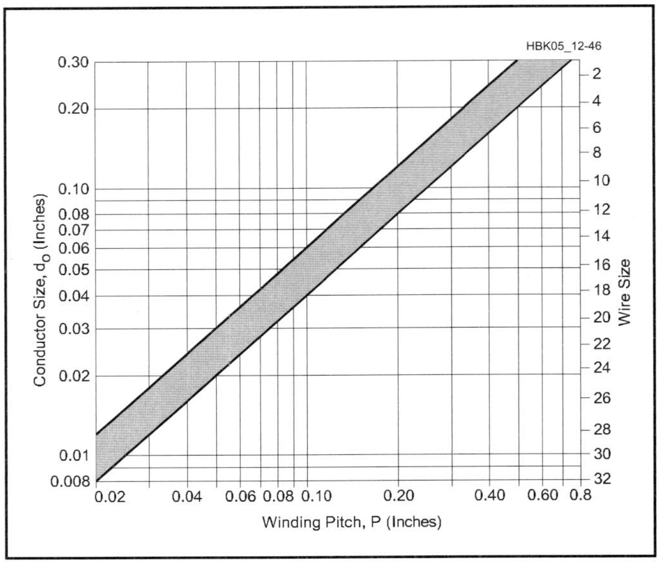

Figure 10.67 — This chart provides the design information of helix conductor size vs winding pitch, P. For example, a winding pitch of 0.047 inch results in a conductor diameter between 0.019 and 0.028 inch (#22 or #24 AWG).

by the equation. At VHF and UHF, however, it is more practical to increase the shield size slightly (that is, increase the selected QU by about 3% before making the calculation). The fringing capacitance at the open-circuit end of the helix is about 0.15D pF (that is, approximately 0.3 pF for a shield 2 inches in diameter). Once the required shield size has been determined, the total number of turns, N, winding pitch, P and characteristic impedance, Z_0, for round and square helical resonators with air dielectric between the helix and shield, are given by:

$$N = \frac{1908}{f_0 D}$$

$$P = \frac{f_0 D^2}{2312}$$

$$Z_0 = \frac{99{,}000}{f_0 D}$$

$$N = \frac{1590}{f_0 S}$$

$$P = \frac{f_0 S^2}{1606}$$

$$Z_0 = \frac{82{,}500}{f_0 S}$$

In these equations, dimensions D and S are in inches and f_0 is in megahertz. The design nomograph for round helical resonators in **Figure 10.65** is based on these formulas.

Although there are many variables to consider when designing helical resonators, certain ratios of shield size to length and coil diameter to length, provide optimum results. For helix diameter, d = 0.55 D or d = 0.66 S. For helix length, b = 0.825D or b = 0.99S. For shield length, B = 1.325 D and H = 1.60 S.

Design of filter dimensions can be done using the nomographs in this section or with computer software. The program *Helical* for designing and analyzing helical filters is available with the online content. Use of the nomographs is described in the following paragraphs.

Figure 10.66 simplifies calculation of these dimensions. Note that these ratios result in a helix with a length 1.5 times its diameter, the condition for maximum Q. The shield is about 60% longer than the helix — although it can be made longer — to completely contain the electric field at the top of the helix and the magnetic field at the bottom.

The winding pitch, P, is used primarily to determine the required conductor size. Adjust the length of the coil to that given by the equations during construction. Conductor size ranges from 0.4 P to 0.6 P for both round and square resonators and are plotted graphically in **Figure 10.67**.

Obviously, an area exists (in terms of frequency and unloaded Q) where the designer must make a choice between a conventional cavity (or lumped LC circuit) and a helical resonator. The choice is affected by physical shape at higher frequencies. Cavities are long and relatively small in diameter, while the length of a helical resonator is not much greater than its diameter. A second consideration is that point where the winding pitch, P, is less than the radius of the helix (otherwise the structure tends to be non-helical). This condition occurs when the helix has fewer than three turns (the "upper limit" on the design nomograph of Figure 10.65).

10.9.6 Helical Filter Construction

The shield should not have any seams parallel to the helix axis to obtain as high an unloaded Q as possible. This is usually not a problem with round resonators because large-diameter copper tubing is used for the shield, but square resonators require at least one seam and usually more. The effect on unloaded Q is minimized if the seam is silver soldered carefully from one end to the other.

Results are best when little or no dielectric is used inside the shield. This is usually no problem at VHF and UHF because the con-

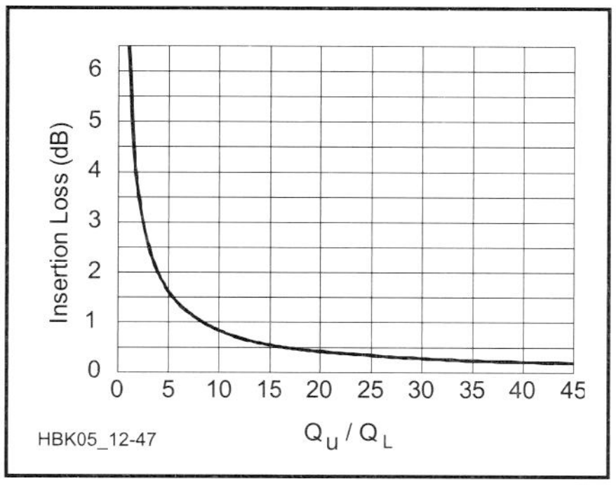

Figure 10.68 — The ratio of loaded (Q_L) to unloaded (Q_U) Q determines the insertion loss of a tuned resonant circuit.

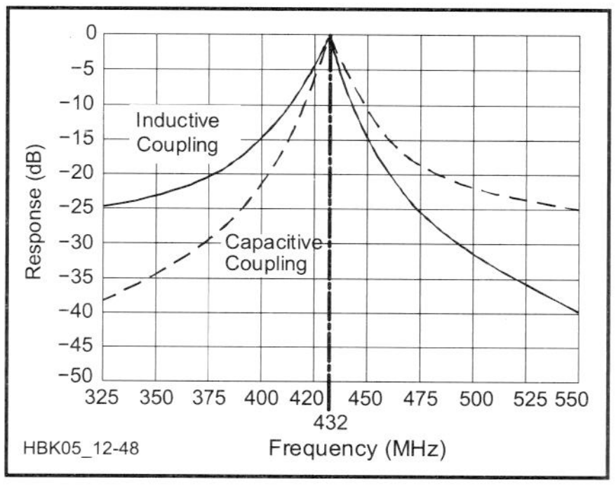

Figure 10.69 — This response curve for a single-resonator 432 MHz filter shows the effects of capacitive and inductive input/output coupling. The response curve can be made symmetrical on each side of resonance by combining the two methods (inductive input and capacitive output, or vice versa).

ductors are large enough that a supporting coil form is not required. The lower end of the helix should be soldered to the nearest point on the inside of the shield.

Although the external field is minimized by the use of top and bottom shield covers, the top and bottom of the shield may be left open with negligible effect on frequency or unloaded Q. Covers, if provided, should make electrical contact with the shield. In those resonators where the helix is connected to the bottom cover, that cover must be soldered solidly to the shield to minimize losses.

10.9.7 Helical Resonator Tuning

A carefully built helical resonator designed from the nomograph of Figure 10.65 will resonate very close to the design frequency. Slightly compress or expand the helix to adjust resonance over a small range. If the helix is made slightly longer than that called for in Figure 10.66, the resonator can be tuned by pruning the open end of the coil. However, neither of these methods is recommended for wide frequency excursions because any major deviation in helix length will degrade the unloaded Q of the resonator.

Most helical resonators are tuned by means of a brass tuning screw or high-quality air-variable capacitor across the open end of the helix. Piston capacitors also work well, but the Q of the tuning capacitor should ideally be several times the unloaded Q of the resonator. Varactor diodes have sometimes been used where remote tuning is required, but varactors can generate unwanted harmonics and other spurious signals if they are excited by strong, nearby signals.

When a helical resonator is to be tuned by a variable capacitor, the shield size is based on the chosen unloaded Q at the operating frequency. Then the number of turns, N and the winding pitch, P, are based on resonance at 1.5 f_0. Tune the resonator to the desired operating frequency, f_0.

10.9.8 Helical Resonator Insertion Loss

The insertion loss (dissipation loss), I_L, in decibels, of all single-resonator circuits is given by

$$I_L = 20 \log_{10}\left(\frac{1}{1-\frac{Q_L}{Q_U}}\right)$$

where

Q_L = loaded Q, and
Q_U = unloaded Q.

This is plotted in **Figure 10.68**. For the most practical cases ($Q_L > 5$), this can be closely approximated by $I_L \approx 9.0\ (Q_L/Q_U)$ dB. The selection of Q_L for a tuned circuit is dictated primarily by the required selectivity of the circuit. However, to keep dissipation loss to 0.5 dB or less (as is the case for low-noise VHF receivers), the unloaded Q must be at least 18 times the Q_L.

10.9.9 Coupling Helical Resonators

Signals are coupled into and out of helical resonators with inductive loops at the bottom of the helix, direct taps on the coil, or a combination of both. Although the correct tap point can be calculated easily, coupling by loops and probes must be determined experimentally.

The input and output coupling is often provided by probes when only one resonator is used. The probes are positioned on opposite sides of the resonator for maximum isolation. When coupling loops are used, the plane of the loop should be perpendicular to the axis of the helix and separated a small distance from the bottom of the coil. For resonators with only a few turns, the plane of the loop can be tilted slightly so it is parallel with the slope of the adjacent conductor.

Helical resonators with inductive coupling (loops) exhibit more attenuation to signals above the resonant frequency (as compared to attenuation below resonance), whereas resonators with capacitive coupling (probes) exhibit more attenuation below the pass-band, as shown for a typical 432 MHz resonator in **Figure 10.69**. Consider this characteristic when choosing a coupling method. The pass-band can be made more symmetrical by using a combination of coupling methods (inductive input and capacitive output, for example).

If more than one helical resonator is required to obtain a desired band-pass characteristic, adjacent resonators may be coupled through apertures in the shield wall between the two resonators. Unfortunately, the size and location of the aperture must be found empirically, so this method of coupling is not very practical unless you're building a large number of identical units.

Since the loaded Q of a resonator is determined by the external loading, this must be considered when selecting a tap (or position of a loop or probe). The ratio of this external loading, R_b, to the characteristic impedance, Z_0, for a ¼ λ resonator is calculated from:

$$K = \frac{R_b}{Z_0} = 0.785\left(\frac{1}{Q_L} - \frac{1}{Q_U}\right)$$

10.10 HF Transmitting Filters

Multi-transmitter stations operate transmitters and receivers in close proximity, either for contesting, special events, DXpeditions and Field Day operation, or emergency and disaster response communications. In such stations, band-pass filters (BPFs) are commonly used to reduce interstation interference from harmonics and broadband noise. Band-pass filters also make receivers less susceptible to overload from nearby transmitters. Remember that when trying to eliminate harmonics or noise, the filter must be installed at the *source*. Once transmitted and radiated, these spurious signals act as in-band signals at the receiver and cannot be filtered out.

This section is based on material from articles by Ethan Miller, K8GU; Edward Wetherhold, W3NQN; and Jeffrey Crawford, KØZR, referenced at the end of this section. The most popular HF BPF designs in use today are the 200 W designs by W3NQN. They can be built from individual components, as kits, and as complete units from several manufacturers. Complete design and building information are available in the two *QST* articles by W3NQN that are referenced at the end of this section. Another set of designs by James Tonne, W4ENE, for lower-power applications to meet the FCC emissions limits in §97.307 is provided in the Projects section of this chapter.

In addition, two publications are important complements to this section. The first is *Managing Interstation Interference* by George Cutsogeorge, W2VJN, available at no cost from Inrad at **www.vibroplex.com/techdocs/INRAD/MII_W2VJN.pdf**. The other is *Low-Band DXing* by John Devoldere, ON4UN. ON4UN's chapter on contest station engineering contains good information on filtering techniques.

10.10.1 Transmitting Filter Basics

There are two places in the station to install transmitting filters: Between the transceiver and amplifier and after the amplifier at the input to the antenna system. They operate in both the receive path and transmit paths (unless separate receive antennas are used). These filters reduce the level of transmitted harmonics and broadband noise and help to protect the receiver from nearby transmitters.

Transmitting filters can be constructed from transmission line stubs or discrete components. At high power levels stubs tend to be less expensive than discrete component band-pass filters. They also have the advantage that they can be designed to work well at harmonically related frequencies.

The two primary performance requirements for transmitting filters are for minimum insertion loss and maximum attenuation at the second harmonic, which is usually the strongest and most troublesome harmonic in a multistation environment along with wideband noise from the transmitter's oscillators and synthesizers.

In practice, filters built from discrete components are usually capable of greater attenuation than stubs although the power-handling capability of a stub-based filter may make it a more appropriate choice. For exciter power levels (i.e., <200 W), discrete filters tend to be quite economical. However, if the transceiver or antenna is not well matched to the filter, the effectiveness of the system is reduced. This situation is very much installation specific and difficult to evaluate in advance.

Insertion loss for transmitting filters should be significantly less than 1.0 dB which is equivalent to approximately 20 percent of the transmitted power. It is recommended to use a vector network analyzer (VNA — see the **Test Instruments and Measurements** chapter) when tuning or adjusting the filters. It is important to measure SWR into the filter, loss through the filter, and attenuation on adjacent bands. Measurements must be made with the filter cover in place to include the effects of component-to-enclosure proximity.

It is important that transmitting filters be operated with a low SWR at the output either looking into an amplifier input or into the antenna system. The filter capacitors are usually the most susceptible to failure by overheating when transmitting into a high SWR load. This can occur when transmitting into the "wrong" antenna or transmitting into the filter on the "wrong" band. An automatic antenna tuner can also cause filter failure by allowing full power to be input to the filter, even if on the wrong band or antenna. If possible, use the transceiver (and amplifier) band output data to automatically switch equipment and antennas to the "right" setting.

BUILDING TRANSMITTING FILTERS

Transmitting filters require sturdy components that can handle high current and high voltage, which increases their size and cost. At and above 100 W, toroidal inductors are generally used because of their self-shielding properties that allows the inductors to be physically close with little interaction, making for a more compact filter. As an example, **Figure 10.70** shows a W3NQN filter constructed by K8GU.

For power levels below 5 W, toroid core sizes of 0.5 inches outer diameter or less are commonly used. At power levels of 150 to 200 W, however, the core size must be much larger to dissipate the heat resulting from core and winding losses without excessive temperature rise. It was once assumed that core saturation was the primary limiting factor in high-power RF applications. However, excessive temperature increases resulting from losses in the winding and core material is the limiting factor. A temperature increase of less than 40 °C is preferred so that in an ambient of 90 °F (typical temperature for a hot day), the core temperature will be not more than 32 + 40 °C = 72 °C, or well below 90 °C.

To minimize time spent adjusting inductor values in the assembled filter, use an LC meter (see the **Test Equipment and Measurements** chapter) to adjust the inductor to the exact value before soldering it into the final assembly. Capacitors should be measured, as well. Capacitors can be combined in parallel to reach a final, exact value. This has the additional benefit of reducing current and heating in the capacitors.

Most transmitting filters at the 100 W level use high-voltage disc ceramic capacitors. To minimize the chance of failure caused by greater-than-anticipated voltages or currents, and to derate the standard dc voltage rating by 50% for RF applications, capacitors with dc ratings of 2, 3, and 4 kV are used. To minimize

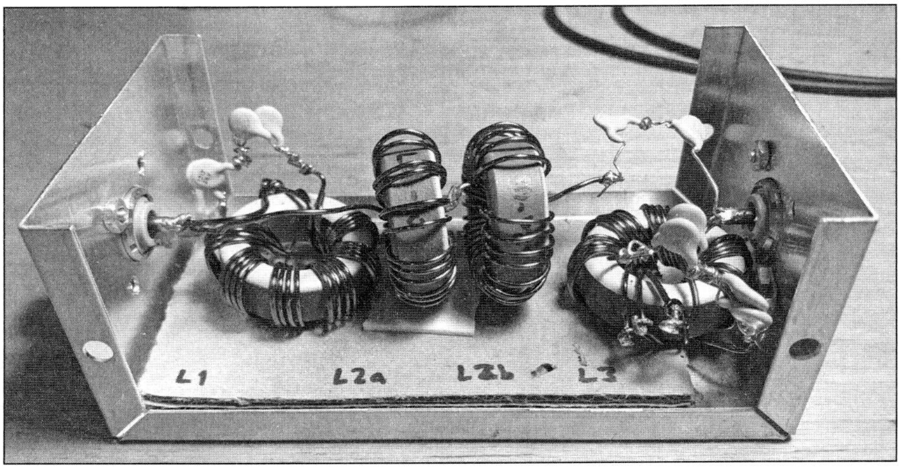

Figure 10.70 — K8GU band-pass Filter.

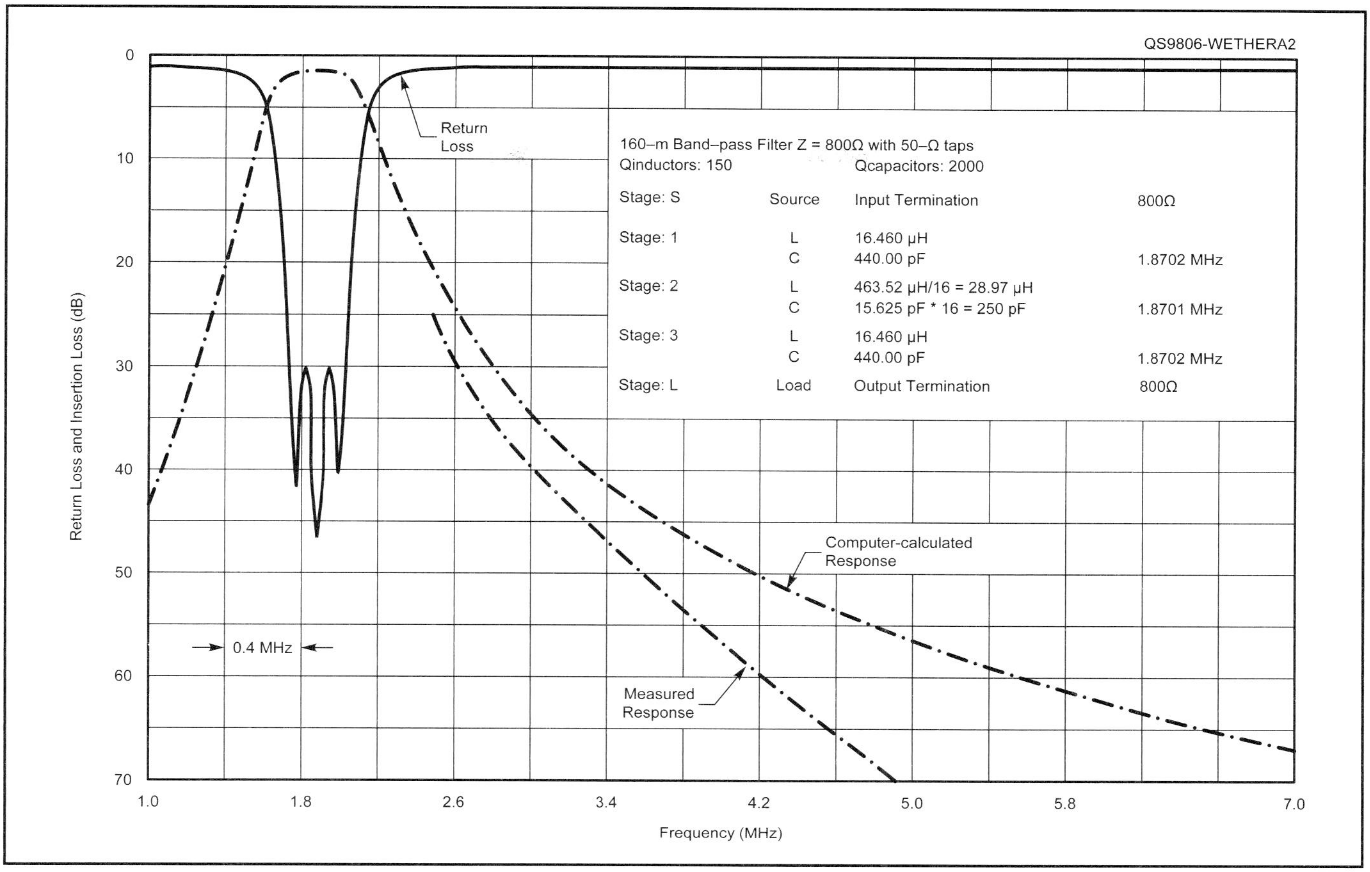

Figure 10.71 — Predicted and as-built performance of a W3NQN 160-meter BPF.

the temperature rise caused by several amps of current, two or more capacitors in parallel can be used. For high currents, CDV-16 capacitors from CDE are rated for 5 A of continuous current at HF.

Adjustable mica compression trimmers (usually made by Arco) were popular components for tunable filters but are no longer available except as surplus and new old stock (NOS). Trimmer capacitors available now are generally low-value (<50 pF, max.) and not rated above 200 V. High-quality Teflon or semi-rigid coax can also be substituted for low-value capacitors and trimmed to the proper value as described in the article "A Coaxial Alternative to low value SBH capacitors" by Bob Henderson, 5B4AGN, at **groups.io/g/TXBPF/files/A%20Coaxial%20Alternative%20to%20low%20value%20SBH%20capacitors.pdf**.

Once the filter has been constructed, the easiest way to complete tuning is to use either a spectrum analyzer with tracking generator or a VNA. Compare the actual behavior to the design's predicted value as in **Figure 10.71** which shows a W3NQN 160 meter BPF. The filter designer may also provide tuning instructions.

Figure 10.72 — A high-power 20 meter BPF assembly by KØZR.

10.10.2 High-Power Transmitting Filters

It is important in a high-power filter design to design carefully and leave little room for error. For example, consider a filter with a dissipative insertion loss of 0.3 dB. With 1500 W applied, 100 W will be dissipated in the filter. This heat will age the components more rapidly, may shift the filter's return loss as a function of duty cycle, and may unnecessarily cause thermal stresses to the filter, leading to possible premature failure. Heat is one of the enemies in high-power filter design.

The voltages and currents encountered in a high-power filter can be very high, especially when the filter is operated with loads far from 50 Ω. Circuit-analysis tools that show conditions for individual components should be used to perform "what if" calculations such as, "what happens to RF currents if the SWR were to be 2:1 instead of 1:1?"

Current-carrying inductors should be wound with #12 or #14 AWG Thermaleze-coated wire. Capacitors can be placed in series when higher breakdown voltages are required, and similarly, placed in parallel to increase the net current capacity of a given capacitor. The article on high-power filters by W2VJN presents a technique of using shorted coaxial stubs. **Figure 10.72** shows a high-power 20 meter BPF constructed by KØZR. Note that it is mounted on a metal plate with the input and output SO-239 connectors at the left and right.

Because of the extra heat dissipation in filters used at high power, it is also important to consider ventilation and cooling. Enclosures should have holes or slots that are big enough to allow for air circulation. After assembling a filter for the first time, it is recommended to do a heat-soak test at full power and measure the temperature of all current-carrying components.

10.10.3 Transmitting Filter References

Baker, A., KG4JJH, "Contest Band-pass Filters," **kg4jjh.com/pdf/Contest Bandpass Filters.pdf**.

Brown, J., K9YC, "Band-pass Filter Survey," Jim Brown, **www.audiosystemsgroup.com/Bandpass FilterSurvey.pdf**.

Crawford, J., KØZR, "High-Power HF Band-Pass Filter Design," *QEX*, Mar. 2018, pp. 3 – 10.

Cutsogeorge, G., W2VJN, "High-Power Harmonic Filters," *NCJ*, Nov. 1998, pp.14 – 15.

Miller, E., K8GU, "Selecting Band-Pass Filters and Switching Hardware," *NCJ*, Mar. 2008, pp. 29 – 31. Additional information at **www.k8gu.com/engineering/w3nqn-tx-filters**.

Wetherhold, E., W3NQN, "Clean Up Your Signals with Band-Pass Filters," *QST*, Part 1 May 1998; Part 2 Jun. 1998.

10.10.4 Diplexer Filter

This section, covering diplexer filters, was written by William E. Sabin, WØIYH. The diplexer is helpful in certain applications, such as frequency mixer terminations.

The terms "diplexer" and "duplexer" are often confused. A duplexer is a device that allows transmitters and receivers to share a common antenna on a single band. A duplexer can be a device such as a circulator or isolator (types of waveguide components that allow power to flow in only one direction) or filters (cavity or lumped-circuit). Diplexers are filters that enable antennas on different bands to be used by a transceiver through a common feed line. Diplexers can also be used to allow transceivers on different bands to share

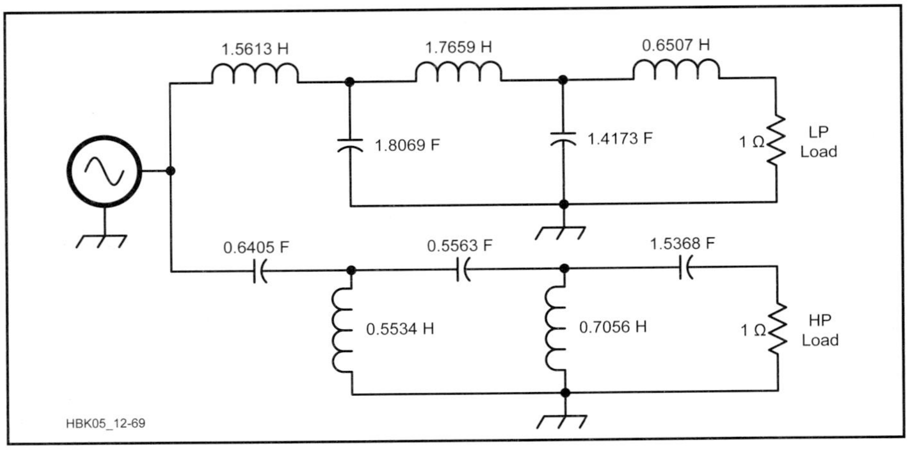

Figure 10.73 — Low-pass and high-pass prototype diplexer filter design. The low-pass portion is at the top, and the high-pass at the bottom of the drawing. See text.

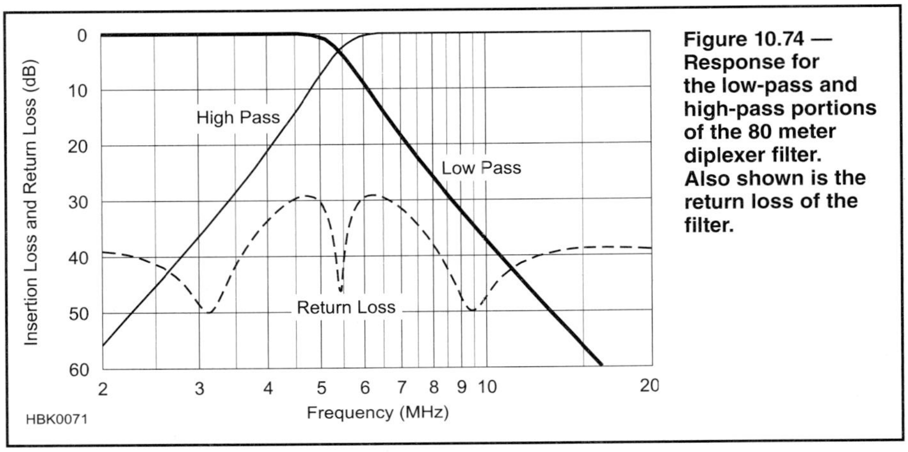

Figure 10.74 — Response for the low-pass and high-pass portions of the 80 meter diplexer filter. Also shown is the return loss of the filter.

a multiband antenna. A triplexer extends the diplexer to three bands.

It is important that a diplexer or triplexer have a constant filter-input resistance that extends to the stop band as well as the passband. Ordinary filters that become highly reactive or have an open or short-circuit input impedance outside the passband may degrade performance of the devices to which they are attached. For example, impedances far from 50 Ω outside the operating frequency range may cause an amplifier to become unstable and generate spurious emissions or oscillate.

Figure 10.73 shows a *normalized* prototype 5-element, 0.1-dB Chebyshev low-pass/high-pass (LP/HP) filter. This idealized filter is driven by a voltage generator with zero internal resistance, has load resistors of 1.0 Ω and a cutoff frequency of 1.0 radian per second (0.1592 Hz). The LP prototype values are taken from standard filter tables.[1] The first element is a series inductor. The HP prototype is found by:

a) replacing the series L (LP) with a series C (HP) whose value is 1/L, and

b) replacing the shunt C (LP) with a shunt L (HP) whose value is 1/C.

For the Chebyshev filter, the return loss is improved several dB by multiplying the prototype LP values by an experimentally derived number, K, and dividing the HP values by the same K. You can calculate the LP values in henrys and farads for a 50 Ω RF application with the following formulas:

$$L_{LP} = \frac{KL_{P(LP)}R}{2\pi f_{CO}};\ C_{LP} = \frac{KC_{P(LP)}}{2\pi f_{CO}R}$$

where

$L_{P(LP)}$ and $C_{P(LP)}$ are LP prototype values,
K = 1.005 (in this specific example),
R = 50 W, and
f_{CO} = the cutoff (–3 dB response) frequency in Hz.

For the HP segment:

$$L_{HP} = \frac{L_{P(HP)}R}{2\pi f_{CO}K};\ C_{HP} = \frac{C_{P(HP)}}{2\pi f_{CO}KR}$$

where $L_{P(HP)}$ and $C_{P(HP)}$ are HP prototype values.

Figure 10.74 shows the LP and HP responses of a diplexer filter for the 80 meter band. The following items are to be noted:

• The 3 dB responses of the LP and HP meet at 5.45 MHz.

• The input impedance is close to 50 Ω at all frequencies, as indicated by the high value of return loss (SWR <1.07:1).

• At and near 5.45 MHz, the LP input reactance and the HP input reactance are conjugates; therefore, they cancel and produce an almost perfect 50 Ω input resistance in that region.

• Because of the way the diplexer filter is derived from synthesis procedures, the transfer characteristic of the filter is mostly independent of the actual value of the amplifier dynamic output impedance.[2] This is a useful feature, since the RF power amplifier output impedance is usually not known or specified.

• The 80 meter band is well within the LP response.

• The HP response is down more than 20 dB at 4 MHz.

• The second harmonic of 3.5 MHz is down only 18 dB at 7.0 MHz. Because the second harmonic attenuation of the LP is not great, it is necessary that the amplifier itself be a well-balanced push-pull design that greatly rejects the second harmonic. In practice this is not a difficult task.

• The third harmonic of 3.5 MHz is down almost 40 dB at 10.5 MHz.

Figure 10.75A shows the unfiltered output of a solid-state push-pull power amplifier for the 80 meter band. In the figure you can see that:

• The second harmonic has been suppressed by a proper push-pull design.

• The third harmonic is typically only 15 dB or less below the fundamental.

The amplifier output goes through our diplexer filter. The desired output comes from the LP side, and is shown in Figure 10.75B. In it we see that:

• The fundamental is attenuated only about 0.2 dB.

• The LP has some harmonic content; however, the attenuation exceeds FCC requirements for a 100 W amplifier.

Figure 10.75C shows the HP output of the diplexer that terminates in the HP load or dump resistor. A small amount of the fundamental frequency (about 1%) is also lost in this resistor. Within the 3.5 to 4.0 MHz band, the filter input resistance is almost exactly the correct 50 W load resistance value. This is because power that would otherwise be reflected back to the amplifier is absorbed in the dump resistor.

Solid state power amplifiers tend to have stability problems that can be difficult to debug.[3] These problems may be evidenced by level changes in: load impedance, drive, gate or base bias, supply voltage, etc. Problems may arise from:

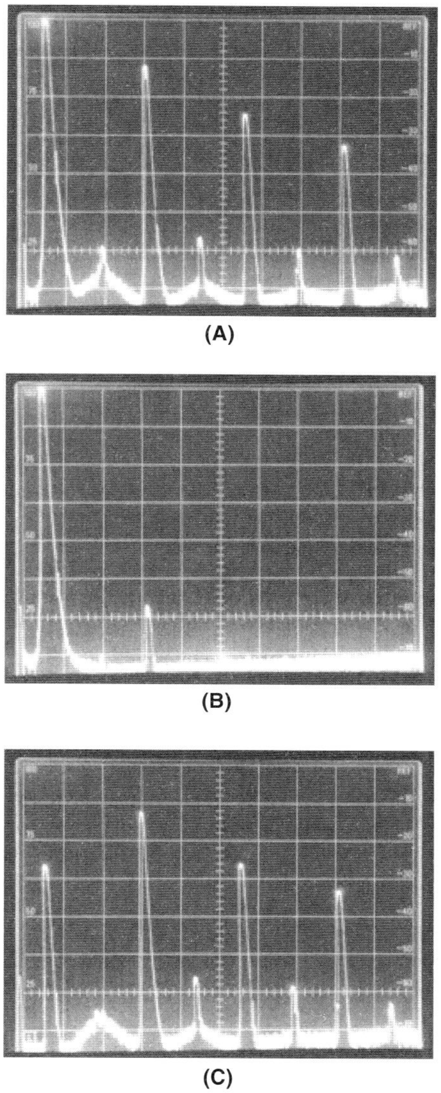

Figure 10.75 — At A, the output spectrum of a push-pull 80 meter amplifier. At B, the spectrum after passing through the low-pass filter. At C, the spectrum after passing through the high-pass filter.

• The reactance of the low-pass filter outside the desired passband. This is especially true for transistors that are designed for high-frequency operation.

• Self-resonance of a series inductor at some high frequency.

• A stop band impedance that causes voltage, current and impedance reflections back to the amplifier, creating instabilities within the output transistors.

Intermodulation performance can also be degraded by these reflections. The strong third harmonic is especially bothersome for these problems.

The diplexer filter is an approach that can greatly simplify the design process, especially for the amateur with limited PA-design experience and with limited home-lab facilities. For these reasons, the amateur homebrew enthusiast may want to consider this solution, despite its slightly greater parts count and expense.

The diplexer is a good technique for narrowband applications such as the HF amateur bands.[4] From Figure 10.74, we see that if the signal frequency is moved beyond 4.0 MHz the amount of desired signal lost in the dump resistor becomes large. For signal frequencies below 3.5 MHz the harmonic reduction may be inadequate. A single filter will not suffice for all the HF amateur bands.

This treatment provides you with the information to calculate your own filters. A *QEX* article has detailed instructions for building and testing a set of six filters for a 120 W amplifier. These filters cover all nine of the MF/HF amateur bands.[5]

You can use this technique for other filters such as Bessel, Butterworth, linear phase, Chebyshev 0.5, 1.0, etc.[6] However, the diplexer idea does *not* apply to the elliptic function types.

The diplexer approach is a resource that can be used in any application where a constant value of filter input resistance over a wide range of passband and stop band frequencies is desirable for some reason. Computer modeling is an ideal way to finalize the design before the actual construction. The coil dimensions and the dump resistor wattage need to be determined from a consideration of the power levels involved.

Another significant application of the diplexer is for elimination of EMI, RFI, and TVI energy. Instead of being reflected and very possibly escaping by some other route, the unwanted energy is dissipated in the dump resistor.[7]

See the discussion "Design Software for LC Filters" at the end of the Passive LC Filters section of this chapter. The software package provided by Jim Tonne, W4ENE, of Tonne Software (**www.tonnesoftware.com**) includes *Diplexer* which greatly simplifies the process of designing diplexer filters. The software is part of the online content for this book.

Notes

[1]Williams, A. and Taylor, F., *Electronic Filter Design Handbook*, any edition, McGraw-Hill.

[2]Storer, J., *Passive Network Synthesis* (McGraw-Hill, 1957), pp. 168 – 170. This book shows that the input resistance is ideally constant in the passband and the stop band and that the filter transfer characteristic is ideally independent of the generator impedance.

[3]Sabin, W. and Schoenike, E., *HF Radio Systems and Circuits* (Noble Publishing, 1998).

[4]Dye, N. and Granberg, H., *Radio Frequency Transistors, Principles and Applications* (Butterworth-Heinemann, 1993) p. 151.

[5]Sabin, W., WØIYH, "Diplexer Filters for the HF MOSFET Power Amplifier," *QEX*, July/Aug. 1999. Also check the ARRL website at **www.arrl.org/qex**.

[6]See note 1. *Electronic Filter Design Handbook* has LP prototype values for various filter types, and for complexities from 2 to 10 components.

[7]Weinrich, R. and Carroll, R., "Absorptive Filters for TV Harmonics," *QST*, Nov. 1968, pp. 10 – 25.

10.11 Filter Projects

The filter projects to follow are by no means the only filter projects in this book. Filters for specific applications may be found in other chapters of this *Handbook* and in its online content. Receiver input filters, transmitter filters, inter-stage filters and others can be extracted from the various projects and built for other applications. Since filters are a first line of defense against *electromagnetic interference* (EMI) problems, additional filter methods appear in the **RF Interference** chapter.

10.11.1 Optimized Harmonic Transmitting Filters

Low-pass filters should be placed at the output of transmitters to ensure that they meet the various regulatory agency requirements for harmonic suppression. These are commonly designed to pass a single amateur band and provide attenuation at harmonics of that band sufficient to meet the requirements. The material presented here by Jim Tonne, W4ENE, is based on material originally published in the September/October 1998 issue of *QEX*. Some 100 W and higher power band-pass filters are discussed in articles referenced by this chapter's previous section HF Transmitting Filters which are included in the online content.

The basic approach is to use a computer to optimize the performance in the passband (a single amateur band) while simultaneously maximizing the attenuation at the second and third harmonic of that same band. When this is done, the higher harmonics will also be well within spec.

The schematic of this filter along with parts values for the 3.5 to 4.0 MHz amateur band is shown in **Figure 10.76**. The responses of that filter are shown in **Figure 10.77**.

Component values for the 160-meter through the 6-meter amateur bands are shown in **Table 10.5**. The capacitors are shown in pF and the inductors in µH. The capacitors are the nearest 5% values; both the nearest 5% and the exact inductor values are shown.

Using the nearest 5% inductor values will result in satisfactory operation. If the construction method is such that exact-value (adjustable) inductors can be used then the "Exact" values are preferred. These values were obtained from the program *SVC Filter Designer* which is available with the online content.

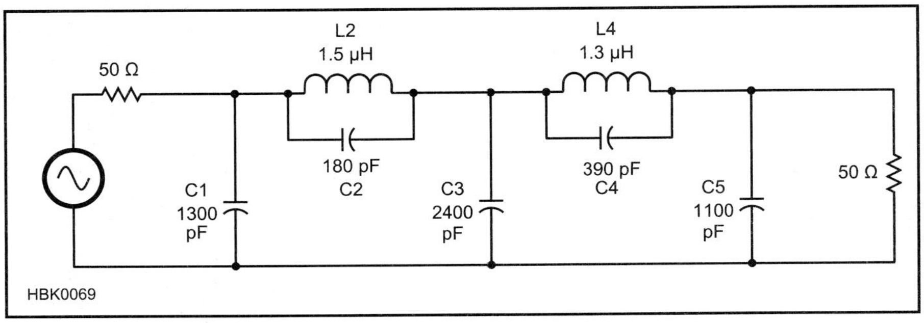

Figure 10.76 — Optimized low-pass filter. This design is for the 80 meter amateur band. It is similar to a Cauer design but the parts values have been optimized as described in the text and in the Sep/Oct 1998 issue of *QEX*.

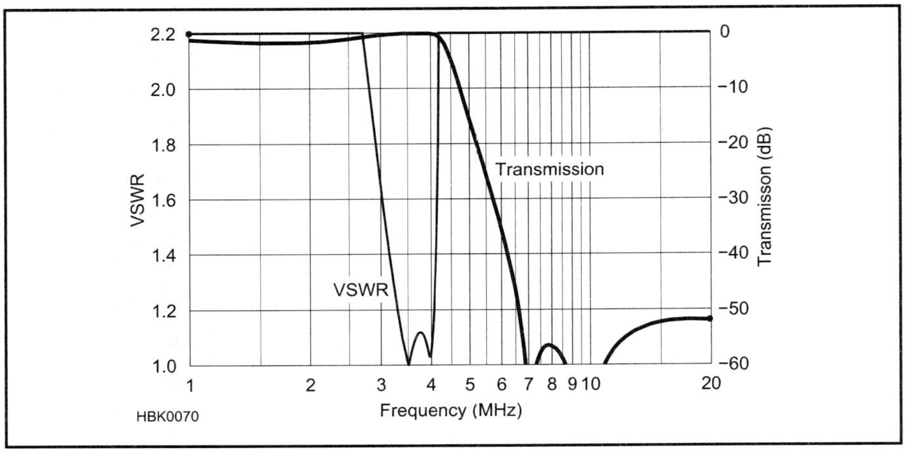

Figure 10.77 — Responses of the filter shown in Figure 10.71. Note the low values of SWR from 3.5 to 4 MHz. At the same time the harmonics are attenuated to meet regulations. Responses for the other amateur bands are very similar except for the frequency scaling.

Table 10.5

Values for the Optimized Harmonic Filters

Band (meters)	*C1 (pF)*	*L2, 5% (µH)*	*L2, Exact (µH)*	*C2 (pF)*	*C3 (pF)*	*L4, 5% (µH)*	*L4, Exact (µH)*	*C4 (pF)*	*C5 (pF)*
160	2400	3.0	2.88	360	4700	2.4	2.46	820	2200
80	1300	1.5	1.437	180	2400	1.3	1.29	390	1100
60	910	1.0	1.029	120	1600	0.91	0.8897	270	750
40	680	0.75	0.7834	91	1300	0.62	0.6305	220	560
30	470	0.56	0.5626	68	910	0.47	0.4652	160	430
20	330	0.39	0.3805	47	620	0.33	0.3163	110	300
17	270	0.30	0.3063	36	510	0.27	0.2617	82	240
15	220	0.27	0.2615	30	430	0.22	0.2245	68	200
12	200	0.24	0.241	27	390	0.20	0.2042	62	180
10	180	0.20	0.2063	24	330	0.18	0.1721	56	150
6	91	0.11	0.108	13	180	0.091	0.0911	30	82

10.11.2 Combline Filters for 50 – 432 MHz

(This project description is condensed from the paper "Combline Filters for VHF and UHF" by Paul Wade, W1GHZ, at the 41st Eastern VHF Conference in 2015. The full paper is available in the online content available with this book.)

RF pollution is rampant at good portable locations on mountaintops and other high places — anywhere accessible is populated with cell phone towers, TV and FM broadcast stations, two-way radio and pager transmitters, and even amateur repeaters. Most of these are high power, producing signals strong enough to seriously overload the VHF and UHF transceivers we use for contest operation or microwave liaison. The advent of broadband MMIC preamps acerbates the problem. The problem often manifests itself as a very high noise level.

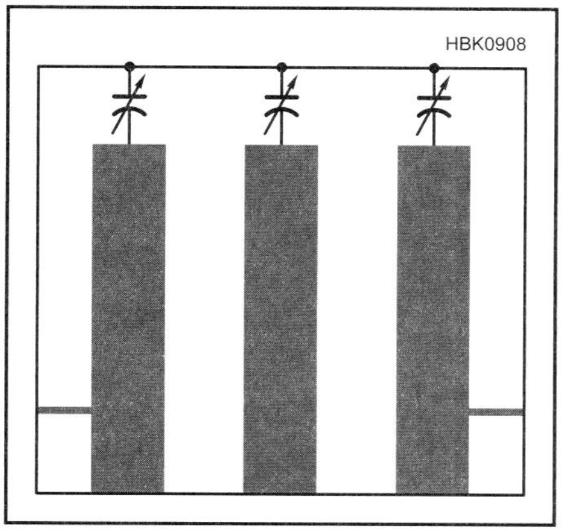

Figure 10.78 — Sketch of combline filter in stripline.

A combline filter uses parallel transmission line resonators less than a quarter-wave long, loaded by capacitance at the open end. This allows tuning over a range of frequencies by varying the capacitance. Typical electrical length of the resonators is between 30 and 60 electrical degrees; a quarter-wavelength is 90 degrees. These filters use stripline construction with tapped input and output coupling, as sketched in **Figure 10.78**, and the design procedure is given in the full paper.

A good, sharp filter must be mechanically robust to stay on frequency, especially for rover work. For low loss, high Q is important, requiring wide striplines with good contact to ground at the bottom, the high-current point. The filter uses all-aluminum construction to prevent dissimilar metal corrosion. All connections are made with #4 tinned solder lugs and stainless-steel hardware, metals that are least likely to interact with aluminum. An inexpensive aluminum enclosure of 220 × 145 × 60 mm was used for the 144 MHz filter in **Figure 10.79**. Dimensions for all of the filters are included in **Table 10.6**. Enclosure model numbers can be searched online to find sources for the boxes. A close match is sufficient.

The full paper includes construction guidelines, including selecting parts, metalworking, and assembly details. Tuning instructions are given and require the use of a sweep generator or a vector network analyzer can be used. The performance curves in **Figure 10.80** are indicative of completed filter performance. If that equipment is not available, the filters can be optimized at one frequency and used as narrowband filters.

10.11.3 Broadcast-Band Rejection Filters

Inadequate front-end selectivity or poorly performing RF amplifier and mixer stages often result in unwanted cross-talk and overloading from adjacent commercial or amateur stations. Two passive receive-only filters are described here — a high-pass, multi-section filter and a simple series wave trap.

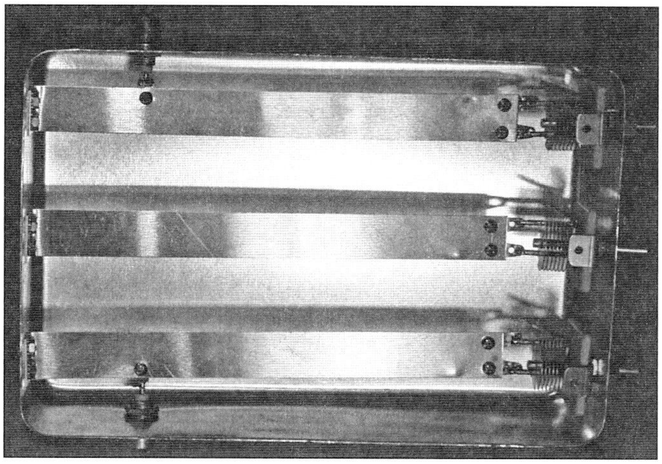

Figure 10.79 — Combline filter for 144 MHz.

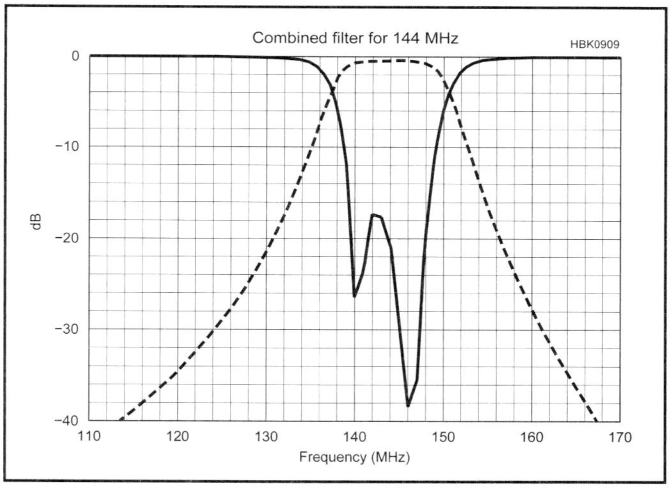

Figure 10.80 — Performance of 144 MHz combline filter. The solid line is Return Loss and the dashed line is Insertion Loss.

Table 10.6
Dimensions for Combline Filter in Stripline

Band (MHz)	*width (mm)*	*Bandwidth (MHz)*	*Strip spacing (mm)*	*BoxModel*	*Strip c to c (mm)*	*Length (mm)*	*Tap point (mm)*	*Width (mm)*	*Capacitor (pF)*	*Depth (mm)*	*Strip*
144	2.5	AC-406	9 in	7 in	2 in	33	44	77	22	24	
222	8	U3879 mid	202	129	54	34	40	74	30	15	
432	11	1590-BB	115	90	30	16	25	41	16	5	
432	13	U3879 sm	176	99	43	29	35	64	15	5	
50	3	AC-1418	8 in	10 in	2.5 in	30	40	70	90	150	

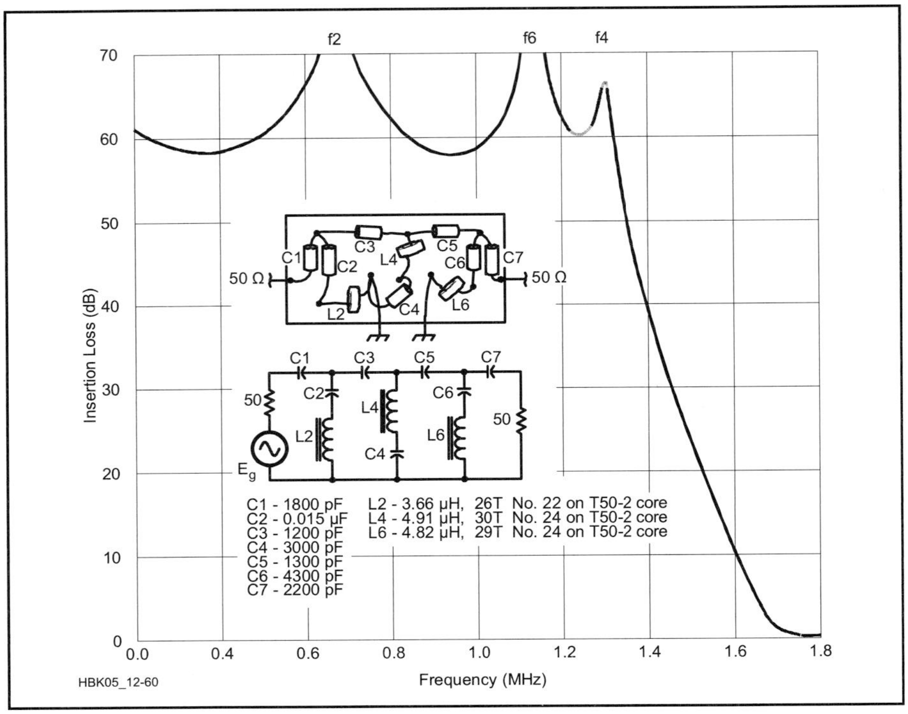

Figure 10.81 — Schematic, layout and response curve of the broadcast band rejection filter.

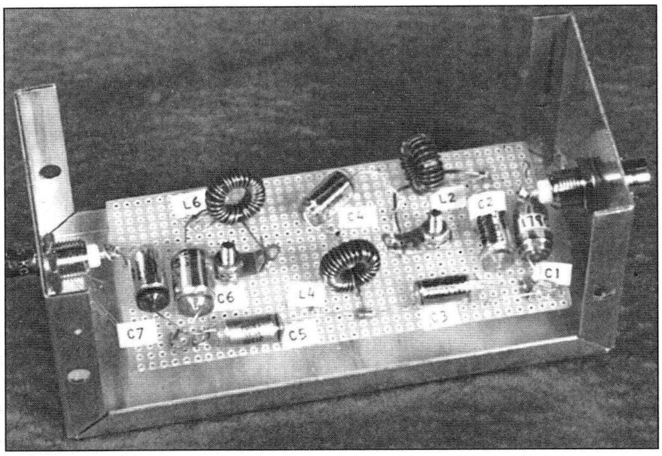

Figure 10.82 — The filter fits easily in a 2 × 2 × 5 inch enclosure. The version in the photo was built on a piece of perfboard.

BROADCAST-BAND REJECTION HIGH-PASS FILTER

The filter shown is inserted between the antenna and receiver. It attenuates the out-of-band signals from broadcast stations but passes signals of interest (1.8 to 30 MHz) with little or no attenuation.

The high signal strength of local broadcast stations requires that the stop-band attenuation of the high-pass filter also be high. This filter provides about 60 dB of stop-band attenuation with less than 1 dB of attenuation above 1.8 MHz. The filter input and output ports match 50 Ω with a maximum SWR of 1.353:1 (reflection coefficient = 0.15). A 10-element filter yields adequate stop-band attenuation and a reasonable rate of attenuation rise. The design uses only standard-value capacitors.

The filter parts layout, schematic diagram, response curve and component values are shown in **Figure 10.81**. The standard capacitor values listed are within 2.8% of the design values. If the attenuation peaks (f2, f4, and f6) do not fall at 0.677, 1.293, and 1.111 MHz, tune the series-resonant circuits by slightly squeezing or separating the inductor windings.

Construction of the filter is shown in **Figure 10.82**. Use polypropylene film-type capacitors. These capacitors are available through Digi-Key and other suppliers. The powdered-iron T50-2 toroidal cores are available through Amidon, Palomar Engineers, and others.

For a 3.4 MHz cutoff frequency, divide the L and C values by 2. (This effectively doubles the frequency-label values in Figure 10.81.) For the 80 meter version, L2 through L6 should be 20 to 25 turns each, wound on T50-6 cores. The actual turns required may vary one or two from the calculated values. Connect capacitors in parallel as needed to achieve the nonstandard capacitor values required for this filter.

The measured filter performance is shown in Figure 10.81. The stop-band attenuation is more than 58 dB. The measured cutoff frequency (less than 1 dB attenuation) is under 1.8 MHz. The measured passband loss is less than 0.8 dB from 1.8 – 10 MHz. Between 10 and 100 MHz, the insertion loss of the filter gradually increases to 2 dB. Input impedance was measured between 1.7 and 4.2 MHz. Over the range tested, the input impedance of the filter remained within the 37 to 67.7 Ω input-impedance window (equivalent to a maximum SWR of 1.353:1).

WAVE TRAP FOR BROADCAST STATIONS

Nearby medium-wave broadcast stations can sometimes cause interference to HF receivers over a broad range of frequencies. A wave trap can catch the unwanted frequencies and keep them out of your receiver.

The way the circuit works is quite simple. Referring to **Figure 10.83**, you can see that it consists essentially of only two components, a coil L1 and a variable capacitor C1. This series-tuned circuit is connected in parallel with the antenna circuit of the receiver. The characteristic of a series-tuned circuit is that the coil and capacitor have a very low impedance (resistance) to frequencies very close to the frequency to which the circuit is tuned. All other frequencies are almost unaffected. If the circuit is tuned to 1530 kHz, for example, the signals from a broadcast station on that frequency will flow through the filter to ground, rather than go on into the receiver. All other frequencies will pass straight into the receiver. In this way, any interference caused in the receiver by the station on 1530 kHz is significantly reduced.

This is a series-tuned circuit that is adjustable from about 540 kHz to 1600 kHz. It is built into a metal box, **Figure 10.84**, to shield it from other unwanted signals and is connected as shown in Figure 10.83. To make the inductor, first make a *former* by winding two layers of paper on the ferrite rod. Fix this in place with black electrical tape. Next, lay one end of the wire for the coil on top of the former, leaving about an inch of wire protruding beyond the end of the ferrite rod. Use several turns of electrical tape to secure the wire to the former. Now, wind the coil along the former, making sure the turns are in a single layer and close together. Leave an

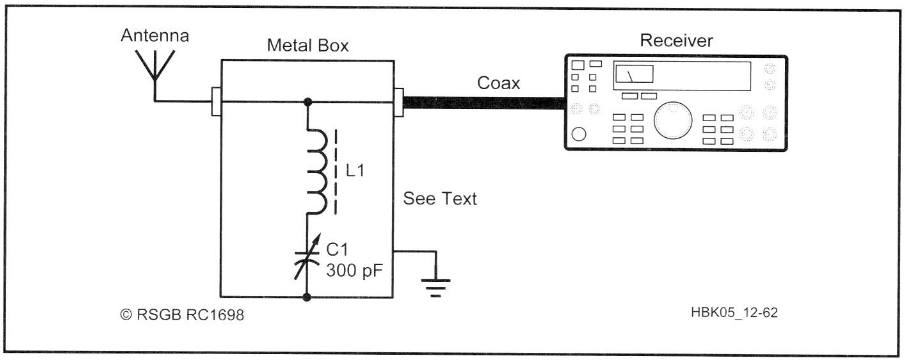

Figure 10.83 — The wave trap consists of a series tuned circuit, which "shunts" unwanted signals on its resonant frequency to ground.

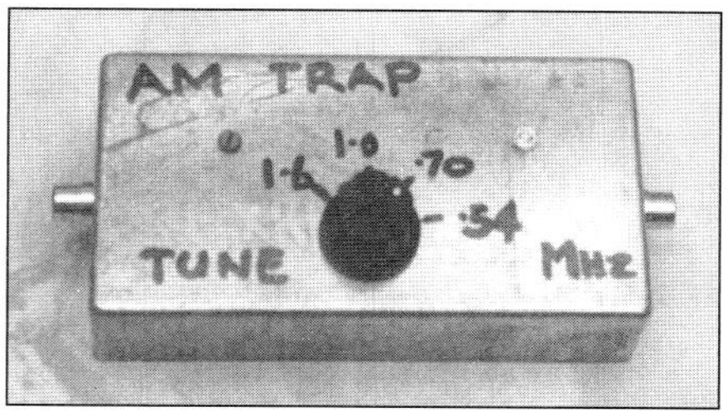

Figure 10.84 — The wave trap can be roughly calibrated to indicate the frequency to which it is tuned.

inch or so of wire free at the end of the coil. Once again, use a couple of turns of electrical tape to secure the wire to the former. Finally, remove half an inch of enamel from each end of the wire.

Alternatively, if you have an old AM transistor radio, a suitable coil can usually be recovered already wound on a ferrite rod. Ignore any small coupling coils. Drill the box to take the components, then fit them in and solder together as shown in **Figure 10.85**. Make sure the lid of the box is fixed securely in place, or the wave trap's performance will be adversely affected by pick-up on the components.

Connect the wave trap between the antenna and the receiver, then tune C1 until the interference from the offending broadcast station is a minimum. You may not be able to eliminate interference completely, but this handy little device should reduce it enough to listen to the amateur bands. Let's say you live near an AM transmitter on 1530 kHz, and the signals break through on your 1.8 MHz receiver. By tuning the trap to 1530 kHz, the problem is greatly reduced. If you have problems from more than one broadcast station, the problem needs a more complex solution.

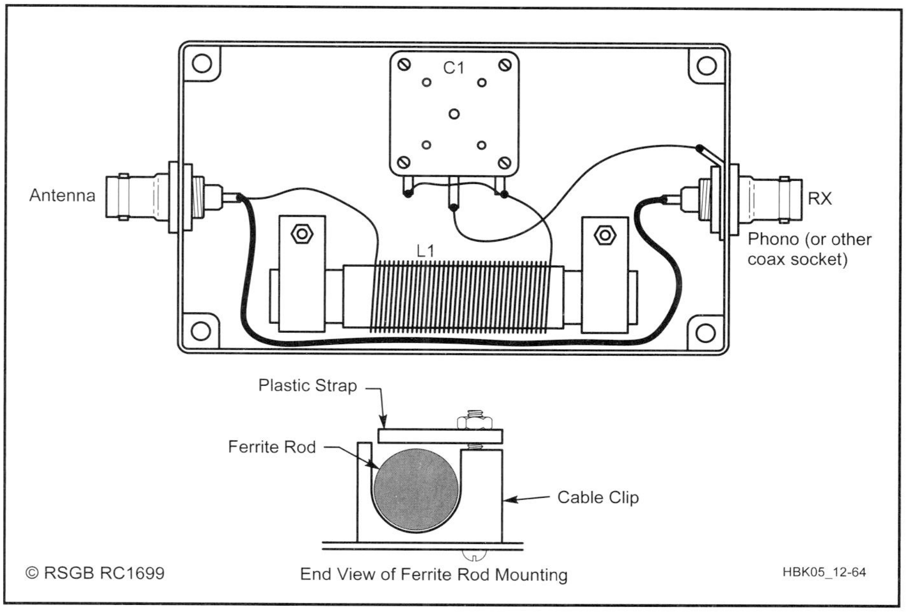

Figure 10.85 — Wiring of the wave trap. The ferrite rod is held in place with cable clips.
C1 — 300 pF polyvaricon variable.
L1 — 80 turns of 30 SWG enameled wire, wound on a ferrite rod.
Associated items: Case (die-cast box), knobs to suit, connectors to suit, nuts and bolts, plastic cable clips.

10.11.4 High-Performance, Low-Cost 1.8 – 54 MHz Low-Pass Filter

The low-pass filter shown in **Figure 10.86** offers low insertion loss, mechanical simplicity, easy construction and operation on all amateur bands from 160 through 6 meters. Originally built as an accessory filter for a 1,500 W 6-meter amplifier, the filter easily handles legal limit power. No complicated test equipment is necessary for alignment. It was originally described by Bill Jones, K8CU, in November 2002 *QST*. The complete original *QST* article for this project is included with the online content. The article supplies complete assembly and alignment drawings.

Although primarily intended for coverage of the 6-meter band, this filter has low

Figure 10.86 — The 1.8 to 54 MHz low-pass filter is housed in a die-cast box.

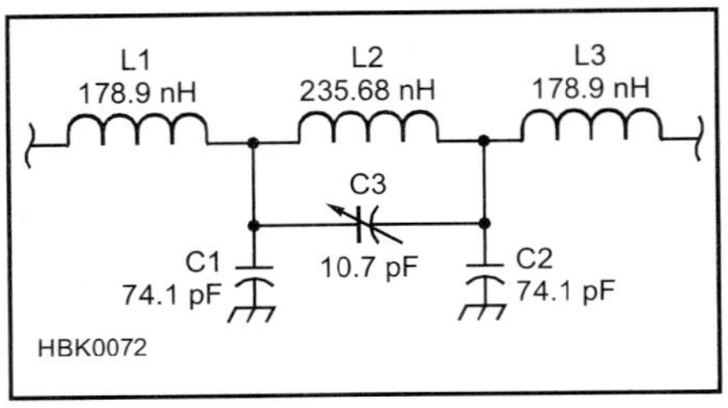

Figure 10.87 — The low-pass filter schematic. See construction details in the article with the online content.

C1, C2 — 74.1 pF. 2 × 2.65 inch brass plate sandwiched with 0.03125 inch thick Teflon sheet. The metal enclosure is the remaining grounded terminal of this capacitor.

C3 — Homemade variable using brass and Teflon.

L1, L3 — 178.9 nH. Wind with ⅛ inch OD soft copper tubing, 3.5 turns, 0.75 inch diameter form, 0.625 inch long, ¼ inch lead length for soldering to brass plate. The length of the other lead to RF connector as required.

L2 — 235.68 nH. Wind with ⅛ inch OD soft copper tubing, 5 turns, 0.75 inch diameter form, 1.75 inches long. Leave ¼ inch lead length for soldering.

insertion loss and presents excellent SWR characteristics for all HF bands. Although harmonic attenuation at low VHF frequencies near TV channels 2, 3, and 4 does not compare to filters designed only for HF operation, the use of this filter on HF is a bonus to 6 meter operators who also use the HF bands. Six meter operators may easily tune this filter for low insertion loss and SWR in any favorite band segment, including the higher frequency FM portion of the band.

ELECTRICAL DESIGN

The software tool used to design this low-pass filter is *Elsie* by Jim Tonne, W4ENE, which is available with the online content. The *Elsie* format data file for this filter, DC54.lct, may be downloaded for your own evaluation from the author's website at **www.realhamradio.com**.

Figure 10.87 is a schematic diagram of the filter. The use of low self-inductance capacitors with Teflon dielectric easily allows legal limit high power operation and aids in the ultimate stop band attenuation of this filter. Capacitors with essentially zero lead length will not introduce significant series inductance that upsets filter operation. This filter also uses a trap that greatly attenuates second harmonic frequencies of the 6 meter band. The parts list for the filter is given in **Table 10.7**.

PERFORMANCE DISCUSSION

Assuming the 6 meter SWR is set to a low value for a favorite part of the band, the worst case calculated forward filter loss is about 0.18 dB. The forward loss is better in the HF bands, with a calculated loss of only 0.05 dB from 1.8 through 30 MHz. The filter cutoff frequency is about 56 MHz, and the filter response drops sharply above this.

Figure 10.88 shows the calculated filter response from 1 to 1000 MHz. The impressive notch near 365 MHz is because of these inherent stray capacitances across each of the coils. Slight variations in each coil will make slightly different tuned traps. This will introduce a stagger-tuned effect that results in a broader notch.

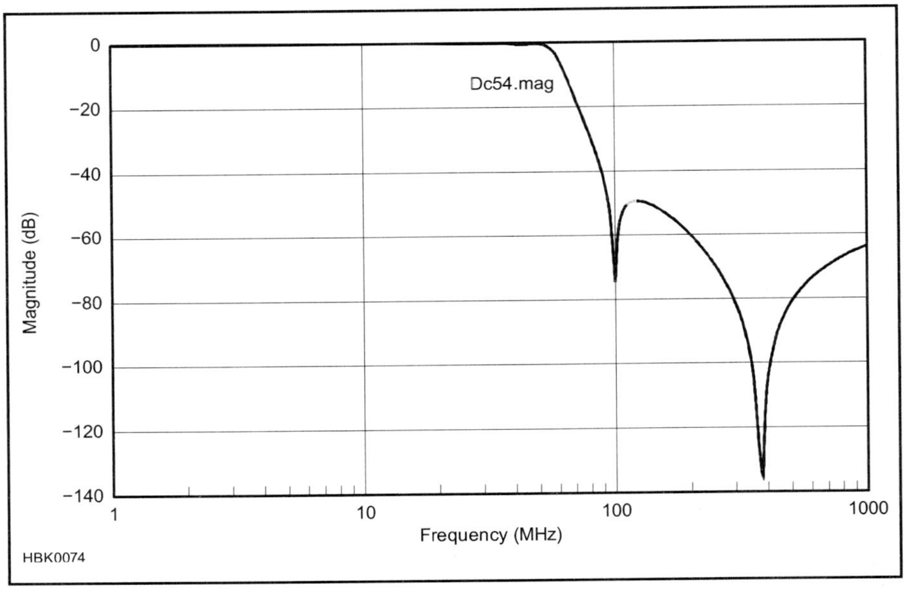

Figure 10.88 — Modeled filter response from 1 to 1000 MHz.

Table 10.7

Low-Pass Filter Parts List

Qty	*Description*
1	Miniature brass strip, 1 × 12 in., 0.032 in. thick (C3)
1	Miniature brass strip, 2 × 12 in., 0.064 in. thick (C1, C2)
5 ft	⅛ inch diameter soft copper tubing
4	¼ × 20 × ½ in. long hex head bolt
4	Plastic spacer or washer, 0.5 in. OD, 0.25 in. ID, 0.0625 in. thick
6	¼ × 20 hex nut with integral tooth lock washer
1	¼ × 20 × 4 in. long bolt
1	¼ × 20 threaded nut insert, PEM nut, or "Nutsert"
1	1 × 0.375 in. diameter nylon spacer. ID smaller than 0.25 in. (used for C3 plunger).
4	Nylon spacer, 0.875 in. OD, 0.25 to 0.34 in. ID, approx. 0.065 in. or greater thickness (used to attach brass capacitor plates).

Aluminum diecast enclosure is available from Jameco Electronics (www.jameco.com) part no. 11973. The box dimensions are 7.5 × 4.3 × 2.4 in. The 0.03125 in. thick Teflon sheet is available from McMaster-Carr Supply Co (www.mcmaster.com), item #8545K21 is available as a 12 × 12 in. sheet.

10.11.5 Band-Pass Filter for 145 MHz

The following project is based on a design from the RSGB *Radio Communication Handbook, 11th edition*. This filter is intended to reduce harmonics and other out-of-band spurious emissions when transmitting and suppress strong out-of-band incoming signals which could overload the receiver. The filter's schematic and response curve are shown in **Figure 10.89**. This filter design is not suitable for use as a repeater duplexer for in-band signals sharing a common antenna — a high-Q cavity resonator is required.

Direct inductive coupling from the proximity of the coils is used between the first two and the last two resonant circuits. C_C performs "top coupling" between the center two sections where a shield prevents stray

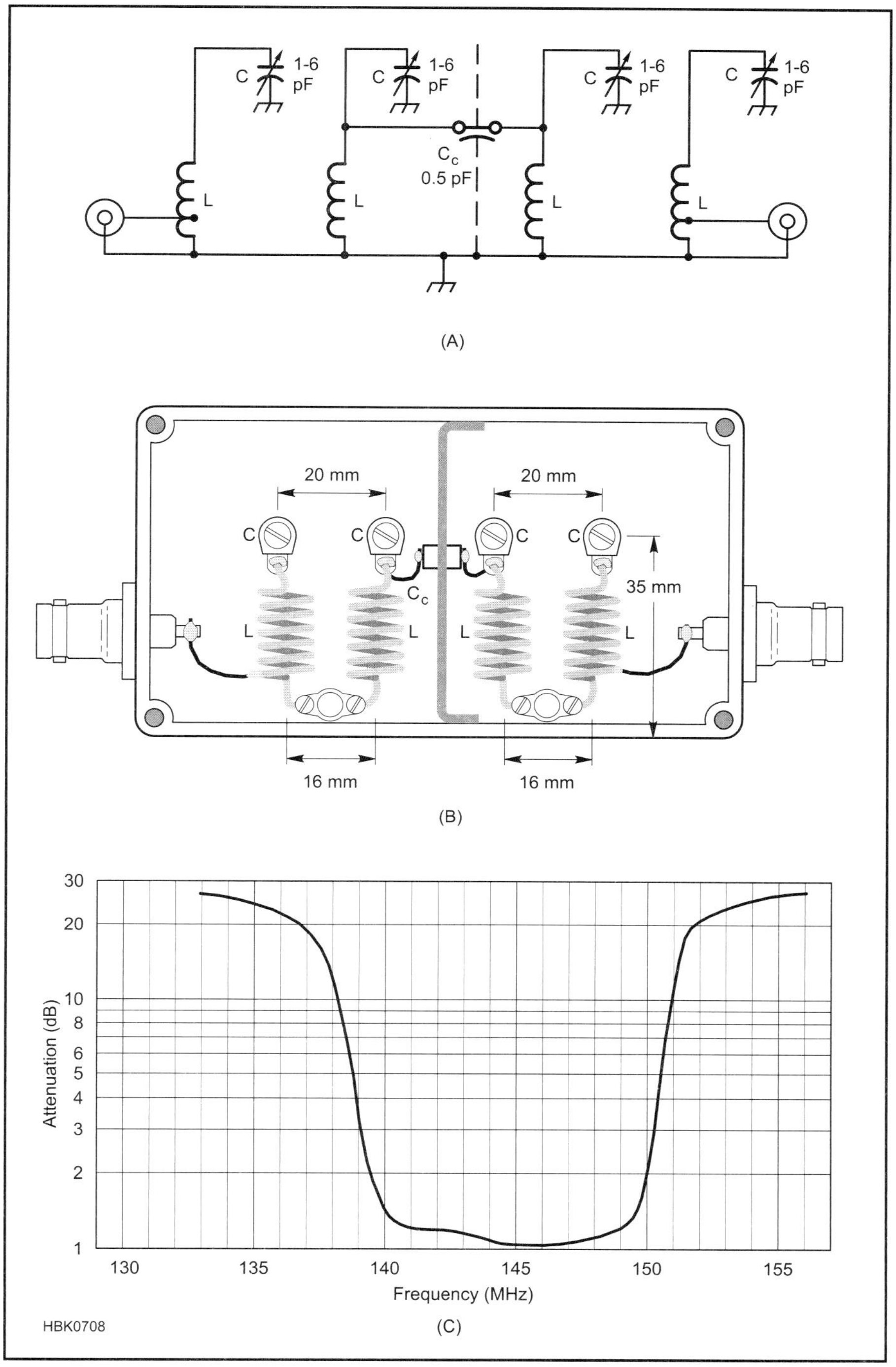

Figure 10.89 — A four-section band-pass filter for 145 MHz for attenuating strong out-of-band signals and reducing harmonics or other out-of-band spurious emissions.

coupling. The input and output connections are tapped down on their respective coils to transform the 50 Ω source and load into the proper impedances for terminating the filter

At VHF, self-supporting coils and mica, ceramic or air-dielectric trimmer capacitors give adequate results for most applications. Piston ceramic trimmers can be used for receiving or for low-power signals of a few watts. At higher power, use an air-variable capacitor. The filter is assembled in a 4 × 2½ × 1½ inch die-cast aluminum box so there is room for either piston or air-variable capacitors.

The band-pass filter is made from four parallel resonant circuits formed by the inductance of the coils and their associated parasitic interturn capacitance described in the **RF Techniques** chapter. A 1-6 pF series capacitor (C in Figure 10.89) connected to ground adjusts the resonant frequency of each coil. A ceramic capacitor should be used for C_C which has a value of 0.5 pF.

The dimensions of the coils are important to control the self-resonant frequency. Each coil is constructed from 6½ turns of #17 AWG solid, bare wire (1.15 mm dia), ⅜ inch (9.5 mm) in diameter. Each turn is spaced 1 wire diameter apart. The original design used British #18 SWG wire which is slightly thicker (1.22 mm) so coils made with #17 AWG wire will have a slightly higher resonant frequency. The taps for the coils are made one turn from the grounded end of the coil as shown in the figure.

Adjustment will be required after assembly to tune out the stray capacitances and inductances. A sweep generator and oscilloscope (see the **Test Equipment and Measurements** chapter) provide the most practical adjustment method. A variable oscillator with frequency counter and a voltmeter with RF probe, plus a good deal of patience, can also do the job.

10.12 References and Bibliography

References

"A Filter Primer," Maxim Integrated Tutorial 733, 2008. **pdfserv.maximintegrated.com/en/an/AN733.pdf**

Bottom, V., *Introduction to Quartz Crystal Unit Design* (Van Nostrand Reinhold Company, 1982).

Butterworth, S., "On the Theory of Filter Amplifiers," *Experimental Wireless and Wireless Engineer*, Oct. 1930, pp. 536 – 541.

Chirlian, P., "Laplace Transforms," *Basic Network Theory* (McGraw Hill, 1969).

Crawford, J. KØZR, "High-Power HF Band-Pass Filter Design," *QEX*, Mar. 2018, pp. 3 – 10.

Cutsogeorge, G., W2VJN, "High-Power Harmonic Filters" *NCJ*, Nov. 1998, pp. 14 – 15.

Darlington, S., "Synthesis of Reactance 4-Poles Which Produce Prescribed Insertion Loss Characteristics," *Journal of Mathematics and Physics*, Sep. 1939, pp. 257 – 353.

Dishal, M., "Top Coupled Band-Pass Filters," *IT&T Handbook*, 4th edition (American Book, Inc, 1956).

Donadio, M., "CIC Filter Introduction" **dspguru.com/dsp/tutorials/cic-filter-introduction**.

Fisk, J., W1DTY, "Helical-Resonator Design Techniques," *QST*, June 1976, pp. 11 – 14.

Geffe, P., *Simplified Modern Filter Design* (John F. Rider, 1963).

Gordon-Smith, D., G3UUR, "Extended Bandwidth Crystal Ladder Filters With Almost Symmetrical Responses," *QEX*, July/Aug. 2011, pp. 36 – 44.

"Group Delay Explanations and Applications," Applied Radio Labs, DN004, **www.radio-labs.com/wp/index.php/design-file/**.

Hayward, W., Campbell, R., and Larkin, R., *Experimental Methods in RF Design*, (ARRL, 2003).

Hogenauer, E., "An Economical Class of Digital Filters For Decimation and Interpolation," *IEEE Transactions on Acoustics, Speech and Signal Processing*, Vol. 29, Apr. 1981 pp.155 – 162.

Lyons, R., "Understanding Cascaded Integrator-Comb Filters," *EE Times*, Mar. 31, 2005.

Lyons, R., *Understanding Digital Signal Processing* 3rd Edition (Pearson, 2011).

Mack, R., W5IFS, "SDR: Simplified" Jan./Feb., Mar./Apr., May/June, July/Aug., and Nov./Dec. *QEX* 2009.

Metcalf, W., "Graphs Speed Interdigitated Filter Design," *Microwaves*, Feb. 1967.

Munson, D., "Fourier Waveform Analysis," *Reference Data for Engineers* 7th Edition, Jordan, E., Ed., (Howard W. Sams & Co., 1985).

Rhea, R., *HF Filter Design and Computer Simulation* (McGraw-Hill, 1995).

Saal, R. "Cauer Elliptic Filters" *The Design of Filters Using the Catalogue of Normalized Low-Pass Filters*, (AEH Telefunken, 1968).

Shepherd, F., "Modern-Network-Theory Design," *Reference Data for Engineers*, 7th Edition, Jordan, E., Ed., (Howard W. Sams & Co., 1985) pp. 9-5 – 9-11.

Silver, W., NØAX, "Hands-On Radio," *QST*, Experiment 50 Mar. 2007, Experiment 51 Apr. 2007, Experiment 87 Apr. 2010, Experiment 88 May 2010. **www.arrl.org/hands-on-radio**

Steder, H., DJ6EV, and Hardcastle, J., G3JIR, "Crystal Ladder Filters for All," *QEX*, Nov./Dec. 2009, pp. 14 – 18.

Weinberg, L., "Network Design by use of Modern Synthesis Techniques and Tables," *Proceedings of the National Electronics Conference*, 1956, Vol. 12.

White, D., *A Handbook on Electrical Filters: (Synthesis, Design and Applications)* (Don White Consultants, Inc., 1980).

Williams, A., *Electronic Filter Design Handbook* (McGraw-Hill, 1981).

Zobel, O., "Theory and Design of Electric Wave Filters," *Bell System Technical Journal*, Jan. 1923.

Zumbahlen, H., "Allpass Filters," Analog Devices, Mini-tutorial MT-202.

Zumbahlen, H., "The Bessel Response," Analog Devices, Mini-tutorial MT-204.

Zumbahlen, H., "Biquadratic (Biquad) Filters," Analog Devices, Mini-tutorial MT-205.

Zumbahlen, H., "The Butterworth Response," Analog Devices, Mini-tutorial MT-224.

Zumbahlen, H., "The Chebyshev Response," Analog Devices, Mini-tutorial MT-206.

Zumbahlen, H., "F_0 and Q in Filters," Analog Devices, Mini-tutorial MT-210.

Zumbahlen, H., "Low-Pass to Band-Pass Filter Transformation," Analog Devices, Mini-tutorial MT-215.

Zumbahlen, H., "Low-Pass to Band-Stop (Notch) Filter Transformation," Analog Devices, Mini-tutorial MT-216.

Zumbahlen, H., "Low-Pass to High-Pass Filter Transformation," Analog Devices, Mini-tutorial MT-217.

Zumbahlen, H., "Multiple Feedback Filters," Analog Devices, Mini-tutorial MT-220.

Zumbahlen, H., "Passive and Active Filtering," Analog Devices, AN-281.

Zumbahlen, H., "Sallen-Key Filters," Analog Devices, Mini-tutorial MT-222.

Zumbahlen, H., "State Variable Filters," Analog Devices, Mini-tutorial MT-223.

Zumbahlen, H., "Twin T Notch Filter," Analog Devices, Mini-tutorial MT-225.

Zumbahlen, H., "Using the Analog Devices Active Filter Design Tool," Analog Devices, AN-649, **https://www.analog.com/media/en/technical-documentation/app-notes/an-649.pdf**.

Zverev, A., *Handbook of Filter Synthesis* (John Wiley and Sons, 1967).

Practical Filter Designs

Antoniazzi, P., IW2ACD, and Arecco, M., IK2WAQ, "Easy Microwave Filters Using Waveguides and Cavities," *QEX*, Sep./Oct. 2006, pp. 37 – 42.

Bartlett, B., VK4UW, and Loftus, J., VK4EMM, "Band-Pass Filters for Contesting," *National Contest Journal*, Jan. 2000, pp. 11 – 15.

Cefalo T., WA1SPI, "Diplexers, Some Practical Applications," *Communications Quarterly*, Fall 1997, pp. 19 – 24.

Cope, P., W2GOM/7, "The Twin-T Filter," *QEX*, July/Aug. 1998, pp. 45 – 49.

Fisher, R., W2CQH, "Combline V.H.F. Band-pass Filters," *QST*, Dec. 1968, pp. 44 – 45.

Fisher, R., W2CQH, "Interdigital Band-pass Filters for Amateur VHF/UHF Applications," *QST*, Mar. 1968, p. 32.

Gordon-Smith, D., G3UUR, "Seventh-Order Unequal-Ripple Low-pass Filter Design," *QEX*, Nov./Dec. 2006, pp. 31 – 34.

Gordon-Smith, D., G3UUR, "Fifth-Order Unequal-Ripple Low-pass Filter Design," *QEX*, Nov./Dec. 2010, pp. 42 – 47.

Hayward, W., W7ZOI, "Extending the Double-Tuned Circuit to Three Resonators," *QEX*, Mar./Apr. 1998, pp. 41 – 46.

Heemstra, F., KT3J, "A Double-Tuned Active Filter with Interactive Coupling," *QEX*, Mar./Apr. 1999, pp. 25 – 29.

Jansson, D., WD4FAB, and Wetherhold, E., W3NQN, "High-Pass Filters to Combat TVI," *QEX*, Feb. 1987, pp. 7 – 8 and 13.

Lau, Z., W1VT, "A Narrow 80-Meter Band-Pass Filter," *QEX*, Sep./Oct. 1998, p. 57.

Lumachi, R., WB2CQM, "How to Silverplate RF Tank Circuits," *73 Amateur Radio Today*, Dec. 1997, pp. 18 – 23.
Moliere, T., DL7AV, "Band-Reject Filters for Multi-Multi Contest Operations," *CQ Contest*, Feb. 1996, pp. 14 – 22.
Rahe, W., DC8NR, "High Performance Audio Speech Low-Pass and CW Band-Pass Filters in SVL Design," *QEX*, July/Aug. 2007, pp. 31 – 39.
Sabin, W., WØIYH, "Diplexer Filters for the HF MOSFET Power Amplifier," *QEX*, July/Aug. 1999, pp. 20 – 26.
Sabin, W., WØIYH, "Narrow Band-Pass Filters for HF," *QEX*, Sep./Oct. 2000, pp. 13 – 17.
Sutcliffe, VK3TCT, "A BPSK band-pass filter," *Amateur Radio*, May 2012, pp. 44 – 46.
Tonne, J., WB6BLD, "Harmonic Filters, Improved," *QEX*, Sep./Oct. 1998, pp. 50 – 53.
Wetherhold, E., W3NQN, "Band-Stop Filters for Attenuating High-Level Broadcast-Band Signals," *QEX*, Nov. 1995, pp. 3 – 12.
Wetherhold, E., W3NQN, "Clean Up Your Signals with Band-Pass Filters," *QST*, Part 1, May 1998; Part 2, June 1998.
Wetherhold, E., W3NQN, "CW and SSB Audio Filters Using 88-mH Inductors," *QEX*, Dec. 1988, pp. 3 – 10.
Wetherhold, E. "CW Filter Combines High Performance and Low Cost" *Radio Handbook*, 23rd Edition, Orr, W., Ed., (Howard W. Sams and Co., 1987) p. 13-4.
Wetherhold, E., W3NQN, "A CW Filter for the Radio Amateur Newcomer," *Radio Communication*, Jan. 1985, pp. 26 – 31.
Wetherhold, E., W3NQN, "Modern Design of a CW Filter Using 88 and 44-mH Surplus Inductors," *QST*, Dec 1980, pp. 14 – 19. See also Feedback in Jan. 1981 *QST*, p. 43.
Wetherhold, E., W3NQN, "Receiver Band-Pass Filters Having Maximum Attenuation in Adjacent Bands," *QEX*, July/Aug. 1999, pp. 27 – 33.
Wetherhold, E., W3NQN, "Second-Harmonic-Optimized Low-Pass Filters," *QST*, Feb. 1999, pp. 44 – 46.

Contents

Chapter 11 — Online Content

Articles

- About FM by Ward Silver, NØAX
- About SSB by Ward Silver, NØAX
- Emissions Designator Table
- Modulation Glossary
- SDR Simplified — Introduction to I and Q by Ray Mack, W5IFS

Chapter 11
Modulation

Radio amateurs use a wide variety of modulations to convey information. This chapter, updated from Alan Bloom, N1AL's original material by Doug Grant, K1DG, explores various characteristics of modulation commonly used by amateurs. Traditional modulation types used for analog signals are discussed, as well as techniques suited for digital transmissions. Image modulations, updated for fast-scan ATV by Jim Andrews, KH6HTV, are also included, as well as pulse modulation as used by amateurs. Practical methods of using these modulations are presented in the chapters on **Receiving** and **Transmitting.** The chapter concludes with a discussion of modulation impairments and suggestions for additional reading.

11.1 Introduction

The purpose of amateur radio transmissions is to send information via radio. The one possible exception to that is a beacon station that transmits an unmodulated carrier for propagation testing. In that case the only information being sent is, "I am transmitting (or not) from this location." One can think of it as a single data bit with two states, *on* or *off*. However, in reality even a beacon station or test transmission must periodically identify with the station call sign! Modulation is what allows the signal to carry the information, no matter how much or little.

To represent the information being sent, the radio signal must periodically change its characteristics (or *state*) in some way that can be detected by the receiver. In the early days of radio, the only way to do that was on-off keying using Morse code. By alternating the on and off states with the proper sequence and rhythm, a pair of highly-skilled operators can exchange textual data at rates up to perhaps 60 WPM.

Later, engineers figured out how to amplify the signal from a microphone and use it to vary the power of the radio signal continuously. Thus was born amplitude modulation, which allowed transmitting voices at full speaking speeds, that is, up to about 200 WPM. That led to analog modes such as television with even faster information rates. Today's digital radio systems are capable of transferring tens of megabits of information per second, equivalent to tens of millions of words per minute.

Modulation is but one component of any transmission mode. For example AM voice and NTSC television (the old analog TV system) both use amplitude modulation, but the transmission protocol and type of information sent are very different. Digital modes also are generally defined not only by the modulation but also by multiple layers of protocol and data coding. Of the three components of a mode (modulation, protocol and information), this chapter will concentrate on the first. The **Digital Protocols and Modes** chapter covers digital transmission protocols and the **Digital Communications** operating chapter with the online content covers practical aspects of operating using the various digital modes.

Types of modulation are often referred to as "analog" or "digital." There is really no distinction between the two. What is usually meant is whether the modulating signal itself is an analog (continuously varying) or digital (some number of fixed states) signal.

The type of modulation is the choice of the radio system designer. The combination of modulation and a protocol create a mode. Modes can be analog or digital, again referring to whether they carry a continuously varying signal or not.

As an example of the difference between mode and modulation, consider HF and VHF packet radio. Both modes use the same AX.25 protocol to control how the data packets are formed and how the stations establish, conduct, and terminate the contact. Both modes also encode data as a pair of audio tones. On HF, however, the audio tones modulate an SSB transceiver to create an AFSK signal. On VHF, the tones modulate an FM transceiver, creating a very different type of signal. Both transceivers (SSB and FM) don't "know" whether the audio tones represent speech or data. If the audio is speech, the result is an "analog" mode. If the audio tones represent data, the result a "digital" mode.

11.1.1 Emission Designators

We tend to think of a radio signal as being "on" a particular frequency. In reality, any modulated signal occupies a band of frequencies. The bandwidth depends on the type of modulation and the data rate. A Morse code signal can be sent within a bandwidth of a couple hundred Hz at 60 WPM, or less at lower speeds. An AM voice signal requires about 6 kHz. For high-fidelity music, more bandwidth is needed; in the United States, an FM broadcast signal occupies a 200-kHz channel. Television signals need about 6 MHz, while 802.11ad, the latest generation of "WiFi" wireless LAN, uses up to 2 GHz for data transmission at its maximum transfer rate.

The International Telecommunication Union (ITU) has specified a system for designating radio emissions based on the bandwidth, modulation type and information to be transmitted. The emission designator begins with the bandwidth, expressed as a maximum of five numerals and one letter. The letter occupies the position of the decimal point and represents the unit of bandwidth, as follows: H = hertz, K = kilohertz, M = megahertz and G = gigahertz. The bandwidth is followed by three to five emission classification symbols, as defined in the table of emission designators available with the online content. The first three symbols are mandatory; the fourth and fifth symbols are supplemental. These designators are found in Appendix 1 of the ITU Radio Regulations, ITU-R Recommendation SM.1138 and in the FCC rules §2.201. More information on emissions designators is also available online at **fccid.io/Emissions-Designator**.

For example, the designator for a CW signal might be 150H0A1A, which means 150 Hz bandwidth, double sideband, digital information without subcarrier, and telegraphy for aural reception. SSB would be 2K5J3E, or 2.5 kHz bandwidth, single sideband with suppressed carrier, analog information, and telephony. The designator for a PSK31 digital signal is 60H0J2B, which means 60 Hz bandwidth, single sideband with suppressed carrier, digital information using a modulating subcarrier, and telegraphy for automatic reception.

Authorized modulation modes for amateur radio operators depend on frequency, license class, and geographical location, as specified in the FCC regulations §97.305. Technical standards for amateur emissions are specified in §97.307. Among other things, they require that no amateur station transmission shall occupy more bandwidth than necessary for the information rate and emission type being transmitted, in accordance with good amateur practice. We will discuss the necessary bandwidth for each type of modulation as it is covered in the following sections.

11.1.2 Bandwidth Definition

The general definition of bandwidth for a tuned circuit or filter are not used for the legal definition of signal bandwidth. The FCC regulations are more concerned with the amount of spectrum a signal consumes. This leads to the concept of *occupied bandwidth*. This is the range of frequencies within which a specified percentage of the total power occurs. A common percentage used is 99%. This means that the total signal power outside the occupied bandwidth must be less than 20 dB less than the total signal power. For a properly-adjusted, low-distortion transmitting system, the occupied bandwidth is determined mainly by the modulation type and filtering and, in the case of digital modulation, the symbol rate.

FCC Rule §97.3(a)(8) provides the legal definition of a signal's bandwidth as "The width of a frequency band outside of which the mean power of the transmitted signal is attenuated at least 26 dB below the mean power of the transmitted signal with the band." The 26 dB limit is equivalent to 1/400th of the signal's power. **Figure 11.1** illustrates this relationship.

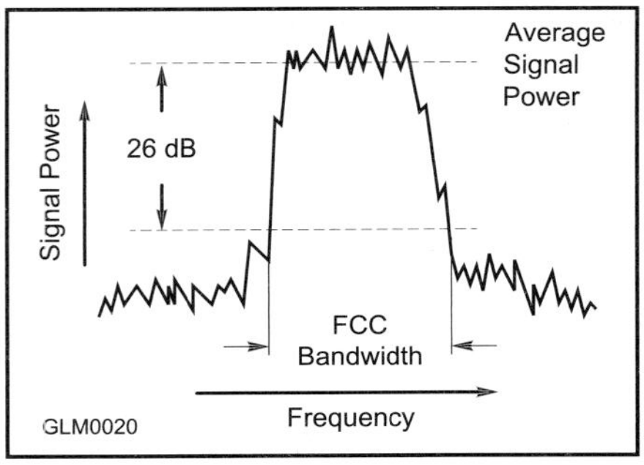

Figure 11.1 — The FCC definition of bandwidth (see text).

11.2 Amplitude Modulation (AM)

Of the various properties of a signal that can be modulated to transmit voice information, amplitude was the first to be used. Not only are modulation and demodulation of AM signals simple in concept, but they are simple to implement as well.

11.2.1 Double-Sideband, Full-Carrier AM

An AM signal is created from two signals; the RF signal that can be transmitted and the modulating signal that will be combined with the RF signal. The RF signal is called the *carrier*, c(t), which is a single-frequency sinusoid at a frequency of f_C.

$c(t) = C \sin(2\pi f_C\, t)$

The modulating signal is represented by m(t) and may be a sine wave or a complex signal like speech. The modulating signal is also referred to as *baseband modulation*.

$m(t) = M \cos(2\pi f_M\, t)$

(The use of sine and cosine are to help identify which signal is which — there is no requirement for the carrier and modulating signal to have a specific phase relationship.)

Amplitude modulation is performed when c(t) is multiplied by m(t). Mathematically, the process of amplitude modulation is easiest to envision if the modulating signal, m(t) is

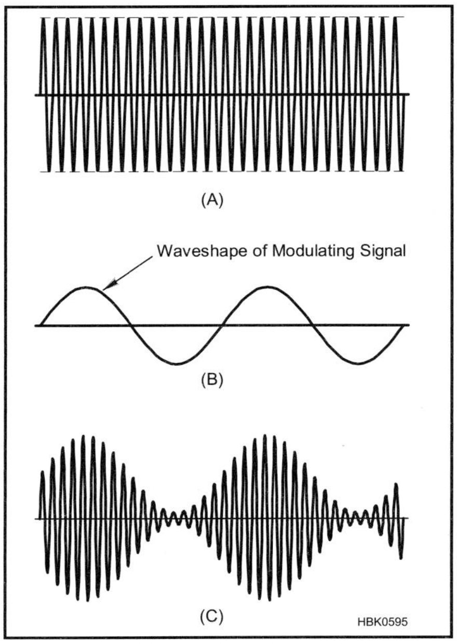

Figure 11.2 — Graphical representation of amplitude modulation. In the unmodulated carrier (A) each RF cycle has the same amplitude. When the modulating signal (B) is applied, the RF amplitude is increased or decreased according to the amplitude of the modulating signal (C). A modulation index of approximately 75% is shown. With 100% modulation the RF power would just reach zero on negative peaks of the modulating signal.

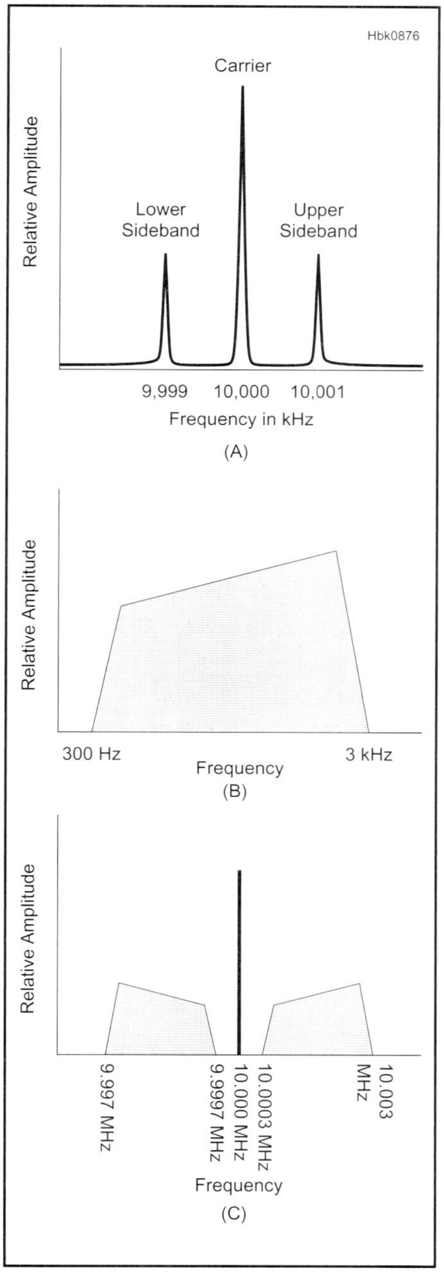

Figure 11.3 — AM with a single 1 kHz tone modulating a 10 MHz carrier (A) and a speech waveform (B) modulating the same carrier in (C).

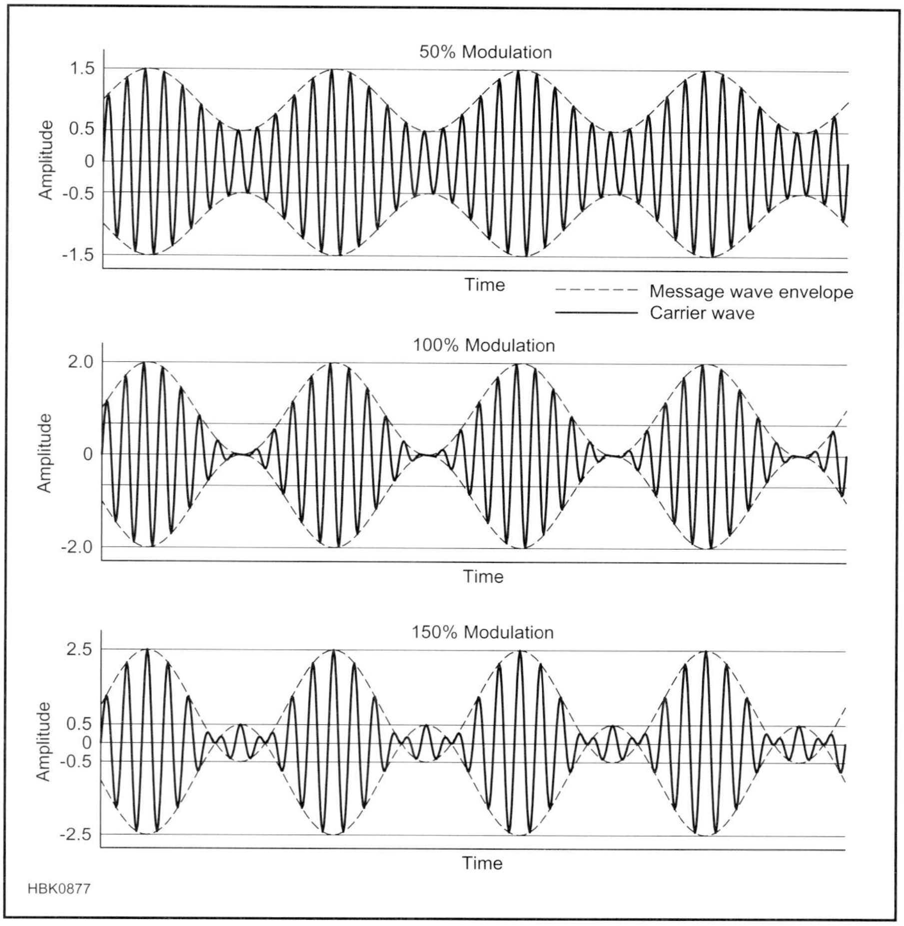

Figure 11.4 — AM waveforms showing 50% modulation (A) and 100% modulation (B). Overmodulation (C) causes distortion of the envelope and reversal of the RF carrier's phase, creating spurious emissions and interference on adjacent channels.

a single audio tone with a frequency of f_M.

$c(t) \times [1 + m(t)] = [1+M \cos (2\pi f_M t)] \times C \sin (2\pi f_C t)$

The constant 1 represents a dc component which is necessary to allow the *envelope* of the AM signal to both increase and decrease as in **Figure 11.2**. Note that if the modulating signal is zero (such as when there is no speech or tone) then the result is just the original carrier, c(t). If M=1, the value of 1 + m(t) can vary from zero to 2 and the signal would just go to zero on the peaks of the modulating signal. Figure 11.2C shows a case very similar to this where the signal's envelope almost goes to zero but not quite.

Mathematically, the resulting AM signal is:

$$C\sin(2\pi f_C t)+\frac{C\times M}{2}\sin(2\pi(f_C+f_M)t)$$

$$+\frac{C\times M}{2}\sin(2\pi(f_C-f_M)t)$$

This expression describes a signal with three components represented by the three terms: first, the carrier; second, an *upper sideband* with a frequency of $f_C + f_M$; and third, a *lower sideband* with a frequency of $f_C - f_M$. This is *double-sideband, full-carrier* (DSB-FC) AM.

If M=1, each sideband's amplitude (C/2) has half the amplitude of the carrier (C). This means each sideband has one-quarter of the carrier's power level or is 6 dB weaker than the carrier. Adding the two sidebands together means that ⅓ of the total signal power is in the sidebands and 2/3 in the carrier. For example, if the carrier power is 100 watts, each sideband will have 100 × ¼ = 25 watts. The total signal power is 100 + 2 × 25 = 150 watts.

AM SPECTRUM

Figure 11.3A shows the spectrum of a 10-MHz signal modulated with a 1-kHz sine wave. The upper sideband is a single frequency at 10 MHz + 1 kHz = 10.001 MHz. The spectrum of the lower sideband is inverted, so it is at 10 MHz – 1 kHz = 9.999 MHz.

When the modulating signal is speech, instead of a single tone, the result is shown in Figure 11.3B and 11.3C. The speech signal is represented by the shaded region in Figure 11.3B extending from 300 Hz to 3 kHz with the higher frequency speech components having a slightly higher amplitude. When the 10 MHz carrier is modulated with the speech signal, the resulting sidebands are shown in Figure 11.3C. Each speech component behaves like a separate tone, creating its own sideband. The set of components in speech create the set of sidebands shown by the shaded regions. Note that the upper and lower sideband spectra are "mirror images" of each other with the lower components closest to the carrier.

With both sidebands and the carrier, the AM signal's total bandwidth is twice the bandwidth of the baseband modulation signal. (A single tone's bandwidth is considered to be the frequency of the tone.) The bandwidth of an AM signal modulated by communications-quality speech with a frequency range of 300 Hz to 3 kHz is 2 × 3 kHz = 6 kHz. Note that the AM signal's bandwidth does not depend on the lower limit of the baseband modulation signal.

AM MODULATION INDEX

If the AM signal's envelope is zero on negative peaks of the modulating signal, that corresponds to 100% modulation. For the case where the modulation signal is zero leaving just the steady carrier signal, that is 0% modulation. Expressed as a value from 0 to 1, this is the *modulation index*:

Modulation index = M / C

Modulation percentage = (M / C) × 100%

Figure 11.4A and 11.4B show the AM signal waveform when M = C/2 (50% modulation or a modulation index of 0.5) and when M = C (100% modulation or a modulation index of 1.0.) If M is greater than C, the envelope of the AM signal can go "below zero" as shown in Figure 11.4C, causing the envelope of the signal to be distorted on negative peaks (and possibly positive peaks depending on the design of the modulator circuit). This is the condition of *overmodulation* and the distortion is known as "flat-topping" because of the shape of the envelope which often exhibits a flattening of peaks during overmodution. This also results in spurious emissions the extend beyond the upper and lower sideband, causing interference on adjacent channels.

11.2.2 Double-Sideband, Suppressed Carrier (DSB-SC) AM

Another way of seeing how an AM signal is constructed is illustrated in **Figure 11.5**. Figure 11.5A shows the carrier and the sidebands from a modulating tone are shown in 11.5B and 11.5C. If you look closely, you can see that the waveforms in Figure 11.5B and 11.5C have slightly different frequencies than the carrier. If the two sidebands are added together, the signal of Figure 11.5D is produced. This is a *double-sideband, suppressed carrier* (DSB-SC) signal and its spectrum is shown in **Figure 11.6**, assuming the same 10 MHz carrier and 300-3000 Hz sidebands.

When the carrier signal is added, the full AM signal is produced in Figure 11.5D. When all of the signals are in-phase, the resulting signal has its maximum amplitude. When all of the signals are out of phase, the resulting signal goes to zero. If the carrier's phase is used as our reference, the phase of each sideband can be viewed as slipping behind (lower sideband) or moving ahead (upper sideband) of the carrier. The sidebands are out of phase with each other at the frequency of the tone so the resulting envelope reproduces the modulating tone's sine wave.

11.2.3 Single-Sideband, Suppressed Carrier (SSB-SC)

As we have seen, since the carrier itself contains no modulation, it does not need to be transmitted, which saves at least 67% of the transmitted power. Since the two sidebands carry identical information, one of them may be eliminated as well, saving half the bandwidth. The result is *single sideband, suppressed carrier* (SSB-SC), which is commonly referred to simply as "SSB." If the lower sideband is eliminated, the result is called *upper sideband* (USB). If the upper sideband is eliminated, you're left with *lower sideband* (LSB). **Figure 11.7** shows the result of removing one of the sidebands in Figure 11.6.

The effect of SSB modulation is that the baseband modulation signal is simply frequency-shifted to the RF carrier frequency (whether the carrier is transmitted or not.)

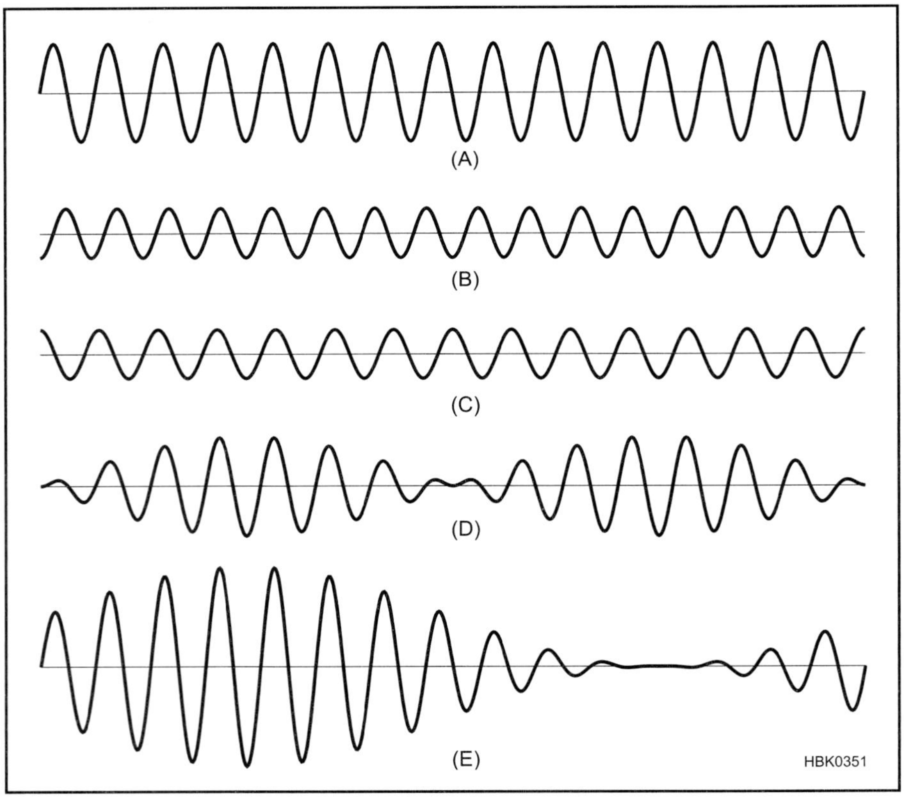

Figure 11.5 — At A is an unmodulated carrier. If the upper (B) and lower (C) sidebands are added together a double-sideband suppressed carrier (DSB-SC) signal results (D). If each sideband has half the amplitude of the carrier, then the combination of the carrier with the two sidebands results in a 100%-modulated AM signal (E). Whenever the two sidebands are out of phase with the carrier, the three signals sum to zero. Whenever the two sidebands are in phase with the carrier, the resulting signal has twice the amplitude of the unmodulated carrier.

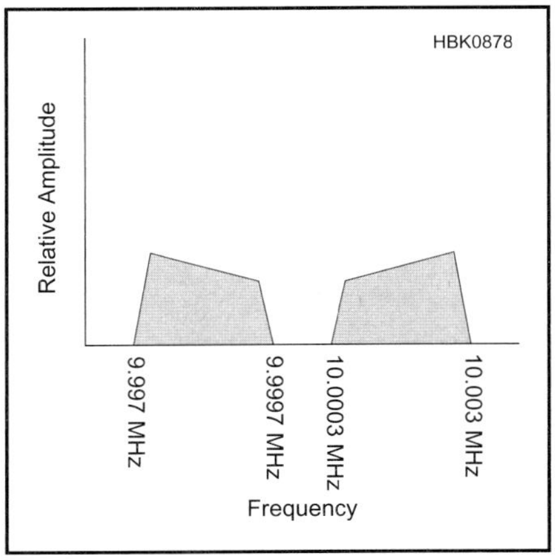

Figure 11.6 — The spectrum of a double-sideband, suppressed-carrier (DSB-SC) signal created from the signal shown in Figure 11.3.

The spectrum of the modulation may be inverted during the shifting process, requiring the demodulation process to re-invert the signal as it is translated back to the baseband frequency range. Aside from that technical consideration, there is no difference between the information in the LSB and USB signals.

The bandwidth of an SSB signal is a little less than half that of an equivalent AM signal. Using the same 300 Hz to 3 kHz speech bandwidth, the SSB signal has a bandwidth of 3000 – 300 = 2700 Hz = 2.7 kHz. In practice, SSB transmitters have an overall audio response from 300 Hz to 2.8 kHz, resulting in 2.5 kHz bandwidth.

Compared to most other analog modulation modes, SSB has excellent power efficiency because the transmitted power is proportional to the modulating signal and no power at all is transmitted during pauses in speech. Another advantage is that the lack of a carrier results in less interference to other stations.

SSB transceivers also are well-suited for the narrowband digital modes. Because an SSB transceiver simply frequency-translates the baseband audio signal to RF in the transmitter and back to audio again in the receiver, digital modulation may be generated and detected at audio frequencies using the sound card in a personal computer.

11.2.4 Amplitude-Modulated On-Off Keying (OOK)

On-off keying (OOK) is a special case of *amplitude-shift keying* (ASK) and is normally used for sending Morse code. For historical reasons dating from the days of spark transmitters, amateurs often refer to this as *continuous wave* (CW) even though the signal is actually keyed on and off.

You can think of OOK as being the same as analog AM that is modulated with a two-level signal that switches between full power and zero power. For example, imagine that the modulation is a 10 Hz square wave, equivalent to sending a series of dits at 24 WPM. A square wave may be decomposed into a sine wave of the same frequency and a theoretically infinite succession of odd harmonics. Recall that the lower and upper sidebands of an AM signal are simply the inverted and non-inverted spectra of the baseband modulation. The RF spectrum therefore contains sidebands at the carrier frequency ±10 Hz, ±30 Hz, ±50 Hz, and so on to plus and minus infinity. This phenomenon is called *key clicks*. Stations listening on nearby frequencies hear a click upon every key closure and opening.

To prevent interference to other stations, the modulation must be low-pass filtered, which slows down the transition times between the on and off states. See **Figure 11.8**. If the transitions are too fast, then excessive key clicks occur. If the transitions are too slow, then at high speeds the previous transition may not have finished before the next one starts, which makes the signal sound mushy and hard to read. Traditionally, filtering was done with a simple resistor-capacitor low-pass filter on the keying line, but using a transition with a raised-cosine shape allows faster transition times without excessive key clicks. Some modern transceivers use DSP techniques to generate such a controlled transition shape.

The optimum transition time, and thus bandwidth, depends on keying speed. It also depends on propagation conditions. When the signal is fading, the transitions must be sharper to allow good copy. **Figure 11.9** gives recommended keying characteristics based on sending speed and propagation. As a compromise, many transmitters use a 5 ms rise and fall time. That limits the bandwidth to approximately 150 Hz while allowing good copy up to 60 WPM on non-fading channels and 35 WPM on fading channels, which covers most requirements.

With any digital system, the number of changes of state per second is called the *baud rate* or the *symbol rate*, measured in *bauds* or

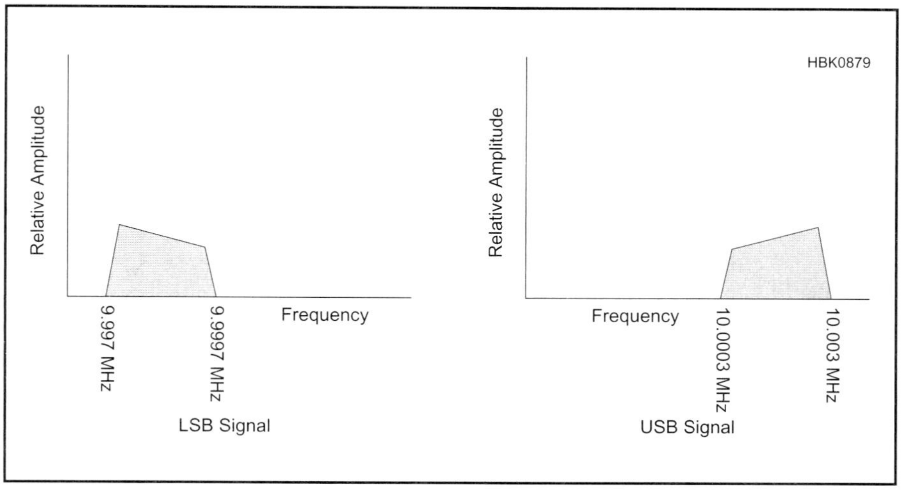

Figure 11.7 — The spectrum of the two single sideband (SSB) signals that could be created from the signal shown in Figure 11.6.

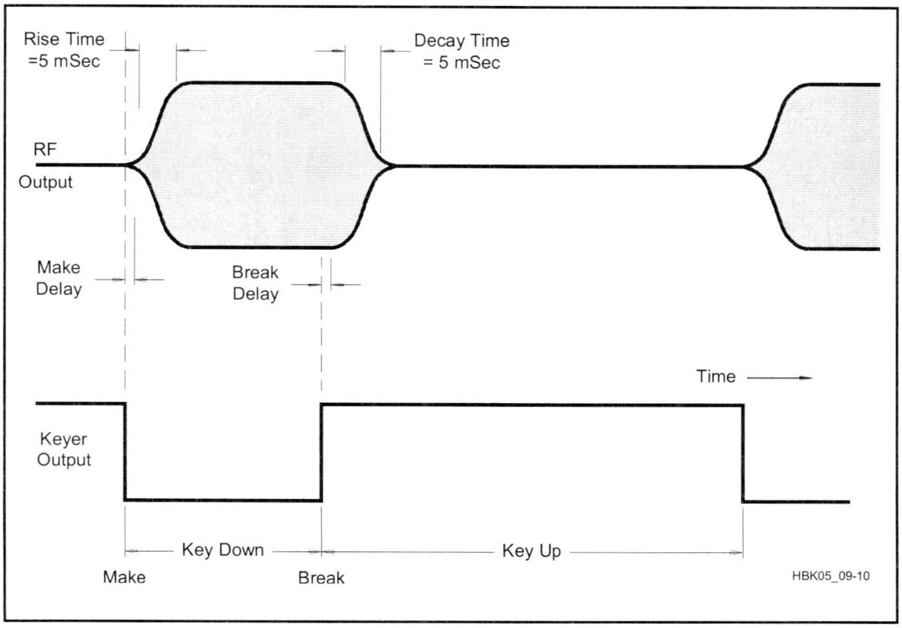

Figure 11.8 — A filtered CW keying waveform. The on-off transitions of the RF envelope should be as smooth as possible while transitioning as quickly as possible. The shape of the RF Output waveform is nearly optimum in that respect.

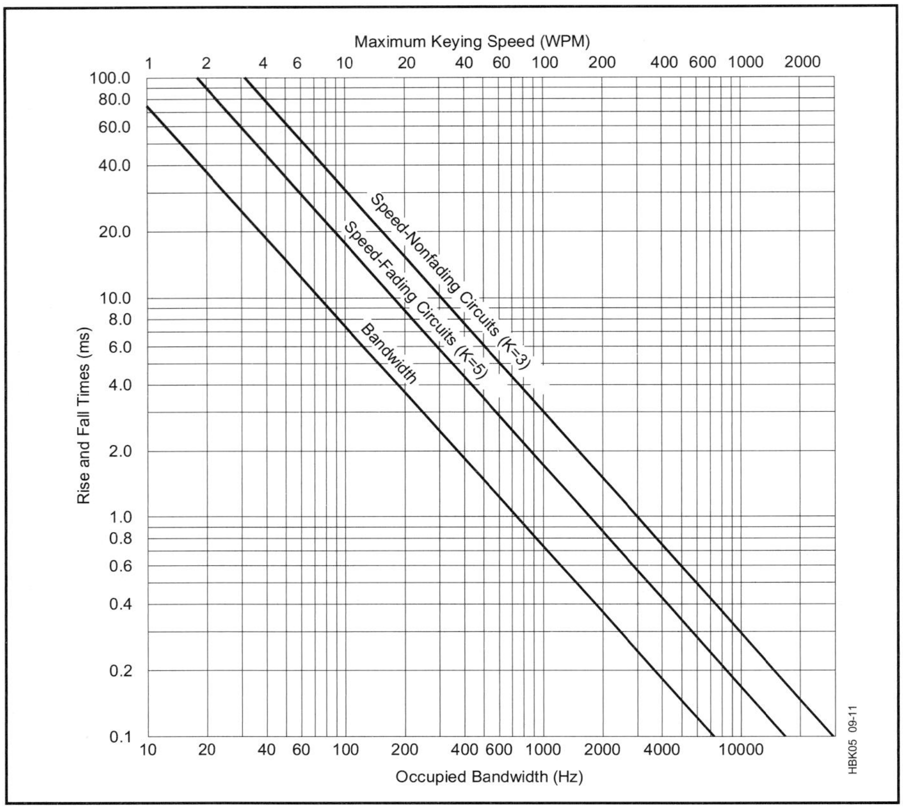

Figure 11.9 — Keying speed vs rise and fall times vs bandwidth for fading and non-fading communications circuits. For example, to optimize transmitter timing for 20 WPM on a non-fading circuit, draw a vertical line from the WPM axis to the K = 3 line. From there draw a horizontal line to the rise/fall axis (approximately 15 ms). Draw a vertical line from where the horizontal line crosses the bandwidth line and see that the bandwidth is about 50 Hz. Harder (more abrupt) keying is required to maintain copying ability in the presence of fading.

symbols per second. Sending a single Morse code dit requires two equal-length states, or symbols: *on* for the length of the dit and *off* for the space between dits. Thus, a string of dits sent at a rate of 10 dits per second has a symbol rate of 20 bauds.

Refer to **Figure 11.10**. A Morse code dash is on for three times the length of a dit. Including one symbol for the off time, the total time to send a dash is four symbols, twice the time to send a dit. For example, at a baud rate of 20 bauds, there are 10 dits per second or 5 dashes per second. The spacing between characters within a word is three symbols, two more than the normal space between dits and dashes. The spacing between words is seven symbols.

For purposes of computing sending speed, a standard word is considered to have five characters, plus the inter-word spacing. On average, that results in 50 symbols per word. From that, the speed in words per minute may be computed from the baud rate:

$$\text{WPM} = \frac{60\ (\text{sec / min})}{50\ (\text{symbols / word})} \times \text{bauds}$$

$$= 1.2 \times \text{bauds} = 2.4 \times \text{dits / sec}$$

Characters in Morse code do not all have the same length. Longer codes are used for characters that are used less frequently while the shortest codes are reserved for the most common characters. For example, the most common letter in the English language, E, is sent as a single dit. In that way, the average character length is reduced, resulting in a faster sending speed for a given baud rate. Such a variable-length code is known as *varicode*, a technique that has been copied in some modern digital modes (see the **Digital Protocols and Modes** chapter).

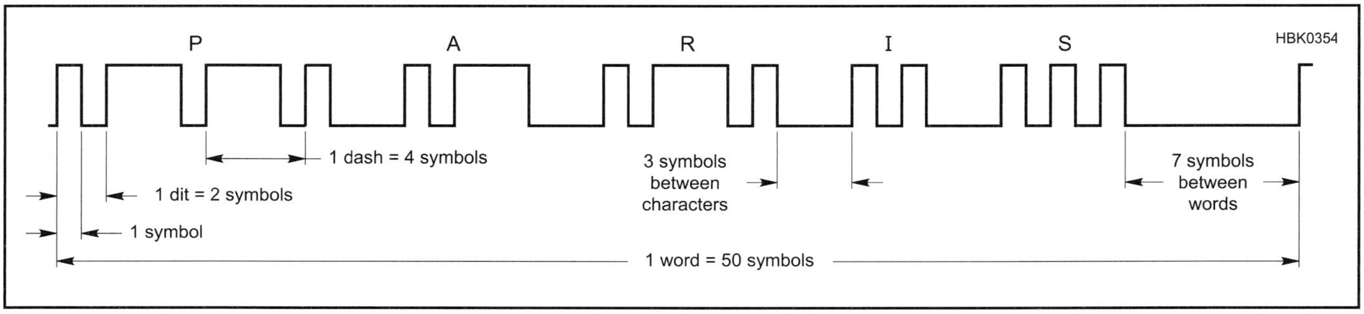

Figure 11.10 — CW timing of the word PARIS, which happens to have a length equal to the standard 50 symbols per word. By programming it multiple times into a memory keyer, the speed may be calibrated simply by counting the number of times the word is completed in one minute.

11.3 Angle Modulation

Angle modulation varies the phase angle of an RF signal in response to the modulating signal. In this context, "phase" means the phase of the modulated RF sine wave with respect to the unmodulated carrier. Angle modulation includes both frequency modulation and phase modulation because any change in frequency results in a change in phase. For example, the way to smoothly ramp the phase from one value to another is to change the frequency and wait. If the frequency is changed by +1 Hz, then after 1 second the phase will have changed by +360°. After 2 seconds, the phase will have changed by +720°, and so on. Change the frequency in the other direction and the phase moves in the opposite direction as well. With sine-wave modulation, the frequency and phase both vary in a sinusoidal fashion. See **Figure 11.11**.

For angle modulation, the modulation signal, m(t) is applied to the frequency or phase of the carrier, not its amplitude:

$\sin(2\pi f_C t + k_P m(t))$ or $\sin(2\pi f_C t \times f_d m(t))$

Mathematically, these are equivalent and most texts work with the second form. The constants k_P and f_d are described below in the discussion on modulation index.

Since any change in frequency results in a change in phase and vice versa, *frequency modulation* (FM) and *phase modulation* (PM) are fundamentally the same. The term *frequency deviation* means the amount the RF frequency deviates from the center (carrier) frequency with a given modulating signal. The instantaneous deviation of an FM signal is proportional to the instantaneous amplitude of the modulating signal. The instantaneous deviation of a PM signal is proportional to the instantaneous *rate of change* of the modulating signal. Since the rate of change of a sine wave is proportional to its frequency as well as its amplitude, the deviation of a PM signal is proportional to both the amplitude and the frequency of the modulating signal.

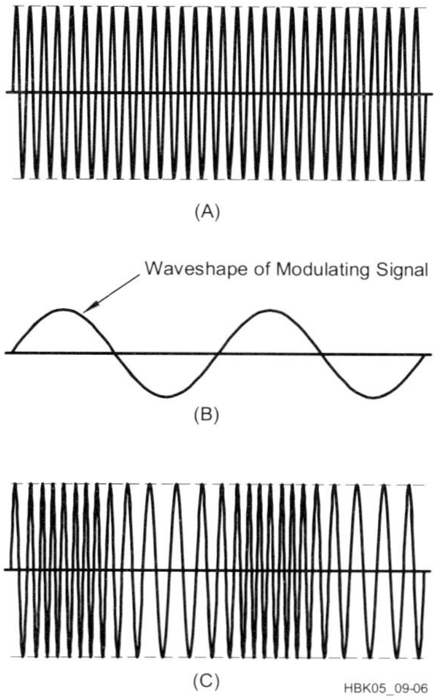

Figure 11.11 — Graphical representation of frequency modulation. In the unmodulated carrier (A) each RF cycle occupies the same amount of time. When the modulating signal (B) is applied, the radio frequency is increased or decreased according to the amplitude and polarity of the modulating signal (C).

11.3.1 Angle-Modulated Modulation Index

The *modulation index* is the ratio of the peak deviation to the highest audio frequency. The *deviation ratio* is the ratio of maximum permitted peak deviation to the maximum permitted modulating frequency. The modulation index is a measure of the maximum amount of phase change (ϕ_{MAX}) the modulation signal can cause. For a single-tone modulating signal:

Modulation index = $A \times f_d / f_M$ (for FM)

Modulation index = $(\phi_{MAX}) = k_P \times A$ (for PM)

where the modulation index is calculated in radians and there are $180/\pi$ radians in the 360° of one complete sine wave cycle (1 radian ≈ 57.3°). In addition,

- A is the amplitude of the modulation signal in volts
- f_M is the frequency of the message signal in hertz
- f_d is the *frequency deviation constant* that represents the sensitivity of the modulator in hertz of deviation per volt of the modulating signal
- $A \times f_d$ is the *peak deviation*
- ϕ_{MAX} is the maximum value of phase change caused by the modulation signal
- k_P is the *phase deviation constant* and is similar to f_d in that it specifies the sensitivity of the phase modulator in radians of phase change per volt of the modulating signal.

For PM, the modulation index doesn't depend on message frequency at all. For an FM signal, m will be larger if the peak deviation gets larger or if f_M gets smaller. For example, loud low-frequency signals can cause the modulation index to become quite large unless the transmitter limits deviation and microphone gain or frequency response.

FCC regulations limit the modulation index at the highest modulating frequency to a maximum of 1.0 for frequencies below 29 MHz. If the audio is low-pass filtered to 3 kHz, then the deviation may be no more than 3 kHz. For that reason, FM transmitters for frequencies below 29 MHz are usually set for 3 kHz deviation while FM transmitters at higher frequencies are typically set for about 5 kHz deviation. The term *narrowband FM* generally refers to deviation of no more than 3 kHz and *wideband FM* refers to deviation greater than 3 kHz.

11.3.2 Angle Modulation Audio Frequency Response

The only difference between FM and PM is the audio frequency response. An FM transmitter with 6 dB/octave pre-emphasis of the modulating signal is indistinguishable from a PM transmitter. A PM transmitter with 6 dB/octave de-emphasis is indistinguishable from an FM transmitter. The reverse happens at the receiver. A frequency detector followed by a 6 dB/octave de-emphasis network acts like a phase detector. It is interesting to note that most VHF and UHF amateur "FM" transceivers should really be called "PM" transceivers due to the pre-emphasis and de-emphasis networks used in the transmitters and receivers respectively.

Most FM and PM transmitters include some kind of audio compressor before the modulator to limit the maximum deviation. Common usage of the term *deviation* is that it refers to the maximum peak deviation allowed by the audio compressor. If the frequency swings a maximum of 5 kHz above the center frequency and 5 kHz below the center frequency, we say the deviation is 5 kHz.

11.3.3 Angle Modulation Bandwidth

The spectrum of an angle-modulated signal is fairly complex. With sine-wave modulation, the RF frequency spectrum from an angle-modulated transmitter consists of the carrier and a series of sideband pairs, each spaced by the frequency of the modulation. For example, with 3 kHz sine-wave modulation, the spectrum includes tones at ±3 kHz, ±6 kHz, ±9 kHz and so on with respect to the carrier. When the modulation index is much less than 1.0, only the first sideband pair is significant, and the spectrum looks similar to that of an AM signal (although the phases of the sidebands are different).

As the modulation index is increased, more and more sidebands appear. Unlike with AM, the carrier amplitude also changes with deviation and actually disappears altogether for certain modulation indices. Because the amplitude of an angle-modulated signal does not vary with modulation, the total power of the carrier and all sidebands is constant. **Figure 11.12** shows several graphs of an FM signal with a 1 kHz modulation signal at different modulation indexes. Note how the sidebands increase and decrease as modulation index changes. All of these spectra have the same total power.

The amplitudes of the carrier and the various sidebands are described by a series of mathematical equations called Bessel functions, which are illustrated graphically in **Figure 11.13**. Note that the carrier disappears when the modulation index equals 2.405 and 5.52. That fact can be used to set the deviation of an FM transmitter. For example, to set 5 kHz deviation, connect the microphone input to a sine-wave generator set for a frequency of 5 / 2.405 = 2.079 kHz, listen to the carrier on a narrowband receiver, and adjust the deviation until the carrier disappears.

11.3.4 Carson's Rule

Unlike amplitude modulation, angle modulation is nonlinear. Recall that with amplitude modulation, the shape of the RF spectrum is the same as that of the modulation spectrum, single-sided with SSB and double-sided with DSB. That is not true with angle modulation. A double-sideband AM signal with audio band-limited to 3 kHz has 6 kHz RF bandwidth. It is easy to see that an FM transmitter with the same audio characteristics but with, say, 5 kHz deviation must have a bandwidth of at least 10 kHz. While you might think that the bandwidth equals twice the deviation, in reality the transmitted spectrum theoretically extends to infinity, although it does become vanishingly small beyond a certain point. As a rule of thumb, approximately 98% of the spectral energy is contained within the bandwidth defined by *Carson's rule*:

$$BW = 2(f_d + f_m)$$

where
f_d = the peak deviation, and
f_m = the highest modulating frequency.

For example, if the deviation is 5 kHz and the audio is limited to 3 kHz, the bandwidth is approximately 2 (5 + 3) = 16 kHz.

Be careful not to "over apply" Carson's Rule. It is just a method of estimating the bandwidth in which a certain amount (98%) of the signal power is contained. This is not as strict a definition as the FCC definition cited earlier in this chapter. Limiting the modulation index and stating the bandwidth in the rules defines a single FM or PM "channel" and how much energy is allowed to be outside that channel. **Figure 11.14** shows a typical 2 meter FM repeater output and compares Carson's Rule and the FCC bandwidth limits. The bandwidths derived from the two rules are "close" but not equivalent. The FCC would judge the signal to be wider than what Carson's Rule predicts. Relying on Carson's Rule is not sufficient to guarantee that there will be no interference to adjacent channels at minimum spacing.

11.3.5 AM Noise and FM Signals

Since the amplitude carries no information, FM receivers are designed to be as insensitive to amplitude variations as possible. Because noise tends to be mostly AM in nature, that results in a quieter demodulated signal. Typically the receiver includes a *limiter*, which is a very high-gain amplifier that causes the signal to clip, removing any amplitude variations, before being applied to the detector. Unlike with AM, as an FM signal gets stronger the volume of the demodulated audio stays the same, but the noise is reduced. Receiver sensitivity is often specified by how much the noise is suppressed for a certain input signal level. For example, if a 0.25 μV signal causes the noise to be reduced by 20 dB, then we say the receiver sensitivity is

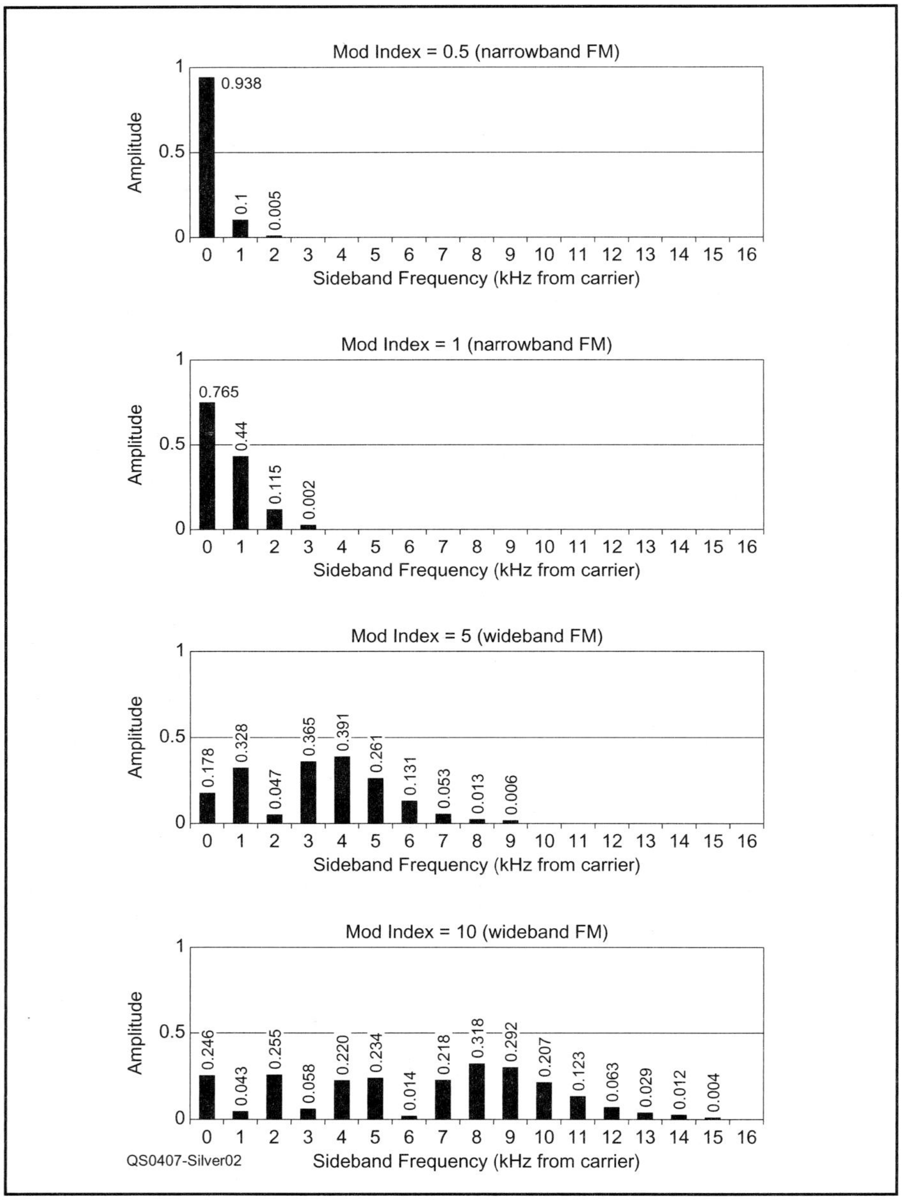

Figure 11.12 — Charts showing one-half of an FM signal's sidebands for a modulating signal frequency of 1 kHz at different modulation indexes.

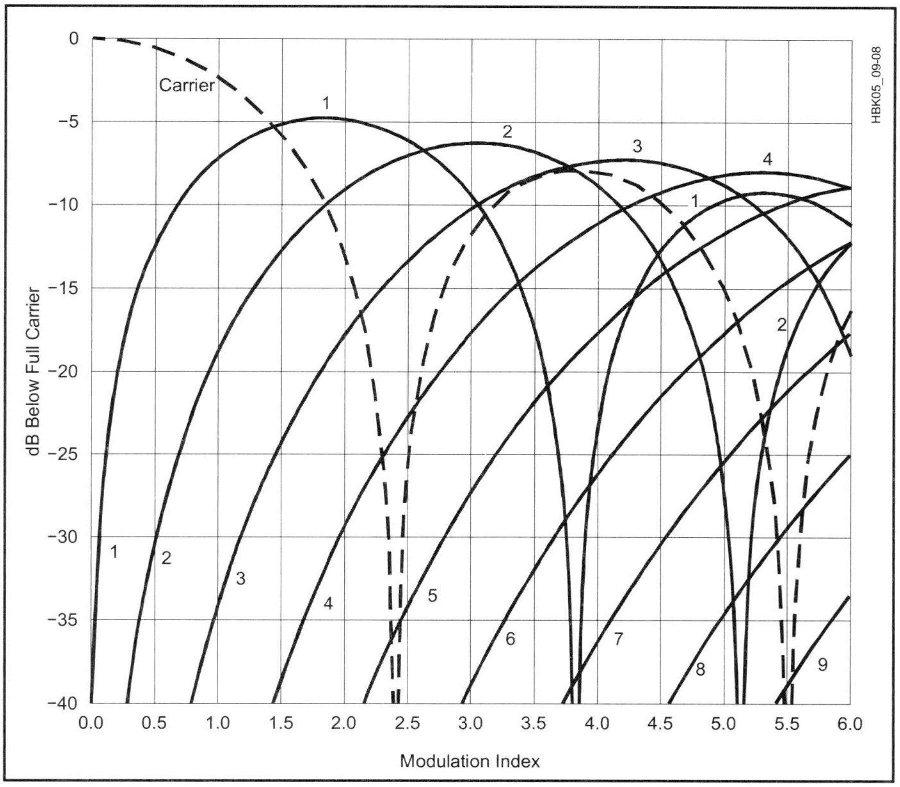

Figure 11.13 — Amplitude of the FM carrier and the first nine sidebands versus modulation index. This is a graphical representation of mathematical functions developed by F. W. Bessel. Note that the carrier completely disappears at modulation indices of 2.405 and 5.52.

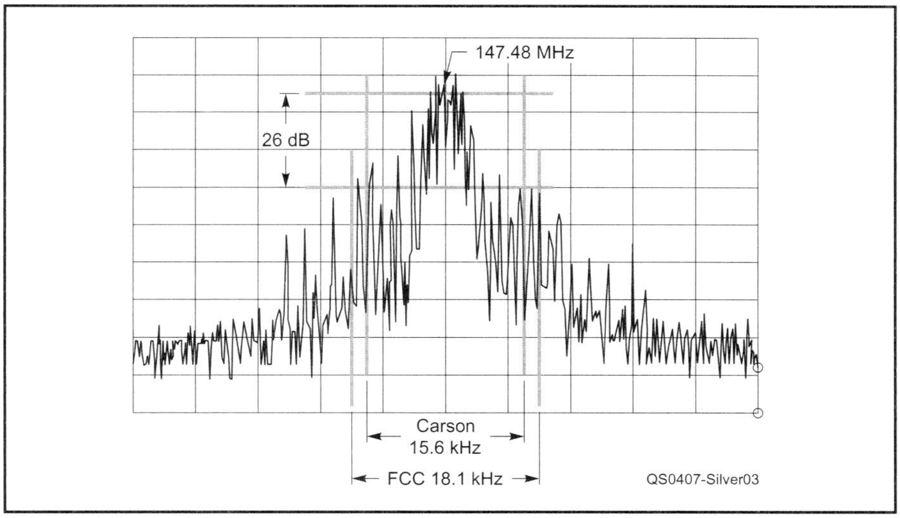

Figure 11.14 — Spectrum analyzer display for a typical 2 meter FM repeater. The bandwidths and amplitudes for Carson's Rule and the FCC's bandwidth definition are overlaid on the spectrum.

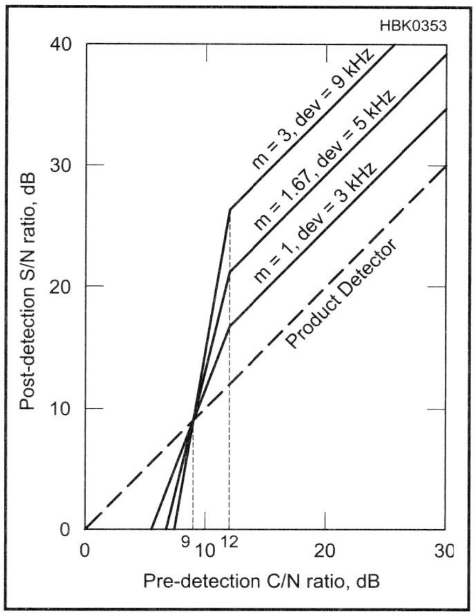

Figure 11.15 — A plot of post-detection signal-to-noise ratio (S/N) versus input carrier-to-noise ratio (C/N) for an FM detector at various modulation indices, m. For each modulation index the deviation is also noted assuming a maximum modulating frequency of 3 kHz. For comparison, the performance of an SSB product detector is also shown.

"0.25 µV for 20 dB of quieting."

The limiter also causes a phenomenon known as *capture effect*. If more than one signal is present at the same time, the limiter tends to reduce the weaker signal relative to the stronger one. We say that the stronger signal "captures" the receiver. The effect is very useful in reducing on-channel interference.

The suppression of both noise and interference is greater the wider the deviation. FM signals with wider deviation do take up more bandwidth and actually have a poorer signal-to-noise (S/N) ratio at the detector output for weak signals but have better S/N ratio and interference rejection for signal levels above a certain threshold. See **Figure 11.15**.

In addition to the noise and interference-reduction advantage, angle-modulated signals share with full-carrier AM the advantages of non-critical frequency accuracy and the continuous presence of a signal, which eases the task of the automatic gain control system in the receiver. In addition, since the signal is constant-amplitude, the transmitter does not need a linear amplifier. Class C amplifiers may be used, which have greater power efficiency.

11.4 FSK and PSK

While they are technically types of angle modulation, *frequency-shift keying* (FSK) and *phase-shift keying* (PSK) are more easily discussed on their own terms. The technique of shifting a signal's frequency or phase to represent the 0 and 1 values of digital data bits is the dominant form of communications outside of amateur radio. Amateurs are making wider use of these techniques as they adopt more digital modes, particularly on the HF and MF bands. (See the **Digital Protocols and Modes** chapter.)

11.4.1 Frequency-Shift Keying (FSK)

FSK was the first digital angle-modulated format to come into common use. It can be thought of as being the same as analog FM that is modulated with a binary (two-level) signal that causes the RF signal to switch between two frequencies. It can also be thought of as equivalent to two on-off-keyed (OOK) signals on two nearby frequencies that are keyed in such a way that whenever one is on the other is off. Just as with OOK, if the transitions between states are instantaneous, then excessive bandwidth occurs — causing interference on nearby channels similar to key clicks. For that reason, the modulating signal must be low-pass filtered to slow down the speed at which the RF signal moves from one frequency to the other.

Although FSK is normally transmitted as a true constant-amplitude signal with only the frequency changing between symbols, it does not have to be received that way. The receiver can treat the signal as two OOK signals and demodulate each one separately. This is an advantage when HF propagation conditions exhibit selective fading — even if one frequency fades out completely the receiver can continue to copy the other, a form of *frequency diversity*. To take advantage of it, wide shift (850 Hz) is normally used. With narrow-shift FSK (170 Hz), the two tones are generally too close to exhibit selective fading.

As previously discussed, selective fading is caused by the signal arriving at the receiver antenna by two or more paths simultaneously. The same phenomenon can cause another signal impairment known as *inter-symbol interference* (ISI) as shown in **Figure 11.16.** If the difference in the two path lengths is great enough, then the signal from one path may arrive delayed by one entire symbol time with respect to the other. The receiver sees two copies of the signal that are time-shifted by one symbol. In effect, the signal interferes with itself. One solution is to slow down the baud rate so that the symbols do not overlap. It is for this reason that symbol rates employed on HF are usually no more than 50 to 100 baud.

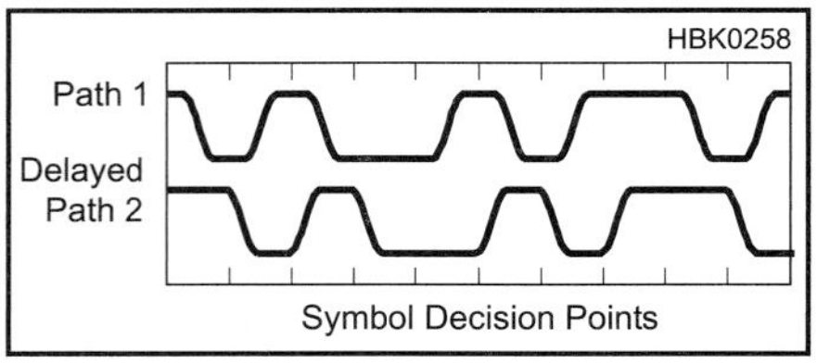

Figure 11.16 — Multipath propagation can cause inter-symbol interference (ISI). At the symbol decision points, which is where the receiver decoder samples the signal, the path 2 data is often opposing the data on path 1.

GAUSSIAN FSK (GFSK)

Gaussian FSK (GFSK) refers to FSK in which the modulating signal has been filtered with a low-pass filter with a Gaussian frequency response. As mentioned before, with any type of angle modulation the spectrum of the modulation is not duplicated at RF but spreads out into an increased bandwidth. For that reason, there is no point in using a modulation filter with a sharp cutoff since the RF spectrum will be wider anyway. A Gaussian filter, with its gradual transition from passband to stopband, has the optimum shape for an angle-modulated digital system.

Figure 11.17 shows band-pass responses that compare the more familiar Butterworth response to Gaussian with an ideal or "brick wall" response as a reference. (See the **Analog and Digital Filtering** chapter for more about filter responses.) While the Butterworth response (with an order n=5) is closer to the ideal filter, the Gaussian filter's smoother transition outside the passband creates fewer products that increase the modulated RF signal's bandwidth.

However, a Gaussian filter also has a gradual transition from symbol-to-symbol in the time domain. The transition is not totally completed by the time of the next symbol, which means that there is some inter-symbol interference (ISI) in the transmitted signal, even in the absence of propagation impairments. With the proper choice of filter bandwidth, however, the ISI is small enough not to seriously affect performance

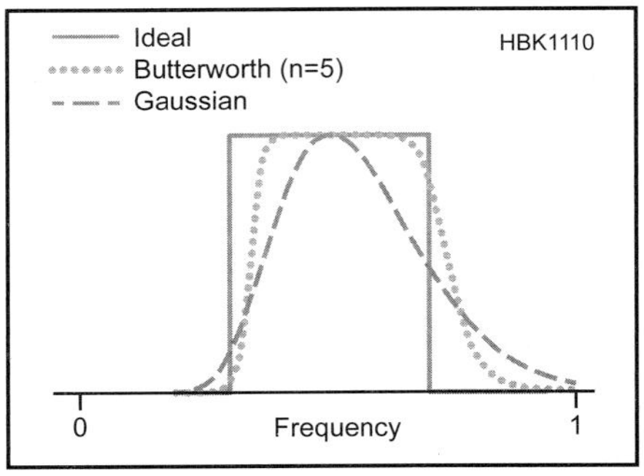

Figure 11.17 — The Gaussian bandpass filter response compared to Butterworth and ideal "brick wall" responses. A Gaussian low-pass filter can be applied to digital data before performing FSK modulation to reduces the resulting RF signal bandwidth.

MULTI-LEVEL FSK

Multi-level FSK (MFSK) is one method to reduce the symbol rate. Unlike with conventional binary FSK, more than two shift frequencies are allowed. For example, with eight frequencies (8-FSK), each symbol can have eight possible states. Since three bits are required to represent eight states, three bits are transmitted per symbol. That means you get three times the data rate without increasing the baud rate. The disadvantage is a reduced signal-to-noise and signal-to-interference ratio. If the maximum deviation is the same, then the frequencies are seven times closer with 8-FSK than with binary FSK and the receiver is theoretically seven times (16.9 dB) more susceptible to noise and interference. However, an IF limiter stage can be used to remove most amplitude variations before the signal arrives at the FSK detector, so the actual increase in susceptibility is less for signal levels above a certain threshold.

With any type of FSK, you can theoretically make the shift as narrow as you like. The main disadvantage is that the receiver becomes more susceptible to noise and interference, as explained above. In addition, the bandwidth is not reduced as much as you might expect. Just as with analog FM, you still get sidebands whose extent depends on the symbol rate, no matter how small the deviation.

11.4.2 Phase-Shift Keying (PSK)

Binary phase-shift keying (BPSK), often referred to simply as phase-shift keying (PSK), can be implemented as a true constant-amplitude angle modulation. An example is *minimum-shift keying* (MSK) which is FSK with a deviation that is at the minimum practical level, taking bandwidth and signal-to-noise ratio into account. That turns out to be a frequency shift from the center frequency of 0.25 times the symbol rate or, using the common definition of frequency shift, a difference between the tone frequencies of 0.5 times the symbol rate. If, on a given symbol, the frequency is shifted to the upper tone, then the phase of the RF signal will change by 0.25 cycles, or 90°, during the symbol period. If the lower tone is selected, the phase shift is –90°.

Thus, MSK may be regarded as either FSK with a frequency shift of 0.5 times the symbol rate, or as *differential phase-shift keying* (DPSK) with a phase shift of ±90°. The binary data can cause the phase to change by either +90 or –90° from one symbol to the next. *Gaussian minimum-shift keying* (GMSK) applies a Gaussian filter to the modulating signal for the same reasons as for GFSK which is described in the preceding section.

The term BPSK is normally understood, however, to refer to phase-shift keying with a 180° phase difference between symbols. To transition from one state to the other, the modulation filter smoothly reduces the amplitude to zero where the polarity reverses (phase changes 180°) before smoothly ramping up to full amplitude again. For that reason, BPSK as usually implemented really should not be considered to be PSK, but rather a form of amplitude-shift keying (ASK) with two modulation amplitudes, +1 and –1. Unlike true angle modulation, it is linear so that the spectrum of the modulation filter is duplicated at RF. The transmitter must use a linear amplifier to prevent distortion and excessive bandwidth similar to the splatter that results from an over-driven SSB transmitter.

11.4.3 Audio Frequency-Shift Keying (AFSK)

Audio frequency-shift keying (AFSK) is the generation of radio-frequency FSK using an audio-frequency FSK signal fed into the modulation input of an SSB transmitter, usually the microphone input. Assume the SSB transmitter is tuned to 14.000 MHz, USB. If the audio signal consists of a sine wave that shifts between 2125 Hz and 2295 Hz (170-Hz frequency shift), then the RF signal is a sine wave shifting between 14.002125 and 14.002295 MHz. The frequency shift and spectral characteristics are theoretically unchanged, other than being translated 14 MHz upward in frequency. The RF signal should be indistinguishable from one generated by varying the frequency of an RF oscillator directly.

This technique works not only for FSK but also for nearly any modulating signal with a bandwidth narrow enough to fit within the passband of an SSB transmitter and receiver. The most common non-voice analog modulation type to use this technique on the amateur bands is FT8, part of the *WSJT-X* software suite. Nearly all narrowband digital signals today are generated in this manner. The audio is usually generated and received by a PC sound card or external audio to USB interface. Dedicated hardware modulator/demodulator (modems) are also used, such as for PACTOR-III and PACTOR-IV.

Whatever the method, it is important to ensure that unwanted interference is not caused by audio distortion or by insufficient suppression of the carrier and unwanted sideband. For example, with the AFSK tone frequencies mentioned above, 2125 and 2295 Hz, the tone harmonics cause no trouble because they fall outside of the transmitter's passband. However, some AFSK modems use 1275 and 1445 Hz (to accommodate 850-Hz shift without changing the 1275-Hz mark frequency). In that case, the second harmonics at 2550 and 2890 Hz must be suppressed since those frequencies are not well-attenuated by the transmitter.

With non-constant-envelope modulation types such as QPSK or the various multi-carrier modes, it is important to set the amplitude of the audio input to the SSB transmitter below the level that activates the transmitter's automatic level control (ALC). That is because the ALC circuit itself generates distortion of signals within the bandwidth of its feedback loop.

11.5 I-Q Modulation

11.5.1 I and Q Components

A sinusoidal wave of any arbitrary amplitude and phase with a frequency of $\omega = 2\pi f$ (f is frequency in Hz) can be represented by the weighted sum of a sine and cosine wave:

$$x(t) = I \cos(\omega t) + Q \sin(\omega t)$$

(See the online **Radio Mathematics** supplement for a list of math tutorials, including trigonometry.) I, the abbreviation for "in-phase" should not be confused with the imaginary number *i*. Q is the abbreviation for "quadrature".

For computational purposes, it is convenient to consider the *in-phase (I)* and *quadrature (Q)* components separately. The components are usually represented as vectors on a *phasor* diagram with the I and Q axes at a 90 degree angle from each other as in **Figure 11.18**. The I component lies on the horizontal I axis and the Q component on the vertical Q axis. The sum of the components is represented by the vector with a magnitude of A at an angle, φ, from the I axis.

Each component oscillates back and forth along its own axis. For example, if Q = 0, then as time increases the I signal oscillates along the I (horizontal) axis, tracing out the path back and forth between I = +1 and I = –1 in a sinusoidal fashion. Conversely, if I = 0, then the signal oscillates along the Q axis.

What if both I and Q are non-zero, for example I = Q = 1? The cosine, $x(t) = \cos(\omega t)$, and sine waves $x(t) = \sin(\omega t)$, are 90° out of phase as shown in **Figure 11.19**. When t = 0, $\cos(\omega t) = 1$ and $\sin(\omega t) = 0$. A quarter cycle later, $\cos(\omega t) = 0$ and $\sin(\omega t) = 1$. Comparing Figure 11.23 with Figure 11.24 you can imagine that the signal (the tip of the vector) is tracing out a circle in the counter-clockwise direction. Similarly, you can imagine that changing ω to –ω results in a signal that circles the origin in the opposite, clockwise direction.

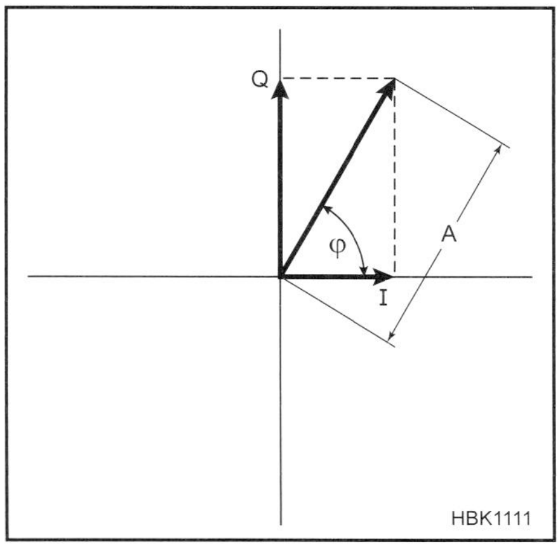

Figure 11.18 — In-phase (I) and quadrature (Q) components of a signal on a phasor diagram.

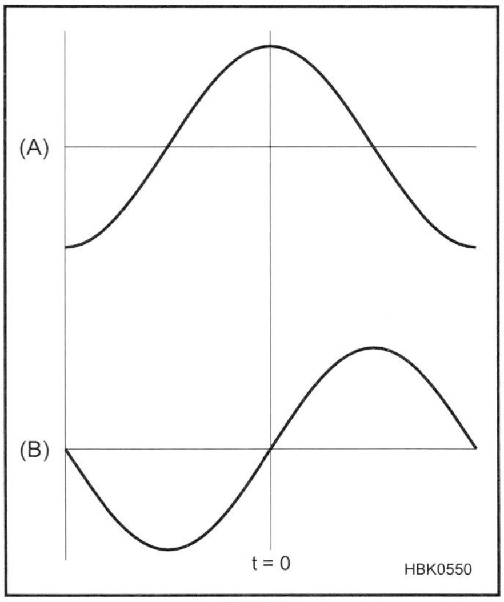

Figure 11.19 — Cosine wave (A) and sine wave (B).

NEGATIVE FREQUENCY

What is negative frequency? Changing the sign of the frequency is equivalent to running time backwards because $(-\omega)t = \omega(-t)$. By examining Figure 11.19A you can see that, because a cosine wave is symmetrical about the time t = 0 point, a negative frequency results in exactly the same signal. That is, $\cos(-\omega t) = \cos(\omega t)$. If, for example, you add a positive-frequency cosine wave to its negative-frequency twin, you get the same signal with twice the amplitude:

$$x(t) = [\cos(\omega t) + \sin(\omega t)] + [\cos(-\omega t) + \sin(-\omega t)] = 2\cos(\omega t)$$

Figure 11.20 illustrates that graphically. Imagine the two vectors rotating in opposite directions, the top one in the counter-clockwise direction of positive frequency. If you mentally add them by placing the tail of one vector on the head of the other, as shown by the dotted line, the result always lies on the I axis and oscillates between +2 and –2 as described by $2\cos(\omega t)$.

That assumes that the phase of the signal is such that it reaches a peak at t = 0. What if instead we had a sine wave, which is zero at t = 0? From Figure 11.24B you can see that running time backwards results in a reversal of polarity, $\sin(-\omega t) = -\sin(\omega t)$. If you add positive and negative-frequency sine waves of the same frequency and amplitude, they cancel, resulting in zero net signal.

That is why we say that a single sinusoidal signal, $\cos(\omega t)$, actually contains two frequencies, $+\omega$ and $-\omega$. It also offers a logical explanation of why a mixer or modulator produces the sum and difference of the frequencies of the two inputs. For example, an AM modulator produces sidebands at the carrier frequency plus and minus the modulating frequency precisely because those positive and negative frequencies are actually already present in the modulating signal.

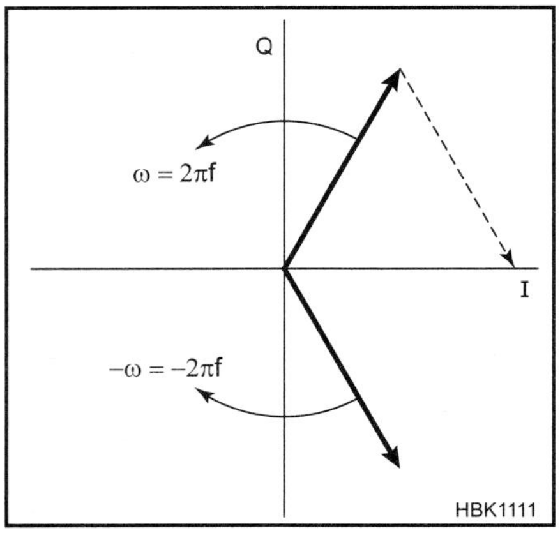

Figure 11.20 — A real signal on the I axis with a single frequency is generated by the sum of I and Q components with positive and negative frequencies.

11.5.2 Analytic Signals

For many purposes, it is useful to separate the portion of the signal that specifies the amplitude and phase (I and Q) from the oscillating part ($\sin(\omega t)$ and $\cos(\omega t)$). For mathematical convenience, the I/Q part is represented by a complex number, $x = I + jQ$. (The **Radio Mathematics** tutorials cover complex numbers.) The oscillating part is also a complex number $e^{-j\omega t} = \cos(\omega t) - j\sin(\omega t)$. (Don't worry if you don't know where that equation comes from — concentrate on the part to the right of the equal sign.) In the equations, $j = \sqrt{}$ of –1. Of course, –1 does not have a real square root (any real number multiplied by itself is positive) so *j*, or any real number multiplied by *j*, is called an *imaginary number*. A number with both real and imaginary parts is called a *complex number*. The total *analytic signal* is a complex number equal to

$$x(t) = xe^{-j\omega t} = (I + jQ)(\cos(\omega t) - j\sin(\omega t))$$

In the above equation, the $\cos(\omega t) - \sin(\omega t)$ portion generally represents an RF carrier, with ω being the carrier frequency (a positive or negative value). The $I + jQ$ part is the modulation. The *scalar* value of a modulated signal (what you would measure with an oscilloscope) is just the real (Re) part of the analytic signal. Using the fact that $j^2 = -1$,

$$Re[x(t)] = Re[(I + jQ)(\cos(\omega t) - j\sin(\omega t))]$$

$$Re[x(t)] = Re[(I\cos(\omega t) + Q\sin(\omega t)) + j(Q\cos(\omega t) - I\sin(\omega t))]$$

$$Re[x(t)] = I\cos(\omega t) + Q\sin(\omega t)$$

Note that if the modulation (I and Q) varies with time, the above equation assumes that the modulated signal does not overlap zero Hz. That is, I and Q have no frequency components greater than ω.

An I/Q phasor diagram, such as Figure 11.18, shows only I and Q (the modulation) and not the oscillating part (the RF carrier). The I/Q vector represents the *difference* in phase and amplitude of the RF signal compared to the unmodulated carrier. For example, if the I/Q vector is at 90°, that means the carrier has been phase-shifted by 90° from what it otherwise would have been. If the I/Q vector is rotating counter-clockwise 10 times per second, then the signal has been increased by 10 Hz from the carrier frequency.

It is worth noting that the modulation can be specified either by the in-phase and quadrature (I and Q) values as shown or alternatively by the amplitude and phase. The amplitude is the length of the I/Q vector in the phasor diagram,

$$A = \sqrt{I^2 + Q^2}$$

The phase is the angle of the vector with respect to the +I axis,

$$\varphi = \arctan\left(\frac{Q}{I}\right)$$

An alternative expression for the modulated analytic signal using amplitude and phase is

$$\begin{aligned} x(t) &= Ae^{-j(\varphi+\omega t)} \\ &= A[\cos(\varphi + \omega t) + j\sin(\varphi + \omega t)] \end{aligned}$$

and for the scalar signal

$$Re[x(t)] = A\cos(\varphi + \omega t)$$

One final comment. So far, we have been looking at signals that consist of a single sinusoidal frequency. In any linear system, anything that is true for a single frequency is also true for a combination of many frequencies. Each frequency is affected by the system as though the others were not present. Since any complicated signal can be broken down into a (perhaps large) number of single-frequency sinusoids, all our previous conclusions apply to multi-frequency signals as well.

11.5.3 I/Q Modulation and Demodulation

An *I/Q modulator* is just a device that controls the amplitude and phase of an RF signal directly from the in-phase (I) and quadrature (Q) components. See **Figure 11.21A**. You can think of an I/Q modulator as a device that converts the analytic signal $I + jQ$ into a scalar signal at some RF frequency. The spectrum of the I/Q signal, both positive and negative frequencies, is translated upward in frequency so that it is centered on the carrier frequency. Thinking in terms of the phasor diagram, any components of the I/Q signal that are rotating counter-clockwise appear above the carrier frequency and clockwise components appear below.

By using two modulators fed with RF sine waves in quadrature (90° out of phase with each other), any amplitude and phase may be obtained by varying the amplitudes of the two modulation inputs. The input labeled I (for in-phase) moves the symbol location horizontally in the constellation diagram and the one labeled Q (for quadrature) moves it vertically. For example, to obtain an amplitude of 1 and a phase angle of –45°, set I to +0.707 and set Q to –0.707.

It is possible to generate virtually any type of modulation using an I/Q modulator. For example, to generate BPSK or on-off keying, simply disconnect the Q input and apply the modulation to I. For angle modulation, such as FM or PM, a waveform generator applies a varying signal to I and Q in such a manner to cause the symbol to rotate at constant amplitude with the correct phase and frequency (rotation rate). Even a multi-carrier signal may be generated with a single I/Q modulator by applying the sum of a number of signals, each representing one carrier, to the I and Q inputs. The phase of each signal rotates at a rate equal to the frequency offset of its carrier from the center.

An *I/Q demodulator* is basically an I/Q modulator in reverse. It generates I and Q signals that represent the in-phase and quadrature components of the incoming RF signal. See Figure 11.21B. Assuming the demodulator's local oscillator is on the same frequency and is in phase with the carrier of the signal being received then the I/Q output of the receiver's demodulator is theoretically identical to the I/Q input at the transmitter end.

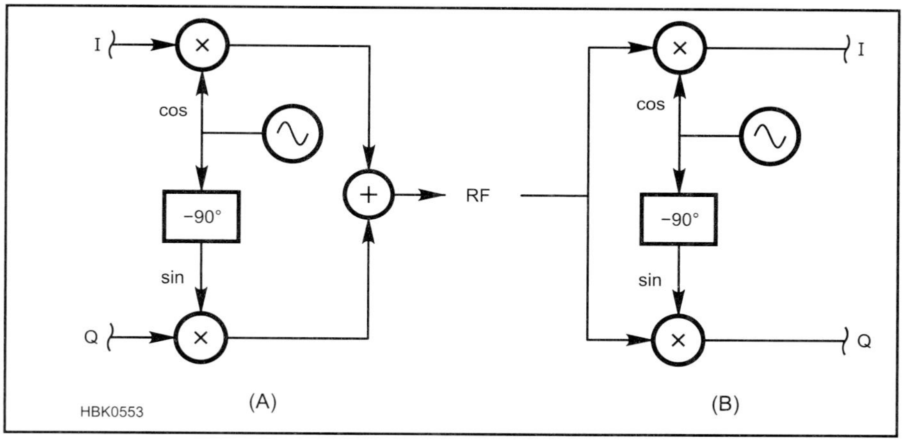

Figure 11.21 — I/Q modulator (A) and demodulator (B). The demodulator is essentially the same circuit as the modulator but operated in the reverse direction.

If the demodulator's local oscillator is not tuned to exactly the same frequency as the one in the transmitter (after down-conversion to the IF in a superhet receiver) then the demodulated signal will rotate in the I/Q plane at a rate equal to the frequency error. Most receivers include a *carrier-recovery* circuit, which phase-locks the local oscillator to the average frequency of the incoming signal to obtain stable demodulation. While that corrects the frequency error, it does not correct the phase, which must be accounted for in some other manner such as by using differential modulation or some kind of symbol-recovery mechanism.

I/Q modulators and demodulators can be built with analog components. The LO could be a transistor oscillator and the 90° phase-shift network could be implemented with coils and capacitors. The circles with the multiplication symbol would be double-balanced mixers. Also required would be trim adjustments to balance the amplitude between the I and Q channels and to adjust the phase shift as close as possible to 90°.

No analog circuit is perfect, however. If the 90° phase-shift is not exactly 90° or the amplitudes of the I and Q channels are not perfectly balanced, you don't get perfect opposite-sideband rejection. The modulator output includes a little bit of signal on the unwanted sideband and the I/Q signal from the demodulator includes a small signal rotating in the wrong direction. If there is a small dc offset in the amplifiers feeding the modulator's I/Q inputs, that shows up as carrier feedthrough. On receive, a dc offset makes the demodulator think there is a small signal at a constant amplitude and phase angle that is always there even when no actual signal is being received. Nor is analog circuitry distortion-free, especially the mixers. Intermodulation distortion shows up as out-of-channel "splatter" on transmit and unwanted out-of-channel responses on receive.

All those problems can be avoided by performing these operations digitally. If the analog I/Q inputs to the modulator are converted to streams of digital numbers with a pair of ADCs, then the mixers, oscillator, phase-shift and summing functions can all be digital. In many systems, the I and Q signals are also generated digitally, so that the digital output signal has perfect unwanted sideband rejection, no carrier feedthrough and no distortion within the dynamic range afforded by the number of bits in the data words. A similar argument holds for a digital demodulator. If the incoming RF signal is first digitized with an ADC, then the demodulation can be done digitally without any of the artifacts caused by imperfections in the analog circuitry.

I/Q modulators and demodulators can also be used as so-called *imageless mixers*. A normal mixer with inputs at f_1 and f_2 produces outputs at f_1, f_2, $f_1 + f_2$, and $f_1 - f_2$. A balanced mixer eliminates the f_1 and f_2 terms but both the sum and difference terms remain, even though normally only one is desired. By feeding an RF instead of AF signal into the input of an SSB modulator, we can choose the sum or difference frequency in the same way as choosing the upper or lower sideband. If the input signal is a sine wave, the Hilbert transformer can be replaced by a simple 90° phase shifter. Similarly, a mixer with the same architecture as an SSB demodulator can be used to down-convert an RF signal to IF with zero image response. Analog imageless mixers are covered in the **Receiving** chapter. They are sometimes used in microwave receivers and transmitters where it is difficult to build filters narrow enough to reject the image response, but they typically only achieve image rejection in the 20-30 dB range. With a digital imageless mixer, the image rejection is "perfect" within the dynamic range of the bit resolution.

11.6 Applications of I/Q Modulation

11.6.1 Quadrature Modulation

Quadrature modulation encodes digital signals using a combination of amplitude and phase modulation. With two types of modulation to work with, it is possible to pack more data bits into each modulation symbol, which allows more throughput for a given bandwidth. Modulation formats in use today have up to 8 or more bits per symbol which would be impractica l for ASK, FSK or PSK alone.

The modulation states of the various digital formats map to positions in the phasor diagram, what is called a *constellation diagram*. The transmitter generates the correct modulation states by placing the correct values on the I and Q inputs to the I/Q modulator. In the receiver, the filtered I and Q values are sampled at the symbol decision times to determine which modulation state they most closely match.

Since quadrature-modulated symbols are defined by both amplitude and phase, the most common way to represent symbol states is on the constellation diagram. **Figure 11.22** shows a *four-level quadrature amplitude modulation* (4-QAM) signal, often referred to simply as QAM. The distance of each state from the origin represents the amplitude. The phase angle with respect to the +I axis represents the phase. The four states shown have phase angles of +45°, +135°, –45° and –135°. Normally, the receiver has no absolute phase information and can only detect phase differences between the states. For that reason, we could just as easily have drawn the four states directly on the I and Q axes, at 0°, +90°, 180° and –90°. Note that each of the four states has the same amplitude, differing only in phase. For that reason, 4-QAM is often referred to as *quadrature phase-shift keying* (QPSK), in the same manner that two-level ASK is normally referred to as BPSK.

Since 4-QAM has four possible states, there are two data bits per symbol. However, QAM is not limited to four states. **Figure 11.23** illustrates a 256-QAM signal. It takes 8 bits to represent 256 states, so 256-QAM

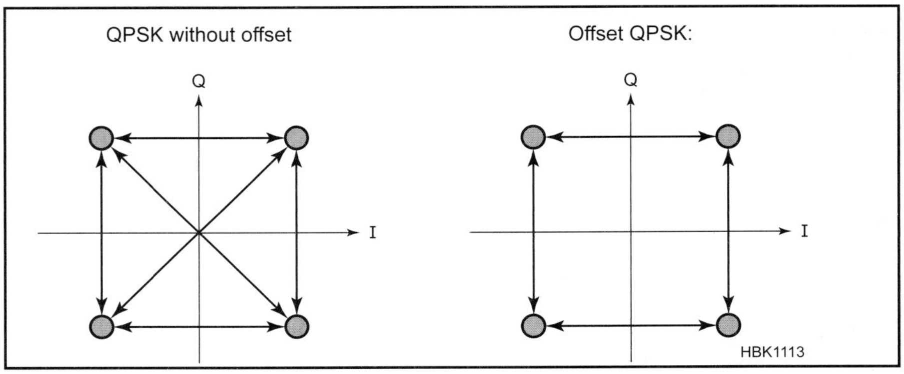

Figure 11.24 — Constellation diagrams of QPSK and OQPSK. Transitions from symbol to symbol for QPSK can pass through the origin where amplifier linearity can be poor. OQPSK avoids this problem with two-step symbol transitions which eliminate direct corner-to-corner transitions.

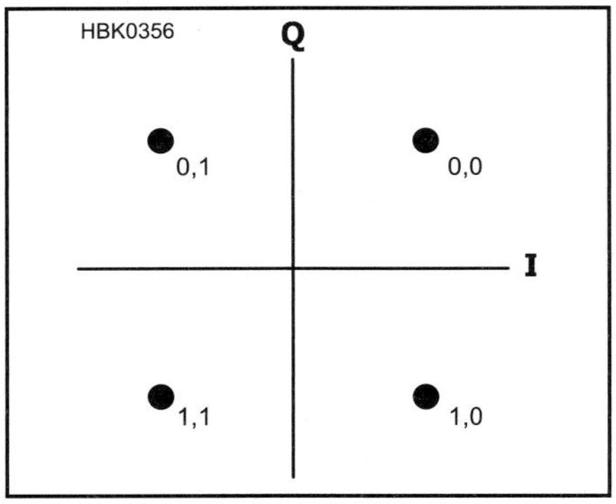

Figure 11.22 — Constellation diagram of a 4-QAM signal, also known as QPSK. The four symbol locations all have the same amplitude and have phase angles of 45°, 135°, –135° and –45°. 4-QAM is a two-bits-per-symbol format. The four symbols are selected by the four possible states of the two data bits, which can be assigned in any order. With the assignment shown, the Q value depends only on the first bit and the I value depends only on the second, an arrangement that can simplify symbol encoding.

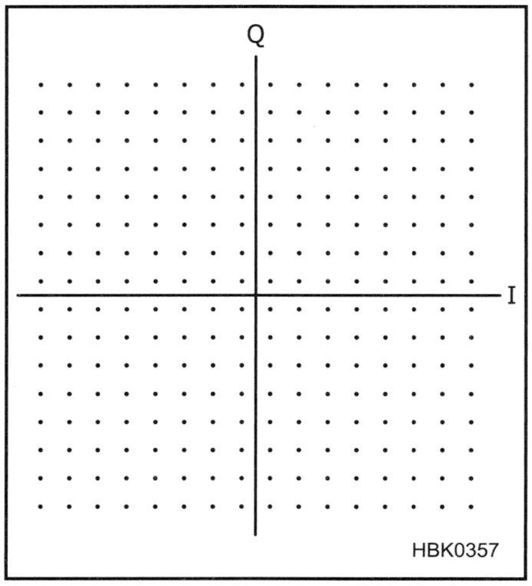

Figure 11.23 — Constellation diagram of a 256-QAM signal. Since an 8-bit number has 256 possible states, each symbol represents 8 bits of data.

packs 8 data bits into each symbol. The disadvantage of using lots of states is that the effect of noise and interference is worse at the receiver. Since, for the same peak power, the states of a 256-QAM signal are 15 times closer together than with 4-QAM, the receiver's decoder has to determine each symbol's location to 15 times greater accuracy. That means the ratio of peak power to noise and interference must be 20 log(15) = 23.5 dB greater for accurate decoding. That is why QAM with a large number of states is normally not used on the HF bands where fading, noise and interference are common. A more common application is digital cable television where the coaxial-cable transmission channel is much cleaner. For example, the European DVB-C standard provides for 16-QAM through 256-QAM, depending on bandwidth and data rate.

One problem with QAM is that whenever the signal trajectory between two symbol states passes through the origin, the signal amplitude momentarily goes to zero. That imposes stringent linearity requirements on the RF power amplifier, since many amplifiers exhibit their worst linearity near zero power. One solution is *offset QPSK* (OQPSK) modulation as shown in **Figure 11.24**. In this case, the symbol transitions of the I and Q channels are offset by half a symbol. That is, for each symbol, the I channel changes state first then the Q channel changes half a symbol time later. That allows the symbol trajectory to sidestep around the origin, allowing use of a higher-efficiency or lower-cost power amplifier that has worse linearity.

Another solution to the zero-crossing problem is called *Pi-over-4 differential QPSK* (π/4 DQPSK). See **Figure 11.25**. This is actually a form of 8-PSK, where the eight symbol locations are located every 45° around a constant-amplitude circle. On any given symbol, however, only four of the symbol locations are used. The symbol location always changes by an odd multiple of 45° (π/4 radians). If the current symbol is located on the I or Q axis,

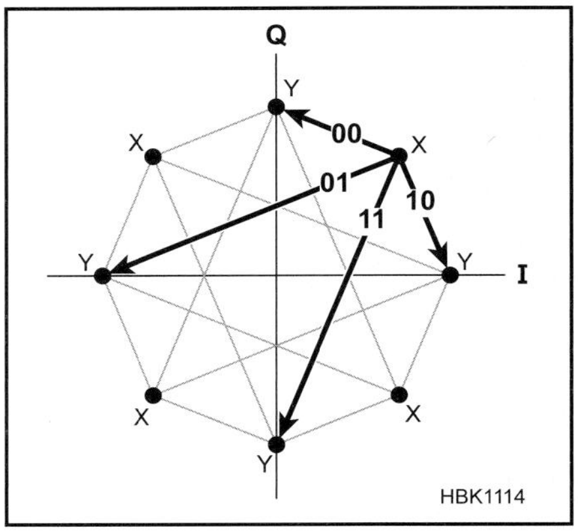

Figure 11.25 — Constellation diagram of a π/4 DQPSK signal. There are eight possible symbol locations. If the current symbol location is at one of the four positions labeled "X" then the next symbol will be at one of the four locations labeled "Y" and vice versa. That guarantees that no possible symbol trajectories (the lines between symbol locations) can ever pass through the origin.

then on the next symbol only the four non-axis locations are available and vice versa. As with OQPSK, that avoids transitions that pass through the origin.

Another advantage of π/4 DQPSK which is shared with other types of differential modulation is that absolute phase doesn't matter. The information is encoded only in the *difference* in the phase of successive symbols. That greatly simplifies the job of the receiver's demodulator.

11.6.2 AM Using I/Q Modulation

How do you make an AM modulator and demodulator using DSP and software? The modulator is easy. Simply add a constant value, representing the carrier, to the audio signal and multiply the result by a sine wave at the carrier frequency, as shown in **Figure 11.26**.

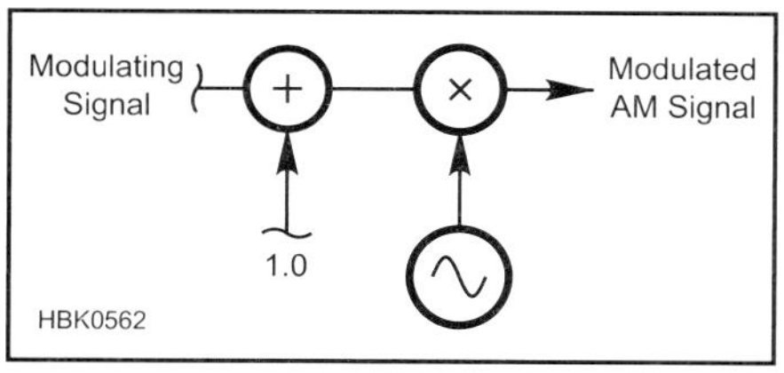

Figure 11.26 — A digital AM modulator

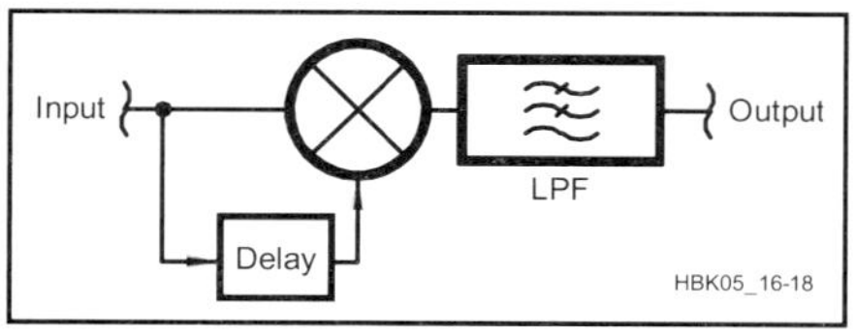

Figure 11.27 — A digital quadrature detector.

Demodulation is almost as easy. We could just simulate a full-wave rectifier by taking the absolute value of the signal, as mentioned previously, and low-pass filter the result to remove the RF energy. If the signal to be demodulated is complex, with I and Q components, then instead of absolute value we take the magnitude

$$A = \sqrt{I^2 + Q^2}$$

The dc bias can be removed by adding a "series blocking capacitor" — a high-pass filter with a suitable cut-off frequency.

A little more elegant way to do it would be to include the AM detector as part of the AGC loop. In the C code snippet shown in **Table 11.1**, the variable "carrier" is the average AM carrier level. It is passed to another subroutine to control the gain.

Note that no "series capacitor" is needed since the audio signal is computed by subtracting the average historical value, carrier, from the magnitude of the current I/Q signal, am. A small fraction of its value is added to the historical value so that the AGC tracks the average AM carrier level. AGC speed is controlled by that fraction. Dividing by $2^{10} = 1024$ gives a time constant of about 1024 clock cycles.

Another type of detector we haven't discussed yet is for frequency modulation. For a scalar signal, the *quadrature detector* shown in **Figure 11.27** is one elegant solution. This is the same circuit whose analog equivalent has been used in millions of FM receivers around the world. In the digital implementation, the delay block is a FIFO buffer constructed from a series of shift registers. Multiplying the signal by a delayed version of itself gives an output with a cosinusoidal response versus frequency. The response crosses zero whenever the carrier frequency is $f = N/(4\tau)$, where N is an odd integer and the delay in seconds is $\tau = n/f_s$, where n is the number of samples of delay and f_s is the sample frequency. As the carrier deviates above and below the zero-crossing frequency the output varies above and below zero, just what we want for an FM detector.

For an I/Q signal, probably the most straightforward FM detector is a phase detector followed by a differentiator to remove the 6 dB per octave rolloff caused by the phase detector. The phase is just

$$\varphi = \arctan\left(\frac{Q}{I}\right)$$

You have to be a little careful since there is a 180° phase ambiguity in the arctangent function. For example,

$$\arctan\left(\frac{1}{1}\right) = \arctan\left(\frac{-1}{-1}\right)$$

Software will have to check which quadrant of the phasor diagram the I/Q signal is in and add 180° when necessary. If there is no arctan function in the library, one can be constructed using a look-up table. Frequency is the derivative of the phase. A differentiator is nothing more than a subtractor that takes the difference between successive samples.

$$f = \frac{\varphi_n - \varphi_{n-1}}{2\pi f_s}$$

where n is the sample number and f_s is the sample rate. It is important to make sure that the difference equation functions properly around 360°. If the phase variable is scaled so that 360° equals the difference between the highest and lowest representable numbers, then standard two's complement subtraction should roll over to the right value at the 360° to 0° transition. Another thing to watch out for is that the derivative of the phase may be a rather small signal, so it might be necessary to carry through all the calculations using long integers or floating-point numbers.

Table 11.1
Software AM Detector

```
static long int carrier;
long int am;
int i, q, signal;

/* Code that generates i and q omitted */
am = (long int)sqrt((long int)i*i + (long int)q*q);
signal = am - carrier;
// Divide signal by 2^10:
carrier += signal >> 10;
// Send audio output to DAC:
```

11.6.3 SSB Using I/Q Modulation

This section walks through the process of generating an upper-sideband signal using an I/Q modulator. See **Figure 11.28**. We'll first describe the mathematics in the following paragraph and then give the equivalent explanation using the phasor diagram.

The modulating signal is a sine wave at a frequency of ω_m radians per second ($\omega_m / 2\pi$

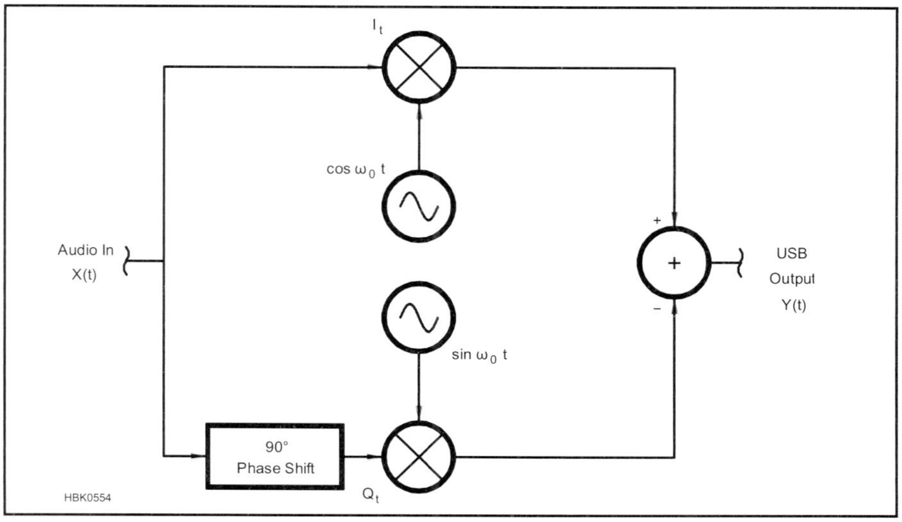

Figure 11.28 — Generating a USB signal with an I/Q modulator.

cycles per second). Because ω_m is a positive frequency the signals applied to the I/Q inputs are $I(t) = \cos(\omega_m t)$ and $Q(t) = \sin(\omega_m t)$. Assume the modulating frequency ω_m is much less than the RF frequency ω. The analytic signal is

$$x(t) = A\cos(\omega_n t) + j\sin(\omega_m t) x \left[\cos(\omega t) + j\sin(\omega t)\right]$$

so that the real, scalar signal that appears at the modulator output is

$$\mathrm{Re}[x(t)] = \cos(\omega_m t)\cos(\omega t) + \sin(\omega_m t)\sin(\omega t)$$

At the moment when t = 0, then $\cos(\omega_m t) = 1$ and $\sin(\omega_m t) = 0$, so the real signal is just $\cos(\omega t)$, the RF signal with zero phase. One quarter of a modulation cycle later $\omega_m t = \pi/2$, so $\cos(\omega_m t) = 0$ and $\sin(\omega_m t) = 1$, and the real signal is now $\sin(\omega t)$, the RF signal with a phase of $+\pi/2$, or +90°. Every quarter cycle of the modulating signal, the RF phase, increases by 90°. That means that the RF phase increases by one full cycle for every cycle of the modulation, which is another way of saying the frequency has shifted by ω_m. This creates an upper sideband at a frequency of $\omega + \omega_m$.

On the phasor diagram, the I/Q signal is rotating counterclockwise at a frequency of ω_m radians per second. As it rotates it is increasing the phase of the RF signal at the same rate, which causes the frequency to increase by ω_m radians per second. To cause the phasor to rotate in the opposite direction, creating a lower sideband, you could change the polarity of either I or Q or you could swap the I and Q inputs.

THE HILBERT TRANSFORMER

For this process to work, the baseband signals applied to the I and Q inputs must be 90° out of phase. That's not hard to do for a single sine wave, but to generate a voice SSB signal, all frequencies in the audio range must be simultaneously phase-shifted by 90° without changing their amplitudes. To do that with analog components requires a broadband phase-shift network consisting of an array of precision resistors and capacitors and a number of operational amplifiers.

To do that with DSP requires a *Hilbert transformer*, an FIR filter with a constant 90° phase shift at all frequencies. Recall that a symmetrical FIR filter has a constant *delay* at all frequencies. That means that the phase shift is not constant — it increases linearly with frequency. It turns out that with an *anti-symmetrical* filter, in which the top half of the coefficients are the negative of the mirror image of the lower half, the phase shift is 90° at all frequencies, which is exactly what we need to generate an SSB signal.

The Hilbert transformer is connected in series with either the I or Q input, depending on whether USB or LSB is desired. Just as with any FIR filter, a Hilbert transformer has a delay equal to half its length, so an equal delay must be included in the other I/Q channel as shown in **Figure 11.29**. It is possible to combine the Hilbert transformer with the normal FIR filter that may be needed anyway to filter the baseband signal. The other I/Q

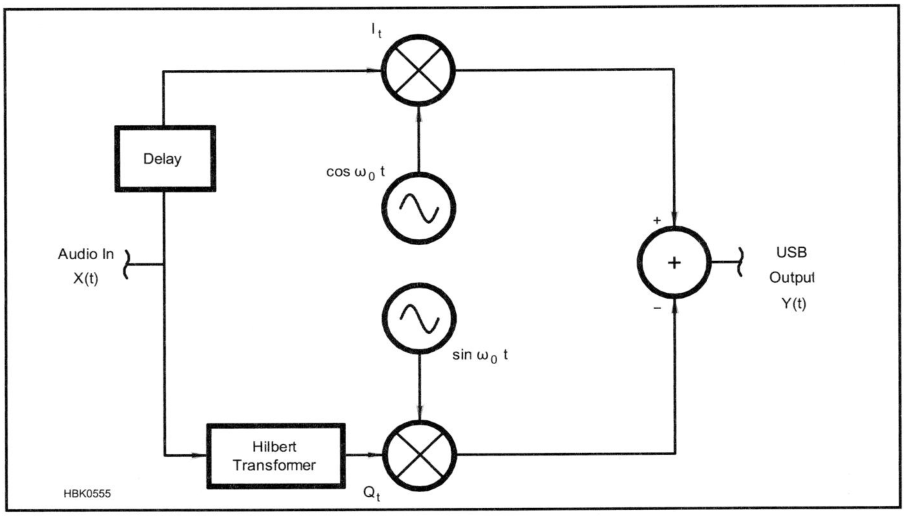

Figure 11.29 — Generating USB for a non-sinusoidal audio or speech with an I/Q modulator.

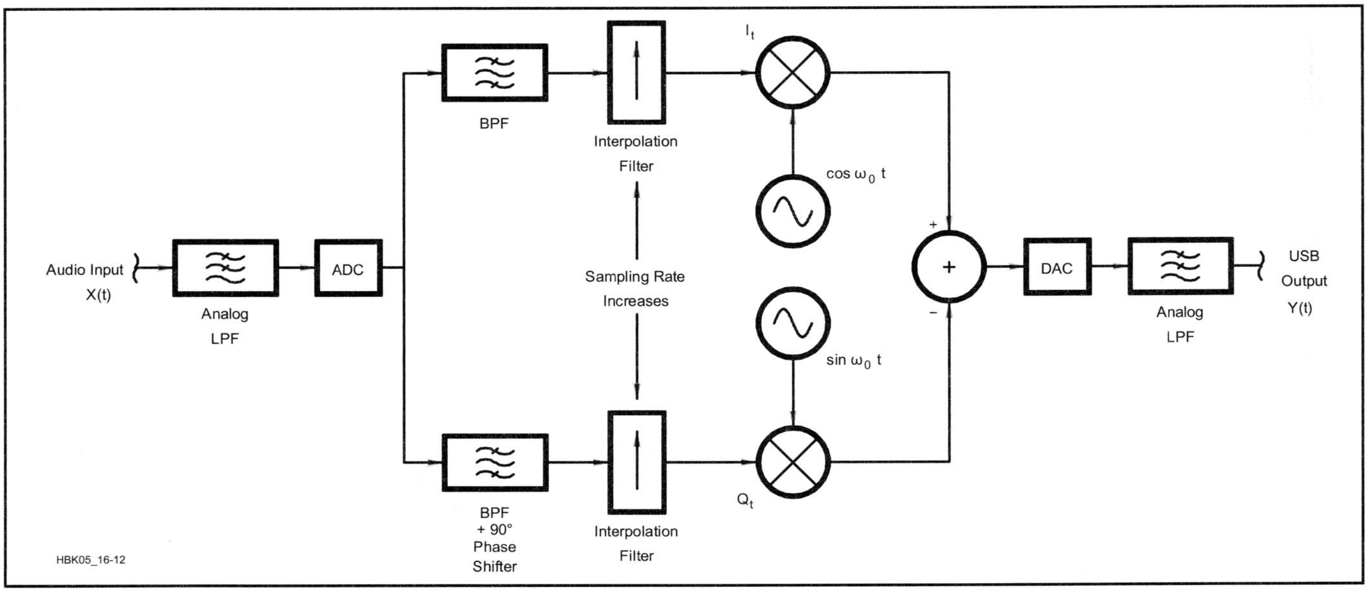

Figure 11.30 — Block diagram of a digital SSB modulator.

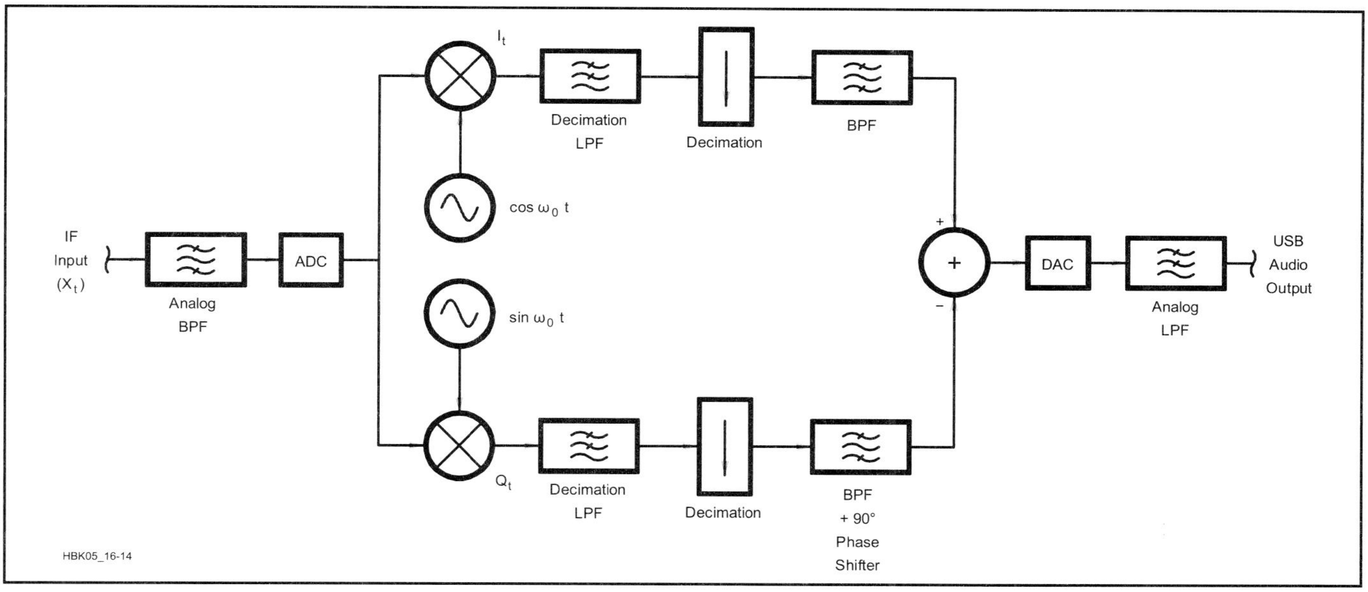

Figure 11.31 — Block diagram of a digital SSB demodulator.

channel then simply uses a similar filter with the same delay but without the 90° phase shift.

Because the RF output of the modulator is normally at a much higher frequency than the audio signal, it is customary to use a higher sample rate for the output signal than for the input. The FIR filters can still run at the lower rate to save processing time, and their output is then up-sampled to a higher rate with an interpolator. It is convenient to use an output sample rate that is exactly four times the carrier frequency because each sample advances the RF phase by exactly 90°. The sequence of values for the sine wave is 0, 1, 0 and –1. To generate the 90° phase shift for the cosine wave, simply start the sequence at the second sample: 1, 0, –1, 0. The complete block diagram is shown in **Figure 11.30**.

This is also known as the "phasing method" of SSB generation. It was popular when SSB first became common on the amateur bands back in the 1950s because suitable crystal filters were expensive or difficult to obtain. The phasing method had the reputation of producing signals with excellent-quality audio, no doubt due to the lack of the phase distortion caused by crystal filters.

A Hilbert transformer may also be used in an SSB demodulator at the receiver end of the communications system. It is basically the same block diagram drawn backwards, as illustrated in **Figure 11.31**.

It is important to note that an ideal Hilbert transformer is impossible to construct because it theoretically has an infinitely-long impulse response. However, with a sufficiently-long impulse response, the accuracy is much better than an analog phase-shift network. Just as with an analog network, the frequency passband must be limited both at the low end as well as the high end. That is, the audio must be bandpass filtered before the 90° phase shift. Actually, the filtering and phase shifting can be combined into one operation using the following method.

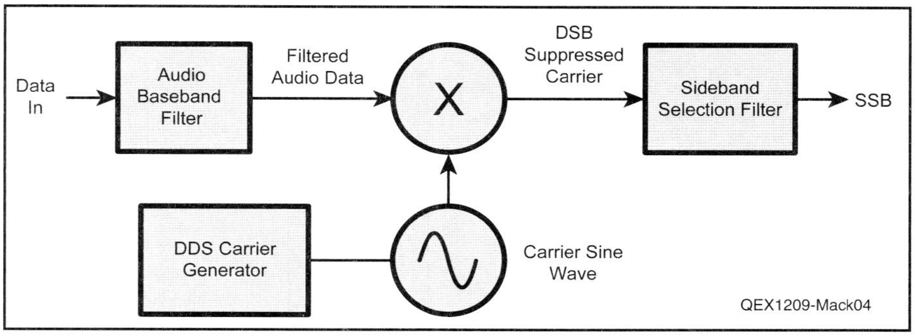

Figure 11.32 — The DSP method of using filters to create a SSB signal is simple and equivalent to the analog filter method.

First design a low-pass FIR filter with a bandwidth one-half the desired audio bandwidth. For example, if the desired passband is 300 to 2700 Hz, the low-pass filter bandwidth should be (2700 – 300)/2 = 1200 Hz. Then multiply the impulse response coefficients with a sine wave of a frequency equal to the center frequency of the desired passband, (2700 + 300)/2 = 1500 Hz in this case. That results in a band-pass filter with the desired 300 – 2700 Hz response. By using sine waves 90° out of phase for the I and Q channels, you end up with two bandpass filters with the same amplitude response and delay but a 90° phase difference at all frequencies. Multiply by a cosine for zero phase and by a sine for a 90° phase shift.

This bears a striking resemblance to the Weaver method, the so-called "third method" of SSB generation (after "filter" and "phasing"). It was originally used back in the late 1950s to eliminate the need for a wide-band audio phase-shift network. It is almost as if there is no such thing as truly new technology, just the same mathematics implemented with different technology!

THE FILTER METHOD IN DSP

As DSP hardware has grown more powerful, it has also become possible to implement the filter method of SSB generation in software. In fact, the DSP filters that can be constructed in software are quite a bit better than the best analog filters. The result is a very simple system as shown in **Figure 11.32** that is equivalent to the analog method. Ray Mack, W5IFS, discusses the various tradeoffs and filter characteristics required of this method in his Sep/Oct 2012 *QEX* SDR Simplified column (included with this book's online content).

Phil Karn, KA9Q, uses a DSP-based filter method for SSB generation because it avoids the need for a Hilbert transformer function. He performs his filtering by converting the signal to the frequency domain with an FFT (see the **DSP and SDR Fundamentals** chapter), multiplies each frequency component by a constant that defines the desired filter shape, and then converts back to the time domain with the inverse FFT. The windowing functions discussed with the FFT material can be applied to this filtering operation as well, with the Kaiser window recommended as supporting smooth trade-offs for sharp response against stop-band attenuation.

Karn has developed a fairly general purpose package in C that down-converts, filters, and detects a range of modulation/demodulates methods. The modulation/demodulation package and code for a WWV emulator is available as a github repository. For more information, see this book's website at **www.arrl.org/arrl-handbook-reference**. It currently demodulates AM, CAM (coherent AM), LSB, USB, FM, ISB (independent sideband) and IQ (straight IQ pass through). The software should work on processors that support C and floating point operations. It's not yet a polished product for the end user, but rather a set of building blocks for the experimenter.

11.6.4 Multi-Carrier Modulation

An effective method to fit more data bits into each symbol is to use more than one separately-modulated signal at a time, each on its own carrier frequency spaced an appropriate distance from the frequencies of the other carriers. An example is multi-carrier FSK. This is not to be confused with MFSK which also uses multiple frequencies, but only one at a time. With multi-carrier FSK, each carrier is present continuously and is frequency-shifted in response to a separate data stream. The total data rate equals the data rate of one carrier times the number of carriers. A disadvantage of multi-carrier FSK is that the resulting signal is no longer constant-amplitude — a linear amplifier must be used. In general, multi-carrier signals using any modulation type on each carrier tend to have high peak-to-average power ratios.

In the presence of selective fading, one or more of the carriers may disappear while the others are still present. An advantage of multi-carrier modulation is that error-correcting coding can use the unaffected carriers to reconstruct the missing data. Also, since each carrier signal is relatively narrowband, propagation conditions are essentially constant within that bandwidth. That makes it easier for the receiver to correct for other frequency-selective propagation impairments such as phase distortion. If a single-carrier signal of the same total bandwidth had been used instead, the receiver would need an adaptive equalizer to correct for the amplitude and phase variations across the transmission channel.

By using multiple carriers each with multiple-bit-per-symbol modulation it is possible to obtain quite high data rates while maintaining the low symbol rates that are required to combat the effects of multi-path propagation on the HF bands. For example, PACTOR-

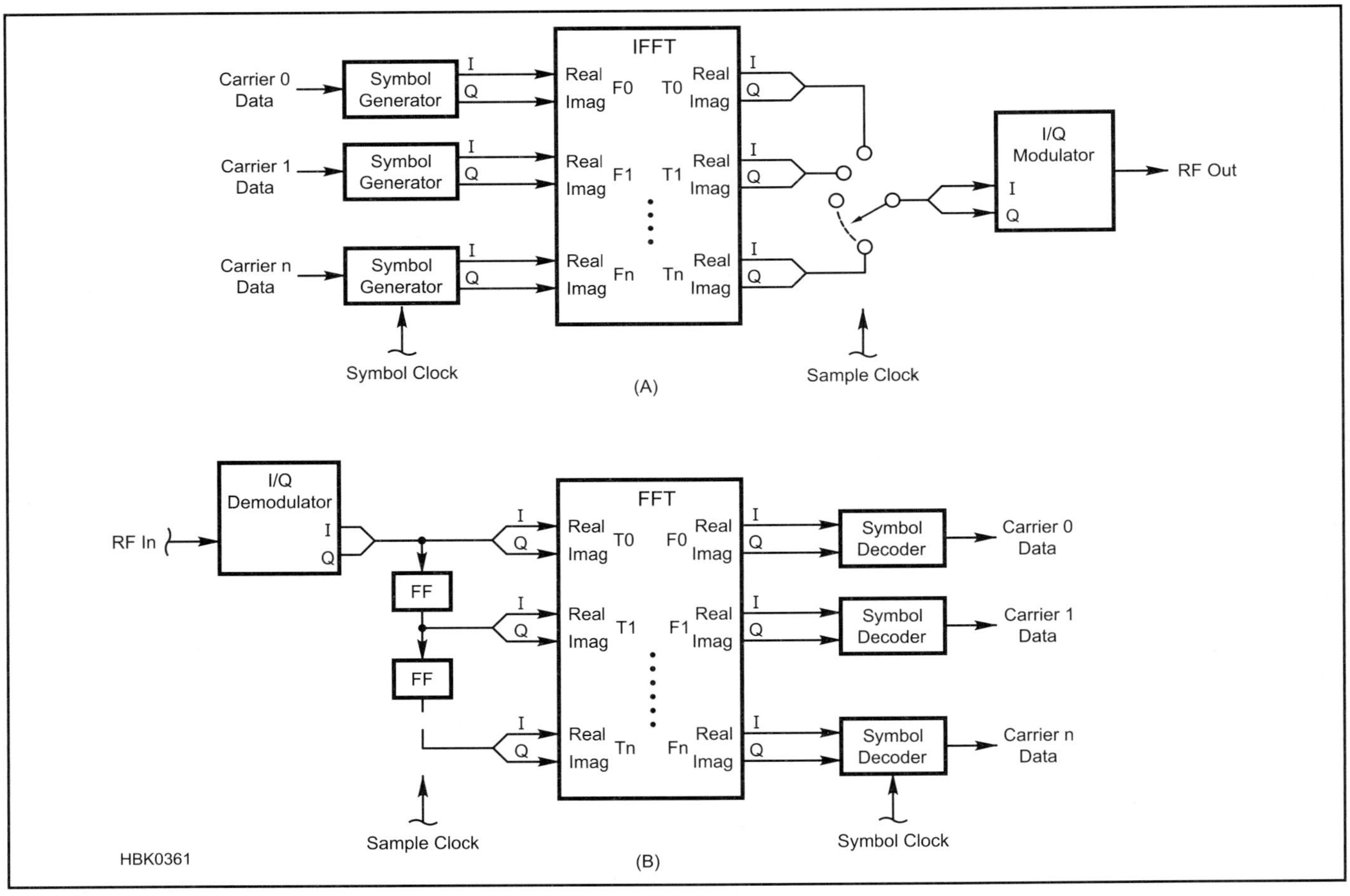

Figure 11.33 — Block diagram of an OFDM modulator (A) and demodulator (B) using the FFT/IFFT technique. The number of carriers is n, and the sample rate is n times the symbol rate. Once per symbol, all the symbol generators in the modulator are loaded with new data and the inverse fast-Fourier transform (IFFT) generates n output samples, which are selected in succession by the switch. In the demodulator, n samples are stored in a shift register (string of flip-flops) for each symbol, then the FFT generates one "frequency" output for each carrier frequency. From the amplitude and phase of each frequency, the symbol decoders can determine the symbol locations and thus the data.

III achieves a raw data rate of 3600 bits per second with a 100-baud symbol rate using 18 carriers of DQPSK. (100 bauds × 2 bits/symbol × 18 carriers = 3600 bits per second.) Similarly, Clover-2000 modulation gets 3000 bits per second with a 62.5-baud symbol rate using eight carriers of 16-DPSK combined with 4-level DASK. (62.5 bauds x (4+2) bits/symbol × 8 carriers = 3000 bits per second.) Decoding is rather fragile using these complex modulation techniques, so PACTOR and Clover include means to automatically switch to simpler, more-robust modulation types as propagation conditions deteriorate.

What is the minimum carrier spacing that can be used without excessive interference between signals on adjacent frequencies? The answer depends on the symbol rate and the filtering. It turns out that it is easy to design the filtering to be insensitive to interference on frequencies that are spaced at integer multiples of the symbol rate. (See the sub-section on filtering and bandwidth later in this chapter.) For that reason, it is common to use a carrier spacing equal to the symbol rate. The carriers are said to be *orthogonal* to each other since each theoretically has zero correlation with the others.

Orthogonal frequency-division multiplexing (OFDM), sometimes called *coded OFDM* (COFDM), refers to the multiplexing of multiple data streams onto a series of such orthogonal carriers. The term usually implies a system with a large number of carriers. In that case, an efficient decoding method is to use a DSP algorithm called the fast Fourier transform (FFT). (See the **DSP and SDR Fundamentals** chapter) The FFT is the software equivalent of a hardware spectrum analyzer. It gathers a series of samples of a signal taken at regular time intervals and outputs another series of samples representing the frequency spectrum of the signal. See **Figure 11.33**. If the length of the series of input samples equals one symbol time and if the sample rate is selected properly, then each frequency sample of the FFT output corresponds to one carrier. Each frequency sample is a complex number (containing a "real" and "imaginary" part) that represents the amplitude and phase of one of the carriers during that symbol period. Knowing the amplitude and phase of a carrier is all the information required to determine the symbol location in the I/Q diagram and thus decode the data. At the transmitter end of the circuit, an inverse FFT (IFFT) can be used to encode the data, that is, to convert the amplitudes and phases of each of the carriers into a series of I and Q time samples to send to the I/Q modulator. See Figure 11.33A.

One advantage of OFDM is high spectral efficiency. The carriers are spaced as closely as theoretically possible and, because of the narrow bandwidth of each carrier, the overall spectrum is very square in shape with a sharp drop-off at the passband edges. One disadvantage is that the receiver must be tuned very accurately to the transmitter's frequency to avoid loss of orthogonality, which causes cross-talk between the carriers.

11.7 Image Modulation

The following section covers the modulation of amateur television communications. Both full-motion (*fast-scan*) and still-frame (*slow-scan*) images can be transmitted. Fast-scan amateur television or ATV is full-motion, commercial broadcast quality, color TV video and audio using very wide channel bandwidths of several MHz. Bandwidth restrictions restrict the use of wide-band fast-scan signals to the 70 cm band and higher frequencies. Amateurs also transmit slow-scan TV (SSTV) on the HF bands below 30 MHz in a SSB voice channel bandwidth.

More detailed information on the operating conventions for amateur television can be found in the **Image Communications** operating chapter in this book's online content. Mesh network video using IP protocols over a wireless network such as AREDN is not considered in this section. See the section on High-Speed Multimedia in the **Digital Protocols and Modes** chapter.

11.7.1 Fast-Scan Television

NTSC BASEBAND VIDEO

NTSC is an abbreviation for the National Television Standards Committee that developed the modulation standard. It was released initially in 1941 (black-and-white only) and again in 1953 (color). It remained the dominant television standard until the late 1990s. The *baseband* portion of the standard describes the video signal, such as is output by a camera or video recorder before it is used to modulate an RF signal. (The current NTSC standard is SMPTE 170M — **www.smpte.org**.)

The complete NTSC video picture consists of 525 *horizontal scan lines* organized in two interlaced *even* and *odd fields* which contain the even-numbered or odd-numbered lines. (*Scanning* refers to moving an electron beam across the screen of a cathode ray tube to create the picture.) The scan lines are organized to display the picture from left to right and top to bottom.

The fields alternate at 60 Hz to display one complete *frame* 30 times per second. (The actual rates are 59.94 fields/sec and 39.97 frames/sec.) Of the 525 lines, only 480 lines are visible. The interlaced horizontal lines are the reason for the video designation for NTSC video as 480i. The complete image shown on a monitor is called a *raster*.

The video signal includes pulses to synchronize the vertical and horizontal scan display circuitry or software in the receiver. See **Figure 11.34A** which defines the basic elements of a single line in an NTSC signal. Figure 11.34B shows one horizontal line of a typical baseband NTSC video waveform in a picture of six vertical color bars. Each line contains six short segments of different colors which are seen as the sequential bursts of color information.

The video *luminance* (brightness) is represented by the signal's amplitude. The waveform's top level of 1 V represents white. The level representing black can be seen as the straight line on the far right of Figure 11.34B following the final color bar segment. Signal amplitudes below this level are termed "blacker than black." The sync pulses and the *front porch* and *back porch* areas that bracket them (see Figure 11.34A) are blacker than black.

The original function of using blacker-than-black signal levels was to turn off or *blank* the electron beam in a cathode ray tube as it returned to the left side or top of the picture without creating any visible lines or artifacts. In computer-type displays there is no electron beam but the pulses are used for timing synchronization.

Each line of video starts with a negative-going, 4.88 µs duration, *horizontal sync pulse* that occurs at a 15.734 kHz rate. The horizontal sync pulse marks the start (left-hand side) of a single horizontal line of video information. A longer, *vertical sync pulse* (not shown) occurs between the interlaced frames of information to reset the display to the top of the screen.

The horizontal sync pulse is followed by an eight-cycle sine wave *color burst* at

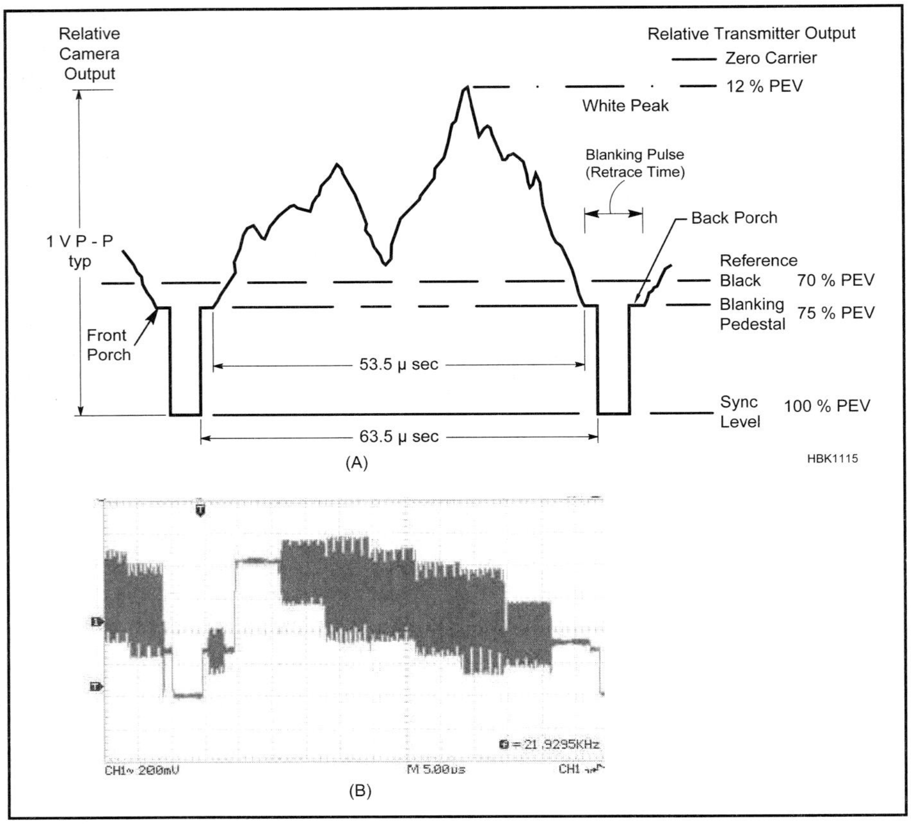

Figure 11.34 — Part A illustrates an ATV waveform, showing the relative camera output as well as the transmitter output RF power during one horizontal line scan for black-and-white TV. (A color camera would generate a "burst" of 8 cycles at 3.58 MHz on the back porch of the blanking pedestal.) Note that "black" corresponds to a higher transmitter power than "white." Part B shows the NTSC video waveform. One horizontal scan line is shown. Vertical scale is 200 mV/div and horizontal scale is 5 μs/div.

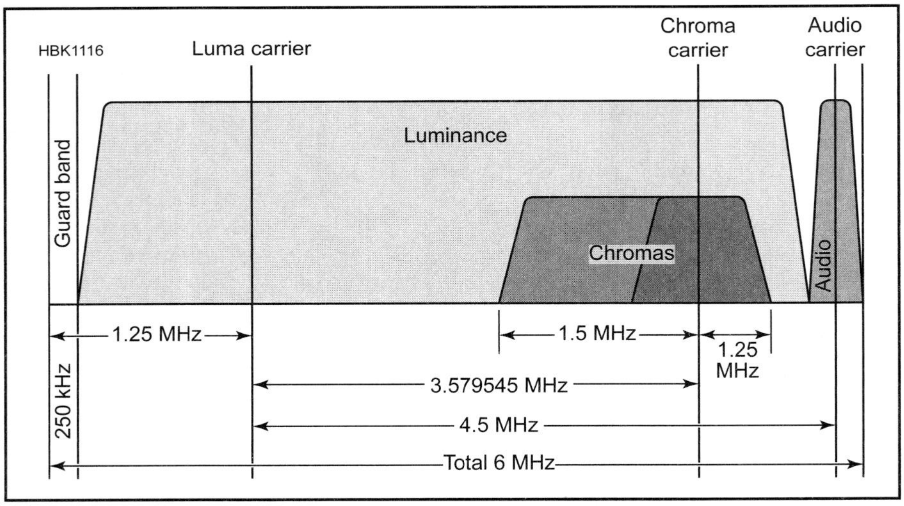

Figure 11.35 — The spectrum of an analog NTSC, VUSB-TV signal.

3.58 MHz (actually 3.579545 MHz) in the middle of the back porch. The color burst is used as a reference frequency for the *chroma* or color information and the internal oscillator used by the receiving display. The video information for this line follows the color burst.

ANALOG FAST-SCAN ATV (ATV)

Although obsolete commercially, there is a lot of ATV activity using the same NTSC modulation standard for US broadcast analog television stations prior to the change to digital TV in 2009. Analog TV receivers can be used to receive and display analog ATV signals.

Figure 11.35 shows the spectrum of an NTSC analog TV full-carrier, double-sideband AM signal filtered to partially remove the lower sideband. The partially-filtered *vestigial sideband* (VSB) extends 1.25 MHz below the carrier frequency. This is called VUSB-TV.

The resulting TV channel is 6 MHz wide to accommodate the composite video and two sub-carriers. One sub-carrier is at 3.58 MHz for the color information and the other is an FM-modulated sub-carrier at 4.5 MHz for the sound. The video bandwidth is limited to 4.2 MHz but the overall TV channel is 6 MHz wide to accommodate both the audio and video information.

In an analog AM-TV transmitter, the RF envelope waveform is inverted from the baseband video waveform. The sync pulses thus become the strongest portion of the envelope and the white levels are the weakest. An AM-TV transmitter is rated like a SSB transmitter with a peak envelope power (PEP) representing the tip of the sync pulses. The video-to-sync ratio must remain constant throughout all of the linear amplifiers in the transmit chain as the video level from the camera changes. To maintain the *sync tips* (bottom of each sync pulse) at 100% of peak power, the modulator usually contains a clamp circuit that also acts as a *sync stretcher* to compensate for amplifier gain compression.

For simplicity in transmitter design, analog ATV stations often transmit an unfiltered full-DSB AM signal, but the normally-removed portion of the lower sideband is unused by the TV receiver. (To conserve spectrum, this practice is discouraged.) The bulk of an NTSC TV signal's energy is contained close to the video carrier and within ±1 MHz of the video carrier frequency. To reduce signal bandwidth, a 6 MHz wide bandpass filter is used after DSB-AM modulation to attenuate the unused lower sideband color and sound subcarrier frequencies by 20-30 dB. This creates a Vestigial Upper Sideband or VUSB-TV signal with a spectrum as shown in Figure 11.35.

With the typical analog TV receiver's 3-dB rolloff at 3 MHz (primarily in the IF filter), up to 480 pixels per horizontal line can be seen. Color bandwidth in a TV set is lower than that, resulting in up to 100 color lines. (This is referred to as *lines of resolution* and should not be confused with the number of horizontal scan lines per frame.)

The PAL analog TV system is an AM system based on NTSC. The SECAM system uses FM color subcarriers. As described previously, FM achieves superior noise and interference suppression for signal levels above a certain threshold, although AM seems to work better for receiver signal levels below about 5 μV.

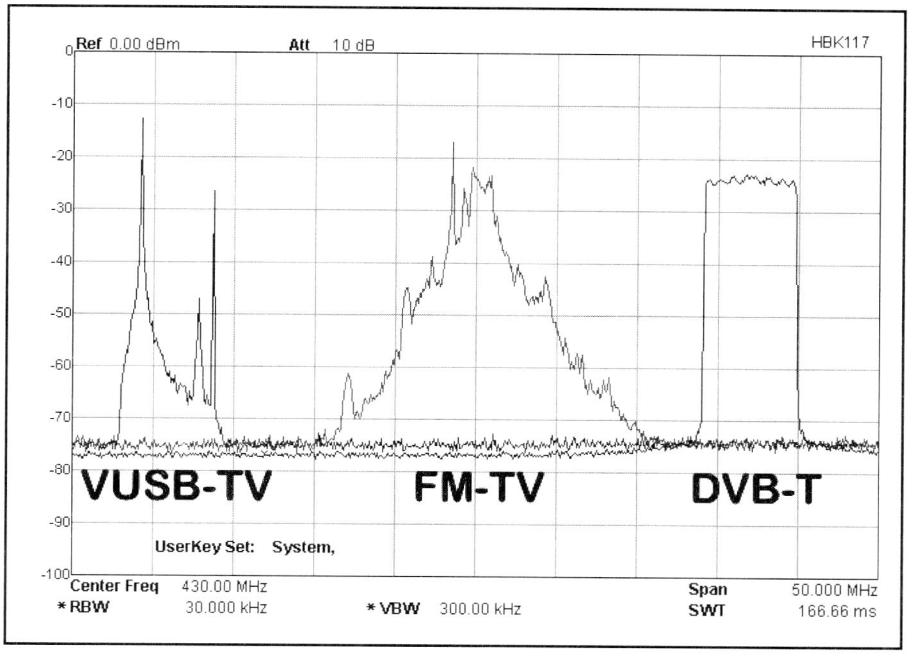

Figure 11.36 — Typical spectra of commonly used TV modulations. Span of 50 MHz shown with vertical scale 10 dB/div and horizontal scale of 5 MHz/div.

FM ATV

On the 33 cm and 23 cm bands and higher frequencies, ATV stations often use FM modulation instead of AM. The main disadvantage is the extremely wide spectrum occupied. **Figure 11.36** shows the spectrum of a typical FM-TV signal with video modulation only. If FM sound sub-carrier(s) are added, the spectrum becomes even broader.

FM ATV in the United States typically uses 4 MHz deviation with NTSC video and a 5.5-MHz sound subcarrier set to 15 dB below the video level. Using Carson's rule, the occupied bandwidth comes out to just under 20 MHz. Because it is so wide, FM-TV is not used on the 70 cm band. Most available FM ATV equipment is made for the 1.2, 2.4 and 10.25-GHz bands.

DIGITAL FAST-SCAN ATV (DATV)

Commercial digital TV (DTV) uses the same 6 MHz-wide channels for analog TV. The 6 MHz-wide over-the-air channels themselves didn't change, just the signals carried in them. Channel 2 is still 54 – 60 MHz, Channel 3 is 60 – 66 MHz, and so forth, up to Channel 51. Channels 2 – 6 have largely been abandoned in the US as broadcast digital TV moved to UHF channels. UHF channels 52 – 69 will be reallocated to other uses.

The US digital broadcast standard is called Advanced Television Standards Committee (ATSC). ATSC uses 8-VSB or 8-level vestigial sideband modulation. This is similar to 256-QAM which means 256-state *quadrature amplitude modulation*. (See the discussion on quadrature modulation earlier in this chapter.) The 256 "states" are combinations of signal phase and amplitude values that represent the 256 different transmitted symbols. This digital format is used in cable TV networks (64-QAM is also used by cable companies). In the case of 8-VSB, the "8" refers to the eight-level baseband DTV signal that amplitude modulates an IF signal. For more information on 8-VSB modulation, see the online article, "What Exactly Is 8-VSB Anyway?" by David Sparano at **www.arrl.org/files/file/Technology/TV_Channels/8_Bit_VSB.pdf** and other references at **www.arrl.org/tv-channels-air-catv**.

The eight levels result in three bits per symbol, resulting in approximately 32 Mb/second raw data rate with a 10.76 Mbaud symbol rate. The net data rate with error correction and other overhead is 19.39 Mb/second.

Digital television signals appear as noise-like signals over their 6 MHz bandwidth. If a digital TV signal interferes with analog radio communications such as FM or SSB, the effect is similar to degraded signal-to-noise ratio. Indeed, a digital TV signal can be thought of as a 6 MHz wide "pile of noise." On a spectrum analyzer, it looks like a "haystack" as shown in Figure 11.36. Of course, the IF bandwidth of the amateur receiver is quite narrow relative to the 6 MHz-wide digital TV channel, so the actual noise level seen by the receiver (assuming no overload or IMD problems) will in part be determined by the receiver's IF bandwidth. Still, the effect on the receiver is what amounts to an elevated noise floor, similar to the effects of wide-spectrum broadband noise.

In Europe and most of the rest of the world, Digital Video Broadcast (DVB) is used. For DVB three basic variations are used depending on what type of signal propagation is available.

DVB-C is used for cable TV where the medium is perfect with strong signal levels and low VSWR, i.e. no reflections. It is not used for DATV as it does not work well when multi-path is present.

DVB-S is for satellite TV where the signals are weak but with little or no multi-path. DVB-S uses MPEG2 audio and video data compression and QPSK modulation with symbol rates up to 20 Mbaud. Most DATV use of DVB-S has been on the 23 cm band since inexpensive set-top demodulation boxes are available for that frequency range.

DVB-T is for terrestrial broadcast TV where the signals could be weak or strong, but where multi-path (what causes "ghost" images on analog TV) is almost always present. Here in the USA, most DATV stations are using DVB-T modulation. In the US, a new modulation standard called ATSC 3.0 is being introduced for broadcast TV. It is an enhanced form of DVB-T and is not backward compatible with ATSC 1.0. DVB-T2 and DVB-S2 are newer versions that will allow 4K high-definition video.

The British Amateur Television Club (BATC) has taken a lead in pushing the bandwidth limits for DTV down far below the commercial standards. They have developed systems in both DVB-S and DVB-T where the bandwidth is less than 1 MHz and still transmit high quality digital video. DATV stations living in major metro areas where the radio spectrum is very crowded are now successfully using narrow-band DVB-T with 2 and 4 MHz bandwidths.

ATV REPEATERS

ATV stations also use repeaters to greatly expand the reach of their signals. Many of the current ATV repeaters are still analog but some are replacing their analog systems or adding digital capability. ATV repeaters are predominately found on the 70 cm and 23 cm bands, with some having capabilities up to the 3 cm (10 GHz) band.

European DATV stations can also use the QO-100 satellite for relaying DATV signals. QO-100 is a stationary satellite covering Europe, Asia, and Africa. The uplink is on the 2.4 GHz band and the downlink on the 10 GHz band. There are some US ATV repeater groups experimenting with DVB-S, ATSC, and ATSC 3.0.

11.7.2 Slow-Scan Television

Despite its name, so-called *slow-scan television* (SSTV) is a method for sending still images, like facsimile. The original monochrome analog SSTV format takes approximately 8 seconds to send one complete frame. The 1500 to 2300 Hz frequency-modulated audio tone resembles that of a fax signal but is sent at a faster rate and includes pulses at 1200 Hz for synchronization.

Color may be sent using any of several methods. The first to be used was the *frame-sequential* method, in which each of the three primary colors (red, green, and blue) is sent sequentially, as a complete frame. A disadvantage of sending complete frames for each color is that you must wait for the third frame to begin before colors start to become correct, and any noise or interference is three times more likely to corrupt the image and risks ruining the image registration (the overlay of the frames), thus spoiling the picture.

In the *line-sequential* method, each line is electronically scanned three times: once each for red, green, and blue. Pictures scan down the screen in full color as they are received, and registration problems are reduced. The Wraase SC-1 modes are examples of early line-sequential color transmission. They have a horizontal sync pulse for each of the color component scans. The major weakness of this method is that if the receiving system gets out of step, it doesn't know which scan represents which color.

Rather than sending color images with the usual RGB (red, green, blue) components, Robot Research decided to use luminance and chrominance signals for their 1200C modes. The first half or two thirds of each scan line contains the luminance information, which is a weighted average of the R, G, and B components. The remainder of each line contains the chrominance signal with the color information. Existing black-and-white equipment could display the B&W-compatible image on the first part of each scan line and the rest would go off the edge of the screen. That compatibility was very beneficial when most people still had only B&W equipment.

The luminance-chrominance encoding makes more efficient use of the transmission time. A 120-line color image can be sent in 12 seconds, rather than the usual 24. Our eyes are more sensitive to details in changes of brightness than color, so the time is used more efficiently by devoting more time to luminance than chrominance. The NTSC and PAL broadcast standards also take advantage of this vision characteristic and use less bandwidth for the color part of the signal. For SSTV, luminance-chrominance encoding offers some benefits, but image quality suffers. It is acceptable for most natural images but looks bad for sharp, high-contrast edges, which are more and more common as images are altered via computer graphics. As a result, all newer modes have returned to RGB encoding.

The 1200C introduced another innovation, called *vertical interval signaling* (VIS). It encodes the transmission mode in the vertical sync interval. By using narrow FSK encoding around the sync frequency, compatibility is maintained. This new signal just looks like an extra-long vertical sync to older equipment.

The Martin and Scottie modes are essentially the same except for the timings. They have a single horizontal sync pulse for each set of RGB scans. Therefore, the receiving end can easily get back in step if synchronization is temporarily lost. Although they have horizontal sync, some implementations ignore them on receive. Instead, they rely on very accurate time bases at the transmitting and receiving stations to keep in step. The advantage of this "synchronous" strategy is that missing or corrupted sync pulses won't disturb the received image. The disadvantage is that even slight timing inaccuracies produce slanted pictures.

In the late 1980s, yet another incompatible mode was introduced. The AVT mode is different from all the rest in that it has *no horizontal sync.* It relies on very accurate oscillators at the sending and receiving stations to maintain synchronization. If the beginning-of-frame sync is missed, it's all over. There is no way to determine where a scan line begins. However, it's much harder to miss the 5-s header than the 300-ms VIS code. Redundant information is encoded 32 times and a more powerful error-detection scheme is used. It's only necessary to receive a small part of the AVT header to achieve synchronization. After this, noise can wipe out parts of the image, but image alignment and colors remain correct.

Digital images may be sent over amateur radio using any of the standard digital modulation formats that support binary file transfer. *Digital SSTV* (DSSTV) is one method of transmitting computer image files, such as JPEG or GIF, as described in an article by Ralph Taggart, WB8DQT, in the Feb 2004 issue of *QST.* This format phase-modulates a total of eight subcarriers (ranging from 590 to 2200 Hz) at intervals of 230 Hz. Each subcarrier has nine possible modulation states. This signal modulation format is known as *redundant digital file transfer* (RDFT) developed by Barry Sanderson, KB9VAK.

Most digital SSTV transmission has switched to using Digital Radio Mondiale (DRM), derived from a system developed for shortwave digital voice broadcasting. The DRM digital SSTV signal occupies the bandwidth between 350 and 2750 Hz. As many as 57 subcarriers may be sent simultaneously, all at the same level. Three pilot carriers are sent at twice the level of the other subcarriers. The subcarriers are modulated using OFDM and QAM, which were described earlier in this chapter. DRM SSTV includes several methods of error correction.

11.8 Spread Spectrum Modulation

A *spread-spectrum* (SS) system is one that intentionally increases the bandwidth of a digital signal beyond that normally required by means of a special spreading code that is independent of the data sequence. There are several reasons for spreading the spectrum in that way.

Spread spectrum was first used in military systems, where the purpose was to encrypt the transmissions to make it harder for the enemy to intercept or jam them. Amateurs are not allowed to encrypt transmissions for the purpose of concealing the information, but reducing interference, intentional or otherwise, is an obvious benefit. The signal is normally spread in such a fashion that it appears like random noise to a receiver not designed to receive it, so other users of the band may not even be aware that an SS signal is present.

Another advantage to spreading the spectrum is that it can make frequency accuracy less critical. In addition, the wide bandwidth means that expensive narrow-bandwidth filters are not required in the receiver. It also provides a measure of frequency diversity. If certain frequencies are unusable because of interference or selective fading, the signal can often be reconstructed using information in the rest of the bandwidth.

There are several ways to spread the spectrum — we will cover the two most common methods below — but they all share certain characteristics. Imagine that the unspread signal occupies a 10 kHz bandwidth and it is spread by a factor of 100. The resulting SS signal is 1000 kHz (1 MHz) wide. Each 10-kHz channel contains 1/100 of the total signal power, or –20 dB. That means that any narrowband stations using one of those 10-kHz channels experience a 20 dB reduction in interference, but also are more likely to be interfered with because of the 100-times greater bandwidth of the SS station's emissions.

How is the spread-spectrum station affected by interference from narrowband stations? In effect, the SS receiver attenuates the signal received on each 10 kHz channel by 20 dB in order to obtain a full-power signal when all 100 channels are added together. That means that the interference from a narrowband station is reduced by 20 dB but, again, the interference is more likely to occur because of the 100-times greater bandwidth of the SS station's receiver.

How is the spread-spectrum station affected by interference from another SS station on the same frequency? It turns out that if the other station is using a different orthogonal spreading code then, once again, the interference reduction is 20 dB for 100-times spreading. That means that many SS stations can share the same channel without interference as long as they are all received at roughly the same signal level. Commercial mobile-telephone SS networks use an elaborate system of power control with real-time feedback to ensure that the signals from all the mobile stations arrive at the base station at approximately the same level.

That scheme works well in a one-to-many (base station to mobile stations) system architecture but would be much more difficult to implement in a typical amateur many-to-many arrangement because of the different distances and thus path losses between each pair of stations in the network. On the HF bands it is not uncommon to see differences in signal levels of 80 to 90 dB or more. (For example, the difference between S1 and 40 dB over S9 is 88 dB, assuming 6 dB per S-unit.) A spread spectrum signal at S9 + 40 dB with a spreading ratio of 100 times would interfere with any other signals below about S9 + 20 dB. It works the same in the other direction as well. The SS signal would experience interference from any other stations that are more than 20 dB louder than the desired signal.

Normally, increasing the bandwidth of a transmission degrades the signal-to-noise (S/N) ratio at the receiver. A 100-times greater bandwidth contains 100 times as much noise, which causes a 20 dB reduction in S/N ratio. However SS receivers benefit from a phenomenon known as processing gain. Just as the receiver is insensitive to other SS signals with different orthogonal spreading codes, so it is insensitive to random noise. The improvement in S/N ratio due to processing gain is:

$$\text{Processing gain} = 10 \times \log\left(\frac{\text{spread bandwidth}}{\text{unspread bandwidth}}\right) \text{dB}$$

That is exactly equal to the reduction in S/N ratio due to the increased bandwidth. The net result is that an SS signal has neither an advantage nor a disadvantage in signal-to-noise ratio compared to the unspread version of the same signal. When someone states that, because of processing gain, an SS receiver can receive signals that are below the noise level (signals that have a negative S/N ratio), that is a true statement. However, it does not imply better S/N performance than could be obtained if the signal were not spread.

11.8.1 Frequency Hopping Spread Spectrum

One simple way to spread the spectrum of a narrowband signal is to repetitively sweep it across the frequency range of the wider spectrum, either continuously or in a series of steps at discrete frequencies. That technique, called *chirp modulation*, can be considered a special case of *frequency-hopping spread spectrum* (FHSS), in which the narrowband signal covers the expanded spectrum by rapidly hopping back and forth from frequency to frequency in a pseudo-random manner. On average, each frequency is used the same percentage of time so that the average spectrum is flat across the bandwidth of the FHSS signal. See **Figure 11.37**.

If the receiver hops in step with the transmitter, using the same pseudo-random sequence synchronized to the one in the transmitter, then the transmitter and receiver are always tuned to the same frequency and the receiver's detector sees a continuous, non-hopped narrowband signal which can be demodulated in the normal way. We say that the signal has been *de-spread*, that is, returned to its normal narrowband form. Synchronization between the receiver and transmitter is one of the challenges in an FHSS system. If the timing of the two sequences differs by even one hop, then the receiver is always tuned to the wrong frequency, unless the same frequency happens to occur twice in succession in the pseudo-random sequence. Any signal with an unsynchronized sequence, or with a different sequence, is reduced in amplitude by the processing gain.

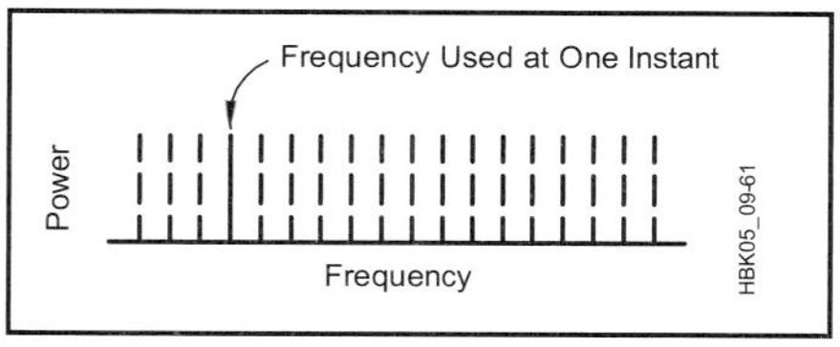

Figure 11.37 — Power versus frequency for a frequency-hopping spread spectrum signal. Emissions jump between discrete frequencies in pseudo-random fashion. Normally the spacing of the frequencies is approximately equal to the bandwidth of the unspread signal so that the average spectrum is approximately flat.

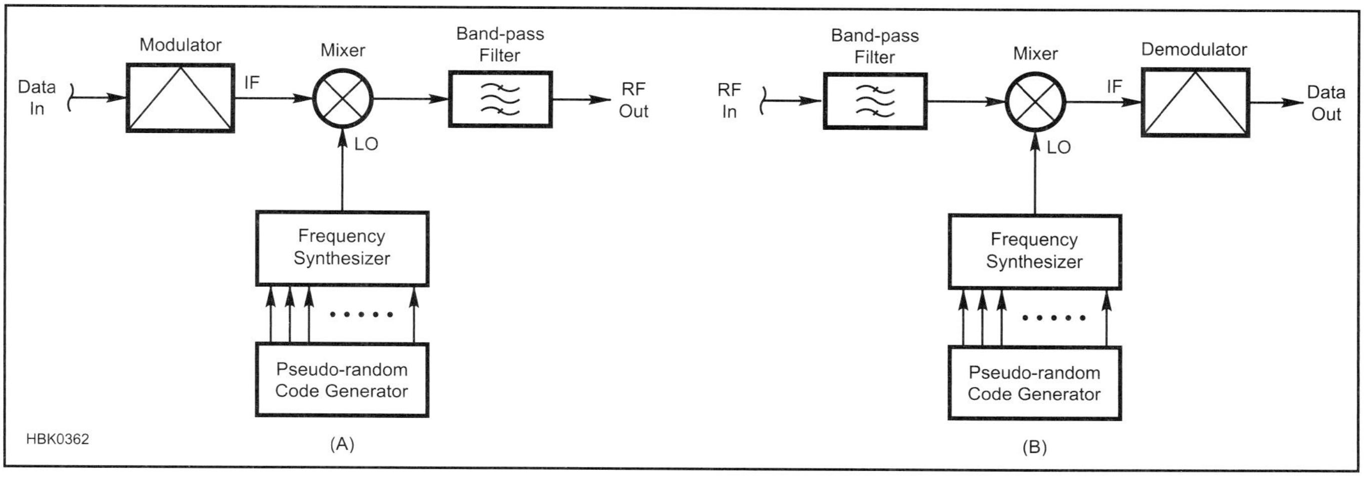

Figure 11.38 — Block diagram of an FHSS transmitter (A) and receiver (B). The receiver may be thought of as a conventional superhet with a local oscillator (LO) that is continually hopping its frequency in response to a pseudo-random code generator. The transmitter has a similar architecture to up-convert a conventionally-modulated intermediate frequency (IF) to a frequency-hopped radio frequency (RF).

There are two types of FHSS based on the rate at which the frequency hops take place. *Slow-frequency hopping* refers to a hop rate slower than the baud rate. Several symbols are sent per hop. With *fast-frequency hopping*, the hop rate is faster than the baud rate. Several hops occur during each symbol. The term *chip* refers to the shortest-duration modulation state in the system. For slow-frequency hopping, that is the baud rate. For fast-frequency hopping it is the hop rate. Fast-frequency hopping can be useful in reducing the effects of multi-path propagation. If the hop period is less than the typical time delay of secondary propagation paths, then those signals are uncorrelated to the main path and are attenuated by the processing gain.

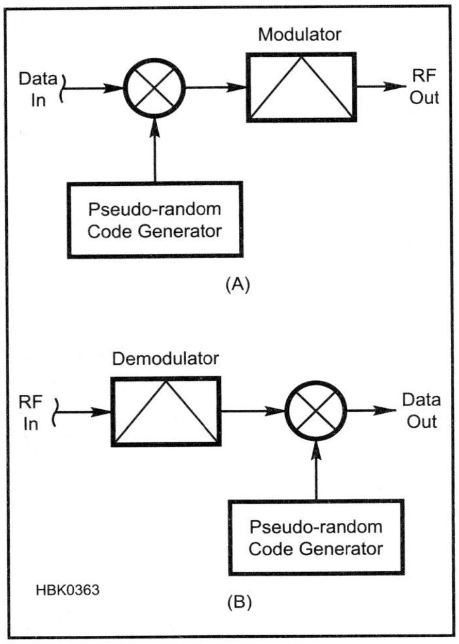

Figure 11.39 — Block diagram of a DSSS transmitter (A) and receiver (B). For BPSK modulation, the modulation has the same format as the bipolar data (±1), so the modulator and demodulator could be moved to the other side of the multiplier if desired.

A diagram of a frequency-hopping spread spectrum system is shown in **Figure 11.38**. In both the transmitter and the receiver, a pseudo-random code generator controls a frequency synthesizer to hop between frequencies in the correct order. In this way, the narrowband signal is first spread by the transmitter, then sent over the radio channel, and finally de-spread at the receiver to obtain the original narrowband signal again. One issue with FHSS is that many synthesizers do not maintain phase coherence over successive frequency hops. That means the basic (non-spread) modulation must be a type that does not depend on phase information. That rules out PSK, QPSK and QAM. That is the reason that modulation types that do not depend on phase, such as FSK and MFSK with non-coherent detection, are frequently used as the base modulation type in FHSS systems.

11.8.2 Direct Sequence Spread Spectrum

Whereas an FHSS system hops through a pseudo-random sequence of frequencies to spread the signal, a *direct-sequence spread spectrum* (DSSS) system applies the pseudo-random sequence directly to the data in order to spread the signal. See **Figure 11.39**. The binary data is considered to be in *polar* form, that is, the two possible states of each bit are represented by –1 and +1. The data bits are multiplied by a higher-bit-rate pseudo-random sequence, also in polar form, which chops the data into smaller time increments called *chips*. The ratio of the chip rate to the data's bit rate equals the ratio of the spread bandwidth to the unspread bandwidth, which is just the processing gain:

$$\text{Processing gain} = 10 \times \log\left(\frac{\text{chip rate}}{\text{bit rate}}\right) \text{dB}$$

Although it doesn't have to, DSSS normally uses a one-bit-per-symbol modulation type such as BPSK. In that case, the modulator and demodulator in Figure 11.39 would consist simply of a mixer, which multiplies the RF local oscillator by the bipolar DSSS modulating signal. Since unfiltered BPSK is constant-envelope, a nonlinear class-C power amplifier may be used for high efficiency. The demodulator in the receiver would also be a mixer, which multiplies the RF signal by a local oscillator to regenerate the DSSS modulating signal. Not shown are additional mixers and filters that would be used in a superheterodyne receiver to convert the received signal to an intermediate frequency before demodulation and decoding.

The unfiltered spectrum has the form of a sinc function. That shows up clearly in the spectrum of an actual DSSS signal in **Figure 11.40A**, which is plotted on a logarithmic scale calibrated in decibels. The humps in the response to the left and right (and additional ones not shown off scale) are not needed for communications and should be filtered out to avoid excessive occupied bandwidth. Figure 11.40B shows the DSSS signal in the presence of noise at the input to the receiver, and Figure 11.40C illustrates the improvement in signal-to-noise ratio of the de-spread narrowband signal.

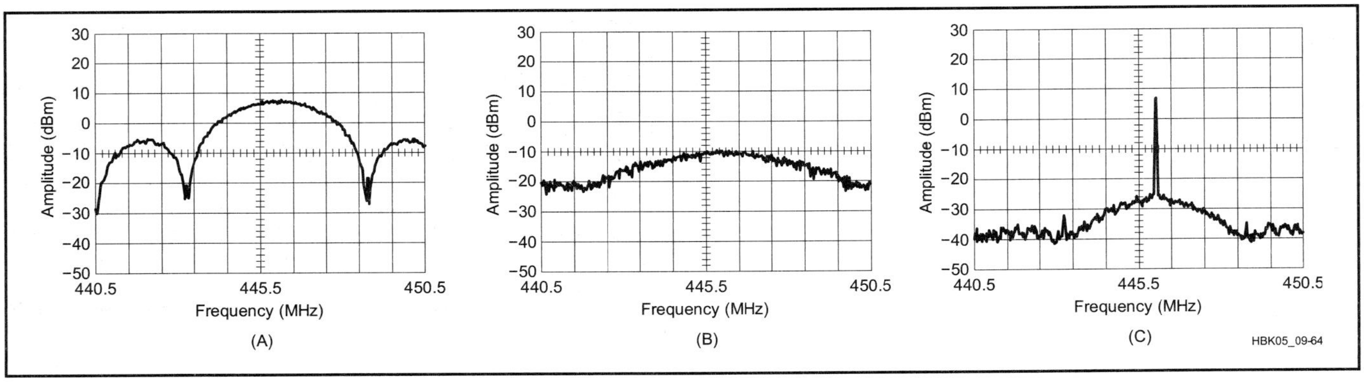

Figure 11.40 — (A) The frequency spectrum of an actual unfiltered biphase-modulated spread spectrum signal as viewed on a spectrum analyzer. In this practical system, band-pass filtering is used to confine the spread spectrum signal to the amateur band. (B) At the receiver end of the line, the filtered spread spectrum signal is apparent only as a 10-dB hump in the noise floor. (C) The signal at the output of the receiver de-spreader. The original carrier — and any modulation components that accompany it — has been recovered. The peak carrier is about 45 dB above the noise floor — more than 30 dB above the hump shown at B. (These spectrograms were made at a sweep rate of 0.1 s/division and an analyzer bandwidth of 30 kHz; the horizontal scale is 1 MHz/division.)

11.8.3 Code-Division Multiple-Access (CDMA)

As mentioned before, to receive an SS signal, the de-spreading sequence in the receiver must match the spreading sequence in the transmitter. The term *orthogonal* refers to two sequences that are coded in such a way that they are completely uncorrelated. The receiver's response to an orthogonal code is the same as to random noise, that is, it is suppressed by a factor equal to the processing gain. One can take advantage of this property to allow multiple SS stations to access the same frequency simultaneously, a technique known as *code-division multiple access* (CDMA). Each transmitter is assigned a different orthogonal code. A receiver can "tune in" any transmitter's signal by selecting the correct code for de-spreading.

If multiple stations want to be able to transmit simultaneously without using spread spectrum, they must resort to either *frequency-division multiple access* (FDMA), where each station transmits on a different frequency channel, or *time-division multiple access* (TDMA), where each transmission is broken up into short time slots which are interleaved with the time slots of the other stations. Compared to TDMA, CDMA has the advantage that it does not require an external synchronization network to make sure that different stations' time slots do not overlap. Compared to both TDMA and FDMA, CDMA has the further advantage that it experiences a gradual degradation in performance as the number of stations on the channel increases. It is relatively easy to add new users to the system. Also CDMA has inherent resistance to interference due to multi-path propagation or intentional jamming. The primary disadvantages of CDMA are the relative complexity and the necessity for accurate power-level control to make sure that unwanted signals do not exceed the level that can be rejected through processing gain.

11.9 Pulse Modulation

Another type of digital amplitude modulation comes under the general category of *pulse modulation*. The RF signal is broken up into a series of pulses, which are usually equally-spaced in time and separated by periods of no signal. We will discuss three types of pulse modulation, PAM, PWM and PPM.

Pulse-amplitude modulation (PAM) consists of a series of pulses of varying amplitude that correspond directly to the amplitude of the modulating signal. See **Figure 11.41B**. The pulses can be positive or negative, depending on the polarity of the signal. The modulating signal can be recovered simply by low-pass filtering the pulse train. The effect of the negative pulses is to reverse the polarity of the RF signal. The result is very similar to a double-sideband, suppressed-carrier signal that is rapidly turned on and off at the pulse rate of the PAM. In other words, the DSBSC signal is periodically sampled at a certain pulse repetition rate.

For the signal to be properly represented, its highest modulating frequency must be less than the *Nyquist frequency*, which is one-half the sample rate. That condition, known as the *Nyquist criterion*, applies not only to PAM but to any digital modulation technique.

There are two variations of PAM that should be mentioned. *Natural sampling* does not hold the amplitude of each pulse constant throughout the pulse as shown in Figure 11.41B, but rather follows the shape of the analog modulation. The *single-polarity method* adds a fixed offset, or pedestal, to the modulating signal before the PAM modulator. As long as the pedestal is greater than or equal to the peak negative modulation, then the RF phase never changes and the signal is equivalent to sampled full-carrier AM rather than DSB-SC.

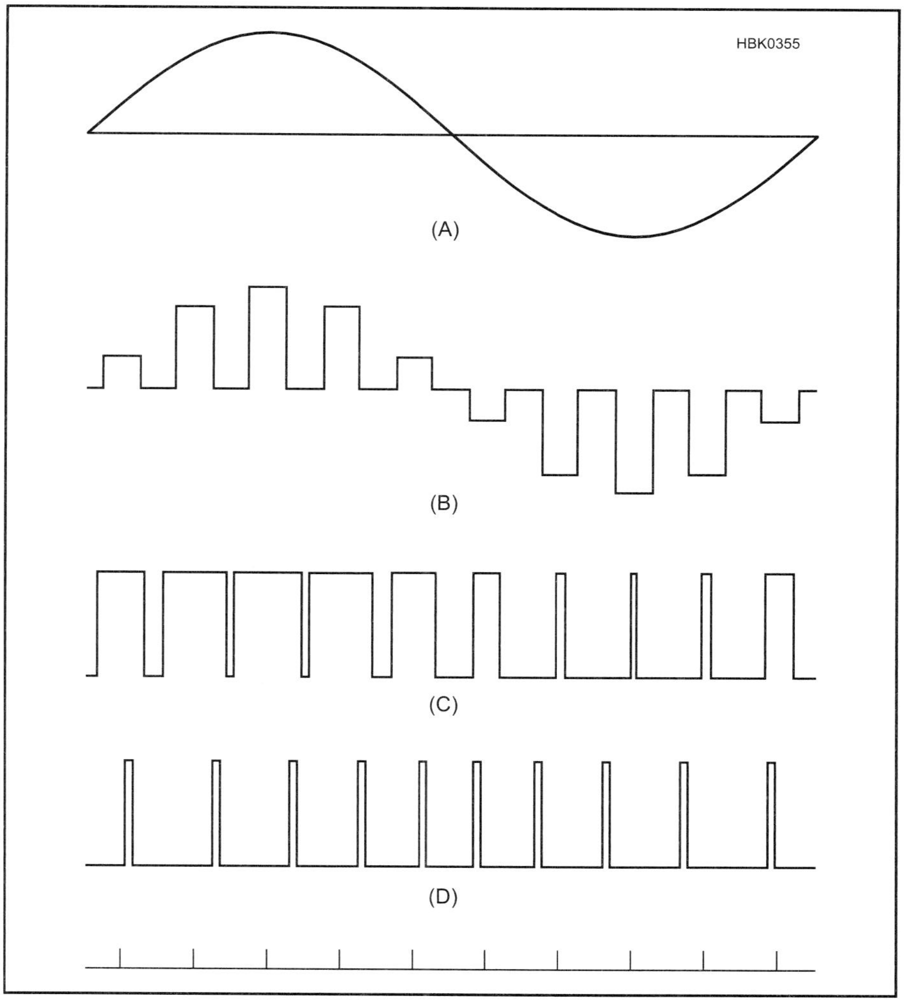

Figure 11.41 — Three types of pulse-code modulation. A sine-wave modulating signal (A) is shown at the top. Pulse-amplitude modulation (B) varies the amplitude of the pulses, pulse-width modulation (C) varies the pulse width, and pulse-position modulation (D) varies the pulse position, proportional to the modulating signal. The tic marks show the nominal pulse times. The RF signal is created by an AM modulator using (B), (C) or (D) as the modulating signal.

PAM is rarely used on the amateur bands because it increases the transmitted bandwidth and adds circuit complexity with no improvement in signal-to-noise ratio for most types of noise and interference. The concept is useful, however, because PAM is similar to the signal generated by the sample-and-hold circuit that is used at the input to an analog-to-digital converter (ADC). An ADC is used in virtually every digital transmitter to convert the analog voice signal to a digital signal suitable for digital signal processing.

Pulse-width modulation (PWM) is a series of pulses whose width varies in proportion to the amplitude of the modulating signal. Figure 11.41C shows the pulses centered on the sample times, but in some systems the sample times may correspond to the leading or trailing pulse edges. With either method, the modulating signal can be recovered by low-pass filtering the pulse train and passing it through a coupling capacitor to remove the dc component.

Pulse-position modulation (PPM), Figure 11.41D, varies the position, or phase, of the pulses in proportion to the amplitude of the modulating signal. With both PWM and PPM, the peak amplitude of the signal is constant. That allows the receiver to be designed to be insensitive to amplitude variations, which can result in a better post-detection signal-to-noise ratio, in a manner similar to analog angle modulation.

11.10 Modulation Bandwidth and Impairments

Most of the previous discussion of the various modulation types has assumed the modulation is perfect. With analog modulation, that means the audio or video modulating signal is perfectly reproduced in the RF waveform without distortion, spurious frequencies or other unwanted artifacts. With digital modulation, the symbol timing and the locations and trajectories in the I/Q constellation are perfectly accurate. In all cases, the RF power amplifier is perfectly linear, if so required by the modulation type, and it introduces no noise or other spurious signals close to the carrier frequency.

In the real world, of course, such perfection can never be achieved. Some modulation impairments are caused by the transmitting system, some by the transmission medium through which the signal propagates, and some in the receiving system. This section will concentrate on impairments caused by the circuitry in the transmitter and, to some extent, in the receiver. Signal impairments due to propagation are covered in detail in the **Propagation of Radio Signals** chapter.

11.10.1 Filtering and Bandwidth of Digital Signals

We have already touched on this topic in previous sections, but let us now cover it a little more systematically. The bandwidth required by a digital signal depends on the filtering of the modulation, the symbol rate and the type of modulation. For linear modulation types such as OOK, BPSK, QPSK and QAM, the bandwidth depends only on the symbol rate and the modulation filter.

As an example, an unfiltered BPSK modulating signal with alternating ones and zeroes for data (10101010...) is a square wave at one-half the symbol rate. Like any square wave, its spectrum can be broken down into a series of sine waves at the fundamental frequency (symbol rate / 2) and all the odd harmonics. If the data consists of alternating pairs of ones and pairs of zeroes (11001100...) then we have a square wave at one-fourth the symbol rate and the spectrum is a series of sine waves at one-fourth the symbol rate and all its odd harmonics. Random data contains energy at all frequencies from zero to half the symbol rate and all the odd harmonics of those frequencies. The harmonics are not needed for proper demodulation of the signal, so they can be filtered out with a low-pass filter with a cutoff frequency of one-half the symbol rate.

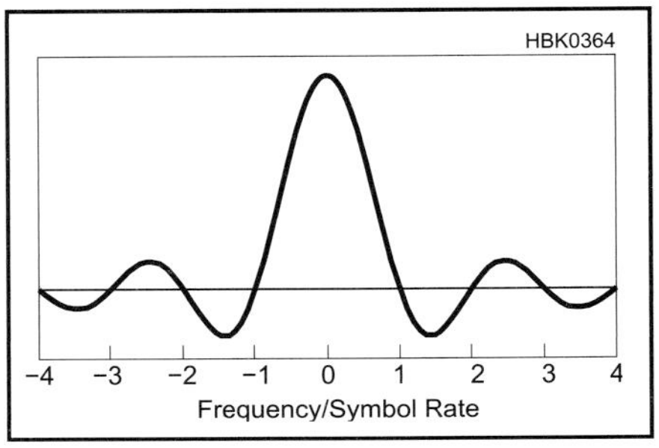

Figure 11.42 — The sinc function, which is the spectrum of an unfiltered BPSK modulating signal with random data, plotted with a linear vertical scale. The center (zero) point corresponds to the RF carrier frequency, To see what the double-sided RF spectrum looks like on a logarithmic (dB) scale, see Figure 11.40A.

With random data, the shape of the unfiltered spectrum is a sinc function,

$$\mathrm{sinc}(f / f_s) = \frac{\sin(\pi f / f_s)}{\pi f / f_s}$$

where f_s is the symbol rate. See **Figure 11.42**. Note that the response is zero (minus infinity dB) whenever f is an integer multiple of the symbol rate. That is why multi-carrier modulation generally uses a carrier spacing equal to the symbol rate.

The previous discussion applies to the baseband signal, before it modulates the RF carrier. BPSK is a double-sideband-type modulation so the baseband spectrum appears above and below the RF carrier frequency, doubling the bandwidth to 2 × (symbol rate / 2) = symbol rate. In reality, since no practical filter has an infinitely-sharp cutoff, the occupied bandwidth of a BPSK signal must be somewhat greater than the symbol rate. That is also true for all other double-sideband linear modulation types, such as OOK and the various forms of QPSK and QAM.

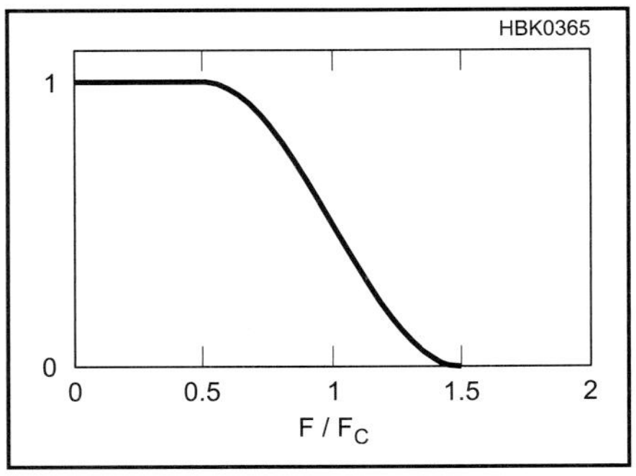

Figure 11.43 — Amplitude versus frequency for a 0.5-alpha, raised-cosine filter. The vertical scale is linear, not logarithmic as would be seen on a spectrum analyzer. The amplitude is 0.5 (–6 dB) at the cutoff frequency, Fc. The amplitude is 1.0 for frequencies less than Fc × (1 – alpha) and is 0.0 for frequencies above Fc × (1 + alpha).

If the low-pass filtering is not done properly, it may slow down the transition between symbols to the point where one symbol starts to run into another, causing *inter-symbol interference* (ISI). A type of filter that avoids that problem is called a *Nyquist filter*. It ensures that each symbol's contribution to the modulating signal passes through zero at the center of all other symbols, so that no ISI occurs. The most common type of Nyquist filter is the *raised-cosine* filter, so-called because the frequency response (plotted on a linear scale) in the passband-to-stopband transition region follows a raised-cosine curve. See **Figure 11.43**. The sharpness of the frequency cutoff is specified by a parameter called *alpha*. If alpha is 1.0, then the transition from passband to stopband is very gradual — it starts to roll off right at zero hertz and finally reaches zero response at two times the nominal cutoff frequency. An alpha of 0.0 specifies an ideal "brick-wall" filter that transitions instantaneously from full response to zero right at the cutoff frequency. Values in the range of 0.3 to 0.5 are common in communications systems.

Unfortunately, if any additional filter is placed before or after the Nyquist filter it destroys the anti-ISI property. In order to allow filtering in both the transmitter and the receiver many systems effectively place half the Nyquist filter in each place. Because the frequency response of each filter is the square root of the response of a Nyquist filter, they are called *root-Nyquist* filters. The *root-raised-cosine* filter is an example. While a Nyquist or root-Nyquist response theoretically could be approximated with an analog filter, they are almost always implemented as digital filters. More information on digital filters appears in the **DSP and SDR Fundamentals** and the **Analog and Digital Filtering** chapters.

As mentioned before, filtering is more difficult with angle modulation because it is nonlinear. The RF spectrum is not a linear transposition of the baseband spectrum as it is with linear modes and Nyquist filtering doesn't work. Old-fashioned RTTY transmitters traditionally just used an R-C low-pass filter to slow down the transitions between mark and space. While that does not limit the bandwidth to the minimum value possible, the baud rate is low enough that the resulting bandwidth is acceptable anyway. In more modern systems, there is a tradeoff between making the filter bandwidth as narrow as possible for interference reduction and widening the bandwidth to reduce inter-symbol interference. For example, the GSM (Global System for Mobile communications) standard, used for some cellular telephone systems, uses minimum-shift keying and a Gaussian filter with a BT (bandwidth symbol-time product) of 0.3. A 0.3 Gaussian filter has a 0.3 ratio of 3-dB bandwidth to baud rate, which results in a small but acceptable amount of ISI and a moderate amount of adjacent-channel interference.

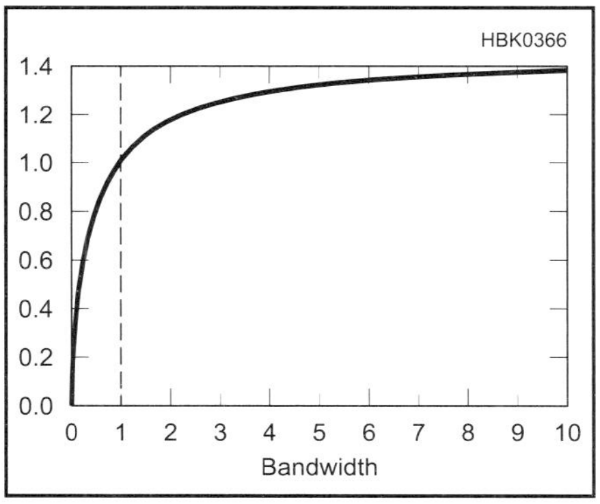

Figure 11.44 — Plot of channel capacity versus bandwidth, calculated by the Shannon-Hartley theorem. The S/N ratio has been selected to be unity at Bandwidth = 1.

CHANNEL CAPACITY

It is possible to increase the quantity of error-free data that can be transmitted over a communications channel by using an error-correcting code. That involves adding additional error-correction bits to the transmitted data. The more bits that are added, the greater the errors that can be corrected. However, the extra bits increase the data rate, which requires additional bandwidth, which increases the amount of noise. For that reason, as you add more and more error-correction bits, requiring more and more bandwidth, you eventually reach a point of limited additional return. In the 1940s, Claude Shannon worked out a formula (called the *Shannon-Hartley theorem*) for the maximum capacity possible over a communications channel, assuming a theoretically-perfect error-correction code:

$$C = B \log_2 \left(1 + \frac{S}{N} \right) \text{bit / s}$$

where

C = the net channel capacity, not including error-correcting bits,

B = the bandwidth in Hz, and

S/N = the signal-to-noise ratio, expressed as a power ratio.

Note that as B increases, N increases in the same proportion. **Figure 11.44** is a plot of channel capacity versus bandwidth based on the formula. As bandwidth is increased, channel capacity increases rapidly until the point where the S/N ratio drops to unity (labeled Bandwidth = 1 in the graph) after which channel capacity increases much more slowly.

11.10.2 Intermodulation Distortion

To minimize distortion of the modulation, each stage in the signal chain must be linear, from the microphone or modem, through all the intermediate amplifiers and processors, to the modulator itself. For the linear modulation types, all the amplifiers and other stages between the modulator and the antenna must be linear as well.

If the modulation consists of a single sine wave, then nonlinearity causes only harmonic distortion, which produces new frequencies at integer multiples of the sine-wave frequency. If multiple frequencies are present in the modulation, however, then *intermodulation distortion* (IMD) products are produced. IMD occurs when a nonlinear amplifier or other device acts as a mixer, producing sum and difference frequencies of all the pairs of frequencies and their harmonics. For example if two frequencies, F1 and F2 > F1, are present, then IMD will cause spurious frequencies to appear at F1 + F2, F2 – F1, 2F1, 2F2, 2F1 – F2, 2F2 – F1, 2F2 – 2F1, 3F1, 3F2, and so on. *Odd-order* products are those that include the original frequencies an odd number of times, such as 3F1, 2F1 + F2, 3F1 – 2F2, and so on. *Even-order* products contain an even number of the original frequencies, such as 2F2, F1 + F2, 3F1 + F2, and so on. If more than two frequencies are present in the undistorted modulation, then the number of unwanted frequencies increases exponentially.

Although intermodulation distortion that occurs before modulation is fundamentally the same as IMD that occurs after the modulator (at the intermediate or radio frequency), the effects are quite different. Consider two frequency components of a modulating signal at, for example, 1000 Hz and 1200 Hz that modulate an SSB transmitter tuned to 14.000 MHz, USB. The desired RF signal has components at 14.001 and 14.0012 MHz. If the IMD occurs before modulation, then the F1 + F2 distortion product occurs at 1000 + 1200 = 2200 Hz. Since that is well within the audio passband of the SSB transmitter, there is no way to filter it out so it shows up at 14.0022 MHz at the RF output. However, if the distortion had occurred after the modulator, the F1 + F2 product would be at 14.001 + 14.0012 = 28.0022 MHz which is easily filtered out by the transmitter's low-pass filter.

That explains why RF speech processors work better than audio processors. A speech processor clips or limits the peak amplitude of the modulating signal to prevent over-modulation in the transmitter. However, the limiting process typically produces considerable intermodulation distortion. If the limiter is located after the modulator rather than before it, then it is an RF signal being clipped, rather than audio. If the RF limiter is followed by a band-pass filter then many of the distortion products are removed, resulting in a less-distorted signal.

If the distortion is perfectly symmetrical (for example equal clipping of positive and negative peaks) then only odd-order products are produced. Most distortion is not symmetrical so that both even and odd-order products appear. However, the odd-order products are of particular interest when measuring the linearity of an RF power amplifier. The reason is that even-order products that occur after the modulator occur only near harmonics of the RF frequency where they are easy to filter out. Odd-order products can fall within the desired channel, where they cause distortion of the modulation, or at nearby frequencies, where they cause interference to other stations.

The informal term for such IMD that interferes with other stations outside the

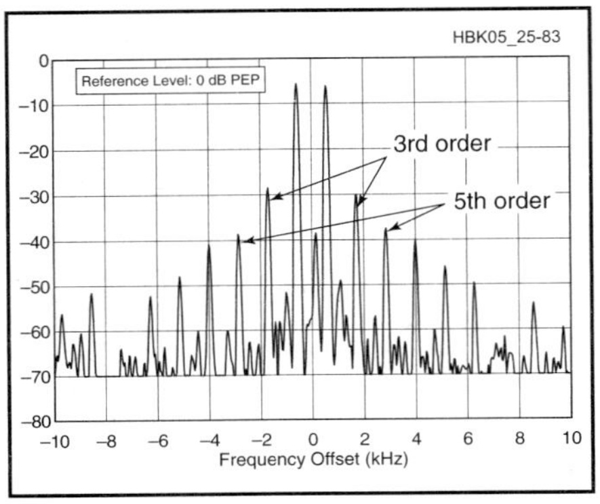

Figure 11.45 — Intermodulation (IMD) products from an SSB transmitter modulated by a pair of audio tones (see text). The two signals at the center are the desired modulation sidebands. The frequencies of the third-order IMD products are separated from the frequencies of the desired sidebands by the tone spacing (1 kHz) and from each other by three times the tone spacing (3 kHz). The fifth-order products are separated from the third-order products by twice the tone spacing (2 kHz) and from each other by five times the tone spacing (5 kHz).

desired channel is *splatter*. It becomes severe if the linear amplifier is over-driven, which causes clipping of the modulation envelope with the resulting odd-order IMD products.

The method commonly used to test the linearity of an SSB transmitter or RF power amplifier is the *two-tone test*. Two equal amplitude audio-frequency tones are fed into the microphone input, the transmitter and/or amplifier is adjusted for the desired power level, and the output signal is observed on a spectrum analyzer. See **Figure 11.45**. If the modulating tones are F1 and F2, the third-order IMD products closest to the modulation sidebands have frequencies of 2F1 – F2 and 2F2 – F1. The difference in frequency between the third-order IMD products is 3(F1 – F2). Similarly, the fifth-order IMD products occur at 3F1 – 2F2 and 3F2 – 2F1, so they are spaced at 5(F1 – F2). As in the figure, if F1 and F2 are 1 kHz apart, the two third-order products are 1 kHz above the upper modulation sideband and 1 kHz below the lower sideband. That is a spacing of 3 × 1 kHz = 3 kHz. The fifth-order products are twice the tone spacing (2 kHz) from the modulation sidebands for a product-to-product spacing of 2 + 1 + 2 kHz = 5 kHz.

11.10.3 Transmitted Bandwidth

We have already discussed the necessary bandwidth for each of the various modulation types. The previous section explained how intermodulation distortion is one phenomenon that can cause unwanted emissions outside of the desired communications channel. Another is failing to properly low-pass filter the modulating signal to the minimum necessary bandwidth. That is especially a concern for linear modulation modes such as SSB, AM, OOK (CW), BPSK and QAM. For angle-modulated modes like FM and FSK, excessive bandwidth can result from simply setting the deviation too high.

There are other modulation impairments that cause emissions outside the desired bandwidth. For example, in an SSB transmitter, if the unwanted sideband is not sufficiently suppressed, the occupied bandwidth is up to twice as large as it should be. Also, an excessively strong suppressed carrier causes particularly-annoying heterodyne interference to stations tuned near that frequency. In some SSB modulators, there is an adjustment provided to optimize carrier suppression.

The carrier suppression may be degraded by a modulating signal that is too low in amplitude. For example, if the signal from the microphone to an SSB transmitter is one-tenth (–20 dB) of the proper amplitude, and if the gain of the RF amplifier stages is increased to compensate, then the carrier suppression is degraded by 20 dB.

The term *adjacent-channel power* (ACP) refers to the amount of transmitted power that falls into an adjacent communications channel above or below the desired channel. Normally the unwanted out-of channel power is worse for the immediately-adjacent channels than for those that are two channels away, the so-called *alternate* channels. ACP is normally specified as a power ratio in dB. It is measured with a spectrum analyzer that can measure the total power within the desired channel and the total power in an adjacent channel, so that the dB difference can be calculated.

The *occupied bandwidth* is the bandwidth within which a specified percentage of the total power occurs. A common percentage used is 99%. For a properly-adjusted, low-distortion transmitting system, the occupied bandwidth is determined mainly by the modulation type and filtering and, in the case of digital modulation, the symbol rate. For example, the IS-54 TDMA format that has been used in some US digital cellular networks has about 30 kHz occupied bandwidth using 24.3-kilosymbols/sec, π/4-DQPSK modulation with a 0.35-alpha root-raised-cosine filter. The GSM cellular standard requires about a 350-kHz occupied bandwidth for its 270.833-kilosymbols/sec, 0.3 Gaussian-filtered MSK signal.

11.10.4 Modulation Accuracy

For analog modes, modulation accuracy is mainly a question of maintaining the proper frequency response across the desired bandwidth with minimal distortion and unwanted signal artifacts. In-band artifacts like noise and spurious signals should not be a problem with any reasonably-well-designed system. Maintaining modulation peaks near 100% for AM signals or the proper deviation for FM signals is facilitated by an audio compressor. It can be either the type that uses a detector and an automatic-gain-control feedback loop to vary the gain in the modulation path or a clipper-type compressor that limits the peak amplitude and then filters the clipped signal to remove the harmonics and intermodulation products that result. SSB transmitters can also use audio speech compression to maintain the proper peak power level although, as explained previously, clipping of the signal before it reaches the modulator can cause unacceptable distortion unless special techniques are used.

For digital signals, there are a number of other possible sources of modulation inaccuracy. For modes that use Nyquist filtering, the cutoff frequency and filter shape must be accurate to ensure no inter-symbol interference (ISI). Fortunately, that is easy to do with digital filters. However any additional filtering in the signal path can degrade the ISI. For example, most HF digital modes use an analog SSB transceiver to up-convert the signal from audio to RF for transmission and to down-convert from RF to audio again at the receiver end. The crystal filters used in the transmitter and receiver can significantly degrade group delay flatness, especially near the edges of the filter passband. That is why most HF digital modes use a bandwidth substantially less than a typical transceiver's passband and attempt to center the signal near the crystal filters' center frequency.

Distortion can also impair the proper decoding of digital signals, especially formats with closely-spaced symbols such as 256-QAM. Any "flat-topping" in the final amplifier causes the symbols at the outermost corners of the constellation to be closer together than they should be. The accuracy of the symbol clock is critical in formats like JT65 that integrate the detection process over a large number of symbols. Perhaps surprisingly, the clock rate on some inexpensive computer sound cards can be off on the order of a percent, which results in a similar error in symbol rate.

The modulation accuracy of an FSK signal is normally characterized by the frequency shift at the center of each symbol time, the point at which the receiver usually makes the decision of which symbol is being received. For other digital formats, the modulation accuracy is normally characterized by the amplitude and phase at the symbol decision points. Amplitude error is typically measured as a percentage of the largest symbol amplitude. The phase

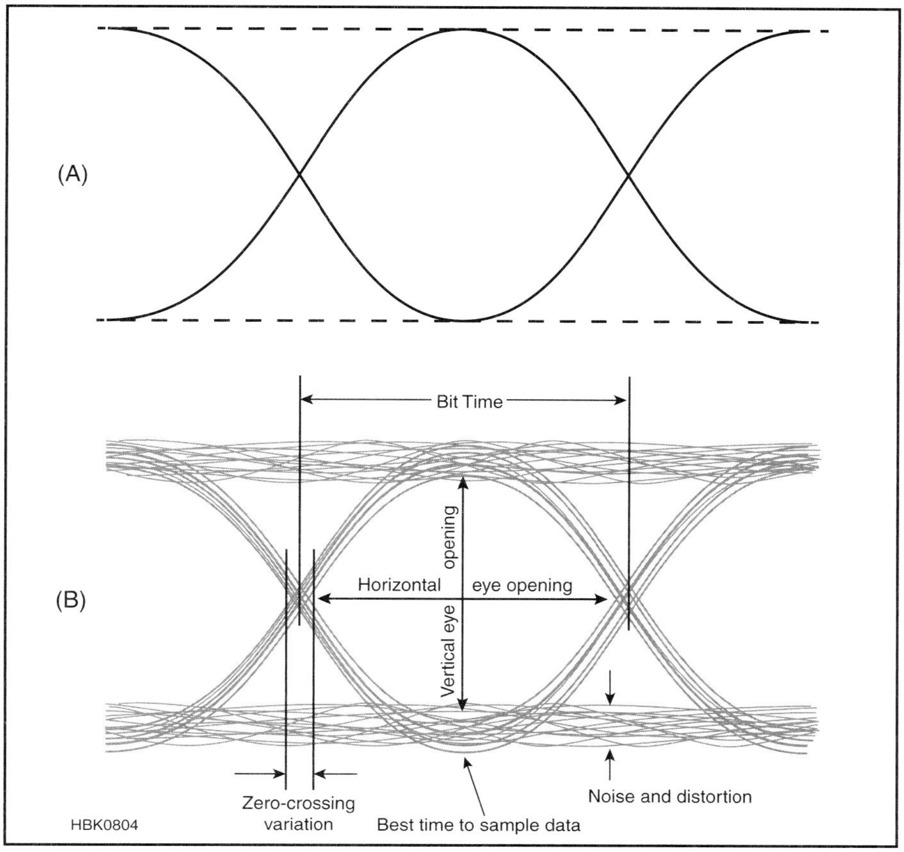

Figure 11.47 — The eye diagram is an oscilloscope-style diagram showing repeated samples of the digital signal as it transitions between points in the constellation. The eye diagram at A is for a perfect, undistorted signal. The eye diagram in B shows a signal with noise, distortion, and timing variations.

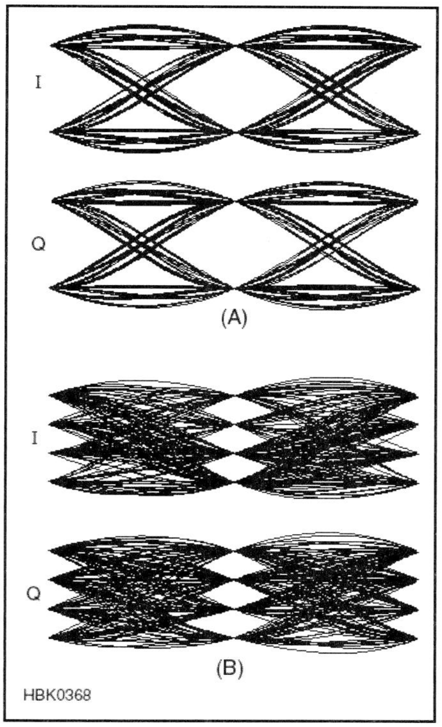

Figure 11.48 — Eye diagrams of the I and Q outputs of a QPSK generator (A) and a 16-QAM generator (B). The "eye" is the empty area between the symbol decision points, visible as the points where all the symbol trajectories come together. The bigger the "eye" the easier the signal is to decode.

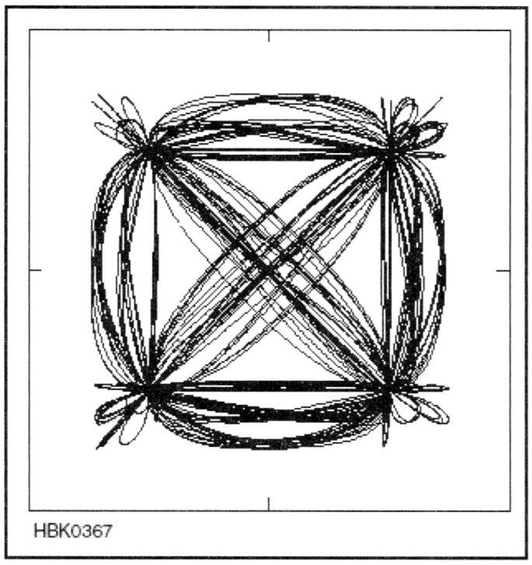

Figure 11.46 — Simulated I/Q constellation display of the trajectory of a QPSK signal over an extended period. A 0.5-alpha raised-cosine filter was used. Because it is a Nyquist filter, the trajectories pass exactly through each symbol location.

error is quoted in degrees. In both cases one can specify either the average RMS value or the peak error that was detected over the measurement period. Amplitude error is the most important consideration for modulation types such as BPSK where the information is encoded as an amplitude. For a constant-envelope format like MSK, phase error is a more important metric.

For formats like QPSK and QAM, both amplitude and phase are important in determining symbol location. A measurement that includes both is called *error vector magnitude* (EVM). It is the RMS average distance between the ideal symbol location in the constellation diagram and the actual signal value at the symbol decision point, expressed as a percentage of either the RMS signal power or the maximum symbol amplitude. This is a measurement that requires specialized test equipment not generally available to home experimenters. However it is possible to estimate the effect on EVM of various design choices using computer simulations.

Previously, we have plotted constellation diagrams with the transitions between symbol locations indicated by straight lines. In an actual system, both the I and Q signals are low-pass filtered which makes the symbol transitions smoother, with no abrupt changes of direction at the symbol locations. **Figure 11.46** shows the actual symbol trajectories for a QPSK signal using a 0.5-alpha raised-cosine filter with random data. Since a raised-cosine is a type of Nyquist filter, the trajectories pass exactly through each symbol location. This is what you would see if you connected the I and Q outputs from a QPSK baseband generator to the X (horizontal) and Y (vertical) inputs of an oscilloscope. It is also what you would see in a receiver at the input to the symbol detector after the carrier and clock-recovery circuits have stabilized the signal. If the symbol trajectories were not accurate, then the dark areas at the symbol points would be less distinct and more spread out. Such a constellation diagram is also a good troubleshooting tool for other purposes, for example to check for amplitude distortion or to see if a faulty symbol encoder is missing some symbols or placing them at the wrong location.

In a system where a root-Nyquist filter is used in both the transmitter and receiver to obtain a net Nyquist response, the transmitter output will not display sharply defined symbol locations as in the figure. In that case, the measuring instrument must supply the missing root-Nyquist filter that is normally in the receiver, in order to obtain a clean display. Professional modulation analyzers normally include a means of selecting the proper matching filter.

Another way to view modulation accuracy is with an *eye diagram*. See **Figures 11.47** and **11.48** In this case the oscilloscope's horizontal axis is driven by its internal sweep generator, triggered by the symbol clock. The two vertical channels of the oscilloscope are connected to the I and Q outputs of the baseband generator. The "eye" is the empty area at the center of the I and Q traces in between the symbol decision points, where all the traces come together. The eye should be as "open" as possible to make the job of the receiver's symbol decoder as easy as possible. The oscilloscope would typically be set for infinite persistence, so that the worst-case excursions from the ideal symbol trajectory are recorded. QPSK has four symbol locations, one in each quadrant, such that I and Q each only have two possible values and there is only one "eye" for each. 16-QAM has 16 symbol locations with four possible locations for I and Q, which forms three "eyes."

11.11 References

"A Brief Introduction to Sigma-Delta Conversion," Intersil Application Note AN9504. **www.renesas.com/us/en/www/doc/application-note/an9504.pdf**

Andrews, J., KH6HTV, *ATV Handbook*, revised 2021, **www.arrl.org/files/file/Technology/pdf/an-55a-atv-handbook-1.pdf**

Costas, J. P., "Poisson, Shannon, and the Radio Amateur," *Proceedings of the IRE*, vol 47, pp 2058-2068, Dec 1959.

"Digital Modulation in Communications Systems — An Introduction" Agilent Technologies, Application Note 1298. **literature.cdn.keysight.com/litweb/pdf/5965-7160E.pdf?id=1817805**

Ford, S., WB8IMY, *Get On the Air with HF Digital* 3rd Edition (ARRL, 2022).

Ford, S., WB8IMY, *ARRL's VHF Digital Handbook* (ARRL, 2008).

Haykin, S., *Digital Communications* (Wiley, 1988).

Langner, J., WB2OSZ, "SSTV Transmission Modes," **www.comunicacio.net/digigrup/ccdd/sstv.htm**.

Mack, R., W5IFS, SDR: Simplified column, *QEX*, from 2009.

"Manual of Transmission Methods – Reference Document," Rohde & Schwarz, 2014, available at **resources.rohde-schwarz-usa.com/c/manual-of-transmissi-2**

Nagle, J., K4KJ, "Diversity Reception: an Answer to High Frequency Signal Fading," *Ham Radio*, Nov 1979, pp. 48 – 55.

Sabin, W., et al, *Single-Sideband Systems and Circuits* (McGraw Hill, 1987).

Shavkoplyas, A., VE3NEA, "CW Shaping in DSP Software," *QEX*, May/Jun. 2006, pp. 3 – 7.

Seiler, T., HB9JNX/AE4WA, et al, "Digital Amateur TeleVision (D-ATV)," proc. 2001 ARRL/TAPR Digital Communications Conference.

Silver, W., NØAX, "About FM," *QST*, Jul. 2004, pp. 38 – 42.

Silver, W., NØAX, "About SSB," *QST*, Jan. 2016, pp. 51 – 54.

Smith, D., KF6DX, "Distortion and Noise in OFDM Systems," *QEX*, Mar./Apr. 2005, pp. 57 – 59.

Smith, D., KF6DX, "Digital Voice: The Next New Mode?," *QST*, Jan. 2002, pp. 28 – 32.

Stanley, J., K4ERO, "Observing Selective Fading in Real Time with Dream Software," *QEX*, Jan./Feb. 2007, pp. 18 – 22.

Taggart, R., WB8DQT, "An Introduction to Amateur Television," *QST*, Apr., May., and Jun. 1993.

Taggart, R., WB8DQT, *Image Communications Handbook* (ARRL, 2002).

Taggart, R., WB8DQT, "Digital Slow-Scan Television," *QST*, Feb. 2004, pp. 47 – 51.

Van Valkenburg, M., et al, *Reference Data for Engineers: Radio, Electronics, Computer and Communications*, 9th Edition (Newnes, 2001). Chapters 23, 24 and 25.

Contents

Chapter 12 — Online Content

Articles
- HF Receiver Testing by Adam Farson, VA7OJ/AB7OJ
- Noise Power Ratio (NPR) Testing on HF Receivers by Adam Farson, VA7OJ/AB7OJ
- Performance Capability of Active Mixers by Dr. Ulrich Rohde, N1UL
- SDR Simplified — Filter Design Program by Ray Mack, W5IFS
- SDR Simplified — Introduction to CIC Filters by Ray Mack, W5IFS
- SDR Simplified — More Filter Activities by Ray Mack, W5IFS
- SDR Simplified — Nyquist Meets Real World by Ray Mack, W5IFS
- VHF and UHF Receivers and UHF and Microwave Techniques

Also see the Online Content for the **Transmitters** and **Transceivers** chapters.

Projects
- 10 GHz preamp PCB template by Zack Lau, W1VT
- A Dual Band Low Noise Amplifier for 2 Meters and 70 Centimeters by Jim Kocsis, WA9PYH
- A High Performance 45 MHz IF Amplifier for an Up-Conversion HF/LF Receiver by Colin Horrabin, G3SBI
- A Long-wave Upconverter by Fred Brown, W6HPH
- A Software-Based Remote Receiver Solution by Martin Ewing, AA6E
- A Software Controlled Radio Preselector by J. Onate, MØWWA and X. de Fortuny
- All Mode 1 kHz to 1.7 GHz SDR Receiver by James Forkin, WA3TFS
- Binaural I-Q Receiver project by Rick Campbell, KK7B
- General Coverage Preselector by George Hirshfield, W5OZF
- Receiver projects from previous editions of the *ARRL Handbook*
- Rock Bending Receiver PCB template by Randy Henderson, WI5W
- Simple SDR Receiver by Michael Hightower, KF6SJ
- Tunable RF Preamplifier Using Varicap Diode by George Steber, WB9LVI
- Universal MMIC Preamp by Paul Wade, W1GHZ

Chapter 12
Receiving

Receivers have traditionally been at the forefront of Amateur Radio technology and certainly are today, even as the long-reigning analog superheterodyne architecture is being overtaken by the digital software defined radio. While commercial designs and devices abound, receiving in the crowded amateur bands is still a demanding application with strong signals immediately adjacent to signals at the noise floor, a variety of natural and man-made interference to reject, and modes ranging from manually sent Morse to the latest experimental modulations. Like the equipment in use today, expect this chapter to change from edition to edition as technology marches on.

In order to address these challenges, this chapter focuses on the functions of receiving and how they are implemented, whether by analog or digital technologies. Metrics of receiver performance are discussed, recognizing that digital receivers behave differently than their established analog counterparts, requiring new measurement definitions and techniques. Receiver home-brew projects from past editions have been collected into a set of projects in the online content accompanying this book.

Radio astronomy is a hobby that is closely related to amateur radio, and there are many people who enjoy both hobbies. A short discussion of the receivers used for radio telescopes addresses arguably the weakest received signals and what techniques are used to receive them.

This chapter's material has been adapted and updated from or provided by a number of authors. A great deal of the receiver material was originally written by Joel Hallas, W1ZR, and the sections on mixers by Dave Newkirk, W9VES, and Rick Karlquist, N6RK. Some sections on SDR functions were adapted from "SDR: Simplified" columns in *QEX* magazine by Ray Mack, W5IFS. Material on SDR receivers, architecture, and noise reduction was contributed by Steve Hicks, N5AC, and Doug Grant, K1DG. Jim Brown, K9YC, provided material on active noise canceling and diversity reception. Bob Allison, WB1GCM, and Adam Farson, VA7OJ/AB4OJ, contributed material on receiver testing. Dr. Richard Russel, AC0UB, provided material from members of the Society of Amateur Radio Astronomy (SARA) to describe receivers in radio telescopes. Additional material was taken from Chapters 10 and 11 in Experimental Methods in RF Design by Hayward, Campbell, and Larkin. The reader interested in the professional perspective and depth of detail is referred to the comprehensive *Communications Receivers — Principles and Design*, 4th Edition by Rohde, Whitaker, and Zahnd.

Coverage of VHF/UHF/microwave receivers from previous editions is included in the online content.

The major subsystems of a radio receiving system are the antenna, the receiver and the information processor. The antenna's task is to provide a transition from an electromagnetic wave in space to an electrical signal that can be conducted on wires. The receiver has the job of retrieving the information content from a particular signal coming from the antenna and presenting it in a useful format to the processor for use.

The processor typically is an operator but can also be an automated system. When you consider that most "processors" require signals in the range of volts (to drive an operator's speaker or headphones, or even the input of an A/D converter), and the particular signal of interest arrives from the antenna at a level of mere microvolts, the basic function of the receiver is to amplify the desired signal by a factor of a million. It must do this, while in the presence of signals many orders of magnitude greater and of completely different characteristics, without distortion of the desired signal or loss of the information it carries.

12.1 Characterizing Receivers

As we discuss receivers we need to characterize their performance, and often their performance limitations, using certain key parameters. The most commonly encountered are as described in the following sections. These are often the key performance parameters, but in many cases there are others that are important to specify, as well. Examples are audio output power, power consumption, size, weight, control capabilities and so forth.

The ARRL Lab has developed an extensive set of standardized tests that it performs on transceivers in support of *QST* Product Reviews.

12.1.1 Receiver Sensitivity and Noise

Sensitivity is a measure of how weak a signal the receiver can extract information from. This generally is expressed at a particular signal-to-noise ratio (SNR) since noise is generally the limiting factor. A typical specification might be: "Sensitivity: 1 μV for 10 dB SNR with 3 kHz bandwidth." The bandwidth is stated because the amount (or power) of the noise, the denominator of the SNR fraction, increases directly with bandwidth. Generally, the noise parameter refers to the noise generated within the receiver, often less than the noise that arrives with the signal from the antenna. (See the **RF Techniques** chapter for a more complete discussion of noise and noise sources.)

The sensitivity of a receiver is a measure of the lowest power input signal that can be received with a specified SNR. In the early days of radio, this was a very important parameter and designers tried to achieve the maximum practical sensitivity. In recent years, device and design technology have improved to the point that other parameters may be of higher importance, particularly in the HF region and below. Sensitivity remains an issue for receiving systems at VHF and above, particularly with respect to noise.

Noise level is as important as signal level in determining sensitivity. This section builds on the discussion of noise in the **RF Techniques** chapter. The most important noise parameters affecting receiver sensitivity are noise bandwidth, noise figure, noise factor, and noise temperature.

Since received noise power is directly proportional to receiver bandwidth, any specification of sensitivity must be made for a particular noise bandwidth. For DSP receivers with extremely steep filter skirts, receiver bandwidth is approximately the same as the filter or operating bandwidth. For other filter types, noise bandwidth is somewhat larger than the filter's 6 dB response bandwidth.

The relationship of noise bandwidth to noise power is one of the reasons that narrow bandwidth modes, such as CW, have a significant signal-to-noise advantage over modes with wider bandwidth, such as voice, assuming the receiver bandwidth is the minimum necessary to receive the signal. For example, compared to a 2400-Hz SSB filter bandwidth, a CW signal received in a 200 Hz bandwidth has a 2400/200 = 12 = 10.8 dB advantage in received noise power. That is the same difference as an increase in transmitter power from 100 to 1200 W.

Table 12.1
Signal Strengths for HF and VHF at Receiver Input Expressed as S-units

	HF Signal Strength		*V/UHF Signal Strength*	
S-unit	*dBm*	*μVRMS for 50Ω load*	*dBm*	*μVRMS for 50Ω load*
0	-127	0.1	-147	0.01
1	-121	0.2	-141	0.02
2	-115	0.4	-135	0.04
3	-109	0.8	-129	0.08
4	-103	1.6	-123	0.16
5	-97	3.2	-117	0.32
6	-91	6.3	-111	0.63
7	-85	12.6	-105	1.26
8	-79	25.1	-99	2.51
9	-73	50.1	-93	5.01
9 + 10 dB	-63	158.3	-83	15.83
9 + 20 dB	-53	500.6	-73	50.06
9 + 30 dB	-43	1583.0	-63	158.30
9 + 40 dB	-33	5005.9	-53	500.59
9 + 50 dB	-23	15830.1	-43	1583.01
9 + 60 dB	-13	50059.3	-33	5005.93

Table 12.2
Typical Noise Levels (Into the Receiver) and Their Source, by Frequency

Frequency Range	Dominant Noise Sources	Typical Level (mV/m)*
LF 30 to 300 kHz	atmospheric	150
MF 300 to 3000 kHz	atmospheric/man-made	70
Low HF 3 to 10 MHz	man-made/atmospheric	20
High HF 10 to 30 MHz	man-made/thermal	10
VHF 30 to 300 MHz	thermal/galactic	0.3
UHF 300 to 3000 MHz	galactic/ thermal	0.2

*The level assumes a 10 kHz bandwidth. Data from *Reference Data for Engineers,* 4th Ed, p. 273, Figure 1.

RECEIVED SIGNAL STRENGTH

The total signal plus noise that is received is often displayed on a Received Signal Strength Indicator (RSSI). Most amateur radio receivers provide an indication of RSSI on the S-meter. **Table 12.1** shows the S-units that are displayed on most radios and their corresponding signal strengths. Different S-unit scales were created for HF radios and those at higher frequencies. The reason for the different scales has a historical basis (see IARU Region 1 Technical Recommendation R.1, 1981). Not all S-meters on amateur radio receivers are calibrated (DeMaw QST article) and, often, the RST report that you give to another station is relative only to signals received by your own station. Even though we use the word, signal, to describe the levels that are seen by the receiver, RSSI has no way of distinguishing between signal and noise and merely shows the sum of both types of energy.

SOURCES OF NOISE

Any electrical component generates a certain amount of noise due to random electron motion. Any gain stages after the internal noise source amplifies the noise along with the signal. Thus, a receiver with no signal input source has a certain amount of noise generated and amplified within the receiver itself.

Connecting an antenna to a receiver introduces any noise external to the receiver that is on the received frequency. **Table 12.2** presents typical levels of external noise in a 10 kHz bandwidth present in the environment from different sources. Descriptions of the usual sources of noise and their properties follow:

Atmospheric noise. This is noise generated within our atmosphere due to natural phenomena. The principal cause is lightning, which sends wideband signals over great distances. All points on the Earth receive this noise, but it is much stronger in some regions than others depending on the amount of local lightning activity. This source is usually the strongest noise source in the LF range and may dominate well into the HF region, depending on the other noises in the region. The level of atmospheric noise tends to drop off by around 50 dB every time the frequency is increased by a factor of 10. This source usually drops in importance by the top of the HF range (30 MHz).

Man-made noise. This source acts in a similar manner to atmospheric noise, although it is more dependent on local activity rather than geography and weather. The sources tend to be sparks from rotating and other kinds of electrical machinery, as well as gasoline engine ignition systems and some types of lighting. In recent years, noise from computing and network equipmensswitch mode power supplies, and appliances has increased significantly in urban and suburban environments. All things being equal, this source, on average, drops off by about 20 dB every time the frequency is increased by a factor of ten. The slower decrease at higher frequencies is due to the sparks having faster rise times than lightning. The effect tends to be comparable to atmospheric noise in the broadcast band, less at lower frequencies and a bit more at HF.

Galactic Noise. This is noise generated by the radiation from heavenly bodies outside our atmosphere. Of course, while this is noise to communicators, it is the desired signal for radio astronomers. This noise source is a major factor at VHF and UHF and is quite dependent on exactly where you point an antenna (antennas for those ranges tend to be small and are often steerable). It also happens that as the Earth turns an antenna can move into positions where it inadvertently is aimed at a noisy area of the galaxy. If the Sun, not surprisingly the strongest signal in our solar system, appears behind a communications satellite, communications is generally disrupted until the Sun is out of the antenna's receiving pattern. Galactic noise occurs on HF, as well: Noise from the planet Jupiter can be heard on the 15 meter band under quiet conditions, for example.

Thermal Noise. Unlike the previous noise sources, this one comes from our receivers. All atomic structures have electrons that move within their structures. This motion results in very small currents that generate small amounts of wideband signals. While each particle's radiation is small, the cumulative effect of all particles becomes significant as the previous sources roll off with increasing frequency. The reason that this effect is called thermal noise is because the electron motion increases with the particle's temperature. In fact the noise strength is directly proportional to the temperature, if measured in terms of absolute zero (0 K). For example, if we increase the temperature from 270 to 280 K, that represents an increase in noise power of 10/270 = 0.037, or about 0.16 dB. Some extremely sensitive microwave receivers use cryogenically-cooled front-end amplifiers to provide large reductions in thermal noise.

Oscillator Noise. As noted in the **Oscillators and Synthesizers** chapter, real oscillators have noise sidebands that extend out on either side of the nominal carrier frequency at low amplitudes. Any such noise is transferred to the received signal and, through reciprocal mixing, create noise products from signals on adjacent channels through the mixer. A good receiver is designed with oscillators that generate noise well below the expected level of received noise.

Noise Measurement Terminology

There are many ways to measure and specify noise, each with its unique name and abbreviation. Most amateurs are familiar with SNR (signal-to-noise ratio) at least in a general sense, but fewer are aware of the need to specify bandwidth and possibly temperature. Communications professionals and receiver designers use a variety of names and methods to specify noise performance. Many of these are explained in Walt Kester's excellent tutorial "Understand SINAD, ENOB, SNR, THD, THD+N, and SFDR so You Don't Get Lost in the Noise Floor" published by Analog Devices as MT-003. See the References section for the complete URL for this document.

NOISE POWER AND SENSITIVITY

There are several related measures that can be used to specify the amount of noise that is generated within a receiver. If that noise approaches, or is within perhaps 10 dB of the amount of external noise received, then it must be carefully considered and becomes a major design parameter. If the internally generated noise is less than about 10 dB below that expected from the environment, efforts to minimize internal noise are generally not beneficial and can, in some cases, be counterproductive.

While the total noise in a receiving system is, as discussed, proportional to received bandwidth, the noise generating elements are generally not. Thus, it is useful to be able to specify the internal noise of a system in a way that is independent of bandwidth. It is important to note that even though such a specification is useful, the actual noise is still directly proportional to bandwidth and any bandwidth beyond that needed to receive signal information reduces SNR.

To evaluate the effect of noise power on sensitivity, refer to the discussion of noise in the **RF Techniques** chapter. The discussion hinges on the value of N_i, the equivalent noise power at the input of a perfect receiver that would result in the same noise output. N_i is generally expressed in dBm:

$$N_i = -198.6 + (10\log_{10}B) + (10\log_{10}T_E)$$

where B is the system bandwidth in Hz and T_E is the equivalent noise temperature expressed in K.

If N_i is greater than the noise generated internally by the receiver, the receiver's sensitivity is limited by the external noise. This is usually the case for HF receivers where atmospheric and man-made noise are much stronger than the receiver's internal noise floor. If N_i is up to, perhaps, 10 dB higher than the receiver's internal noise, then the effect of the receiver's internal circuits on overall system sensitivity must be considered. At VHF and above, noise generated by the system components is more likely to exceed input noise.

EFFECT OF INPUT NOISE

A receiver designer needs to know how strong the expected signals will be to establish the range of signals the receiver is required to handle. One may compare the equivalent internal noise with the expected external noise to determine whether the overall receiver SNR will be determined by external or internal noise. A reasonable design objective is to have the internal noise be less than 10 – 20 dB below the external noise. As noted above, this is related to the frequency of signals we want to receive. Any additional sensitivity does not provide a noticeable benefit to SNR, and may result in reduced dynamic range, as is discussed in the next section.

For frequencies at which external noise sources are strongest, the noise power (and signal power) is also a function of the antenna design. In such cases, the SNR ratio can be improved by using an antenna that picks up more signal and less noise, such as with a directional antenna that can reject noise from directions other than that of the signal. Some antennas improve SNR simply by rejecting noise, such as the Beverage antenna. Signals from a Beverage antenna are usually much weaker than from a conventional antenna but their ability to reject noise from undesired directions creates a net improvement in the SNR at the receiver input.

MINIMUM DETECTABLE (DISCERNIBLE) SIGNAL (MDS)

Also referred to as the *noise floor* of a receiver, the MDS is the strength of the smallest input signal that produces a specified increase in the output power of a receiver. MDS depends on the required SNR (SNR_{MIN} in dB), the system bandwidth (B in Hz), the temperature of the receiver (TE), and the receiver noise figure (NF). (Measurement of MDS is discussed in the **Test Equipment and Measurements** chapter.)

An ideal receiver at room temperature with a bandwidth of 1 Hz has a theoretical MDS of –174 dBm. This is often referred to as the "1 Hz noise floor." As bandwidth, noise figure, and temperature increase, so does MDS as follows:

$$\text{MDS (in dBm)} = -174 + 10\log(B) + NF + SNR_{MIN} + 10\log(T_E/290)$$

For example, if a receiver at room temperature has a bandwidth of 1 kHz and a noise figure of 2.5 dB, for an SNR_{MIN} of 3 dB:

$$\text{MDS} = -174 + 30 + 2.5 + 3 + 0 = -138.5 \text{ dBm}$$

If the receiver's equivalent temperature then increases to 300 K, the MDS also increases to:

$$\text{MDS} = -174 + 30 + 2.5 + 3 + 0.15 = -138.35 \text{ dBm}$$

Many commercial radios achieve an MDS of –135 to –140 dBm in a 500 Hz bandwidth, which is quite good, but at HF the external noise is much higher, making lower MDS specifications irrelevant. As frequency increases through the higher HF bands and into VHF, and external noise decreases, lower MDS becomes more important.

Radio Astronomy Receivers

Arguably the weakest signals that need to be received come from outer space. The radio astronomy field faces very small signals in the presence of noise, much of which originates on Earth and often has a higher signal strength than the desired signal. In the preceding discussion of MDS, units of dBm were used to describe signal strengths. The signal strengths in radio astronomy are so much smaller that a new measure was devised. The Jansky (Jy), named after one of the pioneers of radio astronomy (for more on the history of radio astronomy, see the **Amateur Radio Data Platforms** chapter), represents 10^{-23} mW/m^2/Hz, which is equal to –230 dBm/m^2/Hz. Jy is a measure of power density and bandwidth. This gives a clue about one method radio telescopes use to detect very small signals: they increase the antenna surface area.

A typical Earth-Moon-Earth (EME) transmission with a bandwidth of 3.0 kHz has a received signal strength of –207 dBm. For comparison, your receiver displays S1 when the incoming signal is –121 dBm. Standardizing on an isotropic antenna size ($\lambda^2/4\pi$), the returned EME signal would be equal to about 2,500 Jy. Most strong sources from space are below 2,000 Jy (The Sun is an exception, with received signals greater than 2,500 Jy at 1400 MHz, see the **Propagation of Radio Signals** chapter). The strongest Pulsars have signal strengths of 1 Jy and most others are in the milli-Jy and micro-Jy range.

The MDS of a good receiver is stated above to be around -140 dBm in a 500 Hz bandwidth, or –170 dBm/Hz. For an antenna aperture of 1 m^2 the MDS of a good receiver is equivalent to 1,000,000 Jy. So how do radio astronomers detect signals from space that are more than 60 dB weaker than the MDS of their receivers?

One tool in their arsenal is large directional antennas with high gains. For instance, the 100 m diameter astronomical dish at Green Bank, WV provides an additional 40 dB of gain to the system. Also, an antenna that is pointed up to the sky collects as much as 10 dB less thermal noise than a terrestrial antenna.

Radio astronomy receivers have another potential burden to overcome. Unlike other radio receivers, which can be set to receive signals on distinct frequencies that are regulated to avoid interference, signals from space do not coordinate with terrestrial radio transmitters, so they are often found on the same frequencies that are already in use on the Earth. Again, the highly directional antenna pointed to the sky can serve to attenuate terrestrial signals.

Signal processing adds additional gain to the received signal. Radio astronomy uses averaging over relatively long time periods to enhance signal strength and decrease ambient noise. For this type of processing to work, the signal must be present and unchanging during the averaging period, or the signal must be repetitive and the averaging must be synchronized so that the different parts of the signal that are added together are in phase. This technique is clearly not valid for all types of signals but is a powerful tool to detect signals that are far below the noise level.

Other very high end techniques, such as cooling receiving amplifier hardware to decrease the thermal noise that it adds to the signal, are also used to maximize receiver performance. Even without such extreme solutions, by combining the advantages of large directional antennas and signal processing, radio astronomical measurements can improve the received signal power by up to 90 dB, which more than makes up for the limitations of the receiver hardware.

12.1.2 Preamplifiers and Noise

We are often faced with the requirement of determining the noise figure of a system with multiple stages. In general, adding an amplification stage between the antenna and the rest of the system reduces the equivalent noise figure of the system by the amount of gain of the stage, adding only the noise created by that stage. In general, to improve system sensitivity, the preamp's internal noise must be much lower than the noise generated by the rest of the receiver. Because the noise added by the preamplifier dominates the noise of the entire system, special noise reduction design techniques are used for such preamplifiers, which are called Low Noise Amplifiers (LNAs). (See the discussion of Cascaded Amplifiers in the **RF Techniques** chapter.) Several preamplifier projects are included in the online content for this chapter.

In many cases elements of a receiving system exhibit loss rather than gain. Since they reduce the desired signal, while not changing the noise of following stages, they increase the noise figure at their input by an amount equal to the loss. This is the reason that often a VHF LNA is mounted at the antenna, ahead of a noisy receiver as shown in **Figure 12.1**. For fairly typical values, while the addition of a low noise amplifier does reduce the system's noise figure, as can be observed, the amplifier gain and noise figure of the rest of the receiver can make a big difference.

As an example of this difference, if the coax between the antenna and radio has a loss of 1 dB, and the preamp has a noise figure of 1 dB, the resulting noise figure with the preamp at the radio is 2.5 dB. However, if the LNA is located at the antenna, the noise added by the coax is not amplified by the gain of the LNA, resulting in a significant improvement in received SNR with a reduced NF of 1.04 dB. **Figure 12.2** shows a typical example, modeling the two configurations with ADI-SimRF (see the **Computer Modeling and Simulation** chapter).

Along with noise figure, the preamplifier's effect on distortion and dynamic range must be considered. As discussed in later sections of this chapter, distortion products can be generated by intermodulation from in-band and out-of-band signals. These distortion products can be considered to be another type of noise that can also reduce receiving system sensitivity.

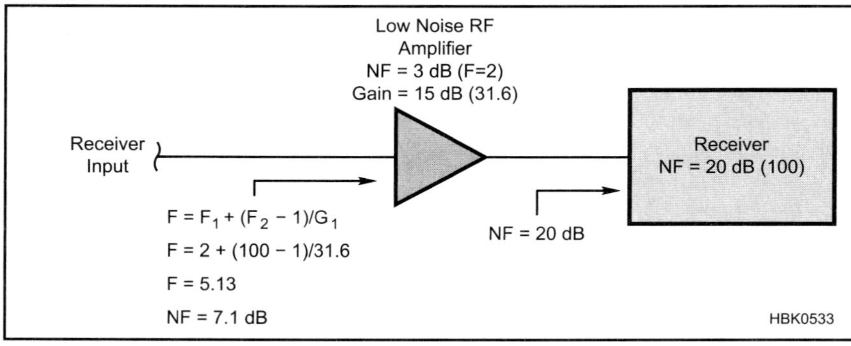

Figure 12.1 — The effect of adding a low-noise preamplifier ahead of a noisy receiver system.

System *spurious-free dynamic range (SFDR)* is the difference between MDS (see previous section) and the highest level of input signal that does not create spurious signals. Although the preamp may increase the amplitude of the desired signal, it also increases the amplitude of the unwanted distortion products even more. For this reason, spurious-free dynamic range of the receiving system can be degraded by adding a preamplifier with a limited dynamic range.

Both dynamic range and noise figure must be considered when choosing a preamp. A thorough discussion of how to evaluate and balance gain and noise figure when choosing a preamp is available in Paul Shuch, N6TX's online article "Quiet! Preamplifier at Work" listed in the References section.

12.1.3 Receiver Selectivity

Selectivity — Selectivity is just the bandwidth discussed above. This is important because to a first estimate it identifies the receiver's ability to separate stations. With a perfect "brick wall" filter in an ideal receiver, stations within the bandwidth are heard, while those outside it aren't detected. The selectivity thus describes how closely spaced adjacent channels can be. With a perfect 3 kHz bandwidth selectivity, and signals restricted to a 3 kHz bandwidth at the transmitter, a different station can transmit 3 kHz across the spectrum. In a less than ideal situation, it is usually necessary to include a guard band between channels.

Note that the word channel is used here in its generic form, meaning the amount of spectrum occupied by a signal, and not defining fixed frequencies, such as an AM broadcast channel. A CW channel is about 300 Hz wide, an SSB channel about 2.5 – 3 kHz wide, and so forth. "Adjacent channel" refers to spectrum immediately higher or lower in frequency. **Table 12.3** shows the bandwidth required for such a "channel" in various services and modes.

Table 12.3

Typical Communications Bandwidths for Various Operating Modes

Mode	*Bandwidth (kHz)*
FM Voice (WBFM Broadcast)	*FM Voice (NBFM Amateur)*
FM Voice (NBFM Amateur)	15
AM Voice (Broadcast)	10
AM Voice (Amateur)	4-6.6
SSB Voice (Amateur)	1.8-3
Digital Voice (Amateur)	1.0-1.2
RTTY (170-850 Hz shift)	0.3-1.5
CW	0.1-0.5

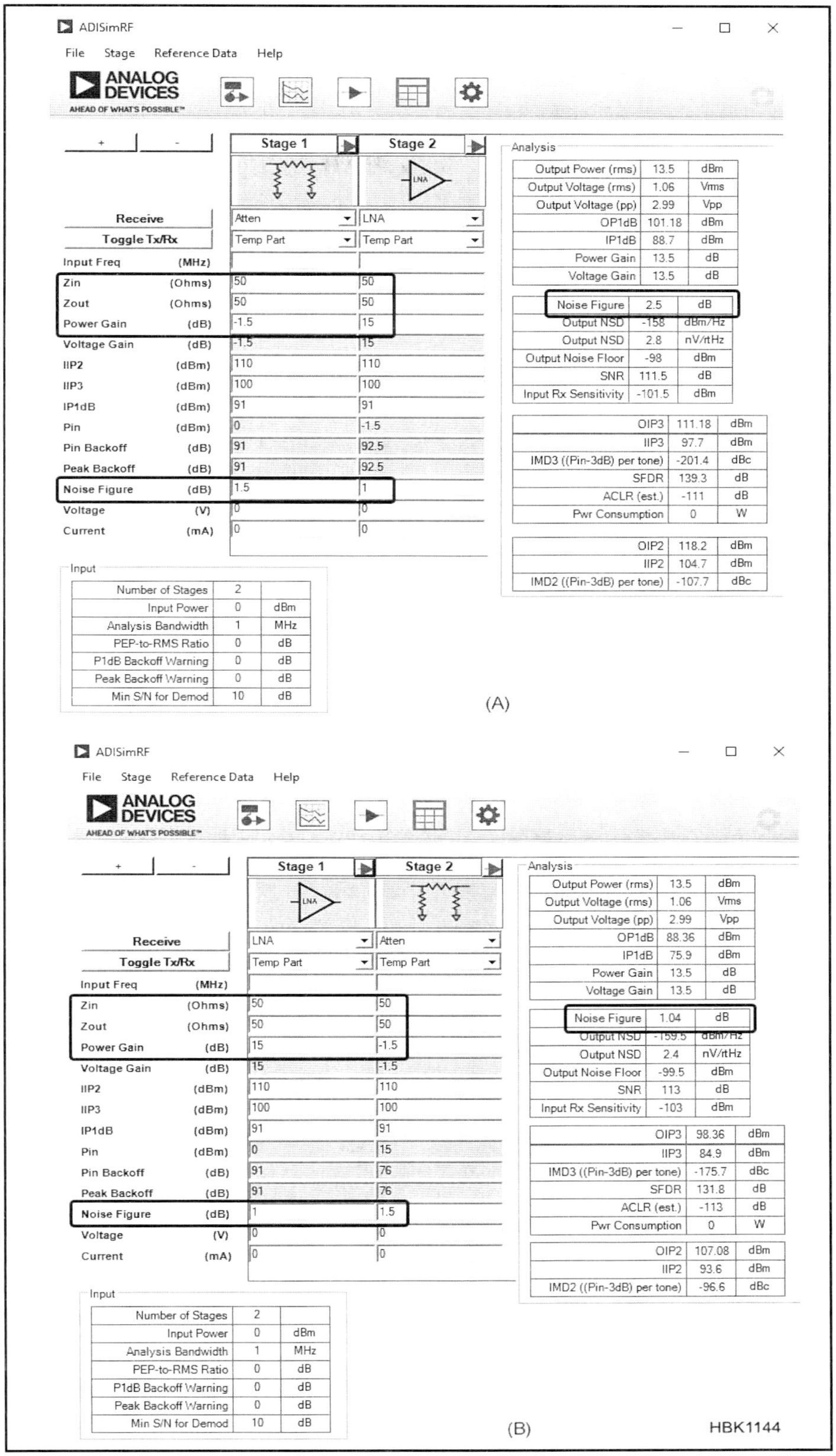

Figure 12.2 — Examples comparing the effect on input NF of placing the preamplifier at different places in the antenna system. Simulation is performed with ADISimRF, where the coax loss is modeled as an attenuator and the preamp is an LNA. To model this problem, you need only enter values for the highlighted fields for each module: Zin, Zout, Power Gain, and NF. The result displayed to the right of each model has the desired result highlighted: Noise Factor. At (A), the coax loss before the preamp results in an input NF to the receiver of 2.5 dB, a reduction in sensitivity. To improve the NF, the preamp can be mounted at the antenna, ahead of the coax cable loss (B), resulting in an input NF to the receiver of 1.04 dB.

Selectivity under linear operating conditions is determined by the receiver's filters. In analog heterodyne designs, the filters are typically crystal filters with a fixed bandwidth. Clever frequency shifting schemes (Passband Tuning or IF Shift, for example) create a continuously adjustable filter by shifting two fixed passbands relative to each other. DSP is also used in both hybrid and direct-sampling SDR designs to provide performance that surpasses that of filters using discrete components. Both types of filters are discussed in the **Analog and Digital Filtering** chapter.

12.1.4 Receiver Dynamic Range

In the case of all real receivers, there is a range of signals that a receiver can respond to linearly without distortion. This is referred to as *dynamic range* and, as is discussed in more detail, can be established based on a number of different criteria. In the most general sense, dynamic range can be defined as the ratio of the strongest to the weakest signal that a system, in this case our receiver, can respond to linearly. This range typically extends from the receiver's noise floor to a level at which intermediate stages overload in some way.

Table 12.2 gives us an idea of how small a signal we might want to receive. The designer must create a receiver that handles signals from below the noise floor to those as strong as the closest transmitter can generate. Most receivers have a specified highest input power that can be tolerated, representing the other end of the range. Usually, the maximum power specified is the power at which the receiver will not be damaged, while a somewhat lower power level is the highest that the receiver can receive without overload and the accompanying degradation of quality of reception of the desired signal.

The type and severity of the overload is often part of the specification. A straightforward example might be a 130 dB dynamic range. The nature of the distortion determines the observed phenomenon. If the weakest and strongest signals are both on the same channel, for example, we would not expect to be able to process the weaker of the two. However, the more interesting case would be with the strong signal in an adjacent channel. In an ideal receiver, we would never notice that the adjacent signal was there. In a real receiver with a finite dynamic range and non-ideal selectivity, there is some level of adjacent channel signal that interferes with reception of the weaker on-channel signal.

Dynamic range is also affected by preamplifiers in the receiver front-end circuitry. (Passive resistive attenuators have less effect on dynamic range as they typically remain linear over a very wide range of signals.) Turning on a preamplifier may worsen various types of dynamic range performance if the preamplifier's dynamic range is less than that of the rest of the receiver. Conversely, in direct-sampling SDR receivers, having the preamp on may improve performance. Dynamic range specifications and measurements must include whether a preamp is on or off.

It is important to note that the behavior of a superheterodyne receiver and an SDR receiver are quite different with respect to dynamic range. The notion of an "intercept point" discussed below is based on the assumption of intermodulation products increasing in a certain way that is characteristic of analog circuitry. This assumption is not valid for direct-sampling SDR equipment. An SDR's input ADC is basically linear up to the point where the instantaneous sum of all signals present at the receiver input exceeds the full-scale range of the input ADC. (See the **DSP and SDR Fundamentals** chapter.) Beyond that point, the receiver generates spurious products based on its software and how that specific ADC responds. Hybrid analog/digital receivers that apply DSP at IF or audio frequencies behave more like traditional analog heterodyne receivers.

ON-CHANNEL DYNAMIC RANGE

The signal you wish to listen to can range from the strongest to the weakest, sometimes changing rapidly with conditions, or in a situation with multiple stations, such as a net. While a slow change in signal level can be handled with manual gain controls, rapid changes require automatic systems to avoid overload and operator discomfort.

Nonlinear Signal Combinations

Although a mixer is often thought of as nonlinear, it is neither necessary nor desirable for a mixer to be nonlinear. An ideal mixer is one that linearly multiplies the LO voltage by the signal voltage, creating two products at the sum and difference frequencies and only those two products. From the signal's perspective, it is a perfectly linear but time-varying device. Ideally a mixer should be as linear as possible.

When a signal is applied to a nonlinear device, however, the output is not just a copy of the input, but can be described as the following infinite series of output signal products:

$$V_{OUT} = K_0 + K_1 \times V_{IN} + K_2 \times V_{IN}^2 + K_3 \times V_{IN}^3 + \ldots + K_N \times V_{IN}^N$$

What happens if AS input V_{IN} consists of two sinusoids at F_1 and F_2, or A × [sin ($2\pi F_1$) × t] and B × [sin ($2\pi F_2$) × t]? Begin by simplifying the notation to use angular frequency in radians/second ($2\pi F = \omega$). Thus V_{IN} becomes $A\sin\omega_1 t$ and $B\sin\omega_2 t$ and equation A becomes:

$$V_{OUT} = K_0 + K_1 \times (A\sin\omega_1 t + B\sin\omega_2 t) + K_2 \times (A\sin\omega_1 t + B\sin\omega_2 t)^2 + K_3 \times (A\sin\omega_1 t + B\sin\omega_2 t)^3 + \ldots + K_N \times (A\sin\omega_1 t + B\sin\omega_2 t)^N$$

The zero-order term, K_0, represents a dc component and the first-order term, $K_1 \times (A\sin\omega_1 t + B\sin\omega_2 t)$, is just a constant times the input signals. The second-order term is the most interesting for our purposes. Performing the squaring operation, we end up with:

$$\text{Second order term} = [K_2A^2\sin^2\omega_1 t + 2K_2AB(\sin\omega_1 t \times \sin\omega_2 t) + K_2B^2\sin^2\omega_2 t]$$

Using the trigonometric identity:

$$\sin á \sin â = 1/2 \{\cos(á - â) - \cos(á + â)\}$$

the product term becomes:

$$K_2AB \times [\cos(\omega_1 - \omega_2)t - \cos(\omega_1 + \omega_2)t]$$

These products are at the sum and difference frequency of the input signals! The signals, originally sinusoids are now cosinusoids, signifying a phase shift. These signals, however, are just two of the many products created by the nonlinear action of the circuit, represented by the higher-order terms in the original series.

In the output of a mixer or amplifier, those unwanted signals create noise and interference and must be minimized or filtered out. This nonlinear process is responsible for the distortion and intermodulation products generated by amplifiers operated nonlinearly in receivers and transmitters.

This is a problem that has been long solved with automatic gain control (AGC) systems. These systems are described in a later section of the chapter, but it is worth pointing out that the measurement and gain control points need to be applied carefully to the most appropriate portions of the receiver to maintain optimum performance. If all control is applied to early stages, the SNR for strong stations may suffer, while if applied in later stages, overload of early stages may occur in the presence of strong stations. Thus, gain control has to be designed into the receiver distributed from the input to the detector.

The next two sections illustrate a frequent limitation of receiver performance — dynamic range between the reception of a weak signal in the presence of one or more strong signals outside of the channel.

BLOCKING GAIN COMPRESSION

A very strong signal outside the channel bandwidth can cause a number of problems that limit receiver performance. *Blocking gain compression* (or "blocking") occurs when strong signals overload the receiver's high gain amplifiers and reduce its ability to amplify weak signals. (Note that the term "blocking" is often used outside Amateur Radio when referring to reciprocal mixing of oscillator noise with strong local signals.)

While listening to a weak signal, all stages operate at maximum gain. If the weak signal were at a level of S0, a strong signal could be at S9 + 60 dB. From the HF definitions in Table 1, the receiver's front-end stages would see a 0.1 µV signal and a 50,000 µV signal into the front end at the same time. A perfectly linear receiver would amplify each signal equally until the undesired signal is eliminated at a later stage in the receiver. However, in practical receivers, after a few stages of full gain amplification, the stronger signal causes amplifier clipping, which reduces the available gain. This is seen as a gradual reduction in gain as the input signal amplitude increases. Gain reduction also reduces the amplitude of the weaker signal, which is perceived to fade as the strong signal increases in amplitude. Eventually, the weaker signal is no longer receivable and is said to have been "blocked," thus the name for the effect.

The ratio in dB between the strongest signal that a receiver can amplify linearly, with no more than 1 dB of gain reduction, and the receiver's noise floor in a specified bandwidth is called the *receiver's blocking dynamic range (BDR)* or the *compression-free dynamic range (CFDR)*. In an analog superheterodyne, BDR is established by the linear regions of the IF amplifiers and mixers. If the receiver employs DSP, the range of the analog-to-digital converter usually establishes the receiver's BDR. The spacing of the signals must also be specified so that any internal filtering is accounted for. Typical signal spacing is 2, 5, 20, and 100 kHz.

A related term is "near-far interference" which is used primarily in the commercial environment to refer to a strong signal causing a receiver to reduce its gain and, along with it, the strength of weak received signals.

INTERMODULATION DISTORTION (IMD)

Blocking dynamic range is the straightforward response of a receiver to a single strong interfering signal outside the operating passband. In amateur operation, we often have more than one interferer. While such signals contribute to the blocking gain compression in the same manner as a single signal described above, multiple signals also result in a potentially more serious problem resulting from *intermodulation (IM) products.* As noted previously, analog superheterodyne and SDR receivers behave quite differently with respect to intermodulation.

If we look again at the equation for two sinusoids in the sidebar on Nonlinear Signal Combinations, we note that there are an infinite number of higher order terms. In general, the coefficients of these terms are progressively lower in amplitude, but they are still greater than zero. Of primary interest is third-order term, $K_3 \times V_{IN}{}^3$, when considering V_{IN} as the sum of two interfering signals (f_1 and f_2) near our desired signal (f0) and within the first IF passband, but outside the operating bandwidth.

$$V_{OUT} = K_3 \times [A\sin(f_1)t+B\sin(f_2)t]^3$$
$$= K_3 \times \{A^3\sin^3(f_1)t + 3A^2B\,[\sin^2(f_1)t \times \sin(f_2)t] + 3AB^2\,[\sin(f_1)t \times \sin^2(f_2)t] + B^3\sin^3(f_2)t\,\}$$

The cubic terms (the first and last terms) result in products at three times the frequency and can be ignored in this discussion. Using trigonometric identities to reduce the remaining $\sin^2$ terms and the subsequent cos( )sin( ) products reveal individual intermodulation (IM) products, recognizing that the signals have cross-modulated each other due to the nonlinear action of the circuit. (Math handbooks such as the CRC Standard Mathematical Tables and Formulae have all the necessary trigonometry identities.)

IM products have frequencies that are linear combinations of the input signal frequencies, written as n(f1) ± m(f2), where n and m are integer values. The entire group of products that result from intermodulation are broadly referred to as *intermodulation distortion* or IMD. The ratio in dB between the amplitude of the interfering signals, f_1 and f_2, and the resulting IM products is called the intermodulation ratio.

If all the higher-order terms in the original equation are considered, n and m can take on any integer value. If the sum of n and m is odd, (2+1, or 3+2, or 3+4, etc.) the result is products that have frequencies near our desired signal, for example, 2(f1) – 1(f2). Those are called odd-order products. Odd-order products have frequencies close enough to those of the original signals that they can cause interference to the desired signal. If the sum of n and m is three, those are third-order IM products or 3IMD. For fifth-order IMD, the sum of n and m is five, and so forth. The higher the order of the IM products, the smaller their amplitude, so our main concern is with 3IMD.

If the two interfering signals have frequencies of $f_0 + \Delta$, and $f_0 + 2\Delta$, where Δ is some offset frequency, we have for the third-order term:

$$V_{OUT} = K_3 \times [A \sin(f_0 + \Delta)t + B \sin(f_0 + 2\Delta)t]^3$$

A good example would be interfering signals with offsets of 2 kHz and 2 × 2 kHz or 4 kHz from the desired frequency, a common situation on the amateur bands.

Discarding the cubic terms and applying the necessary trigonometric identities shows that a product can be produced from this combination of interfering frequencies that has a frequency of exactly f_0 — the same frequency as the desired signal! (The higher-order terms of the sidebar's equation for sinusoidal signals can also produce products at f0, but their amplitude is usually well below those of the third-order products.)

Thus, we have two interfering signals that are not within our operating bandwidth, so we don't hear either by themselves. Yet they combine in a nonlinear circuit and produce a signal exactly on top of our desired signal. If the interfering signals are within the passband of our first IF and are strong enough the IM product can be heard.

As the strength of the interfering signals increases, so do those of the resulting intermodulation products. For every dB of increase in the interfering signals, the third-order IM products increase by approximately 3 dB. Fifth-order IM increases by 5 dB for every dB increase in the interfering signals, but was much smaller to begin with, and so forth. Our primary concern, however, is with the third-order products because they are the strongest and cause the most interference.

IMD DYNAMIC RANGE AND INTERCEPT POINT

As with blocking, discussed above, a

How Phase Noise Affects a Receiver

By Steve Hicks, N5AC

Today's receivers are generally built from one of two primary technologies: mixing or sampling. Some receivers will have components of both. In either case, an oscillator sets the frequency of the mixing or the sampling rate. Let's take a look at how phase noise from an oscillator can affect performance of our receiver, looking first at mixing. *(The effects of reciprocal mixing are produced by both amplitude and phase noise, a combination referred to as "composite noise."– Ed)*

Figure 12.A1 has a block diagram of a typical mixer circuit. The oscillator used with a mixer is called a local oscillator (LO) and the mixing process is also known as heterodyning. Let's assume for the moment that our radio designer selected an off-the-shelf synthesizer IC to use as a local oscillator. The phase noise chart in **Figure 12.A2** represents the phase noise provided by the manufacturer of the synthesizer integrated circuit. Assuming that the phase noise in our design matches the manufacturer's data, we can use the phase noise plot to determine how phase noise affects our receiver. Phase noise generally changes as the oscillator frequency is moved and, in the case of a synthesizer, can also change with the parameters used to lock to a reference signal (An overview of PLL operation can be found in a number of good references — for example: **www.ti.com/lit/an/swra029/swra029.pdf** and in **www.analog.com/media/en/training-seminars/tutorials/MT-086.pdf**). For now, let's assume this plot was taken at 14 MHz and this is where our receiver operates.

The two numbers we use in our example are at an offset of 2 kHz and 10 kHz. Reading the plot, we see that the 2 kHz phase noise is –93 dBc/Hz and the 10 kHz phase noise is –96 dBc/Hz. The measurement units of dBc/Hz indicates the noise that would be present in a 1 Hz bandwidth receiver, the number of dB below the carrier. For example, –93 dBc/Hz indicates that the noise floor would be 93 dB below a carrier signal measured using a 1 Hz bandwidth receiver. To convert this to a 100 Hz bandwidth we might use in a CW filter we must add 10 log (100 Hz/1 Hz) or 20 dB to our phase noise numbers. So at a 2 kHz offset from our carrier, we have –93 dBc + 20 dB = –73 dBc and at 10 kHz, –96 dBc + 20 dB = –76 dBc. In a mixer, this noise is added to all signals present in the receiver.

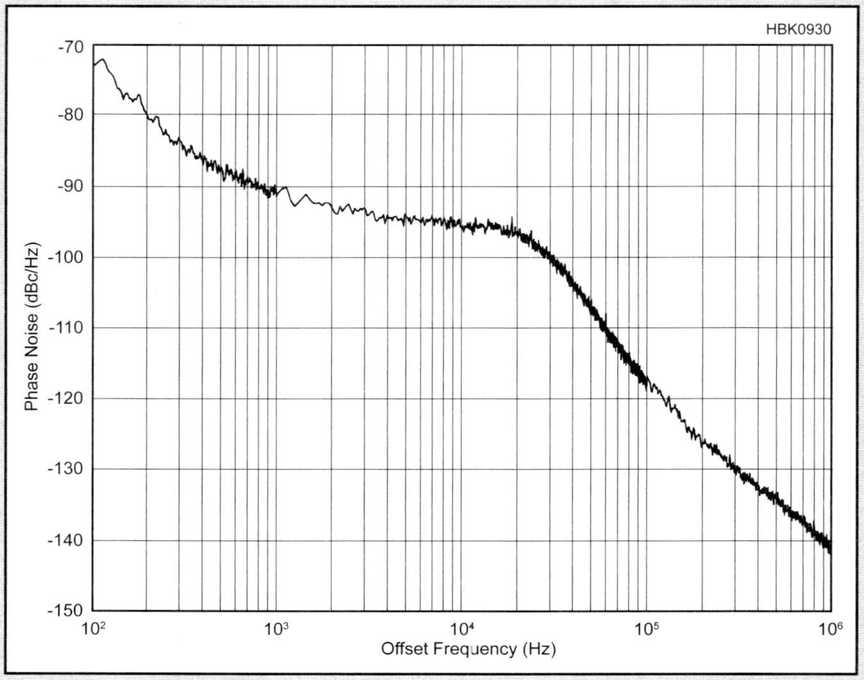

Figure 12.A2 — Plot of phase noise versus frequency separation from the carrier.

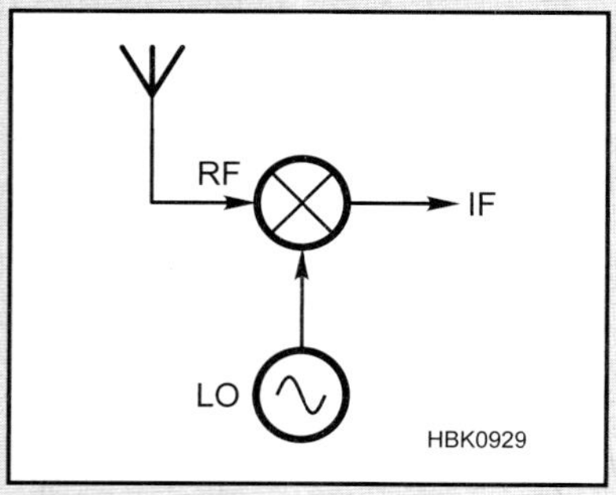

Figure 12.A1 — Typical mixer block diagram.

Looking at **Figure 12.A3**, we show a strong S8 CW signal at 14.000 MHz. S8 corresponds to –79 dBm as shown in the graph. Two kilohertz away is the signal we are trying to receive, an S3 (–109 dBm) CW signal using a 100 Hz CW filter. Using our phase noise calculation, the S8 signal at –79 dBm would have phase noise at a 2 kHz offset of –79 dBm – 73 dB = –152 dBm. This is well below our atmospheric noise floor and does not interfere with our ability to listen to the signal at 14.002 MHz.

Figure 12.A4 shows a signal in the same location that is S9+40 dB or –33 dBm. Using the same math, at 14.002 MHz we experience phase noise of –33 dBm – 73 dB = –106 dBm. Since the signal we are trying to receive is S3 (–109 dBm), our noise floor in 100 Hz is 3 dB above this level, causing us interference while trying to copy the signal. The amount of actual interference received also depends on the required SNR required to successfully copy a signal.

Although it varies by type of oscillator, phase noise generally decreases the further we are from the carrier, meaning that weak signals that are closer to strong signals are more likely to be interfered with than weak signals farther away from a strong carrier. Stations operating in a rural area with modest antennas and no strong neighbors are much less likely to suffer from the effects of phase noise than are multi-multi contest stations with large antennas and local transmitters in the same band. Field Day is another situation where many are familiar with the effects of phase noise. If you've ever operated the sideband station at your local Field Day and heard the noise floor modulated to the sound of the CW station next to you (or vice versa), you've experienced the effects of interference from phase noise.

The effects of phase noise also vary by mode. A CW signal may be copied with a SNR in the –10 to –20 dB range while a sideband signal may require an SNR of >+10 dB. Since sideband signals require a larger filter, we must also run our calculations with this larger filter, recog-

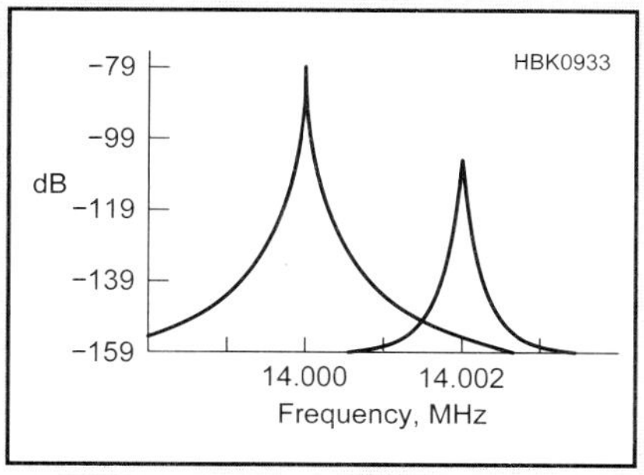

Figure 12.A3 — Spectrum showing adjacent CW signals and the phase noise from each.

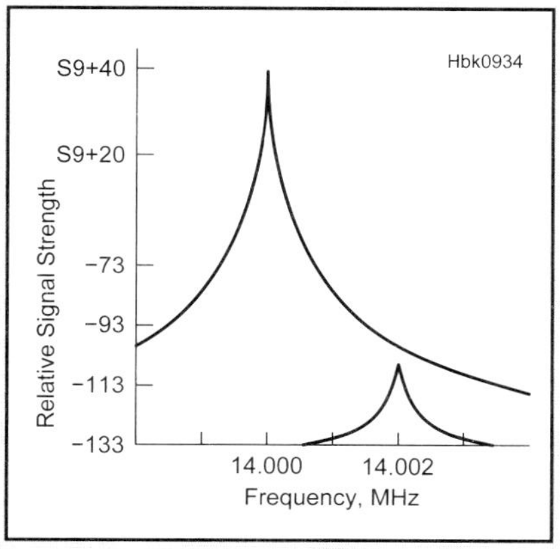

Figure 12.A4 — Same spectrum as Figure 12.A3 but with the left-hand signal replaced with one that is much stronger.

nizing that more noise is present. Both the additional noise and the larger SNR required to copy a sideband signal make it more susceptible to phase noise interference than a CW transmission.

Division is Good

In most local oscillators for superheterodyne receivers, the phase noise of the oscillator benefits from division. This means that the more the oscillator is reduced in frequency through division, the better the resulting phase noise of the oscillator. Typically, we see a 6 dB improvement in phase noise with each division by two of the original oscillator frequency. So a radio that might have good oscillator performance on 20 meters could lose 11 dB or more performance when operating on 6 meters. The superheterodyne receiver designer can take advantage of this by using a good oscillator design and dividing down the oscillator as much as possible, thereby achieving better phase noise.

direct-sampling SDR receiver does not respond to signals in the same way as an all-analog or hybrid analog-DSP IF receiver. For that reason, 3IMD and Intercept Point measurements only apply to analog and hybrid analog-digital receivers. IMD can be created by analog stages between the RF input and ADC input, such as preamps, preselectors, and any analog stages prior to digitization in the ADC device. However, these sources of IMD are typically much less strong than in legacy receivers. (Passive LC filters used as bandpass filters or preselectors can exhibit passive intermodulation (PIM) caused by non-linearities in the components. This can be measured by means of a two-port noise power ratio (NPR, described below) test.)

Third-order IMD dynamic range (3IMD_DR) is the difference between a receiver's MDS and the input level of two interfering signals that create an IMD product on the same frequency that is as strong as a desired signal. (The complete test is described in the **Test Measurements and Instruments** chapter and in the book by Allison listed in the References.) The test is performed with the interfering signals at different spacings (usually 2, 5, and 20 kHz) and with the desired signal at several different levels. The ARRL uses signals at the MDS, at S5 (–97 dBm) and 0 dBm. For *QST* Product Reports, the test is performed on the 3.5, 14, and 50 MHz bands. These conditions must be specified; no single number characterizes 3IMD_DR performance entirely.

This dynamic range is particularly important to the contest and DX community since they often need to copy very weak signals with very strong signals on an adjacent channel or just a few kHz away. Combined with reciprocal mixing, band noise, and spurious emissions from transmitters, 3IMD products can make for very difficult reception. (Note that SDR receivers do not specify this dynamic range because their circuitry does not behave the same as analog superheterodyne receivers, as discussed in the sections below.)

Intercept point describes the IMD performance of an individual stage or a complete receiver. For example, in an analog superheterodyne receiver, third-order IM products increase at the rate of 3 dB for every 1 dB increase in the level of each of the interfering input signals (ideally, but not always exactly true). As the input levels increase, the distortion products seen at the output on a spectrum analyzer could catch up to, and equal, the level of the two desired signals if the receiver did not begin to exhibit blocking as discussed earlier.

The input level at which this occurs is the *input intercept point*. **Figure 12.3** shows the concept graphically and derives an equation

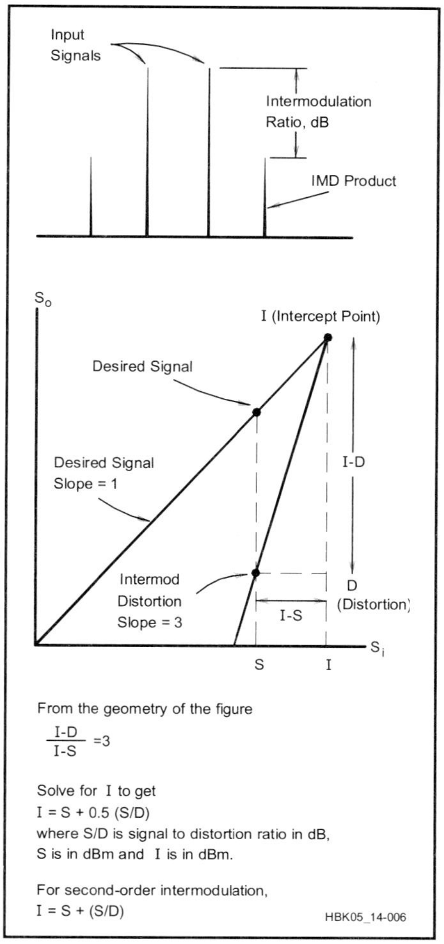

Figure 12.3 — Graphical representation of the third-order intercept concept.

from the geometry that relates signal level, distortion, and intercept point. The intercept point of the most interest in receiver evaluation is that for third-order IM products and is called the *third-order intercept point* or *IP_3*. A similar process is used to get a second-order intercept point for second-order IMD. A higher IP^3 means that third-order IM products are weaker for specific input signal strengths and the operator experiences less interference from IM products due to strong adjacent signals.

These formulas are very useful in designing radio systems and circuits. If the input intercept point (in dBm) and the gain of the stage (in dB) are added, the result is an output intercept point (in dBm). Receivers are specified by input intercept point, referring distortion back to the receive antenna input. Intercept point is a major performance limitation of receivers used in high density contest or DX operations. Keep in mind that we have been discussing this as an effect of two signals, one that is Δ away from our operating frequency and another at twice Δ. In real life, we may be trying to copy a weak signal at f_0, and have other signals at $f_0 \pm$ 500, 750, 1000, 1250…5000 Hz. There are many combinations that produce products at or near our weak signal's frequency.

Note that the products don't need to end up exactly on top of the desired signal to cause a problem; they just need to be within the operating bandwidth. So far, we have been talking about steady carriers, such as would be encountered during CW operation with interference from nearby CW stations. SSB or other wider bandwidth modes with spectrum distributed across a few kHz have signal components that go in and out of a relationship that results in on-channel interference from IMD. This manifests itself as a time-varying synthetic noise floor, composed of all the resulting products across the channel. The difference in this low level "noise" can be dramatic between different receivers, especially when added to phase noise received from other stations and reciprocal mixing inside the receiver!

SPURIOUS-FREE DYNAMIC RANGE (SFDR)

IM products increase with the amplitude of the interfering signals that cause them, and at some point become detectable above the receiver's noise floor. The ratio of the strength of the interfering signals to the noise floor, in dB, is the receiver's *spurious-free dynamic range* or *SFDR*. This is the range of signal strengths over which the receiver does not produce any detectable spurious products. SFDR can be specified for a specific order of IM products; for example, SFDR3 is the SFDR measured for third-order IM products only. The bandwidth for which the receiver's noise floor is measured must also be specified, since smaller bandwidths result in a lower noise floor.

INTERFERENCE FREE SIGNAL STRENGTH (IFSS)

To address the behavior of SDR receivers for which the IP^3 specification is irrelevant, new performance metrics have been proposed. One such measurement is the *interference-free signal strength* or *IFSS*. This measures the largest input signal for which no interference products are produced above the receiver noise floor. The noise floor is specified using ITU-R P.372-16 noise levels. IFSS is like SFDR but does not specify a type of intermodulation product.

Based on the ITU noise levels (see the **RF Techniques** chapter), the IFSS for a given receiver can be much higher on the lower HF bands, where external band noise is high and masks any interfering products. On the upper HF and VHF bands, where band noise is lower, the IFSS can also be lower. Similarly, there can be a difference between the rural, suburban, and urban environments. As such, IFSS measurements ignore distortion products below the noise floor.

IFSS measurements on current receivers show a great deal of variation in the performance over a wide range of signal levels. The best performers show a smooth increase in the distortion product level beginning at the input signal level where the interference products are equal to the noise floor. These radios exhibit predictable performance over a wide range of inputs. Other models show large swings in distortion product levels, even decreasing with increased input over some ranges of signal levels. This indicates a need for better understanding of SDR behavior and design. Nevertheless, IFSS is a useful tool in assessing SDR performance.

RECIPROCAL MIXING DYNAMIC RANGE (RMDR)

All oscillators produce some noise. When the oscillator is used to control a receiver or transmitter, the noise is transferred to signals, causing extra noise in the receiver passband and in the transmitter output signal. In an analog superheterodyne receiver, the noise is transferred to the signals in the mixing process from the local oscillator(s). While all the LOs contribute some noise, in practice the first mixer is the dominant source. In an SDR receiver, noise is transferred primarily from the sampling clock for the input ADC and the DDS signal sources in the receiver. Some noise is also contributed by jitter in the digitizing ADC.

RMDR is measured by increasing a test input signal until the noise in the receive channel increases by a specified amount, usually 3 dB. The difference between that input signal level and the receiver's noise floor (MDS) is the RMDR.

The raised background noise masks weak signals, in effect raising the receiver's noise floor when a nearby strong signal is present to transfer the oscillator noise. For example, if a receiver's noise floor or MDS is –127 dBm and an input signal of –37 dBm causes the 3 dB increase in noise, the receiver's RMDR = –37 – (–127) = 90 dB.

Since oscillator noise is lower farther from the oscillator frequency, the transferred noise has a similar profile. (The discussion of noise in the **Oscillators and Synthesizers** chapter shows the noise profile of typical oscillators.) This requires RMDR to be measured with the strong signal a specified distance from the receiver channel in which noise is being measured. Typical separations are 2, 5, and 20 kHz.

RMDR tends to be the most limiting dynamic range at the close signal spacing of 2 and 5 kHz. Since close-spaced signals are typical of the most demanding receiver environments for contesting and DXing, this is an important measurement to compare when purchasing a transceiver.

NOISE POWER RATIO (NPR)

Another test method applicable to both SDR and superheterodyne receivers, *noise power ratio tests* were originally developed for multiplexed telephone systems. These early systems combined many individual channels on one transmission system, such as cable or radio. Nonlinearities in the system would generate distortion products and

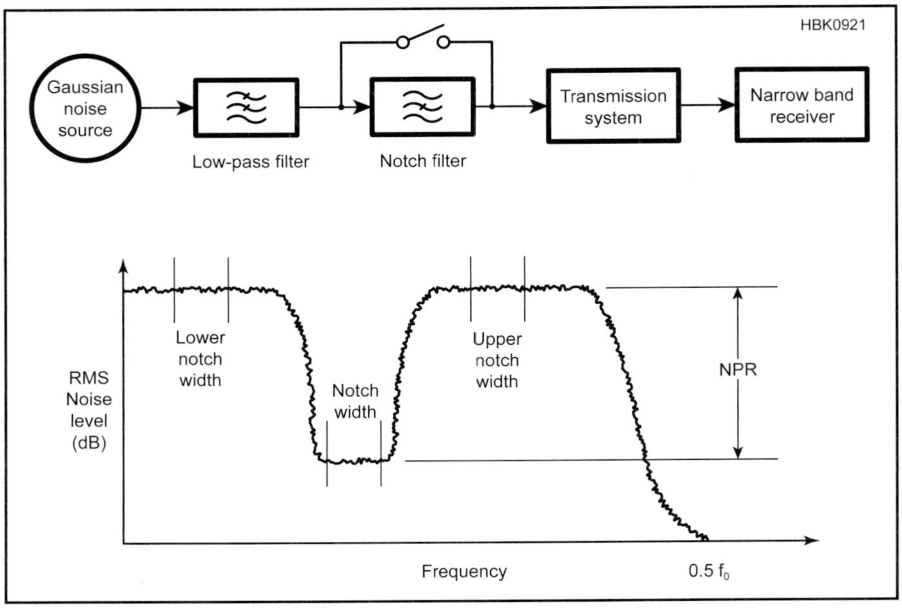

Figure 12.4 — Noise power measurement test set and typical spectrum (after Analog Devices MT-005, see References).

interfere with the channels. To simulate the effect of many active channels, wideband noise was transmitted with one channel removed by a narrow notch filter (see **Figure 12.4).**

The noise simulates the presence of complex speech waveforms in the other channels. Any distortion products from the noise appears as a raised noise floor in the notched channel. As the noise amplitude is raised, the in-channel noise rises, as well.

This is a close analogue to a busy band with multiple channels of communication. Nonlinearities in the receiver generate distortion products in the monitored channel. The ratio of the out-of-channel to in-channel noise is the noise power ratio. The more linear the receiver, the more noise power can be injected for a given amount of increase in the in-channel noise. (NPR is explained in the March/April 2015 *QEX* article, "Noise Power Ratio (NPR) Testing on HF Receivers" by Adam Farson, VA7OJ, and in the Analog Devices tutorial MT-005, "Noise Power Ratio (NPR)" — see the References section.

The above parameters of receiver performance described above can be calculated from the input/output parameters of the individual stages that make up the receiver. Analog Devices distributes a program that can be used for this purpose. See the **Modeling and Simulation** chapter for more information.

ADC OVERLOAD

When overloaded, ADCs behave similarly to amplifiers driven into saturation or clipping. In this condition, the ADC cannot distinguish between an input signal exactly at the limits of its range and a signal that exceeds the input range. In fact, the ADC cannot tell if the input range has been slightly exceeded or grossly exceeded. This is an example of hard clipping.

Most ADCs used in receivers have an output signal to indicate the input range has been exceeded and the converter is in overload or overflow/underflow mode. This normally activates an overflow indicator and can result in the receiver reducing the gain of an amplifier or increasing the attenuation of any stages ahead of the ADC. Alternately, the operator can reduce gain or turn on an attenuator to keep the input signals within range of the ADC.

Overload generally occurs only when a strong out-of-channel signal is present. In-channel signals stimulate AGC action which attenuates the signal at the ADC input. The usual test method is to offset the receiver from the test signal and then increasing the test signal level until ADC overflow occurs.

LATENCY

With respect to a single receiver, latency is the time interval between arrival of an event at the receiver's RF input and the corresponding output event at the receiver's demodulated output. The receiver output can be either an audio signal or some other type of baseband information, such as I/Q signals. The usual RF test signal is a pulse modulated or repetitive burst signal that occurs at a well defined time. A dual-channel oscilloscope is used to measure the time interval between the input and output events.

Latency changes with various detection and filter bandwidth settings. If the receiver includes a waterfall or spectrum display, latency can include the time to display a change after the input changes. A single receiver may show different latencies for audio, data, and visual outputs. This definition applies to only one piece of equipment and does not include network delays for receivers that are operated over a remote-control link.

Latency in an SDR is caused by several factors, including the ADC and DAC conversion times, filter group delay, frequency shifting delay, and demodulation time. Converter latency tends to be consistent, while radio group delays are more variable, depending on mode and filter settings.

Whether latency is considered excessive depends on the mode being used. For example, some digital modes that use ARQ protocols (see the **Digital Protocols and Modes** chapter) have a maximum response time before a data frame is assumed to have been lost. The resulting re-transmissions can slow the system or cause the link to fail completely. Excessive latency when monitoring one's own speech signal can become quite confusing, as well. Hybrid receivers with analog front ends and DSP IF and audio filtering can also exhibit excessive latency if the DSP is slow or underpowered. Latency in analog receivers is mostly insignificant.

TYPICAL PERFORMANCE

Product reviews of receivers and the receiver sections of transceivers published in *QST* now provide the measured dynamic range in the presence of interfering signals with spacings of 2, 5, and 20 kHz. (Details of the test procedures used are given in the **Test Equipment and Measurements** chapter and in the book by Allison in the References.) At 20 kHz spacing, the interfering signal is usually outside of the roofing filter bandwidth of most amateur receivers. Spacing of 2 and 5 kHz represents likely conditions on a crowded band.

A caveat about transceiver testing, particularly receiver performance, is warranted. While the dynamic range of a receiver is certainly important, most modern receivers perform quite adequately. By managing receiver gain (RF gain, attenuation, preamp, or RF amplifier on or off) even a modest receiver performs well, even under crowded conditions. Evaluation of a receiver should include several parameters that affect performance according to the operator's personal interests. THD (distortion), audio hiss, filter adjustment and selection, and noise performance are just a few of the features that matter to an operator. While dynamic range values get a lot of attention, don't buy (or replace) a receiver based on one or two optimized figures. Try to look at the whole picture and consider how you are likely to use the receiver first.

A look at recent top-performing receiver measurements by the ARRL Lab and Sherwood Engineering (**www.sherweng.com/table.html**) indicates that receivers have IMD (or distortion-product) dynamic range with 2 kHz spacing results in the following ranges:

- Direct-sampling SDR: 80 to 110 dB
- Up-converting with VHF IF: 80 to 105 dB
- Down-converting with HF IF: 75 to 110 dB
- Hybrid superheterodyne/DSP architecture: Omni VII, 82 dB (2007); Icom IC-756PRO3, 75 dB (2009)

(Note that SDR distortion products are not necessarily IMD. Receiver block diagrams are provided in the following sections. Data is taken from ARRL Lab tests and the long-term test program conducted by Rob Sherwood, NCØB, at **www.sherweng.com/table.html**.)

As we would expect, the blocking and IMD dynamic range (IMD_DR) performance of a superheterodyne receiver depends on a combination of the early-stage filtering, the linearity of the mixers and amplifiers, and the dynamic range of any ADC used for DSP at the IF or audio stages.

Let's take an example of what this would mean. If we are listening to a signal at S3, for signals to generate a third-order IMD product at the same level in a receiver with a dynamic range of 60 dB, the $f_0 + D$ and $f_0 + 2D$ signals would have to have a combined power equal to S9 +27 dB, or each at S9 +24 dB. This is not unusual on today's amateur bands. On the other hand, if we had an IMD dynamic range of 102 dB, the interfering signals would both have to be S9 +66 dB, which is much less likely. How much dynamic range you need depends in large measure on the kind of operating you do, how much gain your receiving antennas have and the closeness of the nearest station that operates on the same bands as you. Given that transmitters generate both phase noise and IMD products of their own, the dynamic range of most receivers on the market today is quite adequate for even the most demanding situations. Receivers may have other performance issues to consider, such as filter design, AGC reaction, audio distortion, and so forth.

Keep in mind also that it is often difficult to tell whether or not you are hearing internal receiver-generated IMD — it just sounds like there are many more signals than are really present. A good test to assess the source of interference is to first switch off any preamplifiers and noise-blankers or noise-reduction systems that affect the receiver's linearity. Observe the level of the interference (if it's still there) and then switch in some attenuation at the front-end of the receiver. If the level of the interference goes down by more than the level of attenuation (estimate 6 dB per S unit), then at least some of the interference is being generated by non-linearity inside the receiver. Continue to increase attenuation until the interference either goes away or goes down at the same rate as the attenuation is increased. You might be surprised at how much better the band "sounds" when your receiver is operating in its linear region!

12.2 Superheterodyne Receivers

The *heterodyne* receiver combines the input signal with a signal from a *local oscillator* (LO) in a mixer, as discussed previously, to generate the sum and difference frequencies as shown in **Figure 12.5**. The receiver may be designed so the output signal is anything from DC (a so-called direct conversion receiver) to any frequency above or below either of the two frequencies. The major benefit is that most of the gain, bandwidth setting, and processing are performed at a single frequency, simplifying the design dramatically.

By changing the frequency of the LO, the operator shifts which signal at the input appears at the output, along with all its modulated information. In most receivers the mixer output frequency is designed to be an RF signal, either the sum or difference, the other being filtered out at this point. This output frequency is called an *intermediate frequency* or *IF*. The IF amplifier system can be designed to provide the selectivity and other desired characteristics centered at a single fixed frequency.

When more than one frequency conversion process is used, the receiver becomes a *superheterodyne*. A block diagram of a typical superheterodyne receiver is shown in **Figure 12.6.** In traditional form, the RF filter is used to limit the input frequency range to those frequencies that include only the desired sum or difference but not the other — the so-called image frequency. The dotted line represents the fact that in receivers with a wide tuning range, such as a simple AM broadcast receiver that tunes from 500 to 1700 kHz, a more than 3:1 range, the input RF amplifier and filter is often tracked along with the local oscillator. The IF filter is used to establish selectivity — the operating bandwidth required by the information. Circuits from the antenna input through and including the mixer (the first mixer if more than one mixing stage is used) are generally referred to as the receiver's *front-end.*

For reception of suppressed carrier single-sideband voice (SSB) or on-off or frequency-shift keyed (FSK) signals, a second beat *frequency oscillator* or BFO is

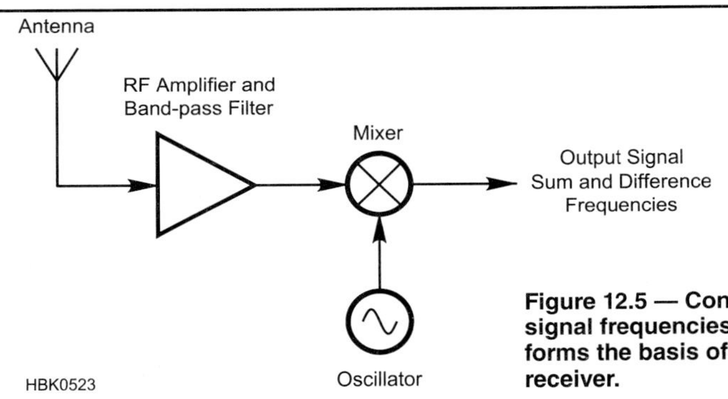

Figure 12.5 — Converting received signal frequencies using a mixer forms the basis of the heterodyne receiver.

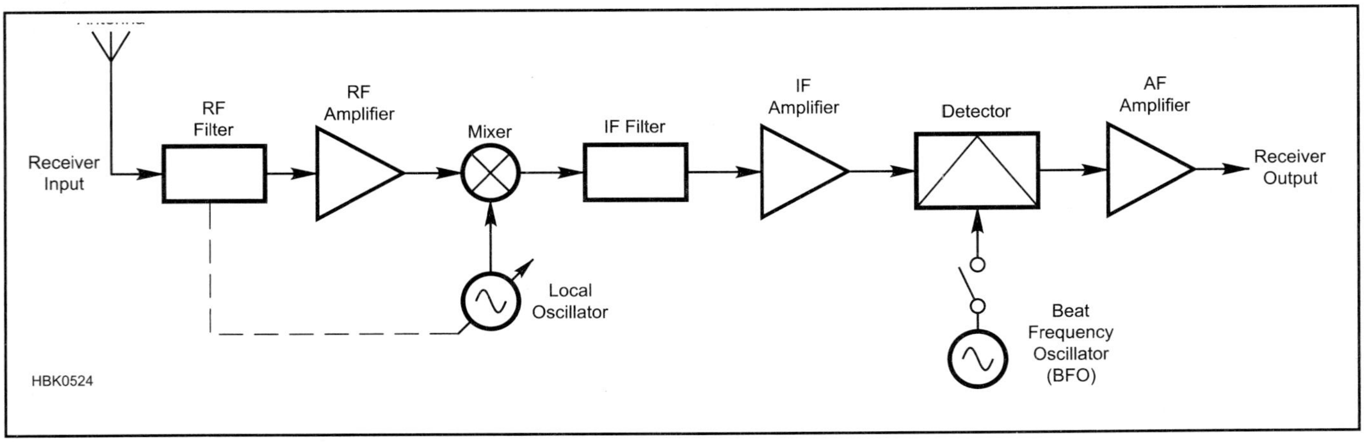

Figure 12.6 — Elements of an analog superheterodyne radio receiver.

employed to provide an audible voice, an audio tone or tones at the output for operator or FSK processing. This is the same as a heterodyne mixer with an output centered at dc, although the IF filter is usually designed to remove one of the output products.

Recent superheterodyne receivers convert the incoming signal to digital form at one of the intermediate frequencies. DSP techniques are then used to control operating bandwidth and demodulate the input signal. See the **DSP and SDR Fundamentals** chapter for more information on these techniques.

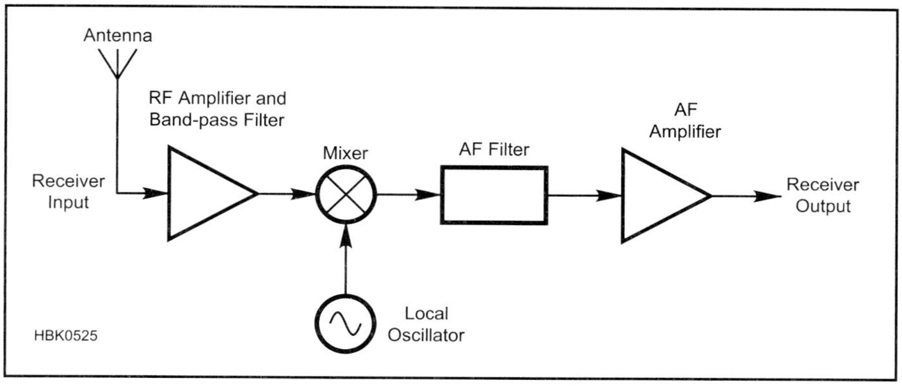

Figure 12.7 — Block diagram of a direct conversion receiver.

12.2.1 The Direct Conversion Receiver

The heterodyne process can occur at different points in the receiver. The simplest form of heterodyne receiver is called a *direct conversion* or *DC* receiver because it performs the translation directly from the signal frequency to the audio output. It is, in effect, just the BFO and detector of the general superheterodyne shown in Figure 12.6. In this case, the detector is often preceded by an RF amplifier with a typical complete receiver shown in **Figure 12.7.** Such a receiver can be very simple to construct, and yet can be quite effective — especially for the ultra-compact low-power consumer-oriented portable station known as the mobile telephone! In fact, given the direct conversion receiver's use in the mobile telephone, it is the most widely used of all receivers.

The basic function of a mixer is to multiply two sinusoidal signals and generate two new signals with frequencies that are the sum and difference of the two original signals. This function can be performed by a linear multiplier, a switch that turns one input signal on and off at the frequency of the other input signal, or a nonlinear circuit such as a diode. (The output of a nonlinear circuit is made up of an infinite series of products, all different combinations of the two input signals, as described in the sidebar on Nonlinear Signal Combinations.) Much more information about the theory, operation and application of mixers may be found in this chapter's section on them.

Figure 12.8 shows the progression of the spectrum of an on-off keyed CW signal through such a receiver based on the relationships described above. (This example is based on the "Rock-Bending Receiver for 7 MHz" which is included in the online projects.) In 12.8C, we include an undesired image signal on the other side of the local oscillator that also shows up in the output of the receiver. Note each of the desired and undesired responses that occur as outputs of the mixer.

Some mixers are designed to be *balanced* to cancel one of the input signals at the output while a *double-balanced mixer cancels both. A double-balanced* mixer simplifies the output filtering job as shown in Figure 12.8D.

Products generated by nonlinearities in the mixing process (see the previous sidebar on Nonlinear Signal Combinations) are heard as intermodulation distortion signals that are discussed later. Note that the nonlinearities also allow mixing with unwanted signals near multiples of local oscillator frequency. These signals, such as those from TV or FM broadcast stations, must be eliminated in the filtering before the mixer since their audio output lands in the desired passband on the output of the mixer.

12.2.2 Superheterodyne Receivers

In many instances, it is not possible to achieve all the receiver design goals with a single-conversion receiver and multiple conversion steps are used, creating the superheterodyne architecture. Traditionally, the first conversion is tasked with removing the RF image signals, while the second generates the IF signal that is amplified and filtered.

The superheterodyne concept was introduced by Major Edwin Armstrong, a US Army artillery officer, just as WW I was ending. He is the same Armstrong who invented frequency modulation (FM) some years later and who held many radio patents between WWI and WWII.

In a superheterodyne receiver, a local oscillator and mixer are used to translate the received signal to an intermediate frequency rather than directly to audio. This provides an opportunity for additional amplification and processing. Then a second mixer is used, as in the DC receiver, to detect the IF signal, translating it to audio. The configuration is shown in Figure 12.6.

An example illustrates how this works. Let's pick a common IF frequency used in a simple AM broadcast radio, 455 kHz. If we want to listen to a 600 kHz broadcast station, the RF stage is set to amplify the 600 kHz signal and the LO is set to 600 + 455 kHz or 1055 kHz. The 600 kHz signal, along with any audio information it contains, is translated to the IF frequency and is amplified and then detected.

Note that we could have also set the local oscillator to 600 – 455 kHz or 145 kHz. By setting it to the sum, we reduce the relative range that the oscillator must tune. To cover the 500 to 1700 kHz with the difference, our LO would have to cover from 45 to 1245, a 28:1 range. Using the sum requires LO coverage from 955 to 2155 kHz, a range of about 2.5:1, which is much easier to implement.

Note that to detect standard AM signals, the receiver's second oscillator, the beat frequency oscillator (BFO), is turned off since the AM station provides its own carrier signal over the air. Receivers designed only for standard AM reception generally don't have a BFO at all.

It's not clear yet that we've gained anything by doing this; so let's look at another example. If we decide to change from listening to the station at 600 kHz and want to listen to another station at, say, 1560 kHz, we can tune the single dial of our superheterodyne to 1560 kHz. The RF stage is tuned to 1560 kHz, the LO is set to 1560 + 455 or 2010 kHz, and now the desired station is translated to our 455 kHz IF where the bulk of our amplification can take place. Note also that with the superheterodyne configuration, selectivity (the ability to separate stations) occurs primarily in the intermediate frequency (IF) stages and is thus the same no matter what frequencies we choose to listen to. This simplifies the design of each stage considerably.

12.2.3 Superheterodyne Bandwidth

Now we will discuss the bandwidth requirements of different operating modes

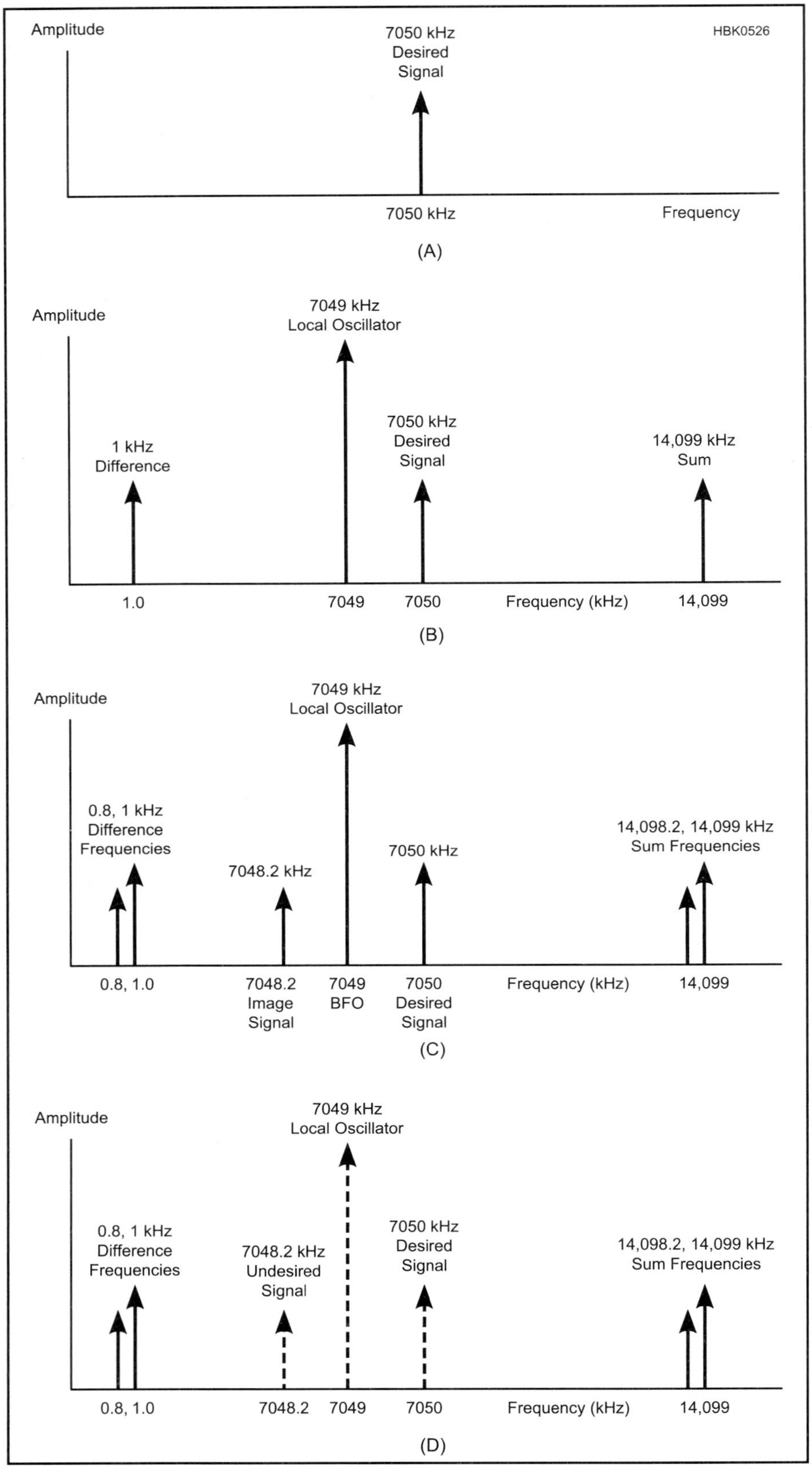

Figure 12.8 — Frequency relationships in a direct conversion receiver. At (A), the desired receive signal from antenna, a 7050 kHz on-off keyed carrier. At (B), the internal local oscillator (LO) and receive frequency relationships. At (C), the frequency relationships of mixer/detector products (not to scale). At (D), the sum and difference outputs from a double balanced mixer (not to scale). Note the mixer inputs that are balanced out so that they cancel in the output (dashed lines).

and how that affects superheterodyne receiver design. One advantage of a superheterodyne is that the operating bandwidth can be established by the IF stages, and further limited by the audio system. It is thus independent of the RF frequency to which the receiver is tuned. It should not be surprising that the detailed design of a superheterodyne receiver is dependent on the nature of the signal being received. We will briefly discuss the most commonly received modulation types and the bandwidth implications of each below. The typical operating modes expected to be encountered by an HF communications receiver are tabulated in **Table 12.3.** (Each modulation type is discussed in more detail in the **Modulation** or **Digital Protocols and Modes** chapters.) The same concerns for receiver bandwidth apply to SDR receivers, although the filters are implemented using DSP techniques.

AMPLITUDE MODULATION (AM)

As shown in Figure 12.8, multiplying (resulting in, modulating) a carrier with a single tone causes the tone to be translated to frequencies of the sum and difference of the two. Thus, if a transmitter were to multiply a 600 Hz tone and a 600 kHz carrier signal, it would generate additional new frequencies at 599.4 and 600.6 kHz. If instead we were to modulate the 600 kHz carrier signal with a band of frequencies corresponding to human speech of 300 to 3300 Hz (the usual range of communication quality voice signals), we would have a pair of information-carrying sidebands extending from 596.7 to 603.3 kHz, as shown in **Figure 12.9**.

Note that the total bandwidth of this AM voice signal is twice the highest modulation signal frequency, or 6600 Hz. If we choose to transmit speech and limited-range music, we might allow modulating frequencies up to 5000 Hz, resulting in a bandwidth of 10,000 Hz or 10 kHz. This is the standard channel spacing that commercial AM broadcasters use in the US (9 kHz is used in Europe). In actual use, transmitters on

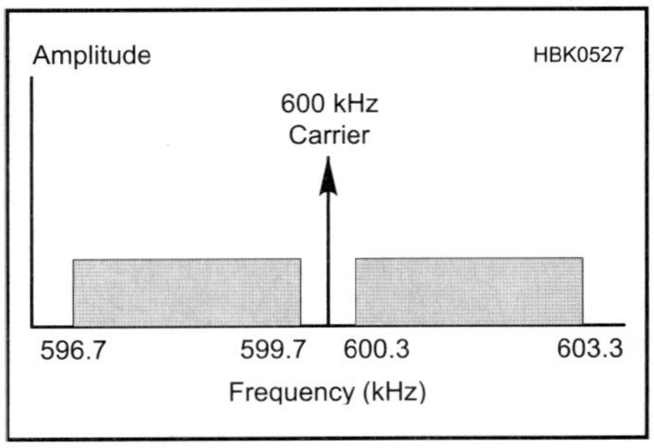

Figure 12.9 — Spectrum of sidebands of an AM voice signal with a carrier frequency of 600 kHz.

adjacent channels are generally geographically separated, so broadcasters can extend some energy into the adjacent channels for improved fidelity. We refer to this as a *narrow-bandwidth* mode.

What does this say about the bandwidth needed for our receiver? If we want to receive the full information content transmitted by a US AM broadcast station, then we need to set the bandwidth to at least 10 kHz. If our receiver has a narrower bandwidth, we will lose the higher frequency components of the transmitted signal, perhaps ending up with a radio suitable for voice but not very good at reproducing music.

On the other hand, what is the impact of having too wide a bandwidth in our receiver? In that case, we are able to receive the full transmitted spectrum, but we also receive some of the adjacent channel information. This sounds like interference and reduces the quality of what we are receiving. If there are no adjacent channel stations, we get increased noise from the additional bandwidth, without more information. The general rule is that the received bandwidth should be matched to the bandwidth of the signal we are trying to receive to maximize SNR and minimize interference.

As the receiver bandwidth is reduced, intelligibility suffers, although the SNR is improved. With the carrier centered in the receiver bandwidth, most amplitude modulated voices are difficult to understand at bandwidths less than around 4 kHz. In cases of heavy interference, full carrier AM can be received as if it were SSB, as described below, with the carrier inserted at the receiver, and the receiver tuned to whichever sideband has the least interference.

SINGLE-SIDEBAND (SSB)

A single-sideband signal contains just one of the AM signal's sidebands and no carrier, as shown in **Figure 12.10**. To receive the SSB signal, the receiver uses a BFO (beat frequency oscillator) to provide a substitute carrier. The BFO oscillator signal is multiplied with the sideband to provide demodulated audio output. The implications are that the bandwidth can be slightly less than half that required for double sideband AM (DSB). The tradeoff is that the BFO signal must be at exactly the right frequency. If the frequency is improperly set, the frequency of the demodulated audio is offset from the original signal. The effect is quite audible even for small frequency errors of a few tens of Hz. (see the References section for a *QST* article "About SSB" by Ward Silver, NØAX, that illustrates this effect, including a video on tuning in an SSB signal).

This results in a requirement for a much more stable receiver design with a much finer tuning system, a more expensive

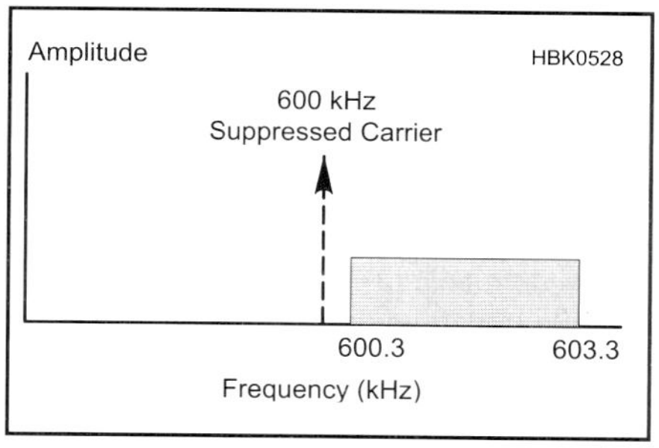

Figure 12.10 — Spectrum of single sideband AM voice signal with the suppressed carrier frequency of 600 kHz.

proposition than the DC receiver. An alternate is to transmit a reduced level carrier, usually called a *pilot* carrier, and have the receiver lock on to it. Note that the pilot carrier need not be of sufficient amplitude to demodulate the signal, just enough to allow a BFO to lock onto it. These alternatives are effective, but tend to make SSB receivers expensive, complex, and most appropriate for the case in which a small number of receivers are listening to a single transmitter.

Note that the bandwidth required to effectively demodulate an SSB signal is less than half that required for the AM signal because the range centered on the AM carrier need not be received. Thus, the communications-quality range of 300 to 3300 Hz can be received in a bandwidth of only 3000 Hz. Early SSB receivers typically used a bandwidth of around 3 kHz, but with the heavy interference frequently found in the amateur bands, it is more common for amateurs to use bandwidths of 1.8 to 2.4 kHz with the corresponding loss of some of the higher- and lower-frequency components of speech.

RADIOTELEGRAPHY (CW)

We have described radiotelegraphy as being transmitted by "on-off keying of a carrier." You might think that since a carrier takes up just a single frequency, the receive bandwidth needed should be almost zero. This is only true if the carrier is never turned on and off. In the case of CW, it is turned on and off quite rapidly. The rise and fall of the carrier results in sidebands extending on either side of the carrier, and they must be received to reconstruct the signal in the receiver.

A rule of thumb is to consider the rise and fall time as about 10% of the pulse width, with the maximum bandwidth as the reciprocal of the quickest of rise or fall time. This results in a bandwidth requirement of about 50 to 200 Hz for the usual CW transmission rates. Another way to visualize this is with the bandwidth being set by a high-Q tuned circuit. Such a circuit continues to "ring" after the input pulse is gone. Thus, too narrow a bandwidth "fills in" between the code elements and acts like a "no bandwidth" full period carrier. This is what is heard if a very narrow crystal filter is used when receiving CW.

DIGITAL MODULATION

(This section contains an introduction to receiver requirements for the digital modes. See the **Modulation** and **Digital Protocols and Modes** chapters and the **Digital Communications** chapter in the *Handbook's* online content for more in-depth treatment.)

The Baudot code, used for teletype communications, and the ASCII code are two popular digital communications codes used by amateurs. They are constructed with sequences of bits. The state of each bit, ON or OFF, is represented by a signal that has one of two distinct frequencies: one designated *mark* and the other designated *space*. Switching between these frequencies is a modulation type called *frequency shift keying* (FSK).

Amateur operators typically use a 170 Hz separation between the mark and space frequencies for the most popular FSK mode, radioteletype, or RTTY, depending on the data rate and local convention; 850 Hz is also sometimes used. The minimum bandwidth required to recover the data is approximately twice the spacing between the tones. Note that the tones can be generated by directly shifting the carrier frequency *(direct FSK),* or by using a pair of 170 Hz spaced audio tones applied to the audio input of an SSB transmitter *(audio FSK* or *AFSK).* Direct FSK and AFSK are indistinguishable to a receiver.

If the standard audio tones of 2125 Hz (mark) and 2295 Hz (space) are used, they fall within the bandwidth of a voice channel and thus a voice transmitter and receiver can be employed without any additional processing needed outside the radio equipment.

If the receiver can shift its BFO frequency appropriately, the two tones can be received through a filter designed for CW reception with a bandwidth of about 300 Hz or wider. Some receivers provide such a narrow filter with the center frequency shifted midway between the tones (2210 Hz) to avoid the need for retuning. The most advanced receivers provide a separate filter for mark and space frequencies, thus maximizing interference rejection and SNR. Using a pair of tones for FSK or AFSK results in a maximum data rate of about 1200 bits per second (baud) over a high-quality voice channel.

Phase shift keying (PSK) can also be used to transmit bit sequences, requiring good frequency stability to maintain the required time synchronization to detect shifts in phase. If the channel has a high SNR, as is often the case at VHF and higher, telephone network data-modem techniques such as Bell 102 and Bell 202 can be used. (FCC §97.307(f) specifies a maximum transmitted symbol rate for each band.)

At HF, the signal is subjected to phase and amplitude distortion as it travels. Noise is also substantially higher on the HF bands. Under these conditions, modulation and demodulation techniques designed for "wire-line" connections become unusable at bit rates of more than a few hundred baud. As a result, amateurs have begun adopting and developing state of the art digital modulation techniques. These include the use of multiple carriers (MFSK, Clover, PACTOR III, etc.), multiple amplitudes and phase shifts (QAM and QPSK techniques), and advanced error detection and correction methods to achieve a net data throughput as high as 3600 baud over a voice-bandwidth channel. Newly developed coding methods for digital voice using QPSK modulation result in a signal bandwidth of less than 1200 Hz. (See **freedv.org** for more information.) Spread-spectrum techniques are also being adopted on the amateur UHF bands but are beyond the scope of this discussion.

The bandwidth required for data communications can be as low as 100 Hz for PSK31 to 1 kHz or more for the faster speeds of PACTOR III and Clover. Beyond having sufficient bandwidth for the data signal, the primary requirements for receivers used for data communications are linear amplitude and phase response over the bandwidth of the data signal to avoid distorting these critical signal characteristics. The receiver must also have excellent frequency stability to avoid drift and frequency resolution to enable the receiver filters to be set on frequency.

FREQUENCY MODULATION (FM)

Another popular voice mode is *frequency modulation* or *FM*. FM can have several variations depending on purpose. Amateur Radio and commercial mobile communications on the shortwave bands use *narrow band FM (NBFM)*. In NBFM, the frequency deviation is limited to around the maximum modulating frequency, typically 3 kHz. The bandwidth requirements at the receiver can be approximated by Carson's Rule of BW = 2 × (D + M), where D is the deviation and M is the maximum modulating frequency. Thus, 3 kHz deviation and a maximum voice frequency of 3 kHz results in a bandwidth of 12 kHz, not much wider than the requirements for broadcast AM. (Additional signal components extend beyond this bandwidth but are not required for voice communications.)

In contrast, broadcast or *wideband FM (WBFM)* occupies a channel width of 150 kHz. Originally, this provided for a higher modulation index, even with 15 kHz audio, that resulted in an improved SNR. However, with multiple channel stereo and sub-channels all in the same allocated bandwidth the deviation is around the maximum transmitted signal bandwidth.

In the US, FCC amateur rules limit wideband FM use to frequencies above 29 MHz. Some, but not all, HF communication receivers provide for FM reception. For proper FM reception, two changes are required in the receiver architecture as shown within the dashed line in **Figure 12.11**. The fundamental change is that the detector must recover information from the frequency variations of the input signal. The most common such detector is called a *discriminator*. The discriminator does not require a BFO, so that is turned off, or eliminated, in a dedicated FM receiver. Amplitude variations convey no information in FM so they are generally eliminated by a *limiter*. The limiter is a high-gain IF amplifier stage that clips the positive and negative peaks of signals above a certain threshold. Since most noise of natural origins is amplitude modulated, the limiting process also strips away noise from the signal.

12.2.4 Superheterodyne FM Receivers

NBFM is the most common mode used on VHF and UHF. **Figure 12.12** is a block diagram of an FM receiver for the VHF/UHF amateur bands. Many FM transmitters are in reality phase-modulated (PM), but aside from frequency response of the transmitted audio, the two types of signals can be received with the same equipment. This section's references to FM include PM signals unless noted otherwise.

FRONT END

A low noise, high gain front end is desirable because of the decreased atmospheric noise level at VHF and above, and because portable gear often uses short rod antennas at ground level. Nonetheless, the possibilities for gain compression and harmonic IMD, multitone IMD and cross modulation are substantial. Therefore, dynamic range is an important design consideration, especially if large, high-gain antennas are used. FM limiting should not be applied until after the IF crystal filter. Because of the high occupancy of the VHF/UHF spectrum by powerful broadcast transmitters and nearby two-way radio services, front-end preselection is necessary, so that a low noise figure can be achieved economically within the amateur band. (See the section on Preselectors earlier in this chapter.)

DOWN-CONVERSION

Down-conversion to the final IF can occur in one or two stages. Usual IFs are in the 5 to 10 MHz region (10.7 MHz is common), but at the higher frequencies, rejection of the image 10 to 20 MHz away can be difficult, requiring considerable preselection, as discussed in the sections below on IF selec-

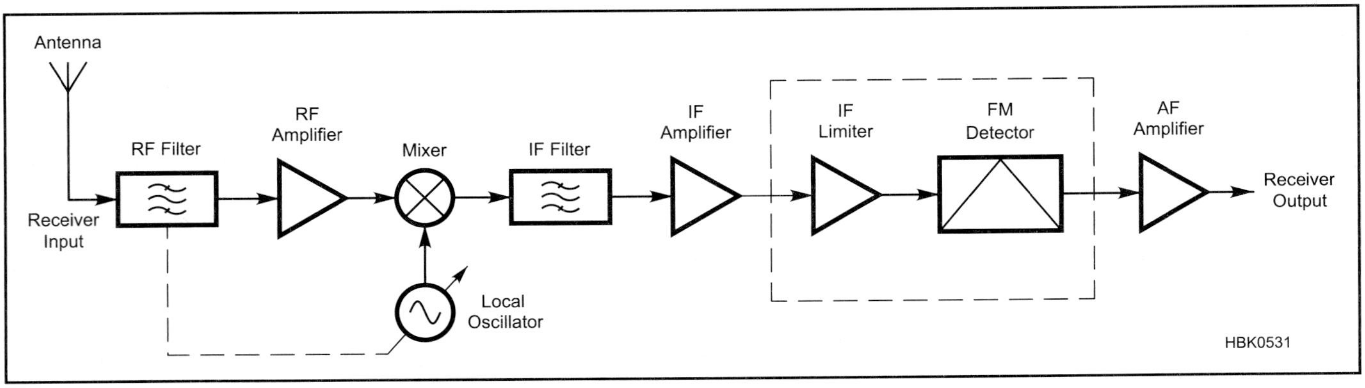

Figure 12.11 — Block diagram of an FM superheterodyne receiver. Changes from an AM/SSB receiver are enclosed by the dashed line.

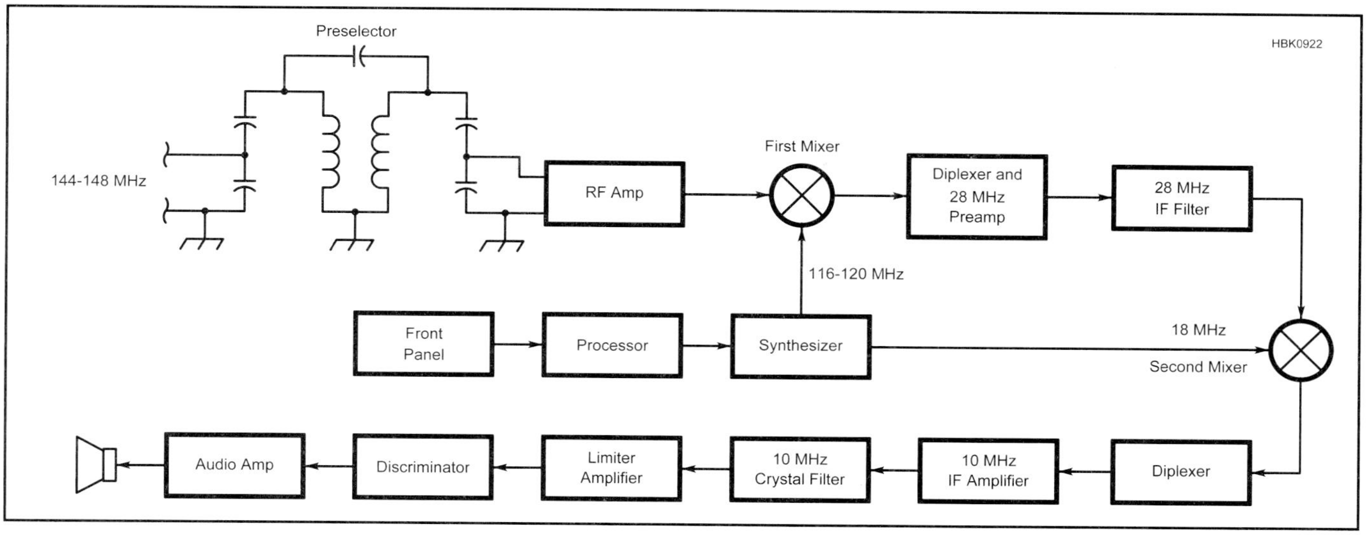

Figure 12.12 — Block diagram of a typical VHF FM receiver using dual down-conversion.

tion and image rejection. At the higher frequencies an intermediate IF in the 30 to 50 MHz region is a better choice. Figure 12.12 illustrates dual down-conversion.

IF FILTERS

The customary peak frequency deviation in amateur FM on frequencies above 29 MHz is 5 kHz and the audio speech band extends to 3 kHz. This defines a maximum modulation index (defined as the deviation ratio) of 5/3 = 1.67. An inspection of the Bessel functions that describe the resulting FM signal shows that this condition confines most of the 300 to 3000 Hz speech information sidebands within a 15 kHz bandwidth. Using filters of this bandwidth, channel separations of 20 or 25 kHz are achievable.

Many amateur FM transceivers are channelized in steps that can vary from 1 to 25 kHz. For low distortion of the audio output (after FM detection), IF filters should have good phase linearity across the bandwidth. This would seem to preclude filters with very steep descent outside the passband, which tend to have very nonlinear phase near the frequency cutoff. However, since the amount of energy in the higher speech frequencies is naturally less, the actual distortion due to this effect may be acceptable for speech purposes. The normal practice is to apply pre-emphasis to increase the gain of higher speech frequencies at the transmitter and then de-emphasis at the return those frequencies to normal amplitudes.

LIMITING AND DEMODULATION

After the IF filter, hard limiting is needed to remove any amplitude modulation components. In a high-quality receiver, special attention is given to any nonlinear phase shift that might result from the limiter circuit design. This is especially important for data receivers, in which phase response must be controlled. In amateur receivers used for speech it may be less important. Also, the *ratio detector* largely eliminates the need for a limiter stage, although the limiter approach is still preferred. FM demodulation is described in the section on Demodulation and Detection.

FM PERFORMANCE FOR WEAK SIGNALS

The noise bandwidth of the IF filter in NBFM is not much greater than twice the audio bandwidth of the speech modulation, and far less than it would be in wideband FM. Therefore, such things as capture effect, threshold effect, and the noise quieting effect, so familiar to wideband FM, still occur, but to a lesser extent. For FM receivers, sensitivity is specified in terms of a SINAD (see the **Test Equipment and Measurements** chapter) ratio of 12 dB. Typical values are –110 to –125 dBm, depending on the low-noise RF pre-amplification that often can be selected or deselected, depending on the strength of the signal environment.

EFFECT OF LO PHASE NOISE

In an FM receiver, LO phase noise is superimposed on phase modulation, creating frequency modulation of the desired signal that reduces the ultimate SNR within the passband. This effect is called *incidental FM (IFM)*. The power density of IFM (W/Hz) is proportional to the phase noise power density (W/Hz) multiplied by the square of the modulating frequency (the familiar parabolic effect in FM). If the receiver uses high-frequency de-emphasis at the audio output (–6 dB per octave from 300 to 3000 Hz, a common practice), the IFM level at higher audio frequencies can be reduced. Ordinarily, as the signal increases the noise would be "quieted" (that is, "captured") in an FM receiver. In this case the signal, with phase noise riding "piggy back," increase in the same proportion as described in this and the **Oscillators and Synthesizers** chapter's discussion of reciprocal mixing. IFM is not a significant problem in modern FM radios, but phase noise can become a concern for adjacent-channel interference.

As the signal becomes large the SNR approaches some final value. A similar ultimate SNR effect occurs in SSB receivers. On the other hand, a perfect AM receiver tends to suppress LO phase noise. (See the reference entry for Sabin.)

FM RECEIVER ICs

A wide variety of special ICs for NBFM receivers may be available. Although many were discontinued from production many years ago, parts may still be available at hamfests or by searching for old stock on the internet. One is an RF amplifier chip (NE/SA5204A, which is no longer manufactured) for 50 Ω input to 50 Ω output with 20 dB of gain. The second chip (NE/SA602A, NXP Semiconductors, **www.nxp.com**) is a front-end device with an RF amplifier, mixer, and LO. The third is an IF amplifier, limiter, and quadrature FM detector (NE/SA602A, NXP Semiconductors, **www.nxp.com**) that includes a very useful logarithmic RSSI output and a "mute" function. The fourth is the LM386, a widely used audio-amplifier chip. A complete FM receiver chip in one package is the MC3371P (no longer manufactured).

The MC13135 (no longer manufactured) features double conversion and two IF amplifier frequencies. This allows more gain on a single chip with less of the cross coupling that can degrade stability. This desirable feature of multiple down-conversions was mentioned previously in this chapter.

Design details and specific parts values can be learned from a careful study of the data sheets and application notes provided by the IC vendors. Amateur designers should learn how to use these data sheets and other information, such as application notes available online from the manufacturers.

12.2.5 Superheterodyne Image Rejection

Now that we have established the range of bandwidths that our receiver needs to pass, we can discuss the selection of the IF frequency at which those bandwidths are established. Selection of IF frequencies is important in receiving weak signals on bands where strong signals are present. If the desired signal is near the noise, a signal at an image frequency could easily be 100 dB stronger. To avoid interference, an image rejection of 110 dB is needed. While some receivers meet that target, the receiver sections of most current amateur transceivers are in the 70 – 100 dB range.

IF IMAGE RESPONSE

As noted earlier, a superheterodyne with a single LO and specified IF can receive two frequencies, the desired frequency, and an undesired frequency. For example, using a receiver with an IF of 455 kHz to listen to a desired signal at 7000 kHz can use an LO of 7455 kHz. However, the receiver also receives a signal at 455 kHz above the LO frequency, or 7910 kHz. This undesired signal frequency, located at twice the IF frequency from the desired signal, is called an *image.*

Images are separated from the desired frequency by twice the IF. The filter preceding the associated mixer must reduce the image signal by the amount of the required *image rejection*. For a given IF, this gets more difficult as the received frequency is increased. For example, with a 455 kHz single conversion system tuned to 1 MHz, the image is at 1.91 MHz, almost a 2:1 frequency ratio and relatively easy to reject with a filter. The same receiver tuned to 30 MHz, would have an image at 30.91 MHz, a much more difficult filtering problem.

While an image that falls on an occupied channel is obviously a problem, additional noise is received even if the image frequency is clear of signals. This is because the atmospheric noise in the image bandwidth is added to the noise of the desired channel, plus any internally generated noise in the RF amplifier stage. When the image response is at the same level as the desired signal response, there is a 3 dB reduction in SNR.

REDUCING IMAGES BY INCREASING IF

An obvious solution to the RF image response is to raise the IF frequency high enough so that signals at twice the IF frequency from the desired signal are sufficiently attenuated by filters preceding the mixer. This can be done with IF stages that operate at 5% to 10% of the highest receiving frequency (1.5 – 3 MHz for a receiver that covers the 3 – 30 MHz HF band). The same concept is used at higher frequencies as well. The FM broadcast band (150 kHz wide channels over 87.9 to 108 MHz in the US) is generally received on superheterodyne receivers with an IF of 10.7 MHz, which places all image frequencies outside the FM band, eliminating interference from other FM stations.

The use of higher frequency tuned circuits for IF selectivity works well for the 150 kHz wide FM broadcast channels, but not so well for the relatively narrow channels encountered on HF or lower, or even for many V/UHF narrowband services. Fortunately, there are three solutions that were commonly used to resolve this problem.

The first, *double conversion*, converts the desired signal to a relatively high IF followed by a second conversion to a lower IF to set the selectivity. This was a popular technique in the 1950s. The success of the double conversion technique led to triple conversion with a very low, highly selective third IF. The *Collins system* of moving a single-range VFO to the second mixer and using switchable crystal oscillators for the first mixer also became popular. A third approach to double conversion is a combination of the two, called pre-mixing, that uses a single variable oscillator range, as with the Collins, but mixes the VFO and LO outside of the signal path before applying them to the first mixer.

These methods are rarely used in current receivers but are commonly encountered in vintage equipment. Further discussion can be found in previous editions of the Handbook.

HIGH-FREQUENCY CRYSTAL LATTICE FILTERS

Commercial quartz-crystal filters with bandwidths appropriate for CW and SSB

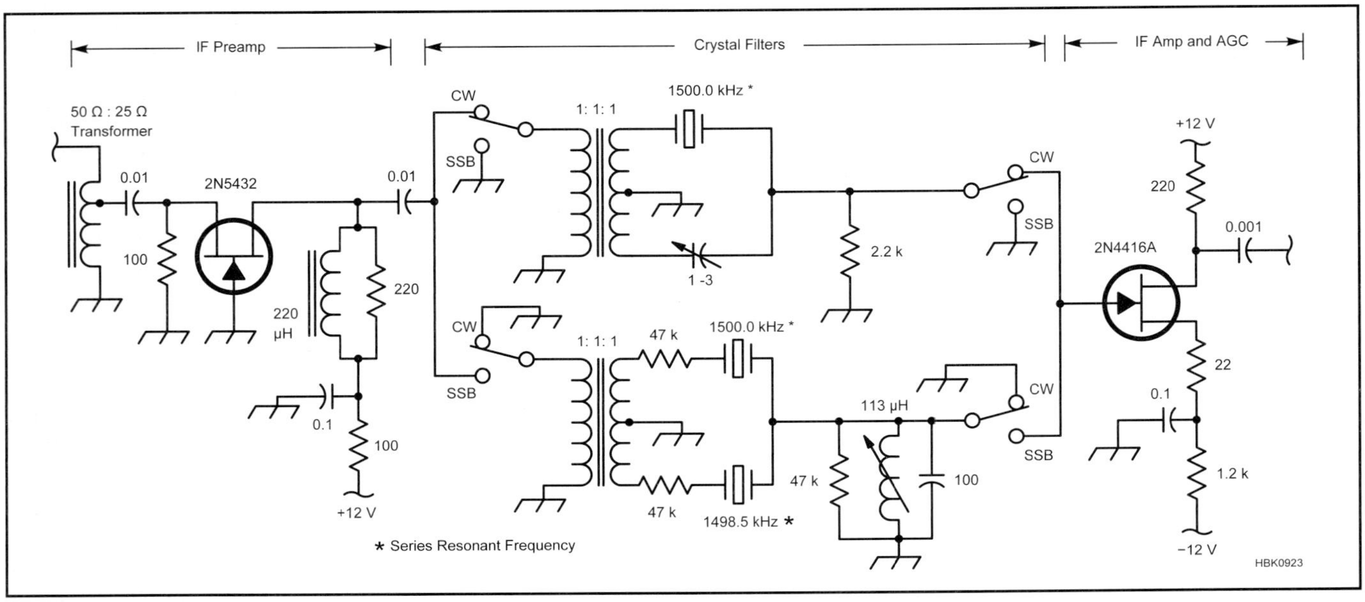

Figure 12.13 — The IF section of a superhet using crystal filters to establish receiver bandwidth.

became available in the 1970s with center frequencies into the 10 MHz range. This allowed a single-conversion receiver (see Figure 12.12) with an IF in the HF range to provide both high image rejection and needed channel selectivity. This single-conversion architecture remains popular among designers of portable and low-power equipment. Crystal filter design is discussed in the **Analog and Digital Filtering** chapter as well as in the online content.

As an example, **Figure 12.13** shows the IF section of a single-conversion superheterodyne using simple filters centered at 1500 kHz. While the simple filters shown are buildable by amateurs, low cost, multiple-section filters with much better performance can be purchased or constructed. Other IF frequencies can be used, depending on crystal or filter availability.

The circuit shown demonstrates the concepts involved and can be reproduced at low cost. Remaining receiver functional blocks such as the AGC circuitry, detectors and BFO, and audio amplifiers and filters can be found elsewhere in the book.

DSP IF FILTERS

Digital signal processing provides a level of filter performance not practical with other technologies. (See the **Analog and Digital Filtering** chapter.) The best IF bandpass filters in amateur receivers are either crystal or mechanical, and they are much better than LC filters. However, none of them are close to the rectangular shaped frequency response of an ideal filter. Filters found in amateur gear tend to have skirts with a 6 to 60 dB response of perhaps 1.4 to 1. That means if we select an SSB filter with a nominal (6 dB) bandwidth of 2400 Hz, the width at 60 dB down is typically 2400 × 1.4 or 3360 Hz. Thus, a signal in the next channel that is 60 dB stronger than the signal we are trying to copy (as often happens) has energy in our IF bandwidth that is just as strong as our desired signal.

DSP filtering approaches the ideal response. **Figure 12.14** shows the ARRL Lab measured response of a DSP bandwidth filter with a 6 dB bandwidth of 2400 Hz. Note how rapidly the skirts drop to the noise level. In addition, while analog filtering generally requires a separate filter assembly for each desired bandwidth, DSP filtering is adjustable — often in steps as narrow as 50 Hz — in both bandwidth and center frequency. In addition to bandpass filtering, the same DSP can provide digital noise reduction and digital notch filtering to remove interference from fixed frequency carriers.

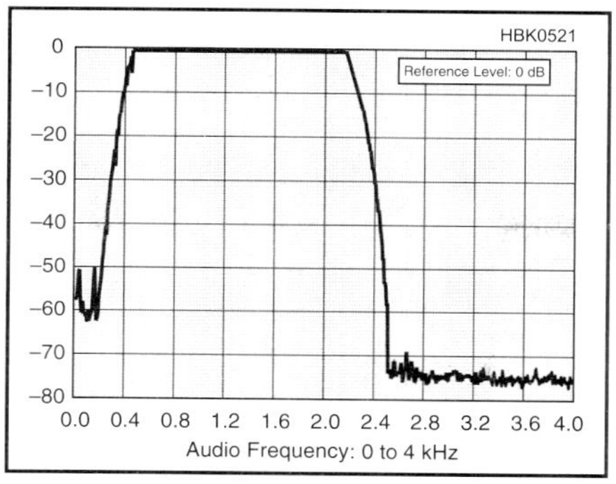

Figure 12.14 — ARRL Lab measured response of an aftermarket 2400 Hz DSP bandpass filter.

UP-CONVERSION AND DOWN-CONVERSION

Current crystal filter technology allows *down-conversion* HF receivers to use an IF in the 4 – 10 MHz range. With a 10 MHz IF and an LO above the signal frequency, a 30 MHz signal would have an image at 50 MHz. This makes image-rejection filtering relatively straightforward, although many receiver IF frequencies tend to be at the lower end of the above range. Still, as is discussed in the next section, they have other advantages.

Many current HF receivers have elected to employ an *up-conversion* architecture. They typically have an IF in the VHF range, perhaps 60 to 70 MHz, making HF image rejection easy. A 30 MHz signal with a 60 MHz IF is an image at 150 MHz. Not only is it five times the signal frequency, but signals in this range (other than perhaps the occasional taxicab) tend to be weaker than some undesired HF signals. Receivers with this architecture have image responses at the upper end of the above range, often with the image rejected by a relatively simple low-pass filter with a cut off at the top of the receiver range.

Another advantage of this architecture is that the local oscillator can cover a wide continuous range, making it convenient for a general coverage receiver. For example, with a 60 MHz IF, a receiver designed for LF through HF would need an LO covering 60 – 90 MHz, a 1.5 – 1 range, easily provided by a number of synthesizer technologies, as described in the **Oscillators and Synthesizers** chapter. The article "A High Performance 45 MHz IF Amplifier" by Colin Horrabin, G3SBI, has been included in the online content as an example of circuits suitable for up-conversion receivers.

The typical up-conversion receiver uses multiple conversions to move signals to frequencies at which operating bandwidth can be established. While crystal filters in the VHF range used by receivers with up-conversion IFs have become available with bandwidths as narrow as around 3 kHz, they do not yet achieve the shape factor of similar bandwidth filters at MF and HF. Thus, these are commonly used as *roofing filters,* discussed in the following section on Superheterodyne Design for Dynamic Range. They are inserted in the IF chain, prior to a conversion to one or more lower IF frequencies at which the operating bandwidth is established. **Figure 12.15** is a block diagram of a typical up-conversion receiver using DSP for setting the operating bandwidth.

THE IMAGE-REJECTING MIXER

Another technique for reduction of image response in receivers is not as commonly encountered in HF receivers as the preceding designs, but it deserves mention because it has some very significant applications. The *image-rejecting* mixer requires phase-shift networks, as shown in **Figure 12.16.** Frequency F_1 represents the input frequency while F_2 is that of the local oscillator. Note that the two 90° phase shifts are applied at different frequencies. The phase shift network following Mixer 1 is at a fixed center frequency corresponding to the IF, while the phase shift network at F_2 must provide the required phase shift as the local oscillator tunes across the band.

If the local oscillator is required to tune over a limited fractional frequency range, this is a very feasible approach. On the other hand, maintaining a 90° phase shift over a wide range can be tricky. The good news is that this approach provides image reduction that is independent of, and in addition to, any other mechanisms such as filters that are employed toward that end.

Additionally, with the ability of DSP

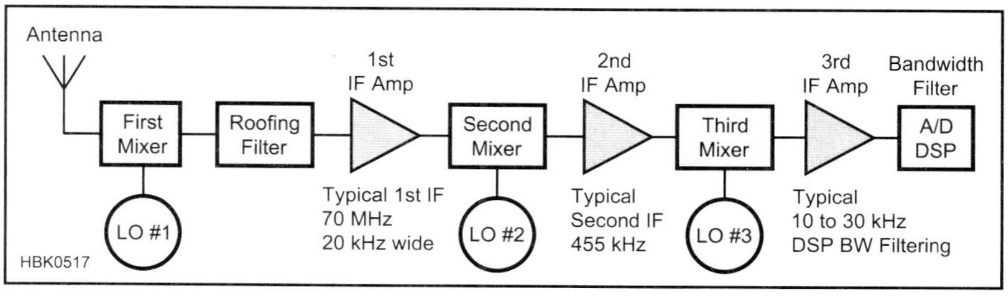

Figure 12.15 — Block diagram of a typical upconverting receiver using DSP for the 3rd IF filter which sets operating bandwidth. Hybrid heterodyne/SDR receivers applying DSP filters at IF frequencies are common.

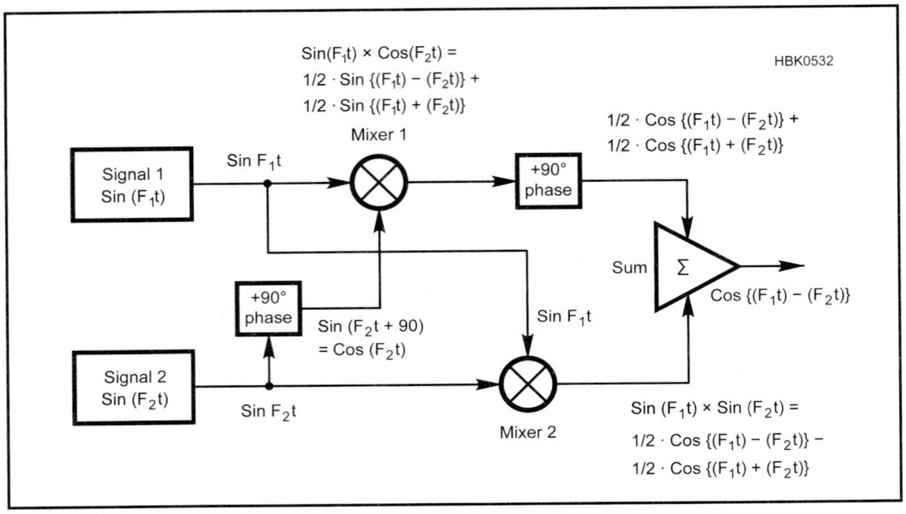

Figure 12.16 — Image rejecting mixer, block diagram, and signal relationships.

components to operate at higher and higher frequencies, the necessary operations seen in Figure 12.16 can be performed in software, which does not depend on precision hardware design to maintain nearly exact phase relationships.

The image-rejecting mixing process has several attractive features:

• It is the only way to provide "single signal" reception with a direct conversion receiver, effectively reducing the audio image. This can make the DC receiver a very good performer, although the added complexity is not always warranted in typical amateur DC applications.

• This option is frequently found in microwave receivers in which sufficiently selective RF filtering can be difficult to obtain. Since they often operate on fixed frequencies, maintaining the required phase shift can be straightforward.

• It is found in advanced receivers that are trying to achieve optimum performance. Even with a high first IF frequency, additional image rejection can be provided.

• In transmitters, the same system is called the *phasing method* of SSB generation. The same blocks run "backwards" — one of the phase shift networks can be applied to the speech band and used to cancel one sideband. This is discussed in the **Transmitting chapter.**

12.2.6 Preselectors

Placing a filter at the receiver's antenna input, called a *preselector,* is a technique that improves both image rejection and overload from out-of-band signals. You can see such a filter at the front-end of several of the receiver block diagrams in this chapter. Preselection to avoid overload is often required with the inexpensive USB "dongle" style SDR receivers that are designed for cable TV reception, and not for strong-signal performance.

There are three types of preselectors:

• Manually-tuned — a tuned circuit or tuned input transformer that is adjusted by the operator for maximum signal level. The tuned circuit usually has several ranges that are selected by the receiver band switch. These were once common, especially on general-coverage receivers. Preselectors were also very popular in European equipment where extremely strong signals from high-power shortwave broadcast (SWBC) stations often caused receiver overload unless attenuated with a filter.

• Switched bandpass — a bank of bandpass filters selected by the band switch or under control of the receiver's controlling microprocessor. Most of the top-performing receivers feature switched bandpass preselectors that do not require manual adjustment.

• Tracking — a continuously-variable tuned circuit or bandpass filter that is controlled by software to have peak response at the frequency of operation. This feature is relatively expensive to implement so it is only available on high-end receivers.

The online content includes a pair of preselector designs. The manual general-coverage preselector from *Ham Radio* magazine is by George Hirshfield, W5OZF. It covers from 0.5 to 30 MHz with a tuning capacitor augmented by fixed capacitors and tapped inductors. A software-controlled switched bandpass preselector covering 1.8 – 30 MHz from *QEX* is included as a project. Designed by Juan Onate, MØWWA and Xavier de Fortuny, the preselector includes several auxiliary features (input protection, attenuators, LNA, and so on) and is controlled by a PIC microcontroller. All software is available from ARRL as described in the article.

12.2.7 Superheterodyne Design for Dynamic Range

In the past, the receivers with the best close-in third order intermodulation distortion and maximum blocking dynamic range were amateur-band-only receivers, such as the primary receiver in the TEN-TEC Orion and Omni series, and the receivers in the Elecraft K2 and K3. A look at a typical block diagram, as shown in **Figure 12.17**, makes it easy to see why. The problems resulting from strong unwanted signals near a desired one are minimized if the unwanted signals are kept out of the places in the receiver where they can be amplified even more and cause the nonlinear effects that we try to avoid.

Note that in Figure 12.17, the only place where the desired and undesired signals all coexist is before the first mixer. If the first mixer and any RF preamp stages have sufficient strong-signal handling capability, the undesired signals are eliminated by the filter immediately behind the first mixer. This HF crystal filter is generally switchable to support desired bandwidths as narrow as 200 Hz. The successive amplifier, mixer and DSP circuits then only deal with the signal we want.

Careful attention to gain distribution among the stages between the filters maintains desired sensitivity, but gains are not so high that undesired products have a chance

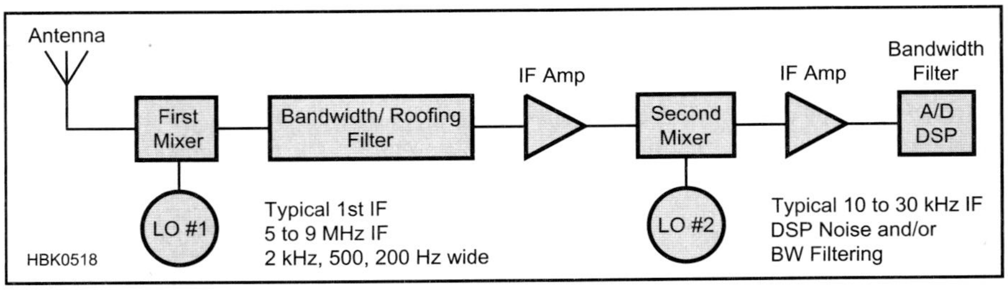

Figure 12.17 — Block diagram of a down-converting amateur band receiver with roofing filters and operating bandwidth filter to improve performance in the presence of strong in-band signals.

to become serious problems. In the Yaesu FT-1000MP, with bandwidths of 20, 6 and 2.5 kHz supplied, and 500 and 300 Hz as accessories, the undesired close-in signals are eliminated before they have an opportunity to cause serious trouble in the DSP stages that follow.

Another variation is found in the Kenwood TS-590S and SG models, which switch between down-conversion on the more crowded "contest bands" (160, 80, 40, 20 and 15 meters) and up-conversion on the remaining bands. In effect, this trades sensitivity for dynamic range.

ROOFING FILTERS

Now look back at a typical modern general-coverage receiver, as shown in Figure 12.15. In this arrangement, a single digital synthesizer, perhaps covering from 70 – 100 MHz, shifts the incoming signals to a VHF IF, often near 70 MHz. A roofing filter at 70 MHz follows the first mixer. The name "roofing filter" comes from the filter's wide bandwidth extending over the narrower operating bandwidth filter, like a house's roof. This arrangement offers simplified local oscillator (LO) design and the possibility of excellent image rejection. Unfortunately, crystal filter technology has only recently been able to produce narrow filters for 70 MHz, and so far, they have much wider skirts than the crystal filters used in Figure 12.17. As technology is rapidly shifting to SDR designs, the current models likely exhibit the peak performance level for upconverting receivers.

Many receivers and transceivers set the roofing filter bandwidth wider than any operating bandwidth and use DSP filtering later in the signal chain to set the final operating bandwidth. For a receiver that receives FM, AM, SSB, and CW, that usually means a roofing filter with a bandwidth of around 20 kHz. With this arrangement, all signals in that 20 kHz bandwidth pass through IF amplifiers and mixers and into the ADC before we attempt to eliminate them using DSP filters. By that time, they have had an opportunity to generate intermodulation products and cause the blocking and IMD problems that we are trying to eliminate. The reference entry for the Hallas Product Review of a Roofing Filter Accessory for the FT-1000MP describes the improvement that a narrow filter provides. Top of the line superheterodyne radios such as the Icom IC-7851 and Yaesu FTDX9000 transceivers feature both general coverage plus HF and VHF roofing filters.

12.3 Software Defined Radio (SDR) Receivers

The Software Defined Radio (SDR) is discussed in the **DSP and SDR Fundamentals** chapter. Repeated here are the sections of that chapter describing SDR architecture and performance issues. This section presents and compares several block-diagram-level concepts for SDR.

12.3.1 Digitizing at IF

The first generation of radios to make use of DSP techniques at RF performed analog-to-digital conversion (ADC) on an IF signal. **Figure 12.18** shows such a design. Placing the ADC after a crystal IF filter improves the blocking dynamic range (BDR) for interfering signals that fall outside the crystal filter bandwidth. As shown, the down-conversion to I/Q format still uses lower-speed ADCs, but often the signal is at a low IF, around 15 kHz. This allows an SSB-bandwidth signal to be contained within the 20 kHz bandwidth of a typical audio codec and avoids errors due to dc offsets in the signal path. With careful design, a receiver with such an architecture can achieve 140 dB or more of BDR (if there are no other limiting factors such as LO phase noise). The third-order dynamic range is like that achieved with a conventional analog architecture since the circuitry up to the crystal filter, including amplifiers and mixers, is the same.

Another advantage of the IF-based approach, compared to directly sampling the RF frequency, is that the ADC does not have

Sampling Rates

Digital audio equipment, that is audio equipment that uses digital sampling at some point in the device, has been around since the 1970s. There are two primary fundamental sampling rates that are used in all digital audio equipment: 44.1 ksps and 48 ksps. From these two rates, all other rates are derived by using factors of these two numbers. In amateur radio, the latter number, 48 kHz, is most often used and is why all our signal chains typically operate at a multiple or sub-multiple of these numbers. The most common processing speeds are 24 ksps, 36 ksps, 48 ksps, 96 ksps and 192 ksps. 48 ksps is by far the most common, but it varies by the manufacturer and radio design.

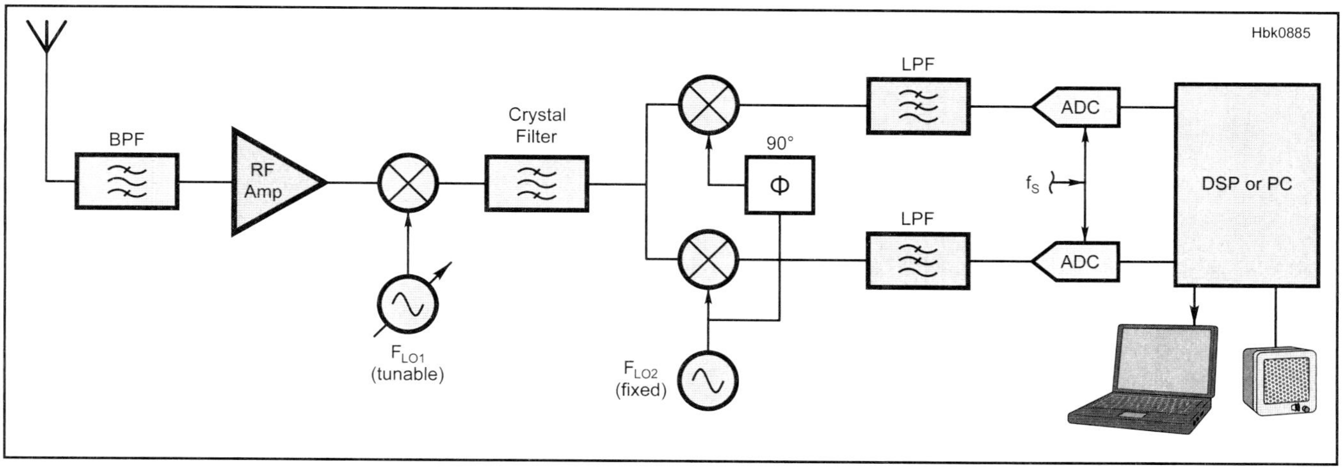

Figure 12.18 — Hybrid superhet/DSP SDR receiver architecture.

to run at such a high sample rate. In fact, because the crystal filter acts as a high-performance, narrow-bandwidth anti-aliasing filter, *undersampling* is possible if the ADC has sufficient sampling bandwidth, which ADCs intended for audio applications generally do not. With bandwidths of a few kHz or less, sample rates in the tens of kHz can be used even though the center frequency of the IF signal is much higher, so long as the ADC's sample-and-hold circuit has sufficient bandwidth.

12.3.2 Direct RF Digitizing

The ultimate SDR architecture is to convert between the analog and digital domains right at the frequency to be transmitted or received, or convert a wide range of frequencies and do all filtering in the digital domain. The receive path of such a design is shown in **Figure 12.19**. In this receiver, the only remaining analog components in the signal chain are a wide-band anti-aliasing filter and an amplifier to improve the noise figure of the ADC. A preselector stage may be used to prevent overload from strong signals far from the selected frequency or frequency band. The local oscillator, mixer, IF filters, AGC, demodulators and other circuitry are all replaced by digital hardware and software. The digital/software implementations of these functions are perfectly stable with time, temperature, and need no adjustments.

It is only recently that low-cost high-speed ADCs have become available with specifications good enough to allow reasonable performance in an RF-sampling communications receiver. Today it is possible to achieve BDR of 130 dB. That is not quite as good as the best analog or hybrid radios, but with every new generation of ADC that becomes available for use in SDR, the performance gap narrows.

It is worth noting that while huge BDR numbers can be measured in the laboratory, the performance achieved in a real-world environment with a receiver connected to an antenna is quite different. Often local noise sources raise the noise floor such that the receiver's full BDR cannot be utilized and other specifications become more important. In many of these specifications (e.g., dynamic range for close-spaced signals), RF-sampling SDRs can provide performance comparable or superior to conventional all-analog and IF-sampling receivers.

Third-order intermodulation distortion dynamic range (3IMD_DR) is not a meaningful specification for this type of radio because it is based on the behavior of analog circuits. In addition, calculation of IP_3 assumes that distortion products increase 3 dB for each 1 dB increase in signal level, which is not always true for an ADC. The level of the distortion products in an ADC tends to be independent of signal level until the signal peak exceeds the ADC's full-scale input, at which point the distortion increases dramatically. It is important to read the data sheet carefully and note the test conditions for the distortion measurements.

There are definite advantages to sampling at RF. For one thing, it saves a lot of analog circuitry. Even though a high-speed ADC is more expensive than an audio converter, the radio may end up being cheaper to build because of the reduced component count and fewer adjustments. Performance is improved in some areas. For example, image rejection is no longer a worry, if the anti-aliasing filter and any preselector filters that may be present are doing their jobs. The dynamic range of an SDR theoretically does not depend on signal spacing. Close-in dynamic range is often better than with a conventional architecture that uses a wide IF filter. With no mixers in the signal chain, intermodulation products that need to be filtered are not created. With no crystal filters in the signal chain, the entire system has a completely linear phase response, which can improve the quality of both analog and digital signals after demodulation.

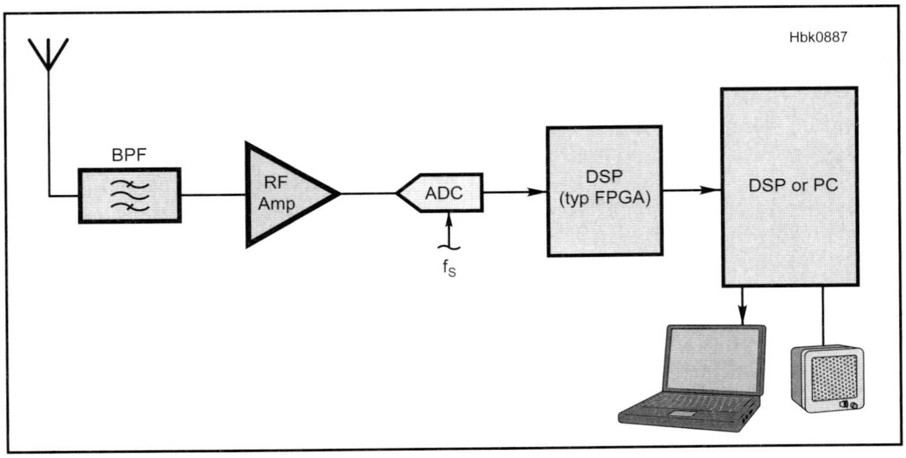

Figure 12.19 — Direct RF-sampling DSP SDR receiver architecture.

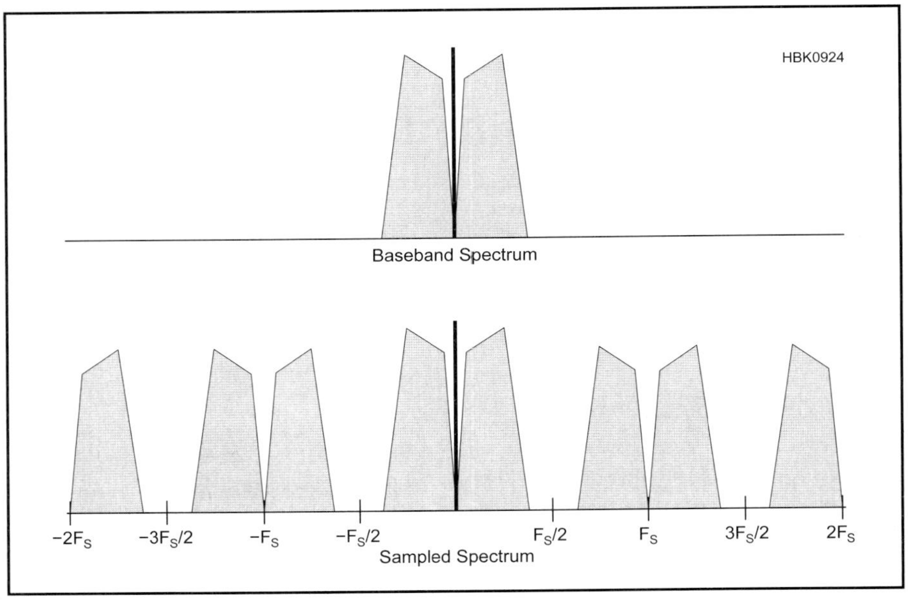

Figure 12.20 — The baseband spectrum of an analog signal before sampling is given in the upper graph, showing both positive and negative frequencies. The lower graph shows the spectrum of the baseband signal after it is sampled by an ideal sampling waveform. Note that there is an upper and lower sideband version of the baseband for each harmonic of the sampling frequency, just as if it were a double sideband suppressed carrier signal. Also note that the bandpass of the analog signal is lower than half the sampling frequency, so there is no overlap, or aliasing, of the harmonic copies of the spectrum with the baseband spectrum.

The biggest challenge with RF sampling is what to do with the torrent of high-speed data coming out of the receiver's ADC. To cover 0 – 54 MHz without aliasing requires a sample rate of at least 120 MHz, and commercial products typically operate the ADC at sample rates well over 200 MHz. That is much faster than a typical microprocessor or programmable DSP can handle. The local oscillator, mixer, and decimator or interpolator must be implemented in digital hardware so that the DSP can send and receive data at a more reasonable sample rate. Digital down-converters (DDC) perform those functions and output a lower-sample-rate digital I/Q signal to the DSP. Stand-alone DDC ICs were available in the past, but the function is now usually integrated with the ADC. It is also possible to implement a DDC in an application specific integrated circuit (ASIC or a field-programmable gate array (FPGA). (See the "Hands-On SDR" *QEX* columns provided in the online content.) Digital up-converters (DUC) do the same conversion in reverse for the transmitter and are available integrated with the digital-to-analog converter (DAC) or can be implemented in an FPGA. Some commercial integrated DDC/DAC products even include the capability to encode several digital modulation formats such as GMSK, QPSK and p/4 DQPSK. To simplify the interface to the digital domain, many high-speed converters now use a standardized serial interface specification called JESD204B, capable of handling up to 12 Gb/s. Code to implement this interface on the digital FPGA is readily available.

Some designers have been successful in repurposing a graphics processing unit (GPU) for this purpose, and some GPU manufacturers now offer FFT libraries to assist in the design process.

12.3.3 Sample-Rate Down-Conversion

Direct RF sampling SDR receivers generally do not use analog mixers to convert RF frequencies to an IF or to baseband. Down-conversion is achieved by the process of *decimation* as described in the **DSP and SDR Fundamentals** chapter. The frequency of a sampled signal can be divided by N by removing every Nth sample from the digital signal data. This creates an alias of the original signal as if it had been sampled at a rate lower than the Nyquist rate. As shown in **Figure 12.20**, this process replicates the original signal or original spectrum at harmonics of the sampling rate. One of the replicated spectra can then be selected with a filter and operated on just as if it had been processed by an analog mixer.

This is one example of *multirate* signal processing which is not nearly as complicated as it sounds. The topic we are interested in now is sample-rate down conversion (decimation). Sample-rate up conversion is another multirate signal processing technique that performs frequency multiplication in much the same way as down-conversion performs frequency division.

If we want to directly receive the 40-meter band, for instance, we would need to sample above 14.6 MHz to satisfy the Nyquist criterion. We do not need a sample rate that fast to manipulate the information received, though, since the bandwidth of the widest signal is on the order of 7 kHz or less. Even if we want to look at the entire band and generate a spectrum display, we only need a sample rate of about 650 kHz since the band is only 300 kHz wide. Any higher received sample rate is a waste of processor resources.

There are two main reasons why we would want to match the sample rate closely to our intended bandwidth. The first reason to lower the sample rate is that the transition

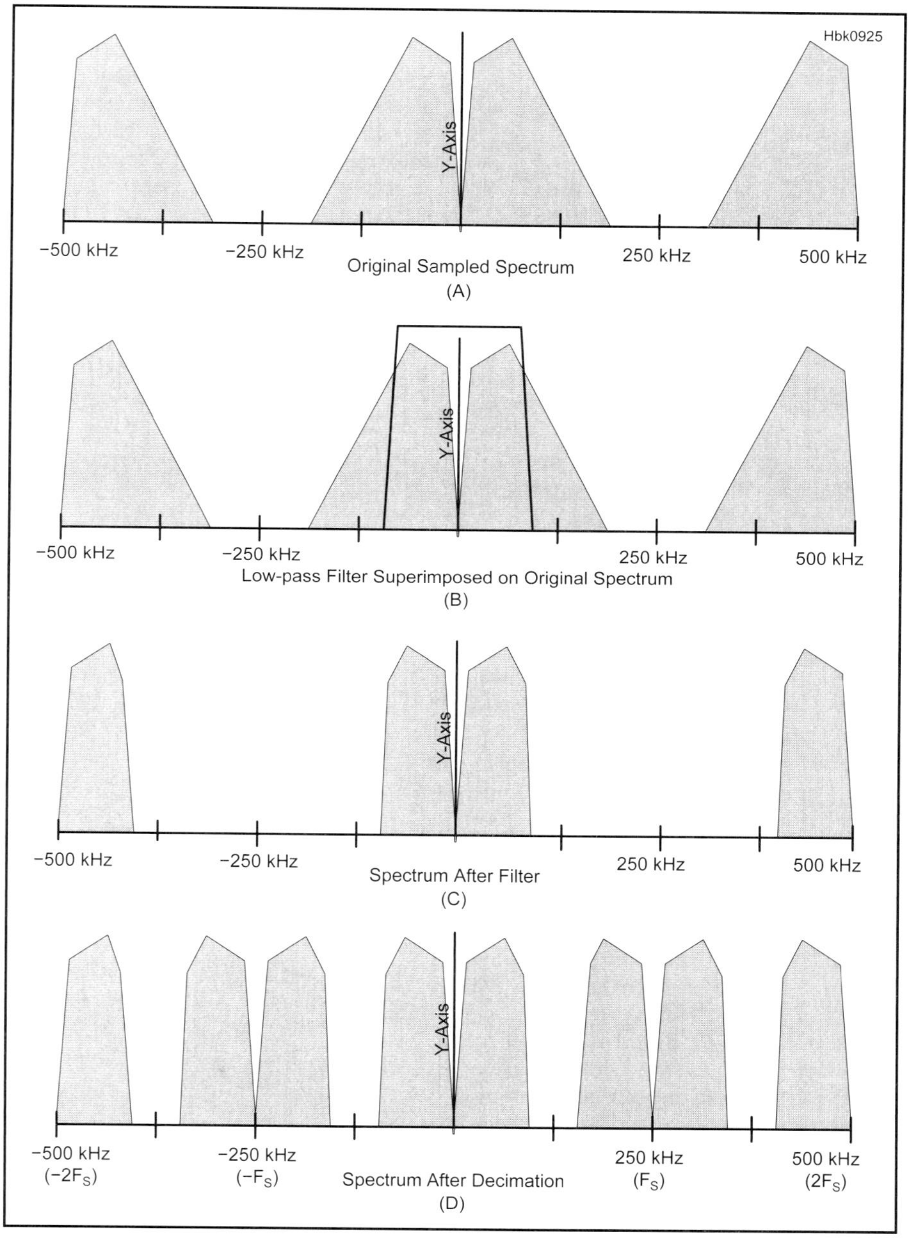

Figure 12.21 — Part A shows the spectrum of the sampled signal. Note that the frequency only extends from –500 kHz to +500 kHz since that is the extent that the math can manage. Part B shows the digital low-pass filter response superimposed on the sampled spectrum. The spectrum of the filtered signal, which is still sampled at 500 kHz, is shown at C. Part D shows the spectrum of the resulting sampled waveform after decimation by 2 (every other sample discarded). This is the spectrum of the original signal as if it was sampled at 250 kHz with an appropriate anti-aliasing filter.

band and ripple of our filters are dependent on the ratio of the filter length (the number of taps) to the sample rate. The second reason is to allow more CPU cycles for processing each sample.

An AM radio is a good example for sample rate conversion. Let's say we want to receive a band of signals through a tuned-circuit input filter with a bandwidth of 80 kHz centered at 590 kHz. If we undersample the input at a rate of 100 kHz, we create a range of signals centered at 90 kHz so that the energy is all within the range of dc to 125 kHz. This is called low-pass filter decimating. (See the *QEX* SDR: Simplified column for July/August 2009 for a more complete discussion of this process.)

The first step in the decimation process is to filter the input signal so that there is no energy above one half the new sample frequency after decimation. A DSP low-pass filter with cutoff at one fourth of the original sample rate can be implemented with a small number of taps. We sampled our AM radio at 500 kHz, so we need 125 kHz cutoff for the low pass filter. We now have a signal that has no energy above 125 kHz and is sampled at 500 kHz. We can now throw away every other sample to create a signal with a sampling rate of 250 kHz, and energy up to 125 kHz. The Nyquist criterion is satisfied with this new signal. **Figure 12.21** shows the spectrum of the process.

The signal at 90 kHz is only about 40 kHz wide (70 – 110 kHz), so it would be nice if we could drop the sample rate even further. It turns out that our signal fits into the band from dc to 62.5 kHz if it is translated down in frequency, which only requires a 125 kHz sample rate. This is *integer bandpass decimating*. If we throw away 3 of every 4 samples from the original data, we can accomplish both frequency translation and sample rate reduction. The energy now spans from 15 – 55 kHz. The signal spectrum is also inverted. In a superheterodyne AM radio, you would need to bandpass filtered the signal before sampling. It is possible to do the band pass filtering digitally with a small number of taps, and then do the decimation.

We succeeded in reducing the sample rate by a factor of 4. We cannot further reduce the sample rate using straight decimation. Each integer sample rate reduction requires that the band limited data fit completely within one of the Nyquist regions for the new sample rates. The next integer sample rate would reduce the sample rate by 5. The new sample rate would be 100 kHz. The data is contained in both the k = 1 (50 – 100 kHz) and k = 2 (100 – 150 kHz) Nyquist bands. **Figure 12.22** shows the overlap. The overlap into the third Nyquist zone prevents further rate reduction. The requirement that all the energy fits into one Nyquist zone limits the usefulness of this technique.

Integer sample rate reduction is a very useful tool because the filters are easily realized in a field programmable gate array (FPGA) or other dedicated hardware.

12.3.4 Decimation and Dynamic Range

When we talk about dynamic range, we are discussing the range of largest to smallest signals that can represented simultaneously. What does dynamic range buy us? Having a high dynamic range allows us to represent or hear weak signals in the presence of large signals. With a lower dynamic range, our receiver may overload in the presence of a large signal, or we may be forced to shift our dynamic range to accommodate the large signals. For example, if we only had 80 dB of dynamic range in our receiver, with the weakest signal possible of –120 dBm and the largest of –40 dBm, this means the strongest signal we can hear is S9+33 dB. If an S9+50 dB (–23 dBm) signal is present in our receiver, we would overload.

To allow the reception of large and small signal, receivers can implement automatic gain control (AGC) in the RF section. The AGC adds attenuation to the receiver, shifting the available dynamic range up. For example, it the AGC adds 20 dB of attenuation, we can now hear signals from –100 dBm to –20 dBm and we have reduced the sensitivity of our receiver by 20 dB. When we try listening to a weak CW signal with a power of –110 dBm, we are now unable to hear it. RF AGC is necessary in some receivers where the instantaneous dynamic range is not sufficient to handle both very strong and very weak signals simultaneously. This should not be confused with audio-derived AGC which is discussed in the section on AGC below.

A common question about direct sam-

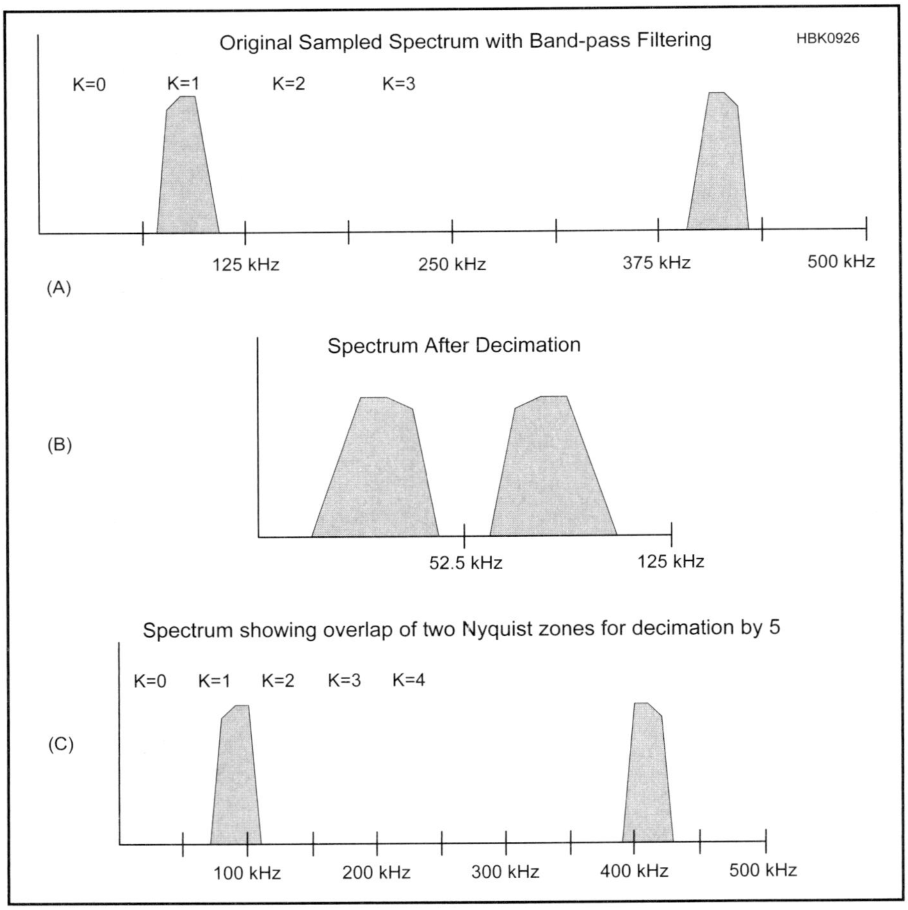

Figure 12.22 — Part A shows the spectrum of a 90 kHz signal that was band pass filtered and then sampled at 500 kHz (only one-half of the spectrum is shown for clarity). The resulting spectrum when the signal is decimated by 4 (3 of every 4 samples discarded) is shown in Part B. Notice that this causes the signal to be aliased and the signal in the first Nyquist zone (k=0) has its frequencies inverted just as a lower sideband signal is inverted. The signal is now sampled at 125 kHz. The initial sampled signal, showing the Nyquist zones that would occur if a decimation by 5 was attempted, is shown at C. Note that input energy exists in both the second and third Nyquist zones.

pling receivers centers around available dynamic range. It is commonly believed that for each bit in an ADC, we can have 6 dB of dynamic range. Often, someone attempts to look at wideband sampled converter (ADC), multiplies the number of bits by 6 dB and states that this is the available dynamic range for narrowband operation. As it turns out, this is an incorrect if not uncommon assumption. In a wideband sampled system where the receiver outputs are narrowband, we use decimation to reduce the bandwidth and increase dynamic range. How does this work?

Decimation was described above in the Sample-Rate Down-Conversion section. When decimating, the digital filter combines multiple samples to produce the output sample stream and, in this combination, we pick up additional bits of resolution in the output samples. This increase in dynamic range is called *processing gain.* Specifically, we typically pick up about one half bit of resolution for each decimation by two. Imagine a sampled system with 16-bits of resolution at 196.608 Msps.If we decimate down to 48 ksps, this represents twelve decimate-by-two operations. This means that we pick up about six bits per sample, increasing our dynamic range by about 36 dB. As a practical matter, the dynamic range of an ADC depends not only on the number of samples in the output, but also the level of the spurs. ADC manufacturers consider any level below the spurs to be tarnished by the spurs and therefore not included in the specified dynamic range. As a first order approximation for the dynamic range that can be achieved, we can take the dynamic range specified by the ADC manufacturer and add the processing gain that we've computed.

Finally, in amateur radio we generally talk about dynamic range in terms of a 500 Hz bandwidth. So, for the purposes of dynamic range in the amateur world, we don't use the entire 48 ksps bandwidth, we look just at the 500 Hz bandwidth as if we had a 500 Hz filter. We can use the formula 3 × $\log_2$ (sampling rate/final bandwidth) to get an approximation of the processing gain from our original sampling rate to a 500 Hz bandwidth receiver. (On a calculator without a log-base-2 ($\log_2$) function, we can calculate this by first calculating 3 × $\log_{10}$ (sampling rate/final bandwidth) and then dividing that result by $\log_{10}$ (2)).

12.3.5 Phase Noise in Sampled Systems

Just like with mixers, analog to digital converters that sample signals and produce a digital output are also susceptible to phase noise. Many of today's older transceivers use superheterodyne receivers, which use mixers and terminate in an ADC operating at baseband. For receiver systems where the ADC is operating in the tens to hundreds of ksps, phase noise introduced by the ADC is rarely an issue. Since phase noise is easily achievable in this frequency range, most designers ensure that phase noise does not impact the design in the ADC. It is much more likely in such a radio that the LO feeding the mixers nearer the antenna would be at fault. With today's technology, the higher we go in oscillator frequency, the more difficult it is to design a low phase noise oscillator.

In an RF sampled system where the ADC is placed either right on the antenna or just behind an amplifier and preselector, we sample at a significantly higher frequency. If we want to design a 160- through 6-meter receiver, we must sample at least double the highest frequency to satisfy the Nyquist sampling theorem. This means we sample faster than 108 MHz. A sampling rate of 122.88 MHz is a common frequency because it divides evenly down to standard audio sampling rates (see the sidebar on sampling rates). In these systems, the local oscillator may be called a clock oscillator and phase noise may be referred to as jitter. While phase noise and jitter are roughly interchangeable, jitter benefits the digital designer's view of the world and discusses a time-domain effect while phase noise benefits the RF designer and discusses a frequency domain effect. Most ADCs used for RF specify both. There is no absolute conversion from jitter directly to phase noise, but there are estimates that make assumptions about the characteristics of the oscillator.

Unlike the superheterodyne system, phase noise in a direct sampling receiver is imparted on the signal during the sampling process at the sampling frequency. What does this mean? While a superheterodyne system may rely on the division of the clock to produce an excellent phase noise signal on the band of interest, signals in a direct sampling system inherit the phase noise of the ADC clock at the sampling frequency. So, if our phase noise is –110 dBc/Hz at 10 kHz for a 122.88 MHz clock, this is the phase noise we have on all sampled frequencies. This makes clock (or LO) selection for the direct sampling receiver even more important. We must ensure that the oscillator's phase noise characteristics are not an issue at any frequency of interest.

More detailed discussions about the effects of phase noise in direct-sampled receivers can be found in Analog Devices application notes AN-741 and AN-756. (See the References section.)

12.4 Mixing and Mixers

This section examines mixers that are used for frequency shifting or conversion in superheterodyne receivers and transmitters. Mixers are often used as modulators and demodulators because they translate information to an RF signal and back again. These translation processes can be thought of as forms of frequency translation or frequency shifting, the function traditionally ascribed to mixers. We'll therefore begin our investigation by examining what a mixer is (and isn't), and what a mixer does.

MULTIPLYING VERSUS ADDING

Mixer is the term for a circuit that shifts one signal's frequency up or down by combining it with another signal. The word mixer is also the name of a device used to blend multiple audio inputs together for recording, broadcast, or sound reinforcement. A radio mixer makes new frequencies from its input signals and an audio mixer does not. In their most basic, ideal forms, both devices have two inputs and one output.

The audio mixer is a combiner that simply adds the instantaneous voltages of the two signals together to produce the output at each point in time (**Figure 12.23**). The radio mixer, on the other hand, multiplies the instantaneous voltages of the two signals together to produce its output signal from instant to instant (**Figure 12.24**). Comparing the output spectra of the combiner and mixer, we see that the combiner's output contains only the frequencies of the two inputs, and nothing else, while the mixer's output contains new frequencies. The process is called heterodyning as used in the superheterodyne receivers described in the preceding sections. The sidebar, "Mixer Math: Mixing as Multiplication," describes this process mathematically. Use of the word "mixer" in this book should be assumed to mean the radio mixer.

The key principle of a radio mixer is that in mixing multiple signal voltages together, it adds and subtracts their frequencies to produce new frequencies. (In the field of signal processing, this process, multiplication in the time domain, is recognized as equivalent to the process of convolution in

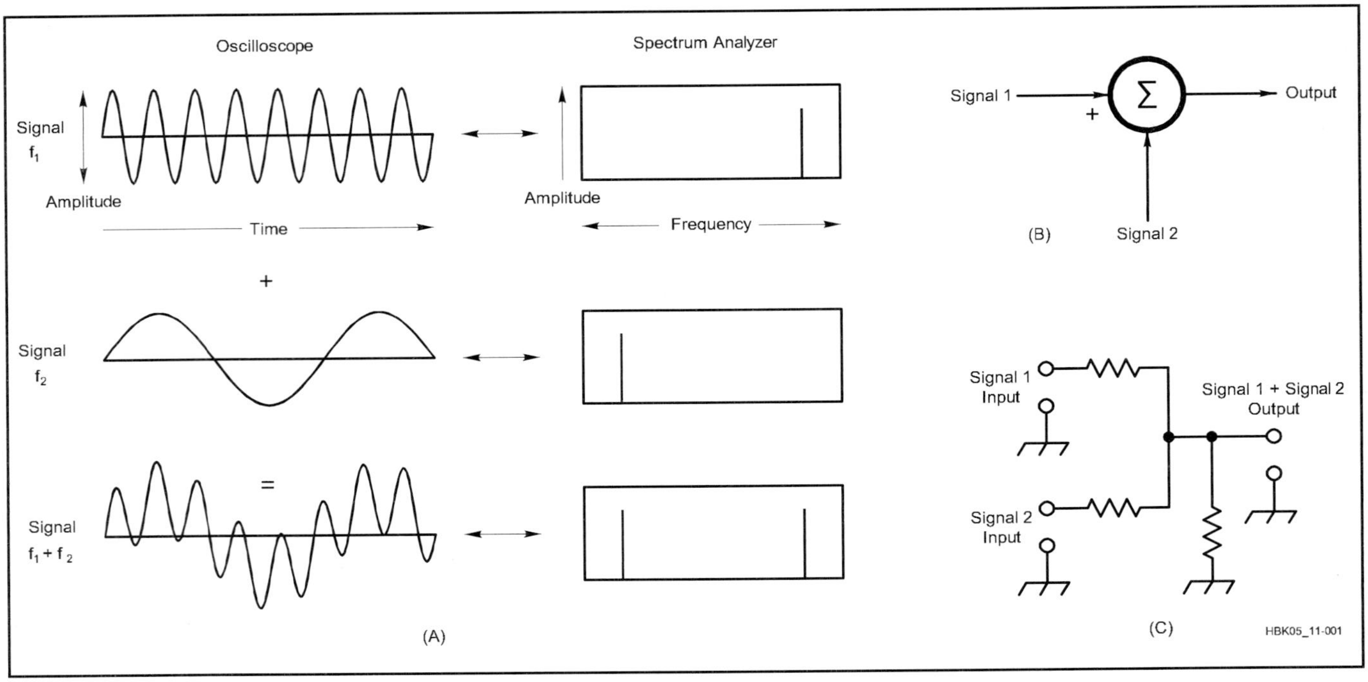

Figure 12.23 — The drawings in (A) illustrate adding, or summing, two sine waves of different frequencies (f_1 and f_2) that combines their amplitudes without affecting their frequencies. Viewed with an *oscilloscope* (a real-time graph of amplitude versus time, such as in the lower left plot), adding two signals appears as a simple superimposition of one signal on the other. Viewed with a spectrum analyzer (a real-time graph of signal amplitude versus frequency, such as in the lower right plot), adding two signals just sums their spectra. The signals coexist on a single cable or wire. All frequencies that go into the adder come out of the adder, and no new signals are generated. Drawing (B), a block diagram of a summing circuit, emphasizes the stage's mathematical operation rather than showing circuit components. Drawing (C) shows a simple summing circuit, such as might be used to combine signals from two microphones. In audio work, a circuit like this is called a mixer, but it does not perform the same function as an RF mixer.

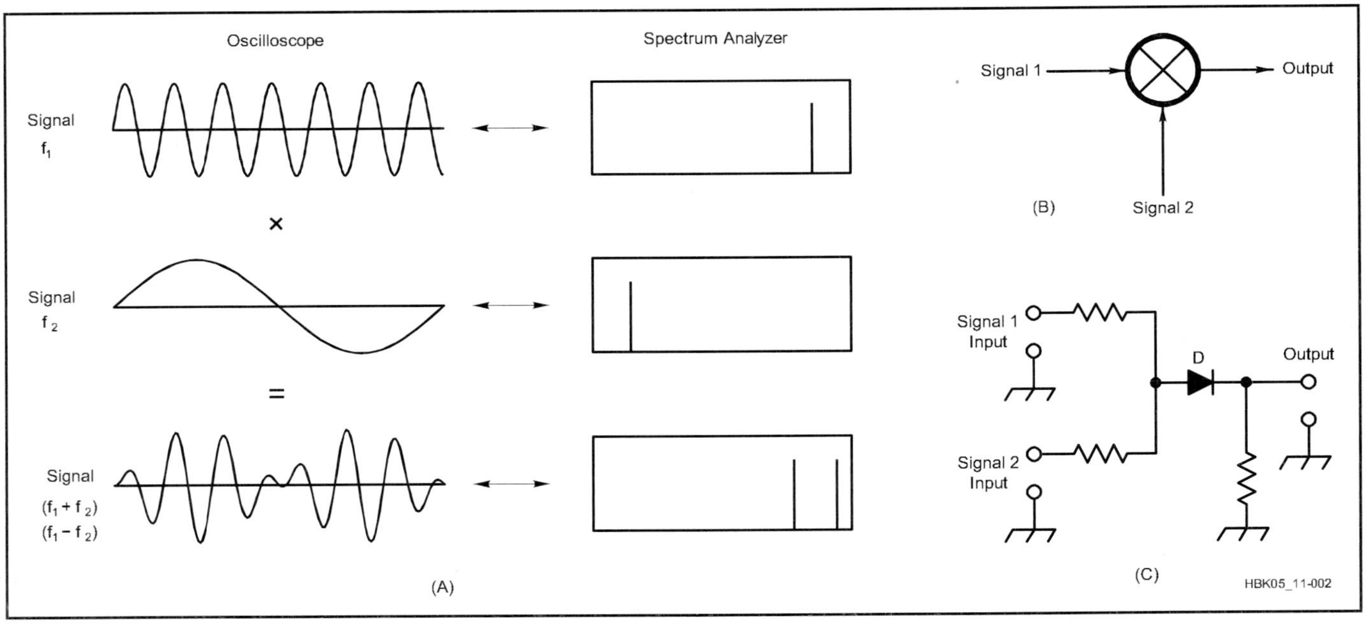

Figure 12.24 — The drawings in (A) illustrate multiplying two sine waves of different frequencies that produces a new output spectrum. Viewed with an oscilloscope, the result of multiplying two sine waves is shown in the lower left graph. A spectrum-analyzer view of the same wave reveals that the original sine wave frequencies disappear entirely and are replaced by two new signals — at the *sum* and *difference* of the original signals' frequencies. Drawing (B) diagrams a multiplier, known in radio work as a mixer. The circled X emphasizes the stage's mathematical operation. Drawing (C) shows a very simple multiplier circuit. The diode, D, does the mixing.

Mixer Math: Quadrature Demodulation

Demodulating an angle-modulated signal merely by multiplying it with a time-delayed copy of itself in a double-balanced mixer results in quadrature demodulation (Figure 12.54). To illustrate this mathematically, for simplicity's sake, we'll represent the mixer's RF input signal as just a sine wave with an amplitude, *A*:

$$A \sin (2 \pi ft)$$

and its time-delayed twin, fed to the mixer's LO input, as a sine wave with an amplitude, *A*, and a time delay of d:

$$A \sin [2 \pi f (t + d)]$$

Setting this special mixing arrangement into motion, we see

$$A\sin(2\pi ft) \times A\sin(2\pi f(t+d)) =$$

$$\frac{A^2}{2}\cos(2\pi fd) - \frac{A^2}{2}\cos(2\pi fd)\cos(2\times 2\pi ft) + \frac{A^2}{2}\sin(2\pi fd)\sin(2\times 2\pi ft)$$

Two of the three outputs — the second and third terms — emerge at twice the input frequency; in practice, we're not interested in these, and filter them out. The remaining term — the one we're after — varies in amplitude and sign according to how far and in what direction the carrier shifts away from its resting or center frequency (at which the time delay, d, causes the mixer's RF and LO inputs to be exactly 90° out of phase — in *quadrature* — with each other). We can examine this effect by replacing f in the RF input and LO input sinusoids with the sum term $f_c + f_s$, where f_c is the center frequency and f_s is the frequency shift. A 90° time delay is the same as a quarter cycle of f_c, so we can restate d as

$$d = \frac{1}{4f_c}$$

The first term of the detector's output then becomes

$$\frac{A^2}{2}\cos(2\pi(f_c + f_s)d)$$

$$= \frac{A^2}{2}\cos\left(2\pi(f_c + f_s)\frac{1}{4f_c}\right)$$

$$= \frac{A^2}{2}\cos\left(\frac{\pi}{2} + \frac{\pi f_s}{2f_c}\right)$$

When f_s is zero (that is, when the carrier is at its center frequency), this reduces to

$$\frac{A^2}{2}\cos\left(\frac{\pi}{2}\right) = 0$$

As the input signal shifts higher in frequency than f_c, the detector puts out a positive dc voltage that increases with the shift. When the input signal shifts lower in frequency than fc, the detector puts out a negative dc voltage that increases with the shift. The detector therefore recovers the input signal's frequency or phase modulation as an amplitude-varying dc voltage that shifts in sign as f_s varies around f_c — in other words, as ac. We have demodulated FM by means of quadrature detection.

the frequency domain. Those interested in this alternative approach to describing the generation of new frequencies through mixing can find more information about it in the many textbooks available on this subject.)

The difference between the mixer we've been describing and any mixer, modulator, or demodulator that you'll ever use is that it's ideal. We put in two signals and got just two signals out. Real mixers, modulators and demodulators, on the other hand, also produce IMD products (discussed earlier) that make their output spectra "dirtier" as well as putting out some energy at input-signal frequencies and their harmonics. Much of the art and science of making good use of multiplication in mixing, modulation, and demodulation goes into minimizing these IMD products and making multipliers perform frequency translation as efficiently as possible.

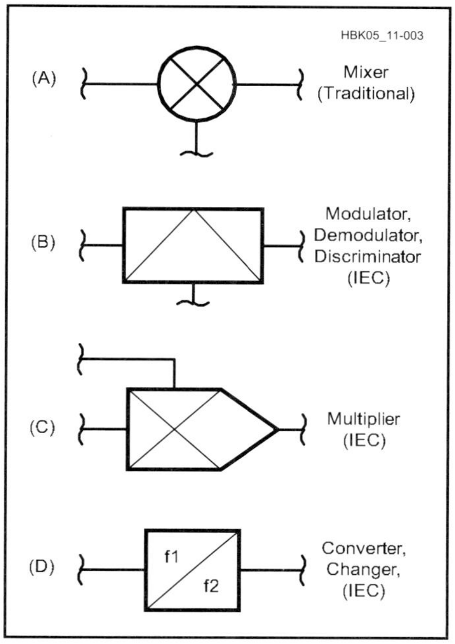

Figure 12.25 — The traditional symbol for a mixer is a circled X (A) although some standards allocate this symbol to a lamp. Current practice is to use one of the three IEC symbols shown at (B), (C), or (D). For the frequency converter or changer symbol at (D), a third connection can be included for the local oscillator. (IEC stands for *International Electrotechnical Commission* and the symbols are published in the IEEE 315A standard.)

12.4.1 Mixers and Distortion

We start with a discussion of the common underlying mechanism of mixers before discussing practical mixer, modulator, and demodulator circuits. **Figure 12.25** shows the block symbol for a traditional mixer along with several IEC symbols for other functions mixers may perform.

NONLINEAR DISTORTION

The mechanism underlying multiplication, mixing, modulation and demodulation is a pretty straightforward thing: Any circuit structure that nonlinearly distorts ac waveforms acts as a multiplier to some degree.

The phrase nonlinear distortion sounds redundant, but isn't. Distortion, an externally imposed change in a waveform, can be linear; that is, it can occur independently of signal amplitude. Consider a radio receiver front-end filter that passes only signals between 6 and 8 MHz. It does this by linearly distorting the single complex waveform corresponding to the wide RF spectrum present at the radio's antenna terminals, reducing the amplitudes of frequency components below 6 MHz and above 8 MHz relative to those between 6 and 8 MHz. The filter's output waveform differs from its input waveform; the waveform has been distorted. But because this distortion occurs independently of signal level or polarity, the distortion is linear. No new frequency components are created; only the amplitude relationships among the wave's existing frequency components are altered. This is amplitude or frequency distortion, and all filters do it, or they wouldn't be filters.

Phase or delay distortion, also linear, causes a complex signal's various component frequencies to be delayed by different amounts of time, depending on their frequency but independently of their amplitude. No new frequency components occur, and amplitude relationships among existing frequency components are not altered. Phase distortion occurs to some degree in all real filters.

The waveform of a nonsinusoidal signal can be changed by passing it through a circuit that has only linear distortion. Only nonlinear distortion can change the waveform of a simple sine wave. It can also produce an output signal whose output waveform changes as a function of the input amplitude, something not possible with linear distortion. Nonlinear circuits often distort excessively with overly strong signals, but the distortion can be a complex function of the input level.

Nonlinear distortion may take the form of harmonic distortion, in which integer multiples of input frequencies occur, or intermodulation distortion (IMD), in which different components multiply to make new ones, as described in previous sections.

Any departure from absolute linearity results in some form of nonlinear distortion, and this distortion can work for us or against us. Any amplifier, including a so-called linear amplifier, distorts nonlinearly to some degree; any device or circuit that distorts nonlinearly can work as a mixer, modulator, demodulator, or frequency multiplier. An amplifier optimized for linear operation mixes, though inefficiently. An amplifier biased for nonlinear amplification may be practically linear over a given tiny portion of its input-signal range. The trick is to use careful design and component selection to maximize nonlinear distortion when we want it (as in a mixer), and to minimize it when we don't. Once we've decided to maximize nonlinear distortion, the trick is to minimize the distortion products we don't want and maximize the products we want.

MINIMIZING UNWANTED DISTORTION PRODUCTS

Ideally, a mixer multiplies the signal at one of its inputs by the signal at its other input but does not multiply a signal at the same input by itself, or multiple signals at the same input by themselves or by each other. (Multiplying a signal by itself — squaring it — generates harmonic distortion, specifically, second-harmonic distortion, by adding the signal's frequency to itself. Simultaneously squaring two or more signals generates simultaneous harmonic and intermodulation distortion.)

Consider what happens when a mixer must handle signals at two different frequencies (f_1 and f_2) applied to its first input, and a signal at a third frequency (f_3) applied to its other input. Ideally, a mixer multiplies f_1 by f_3 and f_2 by f_3 but does not multiply f_1 and f_2 by each other. This produces output at the sum and difference of f_1 and f_3, and the sum and difference of f_2 and f_3, but not the sum and difference of f_1 and f_2. **Figure 12.26** shows that feeding two signals into one input of a mixer results in the same output as if f_1 and f_2 are each first mixed with f_3 in two separate mixers, and the outputs of these mixers are combined. This shows that a mixer, even though constructed with nonlinearly distorting components, behaves as a linear frequency shifter. Traditionally, we refer to this process as mixing and to its outputs as mixing products, but we may also call it frequency conversion, referring to a device or circuit that does it as a converter, and to its outputs as conversion products. If a mixer produces an output frequency that is higher than the input frequency, it is called an up-converter; if the output frequency is lower than the input, a down-converter.

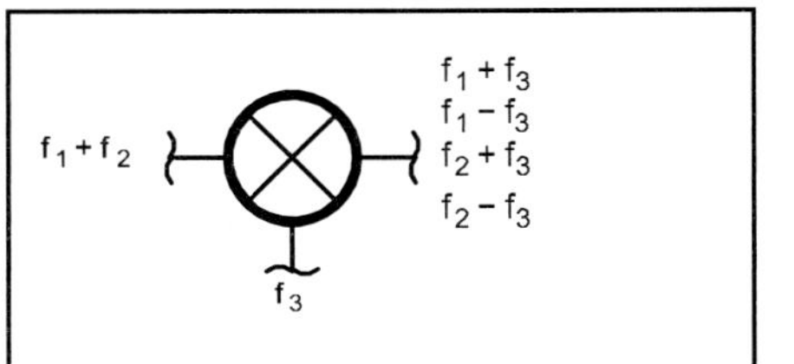

Figure 12.26 — Feeding two signals into one input of a mixer results in the same output as if f_1 and f_2 are each first mixed with f_3 in two separate mixers, and the outputs of these mixers are combined.

Real mixers, however, at best act only as reasonably linear frequency shifters, generating some unwanted IMD products — spurious signals, or spurs — as they go. Receivers are especially sensitive to unwanted mixer IMD because the signal-level range over which they must operate without generating unwanted IMD is often 90 dB or more, and includes infinitesimally weak signals. In a receiver, IMD products so weak that you'd never notice them in a transmitted signal can easily obliterate weak signals. This is why receiver designers apply so much effort to achieving high dynamic range.

The degree to which a given mixer, modulator, or demodulator circuit produces unwanted IMD is often the reason why we use it, or don't use it, instead of another circuit that does its wanted-IMD job as well or even better.

MISCELLANEOUS MIXING PRODUCTS

In addition to desired sum-and-difference products and unwanted IMD products, real mixers also put out some energy at their input frequencies. Some mixer implementations may suppress these outputs — that is, reduce one or both of their input signals by a factor of 100 to 1,000,000 or 20 – 60 dB. This is good because it helps keep input signals at the desired mixer-output sum or difference frequency from showing up at the IF terminal — an effect reflected in a receiver's IF rejection specification. Some mixer types, especially those used in the vacuum-tube era, do not suppress their input-signal outputs significantly.

Input-signal suppression is part of an overall picture called port-to-port isolation. Mixer input and output connections are traditionally called ports. By tradition, the port to which we apply the shifting signal is the local-oscillator (LO) port. By convention, the signals to be frequency-shifted are applied to the RF (radio frequency) port, and the frequency-shifted (product) signals emerge at the IF (intermediate frequency) port. This illustrates the function of a mixer in a superheterodyne receiver: Since it is often impractical to achieve the desired gain and filtering at the incoming signal's RF frequency, a mixer is used to translate the incoming RF signal to an intermediate frequency (IF), where gain and filtering can be applied. The IF may be either lower or higher than the incoming RF signal. Conversely, in a transmitter, the modulated signal may be created at an IF, and then translated in frequency by a mixer to the operating RF frequency.

A mixer may be used in an SDR to convert a range of RF signals into an IF range the SDR can process. It is commonly used when SDR equipment designed for VHF and higher frequencies needs to receive HF signals. In that case an up-converting mixer is required.

Some mixers are bilateral; that is, their RF and IF ports can be interchanged, depending on the application. Diode-based mixers are usually bilateral. Many mixers are not bilateral (unilateral); the popular SA602/612 Gilbert cell IC mixer is an example of this.

It is generally a good idea to keep a mixer's input signals from appearing at its output port because they represent energy that we'd rather not pass to subsequent circuitry. In particular, a receiver that passes this energy back to the RF input port can present spurious emissions that are transmitted by its antenna. FCC regulations limit the emitted power densities of spurious emissions that are created by a receiver in that fashion. It therefore follows that it is usually a good idea to keep a mixer's LO-port energy from appearing at its RF port, or its RF-port energy from making it through to the IF port.

12.4.2 Switching Mixers

Depending on the application, mixers can vary from the extremely simple to the complex. For example, a simple half-wave rectifier (a signal diode, such as a 1N34 [germanium] or a 1N914 [silicon]) can do the job. This is an example of a *switching mixer*, in which mixing occurs as one signal — in this case, the carrier, which turns the diode on and off as its polarity reverses — interrupts the transmission of another (in demodulation of full-carrier AM, the sidebands).

A switch can be thought of as an amplifier toggled between two gain states, off and on, by a control signal. A binary amplifier is not necessary; any device that can be gain-varied in accordance with the amplitude of a control signal can serve as a mixer.

Most modern radio mixers act more like fast analog switches than analog multipliers. In using a mixer as a fast-switching device, we apply a square wave, rather than a sine wave, to its LO input, and feed sine waves, audio, or other complex signals to the mixer's RF input. The RF port serves as the mixer's "linear" input, and therefore must preferably exhibit low intermodulation and harmonic distortion. Feeding a ±1 V square wave into the LO input alternately multiplies the linear input by +1 or −1. Mul-

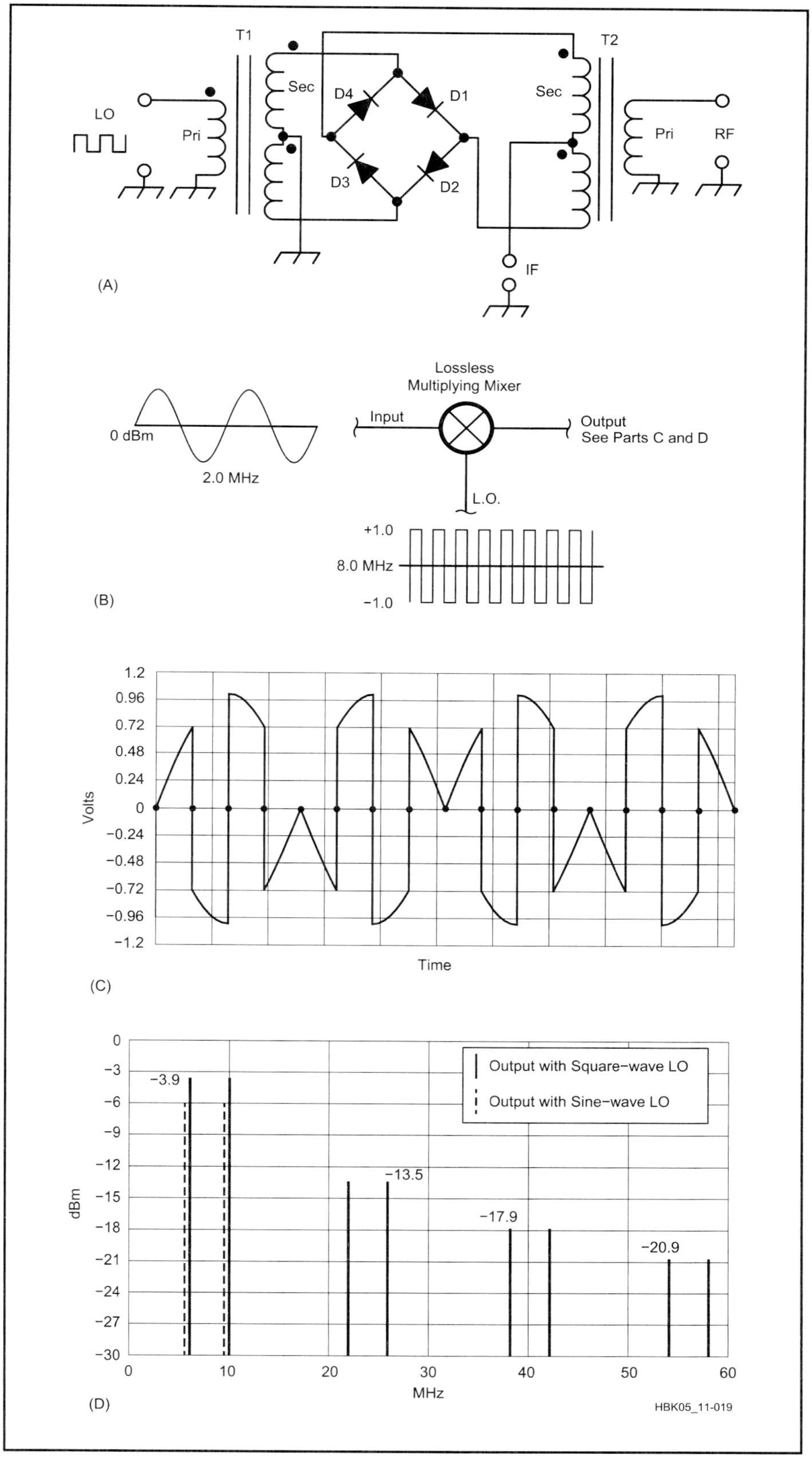

Figure 12.27 — Part (A) shows a general-purpose diode *reversing-switch* mixer. This mixer uses a square-wave LO and a sine-wave input signal. The text describes its action. Part (B) is an ideal *multiplier* mixer. The square-wave LO and a sine-wave input signal produce the output waveform shown in part (C). The solid lines of part (D) show the output spectrum with the square-wave LO. The dashed lines show the output spectrum with a sine-wave LO.

tiplying the RF-port signal by +1 transfers it to the output with no change. Multiplying the RF-port signal by –1 does the same thing, except that the signal inverts (flips 180° in phase). The LO port need not exhibit low intermodulation and harmonic distortion; all it must do is preserve the fast rise and fall times of the switching signal.

REVERSING-SWITCH MIXERS

We can multiply a signal by a square wave without using an analog multiplier at all. All we need is a pair of balun transformers and four diodes (**Figure 12.27A**).

With no LO energy applied to the circuit, none of its diodes conduct. RF-port energy (1) cannot make it to the LO port because there is no direct connection between the secondaries of T1 and T2, and (2) doesn't produce IF output because T2's secondary balance results in energy cancellation at its center tap, and because no complete IF-energy circuit exists through T2's secondary with both of its ends disconnected from ground.

Applying a square wave to the LO port biases the diodes so that half the time, D1 and D2 are on and D3 and D4 are reverse-biased off. This unbalances T2's secondary by leaving its upper wire floating and connecting its lower wire to ground through T1's secondary and center tap. With T2's secondary unbalanced, RF-port energy emerges from the IF port.

The other half of the time, D3 and D4 are on and D1 and D2 are reverse-biased off. This unbalances T2's secondary by leaving its lower wire floating and connects its upper wire to ground through T1's secondary and center tap. With T2's secondary unbalanced, RF-port energy again emerges from the IF port — shifted 180° relative to the first case because T2's active secondary wires are now, in effect, transposed relative to its primary.

A reversing switch mixer's output spectrum is the same as the output spectrum of a multiplier fed with a square wave. This can be analyzed by thinking of the square wave in terms of its Fourier series equivalent, which consists of the sum of sine waves at the square wave frequency and its odd harmonics. The amplitude of the equivalent series' fundamental sine wave is 4/π times (2.1 dB) greater than the amplitude of the square wave. The amplitude of each harmonic is inversely proportional to its harmonic number, so the third harmonic is only one third as strong as the fundamental (9.5 dB below the fundamental), the 5th harmonic is only one fifth as strong (14 dB below the fundamental) and so on. The input signal mixes with each harmonic separately from the others, as if each harmonic were driving its own separate mixer, just as we illustrated with two sine waves in Figure 12.26. Normally, the harmonic outputs are so widely removed from the desired output frequency that they are easily filtered out, so a reversing-switch mixer is just as good as a sine-wave-driven analog multiplier for most practical purposes, and usually better, for radio purposes, in terms of dynamic range and noise.

An additional difference between multiplier and switching mixers is that the signal flow in a switching mixer is reversible (that is, bilateral). It only has one dedicated input (the LO input). The other terminals can be thought of as I/O (input/output) ports, since either one can be the input if the other is the output.

CONVERSION LOSS IN SWITCHING MIXERS

Figure 12.27B shows a perfect *multiplier* mixer. That is, the output is the product of the input signal and the LO. The LO is a perfect square wave. Its peak amplitude is ±1.0 V and its frequency is 8 MHz. Figure 12.27C shows the output waveform (the product of two inputs) for an input signal whose value is 0 dBm and whose frequency is 2 MHz. Notice that for each transition of the square-wave LO, the sine-wave output waveform polarity reverses. There are 16 transitions during the interval shown, at each zero-crossing point of the output waveform. Figure 12.27D shows the mixer output spectrum. The principal components are at 6 MHz and 10 MHz, which are the sum and difference of the signal and LO frequencies. The amplitude of each of these is –3.9 dBm. Numerous other pairs of output frequencies occur that are also spaced 4 MHz apart and centered at 24 MHz, 40 MHz, and 56 MHz and higher odd harmonics of 8 MHz. The ones shown are at –13.5, –17.9, and –20.9 dBm. Because the mixer is lossless, the sum of the output powers must be exactly equal to the power value of the input signal. As explained previously, this output spectrum can also be understood in terms of each of the odd-harmonic components of the square-wave LO operating independently.

If the mixer switched without losses, such as in Figure 12.27A, with diodes that are perfect switches, the results would be mathematically identical to the above example. The diodes would commutate the input signal exactly as shown in Figure 12.27C.

Now consider the perfect multiplier mixer of Figure 12.27B with an LO that is a perfect sine wave with a peak amplitude of ±1.0 V. In this case the dashed lines of Figure 12.27D show that only two output frequencies are present, at 6 MHz and 10 MHz (see also Figure 12.24). Each component now has a –6 dBm level. The product of the 0 dBm sine-wave input at one frequency and the ±1.0V sine-wave LO at another frequency (see equation 6 in this chapter) is the –3 dBm total output.

These examples illustrate the difference between the square-wave LO and the sine-wave LO, for a perfect multiplier. For the same peak value of both LO waves, the square-wave LO delivers 2.1 dB more output at 6 MHz and 10 MHz than the sine-wave LO. An actual diode mixer such as Figure 12.27A behaves more like a switching mixer. Its sine-wave LO waveform is considerably flattened by interaction between the diodes and the LO generator, so that it looks somewhat like a square wave. The diodes have nonlinearities, junction voltages, capacitances, resistances and imperfect parameter matching. (See the **RF Techniques** chapter.) Also, "remixing" of a diode mixer's output with the LO and the input is a complicated possibility. The practical end-result is that diode switching mixers have a conversion loss, from input to each of the two major output frequencies, in the neighborhood of 5 – 6 dB. (Conversion loss is discussed in a later section.)

12.4.3 The Diode Double-Balanced Mixer (DBM)

The diode *double-balanced mixer* (DBM) is standard in many commercial, military, and amateur applications because of its excellent balance and high dynamic range. DBM's can serve as mixers (including image-reject types), modulators (including single- and double-sideband, phase, biphase, and quadrature-phase types) and demodulators, limiters, attenuators, switches, phase detectors and frequency doublers. In some of these applications, they work in conjunction with power dividers, combiners, and hybrids.

THE BASIC DBM

We have already seen the basic diode DBM circuit (Figure 12.27A, though normally with a sine wave LO). In its simplest form, a DBM contains two or more unbalanced-to-balanced transformers and a Schottky-diode ring consisting of 4 × *n* diodes, where n is the number of diodes in each leg of the ring. Each leg commonly consists of up to four diodes.

As we have seen, the degree to which a mixer is *balanced* depends on whether either, neither, or both of its input signals (RF and LO) emerge from the IF port along with mixing products. An unbalanced mixer suppresses neither its RF nor its LO; both are present at its IF port. A single-balanced mixer suppresses its RF or LO, but not both. A double-balanced mixer suppresses its RF and LO inputs. Diode and transformer uniformity in the Figure 12.27A circuit results in equal LO potentials at the center taps of

T1 and T2. The LO potential at T1's secondary center tap is zero (ground); therefore, the LO potential at the IF port is zero.

Balance in T2's secondary likewise results in an RF null at the IF port. The RF potential between the IF port and ground is therefore zero — except when the DBM's switching diodes operate!

The Figure 12.27A circuit normally also affords high RF-IF isolation because its balanced diode switching precludes direct connections between T1 and T2. A diode DBM can be used as a current-controlled switch or attenuator by applying dc to its IF port, albeit with some distortion. This causes opposing diodes (D2 and D4, for instance) to conduct to a degree that depends on the current magnitude, connecting T1 to T2.

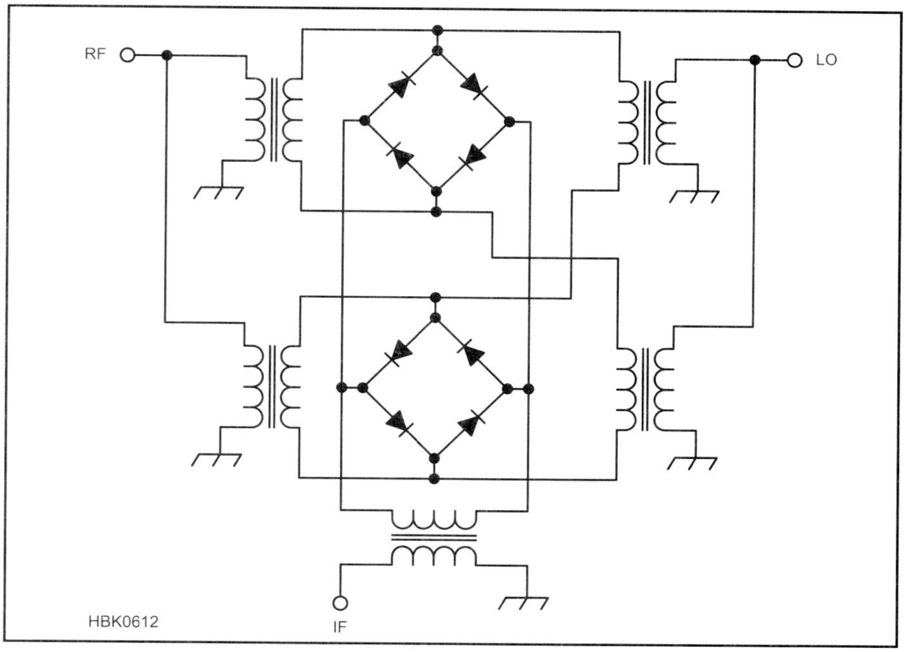

Figure 12.28 — The triple-balanced mixer uses a pair of diode rings and adds an additional balancing transformer to the IF port.

TRIPLE-BALANCED MIXERS

The triple-balanced mixer shown in **Figure 12.28** (sometimes called a "double double-balanced mixer") is an extension **of the** single-diode-ring mixer. The diode rings are fed by power-splitting baluns at the RF and LO ports. An additional balun is added at the IF output. The circuit's primary advantage is that the IF output signal is balanced and isolated from the RF and LO ports over a large bandwidth — commercial mixer IF ranges of 0.5 to 10 GHz are typical.

This type of mixer has higher signal-handling capability and dynamic range (a 1 dB compression point within 3 – 4 dB below LO signal levels) and lower intermodulation levels (by 10 dB or more) than a single-ring mixer. It is used when a very wide IF range is required.

Adding the balancing transformer in the IF output path increases IF-to-LO and IF-to-RF isolation. This makes the conversion process much less sensitive to IF impedance mismatches. Since the IF port is isolated from the RF and LO ports, the three frequency ranges (RF, LO, and IF) can overlap. A disadvantage of IF transformer coupling is that a dc (or low-frequency) IF output is not available, so the triple-balanced mixer cannot be used for direct-conversion receivers.

DIODE DBM COMPONENTS

Commercially manufactured diode DBM's generally consist of a supporting base, a diode ring, two or more ferrite-core transformers commonly wound with two or three twisted-pair wires, encapsulating material, and an enclosure.

Diodes

Hot-carrier (Schottky) diodes are the devices of choice for diode-DBM rings because of their low ON resistance, although ham-built DBM's for noncritical MF/HF use commonly use switching diodes like the 1N914 or 1N4148. The forward voltage drop, V_f, across each diode in the ring determines the mixer's optimum local-oscillator drive level. Depending on the forward voltage drop of each of its diodes and the number of diodes in each ring leg, a diode DBM is often specified by the optimum LO drive level in dBm (typical values are 0, 3, 7, 10, 13, 17, 23, or 27). As a rule of thumb, the LO signal must be 20 dB stronger than the RF and IF signals for proper operation. This ensures that the LO signal, rather than the RF or IF signals, switches the mixer's diodes on and off — a critical factor in minimizing IMD and maximizing dynamic range.

Transformers

From the DBM schematic shown in Figure 12.27, it's clear that the LO and RF transformers are unbalanced on the input side and balanced on the diode side. The diode ends of the balanced ports are 180° out of phase throughout the frequency range of interest. This property causes signal cancellations that result in higher port-to-port isolation. **Figure 12.29A** plots LO-RF and LO-IF isolation versus frequency for Synergy Microwave's CLP-403 DBM, which is specified for +7 dBm LO drive level. Isolations on the order of 70 dB occur at the lower end of the frequency range as a direct result of the balance among the four diode-ring legs and the RF phasing of the balanced ports.

As we learned in our discussion of generic switching mixers, transformer efficiency plays an important role in determining a mixer's conversion loss and drive-level requirement. Core loss, copper loss, and impedance mismatch all contribute to transformer losses. Ferrite in toroidal, bead, balun (multi-hole), or rod form can serve as DBM transformer cores. Radio amateurs commonly use Fair-Rite Mix 43 ferrite (μ = 950) in HF and VHF applications.

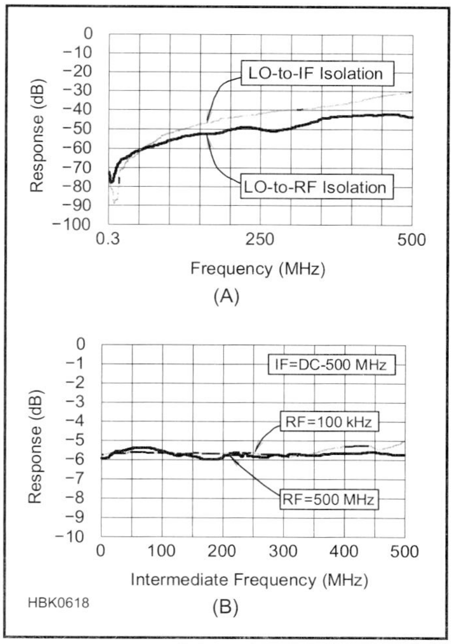

Figure 12.29 — The port-to-port isolation of a diode DBM depends on how well its diodes match and how well its transformers are balanced. (A) shows LO-IF and LO-RF isolation versus frequency and (B) shows conversion loss for a typical diode DBM, the Synergy Microwave CLP-403 mixer. In (B), LO driver level is +7 dBm.

RF transformers combine lumped and distributed capacitance and inductance. The interwinding capacitance and characteristic impedance of a transformer's twisted wires sets the transformer's high-frequency response. The core's permeability (μ) and size, and the number of windings, determine the transformer's lower frequency limit. Covering a specific frequency range requires a compromise in the number of turns used with a given core. Increasing a transformer's core size and number of turns improves its low-frequency response. Cores may be stacked to meet low-frequency performance specs.

Inexpensive mixers operating up to 2 GHz most commonly use twisted trifilar (three-wire) windings made of a wire size between #36 and #32. The number of twists per unit length of wire determines a winding's characteristic impedance. Twisted wires are analogous to transmission lines. The transmission-line effect predominates at the higher end of a transformer's frequency range.

PRACTICAL DIODE DBM'S

Important DBM specifications include conversion loss and amplitude flatness across the required IF bandwidth, variation of conversion loss with input frequency, variation of conversion loss with LO drive, 1 dB compression point, isolation (LO-RF, LO-IF, and RF-IF), intermodulation products, noise figure (usually within 1 dB of conversion loss), port SWR, and dc offset, which is directly related to isolation among the RF, LO, and IF ports. Most of these parameters also apply to other mixer types.

Conversion Loss of Diode DBM's

Figure 12.29B shows conversion loss versus intermediate frequency in a typical DBM. The curves show conversion loss for two fixed RF-port signals, one at 100 kHz and the another at 500 MHz, while varying the LO frequency from 100 kHz – 500 MHz.

Figure 12.30 graphs a diode DBM's simulated output spectrum. Note that the RF input (900 MHz) is –40 dBm and the desired IF output (51 MHz, the frequency difference between the RF and LO signals) is –46 dBm, implying a conversion loss of 6 dB. Very nearly the same value (5 dB) applies to the sum of both signals (RF + LO). We minimize a diode DBM's conversion loss, noise figure and intermodulation by keeping its LO drive high enough to switch its diodes on fully and rapidly. Increasing a mixer's LO level beyond that sufficient to turn its switching devices all the way on merely makes them dissipate more LO power without further improving performance.

Insufficient LO drive results in increased noise figure and conversion loss. IMD also increases because RF-port signals have a greater chance to control the mixer diodes when the LO level is too low.

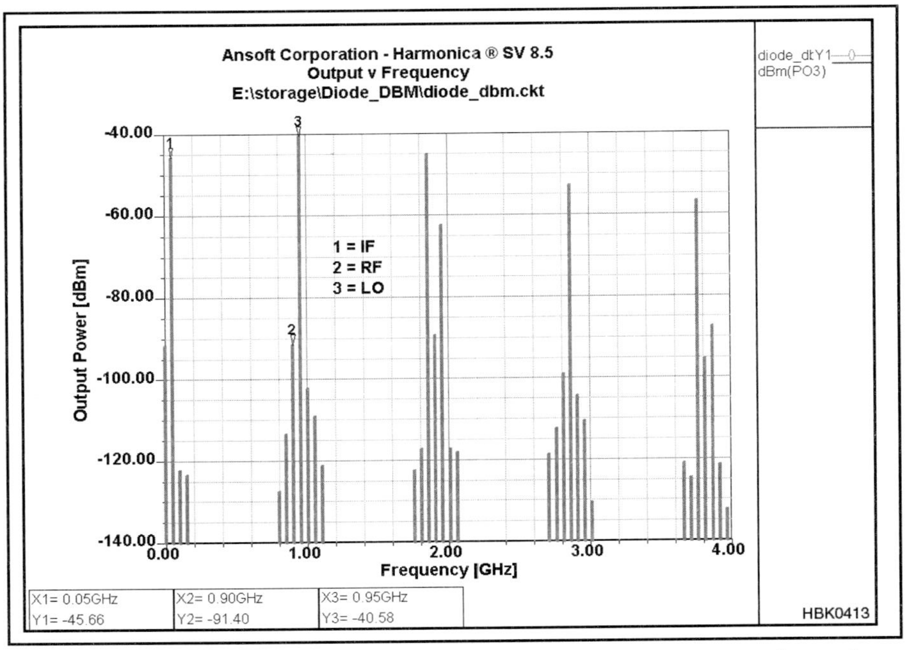

Figure 12.30 — Simulated diode-DBM output spectrum with four LO harmonics evaluated. Note that the desired output products (the highest two products, RF – LO and RF + LO) emerge at a level 5 to 6 dB below the mixer's RF input (–40 dBm). This indicates a mixer conversion loss of 5 to 6 dB. (*Serenade SV8.5* simulation.)

APPLYING DIODE DBM'S

At first glance, applying a diode DBM is easy: We feed the signal(s) we want to frequency-shift (at or below the maximum level called for in the mixer's specifications, such as –10 dBm for the Mini-Circuits SBL-1 and TUF-3, and Synergy Microwave S-1, all popular 7 dBm LO power parts) to the DBM's RF port, feed the frequency-shifting signal (at the proper level) to the LO port, and extract the sum and difference products from the mixer's IF port.

There's more to it than that, however, because diode DBM's (along with most other modern mixer types) are *termination-sensitive. That is, their ports, particularly their IF (output) ports, must be resistively terminated with the proper impedance (commonly 50* W, resistive). A wideband, resistive output termination is particularly critical if a mixer is to achieve its maximum dynamic range in receiving applications. Such a load can be achieved by:

• Terminating the mixer in a 50 W resistor or attenuator pad (a technique usually avoided in receiving applications because it directly degrades the system noise figure);

• Terminating the mixer with a low-noise, high-dynamic-range *post-mixer amplifier* designed to exhibit a wideband resistive input impedance; or

• Terminating the mixer in a *diplexer,* a frequency-sensitive signal splitter that appears as a two-terminal resistive load at its input while resistively dissipating unwanted outputs and passing desired outputs through to subsequent circuitry.

Termination-insensitive mixers are available, but this label can be misleading. Some termination-insensitive mixers are nothing more than a termination-sensitive mixer packaged with an integral post-mixer amplifier. True termination-insensitive mixers are less common and considerably more elaborate. Amateur builders are more likely to use one of the many excellent termination-sensitive mixers available in connection with a diplexer, post-mixer amplifier or both.

Figure 12.31 shows one diplexer implementation. In this approach, L1 and C1 form a series-tuned circuit, resonant at the desired IF, that presents low impedance between the diplexer's input and output terminals at the IF. The high-impedance parallel-tuned circuit formed by L2 and C2 also resonates at the desired IF, keeping desired energy out of the diplexer's 50 W load resistor, R1.

The preceding example is called a *bandpass diplexer.* **Figure 12.32** shows another type: a *high-pass/low-pass diplexer* in which each inductor and capacitor has a reactance of 70.7 Ω at the 3 dB cutoff frequency. It can be used after a "difference" mixer (a mixer in which the IF is the difference between the signal frequency and LO) if the desired IF and its image frequency are far enough apart

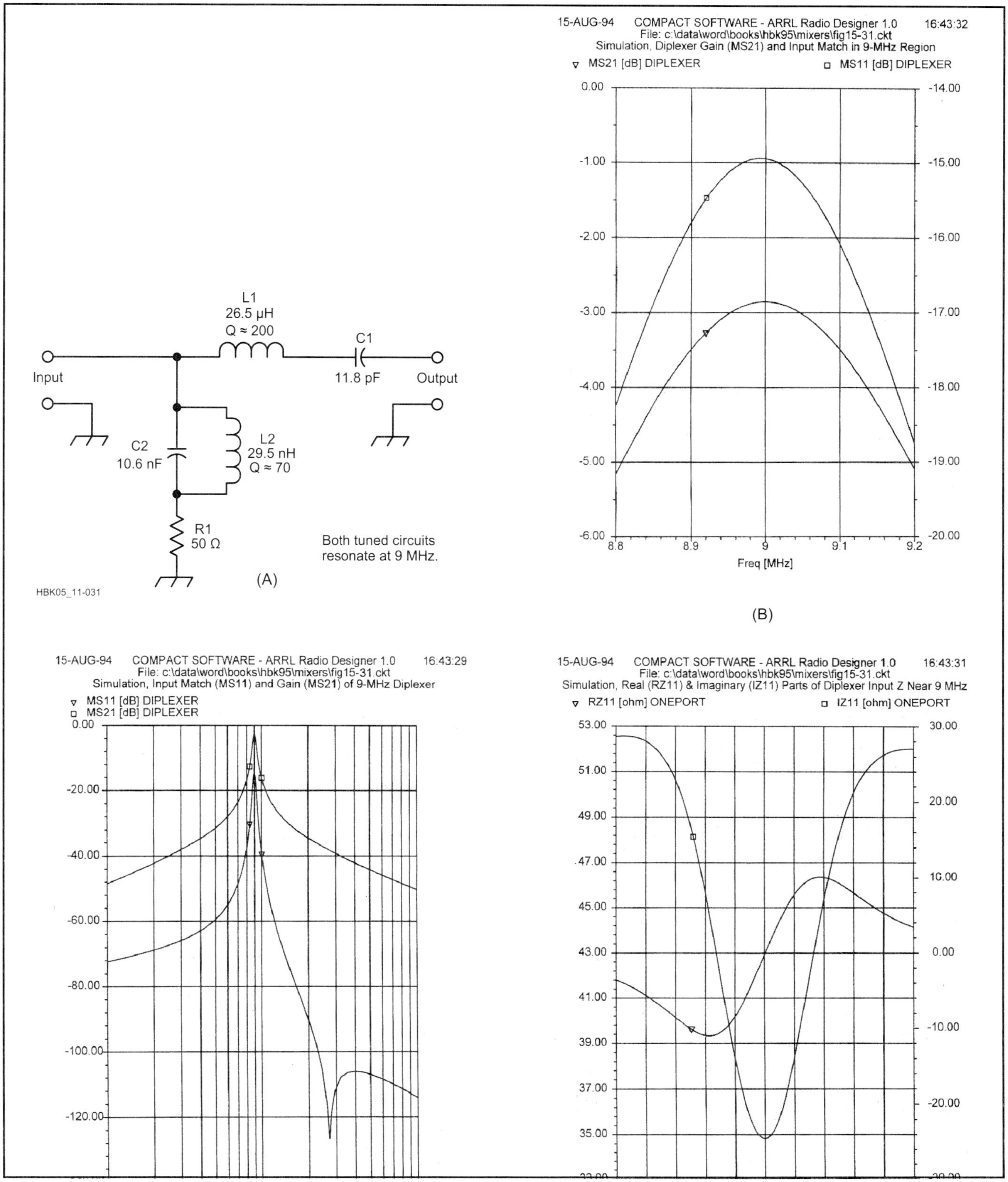

Figure 12.31 — A diplexer resistively terminates energy at unwanted frequencies while passing energy at desired frequencies. This band-pass diplexer (A) uses a series-tuned circuit as a selective pass element, while a high-C parallel-tuned circuit keeps the network's terminating resistor R1 from dissipating desired-frequency energy. Computer simulation of the diplexer's response with *ARRL Radio Designer 1.0* characterizes the diplexer's insertion loss and good input match from 8.8 to 9.2 MHz (B) and from 1 to 100 MHz (C); and the real and imaginary components of the diplexer's input impedance from 8.8 to 9.2 MHz with a 50 Ω load at the diplexer's output terminal (D). The high-C, low-L nature of the L2-C2 circuit requires that C2 be minimally inductive; a 10,000-pF chip capacitor is recommended. This diplexer was described by Rohde and Bucher in *Communications Receivers: Principles and Design, 3rd Edition*.

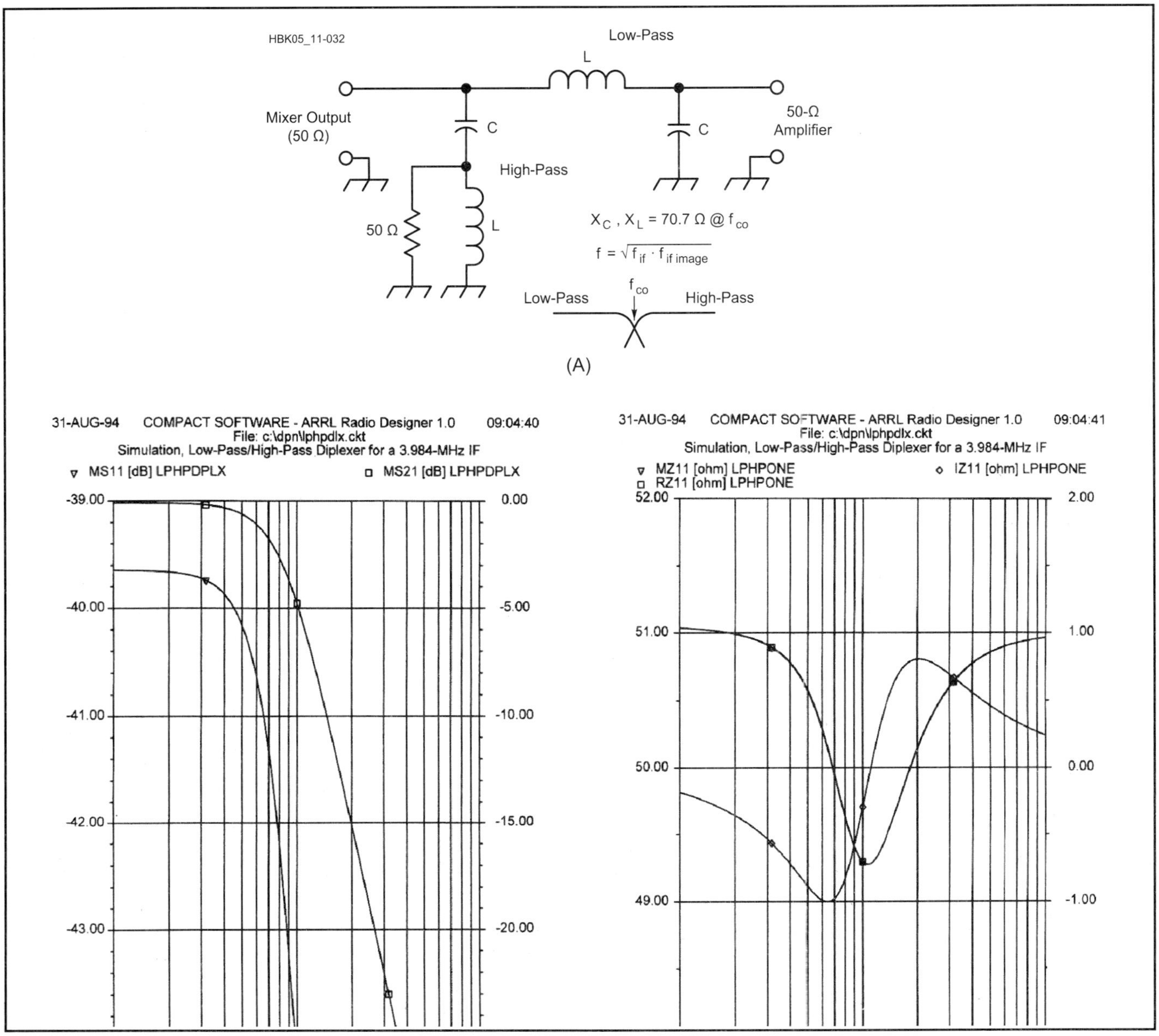

Figure 12.32 — All of the inductors and capacitors in this high-pass/low-pass diplexer (A) exhibit a reactance of 70.7 Ω at its tuned circuits' 3-dB cutoff frequency (the geometric mean of the IF and IF image). B and C show *ARRL Radio Designer* simulations of this circuit configured for use in a receiver that converts 7 MHz to 3.984 MHz using a 10.984-MHz LO. The IF image is at 17.984 MHz, giving a 3-dB cutoff frequency of 8.465 MHz. The inductor values used in the simulation were therefore 1.33 µH (Q = 200 at 25.2 MHz); the capacitors, 265 pF (Q = 1000). This drawing shows idler load and "50 Ω Amplifier" connections suitable for a receiver in which the IF image falls at a frequency *above* the desired IF. For applications in which the IF image falls below the desired IF, interchange the 50 Ω idler load resistor and the diplexer's "50 Ω Amplifier" connection so the idler load terminates the diplexer low-pass filter and the 50 Ω amplifier terminates the high-pass filter.

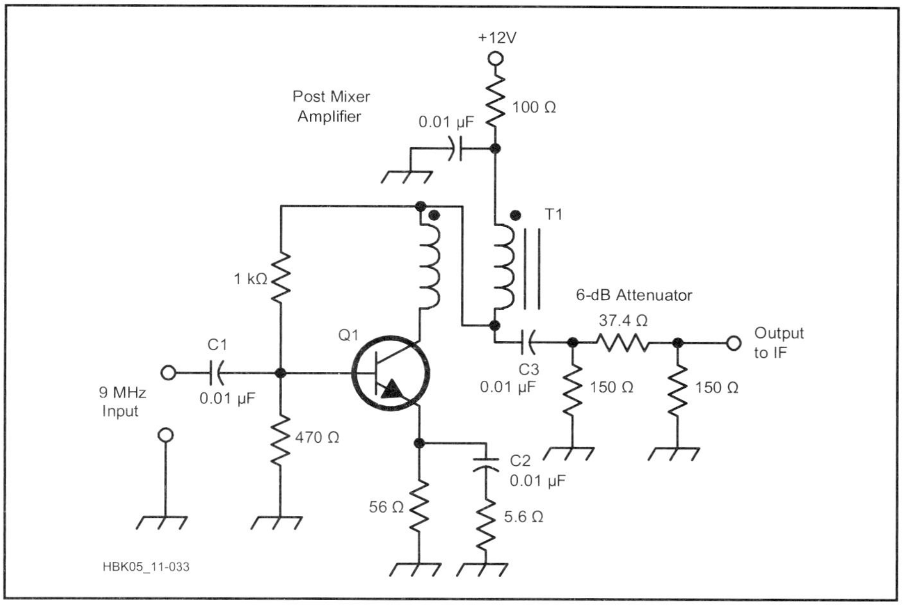

Figure 12.33 — The post-mixer amplifier from Hayward and Lawson's "Progressive Communications Receiver" (November 1981 *QST*). This amplifier's gain, including the 6-dB loss of the attenuator pad, is about 16 dB; its noise figure, 4 to 5 dB; its output intercept, 30 dBm. The 6-dB attenuator is essential if a crystal filter follows the amplifier; the pad isolates the amplifier from the filter's highly reactive input impedance. This circuit's input match to 50 Ω below 4 MHz can be improved by replacing 0.01-µF capacitors C1, C2 and C3 with low-inductance 0.1-µF units (chip capacitors are preferable). Q1 is a TO-39 CATV-type bipolar transistor, f_T = 1 GHz or greater (2N3866, 2N5109, 2SC1252, 2SC1365 or MRF586 suitable.) Use a small heat sink on this transistor. T1 is a broadband ferrite transformer, ≈42 µH per winding: 10 bifilar turns of #28 enameled wire on an FT 37-43 core.

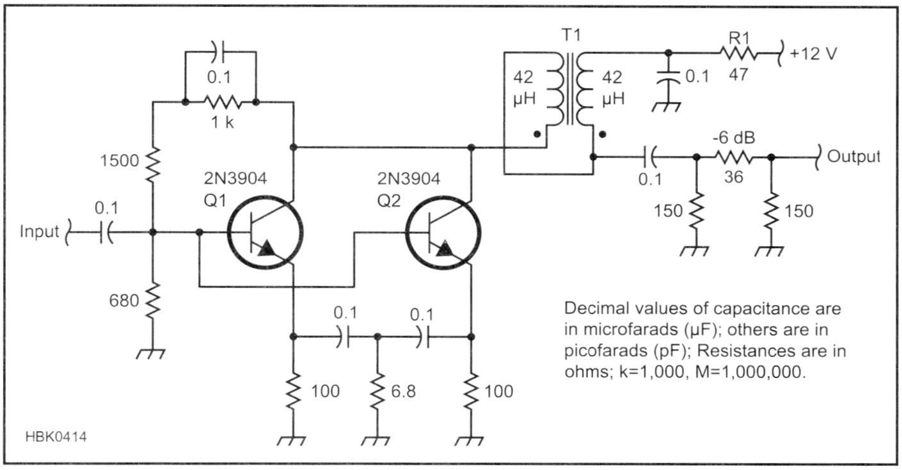

Figure 12.34 — At MF and HF, paralleled 2N3904 BJTs can provide performance comparable to that of the Figure 12.33 circuit with sufficient attention paid to device standing current, here set at »30 mA for the pair. The value of decoupling resistor R1 is critical in that small changes in its value cause a relatively large change in the 2N3904s' bias point. This circuit is part of the "EZ-90 Receiver," described by Hayward, Campbell and Larkin in *Experimental Methods in RF Design.*

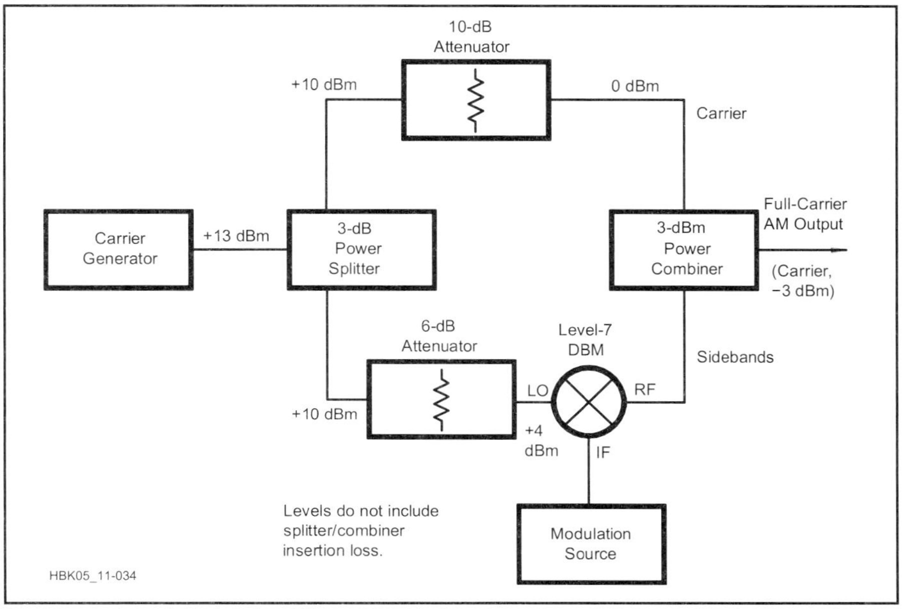

Figure 12.35 — Generating full-carrier AM with a diode DBM. A practical modulator using this technique is described in *Experimental Methods in RF Design*.

so that the image power is "dumped" into the network's 51 Ω resistor. (For a "summing" mixer, a mixer in which the IF is the sum of the desired signal and LO, interchange the 50 Ω idler load resistor and the diplexer's "50 Ω Amplifier" connection.)

Figure 12.33 shows a BJT post-mixer amplifier design made popular by Wes Hayward, W7ZOI, and John Lawson, K5IRK. RF feedback (via the 1 kΩ resistor) and emitter degeneration (the ac-coupled 5.6 Ω emitter resistor) work together to keep the stage's input impedance near 50 Ω and uniformly resistive across a wide bandwidth. Performance comparable to the circuit in Figure 12.33 can be obtained at MF and HF by using paralleled 2N3904s as shown in **Figure 12.34.**

Phase Detection with a DBM

As we saw in our exploration of quadrature detection, applying two signals of equal frequency to a DBM's LO and RF ports produces an IF-port dc output proportional to the cosine of the signals' phase difference (**Figure 12.35).** This assumes that the DBM has a dc-coupled IF port, of course. If it doesn't, and some DBM's don't, phase-detector operation is out. Any dc output offset introduces error into this process, so critical phase-detection applications use low-offset DBM's optimized for this service.

12.4.4 Active Mixers

We've covered diode DBM's in depth because their ease of use in homebrew projects, high performance, and suitability for direct connection into 50 Ω systems makes them attractive to amateur builders. The abundant availability of high-quality manufactured diode mixers at reasonable prices makes them excellent candidates for home construction projects. Although diode DBM's are common in telecommunications, their conversion loss and relatively high LO power requirement have usually driven the manufacturers of high-performance MF/HF Amateur Radio receivers and transceivers to other solutions. Those solutions have generally involved single- or double-balanced FET mixers; MOSFETs in the late 1970s and early 1980s, JFETs from the early 1980s to the present. A comprehensive paper that explores the differences between various forms of active mixers, "Performance Capabilities of Active Mixers," by Ulrich Rohde, N1UL, is included in the online content accompanying this *Handbook.*

Many of the JFET designs are variations of a single-balanced mixer circuit introduced to *QST* readers in 1970. **Figure 12.36** shows the circuit as it was presented by William Sabin in "The Solid-State Receiver," *QST,* July 1970. Two 2N4416 JFETs operate in

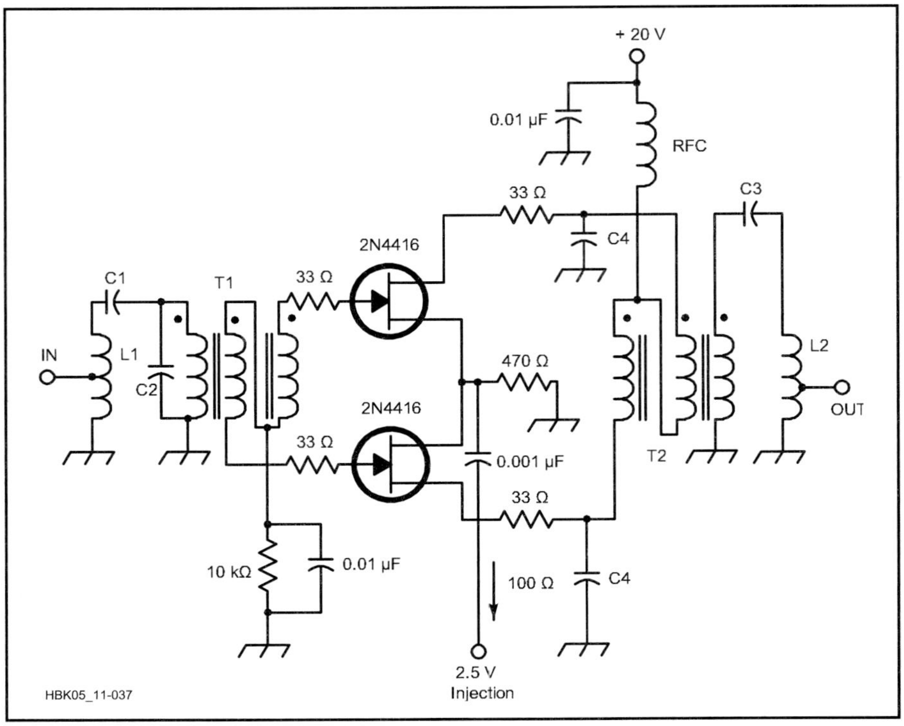

Figure 12.36 — Two 2N4416 JFETs provide highdynamic range in this mixer circuit from Sabin, *QST*, July 1970. L1, C1 and C2 form the input tuned circuit; L2, C3, and C4 tune the mixer output to the IF. The trifilar input and output transformers are broadband transmission-line types.

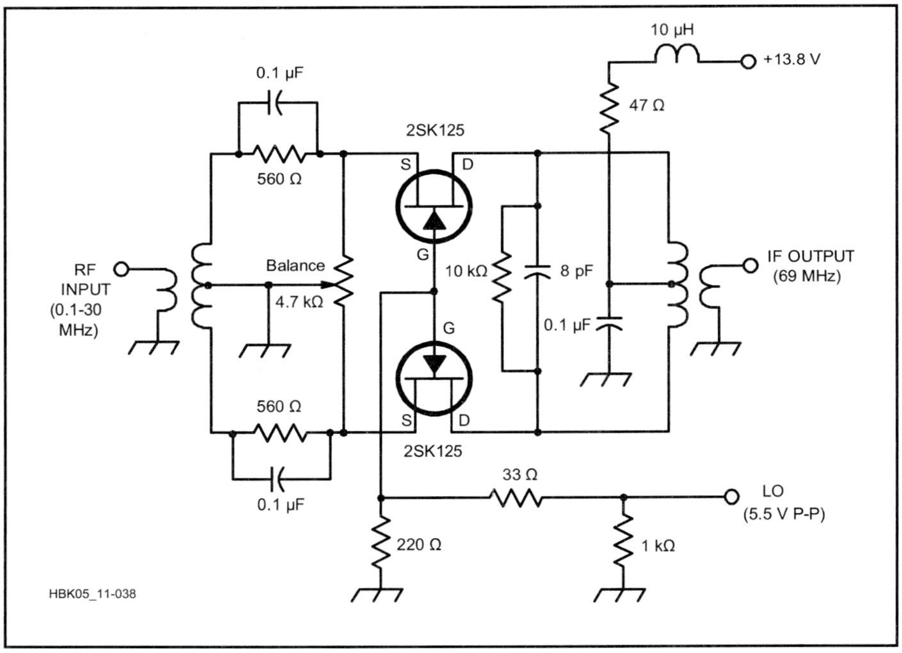

Figure 12.37 — The ICOM IC-765's single-balanced 2SK125 mixer achieves a high dynamic range (per *QST* Product Review, an IP_3 of 10.5 dBm at 14 MHz with preamp off). The first receive mixer in many commercial amateur radio transceiver designs of the 1980s and 1990s used a pair of 2SK125s or similar JFETs in much this way.

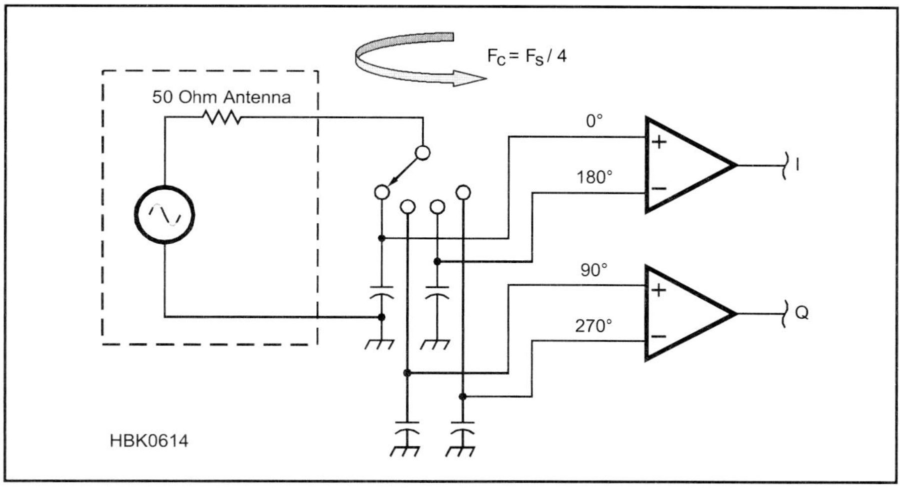

Figure 12.38 — Tayloe detector: The switch rotates at the carrier frequency so that each capacitor samples the signal once each revolution. The 0° and 180° capacitors differentially sum to provide the in-phase (I) signal and the 90° and 270° capacitors sum to provide the quadrature (Q) signal.

Testing Mixer Performance

In order to make proper tests on mixers using signal generators, a hybrid coupler with at least 40 dB of isolation between the two input ports and an attenuator are required. The test set-up provided by DeMaw and Collins in *QST*, January 1981, shown in **Figure 12.C1** is ideal for this.* Two signal generators operating near 14 MHz are combined in the hybrid coupler, then isolated from the mixer under test (MUT) by a variable attenuator. The LO is supplied by a VFO covering 5.0 – 5.5 MHz and applied to the MUT through another variable attenuator. The output is isolated with another attenuator, amplified and applied to a spectrum analyzer for analysis.

Attenuation should be sufficient to provide isolation (minimum of 6 to 10 dB required) and to result in signal levels to the mixer under test (MUT) appropriate for the required testing and as suitable for the particular mixer device.

The 2N5109 amplifier shown may not be sufficient for extremely high intercept point tests as this stage may no longer be transparent (operate linearly) at high signal levels. For stability tests, it is recommended to have a reactive network at the output of the mixer for the sole purpose of checking whether the mixer can become unstable.

The two 14 MHz oscillators must have extremely low harmonic content and very low noise sidebands. A convenient oscillator circuit is provided in **Figure 12.C2**, based on a 1975 *Electronic Design* article.** — *Dr Ulrich L. Rohde, N1UL*

*DeMaw, W1FB, and Collins, ADØW, "Modern Receiver Mixers for High Dynamic Range," QST, Jan 1981, p 19.

**U. Rohde, "Crystal Oscillator Provides Low Noise," *Electronic Design*, Oct 11, 1975.

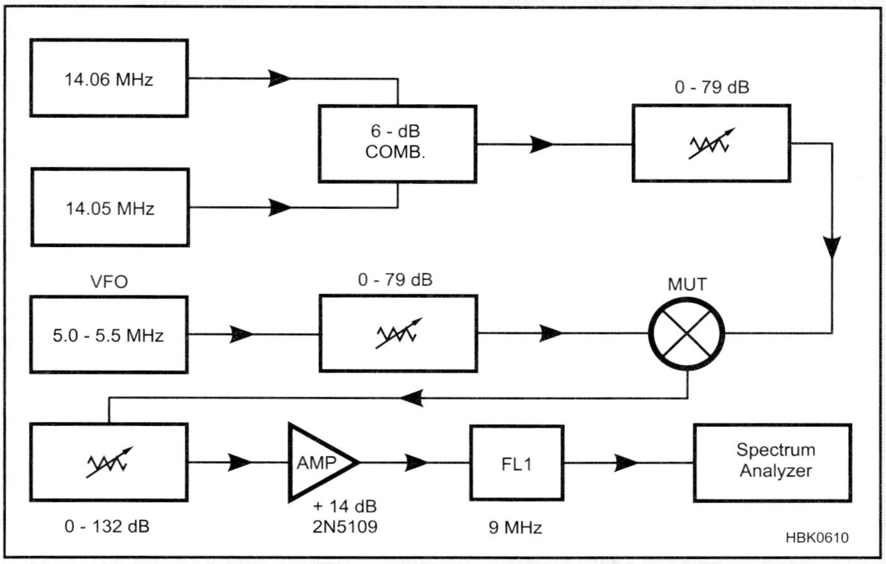

Figure 12.C1 — The equipment setup for measuring mixer performance at HF.

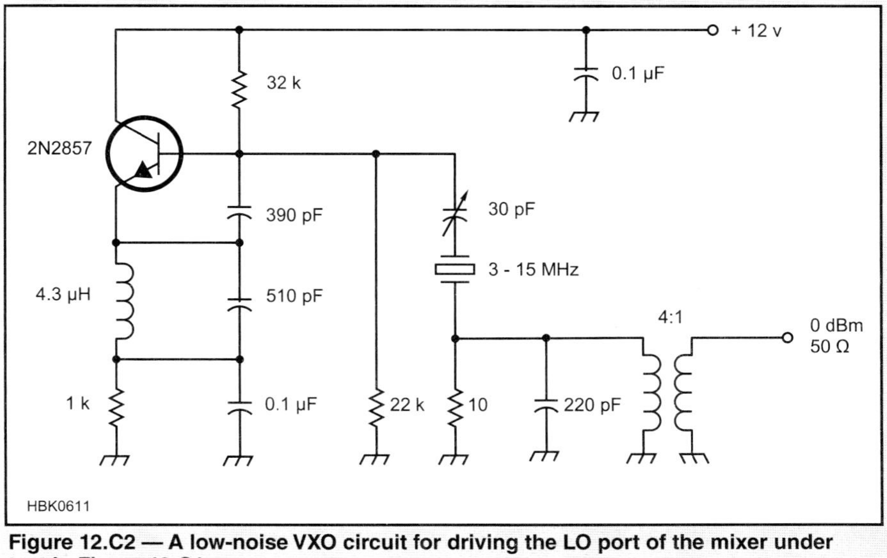

Figure 12.C2 — A low-noise VXO circuit for driving the LO port of the mixer under test in Figure 12.C1.

a common-source configuration, with push-pull RF input and parallel LO drive. **Figure 12.37** shows a similar circuit as implemented in the ICOM IC-765 transceiver. In this version, the JFET's (2SK125s) operate in common-gate, with the LO applied across a 220 Ω resistor between the gates and ground.

12.4.5 The Tayloe Mixer

[The following description of the Tayloe Product Detector (a.k.a. the Tayloe Mixer) is adapted from the July 2002 *QEX a*rticle, "Software-Defined Radio for the Masses, Part 1" by Gerald Youngblood AC5OG, now K5SDR. — Ed.]

The beauty of the Tayloe detector (see reference listings for Tayloe) is found in both its design elegance and its exceptional performance. In its simplest form, you can build a complete quadrature downconverter with only three or four ICs (less the local oscillator) at a cost of less than $10.

Figure 12.38 illustrates a single-balanced version of the Tayloe detector. It can be visualized as a four-position rotary switch revolving at a rate equal to the carrier frequency. The 50 Ω antenna impedance is connected to the rotor and each of the four switch positions is connected to a sampling capacitor. Since the switch rotor is turning at exactly the RF carrier frequency, each capacitor tracks the carrier's amplitude for exactly one-quarter of the cycle and then holds its value for the remainder of the cycle. The rotating switch, therefore samples the signal at 0°, 90°, 180°, and 270°.

As shown in **Figure 12.39**, the 50 W impedance of the antenna and the sampling capacitors form an RC low-pass filter during the period when each respective switch is turned on. Therefore, each sample rep-

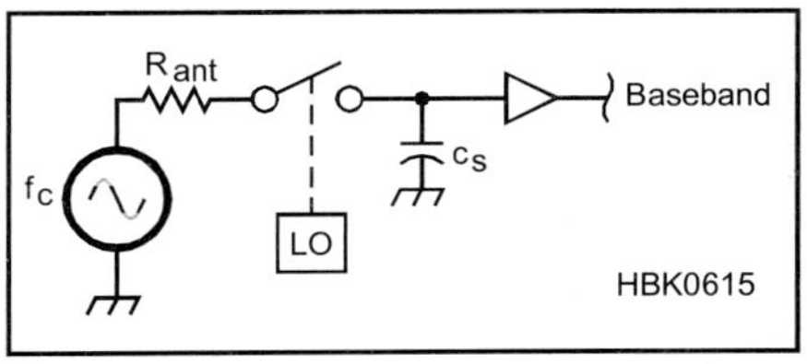

Figure 12.39 — Track-and-hold sampling circuit: Each of the four sampling capacitors in the Tayloe detector form an RC track-and-hold circuit. When the switch is on, the capacitor charges to the average value of the carrier during its respective one-quarter cycle. During the remaining three-quarters cycle, it holds its charge. The local-oscillator frequency is equal to the carrier frequency so that the output is at baseband.

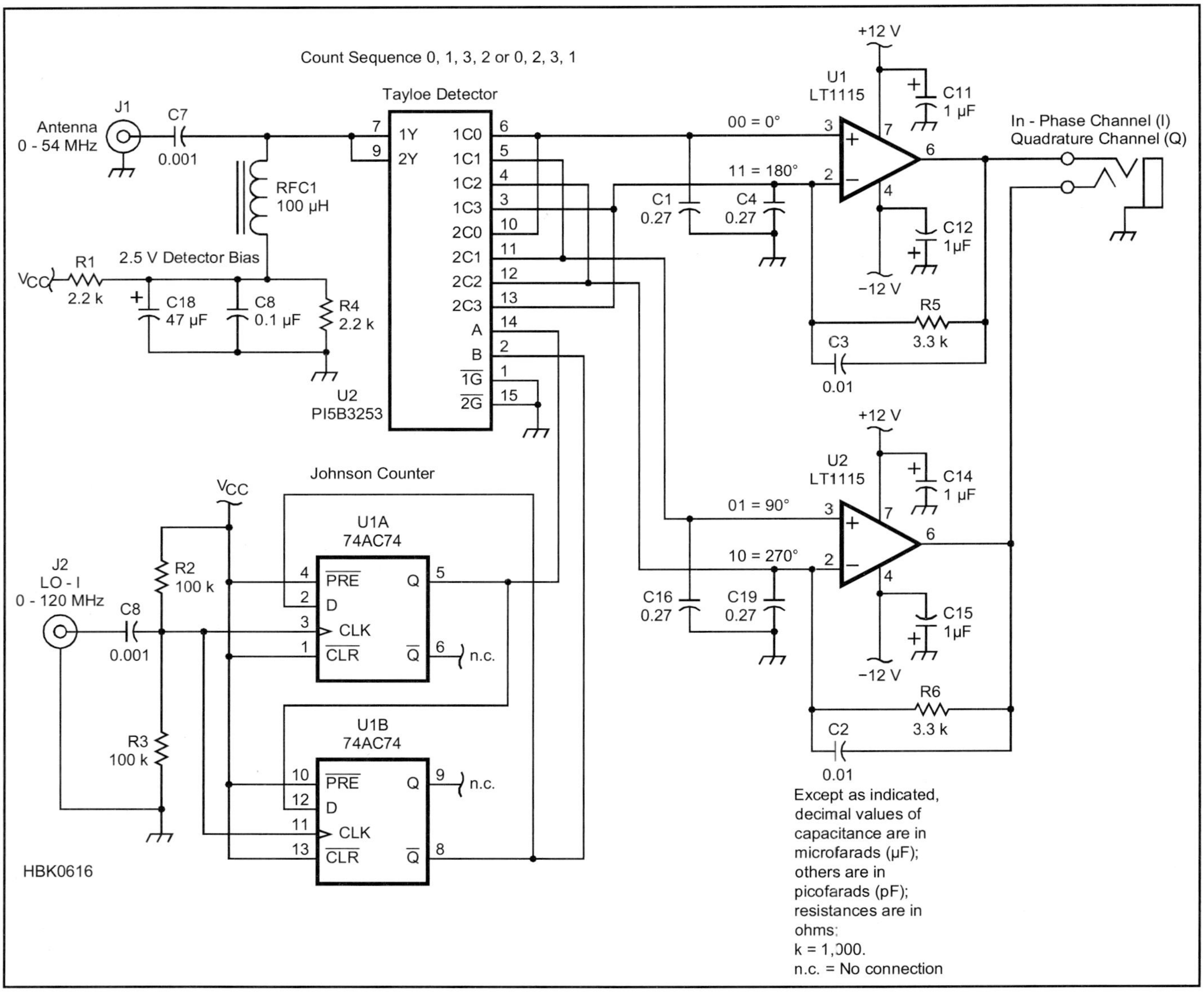

Figure 12.40 — Single-balanced Tayloe detector.

resents the integral or average voltage of the signal during its respective one-quarter cycle. When the switch is off, each sampling capacitor holds its value until the next revolution. If the RF carrier and the rotating frequency are exactly in phase, the output of each capacitor is a dc level equal to the average value of the sample.

If we differentially sum outputs of the 0° and 180° sampling capacitors with an op amp (see Figure 12.38), the output is a dc voltage equal to two times the value of the individually sampled values when the switch rotation frequency equals the carrier frequency. Imagine, 6 dB of noise-free gain! The same would be true for the 90° and 270° capacitors as well. The 0°/180° summation forms the I channel and the 90°/270° summation forms the Q channel of a quadrature down conversion. (See the **Modulation** chapter for more information on I/Q modulation.)

As we shift the frequency of the carrier away from the sampling frequency, the values of the inverting phases are no longer dc levels. The output frequency varies according to the "beat" or difference frequency between the carrier and the switch-rotation frequency to provide an accurate representation of all the signal components converted to baseband.

Figure 12.40 is the schematic for a simple, single-balanced Tayloe detector. It consists of a PI5V331, a 1:4 FET demultiplexer (an analog switch), that switches the signal to each of the four sampling capacitors. The 74AC74 dual flip-flop is connected as a divide-by-four Johnson counter to provide the two-phase clock to the demultiplexer chip. The outputs of the sampling capacitors are differentially summed through the two LT1115 ultra-low-noise op amps to form the I and Q outputs, respectively.

Note that the impedance of the antenna forms the input resistance for the op-amp gain as shown in the equation for gain below. This impedance may vary significantly with the actual antenna. In a practical receiver, a buffer amplifier should be used to stabilize and control the impedance presented to the mixer.

Since the duty cycle of each switch is 25%, the effective resistance in the RC network is the antenna impedance multiplied by four in the op-amp gain formula:

$$G = \frac{R_f}{4R_{ant}}$$

For example, with a feedback resistance, R_f of 3.3 kΩ and antenna impedance, Rant, of 50 Ω, the resulting gain of the input stage is:

$$G = \frac{3300}{4 \times 50} = 16.5$$

The Tayloe detector may also be analyzed as a *digital commutating filter* (see reference by Kossor). This means that it operates as a very-high-Q tracking filter, where the following equation determines the bandwidth:

$$BW_{det} = \frac{1}{\pi \; n \; R_{ant} \; C_s}$$

where n is the number of sampling capacitors, R_{ant} is the antenna impedance and Cs is the value of the individual sampling capacitors. Qdet of the filter is:

$$Q_{det} = \frac{f_c}{BW_{det}}$$

where f_c is the center frequency and BWdet is the bandwidth of the filter.

By example, if we assume the sampling capacitor to be 0.27 µF and the antenna impedance to be 50 W, then BW and Q at an operating frequency of 14.001 MHz are computed as follows:

$$BW_{det} = \frac{1}{\pi \times 4 \times 50 \times (2.7 \times 10^{-7})} = 5895 \text{ Hz}$$

$$Q_{det} = \frac{14.001 \times 10^{-6}}{5895} = 2375$$

The real payoff in the Tayloe detector is its performance. It has been stated that the *ideal* commutating mixer has a minimum conversion loss (which determines noise figure — see the **RF Techniques** chapter) of 3.9 dB. Typical high-level diode mixers have a conversion loss of 6 – 7 dB and noise figures 1 dB higher than the loss. Remarkably, the Tayloe detector has less than 1 dB of conversion loss. How can this be? The reason is that it is not really a mixer but a sampling detector in the form of a quadrature track-and-hold circuit. This means that the design adheres to discrete-time sampling theory, which, while similar to mixing, has its own unique characteristics. Because a track-and-hold maintains the signal value between samples, the signal output never goes to zero. (See the DSP and SDR Fundamentals chapter for more on sampling theory.)

This is where aliasing can be used to our benefit. Since each switch and capacitor in the Tayloe detector samples the RF signal once each cycle, it responds to alias frequencies as well as those within the Nyquist frequency range. In a traditional direct-conversion receiver, the local-oscillator frequency is set to the carrier frequency so that the difference frequency, or IF, is at 0 Hz and the sum frequency is at double the carrier frequency. We normally remove the sum frequency through low-pass filtering, resulting in conversion loss and a corresponding increase in noise figure. In the Tayloe detector, the sum frequency resides at the first alias frequency as shown in Figure 12.41. Remember that an alias is a real signal and appears in the output as if it were a baseband signal. Therefore, the alias adds to the baseband signal for a theoretically lossless detector. In real life, there is a slight loss, usually less than 1 dB, due to the resistance of the switch and aperture loss due to imperfect switching times.

12.4.6 The NE602/SA602/SA612 Gilbert Cell Mixer

Introduced as the Philips NE602 in the mid-1980s, the NXP SA602/SA612 mixer-oscillator IC has become greatly popular with amateur experimenters for receive mixers, transmit mixers, and balanced modulators. The SA602/612's mixer is a *Gilbert cell* multiplier. **Figure 12.42** shows its equivalent circuit. A Gilbert cell consists of two differential transistor pairs whose bias current is controlled by one of the input signals. The other signal drives the differential pairs' bases, but only after being "predistorted" in a diode circuit. (This circuit distorts the signal equally and oppositely to the inherent distortion of the differential pair.)

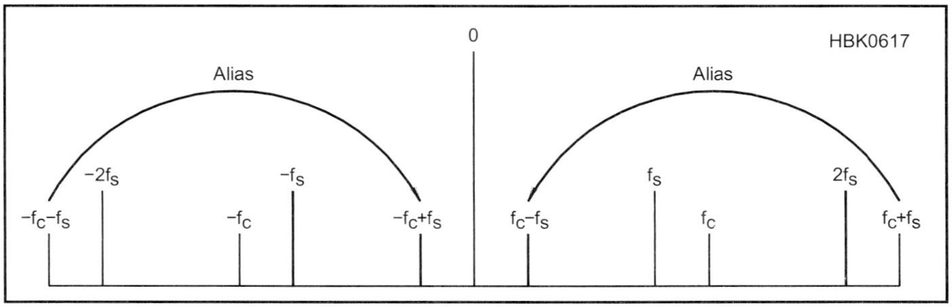

Figure 12.41 — Alias summing on Tayloe detector output: Since the Tayloe detector samples the signal, the sum frequency ($f_c + f_s$) and its image ($-f_c - f_s$) are located at the first alias frequency. The alias signals sum with the baseband signals to eliminate the mixing product loss associated with traditional mixers. In a typical mixer, the sum frequency energy is lost through filtering thereby increasing the noise figure of the device.

The resulting output signal is an accurate multiplication of the input voltages.

SA602/612A VARIANTS

The SA602/612 began life as the NE602/SA602 (SA602A, NXP Semiconductors, **www.nxp.com**). SA-prefixed 602/612 parts are specified for use over a wider temperature range than their NE-prefixed equivalents. Parts without the A suffix have a slightly lower IP_3 specification than their A counterparts. The pinout-identical NE612A and ASA612A (no longer manufactured) cost less than their 602 equivalents due to wider tolerances. All variants of this popular part should work satisfactorily in most "NE602" experimenter projects. The same mixer/oscillator topology, modified for slightly higher dynamic range at the expense of somewhat less mixer gain, is also available in the mixer/oscillator/FM IF chips NE/SA605 (input IP_3 typically –10 dBm, no longer manufactured) and NE/SA615 (input IP_3 typically – 13 dBm, no longer manufactured).

For the parts that are no longer being manufactured, it is often possible to find small numbers of old stock on internet or at hamfests. Some equivalents have been suggested, such as NTE7164 (**www.nteinc.com**), and some amateurs have built Gilbert cell mixers from discrete transistors.

SA602/612 USAGE NOTES

The SA602/612's typical current drain is 2.4 mA; its supply voltage range is 4.5 to 8.0 V. Its inputs (RF) and outputs (IF) can be single- or double-ended (balanced) according to design requirements (**Figure 12.43**). The equivalent ac impedance of each input is approximately 1.5 kW in parallel with 3 pF; each output's resistance is 1.5 kW. **Figure 12.44** shows the use of an NPN transistor at the SA602/612 output to obtain low-impedance drive for a crystal filter; **Figure 12.45** shows how AGC can be applied to an SA602/612.

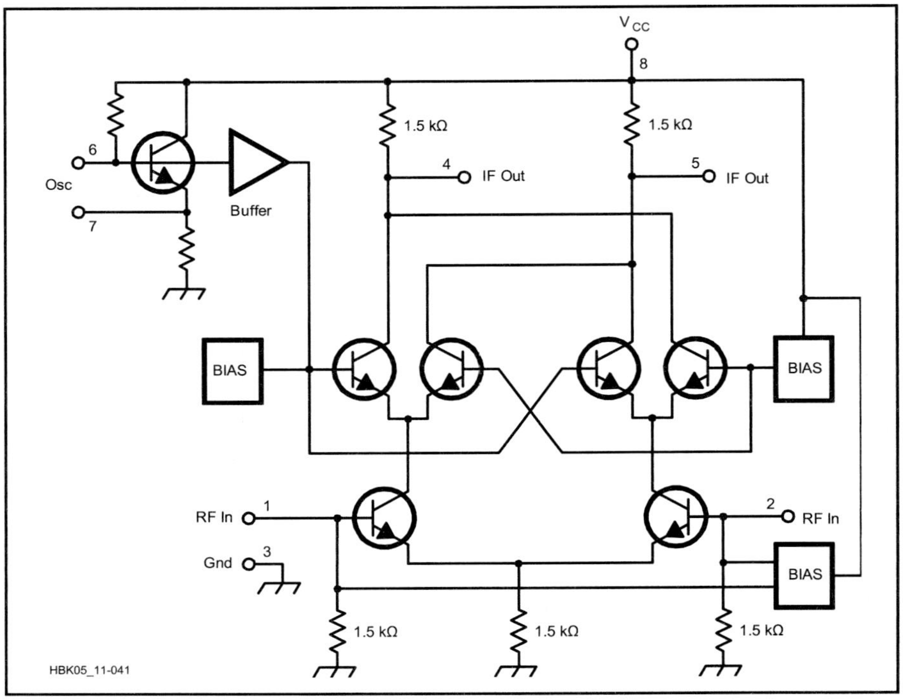

Figure 12.42 — The SA602/612's equivalent circuit reveals its Gilbert-cell origins.

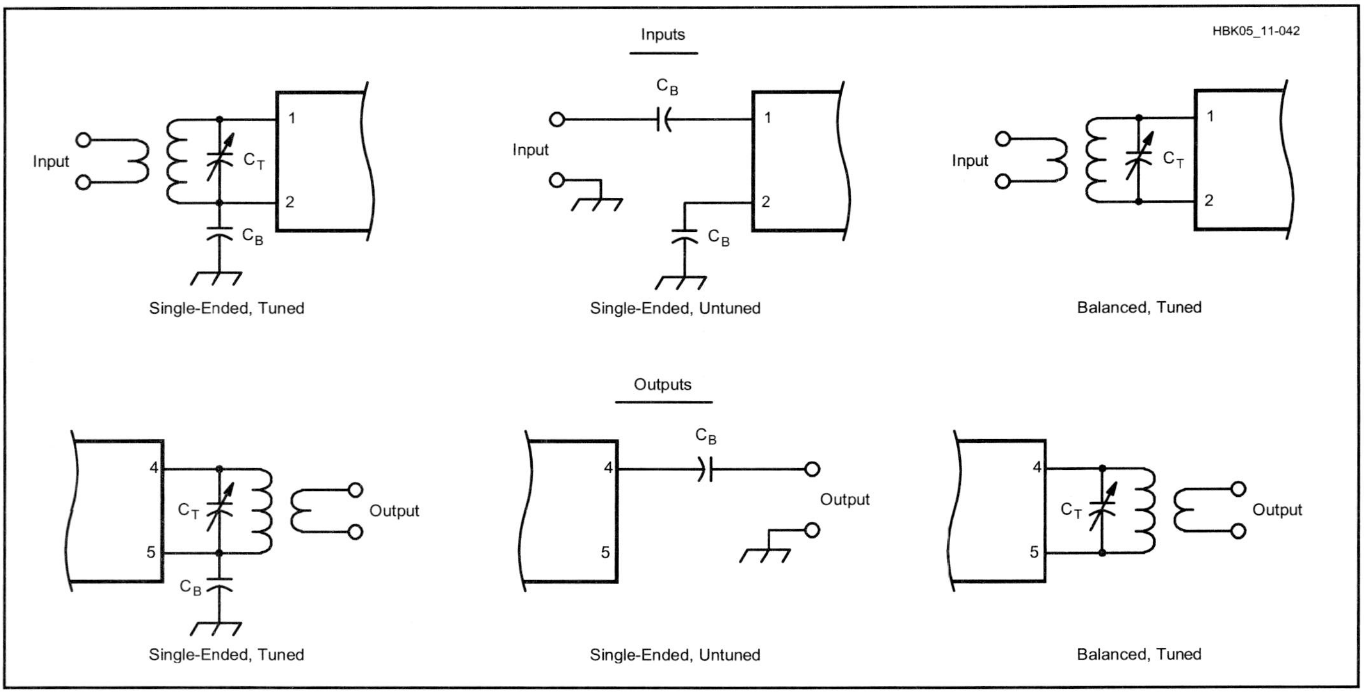

Figure 12.43 — The SA602/612's inputs and outputs can be single- or double-ended (balanced). The balanced configurations minimize second-order IMD and harmonic distortion, and unwanted envelope detection in direct-conversion service. C_T tunes its inductor to resonance; C_B is a bypass or dc-blocking capacitor. The arrangements pictured don't show all the possible input/output configurations; for instance, a center-tapped broadband transformer can be used to achieve a balanced, untuned input or output.

The SA602/612 mixer can typically handle signals up to 500 MHz. At 45 MHz, its noise figure is typically 5.0 dB; its typical conversion gain, 18 dB. Note that in contrast to the diode-based mixers described earlier, which have conversion *loss,* most Gilbert-cell mixers have conversion *gain.* Considering the SA602/612's low current drain, its input IP_3 (measured at 45 MHz with 60 kHz spacing) is usefully good at –15 dBm. Factoring in the mixer's conversion gain results in an equivalent output IP_3 of about 3 dBm.

The SA602/612's onboard oscillator can operate up to 200 MHz in LC and crystal-controlled configurations (**Figure 12.46** shows three possibilities). Alternatively, energy from an external LO can be applied to the chip's pin 6 via a dc blocking capacitor. At least 200 mV P-P of external LO drive is required for proper mixer operation.

The SA602/612 was intended to be used as the second mixer in double-conversion FM cellular radios, in which the first IF is typically 45 MHz, and the second IF is typically 455 kHz. Such a receiver's second mixer can be relatively weak in terms of dynamic range because of the adjacent-signal protection afforded by the high selectivity of the first-IF filter preceding it. When used as a first mixer, the SA602/612 can provide a two-tone third-order dynamic range between 80 and 90 dB, but this figure is greatly diminished if a preamplifier is used ahead of the SA602/612 to improve the system's noise figure.

When the SA602/612 is used as a second mixer, the sum of the gains preceding it should not exceed about 10 dB. An SA602/612 can serve as low-distortion (THD <1%) product detector if overload is avoided through the use of AGC and appropriate attenuation between the SA602/612 and the IF strip that drives it.

The SA602/612 is generally not a good choice for VHF and higher-frequency mixers because of its input noise and diminishing IMD performance at high frequencies. There are applications, however, where 6 dB noise figure and 60 – 70 dB dynamic range performance is adequate. If your target specifications exceed these numbers, you should consider other mixers at VHF and up.

Figure 12.47 shows the schematic of a complete 7 MHz direct-conversion receiver based on the SA602/612 and the widely used LM386 AF power amplifier IC. Such simple product-detector-based receivers sometimes suffer from incidental envelope detection, which causes audio from strong, full-carrier-AM shortwave or medium-wave broadcast stations to be audible regardless of where the receiver LO is tuned. RF attenuation and/or band-limiting the receiver input with a double- or triple-tuned-circuit filter can usually reduce this effect to inaudibility.

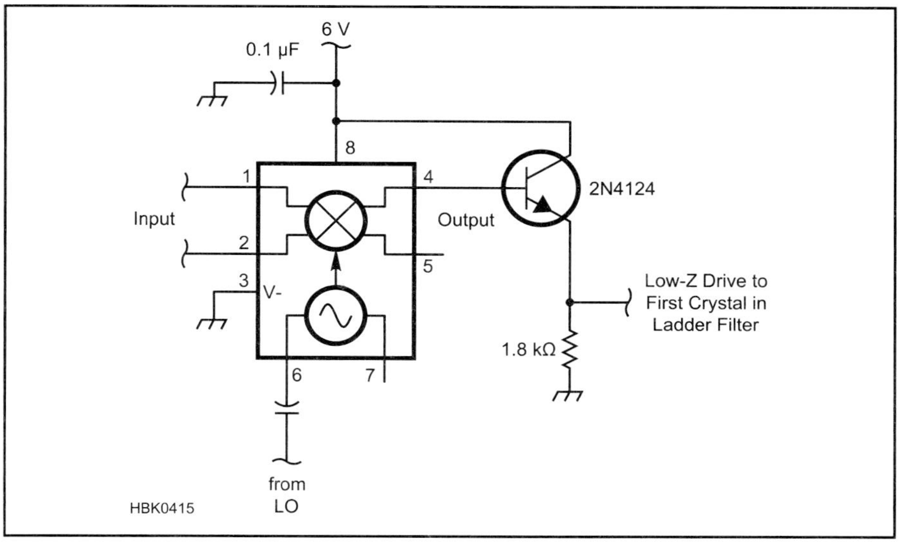

Figure 12.44 — An NPN transistor at the output of an SA602/612 mixer provides power gain and low-impedance drive for a 4.914-MHz crystal filter. A low-reactance coupling capacitor can be added between the emitter and the circuitry it drives if dc blocking is necessary. [Circuit from the Elecraft KX1 transceiver courtesy of Wayne Burdick, N6KR]

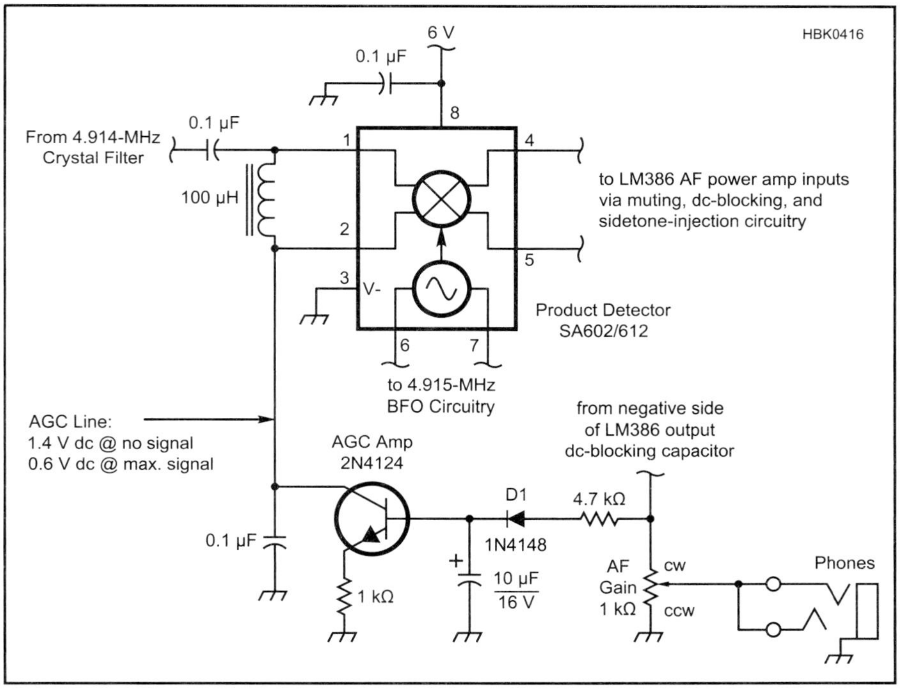

Figure 12.45 — SA602/612 product detector AGC from the Elecraft KX1 transceiver. Designed by Wayne Burdick, N6KR, this circuit first appeared in the Wilderness Radio SST transceiver with an LED used at D1 for simultaneous signal indication and rectification. The selectivity provided by the crystal filter preceding the detector works to mitigate the effects of increasing detector distortion with gain reduction. [Circuit courtesy of Wayne Burdick, N6KR, and Bob Dyer, K6KK]

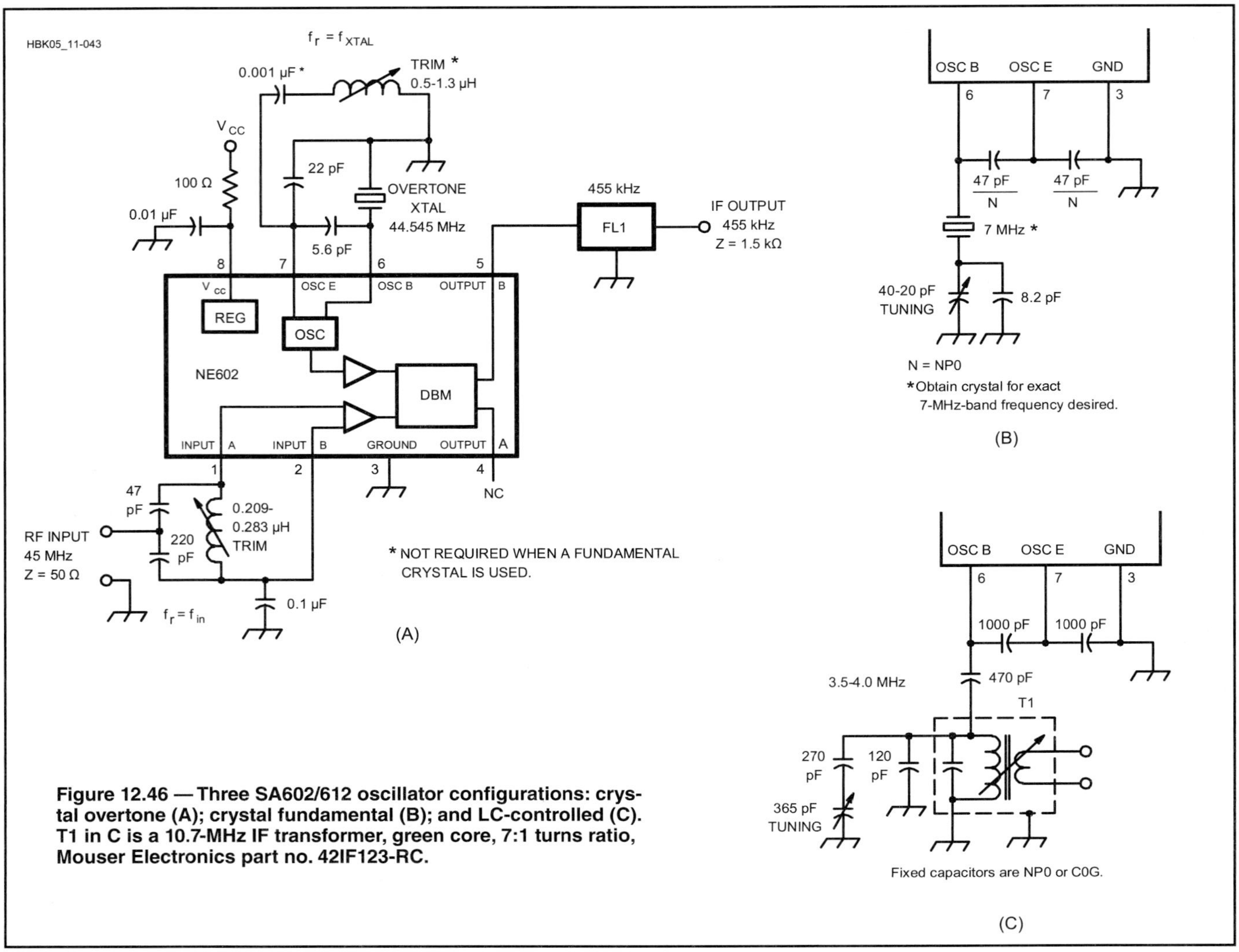

Figure 12.46 — Three SA602/612 oscillator configurations: crystal overtone (A); crystal fundamental (B); and LC-controlled (C). T1 in C is a 10.7-MHz IF transformer, green core, 7:1 turns ratio, Mouser Electronics part no. 42IF123-RC.

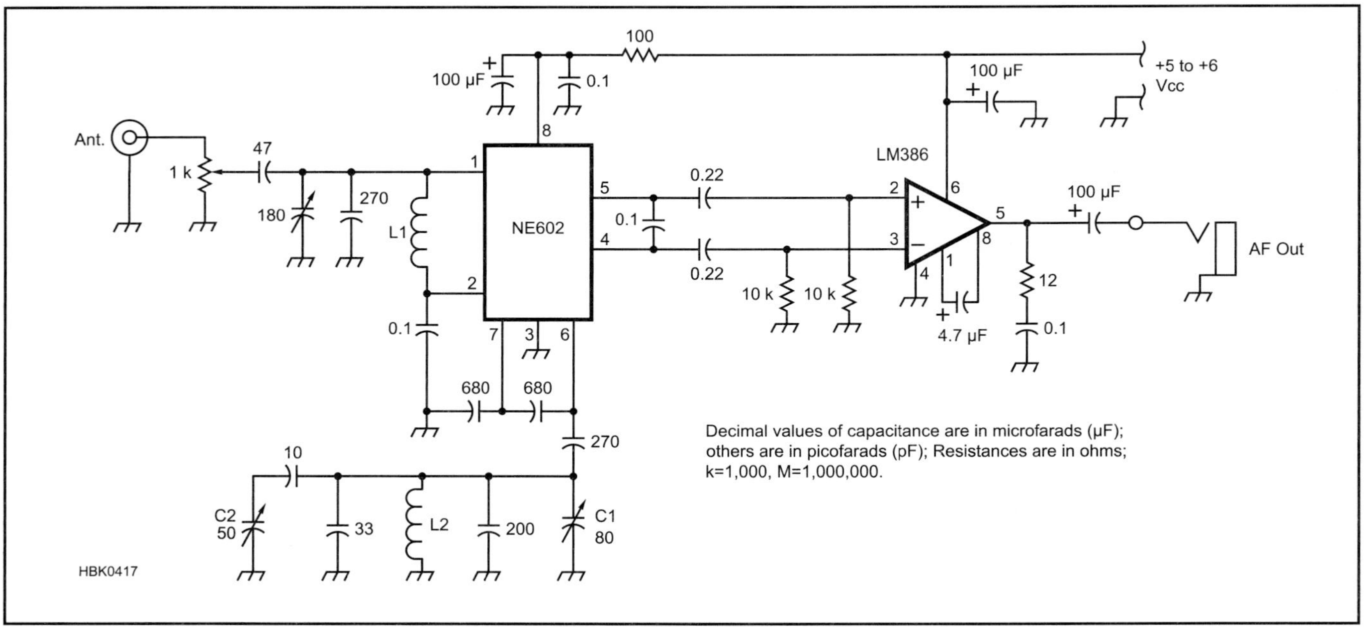

Figure 12.47 — A 7-MHz direct-conversion receiver based on the NE602/SA602/612. Equipped with a stage or two of audio filtering and a means of muting during transmit periods, such a receiver is entirely sufficient for basic Amateur Radio communication at MF and HF. L1 and L2 are 1.2 µH. This receiver is described in greater detail in *Experimental Methods in RF Design.*

12.5 Demodulation and Detection

Translating information from radio form back into its original form — demodulation — is also traditionally called *detection.* If the information signal we want to detect consists merely of a baseband signal frequency-shifted into the radio realm, almost any low-distortion frequency-shifter that works according to the sidebar "Mixer Math: Mixing as Multiplication" can do the job acceptably well.

Sometimes we recover a radio signal's information by shifting the signal back to its original form with no intermediate frequency shifts. This is direct conversion. More commonly, we first convert a received signal to an intermediate frequency so we can amplify, filter and level-control it prior to detection. This is superheterodyne reception. Whatever the receiver type, however, the received signal ultimately makes its way to one last mixer or demodulator (analog or digital) that completes the final translation of information back into audio, video, or a signal form suitable for device control or computer processing.

In a superheterodyne receiver's last translation, the incoming signal is converted back to recovered information by mixing it with one last RF signal. In heterodyne or product detection, that final frequency-shifting signal comes from a BFO. The incoming signal's energy goes into one mixer input port, BFO energy goes into the other, and audio (or whatever form the desired information takes) results. In SDR receivers, the process may involve the digital equivalent of the analog process or some other mathematical process may be used.

12.5.1 Envelope Detection and Full-Carrier AM

If the incoming signal is full-carrier AM and we do not need to hear the carrier as a tone, we can modify this process somewhat. We can use the carrier itself to provide the heterodyning energy in a process called *envelope detection.*

A full-carrier AM signal's modulation envelope corresponds to the shape of the modulating wave. If we can derive from the modulated signal a voltage that varies

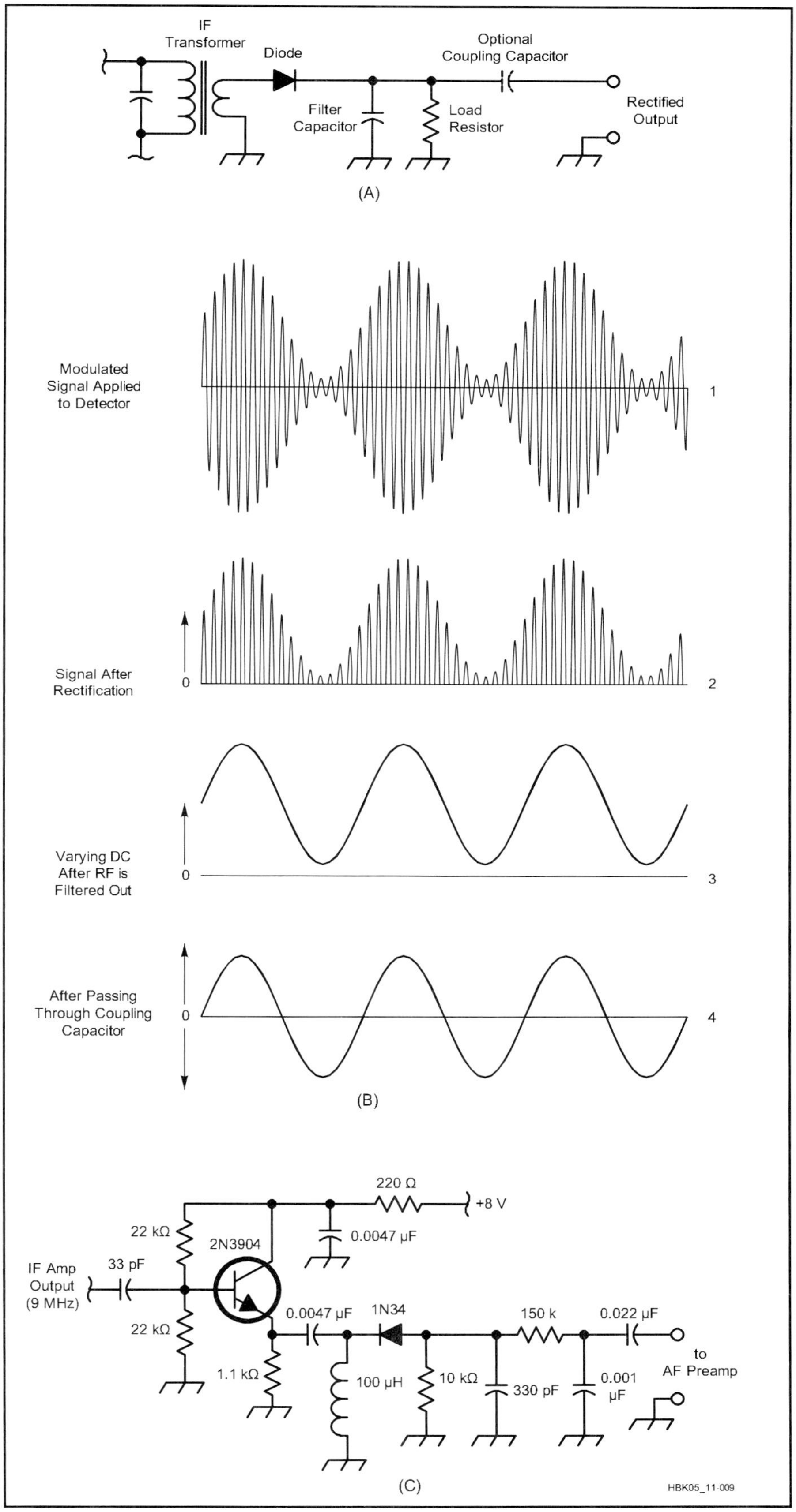

Figure 12.48 — Radio's simplest demodulator, the diode rectifier (A), demodulates an AM signal by acting as a switch that multiplies the carrier and sidebands to produce frequency sums and differences, two of which sum into a replica of the original modulation (B). Modern receivers often use an emitter follower to provide low-impedance drive for their diode detectors (C).

according to the modulation envelope, we have successfully recovered the information present in the sidebands. This process is called envelope detection, and we can achieve it by doing nothing more complicated than half-wave-rectifying the modulated signal with a diode (**Figure 12.48**).

That a diode demodulates an AM signal by allowing its carrier to multiply with its sidebands may jar those long accustomed to seeing diode detection ascribed merely to "rectification." But a diode is certainly nonlinear. It passes current in only one direction, and its output voltage is (within limits) proportional to the square of its input voltage. These nonlinearities allow it to multiply.

Exploring this mathematically is tedious with full-carrier AM because the process squares three summed components (carrier, lower sideband, and upper sideband). Rather than fill the better part of a page with algebra, we'll instead characterize the outcome verbally: In "just rectifying" a DSB, full-carrier AM signal, a diode detector produces

- Direct current (the result of rectifying the carrier);
- A second harmonic of the carrier;
- A second harmonic of the lower sideband;
- A second harmonic of the upper sideband;
- Two difference-frequency outputs (upper sideband minus carrier, carrier minus lower sideband), each of which is equivalent to the modulating waveform's frequency, and both of which sum to produce the recovered information signal; and
- A second harmonic of the modulating waveform (the frequency difference between the two sidebands).

Three of these products are RF signals. Low-pass filtering, sometimes little more than a simple RC network, can remove the RF products from the detector output. A capacitor in series with the detector output line can block the carrier-derived dc component. That done, only two signals remain: the recovered modulation and, at a lower level, its second harmonic — in other words, second-harmonic distortion of the desired information signal.

12.5.2 Detecting AM Signals

Note that the shape of the envelope of the modulated RF signal matches the shape of the modulating signal. (This is only true for full AM including the carrier and both sidebands. The envelope of a single-sideband SSB signal is not an accurate reproduction of the modulating signal, which is why it cannot be recovered by an envelope detector.) That suggests a possible demodulation method. A diode detector puts out a signal proportional to the envelope of the RF signal, recovering the original modulation. See **Figure 12.49**. The capacitor should be large enough so that it filters out most of the RF ripple but not so large that it attenuates the higher audio frequencies.

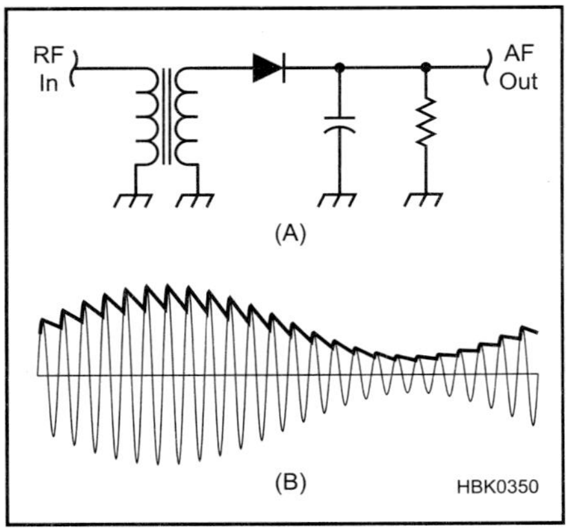

Figure 12.49 — A simple diode-type AM detector, also known as an *envelope detector* (A). The demodulated output waveform has the same shape as the envelope of the RF signal (B). In (B) the thin line is the RF signal modulated with a sine wave and the darker line is the demodulated audio frequency with some residual RF ripple.

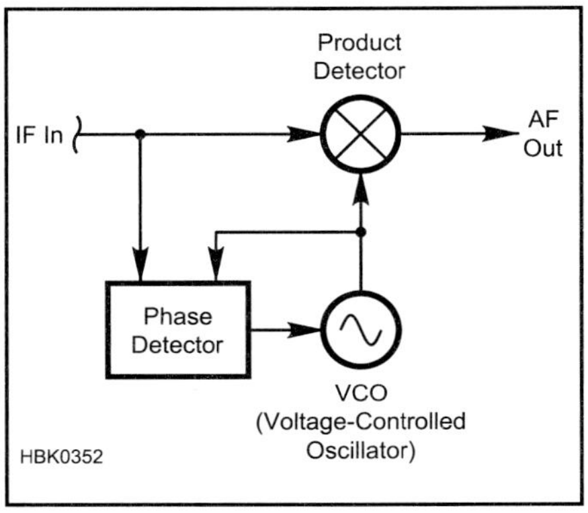

Figure 12.50 — Block diagram of a synchronous detector. The voltage-controlled oscillator (VCO) is part of a phase-locked loop that locks the oscillator to the carrier frequency of the incoming AM or DSBRC signal.

One problem with AM is that if the amplitude of the carrier becomes attenuated for any reason, then the modulation is distorted, especially the negative-going portion near the 100% modulation (zero power) point. This can happen due to a propagation phenomenon called *selective fading.* It occurs when the signal arrives simultaneously at the receive antenna via two or more paths, such as ground wave and sky wave. If the difference in the distance of the two paths is an odd number of half-wavelengths, then the two signals are out of phase. If the amplitudes are nearly the same, they cancel and a deep fade results. Since wavelength depends on frequency, the fading is frequency-selective. On the lower-frequency amateur bands it is possible for an AM carrier to be faded while the two sidebands are still audible.

A solution is to regenerate the carrier in the receiver. Since the carrier itself carries no information about the modulation, it is not necessary for demodulation. The transmitted signal may be a standard full-carrier AM signal or the carrier may be suppressed, resulting in double sideband, suppressed carrier (DSB-SC). An AM detector that regenerates the carrier from the signal is known as a synchronous detector. Often the regenerated carrier oscillator is part of a phase-locked loop (PLL) that locks onto the incoming carrier. See **Figure 12.50.** Synchronous detectors not only reduce the effects of selective fading but also are usually more linear than diode detectors, so they have less distortion. Some commercial short-wave broadcasts include a reduced, but not suppressed, carrier (DSB-RC) to allow operation with PLL-type synchronous detectors.

Since radio amateurs started using single sideband in the 1960s, full-carrier double-sideband AM has become less prevalent in Amateur Radio. It does retain several advantages however. We have already mentioned the simplicity of the circuitry. Another advantage is that, because of the presence of the carrier, the automatic gain control system in the receiver remains engaged at all times, ensuring a constant audio level. Unlike with SSB, there is no rush of noise during every pause in speech. Also tuning is less critical than with SSB. There is no "Donald Duck" sound if the receiver is slightly mistuned. Finally, the audio quality of an AM signal is usually better than that of SSB because of the lack of a crystal filter in the transmit path and the wider filter in the receiver.

SDR AM SQUARE-LAW DETECTOR

(This section is taken from the July/August 2009 QEX column SDR: Simplified by W5IFS.) We need to convert the full-carrier double sideband signal into the baseband information. One of the classic AM demodulation methods is a square-law detector. Very few devices have a true square law response over a large range of signals. The JFET comes close, but has limited range. With DSP we can implement a square law detector with range that is limited only by the size of the data that the DSP can handle. If we use 16-bit data for the input, our mul-

tiplier must have a 32-bit result. This is a very simple and fast operation for the DSP.

The square-law response to a DSB-AM signal with a carrier, sin(y), and two sidebands, cos(x-y) and cos(x+y) is as follows:

$$V_{out} = \left(\sin(y) + ½\cos(x-y) - ½\cos(x+y)\right)^2$$

Applying trig identities and working through the math:

$$V_{out} = \sin(y)^2 + \sin(y)\cos(x-y) - \sin(y)$$
$$\cos(x+y) + \frac{1}{4}\cos(x+y)^2 + \frac{1}{4}\cos(x-y)^2$$
$$V_{out} = \frac{3}{4} - \frac{1}{2}\sin(2y) + \frac{1}{2}\sin(x) +$$
$$\frac{1}{2}\sin(2y-x) - \frac{1}{2}\sin(-x) - \frac{1}{2}\sin(2y+x) +$$
$$\frac{1}{8}\cos(2y+2x) + \frac{1}{8}\cos(2y-2x)$$

Finally:

$$V_{out} = \frac{3}{4} + \sin(x) - \frac{1}{2}\sin(2y) + \frac{1}{2}\sin(2y-x) -$$
$$\frac{1}{2}\sin(2y+x) + \frac{1}{8}\cos(2y+2x) + \frac{1}{8}\cos(2y-2x)$$

(This result has 32 bits, but our filter math should only be 16 bits. We can divide the result by 65536 to get the signal back into range for further operations. In software, this just means we throw away the bottom 16 bits of data before we store the result.)

The signal we want is sin(x), but we have a significant dc component (¾) and a whole bunch of unwanted signals at or close to twice the carrier frequency, 2y. The only real problem is the dc component, since it is close to the desired audio signal. We need to use a bandpass filter to remove the dc and the RF components. Since all but the dc are far removed in frequency from the desired audio, we can use a relatively simple bandpass filter.

12.5.3 Demodulating SSB Signals

A complete discussion of the three main techniques used to receive and demodulate SSB signals, filter, phasing, and Weaver, can be found in the **Modulation** chapter. Examples of practical circuits that implement these methods are included in the receiver projects provided in the online content.

SUPERHETERODYNE RECEIVERS

As with an AM synchronous detector, the carrier is regenerated in a superheterodyne receiver but since only one sideband is present, the synchronous detector's phase-locked loop (described above) does not work. Instead, the detector uses a free-running *beat-frequency oscillator* (BFO). The detector itself is called a product detector because its output is the mathematical product of the BFO and the SSB signal. The BFO must be tuned to the same frequency as the suppressed carrier to prevent distortion of the recovered audio. That is done by carefully tuning the local oscillator (the main tuning dial) such that after conversion to the intermediate frequency, the suppressed carrier aligns with the BFO frequency. Demodulation of SSB signals in an analog heterodyne receiver must be done using a product detector or equivalent technique. Envelope detection does not work because the waveform does not have the shape of the modulating waveform.

SDR RECEIVERS

SDR receivers use I/Q quadrature demodulator architecture for SSB and many other modes. This technique is discussed in the **Modulation** chapter. A block diagram of an SDR demodulator for SSB is shown in **Figure 12.51**. This particular system shows the phasing method of SSB generation in reverse. The excellent tutorial, "Understanding the 'Phasing Method' of Single Sideband Demodulation" by Rick Lyons (**dsprelated.com/showarticle/176.php**) explains the technique in detail. The filter method and the Weaver method can also be implemented with an I/Q architecture, although this is not the usual method used by designers.

A quadrature I/Q demodulator generates the I_1 and Q1 signals which are low-pass filtered to remove high-frequency components, then down-converted using decimation. The I and Q channels are bandpass filtered, with a Hilbert transformer applied to the Q channel as described in the **DSP and SDR Fundamentals** chapter. The resulting streams are then combined so that the Q channel is subtracted from the I channel. This leaves the USB signal, which is con-

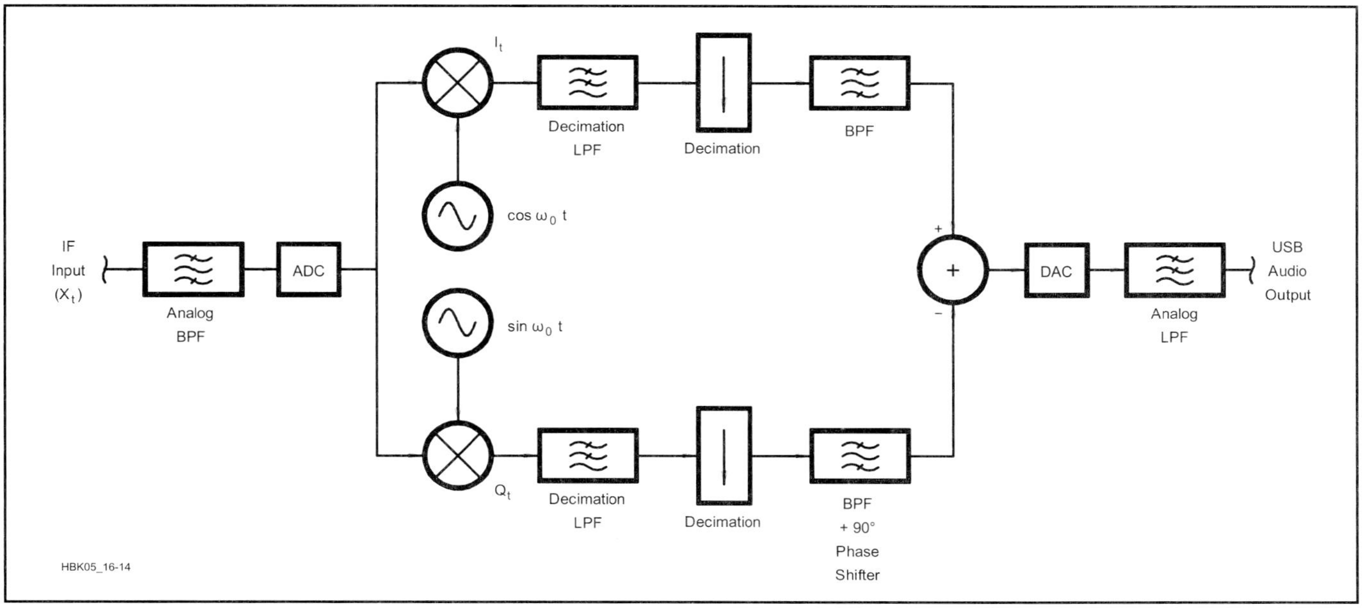

Figure 12.51 — Block diagram of a digital SSB demodulator.

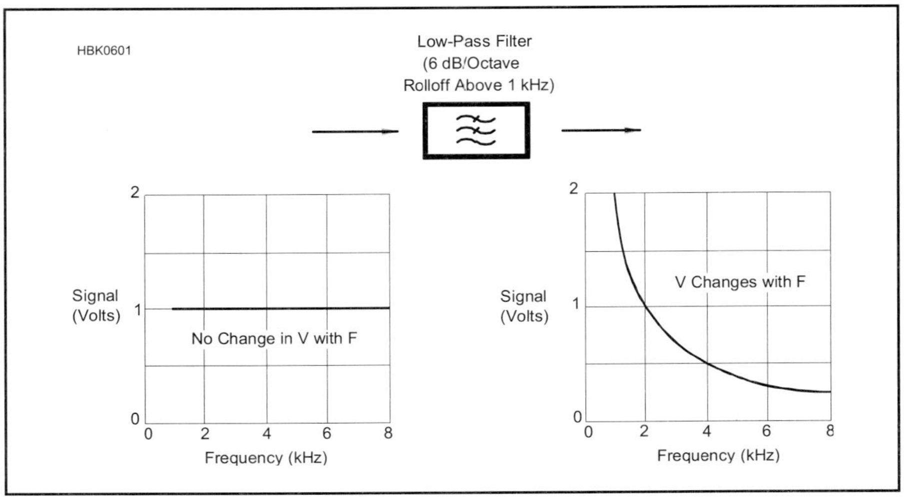

Figure 12.52 — Frequency-sweeping a constant-amplitude signal and passing it through a low-pass filter results in an output signal that varies in amplitude with frequency. This is the principle behind the angle-demodulation process called *frequency discrimination.*

verted to analog audio, and then low-pass filtered.

If an LSB signal is being received, change the summing symbol immediately preceding the DAC to add the I and Q channels together. This produces the LSB signal, which is then converted to audio and filtered as before. In practice, both the LSB and USB signals are available at any time and the digital audio stream can be routed to the DAC and low-pass filter under software control.

12.5.4 Demodulating FM and PM

ANALOG DEMODULATORS

Although angle modulation does generate an infinite number of sidebands, demodulating angle modulation requires little more than turning it into AM and then envelope- or product-detecting it! This is what happens in many of our FM receivers and transceivers, and we can get a handle on this process by realizing that a form of angle-modulation-to-AM conversion is created from linear distortion of the modulation by amplitude-linear circuitry. This happens to angle-modulated signals in any linear circuit that does not have an amplitude-versus-frequency response that's utterly flat out to infinity.

Think of what happens, for example, when we sweep a constant-amplitude signal up in frequency — say, from 1 kHz to 8 kHz — and pass it through a 6 dB per octave filter (**Figure 12.52**). The filter's rolloff causes the output signal's amplitude to decrease as frequency increases. Now imagine that we linearly sweep our constant-amplitude signal *back and forth* between 1 kHz and 8 kHz at a constant rate of 3 kHz per second. The filter's output *amplitude* now var-

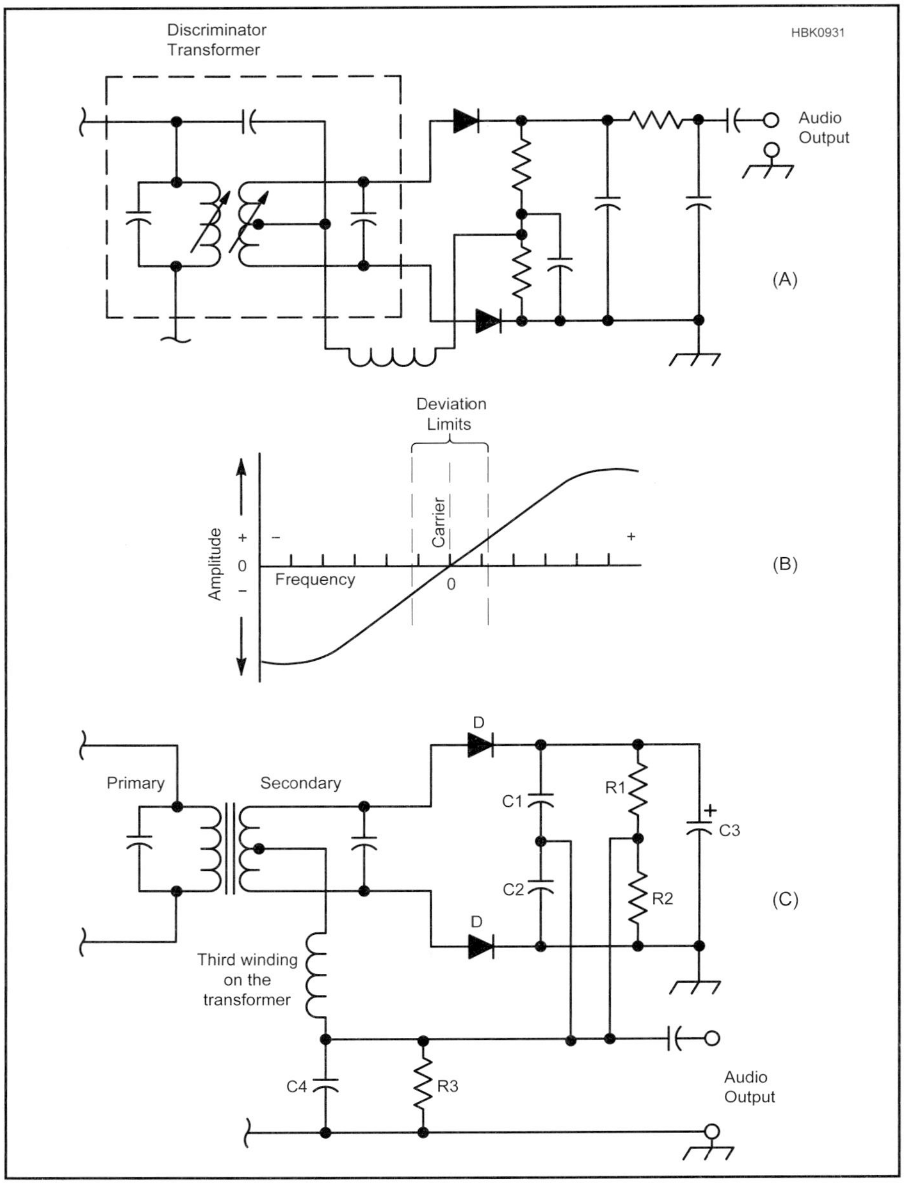

Figure 12.53 — A *discriminator* (A) converts an angle-modulated signal's deviation into an amplitude variation (B) and envelope-detects the resulting AM signal. For undistorted demodulation, the discriminator's amplitude-versus-frequency characteristic must be linear across the input signal's deviation. A *crystal discriminator* uses a crystal as part of its frequency-selective circuitry. The *ratio detector* at (C) operates similarly but has an improved rejection of amplitude modulated noise.

ies cyclically over time as the input signal's *frequency* varies cyclically over time. Right before our eyes, a frequency change turns into an amplitude change. The process of converting angle modulation to amplitude modulation has begun.

Here is what happens whenever an angle-modulated signal passes through circuitry with an amplitude-versus-frequency response that is not flat out to infinity. As the signal deviates across the frequency-response curves of whatever circuitry passes it, its angle modulation is, to some degree, converted to AM, a form of crosstalk between the two modulation types, if we wish to look at it that way. (Variations in system phase linearity also cause distortion and FM to AM conversion, because the sidebands do not have the proper phase relationship with respect to each other and with respect to the carrier.)

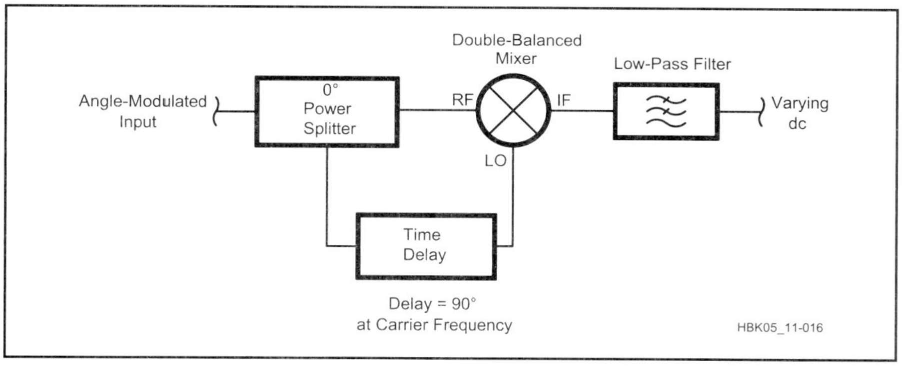

Figure 12.54 — In *quadrature detection*, an angle-modulated signal multiplies with a time-delayed copy of itself to produce a dc voltage that varies with the amplitude and polarity of its phase or frequency excursions away from the carrier frequency. A practical quadrature detector can be as simple as a 0° power splitter (that is, a power splitter with in-phase outputs), a diode double-balanced mixer, a length of coaxial cable λ/4 (electrical) long at the carrier frequency, and a bit of low-pass filtering to remove the detector output's RF components. IC quadrature detectors achieve their time delay with one or more resistor-loaded tuned circuits (Figure 12.55).

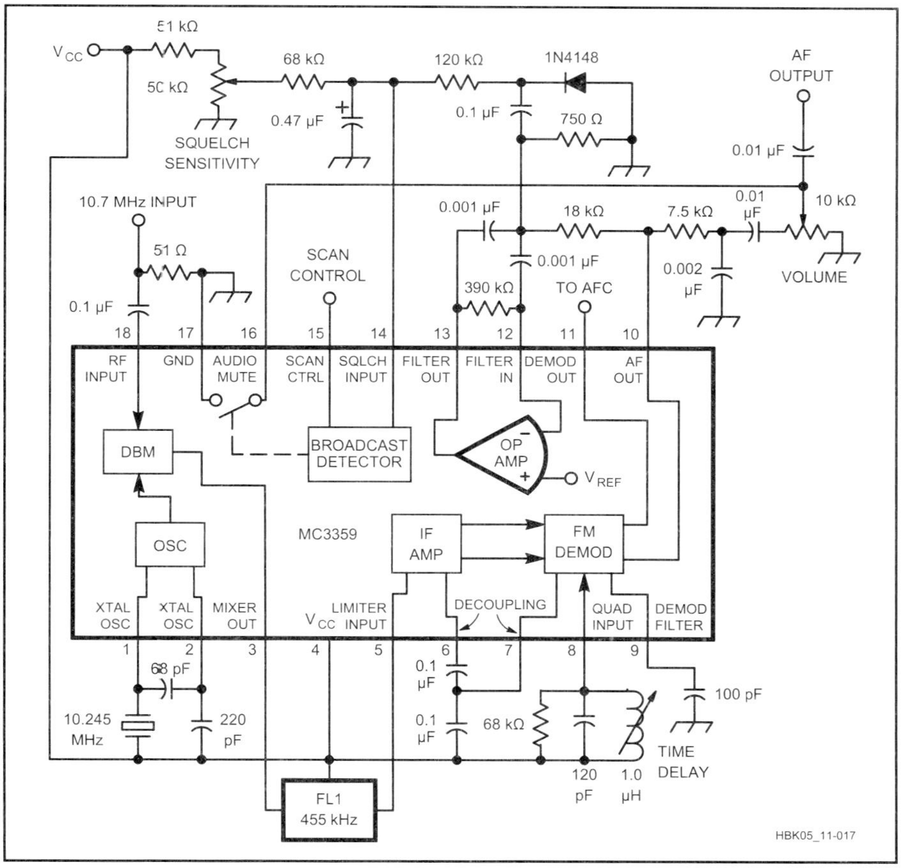

Figure 12.55 — The obsolete Freescale MC3359 (equivalent NTE680 is still in production) is one of many FM subsystem ICs that include limiter and quadrature-detection circuitry. The TIME DELAY coil is adjusted for minimum recovered-audio distortion.

All we need to do to put this effect to practical use is develop a circuit that does this frequency-to-amplitude conversion linearly across the frequency span of the modulated signal's deviation. Then we envelope-demodulate the resulting AM, and we have achieved angle demodulation.

Figure 12.53 shows such a circuit, a Foster-Seeley discriminator, and the sort of amplitude-versus-frequency response we expect from it **(electronics-notes.com/articles/radio/modulation/fm-frequency-demodulation-foster-seeley-detector-discriminator.php).** It's actually possible to use an AM receiver to recover understandable audio from a narrow angle-modulated signal by "off-tuning" the signal so its deviation rides up and down on one side of the receiver's IF selectivity curve. This slope detection process served as an early, suboptimal form of frequency discrimination in receivers not designed for FM. It is always worth trying as a last-resort means of receiving narrowband FM with an AM receiver.

The ratio detector in Figure 12.53C is a variation on the discriminator which is better at rejecting AM noise mixed with the FM signal (**electronics-notes.com/articles/radio/modulation/fm-frequency-demodulation-ratio-detector-discriminator.php**). Note that the diodes are in series in this circuit. A similar phase shift occurs in the third transformer winding as the signal moves away from the frequency of the tuned transformer winding. This causes an imbalance in the transformer secondary and current flows in the third winding. That signal is then filtered by C4 and R3 before being passed to the audio stages of the receiver.

QUADRATURE DETECTORS

It is also possible to demodulate an angle-modulated signal merely by multiplying it with a time-delayed copy of itself in a double-balanced mixer, as shown in **Figure 12.54;** the sidebar, "Mixer Math: Quadrature Demodulation," explains the process mathematically.

An ideal quadrature detector puts out 0 V dc when no modulation is present (with the carrier at f_c). The output of a real quadrature detector may include a small dc offset that requires compensation. If we need the detector's response all the way down to dc, we have it; if not, we can put a suitable blocking capacitor in the output line for ac-only coupling.

Quadrature detection is more common than frequency discrimination in current receivers because it does not require a special discriminator transformer or resonator, and because the necessary balanced-detector circuitry can easily be implemented in IC structures along with limiters and other receiver circuitry. The NXP Semiconductor SA604A and SA614A FM IF ICs (both still in production by NXP Semiconductor, **www.nxp.com**) are examples of this. **Figure 12.55** shows another, the obsolete Freescale (formerly Motorola) Semiconductor MC3359 (equivalent, NTE860, **www.nteinc.com**).

AMPLITUDE LIMITING

FM radio communication systems are superior to AM in their ability to suppress and ignore static, manmade electrical noise and (through a characteristic called *capture effect)* co-channel signals sufficiently weaker than the desired signal. AM noise immunity and capture effect are not intrinsic to angle modulation, however; they must be designed into the angle-modulation receiver in the form of signal amplitude *limiting.*

The amplitude of a quadrature detector's input signal affects the amplitude of a quadrature detector's three output signals. A quadrature detector therefore responds to AM, and so does a frequency discriminator. To achieve FM's excellent amplitude noise immunity, then, these angle demodulators must be preceded by limiter circuitry that removes all amplitude variations from the incoming signal.

SDR ANGLE DEMODULATION

A popular technique for angle demodulation starts with a phase detector as shown in **Figure 12.56.** The angle-modulated signal is down-converted to a convenient frequency and input to a pair of multipliers where it is mixed with a pair of constant frequency LO signals in quadrature. Low-pass filters then remove signals at the sum frequency, leaving only the I and Q baseband signals.

In PM demodulation, the phase comparison is an absolute comparison against the carrier phase. The phase angle of the input signal, relative to the LO center frequency, can be determined from the I and Q signals as:

$$\varphi = \tan^{-1}\left(\frac{V_Q}{V_I}\right)$$

If the signal is a PM signal, the phase signal contains the original modulation and can be filtered and output as audio.

FM demodulation requires additional steps as a special case of PM. The change in angle from the unmodulated carrier is constantly increasing or decreasing and exceeds ±90° as a normal part of operation. For that reason, a simple phase comparator using the recovered carrier will not demodulate FM. FM requires that we measure the rate-of-change of phase and sum those incremental phase changes.

Since frequency is defined as the rate-of-change of phase, it is necessary to differentiate the phase signal to recover the FM modulation signal. Implementing differentiation in DSP involves subtracting each sample from the previous sample. It is necessary to account for the point at which phase passes through 360º and resets to 0º. This can be done through special routines or by scaling of fixed-point data in the computations so that the phase rollover and integer rollover (65535 to 0, for example, in 16-bit math) coincide.

The FM signal will also need a de-emphasis filter (see the **Modulation and Transmitting** chapters) to cancel the high-frequency gain that was added during the FM process by the modulator.

Mixer Math: Mixing as Multiplication

Since a mixer works by means of multiplication, a bit of math can show us how they work. To begin with, we need to represent the two signals we'll mix, A and B, mathematically. Signal A's instantaneous amplitude equals

$$A_a \sin 2\pi f_a t$$

in which A is peak amplitude, f is frequency, and t is time. Likewise, B's instantaneous amplitude equals

$$A_b = A \sin(2\pi f_b t)$$

Since our goal is to show that multiplying two signals generates sum and difference frequencies, we can simplify these signal definitions by assuming that the peak amplitude of each is 1. The equation for Signal A then becomes

$$a(t) = A \sin(2\pi f_a t)$$

and the equation for Signal B becomes

$$b(t) = B \sin(2\pi f_b t)$$

Each of these equations represents a sine wave and includes a subscript letter to help us keep track of where the signals go.

Merely combining Signal A and Signal B by letting them travel on the same wire develops nothing new:

$$a(t) + b(t) = A \sin(2\pi f_a t) + B \sin(2\pi f_b t)$$

As simple as that equation may seem, we include it to highlight the fact that multiplying two signals is a quite different story. From trigonometry, we know that multiplying the sines of two variables can be expanded according to the relationship

$$\sin x \sin y = \frac{1}{2}\left[\cos(x - y) - \cos(x + y)\right]$$

Conveniently, Signals A and B are both sinusoidal, so we can use this relationship to determine what happens when we multiply Signal A by Signal B. In our case, $x = 2\pi f_a t$ and $y = 2\pi f_b t$, so plugging them into the last equation gives us

$$a(t) \times b(t) = \frac{AB}{2}\cos\left(2\pi\left[f_a - f_b\right]t\right) - \frac{AB}{2}\cos\left(2\pi\left[f_a + f_b\right]t\right)$$

Now we see two momentous results: a sine wave at the frequency *difference* between Signal A and Signal B, $2\pi(f_a - f_b)t$, and a sine wave at the frequency *sum* of Signal A and Signal B, $2\pi(f_a + f_b)t$. (The products are cosine waves, but since equivalent sine and cosine waves differ only by a phase shift of 90°, both are called *sine waves* by convention.)

This is the basic process by which we translate information into radio form and translate it back again. If we want to transmit a 1-kHz audio tone by radio, we can feed it into one of our mixer's inputs and feed an RF signal — say, 5995 kHz — into the mixer's other input. The result is two radio signals: one at 5994 kHz (5995 – 1) and another at 5996 kHz (5995 + 1). We have performed modulation.

Converting these two radio signals back to audio is just as straightforward. All we do is feed them into one input of another mixer, and feed a 5995-kHz signal into the mixer's other input. Result: a 1-kHz tone. We have performed demodulation.

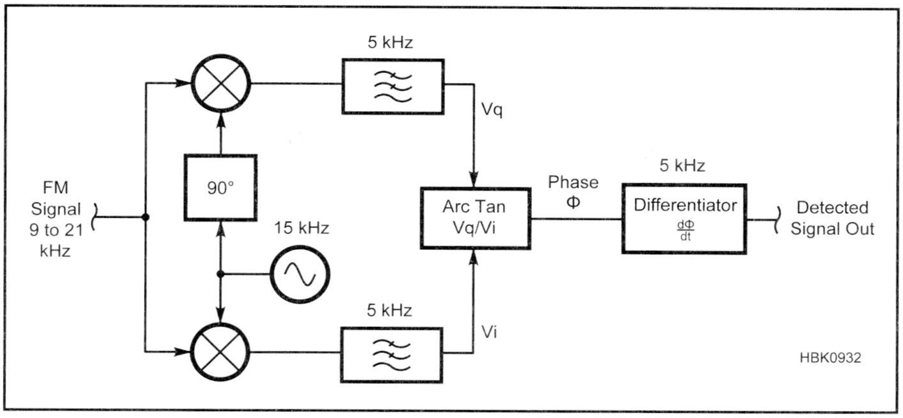

Figure 12.56 — An FM detector using an arctangent phase detector and a differentiator.

12.6 Automatic Gain Control (AGC)

The amplitude of the desired signal at each point in the receiver is generally controlled by the AGC system, although manual control is usually provided as well. Each stage has a distortion versus signal level characteristic that must be known, and the stage input level must not become excessive. The signal being received has a certain signal-to-distortion ratio that must not be degraded too much by the receiver. For example, if an SSB signal has –30 dB distortion products the receiver should not create additional distortion greater than –40 dB with respect to the desired signal. The correct AGC design ensures that each stage gets the right input level. It is often necessary to redesign some stages to accomplish this task.

While this section deals mostly with AGC in the guise of analog circuits, the same function is also implemented digitally in DSP and SDR receivers. The goal of both is the same: to maintain a signal level at all stages of the receiver that is neither too large nor too small so that the various processing systems operate properly. Whether or not the AGC offset and time constant are implemented by an analog component or by a microprocessor output is immaterial. The point is to manage the RF amplifier gain so that the overall receiver behavior is satisfactory.

AGC in the receive signal chain of an SDR should not be confused with AGC in an analog receiver. AGC in the analog receiver will adjust the fixed dynamic range of the receiver up or down, altering the weak and strong signal performance of the radio dynamically.

The effects of an improperly operating AGC system can be either subtle or nearly disabling to a receiver and vary with how the AGC system is constructed. This section attempts to describe the requirements for proper operation and provides some examples of implementation and common AGC failures in terms of analog circuitry which is somewhat easier to describe than software algorithms, noting that similar behaviors exist even in purely software receivers. The interested reader should consider studying the AGC systems of commercial receivers to understand how professional design teams deal with the problem of managing so much gain with such stringent requirements for linearity and distortion.

THE AGC LOOP

Figure 12.57A shows a typical AGC loop that is often used in amateur superheterodyne receivers. The AGC is applied through RF decoupling circuits that prevent the stages from interacting with each other. The AGC amplifier helps to provide enough AGC loop gain so that the gain-control characteristic of Figure 12.57B is achieved. If effective, the AGC system causes the receiver to act as a compression amplifier with lower overall gain for stronger signals.

The AGC action does not begin until a certain signal level, called the AGC threshold, is reached. The THRESHOLD VOLTS input in Figure 12.57A serves this purpose. After that level is exceeded, the audio level increases more slowly than for weaker signals. The audio rise beyond the threshold value is usually in the 5 – 10 dB range. Too much or too little audio rise are both undesirable for most operators.

As an option, the AGC signal to the RF amplifier is offset by the 0.6 V forward drop of the diode so that the RF gain does not start to decrease until larger signals appear. This prevents a premature increase of the receiver noise figure. Also, a time constant of one or two seconds after this diode helps keep the RF gain steady over the short term.

Figure 12.58 is a typical plot of the signal levels at the various stages of a certain ham band receiver using analog circuitry. Each stage has the proper level and a 115 dB change in input level produces a 10 dB change in audio level. A manual gain control could produce the same effect.

AGC PUMPING

AGC *pumping* can occur in receivers in which the AGC measurement point is located ahead of the stages that determine operating bandwidth, such as when an audio filter is added to a receiver externally and outside the reach of the AGC system. If the weak signal is the only signal within the first IF passband, the AGC will cause the receiver to be at maximum gain and optimum SNR. If an interfering signal is within the first IF passband, but outside the audio DSP filter's passband, we will not hear the interfering signal, but it will enter the AGC system and reduce the gain so that we might not hear our desired weak signal. AGC pumping is audible as sudden reductions in signal strength without a strong signal in the passband of the receiver.

AGC pumping is generally not as much of a problem in SDR receivers as it is in superheterodyne receivers, but the phenome-

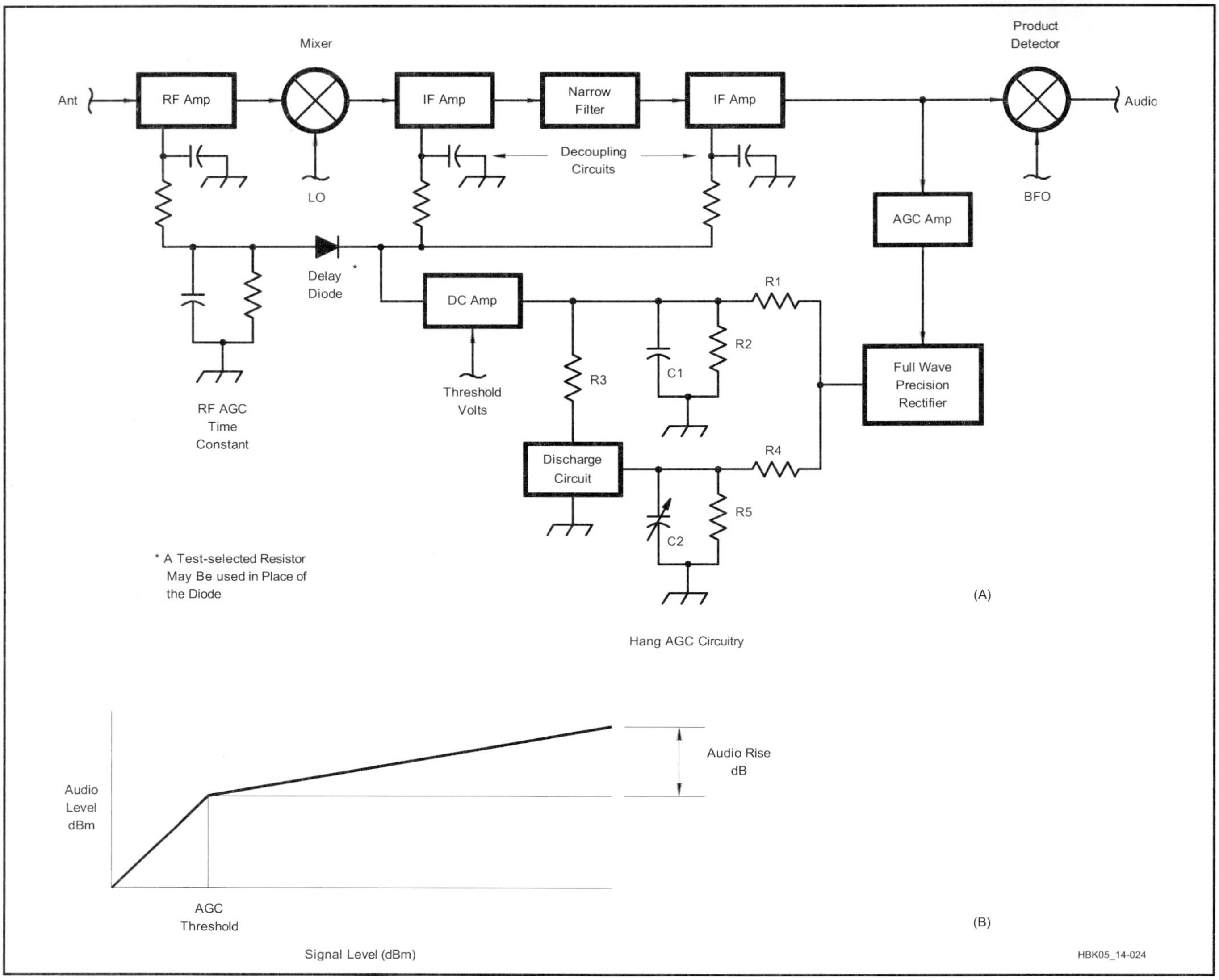

Figure 12.57 — AGC principles. (A): typical superhet receiver with AGC applied to multiple RF and IF stages. (B): audio output level as a function of antenna signal level.

non still exists. The severity depends on the algorithm employed by the SDR and the operating configuration controlled by the user.

AGC TIME CONSTANTS

There are two primary AGC time constants. AGC *attack time* describes the time it takes the AGC system to respond to the presence of a signal. AGC *decay time* describes the response of the AGC system to changes in a signal that is present. The optimum time constants for the AGC system depends on the type of signal being received, the type of operation being conducted, and the operator's preference.

The operator usually has a control that allows for setting the time that it takes the AGC to recover or decay. If we are listening to two relatively loud stations converse, we may set the AGC to slow. Then as each station stops transmitting, the noise floor is slow to rise and we often hear the next station before the noise floor is again heard. In contrast, we may want to set the AGC decay time to be fast, which allows us to hear a weak station immediately after the strong station.

In Figure 12.57A, following the precision rectifier, R1 and C1 set an attack time, to prevent excessively fast application of AGC. One or two msec is a good value for the R1 × C1 product. If the antenna signal suddenly disappears, the AGC loop is opened because the precision rectifier stops conducting. C1 then discharges through R2 and the C1 × R2 product can be in the range of 100 to 200 msec. At some point the rectifier becomes active again, and the loop is closed again.

An optional modification of this behavior is the hang AGC circuit. If we make R2 × C1 much longer, say 3 seconds or more, the AGC voltage remains almost constant until the R5-C2 circuit decays with a switch selectable time constant of 100 to 1,000 msec. At that time R3 quickly discharges C1 and full receiver gain is quickly restored. This type of control is appreciated by many operators because of the lack of AGC pumping that could be caused by modulation, rapid fading, and other sudden signal level changes.

In an SDR receiver, the typical AGC algorithm has a fast-attack and slow-decay that is adjustable. The AGC algorithm has a gain value that it applies to the receiver buffers, which persists from one buffer to the next. If a loud signal appears while the gain is set high, it immediately lowers the gain to prevent a loud sound in our headphones. Once the sound has subsided, the gain is allowed to slowly increase back to its previous value.

AGC LOOP RESPONSE PROBLEMS

If the various stages have the property that each 1 V change in AGC voltage modifies the gain by a constant amount (in dB), the AGC loop is said to be *log-linear* and regular feedback principles can be used to analyze and design the loop. But there are some difficulties that complicate this textbook model. One has already been mentioned. When the signal is rapidly decreasing the loop becomes open and the various capacitors discharge in an open loop manner. As the signal is increasing beyond the threshold, or if it is decreasing slowly enough, the feedback theory applies more accurately.

In SSB and CW receivers, rapid changes are the rule and not the exception. It is important that the AGC loop not overshoot or ring when the signal level rises past the threshold. The idea is to design the ALC loop to be stable when the loop is closed. It obviously will not oscillate when open (during decay time). But the loop must have smooth and consistent transient response when the loop goes from open to closed state.

Another problem involves the narrow bandpass analog IF filters. The group delay of analog filters constitutes a time lag in the loop that can make loop stabilization difficult. Moreover, these filters nearly always have much greater group delay at the edges of the passband, so that loop problems are aggravated at these frequencies. Overshoots and undershoots, called *gulping*, are very common. Compensation networks that advance the phase of the feedback help to offset these group delays. The design problem arises because some of the AGC is applied before the filter and some after the filter. It is a good idea to put as much fast AGC as possible after the filter and use a slower decaying AGC ahead of the filter. The delay diode and RC in Figure 12.57A are helpful in that respect. Complex AGC designs using two or more compensated loops are also in the literature. If a second cascaded narrow filter is used in the IF it is usually a lot easier to leave the second or downstream filter out of the AGC loop at the risk of allowing AGC pumping, as described in the preceding section.

Another problem is that the control characteristic is often not log-linear. For example, dual-gate MOSFETs tend to have much larger dB/V at large values of gain reduction. Many IC amplifiers have the same problem. The result is that large signals cause instability because of excessive loop gain. Variable gain op amps and other similar ICs are available that are intended for gain control loops.

Audio frequency components on the AGC bus can cause problems because the amplifier gains are modulated by the audio and distort the desired signal. A hang AGC circuit (essentially a low-pass filter) can reduce or eliminate this problem.

Finally, if we try to reduce the change in audio levels to a very low value, the required loop gain becomes very large, and stability problems become very difficult. It is much better to accept a 5 – 10 dB variation of audio output.

Because many parameters are involved and many of them are not strictly log-linear, it is best to achieve good AGC performance through an initial design effort and finalize the design experimentally. Use a signal generator, attenuator, and a signal pulser (2 msec rise and fall times, adjustable pulse rate and duration) at the antenna and a synchronized oscilloscope to look at the IF envelope. Tweak the time constants and AGC distribution by means of resistor and capacitor decade substitution boxes. Be sure to test throughout the passband of each filter. The result should be a smooth and pleasant sounding SSB/CW response, even with maximum RF gain and strong signals. Patience and experience are helpful.

12.6.1 Audio-Derived AGC

Some receivers, especially direct-conversion types, use audio-derived AGC. There are problems with this approach as well. At low audio frequencies the AGC control action can be slow to develop. That is, low-frequency audio sine waves take longer to reach their peaks than the AGC time constants. During this time the RF/IF/AF stages can be over-driven. If the RF and IF gains are kept at a low level this problem can be reduced. Also, attenuating low audio frequencies prior to the first audio amplifier should help. With audio AGC, it is important to avoid so-called "charge pump" rectifiers or other slow-responding circuits that require multiple cycles of audio to pump up the AGC voltage. Instead, use a peak-detecting circuit that responds accurately on the first positive or negative half-cycle.

12.6.2 AGC Circuits

Figure 12.59 shows some gain-controllable circuits. Figure 12.59A shows a two-stage 455 kHz IF amplifier with PIN diode gain control. This circuit is a simplified adaptation from a production receiver, the Collins 651S. The IF amplifier section shown is preceded and followed by selectivity circuits and additional gain stages with AGC. The 1.0 µF capacitors aid in loop compensation. The favorable thing about this approach is that the transistors remain biased at their optimum operating point. Right at the point at which the diodes start to conduct, a small increase in IMD may be noticed, but that goes away as diode current increases slightly. Two or more diodes can be used in series if this is a problem, though it seldom is. Another solution is to use a PIN diode that is more suitable for such a low frequency IF. Look for a device with t > 10 / (2πf) where t is the minority carrier lifetime in msec and f is the frequency in MHz.

Figure 12.59B is an audio derived AGC circuit using a full-wave rectifier that responds to positive or negative excursions of the audio signal. The RC circuit follows the audio closely.

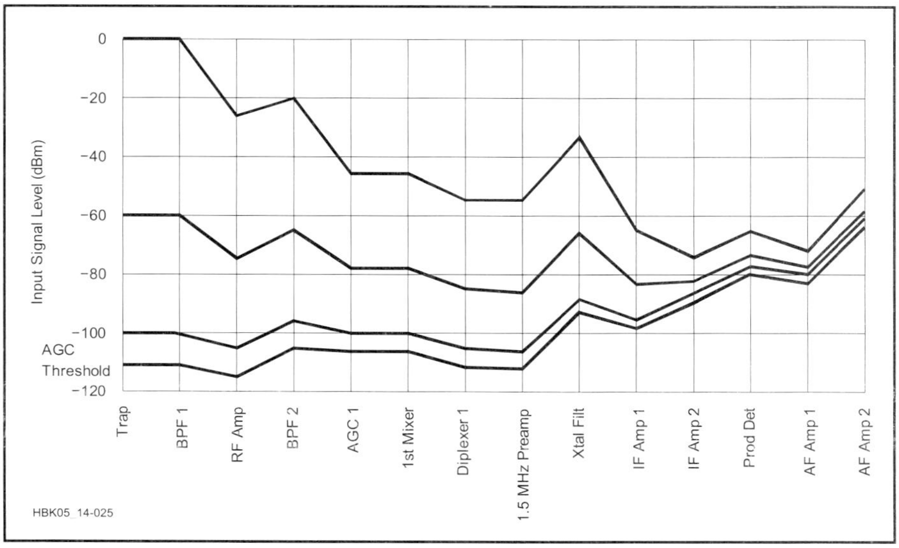

Figure 12.58 — Gain control of a ham-band receiver using AGC. A manual gain control could produce the same result.

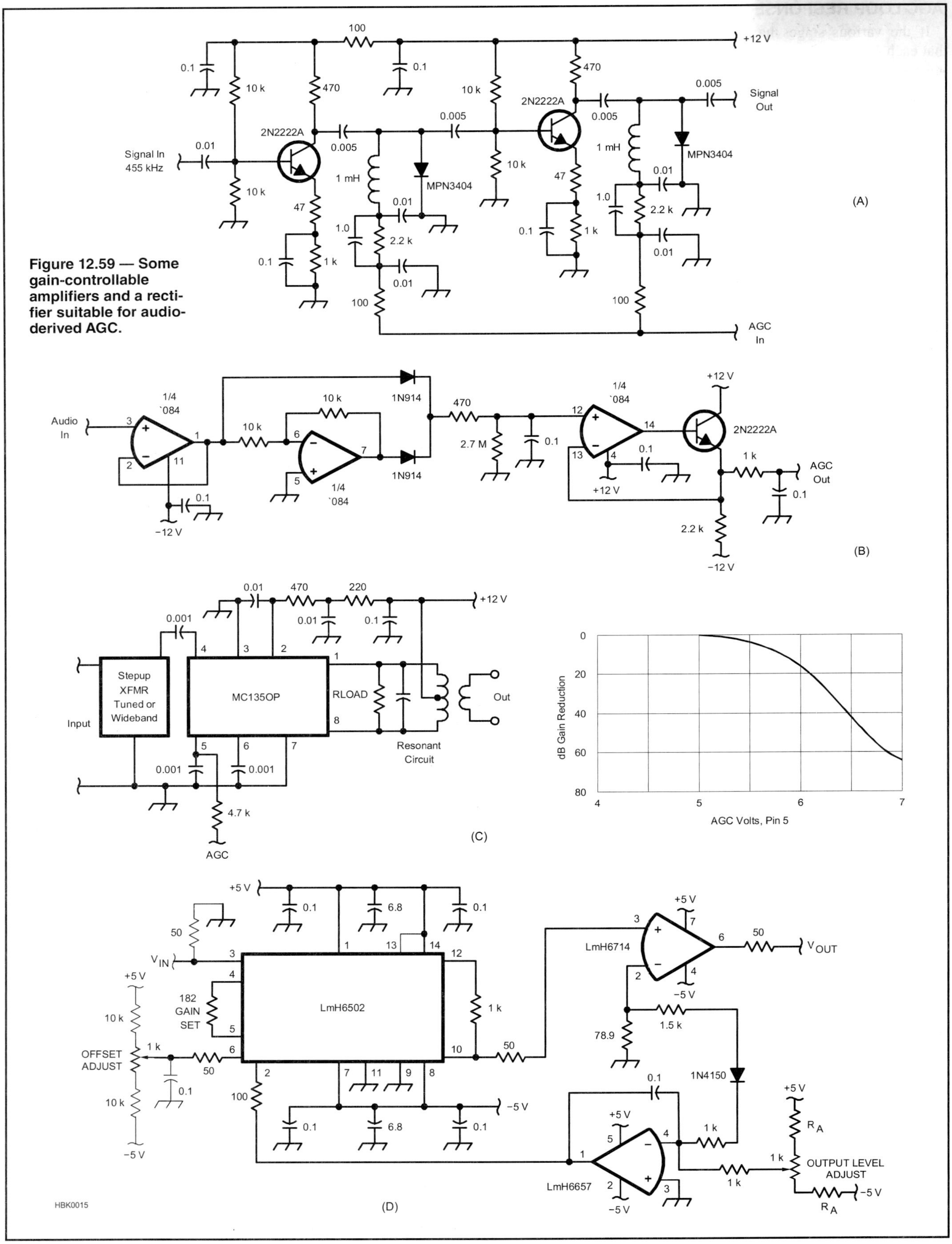

Figure 12.59 — Some gain-controllable amplifiers and a rectifier suitable for audio-derived AGC.

Figure 12.59C shows a typical circuit for the MC1350P RF/IF amplifier (note, this part is no longer manufactured). The graph of gain control versus AGC voltage shows the change in dB/V. If the control is limited to the first 20 dB of gain reduction this chip should be favorable for good AGC transient response and good IMD performance. Use multiple low-gain stages rather than a single high-gain stage for these reasons. The gain control within the MC1350P is accomplished by diverting signal current from the first amplifier stage into a *current sink*. This is also known as the *Gilbert cell multiplier* architecture. Another chip of this type is the NE/SA5209 (note, this part is no longer manufactured). This type of approach is simpler to implement than discrete circuit approaches, such as dual-gate MOSFETs that are now being replaced by IC designs.

Figure 12.59D shows the high-performance TI LMH6502MA voltage-controlled amplifier. It is specially designed for accurate log-linear AGC from 0 – 40 dB with respect to a preset maximum voltage gain from 6 – 40 dB. Its ±3 dB bandwidth is 130 MHz. It is an excellent IF amplifier for high performance receiver or transmitter projects.

Additional info on voltage-controlled amplifier ICs can be found on the Analog Devices web site **www.analog.com/media/en/training-seminars/tutorials/MT-073.pdf**.

12.7 Noise Management

A major problem for those listening to receivers has historically been local impulse noise. For HF and VHF receivers it is often from the sparks of internal combustion engine spark plugs, electric fence chargers, light dimmers, faulty power-line insulators and many other similar devices that put out short duration wide band signals. In the UHF and microwave region, radar systems can cause similar problems.

Additional sources of noise are atmospheric and man-made noise with a variety of different profiles: static crashes, power-line buzz, and the ever-increasing white noise and spurious signals from consumer and industrial electronics, particularly switch mode devices. The capabilities of DSP can be used to combat these diverse types of noise.

Noise canceling by subtracting it from the incoming signals is available as a station accessory using external antennas for sensing and beam-steering. True diversity reception is also available, using the spatial characteristics of the arriving noise signals to discriminate between them and the desired signals. All these techniques provide formidable tools for noise management that were unavailable to amateurs of earlier eras.

12.7.1 The Noise Limiter

The first device used in an early (1930s) attempt to limit impulse noise was called a *noise limiter* or *clipper* circuit, as originally described by H. Robinson, W3LW. This circuit would clip, or limit, noise (or signal) peaks that exceeded a preset limit. The idea was to have the limit set to about as loud as you wanted to hear anything and nothing louder would get through. This was helpful in eliminating the loudest part of impulse noise or even nearby lightning crashes, but it had two problems. First, it didn't eliminate the noise, it just reduced the peak loudness. Second, it also reduced the loudness of loud signals and, in the process, distorted them considerably.

The second problem was fixed shortly thereafter, with the advent of the *automatic noise limiter* (ANL). The ANL automatically set the clipping threshold to that of a loud signal. It thus would adjust itself as the loudness of signals you listened to varied with time. An ANL was easy to implement and became standard equipment on amateur receivers starting in the late 1930s. While ANL circuits are no longer common, simple receivers used today do sometimes incorporate passive clipping circuits to account for their limited AGC ability.

12.7.2 The Noise Blanker

It turned out that improvements in receiver selectivity over the 1950s and beyond, while improving the ability to reduce random noise, made receiver response to impulse noise worse. The reason for this is that a very short duration pulse will be lengthened while going through a narrow filter. This is due to the filter's different phase delay times for the pulse's wide spectrum, resulting in the different frequencies arriving at the filter output at different times. You can demonstrate this in your superheterodyne receiver if it has a narrow crystal filter. Find a frequency with heavy impulse noise and switch between wide and narrow filters. If your narrow filter is 500 Hz or less, the noise pulses will likely be more prominent with the narrow filter. DSP filters, with their superior group delay performance, exhibit less smearing than their analog counterparts.

The noise limiters described so far were all connected to the output of the IF amplifiers. At that point, the noise had passed most of the selectivity of the previous stages and had been widened by the receiver filters. In SSB receivers, since signals vary in strength as someone talks, the usual AGC responds quickly to reduce the gain of a strong signal and then slowly increases it after the signal is no longer there. This means that a short noise pulse that is stronger than the desired signal may reduce the receiver gain for much longer than the pulse lasts.

The solution to this is a noise blanker. An analog *noise blanker* is almost a separate wideband receiver. It takes its input from an early stage in the superheterodyne receiver before much selectivity or AGC has been applied. It amplifies the wideband signal and detects the narrow noise pulses without lengthening them. The still narrow noise pulses are used to shut off or "blank" the receiver at a point ahead of the selectivity and AGC, thus keeping the noise from getting to the parts of the receiver at which the pulses would be extended. In other words, the receiver is shut off or *gated* during the noise pulse.

In addition to an ON/OFF switch, noise blankers include a control labeled THRESHOLD, that adjusts the level of noise that will blank the receiver. If the threshold is set too low, it will blank on signal peaks as well as noise, resulting in distortion of the signal.

An SDR noise blanker can implement the same detect-and-gate technique as an analog receiver. Once the noise impulse is detected, the samples making up the impulse are replaced by an interpolation between the "normal" samples just before and just after the impulse. It is as if the noise impulse was never received.

An alternative technique that also handles longer noise crashes and other large-amplitude, long-duration noise waveforms is described in **Figure 12.60**. The basic idea is to create a filtered average amplitude and watch for incoming signals that exceed the average multiplied by a weighting factor which is how the SDR noise blanker threshold is controlled. When the incoming signal is below this value, it is simply passed along to the rest of the receiver. Above this level, the blanker generates a ramp factor that limits the rate at which the signal can increase. This turns short, sharp impulses into relatively slowly increasing and decreasing ramps, leaving normal signals unaffected.

A well-designed noise blanker can be very effective. Instead of just keeping the noise at the level of the signal as a noise limiter does, the noise blanker can *eliminate* the noise. If the pulses are narrow enough, the loss of desired signal during the time the receiver is disabled is not noticeable and the noise may seem to disappear entirely.

Noise blankers can also create problems, particularly in superheterodyne receivers. The wide-band circuit that detects noise pulses also detects any signals in that bandwidth. If such a signal is strong and has sharp peaks (as voice and CW signals do), the noise blanker will treat them as noise pulses and shut down the receiver accordingly. This causes tremendous distortion and can make it sound as if the strong signal to which the noise blanker is responding is generating spurious signals that cause the distortion. This is less of a problem in SDR receivers. Before you assume that the strong signal is causing problems, turn the noise blanker on and off to check. When the band is full of strong signals, a noise blanker may cause more problems than it solves.

12.7.3 Operating Noise Limiters and Blankers

Many current receivers include both a noise limiter (often labeled NL) and a noise blanker (labeled NB). If your receiver has both, they have separate controls and it is worthwhile to try them both. There are times that one will work better than the other, and other times when the opposite is true, depending on the characteristics of the noise. Then there are times when using both works better than either one by itself. In any case, they can make listening a lot more pleasant. Remember to turn them off when you don't need them, since either type can cause some distortion, especially on strong signals that should otherwise be easy to listen to.

Recognizing that it is difficult for a single noise blanker to work properly with the wide variations of noise pulses, it is common for current receivers to have two noise blankers with different characteristics that are optimized for the different pulse types. One noise blanker is typically optimized for very short pulses and the other for longer pulses. The operator can switch between the blankers to see which works best on the noise at hand.

The usual approach to operating the noise blanker is to activate it, then adjust the THRESHOLD control until the noise is just blanked. You probably need to make occasional adjustments as the noise impulse characteristics change. Again, don't forget to turn the noise blanker off when the noise goes away.

The previous techniques represent the most commonly available to reduce impulse noise. There are a few other solutions as well. Note that we haven't been talking about reducing interference here. By interference, we mean another intended signal encroaching on the channel to which we want to listen. There are several techniques to reduce interference, and some also can help with impulse noise.

Many times, impulse noise is coming from a particular direction. If so, by using a directional antenna, we can adjust the direction for minimum noise. When we think about directional antennas, the giant HF Yagi springs to mind. For receiving purposes, especially on the lower bands such as 160, 80, and 40 meters (where the impulse noise often seems the worst), a small indoor or outdoor receiving loop antenna as described in the *ARRL Antenna Book* can be very effective at eliminating either interfering stations or noise, or both if they happen to be coming from the same direction.

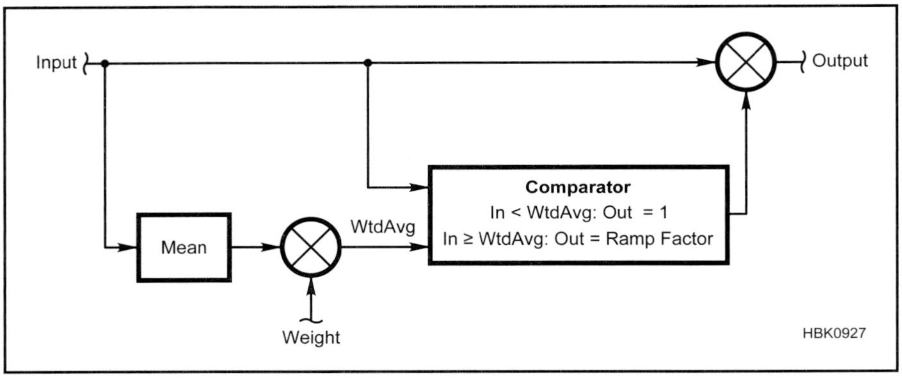

Figure 12.60 — Block diagram of a noise blanker. The input signal is compared to a weighted average. If exceeded, the input signal rise and fall times are limited by a ramp factor. [After *Communications Receivers, 4th ed*]

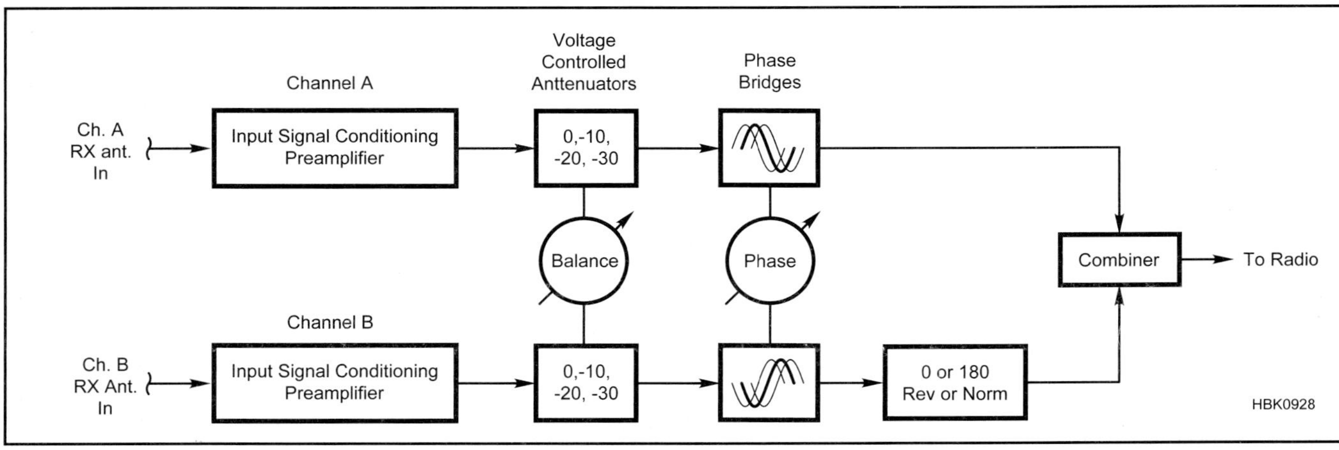

Figure 12.61 — Block diagram of a noise canceling system following the DX Engineering NCC-2. [Courtesy DX Engineering, used by permission]

12.7.4 Noise Canceling

Noise cancellers work by combining signals from a main antenna with the signals from a "sense" antenna and feeding that combination to the receiver. Variable gain stages and phasing networks within the unit must then be carefully adjusted so that the two noise signals are equal in level and 180 degrees out of phase. Noise cancelation is frequency sensitive, so it must be readjusted each time we change frequency. It must also be readjusted for every noise source. Signals coming from the same direction as the noise source are also canceled, but this effect is minimized by placing the sense antenna as close as practical to the noise source. An active noise canceller can greatly reduce a single noise source, but it won't help with more than one source at a time. **Figure 12.61** shows the block diagram of a typical noise canceller and antennas, in this case the DX Engineering NCC-2.

Be careful when using any unit in line with the transceiver output; the carrier detect-driven relay that switches from receive to transmit (that is, bypassing the unit in transmit mode) is known to generate key clicks (transients) in some products. The MFJ 1026 and the DX Engineering NCC-1 and NCC-2 are generally good performers in this regard.

Some transceivers provide connectors where an external preamp or receive antenna can be connected. The click problem can be avoided by connecting the noise canceller at this point. Some SDRs have multiple antenna inputs and the ability to feed them to separate software-defined "receivers," and some software has been written to allow those separate receivers to provide the phasing networks needed to provide noise canceling. In some software, this function is improperly labeled "Diversity Receive." (See the Diversity Reception description below.)

BEAM STEERING

Noise cancellers can also be used to vary the phase of the signals from two receive antennas to vary their directivity. Two vertical antennas, connected to a single receiver by individual feed lines, form a directional antenna. The directional pattern depends on the phase shift provided by the two feed lines (as well as the physical spacing and spatial orientation of the antennas), and it can be adjusted by varying the length of either of the feed lines. Used with two omnidirectional verticals, for example, a noise canceller can vary the phase relationship between the two antennas, accomplishing the same result.

SIGNAL FADING

Most periodic (repetitive) fading is the result of the cancellation of signals from the same source taking slightly different paths, so that one is delayed with respect to the other. The direct and reflected signals cancel when they are nearly equal and nearly 180 degrees out of phase with each other. The frequencies at which this 180-degree phase relationship exists are a function of the time difference; the fading interval is longer for lower frequencies. This fading mechanism is heard as very slow fading on the AM broadcast band and 160 meters, and as much faster "picket fencing" at VHF and UHF.

12.7.5 Diversity Reception

Diversity reception has been used since the earliest days of radio to reduce the effects of signal fading. A receiving diversity system consists of two receivers, each connected to its own antenna and to its own loudspeaker (or opposite ears of stereo headphones). When the two antennas are separated by at least a quarter of a wavelength, the probability of cancellation occurring at both antennas at the same time is low, and the operator copies from the receiver providing the best signal.

Diversity reception is also widely used to listen using different receive antennas having different directivity in the horizontal or vertical plane or aimed in different directions. This use is quite common on the lower HF bands and 160 meters. An operator may listen to the transmit antenna in one ear and a Beverage in the other, or to a loop and a Beverage.

Diversity reception is widely used in consumer FM receivers in vehicles, and in the wireless microphone systems used in live sound and broadcasting. In these applications, a circuit called a "voter" chooses the signal from the receiver having the best quality and switches it to the output. Many VHF and UHF repeater systems have receivers at multiple sites that are relayed back to the main transmit site, where a voter chooses between the best signal. Diversity is also used with receive antennas having different wave polarization; cross-polarization between a receive antenna and the wavefront typically results in a 20 dB loss of gain.

12.7.6 DSP Noise Reduction

DSP noise reduction (often labeled NR on the receiver controls) can look at the statistics of the signal and noise, figure out which is the noise, and then reduce it significantly. These *adaptive filters* cannot quite eliminate the noise and need enough of the desired signal to figure out what's happening, so they won't work if the signal is far below the noise. (See the **DSP and SDR Fundamentals** and **Analog and Digital Filtering** chapters for information on adaptive filters.) Many DSP systems "color" the resulting audio to a degree. Nonetheless, they do improve the SNR of a signal in random or impulse noise.

Noise reduction is designed to decrease uncorrelated noise, such as atmospheric noise and thermal noise. Most noise reduction relies on the knowledge that noise is much more random in nature than the signals we are attempting to demodulate. A typical NR uses a Least Mean Squares (LMS) or similar algorithm to detect correlated signals and reduce the amplitude of anything else in the passband.

How does this work? The algorithm looks for signals that are highly correlated, meaning that a copy of the signal shifted in time resembles the same signal at a different time. These are the signals that are preserved while the other signals in the passband are judged to be noise and are de-emphasized or reduced in volume. Because there are several methods for achieving noise reduction in different classes of noise, some vendors offer more than one type of noise reduction. Each manufacturer's algorithms for noise reduction are generally proprietary, although the *GNU Radio* community's (see the **Transceiver Design Topics** chapter) open-source design approach offers some guidance as to how noise reduction is achieved.

As with noise blankers, receivers frequently offer two or more noise reduction settings that apply different noise reduction algorithms optimized for different conditions. Different combinations and types of noise require the operator to select and adjust the noise reduction system for best performance.

12.7.7 DSP Notch Filtering

An Automatic Notch Filter (ANF) is really the opposite of an NR filter. The ANF looks for the same correlated signals, but instead or preserving these signals, the algorithm attempts to remove these signals. The classic example for which an ANF is useful is when there is a birdie (spur) or undesired CW signal in the passband of a sideband signal. Because the CW signal or spur are much more correlated than the speech we are listening to, these signals are judged to be undesired, and the filter de-emphasizes them. In this way, the NR and ANF filters are cousins, doing the opposite of each other in support of their specific noise reduction goals.

12.8 References and Bibliography

Books

Allison, B., WB1GCM, *Amateur Radio Transceiver Performance Testing* (ARRL, 2013).

Hayward, W., W7ZOI, *Introduction to Radio Frequency Design* (ARRL, 1994).

Hayward, W., Campbell, R., and Larkin, B., *Experimental Methods in RF Design* (ARRL, 2009).

Horowitz, P., Hill, W., *The Art of Electronics,* 2nd Edition (Cambridge University Press, 1989).

Larson, L., *RF and Microwave Circuit Design for Wireless Communication* (Artech House, 1996).

Maas, S., *The RF and Microwave Circuit Design Cookbook* (Artech House, 1998).

Maas, S., *Nonlinear Microwave and RF Circuits,* 2nd Edition (Artech House, 2003).

Maas, S., *Noise in Linear and Nonlinear Circuits* (Artech House, 2005).

McClanning, K., and Vito T., Radio *Receiver Design* (Noble Publishing Corporation, 2001).

RF/IF Designer's Handbook (Scientific Components, 1992).

RF Communications Handbook (Philips Components-Signetics, 1992).

Rohde, U., Newkirk, D., *RF/Microwave Circuit Design for Wireless Applications* (John Wiley & Sons, 2000). (Large number of useful references in the mixer chapter.)

Rohde, U., N1UL, Whitaker, J., and Zahnd, *H., Communications Receivers,* 4th Edition (McGraw-Hill, 2017). (Large number of useful references in each chapter, as well.)

Synergy Microwave Corporation Product Handbook (Synergy Microwave Corporation, 2007).

Terman, F., *Radio Engineers' Handbook* (McGraw-Hill, 1943).

Articles About Mixers

Henderson, B., "Mixers in Microwave Systems", Watkins-Johnson Tech Note, Part 1, Vol. 17 No. 1, 1990; Part 2, Vol. 17 No. 2 1990.

Gilbert, B., "Demystifying the Mixer," (self-published monograph, 1994).

Joshi, S., "Taking the Mystery Out of Double-Balanced Mixers," *QST,* Dec. 1993, pp. 32 – 36.

Kazdan D., "What's a Mixer?" *QST,* Aug. 1992, pp. 39 – 42.

Kearney F., and Frizelle, D., "Complex RF Mixers, Zero-IF Architecture, and Advanced Algorithms," *Analog Dialogue*, 51-02, Analog Devices, **www.analog.com/media/en/analog-dialogue/volume-51/number-1/articles/complex-mixers-zif-architecture-advanced-algorithms-black-magic-next-generation-sdr-transceivers.pdf**.

Kossor, M., "A Digital Commutating Filter," *QEX,* May/June 1999, pp. 3 – 8.

Rohde, U., "Key Components of Modern Receiver Design," Part 1, *QST,* May 1994; Part 2, June 1994; Part 3, Jul. 1994.

Rohde, U., "Testing and Calculating Intermodulation Distortion in Receivers," *QEX,* July 1994, pp. 3 – 4.

Rohde, U., and Poddar, A., "High Intercept Point Broadband, Cost Effective and Power Efficient Passive Reflection FET DBM," *EuMIC Symposium,* 10 – 15 Sep. 2006, UK.

Rohde U., and Poddar, A., "A Unified Method of Designing Ultra-Wideband, Power-Efficient, and High IP3 Reconfigurable Passive FET Mixers," *IEEE International Conference on Ultra-Wide Band,* Sep. 24 – 27, 2006, MA, USA.

Rohde U., and Poddar, A., "Low Cost, Power-Efficient Reconfigurable Passive FET Mixers," 20th *IEEE Canadian Conference on Electrical and Computer Engineering* 2007, 22 – 26 Apr. 2007, British Columbia, Canada.

Tayloe, D., "Letters to the Editor, Notes on 'Ideal' Commutating Mixers (Nov/Dec 1999)," *QEX,* Mar./Apr. 2001, p. 61.

Tayloe, D., "A Low-noise, High-performance Zero IF Quadrature Detector/Preamplifier", Mar. 2003, *RF Design.*

Tayloe, D., US Patent 6,230,000, "A Product Detector and Method Therefore."

Vermusvuori, J., "A Synchronous Detector for AM Transmissions," *QST,* pp. 28 – 33. Uses SA602/612 mixers as phase-locked and quasi-synchronous product detectors.

Wiers, P., "Mirror, Mirror on the Wall — Understanding Image Rejection and Its Impact on Desired Signals," *Analog Dialogue,* 51-08, Analog Devices, **www.analog.com/media/en/analog-dialogue/volume-51/number-3/articles/mirror-mirror-on-the-wall-understanding-image-rejection-and-its-impact-on-desired-signals.pdf**.

Zavrel, R., "Using the NE602," Technical Correspondence, *QST,* May 1990, pp. 38 – 39.

Articles About Receivers

ARRL Lab, "Test Procedures Manual," **www.arrl.org/product-review.**

Brannon, B., "Sampled Systems and the Effects of Clock Phase Noise and Jitter," Analog Devices, AN-756, **www.analog.com/media/en/technical-documentation/application-notes/AN-756.pdf**.

Brannon, B., "Some Recent Development in the Art of Receiver Technology," *Analog Dialogue,* 52-08, Analog Devices, **www.analog.com/media/en/analog-dialogue/volume-52/number-3/a-selected-history-on-receiver-innovations-over-the-last-100-years.pdf**.

DeMaw, D., "What Does My S-meter Tell Me?" *QST,* Jun. 2015 pp. 49-51.

Farson, A., "Noise Power Ratio (NPR) Testing on HF Receivers," *QEX,* Mar. 2015, pp. 20 – 27.

Hallas, J., W1ZR, "International Radio Roofing Filters for the Yaesu FT-1000 MP Series Transceivers," Product Reviews, QST, Feb. 2005, pp. 76.

Harris, J., "What's Up With Digital Downconverters — Part 1 and Part 2," *Analog Dialogue,* 50-07 and 50-11, Analog Devices, **www.analog.com/media/en/analog-dialogue/volume-50/number-3/articles/whats-up-with-digital-downconverters-part-1.pdf and www.analog.com/media/en/analog-dialogue/volume-50/number-4/articles/whats-up-with-digital-downconverters-part-2.pdf**.

Henderson, R., "A Rock Bending Receiver for 7 MHz," *QST*, Aug. 1995, pp. 22 – 25.

Hirshfield, G., "Ham Notebook: General-Coverage Preselector," *Ham Radio,* Oct. 1970, p. 77.

Kester, W., "Understand SINAD, ENOB, SNR, THD, THD+N, and SFDR so You Don't Get Lost in the Noise Floor," Analog Devices, MT-003 Tutorial, 2008, **www.analog.com/media/en/training-seminars/tutorials/MT-003.pdf.**

Kester, W., "Noise Power Ratio (NPR)," Analog Devices, MT-005 Tutorial, 2008, **www.analog.com/media/en/training-seminars/tutorials/MT-005.pdf.**

Lyons, R., "Understanding the 'Phasing Method' of Single Sideband

Demodulation," **www.dsprelated.com/showarticle/176.php.**
Onate, J., Fortuny, X., "A Software-Controlled Preselector," *QEX,* May/June 2008, pp. 11 – 18.
Sabin, W., "Envelope Detection and AM Noise-Figure Measurement," *RF Design,* Nov. 1988, p. 29.
Shuch, P., N6TX, "Quiet! Preamp at Work," *ham radio,* Nov. 1984, pp. 14 – 20, **www.setileague.org/articles/ham/preamp.pdf.**
Silver, W., NØAX, "About FM," *QST,* Jul. 2004, pp. 38 – 42.
Silver, W., "About SSB," *QST,* Jan. 2016, pp. 51 – 54.
Smith, D., "Improved Dynamic Range Testing," *QEX,* Jul./Aug. 2002, pp. 46 – 52.
Smith, P., "Little-Known Characteristics of Phase Noise," Analog Devices, AN-741, **www.analog.com/media/en/technical-documentation/application-notes/5893248557488124036944485577034342 17**

Contents

Chapter 13 — Downloadable Supplemental Content

Articles
- Clean, Punchy, Competitive Contest Audio Without Splatter by Jim Brown, K9YC
- SDR Simplified — Demystifying PID Control Loops by Ray Mack, W5IFS
- SDR Simplified — Noise Reduction and Adaptive Filters by Ray Mack, W5IFS
- Speech Processing: Some New Ideas by Jim Tonne, W4ENE

*Also see the Downloadable Content for the **Receivers** and **Transceivers** chapters*

HF Transmitter and Transceiver Projects
- A Fast TR Switch by Jack Kuecken, KE2QJ
- Designing and Building Transistor Linear Power Amplifiers Parts 1 and 2 by Rick Campbell, KK7B
- MicroT2 — A Compact Single-Band SSB Transmitter by Rick Campbell, KK7B
- MkII — An Updated Universal QRP Transmitter by Wes Hayward, W7ZOI
- TAK-40 SSB/CW Transceiver by Jim Veatch, WA2EUJ
- The Rockmite — A Simple Single-Band CW Transceiver by Dave Benson, K1SWL (article and folder with HEX file)
- The Tuna Tin 2 Today by Ed Hare, W1RFI
- Transmitter and transceiver projects from previous editions of the *ARRL Handbook.*

VHF/UHF Transmitter and Beacon Projects
- A 2-Meter Transmitter for Fox Hunting by Mark Spencer, WA8SME
- A Microwave Transverter Controller by Hamish Kellock, OH2GAQ
- CW Beacon Exciter for 50 MHz by Michael Sapp, WA3TTS (article and parts list)
- Simple Frequency Doublers with High Performance by Paul Wade, W1GHZ
- VHF and UHF CW Beacons by Michael Sapp, WA3TTS
- VHF Open Sources by Rick Campbell, KK7B (article and parts placement)

Chapter 13
Transmitting

As with receiving, analog heterodyne transmitting is rapidly being displaced by SDR techniques. Nevertheless, many of the same functions are required, regardless of whether they are implemented in analog electronics or digitally by software. As such, each function is discussed from both the traditional analog perspective and from the SDR perspective. Subsequent editions will continue to evolve as practices change and standardize on new approaches.

The discussion on microphone selection and audio optimization is by Jim Brown, K9YC, and a set of analog audio signal processing circuits is provided by Jim Tonne, W4ENE. Sections on SDR implementations of basic functions are taken from SDR: Simplified columns in *QEX* by Ray Mack, W5IFS and from material contributed by Steve Hicks, N5AC. Material for the section on PIN Diode RF Switching was contributed by Hans Summers, GØUPL. Bob Allison, WB1GCM, and Adam Farson, VA7OJ/AB4OJ, contributed material on transmitter testing.

The RF power stages of a transmitter are still firmly in the analog camp although design has greatly simplified through integrated circuits and amplifier modules. Circuits for power levels above 100 W (at HF) are covered in the **RF Power Amplifiers** chapter. The **DSP and SDR Fundamentals** chapter has more information on digital techniques and architectures.

Projects included with previous editions have been collected into a set of projects in the online content accompanying this book.

Transmitter technology has advanced in a parallel process similar to that of the technology of receivers. While transmitters are composed of many of the same named blocks as those used in receivers, it's important to keep in mind that there are significant differences. An RF amplifier in a receiver may deal with amplifying picowatts while one in a transmitter may output up to kilowatts. While the circuits may even look similar, the size of the components, especially cooling systems and power supplies, may differ significantly in scale. Still, many of the same principles apply.

Transmitters also make use of many of the same functional blocks as receivers, such as filters and mixers. You will find additional material on these functions covered in the **Receiving** chapter. Oscillators are covered in the **Oscillators and Synthesizers** chapter. Elements of these functions that are pertinent to transmitter design are covered in this chapter but not duplicated.

Transmitters (and transceivers) using vacuum tubes and solid-state matching circuits are likely to present hazardous voltages. At higher power levels RF exposure issues must be considered — review the **Safety** chapter for more information. Techniques for transmitter measurement are covered in the **Test Equipment and Measurements** chapter.

13.1 Characterizing Transmitters

13.1.1 FCC Rules

A survey of the FCC rules in Part 97 shows that most of them are about transmitted signals! Thus, amateurs should have a clear idea of what required their transmitters are expected to satisfy.

There are two important definitions that FCC Part 97 applies to amateur signals (also known as *emissions*):

97.3(a)(8) *Bandwidth.* The width of a frequency band outside of which the mean power of the transmitted signal is attenuated at least 26 dB below the mean power of the transmitted signal within the band. (See Figure 3.16 later in this chapter.)

97.3(a)(43) *Spurious emission.* An emission, or frequencies outside the necessary bandwidth of a transmission, the level of which may be reduced without affecting the information being transmitted.

There are also some important definitions in Part 2.1031 – 2.1060 that apply to all wireless services:

2.202 *Bandwidths.*

(a) *Occupied bandwidth.* The frequency bandwidth such that, below its lower and above its upper frequency limits, the mean powers radiated are each equal to 0.5 percent of the total mean power radiated by a given emission.

(b) *Necessary bandwidth.* For a given class of emission, the minimum value of the occupied bandwidth sufficient to ensure the transmission of information at the rate and with the quality required for the system employed, under specified conditions. Emissions useful for the good functioning of the receiving equipment as, for example, the emission corresponding to the carrier of reduced carrier systems, shall be included in the necessary bandwidth.

The measurement of bandwidth (and other characteristics) is covered by Part 2.1031-2.1060.

Part 97, Subpart D, sets forth a number of technical standards for signals. Part 97.307 — Emission Standards covers all of the signal characteristics from signal quality through symbol rates. Part 97.313 covers transmitter power limits by band. Parts 97.315 and 97.317 set the standards for certification of external RF power amplifiers.

The FCC rules are easily accessible online via the ARRL website at **www.arrl.org/part-97-amateur-radio** and so are not reproduced here.

13.1.2 Performance Measurements

The ARRL Lab has created a series of standardized tests for product review and compliance testing. The entire test program for receivers and transmitters is described in the book *Amateur Radio Transceiver Performance Testing* by Bob Allison, WB1GCM. Important sections are summarized here. The book also addresses a number of other significant parameters that affect various elements of on-the-air performance. The chapter **Test Measurements and Equipment** also discusses transceiver performance.

EMISSION STANDARDS

A spurious signal is any signal (unwanted) other than the intended (fundamental) transmitted signal. Any oscillator will create unwanted products, such as spurs (spurious emissions) and harmonics (multiples of the fundamental frequency). Some consider harmonics to be spurious emissions but they are addressed separately in this book.

Spurs are usually found near or around the fundamental frequency. They can be reduced with appropriate circuit design and, to some extent, with filtering. Harmonics are suppressed to an acceptable level with the use of band-pass filters. An "acceptable level" is one that will not create interference on two times the fundamental frequency (second harmonic), three times the fundamental frequency (third harmonic), and so on.

The single most important standard any transceiver operating in the amateur bands must meet is the FCC Part 97 rules for emissions. According to Part 97.307(d) and 97.307(e), an HF transceiver's emissions must have at least 43 dB of spurious emission and harmonic suppression below 30 MHz. For 30 MHz through 225 MHz, the spurious and harmonic suppression must be 60 dB. (Emission designators are discussed in the **Modulation** chapter.)

CARRIER AND SIDEBAND SUPPRESSION

While there are no FCC specifications for suppression of the carrier and unwanted sideband for SSB signals, the ARRL Lab has established 60 dB of suppression as good practice. 50 dB of suppression is considered a minimum acceptable level.

INTERMODULATION DISTORTION (IMD)

Spurious emissions caused by IMD are definitely of concern, even though they might satisfy the rules on signal bandwidth. For today's transceivers, a third-order product measurement of 30 dB below PEP is typical, 35 dB is considered good, and 25 dB below PEP is mediocre. A clean signal with the lower possible IMD products should be a key consideration for purchasing or operating a transceiver. It is also important to remember that even the best transmitter can be misadjusted to put out an excessively wide signal that creates a lot of interference.

Figures 13.1 and **13.2** show a typical example of a signal with good IMD performance (Figure 13.1) and an unacceptably wide signal (Figure 13.2). The signal in Figure 13.1 has acceptable 3rd-order IMD products and the higher-order products diminish rapidly farther from the main signal. The signal in Figure 13.2, however, has mediocre performance for the 3rd-order products and the higher-order products are far too high out to ±6 kHz from the carrier frequency. This signal would interfere with at least two SSB contacts to each side of it.

Using AFSK to generate multi-tone FSK (MFSK) with multiple simultaneous tones or simultaneous FSK signals in the same transmitter (for example, the fox-and-hound or DXpedition mode of FT8) can result in IMD similar to that produced by the two-tone test method. As with the two-tone method, the IMD can occur from non-linearities in the audio processing path even if the FSK tones are undistorted and not over-driving the audio input.

CW AND FSK KEYING

The shape of a transmitter's keying waveform impacts the quality and effectiveness of its transmissions. By observing the keying waveform, it is possible to assess whether it has characteristics that will affect the quality of the transmitted signal and its impact on adjacent signals.

There are two primary signatures of problems that are can be detected by observing the shape of the keyed waveform. The first is *overshoot* in which the transmitter sends a short high-power transient at the beginning

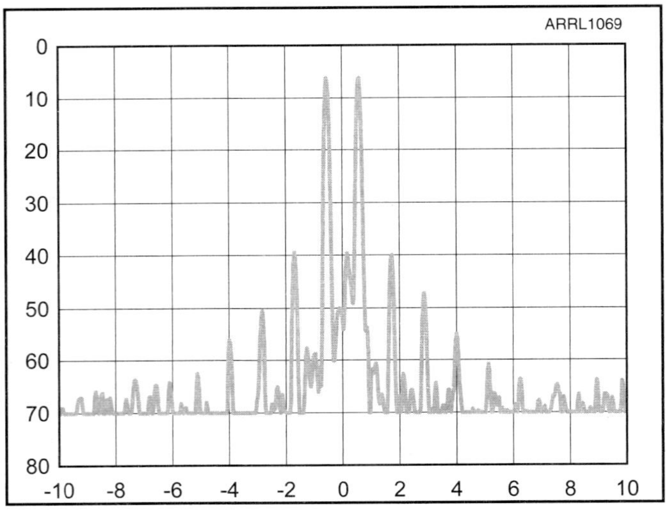

Figure 13.1 — Two-tone IMD test results for a good transmitted signal. The 3rd-order products are close to 40 dB below peak power. Scales are 2 kHz/div (horizontal) and 10 dB/div (vertical).

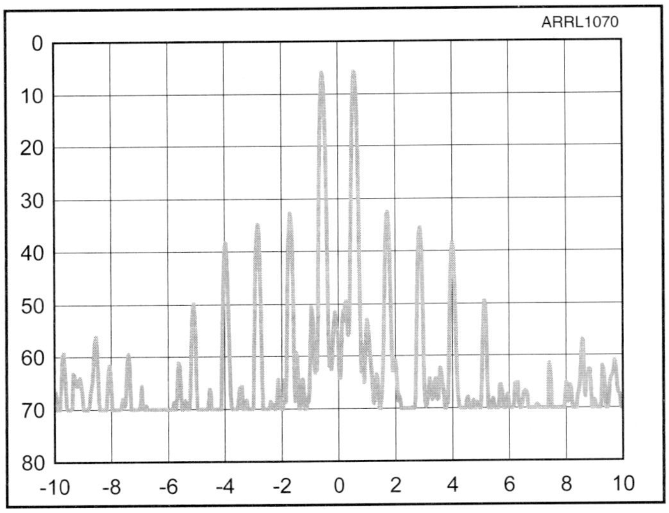

Figure 13.2 — Third-order IMD products are acceptable at 32 dB below PEP but the higher-order products are unacceptably high. Scales are 2 kHz/div (horizontal) and 10 dB/div (vertical).

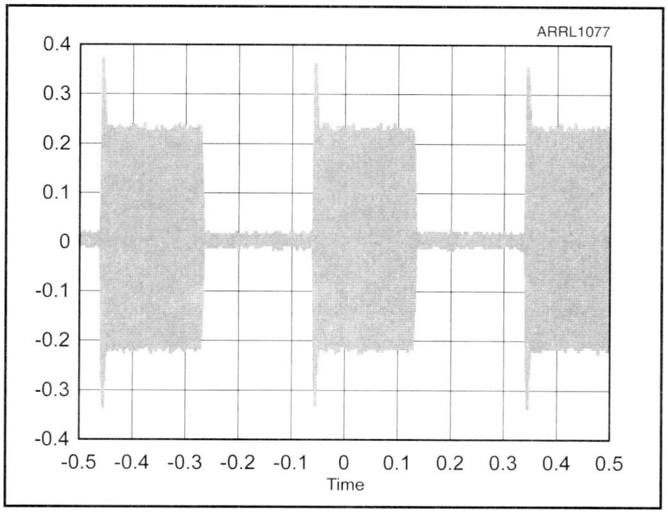

Figure 13.3 — A transient at the leading edge of a keying waveform.

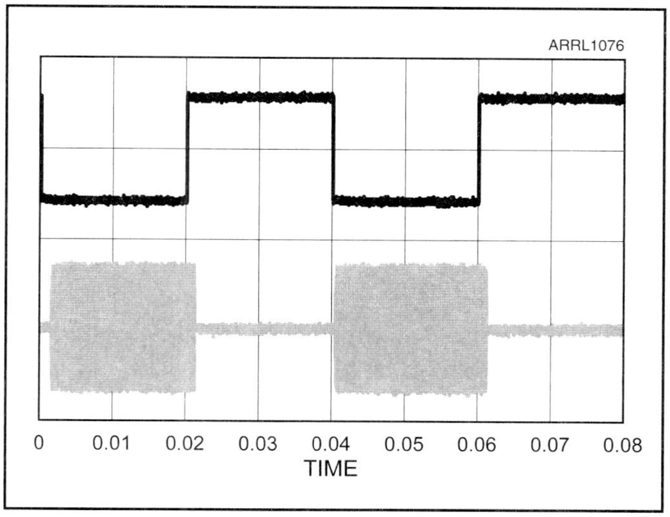

Figure 13.4 — Abrupt, nearly vertical rising and falling edges of "hard" keying.

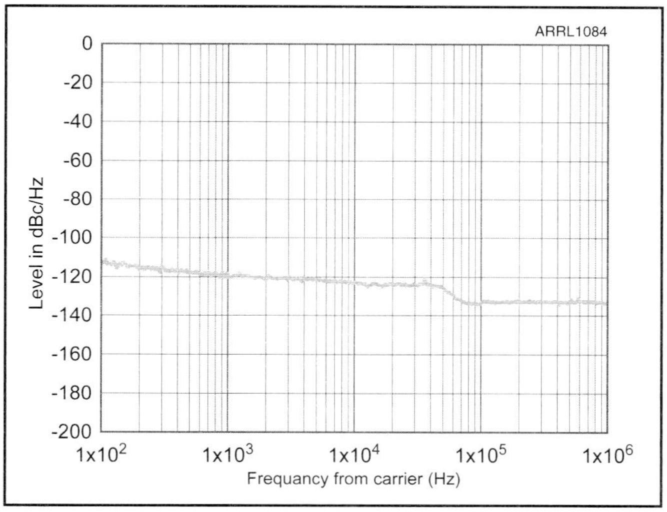

Figure 13.5 — A transmitter with a relatively low composite noise profile. This transmitter is unlikely to cause significant interference on adjacent channels or other bands.

of a transmission (see **Figure 13.3**). This is caused by the power control subsystem not reacting properly to the rapid change in output power. The transient can cause an amplifier's power protection circuitry to activate or trip. It can also cause a key click on adjacent channels. The system at fault can be the ALC system, a software problem, or some other internal power control mechanism.

The second signature is too-rapid rise and fall times or abrupt edges and corners of the envelope as shown in **Figure 13.4**. There may also be discontinuities or artifacts present during the rising and falling edges of the waveform. These create distortion products that extend well beyond the CW signal's necessary bandwidth as key clicks that cause interference to adjacent channels.

FSK keying can also generate spurious products and sidebands outside the intended channel. An abrupt change in tone waveform as the frequency changes creates keying sidebands, just as turning on and off a CW signal does. These artifacts can also disrupt demodulation by the receiver. There are various techniques for shifting between frequencies that minimize these artifacts. (See the **Modulation** chapter.) Direct-FSK which uses a digital signal to shift the transmit frequency directly by changing the frequency of an oscillator can create similar artifacts.

COMPOSITE (PHASE) NOISE

Composite noise is composed of both amplitude and phase noise on the transmitted signal. We are concerned with noise present from 100 Hz to 1 MHz from the carrier frequency of the signal. (See **Figure 13.5**) Composite noise can raise the noise floor of adjacent channels and in severe cases, across an entire band or on multiple bands. (See the discussions on phase noise in this chapter and in the **Transceiver Design Topics** chapter.)

In addition to key clicks, composite noise is also transmitted with CW and FSK signals, so it is appropriate to consider the overall set of keying sidebands output by the transmitter. This parameter is generally measured by keying the transmitter with continuous dits (an approximately square wave keying waveform) at high speed while using a spectrum analyzer to observe the resulting output on frequencies a few kHz above and below the carrier frequency.

ALC OVERSHOOT

The purpose of ALC (Automatic Level Control) is to prevent the various stages in the transmitter from being overdriven. Overdrive can generate out-of-band distortion or cause excessive power dissipation, either in the amplifiers or in the power supply. ALC operates by sampling the peak amplitude of the modulation (the envelope variations) of the output signal and developing a dc gain-control voltage that is applied to an earlier amplifier stage. An ALC voltage can also be developed externally by an RF power amplifier and applied to a transmitter acting as an exciter.

The ALC system will respond to variations in the output signal amplitude. If the response is too slow, output power can be too high until the ALC system regains control. This is called *ALC overshoot*, and the result is distortion of the output signal, usually at the leading or trailing edge of keyed waveforms like CW or of modulations with rapidly changing amplitudes such as speech. The transient at the leading edge of a CW waveform in Figure 13.3 is typical but there can also be artifacts in the rising and falling edges of the waveform. The ALC system may have different time responses on attack and decay as it responds to variations in the signal. Waveform distortion and artifacts created by the ALC system create spurious emissions on adjacent channels.

For digital modes that are not constant-power, ALC action that distorts the waveform can also make it more difficult for a demodulator to recover digital data. For these modes, power output should be reduced to where the ALC system is inactive or the ALC is turned off.

An ancillary function of the ALC system is reduction of drive power (excitation) to a safe level when a load mismatch is detected, typically by an SWR meter at the output of the transmitter or external RF power amplifier.

COMPRESSION AND SPEECH PROCESSING

The purpose of compression and speech processing functions is to increase a voice signal's average power so the output RF signal "sounds louder" to the receiving station. (See the Speech Amplification and Processing section of this chapter.) This can be performed on the baseband audio signal, on an IF signal in the exciter chain, or low-level modulated RF signal before it is amplified. Properly configured, compression should not increase spurious emissions on adjacent channels. Compression should not be used on digital modulation as the resulting distortion products will make demodulation more difficult.

"HOT SWITCHING" AND TURNAROUND

The transmitter's output SEND or KEY control signal operates in conjunction with the RF generation circuits to indicate that the transmitter is in a transmit state and RF output may be present. It is important that the SEND signal becomes active before RF output is generated and stays active until RF output is no longer present. This allows external devices such as preamplifiers and power amplifiers to switch safely between receive and transmit before RF is applied to the antenna system or amplifier input.

If RF output power is present before or after the SEND signal is active, that is called *hot switching*. Hot switching can damage sensitive receiving components or cause switch and relay contacts to be damaged from closing or opening when high voltages or currents are present. Most transmitters have adjustable delay settings to control the time between when a keying signal is asserted and RF is actually generated.

A related metric is *turnaround time* which is the time between when a transmitter stops outputting RF power and a receiver (usually part of an integrated transceiver) can again receive signals. This affects high-speed, full break-in (QSK) CW operation and digital modes that use ARQ protocols in which each transmitted packet is acknowledged by the receiver. (See the **Digital Protocols and Modes** chapter.) The shorter the recovery time, the faster CW speeds and higher packet transmission rates can be.

AUDIO FREQUENCY RESPONSE AND DEVIATION

A transmitter intended for speech modes must be able to transmit audio frequencies from 300 Hz to 3000 Hz although filtering may be applied to reduce this bandwidth under crowded conditions. AFSK digital modes such as FT8 can make use of this entire range. SDR transmitters can have an even wider audio response at the higher end of the range. Low frequency audio components are usually attenuated or "rolled off" below 300 Hz to avoid interfering with the carrier which would cause distortion in the demodulated signal. Many transceivers feature audio equalization on both transmit and receive to accommodate different voices, operating styles, and microphones. This will also affect frequency response and any tests should note the equalization settings. Equalization should be turned off when using digital modulation via the speech processing circuits or functions.

In FM transmitters, deviation is a function of both input audio amplitude and frequency. Insufficient deviation will result in low audio level recovery. Excessive deviation will increase the signal bandwidth and the resulting sidebands can cause interference on adjacent channels. FCC rules (§97.307(f)(1)) also limit angle-modulated signals (both FM and PM) below 30 MHz to a maximum modulation index of 1 to control bandwidth.

LATENCY

Latency of a transmitter is the time interval between an event at the transmitter's modulation input and the corresponding change in RF output at the transmitter's output. The transmitter input can be the CW keying input, the microphone input (used for both voice and AFSK), a USB audio codec, a dedicated data audio input for AFSK, or an FSK keying line. Each is measured by using a dual-channel oscilloscope to observe both the modulation input and RF output. If a USB audio codec is the input, the input to the USB encoding device is the point of input measurement.

Latency in an analog transmitter is generally quite low, limited primarily by delays in any keying sequences that switch filters and amplifiers. An electromechanical relay will create an additional latency of several msec. In an SDR, the processing time in DSP filters dominate the delay.

As with a receiver, whether latency is considered excessive depends on the mode being used. For example, some digital modes that use ARQ protocols (see the **Digital Protocols and Modes** chapter) have a maximum response time before a data frame is assumed to have been lost. The resulting re-transmissions can slow the system or cause the link to fail completely. Excessive latency when monitoring one's own speech or manually sent CW signal can become quite confusing, as well.

13.1.3 CCS, ICAS, and IMS Ratings

Amateur equipment is usually advertised as being rated at some level of ICAS service. There are several related types of operating service. From the RCA Transmitting Tubes manual (TT-4, 1956), the following definitions are obtained:

Continuous Commercial Service (CCS) covers applications involving continuous tube operation in which maximum dependability and long tube life are the primary considerations.

Intermittent Commercial and Amateur Service (ICAS) covers applications in which high tube output is a more important consideration than long tube life. The term "Intermittent Commercial" in this title applies to types of service in which the operating or "on" periods do not exceed 5 minutes each, and are followed by "off" or stand-by periods of the same or greater duration. The term "Amateur Service" covers other applications where operation is of an infrequent or highly intermittent nature, as well as the use of tubes in "amateur" transmitters. ICAS ratings generally are considerably higher than CCS ratings.

Although the ability of a tube to produce greater output power is usually accompanied by a reduction in tube life, the equipment designer may decide that a small tube operated at its ICAS ratings meets the requirements better than a larger tube operated within CCS ratings.

Intermittent Mobile Service (IMS) covers applications in which very high power output for short periods is required from equipment of the smallest practical size and weight. Tube ratings for IMS service are based on the premise that transmitter "on" periods do not exceed 15 seconds each, and are followed by "off" periods of at least 60 seconds duration. In equipment tests, however, maximum "on" periods of not more than 5 minutes each followed by "off" periods of at least 5 minutes are permissible, provided the total "on" time of such test periods does not exceed 10 hours during the life of the tube. Although tubes operated under IMS ratings may have a life of only about 100 hours, the use of these ratings is economically justified where high power must be obtained intermittently from very small tubes.

If equipment is specified as meeting

Continuous Commercial Service requirements, that is exactly how the equipment or device is certified to perform. Very few amateur transmitters or amplifiers, however, are operated that way and if they had to meet the CCS level of service, would be quite expensive.

ICAS, on the other hand, is just a name for the type of service the equipment is expected to provide. The description is intended to apply to use in which the equipment is operated with a 50% duty cycle having equal "on" and "off" periods of 5 minutes or less. Terms like "ICAS 50%" have no standard meaning. That could refer to the original ICAS definition or it could mean a period of 50% duty cycle operation followed by an equal period of no operation at all. The safest choice is to ask the manufacturer directly for their interpretation of ICAS or any similar rating and not assume anything.

Regardless of which rating is used, the intent is to describe the conditions under which the equipment (or devices) may be operated and still be expected to meet the performance specifications. Typically, devices are *de-rated* from their maximum ratings in order to meet the performance specifications but there are many other considerations to take into account when the equipment is designed. For example, equipment designed to operate at a high ambient temperature would have to be de-rated more than equipment operated at a lower temperature.

13.2 Transmitter Architecture

In the following section, "in-band" refers to signal frequencies within the bandwidth of the desired signal. For example, for an upper sideband voice signal with a carrier frequency of 14.200 MHz, frequencies of approximately 14.2003 to 14.203 would be considered in-band. Out-of-band refers to frequencies outside this range, such as on adjacent channels.

13.2.1 Upconverting Heterodyne Architecture

Figure 13.6 shows a traditional superheterodyne architecture for transceivers in which the sideband filter, some amplifiers, and other filters are shared between transmit and receive modes through the use of extensive switching. A limitation of that architecture is that it is difficult to provide operation on frequencies near the first IF. The typical transceiver designer selected a first IF frequency away from the desired operating frequencies and proceeded on that basis.

New amateur bands at 30, 17, and 12 meters were approved at the 1979 ITU World Administrative Radio Conference. The difficulties of managing image rejection on the new bands and the desire for continuous receiver coverage of LF, MF and HF bands (general-coverage receive) required a significant change in the architecture of receivers and transceivers. Thus, the upconverting architecture discussed in the **Receiving** chapter became popular in the 1980s and, with a few notable exceptions, became almost universal in commercial products over the following decade.

The solution was to move to the upconverting architecture shown in **Figure 13.7**. By selecting a first IF well above the highest receive frequency, the first local oscillator can cover the entire receive range without any

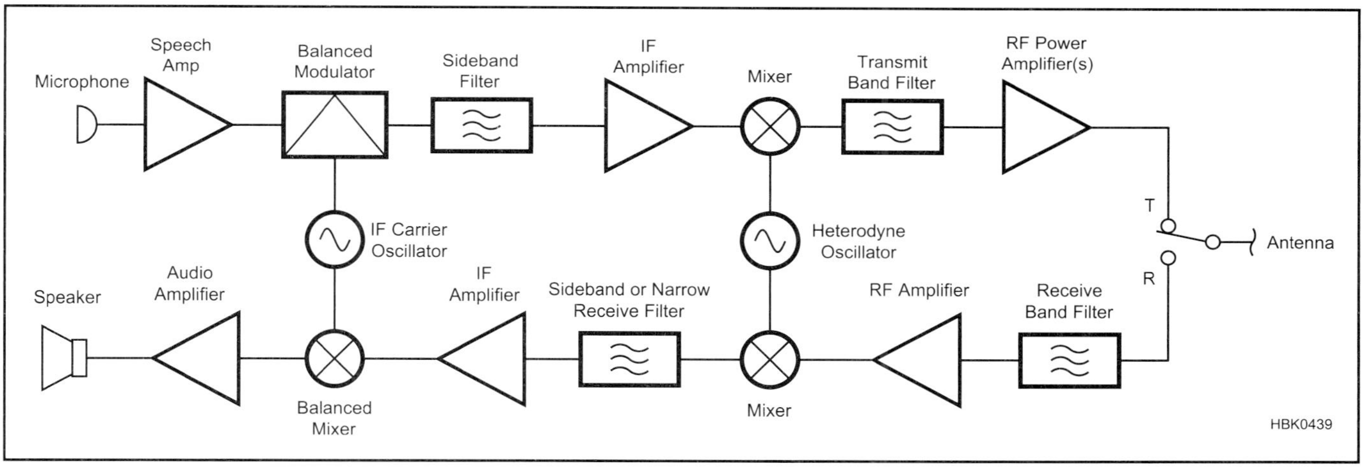

Figure 13.6 — Block diagram of a simple SSB transceiver sharing oscillator frequencies.

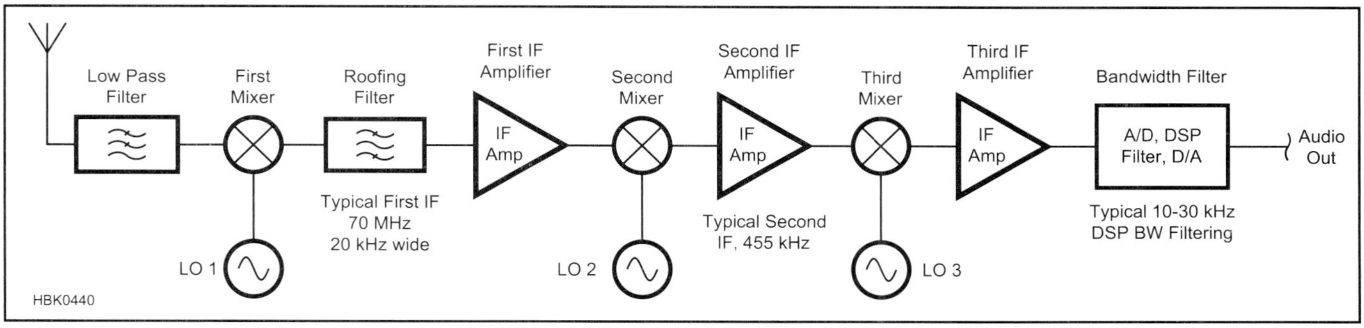

Figure 13.7 — Simplified block diagram of upconverting general coverage transceiver, receiver section shown.

gaps. With the 70 MHz IF shown, the full range from 0 to 30 MHz can be covered by an LO covering 70 to 100 MHz, less than a 1.5:1 range, making it easy to implement with modern PLL or DDS technology. Note that the high IF makes image rejection very easy and, rather than the usual tuned band-pass front end, we can use more universal low-pass filtering. The low-pass filter is generally shared with the transmit side and designed with octave cutoff frequencies to reduce transmitter harmonic content. A typical set of HF transceiver low-pass filter cut-off frequencies would be 1, 2, 4, 8, 16 and 32 MHz.

This architecture offers significant benefits. By merely changing the control system programming, any frequency range or ranges can be provided with no change to the architecture or hardware implementation. Unlike the more traditional transceiver architecture (Figure 13.6), continuous receive frequency coverage over the range is actually easier to provide than to not provide, offering a marketing advantage for those who also like to do shortwave or broadcast listening.

IF FILTERS

The desired IF filter response is shown in **Figure 13.8A**. The reduction of the carrier frequency is augmented by the filter response. It is common to specify that the filter response be down 20 dB at the carrier frequency. Rejection of the opposite sideband should (hopefully) be 60 dB, starting at 300 Hz below the carrier frequency, which is the 300-Hz point on the opposite sideband. The ultimate attenuation should be at least 70 dB. This would represent a very good specification for a high quality transmitter. The filter passband should be as flat as possible (with passband ripple less than 1 dB or so).

Special filters, designated as USB or LSB, are designed with a steeper roll-off on the carrier frequency side, in order to improve rejection of the carrier and opposite sideband. Mechanical filters are available that do this. Crystal-ladder filters (see the **Analog and Digital Filtering** chapter) are frequently called "single-sideband" filters because they also have this property. The steep skirt can be on the low side or the high side, depending on whether the crystals are across the signal path or in series with the signal path, respectively.

Filters require special attention to their terminations. The networks that interface the filter with surrounding circuits should be accurate and stable over temperature. They should be easy to adjust. One very good way to adjust them is to build a narrow-band sweep generator and look at the output IF envelope with a logarithmic amplifier, as indicated in Figure 13.8B. There are three goals:

- The driver stage must see the desired load impedance.
- The stage after the filter must see the desired source (generator) impedance.
- The filter must be properly terminated at both ends.

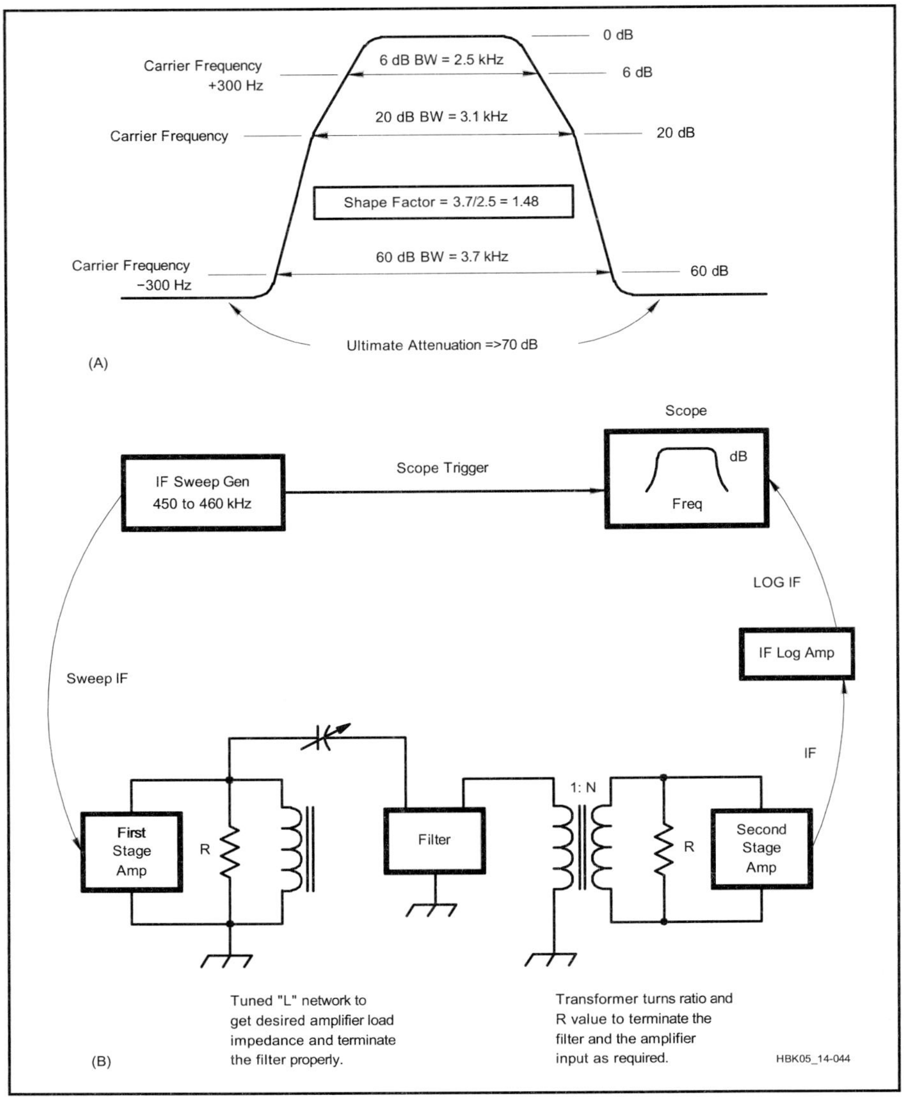

Figure 13.8 — At (A), desired response of a SSB IF filter. At (B), one method of terminating a mechanical filter that allows easy and accurate tuning adjustment and also a possible test setup for performing the adjustments.

Lack of any of these conditions will result in loss of specified filter response. Figure 13.8B shows two typical approaches. This kind of setup is a very good way to make sure the filters and other circuitry are working properly.

Finally, overdriven filters (such as crystal or mechanical filters) can become nonlinear and generate distortion. Thus it is necessary to stay within the manufacturer's specifications. Magnetic core materials used in the tuning networks must be sufficiently linear at the signal levels encountered. They should be tested for IMD separately.

IF Linearity and Noise

Figure 13.9 indicates that after the last SSB filter, whether it is just after the SSB modulator or after the IF clipper, subsequent BPFs are considerably wider. For example, the 70 MHz crystal filter may be 15 to 30 kHz wide. This means that there is a "gray region" in the transmitter in which out-of-band IMD that is generated in the IF amplifiers and mixers can cause adjacent-channel interference.

A possible exception, not shown in Figure 13.9, is that there may be an intermediate IF in the 10 MHz region that also contains a narrow filter.

The implication is that special attention must be paid to the linearity of these circuits. It's the designer's job to make sure that distortion in this gray area is much less than distor-

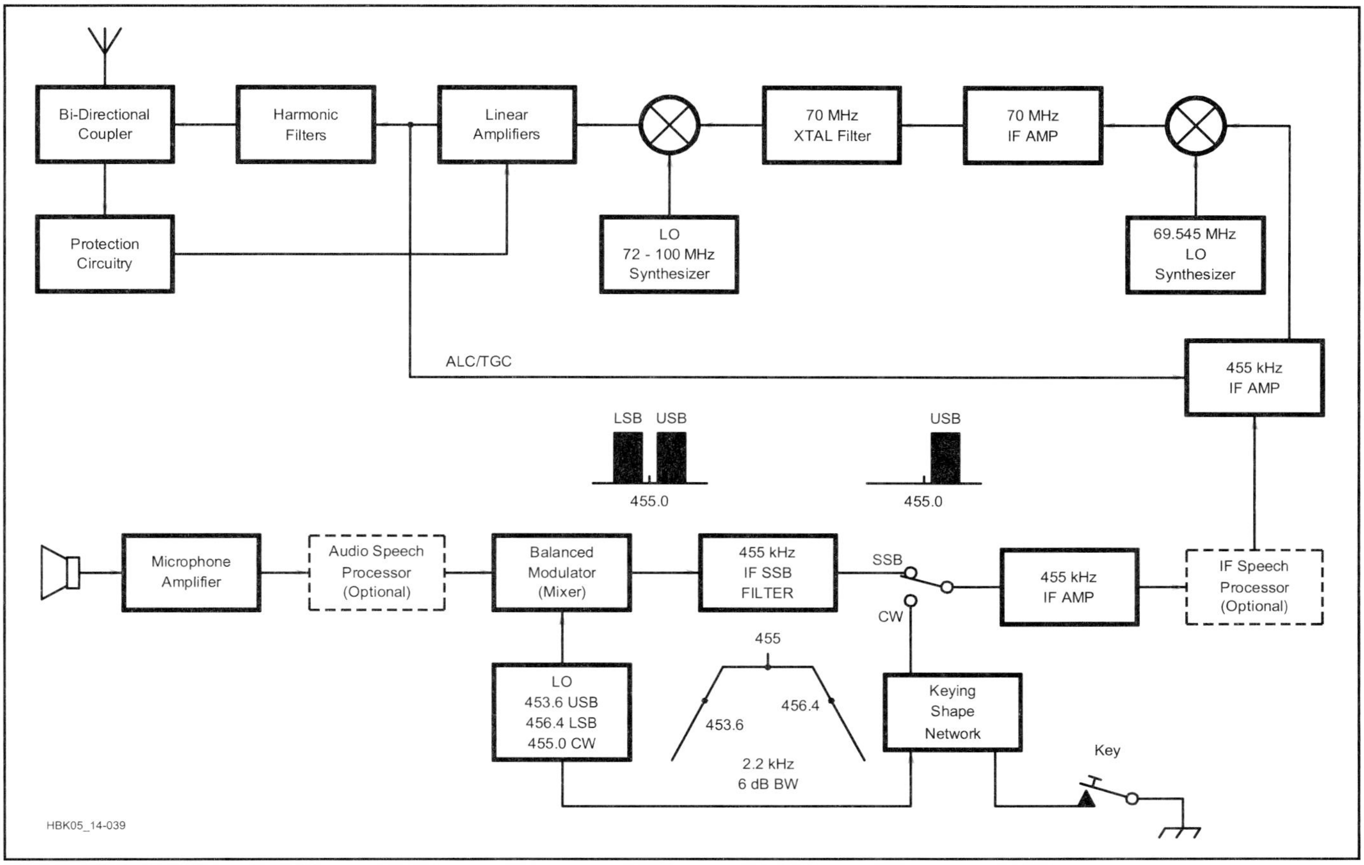

Figure 13.9 — Block diagram of an upconversion SSB/CW transmitter.

tion generated by the PA and also less than the phase noise generated by the final mixer. Recall also that the total IMD generated in the exciter stages is the result of several amplifier and mixer stages in cascade; therefore, each element in the chain must have at least 40 to 50 dB IMD quality. The various drive levels should be chosen to guarantee this. This requirement for multistage linearity is one of the main technical and cost burdens of the SSB mode.

Of interest also in the gray region are additive white, thermal and excess noises originating in the first IF amplifier after the SSB filter and highly magnified on their way to the output. This noise can be comparable to the phase noise level if the phase noise is low, as it is in a high-quality radio. Recall also that phase noise is at its worst on modulation peaks, but additive noise may be (and often is) present even when there is no modulation. This is a frequent problem in co-located transmitting and receiving environments. Many transmitter designs do not have the benefit of the narrow filter at 70 MHz, so the amplified noise can extend over a much wider frequency range.

TRANSMIT MIXER LINEARITY AND NOISE

The last IF and the last mixer LO in Figure 13.9 are selected so that, as much as possible, harmonic IMD products are far enough away from the operating frequency that they fall outside the passband of the low-pass filters and are highly attenuated. This is difficult to accomplish over the transmitter's entire frequency range. It helps to use a high-level mixer and a low enough signal level to minimize those products that are unavoidable. Low-order crossovers that cannot be sufficiently reduced are unacceptable, however; the designer must go back to the drawing board.

13.2.2 SDR Transmitter Architecture

SDR transmitter architecture looks a lot like the SDR receiver architecture as described in the **DSP and SDR Fundamentals** and **Receiving** chapters "turned around." For example, the FFT can be reused as its own inverse to translate back and forth between the time and frequency domains. I/Q modulation looks very much like I/Q demodulation. Substitute a DAC for the ADC to change digital to analog and vice versa. Digital filters just need a data stream and enough clock speed to handle the throughput.

The main question is one of providing enough computing resources and deciding where to make the jump from analog to digital. As the speed and resolution of data converters increases while the cost plummets, the transition between the analog and digital realms is moving ever closer to the antenna. Commercial mainstream transceiver designs being introduced in 2017 are no longer based on the traditional analog superheterodyne architecture. It is only a matter of time before the superheterodyne becomes a legacy technology.

Superheterodyne techniques are still used at points in the signal path, however. **Figure 13.10** shows the progression of transmitter architectures from all-superheterodyne through direct-sampled. If you think they look like the SDR receiver architectures of Figures 8.3 through 8.6 in the **DSP and SDR Fundamentals** chapter, you're right. They are essentially the same processes used for receiving but converting signals to RF instead of vice versa. (See the **Receiving** chapter for a discussion of the architecture pros and

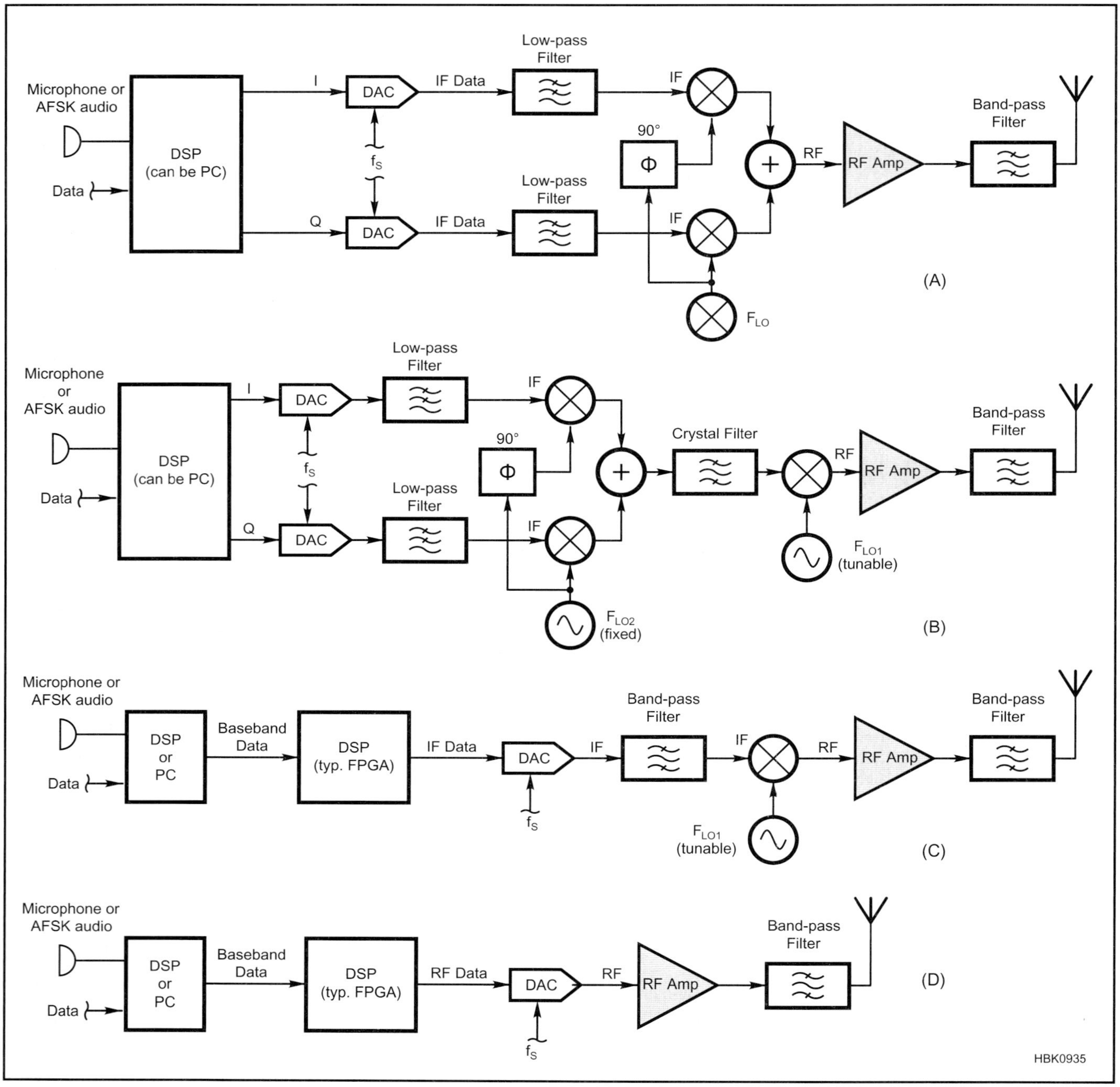

Figure 13.10 — Several SDR transmitter architectures. Direct-from-baseband (A), Hybrid superhet/SDR combination (B), Hybrid DSP-at-IF (C), Direct-sampled (B).

cons.) All of these architectures are in use by amateurs in homebuilt and commercial equipment today.

In any specific transceiver, the receiver and architectures are complementary in that they make use of the same DSP components and convert between analog and digital at roughly the same point in the signal path. The architecture is largely driven by the cost and availability of the data converters and the FPGA or similar DSP computing devices. (Some transceivers are based on generic FPGA parts while others make use of specialized DSP or graphics processors.) Tradeoffs involved with the various architectures are discussed in the **Transceiver Design Topics** chapter.

SDR technology is evolving rapidly so it is premature to make blanket statements about expected levels of performance. The best approach is to read the *QST* Product Reviews, review comparisons such as by Rob Sherwood, NCØB (**www.sherweng.com/table.html**) and Jim Brown, K9YC (**k9yc.com/publish.html**), ask experienced operators for their opinions, and then try the equipment for yourself!

13.3 Modulators

Previous editions of the *Handbook* covered modulators in a general treatment of mixers. Much of that discussion is now included in the **Receiving** chapter where you can find more detailed information on mixer operation.

13.3.1 Amplitude Modulators

You can see how an AM signal is constructed as illustrated in **Figure 13.11**. Figure 13.11A shows the carrier, and the sidebands from a modulating tone are shown in 13.11B and 13.11C. The waveform of an AM signal appears to vary the carrier amplitude but this is not the case. The varying envelope of the AM signal results from the signal's three components — the carrier and the two sidebands — adding together. As the components reinforce and cancel each other, their sum (which appears as the envelope) rises and falls.

If you look closely, you can see that the waveforms in Figures 13.11B and 13.11C have slightly different frequencies than the carrier. If the two sidebands are added together, the signal of Figure 13.11D is produced. This is what the two sidebands look like as waveforms without the carrier. This is a *double-sideband, suppressed carrier* (DSBSC).

When the carrier signal is added, the full AM signal is produced in Figure 13.11D. When all of the signals are in-phase, the resulting signal has its maximum amplitude. When all of the signals are out of phase, the resulting signal goes to zero. If the carrier's phase is used as our reference, the phase of each sideband can be viewed as slipping behind (lower sideband) or moving ahead (upper sideband) of the carrier. The sidebands are out of phase with each other at the frequency of the tone so the resulting envelope reproduces the modulating tone's sine wave.

In the transmitter, the SSB modulator must suppress both the carrier and the unwanted sideband. Carrier suppression is normally accomplished with a *balanced modulator*, a

Mixer Math: Amplitude Modulation

We can easily allow the carrier to be part of the mixer along with the sidebands merely by adding enough *dc level shift* into the information we want to mix so that its waveform never goes negative. In the sidebar "Mixer Math: Mixing as Multiplication" in the **Receiving** chapter, mixer math was kept relatively simple by setting the peak voltage of the input signals directly equal to their sine values. Each input signal's peak voltage therefore varies between +1 and –1, so all we need to do to keep our modulating- signal term (provided with a subscript *m* to reflect its role as the modulating or information waveform) from going negative is add 1 to it. Identifying the carrier term with a subscript c, we can write

$$\text{AM signal} = (1 + m \sin 2\pi f_m t) \sin 2\pi f_c t \qquad \text{(A)}$$

Notice that the modulation ($2\pi f_m t$) term has company in the form of a coefficient, *m*. This variable expresses the modulating signal's varying amplitude — variations that ultimately result in amplitude modulation. Expanding the equation gives us:

$$\text{AM signal} = \sin 2\pi f_c t + \frac{1}{2} m \cos(2\pi f_c - 2\pi f_m)t - \frac{1}{2} m \cos(2\pi f_c - 2\pi f_m)t \qquad \text{(B)}$$

The modulator's output now includes the carrier ($\sin 2\pi f_c t$) in addition to sum and difference products that vary in strength according to m. According to the conventions of talking about modulation, we call the sum product, which comes out at a frequency higher than that of the carrier, the *upper sideband (USB)*, and the difference product, which comes out a frequency lower than that of the carrier, the *lower sideband (LSB)*.

Why We Call It Amplitude Modulation

This process is called *amplitude modulation* because the complex waveform consisting of the sum of the sidebands and carrier varies with the information signal's magnitude (m). Concepts long used to illustrate AM's mechanism may mislead us into thinking that the *carrier* varies in strength with modulation, but careful study of the equation above shows that this doesn't happen. The carrier, $\sin 2\pi f_c t$, goes into the modulator as a sinusoid with an unvarying maximum value of |1|. The modulator multiplies the carrier by the dc level (+1) that we added to the information signal ($m \sin 2\pi f_m t$). Multiplying $\sin 2\pi f_c t$ by 1 merely returns $\sin 2\pi f_c t$. Thus, the carrier's amplitude does not vary as a result of amplitude modulation.

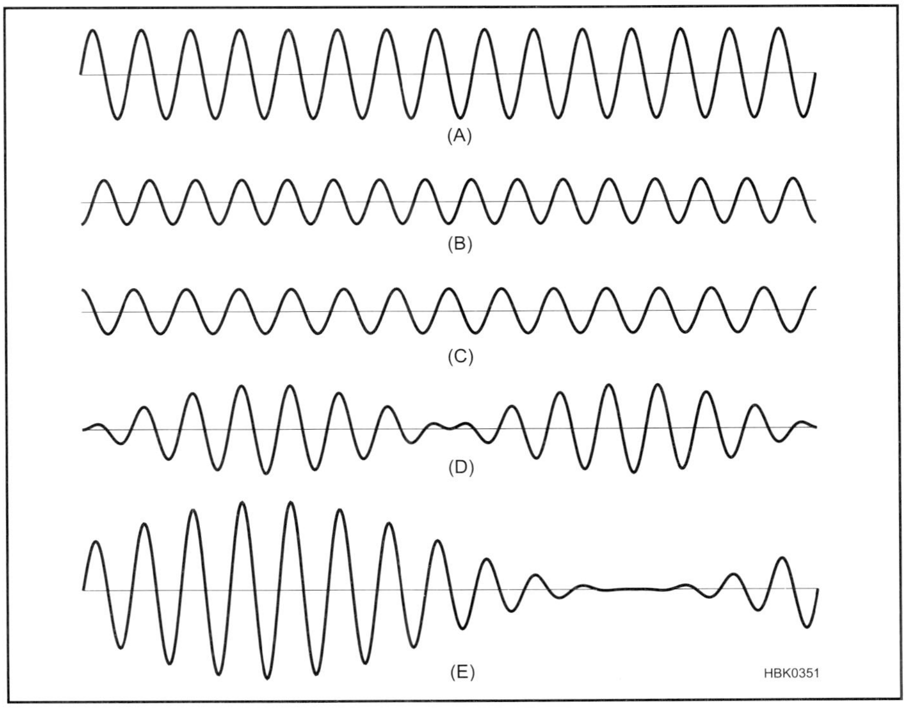

Figure 13.11 — At A is an unmodulated carrier. If the upper (B) and lower (C) sidebands are added together a double-sideband suppressed carrier (DSBSC) signal results (D). If each sideband has half the amplitude of the carrier, then the combination of the carrier with the two sidebands results in a 100%-modulated AM signal (E). Whenever the two sidebands are out of phase with the carrier, the three signals sum to zero. Whenever the two sidebands are in phase with the carrier, the resulting signal has twice the amplitude of the unmodulated carrier.

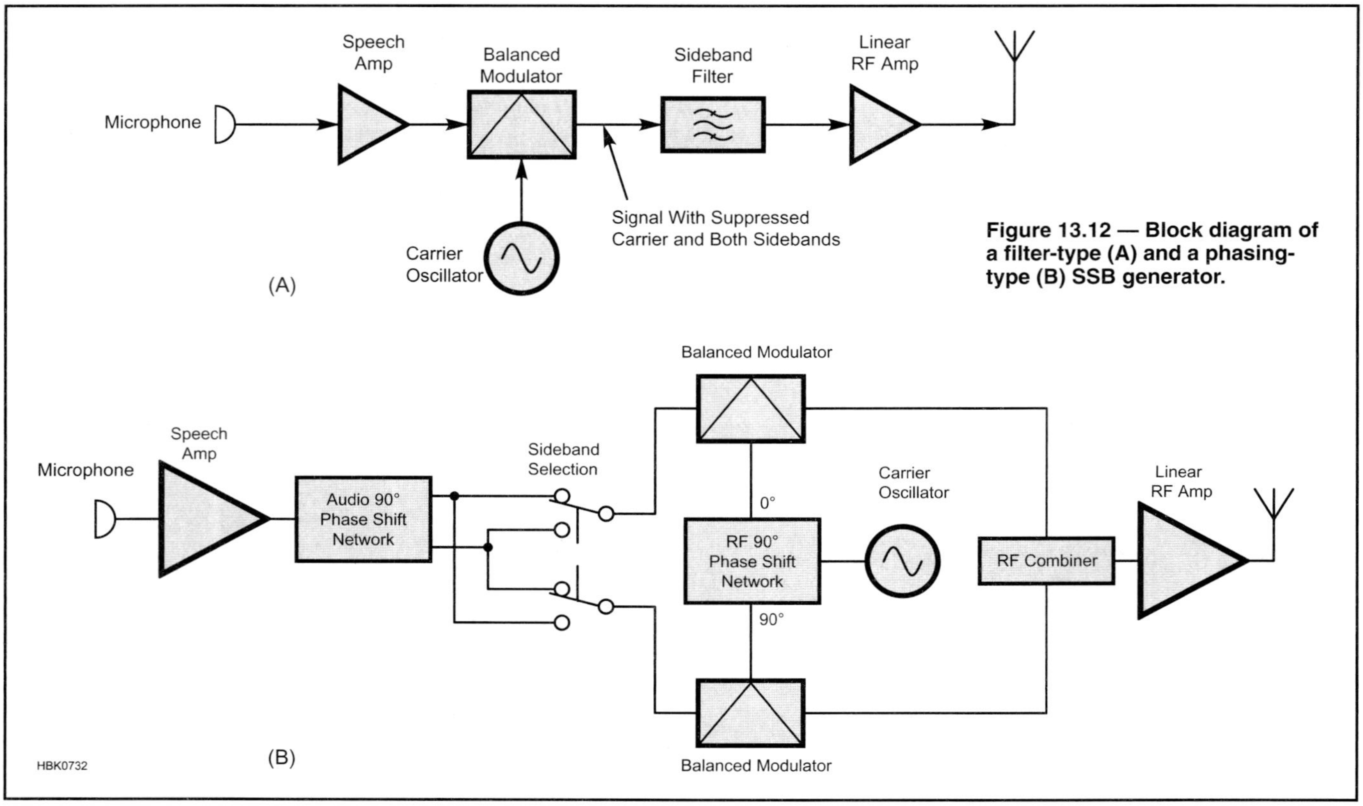

Figure 13.12 — Block diagram of a filter-type (A) and a phasing-type (B) SSB generator.

type of mixer whose output contains the sum and difference frequencies of the two input signals (the modulating signal and the carrier) but not the input signals themselves. There are several ways to eliminate the unwanted sideband, but the most common, shown in **Figure 13.12A**, is to pass the output of the balanced mixer through a crystal filter that passes the wanted sideband while filtering out the unwanted one. This is convenient in a transceiver since the same filter can be used in the receiver by means of a transmit-receive switch.

Another method to generate single sideband is called the *phasing method*. See Figure 13.12B. Using trigonometry, it can be shown mathematically that the sum of the signals from two balanced modulators, each fed with audio signals and RF carriers that are 90° out of phase, consists of one sideband only. The other sideband is suppressed. The output can be switched between LSB and USB simply by reversing the polarity of one of the inputs, which changes the phase by 180°. The phasing method eliminates the need for an expensive crystal filter following the modulator or, in the case of a transceiver, the need to switch the crystal filter between the transmitter and receiver sections. In addition, the audio quality is generally better because it eliminates the poor phase dispersion that is characteristic of most crystal filters. Using analog techniques, designing and building an audio phase-shift network that accurately maintained a 90° differential over a decade-wide frequency band (300 Hz to 3000 Hz) was rather complicated. With modern DSP techniques, the task is much easier.

BALANCED MODULATORS

A balanced modulator is a mixer. Briefly, the IF frequency LO (455 kHz in the example of Figure 13.9) translates the audio frequencies up to a pair of IF frequencies — the LO plus the audio frequency and the LO minus the audio frequency. The balance from the LO port to the IF output causes the LO frequency to be suppressed by 30 to 40 dB. Adjustments are provided to improve the LO null.

The filter method of SSB generation uses an IF band-pass filter to pass one of the sidebands and block the other. In Figure 13.9 the filter is centered at 455.0 kHz. The LO is offset to 453.6 kHz or 456.4 kHz so that the upper sideband or the lower sideband (respectively) can pass through the filter. This creates a problem for the other LOs in the radio, because they must now be properly offset so that the final transmit output's carrier (suppressed) frequency coincides with the frequency readout on the front panel of the radio.

Various schemes have been used to create the necessary LO offsets. One method uses two crystals for the 69.545 MHz LO that can be selected. In synthesized radios, the programming of the microprocessor controls the various LOs. Some synthesized radios use two IF filters at two different frequencies, one for USB and one for LSB, and a 455.0 kHz LO, as shown in Figure 13.9. These radios can be designed to transmit two independent sidebands (ISB) resulting in two separate channels in the spectrum space of the usual AM channel.

The data sheets for balanced modulators and mixers specify the maximum level of audio for a given LO level. Higher audio levels create excessive IMD. The IF filter following the modulator removes higher-order IMD products that are outside its passband but the in-band IMD products should be at least 40 dB below each of two equal test tones. Speech clipping (AF or IF) can degrade this to 10 dB or so, but in the absence of speech processing the signal should be clean, in-band.

AMPLITUDE MODULATION WITH A DBM

We can generate DSB, suppressed-carrier AM with a DBM (double balanced mixer) by feeding the carrier to its RF port and the modulating signal to the IF port. (See the **Receiving** chapter for a detailed discussion of the DBM.) This is a classical *balanced modulator*, and the result — sidebands at radio frequencies corresponding to the carrier signal plus audio and the RF signal minus audio — emerges from the DBM's LO port. If we also want to transmit some carrier along with the sidebands, we can dc-bias the IF port (with a current of 10 to 20 mA) to upset the mixer's balance and keep its diodes from turning all the way off. (This technique is sometimes used for generating CW with a balanced modulator otherwise intended to generate DSB as part of an SSB-generation process.) **Figure 13.13** shows a more elegant approach to generating full-carrier AM with a DBM.

As we saw earlier when considering the many faces of AM, two DBMs, used in conjunction with carrier and audio phasing, can be used to generate SSB, suppressed-carrier AM. Likewise, two DBMs can be used with RF and LO phasing as an image-reject mixer.

AN MC1496P BALANCED MODULATOR

Although it predates the SA602/612, Freescale's MC1496 Gilbert cell multiplier remains a viable option for product detection and balanced modulator service. It has been around for decades and is still one of the best and least expensive. **Figure 13.14** is a typical balanced modulator circuit using the MC1496. The circuit of **Figure 13.15** includes an MC1496-based balanced modulator that is capable of carrier suppression greater than 50 dB. Per its description in Hayward, Campbell, and Larkin's *Experimental Methods in RF Design*, its output with audio drive should be kept to about –20 dBm with this circuit. LO drive should be 200 to 500 mV P-P.

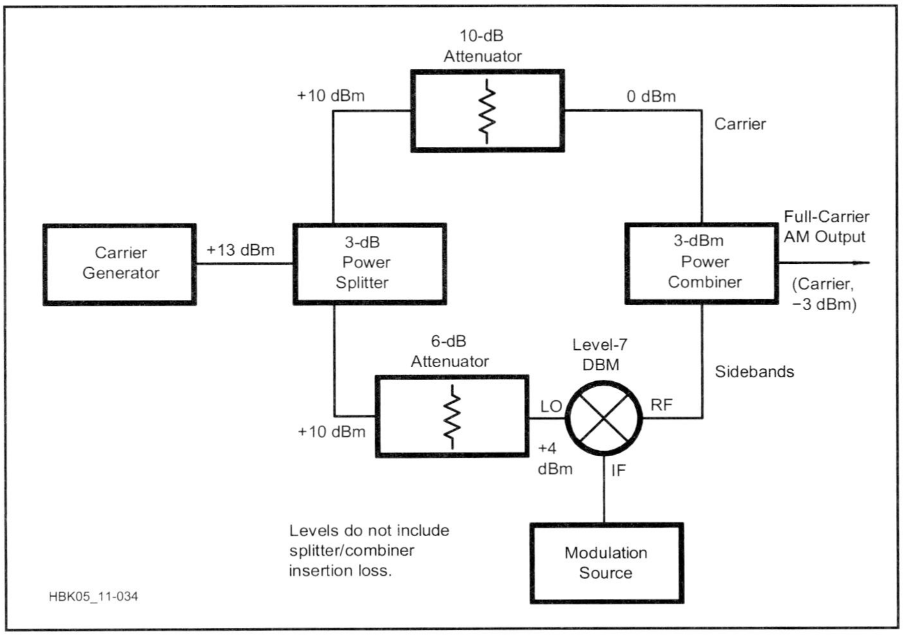

Figure 13.13 — Generating full-carrier AM with a diode DBM. A practical modulator using this technique is described in *Experimental Methods in RF Design*.

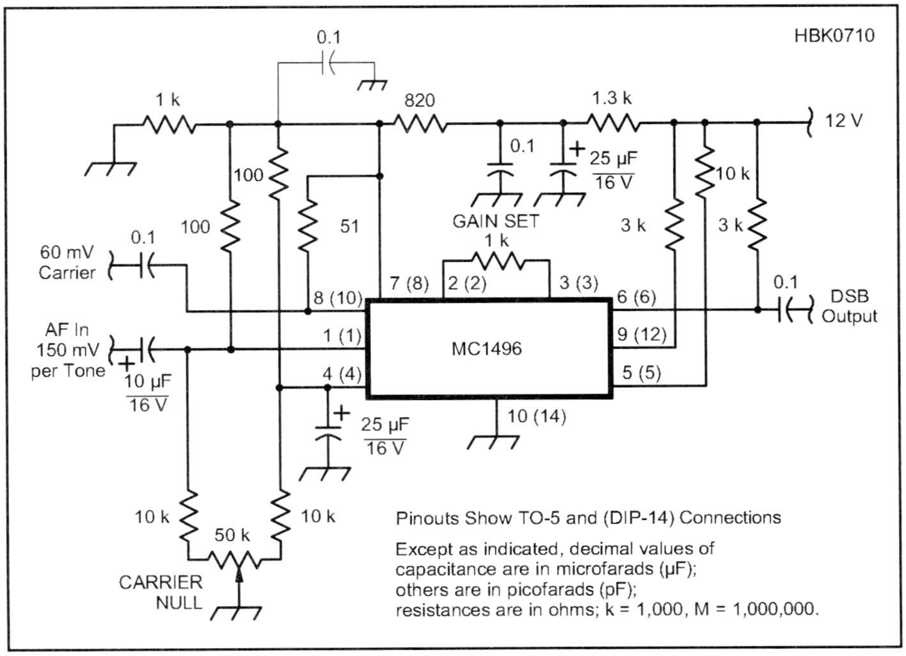

Figure 13.14 — An IC balanced modulator circuit using the MC1496 IC. The resistor between pins 2 and 3 sets the subsystem gain.

SDR SSB GENERATORS

(This section is taken from the SDR: Simplified column by Ray Mack, W5IFS in the September/October 2012 issue of *QEX*.) The structure of this SDR SSB generator software is the same as if it were implemented in analog hardware. **Figure 13.16** shows the block diagram of the system. The program operates in a serial fashion: first an audio baseband filter limits the audio to a band of 300 Hz to 3 kHz. Second, the DDS phase step value to determine the carrier frequency is computed. Following that is the multiplication for the balanced mixer. Finally the undesired sideband is removed. Once the single sideband signal has been created, up-conversion is used to translate it to RF.

The audio band-pass filter response is only useful with 200 taps or more (See **Figure 13.17**). At 100 taps the rejection is only on the order of 12 dB below 100 Hz. Likewise, the opposite sideband filter requires on the order of 700 to 1000 taps to give approximately 60 dB of opposite sideband suppression. The large number of taps also makes the skirts very steep, so that we can use the filter to also further reduce any carrier feed-through. **Figure 13.18** illustrates how steep the skirts can be. If the low-frequency carrier is at 18 kHz, carrier suppression is approximately 52 dB and the unwanted sideband more than that.

One alternative to reduce taps required in the audio filter is to simply use a dc block in the analog portion of the audio chain to set a lower boundary on the frequency. The response will be zero at 0 Hz and rise very rapidly to the frequency we set. This reduces the need for a sharp cutoff in DSP.

The close-in rejection of audio above 3 kHz is 45 dB or more with the 200-tap filter. Additionally, there is almost no energy above 3 kHz in the human voice, so energy in that region will likely be at least 60 dB below the lower frequencies after filtering.

Limiting the higher frequencies allows the use of a 6 kHz wide sideband selection filter instead of the normal 3 kHz filter to get better skirt response. A low-pass or high-pass filter would also work and give approximately the

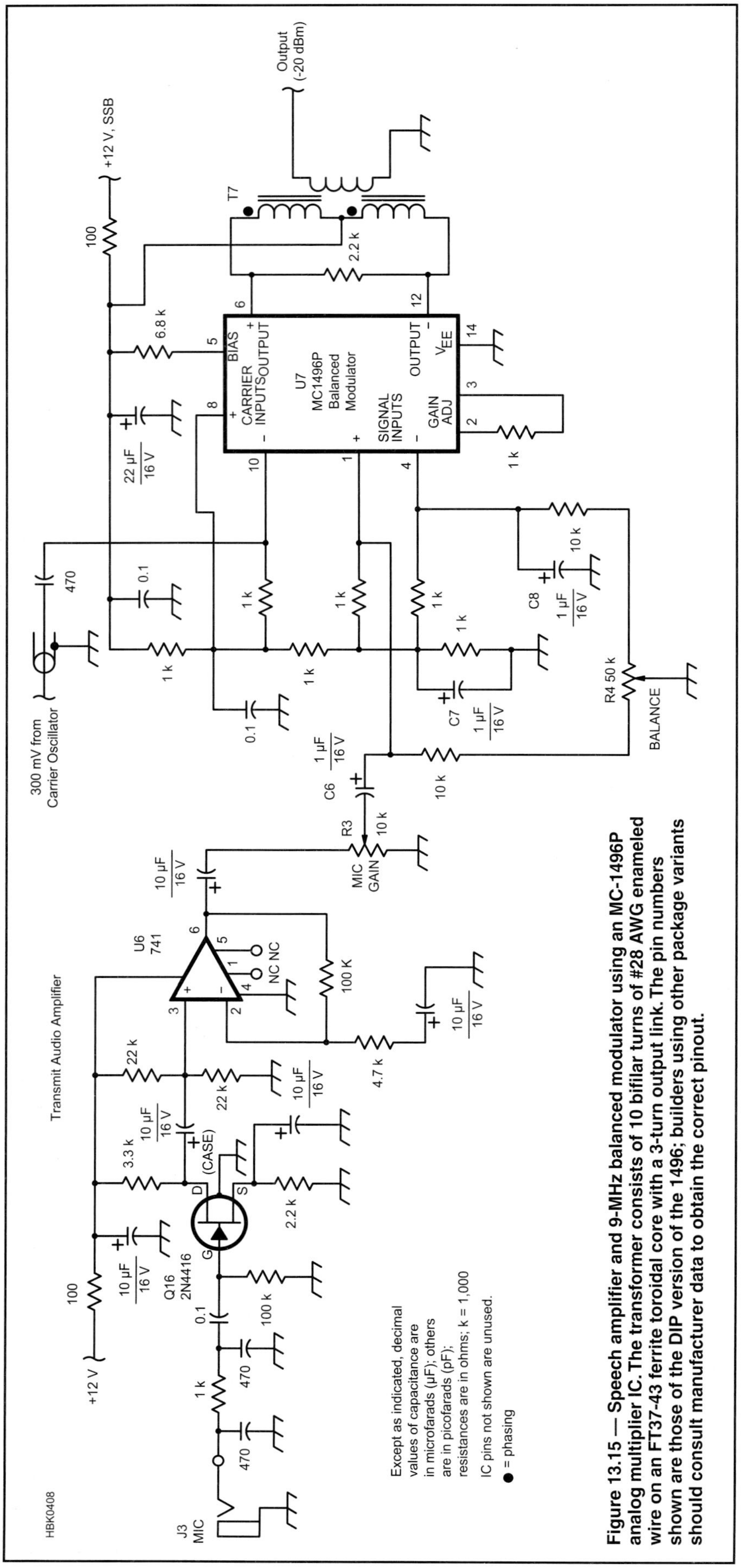

Figure 13.15 — Speech amplifier and 9-MHz balanced modulator using an MC-1496P analog multiplier IC. The transformer consists of 10 bifilar turns of #28 AWG enameled wire on an FT37-43 ferrite toroidal core with a 3-turn output link. The pin numbers shown are those of the DIP version of the 1496; builders using other package variants should consult manufacturer data to obtain the correct pinout.

same skirt response, but we want to be sure to eliminate any residual energy at baseband in the case of a lower sideband transmission. The wider bandwidth limits the lower frequency for our carrier. We want the carrier frequency to be as high as possible in order to limit image response when we up-convert to our final RF signal.

There is a practical limit with respect to the number of taps in the filters. Each tap requires one multiply-accumulate operation, which is a MAC in the DSP world. (MAC also means Media Access Control to a networking hardware person!) The DSP is capable of one MAC for each MHz of clock frequency for each portion of the hardware chain. If voice is digitized at a 48 kHz sample rate, a transmitter filter with 200 taps for audio and 1000 taps for sideband selection will need 57.6 MMACs to do its job. (See the **Receiving** chapter note from KA9Q regarding fast convolution to implement the filter method as a more efficient method if the hardware can support it.) In addition, a large number of taps can create latency. In this case, for a 48 kHz sample stream, 1000 taps represents 20 msec. Depending on how the SDR is implemented, this could add to overall microphone-to-RF latency.

OVERMODULATION

Since the information we transmit using AM shows up entirely as energy in its sidebands, it follows that the more energetic we make the sidebands, the more information energy will be available for an AM receiver to "recover" when it demodulates the signal. Even in an ideal modulator, there's a practical limit to how strong we can make an AM signal's sidebands relative to its carrier, however. Beyond that limit, we severely distort the waveform we want to translate into radio form.

We reach AM's distortion-free modulation limit when the sum of the sidebands and carrier at the modulator output *just reaches zero* at the modulating wave-form's most negative peak (**Figure 13.19**). We call this condition *100% modulation*, and it occurs when *m* in equation A in the sidebar "Mixer Math: Amplitude Modulation" equals 1. (We enumerate *modulation percentage* in values from 0 to 100%. The lower the number, the less information energy is in the sidebands. You may also see modulation enumerated in terms of a *modulation factor* from 0 to 1, which directly equals *m*; a modulation factor of 1 is the same as 100% modulation.) Equation B in the sidebar shows that each sideband's voltage is half that of the carrier. Power varies as the square of voltage, so the power in each sideband of a 100%-modulated signal is therefore $(\frac{1}{2})^2$ times, or ¼, that of the carrier. A transmitter capable of 100% modulation when operating at a carrier power of 100 W

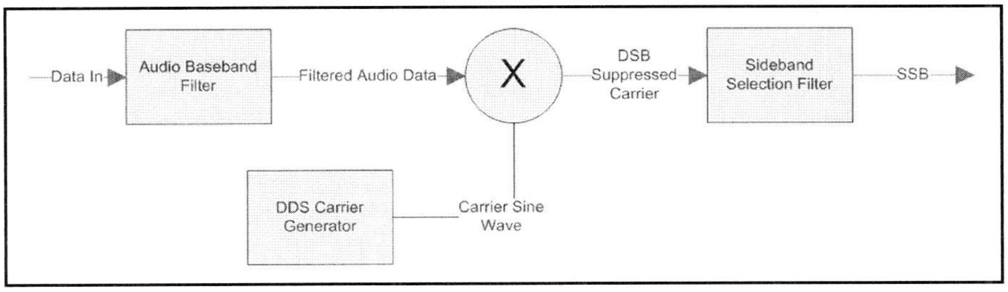

Figure 13.16 — This block diagram shows the software SSB transmit generator using the filter method.

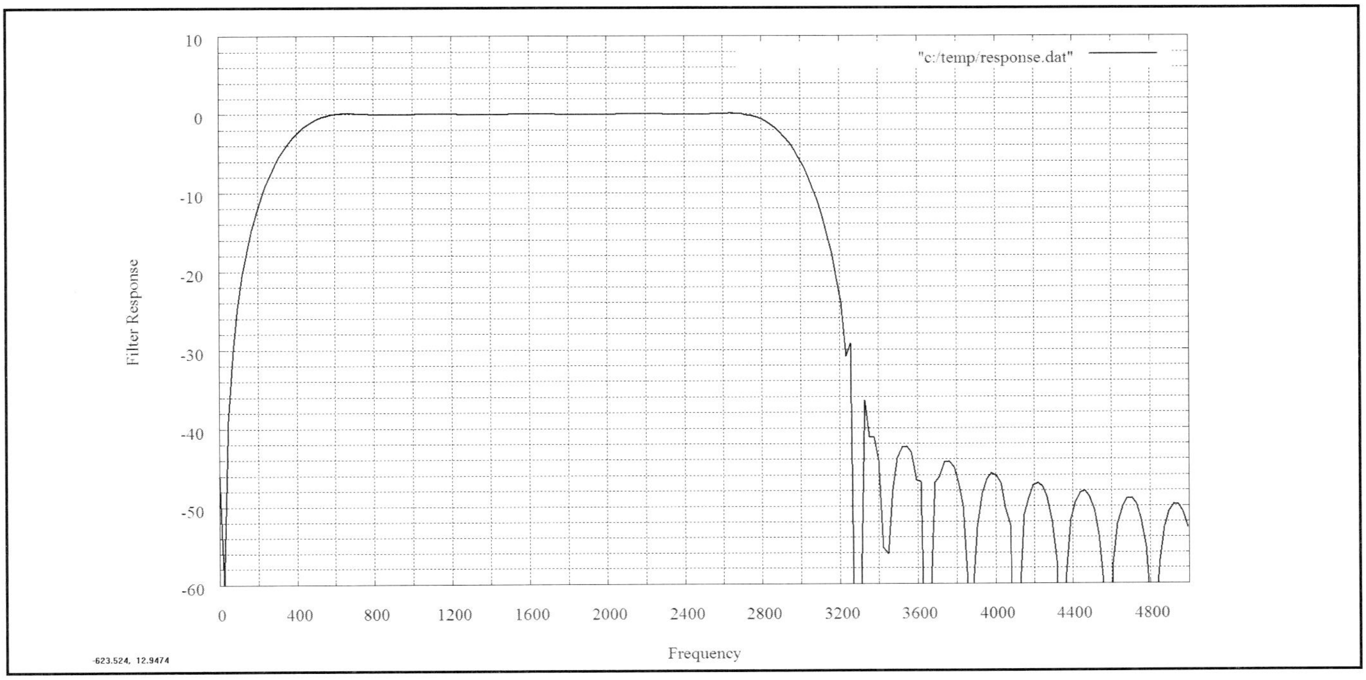

Figure 13.17 — The 200-tap baseband filter response.

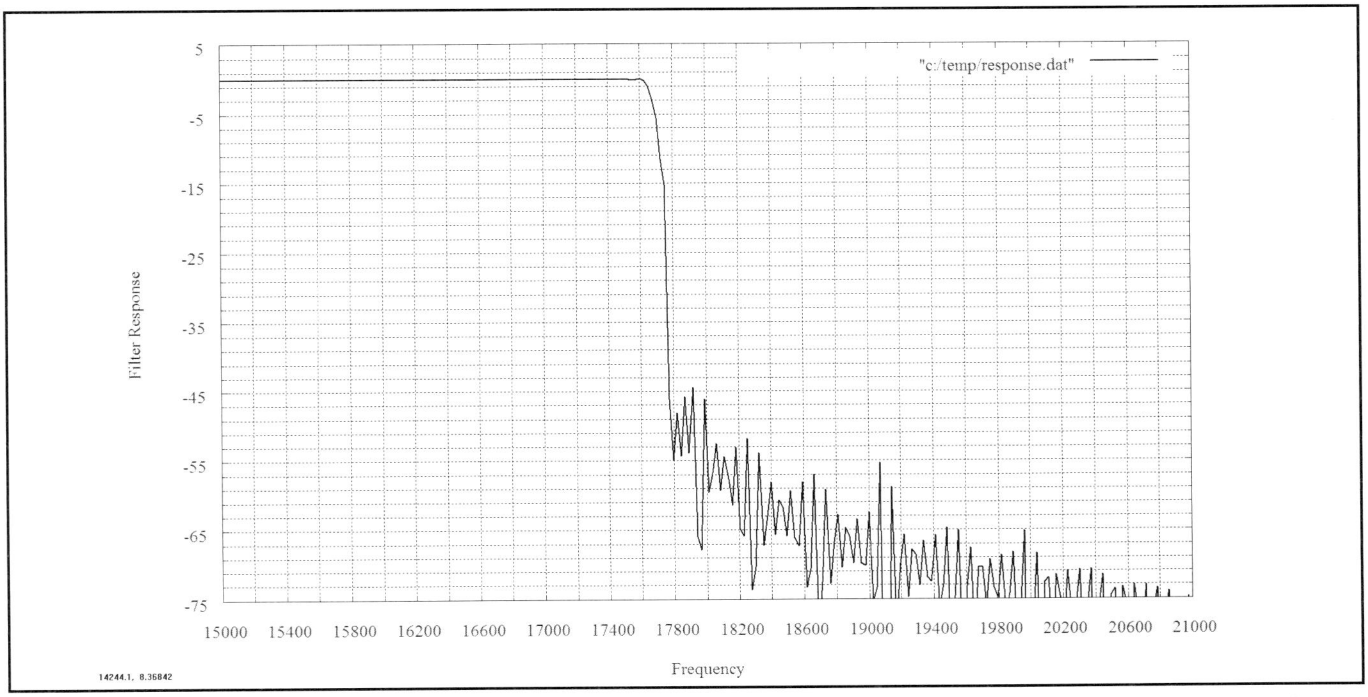

Figure 13.18 — The response of a 700-tap filter, showing a wider frequency view. The filter is 6 kHz wide to allow for a steep skirt on the carrier side. The 6 dB cutoff point is set at 300 Hz away from the low-frequency carrier, which is at 18,000 Hz (see text).

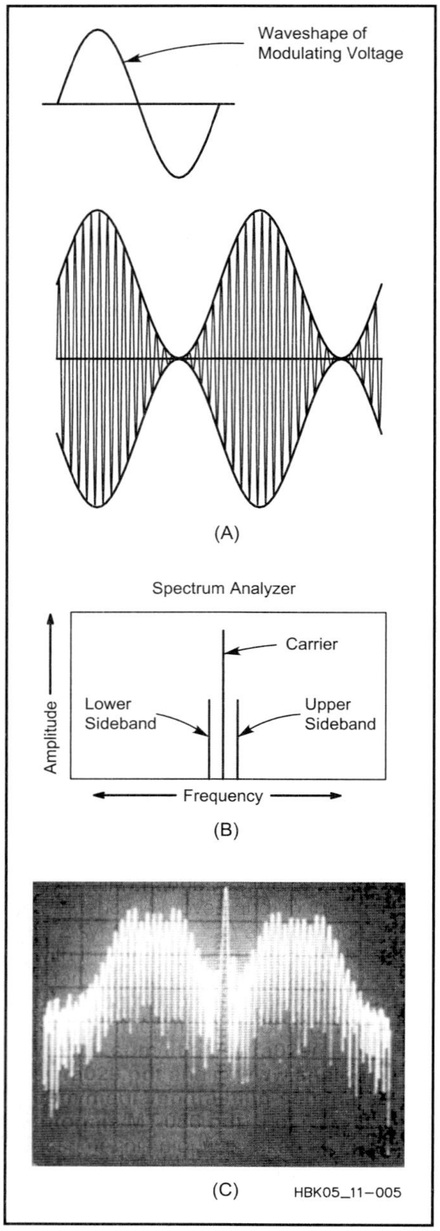

Figure 13.19 — Graphed in terms of amplitude versus time (A), the *envelope* of a properly modulated AM signal exactly mirrors the shape of its modulating waveform, which is a sine wave in this example. This AM signal is modulated as fully as it can be — 100% — because its envelope *just* touches zero on the modulating wave's negative peaks. Graphing the same AM signal in terms of amplitude versus frequency (B) reveals its three spectral components: Carrier, upper sideband and lower sideband. B shows sidebands as single-frequency components because the modulating waveform is a sine wave. With a complex modulating waveform, the modulator's sum and difference products really do show up as bands on either side of the carrier (C).

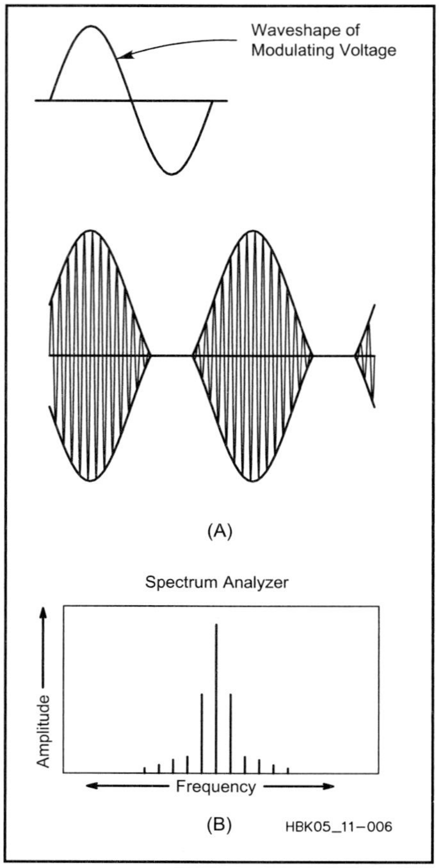

Figure 13.20 — Negative-going overmodulation of an AM transmitter results in a modulation envelope (A) that doesn't faithfully mirror the modulating waveform. This distortion creates additional sideband components that broaden the transmitted signal (B). Positive-going modulation beyond 100% is used by some AM broadcasters in conjunction with negative-peak limiting to increase "talk power" without causing negative overmodulation.

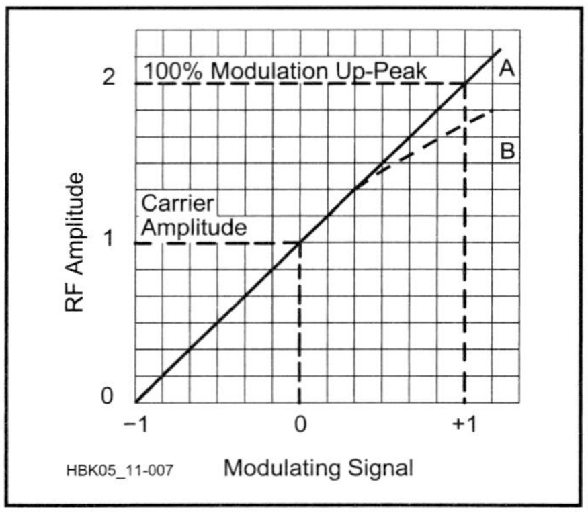

Figure 13.21 — An ideal AM transmitter exhibits a straight-line relationship (A) between its instantaneous envelope amplitude and the instantaneous amplitude of its modulating signal. Distortion, and thus an unnecessarily wide signal, results if the transmitter cannot respond linearly across the modulating signal's full amplitude range (B).

therefore puts out a 150-W signal at 100% modulation, 50 W of which is attributable to the sidebands. (The *peak* envelope power [PEP] output of a double-sideband, full-carrier AM transmitter at 100% modulation is four times its carrier PEP. This is why our solid-state, "100-W" MF/HF transceivers are usually rated for no more than about 25 W carrier output at 100% amplitude modulation.)

One-hundred-percent negative modulation is a brick-wall limit because an amplitude modulator can't reduce its output to less than zero. Trying to increase negative modulation beyond the 100% point results in *overmodulation* (**Figure 13.20**), in which the modulation envelope no longer mirrors the shape of the modulating wave (Figure 13.20A). A negatively overmodulated wave contains more energy than it did at 100% modulation, but some of the added energy now exists as *harmonics of the modulating waveform* (Figure 13.20B). This distortion makes the modulated signal take up more spectrum space than it needs. In voice operation, overmodulation commonly happens only on syllabic peaks, making the distortion products sound like transient noise we refer to as *splatter*.

MODULATION LINEARITY

If we increase an amplitude modulator's modulating-signal input by a given percentage, we expect a proportional modulation increase in the modulated signal. We expect good *modulation linearity*. Suboptimal amplitude modulator design may not allow this, however. Above some modulation percentage, a modulator may fail to increase modulation in proportion to an increase in its input signal (**Figure 13.21**). Distortion, and thus an unnecessarily wide signal, results.

13.3.2 Angle Modulators

Amplitude modulation served as our first means of translating information into radio form because it could be implemented as simply as turning an electric noise generator on and off. (A spark transmitter consisted of little more than this.) By the 1930s, we had begun experimenting with translating information into radio form and back again by modulating a radio wave's angular velocity (frequency or phase) instead of its overall amplitude. The result of this process is *frequency modulation (FM)* or *phase modulation (PM)*, both of which are often grouped under the name *angle modulation* because of their underlying principle.

A change in a carrier's frequency or phase

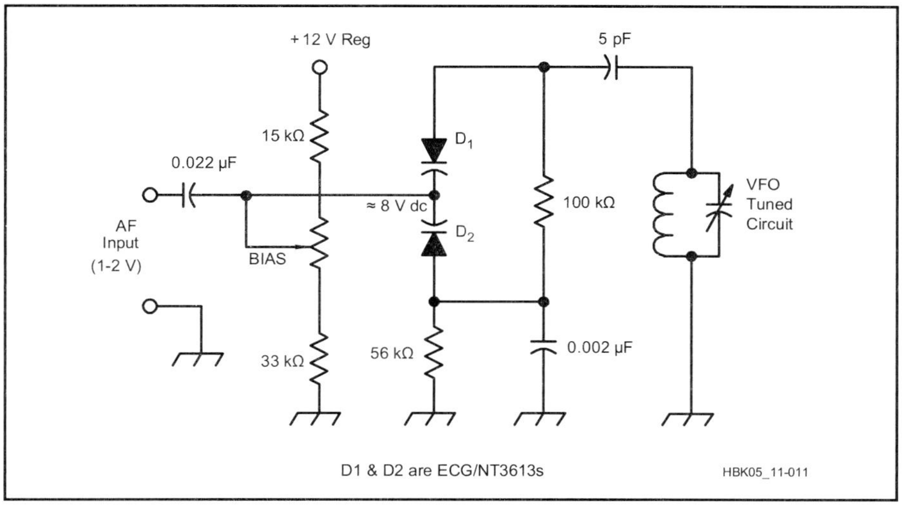

Figure 13.22 — One or more tuning diodes can serve as the variable reactance in a reactance modulator. This HF reactance modulator circuit uses two diodes in series to ensure that the tuned circuit's RF-voltage swing cannot bias the diodes into conduction. D1 and D2 are "30-volt" tuning diodes that exhibit a capacitance of 22 pF at a bias voltage of 4. The BIAS control sets the point on the diode's voltage-versus-capacitance characteristic around which the modulating waveform swings.

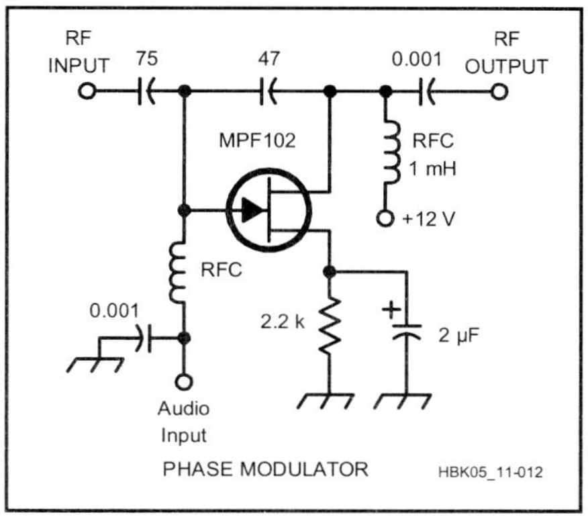

Figure 13.23 — A series reactance modulator acts as a variable shunt around a reactance — in this case, a 47-pF capacitor — through which the carrier passes.

for the purpose of modulation is called *deviation*. An FM signal deviates according to the amplitude of its modulating waveform, independently of the modulating waveform's frequency; the higher the modulating wave's amplitude, the greater the deviation. A PM signal deviates according to the amplitude *and frequency* of its modulating waveform; the higher the modulating wave's amplitude *and/or frequency*, the greater the deviation. See the sidebar, "Mixer Math: Angle Modulation" for a numerical description of these processes.

If you vary a reactance in or associated with an oscillator's frequency-determining element(s), you vary the oscillator's frequency. If you vary the tuning of a tuned circuit through which a signal passes, you vary the signal's phase. A circuit that does this is called a *reactance modulator*, and can be little more than a tuning diode or two connected to a tuned circuit in an oscillator or amplifier (**Figure 13.22**). Varying a reactance through which the signal passes (**Figure 13.23**) is another way of doing the same thing.

The difference between FM and PM depends solely on how, and not how much, deviation occurs. A modulator that causes deviation in proportion to the modulating wave's amplitude and frequency is a phase modulator. A modulator that causes deviation only in proportion to the modulating signal's amplitude is a frequency modulator.

ANGLE MODULATION SIDEBANDS

Although angle modulation produces uncountable sum and difference products, most of them are vanishingly weak in practical systems. They emerge from the modulator spaced from the average ("resting," unmodulated) carrier frequency by integer multiples of the modulating frequency (**Figure 13.24**). The strength of the sidebands relative to the carrier, and the strength and phase of the carrier itself, vary with the degree of modulation — the modulation index. (The *overall* amplitude of an angle-modulated signal does not change with modulation, however; when energy goes out of the carrier, it shows up in the sidebands, and vice versa.) In practice, we operate angle-modulated transmitters at modulation indexes that make all but a few of their infinite sidebands small in amplitude. (A mathematical tool called *Bessel functions* helps determine the relative strength of the carrier and sidebands according to modulation index. The **Modulation** chapter includes a graph to illustrate this relationship.)

Mixer Math: Angle Modulation

An angle-modulated signal can be mathematically represented as

$$f_c(t) = \cos(2\pi f_c t + m \sin(2\pi f_m t))$$

$$= \cos(2\pi f_c t)\cos(m \sin(2\pi f_m t)) - \sin(2\pi f_c t)\sin(m \sin(2\pi f_m t))$$

In this equation, we see the carrier frequency ($2\pi f_c t$) and modulating signal ($\sin 2\pi f_m t$) as in equation A shown in the sidebar Mixer Math: Amplitude Modulation. We again see the modulating signal associated with a coefficient, m, which relates to degree of modulation. (In the AM equation, *m* is the modulation factor; in the angle-modulation equation, *m* is the *modulation index* and, for FM, equals the deviation divided by the modulating frequency.) We see that angle-modulation occurs as the cosine of the sum of the carrier frequency ($2\pi f_c t$) and the modulating signal ($\sin 2\pi f_m t$) times the modulation index (*m*). In its expanded form, we see the appearance of sidebands above and below the carrier frequency.

Angle modulation is a multiplicative process, so, like AM, it creates sidebands on both sides of the carrier. Unlike AM, however, angle modulation creates an *infinite* number of sidebands on either side of the carrier! This occurs as a direct result of modulating the carrier's angular velocity, to which its frequency and phase directly relate. If we continuously vary a wave's angular velocity according to another periodic wave's cyclical amplitude variations, the rate at which the modulated wave repeats *its* cycle — its frequency — passes through an infinite number of values. (How many individual amplitude points are there in one cycle of the modulating wave? An infinite number. How many corresponding discrete frequency or phase values does the corresponding angle-modulated wave pass through as the modulating signal completes a cycle? An infinite number!) In AM, the carrier frequency stays at one value, so AM produces two sidebands — the sum of its carrier's unchanging frequency value and the modulating frequency, and the difference between the carrier's unchanging frequency value and the modulating frequency. In angle modulation, the modulating wave shifts the frequency or phase of the carrier through an infinite number of different frequency or phase values, resulting in an infinite number of sum and difference products.

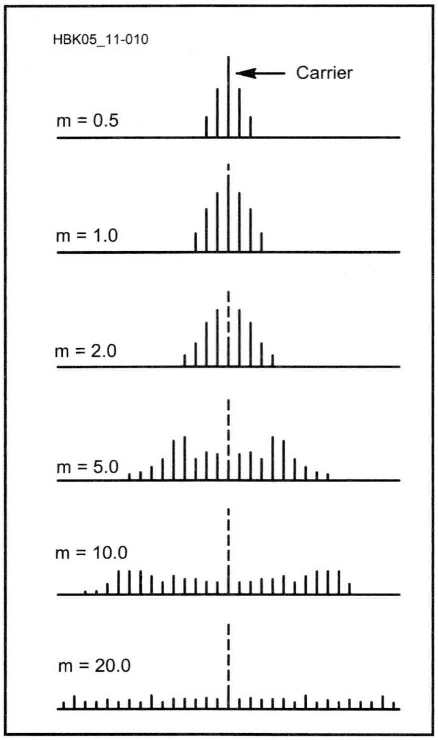

Figure 13.24 — Angle-modulation produces a carrier and an infinite number of upper and lower sidebands spaced from the average ("resting," unmodulated) carrier frequency by integer multiples of the modulating frequency. (This drawing is a simplification because it only shows relatively strong, close-in sideband pairs; space constraints prevent us from extending it to infinity.) The relative amplitudes of the sideband pairs and carrier vary with modulation index, *m*.

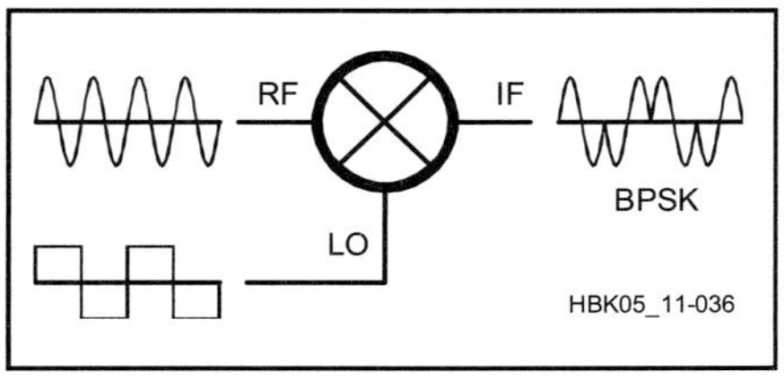

Figure 13.25 — Mixing a carrier with a square wave generates biphase-shift keying (BPSK), in which the carrier phase is shifted 180° for data transmission. In practice, as in this drawing, the carrier and data signals are phase-coherent so the mixer switches only at carrier zero crossings.

Selectivity in transmitter and receiver circuitry further modify this relationship, especially for sidebands far away from the carrier.

BIPHASE-SHIFT KEYING (BPSK) MODULATION WITH A DBM

Back in our discussion of square-wave mixing, we saw how multiplying a switching mixer's linear input with a square wave causes a 180° phase shift during the negative part of the square wave's cycle. As **Figure 13.25** shows, we can use this effect to produce *biphase-shift keying (BPSK)*, a digital system that conveys data by means of carrier phase reversals. A related system, *quadrature phase-shift keying (QPSK)* uses two DBMs and phasing to convey data by phase-shifting a carrier in 90° increments.

DEVIATION AND FREQUENCY MULTIPLICATION

Maintaining modulation linearity is just as important in angle modulation as it is in AM, because unwanted distortion is always our enemy. A given angle-modulator circuit can frequency- or phase-shift a carrier only so much before the shift stops occurring in strict proportion to the amplitude (or, in PM, the amplitude and frequency) of the modulating signal.

If we want more deviation than an angle modulator can linearly achieve, we can operate the modulator at a suitable sub-harmonic — submultiple — of the desired frequency, and process the modulated signal through a series of *frequency multipliers* to bring it up to the desired frequency. The deviation also increases by the overall multiplication factor, relieving the modulator of having to do it all directly. A given FM or PM radio design may achieve its final output frequency through a combination of mixing (frequency shift, no deviation change) and frequency multiplication (frequency shift *and* deviation change).

"TRUE FM"

Something we covered a bit earlier bears closer study: "An FM signal deviates according to the amplitude of its modulating waveform, independently of the modulating waveform's frequency; the higher the modulating wave's amplitude, the greater the deviation. A PM signal deviates according to the amplitude *and frequency* of its modulating waveform; the higher the modulating wave's amplitude *and/or frequency*, the greater the deviation."

The practical upshot of this excerpt is that we can use a phase modulator to generate FM. All we need to do is run a PM transmitter's modulating signal through a low-pass filter that (ideally) halves the signal's amplitude for each doubling of frequency (a reduction of "6 dB per octave," as we sometimes see such responses characterized) to compensate for its phase modulator's "more deviation with higher frequencies" characteristic. The result is an FM, not PM, signal. FM achieved with a phase modulator is sometimes called *indirect FM* as opposed to the *direct FM* we get from a frequency modulator.

We sometimes see claims that one piece of gear is better than another solely because it generates "true FM" as opposed to indirect FM. We can debunk such claims by keeping in mind that direct and indirect FM *sound exactly alike in a receiver* when performed correctly.

CONVEYING DC LEVELS WITH ANGLE MODULATION

Depending on the nature of the modulation source, there *is* a practical difference between a frequency modulator and a phase modulator. Answering two questions can tell us whether this difference matters: Does our modulating signal contain a dc level or not? If so, do we need to accurately preserve that dc level through our radio communication link for successful communication? If both answers are *yes*, we must choose our hardware and/or

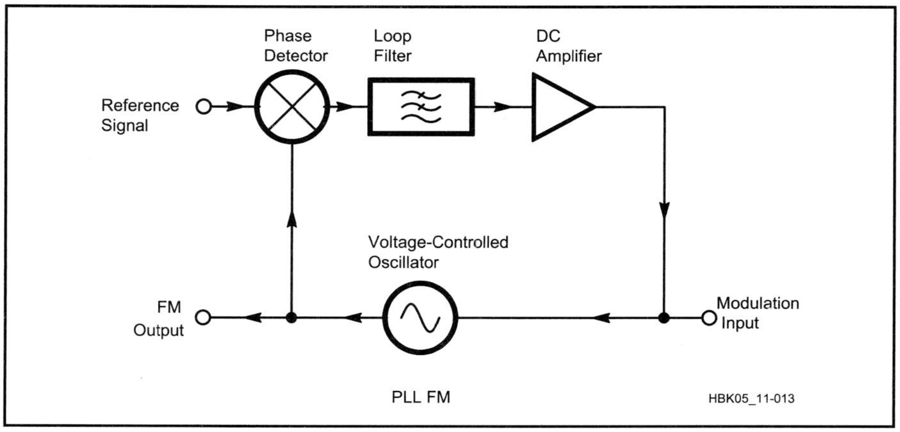

Figure 13.26 — Frequency modulation using a phase-locked loop (PLL).

information-encoding approach carefully, because a frequency modulator can convey dc-level shifts in its modulating waveform, while a phase modulator, which responds only to instantaneous changes in frequency and phase, cannot.

Consider what happens when we want to frequency-modulate a phase-locked-loop-synthesized transmitted signal. **Figure 13.26** shows the block diagram of a PLL frequency modulator. Normally, we modulate a PLL's VCO because it's the easy thing to do. As long as our modulating frequency results in frequency excursions too fast for the PLL to follow and correct — that is, as long as our modulating frequency is outside the PLL's *loop bandwidth* — we achieve the FM we seek. Trying to modulate a dc level by pushing the VCO to a particular frequency and holding it there fails, however, because a PLL's loop response includes dc. The loop, therefore, detects the modulation's dc component as a correctable error and "fixes" it. FMing a PLL's VCO therefore can't buy us the dc response "true FM" is supposed to allow.

We *can* dc-modulate a PLL modulator, but we must do so by modulating the frequency of the loop *reference*. The PLL then adjusts the VCO to adapt to the changed reference, and our dc level gets through. In this case, the modulating frequency must be *within* the loop bandwidth — which dc certainly is — or the VCO won't be corrected to track the shift.

SDR ANGLE MODULATORS

Figure 13.27 shows the block diagram of an FM transmitter using DSP to produce the carrier, and using addition to create true FM from the audio input. An FM signal can be generated directly through the DDS by simply adding or subtracting a small value that corresponds to the audio voltage to the tuning value used for the DDS accumulator. The frequency deviation is adjusted by controlling the gain applied to the audio signal. This method produces true frequency modulation. If the audio signal is used to control the phase accumulation of the DDS, the result is PM.

Angle modulation can also be produced using I/Q modulation as in **Figure 13.28**. A DSB-SC (double-sideband, suppressed-carrier) signal is produced by a balanced modulator or DSP multiplier (see SDR SSB Generators earlier in this chapter). The DSB-SC signal is then added to the carrier signal with a 90º phase difference. The result is a PM signal.

Frequency modulation requires that an integrator (low-pass filter) be applied to the modulating signal. This is because frequency is the time-derivative of phase. By applying the integrated signal to a phase modulator, FM is produced with a deviation that does not depend on signal amplitude — only frequency.

Because of the low-pass filter, speech is usually given a high-frequency boost by a high-pass *pre-emphasis* network. A corresponding *de-emphasis* (low-pass) network must be applied to the recovered modulation in the receiver to restore the original modulating signal's frequency response. This improves intelligibility and signal-to-noise ratio of the received audio.

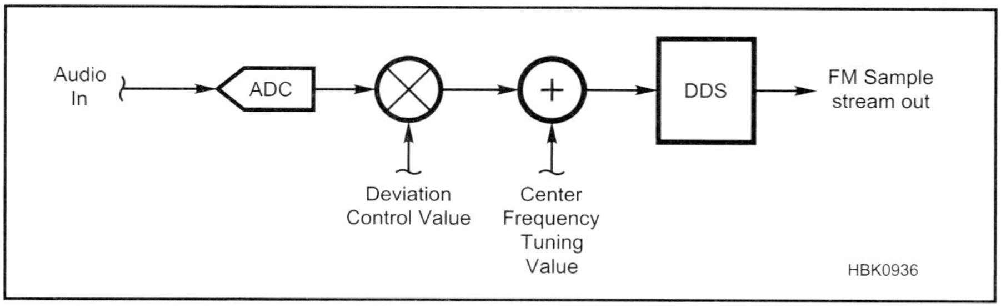

Figure 13.27 — Creating FM by controlling the frequency of a DDS signal source. By controlling the phase step instead of the frequency, PM would be created.

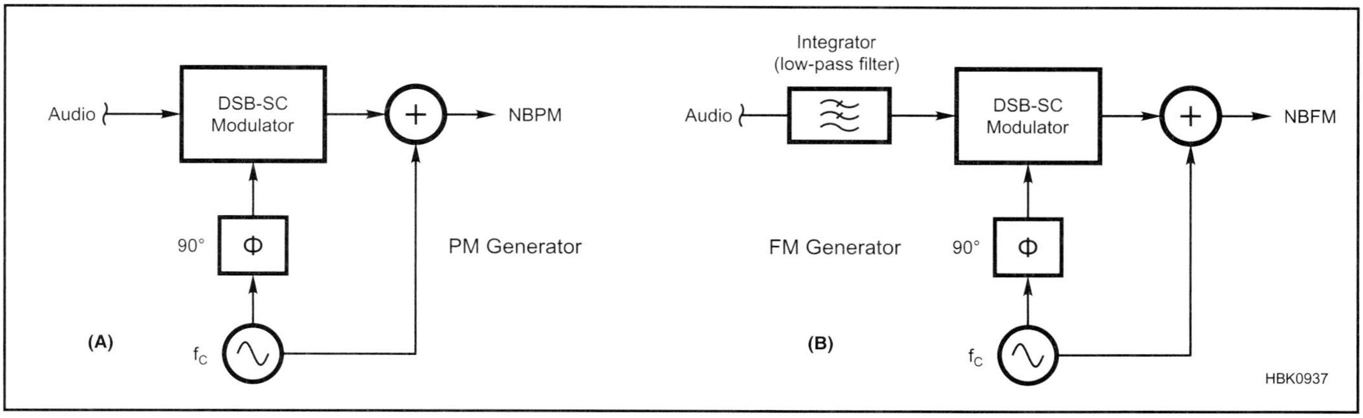

Figure 13.28 — Two methods of using I/Q modulation to produced angle-modulated signals. The block diagram includes a DSB generator (Figure 13.16) with the additional step of mixing (multiplying) the DSB signal with a phase-shifted carrier signal to produce PM (A) or FM (B). FM requires applies an integrator (low-pass filter) to the modulating signal so that output frequency depends only on the amplitude of the modulating signal and not its frequency.

13.4 Transmitting CW

Earlier in this chapter, the importance of shaping the time envelope of the keying pulse of an on-off keyed transmitter is discussed. There are serious ramifications of not paying close attention to this design parameter. The optimum shape of a transmitter envelope should approach the form of a sinusoid raised to a power with a tradeoff between occupied bandwidth and overlap between the successive pulses. This can be accomplished either through filtering of the pulse waveform before modulation in a linear transmitter, or through direct generation of the pulse shape using DSP.

The differences between well-designed and poor pulse shaping can perhaps be best described by looking at some results. The following figures are from recent *QST* product reviews of commercial multimode 100 W HF transceivers. **Figure 13.29** shows the CW keying waveform of a transmitter with good spectrum control. The top trace is the key closure, with the start of the first contact closure on the left edge at 60 WPM using full break-in. Below it is the nicely rounded RF envelope. **Figure 13.30** shows the resultant signal spectrum. Note that the signal amplitude is about 80 dB down at a spacing of ±1 kHz, with a floor of –90 dB over the 10 kHz shown. **Figures 13.31** and **13.32** are similar data taken from a different manufacturer's transceiver. Note the sharp corners of the RF envelope, as well as the time it takes for the first "dit" to be developed. The resulting spectrum is not even down 40 dB at ±1 kHz and shows a floor that doesn't quite make –60 dB over the 10 kHz range. It's easy to see the problems that the latter transmitter will cause to receivers trying to listen to a weak signal near its operating frequency. The unwanted components of the signal are heard on adjacent channels as sharp clicks when the signal is turned on and off, called *key clicks*. Note that even the best-shaped keying waveform in a linear transmitter will become sharp with a wide spectrum if it is used to drive a stage such as an external power amplifier beyond its linear range. This generally results in clipping or limiting with subsequent removal of the rounded corners on the envelope. Trying to get the last few dB of power out of a transmitter can often result in this sort of unintended signal impairment.

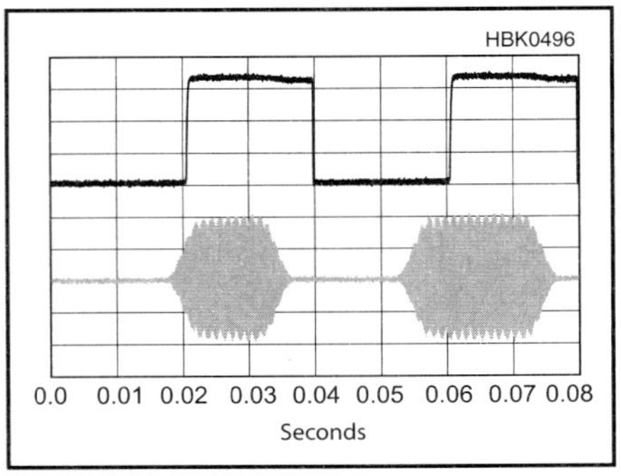

Figure 13.29 — The CW keying waveform of a transmitter with good spectrum control. The top trace is the key closure, with the start of the first contact closure on the left edge at 60 WPM using full break-in. Below that is the nicely rounded RF envelope.

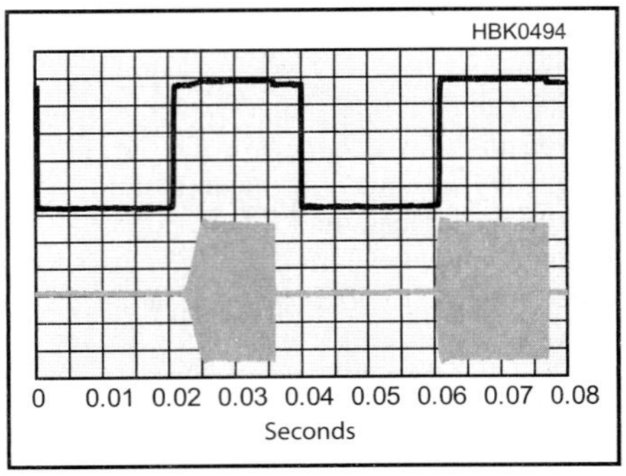

Figure 13.31 — The CW keying waveform of a transmitter with poor spectrum control. The top trace is the key closure, with the start of the first contact closure on the left edge at 60 WPM using full break-in. Note the sharp corners of the RF envelope that result in excessive bandwidth products.

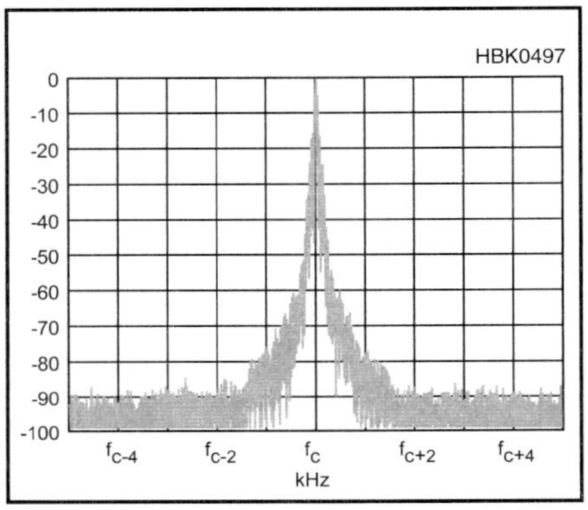

Figure 13.30 — The resultant signal spectrum from the keying shown in Figure 13.29. Note that the signal amplitude is about 80 dB down at a spacing of ±1 kHz, with a floor of –90 dB over the 10 kHz shown.

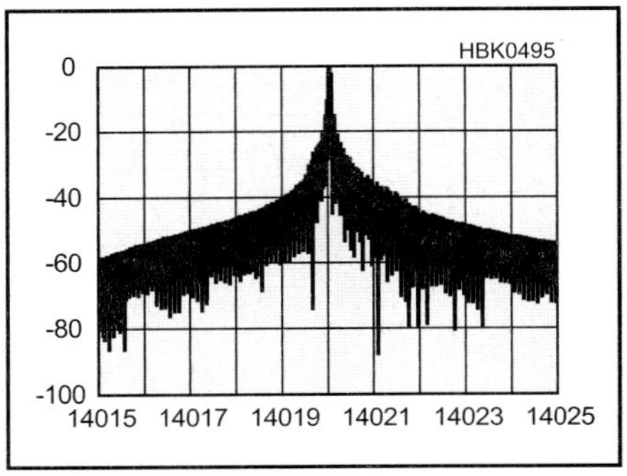

Figure 13.32 — The resultant signal spectrum from the keying shown in Figure 13.31. The resulting spectrum is not even down 40 dB at ±1 kHz and shows a floor that doesn't quite make 60 dB below the carrier across the 10 kHz.

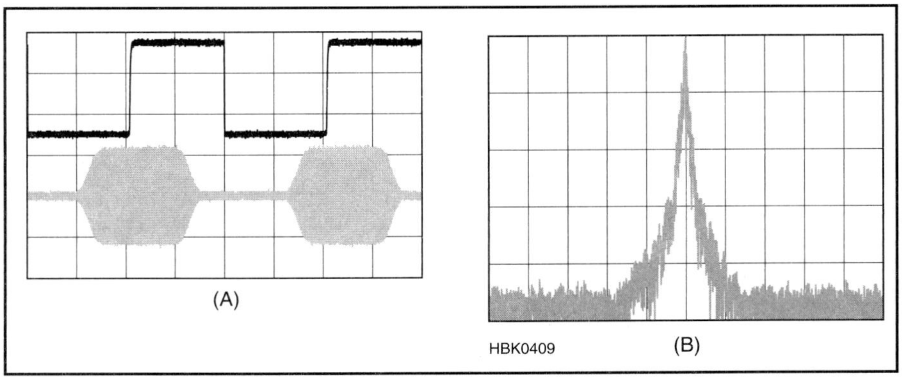

Figure 13.33 — Waveshaping in a CW transmitter often causes a CW signal's RF envelope (lower trace in the amplitude-versus-time display at A) to contain less harmonic energy than the abrupt transitions of its key closure waveform (upper trace in A) suggest should be the case. B, an amplitude-versus-frequency display, shows that even a properly shaped CW signal has many sideband components.

13.4.1 CW Operation

Figure 13.33A closely resembles what we see when a properly adjusted CW transmitter sends a string of dots. Keying a carrier on and off produces a wave that varies in amplitude and has double (upper and lower) sidebands that vary in spectral composition according to the duration and envelope shape of the on-off transitions. The emission mode we call CW is therefore a form of AM. The concepts of modulation percentage and overmodulation are usually not applied to generating an on-off-keyed Morse signal, however. This is related to how we copy CW by ear, and the fact that, in CW radio communication, we usually don't translate the received signal all the way back into its original pre-modulator (*baseband*) form, as a closer look at the process reveals.

In CW transmission, we usually open and close a keying line to make dc transitions that turn the transmitted carrier on and off. See Figure 13.33B. CW reception usually does not entirely reverse this process, however. Instead of demodulating a CW signal all the way back to its baseband self — a shifting dc level — we want the presences and absences of its carrier to create long and short audio tones. Because the carrier is RF and not AF, we must mix it with a locally generated RF signal — from a *beat-frequency oscillator (BFO)* — that's close enough in frequency to produce a difference signal at AF (this BFO can, of course, also be inserted at an IF stage). What goes into our transmitter as shifting dc comes out of our receiver as tone bursts of dot and dash duration.

It so happens that we always need to hear one or more harmonics of the fundamental keying waveform for the code to sound sufficiently crisp. If the transmitted signal will be subject to fading caused by varying propagation — a safe assumption for any long-distance radio communication — we can *harden* our keying by making the transmitter's output rise and fall more quickly. This puts more energy into keying sidebands and makes the signal more copyable in the presence of fading — in particular, *selective fading*, which linearly distorts a modulated signal's complex waveform and randomly changes the sidebands' strength and phase relative to the carrier and each other. The appropriate keying hardness also depends on the keying speed. The faster the keying in WPM, the faster the on-off times — the harder the keying — must be for the signal to remain ear- and machine-readable through noise and fading.

Instead of thinking of this process in terms of modulation percentage, we just ensure that a CW transmitter produces sufficient keying-sideband energy for solid reception. Practical CW transmitters usually do not do their keying with a modulator stage as such. Instead, one or more stages are turned on and off to modulate the carrier with Morse, with rise and fall times set by *R* and *C* values associated with the stages' keying and/or power supply lines. A transmitter's CW *waveshaping* is therefore usually hardwired to values appropriate for reasonably high-speed sending (35 to 55 WPM or so) in the presence of fading. However, some transceivers allow the user to vary keying hardness at will as a menu option. Rise and fall times of 1 to 5 ms are common; 5-ms rise and fall times equate to a keying speed of 36 WPM in the presence of fading and 60 WPM if fading is absent.

The faster a CW transmitter's output changes between zero and maximum, the more bandwidth its carrier and sidebands occupy. See Figure 13.33B. Making a CW signal's keying too hard is therefore spectrum-wasteful and inconsiderate of other stations because it makes the signal wider than it needs to be. Keying sidebands that are stronger and wider than necessary are traditionally called *clicks* because of what they sound like on the air.

Radiotelegraph or CW operation can be easily obtained from the transmitter architecture design shown in Figure 13.9. For CW operation, a carrier is generated at the center of the SSB filter passband. There are two ways to make this carrier available. One way is to unbalance the balanced modulator so that the LO can pass through. Each kind of balanced modulator circuit has its own method of doing this. The approach chosen in Figure 13.9 is to go around the modulator and the SSB filter.

A shaping network controls the envelope of the IF signal to accomplish two things: control the shape of the Morse code character in a way that limits wideband spectrum emissions that can cause interference, and make the Morse code signal easy and pleasant to copy.

13.4.2 RF Envelope Shaping

On-off keying (CW) is a special kind of low-level amplitude modulation (a low signal-level stage is turned on and off). It is special because the sideband power is subtracted from the carrier power, and not provided by a separate "modulator" circuit, as in high-level AM. It creates a spectrum around the carrier frequency whose amplitude and bandwidth are influenced by the rates of signal amplitude rise and fall and by the curvature of the keyed waveform.. For additional information see the article by Sabin on IF signal processing in the References section of this chapter.

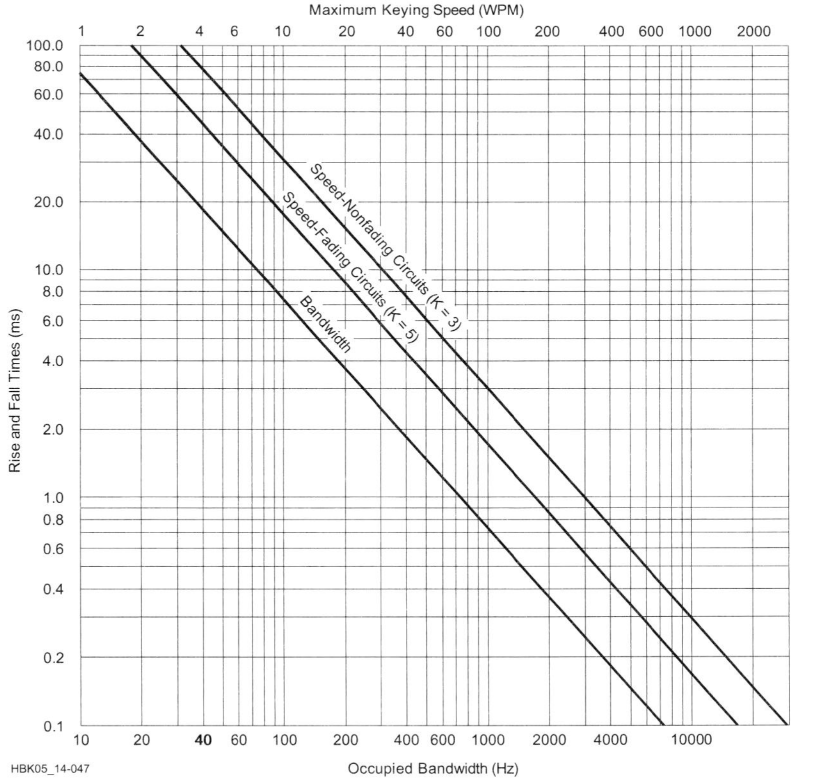

Figure 13.34 — Keying speed versus rise and fall times versus bandwidth for fading and nonfading communications circuits. For example, for transmitter output waveform rise and fall times of approximately 6 ms, draw a horizontal line from 6.0 ms on the rise and fall times scale to the bandwidth line. Then draw a vertical line to the occupied bandwidth scale at the bottom of the graph. In this case the bandwidth is about 130 Hz. Also extend the 6.0 ms horizontal line to the K = 3 line for a nonfading circuit. Finally draw a vertical line from the K = 3 line to the WPM axis. The 6 ms rise and fall time should be suitable for keying speeds up to about 50 WPM in this example.

Now look at **Figure 13.34**. The vertical axis is labeled Rise and Fall Times (ms). For a rise/fall time of 6 ms (between the 10% and 90% values) go horizontally to the line marked Bandwidth. A –20 dB bandwidth of roughly 120 Hz is indicated on the lower horizontal axis. Continuing to the K = 5 and K = 3 lines, the upper horizontal axis suggests code speeds of 30 WPM and 50 WPM respectively.

These code speeds can be accommodated by the rise and fall times displayed on the vertical axis. For code speeds greater than these the Morse code characters become "soft" sounding and difficult to copy, especially under less-than-ideal propagation conditions.

The ITU Classification of Emission Standards for determining necessary bandwidths of signals uses a value of 0.8 for the conversion between baud and WPM and suggests a typical value for K of 5 on an HF channel where the signal is subjected to fading. The bandwidth for a 13 WPM signal would then be:

$$BW = WPM \times 0.8 \times 5 = 10.4 \times 5 = 52 \text{ Hz}$$

For a narrow spectrum and freedom from adjacent channel interference, a further requirement is that the spectrum must fall off very rapidly beyond the –20 dB bandwidth indicated in Figure 13.34. A sensitive narrowband CW receiver that is tuned to an adjacent channel that is only 1 or 2 kHz away can detect keying sidebands that are 80 to 100 dB below the key-down level of a strong CW signal.

An additional consideration is that during key-up a residual signal, called *backwave*, should not be noticeable in a nearby receiver. A backwave level at least 90 dB below the key-down carrier is a desirable goal.

Microprocessor-controlled transceivers manufactured today control CW keying rise- and fall-time through software. The operator generally accesses the keying shape parameter through a menu selection and adjustment process. Three to four ms is a typical value for most transceivers that balances crisp keying characteristics against excessive off-channel artifacts. See *QST* Product Reviews for waveforms and discussions of rise- and fall-time settings.

Homebrew equipment usually relies on analog circuitry to control keying waveforms. **Figure 13.35** is the schematic of one wave-shaping circuit that has been used successfully. A Sallen-Key third-order op amp low-pass filter (0.1 dB Chebyshev response) shapes the keying waveform, produces the rate of rise and fall and also softens the leading and trailing corners just the right amount. The key closure activates the CMOS switch, U1, which turns on the 455-kHz IF signal. At the key-up time, the input to the wave-shaping filter is turned off, but the IF signal switch remains closed for an additional 12 ms.

The keying waveform is applied to the gain control pin of a CLC5523 amplifier IC. This device, like nearly all gain-control amplifiers, has a *logarithmic* control of gain; therefore some experimental "tweaking" of the capacitor values was used to get the result shown in **Figure 13.36A**. The top trace shows the on/off operation of the IF switch, U1. The signal is turned on shortly before the rise of the keying pulse begins and remains on for

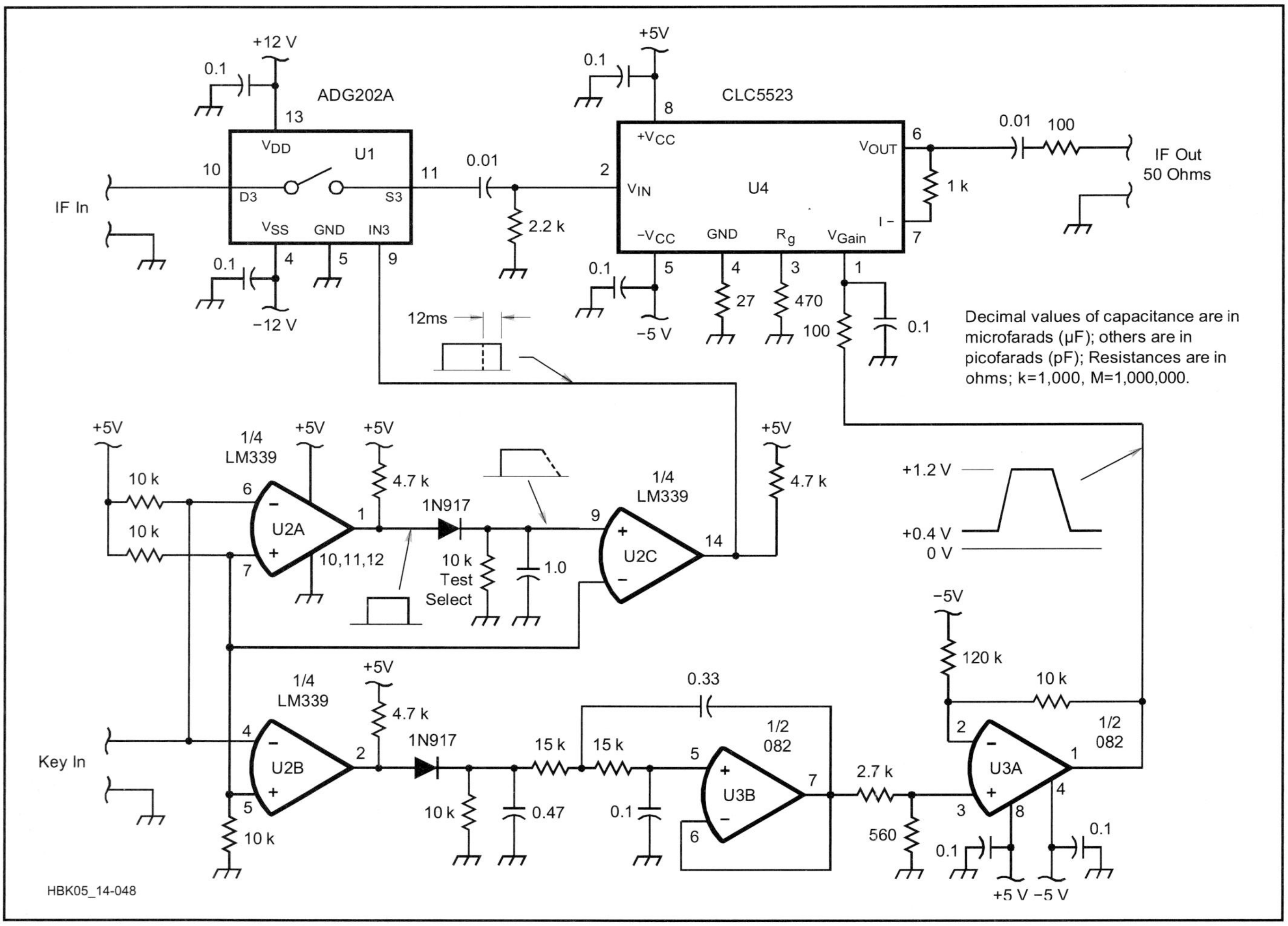

Figure 13.35 — This schematic diagram shows a CW waveshaping and keying circuit suitable for use with an SSB/CW transmitter such as is shown in Figure 13.9.

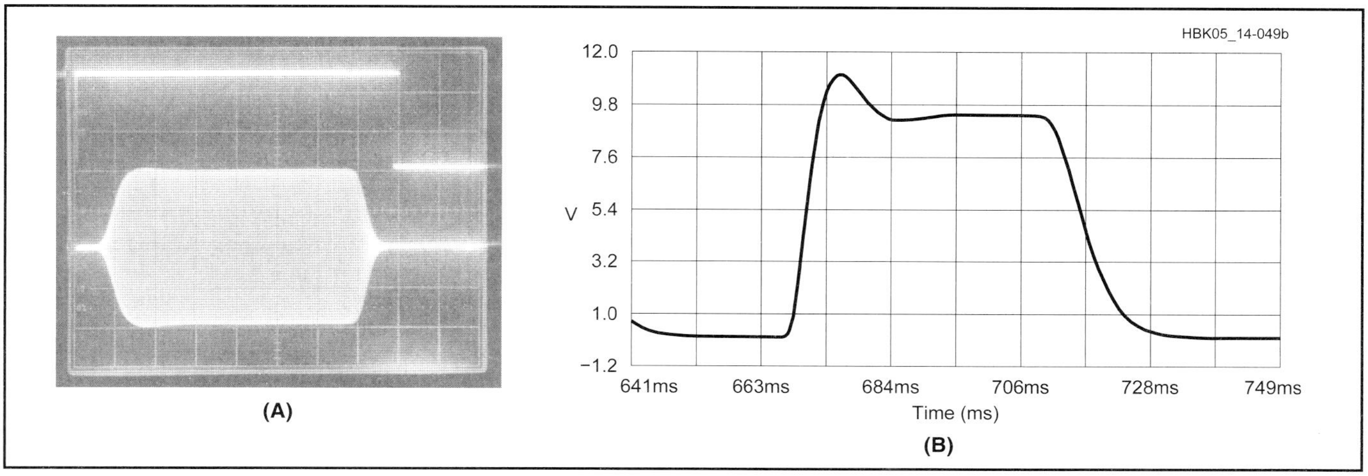

Figure 13.36 — At (A) is the oscilloscope display of the CW waveshaping and keying circuit output. The top trace is the IF keying signal applied to S1 of Figure 13.35. The bottom trace is the transmitter output RF spectrum. At (B) is a *SPICE* simulation of the waveshaping network. When this signal is applied to the logarithmic control characteristic of the CLC5523 amplifier, the RF envelope is modified slightly to the form shown in (A).

about 12 ms after the keying pulse is turned off, so that the waveform falls smoothly to a very low value. The result is an excellent spectrum and an almost complete absence of backwave. Compare this to the factory transmitter waveshapes shown in Figures 13.29 and 13.31. The bottom trace shows the resulting keyed RF output waveshape. It has an excellent spectrum, as verified by critical listening tests. The thumps and clicks that are found in some CW transmitters are virtually absent. The rise and fall intervals have a waveshape that is approximately a cosine. Spread-spectrum frequency-hop waveforms have used this approach to minimize wideband interference.

Figure 13.36B is an accurate *SPICE* simulation of the wave shaping circuit output before the signal is processed by the CLC5523 amplifier. To assist in adjusting the circuit, create a steady stream of 40 ms dots that can be seen on an RF oscilloscope that is looking at the final PA output envelope. It is important to make sure that the excellent waveshape is not degraded on its way to the transmitter output. Single-sideband linear power amplifiers are well suited for a CW transmitter, but they must stay within their linear range, and the backwave problem must be resolved.

When evaluating the spectrum of an incoming CW signal during on-the-air operations, a poor receiver design can contribute problems caused by its vulnerability to a strong but clean adjacent channel signal. Clicks, thumps, front end overload, reciprocal mixing, and other issues can be created in the receiver. It is important to put the blame where it really belongs.

13.4.3 Break-In CW Operation

Most current 100 W class HF transceivers use high-speed relays (with the relay actually following the CW keying) or solid-state PIN diodes to implement full break-in CW. Some RF power amplifiers use high-speed vacuum relays for the TR switching function. See the section on TR Switching later in this chapter for more information about circuits to perform this function. Two projects for adding QSK switching to linear amplifiers are included in the **Station Accessories** online material.

The term *semi-break-in* is used to designate a CW switching system in which closing the key initiates transmission, but switching back to receive happens between words, not between individual dits. Some operators find this less distracting than full break-in, and it is easier to implement with less-expensive relays for the TR switching.

13.5 Transmitting AM and SSB

13.5.1 Amplitude-Modulated Full-Carrier Voice Transmission

A popular form of voice amplitude modulation is called *high-level amplitude modulation.* It is generated by mixing (or modulating) an RF carrier with an audio signal. **Figure 13.37** shows the conceptual view of this. **Figure 13.38** is a more detailed view of how such a voice transmitter would actually be implemented. The upper portion is the RF channel and the lower portion is the audio frequency or AF channel, usually called the *modulator*. The modulator is nothing more than an audio amplifier designed to be fed from a microphone and with an output designed to match the anode or collector impedance of the final RF amplifier stage.

The output power of the modulator is applied in series with the dc supply of the output stage (only) of the RF channel of the transmitter. The level of the voice peaks needs to be just enough to vary the supply to the RF amplifier collector between zero volts, on negative peaks, and twice the normal supply voltage on positive voice peaks. This usually requires an AF amplifier with about half the average power output as the dc input power (product of dc collector or plate voltage times the current) of the final RF amplifier stage.

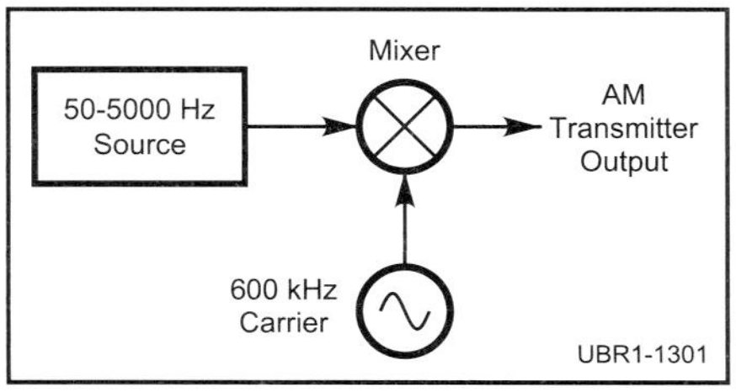

Figure 13.37 — Block diagram of a conceptual AM transmitter.

The output signal, called *full-carrier double-sideband AM*, occupies a frequency spectrum as shown in **Figure 13.39.** The spectrum shown would be that of a standard broadcast station with an audio passband from 50 Hz

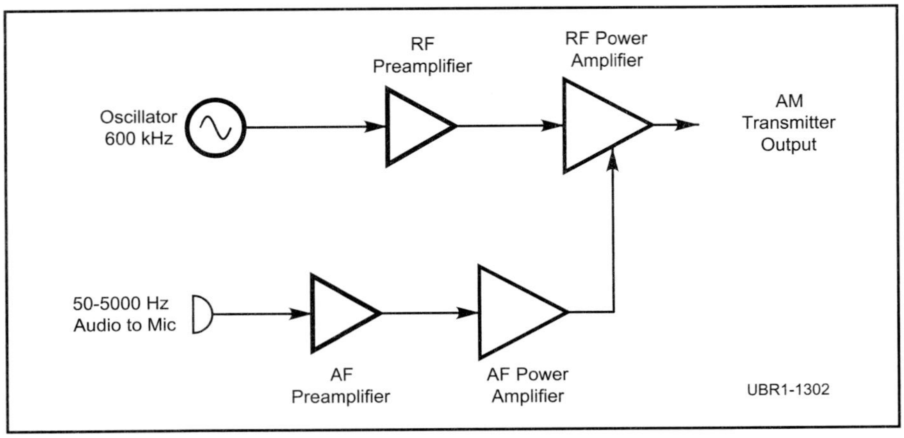

Figure 13.38 — Block diagram of a 600 kHz AM broadcast transmitter.

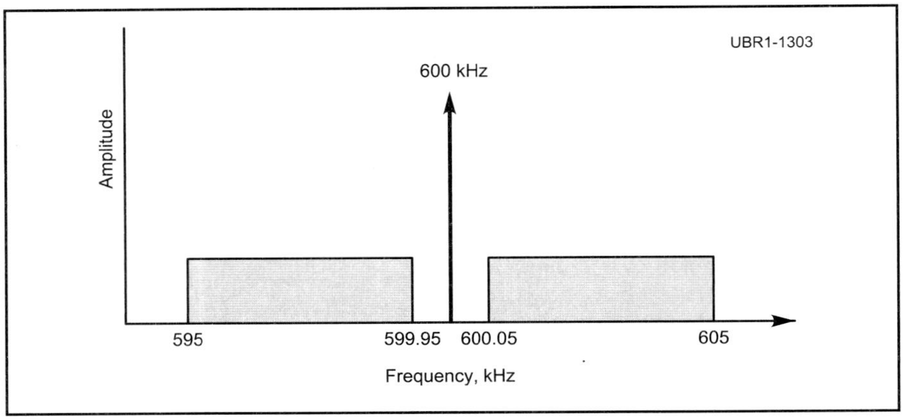

Figure 13.39 — The range of spectrum used by a 600 kHz AM broadcast signal showing sidebands above and below a carrier at 600 kHz.

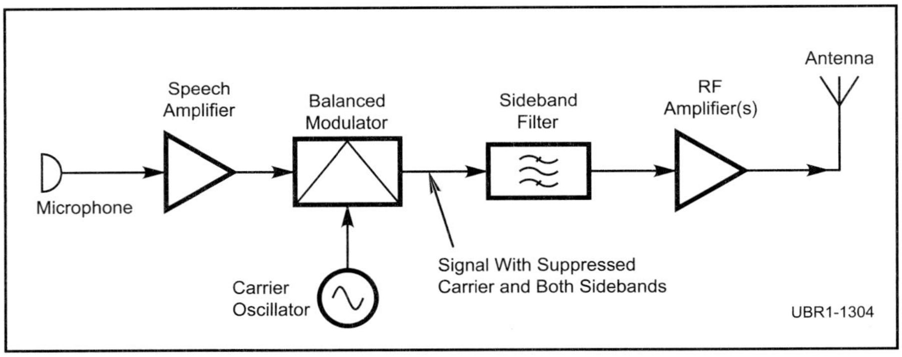

Figure 13.40 — Block diagram of a filter type single-sideband suppressed-carrier (SSB) transmitter.

to 5 kHz. Note that the resulting channel width is twice the highest audio frequency transmitted. If the audio bandwidth were limited to typical "telephone quality speech" of 300 to 3300 Hz, the resulting bandwidth would be reduced to 6.6 kHz. Note also that while a perfect multiplication process would result in just two sidebands and no carrier, this implementation actually provides the sum of the carrier and the sidebands from the product terms. (See the **Receiving** chapter for the mathematical description of signal multiplication.)

Full-carrier double-sideband AM is used in fewer and fewer applications. The spectral and power efficiency are significantly lower than single sideband (SSB), and the equipment becomes quite costly as power is increased. The primary application is in broadcasting — largely because AM transmissions can be received on the simplest and least expensive of receivers. With a single transmitter and thousands of receivers, the overall system cost may be less and the audience larger than for systems that use more efficient modulation techniques. While the PEP output of an AM transmitter is four times the carrier power, none of the carrier power is necessary to carry the information, as we will discuss in the next section.

13.5.2 Single-Sideband Suppressed-Carrier Transmission

The two sidebands of a standard AM transmitter carry (inverted) copies of the same information, and the carrier carries essentially no information. We can more efficiently transmit the information with just one of the sidebands and no carrier. In so doing, we use somewhat less than half the bandwidth, a scarce resource, and also consume much less transmitter power by not transmitting the carrier and the second sideband.

SSB — THE FILTER METHOD

The block diagram of a simple single-sideband suppressed-carrier (SSB) transmitter is shown in **Figure 13.40**. This transmitter uses a balanced mixer as a *balanced modulator* to generate a double sideband suppressed carrier signal without a carrier. That signal is then sent through a filter designed to pass just one (either one, by agreement with the receiving station) of the sidebands.

Depending on whether the sideband above or below the carrier frequency is selected, the signal is called *upper sideband* (*USB*) or *lower sideband* (*LSB*), respectively. The resulting SSB signal is amplified to the desired power level and we have an SSB transmitter. Amateur practice is to use USB above 10 MHz and LSB on lower frequencies. The exception is 60 meter channels, on which amateurs are required to use USB for data and voice signals.

While a transmitter of the type in Figure 13.40 with all processing at the desired transmit frequency will work, the configuration is not often used. Instead, the carrier oscillator and sideband filter are often at an intermediate frequency that is heterodyned to the operating frequency as shown in **Figure 13.41**. The reason is that the sideband filter is a complex narrow-band filter and most manufacturers would rather not have to supply a new filter design every time a transmitter is ordered for a new frequency. Many SSB transmitters can operate on different bands as well,

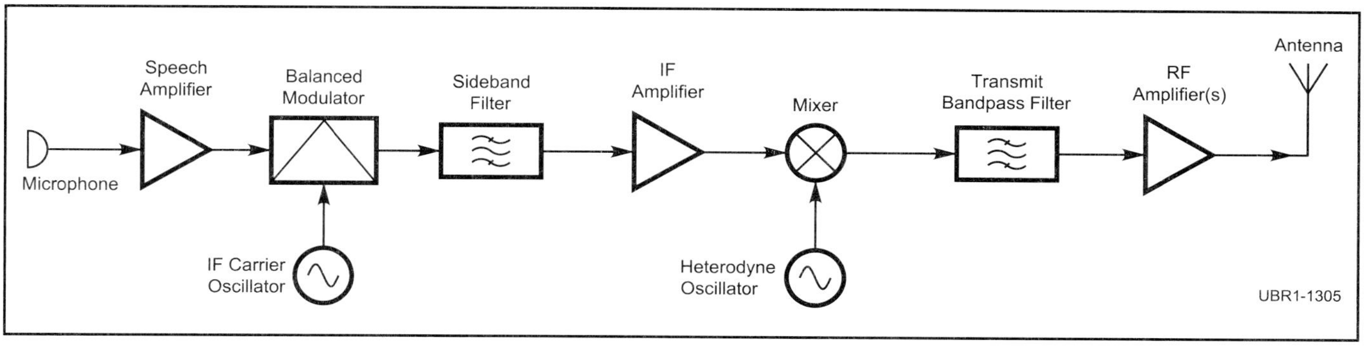

Figure 13.41 — Block diagram of a heterodyne filter-type SSB transmitter for multiple frequency operation.

so this avoids the cost of additional mixers, oscillators and expensive filters.

Note that the block diagram of our SSB transmitter bears a striking resemblance to the diagram of a superheterodyne receiver except that the signal path is reversed to begin with information and produce an RF signal. The same kind of image rejection requirements for intermediate frequency selection that were design constraints for the superhet receiver applies here as well.

The filter method can be implemented by analog circuitry, using high-quality crystal filters to remove the unwanted sideband and a carefully adjusted balanced modulator to eliminate the carrier. An SDR can also implement the filter method by using very sharp DSP filters at a low frequency (carrier frequency less than 100 kHz) and then up-converting the signal to the desired RF frequency. As DSP hardware continues to improve in speed, this method should continue to grow in popularity

SSB — THE PHASING METHOD

Most current transmitters use the method of SSB generation shown in Figure 13.40 to generate the SSB signal. That is the *filter method*, but really occurs in two steps — first a balanced modulator is used to generate sidebands and eliminate the carrier, then a filter is used to eliminate the undesired sideband, and often to improve carrier suppression as well.

The *phasing method* of SSB generation is exactly the same as the image-rejecting mixer described in the **Receiving** chapter. This uses two balanced modulators and a phase-shift network for both the audio and RF carrier signals to produce the upper sideband signal as shown in **Figure 13.42A**. By a shift in the sign of either of the phase-shift networks, the opposite sideband can be generated. This method trades a few phase-shift networks and an extra balanced modulator for the sharp sideband filter of the filter method.

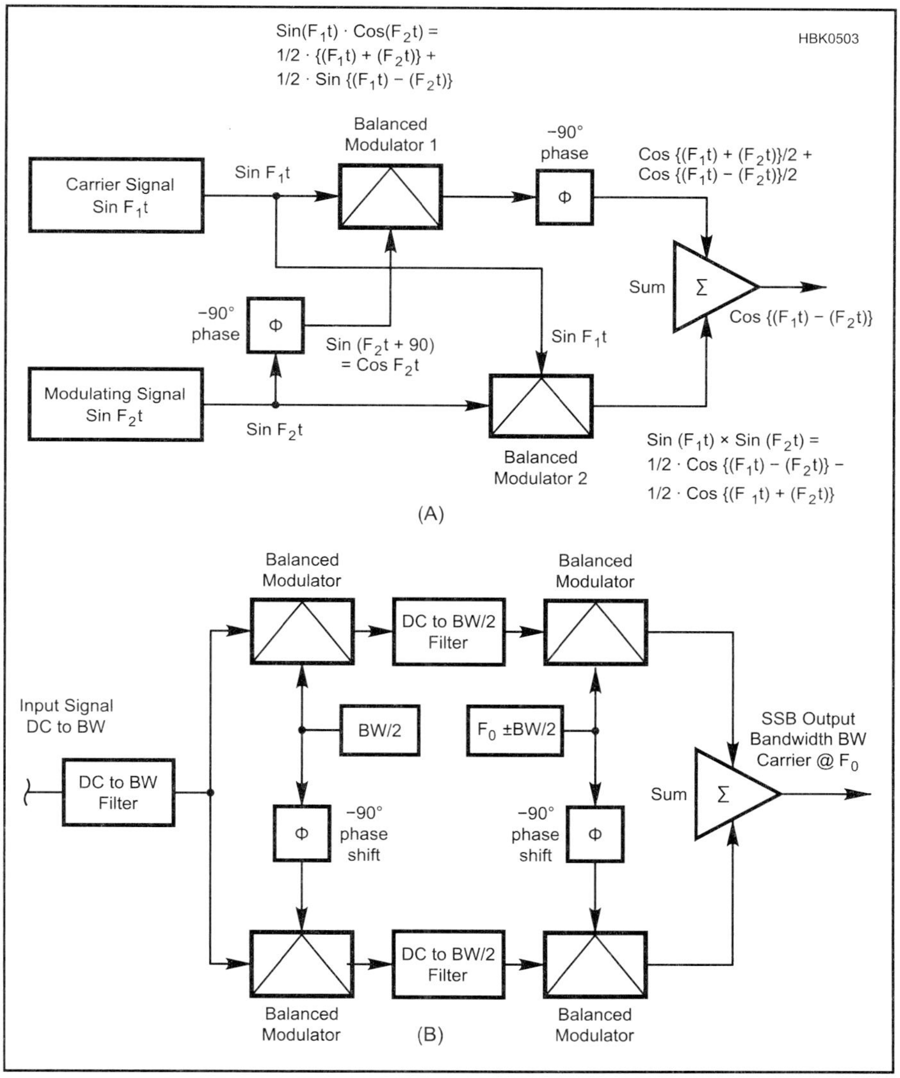

Figure 13.42 — Block diagram of (A) phasing type and (B) Weaver method SSB generators for single-frequency operation.

While it looks deceptively simple, a limitation is in the construction of an analog phase-shift network that will have a constant 90° phase shift over the whole audio range. Errors in phase shift result in less than full carrier and sideband suppression. Nonetheless, there have been some successful examples offered over the years.

The 90° phase shift network is also known as a Hilbert transformer after the mathematical operation it performs. Difficult to implement well in analog form, the phase shift is straightforward to implement digitally. That makes the phasing method a good choice for SSB generation by an SDR transmitter.

SSB — THE WEAVER METHOD

Taking the phasing method one step further, the Weaver method solves the problem of requiring phase-shift networks that must be aligned across the entire audio range. Instead, the Weaver method, shown in Figure 13.42B, first mixes one copy of the message (shown with a bandwidth of dc to BW Hz) with an in-band signal at BW/2 Hz and another copy with a signal at BW/2 Hz that is phase-shifted by –90°. Instead of phase-shifting the message, only the signal at BW/2 Hz must be phase-shifted — a much simpler task!

The output of each balanced modulator is filtered, leaving only components from dc to BW/2. These signals are then input to a second pair of balanced modulators with a more conventional LO signal at the carrier frequency, f_0, offset by +BW/2 for USB and –BW/2 for LSB. The output of the balanced modulators is summed to produce the final SSB signal.

The Weaver method is difficult to implement in analog circuitry, but is well-suited to digital signal processing systems. The Weaver method has become common in DSP-based equipment that generates the SSB signal digitally.

13.6 Transmitting Angle Modulation

13.6.1 Angle-Modulated Transmitters

Transmitters using frequency modulation (FM) or phase modulation (PM) are generally grouped into the category of *angle modulation* since the resulting signals are often indistinguishable. An instantaneous change in either frequency or phase can create identical signals, even though the method of modulating the signal is somewhat different. To generate an FM signal, we need an oscillator whose frequency can be changed by the modulating signal.

We can make use of an oscillator whose frequency can be changed by a "tuning voltage." If we apply a voice signal to the TUNING VOLTAGE connection point, we will change the frequency with the amplitude and frequency of the applied modulating signal, resulting in an FM signal.

The phase of a signal can be varied by changing the values of an R-C phase-shift network. One way to accomplish phase modulation is to have an active element shift the phase and generate a PM signal. In **Figure 13.43**, the drain current through the field-effect transistor is varied with the applied modulating signal, varying the phase shift at the stage's output. Because the effective load on the stage is changed, the carrier is also amplitude-modulated and must be run through an FM receiver-type limiter in order to remove the amplitude variations.

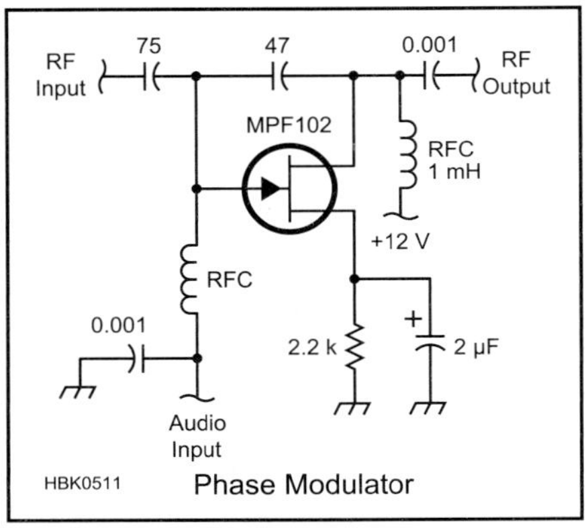

Figure 13.43 — Simple FET-based phase modulator circuit.

FREQUENCY MODULATION TRANSMITTER DESIGN

Frequency modulation is widely used as the voice mode on VHF for repeater and other point-to-point communications. **Figure 13.44** shows the phase-modulation method, also known as *indirect FM*, as used in many FM transmitters. It is the most widely used approach to FM. Phase modulation is performed at a low frequency, say 455 kHz. Prior to the phase modulator, speech filtering and processing perform four functions:

1. Convert phase modulation to frequency modulation (see below).

2. Apply pre-emphasis (high-pass filtering) to the speech audio higher speech frequencies for improved signal-to-noise ratio after de-emphasis (low-pass filtering) of the received audio.

3. Perform speech processing to emphasize the weaker speech components.

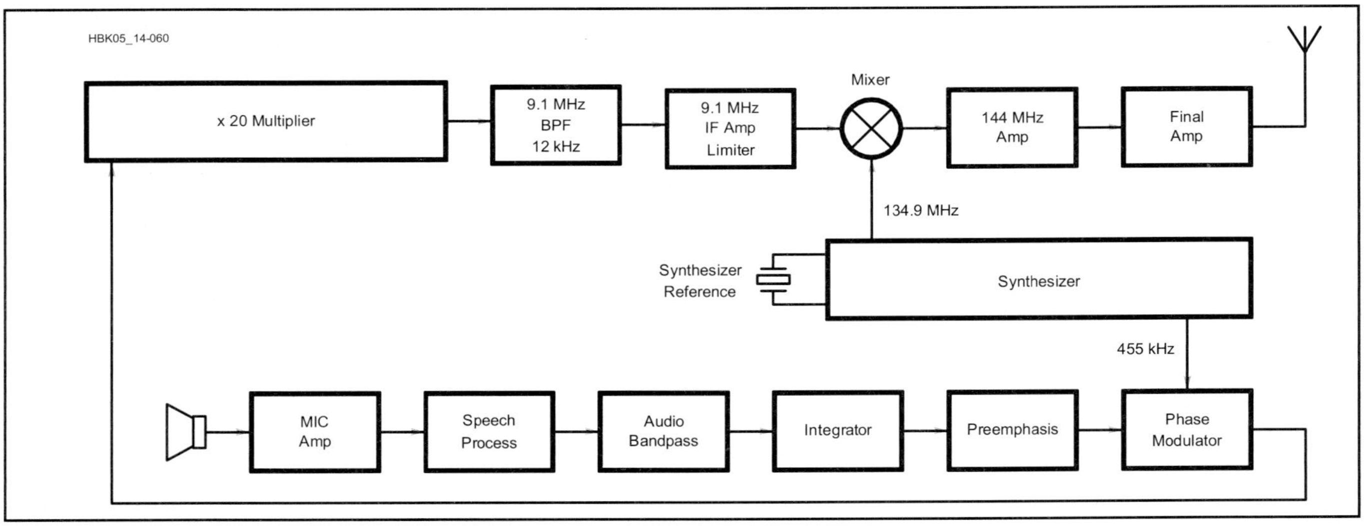

Figure 13.44 — Block diagram of a VHF/UHF NBFM transmitter using the indirect FM (phase modulation) method.

4. Compensate for the microphone's frequency response and possibly also the operator's voice characteristics.

Multiplier stages then move the signal to some desired higher IF and also multiply the frequency deviation to the desired final value. If the FM deviation generated in the 455 kHz modulator is 250 Hz, the deviation at 9.1 MHz is 20 × 250, or 5 kHz.

13.6.2 Frequency Multipliers

Frequency multipliers are frequently used in FM transmitters as a way to increase the deviation along with the carrier frequency. They are composed of devices that exhibit high levels of harmonic distortion, usually an undesired output product. In this case the desired harmonic is selected and enhanced through filtering. The following examples show the way this can be done, both with amplifiers and with passive diode circuits. (Additional discussion of frequency multipliers can be found in Chapter 5.4 of *Experimental Methods in RF Design* and in the "VHF Signal Sources" article by Rick Campbell, KK7B which is included in this book's downloadable supplemental content.)

A passive multiplier using diodes is shown in **Figure 13.45A**. The full-wave rectifier circuit can be recognized, except that the dc component is shorted to ground. If the fundamental frequency ac input is 1.0 V_{RMS}, the second harmonic is 0.42 V_{RMS} or 8 dB below the input, including some small diode losses. This value is found by calculating the Fourier series coefficients for the full-wave-rectified sine wave, as shown in many textbooks.

Transistor and vacuum-tube frequency multipliers operate on the following principle: if a sine wave input causes the plate/collector/drain current to be distorted (not a sine wave) then harmonics of the input are generated. If an output resonant circuit is tuned to a harmonic, the output at the harmonic is emphasized and other frequencies are attenuated. For a particular harmonic the current pulse should be distorted in a way that maximizes that harmonic. For example, for a doubler the current pulse should look like a half-wave rectified sine wave (180° of conduction). A transistor with Class B bias would be a good choice. For a tripler, use 120° of conduction (Class C).

An FET, biased at a certain point, is very nearly a *square-law* device as described in the **Electrical Fundamentals** chapter. That is, the drain-current change is proportional to the square of the gate-voltage change. It is then an efficient frequency doubler that also de-emphasizes the fundamental.

A push-push doubler is shown in Figure 13.45B. The FETs are biased in the square-law region and the BALANCE potentiometer minimizes the fundamental frequency. Note

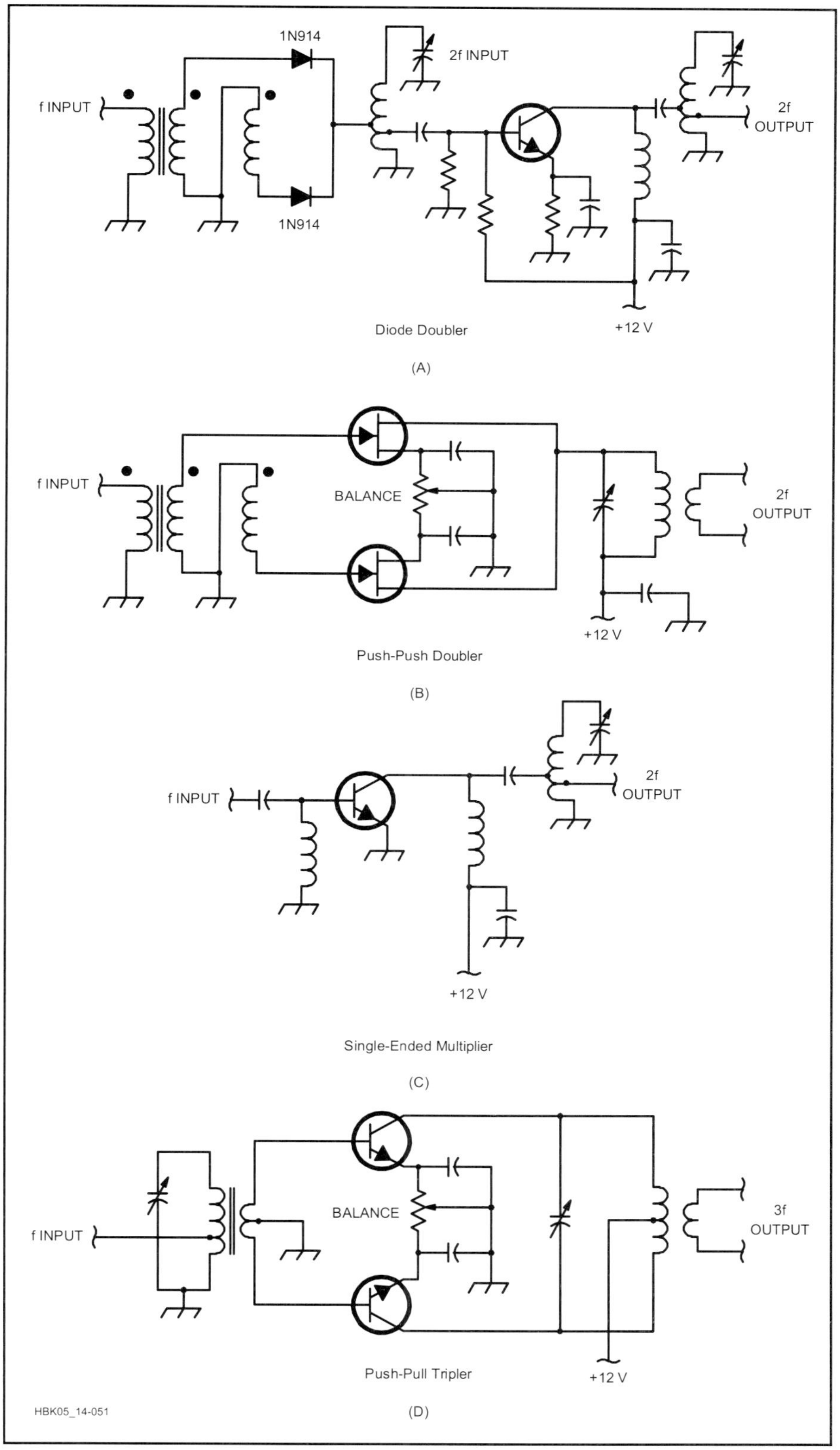

Figure 13.45 — Frequency multipliers. A: diode doubler. B: push-push doubler using JFETS. C: single-ended multiplier using a BJT. D: push-pull tripler using BJTs.

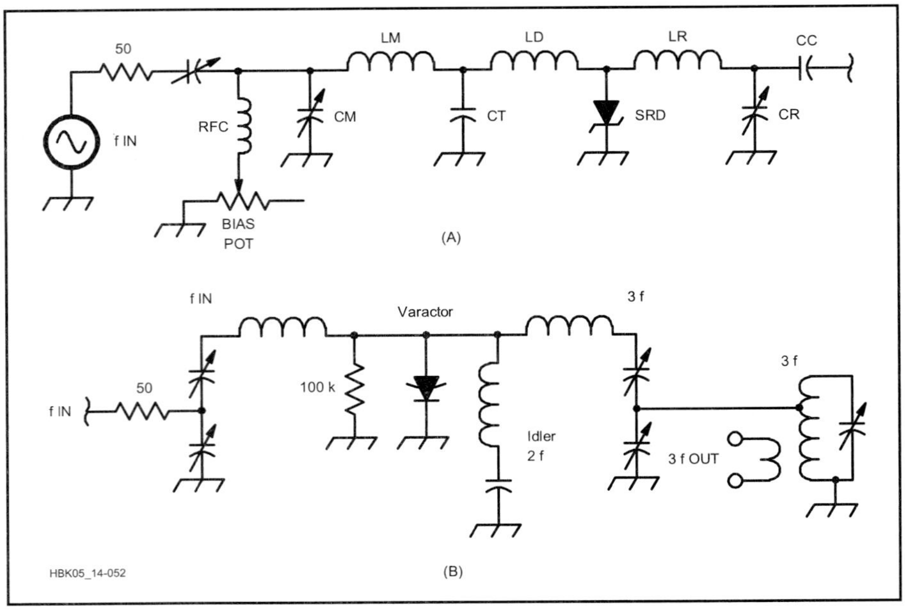

Figure 13.46 — Specialized diode frequency multipliers. A: step-recovery diode multiplier. B: varactor diode multiplier.

that the gates are in push-pull and the drains are in parallel. This causes second harmonics to add in-phase at the output and fundamental components to cancel.

Figure 13.45C shows an example of a single-ended doubler using a bipolar transistor. The efficiency of a doubler of this type is typically 50%, that of a tripler 33%, and of a quadrupler 25%. Harmonics other than the one to which the output tank is tuned will appear in the output unless effective band-pass filtering is applied. The collector tap on L1 is placed at the point that offers the best compromise between power output and spectral purity.

A push-pull tripler is shown in Figure 13.45D. The input and output are both push-pull. The balance potentiometer minimizes even harmonics. Note that the transistors have no bias voltage in the base circuit; this places the transistors in Class C for efficient third-harmonic production. Choose an input drive level that maximizes harmonic output.

The step recovery diode (SRD) shown in **Figure 13.46A** is an excellent device for harmonic generation, especially at microwave frequencies. The basic idea of the SRD is as follows: When the diode is forward conducting, a charge is stored in the diode's diffusion capacitance, and if the diode is quickly reverse-biased, the stored charge is very suddenly released into an LC harmonic-tuned circuit. The circuit is also called a "comb generator" because of the large number of harmonics that are generated. (The spectral display looks like a comb.) A phase-locked loop (PLL) can then lock onto the desired harmonic.

A varactor diode can also be used as a multiplier. Figure 13.46B shows an example. This circuit depends on the fact that the capacitance of a varactor changes with the instantaneous value of the voltage across it, in this case the RF excitation voltage. This is a nonlinear process that generates harmonic currents through the diode. Power levels up to 25 W can be generated in this manner.

Following frequency multiplication, a second conversion to the final output frequency is performed. Prior to this final translation, IF band-pass filtering is performed in order to minimize adjacent channel interference that might be caused by excessive frequency deviation. This filter needs good phase linearity to assure that the FM sidebands maintain the correct phase relationships. If this is not done, an AM component is introduced to the signal, which can cause nonlinear distortion problems in the PA stages. The final frequency translation retains a constant value of FM deviation for any value of the output signal frequency.

The IF/RF amplifiers can be nonlinear Class C amplifiers because the signal in each amplifier contains, at any one instant, only a single value of instantaneous frequency and not multiple simultaneous frequencies whose relationship must be preserved as in SSB. These amplifiers are not sources of IMD, so they need not be "linear." The sidebands that appear in the output are a result only of the FM process. (The spectrum of an FM signal is described by Bessel functions.)

In phase modulation, the frequency deviation is directly proportional to the frequency of the audio signal. (In FM, the deviation is proportional to the audio signal's amplitude.) To make deviation independent of the audio frequency, an audio-frequency response that rolls off at 6 dB per octave is needed. An op-amp low-pass circuit in the audio amplifier accomplishes this function. This process converts phase modulation to frequency modulation.

In addition, audio speech processing helps to maintain a constant value of speech amplitude, and therefore constant IF deviation, with respect to audio speech levels. Pre-emphasis of speech frequencies (a 6 dB per octave high-pass response from 300 to 3000 Hz) is commonly used to improve the signal-to-noise ratio at the receive end. Analysis shows that this is especially effective in FM systems when the corresponding de-emphasis (complementary low-pass response) is used at the receiver. (See reference for Schwartz.) By increasing the amplitude of the higher audio frequencies before transmission and then reducing them in the receiver, high-frequency audio noise from the demodulation process is also reduced, resulting in a "flat" audio response with lower hiss and high-frequency noise.

An IF limiter stage may be used to ensure that any amplitude changes that are created during the modulation process are removed. The indirect-FM method allows complete frequency synthesis to be used in all the transmitter local oscillators (LOs), so that the channelization of the output frequency is very accurate. The IF and RF amplifier stages are operated in a highly efficient Class-C mode, which is helpful in portable equipment operating on small internal batteries.

FM is more tolerant of frequency misalignments between the transmitter and receiver than is SSB. In commercial SSB communication systems, this problem is solved by transmitting a *pilot carrier* with an amplitude 10 or 12 dB below the full PEP output level. The receiver is then phase-locked to this pilot carrier. The pilot carrier is also used for squelch and AGC purposes. A short-duration "memory" feature in the receiver bridges across brief pilot-carrier dropouts, caused by multipath nulls.

In a "direct FM" transmitter, a high-frequency (say, 9 MHz or so) crystal oscillator is frequency-modulated by varying the voltage on a varactor diode. The audio is pre-emphasized and processed ahead of the frequency modulator as for indirect-FM.

13.7 Effects of Transmitted Noise

With receiver sensitivity, selectivity, and linearity having reached extraordinary levels of performance, a reduction in transmitted spurious emissions is clearly in the best interests of all amateurs. It does us no good to spend time and effort creating an exceptional receiver if the channel is filled with transmitted noise and distortion products! (See the article by Grebenkemper in the Reference section.)

In heterodyne transmitters the last mixer and the amplifiers after it are wideband circuits, limited only by the harmonic filters and by any selectivity that may be in the antenna system. Wide-band phase noise transferred onto the transmitted modulation by the last LO (almost always a synthesizer of some kind) can extend over a wide frequency range; therefore LO cleanliness is always a matter of great concern.

The amplifiers after this mixer are also sources of wide-band "white" or additive noise. This noise can be transmitted even during times when there is no modulation, and it can be a source of local interference. To reduce this noise, use a high-level mixer with as much signal output as possible, and make the noise figure of the first amplifier stage after the mixer as low as possible.

SDR transmitters may not have analog mixers but they certainly have data converters and many processes that generate phase noise through clock jitter, non-linear data conversion, and other causes, even rounding errors! The result — heard on the air as transmitted noise — causes the same problems as noise generated by an analog transmitter. **Figure 13.47** is a collection of composite noise (phase noise) and CW keying sideband noise from several current transceivers evaluated by the ARRL Lab during product review assessments. The data is from a single band (14 MHz) and does not represent either the best or worst performance. This chart was initially created in 2014 from that data by Jim Brown, K9YC, to compare the noise performance of several transceivers. (The FT1000MP Mark V Field legacy spectrum shows the improved performance available today.) The data points on the graph represent noise peaks from the measured noise spectra. For the CW sideband noise data, a comparison of averaged and smoothed data is available in K9YC's online paper referenced at the end of this section. The paper also includes data for several older transceiver models.

There is quite a bit of difference between the various models as described in the *QST* Product Reviews. The variation shows why it is important for amateurs to pay attention to noise performance specifications, especially if they plan on operating in the crowded HF bands. With many transceivers now offering upgradable firmware, amateurs are strongly encouraged to install the latest version. This will improve performance on both receive and transmit, as well as help all of us transmit cleaner signals on the air.

Transmitted noise plays the same role in interference as receiver phase noise. The two are ultimately additive and interference becomes a "weakest link" problem in that poor noise performance of the receiver or the transmitter can be the culprit in an interference

Noise Consideration for Antenna Layout

By Steve Hicks, N5AC

Transmitted noise is a point to consider when laying out the antennas at a multi-transmitter site like Field Day or a contest station. For every 10 dB stronger any given signal appears in the receiver, another 10 dB improvement in phase noise performance is required to ensure interference-free operation. A poor antenna layout that produces very strong signals in receivers when local transmitters are operating can render even the best phase noise performance receivers inoperable. George Cutsogeorge, W2VJN's book *Managing Interstation Interference* (available from Inrad (**www.inrad.net**) is an excellent reference for station planning and interference mitigation techniques.

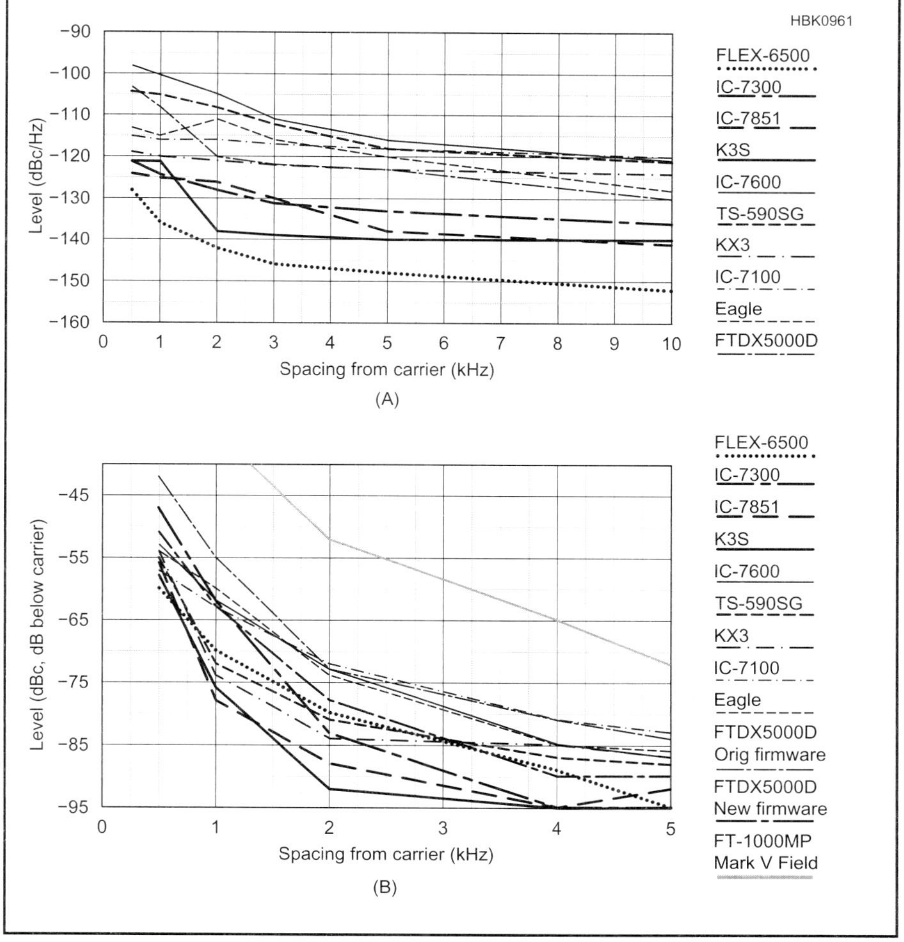

Figure 13.47 — Transmitted composite noise (A) and keying sidebands (B) from a selection of representative modern transmitters measured by the ARRL Lab. These graphs are based on ARRL Lab test data published in *QST* Product Reviews, following graphs originally created by Jim Brown, K9YC

problem. While both amplitude and phase noise contribute to transmitted composite noise, most of the noise contribution is phase noise from the various oscillators and clock signals in the transceiver. (The effects of receiver phase noise are addressed in the **Receiving** chapter.)

In a multiple-transmitter operation such as Field Day or a contest station, a single radio with poor transmitted noise performance can render all of the receivers at the site useless. Similarly, a receiver with poor phase noise performance can suffer from interference issues regardless of how good the performance of co-located transmitters is. In situations where multiple radios are brought to a single location for a joint operation, transmit phase noise is typically more strongly scrutinized since a single poor performer in the transmit phase noise arena can render the whole operation a failure.

Transmitted noise is not just a problem when multiple stations are at one site. Poor noise performance at full power can have an adverse effect on other stations for miles around. In competitive environments, noise from a closely-spaced strong signal can render adjacent channels nearly unusable. Under normal circumstances, transmitted noise raises the noise floor everywhere there is propagation. After decades of receiver improvement, it is important for amateurs to pay closer attention to transmitted noise performance.

The ARRL Lab measures transmitted noise as part of Product Reviews for *QST* magazine. These tests are described in the **Test Measurements and Equipment** chapter and in the book *Amateur Radio Transceiver Performance Testing* by Bob Allison, WB1GCM. Individual transceiver performance is documented in the *QST* Product Review. In addition to creating Figure 13.47, Jim Brown, K9YC has compiled the information from these reviews in an online paper, "Comparison of ARRL Lab Data for Selected Transceivers" which is available at **k9yc.com/publish.html**. Figure 13.47 is typical of the many figures in the paper comparing various noise measurements.

13.8 Microphones and Speech Processing

13.8.1 Frequency Content of Speech

Human speech has content from about 100 Hz to 8 kHz, but only the energy between about 400 Hz and 4 kHz contributes to speech intelligibility. Vocal content below 400 Hz provides "body" to the voice (great for singers and radio announcers), but that low frequency output of the microphone also contains breath pops, room noise, microphone handling noise, wind noise, and reverberation. This low frequency energy can easily be as much as half of the power picked up by the microphone, but it contributes nothing to communications, so it wastes transmitter power.

Likewise, speech content above 3 kHz provides "presence" and helps communications a bit, but the added bandwidth adds noise (and QRM to and from other stations). Most SSB transmitting filters are 2.7 kHz wide, so a well-adjusted radio will align those filters so that they pass audio between 400 Hz and 3.1 kHz. A few radios allow the user to adjust this setting via a configuration menu.

These bandwidth limits for speech communications were established in the earliest days of long distance telephony — they allow what's necessary, but nothing extra. Over more than a century, they have allowed more and more conversations to be packed into less and less bandwidth.

Thus, our first rule is to minimize any part of the audio signal below about 400 Hz, and to not waste bandwidth transmitting sound above 3 kHz. We have several controls over this. First, we can choose a microphone without excess low frequency response. See "Choosing a Microphone" later in this section. Many transceivers provide menu settings to tailor the audio frequency response. Study the manual to understand and choose settings for your radio.

Some newer transceivers make it even easier to tailor the frequency response — they have a built-in octave-band equalizer covering the speech range. (An octave is a 2:1 frequency step). Each band can be set for up to 18 dB of boost or cut in 1 dB steps. A good starting point for most microphones and voices is maximum cut of the three lowest bands (50, 100, 200 Hz), and 3–6 dB cut of the fourth band (400 Hz) leaving all other bands set flat (no boost or cut). Some microphone or voices may benefit from a bit more cut at 400 Hz or from 3-6 dB of cut or boost in the two highest bands. Save these adjustments for when you have a trained listener to advise you.

13.8.2 Dynamics Processing

The loudness of speech varies over a wide range as we speak, sometimes by as much as 20 dB. The audio section of our transmitter must be adjusted so that we never overmodulate, which causes interference on adjacent frequencies (splatter), but keeping audio level as close as practical to 100% modulation makes our signal louder at the other end. Modern transceivers include peak limiters to prevent overmodulation on loudness peaks and compressors to increase the audio gain for quieter parts of speech. When well-designed and carefully adjusted, these circuits work well, increasing our "talk power," but when badly adjusted, cause transmitted audio to be distorted, mushy, dull, and hard to understand.

The strength of sound falls off with distance from the source by inverse square law, just as RF field strength falls off with distance from the antenna. Doubling (or halving) the distance between microphone and mouth causes loudness to change by 6 dB. A boom microphone attached to a headset helps maintain more constant level by keeping the microphone element at a fixed distance.

To adjust these circuits, first set equalization as in the previous section, then set microphone gain so that normal speech causes near 100% modulation with compression disabled. Then set the transceiver display to show compression and adjust compression (or processing) to provide about 10 dB gain reduction on voice peaks. While radios can be set for more compression, few sound very good with more than 10 dB, and too much compression can make speech hard to copy. (On some transceivers, compression may be called "processing.") The combined effect of equalization to eliminate speech content below 400 Hz and using 10 dB of compression is about 13 dB, which is equivalent to multiplying transmitter power by a factor 20!

13.8.3 Types of Microphones

Most common microphones in the amateur station are one of two basic types:

Dynamic microphones operate on the same principle as a loudspeaker, (a coil mounted to a diaphragm is moved by varying magnetic field that surrounds it makes air vibrate) but in reverse (vibration moves a diaphragm in the microphone, generating a voltage in the coil as it moves within the magnetic field). A loudspeaker works pretty well as a microphone, and has been used that way for more than half a century in intercom systems.

Electret condenser microphones are very different — the diaphragm is one plate of a capacitor; a voltage is applied between the two plates (the other being fixed). The source impedance is quite high (megohms), and must be transformed to a lower impedance by a FET follower built into the microphone (so that what it feeds doesn't load down the microphone). The electret capsule is pre-polarized, but the FET follower needs a small positive voltage fed through a load resistor to operate. This is referred to as *bias voltage* and it is applied to the output of the microphone: 8 V dc through a 5.6 kΩ resistor is typical. (Note that if bias voltage is applied to a dynamic microphone element, the result will be muffled, low-volume audio. If the radio's microphone connection supplies bias voltage, a series blocking capacitor of 0.1 to 1 μF should be used to remove the bias voltage.)

Both electrets and dynamic microphones are available with an *omnidirectional* pattern (picks up equally in all directions) or a *cardioid* pattern (picks up better in front of the microphone with a null to the rear). Cardioids can be thought of as "half space" microphones, meaning they pick up sounds from anywhere in front of the microphone but reject sound from all directions to the rear.

Cardioid microphones have an important characteristic called *proximity effect*, which is a very strong bass boost for sound sources very close to the microphone. In addition to making voices "bass heavy," proximity effect magnifies breath pops, wind noise, and handling noise. Virtually all microphones used in live sound are cardioids, and those intended for use by singers have a strong low frequency rolloff that partially compensates proximity effect. Although cardioids reduce room noise pickup, proximity effect generally makes them a poor choice in the ham shack. A microphone designed for radio communications is a better choice.

Cardioids work on the principle of acoustic cancellation between sound reaching the element via front and rear openings of the microphone housing. (Omnidirectional microphones have a single opening). Proximity effect is the result of that process and there being a single front opening and a single rear opening.

An important variation of the cardioid microphone is built with extra openings spaced along the length of the handle, which greatly reduces proximity effect. The ElectroVoice (EV) 664 and 666 were the first popular microphones of this type, which are called "variable-D" (for the variable distant openings), as opposed to "single-D" cardioids with a single rear opening. If you're looking for a good used pro-quality microphone for your ham station, the variable-D EV RE10, 11, 15, 16, 18, 20, and 27, and the Shure SM53 and SM54 are great choices. All but the RE16, 20, and 27 are long discontinued, but dynamic microphones last indefinitely as long as they are not badly mistreated, so buying used from a trustworthy source is a good option.

An omnidirectional microphone, whether dynamic or an electret, or one of the variable-D models listed above, are the best choice for ham radio. They have no proximity effect, so can be used close to the mouth. This minimizes breath pops, while still being close enough to minimize room noise. The soft foam supplied with many microphones is intended to reduce breath pops.

MICROPHONE EQUALIZATION

Beginning in the late 1950s, Shure introduced the model 440, the first microphone designed specifically for SSB transmission. The modern version is the 444D. These are omni-directional mics with the recommended equalization built in — low frequency response rolling off below 400 Hz and a pronounced peak around 3 kHz that compensates for some of the loss in the SSB transmit filter. These are excellent sounding microphones and are primarily desktop models. Most microphones in the Heil line use the same concepts.

13.8.4 Using a Professional or PC Microphone

Pro microphones are balanced and are designed to feed balanced inputs using shielded, twisted pair cables, while ham microphones and rigs use unbalanced wiring. Pro electret microphones cannot easily be used with ham gear (because of the method used to power the balanced microphone's FET follower output stage).

Pro dynamic microphones work well and are easy to connect with ham gear. Their 3-pin XL-connector comes wired for balanced circuits — Pin 1 is the shield, Pins 2 and 3 carry the signal. To connect them to your ham rig with shielded twisted pair, wire the shield to the shell of the microphone connector and connect the signal pair to MIC and MIC RETURN (or MIC GND).

Alternatively, with coaxial cable between the microphone and the radio, wire the cable shield to pin 1 and 3 at the microphone and the center conductor to pin 2. At the radio's microphone connector, wire the center conductor to the radio's MIC input and wire the cable shield both to MIC RETURN and to the connector shell.

Pro microphones generally have uniform ("flat") frequency response. They have more bass response than communications microphones and lack the boost around 3 kHz, so they generally require more equalization. The low frequency rolloff in audio circuits is set by the time constant of interstage coupling capacitors and their resistive load. Using a smaller value capacitor raises the –3dB frequency ($X_C = R$) producing a gentle low frequency rolloff (6 dB/octave). In transmitters that lack an equalizer, raising the –3dB frequency to 1-2 kHz (by reducing those time constants) can provide much of the equalization needed to make a pro microphone produce good communications audio.

Headsets with a microphone attached made for use with computers work well with ham transceivers. **Figure 13.48** illustrates good placement that minimizes breath pops and allows an occasional sip of water or coffee. Most of these microphones are electrets with both headphones and microphone wired to 3.5-mm (⅛-inch) phone plugs (also known as TRS or tip-ring-sleeve plugs). Almost any of these headsets will work with ham transceivers, but some are far more comfortable than others. The Yamaha CM500 and Koss SB-45 and CS-100 are popular with contesters and DXers. All that is needed is a cable adapter to a Foster plug (the round 8-pin connector used on most transceivers) to mate with the 8-pin microphone connector on transceivers that lack a 3.5 mm microphone jack. Adapters for RJ-45 microphone connectors are also available. **Figure 13.49** shows a typical

Figure 13.48 — Using a computer-style headset electret boom microphone. Place the microphone far enough from your mouth to avoid picking up excessive noise from breathing and pops from speech. Balance the placement for normal speech levels while minimizing pickup of room and fan noise. On-the-air testing is important to account for your station circumstances and speaking style.

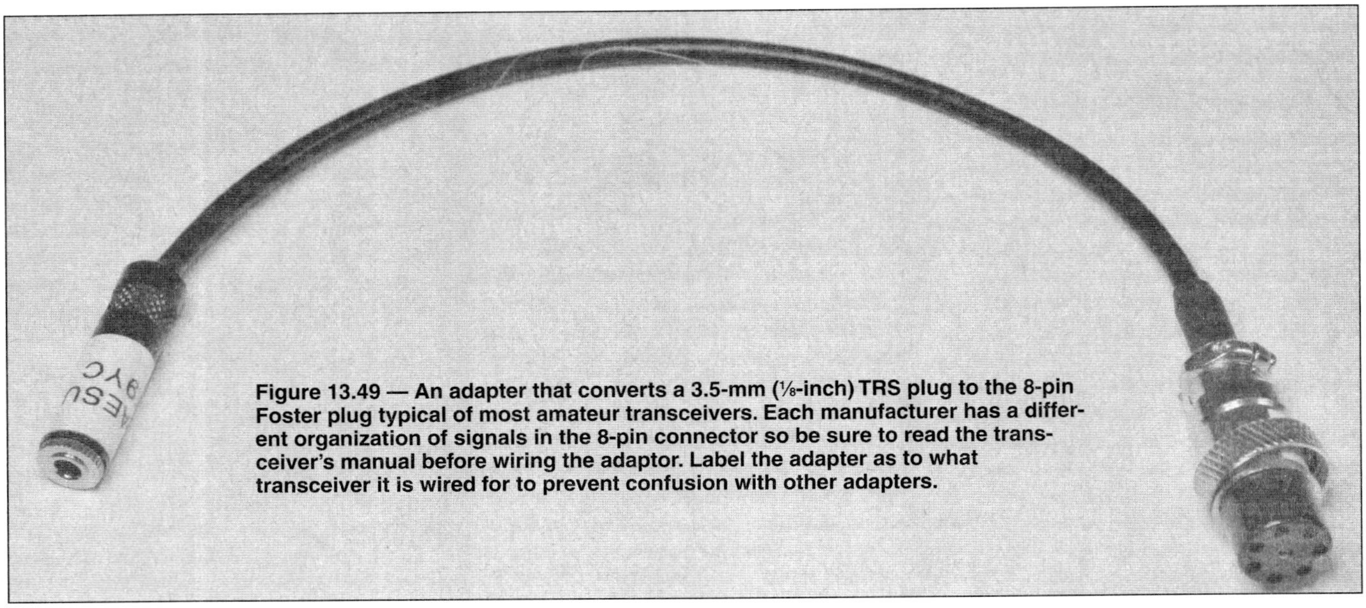

Figure 13.49 — An adapter that converts a 3.5-mm (⅛-inch) TRS plug to the 8-pin Foster plug typical of most amateur transceivers. Each manufacturer has a different organization of signals in the 8-pin connector so be sure to read the transceiver's manual before wiring the adaptor. Label the adapter as to what transceiver it is wired for to prevent confusion with other adapters.

adapter you can build yourself.

To make the adapter, you'll need a cable-mount plug to mate with your radio and a female ⅛-in TRS jack to mate with the TRS plug on the headset. Check the manual for your transceiver for wiring of the microphone connector and label the adapter.

To connect the microphone to the radio, run a single-conductor shielded cable (such as mini-coax or braid-shielded audio cable) from the tip of the TRS jack to the microphone input pin of the Foster plug, connecting the cable shield to the sleeve of the TRS jack and the shell of the Foster plug.

Nearly all modern radios supply bias voltage in the range of 8 V on one pin of the microphone connector; wire a 5.6 kΩ resistor between that pin and the MIC pin. This resistor can have a very low power rating, so it's usually possible to fit it inside the Foster plug. Buy Foster plugs from ham vendors; female TRS jacks can be bought from pro audio vendors such as Full Compass and Sweetwater. (Neutrik part number NYS240BG is typical.)

13.8.5 Optimizing Your Microphone Audio

Summarizing the steps to optimize audio for communications:

1) Set your radio to minimize audio content below 400 Hz and above 3.2 kHz.

2) Keep the distance between microphone and mouth as constant as possible. A boom microphone attached to a headset solves this problem.

3) Get audio gains set right: adjust the mic input of the radio (or of the computer), the output gain of the computer sound card, and/or the line input of the radio.

4) Set processing for an indicated 10 dB on voice peaks.

5) Resist the urge to turn up mic gain or compression — once levels have been set as described here, turning it up louder makes your voice sound *worse*, not better.

You should also make sure the transmitter RF controls are set properly:

6) Tune the RF power amp carefully.

7) Don't overdrive the RF amp, and don't use ALC to set TX power.

8) If your radio requires a nominal 13.8 V dc power supply, make sure the supply is as close to 14 V as possible.

13.8.6 Microphone Pads

(The following material was contributed by Ethan Winer. More design information is available on his website: **ethanwiner.com/gadgets.html**.)

The balanced microphone attenuator or *pad* shown in **Figure 13.50**. is about as basic a circuit as you're likely to encounter. But finding the optimum resistor values can be a challenge for the beginner, and a time consuming nuisance for the more advanced.

Figure 13.51.shows the same general configuration, except this circuit yields a low-end rolloff beginning at a frequency dictated by the resistor and capacitor values. Notice that the designation "R1" does not appear in this diagram. The same component numbering is used throughout the figures to make the

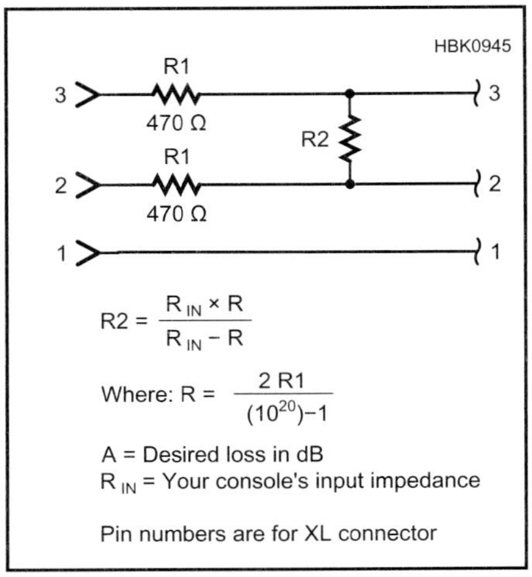

Figure 13.50 — Basic microphone attenuation pad.

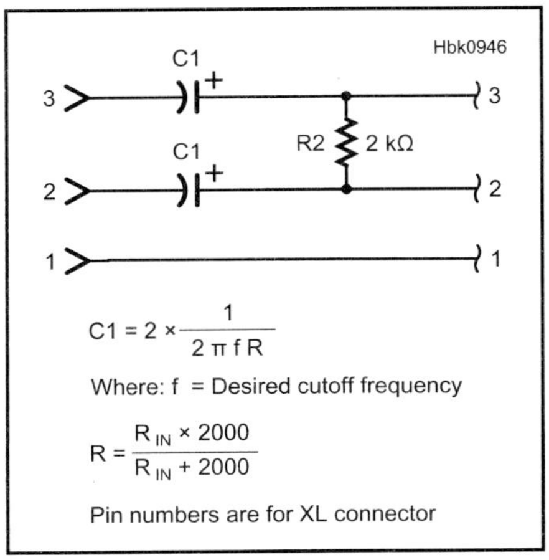

Figure 13.51— Low-frequency rolloff pad.

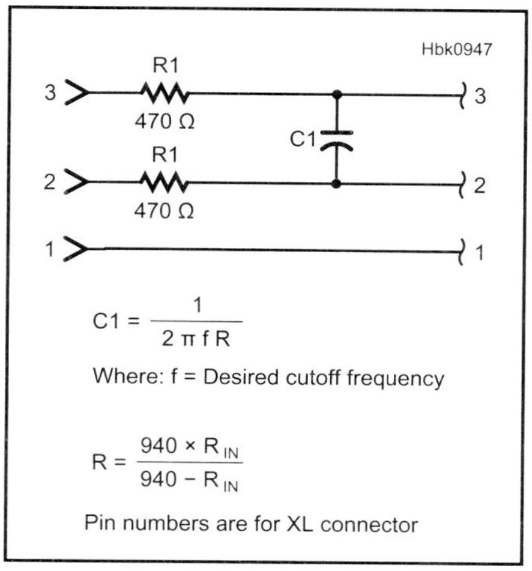

Figure 13.52— High-frequency rolloff pad.

formulas easier to understand. Also note that the low-cut circuit in Figure 51 will not pass bias voltage (also known as *phantom power* in professional audio).

Exchanging the capacitors and resistor in Figure 13.51 results in a high-frequency loss (**Figure 13.52**), while adding an inductor allows adjusting the midrange response for boosting presence or controlling sibilance (**Figures 13.53A** and **13.53B**).

One problem with inductors is their inherent series resistance, which is inevitable due to the wire they're made from. The smaller (physically) a given inductor is built, the higher the resistance since the wire must be a smaller diameter. The Mouser 43LH and 43LJ series inductors minimize this by winding the wire around a ferrite core. At low currents the ferrite core does a great job of increasing the inductance without requiring as many turns of wire, which keeps the series resistance relatively low. Also, these coils are encapsulated in a rugged phenolic case which eliminates the danger of damaging the hair-thin wires during assembly.

For the midrange networks shown in this sidebar you can expect to alter the response by about 4 or 5 dB, though this can be varied by adjusting R2 up or down in value. With the high- and low-end rolloffs, the cutoff slope is 6 dB. per octave beyond the chosen frequency. This rolloff rate is hard to change because additional capacitors and resistors would be needed. That not only complicates the design, but also reduces the signal level overall.

CONSTRUCTION NOTES

The pads can be constructed in the usual metal enclosures. Note that the S3FM connector assembly made by Switchcraft — a 3.5-inch tube fitted with a male XL (or XLR) at one end and a female at the other — has enough room in the middle to build a mike pad, low-end rolloff network, polarity reverser, or any number of other useful little gadgets.

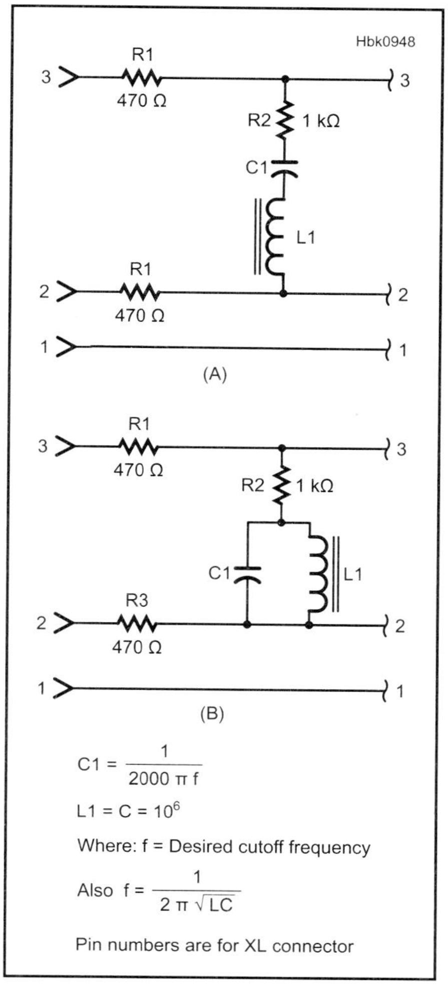

Figures 13.53 — Midrange cut (A) and boost (B) pads.

With space generally at a premium in any small enclosure, you must use the smallest components you can find. For microphone-level signals you can use ⅛-watt resistors. Another space saver is to use low-voltage capacitors. When the component values are large, such as the low-frequency rolloff, tantalum capacitors are the best choice and you may be able to find ultra-miniature types rated as low as 6 or even 3 V. Tantalums cost more than standard electrolytics, but they are smaller and generally higher quality.

Use shrink-fit sleeving over all of the components or otherwise ensure that the components don't short out to one another. The S3FM comes with a plastic-coated cardboard tube that lines the inside of the case, and this prevents any wires from touching the grounded case. Clear shrink tubing lets you see what parts are inside — a definite advantage if you have to take it apart for repair or just to see what's in there. Avoid black electrical tape for insulation; not only can't you see through it, but after a while it can turn into a sticky mess that's difficult to unwrap. Draw the schematic on a small piece of paper, and roll it up to serve as the outer insulation.

13.8.7 Speech Amplification and Processing

AUDIO SIGNAL LEVELS

The output level of most modern microphones ranges from about 10 mV to a few hundred mV when used close to the mouth. Microphone inputs are designed to accept signals in this range and are likely to be overloaded by stronger signals. In addition to microphone inputs, most ham equipment also features line-in inputs that are designed to accept signals between about 100 mV and 2 V.

IMPEDANCE IN AUDIO CIRCUITS

Since solid-state electronics became standard, audio circuits are no longer impedance-matched. Instead, audio circuits are voltage-matched. *Output* circuits have a low source impedance, typically around 100 Ω for pro line level, 300 Ω for consumer line level, and around 200 Ω for most modern microphones. Audio *input* circuits have a high input impedance, typically 10 kΩ for pro line level inputs, 50 kΩ for consumer line level inputs, and at least 1 MΩ for microphone inputs.

Thanks to these impedance relationships, very little current flows in microphone and line circuits. Output stages are constant voltage sources, so they can easily drive multiple inputs in parallel. While output circuits are usually *rated* for a 600 Ω load, loading them with 600 Ω degrades their performance.

Loudspeaker amplifiers are different, in that they *do* provide power. They are still constant voltage sources, and those that drive loudspeakers typically have output impedance of a few tens of milliohms, designed to drive loads in the range of 4-16 Ω. Since they are constant voltage sources, they will deliver four times more power into a 4-Ω load than into a 16-Ω load. The ratio of load impedance to source impedance is the *damping factor*.

Headphone amplifiers also supply power but a lot less of it. Headphones range in impedance from 8 Ω to more than 600 Ω, and most will be pretty loud with only a few volts drive. Most headphone amplifiers can drive more than one pair of headphones in parallel. Most headphone amplifiers have a resistor in series with their output so that they are not damaged when a plug is not inserted properly.

Computer sound cards have stereo *outputs* designed to drive headphones and line level inputs. The sound cards built into many laptop computers have only a mono *input*, often designed only for microphone levels. Most outboard sound cards, and better built-in sound cards, provide both a mono microphone and a stereo line input, or a single stereo input that can be switched to microphone or line level via the computer's operating system or the software that uses the audio signal.

SPEECH PROCESSING

Transmitters generally require audio signals in the volt region and it will be found advantageous to shape the frequency response prior to transmission. ("Amplified" microphones are outside this discussion.) Before being applied to the modulator circuits in the transmitter the audio signal is applied to a clipper or compressor for amplitude control. Such circuits along with their associated filtering increase the average value of signal level into the transmitter and so make the signal sound louder and define the occupied bandwidth.

PREAMPLIFIER

The output of the microphone is amplified and its frequency response shaped in a preamplifier as shown in **Figure 13.54.** R4 allows coarse gain adjustment and it can be a 100 kΩ potentiometer if various microphones are to be accommodated. The low-pass filter has a flat response out to about 3 kHz, including a peak of several dB at 2500 Hz (discussed below). The filters work in concert with the following treble peaking circuit to produce a response very suitable for communications as shown in **Figure 13.55.** The magnitude of the peak at 2500 Hz may be adjusted by simply changing the value of R12. This treble peaking circuit only affects speech signals in the 2500 Hz region.

The response as shown has a rise of 15 dB at 2500 Hz relative to 400 Hz but is nevertheless down 30 dB at 10 kHz. Very low frequencies have been rolled off. The output of the preamplifier is applied to a level control and then to a clipper or compressor for level control.

CLIPPER

After the speech signal has been increased in level and the basic response has been shaped, it may be applied to a clipper to increase its average volume level. At this point it must be pointed out that clipping must be used with great care if the transmitter is operating in the single-sideband mode. SSB transmitters cannot handle clipped waveforms gracefully. Clipping can always be used to catch occasional overshoots resulting from sluggish AGC systems, as an example, but in SSB systems should not be used as a routine method of volume maximization. However, clipping can be used in AM systems with great effect. A clipper followed by a low-pass filter sets the occupied bandwidth of the transmitter, assuming the following stages are operating cleanly.

Figure 13.56 shows the schematic of a speech clipper with its associated filtering. The "preclip" filters are optional but recommended. The "postclip" low-pass filter is a mandatory requirement to limit the transmitted bandwidth.

The 200 Hz high-pass filter ahead of the clipper greatly reduces low-frequency intermodulation distortion. The 3 kHz low-pass filter ahead of the clipper prevents high-frequency audio (sibilants) from being clipped and similarly causing high-frequency intermodulation distortion.

This circuit uses an over-driven op-amp as a clipper. Its sensitivity may be changed by changing the value of R8. The output is reduced by a factor of two (R11 and R12) so the following low-pass filter can handle the signal in a linear mode.

This filter has a sharp cutoff which would ordinarily have overshoots in its transient response when a clipped waveform is applied to its input. This can result in overmodulation, but it has been modified to have a step in its frequency response with the result that the overshoots have been turned into "undershoots" and so are rendered harmless. The overall response of the filtering in this block including the high-pass and both low-pass filters is shown in **Figure 13.57.**

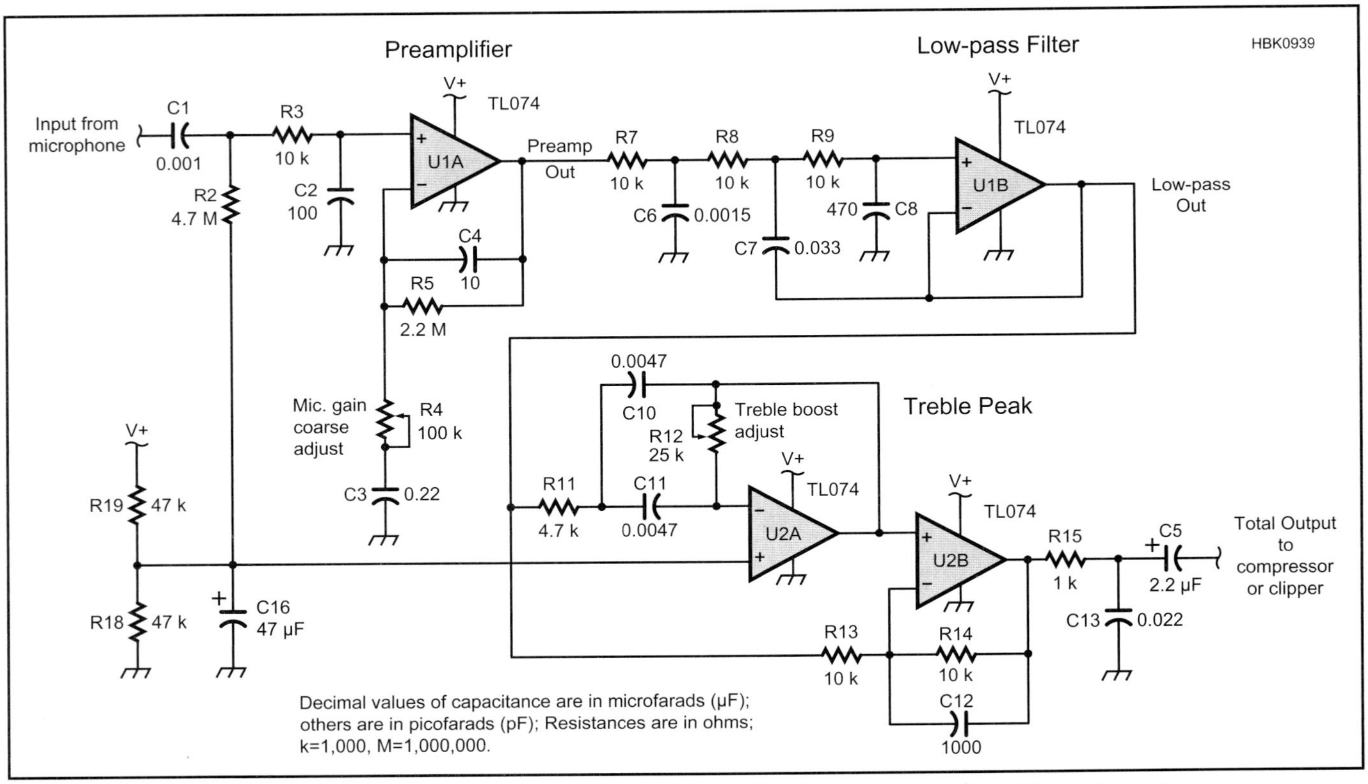

Figure 13.54 — Schematic of a microphone preamplifier with a gain stage, a low-pass filter and a treble peaking stage.

AUDIO COMPRESSOR

Another way to increase the average modulation level is to use automatic gain control (AGC) or a compressor. Such a circuit is shown in **Figure 13.58.** Compressors can be used for any mode of speech transmission. Compression should *never* be used with digital modes using AFSK modulation.

On the output of the op-amp are two LEDs in parallel but with opposite polarities. Audio voltages of either polarity illuminate a photoconductor causing its resistance to decrease. The photoconductor will then adjust the gain of the associated op-amp circuit. As audio levels increase, the LEDs illuminate the photoconductor more, reducing its resistance. This causes the photoconductors to reduce the circuit gain, maintaining a constant audio output level. A second photoconductor monitors the illumination and so shows the degree of compression. The photoconductors (Luna PDV-P8101 or equivalent) are mounted directly on top of the white LEDs (Cree C503D-WAN-CCBEB152 or equivalent). This is shown in the inset for Figure 13.58.

This circuit responds to signal level increases within 2 or 3 milliseconds. After a transient, it increases the gain back to normal in less than 100 milliseconds. The result is a very high average modulation level with far lower distortion than a clipper. Some signal overshoots will escape while gain reduction is underway but time-wise, on a percentage basis, they are quite small. If this block of circuitry is followed by a clipper to catch the overshoots then even those small overshoots

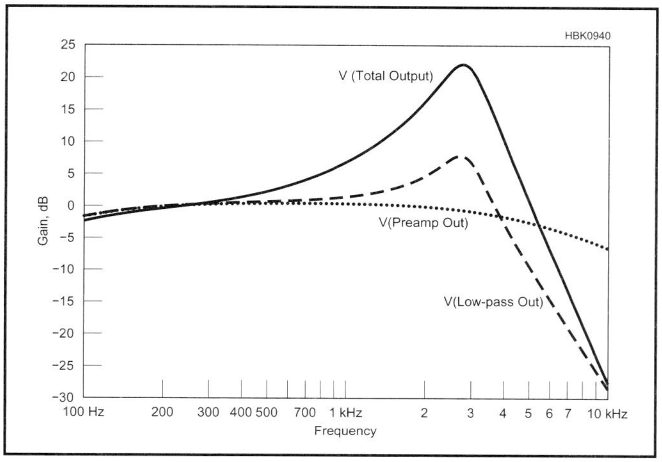

Figure 13.55 — Frequency response at various points within the speech amplifier.

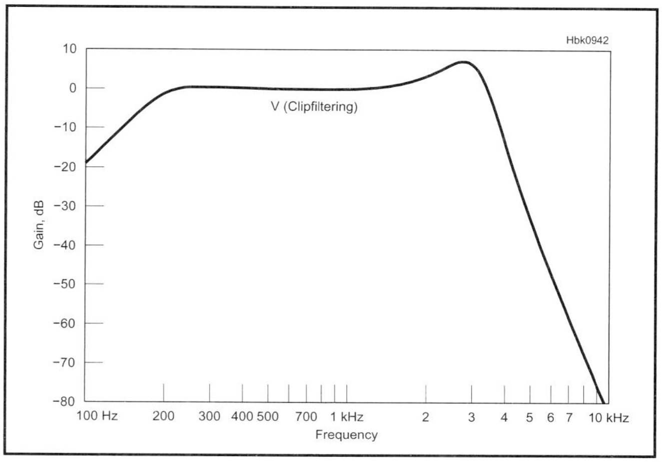

Figure 13.57 — Overall response of the clipper-filtering block.

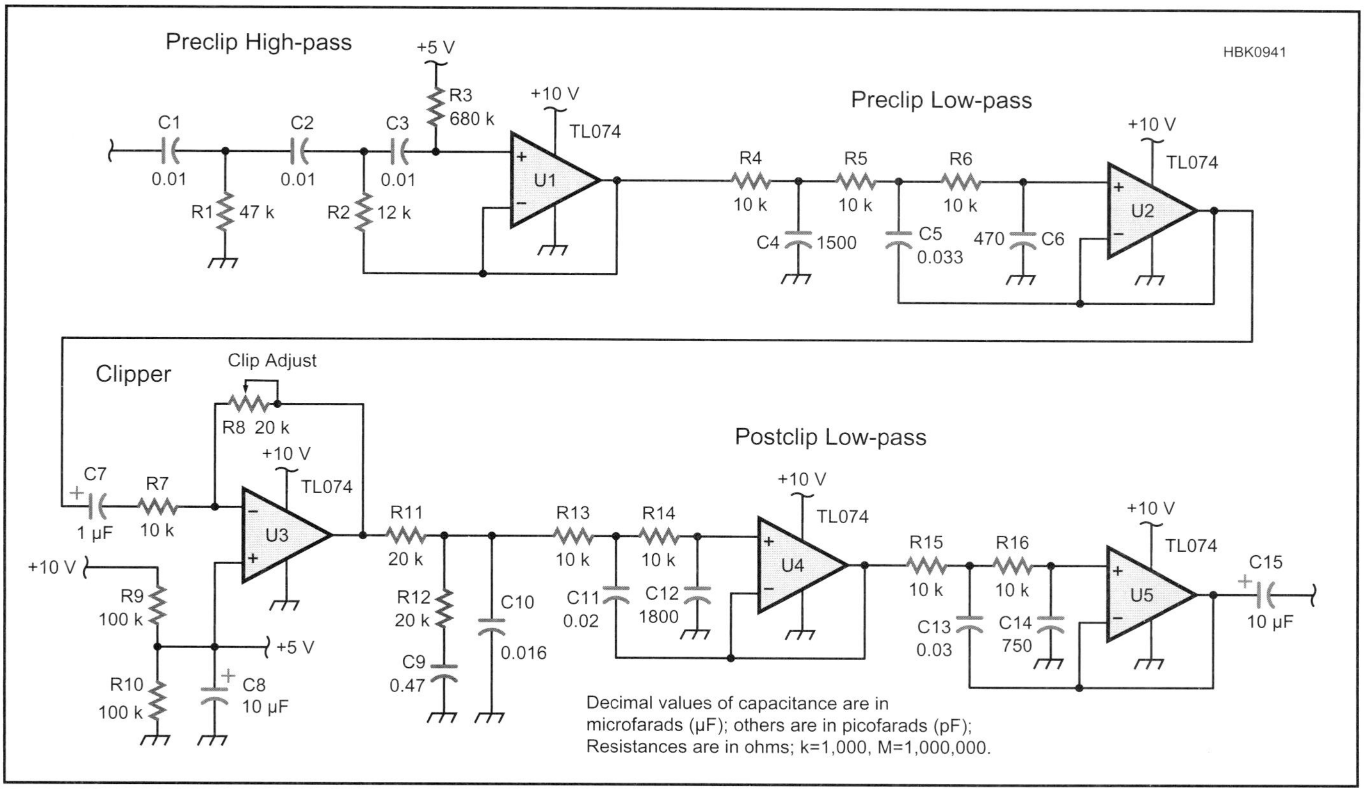

Figure 13.56 — Schematic of the speech clipper and its associated filtering.

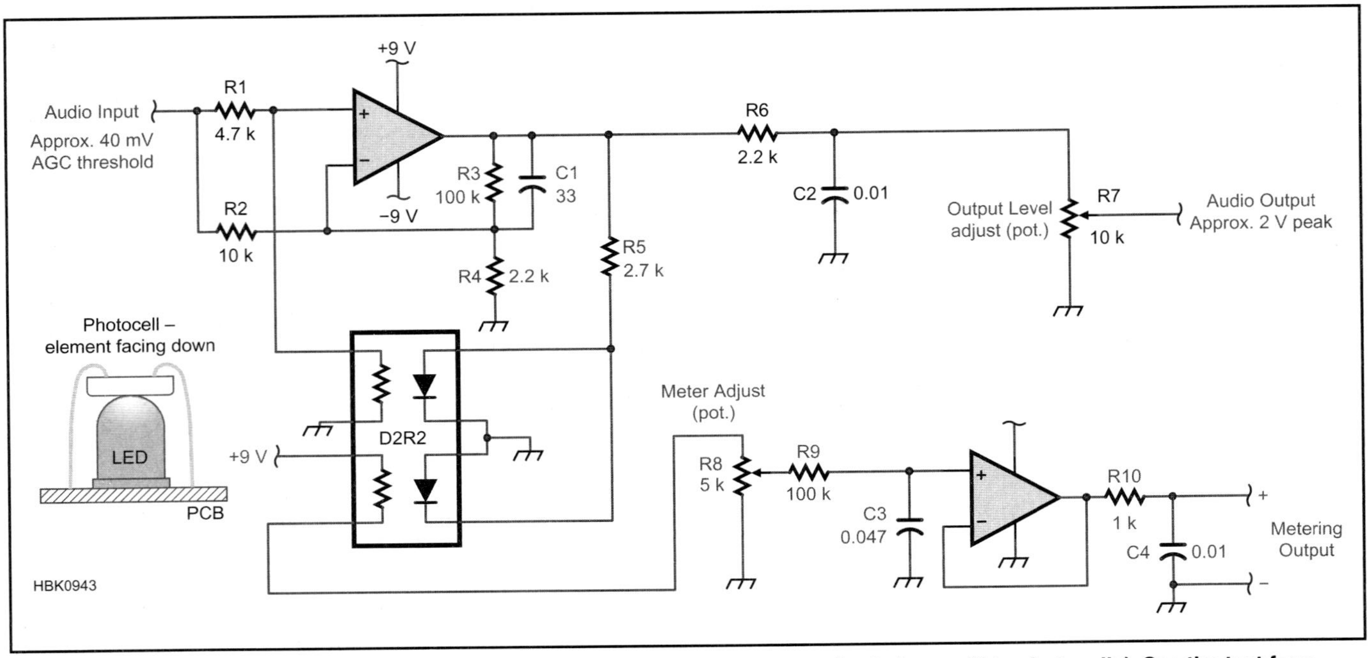

Figure 13.58 — Compressor circuit based on LEDs shining on photoconductors (cadmium sulfide photocells). See the text for a description of how to fabricate D2R2.

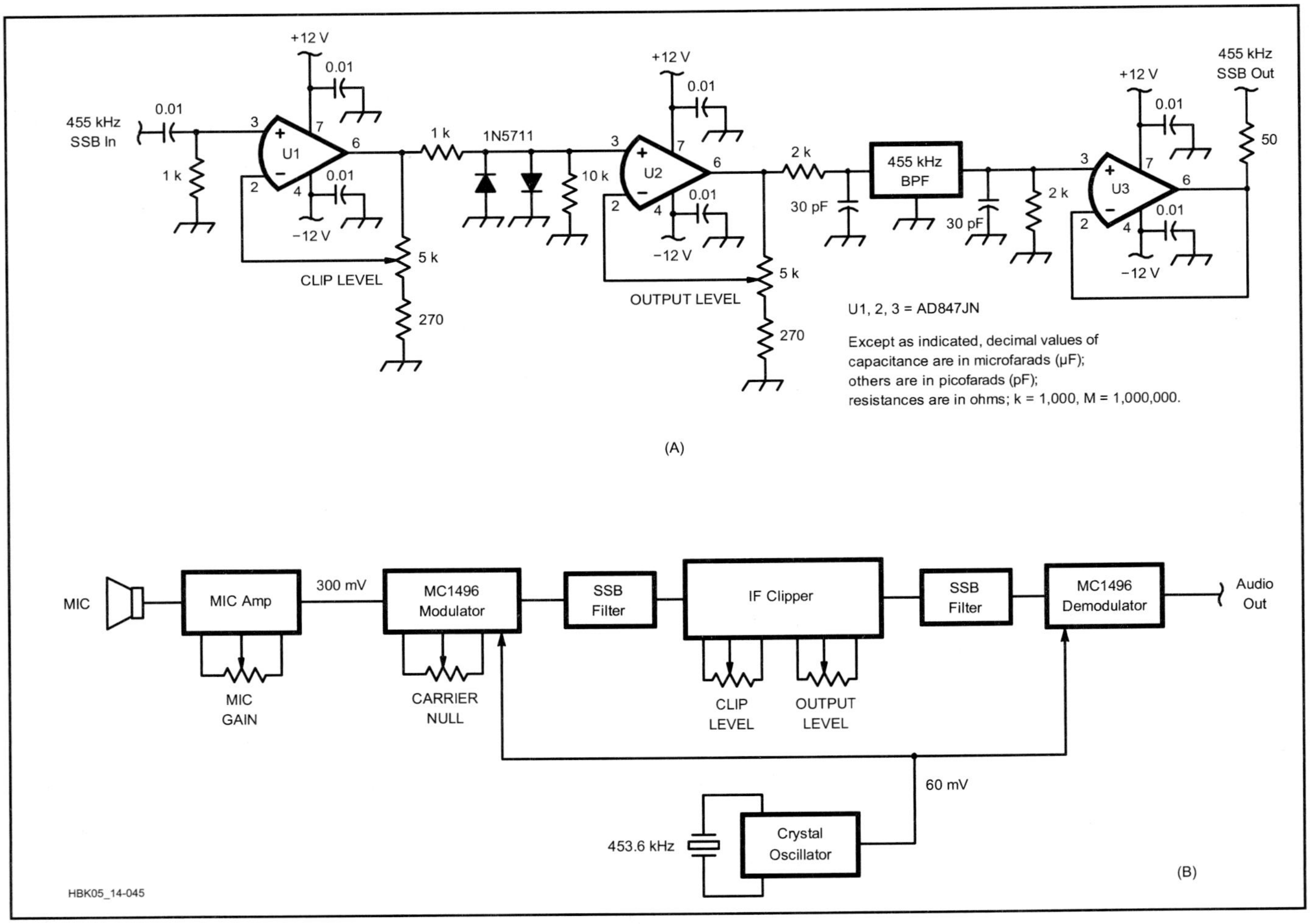

Figure 13.59 — IF speech clipping. At (A), schematic diagram of a 455 kHz IF clipper using high-frequency op amps. At (B) block diagram of an adaptation of the above system to an audio in-audio out configuration.

will be of no concern.

The pre-clip high-pass and pre-clip low-pass filters shown in the clipper block may be used in the compressor; they were not shown here. If used, they must be on the input to the compressor.

IF SPEECH CLIPPER

Audio clipper speech processors can generate a considerable amount of in-band harmonics and IMD (involving different simultaneously occurring speech frequencies). The total distortion detracts somewhat from speech intelligibility. IF clippers (also known as RF speech processors) overcome most of these problems, especially the Hilbert Transform problem. (See Sabin and Schoenike in the References section.)

Figure 13.59A is a schematic diagram of a 455 kHz IF clipper using high-frequency op-amps. 20 dB of gain precedes the diode clippers. A second amplifier establishes the desired output level. The clipping produces a wide band of IMD products close to the IF frequency. Harmonics of the IF frequency are easily rejected by subsequent selectivity. "Close-in" IMD distortion products are band-limited by the 2.5 kHz wide IF filter so that out-of-band splatter is eliminated. The in-band IMD products are at least 10 dB below the speech tones.

Figure 13.59B shows a block diagram of an adaptation of the above system to an audio in-audio out configuration that can be inserted into the mic input of any transmitter to provide the benefits of RF speech processing. These are sometimes offered as aftermarket accessories.

Figure 13.60 shows oscilloscope pictures of an IF clipped two-tone signal at various levels of clipping. The level of clipping in a radio can be estimated by comparing with these photos. Listening tests verify that the IMD does not sound nearly as bad as harmonic distortion. In fact, processed speech sounds relatively clean and crisp. Tests also verify that speech intelligibility in a noise background is improved by 8 dB. (See the article on RF clippers by Sabin in the References section.)

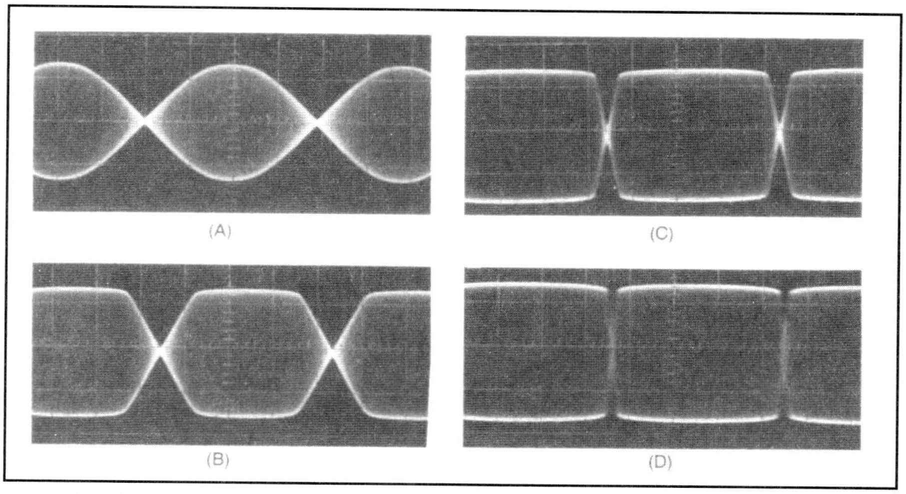

Figure 13.60 — Two-tone envelope patterns with various degrees of RF clipping. All envelope patterns are formed using tones of 600 and 1000 Hz. At A, clipping threshold; B, 5 dB of clipping; C, 10 dB of clipping; D, 15 dB of clipping.

The repeaking effect from band-pass filtering the clipped IF signal occurs, and must be accounted for when adjusting the output level. A two-tone audio test signal or a speech signal should be used. The ALC circuitry (discussed later) will reduce the IF gain to prevent splattering in the power amplifiers. If the IF filter is of high quality and if subsequent amplifiers are clean, the transmitted signal is of very high quality and is very effective in noisy situations and often also in pile-ups.

The extra IF gain implies that the IF signal entering the clipper must be free of noise, hum and spurious products. The cleanup filter also helps reduce the carrier frequency, which is outside the passband.

An electrically identical approach to the IF clipper can be achieved at audio frequencies. If the audio signal is translated to, say 455 kHz, processed as described and translated back to audio, all the desirable effects of IF clipping are retained. This output then plugs into the transmitter's microphone jack. Figure 13.59B shows the basic method. The mic amplifier and the MC1496 circuits have been previously shown and the clipper circuit can be the same as in Figure 13.59A.

The interesting operating principle in all of these examples is that the characteristics of the IF clipped (or equivalent) speech signal do not change during frequency translation, even if translated down to audio and then back up to IF in a balanced modulator.

13.9 Managing Computer Audio

13.9.1 Computer Audio

DIGITAL DYNAMIC RANGE

The number of bits in a digital system (called the *bit depth*) sets the limit for its maximum dynamic range. An ideal 16-bit system would provide 96 dB of dynamic range (the difference between the highest and lowest instantaneous voltage that it can accurately reproduce). The dynamic range of real products is a few dB less. Each additional bit doubles the range of voltage that can be reproduced, and double (or half) the voltage is a change of 6 dB. Thus, a 12-bit system should be capable of a few dB less than 72 dB dynamic range.

DATA RATES

The *bit rate* of a digital system, expressed in kilobits/sec (kbps), sets the limit for its audio bandwidth, which is 90% of half the numerical value, in kHz. Thus, a 48 kHz system can provide about 22 kHz audio bandwidth. When a sound card is used as part of a spectrum display, the maximum displayed bandwidth in kHz is equal to the maximum bit rate in kbps. Most computer sound cards and sound recording software can operate at standard sample rates and bit depth less than their maximum value, and most software that uses a computer sound system can vary these settings either automatically or as desired by the user.

AUDIO DATA COMPRESSION

The size of a sound recording file depends both on its bit depth and bit rate. A 12-bit sound file recorded at 12 kbps provides audio bandwidth of about 5 kHz with about 70 dB of dynamic range and is a good choice for use in amateur radio. Uncompressed sound files (such as AIFF and WAV) require about 16 kB on a hard drive for each second of recording time for each channel of audio recorded. WAV files are the standard for CD-ROM recordings at 16-bit sample width and 44.1 kHz sample rate.

Digital audio signals can be compressed to reduce both the bandwidth needed to transmit them and the size of files needed to contain them. *Lossless compression* (such as ALAC, FLAC, and ZIP) reduces file size somewhat without losing any of the sound information. Lossless systems provide an exact copy of the original. ZIP provides very little compression of uncompressed sound files, but ALAC and FLAC can reduce file size by about one half.

Lossy compression systems like AAC, MP3, and OGG use algorithms to approximate the audio by leaving out parts of the waveform that the ear/brain is unlikely to miss. On playback, the system guesses at what parts were left out and adds them back in. Such systems are referred to as lossy because they do not provide a perfect copy of the original digital signal. These algorithms may compromise digital mode modulation and demodulation.

While file compression can save space on a hard drive, it takes time for the compression and de-compression algorithms to operate. MP3 compression is a great choice for recording a QSO, but a poor choice for recording messages for playback during a contest due to the delay. MP3 compression can reduce file size (and transmission bandwidth) by 75-95% (that is, files and bandwidth between 5% and 25% of their uncompressed size). Since contest messages are short, a lossless format is recommended.

USB AUDIO

It is increasingly common for transceivers to have a USB port for a PC CAT interface and audio input and output. A radio's USB interface is usually a Silicon Labs or FTDI IC. Driver software for the PC to use the radio interface will be available from the radio manufacturer, from the IC supplier, or from the developer of the PC software.

Audio can be transferred over the USB interface as a digital data stream called a USB Audio Codec. This digital interface is recognized by computers and other devices that support the USB standard. Your computer software will treat the USB Audio Codec as an audio playback or recording device for which you select and adjust levels the same as speakers or microphones.

A transceiver's USB interface will use the same data rates (up to 115 kbaud or higher) for PC control and audio transfer so use the highest reliable data rate available to minimize audio latency or dropouts. The radio will also have adjustable input and output levels for USB audio.

USB sound cards are standalone devices that have a standard analog audio interface and a USB port that connects to a host PC. USB sound cards designed for semi-pro audio users and even for gamers are usually good performers. They often perform significantly better than the sound cards built into computers, especially laptops. An inferior quality sound card can degrade the decoding capability of software for digital modes. Some higher-end units have LEDs to indicate safe signal levels — for example, green for good levels, red for overload (also called "digital clip").

Using a USB audio interface can eliminate audio cables between a PC and transceiver. This also reduces the possibility of RFI although USB interfaces can be susceptible to RFI. Use high-quality, shielded USB cables to minimize RFI.

13.9.2 Setting Digital Mode Levels

Many digital modes use PC software to perform audio encoding and decoding. If a USB audio codec is not used, the analog audio is then routed through a sound card to the transceiver's microphone or data audio input on an accessory connector. For USB audio, follow the transceiver and software level adjust instructions. The rest of this section assumes that analog audio is being used.

The transceiver uses the input audio to perform AFSK modulation of a SSB or FM signal. To avoid distorting the modulated audio and generating spurious products in the audio and on adjacent RF channels, it is important to use the right levels for the transceiver's audio input. Similarly, the decoding software in the PC will work best with undistorted audio from the transceiver that is well above the noise floor.

Remember that sound card inputs and outputs will be relatively low-level analog signals and shielded cables are required. Be sure that the shields are properly connected to the outside of the transceiver's metal enclosure. Do not connect the shield to a microphone input ground pin or directly to a circuit board inside the radio. RF current on the shield of these cables will cause RFI if given a path to the internal circuitry. In addition, PC and transceiver enclosures should be bonded to minimize any voltage between them that could result in RFI and noise in the audio circuits. If you do experience RFI, use a ferrite core of the right mix for the frequency range involved. See the **RFI and EMC** chapter for more about managing RF in your station.

SETTING SOUND CARD INPUT LEVEL

This is important to achieving maximum signal-to-noise performance. For sound cards with a digital clip indicator, turn up the input gain until the red light flashes with the loudest signals, then reduce it slightly so that it never flashes. Then set the digital gain in the encode/decode software as directed by the manual.

SETTING SOUND CARD OUTPUT LEVEL

It is just as important to avoid transmitting a distorted or overmodulated signal. There are three steps, all of which are important: (1) making the output of the computer clean, (2) not overdriving the radio's audio input stage, and (3) setting the audio input gain in the radio. There are (at least) four good ways to set output level from the computer, depending on what test equipment is available.

The first method uses an oscilloscope connected to the output of the sound card that feeds the radio. Set the sweep so that you see clearly defined sine waves that make up the signal from the digital program (RTTY, PSK, WSJT modes, and so on) in transmit mode. Because these modulation schemes include multiple frequencies, you probably won't get the display to sync. Increase the output level until you see squaring ("flat-topping") at the top of the sine waves or "spiky" digital distortion, then reduce the output level by half the voltage.

The second method uses an audio spectrum analyzer connected to the output of the sound card. With the digital program in transmit mode and output level set fairly low, note the spectrum lines. With JT65, you should see only the tones that you would see on the WSJT-X or JT65-HF display, roughly 200 Hz wide. With RTTY, you should see only two tones spaced by 170 Hz. With JT9 or PSK, you should only see a single tone. Now, increase levels until you begin to see additional lines spread out from the normal tones. These additional lines are distortion products and will cause interference. Note the *difference* in strength (in dB) between the signal tones and the distortion, then gradually increase output level until you see the difference become smaller (which indicates that the percentage distortion is rising). Now, reduce the output level until the difference is larger (the distortion is much less). With proper adjustment, the distortion products should be at least 40 dB below the tones.

The third method uses an audio voltmeter connected to the output of the sound card. With the digital program in transmit mode, start with very low output, and gradually increase it until you no longer see voltage increasing, then reduce the output level to one-half the maximum measured voltage.

The fourth method uses your ears and headphones connected to the output of the sound card. Start with the output of the sound card set low, gradually increasing it until you hear harshness or sharpness in the tones. That harshness is the distortion products. Now, reduce the output of the sound card until the tones sound half as loud. This works because a change of 6 – 10 dB is perceived as half (or twice) as loud.

With all of these methods, adjust the input gain control of your radio according to the user manual for the radio for transmitting digital modes. Always make sure that any processing in the radio is turned off.

As a final check, ask another amateur who is receiving your signal fairly well to look carefully for sidebands in the received waterfall display (first making sure that the noise blanker is turned off, and that you are not overloading the receiver).

CONNECTING COMPUTER AUDIO TO THE RADIO

To avoid overdriving the radio's audio input, feed the computer sound card to the radio's LINE IN input (if there is one). In older equipment, this may be labeled as a phone patch input, or for use with a hardware RTTY interface. It may be available on an accessory connector.

If the radio has no line input, you'll need to feed the microphone input through a simple voltage divider (often called a "pad" — see the previous section in this chapter) so that you don't overload the microphone preamp. All it takes is two resistors, one in series with the audio path, and one in parallel with the input of the radio. 20 dB (a 10:1 voltage divider) of attenuation should be enough for most radios and calls for a 10:1 ratio between the two resistors. The values are not critical, but 10 kΩ for the series resistor and 1 kΩ for the parallel resistor, or 4.7 kΩ and 470 Ω are good choices. Low-wattage resistors may be used here and can fit within the connectors of the audio cable between the computer and the radio if you use the right connector. RCA and ⅛-inch connectors made by Switchcraft have the most space inside them. Neutrik is also a good brand.

13.10 Voice Operation

13.10.1 Push-To-Talk for Voice

Another advance in amateur station switching followed longstanding practices of aircraft and mobile voice operators who had other things to contend with besides radio switches. Microphones in those services included built-in switches to activate TR switching. Called push-to-talk (PTT), this function is perhaps the most self-explanatory description in our acronym studded environment.

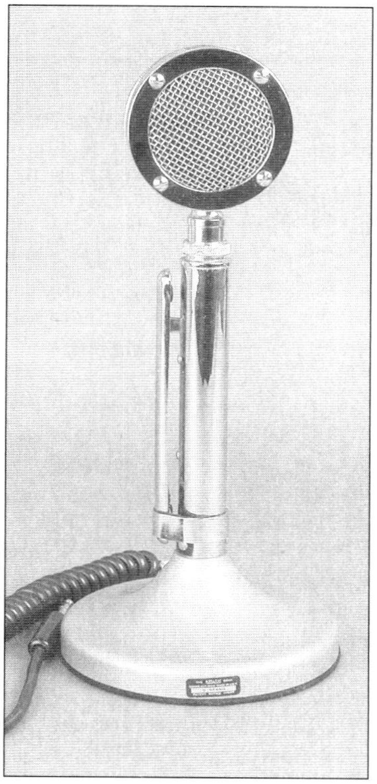

Figure 13.61 — A classic Astatic D-104 mic with PTT stand.

Relays controlled the various switching functions when the operator pressed the PTT switch. Some top-of-the-line transmitters of the period included at least some of the relays internally and had a socket designed for PTT microphones. **Figure 13.61** is a view of the ubiquitous Astatic D-104 microphone with PTT stand, produced from the 1930s to 2004, and still popular at flea markets and auction sites. PTT operation allowed the operator to be out of reach of the radio equipment while operating, permitting "easy chair" operation for the first time.

Modern transceivers include some form of PTT (or "one switch operation"). Relays, diodes, transistors and other components seamlessly handle myriad transmit-receive changeover functions inside the transceiver. Most transceivers have additional provisions for manually activating PTT via a front-panel switch. And many have one or more jacks for external PTT control via foot switches, computer interfaces or other devices.

13.10.2 Voice-Operated Transmit-Receive Switching (VOX)

How about break-in for voice operators? SSB operation enabled the development of voice operated transmit/receive switching, or VOX. During VOX operation, speaking into the microphone causes the station to switch from receive to transmit; a pause in speaking results in switching back to receive mode. Although VOX technology can work with AM or FM, rapidly turning the carrier signal on and off to follow speech does not provide the smooth operation possible with SSB. (During SSB transmission, no carrier or signal is sent while the operator is silent.)

VOX OPERATION

VOX is built into current HF SSB transceivers. In most, but not all, cases they also provide for PTT operation, with switches or menu settings to switch among the various control methods. Some operators prefer VOX, some prefer PTT and some switch back and forth depending on the operating environment.

VOX controls are often considered to be in the "set and forget" category and thus may be controlled by a software menu or by controls on the rear panel, under the top lid or behind an access panel. The following sections discuss the operation and adjustment of radio controls associated with VOX operation. Check your transceiver's operating manual for the specifics for your radio.

Before adjusting your radio's VOX controls, it's important to understand how your particular mic operates. If it has no PTT switch, you can go on to the next section! Some mics with PTT switches turn off the audio signal if the PTT switch is released, while some just open the control contacts. If your mic does the former, you will need to lock the PTT switch closed, have a different mic for VOX, or possibly modify the internal mic connections to make it operate with the VOX. If no audio is provided to the VOX control circuit, it will never activate. If the mic came with your radio, or from its manufacturer, you can probably find out in the radio or mic manual.

VOX Gain

Figure 13.62 shows some typical transceiver VOX controls. The VOX gain setting determines how loud speech must be to initiate switchover, called "tripping the VOX." With a dummy load on the radio, experiment with the setting and see what happens. You should be able to advance it so far that it

Figure 13.62 — The function of VOX controls is described in the text. They require adjustment for different types of operating, so front-panel knobs make the most convenient control arrangement. In some radios, VOX settings are adjusted through the menu system.

switches with your breathing. That is obviously too sensitive or you will have to hold your breath while receiving! If not sensitive enough, it may cause the transmitter to switch off during softly spoken syllables. Notice that the setting depends on how close you are to the microphone, as well as how loud you talk. A headset-type microphone (a "boom set") has an advantage here in that you can set the microphone distance the same every time you use it.

The optimum setting is one that switches to transmit whenever you start talking, but isn't so sensitive that it switches when the microphone picks up other sounds, such as a cooling fan turning on or normal household noises.

VOX Delay

As soon as you stop talking, the radio can switch back to receive. Generally, if that happens too quickly, it will switch back and forth between syllables, causing a lot of extra and distracting relay clatter. The VOX delay control determines how long the radio stays in the transmit position once you stop talking. If set too short, it can be annoying. If set too long, you may find that you miss a response to a question because the other station started talking while you were still waiting to switch over.

You may find that different delay settings work well for different types of operation. For example, in a contest the responses come quickly and a short delay is good. For casual conversation, longer delays may be appropriate. Again, experiment with these settings with your radio connected to a dummy load.

Anti-VOX

This is a control with a name that may mystify you at first glance! While you are receiving, your loudspeaker is also talking to your microphone — and tripping your VOX — even if you aren't! Early VOX users often needed to use headphones to avoid this problem. Someone finally figured out that if a sample of the speaker's audio signal were fed back to the mic input, out-of-phase and at the appropriate amplitude, the signal from the speaker could be cancelled out and would not cause the VOX circuit to activate the transmitter. The ANTI-VOX (called ANTI-TRIP in the photo) controls the amplitude of the sampled speaker audio, while the phase is set by the transceiver design.

As you tune in signals on your receiver with the audio output going to the speaker, you may find that the VOX triggers from time to time. This will depend on how far you turn up the volume, which way the speaker is pointed and how far it is from the mic. You should be able to set the anti-VOX so that the speaker doesn't trip the VOX during normal operation.

Generally, setting anti-VOX to higher values allows the speaker audio to be louder without activating the VOX circuit. Keep in mind that once you find a good setting, it may need to be changed if you relocate your microphone or speaker. With most radios, you should find a spot to set the speaker, microphone and anti-VOX so that the speaker can be used without difficulty.

13.11 Transmitter Power Stages

The functions described so far that process input data and information and result in a signal on the desired output radio frequency generally occur at a low level. The one exception is full-carrier AM, in which the modulation is classically applied to the final amplification stage. More modern linear transmitter systems generate AM in the same way as SSB at low levels, typically between 1 mW and 1 W.

13.11.1 Types of Power Amplifiers

The **RF Power Amplifiers** chapter provides a detailed view of power amplifiers; however, we will take a quick peek here to set the stage for the following discussions. Amplifiers use dc power applied to active devices in order to increase the power or level of signals. As will all real devices, they introduce some distortion in the process, and are generally limited by the level of distortion products. Power amplifiers can be constructed using either solid-state devices or vacuum tubes as the active device. At higher powers, typically above a few hundred watts, vacuum tubes are more frequently found, although there is a clear trend toward solid state at all amateur power levels.

Independent of the device, amplifiers are divided into classes based on the fraction of the input cycle over which they conduct. A sinusoidal output signal is provided either by the *flywheel* action of a resonant circuit or by other devices contributing in turn. The usual amplifier classes are summarized in **Table 13.1**. Moving from Class A toward Class C, the amplifiers become progressively less linear but more efficient. The amplifiers with a YES in the LINEAR column thus are not all equally linear however A, AB or Class B amplifiers can be suitable for operation in a linear transmitter chain. Class C amplifiers can be used only for amplification of signals that do not have modulation information contained in the amplitude, other than on-off keyed signals. Thus class C amplifiers are useful for amplification of sinusoids, CW, FM, or as the nonlinear stage at which high-level AM modulation is employed.

Table 13.1
Characteristics of Transceiver Power Amplifier Classes

Values are typical

Class	*Conduction*	*Linear*	*Efficiency*
A	360°	Yes	30%
AB	270°	Yes	55%
B	180°	Yes	65%
C	90°	No	74%

Recent developments in switching-type amplifiers and in single-band matching network design have created several additional classes. Class D is a switchmode amplifier most often used for high-efficiency audio amplification. Class E and F use tuned output networks that let the amplifying device act like a switch but prevent high voltage and high current at the same time. Class G is similar to a Class B amplifier, but switches between two voltage levels to reduce power dissipation at low signal levels. Class I uses two devices driven with complementary pulse duty cycles to cancel harmonics and follow the input waveform. Class S is a variation on Class D, and Class T uses DSP to optimize pulse widths in a Class D amplifier for better performance. See the references at the end of this chapter for Rosu and for Silver.

13.11.2 Linear Amplifiers

While transmitters at power levels of 1 mW to 1 W have been successfully used for communication across many portions of the spectrum, most communications systems operate with more success at higher powers. The low level stage is usually referred to as an exciter, while higher power is provided by one or more linear amplifier stages as shown in **Figure 13.63**.

The power levels shown at the various points in Figure 13.63 are fairly typical for a high powered amateur station. The 1500 W PEP output represents the legal limit for US amateurs in most bands (200 W PEP on 30 meters and 100 W ERPD on the 60 meter channels are notable exceptions). The first amplifier block may contain more than one stage, while the final output amplifier is often composed of multiple parallel active devices.

Typical power supply requirements for the amplifier stages are noted for a number of reasons. First, while power is rarely an issue at the exciter level, often it is a significant issue at the power levels shown for the amplifiers. The power supplies represent a large portion of the cost and weight of the system as the power increases. Some manufacturers are beginning to use switching-type power supplies for high-power amplifiers, resulting in a major reduction in size and weight.

Note also that a gross amplifier efficiency of about 50% is assumed for the amplifiers, taking into account ancillary subsystems as well as the inefficiency of the active devices in linear mode. The 50% that doesn't result in actual RF output is radiated as heat from the amplifier and must be removed from the amplifier as it is generated to avoid component damage. This represents another cost and weight factor that increases rapidly with power level.

The voltages shown for the supplies are those typical of modern solid state amplifiers. While virtually all commercial equipment now includes solid state amplifiers at the 100 W level, vacuum tube active devices are frequently found at higher levels, although the trend is clearly moving toward solid state. Vacuum tube amplifiers typically operate at voltages in the 2 to 4 kV range, requiring stringent measures be taken to avoid arcing across components. In addition, vacuum tube amplifiers typically dissipate up to 100 W of filament power that must be added to the power supply and heat dissipation planning.

13.11.3 Nonlinear Amplifiers

Nonlinear transmitters are somewhat different in architecture than the linear systems discussed previously. The configuration of a high-level AM modulated transmitter is shown in **Figure 13.64**. Note that none of the upper RF stages (the "RF chain") need to be particularly linear. The final stage must be nonlinear to have the modulation applied. Thus the RF stages can be the more power-efficient Class C amplifiers if desired.

There are some observations to be made here. Note that the RF chain is putting out the full carrier power whenever in transmit mode, requiring a 100% duty cycle for power and amplifier components, unlike the SSB systems discussed previously. This imposes a considerable weight and cost burden on the power supply system. Note also that the PEP output of a 100% modulated AM system is equal to four times the carrier power.

The typical arrangement to increase the power of such a system is to add not only an RF amplifier stage capable of handling the desired power, but also to add additional audio power amplification to fully modulate the final RF stage. For 100% high-level plate modulation, an audio power equal to half the dc input power (plate voltage times plate current of a vacuum tube amplifier) needs to be provided. This arrangement is shown in **Figure 13.65**. In the example shown, the lower level audio stages are provided by those of the previous 50 W transmitter, now serving as an exciter for the power amplifier and as a driver for the modulating stage. This was frequently provided for in some transmitters of the AM era, notably the popular E. F. Johnson Ranger series, which provided special taps on its modulation transformer for use as a driver for higher-power systems.

It is worth mentioning that in those days the FCC US amateur power limit was expressed in terms of dc *input* to the final stage and was limited to 1000 W, rather than the 1500 W PEP *output* now specified. A fully modulated 1000 W dc input AM transmitter would likely have a carrier output of 750 W or 3000 W PEP — 3 dB above our current limit. If you end up with that classic Collins KW-1 transmitter, throttle it back to make it last and stay out of trouble!

13.11.4 Hybrid Amplifiers

Another alternative that is convenient with current equipment is to use an AM transmitter

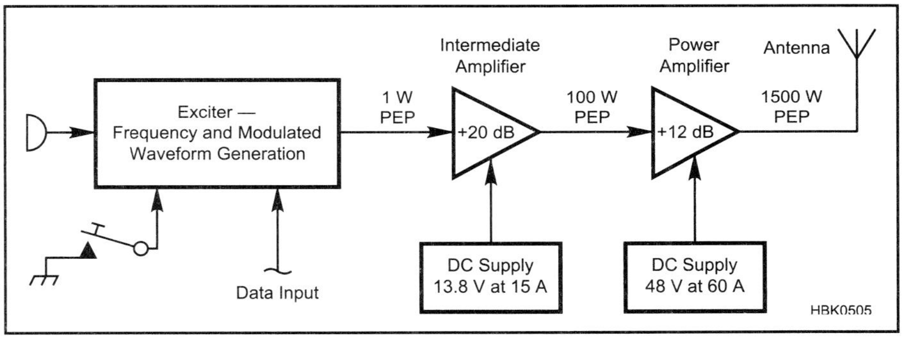

Figure 13.63 — Block diagram of a solid-state linear transmitter chain with multiple amplifier stages.

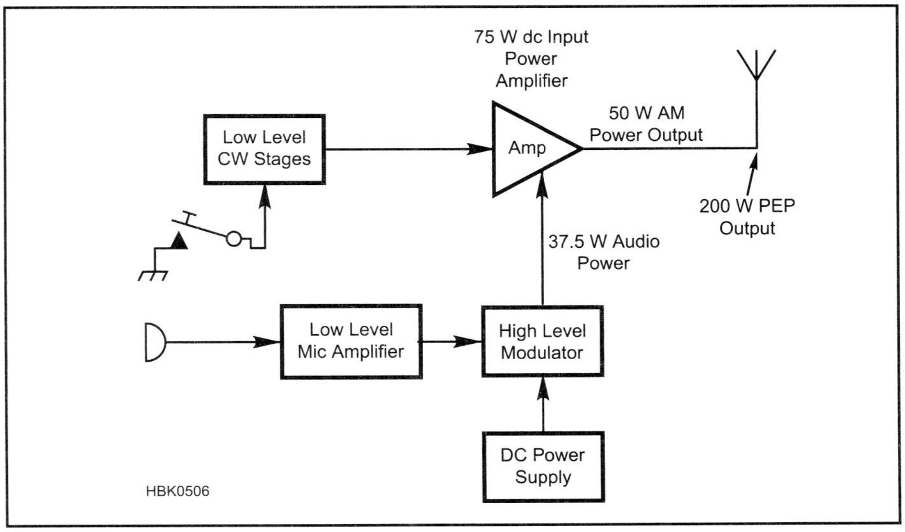

Figure 13.64 — Block diagram of a high level AM modulated transmitter.

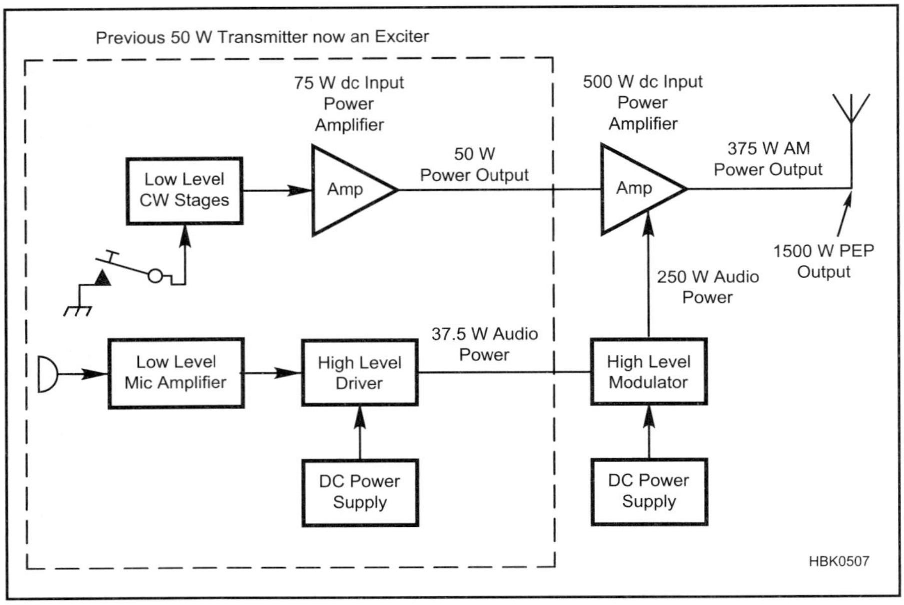

Figure 13.65 — Block diagram of a high level AM modulated transmitter with added output stage.

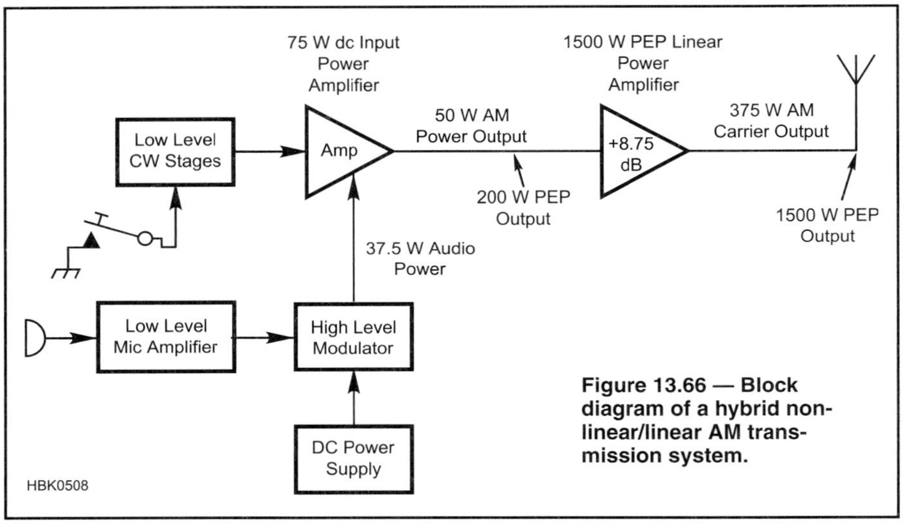

Figure 13.66 — Block diagram of a hybrid non-linear/linear AM transmission system.

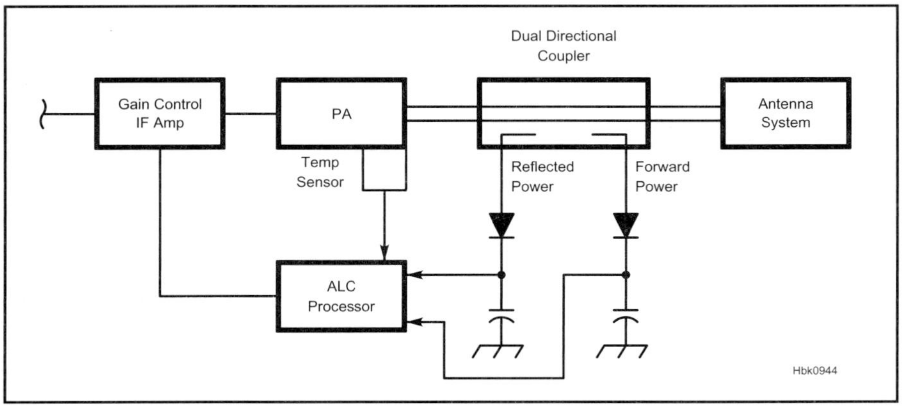

Figure 13.67 — An ALC protection method for a solid-state transmitter using a directional coupler to sense power level.

with a linear amplifier. This can be successful if the relationship that PEP = 4 × Carrier Power is maintained. **Figure 13.66** shows a 1500 W PEP output linear amplifier following a typical 50 W AM transmitter. In this example, t he amplifier would be adjusted to provide a 375 W carrier output with no modulation applied to the exciter. During voice peaks the output seen on a special PEP meter, or using an oscilloscope, should be 1500 W PEP.

Note that during AM operation, the amplifier is producing a higher average power than it would without the carrier being present, as in SSB mode. The duty cycle specification of the amplifier should be checked to be sure it can handle the heavier load. If the amplifier has an RTTY rating, it should be safe to run an AM carrier at 66% of the RTTY output, following the required on and off time intervals.

13.11.5 Automatic Level Control (ALC)

ALC is usually derived from the last stage in a transmitter. This ensures that this last stage will be protected from overload. However, other stages prior to the last stage may not be as well protected; they may generate excessive distortion. It is possible to derive a composite ALC from more than one stage in a way that would prevent this problem. But designers usually prefer to design earlier stages conservatively enough so that, given a temperature range and component tolerances, the last stage can be the one source of ALC. The gain control is applied to an early stage so that all stages are aided by the gain reduction.

Note that ALC should be minimally active with most digital mode transmissions. The modulation of these signals requires linear amplification to preserve the waveform shape and minimize distortion products. ALC action creates distortion as it alters the power level of the signal. Adjust the radio drive levels so that the ALC is at its minimum level of activity — usually shown as the lower bar of a multi-segment LCD meter or a needle position just above zero. (The same caution applies to any form of audio or speech processing if the digital signal is generated by audio tones applied to the transmitter's microphone input.)

Figure 13.67 shows how a dual directional coupler can be used to provide ALC for a solid-state power amplifier (PA). The basic idea is to protect the PA transistors from excessive SWR and dissipation by monitoring both the forward power and the reflected power.

13.11.6 Transmit-Receive (TR) Switching

As the complexity of a transceiver increases, switching between receive and transmit becomes quite complex. In commercially built equipment, this function is usually controlled by a microprocessor that manages any necessary sequencing and interlock functions that would require an excessive amount of circuitry to implement with discrete components. For an example of just how complex TR switching could be in an advanced transceiver, look at the schematic for any modern mid-level or top-of-the-line solid-state transceiver.

Nevertheless, the basic functions of TR switching are well within scope for the amateur building a transceiver. Understanding TR switching will also assist in troubleshooting a more complex commercial radio.

QRP TR SWITCHING

Numerous schemes are popular for switching an antenna between transmitter and receiver functions. But these schemes tend to get in the way when one is developing both simple receivers and low-power transmitters, perhaps as separate projects. A simple relay-based TR scheme is then preferred and is presented here. In this system, used in the MkII Updated Universal QRP Transmitter by Wes Hayward, W7ZOI (see the full article in this book's online supplemental information), the TR relay not only switches the antenna from the receiver to the transmitter, but disconnects the headphones from the receiver and attaches them to a sidetone oscillator that is keyed with the transmitter.

The circuitry that does most of the switching is shown in **Figure 13.68**. A key closure discharges capacitor C1. R2, the 1 kΩ resistor in series with C1, limits the discharge current. Key closure causes Q6 to saturate, causing Q7 to also saturate, turning the relay on. The relay picked for this example has a 700 Ω, 12 V coil with a measured 4 ms pull-in time.

If full break-in TR switching is required, a high-speed reed relay can be used. KE2QJ provided circuits that can be adapted for internal use in a home-built transceiver, although the original purpose was to integrate a stand-alone receiver with a transceiver and linear amplifier. His full article is available in this book's downloadable supplemental information. The References also include an excellent online paper by W8ZR on adding a high-speed vacuum relay QSK switch for full legal-limit QSK operation.

If you already have a receiver and transmitter and want to integrate them under the control of a separate TR switch, the K8IQY "Magic Box" (**www.4sqrp.com/MagicBox.php**) incorporates a number of useful features. This is a microprocessor-controlled design that can handle up to 10 W of transmitter power, switches at up to 50 WPM, and includes an audio sidetone output, as well. Although the kit is no longer offered, complete documentation is available online, including schematics and design information. The design could be extended to handle more transmit power with heavier components and the appropriate circuit changes.

AMPLIFIER-TRANSCEIVER TR SWITCHING

Amateur transceivers intended for use with external amplifiers have a KEY OUT output. This is usually a contact closure while in transmit mode intended to connect to a corresponding KEY IN input on the external amplifier. Check the transceiver and amplifier manuals to find out what they are called on your equipment. A diagram of the proper cabling to connect the transceiver and amplifier will be provided in the manual. Amplifiers may

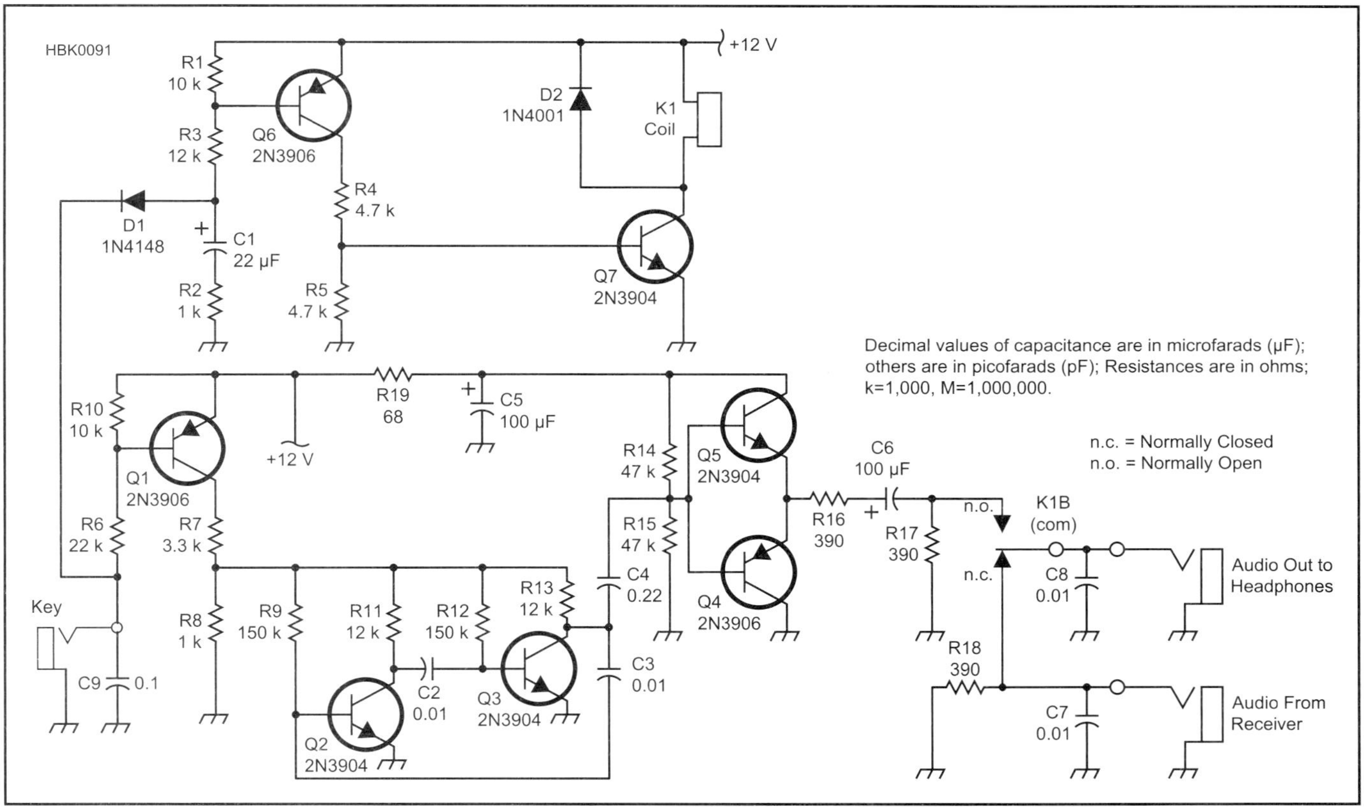

Figure 13.68 — Detailed schematic diagram and parts list for transmit-receive control section and sidetone generator of the universal QRP transmitter. Resistors are ¼ W, 5% carbon film.

C1 — 22 µF, 25 V electrolytic
C2, C3, C7, C8 — 0.01 µF, 50 V ceramic
C4 — 0.22 µF, 50 V ceramic
C5, C6 — 100 µF, 25 V electrolytic
C9 — 0.1 µF, 50 V ceramic
K1 — DPDT 12 V coil relay. An NAIS DS2Y-S-DC12, 700 Ω, 4 ms relay was used in this example.
Q1, Q4, Q6 — 2N3906, PNP silicon small signal transistor
Q2, Q3, Q5, Q7 — 2N3904, NPN silicon small signal transistor

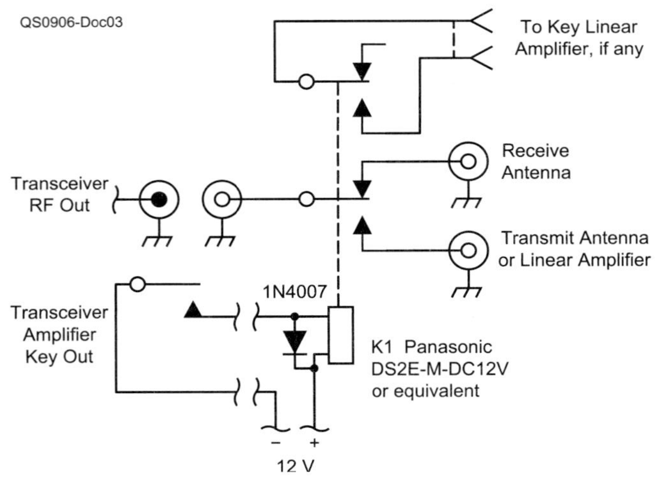

Figure 13.69 — Schematic of an external circuit that allows a modern transceiver to key a linear amplifier with TR switch voltage or current requirements that exceed the transceiver's ratings. As a bonus, it can also be used to allow reception from a low-noise receiving antenna.

also have an output signal that can be used to inhibit RF output from the transceiver until TR switching is complete and the amplifier is ready to operate. This avoids "hot switching" when the transceiver output is active while the amplifier's TR switch is still changing state.

Check the ratings to find out how much voltage and current the transceiver can safely switch, whether by a relay or a solid-state device. Many transceivers have both with the solid-state output used for high-speed switching or full break-in (QSK). Although most amplifiers are compatible with the switching capabilities of current transceivers, the voltage and/or current required to switch the relays in an older linear amplifier may exceed the ratings.

If your amplifier manual doesn't say what the switching voltage is, you can find out with a multimeter or DMM. Set the meter to read voltage of 250 V dc or higher. Connect the positive meter probe to amplifier key jack's center conductor, and connect the negative meter probe to the chassis ground (or the other key jack terminal if it's not grounded). This will tell you what the open circuit voltage is on the amplifier key jack. You may need to try a lower voltage range or switch to read ac voltage.

Now set the meter to read current. Start with a range that can read 1 A dc, and with the leads connected as before, you should hear the amplifier relay close and observe the current needed to operate the TR relay or circuit. Adjust the meter range, if needed, to get an accurate reading.

These two levels, voltage and current, are what the transceiver will be asked to switch. If *either* reading is higher than the transceiver specification, do not connect the transceiver and amplifier together. Doing so will likely damage your transceiver. You will need a simple interface circuit to handle the amplifier's switching voltage and current.

The simple, low-cost relay circuit shown in **Figure 13.69** can be used to key an older amplifier with a modern transceiver. It offers an added benefit: Another potential use of the transceiver KEY OUT jack is to switch to a separate low-noise receive antenna on the lower bands. While most high-end transceivers have a separate receive-only antenna connection built in, many transceivers don't. If you don't need one of the extra functions, just omit that connection.

13.11.7 PIN Diode RF Switching

Many current transceivers and amplifiers use PIN diodes for RF switching. They switch very quickly (as fast as nanoseconds) and do not wear out. They can be destroyed through excess reverse voltage or average forward current, requiring more protection than a simple pair of relay contacts. If high-speed, transparent break-in or digital mode operation is required, the PIN diode is the practical choice. Two practical papers available on-line with a lot of information about PIN diode use are recommended to learn more about switch design — see the References for Huff and Summers. Two industry papers on using PIN diodes are also referenced from Microsemi and Skyworks as well as a general paper by Rosu on RF switches using PIN diodes.

BASIC PIN DIODE OPERATION

A PIN diode is a diode with a wide undoped "intrinsic semiconductor" region between the usual P-type and N-type doped semiconductor layers of a regular diode (See the discussion of PIN diodes in the **Circuits and Components** chapter.) The PIN diode's wide intrinsic layer acts as a storage reservoir for charge. When forward bias current is flowing, the intrinsic layer fills up with electrons. When the diode is reverse biased, it takes time for the stored electrons to empty out of the intrinsic layer. At high enough frequencies, the time available is too short and the diode never "turns off". It remains conducting through the whole RF cycle.

PIN diodes behave like variable resistances. The resistance is determined by the amount of dc forward bias current. They are near-perfect variable resistors which add very little distortion to the signal. For switching, all we want is either a very low resistance (ON), or a very high resistance (OFF).

- To switch the PIN diode OFF, apply a large reverse bias voltage. The voltage should be higher than the peak voltage of the RF being switched. That means high voltage is required for the switch to withstand high power RF voltages when it is turned off.
- To switch the PIN diode ON, apply a forward bias current; the more current, the lower the insertion loss.

The basic element of the PIN diode switch is shown in **Figure 13.70**. The input RF signal passes through C2, then D1, and finally C3 to the output. (RF can flow in either direction — the left-to-right convention is followed here for convenience.) The dc blocking capacitors C2 and C3 are necessary to isolate the dc bias for D1. For the HF range, a 0.1 µF capacitor is satisfactory, with a reactance of approximately 1 Ω at 1.8 MHz. Remember that the capacitor will have to carry all of the RF current so a low-loss, low-inductance type is required for transmitting applications. L1 and L2 are RF chokes that prevent RF from getting into the bias source and switching

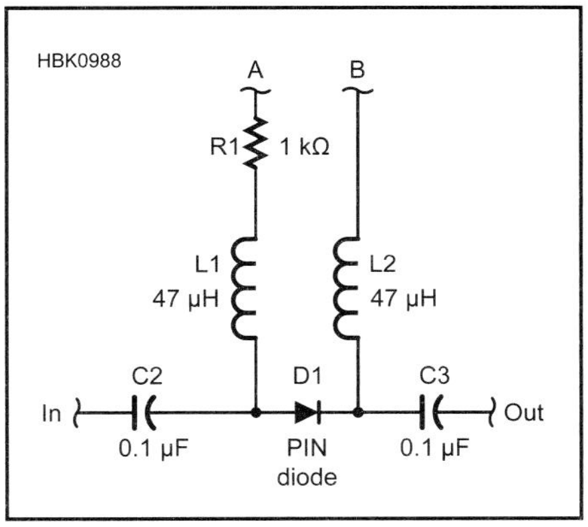

Figure 13.70 — The basic PIN diode RF switch. See text.

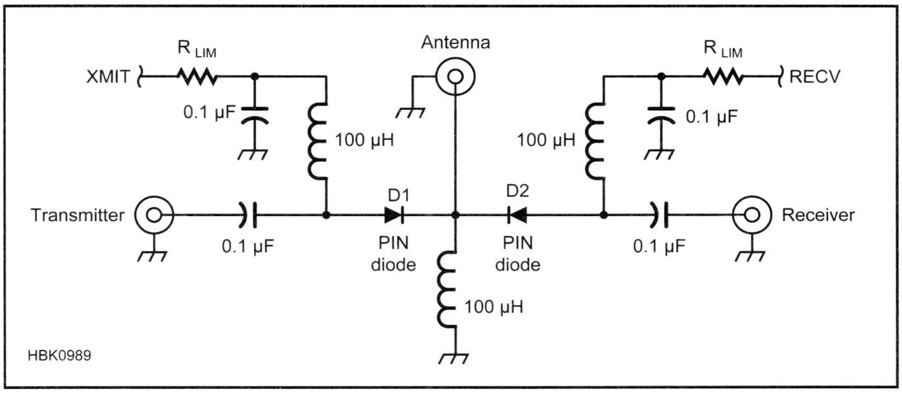

Figure 13.71 — A simple SPDT PIN diode RF TR switch to connect either a transmitter or receiver to an antenna. See text for design information.

circuits. At 1.8 MHz, the reactance of 47 µH is over 500 Ω and should be sufficient in a 50 Ω system.

To switch on D1, we apply a forward bias current through D1. For this, we set A to the supply voltage (assumed to be +12 V here) and B to ground. The current flows through the diode, limited by R1. Neglecting the diode's forward voltage, a 1 kΩ resistor results in 12 / 1000 = 12 mA. This is typical for low-power designs. Higher-power PIN diodes require more bias current to minimize insertion loss.

To switch off D1, we apply a reverse bias voltage by setting A to 0 V and B to the supply voltage. The reverse bias voltage must be higher than the peak voltage of the applied RF with some margin. The reverse voltage must also be high enough that when the diode is off there is enough isolation for the circuit requirements. Complex PIN diode switches can achieve several tens of dB of isolation. 30 dB is typical of single-diode switches like this one.

BASIC SPDT PIN DIODE SWITCH

The switch in Figure 13.70 is a SPST switch and RF current can flow through it in either direction when the PIN diode is forward biased. The switch in **Figure 13.71** is a SPDT switch with the antenna port connected to the switch common point. The diodes are turned on alternately to connect either the transmitter or the receiver to the antenna, but not both at the same time. This is a typical PIN diode TR switch.

When the XMIT signal is ON (a positive voltage), PIN diode D1 is forward biased and conducting RF while D2 is reverse biased and non-conducting. The transmitter is connected to the antenna and the receiver input path is open-circuited. When the RECV signal is ON, the situation is reversed and the receiver is connected to the antenna and the transmitter path open-circuited. The circuit's user must ensure that both XMIT and RECV cannot be on at the same time.

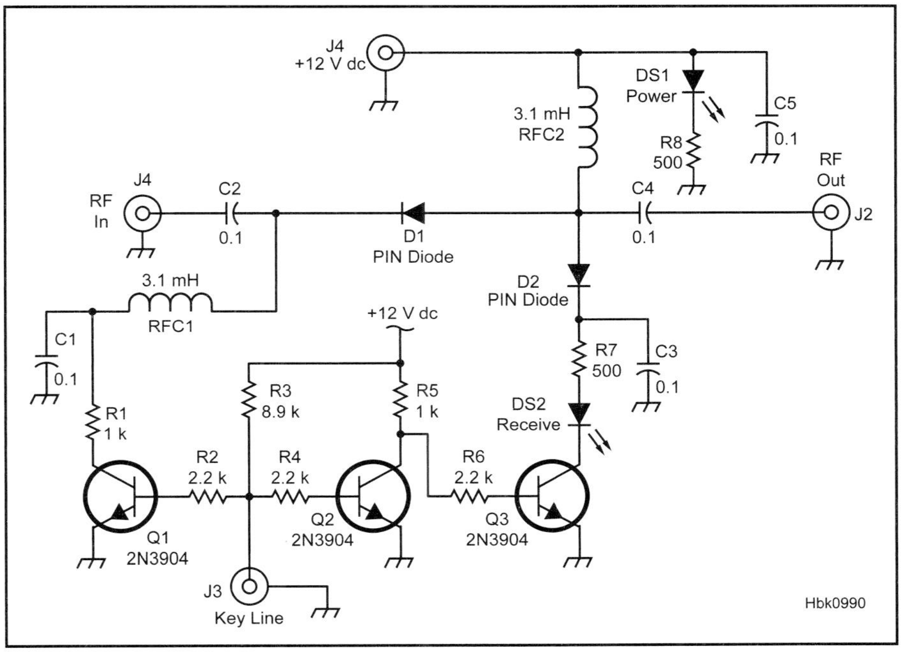

Figure 13.72 — Schematic and parts list for the PIN diode TR switch. Resistors are ¼ W.

C1-C5 — 0.1 µF ceramic capacitor
D1, D2 — PIN diode
D3, D4 — Switching diode, 1N914 or equivalent
DS1, DS2 — LED
J1, J2 — Chassis mount BNC connector
J3, J4 — Chassis mount phono connector
Q1-Q3 — Small signal PNP transistor, 2N3904 or equivalent
RFC1, RFC2 — 3.1 mH RF choke, 225 turns #30 AWG enameled wire wound on a 5⁄16 inch diameter, 5⁄8 inch long plastic tube

Depending on the type of diode being used, isolation between the transmitter and receiver ports will be about 30 dB at 28 MHz and as much as 60 dB at 1.8 MHz if careful construction techniques are followed. If additional isolation is required, multiple diode sections can be added in series or a shunt diode added as in the next section.

The value of R_{LIM} and the dc resistance of the 100 µH RF chokes limit diode current. Typical PIN diodes require 10 to 20 mA of forward bias to reach insertion loss of 0.1 dB at HF. The choice of diode will determine the amount of forward bias current required.

To perform PIN diode switching at high power levels, considerably more details are involved in the design. The online paper by Garland should be studied to learn more about use of PIN diodes at full legal limit power levels.

SERIES-SHUNT PIN DIODE SWITCH

If the transmitter is connected to the antenna at all times while the receiver is switched in and out, the series-shunt circuit of **Figure 13.72** can be used. This circuit was designed by KE2QJ and its complete operation is described in the article included with the downloadable supplemental information. When the receiver is to be connected to the antenna (RF In), series diode D1 is forward biased on and bias is removed from shunt diode D2. (It is not reverse biased.) When the transmitter is keyed (Key Line shorted to ground), forward bias is removed from series diode D1 and shunt diode D2 is forward biased on, connecting the receiver input to ground through C3. The designer reports isolation between the transmitter/antenna and receiver ports to be between 43 and 53 dB depending on bias current level.

REVERSE BIAS

To switch a PIN diode completely off, the reverse bias must be larger than the peak voltage of the applied RF. For a given power level, P, and impedance, Z:

$$V_{PK} = 1.414\ V_{RMS} = 1.414\sqrt{P \times Z}$$

At P = 10 W, V_{PK} = 32 V; at P = 100 W, V_{PK} = 100 V; and at P = 1.5 kW, V_{PK} = 387 V. This assumes SWR = 1:1. At higher SWR, the voltage at the output of a transmitter or amplifier will be higher. (See the **Transmission Lines** chapter.)

The current requirements for reverse bias voltage are very low, consisting of the diode reverse leakage current and any leakage through switching transistors. For two examples of how reverse bias voltage is obtained, see the Reference entries for Summers and Huff. A small dc-to-dc converter is another possible choice.

USING 1N4007 RECTIFIERS AS PIN DIODES

PIN diodes tend to be expensive items. However, the 1N4007 rectifier diode has a peak inverse voltage (PIV) rating of 1000 V. In order to achieve this very high PIV rating, its internal construction is very similar to a "real" PIN diode. Hence the 1N4007 is often referred to as the "Poor man's PIN diode."

Since the 1N4007 is sold as a power rectifier, its RF characteristics are not highly repeatable and its resistance is not linearly controllable. Nevertheless, to act like an ON/OFF switch across the HF range, the 1N4007 costs just a few cents and functions as well as a PIN diode designed for the same application and ratings. See the References for Huff and Summers for more information about 1N4007 performance in this application.

A properly used 1N4007 should not impact the dynamic range or third-order intercept point, IP3 in the receiver. An excellent set of measurements by IN3OTD confirms the 1N4007 is suitable for use at HF as a switching diode. (**www.qsl.net/in3otd/electronics/PIN_diodes/1N4007.html**). His broad conclusions are that a properly biased 1N4007 at HF (1.8 to 30 MHz) has the following characteristics:

• "ON" insertion loss less of than 0.1 dB for 10 mA forward bias current

• "OFF" isolation of at least 30 dB (at 30 MHz)

• IP3 > +50 dBm

Thus, the use of 1N4007 PIN diode switches should result in very little loss of transmitter output power, very little loss of sensitivity on receive, and will not degrade the excellent IP3 (and dynamic range) modern receivers.

13.12 References and Bibliography

See also the list of Books in the **Receiving** chapter references along with a list of articles about Mixers.

Design with PIN Diodes, Skyworks Application Note, **https://www.skyworksinc.com/-/media/SkyWorks/Documents/Products/1-100/Design_With_PIN_Diodes_200312E.pdf**

Garland, J., W8ZR, "Add Full Break-In QSK Keying to your Linear Amplifier," *QST*, June 2015, pp. 37 – 41.

Grebenkemper, J., KI6WX, "Phase Noise and its Effect on Amateur Communications," *QST*, Part 1, Mar. 1988, pp. 14 – 20; Part 2, Apr. 1988, pp. 22 – 25.

Hallas, J., W1ZR, *Basic Radio* (ARRL, 2005).

Hayward, W., Campbell, R., and Larkin, B., *Experimental Methods in RF Design* (ARRL, 2003).

Hayward, W., W7ZOI, "Crystal Oscillator Experiments," Technical Correspondence, *QST*, July 2006, pp. 65 – 66.

Hayward, W., W7ZOI, and DeMaw, D., W1FB, *Solid State Design for the Radio Amateur* (ARRL, 1977), pp. 26 – 27.

Huff, D., W6JL, "Homebrew PIN Diode QRO QSK System," Funkamateur, Mar. 2016, **www.funkamateur.de/tl_files/downloads/hefte/2017/w6jl_improved_qsk_system_mar_2016.pdf**.

Lathi, B., *Modern Digital and Analog Communication Systems*, 4th Edition (Oxford University Press, 2010).

Pozar, D., *Microwave Engineering*, 4th Edition (John Wiley and Sons, 2012).

Rosu, I., YO3DAC/VA3IUL, "RF Power Amplifiers," **https://www.qsl.net/va3iul/RF%20Power%20Amplifiers/RF_Power_Amplifiers.pdf**.

Rosu, I., YO3DAC/VA3IUL, "RF Switches," **www.qsl.net/va3iul/RF_Switches/RF_Switches.pdf**.

Sabin, W., WØIYH, "A 455 kHz IF Signal Processor for SSB/CW," *QEX*, Mar./Apr. 2002, pp. 11 – 16.

Sabin, W., and Schoenike, E., Eds., *Single-Sideband Systems and Circuits* (McGraw-Hill, 1987).

Sabin, W., WØIYH, "RF Clippers for SSB," *QST*, July 1967, pp. 13 – 18.

Silver, W., NØAX, Hands-On Radio, Experiment #174 "Switching Amplifiers," *QST*, July 2017, pp. 58 – 59.

Scarlett, J., KD7O, "A High-Performance Digital-Transceiver Design," Part 1, *QEX*, July/Aug. 2002; Mar./Apr. 2003; and Nov./Dec. 2003.

Schwartz, M., *Information Transmission, Modulation and Noise*, 3rd Edition (McGraw-Hill, 1980).

Summers, H., GØUPL, "50W HF QCX Power Amplifier kit assembly manual," Section 6.6 "Transmit/Receive Switching," **www.qrp-labs.com/50wpa.html**.

The PIN Diode Circuit Designer's Handbook, Microsemi Corp., **www.ieee.li/pdf/essay/pin_diode_handbook.pdf**.

Contents

Chapter 14 — Online Content

Also see the Downloadable Content for the ***Receivers*** *and* ***Transmitters*** *chapters*

Software Defined Radio and DSP Articles and Projects
- Build Your Own IF SDR by Alex Schwarz, VE7DXW and Guy Roels, ON6MU
- Chapter 14 — Audio Oscillator example.grc (GNU Radio design file)
- Digital Signal Processing and GNU Radio Companion by John Petrich, W7FU and Tom McDermott, N5EG
- Digital Signal Processing (DSP) Projects: Examples of GNU Radio and GRC Functionality by John Petrich, W7FU, and Tom McDermott, N5EG
- Hands On SDR — FPGAs by Scotty Cowling, WA2DFI
- Hands On SDR — Introduction by Scotty Cowling, WA2DFI
- Hands On SDR — More on FPGAs by Scotty Cowling, WA2DFI
- Hands On SDR — Sharing Radios on the Network by Scotty Cowling, WA2DFI
- Hands On SDR — Using FPGAs in SDR Designs by Scotty Cowling, WA2DFI
- Hardware Building Blocks for High Performance SDRs by Scotty Cowling, WA2DFI
- SDR Simplified — Demystifying PID Control Loops by Ray Mack, W5IFS
- SDR Simplified — Step One Toward a Working SDR by Ray Mack, W5IFS

Transverter Articles and Projects
- A 222 MHz Transverter for the Yaesu FT-817 by Paul Wade, W1GHZ
- A 222 MHz Transverter for the Yaesu FT-817 — Revisited by Paul Wade, W1GHZ
- Assembling a Microwave Transverter System by Paul Wade, W1GHZ
- Cheap & Simple 1296 MHz Transverter Update by Paul Wade, W1GHZ
- Corrections and Improvements for Simple and Cheap Multi-band Microwave Transverters by Paul Wade, W1GHZ
- Sequencers for Transverter Control by Paul Wade, W1GHZ
- Smart Fool-resistant Conditional Sequencer by Paul Wade, W1GHZ
- The Transverter — An Introduction to a Useful Device by Bill Wageman, K5MAT
- The Transverter by Tom Williams, WA1MBA
- Universal MMIC Preamp by Paul Wade, W1GHZ

Chapter 14
Transceiver Design Topics

This chapter includes high-level material on SDR design. Steve Hicks, N5AC, contributed the sections on signal chains and interfaces. John Petrich, W7FU, updated the introductory material on *GNU Radio* and expanded the list of reference materials. The additional short introduction to FPGA data engines is based on the series of *QEX* columns "Hands-On SDR" by Scotty Cowling, WA2DFI.

Material on transverters was contributed by Paul Wade, W1GHZ, and Wayne Overbeck, N6NB. Material on adding panadapters to older radios was contributed by Ed Krome, K9EK.

There is a great of deal of material that doesn't fit neatly into the categories of receiving, transmitting, DSP/SDR fundamentals, or any of the major radio functions. This information is still required, however, to integrate all of the functions into a single package — the transceiver. There are questions of architecture, interfaces, development tools, and many other topics that must be addressed to build an effective transceiver or communications system. That is the purpose of this general-purpose chapter, to create a home in the *Handbook* for material that would be inappropriate for a chapter focused on a particular aspect of the radios themselves. Amateur Radio is poised at the threshold of great innovation and change — it's a bright future!

14.1 Signal Chains in SDR Transceivers

In an SDR, most work in both transmit and receive signal processing is performed in a computing or processing element that operates on digital samples. **Figure 14.1** shows a simplistic diagram of a transceiver setup for phone operation. Looking first at the transmit side of the radio, the analog microphone audio is converted to digital samples in an analog to digital converter (ADC). These digital samples are passed to the processing element where they enter the *transmit signal chain*. The transmit signal chain typically has a series of functional blocks that perform various functions on the signal, including the modulator. While it is possible to use a purely digital exciter, most radios then pass the digital samples into a digital to analog converter (DAC) which turns the digital signal back into analog form. The resulting analog signal is amplified by an RF power amplifier and transmitted through the antenna.

In the receiver, the direction of flow is the opposite: analog RF data enters through the antenna and is turned into digital signals in an ADC. This data is passed to the receive *signal chain* where it is filtered, demodulated and processed into a signal we can hear. In some digital modes, such as RTTY, the demodulated signals are filtered and turned directly into digital symbols without ever being turned back into analog. For modes that we are expected to hear, the digital samples are sent into a codec where a DAC turns them back into analog for playback through headphones, speakers, or via a line-level output.

The transmit signal chain and receive signal chain are at the heart of an SDR. Let's take a closer look at how they operate.

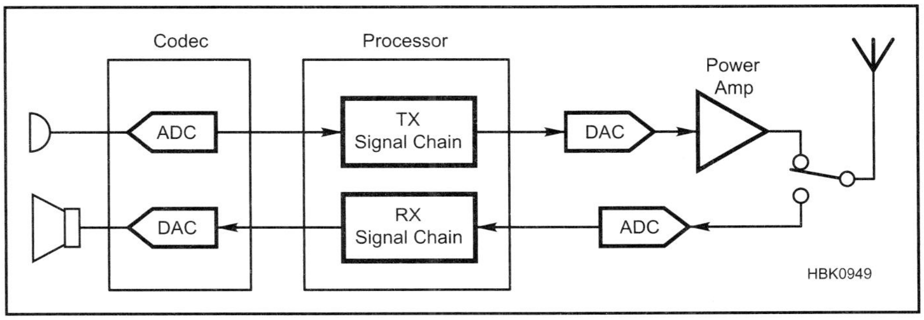

Figure 14.1 — A high-level block diagram of an SDR transceiver configured for phone operation.

14.1.1 Receive Signal Chain

Most receivers will capture more digital samples than are needed for a single receiver. This can happen for many reasons. For example, we might have a wideband sampling receiver that captures a substantial portion of the spectrum and places several receivers in that spectrum. In this case, each individual narrowband receiver only requires a portion of the captured data. When we use the term narrowband receiver, we're talking about a receiver that will be used to demodulate signals in a 50-20,000 Hz slice of the RF spectrum. In another example, the receiver might be designed using an analog or digital mixing technique that reduces the bandwidth to an "audio" range (such as 0-192 kHz). The radio may preserve this larger bandwidth to plot a spectrum display, but will reduce it for narrowband receiver processing.

To understand the next steps, it's important to first understand the relationship between sampling rate and available bandwidth. Generally, in a sampled RF system we will either be using real or complex samples. (See the **DSP and SDR Fundamentals** chapter) *Real samples* will be used for audio input and output. Real samples consist of a single fixed or floating point number that represents each sample.

According to the Nyquist sampling theorem, we can represent at most one half of our sampling rate in bandwidth. For example, if we sample a microphone signal at 48 ksps (real samples), we can represent audio from 0-24 kHz. As a practical matter, the filters we use to process the audio will have a transition band that will further restrict this to something around 20 kHz.

If we sample using the in-phase and quadrature sampling (I/Q) commonly used in RF systems, then each sample consists of a pair of fixed or floating point numbers, one for I and one for Q. This is also commonly referred to as *complex sampling*. In a complex sampled system, sample speed and sampled bandwidth are equal, ignoring filter transition bands. For example, if the sample rate is 48 ksps (complex samples), the samples can represent almost of 48 kHz of RF spectrum (again accounting for filter transition bands). For a better understanding of the reasons behind these general rules there are several excellent references, including Richard Lyons' book *Understanding Digital Signal Processing*, that can be consulted. (See the Reference sections of the **DSP and SDR Fundamentals**, **Receiving**, and **Transmitting** chapters for a comprehensive set of references.

In either situation, the primary need to reduce the bandwidth for the receiver signal chain is driven by the processing resources required to process the receiver signals. The more bandwidth that is captured digitally, the higher the sampling rate and consequently the more processing power required in our signal chain. If we plan to ultimately discard the extra bandwidth in a filter, then we are wasting valuable processing resources by processing a wider bandwidth than we ultimately need.

In any of these cases, our first step is generally to reduce the bandwidth of the incoming signals through decimation. (See the **DSP and SDR Fundamentals** chapter) By decimation, we reduce the number of samples in the receiver by combining multiple samples and producing an output that has fewer samples, but with more bits per sample (greater bit width). After decimation, the receiver will typically have a sampling rate of somewhere between 24 and 48 ksps. We will then process all elements of our signal chain at this rate. While this is not strictly necessary, it is the most convenient. If our sampling rate varies in each block of our signal chain, we will be constantly buffering and resampling, again wasting valuable time and computing resources.

A typical receiver signal chain is shown in **Figure 14.2**, again computing from right to left. (This is the same direction as signal flow in the block labeled "Receive Signal Chain" in Figure 14.1.) Each DSP block in the chain will take input samples, generally in a buffer or block of samples, process the samples using the algorithm for the block, and then pass the resulting samples to the next block. It is a requirement of the receive signal chain that all blocks are processed in less than the time it takes before another buffer of samples arrives.

What does this mean? Let's say that our system is running at 48 ksps and our sample buffers are 512 samples each. We will receive a new buffer every 10.67 ms. If we are not able to process the next sample buffer because we are still processing the last buffer we received, then the buffers will stack up, forming a queue of samples needing processing. Even worse, our DAC at the output of the signal chain has the responsibility for turning a stream of sample buffers into a continuous audio signal. If we "starve" it by not sending a buffer when it has exhausted the previous buffer, it will be forced to play empty audio or repeat previous buffers. If you've ever used digital electronics built this way, you've likely experienced the effects firsthand. Here are some examples of how other engineers dealt with a similar problem:

1. In XM satellite radio, when the satellite signal is briefly lost, some satellite radios play "comfort noise," a low-level hiss that sounds like a normal AM or FM signal that has faded away. This noise is completely synthetic and doesn't represent any actual noise in the receiver. Since the source signal is digital, the receiver knows when the signal has faded and could easily play "blank audio," but it was probably judged that this sound would best convey to the consumer that the radio is working but has lost a signal.

2. Many CD players, when starved for digital data from the disc due to a scratch and the inability to recover the signal, will replay the contents of the last digital buffer, resulting in a repetitive sound reminiscent of a skipping vinyl record player. This sound conveys to traditional (analog) vinyl users that there is a problem with the disc. (Data, however, does not warp when left in the sunlight on a vehicle's back seat!)

It is, therefore, important to budget processing power to ensure all blocks can be processed in the time it takes to collect a new buffer. If our radio has more than one receiver but a single processing element, our requirement will be to process all the DSP blocks for all receivers in the requisite time for a single buffer. While beyond the scope of this discussion, it's worth mentioning that multicore processors, both symmetric and asymmetric, can provide a platform for some unique solutions that stagger or pipeline DSP blocks and receivers and allow spreading the processing requirements across different cores.

PASS-BAND FILTER

Next, we filter the buffered data. The filter removes unneeded signals from the buffer, limiting it to just our bandwidth of interest, called the *receiver pass-band*. Since the samples entering the filter are 48 ksps of I/Q data, we start out with something on the order of 40 kHz of audio bandwidth. If we are using sideband, we're only typically interested in a bandwidth of 2-5 kHz. If we're using CW we might want to reduce the bandwidth down to 50-100 Hz. The filter achieves this reduction in bandwidth by employing DSP filtering techniques. (See the **Analog and Digital Filtering** chapter)

The filter may need to accumulate several buffers of time-domain samples before performing the filter processing depending on how steep or sharp the filter skirts are required to be in order to create the filter's required

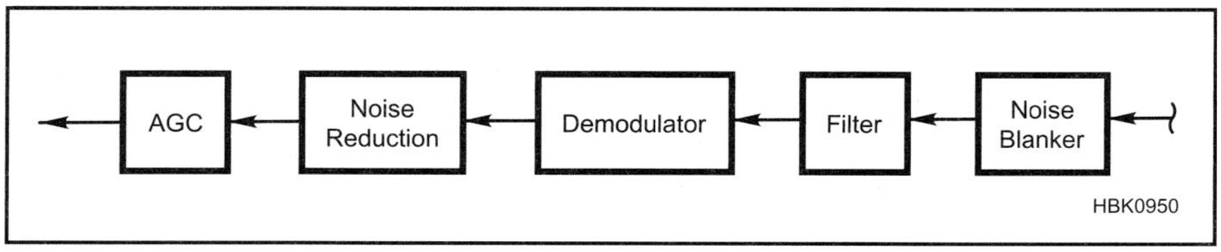

Figure 14.2 — Typical functional blocks that make up a receive signal chain.

shape factor. As mentioned previously, it is also required that once we have accumulated the requisite buffers, we must process the filter within the duration of our buffering. For example, if our buffer size is 512 samples and we are using an overlap-and-save FIR filter with 2048 taps, we will need to accumulate 2048 samples before processing the filter. After this, 1024 samples will be added to the filter for the next processing, since 1024 samples are saved from the last processing. In aggregate, we must be able to process this filter in under 1024/48 ksps = 20.3 ms of CPU time (including all other signal chain operations).

Filters used in receivers can be of many different topologies, including finite impulse response (FIR) and infinite impulse response (IIR). Even within these two broad categories there are plenty of additional variations that affect passband and stop-band ripple, phase distortion (detrimental to some digital modes), group delay, and other parameters. Also, this filter, which is often called the *final filter* since it is the last filter before the signals are presented to the operator, will determine a significant part of the receiver latency (the delay between the input of an RF signal to the output of the data or audio). Why is this? The slower the sampling rate, the more impact collecting samples and filtering them will have on latency.

Some radios allow the operator to make tradeoffs in filter parameters, favoring low latency at one time and favoring the best filtering at another. A sizable FIR filter using 4096 taps will add roughly 1.5 x 4096 / 48 ksps = 128 ms of latency. This is not a large amount for phone conversations, but would be significant for CW contesting.

One of the great triumphs of SDR is that we have eliminated our dependency on needing a collection of electronic or mechanical filters. We now have continuously variable filters that let us tailor our pass-band to meet our operating demands without worrying about the design cost or space for our old electronic and mechanical filter collection. In some radios, we can also adjust the taps in the filter to select better filter shape factor or lower latency as our operations dictate.

DEMODULATOR

The demodulator performs the conversion of on-air signal to the baseband audio signal that we will be hearing. For example, with FM we must convert a fixed-amplitude carrier changing frequency according to the amplitude of the audio into a time-domain amplitude-varying signal that we can understand with our ears and brain. The output of the demodulator is audio that we can hear and understand.

The demodulator block is selected based on the mode being used. When you change the mode of the radio on the front panel, you are switching one demodulator with another for the mode you have just selected. Of course, the filtering and noise mitigation techniques may vary by mode also, and so we may be switching other blocks in the DSP chain based on the requirements of each mode.

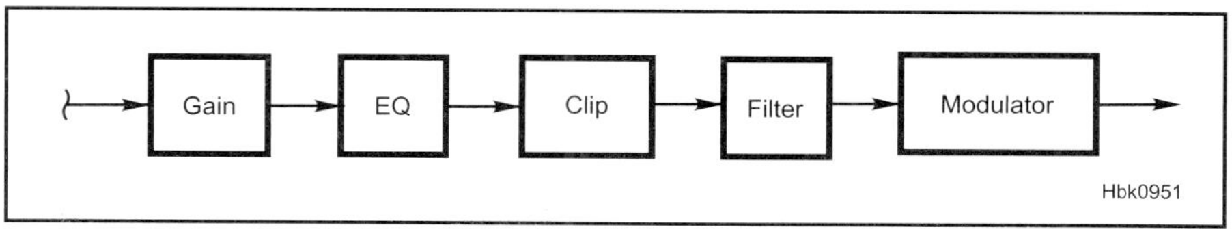

Figure 14.3 — Typical functional blocks that make up a transmit signal chain.

14.1.2 Transmit Signal Chain

For phone, the transmit signal chain will prepare spoken audio for transmission over the air. If our mode is CW or other digital modes, the transmit signal chain will be very different, perhaps just a single or a pair of software oscillators that produce the tones that we will send on air. **Figure 14.3** shows a typical signal chain for phone transmission.

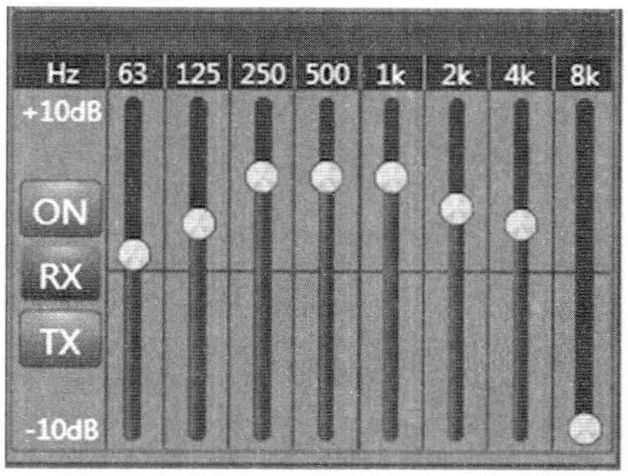

Figure 14.4 — An equalizer control window showing the adjustments for each frequency band of interest.

GAIN

Depending on how the designer has implemented the ability to change output power in the radio, a gain stage may exist at the start of the signal chain that can be adjusted to raise or lower the power of the transmitter. This gain stage could be implemented at other places in the radio as well. A gain stage will multiply all samples in the buffer by an input level value, raising or lowering the level of all audio samples in the buffer.

EQUALIZATION

In a software defined radio, it is easy to isolate frequency bands in an audio buffer and allow each band's relative power to be raised or lowered. This equalization function can be used to tailor a person's voice to produce the best sound on the receiving end. This function is known as equalization or simply EQ. The input to the equalization function is the relative increase or decrease for each band of interest. A typical equalizer panel with vertical sliders for each audio band is shown in **Figure 14.4**. At the output of the equalization stage, we have audio prepared for transmission, but it may exceed required output levels that have been set in the radio. For example, we may independently have set a maximum output power for the radio to prevent overshooting an amplifier input. This is where a clipper becomes useful.

CLIPPER / ALC

A clipper may be employed to reduce the level of any signals that exceed required power levels. It generally does this by checking each sample in the buffer for the maximum allowable level and replacing any samples that exceed the threshold with a corresponding sample of the maximum level, thereby clipping the peak value of that sample.

While clipping can produce undesirable effects, it may be necessary to prevent the radio from operating outside the limits provided by the operator. Automatic level control (ALC) may also be used to reduce signal levels if a downstream amplifier provides voltage to the radio requesting a lowering of the levels.

Generally, a series of meter displays or values derived at different stages in the signal chain will provide the operator with information on undesired effects occurring within the signal chain. This allows the operator to make adjustments to correct any undesired effects. Since we have absolute control over the signal level by altering samples in the radio, it is generally better to set a maximum power out in the radio and allow the signal chain to make the appropriate adjustments than to rely on ALC to adjust the levels when they are too high for an amplifier.

An ALC subsystem that is too active can modulate the transmit signal. This results in distortion that could be prevented by adjusting the proper levels in the signal chain to start with. This is why many digital mode guidelines specify that the ALC system be switched off to avoid unintentional distortion of the signal by power control systems.

FILTER

Since our audio input may contain frequency components well above those required or that comply with good practices, a filter is employed to remove them. For normal SSB operation, we could set the filter to pass audio from 100 Hz–2.8 kHz. For Extended Signal Sideband (ESSB), we may set the filter to pass audio from 50 Hz-4 kHz or more. The transmit filter setting may also be set to comply with communication laws in the country where the transmission is occurring.

The filter will remove the unneeded frequency components of the signal before transmission using similar DSP techniques to those in the receive signal chain filter.

MODULATOR

The modulator will then take the audio signal and create a modulated RF or IF signal required by the operating mode. The modulator takes care of placing the audio in the proper sideband for SSB, producing the carrier for AM and performing frequency modulation for FM. After modulation, the signal is ready to be transmitted.

SUMMARY

Signal chains are the heart of the software defined radio. This overview provides the basics of how a signal chain is put together and the requirements on each of the DSP functions. A typical signal chain in an SDR may have many more blocks including blocks to meter samples, providing valuable feedback to the operator, more audio tailoring and filtering blocks, oscillators for in-air testing, speech compressors, and so on.

14.2 User Interfaces

Many years ago, the knob we used to tune our radios was often attached to a frequency indicator stenciled on a piece of plastic that would rotate into view as the knob was turned. The plastic marker displayed an offset into a band selected by a band switch. The strength of any incoming signal would be displayed on an S meter, a magnetic meter movement excited by the voltage from the AGC circuit. In the 1970s and 1980s, the *liquid crystal display* (LCD) and *light emitting diode* (LED) display came into popularity, and most radio manufacturers adopted some form of display device that used LCD and or LEDs. Both technologies were used to show frequency, mode selections, and simulated S meters without their mechanical counterparts.

These technologies of the past provided only minimal *situational awareness*. We only had a narrow view of the spectrum limited to the instruments at hand showing us the signal strength in our narrow receiver, along with a display of our current frequency. There was no way to see how many signals were on the band, how strong they were, whether the noise floor in the band was rising, or if we might just be listening where some local noise was causing us interference.

With *panadapters* now available on modern transceivers, we can see all the signals across the band. Finding a place to call CQ can be difficult when the bands are busy. With a panadapter, we can glance at the band and notice where there are gaps, quickly tune to an empty frequency, and call CQ. We can also instantly tell where the strong and weak signals are in the band. If we're looking for signals arriving at our location through particular propagation, we can search for signals that look alike or share the similar power levels.

The *waterfall display* provides the same information with a time-historical twist. In the waterfall we can see how signals and noise are changing over time. Imagine working a DX pileup where a few hundred amateurs are all trying to work the same station. We can watch the DX station systematically contact every station and observe the operating patterns. We know where the DX station is listening because we can see each station in the pileup respond to the DX station as they are called. Observing this pattern, we now know where the DX will listen next and can simply drop our call in this location and make the contact. Those enjoying panadapters today lament: how did we survive without this capability?

The tools available today for gaining a situational awareness of the bands and our receiver have dramatically changed. Let's look at each of these tools in some more detail.

14.2.1 The Panadapter Display

The *panadapter*, or spectrum display, shown in **Figure 14.5** displays the amplitude

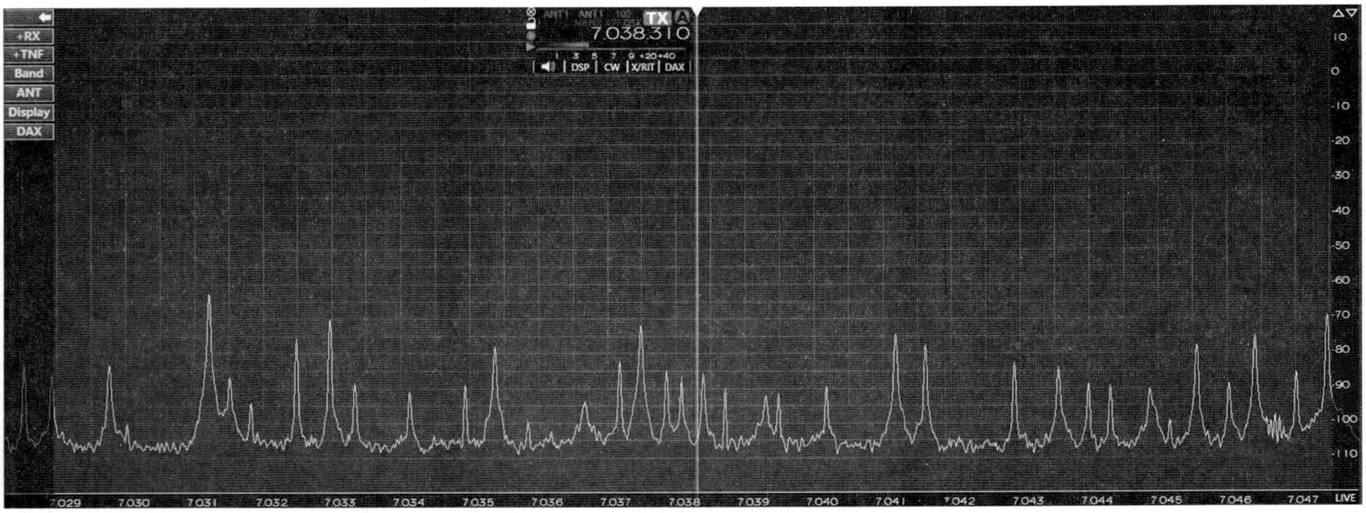

Figure 14.5 — A typical panadapter display showing a number of signals over about 18 kHz of the 40 meter CW segment.

of signals across a range of frequencies. The panadapter's design roots come from spectrum analyzers that were designed in the 1950s and 1960s. These instruments featured a swept receiver that tuned across the band. As the receiver tuned the band, the signal level in the receiver was captured and displayed. The resulting display is a graph with frequency on the x-axis and amplitude on the y-axis.

The spectrum analyzer has proven its worth in many areas, from radio design and troubleshooting to the analysis of on-air interference problems. The almost continuous view of the spectrum allows the operator to quickly understand the spectrum and act accordingly. A spectrum analyzer's controls allow alteration of the bandwidth of the sweep receiver, known as *resolution bandwidth (RBW)*, the *sweep rate* of the receiver, the total bandwidth displayed, and other parameters. (See the **Test Measurements and Equipment** chapter for more details about spectrum analyzers.)

In the absence of a signal, the swept receiver would return the higher of the noise from the input port of the analyzer or the spectrum analyzer's noise floor. This noise floor visualized or painted is partly a function of the bandwidth of the sweep receiver itself: the wider the bandwidth, the more noise was input to the receiver and the higher the noise floor appears. If the noise floor visualized on the sweep was higher than signals of interest, the resolution bandwidth of the receiver could be lowered, reducing the noise let into the receiver and lowering the noise floor of the display.

In the absence of a signal and assuming evenly distributed noise, the noise floor will drop by 3 dB each time the resolution bandwidth is halved, since the total noise in the receiver was cut in half by the bandwidth filter. Typically, as the display is "zoomed in" to show less bandwidth on the screen, the instrument will automatically reduce the receiver bandwidth in the sweep (resolution bandwidth) and the noise floor of the display will drop. The swept receiver approach to a spectrum display is used in radios designed in the early 2000s, such as the Icom 756 Pro-series.

The swept receiver design has some key drawbacks, though. First, time is required for the receiver to settle, a measurement to be taken, and the receiver tuned to the next frequency. This limits the speed at which the receiver can tune and often results in a slow "painting" of the display that lacks the full-motion look of modern panadapters.

The slower sweep rate reduces situational awareness. For example, imagine someone sitting sending a series of CW dits. In the swept design, the receiver will typically hear a signal or no signal as it passes the CW dit signal. We will either see a nice signal out of the noise or nothing, depending on the timing of the dits in relation to our sweep receiver. With a modern panadapter refreshing many times per second, we can see the signal rise and fall with the CW dits, giving us a clear view of what's going on.

The second drawback of the swept receiver design is the tradeoff between sweep speed and resolution bandwidth. As we reduce RBW to see more detail in the spectrum, the sweep rate must slow because the measurement takes longer, primarily because the narrower filter requires more time for signals to pass through and stabilize before the measurement takes place. The more detail we want to see, the slower our display updates, and the lower our situational awareness.

Today's transceiver panadapter displays are built on many of the same principles of the spectrum analyzer with one notable exception: today's receiver panadapters are almost always built from a Fast Fourier Transform or FFT (See the **DSP and SDR Fundamentals** chapter). The FFT accepts a wide bandwidth of time domain samples from an analog-to-digital converter and translates the time-domain signals into the frequency spectrum.

PANADAPTER TYPES

There are two main panadapter types available in current amateur products, the *dependent* and *independent* panadapter. In a dependent panadapter design, the data for the panadapter is from the receiver's IF signal. In this scenario, the panadapter will always show the spectrum to either side of our received frequency, which is shown in the center. For example, if the panadapter dis-play spans 192 kHz, our receive frequency will be in the center and we will see 96 kHz of spectrum on either side of the receiver.

In contrast, an independent display operates independently of the receiver. The panadapter display frequencies can be moved, and the received frequency can be moved without being in the center of the display. This allows showing a DX pileup with the DX station on the left and the pileup on the right, and we can tune our transmitter across the pileup without shifting the spectrum display. Internally, this requires a separate receiver just for the panadapter that is tuned independent of the receiver(s) in the radio.

The FFT used to build the panadapter is always measured in *points* or *bins*, each of which is essentially a collection of receiver channels like our sweep receiver in the spectrum analyzer. The bins are spaced evenly across the range of frequencies output by the FFT. The bandwidth of each bin is determined by the total sampling width presented to the FFT divided by the number of FFT points. For example, if our input is 192 kilosamples per second (abbreviated ksps) and we use a 4096-point FFT, each bin of the FFT would represent a receiver that is 192,000 / 4096 = 46.875 Hz in width. In other words, each point in the display of our FFT represents the signal and noise heard by a receiver with a bandwidth of about 47 Hz. In this way, you can think of the panadapter as a series of equally spaced S meters, laying on their sides continually showing the signal level in each of our FFT bins.

To compute an FFT for a panadapter, the FFT must first collect enough time-domain samples to fill the input registers of the FFT. Using the previous example, a 4096-point FFT required 4096 time-domain input samples in order to create a 4096-point frequency-domain output. For a receiver operating at 192 ksps, the time required to collect these samples is 4096 / 192,000 = 21.3 ms. After this time, the FFT can be computed and a *frame* can be displayed for the operator. The maximum possible frame rate for the display is the reciprocal of the time to fill the input. In this example, 1/21.3ms = 46.875 frames per second (46.875 Hz refresh or update rate). Note that this frame rate is exactly equal to the bin width of the FFT.

Considering a typical sideband signal with a bandwidth of 3 kHz, the panadapter just described would show 3,000 Hz / 46.875 Hz = 64 different points across the audio range of a received signal. Since a sideband signal is the audio in the operator's voice shifted in frequency to the RF spectrum, we can see the energy in the operator's voice at each of 64 different frequencies as the operator talks. Outside of the operator's intended transmit bandwidth we can also observe any RF that is transmitted unintentionally. This allows us to observe the cleanliness of signals on the band in significant detail.

Figure 14.6 shows a transmitted SSB signal that is fairly clean and has little RF outside of the transmitted band of interest. **Figure 14.7** shows a significantly less clean signal displaying splatter caused by an improperly configured amplifier. This capability is particularly useful for assisting another amateur to diagnose a transmitter that has signal anomalies. The visual representation of their transmission can make quick work of identifying intermodulation distortion, splatter, and other problems.

For CW or narrowband digital mode operation, a panadapter with a higher resolution than 47 Hz is desired. There are two ways to achieve higher resolution:

1. An increase in the number of points in the FFT directly reduces the bandwidth of each FFT bin, both lowering the noise floor and increasing the resolution.

2. A reduction in the bandwidth of the samples presented from the ADC also directly affects the resolution. For example, if we simultaneously reduce the ADC bandwidth to 96ksps and increase the FFT to 8192 points, our bin size is now 96,000 / 8192 = 11.7 Hz, a reduction factor of four in the bin size. The noise floor of the narrower bin will decrease by 6 dB as well.

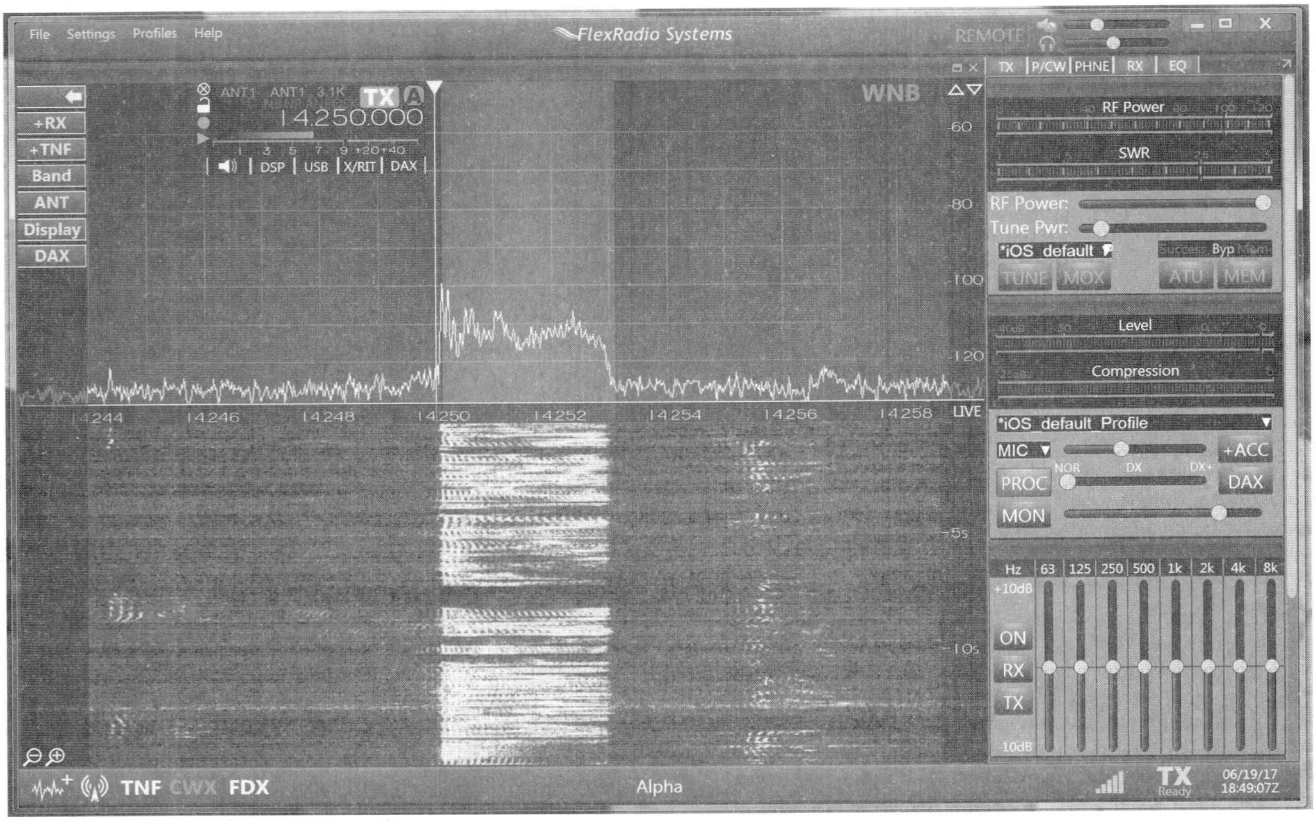

Figure 14.6 — A panadapter display showing a cleanly transmitted SSB signal.

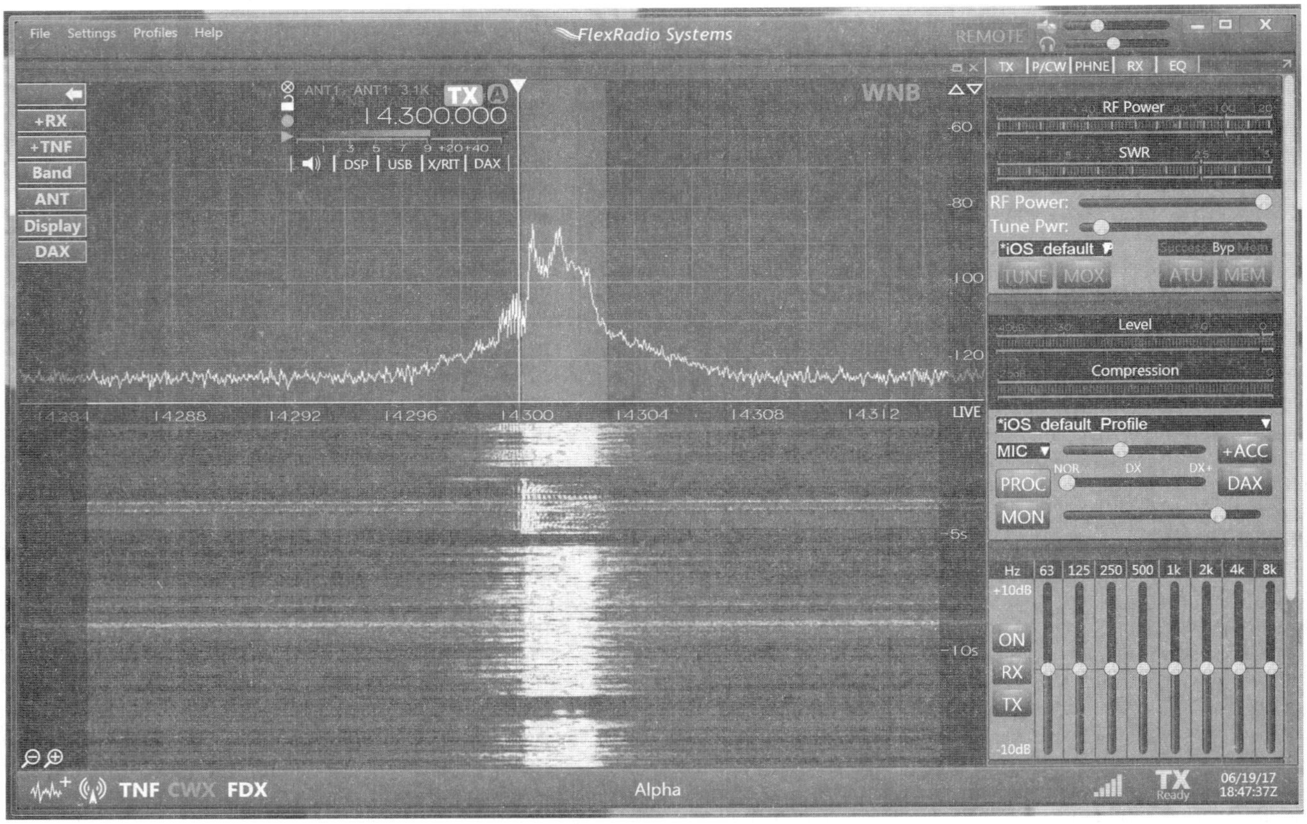

Figure 14.7 — A panadapter display showing an SSB signal with splatter and other distortion products.

The resulting display shows higher resolution, but this comes at a price. First, the frame rate of our FFT is now 11.7 frames per second. This is because we must now collect 8192 / 96,000 = 85.3 ms of data before calculating the FFT. Also, because the FFT is now based on a larger time period of data, fine temporal (moment-to-moment) details are lost in the display. For example, a 60 WPM CW signal is composed of 20 ms dits and 20 ms spaces between elements. A string of dits at this speed could be visible as individual pulses in our first FFT example, but would be somewhat lost in the new FFT since several transitions would occur during each FFT frame. This time limit tends to place practical limits on how fine-grained an FFT can be used in a panadapter. The other practical limit is computational complexity. As the number of points in the FFT is increased, more computing power is required to calculate the FFT that is the basis for the panadapter.

How far we might "zoom in" to a panadapter display is a function of what we want to observe. For SSB with signals approximately 3 kHz wide, we often want to observe many tens of kilohertz of spectrum to get a good view of other signals on the bands. In contrast, with CW we might only want to see a few kHz of bandwidth to determine the exact frequency of each CW signal. With two-tone FSK (RTTY) and PSK (PSK31) signals, it can be helpful to see both components clearly to aid in tuning, but not at such a fine resolution that the display operates slowly or incurs temporal distortion. A properly designed modern panadapter will allow us to make these tradeoffs to get the panadapter tuned "just right" to give us the view of the spectrum we need to help most in our operating mode.

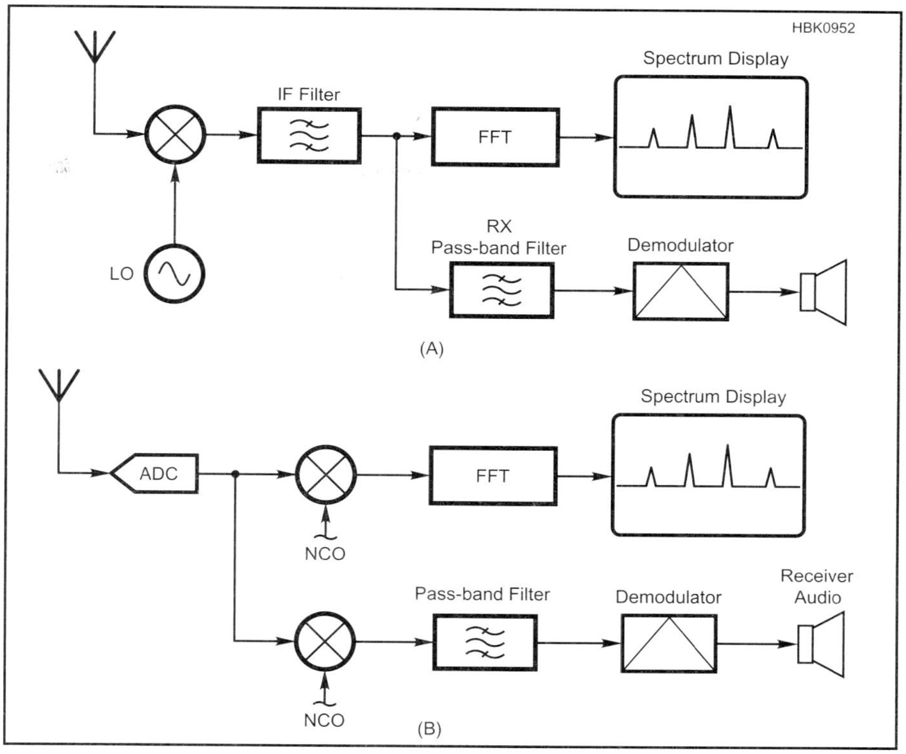

Figure 14.8 — Panadapter architecture. (A) Dependent architecture in which the panadapter and main receiver signal chain share common tuning and filtering mechanisms. The panadapter and receiver are tuned together. Dependent panadapters can be implemented as both superheterodyne and direct sampling receivers. (B) shows an independent architecture in which the panadapter and receiver have their own signal chain. The panadapter and receiver can be tuned independently of each other.

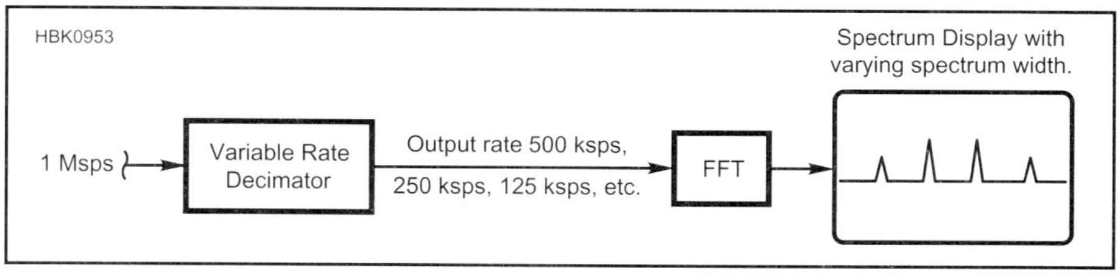

Figure 14.9 — Variable rate decimation.

PANADAPTER CONSTRUCTION

There are several methods for building panadapter displays, but the two key pieces of a panadapter display are the receiver and the FFT. As mentioned earlier, dependent panadapters (**Figure 14.8A**) are derived from the same receiver we use to demodulate and copy stations we are trying to work. In this case, the primary decision is how wide to make the receiver (sampling rate), since this will determine the bandwidth of our panadapter.

In radio systems with dependent panadapters for which the processing of the receiver signal chain is performed on a PC, the designer must be careful to limit the panadapter bandwidth, since all this data will need to be sent to the PC and processed using the PC's main CPU. The dependent panadapter architecture was invented with *PowerSDR* and first shown on the original FlexRadio transceivers such as the FLEX-5000. This panadapter architecture is also used in the HPSDR and several other radios.

With the independent panadapter architecture shown in Figure 14.8B, we see that the receiver and panadapters are separate. This architecture allows tuning the panadapter independently of the receiver and gives rise to more flexibility in setting up the panadapter display and receivers. The other key consideration is the rate of the sample stream from the receiver into the FFT. As discussed above, this affects the bin width of the display along with the refresh rate of the display. It also affects the amount of processing power required to process the raw data from the receiver. The wider the panadapter, the higher the sampling rate data is required to produce the display and the more computing power is required, assuming no loss of data.

To implement a "simple" panadapter display using the independent panadapter architecture, we could use a signal receiver with a bandwidth of 1 MHz, running into a FFT with 16,384 FFT points (the FFT algorithm requires the number of bins to be a power of two, in this case 2^{14}). This would give us bin sizes of 61 Hz. If our radio display has a horizontal width of 600 pixels we can implement a zoom-in and zoom-out feature by simply discarding or using a MAX function across bins we are consolidating into a single pixel. This type of display will also have a consistent noise floor, commensurate with a 61 Hz receiver bandwidth.

To implement a more advanced panadapter using the independent panadapter architecture, we could implement a variable rate decimation function as shown in **Figure 14.9**. The decimator or resampler can take in a sample stream at one sample rate and output a stream at other rates. In this way, we can zoom in and zoom out by varying the decimation rate. For example, let's assume out input rate to the decimator is 1 Mbps. We could output 1 Mbps, 500 kbps, 250 kbps, and so on, down to any rate we needed. This data could then be input to the same FFT function.

The resulting FFT output frame rate will be the input sampling rate divided by the number of FFT points and will have a bin width equal to the sampling rate divided by the number of bins (assuming I/Q input data). This architecture is used in the FLEX-6000 Signature Series transceivers and requires substantially more computing resources than the simple implementation, but it provides a more capable panadapter that can expand to many MHz of spectrum or zoom in to just a few kHz.

PANADAPTER APPLICATIONS

The benefits of a panadapter can be wide-reaching. In the simplest of cases, the instantaneous view of the spectrum can quickly demonstrate where the band is and is not in use. Watching the band for just a few seconds gives the operator an idea for a good place to call CQ. It's always a good idea to check if the band is in use before calling since only one side of a conversation may be audible from your QTH, but this same awareness of the band is difficult to achieve by spinning the dial.

When operating on microwave weak signal bands, it can be hard to find other signals. Rover (mobile) stations using crystal-controlled rigs in very cold or very hot weather may have drifted many kHz from their dial frequency due to the multiplicative effect of crystal error on higher bands and temperature changes in the rover. A panadapter makes quick work of locating these operators. A common methodology is to ask a rover to "send dashes" on a microwave band and then watch the waterfall for the telltale dashes to appear. Because signals deep into the 500 Hz noise floor can be seen on the waterfall display, we can see signals from stations before we can hear them, and this can also be used to assist in pointing antennas. As a fixed station, we can tell the rover operator on a liaison frequency to rotate his antennas as we watch the panadapter for a rise or fall of signal strength. Some panadapters have increased their FFT bins to 2^{17} to reduce their bin size and allow them to see even further into the noise floor.

In contesting, there are often advantages to working stations in a particular order. For example, weaker stations that have reduced signals in the panadapter may be DX stations that would result in a multiplier. Stronger stations are easier to work, and bagging these stations first might be a strategy worthy of consideration. In some sprint-style contests *running*, continually calling CQ on the band and working stations that respond to your CQ, is not permitted. Finding other stations quickly becomes paramount in this type of operating event. With a panadapter, finding stations quickly is simplified to looking for peaks in the panadapter display.

Interference in contests is a frequent problem, and rapid identification and resolution of interference sources is important. An adjacent splattering station is obvious in a panadapter and can allow an operator to make a quick comment to the right station, asking for the other operator to check the transmitting equipment. In some situations, changing frequency may be required, and again the panadapter can help locate where to move. From a competitive standpoint, having a panadapter is akin to having extra mirrors on a race car to see where other competitors are and what they are up to.

For panadapters displayed on a computer monitor, tuning can often be achieved by placing the a cursor on a signal or specific frequency and clicking. Quickly working stations in this manner bypasses the required tuning on traditional receivers. In this situation, the panadapter can provide a significant advantage. This is also true of panadapters on touch-screen devices where tuning can be accomplished just by tapping on a signal seen on the panadapter display.

Interference and noise are common in today's RF landscape. The panadapter offers the same capabilities of a spectrum analyzer by showing instantaneous views of the spectrum to help with isolation and characterization of noise. Some noise is limited to a narrow range of frequencies, while other noise sources may repeat on a specific frequency interval. Knowing the repeating frequency interval of a noise source or the similarities of the noise across the band can aid in diagnosis of the type of noise. Some types of noise drift over time, while others remain fairly constant with time. Other noise is broadbanded and raises the noise floor of the entire band. All of these different characteristics of noise can be easily observed in a panadapter, giving the amateur a thorough view of the type of noise under investigation.

14.2.2 The Waterfall Display

The *waterfall* display is similar to the panadapter in that it has frequency along the x-axis. The y-axis in a waterfall is time rather than amplitude, as shown in **Figure 14.10**. The waterfall display is typically created line-by-line, starting at the top of the display. A new line is added to the display with each unit of time, and all previous lines of the display are shifted down one line. In this way, the waterfall shows what has happened in the spectrum over time. It is this flowing motion of the waterfall that gives the display its name, since it gives the impression of a band of slowly falling water.

Let's build on the earlier analogy of a panadapter as equally spaced S meters lying on their sides. Similarly, the waterfall display can be envisioned as a series of panadapter frames, equally spaced in time and lying on their sides so we view them "from the top." What we see in the waterfall is color or intensity representing the height of the signal (amplitude) in the panadapter. Each line of the waterfall has either different colors or different intensities on each frequency that represent the signal level from the panadapter during that period of time. Typically, brighter or more intense colors represent stronger signals. By looking at the waterfall display, we can see the occupation of frequencies across the spectrum over time. The historical view of the waterfall can assist

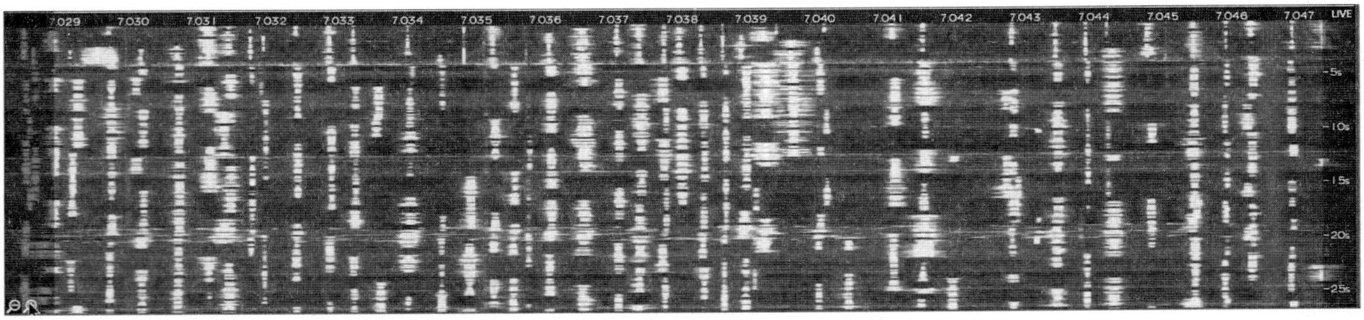

Figure 14.10 — A waterfall display showing the same spectrum as the panadapter in Figure 14.5 with CW and digital signals. More recent data is at the top of the display, and signal strength is indicated by brightness.

in locating a frequency that either has or has not been in use depending on the need.

Like its cousin the panadapter, waterfall displays are also calculated from an FFT, sometimes from the same data used to produce the panadapter. The waterfall also uses the same FFT bins and follows the same rules as the panadapter. The noise floor visible in the waterfall is determined by the bin size of the FFT. The narrower the bins in the FFT, the less noise in the bin and the more weak signals can be observed in the waterfall. In Amateur Radio, we generally refer to the noise floor as the noise observed in a receiver with a 500 Hz bandwidth. If our FFT bin width is 50 Hz, our noise floor would be 10 dB lower, calculated as $10\log_{10}(500/50)$. Similarly, a 5 Hz bin would have a noise floor 20 dB lower than for a 500 Hz receiver bandwidth. To see signals "below the noise floor" is a useful capability in both the panadapter and waterfall displays.

The historical view of a waterfall can provide an additional level of information not always visible in a panadapter. Using the same microwave operating example, assume that we are attempting to hear 10 GHz signal from a microwave rover with the dish antenna slightly off our bearing. A common technique is to ask the rover to send a series of slow CW dashes "please send dashes" using a VHF liaison frequency. By watching the noise floor in the waterfall carefully, we may be able to see a faint line representing those dashes. We can then ask the rover to rotate the dish one way or the other and see if the intensity increases. The final outcome would hopefully be to align the dish where we can hear the rover.

Using an HF example, the waterfall can provide unique insight into the patterns used by a DXpedition station operator. Knowing these patterns and where the DX station is transmitting can help the DXer pick a frequency that is likely to be heard, and worked, by the DX station. For example, by observing the DX station working several stations across a pileup, the DXer can immediately put the information to use and "tailgate" the last contact hoping to work the DX entity before the listening frequency changes. Consistently locating the DX entity on a traditional receiver by spinning the dial is futile in comparison.

COMBINING THE PANADAPTER AND WATERFALL

A common way to show both a panadapter and a waterfall display at the same time is to position the panadapter above the waterfall, both lining up in frequency so that a common frequency axis is used for both. In this way, we can see what's happening now in the panadapter and what happened in the past in the waterfall.

A vast amount of information is presented to the operator with both displays present simultaneously. A short burst of noise heard in the receiver could be anything, but close examination of the panadapter and waterfall can show that it is a signal slowly moving across the spectrum (perhaps an ionosonde) or maybe it's a spread-spectrum signal that is 10 kHz wide moving around the ham bands. The difference will be apparent looking at a wider view of the waterfall and seeing the same signal move around the band. Most operators find that after using a panadapter and waterfall for a while they are not able to easily go back to using a radio without these features, which enhance situational awareness.

14.2.3 User Control Interfaces

The front panel of the radio, referred to here as the *user interface* or simply *interface*, provides for control, meters, status, frequency indicators, and displays such as a panadapter and/or waterfall to visualize the spectrum. Typical controls allow for changing the frequency, mode, mode-specific settings, and other features of the radio.

THICK CLIENTS

The SDR-1000, an amateur SDR designed by Gerald Youngblood, K5SDR, was featured in a 2002 series of *QEX* articles listed in the references for the **DSP and SDR Fundamentals** chapter. The SDR-1000 hardware did not have an interface itself, but relied instead on a computer program as its interface. This first SDR interface, *PowerSDR*, was designed to display the spectrum and control the radio. *PowerSDR* is not only the interface for the SDR-1000, it also performs all of the signal processing required to both modulate and demodulate audio and produce all of the spectrum and other displays in the software. This type of interface is known as a *thick client* in the computer industry. The term thick client is used when substantial work must be done in the client software. **Figure 14.11** shows how

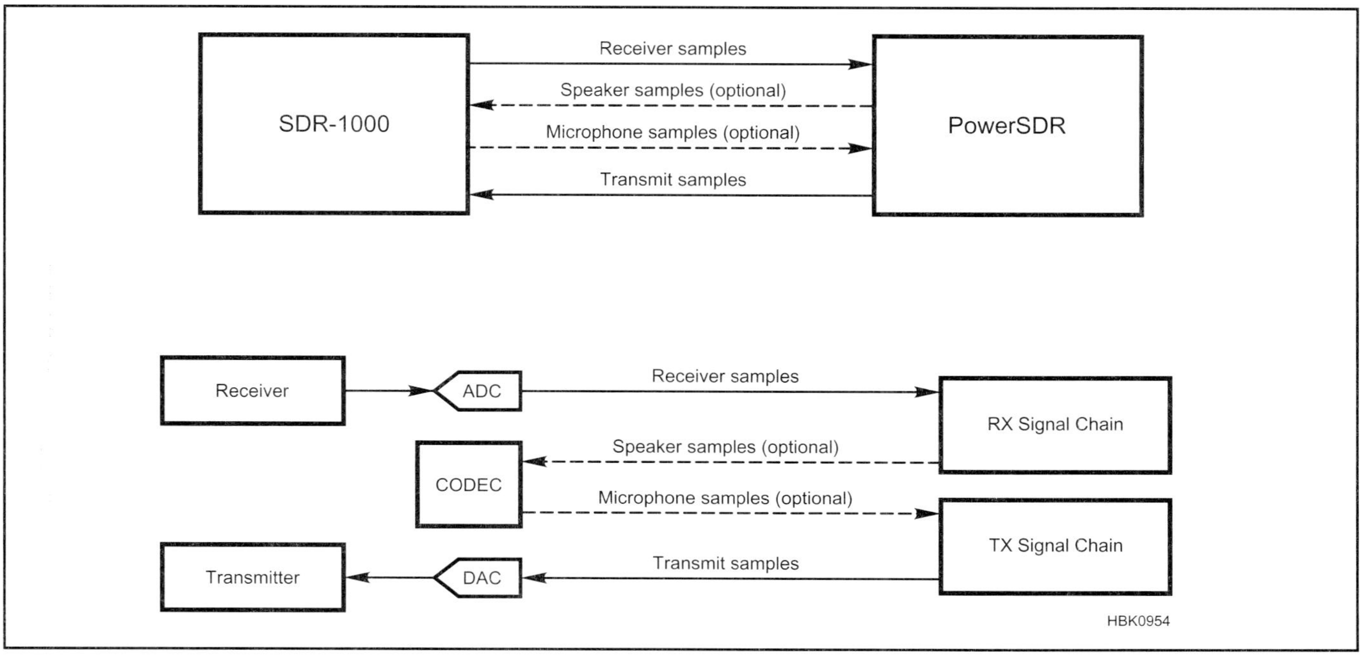

Figure 14.11 — A block diagram of the SDR-1000 showing the main data flows for signal processing.

the signal processing software in *PowerSDR* works with the radio.

There are generally four streams of real-time data that flow between a thick client and the SDR hardware:

1) Receiver samples are time-domain samples collected by the SDR hardware. The hardware itself can be direct-sampling, a superheterodyne (followed by an ADC), or direct conversion. The samples sent to the software for processing are generally in the range of 48 ksps to 384 ksps. Since the samples are often sent as both in-phase and quadrature (I/Q) and are often 16-32 bits each, a 384 ksps stream can contain as much as 25 Mbps of data.

2) The receiver samples are tuned, demodulated, filtered, and ultimately provided to the radio hardware for audio output, usually from headphones or speakers.

3) Microphone samples are captured in the radio hardware and sent to the software for processing and preparation for transmission.

4) Transmit samples are sent from the software back to the radio for transmission when the radio is transmitting.

Some radios require both the speakers and microphone to reside on the computer rather than the radio. In this case, both the speaker and microphone streams are generally not sent to and from the radio, respectively.

Once the interface software has the samples collected from the receiver, it can perform the DSP functions required to create the audio signal. (See the sections on Receive and Transmit Chains earlier in this chapter.)

Similarly, the transmit signal chain commonly has a modulator, filter and limiter. When the radio is transmitting CW or FSK, the SDR software generally synthesizes the waveforms for both of these modes directly without the need for many of the other transmit signal chain elements. For this reason, it is common for there to be a separate signal chain for each mode.

The thick client SDR interfaces such as *PowerSDR* place a higher demand on the computing hardware, since many of the DSP functions are performed in the computer. There is also a stringent set of requirements around timing to ensure that the constant flow of samples on both the receive and transmit sides is not interrupted. If an interruption occurs on the receive side, it will result in audio dropouts. Interruptions on the transmit side can result in discontinuities in the transmit signal. Depending on the hardware involved, discontinuities in the transmit signal can result in high frequency components that cause the transmitted signal to reside outside of the desired transmit passband. There are advantages to the thick client architecture, specifically:

1) Since computing and software requirements for the radio itself are minimal, the radio hardware can be less expensive.

2) Software that runs on the PC is typically easier to develop, test, and debug.

3) As PC hardware has continued to achieve higher price-performance ratios, more power for more capabilities can be added without changing the radio..

The thick client architecture is now common, due in a large part to the availability of open source *PowerSDR* and the ease of adapting it to other radios. More than one HPSDR radio has been designed using this architecture, as have the ANAN radios, adapted from the HPSDR designs.

THIN CLIENTS

In contrast to a thick client interface are *thin client* interfaces. In a thin client, all or most of the work required to modulate, demodulate, and produce the displays for the radio is performed inside the radio hardware. The thin client is used, then, to display the interface and accept control commands, which are passed to the radio for processing. An example of the thin client approach is the FLEX-6000 Signature Series radios by FlexRadio Systems.

In the FLEX-6000, all modulation, demodulation, and radio control are performed in the radio in the *SmartSDR* software running under *Linux*. This reduces the performance requirements on the client system used for the interface. In turn, the interface can easily be run on cellular phones, tablet computers, and other less expensive computing hardware. Key advantages of the thin client architecture are:

1) A wide array of computing hardware can serve as the interface due to the low computing demands.

2) Reduction of network bandwidth between the client and radio due to the processing occurring in the radio. This allows remote operation over low-speed network interfaces making remote connection directly to a cell phone possible, for example.

3) The client software is considerably simpler since it is just responsible for display and control and can, therefore, be easily ported to new computing devices resulting in a wide array of available interfaces and interface platforms.

4) Availability of an applications programming interface (API) due to the necessity of an API for the client. This API can then be used by other amateur applications wishing to use the radio as a server or utility.

14.2.4 Panadapters for Older Radios

Panadapters can also be added to an older analog or analog-digital hybrid transceiver. The concept is straightforward: use an SDR receiver to monitor the radio's IF spectrum while the radio operates normally. The radio provides the front-end preselector, band-pass filtering, and initial IF filtering. The SDR and software displays signals in the IF passband. As the radio is tuned, signals move across the display. This is an example of *dependent architecture* as described in this section (see Figure 14.8). Panadapter software demodulates signals independently of the radio, so different signals can be monitored simultaneously on the radio and the panadapter.

Adding a panadapter to an older radio requires 1) an SDR device with display software that will cover the receiver's IF; and 2) a method of interfacing the SDR to the IF of the radio. SDR devices come in all capabilities and price ranges. The two most common and least expensive types are:

1) An external commutating mixer (also

Software for External Panadapters

Software for display and demodulation is dependent on SDR type and PC operating system. All listed titles below have versions available for multiple operating systems. The following packages support the Windows operating system:

- *HDSDR*: (**hdsdr.de**) Powerful, many customizable features. Can record RF or AF on a timer. Mild learning curve. Widely used.
- *SDR Sharp*: Powerful. Easy to use. Lots of add-ons and customizations. The port driver *Zadig* (included with the distribution) is required for using *SDR#* with RTL-SDR. Specific instructions on how to properly load *SDR#* on the **RTL-SDR.com** website must be followed exactly. (**sdrsharp.com** redirects the browser to AirSpy website where *SDR#* software is available under downloads)
- *SDRuno*: (**sdrplay.com**) Supports all *SDRplay* and other packages.
- *SDR Console*: (**sdr-radio.com**) Extremely powerful, multiple windows. More advanced learning curve.
- *RTL-SDR*: (**rtl-sdr.com**) Supports the RTL2832+R820. Detailed installation instructions, much information.
- *Rocky*: (**dxatlas.com**) For SoftRock (or other I/Q type devices). Simple and works well.

Two packages for the Android operating system include *RF Analyzer* and *SDR Touch*. Both are designed for use with RTL-SDR connected to the Android device by an OTG ("On-The-Go") cable. More software is available from Google Play.

called a Tayloe mixer — see the Receiving chapter) such as a SoftRock. These use a fixed or variable LO to switch a multiplexer IC that samples the IF signal in quadrature (every 90 degrees). The result is a pair of audio frequency I and Q signals that are fed to a computer soundcard for the display and demodulation software.

Limited by the performance of inexpensive components and the requirement for an LO that usually runs at four times the IF, these are generally limited to a maximum IF frequency of 30 MHz. An additional limitation is the sample width and quality of the soundcard. Built-in PC soundcards are often limited to 48 kHz, so 24 kHz is the limit of the displayable RF bandwidth. Higher performance external or add-in soundcards are available with sample rates up to 192 kHz with 24-bit resolution.

2) USB "dongles" such as the FunCube Dongle Pro+, a USB plug-in with a 0.15 to 1900 MHz RF range and 192 kHz bandwidth. These have automatically selected internal RF filters and offer surprisingly good performance. Other popular types are SDRPlay, AirSpy, and HackRF. For an IF higher than 25 MHz, the RTL2832U+R820 European TV tuner dongle, commonly called RTL-SDR, includes software that converts it into a 24 – 1700 MHz receiver of 2 MHz bandwidth and 8-bit resolution. These have no internal filtering but if applied properly, work OK for IF monitoring.

Interfacing to the IF requires modifications to the radio. Making the modifications to add the panadapter could void the warranty and can damage the radio if done incorrectly. Refer to a skilled transceiver technician if you are not comfortable making the necessary modifications on your own.

Adding an IF monitoring tap is not a simple task. Modern analog radios have low-impedance circuitry in the IF, so simply adding a 50-Ω device in parallel with the existing IF circuitry will be detrimental to stage gain and balance. Some SDR devices are known to generate switching noise from their input port where it is injected into the receiver's IF, affecting all signals received by the main radio.

SDR Panadapter Hardware

- JFET buffer amps from **SDR-Kits.net** and **huprf.com**
- Funcube: FCD Pro+ from **funcube-dongle.com**
- SDRplay: (**sdrplay.com**) Several versions available; 10 kHz to 2 GHz with up to 10 MHz bandwidth.
- Airspy: **airspy.com**
- HackRF: (**greatscottgadgets.com**) Frequencies to 6 GHz. Half duplex transceiver. Open source.
- RTL2832+R820 (**rtl-sdr.com**) Also called RTL-SDR. Available from Amazon, Nooelec, other sources.
- SoftRock: (**fivedash.com**) simple kits and pre-built.

There are several ways of tapping IF signal out of the radio. The radios for which it's easiest to add the panadapter have buffered IF outputs already built in. Examples include the Icom IC-R7000 receiver and Elecraft K2 and K3 transceivers. Simply connect the SDR device input to this output port. Some radios add a dc voltage to the IF output for supplying power to an external device. If the input of the SDR device you are using is dc-coupled, a dc blocking capacitor must be added. A value of 0.01 µF will work in most cases. Use a ceramic capacitor with low lead inductance to minimize reactance at VHF and higher frequencies.

It may be possible to find a point in the IF signal path that is buffered. The Kenwood TS-430S and TS-700SP transceivers both have splitters and buffers in the IF path that feed separate FM demodulator circuits. (In these radios, a signal was always present at the buffer output, regardless of mode selected, but that may not be true of all radios. Check the in-depth User's or Service Manual for the transceiver.) An SDR connected to that point should have little effect on the overall receiver performance, although dc blocking may still be required. Careful testing before and after adding the separate IF output is required.

Another widely used method of interfacing is to tap into the IF with a high impedance buffer amplifier. An excellent design is available from SDR-Kits (**www.sdr-kits.net**) utilizing a JFET input buffer (100 kΩ input impedance) followed by a buffer amplifier and low-pass filter to drive the SDR device. These interfaces provide reverse isolation to keep SDR noise out of the IF chain. Information from the interface's original designer (G4HUP) is maintained on the SDR-Kits website.

It is also very useful to remove dc power or disable the IF output during transmit to prevent spurious audio noise and display artifacts during transmit. A source of dc in the radio that turns off during transmit or a control signal to enable and disable the IF output is useful. See the information from G4HUP on the SDR-Kits website for suggestions.

Examples of radios evaluated by the author include:

- Icom IC-R7000: An easy radio for adding a panadapter. The buffered 10.7 MHz IF is brought out to a jack on the back panel. No buffer amp required, but adding a dc blocking capacitor is required.
- Kenwood TS-430S: 8.83 MHz IF. Worked well with SoftRock.
- Kenwood TS-700SP: A buffered path in the 10.7 MHz IF, similar to that in the TS-430S.
- Yaesu FT-817: No buffered path in the 68.33 MHz first IF so it requires a high-impedance buffer such as the G4HUP "PAT." There is a version with a 70 MHz low-pass filter that removes artifacts before the following SDR. If the IF is tapped before the first IF bandpass filter, the SDR input bandwidth is approximately ±75 kHz from the center operating frequency. After the first IF filter the SDR input has a bandwidth of ±10 kHz.

14.3 Configuration and Control Interfaces

Computer control interfaces for radios play a key role in use of our radios beyond all but the simplest of operation. These interfaces allow us to check the status of our radios, change their settings, tune them, and operate from a distance. Logging programs often use them during contests to automatically acquire frequency and other data for the contest log. In combination with logging programs, the interfaces can be used to automatically select antennas, turn rotators, and control other equipment in the station.

Each radio manufacturer has selected or designed an interface to work with their radios. Typically a manufacturer will use a common interface across their product lines. The user's manual and service manual contain complete information on the commands and data that are exchanged via the interface protocol.

14.3.1 Icom CI-V

Icom's control interface, CI-V, named as the fifth (V) Communications Interface (CI), is a serial "one-wire" protocol that uses the RS-232 electrical interface (generally with 5 V rather than the RS-232 standard ±12 V), but allows multi-drop or multiple transceivers on a single CI-V communications line.

By using CSMA/CD (carrier sense, multiple access, collision detect), multiple radios can avoid using the communications interface at the same time, detecting when a collision occurs and retransmitting data. This is the same technique used in 10Base2 Ethernet in which coaxial cable was used and multiple computers used the same communications bus.

In CI-V, each radio has an address (which can be changed) and the computer can communicate with any of the radios on the CI-V bus. The CI-V protocol allows tuning the radio, changing modes, selecting memory channels, and other functions. CI-V generally treats the RS-232 data as bytes and references commands using their hexadecimal values. It is essentially a binary communications protocol. CI-V is also used in newer Ten-Tec radios, but with additional commands specific to Ten-Tec transceivers.

14.3.2 CAT (Computer Aided Transceiver) Interface

CAT is a control interface based on the RS-232 electrical standard. Most manufacturers such as Kenwood, Yaesu, Elecraft, and FlexRadio implement CAT via RS-232 or via a USB interface that emulates RS-232 signals. The protocol used by each manufacturer is proprietary to that line of equipment. Unlike CI-V, CAT is a point-to-point protocol, meaning the expectation is that there is a single computer controlling a single radio on the line. CSMA/CD protocols are no longer required, but an additional communications port must be used for each radio. CAT is not a binary communications interface, but uses ASCII characters. As with CI-V, each radio implements only commands that are meaningful to that radio.

With the advent of USB, many radios now make their CAT interfaces available over USB in the form of a USB communications port. When the USB cable is plugged into a computer, the USB protocol recognizes the device as a communication port and adds a new port to the computer. In this way, many radios can be added to a single computer using a USB hub.

As Ethernet and the Internet have been added to our computing palette, CAT has been implemented over TCP/IP, an Internet standard communications protocol pair. For example, with the FLEX-6000's *SmartSDR* CAT program, a TCP/IP port can be added that responds to CAT commands. The base protocol remains the same, but the communications transport mechanism is TCP/IP over Ethernet rather than RS-232. This allows a relatively simple conversion for programs attempting to access radios over Ethernet rather than either RS-232 or USB.

FlexRadio Systems, although adopting CAT to allow for interoperation with legacy software such as existing loggers and digital mode programs, also added a new computer control interface, the *SmartSDR* API. CAT is restricted to commands relating to two variable frequency oscillators (VFO), typically called VFO A and VFO B. Since the FLEX-6000 can currently have up to eight slice receivers (synonymous with a VFO for the purposes of controlling the radio), the CAT standard could not sufficiently provide for control of the radio. The SmartSDR API is also based on ASCII commands and allows full control of the radio over a TCP/IP port or using a Microsoft *Windows* .NET DLL interface. FlexRadio's *SmartSDR* thin client running on a *Windows* PC achieves all of its display and control capabilities using the *SmartSDR* API.

14.4 SDR Design Tools

A description of how SDR works and the opportunities it presents are interesting but not of much use unless the reader can acquire and use the necessary tools to work with SDR technology. This section discusses *GNU Radio*, a popular software development environment for working with digital signal processing (DSP) and applying it to SDR. A short introduction to working with FPGA technology is also included. DSP software is key to working effectively with SDR at the level to which amateurs are accustomed with traditional analog technology and a soldering iron.

14.4.1 GNU Radio

INTRODUCTION TO *GNU RADIO*

The ability to create and work with software for SDR projects is a new skill for many hams. Professional programmers have many options to pursue. For hams without a programming background, software packages are now available that create a gateway for ordinary hams to build custom DSP software. Software authoring has been vastly simplified. The software packages feature graphic user interfaces (GUI) and offer useful DSP functions in DSP source code libraries.

The ham experimenter has choices of graphically based DSP software packages: the proprietary packages *Matlab / Simulink* by MathWorks and *LabVIEW* by National Instruments, or the open source *GNU Radio / GNU Radio Companion* (*GRC*) DSP library package. This section introduces the combination of *GNU Radio* and *GRC* package as a powerful, "ham friendly" development environment for SDR projects. The material is based on a pair of *QEX* articles by John Petrich, W7FU (see the *GNU Radio* references), which are included in the online supplemental information for this book.

GNU Radio was conceived as a development tool intended for advanced educational programs and commercial product development labs. With a little effort, hobbyists can use these tools as well. *GNU Radio* can be used as a simulation tool to study DSP without being connected to an SDR. *GNU Radio* can also be used to develop practical and operationally satisfying DSP software for SDR transmitters, receivers, and transceivers as well as test equipment.

To make DSP authoring accessible for all levels of users, the *GNU Radio* developers include the user option of *GNU Radio Companion (GRC)*. *GRC* is the graphical DSP library interface, an integral part of *GNU Radio*. The foundational *GNU Radio* DSP library is accessed with character-based command line entries. *GRC* is a graphical overlay to the *GNU Radio* DSP library with the same full functionality as the command line approach. *GRC* emulates the command line entries and represents each DSP function as a graphical icon without command line entries. (For the sake of completeness, a very few of the *GRC* DSP parameters are expressed as simple Python code commands. These Python commands are covered in the section **Learning Resources: DSP Competency and *GNU Radio***)

An arrangement of DSP icons and digital data links is termed a *flow graph*. When executed, the *GRC* flow graph is compiled to a Python code file, which itself is a wrapper for C++ executable DSP code. This artful layered construction, like a set of nesting dolls, provides a high level, user friendly, graphical interface for DSP authoring with the speed and efficiency of lower level code for DSP process speed and efficiency. More information from the developers about the construction of this DSP library environment is available at **wiki.gnuradio.org/index.php/What_is_GNU_Radio%3F**

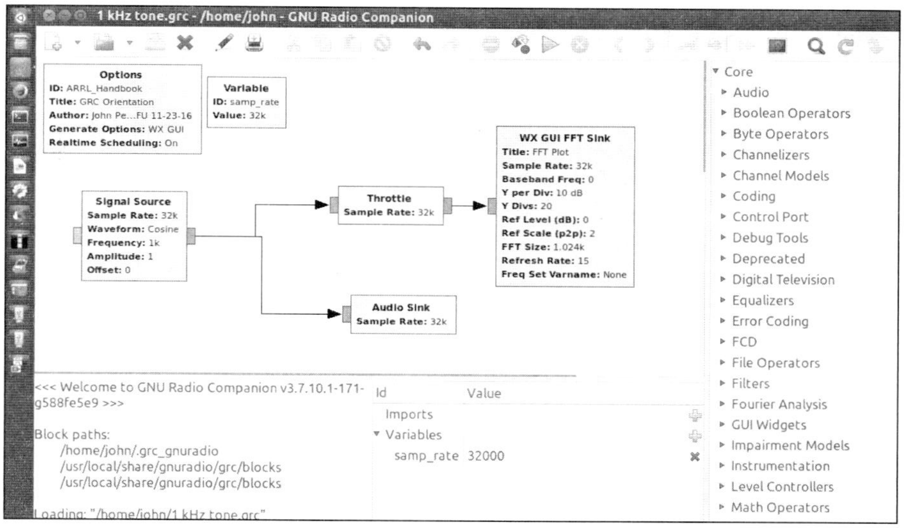

Figure 14.12 — The development window user interface for *GNU Radio* functional blocks.

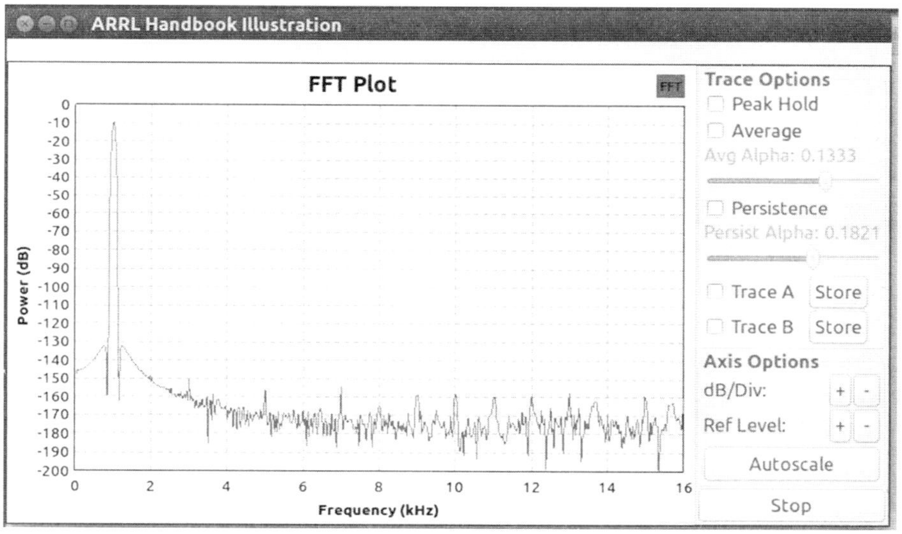

Figure 14.13 — The user interface window showing an operating 1 kHz audio tone FFT.

GNU RADIO AND THE GRC TOOL

GRC, the easy to use graphical interface, makes this DSP library particularly appealing. Reduced to the basics, the step-by-step design of a functional DSP begins with envisioning or drawing a block diagram pattern of a traditional analog signal processing circuit. Once the general signal processing pattern is conceived, select the desired DSP function(s) from the DSP library (filter, mixer, oscillator, amplifier, etc.), arrange the DSP function(s) as graphical blocks in the work screen according to the design pattern, connect the blocks with an appropriate type of data link, add the DSP parameters in each block, and execute the program. The details of constructing a DSP flow graph are covered in the **Success with *GNU Radio* Tools** section and in the online tutorials referenced in the section **Learning Resources: DSP Competency and *GNU Radio***. The process is straightforward.

Let's examine the *GRC* work environment and flow graph composition at the highest level or block level. The *GRC* work environment consists of four parts: the work area, the control bar, the library, and the console. **Figure 14.12** illustrates the *GRC* user graphic user interface (GUI). The DSP is authored in the work area in the center of the GUI screen. The *GRC* control bar is located across the top of the work area. The *GNU Radio* DSP library runs along the right margin of the screen. The operating console is visible at the very bottom of the work environment and becomes active when a DSP function is executed.

Centered in the work area are six DSP blocks, which in aggregate constitute a DSP flow graph. The flow graph implements, in DSP, a 1 kHz audio tone output to both an audio sink (computer speaker) and to an FFT panadapter display (on screen). **Figure 14.13** is the FFT display output once the flow graph has been executed. Note the signal spike is accurately centered at 1 kHz on the FFT display and the 1 kHz tone would be heard via the computer sound card.

Continuing to the parameter level, one step below the block level, each DSP library item is displayed as a rectangular shaped "block" with input / output ports. Each block must be "opened" with a mouse right-click to gain access to the parameter fields that define that DSP function. Each parameter has a field in which the values are either added or selected from a menu.

Figure 14.14 displays the parameter screen for a low-pass filter DSP block. Note how the digital data type (*FIR Type*), sample rate, filter bandwidth and filter shape, and other parameters can be selected or specified. Some parameter values are intuitive, direct carry-overs from analog design. For example, the bandwidth and shape factor parameters for a DSP filter block are specified in kHz in the same way an analog filter is specified. Other parameter values are not as intuitive and refer to parameters unique to DSP. Examples of these parameters are the DSP *Sample Rate* and *FIR Type* choices. More detail about the most commonly used DSP blocks and appropriate parameter choices are explained in

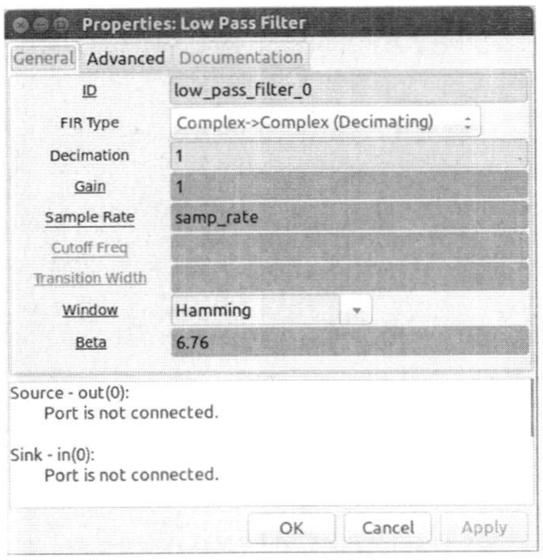

Figure 14.14 — A screen showing how parameters are managed for a typical DSP filter block.

greater detail in the section below, **The Rest of the Story: Basic DSP Theory Applied to a Flow Graph Segment**.

Figure 14.15 depicts the more complex flow graph of an SDR broadcast FM receiver. This flow graph uses an RTL-SDR as the receiver source. Other SDR receiver hardware can be substituted easily as explained in the section ***GNU Radio*, Radio Hardware, and UHD**.

The RTL-SDR is represented by the *RTL-SDR Source* block on the left. The *Audio Sink* on the right outputs the audio sound. The *WX GUI FFT Sink* in the center generates a panadapter display. The chain of DSP blocks in the center of the work screen perform the filtering and signal demodulation functions for this radio. Compare this flow graph with a block diagram for a simple analog FM receiver to appreciate the fundamental similarity. **Figure 14.16** illustrates the operating receiver GUI that appears when the flow graph is executed. Shown are the radio tuning controls for frequency and audio output, and the FFT panadapter display.

SUCCESS WITH *GNU RADIO* TOOLS

Just as success at woodworking comes from competency with woodworking tools, success with *GNU Radio* comes from learning and using its tools. Viewing a flow graph is the first step is learning the tools. Success with *GNU Radio* inevitably proceeds to DSP competency as well.

An important part of success with *GNU Radio* comes from an organized and disciplined authoring approach. Lacking organization and discipline, flow graph construction can easily become a chaotic, non-functional mess. This section describes one general disciplined approach. Following a description of the general approach, a detailed example flow graph creates entry level DSP competency through flow graphs construction.

DSP competency for new users is not necessary for initial success. Many of the automatic default settings in *GRC* are adequate for entry level flow graphs. Ultimately, DSP competency comes from incrementally constructing more advanced DSPs with simple flow graphs.

GENERAL APPROACH: THE FOUR STEPS TO FLOW GRAPH SUCCESS

The disciplined approach involves following a pattern of four basic steps. These basic steps are useful at getting started the first time with DSP flow graph construction and as an efficient approach to construct more advanced and complex flow graphs.

How does one begin a DSP flow graph using the *GNU Radio GRC* graphical interface? Four steps are required to construct any flow graph: the "block" step, the "parameter" step, the "logical interconnection" step, and the "DSP" step. Each step is explained below.

Step 1 — the Blocks: Arrange the appropriate DSP blocks from the *GRC* library to form a diagram of the desired circuit in the work area. See how the functional blocks are arranged in the example flow graph in the work area of **Figure 14.17**.

Step 2 — the Parameters: To perform the desired DSP function, each DSP block re-

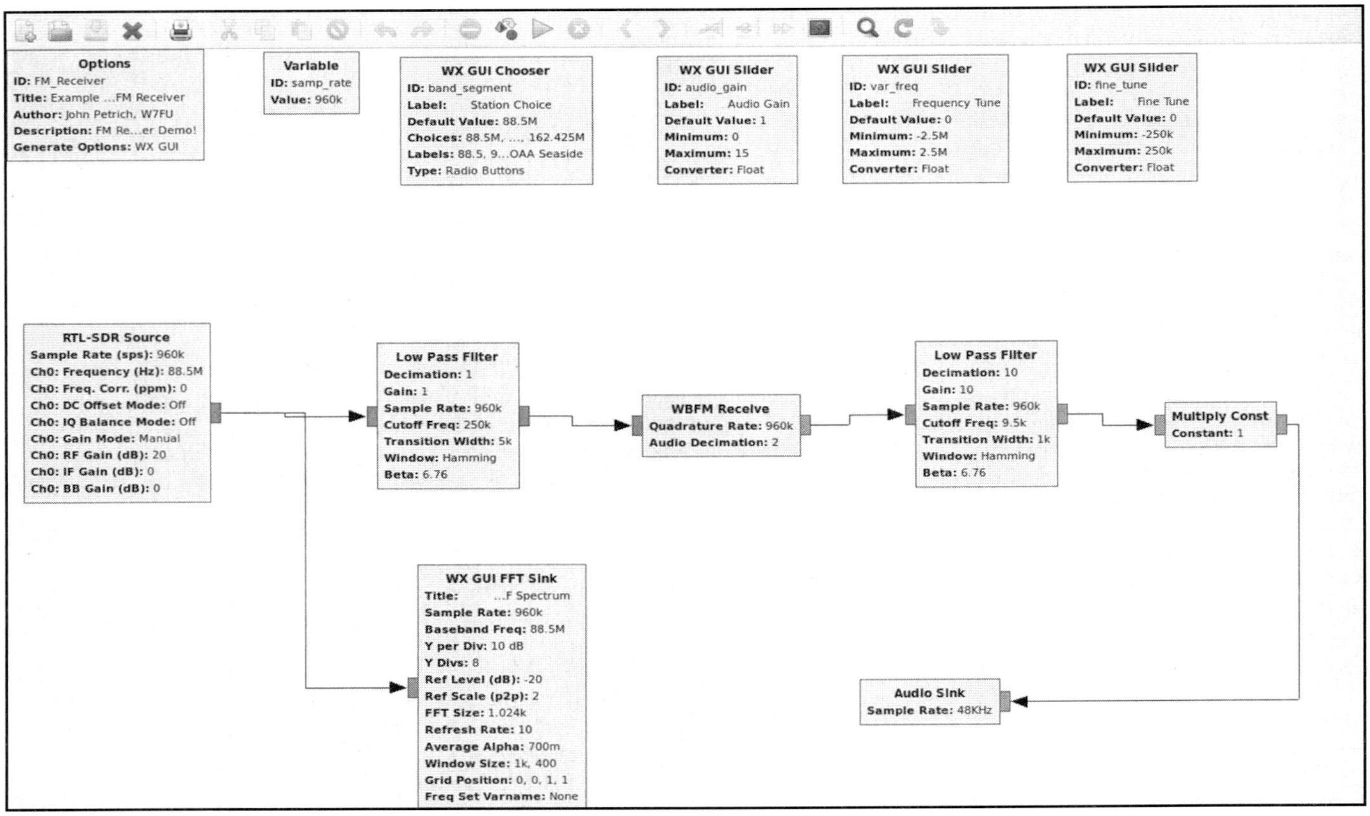

Figure 14.15 — Flow graph for a broadcast FM receiver showing the various function blocks involved.

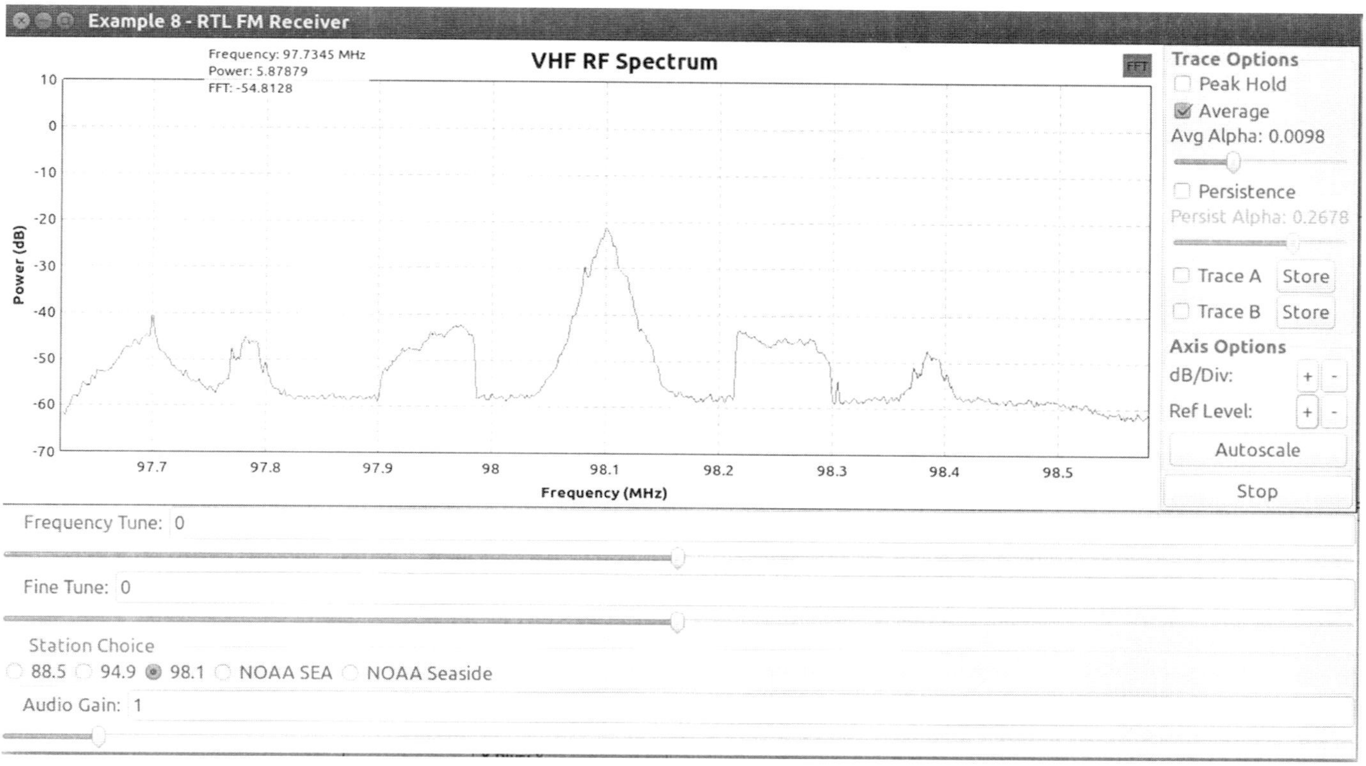

Figure 14.16 — The user interface window for the operating FM broadcast receiver created from the flow graph in Figure 14.15.

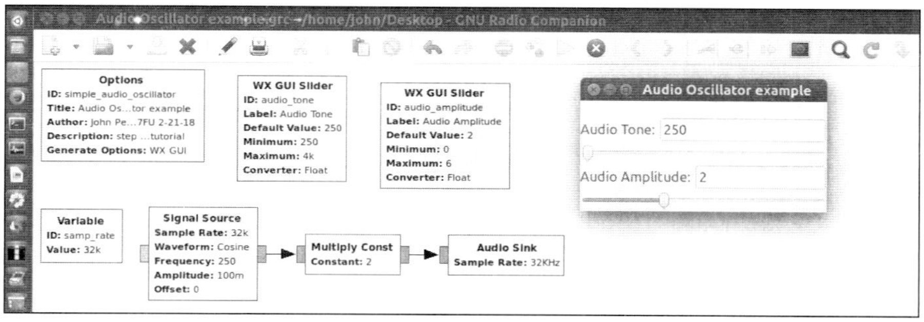

Figure 14.17 — The flow graph for the audio oscillator.

quires user-specified parameters. In the parameter step, right-click each block, select *Properties*, and enter the parameter values in the available fields. Each block in the flow graph diagram is opened in turn, and the desired parameters are entered in each block. Typical parameters are familiar values such as gain, frequency, bandwidth, and so on, as well as some less familiar values such as *Sample Rate* and *Output Type*. The *Sample Rate* and *Output Type* parameter values are unique to DSP as discussed below. These DSP parameters are set by default in many flow graphs and automatically propagated into every DSP block. For one to meaningfully change these default parameters, a basic understanding of DSP theory is required as described in Step 4.

Step 3 — the Logical Interconnections: Each DSP block must be connected to the other blocks by a data link arrow in a logical manner. The data links begin with the "input" or *Source* location in the flow graph and end with the "output" or *Sink* in the flow graph.

Step 4 — DSP Parameters: It is in this fourth and final step that authoring a flow graph involves DSP issues. Let's begin with the special *Variable* block located in the left-hand corner of the work area. That block is automatically present whenever *GRC* is opened for programming. The *Variable* block is the location for the base line *Sample Rate* parameter for each block in the DSP. The default parameter is 32,000 samples per second (SPS). That default parameter is automatically propagated to the *Sample Rate* field in each DSP block in the flow graph work area. The default *Sample Rate* value is a good starting place for many DSP flow graphs and can be used as a default for many designs. Advanced DSP designs may require different sample rates and adjustments to the sample rate parameter in the DSP blocks.

Another default DSP parameter is the *Output Type* parameter for each DSP block. The default parameter is termed *Complex* or I and Q. This default parameter is useful for many but not all radio-related DSPs. The flow graph example described below illustrates a flow graph where the default *Output Type* must be changed for the simple flow graph to function. (See the **Modulation** chapter for more information about complex signals and modulation.)

FLOW GRAPH EXECUTION

In practice, the above outlined flow graph is created in the *work area* of the *GRC*. To operationalize the flow graph, select the *Execute flow graph* control (a green arrow on the screen) from the tool bar above the *work area*. After the software compiles, the DSP will spring to life. The operational GUI appears on the screen with the control functions displayed as shown on the right side of the screen in Figure 14.17.

SPECIFIC FLOW GRAPH EXAMPLE: SIMPLE SINGLE-TONE AUDIO OSCILLATOR

Step 1 — the Blocks: Figure 14.17 is a screen capture of the audio oscillator flow graph. The operating GUI is superimposed on the right side of the figure. The operating GUI only appears only after the *Execute flow graph* is selected from the tool bar.

The flow graph blocks are selected from the library off the right-hand edge depicted in Figure 14.12 (not Figure 14.17) and placed in the work area. At a glance, the flow graph block diagram is constructed to look essentially the same as a traditional analog radio block diagram with the same components. The diagram and component content reflect standard signal processing practice. A traditional circuit block diagram for a simple sine wave oscillator with audio output would consist of series of blocks linked by lines: an oscillator block, a frequency control, an audio amplifier, an audio volume control, all connected one after the other, to an audio output block.

In this audio oscillator example, the signal processing pattern is the same for the traditional analog and the *GNU Radio* digital implementation. With a DSP, the interconnections between blocks are no longer signal links (wires in traditional circuits) but digital data links, as we will see in Step 3 below.

Step 2 — the Parameters: Most of the parameters for each DSP block are visible in Figure 14.17. Examine the parameters of each block with a right-click on the block and select *Properties*. Right-click the *QT GUI Range* control block with the *ID audio_tone* located at the top of the work space. This block is the control block for the oscillator frequency or audio tone. Note the block ID, *audio_tone*, and the tone's tuning range from 250 Hz to 4 kHz in 1 Hz steps. This block connects to and controls the *Signal Source* DSP block output frequency, which is the audio oscillator tone in this example.

The *ID audio_tone* comprises the software link between the control block and the *Signal Source* DSP block. The term *audio_tone* is entered into both the *Signal Source frequency* parameter field and the *QT GUI Range ID* field, thereby connecting the two blocks in software.

In the same manner, the audio volume control is linked via software to the audio amplifier. Select *Properties* in the *QT GUI Range* control block with the *ID audio_amplitude*. The multiplication values range from 0 to 10. These numbers are factors that tell the *Multiply Const* DSP block to multiply the digital data stream amplitude for audio volume control. The ID *audio_amplitude* denotes the software link to the *Multiply Const* DSP block. That ID is entered in the *constant* parameter field in the *Multiply Const* DSP block. The two blocks are linked in the same way as the audio tone frequency is linked to the audio oscillator. The *Sample Rate* value from the *Variable* block, 32k, is automatically propagated to the other DSP blocks but not to the two control blocks.

Step 3 — the Logical Interconnections: To complete the basic DSP flow graph, each block is logically interconnected to the others. Additional attention to details are important in this step. Not appreciated in the screen capture is how the *Multiply Const* DSP block became configured to have only single input and single output ports. This block and many other blocks from the DSP library are pre-configured for multiple input and output ports. When placed in the *GRC* work area, pre-configured blocks from the library are reconfigured for the desired number of ports to fit the flow graph. To reconfigure the number of ports, right-click the block, select *Properties,* and enter a value for the number of desired ports in the appropriate port parameter field. In this example, a value of 1 was entered in the *Input Port* field of the *Multiply Const* DSP block. Select OK, the window closes, and the block appears on screen ready to use with a single input and single output port.

Once the block ports are configured, the DSP blocks must be linked with the data stream. To form these data stream links, left-click on the output port of the *Signal Source* DSP block, followed by a left-click on the input port of the *Multiply Const* DSP block. A data link arrow appears, and those two blocks are digitally connected. The output of the *Multiply Const* DSP block connects in a similar manner to the input port of the *Audio Sink* DSP block.

The flow graph is almost complete. The two control blocks have no visible data link arrows. They are automatically connected via the software *ID* parameters described above. Next, we consider the DSP parameter step.

Step 4 — DSP parameters: The audio oscillator flow graph functions fine with the default *Sample Rate* parameter in conjunction with modern computer sound cards. No change is required for the *Sample Rate* parameter. The *Output Type* parameter is different and must be changed from the default setting for this flow graph to function.

The *Output Type* parameter for each DSP block is displayed in a drop-down menu, found by selecting *Properties* for that DSP block. The *Output Type* is also visually displayed in the flow graph by the color of the input and output port tabs on the blocks. (This color code is a useful bug fix hint with advanced DSP programing and not central for this description. The change in color is only visible when viewing the screen capture or flow graph on a computer screen.) The desired *Output Type* for this flow graph is "Float" (digital data using floating-point values). The *Output Type* parameter for each DSP block in this example must be the same: "Float." Right-click and select *Properties* for each DSP block in the flow graph, navigate to the *Output Type* parameter, and select "Float" from the drop-down menu. The two control blocks don't require any DSP parameters.

Once all steps are completed, the flow graph is ready to be executed. Upon execution, a new GUI appears on the screen with the controls displayed. That GUI screen is superimposed in the upper right-hand corner of the screen capture in Figure 14.17. Listen for the 250 Hz audio tone in the speaker. Adjust the audio volume and audio frequency with the GUI controls. You can also experiment with adjusting the sample rate in the *Variable* block or parameter values in any block and observe the effect these changes have on the operation of the flow graph. (The GRC flow graph for this example is included in the downloadable supplemental package for this book. The file, Chapter 14 - Audio Oscillator example.grc, can be executed in the *GNU Radio* application.)

THE REST OF THE STORY: BASIC DSP THEORY APPLIED TO A FLOW GRAPH

To better understand this DSP flow graph example and to aid authoring more advanced DSP flowgraphs, basic information about the relationship of sample rate, data stream output type, and DSP is important. (See the **DSP and SDR Fundamentals** chapter for a detailed discussion of the analog/digital conversion process.)

Sampling is a foundational aspect of DSP. DSP comes into action in the A/D converter where an analog signal is converted to a digital data stream. The DSP acts to manipulate the series of data points which form a replica of the sinusoid signal as stream of digital data. The greater the number of samples, the more closely the replicated digital data corresponds to the original sinusoid signal. After digital processing, the digital data stream can be reconstructed to generate the familiar analog sound or radio signals with a D/A converter.

Connecting the sampling concept to a functioning flow graph, each DSP block in the flow graph must function with a compatible sample rate parameter. *GRC* automatically enters the default sample rate for each DSP block, which is fine for many DSP processes. Advanced DSP techniques using *GNU Radio and* GRC call for a wide range of sample rate parameters beyond the default value. The sample rate parameters can also be individually adjusted for each DSP block in the flow graph. These sample rate changes are accomplished using *Resampler* DSP blocks. Sample rate adjustments within a single flow graph are termed "multi-rate" sampling, a common practice with advanced DSPs such as those for SDR applications.

Another foundational concept regarding DSP is to understand that the sole purpose of each DSP block is to mathematically manipulate the digital data stream. Fortunately, with *GNU Radio* and the *GRC* graphical approach to programming, the user is not required to understand or program the underlying mathematics. The DSP functions in *GNU Radio* DSP library are preloaded with the necessary mathematical algorithms to perform their intended DSP function.

The data stream math algorithms are unique in each DSP block. The operation can be single purpose, such as for the *Multiply Constant* DSP block, which functions as an amplifier or attenuator by simple multiplication of the digital data stream amplitude. Other DSP blocks provide filtering or mixing functions and frequently combinations of these functions. All operations are performed mathematically based on algorithms programed by the developers into each DSP block.

Not only is the data stream manipulated in a mathematically determined manner, but the mathematical structure of the digital data stream has an impact on how a flow graph can function. Return to the flow graph in Figure 14.17 where the *Output Type* DSP parameter reflects the structure of the digital data stream that communicates among the DSP blocks. The mathematically processed digital data stream *Output Type* can be customized as necessary for advanced DSP processes.

As a practical matter, the default *Output Type* digital data structure for most *GNU Radio* DSP blocks is set by default to "Complex" or I and Q data. The default setting "Complex" is the preferred data type for many, if not most, ordinary radio related DSP applications. Some DSP flow graphs, such as those employing phasing techniques and the audio oscillator example in this case, require an *Output Type* that is different from the default "Complex." (Some SDR hardware digital data output types are derived from different math conversions and require a non-default data stream structure conversion in the flow graph, as well.) In other words, one type of digital data stream doesn't fit all. The *Output Type* data stream parameter can be quite specific for the some SDR hardware and quite specific for some DSP process. Mastering the details of *Output Type* for your flow graph constructions will come with more in-depth understanding of DSP fundamentals. For further information on sampling rate choices and *Output Type* digital data streams, review the information on the GNU Radio website.

LEARNING RESOURCES: DSP COMPETENCY AND *GNU RADIO*

This introduction to *GNU Radio*, *GRC* and flow graph construction presents DSP at the high level of blocks and parameters, consistent with *GNU Radio*'s high-level graphical environment. But that is only the beginning of a journey into the world of DSP. The **DSP and SDR Fundamentals** and the **Receiving** chapters contain much useful information about understanding DSP. That information offers important background and context to more in-depth understanding of DSP to apply to *GNU Radio*, although not specific to *GNU Radio*.

DSP competency is readily learned using *GNU Radio* as a tool for experiments and simulation. The online educational resources specific to *GNU Radio* are great learning resources. Detailed examples of DSP applications and parameter selection, awkward and sometimes confusing in a text format, are much more easily understood via online video tutorials. The Ettus Corporation-sponsored series of DSP tutorials (**files.ettus.com/tutorials/labs/Lab_1-5.pdf**) and the related video series (**www.youtube.com/playlist?list=PL618122BD66C8B3C4**) are highly recommended for this learning format.

For a more academic treatment of DSP consider both the **DSP and SDR Fundamentals** chapter in this *Handbook* and the DSP texts by Richard Lyons. The Lyons texts provide a particularly "ham friendly," intuitive approach to understanding DSP. (See the ***GNU Radio* References** section.)

INSTALLING *GNU RADIO*

The very first step for a new user is to review the *GNU Radio* web portal for authoritative information (**gnuradio.org/redmine/projects/gnuradio/wiki**). The *GNU Radio* developers have published a rich source of detailed information to help the new user get started using *GNU Radio*. From that portal page one can link to installation instructions, tutorials, latest news, and in-depth information about the software development process. Refer to it frequently, since the content changes with the ongoing software development.

To get started with *GNU Radio*, a particularly appealing option is to use the *GNU Radio Live SDR Environment* (**wiki.gnuradio.org/index.php/GNU_Radio_Live_SDR_Environment**). The *Live SDR Environment* allows the user to implement *GNU Radio* without installing *GNU Radio* on the computer hard drive. The latest versions of the *Ubuntu* operating system and *GNU Radio* software are downloaded from the Internet to a storage medium (for example, DVD or USB memory stick) as an executable file. Upon executing from the medium, the *Live Environment* opens in a complete *Ubuntu* operating system with the *GNU Radio* application pre-installed. *GNU Radio* and *GRC* can be immediately executed to run at essentially full capability.

To install *GNU Radio* directly to a computer hard drive, installation from the official Source with the "*build-gnuradio script*" is far and away the most dependable and highly recommended approach (**wiki.gnuradio.org/index.php/InstallingGRFromSource**). Installing *GNU Radio* from the Source with the *build-gnuradio* script automatically installs the most recent *GNU Radio* version, all add ons, and the latest driver for compatible SDR hardware, the *Universal Hardware Driver* (*UHD*). The *build-gnuradio* script is a preferred method for initial installation and subsequent updates of *GNU Radio* and UHD as needed. Step by step "ham friendly" help with the installation process is also available at **www.w7fu.com**

GETTING THE MOST OUT OF *GNU RADIO*

Support for *GNU Radio* is available online via the "discuss-gnuradio" message reflector (**lists.gnu.org/mailman/listinfo/discuss-gnuradio**). Questions can be posted on the reflector, and the *GNU Radio* developers respond via email. The reflector is very active and a rich source of information.

The most efficient and user-friendly approach to get full benefit of what *GNU* Radio has to offer is to implement *GNU Radio* in the Linux *Ubuntu* operating system. The *GNU Radio* developers develop and maintain *GNU Radio* in *Ubuntu*. Potential cross-platform bugs and incompatibilities are avoided by adopting the same *Ubuntu* operating system that the developers use. *GNU Radio* is reported to operate on Windows and Mac OSX, but there are challenges with full implementation with these operating systems, which make them best avoided by beginners.

Computer users unfamiliar with Linux operating systems shouldn't be intimidated by using the *Linux Ubuntu* operating system. *Ubuntu* has the look and feel and straight forward ease of use of MS Windows. The *Ubuntu* operating system is open source, downloadable via the internet, and can easily be installed in a separate hard drive partition if necessary. Installing *Ubuntu* via the internet is straightforward and quick. The official *Ubuntu* website provides the latest downloadable *Ubuntu* versions and instructions for using *Ubuntu* (**www.ubuntu.com/download/desktop**).

GNU Radio is a dynamic application with version updates on an irregular but frequent basis. It is important to keep abreast of the changes via the web portal and learn to update the application re-using the *build-gnuradio* script as necessary. Follow the "Latest news" for updates and other information (**gnuradio.org/redmine/projects/gnuradio/wiki**).

For serious DSP authoring and most SDR projects, avoid installing the Linux "distribution" versions of *GNU Radio*. The "distribution" versions of *GNU Radio* are built into the *Ubuntu* operating version at the time of that operating system's release. These "distribution" versions can be installed easily from within the *Ubuntu* version. Because *GNU Radio* is built into the *Ubuntu* version at the time of initial release, these *GNU Radio* versions are typically out-of-date, have outdated *UHD* drivers, and are installed differently within the file system from the versions downloaded from the official Source. The UHD might not work with newer SDR hardware. Updating and adding DSP blocks to distribution versions is not possible. For all of these reasons it is best to avoid installing a "distribution" version of *GNU Radio*. Instead install a full installation of the current *GNU* Radio version.

A capable computer is necessary to use *GNU Radio* to its full real-time DSP potential. The developers recommend a dual-core CPU at the i3 level or above to users. Single-core and early dual-core CPUs lack the speed and capacity for many ham DSP designs.

GNU RADIO, RADIO HARDWARE, AND UHD

To implement real-time SDR radio projects with *GNU Radio*, it is necessary to use compatible SDR hardware. The *Universal Hardware Driver* (*UHD*) is a dedicated software package to interface *GNU Radio* and compatible SDR hardware. Of late, newer generation SDRs have appeared that interface with a multi-layered PPA system of *GNU Radio* packages or a suite of applications. The documentation for compatible hardware will refer to the *GNU Radio UHD* capability or to a PPA *GNU Radio* package. Given the popularity of *GNU Radio* with the DSP developer community, *UHD* or the *GNU Radio* PPA is commonly specified for modern SDR hardware.

The latest version of *UHD* is installed automatically whenever *GNU Radio* is installed via the *build-gnuradio* script from Source. The *GNU Radio* PPA requires a separate and additional installation. The ever-expanding list of compatible hardware is published at **wiki.gnuradio.org/index.php/Hardware.** Searches of the internet and close reading of the specifications can sometimes find *UHD*-compatible SDR hardware that hasn't yet reached the official list.

GNU RADIO REFERENCE MATERIAL

J. Petrich, W7FU, "Digital Signal Processing and GNU Radio Companion," Jul/Aug 2017, *QEX*, pp. 41-46.

J. Petrich, W7FU, T. McDermott, N5EG, "Digital Signal Processing (DSP) Projects: Examples of GNU Radio and GRC Functionality," Sep/Oct 2014, *QEX*, pp. 25 – 30.

R. G. Lyons, "Understanding Digital Signal Processing" ISBN 0-201-63467-8

"GNU Radio Workshop," Microwave Update, October 2019, North Texas Microwave Society (**microwaveupdate.org**)

www.w7fu.com — step by step information on installing and maintaining *GNU Radio*, with examples of ham radio oriented SDR flow graphs.

www.reddit.com/r/GNURadio — a very active *GNU Radio* development community.

Suggested reading: **wiki.gnuradio.org/index.php/SuggestedReading**

14.4.2 FPGA Data Engines

What actually implements the DSP functional blocks referenced in the previous chapters? In the early days of amateur SDR and DSP, the only suitable hardware platform with a reasonable cost was a high-end PC or specialized DSP or graphics processing microprocessors (microcontroller units – MCU). These are still used today but after some years in transition, amateur SDR is largely implemented by FPGAs — Field-Programmable Gate Arrays.

An MCU executes instructions from a predefined instruction set in sequential order; the hardware is fixed, but the sequence of instructions is programmable. An FPGA, on the other hand, has no fixed instruction set or sequence of instructions, operates on data in parallel, and has programmable logic and interconnections. Software defined radio (SDR) implementations benefit greatly from the FPGA parallel hardware architecture, since they can be configured to implement the required math operations directly in hardware at very high speed. FPGA ICs are now available inexpensively and with a sufficient number of gates (in the 10s of 1000s) to implement high-performance SDR functions.

The FPGA as typically used by non-professional developers is part of a general-purpose computing unit, also known as a *data engine*. Support ICs are included in the package to allow the developer to interact with the FPGA and program it. Other data ports are available, depending on what applications the data engine is designed to handle. When combined with external RF hardware and either an on-board user interface or the ability to stream data to an external user interface, the result is an effective SDR development environment. The example in **Figure 14.18** combines the SDRStick receiver and transmitter with a BEMICROCV-A9 data engine. An Ethernet port provides enough bandwidth to implement the user interface and display on a PC. This is a thin-client implementation as described in the previous section on User Interfaces.

In a series of Hands-On SDR columns in *QEX* magazine beginning in 2014, Scotty Cowling, WA2DFI, described several data engines and the software development environment needed to program the FPGAs. Several of the key columns dealing with FPGA development are included with the online supplemental information for this book.

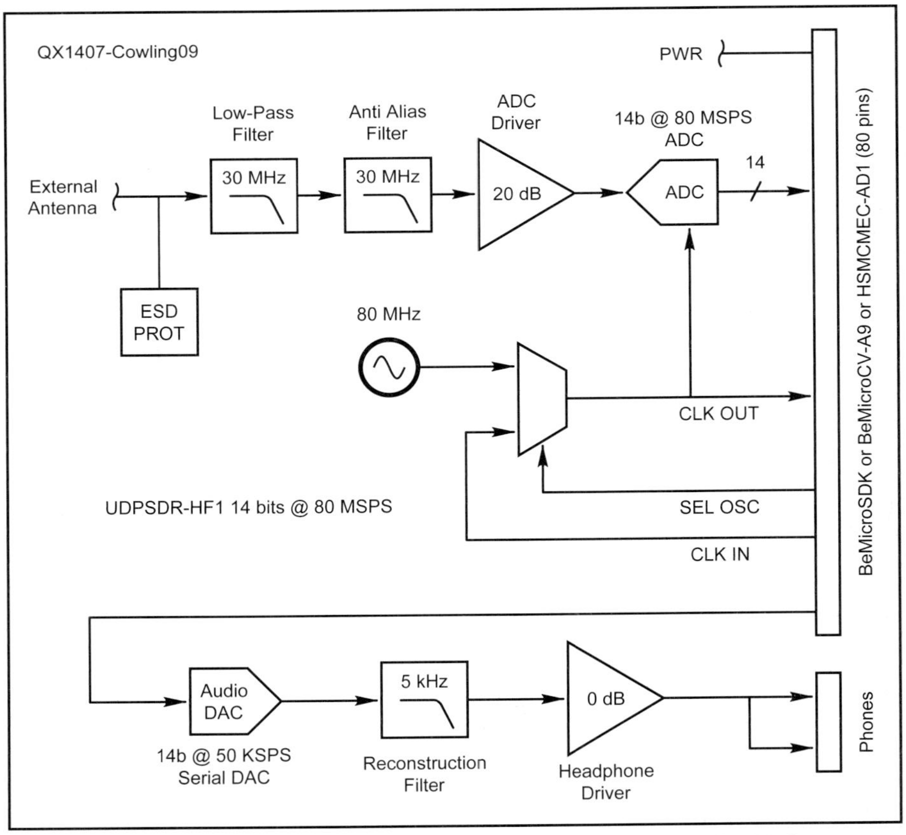

Figure 14.18 — An SDR development system based on the BEMICROCV-A9 data engine and separate SDRStick receive and transmit modules.

14.5 Transverters

The material in this section is largely taken from material provided by Paul Wade, W1GHZ, and Wayne Overbeck, N6NB. Along with downloadable materials in the supplemental information package, the print and online articles in the references provide much more detail and technical depth.

Transverters allow an HF or VHF transceiver to operate on a band it doesn't support by converting both the transmitted and received signal. As a result, the full features of the transceiver are available on the converted band. At one time, building and installing transverters presented a serious challenge for all but the most experienced and technically adept amateurs, but today's off-the-shelf transverters are almost as easy to install as a commercial transceiver. For those who prefer the challenge of building their own transverters from kits or from scratch, those options offer a real sense of accomplishment, as well.

Non-repeater operation on the VHF/UHF/microwave bands is usually CW or SSB, requiring a multimode transceiver. (FT8, JT65, and similar digital modes use the SSB or SSB DATA features.) Some transceivers cover 6 meters, 2 meters, and 70 cm in addition to HF. Although some VHF/UHF transceivers such as the TS-2000 and IC-9700 cover 23 cm (1296 MHz), multimode radios that cover 222 – 225 MHz, 33 cm (902 MHz), and the bands above 2.3 GHz are not available as of early 2020. Wideband SDR receivers that cover from VHF through several GHz are widely available, but companion transmitters are just starting to become available. That limits the equipment choices to surplus equipment that can be modified — and transverters.

Similarly, for the amateur MF and LF bands at 630 meters (472 – 479 kHz) and 2200 meters (136 kHz), commercial transceivers covering these bands are uncommon. Surplus equipment can be adapted to amateur use, but availability is quite variable and the equipment is likely to be designed for operation on fixed channels. As a result, transverters are likely to be used on those bands for the foreseeable future.

Commercial transverters such as those in **Figure 14.19** are available as finished packages for 6 meters, 2 meters, 222 MHz, 432 MHz, 902 MHz, 1296 MHz, 2304 MHz, 3456 MHz, 5760 MHz, and 10 GHz. Recently, transverters have also become available for much higher frequency bands, including 24 GHz, 47 GHz and 76 GHz. Manufacturers that offer complete transverters include DB6NT (**shop.kuhne-electronic.com**); Down East Microwave (**www.downeastmicrowave.com**); Q5 Signal (**www.q5signal.com**); and SG-Lab (**www.sg-lab.com**). A number of small shops also produce kits or finished prod-

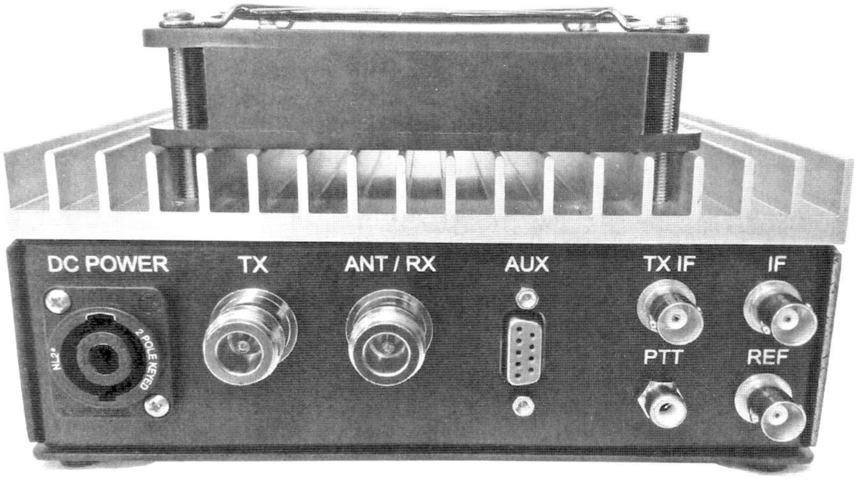

(A)

(B)

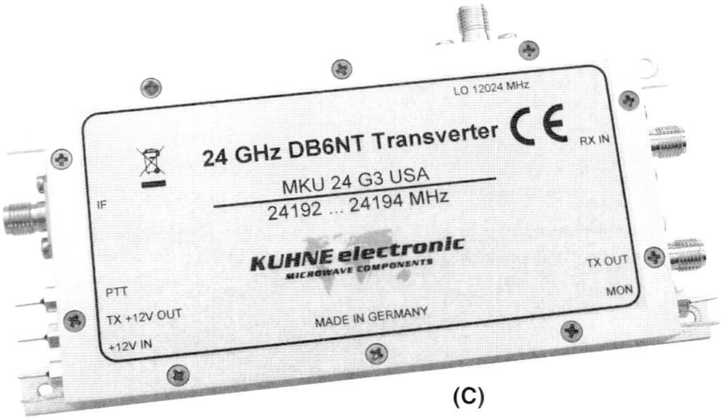

(C)

Figure 14.19 — The Q5 Signal 222 MHz transverter (28 MHz IF) is shown at A. At B is the SG-Lab 2.3 GHz transverter (430 MHz IF), and at C is the Kuhne MKU 24 G3 transverter for 24 GHz (432 MHz IF).

ucts as single PC boards that can be mounted in an enclosure and integrated into a system. For the kits and single-board products, an internet search may turn up vendors.

Becoming a member of a VHF+ club is also a good way to find used and surplus equipment along with information about vendors and their products. These groups are typically great for mentoring and technical advice on transverters, as well as tips on operating.

14.5.1 Transverter Basics

A transverter is both a receive converter and a transmit converter for a conventional HF or VHF transceiver, usually operating on 10 meters or 2 meters. The HF/VHF transceiver acts similarly to the intermediate frequency (IF) sections of a conventional superheterodyne radio and is referred to as the *IF radio*. For those using a full-featured commercial transceiver, the various receive and transmit functions are also available on the converted band, creating a high-performance VHF/UHF/microwave (or MF/LF) transceiver. Some transceivers also have the ability to display the converted frequency so that the operator does not have to convert the lower frequency display to the higher band. (From this point on, discussion will assume the transverter is supporting a VHF/UHF/microwave band.)

An IF radio operating on the 28 MHz or 144 MHz band will cover either 2 or 4 MHz of the band that is being converted if the operator tunes from 28 to 30 MHz or 144 to 148 MHz. That is adequate to cover the entire weak signal portion of a VHF/UHF/microwave band. However, the amateur bands above 400 MHz are much wider than the 10 meter and 2 meter bands, so the entire band will not be covered. For FM repeater coverage of 432, 902, or 1296 MHz, a dedicated mobile or handheld FM transceiver is needed, since FM activity may cover a span of 20 MHz or more.

The 28 MHz band is commonly used as an IF for 50 MHz, 70 MHz (in Europe), and 144 MHz transverters, and sometimes for transverters on higher bands. The most common IF band for the higher bands is 144 MHz. For transverters on higher bands, 432 MHz is becoming more popular as an IF both to provide coverage of more of the band and to achieve better technical performance.

Even older transceivers and single-band 10 meter and 11 meter transceivers can provide reasonably good performance as IF radios. Low-power transceivers such as the Yaesu FT-817/818 are widely used because most transverters are now designed for up to 5 W of IF drive, nicely matching the output power of such transceivers.

14.5.2 Integrating a Transverter with a Transceiver

Figure 14.20 shows the two most common ways of connecting a transverter and transceiver IF radio. Some transceivers have dedicated low-level transmit outputs as in Figure 14.20A, often available through an auxiliary connector, that can be connected directly to a transverter regardless of the power available at the main antenna connector. (A dummy load may have to be connected during transverter operation.)

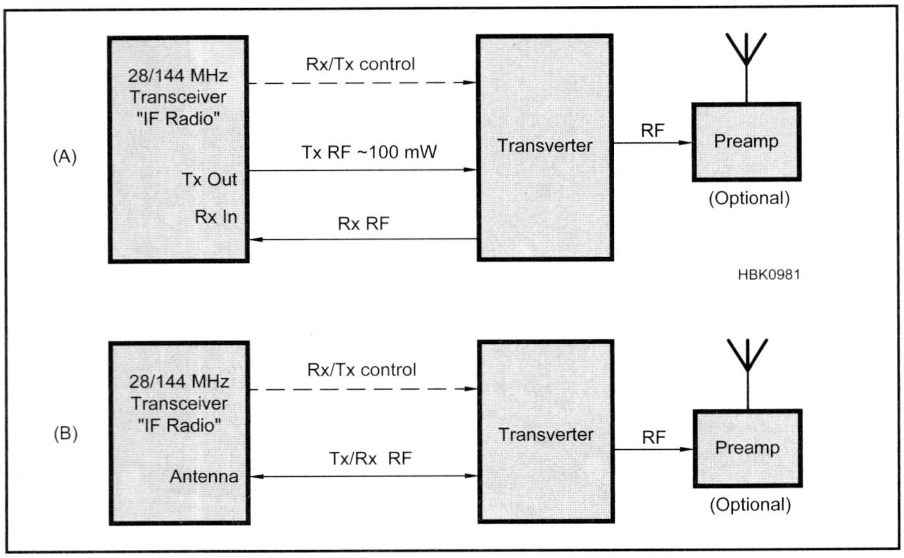

Figure 14.20 — The illustration at A shows a transceiver with receiver input and separate low-power transmit output connections. At B, the RF path is through a direct connection to the transceiver's main antenna connector, requiring the RF power output to be reduced to 5 W for most modern transverters.

If the transverter is rated to handle the power, a simpler approach is to turn the transceiver output power down to 5 W and connect its antenna connector to the IF port on the transverter that handles both transmit and receive signal paths. The transverter internally switches the IF port between the transmit and receive sections of the transverter. Most modern transverters have built-in attenuators that can handle up to 5 W of transmitter power from the transceiver. That's one reason why 5-W transceivers are so popular for use with transverters. This method is shown in Figure 14.20B. Of course, the operator has to remember to reduce the transmitter power of a higher power transceiver from the full output to 5 W when switching from "straight through" operation on 10 meters or 2 meters to transverter operation. Be aware that some transceivers transmit a full-power transient at the beginning of a transmission before the ALC system can control the output power. If this is the case, turning the power down manually is probably a poor option. **Table 14.1** shows some values for attenuators you can build to reduce transmit power.

If the transverter does not have the option to operate directly with the transceiver's main RF port, a TX/RX switch is required. Switching is controlled either by RF sensing or by the transceiver's transmit control line, such as a grounded PTT line or a positive-true TX ON signal. Performing switching control through the IF radio's RF port does eliminate one cable and connector set.

Optional preamps and power amplifiers may be mounted directly at the antenna for best performance. Because feed line losses are so high at microwave frequencies, many amateurs mount transverters and amplifiers in a weatherproof enclosure at the antenna and only send the IF signals, dc power, and any control lines from the transceiver.

Transverter users must also consider frequency stability, especially on the microwave bands. Most transverters are now designed to accept a highly stable 10 MHz "reference signal" from a *GPS-disciplined oscillator* or a *rubidium oscillator* (see the **Oscillators and Synthesizers** chapter). Commercial transverters usually have built-in local oscillators (LO) that are designed to be fairly stable. However, the best LO circuits will drift several kilohertz or more on microwave bands during warmup. Even after a long warmup, an internal LO may drift too much for modern digital modes. Most amateurs who want good performance will add an external GPSDO or Rubidium reference to "lock" their transverters on the correct frequency. One outboard reference signal can be fed into a distribution amplifier to serve several transverters.

Table 14.1
Attenuator Component Values

IF Power	10 W	6W	½ – 3 W	< 250 mW
First attenuator	23 dB	20 dB	18 dB	6 dB
R1	50 @ 8W	50 @ 4W	50 @ 2W	50
R2	27 * 4	43 * 4	56 * 4	100
R3	1.5k * 4	1k * 4	560 * 3 or 750 * 4	39
R4	56	62	62	150

Notes:
*4 means four resistors in parallel; * 3 means three resistors in parallel.
See Figure 14.25 for the attenuator schematic.
All resistances are in ohms and all resistors are ¼ W unless noted otherwise.

14.5.3 Basic Transverter Structure

Most radio amateurs now buy commercial transverters and install them as just described. This section is for those who wish to understand the inner workings of their transverters — or perhaps wish to build their own transverters.

The block diagrams of two typical transverter structures are shown in **Figures 14.21** and **14.22**. The transverter in Figure 14.21 has separate receive and transmit paths. A low-level output, typically 100 mW (20 dBm) or less, from the IF radio (labeled IF IN) is mixed with the transceiver's LO. The desired VHF mixing product is selected with the VHF band-pass filter and amplified by one or more low-power amplifier modules or ICs. This architecture requires the IF radio to supply a separate receiver input (typically labeled RX IN) and low-power transmit output (usually referred to as a "transverter output.")

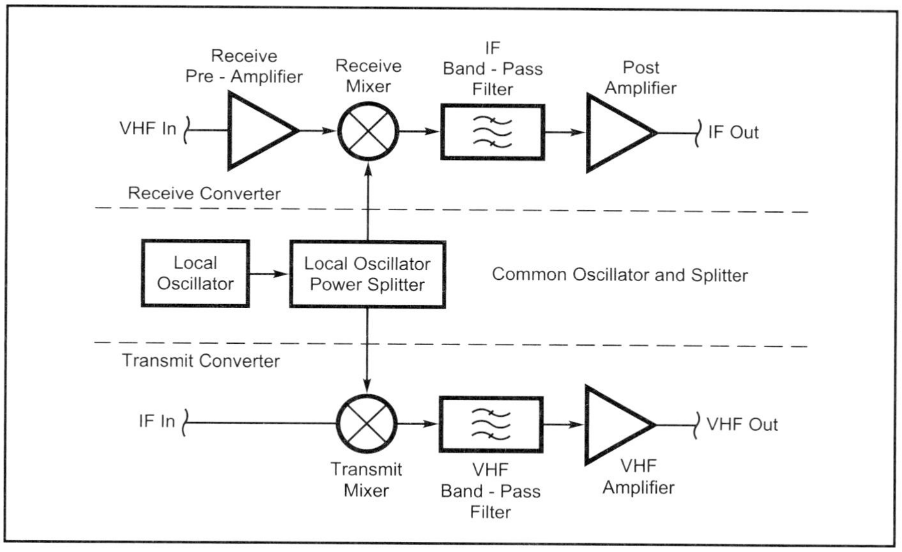

Figure 14.21 — Block diagram for a typical transverter. The upper chain is the receiving converter and the lower chain is the transmitting converter.

In Figure 14.22, the transceiver uses only a single *bilateral* mixer that converts signals flowing in either direction as transmit or receive. This allows the IF transceiver to operate as it normally does, with its internal transmit-receive (TR) switching routing signals. In the case of the 222 MHz transceiver, it is operating on the 24 MHz band. The transceiver output power must be attenuated or turned down to an acceptable level to avoid damaging the transverter's input circuits.

14.5.4 Mixers

The heart of a transverter is a mixer, sometimes called a "frequency mixer" or a "frequency changer," a nonlinear device that combines two signals at different input frequencies to produce new frequencies at the mixer output. These new output frequencies are the sum and difference of the original input frequencies. Mixers are discussed in detail in this book's **Receiving** chapter.

In the 222 MHz transverter of Figure 14.22, the mixer combines input signals with a 198 MHz local oscillator. To transmit, a 24-MHz signal is applied at the IF port from the transceiver; the sum frequency is 222 MHz, which is amplified and sent to the antenna. To receive, 222 MHz signals from the antenna combine with the local oscillator and the difference frequency, 24 MHz, is sent to the IF port on the 12 meter band.

A common choice for a transverter mixer is an inexpensive, double-balanced diode mixer. A high-level version will provide for better dynamic range (see the **Receiving** chapter). A wide dynamic range is important for rejecting out-of-band signals from other services, particularly TV, FM, and commercial/public safety signals, which are often present at the elevated locations common for portable operating.

14.5.5 Preamplifiers

Double-balanced mixers typically have a relatively high noise figure, so it is almost always necessary to use a low-noise preamp ahead of the mixer. The preamp can be a dis-

Figure 14.22 — Block diagram of a 222 MHz transverter.

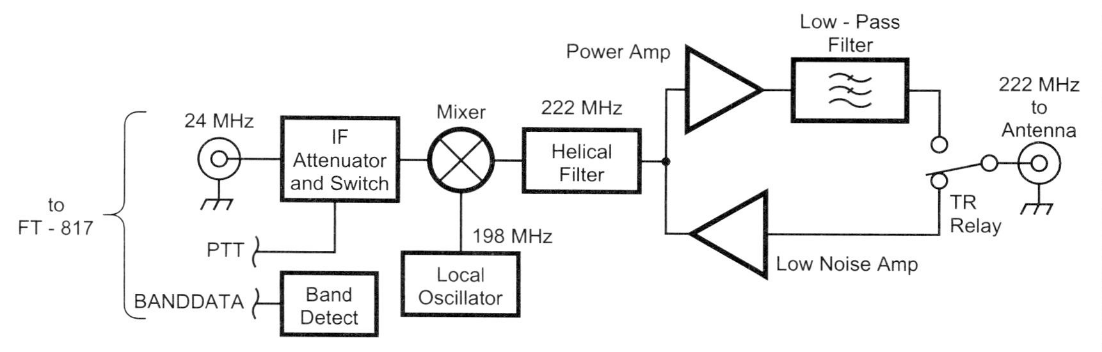

crete semiconductor circuit, or an MMIC RF amplifier can be used. If a preamp is used, a band-pass filter should be used at the input. (See the **Receiving** chapter for more information on preamps, including some discrete component design references.) A wide-band preamp design useful up to 1296 MHz is provided in this book's downloadable supplemental information. (See "Universal MMIC Preamp" by Paul Wade, W1GHZ)

External preamplifiers are often mounted directly at the antenna with power supplied via the main feed line. An RF sensing circuit or a separate control line from the transceiver or transverter is required to route transmit RF around the preamplifier. Some transverters can be mounted at the antenna as well, and may include a preamplifier. Both techniques improve the noise figure (see the **Receiving** chapter) of the receiving chain and also reduce the antenna system cost by allowing less-expensive coax to be used.

14.5.6 Filters

Transverters typically use a filter before and after the mixer. Comb filters, helical filters, and lumped LC filters are all used to provide band-pass responses. A low-pass filter is generally sufficient for the IF input and output, since the IF radio's filters will be used. Another low-pass filter at the output of the transverter power amplifier path reduces harmonics. A band-pass filter at the antenna port can serve the same function and keeps spurious signals from being transmitted on the air. It also helps prevent overload or intermodulation from strong out-of-band signals on receive.

Regardless of filter type, both filters and mixers behave much better when they are terminated in a load impedance that is constant over a wide bandwidth. Filters also behave better when their inputs and outputs are terminated by the impedance for which they were designed. A 3-dB pad made up of chip resistors at filter and mixer inputs and outputs allows each to achieve a reasonable impedance match, which makes both circuits perform more predictably.

Filter designs are available in the **Analog and Digital Filtering** chapter or can be designed using the *ELSIE* software package that is included with this book's downloadable supplemental information. If amplifier modules are used, some models are less stable with capacitor-input filters, so the inductor-input designs are recommended.

Microstrip line and cavity filters are most often used in amateur microwave transverters. Microstrip line filters consist of patterns of printed circuits that are frequency selective. Cavity filters are usually constructed from small plumbing parts that are soldered to circuit boards and then tuned with setscrews. Waveguide filters are also used, most notably at 10 GHz and higher. These consist of a piece of waveguide with metal rods and irises inserted. Tuning is performed with screws.

If portable operation is planned at hilltop sites where commercial or government transmitters are in operation, you should consider using external band-pass filters to reject the extremely strong signals that will be present. Be aware that noise from these transmitters (and possibly intermodulation products) may be quite strong compared to weaker amateur signals.

14.5.7 Local Oscillator

A good local oscillator (LO) is usually the hardest part of any transverter design. It must provide a clean signal, because any undesired frequencies generated by the LO can become a source of "birdies" in the receiver. It must also be stable and have low phase noise. Low phase noise is important, as a noisy local oscillator will raise the noise floor of the receiver, particularly in the presence of strong undesired signals, either out-of-band or in-band. Above 50 MHz, the interesting signals are usually the weak ones, so a low noise floor gets them in the log.

A crystal oscillator is the most obvious LO solution (see the **Oscillators and Synthesizers** chapter). Good crystals have become more expensive and harder to find, except those made for frequencies used in computer hardware — those are cheap and produced in high volume. Oscillator modules are the usual choice, having all of the circuitry optimized for the crystal and temperature stability. Packaged oscillators don't have any provisions for frequency adjustment, but because they are inexpensive, several can be ordered and the module closest to the desired frequency used. In addition, many wide-band SDR transceivers also have transverter outputs that can operate over a wide range of frequencies, allowing more choices for the LO frequency.

The oscillator frequency is usually not high enough to use directly, so it must be multiplied. **Figure 14.23** shows a typical scheme for generating a 10 GHz signal from a VHF crystal oscillator. For this oscillator to be reasonably temperature stable, a temperature stabilized miniature oven is usually used. After warmup, a simple ovenized 200 MHz oscillator stays within a few tenths of a hertz over the time that typical contacts are made. The 200-MHz oscillator is multiplied by approximately 50 to achieve an LO for a 10.386 GHz radio. One hertz of variation in the crystal oscillator will result in about 50 Hz change at 10 GHz, a noticeable change that may cause problems on some digital modes. If the LO is to drive an 80 GHz or 145 GHz radio, the sensitivity to drift is ten times more critical. (See K7MDL's "Multiband Microwave Local Oscillator" paper in the References at the end of this section for an approach to developing a stable signal source that is designed for transverter operation.)

While multiplier circuits can be difficult to design and adjust, the oscillator modules used for digital circuits have a square wave output. These oscillators have very fast rise and fall times, so they are rich in odd harmonics. Separating a third or fifth harmonic with a narrow filter or diplexer and amplifying it with an MMIC produces satisfactory output.

Microwave LOs in commercial and scientific converters are usually phase locked oscillators or PLOs. These items cost over a thousand dollars if purchased new. However, these PLO "bricks" are sometimes available at flea markets and auction sources at low cost. The negatives of using a surplus PLO include the effort to retune them, acquisition of a special crystal, greater weight, and the need for –20 V at 1 A of power that most of them require. The positives include excellent phase noise, great stability after warm up, and low cost. Synthesized LO modules with low phase noise are also available, such as the digiLO from **g5systems.com**. Most of these can also be locked to a GPS-disciplined 10 MHz reference for very accurate and stable frequency control.

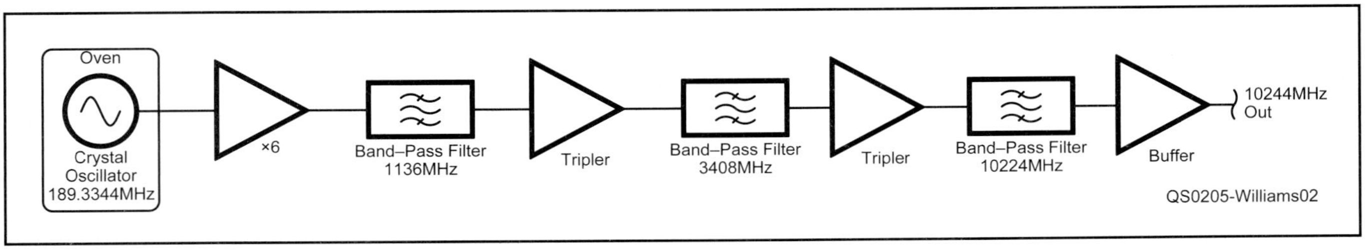

Figure 14.23 — A local oscillator multiplier chain for a 10 GHz transverter.

A high-stability TCXO is highly recommended in the IF radio for use with transverters. For the microwave bands, LO oscillator drift and microphonics can be a problem. It is important to keep the IF radio and the transverter at a constant temperature and avoid vibration. As noted earlier, most modern transverters have a provision for an external frequency reference that will lock the transverter on frequency with an accuracy of plus or minus only a few hertz even at 10 GHz.

14.5.8 Transceiver Power Output Control

Assuming your transceiver can be set to output ½ W or less, perhaps through a dedicated transverter output port, ordinary ¼ W resistors can be used as an input attenuator which may be fixed or variable. Be sure to test your attenuator design before using it between the transceiver and transverter. In the 222 MHz transverter shown in Figure 14.22, two attenuators are used in series, one fixed and one variable. PIN diodes are used to bypass the variable attenuator on receive, but a smaller fixed attenuator is left in the line to protect the mixer in case of a switching failure.

Most modern commercial transverters have built-in attenuators so they can handle the full output power of a 5-W transceiver. If your transmitter outputs 100 W, however, it's up to you to remember to turn the power down every time you switch from straight-through HF or 2 meter operation to transverter operation. If you forget to reduce the power in the heat of battle to try to work a new multiplier, you may not only miss the multiplier but also fry the input circuitry of your transverter and set yourself up for a costly repair.

Another option for power control is to use an external dc voltage to control the transceiver's ALC function. For most transceivers, a negative dc voltage applied to the ALC input can reduce the output power to nearly zero. The ALC input is a high-impedance circuit so little current is required. **Figure 14.24** shows a simple circuit that will work with most transceivers. The Inverting DC-DC Converter project in the downloadable supplemental information for the **Power Sources** chapter will also work and can be connected to the radio's power supply.

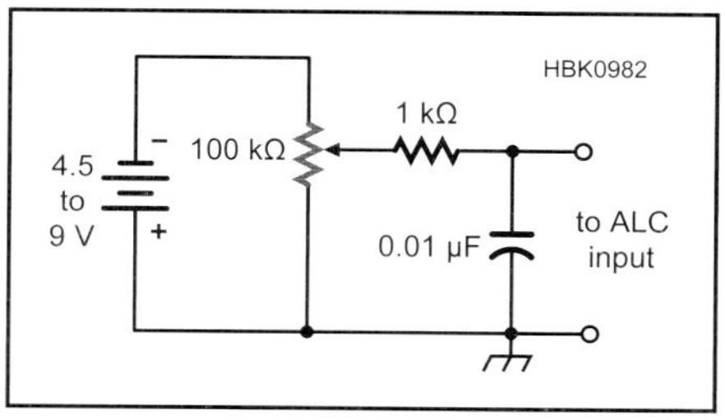

Figure 14.24 — Negative voltage ALC power control circuit. Note the battery polarity and chassis connection. The 0.01 µF capacitor filters out any RF that might be present, and the 1 kΩ resistor limits current into the ALC input.

14.5.9 Power Amplification

Transverter modules commercially available as kits have an output of 10 to 100 mW (10 to 20 dBm) and usually require additional amplification to be able to fully drive a stand-alone solid-state amplifier to the 10 to 50 W level. RF amplifier modules designed for commercial service are available to amplify the low-level signals to useful power levels.

The power level at the output of a mixer/buffer might be as low as 100 µW (–10 dBm) and must be increased to develop useful output power levels. MMICs are an easy and inexpensive way to raise the power level to 20 – 100 mW (+13 dBm to +20 dBm) required to drive higher power devices or modules. Gain at and above UHF has become much less expensive in recent years due to consumer devices operating at these frequencies.

Hybrid modular amplifiers can be used to easily reach the 10 – 25 W level. While they seem a bit expensive, they are easy to use, and usually the only other components required are a few capacitors and chokes. Higher power modules require more current and heat-sinking, increasing the size and weight of the transverter. Amplifier modules made by Mitsubishi /Toshiba (**www.mitsubishielectric.com/semiconductors/products/hf/sirfpowermod/index.html**) are very common in amateur VHF/UHF FM transceivers and can also be used for transverters. Remember that the module must be capable of linear operation for use with SSB and AFSK modulation. Vendors such as RF Parts (**www.rfparts.com**) can supply modules in small quantities. (Also see the RF Power Amplifier Modules listing in the **Component Data and References** chapter.)

14.5.10 TR Switching and Sequencers

Just as in a transceiver, the transmit-receive (TR) functions must be switched at the appropriate time. The simplest approach is to use a relay activated by the transceiver's PTT circuit to switch the input and output circuits. If the timing of the PTT and RF output from the transceiver is well-behaved and some protective attenuation is provided at the receiver connections, this approach may work fine.

However, applying RF from a transmitter before the receive circuits have been disconnected can destroy a sensitive receive preamplifier or mixer. Many commercial transceivers can transmit power immediately upon being activated, before any external circuits can switch from transmit to receive. The unwanted power can burn out mixers and preamps, as well as coaxial relays that are not meant to hot-switch RF power — it doesn't take much power to damage a relay contact while it is switching. Another common problem is that some transceivers output a high-power transient until the ALC loop regains control of the transmitter power. This is addressed by using a sequencer. Several manufacturers such as Kuhne Electronic sell highly reliable sequencers for a modest price to address these problems. If you use one of these and carefully follow the directions, you should be fine.

Another pitfall in the TR switching process is caused by the relay itself. The coil that activates a relay may send a powerful voltage spike back into the transverter when it is de-energized, with devastating consequences for the low-level circuitry. Read the instructions from the transverter manufacturer carefully. If it says to add a protective diode across the relay coil, follow those instructions literally, observing the recommended polarity. (A small bypass capacitor across the diode will keep it from generating harmonics of any RF that is present.) If it doesn't say to take that precaution, consider doing it anyway.

Some approaches involve trying to delay the transceiver output, either by intercepting the PTT line from the microphone or keying devices or by controlling a Transmitter Inhibit input to the transceiver. Both of these approaches work well but need additional connections to the transceiver, are model-specific, and present additional points of possible failure.

A simpler approach is to have a single connection from transceiver to transverter, with PTT voltage added to the IF transceiver output feed line. A microcontroller can then monitor the switching operation. The article "A Smart Fool-resistant Conditional Sequencer" by Paul Wade, W1GHZ (see this book's downloadable supplemental information for the complete article) describes the Arduino-based controller (an Arduino Pro Trinket) suitable for adaptation to many transverter circuits. For example, an RF sensor can detect any transmitted power and switches the input circuit to a safe state to absorb unwanted power until the sequencer finishes all switching.

In **Figure 14.25**, the PIN diode switching circuit is controlled by the signal lines TXE (Transmit Enable) and RXD (Receive Disable). Both controlling signals for the PIN switch are asserted High. To receive, both are de-asserted. RXD is asserted, but not TXE, for the SAFE state, and then TXE is asserted along with RXD to transmit. TXE should never be asserted without RXD, a combination that would enable both transmitter and receiver simultaneously.

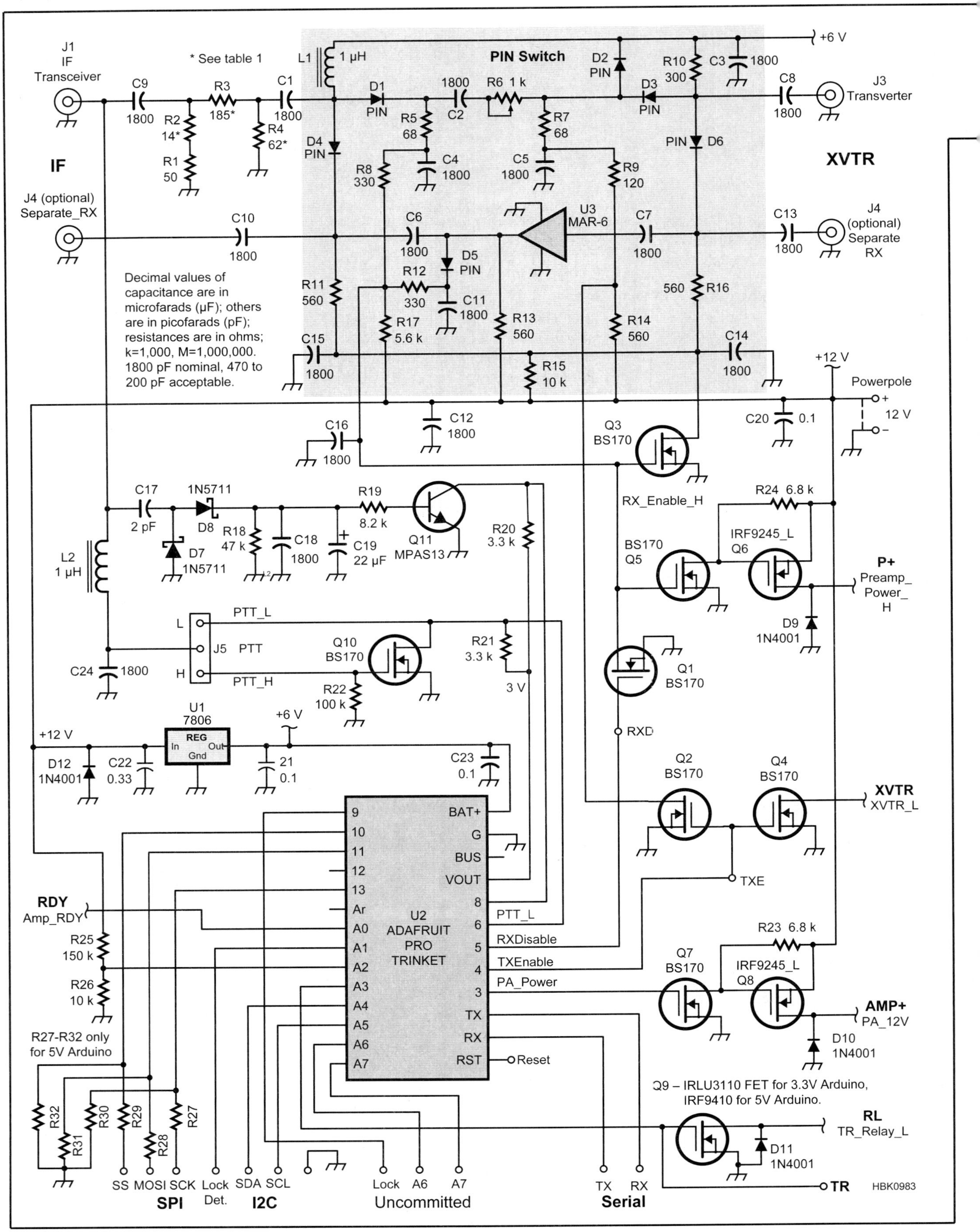
J1
IF
Transceiver
* See table 1
C9
1800
R3
185*
C1
1800
R2
14*
R1
50
R4
62*
IF
L1
1 μH
PIN Switch
D1
PIN
1800
C2
R6 1 k
D2
PIN
D3
PIN
R10
300
C3
1800
+6 V
C8
1800
J3
Transverter
R5
68
R7
68
D4
PIN
PIN
D6
XVTR
R8
330
C4
1800
C5
1800
R9
120
J4 (optional)
Separate_RX
C10
1800
C6
1800
U3
MAR-6
C7
1800
C13
1800
J4
(optional)
Separate
RX
D5
PIN
R12
330
C11
1800
R11
560
560
R16
Decimal values of capacitance are in microfarads (μF); others are in picofarads (pF); resistances are in ohms; k=1,000, M=1,000,000. 1800 pF nominal, 470 to 200 pF acceptable.
R17
5.6 k
R13
560
R14
560
C15
1800
C14
1800
R15
10 k
+12 V
Powerpole
+
12 V
–
C16
1800
C12
1800
C20
0.1
Q3
BS170
RX_Enable_H
C17
2 pF
1N5711
D8
R19
8.2 k
Q11
MPAS13
R20
3.3 k
R24 6.8 k
L2
1 μH
D7
1N5711
R18
47 k
C18
1800
C19
22 μF
BS170
Q5
IRF9245_L
Q6
P+
Preamp_
Power_
H
D9
1N4001
L
J5 PTT
H
PTT_L
PTT_H
Q10
BS170
R21
3.3 k
3 V
C24
1800
R22
100 k
Q1
BS170
RXD
U1
7806
+12 V
REG
In
Out
Gnd
+6 V
D12
1N4001
C22
0.33
21
0.1
C23
0.1
Q2
BS170
Q4
BS170
XVTR
XVTR_L
9
10
11
12
13
Ar
A0
A1
A2
A3
A4
A5
A6
A7
BAT+
G
BUS
VOUT
8
6
5
4
3
TX
RX
RST
U2
ADAFRUIT
PRO
TRINKET
TXE
PTT_L
RXDisable
TXEnable
PA_Power
RDY
Amp_RDY
R25
150 k
R26
10 k
R23 6.8 k
Q7
BS170
IRF9245_L
Q8
AMP+
PA_12V
D10
1N4001
Reset
R27-R32 only
for 5V Arduino
Q9 – IRLU3110 FET for 3.3V Arduino,
IRF9410 for 5V Arduino.
R32
R31
R30
R29
R28
R27
RL
TR_Relay_L
D11
1N4001
SS
MOSI
SCK
SPI
Lock
Det.
SDA
SCL
I2C
Lock
A6
A7
Uncommitted
TX
RX
Serial
TR
HBK0983

Figure 14.25 — A processor-controlled sequencer. The PIN switch circuitry at the top is controlled by the Arduino processor through the TXE (Transmit Enable) and RXD (Receive Disable) control lines.

14.5.11 Transverter Systems

Building a separate transverter for each band has been the norm for decades, but recent technological advances are making multiband transverters possible. For example, Down East Microwave is in the final pre-production stages (as of November 2019) of producing a six-band transverter covering 902, 1296, 2304, 3456, 5760, and 10368 MHz requiring only a single 144-MHz IF radio. Wideband SDR-based radios are also becoming available, moving transceiver performance to the UHF and microwave bands, with transverter opportunities extending into the millimeter-wave spectrum.

Another approach is taken by Wayne Overbeck, N6NB, in creating his "Ten-Band Toolbox Transverter" system designed for roving in VHF+ contests. (The full article describing the transverter system is available online; see the References list at the end of this section.) The system consists of two packages: an IF radio and control unit which goes inside the vehicle (see **Figure 14.26**) and a transverter package designed for mounting on top of the vehicle, **Figures 14.27** and **14.28**. The transverter package is integrated with antennas so that the whole package is easy to install and remove.

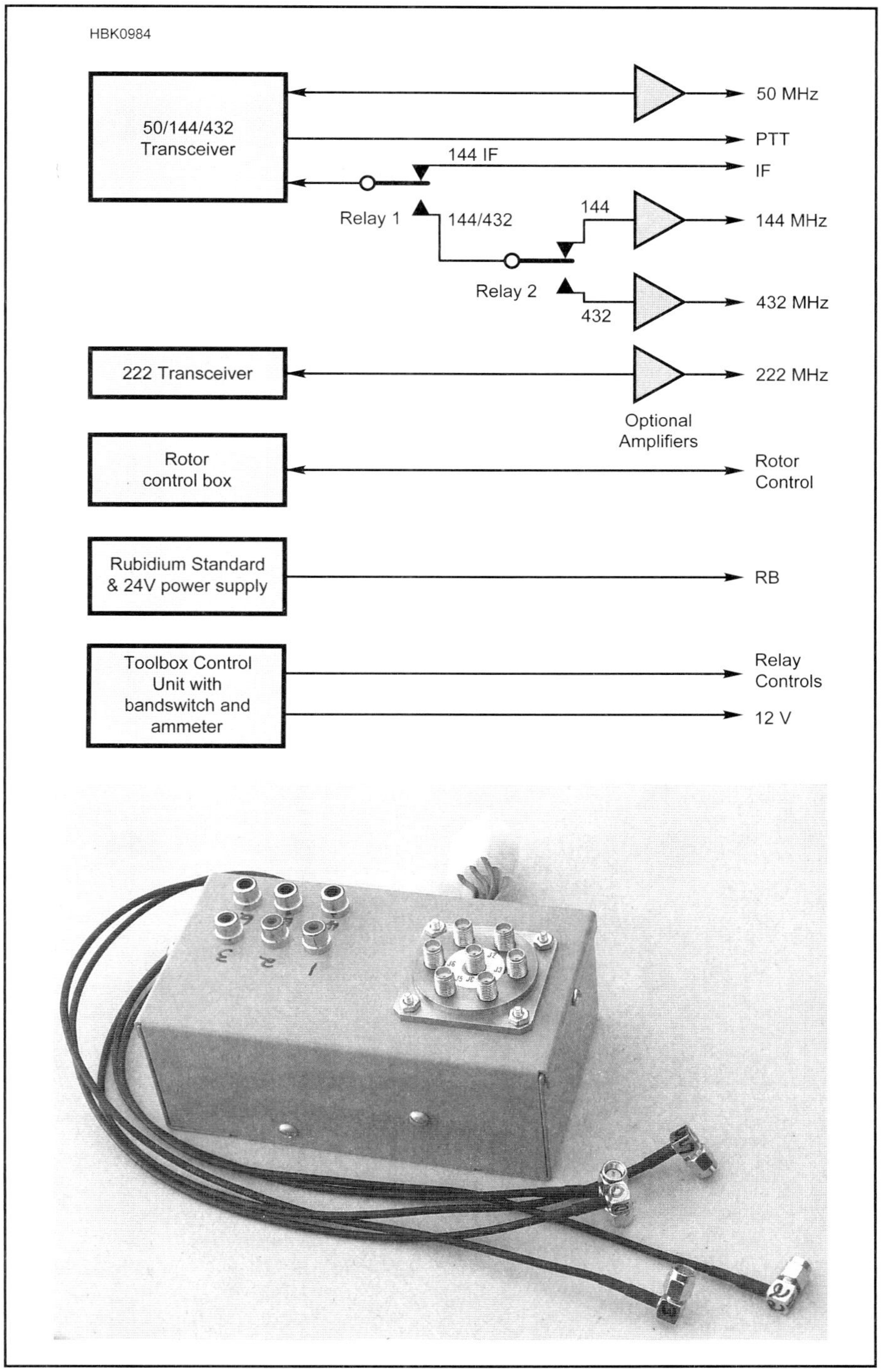

Figure 14.26 — Control unit for multiband transverter system by N6NB. The switch enclosure goes inside the toolbox or other microwave equipment box. It switches the IF, PTT line, and rubidium standard reference signal through the six bands from 902 MHz to 10 GHz.

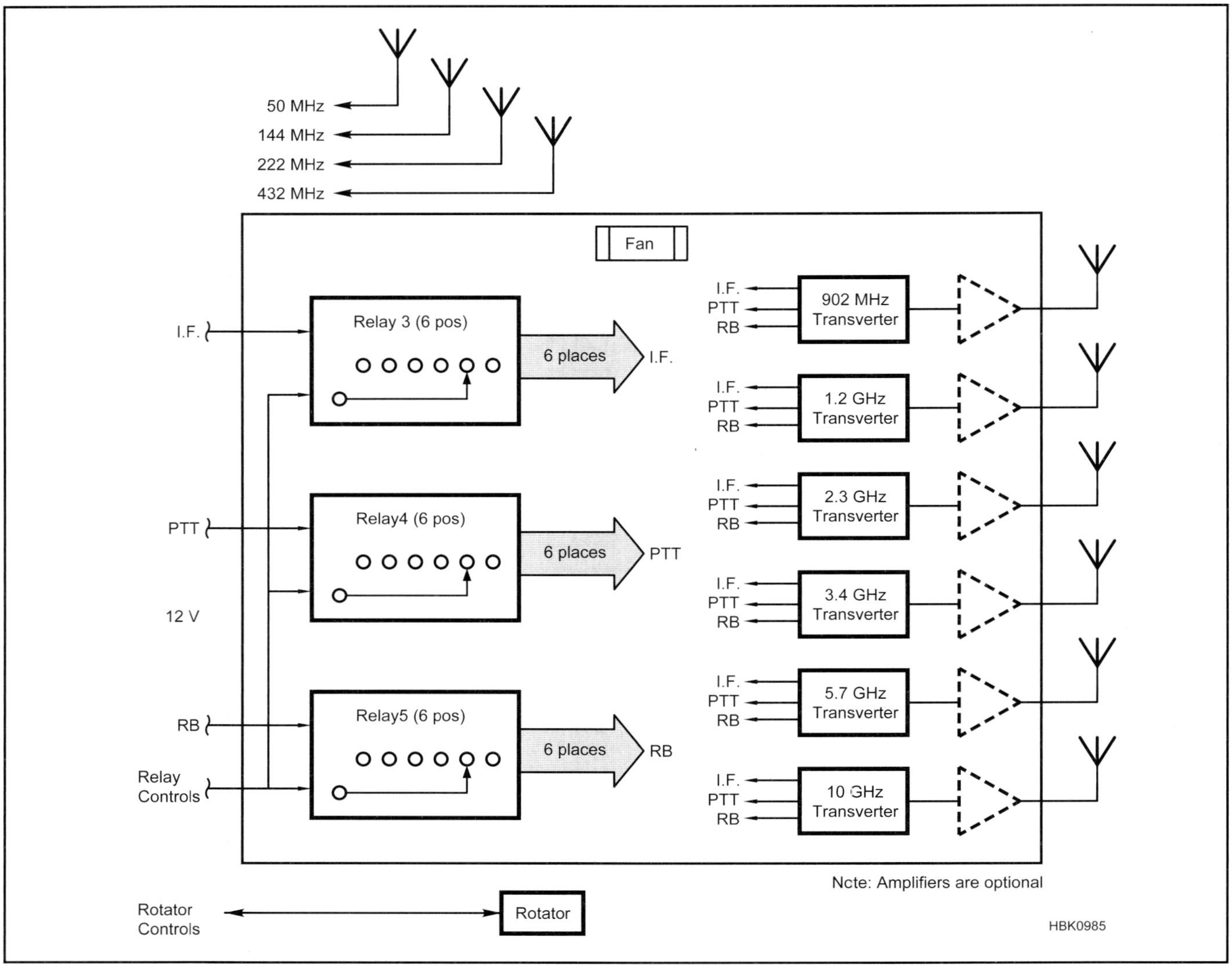

Figure 14.27 — Block diagram of integrated transverter package.

This package can be used as a fixed or mobile/rover station and covers all bands from 50 MHz to 10 GHz with a single transceiver and a group of transverters. A six-position or eight-position rotary switch is mounted in a small box to serve as a band switch. One set of its contacts is used to switch the PTT control line (the TR switching line) between transverters. Another set of contacts switches a dc voltage (usually either 12 V or 28 V) to control a six-position SMA relay (available on the surplus market) that routes the 144 MHz IF signals between transverters. The result is a multiband station with rapid single-knob band switching.

If additional switch positions are available, the relay can also be configured to transfer the IF radio's output directly to an antenna for straight-through operation on 144 or 432 MHz. Sometimes a 222 MHz transverter is also connected to this package, although that often requires switching the IF rig from 144 MHz to 28 MHz, the usual IF for a 222 MHz transverter. That's why many VHF operators have a separate transceiver and transverter for 222 MHz.

Figure 14.28 — Seven transverter packages designed to be mounted on the roof of the vehicle with antennas attached.

14.5.12 Transverter References

2200 and 630 Meter Transmit Converters and Amplifiers

njdtechnologies.net/monitor-sensors-630-meter-and-2200-meter-linear-transverters

Wilson, M., K1RO, "Monitor Sensors TVTR1 630-Meter Transverter," *QST*, July 2018, pp. 44 –48.

Wilson, M., K1RO, "Monitor Sensors TVTR2 2200-Meter Transverter," *QST*, Oct. 2019, pp. 47 – 49.

www.monitorsensors.com/ham-radio

www.n1bug.com/lfmf

Articles and Papers

Kostro, S., N2CEI, "Multiband VHF/UHF and Microwave Transverters, Rev C," Down East Microwave, Inc., **www.pnwvhfs.org/conference/2018/pdf/Steve-N2CEI-Multiband-Transverters- RevC.pdf**.

Lau, Z., KH6CP/W1VT, "Home-Brewing a 10-GHz SSB/CW Transverter," *QST*, Part 1, May 1993; Part 2, June 1993.

Lau, Z., KH6CP/W1VT, "RF" column, *QEX*.

Lewis, M., K7MDL, "Multiband Microwave Local Oscillator," **pnwvhfs.org/conference/2019/pdf/Multiband-LO- Project-by-Mike-K7MDL.pdf**.

Lewis, M., K7MDL, "Multi-Band VHF+ Local Oscillator, Decoder, and Remote Antenna Switch," *2019 Central States VHF Conference Proceedings*.

Overbeck, W., N6NB, "10-Band Toolbox Stations for Roving," **n6nb.com/toolbox.htm**.

Overbeck, W., N6NB, "Small 10-Band Rover Stations that a Septuagenarian Can Lift and Roll Aboard a Plane," **n6nb.com/small_10_band_rovers_with_up_to_seven_transverters.pdf**.

w1ghz.org, W1GHZ Microwave page.

Wade, P., W1GHZ, "432 MHz Transverter for an SDR," *2019 Central States VHF Conference Proceedings*.

Wade, P., W1GHZ, "2304 and 5760 GHz Transverters," *2016 Microwave Update Proceedings*, pp. 104 – 106 and 110 – 122.

www.downeastmicrowave.com, DEMI Knowledge Base.

Contents

Chapter 15 — Online Content

Articles and Data Tables

- ASCII Character Code Table
- Digital Modes — Lowest Permitted Amateur Frequency
- High-Speed Wireless Networking in the UHF and Microwave Bands by David Bern, W2LNX, and Keith Elkin, KB3TCB
- ITA2 Character Code Table
- Legacy Digital Modes
- Operating Tips for Digital Voice Using *FreeDV* by Mel Whitten, KØPFX
- The FT4 and FT8 Communication Protocols by Steve Franke, K9AN, Bill Somerville, G4WJS, and Joe Taylor, K1JT
- Varicode tables for PSK31, MFSK16 and DominoEX

Chapter 15

Digital Protocols and Modes

Since adapting radioteletype (RTTY) after World War II, radio amateurs have been creating new digital modes to communicate both as "chat-style" keyboard-to-keyboard operation or using packetized network-compatible protocols. Many new protocols have been entirely implemented using software and a computer sound card, making tools for experimentation and implementation easily available worldwide via the internet. SDR technology allows digital protocols and modes to be implemented directly without an intervening audio modulation/demodulation step. It is an exciting period for amateur radio with frequent innovations and improvements.

This chapter will focus on the protocols for transferring various data types, focusing on the processes by which data is encoded, compressed and error checked, packaged and exchanged. Joe Taylor, K1JT, and Steve Franke, K9AN, updated the sections on *WSJT-X* modes. HF Digital Voice was updated by Mel Whitten, KØPFX. The sections on VARA and ARDOP were contributed by Steve Stroh, N8GNJ. Packet radio and APRS material was updated by John Langner, WB2OSZ and Lynn Deffenbaugh, KJ4ERJ. Al Francisco, K7NHV updated the section on Winlink. The section on D-STAR was rewritten by John Davis, WB4QDX and Ed Woodrick, WA4YIH. DMR information was updated by John Burningham, W2XAB. Cory Sickles, WA3UVV updated the System Fusion section. The new section on IP Microwave Networking was contributed by Orv Beach, W6BI. Ken Humbertson, WØKAH updated the sidebar on *fldigi* software.

Modulation methods are covered in the **Modulation** chapter, and the **Digital Communications** chapter available with the online content discusses the practical considerations of operating using these modes.

There is a broad array of digital modes to service various needs with more coming. The most basic modes simply encode text data for transmission in a keyboard-to-keyboard chat-type environment. These modes may or may not include any mechanism for error detection or correction. The second class of modes are generally more robust and support more sophisticated data types by structuring the data sent and including additional error-correction information to properly reconstruct the data at the receiving end. The third class of modes discussed will be networking modes with protocols often the same or similar to versions used on the Internet and computer networks.

15.1 Digital "Modes"

The ITU uses *Emission Designators* to define a "mode" as demonstrated in the **Modulation** chapter. These designators include the bandwidth, modulation type and information being sent. This system works well to describe the physical characteristics of the modulation, but digital modes create some ambiguity because the type of information sent could be text, image or even the audio of a CW session. As an example, an FM data transmission of 20K0F3D could transmit spoken audio (like FM 20K0F3E or 2K5J3E) or a CW signal (like 150H0A1A). These designators don't help identify the type of data supported by a particular mode, the speed that data can be sent, if it's error-corrected, or how well it might perform in hostile band conditions. Digital modes have more characteristics that define them and there are often many variations on a single mode that are optimized for different conditions. We'll need to look at the specifics of these unique characteristics to be able to determine which digital modes offer the best combination of features for any given application.

15.1.1 Symbols, Baud, Bits, and Bandwidth

The basic performance measure of a digital mode is the *data rate*. This can be measured a number of ways and is often confused. Each change of state on a transmission medium defines a *symbol* and the *symbol rate* is also known as *baud*. (While commonly used, "baud rate" is redundant because "baud" is already defined as the rate of symbols/second.) Modulating a carrier increases the frequency range, or bandwidth, it occupies. The FCC currently limits digital modes by symbol rate on the various bands as an indirect (but easily measurable) means of controlling bandwidth.

The *bit rate* is the product of the *symbol rate* and the number of bits encoded in each symbol. In a simple two-state system like an RS-232 interface, the bit rate will be the same as baud. More complex waveforms can represent more than two states with a single symbol so the bit rate will be higher than the baud. For each additional bit encoded in a symbol, the number of states of the carrier doubles. This makes each state less distinct from the others, which in turn makes it more difficult for the receiver to detect each symbol correctly in the presence of noise. A V.34 modem may transmit symbols at a baud rate of 3420 baud and each symbol can represent up to 10 discrete states or bits, resulting in a *gross* (or *raw*) *bit rate* of 3420 baud × 10 or 34,200 bits per second (bit/s). Framing bits and other overhead reduce the *net bit rate* to 33,800 bit/s.

Bits per second is abbreviated here as *bit/s* for clarity but is also often seen as *bps*. Bits per second is useful when looking at the protocol but is less helpful determining how long it

Table 15.1
Data Rate Symbols and Multipliers

Name	*Symbol*	*Multiplier*
kilobit per second	kbit/s or kbps	1000 or 10^3
Megabit per second	Mbit/s or Mbps	1,000,000 or 10^6
Gigabit per second	Gbit/s or Gbps	1,000,000,000 or 10^9

takes to transmit a specific size file because the number of bits consumed by overhead is often unknown. A more useful measure for calculating transmission times is *bytes per second* or *Bps* (note the capitalization). Although there are only eight bits per byte, with the addition of start and stop bits, the difference between bps and Bps is often tenfold. Since the net bit rate takes the fixed overhead into account, Bytes per second can be calculated as $bps_{net}/8$. Higher data rates can be expressed with their metric multipliers as shown in **Table 15.1**.

Digital modes constantly balance the relationship between symbol rate, bit rate, bandwidth and the effect of noise. The Shannon-Hartley theorem demonstrates the maximum channel capacity in the presence of Gaussian white noise and was discussed in the **Modulation** chapter in the Channel Capacity section. This theorem describes how an increased symbol rate will require an increase in bandwidth and how a reduced signal-to-noise ratio (SNR) will reduce the potential *throughput* of the channel.

15.1.2 Error Detection and Correction

Voice modes require the operator to manually request a repeat of any information required but not understood. Using proper phonetics makes the information more easily understood but takes longer to transmit. If 100% accuracy is required, it may be necessary for the receiver to repeat the entire message back to the sender for verification. Computers can't necessarily distinguish between valuable and unnecessary data or identify likely errors but they offer other options to detect and correct errors.

ERROR DETECTION

The first requirement of any accurate system is to be able to detect when an error has occurred. The simplest method is *parity*. With 7-bit ASCII data, it was common to transmit an additional 8th parity bit to each character. The parity bit was added to make the total number of 1 bits odd or even. The binary representation for an ASCII letter Z is 1011010. Sent as seven bits with even parity, the parity bit would be 0 because there are already an even number of 1 bits and the result would be 01011010. The limitation of parity is that it only works with an odd number of bit inversions. If the last two bits were flipped to 01011001 (the ASCII letter Y), it would still pass the parity check because it still has an even number of bits. Parity is also rarely used on 8-bit data so it cannot be used when transferring binary data files.

Checksum is a method similar to the "check" value in an NTS message. It is generally a single byte (8-bit) value appended to the end of a packet or frame of data. It is calculated by adding all the values in the packet and taking the least significant (most unique) byte. This is a simple operation for even basic processors to perform quickly but can also be easily mislead. If two errors occur in the packet of equal amounts in the opposite direction (A becomes B and Z becomes Y), the checksum value will still be accurate and the packet will be accepted as error-free.

Cyclic redundancy check (CRC) is similar to checksum but uses a more sophisticated formula for calculating the check value of a packet. The formula most closely resembles long division, where the quotient is thrown away and the remainder is used. It is also common for CRC values to be more than a single byte, making the value more unique and likely to identify an error. Although other error detection systems are currently in use, CRC is the most common.

ERROR CORRECTION

There are two basic ways to design a protocol for an error correcting system: *automatic repeat request (ARQ)* and *forward error correction (FEC)*. With ARQ the transmitter sends the data packet with an error detection code, which the receiver uses to check for errors. The receiver will request retransmission of packets with errors or sends an acknowledgement (ACK) of correctly received data, and the transmitter re-sends anything not acknowledged within a reasonable period of time.

With forward error correction (FEC), the transmitter encodes the data with an *error-correcting code (ECC)* and sends the encoded message. The receiver is not required to send any messages back to the transmitter. The receiver decodes what it receives into the "most likely" data. The codes are designed so that it would take an "unreasonable" amount of noise to trick the receiver into misinterpreting the data. It is possible to combine the two, so that minor errors are corrected without retransmission, and major errors are detected and a retransmission requested. The combination is called *hybrid automatic repeat request (hybrid ARQ)*.

There are many error correcting code (ECC) algorithms available. Extended Golay coding is used on blocks of ALE data, for example, as described in the section below on G-TOR. In addition to the ability to detect and correct errors in the data packets, the modulation scheme allows sending multiple data streams and interleaving the data in such a way that a noise burst will disrupt the data at different points.

15.1.3 Data Representations

When comparing digital modes, it is important to understand how the data is conveyed. There are inherent limitations in any method chosen. PSK31 might seem a good choice for sending data over HF links because it performs well, but it was only designed for text (not 8-bit data) and has no inherent error correction. It is certainly possible to use this modulation scheme to send 8-bit data and add error correction to create a new mode. This would maintain the weak signal performance but the speed will suffer from the increased overhead. Similarly, a digital photo sent via analog SSTV software may only take two minutes to send, but over VHF packet it could take 10 minutes, despite the higher speed of a packet system. This doesn't mean SSTV is more efficient. Analog SSTV systems generally transmit lower resolution images with no error correction. Over good local links, the VHF packet system will be able to deliver perfect images faster or of higher quality.

TEXT REPRESENTATIONS

Morse code is well known as an early code used to send text data over a wire, then over the air. Each letter/number or symbol is represented with a varying length code with the more common letters having shorter codes. This early *varicode* system is very efficient and minimizes the number of state changes required to send a message.

The Baudot code (pronounced "bawd-OH") was invented by Émile Baudot and is the predecessor to the character set currently known more accurately as International Telegraph Alphabet No 2 (ITA2). This code is used for radioteletype communications and contains five bits with start and stop pulses. This only allows for 2^5 or 32 possible characters to be sent, which is not enough for all 26 letters plus numbers and characters. To resolve this, ITA2 uses a LTRS code to select a table of upper case (only) letters and a FIGS code to select a table of numbers, punctuation and special symbols. The code is defined in the ITA2 codes table with the online content.

Early computers used a wide variety of alphabetic codes until the early 1960s until the advent of the American Standard Code for Information Interchange or ASCII (pronounced "ESS-key"). At that point many computers standardized on this character set to allow simple transfer of data between machines. ASCII is a 7-bit code which allows for 2^7 or 128 characters. It was designed without the *control characters* used by Baudot for more reliable transmissions and the letters appear in English alphabetical order for easy sorting. The code can be reduced to only six bits and still carry numbers and uppercase letters. Current FCC regulations provide that amateur use of ASCII conform to ASCII as defined in ANSI standard X3.4-1977. The international counterparts are ISO 646-1983 and International Alphabet No. 5 (IA5) as published in ITU-T Recommendation V.3. A table of ASCII characters is presented as "ASCII Character Set" with the online content.

ASCII has been modified and initially expanded to eight bits, allowing the addition of foreign characters or line segments. The different extended versions were often referred to as *code pages*. The IBM PC supported code page 437 which offers line segments, and English *Windows* natively supports code page 1252 with additional foreign characters and symbols. All of these extended code pages include the same first 128 ASCII characters for backward compatibility. In the early 1990s efforts were made to support more languages directly and *Unicode* was created. Unicode generally requires 16 bits per character and can represent nearly any language, becoming the standard for internet, web pages, and software.

More recent schemes use varicode, where the most common characters are given shorter codes (see **en.wikipedia.org/wiki/Prefix_code**). Varicode is used in PSK31and MFSK to reduce the number of bits in a message. Although the PSK31 varicode contains all 128 ASCII characters, lower-case letters contain fewer bits and can be sent more quickly. Tables of PSK and MFSK varicode characters are included with the online content.

IMAGE DATA REPRESENTATIONS

Images are generally broken into two basic types, *raster* and *vector*. Raster or *bitmap* images are simply rows of colored points known as *pixels* (picture elements). Vector images are a set of drawing primitives or instructions. These instructions define shapes, placement, size and color. Similar coding is used with plotters to command the pens to create the desired image.

Bitmap images can be stored at various *color depths* or bits per pixel indicating how many colors can be represented. Common color depths are 1-bit (2 colors), 4-bit (16 colors), 8-bit (256 colors), 16-bit (65,536 colors also called *high color*) and 24-bit (16 million also called *true color*). True color is most commonly used with digital cameras and conveniently provides eight bits of resolution each for the red, green and blue colors. Newer scanners and other systems will often generate 30-bit (2^{30} colors) and 36-bit (2^{36} colors). Images with 4- and 8-bit color can store more accurate images than might be obvious because they include a *palette* where each of their 16 or 256 colors respectively are chosen from a palette of 16 million. This palette is stored in the file which works well for simple images. The GIF format only supports 256 colors (8-bit) with a palette and lossless compression.

Digital photographs are raster images at a specific resolution and color depth. A typical low resolution digital camera image would be 640×480 pixels with 24-bit color. The 24-bit color indicated for each pixel requires three bytes to store (24 bits / 8 bits per byte) and can represent one of a possible 2^{24} or 16,777,216 colors with eight bits each for red, green and blue intensities. The raw (uncompressed) storage requirement for this image would be 640×480x3 or 921,600 bytes. This relatively low resolution image would require significant time to transmit over a slow link.

Vector images are generally created with drawing or CAD packages and can offer significant detail while remaining quite small. Because vector files are simply drawn at the desired resolution, there is no degradation of the image if the size is changed and the storage requirements remain the same at any resolution. Typical drawing primitives include lines and polylines, polygons, circles and ellipses, Bézier curves and text. Even computer font technologies such as TrueType create each letter from Bézier curves allowing for flawless scaling to any size and resolution on a screen or printer.

Raster images can be resized by dropping or adding pixels or changing the color depth. The 640×480×24-bit color image mentioned above could be reduced to 320×240×16-bit color with a raw size of 153,600 bytes — a significant saving over the 921,600 byte original. If the image is intended for screen display and doesn't require significant detail, that size may be appropriate. If the image is printed full sized on a typical 300 dpi (dots per inch) printer, each pixel in the photo will explode to nearly 100 dots on the printer and appear very blocky or pixilated.

AUDIO DATA REPRESENTATIONS

Like images, audio can be stored as a sampled waveform or in some type of primitive format. Storing a sampled waveform is the most versatile but can also require substantial storage capacity. MIDI (musical instrument digital interface, pronounced "MID-ee") is a common music format that stores instrument, note, tempo and intensity information as a musical score. There are also voice coding techniques that store speech as *allophones* (basic human speech sounds).

As with images, storage of primitives can save storage space (and transmission times) but they are not as rich as a high quality sampled waveform. Unfortunately, the 44,100 Hz 16-bit sample rate of an audio compact disc (CD) requires 176,400 bytes to store each second and 10,584,000 bytes for each minute of stereo audio.

The Nyquist-Shannon sampling theorem states that perfect reconstruction of a signal is possible when the sampling frequency is at least twice the maximum frequency of the signal being sampled. With its 44,100 Hz sample rate, CD audio is limited to a maximum frequency response of 22,050 Hz. If only voice-quality is desired, the sample rate can easily be dropped to 8000 Hz providing a maximum 4000 Hz frequency response. (See the **DSP and SDR Fundamentals** chapter for more information on sampling.)

The *bit depth* (number of bits used to represent each sample) of an audio signal will determine the theoretical dynamic range or signal-to-noise ratio (SNR). This is expressed with the formula SNR = (1.761 + 6.0206 × bits) dB. A dynamic range of 40 dB is adequate for the perception of human speech. **Table 15.2** compares the audio quality of various formats with the storage (or transmission) requirements.

VIDEO DATA REPRESENTATIONS

The most basic video format simply stores a series of images for playback. Video can place huge demand on storage and bandwidth because 30 frames per second is a common rate for smooth-appearing video. Rates as low

Table 15.2
Typical Audio Formats

Audio Format	*Bits per Sample*	*Dynamic Range*	*Maximum Frequency*	*kbytes per Minute*
44.1 kHz stereo	16	98	22.050 kHz	10,584
22 kHz mono	16	98	11 kHz	2,640
8 kHz mono	8	50	4 kHz	960

as 15 frames per second can still be considered "full-motion" but will appear jerky and even this rate requires substantial storage. The most popular full-motion video formats are MP4, MOV, and WMV. GIF can also support limited animation.

15.1.4 Compression Techniques

There are a large variety of compression algorithms available to reduce the size of data stored or sent over a transmission medium. Many techniques are targeted at specific types of data (text, audio, images or video). These compression techniques can be broken into two major categories: *lossless compression* and *lossy compression*. Lossless algorithms are important for compressing data that must arrive perfectly intact but offer a smaller compression ratio than lossy techniques. Programs, documents, databases, spreadsheets would all be corrupted and made worthless if a lossy compression technique were used.

Images, audio and video are good candidates for lossy compression schemes with their substantially higher compression rates. As the name implies, a lossy compression scheme deliberately omits or simplifies that data to be able to represent it efficiently. The human eye and ear can easily interpolate missing information and it simply appears to be of lower quality.

Compression of a real-time stream of data such as audio or video is performed by software or firmware *codecs* (from *co*der-*dec*oder). Codecs provide the real-time encoding or decoding of the audio or video stream. Many codecs are proprietary and have licensing requirements. Codecs can be implemented as operating system or application plug-ins or even as digital ICs, as with the P25 IMBE and D-STAR AMBE codecs. The use of proprietary codecs is controversial in amateur radio, open-source software, and internet systems. Open-source codecs such as *CODEC2* (see the section on Digital Voice) are an active area of development both in amateur radio and the open-source community.

LOSSLESS COMPRESSION TECHNIQUES

One of the earliest lossless compression schemes is known as *Huffman coding*. Huffman coding creates a tree of commonly used data values and gives the most common values a lower bit count. Varicode is based on this mechanism. In 1984, Terry Welch released code with improvements to a scheme from Abraham Lempel and Jacob Ziv, commonly referred to as *LZW* (Lempel-Ziv-Welch). The Lempel-Ziv algorithm and variants are the basis for most current compression programs and is used in the GIF and optionally TIFF graphic formats. *LZW* operates similarly to Huffman coding but with greater efficiency.

The actual amount of compression achieved will depend on how redundant the data is and the size of the data being compressed. Large files will achieve greater compression rates because the common data combinations will be seen more frequently. Simple text and documents can often see 25% compression rates. Spreadsheets and databases generally consist of many empty cells and can often achieve nearly 50% compression. Graphic and video compression will vary greatly depending on the complexity of the image. Simple images with solid backgrounds will compress well where complex images with little recurring data will see little benefit. Similarly, music will compress poorly but spoken audio with little background noise may see reasonable compression rates.

Run length encoding (*RLE*) is a very simple scheme supported on bitmap graphics (*Windows* BMP files). Each value in the file is a color value and "run length" specifying how many of the next pixels will be that color. It works well on simple files but can make a file larger if the image is too complex.

It is important to note that compressing a previously compressed file will often yield a larger file because it simply creates more overhead in the file. There is occasionally some minor benefit if two different compression algorithms are used. It is not possible to compress the same file repeatedly and expect any significant benefit. Modern compression software does offer the additional benefit of being able to compress groups of files or even whole directory structures into a single file for transmission.

The compression mechanisms mentioned above allow files to be compressed prior to transmission but there are also mechanisms that allow near real-time compression of the transmitted data. Winlink uses a compression scheme called B2F and sees an average 44% improvement in performance since most Winlink data is uncompressed previously. There is a slight delay (latency) as a result of this compression but over a slow link, the additional latency is minimal compared to the performance gain. (See the Winlink section below.)

LOSSY COMPRESSION SCHEMES

Lossy compression schemes depend on the human brain to "recover" or simply ignore the missing data. Since audio, image and video data have unique characteristics as perceived by the brain, each of these data types have unique compression algorithms. New compression schemes are developed constantly to achieve high compression rates while maintaining the highest quality. Often a particular file format actually supports multiple compression schemes or is available in different versions as better methods are developed. In-depth discussion of each of these algorithms is beyond the scope of this book but there are some important issues to consider when looking at these compression methods.

Lossy Audio Compression

Audio compression is based on the psychoacoustic model that describes which parts of a digital audio signal can be removed or substantially reduced without affecting the perceived quality of the sound. Lossy audio compression schemes can typically reduce the size of the file 10- or 12-fold with little loss in quality. The most common formats currently are MP3, WMA, AAC, Ogg Vorbis and ATRAC.

Lossy Image Compression

The Joint Photographic Experts Group developed the JPEG format (pronounced "JAY-peg") in 1992 and it has become the primary format used for lossy compression of digital images. It is a scalable compression scheme allowing the user to determine the level of compression/quality of the image. Compression rates of 10-fold are common with good quality and can be over 100-fold at substantially reduced quality. The JPEG format tends to enhance edges and substantially compress fields of similar color. It is not well suited when multiple edits will be required because each copy will have generational loss and therefore reduced quality.

Lossy Video Compression

Because of the massive amount of data required for video compression it is almost always distributed with a lossy compression scheme. Lossless compression is only used when editing to eliminate generational loss. Video support on a computer is generally implemented with a codec that allows encoding and decoding the video stream. Video files are containers that can often support more than one video format and the specific format information is contained in the file. When the file is opened, this format information is read and the appropriate codec is activated and fed into the data stream to be decoded.

The most common current codecs are H.261 (video conferencing), MPEG-1 (used in video CDs), MPEG-2 (DVD Video), H.263 (video conferencing), MPEG-4, DivX, Xvid, H.264, Sorenson 3 (used by *QuickTime*), WMV (Microsoft), VC-1, RealVideo (RealNetworks) and Cinepak.

The basic mechanism of video compression is to encode a high quality "key-frame" that could be a JPEG image as a starting image. Successive video frames or "inter-frames"

fldigi

By Ken Humbertson, WØKAH, and Jeff Coval, ACØSC

You may have heard of the free digital mode software *fldigi* by David H. Freese Jr., W1HKJ. It is very popular for use in emergency communications, for example, because of its multimode capabilities and its ability to work with a companion program *flmsg* that generates standard message forms to be transmitted using *fldigi*. The software supports numerous modes, including CW, as well as variations of tones, bandwidth, baud rates, number of bits, and other variations for many modes. Versions of *fldigi* are available for Windows, Linux, Unix, and OS X. See **www.w1hkj.com** for information on downloading the files.

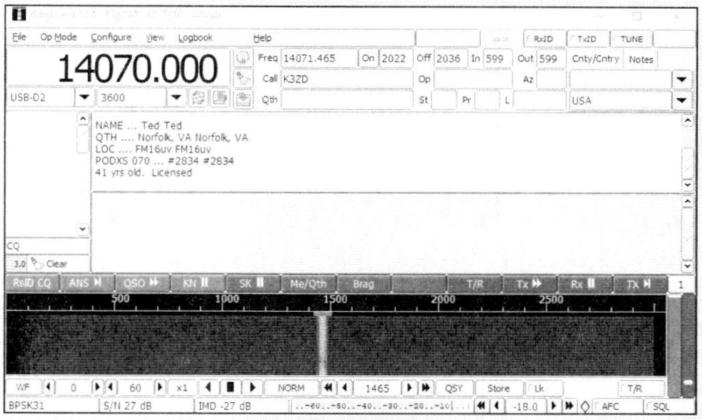

Figure 15.A — Part of a contact being conducted using BPSK-31.

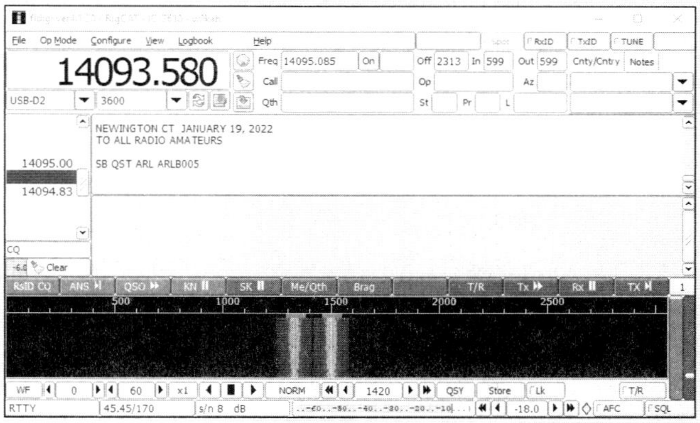

Figure 15.B — A W1AW bulletin being sent using RTTY-45.

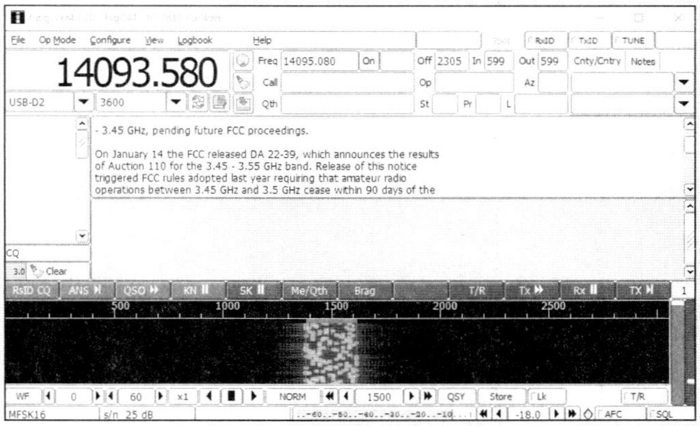

Figure 15.C — A W1AW bulletin being sent using MFSK-16.

If you have a computer with a sound card and microphone, you can begin using *fldigi* to receive with no additional hardware. A quick example would be to tune to 14.070 MHz or 21.070 MHz USB during the day. Set the radio speaker volume to a normal listening level. Start *fldigi* on the computer with sound card and microphone connected and you should see a waterfall display similar to **Figure 15.A** when PSK signals are present.

If your rig has computer control capability, *fldigi* can likely interface with it to display frequency and mode, as well as control the radio from the program. In Figure 15.A, an ICOM IC-7610 is controlled by *fldigi* and thus shows the current operating frequency and mode of 14070.000 kHz USB, the actual contact frequency of 14071.465 (kHz) (in the Feq window, upper center just right of dial frequency), and the current mode of BPSK31 (lower left corner).

When you see multiple signals in the waterfall, *fldigi* will decode whichever one you choose when you place the mouse cursor over a signal of interest and line up the two vertical red lines on your screen with the sides of the signal, as shown near 1500 Hz. When you change modes or bandwidths within a mode, the spacing of the red lines will adjust to the new bandwidth. When selecting **RTTY-45** under the **OP MODE** tab, you will notice that the red line spacing increases to match the familiar RTTY tone shift of 170 Hz. Simply place the two red lines over the signal you wish to decode and the program does the rest.

BPSK-31 and RTTY-45 are both modes that are well established and remain popular. However, they both need a fairly strong signal for error-free copy because they have no error correction. MFSK-16 is a mode that is popular among emcomm operators because of its robust error correction. W1AW sends digital bulletins in BPSK-31, MFSK-16, and RTTY-45, so you have the opportunity to copy the bulletins on all three modes. **Figures 15.B** and **15.C** show a portion of a decoded ARRL digital bulletin from January 18, 2022 in RTTY-45 and MFSK16, respectively.

Actions in *fldigi* are performed by action buttons that invoke *macros* — text scripts that control the program. For example, to call CQ use the mouse to click on an unoccupied spot in the waterfall display. This shifts the modulating tones of the signal to that offset within your receive bandwidth. Click the CQ action button on the *fldigi* display. The transmitted text is displayed in red above the waterfall display. When the text is finished, *fldigi* will return to receive mode. The tutorial "Beginner's Guide to Fldigi" at **www.w1hkj.com/beginners.html** is recommended and the program has an extensive Help file. An excellent guide to understanding the technical details of the different modes is also available at **www.w1hkj.com/modes/index.htm**.

The program can be used with an internal sound card or external sound card adapter such as the Tigertronics Signalink USB (**www.tigertronics.com**) or West Mountain Radio Rigblaster series (**www.westmountainradio.com**). Setting up a sound card to use *fldigi* may require manipulation of the audio device configuration for your computer's operating system. Follow the instructions in the *fldigi* manuals and the manufacturer's manuals if you are using an external adapter. Many radios now have either a direct digital data connector (typically a 6-pin mini-DIN connector) or an internal USB soundcard. These are the recommended digital data interfaces. A digital data interface that allows you to connect the sound card to your radio's microphone input is also an option. (See the **Assembling a Station** and **Digital Communications** chapters.)

Fldigi is frequently updated to include new modes that are being developed. The program is a user-friendly way to become active on the digital modes, a rapidly expanding aspect of amateur radio.

contain only changes to the previous frame. After some number of frames, it is necessary to use another key-frame and start over with inter-frames. The resolution of the images, frame rate and compression quality determine the size of the video file.

BIT RATE COMPARISON

Table 15.3 provides an indication of minimum bit rates required to transmit audio and video such that the average listener would not perceive them significantly worse than the standard shown.

Table 15.3
Audio and Video Bit Rates

Audio	
8 kbit/s	Telephone quality audio (using speech codecs)
32 kbit/s	AM broadcast quality (using MP3)
96 kbit/s	FM broadcast quality (using MP3)
224-320 kbit/s	Near-CD quality, indistinguishable to the average listener (using MP3)
Video	
16 kbit/s	Videophone quality (minimum for "talking head")
128-384 kbit/s	Business videoconference system quality
1.25 Mbit/s	VCD (video compact disk) quality
5 Mbit/s	DVD quality
10.5 Mbit/s	Actual DVD video bit rate

15.1.5 Compression vs. Encryption

There is some confusion about compression being a form of encryption. It is true that a text file after compression can no longer be read unless uncompressed with the appropriate algorithm. In the United States the FCC defines encryption in part 97.113 as "messages encoded for the purpose of obscuring their meaning." Compressing a file with ZIP or RAR (common file compression methods) without password protection and transmitting it over the air is simply an efficient use of spectrum and time and is not intended to "obscure its meaning."

As amateur digital modes interact more with internet-based services, the issue arises because many of these services utilize encryption of various types. For example, traffic with any site accessed using the Secure Hypertext Protocol where the URL begins "https://" is encrypted. Banks and other retailers may encrypt their entire transactions to insure confidentiality of personal data. Other systems as benign as e-mail may simply encrypt passwords to properly authenticate users. The FCC has offered no additional guidance on these issues.

15.2 Unstructured Digital Modes

The first group of modes we'll examine are generally considered "sound card modes" for keyboard-to-keyboard communications. Because each of these modes is optimized for a specific purpose by blending multiple features, they often defy simple categorization.

15.2.1 Radioteletype (RTTY)

Radioteletype (*RTTY*) consists of a *frequency shift keyed* (*FSK*) signal that is modulated between two carrier frequencies, called the *mark* frequency and the *space* frequency. The protocol for amateur RTTY calls for the mark carrier frequency to be the higher of the two in the RF spectrum. The difference between the mark and space frequencies is called the *FSK shift*, usually 170 Hz for an amateur RTTY signal.

At the conventional data speed of 60 WPM, binary information modulates the FSK signal at 22 ms per bit, or equivalent to 45.45 baud. Characters are encoded into binary 5-bit Baudot coded data. Each character is individually synchronized by adding a start bit before the 5-bit code and by appending the code with a stop bit. The start bit has the same duration as the data bits, but the stop bit can be anywhere between 22 to 44 ms in duration. The stop bit is transmitted as a mark carrier, and the RTTY signal "rests" at this state until a new character comes along. If the number of stop bits is set to two, the RTTY signal will send a minimum of 44 ms of mark carrier before the next start bit is sent. A start bit is sent as a space carrier. A zero in the Baudot code is sent as a space signal and a one is sent as a mark signal. **Figure 15.1** shows the character D sent by RTTY.

BAUDOT CODE

The Baudot code (see the online content) is a 5-bit code; thus, it is capable of encoding only 32 unique characters. Since the combination of alphabets, decimal numbers and common punctuations exceeds 32, the Baudot code is created as two sets of tables. One table is called the LTRS Shift, and consists mainly of alphabetic characters. The second table is called the FIGS Shift, and consists mainly of decimal numerals and punctuation marks. Two unique Baudot characters called LTRS and FIGS are used by the sender to command the decoder to switch between these two tables.

Mechanical teletypewriter keyboards have two keys to send the LTRS and FIGS characters. The two keys behave much like the caps lock key on a modern typewriter keyboard. Instead of locking the keyboard of a typewriter to upper case shift or lower case shift, the two teletypewriter keys lock the state of the teletypewriter into the LTRS table or the FIGS table. LTRS and FIGS, among some other characters such as the space character, appear in both LTRS and FIGS tables so that you can send LTRS, FIGS and shift no matter which table the encoder is using.

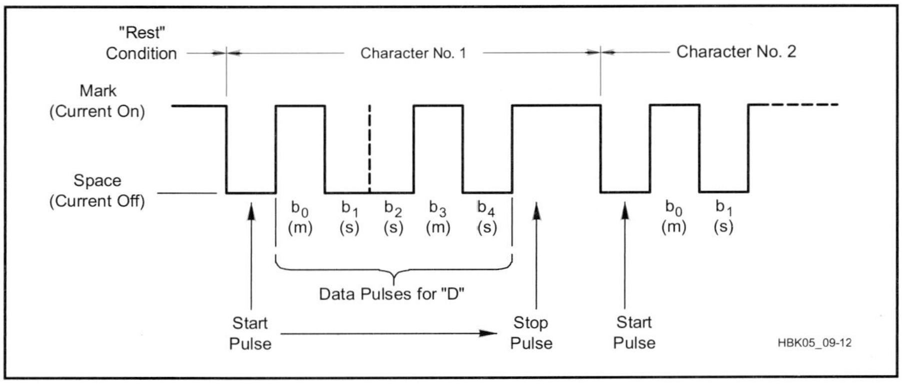

Figure 15.1 — The character "D" sent via RTTY.

To send the letter Q, you need to first make sure that the decoder is currently using the LTRS table, and then send the Baudot codeword for Q, 17 (hexadecimal or "hex" value). If the same hex 17 is received when the decoder is in the FIGS shift, the number 1 will be decoded instead of the Q. Modern software does away with the need for the operator to manually send the LTRS and FIGS codes. When the operator sends a character from the LTRS table, the software first checks to make sure that the previous character had also used a character from the LTRS table; if not, the software will first send a LTRS character.

Noise can often cause the LTRS or FIGS character to be incorrectly received. This will cause subsequent characters to decode into the wrong Baudot character, until a correct LTRS or FIGS is received (see also USOS below). Instead of asking that the message be repeated by the sender, a trick that many RTTY operators use is to observe that on a standard QWERTY keyboard, the Q key is below the 1 key in the row above it, the W key is below the 2 key, and so on. Q and 1 happen to share the same Baudot code; W and 2 share the same Baudot code, and so forth. Given this visual aid, a printed UE can easily be interpreted by context to 73 and TOO interpreted as 599.

If the sender uses one stop bit, an RTTY character consists of a total of seven bits after adding the start and stop bits. If the sender uses 1.5 stop bits, each RTTY character has a total length of 7.5 bits. The least significant bit of the Baudot code follows the start bit of a character.

INVERTED SIGNALS

There are times when the sender does not comply with the RTTY standard and reverses the mark and space order in the spectrum. This is often called an "inverted" signal or "reverse shift." Most RTTY modulators and demodulators have a provision to reverse the shift of an inverted signal.

DEMODULATION AND DECODING

The most common way to decode an RTTY signal is to use a single sideband (SSB) receiver to first translate the two FSK carriers into two audio tones. If the carriers are 170 Hz apart, the two audio tones (called the *tone pair*) will also be 170 Hz apart. The RTTY demodulator, in the form of a terminal unit (TU) or a software modem (modulator-demodulator), then works to discriminate between the two audio tones. Some packet TNCs (terminal node controllers) can be made to function as RTTY demodulators, but often they do not work as well under poor signal-to-noise conditions because their filters are matched to packet radio FSK shifts and baud rate instead of to RTTY shift and baud rate.

As long as the tone pair separation is 170 Hz, the frequencies of the two audio tones can be quite arbitrary. Many TUs and modems are constructed to handle a range of tone pairs. If reception uses a lower sideband (LSB) receiver, the higher mark carrier will become the lower of the two audio tones. If upper sideband (USB) is used, the mark carrier will remain the higher of the tone pair. It is common to use 2125 Hz as the mark tone and 2295 Hz as a space tone. Since the mark tone (2125 Hz) is lower in frequency, the receiver will be set to LSB. In general, modem tone pairs can be "reversed" (see the Inverted Signals section), so either an LSB or a USB receiver can be used. Moreover, the tone pairs of many modems, especially software modems, can be moved to place them where narrowband filters are available on the receiver.

In the past, audio FSK demodulators were built using high-Q audio filters followed by a "slicer" to determine if the signal from the mark filter is stronger or weaker than the signal that comes out of the space filter. To counter selective fading, where ionospheric propagation can cause the mark carrier to become stronger or weaker than the space carrier, the slicer's threshold can be designed to adapt to the imbalance. Once "sliced" into a bi-level signal, the binary stream is passed to the decoder where start bit detection circuitry determines where to extract the 5-bit character data. That data is then passed to the Baudot decoder, which uses the current LTRS or FIGS state to determine the decoded character. The mark and space transmitted carriers do not overlap, although this can occur after they pass through certain HF propagation conditions. Sophisticated demodulators can account for this distortion.

A software modem performs the same functions as a hardware terminal unit, except that the software modems can apply more sophisticated mathematics that would be too expensive to implement in hardware. Modern desktop computers have more than enough processing speed to implement an RTTY demodulator. Software modems first convert the audio signal into a sequence of binary numbers using an analog-to-digital converter which is usually part of an audio chip set either on the motherboard or sound card. Everything from that point on uses numerical algorithms to implement the demodulation processes.

FSK VS AFSK MODULATION

An RTTY signal is usually generated as an F1B or F2B emission. F1B is implemented by directly shifting an RF carrier between the two (mark and space) frequencies. This method of generating FSK is often called *direct FSK*, or *true FSK*, or simply FSK. F2B is implemented by shifting between two audio tones, instead of two RF carriers. The resultant audio is then sent to an SSB transmitter to become two RF carriers. This method of first generating an audio FSK signal and then modulating an SSB transmitter to achieve the same FSK spectrum is usually called AFSK (audio frequency shift keying).

AFSK can be generated by using either an upper sideband (USB) transmitter or a lower sideband (LSB) transmitter. With a USB transmitter, the mark tone must be the *higher* of the two audio tones in the audio FSK signal. The USB modulator will then place the corresponding mark carrier at the higher of the two FSK carrier frequencies. When LSB transmission is used, the mark tone must be the *lower* of the two audio tones in the audio FSK signal. The LSB modulator will then place the corresponding mark carrier at the higher of the two FSK carrier frequencies. As when receiving, the actual audio tones are of no importance. The important part of AFSK is to have the two audio tones separated by 170 Hz, and have the pair properly flipped to match the choice of USB or LSB transmission. (See the **Modulation** chapter for more information on USB and LSB modulation and the relationship between modulating frequency and transmitted signal frequency.)

When using AFSK with older transceivers, it is wise to choose a high tone pair so that harmonic by-products fall outside the passband of the transmitter. Because of this, a popular tone pair is 2125 Hz/2295 Hz. Most transceivers will pass both tones and also have good suppression of the harmonics of the two tones. Not all transceivers that have an FSK input are FSK transmitters. Some transceivers will take the FSK keying input and modulate an internal AFSK generator, which is then used to modulate an SSB transmitter. In this case, the transmitter is really operating as an F2B emitter. This mode of operation is often called "keyed AFSK."

"SPOTTING" AN RTTY SIGNAL

By convention, RTTY signals are identified by the frequency of the mark carrier on the RF spectrum. "Spotting" the suppressed carrier frequency dial of an SSB receiver is useless for someone else unless they also know whether the spotter is using upper or lower sideband and what tone pair the spotter's demodulator is using. The mark and space carriers are the only two constants, so the amateur RTTY standard is to spot the frequency of the mark carrier.

DIDDLE CHARACTERS

In between the stop bit of a preceding character and the start bit of the next character, the RTTY signal stays at the mark frequency. When the RTTY decoder is in this "rest" state, a mark-to-space transition tells a decoder that the start of a new character has arrived. Noise that causes a start bit to be misidentified can cause the RTTY decoder to fall out of sync. After losing sync, the decoder will use subsequent data bits to help it identify the location of the next potential start bit.

Since all mark-to-space transitions are potential locations of the leading edge of a start bit, this can cause multiple characters to be incorrectly decoded until proper synchronization is again achieved. This "character slippage" can be minimized somewhat by not allowing the RTTY signal to rest for longer than its stop bit duration. An idle or *diddle character* (so called because of the sound of the demodulated audio from an idle RTTY signal sending the idle characters) is inserted immediately after a stop bit when the operator is not actively typing. The idle character is a non-printing character from the Baudot set and most often the LTRS character is used. Baudot encodes a LTRS as five bits of all ones making it particularly useful when the decoder is recovering from a misidentified start bit.

An RTTY diddle is also useful when there is selective fading. Good RTTY demodulators counter selective fading by measuring the amplitudes of the mark and space signals and automatically adjusting the decoding threshold when making the decision of whether a mark or a space is being received. If a station does not transmit diddles and has been idle for a period of time, the receiver will have no idea if selective fading has affected the space frequency. By transmitting a diddle, the RTTY demodulator is ensured of a measurement of the strength of the space carrier during each character period.

UNSHIFT-ON-SPACE (USOS)

Since the Baudot code aliases characters (for example, Q is encoded to the same 5-bit code as 1) using the LTRS and FIGS Baudot shift to steer the decoder, decoding could turn into gibberish if the Baudot shift characters are altered by noise. For this reason, many amateurs use a protocol called *unshift-on-space* (USOS). Under this protocol, both the sender and the receiver agree that the Baudot character set is always shifted to the LTRS state after a space character is received. In a stream of text that includes space characters, this provides additional, implicit, Baudot shifts.

Not everyone uses USOS. When used with messages that have mostly numbers and spaces, the use of USOS causes extra FIGS characters to be sent. A decoder that complies with USOS will not properly decode an RTTY stream that does not have USOS set. Likewise, a decoder that has USOS turned off will not properly decode an RTTY stream that has USOS turned on.

OTHER FSK SHIFTS AND RTTY BAUD RATE

The most commonly used FSK shift in amateur RTTY is 170 Hz. However, on rare occasions stations can be found using 425 and 850 Hz shifts. The wider FSK shifts are especially useful in the presence of selective fading since they provide better frequency diversity than 170 Hz.

Because HF packet radio uses 200 Hz shifts, some TNCs use 200 Hz as the FSK shift for RTTY. Although they are mostly compatible with the 170 Hz shift protocol, under poor signal to noise ratio conditions these demodulators will produce more error hits than a demodulator that is designed for 170 Hz shift. Likewise, a signal that is transmitted using 170 Hz shift will not be optimally copied by a demodulator that is designed for a 200 Hz shift.

To conserve spectrum space, amateurs have experimented with narrower FSK shifts, down to 22.5 Hz. At 22.5 Hz, optimal demodulators are designed as *minimal shift keyed* (*MSK*) instead of frequency shift keyed (FSK) demodulators.

SOME PRACTICAL CHARACTERISTICS

When the demodulator is properly implemented, RTTY can be very resilient against certain HF fading conditions, namely when selective fading causes only one of the two FSK carriers to fade while the other carrier remains strong. However, RTTY is still susceptible to "flat fading" (where the mark and space channels both fade at the same instant). There is neither an error correction scheme nor an interleaver (a method of rearranging — interleaving — the distribution of bits to make errors easier to correct) that can make an RTTY decoder print through a flat and deep fade. The lack of a data interleaver, however, also makes RTTY a very interactive mode. There is practically no latency when compared to a mode such as MFSK16, where the interleaver causes a latency of over 120 bit durations before incoming data can even be decoded. This makes RTTY attractive to operating styles that have short exchanges with rapid turnarounds, such as in contests.

Although RTTY is not as "sensitive" as PSK31 (when there is no multipath, PSK31 has a lower error rate than RTTY when the same amount of power is used) it is not affected by phase distortion that can render even a strong PSK31 signal from being copied. When HF propagation conditions deteriorate, RTTY can often function as long as sufficient power is used. Tuning is also moderately uncritical with RTTY. When the signal-to-noise ratio is good, RTTY tuning can be off by 50 Hz and still print well.

15.2.2 PSK31 and Variants

PSK31 is a family of modes that uses differentially encoded varicode (see the next section), envelope-shaped phase shift keying. BPSK31, or binary PSK31, operates at 31.25 bit/s (one bit every 32 ms). QPSK31, or quadrature PSK31, operates at 31.25 baud. Each symbol consists of four possible quadrature phase change (or *dibits*) at the signaling rate of one dibit every 32 ms. QPSK31 sends a phase change symbol of 0°, 90°, 180° or 270° every 32 ms.

Characters that are typed from the keyboard are first encoded into variable-length varicode binary digits. With BPSK31, the varicode bits directly modulate the PSK31 modulator, causing a 180° phase change if the varicode bit is a 0 and keeping a constant phase if the varicode bit is a 1. With QPSK31, the varicode bits first pass through two convolution encoders to create a sequence of bit pairs (dibits). Each dibit is then used to shift the QPSK31 modulator into one of four different phase changes.

PSK63 is a double-clock-rate version of a PSK31 signal, operating at the rate of one symbol every 16 ms (62.5 symbols per second). PSK125 is a PSK31 signal clocked at four times the rate, with one symbol every 8 ms (125 symbols per second). Although PSK63 and PSK125 are both in use, including the binary and quadrature forms, the most popular PSK31 variant remains BPSK31.

Most implementations of PSK31 first generate an audio PSK31 signal. The audio signal is then used to modulate an SSB transmitter. Since BPSK31 is based upon phase reversals, it can be used with either upper sideband (USB) or lower sideband (LSB) systems. With QPSK31 however, the 90° and 270° phase shifts have to be swapped when using LSB transmitters or receivers.

PSK31 VARICODE

Characters that are sent in PSK31 are encoded as varicode, a variable length prefix code as varicode. As described earlier, characters that occur more frequently in English text, such as spaces and the lower-case e, are encoded into a fewer number of bits than characters that are less frequent in English, such as the character q and the upper case E.

PSK31 varicode characters always end with two bits of zeros. A space character is sent as a one followed by the two zeros (100); a lower case e is sent as two ones, again terminated by the two bits of zero (1100). None of the varicode code words contain two consecutive zero bits. Because of this, the PSK31 decoder can uniquely identify the boundary between characters. A special "character" in PSK31 is the idle code, which consists of nothing but the two prefix bits. A long pause at the keyboard is encoded into a string of even numbers of zeros. The start of a new character is signaled by the first non-zero bit received after at least two consecutive zeros.

CONVOLUTION CODE

As described earlier, QPSK31 encodes the varicode stream with two convolution encod-

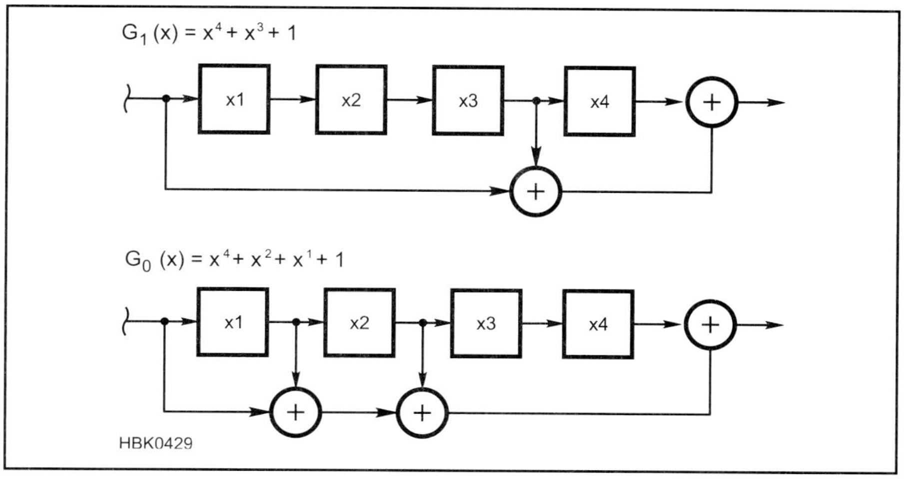

Figure 15.2 — QPSK31 convolution encoders.

ers to form the dibits that are used to modulate the QPSK31 generator. Both convolution encoders are fourth-order polynomials, shown in **Figure 15.2**.

The varicode data is first inverted before the bits are given to the convolution encoders. The first polynomial generates the most significant bit and the second polynomial generates the least significant bit of the dibit. Since we are working with binary numbers, the GF(2) sums shown in the above figure are exclusive-OR functions and the delay elements x1, x2, x3 and x4 are stages of binary shift registers. As each inverted varicode bit is available, it is clocked into the shift register and the two bits of the dibit are computed from the shift register taps.

MODULATION

PSK31 uses both differential phase shift modulation and envelope modulation to maintain its narrow-band characteristics. The most common way to generate an envelope-shaped PSK31 signal is to start with baseband in-phase (I) and quadrature (Q) signals. Both I and Q signals settle at either a value of +1 or a value of –1. When no phase transition is needed between symbols, the I and Q signals remain constant (at their original +1 or –1 values). See the **Modulation** chapter for information on creating I and Q signals.

To encode a 180° transition between symbols, both I and Q signals are slewed with a cosinusoidal envelope. If the in-phase signal had a value of +1 during the previous symbol, it is slewed to –1 cosinusoidally. If the in-phase signal had a value of –1 in the previous symbols, it is slewed cosinusoidally to +1. The quadrature signal behaves in a likewise manner. To encode a 90° phase shift in QPSK31, only the in-phase signal is slewed between the two symbols, the quadrature signal remains constant between the two symbols. To encode a 270° phase shift in QPSK31, only the quadrature signal is slewed between the two symbols; the in-phase signal remains constant between the two symbols.

The envelope of the real signal remains constant if there is no phase change. When the signal makes a 180° phase transition, the amplitude of a PSK31 signal will drop to zero in between the two symbols. The actual phase reversal occurs when the signal amplitude is zero, when there is no signal energy. During the 90° and 270° phase shifts between symbols of QPSK31, the amplitude of the signal does not reach zero. It dips to only half the peak power in between the two symbols.

To provide a changing envelope when the operator is idle at the keyboard, a zero in the varicode (remember that an idle varicode consists of two 0 bits) is encoded as a 180° phase change between two BPSK31 symbols. A 1 in the varicode is encoded as no phase change from one BPSK31 symbol to the next symbol. This changing envelope allows the receiver to extract bit timing information even when the sender is idle. Bit clock recovery is implemented by using a comb filter on the envelope of the PSK31 signal.

The convolution code that is used by QPSK31 converts a constant idle stream of zeros into a stream of repeated 10 dibits. To produce the same constant phase change during idle periods as BPSK31, the QPSK31 10 dibit is chosen to represent the 180° phase shift modulating term. **Table 15.4** shows the QPSK31 modulation.

To produce bit clocks during idle, combined with the particular convolution code that was chosen for QPSK31, results in a slightly sub-optimal (non-Gray code) encoding of the four dibits. At the end of a transmission, PSK31 stops all modulation for a short period and transmits a short unmodulated carrier. The intended function is as a squelch mechanism.

DEMODULATION AND DECODING

Many techniques are available to decode differential PSK, where a reference phase is not present. Okunev has a good presentation of the methods (reference: Yuri Okunev, *Phase and Phase-difference Modulation in Digital Communications* (1997, Artech House), ISBN 0-89006-937-9).

As mentioned earlier, there is sufficient amplitude information to extract the bit clock from a PSK31 signal. The output of a differential-phase demodulator is an estimate of the phase angle difference between the centers of one symbol and the previous one. With BPSK31, the output can be compared to a threshold to determine if a phase reversal or a non-reversal is more likely. The decoder then looks for two phase reversals in a row, followed by a non-reversal, to determine the beginning of a new character. The bits are gathered until two phase reversals are again seen and the accumulated bits are decoded into one of the characters in the varicode table.

QPSK31 decoding is more involved. As in the BPSK31 case, the phase difference demodulator estimates the phase change from one bit to another. However, one cannot simply invert the convolution function to derive the data dibits. Various techniques exist to decode the measured phase angles into dibits. The *Viterbi algorithm* is a relatively simple algorithm for the convolution polynomials used in QPSK31. The estimated phase angles can first be fixed to one of the four quadrature angles before the angles are submitted to the Viterbi algorithm. This is called a hard-decision *Viterbi decoder*.

A soft-decision Viterbi decoder (that is not that much more complex to construct) usually gives better results. The soft-decision decoder uses arbitrary phase angles and some measure of how "far" an angle is away from one of the four quadrature angles. Error correction occurs within the trellis that implements the Viterbi algorithm. References to convolution code, trellis and the Viterbi algorithm can be found at **en.wikipedia.org/wiki/Convolutional_code**.

Table 15.4
QPSK31 Modulation

Dibit	*Phase Change*
00	0°
01	90°
10	180°
11	270°

15.2.3 MFSK16

MFSK16 uses M-ary FSK modulation with 16 "tones" (also known as 16-FSK modulation), where only a single tone is present at any instant. MFSK16 has a crest factor of 1, with no wave shaping performed on the data bits. The tone centers of MFSK16 are separated by

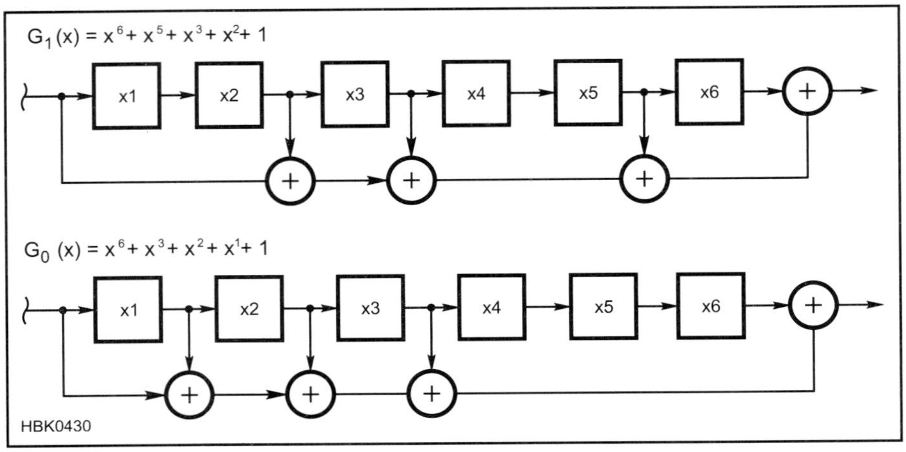

Figure 15.3 — MFSK16 convolution encoders.

15.625 Hz. Data switches at a rate of 15.625 baud (one symbol change every 64 ms).

Characters are first encoded into the MFSK16 varicode, creating a constant bit stream whose rate is one bit every 32 ms. This results in a similar, although not identical, character rate as PSK31. The difference is due to the different varicode tables that are used by the two modes.

This bit stream is clocked into a pair of 6th-order convolution encoders. The result is a pair of bits (dibit) for each varicode bit, at a rate of one dibit every 32 ms. Consecutive pairs of dibits are next combined into sets of four bits (*quadbit* or *nibble*). Each nibble, at the rate of one per 64 ms, is then passed through an interleaver. The interleaver produces one nibble for each nibble that is sent to it. Each nibble from the interleaver is then Gray-coded and the resultant 4-bit Gray code is the ordered index of each 16 tones that are separated by 15.625 Hz. The result is a 16-FSK signal with a rate of one symbol per 64 ms. Since the symbol time is the reciprocal of the tone separation, a 16-point fast Fourier transform (FFT) can be conveniently used as an MFSK16 modulator.

MFSK16 VARICODE

Although the varicode table that is used by MFSK16 (see the online content) is not the same as the one used by PSK31, they share similar characteristics. Please refer to the PSK31 section of this chapter for a more detailed description of varicode. Unlike PSK31 varicode, MSK16 varicode encodings can contain two or more consecutive zero bits, as long as the consecutive zeros are at the tail of a code word. Character boundaries are determined when two or more consecutive zeros are followed by a one.

CONVOLUTION CODE

As described earlier, MFSK16 encodes the varicode stream with two convolution encoders to form the dibits that are passed on to the interleaver. Both convolution encoders are sixth order polynomial, shown in **Figure 15.3**.

The first polynomial generates the most significant bit and the second polynomial generates the least significant bit of the dibit. Since we are working with binary numbers, the GF(2) sums shown in the above figure are exclusive-OR functions and the delay elements x1, x2, x3, x4, x5 and x6 are stages of binary shift registers. As each varicode bit is available, it is clocked into the shift register and the two bits of the dibit are computed from the shift register taps.

INTERLEAVER

HF fading channels tend to generate "burst" errors that are clumped together. An interleaver is used to permute the errors temporally so that they appear as uncorrelated errors to the convolution decoder. MFSK16 uses 10 concatenated stages of an IZ8BLY Diagonal Interleaver. More information on the interleaver can be found at **www.qsl.net/zl1bpu/MFSK/Interleaver.htm**. **Figure 15.4** illustrates how a single IZ8BLY interleaver spreads a sequence of bits.

The bits enter the IZ8BLY interleaver in the order 0, 1, 2, 3, 4,... and are passed to the output in the order 0, 5, 10, 15, 8, 13, ... (shown in the diagonal boxes). In MFSK16, the output of one interleaver is sent to the input of a second interleaver, for a total of 10 such stages. Each stage spreads the bits out over a longer time frame.

This concatenated 10-stage interpolator is equivalent to a single interpolator that is 123 bits long. **Figure 15.5** shows the structure of the single interpolator and demonstrates how four consecutive input bits are spread evenly over 123 time periods.

An error burst can be seen to be spread over a duration of almost two seconds. This gives MFSK16 the capability to correct errors over deep and long fades. While it is good for correcting errors, the delay through the long interleaver also causes a long decoding latency.

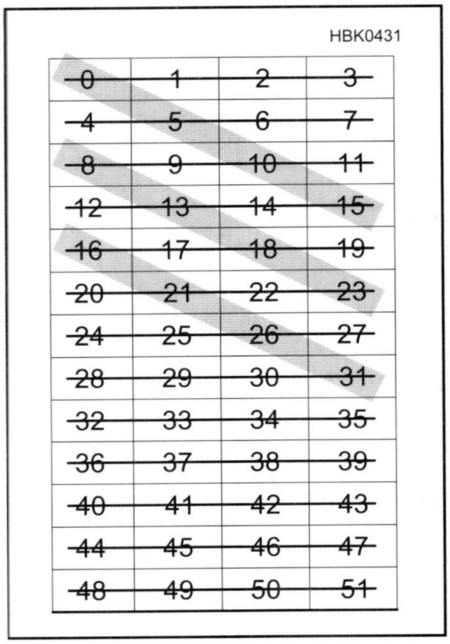

Figure 15.4 — IZ8BLY interleaver.

GRAY CODE

The Gray code creates a condition where tones that are next to one another are also different by a smaller *Hamming* distance. This optimizes the error correction process at the receiver. References to Gray code and Hamming distance can be found at **en.wikipedia.org/wiki/Gray_code** and **en.wikipedia.org/wiki/Hamming_distance**.

DEMODULATION AND DECODING

A 16-point FFT can be used to implement a set of matched filters for demodulating an MFSK16 signal once the input waveform is properly time aligned so that each transform is performed on an integral symbol and the signal is tuned so that the MFSK16 tones are perfectly centered in the FFT bins. A reference to a matched filter can be found at **en.wikipedia.org/wiki/Matched_filter**.

The 16 output bins from an FFT demodulator can first be converted to the "best" 4-bit index of an FFT frequency bin, or they can be converted to a vector of four numerical values representing four "soft" bits. The four bits are then passed through an MFSK16 de-interleaver. In the case of "soft decoding," the de-interleaver would contain numerical values rather than a 0 or 1 bit.

The output of the de-interleaver is passed into a convolution decoder. The Gray code makes sure that adjacent FFT bins also have the lowest Hamming distance; i.e., the most likely error is also associated with the clos-

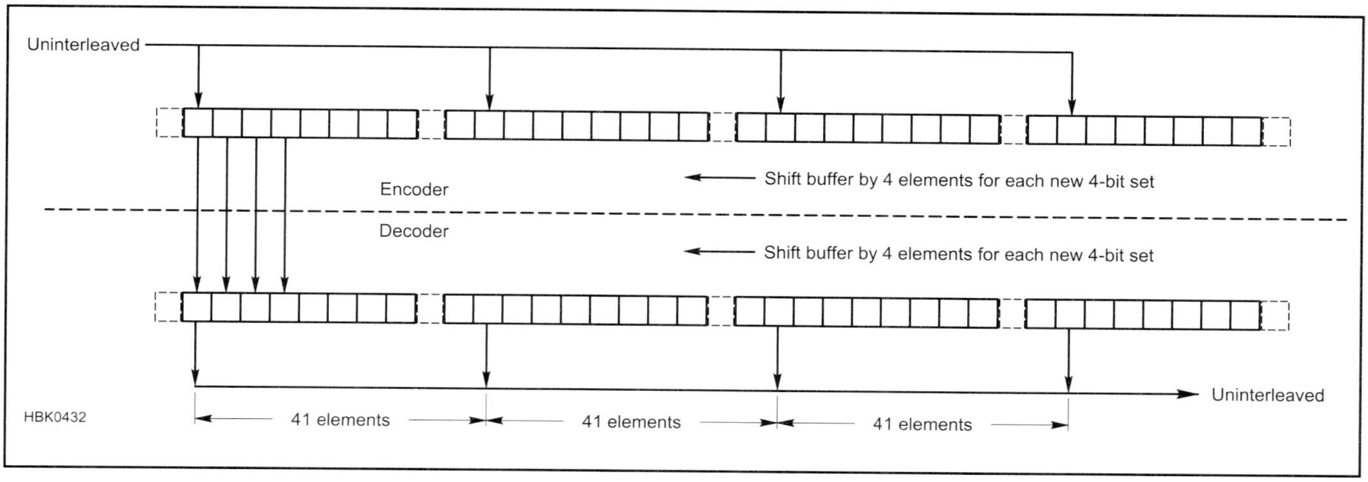

Figure 15.5 — Bit spreading through interpolation.

est FFT bin. The hard or the soft Viterbi Algorithm can be used to decode and correct errors.

15.2.4 DominoEX

DominoEX is a digital mode with MFSK (multi-frequency shift keying) designed for simplex chats on HF by Murray Greenman, ZL1BPU. It was designed to be easier to use and tune than other similar modes, offer low latency for contesting or other quick exchange situations, offer reliable copy down into the noise floor, and work well as an NVIS (near-vertical incidence skywave, see the **Propagation of Radio Signals** chapter) mode for emergency communications.

Generally MFSK requires a high degree of tuning accuracy and frequency stability and can be susceptible to multipath distortion. DominoEX specifically addresses these issues. To avoid tuning issues, IFK (*incremental frequency keying*) is used. With IFK, the data is represented not by the frequency of each tone, but by the frequency difference between one tone and the next. It also uses offset incremental keying to reduce inter-symbol interference caused by multipath reception. These techniques provide a tuning tolerance of 200 Hz and a drift of 200 Hz/minute. DominoEX also features an optional FEC mode that increases latency but provides communications over even more difficult channels. More information can be found online at **www.qsl.net/zl1bpu/MFSK/DEX.htm**.

DominoEX uses M-ary FSK modulation with 18 tones in which only a single tone is present at any instant. Information is sent as a sequence of separate 4-bit ("nibble") symbols. The value of each nibble is represented during transmission as the position of the single tone.

The position of the tone is computed as the difference of the current nibble value from the nibble value of the previously transmitted symbol. In addition, a constant offset of 2 is applied to this difference. Because there are 18 tones, any possible 4-bit value between 0 and 15 can be represented, including the offset of 2.

The additional offset of 2 tone positions ensures that a transmitted tone is separated from the previously transmitted tone by at least two tone positions. It is thus impossible for two sequential symbols to result in the same tone being transmitted for two sequential tone periods. This means sequential tones will always be different by at least two positions, an important consideration in maintaining sync.

This minimum separation of successive tones of incremental frequency keying (IFK) in DominoEX reduces the inter-symbol distortion that results from a pulse being temporally smeared when passing through an HF channel. The double-tone spacing of DominoEX modes (see **Table 15.5**) further reduces inter-symbol distortion caused by frequency smearing.

Incremental frequency keying allows the DominoEX nibbles to immediately decode without having to wait for the absolute tone to be identified. With MFSK16, a tone cannot be uniquely identified until the lowest and highest of the 16 tones have passed through the receiver. This contributes to the decoding latency of MFSK15. There is no such latency with IFK.

Since IFK depends upon frequency differences and not absolute tone frequencies, DominoEX tolerates tuning errors and drifting signals without requiring any additional automatic frequency tracking algorithms.

Like MFSK16, the DominoEX signal is not wave-shaped and has constant output power. The baud rates for DominoEX are shown in Table 15.5. The tone spacings for DominoEX 11, DominoEX 16 and DominoEX 16 have the same values as their baud rates. The tone spacings for DominoEX 8, DominoEX 5 and DominoEX 4 are twice the value of their baud rates.

Unlike PSK31 and MFSK16, characters in DominoEX are encoded into varicode nibbles instead of encoding into varicode bits. The DominoEX varicode table can be found with the online content.

BEACON MESSAGE

Instead of transmitting an idle varicode symbol when there is no keyboard activity, DominoEX transmits a "beacon" message from an alternate set of varicode (SECVAR columns in the DominoEX varicode table). This user-supplied repeating beacon message is displayed at the receiving station when the sending station is not actively sending the primary message. On average, the character rate of the beacon channel is about half of the character rate of the primary channel.

Table 15.5
Comparison of DominoEX Modes

Mode	*Baud/ (sec)*	*BW (Hz)*	*Tones*	*Speed (WPM)*	*FEC (WPM)*	*Tone Spacing*
DominoEX 4	3.90625	173	18	~25	~12	Baud rate ×2
DominoEX 5	5.3833	244	18	~31	~16	Baud rate ×2
DominoEX 8	7.8125	346	18	~50	~25	Baud rate ×2
DominoEX 11	10.766	262	18	~70	~35	Baud rate ×1
DominoEX 16	15.625	355	18	~100	~50	Baud rate ×1
DominoEX 22	21.533	524	18	~140	~70	Baud rate ×1

FORWARD ERROR CORRECTION (FEC)

When FEC is turned off, DominoEX has very low decoding latency, providing an interactive quality that approaches RTTY and PSK31. The first character is decoded virtually instantly by the receiver after it is transmitted. Because of that, FEC is not usually used even though it is available in most software that implements DominoEX.

DominoEX FEC is similar to the FEC that is used in MFSK16. When FEC is on, each 4-bit IFK symbol that is decoded by the receiver is split into two di-bits. The dibits enter an identical convolution coder to that used by MFSK16. However, instead of a 10-stage IZ8BLY interleaver (see Figure 15.4), only 4 cascaded stages of the basic 4-bit interleaver are present in DominoEX. In the presence of long duration fading, the performance of the shortened interleaver is moderately poor when used with DominoEX 16 and DominoEX 22. However, the interleaver is quite efficient in countering fading when FEC is used with DominoEX 4, DominoEX 5 and DominoEX 8, with their longer symbol periods.

Since DominoEX works in dibit units rather than nibble units when FEC is turned on, it also switches to using the same binary varicode used by MFSK16 instead of using the nibble-based varicode. DominoEX does not implement Gray code as is used by MFSK16 FEC.

Even without FEC, DominoEX works well under many HF propagation conditions, including the ITU NVIS ("Mid-latitude Disturbed") propagation profile. However, there are conditions where DominoEX is not usable unless FEC is switched on, specifically the CCIR Flutter and ITU High Latitude Moderate Conditions profiles. DominoEX modes, especially those with tone spacings that are twice the baud rate, are very robust even under these extreme conditions once FEC is switched on.

DominoEX performance charts (character error rates versus signal-to-noise ratios) are included as the HTML document "DominoEX Performance" with the online content and online at **www.w7ay.net/site/Technical/DominoEX/Measurements/**.

CHIP64/128

Chip64 and Chip128 modes were released in 2004 by Antonino Porcino, IZ8BLY. The modes were tested on the air by IZ8BLY, Murray Greenman, ZL1BPU, (who also contributed in the design of the system), Chris Gerber, HB9BDM, and Manfred Salzwedel, OH/DK4ZC. According to IZ8BLY, "The design of this new digital mode served to introduce the spread spectrum technology among radio amateurs by providing a communication tool to experiment with. Its purpose was to prove that it's possible to take advantage of the spread spectrum techniques even on the HF channels, making the communication possible under conditions where traditional narrowband modes fail.

"Among the different possible implementations of spread spectrum, Chip64 uses the so-called Direct Sequence Spread Spectrum (DSSS). In a DSSS transmission, the low speed signal containing the data bits to be transmitted is mixed (multiplied) with a greatly higher speed signal called code. The result of this mixing operation, called *spreading*, is a high-speed bit stream which is then transmitted as a normal DBPSK. Indeed, a DSSS signal looks like nothing else than wideband BPSK.

"The system proved to be efficient and we found it comparable to the other modern digital modes. Being totally different in its architecture, it shows better performance during certain circumstances, while in others it shows no actual gain. In particular, it performs better under multipath where normal BPSK can't track arriving symbols, but in quiet environments it doesn't show any improvement over plain BPSK. This is expected because of the losses that occur due to the imperfect autocorrelation of the codes."

Chip64 has a total data rate of 37.5 bit/s and the more robust Chip128 is 21.09 bit/s. Both use the same varicode used by MFSK15. The software can be downloaded from **antoninoporcino.xoom.it/Chip64/** and more information is available at **www.arrl.org/technical-characteristics**. Spread spectrum is discussed in the **Modulation** chapter, as well.

15.2.5 THROB

Throb is an experimental mode written by Lionel Sear, G3PPT, and gets the name from the "throbbing" sound it makes on the air. It uses either single tones or pairs from a possible nine tones spaced 8 or 16 Hz apart, resulting in a bandwidth of 72 or 144 Hz, respectively. It has three transmission speeds — 1, 2 and 4 throbs/s — resulting in data rates of 10, 20 and 40 WPM, respectively. The 1 and 2 throb/s speeds use a tone spacing of 8 Hz for a 72 Hz bandwidth and the 4-throb/s speed uses a spacing of 16 Hz for a 144 Hz bandwidth. It is implemented as a standalone application or included in a multimode package such as *MixW* (**www.mixw.net**).

15.2.6 MT63

MT63 is a mode developed by Pawel Jalocha, SP9VRC. MT63 is very complex with wide bandwidth, low speed and very high noise immunity. By using 64 different modulated tones, MT63 includes a large amount of extra data in the transmission of each character, so that the receiving equipment can work out, with no ambiguity, which character was sent, even if 25% of the character is obliterated. MT63 also features a secondary channel that operates simultaneously with the main channel that can be used for an ID or beacon.

MT63 likely has the most extensive error correction and can be quite processor intensive. It uses a Walsh function that spreads the data bits of each 7-bit ASCII character across all 64 of the tones of the signal spectrum and simultaneously repeats the information over a period of 64 symbols within any one tone. This coding takes several seconds. The combination of time domain (temporal) and frequency domain (spectral) interleaving results in superb impulse noise rejection. At the same time, in the frequency domain, significant portions of the signal can be masked by unwanted noise or other transmissions without any noticeable effect on successful reception.

On each of the 64 tones, the transmission data rate is fairly slow, which suits the nature of ionospheric disturbances. Despite the low data rate, good text speed is maintained because the text is sent on many tones at once. The system runs at several different speeds, which can be chosen to suit conditions but 100 WPM is typical of the MT63-1K mode. Although the 1 kHz bandwidth mode is typical, MT63 can also run at 500 Hz and 2 kHz bandwidth where the tone spacing and baud rate are halved or doubled and the throughput is halved or doubled, respectively.

Tuning of MT63 modes is not critical. This is because the mode can use FEC techniques to examine different combinations of the 64 tones that calculate the correct location within the spectrum. As an example, MT63-1K will still work if the decoder is off-frequency by as much as 100 Hz. MT63-2K requires even less precision and can tolerate an error of 250 Hz.

The incredible noise immunity comes at a price beyond the large bandwidth required. There is a large latency caused by the error correction and interleaving process. Quick-turnaround QSOs are not possible because there is a several second delay between typing the last character and it being transmitted.

Without confirming each transmission with some type of ARQ mode, there is no more robust digital mode than MT63.

15.2.7 Olivia

Olivia is an MFSK-based protocol designed to work in difficult (low signal-to-noise ratio plus multipath propagation) conditions on the HF bands. The signal can still be copied accurately at 10 dB below the noise floor. Olivia was developed in 2003 by Pawel Jalocha, SP9VRC, and performs well for digital data transfer with white noise, fading and multipath, polar path flutter and auroral conditions.

Olivia transmits a stream of 7-bit ASCII characters. The characters are sent in blocks of five with each block requiring two seconds to transmit. This results in an effective data rate of 2.5 characters/second or 150 characters/minute. A transmission bandwidth of 1000 Hz and the baud rate of 31.25 MFSK tones/second, also known as *Olivia 1000/32*, is the most common. To adapt to different propagation conditions, the number of tones and the bandwidth can be changed and the time and frequency parameters are proportionally scaled. The number of tones can be 2, 4, 8, 16, 32, 64, 128 or 256 and the bandwidth can be 125, 250, 500, 1000 or 2000 Hz.

The Olivia is constructed of two layers: the lower, modulation and FEC code layer is a classical MFSK while the higher layer is an FEC code based on Walsh functions. More detail on Walsh functions is available online at **en.wikipedia.org/wiki/Walsh_function**.

Assuming Olivia 1000/32 is being used, in the first layer the orthogonal functions are cosine functions, with 32 different tones. Since only one of those 32 tones is being sent at a time, the demodulator measures the amplitudes of all the 32 possible tones and identifies the tone with the highest amplitude. In the second layer every ASCII character is encoded as one of 64 possible Walsh functions. The receiver again measures the amplitudes for all 64 vectors and selects the greatest as the true value.

To avoid simple transmitted patterns (like a constant tone) and to minimize the chance for a false lock at the synchronizer, the characters encoded into the Walsh function pass through a scrambler and interleaver. The receiver synchronizes automatically by searching through time and frequency offsets for a matching pattern.

More information can be found online at **n1su.com/olivia** and Olivia is supported in a number of digital multimode packages such as *MixW*, *MultiPSK* and *Ham Radio Deluxe*.

15.3 Fuzzy Modes

There is a group of modes referred to as "fuzzy modes" because although they are machine generated and decoded, they are designed to be human-read. These include facsimile (fax), slow-scan TV (SSTV) and Hellschreiber.

15.3.1 Facsimile (WEFAX)

Also known as Weatherfax, HF-FAX, Radiofax, and Weather Facsimile, WEFAX was developed as a mechanically transmitted technology where the source material was placed on a spinning drum and scanned line by line into an electrical signal which would be transmitted by wire or over the air. It is important that the receiving station have their drum spinning at the correct speed in order to correctly recreate the image. A value known as the *index of cooperation* (*IOC*) must also be known to decode a transmission. IOC governs the image resolution and is the product of the total line length and the number of lines per unit length divided by π. Most fax transmissions are sent with LPM (RPM) at 120 and an IOC of 576.

Facsimile is generally transmitted in single sideband with a tone of 1500 Hz representing black and 2300 Hz representing white. The *automatic picture transmission* (*APT*) format is used by most terrestrial weather facsimile stations and geostationary weather satellites. It features a start tone that triggers the receiving system, originally used to allow the receiving drum to come up to speed. It also includes a phasing signal with a periodic pulse that synchronizes the receiver so the image appears centered on the page. A stop tone, optionally followed by black, indicates the end of the transmission. The APT format is shown in **Table 15.6**.

Stations with Russian equipment sometimes use RPM 60 or 90 and sometimes an IOC of 288. Photofax transmissions such as those from North Korea use RPM 60 and an IOC 352 with gray tones, and satellite rebroadcast use also RPM 120 IOC 576, with gray tones (4 or more bit depth). For software decoding of weather fax images it is best to decode with Black and White (2-bit depth).

Table 15.6
Facsimile Automatic Picture Format

Signal	*Duration*	*IOC576*	*IOC288*	*Remarks*
Start Tone	5 s	300 Hz	675 Hz	200 Hz for color fax modes
Phasing Signal	30 s			White line interrupted by black pulse
Image	Variable	1200 lines	600 lines	At 120 LPM
Stop Tone	5 s	450 Hz	450 Hz	
Black	10 s			

15.3.2 Slow-Scan TV (SSTV)

Slow-Scan TV or SSTV is similar to facsimile where a single image is converted to individual scanned lines and those lines sent as variable tones between 1500 and 2300 Hz. Modern systems use computer software and a sound card to generate and receive the required tones. (Some SSTV communication uses purely digital protocols, in which the picture content is sent as digital data and not directly represented in the modulation scheme.)

There are a number of different SSTV "modes" that define image resolution and color scheme. A color image takes about 2 minutes to transmit, depending on mode. Some black and white modes can transmit an image in under 10 seconds. More information about SSTV may be found in the **Image Communications** chapter with the online content.

15.3.3 Hellschreiber, Feld-Hell, or Hell

Hellschreiber is a facsimile-based mode developed by Rudolph Hell in the 1920s. The name is German and means "bright writer" or "light writer" and is a pun on the inventor's name. In Hellschreiber, text is transmitted by dividing each column into seven pixels and transmitting them sequentially starting at the lowest pixel. Black pixels are transmitted as a signal and white as silence at 122.5 bit/s (about a 35 WPM text rate). Originally the text was printed on continuous rolls of paper so the message could be any length.

Even though each pixel is only transmitted once, they are printed twice, one below the other. This compensates for slight timing errors in the equipment that causes the text to slant. If properly in sync, the text will appear as two identical rows, one below the other or a line of text in the middle with chopped lines top and bottom. Regardless of the slant, it is always possible to read one copy of the text. Since the text is read visually, it can be sent in nearly any language and tends to look like an old dot matrix printer. More information can be found online at **www.qsl.net/zl1bpu/HELL/Feld.htm** and Randy, K7AGE, has a great introduction to Hellschreiber available on YouTube at **www.youtube.com/watch?v=yR-EmyEBVqA**.

15.4 Structured Digital Modes

This group of digital modes has more structured data and as a result provides more robust connections and better weak signal performance. Each of these modes bundles data into packets or blocks that can be transmitted and error checked at the receive end.

15.4.1 WSJT-X Modes

All modes described in Sections 15.4.1 through 15.4.4 are available in a multi-author, open-source software package called *WSJT-X*. Some are also available in other programs derived from *WSJT-X*. These modes are designed around structured messages that optimize the exchange of minimal contact information including call signs, signal reports, Maidenhead grid locators, and acknowledgments, even at very low signal-to-noise ratios. These modes were pioneered by Joe Taylor, K1JT, starting in 2001 when home computers were first widely equipped with sound cards and fast enough to do significant signal processing. The first modes were optimized for use over especially difficult paths such as meteor scatter and Earth-Moon-Earth (EME, or "moonbounce") on VHF and higher bands. More recently, several of the modes have become very popular for LF, MF, and HF DXing, as well.

The *WSJT-X* structured modes divide naturally into two groups, *slow* and *fast*. Slow modes send each message frame once per transmission interval, while fast modes send the message frame repeatedly, as many times as will fit into the transmission interval. The modes intended for making two-way QSOs use message formats based on compressed, fixed-length packets of 72 or 77 information bits. Modes called WSPR and FST4W allow individual stations to transmit a call sign, location and power level using 50-bit packets. As outlined in this chapter's section "Error Detection and Correction," additional bits are added to the message frames to effect error detection and correction. Details of the mathematical encoding schemes and modulation types differ from mode to mode. In all cases the forward error correction (FEC) algorithm is strong enough that false decodes are very rare: a receiving operator nearly always sees exactly the message that was transmitted, or nothing is displayed at all. All of the modes implemented in *WSJT-X* use *constant-envelope* waveforms, so they can be amplified by efficient nonlinear amplifiers without generating intermodulation products. Transmissions have a fixed duration, for example 15 seconds or one minute, and their start times are synchronized with Coordinated Universal Time (UTC) as maintained by the computer's operating system.

15.4.2 WSJT-X Message Compression

The information packets normally consist of two 28-bit call signs and a 15-bit field for a grid locator, signal report, acknowledgment, or "73". Additional bits can flag message formats containing arbitrary text (up to 13 characters), nonstandard call signs, contest exchanges, or the like. The basic aim is to compress the most common message formats used for minimal contacts into fixed-length packets. The older modes (JT4, JT9, and JT65) use 72-bit message payloads, while FST4, FT4, FT8, and Q65 use 77 bits and the propagation-probing modes WSPR and FST4W use 50.

A standard amateur call sign consists of a one- or two-character prefix, at least one of which must be a letter, followed by a decimal digit and a suffix of one to three letters. Within these rules, the number of possible call signs is equal to 37×36×10×27×27×27, or somewhat over 262 million. (The numbers 27 and 37 arise because in the first and last three positions a character may be absent, or a letter, or perhaps a digit.) Since 2^{28} is more than 268 million, 28 bits are enough to encode any standard call sign uniquely. Similarly, the number of 4-character Maidenhead grid locators on Earth is 180×180 = 32,400, which is less than 2^{15} = 32,768; thus, a grid locator requires 15 bits. Some otherwise unused values in the 28-bit fields are used for special message components such as CQ, DE, and QRZ, and some values of the 15-bit field not needed for grid locators are used for signal reports, acknowledgments, and the like. Structured messages using these 28-bit and 15-bit fields underlie the efficient, reliable exchange of basic and essential information for minimal station-to-station contacts.

15.4.3 WSJT-X Slow Modes

The slow modes in *WSJT-X* (versions 2.4.0 and later) are called FST4, FST4W, FT4, FT8, JT4, JT9, JT65, Q65, and WSPR. Each mode uses *continuous-phase frequency-shift keying* (FSK), with waveforms normally generated at audio frequency and transmitted as upper sideband by a standard SSB transceiver. Numbers in the mode names indicate the number of distinct tone frequencies used. FST4 was designed for the LF, MF and HF bands; JT4 and JT65 for EME on VHF, UHF, and microwave bands; JT9 for MF and lower HF bands; FT8 for multi-hop sporadic E at 50 MHz; FT4 for fast contest QSOs on HF and 50 MHz; and FST4W and WSPR for probing potential propagation paths using low-power transmissions. Not surprisingly, amateurs have discovered many other ways in which each mode can be used. On the HF bands, worldwide contacts are possible with these modes using power levels of a few watts (or even milliwatts) and compromise antennas. Reliable two-way contacts can be made with signal levels far too weak to be heard, 10 to 15 dB below the minimum levels required for Morse-coded CW received and decoded by ear.

Table 15.7A provides a brief summary of essential parameters for the *WSJT-X* slow modes. Column 1 gives the mode name, column 2 the message payload size and number of cyclic redundancy check (CRC) bits, and column 3 the type of forward error correction used. Each code maps a sequence of k information symbols into a longer sequence of n transmitted symbols. The results are called (n, k) *block codes*. After encoding, digital message information is modulated onto a carrier so that transfer can take place over a radio channel. The basic unit of transmitted data is called a *channel symbol*. Parameter Q (column 5 in the table) is the *alphabet size,* the number of different symbols used. Q is also the number of distinct waveforms used for conveying information. For frequency-shift keying the waveforms are sinusoids at different frequencies. Additional columns in Table 15.7A specify the keying rate, occupied bandwidth, fraction of transmitted energy devoted to synchronization, duration of transmitted waveform, and threshold signal-to-noise ratio for reliable decoding of each mode. Here and elsewhere in this section, signal-to-noise ratios are measured in a standard reference bandwidth of 2500 Hz.

Figure 15.6 illustrates the appearance of each of the *WSJT-X* slow modes in a typical waterfall-type spectral display. For comparison, this collection of simulated signals also includes an unmodulated carrier and a 25 WPM CW signal. The signals were generated with a key-down signal-to-noise ratio of 0 dB, thus simulating typical over-the-air reception of moderately weak, barely audible signals. WSPR has the narrowest occupied bandwidth of the basic *WSJT-X* modes, 5.9 Hz, while JT65 is the widest at 177.6 Hz. JT4, JT9, and JT65 use one-minute timed sequences of transmission and reception, synchronized with UTC, while WSPR uses two-minute sequences, FT8, 15 seconds, and FT4, 7.5 seconds. FST4, FST4W, and Q65 offer a wide range of sequence lengths. The following paragraphs give further details for each of the slow modes and describe their typical uses.

FT8

The FT8 protocol was developed by Steven Franke, K9AN, and K1JT. The mode became popular soon after its introduction in early

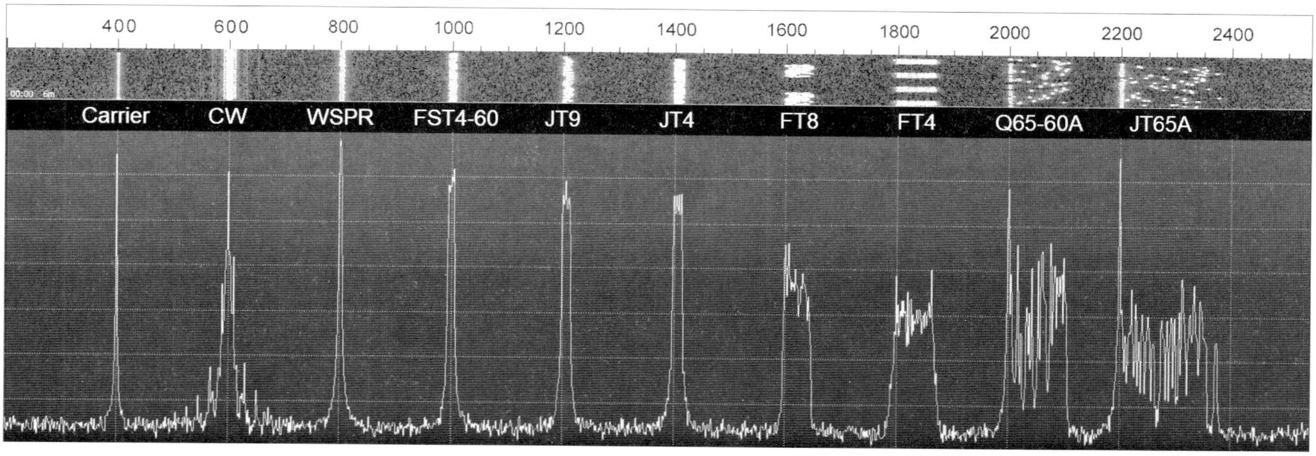

Figure 15.6 — Spectrograms of simulated signals for an unmodulated carrier, a 25 WPM CW signal, and the *WSJT-X* slow modes WSPR, JT9, JT4, FT8, FT4, QRA64, and JT65. All signals have S/N = 0 dB in a 2500 Hz reference bandwidth. Horizontal scale is frequency in Hz; vertical axis represents time over the displayed interval of about 50 seconds.

summer of 2017, and is now one of the most widely used modes of any type in amateur radio. FT8 was designed for circumstances where signals may be weak and fading, so that quick completion of reliable, confirmable QSOs is especially desirable. The terse message exchanges of call signs, locators, signal reports, and acknowledgments serve well for those engaged in country-counting, award-chasing, and the like. The short, 15-second timed sequences for transmission and reception mean that basic sensitivity for steady signals is several dB worse than the modes with one-minute transmissions. Message packets include 77 information bits and a 14-bit CRC. Three of the information bits define special message types, some of which include several subtypes. These are used for such purposes as free text and contest exchanges. A detailed description of the 77-bit message format and the FT8 protocol was published in *QEX* in 2020 (see References and Bibliography). FT8 uses a (174,91) *low-density parity check* (LDPC) code for forward error correction, and the 14-bit CRC ensures a very low false decode rate. FT8 modulation uses Gaussian filtered 8-tone FSK (8-GFSK) at 6.25 baud, and accomplishes frame synchronization with three 7 × 7 Costas arrays.

One special message subtype is used for *FT8 DXpedition Mode* (also called *Fox and Hounds*), which enables a much-sought-after station to make QSOs at high rates. In this mode *WSJT-X* allows the rare station to transmit up to five FT8 signals simultaneously, thereby conducting as many as five minimal contacts at once. The special multi-signal transmissions are subject to undesirable spectral broadening caused by intermodulation between the signals. This is true even when care is taken to avoid overdriving the transmitter audio chain because most modern transmitters produce intermodulation products at levels that can cause significant interference to weak-signal digital modes. Multi-signal transmissions are therefore not suitable for general use or for use in contests — situations where good engineering practice and common courtesy dictate that any individual station use only the minimum bandwidth necessary. Multi-signal transmission is intended only for use by DXpeditions who will operate on frequency segments that do not overlap with those

Table 15.7
Parameters of Structured Modes
A — Structured Slow Modes

Mode	*FEC Type*	*(n, k)*	*Q*	*Modulation Type*	*Keying Rate*	*Bandwidth (Hz)*	*Sync Energy (s)*	*Transmit Duration (dB)*	*S/N Threshold*
FT4	LDPC	(174,91)	4	4-GFSK	20.833	83.3	0.16	5.04	−17.5
FT8	LDPC	(174,91)	8	8-GFSK	6.25	50.0	0.27	12.6	−20.8
JT4	Convolutional	(206,72)	2	4-FSK	4.375	17.5	0.50	47.1	−23
JT9	Convolutional	(206,72)	8	9-FSK	1.736	15.6	0.19	49.0	−27
JT65	Reed Solomon	(63,12)	64	65-FSK	2.692	177.6	0.50	46.8	−25
QRA64	Q-ary Repeat Accumulate	(63,12)	64	64-FSK	1.736	111.1	0.25	48.4	−26
WSPR	Convolutional	(162,50)	2	4-FSK	1.465	5.9	0.50	110.6	−31

B — Structured Fast Modes

Mode	*FEC Type*	*(n, k)*	*Q*	*Modulation Type*	*Keying Rate*	*Bandwidth (Hz)*	*Sync Energy*	*Transmit Duration (s)*
JT9E	Convolutional	(206,72)	8	9-FSK	25	225	0.19	3.4
JT9F	Convolutional	(206,72)	8	9-FSK	50	450	0.19	1.7
JT9G	Convolutional	(206,72)	8	9-FSK	100	900	0.19	0.85
JT9H	Convolutional	(206,72)	8	9-FSK	200	1800	0.19	0.425
MSK144	LDPC	(128,90)	2	OQPSK	2000	2400	0.11	0.072

in use for regular QSOs, to avoid causing unnecessary QRM.

Recognizing a need expressed by the VHF+ contest community, the WSJT-X development team added the ability to send band-change messages as one of the Tx# messages. This change became effective with version 2.5.3 in Dec 2021. The capability is described in the WSJT-X User Guide at **wsjt.sourceforge.io/wsjtx.html**. Read the Tx Macros section under Settings for complete information.

Summarizing, the messages now support the use of band abbreviations ABCD… etc. that represent 50 MHz, 144 MHz, 222 MHz, 432 MHz, and so forth. The band abbreviation is followed by a mode abbreviation: V (SSB Voice), W (CW), or 8 (FT8). For example, the message $DXCALL BV 228 means, "Send the call sign of the station being worked and request they QSY to 2 meter voice on 144.228 MHz." The station being queried would presumably send an acknowledgment message, although that is not required. Macros can be preprogrammed with a message that is selected from the Tx5 drop-down menu. Text can also be entered into the Tx5 message at any time.

For more information about using macros to move stations, read the article, "Moving on up with FT8" by Phil Miguelez, WA3NUF, with contributions by Joe Taylor, K1JT, in the November 2021 issue of the Mt. Airy V.H.F. Radio Club's newsletter, "Cheese sBits" at **www.packratvhf.com** in the "NEWSLETTER" section. The codes and techniques may change as operating practices and the WSJT-X software evolve.

FT4

FT4 was designed for use in contests where, under certain circumstances, it is worthwhile to trade a few dB of sensitivity for higher QSO rates. It uses the same 77-bit messages and (174,91) error-correcting code as FT8, but with transmit/receive sequences of just 7.5 seconds. Although twice as fast as FT8, with only 4 tones the occupied bandwidth of FT4 is only 83 Hz, less than twice that of FT8. The decoding threshold is –17.5 dB, 3.3 dB higher than that of FT8. FT4 uses Gaussian filtered 4-tone FSK (4-GFSK) waveforms at 20.833 baud, and frame synchronization uses four 4 × 4 Costas arrays. A *QEX* article describing how FT8 and FT4 work is part of the online information for this chapter.

FST4 and FST4W

FST4 was designed for weak-signal QSOs on LF, MF, and HF bands. It uses 77-bit messages, a 24-bit CRC and a (240,101) LDPC code for error detection and correction. Modulation is Gaussian filtered 4-tone FSK (4-GFSK) with 120 message-bearing channel symbols and 40 symbols for frame synchronization; a total of 160 channel symbols. A companion mode named FST4W (the W is for "WSPR-like") is intended for use as a propagation probe, like the popular WSPR mode. FST4W uses 50-bit messages and a (240,74) LDPC code. FST4 and FST4W use the same modulation format and frame-synchronization scheme so these signals can be detected and synchronized using the same algorithms. FST4 and FST4W offer several different keying rates. FST4 supports transmission lengths of 15, 30, 60, 120, 300, 900, and 1800 seconds and FST4W supports transmission lengths of 120, 300, 900, and 1800 seconds. In general, longer transmission lengths (lower keying rates) provide proportionally higher sensitivity, narrower bandwidth, and decreasing tolerance for Doppler frequency spread. In practice, the transmit duration in seconds is appended to the mode designator to describe a particular variant. For example, FST4-60 refers to FST4 with 60 second transmissions. On channels where Doppler spread is small compared to the keying rate, FST4 and FST4W are the most sensitive weak-signal modes implemented in *WSJT-X*.

JT4

In JT4 mode, each channel symbol carries one information bit (the most significant bit) and one synchronizing bit. The sync bits are defined by a pseudo-random binary sequence known to the software at both transmitting and receiving ends. FEC is accomplished using a strong convolutional code designed by NASA for deep space use. After the sync pattern is recognized and removed by the receiving software, the remaining signal amounts to a 2-tone FSK signal. Submodes JT4A through JT4G have tone spacings at increasing multiples 1, 2, 4, 9, 18, 36, and 72 times the basic keying rate, 4.375 baud. The wider submodes have proven very useful on propagation paths with large Doppler spread. For example, JT4F is frequently used for Earth-Moon-Earth (EME) communication on the 10 GHz band.

JT9

The JT9 mode uses eight tone frequencies to convey message information, one additional tone for synchronization, and the same convolutional code as JT4. The message frame includes 69 data symbols and 16 synchronizing symbols. Submodes JT9A-H have tone spacings at multiples 1, 2, 4, 8, 16, 32, and 64 times the 1.736-baud keying rate. JT9A (often called simply JT9) uses less than 10% the bandwidth of JT65, and for steady, undistorted signals is about 2 dB more sensitive. These characteristics made JT9 popular on the 630 meter band, but its use there and on 2200 meters is now being supplanted by FST4.

JT65

The JT65 protocol is the oldest of the weak-signal structured modes; a detailed description was published in *QEX* in 2005 (see References and Bibliography). The protocol uses a (63,12) Reed-Solomon code with alphabet size $Q = 64$. Modulation uses one synchronizing tone and 64 data tones, with data and sync information interspersed according to a pseudo-random sync pattern. In *WSJT-X* decoding is accomplished using the Franke-Taylor soft decision algorithm (see References and Bibliography). Special features can be used to convey an EME-style "OOO" signal report and short messages interpreted as RO, RRR, and 73. Submodes JT65B and JT65C use tone spacings 2 and 4 times larger than JT65A. JT65 is widely used for EME on VHF and higher bands. It was popular for low-power DXing at MF and HF, as well, but has been largely superseded on those bands by FT8.

Q65

Q65 is a jack-of-all trades mode intended for use on VHF, UHF, and microwave bands for EME and other extreme weak-signal paths. On paths with Doppler spread more than a few Hz, the weak-signal performance of Q65 is the best among all *WSJT-X* modes. Its error-correcting code was designed by Nico Palermo, IV3NWV (see References and Bibliography). Q65 uses 65-FSK with the lowest tone dedicated to synchronization. The message frame includes 63 data symbols and 22 sync symbols. Five different keying rates are available, yielding T/R sequence lengths of 15, 30, 60, 120, and 300 s. Submodes A through E provide tone spacings 1, 2, 4, 8, and 16 times the keying rate. Suitable choices of sequence length and tone spacing make Q65 a highly effective mode for tropospheric scatter, ionospheric scatter, rain scatter, trans-equatorial propagation (TEP), and EME on VHF and higher bands, as well as for other types of fast-fading signals. As a few particular examples, submodes Q65-15C and Q65-30A are widely used for TEP and ionospheric scatter, respectively, on the 50 MHz band; Q65-60C for EME on 1296 MHz; and Q65-60D for both rain scatter and EME at 10 GHz.

WSPR

The WSPR mode was designed as a propagation probe rather than for making two-way contacts. It uses the same convolutional code as JT4 but differs from the other structured modes in *WSJT-X* by using message lengths of 50 information bits and two-minute T/R sequences. Message packets normally include a 28-bit call sign, a 15-bit grid locator, and 7 bits to convey transmitter power in dBm (decibels above one milliwatt). Alternative formats can convey a compound call sign and/or a 6-digit grid locator, using a two-transmission sequence. Typical WSPR usage was described in *QST* in 2010 (see References and Bibliography).

15.4.4 WSJT-X Fast Modes

The fast modes in *WSJT-X* aim to take advantage of brief propagation enhancements that bring a signal up to useful levels for a very short time. Keying rates and occupied bandwidths are much larger than for the slow modes so that full messages can be conveyed in these very short intervals. **Table 15.7B** lists essential parameter values for the *WSJT-X* fast modes. The last column gives the time required to transmit a message once at the specified keying rate. In these modes the transmitted information is repeated for the full duration of a T/R sequence.

JT9 Fast

Fast submodes JT9E-H differ from their slow JT9 counterparts by using much higher keying rates and wider tone spacings. Otherwise, the coding, modulation, and synchronization schemes are the same as for the slow JT9 modes. JT9 fast modes have proven useful for such propagation types as ionospheric scatter and weak multi-hop sporadic E on the 6 meter band.

MSK144

Messages in MSK144 contain 77 information bits and a 13-bit CRC, and use the same types of user message as FT8 and FT4. Forward error correction is accomplished with a (128, 90) LDPC code. Two 8-bit synchronizing sequences are added to make a message frame 144 bits long. Modulation uses *offset quadrature phase-shift keying* (OQPSK) at 2000 baud, so the frame duration is 144/2000 = 0.72 second. With the advantages of strong error correction and an effective character transmission rate around 250 characters per second (see References and Bibliography), MSK144 has become the dominant mode for amateur meteor-scatter contacts.

15.4.5 HF Digital Voice

AOR

In 2004, AOR Corporation introduced its HF digital voice and data modem, the AR9800. Digital voice offers a quality similar to FM with no background noise or fading as long as the signal can be properly decoded. The AR9800 can alternatively transmit binary files and images. AOR later released the AR9000 which is compatible with the AR9800 but less expensive and only supports the HF digital voice mode.

The AR9800 uses a protocol developed by Charles Brain, G4GUO. The protocol uses the AMBE (Advanced Multi-Band Excitation) codec from DVSI Inc. to carry voice. It uses 2,400 bit/s for voice data with an additional 1,200 bit/s for Forward Error Correction for a total 3,600 bit/s data stream. The protocol is detailed below:

- Bandwidth: 300 – 2500 Hz, 36 carriers
- Symbol rate: 20 ms (50 baud)
- Guard interval: 4 ms
- Tone steps: 62.5 Hz
- Modulation method: 36 carriers: DQPSK (3.6K)
- AFC: ±125 Hz
- Error correction: voice: Golay and Hamming
- Video/data: convolution and Reed-Solomon
- Header: 1 s; 3 tones plus BPSK training pattern for synchronization
- Digital voice: DVSI AMBE2020 coder, decoder
- Signal detection: automatic digital detect, automatic switching between analog mode and digital mode
- Video compression: AOR original adaptive JPEG

AOR has more information online at **www.aorusa.com/others/ard9800.html**.

FREEDV

FreeDV (initial release Dec 2012) is available for *Windows*, *Linux*, and *MacOS* clients, and allows any SSB radio to be used for low-bit-rate digital voice. Since its release, *FreeDV* has added new modes to choose from based on current band conditions. In addition, new features have been added to make set up and operation easier including simultaneous decoding for all HF modes. PSK Reporter integration provides frequency, call sign, and grid square on a map-based world view of stations being copied.

FreeDV speech and call sign data is compressed, which then modulates a mode-dependent OFDM/QPSK signal which is then applied to the microphone input of an SSB transceiver. On receive, the signal is received as SSB, demodulated to audio, then further demodulated, and decoded by *FreeDV* running on a PC.

FreeDV is entirely open source — even the voice codec. This makes it unique among amateur radio digital voice systems which typically rely on a proprietary voice codec that is not available to hams for experimentation.

FreeDV was coded from scratch by David Witten, KDØEAG (GUI, architecture), and David Rowe, VK5DGR (*Codec2*, modem implementation, integration). Recently, additional development was coded by Mooneer Salem, K6AQ (simultaneous mode receive, PSK Reporter support, Flex 6000-series Teensy dongle support, various usability enhancements). The *FreeDV* design and user interface is based on *FDMDV* (2007), developed by Francesco Lanza, HB9TLK. The large team of reviewers and beta testers who have supported the development of *FreeDV* are credited on the **freedv.org** home page.

Key *FreeDV* Features

- Waterfall, spectrum, scatter, and audio oscilloscope displays
- Adjustable squelch
- TX attenuation control (drive level)
- Fast/slow SNR estimation
- Microphone and speaker signal audio equalizer
- Control of transmitter HamLib and PTT via RS-232 levels

Table 15.8
Implementations of *FreeDV*

Mode	*Codec*	*Modem*	*RF BW*	*Raw bits/s*	*FEC*	*Text bits/s*	*SNR min*	*Multipath*
1600	Codec2 1300	14 DQPSK + 1 DBPSK pilot	1125	1600	Golay (23,12)	25	4	poor
700C	Codec2 700C	14 carrier coherent QPSK + diversity	1500	1400	- -		2	good
700D	Codec2 700C	17 carrier coherent OFDM/QPSK	1000	1900	LDPC (224,112)	25	–2	fair
700E	Codec2 700C	21 carrier coherent OFDM/QPSK	1500	3000	LDPC (112,56)	25	1	good
2020	LPCNet 1733	31 carrier coherent OFDM/QPSK	1600	3000	LDPC (504,396)	22.2	2	good

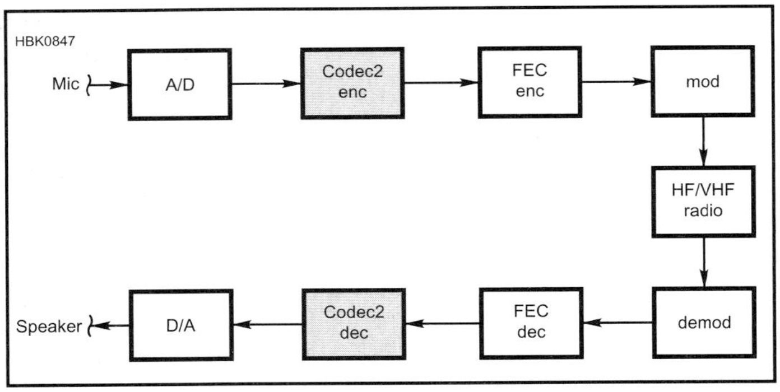

(A)

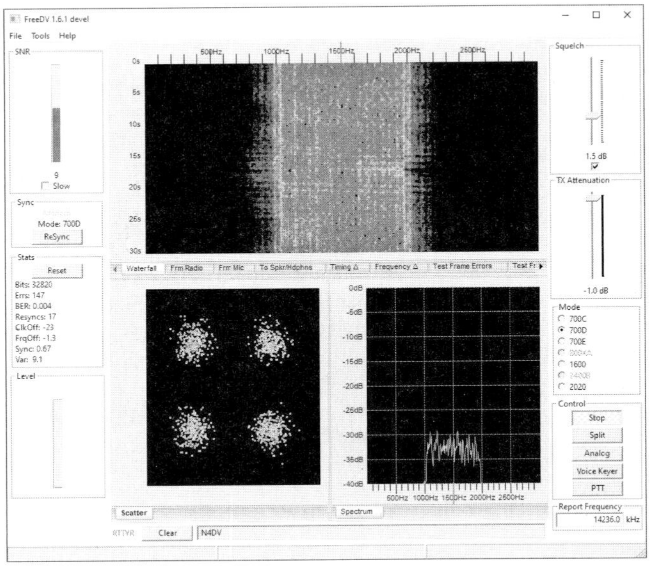

(B)

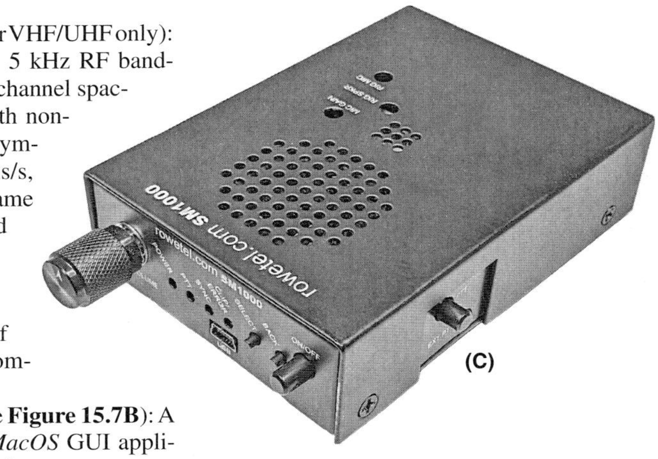

(C)

Figure 15.7 — Typical *FreeDV* system architecture (A), *FreeDV* 1.61 screenshot during reception (B), and *FreeDV* SM1000 standalone adapter — no host PC needed (C).

• Works with one (receive only) or two (transmit and receive) sound cards

(For example, a built-in sound card and USB headphones)

FreeDV Overview

There are five HF implementations of *FreeDV* summarized in **Table 15.8**. The characteristics of each variation are as follows:

FreeDV 1600: 1,275 bit/s voice coding; with 300 bits/s FEC using original FDMDV modem and 75 Hz carrier spacing. Pilot carrier with 2X power useful for tuning and robust synchronization. Good voice quality. 4000 Hz speech BW.

FreeDV 700(C): 700 bit/s voice coding; with no FEC. Improves performance in fast-fading channels using a simple transmit diversity scheme that transmits the signal twice at two different frequencies. Similar voice quality to mode 1600 but at one-half the bit rate. 4000 Hz speech BW.

FreeDV 700(D) (default mode): 700 bit/s voice coding; with Low Density Parity Check (LDPC) FEC. Sensitive to tuning, must be within ±60 Hz of the transmit frequency. Most robust mode. Error free at –2 dB SNR approximates SSB quality. 4000 Hz speech BW.

FreeDV 700(E): 700 bit/s voice encoding; with shorter 80 ms frame size to reduce latency and sync time. Best in multi-path fast fading channels with up to 4 Hz Doppler spread and 6 ms delay spread.

FreeDV 2020: Experimental codec based on the LPCNet neural net (deep learning) synthesis engine developed by Jean-Marc Valin. Capable of providing 8 kHz audio BW in a narrow 1600-Hz bandwidth. Highly experimental, believed to be the first use of neural net vocoders in a real world over-the-air system.

FreeDV 800XA (Note: for VHF/UHF only): 700 bit/s codec plus 100 bits/s for sync. 4FSK modulation suitable for Class C power amplifiers.

FreeDV 2400B (Note: for VHF/UHF only): 2,400 bit/s encoding with 5 kHz RF bandwidth requiring 6.25 kHz channel spacing. 4FSK modulation with non-coherent demodulation. Symbol rate of 1,200 symbols/s, tone spacing 1200 Hz, frame period 40 ms, unique word 16 bits/frame. Codec2 at 1300 with 52 bits/frame and 28 bits/frame spare.

FreeDV is composed of two primary software components:

1) The *FreeDV* GUI (see **Figure 15.7B**): A *FreeDV Windows/Linux/MacOS* GUI application. The *FreeDV* GUI software runs on any PC and requires two sound cards, for example a USB rig interface and a USB headset.

2) The *FreeDV* API: An open source, C-callable software library, that has interfaces for audio and modem samples. The API can be linked into third-party SDR programs that wish to support *FreeDV*.

A standalone hardware all open-source implementation of *FreeDV* (SM1000) is also available that does not require a PC host. Software/hardware developed by David Rowe, VK5DGR, with CAD PCB and mechanical enclosure by Rick Barnich, K8BMA. This small (see Figure 15.7C) adapter is excellent for portable and mobile work. Includes a built-in microphone and speaker with PTT switch. Easily connects to a SSB radio with RJ-45 cable or 3.5 mm plugs.

FreeDV Architecture

Figure 15.7A shows a typical *FreeDV* system using 1600 mode. Voice signals from a microphone are sampled by the ADC, typically at 8 kHz with 16-bit resolution. A speech encoder (*Codec2*) then compresses the speech to a low bit rate, for example 1,300 bit/s. FEC bits are then added to protect against errors encountered in the channel. This may increase the bit rate to 1,600 bit/s. A modulator (labeled "mod") then converts the bit stream to tones that can be passed via a radio channel. Over the channel the digital speech signal will encounter noise and other impairments such as fading and frequency offsets.

On the receive side, the demodulator (labeled "demod") then converts the tones back into a bit stream. The FEC decoder attempts to correct channel errors before passing the payload compressed voice information to the voice decoder (labeled "codec 2 dec"). The output of the voice decoder is an 8 kHz, 16-bit/sample sequence that is converted back into analog speech by the DAC and played through a speaker or headphones.

The screenshot in Figure 15.7B shows the software in operation using the 700D mode.

A waterfall display at the top shows the individual carriers in the signal. At the bottom right is a spectrum display of the signal. The constellation diagram of the QPSK signal is shown at lower left to give the operator an idea of the signal-to-noise performance along with the SNR meter at the upper left. More operating controls and status values are also displayed. More details and the software can be found at **freedv.org** as well as written and video setup guides.

15.4.6 ALE

Automatic link establishment (*ALE*) was created as a series of protocols for government users to simplify HF communications. The protocol provides a mechanism to analyze signal quality on various channels/bands and choose the best option. The purpose is to provide a reliable rapid method of calling and connecting during constantly changing HF ionospheric propagation, reception interference and shared spectrum use of busy or congested HF channels. It also supports text messages with a very robust protocol that can get through even if no voice-quality channel can be found.

Each radio ALE station uses a call sign or address in the ALE controller. When not actively in communication with another station, each HF SSB transceiver constantly scans through a list of frequencies, listening for its call sign. It decodes calls and soundings sent by other stations, using the bit error rate to store a quality score for that frequency and sender call sign.

To reach a specific station, the caller simply enters the call sign, just like dialing a phone number. The ALE controller selects the best available frequency and sends out brief digital selective calling signals containing the call signs. When the distant scanning station detects the first few characters of its call sign, it stops scanning and stays on that frequency. The two stations' ALE controllers automatically handshake to confirm that a link of sufficient quality is established and they are ready to communicate.

When successfully linked, the receiving station which was muted will typically emit an audible alarm and visual alert for the receiving operator of the incoming call. It also indicates the call sign of the linked station. The operators then can talk in a regular conversation. At the conclusion of the QSO, one of the stations sends a disconnect signal to the other station, and they each return their ALE stations to the scanning mode. Some military / commercial HF transceivers are available with ALE available internally. Amateur Radio operators commonly use the PCALE soundcard software ALE controller, interfaced to a ham transceiver via rig control cable and multi-frequency antenna.

Table 15.9
ALE Tones

Frequency	*Data*
750 Hz	000
1000 Hz	001
1250 Hz	011
1500 Hz	010
1750 Hz	110
2000 Hz	111
2250 Hz	101
2500 Hz	100

The ALE waveform is designed to be compatible with the audio passband of a standard SSB radio. It has a robust waveform for reliability during poor path conditions. It consists of 8-ary frequency-shift keying (FSK) modulation with eight orthogonal tones, a single tone for a symbol. These tones represent three bits of data, with least significant bit to the right, as shown in **Table 15.9**.

The tones are transmitted at a rate of 125 tones per second, 8 ms per tone. The resultant transmitted bit rate is 375 bit/s. The basic ALE word consists of 24 bits of information. Details can be found in Federal Standard 1045, at **hflink.com/standards/FED_STD_1045A.pdf**.

It would require a lot of time for the radio to go through the sequence of calling a station on every possible frequency to establish a link. Time can be decreased by using a "smarter" way of predictive or synchronized linking. With *Link Quality Analysis* (LQA), an ALE system uses periodic sounding and linking signals between other stations in the network to stay in touch and to predict which channel is likely to support a connection to the desired station at any given time. Various stations may be operating on different channels, and this enables the stations to find and use a common open channel.

The *PCALE* software developed by Charles Brain, G4GUO, is available for download at **hflink.com/software**. Much more ALE information and real-time data is available online at **hflink.com**.

15.5 Network Modes and Systems

This section describes both digital modes and systems that use them to build computer-to-computer networks. Even though communication using these modes may not involve the creation of a network, the modes are referred to as "networking modes" because of their structure. In systems such as D-STAR and System Fusion, network communication over the air uses a specific mode. Other systems such as Winlink may use several different modes to communicate. The features of these systems are described along with the modes and protocols used to implement communications within the network.

15.5.1 OSI Networking Model

The Open Systems Interconnection Model or *OSI Model* is an abstract description for computer network protocol design. It defines seven different *layers* or functions performed by a *protocol stack*. In the OSI model, the highest level is closest to the user and the lowest is closest to the hardware required to transport the data (network card and wire or radio). The seven layers are described in **Table 15.10**. The modes examined previously implemented the protocols as a *monolithic stack* where all the functions are performed inside a single piece of code. The modes described in this section implement networking features in a more modular fashion. This allows greater flexibility (and complexity) when mixing features.

As a data packet moves through these layers the header or preamble is removed and any required action performed before the data is passed to the next layer, much like peeling away layers of an onion until just the basic clean data is left. The OSI model does not define any interfaces between layers; it is just a conceptual model of the functions required. Real-world protocols rarely implement each layer individually and often span multiple layers.

Table 15.10
OSI Seven Layer Networking Model

7 — Application Layer	End-user program or "application" that uses the network
6 — Presentation Layer	The format of data after transfer (code conversion, encryption)
5 — Session Layer	Manages the transfer process
4 — Transport Layer	Provides reliable data transfer to the upper layers
3 — Network Layer	Controls data routing
2 — Data Link Layer	Provides error detection and flow control
1 — Physical Layer	Signal used on the medium—voltage, current, frequency, etc.

Table 15.11
Networking Protocols in the OSI Model

Layer	*Examples*	*IP Protocol Suite*
7 — Application		NNTP, DNS, FTP, Gopher, HTTP, DHCP, SMTP, SNMP, TELNET
6 — Presentation	ASCII, EBCDIC, MIDI, MPEG	MIME, SSL
5 — Session	Named Pipes, NetBIOS, Half Duplex, Full Duplex, Simplex	Sockets, Session establishment in TCP
4 — Transport		TCP, UDP
3 — Network	AX.25	IP, IPsec, ICMP, IGMP
2 — Data Link	802.3 (Ethernet), 802.11a/b/g/n MAC/LLC, ATM, FDDI, Frame Relay, HDLC, Token Ring, ARP (maps layer 3 to layer 2 address)	PPP, SLIP, PPTP, L2TP
1 — Physical	RS-232, T1, 10BASE-T, 100BASE-TX, POTS, DSL, 802.11a/b/g/n, Soundcard, TNC, Radio	

This description is by no means exhaustive and more information can be found online and in every networking textbook. **Table 15.11** shows the placement of commonly recognized protocols within the OSI layered structure.

15.5.2 Connected and Connectionless Protocols

The protocols discussed to this point have been *connectionless* meaning they don't establish a connection with a specific machine for the purpose of transferring data. Even with packetized modes like FSK441 with a destination call sign specified, the packet is transmitted and it's up to the user to identify they are the intended recipient. In a packet-switched network, connectionless mode transmission is a transmission in which each packet is prepended with a header containing a destination address to allow delivery of the packet without the aid of additional instructions. A packet transmitted in a connectionless mode is frequently called a *datagram*.

In *connection-oriented protocols*, the stations about to exchange data need to first declare to each other they want to "establish a connection". A connection is sometimes defined as a logical relationship between the peers exchanging data. Connected protocols can use a method called *automatic repeat request* (*ARQ*) to insure accurate delivery of packets using *acknowledgements* and *timeouts*. This allows the detection and correction of corrupted packets, misdelivery, duplication, or out-of-sequence delivery of the packets.

Connectionless modes can have error correction and detection included by a higher layer of the protocol but they have no mechanism to request a correction. An advantage of connectionless mode over connection-oriented mode is that it has a low data overhead. It also allows for *multicast* and *broadcast* (net-type) operations, which may save even more network resources when the same data needs to be transmitted to several recipients. In contrast, a connected mode is always *unicast* (point-to-point).

Another drawback of the connectionless mode is that no optimizations are possible when sending several frames between the same two peers. By establishing a connection at the beginning of such a data exchange the components (routers, bridges) along the network path would be able to pre-compute (and hence cache) routing-related information, avoiding re-computation for every packet.

Many network modes incorporate both types of protocol for different purposes. In the Internet TCP/IP protocol, TCP is a connection-oriented transport protocol where UDP is connectionless.

15.5.3 The Terminal Node Controller (TNC)

Hardware *terminal node controllers* (TNCs) such as the venerable TAPR TNC 2 are now mostly legacy devices. Their limited processing power, modem ICs, and limited interface options have been largely rendered obsolete by the use of audio interfaces coupled with software on a more powerful host computer. Coupled with a simple audio interface and software such as *Dire Wolf*, a software modem running on a Raspberry Pi is much more capable than a hardware-based TNC.

A TNC is actually a computer that contains the protocols implemented in firmware and a *modem* (modulator/demodulator). The TNC generally connects to a PC as a serial or USB device on one side and to the radio with appropriate audio and PTT cables on the other. Most of the newer rigs have dedicated data connections available that feature audio lines with fixed levels that are unaffected by settings in the radio. These jacks make swapping mike cables unnecessary when switching between voice and digital modes. Bypassing internal audio processing circuitry eliminates a number of issues that can cause problems with digital modes and makes the use of digital modes more reproducible/reliable by eliminating a number of variables when configuring equipment. These same data jacks are recommended when using a computer sound card.

Although many of the modes discussed can use a computer sound card to generate the required modulation and a separate mechanism to support push-to-talk (PTT), a TNC offers some advantages:

• TNC hardware can be used with any computer platform.

• A computer of nearly any vintage/performance level can be used.

• Data transmission/reception is unaffected by computer interruptions from virus checkers or other "inits."

• Initialization settings are held internal to the TNC and can easily be reset as needed — once working, they stay working.

• Virtually eliminates the computer as a problem/failure point.

• Offers features independent of the computer (digipeat, BBS, APRS beacon, telemetry, weather beacon, and so forth).

The majority of TNCs are designed for 300 or 1200-bit/s packet and implement the Bell 103 or Bell 202 modulation respectively. A *multimode communications processor* (*MCP*) or *multi-protocol controller* (*MPC*) may offer the capability to operate RTTY, CW, AMTOR, PACTOR, G-TOR, Clover, fax, SSTV and other modes in addition to packet. Some of these modes are only available in TNC hardware because the real-time operating system in the TNC provides a more reliable platform to implement the mode and it also helps protect proprietary intellectual property.

KISS-Mode TNCs have become popular. These devices simply provide the modem and filters to implement the baseband signals for a type of digital modulation. They rely on the computer software to generate the appropriate packet protocol and complete the mode. This means the software must be written specifically to support these TNCs by creating the entire AX.25 packet with the data embedded, rather than simply sending the data to the TNC expecting the TNC to frame the packet. By leaving the TNC to handle only the baseband signal generation and data recovery, much simpler, smaller and less expensive designs are possible while still retaining the platform independence and robustness of a separate TNC.

Modern software-based TNCs such as *Dire Wolf* provide a virtual KISS connection, typically as a TCP/IP socket. Functionally, that is equivalent to an application such as APRS, running on a host computer, connecting over a serial link to a KISS TNC.

15.5.4 PACTOR-I

PACTOR, now often referred to as PACTOR-I, is an HF radio transmission system developed by German amateurs Hans-Peter Helfert, DL6MAA, and Ulrich Strate, DF4KV. It was designed to overcome the shortcomings of AMTOR and packet radio. It performs well under both weak-signal and high-noise conditions. PACTOR-I has been overtaken by PACTOR-II and PACTOR-III but remains in use.

Table 15.13
PACTOR Control Signals

Code	*Chars (hex)*	*Function*
CS1	4D5	Normal acknowledge
CS2	AB2	Normal acknowledge
CS3	34B	Break-in (forms header of first packet from RX to TX)
CS4	D2C	Speed change request

All control signals are sent only from RX to TX

Table 15.14
PACTOR Initial Contact

Master Initiating Contact

Size (bytes)	1	8	6
Content	/Header	/SLAVECAL	/SLAVECAL/
Speed (baud)	100	100	200

Slave Response

The receiving station detects a call, determines mark/space polarity, and decodes 100 baud and 200-bd call signs. It uses the two call signs to determine if it is being called and the quality of the communication path. The possible responses are:

First call sign does not match slave's call sign (Master not calling this slave)	none
Only first call sign matches slave's call sign (Master calling this slave, poor communications)	CS1
First and second call signs both match the slaves (good circuit, request speed change to 200 baud)	CS4

TRANSMISSION FORMATS

All packets have the basic structure shown in **Figure 15.8**, and their timing is as shown in **Table 15.12**.

• Header: Contains a fixed bit pattern to simplify repeat requests, synchronization and monitoring. The header is also important for the Memory ARQ function. In each packet that carries new information, the bit pattern is inverted.

• Data: Any binary information. The format is specified in the status word. Current choices are 8-bit ASCII or 7-bit ASCII (with Huffman encoding). Characters are not broken across packets. ASCII RS (hex 1E) is used as an IDLE character in both formats.

• Status word: See Table 15.12

• CRC: The CRC is calculated according to the CCITT standard, for the data, status and CRC.

The PACTOR acknowledgment signals are shown in **Table 15.13**. Each of the signals is 12-bits long. The characters differ in pairs in eight bits (Hamming offset) so that the chance of confusion is reduced. If the CS is not correctly received, the TX reacts by repeating the last packet. The request status can be uniquely recognized by the 2-bit packet number so that wasteful transmissions of pure RQ blocks are unnecessary.

The receiver pause between two blocks is 0.29 s. After deducting the CS lengths, 0.17 s remains for switching and propagation delays so that there is adequate reserve for DX operation.

Table 15.12
PACTOR Timing

Object	*Length (seconds)*
Packet	0.96 (200 baud: 192 bits; 100 baud: 96 bits)
CS receive time	0.29
Control signals	0.12 (12 bits at 10 ms each)
Propagation delay	0.17
Cycle	1.25

CONTACT FLOW

In the listen mode, the receiver scans any received packets for a CRC match. This method uses a lot of computer processing resources, but it's flexible.

A station seeking contacts transmits CQ packets in an FEC mode, without pauses for acknowledgment between packets. The transmit time length, number of repetitions and speed are the transmit operator's choice. (This mode is also suitable for bulletins and other group traffic.) Once a listening station has copied the call, the listener assumes the TX station role and initiates a contact. Thus, the station sending CQ initially takes the RX station role. The contact begins as shown in **Table 15.14**.

With good conditions, PACTOR's normal signaling rate is 200 baud, but the system automatically changes from 200 to 100 baud and back, as conditions demand. In addition, Huffman coding can further increase the throughput by a factor of 1.7. There is no loss of synchronization speed changes; only one packet is repeated.

When the RX receives a bad 200-baud packet, it can acknowledge with CS4. TX immediately assembles the previous packet in 100-baud format and sends it. Thus, one packet is repeated in a change from 200 to 100 baud.

The RX can acknowledge a good 100-baud packet with CS4. TX immediately switches to 200 baud and sends the next packet. There is no packet repeat in an upward speed change.

The RX station can become the TX station by sending a special change-over packet in response to a valid packet. RX sends CS3 as the first section of the changeover packet.

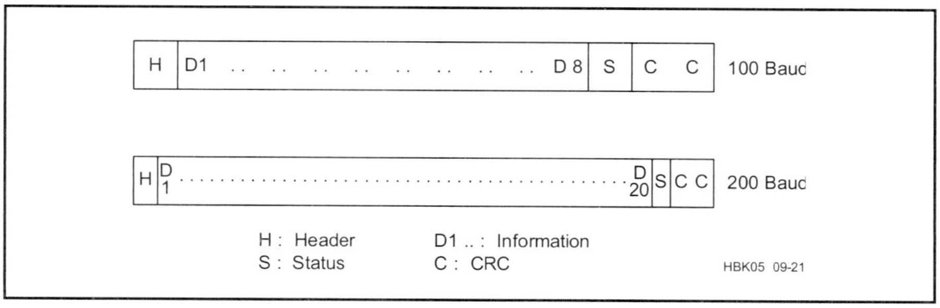

Figure 15.8 — PACTOR packet structure.

This immediately changes the TX station to RX mode to read the data in that packet and responds with CS1 and CS3 (acknowledge) or CS2 (reject).

PACTOR provides a sure end-of-contact procedure. TX initiates the end of contact by sending a special packet with the QRT bit set in the status word and the call of the RX station in byte-reverse order at 100 baud. The RX station responds with a final CS.

15.5.5 PACTOR-II

This is a significant improvement over PACTOR-I, yet it is fully compatible with the older mode. PACTOR-II uses 16PSK to transfer up to 800 bit/s at a 100 baud rate. This keeps the bandwidth less than 500 Hz.

PACTOR-II uses digital signal processing (DSP) with Nyquist waveforms, Huffman and Markov compression and powerful Viterbi decoding to increase transfer rate and sensitivity into the noise level. The effective transfer rate of text is over 1200 bit/s. Features of PACTOR II include:

- Frequency agility — it can automatically adjust or lock two signals together over a ±100 Hz window.
- Powerful data reconstruction based upon computer power — with over 2 Mbyte of available memory.
- Cross correlation — applies analog Memory ARQ to acknowledgment frames and headers.
- Soft decision making — Uses artificial intelligence (AI), as well as digital information received to determine frame validity.
- Extended data block length — when transferring large files under good conditions, the data length is doubled to increase the transfer rate.
- Automatic recognition of PACTOR-I, PACTOR-II and so on, with automatic mode switching.
- Intermodulation products are canceled by the coding system.
- Two long-path modes extend frame timing for long-path terrestrial and satellite propagation paths.

This is a fast, robust mode that has excellent coding gain as well. PACTOR-II stations acknowledge each received transmission block. PACTOR-II employs computer logic as well as received data to reassemble defective data blocks into good frames. This reduces the number of transmissions and increases the throughput of the data.

15.5.6 PACTOR-III and PACTOR-IV

PACTOR-III

PACTOR-III is a software upgrade for existing PACTOR-II modems that provides a data transmission mode for improved speed and robustness. Both the transmitting and receiving stations must support PACTOR-III for end-to-end communications using this mode.

PACTOR-III's maximum uncompressed speed is 2722 bit/s. Using online compression, up to 5.2 kbit/s is achievable. This requires an audio passband from 400 Hz to 2600 Hz (for PACTOR-III speed level 6). On an average channel, PACTOR-III is more than three times faster than PACTOR-II. On good channels, the effective throughput ratio between PACTOR-III and PACTOR-II can exceed five. PACTOR-III is also slightly more robust than PACTOR-II at their lower SNR edges.

The ITU emission designator for PACTOR-III is 2K20J2D. Because PACTOR-III builds on PACTOR-II, most specifications like frame length and frame structure are adopted from PACTOR-II. The only significant difference is PACTOR III's multitone waveform that uses up to 18 carriers while PACTOR-II uses only two carriers. PACTOR-III's carriers are located in a 120 Hz grid and modulated with 100 symbols per second DBPSK or DQPSK. Channel coding is also adopted from PACTOR-II's Punctured Convolutional Coding.

PACTOR-III Link Establishment

The calling modem uses the PACTOR-I FSK connect frame for compatibility. When the called modem answers, the modems negotiate to the highest level of which both modems are capable. If one modem is only capable of PACTOR-II, then the 500 Hz PACTOR-II mode is used for the session. With the MYLevel (MYL) command a user may limit a modem's highest mode. For example, a user may set MYL to 1 and only a PACTOR-I connection will be made, set to 2 and PACTOR-I and II connections are available, set to 3 and PACTOR-I through III connections are enabled. The default MYL is set to 2 with the current firmware and with PACTOR-III firmware it will be set to 3. If a user is only allowed to occupy a 500 Hz channel, MYL can be set to 2 and the modem will stay in its PACTOR-II mode.

The PACTOR-III Protocol Specification is available online at **www.scs-ptc.com/pactor.html**. More information can also be found online at **www.arrl.org/technical-characteristics**.

PACTOR-IV

PACTOR-IV is 1.5 – 3 times faster than PACTOR-III and has 10 speed levels. Its emission designator is 2K20J2D or 2K40J2D (at a symbol rate up to 1800 symbols per second), with a maximum speed of 5513 bps.

Because PACTOR-IV uses 1800 symbols per second (baud), and current FCC rules permit a maximum of 300 baud on amateur radio bands below 28 MHz, PACTOR-IV is not allowed for use by US amateur radio operators. The FCC has allowed a number of Special Temporary Authority (STA) usages of PACTOR-IV during emergency situations such as during the aftermath of major hurricanes. Outside the US, PACTOR-IV is used by individual amateur radio operators and Winlink stations.

15.5.7 VARA

VARA was created by Jose Alberto Nieto Ros, EA5HVK as a robust and fast "sound card" data mode for reliable file transfers. VARA is most typically used by Winlink user stations and Winlink Remote Mail Servers (RMS) for faster and more reliable file transfers than previous methods such as WINMOR and packet radio. (See the Winlink section later in this chapter.)

VARA is not compatible with any other data communications mode (such as packet radio); VARA stations can only communicate with VARA stations. VARA HF 4.0 has been adopted by the Amateur Radio Safety Foundation for the worldwide Winlink Amateur Radio email system as a co-equal (with ARDOP) on HF. (**www.winlink.org/content/vara_hf_40_network_cutover_june_30_2020**) VARA FM has been similarly adopted for Winlink access on VHF/UHF. Winlink *Radio Mail Server* (RMS) stations accept connections from users using VARA, and the *Winlink Express* client software includes direct compatibility with VARA HF and VARA FM software. For example, the *Winlink Express* software has specific sections for easy integration with VARA software.

VARA achieves its robustness and speed by incorporating a number of techniques:

- Orthogonal Frequency Division Multiplexing (OFDM) generates multiple subcarriers within the audio signal. (See **Figure 15.9**.)
- Varying modulation methods (Modulation Index) – FSK through 256 QAM depending on mode and quality of channel between stations.
- Robust handshake between transmitting and receiving station to negotiate best possible speed on each transmission.
- Huffman data compression.
- Turbo Codes Forward Error Correction.

Of these techniques, the most significant is OFDM, which uses "independent subcarriers." (See the Multi-Carrier Modulation section of the **Modulation** chapter.) Subcarriers are independently managed within the transmitted audio signal. If one or more subcarriers are damaged in transmission such as a noise burst at a specific audio frequency, other subcarriers are unaffected.

Audio levels are a consistent issue with "sound card" modes and VARA addresses this by including a "Tune" function to adjust

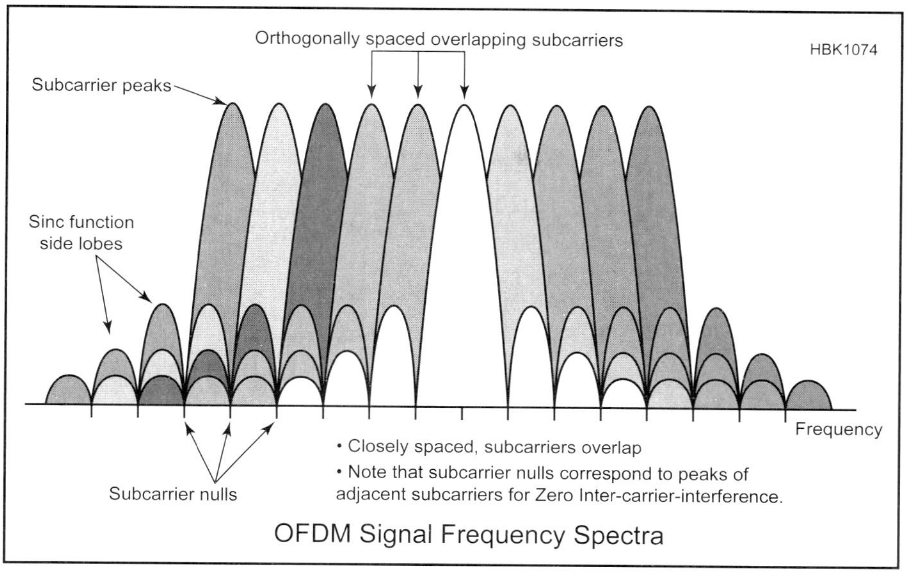

Figure 15.9 — The orthogonal frequency domain multiplexing (OFDM) structure of VARA modulation.

transmitted audio levels into the radio, and an "Autotune" function to test audio levels with another VARA station. If audio levels are too high or too low, an error message is displayed. VARA also includes a "VU Meter" to monitor the receive audio level (from the radio) into the audio interface.

If a US call sign is used, VARA will transmit a CW identification. When a non-US call sign is used, CW identification is optional.

VARA RESOURCES

All versions of the software, along with specification documents, and guides to configuring and using VARA are maintained at the author's website, **rosmodem.wordpress.com**. Videos and presentation are also available. Besides those linked at the author's website, there are numerous videos available on YouTube that offer presentations on VARA, and demonstrate setup, troubleshooting, and operation of all three VARA variants. For discussion of VARA HF and VARA FM, there is an independent email list that is quite active (1,000+ subscribers) at **groups.io/g/VARA-MODEM** (vara-modem@groups.io).

VARA IMPLEMENTATIONS

There are three variants of VARA: VARA HF, VARA FM, and VARA SAT. Each variant is different software (not three modes of the same software). Tables showing the symbol rate, number of carriers, modulation, and net bit rate for all levels of the three VARA variants is available in the online document *VARA 4.3 quick guide.pdf*. The document is available from several repositories — an online search for the name will obtain the document from the repository.

VARA HF

VARA HF is optimized for narrow, noisy channels of the HF bands, and has three modes:

VARA HF 2750 (Tactical) is optimized for commercial and military HF radios that can use wider channels (2750 Hz) to achieve data transfer speeds up to 8,489 bps.

VARA HF 2300 (Standard) is optimized for typical amateur radio HF radios (2300 Hz channels) to achieve data transfer speeds up to 7,050 bps.

VARA HF 500 (Narrow) is optimized for use on HF bands where channel width is limited to 500 Hz. VARA HF 500 can achieve data transfer speeds up to 1,543 bps.

VARA FM

VARA FM is optimized for VHF/UHF radios and the wider, and quieter, channels available on the VHF/UHF bands (and implemented in VHF/UHF FM transceivers). VARA FM has two modes — Wide and Narrow.

VARA FM Wide is optimized for VHF/UHF radios that have a "flat audio" interface. This is usually a 6-pin Mini-DIN Data connector providing connection to the transmit modulator and receive discriminator (**www.soundcardpacket.org/7miniDIN.aspx**). This connection has been traditionally used for packet radio 9,600 bps Frequency Shift Keying (FSK) modulation.

For maximum data throughput, VARA FM Wide also requires a wide-bandwidth audio interface. For example, the Masters Communications series of Digital Radio Adapters offers various combinations of bandwidth and connectivity. (**www.masterscommunications.com/products/radio-adapter/dra/dra-features.html**)

If both the transmitting and receiving radios are configured for VARA FM Wide, and both radios have wide-bandwidth audio interfaces, data transfer speeds up to 25,210 bps are possible. Note the significant speed difference between typical "fast" packet radio (9,600 bps) versus VARA FM Wide (25,210 bps) — more than 2.5 times faster.

VARA FM Narrow is optimized for more typical VHF/UHF radios that do not have a "flat audio" interface, connecting via the microphone input and speaker output where CTCSS filtering and pre-emphasis and de-emphasis processing occurs in the audio path.

An advantage of VARA FM Narrow is that almost any audio interface is suitable, including radios with integrated audio interfaces that are not capable of VARA FM Wide. Some audio interfaces that cannot achieve full VARA FM Wide speeds *can* achieve speeds greater than VARA FM Narrow; thus, experimentation is warranted when configuring VARA FM on a particular radio/audio interface combination. A radio using VARA FM Narrow can achieve data transfer speeds up to 12,750 bps (depending on channel conditions).

A compelling feature of VARA FM is that VARA FM Wide and VARA FM Narrow are fully (and automatically) interoperable, unlike 1,200 bps and 9,600 bps packet radio. During the initial connection between two VARA FM stations, the two stations negotiate their fastest common speed. If a station configured for VARA FM Wide connects to a station configured for VARA FM Narrow, the "Wide" station "downshifts" to VARA FM Narrow for data transfer with the "Narrow" station.

Other features of VARA FM include:

- Can connect via one or two VARA FM stations acting as digipeaters (VARA digipeaters perform the same function as packet radio digipeaters but are not compatible with packet radio signals.)
- Can connect via FM voice repeaters. (Repeaters can be "optimized" for VARA but the details are not publicly available.)
- Optimizations for use with portable radios, including TX delay to accommodate slow transmit/receive switching.
- Some VHF/UHF transceivers have poor audio response at low frequencies and VARA FM Narrow mitigates that issue by "shifting the frequency range middle frequency up 200 Hz."
- Ping function to verify connectivity between two VARA FM stations. "Ping" used in VARA FM is a functional description to perform a simple connectivity test. VARA FM is not compatible with the Ping function using TCP/IP over the air.
- Monitor Mode displays all VARA FM stations heard, including their SNR.
- Transmit Delay (TXD) is automatically optimized.

VARA SAT

VARA SAT is based on VARA HF with optimizations for satellite operations, including the unique aspects of QO-100 such as significant latency. (QO-100 is the amateur payload on the geosynchronous Es'hail-2 satellite over Europe and Africa.) Optimizations in VARA SAT include variable transmit power, delays in equipment key-up (software defined transceivers, amplifier key-up, transverter inline, etc.), transponder operation, and adjustments for Doppler effects.

INTEGRATING VARA WITH OTHER SOFTWARE

VARA provides interoperability with other software using the Telnet protocol via TCP/IP socket 8300 for commands and socket 8301 for the data stream. The author provides a specification document, *VARA Protocol Native TNC Commands*, at **rosmodem.wordpress.com** to describe the various parameters and variables that can be sent and queried to interoperate with VARA.

VARA also provides a "KISS" Interface so that software designed to work with packet radio TNCs (and software implementation of TNCs) that offer a KISS interface can use VARA. (See the Packet Radio section of this chapter.) The author provides a specification document, *VARA KISS Interface*, to describe the unique aspects of the interface.

A unique aspect of VARA's KISS interface is that when other software (such as APRS) transmits (via KISS) an AX.25 Unnumbered Information (UI) frame, such transmissions are transmitted by VARA in a "robust VARA Broadcast frame" (114 bytes).

In VARA HF, KISS frames are transmitted using "VARA HF 500 Hz Level 5" — Symbol Rate 94, 2 carriers, 4PSK, data rate of 177 bps. "VARA HF 500 Hz Level 5" can be received by VARA HF 2750, VARA HF 2300, and VARA HF 500.

Note that if VARA HF is configured to work with a Winlink software client, the VARA HF's KISS interface is disabled. The author states, "This is to disable use of APRS by Winlink gateways operating under automatic control. Winlink gateways have position and frequencies published on the Winlink web site, so a position beacon is unnecessary."

In VARA FM, KISS frames are transmitted using "VARA FM Narrow Level 1" — Symbol Rate 42, 14 carriers, 4PSK, data rate of 549 bps. "Narrow Level 1" is presumably used for maximum compatibility with all VARA FM stations, including those stations that have not purchased a registration key and thus can only operate at "Level 1." This detail in the author's document "VARA Kiss Interface.pdf" "Rev, October 21th 2021" is ambiguous.

In VARA SAT, KISS frames are transmitted using "VARA SAT Level 3" — Symbol Rate 94, 20 carriers, FSK, data rate of 175 bps.

VARA-COMPATIBLE SOFTWARE

VARA's author provides two utility programs specifically for use with VARA:

- *VARA Chat* — Chat and file transfers, and
- *VARA Terminal* — Dumb terminal for use with Bulletin Board Systems.

Other software that directly supports VARA (specifically mentioned in various VARA documentation) includes *BPQ32, JNOS, Winlink Express, RMS Packet, RMS Trimode,* and *vARIM*.

LIMITATIONS OF VARA

- VARA is, essentially, a file transfer protocol. It does not have basic network functionality inherent in packet radio such as "routing" (with the exception of digipeater functionality).
- VARA is proprietary software; thus, any bug fixes, updates, and improvements are only available from the author. As described in the author's *VARA HF Modem Specification Revision 1.0.0 Oct30 2017,* VARA is a proprietary system developed by Jose Alberto Nieto Ros, EA5HVK and can be used under shareware license.
- VARA runs only on Windows but there have been some projects to run VARA FM on Linux using *WINE* (Windows emulation) software. (**k6eta.com/linux/installing-rms-express-on-linux-with-wine**)
- Full VARA functionality requires the user purchase a "registration key" for each unique station call sign. (There may be up to 15 SSIDs per call sign.) The registration key is purchased from within VARA HF, VARA FM, or VARA SAT software. One registration key will work with all three VARA variants. If a registration key is not purchased, VARA will only function at "Level 1" of each VARA variant. Example, VARA FM Level 1 is 566 bps (Wide) and 549 bps (Narrow) versus (with a registration key) 25,210 bps (Wide) and 12,750 bps (Narrow).
- Some updates of VARA software are not compatible with previous versions. For maximum compatibility within a group of VARA stations, the same version of VARA software should be used.

15.5.8 Amateur Radio Digital Open Protocol (ARDOP)

Development on ARDOP began in 2015 as a replacement for the earlier Winlink Message Over Radio (WINMOR) "transport" protocol used in Winlink clients and servers. (WINMOR is described in the paper "Legacy Digital Modes" in this chapter's online content.) Compatibility with WINMOR was not a requirement for ARDOP, and development of WINMOR has ceased.

The overall goal of ARDOP is to create a radio modem that is sufficiently capable that expensive, proprietary hardware radio modems would not generally be required except for the most demanding requirements (highest possible speeds). The key aspect of ARDOP is "Open;" the stated goal of ARDOP was, "The specification and the protocol will be released to the public domain." That goal has been met.

The details of the ARDOP protocol are well defined in *Amateur Radio Digital Open Protocol (ARDOP) Specification*, by Rick Muething, KN6KB. The current revision is 0.3.1, Mar 25, 2015.

There is an active ARDOP Support Mailing List with nearly 1,500 subscribers as of early 2022 (**ardop.groups.io/g/main**) that supports three active specialized subgroups with their own active mailing lists: Applications, Developers, and Users.

ARDOP REQUIREMENTS AND GOALS

High level requirements for ARDOP:

- Hardware compatibility — accommodate frequency accuracy of typical HF and VHF/UHF radios, majority of audio interfaces ("Sound cards"), all popular methods of Push-To-Talk (PTT) control, and flexible host interface (TCP/IP socket or serial).
- Use channel width of 200, 500, 1000, and 2000 Hz (negotiated between client and server).
- Optimize choices of channel size, modulation method, forward error correction (FEC) to maximize the fastest possible error-free data transfers.
- Compliance with the FCC symbol rate limit of 300 baud below 28 MHz (§97.307(f)(3)).
- Strong resistance to multipath distortion.
- Listen-before-transmit mechanism to avoid causing interference with a communication in progress.
- Agnostic implementation — software on host computer ("Sound card" mode) or dedicated processor/DSP (real-time or no operating system).
- Usable for both HF (single sideband, AFSK) radios and VHF/UHF (FM) radios.
- Both Forward Error Correction (FEC) and Automatic Retry Request (ARQ) for error-free data transfer (reliable transport).
- Automatic adjustment of timing parameters, including for use with VHF/UHF voice repeaters, round trip timing, latencies in the operating system (if any), TX delay, networking, etc.

Two interesting goals of ARDOP are to support both file transfers between two stations using ARQ, and a broadcast/multicast mode using FEC.

ARDOP IMPLEMENTATION

There are two stable "code" implementations of ARDOP:

• ARDOPc v1 by John Wiseman G8BPQ, for Linux (**www.cantab.net/users/john.wiseman/Documents/ARDOPC.htm**)

• WIN_ARDOP v1 by Rick Muething, for Windows (**ardop.groups.io/g/users/files**)

Progress on ARDOP development beyond these two v1.0 implementations appears to be stalled, possibly due to the rapid evolution of VARA as a competitor mode to ARDOP. (**github.com/la5nta/pat/issues/148#issuecomment-483914337**)

ARDOP has been implemented in the following Winlink software:

Pat is "Winlink anywhere," a Winlink client developed as an open-source project with versions available for Windows, Mac OS, and Linux. (**getpat.io**) The Linux version of *Pat* is popular because it has been ported to Raspberry Pi's Linux and is now included with many amateur radio software distributions for that inexpensive, capable computer. Developers for *Pat* use GitHub for *Pat* development. (**github.com/la5nta/pat**) *Pat* is unaffiliated with the Winlink Development Team.

Winlink Express (WE, **winlink.org/WinlinkExpress**) is the "preferred Winlink radio email client" because it "does it all," including many radio access methods other than ARDOP. WE is developed and supported by the Winlink Development Team. *Winlink Express* only operates on Windows (newer than Windows XP). (See the Winlink section later in this chapter.)

RMS Trimode (**www.winlink.org/content/rms_trimode**) is software to host a Winlink Radio Mail Server (RMS) — HF gateway stations in the worldwide Winlink system. ARDOP is one of several supported radio connection methods in *RMS Trimode*, including proprietary modes that require hardware modems.

ARDOP HARDWARE MODEM — TNC-PI9K6

The ARDOP goal of "Agnostic implementation ... dedicated processor/DSP (real-time or no operating system" was realized. John Wiseman, G8BPQ implemented ARDOP (**www.cantab.net/users/john.wiseman/Documents/ARDOPTeensy.html**) on a Teensy 3.6 microcontroller (**www.pjrc.com/store/teensy36.html**) that was the basis for the Coastal Chipworks TNC-Pi9K6 "TNC" HAT (Hardware Attached on Top) board (**www.tnc-x.com/TNCPi9k6.htm**) for a Raspberry Pi computer. Although Coastal Chipworks is no longer in business, another version of the TNCPi-9k6 is available (**www.etsy.com/listing/775608115/tnc-pi9k6-raspberry-pi-hat-wteensy-32**) from the West Valley Amateur Radio Club (WVARC, Sun City, Arizona; **westvalleyarc.com**). It is unknown if the WVARC version of the TNC-Pi9k6 is compatible with G8BPQ's implementation on ARDOP on the Coastal Chipworks version of the TNC-Pi9k6.

ARDOP PERFORMANCE COMPARISON

The undated paper by Tom Whiteside, N5TW: "A comparison of Winlink digital mode performance based on simulation results using the Teensy IONOS Simulator" (**winlink.org/sites/default/files/downloads/a_winlink_digital_mode_performance_comparison_based_on_the_ionos_hf_vhf_channel_simulator_-_may_18_2020.pdf**) gives some perspective of ARDOP's overall performance compared to other data modes:

"Executive Summary:

"Hardware SCS modems running PACTOR 2,3 and 4 were evaluated as were sound card modes WINMOR, ARDOP and VARA across a variety of channel models and across a range of signal to noise conditions for HF. VARA testing included a new VARA 500 500 Hz mode. VARA FM and AX.25/FX.25 packet were simulated for a VHF channel.

"Spoiler alert: The SCS modems did very well as you would expect. The much less expensive VARA HF did especially well across the range of conditions tested. The SCS modems and VARA could run long test cases without losing a connection while ARDOP and WINMOR were slow and unable to run many of the cases. VARA FM crushed AX.25/FX.25 VHF cases.

...

"WINMOR and ARDOP were really breakthrough modes when they were first created at a time when computer/sound card technology was much less advanced than today, but their performance lags the other modes by a great deal. They were also much less reliable across the multipath cases with incredibly low rates and frequent connection drops."

15.5.9 Packet Radio

Amateur Packet Radio began in the late 1970s when groups of Canadian hams started experimenting with a new mode of communication. American hams had a slow start because the FCC did not authorize use of ASCII until 1980. Tucson Amateur Packet Radio (TAPR; **tapr.org**) refined the protocol and called it AX.25 because it was based on the commercial X.25 standard.

Packet radio offers several advantages over non-structured, character-based protocols:

• Transmissions are quick bursts (packets) rather than single characters

• Source and destination routing information in each packet.

• Error checking to detect corruption

• Connection-oriented protocol for guaranteed delivery or notification of failure (*reliable transport*)

• Digital repeaters (*digipeaters*) to extend range.

• Multiple packet radio communications can use the same frequency concurrently.

PACKET RADIO OVERVIEW

Packet radio is a method of exchanging data between stations that is most often used for direct, keyboard-to-keyboard connections and the transfer of small files between stations, either between two live operators or between an operator and a bulletin board system (BBS) or email servers.

A network of packet stations is built using the AX.25 protocol. The AX.25 protocol specification (version 2.2, July 1998) can be found on the TAPR website at **www.tapr.org/pdf/AX25.2.2.pdf**.

Despite its low data rate, Bell 202 modulation at 1,200 bps (bits/sec) has remained the standard for VHF/UHF operation in most areas because it is inexpensive and performs reliably over a standard FM voice channel. 9,600 bps is also popular on the VHF/UHF bands, although it is more technically demanding and requires about 5 kHz of bandwidth for higher speed data. This audio connection is usually available as a separate output on an accessory or data connector. At HF frequencies, Bell 103 modulation is used, due to the FCC rule (§97.307(f)(3)) limiting a data signal's symbol rate to 300 bauds below 28 MHz. (1200 bauds may be used on 10 meters.) As of early 2022, FCC action is pending on action to change the rules regarding maximum data symbol rates.

In the US, 145.01, 145.03, 145.05, 145.07, and 145.09 MHz.are the usual frequencies used for packet radio. Packet can be used on VHF/UHF channels as a transport method for applications like *Winlink Express* which require reliable transport (error-free) of small data files and messages.

TERMINAL NODE CONTROLLER (TNC)

The AX.25 protocol was first implemented with specialized hardware called a *Terminal Node Controller* (TNC), composed of a modem and a microprocessor. TAPR offered a complete kit, the TNC-1, in 1983, followed by the TNC-2 which was smaller and cheaper to manufacture. The design was then licensed to multiple manufacturers.

In the late 1980s, numerous TNC variations became available. Manufacturers added higher speed modems (2,400 and 9,600 bps), built-in bulletin board system software (BBS), transmission of telemetry and weather station data, and additional modes such as weather satellite image reception. A slide show history of TNC evolution can be found at **github.com/wb2osz/direwolf-presentation**.

KISS

The original text-based user interface was designed for operators communicating between dumb terminals. This interface was not well suited for communication with a computer application. Commands, responses, and data were all combined in a single stream. The format varied by manufacturer and depended on the operating mode of the TNC.

The KISS protocol (Keep It Simple, Stupid; **www.ka9q.net/papers/kiss.html**) was invented in 1986 and better suited for communicating with applications. Most of the protocol handling is moved to the host PC. The TNC does little more than converting between the KISS protocol and the frames (aka "packets," see section below) sent over the air. KISS-only TNCs are the standard today because the connection-oriented functions of the protocol are moved to the computer or not needed at all as with APRS (see the next section).

KISS is very simple and doesn't provide any information about the TNC status or what is happening on the radio channel. For example, an application might send a dozen packets to a TNC to be transmitted. Since KISS does not support any form of flow control, there is no standard way for the application to know whether they have been transmitted, are still in the transmit queue, waiting because of a busy channel, or discarded.

SOFTWARE TNC

The physical TNC device can be eliminated and replaced by software on a host PC with a sound card, similar to *WSJT-X*. The "software TNC" approach is very flexible. When access to a TNC is available over a computer network (Ethernet, WiFi), rather than being constrained by direct RS-232 cable connection, it is known as a *network TNC*.

As shown in **Figure 15.10**, even a single-board Raspberry Pi computer can run multiple digipeaters at the same time, send telemetry data from a weather station, and act as an internet gateway (IGate), connecting your local radio network to others around the world over the internet.

The early "soundcard modem" software produced poor results, giving this approach a bad reputation. At one time, these applications needed special hardware for acceptable results but that is no longer true with today's computers.

Currently, two software TNC implementations, *UZ7HO SoundModem* (**uz7.ho.ua/packetradio.htm**) and *Dire Wolf* (**github.com/wb2osz/direwolf**), outperform all of the hardware modem ICs. The test protocols and results are available at **github.com/wb2osz/direwolf/blob/dev/doc/WA8LMF-TNC-Test-CD-Results.pdf**.

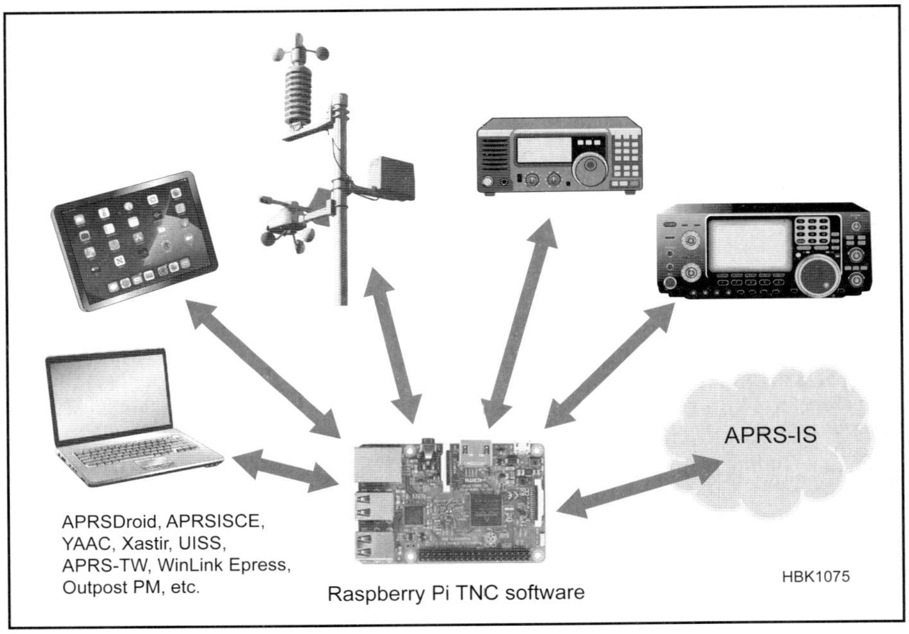

Figure 15.10 — A network TNC is not constrained by the traditional RS-232 interface to a PC and offers many interfacing possibilities with multiple applications and devices at the same time.

New Packet Radio

An IP-based high-speed version of packet radio, New Packet Radio, was created by F4HDK in 2019 (**hackaday.io/project/164092-npr-new-packet-radio**). Operating on 70 cm, the protocol supports 50 – 500 kbps and fully bi-directional links. Due to symbol rate restrictions in the US on 70 cm, a slower configuration of the protocol is available. The local interface is Ethernet so no special host software is required. The modem software is open-source, as well.

DIGIPEATERS & NETWORKING

Packet radio range can be extended by using digital repeater stations (digipeaters). Unlike voice repeaters that simultaneously receive on one frequency and transmit on another, digipeaters usually use "store and forward" on a single radio frequency. Packets are received, held in memory, and retransmitted when the frequency is clear. Digipeaters are used extensively with APRS (Automatic Packet Reporting System) covered below. Traditional full-featured TNCs included digipeating capability. A KISS-only TNC requires a separate software application to perform the digipeating function.

The original digipeating process required the user to understand the possible communication paths among the digipeaters and manually construct a path to be taken by the packets. Various networking schemes such as NET/ROM, BPQ, Xrouter, *NOS, Flex, TheNET, and ROSE were devised to solve the routing problem. The Terrestrial Amateur Radio Packet Network (TARPN; **tarpn.net**) is now vigorously promoting packet networking and provides detailed instructions on how to build a node.

AX.25 FRAMES AND ERROR DETECTION

AX.25 exchanges data as HDLC frames (High-Level Data Link Control protocol) that contain routing, identification, data, and error detection information. There are three general types of AX.25 frames:

a) Information frame (I frame);
b) Supervisory frame (S frame); and
c) Unnumbered frame (U frame).

Each frame is made up of several smaller groups, called fields. **Figure 15.11** illustrates the three basic types of frames. Note that the first bit to be transmitted is on the left side. FCS is the Frame Check Sequence field and PID is the Protocol Identifier field.

Packet radio uses the Frame Check Sequence to include a two-byte checksum in every frame. The checksum is generated using the CRC-CCITT polynomial and is sent low-byte first. If the checksum indicates an error, the frame is discarded. In AX.25 connected mode a retransmission request is made as in other ARQ modes. The FX.25 protocol extension adds forward error correction (FEC) to the protocol while retaining complete in-

Flag	Address	Control	Info	FCS	Flag
01111110	112/224 Bits	8/16 Bits	N*8 Bits	16 Bits	01111110

(A) U and S frame construction

Flag	Address	Control	PID	Info	FCS	Flag
01111110	112/224 Bits	8/16 Bits	8 Bits	N*8 Bits	16 Bits	01111110

(B) Information frame construction

HBK1076

Figure 15.11 — The three types of frames or packets used in AX.25 networks. Unnumbered (U) frames are used for commands and responses to commands such as requests to connect and disconnect stations. Supervisory (S) frames are used to acknowledge or reject information frames. The Information (I) frames are used to exchange data.

teroperability with regular AX.25. **github.com/wb2osz/direwolf/blob/master/doc/AX25_plus_FEC_equals_FX25.pdf**.

TCP/IP

TCP/IP (Transmission Control Protocol/Internet Protocol) is a standardized set of protocols that form the backbone of the internet and are compatible with existing network technologies and applications. The first widely used implementation of TCP/IP for amateur radio was *NOS*, created by Phil Karn, KA9Q. *NOS* was influential in the development of TCP/IP in amateur radio because he wrote *NOS* in (highly portable) C programming language and made the *NOS* source code publicly available (what we now call open source). *NOS* and its follow-ons are now essentially obsolete as "TCP/IP stacks" are available as modules to be added into software. TCP/IP capability is an integral part of Linux and other operating systems.

The only other item necessary to use TCP/IP packet software is your own static IP address in Network 44, termed AMPRNet (AMateur Packet Radio Network). Individual IP Address Coordinators assign addresses to new TCP/IP users in internet network 44 (44.0.0.0/9 and 44.128.0.0/10) based on physical location worldwide. More information can be found on the AMPRNet website at **portal.ampr.org**.

With globally unique addresses, routing could be performed over the internet when necessary. A big advantage of using the standard TCP/IP routing and data transport protocols is that we can use ordinary internet applications for e-mail, file transfer, remote login, web browsers, Voice over IP (VoIP), routing, and so on rather than reinventing our own equivalent.

The additional overhead for TCP/IP makes it painful over AX.25 when using 1,200 or 9,600 bps modems, so it has not reached its full potential. Perhaps we will see a resurgence if higher speed (56 kbps+) data radios become available.

15.5.10 APRS

The Automatic Packet Reporting System (APRS) was developed by Bob Bruninga, WB4APR, SK, as a real time local tactical information resource for emergencies, public service, and everyday use. It is not a vehicle tracking system. APRS data is intended for integration with maps and other forms of information display.

Besides reporting positions of user stations and vehicles, packets may include weather reports, telemetry, and messages to an individual or bulletins to groups (similar to mobile text messages) with queries and responses. APRS messages support the "Ham Radio of Things" (HRoT, the amateur radio equivalent of the Internet of Things or IoT). An image format called Packet Compressed Sensing Imaging (**maqifrnswa.github.io/PCSI**) has also been defined to send images when individual packets may be lost or received out of order.

APRS uses the same AX.25 protocol and TNCs as packet radio (see the previous section), but in a different way. Rather than one-to-one data transfer, it usually operates as a one-to-many transmission without the expectation of reliable data transport. In this way, APRS uses the redundancy of many stations in the network to deliver messages.

APRS STATIONS AND BEACONS

Traditional full featured TNCs can transmit fixed location beacons and serve as digipeaters. Depending on the model, they might also send telemetry, data from a weather station, or a location from a GPS receiver. KISS-only TNCs rely on a host computer with applications to provide the functionality. Software TNCs have varying degrees of functionality (GPS tracker, digipeater, IGate, DTMF decoding, etc.) built in so they can provide services autonomously without additional applications.

Stations are generally set to transmit or *beacon* their position and other information periodically. Fixed stations should beacon infrequently to avoid crowding the channel with repetitive information. Moving mobile stations can receive their coordinates from a GPS and should report more frequently. "Smart beaconing" automatically adjusts beaconing to be more frequent at higher speeds and when making sharp changes in direction.

APRS station names are usually a call sign followed by an optional number, up to 15, called the *substation identifier* (SSID). This allows multiple stations to be operated using the same base call sign. An APRS *beacon* transmission will generally contain a call sign with SSID, position report (lat/long), heading, speed, altitude, display icon type, status text and routing information. It may also contain antenna/power information, weather data, or a short text comment.

The SSID has historically provided information about the type of station. For example, a call sign of N7SS-9 would generally indicate a mobile station. Since there are several systems in use, the SSID information is not definitive. Tactical call signs can also be used if the assigned call sign is included as part of the status text.

Some VHF/UHF transceivers (both handheld and mobile) have a built in TNC, a display, often a GPS receiver, and other features for APRS. This is very convenient because everything is in a single package. They maintain lists of stations heard and calculate the distance away and direction. They can also be used to send messages to others and receive them. Some dual-receiver models can use the frequency, offset, and tone information in a packet and switch voice operation to that channel. This might be a request to contact a person listening or an Object advertising a repeater. These radios include a method of displaying and entering messages without requiring the use of a computer meaning their user can directly respond to a message sent.

In North America the standard APRS frequency is 144.39 MHz. See **www.aprs.org** for more information about using APRS and the network structure. The APRS protocol specification (August 2000) can be found at **www.aprs.org/doc/APRS101.PDF**. Many enhancements since then have less formal documentation — see **www.aprs.org/aprs11.html** and **www.aprs.org/aprs12.html**.

APRS DIGIPEATERS

Although APRS is based on packet radio, APRS packets spread in all directions through the generic network of digipeaters rather than flowing toward a specific pre-determined destination. As the packet moves through the network, it often encounters an IGate station that will relay the packet to an APRS-IS server.

The sending station determines the maximum number of times the packet can be relayed with a PATH value that sets the maximum number of "hops." (A hop is a relay between digipeaters or between a digipeater and an internet gateway or IGate.) This count is decremented each time the information is repeated in order to limit the number of hops, preventing data from being relayed too far from the originating station. APRS digipeaters remember what they transmitted in the last half minute and will not send a duplicate. This prevents the same packet from bouncing back and forth between digipeaters, as well. Digipeating for APRS is covered in much more detail at **www.aprs.net.au/vhf/aprs-digipeaters-101**.

A local group should be able to offer guidance on how the PATH should be set to avoid overloading the local network. Listening to packet signals on the national APRS frequency of 144.390 MHz will provide some idea how busy the channel is. In much of the country the APRS network is well developed and a beacon doesn't need more than one or two hops to cover a wide area and find an IGate.

APRS SOFTWARE

Many applications are available to display and generate APRS information. Most involve a map to show locations of objects and capability to send/receive APRS messages, telemetry, weather reports, and function as digipeaters and IGates. Some of the most commonly used modern applications are:

APRSISCE/32 — **aprsisce.wikidot.com**

PinPoint APRS — **www.pinpointaprs.com**

SARTrack — **sartrack.co.nz** which is specifically intended for Search and Rescue and Emergency Management

UISS — **www.qsl.net/on6mu/uiss.htm** designed for packet communication with the International Space Station (ISS), and compatible satellites.

UI-View — **www.ui-view.net** (popular but no longer maintained)

Xastir — **xastir.org/index.php/Main_Page**

Yet Another APRS Client (YAAC) — **www.ka2ddo.org/ka2ddo/YAAC.html**

The app *APRS Telemetry Watcher* (APRS-TW; **aprstw.blandranch.net**) is different than other map-centric applications. Its purpose is to monitor APRS telemetry and alert the operator to predefined conditions or events, such as low battery voltage.

APRS-IS AND INTERNET GATEWAYS (IGATE)

The APRS Internet Service (APRS-IS; **aprs-is.net**) is an internet-based network which interconnects various APRS radio networks around the world and in space. Internet gateway (IGate) stations provide a two-way link between radio networks and the APRS-IS servers. This allows information from one isolated radio network to be routed to another radio network around the world. It is highly recommended that all IGate stations be bi-directional if possible. Restrictions vary by country. Receive-only IGate stations break the intended two-way communication.

The online mapping applications at **aprs.fi** and **www.aprsdirect.com** gather information from APRS-IS and display it on maps accessible over the internet. The **findu.com** website offers historical weather data, a message display, and a large list of other queries. The IGate stations do not send information directly to these websites, they are only clients of the APRS-IS network.

APRS INTEGRATION WITH OTHER TECHNOLOGIES

The Automatic Identification System (AIS; **www.navcen.uscg.gov/?pageName=aismain**) is an international tracking system for ships in coastal waters using two dedicated channels in the VHF FM Marine Radio band. AIS transponder equipment is required by law in commercial ships and is recommended for private vessels. Messages can contain position, speed, course, name, destination, status, dimensions, destination, ETA, and many other types of information. Due to its similarity to APRS, it is not hard to write a converter to translate AIS sentences to APRS format. (**github.com/wb2osz/direwolf/blob/master/doc/AIS-Reception.pdf**) This allows information about ships to be merged with APRS data and displayed on APRS mapping applications.

With billions of computers and mobile phones (handheld computers) all connected by the Internet, the large growth is expected from the "Internet of Things." What is a "thing?" It could be a temperature sensor, garage door opener, motion detector, flood water level, smoke alarm, antenna rotator, coffee maker, lights, station thermostat, or just about anything you might want to monitor or control. There exists a global APRS network with numerous digipeaters and IGate stations joined by APRS-IS servers. This is also referred to as the "APRS of Things" or the "Ham Radio of Things" (HRoT; **github.com/wb2osz/hrot**) The network is very lightly used and available for experimentation and new applications.

Two different and parallel e-mail gateways allow APRS messages to be converted to short e-mail or text messages as documented at **www.aprs-is.net/email.aspx**. When entering a message, "EMAIL" or "EMAIL-2" is used for the addressee. The message body must then contain the actual e-mail address followed by a space and very short message. If the message is picked up by an IGate it will be sent to the mail server and out to the recipient, with a confirmation APRS message. For dozens of other messaging options, check **aprs.org/aprs-messaging.html**.

There is connectivity between APRS and the Winlink system via APRSLink. APRSLink monitors all APRS traffic gated to the internet, worldwide, and watches for special commands that allow APRS users to:

• Read short e-mail messages sent to their callsign@winlink.org

• Send short e-mail messages to any valid e-mail address or Winlink user

• Perform e-mail related maintenance (see commands below)

• Be notified of pending Winlink e-mail via APRS message

• Query APRSLink for information on the closest Winlink RMS packet station

Attention must be paid to the APRS traffic generated when using these features but they can be very handy. Details on the APRSLink system are available at **winlink.org/APRSlink**.

APRS position reports are also available from GPS-equipped D-STAR radios via DPRS gateways (see the **Repeater Systems** chapter). The DPRS specification defines how to translate the continuous stream of GPS position reports in the D-STAR DV stream to individual APRS packets. This definition restricts the number of APRS packets generated to prevent overrunning a local RF network because the D-STAR radios continuously stream new position reports while the user is talking. Most D-STAR repeaters have DPRS IGates running on the gateway providing the D-STAR radio positions to APRS-IS. Those positions can be gated to local RF by APRS IGates if the APRS IGate sysop enables this feature. The BrandMeister DMR Server also provides an interface from DMR radios to the APRS network.

15.5.11 Winlink

Winlink (**www.winlink.org**) is a worldwide radio messaging system that takes advantage of the internet where possible. It does this in order to allow the end-user more functionality on the crowded radio spectrum. The system provides radio interconnection services including email with attachments, position reporting, graphic and text weather bulletins, emergency/disaster relief communications, and relaying messages. This section is an overview of the Winlink system — more information is available at **www.winlink.org/content/winlink_book_knowledge**, including the Winlink FAQ.

Winlink has been assisting the maritime and RV communities continuously for many years. Emergency communications (emcomm) communities also make use of the system and the Winlink development team has responded by adding features that make the system more reliable and redundant. The role of Winlink in emergency communications is to add another tool in the toolkit of volunteer services deploying emergency communications in their communities.

SYSTEM ARCHITECTURE

The Winlink network structure is explained in "Overview of the Winlink "Hybrid" Network" available on the Winlink website. The system is designed around *Central Message Servers* (CMS) hosted on Amazon Web Services (AWS) in multiple locations for reliability. These ensure that the system will remain operational should any piece of the internet become inoperative. Individual *Radio Message Servers* (RMS) transfer messages between the CMS (via internet) and system users (via radio). Users may also communicate directly with the CMS via the internet. The system has three basic configurations: Conventional (by far the most widely used), Radio-Only, and Peer-to-Peer. *Winlink Express* software manages how each user accesses the system. An individual message may have several destinations that include radio-to-radio addresses as well as radio-to-internet email addresses, allowing complete flexibility.

All messages exchanged within the Winlink system, including directly via the internet, are compressed according to the open-source B2F protocol to minimize transmission time. To comply with FCC rules regarding encryption, within Winlink and in all compatible user software, the modem hardware or software that handles radio modes are always used without any native or proprietary capability to compress, obscure, or encrypt message content. (See **winlink.org/B2F**.)

CONVENTIONAL CONFIGURATION

In the Conventional configuration shown in **Figure 15.12**, RMS stations connect to the CMS via the internet. The set of CMS "in the cloud" act as a hub, routing messages to and from the RMS. The RMS stations act as gateways, providing endpoint connectivity to users via HF or VHF and are located worldwide.

A number of modes can be used to connect with RMS stations, including Packet (VHF), Pactor (HF), Robust Packet, ARDOP, VARA (HF and VHF), and Iridium Go. (These protocols are discussed elsewhere in this section except for Iridium Go.) Users can also connect directly to the CMS servers over the internet via the Telnet protocol. Several slower legacy modes are sometimes used, but are not recommended in order to reduce congestion on the radio frequencies used by Winlink stations.

Because each RMS can access an identical copy of the message it does not matter which one is used to exchange messages. In addition to individual email messages, each RMS stores or has access to hundreds of text-based or graphic weather products, and each can relay the user's position to a web-based view of reporting users and interoperates with the APRS system.

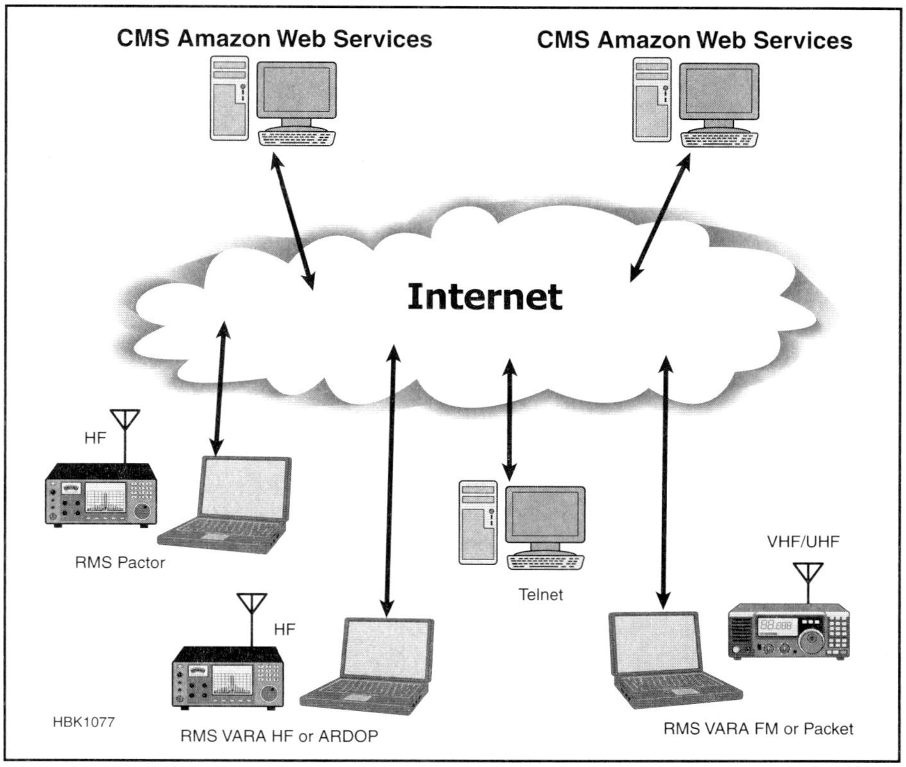

Figure 15.12 — Winlink system. Central Message Servers (CMS) are hosted on Amazon Wireless Services (AWS) servers.

RADIO-ONLY CONFIGURATION

In case of an internet outage, the Radio-Only configuration allows traffic to be exchanged directly between RMS stations. Because the CMS are not available in this configuration, each user must designate at least one *Message Pickup Station* (MPS). (Up to three MPSs may be designated by a user.) When a message is sent to that user, it will be automatically forwarded from one RMS to another until it reaches the recipient's chosen MPS. Because all traffic is over the air, message handling will take longer than in the Conventional configuration and internet email addresses are not available.

Radio-only forwarding is an option that can be selected when you generate a message. It means that message will not be sent over the internet, even if the internet is available. Any message sent by radio-only forwarding will reach each MPS and will be picked up when the recipient next connects to that station.

PEER-TO-PEER CONFIGURATION

Finally, a direct radio-to-radio configuration (Peer-to-Peer) allows stations to exchange messages directly without going through an RMS station or CMS. This requires both stations to be able to communicate using the same protocol on a single frequency. Internet and message relay functions are not available in this configuration.

MAXIMIZING DATA RATE

One of the most important objectives in the eyes of the Winlink development team was to reduce the use of the HF spectrum to only that required to exchange messages with a user and to do that at full "machine" speeds. The HF spectrum is very crowded, so by limiting the forwarding of messages between Winlink RMS stations to the internet, a great deal of radio air time is eliminated, making the time and spectrum available to others for message handling or other operations. Many RMS stations are limited to higher speed modes and have a much higher ratio of traffic minutes and message counts to connect times than do the RMS stations that operate at slower speeds. In other words, the amount of traffic that is passed with a higher speed connection is much greater for an equivalent amount of connect time with approximately the same number of connections.

The fastest HF mode is PACTOR IV, but it is not normally available in the United States because the protocol's data symbol rate is greater than is permitted below 30 MHz by FCC rules. Occasionally, PACTOR IV is temporarily authorized by the FCC during disaster responses to clear HF traffic. It is used regularly by stations outside the United States.

CLIENT ACCESS TO WINLINK

Winlink supports a clean, simple interface to the internet SMTP email system through the *Winlink Express* software available at **winlink.org/WinlinkExpress**. You can obtain a Winlink email address by using *Winlink Express* to connect to the system either over the radio or through Telnet. There is no charge, although a $25 donation (one time) is suggested. You can't obtain a Winlink address over the web.

Any message sent or received may include multiple recipients and multiple binary attachments. Mail addressed to a Winlink address from a non-Winlink email address, however, must be known to the radio user or the message will be rejected. Spam would clog the system otherwise. The system maintains a whitelist of non-Winlink email addresses you've sent mail to recently and you will receive messages sent to you from addresses on that list. You can also add email addresses to your account at **winlink.org/content/how_manage_your_whitelist_spamcontrol**.

In case of an emergency where local services are interrupted, the system may be used by non-amateur groups as an alternative to normal email. Connecting to any of the RMS can automatically allow a local amateur station to pass emergency traffic. Winlink uses no external source for sending or receiving internet email. It is a stand-alone function which interacts directly with the internet rather than through any external internet service provider.

Airmail is independently developed, distributed and supported by Jim Corenman, KE6RK. It is the oldest program for sending and receiving messages using Winlink. Once connected to a Winlink station, message transfer is completely automatic. On the amateur bands, *Airmail* can transfer messages automatically with other stations using *Airmail* and any station supporting Winlink.

When used with Winlink, *Airmail* also contains position reporting capabilities, and a propagation prediction program to determine which of the participating Winlink stations will work from anywhere on Earth. To obtain a copy of *Airmail*, including the installation and operating instructions, download the program from **www.siriuscyber.net/ham**.

When web access is available, Winlink messages can be retrieved or sent via a web interface (webmail). The web browser access is limited to text-based messages without the use of bulletins or file attachments. Password-protected access to your Winlink mailbox is available at **www.winlink.org/webmail**.

Winlink can be accessed while directly connected through the internet to one of the CMS servers. This method of obtaining messages over the internet allows multiple attachments, catalog bulletins, and all other Winlink services normally available over radio channels, but at internet speeds.

Both *Winlink Express* and *Airmail* support Telnet. This allows regular use of the system with the same software configuration at high speed via internet, without using RF bandwidth and provides a full-featured mechanism to access Winlink. An Emergency Operations Center (EOC) with internet access can easily use Telnet with RF as a fallback. Similarly, an RV or marine user can use the software from an internet café or home via Telnet and switch the software back to radio when mobile, with no change in functionality.

WINLINK FEATURES

Bulletins

To address the needs of mobile users for near real-time data, Winlink uses an "on-demand" bulletin distribution mechanism. Users select requested bulletins from an available "catalog" list managed in *Airmail* or *Winlink Express*. When bulletin requests are received by an RMS, a fresh locally cached copy of the requested bulletin is delivered. If no fresh locally-cached version is available, the RMS accesses the internet and finds the bulletin which is then downloaded to the RMS and sent to the user.

The global catalog includes hundreds of weather, propagation, and information bulletins, including instructions for using the system, world news, and piracy reports. All Winlink RMS stations support a single global catalog which ensures users can access any bulletin from any RMS. Bulletins can contain basic text, graphic fax or satellite images, binary or encoded files like GRIB or WMO weather reports. Local processing is used to re-process images to sizes suitable for HF PACTOR transmission. The system prevents bulletin duplication and automatically purges obsolete time-sensitive weather bulletins and replaces them with the current version.

The system also has the ability to contain bulletins with attachment information which is local to each RMS. This is especially useful for the non-public emcomm RMS which may house valuable procedural information needed by specific agencies in any community emergency.

The Winlink network administrator may post notices that are delivered to all individual Winlink users as a private message. This is a valuable tool for notifying users of system changes, outages, software upgrades, emergencies, etc.

Attachments

Multiple binary or text-based file attachments of any type or number may be attached to a message by simply selecting the file to be sent from a Windows selection dialog in *Airmail* or *Winlink Express*. Email message attachments sent through the Winlink system must be limited in size. Users can set the maximum file size. When using the default B2F format, the protocol chosen by the user usually determines the file size of an attachment. A user may also turn off the ability to receive file attachments. Certain file attachment types are blocked from the system to protect the user from viruses and malware.

Standard Form Templates

The U.S. government Federal Emergency Management Agency (FEMA) publishes standard forms which must be used by local agencies for official communications, such as to request resources. To repeatedly transmit form graphics over the radio would be a great waste of limited spectrum. To avoid this, the *Winlink Express* software includes a regularly updated library of forms — called *templates* — with each copy of the software. Using templates means the radio traffic only consists of the form name and the text used to fill out the form. This greatly reduces transmission time for the message. Since the same template is available to the recipient of the message, the completed form appears to the recipient exactly as it was filled in by the originator.

If the recipient software doesn't include a specific template, the text is transmitted in human-readable form. This means the text of the message transmitted can easily be read by the recipient even though it isn't displayed in the original form.

Position Reporting

The bulletin services mentioned above is beyond normal messaging, but Winlink also provides rapid position reporting from anywhere in the world. This facility is interconnected with the APRS, ShipTrak, and YotReps networks. It supports weather reporting from cruising yachts at sea and an interconnection with the YotReps network which is used by government forecasters for weather observations in parts of the world where no others are available. It also allows the maritime user to participate in the National Weather Service's NOAA MAROB, a voluntary marine observation reporting program.

To ensure equitable access to the system, individual users are assigned daily time limits on HF frequencies by RMS operators. The default time per any 24-hour period is 30 minutes, however, the user may request more time from the RMS operator should it be needed. The time limit is individual to each RMS station. Utilization of higher speed modes such as PACTOR III and PACTOR IV are a great timesaver.

Network Maintenance

The system has a number of other secondary network maintenance features. Extensive traffic reports are collected, the state of individual RMS stations is monitored and reported if one becomes inactive. Daily backups are performed automatically at all RMS and CMS to ensure the system integrity. Security is maintained through the vigorous updating of virus definitions and automatic virus screening for internet mail and files. The system has the ability to block any user by both radio (by frequency band) and internet (by email address) to prevent abuse of the system.

15.5.12 D-STAR

D-STAR (Digital Smart Technologies for Amateur Radio) is a digital voice and data protocol specification developed from research by the Japan Amateur Radio League (JARL) into digital technologies for amateur radio in 2001. While other digital on-air technologies for other services are being used by amateurs, D-STAR was one of the first on-air protocols designed specifically for amateur service use to be widely deployed. As of early 2022, radios supporting D-STAR are sold by Icom and Kenwood.

D-STAR is a bit-streaming protocol that encapsulates digitized voice and low speed data (DV) or higher speed Ethernet data (DD). D-STAR radios can communicate either simplex or via repeaters. D-STAR equipment is available for the 2 meter (VHF), 70 cm (UHF) and 23 cm (1.2 GHz) bands.

D-STAR's GMSK modulation similarly to FM modulation: Only one station can talk at a time. Since Forward Error Correction (FEC) is included in D-STAR. FEC will correct bit errors in the data stream. This provides a higher sound quality and slightly greater range than FM modulation. With FM signals, signal quality gradually degrades as signal strength decreases. With D-STAR, the receiving station hears perfect signals down to low signal levels before the signal is no longer received.

An overview of the D-STAR system, "D-STAR — For the Second Century of Amateur Radio" is available at **www.icomamerica.com/en/downloads/DownloadDocument.aspx?Document=366**. A detailed discussion of the D-STAR specification is contained in a three-part *QEX* article by John Gibbs, KC7YXD, in the Jul/Aug, Sep/Oct, and Nov/Dec 2003 issues. These articles are available at **www.icomamerica.com/en/products/amateur/dstar/dstar/default.aspx** in the Downloads area. More information about D-STAR repeater systems may be found in the **Repeater Systems** chapter.

DD AND DV MODE

The D-STAR protocol specifies two modes; Digital Voice (DV) and Digital Data (DD). The DV mode provides both voice and low speed data channel on 2 meters, 70 cm and 23 cm over a 4,800-bps (bit/s) data stream. The high-speed Digital Data (DD) mode operates at 128 kbps on the 23 cm band. The user's amateur radio call sign is included in all transmissions for signal routing and identification.

The DV data stream is broken down to three main packages: voice, forward error correction (FEC) and data. (See **Figure 15.13A**.) The largest portion of the data stream is the voice package, which is a total of 3,600 bits/s with 1,200 bps dedicated to forward error correction, leaving 1,200 bps for data. This additional data contains various data flags as well as the data header, leaving about 950 bps available for either GPS or serial data. This portion of the data stream does not provide any type of error correction, which has been overcome by implementing error correction in the application software. A second data option has been made within the DV protocol in which the serial data rate is 3,480 bps and utilizes the entire channel.

The focus of D-STAR's design was the most efficient way to conserve RF spectrum. While D-STAR's occupied bandwidth is 6.25 kHz, tests reveal a band plan of 10 kHz spacing is adequate to incorporate the D-STAR signal as well as provide space for channel guards.

In later models of D-STAR radios, a higher speed data protocol was implemented in DV mode which "borrows" some of the voice bits for data transmission of pictures and files at a maximum rate of 3,600 bits/s increasing data throughput.

In addition to DV mode, the D-STAR protocol outlines the high-speed Digital

Voice Mode Message Transfer Details

Radio Header								Data						
			ID											
Bit Sync	Frame Sync	Flag	Destination Repeater Callsign	Departure Repeater Callsign	Companion Callsign	Own Callsign	PFCS	Voice Frame	Data Frame	Voice Frame	Data Frame	- -	Voice Frame	Last Frame
64bit	15bit	1byte	8byte	8byte	8byte	8byte	2byte	48bit	48bit	48bit	48bbit		48bit	48bit

(A)

Data Mode Message Transfer Details

Radio Header								Data					FCS
			ID						MAC Header				
Bit Sync	Frame Sync	Flag	Destination Repeater Callsign	Departure Repeater Callsign	Companion Callsign	Own Callsign	PFCS	ELen	SA	DA	Type	Data Frame	CRC
64bit	15bit	1byte	8byte	8byte	8byte	8byte	2byte	2byte	6byte	6byte	2byte	46-1500byte	4byte

(B)

Figure 15.13 — D-STAR protocol frame structure for DV (A) and DD (B).

Data (DD) mode. (See Figure 15.13B.) This higher speed data, 128 kbps, is available only on the 23 cm band because it requires a 150 kHz bandwidth, only available at 23 cm in world-wide band plans. Unlike the DV mode repeaters, the DD mode module operates as an "access point" operating in half duplex, switching quickly on a single channel.

As with the DV mode, there is a portion of the data stream used for signal identification with the data header as well as various system flags and other D-STAR related items. Once this portion of the data stream is taken into consideration, the 128 kbps is reduced to approximately 100 kbps — still more than double a dial-up connection speed with significant range.

Another consideration is the data rate specified at 128 kbps is the gross data rate. Therefore, the system developers are challenged by the area coverage/potential user issue. This means the higher the elevation of the system, the more potential users and the slower the system will become as all the users split the data bandwidth. Finally, there is an issue from the days of packet radio. While technically, the opportunity for "hidden transmitter" issues does exist and collisions do occur, the TR switching is very fast and this effect is handled by TCP/IP as it is for WiFi access points.

The simplex channel eliminates the need for duplexers at a repeater site if only the DD mode system is installed. It is still recommended to have filtering, such as a band-pass filter, in place to reduce possible interference from other digital sources close to the 23 cm band as well as to reduce RF overload from nearby RF sources. The Icom 23cm DD repeater module operates at an output of 10 watts.

Radios currently providing DV mode data service use a serial port for low-speed data (1,200 bps), while the DD mode radio offers a standard Ethernet connection for high-speed (128 bps) connections, to allow easy interfacing with computer equipment. The DD mode Ethernet jack allows two radios to act as an Ethernet bridge without any special software support required. This allows standard file sharing, FTP, TELNET, HTTP/web browsing, IRC chat or even remote desktop connections to function as if connected by Ethernet.

D-STAR VOICE CODEC

The AMBE voice codec (encoder/decoder, also referred to as a *vocoder*) for D-STAR uses a voice data rate (2,400 bps) optimized for use by public service and satellite communications services. These characteristics made it suitable for amateur radio, as well. The AMBE codec is proprietary to Digital Voice Systems, Inc. (DVSI). AMBE stands for Advanced Multi-Band Excitation and as firmware, it is embedded in ICs for radio and other communications equipment.

There are several AMBE products with varying rates of voice compression. JARL tested the different rates and determined that one of the lowest rates would be acceptable to amateur radio operators. While there has been considerable discussion about the proprietary nature of AMBE, at the time of development of D-STAR, there were no open protocols available that could be implemented in a handheld radio with the processors available at that time, at a reasonable cost. AMBE was already in service in many applications, including P25 and later, DMR.

D-STAR REPEATERS

A D-STAR repeater system consists of at least one RF module and a controller. While any combination of RF modules can be installed, typically a full system includes the three voice modules (2 meters, 70 cm and 23 cm) and the 23 cm DD mode module shown in **Figure 15.14**. Repeaters only retransmit the data stream and do not convert the voice data back to analog form.

A computer with dual Ethernet ports, running the Gateway software, is required for internet access to the global network. An additional server, as shown in the diagram, can be incorporated for local hosting of e-mail, chat, FTP, web, and other services.

In a D-STAR system installation, the standard repeater components (cavities, isolators, antennas, and so forth) are not shown but are required as with any analog system. Some groups have removed analog gear and replaced it with D-STAR components on the same frequency with no additional work beyond connecting the power and feed lines.

GATEWAY CONFIGURATION

In DV mode, user registration is not required but allows for additional capabilities such as repeater linking/unlinking and access via hotspots (discussed in the **Repeater Systems** chapter). An unregistered user may access a local repeater and if it is linked to another repeater or reflector the user will be included in the local conversation as well as the transmission "sent" to other linked repeaters or reflectors. Only registered users may link and unlink the repeater. Users accessing the D-STAR network through hotspots must be registered to connect to repeaters or reflectors.

In a Gateway configuration, *all* DD mode users must be registered in the network. This provides the DD mode sysop a layer of authorization, meaning that if someone wants to use a DD mode system, and they have not received authorization to use the gateway, their DD mode access will be denied. Any gateway registered user, on the common network, can use any DD mode system, even if the registration was not made on that system. While we are not able to use encryption in the amateur radio service, security can be implemented in standard software or consumer routers and firewalls.

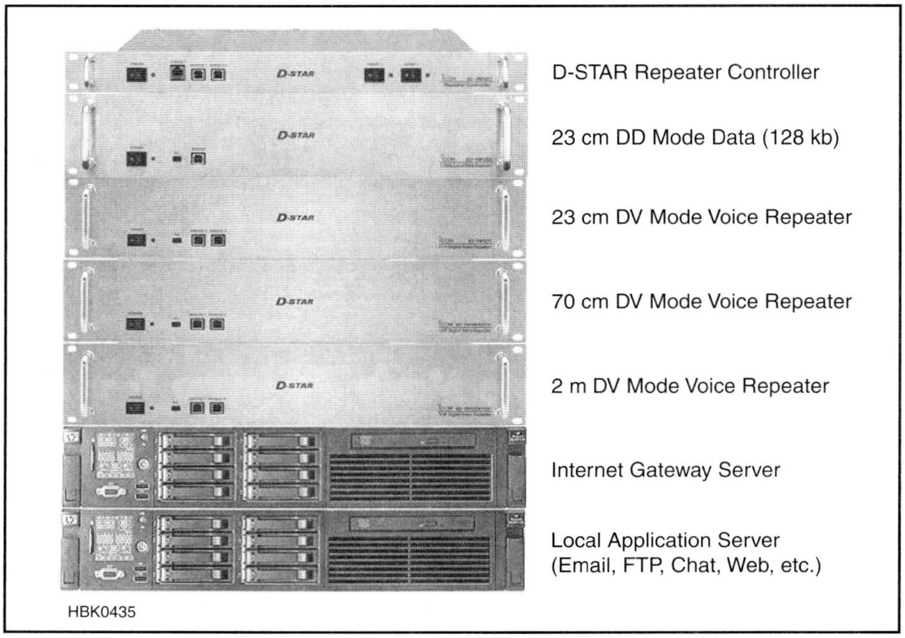

Figure 15.14 — Full D-STAR repeater system showing the controller and complement of DV and DD repeater modules.

15.5.13 APCO Project 25 (P25)

Project 25 (P25) or APCO-25 is a suite of standards for digital radio communications for use by public safety agencies in North America. P25 was established by Association of Public-Safety Communications Officials (APCO) to address the need for common digital public safety radio communications standards for first responders and Homeland Security/emergency response professionals. In this regard, P25 fills the same role as the European Tetra protocol, although not interoperable with it.

P25 has been developed in phases with Phase 1 completed in 1995. P25 Phase 1 radio systems operate in 12.5 kHz analog, digital or mixed mode. Phase 1 radios use Continuous 4-level FM (C4FM) modulation for digital transmissions at 4,800 baud and two bits per symbol, yielding 9,600 bit/s total channel throughput. In the case of data transmission, data packets basically consist of a header, containing overhead information, followed by data. In the case of digitized voice transmission, after the transmission of a header containing error protected overhead information, 2,400 bit/s is devoted to periodically repeating the overhead information needed to allow for late entry (or the missed reception of the header).

After evaluating several candidates, Project 25 selected the IMBE (Improved MultiBand Excitation) vocoder from DVSI for Phase 1, operating at 4,400 bit/s. An additional 2,800 bit/s of forward error correction is added for error correction of the digitized voice. An 88.9-bit/s low-speed data channel is provided in the digitized voice frame structure and several forms of encryption are supported.

P25 Phase 2 requires a further effective bandwidth reduction to 6.25 kHz by operating as two-slot TDMA (Time Division Multiple Access) with support for trunking systems. This provides two channels from the same 12.5 kHz spectrum when working through a repeater to provide time synchronization. The TDMA timing requirements does limit range to a published 35 km but this is plenty for public safety users. It uses the newer AMBE (Advanced MultiBand Excitation) codec from DVSI to accommodate the bit rate reduction to 4,800 bit/s. Phase 2 radios will still offer backward compatibility with Phase 1. As of 2022, Phase 3 access to the 700 MHz band in the US has been implemented.

P25 radios offer a number of features of interest to public service agencies such as an emergency mode, optional text messaging, over the air programming/deactivation. More P25 information can be found at **www.apcointl.org/technology/interoperability/project-25**.

15.5.14 Yaesu System Fusion

Yaesu released a specification for System Fusion in 2013, becoming the most recent DV methodology. System Fusion supports digital voice and data in a 12.5 kHz narrow-band channel at 9,600 bps, using C4FM modulation over VHF (144-148 MHz) and UHF (440-450 MHz). System Fusion supports three modes of operation: voice full rate (Voice FR) mode, data full rate (Data FR) mode, or voice/data (V/D) mode. Voice FR mode is typically displayed on System Fusion transceivers as "VW," with voice/data mode appearing as "DN." The two full rate modes use the entire 9,600 bps channel for their respective voice or data payloads, whereas the V/D mode splits the channel into two 4,800 bps payloads, with voice information on one and data on the other.

The voice modes also include robust forward error correction (FEC) and utilize the AMBE+2 (Advanced MultiBand Excitation) CODEC from DVSI. In V/D mode, the voice and FEC data use the AMBE+2 "Enhanced Half-Rate" mode where voice and FEC payloads consume individual 3,600 bps totals, with header, synchronization, and routing data consuming much of the remainder of the 4,800 bps available. The data component of the V/D mode can be used for features like GPS information, in order to provide APRS-like functionality. The Voice FR mode encodes using the AMBE+2 "Enhanced Full-Rate" mode, where voice information and FEC consume 7,200 bps. The Voice FR mode provides the highest quality voice exchange.

Each transmission comprises 960 byte (100 ms) packets. Each transmission begins with a header packet (HC), followed by communication channel packets (CC), and ends with a terminator packet (TC). The HC and TC packets synchronization information, source, and destination call signs, plus uplink and downlink call signs, as used for routing. In the V/D mode, each of the CC packets contain 40 bytes of frame sync, then 200 bytes of frame information, followed by 72 bytes of interleaved voice and data frames, for a total of 960 bytes. In the Voice FR mode, the first CC packet contains some overflow call sign information, but is generally segmented into 144-byte frames with voice information.

Yaesu repeaters offer the System Fusion environment with an "Automatic Mode Select" (AMS), operating in both FM analog and System Fusion digital modes. The repeaters can be configured in a combination of modes. While the repeaters are capable of operation in a purely analog mode, this prevents digitally equipped users from taking advantage of the enhancements that digital operation offers. With a similar discouragement toward digital enhancements, the repeaters can be configured in such a way as to allow analog or digital reception, but forcing the output to analog only. While this configuration allows both analog and digital users to coexist, it does so in a constrained fashion, negating the digital advantages of a white noise and hiss free experience, plus strips away GPS and call sign information. Operating in purely digital mode is also possible, offering enhancements for digital users, but effectively "locking out" analog FM operations, as with other DV methodologies. This is typically implemented when there is more than one System Fusion repeater operating in a given area.

More interesting (and common) is the hybrid AMS configuration, allowing analog FM in to analog FM out and digital in to digital out. In this configuration, analog users are not suddenly "disconnected" from a repeater and other analog-only capable operators. Also, digital users are free to take full advantage of the enhancements available with their transceivers. With the users' System Fusion transceivers configured in AMS, digital users can hear an analog call placed on the repeater, in between transmission exchanges. The transceivers will switch to analog, automatically allowing them to communicate with the analog station. This furthers an environment of inclusion for both "camps" to enjoy and allows analog users to upgrade to System Fusion digital at a time of their choosing, rather than forcing a transition, all at once.

In order for analog FM transceivers to remain quiet during digital transmissions on the selected receive frequency, enabling the tone squelch feature to match the repeater's transmitted continuous tone coded subaudible squelch (CTCSS) or digital coded squelch (DCS) signal, is required. Watching for a visual "channel busy" indicator on the analog users' transceivers or enabling a Busy Channel Lockout feature will prevent accidental interference, when digital communications are taking place.

15.5.15 Digital Mobile Radio — DMR

Digital Mobile Radio (DMR) is divided into three tiers, Tier 1 is FDMA (Frequency Division Multiple Access), Tier 2, is TDMA (Time Division Multiple Access), and Tier 3 is TDMA Trunking. Amateurs utilize Tier 2.

Digital Mobile Radio (DMR) is a worldwide standard defined by the European Telecommunications Standards Institute (ETSI) based on Motorola Solutions MOTOTRBO Tier 2 and is used in commercial products from a number of manufacturers. Tier 2 and Tier 3 transmissions fit in a standard 12.5 kHz narrowband channel structured as two-time-slot TDMA. The two channels result in a spectral efficiency of 6.25 kHz per channel making it ultra-narrowband.

DMR uses 4FSK modulation at 4,800

symbols/second, corresponding to 9,600 bps (bits/second) for both time slots. Each DMR timeslot is 30 msec long with 2.5 msec of guard time between time slots resulting in a 27.5 msec frame of 264 bits; 108-bit payload; 48-bit SYNC or embedded signaling; and a second 108-bit payload for a total of 216 bits of payload per frame.

While not specified in the ETSI DMR standard, manufacturers have followed Motorola Solutions' lead and utilize the DVSI AMBE+2 codec to encode and decode the voice audio. The vocoder compresses 60 msec of audio with FEC (forward error correction) into 216 bits of data for transmission.

Since a user radio (as opposed to a repeater) only transmits during a single time slot for 27.5 msec of each 60 msec time period, the result is an approximately 50% transmit duty cycle, unlike an FM or FDMA-based digital mode transmitter. This translates into longer battery life on handheld radios. It should be noted that typical transmit power measurements are not possible for TDMA transmitters.

Synchronization of TDMA transmitters is critical. When using a repeater, the repeater is the time source for synchronization of the user radios. During simplex communications the first transmitter becomes the time source for synchronization. The critical nature of DMR TDMA limits the practical range between user radios and repeaters to about 50 miles because of synchronization. Simplex range between user radios does not have any distance limitation. For amateur DMR repeaters the 50-mile limitation is only an issue if users are active on both time slots at the same time and one is near the repeater and the other is beyond the 50-mile limit.

The original Tier 2 specification for DMR covered conventional simplex or repeater operation. In 2013, the Tier 3 standard for trunking operation was approved by ETSI. While the ETSI standards allow for interconnection of repeaters over IP networks, beaconing by repeaters allows roaming by user radios properly configured in a homogeneous network of repeaters.

Like many digital modes, the DMR specification only covers the air interface to allow the radios to interoperate. It does not cover the network linking protocols, leaving that to the manufacturer. Motorola Solutions uses their IP Site Connect (IPSC) protocol to interconnect repeaters over IP networks. These IPSC networks have been expanded using gateway/conference bridging software on servers to expand networks worldwide. The main gateway/conference bridging products are RavenNet's cBridge (sold by Rayfield Communications) and TL-NET (sold by Bridgecomm Systems), DMRplus (IPSC2), and Brandmeister. The cBridge/TL-NET product supports only MOTOTRBO repeaters and analog interfaces. Brandmeister and DMRplus support a variety of other repeater manufacturers as well as hotspots.

Table 15.15
Wireless Networking Frequencies

airMAX	*Ubiquiti*	*ISM*	*Amateur*
M900 900 MHz	902 – 928	902 – 928	902 – 928
M2 2.4 GHz	2402 – 2462	2400 – 2500	2390 – 2459
M3 3 GHz[1]	3370 – 3730		3300 – 3500[3]
M5 5 GHz	5725 – 5850	5725 – 5875[2]	5650 – 5925

[1]For export from USA
[2]U-NII: 5150 – 5350, 5470 – 5825 MHz
[3]ARRL Band Plan

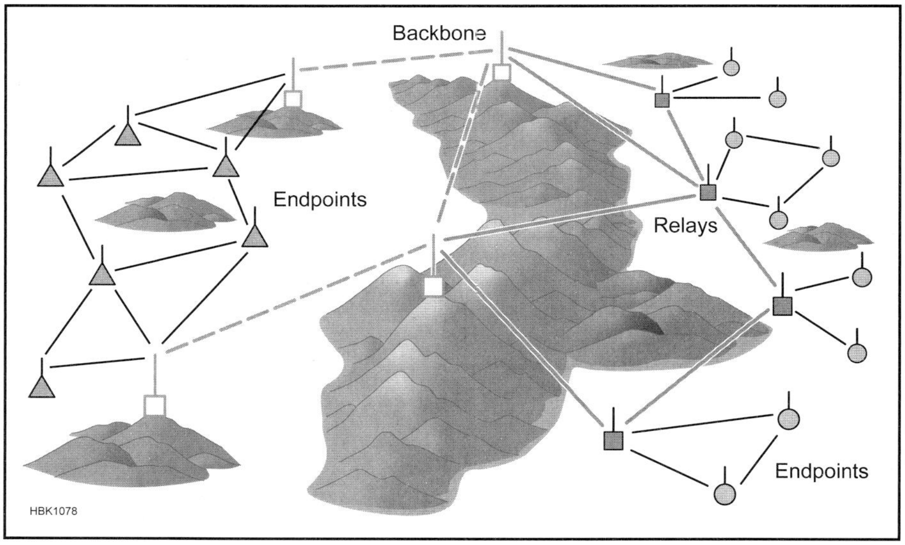

Figure 15.15 — A mesh network is created by connecting the individual user stations (endpoints) either directly or through relay nodes that communicate through backbone links. In this way an entire region can be connected using standard networking technology.

Conferencing *bridges* are used to connect groups of users. DMR uses the term *Talk Group* (TG) to identify a specific bridge. System Fusion uses the term *Room* and D-STAR uses the term *Reflector*. The amateur use of these conference bridges also allows for interconnecting other digital and analog networks together. Many amateur DMR networks utilize one time slot for local traffic and the other for wider area communications. (See the **Repeater Systems** chapter for more information about bridges in these networks.)

15.5.16 IP-Based Microwave Networking

Implementing an IP-based network using amateur radio allows implementation of any IP-based service (subject to the restrictions of FCC Part 97) on amateur frequencies. Microwave bands are used because wide-bandwidth signals are permitted on amateur frequencies above 902 MHz (33 cm). Some examples of services that can be supported over an IP-based network include:

- Keyboard to keyboard (text)
- Voice (VOIP phone systems, chat servers)
- Video (video chat, webcams)
- Email/messaging
- Document editing/management
- File sharing services
- Web servers
- Repeater linking
- Games

NETWORK TOPOLOGIES

There are two primary network topologies used to implement amateur radio microwave IP networks. Amateur Radio Emergency Data Network (AREDN) can create a peer-to-peer mesh or be used in point-to point or point-to-multipoint topologies while HamWAN, HamNet, and Mi6WAN implement star topologies. **Figure 15.15** illustrates the basic mesh network components.

All are TCP/IP-based and the same applications and services can be supported regardless of which networking technology is used. The

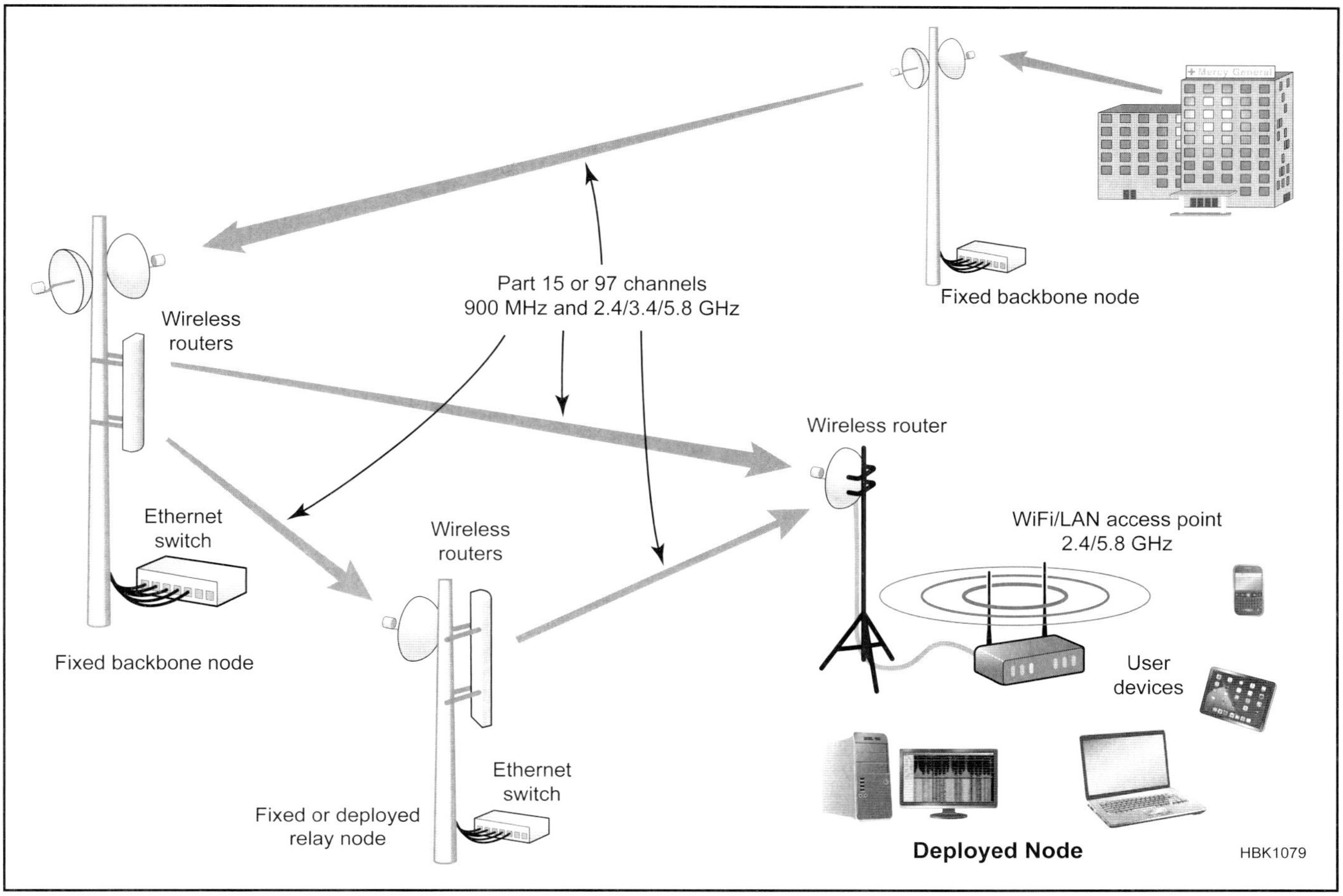

Figure 15.16 — The components of a typical AREDN network.

systems adapt commercial hardware operating in the amateur allocations of microwave frequency bands as shown in **Table 15.15**.

Additional information on amateur networking is available at the following websites:

- AREDN: **www.arednmesh.org**
- HamWAN: **hamwan.org**
- Hamnet (European): **hamnetdb.net**
- Mi6WAN (Central Michigan): **w8cmn.net**

AREDN (AMATEUR RADIO EMERGENCY DATA NETWORK)

AREDN develops software for the Ubiquiti, Mikrotik, TP-Link and Gl.iNet series of wireless routers, in the 900 MHz, 2.4 GHz, 3.4 GHz and 5.8 GHz amateur bands. (See **www.arednmesh.org/content/supported-platform-matrix** for the current list of supported devices.) The stock firmware in the routers is replaced to become a fully functional node in a peer-to-peer network. (See **Figure 15.16**) The map of a typical AREDN network in Western Oregon is shown in **Figure 15.17**.

In an AREDN network, nodes are self-discovering, self-configuring, and self-advertising. This greatly simplifies the establishment of a network since minimal networking knowledge is required to set up a node. For complete information on AREDN network design and technologies, an extensive set of documentation is available at **arednmesh.readthedocs.io**.

A typical AREDN endpoint node consists of a wireless router to connect to the network and a wireless data access point, usually sup-

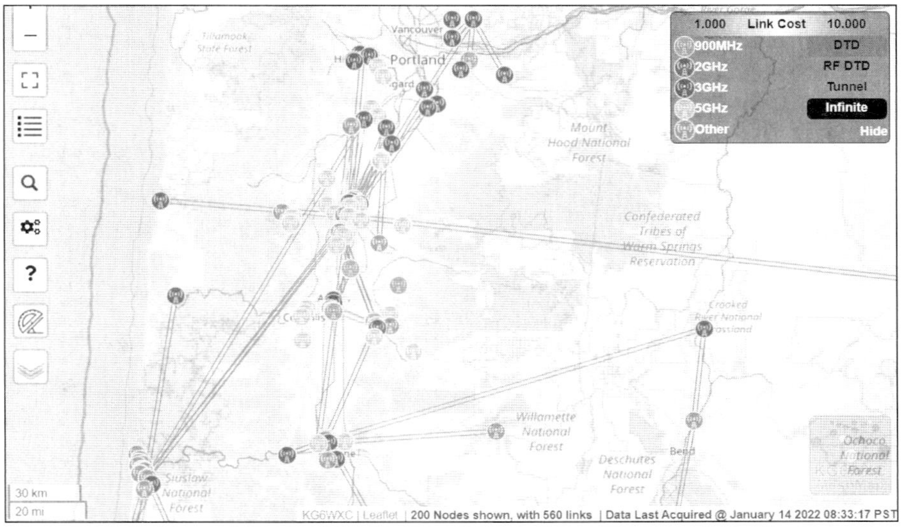

Figure 15.17 — The Western Oregon AREDN network as of early January.

Figure 15.18 — A backbone site (right tower) in Southern California. User access points for 2 and 5 GHz are shown, along with a 5 GHz county backbone link. A 3 GHz local backbone node and a Part 15 wireless access point are also shown. (Photo courtesy Orv Beach, W6BI)

porting both WiFi and Ethernet LAN connections. The access point provides connectivity that is the same as in a typical home network and many AREDN nodes are part of a ham's home network. Relay nodes and backbone nodes consist of two or more wireless routers connected together with an Ethernet switch. No other equipment is required. Installed in advantageous locations, backbone nodes support long-distance links between relay nodes in different areas. **Figure 15.18** shows the K6PVR backbone site in the mountains of Southern California.

The current firmware also includes tunneling capability to allow RF "islands" to be connected through the internet. In this configuration, individual groups of nodes (see **Figure 15.19**) communicate with each and one or more has regular internet access. Tunneling protocols are then used to connect the groups of nodes together as if they were using a purely RF system of relays and backbone links.

HAMWAN

HamWAN utilizes the international version of MikroTik wireless hardware. It's based on a star network topology in which all of the user nodes (also known as *client nodes*) connect directly to a central node, also known as a *cell site* or *distribution node*. Each distribution node communicates over an *uplink path* to a *ring node* connected in a backbone called a *data ring*. (See **Figure 15.20**) Each ring node consists of at least two wireless routers supporting the backbone and one providing the link to the distribution nodes, similarly to the AREDN backbone nodes. A current implementation of the ring-and-distribution network in the Western Washington area is shown in **Figure 15.21**.

HamWAN distribution nodes communicate to the ring nodes over a dedicated link at 3.4 or 5.8 GHz. HamWAN client nodes use MikroTik Nv2 command protocols which support *time division multiple access* (TDMA) technology. HamWAN also uses digital certificates to authenticate users. Unlike AREDN, HamWAN requires pre-coordination to be arranged between the nodes and more extensive network knowledge to configure the nodes. Complete documentation of the HamWAN network is available at **hamwan.org**.

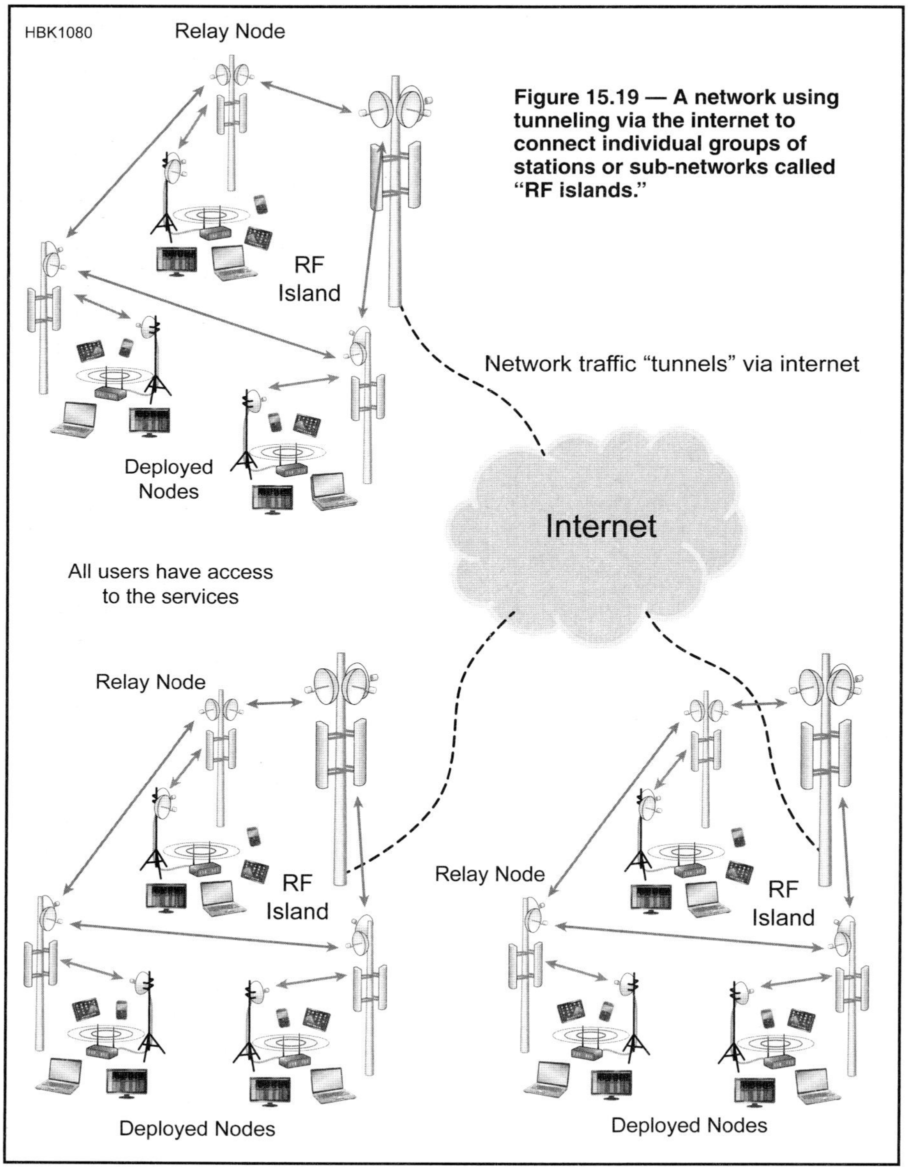

Figure 15.19 — A network using tunneling via the internet to connect individual groups of stations or sub-networks called "RF islands."

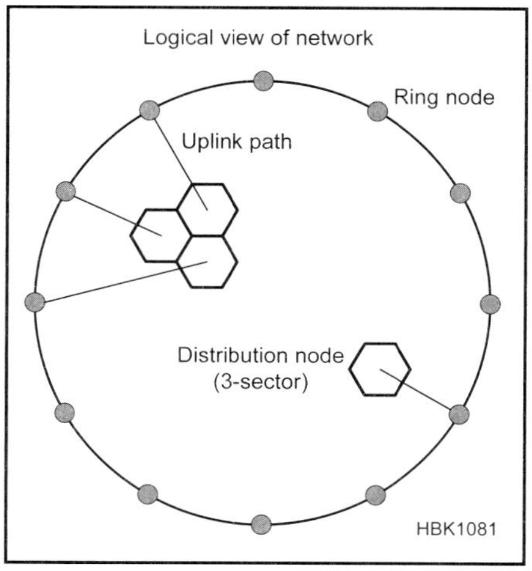

Figure 15.20 — The HamWAN ring network topology is built from distribution nodes that serve a small area and a backbone of ring nodes that connect all of the distribution nodes together. (Graphic from hamwan.org)

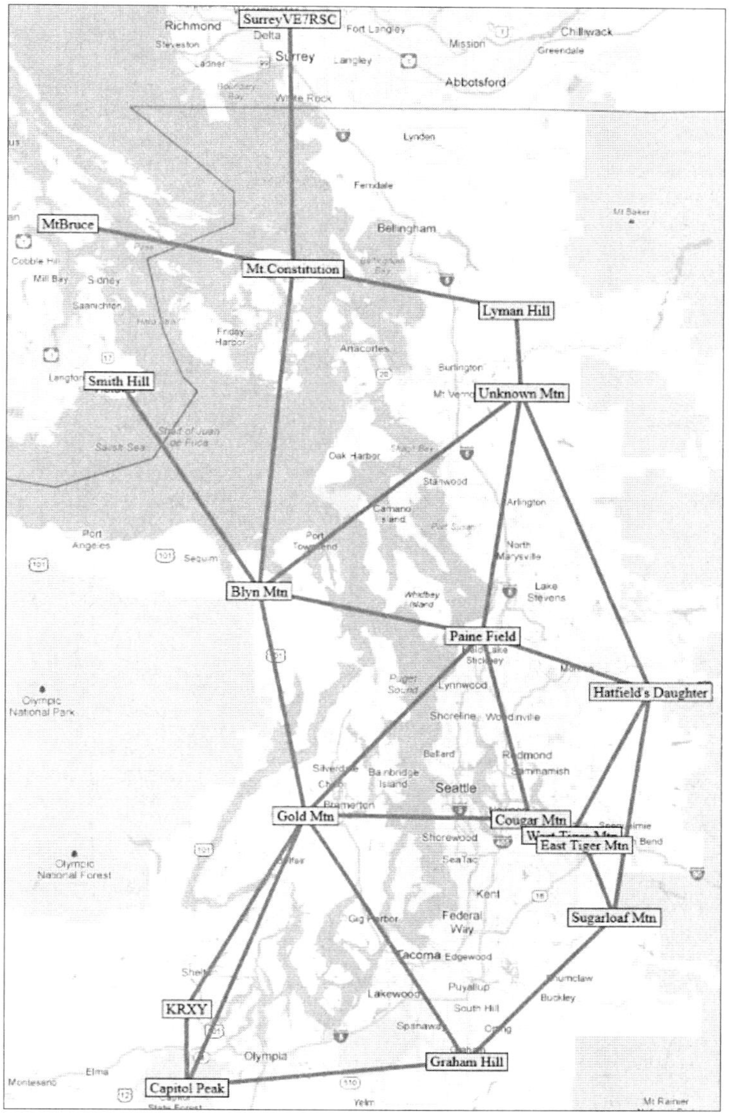

Figure 15.21 — The Puget Sound Data Ring is an HamWAN network in the Western Washington area. (Graphic from hamwan.org)

USE OF FCC PART 15 OR COMMERCIAL CHANNELS

Many amateur networks operating in the microwave spectrum use Part 15 channels for their backbones between groups of nodes which communicate between each other on Part 97 channels. (See Figure 15.16) Other networks are built purely of nodes using only Part 97 channels. There are pros and cons to each approach:

Part 97-only networks use channels that are much less crowded and often have a lower noise floor and thus potentially higher throughput or longer links. However, Part 97 rules prohibit encryption so protocols such as TLS (Transport Layer Security) and SSL (Secure Sockets Layer) cannot be used. Websites that require the use of HTTPS (Secure HTTP) cannot be accessed. Part 15 and other commercial channels are free of rules prohibiting encryption but the channels can be busy with commercial traffic from WISPS (Wireless Intranet Service Providers) operating on these channels

If you can "hitch a ride" on an existing commercial network as a backbone link, the only requirement for supporting an AREDN-based network is that it provides VLAN 2 from point to point. VLAN 2 is the VLAN (Virtual Local Area Network) that AREDN uses to carry DtD (Device to Device) traffic between nodes.

15.6 Digital Mode Table

Table 15.16 is a summary of common digital modes used by amateurs as of early 2013 and their primary characteristics. The following text is intended to make high-level comparisons. For more information on these modes see the earlier sections of this chapter. This table and a detailed listing showing the lowest permitted frequency for a number of digital mode variants are also available as a PDF file with the online content.

Many modes have a number of variants, with the most common shown in the table. High reliability, low data rate modes are more common on HF. Higher data rates are available on VHF/UHF, where the bands have less noise and require less effort to make them reliable. Some modes have very specific intended uses like the meteor scatter, moonbounce and beaconing modes created by Joe Taylor, K1JT as *WSJT* (see the section Structured Digital Modes).

Table 15.16

Digital Modulation Modes and Formats Used in Amateur Radio

See the text of this chapter for definitions and abbreviations

Lowest Permitted Frequency is the lowest amateur frequency at which this emission may be used according to Part 97.307 limits on symbol rate and bandwidth. If no frequency is listed, the emission is permitted in all data segments.

See also the online PDF file "Digital Modes - Lowest Permitted Amateur Frequency" in the online content accompanying this book.

Mode or Format Name	*Developer*	*Principal Freq*	*Principal Application*	*Lowest Permitted Frequency*	*Data Rate (bits/sec)*	*Bit rate (bits/sec)*
ALE	MIL-STD-188-141, FED-STD-1045	HF	Data		<375	375
AMTOR-A	G3PLX	HF	Data		53	114
AMTOR-B	G3PLX	HF	Data		57	114
AOR AMBE	AOR Corp	HF	Voice, Data		2400	3600
APCO P25	APCO	VHF	Voice	50 MHz	6800	9600
ARDOP	KN6KB	HF	Winlink email		25-167	46-2296
Chip64	IZ8BLY	HF	Keyboarding		21.1/37.5	300
CLOVER-II	Hal Comm.	HF	Data		< 37.5-750	62.5-750
CLOVER-2000	Hal Comm.	HF	Data		108-1994	500-3000
DMR	Motorola	VHF	Voice	50 MHz		9600
Domino	ZL2AFP	HF	Keyboarding		31/44/62?	31/44/62
DominoEX	ZL1BPU	HF	Keyboarding		(<15.63)-86.13	15.63-86.13
D-Star (DV)	JARL	VHF	Voice & data	28 MHz	D:960 V:2400	4800
D-Star (DD)	JARL	UHF	Data	902 MHz	72k-124k	128000
Facsimile		HF	Image			
FDMDV	G3PLX/HB9TLK	HF	Voice		1450	1450
FSK441	K1JT	VHF	Meteor scatter	28 MHz	882	882
FFST4*	K9AN, G4WJS, K1JT	LF-MF	Weak signal		1.49	6.18
FST4W*	K9AN, G4WJS, K1JT	LF-MF	Weak signal beacon		0.457	2.92
FT4	K9AN, K1JT	HF	Contests		15.3	41.67
FT8	K9AN, K1JT	HF	DXing		6.1	18.75
G-TOR	Kantronics	HF	Data		35/75/115	80/160/240
Hellschreiber (Feld)	Rudolf Hell	HF	Keyboarding		2.5 char/s	122.5
JT4*	K1JT	V/UHF	Moonbounce		1.64	8.75
JT9*	K1JT	MF-VHF	Weak signal		1.47	5.21
JT65*	K1JT	V/UHF	Moonbounce		1.54	16.1
MFSK16	ZL1BPU/IZ8BLY	HF	Keyboarding, Data		31.25	62.5
MSK144	K9AN, K1JT	VHF	Meteor scatter	50 MHz	1069.4	2000
MT63	SP9VRC	HF	Keyboarding		35/70/140	320/640/1280
Olivia	SP9VRC	HF	Keyboarding		8.75/17.5	78.13/156.25
Packet (Bell202)		VHF	Data	28 MHz	<1200	1200
PACTOR-I	DL6MAA/DF4KV	HF	Data, Winlink email		51.2/128	100/200
PACTOR-II	Spec. Comm. Sys.	HF	Data, Winlink email		100-700	200-800
PACTOR-III	Spec. Comm. Sys.	HF	Data, Winlink email		85-2722	200-3600
PSK31	G3PLX	HF	Keyboarding		31.25	31.25
Q65*	IV3NWV, K1JT	V/UHF	Scatter, Moonbounce		1.51	10.0
QPSK31	G3PLX	HF	Keyboarding		31.25	62.5
PSK63/125		HF	Keyboarding		62.5/125	62.5/125
QPSK62/126		HF	Keyboarding		62.5/125	125/250
Q15X25	SP9VRC	HF	Data		300/1200/2400	2500
RTTY (Baudot)		HF	Keyboarding, contests		30.3	45.45
SSTV (traditional)	WØORX	HF	Image		8 s/frame	
SSTV Martin M1	Martin Emmerson	HF	Image		114 s/frame	
SSTV Scottie S1	Eddie Murphy	HF	Image		110 s/frame	
System Fusion	Yaesu	VHF	Voice, data	50 MHz		9600
Throb	G3PPT	HF	Keyboarding		10/20/40 wpm	
VARA	EA5HVK	HF-VHF	Winlink email		18 – 12,750	
WSPR (MEPT-JT)	K1JT	HF-VHF	Weak signal beacon		0.45	2.93

* A number of submodes are available.

There can be a significant difference between data rate and bit rate in high reliability modes. The data rate is the amount of user data transmitted where the bit rate includes the packet and error correction overhead. Some of these rates are approximate and can vary based on conditions and the variant of the mode used. The symbol rate is a function of the modulation scheme and how many simultaneous carriers are used to transmit the data.

The error handling mechanism is critical to note when sending data. Some modes include no error handling which means errors are not detected and must be addressed in a higher protocol layer (as in AX.25 packet). Forward error correction (FEC) makes a best effort to address errors in real time as part of the data sent in each packet. FEC can add substantial overhead to each packet and does not guarantee error-free delivery but does make the mode robust in high noise environments. Automatic Retry Request (ARQ) can guarantee error-free delivery of data but has no ability to actually correct received data. The combination of FEC and ARQ allows minor errors to be corrected in real time with major errors generating a retry request.

Symbol rate (baud)	*Modulation ("N-" means multi-carrier)*	*Error handling*
125	8FSK	FEC
114	FSK	ARQ
114	FSK	FEC
50	36-QPSK	FEC
4,800	4FSK/QPSK	FEC
300	MFSK, MPSK	FEC
300	DBPSK-DSSS	FEC
31.25	4-(2-16)DPSK/(2-4)DASK	FEC/FEC+ARQ
62.5	8-(2-16)DPSK/(2-4)DASK	FEC/FEC+ARQ
4800	4FSK/QPSK/C4FM	FEC
7.8/11/15.6	16FSK	None
3.9-21.5	18FSK	None/FEC
4800	0.5 GMSK/QPSK/4PSK	FEC
128000	0.5 GMSK/QPSK/4PSK	FEC
120 lpm	FM, 1500-2300 Hz	None
50	15-QPSK	None
441	4FSK	None
3.09	4-GFSK	FEC + CRC
1.46	4-GFSK	FEC + CRC
20.833	4-GFSK	FEC + CRC
6.25	8-GFSK	FEC + CRC
100/200/300	FSK, 170/200 Hz shift	FEC + ARQ
122.5	ASK	None
4.375	4FSK	FEC
1.736	9FSK	FEC
2.69	65FSK	FEC
15.625	16FSK	FEC
2000	OQPSK	FEC + CRC
5/10/20	64-DPSK	FEC
15.63/31.25	32-FSK	FEC
1200	FSK	ARQ
100/200	FSK, 200 Hz shift	ARQ
100	2-DBPSK/PI-4DQPSK/8,16DPSK	FEC + ARQ
100	(2-18)-DBPSK/DQPSK	FEC + ARQ
31.25	BPSK	None
1.667	65FSK	FEC + CRC
31.25	QPSK	FEC
62.5/125	BPSK	None
62.5/125	QPSK	FEC
83.33	15-QPSK	FEC + ARQ
45.45	FSK, 170 Hz shift	None
120-line B/W	FM, 1200-2300 Hz	None
240-line RGB	FM, 1200-2300 Hz	None
240-line RGB	FM, 1200-2300 Hz	None
4800	4FSK/QPSK/C4FM	FEC
1/2/4	9FSK/2-9FSK	None
23-94	OFDM,MFSK, MPSK, QAM	FEC
1.46	4FSK	FEC

15.7 References and Bibliography

Bern, D., W2LNX, and Elkin, K., KB3TCB, "High-Speed Wireless Networking in the UHF and Microwave Bands," *QST*, Nov. 2015, pp. 33 – 37.

Bern, D., W2LNX, and Elkin, K., KB3TCB, "High-Speed Wireless Networking in the UHF and Microwave Bands," **files.tapr.org/meetings/DCC_2014/DCC2014-High-Speed-Wireless-Networking-UHF-Microwave-Bands-W2LNX-presentation.pdf**.

Bombardiere, A., K2MO, "A Quick-Start Guide to ALE400 ARQ FAE," *QST*, June 2010, pp. 34 – 36.

Ford, S., WB8IMY, *Get on the Air with HF Digital*, 3rd Edition (ARRL, 2022).

Ford, S., WB8IMY, *HF Digital Handbook*, 4th Edition (ARRL, 2007).

Ford, S., WB8IMY, *VHF Digital Handbook* (ARRL, 2008).

Franke, S., K9AN, and J. Taylor, K1JT, "Open Source Soft-Decision Decoder for the JT65 (63,12) Reed-Solomon Code," *QEX*, May/June 2016, pp. 8 – 17.

Franke, S., K9AN, and Taylor, J., K1JT, "The MSK144 Protocol for Meteor-Scatter Communication," *QEX*, Sep./Oct. 2018, pp. 30 – 36.

Franke, S., Somerville, B., and Taylor, J., "The FT4 and FT8 Communication Protocols," *QEX*, July/Aug. 2020, pp. 7 – 17.

Lathi, B., *Modern Digital and Analog Communication Systems* (Oxford University Press, 1998).

"Manual of Transmission Methods – Reference Document," Rohde & Schwarz, 2014, available at **www.udxf.nl/R&S-Manual-of-Transmissing-Methods.pdf**.

Palermo, N., IV3NWV, "Q-ary Repeat-Accumulate Codes for Weak Signals Communications."

Popiel, G., KW5GP, *High-Speed Multimedia for Amateur Radio* (ARRL, 2016).

Sklar, B., *Digital Communications, Fundamentals and Applications*, Prentice Hall, 1988.

Yaesu System Fusion Standard, "Amateur Radio Digital Standards," v1.0 January 15, 2013, rev 1.01 April 18, 2013, available for download from **www.yaesu.com**.

Taylor, J., K1JT, "The JT65 Communications Protocol," *QEX*, Sep./Oct. 2005, p. 3.

Taylor, J., K1JT, and Walker, B., W1BW, "WSPRing Around the World," *QST*, Nov. 2010, p. 30.

Taylor, J., K1JT, "WSJT: New Software for VHF Meteor-Scatter Communication," *QST*, Dec. 2001, pp. 36 – 41.

Contents

Chapter 16 — Online Content

Articles

- A Simple Sensor Package for High Altitude Ballooning by John Post, KA5GSQ
- APRS Unveiled by Bob Simmons, WB6EYV
- APRS with a Smartphone by Pat Cain, KØPC
- ARRL Education and Technology Program Space/Sea Buoy by Mark Spencer, WA8SME
- Fox-1 Satellite Telemetry – Part 1: On the Satellite by Burns Fisher, W2BFJ
- Fox-1 Satellite Telemetry – Part 2: FoxTelem by Chris Thompson, AC2CZ
- Touching Near Space on a Budget by Paul Verhage, KD4STH
- High Altitude Platforms – folder with a collection of Powerpoint presentations and PDF articles by Paul Verhage, KD4STH

Chapter 16

Amateur Radio Data Platforms

This chapter addresses the increasing use of amateur communication as a key element of scientific experimentation. The use of amateur means to collect and track data focuses on high-altitude balloons, which are the most popular platform used for this purpose today. Other types of platforms, such as drones and rockets, are also used for experimenting and enjoying operating them. Amateur Radio plays an important role in supporting their use.

The chapter begins with a discussion of what types of sensors and transducers are used to collect data on these platforms. Navigation and telemetry streams are also fundamental to collecting data and operating the platform. Several types of telemetry and location data are covered here along with a review of how the information is transmitted from the platform. (See the **Digital Protocols and Modes** for descriptions of the modes themselves.)

Future editions will continue to expand coverage to additional technologies, platforms, and applications. Material in this edition was updated by Paul Verhage, KD4STH and Bill Brown, WB8ELK.

Amateurs conducting science experiments and operating mobile craft frequently use amateur radio for their data links and control signals, even as they cross continents and entire oceans! Such *platforms* include balloons and multi-rotor copters. Others are land-based (such as weather stations, animal tracking, or robots) or marine (ocean or river buoys or rovers). CubeSats (**www.cubesat.org**) also use Amateur Radio control and data links.

Building these platforms combines several technical fields: using sensors and data acquisition systems to measure events and phenomena, mechanical and electrical engineering to construct the platform, 3-dimensional navigation, and the data link and associated radio technologies. These hybrids are attracting the experimentalists and scientists to amateur radio, just as they were attracted at the dawn of the wireless age.

Support of scientific experimentation is hardly a new aspect of amateur radio — amateurs have supported science almost since the beginning when Tom Mix, 1TS, accompanied the explorer MacMillan to the Arctic aboard the *Bowdoin* in 1923. Amateurs assisted the Naval Laboratory in listening tests that helped establish the existence of the ionosphere. The story continued with Grote Reber, W9GFZ, and radio astronomy in the 1930s, broad participation by hams during the International Geophysical Year of 1957 – 1958, wildlife tracking, propagation reporting, satellite construction, and numerous other instances. Recently, hundreds of amateurs participated in the Solar Eclipse QSO Party (**hamsci.org**) to observe the effect of a total solar eclipse on HF propagation. There is no doubt that fulfillment of FCC Part 97.1 is thriving as hams continue to "improve the radio art" by adapting technology to new uses.

This chapter is organized in four parts: Platform Overview, Sensors, Telemetry and Navigation, and Platform Design. The goal is to cover the engineering necessary to assemble an effective platform. In recognition of the rapid innovation in these activities, this chapter will change in future editions. Expect to see new digital protocols and miniature telemetry and audio-image transmitters. Improvements will be forthcoming in portable-mobile power sources and antennas.

Although the *ARRL Handbook* strives to be complete, covering all aspects of these efforts is beyond the scope of this book. The referenced magazines and websites in the References and Bibliography section at the end of this chapter provide additional material and the latest updates on platforms and technology.

16.1 Platform Overview

Automated platforms are found in many locations carrying out a multitude of functions. They are often the solution to data collection in locations where it's not practical for humans to operate data collection devices for reasons such as safety or extensive time commitment requirements. Aside from data collection, automated platforms are also pressed into service as transponders, repeaters, and beacons. In instances such as these, automated platforms are well served by amateur radio because of the wireless communication requirements these platforms need to telemeter the data they generate, report their location and status, or forward analog and digital radio communications.

Although the primary platforms used today involve high-altitude balloon-borne experiments, similar considerations apply to other fixed and mobile terrestrial and marine experiment platforms.

16.1.1 Platform Structure

The generic structure of a remote sensing platform is shown in **Figure 16.1**. Along with the power source, there are five separate functions:

1) Sensor data or image acquisition — conversion of analog data into digital format and acquisition of still images or video

2) GPS or navigation data — acquisition of location data in digital form

3) Integration of sensor and location data — collection of all data to be stored and/or transmitted to the ground station

4) Protocol engine — packaging and encoding of data for transmission

5) Amateur transmitter — generates the digitally modulated RF signal

These functions can be implemented by separate modules, or everything can be performed by a single microcontroller-based module such as one of the APRS trackers. The choice is completely up to the platform designer and varies with the requirements for the particular mission. For example, **Figure 16.2** shows a two-module solution in which everything except GPS location is provided by the single MMT module. The ATV payload described later uses a separate controller to integrate the video and GPS data for transmission as part of the overall audio-video signal. The combinations are endless! The websites listed in the following section on high-altitude platforms (balloons) are good places to begin looking for the right subsystems for your mission.

DATALOGGERS

Dataloggers are standalone microprocessor-based devices that acquire or "log" digital and analog data on a pre-determined schedule, or when prompted by a pre-determined event. They can also acquire digital data from avionics and navigation equipment. The data is stored or transmitted for later analysis. Dataloggers are available as commercial products, or they can be home-built. Commercial dataloggers have the advantage of being durable and standardized. On the other hand, they may not be customizable to the extent desired for a science flight into the stratosphere. Popular microcontrollers such as the BASIC Stamp (Parallax), PICAXE, Arduino, and Raspberry Pi have made it much easier to create custom dataloggers that are also capable of operating experiments as well as just recording data.

TEMPERATURE RANGES AND HUMIDITY

The platforms discussed in this chapter are often used outdoors or in unprotected environments. As a result, the electronics and other sub-systems can be subjected to extreme temperatures for long periods of time. When selecting components, modules, or other equipment, make sure they are properly rated for the intended use.

The four most common temperature specifications for electronics and electro-mechanical systems are as follows:

Commercial: 0 °C to 85 °C
Industrial: –40 °C to 100 °C
Automotive: –40 °C to 125 °C
MIL-SPEC (or MIL-STD): –55 °C to 125 °C

Components meeting Commercial Specifications will work well for fixed stations residing indoors. However, when designing outdoor fixed stations, the amateur must consider using components that at least meet industrial specifications.

A closely related specification is for relative humidity (RH), which is specified in percent and condensing or non-condensing. RH can range from 0% in the desert or upper atmosphere to 100% from rain or fog or in water-borne environments. "Condensing" means that liquid water forms on surfaces of the device or component directly from the air. If your platform will be subjected to condensing humidity or direct water spray or splash, consult the equipment manufacturer for the best methods of protecting the platform components.

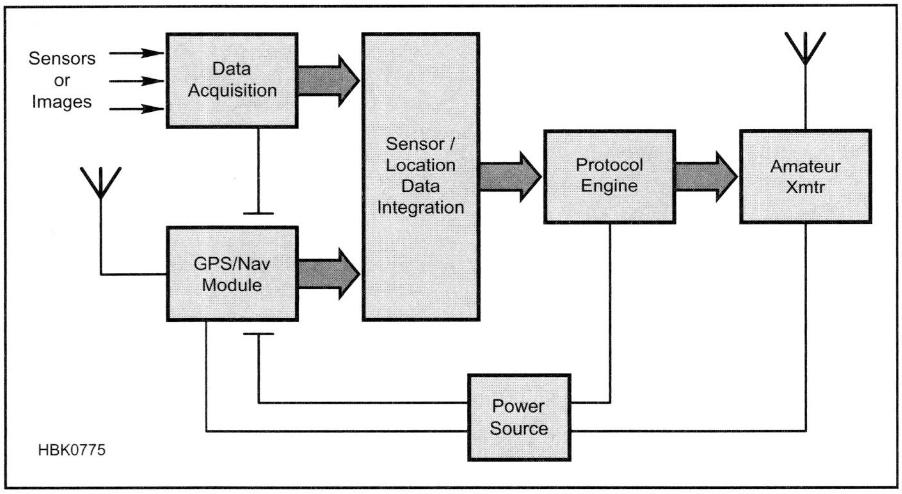

Figure 16.1 — The basic structure of a remote sensing platform using Amateur Radio for the telemetry link.

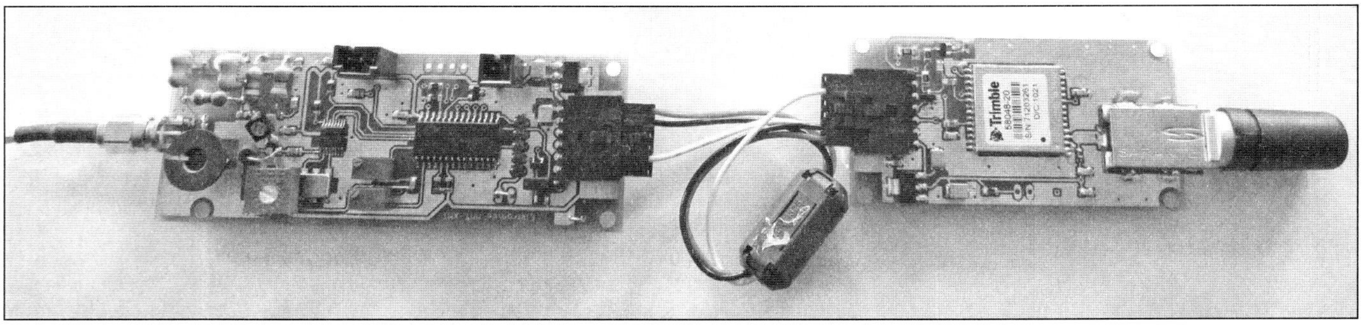

Figure 16.2 — A two-module payload consisting of a GPS receiver module (right) and WB8ELK's MMT (Multi-Mode Transmitter) on the left.

16.1.2 Types of Platforms

HIGH ALTITUDE BALLOONS

High altitude balloons are most frequently latex (mixed with neoprene) weather balloons designed to carry automated platforms weighing up to 12 pounds (heavier weights are possible) into near space. Altitudes above 60,000 feet (flight level 600) and below 328,000 feet are often referred to as *near space*. This region of the atmosphere has conditions that are closer to those found in Earth orbit than to Earth's surface.

Altitudes in excess of 100,000 feet are accessible using large weather balloons and lightweight platforms. Heavier platforms rely on polyethylene skinned balloons called *zero pressure balloons*. These balloons are vented to the atmosphere and do not develop an internal pressure above that of the surrounding atmosphere at any height (**www.eoss.org/faq/zero_pressure**).

The need for long duration flights across continents, oceans, and even circumnavigating the planet has necessitated the development of *super-pressure balloons*. These balloons have skins of high-tech plastics and are sealed airtight. This prevents their lifting gas from escaping and shortening their flight times. While zero pressure and super pressure balloons are traditionally in the realm of professionals, amateurs have been making use of sealed Mylar foil party balloons as well as larger, custom-built, super-pressure envelopes using a special plastic film.

To be successful, the total weight of the payload should be a half-ounce or less. A handful of amateurs have developed incredibly lightweight payloads, and most are totally solar-powered. In one instance, a UK-launched Mylar party balloon remained aloft for months and circled the globe 9 times. More recently, a balloon launched from Atlanta, Georgia, has stayed aloft over a year and has circled the globe more than 20 times.

A number of conditions are commonly measured or monitored, with a microprocessor-based datalogging system to acquire the data and convert it to digital form. The data is stored in onboard memory for recovery or transmitted to ground stations as a telemetry stream. The sensor data can also be integrated with the GPS data for transmission via APRS. Some APRS trackers can acquire analog and digital sensor data and integrate it into the APRS data messages.

Standalone experiments can also be attached to the balloon, then located and recovered after landing. Data collected by sensors carried by the weather balloon can then be analyzed and correlated to the balloon's altitude. Thus, it is extremely important that the payload be able to transmit or store position and altitude data during the flight while collecting data.

You can find a considerable amount of information about balloons and experimental platforms suitable for balloons at **arhab.org**, **wb8elk.com**, **eoss.org**, **ansr.org**, and **nearsys.com**. See also the References and Bibliography section at the end of this chapter.

BalloonSats

BalloonSats are packages with standalone high-altitude experiments. They do not include a tracker, since it is assumed they will be carried by operational platforms that include trackers. BalloonSats are an excellent way to introduce school students to near space exploration. They are simple systems that permit students to perform experiments in the space-like environment of near space.

UNMANNED AERIAL VEHICLES (UAV)

UAV development goes back centuries. However, it was the development of lightweight electronic autopilots that turned the UAV into the practical and affordable platform of today. With the commercial marketing of the four-bladed quadcopter and the six-bladed hexacopter with rechargeable lithium-polymer batteries, amateurs gained access to practical UAVs. The UAV has severe altitude and weight limits that balloons do not. However, the UAV provides a platform for imaging and datalogging at higher resolutions, over more restricted regions, with higher data immediacy and quicker turnaround time. The popularity of this platform and the potential risks posed to conventional aviation have resulted in registration requirements and flight restrictions.

AMATEUR ROCKETRY

Experiments with amateur rocketry began in the late 1920s in Germany. By the 1950s, amateur rocketry had become a growing hobby for post-war America, especially after the launch of Sputnik 1 in 1957.

With the introduction of modern composite propellants and lightweight digital electronics, serious data collection is now possible. Unlike balloons and UAVs, rocket-based platforms must account for the rapid accelerations and tight dimensions inherent in rocketry. Because of their high accelerations, the rocket makes data collection possible at high altitudes but only over very short time frames.

ROBOTICS

Robots can be described as either *autonomous* or *semiautonomous*, depending on the amount of human input they require. Among the most famous robots are the series of rovers NASA has sent to explore the surface of Mars. Inexpensive microcontrollers, which are a form of programmable logic, have brought robotics to the amateur level. Robots often have fewer weight limits than balloons, rockets, or UAVs. They can also carry platforms to desired locations and remain on site while their platform performs its mission. Finally, robots usually place fewer power limits on their platforms than balloons, rockets, and UAVs. However, robots tend to cover the ground at slow speeds and limit their platforms to recording data from the ground level.

FIXED AND FLOATING PLATFORMS

Fixed platforms are any data collection stations that remain in one location. This can be due to their being mounted to a pole like a weather station or anchored to the seabed like a buoy. Typically, fixed platforms remain in place for long duration data collecting or imaging. As such, they require protection from the elements. They also require occasional maintenance from a human who goes onsite to service the fixed platform.

Some are floating buoys that may or may not be anchored. Anchored buoys can be treated as a fixed station. Free-floating buoys or marine rover-style platforms travel on the body of water either under self-contained power or by following the winds and currents.

16.2 Sensors

Inexpensive sensors and microcontrollers combined with amateur radio create opportunities for the amateur to perform experiments in environments that are otherwise inaccessible for one reason or another. Many interesting regions of the Earth, including extremely high altitudes in the atmosphere or the distant ocean, fall into this category. Hams can be instrumental in helping both amateurs and professionals explore these environments.

These platforms enable *remote sensing* — observing or measuring an object or event without a human being actually being in contact with the condition being measured. Data from the measurement is then stored on the platform for eventual collection after recovery or transmitted to a ground station for recording and analysis (telemetry). Examples of parameters that are measured by amateur remote sensing platforms include temperature, pressure (air and water/fluid), humidity, ozone and other gasses, acceleration, and light.

A sensor is a device that reacts to a specific condition of interest, such as temperature or pressure, and produces a predictable output in response. The first step is to select the appropriate sensor or sensors for the parameter of interest and a means of converting sensor outputs to digital data, usually by connecting the outputs to a microcontroller or analog to digital converter IC. Sensors and their associated *signal conditioning* circuits are the "front end" of remote sensing. Analogous to the speech and video circuits of traditional amateur transmitters, the same care in their design is required if quality results are to be obtained.

This section divides sensors into the following four types of outputs: *resistance-based, current-based, voltage-based,* and *digital.* In addition, four common types of sensor outputs are discussed here. Not all of these outputs are directly readable by a microcontroller. However, methods exist to convert the output of these sensors into a form that can be interfaced to a microcontroller. (See the Analog/Digital Conversion section of the **DSP and SDR Fundamentals** chapter.)

16.2.1 Resistance-Based Sensors

Resistance-based sensors change an internal resistance in response to the environmental variable they measure. An example includes the photocell, which is constructed of the chemical cadmium sulfide (CdS), a semiconductor that produces electrons and holes when irradiated by light. The production of free electrons and holes reduces the resistance of CdS when it is exposed to light.

In many cases, the change in resistance in response to changes in the measured condition is small. Therefore, sensor manufacturers often incorporate additional circuitry with the sensing element to convert this changing resistance into a more easily measured change in voltage. However, resistive-type sensors (without signal conditioning) are still available and quite useable.

Resistance-based sensors do not create a signal that a microcontroller can measure directly. Instead, the resistance of the sensor is used to vary the voltage from a regulated voltage source. A simple and popular circuit capable of converting a changing resistance into a changing voltage is the voltage divider as described in the **Electrical Fundamentals** chapter.

The voltage divider circuit of two resistors is shown in **Figure 16.3**. One resistor is fixed in resistance (R_F), and the other is the sensor and therefore variable in resistance (R_V). The current through the voltage divider circuit is variable. It increases as the resistance of the variable sensing element decreases and vice-versa. However, the sum of the two voltage drops is always equal to the supply voltage. The preferred arrangement of the two resistors depends on the response of the sensing resistor to the condition to which it responds — temperature, humidity, illumination, and so on.

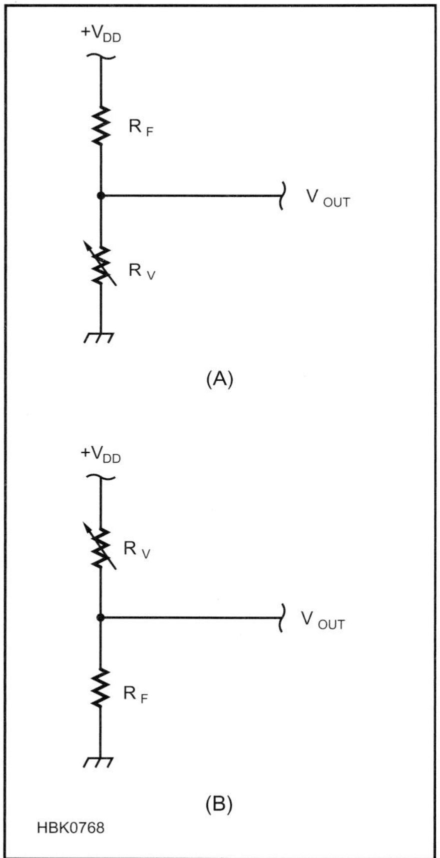

Figure 16.3 — The voltage divider circuit, a series circuit of two resistors. The orientation of resistors in A is preferable when the sensing resistor increases resistance in response to an increasing condition. Use B when the sensing resistor decreases resistance in response to an increasing condition. This results in the output voltage increasing with the increasing condition.

A microcontroller connected to the voltage divider circuit digitizes the voltage across the resistor connected to ground. This permits the design of resistance-based sensors into circuits that produce changing voltages which follow the change in the condition. For example, the resistance of photocells decreases as the light intensity increases. A microcontroller digitizing the voltage across a photocell connected as R_V in Figure 16.3A will observe V_{OUT} *increasing* as light intensity *decreases*. If however, the photocell is connected as R_V in Figure 16.3B, V_{OUT} *increases* as the light intensity *increases*. The latter case is easier to understand and work with than having output voltage and the sensed condition varying in opposite directions (or inversely proportional).

There are two equations that describe the output of the voltage divider circuit. The first describes the voltage drop across the variable resistor, and the second describes the voltage drop across the fixed resistor.

For Figure 16.3A:

$$V_{OUT} = +V\ (R_V/(R_F + R_V))$$

For Figure 16.3B:

$$V_{OUT} = +V\ (R_F/(R_F + R_V))$$

OPTIMIZING R_F

The equations above show that the value selected for R_F has a large impact on the range of output voltages created by the voltage divider circuit. The precision of the sensor output is greatest when the voltage range of V_{OUT} is maximized. The value of R_F that generates the maximum range is the geometric mean of the sensor's highest expected resistance (R_H) and lowest expected resistance (R_L). The equation for calculating the best fixed resistor value (R_F) in a voltage divider circuit is:

$$R_f = \sqrt{R_H R_L}$$

The maximum range for V_{OUT} of the voltage divider circuit is thus equal to 1/3 of the supply voltage, V_{DD}. Furthermore, the voltage range is centered at the midpoint of the supply voltage. The following three equations calculate the minimum voltage, maximum voltage, and voltage range of an optimized voltage divider.

$$V_{MIN} = V_{DD}/3$$

$$V_{MAX} = 2V_{DD}/3$$

$$Range = V_{MIN} - V_{MAX} = V_{DD}/3$$

16.2.2 Current-Based Sensors

Some types of sensors generate or change output current in response to the environmental condition they are measuring. Examples include the photodiode, solar cell, and light-emitting diode (LED). All three of these devices are similar, although not used in similar ways. When a photon of light is absorbed, its energy gives an electron in the device enough energy to jump across the PN junction. The electron creates a measurable current from the sensor.

The LED is one of the most surprising

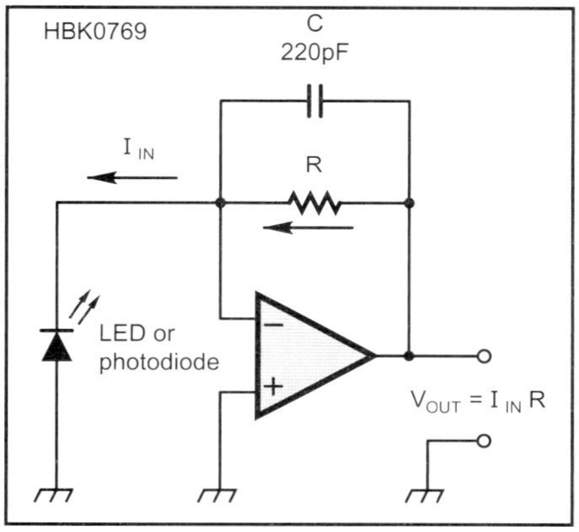

Figure 16.4 — The transimpedance amplifier converts current at its input to voltage at the output by balancing current through R with the input current.

current-based sensors. While the photodiode is sensitive to a wide range of frequencies, the LED is most sensitive to light at the wavelength it emits when forward biased. This makes the LED a very inexpensive spectrally sensitive photometer. (See the References and Bibliography section entry for Mims for a description of the LED responses.)

A current-based sensor can provide useful data when connected to a digital multimeter (DMM) set to measure milliamps of current. However, this is not a suitable configuration for a microcontroller with the capability to digitize voltage. Therefore, it is necessary to find a way to convert changing current into a changing voltage. Two popular ways to accomplish this are to use a transimpedance amplifier or by measuring the charging time of a capacitor.

THE TRANSIMPEDANCE AMPLIFIER

The transimpedance amplifier in **Figure 16.4** is a popular op amp circuit that converts input current into an output voltage. The feedback resistor, R, sets the gain of the transimpedance amplifier. The output voltage is given by the following equation:

$$V_{OUT} = I_{IN} \times R$$

The capacitor, C, reduces gain at high frequencies above 1/RC, acting as a low-pass filter to reduce noise. A generally useful value is 220 pF with the usual values of R for LED light-sensing. You will have to take the bandwidth of your measurement into account when choosing the value of C.

It is important that the value selected for the feedback resistor does not result in amplifier saturation for high sensor output levels. In those circumstances, data is lost when the sensor output is too high and the amplifier saturates.

USING CAPACITOR CHARGE TIME

A second method to digitize the current from a sensor is to measure the length of time required for a current to charge or discharge a capacitor to a certain voltage. One example can be found in the book *Earth Measurements* by Parallax (**www.parallax.com**, manufacturer of the BASIC Stamp microcontrollers). Here, the BASIC Stamp initially charges a capacitor. The Stamp then measures the length of time required for the capacitor to discharge due to the current entering it from the current-based sensor. The capacitor and resistor values are selected according to the expected current output of the sensor. The book uses the circuit in **Figure 16.5** to measure the current output of a photodiode or LED.

The program shown in **Table 16.1** (*Earth Measurements*, Program 4.2) was written to use the schematic in Figure 16.5. It assumes the circuit connects to the BASIC Stamp via I/O pin 6. Change the I/O reference to another pin as needed by your circuit.

The program reports the light intensity once per second. It begins by charging the capacitor to the same potential as the supply voltage through the use of the HIGH 6 command. Afterward, the reverse current emitted by the LED, due to its exposure to light, discharges the capacitor. The changing potential of the capacitor makes the voltage drop across the LED appear to decrease from its start at +5 V. Any voltage above 1.4 V is treated as a logic high by the BASIC Stamp. Therefore, as reverse current from the LED brings the capacitor voltage lower, the voltage across the LED eventually becomes lower than 1.4 V and a logic low.

The RCTime command counts the time (in units of 2 μs) required for I/O pin 6 to change from a logic high (above 1.4 V) to a logic low (below 1.4 V). The result in units of 2 μs is stored in the variable RCT. The greater the

Table 16.1
BASIC Stamp Program

This BASIC Stamp program is used with the circuit of Figure 16.5 for measuring light intensity.

```
RCT VAR Word
Light VAR Word
HIGH 6

Loop:
      RCTIME 6,1,RCT
      HIGH 6
      Light = 65535/RCT
      DEBUG "Light Intensity: ", DEC Light
      PAUSE 1000
      GOTO Loop
```

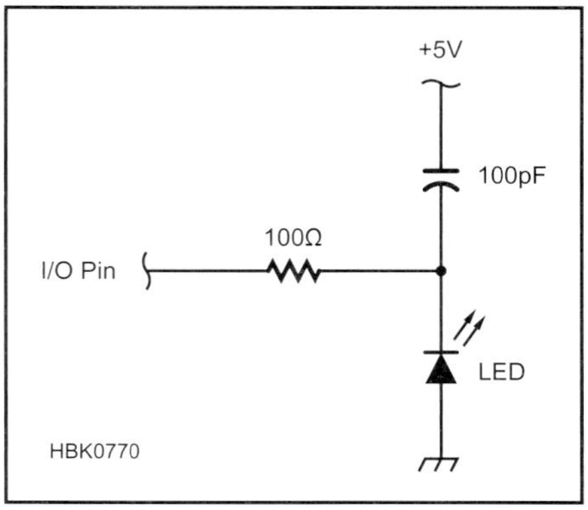

Figure 16.5 — The circuit recommended by Parallax for digitizing the current output of a photodiode or LED. See text for more information about using this circuit for current measurement.

intensity of the light shining on the LED, the faster the capacitor discharges and the smaller the value stored in the variable RCT. The value in RCT is then divided into 65535 to invert the relationship and is stored in the variable LIGHT, which contains a value directly proportional to light intensity.

16.2.3 Voltage-Based Sensors

Some types of sensors change their voltage output in response to the condition they are measuring. Examples include the LM355 temperature sensor, Honeywell's HIH-4000 relative humidity sensor, and Microdyne's MPS-3138 pressure sensor. These devices are typically current or resistance-based sensors along with circuitry to amplify and condition the output into a useable voltage change.

The voltage-based sensors easiest to interface are those that include signal amplification and correction on the chip. The result can be a ratiometric output that is linear and proportional to the supply voltage. The conversion factor for the sensor needed to convert its voltage into the environmental condition being measured is documented in the device's datasheet.

Pressure sensors can be used as altitude sensors for an airborne platform if a digital solution, such as GPS, is not available. Absolute pressure is preferred, although it must be calibrated against ground barometric pressure before launch and, if the flight covers long distance, requires additional corrections based on local pressure data.

16.2.4 Capacitance-Based Sensors

Capacitance-based sensors use changes in capacitance as their primary means of measurement. Capacitance between two electrodes of known area depends on the distance between the electrodes and the dielectric constant of the insulating material separating them. Any process that changes either separation or dielectric constant can be then be sensed as a change in the capacitance. Parameters that are sensed in this manner include motion, moisture, fluid and material level, chemical composition, and acceleration.

Sensors based on capacitance are rarely supplied without signal conditioning and linearization. Many have digital outputs that supply the measurement as a digital value. Another option is to have the sensor's capacitance control the frequency of an oscillator, which can then be read by a digital circuit. For more information on capacitive sensing, the excellent overview at **www.capsense.com/capsense-wp.pdf** is recommended.

16.2.5 Sensor Calibration

Sensors come in two basic configurations: *sensing elements* and *conditioned sensors*. The voltage divider discussed earlier is an example of a sensing element. There are no electronics associated with the divider — the package contains only the two resistors and the necessary connecting wires or terminals. Conditioned sensors contain electronic circuitry that operates on the signal from the sensing element before it is made available externally. The circuits usually regulate power applied to the sensor and also *linearize* the data so that a linear range of measurements is represented by a linear change in output voltage.

All sensing elements and some conditioned sensors require a calibration equation to convert the output signal into the parameter value the sensor is measuring. In some cases, the equation is simple and linear as in the LM335 temperature sensor. In other cases, the equation may be complicated, such as for the thermistor and photocell when used in a voltage divider circuit.

It is important to understand the range over which a sensor will be measuring a condition before attempting to calibrate it. The calibration equation is usually more accurate when based on the interpolation of measurements than when based on the extrapolation of measurements. There are exceptions to this rule. For example, the calibration equations of linear sensors can be just as accurate when extrapolated, as long as the maximum operating conditions of the sensor are not exceeded. Otherwise, is it best to expose the sensor to the entire range of expected environmental conditions while collecting measurements of its output to create the calibration equation.

The ham can easily create some of these conditions, such as temperature, on the bench top. High temperatures can be created with the use of heat lamps and low temperatures created with the use of dry ice packed in Styrofoam coolers. Other conditions might need to be simulated. For example, light intensity is easily changed by changing the distance between the sensor and a fixed light source. Recall however that light intensity decreases by a factor of $1/r^2$ when using this method to create the calibration curve of a sensor.

The spreadsheet is a powerful tool for creating calibration equations. To create the calibration equation, carefully measure the output of the sensor as the environmental condition is varied. Enter the readings and distance into a spreadsheet. In the next column, calculate the intensity of the source, based solely on its distance from the sensor. Graph the results so that the independent variable (X axis) is the distance and the dependent variable (Y axis) is the intensity. Then select the function to create a regression line from the data in the chart.

16.2.6 Digital Sensor Protocols

Some types of sensor outputs are in digital form. These sensors communicate their data as a serial protocol in which data is exchanged as a series of bits over one or more circuits. Data can be transmitted synchronized to an external timing signal (*synchronous protocol*) or synchronized to special signals embedded within the data being transmitted (*asynchronous protocol*).

Examples of synchronous serial data protocols include 1-Wire, Inter-Integrated Circuit (I2C), and Serial Peripheral Interface (SPI). Examples of asynchronous serial data transmission include USB and the RS-232 (COM) ports on PCs. These serial protocols can transfer measured data to a microprocessor without additional conditioning.

Another type of digital sensor is one in which an event's detection is signaled as a voltage pulse or as a switch closure. For example, the detection of ionizing radiation by Geiger counters is signaled by voltage pulses created when ionizing radiation passes through a Geiger-Muller tube. These signals require additional processing, such as by a counter or register circuit that is often implemented by a microprocessor.

SYNCHRONOUS SENSOR DATA PROTOCOLS

The following protocols are by no means the only ones used by sensors, but they are the ones amateurs are most likely to use or encounter. The manufacturer websites mentioned below have numerous resources to support design and development with devices that support these protocols.

1-Wire

1-Wire is a communication system developed by Dallas Semiconductor (now part of Maxim Electronics — **www.maximintegrated.com**) to enable communication between two or more integrated circuits. Devices on a 1-Wire network are daisy-chained together on a single-wire bus, called a *microlan*. (See **Figure 16.6**.) One device acts as the main device, and it controls communication between itself and the subnode devices connected to the microlan. Some available 1-Wire devices include:

- Temperature sensor: (MAX51826)
- EEPROM memory (DS24B33)
- Low voltage sensor (DS25LV02)

Since a microlan may not include a separate power wire, many devices attached to the microlan include a small capacitor in their design. The capacitor provides *parasitic*

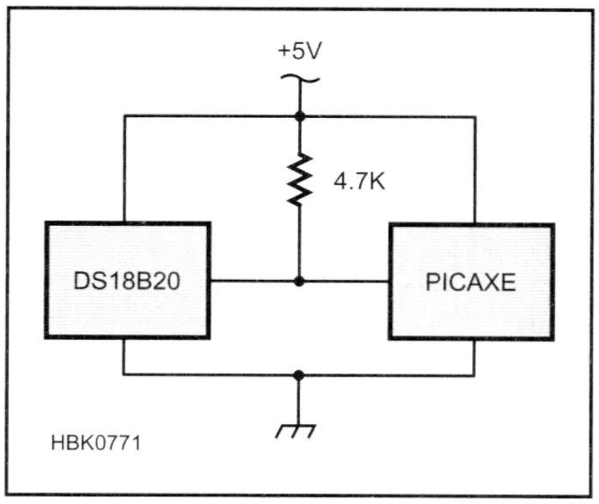

Figure 16.6 — The DS18B20 is a 1-Wire temperature sensor. In this circuit, the device does not use parasitic power and is connected to a 5 V source. A PICAXE microcontroller communicates with this device using the READTEMP or READTEMP12 command.

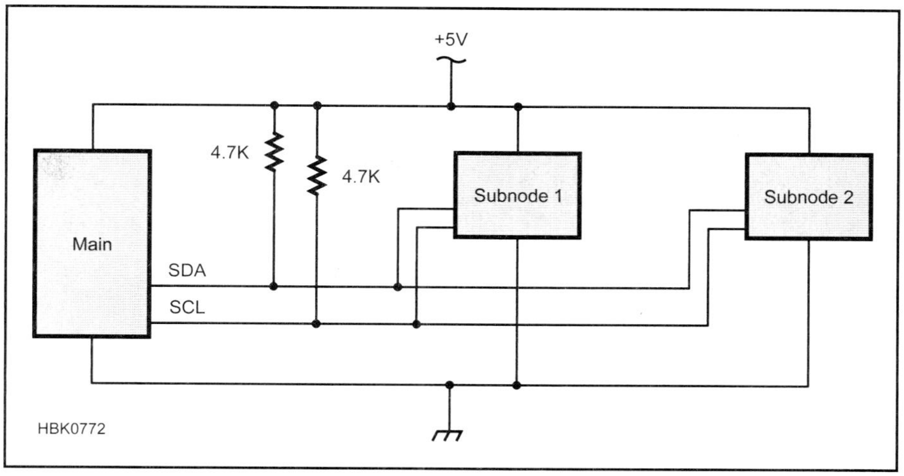

Figure 16.8 — An example of a main and two subnode ICs connected via an I2C network.

Figure 16.7 — A comparison between an iButton and a nickel.

power to the device while communications are taking place. The capacitor is necessary because communication requires the voltage on the single wire to alternate between power and ground. Without some temporary power source, devices would lose power during communications.

The main device communicates with each subnode device by transmitting the subnode's address over the microlan prior to other commands. Because multiple devices can be connected to a microlan, each device must have a unique address to avoid confusion. Subnode addresses are laser etched into 1-Wire devices. Alternatively, if a single 1-Wire device is attached to a microlan, communication on the network can ignore addressing altogether.

1-Wire is a two-way communication protocol. The main device begins communication by sending the subnode device's address and then commands over the network. Only the device with the address in the message will respond to the commands.

iButtons

An iButton is a sealed 1-Wire device resembling a thick watch battery (see **Figure 16.7**). iButtons include memory and a lithium battery. The memory contains the ID of the device and can often be used to store data. The battery permits an iButton to operate independently of a microlan.

iButton devices download their stored data when connected to a microlan. The microlan connection is made by pressing the iButton device against a 1-Wire receptor. 1-Wire receptors are available for integration into microcontroller projects. Some available iButton devices include the following:

- Time and temperature loggers (DS1920 Thermochron)
- Time, temperature, and humidity data loggers (DS1923 Hygrochron)

The amateur may be interested in the ongoing development of a 1-Wire weather station. Consult Maxim Integrated (**www.maximintegrated.com**) for the latest information concerning 1-Wire devices, including iButtons.

I2C

Inter-Integrated Circuit or I2C is a communication method developed by Phillips to enable communication between two or more ICs. Devices on an I2C network are daisy-chained together on a two-wire bus as in **Figure 16.8**. One device acts as the main device and controls communications between itself and the subnode devices connected to the network. The I2C network is described in detail at **www.i2c-bus.org** and in the application notes supplied by manufacturers of devices that use it.

The first connection in the I2C network is the serial data (SDA) line. This line carries subnode device addresses, data, and instructions between devices. The second line is the serial clock (SCL) line. This connection provides timing pulses to synchronize the data sent from the sending IC (main) to the receiving IC (subnode). In an I2C network, the SDA and SCL lines are pulled up to +5 V by pull-up resistors. A value of 4.7 kΩ works well.

The main device communicates with each subnode device by transmitting an address over the I2C bus prior to other commands. Because multiple devices can be connected to an I2C network, each device must have a unique address to avoid confusion. Subnode addresses may be designed into the IC or may be externally configured for an IC by connecting a combination of address pins to +5 V and ground.

I2C is a two-way communication protocol. The main device begins communication by sending the subnode device's address and then commands over the network. Only the device receiving its address in the message will respond to the commands. Serial data can be sent in either in fast (400 kHz) or slow (100 kHz) mode. The main device sends commands and memory addresses in either one byte or one word (two bytes) long commands. Some available I2C devices include the following:

- Memory: the 24LCxxx series of I2C memory.
- Real-time clocks: DS1307
- Analog to digital converters: LTC2903 (12 bit), AD7991 (12 bit), and MCP3421 (18 bit)

SPI and Microwire

Serial Peripheral Interface or SPI is a communication method developed by Motorola (now Freescale) to enable communication between two or more ICs. Devices on a SPI network are daisy-chained together on a two- or three-wire bus. (See **Figure 16.9**.) Like I2C, one device is the main that controls communications between it and the subnodes connected to the network. The Microwire network originally developed by National Semiconductor (now Texas Instruments) is essentially a subset of SPI. Microchip (manufacturer of the PIC processor family) has published an overview and tutorial about SPI at **ww1.microchip.com/downloads/en/DeviceDoc/spi.pdf**.

Two lines of the SPI bus are used to trans-

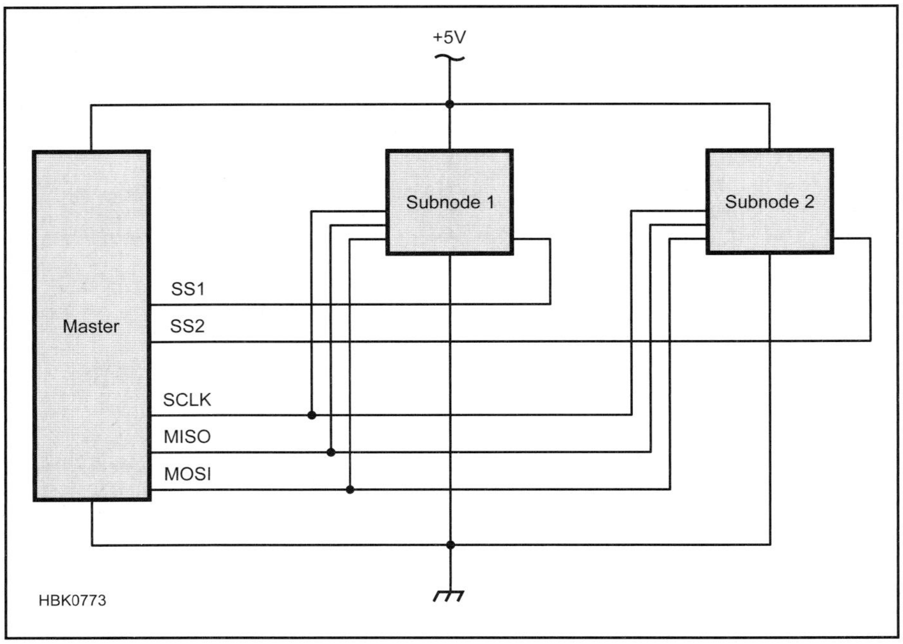

Figure 16.9 — An example of a main and two subnode ICs connected via an SPI network.

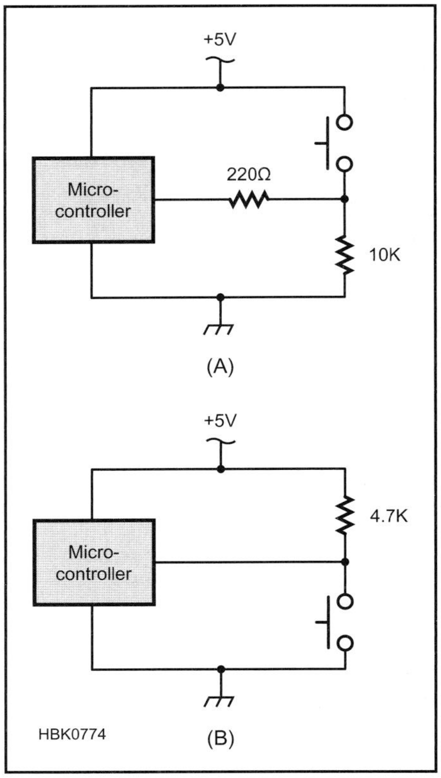

Figure 16.10 — (A) This circuit produces a logic high signal, typically 5 V, when an event is detected, represented as a switch closure. (B) This circuit produces a logic low signal, typically ground or 0 V, when an event is detected.

mit data and instructions: MOSI (main out/subnode in) and MISO (main in/subnode out). In some cases, the MISO and MOSI lines can be combined into a single shared line. The third line of the bus is the serial timing clock line (SCLK), which provides timing pulses to synchronize the data sent between the main device and the subnode. None of these lines requires being pulled high by a resistor.

The main device communicates with the subnodes by activating each device's Subnode Select (SS) line. To avoid confusion, each subnode must have a unique connection to the main device. This is a major difference between I2C and SPI. An I2C network requires only two communication lines between devices, while an SPI network requires two or three communication lines in addition to an SS line between the main and each subnode. A large number of subnodes require a large number of dedicated SS lines between the main device and the subnodes.

SPI is also a two-way communication protocol. The main device begins communications by activating the subnode device's SS line. Only the device with the activated SS line will respond to the commands. Serial data is sent as fast as the main device pulses the SCLK line. The number of bytes in each transmission between the main device and the subnode is limited by the design of the subnode rather than to eight or 16 bits. Some available SPI devices include the following:

• Analog to digital converters: MAX186 (12 bit resolution)
• Temperature sensor: LM74
• Hall effect sensor: MLX90363
• Pressure sensor: MPL115A1
• Memory: AT25010B

Note that the popular Dallas Semiconductor DS1620 Temperature Sensor uses a three-wire interface similar to SPI.

ASYNCHRONOUS SENSOR DATA PROTOCOLS

Asynchronous communication is any form of communication that does not use a clock signal to maintain timing between the sender and the receiver. A message begins with a start signal that allows the receiver to synchronize with the transmitter's message. The rest of the communication follows at a predefined rate in bits of data per second or baud. (See the **Digital Protocols and Modes** chapter for a discussion of data rate.) As long as the sender and receiver use equally accurate clocks, they will transmit and receive the same bits of data.

Some sensors supply their output data using RS-232 and USB ports. The data is transmitted as a stream of characters controlled by a protocol developed by the manufacturer. USB devices often conform to certain classes of data objects so that generic device drivers can be used to acquire data from the sensor. Control and configuration protocols that allow the user to interact with the sensor are usually proprietary.

Time-independent serial devices produce a change in output voltage only at the detection of an event. The primary example is the *event counter*. The simplest event counters detect the closure of a switch, which can be useful for detecting the presence of wildlife. Game cameras use switches in this way to trigger a camera to record an image of wildlife. Thermostats and thermal switches are another example.

Switches can be used to signal a microcontroller by two different methods. In the first, called *active low*, the switch connects a microcontroller I/O pin to ground at the detection of an event. When the event is not present, the I/O pin is connected to positive voltage or pulled up to a positive voltage by a pull-up resistor. In the second method, called *active high*, the switch connects a microcontroller I/O pin to positive voltage at the detection of an event. When the event is not present, the I/O pin is connected to ground. Schematics for both of these switch circuits are shown in **Figure 16.10**.

An example of a sensor that produces asynchronous output is the Geiger counter. The output of the RM-60 Geiger counter from Aware Electronics' RM-60 (**www.aw-el.com**) maintains a 5 V level until ionizing radiation is detected. Then the output voltage drops to 0 V for 20 μs. The amount of radiation detected by a RM-60 Geiger counter is measured by counting the number of pulses emitted by the sensor over a fixed period.

Other event counters can be modified for microcontroller use if they use an LED indicator or piezoelectric annunciator. When an

LED is illuminated, greater than 1.4 V appears across its terminals. Wires soldered to the LED can be connected to ground and one of a microcontroller's I/O pins to permit the microcontroller to count the number of LED flashes. Care in counting the number of flashes is necessary, since some inexpensive sensors may output several pulses each time the event is detected or in the case of contact bounce for a switch closure.

16.2.7 Powering Sensors

The output of sensing element sensors is typically very sensitive to power supply voltage and noise. Any changes in power supply voltage on the voltage divider also appear, proportionally reduced, at the output of the voltage divider. This includes noise, transients, slowly dropping battery voltage — any change in the sensing element's supply voltage. The sensing element user must provide clean, filtered, regulated power to the sensor to avoid contaminating the sensor output voltage.

Loading of the sensing element is also an issue for the designer to deal with. A high-impedance sensing element will output erroneous voltages if connected to a load impedance that is too low. Be sure you know what the sensing element's ratings are!

Conditioned sensors are far less sensitive to noise and power supply variations. Some kind of voltage regulator circuit is included to make sure the electronics operate with a "clean" supply. The conditioning electronics, which often include laser-trimmed calibration circuitry, assume clean, well-regulated dc voltage from a power supply. They are much less sensitive to the effects of output loading, although there are usually limits as to the amount of capacitance they can tolerate at the output, such as from a long run of wire.

In portable or mobile platforms, power is usually supplied by a battery pack. Make sure you have fresh, fully charged batteries before heading out to launch the platform for the experiment. Take into account the gradual reduction in voltage from the battery pack as its charge is consumed — it's awfully hard to swap out batteries with a balloon that is in flight! In the quest to save weight in these platforms, make sure you still have enough capacity in the battery pack voltage (see the **Power Sources** chapter) so that the experiment won't run out of power during its mission.

16.3 Navigation Data and Telemetry

Navigation data allows the sensor measurements to be combined with geographical data, which is important for correlating data to location (including altitude). A final step involves using amateur radio to either transmit the collected data to a ground station as telemetry or to track and recover a remote sensing payload for later data extraction. Since the most common use of this data is for weather balloons and other near-space missions, that context will be used.

It is important to note that transmitting data as ASCII characters (7- or 8-bit) is preferred to more compact binary formats. ASCII characters have the advantage of being human-readable so that even raw data can be inspected and used. At the low data rates of most amateur remote sensing, little overall throughput is lost by using ASCII characters. The ability to read the raw data stream directly is often invaluable during troubleshooting, as well.

16.3.1 Dead Reckoning

If a digital navigation data source such as GPS is not available, it is also possible to estimate platform position, including altitude, by the process of dead reckoning. In dead reckoning, navigation (or tracking) depends on determining a known position — called a *fix* — and then calculating subsequent positions from the platform speed and direction.

Direction data can be obtained from compass sensors that output direction as an analog voltage or digitally encoded signal. Altitude can be calculated based on ground barometric pressure and absolute pressure readings from the platform.

Obtaining accurate ground speed data is difficult for mobile platforms such as balloons or water-borne instruments, which move with the wind or current. If some other form of position tracking is available, it is possible to infer ground speed, although rarely accurately.

16.3.2 GPS Data

As currently practiced, a GPS (Global Positioning System) module is the usual means of acquiring navigation data, which is then transmitted as a telemetry stream using the Automatic Packet Reporting System (APRS). Thus, the two are combined in this section.

Depending on the model, GPS receivers produce a number of *navigation sentences*, such as the GPGGA and GPRMC sentences described below. GPS sentences are human-readable text with information in fields separated by commas. Below is a brief description of the two more important GPS sentences (when it comes to high altitude ballooning) and their fields.

THE GPGGA SENTENCE

The GPGGA sentence is the Global Positioning System Fixed Data sentence. A typical GPGGA sentence from a balloon-based GPS looks like this.

```
$GPGGA,153919.00,4332.2076,N,
11608.6666,W,1,08,1.1,13497.1
,M,18.3,M,,*78
```

There are 13 fields in the GPGGA sentence following the sentence identifier, "$GPGGA."

The fields from left to right are as follows.

1) Time in UTC (hours, minutes, seconds)

2) Latitude North (degrees and decimal minutes — note that there is no separator between degrees and minutes)

3) N (north)

4) Longitude West (degrees and decimal minutes — note that there is no separator between degrees and minutes)

5) W (west)

6) GPS Quality Indicator (0 = no GPS fix, 1 = GPS fix, and 2 = differential GPS fix)

7) Number of Satellites (number of satellites detected — not all of them may be used in determining the position)

8) Dilution of Horizontal Position (or DOHP, which is an indication of how precise the fix is — the closer to 1.0 the better)

9) Altitude (in meters)

10) M (meters)

11) Geoidal Separation (the difference in the actual height and a mathematic description of the height of an idealized Earth's surface in meters)

12) M (meters)

13) Checksum (result of exclusive OR-ing the sentence and used to verify that the text is not corrupted)

THE GPRMC SENTENCE

The GPRMC sentence is the Recommended Minimum Specific GPS/Transit Data sentence. A typical GPRMC sentence from a balloon-based GPS looks like this.

```
$GPRMC,153924.00,A,4332.2317,
N,11608.6330,W,24.4,46.3,2310
99,16.1,E*7E
```

There are 12 fields in the GPRMC sentence following the sentence identifier, "$GPRMC." The fields from left to right are as follows:

1) Time in UTC (hours, minutes, seconds)

2) Navigation warning (A = okay and V = warning)

3) Latitude North (degrees and decimal minutes — note that there is no separator between degrees and minutes)

4) N (north)

5) Longitude West (degrees and decimal minutes — note that there is no separator between degrees and minutes)

6) W (west)

7) Speed (in knots)

8) Heading (in degrees true north)

9) Date (day, month, and year — note that there is no separation between them)

10) Magnetic Variation (number of degrees)

11) Direction of magnetic variation (E = east and W = west)

12) Checksum (result of exclusive ORing the sentence and used to verify that the text is not corrupted)

16.3.3 Automatic Packet Reporting System (APRS)

Most mobile platforms include an APRS station in order to follow the platform's position and altitude throughout a mission to the edge of space. The APRS position reports, usually containing GPS data as described above, can be used directly to locate the position of the platform for recovery or tracking. (For more details about APRS, see the **Digital Protocols and Modes** chapter.)

There is a large network of dedicated ground stations and digipeater and Internet gateway stations operating on the US national APRS frequency of 144.390 MHz (144.800 and other frequencies are used elsewhere in the world). Thanks to this network, the platform's position will be plotted onto a map in near real-time. Two popular websites to view the maps are at **aprs.fi** and **findu.com**. These sites are databases of APRS packets received and routed through APRS Internet gateways.

Chase crews collect a platform's APRS data directly over amateur radio or over the Internet using the APRS maps. The platform or payload can then be recovered based on this position data. Later the data is correlated with other sensor data and images that are stored in on-board memory.

There are a number of APRS "trackers" that combine a low-power GPS module with a VHF transmitter and microprocessor that creates the APRS message packets. For example, Byonics (**www.byonics.com**) makes a number of APRS tracking and telemetry products, including the Micro-Trak RTG FA High Altitude Combo, which contains an altitude-certified GPS for balloon payloads. The RPC-Electronics (**www.rpc-electronics.com**) RTrak-HAB - High Altitude APRS Tracker Payload is specially made for high-altitude ballooning, as well.

The tracker combination built by WB8ELK is shown in Figure 16.2. On the right is a GPS module that creates the GPS sentences discussed previously. On the left is the MMT (Multi-Mode Transmitter) that creates and transmits the APRS packets.

APRS POSITION DATA

A simple APRS tracker can generate a stream of useful navigation data for a mission. The data begins at the GPS receiver where two navigation sentences are generated. The sentences are then combined to create a position report in the required APRS format. Like GPS sentences, the raw APRS packets are also readable text that is easily interpreted.

An APRS position report uses a combination of commas and slashes as field delimiters. An example of an APRS report from a balloon mission looks like this:

```
13:37:23 UTC: KD4STH-8>APT311,WIDE1-2,qAS,KC0QBU,133721h3836.39N/09500.51W>160/031/A=049114
```

There are 12 fields in the APRS report. The fields from left to right are as follows:

1) Time in UTC (hours, minutes, seconds)

2) Call sign and SSID

3) Routing Information

4) GPS Time (time in UTC — note there is no separator between hours, minutes, and seconds)

5) h (hours)

6) Latitude North (degrees and decimal minutes — note that there is no separator between degrees and minutes)

7) N/ (north)

8) Longitude West (degrees and decimal minutes — note that there is no separator between degrees and minutes)

9) W> (west)

10) Heading (in degrees from true north)

11) Speed (in knots)

12) A= (altitude equals)

13) Altitude (feet)

There are other formats and fields that may be present. Additional telemetry fields can be added in a variety of formats following the altitude data. For more information about the APRS reports, see "APRS Formats Used in Edge of Space Sciences," at **www.eoss.org/aprs/aprs_formats_eoss**.

APRS CONFIGURATION

High altitude APRS trackers should be programmed to provide altitude data and not to use Smart Beaconing. Altitude data is a fascinating datum and useful for determining when a platform has landed. Smart Beaconing prevents APRS trackers from transmitting their position while the tracker is not changing its speed or direction. At landing, it's important that the tracker continue to announce its position on a regular schedule. This is particularly important if the tracker was out of range of the chase crew at the time of landing.

The horizon for high altitude balloons and larger rockets can be hundreds of miles away. Therefore, their transmission footprint can cover tens of thousands of square miles. To prevent high altitude APRS trackers from interfering with orderly use of APRS, the high-altitude ballooning community recommends programming an APRS tracker with the following settings if a frequency of 144.390 MHz is used. (Thanks, Jerry Gable, KF7MVY)

1. Path Recommendations: Use no Path or set Path to WIDE2-1. Never use a two-part path such as the common WIDE2-1, WIDE-1 nor use WIDE2-2 or WIDE3-3.

2. Transmit Rate: Limit the transmit rate to once per 60 seconds during ascent and no less than 30 seconds during descent.

If possible or practical, you may want to use a frequency other than 144.390 MHz.

16.3.4 Satellite Telemetry

Amateur satellites use a variety of modulation methods and data encoding to construct and transmit the stream of data coming from the satellite. Bit rates vary from 1200 to 9600 bps and modulation types of CW, PSK, BPSK, FSK, and AFSK are common. Each satellite also uses a custom scheme to encode the data, often using the AX.25 packet radio protocol.

Decoding satellite data would be very challenging except that the team building the satellite usually publishes a description of the telemetry stream and provides software for receiving and decoding the data. In this way, individual amateurs can collect telemetry data for the satellite's operational team.

Information on particular satellites is often available at one of the AMSAT websites such as **www.amsat.org** (AMSAT North America) or **amsat-uk.org/satellites/telemetry/** (AMSAT UK). Satellite status, including whether telemetry decoder software is available, can be found at DK3WN's website (**www.dk3wn.info/p/?page_id=29535**). Additional information for CubeSats launched by universities or other private groups is usually available on a web page provided by the satellite's sponsor.

To decode satellite telemetry reliably at the higher bit rates, particularly 9600 bps, you will need to provide the full bandwidth audio from the receiver or use packet radio TNCs that are rated for 9600 baud operation.

The Fox-1 satellite (AO-85, see **www.amsat.org/status** for the satellite's current status) has both a low-speed 200 bps DUV (data under voice) and a high-performance 9600 bps FSK telemetry stream. *FoxTelem* software is available for decoding the information. As an example of current best practices for satellite data, a two-part article from *AMSAT Journal* on the Fox-1 telemetry system is available as a PDF on the downloadable supplemental information. (See the reference listings for Burns Fisher, W2BFJ, and Chris Thompson, AC2CZ.) Numerous other satellites use the same or similar telemetry formats and modulations.

16.3.5 Non-Licensed Telemetry Transmissions

There are many low-power data links operating in the unlicensed 915 MHz and 2.4 GHz bands. Typically, these are intended to be used for short-range applications but with the balloon payload at great altitude, the range of these devices is much longer, particularly if a high-gain Yagi antenna is used to track the payload. (See the **Antennas** chapter for information on VHF and UHF beams.)

Many of the data link modules use a standard two-way protocol such as Zigbee and have direct analog and digital inputs and outputs. Some modules support Ethernet and Bluetooth interfaces, offering even more options for modules that can be assembled into the payload.

It is also important to note that unlicensed transmitters operating under FCC Part 15 rules are also subject to certain restrictions such as field strength. In addition, the type of antenna may be fixed and even required to be attached to the transmitter permanently. These and other restrictions are required in order to limit the range of these devices. Amateurs are used to modifying and adjusting their equipment, and this may not be allowed for some of these devices! Be sure to obtain the full documentation for any unlicensed device you plan on using and be sure you can use it in the way you expect.

16.3.6 Other Telemetry Digital Modes

The usual method of communication from airborne and other remotely located platforms is via the APRS network. APRS messages are packaged in X.25 packets and usually transmitted as FSK or PSK modulation on FM transmissions. This works well and takes advantage of the extensive ground network of APRS digipeaters and servers.

Along with APRS, other subsystems are popular in the amateur balloon community. These include Weak Signal Propagation Reporter (WSPR), Hellschreiber, DominoEX, Contestia, JT9, PSK31, RTTY, CW (especially for fox hunting a lost near space balloon), FM for voice repeaters, and imagery through ATV and SSTV. (All of these digital modes are described in detail in the **Digital Protocols and Modes** chapter.)

WSPR

WSPR, part of the *WSJT-X* software suite (see the **Digital Protocols and Modes** chapter), is a simple location reporting method using HF and therefore a practical tracking data system for long duration balloon flights. WSPR takes a bit less than two minutes for one transmission.

The format of a WSPR report specified in the protocol definition is limited to: call sign, transmitter location (using the Maidenhead grid locator system), and transmitter power (in dBm). To send telemetry, amateurs have developed a way to encode altitude, voltage, temperature, and a six-digit grid locator by sending a second WSPR transmission that uses a call sign beginning with a "0" or a "Q," which are prefixes not used in Amateur Radio. The telemetry is embedded in the call sign field and power levels, and one method also uses the grid square. There are several protocols in use and undergoing development as of early 2018. You can find information on them by doing an Internet search for "wspr telemetry."

Figure 16.11 shows the track of WB8ELK's ballon that took six trips around the world in 75 days. The track is based on WSPR.net reports and displayed on the UK High Altitude Society's (UKHAS) **tracker.habhub.org** website, which features tracking of many balloons that are flying at any given time.

WSPR has a similar advantage to APRS in that there is a distributed network of ground stations around the world that relay reports to a centralized web server, which can be viewed on the **wsprnet.org** website. The data from a long duration balloon can be picked up from many thousands of miles away using a very low power HF transmitter in the 10 to 25-mW range.

JT9

JT9, part of the *WSJT-X* software suite (see the **Digital Protocols and Modes** chapter), is also very effective for low-power HF telemetry and has been used for long duration, high altitude balloon platforms. It takes less than a minute to send a transmission. It does allow a free-form transmission in which telemetry can be embedded, however the amount of information is very limited. There is no distributed worldwide ground station network that exists, as there is for the APRS and WSPR modes.

HELLSCHREIBER

Hellschreiber ("light writer" in German) is an HF transmission method that can send balloon location and other data as a fax image. The characters are formed with an accurately timed sequence of on/off keying that forms an image of the data in a received waterfall-style display. This makes it an ideal mode for poor reception conditions, since human eyes are very good at retrieving information from a noisy image.

ASCII RTTY

ASCII RTTY is very easy to implement with a small microcontroller. It uses two tones and speeds similar to Baudot RTTY but uses the full ASCII character set to send telemetry. Speeds range from 45 baud for HF systems. VHF and UHF FM transmitters can send RTTY at 100 to 300 baud.

CONTESTIA AND OLIVIA

Contestia and Olivia are MFSK modes with excellent FEC correction. They can be used for HF, VHF, and UHF payloads. Information about these modes and how to encode them in a microcontroller can be found here: **ukhas.org.uk/guides:olivia_and_contestia**

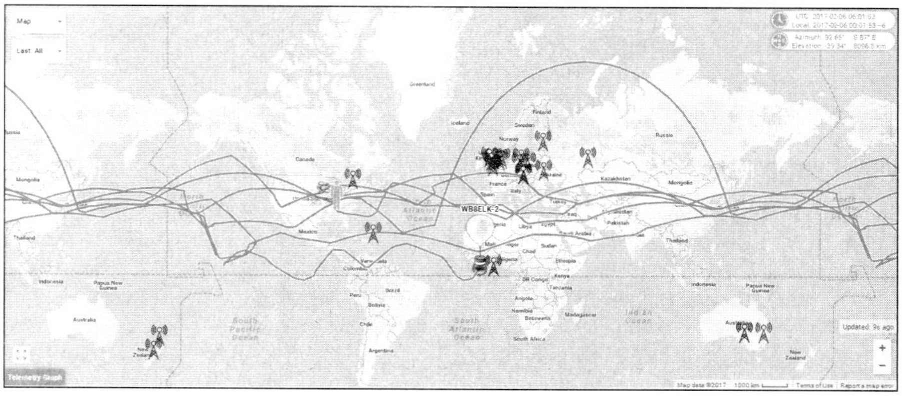

Figure 16.11 — The track of a balloon launched by WB8ELK that managed 6 trips around the world in 75 days. The image was generated from WSPR reports collected by the UKHAS website (tracker.habhub.org).

DOMINOEX

DominoEX is designed for very weak signals using an 18-tone sequence that can send balloon data using a low power signal on either HF, VHF, or UHF. It uses Offset Incremental Frequency Keying (IFK+), which uses the frequency difference between tones rather than the absolute frequency of each tone. As a result, it is very tolerant of frequency drift.

PSK31

PSK31 is a phased shift keyed mode primarily for HF payloads that requires a stable transmitter and can provide a 31.25 baud data rate.

CW (MORSE CODE)

CW is likely the easiest mode to implement with a small microcontroller performing On/Off Keying or OOK. It is primarily used as a backup transmitter for direction-finding a lost balloon payload. With the addition of a small GPS, some balloon groups actually send the position and altitude via Morse code so that the chase crew can decode the position without using a decoding program on a laptop computer.

SATELLITE COMMUNICATION

Some missions use satellite communications as a backup and telemetry-message system. These are unlicensed communication methods that charge a subscription fee. The two satellite communications services used today are Spot (for position tracking below 20,000 feet altitude — **www.findmespot.com**) and Iridium modems (for sending messages —such asthe RockBlock Mk2 modem (**www.sparkfun.com/products/13745**). There are no format standards for data transmitted using satellite or mobile phone systems.

16.3.7 Receiving and Relaying Telemetry

Most digital modes can be readily received by a modified version of the free software package *FLdigi* called *dl-FLdigi* which can be downloaded from **ukhas.org.uk/projects:dl-fldigi**. This software receives and decodes the telemetry, uses a checksum to ensure the accuracy of the received data and sends that data to a server. The position report can then be viewed at: **tracker.habhub.org** which is maintained by the UK High Altitude Society (UKHAS). The site can display telemetry from any platform but has a number of features specifically designed for high altitude balloon platforms, both Amateur Radio and license-free transmitters.

The format for generating telemetry for use with **tracker.habhub.org** is comma-separated values (CSV) of ASCII characters as follows:

$$CALLSIGN,
sentence_id,
time,latitude,
longitude,
altitude,
speed (optional),
bearing (optional),
internal temperature (optional),
*CHECKSUM\n

Although these parameters can be changed with the exception of $$ to start and a checksum at the end. More information on the telemetry formats can be found at **ukhas.org.uk/communication:protocol**.

16.4 Payloads

16.4.1 VHF/UHF/Microwave Payloads

NEAR SPACE TRACKER

A very popular configuration for high-altitude payloads is the Near Space Tracker, consisting of avionics to acquire and record or transmit data, an APRS tracker module coupled with a GPS receiver for position data, and a simple dipole or omnidirectional antenna. **Figure 16.12** shows a typical physical layout for such a package. The APRS tracker module consists of a *terminal node controller* (TNC), and a transceiver module. It is advisable to use a complete APRS tracker for balloons rather than a separate TNC and radio. Using a combined system reduces the complexity, weight, and battery needs of the tracker.

A complete construction article is included in this book's downloadable supplemental content, titled "Touching Near Space on a Budget," by Paul Verhage, KD4STH. The assembled tracker is shown in **Figure 16.13**. It uses an inexpensive insulated lunch cooler to hold the electronics, battery, and antenna.

Tracker packages such as this are excellent experiments not only for hams but for students and other groups when assisted by a licensed amateur to allow the use of the APRS module. Such experiments are a good way to introduce students to amateur radio and can often be supported by a local radio club.

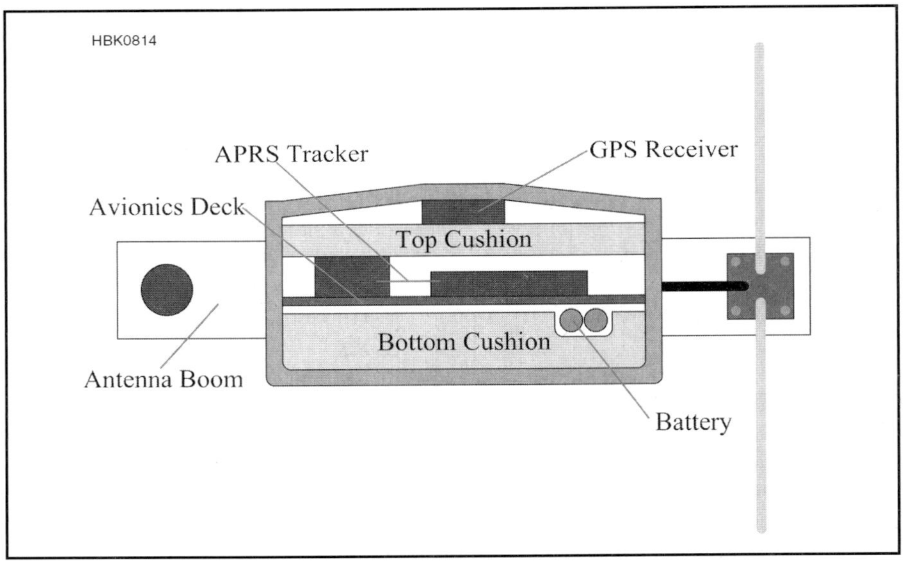

Figure 16.12 — A cut-away graphic showing an idealized Near Space Tracker including an APRS tracker, GPS receiver, an avionics package, batteries, and a dipole antenna.

Key considerations for trackers are their weight (less is better) and time of operation (or how long will they operate on a fully charged battery). The lighter the tracking system, the more available weight for science payloads. It also means a given balloon will reach a higher altitude before bursting (on account of the reduced initial volume of lifting gas). Since it can take an hour to fill a launch a balloon, ninety minutes to reach peak altitude, forty-five minutes to descend, and a few hours to recover, a tracker should be

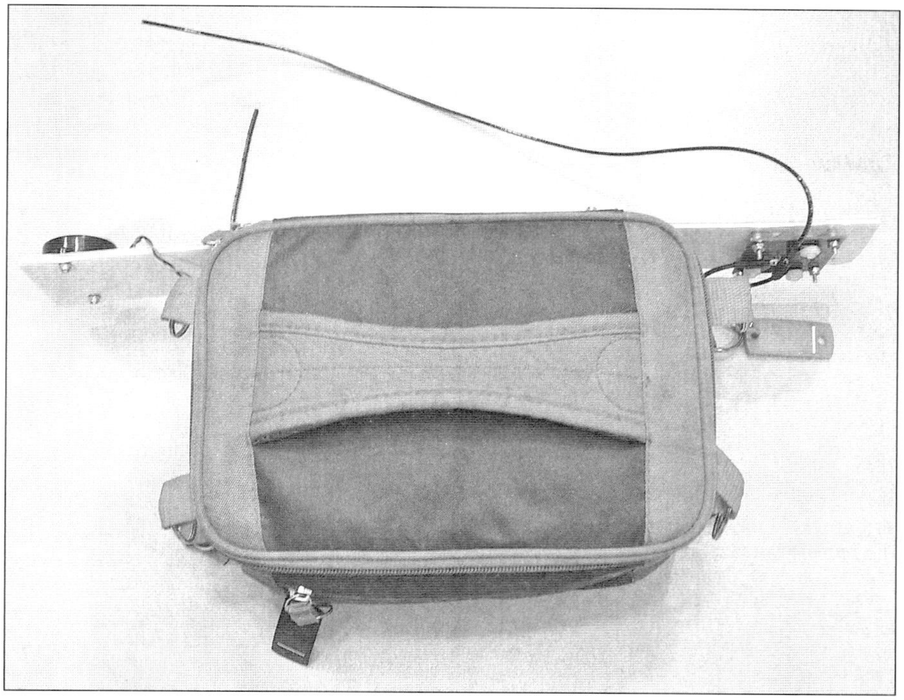

Figure 16.13 — The complete Near Space Tracker assembled in an insulated fabric lunch cooler.

capable of running for at least six hours on a set of batteries.

APRS trackers such as the Byonic TinyTrak 3, Argent Data Systems Tracker3, Tracksoar, and others are software configurable. This means the tracker's behavior can be pre-programmed using configuration software provided for the tracker. An APRS tracker is configured for settings such as Callsign, SSID, Symbol, Smart Tracking, Transmit Times, Time Slotting, and Status Messages. (See the previous section on APRS.)

A tracker producing a 300 mW signal is adequate for a balloon flight. Even an APRS transmitter producing only 25 mW such as the Skytracker (**wb8elk.com**) can be used as a backup tracker. Even at this low power it is comparable to higher power transmitters down to about 1500 feet AGL (above ground level). This means a battery with a capacity of several hundred mAhrs will be sufficient for most balloon flights (a 2200 mAh is even better, since it permits an overnight recovery).

Since a 5-V LM2940T-5 voltage regulator has a drop-out of 0.5 V, a two-cell (2S) rechargeable LiPo battery can provide more than enough power for most avionics on a balloon flight. In addition, these batteries are commonly available from hobby stores where they are sold for RC racing cars.

A vertical dipole is the recommended antenna for near space tracking modules using APRS. They are simple to construct, light weight, and are not direction sensitive with respect to chase vehicle antennas (which tend to be mag-mounted vertical whips).

GPS RECEIVERS

GPS receivers are a vital component for APRS trackers. To prevent their use in guided missiles, the Coordinating Committee for Multilateral Export Controls requires GPS receivers to stop creating position data when they are moving faster than 1,000 knots at an altitude above 59,000 feet. Some companies have taken this as an OR condition and not an AND condition as it was meant. Therefore, amateurs must verify their selected GPS receiver will operate above 59,000 feet. Amateurs can find a list of appropriate receivers through an internet search.

GPS signals are weak signals. This means it can be easy to block the radio signals that GPS receivers depend upon. Experiments have shown that several inches of Styrofoam will not block a GPS signal, but that a layer of aluminized Mylar will. It's important that amateurs test their APRS tracker prior to launch after they have finished constructing their airframe.

SIMPLEX REPEATER

Readers should be aware that a high-altitude balloon at 100,000 feet has a radio line-of-sight horizon of more than 400 miles. The formula for radio signal line-of-sight in miles can be calculated as follows:

$$\text{Distance (in miles)} = 1.41 \times \sqrt{\text{height (in feet)}}$$

where H is the height in feet.

Since antenna height is so important for operating on VHF and UHF, imagine having an antenna that is 19 miles high. If you could fly a repeater to that altitude, two ground stations 800 miles apart could communicate with each other through the repeater. One simple way to do this is to fly a single handheld radio operating on 2 meters or the 70 cm band. Connecting a voice recorder and playback device to the handheld radio creates a *simplex repeater*. One such device is the Argent Data ADS-SR1. A discontinued Radio Shack simplex repeater module can be sometimes found online.

It takes some practice and patience to get the hang of a simplex repeater conversation, but this provides a very simple way to make some very exciting contacts over a multi-state region using minimal equipment on the ground.

CROSSBAND REPEATER

If you use two handheld radios, one on 2 meters and one on 70 cm, you can build a crossband repeater payload. (Some handheld radios can also operate as crossband repeaters by themselves.) You'll need to build an audio level control to adjust the audio between the two radios and also provide a PTT control.

Although more complicated, heavier, and more expensive than the simplex repeater, this does provide a real-time repeater without having to worry about flying large filters to prevent desense. The input is usually on the 2 meter band with the output on the 70 cm band. Although you can set it up the other way around, the 3rd harmonic of the 2 meter transmit can cause desense issues with the 70 cm receiver.

STILL IMAGES

A great addition to any balloon flight is the ability to actually receive live images during the flight. From 100,000 feet you can clearly see a spectacular view of the blackness of space and the curve of the Earth, since the balloon is above 99 percent of the atmosphere. **Figure 16.14** shows a photo taken from by a balloon-launched camera. Suitable lightweight cameras are available in thumb drive (USB) formats and helmet- or bike-cams designed to be used while being worn.

SSTV has been flown since at least 1998 and is still used with 2-meter FM transmitters by using a SSTV module such as SSTVcam by Argent Data (**www.argentdata.com**). The SSTV images are decoded using sound cards and software on a PC. The ability to use software means amateurs are free to use virtually any SSTV mode they prefer. The Argent system is configurable to produce images in the following four modes, Robot 36, Robot 72, Scottie 1, and Scottie 2. Scottie 2 will transmit one image in 71 seconds. Robot 36 is somewhat quicker although with some loss of

Figure 16.14 — A balloon carrying a camera payload was launched from the Dayton Hamvention in 2010 by Bill Brown, WB8ELK. This picture was obtained a few minutes later from an altitude of about 1000 feet.

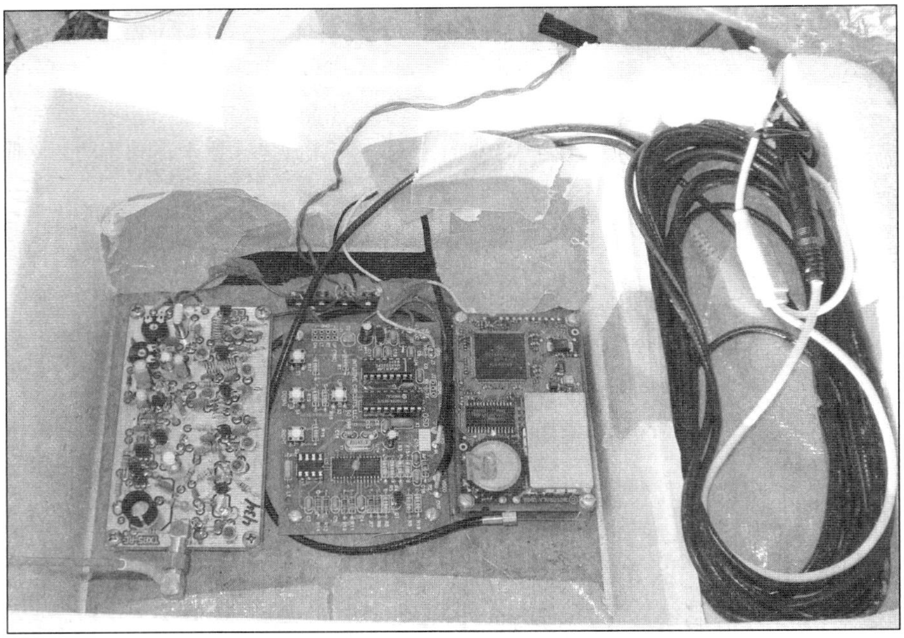

Figure 16.15 — This payload consists of a GPS receiver (right), payload controller (center), and an ATV transmitter (left). Batteries and cables are in the far-right compartment and the entire platform is contained in a Styrofoam enclosure.

resolution. The Scottie modes are most commonly used for balloon flights. (See the downloadable **Image Communications** chapter)

There are several programs to decode the SSTV audio signal and display the images on a computer screen. *MMSSTV*, *MixW*, *MultiPSK*, and *Ham Radio Deluxe* (DM780) are a few programs that can be used to capture and view SSTV images.

VIDEO

There is nothing like watching a live video camera view from a flight to the edge of space. Typically, a 1 W to 3 W AM-modulated ATV transmitter on either 434 or 439.25 MHz is used for best results. You can also use FM ATV transmitters on higher frequencies, such as the 23 cm and 13 cm bands, but the path loss will be much higher on those frequencies, which will limit your maximum downrange reception distance. There are many lightweight video cameras that can be used. (See the **Image Communications** chapter with the downloadable supplemental material for more information about analog and digital ATV.)

Analog and digital methods are used to transmit video. In digital mode, Raspberry Pi computers can produce the video, and 5.8 GHz modems are used to transmit the video. UHF can also be used to transmit digital video. By using video overlays, GPS data can be superimposed on images as a backup tracking method.

A variety of transmitters can be found at these websites: **www.hamtv.com**, **www.hamtv.com/videolynx.html#VM70X**, **kh6htv.com**, and **www.hides.com.tw/product_eng.html**.

Figure 16.15 shows a typical ATV payload in the insulating Styrofoam box enclosure. On the left is the low-power 70 cm transmitter. In the middle is the microprocessor-based controller. The GPS receiver module is on the right. The batteries and cables are placed in the separate compartment at the far right. Note that the three electronics boards are mounted over a common PCB ground-plane to provide mechanical stability and to minimize RFI from the transmitter. The antenna for the ATV link hangs below the package.

On UHF, Little Wheel or Big Wheel omnidirectional antennas have been a staple of high altitude weather balloons. The balloon group Project Traveler has a webpage on making a Little Wheel antenna for high altitude balloon flights at **www.projecttraveler.org/index.php/how-to-s/6-little-wheel-antenna-for-70cm-atv**. PC Electronics (**www.hamtv.com/wheel.html**) carries the Olde Antenna Labs line of "wheel" antennas for various bands as well as video camera modules, and ATV transmitters and receivers.

Analog video will produce P5 (high-quality) signals early in the flight while the balloon's altitude is relative low. At high altitudes, it's not uncommon to receive P1 (low-quality) signals on the ground, even from below the balloon.

Remember that the power requirements for a continually operating 1 W TV transmitter will be much higher than an APRS or audio repeater payload. You'll typically need at least 12 V with an Ah rating sufficient to allow for at least three hours of operating time. High capacity, four-cell, RC racing car batteries are a good option for powering ATV transmitters.

You will need a good antenna on the ground, an ATV downconverter, and an analog TV receiver. If you are flying in an area where horizontal polarization is used for local ATV activity, the Big Wheel antenna is a good option for the payload's ATV antenna. It provides good coverage at the horizon as well as underneath the payload. You can also use a vertical antenna, but there will be a null directly underneath a vertical radiator.

16.4.2 HF Payloads

The RF range of a high-altitude balloon at peak altitude is limited to about 450 miles when using VHF and UHF. Some balloon groups have flown transmitters on the HF bands with reception reports many thousands of miles away. It's a great way to include amateur radio operators far outside your local region.

There are several digital modes that can be programmed into a small microcontroller without having to invoke floating point math (see **www.elktronics.com** for an example of a multi-mode HF balloon transmitter). Morse code, RTTY, PSK31, DominoEX, and Hellschreiber have all been successfully flown as well.

Transmit power levels under 1 W will work well due to the weak signal advantage of some of these digital modes. DominoEX, Hellschreiber, and PSK31 are particularly good for very weak signal reception. WSPR will also provide tracking information.

RESTRICTIONS ON UNATTENDED OPERATION

Note that unattended "beacon" operation and stations making transmission on their own are restricted to the automatic control band segments when located in areas under FCC jurisdiction. There should be a way to turn the HF transmitter on or off under remote control. (Being able to turn the transmitter on and off does not constitute full control via the remote link, so the transmitter is still considered to be operating under automatic control.) There are a number of inexpensive and lightweight UHF handheld radios that can be used as a control receiver along with a DTMF decoder board.

Be sure to comply with the requirements of FCC Part 97.221 – Restricted Operation, which reads as follows:

§97.221 *Automatically controlled digital station.*

(a) This rule section does not apply to an auxiliary station, a beacon station, a repeater station, an earth station, a space station, or a space telecommand station.

(b) A station may be automatically controlled while transmitting a RTTY or data emission on the 6 m or shorter wavelength bands, and on the 28.120-28.189 MHz, 24.925-24.930 MHz, 21.090-21.100 MHz, 18.105-18.110 MHz, 14.0950-14.0995 MHz, 14.1005-14.112 MHz, 10.140-10.150 MHz, 7.100-7.105 MHz, or 3.585-3.600 MHz segments.

(c) Except for channels specified in §97.303(h), a station may be automatically controlled while transmitting a RTTY or data emission on any other frequency authorized for such emission types provided that:

(1) The station is responding to interrogation by a station under local or remote control, and

(2) No transmission from the automatically controlled station occupies a bandwidth of more than 500 Hz.

16.5 High Altitude Balloon Platforms

The reader will find plenty of near space information downloadable by using search terms such as "near space," "ARHAB," or "BalloonSats." Persons and organizations planning a near space event will find a series of papers and presentations on the **nearsys.com** website in its "Other People's Helium" section (**www.nearsys.com/arhab/ophe/ophe.htm**). The free e-book *BASIC Stamp Near Space* (see the References section) covers every aspect of high altitude ballooning. It is written for the amateur who wants to begin a near space program from scratch or who is looking for new ideas. Finally, the "Near Space" column in *Nuts and Volts* magazine contains articles on designing and using microcontrollers for high altitude balloon projects.

Here are a few websites with a great deal of information about amateur radio high altitude ballooning (ARHAB):

- Amateur Radio High Altitude Ballooning (ARHAB) — **www.arhab.org**
- Near Space — **nearsys.com**
- Edge of Space Sciences — **www.eoss.org**
- WB8ELK Balloons — **www.wb8elk.com**
- UK High Altitude Society — **www.habhub.org**
- Great Plains Super Launch — **www.superlaunch.org**

16.5.1 FAA Requirements

The regulating agency for unmanned free balloons is the Federal Aviation Administration (FAA). The applicable regulation for unmanned free balloons is the Federal Aviation Regulation Part 101 (FAR 101) or Title 14: Aeronautics and Space, 14 CFR 101.

FAR 101 Section 101.1 applies to any unmanned free balloon if it falls under one of these four conditions:

1. The unmanned free balloon carries a payload weighing more than four pounds and has a weight/size ratio greater than three ounces per square inch (this "surface density" is measured on the side of a payload with the smallest area).

2. The unmanned free balloon carries a single payload weighing more than six pounds.

3. The unmanned free balloon carries a payload of two or more packages weighing a combined 12 pounds.

4. The unmanned free balloon uses a rope or line to connect the payload to the balloon, which requires more than 50 pounds of force to separate the payload form the balloon.

Amateurs are strongly encouraged not to exceed the four limitations outlined above. When any of these limitations are exceeded, the balloon flight must meet additional requirements. These requirements are explained in Subpart D and include the following.

A) Limitations on launch site selection

B) Limitations regarding cloud cover at the time of launch

C) Requirement to add payload cutdown devices

D) Requirement to add balloon termination devices

E) Requirement to add a radar reflector to the balloon train

F) Requirement to increase the visibility of the balloon train

G) Additional pre-launch notification requirements

H) The requirement to regularly report the balloon's position

Section 101.3 states that one can request a waiver for a balloon flight that cannot meet the requirements of FAR 101. Anyone requiring a waiver must complete an FAA Form 7711-2, Application for Certificate of Wavier or Authorization.

Section 101.5 states that an unmanned free balloon cannot be launched from a restricted or prohibited area without permission from the controlling agency, or an agency that uses that restricted or prohibited area.

Section 101.7 states that you cannot operate a balloon in a manner that creates a hazard to people or their property. For example, a balloon cannot be launched if during its flight objects will be dropped in such a way that can harm or injure people or their property.

The maximum weight per payload is 6 pounds for a total of 12 pounds for the platform. Launching additional weight requires getting special permission from the FAA. This doesn't mean that you can fly 6-pound lead weights. You have to make sure that the density of your payload will not inflict damage to others, and you also need to protect all those expensive electronics that you have packed inside.

16.5.2 Balloon Platform Environment

The near space environment has environmental conditions that are in many ways closer to those found in outer space than Earth's surface, as **Table 16.2** illustrates (LEO stands for Low Earth Orbit).

HORIZONTAL DEPRESSION

For an observer at the Earth's surface, the visual horizon forms a horizontal line, 90 degrees from the zenith (the point overhead) for all azimuths (directions). As height increases, the horizon is determined by the point at which a line from the observer's eye is tangent to the Earth's surface. (The radio horizon is a bit more distant, as explained in the chapter on **Propagation of Radio Signals**.) As an observer's altitude increases, as shown in **Figure 16.16**, that tangent point becomes more distant due to the curvature of the Earth. The angle from the zenith to the tangent line also increases, therefore the horizon appears lower to the observer. This lowering of the horizon is called *horizontal depression*.

ATMOSPHERIC DENSITY AND PRESSURE

Air pressure, and therefore density, decreases as altitude increases because the weight of air still above that level decreases. The scientific way of specifying changes in atmospheric density is *scale height*. Similar to a capacitor discharging, scale height is the change in height by which the density of the atmosphere decreases by a factor of e (which equals 2.718). Scale height depends on factors such as acceleration of gravity, the average atomic mass of atmospheric gases, and temperature. Scale height is a characteristic of every planet's atmosphere and a useful measurement for astrodynamics and astronomy. The scale height of Earth's atmosphere is 4.9 miles. Therefore, for every 4.9-mile increase in height, Earth's atmosphere is 1/e or 37% as dense. A simple rule of thumb is that air pressure drops by 50% per 18,000-foot increase in altitude.

Air pressure (see **Figure 16.17**) is a concern in near space experiments for keeping organisms alive, high voltage electrical circuits that might arc over without the insulation provided by the air, and where contact with the air is a factor in keeping devices cool.

Table 16.2
Balloon Environment Summary

	Sea Level	*Near Space*	*LEO*
Pressure	1013 mb	60 mb to 5 mb	0 mb
Temperature	59 °F	–60 °F to 20 °F	undefined
Gravity	32.2 ft/sec^2	31.09 ft/s^2	28.9 ft/s^2
Distance to Horizon*	3 miles for 6' tall adult	300 miles to 400 miles	1260 miles
Horizontal Depression**	0 degrees	5 degrees	18 deg
Cosmic Rays	8 counts/min	800 counts/min	?
UV Radiation	UV-A, small amount UV-B	UV-A and UV-B	UV-A,B, & C

* Distance to horizon can be calculated by the formula distance (miles) = sqrt(height(feet) * 1.5)
** See text

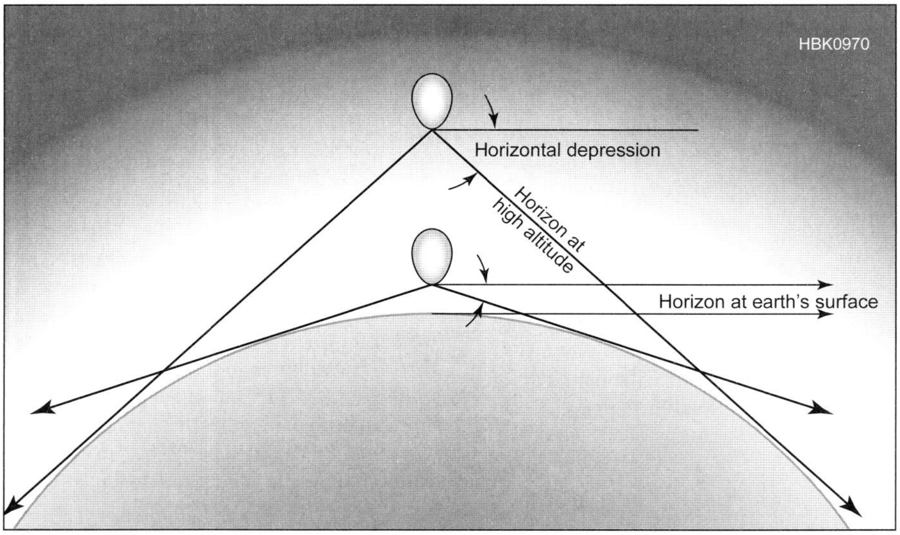

Figure 16.16 — In this highly exaggerated view, the higher balloon has a more distant horizon than at lower altitudes. This depresses the higher altitude balloon's horizon as altitude increases.

TEMPERATURE

The atmosphere consists of four layers, two of which are observable in the **Figure 16.18** chart. The lowest layer is the *troposphere*, which includes all our weather. It's heated by its contact with the ground (which is heated by sunlight) and not directly by sunlight. The higher in the troposphere, the farther from the warm ground and therefore the colder the air.

The second atmospheric layer is the *stratosphere*, which grows warmer from energy absorbed by ozone molecules in the layer. As ozone blocks solar ultraviolet from reaching the ground, the energy of the ultraviolet photons warms the ozone molecules. The higher one climbs into the stratosphere, the more ultraviolet there is to block and therefore the warmer the air.

The *tropopause* is the boundary between the troposphere and stratosphere. Sensors often show the air temperature remaining constant within this transition layer. Figure 16.18 shows a typical temperature profile measured during a balloon experiment.

Temperature is a factor in how fast chemical processes operate. Low temperatures are a concern for the voltage and current output of batteries. Low temperature can also increase viscosity to the point that lubricated mechanical systems will seize. Another issue is that some items (such as plastic zip ties) become brittle in the cold and may not perform as expected in near space.

ALTITUDE AND TEMPERATURE RATING

Many GPS receiver modules will not work above 60,000 feet. When choosing a GPS receiver, make sure the datasheet specifies a maximum altitude above the expected maximum altitude of the balloon. Some popular modules known to work at stratospheric altitudes are those made by Trimble, Garmin, u-Blox, and Inventek, as well as high-altitude modules offered by Byonics and Argent Data.

Instrument modules designed for high altitude experiments should have a sufficient rating for conditions found at high altitude. Modules and other items made for general purpose use may not have an adequate altitude and temperature rating. That does not mean they will not work when those ratings are exceeded, but they may become unreliable or

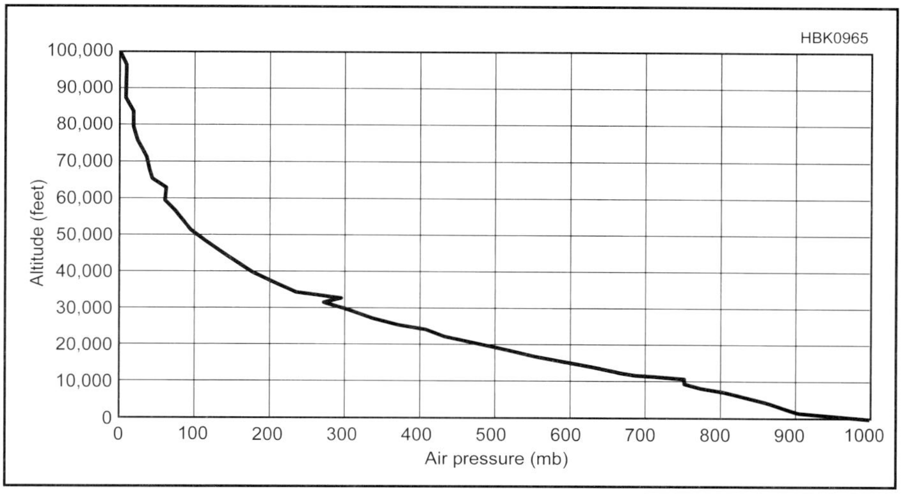

Figure 16.17 — A typical change of air pressure versus altitude.

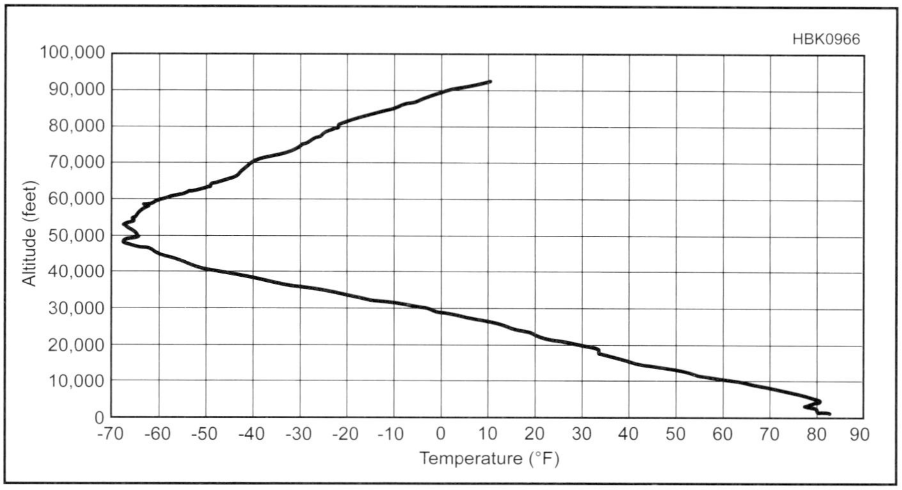

Figure 16.18 — Typical air temperature profile versus altitude, showing the troposphere (up to 50,000 feet) and the stratosphere (above 50,000 feet).

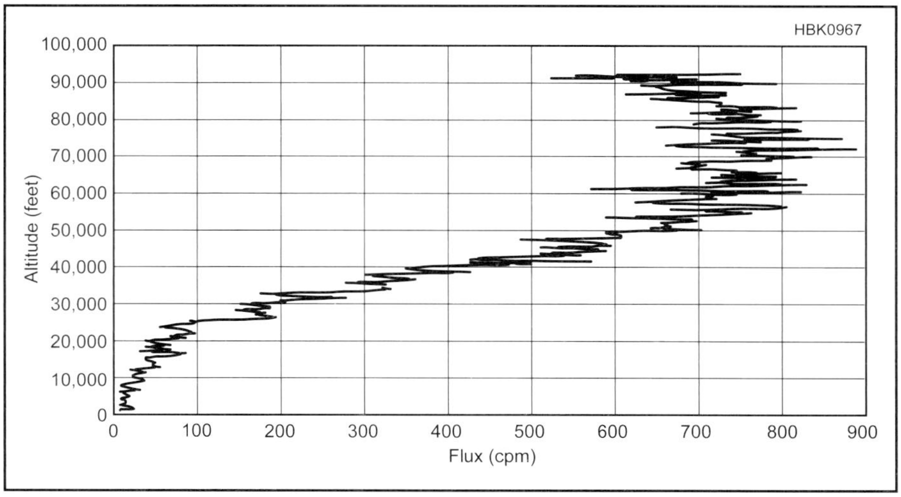

Figure 16.19 — Cosmic rays are encountered more frequently with increasing altitude in this typical balloon experiment.

fail to meet their performance specifications.

Since most consumer electronics and integrated circuits are rated for 0 to 70 ºC, you may want to find devices that are rated for the industrial temperature range (–40 to 100 ºC) or the automotive temperature range (–40 to 125 ºC) for near space payloads. Forums and user communities involved with high altitude ballooning will often have recommendations for specific models of equipment that perform well in these extreme conditions (see the preceding section on Temperature and Relative Humidity).

Cosmic rays are energetic subatomic particles. Most of this population of particles are protons, or the nuclei of hydrogen atoms. A smaller population of heavier nuclei can also be found in cosmic rays. When atomic particles enter the upper atmosphere, they are called *primary cosmic rays*.

When primary cosmic rays collide with molecules of oxygen and nitrogen, they break apart the atoms and create a shower of *secondary cosmic rays*. These secondary cosmic rays contain energetic X-rays and other subatomic particles such as neutrons, pions, and muons. Many of the secondary cosmic rays decay and lose energy though further collisions. The result of the multiple collisions is a peak in cosmic ray flux at around 62,000 feet (called the Pflotz Line). Depending on the type of radiation sensor used, at the Pflotz line the cosmic ray flux can be one hundred times greater than at sea level. (See **Figure 16.19**.)

Cosmic ray strikes can affect sensitive electronics. However, this should only be a concern in electronic devices with Very Large Scale Integration (VLSI) integrated circuits. This includes memories, processors, and advanced processing units. Smaller scale ICs have larger logic and circuit structures, making it more difficult for a cosmic ray to "flip a bit."

High-speed avionics may be at a slightly greater risk from cosmic-ray induced *single-event upsets* (SEU) during a near space flight. In fact, NASA research has found that avionics at 100,000 feet may experience an SEU once per 2.3 hours on average. Chances are the amateur will not have access to radiation-hardened electronics; in fact, it may raise suspicions if an amateur were to purchase such electronics! Advanced avionics uses software to detect and correct SEUs, but this may not be practical for amateurs either. The amateur's simplest solution is to design their avionics using electronics with large memory cells, and therefore not to use VLSI integrated circuits.

16.5.3 Balloon Platform Physical Design

Figure 16.20 shows a pair of typical balloon platforms. A weather balloon, once filled, is around seven feet tall. It is attached to the parachute and payloads with a *load line*. The load line is cut between 15 and 20 feet long so that the burst balloon's remains don't collapse on top of the parachute. Doing so may prevent the parachute from opening properly and slowing down the payload. Just after burst, the balloon's remains will fall slower than the parachute, which places the balloon above the parachute during the early descent.

There is however a risk of entanglement between the burst balloon and the parachute shroud lines or tracking antenna during late descent. This risk rarely is a problem as long as the load line is long enough to suspend the burst balloon below the tracking modules.

The parachute's diameter is chosen based on the payload's total weight. The goal is to provide a low-speed landing no greater than 5 meters/second, 1,000 feet/minute, or 10 miles per hour. A parachute six feet in diameter is usually large enough for a 12-pound payload.

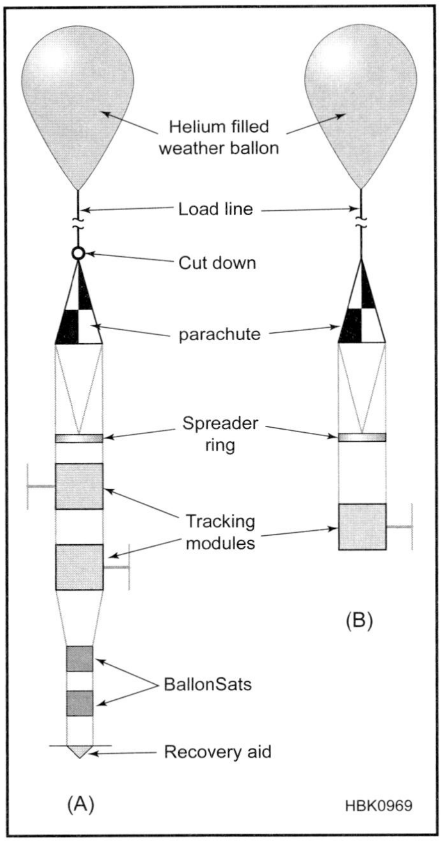

Figure 16.20 — (A) An idealized balloon platform showing the various components. (B) A simplified balloon platform with only the minimum necessary components.

Below the parachute and part of its shroud lines is a plastic ring called the *spreader ring*. The spreader ring is usually a plastic ring one foot in diameter loosely suspended between the parachute's shroud lines. It keeps the shroud lines from touching and wrapping around each other during the ascent. As long as the parachute shroud lines do not twist, the parachute will open properly soon after the balloon bursts (video records indicate this can take as little as one second). If the shroud lines are allowed to twist, the payload's descent is not properly braked. A descent from 70,000 feet can take as little as seven minutes in such cases. Furthermore, an improperly operating parachute can tangle up in payload antennas or even the payload itself.

A *cutdown device* is optional on a balloon flight meeting the strictest limits of FAR 101. The cutdown separates the parachute and payload from the balloon and load line on command or when conditions such as balloon burst or time of flight requirements are met. Most, if not all cutdown systems in use today use a series of DTMF tones sent over VHF or UHF FM channels. Typically, a DTMF decoder chip is connected to a VHF or UHF receiver on the payload, and the commands are sent as audio tones from a transmitter on the ground on the command frequency.

DTMF tone control permits the user to transmit a cutdown command followed by its password. Password protection for balloon cutdown ensures no false activation from DTMF tones intended for another purpose or payload. The cutdown is designed to either melt the load line with a coil of nichrome wire or slice the load line with a sharp blade.

Initiating cutdown starts the descent of the payload and allows the balloon to continue rising until it reaches its burst altitude. A cutdown that operates after the balloon bursts separates the balloon remains from the parachute. Separating the balloon fragments and load line from the parachute makes the descent less chaotic and protects the payload from the shock and acceleration of initial descent.

Following cutdown, the release of a near space balloon generates shaking and twisting for the attached platforms; however, the descent after balloon burst is significantly more traumatic. In fact, it's very evident when the balloon burst in the accelerometer chart shown in **Figure 16.21**.

Any object that places mass away from its center (has a lever arm) is subject to damage during balloon burst. Any object that is significantly denser than the rest of the platform can break free during balloon burst and become an impact hazard inside the platform or even escape the platform completely. Mounting methods that become brittle in the cold cannot be used to restrain dense or heavy objects on the platform.

ENCLOSURE AND INSTALLATION

The platform design must take into account the low temperature and low pressure of near space, the shock and vibration of the balloon burst, and to a much lesser extent, increased cosmic radiation.

A Styrofoam box is one of the most common enclosures. The foam is very light, provides insulation against the extreme temperatures encountered during flight for the payload electronics, and helps reduce the impact of landing. Even on the hottest summer day on the ground, it can be approximately –60 °F at 50,000 feet above the Earth. Most battery types do not work well at these temperature extremes, which are also outside the specification range of most electronic components. Fortunately, a Styrofoam box will help keep the internal temperatures well above those brutal outside conditions.

Covering the exterior of the Styrofoam housing with black tape (such as black packing tape) creates a passive heating system for the enclosed components. A second heat source is a chemical hand warmer. Hand warmers produce heat from the oxidation of iron and can produce heat for a two-hour flight. Start hand warmers an hour before launch to ensure the items they are heating get thoroughly warm. The ability of hand warmers to produce heat will degrade during the ascent (less oxygen availability), but the initial warmth and gradually decreasing heat output of the hand warmers will keep items about 10 degrees warmer throughout the flight.

Another technique is to mount the electronics and batteries on a foam-core board and wrap everything with three layers of small-cell bubble wrap. The insulation and trapped sunlight will keep the electronics warm.

Components inside the balloon platform can be protected from shock using foam rubber (foamed urethane). In all cases, every component must be mounted to the platform using fasteners that are strong enough to hold several times the weight of the component being immobilized. In addition, the fasteners cannot be constructed of materials that become brittle due to cold. Therefore, do not use plastic zip ties to restrain heavy or dense objects for high altitude balloon flights.

Any instrument requiring high voltage should receive additional electrical insulation, such as a coating of silicone glue. Items that may expand in volume when exposed to low air pressure should be evaluated on the ground (a vacuum chamber is ideal for this purpose) before being sent into near space. Devices requiring air cooling should be replaced with items not requiring air cooling, or perhaps given larger heat sinks.

BATTERIES

Lithium-based batteries are strongly recommended as a high-altitude power source as they

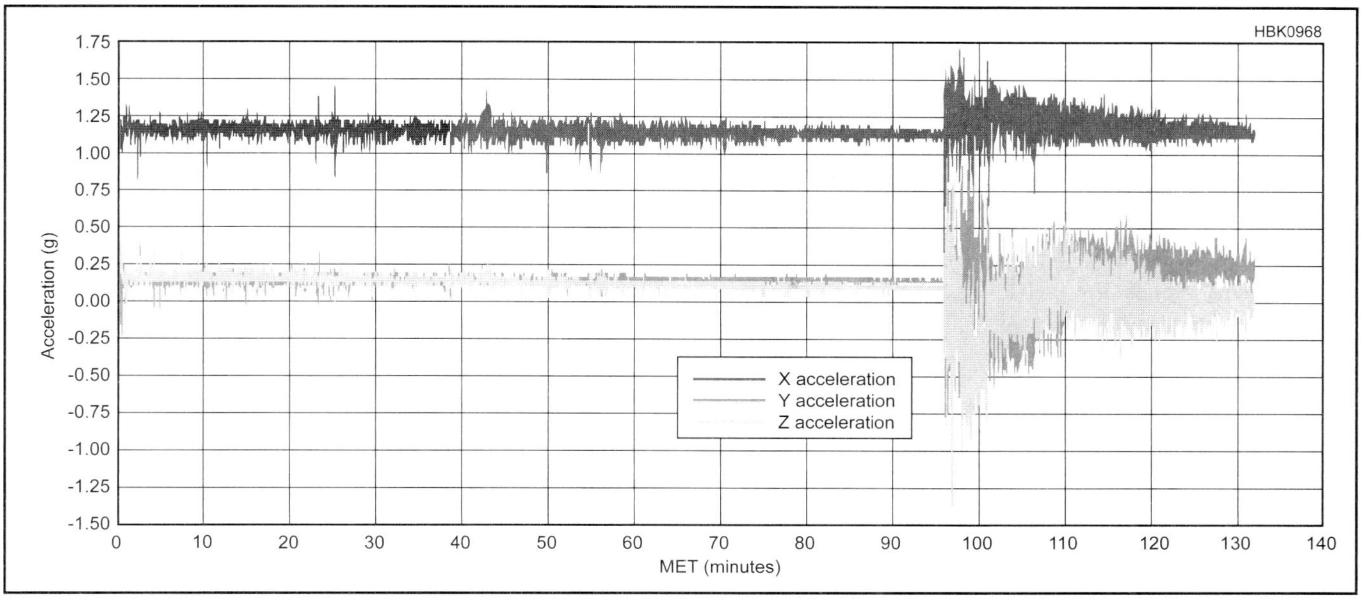

Figure 16.21 — The acceleration profile of a typical balloon experiment clearly shows the balloon bursting very clearly at about 95 minutes.

tend to handle the cold temperatures without excessive voltage drop. Batteries may be incorporated into the main enclosure, or they can be included in a separate container. The weight of a separate battery box below the main enclosure helps to stabilize the entire platform.

A practical source of lithium batteries is an RC hobby shop. Racing car batteries come in several voltages and capacities. Since many APRS trackers are designed to operate on 5 V, select a two- or three-cell battery pack (7.2 V or 11.1 V). If your tracker electronics has a low dropout voltage regulator, then your tracker can operate with a two-cell battery pack. The Eveready L91 AA lithium battery is popular for use with amateur radio high-altitude balloon platforms.

Not only do you have a choice of battery voltages with RC racing car batteries, you also have a choice of capacities. Measure the current draw of your APRS tracker and avionics and multiply that value by 24 hours. That will give a rough idea of the minimum battery capacity for your tracker.

Batteries must have sufficient capacity to operate the tracking system for its typical three-hour near-space flight plus additional time spent on the ground prior to launch and awaiting recovery. For example, it's nice to have extra reserve to make sure the tracker produces position reports overnight should it get lost during descent.

PREDICTING FLIGHT PATHS AND TRACKING

When flying a high-altitude balloon, it is a good idea to run a flight prediction a few days before launch. Repeat the prediction the night before a flight. The goal is to be sure the balloon doesn't land in a densely populated area or where making a successful recovery will be difficult. There are two popular downloadable prediction programs that can be accessed at the **www.arhab.org** website.

Amateur radio is a popular way to monitor the progress of a balloon's flight. The minimum system required to track a weather balloon flight consists of a GPS receiver that is certified to operate above 60,000 feet, an APRS TNC (terminal node controller), and a 2 meter FM transmitter. (A radar reflector is often included to increase the platform's visibility to air traffic control systems.)

The usual APRS configuration is to transmit a position report once a minute with the recommended path set to WIDE2-1. A power level below 1 W is quite sufficient, and many systems work quite well with just 200 mW.

Note that time and altitude data can be used to calculate the ascent rate of the weather balloon as a function of altitude. In addition, the same information can be used to calculate the descent rate of the parachute. Since a parachute's descent rate is a function of air drag, which is controlled by air density, the parachute's descent speed during descent can be used to estimate air density as a function of altitude. Note also that since a weather balloon is captive to the wind, measurements of altitude, speed, and heading are measurements of wind speed and direction at specific altitudes.

Including a second tracker may be an attractive form of insurance for mobile automated platforms such as weather balloons. If an APRS tracker were to fail at an altitude of 100,000 feet, the weather balloon can still be tracked and recovered using the second tracker. Redundant APRS trackers will need to use different SSIDs. In addition, they need to be time slotted so that their transmissions do not occur at the same time and jam each other. Preferably, redundant trackers will transmit on the same frequency as the primary tracker, which simplifies tracking for the people in search of the automated platform.

Finally, a *recovery aid* is a backup recovery method that's most useful after the parachute and its payload have landed. A recovery aid can include systems such as a loud audio beeper, strobe lights, or a fox-hunting (direction finding) transmitter. These help a chase crew recover a balloon platform after it has landed in trees or brush or tall crops where a GPS position is not enough to locate the platform.

APRS TELEMETRY

APRS is most popular and often used on short duration balloon flights (lasting for only a few hours), and where line of sight communication is possible between the chase crew and digipeaters. The use of IGates permits unlicensed individuals to participate in balloon chases by using a smartphone or tablet.

The simplest APRS telemetry string useful for balloon flights includes an identification, time of message, and the balloon's three-dimensional position, the balloon's course and speed. Such a string looks like the examples in the APRS message format specification at **www.eoss.org/aprs/aprs_formats_eoss** and in the APRS-IS.net specification at **www.aprs-is.net/q.aspx**. (See the preceding section on APRS for a sample string.)

16.6 Unmanned Aerial Vehicles (UAVs)

16.6.1 General UAV Platform Requirements

Drones or UAVs and their control systems are grouped in the category of *Unmanned Aerial Systems* (UAS). The regulating agency for UAS is the Federal Aviation Administration (FAA). For the purposes of Part 107, the FAA does not consider UAVs to be the same as model airplanes. One reason is that model airplanes are considered to be simple systems incapable of flying autonomously. Therefore, a model airplane must remain within sight of the pilot to be controlled. The applicable regulation for UAS is Federal Aviation Regulation Part 107 (FAR 107) or Title 14: Small Unmanned Aircraft Systems (14 CFR 107). Part 107 applies to pilots flying UAVs for a business; hobbyists have different requirements. By becoming a licensed remote pilot, however, additional privileges become available to the pilot.

This chapter will focus on the requirement for people flying UAVs as a hobby or for education. For these people, the following stipulations apply:

1) You do not need to be certified as a remote pilot to fly your UAV if you are flying for a hobby or pleasure.

2) You must register your UAV if it weighs between 0.55 and 55 pounds (this requirement has changed several times since it was first implemented).

3) You cannot fly your UAV within five miles of an airport without first contacting the airport and/or *air traffic control* (ATC). Note that some small airports do not have an ATC. In those cases, you'll need to contact the airport manager.

4) Your UAV must always yield right of way to manned aircraft.

5) You must keep your UAV within sight, or in visual line-of-sight.

6) Your UAV must weigh less than 55 pounds.

7) You must follow any community-based safety guidelines.

8) You cannot fly a UAV higher than 400 feet above the ground.

9) You cannot fly a UAV over people or stadiums.

10) You cannot fly a UAV over people who are not a part of the drone operation.

11) You cannot fly a UAV over moving vehicles.

12) You cannot fly a UAV while under the influence of drugs or alcohol.

13) You cannot fly a UAV over or around sensitive property or infrastructure.

14) You cannot use a UAV to spy on or monitor people when they have a reasonable expectation of privacy.

UAV pilots can find additional information at **www.faa.gov/uas** and from the Academy of Model Aeronautics (AMA) at **www.modelaircraft.org**.

UAV pilots are strongly encouraged to download the app *B4UFly* to their mobile phones prior to flying a UAV. The app shows airports and other restricted sites close to your location, or in other words, locations near which you are not allowed to fly a UAV.

16.6.2 UAV Data and Navigation Subsystems

Commercial UAVs require the same digital radio controllers used for traditional radio-controlled airplanes. These controllers can come with the UAV or may be purchased separately. A separate purchase is more common for advanced UAV pilots who are assembling their UAV from parts. Since the radio controller is operating a multirotor aircraft rather than a fixed wing aircraft, the aircraft pilot must configure the radio controller for UAV operation. Otherwise, the operation of the radio controller for a UAV is nearly identical to that of a fixed-wing aircraft.

An outdoor multirotor UAV is so complex that a human pilot cannot directly operate one by controlling each of the UAV's rotors. Instead, many UAVs carry an onboard flight controller that incorporates a GPS receiver and inertial navigation components such as accelerometers, gyroscopes, and a compass (magnetometer). The flight controller makes adjustments to the UAV's rotors based on joystick input and internal conditions. Combining all these elements into a single flight controller makes it so much easier to fly a UAV — the pilot only needs to push joysticks to control the thrust, pitch, roll, and yaw of the UAV.

Many onboard UAV flight controllers permit the pilot to set the UAV's flight characteristics and behaviors by changing the flight computer's settings through RC switch settings or even software. For example, a UAV can be configured to hold altitude (remain at the same altitude when the thrust joystick is left centered), fly in the direction the joysticks are pushed (regardless of the UAV's heading), and even programmed to fly a predefined course without further joystick input.

The most popular data transmitted from a UAV is video imagery (called *first person view* or FPV). Some UAVs transmit helpful navigational data such as altitude, heading, and speed. In addition, battery condition is transmitted in some UAVs to help the pilot manage the UAV's flight. Battery condition is frequently indicated by onboard LED colors.

To be a successful platform, UAVs are not usually required to transmit other types of data as a telemetry stream, because of the immediacy of the data collected. In other words, a drone can rapidly reach operational altitude, record data, then return to the launch site in a few minutes or less. Therefore, unlike weather balloons, data from UAVs is rapidly accessible and remains timely when it's downloaded.

However, if a UAV is going to transmit data, the pilot must test for the compatibility between the data transmitter and the radio controller and the onboard GPS. A pilot cannot safely operate a UAV if an onboard data transmitter interferes with the reception of RC commands, or if it interferes with the UAV's GPS receiver.

Testing for interference between the data transmitter and the UAV must be performed outdoors and with the UAV propellers removed. The pilot can then safely monitor the UAV's status LEDs or other status transmissions while the data transmitter operates onboard the UAV.

One additional safety test the pilot should perform is a measure of the UAV's center of gravity. UAV flight controllers have expectations about the UAV's center of gravity. Adding a payload to a UAV that changes that center of gravity substantially will make the UAV difficult or even impossible to control. If a payload will be added to a UAV, the pilot must determine the center of gravity before and after the modification.

16.6.3 UAV Platform Configuration

Most pilots purchase their UAV *ready to fly* (RTF) or *almost ready to fly* (ATF). Therefore, in the vast majority of the cases there is no need to worry about mounting or protecting the UAV's flight subsystems.

If a pilot is purchasing a UAV in parts and assembling it, then follow the assembly directions to construct the UAV. Directions will show the proper way to mount the UAV's flight subsystems. The proper placement is important in order to maintain the UAV's correct *center of gravity* (CG). When the CG is not located as the UAV's flight controller expects, unpredictable flights and even crashes can result.

16.6.4 UAV Electronic Subsystems

Prior to installing a new electronic system on a UAV, the pilot must ensure RFI from the payload cannot interfere with the UAV's GPS receiver. This can be an issue with some digital cameras, for example, where the crystal-controlled clock inside the camera can emit enough RF to prevent a GPS from gaining a proper lock or even worse, losing satellite lock during flight. Therefore, the UAV should be left on the ground while its new electronics payload is tested in place on the UAV and then tested around the UAV. Since many commercial UAVs are capable of reporting their GPS status through LEDs and the radio controller, the remote pilot should monitor them for a while to ensure GPS lock remains steady as the new payload electronics is tested.

One useful electronic subsystem for a UAV is a voice or digital repeater. (See the preceding section on VHF/UHF Payloads.) The UAV's ability to climb to an altitude of several hundred feet with a payload makes it a rapid-to-deploy antenna tower. The limitations of a UAV-based antenna tower include the following:

- Limited weight — usually no more than one pound
- Limited time aloft — upwards of 30 minutes with new batteries and less as the payload becomes heavier
- Limited authority — poor authority means a UAV cannot prevent it from drifting with the wind, if the wind blows too strongly
- Limited to flying during daylight hours
- Limited antenna orientations — antennas must be placed where they do not interfere with the UAV's propellers
- Limited to only flying above people who are not involved with operation of the UAV

A second useful electronic subsystem is FPV or first person view video. (See the preceding section on Image and Video payloads.) A UAV flown with FPV adds a synergy to search and rescue (SAR). Hams can use the UAV to rapidly scan long distances, fly over steep terrain, and even peer over fences and into steep canyons. These abilities can reduce search times and the risk hams experience climbing into and out of potentially hazardous terrain. A thermal imaging camera is another powerful tool to add to a UAV-enhanced SAR.

Complete FPV systems containing the video camera and the transmitter are readily available at minimal cost. They use the unlicensed 5.8 GHz radio frequencies and transmit with a power of hundreds of milliwatts. In addition, the systems are designed to be lightweight and not to interfere with the operation of the UAV. However, the remote pilot will need an observer to provide assistance to the pilot who is flying the UAV.

A non-electronic application of UAVs is carrying one end of a lightweight string over a tree or structure so that an antenna can be pulled up afterward. This permits an operator to place an antenna with more precision than is usually possible with a sling shot or "antenna launcher."

16.6.5 Powering the UAV

The motors of a UAV have a very high current demand. Twenty or more amps per motor isn't uncommon. The heavy current demands of motors (four of them on a quadcopter and six on a hexacopter) and the requirement for a lightweight quadcopter means a high capacity, high current draw, and a lightweight battery is required. These requirements and the ability to recharge the flight battery are best met with lithium polymer (LiPo) batteries. Read the recommendations for your UAV before purchasing new batteries.

16.7 Rockets

16.7.1 General Rocket Platform Requirements

Like amateur radio, amateur rocketry is a self-regulated activity. Since model rockets use the national air space, the ultimate regulating agency for model rockets is the Federal Aviation Administration (FAA). The applicable FAA regulation for model rockets is the Federal Aviation Regulation Part 101 (FAR 101) or Title 14: Aeronautics and Space, 14 CFR 101. In addition, the National Fire Protection Association (NFPA) creates safety codes that keep rocket motors safe to transport and use. NFPA 1127 - Code for High Power Rocketry Scope covers the construction, propellant mass, and reliability of rocket motors for amateur use. Rocket motor manufacturers follow these codes and as long as amateurs purchase their motors from reputable dealers, this will not be a concern when flying model rockets. The National Association of Rocketry (NAR) and Tripoli Rocketry Association (TRA) are the self-regulating organizations for model rocketry. Therefore, it's strongly encouraged that you join either one of these organizations prior to launching model or amateur rockets.

16.7.2 Suitable Rocket Data and Navigation Subsystems

Amateur radio is seldom used with low-power model rocket flights. This is mostly because of the extreme weight limits placed on payloads by small model rockets and their limited range. Amateur radio does however appear in some high power amateur rockets as an aid to recovering rockets after flight. There are several commercial products available for this purpose.

Radio tracking products for rockets are available in licensed and unlicensed frequencies. They also appear as either APRS (onboard rocket GPS receivers) or radio direction finding (fox hunting) products. The only other tracking system is an audio beacon and therefore does not rely on amateur radio.

Amateur radio is also making an appearance in model rocketry for purposes of telemetry. Data typically includes acceleration, velocity, and altitude. UHF frequencies are preferable for rocket telemetry in order to keep the antenna short and lightweight. The antenna's orientation is necessarily vertical due to the construction of the rocket.

16.7.3 Rocket Platform Configuration

All electronics mounted inside an amateur rocket will experience very high accelerations at lift off, over 10 g of acceleration in some cases. (1 g is equivalent to the force of gravity at the Earth's surface.) Therefore, all electronics must be securely mounted to the rocket's airframe in a way that will prevent them from coming loose. This is especially true for batteries, which can be denser than electronics. In additional to acceleration, rocket payloads can experience mechanical shock and rapid shaking in multiple directions.

Most model rockets can't climb high enough to experience significant changes in temperature and pressure. Therefore, their electronics payloads do not have to be protected from these conditions. However, there are some high power amateur rockets that do climb to over 10,000 feet and can therefore experience rapid changes in air pressure and air temperature. Unlike high altitude weather balloons, the electronics reach high altitudes so fast (compared to a balloon) that the payload does not cold soak long enough to become completely chilled.

Sensors onboard rockets experience rapid changes in the physical conditions they are measuring. This can result in noisy reading from the sensors that must be accounted for. Sensors used to control the operation of the rocket, such as for determining peak altitude for parachute deployment, must have their outputs filtered in real-time to be useful. Other data can be filtered after the rocket is recovered and its data downloaded.

Antennas on rockets must be small and lightweight and they must not interfere with rocket staging or parachute deployment (which involves ejecting the rocket nose cone). These requirements make a UHF dipole a useful antenna. The ends of the antenna must be secured to the rocket body so they don't vibrate in the air flowing over the rocket body. Or the antenna must be secured inside the rocket body. Some radio trackers use flexible antennas and whips. Bird trackers make popular radio trackers for rockets since they have similar requirements.

As with UAVs, any electronic devices included in a rocket's payload should be tested for interference — especially when a GPS receiver is used for the rocket's maximum altitude determination.

Since rocket flights tend to be short, a high capacity battery is not typically required. Form factor (shape and dimensions) and weight are often the important considerations. This means alkaline batteries are often acceptable for rocket electronics.

16.8 Robotics

16.8.1 General Robotics Platform Requirements

Robots are defined as systems capable of acquiring sensory input, making decisions based on that input and the robot's current state, then acting based on the results of that decision. A robot's need to combine sensory data and current status means most robots include some type of programmable microcontroller.

Being able to act doesn't mean a robot must be very mobile. Industrial robots, for example, are typically stationary and only able to operate within defined boundaries. A robot's method of acting can be electrical (such as motors and servos), pneumatic (air pressure), or hydraulic (liquid pressure). Each has its own abilities, requirements, and interfacing requirements with a robot controller.

Robots span a range of autonomies. A robot can be completely autonomous and capable of operating without any human intervention for long periods of time, or a robot can require regular, predictable human intervention (such as a planetary rover). When a robot requires human input, that communication can be wireless or via a tether containing wire or fiber optics.

The many types of sensors, the many internal states, the many types of control systems, the many levels of limitation of mobility, many mechanical systems, and many ways to communicate between humans and robots make it impossible to define a general platform for robotics. Each robot is typically as unique as the person or people who created it.

Unless robots interact with the public, there is no regulation controlling their amateur use. This is changing with the advent of self-driving cars and trucks.

16.8.2 Suitable Robotics Data and Navigation Subsystems

Robots confined to traveling in small areas typically use inertia navigation methods or position measurement systems to navigate. Typical sensors include:

- Accelerometers
- Magnetometers
- Encoders
- Range Finders
- Light Detectors
- Line Detectors
- Laser Range Finders
- Limit Switches

Robots traveling over large distances or regions can combine GPS-based navigation systems into onboard inertia and local position measurement systems.

APRS is a good method for robots (such as the Mars Rover) that send navigation or sensory data wirelessly to a human operator/monitor. In addition, a human operator can send instructions back to the robot via APRS messages.

Robots that transmit imagery can rely on either slow scan (SSTV) or amateur television (ATV). SSTV is a suitable method when conditions are slow to change or when only low power levels are available to send imagery. ATV is important where the robot is traveling at speeds high enough that human intervention is needed constantly. (See the preceding sections on Image and Video payloads.)

Although robotics projects using amateur radio are currently few, one of the more successful as of early 2018 is the HF Voyager project (**www.jrfarc.org/hf-voyager**). An autonomous "wave glider," HF Voyager operates under automatic control, making contacts with the call sign KH6JF/mm on 20 meters using FT8 and PSK-31 as the primary operating modes. It can also use WSPR in times of poor propagation.

A possible solution to long-range communication between humans and a robot might be through satellites. An iridium modem would permit the onboard robot controller to send and receive data from virtually anywhere in the world.

Finally, there are unlicensed radio bands, such as ISM, that amateur robotics engineers are using for their short-range robotics projects. These bands permit human-robot communications in both imagery and data.

16.8.3 Robotics Platform Configuration

Small, indoor robots typically have few limitations (other than those imposed for competitions). The roboticist can replace their batteries frequently, the robots don't experience extreme high or low temperatures, no rain falls on them, and small and lightweight motors suffice for making them mobile.

A roboticist designing a large, outdoor robot will need to balance the robot design with many of the limitations mentioned above. This starts with a strong but lightweight robot body (often made from aluminum). Weatherproof enclosures are required to protect the onboard electronics. Motors need to be sealed where dust and rain are issues. The motors must be powerful enough to make the robot mobile at reasonable speeds. The robot's battery design must have the capacity to operate the robot's electronics and motors without excessively weighing down the robot. Finally, depending on the speed of the robot, the design might call for physical shock protection.

Most robots use flexible antennas to communicate with human operators. The antenna is typically mounted at a high location on the robot where it can't interfere with cameras and robotic arms.

Robots that operate without direct human contact for long periods of time will need to use solar charging in order to keep their batteries topped off. Solar charging systems need to monitor the battery level and then act to recharge batteries at the appropriate voltage and current. The high power demands of robots means they can never operate solely on solar power and without batteries. This also means robots will need down time at some point in order to recharge their batteries.

Lithium polymer (LiPo) batteries are good batteries for robots since they are light weight, can operate across a large range of temperatures, and are capable of meeting high current demands.

16.9 Fixed Stations

16.9.1 General Fixed Station Platform Requirements

Fixed stations such as weather stations are found in locations where environmental conditions are hostile to humans, or where it's not practical or safe to record data. Moored or drifting buoys are included in this classification. Therefore, fixed stations must be constructed to handle extreme weather conditions. And their data subsystems are critically important if they can't be tended on a regular basis.

The FCC is the agency responsible for the telemetry requirements of fixed stations that use radios to communicate. The data transmission requirements for fixed site data collectors and dataloggers (such as moored buoys) are simpler than they are for drifting buoys. This is because a drifting buoy can be carried by ocean currents into international waters or the coastal waters of other countries. Since each country has its own rules governing the use of amateur radio, drifting buoy designers and builders must become acquainted with the communication rules of other countries to insure their drifting buoy's don't violate amateur radio laws.

When the need is to monitor weather conditions continuously on land, a weather station is the preferred solution. An amateur must mount a weather station above the level of obstacles (such as buildings and trees) that will interfere with the flow of wind. This means many weather stations can be found mounted above roofs and on tall masts or radio towers.

When it comes to monitoring conditions in bodies of water, a buoy is used. Buoys can be of two types — those remaining stationary by mooring them to the seabed or those free to drift with the ocean and wind currents. Moored buoys have been used by the National Oceanographic and Atmospheric Administration (NOAA) since 1951, and drifting buoys have been in use since 1979 (Wikipedia). At this time, the editors are unaware of any communication requirements from NOAA.

Anchored or fixed datalogging stations should not be anchored or fixed in locations where people have a reasonable expectation of privacy or their presence will interfere with navigation or right of way.

16.9.2 Suitable Fixed Station Data and Navigation Subsystems

The needs of data and navigation systems for weather stations and buoys are similar to those mentioned in previous sections. The data needs are addressed based on the distance between the station and the amateur recording the data.

For short distances, such as from the roof to the interior of a house, a wire connection between the station and a PC or display often suffices. Where it's not practical to route a communication cable into the house, then an unlicensed radio system will often work and is usually available with the fixed station. Otherwise, where long distance, line-of-sight communication is needed, APRS is one of the best methods available. In fact, there are many weather stations with integrated APRS capability available for purchase.

For fixed stations located long distances from the data user, HF modes are often the best solution. This includes the WSPR, Hellschreiber, RTTY, CW, and JT9 modes.

Satellite communication methods such as Spot (**findmespot.com**) and iridium modems are a possible solution for long distance fixed and mobile stations. A Spot tracker only reports its position; therefore it is best used for reporting the position of mobile stations such as buoys. An iridium satellite modem will telemeter data, unlike the Spot. Both of these satellite systems require a subscription. Therefore the amateur must determine whether or not the value of the data justifies the cost of a subscription prior to placing the fixed station into operation.

16.9.3 Fixed Station Platform Configuration

WEATHER STATIONS

Commercial weather stations use components capable of handling hot and cold temperatures and house most of them inside of an enclosure. Weather sensors, on the other hand, must be exposed to the elements in order to collect their data.

Pressure sensors can be left inside the house (if it's close to the weather station) or inside the enclosure since the enclosure is not air tight. Relative humidity sensors must be exposed to the air, while at the same time, protected from rain and condensation. One solution is to house the relative humidity (RH) sensor inside of a housing with vents cut in its sides. Temperature sensors (such as the LM335) must be exposed to the air, but not to sunlight because solar absorption will warm the sensor above ambient air temperature. A sun shield placed over the temperature sensor usually suffices. Alternatively, a temperature sensor can be mounted to the underside of the weather station enclosure.

Two of the more difficult weather sensors are the anemometer and wind vane. They are often made from position sensors that can rotate 360 degrees without needing to unwind (for the wind vane) or with generators (for the anemometer). These are mechanical systems that need protection from moisture while still remaining free to rotate continuously without impediment. Other systems that are less difficult to protect include the hot wire anemometer and the ultrasonic anemometer and wind vane.

BUOYS

Drifting buoys are used to track ocean currents, water, and weather conditions (temperature, pressure, RH, water temperature, salinity). A moored buoy can report on wind speed and direction, while a drifting buoy can't report with the same level of accuracy due to its movement.

Buoys must have an overall density less than the density of water (1.0 grams/cc) if they are to float. A lower density will raise the top of the buoy higher above the water than for a higher density buoy. Be careful — density that is too low can result in an unstable buoy that tips over. Therefore, buoys need ballast so they can maintain an upright configuration in rolling waters and waves. The ballast placed in a buoy must be located below the buoy's desired center of gravity. Any ballast must be securely attached to the buoy so that it can't shift its position when the buoy is tilted by waves. Finally, the ballast attachment cannot compromise the watertightness of the buoy.

All buoy components must be protected from fresh or salt water. Therefore, buoys must be constructed with water-tight seals.

This can be a challenge for buoy antennas, which are often located on top of the buoy. These antennas will be exposed to fresh or salt water on a regular basis. Therefore, buoy antennas require insulation to keep salt water off of their metal components, including connectors. Be careful in the selection of insulating materials — cold water can make some materials stiff or brittle. Select insulating materials that maintain their strength and flexibility when they get cold.

Insulation slows changes in temperature but can't stop those changes. Since buoys can operate for days or weeks in cold conditions, insulation such as Styrofoam is less effective in protecting electronics. Two options are to add heaters to the buoy or to use electronics that are designed to operate in cold conditions.

16.9.4 Powering the Fixed Station Platform

Indoor and some outdoor fixed stations may have access to household or commercial electrical power. If so, there will be no power concerns for these stations. Other outdoor fixed stations require batteries that must function in extreme conditions unless heating and cooling is provided for the batteries. In addition, batteries may be exposed to damp conditions if they can't be hermetically sealed in protective containers. The requirement to seal batteries is critical for buoys since they are exposed to salt water.

If a fixed station must operate for a good part of a year without maintenance, then the amateur should give consideration to using solar power to recharge the station's batteries. The solar cells must be encapsulated to protect them from moisture.

16.10 References and Bibliography

The ARRL Operating Manual, 11th edition, (ARRL, 2016).

The ARRL Antenna Book, 24th edition, (ARRL, 2019).

Britain, K., WA5VJB, "Cheap Yagis," **www.wa5vjb.com/yagi-pdf/cheapyagi.pdf**.

Buchmann, I., *Batteries in a Portable World* (Cadex, 2011).

Cain, P., KØPC, "APRS with a Smartphone," *NCJ*, Sep./Oct. 2012, pp. 16-17.

Fisher, B., W2BFJ, "Fox-1 Satellite Telemetry – Part 1: On the Satellite," *AMSAT Journal*, Nov./Dec. 2015, pp. 9-12.

Horzepa, S., WA1LOU, *APRS: Moving Hams on Radio and the Internet*, (ARRL, 2004 — out of print).

Mims, F., "Sun Photometer with Light-emitting Diodes as Spectrally Selective Filters," Applied Optics, Vol. 31, No. 33, 20 Nov. 1992, pp. 6965-6967.

Post, J., KA5GSQ, "A Simple Sensor Package for High Altitude Ballooning," *QEX*, May/Jun. 2012, pp. 10-19.

Simmons, B., WB6EYV, "APRS Unveiled," *QEX*, Nov./Dec. 2012, pp. 19-23.

Spencer, M., WA8SME, "ARRL Education and Technology Program Space/Sea Buoy," *QST*, May 2012, pp. 33-35.

Thompson, C., AC2CZ, "Fox-1 Satellite Telemetry – Part 2: FoxTelem," *AMSAT Journal*, Jan./Feb. 2016, pp. 7-9.

Verhage, Paul, KD4STH, *BASIC Stamp Near Space* (see the References section), **nearsys.com/pubs/book/index.htm**.

Williams, D., AJ5W, "APRS and High-Altitude Research Balloons," *QST*, Aug. 2007, pp. 48-49.

CQ VHF Articles (out of print)

Benton, D., KE5URH, "Tulsa (OK) Technology Center's Innovative Approach to Training Future Engineers," *CQ VHF*, Fall 2008, p. 32.

Brown, B., WB8ELK, "Up in the Air: Kentucky Space Balloon-1," *CQ VHF*, Spring 2009, p. 57.

Brown, B., WB8ELK, "Up in the Air: HF Balloon Tracking," *CQ VHF*, Winter 2011, p. 63.

Dean, T., KB1JIJ, "ATV: W2KGY BalloonSat Payload," *CQ VHF*, Fall 2010, p. 68.

Ferguson, D., AI6RE, "Transatlantic Balloon Flight 2012, CNSP-18 K6RPT-12," *CQ VHF*, Winter 2013, p. 8.

Helm, M., WC5Z, "A Low Cost 70 cm Tracking Beacon for Rocket or Balloon Payloads," *CQ VHF*, Fall 2010, p. 31.

Whitham, R., VE3NSA and Cieszecki, J. VE4CZK, "The WinCube Project," *CQ VHF*, Winter 2009, p. 12.

An index to all of the articles in *CQ VHF*, including the "Up In The Air" column on balloons, has been compiled by KB9MWR at **www.qsl.net/kb9mwr/files/ham/cq-vhf.html**.

Nuts and Volts Magazine Articles by Paul Verhage, KD4STH

"Near Space: Approaching the Final Frontier," Mar. 2013, p. 68.

"Near Space: Using the Nearspace Simple Flight Computer," Jan. 2013, p. 72.

"Near Space: Approaching the Final Frontier," Nov. 2012, p. 14.

"Near Space: A New BalloonSat Airframe Design," Jul. 2012, p. 68.

"Near Space: The NearSpace UltraLight — The Everyman's Flight Computer," Jan. 2011, p. 67.

"Near Space: Your Own Micro Datalogger," May 2009, p. 80.

Verhage continues to write the "Near Space" column as of early 2018.

Contents

Articles
- A Tube Tester for High Power Transmitting Tubes by John Mathis, WA5FAC, and Max Landey, KM4UK
- Amplifier Overshoot-Drive Protection by Phil Salas, AD5X
- Amplifier Projects from Previous *ARRL Handbook* Editions
- Design Example — HF Amplifier using 8877 Vacuum Tube by John Stanley, K4ERO
- Design Example — MOSFET Thermal Design by Dick Frey, K4XU
- Designing to Avoid Interactive Tuning and Load Adjustments by John Stanley, K4ERO
- Determining a Transistor's Power Rating (APT Application Note) by Dick Frey, K4XU

HF Amplifier Projects
- 3CX1500D7 RF Linear Amplifier by Jerry Pittenger, K8RA (including PCB layout, Pi-L values spreadsheet, etc.)
- Base diagrams and operating values for popular transmitting tubes
- The Everyham's Amp by John Stanley, K4ERO (including files with construction notes, layouts, use of different tubes, etc.)

VHF Amplifier Projects
- 144 MHz Amplifier Using the 3CX1200Z7 by Russ Miller, N7ART
- A 6 Meter Kilowatt Amplifier by Dick Stevens, W1QWJ
- Build a Linear 2 Meter, 80 W All Mode Amplifier by James Klitzing, W6PQL
- High-Performance Grounded-Grid 220-MHz Kilowatt Linear by Robert Sutherland, W6PO
- Simple Broadband Solid-State Power Amplifiers by Paul Wade, W1GHZ

UHF/Microwave Amplifier Projects
- 2 Watt RF Power Amplifier for 10 GHz by Steven Lampereur, KB9MWR
- 432 MHz 3CX800A7 Amplifier by Steve Powlishen, K1FO
- 2304 MHz 70 W Rover Amplifier by Bill Koch, W2RMA, and John Brooks, N9ZL
- A High-Power Cavity Amplifier for the New 900-MHz Band by Robert Sutherland, W6PO
- A Quarter-Kilowatt 23-cm Amplifier Parts 1 and 2 by Chip Angle, N6CA

Software and Tools
- 135 Degree Pi Network Calculator Spreadsheet by John Stanley, K4ERO
- *MATCH.EXE* software (for use with Tuned (Resonant) Networks discussion)

The following programs are available in the separate Software folder
- *TubeCalculator* by Bentley Chan and John Stanley, K4ERO, for analysis of operation of popular high power transmitting tubes.
- *PI-EL* by Jim Tonne, W4ENE, for design and analysis of pi and pi-L networks for transmitter output.
- *MeterBasic* by Jim Tonne, W4ENE, for design and printing of custom analog meter scales.

Chapter 17

RF Power Amplifiers

Amateur Radio operators typically use a very wide range of transmitted power — from milliwatts to the full legal power limit of 1.5 kW. This chapter covers RF power amplifiers beyond the 100 – 150 W level of the typical transceiver. The sections on tube-type amplifiers, including a sidebar on a design method to avoid interactive tuning, were prepared by John Stanley, K4ERO. The sections on solid-state amplifiers were contributed by Richard Frey, K4XU, along with an overview of the new "pallet" amplifier modules. Roger Halstead, K8RI, contributed material on amplifier tuning and the use of surplus components in amplifier construction that appears with the online content. The section on Pallet Amplifiers was contributed by Jim Klitzing, W6PQL. The section on Solid-State Amplifier IMD was contributed by Rob Sherwood, NCØB; Dick Frey, K4XU; Warren Pratt, NRØV; and Tom Thompson, WØIVJ, who also contributed the material on Adaptive-Predistortion. Amplifier projects using tubes, transistors, and integrated circuits that appeared in previous editions of *The ARRL Handbook* are included with the online content.

17.1 High Power, Who Needs It?

There are certain activities where higher power levels are almost always required for the greatest success — contesting and DXing, for example. While there are some outstanding operators who enjoy being competitive in spite of that disadvantage, the high-power stations usually have the biggest scores and get through the pileups first. On the VHF and higher bands, sometimes high power is the only way to overcome path loss and poor propagation.

Another useful and important area where higher power may be needed is in net operations. The net control station needs to be heard by all potential net participants, some of whom may be hindered by a noisy location or limited antenna options or operating mobile. This can be crucial to effective emergency communications. General operation also benefits from the availability of high power when conditions demand it to maintain contact.

How do you decide that you need amplification? As a rule of thumb, if you have a good antenna and hear a lot of stations that don't seem to hear you, you probably need to run more power. If you operate on bands where noise levels are high (160 and 75/80 meters) or at times when signals are weak then you may find that running the legal limit makes operations more enjoyable. On the other hand, many stations will find that they can be heard fine with the standard 100 W transmitter or even with lower power.

Power requirements also depend on the mode being used. Some digital modes, such as PSK31, work very well with surprisingly low power. CW is more power efficient than SSB voice. Least effective is full carrier AM, which is used by vintage equipment lovers. Once you have determined that higher power will enhance your operations, you should study the material in this chapter no matter whether you plan to build or buy your amplifier.

Note that many power amplifiers are capable of exceeding the legal limit, just as most automobiles are capable of exceeding posted speed limits. That does not mean that every operator with an amplifier capable of more than 1.5 kW output is a scofflaw. Longer life for the amplifying devices and other amplifier components as well as a cleaner signal result from running an amplifier below its maximum rating. However, just as with automobiles, that extra capability also presents certain temptations. Remember that FCC rules require you to employ an accurate way to determine output power, especially when you are running close to the limit.

Safety First!

YOU CAN BE KILLED by coming in contact with the high voltages inside a commercial or homebrew RF amplifier. Please don't take foolish chances. Remember that you cannot go wrong by treating each amplifier as potentially lethal! For a more thorough treatment of this all-important subject, please review the applicable sections of the **Power Sources** and the **Safety** chapters in this *Handbook*.

High-power amplifiers generate strong RF fields, especially near a tube or transistor and around the output circuits. High levels of RF at frequencies of 50 MHz through microwaves are particularly hazardous. Avoid exposing yourself to intense RF fields when adjusting or measuring energized equipment and follow the RF exposure guidelines for yourself and others. See the **Safety** chapter for more information.

17.2 Types of Power Amplifiers

Power amplifiers are categorized by their power level, intended frequencies of operation, device type, class of operation and circuit configuration. Within each of these categories there are almost always two or more options available. Choosing the most appropriate set of options from all those available is the fundamental concept of design.

17.2.1 Why a "Linear" Amplifier?

The amplifiers commonly used by amateurs for increasing their transmitted power are often referred to as "linears" rather than amplifiers or linear amplifiers. What does this mean and why is it important?

The active device in amplifiers, either tube or transistor, is like a switch. In addition to the "on" and "off" states of a true switch, the active device has intermediate conditions where it presents a finite value of resistance, neither zero nor infinity. As discussed in more detail in the **RF Techniques** chapter, active devices may be operated in various *classes of operation*. Class A operation never turns the device fully on or off; it is always somewhere in between. Class B turns the device fully off for about half the time, but never fully on. Class C turns the device off for about 66% of the time, and almost achieves the fully on condition. Class D switches as quickly as possible between the on and off conditions. Other letters have been assigned to various rapid switching methods that try to do what Class D does, only better. Class E and beyond use special techniques to ensure that high voltage and current do not occur during switching.

During the operating cycle, the highest efficiency is achieved when the active device spends most of its time in the on or off condition and the least in the resistive condition. For this reason, efficiency increases as we go from Class A to B to C to D.

A *linear amplifier* is one that produces an output signal that is identical to the input signal, except that it is stronger. Not all amplifiers do this. Linear amplifiers use Class A, AB or B operation. They are used for modes such as SSB where it is critical that the output be a close reproduction of the input.

The Class C amplifiers used for FM transmitters are *not* linear. A Class C amplifier, properly filtered to remove harmonics, reproduces the frequencies present in the input signal, but the *envelope* of the signal is distorted or even flattened completely. (See the **Modulation** chapter for more information on waveforms, envelopes and other signal characteristics.)

An FM signal has a constant amplitude, so

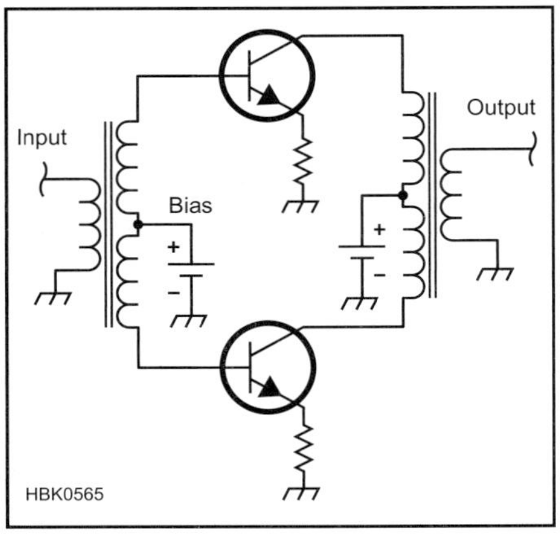

Figure 17.1 — This simple circuit can operate in a linear manner if properly biased.

it carries no information in the envelope. A CW signal does carry information in the amplitude variations. Only the on and off states must be preserved, so a Class C amplifier retains the information content of a CW signal. However, modern CW transmitters carefully shape the pulses so that key clicks are reduced to the minimum practical value. A Class C amplifier will distort the pulse shape and make the key clicks worse. Therefore, except for FM, a linear amplifier is recommended for all amateur transmission modes.

Some digital modes, such as RTTY using FSK, are a form of FM and can also use a nonlinear Class C, D or E amplifier. If these signals are not clean, however, a Class C amplifier may make them worse. Also, Class C or even D and E can be used for very slow CW, for very simple low-power CW transmitters or on uncrowded bands where slightly worse key clicks are not so serious. After all, Class C was used for many years with CW operation.

Class of operation as it relates to tube-type amplifier design is discussed in more detail in a later section of this chapter.

ACHIEVING LINEAR AMPLIFICATION

How is linear amplification achieved? Transistors and tubes are capable of being operated in a linear mode by restricting the input signal to values that fall on the linear portion of the curve that relates the input and output power of the device. Improper bias and excessive drive power are the two most common causes of distortion in linear amplifiers. All linear amplifiers can be improperly biased or overdriven, regardless of the power level or whether transistors or tubes are used.

Figure 17.1 shows a simple circuit capable of operating in a linear manner. For linear operation, the bias on the base of the transistors must be such that the circuit begins to produce an output signal even with very small input signal values. As shown in **Figure 17.2**, when properly adjusted for linear operation, the amplifier's output signal faithfully tracks the input signal. Without proper bias on the transistors, there will be no output signal until the input voltage goes above a threshold. As shown in Figure 17.2B, the improperly adjusted amplifier suddenly switches on and produces output when the input signal reaches 0.5 V.

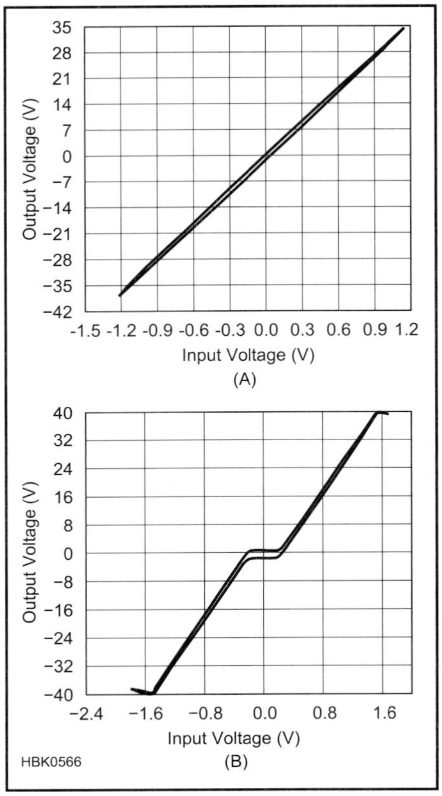

Figure 17.2 — Input versus output signals from an amplifier, as observed with the X-Y display on an oscilloscope. At A, an amplifier with proper bias and input voltage. At B, the same amplifier with improper bias and high input voltage.

Some tubes are designed for *zero bias* operation. This means that an optimum bias current is inherent in the design of the tube when it is operated with the correct plate voltage. Other types of tubes and all transistors must have bias applied with circuits made for that purpose.

All amplifiers have a limit to the amount of power they can produce, even if the input power is very large. When the output power

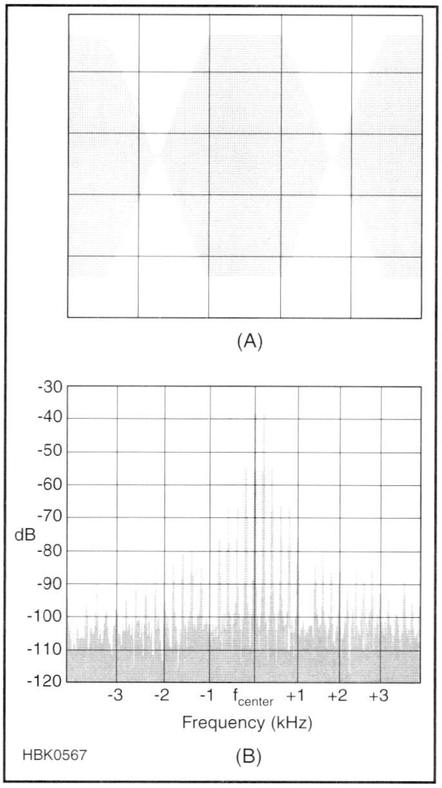

Figure 17.3 — Improper operation of a linear using a two-tone test will show peak clipping on an oscilloscope (A) and the presence of additional frequencies on a spectrum analyzer (B). When these patterns appear, your "linear" has become a "nonlinear" amplifier. Users of adjacent channels will not be happy. More information on transmitter testing may be found in the Test Equipment and Measurements chapter.

runs up against this upper limit (that is, when additional drive power results in no more output power), *flat topping* occurs and the output is distorted, as shown in **Figure 17.3**. Many amplifiers use *automatic level control* (ALC) circuits to provide feedback between the amplifier and the companion transceiver or transmitter. When adjusted properly, ALC will control the transmitter output power, preventing the worst effects of overdriving the amplifier. Even with ALC, however, overdriving can occur.

17.2.2 Solid State vs Vacuum Tubes

With the exception of high-power amplifiers, nearly all items of amateur equipment manufactured commercially today use solid state (semiconductor) devices exclusively. Semiconductor diodes, transistors and integrated circuits (ICs) offer several advantages in designing and fabricating equipment. Solid state equipment is smaller, offers broadband (no-tune-up) operation, and is easily manufactured using PC boards and automated (lower cost) processes.

Based on all these facts, it might seem that there would be no place for vacuum tubes in a solid state world. Transistors and ICs do have significant limitations, however, especially in a practical sense. Individual present-day transistors cannot generally handle the combination of current and voltage needed nor can they safely dispose of the amount of heat dissipated for RF amplification to high power levels. Pairs of transistors, or even pairs of pairs, are usually employed in practical power amplifier designs at the 100 W level and beyond. Sometimes various techniques of power combination from multiple amplifiers must be used.

Tube amplifiers can be more economical to build for a given output power. Vacuum tubes operate satisfactorily at surface temperatures as high as 150-200 °C, so they may be cooled by simply blowing sufficient ambient air past or through their relative large cooling surfaces. The very small cooling surfaces of power transistors should be held to 75-100 °C to avoid drastically shortening their life expectancy. Thus, assuming worst-case 50 °C ambient air temperature, the large cooling surface of a vacuum tube can be allowed to rise 100-150 °C above ambient, while the small surface of a transistor must not be allowed to rise more than about 50 °C.

Furthermore, RF power transistors are much less tolerant of electrical abuse than are most vacuum tubes. An overvoltage spike lasting only microseconds can — and is likely to — destroy RF power transistors. A comparable spike is unlikely to have any effect on a tube. So the important message is this: designing with RF power transistors demands caution to ensure that adequate thermal and electrical protection is provided.

Even if one ignores the challenge of the RF portions of a high-power solid state amplifier, there is the dc power supply to consider. A solid state amplifier capable of delivering 1 kW of RF output might require regulated (and transient-free) 50 V at more than 40 A. Developing that much current can be challenging. A vacuum tube amplifier at the same power level might require 2000 to 3000 V, unregulated, at less than 1 A.

At the kilowatt level, the vacuum tube is still a viable option for amateur constructors because of its cost-effectiveness and ease of equipment design. Because tube amplifiers and solid state amplifiers are quite different in many ways, we shall treat them in different sections of this chapter. Also, the author of the solid state section presents a slightly different perspective on the tube-vs-solid state discussion.

Care and Feeding of Power Grid Tubes Available

The classic handbook of power tube design and maintenance from the Eimac Corporation, *Care and Feeding of Power Grid Tubes*, is once again available, courtesy of Communications and Power Industries (**cpii.com** — Eimac is a division of CPI). The book consists of six PDF sections covering all phases of tube operation and design. It can be downloaded from **www.arrl.org/engineering-references** or **www.arrl.org/arrl-handbook-reference**. (A hard copy of the book is available on request from CPI.) Additional references are downloadable from **cpii.com/library.cfm/9**. The ARRL thanks CPI for making this important reference available to radio amateurs!

17.3 Vacuum Tube Basics

The term *vacuum tube* describes the physical construction of the devices, which are usually tubular and have a vacuum inside. The British call them electron valves which describes the operation of the devices, since they control the flow of electrons, like a water valve controls the flow of water.

17.3.1 Thermionic Emission

Metals are electrical conductors because the electrons in them readily move from one atom to the next under the influence of an electrical field. It is also possible to cause the electrons to be emitted into space if enough energy is added to them. Heat is one way of adding energy to metal atoms, and the resulting flow of electrons into space is called *thermionic emission*. As each electron leaves the metal surface, it is replaced by another provided there is an electrical connection from outside the tube to the heated metal.

In a vacuum tube, the emitted electrons hover around the surface of the metal unless acted upon by an electric field. If a positively charged conductor is placed nearby, the electrons are drawn through the vacuum and arrive at that conductor, thus providing a continuous flow of current through the vacuum tube.

17.3.2 Components of a Vacuum Tube

A basic vacuum tube contains at least two parts: a *cathode* and a *plate*. The electrons are emitted from the *cathode*. The cathode can be *directly heated* by passing a large dc current through it, or it can be located adjacent to a heating element (*indirectly heated*). Although ac currents can also be used to directly heat cathodes, if any of the ac voltage mixes with the signal, ac hum will be introduced into the output. If the ac heater supply voltage can be obtained from a center tapped transformer, and the center tap is connected to the signal ground, hum can be minimized.

The difficulty of producing thermionic emission varies with the metal used for the cathode, and is called the "work function" of that metal. An ideal cathode would be made of a metal with a low work function that can sustain high temperatures without melting. Pure tungsten was used in early tubes as it could be heated to a very high temperature. Later it was learned that a very thin layer of thorium greatly increased the emission. Oxides of metals with low work functions were also developed. In modern tubes, thoriated-tungsten is used for the higher power tubes and oxide-coated metals are commonly used at lower power levels.

Filament voltage is important to the proper operation of a tube. If too low, the emission will not be sufficient. If too high, the useful life of the tube will be greatly shortened. It is important to know which type of cathode is being used. Oxide-coated cathodes can be run at 5 to 10%. Tube manufacturers specify an allowable range of filament voltages for proper operation and maximum tube life — follow those recommendations. Tubes should never be operated with filament voltages above the allowable value. Reducing filament voltage below the specified range in hopes of extending tube life is definitely not recommended for tubes with oxide-coated cathodes. In addition, low filament voltage can cause distortion and spurious emissions for any type of tube. Tube failures from low filament emission in amateur service are rare. For more information, see the Reference section entry for T. Rauch, W8JI, on filament voltage management.

Every vacuum tube needs a receptor for the emitted electrons. After moving though the vacuum, the electrons are absorbed by the *plate,* also called the *anode.* This two-element tube — anode and cathode — is called a *diode*. The diode tube is similar to a semiconductor diode: it allows current to pass in only one direction. If the plate goes negative relative to the cathode, current cannot flow because electrons are not emitted from the plate. Years ago, in the days before semiconductors, tube diodes were used as rectifiers.

TRIODES

To amplify signals, a vacuum tube must also contain a control *grid*. This name comes from its physical construction. The grid is a mesh of wires located between the cathode and the plate. Electrons from the cathode pass between the grid wires on their way to the plate. The electrical field that is set up by the voltage on these wires affects the electron flow from cathode to plate. A negative grid voltage repels electrons, blocking their flow to the plate. A positive grid voltage enhances the flow of electrons to the plate. Vacuum tubes containing a cathode, a grid and a plate are called *triode* tubes (*tri* for three components). See **Figure 17.4**.

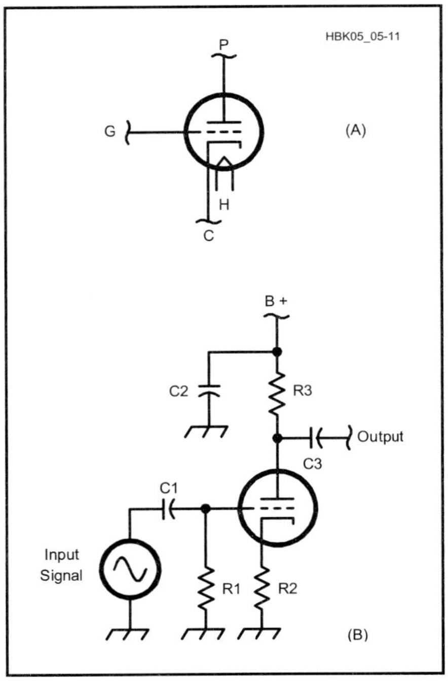

Figure 17.4 — Vacuum tube triode. (A) Schematic symbol detailing heater (H), cathode (C), grid (G) and plate (P). (B) Audio amplifier circuit using a triode. C1 and C3 are dc blocking capacitors for the input and output signals to isolate the grid and plate bias voltages. C2 is a bypass filter capacitor to decrease noise in the plate bias voltage, B+. R1 is the grid bias resistor, R2 is the cathode bias resistor and R3 is the plate bias resistor. Note that although the cathode and grid bias voltages are positive with respect to ground, they are still negative with respect to the plate.

The input impedance of a vacuum tube amplifier is directly related to the grid current. Grid current varies with grid voltage, increasing as the voltage becomes more positive. When the grid voltage is negative, no grid current flows and the input impedance of a tube is nearly infinite. When the grid is driven positive, it draws current and thus presents a lower input impedance, and requires significant drive power. The load placed across the plate of the tube strongly affects its output power and efficiency. An important part of tube design involves determining the optimum load resistance. These parameters are plotted as *characteristic curves* and are used to aid the design process. **Figure 17.5** shows an example.

Since the elements within the vacuum tube are conductors that are separated by an insulating vacuum, the tube is very similar to a capacitor. The capacitance between the cathode and grid, between the grid and plate, and between the cathode and plate can be large enough to affect the operation of the amplifier at high frequencies. These capacitances, which are usually on the order of a few picofarads, can limit the frequency response of an amplifier and can also provide signal feedback paths that may lead to unwanted oscillation. Neutralizing circuits are sometimes used to prevent such oscillations. Techniques for neutralization are presented later in this chapter.

TETRODES

The grid-to-plate capacitance is the chief source of unwanted signal feedback. Therefore tubes were developed with a second grid,

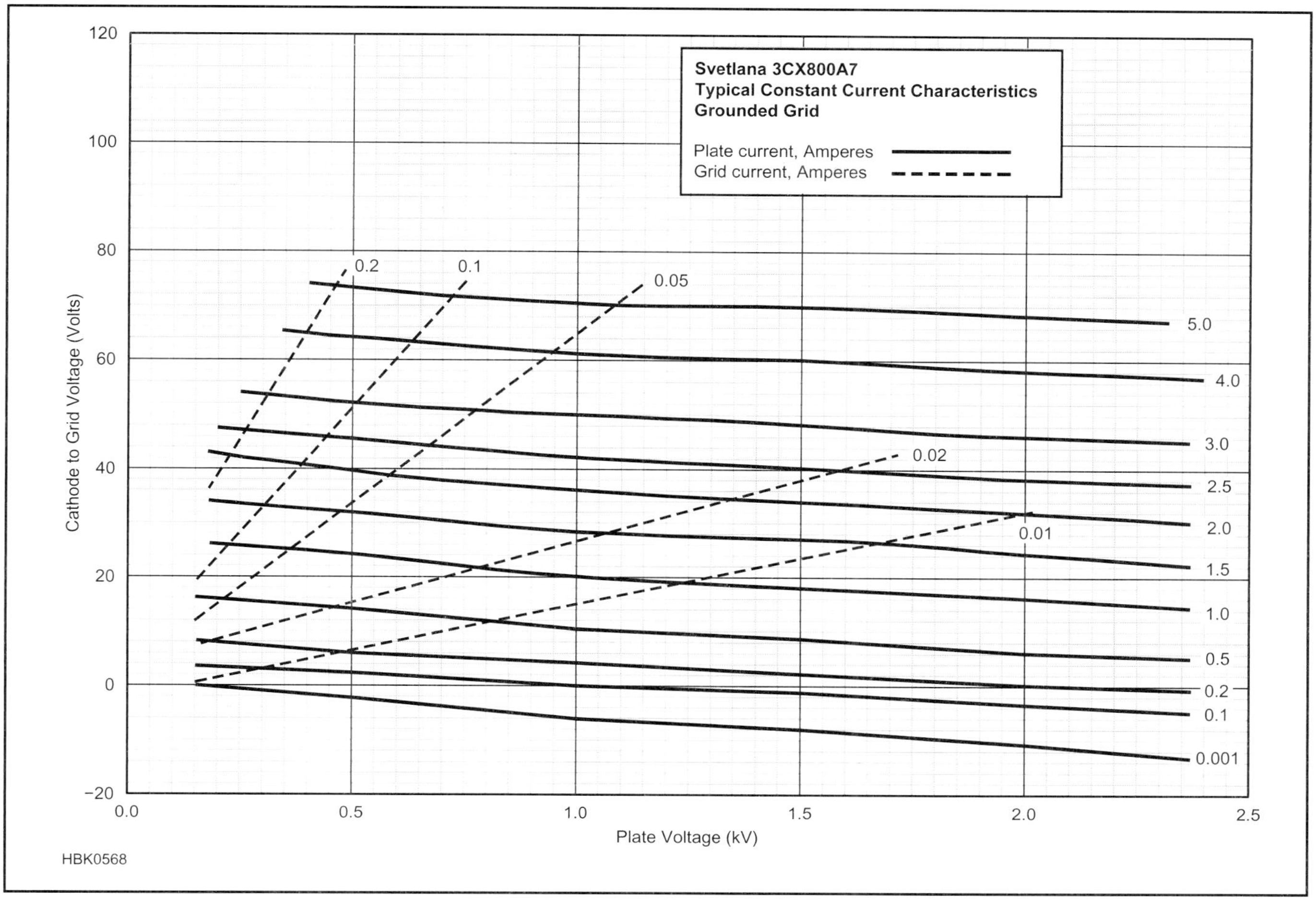

Figure 17.5 — Characteristic curves for a 3CX800A7 triode tube. Grid voltage is plotted on the left, plate voltage along the bottom. The solid lines are plate current, the dashed lines are grid current. This graph is typical of characteristic curves shown in this chapter and used with the *TubeCalculator* program described in the text and available with this book's online content.

called a *screen grid*, inserted between the original grid (now called a *control grid*) and the plate. Such tubes are called tetrodes (having four elements). See **Figure 17.6**. This second grid is usually tied to RF ground and acts as a screen between the grid and the plate, thus preventing energy from feeding back, which could cause instability. Like the control grid, the screen grid is made of a wire mesh and electrons pass through the spaces between the wires to get to the plate.

The screen grid carries a high positive voltage with respect to the cathode, and its proximity to the control grid produces a strong electric field that enhances the attraction of electrons from the cathode. The gain of a tetrode increases sharply as the screen voltage is increased. The electrons accelerate toward the screen grid and most of them pass through the spaces and continue to accelerate until they reach the plate. In large tubes this is aided by careful alignment of the screen wires with the grid wires. The effect of the screen can also be seen in the flattening of the tube curves. Since the screen shields the grid from the plate, the plate current vs plate voltages becomes almost flat, for a given screen and grid voltage. **Figure 17.7** shows characteristic curves for a typical tetrode, and curves for many more tube types are available with this book's online content.

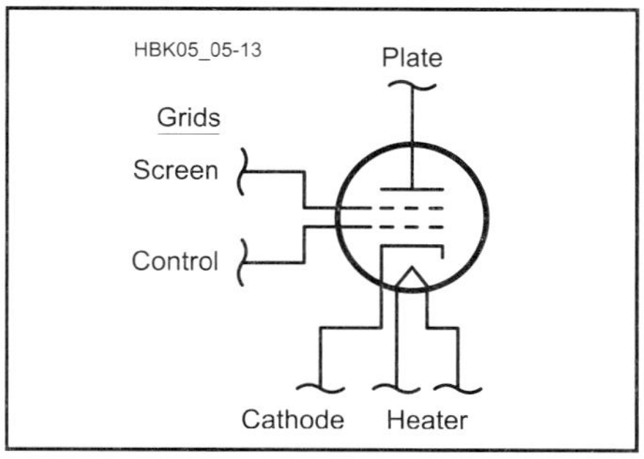

Figure 17.6 — Vacuum tube tetrode. Schematic symbol detailing heater (H), cathode (C), the two grids: control and screen, and the plate (P).

A special form of tetrode concentrates the electrons flowing between the cathode and the plate into a tight beam. The decreased electron-beam area increases the efficiency of the tube. *Beam tetrodes* permit higher plate currents with lower plate voltages and large power outputs with smaller grid driving power. The 6146 is an example of a beam power tube.

PENTODES

Another unwanted effect in vacuum tubes is the so-called *secondary emission*. The electrons flowing within the tube can have so much energy that they are capable of dislodging electrons from the metal atoms in the grids and plate. Secondary emission can cause a grid, especially the screen grid, to lose more electrons than it absorbs. Thus while a screen usually draws current from its supply, it occasionally pushes current into the supply. Screen supplies must be able to absorb as well as supply current.

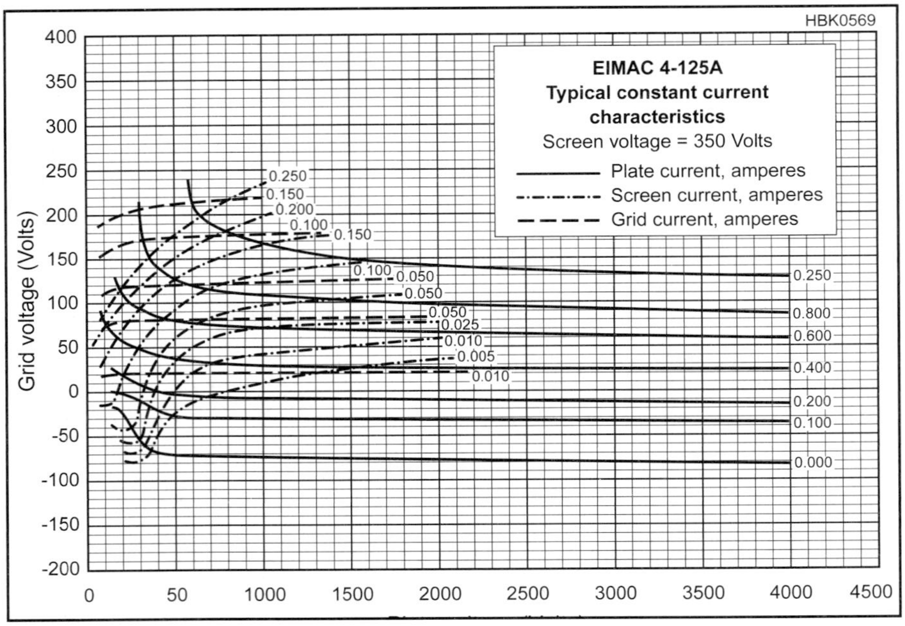

Figure 17.7 — Characteristic curves for a 4-125 tetrode tube.

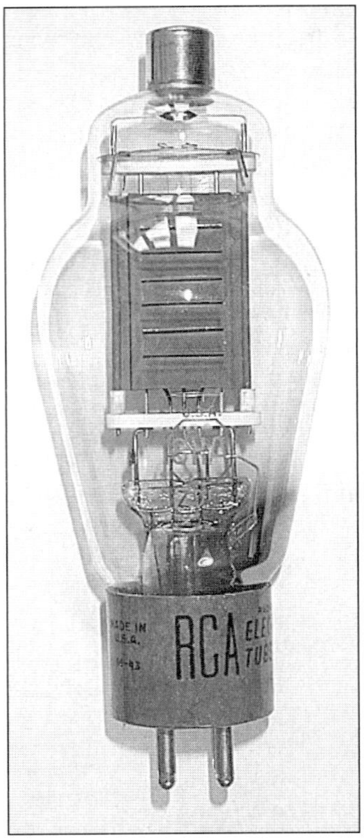

Figure 17.8 — The RCA 811A is an example of a transmitting triode with a glass envelope. [Photo courtesy the Virtual Valve Museum, www.tubecollector.org]

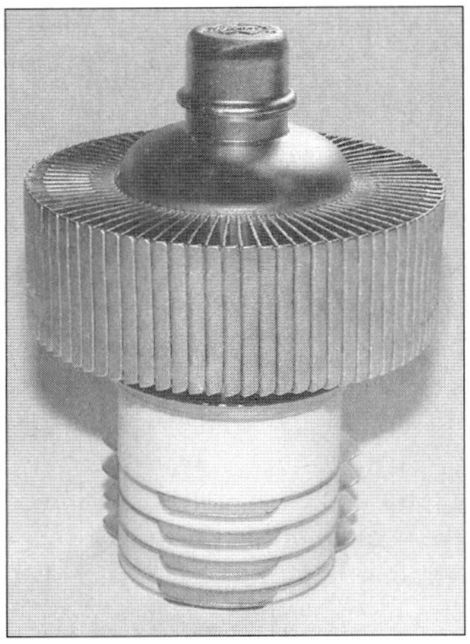

Figure 17.9 — Modern power tubes, such as this 4CX1000A tetrode, tend to use metal and ceramic construction.

A third grid, called the *suppressor grid*, can be added between the screen grid and the plate. This overcomes the effects of secondary emission in tetrodes. A vacuum tube with three grids is called a *pentode* (penta- for five elements). The suppressor grid is connected to a low voltage, often to the cathode.

17.3.3 Tube Nomenclature

Vacuum tubes are constructed with their elements (cathode, grid, plate) encased in an envelope to maintain the vacuum. Tubes with glass envelopes, such as the classic transmitting tube shown in **Figure 17.8**, are most familiar. Over time, manufacturers started exploring other, more rugged, methods for making high power transmitting tubes. Modern power tubes tend to be made of metal parts separated by ceramic insulating sections (**Figure 17.9**).

Because of their long history, vacuum tube types do not all follow a single logical system of identification. Many smaller tubes types begin with an indication of the filament voltage, such as the 6AU6 or 12AT7. Other tubes such as the 811 (Figure 17.8) and 6146 were assigned numbers in a more or less chronological order, much as transistors are today. Some glass envelope power tubes follow a numbering system that indicates number of tube elements and plate dissipation — 3-500Z and 4-1000A are two common examples in amateur circles.

Some power tubes follow the 3CX and 4CX numbering system. The first number indicates a triode (3) or tetrode (4) and the C indicates ceramic/metal construction. The X indicates cooling type: X for air, W for water and V for vapor cooling. The cooling type is followed by the plate dissipation. (The tubes that amateurs use typically have three or four numbers indicating plate dissipation; those used in commercial and broadcast service can have much higher numbers.) Thus a 4CX250 is a ceramic, air cooled 250 W tetrode. A 3CX1200 is a ceramic, air cooled, 1200 W triode. Often these tubes have additional characters following the plate dissipation to indicate special features or an upgraded design. For example, a 4CX250R is a special version of the 4CX250 designed for AB1 linear operation. A 4CX1500B is an updated version of the 4CX1500A.

During the heyday of tube technology, some tube types were developed with several tubes in the same glass envelope, such as the 12AX7 (a dual triode). Except for a very few devices used in the specialty audio market, tubes of this type are no longer made.

17.3.4 Tube Mounting and Cooling Methods

Most tubes mount in some kind of socket so that they can be easily replaced when they reach the end of their useful life. Connections to the tube elements are typically made through pins on the base. The pins are arranged or keyed so that the tube can be inserted into the socket only one way and are sized and spaced to handle the operating voltages and currents involved. Tubes generally use a standard base or socket, although a great many different bases developed over the years. Tube data sheets show pinouts for the various tube elements, just like data sheets for ICs and transistors. Base diagrams of some popular transmitting tubes are provided in the online content.

For transmitting tubes, a common arrangement is for filament and grid connections to be made through pins in the main base, while the plate connection is made through a large pin or post at the top of the tube for easier connection to the high voltage supply and tank circuit. This construction is evident in the examples shown in Figures 17.8 and 17.9. To reduce stray reactances, in some older glass tubes the grid used a separate connection.

Heat dissipation from the plate is one of the major limiting factors for vacuum tube power amplifiers. Most early vacuum tubes were encased in glass, and heat passed through it as infrared radiation. If more cooling was needed, air was simply blown over the outside of the glass. Modern ceramic tubes suitable for powers up to 5 kW are usually cooled by forcing air directly through an external anode. These tubes require a special socket that allows free flow of air. The large external anode and cooling fins may be seen in the example in Figure 17.9. Conduction through an insulating block and water cooling are other options, though they are not often seen in amateur equipment. Practical amplifier cooling methods are discussed in detail later in this chapter.

17.3.5 Vacuum Tube Configurations

Just as the case with solid state devices described in the **Circuits and Components** chapter, any of the elements of the vacuum tube can be common to both input and output. A common plate connection — called a cathode follower and similar to the emitter follower — was once used to reduce output impedance (current gain) with little loss of voltage. This application is virtually obsolete.

Most modern tube applications use either the common cathode or the common grid connection. Figure 17.4B shows the common cathode connection, which gives both current and voltage gain. The common grid (often called *grounded grid*) connection shown in **Figure 17.10** gives only voltage gain. Thus the common cathode connection is capable of 20 to 30 dB of gain in a single stage, whereas the grounded grid connection typically gives 10 to 15 dB of gain.

The input impedance of a grounded grid stage is low, typically less than a few hundred ohms. The input impedance of a grounded cathode stage is much higher.

17.3.6 Classes of Operation in Tube Amplifiers

Class of operation was discussed briefly in the previous section describing the need for linear operation of RF power amplifiers for most Amateur Radio modes except FM. The class of operation of an amplifier stage is defined by its conduction angle, the angular portion of each RF drive cycle, in degrees, during which plate current flows. The conduction angle is determined by the bias on the device, and to a lesser extent on the drive level. These, in turn, determine the amplifier's efficiency, linearity and operating impedances. Refer to **Figure 17.11** for the following discussion.

Class A is defined as operation where plate current is always flowing. For a sine wave this means during 360° of the wave. Class A has the best linearity, but poor efficiency.

Class B is when the bias is set so that the tube is cut off for negative input signals, but current flows when the input signal is positive. Thus for a sine wave, conduction occurs during 1/2 of the cycle, or 180°. Class B is linear only when two devices operate in push-pull, so as to provide the missing half of the wave, or when a tuned circuit is present to restore the missing half by "flywheel" action (discussed later in this chapter).

Class AB is defined as operation that falls between Class A and Class B. For a sine wave, the conduction angle will be more than 180°, but less than 360°. In practice Class AB amplifiers usually fall within the gray area shown in the center area of the graph. Like Class B, a push pull connection or a tuned circuit are needed for linear operation. Class AB is less efficient that class B, but better than Class A. The linearity is better than class B but worse than class A.

Class AB vacuum tube amplifiers are further defined as class AB1 or AB2. In class AB1, the grid is not driven positive, so no grid current flows. Virtually no drive power is required. In Class AB2, the grid is driven positive at times with respect to the cathode and some grid current flows. Drive power and output both increase as compared to AB1. Most linear amplifiers used in the Amateur service operate Class AB2, although for greater linearity some operate Class AB1 or even Class A.

Class C is when conduction angle is less than 180°—typically 120° to 160° for vacuum tube amplifiers or within the gray area to the left in Figure 17.11. The tube is biased well

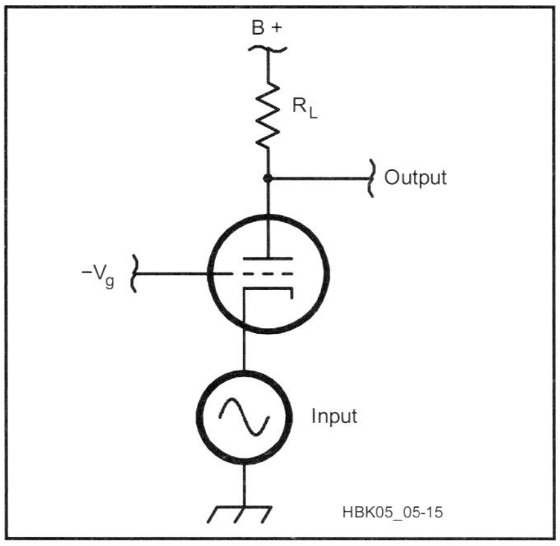

Figure 17.10 — Grounded grid amplifier schematic. The input signal is connected to the cathode, the grid is biased to the appropriate operating point by a dc bias voltage, $-V_G$, and the output voltage is obtained by the voltage drop through R_L that is developed by the plate current, I_P.

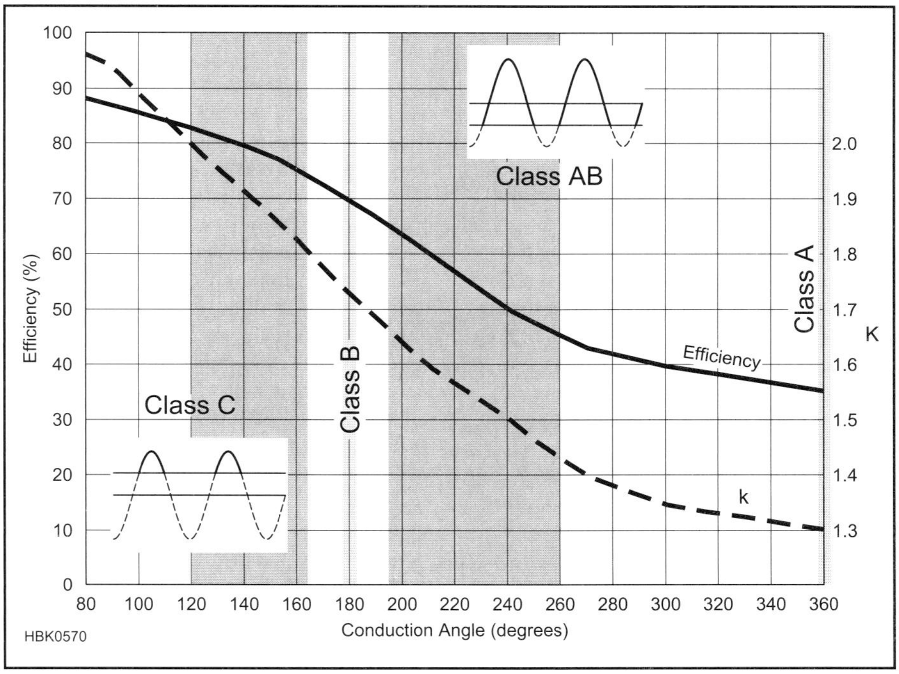

Figure 17.11 — Efficiency and K for various classes of operation. Read the solid line to determine efficiency. Read the dashed line for K, which is a constant used to calculate the plate load required, as explained in the text.

The Flywheel Effect

The operation of a resonant tank circuit (see the **Oscillators and Synthesizers** chapter for a discussion of resonant LC circuits) is sometimes referred to as the "flywheel" effect. A flywheel does illustrate certain functions of a resonant tank, but a flywheel alone is non-resonant; that is, it has no preferred frequency of operation.

A better analogy for a resonant tank — although not exact — is found in the balance wheel used in mechanical watches and clocks in which a weighted wheel rotates back and forth, being returned to its center position by a spiral spring, sometimes called a hairspring. The balance wheel stores energy as inertia and is analogous to an inductor. The hairspring also stores energy and is similar in operation to a capacitor. The spring has an adjustment for the frequency of operation or resonance (1 Hz). An escapement mechanism gives the wheel a small kick with each tick of the watch to keep it going.

A plot of the rotational (or radial) velocity of the balance wheel will give a sine function. It thus converts the pulses from the escapement into smooth sinusoidal motion. In a similar way, pulses of current in the plate of a tube are smoothed into a sine wave in the tank circuit.

The plate (or collector or drain) voltage is a sine wave, even though the plate current is made up of pulses. Just as the escarpment mechanism must apply its kicks at just the right time, the frequency at which current pulses are added to the tank must match the natural resonant frequency of the tank, which is adjusted with the "plate tune" capacitor.

Efficient operation of the tank itself occurs when the losses are small compared to the energy transferred to the output. In a similar manner, lowering the losses in a balance wheel by using jeweled bearings makes the watch more efficient. The amount of energy stored in the balance wheel should also be kept as low as practical since excess "oscillation" of the wheel wastes energy to the air and bearings. Likewise, the "circulating currents" in the tank must be limited to reduce heating from the inevitable losses in the components.

Figure 17.A1 — A balance wheel and hairspring in a mechanical clock illustrate the flywheel effect in tank circuits.

beyond cutoff when no drive signal is applied. Output current flows only during positive crests in the drive cycle, so it con-sists of relatively narrow pulses at the drive frequency. Efficiency is high, but nonlinear operation results. Class C amplifiers always use tuned circuits at the input and output. Attempts to achieve extreme efficiency with very narrow pulses (small conduction angles) require very high drive power, so a point of diminishing returns is eventually reached.

Classes D through H use various switched mode techniques. These are used almost exclusively with solid state circuits.

17.3.7 Understanding Tube Operation

Vacuum tubes have complex current transfer characteristics, and each class of operation produces different RMS values of RF current through the load impedance. As described earlier, tube manufacturers provide characteristic curves that show how the tube behaves as operating parameters (such as plate and grid voltage and current) vary. See Figures 17.5 and 17.7 for examples of characteristic curves for two different transmitting tubes. The use of tube curves provides the best way to gain insight into the characteristics of a given tube. Before designing with a tube, get a set of these curves and study them thoroughly.

Because of the complexity and interaction among the various parameters, computer-aided design (CAD) software is useful in analyzing tube operation. One such program, *TubeCalculator*, is available with this book's online content, along with curves for many popular transmitting tubes. This software makes it easier to do analysis of a given operation with the tube you have chosen. **Figure 17.12** shows a *TubeCalculator* screen with an example of "constant current" curves for a typical tube used in high power amplifiers and tables showing values for the various operating parameters. The curves shown have grid voltage on the vertical axis and plate voltage on the horizontal axis. Older tubes and some newer ones use a slightly different format in which plate voltage is plotted on the horizontal axis and plate current on the vertical. *TubeCalculator* allows analysis using either type.

The tube is the heart of any amplifier. Using the software to arrive at the desired operating parameters is a major step toward understanding and designing an amplifier. The second most important part of the design is the tuning components. Before they can be designed, the required plate load resistance must be determined and *TubeCalculator* will do that. In addition it will give insight into what happens when a tube is under driven or over driven, when the bias is wrong, or when the load resistance incorrect.

If a tube is to be used other than with nominal voltages and currents, analysis using the tube curves is the only solution short of trial and error. Trial and error is not a good idea because of the high voltages and currents found in high power amplifiers. It's best to conduct your analysis using curves and software, or else stick to operating the tube very close to the voltages, currents, drive levels and load values specified by the manufacturer.

ANALYZING OPERATING PARAMETERS

Characteristic curves allow a detailed look at tube operation as voltages and currents vary. For example, you can quickly see how much negative grid voltage is required to set the plate current to any desired value, depending also on the plate voltage and screen voltages. You can also see how much grid voltage is needed to drive the plate current to the maximum desired value.

With RF power amplifiers, both the grid voltage and the plate voltage will be sinusoidal and will be 180° out of phase. With constant current curves, an operating line can be drawn that will trace out every point of the operating cycle. This will be a straight line connecting two points. One point will be at the intersection of the peak plate voltage and the peak negative grid voltage. The other point will be at the intersection of the peak positive grid voltage and the minimum plate voltage.

If we plot the plate current along this line as a function of time, it will be seen that it changes in a nonlinear fashion; the exact shape of which depends on the class of operation. For class AB2 operation, which is the most commonly used in linear RF power amplifiers, it will look something like the plot shown in **Figure 17.13**. This rather complicated waveform is not easily evaluated using simple formulas, but it can be analyzed by taking the current values vs time and apply-

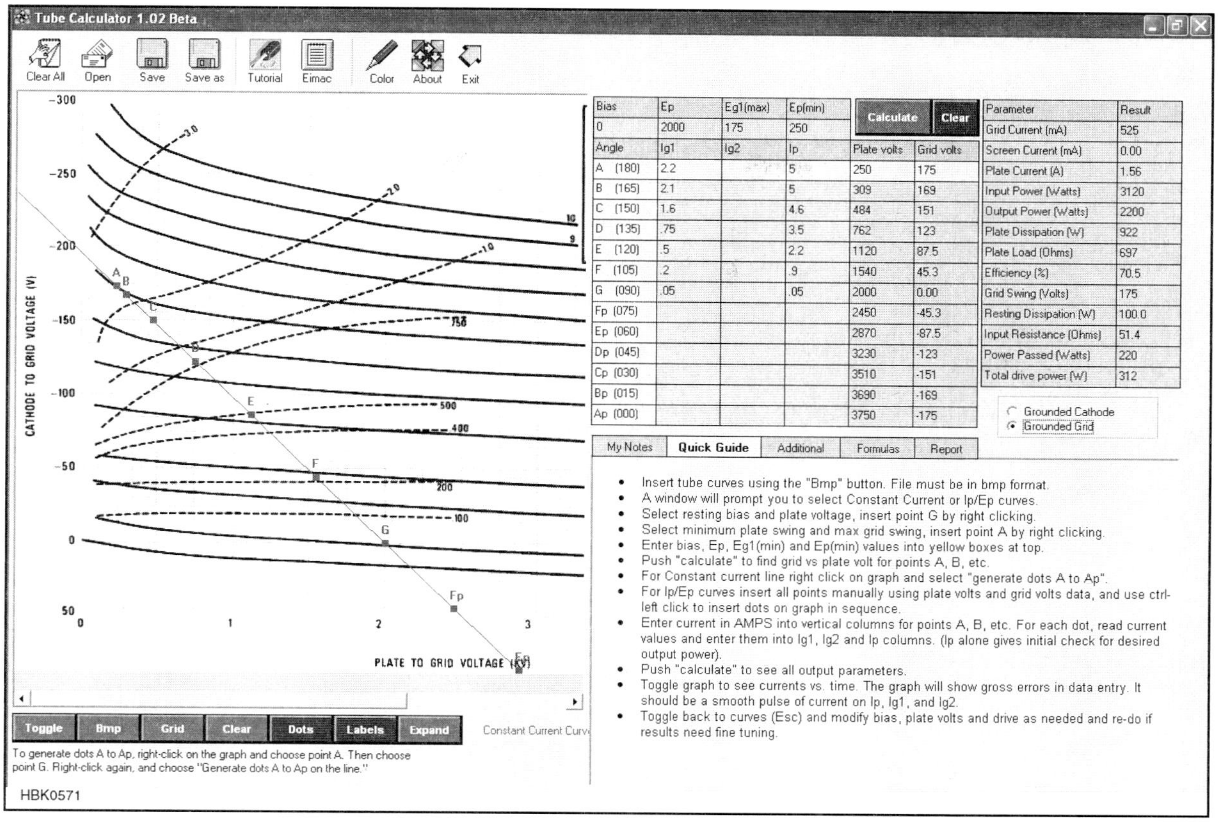

Angle	Ig1	Ig2	Ip	Plate volts	Grid volts
A (180)	2.2		5	250	175
B (165)	2.1		5	309	169
C (150)	1.6		4.6	484	151
D (135)	.75		3.5	762	123
E (120)	.5		2.2	1120	87.5
F (105)	.2		.9	1540	45.3
G (090)	.05		.05	2000	0.00
Fp (075)				2450	-45.3
Ep (060)				2870	-87.5
Dp (045)				3230	-123
Cp (030)				3510	-151
Bp (015)				3690	-169
Ap (000)				3750	-175

Figure 17.12 — Main screen of *TubeCalculator* program (available with this book's online content). A characteristic curve plot is loaded in the window on the left side of the screen.

ing averaging techniques. This method was developed by Chaffee in 1936 and popularized by Eimac, the company that developed many of the power tubes used today.

With *TubeCalculator*, data can be extracted from the curves and can be converted into many useful operating parameters such as input power, output power, grid power required, grid and plate dissipation and required load resistance.

MANUAL METHODS FOR TUBE PARAMETER SELECTION

For those not wishing to use a computer for design, most tube manufacturers will supply a table of typical operating values. (See the base diagrams and operating values material in the online content.) These values have already been determined both from the tube characteristics and actual operational tests. As long as your proposed operation is close to the typical values in terms of voltages and currents, the typical values can provide you the desired load resistance and expected output power and efficiency. The typical operating parameters may also include a suggested optimum load resistance. If not, we can use well established "rules of thumb."

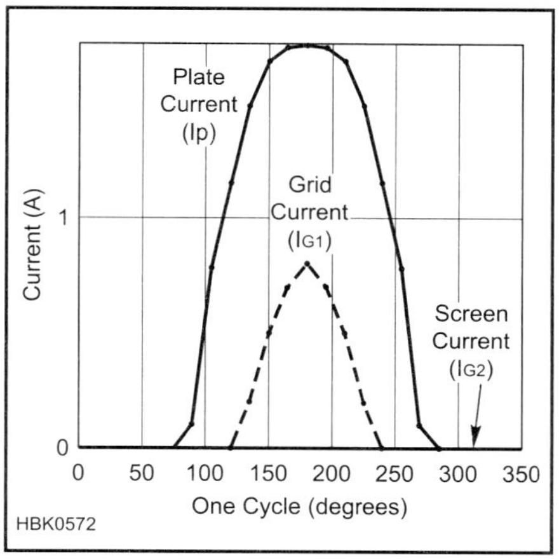

Figure 17.13 — Class AB2 plate and grid current over one cycle as plotted by the *TubeCalculator* program. This plot is for a triode, so there is no curve for screen (G2) current.

The optimum load resistance for vacuum-tube amplifiers can be approximated by the ratio of the dc plate voltage to the dc plate current at maximum signal, divided by a constant appropriate to each class of operation. The load resistance, in turn, determines the maximum power output and efficiency the amplifier can provide. The approximate value for tube load resistance is

$$R_L = \frac{V_P}{K \times I_P} \qquad (1)$$

where

R_L = the appropriate load resistance, in ohms,

V_P = the dc plate potential, in V,

I_P = the dc plate current, in A, and

K = a constant that approximates the RMS current to dc current ratio appropriate for each class. For the different classes of operation: Class A, K ≈ 1.3; Class AB, K ≈ 1.5-1.7; Class B, K ≈ 1.57-1.8; Class C, K ≈ 2. The way in which K varies for different conduction angles is shown in Figure 17.11 (right scale).

Once we determine the optimum load resistance value for the tube(s) to be used we are ready to design the output networks for the amplifier. After tube selection, this is the most important part of the total design.

17.4 Tank Circuits

Usually we want to drive a transmission line, typically 50 Ω, with the output of our amplifier. An output network is used to transform that 50 Ω impedance to the optimum load resistance for the tube. This transformation is accomplished by resonant output networks which also serve to reduce harmonics to a suitable level. The **Radio Fundamentals** chapter of this *Handbook* gives a detailed analysis of the operation of resonant circuits. We summarize here only the most important points.

Resonant circuits have the ability to store energy. Capacitors store electrical energy in the electric field between their plates; inductors store energy in the magnetic field induced by the coil winding. These circuits are referred to as *tank circuits* since they act as storage "tanks" for RF energy. This energy is continuously passed back and forth between the inductive storage and the capacitive storage. It can be shown mathematically that the "alternating" current and voltage produced by this process are sinusoidal in waveform with a frequency of

$$f = \frac{1}{2\pi\sqrt{LC}} \qquad (2)$$

which, of course, is the resonant frequency of the tank circuit.

17.4.1 Tank Circuit Q

In order to quantify the ability of a tank circuit to store energy, a quality factor, Q, is defined. Q is the ratio of energy stored in a system during one complete RF cycle to energy lost.

$$Q = 2\pi \frac{W_S}{W_L} \qquad (3)$$

where

W_S = is the energy stored, and

W_L = the energy lost to heat and the load.

A load connected to a tank circuit has exactly the same effect on tank operation as circuit losses. Both consume energy. It just happens that energy consumed by circuit losses becomes heat rather than useful output. When energy is coupled out of the tank circuit into a load, the loaded Q (Q_L) is:

$$Q_L = \frac{X}{R_{Loss} + R_{Load}} \qquad (4)$$

where R_{Load} is the load resistance. Energy dissipated in R_{Loss} is wasted as heat. And X represents the reactance of the inductor or the capacitor, assumed to be equal at resonance. Ideally, all the tank circuit energy should be delivered to R_{Load}. This implies that R_{Loss} should be as small as possible.

17.4.2 Tank Circuit Efficiency

The efficiency of a tank circuit is the ratio of power delivered to the load resistance (R_{Load}) to the total power dissipated by losses (R_{Load} and R_{Loss}) in the tank circuit. Within the tank circuit, R_{Load} and R_{Loss} are effectively in series and the circulating current flows through both. The power dissipated by each is proportional to its resistance. The loaded tank efficiency can, therefore, be defined as

$$\text{Tank Efficiency} = \frac{R_{Load}}{R_{Load} + R_{Loss}} \times 100 \qquad (5)$$

where efficiency is stated as a percentage. The loaded tank efficiency can also be expressed as

$$\text{Tank Efficiency} = \left(1 - \frac{Q_L}{Q_U}\right) \times 100 \qquad (6)$$

where

Q_L = the tank circuit loaded Q, and

Q_U = the unloaded Q of the tank circuit.

For practical circuits, Q_U is very nearly the Q of the coil with switches, capacitors and parasitic suppressors making a smaller contribution. It follows, then, that tank efficiency can be maximized by keeping Q_L low which keeps the circulating current low and the I^2R losses down. Q_U should be maximized for best efficiency; this means keeping the circuit losses low. With a typical Q_L of 10, about 10% of the stored energy is transferred to the load in each cycle. This energy is replaced by energy supplied by the tube. It is interesting to contemplate that in a typical amplifier which uses a Q_L of 10 and passes 1.5 kW from the tube to the output, the plate tank is storing about 15 kW of RF energy. This is why component selection is very important, not only for low loss, but to resist the high voltages and currents.

Resonant circuits are always used in the plate circuit. When the grid is used as the input (common cathode), both matching and a tuned circuit may be used or else a low impedance load is connected from grid to ground with a broad matching transformer or network. The "loaded grid" reduces gain, but improves stability. In grounded grid operation, a tuned circuit may not be needed in the cathode circuit, as the input Z may be close to 50 Ω, but a tuned network may improve the match and usually improves the linearity. These resonant circuits help to ensure that the voltages on grid and plate are sine waves. This wave-shaping effect is the same thing as harmonic rejection. The reinforcing of the fundamental frequency and rejection of the harmonics is a form of filtering or selectivity.

The amount of harmonic suppression is dependent upon circuit loaded Q_L, so a dilemma exists for the amplifier designer. A low Q_L is desirable for best tank efficiency, but yields poorer harmonic suppression. High Q_L keeps amplifier harmonic levels lower at the expense of some tank efficiency. At HF, a compromise value of Q_L can usually be chosen such that tank efficiency remains high and harmonic suppression is also reasonable. At higher frequencies, tank Q_L is not always readily controllable, due to unavoidable stray reactances in the circuit. Unloaded Q_U can always be maximized, however, regardless of frequency, by keeping circuit losses low.

17.4.3 Tank Output Circuits

THE PI NETWORK

The pi network with the capacitors to ground and the inductor in series is commonly used for tube type amplifier matching. This acts like a low pass filter, which is helpful for getting rid of harmonics. Harmonic suppression of a pi network is a function of the impedance transformation ratio and the Q_L of the circuit. Second-harmonic attenuation is approximately 35 dB for a load impedance of 2000 Ω in a pi network with a Q_L of 10. In addition to the low pass effect of the pi network, at the tube plate the third harmonic

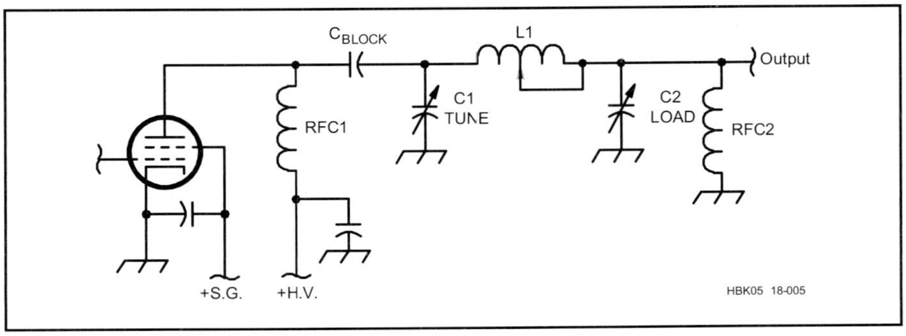

Figure 17.14 — A pi matching network used at the output of a tetrode power amplifier. RFC2 is used for protective purposes in the event C_{BLOCK} fails.

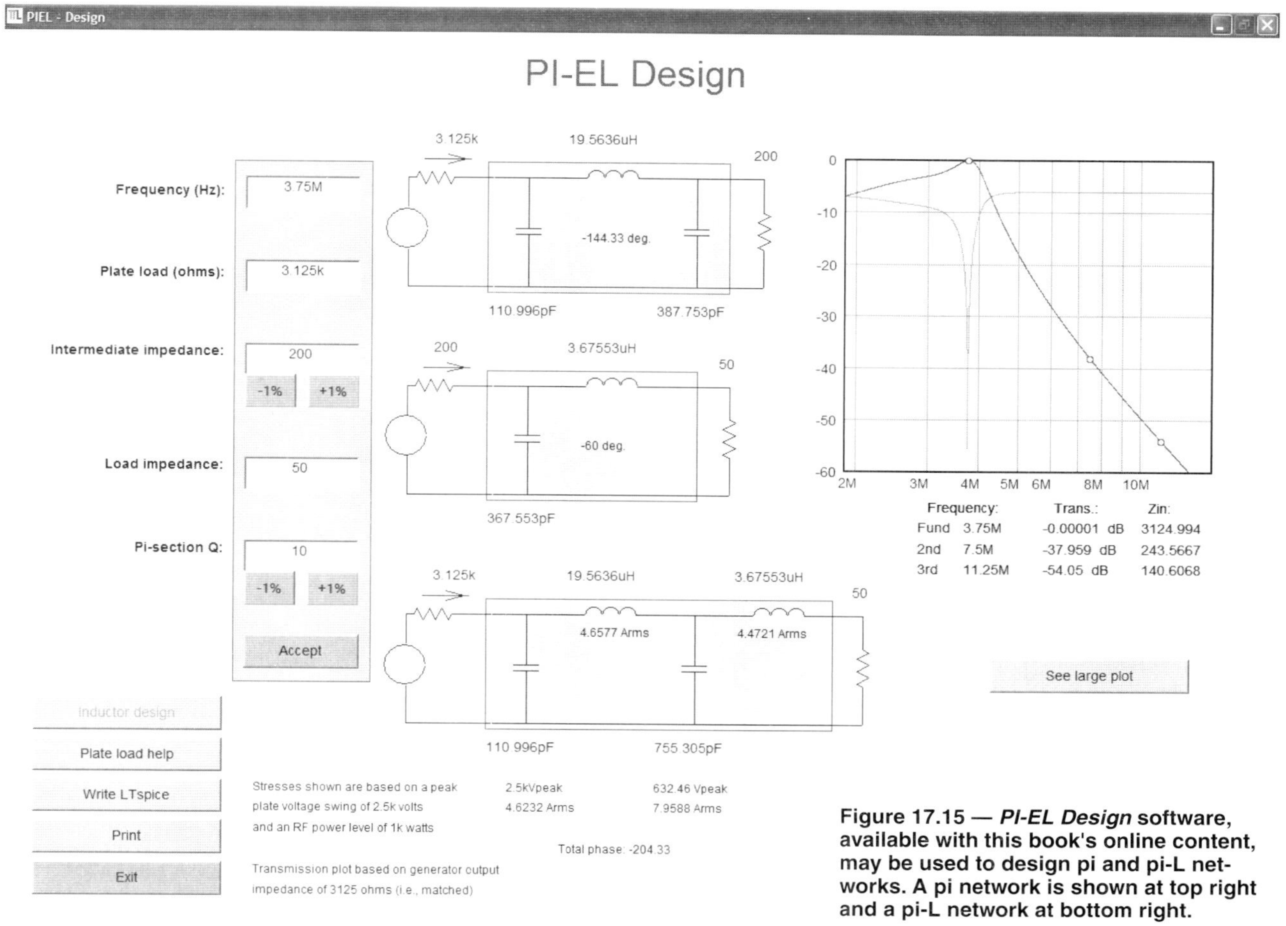

Figure 17.15 — *PI-EL Design* software, available with this book's online content, may be used to design pi and pi-L networks. A pi network is shown at top right and a pi-L network at bottom right.

is already typically 10 dB lower and the fourth approximately 7 dB below that. A typical pi network as used in the output circuit of a tube amplifier is shown in **Figure 17.14**. The **Analog and Digital Filtering** chapter describes harmonic filters that can also greatly reduce harmonics. These are typically not switched but left in the circuit at all times. With such a filter, the requirements for reducing the harmonics on the higher bands with the amplifier pi network is greatly reduced.

The formulas for calculating the component values for a pi network, for those who wish to use them, are included with this book's online content, along with tabular data for finished designs. The input variables are desired plate load resistance, output impedance to be matched (usually 50 Ω), the desired loaded Q (typically 10 or 12) and the frequency. With these inputs one can calculate the values of the components C1, L1 and C2. These components are usually referred to as the plate tune capacitor, the tank inductor and the loading capacitor. In a multi-frequency amplifier, the coil inductance is changed with a band switch and the capacitors are adjusted to the correct value for the band in question.

Tank circuit component values are most easily found using computer software. The program *PI-EL Design* by Jim Tonne, W4ENE, is available with this book's online content and is illustrated here. With this software, all of the components for a pi or pi-L network (described in the next section) can be quickly calculated. Since there are so many possible variables, especially with a pi-L network, it is impractical to publish graphical or tabular data to cover all cases. Therefore, the use of this software is highly recommended for those designing output networks. The software allows many "what-if" possibilities to be quickly checked and an optimum design found.

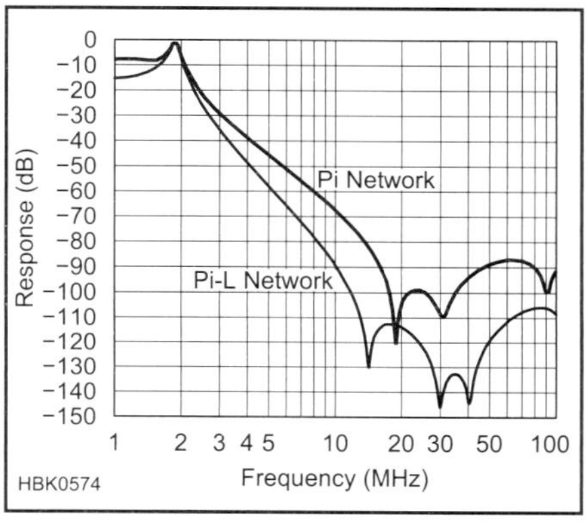

Figure 17.16 — Relative harmonic rejection of pi and pi-L circuits

THE PI-L NETWORK

There are some advantages in using an additional inductor in the output network, effectively changing it from a pi network to a pi-L network as shown in the bottom right corner of **Figure 17.15**. The harmonic rejection is increased, as shown in **Figure 17.16**, and the component values may become more convenient. Alternatively, the Q can be reduced to lower losses while retaining the same harmonic rejection as the simple pi. This can reduce maximum required tuning capacitance to more easily achievable values.

With a pi-L design, there are many more options than with the simple pi network. This is because the intermediate impedance can take on any value we wish to assign between

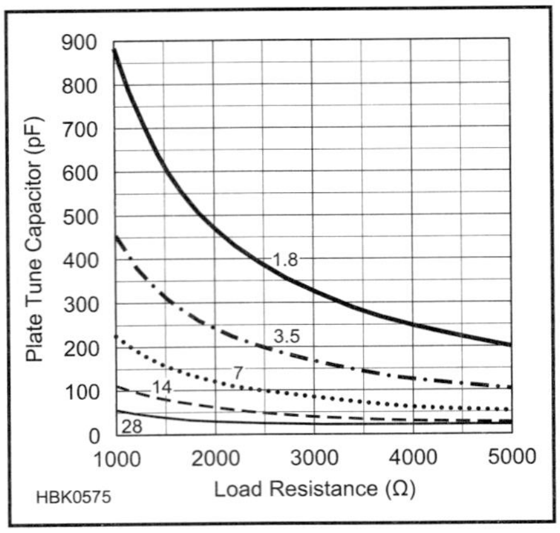

Figure 17.17 — Plate tuning capacitor values for various bands and values of load resistance. Figures17.17 through 17.19 may be used for manual design of a tank circuit, as explained in the text.

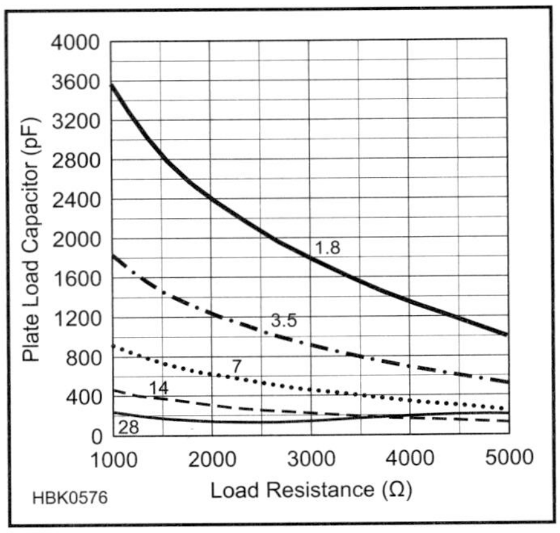

Figure 17.18 — Plate loading capacitor values for various bands and values of load resistance.

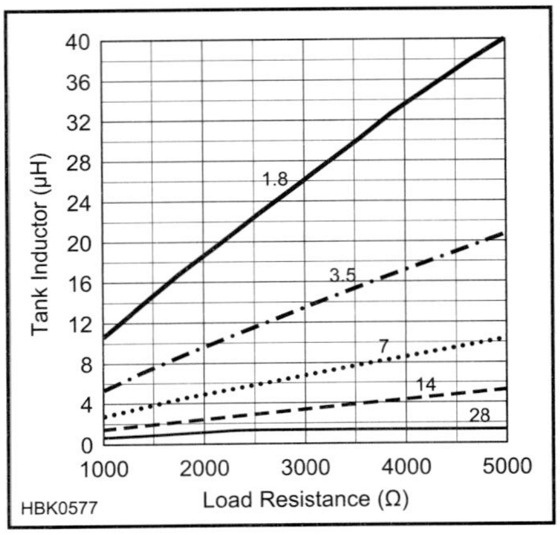

Figure 17.19 — Tank inductor values for various bands and values of load resistance.

the output impedance (usually 50 Ω) and the desired plate resistance. This intermediate impedance need not be the same for each frequency, providing the possibility of further optimizing the design. For that reason, it is especially desirable that the software be used instead of using the chart values when a pi-L is contemplated. Using this software, one can quickly determine component values for the required load resistance and Q values for any frequency as well as plotting the harmonic rejection values. Even the voltage and current ratings of the components are calculated.

Further analysis can be done using various versions of *SPICE*. A popular *SPICE* version is *LTspice* available from Linear Technologies and downloadable for free on their website, **www.linear.com**. The *PI-EL Design* software mentioned above generates files for *LTspice* automatically. *PI-EL Design* assumes that the blocking capacitor has negligible reactance and the RF choke has infinite reactance. There are times when these assumptions may not be valid. The effect of these components and changes to compensate for them can easily be evaluated using *LTspice*. It can also evaluate the effects of parasitic and stray effects, which all components have. The deep nulls in the response curves in Figure 17.16 are caused by the stray capacitance in the inductors. These can be used to advantage but, if not understood, can also lead to unexpected results. See the **RF Techniques** chapter for more information.

MANUAL METHODS FOR TANK DESIGN

For those who wish to try designing a pi network without a computer, pi designs in chart form are provided. These charts (**Figure 17.17** to **Figure 17.19**) give typical values for the pi network components for various bands and desired plate load resistance. For each value of load resistance the component values can be read for each band. For bands not shown, an approximate value can be reached by interpolation between the bands shown. It's not necessary to be able to read these values to high precision. In practice, unaccounted-for stray capacitance and inductance will likely make the calculated values only an approximate starting point. For those desiring greater precision, this data is available in tabular form with this book's online content along with the formulas for calculating them.

Several things become obvious from these charts. The required capacitance is reduced and the required inductance increased when a higher load resistance is used. This means that an amplifier with higher plate voltage and lower plate current will require smaller capacitor values and larger inductors. The capacitors will, of course, also have to withstand higher voltages, so their physical size may or may not be any smaller. The inductors will have less current in them so a smaller size wire may be used. It will be obvious that for an amplifier covering many bands, the most challenging parts of the design are at the frequency extremes. The 1.8 MHz band requires the most inductance and capacitance and the 28 MHz band requires the least. Often, the output capacitance of the tube plus the minimum value of the plate capacitor along with assorted stray capacitance will put a lower limit on the effective plate capacitance that can be achieved. These charts assume that value to be 25 pF and the other components are adjusted to account for that minimum value. An inevitable trade-off here is that the Q of the tank will be higher than optimum on the highest bands.

Table 17.1
Pi-L Values for 1.8 MHz

Plate Load Resistance (Ω)	*Plate Tune Capacitor (pF)*	*Plate Inductor (µH)*	*Plate Load Capacitor (pF)*	*Output Inductor (µH)*	*Intermediate Resistance (Ω)*	*Loaded Q (Q_L)*	*Harmonic Attenuation (3rd/5th, dB)*
1000	883	10.5	3550	None (pi)	None (pi)	12	30/42
1000	740	15	2370	7.7	200	12	38/54
1000	499	21.5	1810	7.7	200	8	35/51
2200	376	26.4	1938	7.7	200	12	39/55
2200	255	37.7	1498	7.7	200	8	36/52
5000	180	50.4	1558	7.7	200	12	39/55
5000	124	71	1210	7.7	200	8	36/53
5000	114	83	928	11.7	400	8	39/55

Designing to Avoid Interactive Tune and Load Adjustments

It can be annoying when tuning a tube amplifier if adjusting the loading control requires that we also re-dip the plate current. This happens when the loading capacitor changes the reactance seen by the final tube, rather than just the resistive part of its load. Setting the output power to a chosen value may require numerous readjustments of both controls.

The plate tuning capacitor is directly across the load seen by the tube, hence it only tunes for resonance. The loading capacitor is seen by the tube through the pi network and the network's phase shift will determine how the plate load impedance changes when the loading control is adjusted. For the loading capacitor to change only the resistive part of the tube load, choose a phase shift value for the pi network that is an odd multiple of 45°. Since a pi network with 45° phase shift will have very low Q, the best choice is a network with 135° phase shift.

The Smith chart in **Figure 17.A2** illustrates this effect. The solid line with dots shows the effect of adjusting the plate tuning capacitor. When close to resonance, it does not change the resistive part of the load seen by the tube, but does tune out the reactive part as shown by the fact that it crosses the horizontal resistance axis at a right angle. When the load is purely resistive at resonance, we get a dip in the plate current.

The thick dashed curve shows what happens when we adjust the loading capacitor with a 135° pi network. The resistive part of the tube load changes as shown by the curve moving along the horizontal axis, but the resonance is largely unaffected. The thin dashed line shows an example of what happens when we change the load capacitor with a network with a phase shift that is not 135°. Both the resistive part and the reactive part of the tube load change, making it necessary to retune the plate capacitor to resonance.

The phase shift of a pi network is determined by the Q of the network and the impedance transformation ratio. The value of Q that will give us the "magic" 135° phase shift will thus depend on the plate load to output impedance ratio. For a 50 Ω load at the output of the pi network, the desired Q value can be determined from the upper curve of **Figure 17.A3**. Putting this value of Q into the pi-L design program will generate the pi network component values desired. For example, for 2500 Ω plate load we see that the desired Q of the tank will be 12.4, a reasonable Q to choose that is close to the commonly used value of 10. Thus, with a slight adjustment of the design Q we can get to a "non interactive loading control" design.

As we can see on the upper curve of Figure 17.A3, with very high or low plate load values, the excursion from the normally used Q value of 10 may be excessive and a fully non-interactive tuning solution not practical with the simple pi network. For a plate load of 4000 Ω we see that a Q of 16.4 would be required. This is somewhat high and would lead to excessive losses in the tank. An alternate approach would be to use the pi-L network, and allow the intermediate value to be 100 Ω. The lower curve shows this allows the pi-L network to be designed using a Q value of 11.2, which is more acceptable.

A further discussion of this method, with formulas, a spreadsheet calculator, and more Smith chart graphics illustrating how it works is available with this book's online content.

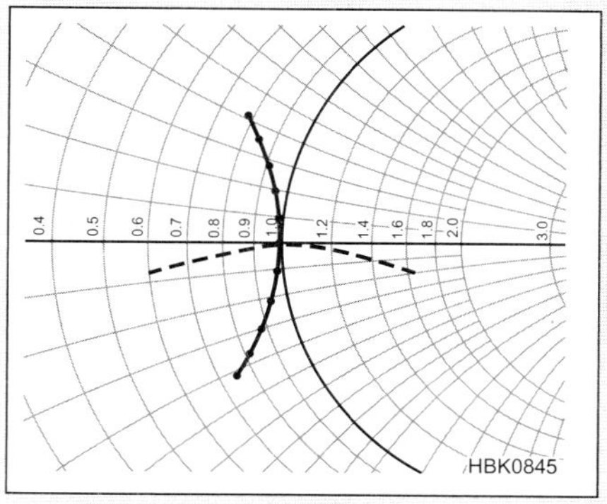

Figure 17.A2 — Effect of plate and load capacitors for various phase shift values through the pi network.

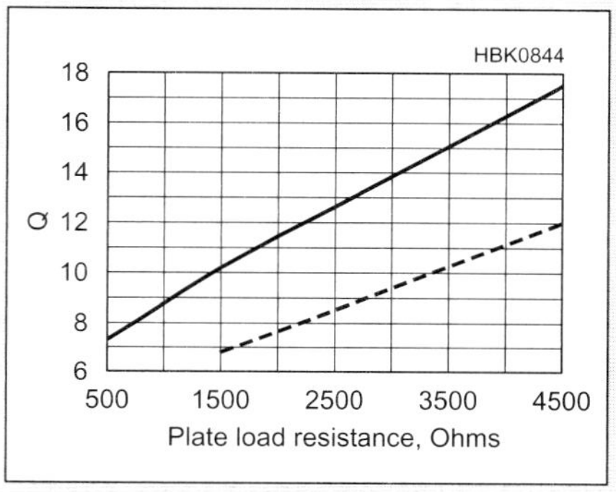

Figure 17.A3 — Pi network Q versus plate load resistance for non-interactive tuning. Use the upper curve for a 50 Ω simple pi network and the lower curve for a pi-L network with a 100 Ω intermediate resistance valu

When tuning to the lower frequency bands, the maximum value of the capacitor, especially the loading capacitor, will be a limiting factor. Sometimes, fixed mica or ceramic transmitting rated capacitors will be switched in parallel with the variable capacitor to reach the total value required.

COMPONENT SELECTION FOR THE PI-L NETWORK

For those wishing to use manual methods to design a pi-L network, there are lookup tables with this book's online content along with the mathematical formulas from which they are derived. **Table 17.1** is an abbreviated version that shows the general trends. Values shown are for 1.8 MHz. Other bands can be approximated by dividing all component values by the frequency ratio. For example, for 18 MHz, divide all component values by 10. For 3.6 MHz, divide values by two, and so on.

For a given Q_L, the pi-L circuit has better harmonic rejection than the pi circuit. This allows the designer to use a lower Q_L, resulting in lower losses and lower capacitor values, which is an advantage on the lower frequencies. However, the inductor values will be higher, and the lower capacitor values may be unachievable at higher frequencies. With the pi-L circuit there are many more variables to work with and, thus, one can try many different possibilities to make the circuit work within the limits of the components that are available.

PROBLEMS AT VHF AND HIGHER FREQUENCIES

As the size of a circuit approaches about 5% of a wavelength, components begin to seriously depart from the pure inductance or capacitance we assume them to be. Inductors begin to act like transmission lines. Capacitors often exhibit values far different from their marked values because of stray internal reactances and lead inductance. Therefore, tuned circuits are frequently fabricated in the form of striplines or other transmission lines in order to circumvent the problem of building "pure" inductances and capacitances. The choice of components is often more significant than the type of network used.

The high impedances encountered in VHF tube-amplifier plate circuits are not easily matched with typical networks. Tube output capacitance is usually so large that most matching networks are unsuitable. The usual practice is to resonate the tube output capacitance with a low-loss inductance connected in series or parallel. The result can be a very high-Q tank circuit. Component losses must be kept to an absolute minimum in order to achieve reasonable tank efficiency. Output impedance transformation is usually performed by a link inductively coupled to the tank circuit or by a parallel transformation of the output resistance using a series capacitor.

Since high values of plate load impedance call for low values of plate tuning capacitance, one might be tempted to add additional tubes in parallel to reduce the required load impedance. This only adds to the stray plate capacitance, and the potential for parasitic oscillations is increased. For these reasons, tubes in parallel are seldom used at VHF. Push-pull circuits offer some advantages, but with modern compact ceramic tube types, most VHF amplifiers use a single tube along with distributed type tuned networks. Other approaches are discussed later in this chapter.

"COLD TUNING" AN AMPLIFIER

Because of the high voltage and current involved, as well as the danger of damaging an expensive tube or other component, it is prudent to "cold tune" an amplifier before applying power to it. This can actually be done early in the construction as soon as tank components and the tube are in place. Cold tuning requires

Amplifier Tuning

For commercial amplifiers, "Read the Manual" should be your guide as the following procedures may not be exactly the same as the manufacturer's directions and can be quite different in some cases. When tuning any amplifier, monitor grid currents closely and do not exceed the specified maximum current as those are the easiest elements of the tube to damage.

In all cases, the last tuning adjustments should be made at full power output, not at reduced power, because the characteristics of the tube change with different power levels. Operating the amplifier at high power after tuning at low power can result in spurious emissions or over-stressing the output network components.

Pre-tuning Preparation

1. Begin by making sure you have all band-switching controls set properly. If the amplifier TUNE and LOAD controls (sometimes referred to as PLATE TUNE and OUTPUT, respectively) have recommended settings on a particular band, start at those settings. If your amplifier can be set to a TUNE mode, do so. If you don't have instructions or pre-set values, set the TUNE control for maximum capacitance.
2. Set the initial amount of drive (input power) from the exciter — read the amplifier manual or check the tube's specifications if a manual is not available. The exciter output should be one-half or more of full power so that the exciter's ALC systems function properly.

Grounded-Grid Triode Amplifiers

Triodes used in currently manufactured amplifiers include the 811A, 572B, 3-500Z, and 3CX-type tubes. Tuning a triode-based, grounded-grid amplifier is the simplest: Tune for maximum or legal-limit output power without exceeding the tube ratings, particularly grid current. The typical procedure is:

1. While monitoring grid current, adjust TUNE for a dip in the plate current which should correspond to a peak in output power.
2. Increase drive until the plate current equals about one-quarter to one-half of the target current (depending on the tube and grid bias) while monitoring the output on a wattmeter or the internal power meter.
3. Adjust the TUNE control then advance the LOAD control for maximum output.
4. Repeat the sequence of peaking TUNE then increasing LOAD until no more output power can be obtained without exceeding the ratings for the tube or the legal power limit is reached.
5. If necessary, increase drive and re-peak both the TUNE and LOAD controls.

Tetrode Amplifiers

Amplifiers that use tetrodes such as the 4CX800A/GU-74B, 4CX1000 employ a different tuning procedure in which tuning for maximum power may result in a destroyed control grid or screen grid! Procedures vary depending not only on the ratings of the tetrode, but the voltages as well.

A grid-driven tetrode (or pentode) amplifier operating near its designed output power uses the TUNE control for peaking output power and the LOAD control for increasing, but not exceeding, the maximum allowable screen current. Generally, the first part of tetrode amplifier tuning is the same as for a triode amplifier with both the TUNE and LOAD controls adjusted for maximum power output while monitoring screen and control grid current.

1. After the initial tuning step, the LOAD control is used to peak the screen current. Maximum power should coincide with maximum screen current. The screen current is just a better indicator.
2. As with the triode amplifier, if drive needs to be increased, readjust the TUNE and LOAD controls.

After Tune-Up

For both triode and tetrode amplifiers, once the procedures above have been completed, try moving the TUNE control a small amount higher or lower and re-peaking the LOAD control. Also known as "rocking" the TUNE control, this small variation can find settings with a few percent more output power or better efficiency.

Once tuning has been completed, it is a good idea to mark the settings of the TUNE and LOAD controls for each band. This reduces the amount of time for on-the-air or dummy-load tuning, reducing stress to the tube and interference to other stations. Usually, a quick "fine-tune" adjustment is all that is required for maximum output. The set of markings also serves as a diagnostic tool, should the settings for maximum power suddenly shift. This indicates a change in the antenna system, such as a failing connector or antenna.

What Does Tuning Really Do?

What is actually happening when the TUNE and LOAD controls are operated this way? First, remember that the pi network acts as an adjustable impedance matching circuit between the tube's high impedance (from a few hundred to several kiloohms) to the antenna system's low impedance (typically under 100 ohms). It acts like two L networks back-to-back, changing the impedance in two steps. The TUNE capacitor and a portion of the inductor form one L network while the LOAD capacitor and the remainder of the inductor form the other L network.

By adjusting the two controls, you adjust impedance presented to the tube. Setting the LOAD control to maximum capacitance and adjusting the TUNE control for minimum plate current presents the maximum impedance or lightest load to the tube.

The next steps of decreasing LOAD capacitance and re-adjusting the TUNE control for minimum plate current ("dipping the plate") steadily reduce the impedance presented to the tube (increasing the load), which increases power output. The goal is to increase the LOAD capacitance while keeping plate current minimized, until the desired output power is reached, and the tube is operated at "full load."

The math behind how the pi network changes the impedance and the tube current with power output is explained on the web page **www.edn.com/pi-network-and-dipping-the-final**.

some test equipment, but is not difficult or time consuming. Only if you have a problem in getting the tuning right will it take much time, but that is exactly the case in which you would not want to turn on the power without having discovered that there is a problem. With cold tuning, you can also add and remove additional components, such as the RF choke, and see how much it affects the tuning.

There are always stray capacitances and inductances in larger sized equipment, so there is a good chance that your carefully designed circuits may not be quite right. Even with commercial equipment, you may want to become aware of the limitations of the tuning ranges and the approximate settings for the dials for each band. The equipment manual may provide this information, but what if the amplifier you have is a bit out of calibration? In all of these cases, cold tests provide cheap insurance against damage caused by bad tuning and, at the same time, give you practice in setting up.

The first step is to ensure that the equipment is truly cold by removing the power plug from the wall. Since you will be working around the high voltage circuits, you may want to remove fuses or otherwise ensure that power cannot come on. In a well-designed amplifier there will be interlocks that prevent turn on and, perhaps, also short out the plate voltage. If these are in a place that affects the RF circuits, they may have to be temporarily removed. Just be sure to put them back when done.

The adjustment of the plate circuit components is the most important. The easiest way to check those is to attach a resistor across the tube from plate to ground. The resistor should be the same value as the design load resistance and must be non-inductive. Several series resistors in the 500 Ω range, either carbon film or the older carbon composition type will work, but *not* wire wound. The tube must remain in its socket and all the normal connections to it should be in place. Covers should be installed, at least for the final tests, since they may affect tuning.

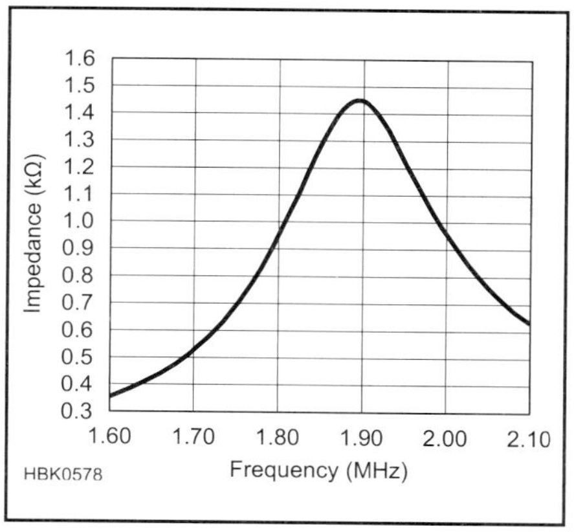

Figure 17.20 — Impedance as would be measured at the plate of an amplifier tube with the output network tuned to 1.9 MHz and the output terminated in its proper load.

Connect a test instrument to the 50 Ω output. This can be an antenna bridge or other 50 Ω measuring device. Examples of suitable test equipment would be a vector network analyzer, preferably with an impedance step up transformer, a vector impedance meter or an RX meter, such as the Boonton 250A. If the wattage of the resistor across the plate can stand it, it can even be a low power transmitter.

Select the correct band and tune the plate and load capacitors so that the instrument at the output connector shows 50 Ω or a low SWR at the 50 Ω point. Since this type of circuit is bilateral, you now have settings that will be the same ones you need to transform the 50 Ω load to look like the resistive load you used at the tube plate for the test. If you have an instrument capable of measuring relatively high impedances at RF frequencies, you can terminate the output with 50 Ω and measure the impedance on the plate. It will look something like **Figure 17.20**.

A similar test will work for the input, although it is sometimes more difficult to know what the input impedance at the tube will be. In fact, it will change somewhat with drive level so a low power cold test may not give a full picture. However, just like the case of the plate circuit, you can put your best estimate of the input impedance at the grid using a non-inductive resistor. Then, tune the input circuits for a match.

An old timer's method for tuning these circuits, especially the plate circuit, is to use a dip meter. These are getting pretty hard to find these days and, in any case, only show resonance. You won't know if the transformation ratio between plate and output is correct, but it is better than nothing. At least, if you can't get a dip at the proper frequency, you will know that something is definitely wrong.

17.5 Transmitting Tube Ratings

17.5.1 Plate, Screen, and Grid Dissipation

The ultimate factor limiting the power-handling capability of a tube is often (but not always) its maximum plate dissipation rating. This is the measure of how many watts of heat the tube can safely dissipate, if it is cooled properly, without exceeding critical temperatures. Excessive temperature can damage or destroy internal tube components or vacuum seals — resulting in tube failure. The same tube may have different voltage, current and power ratings depending on the conditions under which it is operated, but its safe temperature ratings must not be exceeded in any case! Important cooling considerations are discussed in more detail later in this chapter.

The efficiency of a power amplifier may range from approximately 25% to 85%, depending on its operating class, adjustment and circuit losses. The efficiency indicates how much of the dc power supplied to the stage is converted to useful RF output power; the rest is dissipated as heat, mostly by the plate. The *TubeCalculator* program will calculate the dissipation of the plate as one of its outputs. Otherwise, it can be determined by multiplying the plate voltage (V) times the plate current (A) and subtracting the output power (W).

For a class AB amplifier, the resting dissipation should also be noted, since with no RF input, *all* of the dc power is dissipated in the plate. Multiply plate voltage times the resting plate current to find this resting dissipation value. Screen dissipation is simply screen voltage times screen current. Grid dissipation is a bit more complicated since some of the power into the grid goes into the bias supply, some is passed through to the output (when grounded grid is used) and some is dissipated in the grid. Some tubes have very fragile grids and cannot be run with any grid current at all.

Almost all vacuum-tube power amplifiers in amateur service today operate as linear amplifiers (Class AB or B) with efficiencies of approximately 50% to 65%. That means that a useful power output of approximately 1 to 2 times the plate dissipation generally can be achieved. This requires, of course, that the tube is cooled enough to realize its maximum plate dissipation rating and that no other tube rating, such as maximum plate current or grid dissipation, is exceeded.

Type of modulation and duty cycle also influence how much output power can be achieved for a given tube dissipation. Some types of operation are less efficient than others, meaning that the tube must dissipate more heat. Some forms of modulation, such as CW or SSB, are intermittent in nature, causing less average heating than modulation formats in which there is continuous transmission (RTTY or FM, for example).

Power-tube manufacturers use two different rating systems to allow for the variations in service. CCS (Continuous Commercial Service) is the more conservative rating and is used for specifying tubes that are in constant use at full power. The second rating system is based on intermittent, low-duty-cycle operation, and is known as ICAS (Intermittent Commercial and Amateur Service). ICAS ratings are normally used by commercial manufacturers and individual amateurs who wish to obtain maximum power output consistent with reasonable tube life in CW and SSB service. CCS ratings should be used for FM, RTTY and SSTV applications. (Plate power transformers for amateur service are also rated in CCS and ICAS terms.).

MAXIMUM RATINGS

Tube manufacturers publish sets of maximum values for the tubes they produce. No maximum rated value should ever be exceeded. As an example, a tube might have a maximum plate-voltage rating of 2500 V, a maximum plate-current rating of 500 mA, and a maximum plate dissipation rating of 350 W. Although the plate voltage and current ratings might seem to imply a safe power input of 2500 V × 500 mA = 1250 W, this is true only if the dissipation rating will not be exceeded. If the tube is used in class AB2 with an expected efficiency of 60%, the maximum safe dc power input is

$$P_{IN} = \frac{100 P_D}{100 - N_D} = \frac{100 \times 350}{100 - 60} = 875 \text{ W}$$

17.5.2 Tank Circuit Components

CAPACITOR RATINGS

The tank capacitor in a high-power amplifier should be chosen with sufficient spacing between plates to preclude high-voltage breakdown. The peak RF voltage present across a properly loaded tank circuit, without modulation, may be taken conservatively as being equal to the dc plate voltage. If the dc supply voltage also appears across the tank capacitor, this must be added to the peak RF voltage, making the total peak voltage twice the dc supply voltage. At the higher voltages, it is usually desirable to design the tank circuit so that the dc supply voltages do not appear across the tank capacitor, thereby allowing the use of a smaller capacitor with less plate spacing. Capacitor manufacturers usually rate their products in terms of the peak voltage between plates. Typical plate spacings are given in **Table 17.2**.

Table 17.2
Typical Tank-Capacitor Plate Spacings

Spacing Inches	*Peak Voltage*	*Spacing Inches*	*Peak Voltage*	*Spacing Inches*	*Peak Voltage*
0.015	1000	0.07	3000	0.175	7000
0.02	1200	0.08	3500	0.25	9000
0.03	1500	0.125	4500	0.35	11000
0.05	2000	0.15	6000	0.5	13000

Output tank capacitors should be mounted as close to the tube as possible to allow short low inductance leads to the plate. Especially at the higher frequencies, where minimum circuit capacitance becomes important, the capacitor should be mounted with its stator plates well spaced from the chassis or other shielding. In circuits in which the rotor must be insulated from ground, the capacitor should be mounted on ceramic insulators of a size commensurate with the plate voltage involved and — most important of all, from the viewpoint of safety to the operator — a well-insulated coupling should be used between the capacitor shaft and the knob. The section of the shaft attached to the control knob should be well grounded. This can be done conveniently by means of a metal shaft bushing at the panel.

COIL RATINGS

Tank coils should be mounted at least half their diameter away from shielding or other large metal surfaces, such as blower housings, to prevent a marked loss in Q. Except perhaps at 24 and 28 MHz, it is not essential that the coil be mounted extremely close to the tank capacitor. Leads up to 6 or 8 inches are permissible. It is more important to keep the tank capacitor, as well as other components, out of the immediate field of the coil.

The principal practical considerations in designing a tank coil usually are to select a conductor size and coil shape that will fit into available space and handle the required power without excessive heating. Excessive power loss as such is not necessarily the worst hazard in using too-small a conductor. It is not uncommon for the heat generated to actually unsolder joints in the tank circuit and lead to physical damage or failure. For this reason it's extremely important, especially at power levels above a few hundred watts, to ensure that all electrical joints in the tank circuit are secured mechanically as well as soldered.

Table 17.3 shows recommended conductor sizes for amplifier tank coils, assuming loaded tank circuit Q of 15 or less on the 24 and 30 MHz bands and 8 to 12 on the lower frequency bands. In the case of input circuits for screen-grid tubes where driving power is quite small, loss is relatively unimportant and almost any physically convenient wire size and coil shape is adequate.

Table 17.3
Copper Conductor Sizes for Transmitting Coils for Tube Transmitters

Power Output (W)	*Band (MHz)*	*Minimum Conductor Size*
1500	1.8-3.5	10
	7-14	8 or ⅛"
	18-28	6 or ³⁄₁₆"
500	1.8-3.5	12
	7-14	10
	18-28	8 or ⅛"
150	1.8-3.5	16
	7-14	12
	18-28	10

*Whole numbers are AWG; fractions of inches are tubing ODs.

The conductor sizes in Table 17.3 are based on experience in continuous-duty amateur CW, SSB and RTTY service and assume that the coils are located in a reasonably well ventilated enclosure. If the tank area is not well ventilated and/or if significant tube heat is transferred to the coils, it is good practice to increase AWG wire sizes by two (for example, change from #12 to #10) and tubing sizes by ¹⁄₁₆ inch.

Larger conductors than required for current handling are often used to maximize unloaded Q, particularly at higher frequencies. Where skin depth effects increase losses, the greater surface area of large diameter conductors can be beneficial. Small-diameter copper tubing, up to ⅜ inch outer diameter, can be used successfully for tank coils up through the lower VHF range. Copper tubing in sizes suitable for constructing high-power coils is generally available in 50 foot rolls from plumbing and refrigeration equipment suppliers. Silver-plating the tubing may further reduce losses. This is especially true as the tubing ages and oxidizes. Silver oxide is a much better conductor than copper oxide, so silver-plated tank coils maintain their low-loss characteristics even after years of use. (There is some debate in amateur circles about the benefits of silver plating.).

At VHF and above, tank circuit inductances do not necessarily resemble the familiar coil. The inductances required to resonate tank circuits of reasonable Q at these higher frequencies are small enough that only strip lines or sections of transmission line are practical. Since these are constructed from sheet metal or large diameter tubing, current-handling capabilities normally are not a relevant factor.

17.5.3 Other Components

RF CHOKES

The characteristics of any RF choke vary with frequency. At low frequencies the choke presents a nearly pure inductance. At some higher frequency it takes on high impedance characteristics resembling those of a parallel-resonant circuit. At a still higher frequency it goes through a series-resonant condition, where the impedance is lowest — generally much too low to perform satisfactorily as a shunt-feed plate choke. As frequency increases further, the pattern of alternating parallel and series resonances repeats. Between resonances, the choke will show widely varying amounts of inductive or capacitive reactance.

In most high-power amplifiers, the choke is directly in parallel with the tank circuit, and is subject to the full tank RF voltage. See **Figure 17.21A**. If the choke does not present a sufficiently high impedance, enough power will be absorbed by the choke to burn it out. To avoid this, the choke must have a sufficiently high reactance to be effective at the lowest frequency (at least equal to the plate load resistance) and yet have no series resonances near any of the higher frequency bands. A resonant-choke failure in a high-power amplifier can be very dramatic and damaging!

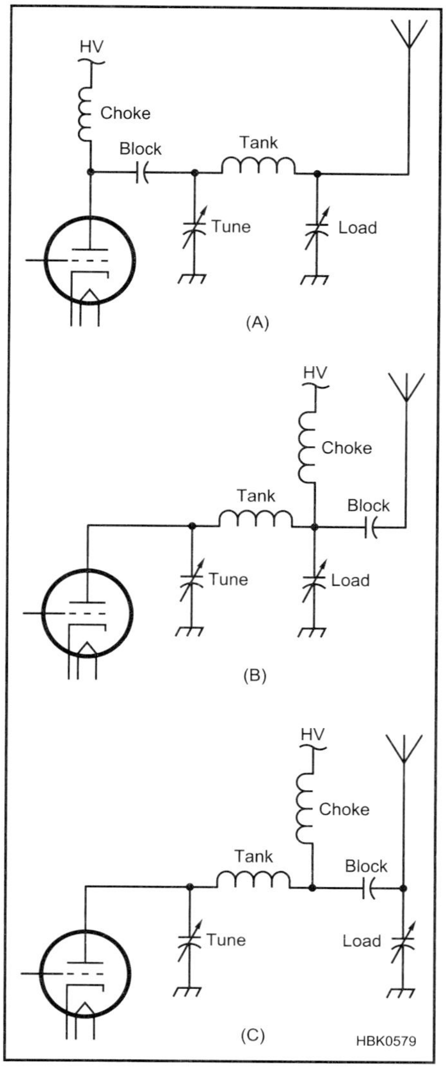

Figure 17.21 — Three ways of feeding dc to a tube via an RF choke. See text for a discussion of the tradeoffs.

An RF choke performs best well below its self-resonant frequency (SRF) but it can be used throughout the range for which it has an acceptably high impedance. The choke's impedance will be inductive below the SRF and capacitive above the SRF. (See the discussion of RF Chokes in the **RF Techniques** chapter.) In the plate circuit, resonances may produce very high voltages at one or more points along the coil. This can cause arcing and damage to the choke. These chokes are often specially wound to minimize the effect of resonances but this must be done to suit the exact application since distributed capacitance to the enclosure and other components affects the SRF.

W8JI has written an informative paper about SRF and plate chokes which can be viewed at **www.w8ji.com/rf_plate_choke.htm**. His companion web page on inductors used in high-power RF circuits (**www.w8ji.com/loading_inductors.htm**) is also very informative.

Thus, any choke intended for shunt-feed use should be carefully investigated. The best way would be to measure its reactance to ground with an impedance measuring instrument. If the dip meter is used, the choke must be shorted end-to-end with a direct, heavy braid or strap. Because nearby metallic objects affect the resonances, it should be mounted in its intended position, but disconnected from the rest of the circuit. A dip meter coupled an inch or two away from one end of the choke nearly always will show a deep, sharp dip at the lowest series-resonant frequency and shallower dips at higher series resonances.

Any choke to be used in an amplifier for the 1.8 to 28 MHz bands requires careful (or at least lucky!) design to perform well on all amateur bands within that range. Most simply put, the challenge is to achieve sufficient inductance that the choke doesn't "cancel" a large part of tuning capacitance at 1.8 MHz. At the same time, try to position all its series resonances where they can do no harm. In general, close wind enough #20 to #24 magnet wire to provide about 135 μH inductance on a ¾ to 1-inch diameter cylindrical form of ceramic, Teflon or fiberglass. This gives a reactance of 1500 Ω at 1.8 MHz and yet yields a first series resonance in the vicinity of 25 MHz. Before the advent of the 24 MHz band this worked fine. But trying to "squeeze" the resonance into the narrow gaps between the 21, 24 and/or 28 MHz bands is quite risky unless sophisticated instrumentation is available. If the number of turns on the choke is selected to place its first series resonance at 23.2 MHz, midway between 21.45 and 24.89 MHz, the choke impedance will typically be high enough for satisfactory operation on the 21, 24 and 28 MHz bands. The choke's first series resonance should be measured very carefully as described above using a dip meter and calibrated receiver or RF impedance bridge, with the choke mounted in place on the chassis.

Investigations with a vector impedance meter have shown that "trick" designs, such as using several shorter windings spaced along the form, show little if any improvement in choke resonance characteristics. Some commercial amplifiers circumvent the problem by band switching the RF choke. Using a larger diameter (1 to 1.5 inches) form does move the first series resonance somewhat higher for a given value of basic inductance. Beyond that, it is probably easiest for an all-band amplifier to add or subtract enough turns to move the first resonance to about 35 MHz and settle for a little less than optimum reactance on 1.8 MHz.

However, there are other alternatives. If one is willing to switch the choke when changing bands, it is possible to have enough inductance for 1.8 to 10 MHz, with series resonances well above 15 MHz. Then for 14 MHz and above, a smaller choke is used which has its resonances well above 30 MHz. Providing an extra pole on the band switch is, of course, the trade-off. This switch must withstand the full plate voltage. Switches suitable for changing bands for the pi network would handle this fine.

Another approach is to feed the high-voltage dc through the main tank inductor, putting the RF choke at the loading capacitor, instead of at the tube. (See Figure 17.21B) This puts a much lower RF voltage on the choke and, thus, not as much reactance is required for satisfactory rejection of the RF voltage. However, this puts both dc and RF voltages on the plate and loading capacitors which may be beyond their ratings. The blocking capacitor can be put before the loading capacitor, as in Figure 17.21C. This removes the dc from the loading capacitor, which typically has a lower voltage rating than the plate capacitor, but puts high current in the blocker.

Yet another method involves using hollow tubing for the plate tank and passing the dc lead through it. This lowers the RF voltage on the choke without putting dc voltage on the tuning components. This method works best for higher power transmitters where the tuning inductor can be made of ⅛ inch or larger copper tubing.

BLOCKING CAPACITORS

A series capacitor is usually used at the input of the amplifier output circuit. Its purpose is to block dc from appearing on matching circuit components of the antenna. As mentioned in the section on tank capacitors, output-circuit voltage requirements are considerably reduced when only RF voltage is present.

To provide a margin of safety, the voltage rating for a blocking capacitor should be at least 25% to 50% greater than the dc voltage applied. A large safety margin is desirable, since blocking capacitor failure can bring catastrophic results. The worse case is when dc is applied to the output of the transmitter and even to the antenna, with potentially fatal results. Often an RF choke is placed from the RF output jack to ground as a safety backup. A shorted blocker will blow the power supply fuse.

To avoid affecting the amplifier's tuning and matching characteristics, the blocking capacitor should have a low impedance at all operating frequencies. If it presents more than 5% of the plate load resistance, the pi components should be adjusted to compensate. Use of a *SPICE* analysis provides a useful way to see what adjustments might be required to maintain the desired match.

The capacitor also must be capable of handling, without overheating or significantly changing value, the substantial RF current that flows through it. This current usually is greatest at the highest frequency of operation where tube output capacitance constitutes a significant part of the total tank capacitance. A significant portion of circulating tank current, therefore, flows through the blocking capacitor. When using the connection of the RF choke shown in Figure 17.21C, the entire circulating current must be accommodated.

Transmitting capacitors are rated by their manufacturers in terms of their RF current-carrying capacity at various frequencies. Below a couple hundred watts at the high frequencies, ordinary disc ceramic capacitors of suitable voltage rating work well in high-impedance tube amplifier output circuits. Some larger disk capacitors rated at 5 to 8 kV also work well for higher power levels at HF. For example, two inexpensive Centralab type DD-602 discs (0.002 µF, 6 kV) in parallel have proved to be a reliable blocking capacitor for 1.5-kW amplifiers operating at plate voltages to about 2.5 kV. At very high power and voltage levels and at VHF, ceramic "doorknob" transmitting capacitors are needed for their low losses and high current handling capabilities. When in doubt, adding additional capacitors in parallel is cheap insurance against blocking capacitor failure and also reduces the impedance. So-called "TV doorknobs" may break down at high RF current levels and should be avoided.

The very high values of Q_L found in many VHF and UHF tube-type amplifier tank circuits often require custom fabrication of the blocking capacitor. This can usually be accommodated through the use of a Teflon "sandwich" capacitor. Here, the blocking capacitor is formed from two parallel plates separated by a thin layer of Teflon. This capacitor often is part of the tank circuit itself, forming a very low-loss blocking capacitor. Teflon is rated for a minimum breakdown voltage of 2000 V per mil of thickness, so voltage breakdown should not be a factor in any practically realized circuit. The capacitance formed from such a Teflon sandwich can be calculated from the information presented elsewhere in this *Handbook* (use a dielectric constant of 2.1 for Teflon). In order to prevent any potential irregularities caused by dielectric thickness variations (including air gaps), Dow-Corning DC-4 silicone grease should be evenly applied to both sides of the Teflon dielectric. This grease has properties similar to Teflon, and will fill in any surface irregularities that might cause problems.

Finally, it is also possible to place the plate at ground potential and the filament or cathode at negative HV or B-. This has the benefit of placing the RF choke at a low-impedance part of the circuit, greatly reducing the effects of self-resonance. Blocking capacitors are still required on the input and grid-bias circuits but they carry much less RF current than when in the output plate circuit. The filament transformer will have to have sufficient insulation to withstand the HV across it, as well. While rarely seen in amateur amplifiers, this technique was used in AM broadcast transmitters and is discussed more at **forums.qrz.com/index.php?threads/the-case-for-a-grounded-b-other-innovations-in-the-design-of-a-vacuum-tube-rf-power-amplifier.771103/**.

17.6 Sources of Operating Voltages

17.6.1 Tube Filament or Heater Voltage

A power vacuum tube can use either a directly heated filament or an indirectly heated cathode. The filament voltage for either type should be held within 5% of rated voltage. Because of internal tube heating at UHF and higher, the manufacturers' filament voltage rating often is reduced at these higher frequencies. The de-rated filament voltages should be followed carefully to maximize tube life.

Series dropping resistors may be required in the filament circuit to attain the correct voltage. Adding resistance in series will also reduce the inrush current when the tube is turned on. Cold tungsten has much lower resistance than when hot. Circuits are available that both limit the inrush current at turn on and also regulate the voltage against changes in line voltage.

The voltage should be measured with a true RMS meter at the filament pins of the tube socket while the amplifier is running. The filament choke and interconnecting wiring all have voltage drops associated with them. The high current drawn by a power-tube heater circuit causes substantial voltage drops to occur across even small resistances. Also, make sure that the plate power drawn from the power line does not cause the filament voltage to drop below the proper value when plate power is applied.

Thoriated filaments lose emission when the tube is overloaded appreciably. If the overload has not been too prolonged, emission, sometimes, may be restored by operating the filament at rated voltage, with all other voltages removed, for a period of 30 to 60 minutes. Alternatively, you might try operating the tube at 20% above rated filament voltage for five to ten minutes.

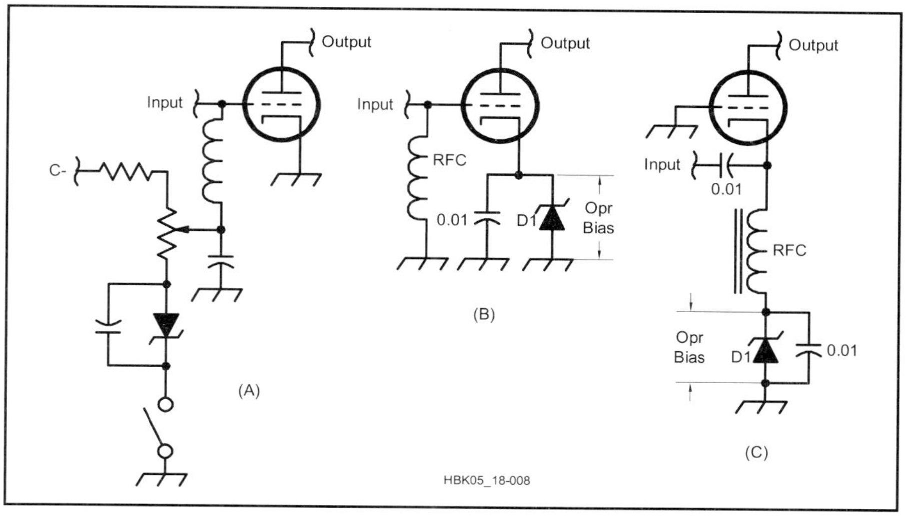

Figure 17.22 — Various techniques for providing operating bias with tube amplifiers.

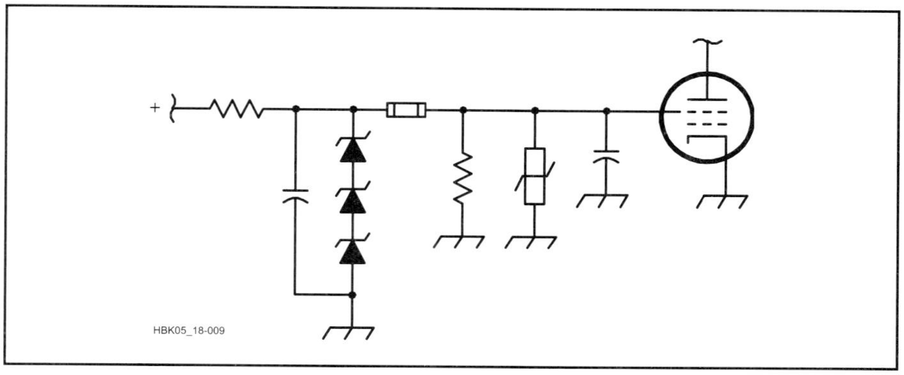

Figure 17.23 — A Zener-regulated screen supply for use with a tetrode. Protection is provided by a fuse and a varistor.

17.6.2 Vacuum-Tube Plate Voltage

DC plate voltage for the operation of RF amplifiers is most often obtained from a transformer-rectifier-filter system (see the **Power Sources** chapter) designed to deliver the required plate voltage at the required current. It is not unusual for a power tube to arc over internally (generally from the plate to the screen or control grid) once or twice, especially soon after it is first placed into service. The flashover by itself is not normally dangerous to the tube, provided that instantaneous maximum plate current to the tube is held to a safe value and the high-voltage plate supply is shut off very quickly.

A good protective measure against this is the inclusion of a high-wattage power resistor in series with the plate high-voltage circuit. The value of the resistor, in ohms, should be approximately 10 to 15 times the no-load plate voltage in KV. This will limit peak fault current to 67 to 100 A. The series resistor should be rated for 25 or 50 W power dissipation; vitreous enamel coated wire-wound resistors have been found to be capable of handling repeated momentary fault-current surges without damage. Aluminum-cased resistors (Dale) are not recommended for this application. Each resistor also must be large enough to safely handle the maximum value of normal plate current; the wattage rating required may be calculated from $P = I^2R$. If the total filter capacitance exceeds 25 µF, it is a good idea to use 50 W resistors in any case. Even at high plate-current levels, the addition of the resistors does little to affect the dynamic regulation of the plate supply.

Since tube (or other high-voltage circuit) arcs are not necessarily self-extinguishing, a fast-acting plate overcurrent relay or primary circuit breaker is also recommended to quickly shut off ac power to the HV supply when an arc begins. Using this protective system, a mild HV flashover may go undetected, while a more severe one will remove ac power from the HV supply. (The cooling blower should remain energized, however, since the tube may be hot when the HV is removed due to an arc.) If effective protection is not provided, however, a "normal" flashover, even in a new tube, is likely to damage or destroy the tube, and also frequently destroys the rectifiers in the power supply as well as the plate RF choke. A power tube that flashes over more than about 3 to 5 times in a period of several months likely is defective and will have to be replaced before long.

17.6.3 Grid Bias

The grid bias for a linear amplifier should be highly filtered and well regulated. Any ripple or other voltage change in the bias circuit modulates the amplifier. This causes hum and/or distortion to appear on the signal. Since most linear amplifiers draw only small amounts of grid current, these bias-supply requirements are not difficult to achieve.

Fixed bias for class AB1 tetrode and pentode amplifiers is usually obtained from a variable-voltage regulated supply. Voltage adjustment allows setting bias level to give the desired resting plate current. **Figure 17.22A** shows a simple Zener-diode-regulated bias supply. The dropping resistor is chosen to allow approximately 10 mA of Zener current. Bias is then reasonably well regulated for all drive conditions up to 2 or 3 mA of grid current. The potentiometer allows bias to be adjusted between Zener and approximately 10 V higher. This range is usually adequate to allow for variations in the characteristics of different tubes. Under standby conditions, when it is desirable to cut off the tube entirely, the Zener ground return is interrupted so the full bias supply voltage is applied to the grid.

In Figure 17.22B and C, bias is obtained from the voltage drop across a Zener diode in the cathode (or filament center-tap) lead. Operating bias is obtained by the voltage drop across D1 as a result of plate (and screen) current flow. The diode voltage drop effectively raises the cathode potential relative to the grid. The grid is, therefore, negative with respect to the cathode by the Zener voltage of the diode. The Zener-diode wattage rating should be twice the product of the maximum cathode current times the rated Zener voltage. Therefore, a tube requiring 15 V of bias with a maximum cathode current of 100 mA would dissipate 1.5 W in the Zener diode. To allow a suitable safety factor, the diode rating should be 3 W or more. The circuit of Figure 17.22C illustrates how D1 would be used with a cathode driven (grounded grid) amplifier as opposed to the grid driven example at B.

In all cases, the Zener diode should be bypassed by a 0.01-µF capacitor of suitable voltage. Current flow through any type of diode generates shot noise. If not bypassed, this noise would modulate the amplified signal, causing distortion in the amplifier output.

17.6.4 Screen Voltage For Tubes

Power tetrode screen current varies widely with both excitation and loading. The current may be either positive or negative, depending on tube characteristics and amplifier operating conditions. In a linear amplifier, the screen voltage should be well regulated for all values of screen current. The power output from a tetrode is very sensitive to screen voltage, and any dynamic change in the screen potential can cause distorted output. Zener diodes are commonly used for screen regulation.

Figure 17.23 shows a typical example of a regulated screen supply for a power tetrode amplifier. The voltage from a fixed dc supply is dropped to the Zener stack voltage by the current-limiting resistor. A screen bleeder resistor is connected in parallel with the Zener stack to allow for the negative screen current developed under certain tube operating conditions. Bleeder current is chosen to be roughly 10 to 20 mA greater than the expected maximum negative screen current, so that screen voltage is regulated for all values of current between maximum negative screen current and maximum positive screen current. For external-anode tubes in the 4CX250 family, a typical screen bleeder current value would be 20 mA. For the 4CX1000 family, a screen-bleeder current of 70 mA is required.

Screen voltage should never be applied to a tetrode unless plate voltage and load also are applied; otherwise, the screen will act like an anode and will draw excessive current. Perhaps the best way to insure this is to include logic circuits that will not allow the screen supply to turn on until it senses plate voltage. Supplying the screen through a series-dropping resistor from the plate supply affords a measure of protection, since the screen voltage only appears when there is plate voltage. Alternatively, a fuse can be placed between the regulator and the bleeder resistor. The fuse should not be installed between the bleeder resistor and the tube because the tube should never be operated without a load on the screen. Without a load, the screen potential tends to rise to the anode voltage. Any screen bypass capacitors or other associated circuits are likely be damaged by this high voltage.

In Figure 17.23, a varistor is connected from screen to ground. If, because of some circuit failure, the screen voltage should rise substantially above its nominal level, the varistor will conduct and clamp the screen voltage to a low level. If necessary to protect the varistor or screen dropping resistors, a fuse or overcurrent relay may be used to shut off the screen supply so that power is interrupted before any damage occurs. The varistor voltage should be approximately 30% to 50% higher than normal screen voltage.

17.7 Tube Amplifier Cooling

Vacuum tubes must be operated within the temperature range specified by the manufacturer if long tube life is to be achieved. Tubes having glass envelopes and rated at up to 25 W plate dissipation may be used without forced-air cooling if the design allows a reasonable amount of convection cooling. If a perforated metal enclosure is used, and a ring of ¼ to ⅜-inch-diameter holes is placed around the tube socket, normal convective airflow can be relied on to remove excess heat at room temperatures.

For tubes with greater plate dissipation ratings, and even for very small tubes operated close to maximum rated dissipation, forced-air cooling with a fan or blower is needed. Most manufacturers rate tube-cooling requirements for continuous-duty operation. Their literature will indicate the required volume of airflow, in cubic feet per minute (CFM), at some particular back pressure. Often, this data is given for several different values of plate dissipation, ambient air temperature and even altitude above sea level.

One extremely important consideration is often overlooked by power-amplifier designers and users alike: a tube's plate dissipation rating is only its maximum potential capability. The power that it can actually dissipate safely depends directly on the cooling provided. The actual power capability of virtually all tubes used in high-power amplifiers for amateur service depends on the volume of air forced through the tube's cooling structure.

17.7.1 Blower Specifications

This requirement usually is given in terms of cubic feet of air per minute (CFM), delivered into a back pressure, representing the resistance of the tube cooler to air flow, stated in inches of water. Both the CFM of airflow required and the pressure needed to force it through the cooling system are determined by ambient air temperature and altitude (air density), as well as by the amount of heat to be dissipated. The cooling fan or blower must be capable of delivering the specified airflow into the corresponding back pressure. As a result of basic air flow and heat transfer principles, the volume of airflow required through the tube cooler increases considerably faster than the plate dissipation, and back pressure increases even faster than airflow. In addition, blower air output decreases with increasing back pressure until, at the blower's so-called "cutoff pressure," actual air delivery is zero. Larger and/or faster-rotating blowers are required to deliver larger volumes of air at higher back pressure.

Values of CFM and back pressure required to realize maximum rated plate dissipation for some of the more popular tubes, sockets and chimneys (with 25 °C ambient air and at sea level) are given in **Table 17.4**. Back pressure is specified in inches of water and can be

Table 17.4

Specifications of Some Popular Tubes, Sockets and Chimneys

Tube	*CFM*	*Back Pressure (inches)*	*Socket*	*Chimney*
3-500Z	13	0.13	SK-400, SK-410	SK-416
3CX800A7	19	0.50	SK-1900	SK-1906
3CX1200A7	31	0.45	SK-410	SK-436
3CX1200Z7	42	0.30	SK-410	—
3CX1500/8877	35	0.41	SK-2200, SK-2210	SK-2216
4-400A/8438	14	0.25	SK-400, SK-410	SK-406
4-1000A/8166	20	0.60	SK-500, SK-510	SK-506
4CX250R/7850	6.4	0.59	SK602A, SK-610, SK-610A SK-611, SK-612, SK-620, SK-620A, SK-621, SK-630	
4CX400/8874	8.6	0.37	SK1900	SK606
4CX400A	8	0.20	SK2A	—
4CX800A	20	0.50	SK1A	—
4CX1000A/8168	25	0.20	SK-800B, SK-810B, SK-890B	SK-806
4CX1500B/8660	34	0.60	SK-800B, SK-1900	SK-806
4CX1600B	36	0.40	SK3A	CH-1600B

These values are for sea-level elevation. For locations well above sea level (5000 ft/1500 m, for example), add an additional 20% to the figure listed.

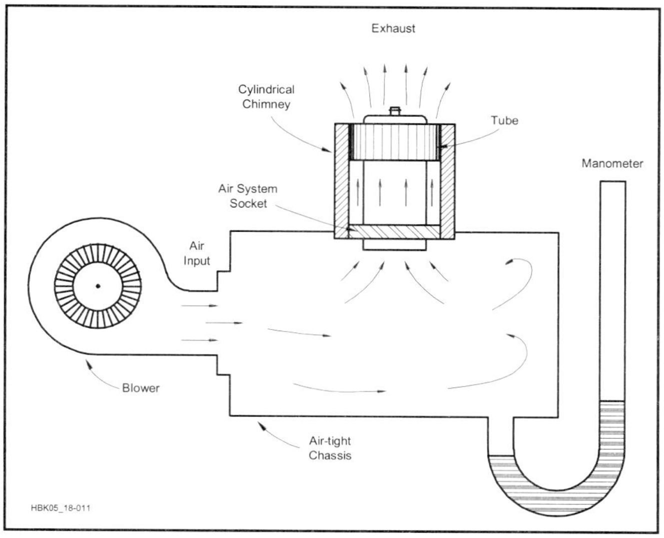

Figure 17.24 — Air is forced into the chassis by the blower and exits through the tube socket. The manometer is used to measure system back pressure, which is an important factor in determining the proper size blower.

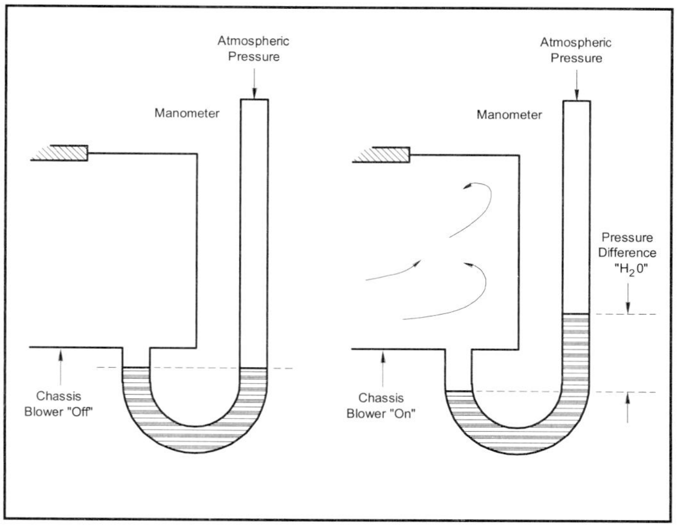

Figure 17.25 — At A the blower is "off" and the water will seek its own level in the manometer. At B the blower is "on" and the amount of back pressure in terms of inches of water can be measured as indicated.

Table 17.5

Blower Performance Specifications

Wheel Dia	*Wheel Width*	*RPM*	*Free Air CFM*	*CFM for Back Pressure (inches)* 0.1	0.2	0.3	0.4	0.5	*Stock No.*
2"	1"	3340	12	9	6	—	—	—	1TDN2
2 15/16"	1½"	3388	53	52	50	47	41	23	1TDN5
3"	1⅞"	3036	50	48	44	39	32	18	1TDN7
3"	1⅞"	3010	89	85	78	74	66	58	1TDP1
3 15/16"	1 15/16"	3016	75	71	68	66	61	56	1TDP3
3¾"	1⅞"	2860	131	127	119	118	112	105	1TDP5

Representative sample of Dayton squirrel cage blowers. More information and other models available from Grainger Industrial Supply (**www.grainger.com**).

measured easily in an operational air system as indicated in **Figures 17.24** and **17.25**. The pressure differential between the air passage and atmospheric pressure is measured with a device called a *manometer*. A manometer is nothing more than a piece of clear tubing, open at both ends and fashioned in the shape of a "U." The manometer is temporarily connected to the chassis and is removed after the measurements are completed. As shown in the diagrams, a small amount of water is placed in the tube. At Figure 17.25A, the blower is "off" and the water seeks its own level, because the air pressure (ordinary atmospheric pressure) is the same at both ends of the manometer tube. At B, the blower is "on" (socket, tube and chimney in place) and the pressure difference, in terms of inches of water, is measured. For most applications, a standard ruler used for measurement will yield sufficiently accurate results.

Table 17.5 gives the performance specifications for a few of the many Dayton blowers, which are available through Grainger Industrial Supply (**www.grainger.com**). Other blowers having wheel diameters, widths and rotational speeds similar to any in Table 17.5 likely will have similar flow and back pressure characteristics. If in doubt about specifications, consult the manufacturer. Tube temperature under actual operating conditions is the ultimate criterion for cooling adequacy and may be determined using special crayons or lacquers that melt and change appearance at specific temperatures. The setup of Figure 17.25, however, nearly always gives sufficiently accurate information.

17.7.2 Cooling Design Example

As an example, consider the cooling design of a linear amplifier to use one 3CX800A7 tube to operate near sea level with the air temperature not above 25 °C. The tube, running 1150 W dc input, easily delivers 750 W continuous output, resulting in 400 W plate dissipation ($P_{DIS} = P_{IN} - P_{OUT}$). According to the manufacturer's data, adequate tube cooling at 400 W P_D requires at least 6 CFM of air at 0.09 inches of water back pressure. In Table 17.5, a Dayton no. 1TDN2 will do the job with a good margin of safety.

If the same single tube were to be operated at 2.3 kW dc input to deliver 1.5 kW output (substantially exceeding its maximum electrical ratings!), P_{IN} would be about 2300 W and $P_D \approx 800$ W. The minimum cooling air required would be about 19 CFM at 0.5 inches of water pressure — doubling P_{DIS}, more than tripling the CFM of air flow required and increasing back pressure requirements on the blower by a factor of 5.5!

However, two 3CX800A7 tubes are needed to deliver 1.5 kW of continuous maximum legal output power in any case. Each tube will operate under the same conditions as in the single-tube example above, dissipating 400 W. The total cooling air requirement for the two tubes is, therefore, 12 CFM at about 0.09 inches of water, only two-thirds as much air volume and one-fifth the back pressure required by a single tube. While this may seem surprising, the reason lies in the previously mentioned fact that both the airflow required by a tube and the resultant back pressure increase much more rapidly than P_D of the tube. Blower air delivery capability, conversely, decreases as back pressure is increased. Thus, a Dayton 1TDN2 blower can cool two 3CX800A7 tubes dissipating 800 W total, but a much larger (and probably noisier) no. 1TDN7 would be required to handle the same power with a single tube.

In summary, three very important considerations to remember are these:

• A tube's actual safe plate dissipation capability is totally dependent on the amount of cooling air forced through its cooling system. Any air-cooled power tube's maximum plate dissipation rating is meaningless unless the specified amount of cooling air is supplied.

• Two tubes will always safely dissipate a given power with a significantly smaller (and quieter) blower than is required to dissipate the same power with a single tube of the same type. A corollary is that a given blower can virtually always dissipate more power when cooling two tubes than when cooling a single tube of the same type.

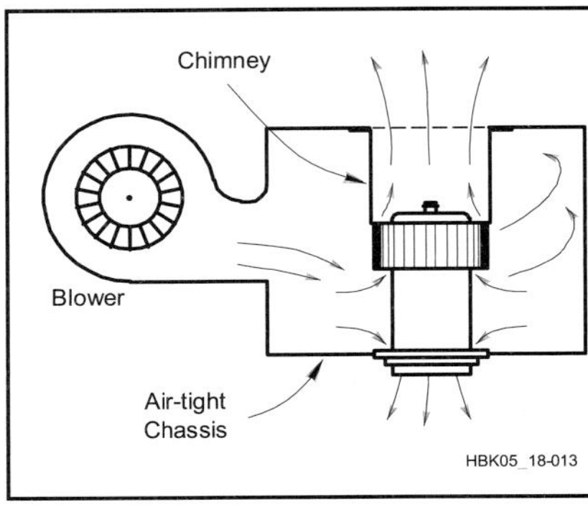

Figure 17.26 — Anode compartment pressurization may be more efficient than grid compartment pressurization. Hot air exits upwards through the tube anode and through the chimney. Cool air also goes down through the tube socket to cool tube's pins and the socket itself.

• Blowers vary greatly in their ability to deliver air against back pressure so blower selection should not be taken lightly.

17.7.3 Other Considerations

A common method for directing the flow of air around a tube involves the use of a pressurized chassis. This system is shown in Figure 17.24. A blower attached to the chassis forces air around the tube base, often through holes in its socket. A chimney is used to guide air leaving the base area around the tube envelope or anode cooler, preventing it from dispersing and concentrating the flow for maximum cooling.

A less conventional approach that offers a significant advantage in certain situations is shown in **Figure 17.26**. Here the anode compartment is pressurized by the blower. A special chimney is installed between the anode heat exchanger and an exhaust hole in the compartment cover. When the blower pressurizes the anode compartment, there are two parallel paths for airflow: through the anode and its chimney, and through the air system socket. Dissipation, and hence cooling air required, generally is much greater for the anode than for the tube base. Because high-volume anode airflow need not be forced through restrictive air channels in the base area, back pressure may be very significantly reduced with certain tubes and sockets. Only airflow actually needed is bled through the base area. Blower back pressure requirements may sometimes be reduced by nearly half through this approach.

Table 17.4 also contains the part numbers for air-system sockets and chimneys available for use with the tubes that are listed. The builder should investigate which of the sockets listed for the 4CX250R, 4CX300A, 4CX1000A and 4CX1600A best fit the circuit needs. Some of the sockets have certain tube elements grounded internally through the socket. Others have elements bypassed to ground through capacitors that are integral parts of the sockets.

Depending on your design philosophy and tube sources, some compromises in the cooling system may be appropriate. For example, if glass tubes are available inexpensively as broadcast pulls, a shorter life span may be acceptable. In such a case, an increase of convenience and a reduction in cost, noise, and complexity can be had by using a pair of "muffin" fans. One fan may be used for the filament seals and one for the anode seal, dispensing with a blower and air-system socket and chimney. The airflow with this scheme is not as uniform as with the use of a chimney. The tube envelope mounted in a cross flow has flow stagnation points and low heat transfer in certain regions of the envelope. These points become hotter than the rest of the envelope. The use of multiple fans to disturb the cross airflow can significantly reduce this problem. Many amateurs have used this cooling method successfully in low-duty-cycle CW and SSB operation but it is not recommended for AM, SSTV or RTTY service.

The true test of the effectiveness of a forced air cooling system is the amount of heat carried away from the tube by the air stream. The power dissipated can be calculated from the airflow temperatures. The dissipated power is

$$P_D = 0.543\ Q_A\ (T_2 - T_1) \qquad (7)$$

where

P_D = the dissipated power, in W,

Q_A = the air flow, in CFM (cubic feet per minute),

T_1 = the inlet air temperature, °C (normally quite close to room temperature), and

T_2 = the amplifier exhaust temperature, °C.

The exhaust temperature can be measured with a cooking thermometer at the air outlet. The thermometer should not be placed inside the anode compartment because of the high voltage present.

17.8 Vacuum Tube Amplifier Stabilization

Purity of emissions and the useful life (or even survival) of a tube depend heavily on stability during operation. Oscillations can occur at the operating frequency, or far from it, because of undesired positive feedback in the amplifier. Unchecked, these oscillations pollute the RF spectrum and can lead to over-dissipation and subsequent failure. Each type of oscillation has its own cause and its own cure.

17.8.1 Amplifier Neutralization

An RF amplifier, especially a linear amplifier, can easily become an oscillator at various frequencies. When the amplifier is operating, the power at the output side is large. If a fraction of that power finds its way back to the input and is in the proper phase, it can be re-amplified, repeatedly, leading to oscillation. An understanding of this process can be had by studying the sections on feedback and oscillation in the **Radio Fundamentals** chapter. Feedback that is self-reinforcing is called "positive" feedback, even though its effects are undesirable. Even when the positive feedback is insufficient to cause actual oscillation, its presence can lead to excessive distortion and strange effects on the tuning of the amplifier and it, therefore, should be eliminated or at least reduced. The deliberate use of "negative" feedback in amplifiers to increase linearity is discussed briefly elsewhere in this chapter.

The power at the output of an amplifier will couple back to the input of the amplifier through any path it can find. It is a good practice to isolate the input and output circuits of an amplifier in separate shielded compartments. Wires passing between the two compartments should be bypassed to ground if possible. This prevents feedback via paths external to the tube.

However, energy can also pass back through the tube itself. To prevent this, a process called neutralization can be used. Neutralization seeks to prevent or to cancel out any transfer of energy from the plate of the tube back to its input, which will be either the grid or the cathode. An effective way to neutralize a tube is to provide a grounded shield between the input and the output. In the grounded grid connection, the grid itself serves this purpose. For best neutralization, the grid should be connected through a low inductance conductor to a point that is at RF ground. Ceramic external tube types may have multiple low inductance leads to ground to enhance the shielding effect. Older glass type tubes may have significant inductance inside the tube and in the socket, and this will limit the effectiveness of the shielding effect of the grid. Thus, using a grounded grid circuit with those tube types does not rule out the need for further efforts at neutralization, especially at the higher HF frequencies.

When tetrodes are used in a grounded cathode configuration, the screen grid acts as an RF shield between the grid and plate. Special tube sockets are provided that provide a very low inductance connection to RF ground. These reduce the feed through from plate to grid to a very small amount, making the effective grid-to-plate capacitance a tiny fraction of one picofarad. If in doubt about amplifiers that will work over a large frequency range, use a network analyzer or impedance measuring instrument to verify how well grounded a "grounded" grid or screen really is. If at some frequencies the impedance is more than an ohm or two, a different grounding configuration may be needed.

For amplifiers to be used at only a single frequency, a series resonant circuit can be used at either the screen or grid to provide nearly perfect bypassing to ground. Typical values for a 50 MHz amplifier are shown in **Figure 17.27**.

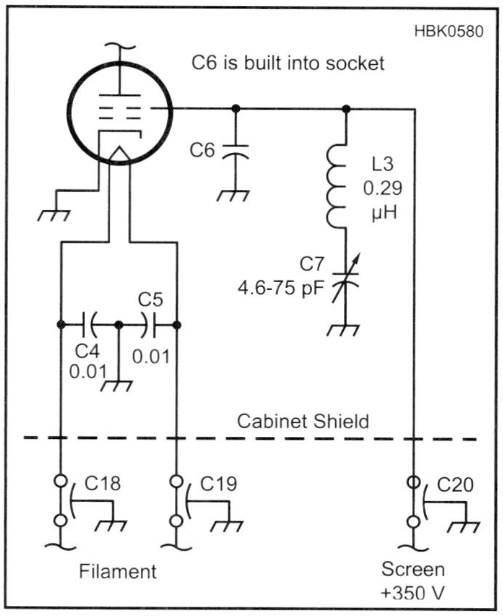

Figure 17.27 — A series-resonant circuit can be used to provide nearly perfect screen or grid bypassing to ground. This example is from a single-band 50 MHz

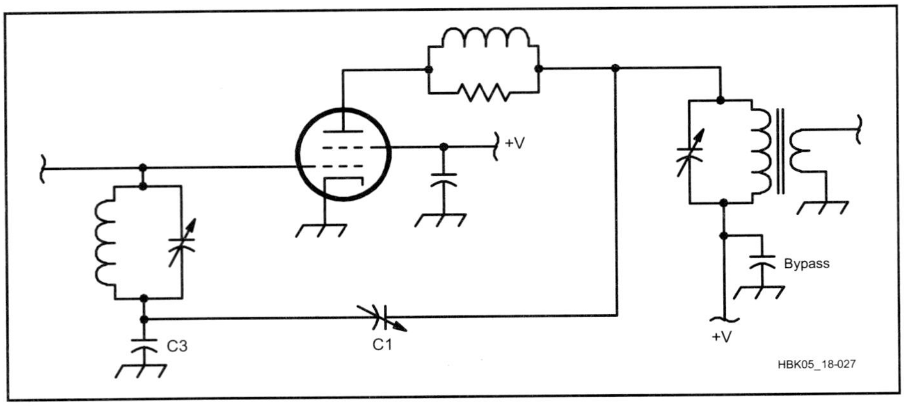

Figure 17.28 — A neutralization circuit uses C1 to cancel the effect of the tube internal capacitance.

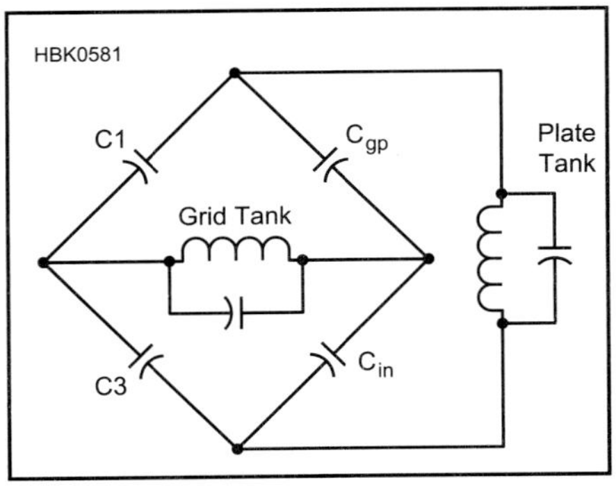

Figure 17.29 — The "bridge neutralization" circuit of Figure 17.28 redrawn to show the capacitance values.

For some tubes, at a certain frequency, the lead inductance to ground can just cancel the grid-to-plate capacitance. Due to this effect, some tube and socket combinations have a naturally self-neutralizing frequency based on the values of screen inductance and grid-to-plate capacitance. For example, the "self-neutralizing frequency" of a 4-1000 is about 30 MHz. This effect has been utilized in some VHF amplifiers.

BRIDGE NEUTRALIZATION

When the shielding effect of a grid or screen bypassed to ground proves insufficient, other circuits must be devised to cancel out the remaining effect of the grid-to-plate or grid-to-cathode capacitance. These, in effect, add an additional path for negative feedback that will combine with the undesired positive feedback and cancel it. The most commonly used circuit is the "bridge neutralization" circuit shown in **Figure 17.28**. This method gets its name from the fact that the four important capacitance values can be redrawn as a bridge circuit, as shown in **Figure 17.29**. Clearly when the bridge is properly balanced, there is no transfer of energy from the plate to the grid tanks. Note that four different capacitors are part of the bridge. C_{gp} is characteristic of the chosen tube, somewhat affected by the screen or grid bypass mentioned earlier. The other components must be chosen properly so that bridge balance is achieved. C1 is the neutralizing capacitor. Its value should be adjustable to the point where

$$\frac{C1}{C3} = \frac{C_{gp}}{C_{IN}} \qquad (8)$$

where

C_{gp} = tube grid-plate capacitance, and
C_{IN} = tube input capacitance.

The tube input capacitance must include all strays directly across the tube. C3 is not simply a bypass capacitor on the ground side of the grid tank, but rather a critical part of the bridge. Hence, it must provide a stable value of capacitance. Sometimes, simple bypass capacitors are of a type which change their value drastically with temperature. These are not suitable in this application.

Neutralization adjustment is accomplished by applying energy to the output of the amplifier, and measuring the power fed through the input. Conversely, the power may be fed to the input and the output power measured. *with the power off*, the neutralization capacitor C1 is adjusted for minimum feed through, while keeping the output tuning circuit and the input tuning (if used) at the point of maximum response. Since the bridge neutralization circuit is essentially broad band, it will work over a range of frequencies. Usually, it is adjusted at the highest anticipated frequency of operation, where the adjustment is most critical.

BROADBAND TRANSFORMER

Another neutralizing method is shown in **Figure 17.30**, where a broadband transformer provides the needed out of phase signal. C4 is adjusted so that the proper amount of negative feedback is applied to the input to just cancel the feedback via the cathode to plate capacitance. Though many 811A amplifiers have been built without this neutralization, its use makes tuning smoother on the higher bands. This circuit was featured in June 1961 *QST* and then appeared in the *RCA Transmitting Tube Handbook*. Amplifiers featuring this basic circuit are still being manufactured in 2009 and are a popular seller. Many thousands of hams have built such amplifiers as well.

An alternate method of achieving stable operation is to load the grid of a grounded cathode circuit with a low value of resistance. A convenient value is 50 Ω as it provides a match for the driver. This approach reduces the power being fed back to the grid from the output to a low enough level that good stability is achieved. However, the amplifier gain will be much less than without the grid load. Also, unlike the grounded grid circuit, where

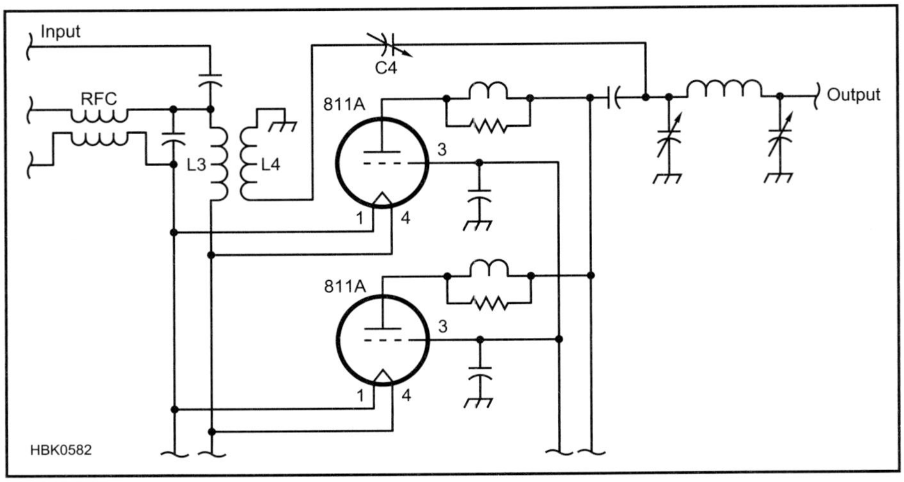

Figure 17.30 — In this neutralizing method, a broadband transformer (L3, L4) provides the needed out-of-phase signal. L3 is 6 turns of #14 wire close wound, ½ inch diameter. L4 is 5 turns of insulated wire over L3. C4 is 6 pF with 0.06-inch spacing. This circuit was originally featured in June 1961 *QST* and is still found in modern amplifiers using 811A tubes.

much of the power applied to the input feeds through to the output, with this "loaded grid" approach, the input power is lost in the load, which must be able to dissipate such power. Distortion may be low in that the driver stage sees a very constant load. In addition, no tuning of the input is required.

17.8.2 VHF and UHF Parasitic Oscillations

RF power amplifier circuits contain parasitic reactances that have the potential to cause so-called parasitic oscillations at frequencies far above the normal operating frequency. Nearly all vacuum-tube amplifiers designed for operation in the 1.8 to 29.7 MHz frequency range exhibit tendencies to oscillate somewhere in the VHF-UHF range — generally between about 75 and 250 MHz depending on the type and size of tube. A typical parasitic resonant circuit is shown in **Figure 17.31**. Stray inductance between the tube plate and the output tuning capacitor forms a high-Q resonant circuit with the tube's C_{OUT}. C_{OUT} normally is much smaller (higher X_C) than any of the other circuit capacitances shown. The tube's C_{IN} and the tuning capacitor C_{TUNE} essentially act as bypass capacitors, while the various chokes and tank inductances shown have high reactances at VHF. Thus, the values of these components have little influence on the parasitic resonant frequency.

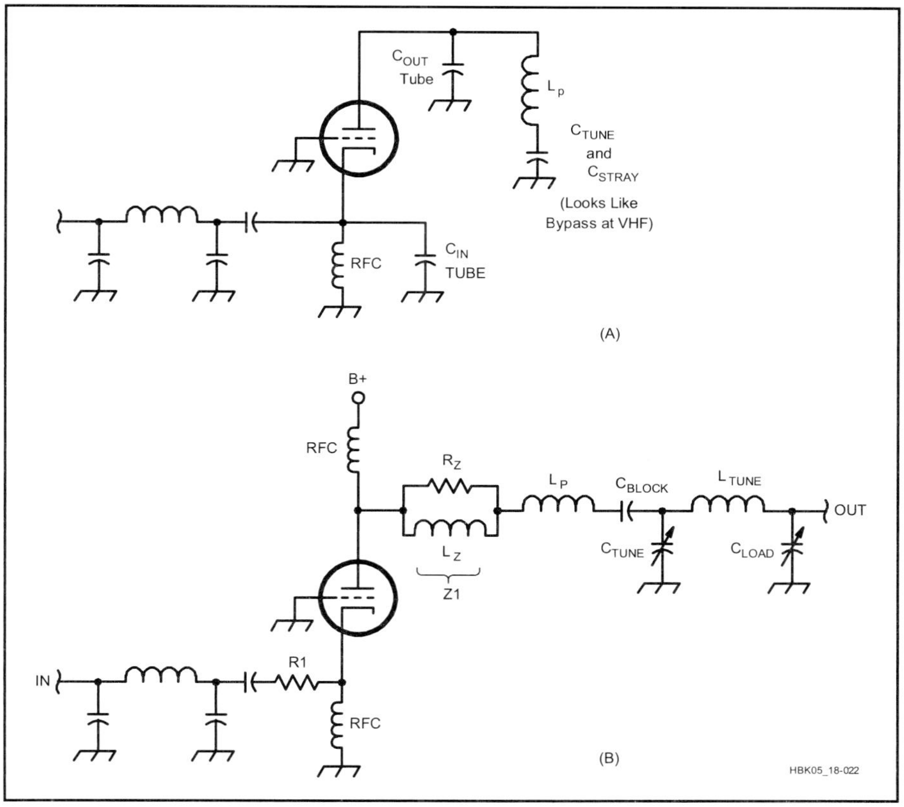

Figure 17.31 — At A, typical VHF/UHF parasitic resonance in plate circuit. The HF tuning inductor in the pi network looks like an RF choke at VHF/UHF. The tube's output capacitance and series stray inductance combine with the pi-network tuning capacitance and stray circuit capacitance to create a VHF/UHF pi network, presenting a very high impedance to the plate, increasing its gain at VHF/UHF. At B, Z1 lowers the Q and therefore gain at parasitic frequency.

Oscillation is possible because the VHF resonant circuit is an inherently high-Q parallel-resonant tank that is not coupled to the external load. The load resistance at the plate is very high and thus, the voltage gain at the parasitic frequency can be quite high, leading to oscillation. The parasitic frequency, f_r, is approximately:

$$f_r = \frac{1000}{2\,\pi\sqrt{L_P C_{OUT}}} \quad (9)$$

where

f_r = parasitic resonant frequency in MHz,

L_P = total stray inductance between tube plate and ground via the plate tuning capacitor (including tube internal plate lead) in µH, and

C_{OUT} = tube output capacitance in pF.

In a well-designed HF amplifier, L_P might be in the area of 0.2 µH and C_{OUT} for an 8877 is about 10 pF. Using these figures, the equation above yields a potential parasitic resonant frequency of

$$f_r = \frac{1000}{2\,\pi\sqrt{0.2 \times 10}} = 112.5 \text{ MHz}$$

For a smaller tube, such as the 3CX800A7 with C_{OUT} of 6 pF, f_r = 145 MHz. Circuit details affect f_r somewhat, but these results do, in fact, correspond closely to actual parasitic oscillations experienced with these tube types. VHF-UHF parasitic oscillations can be prevented (*not* just minimized!) by reducing the loaded Q of the parasitic resonant circuit so that gain at its resonant frequency is insufficient to support oscillation. This is possible with any common tube, and it is especially easy with modern external-anode tubes like the 8877, 3CX800A7 and 4CX800A.

PARASITIC SUPPRESSORS

Z1 of Figure 17.31B is a parasitic suppressor. Its purpose is to add loss to the parasitic circuit and reduce its Q enough to prevent oscillation. This must be accomplished without significantly affecting normal operation. L_z should be just large enough to constitute a significant part of the total parasitic tank inductance (originally represented by L_P), and located right at the tube plate terminal(s). If L_z is made quite lossy, it will reduce the Q of the parasitic circuit as desired.

The inductance and construction of L_z depend substantially on the type of tube used. Popular glass tubes like the 3-500Z and 4-1000A have internal plate leads made of wire. This significantly increases L_P when compared to external-anode tubes. Consequently, L_z for these large glass tubes usually must be larger in order to constitute an adequate portion of the total value of L_P. Typically a coil of 3 to 5 turns of #10 wire, 0.25 to 0.5 inches in diameter and about 0.5 to 1 inches long is sufficient. For the 8877 and similar tubes it usually is convenient to form a "horseshoe" in the strap used to make the plate connection. A "U" about 1-inch wide and 0.75 to 1 inch deep usually is sufficient. In either case, L_z carries the full operating-frequency plate current; at the higher frequencies this often includes a substantial amount of circulating tank current, and L_z must be husky enough to handle it without overheating even at 29 MHz. **Figure 17.32** shows a typical parasitic suppressor.

Regardless of the form of L_z, loss may be introduced as required by shunting L_z with one or more suitable non-inductive resistors. In high-power amplifiers, two composition or metal film resistors, each 100 Ω, 2 W, connected in parallel across L_z usually are adequate. For amplifiers up to perhaps 500 W a single 47 Ω, 2 W resistor may suffice. The resistance and power capability required

Figure 17.32 — Typical parasitic suppressor.

to prevent VHF/UHF parasitic oscillations, while not overheating as a result of normal plate circuit current flow, depend on circuit parameters. Operating-frequency voltage drop across L_Z is greatest at higher frequencies, so it is important to use the minimum necessary value of L_Z in order to minimize power dissipation in R_Z.

The parasitic suppressors described above very often will work without modification, but in some cases it will be necessary to experiment with both L_Z and R_Z to find a suitable combination. Some designers use nichrome or other resistance wire for L_Z.

In exceptionally difficult cases, particularly when using glass tetrodes or pentodes, additional parasitic suppression may be attained by connecting a low value resistor (about 10 to 15 Ω) in series with the tube input, near the tube socket. This is illustrated by R1 of Figure 17.31B. If the tube has a relatively low input impedance, as is typical of grounded-grid amplifiers and some grounded-cathode tubes with large C_{IN}, R1 may dissipate a significant portion of the total drive power.

TESTING TUBE AMPLIFIERS FOR VHF-UHF PARASITIC OSCILLATIONS

Every high-power amplifier should be tested, before being placed in service, to insure that it is free of parasitic oscillations. For this test, nothing is connected to either the RF input or output terminals, and the band switch is first set to the lowest-frequency range. If the input is tuned and can be band switched separately, it should be set to the highest-frequency band. The amplifier control system should provide monitoring for both grid current and plate current, as well as a relay, circuit breaker or fast-acting fuse to quickly shut off high voltage in the event of excessive plate current. To further protect the tube grid, it is a good idea to temporarily insert in series with the grid current return line a resistor of approximately 1000 Ω to prevent grid current from soaring in the event a vigorous parasitic oscillation breaks out during initial testing.

Apply filament and bias voltages to the amplifier, leaving plate voltage off and/or cutoff bias applied until any specified tube warm-up time has elapsed. Then apply the lowest available plate voltage and switch the amplifier to transmit. Some idling plate current should flow. If it does not, it may be necessary to increase plate voltage to normal or to reduce bias so that at least 100 mA or so does flow. Grid current should be zero. Vary the plate tuning capacitor slowly from maximum capacitance to minimum, watching closely for any grid current or change in plate current, either of which would indicate a parasitic oscillation. If a tunable input network is used, its capacitor (the one closest to the tube if a pi circuit) should be varied from one extreme to the other in small increments, tuning the output plate capacitor at each step to search for signs of oscillation. If at any time either the grid or plate current increases to a large value, shut off plate voltage immediately to avoid damage! If moderate grid current or changes in plate current are observed, the frequency of oscillation can be determined by loosely coupling an RF absorption meter or a spectrum analyzer to the plate area. It will then be necessary to experiment with parasitic suppression measures until no signs of oscillation can be detected under any conditions. This process should be repeated using each band switch position.

When no sign of oscillation can be found, increase the plate voltage to its normal operating value and calculate plate dissipation (idling plate current times plate voltage). If dissipation is at least half of, but not more than its maximum safe value, repeat the previous tests. If plate dissipation is much less than half of maximum safe value, it is desirable (but not absolutely essential) to reduce bias until it is. If no sign of oscillation is detected, the temporary grid resistor should be removed and the amplifier is ready for normal operation.

LOW-FREQUENCY PARASITIC OSCILLATIONS

The possibility of self-oscillations at frequencies lower than VHF is significantly lower than in solid state amplifiers. Tube amplifiers will usually operate stably as long as the input-to-output isolation is greater than the stage gain. Proper shielding and dc-power-lead bypassing essentially eliminate feedback paths, except for those through the tube itself.

On rare occasions, tube-type amplifiers will oscillate at frequencies in the range of about 50 to 500 kHz. This is most likely with high-gain tetrodes using shunt feed of dc voltages to both grid and plate through RF chokes. If the resonant frequency of the grid RF choke and its associated coupling capacitor occurs close to that of the plate choke and its blocking capacitor, conditions may support a tuned-plate tuned-grid oscillation. For example, using typical values of 1 mH and 1000 pF, the expected parasitic frequency would be around 160 kHz.

Make sure that there is no low-impedance, low-frequency return path to ground through inductors in the input matching networks in series with the low impedances reflected by a transceiver output transformer. Usually, oscillation can be prevented by changing choke or capacitor values to insure that the input resonant frequency is much lower than that of the output.

17.8.3 Reduction of Distortion

As mentioned previously, a common cause of distortion in amplifiers is over drive (flat topping). The use of automatic level control (ALC) is a practical way of reducing the ill effects of flat topping while still being assured of having a strong signal. This circuit detects the voltage applied to the input of the amplifier. Other circuits are based on detecting the

Overshoot and Overdrive Protection

The ALC and power control circuits of numerous transceivers allow short excess power transients at the beginning of transmission. While tube amplifiers are fairly tolerant of short overloads, solid-state amplifiers are not. To avoid damaging your amplifier input, some kind of protection is necessary. The article "Amplifier Overshoot — Drive Protection" by Phil Salas, AD5X, shows how to use a gas-discharge tube to limit short over-power pulses. It is available as part of the online information for this book.

onset of grid current flow. In either case, when the threshold is reached, the ALC circuit applies a negative voltage to the ALC input of the transceiver and forces it to cut back on the driving power, thus keeping the output power within set limits. Most transceivers also apply an ALC signal from their own output stage, so the ALC signal from the amplifier will add to or work in parallel with that. See **Figure 17.33** for a representative circuit.

Some tube types have inherently lower distortion than others. Selection of a tube specifically designed for linear amplifier service, and operating it within the recommended voltage and current ranges is a good start. In addition, the use of tuned circuits in the input circuits when running class AB2 will help by maintaining a proper load on the driver stages over the entire 360° cycle, rather than letting the load change as the tube begins to draw grid current. Another way to accomplish this is with the "loaded grid," the use of a rather low value of resistance from the grid to cathode. Thus, when grid current flows, the change in impedance is less drastic, having been swamped by the resistive load.

For applications where the highest linearity is desired, operating class A will greatly reduce distortion but at a high cost in efficiency. Some solid state amateur transceivers have provision for such operation. The use of negative feedback is another way of greatly reducing distortion. High efficiency is maintained, but there is a loss of overall gain. Often, two stages of gain are used and the feedback applied around both stages. In this way, gain can be as high as desired, and both stages are compensated for any inherent nonlinearities. Amplifiers using RF negative feedback can achieve values of intermodulation distortion (IMD) as much as 20 dB lower than amplifiers without feedback.

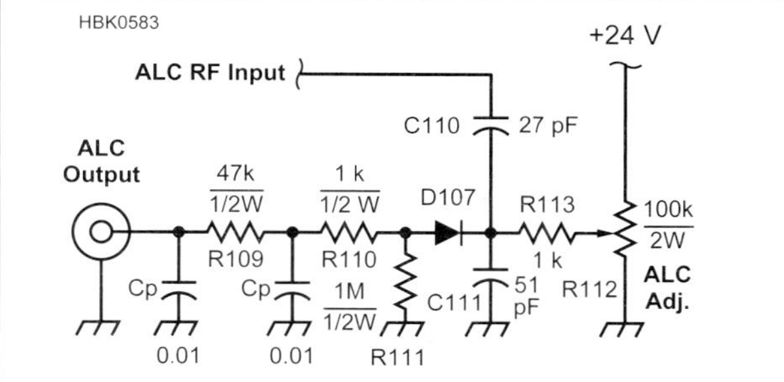

Figure 17.33 — An amplifier ALC circuit can be used to automatically limit the drive power from the transceiver to a safe level.

It must be remembered that distortion tends to be a cumulative problem, with each nonlinear part of the transmission chain adding its part. It is not worthwhile to have a super clean transceiver if it is followed by an amplifier with poor linearity. In the same way, a very good linear will look bad if the transceiver driving it is poor. It is even possible to have a clean signal out of your amplifier, but have it spoiled by a ferrite core tuner inductor or balun that is saturated.

Distortion in a linear amplifier is usually measured with a spectrum analyzer while transmitting a two tone test. If the spectrum analyzer input is overloaded, this can also produce apparent distortion in the amplifier. Reducing the level so that the analyzer is not clipping the input signal is necessary to see the true distortion in the amplifier chain. Use of test equipment for various types of measurements is covered in the **Test Equipment and Measurements** chapter.

17.9 MOSFET Design for RF Amplifiers

There are two general classes of MOSFETs: high and low frequency designs. (See the **Circuits and Components** chapter for MOSFET basics.) The low frequency types are generally optimized for high volume commercial switching applications: computer power supplies, motor controllers, inverters, and so on. They have molded plastic packages, the die are made with aluminum top side metallization, they have maximum junction temperature ratings of 150 to 175°C. Most have polysilicon gate conductors. Polysilicon is easy to manufacture consistently and it's cheap. This works very well for applications up to 200 kHz, but the gate losses start to increase dramatically when they are used at higher frequencies.

A MOSFET gate is essentially a capacitor, but its folded structure is long and skinny. The gate in a 500 W device may be more than a meter long! ("Meter" is not a misprint.) If its conductor material is a lossy material like polysilicon, the gate becomes a long distributed RC network. If an RF signal does make it all the way to the end it will be attenuated and no longer be in phase with the start. This effectively reduces the useful area of the device as frequency increases. It takes a lot of RF current to feed the gate capacitance: $I = CVf$, where C is the gate capacitance, V is the peak gate voltage, and f is the operating frequency. If the gate capacitance is 500 pF and is being driven to 10 V at 30 MHz, the gate current is 150 mA RMS. While the current is directly proportional to the frequency, the power loss is I^2R. If the gate's top conductor is not low loss, it will fail due to I^2R losses at the gate bond pad (the metallization melts) when used at frequencies much higher than it was designed for.

There are two MOSFET manufacturers that use a metal gate instead of polysilicon for their switchmode devices, IXYS and Microsemi. While the die of these devices are quite capable of HF operation, their packaging (usually TO-247 or TO-264) does not provide an optimum HF layout. Because these are aimed at the switchmode market, their drain terminal is on the mounting surface and the source bond-wire length adds gain-killing degeneration at higher frequencies. But on the other hand, they are cheap in terms of cost per watt of dissipation and are acceptable for single-band designs through 20 meters.

When these same metal-gate MOSFET die are placed in packages that are specifically for RF use, the source is often connected to the mounting surface of the package. The source bond wires are thus short, which improves the available gain at all frequencies. This is very convenient because the source is grounded in most RF power amplifier circuits. It also eliminates the need for a mounting insulator, which in turn improves the power dissipation capability.

In MOSFETs specifically designed for RF, the main distinguishing feature is the gate structure. The channels are "shorter" (there is less distance between the gate and source) which reduces the transit time for electrons. As the active area of a device is increased by making the channel "wider," its power dissipation capability is increased. At the same time, the intrinsic (inter-electrode) capacitances also get bigger. A larger device is more difficult to use at higher frequencies because the input impedance (mostly gate capacitance) becomes ever smaller, which makes it harder to drive. In order to solve the gate loss problem mentioned earlier, the long skinny gate is folded into a comb shape. (See **Figure 17.34**.) The gate signal now only has to travel to the end of each finger. The highest frequency designs use multiple combs with shorter fingers. Several of these comb structures are arrayed on the die and are connected in parallel when the die is wire-bonded in the package.

The top metallization for RF parts is either aluminum or gold. Aluminum is cheaper but gold is best because it has a higher operating temperature rating, up to 225 °C, and it is immune to power cycling failures due to its excellent ductility. The downside is that it is more expensive and the devices are much harder to manufacture because gold likes to dissolve into silicon. Its higher temperature rating means you can get more power from a small device, which offsets their higher cost somewhat.

17.9.1 LDMOS versus VDMOS

So far all the devices discussed are vertical MOSFETs, or VDMOS. Their gate and source electrodes are on the top surface of the die and the drain is on the bottom. For RF applications there is another type, LDMOS. This is a lateral device. Here the MOS structure is laid on edge and all the electrodes are on the top side of the die. Vertical p+ source connections are made through the die to make the bottom side of the die a source contact in order to get the optimum "common source" configuration. The channel area is low so the capacitances are smaller, especially the feedback capacitance, C_{GD}. However, the operating voltage capability is also low. There are none rated for more than 50 V operation. The gates are particularly sensitive to ESD and overdrive. However, they have spectacular high frequency capability and gain, and reasonable ruggedness. Your mobile phone would not work without LDMOS technology.

New RF amplifier designs are using LDMOS to replace bipolar transistors, which manufacturers are no longer making. The ability of LDMOS to operate at lower voltages is well suited for 12 V operation and, with its high gain and frequency response, providing 6 meter capability is simple. The downside is that these devices are not as linear as the bipolar transistor they replace.

Bipolar transistors need emitter ballasting resistors in each tiny bipolar cell so they can be paralleled in the die. The resistors also provide negative feedback, which improves linearity. MOSFETs do not require source resistors: paralleled cells will naturally share the load because their ON resistance has a positive temperature coefficient that prevents thermal runaway. As a result, LDMOS amplifiers require more negative feedback to provide comparable IMD performance, which offsets their higher gain advantage.

Regardless of the device type, the packaging is particularly important to an RF device. It must have low parasitics (see the **RF Techniques** chapter) and superior thermal qualities. The package insulator is made of ceramics, beryllia BeO (which is toxic), and/or alumina Al_2O_3, for high temperature capability. The conductors are gold-plated copper or Kovar, and the base flanges are often copper-tungsten or copper-molybdenum. The package is the major determining factor in the cost of an RF part. Parasitic inductance introduced by gate and source bond wires limits the ultimate frequency capability of a VDMOS part. LDMOS parts have gate and drain bond wires. LDMOS devices have an advantage in terms of frequency and package cost because they are free of the gain degen-

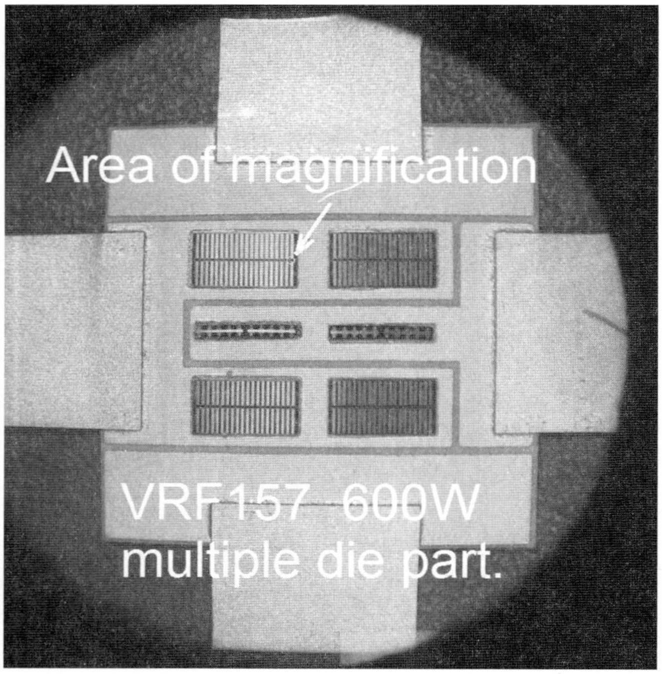

(A)

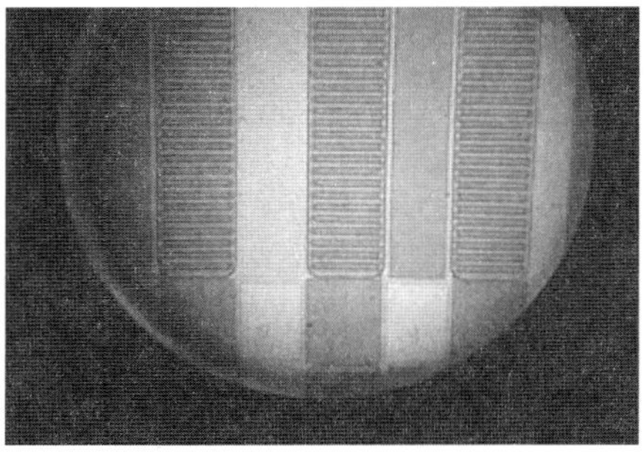

(B)

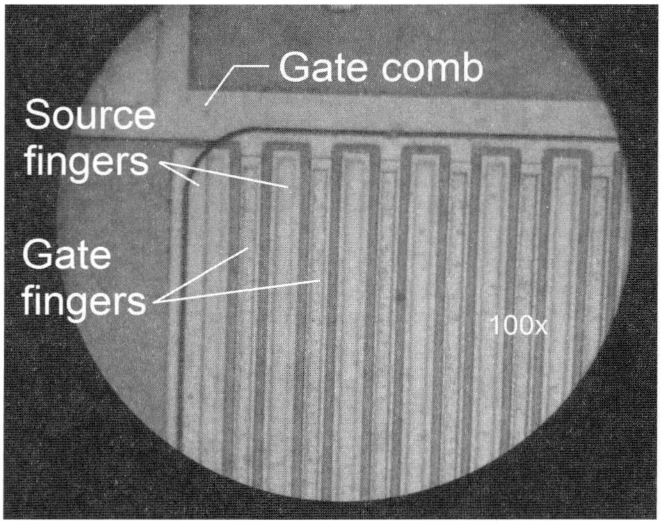

(C)

Figure 17.34 — A shows the layout of a multiple-die RF MOSFET (VRF157). B illustrates the comb structure of the gate. C is a closeup of the gate showing the interleaved source and gate finger structure. [Dick Frey, K4XU, photos]

eration caused by source wire inductance and their die may be soldered directly to a metal mounting flange.

17.9.2 Designing Amplifiers with MOSFETs

Designing an amplifier requires a systems approach. You will need to consider how much power supply is required, as well as the cooling and control systems needed to keep it happy. If you have the transistors and want to build them into an amplifier, the design procedure is a little different. The place to start in any case is with its transistor's data sheet. This will show the voltage and power capabilities, and from these the circuit requirements can be calculated and the cooling system defined.

VOLTAGE RATINGS

Designing an amplifier with MOSFETs requires knowledge of the part being used. Generally, the cheap plastic-packaged switch-mode parts will be best for single-band operation. Switchmode parts are rated by their drain breakdown voltage (BV_{DSS}), ON resistance (R_{DS}), and power dissipation. They are available in voltage ratings from as little as 5 V to over 1200 V. RF parts are sold by operating voltage, V_{DD}, power dissipation and frequency capability.

Select the part to suit your power supply requirements. A 500 V MOSFET will not work on a 12 V supply and will barely work at 50 V. This is because the MOSFET's intrinsic capacitances are higher at low voltage. Between the drain and source of every MOSFET is a parasitic "body diode" as shown in **Figure 17.35**. It's too slow to rectify RF, but like any diode, its capacitance changes with reverse bias voltage. This relationship is always given in the device's C-V curves on its data sheet. (See **Figure 17.36**.) Data sheets can be found on manufacturer or distributor websites or perform an Internet search for the part number and "data sheet."

In class AB operation, a MOSFET works best when operated at a little less than one-half of its rated breakdown voltage, BV_{DSS}. The drain voltage will swing up to 2 or even 3.562 times (for class E) the supply voltage. Under high VSWR, the drain voltage can be somewhat higher still. The RF voltage breakdown of a MOSFET is typically 20% higher than its data sheet value but is hard to specify reliability, so RF devices are rated by their dc operating voltages rather than BV_{DSS}. This takes into account the requirement for operating overhead.

RF parts are usually rated at a specific operating voltage such as 13.5 V, 28 V or 50 V. Originally these were common battery voltages for civilian and military vehicles and the tradition persists. The devices are optimized for their operating voltage. Choosing the operating voltage is a matter of considering many different parameters, not just the breakdown voltage.

THERMAL DESIGN

Suffice it to say that the thermal design of a high power transistor PA is often as challenging as the electrical design. It can be done "by the numbers" but the tricky part is making it fit into the available space. A thermal design example and an Advanced Power Technology application note by the author are provided with this book's online content.

A word of caution is in order: tubes used in power amplifiers have a great deal of "headroom" in their specifications and are quite forgiving of momentary operator errors. RF power transistors, because they are more expensive in terms of cost per watt, are specified much closer to their limits. These limits must be observed at all times. Even though the data sheet says the device can do X watts, the designer must observe the requirements for proper cooling in order to reach this level in practice. In addition, manufacturers rate the power dissipation in theoretical terms. You will be lucky to achieve half of it.

As with tubes, there are CW ratings and SSB ratings. For transistors, the ratings are based on average power. The difference is simply the size of the heat sink required, as the peak power is the same for each. Each transistor has a thermal rating expressed as $R_{\theta JC}$, the thermal resistance from the transistor junction to the bottom of its case. Since the device has an upper junction temperature limit, somewhere between 150 and 200 °C, the power dissipation is determined by the difference between the junction and the case temperature:

$$P_d = (T_J - T_C) / R_{\theta JC}$$

where P_d is the available power dissipation, T_J is junction and T_C is case temperature. What this says is that without any cooling, the transistor's case will be almost the same as the junction temperature so its power dissipation capability is nearly zero. When placed on a heat sink, the case will be cooler and it then has power dissipation capability. It follows that the better the heat sink, the more power can be dissipated by the transistor.

There is another thermal consideration: the thermal resistance between the case of the transistor and the heat sink, $R_{\theta CS}$. Even if the base of the transistor and the heat sink are flat and smooth, microscopic air gaps still exist. These do not conduct heat and so reduce the net effectiveness of the heat sink. The solution is to use a thermal interface compound or *thermal grease*. The simplest and best is silicone oil loaded with zinc oxide powder. The oil does most of the work: the powder thickens it to a paste so it doesn't run out of the joint. It is applied as a very thin coat between the heat sink and device. When using thermal grease, always wiggle the transistor around on the sink to insure a minimum of grease between the part and the sink. Remember, thermal grease is not a good thermal conductor, it's just much better than air. Use it sparingly!

Most commercial high power broadcast amplifiers are cooled with circulated water,

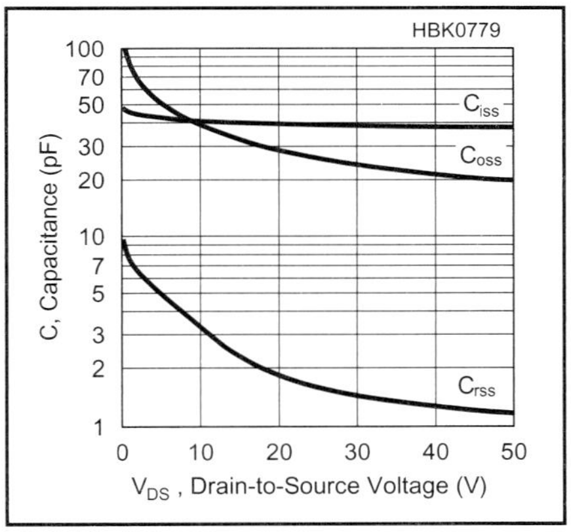

Figure 17.36 — The capacitance versus voltage (C-V) curves for the Microsemi VRF151 RF MOSFET. (Illustration courtesy Microsemi Corp.)

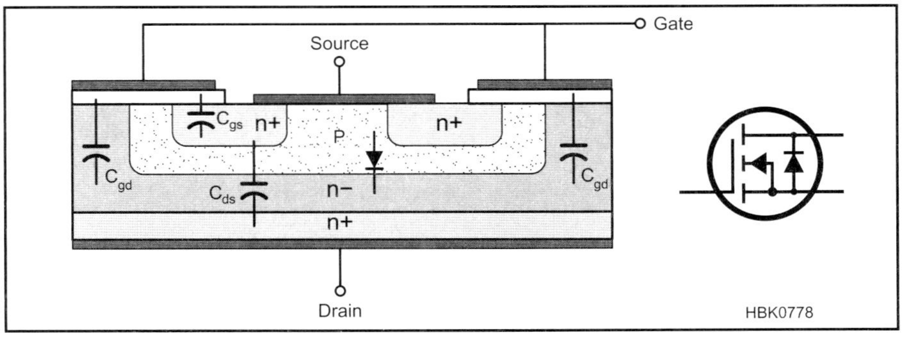

Figure 17.35 — All MOSFETs have parasitic capacitances as shown and a body diode between the drain and source in the cross-section and schematic symbol. The diode is shown for an N-channel device. It is reversed for P-channel devices.

or a water-glycol mix if the minimum ambient temperature will be below 0 °C. While it has yet to be introduced to the amateur market, liquid cooling has great potential. The advances in plastic fittings, small pumps and heat exchangers driven by the high-performance computer market have great potential benefits to the amateur high power amplifier. Water is more than four times better than air for absorbing and moving heat. **Figure 17.37** illustrates both open- and closed-loop cooling systems. But regardless of where it is moved to, the heat must still be dissipated. It can warm the air in the shack, heat the rest of the house, or warm the septic tank.

17.9.3 The Transistor Data Sheet

Regardless of manufacturer, all data sheets contain the same basic information. The following should help make sense of what can be very confusing to a first time user. Transistor specifications rely heavily on several ideal conditions that cannot happen in practice but since it is a common practice by all manufacturers, the numbers are very useful for comparing different devices. As long as you have the part number (and it is not a custom part), the corresponding data sheet can be found easily by searching for the part number and "data sheet."

The Microsemi VRF151 N-channel enhancement-mode VDMOS transistor will be used as an example. It is used in the 250 W broadband amplifier project presented in this chapter. The following discussion assumes the reader has obtained a data sheet for this part (see the company website at **www.microsemi.com**) and can refer to it.

MINIMUM, TYPICAL AND MAXIMUM

All of the specification parameters which the manufacturer guarantees are subject to either a minimum value or a maximum value. This is the worst case. As in an automobile, there is a maximum safe stopping distance and a minimum gas tank capacity. Most parameters also have a *typical* value that is generally representative of typical performance than the specified minimum or maximum. Some quantities have both upper and lower limits. Every parameter has specific test conditions under which it is measured.

ABSOLUTE MAXIMUM RATINGS

These are all dc ratings, easily verified with a variable power supply and a multimeter. If the manufacturer finds that any of these have been exceeded, any warranty claims are voided.

V_{DSS} is the maximum drain to source voltage rating, with the gate is shorted to the source. Think of the device as a high voltage Zener diode. As soon as it draws any current, power is dissipated, and temperature rises very quickly. Damage occurs either from puncturing through the junction or by arcing over the edge of the die.

Maximum drain current, I_D, is the current that will cause the device to dissipate its maximum rated power when fully turned on. Every device has an ON resistance called $R_{DS(ON)}$. The power dissipated is due to $I_D^2 \times 2.5\ R_{DS(ON)}$. The factor of 2.5 accounts for $R_{DS(ON)}$ having a positive temperature coefficient which causes it to roughly double by the point at which the junction temperature is 200 °C.

V_{GS} is the maximum gate to source voltage. The gate is essentially a capacitor, with a SiO_2 dielectric perhaps 400 to 1000 Angstroms (10^{-10} m) thick. The limit is lower on LDMOS than VDMOS, and cannot be exceeded without destroying the device. Because LDMOS devices have much lower V_{GS} ratings and thinner dielectrics, some LDMOS manufacturers build in diode protection to make them less susceptible to electrostatic discharge, ESD.

P_D is the maximum power dissipation of the device under theoretical conditions: The bottom of the case at 25 °C and the junction at its maximum temperature, T_{jmax}. If not stated directly, the thermal resistance $R_{\theta JC}$ is equal to $P_D / (T_{jmax} - 25°C)$.

Storage temperature is straightforward. Maximum T_J is the junction temperature above which things start to come unsoldered, or the reliability seriously impaired, or smoke emitted.

STATIC ELECTRICAL CHARACTERISTICS

V_{DSS} is specified again, this time showing the measurement conditions, the guaranteed minimum, and the typical production values.

$V_{DS(ON)}$ or sometimes $R_{DS(ON)}$ is the minimum resistance between drain and source that is obtained when the device is fully ON and carrying half the rated current. It is more commonly specified on switchmode parts, but it is of particular importance in high-efficiency saturated modes like class D and E because it limits the maximum obtainable efficiency.

I_{DSS} is the maximum drain current flowing with the gate shorted to the source. In an N-channel enhancement device one must apply a positive voltage to the gate to turn it on. The VRF151 will not conduct more than 1 mA at 100 V by itself without gate bias. This is a leakage current. In a perfect device it is zero.

Similarly, I_{GSS} is gate leakage current with the $V_{DS} = 0$. It represents a resistor in parallel with the gate capacitor. In this case, 1 µA at 20 V is 20 MΩ. This does not sound like much but if the drain has voltage on it and there is no resistor across the gate, this leakage current will cause the gate to charge from the drain, eventually turning the device fully ON with bad consequences.

Forward transconductance, g_{fs}, is the dc gain of the device expressed in terms of change in drain current per change in gate volts measured at a particular drain current. It has only a mild relationship with RF gain and too high a g_{fs} can cause bias instability and/or parasitics.

V_{TH} is the gate threshold specification. When V_{DS} is 10 V, V_{GS} of no less than 2.9 V and no more than 4.4 V will cause 100 mA of drain current to flow. A typical Class AB quiescent bias condition is 100 mA. Of all the parameters, this one has the widest window. Enough variation exists so that manufacturers sort parts into "bins" of values across the range, and assign letter codes to each which are marked on the package. This allows one to make matched pairs within a device type.

THERMAL CHARACTERISTICS

$R_{\Theta JC}$ is equal to $P_{Dmax} / (T_{j\,max} - 25°C)$. Specifying the first two parameters ($R_{\Theta JC}$ and P_{Dmax}) generates the third ($T_{j\,max}$). In this

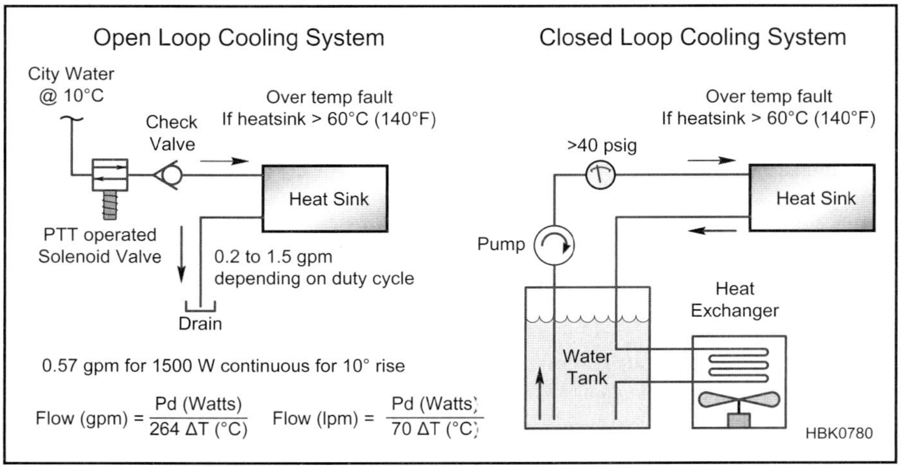

Figure 17.37 — Open- and closed-loop water cooling systems for solid-state amplifiers.

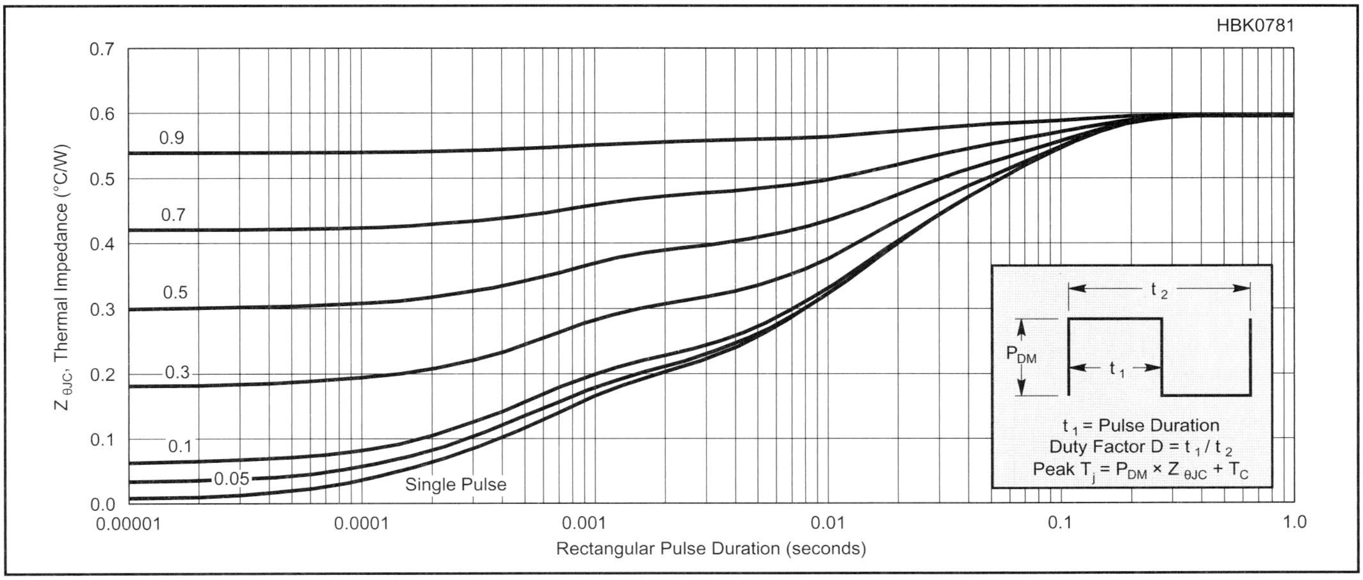

Figure 17.38 — Dynamic thermal impedance for the Microsemi VRF151 RF MOSFET. (Illustration courtesy Microsemi Corp.)

sense specifying $T_{j\ max}$ is redundant but it is often reiterated for those who are looking for the particular parameter.

Because thermal performance is just as important as RF performance, much of the information in the data sheet is concerned with it. In addition to the static thermal impedance, $R_{\Theta JC}$, there is a dynamic thermal impedance $Z_{\Theta JC}$. Transistor packages have thermal mass so the die temperature does not change instantaneously. For pulsed operation, the effective $R_{\Theta JC}$, can be much lower than it is for steady-state operation.

The dynamic thermal impedance is shown in **Figure 17.38** (Figure 5 of the VRF151 data sheet). It shows how the device can be used for pulse operation at higher power than it can on CW because the effective thermal impedance is lower for short pulses. This has some application to SSB, but there are other constraints on peak power that boil down to just allowing a smaller heat sink when used only for intermittent service — a bad practice for amateur amplifiers!

DYNAMIC CHARACTERISTICS

Somewhat of a misnomer, these are the parasitic capacitances between the gate, drain and gate. Except for the gate capacitance which is a fixed value based on device dimensions, the other two are a function of the drain voltage, just like varactors (voltage-variable capacitors). Because it is rather difficult to measure these parameters, the parameters are defined in a common-source configuration. C_{ISS} is the gate-to-source capacitance with the drain ac-shorted to the source. It is actually C_{GS} in parallel with C_{GD}. The three parameters are usually measured at the specified drain supply voltage with V_{GS} at zero. They are also usually displayed as in Figure 17.36 (Figure 3 of the data sheet), a graph of C vs drain voltage.

This voltage-varying capacitance is one of the causes of IMD in a transistor. Its effect is to impart some phase modulation (PM) on the signal. PM can be observed as unequal IMD products on each side of the carrier pair in a two-tone test. PM distortion generates pairs of odd-order carriers like AM distortion but the high side carriers are –180° out of phase with the AM products. This causes them to reduce the level of amplitude products on the high side of the carrier pair and increases them on the lower side. (See the reference list entry for Sabin and Schoenike, *Single Sideband Systems and Circuits*.)

TRANSFER CHARACTERISTICS

As discussed above, g_{fs} describes the gain of a MOSFET as the change in drain current per change in gate voltage: $\Delta I_D / \Delta V_{GS}$. This means that g_{fs} is the slope of the transfer curve. The transfer curve for the VRF151 is shown in **Figure 17.39** (the data sheet's Figure 2). There are three curves in this graph, depicting the transfer characteristic at three different temperatures.

The three curves cross each other at the "thermal neutral point." This is where the temperature coefficient of V_{TH} changes from negative to positive and it is usually at a current much higher than the part normally operates. This explains why MOSFETs must have thermally compensated gate bias. Below the crossover point, where the part would be biased for class AB, the temperature coefficient of V_{TH} is negative. This means the gate voltage required for a given current goes down as the part heats up. Without thermal bias compensation we can have thermal runaway. The part will usually melt before it reaches the crossover point.

The transfer curve also shows that the gain of the device is not constant. It is quite low at low current and increases to a nominal value over its linear range and then the curves flatten out at higher current as the part saturates. This demonstrates why very low distortion amplifiers are usually operated in class A.

FUNCTIONAL CHARACTERISTICS

This section is where the RF performance is specified — how much gain at what frequency, IMD performance, and ruggedness. While gain and IMD are easily understood, ruggedness is more difficult and it is very

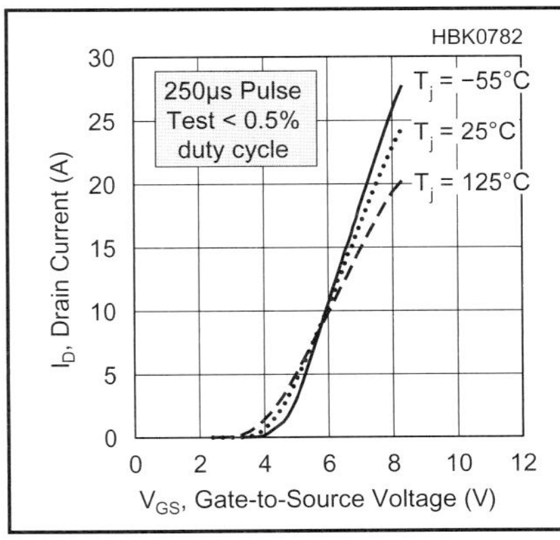

Figure 17.39 — Transfer characteristics for the Microsemi VRF151 RF MOSFET. (Illustration courtesy Microsemi Corp.)

poorly defined by most manufacturers. The VRF151 has a ruggedness specification at all phase angles of a 30:1 VSWR when putting out 150 W PEP at 30 MHz. The test circuit is shown on page 4 in the data sheet. Nothing is said about how long the test takes or how well the device is cooled. If the test time and cooling conditions are not given, the specification is inadequate to guarantee that a design will remain within the device power limits. In the author's experience, 90% of all VSWR-related failures are due to over-dissipation. This says that limiting the amplifier's drain current is a simple and effective means for VSWR protection.

DATA SHEET EXTRAS

Test circuits are usually provided so customers can duplicate the test conditions for gain, IMD and output power. Sometimes circuit layouts and complete part lists are also provided. Note however, that while most high power parts are usually used in push-pull circuits, parts in data sheets are always tested one at a time in single-ended circuits. The last page of the data sheet gives the mechanical outline and sometimes mounting information. The VRF151 is "binned" (sorted into similarly performing groups) for V_{TH} as it exits the final testing and the bin letter code is marked on the package. This allows very close matching of gate threshold which is important when used in a push-pull circuit with a common bias supply for both parts. Most designers use separate bias adjustments regardless of matching, but it insures a measure of gain matching also, especially if the parts are from the same lot date code.

17.9.4 Summary Observations

MOSFETs are not perfect devices. g_{fs}, V_{TH}, BV and R_{DS} are all affected by die temperature. g_{fs} goes down with increasing die temperature. While this might cause the output power to sag a bit as the amplifier gets hotter, it is generally a benefit. MOSFETs can be paralleled without requiring source resistors because as the one carrying more current heats up, it loses gain and its resistance goes up, allowing the others to share the load. R_{DS} rises with temperature, a useful trait in hard-switching applications using paralleled devices. Breakdown voltage, BV_{DSS}, increases with temperature and so is usually not a concern in a part that heats up as it is being used. It has implications in cases of extreme cold, as in satellites.

One caveat when paralleling MOSFETs: always be mindful that they will easily oscillate at UHF. The device's inter-electrode capacitances and bonding wire inductances conspire to form a cross-coupled multivibrator. The solution is to place a small resistor in series with the gate leads, 3.3 Ω will usually suffice. This lowers the circuit Q enough to prevent oscillation from starting. [ref: Motorola *Engineering Bulletin 104*, et al.]

17.10 Solid State RF Amplifiers

17.10.1 Solid State vs Vacuum Tubes

Solid state amplifiers have become the norm in transceivers, but their use in external high power amplifiers has not. The primary reason is economic. It is more expensive to generate a kilowatt or more with transistors because they are smaller and have less dissipation capability. This is changing as the broadcast and industrial RF industry converts to solid state, making RF power transistors available for amateur use at lower prices. A number of legal-limit solid state Amateur Radio amplifiers are available.

17.10.2 Classes of Operation

This topic applies to transistors as well as tubes, and it was covered earlier in this chapter and also in the **RF Techniques** chapter. In communications amplifiers, Class A is used mainly for driver stages where linearity is desired and efficiency is not a concern.

Class B is usually passed over in favor of the more linear Class AB. Class AB offers RF amplifiers increased linearity, mainly in less crossover distortion, for a very small (perhaps 1% or 2%) reduction in efficiency. It is the most commonly used class of operation for linear power amplifiers that must cover a wide range of frequencies. Broadband solid state Class AB amplifiers typically achieve 50 to 60% efficiency.

Class C is used where efficiency is important and linearity and bandwidth or harmonics are not. FM transmitters are the most common application in communications. Single-band tuned amplifiers can be as much as 80% efficient. However, in a single-ended amplifier they require a tank circuit. Class C amplifiers are *not* suitable for on-off keyed modes like CW without extensive pre-distortion of the driving signal in order to prevent key clicks.

Class D and E are most efficient, up to 95% in practical circuit applications. But both of these both require a narrow band tuned tank circuit to achieve this efficiency. They are not linear; their output is essentially either on or off. They can be used quite effectively for linear amplification by the process of EER (envelope elimination and restoration) but it is always in a single-band circuit. On-off keying can be employed if the power supply is keyed with a properly shaped envelope. EER is difficult to do well and requires very complex circuitry. Without careful system design, EER results in poor SSB performance.

17.10.3 Modeling Transistors

The design method using performance curves that was detailed earlier in this chapter is more applicable to vacuum tube amplifier design than solid state. The most common method used with solid state is electronic design analysis (EDA, also called computer-aided design, or CAD) using electronic models. *SPICE* or S-parameter models are available for some high power transistors, and simple amplifiers can be readily designed with the aid of an appropriate analysis program. (See the **Computer Modeling and Simulation** chapter for more information on *SPICE* and related modeling techniques.)

Full-featured circuit design and analysis programs are expensive, and the resulting designs are only as good as the accuracy of the transistor models they use. A complicating factor is that any design relies heavily on models for all the passive components in the circuit. While passive part models can be obtained for some commercial components, many others — such as ferrite-loaded transformers — must also be designed before the circuit can be modeled. It is not unusual for the electronic design to take much longer and cost more than the benefits are worth.

As detailed in the **Computer Modeling and Simulation** chapter, many of the EDA vendors offer free or inexpensive "student versions" of their products. These are fully capable up to a certain level of circuit complexity. Although they usually are not big enough to analyze a whole amplifier, student versions are still particularly useful for looking at parts of the whole.

Electronic design is very useful for getting the circuit design "in the ballpark." The design will be close enough that it will work when built, and any necessary fine tuning can be done easily once it is constructed. Computer modeling is very useful for evaluating the stresses on the circuit's passive components so they can be properly sized. Another very helpful use of CAD is in the development of the output filters. *SVC Filter Designer* by Jim Tonne, W4ENE, available with this book's online content, is exceptional in this regard.

17.10.4 Impedance Transformation — "Matching Networks"

Aside from the supply voltage, there is little difference between the operation of a tube amplifier and a transistor amplifier. Each amplifies the input signal, and each will only work into a specific load impedance. In a tube amplifier, the proper plate load impedance is provided by an adjustable pi or pi-L plate tuning network, which also transforms the impedance down to 50 Ω.

A single-transistor amplifier can be made in the same way, and in fact most single-band VHF amplifier "bricks" are. A tuned matching network provides the proper load impedance for the transistor and transforms it up to 50 Ω. The major difference is that the proper load impedance for a transistor, at any reasonable amount of power, is much *lower* than 50 Ω. For vacuum tubes it is much *higher* than 50 Ω.

BROADBAND TRANSFORMERS

Broadband transformers are often used in matching to the input impedance or optimum load impedance in a power amplifier. Unlike the tuned matching circuits, transformers can provide constant impedance transformation over a wide range of frequency without tuning. Multi-octave power amplifier performance can be achieved by appropriate application of these transformers. The input and output transformers are two of the most critical components in a broadband amplifier. Amplifier efficiency, gain flatness, input SWR, and even linearity all are affected by transformer design and application.

There are two basic RF transformer types, as described in the **RF Techniques** and **Transmission Lines** chapters: the conventional transformer and the transmission line transformer. More information on RF transformers is included with this book's online content.

The conventional transformer is wound much the same way as a power transformer. Primary and secondary windings are wound around a high-permeability core, usually made from a ferrite or powdered-iron material. Coupling between the secondary and primary is made as tight as possible to minimize leakage inductance. At low frequencies, the coupling between windings is predominantly magnetic. As the frequency rises, core permeability decreases and leakage inductance increases; transformer losses increase as well.

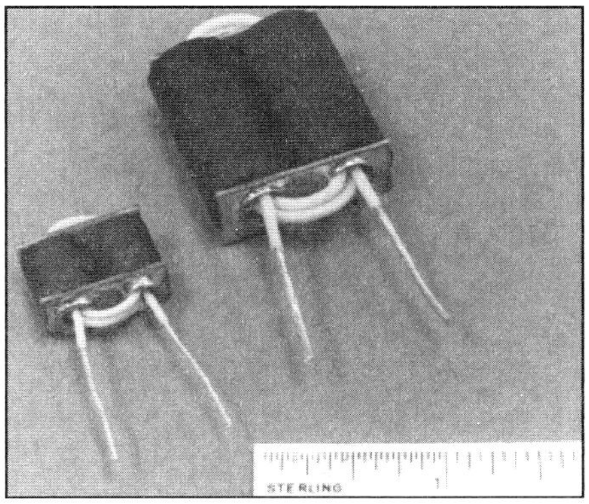

(A)

(B)

Figure 17.40 — The two methods of constructing the transformers outlined in the text. At A, the one-turn loop is made from brass tubing; at B, a piece of coaxial cable braid is used for the loop.

Typical examples of conventional transformers are shown in **Figure 17.40**. In Figure 17.40A, the primary winding consists of brass or copper tubes inserted into ferrite sleeves. The tubes are shorted together at one end by a piece of copper-clad circuit board material, forming a single turn loop. The secondary winding is threaded through the tubes. Since the low-impedance winding is only a single turn, the impedance transformation ratio is limited to the squares of integers — 1, 4, 9, 16 and so on.

The lowest effective transformer frequency is determined by the inductance of the one-turn winding. It should have a reactance, at the lowest frequency of intended operation, at least four times greater than the impedance it is connected to. The coupling coefficient between the two windings is a function of the primary tube diameter and its length, and the diameter and insulation thickness of the wire used in the high-impedance winding. High impedance ratios, greater than 36:1, should use large-diameter secondary windings. Miniature coaxial cable (using only the braid as the conductor) works well. Another use for coaxial cable braid is illustrated in Figure 17.40B. Instead of using tubing for the primary winding, the secondary winding is threaded through copper braid. Because of the increased coupling between the primary and secondary of the transformer made with multiple pieces of coax, leakage reactance is reduced and bandwidth performance is increased.

The cores used must be large enough so the core material will not saturate at the power level applied to the transformer. Core saturation can cause permanent changes to the core permeability, as well as overheating. Transformer nonlinearity also develops at core saturation. Harmonics and other distortion products are produced — clearly an undesirable situation. Multiple cores can be used to increase the power capabilities of the transformer.

Transmission line transformers are similar to conventional transformers, but can be used over wider frequency ranges. In a conventional transformer, high-frequency performance deterioration is caused primarily by leakage inductance, the reactance of which rises with frequency. In a transmission line transformer, the windings are arranged so there is tight capacitive coupling between the two. A high coupling coefficient is maintained up to considerably higher frequencies than with conventional transformers.

The upper frequency limit of the transmission line transformer is limited by the length of the lines. As the lines approach ¼ wavelength, they start to exhibit resonant line effects and the transformer action becomes erratic.

MATCHING NETWORKS AND TRANSFORMERS

The typical tube amplifier tank circuit is an impedance transforming network in a pi or pi-L configuration. With reasonable loaded Q, it also functions as a low-pass filter to reduce the output signal harmonic levels below FCC minimums.

If a transistor amplifier uses a broadband transformer, it must be followed by a separate low-pass filter to achieve FCC harmonic suppression compliance. This is one reason broadband transistor amplifiers are operated in push-pull pairs. The balance between the circuit halves naturally discriminates against even harmonics, making the filtering job easier, especially for the second harmonic. The push-pull configuration provides double the power output when using two transistors, with very little increase in circuit complexity or component count. Push-pull pairs are also easier to match. The input and output impedance of a push-pull stage is twice that of a single-ended stage. The impedance is low, and raising it usually makes the matching task easier.

The transistor's low-impedance operation provides the opportunity to use a simple broadband transformer to provide the transformation needed from the transistor's load impedance up to 50 Ω. This low impedance also swamps out the effects of the device's output capacitance and, with some ferrite loading on the transformer, the amplifier can be made to operate over a very wide bandwidth without adjustment. This is not possible with tubes.

On the other hand, a tube amplifier with its variable output network can be adjusted for the actual output load impedance. The transistor amplifier with its fixed output network cannot be adjusted and is therefore much less forgiving of load variations away from 50 Ω. Circuits are needed to protect the transistor from damage caused by mismatched loads. These protection circuits generally operate "behind the scenes" without any operator intervention. They are essential for the survival of any transistor amplifier operating in the real world.

CALCULATING PROPER LOAD IMPEDANCE

The proper load impedance for a single transistor (or a tube for that matter) is defined by $R = E^2/P$. Converting this from RMS to peak voltage, the formula changes to $R = E^2/2P$. If two devices are used in push-pull (with twice the power and twice the impedance) the formula becomes $R = 2E^2/P$.

There is a constraint. Transformer impedance ratios are the square of their turns ratios. Those with single turn primaries are limited to integer values of 1, 4, 9, 16 and so on. Real-world transformers quickly lose their bandwidth at ratios larger than 25:1 due to stray capacitance and leakage inductance. A design solution must be found which uses one of these ratios. We will use the ubiquitous 100 W, 12 V transceiver power amplifier as an example. Using the push-pull formula, the required load impedance is $2 \times 12.5^2/100 = 3.125\ \Omega$. The required transformer impedance ratio is $50/3.125 = 16$, which is provided by a turns ratio of 4:1.

Transistor manufacturers, recognizing the broadband transformer constraint, have developed devices that operate effectively using automotive and military battery voltages and practical transformer ratios: 65 W devices for 12 V operation and 150 W devices for 48 V. Bigger 50 V devices have been designed (for example, the MRF154) that will put out 600 W, but practical transformer constraints limit 50 V push-pull output power to 900 W. There are higher voltage devices developed for the industrial markets and they are gradually finding their way into amateur designs. Being able to adjust the impedance to a convenient value for a common transformer turns ratio by adjusting the operating voltage is a powerful design option. These device and transformer limitations on output power can also be overcome by combining the outputs of several PA modules.

17.10.5 Combiners and Splitters

With some exceptions, practical solid state amplifiers have an upper power limit of about 500 W. This is a constraint imposed by the available devices and, to some extent, the ability to cool them. As devices are made more powerful by increasing the area of silicon die, the power density can become so high that only water cooling can provide adequate heat removal. Large devices also have large parasitic capacitances that make securing a broadband match over several octaves very difficult. By building a basic amplifier cell or "brick" and then combining several cells together, transmitter output powers are only limited by the complexity. Combiners and splitters have losses and add cost, so there are practical limits.

Broadband combiners usually take the form of an N-way 0° hybrid followed by an N:1 impedance transformer. The square ratio rule applies here too because the output impedance of a broadband combiner with N input ports is 50/N Ω — so combiners are

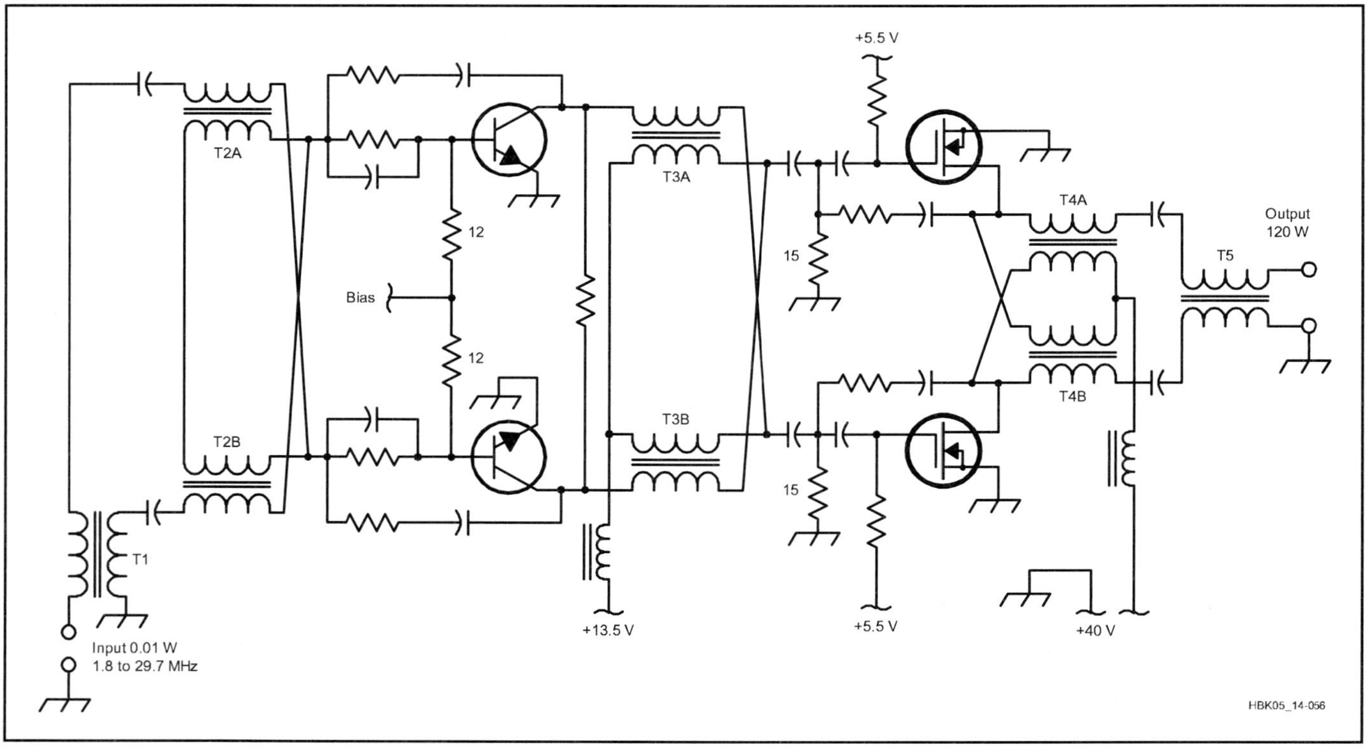

Figure 17.41 — Typical use of transmission line transformers as baluns and combiners in solid state power amplifiers.

usually 2, 4, 9 and 16-way. The higher the ratio, the lower the bandwidth will be.

Many types of combiners have been developed. The most common is the 4-way. It is easy to construct and has very good bandwidth. Most of the commercial "1 kW" broadband amplifiers use a 4-way combiner to sum the output of four 300 W push-pull modules operating on 48 V. Every combiner has loss. It may only be a few percent, but this represents a considerable amount of heat and loss of efficiency for a kilowatt output. This is a case where 4 × 300 does not make 1200.

The combiner approach to make a 1 kW solid state amplifier uses a large number of individual parts. A comparable 1 kW tube amplifier requires relatively few. This makes the high-powered solid state unit more expensive and potentially less reliable.

There is an alternative. If we had higher voltage transistors, we could use the same output transformer network configuration to get more output because power rises with the square of the operating voltage. There are high voltage transistors that can operate on 200 V or more. The problem is that these transistors must be capable of handling the corresponding higher power dissipation. The downside of making bigger, more powerful devices is an increase in parasitic capacitance. These bigger transistors become harder to drive and to match over broad bandwidths. However, the circuit simplicity and elimination of the combiner and its losses makes the higher voltage approach quite attractive.

TRANSMISSION LINE TRANSFORMERS AND COMBINERS

Figure 17.41 illustrates, in skeleton form, how transmission-line transformers can be used in a push-pull solid state power amplifier. The idea is to maintain highly balanced stages so that each transistor shares equally in the amplification in each stage. The balance also minimizes even-order harmonics so that low-pass filtering of the output is made much easier. In the diagram, T1 and T5 are current (choke) baluns that convert a grounded connection at one end to a balanced (floating) connection at the other end, with a high impedance to ground at both wires. T2 transforms the 50 Ω generator to the 12.5 Ω (4:1 impedance) input impedance of the first stage. T3 performs a similar step-down transformation from the collectors of the first stage to the gates of the second stage. The MOSFETs require a low impedance from gate to ground. The drains of the output stage require an impedance step up from 12.5 Ω to 50 Ω, performed by T4. Note how the choke baluns and the transformers collaborate to maintain a high degree of balance throughout the amplifier. Note also the various feedback and loading networks that help keep the amplifier frequency response flat.

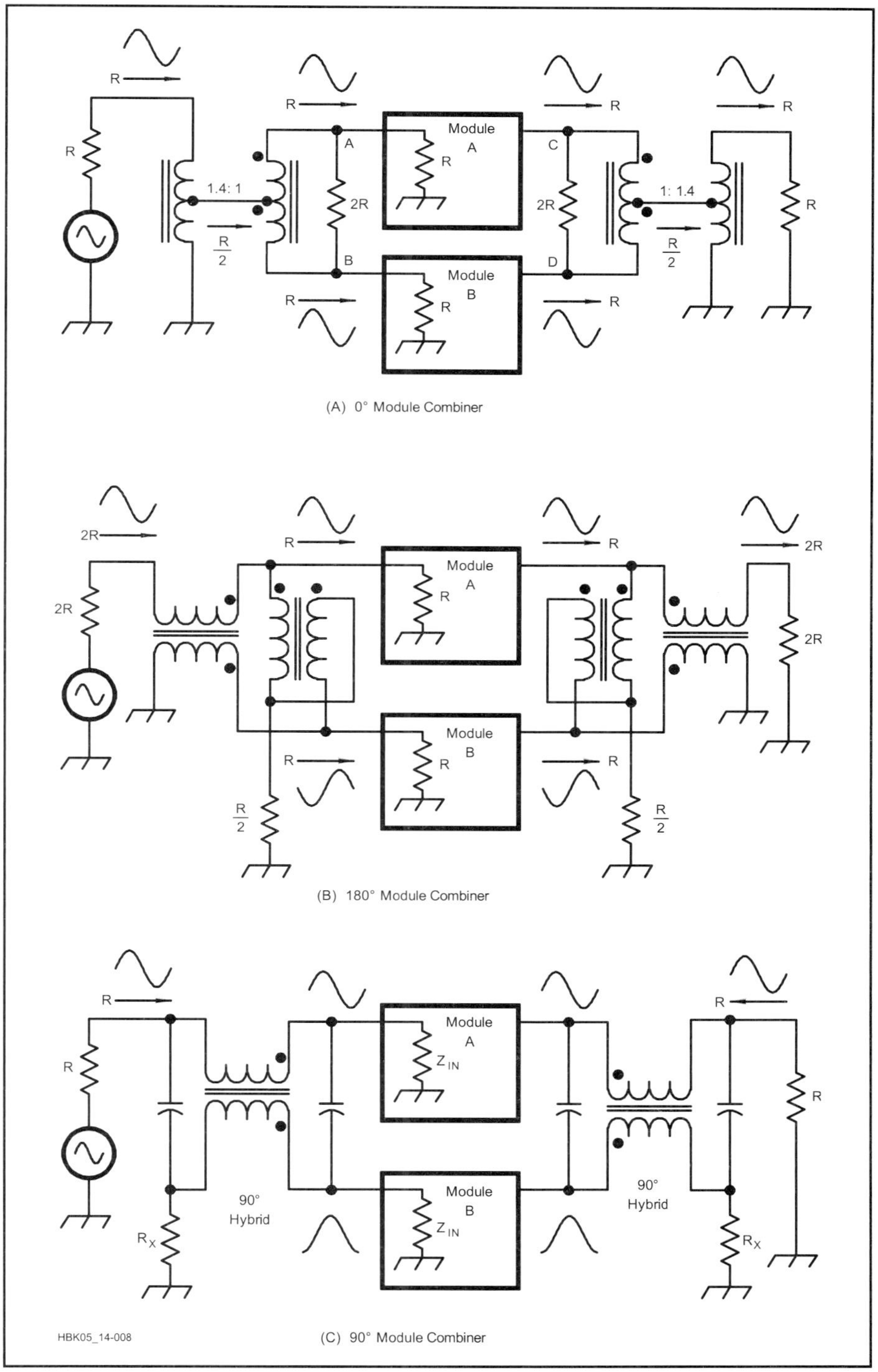

Figure 17.42 — Three basic techniques for combining modules.

Three methods are commonly used to combine modules: parallel (0°), push-pull (180°) and quadrature (90°). In RF circuit design, the combining is often done with special types of "hybrid" transformers called splitters and combiners. These are both the same type of transformer that can perform either function. The splitter is at the input, the combiner at the output.

Figure 17.42 illustrates one example of each of the three basic types of combiners. In a 0° hybrid splitter at the input the tight coupling between the two windings forces the voltages at A and B to be equal in amplitude and also equal in phase if the two modules are identical. The 2R resistor between points A and B greatly reduces the transfer of power between A and B via the transformer, but only

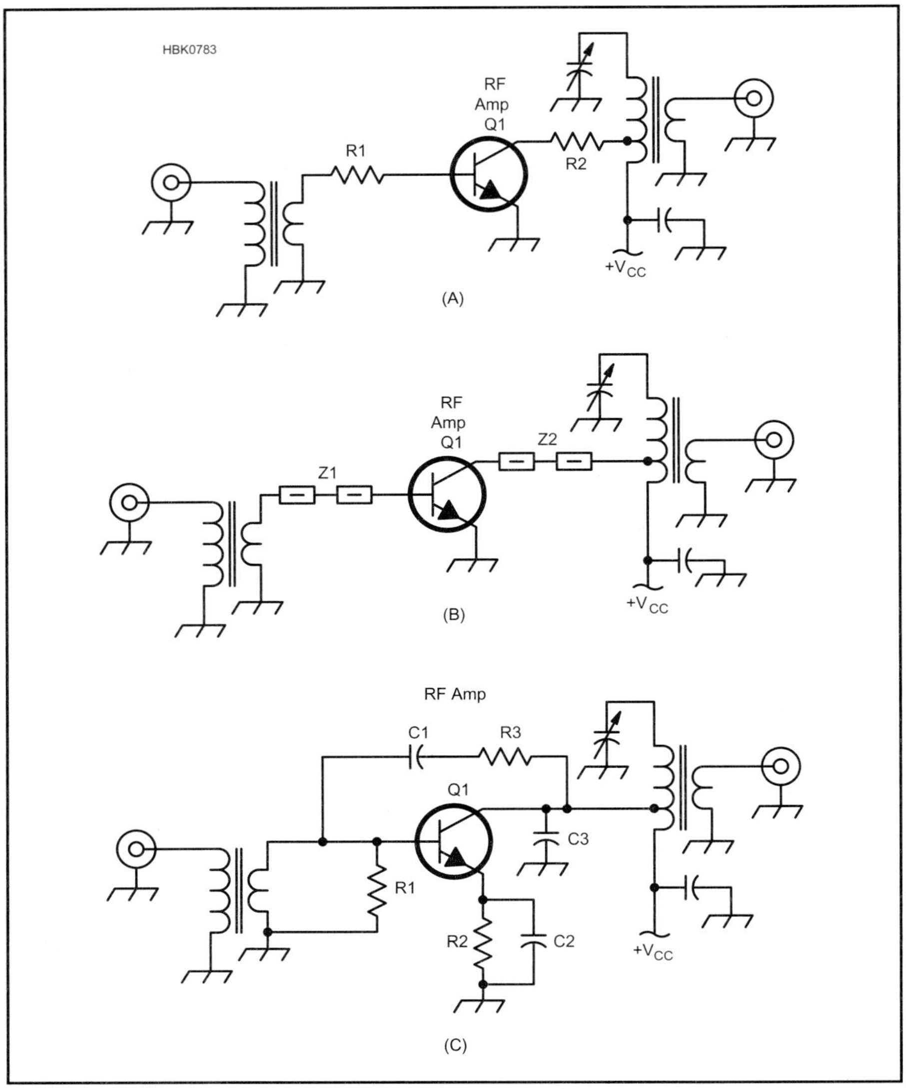

Figure 17.43 — Suppression methods for VHF and UHF parasitics in solid state amplifiers. At A, small base and collector resistors are used to reduce circuit Q. B shows the use of ferrite beads to increase circuit impedance. In circuit C, C1 and R3 make up a high-pass network to apply negative feedback.

if the generator resistance is closely equal to R. The output combiner separates the two outputs C and D from each other in the same manner, if the output load is equal to R, as shown. No power is lost in the 2R resistor if the module output levels are identical. This section covers the subject of combiners very lightly. We suggest that the reader consult the considerable literature for a deeper understanding and for techniques used at different frequency ranges.

17.10.6 Amplifier Stabilization

Purity of emissions and the useful life (or even survival) of the active devices in a tube or transistor circuit depend heavily on stability during operation. Oscillations can occur at or away from the operating frequency because of undesired positive feedback in the amplifier. Unchecked, these oscillations pollute the RF spectrum and can lead to tube or transistor over-dissipation and subsequent failure. Each type of oscillation has its own cause and its own cure.

In a linear amplifier, the input and output circuits operate on the same frequency. Unless the coupling between these two circuits is kept to a small enough value, sufficient energy from the output may be coupled in phase back to the input to cause the amplifier to oscillate. Care should be used in arranging components and wiring of the two circuits so that there will be negligible opportunity for coupling external to the tube or transistor itself. A high degree of shielding between input and output circuits usually is required. All RF leads should be kept as short as possible and particular attention should be paid to the RF return paths from input and output tank circuits to emitter or cathode.

In general, the best arrangement using a tube is one in which the input and output circuits are on opposite sides of the chassis. Individual shielded compartments for the input and output circuitry add to the isolation. Transistor circuits are somewhat more forgiving, since all the impedances are relatively low. However, the high currents found on most amplifier circuit boards can easily couple into unintended circuits. Proper layout, the use of double-sided circuit boards (with one side used as a ground plane and low-inductance ground return), and heavy doses of bypassing on the dc supply lines often are sufficient to prevent many solid state amplifiers from oscillating.

PARASITIC OSCILLATIONS

In low-power solid state amplifiers, parasitic oscillations can be prevented by using a small amount of resistance in series with the base or collector lead, as shown in **Figure 17.43A**. The value of R1 or R2 typically should be between 10 and 22 Ω. The use of both resistors is seldom necessary, but an empirical determination must be made. R1 or R2 should be located as close to the transistor as practical.

At power levels in excess of approximately 0.5 W, the technique of parasitic suppression shown in Figure 17.43B is effective. The voltage drop across a resistor would be prohibitive at the higher power levels, so one or more ferrite beads placed over connecting leads can be substituted (Z1 and Z2). A bead permeability of 125 presents a high impedance at VHF and above without affecting HF performance. The beads need not be used at both circuit locations. Generally, the terminal carrying the least current is the best place for these suppression devices. This suggests that the resistor or ferrite beads should be connected in the base lead of the transistor.

C3 of Figure 17.43C can be added to some power amplifiers to dampen VHF/UHF parasitic oscillations. The capacitor should be low in reactance at VHF and UHF, but must present a high reactance at the operating frequency. The exact value selected will depend upon the collector impedance. A reasonable estimate is to use an X_C of 10 times the collector impedance at the operating frequency. Silver-mica or ceramic chip capacitors are suggested for this application. An additional advantage is the resultant bypassing action for VHF and UHF harmonic energy in the collector circuit. C3 should be placed as close to the collector terminal as possible, using short leads.

The effects of C3 in a broadband amplifier are relatively insignificant at the operating frequency. However, when a narrow-band collector network is used, the added capacitance of C3 must be absorbed into the network design in the same manner as the C_{OUT} of the transistor.

LOW-FREQUENCY PARASITIC OSCILLATIONS

Bipolar transistors and MOSFETs exhibit a rising gain characteristic as the operating frequency is lowered. To preclude low-frequency instabilities because of the high gain, shunt and degenerative feedback are often used. In the regions where low-frequency self-oscillations are most likely to occur, the feedback increases by nature of the feedback network, reducing the amplifier gain. In the circuit of Figure 17.43C, C1 and R3 provide negative feedback, which increases progressively as the frequency is lowered. The network has a small effect at the desired operating frequency but has a pronounced effect at the lower frequencies. The values for C1 and R3 are usually chosen experimentally. C1 will usually be between 220 pF and 0.0015 µF for HF-band amplifiers while R3 may be a value from 51 to 5600 Ω.

R2 of Figure 17.43C develops emitter degeneration at low frequencies. The bypass capacitor, C2, is chosen for adequate RF bypassing at the intended operating frequency. The impedance of C2 rises progressively as the frequency is lowered, thereby increasing the degenerative feedback caused by R2. This lowers the amplifier gain. R2 in a power stage is seldom greater than 10 Ω, and may be as low as 1 Ω. It is important to consider that under some operating and layout conditions R2 can cause instability. This form of feedback should be used only in those circuits in which unconditional stability can be achieved.

R1 of Figure 17.43C is useful in swamping the input of an amplifier. This reduces the chance for low-frequency self-oscillations, but has an effect on amplifier performance in the desired operating range. Values from 3 to 27 Ω are typical. When connected in shunt with the normally low base impedance of a power amplifier, the resistors lower the effective device input impedance slightly. R1 should be located as close to the transistor base terminal as possible, and the connecting leads must be kept short to minimize stray reactances. The use of two resistors in parallel reduces the amount of inductive reactance introduced compared to a single resistor.

17.10.7 "Pallet" Amplifiers

[This section, written by Jim Klitzing, W6PQL, explains pallet construction and how pallets are used in an amplifier system. More information is available at the author's website: ***www.w6pql.com****.]*

In the world of RF power, a "pallet" is a fully functional solid-state power amplifier (SSPA) module intended to be used as a sub-assembly in a high-power amplifier system. A "pallet amplifier" is a complete RF amplifier based on one or more pallet modules. **Figure 17.44** shows a pallet designed by W6PQL for HF amplifiers. Modules and amplifiers are available for the MF/HF, VHF, UHF, and microwave amateur bands.

Figure 17.44 — Typical HF amplifier "pallet" designed by Jim Klitzing, W6PQL (www.w6pql.com/index.htm). This design is capable of 1 kW output power and uses the NXP BLF188XR LDMOS amplifier transistor module. [Jim Klitzing, W6PQL, photo]

THE PALLET MODULE

A typical pallet amplifier module consists of at least the following components:

• The RF transistor itself — The most common type in use today for high-power applications is the LDMOS transistor, usually two identical devices inside the same ceramic or plastic package and intended to be used as a push-pull amplifier. An amplifier so configured has the advantage of suppressing even harmonics, making the design of an output filter less complex. Typical device part numbers are BLF188XR, ART1K6, MRF1K50, MRFX1K80, and many others. The pallet in Figure 17.44 is based on the NXP BLF188XR.

• A copper *heat spreader* base — A heat spreader distributes heat from the device transistor(s) over a larger area. The base of the spreader is then bolted to an even larger aluminum heat sink where the heat is drawn away with airflow provided by fans.

• Input and output matching circuits — The transistor module's input is normally fed with a broadband center-tapped transformer that will need to handle only a few watts of drive power. However, the output circuit must be able to handle 300 W to 2 kW as of early 2022. This level of power involves many amps of RF current and high RF voltages. Capacitors, coaxial or ferrite transformers, and coupling capacitors must be adequately rated to survive these conditions.

• External stabilization components — Most pallet modules have a gain of 30 dB at 2 meters, and much higher gain at HF and below. If this gain is not limited with stabilizing components, there will be spurious signals in the output stream. In the worst case, a module can oscillate at high power levels that ensure the transistors will be quickly destroyed along with other nearby components.

• An adjustable bias supply — An LDMOS transistor is usually an enhancement mode device. Such a transistor designed for 50 V operation may only require a positive bias voltage of 2 – 3 V applied to the gates to conduct. Without bias the device is cut off and will not conduct, though RF drive itself can provide some stimulus.

Transistor modules are either bolted or soldered to the heat spreader. The direct bolt-down method requires heat sink compound to be used on the module where it mates to the copper spreader. This allows for easier device replacement but is less robust than soldering the device directly to the spreader. A properly managed flow-solder process ensures the best possible RF connection to the circuit common ground plane and maximum heat flow into the spreader. Soldering can be performed in a home workshop using a hotplate.

TRANSISTOR BIAS

The pallet's bias supply must be well-regulated and adjustable; once the drain is conducting, an upward change of just a few dozen millivolts on the gates can cause the LDMOS idling current to rise from of a couple of amps to full saturation.

Bias voltage should also be temperature compensated. As the LDMOS transistors heats up during operation, the idling current will drift up as well. If uncompensated, drain current will rise quickly depending on how hot the transistor junctions are. This effect can cascade into thermal runaway.

Bias supply control can be performed by active or passive circuits. The active method uses a voltage regulator and a variable resistor. The passive method uses a Zener diode instead of the active regulator and has the advantages of being simpler to implement and less susceptible to RFI. With either method, a semiconductor PN junction or a thermistor can be employed as the temperature-sensing element.

RF DECOUPLING AND DC POWER

Most high-power LDMOS devices are designed for 50 V dc power. Newer transistors are designed for 55 – 65 V. In order to get to the kW output level or higher at 50 V, the required drain current can be 30 amps or more depending on the efficiency of the design. Higher drain voltages allow full power output at lower current levels.

To route high current to the transistor drains, some designers route dc power through the RF transformers at a properly bypassed RF-neutral winding tap. An alternate method is to route dc power separately to the drain connections through rf chokes bypassed at the dc connection. This second method reduces the additional IR losses in the RF transformers caused by the high currents. The first method is simpler (fewer parts) and the second lets the RF transformers run cooler. Both methods work well and are illustrated in **Figure 17.45**.

POWER LIMITATIONS

LDMOS transistors are available at up to 2 kW rated output in a single push-pull package. This makes a legal-limit HF pallet amplifier using a single device a real possibility.

At frequencies above HF, single-band pallets are the norm. IR losses mount with upward frequency changes and currently available components for a single-device pallet limit output to about 1200 W continuous at 2 meters. Some high-end pallets can easily develop 1500 W in SSB and CW modes where the duty cycle is low.

For FM or digital modes where the duty cycle can be very high, above 1200 W the transformers can over-heat with temperatures high enough to melt the solder holding them in place. One can mitigate this a bit with additional airflow across the pallet, but 1200 W is about the limit. To get to 1500 W in digital modes, two separate pallets or a dual-device pallet and combiners are often used.

At 1296 MHz and above, the maximum single-device pallet output is usually lower and you may need as many as 4 devices (properly combined) for a 1500 W EME amplifier.

There are also limitations on some of the other components in the design and these must be considered. RF and dc voltages and currents are higher, so two possible responses are:

- Increasing the size of the transformer core and wires, and

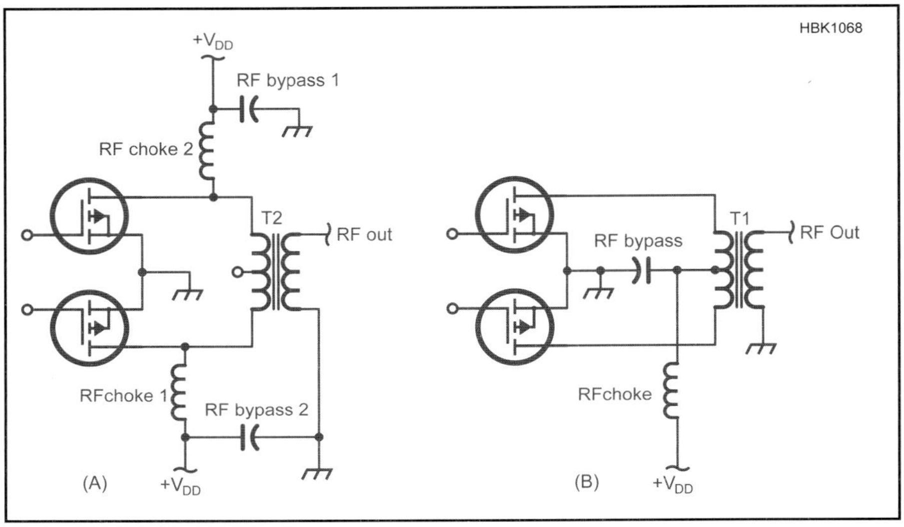

Figure 17.45 — Two methods of feeding dc power to the RF transistors. At (A), power is supplied separately to each transistor through an RF choke. At (B), power is supplied through the shared primary winding center-tap. See text for a comparison of the two methods.

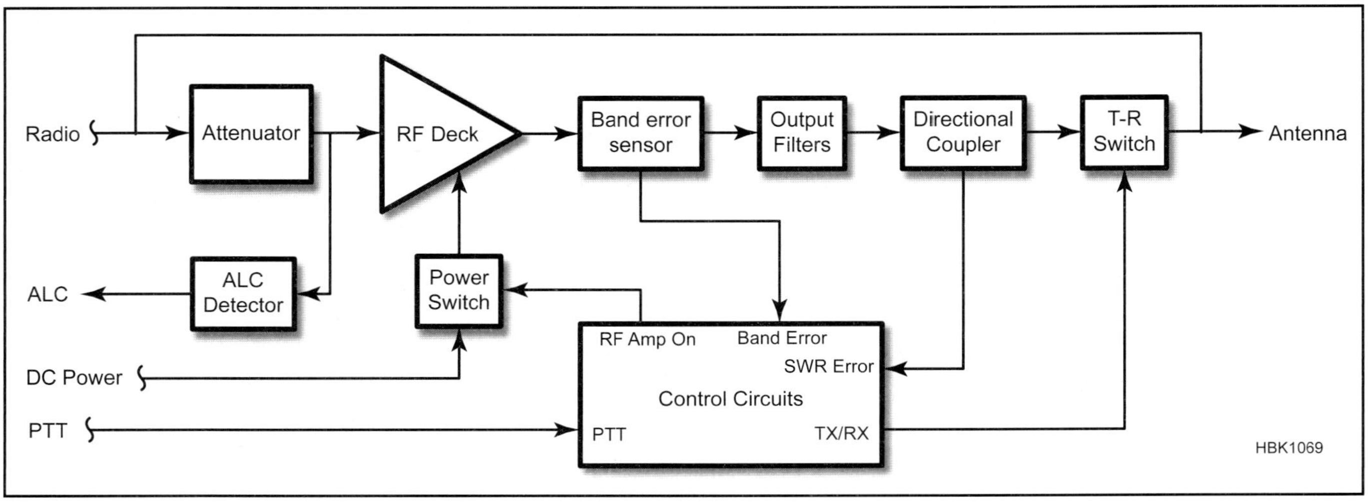

Figure 17.46 — A block diagram of a complete pallet-based RF power amplifier.

• Using matching capacitors rated at higher voltages and currents.

Capacitor current limits can be overcome by using two or more capacitors of the same type in parallel. For example, if one needs a 300 pF capacitor, using two 150 pF parts in parallel will do, as will using three 100 pF parts, etc. This method is currently employed in many designs.

BANDWIDTH AND DISTORTION

Pallet amplifiers designed for use on HF frequencies are normally very broadband, requiring no tuning into a 50 Ω load, and can cover several octaves if the transformers are designed properly. The no-tune feature does require well-matched antennas to facilitate the best power transfer — we give up the "plate" and "load" tuning adjustments associated with tube-type amplifiers for this convenience.

To achieve no-tune broadband performance some compromises are usually made. One of the most notable is efficiency; if one were to optimize for efficiency only, the result would be more narrow band. At VHF and above the narrower bandwidth approach is the most common, and resulting efficiencies are often at least 10% higher than at HF.

If the pallet amplifier is properly biased and not over-driven, typical IMD3 is –30 to –35 dBc. When not driven higher than the 1dB compression point it will be difficult for listeners to tell whether the amplifier is a solid-state pallet design or some other type. Linearity of solid-state RF power amplifiers is discussed the following section.

PALLET AMPLIFIER SYSTEM DESIGN

A useable pallet amplifier involves a complete design, including associated protective and matching circuits, the output filter, power supply, and indicator/monitoring/control circuits. **Figure 17.46** is a block diagram showing the different functions and their relationship. The complete system consists of at least the following parts:

- RF deck (pallet)
- Antenna transmit-receive (T-R or changeover) switch
- Output filtering for harmonic suppression
- Cooling system
- Protective circuits for band selection errors (for multi-band systems), high SWR, and excessive temperature
- Band-select system (for multi-band systems)
- High current switches for dc power
- Control system for the system components
- Power monitors for forward and reflected power
- Metering for voltage and current
- RF attenuator to limit input power
- Power supply
- A suitable cabinet to house all components

Antenna T-R Switch

The antenna T-R switch has two important jobs to do; in its default (unpowered) state, it connects the antenna to the radio. When switched to the transmit state, it will place the amplifier in the transmit path between the driving radio and the antenna.

The most common antenna switch is usually an electro-mechanical device (a relay or relays) but could also be a solid-state switch of some type (typically a PIN diode switch). The mechanical types can be general-pur-pose relays or coaxial types; general-purpose relays are often used at HF and coaxial relays are more commonly found at VHF and above.

For systems with a separate input and output relay, the input relay needs to handle no more power than the driving radio can produce, but the output relay must be more robust and able to handle as much as 1.5 kW. Some relays have two poles and are capable of handling both receive and transmit switching if the isolation between the poles is adequate. A good rule of thumb is to have at least 10 dB more isolation than the amplifier has gain; this is necessary to avoid oscillation caused by positive feedback from leakage of the amplifier output to the input.

Output Filtering

Output filtering for single-band systems can consist of a single bandpass filter or a low-pass filter (LPF), the latter being the most common. **Figure 7.47** shows two examples of amplifier output filters. Figure 7.47A is a multi-band HF filter assembly and Figure 7.47B is a single-band low-pass filter for a 220 MHz amplifier.

For multi-band HF amplifier systems, the filter usually consists of several separate LPFs which are selected according to the band in use. For example, one LPF section with adequate harmonic suppression above 24 MHz can be used for the 14, 18 and 21 MHz bands. It is typical for the LPF to have 5 or 6 different selectable sections with different cut-off frequencies to cover the full range of 160 through 6 meters. These can be assembled on one PC board and selected using general-purpose relays capable of handling the high RF voltages and currents.

Cooling System

The RF deck on its copper heat spreader should be mounted to a heat sink (aluminum extrusion is widely used) with a thin coating

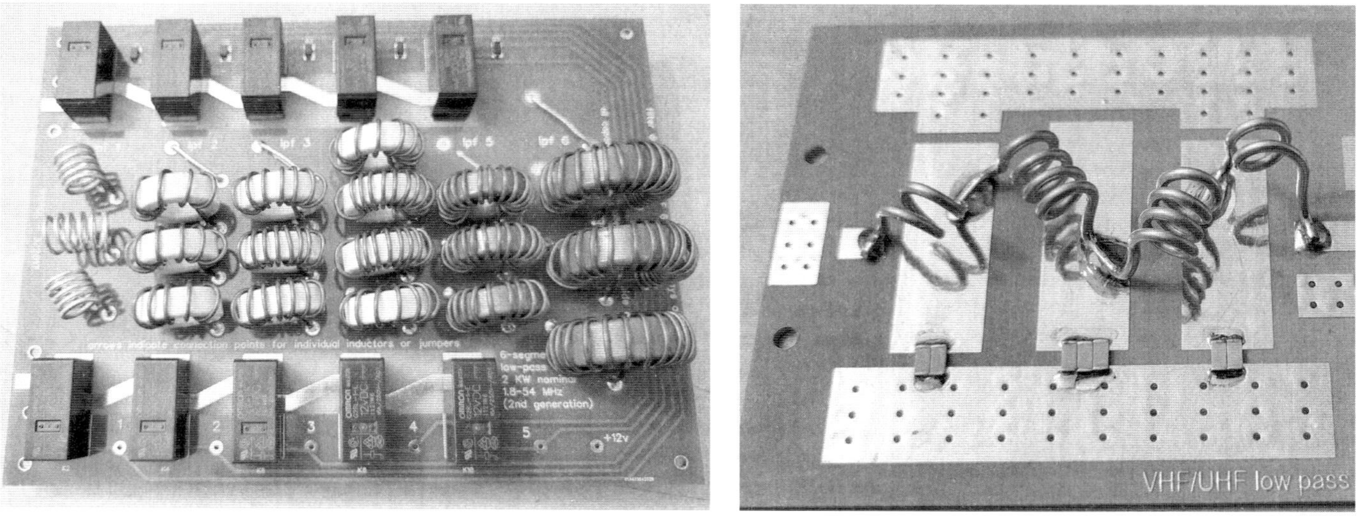

(A) (B)

Figure 17.47 — Output LPF filters for a multi-band HF amplifier (A) and a VHF single-band amplifier (B).

of heat sink compound covering the surface of the copper base where it mates to the heat sink. A good rule of thumb for cooling is six pounds of aluminum heatsink with a 0.375-inch base under the fins and 80 – 120 cfm airflow per kilowatt of power generated.

Some builders have fitted the copper base with water cooling. While this method is quiet and very efficient it also has some disadvantages for convenience and adds additional complexity. It is seldom necessary unless the unit will be operated in an extremely hot environment.

Thermal dynamics is a science all its own and some experimentation may be in order; for example, if the distance between the output of the fans and the heat sink fins is too close, the back pressure to the fans will be high and restrict the airflow. You can evaluate this before you build, changing the distance and monitoring the airflow on the other side of the heat sink to see this effect. About 3/4 inch seems to be the correct distance in most cases. The warm exhaust air must also be directed out of the cabinet in such a manner that it does not flow back over the top or sides of the heat sink and dump heat back into the system. One method which works well is to pull cool air in from the bottom rear of the cabinet and then expel the warm air out the top of the front.

Protective Circuits

Human error is a certainty as is the occasional unexpected act of nature. We get distracted and forget to reconnect our antenna to the radio or amplifier after doing some sort of test, moving equipment around, or perhaps an animal chews on the feed line. It happens and whatever the cause we don't want all that RF energy coming back to overload circuits and cause components to fail.

A control system relies on an SWR sensor to lock out the amplifier when reflected power gets close to damage levels. A simple lockout for a kW amplifier should activate around 100 W of reflected power. Some sensors are even more sophisticated and can sense 2:1 SWR or higher with just a dozen watts of forward power but the simple lockout sensor is adequate, provided it is fast enough to prevent damage. Such a sensor depends on sampling reflected power and is usually part of a dual directional detector monitoring both forward and reflected power levels. See **Figure 17.48** for an example circuit.

A high-temperature lockout sensor should be able to trigger the control system if the amplifier cooling system cannot maintain safe operating temperatures. Unsafe conditions can be caused by hot weather, blocked exhaust ports, or something else causing the amplifier to overheat. Today's modern LDMOS devices are tough but letting things cool down until safe operating temperatures are reached again will prolong the operating life of those and many other components. A small thermistor attached to the main heat sink makes a good sensor. Thermistors are not rapid responders, and this is a good match for the gradual changes in heat sink temperature. Fast sensors tend to make the fans 'stutter' (turn on and off rapidly) and this can be an unnecessary annoyance.

Another protective circuit which is good to have with multi-band (HF) systems is a band-select error sensor. If you transmit on 20 meters with the band switch on 40 meters, the output LPF will reflect all the power to the RF deck. Since the SWR sensor is on the output side of the LPF it won't see this error, so you'll need a separate reverse power sensor between the RF deck and LPF to catch this condition. This sensor would not be needed if the amplifier is fitted with a band-select decoder allowing the radio to control the amplifier band. The sensor threshold is set to a high enough level not to trip on harmonic energy reflected from the LPF.

High-current Switch

A switch for disconnecting the RF deck from the 50 V power connection during receive is another safety feature. This disconnect absolutely shuts off the transistor and helps to ensure stability. For example, during receive the antenna switch will leave the pallet with no load and if left active in that state it could be unstable.

The switch can be a relay, but with power MOSFETs the power can safely be switched on or off in less than 1 msec. In the ON state, the device forward resistance is just a few milliohms so there is very little IR loss (½ V or less). These devices can also handle 35 amps or more of current at voltages higher than 50 V and can be paralleled for additional current capacity if necessary. **Figure 17.49** shows a sample current switch circuit.

Control System

With so many components making up the amplifier system, an effective method of coordinated control is necessary. For the system to work properly some functions need to happen in the proper order. This part of the control system is called a *sequencer*. Other things (like fans) can operate independently. The control system circuits are usually consolidated onto one PC board with the appropriate sensor inputs and control outputs. (A sample control system schematic and PC board layout can be seen on the author's website.)

Everything requiring coordination begins with the radio signaling the amplifier it is going into the transmit state. The channel for this is the PTT connection between the radio and the amplifier; this is usually a control line being connected to ground. When this happens, the following should take place in this order:

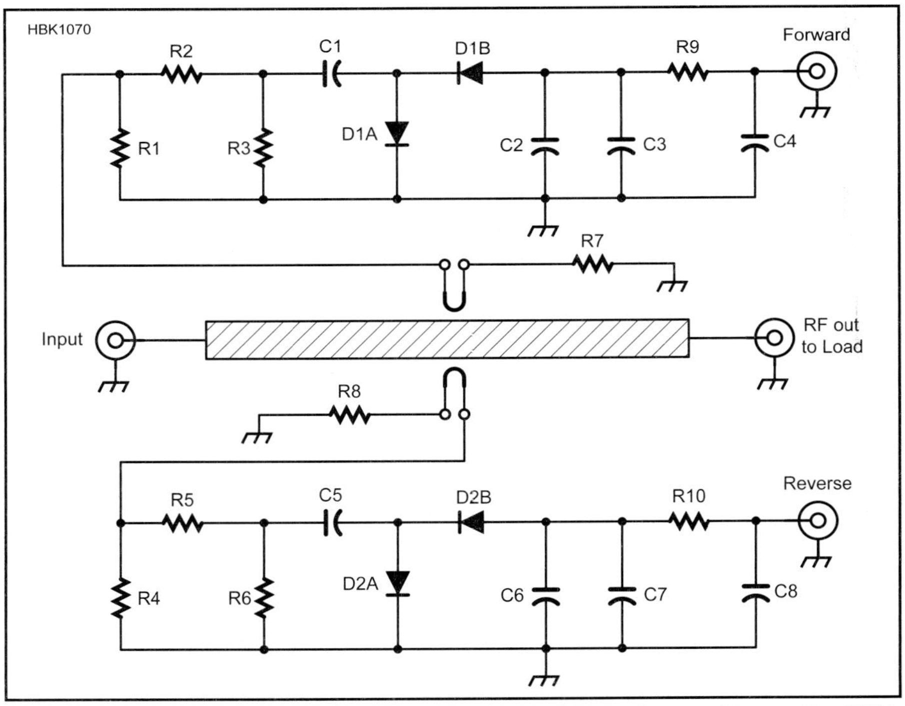

Figure 17.48 — Basic schematic for a dual directional RF coupler, used to monitor SWR in an RF amplifier.

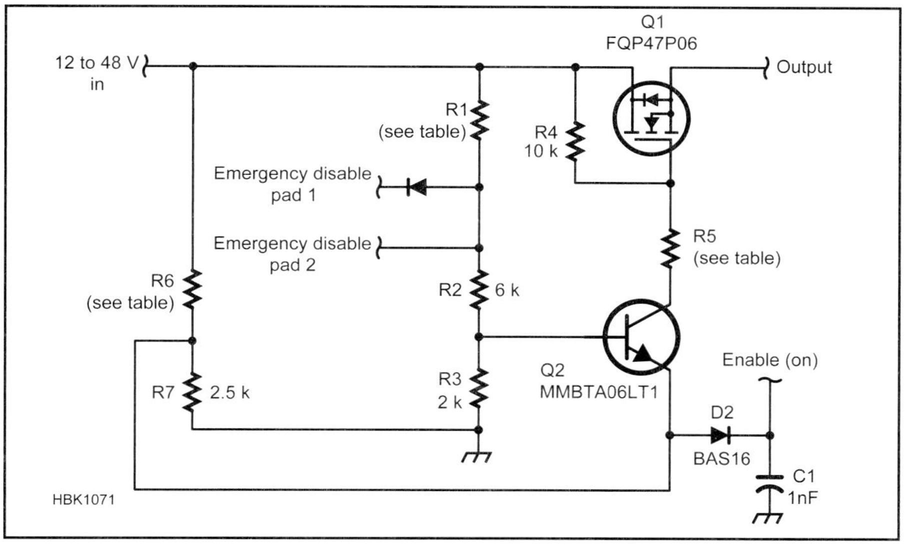

Figure 17.49 — Schematic for a high-current dc switch using a power MOSFET.

1. The fans should be switched on and the change-over relays in the amplifier should switch into the transmit state. Failure to do so would cause them to be hot-switched which shortens the life of the relay contacts and often triggers a high-SWR lockout during the time the relays are settling. The duration of settling is about 20 – 30 msec with general-purpose relays.

2. 30 – 50 msec after Step 1 the high-current dc switch feeding power to the RF deck should be turned on.

3. 30 – 50 msec after Step 2 the bias should be turned on, allowing the amplifier to be fully activated. At this time, transmit inhibit (if used on the driving radio) should be released.

The three events listed above should be switched off in reverse order when the PTT signal from the radio is deactivated, allowing the receiver to function again.

One of the independent safety functions is the heat sensor. It should monitor the temperature of the heat sink continuously and signal the fans to turn on if the temperature rises to moderate levels. If the temperature rises to exceed safe operating levels, the sensor should signal the system to drop out of the transmit path and keep the fans on. This can be done by disabling the PTT function, allowing the sequencer to shut the system down gracefully. Resetting this lockout can be allowed to happen automatically once safe temperatures are reached. Some hysteresis should be built into the mechanism to keep this control from chattering on and off at the temperature thresholds.

A reflected power sensor must trigger an immediate lockout if reflected power exceeds safe levels. Damage can happen quickly with this type of fault so there should be no sequencing delays for this lockout; the control mechanism should immediately command the high current switch to remove power from the RF deck and at the same time tell the sequencer to begin removing the system from the transmit path.

Lastly, some mechanism should be provided to place the amplifier online or offline. An easy way to accomplish this is to use a front panel switch to interrupt power to the sequencer mechanism, or just use the switch to disconnect the PTT input.

Overdrive Protection and ALC

Overdriving an amplifier can cause problems, the most serious of which is damaging the power transistors. One way to provide protection is to use ALC feedback to the driving radio, but this must be done properly to avoid distortion. Whether using a tube or solid-state amplifier, if ALC is only used to control output levels it will cause an amplifier to be non-linear in its response and thus create distortion.

ALC can be employed to prevent overdrive. For example, if an amplifier reaches full linear output with 50 W of drive, setting the ALC to begin reducing drive at 50 to 55 W will prevent damage to the amplifier. In this case, reducing the radio's maximum drive power to just below the point where ALC begins to assert control will get the job done correctly. This is difficult to do on a multi-band amplifier because of different drive levels needed for each band, but if the operator is using a radio capable of controlling power output on each band it will work very well.

An even better way of protecting the amplifier from overdrive damage is to use an input attenuator. Should the radio be capable of a maximum output of 100 W and the amplifier only requires 50 W to reach full output, then an input attenuator reducing the radio's full output power to 80 W or less will provide enough protection to avoid damage. Some accidental overdrive will not damage the transistors, but one must still be aware an overdriven amplifier will cause distortion and unnecessary splatter no matter what method of damage protection is used.

Monitoring Power Levels

Monitoring forward and reflected power is useful for helping the operator make proper adjustments to drive levels and watch for trouble in the antenna system. A good example might be operating in bad weather when high winds, water leakage, or an ice storm might be affecting the antenna; if so, reflected power levels will alert the operator to problems. Adjusting a manual antenna tuner for a good match is also made easy with power monitoring.

The monitoring displays or meters can get their signal information from power detectors; these detectors are usually placed at the output of the LPF with the detectors feeding both a power monitor and the control system at the same time.

While meters are suitable if properly interpreted, peak power is best monitored using peak-reading displays. SSB has lower average power than some of the other modes, but peak power levels are quite a bit higher and should be monitored for compliance with power limit regulations. An LED bar graph display, common on transceivers and amplifiers responds nearly instantly to changing levels and shows peak readings clearly. (Peak and average power measurements are also discussed in the **Test Equipment and Measurements** chapter.)

Power Supplies

Both linear and switching power supplies have advantages and disadvantages. A linear power supply is much heavier due to its iron transformer making the voltage conversion from the AC mains at 50 – 60 Hz. A power transformer for a legal-limit amplifier can weigh 50 pounds or more. The main advantage of linear supplies is the absence of RF noise.

The switching power supply is now the most widely used. A supply equivalent to the linear type might weigh 10 pounds or less and be as efficient as 90%. (See the **Power Sources** chapter for more about switchmode power conversion.) Most supplies are properly shielded and filtered to minimize interference generated by the switching waveforms. Still, it is always good practice to use ferrite chokes on all the power leads going into and out of the power supply enclosure. If this is done properly, even sensitive receivers will be unaffected.

17.11 Solid-State Amplifiers and Intermodulation Distortion

Trade-offs are a fact of life, and we certainly see them in amateur transceivers and linear RF amplifiers (referred to just as "amplifiers" elsewhere in this section). Solid-state transmitters and amplifiers are more convenient to operate with no tuning and automatic band-changing. However, tube transmitters with 6146A tube finals and negative feedback produced much "cleaner" signals with less IMD (intermodulation products — often called "splatter"). The difference is significant since a SSB signal today can be at least twice as wide as for the top tube transmitters such as a Collins 32S-3 or Yaesu FT-102 (when measured where the signal's sideband amplitudes have decreased to 60 dB below their peak value).

The first step, of course, must be to adjust the transceiver so the output signal is clean, with a minimum of IMD products. However, an excessively "wide" signal cannot be blamed solely on transceiver-generated IMD. An amplifier will contribute its own IMD products, and today's solid-state amps produce IMD that can be 10 to 20 dB greater than a tube amp of equivalent power. Why does this happen, and what are the remedies?

17.11.1 Solid-State Device Linearity

QST Product Reviews publish input/output linearity graphs for amplifiers. **Figure 17.50** is an example of a modern LDMOS amp with a relatively straight linearity curve. **Figure 17.51** is an example of an LDMOS amp with a curve that begins to flatten out just above half power. For a clean SSB signal with a minimum of IMD products, this amp should be run at half power. We have to share our crowded bands, and driving an amplifier into its non-linear regions isn't being a good neighbor.

With all the advances in solid-state technology, why are today's MOSFETs worse, from the perspective of amplifier linearity, than the previous generations of bipolar transistors? (See this chapter's section "MOSFET Design for RF Power Amplifiers".)

Bipolar RF power transistor circuits are inherently more linear than MOSFETS because of their thermal characteristics. Bipolar transistors cannot be directly paralleled because the hottest one — even on a perfectly designed IC die — will have increased gain, consume more of the total current, and eventually overheat and destroy itself. The technique, called *emitter ballasting*, deposits a low-value metal resistor in series with each emitter where it connects to the emitter buss of the device.

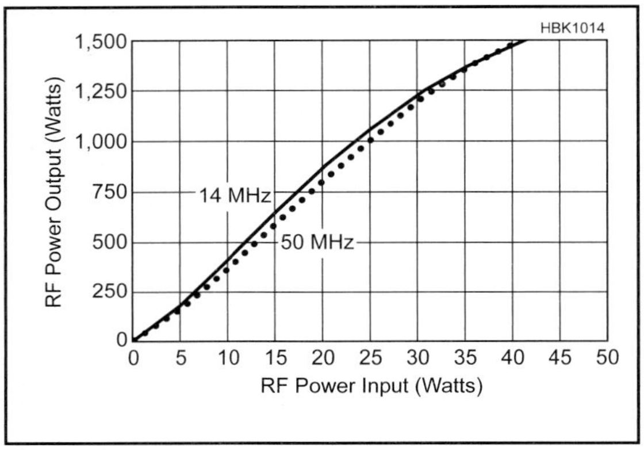

Figure 17.50 — A relatively straight linearity curve for a legal-limit LDMOS solid-state power amplifier.

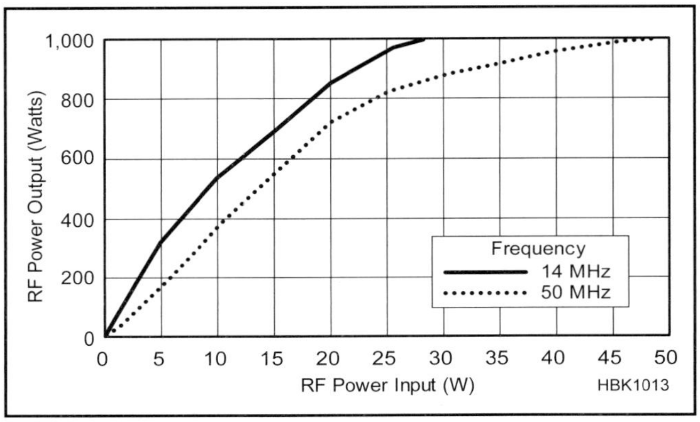

Figure 17.51 — Linearity curve for a 1 kW LDMOS solid-state amplifier that begins to flatten out (becomes nonlinear) at around half of its rated power.

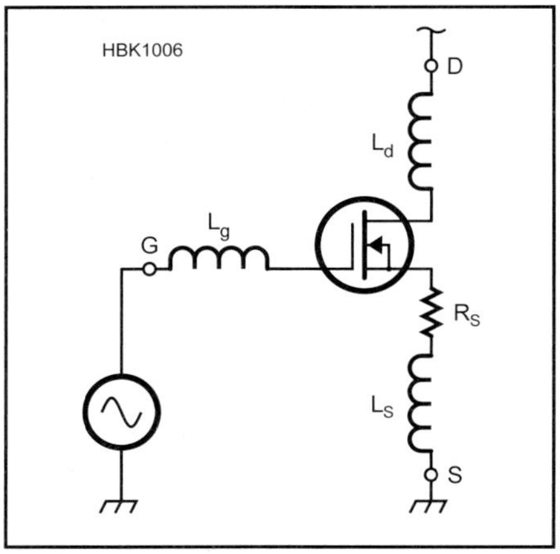

Figure 17.52 — Schematic of a MOSFET connected as a common-source amplifier showing parasitic lead inductance and the intrinsic source resistance.

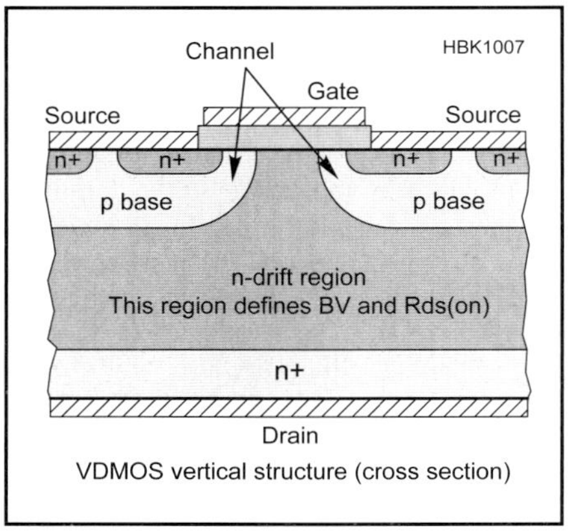

Figure 17.53 — Cross-section of a VDMOS transistor showing the vertical structure with the drain at the bottom which is connected to the device package.

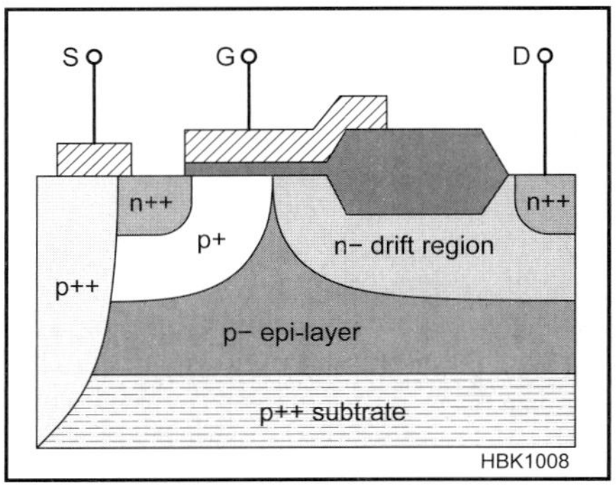

Figure 17.54 — Cross-section of an LDMOS transistor showing the lateral (horizontal) structure with a vertical connection ("sinker") between the source terminal and device substrate.

This solves the thermal problem by adding negative resistive feedback but reduces the available gain by about 3 dB.

Conversely, MOSFET (LDMOS and VDMOS) cells can easily be paralleled (if their size is chip-scale) because the hottest one becomes more resistive via the intrinsic positive temperature coefficient of its drain resistance. This reduces gain and shifts current to the other MOSFETs, stabilizing the transistors thermally.

COMMON-SOURCE AMPLIFIERS

In a common-source amplifier (see **Figure 17.52**), the source is grounded, the drive is applied to the gate, and the output is taken at the drain. Parasitic elements inside the device, such as bond wires connecting the die elements to the package terminals, cannot be ignored. Any impedance in series with the source causes a voltage that subtracts from the drive signal producing negative feedback. Resistance gives feedback that is constant with frequency, and inductance gives feedback that increases with frequency. Parasitic inductance in series with the gate and/or drain can usually be managed by the matching circuits. In general, negative feedback will improve device linearity at the expense of circuit gain.

VDMOS TRANSISTORS

In V(vertical)DMOS parts, the back side of the die is the drain as shown in **Figure 17.53**. The reason LDMOS transistors are more powerful (i.e. much lower thermal resistance) per die size than their VDMOS cousins is that the die does not need to be electrically isolated from its package mounting flange. It can be soldered directly to a thermally and electrically conductive header, which is the source that we wanted to run at ground potential anyway in the common-source configuration. However, it is now impossible to isolate the source enough to introduce some negative feedback there.

The VDMOS gate and source are wirebonded to the package terminals. The die is attached to an electrically insulating BeO (beryllium oxide) substrate that is brazed to the package flange. The thermal resistance of this substrate can easily double the device's junction-to-case thermal resistance ($R_{\theta jc}$), reducing its available power dissipation by half.

LDMOS TRANSISTORS

In L(lateral)DMOS devices, all the electrodes are laid out flat rather than stacked vertically, so the gates and drains are edge-on, giving very low gate-to-drain feedback and very high gain. (See **Figure 17.54**.) The sources are connected to the back side of the die using p-well "sinkers" (at left in the figure), so bond wires for the source connections are no longer needed. The die is only 2 or 3 mils thick, so there is virtually no negative feedback ("source degeneration").

A diffused p++ sinker through the p– epitaxial layer is used to connect the source to the back of the substrate. Each source is thus reliably connected to the back side of the wafer, which is then attached directly to the package flange. The gate and drain are wire-bonded to the package terminals.

LDMOS parts are relatively low voltage because the breakdown voltage of the drain-source junction is a function of doping resistivity and the distance between electrodes. In a lateral device, more distance requires more expensive die area, reduces gain, and lowers efficiency. If the p-well sinkers connecting the individual cell sources to the back side of the die could be made more resistive, there would be an improvement in linearity, just as in the bipolar devices.

In today's LDMOS devices there is a patented high-pass network on the gates that keeps the gain from being almost infinite on lower frequencies, such as 160 meters. It uses a high-value but somewhat-nonlinear doped silicon resistor paralleled by a capacitor, which acts as a divider against the shunt capacitance of the gate. The resistor is needed for biasing the gate. This explains why LDMOS gates are rather intolerant of overdrive in addition to being another source of nonlinearity that is built-in to the device itself.

LDMOS devices are being used for new amateur amplifier designs because they are the dominant commercial technology and thus, the most cost-effective. Amateurs should be thankful there are plenty of high-power RF applications to keep the price of transistors down and availability high.

17.12 Adaptive Predistortion

17.12.1 Predistortion Overview

Predistortion is applied today in many commercial applications to reduce IMD and thereby reduce the bandwidth consumed by transmissions and conserve frequency spectrum. Some current applications include cell phone transmissions, digital TV transmissions, and digital FM broadcasts. While one offshore amateur transceiver manufacturer has provided predistortion since 2014, several newer amplifiers have built-in samplers for predistortion, and other manufacturers have discussed providing predistortion, but this capability has not yet been widely deployed in amateur radio. This is at least partially because the challenges of applying this technology in an amateur radio environment are different than those of commercial environments and new and innovative solutions are often required.

When configuring predistortion for these commercial applications, the antenna and its load impedance, the amplifier gain, phase, and other characteristics, as well as the specific frequency or frequency range of operation are all known in advance. The frequency range is typically a small percentage of the absolute frequency. The situation is very different for amateur radio — amateurs use a wide variety of antennas presenting different load impedances to the amplifier, a wide variety of amplifiers using different technologies and with very different characteristics, and the frequency range of many transceivers and power amplifiers spans from 1.8 MHz to 54 MHz, the highest frequency being thirty times the lowest!

There are certainly other less technical approaches to reducing bandwidth by reducing IMD. These are usually less effective and less convenient compared to predistortion. For example, an amplifier can be operated in its most linear region as much as possible. This usually means operating at power levels well below the maximum rated power. Another approach is to obtain a Class A amplifier; however, these are very inefficient and not common in amateur radio.

The sections that follow present a brief look at predistortion. There are many variations on predistortion, and this material relates approaches that have been demonstrated to work for amateur radio applications.

17.12.2 IMD — The Cause and a Solution

The human voice has two key characteristics that result in the production of IMD in operating modes such as SSB and AM: (1) the amplitude of the signal varies, and (2) the signal contains multiple frequencies. If an amplifier has perfect gain linearity and constant phase shift as the signal amplitude varies, then, the output of the amplifier is a perfect amplified replica of the input. However, if either of these conditions is not met, the amplifier exhibits intermodulation, forming products at new frequencies from the existing frequencies in the input signal. Some of these new frequencies fall within the pass-band of the signal, producing distortion, while others fall outside the pass-band making them spurious emissions, referred to as splatter, which may interfere with communications on nearby frequencies.

Predistortion compensates for this by inserting an additional predistorter stage that modifies the input signal to the amplifier to cancel the distortions the amplifier introduces. **Figure 17.55** illustrates how a predistorter compensates for gain non-linearity.

The predistorter must "know" the gain and phase characteristics of the amplifier to apply compensating predistortion to the signal. The solution is to measure and respond to the characteristics of the amplifier using hardware and software in the transceiver. By doing that repeatedly, the system can automatically adapt to changing bands, temperatures, voltages, and other conditions that may impact the characteristics of the amplifier. This is called *adaptive predistortion*.

Figure 17.56 illustrates the gain characteristics of the same MOSFET amplifier on two different bands. The horizontal axis shows input signal amplitude with a maximum output of 100 W PEP into a 50 Ω load. As can be observed, the required predistortion correction is very different in the two cases.

17.12.3 Predistortion System Design

Figure 17.57 displays a simplified block diagram of a predistortion solution implemented for a direct-sampling radio. The low-level transmitter at the upper left is implemented in software and implements typical functions such as audio processing, band-pass filtering, modulation, and ALC. It produces a fully modulated and filtered baseband signal, the 0 Hz IF. (See the **SDR and DSP Fundamentals** chapter for more information.) Its output goes to the predistorter software that modifies the signal's amplitude and phase. The result is then up-converted to the RF frequency and DAC sample rate and then converted to an analog RF signal by the DAC.

To measure the amplifier's characteristics, a sample of its output is needed. This is obtained with a sampler that can be a resistive or capacitive divider, or perhaps a directional coupler. The sample is digitized by the receiver ADC and down-converted to baseband through the receive path. The predistortion correction required (both gain and phase) is calculated by comparing the down-converted amplifier output with the amplifier input signal just before up-conversion. The result of that calculation — the needed correction information — goes to the predistorter.

17.12.4 Implementing Predistortion

Digital signals are frequently processed in quadrature (*I*/*Q*) form. (See the **Modulation** chapter for more information.) In direct-sampling radios, this is true of the signals down-converted to baseband for receive and of those to be up-converted for transmit. For quadrature digital signals, each signal sample is represented by two numbers, *I*, the In-phase component, and *Q*, the Quadrature component. *Q* differs from *I* in that each frequency component is phase-shifted by 90 degrees. Quadrature signals provide a very convenient way to represent both the magnitude and the phase of a signal. Since *I* and *Q* are in quadrature, the magnitude of each signal sample is

$$M = \sqrt{I^2 + Q^2}$$

and the phase is conveyed by the ratio of *I* and *Q*, specifically $\phi = \tan^{-1}(Q / I)$.

As shown in Figure 17.57, to calculate the required correction, each input sample to the amplifier is compared with the corresponding output sample. For the magnitude

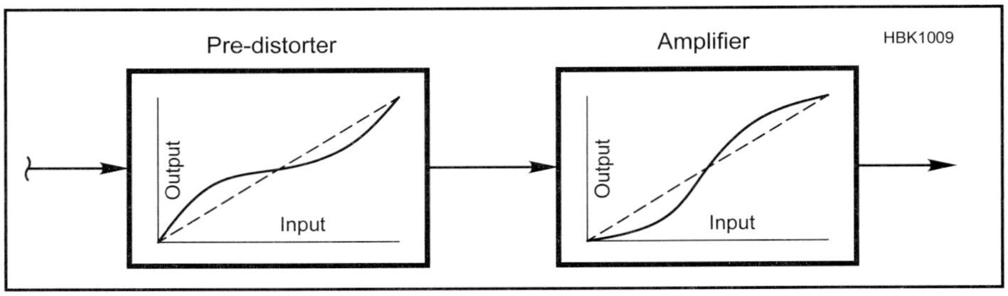

Figure 17.55 — Gain compensation by predistortion.

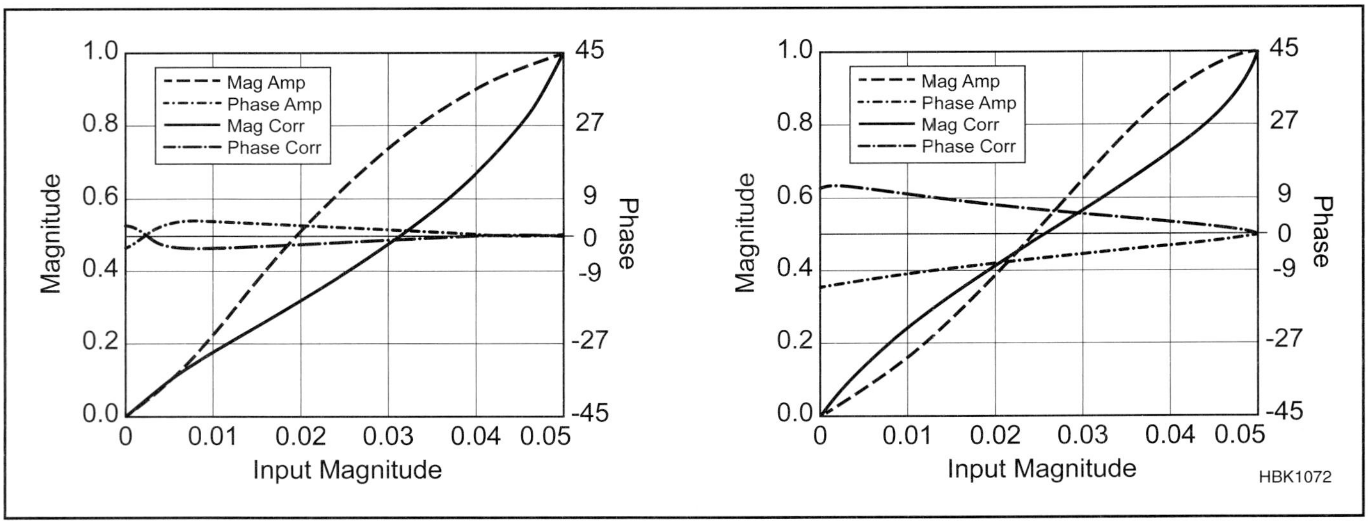

Figure 17.56 — Amplifier characteristics on two bands, 160 meters (left) and 10 meters (right). Dotted lines show the amplifier's magnitude and phase response at different amplitudes. Solid lines show the compensating magnitude and phase predistortion to correct for the amplifier's response. Note the differences between the amplifier characteristics on these two bands.

level of each output sample, the gain correction factor, G, and the phase correction factor, φ, are calculated. After some algebraic manipulation,

$$G = S(\sqrt{I_{in}^2 + Q_{in}^2} \;/\; (\sqrt{I_{out}^2 + Q_{out}^2})$$

$$\varphi = tan^{-1}((\,-I_{in} \times Q_{out} + Q_{in} \times I_{out}\,) \;/\; (I_{in} \times I_{out} + Q_{in} \times Q_{out}))$$

where I_{in} and Q_{in} are the amplifier input sample, I_{out} and Q_{out} are the amplifier output sample, and S is a scale factor to compensate for the amplifier's gain and for the attenuation in the amplifier output feedback path.

Assuming the correction for all output sample values has been calculated, the correction is then applied. The magnitude of the predistorter input sample is first calculated, the corresponding values of G and φ are found, and these values are applied to the samples according to the equations below. G, the gain correction factor, is a multiplier and φ, the phase correction factor, shifts the phase by changing the ratio of I_{pd} and Q_{pd}, the resulting predistorted values of I and Q.

$$I_{pd} = G(I\cos(\varphi) - Q\sin(\varphi))$$

$$O_{pd} = G(I\sin(\varphi) - Q\cos(\varphi))$$

With regard to the phase correction, only the values of sin (φ) and cos (φ) are used to calculate I_{pd} and Q_{pd}. This means that, unless φ is to be used for some other purpose, e.g., a real-time plot of amplifier characteristics, it need not be explicitly calculated. Only sin (φ) and cos (φ) need to be calculated and stored.

There are several alternatives for the storage and retrieval of the correction factors. Two common approaches are table look-up and polynomial form. In using table look-up, it is normally not practical to store values of G and φ for each possible magnitude value, M. When retrieving values for correction, then the problem is what table values to use for a given M. Possible solutions include (1) choosing those at the nearest available value of M, or (2) interpolating between table values. In using a polynomial form, polynomials that are good fits to the G and φ versus M data are generated. Then for correction, these polynomials are evaluated at the signal values of M.

17.12.5 Practical Amateur Radio Implementation

When designing a robust predistortion solution for amateur radio, the expression "The devil is in the details" is to be revered. Some of these implementation challenges are discussed here.

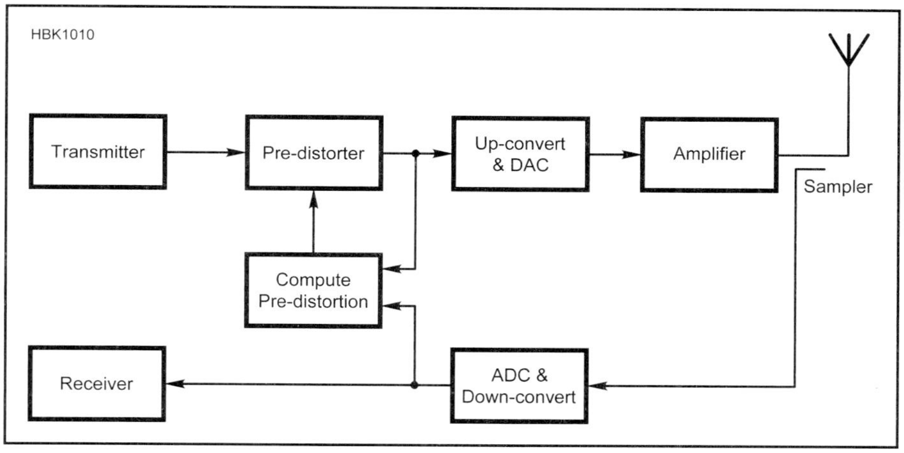

Figure 17.57 — Block diagram of an adaptive predistortion system for a direct-sampling radio.

CORRECTION BANDWIDTH

The preceding section discussed how to modify the gain and phase of the signal to achieve correction. It did not discuss what bandwidth is required for (1) the amplifier output feedback signal, and (2) the corrected upconverted signal driving the DAC. Some technical details are beyond our scope here; however, basically, if the predistorter can't "see" IMD products of a certain order, it can't correct them. This implies that the feedback bandwidth must at least accommodate the highest order products to be corrected. Likewise, if the output of the predistorter cannot convey IMD products of a certain order to the DAC, they cannot be corrected. This implies that the upconversion path must at least accommodate products to be corrected.

In practice, feedback and upconversion bandwidths of 48 kHz are generally adequate for SSB and AM operation. A frequently asked question is "Can a predistorter be implemented in an external box with the correction being applied to the microphone input signal?" The requirement for correction bandwidth is one of the reasons this is not practical since the microphone input signal path is limited by the band-pass filter. (Other reasons include lack of *I-Q* input for phase correction and the fact that other processing, such as compression, is often present in the microphone signal path.)

SAMPLE SYNCHRONIZATION

As discussed above, to calculate the correction factors, samples of the amplifier input and output that correspond in time are compared. I.e., each output sample is compared with the input sample that caused it. However, looking at the simplified diagram in Figure 17.57, it is evident that the feedback path from the amplifier output is much longer, having more delay, than that from the input. Furthermore, the amplifier itself introduces some amount of delay. Also, depending upon the communication channels, the delay from the amplifier output may not always be consistent.

DOWNCONVERSION FILTER RESPONSE

A second problem arises in that the samples from the amplifier output go through a downconversion process which includes the process of decimation to reduce the sample rate. Decimation is implemented in digital filters, which have time-domain overshoot and undershoot which can distort the signal.

A practical solution for both these problems is to provide a second downconversion and communication path, identical to the first. The samples going into the transmit DAC, i.e., the amplifier input samples, are then intercepted and processed through this second downconversion before comparison with the amplifier output samples. This ensures identical filter responses and identical downconversion and communication delays. Still remaining is some added delay in the DAC, amplifier, and ADC. This can be compensated for in software, at baseband, by resampling the feedback signals.

SCALE FACTOR (S)

For a given set of operating conditions and resulting correction solution, S is calculated to maintain precisely the same peak output power as the amplifier has demonstrated prior to the application of the new solution. This calculation can be done from the same data used to calculate the correction. In the event that a peak-power event is not present in the data, the data can be extrapolated to provide a good estimate.

CONTROLLING THE FEEDBACK LEVEL

To obtain optimum IMD reduction and overall performance and considering that the operator may adjust station power output from a few watts to 1.5 kW, an automatic mechanism to adjust the feedback level into the ADC should be provided. To the extent possible, the feedback level should be automatically adjusted to (1) maximize use of the full dynamic range of the ADC, but (2) take into account that passive IMD in the receiver filters and input circuits may increase at very high signal levels. Any passive IMD in the feedback circuit will increase the IMD of the transmitted signal. The typical result would be that the ADC is driven to within just a few dB of its clipping level. This control is easily done by monitoring the ADC peak output level and using that information to adjust a preceding attenuator.

OVER-DRIVEN AMPLIFIERS

Amplifiers that are being over-driven, to the point of having a flat or decreasing output with increasing input, present a special problem in the calculation of magnitude correction. To calculate this correction, the question to be answered is essentially "What is the input signal level that produces each particular output signal level?" However, if there is a range of high input levels for which the output is always the same, the answer to this question is indeterminate. This case can be handled by slightly modifying the measured amplifier response such that a solution is always available. Note that this only happens prior to the application of a predistortion solution; once predistortion is operating, the amplifier characteristic should never have this problem.

17.12.6 Adapting the Solution

In an amateur radio setting, having an adaptive predistortion solution that modifies itself as conditions change is nearly essential. Predistortion requires a high degree of precision since it changes the levels of very small IMD products that, with predistortion applied, may be 40 dB to 70 dB down from the desired signal. Therefore, to obtain best results, it is very important to repeatedly adjust the predistortion solution for even small changes in operating conditions. The solution must adapt to things like equipment temperature changes, small frequency changes, and certainly major events such as band changes and antenna changes.

Fortunately, while applying predistortion, the amplifier's output and input can still be measured and compared. A "next" solution can be calculated while a current solution is in use. Finally, there must be a transition from the current solution to this next solution. This adaptation process continues repeatedly.

Care must be taken in the transition from one solution to another. Considering that predistortion changes the magnitude and phase of the sample stream, an abrupt change in the solution would create a discontinuity in the waveform conveyed by the sample stream. This discontinuity would create a momentary broadening of the frequency spectrum and perhaps an audible "click." The solution is to gradually transition, over a period of a few milliseconds, from one solution to the next.

17.12.7 Memory Effects

Memory effects are present in an amplifier when the amplifier's output at the present time is affected by not only the present input but also by previous inputs. All of our amateur radio amplifiers have memory effects; however, depending upon technologies used and the design, some have more severe memory effects than others. The ability of amplifier components to store energy causes memory effects. For example, if the envelope waveform of the signal is at a high level for a brief time, solid state devices store thermal energy, i.e., they heat up, and this changes the device characteristics such as gain. The amplification of the next time-slice of signal will be affected by these characteristics. The same type of effect happens with voltages and currents. For example, the drain voltage power supply may sag due to a strong burst of signal and the voltage will have some non-zero recovery time. So, the following signal will see a different drain voltage depending upon the strength of the preceding signal. These effects have durations in the microsecond and millisecond range.

The issue for predistortion is that the gain and phase characteristics of the amplifier vary depending upon previous signals. Calculating corrections, when the amplifier characteristics are rapidly changing, presents a challenge. Collecting a set of samples and plotting amplifier output versus amplifier input, in the presence of memory effects, the samples do not line up perfectly along a curve. There is some scattering around the centerline of the curve.

A reasonably simple solution which is generally effective in amateur radio is to do a best fit of the available samples to gain and phase curves and use those curves to calculate correction. Much more advanced solutions are also available. These involve creating a predistortion solution based upon not only the input at the current time, but also based upon previous input values. To apply the solution, both the current input and previous inputs are considered. The Volterra series serves as an excellent mathematical model for such nonlinear time-varying systems. However, many other simplified models that are more practical to implement have been developed. (The Volterra series is used for modeling non-linear systems when including memory effects is important. See Kenington, P., *High-Linearity RF Amplifier Design*, (Artech House, 2000) pp.

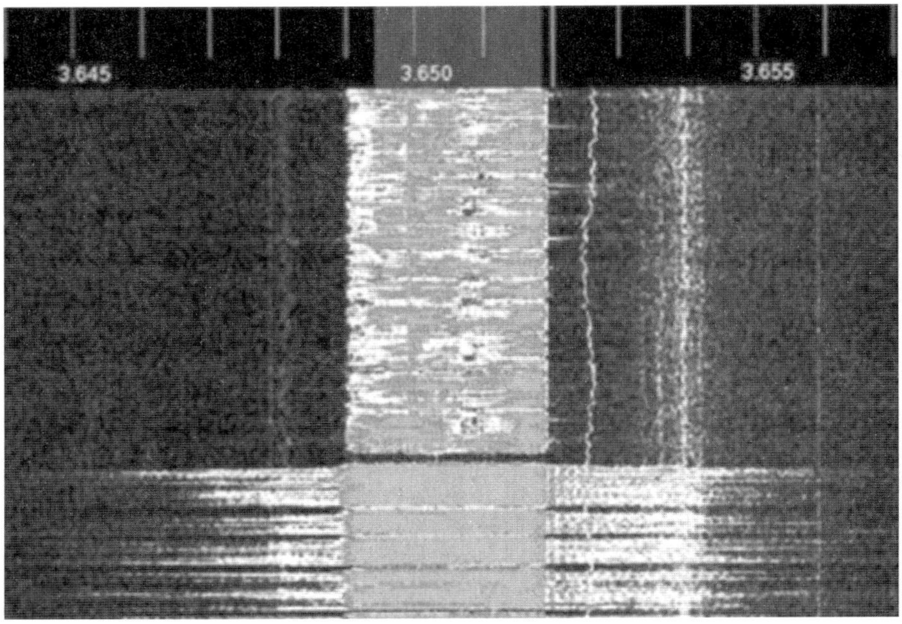

Figure 17.58 — Waterfall display showing two SSB signals without (bottom) and with (top) predistortion.

83 – 84; and Ghannouchi, F., Hammi, O., and Helaoui, M., *Behavioral Modeling and Predistortion of Wideband Wireless Transmitters*, (John Wiley & Sons Ltd, 2015) pp. 38 – 40.

17.12.8 On-the-Air Effects of Predistortion

The waterfall display in **Figure 17.58** shows a typical SSB signal (bottom) without predistortion followed by a very clean transmission using predistortion (top). The IMD produced without predistortion (at the bottom in the figure) creates a signal consuming about three times the bandwidth. Both stations were running legal-limit amplifiers.

17.13 References and Bibliography

Additional references are listed in "RF Power Amplifier References" which is included in the online material.

ARRL members can download many *QST* articles on amplifiers from: **www.arrl.org/arrl-periodicals-archive-search**. Enter the keyword "amplifiers" in the "Title/Keywords:" search window.

VACUUM TUBE AMPLIFIERS

Badger, G., W6TC, "The 811A: Grandfather of the Zero Bias Revolution," *QST*, Apr. 1996, pp. 51 – 53.

Care and Feeding of Power Grid Tubes, **www.cpii.com/library.cfm/9**

EIMAC tube performance computer, **www.cpii.com/library.cfm/9**

Measures, R., AG6K, "Improved Parasitic Suppression for Modern Amplifier Tubes," *QST*, Oct. 1988, pp. 36 – 38, 66, 89.

Orr, W., W6SAI, *Radio Handbook*, 22nd Ed (Howard W. Sams & Co, Inc, 1981).

RCA Transmitting Tubes Technical Manual TT-5, Radio Corporation of America, 1962.

Rauch, T., W8JI, "Filament Voltage life," **www.w8ji.com/filament_voltage_life.htm**.

Rauch, T., W8JI, "Neutralizing Amplifiers," **www.w8ji.com/neutralizing_amplifier.htm**.

Rauch, T., W8JI, "Loading Inductors," **www.w8ji.com/loading_inductors.htm**.

Rauch, T. W8JI, "RF Plate Chokes," **www.w8ji.com/rf_plate_choke.htm**.

Reference Data for Radio Engineers, ITT, Howard W. Sams Co, Inc.

Terman, *Electronic and Radio Engineering* (McGraw-Hill).

Wingfield, E., "New and Improved Formulas for the Design of Pi and Pi-L Networks," *QST*, Aug. 1983, pp. 23 – 29. (Feedback, *QST*, Jan. 1984, p. 49.).

Wingfield, E., "A Note on Pi-L Networks," *QEX*, Dec 1983, pp. 5 – 9.

ADDITIONAL VACUUM TUBE AMPLIFIER PROJECTS

Heck, S. LAØBY/DF9PY, "Description of a 144 MHz high power amplifier for 2 x 4CX250B."

Knadle, R., K2RIW, "A Strip-line Kilowatt Amplifier 432 MHz," *QST*, Apr. 1972, pp. 49 – 55.

Meade, E., K1AGB, "A High-Performance 50-MHz Amplifier," *QST*, Sep. 1975, pp. 34 – 38.

Meade, E., K1AGB, "A 2-KW PEP. Amplifier for 144 MHz," *QST*, Dec. 1973, pp. 34 – 38.

Peck, F., K6SNO, "A Compact High-Power Linear," *QST*, June 1961, pp. 11 – 14.

Additional projects are available with this book's online content.

SOLID STATE AMPLIFIERS

References for RF Power Amplifiers and Power Combining

Classic Works in RF Engineering: Combiners, Couplers, Transformers and Magnetic Materials, Edited by Myer, Walker, Raab, Trask. (Artech House, June 2005).

Abulet, Mihai, *RF Power Amplifiers* (Noble Publishing, 2001).

Pozar, *Microwave Engineering,* 3rd Ed. (Wiley, 2004).

Sevick, J., *Transmission Line Transformers,* 4th Ed., (Noble Publishing, 2001).

White, Joseph F., *High Frequency Techniques: An Introduction to RF and Microwave Engineering* (Wiley, 2004).

Articles

Davis and Rutledge, "Industrial Class-E Power Amplifiers With Low-cost Power MOSFETs and Sine-wave Drive" Conf. Papers for RF Design '97, Santa Clara, CA, Sep. 1997, pp. 283 – 297.

Frey, R., "A 50 MHz, 250 W Amplifier using Push-Pull ARF448A/B," APT Application Note APT9702.

Frey, R., "Low Cost 1000 Watt 300 V RF Power Amplifier for 27.12 MHz," APT Application Note APT9701.

Frey, R., "A push-pull 300 watt amplifier for 81.36 MHz," *Applied Microwave and Wireless,* vol 10 no 3, pp. 36 – 45, Apr. 1998.

Frey, R., "Push-Pull ARF449A/B Amplifier for 81.36 MHz," APT Application Note APT9801.

Granberg, H., "Broadband Transformers and Power Combining Techniques for RF," AN-749, Motorola Semiconductor Products Inc.

Granberg, H. O., WB2BHX, "One KW — Solid-State Style, Part 1," Apr. 1976, *QST*, p. 11.

Granberg, H. O., WB2BHX, "One KW — Solid-State Style, Part 2," May 1976, *QST*, p. 28.

Granberg, Helge, K7ES/OH2ZE, "Build This Solid-State Titan, Part 1," June 1977, *QST*, p. 27.

Granberg, Helge, K7ES/OH2ZE, "Build This Solid-State Titan, Part 2," July 1977, *QST*, p. 11.

Granberg, Helge, K7ES/OH2ZE, "Printed Line Techniques Applied to VHF Amplifier Design," Sep. 1979, *QST*, p. 11.

Granberg, Helge, K7ES/OH2ZE, "MOSFET RF Power: An Update - Part 1," Dec. 1982, *QST*, p. 13.

Granberg, Helge, K7ES/OH2ZE, "MOSFET RF Power: An Update - Part 2," Jan. 1983, *QST*, p. 30.

Hejhall, R. "Systemizing RF Power Amplifier Design," Motorola Semiconductor Products, Inc., Phoenix, AZ, Application note AN282A.

Hilbers, A.H., "Design of HF wideband power transformers," Philips Semiconductors, ECO 6907, Mar. 1998.

Hilbers, A.H., "Design of HF wideband power transformers Part II," Philips Semiconductors, ECO 7213, Mar. 1998.

Hilbers, A.H., "Power Amplifier Design," Philips Semiconductors, Application Note, Mar. 1998.

Klitzing, J, W6PQL, "Build a Linear 2 Meter 80 W All Mode Amplifier," *QST*, May 2013, pp 30 – 34. Feedback, July 2013 *QST*, p. 40.

Trask, C., N7ZWY, "Designing Wide-band Transformers for HF and VHF Power Amplifiers," *QEX,* Mar./Apr. 2005, pp. 3 – 15.

Contents

Chapter 18

Repeater Systems

For decades, FM has been a mainstay of amateur radio operation. FM and repeaters fill the VHF and UHF bands, and most hams have at least one handheld or mobile FM radio. Thousands of repeaters throughout the country provide reliable communication for amateurs operating from portable, mobile, and home stations.

Digital voice systems are now beginning to dominate new installations. Along with the well-established D-STAR (now offered by Kenwood as well as Icom), Yaesu's System Fusion and Digital Mobile Radio (DMR) are two popular new systems. DMR makes extensive use of the internet to build hybrid systems similar to the trunking systems used by public service agencies. A whole new era of repeater-based communications is at hand.

The repeater overview material was written by Gary Pearce, KN4AQ. FM analog repeaters are described by Paul Danzer, N1II. The section on D-STAR, originally by Texas Interconnect Team members Jim McClellan, N5MIJ, and Pete Loveall, AE5PL, was updated by John Davis, WB4QDX, and Ed Woodrick, WA4YIH. John Burningham, W2XAB, continues to maintain the DMR section, and Cory Sickles, WA3UVV, has updated the System Fusion and P25 sections.

18.1 Amateur Repeater History

Few hams today don't operate at least some VHF/UHF FM, and for many hams, FM *is* Amateur Radio. That wasn't always the case.

Until the late 1960s, the VHF and UHF amateur radio bands were home to a relatively small number of highly skilled operators who used CW and SSB for long distance communication and propagation experiments. This operation used just a small fraction of the spectrum available at 50 MHz and above. A somewhat larger number of hams enjoyed low power, local operation with AM transceivers on 6 and 2 meters. Our spectrum was underutilized, while public safety and commercial VHF/UHF two-way operation, using FM and repeaters, was expanding rapidly.

The business and public-safety bands grew so rapidly in the early 1960s that the FCC had to create new channels by cutting the existing channels in half. Almost overnight, a generation of tube-type, crystal-controlled FM equipment had to be replaced with radios that met the new channel requirements. Surplus radios fell into the hands of hams for pennies on the original dollar. This equipment was designed to operate around 150 MHz and 450 MHz, just above the 2 meter and 70 cm amateur bands. It was fairly easy to order new crystals and retune them to operate inside the ham bands. Hams who worked in the two-way radio industry led the way, retuning radios and building repeaters that extended coverage. Other hams quickly followed, attracted by the noise-free clarity of FM audio, the inexpensive equipment, and the chance to do something different.

That initial era didn't last long. The surplus commercial equipment was cheap, but it was physically large, ran hot, and consumed lots of power. By the early 1970s, manufacturers recognized an untapped market and began building solid-state equipment specifically for the amateur radio FM market. The frequency synthesizer, perfected in the mid-1970s, eliminated the need for crystals. The stage was set for an explosion that changed the face of amateur radio. Manufacturers have added plenty of new features to equipment over the years, but the basic FM operating mode remains the same.

In the 1980s, hams experimenting with data communications began modulating their FM radios with tones. *Packet radio* spawned a new system of *digipeaters* (digital repeaters).

Digitized audio has been popular since audio compact discs (CDs) were introduced in the 1980s. In the 1990s, technology advanced enough to reduce the bandwidth needed for digital audio, especially voice, to be carried over the internet and narrowband radio circuits. The first digital-voice public safety radio systems (called APCO-Project 25) appeared, and a variety of Internet voice systems for conferencing and telephone-like use were developed.

Hams are also using VHF/UHF digital voice technology. The Japan Amateur Radio League (JARL) developed a true ham radio standard called D-STAR, a networked VHF/UHF repeater system for digital voice and data that is beginning to make inroads around the world. Yaesu has developed a hybrid analog/digital system called System Fusion. Another digital voice system, generically known as DMR (Digital Mobile Radio) and by the Motorola trade name MOTOTRBO, is being adapted from commercial service for the ham bands. Hams are also using surplus Project 25 (or just P25) radios.

18.2 Repeater Overview

Amateurs learned long ago that they could get much better use from their mobile and portable radios by using an automated relay station called a *repeater*. Home stations benefit as well — not all hams are located near the highest point in town or have access to a tall tower. A repeater, whose basic idea is shown in **Figure 18.1**, can be more readily located where the antenna system is as high as possible and can therefore cover a much greater area.

18.2.1 Types of Repeaters

The most popular and well-known type of amateur repeater is an FM voice system on the 144 or 440 MHz bands. Amateurs operate many repeaters on the 29, 50, 222, 902, and 1240 MHz bands as well, but 2 meters and 70 cm are the most popular. Tens of thousands of hams use mobile and handheld radios for casual ragchewing, emergency communications, public service activities, or just staying in touch with their regular group of friends during the daily commute.

While the digital voice modes are gaining ground, FM is still the most popular mode for voice repeaters. Operations are *channelized* — all stations operate on specific, planned frequencies, rather than the more or less random frequency selection employed in CW and SSB operation. In addition, since the repeater receives signals from mobile or fixed stations and retransmits these signals simultaneously, the transmit and receive frequencies are different, or *split.* Direct contact between two or more stations that listen and transmit on the same frequency is called operating *simplex.*

Individuals, clubs, emergency communications groups, and other organizations all sponsor repeaters. Anyone with a valid amateur license for the band can establish a repeater in conformance with the FCC rules. No one owns specific repeater frequencies, but nearly all repeaters are *coordinated* to minimize repeater-to-repeater interference. Frequency coordination and interference are discussed later in this chapter. Operational aspects are covered in more detail in *The ARRL Operating Manual.* Special operating notes for the digital repeaters are included in the sections on each mode.

18.2.2 Advantages of Using a Repeater

When we use the term *repeater* we are almost always talking about transmitters and receivers on VHF or higher bands, where radio-wave propagation is normally line of sight. Don't take "line of sight" too literally. VHF/UHF radio signals do refract beyond what you can actually see on the horizon, but the phrase is a useful description. (See the **Propagation of Radio Signals** chapter for more information on these terms.)

We know that the effective range of VHF and UHF signals is related to the height of each antenna. Since repeaters can usually be located at high points, one great advantage of repeaters is the extension of coverage area from low-powered mobile and portable transceivers.

Figure 18.2 illustrates the effect of using a repeater in areas with hills or mountains. The same effect is found in metropolitan areas, where buildings provide the primary blocking structures.

Siting repeaters at high points can also have disadvantages. Since most repeaters have co-channel neighbors (other repeaters operating on the same channel) less than 150 miles away, there may be times when your transceiver can receive both. But since it operates FM, the *capture effect* usually ensures that the stronger signal will capture your receiver and the weaker signal will not be heard — at least as long as the stronger repeater is in use.

It is also simpler to provide a very sensitive receiver, a good antenna system, and a slightly higher power transmitter at just one location — the repeater — than at each mobile, portable, or home location. A superior repeater system compensates for the low power (5 W or less) and small, inefficient antennas that many hams use to operate through them. The repeater maintains the range or coverage we want, despite our equipment deficiencies. If both the handheld transceiver and the repeater are at high elevations, for example, communication is possible over great distances, despite the low output power and inefficient antenna of the transceiver (see **Figure 18.3**).

Repeaters also provide a convenient meeting place for hams with a common interest.

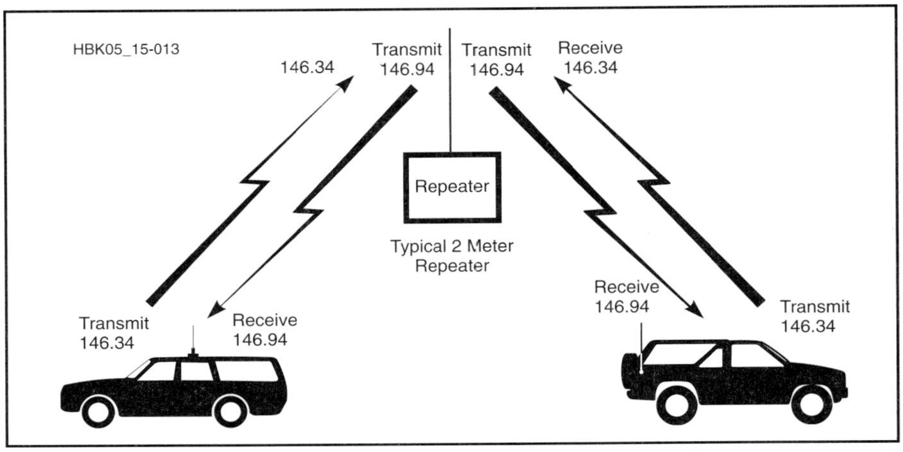

Figure 18.1 — Typical 2 meter repeater, showing mobile-to-mobile communication through a repeater station. Usually located on a hill or tall building, the repeater amplifies and retransmits the received signal on a different frequency.

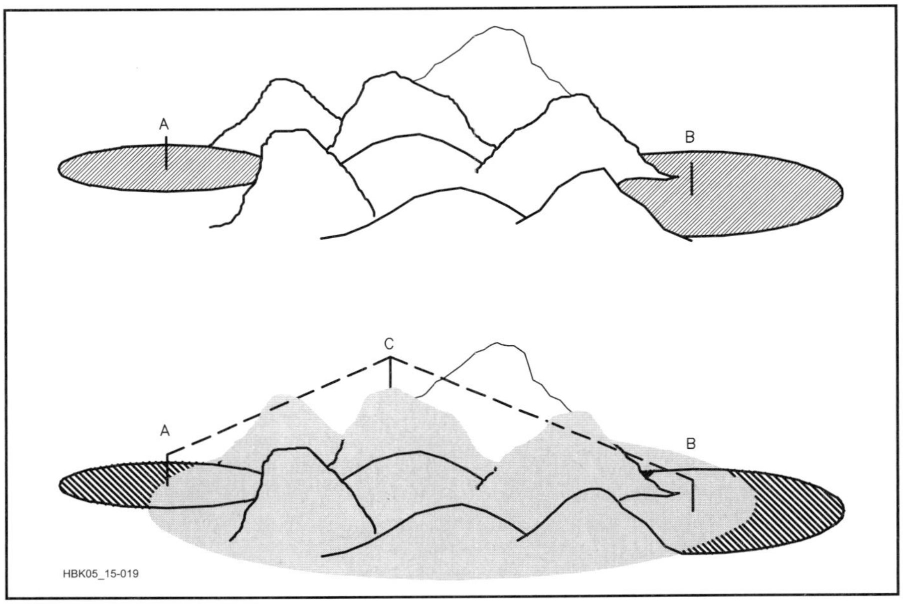

Figure 18.2 — In the upper diagram, stations A and B cannot communicate because their mutual coverage is limited by the mountains between them. In the lower diagram, stations A and B can communicate because the coverage of each station falls within the coverage of repeater C, which is on a mountaintop.

It's usually geographic — your town, or your club. A few repeaters are dedicated to a particular interest such as DX or passing traffic. Operation is channelized, and usually in any area you can find out which channel — or repeater — to pick to ragchew, get highway information, or whatever your need or interest is. The conventional wisdom is that you don't have to tune around and call CQ to make a contact, as on the HF bands. Simply call on a repeater frequency — if someone is there and they want to talk, they will answer you. But with a few dozen repeaters covering almost any medium-size town, you probably use the scan function in your radio to seek activity.

EMERGENCY OPERATIONS

When there is a weather-related emergency or a disaster (or one is threatening), most repeaters in the affected area immediately spring to life. Emergency operation and traffic always take priority over other ham activities, and many repeaters are equipped with emergency power sources just for these occasions (see **Figure 18.4**).

Almost all amateur radio emergency organizations use repeaters to take advantage of their extended range, uniformly good coverage, and visibility. Most repeaters are well known — everyone active in an area with suitable equipment knows the local repeater frequencies.

18.2.3 Digital Voice Repeater Systems

Digital voice (DV) transmits digitized audio speech as a digital data stream over the RF or wired media. The process of converting between analog voice and the digital data stream utilizes a vocoder (voice encoder). A package of software or firmware that performs the encoding and decoding of the speech is called a codec (coder-decoder). The transmitted DV data stream is encapsulated inside a protocol that handles the addressing and communication management. For a sender and receiver to communicate, both must be using the same vocoder and protocol.

See the **Digital Protocols and Modes** chapter for more information about amateur radio DV modes. The websites listed in the References and Bibliography section of this chapter also contain information about the DV modes and their repeater systems.

DIGITAL VS ANALOG VOICE

If you are used to operating on analog FM repeaters, you will have noticed that the audio quality degrades gradually. As a station's signal into the repeater gets weaker there is an increase in noise bursts mixed with the audio until the signal gets so weak that the station can no long access the repeater, or you cannot understand the audio because of noise. Similarly, as you move further from the repeater you will start hearing the same noise as the repeater's transmitted signal gets weaker. A combination of a station's weak signal into a repeater and a repeater's weak signal to the listener degrades usability faster.

All the major DV technologies include *forward error correction* (FEC), which can correct limited errors, slightly extending the usable range and improving communication quality. Better quality receivers can operate at a lower noise floor, and higher power transmitters and higher gain antenna systems will also extend coverage of both analog and digital systems.

Figure 18.3 — In the Rocky Mountain west, handheld transceivers can often cover great distances, thanks to repeaters located atop high mountains. [Photo courtesy Rachel Witte, KCØETU, and Bob Witte, KØNR]

For DV repeaters the audio quality remains the same for both input and output to the edge of the coverage range. At that point, the audio becomes broken (missing portions of the speech) caused by lost packets in the digital protocol. Further signal degradation causes complete signal loss. (The same effect is seen on digital television [DTV] as the image becomes pixilated and finally is dropped completely.) The internet can also drop packets used for transferring the audio data stream between repeaters and bridges, causing the same broken audio affect.

DIGITAL VOICE SYSTEMS

DV repeaters can operate as standalone stations but are usually linked together using the internet to form a repeater system. The interface between a repeater and the internet is usually referred to as a gateway. Each type of system then uses a proprietary protocol specific to that technology to communicate information between the repeater gateways. Operation of these repeater systems and descriptions of the user functions and features are discussed in separate sections later in this chapter.

Once digitized, the voice information in a DV system remains in digital form until it is converted to audio for the user to hear. Repeater systems only retransmit the digital data stream. They do not process the voice

Figure 18.4 — Repeaters provide communication over a wide area to support disaster relief, emergency communication, and public service. They allow portable and mobile stations to communicate effectively with each other and with served agencies and event management. The operators shown here (F-B AE5MT, KE5YBC, and KE5NCR) are supporting an Oklahoma City marathon, getting valuable training while providing a service to the community. (Photo by Frank Tassone, KE5KQL)

Hotspots — What Are They?

A "hotspot" is an interface constructed with a very low power transceiver that interconnects to the internet through an attached computer that is external or integrated to the hotspot. This is similar to a Wi-Fi access point. The user's radio communicates with the hotspot transceiver, which then interacts with the desired repeater via the internet, performing any necessary translations between the digital voice (DV) protocols.

Some hotspots currently available include the DV4mini, openSPOT, DVMEGA, MMDVM, and Micro Node Nano-DV. Most hotspots support DMR, D-STAR, System Fusion, NXDN, and P25 protocols. Hotspots for DMR are supported by both the Brandmeister and DMRplus networks and offer connectivity to other networks.

A hotspot also allows a user outside the coverage area of a repeater to access parts of the network using their radio. The hotspot then connects to the desired repeater via the internet, similarly to EchoLink.

information as analog audio. Digital data such as IDs, addressing or routing information, user data messages, and other data structures may or may not be relayed with the digital voice, depending on the type of system and the configuration of the repeaters and gateway equipment.

D-STAR and System Fusion protocols both include the amateur's call sign in the transmitted digital data. DMR utilizes a Radio ID number for users and repeaters that amateurs must obtain from DMR-MARC (Motorola Amateur Radio Club), which serves as a central registry for worldwide amateur users.

DIGITAL VOICE SYSTEM STANDARDS

The primary DV technologies utilized for repeater systems in the amateur VHF/UHF bands are D-STAR (Digital Smart Technologies for Amateur Radio), DMR (Digital Mobile Radio), System Fusion, and P25. Both D-STAR and System Fusion were designed specifically for the amateur radio market. Developed by the European Telecommunications Standards Institute (ETSI) for government and commercial users, DMR has been adapted for amateur radio use. P25 is an adaptation of the APCO Project 25 standard.

D-STAR is supported by Icom and Kenwood according to the standard originally developed for the Japanese Amateur Radio League (JARL). Yaesu is the only manufacturer for System Fusion. There are over a dozen manufacturers producing DMR compatible equipment including Motorola Solutions, Hytera, Broadcom, Vertex Standard, TYT, Connect Systems, BFDX, Kirisun, and Kenwood.

18.2.4 Non-Voice Repeater Uses

In addition to FM voice, there are several other types of ham radio repeaters.

ATV (amateur TV) — ATV repeaters are used to relay wideband television signals in the 70 cm and higher bands. They are used to extend coverage areas, just like voice repeaters. ATV repeaters are often set up for *cross-band* operation, with the input on one band (say, 23 cm) and the output on another (say, 70 cm). More information on ATV repeaters may be found in the **Image Communications** chapter in the online content.

Digipeaters — Digital repeaters are used primarily for packet communications, including APRS (the Amateur Packet/position Reporting System). They can use a single channel (single port) or several channels (multi-port) on one or more VHF and UHF bands. See the **Digital Protocols and Modes** chapter and the **Digital Communications** chapter in the *Handbook's* online content for details of these systems.

Multi-channel (wideband) — Amateur satellites are the best-known examples. Wide bandwidth (perhaps 50 to 200 kHz) is selected to be received and transmitted so all signals in bandwidth are heard by the satellite (repeater) and retransmitted, usually on a different VHF or UHF band. See the **Space Communications** chapter in the *Handbook's* online content for details.

18.2.5 Repeaters and FCC Regulations

Repeaters are specifically authorized by the FCC rules. For a brief period when the repeater concept was new in the amateur service, the FCC required special repeater licenses identified by a "WR" call sign prefix and a fairly complex application process. While that complexity is gone, repeaters are still restricted to certain band segments and have lower maximum power limits. But most of the "rules" that make our repeater systems work are self-imposed and voluntary. Hams have established frequency coordination, band plans, calling frequencies, digital protocols and rules that promote efficient communication and interchange of information.

FCC rules on prohibited communication have also been somewhat relaxed, allowing hams to communicate with businesses, and allowing employees of emergency-related agencies and private companies to participate in training and drills while "on the clock." There are significant restrictions to this operation, so for the latest rules and how to interpret them, see *QST* and the ARRL website, **www.arrl.org**.

FREQUENCY COORDINATION AND BAND PLANS

Since repeater operation is channelized, with many stations sharing the same frequency pairs, the amateur community has formed coordinating groups to help minimize conflicts between repeaters and among repeaters and other modes. Over the years, the VHF and UHF bands have been divided into repeater and non-repeater sub-bands. These frequency-coordination groups maintain lists of available frequency pairs in their areas (although in most urban areas, there are no "available" pairs on 2 meters, and 70 cm pairs are becoming scarce). A complete list of frequency coordinators, band plans and repeater pairs is included in the *ARRL Repeater Directory*.

Each VHF and UHF repeater band has been subdivided into repeater and nonrepeater channels. In addition, each band has a specific *offset* — the difference between the transmit frequency and the receive frequency for the repeater. While most repeaters use these standard offsets, others use "oddball splits." These nonstandard repeaters are generally also coordinated through the local frequency coordinator. **Table 18.1** shows the standard frequency offsets for each repeater band.

FM repeater action isn't confined to the VHF and UHF bands. There are a large handful of repeaters on 10 meters around the US and the world. "Wideband" FM is permitted only above 29.0 MHz, and there are four band-plan repeater channels (outputs are 29.62, 29.64, 29.66 and 29.68 MHz), plus the simplex channel 29.60 MHz. Repeaters on 10 meters use a 100 kHz offset, so the corresponding inputs are 29.52, 29.54, 29.56 and 29.58 MHz. During band openings, you can key up a repeater thousands of miles away, but that also creates the potential for interference generated when multiple repeaters are keyed up at the same time. CTCSS would help reduce the problem, but not many 10 meter repeater owners use it and too many leave their machines on "carrier access."

Table 18.1
Standard Frequency Offsets for Repeaters

Band	*Offset*
29 MHz	–100 kHz
52 MHz	Varies by region –500 kHz, –1 MHz, –1.7 MHz
144 MHz	+ or –600 kHz
222 MHz	–1.6 MHz
440 MHz	+ or –5 MHz
902 MHz	12 MHz
1240 MHz	12 MHz

18.3 FM Voice Repeaters

Repeaters normally contain at least the sections shown in **Figure 18.5**. Repeaters consist of a receiver and transmitter plus two more special devices. One is a *controller*, which routes the audio between the receiver and transmitter, keys the transmitter, and provides remote control for the repeater licensee or designated control operators.

The second device is the *duplexer*, which lets the repeater transmit and receive on the same antenna. A high power transmitter and a sensitive receiver, operating in close proximity within the same band and using the same antenna, present a serious technical challenge. You might think the transmitter would just blow the receiver away. But the duplexer keeps the transmit energy out of the receiver with a series of tuned circuits. Without a duplexer, the receiver and transmitter would need separate antennas, and those antennas would need to be 100 or more feet apart on a tower. Some repeaters do just that, but most use duplexers. A 2 meter duplexer is about the size of a two-drawer filing cabinet (see **Figure 18.6**).

Receiver, transmitter, controller, and duplexer: the basic components of most repeaters. After this, the sky is the limit on imagination. As an example, a remote receiver site can be used to extend coverage (**Figure 18.7**).

The two sites can be linked either by telephone ("hard wire") or a VHF or UHF link. Once you have one remote receiver site, it is natural to consider a second site to better hear those "weak mobiles" on the other side of town (**Figure 18.8**). Some of the stations using the repeater are on 2 meters while others are on 70 cm? Just link the two repeaters! (See **Figure 18.9**).

For even greater flexibility, you can add an auxiliary receiver, perhaps for a NOAA weather channel (**Figure 18.10**).

The list goes on and on. Perhaps that is why so many hams have put up repeaters.

18.3.1 FM Repeater Operation

There are almost as many operating procedures in use on repeaters as there are repeaters. Only by listening can you determine the customary procedures on a particular machine. A number of common operating techniques are found on many repeaters, however.

One such common technique is the transmission of *courtesy tones*. Suppose several stations are talking in rotation — one following another. The repeater detects the end of a transmission of one user, waits a few seconds, and then transmits a short tone or beep. The next station in the rotation waits until the beep before transmitting, thus giving any other station wanting to join in a brief period to transmit their call sign. Thus the term *courtesy tone* — you are politely pausing to allow other stations to join in the conversation.

Another common repeater feature that encourages polite operation is the *repeater timer*. A 3-minute timer is actually designed to comply with an FCC rule for remotely controlled stations, but in practice the timer serves a more social function. Since repeater operation is channelized — allowing many

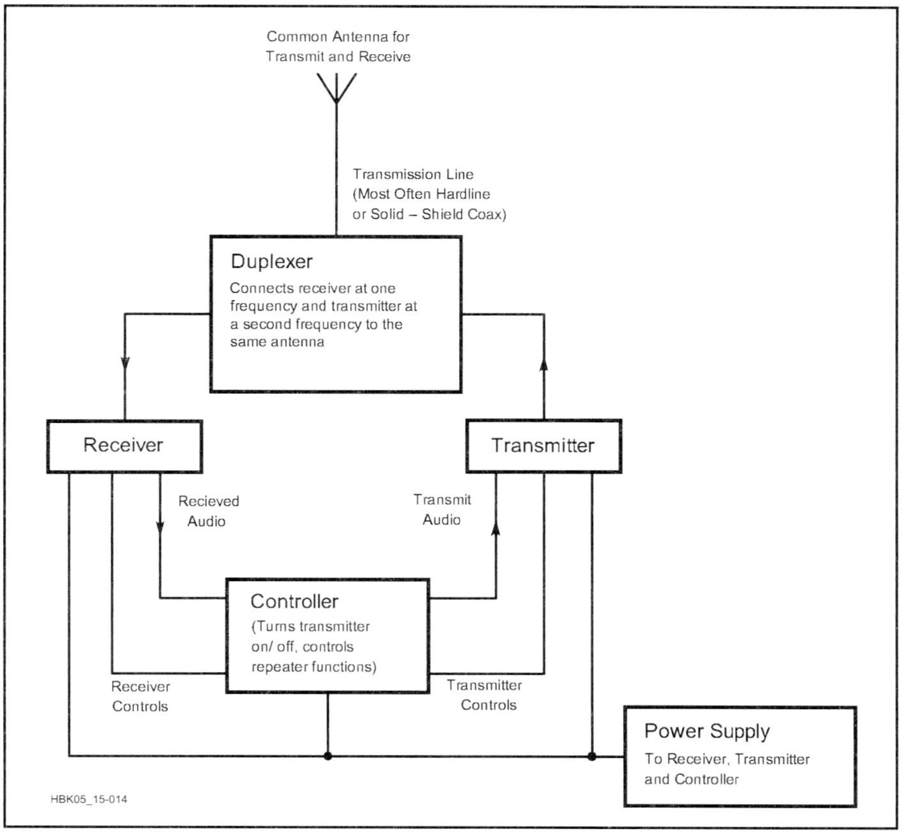

Figure 18.5 — The basic components of a repeater station. In the early days of repeaters, many were home-built. Today, most are commercial, and are far more complex than this diagram suggests.

Figure 18.6 — The W4RNC 2 meter repeater includes the repeater receiver, transmitter, and controller in the rack. The large object underneath is the duplexer. [Photo courtesy Gary Pearce, KN4AQ]

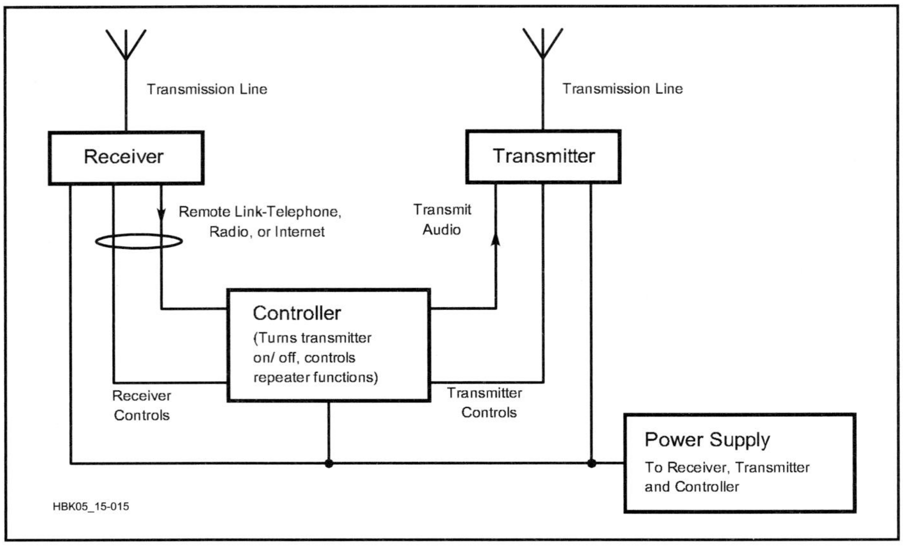

Figure 18.7 — Separating the transmitter from the receiver can extend the repeater's coverage area. The remote receiver can be located on a different building or hill, or consist of a second antenna at a different height on the tower.

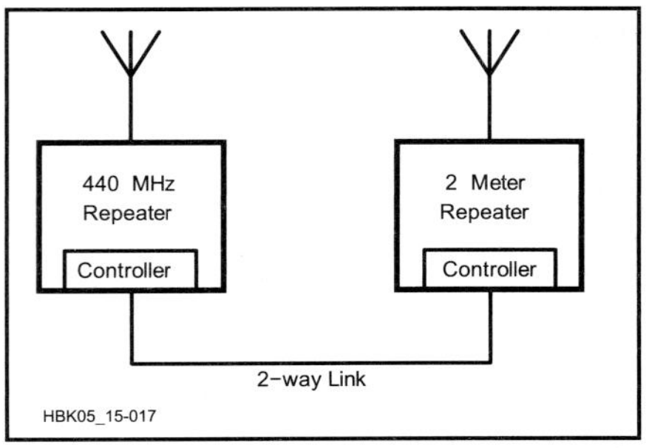

Figure 18.9 — Two repeaters using different bands can be linked for added convenience.

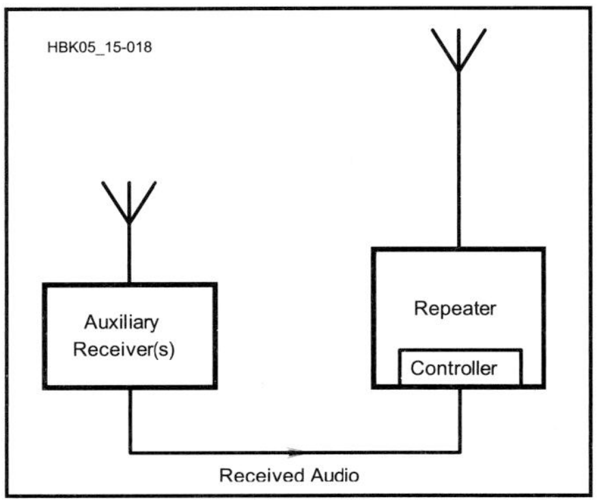

Figure 18.10 — For even greater flexibility, you can add an auxiliary receiver.

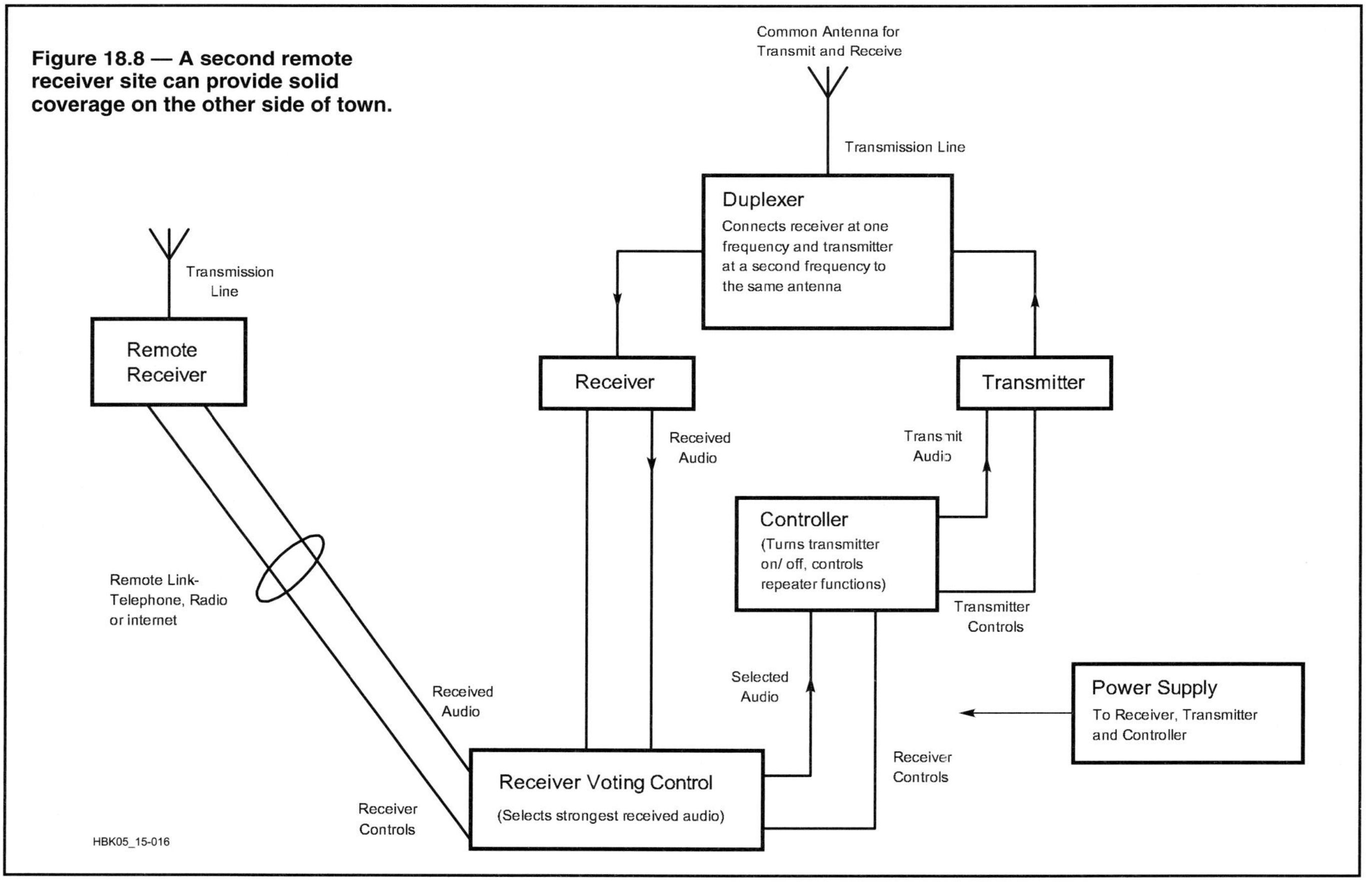

Figure 18.8 — A second remote receiver site can provide solid coverage on the other side of town.

stations to use the same frequency — it is polite to keep your transmissions short. If you forget this little politeness, many repeaters simply cut off your transmission after 2 or 3 minutes of continuous talking. If the repeater *times out*, it remains off the air until the station on the input frequency stops transmitting.

A general rule, in fact law — both internationally and in areas regulated by the FCC — is that emergency transmissions always have priority. These are defined as relating to life, safety, and property damage. Many repeaters are voluntarily set up to give mobile stations priority, at least in checking into the repeater. If there is going to be a problem requiring help, the request will usually come from a mobile station. This is particularly true during rush hours; some repeater owners request that fixed stations limit their use of the repeater during these hours.

Some parts of the country have one or more *closed repeaters*. These are repeaters whose owners wish, for any number of reasons, to not make them available for general use. Often they require transmission of a *subaudible* or *CTCSS* tone (discussed later). Not all repeaters requiring a CTCSS tone are closed — many open repeaters use tones to minimize interference among machines in adjacent areas using the same frequency pair. Other closed repeaters require the transmission of a special tone sequence to turn on. It is desirable that all repeaters, including generally closed repeaters, be made available at least long enough for the presence of emergency information to be made known.

Repeaters have many uses. In some areas they are commonly used for formal traffic nets, replacing or supplementing HF nets. In other areas they are used with tone alerting for severe weather nets. Even when a particular repeater is generally used for ragchewing, it can be linked for a special purpose. As an example, ARRL volunteer officials may hold periodic section meetings across their state with linked repeaters allowing both announcements and questions directed back to them.

One of the most common and important uses of a repeater is to aid visiting hams. Since repeaters are listed in the *ARRL Repeater Directory* and other directories, hams traveling across the country with mobile or handheld radios often check into local repeaters asking for travel route, restaurant, or lodging information. Others just come on the repeater to say hello to the local group. In most areas courtesy prevails — the visitor is given priority to say hello or get the needed help.

Detailed information on repeater operating techniques is included in a full chapter of the *ARRL Operating Manual*.

18.3.2 Home, Mobile, and Handheld Equipment

There are many options available in equipment used on repeaters. A number of these options are shown in **Figure 18.11**.

HANDHELD TRANSCEIVERS

A basic handheld radio with an output power of 500 mW to 5 W can be used almost anywhere — in a building, walking down a street, or in a car.

Several types of antennas can be used in the handheld mode. The smallest and most convenient is a rubber flex antenna, known as a "rubber duck," a helically wound antenna encased in a flexible tube. Unfortunately, to obtain the small size the use of a wire helix or coil produces a very low efficiency.

A quarter-wave whip, which is about 19 inches long for the 2-meter band, is a good choice for enhanced performance. The rig and your hand act as a ground plane, and a reasonably efficient result is obtained. A longer antenna, consisting of several electrical quarter-wave sections in series, is also commercially available. Although this antenna usually produces extended coverage, the mechanical strain of 30 or more inches of antenna mounted on the radio's antenna connector can cause problems. After several months, the strain may require replacement of the connector.

Most newer handheld radios are supplied with lithium-ion (Li-ion) batteries. These high-capacity batteries are lightweight and allow operation for much longer periods than the classic NiCd battery pack. Charging is accomplished either with a "quick" charger in an hour or less or with a trickle charger overnight. See the **Power Sources** chapter for more information on batteries.

Power levels higher than 7 W may cause a safety problem on handheld units, since the antenna is usually close to the operator's head and eyes. See the **Safe Practices** chapter for more information.

For mobile operation, an external antenna provides much greater range than the "rubber duck" as discussed in the following Mobile Equipment section. A power cord plugs into the vehicle cigarette lighter so that the battery remains charged, and a speaker-microphone adds convenience. In addition, commercially available "brick" amplifiers can be used to raise the output power level of the handheld radio to 50 W or more. Many hams initially go this route, but soon grow tired of frequently connecting and disconnecting all the accessories from the handheld radio and install a permanent mobile radio.

MOBILE EQUIPMENT

Compact mobile transceivers operate from 11-15 V dc and generally offer several transmit power levels up to about 50 W. They can operate on one or more bands. Most common are the single-band and dual-band transceivers. "Dual-band" usually means 2 meters and 70 cm, but other combinations are available, as are radios that cover three or more bands.

Mobile antennas range from the quick and easy magnetic mount to permanently mounted assemblies. The four general classes of mobile antennas shown in the center section of Figure 18.11 are the most popular choices. Before experimenting with antennas for your vehicle, there are some precautions to be taken.

Through-the-glass antennas: Rather than trying to get the information from your dealer or car manufacturer, test any such antenna first using masking tape or some other temporary technique to hold the antenna in place. Some windshields are metalicized for defrosting, tinting, and car radio reception. Having this metal in the way of your through-the-glass antenna will seriously decrease its efficiency.

Magnet-mount or "mag-mount" antennas are convenient if your car has a metal roof or trunk. The metal also serves as the ground plane. They work well, but are not a good long-term solution. Eventually they'll scratch the car's paint, and the coax run through a door can be subject to flexing or crimping and can eventually fail.

Through-the-body antenna mounting: Most hams are reluctant to drill a hole in their car roof or trunk, unless they intend to keep the car for a long time. This mounting method provides the best efficiency, however, since the metal body panel serves as a ground plane. Before you drill, carefully plan and measure how you intend to get the antenna cable past headliners or trim to the radio. Be especially careful of side-curtain air bags. Commercial two-way shops can install antennas and power cables, usually for a reasonable price.

Trunk lip and clip-on antennas: These antennas are good compromises. They are usually easy to mount and they perform acceptably. Cable routing must be planned. If you are going to run more than a few watts, do not mount the antenna close to one of the car windows — a significant portion of the radiated power may enter the car interior.

More information on mobile equipment may be found in the **Assembling a Station** and **Antennas** chapters.

HOME STATION EQUIPMENT

Most "base station" FM radios are actually mobile rigs, powered either from rechargeable batteries or ac-operated power supplies. Use of batteries has the advantage of providing back-up communications ability in the event of a power interruption. Some HF transceivers designed for fixed-station

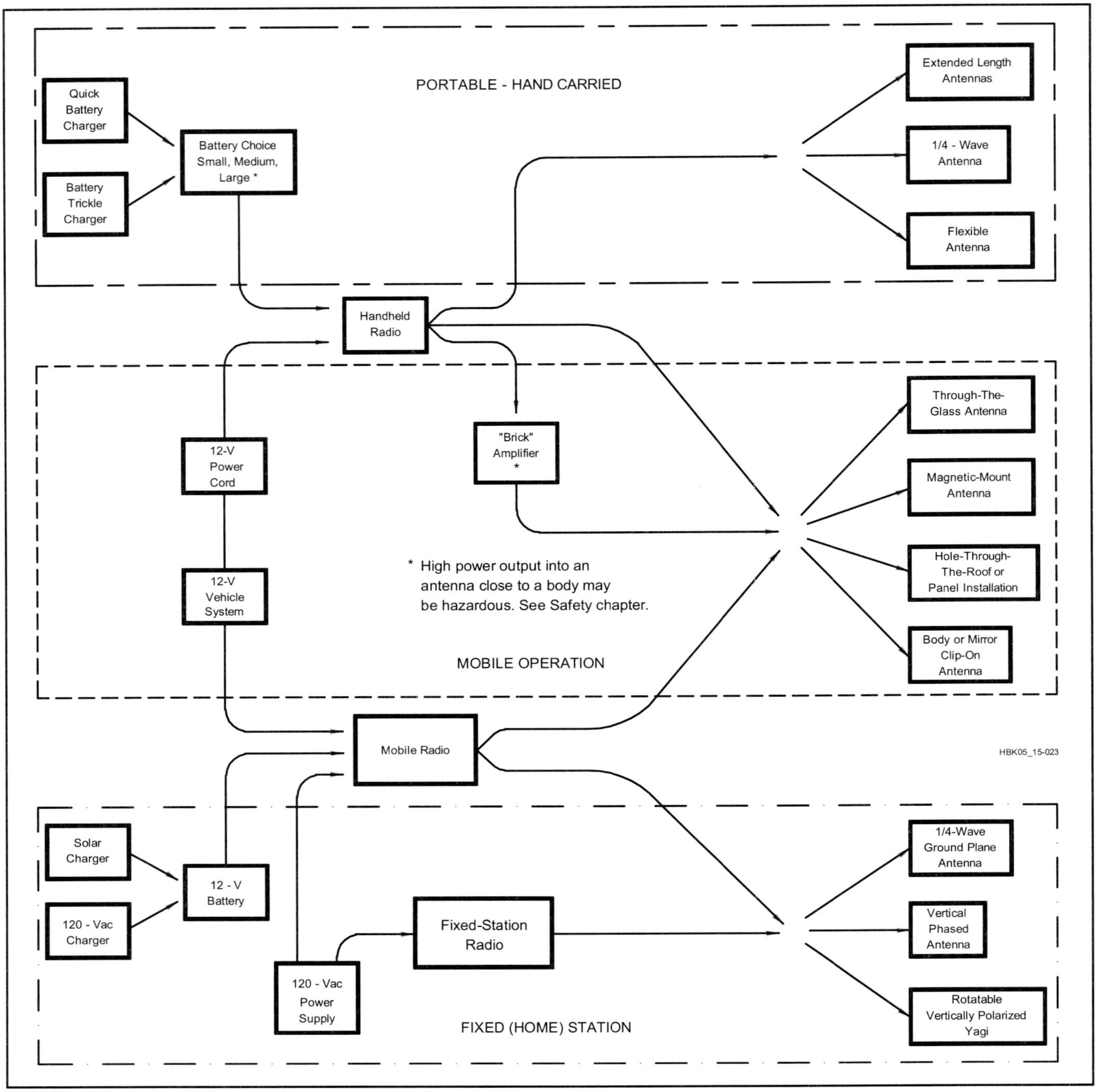

Figure 18.11 — Equipment choices for use with repeaters are varied. A handheld transceiver is perhaps the most versatile type of radio, as it can be operated from home, from a vehicle and from a mountaintop.

use also offer operation on the VHF or VHF/UHF bands, using SSB and CW in addition to FM. Using them means that you will not be able to monitor a local FM frequency while operating HF.

The general choice of fixed-location antennas is also shown in Figure 18.11. Most hams use an omnidirectional vertical, but if you are in an area between two repeaters on the same channel, a rotatable Yagi may let you pick which repeater you will use without interfering with the other repeater. Vertical polarization is the universal custom, since it is easiest to accomplish in a mobile installation. VHF/UHF SSB operation is customarily horizontally polarized. An operator with a radio that does both has a tough choice, as there can be a serious performance hit between stations using cross-polarized antennas.

Both commercial and homemade $\frac{1}{4}$-λ and larger antennas are popular for home use. A number of these are shown in the **Antennas** chapter. Generally speaking, $\frac{1}{4}$-λ sections may be stacked up to provide more gain on any band. As you do so, however, more and more power is concentrated toward the horizon. This may be desirable if you live in a flat area (see **Figure 18.12**).

18.3.3 Coded Squelch and Tones

Squelch is the circuit in FM radios that turns off the loud rush of noise with no signal present. Most of the time, hams use *noise squelch*, also called *carrier squelch*, a squelch circuit that lets any signal at all come through. But there are ways to be more selective about what signal gets to your speaker or keys up your repeater. That's generically known as *coded squelch*, and more than half of the repeaters on the air require you to send coded squelch to be able to use the repeater.

CTCSS

The most common form of coded squelch has the generic name *CTCSS* (Continuous Tone Coded Squelch System), but it is better known by the nickname "tone." Taken from the commercial services, subaudible tones are generally not used to keep others from using a repeater but rather are a method of minimizing interference from users of the same repeater frequency. CTCSS tones are sine-wave audio tones between 67 and 250 Hz that are added to the transmit audio at a fairly low level. They are *sub*audible only because your receiver's audio circuit is supposed to filter them out. A receiver with CTCSS will remain silent to all traffic on a channel unless the transmitting station is sending the correct tone. Then the receiver sends the transmitted audio to its speaker.

For example, in **Figure 18.13** a mobile station on a hill is nominally within the normal coverage area of the Jonestown repeater (146.16/76). The Smithtown repeater, also on the same frequency pair, usually cannot hear stations 150 miles away. The mobile is on a hill and so it is in the coverage area of both Jonestown and Smithtown. Whenever the mobile transmits both repeaters hear it.

The common solution to this problem, assuming it happens often enough, is to equip the Smithtown repeater with a CTCSS decoder and require all users of the repeater to transmit a CTCSS tone to access the repeater. Thus, the mobile station on the hill does not come through the Smithtown repeater, since he is not transmitting the required CTCSS tone.

Table 18.2 shows the available CTCSS tones. Most radios built since the early 1980s have a CTCSS encoder built in, and most radios built since the early 1990s also have a CTCSS *decoder* built in. Newer radios have a "tone scan" feature that will hunt the tone, *if* the repeater output includes a tone. Most repeaters that require tone also transmit their tone, but they don't have to. Listings in the *ARRL Repeater Directory* include the CTCSS tone required, if any.

If your local repeater sends a CTCSS tone, you can use your decoder to monitor just that repeater and avoid hearing the co-channel neighbor, intermod, or the annoying fizzes of nearby consumer electronics. Radios typically store CTCSS frequency and mode in their memory channels.

DIGITAL-CODED SQUELCH (DCS)

A newer form of coded squelch is called *DCS* (Digital-Coded Squelch). DCS appeared in commercial service because CTCSS didn't provide enough tones for the many users. DCS adds another 100 or so code options.

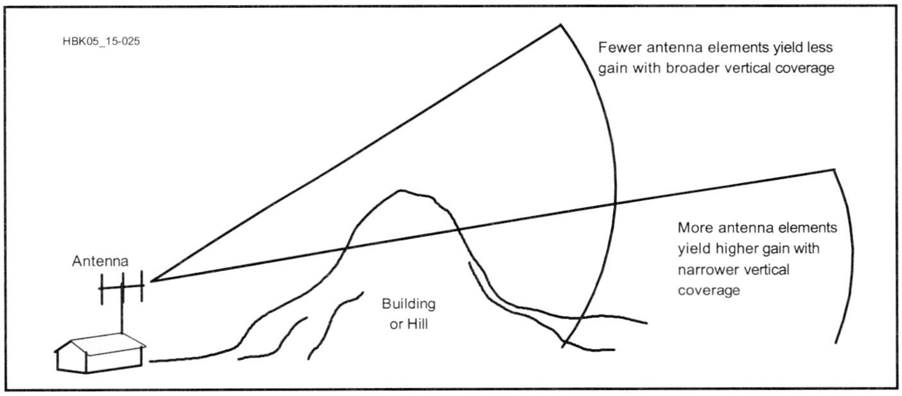

Figure 18.12 — As with all line-of-sight communications, terrain plays an important role in how your signal gets out.

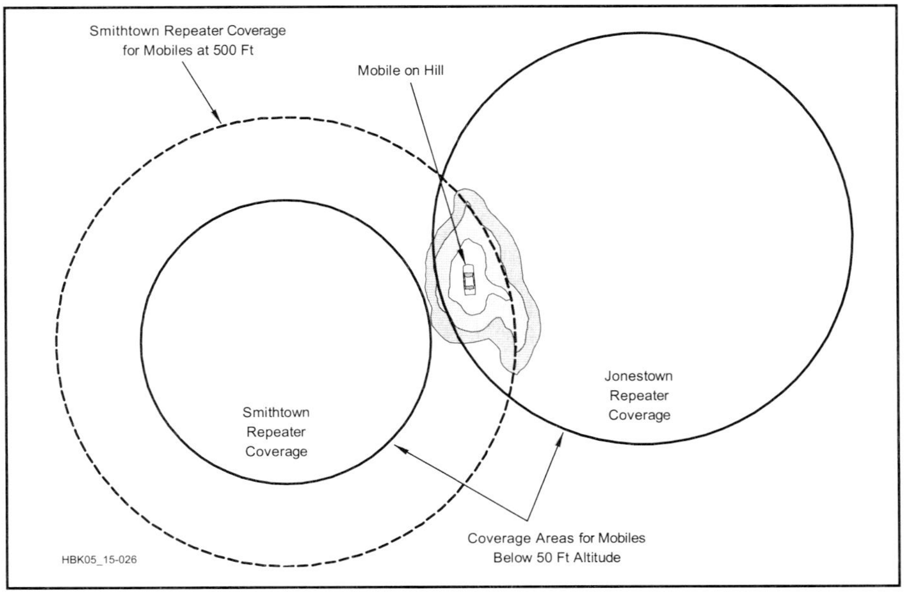

Figure 18.13 — When two repeaters operate on the same frequencies, a well-situated operator can key up both repeaters simultaneously. A directional antenna may help.

Table 18.2

CTCSS Tone Frequencies

The purpose of CTCSS is to reduce cochannel interference during band openings. CTCSS equipped repeaters would respond only to signals having the CTCSS tone required for that repeater. These repeaters would not respond to weak distant signals on their inputs and correspondingly not transmit and repeat to add to the congestion. The standard ANSI/EIA frequency codes, in hertz, are as follows:

	114.8	177.3
67.0	118.8	179.9
69.3	123.0	183.5
71.9	127.3	186.2
74.4	131.8	189.9
77.0	136.5	192.8
79.7	141.3	199.5
82.5	146.2	203.5
85.4	151.4	206.5
88.5	156.7	210.7
91.5	159.8	218.1
94.8	162.2	225.7
97.4	165.5	229.1
100.0	167.9	233.6
103.5	171.3	241.8
107.2	173.8	250.3
110.9		254.1

DCS started showing up in amateur radio transceivers several years ago and is now a standard feature in most new radios. Open repeaters generally still use CTCSS rather than DCS, since many older radios still in use don't have DCS.

DTMF

In the days before widespread use of cell phones, one of the most attractive features of repeaters was the availability of autopatch services that allowed a mobile or portable station to use a standard telephone DTMF (dual-tone multi-frequency, or Touch-Tone) key pad to connect the repeater to the local telephone line and make outgoing calls.

Although autopatches see less use today, DTMF key pads are still used for sending control signals. DTMF can also be used as a form of squelch to turn a receiver on, though it's more often used to control various functions such as autopatch and talking S meters. Some repeaters that require CTCSS have a DTMF "override" that puts the repeater into carrier-squelch mode for a few minutes if you send the proper digits. Other applications for DTMF tones include controlling linked repeaters, described in a later section.

Table 18.3 shows the DTMF tones. Some keyboards provide the standard 12 sets of tones corresponding to the digits 0 through 9 and the special signs # and *. Others include the full set of 16 pairs, providing special keys A through D. The tones are arranged in two groups, usually called the low tones and high tones. Two tones, one from each group, are required to define a key or digit. For example, pressing 5 will generate a 770-Hz tone and a 1336-Hz tone simultaneously.

The standards used by the telephone company require the amplitudes of these two tones to have a certain relationship.

18.3.4 Narrowbanding

We noted in the previous section that in most urban areas, there are no "available" frequencies for new repeaters. And you might recall from our short history section at the beginning of this chapter that the amateur radio FM boom began when the business and public-safety services ran out of spectrum and had to buy all new radios. Their solution to overcrowding, mandated by the FCC, was to reduce the spectrum used for each channel. In the 1960s, that meant reducing the modulation ("deviation") of FM signals from 15 kHz to 5 kHz and splitting each channel in two. Hams inherited the 15 kHz deviation radios (called "wideband" at the time), but soon adopted the 5 kHz "narrowband" standard.

History is repeating itself. Our spectrum neighbors are again out of room, and the FCC is again requiring them to reduce deviation from 5 kHz (now called "wideband") to 2.5 kHz (the new "narrowband").

Will hams follow suit and create space for more repeaters in our own crowded bands? So far, the answer is "no." While most amateur radio FM equipment built in the past decade has a "narrow" option that reduces the deviation and incorporates tighter receive filters, we still have a lot of legacy equipment in the field, and no corporate or municipal budget to draw on to replace it. Most of our repeaters are still made of old hardware converted from commercial service. Few repeater councils have seriously considered adjusting frequency coordination to accommodate narrowbanding. No one expects the FCC to require hams to adopt narrowbanding.

What *is* happening is the placement of D-STAR digital repeaters, which are especially "narrow" already, in between the channels occupied by analog FM repeaters. Since there is still some spectrum overlap, the D-STAR repeaters must also be a good distance away from their new adjacent-channel neighbors — 30 to 50 miles — to reduce the field strength of all the signals involved.

To help you understand how this all works, we'll explain that the terminology "5 kHz" and "2.5 kHz" deviation for analog FM signals is misleading. It refers to the peak frequency shift a modulated signal takes *in one direction* from the center frequency. But the FM signal moves both up and down from that center, and has some sidebands as well. The actual spectrum used by a "5 kHz" FM signal peaks at 16 kHz and the "narrow" 2.5 kHz signal hits 11 kHz on peaks — not much of a savings. The digital D-STAR signal is about 6.25 kHz wide. The digital signals fill their spectrum completely, all the time, and don't vary with modulation.

Narrowbanding may become a voluntary practice in amateur radio, though *your* use would be "mandated" by remaining compatible with narrowbanded repeaters. It's not on the horizon as of this edition of the *Handbook*.

18.3.5 Linked Repeaters

Most repeaters are standalone devices, providing their individual pool of coverage and nothing else. But a significant number of repeaters are linked — connected to one or more other repeaters. Those other repeaters can be on other bands at the same location, or they can be in other locations, or both. Linked repeaters let users communicate between different bands and across wider geographic areas than they can on a single repeater. **Figure 18.14** shows an example.

There are many ways to link repeaters. Repeaters on the same tower can just be wired together, or they may even share the same controller. Repeaters within a hundred miles or so of each other can use a radio link — separate link transmitters and receivers at each repeater, with antennas pointed at each other. Multiple repeaters, each still about 100 miles apart, can "daisy-chain" their links to cover even wider territory. There are a few linked repeater systems in the country that cover several states with dozens of repeaters, but most radio-linked repeater systems have more modest ambitions, covering just part of one or two states.

Repeater linking via the internet makes it possible to tie repeaters together around the world and in nearly unlimited number. We'll talk more about internet linking in the next section.

There are several ways linked repeaters can be operated, coming under the categories of *full-time* and *on demand*. Full-time linked repeaters operate just as the name implies — all the repeaters in a linked network are connected all the time. If you key up one of them, you're heard on all of them, and you can talk to anyone on any of the other repeaters on the network at any time. You don't have to do anything special to activate the network, since it's always there.

In an on-demand system, the linked repeaters remain isolated unless you take some action, usually by sending a code by DTMF digits, to connect them. Your DTMF sequence may activate all the repeaters on the network, or the system may let you address just one specific repeater, somewhat like dialing a telephone. When you're finished, another DTMF code drops the link, or a timer may handle that chore when the repeaters are no longer in use.

Table 18.3
Standard Telephone (DTMF) Tones

	Low Tone Group		*High Tone Group*	
	1209 Hz	*1336 Hz*	*1477 Hz*	*1633 Hz*
697 Hz	1	2	3	A
770 Hz	4	5	6	B
852 Hz	7	8	9	C
941 Hz	*	0	#	D

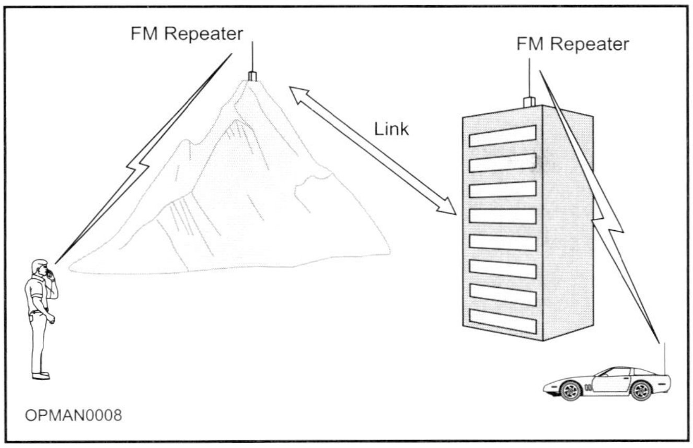

Figure 18.14 — Repeater linking can greatly expand VHF/UHF communication distances. Repeater links are commonly made via dedicated radio hardware or via the Internet.

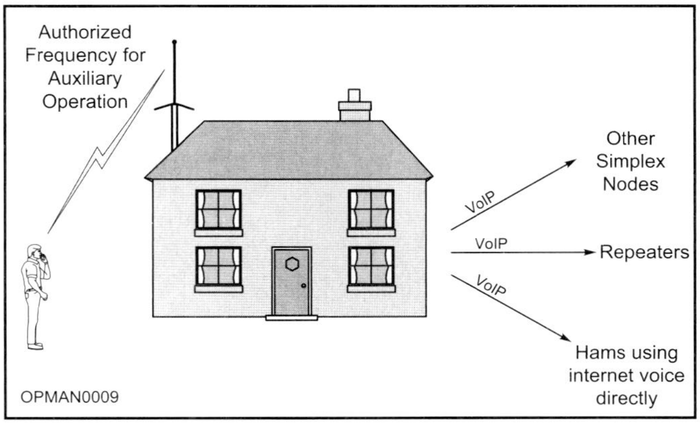

Figure 18.15 — A diagram of a VoIP simplex node. If a control operator is not physically present at the station location and the node is functioning with wireless remote control, the control link must follow the rules for *auxiliary* operation.

INTERNET LINKING

The internet and *Voice over Internet Protocol* (VoIP) has expanded repeater linking exponentially, making worldwide communication through a local repeater commonplace. Two internet linking systems, IRLP and EchoLink, have reached critical mass in the US and are available almost everywhere. The D-STAR and DMR digital systems also have a significant internet linking component. A brief overview is included here; more information may be found in the **Digital Communications** online chapter.

IRLP

The internet Radio Linking Project (IRLP) is the most "radio" based linking system. User access is only via radio, using either simplex or repeater stations, while linking is done using VoIP on the internet. IRLP system operators establish a node by interfacing their radio equipment to a computer with an internet connection and then running IRLP software. Once that's set up, repeater users send DTMF tones to make connections, either directly to other individual repeaters or simplex nodes (**Figure 18.15**), or to *reflectors* — servers that tie multiple nodes together as one big party line.

The direct connections work like on-demand linked repeaters. You dial in the node number you want to connect to and access code (if required), and you are connected to the distant repeater. Once connected, everyone on both ends can communicate. When finished, another DTMF sequence takes the link down. Someone from a distant repeater can make a connection to you as well.

Reflectors work like a hybrid between on-demand and full-time linked repeaters. You can connect your local repeater to a reflector and leave it there all day, or you can connect for a special purpose (like a net) and drop it when the event is over.

EchoLink

EchoLink requires a PC with sound card and appropriate software. It allows repeater connections like IRLP, and it has Conference Servers, similar to IRLP reflectors that permit multiple connections. The big difference is that EchoLink allows individuals to connect to the network from their computers, without using a radio.

The EchoLink conference servers all have more or less specific functions. Some are just regional gathering places, while some are region, topic, or activity based (SKYWARN and National Hurricane Center Nets, Jamboree on the Air, and so on).

You can connect your EchoLink-enabled computer to your base station radio fairly easily through a sound card and create an on-air node. Don't pipe EchoLink to a local repeater without permission from the repeater owner, though.

If you decide to create a full-time link from a computer to a repeater, consider using a dedicated UHF link frequency rather than just a base station on the repeater input. This applies to IRLP connections as well. Of course, the internet is infrastructure dependent, and both power and internet access can be interrupted during storms or other disasters.

18.4 D-STAR Repeater Systems

18.4.1 D-STAR System Overview

A D-STAR repeater system consists of one or more RF modules and a controller. This configuration allows for fixed, handheld, or mobile communications on one or more of the attached RF modules. While there are D-STAR systems with a single RF module, there are a number of "full stack" repeaters that have one of each type of module connected, 2 m voice, 70 cm voice, 23 cm voice, and 23 cm data.

An overview of the D-STAR system, "D-STAR — For the Second Century of Amateur Radio" is available at **www.icomamerica.com/en/downloads/DownloadDocument.aspx?Document=366**. A detailed discussion of the D-STAR specification is contained in a three-part *QEX* article by John Gibbs, KC7YXD, in the Jul/Aug, Sep/Oct, and Nov/Dec 2003 issues. These articles are available at **www.icomamerica.com/en/products/amateur/dstar/dstar/default.aspx** in the Downloads area. See the **Digital Protocols and Modes** chapter for more information about the D-STAR protocol.

D-STAR NETWORKS

To communicate within the D-STAR international network, D-STAR controllers are commonly connected to a gateway computer with *CentOS*, the Icom gateway software, and the additional modules *dplus* (by Robin Cutshaw, AA4RC) and *DStarMonitor* (by Pete Loveall, AE5PL). These additional third-party, amateur-developed modules greatly enhance the functionality of the D-STAR environment and are now considered a base part of the system. (See the section User-Created Features and Tools later in this section.)

Operation within Japan and outside of Japan varies due to differences in regulatory requirements. Inside Japan, gateway computers only run the Icom gateway software. Outside of Japan, additional software is installed to provides enhanced functionality.

Outside of Japan, *dplus* or its derivatives are used to implement repeater linking. Linking repeaters connects RF modules together so that anything heard on one is heard on the other. A link is a semi-permanent connection during which all traffic that is heard on one module is echoed to a destination system. Users hearing a call on linked repeaters do not need to program their radios to respond. Talking with a remote user is the same as talking with a local user.

D-STAR STATION ROUTING

Station routing is used to direct the controller and gateway to send a user's signal to a specific RF module anywhere in the world or to the specific RF module from which another user was last heard. In this situation, only the user's radio is configured to access the remote system. Users who want to respond to a remote user's call must configure their radio to respond to the remote user. This is simplified by the RX CS function on D-STAR radios. The RC CS function automatically configures the radio to the correct response route. Station routing only works between two repeaters and requires all stations participating in a contact to have their radios programmed appropriately.

D-STAR TRUST SERVER

A vital part of a D-STAR system is the gateway server, which networks a single system into a D-STAR network via a *trust server*. The trust server provides a central, master database to look up users and their associated system. The trust server database consists of a list of registered users, the last repeater on which a user was heard, and a list of the D-STAR repeaters. It periodically updates each registered gateway server.

This allows amateur radio operators to respond to calls made to them, regardless of their location on the D-STAR network. Currently, the global D-STAR trust server is maintained by a group of dedicated D-STAR enthusiasts from the Dallas-based Texas Interconnect Team (**www.k5tit.org**).

18.4.2 D-STAR Station IDs

In D-STAR, each station is identified with a *user station ID*. These are usually call signs consisting of six upper-case alphanumeric characters (space padded to the right). Before utilizing D-STAR, the radio must be programmed with a station ID ("My Call Sign"). User call signs are left-justified with spaces to the right. Additional characters are not recommended. If you have multiple radios, the same call sign should be used on all radios.

Repeater modules require a unique call sign different from other users or repeater modules on the network. While seventh and eighth characters of the call sign fields are not used for user station IDs, they are used for repeater module IDs. They are used for repeater module designation to denote which module is being utilized and/or *dplus* functions.

A common implementation of module designation is:

A — 1.2 GHz voice or data
B — 440 MHz voice
C — 146 MHz voice

The position of the module identifier can be the seventh or eighth character, depending on what the command is doing. For example, WD4STR•B or W1AW•••B are examples of repeater call signs (• represents a space).

18.4.3 Configuration of Station ID Fields

The D-STAR protocol can be used for simplex communication or repeater operation depending on how the radio is configured. In the following examples, a typical D-STAR radio screen is shown with abbreviations for the four station IDs. (UR is URCALL, MY is MYCALL, R1 is RPT1, and R2 is RPT2.)

When used for simplex communication, D-STAR radios function similarly to analog radios. To call a station using simplex:

- Set UR to CQCQCQ.
- Set RPT1 is not used
- Set RPT2 is not used
- Set MY to the calling station's ID.

RPT1 and RPT2 are disabled in simplex mode.

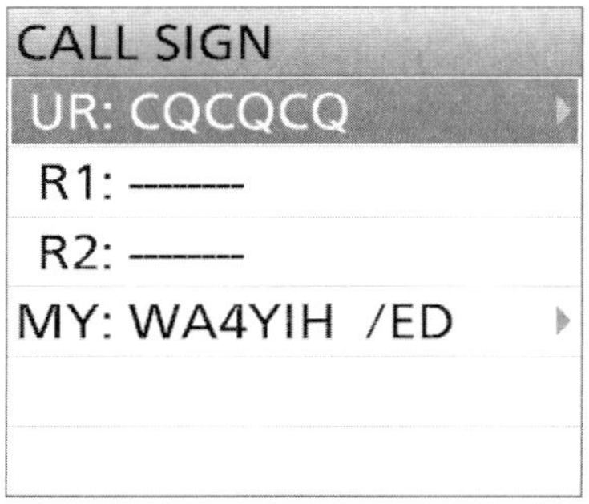

D-STAR Simplex Configuration

Radio programming changes slightly for repeater operation. To operate through a repeater, specify the repeater's station ID as well as the frequency of the repeater. To then call a station using the repeater, configure the radio with four station IDs:

- Set UR to CQCQCQ.
- Set RPT1 to a local gateway station ID.
- Set RPT2 to a local gateway station ID.
- Set MY to the calling station's ID.

D-STAR Repeater Configuration

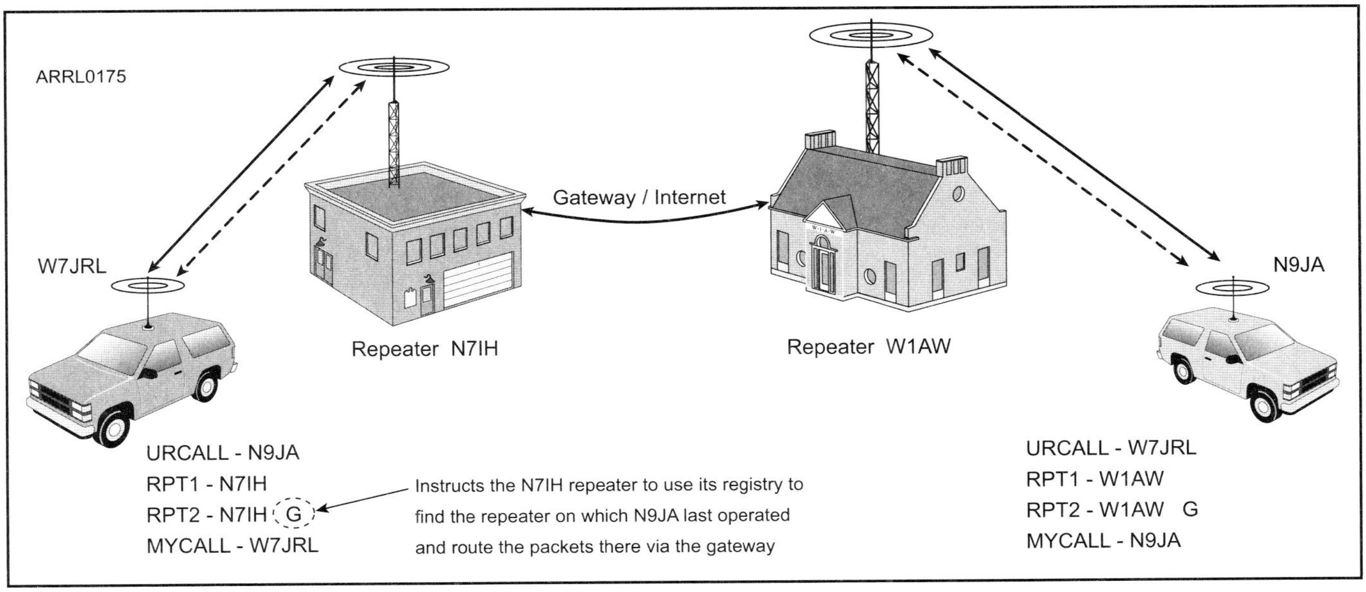

Figure 18.16 — Configuring D-STAR radios for repeater operation. (*Courtesy of the D-STAR Calculator dstarinfo.com/dstar-web-calculator.aspx.*)

In this example, the repeater module WD4STR B (on 440.55 MHz) is being used. The WD4STR repeater's gateway is addressed as WD4STR G. Figure 18.16 illustrates how D-STAR users would communicate using this programming to access a repeater system.

18.4.4 DR Mode

DR Mode effectively hides the relatively complex configuration of the URCALL, RPT1, RPT2, and MYCALL fields, providing a "No programming required" approach. (This is a mode of operation, not a digital transmission mode.) The use of DR mode is enhanced with increased memory space and GPS in the radio. After initially entering a calling station ID, in DR mode only two fields must be entered, TO and FROM. The first action is to choose a nearby repeater in the FROM field.

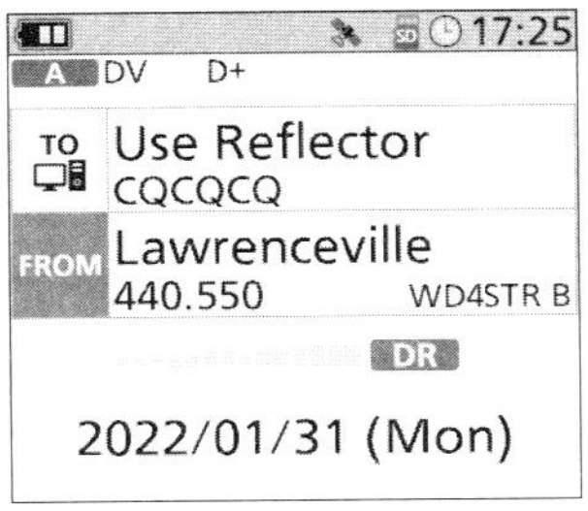

D-STAR Use Reflector

The FROM field presents the options of **Repeater List**, **Nearest Repeater**, and **TxHistory**. The repeater list provides a selection of the repeaters that are preprogrammed into the radio. (All D-STAR radios after 2012 come with a set of D-STAR repeaters preprogrammed into the radio from the factory. This database will need to be updated regularly with the latest changes and additions by downloading the latest copy from **dstarinfo.com/RepeaterDownloads.aspx**.)

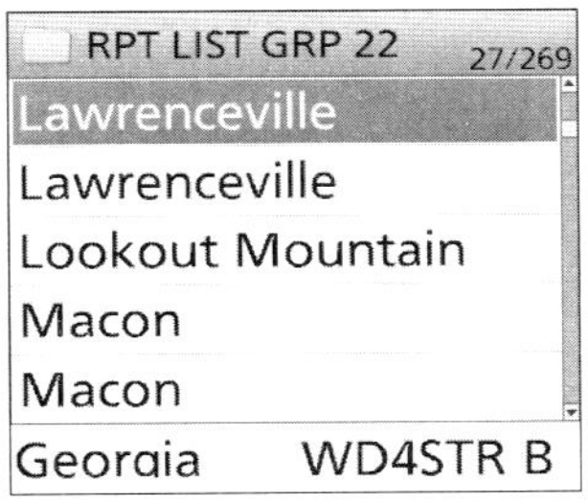

D-STAR Repeater List

Once a repeater is selected, all information needed will have been automatically entered into the radio. To simplify a long list of repeaters, **Nearest Repeater** will provide only a list of the nearest repeaters. This is determined by the GPS location of the radio and geolocation information for each repeater.

Next, use the TO field to select the function desired. For most contacts, select **Use Reflector - CQCQCQ** (**Use Repeater – CQCQCQ** in older radios). Figure 18.16 shows how the reflector is used.

Note: A common mistake is to use **CQCQCQ**. This is not to be confused with **Use Reflector - CQCQCQ**. Calls using the basic **CQCQCQ** will only be heard on the local repeater. Unless linking or unlinking a repeater, select **Use Reflector – CQCQCQ.**

The **Reflector – Link to Reflector** feature can be used to link a repeater to a reflector. To request a repeater or a hotspot to link to a reflector, choose **Reflector – Link to Reflector** and choose the reflector that you want to link to. Press the PTT for about a second and, if accepted, the repeater will link to the specified repeater or reflector.

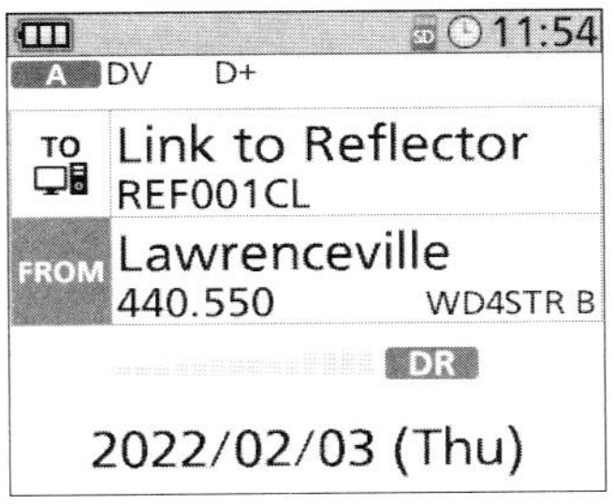

D-STAR Link to Reflector Configuration

18.4.5 D-STAR Reflectors

In addition to linking RF modules between repeaters, *dplus* created a network of virtual repeaters called *reflectors*. Reflectors allow multiple physical repeaters to connect to a single location on the network. This allows all users listening to a linked RF module to hear and talk to other users connected via linked repeaters on the same reflector. When a user transmits on one repeater, that transmission is heard on all connected repeaters. The reflector is software installed in a data center with high-speed internet connectivity that allows many repeaters to connect at once.

D-STAR reflectors have a status page that can be accessed from the internet at **[reflector ID].dstargateway.org** (i.e. **REF001.dstargateway.org**). Note that the module designator is not used in the URL. The status page has a list of repeaters that are connected to each port on the reflector.

Other types of reflectors have been created

since the introduction of dplus. These reflector IDs start with designators such as XRF, DCS, or XLX. Each different reflector system may have slightly different features but are essentially the same.

18.4.6 User Registration

Registration refers to the listing of a user call sign on the D-STAR network. This allows a user to be called from remote repeaters and to access D-STAR network functions such as linking repeaters or accessing the D-STAR network from the internet. Historically, registration was performed by accessing a local repeater. This caused problems because an administrator was required to approve the registration. To remedy this issue, a centralized registration system has been created at **regist.dstargateway.org**.

Follow the instructions, especially entering a call sign in upper case and remembering the password used. Once complete, an email notification will be sent in a few hours with additional instructions to complete registration. Registration is then accepted by all D-STAR systems — do not attempt to register anywhere else.

18.4.7 D-STAR Data Modes

In D-STAR, there are two modes: DV (Digital Voice) and DD (Digital Data). The DV mode provides voice and slow-speed data. The DD mode provides a data throughput of 128 kbps and is only available at 1.2 GHz and above because it has a 300 kHz bandwidth. (See the **Digital Protocols and Modes** chapter section on D-STAR for more information about DV and DD modes.)

DV MODE

DV utilizes a 4,800-bps data stream with an RF signal half the width of a standard analog FM signal. Within the signal, 2,400 bps is used for voice, 1,200 bps is used for forward error correction (FEC) of the codec data, and the remaining 1,200 bps is used for non-voice data.

Before the introduction of the Icom ID-51 Plus2, serial data was available to the user at an effective rate of about 960 bps. Since the introduction of the Icom ID-51 Plus2 and the Kenwood TH-D74, a second data option has been made within the DV protocol in which the serial data rate is 3,480 bps and utilizes the entire channel.

In DV slow speed data mode, both voice and data can be transported at the same time. In DV high speed data mode, voice is not available.

DD MODE

The DD mode provides 128 kbps data throughput and utilizes 150 kHz of RF spectrum, which FCC regulations allow at 1.2 GHz and above. The only radios that support the DD mode are the original Icom ID-1 and Icom ID-9700, as well as the 1.2 GHz DD repeater modules.

In radios supporting DD, the data stream is implemented as Ethernet. A collection of D-STAR radios on the same frequency effectively emulates an Ethernet hub. The radios can also operate through a D-STAR DD repeater. Because DD is implemented as Ethernet, applications do not require special configurations to communicate with each other. A standard application such as email can function without any modification using the D-STAR DD mode for connectivity.

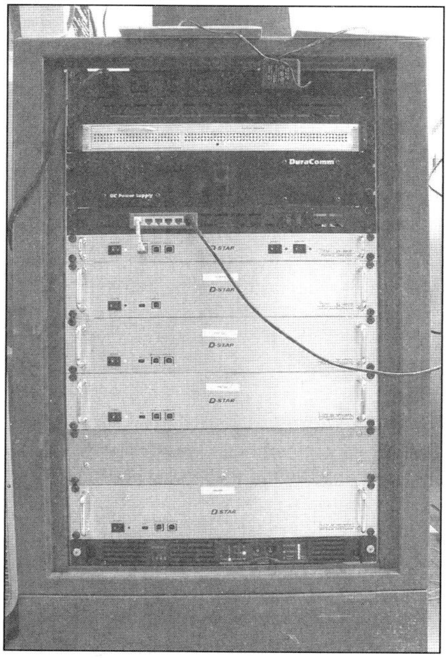

Figure 18.17 — A full rack of D-STAR equipment on the bench of Jim McClellan N5MIJ. Top to bottom: ICOM IP-RPC2 controller, ID-RP2V 1.2-GHz voice repeater, ID-RP4000 440 MHz voice and data repeater, a blank panel, and an ID-RP2000 146 MHz voice and data repeater.

18.4.8 D-STAR Repeaters

Icom has provided three generations of D-STAR repeaters. See **icomamerica.com/en/amateur** for the current set of equipment. The first generation (2003) was short-lived and very limited in distribution, consisting of a 23 cm DD module, a 23 cm DV / Controller module, and a microwave link for connecting repeaters together.

The second, widely deployed generation (2006, updated in 2020) consists of a controller module, plus repeater modules for 2 meter DV, 70 cm DV, 23 cm DV, and 23 cm DD. The DD module is not a repeater — it operates simplex with a single antenna connection. See **Figure 18.17** for a second-generation repeater "stack."

In 2020 Icom released a completely updated set of repeaters. This third-generation repeater stack has integrated controllers in each RF module that can be activated. The new repeaters support FM and D-STAR and are based on RF direct sampling and a FPGA platform.

The RF modules and a controller can operate as a stand-alone repeater system without a gateway. A user on the 2 meter repeater can talk with other users in the 2 meter repeater. Using station routing, users can talk across the modules, i.e. a 2 meter user can talk to a 23 cm user. To communicate with the worldwide D-STAR network, a gateway computer is required. **Figure 18.18** shows a block diagram of the D-STAR system, including repeaters, reflectors, and hotspots.

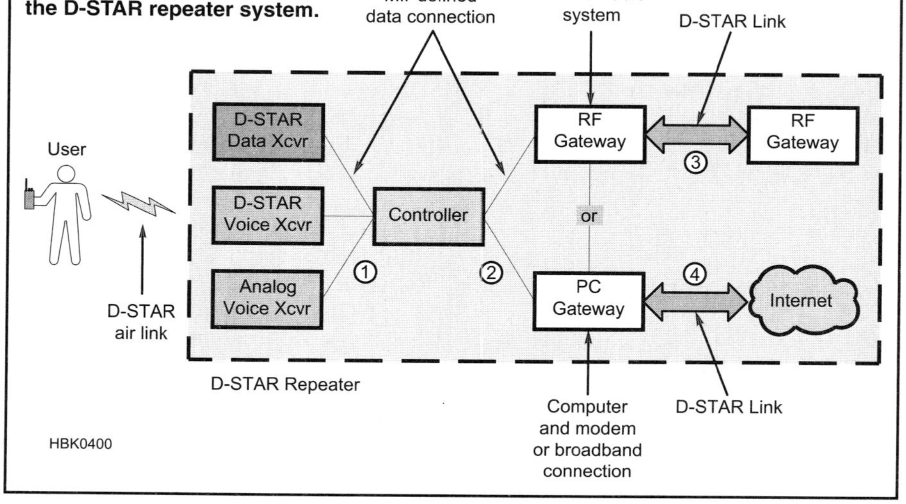

Figure 18.18 — Block diagram showing the components of the D-STAR repeater system.

D-STAR Network Overview

The D-STAR specification defines the repeater controller/gateway communications and the general D-STAR network architecture. The diagram shown here as Figure 18.A1 is taken from the English translation of the D-STAR specification:

The Comp. IP and Own IP are shown for reference as if this were a DD communications. As they do not change and are not passed as part of the D-STAR protocol, they can safely be ignored for the purposes of the following explanation.

Headers 1 through 4 are W$1QQQ calling W$1WWW. Headers 5 through 8 are W$1WWW calling W$1QQQ. Note that "Own Callsign" and "Companion Callsign" are never altered in either sequence. The "Destination Repeater Callsign" and the "Departure Repeater Callsign" are changed between the gateways. This is so the receiving gateway and repeater controller know which repeater to send the bit stream to. It also makes it easy to create a "One Touch" response as ICOM has done by simply placing the received "Own Callsign" in the transmitted "Companion Callsign", the received "Destination Repeater Callsign" in the transmitted "Departure Repeater Callsign", and the received "Departure Repeater Callsign" in the transmitted "Destination Repeater Callsign."

Use of the "special" character "/" at the beginning of a call sign indicates that the transmission is to be routed to the repeater specified immediately following the slash. For instance, entering "/K5TIT B" in the "Companion Callsign" would cause the transmission to be routed to the "K5TIT B" repeater for broadcast. Using the above example, W$1QQQ would put "/W$1SSS" in the "Companion Callsign" for the same sequence 1 through 4 to occur. At the W$1VVV gateway, however, the "/W$1SSS" in the "Companion Callsign" would be changed to "CQCQCQ." All stations within range of W$1SSS would see the transmission as originating from W$1QQQ and going to CQCQCQ just as if that station were local (but the "Departure Repeater Callsign" would be "W$1VVV G" and the "Destination Repeater Callsign" would be "W$1SSS"). Replying would still be done the same way as before, since the received "Companion Callsign" is ignored when programming the radio to reply.

Every "terminal" (station) has an IP address assigned to it for DD purposes. The address is assigned from the 10.0.0.0/8 address range. The D-STAR gateway is always 10.0.0.2. The router to the internet is always 10.0.0.1. The addresses 10.0.0.3–31 are reserved for local-to-the-gateway (not routable) use. This makes possible the ability to send Ethernet packets to another station by only knowing that terminal's IP address, and the remote station can directly respond based solely on IP address. This is because the gateway software can correlate IP address with call sign and ID. This makes it possible to route DD Ethernet packets based on the "Companion Callsign," or based on IP address with "Companion Callsign" set to "CQCQCQ." — *Pete Loveall, AE5PL*

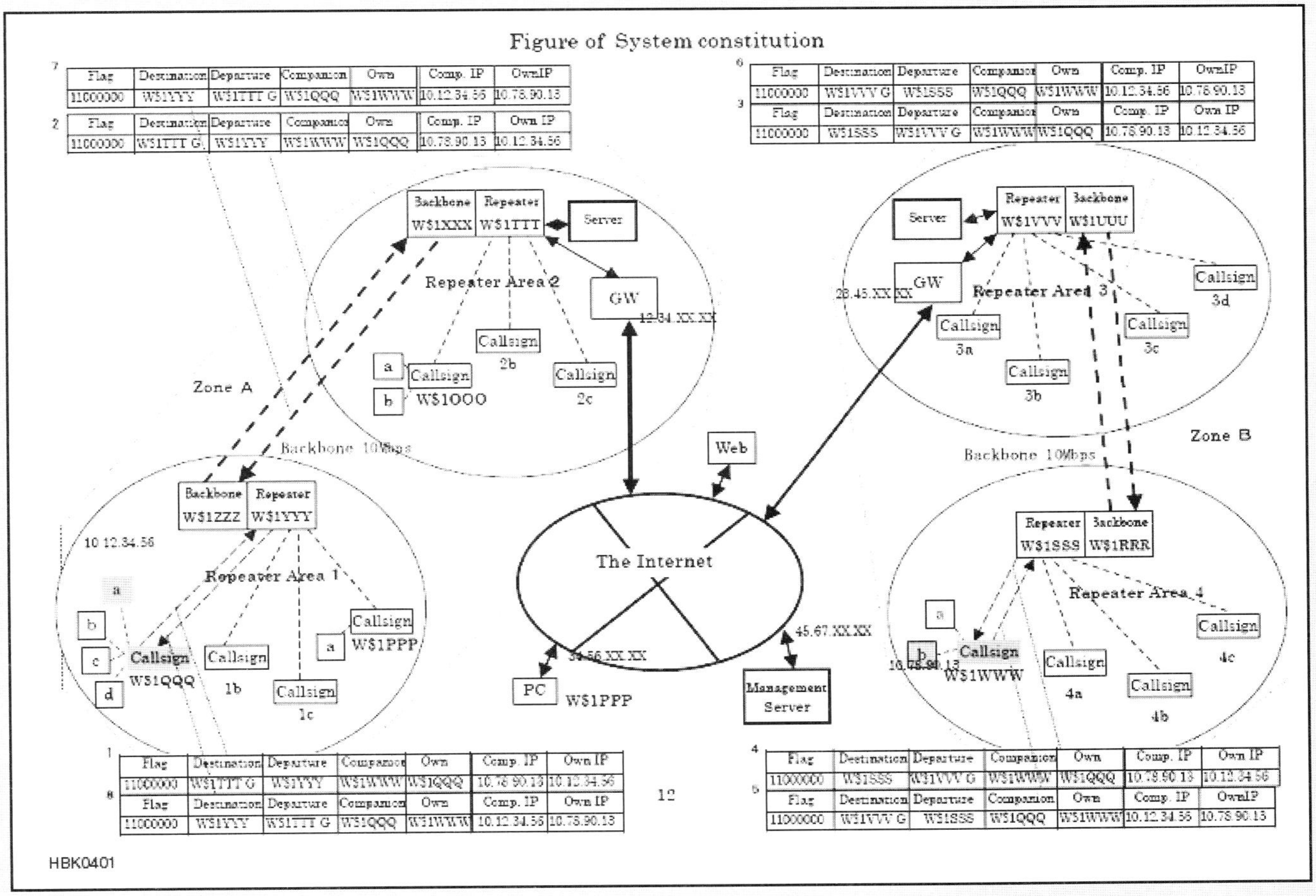

Figure 18.A1 — A D-STAR system overview.

18.4.9 D-STAR Gateways

Internet access from a repeater's gateway computer is required to talk to the reflectors, other repeaters, and the trust server. This access does not require a fixed address but does require inbound connectivity. This allows the trust server and other repeaters to reach the gateway server. Latency on the circuit should be low, to support the VOIP UDP connections. (See the Icom Gateway Control Software RS-RP3C Set Up Instructions [for Revision 3] documentation for minimum computer requirements and protocol port assignments.)

Two applications are often present on the gateway computer along with the Icom gateway software: *dplus* creates the network of *reflectors* that link repeaters, and *DSTAR-Monitor* relays APRS or NMEA type data to the APRS internet network. Both are described in the following section.

18.5.10 User-Created Features and Tools

D-STAR has benefited from a strong user community developing hardware, products, and applications over the last decade. The D-STAR architecture, utilizing the gateway computer running *CentOS*, a Linux distribution, allows users to develop applications and add features and additional functionality to D-STAR.

The following are a sample of the developments taking place. Many individuals and groups have created similar products. Many of these products have been extended to include DMR and other digital voice systems The two websites with the most current D-STAR information are **www.dstarinfo.com** and **www.dstarusers.org**.

D-STAR SOFTWARE APPS

The *dplus* application, developed by Robin Cutshaw, AA4RC, creates reflectors which function as "conference bridges" linking repeaters together via the internet. Users may link repeaters to reflectors with commands from a D-STAR radio to create networks of linked repeaters. *dplus* reflectors are designated as REFnnnx (ex. REF030C for reflector 030 and the letter A, B, C, or D designating a separate reflector module). Nearly 100 *dplus* reflectors are in operation worldwide. Developers of other reflectors designated as DCS, XRF, and XRX also link repeaters and users worldwide.

DStarMonitor (by Pete Loveall, AE5PL) creates a DPRS-to-APRS gateway that captures DPRS packets (APRS packets sent via D-STAR, see the **Digital Protocols and Modes** chapter) and relays them to the APRS network. Although it doesn't natively appear on the APRS RF network (unless IGated specifically), it does appear in all the online APRS tools. DPRS has been implemented in software with *DStarMonitor* running on many D-STAR gateways and *DPRS Interface* available for most computers at **www. aprs-is.net/dprs**.

Dan Smith, KK7DS, developed *D-RATS* with public service users in mind. In addition to chat, *D-RATS* supports file/photo downloads, bulletins, forms, email, and a sophisticated mapping capability. It also has a "repeater" functionality that allows sharing a D-STAR radio data stream over a LAN. The repeater feature is interesting in an environment where the radio would be physically separated from the software user, as in an emergency operations center. Current developments and downloads are available at **groups.io/g/d-rats** and **DSTARInfo.com**.

DSTARRepeater and *ircDDBGateway*, created by Jonathan Naylor, G4KLX, implement D-STAR functions on non-Icom repeater hardware (**github.com/g4klx/DStarRepeater**). A version of this software runs on the Raspberry Pi, which is the basis for many of the hotspots now in use.

PI-STAR by Andy Taylor, MWØMWZ runs on a Raspberry Pi (**www.pistar.uk**). It is a general-purpose digital voice package for several repeater systems, and includes an integrated UI for monitoring and controlling the *DSTARRepeater* and *ircDDBGateway* software.

D-STAR HOTSPOTS

One of the first hardware D-STAR projects, also developed by AA4RC, is the DV Dongle (**www.opendstar.org/tools**), which contains the AMBE codec chip and a USB interface to allow a computer user to talk with other D-STAR voice users much the way EchoLink users can connect to the system with a PC. The *DV Tool* software is written in Java to provide cross-platform support.

AA4RC also created the Digital Voice Access Point (DVAP). This device does not include the AMBE codec but uses a GMSK transceiver instead. The DVAP is a portable access point for D-STAR radios to access the D-STAR network via the internet.

Satoshi Yasuda, 7M3TJZ/AD6GZ, created a D-STAR node adapter or hotspot that connects to a standard simplex FM radio and allows a nearby D-STAR radio user to access the D-STAR network. It can connect to repeaters and reflectors worldwide, even though no local D-STAR repeater may be available. The system requires that the FM radio be accurately configured for a clean signal and loses the D-STAR benefit of only consuming 6.25 kHz of bandwidth. More information can be found online at **www.d-star.asia**.

Most current devices (as of early 2022) use the AMBE+2 chip, an enhanced version of the vocoder from DVSI. The AMBE+2 vocoder not only supports D-STAR digital but is also compatible with APCO Project 25 (P25), DMR, and Yaesu System Fusion, allowing hotspots to support D-STAR, DMR, P25, and System Fusion in one device.

18.5 Digital Mobile Radio (DMR)

The following sections are contributed by John Burningham, W2XAB, from "Amateur Radio Guide to Digital Mobile Radio" (see Reference section).

Digital Mobile Radio (DMR) was developed by the European Telecommunications Standards Institute (ETSI) and is used worldwide by professional mobile radio users. It supports both fully digital and dual-mode (analog/digital) operation.

Amateurs have implemented DMR with over 8,437 repeaters and 181,423 users registered worldwide as of September 2021. A majority of the repeaters are interconnected via the internet. There are a number of US amateur international, regional, and state networks. Most are interconnected with the cBridge/TL-NET, Brandmeister, and DMRplus based networks.

Used commercial DMR gear is available, but new DMR radios are now available with street prices within the range of a typical ham budget ($100 to over $2,000). DMR mobile and handheld radios are referred to as "user radios" to distinguish them from repeaters.

18.5.1 DMR Standards

DMR is divided into three tiers. Tier I is a single channel specification originally for the European unlicensed dPMR446 service. It is a single-channel FDMA 6.25 kHz bandwidth; the standard supports peer-to-peer (mode 1), repeater (mode 2) and linked repeater (mode 3) configurations. The use of the Tier I standard has been expanded into radios for use in other than the unlicensed dPMR446 service.

Tier II is a 2-slot TDMA 4FSK 12.5 kHz wide peer-to-peer and repeater mode specification, resulting in a spectrum efficiency of 6.25 kHz per channel. Each time slot can be either voice and/or data, depending upon system needs. Most amateur radio implementations of DMR are using voice on both time slots.

Tier III builds upon Tier II, adding trunking operation involving multiple repeaters at a single site. Not all manufacturers' trunking implementation is compatible with Tier III. Vendor-specific protocols have expanded trunking to multiple-site operations. Any Tier III-capable radio will also work on Tier II systems but neither will work on Tier I.

DMR Tier II is being implemented in amateur networks with commercial DMR repeaters along with homebrew repeaters using MMDVM (Multi-Mode Digital Voice Modem) interfaces. The IP-based protocols used by different repeater manufacturers are not compatible. It is doubtful the equipment manufacturers will ever standardize for business reasons, but conversion (a "bridge") between the multiple vendor protocols is possible. Any DMR (Tier II) user (mobile or handheld) radio will work on any Tier II system, although some manufacturers offer proprietary features for their infrastructure.

DMR VOCODER

The current implementation of DMR utilizes the DSVI AMBE+2 vocoder by agreement of the manufacturers; it is not specified in the ETSI standard. Many of the radio manufacturers have implemented the vocoder in licensed software, while others use a DSVI IC. The AMBE+2 vocoder is a more recent technology than the AMBE-2000/2020 vocoder used by D-STAR.

DMR NETWORKS

Amateur DMR networks operate the same from the end user perspective. The amateur DMR networks are built on servers that support gateway interconnects and conference bridging. These networks can be divided into the Motorola Solution's MOTOTRBO-specific infrastructure (cBridge/TL-NET) and infrastructure that supports different vendor IP protocols (Brandmeister, DMPplus). Not all the amateur DMR repeaters are connected to the wide area networks; some are standalone either because they have yet to obtain an ISP connection at their repeater site, or because the repeater is only intended for local communications. Some repeater systems are operating in dual-mode (analog/digital), which allows the repeater to support both digital and legacy analog users. MOTOTRBO repeaters operating in dual-mode do not support interconnection via the internet using their proprietary IPSC protocol (see below).

Some hams have installed DMR repeaters in a vehicle, using 3G/4G/5G cellular wireless services for internet access. Others have implemented remote bases to interconnect to other networks or radios; it is important to remember that the wide area networks typically have policies restricting interconnected traffic, but what is implemented locally and stays local is acceptable. FCC regulations regarding commercial and business traffic must be followed when transferring content to and from the internet.

TWO-SLOT TDMA

DMR Tier II/Tier III occupies a 12.5 kHz bandwidth shared between two channels using Time-Division Multiple Access (TDMA), resulting in a spectrum efficiency of 6.25 kHz per channel. Each time slot can carry either voice and/or data depending on system design. The time slots are called Time Slot 1 (TS1) and Time Slot 2 (TS2). You can think of the two time slots as separate channels.

For the amateur, this means one repeater allows two separate channels at the same time. Currently most amateur DMR repeater system implementations utilize both channels for voice and some limited text messaging. Normally one time slot is used for wide-area and the second for local and regional.

For repeater operators, a single two-slot TDMA repeater offers a significant savings over two stand-alone repeaters to obtain two separate communication channels, as only one repeater and one antenna system is required.

The two-slot TDMA implemented in DMR uplinks (portable/mobile to repeater) uses a 30-ms window for each time slot; the 30-ms is further divided into a 27.5-ms frame and a 2.5-ms gap. This means when transmitting, your transmitter is only turned on for 27.5 ms every 60 ms, resulting in about a 40% battery savings on transmit.

The DMR repeater transmits a continuous data stream even if only one time slot is being used; the 2.5-ms uplink gap is replaced with a CACH burst (Common Announcement Channel) that is used for channel management and low-speed signaling.

The 27.5-ms frame consists of a total of 264 bits; a 108-bit payload, 48-bit SYNC or embedded signaling, and a second 108-bit payload for a total of 216 bits of payload per frame. The vocoder must compress 60 ms of audio with FEC (forward error correction) into 216 bits of data for transmission. The 2.5 ms gap is used as guard time to allow for PA ramping (turn-on time) and propagation delay.

IPSC AND BRIDGES

IP Site Connect (IPSC) is a vendor-specific repeater feature offered by some manufacturers. Note that MOTOTRBO repeaters will only interconnect over the internet with other MOTOTRBO repeaters because it is not part of the ETSI specifications and the manufacturers don't want to interconnect their infrastructures. IPSC is not part of the current ETSI standards.

The Motorola Solutions MOTOTRBO IPSC implementation allows up to 15 MOTOTRBO repeaters operating in DMR mode to be connected on a fully meshed IPv4 network, with one of the repeaters or a bridge serving as a Master and the others as Peers. Any traffic originating on one of the interconnected repeaters is relayed over the IP network to each of the other repeaters. The Peers will first establish a connection with the Master and obtain the database of the other Peers along with their IPv4 and port addresses.

The more repeaters in this fully meshed IPSC network, the more IP network bandwidth is required for each repeater. To expand beyond the limits of a basic IPSC network requires the utilization of a bridge to interconnect the different IPSC networks. Rayfield Communications (c-Bridge) and BridgeCom Systems (TL-Net) are the current commercial preference in North America. In the European market, SmartPTT is common. These bridges require static IPv4 addresses and larger network bandwidths than individual repeaters. Besides the commercial bridging products, Brandmeister and DMRplus are available.

18.5.2 DMR Structure

TALK GROUPS (TG)

Talk Groups (TG) are a way for groups of users to share a time slot without distracting and disrupting other users of the time slot. It should be noted that only one TG can be using a time slot at one time on a repeater. If your radio is not programmed to listen to a TG, you will not hear traffic of that TG. In this regard it is similar to coded squelch, discussed earlier in this chapter. Talk Groups can be considered as conference bridges that are used to interconnect multiple users; other technologies use the term *reflectors*, or *rooms*.

The MOTOTRBO-based network supports many Talk Groups on TS1, including World Wide (TG1), North America (TG3), World Wide English (TG13), and DMR-Plus USA (TG133). TS2 is for local, state, and regional traffic. Check with your local repeater operator to find out what Talk Groups/Time Slots are available on a repeater. Other networks such as Brandmeister and DMRplus have their own assignment of TGs. While many TGs are shared between different networks, this is not true for all TGs, even if the different networks are using the same TG number.

There are TGs implemented for individual states and regions on many networks. Some TGs are available all the time, others only at preprogrammed times, or they require a local user to activate PTT for the TG to enable it for a programmed time. Since only one TG can be transmitting at a time on a time slot, many systems will disable other TGs when a local user is active on a different TG on the time slot. Be courteous and try to use TGs that tie up the fewest number of repeaters if you are going to have a long contact. Further information about specific Talk Groups can be found on the DMR-MARC, Brandmeister, DMRplus, and regional group websites.

ZONES

User DMR radios support *zones*. Zones are a way to organize channels, much like file folders or directories on your computer. A zone is just a grouping of individual channels. Some model radios may limit the number of channels per zone and the number of zones allowed.

You could program zones for local channels (DMR or analog), another zone for a neighboring state, and a zone for business and government channels. For example, you could program a zone to include all of the NWS Weather Channels. If you do program non-amateur channels in your radio, make sure they are receiver-only unless you are licensed to use them as required by FCC 90.427(b).

ENCRYPTION AND DMR

The DMR standard also supports private calls (one-to-one), encryption, and data. Private calls are not allowed by most of the amateur networks, and many consider private calls inappropriate for amateur radio. Private calls can tie up large number of repeaters across the network and can't be heard by other users. Encryption is not legal for amateur radio in the US and in most other parts of the world. Data and text messaging is supported on some networks.

18.5.3 DMR Channels

On a DMR radio, a channel is a combination of frequency, Color Code (CC), time slot (TS), and TG. A single repeater may occupy six or more programmed channels, depending on the number of TGs available.

COLOR CODES

DMR repeaters use CCs like analog repeaters use CTCSS or DCS. To access a repeater, you must program your radio to use the same CC as the repeater. There are 16 different CCs (CC0-CC15). The use of Color Codes is not optional on DMR systems. If your Color Code is not set correctly, you will not be able to access the repeater. Many repeater councils are assigning CCs with frequency assignments.

ADMIT CRITERIA

The Admit Criteria determines when your radio can transmit. The recommended setting for repeater channels is COLOR CODE FREE; this configures your radio to work with your own digital system. You should configure the radio in Call Criteria to FOLLOW ADMIT CRITERIA. Simplex channels should be configured as ALWAYS for both Admit Criteria and ALWAYS or FOLLOW TX in Call Criteria.

18.5.4 DMR Equipment

USER RADIOS

There are many sources of new and used DMR radios. Presently a majority of them are professional (commercial) radios marketed primarily to commercial radio users, but some include features for the amateur market. If you want to purchase a new DMR radio for ham use, you can easily find a dealer. Some dealers are "ham friendly" and will offer reasonable discounts to hams. Check with other DMR users or on DMR-related websites for further information.

You can also search on eBay and other online flea markets for both new and used radios. Larger hamfests may also have DMR dealers or sellers in their flea markets or vendor areas. Here are a few things you need to know before buying a DMR radio:

New or Used — For used DMR radios, it is buyer beware! Remember that you will not be able to repair a non-working DMR radio unless you have the technical skills and necessary test equipment, and that test equipment can cost hundreds of times the cost of the radio. The street price for new DMR radios is $100 to $2,000. Used, higher quality, name-brand radios, such as Motorola or Hytera, typically sell for more than brand-new radios from newer entrants into the DMR market. Higher priced radios usually have more features, are better constructed, and can handle more abuse than less expensive radios. For the average amateur, one of the new lower-cost radios is a good initial purchase.

Some of the newer DMR radios are Android-based, such as Motorola's ION (street price of under $2,000) and RFinder B-1 (street price of $1,000).

VHF, UHF, or 900 MHz — UHF is the most commonly used band for DMR, but because military radar takes priority in some US areas, VHF repeaters may be the local choice. There are only a few amateur 902 – 928 MHz DMR repeaters in the US. If you are purchasing UHF equipment, make sure it covers the amateur band (420 – 450 MHz) from the factory.

Programming Software — Some manufacturers supply programming software at no cost. Motorola Solutions charges approximately $175 for a three-year subscription (which covers all their models within a region) to their software and updates. Many DMR radios typically do not allow keyboard programming because they are sold in the professional market. If a vendor charges for the programming software, do not ask another ham to bootleg a copy for you. If you have a legal copy, you may program radios for others, but you cannot legally distribute the software.

Programming Cable — Some radios use standard USB cables for programming, while others use cables that can cost more than $80.

Number of Channels — Some radios have as few as two channels, while others have as more than 1,000 channels. You will need a channel for each combination of frequency, CC, TS, and TG. You can easily use six or more memory channels for each DMR repeater.

Display or Non-Display — Some radios have only a channel selector knob, while others have displays (monochrome or color) that show TG and ID information. Some displays only show channel numbers.

Visually Impaired Operators — Consideration must be given to the channel selection knob on the radios. Most of the non-display models have channel selection knobs that have fixed stops instead of 360-degree continuous rotation to allow the operator to find channel one. Some LCD display models also have fixed stops on the channel-selector knob; these include some Hytera and CSI radios. Many models offer programmable voice announcements.

DTMF Keypad — Some radios have a 12-button DTMF keypad. MOTOTRBO repeaters support an optional proprietary autopatch feature (Digital Telephone Interconnect) that only works with MOTOTRBO radios.

GPS — GPS is available on some models, but DMR does not natively support APRS (Amateur Packet Reporting System). On professional networks, one of the time slots is typically allocated for location reporting and is interconnected to server-based dispatch applications. GPS will shorten battery life if it is enabled. FCC identification requirements must be complied with if transmitting GPS or other data.

Bluetooth and Wi-Fi — Some higher-end radios have Bluetooth built in for wireless headsets and programming. Wi-Fi is also available on some models. This is a great feature at work and home to listen without bothering others. Some radios with Bluetooth support data and programming via the Bluetooth wireless connection to the radio. Some models have Bluetooth adapters available as options. Bluetooth and Wi-Fi will shorten battery life if enabled.

Analog —Most radio models support analog FM. Current FCC rules require narrowband for most commercial/government services. For DMR radios from some manufacturers, this requires a programming entitlement key or a different version of the programming software if you require wideband FM, which is still used on many amateur analog repeaters.

External Antenna on Portable — Not all portable radios support the connection of an external antenna, except for testing and alignment purposes. Using an adapter to connect an external antenna can place undue stress on the portable antenna connector, which may result in premature equipment failure and expensive repair. If you are going to use an external antenna adapter, use an adapter cable that uses miniature coaxial cable to reduce stress on the radio's connector.

Portable (Handheld) or Mobile — Portable models are typically available in the 2 – 5 W range. Mobile radios are available with a maximum of 10 – 45 W. A portable is recommended as a first DMR radio unless you live beyond the handheld coverage of your local DMR repeater. If you spend significant time in your vehicle commuting, you will find a mobile radio a good investment. Mobile radios can also be used as a base station with the addition of an external power supply.

External Amplifier — Many external amplifiers will not work with DMR radios unless they are specifically designed to meet the fast-switching requirements of TDMA on DMR.

CODE PLUGS

A *code plug* is simply a radio's configuration file. The user programs their code plug from scratch or starts with one made available by a local group. This file is uploaded to the radio and should also be saved on your computer as a backup. You can also download the code plug from a radio to modify it. Building a code plug can take many hours, especially if you want to program hundreds of channels. The code plug can also contain a contact list of Radio IDs, call signs, and names to be displayed. You can find copies of configured code plugs online for different models of radio; check first with your local group. All DMR radios support a limited number of entries in the Contact List — there are over 200,000 individual radio IDs currently assigned.

18.5.5 DMR Operation

SIMPLEX

On the professional side of DMR, *Talk-Around* refers to operating simplex on a repeater output channel. Doing so allows for direct communication while still being able to hear the repeater. The benefit of operating simplex on a repeater output channel is that it allows users to directly contact other users listening on the repeater output frequency. Amateurs typically use dedicated simplex channels so as not to interfere with repeaters. The amateur DMR community has published a list of recommended simplex frequencies to be used instead of operating simplex on repeater outputs. **Table 18.4** shows recommended simplex DMR frequencies and configuration.

Avoid creating conflict with non-DMR analog users. Do not use 146.520 or 446.000, the national analog simplex channels. Avoid repeater inputs and outputs, locally used non-DMR simplex channels, satellite sub-bands, and any other frequencies that could disrupt other amateur communications.

ACCESSING A DMR REPEATER

To access a DMR repeater, set the frequency, CC, TS, and TG. When a transmitter is keyed (pressing PTT), an access-request signal is sent to the repeater and the repeater responds to permit transmitting. If a repeater's acknowledgement is not received, the radio will stop transmitting and a negative confirmation tone will be heard. This is one of the advantages of TDMA: allowing bidirectional communications between user radio and the repeater when transmitting. The repeater can also signal the radio to stop transmitting if there is contention on the network because more than one station is transmitting at a time.

Not all DMR repeaters are interconnected on the internet. Internet connectivity may not be available at the repeater site, or not available at a reasonable cost. Some repeater operators may just prefer to keep their repeater for local usage only without connecting to the larger regional and worldwide networks.

OPERATING USING DMR

If you are unsure of "DMR etiquette," spend some time listening to others operate. Be considerate and learn the preferred operating style. A good practice for learning any new mode, system, or protocol: listen, listen, listen.

To place a call to another station, or to make a general call, announce your Talk Group because some users may be scanning or have radios without a display. Avoid calling CQ.

When you are talking on one of the wide area TGs, hundreds of repeaters will be tied up. If you are unable to move to a more localized TG, be considerate of the other users on the network. When one TG is active, other TGs on the same time slot will be blocked. Leave time between transmissions so others can break in. Remember that emergency traffic always has priority over all other traffic.

Table 18.4

Recommended Simplex DMR Frequencies and Configuration

	Frequency (MHz)
UHF	441.000, 446.500, 446.075, 433.450
VHF	145.790, 145.510

Channel configuration: TG99 / CC1 / TS1 / Admit Criteria: Always / In Call Criteria: TX or Always+

18.6 System Fusion

Yaesu released the initial specification for System Fusion in 2013. System Fusion supports digital voice and data in a 12.5 kHz narrow-band channel at 9,600 bps, using C4FM modulation over VHF (144 – 148 MHz) and UHF (430 – 450 MHz). System Fusion's low-level modulation and packet structure are discussed in the **Digital Protocols and Modes** chapter. Unless noted otherwise, references to "System Fusion" in this section apply to all versions of System Fusion equipment.

All System Fusion repeaters are configurable for VHF (144 – 148 MHz) or UHF (440 – 450 MHz) operation. VHF and UHF operation can be selected as required. All repeater models are configured through a touch-screen interface to set call sign, input frequency, output frequency, CTCSS or DCS setting, power level, mode selection, and so on. By using the lowest (5 W) power setting and a suitable attenuator, a variety of power amplifiers can be used.

Firmware upgrades are available for transceivers, including discontinued models manufactured prior to the release of System Fusion II in 2017. Specific support levels of DG-ID memories (see below) and accessibility vary by transceiver. All updated transceivers can access and control unrestricted WIRES-X nodes.

CALL SIGN IDENTIFICATION

Call sign identification is embedded within the digital transmission packets. When powered up for the first time, a System Fusion transceiver prompts users to enter their call sign. This establishes station identification. There is no requirement for advance call sign registration or a subscriber ID in order to get on the air.

18.6.1 System Fusion Versions

SYSTEM FUSION I

The original DR-1 repeater was produced in 2014 for a beta testing program. The DR-1X repeater was then released as the production model. There is a significant difference between the internal controllers in the DR-1 and DR-1X. Note that while many DR-1 (beta) repeaters are installed and in daily use as standalone assets, they were not designed to be directly connected to external controller and networking interfaces in the same manner as the DR-1X (production). System Fusion I is supported by DR-1 and DR-1X repeaters.

SYSTEM FUSION II

System Fusion II adds new features and configuration options and is supported by DR-2X repeaters and enhanced controllers for DR-1X repeaters. For a time, Yaesu offered a DR-1X trade-in program toward the purchase of a DR-2X. The traded-in repeaters were then refurbished, enhanced, and sold as DR-1XFR models. Note that during the upgrade, no changes were made to the exterior and labeling. The most reliable way to determine the difference between a DR-1X and DR-1XFR is by examining the firmware versions.

DR-2X repeaters were released in 2017, along with an internal controller upgraded from DR-1X repeaters. DR-2X repeaters also provide intelligent thermal control for higher duty-cycle operation at full power. DR-2X repeaters feature a second receiver and additional over-the-air controls. The DR-2X can also be commanded to transmit on a second (alternate), pre-assigned frequency using an additional RF connector.

DSQ (Digital Squelch) has been renamed DG-ID (Digital Group Identification). It provides a means of adding selective access for individual repeaters, repeater groups, and WIRES-X nodes. The DG-ID mantissa is 00 ~ 99, with 00 used as a "hear all" setting.

DP-ID (Digital Personal Identification) is transceiver-specific and may be used to designate authorized control operators on a given DR-2X, by inclusion within that repeater's registry. DP-ID can also be used on an ad-hoc basis to aggregate sets of radios within a given DG-ID designation.

18.6.2 System Fusion Modes

System Fusion supports three modes of operation: voice full rate (Voice FR) mode, data full rate (Data FR) mode, or voice/data (V/D) mode. Voice FR mode is typically displayed on System Fusion transceivers as "VW" and Voice/Data as "DN." The two full-rate modes use the entire 9,600 bps channel for their respective voice or data payloads, whereas the V/D mode splits the channel into two 4,800 bps streams, with voice information on one and data on the other. Data is currently limited to image files and internally generated data such as GPS, call sign, and routing elements. There is no external access to the data stream.

System Fusion repeaters can be configured in a combination of modes. While the older DR-1X repeaters are capable of operation in a purely analog mode, this prevents digitally equipped users from accessing the repeater. The repeaters can also be configured to allow analog or digital reception, while forcing the output to analog only. While this configuration allows both analog and digital users to coexist, it does so in a constrained fashion. Analog-output mode strips away GPS, call sign, and other information from a digital-mode input signal. Operating in purely digital mode is also possible, effectively locking out analog FM operation. This is typically implemented when there is more than one System Fusion repeater operating in a given area.

AMS CONFIGURATION

The hybrid Automatic Mode Select (AMS) operates in both FM analog and System Fusion digital modes. This supports both analog FM into analog FM out, and digital into digital out. In this configuration, analog users are not suddenly "disconnected" from a repeater and other analog-only capable operators. Also, digital users are free to take full advantage of available features.

When configured in AMS, digital users can hear an analog call placed on the repeater in between transmission exchanges. The transceivers will switch to analog, automatically allowing them to communicate with the analog station. This allows both analog and digital users to share a repeater and does not require all users to switch to digital simultaneously.

Analog FM users can avoid hearing digital transmissions by enabling their transceiver's TONE SQUELCH feature to match the repeater's transmitted continuous tone coded subaudible squelch (CTCSS) or digital coded squelch (DCS) signal. Watching for a visual "channel busy" indicator on the analog users' transceivers or enabling the BUSY CHANNEL LOCKOUT feature will prevent accidental interference when digital communications are taking place.

Repeater networking options allow remote nodes or "points of presence" in cases where internet connectivity is not locally available. This enables systems without internet service to be integrated into a repeater network. Further, digital and analog signaling is supported throughout the WIRES-X (Wide-coverage Internet Repeater Enhancement System) networking protocols.

18.6.3 System Fusion Network

WIRES-X NETWORK OPERATION

WIRES-X nodes normally consist of a transceiver, a Windows-based computer executing the WIRES-X node software, and the WIRES-X interface. Alternately, a "local" configuration is possible, using the repeater as a replacement for the transceiver. With the acquisition of the HRI-200 WIRES-X interface, a node registration is required in order to enable secure communications to the network. A WIRES-X network map is presented in **Figure 18.19**.

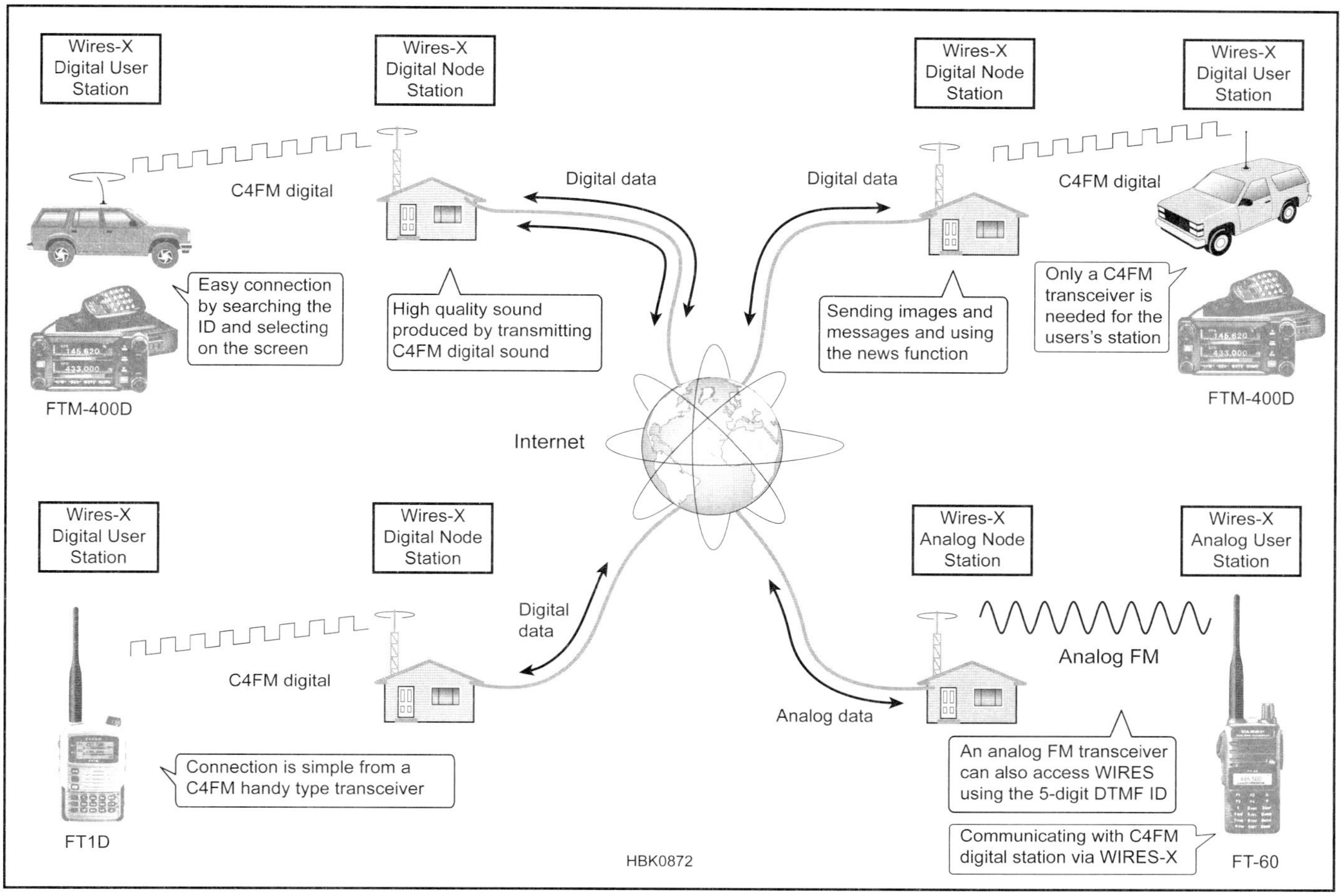

Figure 18.19 — System Fusion I configuration. (*Graphic courtesy of Yaesu.*)

WIRES-X presently supports nodes and "rooms." Each is assigned a number, although rooms can have alphanumeric identifiers ("America Link," "Keystone-Wide," and so on). Compatible digital transceivers can select rooms and nodes with scrollable menus. Analog FM transceivers operating through analog WIRES-X nodes can access rooms and other nodes through use of the numeric identifiers, using the DTMF pad on a suitable microphone or portable radio. (See IMRS section below.)

Transceivers with enhanced feature sets allow for the adhoc selection of different rooms. In a digital setting, room lists may be scrolled through on the display and selected, as desired. In an analog setting, rooms are not listed but may be selected through a direct entry of the desired room's numeric identifier. Although not as popular, mixed rooms — accommodating both analog and digital audio — are supported.

WIRES-X remote nodes can be used as portals with the network or as a means to network a given System Fusion repeater. Remote nodes are supported by all variants of the DR-2X and DR-1X, including the beta DR-1. Remote nodes may make use of the HRI-200 WIRES-X interface or certain mobile, or portable transceiver models can be used as a PDN (Personal Digital Node) without the need for the HRI-200. See the Yaesu product specifications to determine which models support PDN operation.

While an HRI-200 equipped node can be set for analog or digital operation, a PDN is by definition digital-only. With PDN, the DP-ID of the transceiver serves as the node registration identifier.

INTERNET-LINKED MULTIPLE REPEATER SYSTEM (IMRS)

IMRS is a networking structure supported by System Fusion II DR-2X controllers. While WIRES-X is server-based, IMRS utilizes a peer-to-peer architecture using TCP/IP. IMRS allows a number of repeaters to be interconnected and accessed via DG-ID entries assigned to individual repeaters, depending on how the overall repeater group's system administrator configures the repeater network. Some examples: A DG-ID of 01 may be used to access only a local DR-2X. A DG-ID of 11 may access a group of DR-2X repeaters within a county. A DG-ID 21 may access DR-2X repeaters across a region, state, province, and so on.

In System Fusion II, WIRES-X nodes, through the use of DG-ID, can be used to bridge normally non-connected IMRS groups together, or to add the users of a room into an IMRS group. The DR-2X can support IMRS or direct connection of the HRI-200 for WIRES-X, but not both at the same time, due to shared signal lines. An IMRS-enabled repeater with a remote WIRES-X node avoids this limitation.

18.6.4 Third-Party Enhancements and Options

External repeater controllers can be attached via the repeater signal and control interface connections on rear panel HD-15 and 10-pin Mini-DIN connectors. Along with external repeater controllers, analog transceivers with suitable "flat audio" connections (no pre/de-emphasis) may also be attached to further support analog linking or digital voice to analog communications conversions. Adapter cable sets are available that enable a microcontroller to read and assert the correct sequence of signals required for Automatic Mode Select configuration.

Following publication of the Common

Air Interface (CAI) specification for System Fusion, a number of third-party additions have been offered for the repeaters. Protocol converters, typically operating in conjunction with Raspberry Pi3/4 and Arduino single-board computers, may allow the support of other DV functions with the DR-1X. DR-2X and DR-1XFR design enhancements may not support a given third-party protocol converter. Interfacing with analog repeater controllers is unaffected.

Other products in the "hot spot" category allow System Fusion repeaters and transceivers to connect through alternate networking schemes. As such, it is possible to use a System Fusion transceiver and network interface to communicate with other DV methodologies, while staying completely within the digital domain.

18.7 APCO Project 25 (P25)

P25 (or APCO Project 25) is a digital voice system designed for public safety (police, fire, EMS, and so on). It was developed in the 1990s to update the FM infrastructure. After about 10 years, the first P25 radios were retired and acquired by hams, who built P25 repeaters around the country. P25 was the first commonly adopted digital voice methodology used by radio amateurs.

The Project 25 Network Exchange (P25NX) designed by NX7Y allows owners of Motorola Quantar P25-enabled repeaters to inexpensively and easily network their repeaters. Established as a worldwide network, there are now four talkgroups: World Wide, North America, Europe, and Pacific, plus three tactical channels within each talkgroup. Efforts are also being undertaken to allow alternative P25 capable repeaters to be networked with P25NX.

18.8 References and Bibliography

Burningham, J., W2XAB, *Amateur Radio Guide to Digital Mobile Radio (DMR)*, **w2xab@arrl.net**, downloadable copies available at **guide.k4usd.org**.

Digital Mobile Radio Association (Professional DMRA) — **www. dmrassociation.org**.

DMR Communications Interconnect Group (DCI) — **www.trbo.org**.

DMR IDs and database — **radioid.net**.

DMR-MARC (Motorola Amateur Radio Club) — **www.dmr-marc.net**.

DMR-MARC Canada — **www.va3xpr.net/dmr-marc-canada**.

D-STAR Users website — **www.dstarusers.org**.

Japan Amateur Radio League, *Translated D-STAR System Specification*, **www.jarl. com/d-star**.

Loveall, P., AE5PL, "D-STAR Uncovered," *QEX*, Nov./Dec. 2008, pp. 50 – 56.

Moxon, L., K1KRC, "Discovering D-STAR," *QST*, Sep. 2010, p. 72.

Pearce, G., KN4AQ, "Operating D-STAR," *QST*, Sep. 2007, pp. 30 – 33.

Pearce, G., KN4AQ, "From Analog to D-STAR," *QST*, May 2012, pp. 36 – 39.

Texas Interconnect Team website — **www.k5tit.org**

The Repeater Builder's Technical Information Page — **www.repeater-builder.com**

Wilson, M., K1RO, ed., *The ARRL Operating Manual*, 12th Edition (ARRL, 2019). Section 1.4 covers FM, Repeaters, Digital Voice, and Data.

Witte, B., KØNR, "TRBO Hits the Amateur Bands," *CQ VHF*, Spring 2012, pp. 72 –74.

Contents

Chapter 19 — Online Content

Articles

- Build a Homebrew Radio Telescope by Mark Spencer, WA8SME
- F-Region Propagation and the Equatorial Ionosphere Anomaly by Jim Kennedy, K6MIO/KH6
- Gray Line Propagation, or Florida to Cocos (Keeling) on 80m by Ed Callaway, N4II
- Hands On Radio: Recording Signals by Ward Silver, NØAX
- The New Sunspot Numbers by Carl Luetzelschwab, K9LA
- The Penticton Solar Flux Receiver by John White, VA7JW, and Ken Tapping
- The Reverse Beacon Network by Pete Smith, N4ZR and Ward Silver, NØAX
- The Solar Eclipse QSO Party by Ward Silver, NØAX
- Upper Level Lows and 6-Meter Sporadic E by Joe Dzekevich, K1YOW
- What to Expect During the Rising Years of Solar Cycle 25 by Frank Donovan, W3LPL

Chapter 19
Propagation of Radio Signals

Radio waves are what amateur radio is all about, so it's very important for amateurs to understand how they *propagate*, or travel, from point to point. In addition, we need to understand how events ranging from local weather to solar activity affect radio wave propagation. Learning about these things makes us more effective operators and station builders.

This chapter has been comprehensively updated from previous editions by Carl Luetzelschwab, K9LA, with additional input from Frank Donovan, W3LPL, Ward Silver, NØAX, Ethan Miller, K8GU, and Hermann Schumacher, DF2DR. The material provides a basic understanding of the principles of electromagnetic radiation, the structure of the Earth's atmosphere, and solar-terrestrial interactions necessary for a working knowledge of radio propagation. The section on VHF/UHF mobile propagation was contributed by Alan Bloom, N1AL. More detailed discussions and the underlying mathematics of radio propagation physics can be found in the references listed at the end of this chapter.

19.1 Fundamentals of Radio Wave Propagation

Radio belongs to a family of electromagnetic radiation that includes infrared (radiation heat), visible light, ultraviolet, X-rays, and the even shorter-wavelength gamma and cosmic rays. Radio has the longest wavelength of this group and thus the lowest frequency frequency (see **Table 19.1**).

Electromagnetic radio waves are composed of an interrelated electric and magnetic field. The electric and magnetic components are oriented at right angles to each other and are also perpendicular to the direction of travel. The polarization of a radio wave is usually designated the same as the orientation of its electric field. This relationship can be visualized in **Figure 19.1**. Unlike sound waves or ocean waves, electromagnetic waves need no propagating medium, such as air or water. This property enables electromagnetic waves to travel through the vacuum of space.

A radio wave far enough from its source to appear flat is called a *plane wave*. From here on, we will be discussing plane waves. The path taken by a plane wave is called a *ray*. Traced from its source to any point on a spherical surface, such as the Earth, the ray will follow a straight line on an *azimuth-equidistant* map projection with the source at its center. (See the sidebar Azimuth-Equidistant Maps.)

19.1.1 Velocity

Radio waves, like all forms of electromagnetic radiation, travel nearly 300,000 kilometers (186,400 miles) per second in a vacuum. The speed of a radio wave is always the product of wavelength and frequency, whatever the medium. That relationship can be stated simply as:

$$c = f \lambda$$

where

c = speed in meters/second,

f = frequency in Hz, and

λ = wavelength in meters.

The wavelength (λ) of any radio frequency can be determined from this simple formula by rearranging the above equation to $\lambda = c/f$. For example, in free space the wavelength of a

Table 19.1
The Electromagnetic Spectrum

Radiation	*Frequency*	*Wavelength*
Radio	10 kHz – 300 GHz	30 km – 1 mm
Infrared	300 GHz – 428.6 THz	1 mm – 700 nm
Visible light	428.6 THz – 750 THz	700 nm – 400 nm
Ultraviolet	750 THz – 3×10^3 THz	400 nm – 100 nm
Extreme Ultraviolet	3×10^3 THz – 3×10^4 THz	100 nm – 10 nm
"Soft" X-ray	3×10^4 THz – 3×10^5 THz	10 nm – 1 nm
"Hard" X-ray	3×10^5 THz – 3×10^6 THz	1 nm – 0.1 nm

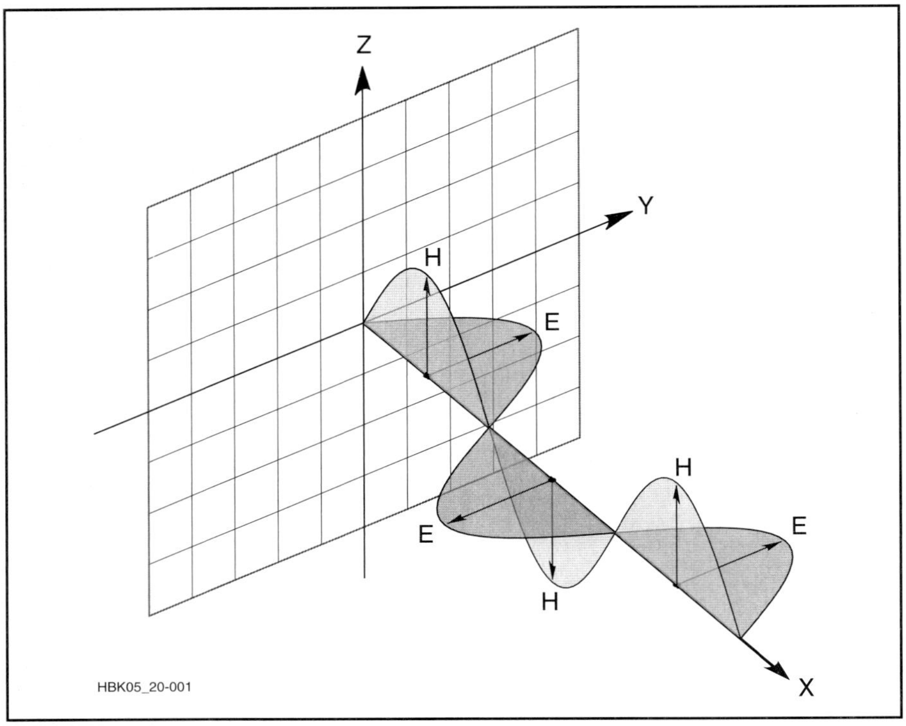

Figure 19.1 — Electric and magnetic field components of the electromagnetic wave. The polarization of a radio wave is the same direction as the plane of its electric field.

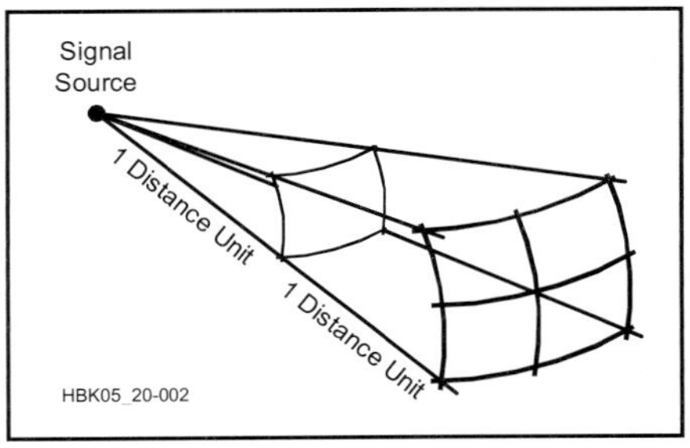

Figure 19.2 — Radio energy disperses as the square of the distance from its source. For the change of one distance unit shown the signal's power per unit of area is only one quarter as strong. Each spherical section has the same surface area.

30 MHz radio signal is thus 10 meters. A simplified equation in metric units is λ in meters = 300 divided by the frequency in MHz. Alternately, λ in feet = 984 divided by the frequency in MHz. Wavelength decreases in other media because the propagating speed is slower.

Radio waves travel more slowly through any other medium than the vacuum of free space. The decrease in speed through the atmosphere is so slight that it is usually ignored, but sometimes even this small difference is significant. The speed of a radio wave in a piece of wire, by contrast, is about 95% of that in free space, and the speed can be even slower in other media.

The reduction of velocity is expressed as the *velocity factor* (VF), which is a value between 0 and 1 representing the medium's fraction of free-space velocity. For example, if the wave travels in RG-213 coaxial cable at 66% of the free-space velocity, the VF for RG-213 is 0.66. Velocity factor must be taken into consideration in antenna designs, in transmission line designs, and in other applications.

19.1.2 Free-Space Attenuation

Free-space attenuation results from the spherical spreading of radio energy as it travels from its source (see **Figure 19.2**). Attenuation grows rapidly with distance because signal strength decreases with the square of the distance from the source. (The signal's field strength in V/m decreases linearly with distance, and its power density in W/m^2 decreases with the square of the distance.) In free space, if the distance between transmitter and receiver is increased from 1 kilometer to 10 kilometers (0.6 to 6 miles), the signal power will be reduced by a factor of 100 (20 dB). Attenuation increases with frequency as well (for example, doubling the frequency results in 6 dB more free-space path loss). Free-space attenuation (path loss) is

$$L_{fs} = 32.45 + 20 \log d + 20 \log f$$

where

L_{fs} = free space path loss in dB,
d = distance in kilometers, and
f = frequency in MHz.

Enhancements (especially at VHF and above) can increase signal strength at the receiver due to multi-path and scattering. Thus, the L_{fs} equation above may not be valid for actual terrestrial conditions.

19.1.3 Losses of Propagation

Radio energy is also lost during refraction, scattering, reflection, and knife-edge diffraction — the very phenomena that can allow long-distance propagation. Indeed, any form of useful propagation is accompanied by attenuation. This may vary from the slight losses encountered by refraction from sporadic E clouds near the maximum usable frequency, to the more considerable losses involved with tropospheric forward scatter (not enough ionization for refraction or reflection, but enough to send weak electromagnetic waves off into varied directions) or D region absorption in the lower HF bands.

Free-space attenuation is a major factor governing signal strength, but radio signals undergo other losses. Energy is lost to *absorption* when radio waves travel through media other than a vacuum. Radio waves propagate through the ionosphere or solid material (like a wire) by exciting electrons, which then reradiate energy at the same frequency. The amount of radio energy lost depends on the characteristic of the medium and on the frequency. In many circumstances, total losses can become so great that radio signals become too weak for communication (they are below the sensitivity of a receiver or receiving system).

19.1.4 Refraction

Electromagnetic waves travel in straight lines until they are deflected by something. Radio waves are *refracted*, or bent, slightly when traveling from one medium to another. *Refraction* is the bending of a ray as it passes from one medium to another at an angle. The apparent bending of a pencil partially immersed in a glass of water demonstrates this principle quite dramatically. The mediums should extend much more than one wavelength around the region in which refraction takes place. Radio waves behave no differently from other familiar forms of electromagnetic radiation in this regard.

Refraction is caused by a change in the velocity of a wave when it crosses the boundary between one propagating medium and another. If this transition is made at an angle, one portion of the wavefront slows down (or

Azimuth-Equidistant Maps

We can better understand several aspects of long-distance propagation if you become accustomed to thinking of the Earth as a ball. This is easy if you use a globe frequently. A flat map of the world, of the *azimuthal-equidistant* projection type, is a useful substitute.

Also known as great circle maps, these are a projection of the Earth's surface that make all great circle paths straight lines from a given location, which is at the center of the map. The outer perimeter of the map is usually 20,000 kilometers (12,427 miles or halfway around the world), but some mapping programs allow other distances.

An azimuth-equidistant map centered on Newington, Connecticut, is shown in **Figure 19.A**. NS6T provides an online service at **ns6t.net/azimuth/azimuth.html** to generate these maps centered on any location. The ARRL World Map is an azimuth-equidistant map centered on Wichita, Kansas. The program *DX Atlas* by VE3NEA (**www.dxatlas.com**) also presents azimuth-equidistant maps, and it offers the option of adding the terminator (the gray line), the auroral oval, and ionospheric parameters.

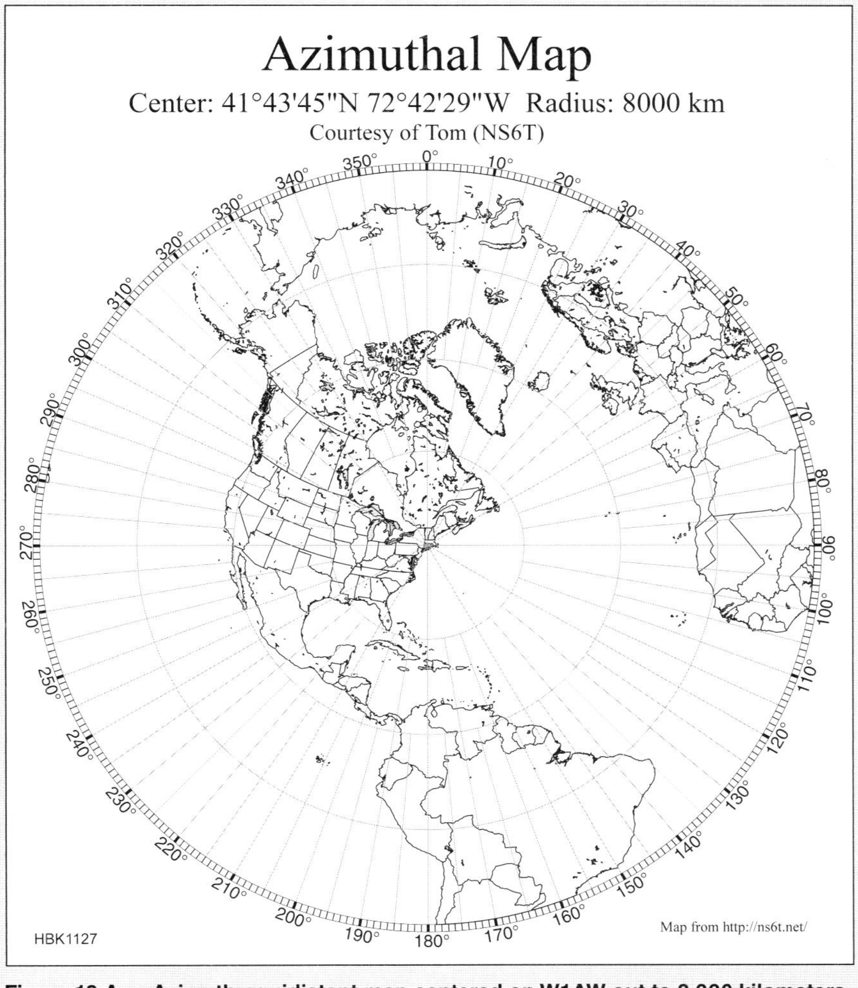

Figure 19.A — Azimuth-equidistant map centered on W1AW out to 8,000 kilometers. (Map provided by ns6t.net/azimuth/azimuth.html)

Optics and Radio Propagation

The four terms *reflection*, *refraction*, *scatter*, and *diffraction* were terms used in optics long before the radio age began. Radio propagation is nearly always a mix of these phenomena, and it may not be easy to identify or separate them while they are happening when we are on the air. This book tends to rely on the words *bending* (refraction) and *scattering* in its discussions, with appropriate modifiers as needed. The important thing to remember is that any alteration of the path taken by energy radiated from an antenna is almost certain to affect on-the-air results.

speeds up) before the other, thus bending the wave slightly. This is shown schematically in **Figure 19.3**.

The amount of bending increases with the ratio of the *refractive indexes* of the two media. The refractive index, n, is simply the velocity of a radio wave in free space divided by its velocity in the medium.

19.1.5 Reflection

Reflection occurs at any boundary between materials of differing dielectric constant when the extent of the materials is on the order of at least one wavelength at the frequency considered. (The surface should extend at least one wavelength from the point at which reflection takes place.) Familiar examples with light are reflections from water surfaces and window panes. Both water and glass are transparent for light, but their dielectric constants are very different from that of air. Light waves, being very short, seem to bounce off both surfaces. Radio waves, being much longer, are practically unaffected by glass, but will be at least partially reflected by a large body of conductive material such as soil or water.

The degree of bending of radio waves at boundaries between air masses which have different indexes of refraction increases with the radio frequency. When the ratio of the refractive indexes of two media is great enough, radio waves can be reflected, just like light waves striking a mirror.

At amateur frequencies above 30 MHz, reflections from a variety of large objects, such as water towers, buildings, airplanes, mountains, and the like, can provide a useful means of extending over-the-horizon paths several hundred kilometers. Two stations need only beam toward a common reflector, whether stationary or moving. A metal surface works well as a reflector if it is several wavelengths in diameter.

Maximum range is limited by the radio line-of-sight distance of both stations to the reflector and by reflector size and shape. The reflectors must be many wavelengths in size and ideally have flat surfaces. Large airplanes make fair reflectors and may provide the best opportunity for long-distance contacts. The calculated limit for airplane reflections is 900 kilometers (560 miles), assuming the largest jets fly no higher than 12,000 meters (40,000 ft), but actual airplane reflection contacts are likely to be considerably shorter.

Reflections from the ionosphere can occur when the plasma frequency (which is proportional to the square root of electron density) is much, much greater than the operating frequency. A good example is sporadic E propagation, since it occurs from a very thin layer

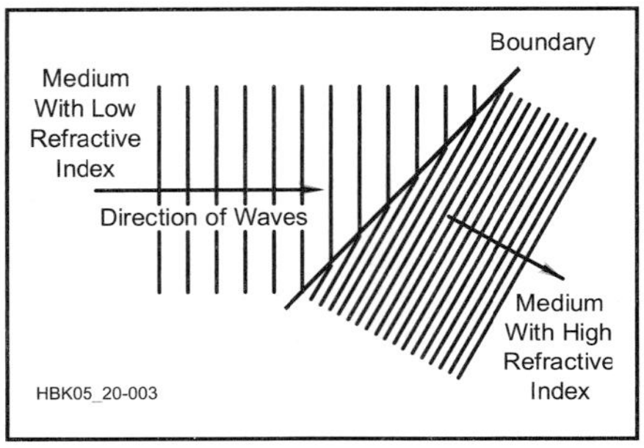

Figure 19.3 — Radio waves are refracted as they pass at an angle between dissimilar media. The lines represent the crests of a moving wave front and the distance between them is the wavelength. The direction of the wave changes because one end of the wave slows down before the other as it crosses the boundary between the two media. The wavelength is simultaneously shortened, but the wave frequency (number of crests that pass a certain point in a given unit of time) remains constant.

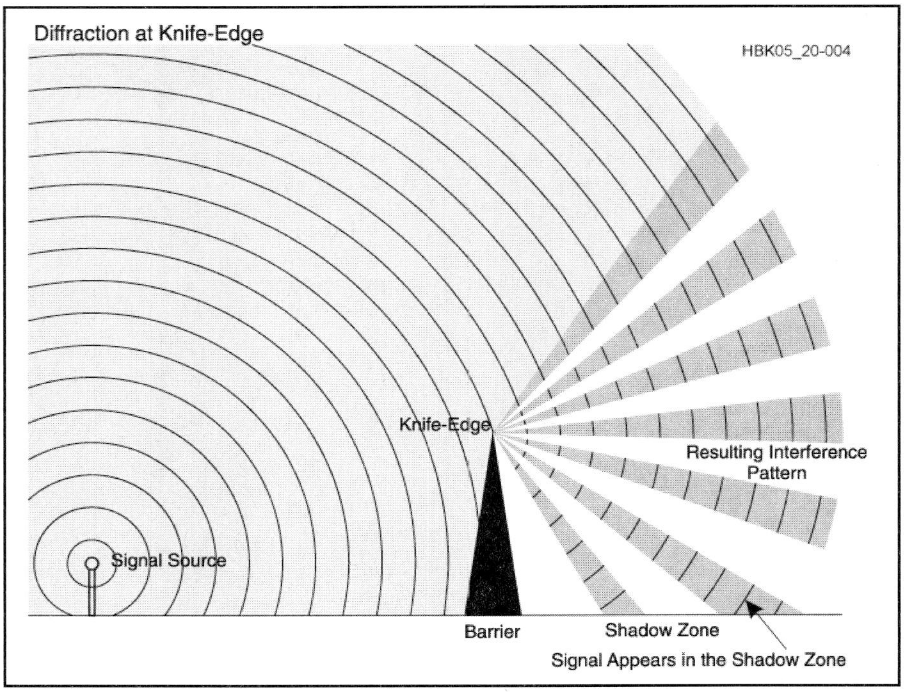

Figure 19.4 — VHF and UHF radio waves, light, and other waves are diffracted around the sharp edge of a solid object that is large in terms of wavelengths. Diffraction results from interference between waves right at the knife-edge and those that are passing above it. Some signals appear behind the knife-edge as a consequence of the interference pattern. Hills or mountains can serve as natural knife-edges at radio frequencies.

(a couple kilometers) that isn't thick enough to return signals to Earth via refraction.

In material of a given electrical conductivity, long waves penetrate deeper than short ones, and so require a thicker conductive region for good reflection. Thin metal, which is highly conductive, is a good reflector of even long-wavelength radio waves. The Earth is a rather lossy reflector of radio signals. With poorer conductors, such as the Earth's crust, long waves may penetrate quite a few feet below the surface. Soil conductivity ranging from poor (sand and rock) to high (wet, rich soil) affects reflection. Water will also reflect radio waves depending on its conductivity. Distilled water is an insulator but salt- and brackish water have high conductivity. The polarization of the reflected wave also affects reflection losses.

19.1.6 Scattering

The direction of radio waves can also be altered through *scattering*. Scatter is the redirection of an electromagnetic wave in many directions from a medium when the volume of material from which the scattering takes place is much smaller than one wavelength at the frequency being considered. Scatter inherently results in loss.

The effect seen by a beam of light attempting to penetrate fog is a good example of light-wave scattering. Even on a clear night, a highly directional searchlight is visible due to a small amount of atmospheric scattering perpendicular to the beam. Radio waves are similarly scattered when they encounter randomly arranged objects of wavelength size or smaller, such as masses of electrons or water droplets. When the density of scattering objects becomes great enough, they behave more like a propagating medium with a characteristic refractive index.

19.1.7 Knife-Edge Diffraction

Radio waves can also pass behind solid objects with sharp edges, such as a mountain range, by *knife-edge diffraction*. This is a common natural phenomenon that affects light, sound, radio, and other coherent waves, but it is difficult to comprehend. **Figure 19.4** depicts radio signals approaching an idealized knife-edge. The portion of the radio waves that strike the base of the knife-edge is entirely blocked, while that portion passing several wavelengths above the edge travel on relatively unaffected. It might seem at first glance that a knife-edge as large as a mountain, for example, would completely prevent radio signals from appearing on the other side but that is not quite true. Something quite unexpected happens to radio signals that pass just over a knife-edge.

Normally, radio signals along a wave front interfere with each other continuously as they propagate through unobstructed space, but the overall result is a uniformly expanding wave. When a portion of the wave front is blocked by a knife-edge, the resulting interference pattern is no longer uniform. This can be understood by visualizing the radio signals right at the knife-edge as if they constituted a new and separate transmitting point, but in-phase with the source wave at that point. The signals adjacent to the knife-edge still interact with signals passing above the edge, but they cannot interact with signals that have been obstructed below the edge. The resulting *interference pattern* no longer creates a uniformly expanding wave front, but rather appears as a pattern of alternating strong and weak bands of waves that spread in a nearly 180° arc behind the knife-edge.

The crest of a range of hills or mountains 50 to 100 wavelengths long can produce knife-edge diffraction at UHF and microwave frequencies. Hillcrests that are clearly defined and free of trees, buildings, and other clutter make the best knife-edges, but even rounded hills may serve as a diffracting edge. Alternating bands of strong and weak signals, corresponding to the interference pattern, will appear on the surface of the Earth behind the mountain, known as the *shadow zone*. The phenomenon is generally reciprocal, so that two-way communication can be established under optimal conditions. Knife-edge diffraction can make it possible to complete paths of 100 kilometers or more that might otherwise be entirely obstructed by mountains or seemingly impossible terrain.

19.1.8 Surface Wave

In general, an electromagnetic *surface wave* travels along an interface between two media with different dielectric constants or refractive indexes. At HF and lower frequencies, the most common type of surface wave travels along the surface of the Earth and is one aspect of what amateurs refer to as "ground wave" as discussed in the next section.

Surface waves can also travel along an interface between media with two different dielectric constants such as layers of air with different temperatures and humidity or insulating layers with different dielectric constants. At microwave frequencies, surface waves travel along the interface between a dielectric (such as air or plastic) and a conducting surface, usually metal.

A radio wave could be traveling in contact with the ground as a surface wave. As the frequency is raised, the distance over which surface waves can travel without excessive energy loss becomes smaller and smaller. The surface wave can provide coverage up to about 100 miles in the standard AM broadcast band during the daytime, but attenuation is high.

Power from a surface wave is typically not transferred to either medium. The wave's E and H fields become weaker exponentially with perpendicular distance from the interface. This property is called *evanescence*. If one or both of the mediums are lossy or leaky, some power can be transferred from the surface wave, and it will become weaker with increasing distance traveled along the interface.

19.1.9 Ground Wave

Ground wave propagation includes two types of radio wave propagation. One type is surface waves (as discussed above) that travel along the Earth's surface. The second is diffracted waves that are bent around the curvature of the Earth by a special form of diffraction that primarily affects longer-wavelength, vertically polarized radio waves.

Ground wave propagation is most apparent in the 3.5 MHz (80 meter), 1.8 MHz (160 meter), 475 kHz (630 meter), and 137 kHz (2200 meter) amateur bands, where practical ground-wave distances may extend beyond 200 kilometers (120 miles). It is also the primary mechanism used by AM broadcast stations in the medium-wave bands. The term ground wave is often mistakenly applied to any short-distance communication, but the actual mechanism is unique to the longer-wave bands.

Ground wave is most useful during the day at 1.8 and 3.5 MHz, when D region absorption makes skywave propagation more difficult. Vertically polarized antennas with excellent ground systems provide the best results. Ground-wave losses are reduced considerably over saltwater and are highest over dry and rocky land. As can be seen from **Figure 19.5**, the range increases with decreasing frequency. The results in this figure are for a typical 100-watt station using a vertical over average ground in a rural noise environment. Your results will vary depending on your power level, antenna, ground conditions, and local noise level.

19.1.10 The Space or Direct Wave

Propagation between two antennas situated within line of sight of each other is shown in **Figure 19.6**. Energy traveling directly between the antennas is attenuated to about the same degree as in free space. Unless the antennas are very high or quite close together, an appreciable portion of the energy is reflected from the ground. This reflected wave combines with direct radiation to affect the actual signal received.

In most communication between two stations on the ground, the angle at which the wave strikes the ground will be small. For a horizontally polarized signal, such a reflection reverses the phase of the wave. If the distances traveled by both parts of the wave were the same, the two parts would arrive out of phase, and would therefore cancel each other. The ground-reflected ray in Figure 19.6 must travel a little farther, so the phase difference between the two depends on the lengths of the paths, measured in wavelengths. The wavelength in use is important in determining the useful signal strength in this type of communication. Let's analyze an example of why the wavelength in use is important.

If the difference in path length is 3 meters, the phase difference with 160 meter waves would be only $360° \times 3/160 = 6.8°$. This is a negligible difference from the 180° shift caused by the reflection, so the effective signal strength over the path would still be very small because of cancellation of the two waves. But with 6 meter radio waves the phase length would be $360° \times 3/6 = 180°$. With the additional 180° shift on reflection, the two rays would add. Thus, the space wave is a negligible factor at low frequencies, but it can be increasingly useful as the frequency is raised. It is a dominant factor in local amateur communication at 50 MHz and higher.

Interaction between the direct and reflected waves is the principal cause of *mobile flutter* observed in local VHF communication between fixed and mobile stations. The flutter effect decreases once the stations are separated enough so that the reflected ray becomes inconsequential. The reflected energy can also confuse the results of field-strength measurements during tests on VHF antennas.

As with most propagation explanations, the space-wave picture presented here is simplified, and practical considerations dictate modifications. There is always some energy loss when the wave is reflected from the ground. Further, the phase of the ground-reflected wave is not shifted exactly 180°, so the waves never

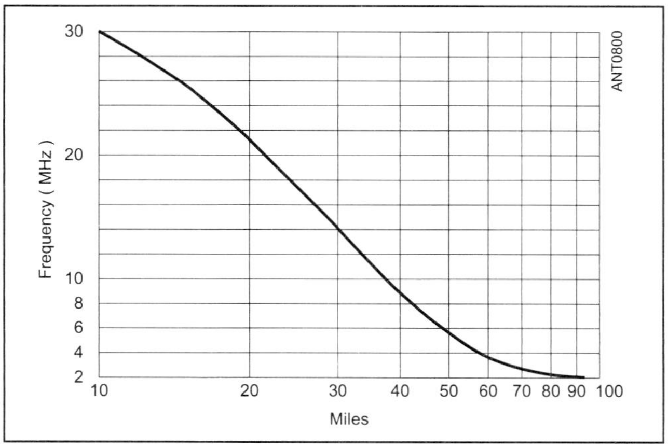

Figure 19.5 — Typical HF ground-wave range as a function of frequency.

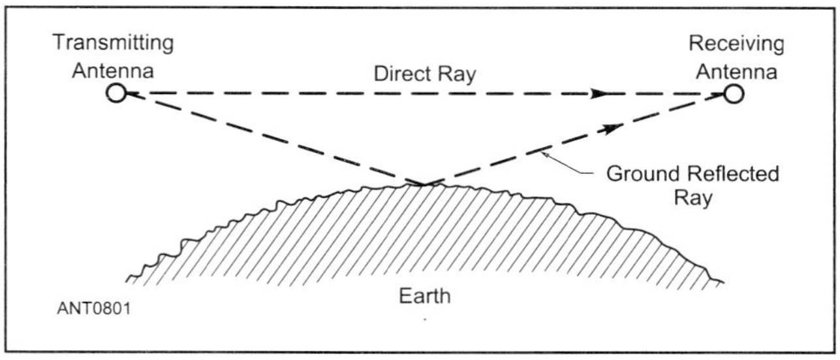

Figure 19.6 — The ray traveling directly from the transmitting antenna to the receiving antenna (direct wave) combines with a ray reflected from the ground (ground reflected ray) to form the space wave. For a horizontally polarized signal, a reflection as shown here reverses the phase of the ground-reflected ray.

cancel completely. At UHF, ground-reflection losses can be greatly reduced or eliminated by using highly directive antennas. By confining the antenna pattern to something approaching a flashlight beam, nearly all the energy is in the direct wave. The resulting energy loss is low enough that microwave relays, for example, can operate with moderate power levels over many miles. Thus, we see that, while the space wave is inconsequential below about 20 MHz, it can be a prime asset in the VHF realm and higher.

19.1.11 Polarization

ANTENNA POLARIZATION SELECTION

If effective communication over long distances were the only consideration, we might be concerned mainly with radiation of energy at the lowest possible angle above the horizon. As an example, our 1.8 and 3.5-MHz bands are well-suited for short-distance communication because they serve that purpose with antennas that are not difficult or expensive to put up. Out to a few hundred miles, simple wire antennas for these bands do well, even though their radiation is mostly at high angles above the horizon. (See the section on NVIS Communication.) Vertical antenna systems are generally better for long-distance use on these frequencies, although they require extensive ground systems for good performance.

Horizontal antennas that radiate well at low angles are most easily erected for 7 MHz and higher frequencies — horizontal wires and arrays are almost standard practice for work on 7 through 29.7 MHz. Vertical antennas, such as a single omnidirectional antenna of multiband design, are also used in this frequency range. An antenna of this type may be a good solution to the space problem for a city dweller on a small lot, or even for the resident of an apartment building.

High-gain antennas are almost always used at 50 MHz and higher frequencies, and most of them are horizontal. The principal exception is mobile communication with FM through repeaters. The height question is answered easily for VHF enthusiasts — the higher the better.

The theoretical and practical effects of height above ground at HF are treated in detail in the *ARRL Antenna Book* chapter **Effects of Ground**. Note that it is the electrical height in *wavelengths* that is important — a good reason to think in the metric system, rather than in feet and inches, since our bands are generally referred to in terms of meters.

In working locally on any amateur frequency band, best results will be obtained with the same polarization at both stations, except on rare occasions when polarization rotation is caused by terrain obstructions or reflections from buildings. Where such a shift is observed, mostly above 100 MHz or so, horizontal polarization tends to work better than vertical. This condition is found primarily on short paths, so it is not too important.

19.2 The Sun and Solar Activity

The Sun is important for amateur radio in two major ways:

1) The Sun emits electromagnetic radiation at appropriate wavelengths to ionize the Earth's atmosphere. This gives us long distance contacts via the ionosphere. Without this energy, we would be limited to line-of-sight propagation and ground wave propagation.
2) The Sun can be quiet or stormy. When the Sun is quiet, it doesn't disrupt the ionosphere. When the Sun becomes stormy, it can disrupt the ionosphere to a level that degrades propagation.

The goal of this section is not to make you a solar scientist. The goal is to give you a basic understanding of the Sun and how it affects propagation. For those who want to learn more about the Sun itself, here is a website that gives much more detail: **www.nasa.gov/mission_pages/sunearth/science/solar-anatomy.html**. Conditions related to solar phenomena that affect radio propagation are measured and reported by the National Oceanic and Atmospheric Administration (NOAA) Space Weather Prediction Center (SWPC: **www.swpc.noaa.gov**). The SWPC is an excellent resource for learning more about how solar behavior affects amateur radio.

Observing Solar Activity

The online content includes articles about observing solar phenomena, including a *QST* article by Mark Spencer, WA8SME, on building your own radio telescope. Numerous online sources such as **solar-center.stanford.edu/observe** explain how to view the Sun for yourself. When observing the Sun, be sure to follow safe viewing practices as described in the online *Sky and Telescope* article "Safe Solar Observing" by Jeff Medkeff (**skyandtelescope.org/observing/celestial-objects-to-watch/safe-solar-observing/**).

19.2.1 The "Quiet" Sun

In this section we describe the activity of the so-called quiet Sun, meaning those times when the Sun is not doing anything more spectacular than acting like a "normal" thermonuclear ball of fusing matter. However, activity on the surface of the Sun is changing continually. The Sun and its effects on earthly propagation can be described in statistical terms — that's what the 11-year solar cycle does as described below.

Because of the continual changes in solar activity, there are continual changes in the state of the Earth's ionosphere and resulting changes in propagation conditions. A short-term burst of solar activity may trigger unusual propagation conditions here on Earth lasting for less than an hour. In more active conditions, described below, activity may result in changes to conditions that last for days at a time. However, on any particular day and time, you are likely to experience very different conditions compared to what a long-term average would suggest.

There are also daily and seasonal variations in the Earth's ionized regions resulting from changes in the amount of ultraviolet and X-ray radiation received from the Sun. The 11-year solar cycle affects propagation conditions because there is a direct correlation between sunspot activity and ionization.

THE SOLAR WIND

The Sun is constantly ejecting material from its surface in all directions into space, making up the *solar wind*. Under relatively quiet solar conditions the solar wind blows around 250 miles per second (400 kilometers per second) — 900,000 miles per hour. Although the velocity is high, the density of the material in the solar wind is very small by the time it has been spread out into interplanetary space. Scientists calculate that the density of the particles in the solar wind is less than that of the best vacuum they've ever achieved on Earth. From **www.swpc.noaa.gov/phenomena/solar-wind**, we see that the solar wind continuously flows outward from the Sun and consists mainly of protons and electrons in a state known as a *plasma*. The solar magnetic field is embedded in the plasma and flows outward with the solar wind.

Different regions on the Sun produce solar

wind of different speeds and densities. Coronal holes produce solar wind of high speed, ranging from 500 to 800 kilometers per second. The north and south poles of the Sun have large, persistent coronal holes, so high latitudes are filled with fast solar wind. In the equatorial plane, where the Earth and the other planets orbit, the most common state of the solar wind is the slow speed wind, with speeds of about 400 kilometers per second. This portion of the solar wind forms the equatorial current sheet. Coronal mass ejections (discussed later) can increase the solar wind speed to 3,000 kilometers per second.

The location of the Earth with respect to the current sheet is important because space weather impacts are highly dependent on the solar wind speed, the solar wind density, and the direction of the magnetic field embedded in the solar wind.

Each of the elements mentioned above play a role in space weather. High speed winds bring geomagnetic storms (described below), while slow speed winds bring calm space weather. Corotating interaction regions and to a lesser extent current sheet crossings can also cause geomagnetic disturbances.

THE INTERPLANETARY MAGNETIC FIELD (IMF)

The Sun's magnetic field isn't confined to the immediate vicinity of our star. The solar wind carries it throughout the solar system. Out among the planets we call the Sun's magnetic field the *interplanetary magnetic field* (IMF). Visit **www.spaceweatherlive.com/en/help/the-interplanetary-magnetic-field-imf.html** for more information.

During solar minimum, the magnetic field of the Sun looks similar to Earth's magnetic field. It looks a bit like an ordinary bar magnet with closed lines close to the equator and open field lines near the poles. Scientist call this orientation of a magnetic field a dipole. The dipole field of the Sun is about as strong as a magnet on a refrigerator (around 50 gauss). The magnetic field of the Earth is about 100 times weaker.

Around solar maximum, when the Sun reaches maximum activity, many sunspots are visible on the visible solar disk. Sunspots are the location of intense magnetic fields which cause ionized material to move along field lines. These fields are often hundreds of times stronger than the surrounding dipole field. This causes the magnetic field around the Sun to be very complex, with many disturbed field lines.

THE EARTH'S GEOMAGNETIC FIELD (GMF)

The Earth's magnetic field, called the *geomagnetic field* (GMF), forms a bubble around our planet called the *magnetosphere*, which deflects solar wind gusts. (Mars, which does not have a protective magnetosphere, has lost much of its atmosphere as a result of solar wind erosion.) Earth's magnetic field and the interplanetary magnetic field come into contact at the *magnetopause*: a place where the magnetosphere meets the solar wind.

Geomagnetic activity is monitored by devices known as *magnetometers*. These may be as simple as a magnetic compass rigged to record its movements. A worldwide network of magnetometers constantly monitors the Earth's magnetic field, because it varies with location. Small variations in the geomagnetic field are scaled to two measures known as the K- and A- indexes, which are described below.

A *geomagnetic storm* is a major disturbance of Earth's magnetosphere that occurs when there is a very efficient exchange of energy from the solar wind into the space environment surrounding Earth. These storms result from variations in the solar wind that produce major changes in the currents, plasmas, and fields in Earth's magnetosphere (from **www.swpc.noaa.gov/phenomena/geomagnetic-storms**).

K-INDEX

The *K-index* provides an indication of geomagnetic activity on a finite logarithmic scale of 0-9, updated every 3 hours. Very quiet conditions are reported as 0 or 1, while geomagnetic storm levels begin at 5 as described later. The K-index reflects readings of the Earth's geomagnetic field at, for example, Boulder, Colorado, over the 3 hours just preceding the data publication. It is the nearest thing to current data on radio propagation available. With new data every 3 hours, the K-index trend is important. For HF propagation, rising is bad news; falling is good, especially related to propagation on paths involving latitudes above 30° north. Because this is a single-location reading of geomagnetic activity, it may not correlate closely with conditions in other areas.

The K-index is also a timely clue to aurora possibilities. Values of 4, and rising, warn that conditions associated with auroras and degraded HF propagation are present in the Boulder area at the time of the bulletin's preparation.

The *planetary K-index, K_p,* is a preliminary value that is updated every minute by the NOAA SWPC with an estimate of the measured K_p of the past 3 hours based on eight ground-based magnetometers at sub-auroral and upper mid-latitude locations. Like the K-index, the estimated 3-hour planetary K_p-index ranges from 0 to 9. It is important to understand that this K_p-index isn't a forecast or an indicator of the current conditions, it always shows the K_p-value that was observed during a certain period.

A-INDEX

Daily geomagnetic conditions are also summarized by the open-ended linear *A-index*, which corresponds roughly to the cumulative K-index values (it's the daily average of the eight K-indexes after converting the K-indexes to a linear scale). See the table at **www.swpc.noaa.gov/products/station-k-and-indices** to translate the logarithmic K-indexes to the linear a-index. Remember the small case 'a' is the linear equivalent of the 3-hour K-index. The capital 'A'-index is the average of the eight 3-hour K-indexes.

The A-index is a daily figure for the state of activity of the Earth's magnetic field. The A-index tells you mainly how yesterday was, but it is very revealing when charted regularly because geomagnetic disturbances nearly always recur at approximately four-week intervals (see the section on 27-Day Recurrence). The A-index commonly varies between 0 and 30 during quiet to active conditions, and up to 100 or higher (the maximum is defined as 400) during geomagnetic storms.

Earth's magnetic field points north at the magnetopause. If the IMF points south — a condition scientists call "southward Bz" (a negative value for Bz) — then the IMF can partially cancel Earth's magnetic field at the point of contact and couple disruptive energy into the Earth's magnetic field. (Material in this section is excerpted from **spaceweather.com/glossary/imf.html**.)

For a geomagnetic storm to develop it is vital that the direction of the interplanetary magnetic field (Bz) turn southward. Persistent values of –10nT and lower (more negative) are good indicators that a geomagnetic storm could develop, but the lower this value goes, the better it is for auroral activity. Only during extreme events with high solar-wind speeds is it possible for a geomagnetic storm (K_p of 5 or higher) to develop with a northward Bz.

SOLAR ULTRAVIOLET RADIATION (UV)

Solar ultraviolet (UV) radiation creates small concentrations of ozone (O_3) molecules between 10 and 50 kilometers (6 and 30 miles) above the Earth's surface. Most UV radiation is absorbed by this process and never reaches the Earth. At even higher altitudes, *ionizing EUV* (Extreme UV with wavelengths of 10 to 100 nanometers or nm) and X-ray radiation partially ionize atmospheric gases. Electrons freed from gas atoms eventually recombine with positive ions to recreate neutral gas atoms, but this process of *recombination* takes some time. In the low-pressure environment at the highest altitudes, atoms are spaced far apart, and the gases may remain ionized for many hours. At lower altitudes, recombination happens rather quickly, and only constant radiation can keep any appreciable portion of the gas ionized.

SOLAR FLUX

Another method of gauging solar activity is the *solar flux*, which is a measure of the intensity of 2800 MHz (10.7 cm) radio noise coming from the Sun. Solar flux is a measure of energy received per unit time, per unit area, per unit frequency interval. (10^{-22} W m^{-2} Hz^{-1}). One solar flux unit equals 10^{-22} joules per second per square meter per hertz.

Values of solar flux at 10.7 cm are measured daily at The Dominion Radio Astrophysical Observatory, Penticton, British Columbia. Data have been collected in Penticton since 1991. (Prior to June 1991, the Algonquin Radio Observatory, Ontario, made the measurements.) Measurements are also made at other observatories around the world, at several frequencies. With some variation, the daily measured flux values increase with increasing frequency of measurement, to at least 15.4 GHz.

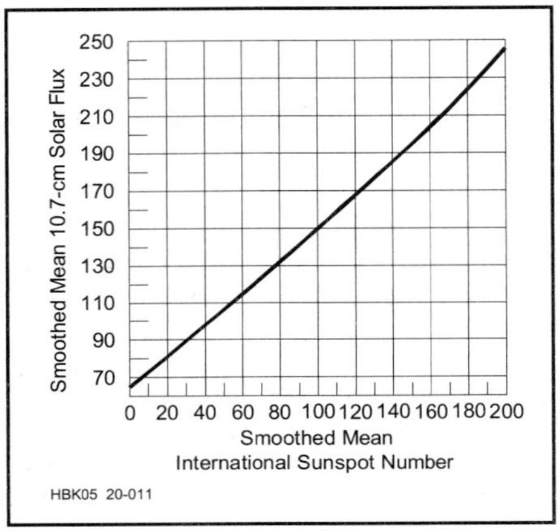

Figure 19.7 — Approximate conversion between solar flux and sunspot number. Note that this is for smoothed values of Solar Flux and Sunspot Numbers.

The daily 2800 MHz Penticton value is sent to the NIST in Boulder, Colorado, where it is incorporated into WWV propagation bulletins. Solar flux, just like a sunspot number, is a proxy (substitute) for the true ionizing radiation at much shorter wavelengths, as solar flux at 2800 MHz does not have enough energy to ionize any atmospheric constituent. The relationship between solar flux and sunspots numbers is discussed later in this section.

The smoothed 2800 MHz radio flux is an indication of the intensity of ionizing UV and X-ray radiation and provides a convenient alternative to sunspot numbers. Solar flux values commonly vary on a scale of 60 – 300 and can be related to sunspot numbers, as shown in **Figure 19.7** (note that this is only valid for converting between smoothed values). The Penticton observatory measures the 2800 MHz solar flux three times a day (centered on local noon). (A supplemental article, "The Penticton Solar Flux Receiver" from February 2013 *QST* is part of this book's online content.) NIST radio station WWV broadcasts the latest solar-flux index at 18 minutes after each hour; WWVH does the same at 45 minutes after the hour. The information is updated every three hours.

19.2.2 The 11-Year Solar Cycle

The density of ionospheric regions depends on the amount of solar radiation reaching the Earth, but solar radiation is not constant. Variations result from daily and seasonal motions of the Earth, the Sun's own 27-day recurrence, and the 11-year cycle of solar activity as described in the following sections. The solar cycle generally has four phases: the solar minimum phase, rising phase, solar maximum phase, and declining phase.

One visual indicator of both the Sun's recurrence and the solar cycle is the periodic appearance of sunspots, which have been observed continuously since the mid-18th century. (The terms "solar cycle" and "sunspot cycle" are often used interchangeably, although the former is more general and so is used in this chapter.)

Until the advent of satellites, solar UV and X-ray radiation could not be measured directly, because they were almost entirely absorbed in the upper atmosphere during the process of ionization. The sunspot number provided the most convenient approximation of general solar activity.

The sunspot number is not a simple count of the number of visual spots, but rather the result of a somewhat complicated formula that takes into consideration size, number, and grouping. (See the following sections on sunspot numbers.) On average, the number of sunspots reaches a maximum every 11 years, but the period has varied between 7 and 17 years. Cycle 19 peaked in 1958, with a smoothed sunspot number of 201, the highest recorded to date. **Figure 19.8** shows the smoothed sunspot numbers (these values are the new Version 2.0 sunspot data) for the past six cycles.

During the peak of the 11-year solar cycle, average solar radiation increases along with the number of flares and sunspots. The ionosphere becomes more intensely ionized consequently, resulting in higher critical frequencies, particularly in the F_2 region. The possibilities for long-distance communications are considerably improved during solar maximum phase, especially in the higher-frequency bands.

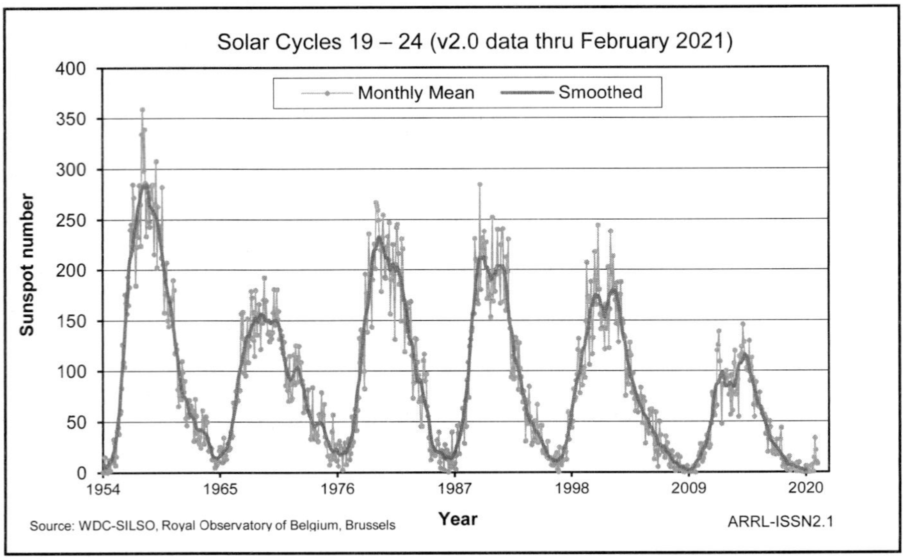

Figure 19.8 — Smoothed sunspot numbers (SSN) through February 2021 for solar cycles 19 to 25.

TRENDS IN SOLAR CYCLES

If one looks at all 24 recorded solar cycles, three characteristics stand out (see **Figure 19.9**). First, there is a cyclic nature to the maximum smoothed sunspot numbers. Second, there have been three high cycle periods of 50 years or so (consisting of several solar cycles) and two low cycle periods of 50 years or so (again, consisting of several solar cycles — the Dalton minimum and the Gleissberg minimum), and the Earth appears to be headed into a third low cycle period. Third, the Earth recently experienced a period of unusually high solar activity known as the Modern Maximum, which ended in 2002. This has allowed excellent worldwide propagation on the higher frequency bands and provided great enjoyment for radio amateurs.

If one looks at solar cycles prior to recorded history through various proxies for solar activity (for example, carbon-14 in trees and beryl-

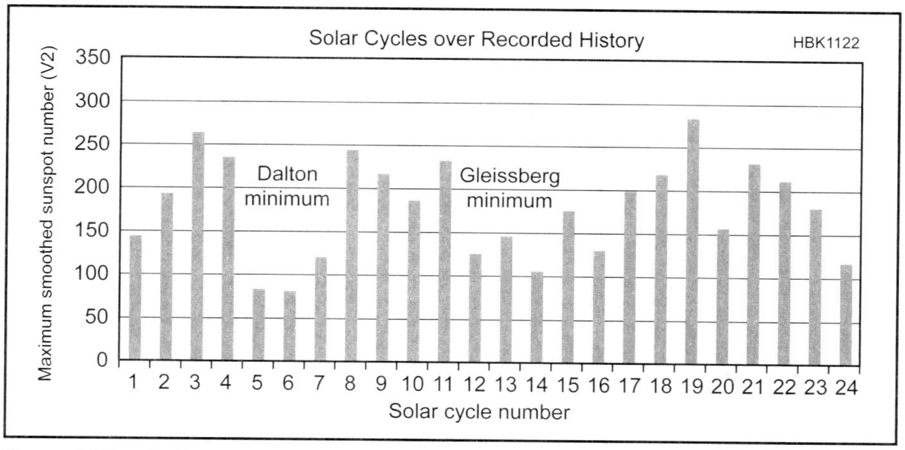

Figure 19.9 — Solar cycles historically

lium-10 in ice cores), one sees extended periods of very low solar activity referred to as Grand Minima (see **Figure 19.10**). The Maunder Minimum between 1645 and 1715 AD is one of the best known. When the Sun is at solar minimum, the Sun's magnetic field is weak, and this lets in galactic cosmic rays that result in more carbon-14 deposits in tree rings and more beryllium-10 deposits in ice cores. It's likely that we'll again enter one of these extended periods, but trying to predict when this will happen is, at best, a wild guess.

With respect to solar minimum periods between solar cycles, historical data shows great variation. The average length of solar minimum, for example defined as the number of months in which the smoothed sunspot number is below 20, is around 37 months. Using this definition, the shortest minimum was 17 months (between Cycles 1 and 2) and the longest minimum was 96 months (between Cycles 5 and 6). The minimum period between Cycle 23 and Cycle 24 turned out to be 56 months. Interestingly, up until the minimum between Cycle 23 and Cycle 24, in living memory, the Earth has experienced minimum periods of approximately 24 months — much shorter than the average. This may lead one to believe this recent solar minimum period was unusual. But historical data, with its great variation, says otherwise. (There is evidence that we may have missed a small solar cycle way back in the 1790 time frame. See **k9la.us/Did_We_Lose_a_Solar_Cycle.pdf**.)

19.2.3 Sunspots

Sunspots are cooler areas on the Sun's surface associated with high magnetic activity. Active regions adjacent to sunspot groups, called *plages* (French for 'beach'), can produce great flares and sustained bursts of electromagnetic radiation in the radio through X-ray spectrum. The white plages are where copious amounts of EUV radiation are emitted, which ionizes the F_2 region of the ionosphere. Since 1948 it has been well known that radio propagation phenomena vary with the number and size of sunspots, and with the position of sunspots on the surface of the Sun. (See the Reference entry for Phillips.)

Individual sunspots may vary in size, appearance, and duration — and may even disappear totally — within a single day. In general, larger active areas persist through several rotations of the Sun. Some active areas have been identified over periods of about a year.

SMOOTHED SUNSPOT NUMBERS

Sunspot data are averaged or smoothed to remove the effects of short-term changes. For example, if a solar cycle is plotted in terms of the daily sunspot number or the monthly mean sunspot number, the resulting curve would be very spiky. Thus, it would be somewhat difficult to ascertain when the solar cycle started, whether it had one or two peaks, and when it ended. Additionally, the sunspot values that should be used for correlating propagation conditions are *Smoothed Sunspot Numbers* (also known as R_{12}), often called 12-month running average values. Data for 13 consecutive months are required to determine a smoothed sunspot number. Sometimes you'll see SSN used as the abbreviation for Smoothed Sunspot Number. Historically SSN has simply meant Sun Spot Number — not a smoothed value. It is best to use R_{12} as the abbreviation for the smoothed sunspot number.

Long-time users have found that the upper HF bands are reliably open for propagation only when the average number of sunspots is above certain minimum levels. For example, between mid-1988 to mid-1992 during Cycle 22, the SSN stayed higher than 100. The 10 meter band was open then almost all day, every day, to some part of the world. However, by mid-1996, few if any sunspots showed up on the Sun and the 10 meter band consequently was rarely open. Even 15 meters, normally a workhorse DX band when solar activity is high, was closed most of the time during the low point in Cycle 22. This behavior is typical for solar cycles and was repeated for Cycles 23 and 24. So far as propagation on the upper HF bands is concerned, the higher the sunspot number, the better the conditions.

Each smoothed number is an average of 13 monthly means, centered on the month of concern. The 1st and 13th months are given a weight of 0.5. A monthly mean is simply the sum of the daily values of the International Sunspot Number (ISN) for a calendar month, divided by the number of days in that month. We would commonly call this value a monthly average.

This may all sound very complicated, but an example should clarify the procedure. Suppose we wished to calculate the smoothed sunspot number for June 1986. We would require monthly mean values for six months prior and six months after this month, or from December 1985 through December 1986. The

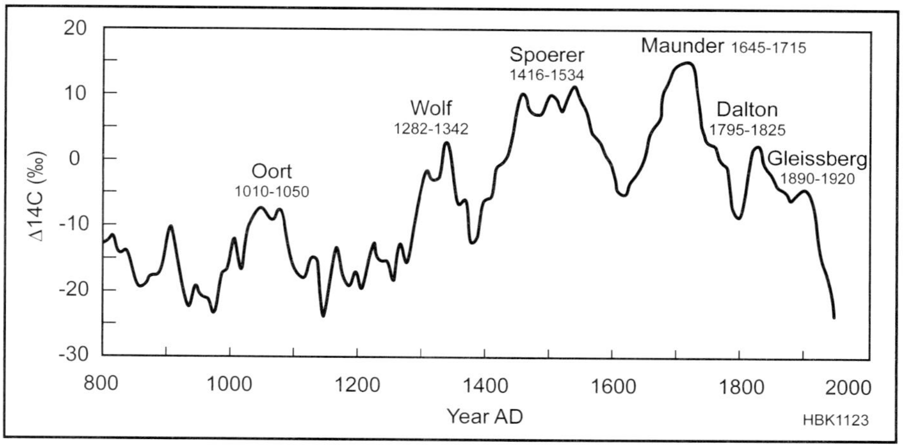

Figure 19.10 — Solar activity and carbon-14 levels from tree rings and other sources of data.

Cycle 25 Status

(This information was prepared in March 2022 by Carl Luetzelschwab, K9LA)

In December 2019, the smoothed sunspot number numerically minimized. This has been heralded as the end of Cycle 24 and the start of Cycle 25. But we have to remember that solar cycles overlap. In other words, we simultaneously see sunspots from the old cycle and the new cycle for a period of time.

Figure 19.B is the latest data for the new Cycle 25. It uses the latest monthly mean sunspot number from February 2022. The smoothed sunspot value is more heavily averaged than the monthly mean value to take out the spikiness of the monthly mean data. The smoothed calculation uses six months of data before and after the target month — thus the latest smoothed sunspot number is six months behind the latest monthly mean sunspot number. (The latest smoothed sunspot number in this graph is from August 2021.)

What's obvious so far is that Cycle 25 appears to be tracking the small Cycle 24, and not the average Cycle 23 nor the moderately big Cycle 21 that resulted in excellent 6-meter propagation. The next six to twelve months will likely give us a better indication of where Cycle 25 is headed. Regardless of where it heads, we should have good worldwide HF propagation for about five years through at least 2026. A good source of forecasting data and predictions for the solar cycle is the Solar Cycle Progression and Forecast at the Marshall Space Flight Center (**www.nasa.gov/msfcsolar**).

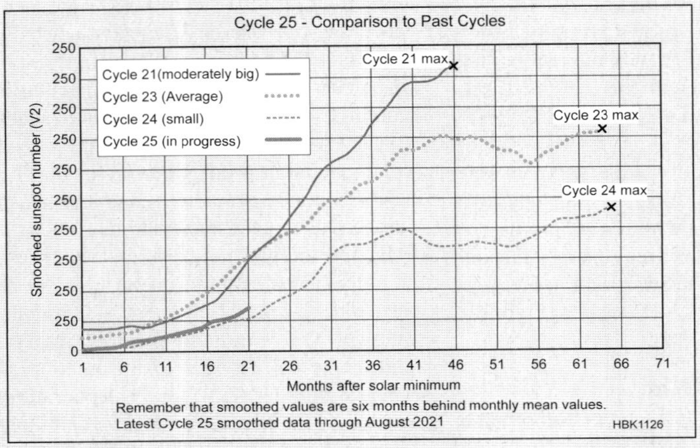

Figure 19.B — Smoothed sunspot numbers for Cycle 25 compared to three other cycles.

monthly mean ISN values for these months (again, these are Version 1.0 values) are:

Dec	85	17.3	Jul	86	18.1
Jan	86	2.5	Aug	86	7.4
Feb	86	23.2	Sep	86	3.8
Mar	86	15.1	Oct	86	35.4
Apr	86	18.5	Nov	86	15.2
May	86	13.7	Dec	86	6.8
Jun	86	1.1			

First, we find the sum of the values, but using only one-half the amounts indicated for the 1st and 13th months in the listing. This value is 166.05. Then we determine the smoothed value by dividing the sum by 12: 166.05/12 = 13.8. (Values beyond the first decimal place are not warranted.) Thus, 13.8 is the smoothed sunspot number for June 1986. From this example, you can see that the smoothed sunspot number for a particular month cannot be determined until six months afterwards.

Generally, the plots we see of sunspot numbers are averaged data. As already mentioned, smoothed numbers make it easier to observe trends and see patterns, but sometimes this data can be misleading. The plots tend to imply that solar activity varies smoothly, indicating, for example, that at the onset of a new cycle the activity just gradually increases. But this is definitely not so. On any given day, significant changes in solar activity can take place within hours, causing sudden band openings at frequencies well above the MUF values predicted from smoothed sunspot number curves. The durations of such openings may be brief, or they may recur for several days running, depending on the nature of the solar activity.

Smoothed numbers are also used in our propagation predictions. When the model of the ionosphere was developed for propagation predictions, scientists determined that the best (acceptable) correlation between what the Sun was doing and what the ionosphere was doing is to use smoothed sunspot numbers and monthly median ionospheric parameters. Median implies a 50% probability and thus our predictions are statistical in nature — we do not have daily predictions.

THE NEW SUNSPOT NUMBERS

Beginning in September 2011, there have been four Sunspot Number Workshops (sponsored by the US National Solar Observatory, the Royal Observatory of Belgium, and the US Air Force Research Laboratory) discussing the quality of the sunspot data. The last workshop reviewed the corrected time series of sunspot numbers from 1610 to the present, and the participants reached an agreement to publish the new data. The old data is designated Version 1.0 (V1.0), while the new data is designated Version 2.0 (V2.0). For more details on this effort, an explanation is provided by K9LA on his website (**k9la.us/Apr16_NEW_Sunspot_Numbers.pdf**). This paper also shows V1.0 vs V2.0 smoothed data (Figure 3 in the paper) from 1950 through 2015, and shows the difference in recent years. V2.0 smoothed values are on average 1.4 times the V1.0 smoothed values for this period.

Going forward, all sunspot numbers are V1.0 values unless otherwise noted as V2.0 data. A discussion of the impact of the new sunspot numbers with respect to propagation predictions will be addressed later.

19.2.4 27-Day Recurrence

Sunspot observations also reveal that the Sun rotates about its own axis. The Sun is composed of extremely hot gases and does not turn uniformly (it is essentially a fluid). At the equator, the period is just over 25 days, but it approaches 35 days at the poles. Sunspots and active regions that affect the Earth's ionosphere, which appear almost entirely within 35° of the Sun's equator, take about 26 days to return to the same position on the solar disk where they affect the Earth. After accounting for the Earth's movement around the Sun, it takes about 27 days for the apparent recurrence of solar activity.

Active regions must face the Earth in the proper orientation to have an impact on the ionosphere. This orientation is referred to as *geo-effective*. They may face the Earth only once before rotating out of view, but the larger active regions often persist for several solar rotations. The net effect is that solar activity from the larger active regions often recurs in 27-day cycles, even though the active regions themselves may last for several solar rotations.

19.2.5 The "Active" Sun

This section describes some of the ways in which strong disruptions to normal solar activity can affect radio wave propagation here on Earth. As far as amateur HF skywave propagation is concerned, the results of these disturbances on the active Sun are not generally beneficial. The most important solar disturbances affecting HF propagation are solar flares, coronal holes, and coronal mass ejections. (See the later section Propagation in Disturbed Conditions.) The occurrence of these conditions and the resulting changes to propagation are monitored and reported on the NIST broadcast stations WWV and WWVH.

NOAA CLASSIFICATION SCALES

In March 2002, NOAA changed the format of the WWV alert (at 18 minutes past the hour) to better align it to the current understanding of disturbances to propagation. NOAA assigns disturbances to propagation into three categories: Radio Blackouts (abbreviated R), Solar Radiation Storms (abbreviated S), and Geomagnetic Storms (abbreviated G). All the details are at **www.swpc.noaa.gov/noaa-scales-explanation**.

NOAA reports all three categories of propagation disturbance on a scale of 1 to 5, with 1 being a minor disturbance and 5 being an extreme disturbance. Although this chapter's discussion of geomagnetic storms, solar radiation storms, and radio blackouts gives you the fundamentals, reviewing the subject matter at the following website will give you a much deeper understanding of the level of impact of each and how and what they impact: See pages 8 – 10 of **www.swpc.noaa.gov/sites/default/files/images/u33/swx_booklet.pdf**.

SOLAR FLARES

Solar flares are large eruptions of electromagnetic radiation from the Sun lasting from minutes to hours. Their intensity and frequency of occurrence follow the 11-year solar cycle. The sudden outburst of electromagnetic energy travels at the speed of light, therefore any effect upon the sunlit side of Earth's exposed outer atmosphere occurs at the same time the event is observed. The increased level of X-ray and extreme ultraviolet (EUV) radiation results in ionization in the regions of the ionosphere on the sunlit side of Earth. (Material in this section is excerpted from **www.swpc.noaa.gov/phenomena/solar-flares-radio-blackouts**.)

Under normal conditions, high frequency (HF) radio waves support communication over long distances by refraction via the upper regions of the ionosphere. When a strong enough solar flare occurs, detrimental ionization is produced in the lower, more dense regions of the ionosphere (the D region), and radio waves that interact with electrons in that region lose energy due to the more frequent collisions that occur in the higher density environment of the D region.

This can cause HF radio signals propagating through the ionosphere on the daylight side of the Earth to become degraded or completely absorbed on the daylight side of Earth. This results in a radio blackout — the absence of HF communication, primarily impacting the 3 to 30 MHz band. This is the Radio Blackout disturbance categorized by NOAA

Very strong solar flares can also eject very high energy protons that cause increased D-region absorption of radio signals propagating through polar regions. This is the Solar Radiation Storm disturbance categorized by NOAA.

Solar flares usually take place in active regions, which are areas on the Sun marked by the presence of strong magnetic fields — typically associated with sunspot groups. As these magnetic fields evolve, they can reach a point of instability and release energy in a variety of forms. These include electromagnetic radiation, which is observed as solar flares.

Solar flare intensities cover a large range and are classified in terms of peak emission in the 0.1 – 0.8 nm spectral band (hard x-rays) of the NOAA/GOES XRS (X-ray Sensor). The X-ray flux levels start with the "A" level (nominally starting at 10^{-8} W/m^2). The next level, ten times higher, is the "B" level ($\geq 10^{-7}$ W/m^2); followed by "C" flares (10^{-6} W/m^2), "M" flares (10^{-5} W/m^2), and finally "X" flares (10^{-4} W/m^2).

Radio Blackouts and Solar Radiation Storms are classified using a five-level NOAA Space Weather Scale, directly related to the flare's max peak in hard X-rays reached or expected. The NOAA Space Weather Prediction Center (SWPC) currently forecasts the probability of C, M, and X-class flares and the probability of an R1 to R3 or greater event as part of the NOAA three-day forecast. Only M- and X-class flares affect HF propagation to a meaningful degree; the weaker M-class flares cause brief blackouts that are of little significance and usually aren't even noticed. The effects of C-class flares are difficult to detect.

CORONAL HOLES

Coronal holes appear as dark areas in the solar corona in extreme ultraviolet (EUV) and soft x-ray solar images. They appear dark because they are cooler, less dense regions than the surrounding plasma and are regions of open, unipolar magnetic fields. This open, magnetic field line structure allows the solar wind to escape more readily into space, resulting in streams of relatively fast solar wind, and is often referred to as a high-speed stream in the context of analysis of structures in interplanetary space. (Material in this section is excerpted from **www.swpc.noaa.gov/phenomena/coronal-holes**.)

Coronal holes can develop at any time and location on the Sun but are most common and persistent during the solar minimum phase. The more persistent coronal holes can sometimes last through several solar rotations (see the preceding section on 27-Day Recurrence). Coronal holes are most prevalent and stable at the solar north and south poles, but these polar holes can grow and expand to lower, more geo-effective, solar latitudes. It is also possible for coronal holes to develop in isolation from the polar holes, or for an extension of a polar hole to split off and become an isolated structure. Persistent coronal holes are long-lasting sources of high-speed solar wind streams.

CORONAL MASS EJECTIONS (CME)

Coronal Mass Ejections (CMEs) are large expulsions of plasma and magnetic field from the Sun's corona. Their intensity and frequency of occurrence tends to follow the solar cycle, although some of the most extremely intense CMEs have occurred outside the solar maximum phase of the solar cycle.

A CME can eject billions of tons of coronal material and carry an embedded magnetic field (frozen in flux) that is stronger than the background solar wind interplanetary magnetic field (IMF) strength. CMEs travel outward from the Sun at speeds ranging from slower than 250 kilometers per second (km/s) to as fast as nearly 3,000 km/s. Most CMEs take 24 to 36 hours to reach Earth, but the fastest take as little as 15 – 18 hours. Slower CMEs can take several days to arrive, or they may not escape the gravity of the Sun at all. They expand in size as they propagate away from the Sun, and larger CMEs can reach a size comprising nearly a quarter of the space between Earth and the Sun by the time they reach our planet. (Material in this section is excerpted from **www.swpc.noaa.gov/phenomena/coronal-mass-ejections**.)

CMEs are often associated with the sudden release of electromagnetic energy in the form of a solar flare, which typically accompanies the explosive acceleration of plasma away from the Sun — the CME. These types of CMEs usually take place from areas of the Sun with localized fields of strong and stressed magnetic flux, such as active regions associated with sunspot groups. CMEs travelling faster than the background solar wind speed can generate a shock wave. These shock waves can accelerate charged particles ahead of them, causing increased radiation storm potential or intensity.

19.3 Sky-Wave or Ionospheric Propagation

The Earth's atmosphere, which reaches to more than 600 kilometers (370 miles) altitude, is usually divided into a number of regions based on a transitioning characteristic of the atmosphere, such as temperature. For propagation purposes, the important regions are shown in **Figure 19.11**. The weather-producing *troposphere* lies between the surface and an average altitude of 10 kilometers (6 miles). Between 10 and 50 kilometers (6 and 30 miles) are the *stratosphere* and the embedded *ozonosphere*, where ultraviolet-absorbing ozone reaches its highest concentrations. About 99% of atmospheric gases are contained within these two lowest regions.

Above 50 kilometers to about 600 kilometers (370 miles) is the *ionosphere*, notable for its effects on radio propagation. At these altitudes, atomic oxygen, molecular oxygen, molecular nitrogen, and nitric oxide predominate under very low pressure and are the important species to consider for propagation. High-energy solar EUV and X-ray radiation ionize these constituents, creating a broad region where ions are created in relative abundance. The ionosphere is subdivided into distinctive D, E, and F regions.

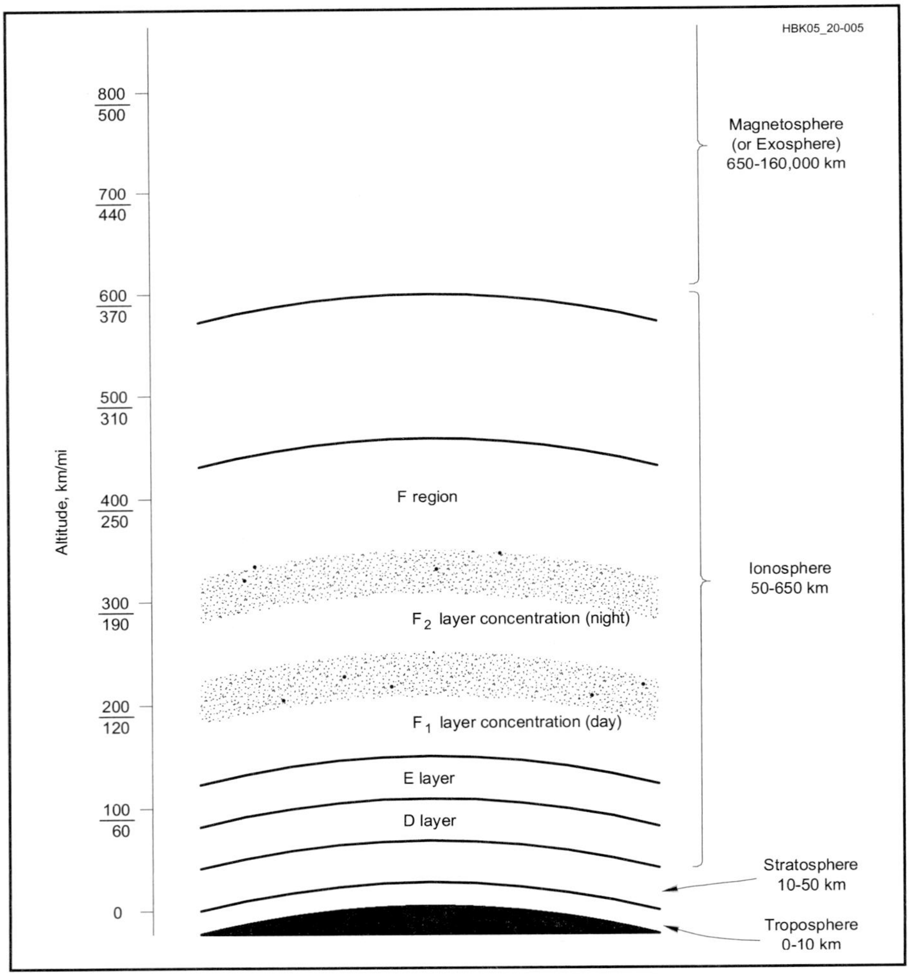

Figure 19.11 — Regions of the lower atmosphere and the ionosphere.

19.3.1 Structure of the Geomagnetic Field

The north and south *magnetic poles* are the points on the surface of Earth at which the planet's magnetic field points vertically downward for the north pole and vertically upward for the south pole. In other words, if a magnetic compass needle is allowed to rotate in three dimensions, it will point straight down at the north magnetic pole and straight up at the south magnetic pole. There is only one location in each hemisphere where this occurs, near (but distinct from) the *geographic* north and south poles.

Related to the magnetic poles are the *geomagnetic poles*. These are the poles of an ideal dipole model of the Earth's magnetic field that most closely fits the Earth's actual magnetic field. The geomagnetic poles are the centers of the auroral ovals. **Figure 19.12** shows the location of the north magnetic pole and how it has drifted much more than the geomagnetic north pole over the years. The figure also shows how the location of the north geomagnetic pole has drifted very little during the same period.

GEOMAGNETIC EQUATOR

The geomagnetic equator is the great circle of the Earth whose plane is perpendicular to the axis of the geomagnetic field. **Figure 19.13** shows the geomagnetic equator (the dashed line) on a Mercator projection of Earth as of 1990. The geomagnetic equator is important for trans-equatorial propagation (TEP) and will be discussed later. This is also called the *magnetic dip equator.*

19.3.2 The Ionosphere

The ionosphere plays a basic role in long-distance communications in all the amateur bands from 1.8 MHz to 50 MHz. The effects of the ionosphere are less apparent at the very high frequencies (30 – 300 MHz), but they persist for terrestrial communications at least through 432 MHz. The basic physics of ionospheric propagation was largely worked out by the 1930s, yet both amateur and professional experimenters made further discoveries throughout the 1930s, 1940s, and 1950s, and continue to make important new discoveries to this day. Sporadic E (the term "layer" is appropriate here as the concentration of electrons is very thin — only a couple kilometers), aurora, trans-equatorial, meteor scatter, and several types of field-aligned scattering were among many additional ionospheric phenomena that required explanation and are still not fully understood.

REGION VERSUS LAYER

The commonly used term "layer" could lead to the erroneous assumption that the ionosphere consists of distinct thin sheets separated by emptiness in between. This is not so. The ionosphere's electronic density versus altitude varies continously with definite peaks and inflection points that define the D, E, F_1, and F_2 regions. (F_1 and F_2 are often labeled F1 and F2 — this chapter will use the subscripted version). These notations are equivalent and refer to the same region. Studies have shown that the height of the various regions may also vary by latitude. On-going ionospheric physics research seeks to better understand the geographic and day-to-day variability of the ionosphere.

IONIZATION PROCESSES

Extreme Ultraviolet (EUV) radiation (wavelengths of 10 to 100 nanometers or nm) from the Sun is the primary cause of ionization in the highest region of the ionosphere (the F_2 region), the region most important for long

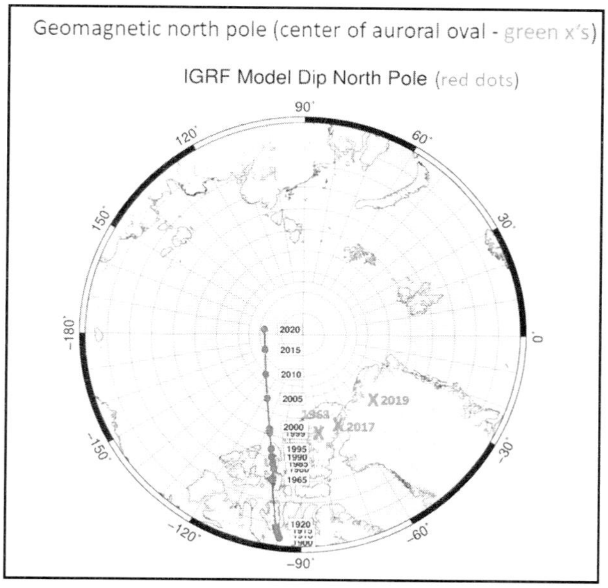

Figure 19.12 — Geomagnetic poles. The location of the north magnetic pole is shown as dots. The geomagnetic north pole locations are labeled with an X.

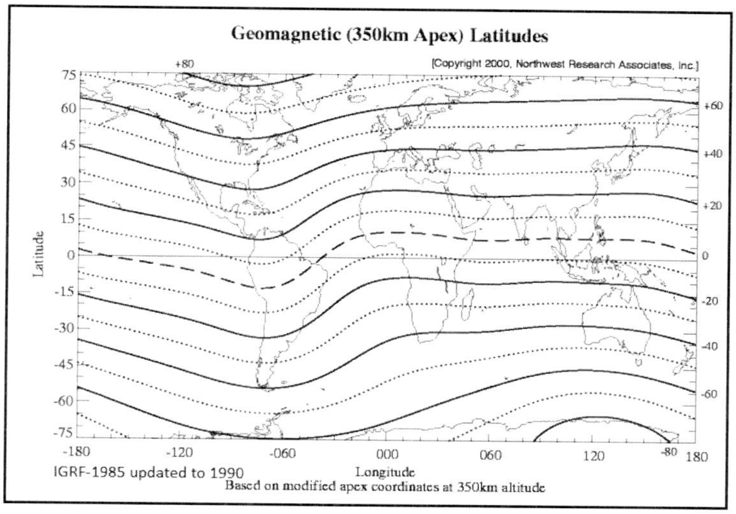

Figure 19.13 — Geomagnetic equator (the dashed line) on a Mercator projection of Earth as of 1990 (thanks Northwest Research Associates).

distance HF propagation. However, there are many other wavelengths of solar radiation as well, including both hard and soft x-rays (0.1 to 1 nm and 1 to 10 nm that contribute to the E region and the D region, respectively), gamma rays, and cosmic rays. The radiated energy breaks up or *photoionizes* atoms and molecules of atmospheric gases into free electrons and positively charged ions. The degree of ionization does not increase uniformly with distance from the Earth's surface. Instead, there are relatively dense regions of ionization, each quite thick and more or less parallel to the Earth's surface, at fairly well-defined intervals outward from about 60 to 400 kilometers (100 to 650 miles). These distinct regions are formed due to complex photochemical reactions of the various types of solar radiation with oxygen, ozone, nitrogen, and nitrous oxide in the rarefied upper atmosphere.

Ionization is not constant within each region. In the E and F_2 region, it tapers off gradually on either side of the maximum. In the D and F_1 regions ionization for all intents and purposes consists of inflection points in the level of ionization with altitude as shown in **Figure 19.14**. The total ionizing energy from the Sun reaching a given point, at a given time, is never constant, so the height and intensity of the ionization in the various regions will also vary. Thus, the practical effect on long-distance communication is an almost continuous variation in signal level, related to the time of day, the season of the year, the distance between the Earth and the Sun, and both short-term and long-term variations in solar activity. It is possible to plan antenna designs, particularly the choosing of antenna heights, to exploit known propagation characteristics.

19.3.3 Ionospheric Refraction

The degree of bending (refraction) of a wave path in an ionized region depends on the density of the ionization and the length of the wave (inversely related to the square of its frequency). The bending at any given frequency or wavelength will increase with increased ionization density and will bend away from the region of most-intense ionization. For a given ionization density, bending increases with longer wavelengths (that is, bending decreases with higher frequencies).

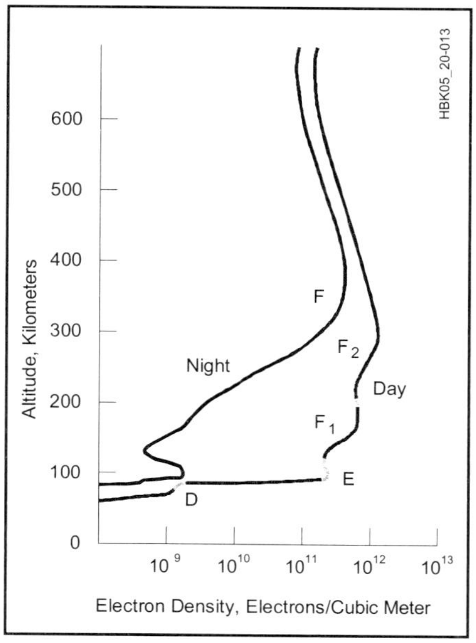

Figure 19.14 — Typical electron densities for the various ionospheric regions.

VIRTUAL HEIGHT

Although refraction is the primary mechanism of ionospheric propagation, it is usually more convenient to think of the process as a reflection. The *virtual height* of an ionospheric region is the equivalent altitude of a reflection that would produce the same effect as the actual refraction. The virtual height of any ionospheric region can be determined using an ionospheric sounder, or *ionosonde*, a sort of vertically oriented radar. The ionosonde sends pulses that sweep over a wide frequency range, generally from 2 MHz to 20 MHz or higher, straight up into the ionosphere. The frequencies of any echoes are recorded against time and then plotted as distance on an *ionogram*. **Figure 19.15** depicts a simple vertical incidence ionogram. (Ionospheric sounding is described in the following sections.)

CRITICAL FREQUENCY

The highest frequency that returns signals from the E and F regions at vertical incidence is known as the *vertical incidence* or *critical frequency*. When the frequency of a vertically incident signal is raised above the critical frequency of an ionospheric region, that portion of the ionosphere is unable to refract the signal back to Earth.

The critical frequency is a function of ion density. The higher the ionization at a particular altitude, the higher becomes the critical frequency. Strictly speaking, the critical frequency is the term applicable to the peak electron density of a region. Physicists relate any electron density in any part of the ionosphere to a plasma frequency, because technically gases in the ionosphere are in a plasma, or partially ionized state. F region critical frequencies commonly range from about 1 MHz and to occasionally as high as 15 MHz during periods of intense ionization.

19.3.4 Real-Time Sounding of the Ionosphere

For many years scientists have *sounded* the ionosphere to determine its communication potential at various elevation angles and frequencies. The word "sound" stems from an old idea — one that has nothing to do with the audio waves that we can hear as "sounds." Long ago, sailors sounded the depths beneath their boats by dropping weighted ropes, calibrated in fathoms, into the water. In a similar fashion, the instrument used to probe the height of the ionosphere is called an *ionosonde*, or ionospheric sounder. It measures distances to various regions by launching an electromagnetic wave directly up into the ionosphere.

VERTICAL-INCIDENCE SOUNDERS

Most ionosondes are *vertical-incidence sounders*, bouncing their signals perpendicularly off the various ionized regions above it by launching signals straight up into the ionosphere. The ionosonde frequency is swept upwards until echoes from the various ionospheric regions disappear, meaning that the critical frequencies for those regions have been exceeded, causing the waves to disappear into space.

Figure 19.15 shows the results of this refraction in a highly simplified ionogram for a typical vertical-incidence sounder. The vertical axis is the virtual height, not the true height attained by the echoes. That's because the virtual height assumes the ray always travels at the speed of light and is simply calculated from the up-down time of flight of the echo. In the real world, though, the ray slows down as it encounters a higher electron density (the group velocity decreases). To determine the electron density profile versus true height, the virtual height traces are used in software such as *POLAN* (POLynomial Analysis) or *ARTIST* (Automatic Real-Time Ionogram Scaler with True height).

Figure 19.16 shows an example of an actual ionogram from the vertical incidence Lowell Digisonde at Millstone Hill in Massachusetts, owned and operated by the Massachusetts Institute of Technology. This ionogram was made on June 18, 2000, and shows the conditions during a period of very high solar activity. The black-and-white rendition in Figure 19.16 of the actual color ionogram unfortunately loses some information. However, you can still see that a real ionogram is a lot more complicated looking than the simple simulated one in Figure 19.15. A map of active ionosonde stations is available at **www.digisonde.com/**

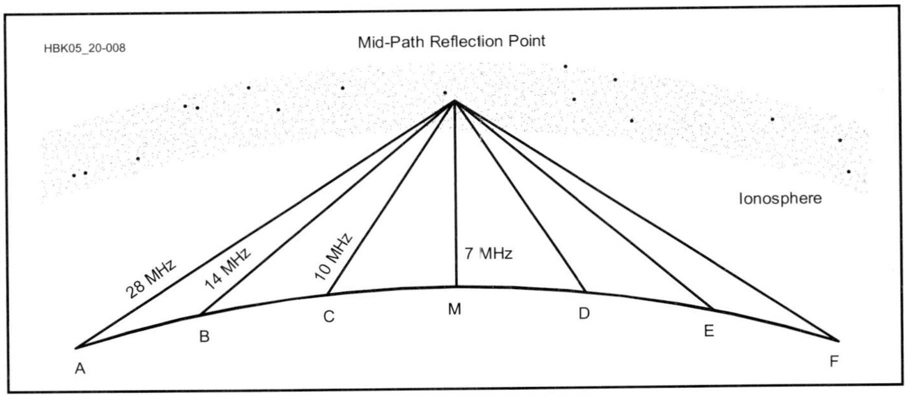

Figure 19.17 — The relationships between critical frequency, maximum usable frequency (MUF), and skip zone can be visualized in this simplified, hypothetical case. The critical frequency is 7 MHz, allowing frequencies below this to be used for short-distance ionospheric communication by stations in the vicinity of point M. These stations cannot communicate by the ionosphere at 14 MHz. Stations at points B and E (and beyond) can communicate because signals at this frequency are refracted back to Earth when they encounter the ionosphere at an oblique angle of incidence. At greater distances, higher frequencies can be used because the MUF is higher at the larger angles of incidence (low launch angles). In this figure, the MUF for the path between points A and F, with a small launch angle, is shown to be 28 MHz. Each pair of stations can communicate at frequencies at or below the MUF of the path between them, but not below the LUF — see text.

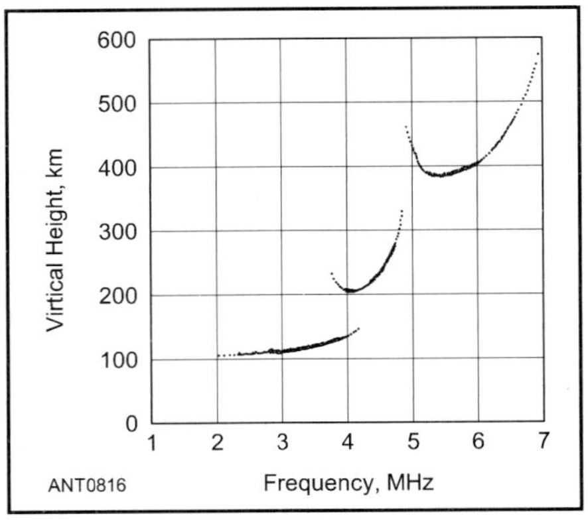

Figure 19.15 — Very simplified ionogram from a vertical-incidence sounder. The lowest trace is for the E region, the middle trace for the F_1 region, and the upper trace for the F_2 region.

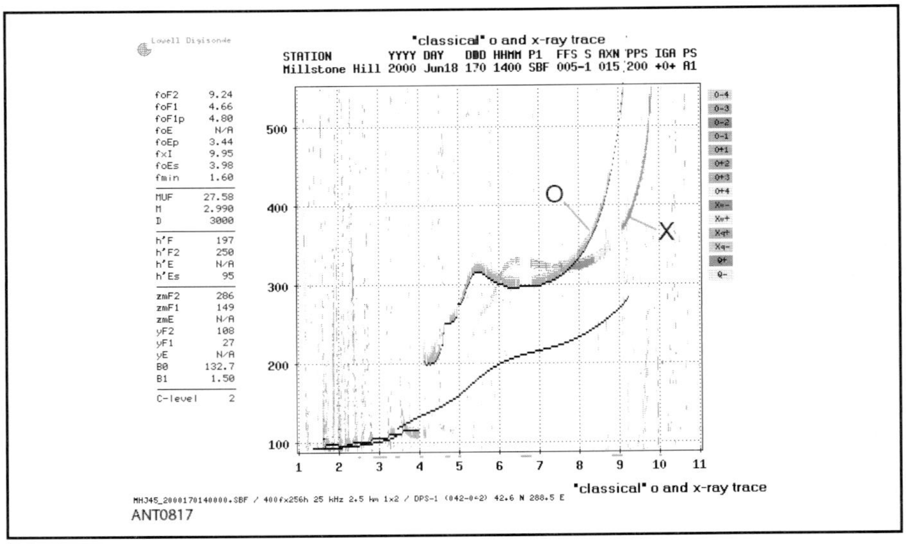

Figure 19.16 — Actual vertical-incidence ionogram from the Lowell Digisonde, owned and operated at Millstone Hill in Massachusetts by MIT (Massachusetts Institute of Technology). The ordinary (o) and extraordinary (x) traces are shown for heights greater than about 300 kilometers. At the upper left are listed the computer-determined ionospheric parameters, such as f_oF_2 at 9.24 MHz and f_oF_1 at 4.66 MHz.

index.html#stationmap-section. Real-time ionograms can be found online at the Digital Ionogram Database, sponsored by the University of Massachusetts Lowell (**giro.uml.edu/DIDBase/**).

The effects of noise and interference from other stations are shown by the many speckled dots appearing in the ionogram. The critical frequencies for various ionospheric regions are listed numerically at the left-hand side of the plot, and the signal amplitudes are color-coded by the color bars at the right-hand side of the plot. The X-axis is the frequency, ranging from 1 to 11 MHz.

19.3.5 Maximum and Lowest Usable Frequencies

A signal entering an ionospheric region above its critical frequency will not be reflected and may be lost to space. However, a signal above the critical frequency may be returned to Earth if it enters the region at an *oblique angle*, rather than at vertical incidence. This is fortunate because it permits two widely separated stations to communicate on significantly higher frequencies than the critical frequency (see **Figure 19.17**). This creates *skip propagation* in which the signal travels from the Earth, into the ionosphere, and back to the Earth where it can be received.

Lower frequency signals can be returned to Earth at higher incident angles, whereas a higher frequency signal at the same angle will not. The higher frequency signal will need to enter the ionosphere at a lower angle if it is to be returned to Earth. The lower angle also means it will be returned to Earth at a more distant point. The combination of ionization, angle of incidence, and fixed station locations creates ranges of frequencies for which communication is possible. This is discussed in the following sections.

MAXIMUM USABLE FREQUENCY (MUF)

The highest frequency supported by the ionosphere for reliable communications between two stations is the *maximum usable frequency* (MUF) for that path. If the separation between the stations is increased, a still higher frequency can be supported at lower launch angles. The MUF for this longer path is higher than the MUF for the shorter path because more refraction can occur for radio waves that encounter the ionosphere at more oblique angles. When the distance is increased to the maximum one-hop distance, the launch angle of the signals between the two stations is zero (that is, the ray path is tangential to the Earth at the two stations) and the MUF for this path is the highest that can be supported by that region of the ionosphere at that location, although antennas producing adequate radiation at very low angles may not be feasible at many locations. This maximum distance is about 4,000 kilometers (2500 miles) for the F_2 region and about 2,000 kilometers (1,250 miles) for the E region (see **Figure 19.18**). This is illustrated more completely in **Figure 19.19**, which gives maximum single-hop distances for E and F region propagation at various wave angles.

The MUF is a function of path, time of day, season, location, solar UV and X-ray radiation levels, and ionospheric disturbances. For vertically incident waves, the MUF is the same as the critical frequency. The seasonal variation of the MUF is due to the ratio of atomic oxygen to molecular nitrogen (O/N_2) in the atmosphere. Atomic oxygen is critical for electron production, whereas molecular nitrogen is critical for recombination. During the winter months in the northern hemisphere, the ratio of O/N_2 is higher, resulting

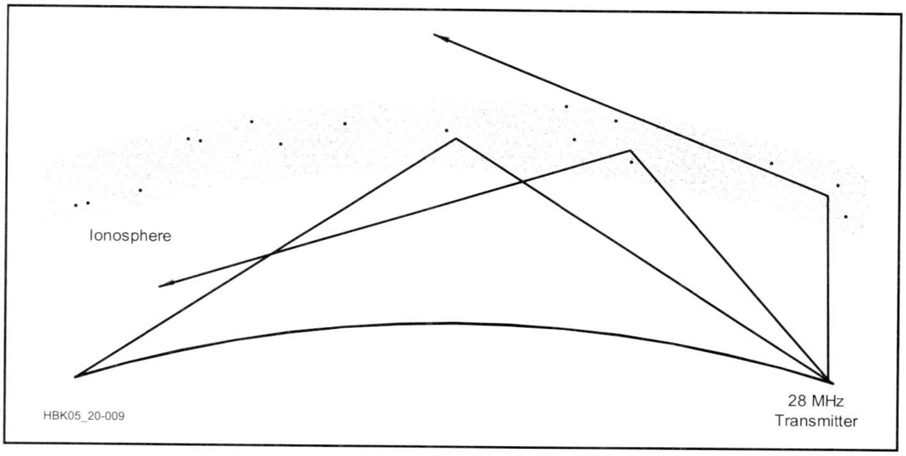

Figure 19.18 — Signals at the MUF propagated at a low angle to the horizon provide the longest possible one-hop distances. In this example, 28 MHz signals entering the ionosphere at higher angles are not refracted enough to bring them back to Earth.

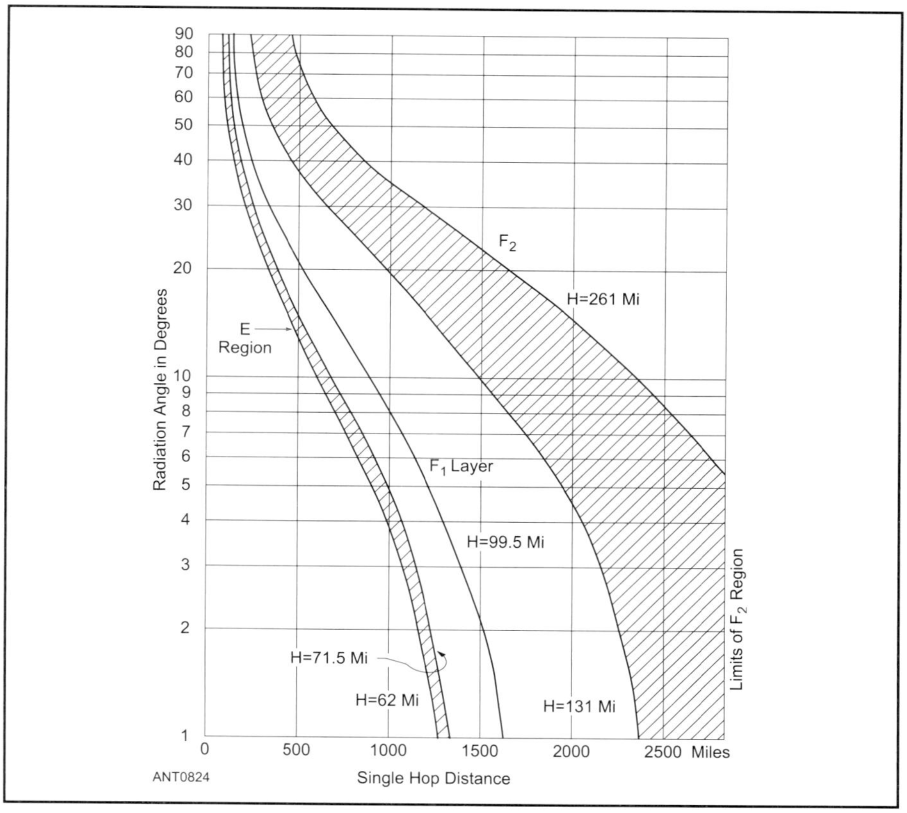

Figure 19.19 — Distance plotted against wave angle (one-hop transmission) for the nominal range of heights for the E, F_1, and F_2 regions.

in higher daytime MUFs.

Precisely speaking, a maximum usable frequency or MUF is defined for communication between two specific points on the Earth's surface, for the conditions existing at the time, including the minimum elevation angle that the station can launch at the frequency in use. (This practical form of MUF is sometimes called the *operational MUF*). At the same time and for the same conditions, the MUF from either of these two points to a third point may be different.

LOWEST USEABLE FREQUENCY (LUF)

The MUF at any time on a particular path is just that — the *maximum* usable frequency. Frequencies below the MUF will also propagate along the path, but ionospheric absorption and noise at the receiving location (due to man-made noise and/or noise from local or distant thunderstorms) may make the received signal-to-noise ratio too low to be usable. This happens because signal absorption increases proportionally to the square of the decrease in frequency. In this case, the frequency is said to be below the *lowest usable frequency* (LUF). The frequency nearest the point where reception became unusable would be the LUF. This occurs most frequently below 10 MHz, where atmospheric and man-made noises are most troublesome.

The LUF can be lowered somewhat by the use of high power and directive antennas, or through the use of communication modes that permit reduced receiver bandwidth or are less demanding of SNR — CW or PSK31 instead of SSB, for example. Digital modes such as JT65 and FT8 that use coding techniques to reject noise can also extend the LUF below what is available to analog modes. This is not true of the MUF, which is limited by the physics of ionospheric refraction, no matter how high your transmitter power or how narrow your receiver bandwidth.

The LUF can be higher than the MUF. When the LUF is higher than the MUF, there is no frequency that supports communication on the particular path at that time.

For example, when solar activity is very high at the peak of a solar cycle, the LUF often rises higher than 14 MHz on the morning Eastern US-to-Europe path on 20 meters. Just before sunrise in the US, the 20 meter band will be first to open to Europe, followed shortly by 15 meters, and then 10 meters as the Sun rises further. By mid-morning, however, when 10 and 15 meters are both wide open, 20 meters will become very marginal to Europe, even when both sides are running maximum legal power levels. By contrast, stations on 10 meters can be worked readily with a transmitter power of only 1 or 2 W, indicating the wide range between the LUF and the MUF.

19.3.6 Skip Zone and Skip Distance

Figure 19.20 shows we can communicate with the point on the Earth labeled "A" (where Wave #3 arrives), but not any closer to our transmitter site. When the critical angle is less than 90° (that is, directly overhead) there will always be a region around the transmitting site where an ionospherically propagated signal cannot be heard, or is heard weakly.

The area between the outer limit of the ground-wave range and the inner edge of signals returning from the ionosphere is the *skip zone*. The outer edge of the skip zone is the distance between the originating site and the beginning of the ionospheric return and is called the *skip distance*. This terminology should not be confused with ham jargon such as "skip is in," referring to the fact that a band is open for skywave propagation. Since both skip distance and maximum ground-wave distance change with frequency, the size of the skip zone also changes with frequency.

The size of the skip zone is closely related to MUF. When two stations are unable to communicate with each other on a particular frequency because the ionosphere is unable to refract the signal enough from one to the other through the required angle — that is, the operating frequency is above the MUF — the stations are in the skip zone for that frequency. Stations within the skip zone may be able to work each other on a lower frequency, or by ground wave, or by other mechanisms if they are close enough. There is no skip zone at frequencies below the MUF for those paths.

19.3.7 Multi-Hop Propagation

As mentioned previously in the discussion about Figure 19.20, the Earth itself can act as a reflector for radio waves, resulting in multiple hops. Thus, a radio signal can be reflected from the reception point on the Earth back into the ionosphere, reaching the Earth a second time at a still more-distant point. This effect is illustrated in **Figure 19.21**, where a single ionospheric region is depicted, although this time we show both the region and the Earth beneath it as curved rather than flat. The wave identified as "Critical Angle" travels from the transmitter via the ionosphere to point A, in the center of the drawing, where it is reflected upwards and travels through the ionosphere to point B, at the right. This shows a two-hop signal.

As in the simplified case in Figure 19.20, the distance at which a ray eventually reaches

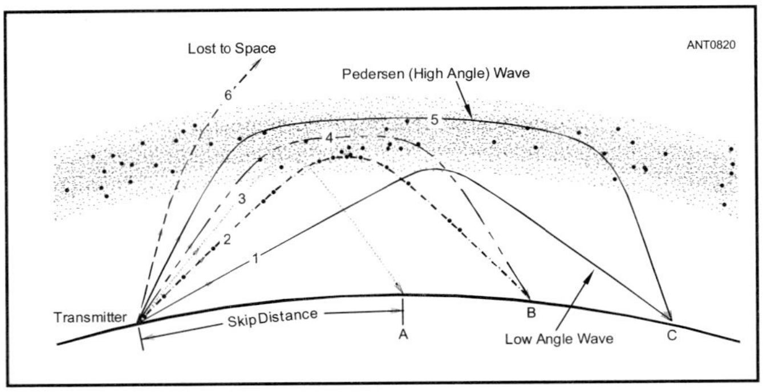

Figure 19.20 — Very simplified smooth-Earth/ionosphere diagram showing how the ground range from transmitter to receiver can vary as the elevation angle is gradually raised. The Pedersen wave, launched at a relatively high angle, has the same ground range as the low-angle wave #1. It may be slightly weaker and less stable, having traveled for a long distance in the ionosphere.

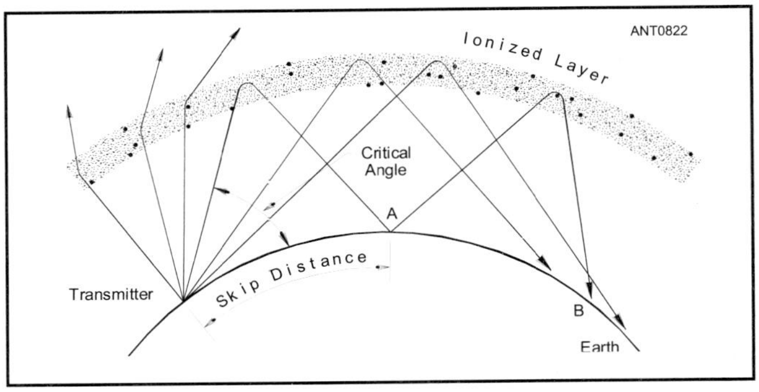

Figure 19.21 — Behavior of waves encountering a simple curved ionospheric region over a curved Earth. Rays entering the ionized region at angles above the critical angle are not bent enough to be returned to Earth and are lost to space. Waves entering at angles below the critical angle reach the Earth at increasingly greater distances as the launch angle approaches the horizontal. The maximum distance that may normally be covered in a single hop is 4,000 kilometers for low elevation angle signals at the upper end of HF. Greater distances are covered with multiple hops. Lower frequencies have shorter maximum hop distances due to more bending in the ionosphere.

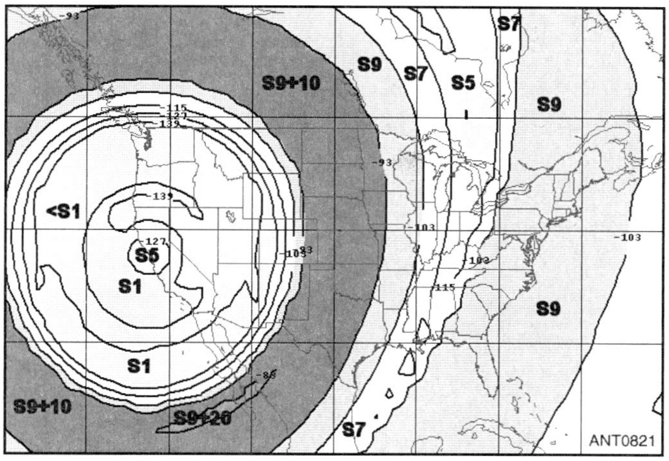

Figure 19.22 — Modified *VOAAREA* plot for 21.2 MHz from San Francisco to the rest of the US, annotated with signal levels in S units, as well as signal contours in dBW (dB below a watt). Antennas are assumed to be 3-element Yagis at 55 feet above flat ground; the transmitter power is 1,500 W; the month is November with SSN=50, a moderate level of solar activity, at 22 UTC. The most obvious feature is the large "skip zone" centered on the transmitter in San Francisco, extending almost one third of the distance across the US.

the Earth depends on the launch elevation angle at which it left the transmitting antenna.

The information in Figure 19.21 is greatly simplified. On actual communication paths the picture is complicated by many factors. One is that the transmitted energy spreads over a considerable area after it leaves the antenna. Even with an antenna array having the sharpest practical beam pattern, there is what might be described as a *cone of radiation* centered on the wave lines (rays) shown in the drawing. The reflection/refraction in the ionosphere is also highly variable and is the cause of considerable spreading and scattering.

Under some conditions it is possible for as many as four or five signal hops to occur over a radio path, as illustrated by the oblique ionogram in Figure 19.19. But no more than two or three hops is the norm. In this way, HF communication can be conducted over many thousands of miles.

An important point should be recognized with regard to signal hopping. A significant loss of signal occurs with each hop — especially at lower HF frequencies. The D and E regions of the ionosphere absorb energy from signals as they pass through, and the ionosphere tends to spread and scatter the radio energy in various directions, rather than confining it in a tight bundle. The roughness of the Earth's surface also scatters the energy at the Earth reflection point.

Assuming that both waves do reach point B in Figure 19.21, the low-angle wave will contain more energy at point B. This wave passes through the lower regions just twice, compared to the higher-angle route, which must pass through these regions four times, plus encountering an Earth reflection. Measurements indicate that although there can be great variation in the relative strengths of the two signals — the one-hop signal will generally be from 7 to 10 dB stronger, although many amateurs may not be capable of transmitting at these low angles. The nature of the terrain at the mid-path reflection point for the two-hop wave, the angle at which the wave is reflected from the Earth, and the condition of the ionosphere in the vicinity of all the refraction points are the primary factors in determining the signal-strength ratio.

The loss per hop becomes significant at greater distances. It is because of these losses that no more than four or five propagation hops are useful; the received signal becomes too weak to be usable over more hops. Although modes other than signal hopping also account for the propagation of radio waves over thousands of miles, backscatter studies of actual radio propagation have displayed signals with as many as five hops. So, the hopping mode is arguably the most prevalent method for long-distance communication.

Figure 19.22 shows another way of looking at propagation — at a *geographic area*. Figure 19.22 shows 15 meter signal levels across the US as they propagate from a transmitting station in San Francisco. This simulation of propagation conditions is for the month of November, with a medium level of solar activity (SSN = 50) at 22 UTC. Figure 19.22 was created using the *VOAAREA* software program, part of the *VOACAP* software suite. Transmitter power is assumed to be 1,500 W, with 3-element Yagis, 55 feet high, at the transmitter and at each receiving location.

From the transmitter out to about 50 miles, signals are moderate, at about S-5 on an S meter. Beyond that coverage area to almost ⅓ of the way across the country (to Colorado), there is a large and distinctive skip zone, where only very weak signals return to Earth (S-1 or less). Beyond Colorado, signals rapidly build up to S-9+10 dB across the middle of the US, falling to S-9 and then to S-7 in the vicinity of Chicago, Illinois. Beyond Chicago, the signals drop to S-5 in a swath from Michigan and part of Ohio down to Alabama. All along the US East Coast, signals come back strong at S-9.

The reason why the signals in Figure 19.22 drop down to S-5 in the Midwest is that the necessary elevation angles to cover this region in a single F_2 hop are extremely low even at a moderate level of solar activity. To achieve launch angles as low as 1 degree requires either very high antenna heights or a high mountain-top location. Beyond the Midwest, out to the US East Coast, two F_2 hops are required, with higher elevation angles and hence greater antenna gain for moderate antenna heights.

19.3.8 NVIS Propagation

The purpose of *near vertical incidence sky-wave* (*NVIS*) propagation is to bridge the gap between where ground wave is too weak and where the skip zone ends. Because NVIS communication operates with high angles of incidence to the ionosphere, signals return to Earth relatively close to the transmitter. By using frequencies close to and below the critical frequency and antennas that produce radiation at high elevation angles, NVIS operation exhibits no skip zone and communications can be maintained over these relatively short distances. (See the previous section on Ionospheric Refraction for an explanation of critical frequency.)

Frequency selection is important since absorption in the ionosphere increases as frequency decreases. By operating close to the critical frequency, absorption is minimized. This requires knowing the critical frequency, which is continually changing from day to day and hour to hour. Propagation modeling software can be used to predict the critical frequency for propagation between two locations. You can also use real-time ionosonde data as discussed previously if the data is representative of your location and the desired communication path. You should expect to change bands as conditions change to remain close to the critical frequency.

A desirable NVIS antenna radiation pattern is one in which most of the radiation is upward at elevation angles greater than 45 degrees. This is referred to as *NVIS directivity*, defined as the average directivity for elevation angles between 70° and 90°. To maximize NVIS directivity, the most common antenna is a λ/2

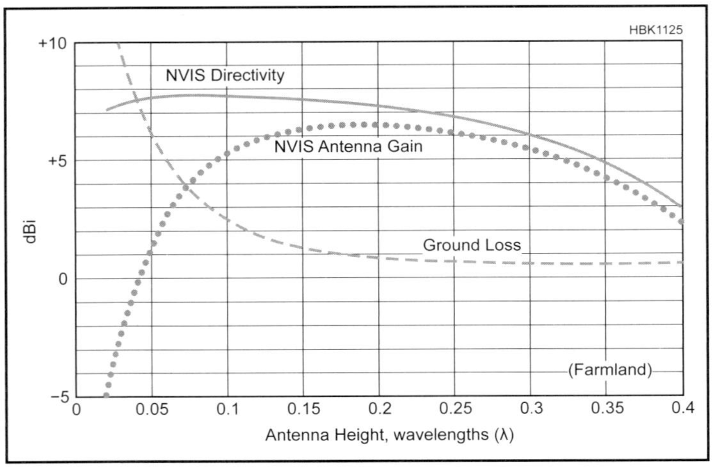

Figure 19.23 — The relationship between antenna height, ground loss, and NVIS directivity at 5.4 MHz. The same general relationship exists on the lower HF bands and for different types of soil. (Graph from Witvliet et al, see References.)

dipole mounted approximately 0.2 λ above the ground. Lower heights incur greater ground losses, and larger heights result in more low-angle radiation such that NVIS performance begins to suffer.

Figure 19.23 summarizes the relationship of antenna height, NVIS directivity, and ground loss at 5.4 MHz. NVIS directivity varies only slowly with height with an optimum at 0.09 λ, whereas the antenna gain has a distinct optimum at 0.19 λ, sharply decreasing at low heights due to excessive ground loss. The same general relationship exists for the other amateur bands below 10 MHz, leading to the overall recommendation for antenna heights of approximately 0.2 λ for best NVIS performance. (This paragraph and figure are from the referenced 2015 article by Witvliet and others — see below.) In the December 2005 issue of *QST*, Dean Straw, N6BV, used the *VOACAP* propagation prediction program to analyze a variety of NVIS paths centered on San Francisco. His analysis showed area coverage maps (signal strength contours versus distance from the transmitter) and elevation patterns of antennas at various heights. His analysis allowed him to formulate a very nice summary: *As a rule-of-thumb, for ham band NVIS, I would recommend that 40 meters be used during the day, 80 meters during the night.* Additionally, during a winter night near solar minimum, changing to 160 meters may be necessary.

Another way to summarize this would be to say that 40 meters (7 MHz) and 60 meters (5 MHz) are good for NVIS propagation out to a couple hundred miles during the day, with 40 meters better at solar maximum and 60 meters better at solar minimum. Likewise, 80 meters (4 MHz, aka 75 meters) and 160 meters (1.9 MHz) are good for NVIS propagation out to several hundred miles during the night, with 80 meters better at solar maximum and 160 meters better at solar minimum (especially during the winter months). This summary takes into account both the MUF and ionospheric absorption.

Along with N6BV's *QST* article, there are several excellent articles that apply to the amateur radio use of NVIS. Ed Farmer, AA6ZM, contributed an overview of NVIS techniques, including an analysis of antenna type and pattern with height, in January 1995 *QST*, which is also listed in the References section. Perhaps the most complete current article on NVIS is from 2015: "Near Vertical Incidence Skywave Propagation: Elevation Angles and Optimum Antenna Height for Horizontal Dipole Antennas" by Witvliet and others in the *IEEE Antennas and Propagation Magazine*. It presents a comprehensive discussion of propagation, coverage, and antenna systems. Tom Kamp, DF5JL, also studied NVIS on the 60 meter band in Germany. See the References entry for his article in *CQ DL*. NVIS antenna systems are also discussed extensively in the *ARRL Antenna Book* chapter on HF Antenna System Design.

19.3.9 Propagation in Disturbed Conditions

So far, we have discussed the Earth's ionosphere when conditions at the Sun are undisturbed. Unfortunately, events on the Sun can disrupt propagation on Earth, especially during the roughly eight to nine years removed from the solar minimum phase. These events cause three types of disturbances to propagation on Earth: radio blackouts, solar radiation storms, and geomagnetic storms. These three categories were defined by NOAA in early 2002 when they changed their format for solar disturbances to better align with the current understanding of these disturbances to propagation. These and related terms, along with the scales for reporting solar disturbances, were described earlier in the section "The Sun and Solar Activity."

GEOMAGNETIC STORMS

Geomagnetic storms are generally caused by coronal mass ejections (CME) and high-speed streams in the solar wind originating from coronal holes. It usually takes 24 to 36 hours for a CME or the effects of a coronal hole to reach and impact the Earth's ionosphere, so we generally have ample warning of the impending disturbance from satellite observations.

CMEs and coronal holes can result in increased geomagnetic field activity, which we see as higher A- and K-indexes. This can cause electrons that are trapped in the Earth's magnetosphere to precipitate into the auroral zones, resulting in increased D and E region ionization. Generally, a similar effect is seen in both auroral zones since the precipitating electrons come from within the Earth's magnetosphere (they don't come directly from the Sun). Thus, as the K-index goes to much higher levels (7 or above), check for auroral VHF propagation. Storm levels range from a K value of 5 (minor) all the way to 9 (extreme). Extreme storms only occur for a few days during a typical solar cycle. (Remember that the A-index reflects yesterday's geomagnetic activity.) See **Table 19.2** for the latest NOAA descriptions of geomagnetic storms.

An elevated K-index would also be a good time to check for skewed paths on the low bands — especially 160 meters — since the amount of refraction by a given electron density profile is inversely proportional to the square of the frequency (the lower the frequency, the more the bending).

Additionally, when the K-index is high, F_2 region ionization generally is depleted at middle and high latitudes, thus moving down in frequency may minimize the impact to propagation. F_2 region ionization can also be enhanced at the low latitudes during geomagnetic storms, so check for enhanced low latitude (equatorial) propagation.

SOLAR RADIATION STORMS AND RADIO BLACKOUTS

Solar radiation storms and radio blackouts are caused by X-class and strong M-class solar flares. When a large solar flare erupts from the Sun's surface, it can launch out into space a wide spectrum of electromagnetic energy. Since electromagnetic energy travels at the speed of light, the first indication of a solar

Table 19.2
Geomagnetic Storms

Scale	*Description*	*Effect on HF Radio*	*Physical measure*	*Average Frequency (1 cycle = 11 years)*
R 5	Extreme	Complete HF (high frequency) radio blackout on the entire sunlit side of the Earth lasting for a number of hours. This results in no HF radio contact with mariners and en route aviators in this sector.	X20	Less than 1 per cycle
R 4	Severe	HF radio communication blackout on most of the sunlit side of Earth for one to two hours. HF radio contact lost during this time.	X10	8 per cycle
R 3	Strong	Wide area blackout of HF radio communication, loss of radio contact for about an hour on sunlit side of Earth.	X1	175 per cycle
R 2	Moderate	Limited blackout of HF radio communication on sunlit side, loss of radio contact for tens of minutes.	M5	350 per cycle
R 1	Minor	Weak or minor degradation of HF radio communication on sunlit side, occasional loss of radio contact.	M1	2000 per cycle

(Data from www.swpc.noaa.gov/noaa-scales-explanation)

flare reaches the Earth in about eight minutes. A large flare shows up as an increase in visible brightness near a sunspot group, accompanied by increases in UV and X-ray radiation and high levels of Sun noise in the VHF radio bands. It is the X-ray radiation that results in radio blackouts for signals propagating across the daytime side of the Earth due to increased D region absorption, and this is called a *sudden ionospheric disturbance* (SID) or a *radio blackout*. The lower frequencies are affected for the longest period. In extreme cases, nearly all background noise will be gone as well. SIDs may last from minutes to a few hours, after which ionospheric conditions return to their pre-blackout conditions.

A large solar flare can also release matter into space, mainly in the form of very energetic particles. These cause *solar radiation storms*, whereby increased absorption in the polar cap (that area inside the auroral oval) degrades over-the-pole propagation paths. This is called a *polar cap absorption* (PCA) event. A PCA event may last for days, dramatically affecting transpolar HF propagation. An interesting fact with respect to PCAs is that they do not necessarily affect the northern and southern polar regions similarly. Thus, if the short path between two points is degraded over one pole, the long path may still be available over the other pole.

19.3.10 E Region Propagation

The lowest portion of the ionosphere useful for long-distance communication by amateurs, the *E region* lies between 90 and 150 kilometers (60 and 90 miles) altitude, but a narrower region centered at 95 to 120 kilometers (60 to 70 miles) is more important for radio propagation. In the E region, nitric oxide and molecular oxygen are ionized by short-wavelength UV and long-wavelength X-ray radiation (so-called soft X-rays). Normally, the E region exists primarily during daylight hours, because, like the D region, it requires a constant source of ionizing radiation. Recombination is not as fast as in the denser D region and absorption is much less. The E region has a daytime critical frequency that varies between 3 and 4 MHz with the solar cycle. At night, the normal E region decays to a minimum critical frequency of 0.3 – 0.5 MHz, which is still enough to refract low elevation angle 1.8 MHz signals.

DAYTIME AND NIGHTTIME E REGION

In the E region, at intermediate atmospheric density, ionization varies with the Sun angle above the horizon (the solar zenith angle), but solar EUV radiation is not the sole ionizing agent. Solar X-rays, meteors, and meteoroids entering this portion of the Earth's atmosphere also play a part. Ionization increases rapidly after sunrise, reaches maximum around noon local time, and drops off quickly after sundown. The minimum is after midnight, local time. As with the D region, the E region absorbs wave energy in the lower-frequency amateur bands when the Sun angle is high,

The E region plays a small role in propagating HF signals but can be a major factor limiting propagation during daytime hours. Its usual critical frequency of 3 to 4 MHz, with an M-factor of about 5, suggests that single-hop E region skip might be useful between 5 and 20 MHz at distances up to 2,300 kilometers (1,400 miles). In practice this is not the case, because the potential for E region skip is severely limited by D region absorption. Signals radiated at low angles at 7 and 10 MHz, which might be useful for the longest-distance contacts, are largely absorbed by the D region. Only high-angle signals pass through the D region at these frequencies, but high-angle E region skip is typically limited to 1,200 kilometers (750 miles) or so. Signals at 14 MHz penetrate the D region at lower angles at the cost of some absorption, but the casual operator may not be able to distinguish between signals propagated by the E region or higher-angle F region propagation.

E REGION BLANKETING

On 20 meters during the summer, on 40 and 30 meters around mid-day, and on 160 meters mostly during the evening hours or well into the night in summer, the E region can block a signal from getting to the higher F region that offers longer hops. This is known as *E region blanketing*, in reference to the E region acting as a blanket that nothing can get through. This is often mistakenly considered to be ionospheric absorption from the D region.

E region blanketing is the primary reason why long-distance propagation is uncommon during midday on the 40- and 30 meter bands and during the summer on 20 meters. D region absorption — to a much lesser degree — also degrades midday long-distance propagation on 40 and 30 meters, mainly by absorbing most low angle, multi-hop, E region propagation.

AURORA

If the orientation of the magnetic field from a powerful CME is aligned southward (opposite to that of the Earth's magnetic field) and if the magnetic field strength is greater than about 5 nanoteslas and *if both conditions* persist for about two hours or more, the magnetic bubble can partially collapse and the particles normally trapped there can be deposited into the Earth's upper atmosphere. This produces a visible or radio *aurora*. An aurora is visible if the time of entry is after dark.

The visible aurora is an optical signature of the plasma curtain capable of refracting radio waves in the range of frequencies above about 20 MHz. D-region absorption increases between midnight and sunrise on lower frequencies during auroras. The exact frequency ranges depend on many factors: time, season, position with relation to the Earth's auroral regions, and the level of solar activity at the time, to name a few.

Radar signals as high as 3 GHz have been scattered by the *aurora borealis* or northern lights (*aurora australis* in the Southern Hemisphere), but amateur aurora contacts are common only from 28 through 432 MHz. By pointing directional antennas generally northward toward the center of aurora activity, oblique paths between stations up to 2,300 kilometers (1,400 miles) apart can be completed (see **Figure 19.24**). High power and large antennas are not necessary. Stations with small Yagis and as little as 10 W output have used auroras on frequencies as high as 432 MHz, but contacts at 902 MHz and higher are exceedingly rare. Auroral propagation works just as well in the Southern Hemisphere, in which case antennas must be pointed generally southward.

In addition to scattering radio signals, auroras have other effects on worldwide radio propagation. Communication below 20 MHz is disrupted in high latitudes, primarily by absorption, and is especially noticeable over polar and near-polar paths. Signals on the AM broadcast band through the 40-meter band late in the afternoon may become weak and watery. The 20-meter band may close down altogether. Satellite operators have also noticed that 144 MHz downlink signals are often weak and distorted when satellites pass near the polar regions. At the same time, the MUF in equatorial regions may temporarily rise dramatically, including an enhancement of transequatorial paths at frequencies as high as 50 MHz.

Auroras occur more often around the spring and fall equinoxes (March-April and September-October), but auroras may appear in any month. Aurora activity generally peaks about two years before and after solar cycle maximum (the "before" generally due to CMEs and the "after" generally due to coronal hole high-speed wind streams). There is a marked diurnal swing in the number of auroras. Favored times are late afternoon and early evening, late evening through early morning, and early afternoon, in about that order. Major auroras often start in early afternoon and carry through to early morning the next day

19.3.11 F Region Propagation

Most of our long-distance communication capability stems from the tenuous outer reaches of the Earth's atmosphere known as the F region. At heights above 100 miles, ions and electrons recombine more slowly, allowing the region to hold its ability to reflect wave energy back to Earth well into the night.

The F region, from 150 kilometers (90 miles) to over 400 kilometers (250 miles) altitude, is by far the most important for long-distance HF communications. F-region oxygen atoms are ionized primarily by ultraviolet radiation. During the day, ionization reaches maxima in two distinct regions, called F_1 and F_2. The F_1 region forms between 150 and 250 kilometers (90 and 160 miles), is most prevalent in the summer months, and disappears at night. The F_2 region extends above 250 kilometers (160 miles), with a peak of ionization around 300 kilometers (190 miles). At night, F-region ionization collapses into one broad region at 300 – 400 kilometers (190 – 250 miles) altitude. Ions recombine very slowly at these altitudes because atmospheric density is relatively low. Maximum ionization levels change significantly with time of day, season, and year of the solar cycle.

The region's height may be from 160 to more than 500 kilometers (100 to over 310 miles), depending on the season of the year, the latitudes, the time of day, and, most capricious of all, what the Sun has been doing in the last few minutes and in perhaps the last three days before the attempt is made. The MUF between Eastern US and Europe, for example, has been anything from 7 to 70 MHz, depending on the conditions mentioned above, plus the point in the long-term solar-activity cycle at which the check is made. The MUF between North America and Europe frequently drops below 7 MHz during the fall and winter during solar minimum years, mainly because of the electron depletion effects of the mid-latitude trough, which is discussed later.

During a summer day, the F region may split into two, the F_1 region and the F_2 region. The lower and weaker F_1 region, about 160 kilometers (100 miles) up, has only a minor role, often blanketing access to the F_2 region during summer daytime hours. At night the F_1 region disappears and the F_2 region height drops somewhat.

Propagation information tailored to amateur needs is published by the ARRL via propagation bulletins that are issued every Friday — they are archived at **arrl.org/w1aw-bulletins-archive-propagation**. Finally, solar and geomagnetic field data, transmitted hourly and updated eight times daily, are given in brief bulletins carried by the US Time Standard stations, WWV (18 minutes past the hour) and WWVH (45 minutes past the hour), and also on internet websites. More information on these services is provided later.

F_1 REGION

The daytime F_1 region is not very important to HF communication. In reality it is an inflection point in the electron density profile, not a peak. It exists only during daylight hours and is largely absent in winter. Radio signals below 10 MHz are not likely to reach the F_1 region, because they are either absorbed by the D region or refracted by the E region.

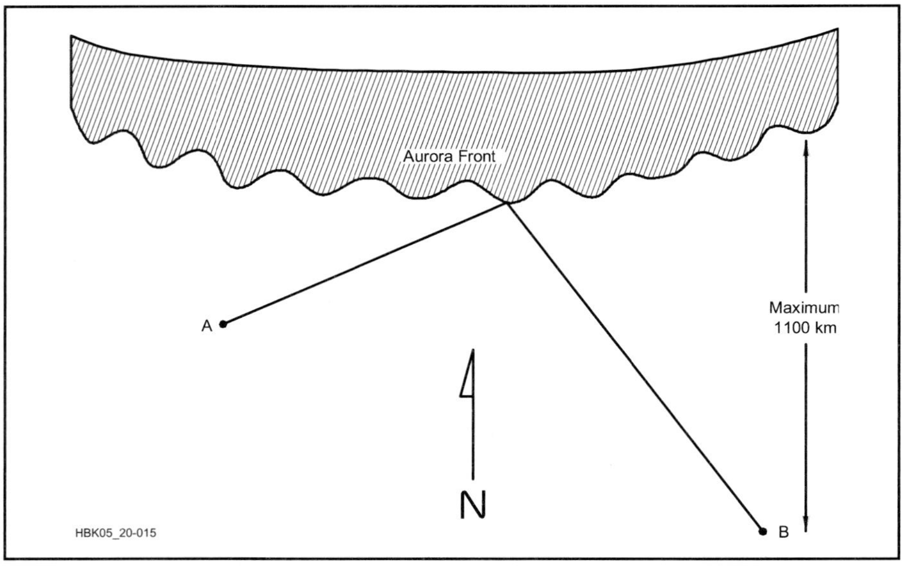

Figure 19.24 — Point antennas generally north to make oblique long-distance contacts on 28 through 432 MHz via aurora scattering. Optimal antenna headings may shift considerably to the east or west depending on the location of the aurora. This necessitates moving your antenna azimuth direction for best propagation and continuing to do so as the aurora progresses.

Signals higher than 20 MHz that pass through both of the lower ionospheric regions are likely to pass through the F_1 region as well, because the F_1 MUF rarely rises above 20 MHz. Absorption diminishes the strength of any signals that continue through to the F_2 region during the day. Some useful F_1 region refraction may take place between 10 and 20 MHz during summer days, yielding paths as long as 3,000 kilometers (1,900 miles), but these would be practically indistinguishable from F_2 skip. The F_1 region often blankets F_2 region daytime 20 meter propagation during the summer, preventing most mid-day long distance propagation. As with E region blanketing, this is often mistaken for absorption.

F_2 REGION

The F_2 region forms between 250 and 400 kilometers (160 and 250 miles) during the daytime and persists throughout the night as a single consolidated F region 50 kilometers (30 miles) higher in altitude. Typical ion densities are the highest of any ionospheric region, with the possible exception of some unusual E region phenomena. In contrast to the other ionospheric regions, F_2 ionization varies considerably with time of day, season, and position in the solar cycle, but it is never altogether absent. These two characteristics make the F_2 region the most important for long-distance HF communications.

The F_2 region MUF is nearly a direct function of ultraviolet (UV) solar radiation, which in turn closely follows the solar cycle. During the lowest years of the cycle, the daytime MUF may climb above 14 MHz for only a few hours a day. In contrast, the MUF may rise beyond 50 MHz during peak years of strong solar cycles and stay above 14 MHz throughout the night. The virtual height of the F_2 region averages 330 kilometers (210 miles) but varies between 200 and 400 kilometers (120 and 250 miles). Maximum one-hop distance is about 4,000 kilometers (2,500 miles). Near vertical incidence skywave propagation just below the critical frequency provides reliable coverage out to 200 – 300 kilometers (120 – 190 miles) with no skip zone. It is most often observed on 7 MHz during the day.

The extremely high-angle *Pedersen ray* can create effective single-hop paths of 5,000 to 12,000 kilometers under certain conditions, but most operators will not be able to distinguish Pedersen ray paths from normal F region propagation. A Pedersen ray path follows the contour of the Earth near the height of the maximum F_2 region electron density, and it requires a fairly stable ionosphere. Pedersen ray paths are most evident over high-latitude east-west paths at frequencies near the MUF. They appear most often about noon local time at mid-path when the geomagnetic field is very quiet. Pedersen ray propagation may be responsible for 50 MHz paths between the US Northeast and Western Europe, for example, when ordinary MUF analysis could not explain the 5,000-kilometer contacts (see part E in **Figure 19.25**).

At any given location on Earth, in general both F_2 region ionization and MUF at that point build rapidly at sunrise, usually reaching a maximum after local noon, and then decrease to a minimum at night prior to the next sunrise. Depending on the season, the MUF is generally highest on paths directed within 20 degrees of the Equator and lower toward the poles. For this reason, transequatorial paths may be open at a particular frequency when all other paths are closed, especially when TEP propagation is available during late afternoon and evening.

In contrast to all the other ionospheric regions, daytime ionization in the winter F_2 region averages four times the level of the summer at the same period in the solar cycle, doubling the MUF. This so-called *winter anomaly* is caused by a seasonal increase in the ratio of atoms to molecules at F_2 region heights (atoms are instrumental in the production of electrons, whereas molecules are instrumental in the loss of electrons). Winter daytime F_2 conditions are much superior to those in summer, because the MUF is much higher.

Propagation modes called *above-the-MUF* can also come into play. Normally we believe the MUF must be equal to or greater than the operating frequency for propagation to occur. This assumes pure refraction. But the MUF can be below the operating frequency — especially for chordal hop modes such as TEP — with additional loss incurred due to the scatter process. The *VOACAP* propagation prediction program includes this above-the-MUF mode, and it will be discussed later.

TRANSEQUATORIAL PROPAGATION

Discovered by amateur radio operators in 1947, *transequatorial propagation* (commonly abbreviated TE or TEP) supports propagation between 5,000 and 8,000 kilometers (3,100 and 5,000 miles) across the magnetic equator from 28 MHz to occasionally as high as 432 MHz. (See the Bibliography entry for *Beyond Line of Sight* by Pocock.) Stations attempting TE contacts must be nearly equidistant from the geomagnetic equator. Many contacts have been made at 50 and 144 MHz between Europe and South Africa, Japan and Australia, and the Caribbean region and South America. Fewer contacts have been made on the 222 MHz band, and TE signals have been heard at 432 MHz.

The ionosphere over equatorial regions — but not directly over the Equator — is higher, thicker, and denser than elsewhere. Because of its more constant exposure to solar radiation, the equatorial belt has high nighttime-MUF possibilities. The potential MUF varies with solar activity, but not to the extent that conventional F-region propagation does.

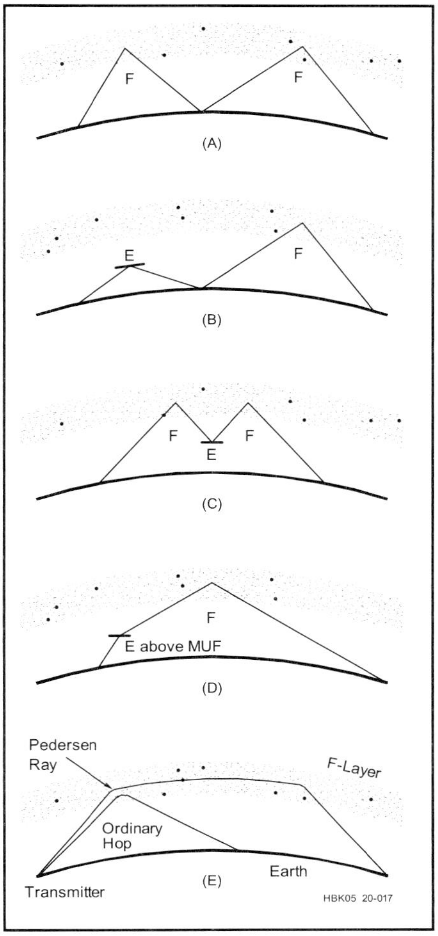

Figure 19.25 — Multihop paths can take many different configurations, including a mixture of E and F region hops. (A) Two F region hops. Five or more consecutive F region hops are possible. (B) An E region hookup to the F region. (C) A top-side E region reflection can shorten the distance of two F region hops. (D) Refraction in the E region above the MUF is insufficient to return the signal to Earth, but it can go on to be refracted in the F region. (E) The Pedersen ray, which originates from a signal launched at a relatively high angle above the horizon into the E or F region, may result in a single-hop path, 5,000 kilometers (3,100 miles) or more. This is considerably farther than the normal 4,000 kilometer (2,500 mile) maximum F-region single-hop distance, where the signal is launched at a very low takeoff angle. The Pedersen ray can easily be disrupted by any sort of ionospheric gradient or irregularity. Not shown in this figure is a chordal hop — since chordal hops are most prevalent in the equatorial ionosphere, refer to Figure 19.26.

TE propagation depends on bulges of intense F_2 region ionization on both sides of the geomagnetic equator. This field-aligned ionization forms shortly after sunset via a process known as the *fountain effect* in an area 100-200 kilometers (60 – 120 miles) north and south of the geomagnetic equator and 500 – 3,000 kilometers (310 – 1,900 miles) wide. It moves west with the setting Sun. The MUF may increase to twice its normal level 15 degrees on either side of the geomagnetic equator (see **Figure 19.26**). For an overview of trans quatorial propagation, read *F-Region Propagation and the Equatorial Ionosphere Anomaly* included with the supplemental content.

The TE range is usually within about 4,000 kilometers (2,500 miles) on either side of the geomagnetic equator. The Earth's magnetic axis is tilted with respect to the geographical axis, so the TE belt appears as a curving band on conventional flat maps of the world (see **Figure 19.27**). As a result, TE has a different latitude coverage in the Americas from that from Europe to Africa.

Unfortunately for most continental US stations, the *geomagnetic equator* dips south of the geographic equator in the Western Hemisphere, as shown in Figure 19.27. Primary northern TEP areas are the Virgin Island, Haiti, the Dominican Republic, Jamaica, Mexico, and much less frequently Florida south of Miami. TE contacts from the southeastern part of the country may be possible with Argentina and Chile.

Transequatorial propagation peaks between 5 p.m. and 10 p.m. during the spring and fall equinoxes, especially during the peak years of the solar cycle. The lowest probability is during the summer and winter solstices. Quiet geomagnetic conditions are required for TE to form. High power and large antennas are not required to work TE, as VHF stations with 100 W and single long Yagis have been successful.

Within its optimum regions of the world, the TE mode extends the usefulness of the 50-MHz band far beyond that of conventional F-region propagation, since the practical TE MUF can be up to 1.5 times that of normal F2 MUF based on analysis with ray tracing. During unusually high periods of F2 region ionization on both sides of the geomagnetic equator (high solar activity) and perhaps aided by geomagnetic field activity, TEP on 144 MHz and 432 MHZ has been observed around the equinoxes for several hours after sunset. Both its seasonal and diurnal characteristics are extensions of what is considered normal for 50-MHz propagation. In that part of the Americas south of about 20 degrees North latitude, the existence of TE affects the whole character of band usage, especially in years of high solar activity. More northerly stations frequently access 50 MHz TEP via an additional E_s hop as discussed in the section on sporadic E. TE propagation is also discussed in the paper "Trans-Equatorial Propagation" by Carl Luetzelschwab, K9LA, on his website, **k9la.us**.

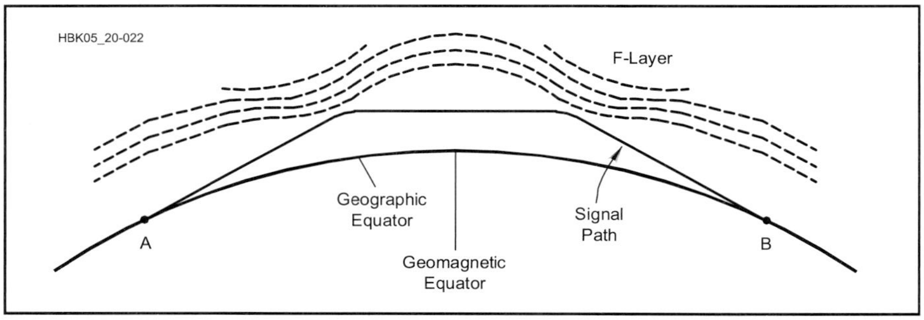

Figure 19.26 — Cross-section of a transequatorial signal path, showing the effects of ionospheric bulging and a double refraction or chordal hop above the normal MUF.

MULTI-HOP F REGION PROPAGATION

Most HF communication beyond 4,000 kilometers (2,500 miles) takes place via multiple ionospheric hops. Radio signals are reflected from the Earth back toward the ionosphere for additional ionospheric refractions. A series of ionospheric refractions and ter-

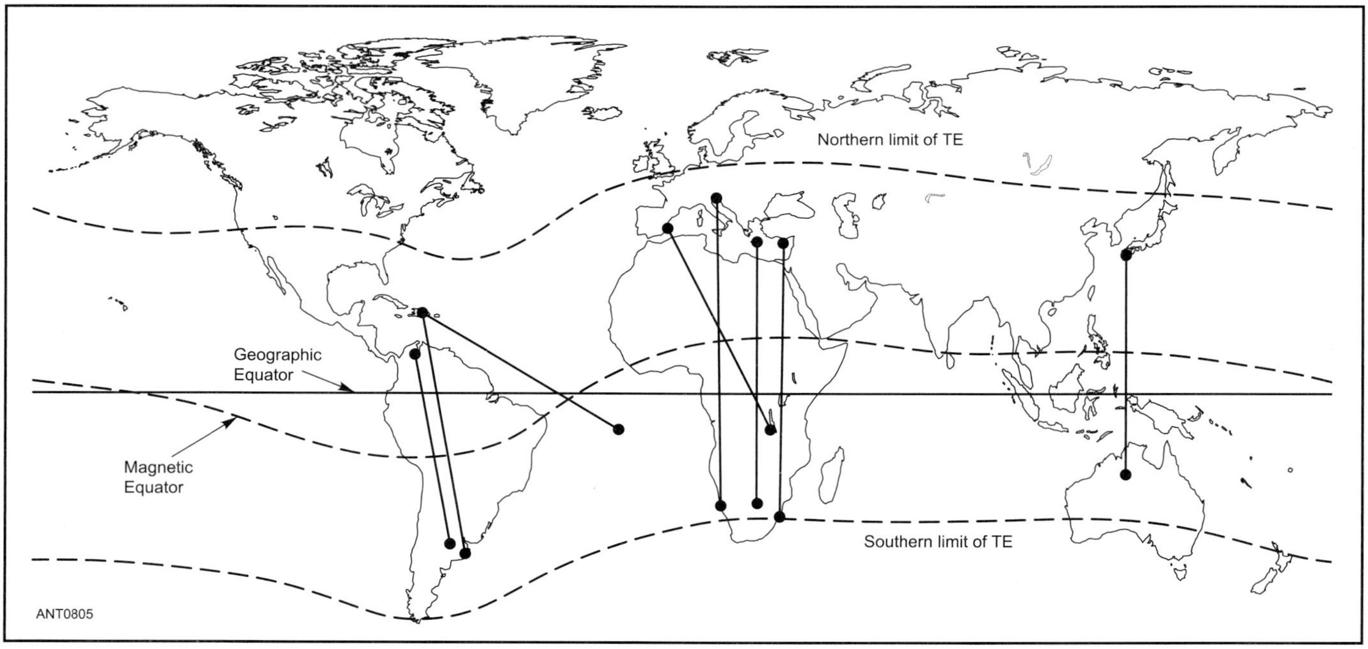

Figure 19.27 — Transequatorial propagation takes place between stations equidistant across the geomagnetic equator. Distances up to 8,000 kilometers (5,000 miles) are possible on 28 through 432 MHz. Note the geomagnetic equator is considerably south of the geographic equator in the Western Hemisphere.

restrial reflections commonly create paths halfway around the Earth. Each hop involves additional attenuation and absorption, so the longest-distance signals tend to be the weakest. Even so, it is possible for signals to be propagated completely around the world and arrive back at their originating point. Multiple reflections within the F region may at times bypass ground reflections altogether, creating what are known as *chordal hops* (see Figure 19.26), with lower total attenuation. It takes a radio signal about 0.14 seconds to make an around-the-world trip.

Multi-hop paths can take on many different configurations, as shown in the examples of Figure 19.25. E region (especially sporadic E) and F region hops may be mixed. In practice, multi-hop signals arrive via many different paths, which often increases the problems of fading. Analyzing multi-hop paths is complicated by the effects of D- and E-region absorption, possible reflections from the tops of sporadic E regions, disruptions in the auroral zone, and other phenomena.

In general, when a band is opening and closing as the MUF changes, extremely low elevation angles are dictated (less than 5 degrees). During the main part of the opening, the elevation angle is higher — generally in the range of 5 to 20 degrees. This is why stations with extremely high antennas (which have higher gain at lower takeoff angles) perform better at band openings and closings.

F-REGION LONG PATH

Most HF communication takes place along the shortest great-circle path between two stations. Short-path propagation is always less than 20,000 kilometers (12,000 miles) — halfway around the Earth. Nevertheless, it may be possible at times to make the same contact in exactly the opposite direction via the *long path*. The long-path distance will be 40,000 kilometers (25,000 miles) minus the short-path length. Signal strength via the long path is usually considerably less than the more direct short path. When both paths are open simultaneously, there may be a distinctive sort of echo on received signals. The time interval of the echo represents the difference between the short-path and long-path distances.

Sunlight is a required element in long-haul communication via the F region above about 10 MHz, but due to recombination being a slow process after sunset, propagation may be supported in the dark ionosphere for many hours after sunset. This fact tends to define long-path timing and antenna aiming. Both are essentially the reverse of the "normal" for a given circuit. We know also that salt-water paths work better than overland ones. This can be significant in long-path work.

There is sometimes a great advantage to using the long path when it is open because signals can be stronger and fading less troublesome, or because fewer interfering signals lie along the path between the stations. There are times when the short path may be closed or disrupted by E region blanketing (described earlier), D region absorption, or F region gaps, especially when operating just below the MUF. Long paths that predominantly cross the night side of the Earth, for example, are sometimes useful because they generally avoid blanketing and absorption problems. Daylight-side long paths may take advantage of higher F region MUFs that occur over the sunlit portions of the Earth. If there is knowledge of this potential at both ends of the circuit, long-path communication may work very well. Cooperation is almost essential, because both the aiming of directional antennas and the timing of the attempts must be right for any worthwhile result.

F-REGION GRAY-LINE

Gray-line paths can be considered a special form of long-path propagation that take into account the unusual ionospheric configuration along the twilight region between night and day. The gray line, as the twilight region is sometimes called, extends completely around the world. Astronomers call this the *terminator*. It is not precisely a line, for the distinction between daylight and darkness is a gradual transition due to atmospheric scattering. Notice that on one side of the Earth, the gray line is coming into daylight (sunrise), and on the other side it is coming into darkness (sunset). In addition, the upper regions of the ionosphere are illuminated longer than lower regions; they come into sunlight first and enter darkness last.

The ionosphere undergoes a significant transformation between night and day. As day begins, the highly absorbent D and E regions are recreated, while the F region MUF rises from its pre-dawn minimum. At the end of the day, the D and E regions quickly disappear, while the F region MUF continues its slow decline from late afternoon. For a brief period just along the gray-line transition, the D and E regions are not well formed, yet the F_2 MUF usually remains higher than 5 MHz. This would normally provide a special opportunity for stations at 1.8 and 3.5 MHz except that:

- ionospheric absorption is still high on these bands in this twilight region from an understanding of the physics of absorption, and
- the electron density gradient across the gray line (also known as the terminator) skews RF away from the day ionosphere into the night ionosphere (shown by ray tracing studies taking into account the Earth's magnetic field and the electron-neutral collision frequency).

What this means is that contrary to decades of popular belief, propagation along the terminator on the low bands is not efficient, and it is not possible for RF on these bands to even follow the terminator. This is supported by the fact that there has never been a report of enhanced propagation on these bands along a short path that is aligned with the gray line.

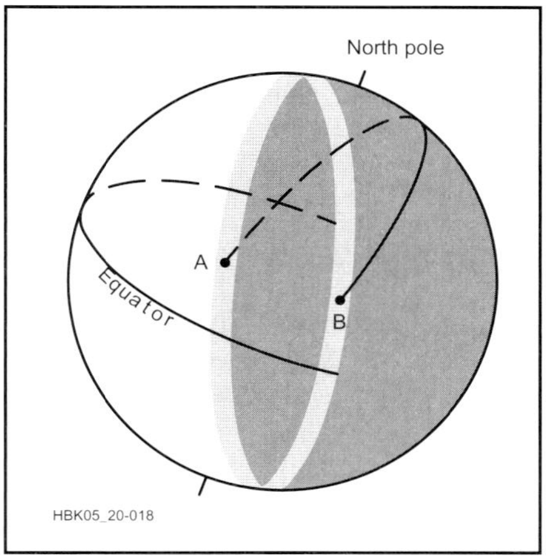

Figure 19.28 — The gray line encircles the Earth, but the tilt at the equator to the poles varies over 46 degrees with the seasons. Long-distance contacts on 1.8 MHz and 3.5 MHz at more than halfway around the Earth can be enabled by the gray line, putting both ends of the path near darkness and allowing RF to cut across the dark ionosphere with minimal absorption. The strength of the signals indicates that multiple Earth-ionosphere hops are not the likely mode of propagation, since losses in many such hops would be prohibitive. Ducting in the electron density valley above the E region peak in the dark ionosphere is the likely mechanism.

Thus, what's important about the gray line is that it puts both ends of the path very near darkness (see **Figure 19.28**). What really appears to be happening is that the RF is taking a shortcut across the dark ionosphere. This requires a skew point to join the two great circle paths in the dark ionosphere out of each end of the path, and the auroral zone is the likely area for the skew. An investigation of this phenomena is contained in N4II's article "Gray Line Propagation, or Florida to Cocos (Keeling) on 80m" at **www.arrl.org/files/file/QEX%20Binaries/2016/Callaway.pdf**.

To take advantage of these paths, you need to understand the mechanics of gray line. The gray line generally runs north-south, but it varies by 23 degrees either side of true north as measured at the equator over the course of the year. This variation is caused by the tilt in the Earth's axis. The gray line is exactly north-south through the poles at the equinoxes (March 21 and September 21) and is at its 23-degree extremes on June 21 and December 20. Over a one-year period, the gray line crosses a 46-degree sector of the Earth north and south of the equator, providing optimum paths to slightly different parts of the world each day. Many commonly available computer programs plot the gray line on a flat map or globe.

To an observer on the Earth, the direction of the terminator is always at right angles to the direction of the Sun at sunrise or sunset. It is important to note that, except at the equinoxes, the gray-line direction will be different at sunrise from that at sunset. This means you can work different areas of the world in the evening than you worked in the morning. Finally, it isn't necessary to be located inside the twilight zone in order to take advantage of gray-line propagation.

A number of web-based applications to calculate worldwide sunrise and sunset times are available online. The software package *DXATLAS* (**www.dxatlas.com/DxAtlas**) includes the gray line with an azimuthal-equidistant map and several logging software packages. For an online gray line map, visit **dx.qsl.net/propagation/greyline.html**.

ENHANCEMENTS NEAR SUNRISE AND SUNSET

There appear to be two mechanisms for signal enhancements around sunrise and sunset. When the path is approximately parallel to the terminator, a signal enhancement can occur when the eastern end is around sunrise or the western end is around sunset. When the path is approximately along the terminator, a signal enhancement can occur when one end of the path is at sunrise and the other end of the path is at sunset (or vice versa).

Propagation Summary, by Band

LOW FREQUENCY (LF) BANDS AND MEDIUM FREQUENCY (MF) BANDS

135.7 – 137.8 kHz (2200 meters) and 472-479 kHz (630 meters)

See the section Propagation Below 1.8 MHz.

1.8 – 2.0 MHz (160 meters)

160 meters suffers from daytime D region absorption. Daytime communication is limited to ground-wave coverage and a single E hop out to about 1,500 kilometers for well-equipped stations (running the full legal limit, quarter-wave verticals with a good ground system, and a low noise receiving environment). At night, the D region quickly disappears and worldwide 160-meter communication becomes possible via F_2 region propagation. Atmospheric and man-made noise limits propagation. Tropical and mid latitude thunderstorms cause high levels of static in summer, making winter evenings the best time to work DX at 1.8 MHz. A proper choice of receiving antenna (Beverage, 4-square, small loop) can often significantly reduce the amount of received noise to improve the signal-to-noise ratio.

HIGH FREQUENCY (HF) BANDS (3 – 30 MHz)

A wide variety of propagation modes are useful on the HF bands. The 80-meter band shares many daytime characteristics with 160 meters. The 60-meter band is better for daytime communications than the 160- and 80 meter bands and is much more similar to the 40 meter band. The transition between bands primarily useful at night or during the day appears around 10 MHz. Most long-distance contacts are made via F_2 region skip. Above 21 MHz, more exotic propagation, including TE, sporadic E, aurora, and meteor scatter, begins to be practical.

3.5 – 4.0 MHz (80 meters for the lower end, 75 meters for the higher end)

The lowest HF band is similar to 160 meters in many respects. Daytime absorption is significant, but not quite as extreme as at 1.8 MHz. At night, signals are often propagated halfway around the world. As at 1.8 MHz, atmospheric noise is a nuisance, especially during the summer, making winter the most attractive season for the 80/75 meter DXer.

5.3 – 5.4 MHz (60 meters)

The distance covered during daytime propagation is very similar to the 40-meter band. At night, worldwide propagation is possible in spite of the relatively low power limit. Signal strengths will typically be higher than on 80 meters but not as high as on 40 meters.

7.0 – 7.3 MHz (40 meters)

The popular 40-meter band supports daytime NVIS propagation and E region propagation out to about 500 miles during most daylight hours. D region absorption is much less severe than on the 160- and 80-meter bands except at low elevation angles, so short-distance skip via the E and F regions is possible. During the day, a typical station can cover a radius of approximately 800 kilometers (500 miles). At night, reliable worldwide communication via F_2 is common on the 40-meter band.

Atmospheric noise is much less troublesome than on 160 and 80 meters, and 40-meter DX signals are often of sufficient strength to override even high-level summer static. For these reasons, 40 meters is the lowest-frequency amateur band considered reliable for DX communication in all seasons. Even during the lowest point in the solar cycle, 40 meters is usually open for worldwide DX throughout the night.

10.1 – 10.15 MHz (30 meters)

The 30-meter band is unique because it shares characteristics of both daytime and nighttime bands. D region absorption is not a significant factor. Communication up to 3,000 kilometers (1,900 miles) is typical during the daytime, and this extends halfway around the world via all-darkness paths. The band is generally open via F_2 on a 24-hour basis, but during a solar minimum the MUF on many DX paths often drops below 10 MHz shortly after sunset. DX paths may drop below 10 MHz at night. Under these conditions, 30 meters adopts the characteristics of the daytime bands at 14 MHz and higher.

14.0 – 14.35 MHz (20 meters)

The 20-meter band is traditionally regarded as the amateurs' primary long-haul DX favorite. Regardless of the 11-year solar cycle, 20 meters can be depended on for at least a few hours of worldwide F_2 propagation during the day. During solar-maximum periods, 20 meters will often stay open to distant locations throughout the night. Skip distance is usually appreciable and is always present to some degree. Daytime E region and sporadic E propagation from sunrise through midnight propagation may be detected along very short paths. Atmospheric noise is not a serious consideration, even in the summer. Because of its popularity, 20 meters tends to be very congested during the daylight hours.

18.068 – 18.168 MHz (17 meters)

The 17-meter band is similar to the 20 meter band in many respects, but the effects of fluctuating solar activity on F_2 propagation are more pronounced. During the years of high solar activity, 17 meters is reliable for daytime and early-evening long-range communication, often lasting well after sunset. During moderate years, the band may open only during sunlight hours and close shortly after sunset. At solar minimum, 17 meters will open to middle and equatorial latitudes, but only for short periods during midday on north-south paths.

21.0 – 21.45 MHz (15 meters)

The 15-meter band has long been considered a prime DX band during solar cycle maxima, but it is sensitive to changing solar activity. During peak years, 15 meters is reliable for daytime F_2 region DXing and will often stay open well into the night. During periods of moderate solar activity, 15 meters is basically a daytime-only band, closing shortly after sunset. During solar minimum periods, 15 meters may not open at all except for infrequent north-south transequatorial circuits. Sporadic E is observed occasionally in early summer and mid-winter, although the effects are not as pronounced as on the higher frequencies.

24.89 – 24.99 MHz (12 meters)

This band offers propagation that combines the best of the 10- and 15-meter bands. Although 12 meters is primarily a daytime band during low and moderate sunspot years, it may stay open well after sunset during the solar maximum. During years of moderate solar activity, 12 meters opens to the low and middle latitudes during the daytime hours, but it seldom remains open after sunset. Periods of low solar activity seldom cause this band to go completely dead, except at higher latitudes. Occasional daytime openings, especially in the lower latitudes, are likely over north-south paths. The main sporadic E season on 24 MHz lasts from late spring through summer and short openings may be observed in mid-winter.

28.0 – 29.7 MHz (10 meters)

The 10-meter band is well known for extreme variations in characteristics and a variety of propagation modes. During solar maxima, long-distance F_2 propagation is so efficient that very low power can produce strong signals halfway around the globe. DX is abundant with modest equipment. Under these conditions, the band is usually open from sunrise to a few hours past sunset. During periods of moderate solar activity, 10 meters usually opens only to low and transequatorial latitudes around noon. During the solar minimum, there may be no F_2 propagation at any time during the day or night.

Sporadic E is fairly common on 10 m, especially May through August, although it may appear at any time. Short skip, as sporadic E is sometimes called on the HF bands, has little relation to the solar cycle and occurs regardless of F region conditions. It provides single-hop communication from 300 to 2,300 kilometers (190 to 1,400 miles) and multiple-hop opportunities of 4,500 kilometers (2,800 miles) and farther.

Ten meters is a transitional band in that it also shares some of the propagation modes more characteristic of VHF. Meteor scatter, aurora, and auroral E provide the means of making contacts out to 2,300 kilometers (1,400 miles), and TEP propagation often supports propagation to 5,000 miles and farther, but these modes often go unnoticed at 28 MHz. Techniques similar to those used at VHF can be very effective on 10 meters, as signals are usually stronger and more persistent. These exotic modes can be more fully exploited, especially during the solar minimum when F_2 DXing has waned.

VERY HIGH FREQUENCY (VHF) BANDS (30 – 300 MHz)

A wide variety of propagation modes are useful in the VHF range. F-region skip appears on 50 MHz during solar cycle peaks. Sporadic E and several other E-region phenomena are most effective in the VHF range. Still other forms of VHF ionospheric propagation, such as field-aligned irregularities (FAI) and transequatorial propagation (TE), are rarely observed at VHF. Tropospheric propagation, which is not a factor at HF, becomes increasingly important above 50 MHz.

50 – 54 MHz (6 meters)

The lowest amateur VHF band shares many of the characteristics of both lower and higher frequencies. In the absence of any favorable ionospheric propagation conditions, well-equipped 50 MHz stations work regularly over a radius of 300 kilometers (190 miles) via tropospheric scatter, depending on terrain, power, receiver capabilities, and antenna. Weak-signal troposcatter allows the best stations to make 500 kilometers (310 mile) contacts nearly any time. Weather effects may extend the normal range by a few hundred kilometers, especially during the summer months, but true tropospheric ducting is rare.

During the peak of the 11-year solar cycle (especially during the winter months), worldwide 50 MHz DX is possible via the F_2 region during daylight hours. F_2 backscatter provides an additional propagation mode for contacts as far as 4,000 kilometers (2,500 miles) when the MUF is just below 50 MHz. TE paths as long as 8,000 kilometers (5,000 miles) across the magnetic equator are common around the spring and fall equinoxes of peak solar cycle years.

Sporadic E is probably the most common and certainly the most popular form of propagation on the 6-meter band. Single-hop E-skip openings may last many hours for contacts from 600 to 2,300 kilometers (370 to 1,400 miles), primarily during the spring and early summer. Multiple-hop E_s provides transcontinental contacts mostly during June and July in the northern hemisphere at distances of 10,000 kilometers or more, and contacts between the US and South America, Europe and Japan via multiple-hop E-skip occur nearly every summer.

Other types of E region ionospheric propagation make 6 meters an exciting band. Maximum distances of about 2,300 kilometers (1,400 miles) are typical for all types of E-region modes. TEP propagation is frequent for stations within range of the geomagnetic equator and for stations farther away when sporadic E propagation couples to TEP. Propagation via FAI often provides additional hours of contacts immediately following sporadic E events. Auroral propagation often makes its appearance in late afternoon when the geomagnetic field is disturbed. Closely related auroral E propagation may extend the 6-meter range to 4,000 kilometers (2,500 miles) and sometimes farther across the northern states and Canada and Alaska and occasionally to Scandinavia, usually after midnight. Meteor scatter provides reliable contacts almost every day around sunrise and especially during one of the dozen or so prominent annual meteor showers.

144 – 148 MHz (2 meters)

Ionospheric effects are significantly reduced at 144 MHz, but they are far from absent. F-region propagation is unknown except for TE, which is responsible for the current 144 MHz terrestrial DX record of nearly 8,000 kilometers (5,000 miles). Sporadic E occurs as high as 144 MHz less than a tenth as often as at 50 MHz, but the usual maximum single-hop distance is the same, about 2,300 kilometers (1,400 miles). Multiple-hop sporadic E contacts greater than 3,000 kilometers (1,900 miles) have occurred from time to time across the continental US, as well as across Southern Europe.

Auroral propagation is quite similar to that found at 50 MHz, except that signals are weaker and more Doppler-distorted. Auroral E contacts are rare. Meteor-scatter contacts are limited primarily to the periods of the great annual meteor showers and require much patience and operating skill. Contacts have been made via FAI on 144 MHz, but its potential has not been fully explored.

(Continued on next page.)

(Continued from previous page.)

Tropospheric effects improve with increasing frequency, and 144 MHz is the lowest VHF band at which terrestrial weather plays an important propagation role. Weather-induced enhancements may extend the normal 300- to 600-kilometer (190- to 370-mile) range of well-equipped stations to 800 kilometers (500 miles) and more, especially during the summer and early fall. Tropospheric ducting extends this range to 2,000 kilometers (1,200 miles) and farther over the continent and at least to 4,000 kilometers (2,500 miles) over some well-known all-water paths, such as that between California and Hawaii.

222 – 225 MHz (135 cm)

The 135 cm band shares many characteristics with the 2-meter band. The normal working range of 222 MHz stations is nearly as far as comparably equipped 144 MHz stations. The 135 cm band is slightly more sensitive to tropospheric effects, but ionospheric modes are more difficult to use. Auroral and meteor-scatter signals are somewhat weaker than at 144 MHz, and sporadic E contacts on 222 MHz are extremely rare. FAI and TE may also be well within the possibilities of 222 MHz, but reports of these modes on the 135 cm band are uncommon. Increased activity on 222 MHz will eventually reveal the extent of the propagation modes on the highest of the amateur VHF bands.

ULTRA-HIGH FREQUENCY (UHF) BANDS (300 – 3000 MHz) AND HIGHER

Tropospheric propagation dominates the bands at UHF and higher, although some forms of E region propagation are still useful at 432 MHz. Above 10 GHz, atmospheric attenuation increasingly becomes the limiting factor over long-distance paths. Reflections from airplanes, mountains and other stationary objects may be useful adjuncts to propagation at 432 MHz and higher.

420 – 450 MHz (70 cm)

The lowest amateur UHF band marks the highest frequency on which ionospheric propagation is commonly observed. Auroral signals are weaker and more Doppler distorted; the range is usually less than at 144 or 222 MHz. Meteor scatter is much more difficult than on the lower bands, because bursts are significantly weaker and of much shorter duration. Although sporadic E and FAI are unknown as high as 432 MHz and probably impossible, TE may be possible.

Well-equipped 432 MHz stations can expect to work over a radius of at least 300 kilometers (190 miles) in the absence of any propagation enhancement. Tropospheric refraction is more pronounced at 432 MHz and provides the most frequent and useful means of extended-range contacts. Tropospheric ducting supports contacts of 1,500 kilometers (930 miles) and farther over land. The current 432 MHz terrestrial DX record of more than 4,000 kilometers (2,500 miles) was accomplished by ducting over water.

902 – 928 MHz (33 cm) and Higher

Ionospheric modes of propagation are nearly unknown in the bands above 902 MHz. Auroral scatter may be just within amateur capabilities at 902 MHz, but signal levels will be well below those at 432 MHz. Doppler shift and distortion will be considerable, and the signal bandwidth may be quite wide. No other ionospheric propagation modes are likely, although high-powered research radars have received echoes from auroras and meteors as high as 3 GHz.

Almost all extended-distance work in the UHF and microwave bands is accomplished with the aid of tropospheric enhancement. The frequencies above 902 MHz are very sensitive to changes in the weather. Tropospheric ducting occurs more frequently than in the VHF bands, and the potential range is similar. At 1296 MHz, 2,000-kilometer (1,200 mile) continental paths and 4,000-kilometer (2,500 mile) paths between California and Hawaii have been spanned many times. Contacts of 1,000 kilometers (620 miles) have been made on all bands through 10 GHz in the US and over 1,600 kilometers (1,000 miles) across the Mediterranean Sea. Well-equipped 903 and 1296 MHz stations can work reliably up to 300 kilometers (190 miles), but normal working ranges generally shorten with increasing frequency.

Other tropospheric effects become evident in the GHz bands. Evaporation inversions, which form over very warm bodies of water, are usable at 3.3 GHz and higher. It is also possible to complete paths by scattering from rain, snow, and hail in the lower GHz bands. Above 10 GHz, attenuation caused by atmospheric water vapor and oxygen becomes the most significant limiting factor in long-distance communication.

19.4 VHF/UHF Non-Ionospheric Propagation

All radio communication involves propagation through the troposphere for at least part of the signal path. Radio waves traveling through the lowest part of the atmosphere are subject to refraction, scattering, and other phenomena, much like ionospheric effects. Tropospheric conditions are rarely significant below 30 MHz, but they are very important at 50 MHz and higher. Much of the long-distance work on the VHF, UHF, and microwave bands depends on some form of tropospheric propagation. Instead of watching solar activity and geomagnetic indices, those who use tropospheric propagation are much more concerned about terrestrial weather as opposed to space weather. General characteristics of the amateur bands above 30 MHz are summarized in the previous sidebar **Propagation Summary — By Band**.

19.4.1 Line of Sight

At one time it was thought that communications in the VHF range and higher would be restricted to line-of-sight paths. Although this has not proven to be the case even in the microwave region, the concept of line of sight is still useful in understanding tropospheric propagation. In the vacuum of space or in a completely homogeneous medium, radio waves do travel essentially in straight lines, but these conditions are almost never met in terrestrial propagation.

Radio waves traveling through the troposphere are ordinarily refracted slightly earthward. The normal drop in temperature, pressure, and water-vapor content with increasing altitude changes the index of refraction of the atmosphere enough to cause refraction. Under average conditions, radio waves are refracted toward Earth enough to make the horizon appear 15 percent farther away than the visual horizon. Under unusual conditions, tropospheric refraction may extend this range significantly.

A simple formula can be used to estimate the distance to the radio horizon under average conditions:

$$d = \sqrt{2h}$$

where
d = distance to the radio horizon, miles, and
h = height above average terrain, ft.

and

$$d = \sqrt{17h}$$

where
d = distance to the radio horizon, kilometers, and
h = height above average terrain, m.

The distance to the radio horizon for an antenna 30 meters (98 feet) above average terrain is thus 22.6 kilometers (14 miles). A station on top of a 1,000-meter (3,280 foot) mountain has a radio horizon of 130 kilometers (80 miles).

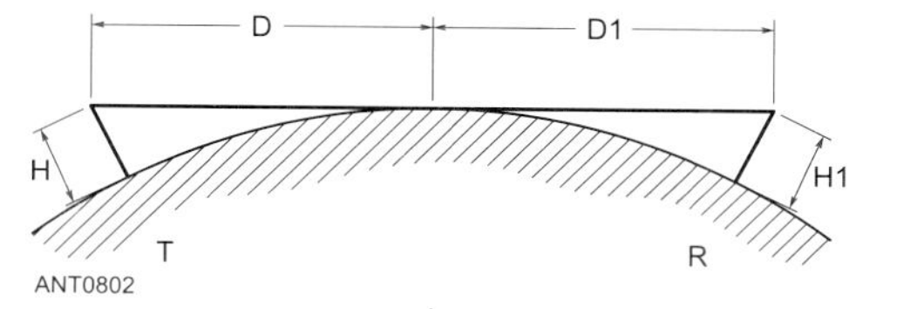

Figure 19.29 — The distance D to the horizon from an antenna of height H is given by equations in the text. The maximum line-of-sight distance between two elevated antennas is equal to the sum of their distances to the horizon as indicated here.

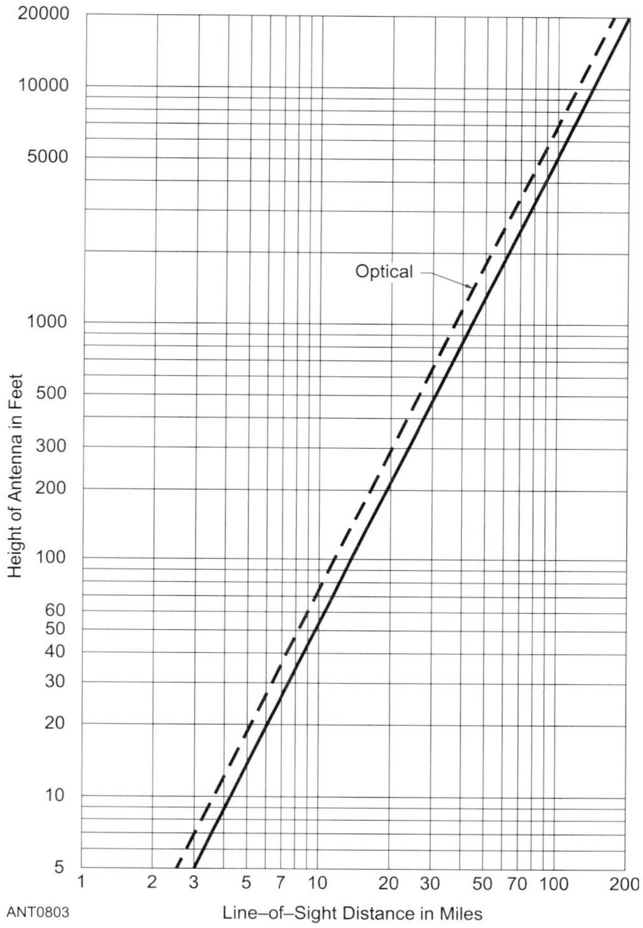

Figure 19.30 — Distance to the horizon from an antenna of given height. The solid curve includes the effect of atmospheric refraction. The optical line-of-sight distance is given by the broken curve.

19.4.2 Beyond Line of Sight

From Figure 19.6 it appears that use of the space wave depends on direct line of sight between the antennas of the communicating stations. This is not literally true, although that belief was common in the early days of amateur communication on frequencies above 30 MHz. When equipment became available that operated more efficiently and after antenna techniques were improved, it soon became clear that VHF waves were actually being bent or scattered in several ways, permitting reliable communication beyond visual distances between the two stations. This was found true even with low power and simple antennas. The average communication range can be approximated by assuming the waves travel in straight lines, but with the Earth's radius increased by one-third. The distance to the *radio horizon* is then given as

$$D_{miles} = 1.415 \sqrt{H_{feet}}$$

or

$$D_{km} = 4.124 \sqrt{H_{meters}}$$

where H is the height of the transmitting antenna, as shown in **Figure 19.29**.

The formula assumes that the Earth is smooth out to the horizon, so any obstructions along the path must be taken into consideration. For an elevated antenna, the communication distance is equal to D + D1, that is, the sum of the distances to the horizon of both antennas. Radio horizon distances are given in graphic form in **Figure 19.30**. Two stations on a flat plain, one with its antenna 60 feet above ground and the other 40 feet, could be up to about 20 miles apart for strong-signal line-of-sight communication (11 + 9 miles). The terrain is almost never completely flat, however, and variations along the way may add to or subtract from the distance for reliable communication. Remember that energy is absorbed, reflected, or scattered in many ways in nearly all communication situations. The formula or the chart will be a good guide for estimating the potential radius of coverage for a VHF FM repeater, assuming the users are mobile or portable with simple, omnidirectional antennas. Coverage with optimum equipment, high-gain directional arrays, and weak-signal analog or digital modes is quite a different matter. A much more detailed method for estimating coverage on frequencies above 50 MHz is given later in this chapter.

For maximum use of the ordinary space wave, it is important to have the antenna as high as possible above nearby buildings, trees, wires, and surrounding terrain. A hill that rises above the rest of the countryside is a good location for an amateur station of any kind,

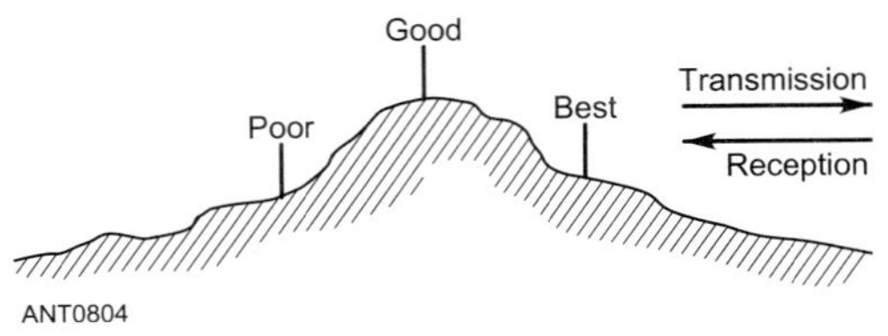

Figure 19.31 — Propagation conditions are generally best when the antenna is located near the top of a downslope facing the distant station. Communication is poor when there is a sharp rise immediately in front of the antenna in the direction of communication.

and particularly so for extensive coverage on the frequencies above 50 MHz. That highest point is not necessarily the best location for the antenna, though. In the example shown in **Figure 19.31**, the hilltop would be a good site in all directions. But if maximum performance to the right is the objective, a downslope in terrain results in more energy at lower elevation angles. This would involve a trade-off with reduced coverage in the opposite direction. Conversely, an antenna situated on the left side, lower down the hill, might do well to the left, but almost certainly would be inferior in performance to the right. For more information about the effect of terrain on ground reflections and elevation patterns, see the discussion of the Fresnel zone in the *ARRL Antenna Book* chapter on HF Antenna System Design.

19.4.3 Long-Distance Propagation

The wave energy of VHF stations does not simply disappear once it reaches the radio horizon. It is scattered, but it can be heard to some degree for hundreds of miles, well beyond line-of-sight range. Everything on Earth, and in the regions of space up to at least 100 miles, is a potential forward-scattering agent. Propagation via various phenomena in the troposphere is often referred to as simply "tropo."

WEATHER EFFECTS AT VHF/UHF

Varied weather patterns over most of the Earth's surface can give rise to boundaries between air masses of very different temperature and humidity characteristics. These boundaries can be anything from local anomalies to air-circulation patterns of continental proportions. The boundaries create opportunities for refraction due to the different indexes of refraction for the different regions.

There is a diurnal effect in temperate climates. At sunrise the air aloft is warmed more rapidly than that near the Earth's surface, and as the Sun lowers late in the day, the upper air is kept warm, while the ground cools. In fair, calm weather such as sunrise and sunset *temperature inversions* create boundaries in the atmosphere that improve signal strength over paths beyond line of sight as much as 20 dB over levels prevailing during the hours of high sun. Operating during hours when the diurnal inversion is present may also extend the operating range for a given strength by some 20 to 50 percent.

Under stable weather conditions, large air masses can retain their characteristics for hours or even days at a time (see **Figure 19.32**). Stratified warm dry air over cool moist air, flowing slowly across the Great Lakes region to the Atlantic Seaboard, can provide the medium for east-west communication on 144 MHz and higher amateur frequencies over as much as 1,200 miles. More common, however, are communication distances of 400 to 600 miles under such conditions.

A similar inversion along the Atlantic Seaboard as a result of a tropical storm air-circulation pattern may bring VHF and UHF openings extending from the Maritime Provinces of Canada to the Carolinas. Propagation across the Gulf of Mexico, sometimes with very high signal levels, enlivens the VHF

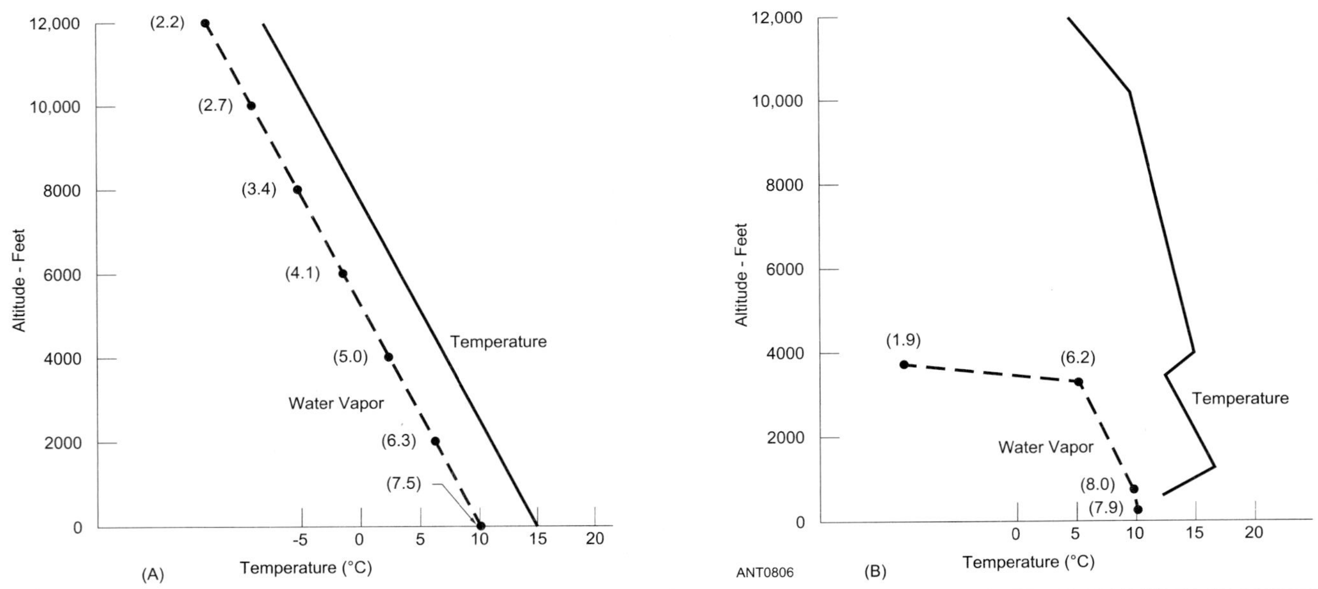

Figure 19.32 — Upper air conditions that produce extended-range communication on the VHF bands. At the left is shown the US Standard Atmosphere temperature curve. The humidity curve (dotted) is what would result if the relative humidity were 70%, from ground level to 12,000 feet elevation. There is only slight refraction under this standard condition. A sounding that is typical of marked refraction of VHF waves appears at the right. Figures in parentheses are the "mixing ratio" — grams of water vapor per kilogram of dry air. Note the sharp break in both curves at about 3,500 feet.

bands in coastal areas from Florida to Texas. The California coast, from below the San Francisco Bay Area to Mexico, experiences a similar propagation aid during the warmer months. Tropical storms moving west across the Pacific below the Hawaiian Islands may provide a transpacific long-distance VHF medium. Amateurs first exploited this on 144, 220, and 432 MHz in 1957. It has been used fairly often in the summer months since, although not yearly.

The examples of long-haul work cited above may occur infrequently, but lesser extensions of the minimum operating range are available almost daily. Under minimum conditions there may be little more than increased signal strength over paths that are workable at any time.

There are other short-range effects of local atmospheric and topographical conditions. Known as *subsidence*, the flow of cool air down into the bottom of a valley, leaving warm air aloft, is a familiar summer-evening pleasure. The daily inshore-offshore wind shift along a seacoast in summer sets up daily inversions that make coastal areas highly favored as VHF sites.

Tropospheric effects can show up at any time, in any season. Late spring and early fall are the most favored periods, although a winter warming trend can produce strong and stable inversions that work VHF magic almost equal to that of the more familiar spring and fall events.

Regions where the climate is influenced by large bodies of water enjoy the greatest degree of tropospheric bending. Hot, dry desert areas see little of it, at least in the forms described above.

TROPOSPHERIC DUCTING

Tropospheric propagation of VHF and UHF waves can influence signal levels at all distances from purely local to something beyond 4,000 kilometers (2,500 miles). The outer limits are not well known. At the risk of oversimplification, we will divide the modes into two classes — extended-local and long-distance. This concept must be modified depending on the frequency under consideration, but in the VHF range the extended-local effect gives way to a form of propagation much like that of microwaves in a waveguide, called *ducting*. The transition distance is ordinarily somewhere around 200 miles. The difference lies in whether the atmospheric condition producing the bending is localized or continental in scope. Remember, we're concerned here with frequencies in the VHF range, and perhaps up to 500 MHz. At 10 GHz, for example, the extent of the required atmospheric conditions producing ducting are smaller.

VHF propagation beyond a few hundred miles probably required more than one weather system. On long-distance paths over the ocean (two notable examples are California to Hawaii and Ascension Island to Brazil), propagation is likely to be between two atmospheric regions. On such circuits, the communicating station antennas must be in the duct, or capable of propagating strongly into it. Here again, we see that the positions and radiation angles of the antennas are important. As with microwaves in a waveguide, the low-frequency limit for the duct is critical. In long-distance ducting it is also very variable. Airborne equipment has shown that duct capability exists well down into the HF region in the stable atmosphere west of Ascension Island. Some contacts between Hawaii and Southern California on 50 MHz are believed to have been by way of tropospheric ducts. Probably all contact over these paths on 144 MHz and higher bands is because of duct propagation.

Long-distance communication using tropospheric modes is possible to some degree on all amateur frequencies from 50 MHz to at least 10 GHz. For forecasts of tropospheric ducting possibilities, visit **www.dxinfocentre.com/tropo.html**.

POLARIZATION FACTORS ABOVE 50 MHZ

Horizontal systems are popular, in part because they tend to reject man-made noise, much of which is vertically polarized. There is some evidence that vertical polarization shifts to horizontal in hilly terrain more readily than horizontal shifts to vertical. With large arrays, horizontal systems may be easier to erect, and they tend to give higher signal strengths over irregular terrain, if any difference is observed.

Practically all work with VHF mobiles is now handled with vertical systems. For use in a VHF repeater system, the vertical antenna can be designed to have gain without losing the desired omnidirectional quality. In the mobile station, a small vertical whip has obvious aesthetic advantages. Often, a telescoping whip used for broadcast reception can be pressed into service for 144 MHz FM. A car-top mount is preferable, but the broadcast whip is a practical compromise. Tests have shown that horizontal polarization can give a slightly larger service area, but mechanical advantages of vertical systems have made them the almost unanimous choice in VHF FM communication.

19.4.4 Reliable VHF Coverage

EFFECTS OF TERRAIN AT VHF/UHF

The coverage figures derived from the above procedure are for smooth terrain. What of stations in mountainous country? Although an open horizon is generally desirable for the VHF station site, mountain country should not be considered hopeless. Help for the valley dweller often lies in the optical phenomenon known as *knife-edge diffraction* as described earlier. A flashlight beam pointed at the edge of a partition does not cut off sharply at the partition edge, but is diffracted around it, partially illuminating the shadow area. A similar effect is observed with VHF waves passing over ridges; there is a shadow effect, but not a complete blackout. If the signal is strong where it strikes the mountain range, it will be heard well at the bottom of a valley on the far side. (See the *ARRL Antenna Book*'s chapter on **The Effects of Ground** for a more thorough discussion of the theory of diffraction.)

This is familiar to all users of VHF communications equipment who operate in hilly terrain. Where only one ridge lies in the way, signals on the far side may be almost as good as on the near side. Under ideal conditions (a very high and sharp-edged obstruction near the midpoint of a long-enough path so that signals would be weak over average terrain), knife-edge diffraction may yield signals even stronger than would be possible with an open path.

The obstruction must project into the radiated beam of the antennas used. Often mountains that look formidable to the viewer are not high enough to have an appreciable effect, one way or the other. Since the normal radiation pattern from a VHF array is several degrees above the horizontal, mountains that are less than about three degrees above the horizon, as seen from the antenna, are missed by the radiation from the array. Moving the mountains out of the way would have substantially no effect on VHF signal strength in such cases.

Rolling terrain, where obstructions are not sharp enough to produce knife-edge diffraction, still does not exhibit a complete shadow effect. There is no complete barrier to VHF propagation — only attenuation, which varies widely as the result of many factors. Thus, even valley locations are usable for VHF communication. Good antenna systems, preferably as high as possible, the best available equipment, and above all the willingness and ability to work with weak signals may make outstanding VHF work possible, even in sites that show little promise to casual inspection.

The availability of detailed terrain data for much of the world enables software to make good predictions about coverage of VHF, UHF, and microwave signals. The software is freely available to amateurs and can run online in a web browser without having to download and install software or large databases on a PC.

Radio Mobile Online is the online version of the popular RF propagation tool *Radio Mobile* by Roger Coudé, VE2DBE (**www.ve2dbe.com/rmonline.html**). It uses digital terrain information and a mathematical model to simulate coverage between two fixed sites (such as a radio link) or between a fixed site

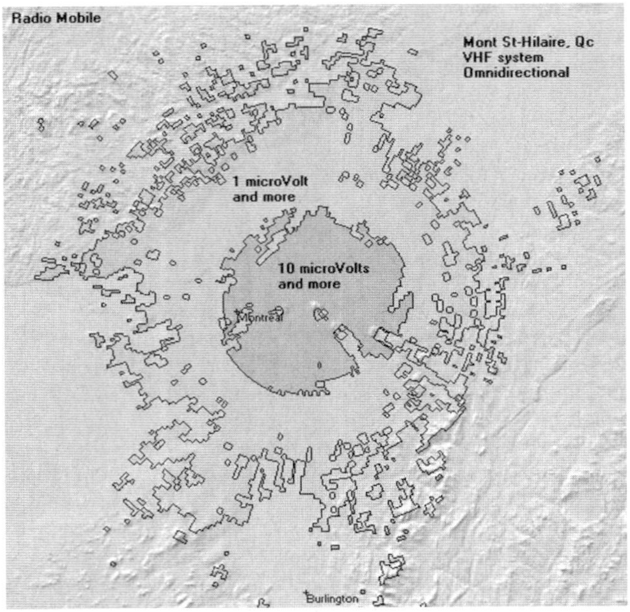

HBK1128 (A)

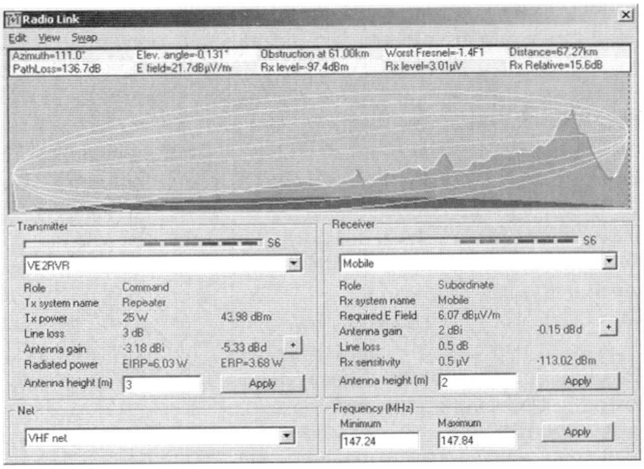

(B)

Figure 19.33 — *Radio Mobile* signal-strength contour map (A) and a point-to-point link map (B). Both figures are from www.ve2dbe.com/rme.html.

and a mobile (such as for repeater coverage). The digital terrain information comprises two databases totaling 198 gigabytes: elevation and land cover. Its online model works with amateur bands from 30 MHz to 25 GHz. The information for both stations (whether fixed/fixed or fixed/mobile) is comprehensive. The model is based on the public-domain ITM model augmented with the author's decades of experience in the mobile communications field. To use the service, you must first create a personal account (free) and your calculations will be saved on the system's server for online access.

A tutorial on *Radio Mobile Online* is available from KØLWC — search **YouTube.com** for "Radio Mobile Tutorial — How To Predict Ham Radio Coverage." Two sample outputs from *Radio Mobile Online* are shown in **Figure 19.33**: a coverage signal-strength contour map (A) and a point-to-point link map (B). Both figures are from **www.ve2dbe.com/rme.html**.

The AREDN network (**www.arednmesh.org**) operates Ubiquiti microwave wireless networking equipment with customized firmware to support amateur radio operation. Nodes in the network are situated to offer coverage to individual stations, relay stations, and stations deployed in disaster response. To support network node setup and site selection, the Ubiquiti *AirLink* planning tool software is used (**link.ubnt.com**).

After creating a free account, the simplest way to get started is to enter your own address, zoom out on the map, and select a nearby location. The link profile will automatically be generated as shown in **Figure 19.34** which

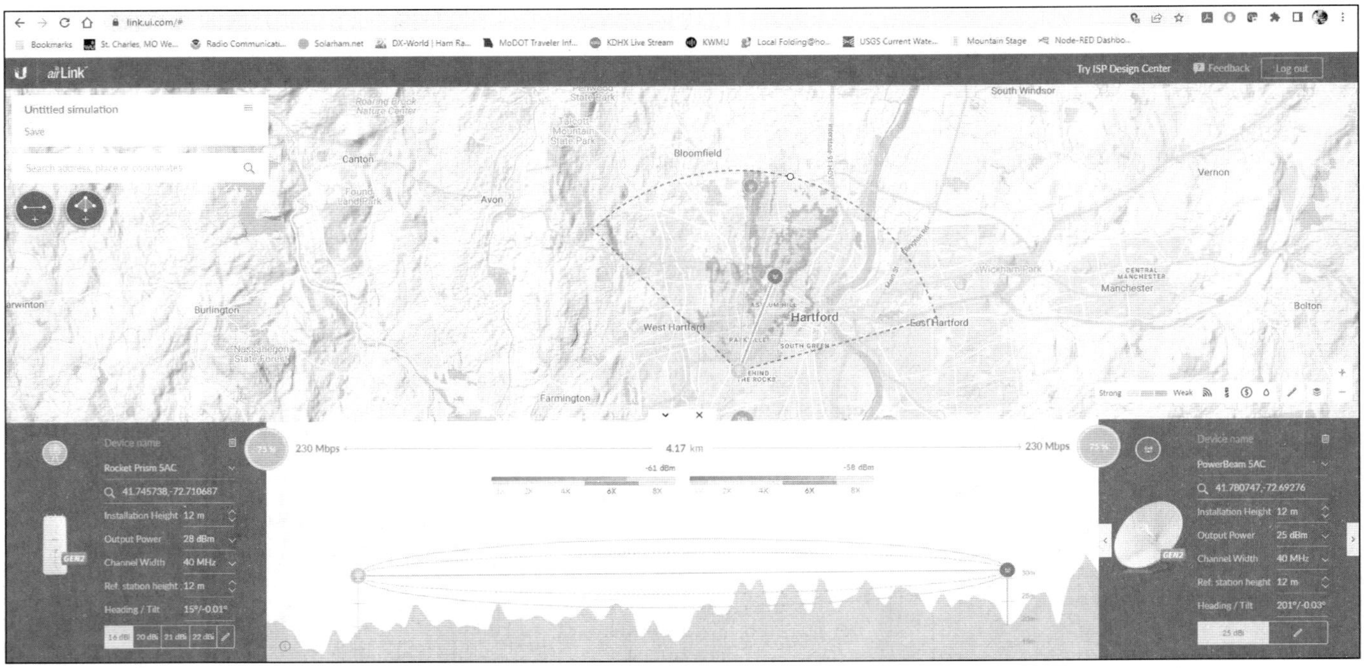

Figure 19.34 — A coverage map for 5 GHz wireless links from the Ubiquiti AirLink coverage and site evaluation tool at link.ui.com. The map and link profile show coverage between ARRL headquarters in Newington, CT, and nearby Hartford, CT.

shows coverage between ARRL headquarters in Newington and nearby Hartford, CT over a 5 GHz link. A step-by-step tutorial is available at **beetledigital.com/how-to-use-the-ubiquiti-airlink-tool-to-design-and-plan-wireless-network-links**.

These are two popular and low-cost tools available to amateurs for use in the VHF, UHF, and microwave range. The AREDN group has developed a very useful guide to wireless network design at **arednmesh.readthedocs.io/en/stable/arednNetworkDesign/network_modeling.html**. It discusses other packages, services, and techniques that you may find valuable. This is an active group and regularly updates its website, so this is a good source for coverage and propagation planning and site evaluation.

19.5 Propagation Predictions for HF Operation

19.5.1 The Big Picture Overhead

There are about 150 vertical-incidence ionosondes around the world. Ionosondes are located on land, even on a number of islands. There are gaps in sounder coverage, however, mainly over large expanses of open ocean. The compilation of all available vertical-incidence data from the worldwide network of ionospheric sounders results in global f_oF_2 maps, such as the map shown in **Figure 19.35**, a simulation from the highly sophisticated *PropLab Pro* computer program (**spacew.com** — listed under Geophysical Software).

This simulation is for 1900 UTC, about four hours before East Coast sunset on September 15, 1998, with a high level of solar activity (SSN of 85) and a planetary A_p index of 5, indicating quiet geomagnetic conditions. (These conditions are typical of what is expected as we move into the peak years of solar Cycle 25.) The contours of f_oF_2 peak over the South American longitudes at around 12 MHz. If you look carefully, you'll see two peaks, or humps, on either side of the geomagnetic equator, at about –12 degrees geographic latitude in the South American sector.

These two "humps" in f_oF_2 form what is known as the *equatorial anomaly* and are caused by upwelling "fountains" of high electron concentration located in daylight areas about ±20 degrees from the geomagnetic equator. The equatorial anomaly is important in transequatorial propagation. Those LU stations in Argentina that you can hear on 28 MHz from the US in the late afternoon, even during low portions of the solar cycle when other stations to the south are not coming through, are benefiting from transequatorial propagation, a form of *chordal hop* propagation, because signals going through this area remain in the ionosphere without lossy intermediate hops to the ground.

From records of f_oF_2 profiles, the underlying electron densities along a path can be computed. And from the electron density profiles computerized "ray tracing" may be done throughout the ionosphere to determine how a wave propagates from a transmitter to a particular receiver location. *PropLab Pro* can do complex ray tracings that explicitly include the effect of the Earth's magnetic field, even taking into effect ionospheric stormy conditions.

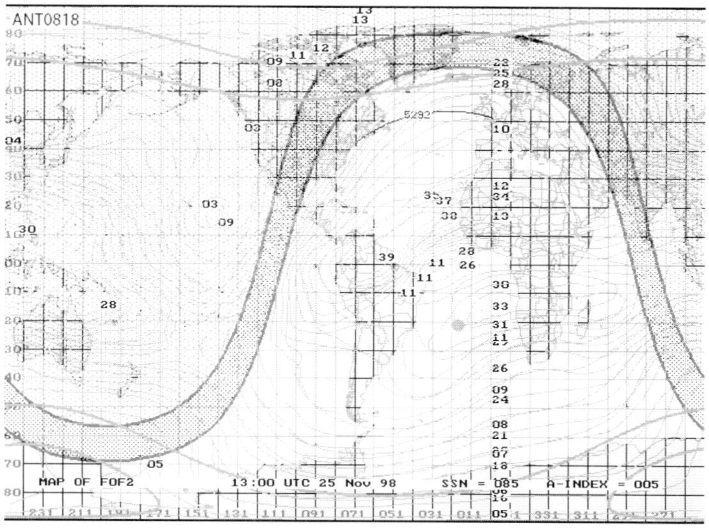

Figure 19.35 — Computer simulation of the f_oF_2 contours for 15 September 1998 at 1900 UTC, for an SSN of 85 and a quiet planetary Ap index of 5. Note the two regions of high f_oF_2 values in the South American sector (they can be seen better in a plot of MUF contours). These are the "equatorial anomalies," regions of high electronic density in the F_2 region that often allow chordal-hop north-south propagation. See also Figure 19.5. (*PropLab Pro V3* simulation courtesy of K9LA)

19.5.2 MUF Prediction

F region MUF prediction is key to forecasting HF communications paths at particular frequencies, dates, and times, but forecasting is complicated by several variables. Solar radiation varies over the course of the day, season, year, and solar cycle. Additionally, the ionization at any given point in the world depends on geomagnetic field activity and events in the lower atmosphere coupling up to the ionosphere. These regular intervals provide the main basis for prediction, yet recurrence is far from reliable. In addition, forecasts are predicated on a quiet geomagnetic field, but the condition of the Earth's magnetic field is most difficult to predict weeks or months ahead. For professional users of HF communications, uncertainty is a nuisance for maintaining reliable communications paths, while for many amateurs it provides an aura of mystery and chance that adds to the fun of DXing. Nevertheless, many amateurs want to know what to expect on the HF bands to make best use of available on-the-air time, plan contest strategy, ensure successful net operations, or engage in other activities.

The online supplement "Frequency Selection for HF Operation" will help you make the best use of your operating time. You may also find an article that appeared in the Summer 2008 issue of *CQ VHF* to be helpful with your 6-meter endeavors: "Predicting 6-meter F2 Propagation," by Carl Luetzelschwab, K9LA.

19.5.3 Solar and Geophysical Data

An amateur radio operator interested in long-distance QSOs on any frequency benefits from being aware of where we are in a solar cycle (solar data), the current state of the Sun's activity with respect to disturbances (geophysical data), and the fundamentals of propagation.

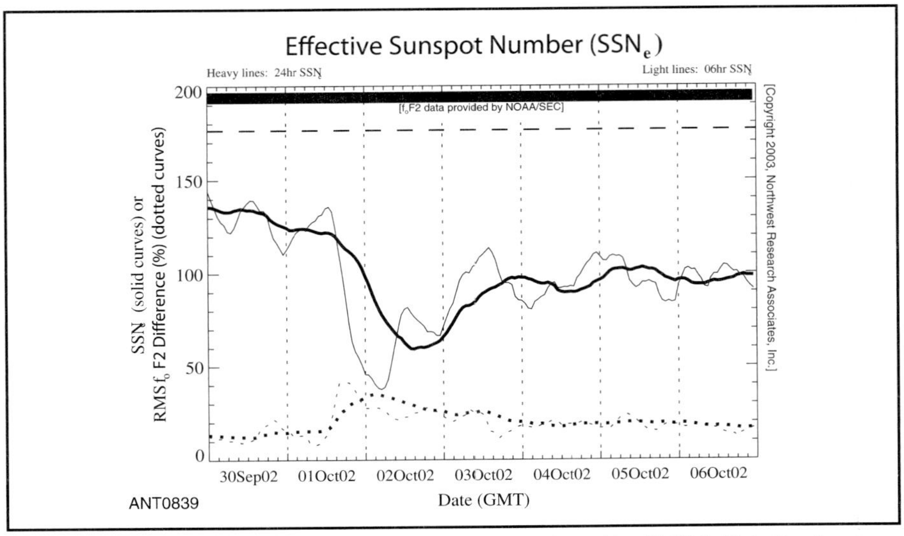

Figure 19.36 — Effective Sunspot Number (SSN_e) produced by NWRA. Note the large drop in effective SSN due to a geomagnetic storm commencing Oct 1, 2002.

For solar and geophysical data, the internet can be very helpful. For example, the Space Weather Prediction Center (SWPC) website at **www.swpc.noaa.gov** can supply pretty much everything you need to know either directly or through links. Sites providing information on more specific propagation topics are referenced in the appropriate sections. Another source of useful information about solar disturbances is **www.spaceweather.com**.

The SWPC home page includes indexes for radio blackouts, solar radiation storm impacts, and geomagnetic storm impacts. Three videos are provided of recent X-ray activity (for assessment of solar radiation storms and radio blackouts), recent coronal mass ejections (for assessment of geomagnetic storms), and predicted visible aurora (which gives a general indication of radio aurora conditions). Plots of X-ray flux (for assessment of solar radiation storms and radio blackouts), proton flux (for assessment of solar radiation storms), and recent K-index activity (for assessment of geomagnetic storms) are also provided.

Under these six images are numerous links in three major categories of additional data: About Space Weather, Products and Data, and Dashboards for specific interests. The Radio dashboard is most suited to amateur radio.

There are also links to other sources of space weather information not on the SWPC website. For example, entering a specific topic (such as STEREO) in the search area in the upper right-hand corner of the home page brings up many links to NASA's STEREO mission (which allows us to look at the backside of the Sun).

Some other useful websites are: **dx.qsl.net/propagation**, **www.solen.info/solar**, **www.solarham.net**, and **hfradio.org/propagation.html**. You may also access propagation information via your preferred spotting network server. Use the command SH/WWV/*n*, where *n* is the number of reports you wish to see (five is the default).

Another excellent method for obtaining an "equivalent sunspot number" (SSN_e) is to go to the Space Weather site of Northwest Research Services at **spawx.nwra.com/spawx/ssne24.html**. NWRA compares real-time ionospheric sounder data around the world with predictions using various levels of SSN looking for the best match. They thus "back into" the actual effective sunspot number. Note that this is necessarily a best fit of ionospheric sounder data to an equivalent sunspot number — it's not a perfect fit for all the data due to the dynamic hour-to-hour variability of the worldwide F_2 region. **Figure 19.36** is a typical NWRA graph that covers the week ending October 6, 2002. This graph was selected as it clearly shows the effects of a typical space weather event: in this case, a sudden decrease in SSNe after a geomagnetic storm depressed SSNe by more than 50 percent.

GEOPHYSICAL DATA FROM WWV AND WWVH

The standard time station WWV (Fort Collins, Colorado) transmits on 2.5, 5, 10, 15 and 20 MHz, and WWVH (Kauai, Hawaii) transmits on 2.5, 5, 10 and 15 MHz. Both of these stations are popular for propagation monitoring. They transmit 24 hours a day. Daily monitoring of these stations for signal strength and quality can quickly provide a good basic indication of propagation conditions. In addition, each hour they broadcast the geomagnetic A- and K-indexes, the 2800 MHz (10.7 cm) solar flux, and a short forecast of conditions for the next day. These are heard on WWV at 18 minutes past each hour and on WWVH at 45 minutes after the hour. This is the same information published by SWPC and **spaceweather.com**. The K-index is updated every three hours, while the A-index and solar flux are updated after 2100 UTC. These data are useful for making predictions on home computers, especially when averaged over several days of solar flux observations. For more details about the broadcasts, visit **www.nist.gov/pml/time-and-frequency-division/radio-stations/wwv/wwv-and-wwvh-digital-time-code-and-broadcast**.

A word of caution — it's easy to be overwhelmed with all this data. In fact, much of it has little direct relevance to radio propagation — but it certainly is colorful and interesting to look at! In the authors' opinions, just knowing the level and trends of the 10.7 cm solar flux and sunspot number, along with the level and trend of the K-index, will give a good picture of HF propagation. The more esoteric parameters may serve to aid in the analysis of a particular propagation observation.

19.5.4 MUF Forecasts

LONG-RANGE

Long-range forecasts several months ahead provide only the most general form of prediction. For many years, ARRL published charts similar to **Figure 19.37** to forecast average propagation for a one-month period over specific paths. These charts are no longer published, but customizable charts are available online from **www.voacap.com/hf/**. Charts like the one in Figure 19.37 assumed a single average solar flux value for the entire month, and they assumed that the geomagnetic field is undisturbed.

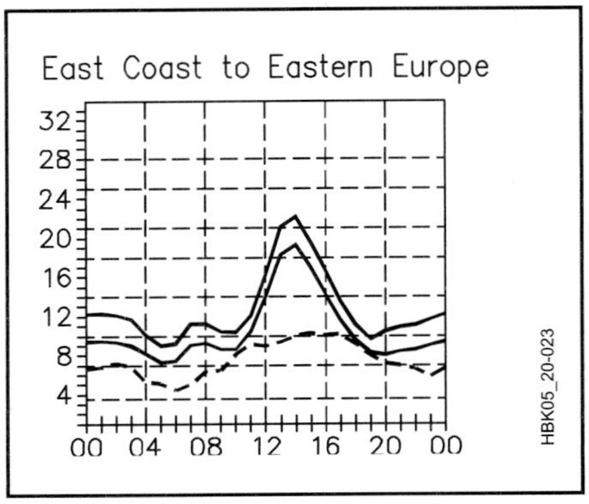

Figure 19.37 — For many years, ARRL published propagation prediction charts like this one for East Coast to Eastern Europe for April 2001. An average 2800 MHz (10.7 cm) solar flux of 159 was assumed for the month. On 10% of the days, the highest frequency propagated is predicted to be at least as high as the uppermost curve (the Highest Possible Frequency, or HPF, approximately 33 MHz), and for 50% of the days as high as the middle curve, the MUF. The lowest curve shows the Lowest Usable Frequency (LUF) for a 1,500 W CW transmitter.

The uppermost curve in Figure 19.37 shows the highest frequency that will be propagated on at least 10% of the days in the month. The given values might be exceeded considerably on a few rare days. On at least half the days, propagation should be possible on frequencies as high as the middle curve. Propagation will exceed the lowest curve on at least 90% of the days. The exact MUF on any particular day cannot be determined from these statistical charts, but you can determine when you should start monitoring a band to see if propagation actually does occur that day — particularly at frequencies above 14 MHz.

SHORT-RANGE

Short-range forecasts of a few days ahead are marginally more reliable than long-range forecasts because underlying solar indexes and geomagnetic conditions can be anticipated with greater confidence. The tendency for disturbances caused by solar phenomena to recur at 27-day intervals may enhance short-term forecasts (see the section 27-Day Recurrence). Daily forecasts may not be any better, as the ionosphere does not instantly react to small changes in solar flux (or sunspot number) and geomagnetic field indexes. Regardless of these limitations, it is always good to know the current solar and geophysical data, as well as understanding warnings provided by observations of the Sun in the visual to X-ray range.

Table 19.3

Features and Attributes of Several Currently Available Propagation Prediction Programs

	ASAPS V. 7	*VOACAP Windows*	*W6ELProp V. 2.70*	*PropLab Pro 3.1*
User Friendliness	Good	Good	Good	Good
Operating System	*Windows*	*Windows*	*Windows*	*Windows*
Uses K-index	No	No	Yes	Yes
User library of QTHs	Yes/Map	Yes	Yes	No
Bearings, distances	Yes	Yes	Yes	Yes
MUF calculation	Yes	Yes	Yes	Yes
LUF calculation	Yes	Yes	No	Yes
Wave angle calculation	Yes	Yes	Yes	Yes
Vary minimum wave angle	Yes	Yes	Yes	Yes
Path regions and hops	Yes	Yes	Yes	Yes
Multipath effects	Yes	Yes	No	Yes
Path probability	Yes	Yes	Yes	Yes
Signal strengths	Yes	Yes	Yes	Yes
S/N ratios	Yes	Yes	Yes	Yes
Long path calculation	Yes	Yes	Yes	No
Antenna selection	Yes	Yes	Indirectly	Yes
Vary antenna height	Yes	Yes	Indirectly	Yes
Vary ground characteristics	Yes	Yes	No	No
Vary transmit power	Yes	Yes	Indirectly	Yes
Graphic displays	Yes	Yes	Yes	2D/3D
UT-day graphs	Yes	Yes	Yes	Yes
Area Mapping	Yes	Yes	Yes	Yes
Documentation	Yes	Online	Yes	Yes
Price class	$AUD385[1]	free[2]	free[3]	$240[4]

[1]See **www.sws.bom.gov.au/Products_and_Services/1/2**. Price as of 2017, subject to change and exchange rate variation.
[2]*VOACAP* available at **www.its.bldrdoc.gov/resources/radio-propagation-software/high-frequency/high-frequency-propagation-models.aspx**
[3]*W6ELProp*, not actively supported, see **www.qsl.net/w6elprop**
[4]*PropLab Pro V3*, see **www.spacew.com/proplab/**

19.5.5 HF Propagation Prediction Software

Like predicting the weather, predicting HF propagation — even with the best computer software available — is not an exact science. The processes that cause a signal to propagate from one point on the Earth to another are enormously complicated and subject to an incredible number of variables. Experience and knowledge of propagation conditions (as related to solar activity, especially unusual solar activity such as flares or coronal mass ejections) are needed when you actually get on the air to check out the bands. Keep in mind, too, that prediction software is written mainly to calculate propagation for great-circle paths via the F region. Scatter, skew-path, auroral, and other such propagation modes may provide contacts when computer predictions indicate no contacts are possible.

Modern programs are designed for quick-and-easy predictions of propagation parameters. See **Table 19.3** for a listing of a number of popular PC programs. (An online description of propagation prediction software is also available at **astrosurf.com/luxorion/qsl-review-propagation-software.htm**.)

The basic input information required is the smoothed sunspot number (R_{12}) or smoothed solar flux (F_{12}), the date (month and day), and the latitudes and longitudes at the two ends of the radio path. The latitude and longitude, of course, are used to determine the great-circle radio path. Most commercial programs tailored for ham use allow you to specify locations by the call sign. The date is used to determine the latitude of the Sun, and this, with the smoothed sunspot number (or smoothed 10.7 cm solar flux converted to a smoothed sunspot number), is used to determine the properties of the ionosphere at critical points on the path.

A very powerful package, *VOACAP* is based on *IONCAP*, short for Ionospheric Communications Analysis and Prediction. It was written by an agency of the US government and has been under development for several decades. The *IONCAP* program has a well-deserved reputation for being difficult to use, since it came from the world of FORTRAN punch cards and mainframe computers.

VOACAP is a version of *IONCAP* adapted to Voice of America predictions, but this one includes a sophisticated Windows interface. The Voice of America (VOA) started work on *VOACAP* in the early 1990s and continued for several years before funding ran out. The program was maintained by a single, dedicated computer scientist, Greg Hand, at NTIA/ITS (Institute for Telecommunication Sciences), an agency of the US Department of Commerce in Boulder, Colorado. More information about *VOACAP* and an online app tailored for amateur radio are available from **www.voacap.com**.

ONLINE PREDICTION SERVICES

As mentioned earlier, *VOACAP* includes the above-the-MUF mode mentioned earlier in the section on HF Scatter modes. If a path can withstand a bit more loss, the QSO may be completed although the MUF is *below* the operating frequency. And with FT8 having the ability to decode signals farther down into the noise than CW, it essentially says the MUF can be even lower than for CW. FT8 will certainly help in the summer sporadic E season as well.

Another propagation prediction website, **soundbytes.asia/proppy/**, is produced in collaboration with the RSGB's Propagation Studies Committee. Both websites have instructions on how to input your data to generate the desired prediction.

Table 19.4
Popular Beacon Frequencies

(MHz)	*Comments*
14.100, 18.110, 21.150, 24.930, 28.200	Northern California DX Foundation beacons
28.2-28.3	Several dozen beacons worldwide
50.0-50.1	Most US beacons are within 50.06-50.08 MHz
70.03-70.13	Beacons in England, Ireland, Gibraltar and Cyprus

PC PREDICTION SOFTWARE

Table 19.3 contains information about four available software packages for the home PC. These programs generally allow predictions from 3 to 30 MHz. Unfortunately, prediction programs are not available on 160 meters (because of an incomplete understanding of the lower ionosphere and ducting mechanisms that contribute to propagation on that band) and on 6 meters (because of an incomplete understanding of openings when the MUF is not predicted to be high enough). As a general guideline, you should look for 160 meter openings when the path to your target station is in darkness and around sunrise/sunset, and you should look for 6 meter F_2 openings in the daytime during winter months near solar maximum.

19.5.6 Beacons

Automated *beacons* in the higher amateur bands can also be useful adjuncts to propagation watching. Beacons are ideal for this purpose because most are designed to transmit 24 hours a day.

One of the best organized beacon systems is the International Beacon Project, sponsored by the Northern California DX Foundation (NCDXF) and International Amateur Radio Union (IARU). The beacons operate 24 hours a day at 14.100, 18.110, 21.150, 24.930 and 28.200 MHz. Eighteen beacons on five continents transmit in successive 10-second intervals (each beacon transmits once every 3 minutes). More on this system can be found at the Northern California DX Foundation website **www.ncdxf.org** (click on the *Beacons* link on the left).

A network of automated SDR receivers called the Reverse Beacon Network (**www.reversebeacon.net**) has been created to monitor the HF CW sub-bands for signals that are automatically decoded by *CW Skimmer* and *RTTY Skimmer* software by VE3NEA (**www.dxatlas.com**). The network logs signal strength and location for the received signals, providing real-time information on HF propagation.

A list of many 28 MHz beacons can be found on the 10-10 International website, **www.ten-ten.org** (look for Beacons under the Resources link). Beacons often include location as part of their automated message, and many can be located from their call sign. Thus, even casual scanning of beacon sub-bands can be useful. **Table 19.4** provides the frequencies where beacons useful to HF propagation are most commonly placed.

There are also many beacons on VHF and higher bands. "G3USF's Worldwide List of 50 MHz Beacons" may be found at **www.keele.ac.uk/depts/por/50.htm**. Information on North American beacons on 144 MHz and up is maintained by Ron Klimas, WZ1V, at **www.newsvhf.com/beacons2.html**.

19.6 VHF/UHF Mobile Propagation

Most amateurs are aware that radio signals in free space obey the inverse-square law: the received signal power is inversely proportional to the square of the distance between the transmitting and receiving antennas. That law applies *only* if the transmitting and receiving antennas have an obstruction-free radio path between them.

Imagine two operators using hand-held 144 MHz radios, each standing on a mountaintop so that they have a direct line of sight. How far apart can they be and still maintain reliable communications? Assume 5 W transmitter power, 0 dBi antenna gain (2.15 dB worse than a dipole), 5 dB receiver noise figure, 10 dB S/N ratio, and 12 kHz receiver bandwidth.

Ask experienced VHF mobile operators that question, and you'll generally get guesses in the range of 20 to 30 miles because their experience tells them that's about the most you can expect when not communicating through a repeater. The correct answer, however, is over 9,500 kilometers (5,938 miles) based on the parameters given in the previous paragraph! This also explains how some operators have been able to work the amateur radio station on the International Space Station using only a handheld transceiver.

The discrepancy is explained by the fact that the line-of-sight scenario is not realistic for a mobile station located close to ground level. At distances greater than a few miles, there usually is no line of sight — the signal is reflected at least once on its journey from transmitter to receiver. As a result, path loss is typically proportional to distance to the third or fourth power, not the second power as the inverse-square law implies.

19.6.1 Rayleigh Fading

Not only is the signal reflected, but it usually arrives at the receiver by several different paths simultaneously (see **Figure 19.38**). Because the length of each path is different, the signals are not in phase. If the signals on two paths happen to be 180 degrees out of phase and at the same amplitude, they will cancel. If they are in phase, then their amplitudes will add. As the mobile station moves about, the phases of the various paths vary in a random fashion. However, they tend to be uncorrelated over distances greater than λ/4 or so, which is about 20 inches on the 2 meter band. That is why if the repeater you are listening to drops out when you are stopped at a traffic light, you can often

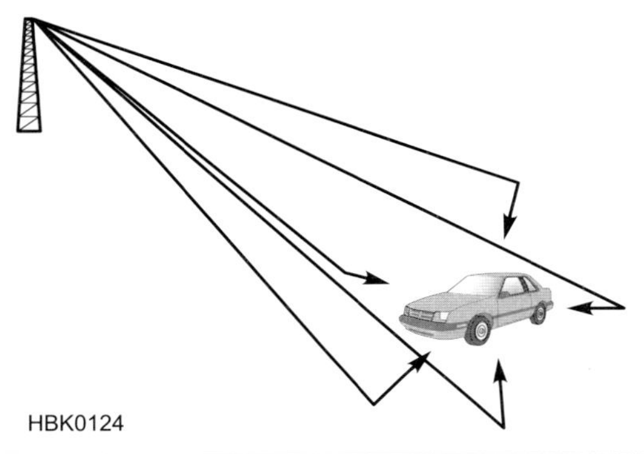

Figure 19.38 — At distances greater than a few miles, there normally is no line-of-sight path to the mobile station. Each path experiences at least one reflection. (Because the radio paths are reciprocal, propagation is the same for both receiving and transmitting.)

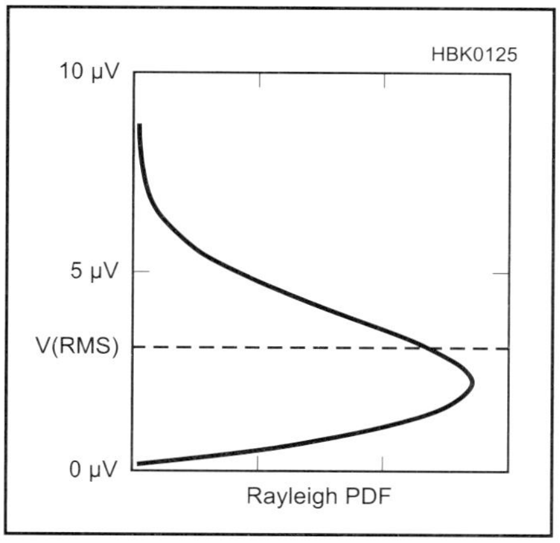

Figure 19.39 — The horizontal axis is the Rayleigh probability density function, which is the relative probability of the different signal levels shown on the vertical axis. For this graph, an RMS value of 3.16 µV has been selected to represent a typical average signal level in a marginal-coverage area.

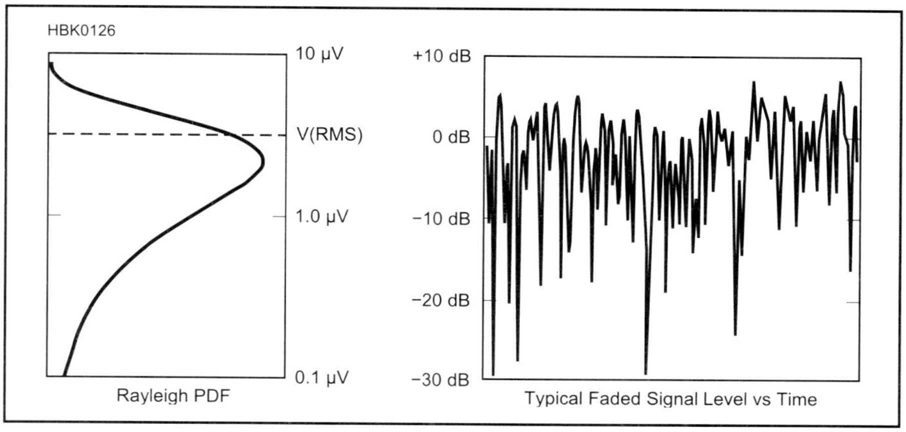

Figure 19.40 — The graph at the left is the same as Figure 19.31 except that the vertical axis is logarithmic (proportional to dB). At the right is a typical Rayleigh-faded signal of the same RMS voltage level plotted using the same vertical scale.

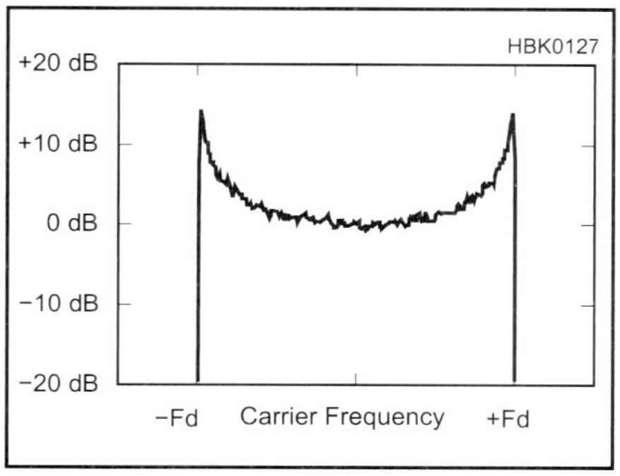

Figure 19.41 — Rayleigh fading spectrum of a CW (unmodulated) signal. The maximum deviation (F_d) from the center frequency is typically less than 10 – 15 Hz on the 144 MHz band.

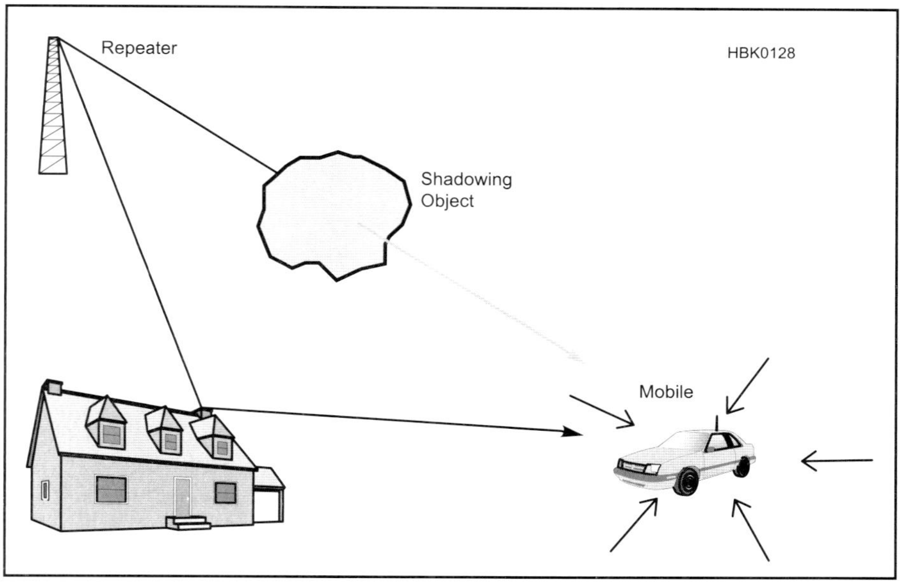

Figure 19.42 — When multipath propagation occurs, each main path is typically Rayleigh-faded by multiple reflectors close to the mobile station.

get it back again by creeping forward a few inches.

Because there are typically dozens of paths, it is rare for their amplitudes and phases to be such that they all cancel perfectly. Fades of 20 to 30 dB or more are common. The range of signal strengths has a *Rayleigh* distribution, named after the physicist/mathematician who first derived the mathematical formula. That is why the phenomenon is called *Rayleigh fading*. **Figure 19.39** shows the relative probability of various signal strengths. **Figure 19.40** is the same graph plotted on a logarithmic (dB) scale, along with a typical plot of signal strength versus time as the mobile station moves down the road.

The closer a reflecting object is to the antenna, the smaller is the path loss. A reflector that is close to the transmitter or receiver antenna gives a much stronger signal than one located halfway between. Even a weak reflector, such as a tree branch or telephone pole, is significant if it is close to the mobile station. Because there are many such close-in reflectors, many rays arrive from all directions.

Rays arriving from in front of a forward-moving vehicle experience a positive Doppler frequency shift, and rays from the rear have a negative Doppler shift. Those from the sides are somewhere in-between, proportional to the cosine of the angle of arrival. The received signal is the sum of all those rays, which results in *Doppler spreading* of the signal as illustrated in **Figure 19.41**. At normal vehicle speeds on the 2-meter band, the Doppler spread is only plus and minus 10 or 15 Hz, calculated from:

Doppler frequency = F_c v/c

where

F_c = the carrier frequency,
v = the vehicle speed, and
c = the speed of light.

Use the same units for v and c. On an FM voice signal the only effect is a slight distortion of the audio, but Doppler spreading can severely affect digital signals, as will be discussed later.

19.6.2 Multipath Propagation

In addition to scattering by local reflectors, it is not uncommon also to have more than one main radio path caused by strong reflectors, such as large metal buildings located some distance away (see **Figure 19.42**). Each main path typically does not reach the mobile station directly but is separately Rayleigh-

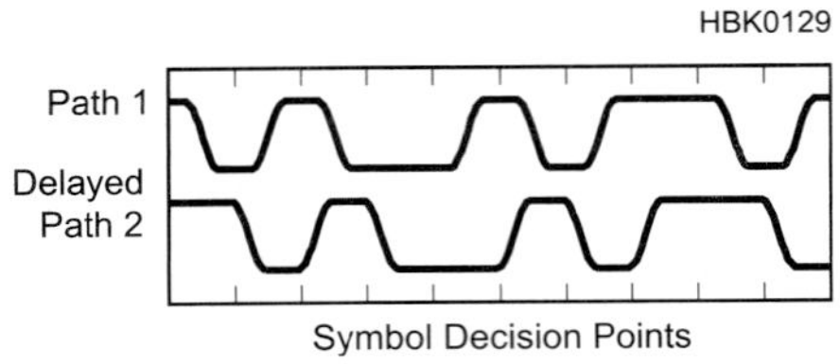

Figure 19.43 — Multipath propagation can cause intersymbol interference (ISI). At the symbol decision points, where the receiver decoder samples the signal, the path 2 data is often opposing the data on path 1.

faded by the local reflectors.

As the mobile station moves around, the shadowing of various paths by intervening hills, buildings, and other objects causes the average signal level to fade in and out, but at a much slower rate than Rayleigh fading. This is called *shadowing* or *slow fading*. It is also called *log-normal fading* because the distribution of average signal levels tends to follow a log-normal curve. That means that the *logarithm* of the signal level (on a dB scale, if you will) has a *normal* distribution (the famous bell-shaped curve). This effect typically causes the average signal level on each path to vary plus and minus 10 – 20 dB (at two standard deviations) from the mean value. This is in addition to the signal variation due to Rayleigh fading.

Under weak-signal conditions Rayleigh fading causes the signal to drop out periodically, even though the average signal level would be high enough to maintain reliable communications if there were no fading. *Picket fencing*, as such rapid periodic dropout is called, is a common occurrence when traveling in a weak-signal area at highway speeds. With analog modulation, normally the only solution is to stop at a location where the signal is strong. Moving the vehicle forward or backward a few inches is often enough to change an unreadable signal to solid copy.

Another possible solution is to employ *diversity reception*. If two (or more) mobile receiver antennas are spaced a half-wavelength or more apart, their Rayleigh fading will be almost entirely uncorrelated. That means it is relatively rare for both to experience a deep fade at the same time. The receiving system must have circuitry to determine which of the antennas has a stronger signal at any given time and to automatically combine the signals using some scheme that minimizes the probability of signal dropout. (One engineering text that has a fairly readable discussion of fading is *Cellular Radio Performance Engineering* by Asha Mehrotra; see the "References and Bibliography" section at the end of this chapter.)

One low-tech scheme is to use stereo headphones with each channel connected to a separate receiver and antenna. That method works better with linear modulation (AM, SSB, CW) than with FM because of the noise burst that occurs when the FM signal drops out.

Diversity antennas can also be used at a repeater site. The conditions are different because most repeater antennas are located in the clear with few local reflectors. The diversity antennas must be located much farther apart, typically on the order of 10 to 20 wavelengths, for the fading to be uncorrelated.

FADING AND DIGITAL SIGNALS

Digitally modulated signals, such as those used in the cellular telephone industry, use several techniques to combat Rayleigh fading. One is *error-correcting coding*, which adds redundant bits to the transmitted signal in a special way such that the receiver can "fill in" missing bits to obtain error-free reception. That technique does not work if the signal drops out completely for longer than a few consecutive bits. The solution often employed to counter longer signal drops is called *interleaving*. This takes data bits that represent points close together in time in the original voice signal and shuffles them into several different time slots for transmission. That way, a single brief dropout does not affect all the bits for that time period. Instead, the lost data are scattered over several different time periods. Since only a few bits are missing at any one point, the error-correcting decoder in the receiver has enough information to reconstruct the missing information.

As the signal gradually gets weaker, the error correction in a digital receiver produces nearly perfect voice quality until the signal gets too weak for the decoder to handle. At that point reception stops abruptly and the call is dropped. With an analog signal, reception gets scratchier as the signal gets weaker; this gives advance warning before the signal drops out completely.

The main paths of a multipath-faded signal can differ in length by several miles. A 10 kilometers (6 mile) difference in path length results in a difference in propagation delay of over 30 µs. While that is not noticeable on an FM voice signal, it can wreak havoc with digital signals by causing *intersymbol interference* (*ISI*) (see **Figure 19.43**). If the delay difference is on the order of one symbol, the receiver sees adjacent symbols superimposed. In effect, the signal interferes with itself.

One solution is to use an *equalizer*. This is a special type of digital filter that filters the demodulated signal to remove the delay differences so that multiple paths become time-aligned. Since the path characteristics change as the mobile station moves around, a special *training sequence* is sent periodically so that the receiver can re-optimize the filter coefficients based on the known symbols in the training sequence.

ISI is an even bigger problem at HF than at VHF/UHF because the path lengths are much greater. It is not unusual to have path length differences up to 3,000 kilometers (1,800 miles), which correspond to propagation delay differences of up to 10 ms. That is why digital modes on HF with symbol periods of less than about 10 to 20 ms (that is, with symbol rates greater than 50 to 100 baud) are not very practical without some method of equalization.

There is much more detailed information available about mobile radio propagation, but unfortunately most of it is written at an engineering level. For example, Keysight Technologies (**keysight.com**) offers fading simulator software that operates in conjunction with their N5182B MXG X-Series RF Vector Signal Generator.

For those who would like to explore the subject further, a *Mathcad* file with equations and explanatory text related to Rayleigh fading is available at **www.arrl.org/qst-in-depth**. Look in the 2006 section for Bloom0806.zip. The file was used to generate some of the graphics in this section. For those without access to *Mathcad*, a read-only PDF version of the file is available in the same directory.

19.7 Special Propagation Modes and Topics

19.7.1 *WSJT-X* and WSPR

WSJT-X (Weak Signal Communication, see **wsjt.sourceforge.io/wsjtx.html**) is a suite of digital protocols optimized for weak signal communications on HF and VHF/UHF. See the **Digital Protocols and Modes** chapter's section on Structured Digital Modes and this chapter's section on E Region Propagation. This software provides a wide variety of modes tailored for specific types of propagation. For example, FT8 is optimized for use at HF and 50 MHz; JT65 for EME on the VHF and UHF bands; JT4 for microwave bands; and JT9 especially for LF and MF. Any of these modes can be used on the HF bands to make worldwide contacts with a few watts and compromise antennas.

The WSPR mode implements a protocol that is specially designed for probing potential propagation paths with low power transmissions. WSPR transmissions carry a station's call sign, Maidenhead grid locator, and transmitter power in dBm. The program can decode signals with a signal-to-noise ratio as low as –28 dB in a 2500 Hz bandwidth. For more details on the WSPR mode and its capabilities, visit **www.wsprnet.org** and **physics.princeton.edu/pulsar/k1jt/wspr.html**.

19.7.2 Sporadic E

Short skip, long familiar on the 10-meter band during the summer months, affects the VHF bands as high as 222 MHz. *Sporadic E*, as this phenomenon is properly called, commonly propagates 28, 50, and, less frequently, 144 MHz radio signals between 500 and 2,300 kilometers (300 and 1,400 miles). Signals are apt to be exceedingly strong, allowing even modest stations to make E_s contacts. At 21 MHz, the skip distance may only be a few hundred kilometers. During the most intense E_s events, skip may shorten to less than 200 kilometers (120 miles) on the 10 meter band, with no skip zone at all on 15 meters. The first confirmed 220 MHz E_s contact was made in June 1987, but such contacts are likely to remain very rare. Much of what is known about E_s has come as the result of amateur pioneering in the VHF range.

Note that sporadic E can also be written as *E-subscript-s* (E_s) but is often written simply as Es (pronounced E ess). Sporadic E results from ionization at the lower boundary of the E-region, but of different origin and communication potential from the E-region that affects mainly our lower amateur frequencies. *[The amateur and professional literature is not strict about the name, with sporadic E and E_s used regularly — Ed.]*

CAUSES OF SPORADIC E

Efforts to predict mid-latitude E_s have not been successful, probably because its causes are complex and not yet well understood. Studies have demonstrated that thin and unusually dense patches of ionization in the E region, between 100 and 110 kilometers (60 and 70 miles) altitude and 10 to 100 kilometers (6 to 60 miles) in extent, are responsible for most E_s reflections. Note that "patch" is a more accurate term than "cloud" in describing the thin regions of ionization that create E_s propagation, but the term "cloud" is far more commonly used.

The formative mechanism for mid-latitude E_s is believed to be wind shear. This explains ambient ionization (believed to be meteoroid ablation — dust from meteors) being distributed and compressed into a thin, high-density region (a couple kilometers thick), without the need for production of extra ionization. Neutral winds of high velocity, flowing in opposite directions at slightly different altitudes, produce wind shears. In the presence of the Earth's magnetic field, the ions are collected at an altitude of about 120 kilometers, slowly drifting down to about 95 kilometers, forming a thin, over-dense layer. Data from rockets entering E_s regions confirm the electron density, wind velocities, and height parameters. The patches of ionization last only minutes to hours and are distributed randomly. They vary in density and, in the middle latitudes of the Northern Hemisphere, move at up to 100 mph, mostly from south to north along magnetic field lines. For a more detailed technical understanding of sporadic E, see the Reference entry for Whitehead.

WHEN SPORADIC E OCCURS

Sporadic E at mid latitudes (roughly 15 to 45 degrees) may occur at any time of the year, but it is most common in the Northern Hemisphere during May, June, and July, with a less intense season from the end of December to early January, near the summer and winter solstices. Its appearance is independent of the solar cycle. E_s propagation is most likely to occur from 9 AM to noon local time and again early in the evening between 5 PM and 8 PM, but can occur any time between dawn and midnight. Mid-latitude E_s events may last only a few minutes or can persist for many hours. In contrast, E_s is an almost constant feature near the auroral oval at night and the geomag-

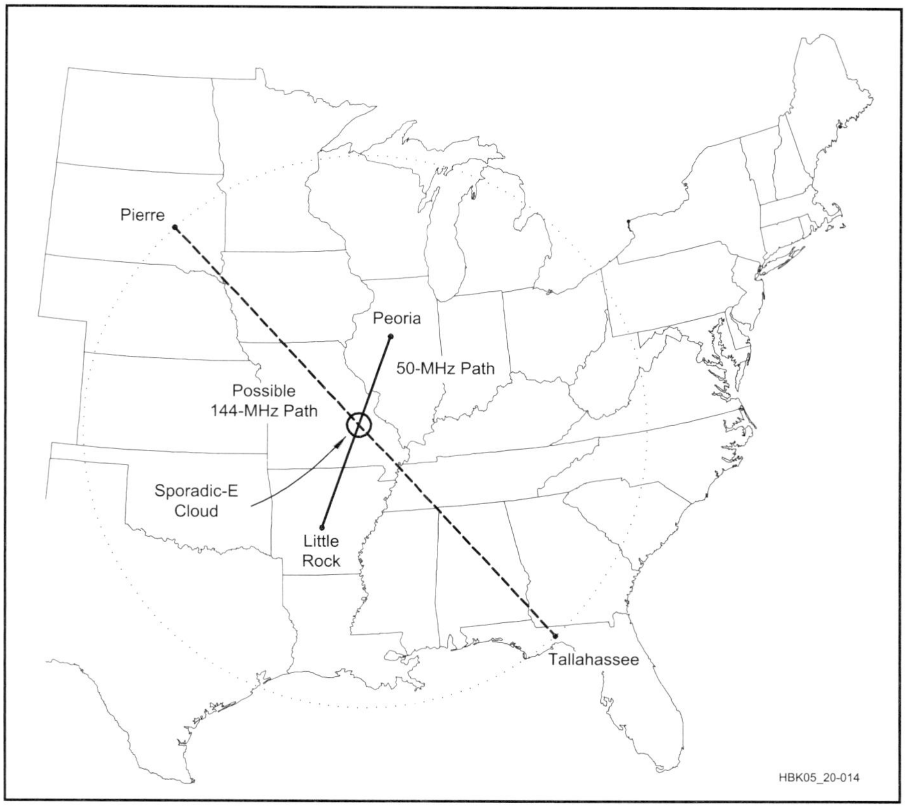

Figure 19.44 — 50 MHz sporadic E contacts of 700 kilometers (435 miles) or shorter (such as between Peoria and Little Rock) indicate that the MUF on longer paths is above 144 MHz. Using the same sporadic E region reflecting point, 144 MHz contacts of 2,200 kilometers (1,400 miles), such as between Pierre and Tallahassee, should be possible.

netic equator during the day.

At the peak of the long E_s season, ionization becomes extremely dense and widespread. This extends the usable range from the more common "single-hop" maximum of about 1,400 miles to longer "multi-hop" distances discussed below. With 50-MHz techniques and interest improving in recent years, it has been shown that distances considerably beyond 2,500 miles can be covered, especially during June and July in the Northern Hemisphere. The evolution of the digital FT8 mode has also opened up new possibilities on 50 MHz.

The seasons and distribution in the Southern Hemisphere are not so well known. Australia and New Zealand seem to have conditions much like those in the US, but with the timing of the seasons reversed, of course.

RANGE OF SPORADIC E

Sporadic E clouds exhibit an MUF that can rise from 28 MHz through the 50 MHz band and higher in just a few minutes. When the skip distance on 28 MHz is as short as 400 or 500 kilometers (250 or 310 miles), it is an indication that the MUF has reached 50 MHz for longer paths at low launch angles. Contacts at the maximum one-hop E_s distance, about 2,300 kilometers (1,400 miles), should then be possible at 50 MHz. *E-skip* (yet another term for sporadic E) contacts as short as 700 kilometers (435 miles) on 50 MHz, in turn, may indicate that 144 MHz contacts in the 2,300-kilometer (1,400-mile) range can be completed (see **Figure 19.44**). Sporadic E openings occur about a tenth as often at 144 MHz as they do at 50 MHz and for much shorter periods.

Sporadic E can also have a detrimental effect on HF propagation by masking or blanketing the F_2 region from below. HF signals may be prevented from reaching the higher levels of the ionosphere and the possibilities of long F_2 skip. Reflections from the tops of E_s patches can also have a masking effect, but they may also lengthen the F_2 propagation path with a top-side intermediate hop that never reflects from the Earth.

E_s has supported contacts on 28 and 50 MHz beyond 4,000 kilometers, up to 10,000 kilometers (6,200 miles), and more than 3,000 kilometers (1,900 miles) on 144 MHz. The term "multi" hop is probably not technically accurate, since it is likely that cloud-to-cloud paths are involved. This is an example of "above-the-MUF" propagation involving chordal hops between the patches of ionization.

There may also be "no-hop" E_s. On very rare occasions, the very high ionization density produces critical frequencies up to the 50-MHz range, with no skip distance at all. It is often said that the E_s mode is a great equalizer. With the reflecting region practically overhead, even a simple dipole close to the ground may do as well over a few hundred miles as a large, stacked antenna array designed for low-angle radiation. It's a great mode for low power and simple antennas on 28 and 50 MHz and a good reason to have high-angle antennas available.

E_s can also "link-up" with other modes. For example, those in the northern continental US normally can't take advantage of TEP (Trans-Equatorial Propagation — see the F Region Propagation section) because they are too far north. But E_s can provide a link to TEP. This hybrid mode is most likely to be experienced on our 50 MHz band. E_s can also provide a link to normal F_2 propagation.

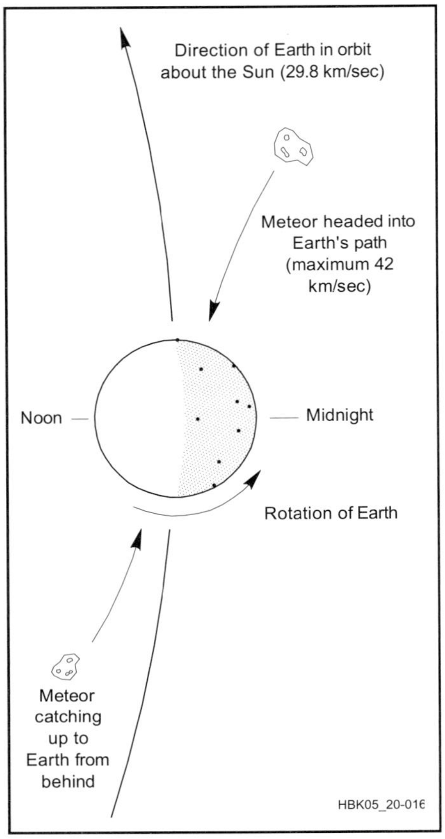

Figure 19.45 — The relative velocity of meteors that meet the Earth head on is increased by the rotational velocity of the Earth in orbit. Fast meteors strike the morning side of the Earth because their velocity adds to the Earth's rotational velocity, while the relative velocity of meteors that "catch up from behind" is reduced.

19.7.3 Meteor Scatter

Contacts between 800 and 2,300 kilometers (500 and 1,400 miles) can be made at 28 through 432 MHz via reflections from the ionized trails left by meteors as they travel through the atmosphere. The kinetic energy of meteors no larger than grains of rice are sufficient to ionize a column of air 20 kilometers (12 miles) long in the E region. The particle itself evaporates and never reaches the ground, but the ionized column may persist for a few seconds to a minute or more before it dissipates. This is enough time to make very brief contacts by reflections from the ionized trails. Millions of meteors enter the Earth's atmosphere every day, but few have the required size, speed, and orientation to the Earth to make them useful for meteor-scatter propagation.

Radio signals in the 30 to 100 MHz range are reflected best by meteor trails, making the 50 MHz band prime for meteor-scatter work. The early morning hours around dawn are usually the most productive, because the morning side of the Earth faces in the direction of the planet's orbit around the Sun. The relative velocity of meteors that head toward the Earth's morning side are thus increased by up to 30 km/sec, which is the average rotational speed of the Earth in orbit (see **Figure 19.45**). The maximum velocity of meteors in orbit around the Sun is 42 km/sec. Thus when the relative velocity of the Earth is considered, most meteors must enter the Earth's atmosphere somewhere between 12 and 72 km/sec.

Meteor contacts ranging from a second or two to more than a minute can be made nearly any morning at 28 or 50 MHz. Meteor scatter contacts at 144 MHz and higher are more difficult because reflected signal strength and duration drop sharply with increasing frequency. A meteor trail that provides 30 seconds of communication at 50 MHz will last only a few seconds at 144 MHz, and less than a second at 432 MHz.

Meteor scatter opportunities are somewhat better during July and August for North America because the average number of

Table 19.5
Major Annual Meteor Showers

Name	*Peak Dates*	*Approximate Rate (meteors/hour)*
Quadrantids	Jan 3	50
Arietids	Jun 7-8	60
Perseids	Aug 11-13	80
Orionids	Oct 20-22	20
Geminids	Dec 12-13	60

meteors entering the Earth's atmosphere peaks during those months. The best times are during one of the great annual *meteor showers*, when the number of useful meteors may increase 10-fold over the normal rate of 5 to 10 per hour (see **Table 19.5**). A meteor shower occurs when the Earth passes through a relatively dense stream of particles, thought to be the remnants of a comet in orbit around the Sun. The most-productive showers are relatively consistent from year to year, although several can occasionally produce great storms.

A technique allowing even smaller VHF/UHF stations to take advantage of meteor scatter and other modes (ionospheric scatter and even moonbounce) is available as the MSK144 mode, which is part of the *WSJT-X* software package, allowing meteor scatter communication to as much as 10 dB below the received noise level.

19.8 References and Bibliography

Barter, A., G8ATD, ed., *International Microwave Handbook*, 2nd edition (Potters Bar: RSGB, 2008). Includes a chapter on microwave propagation.

Barter, A., G8ATD, ed., *VHF/UHF Handbook*, 2nd edition (Potters Bar: RSGB, 2008). Includes a chapter on VHF/UHF propagation.

Bean, B., and Dutton, E., *Radio Meteorology* (New York: Dover, 1968).

Chen, C., "Attenuation of Electromagnetic Radiation by Haze, Fog, Clouds, and Rain," USAF, R-1694-PR, April 1975, **www.rand.org/content/dam/rand/pubs/reports/2006/R1694.pdf**.

Davies, K., *Ionospheric Radio* (Peter Peregrinus Ltd, 1990). Excellent though highly technical text on propagation.

Donovan, F., W3LPL, "What to Expect During the Rising Years of Solar Cycle 25," *QST*, May 2021, pp. 57 – 59.

Farmer, E., AA6ZM, "A Look at NVIS Techniques," *QST*, Jan 1995, p. 39 – 42.

Gosling, J., "The Solar Flare Myth in Solar-Terrestrial Physics," Solar System Plasma Physics: Resolution of Processes in Space and Time, Yosemite National Park, CA, Feb. 2 – 5, 1993.

Hunsucker, R., Hargreaves, J., *The High Latitude Ionosphere and Its Effect on Radio Propagation* (Cambridge University Press, 2003). Highly technical, but with an excellent chapter on fundamental physics of the ionosphere.

Jacobs, G., Cohen, T., Rose, R., *The New Shortwave Propagation Handbook* (CQ Communications, Inc., 1995).

Kamp, T., DF5JL, "60-m-Band: Chancen für NVIS," *CQ DL*, Feb. 2017, pp. 48 – 50.

Kift, F., "The propagation of high frequency radio waves to long distances," *Proceedings of the IEE*, Vol.107, Iss. 32, Mar. 1960, pp. 127 – 140.

Luetzelschwab, C., K9LA, **k9la.us**. Many articles about the ionosphere and propagation as they relate to 160 meters, HF, VHF, contesting, and more.

Maanen, E., PA3DES; Witvliet, B., PA5BW; Visser, G. PAØSIR; Westenberg, A., PAØA, "Elevation Angle Measurements During a Local Contest," *QST*, Jan. 2006, p. 28 – 30.

McNamara, L., *Radio Amateur's Guide to the Ionosphere* (Krieger Publishing Company, 1994). Excellent, quite-readable text on HF propagation.

Mehrotra, A., *Cellular Radio Performance Engineering* (Artech House, 1994). See Chapter 4 "Propagation" and the introduction to Chapter 3 "Cellular Environment." Diversity reception techniques are covered in Chapter 8.

Muldrew, D., and Maliphant, R., "Long Distance One-Hop Ionospheric Radio Wave Propagation," *Journal of Geophysical Research*, Vol. 67, Iss. 5, pp. 1805 – 1815.

Phillips, M., "Comparative Correlations of f^0F2 with 'Ionospheric Sunspot-Number' and Ordinary Sunspot-Number," *Journal of Geophysical Research*, Vol. 53, Iss. 1, pp. 79 – 80.

Straw, D., N6BV, "What's the Deal About 'NVIS'?" *QST*, Dec. 2005, p. 38.

White, J., VA7JW, and Tapping, K., "The Penticton Solar Flux receiver," *QST*, Feb. 2013, pp. 39 – 45.

Whitehead, D., "Sporadic E—A Mystery Solved?" *QST*, part 1, Oct. 1997; Part 2, Nov. 1997.

Zawdie, K., Drob, D., Siskind, D., and Coker, C., "Calculating the absorption of HF radio waves in the ionosphere," *Radio Science*, 2017, Vol. 52, pp. 767 – 783, **https://doi.org/10.1002/2017RS006256**.

Contents

Chapter 20 — Online Content

Articles

- A Commercial Triplexer Design by George Cutsogeorge, W2VJN
- Chokes and Isolation Transformers for Receiving Antennas by Jim Brown, K9YC
- Don't Blow Up Your Balun by Dean Straw N6BV
- HF Yagi Triplexer Especially for ARRL Field Day by Gary Gordon, K6KV
- Legacy Wound-Coax Ferrite Chokes
- Measuring Ferrite Chokes by Jim Brown, K9YC
- Measuring Isolation Between Radios by George Cutsogeorge, W2VJN
- Microwave Plumbing by Paul Wade, W1GHZ
- Multiband Operation with Open-wire Line by George Cutsogeorge, W2VJN
- Optimizing the Performance of Harmonic Attenuation Stubs by George Cutsogeorge, W2VJN
- Optimizing the Placement of Stubs for Harmonic Suppression by Jim Brown, K9YC
- Radio Frequency (RF) Surge Suppressor Ratings for Transmissions into Reactive Loads by Gene Hinkle, K5PA
- RF Connectors and Transmission Lines
- Simple Splice for ⅞ Inch Heliax by Ott Fiebel, W4WSR
- Splicing Window Line by Joel Hallas, W1ZR
- Transmission Lines in Digital Circuits
- Using *TLW* to Design Impedance Matching Networks by George Cutsogeorge, W2VJN

Software and Tools

- *MATCH.EXE* software (for use with Tuned (Resonant) Networks discussion)

The following program is available in the separate Software folder

- *jjSmith* from Tonne Software

Chapter 20
Transmission Lines

RF power is rarely generated right where it is used. A transmitter and the antenna it feeds are a good example. The most effective antenna installation is outdoors and clear of ground and energy-absorbing structures. The transmitter, however, is most conveniently installed indoors, where it is out of the weather and is readily accessible. A *transmission line* is used to convey RF energy from the transmitter to the antenna. A transmission line should transport the RF from the source to its destination with as little loss as possible. As an added benefit, transmission lines permit greater distances between radiated RF energy and people, which makes it likely that excessive human exposure will not occur (see the **Safe Practices** chapter). Additional material on the topics covered in this chapter is available in the *ARRL Antenna Book*.

20.1 Transmission Line Basics

There are three main types of transmission lines used by radio amateurs: coaxial, open-wire and waveguide. The most common type is the *coaxial* line, usually called *coax*, shown in various forms in **Figure 20.1**. Coax is made up of a center conductor, which may be either stranded or solid wire, surrounded by a concentric outer conductor with a *dielectric* center insulator between the conductors. The outer conductor may be braided shield wire or a metallic sheath. A flexible aluminum foil or a second braided shield is employed in some coax to improve shielding over that obtainable from a standard woven shield braid. When the outer conductor is made of solid aluminum or copper, the coax is referred to as *hardline*.

The second type of transmission line uses parallel conductors, side by side, rather than the concentric ones used in coax. Typical examples of such *open-wire* lines are 300 Ω TV ribbon line or *twin-lead* and 450 Ω ladder line (sometimes called *window line*), also shown in Figure 20.1. Although open-wire lines are enjoying a sort of renaissance in recent years because of their inherently lower losses in simple multiband antenna systems, coaxial cables are far more prevalent because they are much more convenient to use.

The third major type of transmission line is the *waveguide*. While open-wire and coaxial lines are used from power-line frequencies to well into the microwave region, waveguides are used at microwave frequencies only. Waveguides are covered at the end of this chapter.

20.1.1 Fundamentals

In either coaxial or open-wire transmission line, currents flowing in the two conductors travel in opposite directions as shown in Figs 20.1E and 20.1I. When the physical spacing between the two parallel conductors in an open-wire line, S, is small in terms of wavelength, the phase difference between the currents is very close to 180°. When the two currents also have equal amplitudes, the field generated by each conductor cancels that generated by the other, and the line does not radiate energy, even if it is many wavelengths long.

The equality of amplitude and 180° phase difference of the currents in each conductor in an open-wire line determine the degree of radiation cancellation. When the currents are, for some reason unequal, or when the phase difference is not 180°, the line radiates energy. How such imbalances occur and to what degree they can cause problems is covered in more detail later.

In contrast to an open-wire line, the outer conductor in a coaxial line acts as a shield, confining RF energy within the line as shown in Figure 20.1E. Because of *skin effect* (see the **RF Techniques** chapter), current flowing in the outer conductor of a coax does so on the inner surface of the outer conductor. The fields generated by the currents flowing on the outer surface of the inner conductor and on the inner surface of the outer conductor cancel each other out, just as they do in open-wire line.

VELOCITY FACTOR

In free space, electrical waves travel at the speed of light, or 299,792,458 meters per second, or 983,569,082 feet per second. The length of a wave in space may be related to frequency as wavelength = λ = velocity/frequency. Thus, the wavelength of a 1 Hz signal is 983,569,082 ft. Changing to a more useful expression gives:

$$\lambda = 983.6/f \qquad (1)$$

where
  λ = wavelength, in feet, and
  f = frequency in MHz.

Thus, at 14 MHz the wavelength is 70.25 ft.

Wavelength (λ) may also be expressed in electrical degrees. A full wavelength is 360°, λ/2 is 180°, λ/4 is 90°, and so forth.

Waves travel slower than the speed of light in any medium denser than a vacuum, such as in free space. A transmission line may have an insulator that slows the wave travel down. The actual velocity of the wave is a function of the dielectric characteristic of that insulator. We can express the variation of velocity as the *velocity factor* for that particular type of dielectric — the fraction of the wave's velocity of propagation in the transmission line compared to that in free space. The velocity factor is related to the dielectric constant of the material in use.

$$VF = 1/\sqrt{\varepsilon} \qquad (2)$$

where
  VF = velocity factor, and
  ε = dielectric constant.

So, the wavelength in a real transmission line, in feet, becomes:

$$\lambda = (983.6/f) \times VF \qquad (3)$$

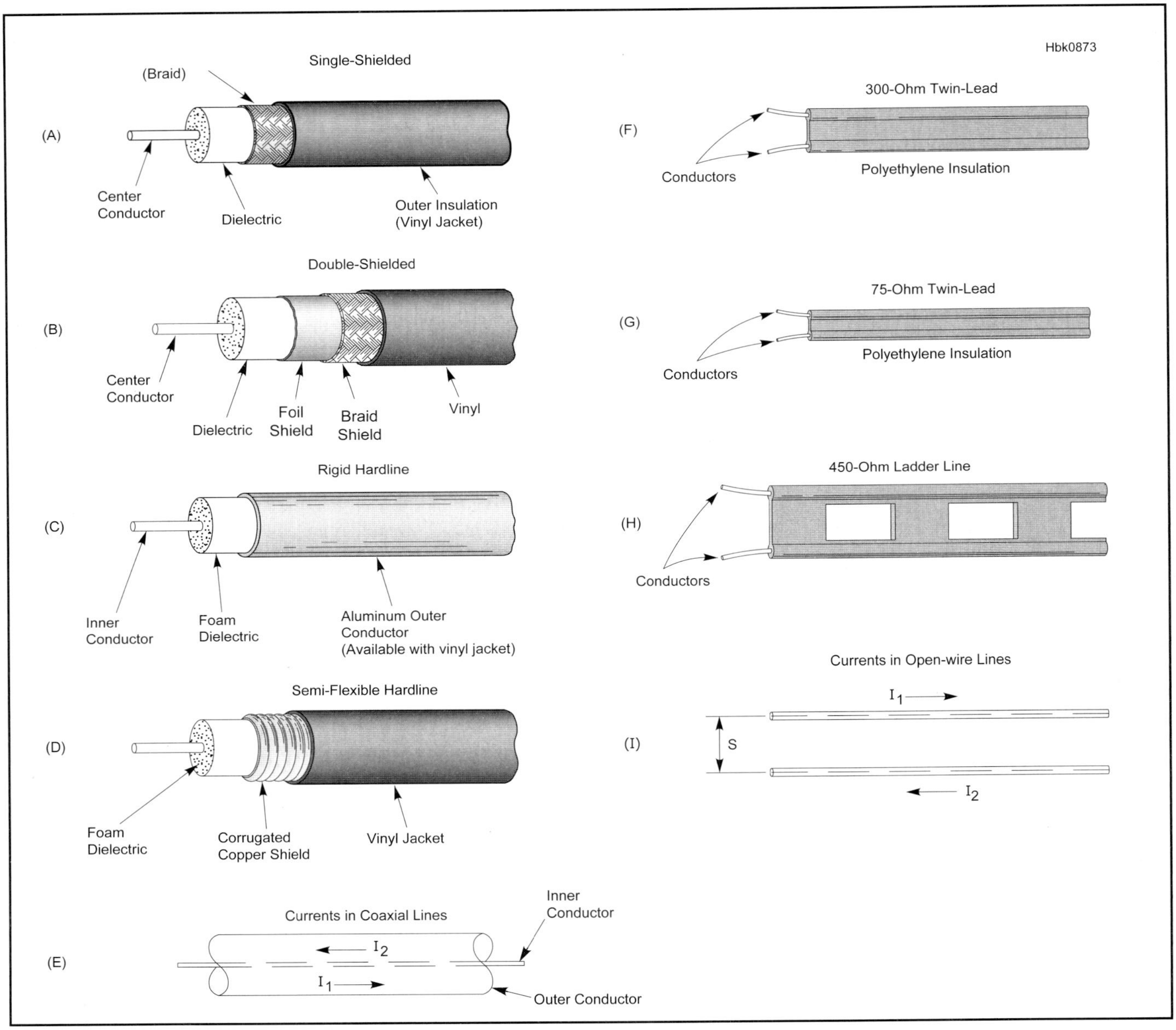

Figure 20.1 — Common types of transmission lines used by amateurs. Coaxial cable, or "coax," has a center conductor surrounded by insulation. The second conductor, called the shield, cover the insulation and is, in turn, covered by the plastic outer jacket. Various types are shown at (A), (B), (C), and (D). The currents in coaxial cable (E) flow on the outside of the center conductor (I_2) and the inside of the outer shield (I_1). Open-wire line ((F), (G), and (H)) has two parallel conductors separated by insulation. In open-wire line, the current flows in opposite directions on each wire (I). Articles on splicing hard line and window line are included in the supplmental information available online.

As an example, many coax cables use polyethylene dielectric over the center conductor as the insulation. The dielectric constant for polyethylene is 2.3, so the VF is 0.66. Thus, wavelength in the cable is about two-thirds of what it is in free space.

The VF and other characteristics of many types of lines, both coax and twin lead, are shown in the table "Nominal Characteristics of Commonly used Transmission Lines" in the RF Connectors and Transmission Lines document in the online content for this chapter.

The VF of any transmission line is not constant. It varies with frequency, although a nominal VF is given in tables of specifications. At the lowest audio frequencies, VF is quite a bit lower than the nominal value, rising rapidly throughout the audio spectrum, continuing to increase through the radio spectrum until it reaches a nearly constant value in the VHF range. It is this constant value that is computed by simple equations for VF that don't take frequency into account.

For most cables, VF at 2 MHz is typically 1 – 2% lower than the nominal value. In other words, an actual transmission line is 1 – 2% longer electrically than using the nominal value that the simplified equation for VF predicts. Because of this change, the physical length of a stub for the lower frequency bands is 1 – 2% shorter than predicted from the specified value. Always measure the actual VF of cables at the frequency of use before calculating electrical lengths for stubs or other length-sensitive uses.

For example, when using software to transform antenna measurements made in the station to the actual impedance at the feed point, the variability of VF must be applied to data for antennas for the lower bands (14 MHz and below). AC6LA's free *Zplots* utility software (**ac6la.com/zplots1.html**) plots VF, Z_0, and attenuation versus frequency from measurements of a known length of a transmission made with the far end open and again with the far end shorted. **Figure 20.2B** is a plot of VF and attenuation computed by *ZPlots* from such measurements on a 176-foot length of RG-11 cable with a #14 AWG solid copper center conductor, a foam dielectric, and a copper braid shield. This behavior and general curve shape are typical of all transmission lines as predicted by fundamental transmission line equations. Exact values for each line differ based on their physical dimensions and their dielectric.

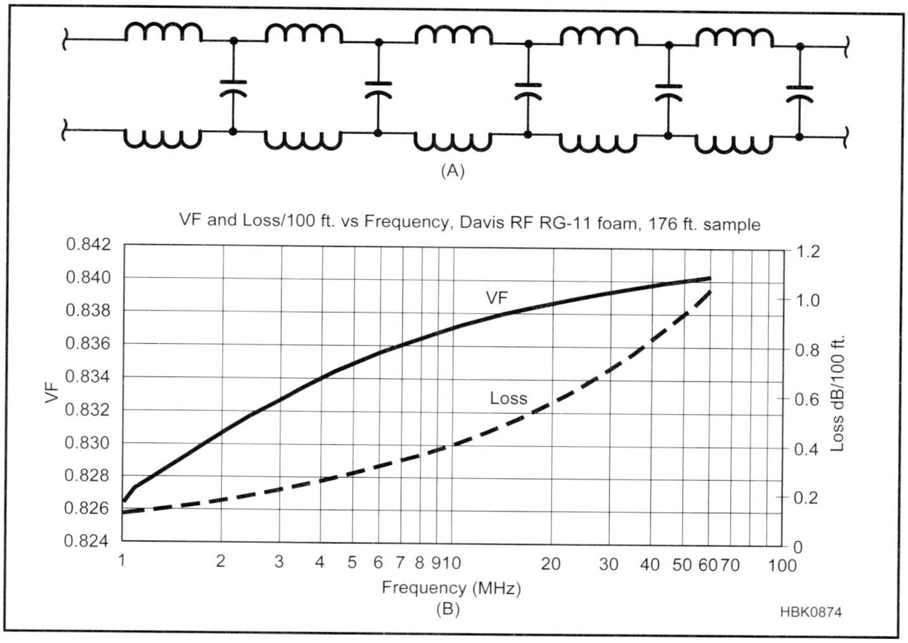

Figure 20.2 — Equivalent circuit (A) of an infinitely long lossless transmission line using lumped circuit constants. The actual performance of transmission line varies with frequency as shown in (B). For precision uses, the exact value for Z_0, VF, and loss should be determined by measuring the cable.

CHARACTERISTIC IMPEDANCE

A perfectly lossless transmission line may be represented by a whole series of small inductors and capacitors connected in an infinitely long line, as shown in Figure 20.2A. (We first consider this special case because we need not consider how the line is terminated at its end, since there is no end.)

Each inductor in Figure 20.2A represents the inductance of a very short section of one wire and each capacitor represents the capacitance between two such short sections. The inductance and capacitance values per unit of line depend on the size of the conductors and the spacing between them. Each series inductor acts to limit the rate at which current can charge the following shunt capacitor, and in so doing establishes a very important property of a transmission line: its *surge impedance*, more commonly known as its *characteristic impedance*. This is usually abbreviated as Z_0,

$$Z_0 \approx \sqrt{L/C}$$

where L and C are the inductance and capacitance per unit length of line.

The characteristic impedance of an air-insulated parallel conductor line, neglecting the effect of the insulating spacers, is given by

$$Z_0 = \frac{120}{\sqrt{\varepsilon}} \cosh^{-1} \frac{S}{d} \qquad (4)$$

where

Z_0 = characteristic impedance,

S = center-to-center distance between the conductors, and

d = diameter of conductors in the same units as S.

When $S \gg d$, the approximation $Z_0 = 276 \log_{10}(2S/d)$ is accurate. When $S < 2d$, the approximations for Z_0 are significantly higher than the correct values, such as is often the case when wires are twisted together to form a transmission line for impedance transformers. A more complete discussion of parallel conductor characteristic impedance for very small spacings is given in the *ARRL Antenna Book*.

The characteristic impedance of an air-insulated coaxial line is given by

$$Z_0 = 138 \log_{10}\left(b/a\right) \qquad (5)$$

where

Z_0 = characteristic impedance,

b = inside diameter of outer conductors, and

a = outside diameter of inner conductor

It does not matter what units are used for S, d, a, or b, as long as they are the *same* units. A line with closely spaced, large conductors has a low characteristic impedance, while one with widely spaced, small conductors has a relatively high characteristic impedance. Practical open-wire lines exhibit characteristic impedances ranging from about 200 to 800 Ω, while coax cables have Z_0 values between 25 and 100 Ω. Except in special instances, coax used in Amateur Radio has an impedance of 50 or 75 Ω.

All practical transmission lines exhibit some power loss. These losses occur in the resistance that is inherent in the conductors that make up the line, and from leakage currents flowing in the dielectric material between the conductors. We'll next consider what happens when a real transmission line, which is not infinitely long, is terminated in an actual load impedance, such as an antenna.

It should be noted that Z_0 also varies with frequency and below VHF is complex. That means it is not a pure resistance and is slightly capacitive. The *TLW* program (Transmission Lines for Windows by N6BV), included with the *ARRL Antenna Book*, provides Z_0 data for most commonly used cables, and can plot voltage and current along the line. These plots clearly show standing waves on a line at 2 MHz when the termination is only resistive. The mismatch is small and the effect on attenuation is insignificant, but it clearly shows up in carefully made measurements of long cable lengths over a range of frequencies as a small ripple in attenuation values.

The simplified calculations given in this section make simplifying assumptions about conductor and insulation characteristics that give useful values at MF through low VHF. At lower and higher frequencies, the actual values become significantly different.

20.1.2 Matched and Mismatched Lines

Real transmission lines have a definite length and are connected to, or *terminate* in, a load (such as an antenna), as illustrated in **Figure 20.3A**. When the load is a pure resistance whose value equals the characteristic impedance of the line, the line is said to be *matched*. To current traveling along the line, such a load at the end of the line appears as though it were still more transmission line of the same characteristic impedance. In a matched transmission line, energy travels along the line from the source until it reaches the load, where it is completely absorbed (or radiated when the load is an antenna).

MISMATCHED LINES

Assume now that the line in Figure 20.3B is terminated in an impedance Z_a which is not equal to Z_0 of the transmission line. The line is now a *mismatched* line. Energy reaching the end of a mismatched line is not fully absorbed by the load impedance. Instead, part of the energy is reflected back toward the source. The amount of reflected versus absorbed energy depends on the degree of mismatch between the characteristic impedance of the line and the load impedance connected to its end.

The reason why energy is reflected at a discontinuity of impedance on a transmission line can best be understood by examining some limiting cases. First, consider the rather extreme case where the line is shorted at its end. Energy flowing to the load encounters the short at the end, and the voltage at that point goes to zero, while the current rises to a maximum. Since the short circuit does not dissipate any power, the energy is all *reflected* back toward the source generator.

When the short at the end of the line is replaced with an open circuit, the opposite happens. Here the voltage rises to maximum, and the current, by definition, goes to zero. The phase reverses, and again all energy is reflected back towards the source. By the way, if this sounds to you like what happens at the end of a half-wavelength dipole antenna, you are quite correct. However, in the case of an antenna, energy traveling along the antenna is lost by radiation on purpose, whereas a good transmission line loses little energy to radiation because the fields from the conductors cancel outside the line.

For load impedances falling between the extremes of short and open circuit, the phase and amplitude of the reflected wave vary. The amount of energy reflected and the amount of energy absorbed in the load depends on the difference between the characteristic impedance of the line and the impedance of the load at its end.

What actually happens to the energy reflected back down the line? This energy may encounter another impedance discontinuity, this time at the source. Reflected energy flows back and forth between the mismatches at the source and load. After a few such journeys, the reflected wave diminishes to nothing, partly as a result of finite losses in the line, but mainly because of partial absorption at the load each time it reaches the load. In fact, if the load is an antenna, such absorption at the load is desirable, since the energy is radiated into the air.

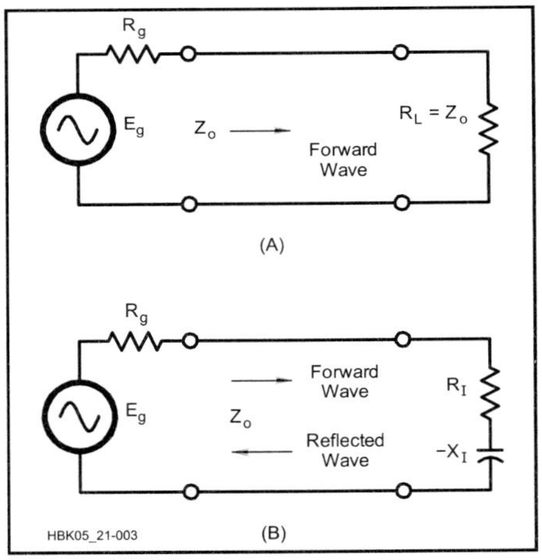

Figure 20.3 — At (A), the coaxial transmission line is terminated with resistance equal to its Z_0. All power is absorbed in the load. At (B), coaxial line is shown terminated in an impedance consisting of a resistance and a capacitive reactance. This is a mismatched line, and a reflected wave is returned back down the line toward the generator. The reflected wave adds to the forward wave, producing a standing wave on the line. The amount of reflection depends on the difference between the load impedance and the characteristic impedance of the transmission line.

When a continuous RF voltage is applied to the terminals of a transmission line, the voltage at any point along the line consists of a vector sum of voltages, the composite of waves traveling toward the load and waves traveling back toward the source generator. The sum of the waves traveling toward the load is called the *forward* or *incident* wave, while the sum of the waves traveling toward the generator is called the *reflected wave*.

20.1.3 Reflection Coefficient and SWR

In a mismatched transmission line, the ratio of the voltage in the reflected wave at any one point on the line to the voltage in the forward wave at that same point is defined as the *voltage reflection coefficient*. This has the same value as the current reflection coefficient. The reflection coefficient is a complex quantity (that is, having both amplitude and phase) and is generally designated by the Greek letter ρ (rho), or sometimes in the professional literature as Γ (Gamma). The relationship between R_L (the load resistance), X_L (the load reactance), Z_0 (the line characteristic impedance, whose real part is R_0 and whose reactive part is X_0) and the complex reflection coefficient ρ is

$$\rho = \frac{Z_L - Z_0}{Z_L + Z_0} = \frac{(R_L \pm jX_L) - (R_0 \pm jX_0)}{(R_L \pm jX_L) + (R_0 \pm jX_0)} \quad (6A)$$

For most transmission lines the characteristic impedance Z_0 is almost completely resistive, meaning that $Z_0 = R_0$ and $X_0 \cong 0$. The magnitude of the complex reflection coefficient in equation 6A then simplifies to:

$$|\rho| = \sqrt{\frac{(R_L - R_0)^2 + X_L^2}{(R_L + R_0)^2 + X_L^2}} \quad (6B)$$

For example, if the characteristic impedance of a coaxial line is 50 Ω and the load impedance is 120 Ω in series with a capacitive reactance of –90 Ω, the magnitude of the reflection coefficient is

$$|\rho| = \sqrt{\frac{(120-50)^2 + (-90)^2}{(120+50)^2 + (-90)^2}} = 0.593$$

Note that if R_L in equation 6A is equal to R_0 and X_L is 0, the reflection coefficient, ρ, is 0. This represents a matched condition, where all the energy in the incident wave is transferred to the load. On the other hand, if R_L is 0, meaning that the load is a short circuit and has no real resistive part, the reflection coefficient is 1.0, regardless of the value of R_0. This means that all the forward power is reflected since the load is completely reactive.

The concept of reflection is often shown in terms of the *return loss* (RL), which is given in dB and is equal to 20 times the log of the reciprocal of the reflection coefficient (Even though negative values generally represent losses, the minus sign makes the numerical result positive since the name of the result is Return Loss and a negative loss would then be a gain).

$$\mathrm{RL(dB)} = -10\log\left(\frac{P_r}{P_f}\right) = -20\log(\rho) \qquad (7)$$

In the example above, the return loss is –20 log (0.593) = 4.5 dB.

If there are no reflections from the load, the voltage distribution along the line is constant or *flat*. A line operating under these conditions is called either a *matched* or a *flat* line. If reflections do exist, a voltage *standing-wave* pattern results from the interaction of the forward and reflected waves along the line. For a lossless transmission line at least λ/4 long, the ratio of the maximum peak voltage anywhere on the line to the minimum value anywhere along the line is defined as the *voltage standing-wave ratio*, or VSWR. (The line must be λ/4 or longer for the true maximum and minimum to be created.) Reflections from the load also produce a standing-wave pattern of currents flowing in the line. The ratio of maximum to minimum current, or ISWR, is identical to the VSWR in a given line.

In amateur literature, the abbreviation SWR is commonly used for standing-wave ratio, as the results are identical when taken from proper measurements of either current or voltage. Since SWR is a ratio of maximum to minimum, it can never be less than one-to-one. In other words, a perfectly matched line has an SWR of 1:1. The SWR is related to the magnitude of the complex reflection coefficient and vice versa by

$$\mathrm{SWR} = \frac{1+|\rho|}{1-|\rho|} \qquad (8A)$$

and

$$|\rho| = \frac{\mathrm{SWR}-1}{\mathrm{SWR}+1} \qquad (8B)$$

The definitions in equations 8A and 8B are valid for any line length and for lines that are lossy, not just lossless lines longer than λ/4 at the frequency in use. Very often the load impedance is not exactly known, since an antenna usually terminates a transmission line. The antenna impedance may be influenced by a host of factors, including its height above ground, end effects from insulators, and the effects of nearby conductors. We may also express the reflection coefficient in terms of forward and reflected power, quantities that can be easily measured using a directional RF wattmeter. The reflection coefficient and SWR may be computed as

$$|\rho| = \sqrt{\frac{P_r}{P_f}} \qquad (9A)$$

and

$$\mathrm{SWR} = \frac{1+\sqrt{\frac{P_r}{P_f}}}{1-\sqrt{\frac{P_r}{P_f}}} \qquad (9B)$$

where

P_r = power in the reflected wave, and

P_f = power in the forward wave.

If a line is not matched (SWR > 1:1) the difference between the forward and reflected powers measured at any point on the line is the net power going toward the load from that point. The forward power measured with a directional wattmeter (often referred to as a reflected power meter or reflectometer) on a mismatched line always appear greater than the forward power measured on a matched line with a 1:1 SWR.

The software program *TLW*, solves these complex equations. The characteristics of many common types of transmission lines are included in the software so that real antenna matching problems may be easily solved. Detailed instructions on using the program are included with it. *TLW* was used for the example calculations in this chapter.

An interactive online animated graphic that shows the relationships between SWR, Reflection Coefficient, Return Loss, Mismatch Loss, and the load (antenna) impedance (for a feed line impedance of 50 Ω) and then displays the incident and reflected waves, the standing wave, and the transmitted wave. This animation can be found at **www.bessernet.com/node/138**. (Note: JavaScript must be enabled in your browser for the animation to play. Make sure you are using a modern browser and all security patches are up to date). **Figure 20.4** is a screen capture from this animation for an SWR of 3.0. It demonstrates a model of the RF signal traveling down the feed line and part of it being reflected at an impedance mismatch at the antenna feed point. This highly instructive animation allows you to enter any of those parameters and then see what happens to a transmitted signal. Try entering an SWR value, press Enter and then observe what happens along the transmission line and at the antenna.

20.1.4 Losses in Transmission Lines

A transmission line exhibits a certain amount of loss, caused by the resistance of the conductors used in the line and by dielectric losses in the line's insulators. The *matched line loss* for a particular type and length of transmission line, operated at a

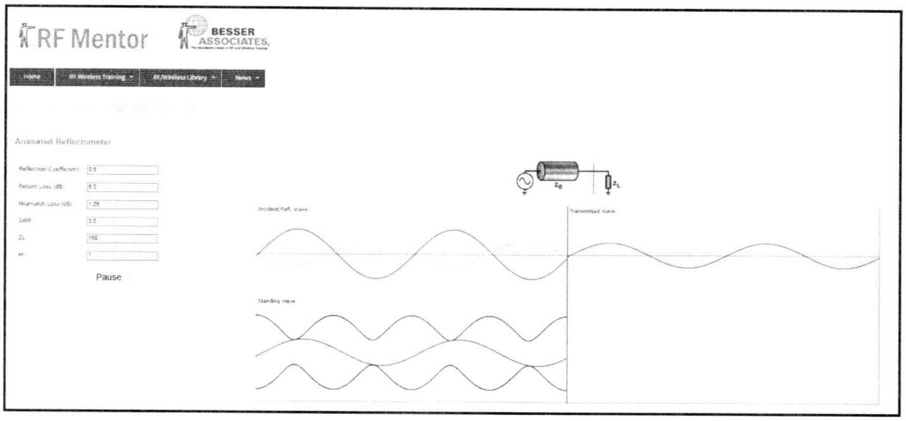

Figure 20.4 — A screen capture from the animation that can be found at www.bessernet.com/node/138 (note: JavaScript must be enabled in your browser). This animation shows an SWR of 3.0:1. The left side of the graph represents the feed line and the right side, the antenna. The vertical line separating the halves represents the impedance mismatch. In the upper left frame, you see the solid forward waveform traveling from left-to-right and the dashed, reflected, waveform traveling from right-to-left. The lower left frame illustrates the envelope of the standing wave, and the line inside it is the vector sum of the forward and reflected waves frozen at one point in time. The solid line in the upper right half of the screen is the waveform that reaches the antenna and radiates into the air. (Image courtesy of Besser Associates/Rex Frobenius, KJ6JFE)

particular frequency, is the loss when the line is terminated in a resistance equal to its characteristic impedance. The loss in a line is lowest when it is operated as a matched line.

Line losses increase when SWR is greater than 1:1. Each time energy flows from the generator toward the load, or is reflected at the load and travels back toward the generator, a certain amount is lost along the line. The net effect of standing waves on a transmission line is to increase the average value of current and voltage, compared to the matched line case. An increase in current raises I^2R (ohmic) losses in the conductors, and an increase in RF voltage increases E^2/R losses in the dielectric. Line loss rises with frequency, since the conductor resistance is related to skin effect, and also because dielectric losses rise with frequency.

Matched-line loss (ML) is stated in decibels per hundred feet at a particular frequency. The matched-line loss per hundred feet versus frequency for a number of common types of lines, both coaxial and open-wire balanced types, is shown graphically and as a table in the RF Connectors and Transmission Lines material in the online content for this chapter. For example, RG-213 coax cable has a matched-line loss of 2.5 dB/100 ft at 100 MHz. Thus, 45 ft of this cable feeding a 50 Ω load at 100 MHz would have a loss of 2.5 dB × (45 ft / 100 ft) = 1.13 dB.

When a line is not matched, standing waves cause additional loss beyond the inherent matched-line loss for that line, resulting in Total Mismatched Line Loss (TMLL).

$$\mathrm{TMLL} = 10\ \log\left[\frac{a^2 - |\rho^2|}{a\left(1-|\rho^2|\right)}\right] \qquad (10)$$

where

$a = 10^{ML/10}$

The effect of SWR on line loss is shown graphically in **Figure 20.5**. The horizontal axis is the SWR at the load end of the line. The family of curves is the matched loss of the length of transmission line in use. From the SWR value on the horizontal axis, proceed vertically to the curve representing the feed line's matched loss (loss with SWR of 1:1). At the intersection, the TMLL can be read on the vertical axis.

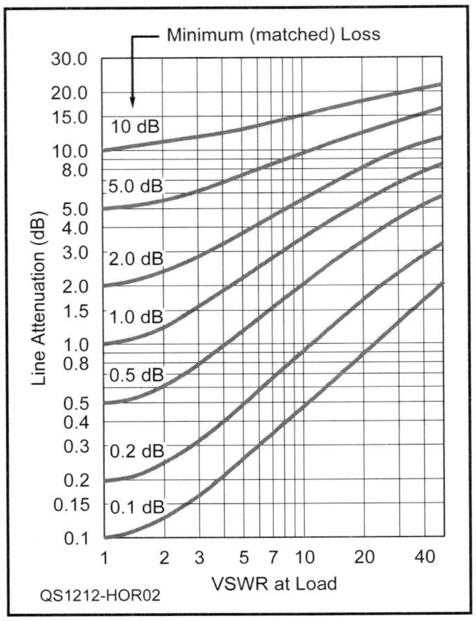

Figure 20.5 —Total insertion loss in a transmission line terminated in a mismatch. To use the chart, start with the SWR at the load. Find that value on the horizontal axis. From the SWR value on the horizontal axis, proceed vertically to the curve representing the feed line's matched loss (loss with SWR of 1:1). At the intersection, the total loss can be read on the vertical axis. (Courtesy of Joe Reisert, W1JR; see reference)

MEASURING TRANSMISSION LINE LOSS

The most obvious method to measure ML is to use a calibrated wattmeter and dummy load. With the wattmeter at the input to the line and the dummy load at the output, apply power to the line and measure forward power, P_{IN}. (With the dummy load attached, there should be no reflected power.) Remove power and connect the input of the line directly to the power source. Connect the wattmeter between the output of the line and the dummy load. Apply the same amount of power and read forward power at the dummy load, P_{OUT}. The loss in the line is equal to 10 log (P_{OUT} / P_{IN}).

Without a wattmeter, loss can be measured by using a calibrated mismatch. Assuming a 50 Ω system, select a non-inductive resistor between 150 Ω (3:1 SWR) and 270 Ω (5.4:1 SWR). Convert the resistor's expected SWR to RL, using the table "Reflection Coefficient, Attenuation, SWR and Return Loss" in the online content for this book. For example, a 220 Ω resistive load results in a 4.4:1 SWR which is a return loss of 4.0 dB. Connect the resistor to the output of the line. Make sure the resistor leads are very short so that they do not add a significant amount of inductance. Measure SWR at the input to the line and convert to return loss. The line loss is the difference between return loss at the line input and return loss of the load. For example, with the 220 Ω load (4.0 dB RL) and 100 feet of RG-58 coax at 10 MHz, the input SWR might be 3.0:1 (RL = 6.0 dB). The line loss at this frequency is 6.0 dB – 4.0 dB = 2.0 dB.

Some methods use an open or short circuit at the load end of the line (an infinite SWR and RL = 0) to measure line loss. Most amateur instrumentation is not well-calibrated at high SWR and give an unreliable reading for SWR and RL. Using a moderate mismatch improves the accuracy of the final result. You can replace the feed line with in-line attenuators to check this more accurate method with the known amounts of loss. See the Reference entry for Stensby describing another method of calculating line loss by using measurements of mismatched impedance values.

20.1.5 Maximum Voltage and Current with SWR

It is useful to know what maximum voltage occurs on a transmission line with an SWR > 1:1. For example, if the line carries the full legal power limit, voltage breakdown is a consideration if the SWR becomes significant, such as at a band edge far from the frequency of minimum SWR. Similarly, an open or short circuit creates high voltage at points along the line.

The following equation assumes P is the net input power, which on a directional wattmeter equals $P_f - P_r$, and none of the reflected power is absorbed by the source (such as a transmitter's output stage) or by line loss. In such a case, power is continually "pumped" into the line. If the termination is short or open, power is only dissipated by line loss and voltage can become very high. (See the *ARRL Antenna Book*, 24th edition for related versions of this calculation.)

$$E_{max} = \sqrt{P \times Z_0 \times SWR}\,;\, I_{max} = \frac{E_{max}}{Z_0} \qquad (11A)$$

$$E_{min} = \frac{E_{max}}{SWR; I_{min}} = \frac{I_{max}}{SWR} \qquad (11B)$$

If $P_f - P_r$ = 100 W of power is applied to a 50 Ω line with an SWR at the load of 4:1, the maximum voltage is:

$$E_{max} = \sqrt{100 \times 50 \times 4} = 141\ \mathrm{V}$$

In practice, when SWR is very high, line losses and transmitter output amplifier protection circuitry limit the peak line voltage caused by standing waves to a lower value. E_{max} and E_{min} are RMS values that would be measured with an ordinary RF voltmeter. E_{max} can be converted to an *instantaneous peak voltage* by multiplying by 1.414 and doubling that to find the instantaneous peak-to-peak voltage.

20.2 Transmission Lines — Practical Considerations

Making the best choice for a particular installation requires balancing the properties of the three common types of feed lines used by amateurs — coaxial, parallel-conductor or open-wire, and hardline — along with cost. The primary electrical considerations for feed line are characteristic impedance and loss. Mechanical concerns include weight, suitability for exposure to weather or soil, and interaction with other cables and conductors. When evaluating cost, be sure to include the costs of connectors and any auxiliary items such as baluns and waterproofing materials.

The entire antenna system, composed of the feed line, tuners or matching networks and the antenna itself, must be considered when evaluating what type of line to use. Along with loss, the effects of SWR on maximum voltage in the system must be considered if high power is used, especially if high SWR is anticipated.

Multiband antenna systems, such as non-resonant wire antennas, can present a challenge because of the range of SWR values and the wide range of frequencies. As an example of the considerations involved, the article "Multiband Operation with Open-Wire Line" is included with the online content. By following the general process of modeling or calculation and evaluation with different types of feed line, the best choices can be made.

20.2.1 Effect of Loss

For most types of line and for modest values of SWR, the additional line loss due to SWR is of little concern. As the line loss increases at higher frequencies, the total line loss (the sum of matched-line loss and additional loss due to SWR) can be surprisingly large at high values of SWR.

For example, RG-213 solid-dielectric coax cable exhibits a matched-line loss at 28 MHz of 1.14 dB per 100 ft. A 250 ft length of this cable has a matched-line loss of 1.14 × 250/100 = 2.86 dB. If we measure the SWR *at the load (antenna)* as 6:1, the total mismatched line loss from equation 10 is 5.32 dB.

However, we rarely measure SWR at the antenna. Rather, the most common place to measure SWR is at the transmitter. This almost always underestimates the true SWR because of the loss in the line. In the above example, we determine the forward power, P_f, with no loss. The signal travels to the antenna with a feed line loss of 2.86 dB and then some of it reflects back toward the transmitter. The reflection travels back to the transmitter, losing another 2.86 dB in the coax. Thus, we measure the reflected power, P_r, after it has lost 5.72 dB compared to the forward power measurement, and the SWR calculated from those two measurements, as in equation 9B, is only 2.2:1.

On the other hand, if expensive 7/8 inch diameter 50 Ω hardline cable is used at 28 MHz, the matched-line loss is only 0.19 dB/100 ft. For 250 ft of this hardline, the matched-line loss is 0.475 dB, and the additional loss due to a 6:1 SWR is 0.793 dB. Thus, the total loss is 1.27 dB.

At the upper end of the HF spectrum, when the transmitter and antenna are separated by a long transmission line, the use of bargain coax may prove to be a very poor cost-saving strategy. Adding a 1,500 W linear amplifier (providing 8.7 dB of gain over a 200 W transmitter), to offset the loss in RG-58A compared to hardline, would cost a great deal more than higher-quality coax. Furthermore, no transmitting amplifier can boost receiver sensitivity. Loss in the line has the same effect as putting an attenuator in front of the receiver.

At the lower end of the HF spectrum, say 3.5 MHz, the amount of loss in common coax lines is less of a problem for the range of SWR values typical on this band. For example, consider an 80-meter dipole cut for the middle of the band at 3.75 MHz. It exhibits an SWR of about 6:1 at the 3.5 and 4.0 MHz ends of the band. At 3.5 MHz, 250 ft of RG-58A small-diameter coax has an additional loss of 2.1 dB for this SWR, giving a total line loss of 4.0 dB. If larger-diameter RG-213 coax is used instead, the additional loss due to SWR is 1.3 dB, for a total loss of 2.2 dB. This is an acceptable level of loss for most 80-meter operators.

The loss situation gets dramatically worse as the frequency increases into the VHF and UHF regions. At 146 MHz, the total loss in 250 ft of RG-58A with a 6:1 SWR at the load is 21.4 dB, 10.1 dB for RG-213A, and 2.7 dB for 7/8 inch, 50 Ω hardline. At VHF and UHF, a low SWR is essential to keep line losses low, even for the best coaxial cable. The length of transmission line must be kept as short as practical at these frequencies.

Table 20.1 lists some commonly used coax cables showing feet per dB of loss vs frequency. The table can help with the selection of coax cable by comparing lengths that result in 1 dB of loss for different types of cable. The larger the value in the table, the less loss in the cable per unit length. (See also Tables 23.3 and 23.4 in the *ARRL Antenna Book* for information on more types of cable.) Determine the length of line your installation requires and the maximum acceptable amount of line loss in dB. Divide the total line length by the maximum acceptable loss to calculate the minimum acceptable length of line with 1 dB of loss. From the table, select a cable type that has a length per 1 dB of loss that is greater than the minimum acceptable length.

Example: An installation requires 400 feet of feed line at 14 MHz with a maxi-

Transmission Lines for Microwave Frequencies

While low-loss waveguide is generally used to carry microwave frequency signals for long distances, *semi-rigid* coaxial cable — essentially miniature hardline — is used for connections inside and between pieces of equipment in the shack. Working with this type of cable requires special techniques, addressed in the online article "Microwave Plumbing" with this book's online content. More information on microwave construction techniques is available in the **Construction Techniques** chapter and in the "Microwavelengths" columns by Paul Wade, W1GHZ in *QST*, available on-line to ARRL members.

Table 20.1
Length in Feet for 1 dB of Matched Loss

MHz	*1.8*	*3.6*	*7.1*	*14.2*	*21.2*	*28.4*	*50.1*	*144*	*440*	*1296*
RG-58	179	122	83	59	50	42	30	18	9	5
RG-8X	257	181	128	90	74	63	47	27	14	8
RG-213	397	279	197	137	111	95	69	38	19	9
LMR-400	613	436	310	219	179	154	115	67	38	21
9913	625	435	320	220	190	155	110	62	37	20
EC4-50				290		202		87		26
EC5-50				787		548		239		75
450 Ω OWL*	1065	758	547	391	322	279	213			
600 Ω OWL**	4550	3030	2130	1430	1150	980	715			

*Wireman #551, 400 Ω characteristic impedance
**Conductors #12 AWG

mum acceptable value of 3 dB of loss. This requires cable with a minimum of 400 / 3 = 133 feet per dB of loss. Find a cable in Table 20.1 that shows more than 133 feet for 1 dB of loss at 14.2 MHz. RG-213 has the highest acceptable loss at this frequency: 137 ft / dB of loss.

Don't forget that you can combine types of cable and accessories to lower the total system cost while still meeting performance requirements. For example, it is common to use a single low-loss cable from the shack to a distant tower with multiple antennas. At the tower, an antenna switch then selects short runs of less-expensive cable to the various antennas.

Using 75 Ω Hardline in 50 Ω Systems

Surplus CATV hardline is usually 75 Ω but can be used in 50 Ω systems without special impedance matching techniques. Make the 75 Ω line some integer multiple of λ/2 at the operating frequency so that the impedance at the load end of the line is reproduced at the input to the line. This can also be done when feeding a 20/15/10 meter beam, for example, by making the 75 Ω line 2 λ on 20 meters so that it is 3 λ long on 15 and 4 λ long on 10 meters. If the load is 50 Ω, the SWR in the line is still 75/50 = 1.5:1 but the simplicity often outweighs the minimal extra loss on the HF and lower VHF bands.

20.2.2 Feed Line Construction and Installation

Either coaxial cable or open-wire feed line is used to connect a radio to an antenna. There are pros and cons for each type of feed line. Coax is the common choice because it is readily available, its characteristic impedance is close to that of a center-fed dipole, and it may be easily routed through or along walls and among other cables. Where a very long feed line is required or the antenna is to be used at frequencies for which the feed point impedance is high, the increased RF loss and low working voltage (compared to that of open-wire line) make it a poor choice. Hardline is the preferred choice at VHF and higher frequencies when the line losses for flexible coax would be too high. It is also often used for very long line lengths at HF.

Note that most traditional RG-type designations are no longer listed under MIL-spec and may not exactly match that cable type's construction and characteristics. For example, cable advertised as RG-213 is actually "RG-213 Type" and may have characteristics quite different from the original RG-213 specification. Use the manufacturer's part number to determine the actual performance specifications.

A list of commonly used coaxial cable types is included in the online material for this chapter, in a document entitled, "RF Connectors and Transmission Line Information." This document also lists the different coaxial connector types and has pictorial instructions for installing them.

COAXIAL CABLE CONSTRUCTION

Coaxial cable is mechanically much easier to use than open-wire line. Because of the excellent shielding afforded by its outer shield layer, coax can be installed along a metal tower leg or taped together with numerous other cables and exhibit virtually no interaction or cross talk. Coax can be used with a rotatable antenna without worrying about shorting or twisting the conductors, which might happen with an open-wire line.

Coaxial cable used in the amateur radio service is, for the most part, made with solid polyethylene (PE), extended or "foamed" polyethylene (FPE), or solid Teflon (TFE) center insulation. Teflon dielectric coax is often used at VHF and UHF frequencies due to its low loss characteristics.

Class 2 PVC (P2) noncontaminating outer jackets are designed for long-life outdoor installations. Class 1 PVC (P1) outer jackets are not recommended for outdoor installations. Coax and hardline can be buried underground, especially if run in plastic piping (with suitable drain holes) so that ground water and soil chemicals cannot easily deteriorate the cable. A cable with an outer jacket of polyethylene (PE) rather than polyvinyl chloride (PVC) is recommended for direct-bury installations.

Respect coax's power-handling ratings. Cables such as RG-58 and RG-59 are suitable for power levels up to 300 W with low SWR. RG-8X cable can handle higher power and there are a number of variations of this type of cable. For legal-limit power or moderate SWR, use the larger diameter cables that are 0.4 inches in diameter or larger.

Table 20.2
Minimum Bend Radii of Common Types of Coaxial Cable

Cable Type	*Diam (in)*	*Manufacturer's Min Bend Radius (in)*	
		Single	*Repeated*
RG-58	0.20	1.0	2.0
RG-8X	0.24	2.4	
RG-213	0.41	4.5	
LMR-400	0.41	1.0	4.0
9913	0.41	4.0	
EC4-50	0.63	2.8	4.7
EC5-50	1.1	4.0	10.0

Subminiature cables, such as RG-174, are useful for very short lengths at low power levels, but the high RF losses associated with these cables make them unsuitable for most uses as antenna feed lines. In these applications, a PTFE (Teflon) insulated cable is a better choice.

BENDING COAXIAL CABLE

Bending coax is acceptable as long as the radius of the bend is larger than the specified minimum bend radius. For example, a common minimum bend radius specification for RG-8 is 4 inches (8 times the cable diameter). Coax with more rigid shield materials has a larger bend radius. **Figure 20.6** shows how bend radius is measured. **Table 20.2** lists some common types of coax and their manufacturer's recommended minimum bend radii.

Bending the coax tighter than the minimum bend radius can cause impedance "bumps" in the line by distorting the geometry of the conductors. It can also cause the center conductor to migrate through the plastic insulation and eventually short to the outer shield. **Figure 20.7** shows what happens inside a coaxial cable when it is bent too tightly. This figure shows a section of hard line with the outer surface cut away. The bend has caused the center conductor

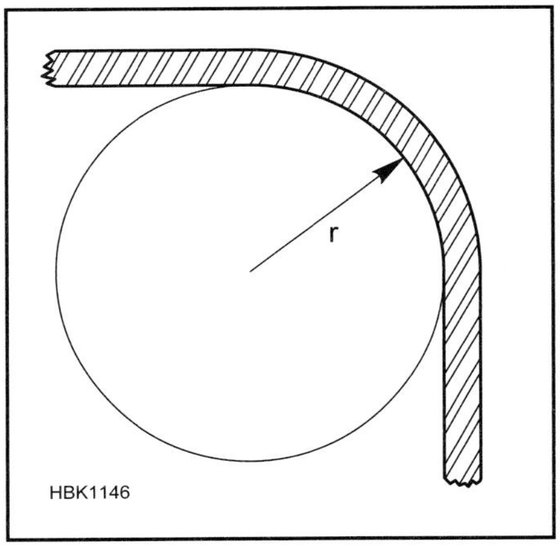

Figure 20.6 — Cable bend radius is determined as if the cable was wrapped around a cylinder with radius, r.

Figure 20.7 — What happens inside coaxial cable when bends are tighter than the minimum bend radius. (Photo courtesy of Greg Lapin, N9GL)

to push close to one side of the shield, compressing the insulation. Even if there is no short circuit between the center conductor and the shield, the impedance of the cable at that point is sharply shifted, resulting in reflections of RF signals that can greatly increase SWR. If this should happen, it may not be immediately evident from the outside of the cable. See the **Measurement and Test Equipment** chapter for ways to detect and locate this type of damage.

The forces pushing on the center conductor from coiling are aggravated by self-heating from cable loss, a direct function of the amount of power applied and SWR. RG-8X is not rated for 1500 W but lots of amateurs use it successfully at that power level. RG-8X gets warm to the touch at 1500 W. Increasing internal temperature softens the foam, facilitating center conductor migration. Tight radius bends taken together with heating are a recipe for impedance shifting and eventual short circuit. Tightly coiled baluns used outdoors can be heated by the sun, in addition to self-heating. A balun made with coiled coaxial cable used outside has higher probability of shorting out over time, particularly when used at high power.

Avoiding center conductor migration is easy: avoid sharp bends, particularly at high power. Use solid dielectric coax to make tightly coiled coaxial baluns and, if stubs must be coiled up, use solid dielectric coax for those too.

BURYING COAXIAL CABLE

There are several reasons why you might choose to bury coax. One is that direct burial cable is virtually free from storm and UV damage, and usually has lower maintenance cost than cable that is out in the open. Another reason might be aesthetics; a buried cable is acceptable in almost all communities. Also, being underground reduces common-mode feed line current on the outside of the shield, helping to reduce inter-station interference and RFI. Buried cable is also less susceptible to lightning.

Although any cable can be buried, a cable that is specifically designed for direct burial will have a longer life. The best cable to use is one that has a high-density polyethylene jacket because it is both nonporous and tolerates a relatively high amount of compressive loads. "Flooded" direct burial cables contain an additional moisture barrier of non-conductive grease under the jacket; this allows the material to leak out, thus "healing" small jacket penetrations. (These can be messy to work with when installing connectors.)

Here are some direct burial guidelines:

1) Because the outer jacket is the cable's first line of defense, any steps that can be taken to protect it goes a long way toward maintaining the internal quality of the cable.

2) Bury the cable in sand or finely pulverized dirt, without sharp stones, cinders, or rubble. If the soil in the trench does not meet these requirements, tamp four to six inches of sand into the trench, lay the cable and tamp another six to eleven inches of sand above it. A pressure-treated board placed in the trench above the sand prior to backfilling provides some protection against subsequent damage that could be caused by digging or driving stakes.

3) Lay the cable in the trench with some slack. A tightly stretched cable is more likely to be damaged as the fill material is tamped.

4) Examine the cable as it is being installed to be sure the jacket has not been damaged during storage or by being dragged across over sharp edges.

5) You may want to consider burying the cable in plastic pipe or conduit. Be careful to drill holes in the bottom of the pipe at all low spots so that any moisture can drain out. While PVC pipe or conduit provides a mechanical barrier, water incursion is practically guaranteed — you can't keep it out. It leaks in directly or condenses from moisture in the air. Use the perforated type so that any water drains out harmlessly. Plastic drain pipe with drain holes also works well.

6) It is important that direct burial is below the frost line to avoid damage by the expansion and contraction of the earth during freezing and thawing of the soil and any water surrounding the buried cables.

OPEN-WIRE LINE

The most common open-wire transmission lines are *ladder line* (also known as *window line*) and *twin-lead*. Since the conductors are not shielded, two-wire lines are affected by their environment. Use standoffs and insulators to keep the line several inches from structures or other conductors. Ladder line has very low loss (twin-lead has a little more), and it can stand very high voltages (created by high SWR) as long as the insulators are clean. Twin-lead can be used at power levels up to 300 W and ladder line to the full legal power limit. For direction on making one's own open-wire line, see the reference for the *QST* article by Robert Zavrel, W7SX. This is very practical, especially for lines that must handle high power or high voltage.

The characteristic impedance of open-wire line varies from 300 Ω for twin-lead to 450 to 600 Ω for most ladder and window line. The common 450 Ω window line with plastic insulation and 1-inch spacing has a characteristic impedance of approximately 360 to 405 Ω and velocity factor (VF) of approximately 91%, depending on the manufacturer and materials used. The solid plastic insulation also means that rain, snow, or ice can affect the line's Z_0 and VF, typically dropping both by about 3% when the line is wet, according to a paper by Wes Stewart, N7WS (**users.triconet.org/wesandlinda/ladder_line.pdf**). This variability suggests that if precise characteristics are important, Z_0 and VF should be measured for the line to be used.

When used with half wave dipoles, the resulting moderate to high SWR requires an impedance matching unit at the transmitter. The low RF losses of open-wire lines make this an acceptable situation on the HF bands.

Ladder line is available with both solid copperweld and stranded copper conductors. The solid conductor types tend to be less expensive but can break if flexed repeatedly, such as from being blown around in the wind. For that reason, stranded conductor ladder line is preferred. If solid conductor ladder line is used, be sure to provide adequate mechanical support to provide stress relief and protection against flexing. To minimize movement due to wind, twist ribbon or open-wire/window line once every two or three feet. This also helps balance the effect of capacitance to nearby objects and surfaces.

CONTROL CABLES

In addition to coaxial cables, most towers have some sort of control cable for rotators, antenna switches, or other accessories. The manufacturer should provide the size that is necessary and again, you should follow their specifications.

Multi-conductor cables are not usually as waterproof as coaxial cable. The jacket is usually just a plastic sleeve around the inner wires. If the jacket is nicked or cut, water can easily get in and collect at the lowest point of the cable. When the water does not leak out, it fills the cable jacket all the way into the station where it runs out the unsealed end of the cable. For this reason, it is common to make a drip loop in the cable where it enters the station and make a small slice or hole in the jacket to allow any accumulated water to leak out.

In the case of rotator cables, some rotators are sensitive to voltage drop so bigger sizes should be used. For really long runs, some amateurs use THHN house wire or UF-Romex, (with the motor start capacitor installed at the rotator) from the local hardware store to get reasonably-priced bigger wire. Only the motor and solenoid (if used) conductors typically require the larger wire.

For medium-length runs of buried cable, multi-conductor irrigation system control cable can be used in place of standard light-duty rotator cable. If the cable has extra conductors, pairs of conductors can be doubled to provide for the higher current brake connections. For example, four conductors in a 10-conductor irrigation cable with #18 conductors can be doubled up in pairs to create two heavy conductors for the brake solenoid circuits. Irrigation cables generally have solid wires, requiring different terminals and splicing techniques than for stranded wire.

20.2.3 Weatherproofing Coaxial Cable

Most manufacturers use some type of feed point system that accepts a PL-259 or N connector. Some antennas require you to split the coax and attach the shield and center conductor to machine screws attachment points on the driven element. The exposed end of the coax is very difficult to seal; indeed, it's nearly impossible. Water will wick down the outer shield and into your shack unless you take great pains to weatherproof it. Coating the entire pigtail and attachment terminals with Liquid Electrical Tape or some other conformal sealant is a good approach, although UV will degrade such coatings over time. Another approach for HF beams is to use a "Budwig HQ-1" style insulator with the integral SO-239 and wires for connecting to the terminals. As always, follow the manufacturer's directions.

With many beam antennas, the feed point is out of reach from the tower and should be connected to a jumper just long enough to reach from the feed point to the antenna mast. That way, the feed line connection and waterproofing can be done at the most convenient location. If you ever have to remove the antenna in the future you can just disconnect the jumper and lower the antenna.

When weatherproofing an RF connector, use high-quality electrical tape, such as 3M Scotch 33+ or Scotch 88 (same as 33+ but 1.5 mil thicker). Avoid inexpensive utility tape. Before weatherproofing, tighten the connector (use pliers carefully to seat threaded connectors — hand-tight isn't good enough), and apply two wraps of tape around the joint.

When you're done making a tape wrap, sever the tape with a knife or tear it very carefully. (Do not stretch the tape until it breaks. This invariably leads to "flagging," where the end of the tape loosens and blows around in the wind.) Let the tape relax before applying the next layer and finishing the wrap. See **Figure 20.8**.

Begin by applying two wraps of electrical tape around the joint. Next put a layer of butyl rubber vapor wrap over the joint. (3M Butyl Mastic Tape 2212 is one such material. This butyl rubber tape is also usually available in the electrical section of

Smith Chart Software

The standard way of visualizing transmission line and impedance matching mechanics is by using a Smith chart. (If you are unfamiliar with the Smith chart, recent editions of this book and the *ARRL Antenna Book* include a detailed tutorial on the Smith chart, either in print or with the online information.) Yesterday's paper charts, however, have been replaced by interactive computer software such as the easy-to-use *jjSmith* (*Windows* only) by Jim Tonne, W4ENE, available with the online content and *SimSmith* (**www.ae6ty.com/Smith_Charts.html**) by Ward Harriman, AE6TY. *SimSmith* is written in Java and runs on a number of operating systems. Learning about the Smith chart is a great aid to understanding the mechanics of transmission lines and impedance matching.

Where to Place Stubs Used as Harmonic Filters

A quarter-wave ($\lambda/4$) shorted stub makes an effective harmonic filter that is able to handle high power without expensive components. As described in this section, it is easy to build without special components or construction techniques. In order to get the best performance from the stub filter, however, it is important to install the stub at an appropriate location in the transmission line.

This type of stub works by presenting an open circuit at the fundamental frequency and a short circuit on the harmonics where it is a multiple of one-half wavelength ($\lambda/2$). For the short circuit to be most effective, the stub should be placed at a location in the transmission line where impedance is high *at the harmonic frequency*. The difference in performance can be dramatic. If the stub is connected at a low-impedance point, attenuation of the harmonic can be less than 10 dB. On the other hand, connected at a high-impedance point, attenuation can be well in excess of 30 dB. Without recognizing this dependence on impedance at the point of connection, applying stub filters results in erratic results.

A complete discussion of how to determine the optimum location for a stub is beyond the scope of this chapter but two *National Contest Journal* articles on stub placement by George Cutsogeorge, W2VJN, and Jim Brown, K9YC, have been added to the supplemental files supplied with the online content. (See the Table of Contents page at the beginning of this chapter.) A convenient rule of thumb that results in good, if not optimum, filter performance is to first determine the output circuit of the amplifier or transmitter. In almost all tube amplifiers, the output circuit is either a Pi or a Pi-L network. (See the chapter on RF Power Amplifiers.) Solid-state transmitters usually have a low-pass LC filter at the output. Determine whether the output component of the network or filter is a shunt (parallel) capacitor or a series inductor. (Ignore protective RF chokes or similar components.)

- If the output component is a series inductor, such as in a low-pass filter or a Pi-L network, the output impedance at the frequency of the harmonic to be attenuated is probably higher than 50 Ω. The stub is attached at the equipment output.
- If the output component is a shunt capacitor, such as in a Pi network, the output impedance at the frequency of the harmonic to be attenuated is probably lower than 50 Ω. The nearest high-impedance point, as a result, is $\lambda/4$ from the output. Connect a $\lambda/4$ jumper to the output and connect the stub at the junction of the jumper and the main feed line.

This simple rule-of-thumb procedure is unlikely to result in the best performance but it is a good substitute if measuring the feed line impedance directly is not practical. Both of the referenced articles go into some detail about feed line impedance measurement and determining the optimum location for the stub taking into account the antenna impedance and length of the main feed line. These are important considerations for high-performance station design to manage interstation interference. (Also see the reference section entry for *Managing Interstation Interference* by George Cutsogeorge, W2VJN.) In addition, the article by Jim Brown, K9YC, discusses placement of stub harmonic filters for receiving.

Commercial Triplexer Design Example

Provided with the online content, the article "Commercial Triplexer Design" by George Cutsogeorge, W2VJN, discusses the issues encountered in adapting a popular *QST* article on a Field Day-style triplexer to become a commercial product. The referenced *QST* article by Gary Gordon, K6KV, is provided as well.

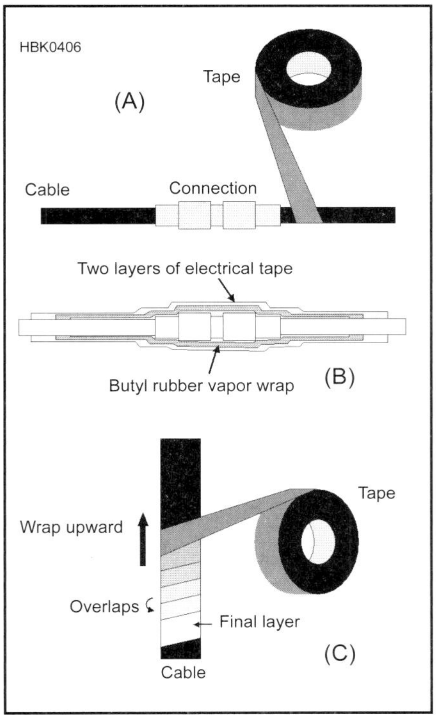

Figure 20.8 — Waterproofing a connector in three steps. At (A), cover the connectors with a layer of good-quality electrical tape. (B) shows a layer of butyl rubber vapor wrap between the two layers of electrical tape. (C) shows how to wrap tape on a vertical cable so that the tape sheds water away from the connection. (Drawing (C) reprinted from *Circuitbuilding for Dummies*, courtesy of Wiley Press)

hardware and home improvement stores.) Finally, add two more layers of regular tape over the vapor wrap, creating a professional-quality joint that will never leak. Finally, if the coax is vertical, be sure to wrap the final layer so that the tape is going up the cable. In that way, the layers act like roofing shingles, shedding water off the connection. Wrapping it top to bottom guides water between the layers of tape.

An alternative method, suggested by K4ZA, begins with a wrap of "military grade" Teflon tape — a thread wrapping tape thicker than what you'll find at your local hardware store. (McMaster-Carr #6802K44) Over that, install a layer of Scotch 130C (liner-less rubber sealing tape), using a 50% wrap (half the tape width is overlapped on each turn). Cover that with a layer of either Scotch 33+ or Scotch 88. Taken apart, 20-year-old joints have revealed connectors with like-new appearance.

A recent product for coax joints is shrink-fit tubing impregnated with hot-melt glue along the inside. As you apply heat to the shrink-fit tubing, it shrinks while the glue melts and oozes inside between the fitting and the tubing. It not only keeps the tubing from slipping, but it also fills in the voids in the joint and provides an additional seal. It's an expensive alternative (approximately $1 per inch) but is very simple to use and remove, if necessary.

Do not use silicone sealant that gives off acetic acid (a vinegary smell) and absorbs water when curing. Acid and water will migrate into the connection causing problems later. Use only aquarium-type sealants or Dow-Corning 3145 for reliable connections. Be aware that once cured, silicone sealants are very hard to remove from connectors — practically impossible.

20.3 The Transmission Line as Impedance Transformer

If the complex mechanics of reflections, SWR and line losses are put aside momentarily, a transmission line can very simply be considered as an impedance transformer. A certain value of load impedance, consisting of a resistance and reactance, at the end of the line is transformed into another value of impedance at the input of the line. The amount of transformation is determined by the electrical length of the line, its characteristic impedance, and by the losses inherent in the line. The input impedance of a real, lossy transmission line is computed using the following equation

$$Z_{in} = Z_0 \times \frac{Z_L \cosh(\eta \ell) + Z_0 \sinh(\eta \ell)}{Z_L \sinh(\eta \ell) - Z_0 \cosh(\eta \ell)} \quad (12)$$

where

Z_{in} = complex impedance at input of line = $R_{in} \pm j\,X_{in}$,

Z_L = complex load impedance at end of line = $R_L \pm j\,X_L$,

Z_0 = characteristic impedance of line = $R_0 \pm j\,X_0$,

η = complex loss coefficient = $\alpha + j\,\beta$,

α = matched line loss attenuation constant, in nepers/unit length (1 neper = 8.688 dB, so divide line loss in dB per unit length by 8.688),

β = phase constant of line in radians/unit length (multiply electrical length in degrees by 2π radians/360 degrees),

ℓ = electrical length of line in same units of length as used for α.

This and other complex equations describing the electrical behavior of transmission lines were traditionally solved through the use of Smith charts. (Smith charts are discussed in the article "Smith Chart Supplement" with the online content. Many references to Smith charts and their use may be found on-line.) While the Smith chart is an extremely effective way of visualizing and understanding the transmission line, using it directly in design has been replaced by software. Programs such as those mentioned in the sidebar "Smith Chart Software" can perform the numerical calculations and display the results on a Smith chart. The software *TLW* provided with the *ARRL Antenna Book* also solves problems of this nature, although without Smith chart graphics.

20.3.1 Transmission Line Stubs

The impedance transformation properties of a transmission line are useful in a number of applications. If the terminating resistance is zero (that is, a short) at the end of a low-loss transmission line which is less than a quarter wave long, the input impedance consists of a reactance, which is given by a simplification of equation 12.

$$X_{in} \cong Z_0 \tan \ell \quad (13)$$

If the line termination is an open circuit, the input reactance is given by

$$X_{in} \cong Z_0 \cot \ell \quad (14)$$

The input of a short (less than $\lambda/4$) length of line with a short circuit as a terminating load appears as an inductance, while an open-circuited line appears as a capacitance. This is a useful property of a transmission line, since it can be used as a low-loss inductor or capacitor in matching networks. Such lines are often referred to as stubs.

A line that is electrically $\lambda/4$ long is a special kind of a stub. When a $\lambda/4$ line is short circuited at its load end, it presents an open circuit at its input. Conversely, a $\lambda/4$ line with an open circuit at its load end presents a short circuit at its input. Such a line inverts the impedance of a short or an open circuit at the frequency for which the line is λ/4 long. This is also true for frequencies that are odd multiples of the λ/4 frequency. However, for frequencies where the length of the line is λ/2, or integer multiples thereof, the line duplicates the termination at its end.

20.3.2 Transmission Line Stubs as Filters

The impedance transformation properties of stubs can be put to use as filters. For example, if a shorted line is cut to be λ/4 long at 7.1 MHz, the impedance looking into the input of the cable is an open circuit. The line has no effect if placed in parallel with a transmitter's output terminals. However, at twice the *fundamental* frequency, 14.2 MHz, that same line is now λ/2, and looks like a short circuit. The line, often dubbed a *quarter-wave stub* in this application, acts as a trap for not only the second harmonic, but also for higher even-order harmonics, such as the fourth or sixth harmonics.

This filtering action is extremely useful in multitransmitter situations, such as Field Day, emergency operations centers, portable communications facilities and multioperator contest stations. Transmission line stubs can operate at high power where lumped-constant filters would be expensive. Using stub filters reduces noise, harmonics and strong fundamental signals from the closely spaced antennas that cause overload and interference to receivers. (For information on determining isolation between radios and filter requirements, see the online article "Measuring Receiver Isolation" by W2VJN in the online content.) Vector impedance analyzers (VIA) are discussed in the **Test Equipment and Measurements** chapter.

Quarter-wave stubs made of good quality coax, such as RG-213, offer a convenient way to lower transmitter harmonic levels. Despite the fact that the exact amount of harmonic attenuation depends on the impedance (often unknown) into which they are working at the harmonic frequency, a quarter-wave stub typically yields 20 to 25 dB of attenuation of the second harmonic when placed directly at the output of a transmitter feeding common amateur antennas. Remember to account for the length added by the required T adapter. If a measurement is made with and without the T adapter, the average of the two measurements should be very close to correct.

This attenuation may be a bit higher when the output impedance of the plate tuning network in an amplifier increases with frequency, such as for a pi-L network. Attenuation may be a bit lower if the tuning network's output impedance decreases with frequency, such as for a pi network.

Because different manufacturing runs of coax have slightly different velocity factors, a quarter-wave stub is usually cut a little longer than calculated, and then carefully pruned by snipping off short pieces, while using an antenna analyzer to monitor the response at the fundamental frequency. Because the end of the coax is an open circuit while pieces are being snipped away, the input of a λ/4 line shows a short circuit exactly at the fundamental frequency. Once the coax has been pruned to frequency, a short jumper is soldered across the end, and the response at the second harmonic frequency is measured. **Figure 20.9** shows how to connect a shorted stub to a transmission line and **Figure 20.10A** shows a typical frequency response.

The shorted quarter-wave stub shows low loss at 7 MHz, and also at the third harmonic on 21 MHz. It nulls 14 MHz, 28 MHz, and all other even harmonics. It can be used for a 21 MHz transmitter as well, and reduces any spurious emissions such as phase noise and wideband noise that might cause interference to receivers operating on 14 or 28 MHz.

The open-circuited quarter-wave stub has a low impedance at the fundamental frequency, so it must be used at two times the frequency for which it is cut. For example, a quarter-wave open stub cut for 3.5 MHz presents a high impedance at 7 MHz where it is λ/2 long. It presents a high impedance at those frequencies where it is a multiple of λ/2, or 7, 14 and 28 MHz. It would be connected in the same manner as shown in Figure 20.9, and the frequency plot is shown in Figure 20.10B.

This open stub can protect a receiver operating on 7, 14, 21, or 28 MHz from interference by a 3.5 MHz transmitter. It also has nulls at 10.5, 17.5 and 24.5 MHz, the 3rd, 5th, and 7th harmonics. The length of a quarter-wave stub may be calculated as follows:

$$L_e = \frac{VF \times 983.6}{4f} \quad (15)$$

where

L_e = length in ft,

VF = propagation constant for the coax in use, and

f = frequency in MHz.

For the special case of RG-213 (and any similar cable with VF = 0.66), equation 15 can be simplified to:

$$L_e = 163.5 / f \quad (16)$$

where

L_e = length in ft, and

f = frequency in MHz.

Table 20.3 solves this equation for the major contesting bands where stubs are often used. The third column shows how

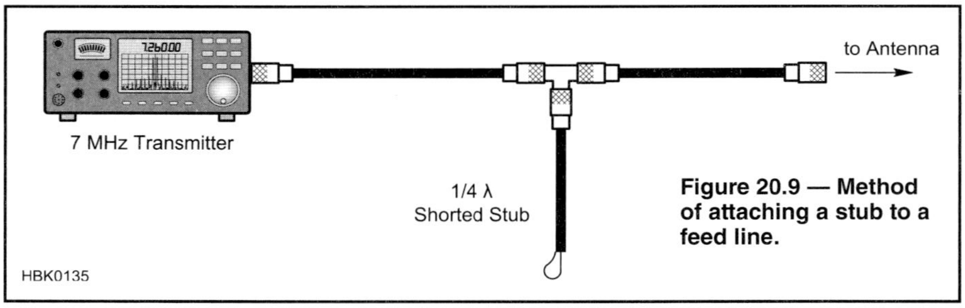

Figure 20.9 — Method of attaching a stub to a feed line.

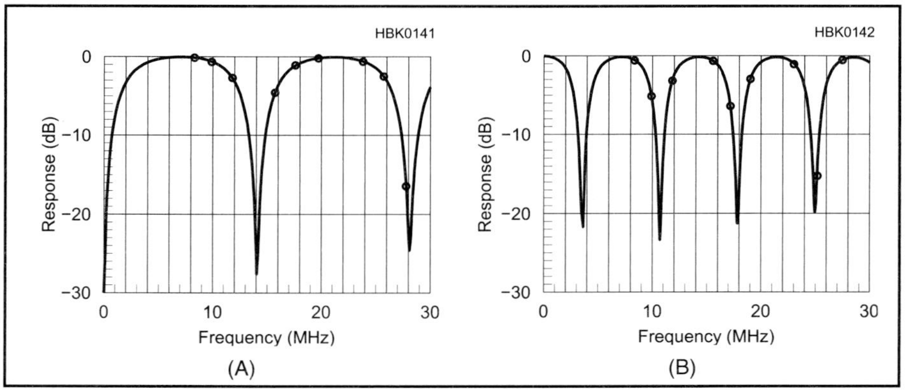

Figure 20.10 — Frequency response with a shorted stub (A) and an open stub (B).

much of the stub to cut off if the desired frequency is 100 kHz higher in frequency. For example: To cut a stub for 14.250 MHz, reduce the overall length shown by 2.5 × 1 inches, or 2.5 inches. There is some variation in dielectric constant of coaxial cable from batch-to-batch or manufacturer-to-manufacturer, so it is always best to measure the stub's fundamental resonance before proceeding.

Some analyzer software can measure phase shift and indicate how much needs to be trimmed to achieve the desired amount. Be cautious when using this information to make large changes in line length. Because VF varies with frequency as described earlier, stubs should be measured as close as practical to the frequency at which they are intended to operate. The highest accuracy is achieved by computing the frequency nearest the operating frequency at which the stub should look like a short circuit, then trimming it to present the short circuit at that frequency.

Table 20.3

Quarter-Wave Stub Lengths for the HF Contesting Bands

Freq (MHz)	*Length (L_e)**	*Cut off per 100 kHz*
1.8	90 ft, 10 in	57⅜ in
3.5	46 ft, 9 in	15½ in
7.0	23 ft, 4 in	4 in
14.0	11 ft, 8 in	1 in
21.0	7 ft, 9 in	7/16 in
28.0	5 ft, 10 in	¼ in

*Lengths shown are for RG-213 and any similar cable, assuming a 0.66 velocity factor (L_e = 163.5/f). See text for other cables.

CONNECTING STUBS

Stubs are usually connected in the antenna feed line close to the transmitter. They may also be connected on the antenna side of a switch used to select different antennas. Some small differences in the null depth may occur for different positions.

To connect a stub to the transmission line it is necessary to insert a coaxial T (as shown in Figure 20.9). If a female-male-female T is used, the male can connect directly to the transmitter while the antenna line and the stub connect to the two females. It should be noted that the T inserts a small additional length in series with the stub that lowers the resonant frequency. The additional length for an Amphenol UHF T is about 3/8 inch. This length is negligible at 1.8 and 3.5 MHz, but on the higher bands it should not be ignored.

MEASURING STUBS WITH ONE-PORT METERS

Many of the common measuring instruments used by amateurs are *one-port devices*, meaning they have one connector at which the measurement, typically VSWR, is made. Probably the most popular instrument for this type of work is the antenna analyzer, available from a number of manufacturers.

To test a stub using an antenna analyzer, connect the stub to the meter by itself and tune the meter for a minimum impedance value, ignoring the VSWR setting. It is almost impossible to get an accurate reading on the higher HF bands, particularly with open stubs. In one test, when a quarter-wave open stub cut for 20 meters was nulled on an MFJ-259 SWR analyzer, the frequency of the very broad null was 14.650 MHz. A recheck with a professional quality network analyzer measured 14.018 MHz. (Resolution on the network analyzer was about ±5 kHz.) Running the same test on a quarter-wave shorted stub gave a measurement of 28.320 MHz on the MFJ-259 and 28.398 MHz on the network analyzer. These inaccuracies are typical of amateur instrumentation that may cost thousands of dollars less than professional devices. Be aware of the potential inaccuracies when using inexpensive instruments for precise measurements.

Other one-port instruments that measure phase can be used to get a more accurate reading. The additional length added by the required T adapter must be accounted for. If the measurement is made without the T and then with the T, the average of the two measurements should be close to correct.

MEASURING STUBS WITH TWO-PORT INSTRUMENTS

A two-port measurement is made with a signal generator and a separate detector. A T connector is attached to the generator with the stub connected to one side. The other side is connected to a cable of any length that goes to the detector. The detector should present a 50 Ω load to the cable. This is how a network analyzer is configured, and it is similar to how the stub is connected in actual use. If the generator is calibrated accurately, the measurement can be very good. There are a number of ways to do this without buying an expensive piece of lab equipment.

An antenna analyzer can be used as the signal generator. Measurements are quite accurate if the detector has 30 to 40 dB dynamic range. Two setups were tested for accuracy. The first used a digital voltmeter with a diode detector. (A germanium diode,1N34A or equivalent, gives the best dynamic range.) Tests on open and shorted stubs at 14 MHz returned readings within 20 kHz of the network analyzer. Another test was run using an oscilloscope as the detector with a 50 Ω load on the input. This test produced results that were essentially the same as the network analyzer.

A noise generator can be used in combination with a receiver as the detector. (An inexpensive noise generator kit is available from Elecraft, **www.elecraft.com**.) Set the receiver for 2 – 3 kHz bandwidth and turn off the AGC. An ac voltmeter connected to the audio output of the receiver serves as a null detector. The noise level into the receiver without the stub connected should be just at or below the limiting level. With the stub connected, the noise level in the null should drop by 25 or 30 dB. Connect the UHF T to the noise generator using any necessary adapters. Connect the stub to one side of the T and connect the receiver to the other side with a short cable. Tune the receiver around the expected null frequency. After locating the null, snip off pieces of cable until the null moves to the desired frequency. Accuracy with this method is within 20 or 30 kHz of the network analyzer readings on 14 MHz stubs.

A *vector network analyzer,* or VNA, is another type of two-port device. VNAs can perform both reflection measurements as a one-port device and transmission measurements through a device or system at two locations. Once only available as expensive lab instruments, a number of inexpensive VNA designs that use a PC to perform display and calculations are now available to amateurs. The use of VNAs is discussed in the **Test Equipment and Measurements** chapter of this book.

STUB COMBINATIONS

A single stub gives 20 to 30 dB attenuation in the null. If more attenuation is needed, two or more similar stubs can be combined. Best results are obtained if a short coupling cable is used to connect the two stubs rather than connecting them directly in parallel. The stubs may be cut to the same frequency for maximum attenuation, or to two slightly different frequencies such as the CW and SSB frequencies in one band. Open and shorted stubs can be combined together to attenuate higher harmonics as well as lower frequency bands.

An interesting combination is the parallel connection of two λ/8 stubs, one open and the other shorted. The shorted stub acts as an inductor and the open stub as a capacitor. Their reactances are equal and opposite, forming a resonant circuit. The null depth with this arrangement is a bit better than a single quarter-wave shorted stub. This presents some possibilities when combinations of stubs are used in a band switching system.

20.3.3 Project: A Field Day Stub Assembly

Figure 20.11 shows a simple stub arrangement that can be useful in a multi-transmitter Field Day station. The stubs reduce out-of-band noise produced by the transmitters that would cause interference to the other stations — a common Field Day problem where the stations are quite close together. This noise cannot be filtered out at the receiver and must be removed at the transmitter. One stub assembly would be connected to each transmitter output and manually switched for the appropriate band.

Two stubs are connected as shown. The two-relay selector box can be switched in four ways. Stub 1 is a shorted quarter-wave 40-meter stub. Stub 2 is an open quarter-wave 40-meter stub. Operation is as shown in **Table 20.4**.

The stubs must be cut and tuned while connected to the selector relays. RG-213 may be used for any amateur power level and provides 25 to 30 dB reduction in the nulls. For power levels under 500 W or so, RG-8X may be used. It provides a few dB less reduction in the nulls because of its higher loss than RG-213.

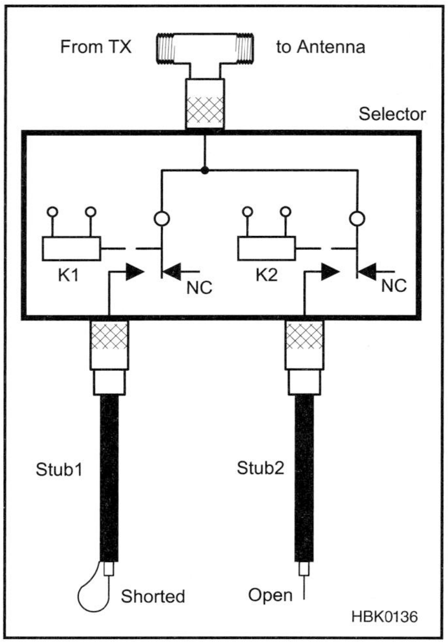

Figure 20.11 — Schematic of the Field Day stub switching relay control box. Table 20.4 shows which relays should be closed for the desired operating band.

Table 20.4
Stub Selector Operation
See Figure 20.11 for circuit details.

Relay K1 Position	*Relay K2 Position*	*Bands Passed (meters)*	*Bands Nulled (meters)*
Open	Open	All	None
Energized	Energized	80	40, 20, 15, 10
Energized	Open	40, 15	20, 10
Open	Energized	20, 10	40, 15

20.4 Matching Impedances in the Antenna System

Only in a few special cases is the antenna impedance the exact value needed to match a practical transmission line. In all other cases, it is necessary either to operate with a mismatch and accept the SWR that results, or else to bring about a match between the line and the antenna.

When transmission lines are used with a transmitter, the most common load is an antenna. When a transmission line is connected between an antenna and a receiver, the receiver input circuit is the load, not the antenna, because the power taken from a passing wave is delivered to the receiver.

Whatever the application, the conditions existing at the load, and *only* the load, determine the reflection coefficient, and hence the SWR, on the line. When the load is purely resistive and equal to the characteristic impedance of the line, there are no standing waves. When the load is not purely resistive, or is not equal to the line Z_0, there are standing waves. No adjustments can be made at the input end of the line to change the SWR at the load. Neither is the SWR affected by changing the line length. The SWR can appear to decrease, even though it does not change in reality, when measured at the input of a lossy line, as described previously.

20.4.1 Conjugate Matching

Technical literature sometimes uses the term *conjugate match* to describe the condition where the impedance seen looking toward the load from any point on the line is the complex conjugate of the impedance seen looking toward the source. This means that the resistive and reactive magnitudes of the impedances are the same, but that the reactances have opposite signs. The complex conjugate of $20 + j\,75$ is $20 - j\,75$. The complex conjugate of a purely resistive impedance, such as $50 + j\,0\ \Omega$, is the same impedance, $50 + j\,0\ \Omega$. A conjugate match is necessary to achieve the maximum possible power transfer from most signal sources.

For example, if 50 feet of RG-213 is terminated in a $72 - j\,34\ \Omega$ antenna impedance, the transmission line transforms the impedance to $35.9 - j\,21.9\ \Omega$ at its input. (The *TLW* program is used to calculate the impedance at the line input.) To create a conjugate match at the line input, a matching network would have to present an impedance of $35.9 + j\,21.9\ \Omega$. The system would then become resonant, since the $\pm j\,21.9\ \Omega$ reactances would cancel, leaving $35.9 + j\,0\ \Omega$. A conjugate match is not the same as transforming one impedance to another, such as from $35.9 - j\,0\ \Omega$ to $50 + j\,0\ \Omega$. An additional impedance transformation network would be required for that to occur.

Conjugate matching is often used for small-signal amplifiers, such as preamplifiers at VHF and above, to obtain the best power gain. The situation with high-power amplifiers is complex and there is considerable discussion as to whether conjugate matching delivers the highest efficiency, gain and power output. Nevertheless, conjugate matching is the model most often applied to impedance matching in antenna systems.

The concept of conjugate impedance matching is explored in more detail by Lou Ernst, WA2GKH, in an online tutorial on this book's web page. The PDF presentation and *Excel* spreadsheet tool are available in the Software section of the web page at **arrl.org/arrl-handbook-reference**.

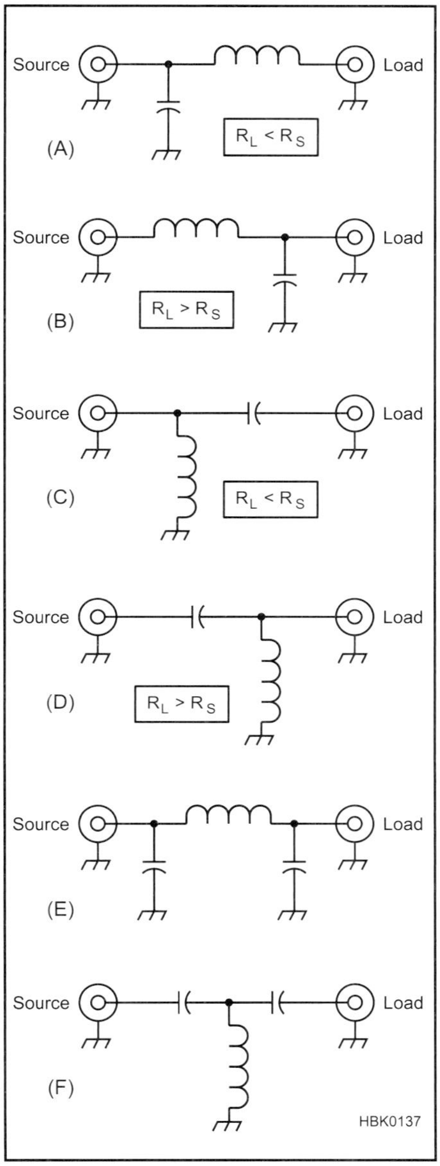

Figure 20.12 — Matching network variations. (A) through (D) show L networks. (E) is a π network, equivalent to a pair of L networks sharing a common series inductor. (F) is a T network, equivalent to a pair of L networks sharing a common parallel inductor.

20.4.2 Impedance Matching Networks

When all of the components of an antenna system — the transmitter, feed line, and antenna — have the same impedance, almost all of the power generated by the transmitter is transferred to the antenna and SWR is 1:1. This is rarely the case, however, as antenna feed point impedances vary widely with frequency and design. This requires some method of *impedance matching* between the various antenna system components.

Many amateurs use an *impedance-matching unit* or "*antenna tuner*" between their transmitter and the transmission line feeding the antenna. (This is described in a following section.) The antenna tuner's function is to transform the impedance, whatever it is, at the transmitter end of the transmission line into the 50 Ω required by the transmitter. Remember that the use of an antenna tuner at the transmitter does not tune the antenna, reduce SWR on the feed line, or reduce feed line losses!

Some matching networks are built directly into the antenna (for example, the gamma and beta matches) and these are discussed in the **Antennas** chapter and in *The ARRL Antenna Book*. Impedance matching networks made of fixed or adjustable components can also be used at the antenna and are particularly useful for antennas that operate on a single band.

Remember, however, that impedance can be transformed anywhere in the antenna system to match any other desired impedance. A variety of techniques can be used as described in the following sections, depending on the circumstances.

An electronic circuit designed to convert impedance values is called an *impedance matching network*. The most common impedance matching network circuits for use in systems that use coax cable are:

1) The low-pass L network.
2) The high-pass L network.
3) The low-pass Pi network.
4) The high-pass T network.

Basic schematics for each of the circuits are shown in **Figure 20.12**. As shown in **Table 20.5**, the L networks can be reversed if matching does not occur in one direction. L networks are the most common for single-band antenna matching. The component in parallel is the *shunt* component, so the L networks with the shunt capacitor or inductor at the inputs (Figs 20.12A and 20.12C) are *shunt-input* networks and the others are *series-input* networks.

Impedance matching circuits can use fixed-value components for just one band when a particular antenna has an impedance that is too high or low, or they can be made to be adjustable when matching is needed on several bands, such as for matching a dipole antenna fed with open-wire line.

Additional material by Bill Sabin, WØIYH, on matching networks can be found in the online content along with his program *MATCH*. The supplemental article "Using *TLW* to Design Impedance Matching Networks" by W2VJN is also in the online content.

DESIGNING AN L NETWORK

The L network, shown in Figures 20.12A through 20.12D, only requires two components and is a particularly good choice of matching network for single-band antennas. The L network is easy to construct so that it can be mounted at or near the feed point of the antenna, resulting in a low SWR on the transmission line to the shack. (Note that L networks as well as Pi and T networks can easily be designed with the *TLW* software.)

To design an L network, both the source and load impedances must be known. Let us assume that the source impedance, R_S, is 50 Ω, representing the transmission line to the transmitter, and that the load is an arbitrary value, R_L.

First, determine the circuit Q.

$$Q = \sqrt{\frac{R_L}{50} - 1} \qquad (17)$$

Next, select the type of L network you want from Figure 20.12. Note that the parallel component is always connected to the higher of the two impedances, source or load. Your choice should take into account whether either the source or load require

Table 20.5
Network Performance

Figure 20.12 Section	*Circuit Type*	*Match*	*Harmonic Attenuation?*
(A)	Low-pass L network	Lower	Fair to good
(B)	Reverse Low-pass L network	Higher	Fair to good
(C)	High-pass L network	Lower	No
(D)	Reverse high-pass L network	Higher	No
(E)	Low-pass Pi network	Lower and Higher	Good
(F)	High-pass T network	Lower and higher	No

a dc ground (parallel-L or shunt-L) and whether it is necessary to have a dc path through the network, such as to power a remote antenna switch or other such device (parallel-C or shunt-C). Once you have selected a network, calculate the values of X_L and X_C:

$$X_L = Q\,R_S \tag{18}$$

and

$$X_C = \frac{R_L}{Q} \tag{19}$$

As an example, we design an L network to match a 300 Ω antenna (R_L) to a 50 Ω transmission line (R_S). $R_L > R_S$ so we can select either Figure 20.12B or Figure 20.12D. The network in B is a low-pass network and attenuates harmonics, so that is the usual choice.

$$Q = \sqrt{\frac{300}{50} - 1} = 2.236$$

$$X_L = 50 \times 2.236 = 112\ \Omega$$

$$X_C = \frac{300}{2.236} = 134\ \Omega$$

If the network is being designed to operate at 7 MHz, the actual component values are:

$$L = \frac{X_L}{2\pi f} = 2.54\ \mu H$$

$$C = \frac{1}{2\pi\ f\ X_C} = 170\ pF$$

The components could be fixed-value or adjustable. By running the *TLW* software, additional information is obtained: At 1500 W there is 942 V peak across the capacitor and 5.5 A flowing in the inductor.

The larger the ratio of the impedances to be transformed, the higher Q becomes. High values of Q (10 or more) may result in impractically high or low component values. In this case, it may be easier to design the matching network as a pair of L networks back-to-back that accomplish the match in two steps. Select an intermediate value of impedance, R_{INT}, the geometric mean between R_L and R_S:

$$R_{INT} = \sqrt{R_L R_S}$$

Construct one L network that transforms R_L to R_{INT} and a second L network that transforms R_{INT} to R_S.

20.4.3 Matching Antenna Impedance at the Antenna

This section describes methods by which a network can be installed at or near the antenna feed point to provide matching to a transmission line. Having the matching system at the antenna rather than down in the shack at the end of a long transmission line does seem intuitively desirable, but it is not always very practical, especially in multiband antennas.

RESONATING THE ANTENNA

If a highly reactive antenna can be tuned to resonance, even without special efforts to make the resistive portion equal to the line's characteristic impedance, the resulting SWR is often low enough to minimize additional line loss. For example, the multiband 100 ft long flat-top antenna in **Figure 20.13** has a feed point impedance of 4.18 – *j* 1590 Ω at 1.8 MHz. Assume that the antenna reactance is tuned out with a network consisting of two symmetrical inductors whose reactance is + *j* 1590/2 = *j* 795 Ω each, with a Q of 200. These inductors are 70.29 µH coils in series with inherent loss resistances of 795/200 = 3.98 Ω. The total series resistance is thus 4.18 + 2 × 3.98 = 12.1 Ω. If placed in series with each leg of the antenna at the feed point, as in the figure, the antenna reactance and inductor reactance cancel out, leaving a purely resistive impedance at the antenna feed point.

If this tuned system is fed with 50 Ω coaxial cable, the SWR is 50/12.1 = 4.13:1, and the loss in 100 ft of RG-213 cable would be 0.48 dB. The antenna's radiation efficiency is the ratio of the antenna's radiation resistance (4.18 Ω) to the total feed point resistance including the matching coils (12.1 Ω), so efficiency is 4.18/12.1 = 34.6% which is equivalent to 4.6 dB of loss compared to a 100% efficient antenna. Adding the 0.48 dB of loss in the line yields an overall system loss of 5.1 dB. Compare this to the loss of 26 dB if the RG-213 coax is used to feed the antenna directly, without any matching at the antenna. The use of moderately high-Q resonating inductors has yielded almost 21 dB of "gain" (that is, less loss) compared to the case without the inductors. The drawback of course is that the antenna is now resonant on only one frequency, but it certainly is a lot more efficient on that frequency!

THE QUARTER-WAVE TRANSFORMER OR "Q" SECTION

The range of impedances presented to the transmission line is usually relatively small on a typical amateur antenna, such as a dipole or a Yagi when it is operated close to resonance. In such antenna systems, the impedance-transforming properties of a λ/4 section of transmission line are often utilized to match the transmission line at the antenna.

If the antenna impedance and the characteristic impedance of a feed line to be matched are known, the characteristic impedance needed for a quarter-wave matching section of low-loss cable is expressed by another simplification of equation 12.

$$Z = \sqrt{Z_L Z_0} \tag{20}$$

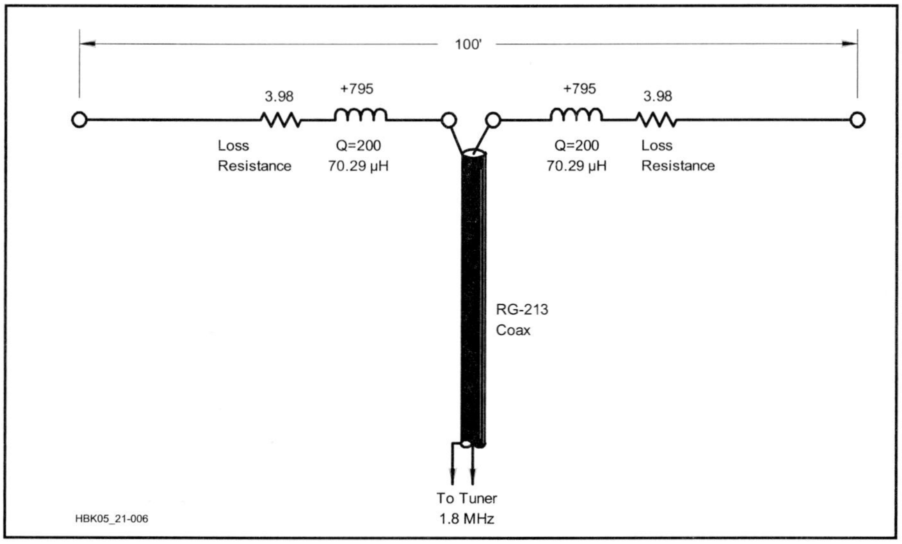

Figure 20.13 — The efficiency of a short dipole can be improved at 1.8 MHz with a pair of inductors inserted symmetrically at the feed point. Each inductor is assumed to have a Q of 200. By resonating the dipole in this fashion the system efficiency, when fed with RG-213 coax, is about 21 dB better than using this same antenna without the resonator. The disadvantage is that the formerly multiband antenna can only be used on a single band.

where

Z = characteristic impedance needed for matching section,

Z_L = antenna impedance, and

Z_0 = characteristic impedance of the line to which it is to be matched.

Such a matching section is called a *synchronous quarter-wave transformer* or a *quarter-wave transformer.* The term synchronous is used because the match is achieved only at the length at which the matching section is exactly λ/4 long. This limits the use of λ/4 transformers to a single band. Different λ/4 lengths of line are required on different bands.

Figure 20.14 shows one example of this technique to feed an array of stacked Yagis on a single tower. Each antenna is resonant and is fed in parallel with the other Yagis, using equal lengths of coax to each antenna called *phasing lines.* A stacked array is used to produce not only gain, but also a wide vertical elevation pattern, suitable for coverage of a broad geographic area. (See *The ARRL Antenna Book* for details about Yagi stacking.)

The feed point impedance of two 50 Ω Yagis fed with equal lengths of feed line connected in parallel is 25 Ω (50/2 Ω); three in parallel yield 16.7 Ω; four in parallel yield 12.5 Ω. The nominal SWR for a stack of four Yagis is 4:1 (50/12.5). This level of SWR does not cause excessive line loss, provided that low-loss coax feed line is used. However, many station designers want to be able to select, using relays, any individual antenna in the array, without having the load seen by the transmitter change. (Perhaps they might wish to turn one antenna in the stack in a different direction and use it by itself.) If the load changes, the amplifier must be retuned, an inconvenience at best.

Example: To match a 50 Ω line to a Yagi stack consisting of two antennas fed in parallel, resulting in a 25 Ω load, the quarter-wave matching section would require a characteristic impedance of

$$Z = \sqrt{50 \times 25} = 35.4\ \Omega$$

RG-83 is a 35 Ω cable suitable for this application. A transmission line with a characteristic impedance of 35 Ω could be closely approximated by connecting two equal λ/4 sections of 75 Ω cable (such as RG-11A) in parallel to yield the equivalent of a 37.5 Ω cable. Three Yagis fed in parallel would require a λ/4 transformer made using a cable having a characteristic impedance of

$$Z = 50 \times 16.7 = 28.9\ \Omega$$

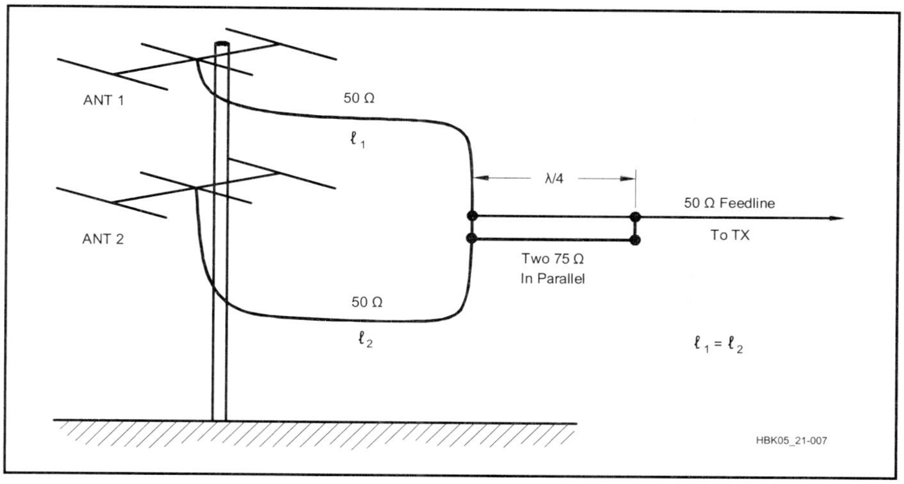

Figure 20.14 — Array of two stacked Yagis, illustrating use of ¼ λ matching sections. At the junction of the two equal lengths of 50 Ω feed line the impedance is 25 Ω. This is transformed back to 50 Ω by the two paralleled 75 Ω, ¼ λ lines, which together make a net characteristic impedance of 37.5 Ω. This is close to the 35.4 Ω value computed by the formula.

This is approximated by using a λ/4 section of 50 Ω cable in parallel with a λ/4 section of 75 Ω cable, yielding a net impedance of 30 Ω, quite close enough to the desired 28.9 Ω. Four Yagis fed in parallel would require a λ/4 transformer made up using cable with a characteristic impedance of 25 Ω, easily created by using two 50 Ω cables in parallel.

The 100-foot flat-top example in the previous section with the two resonating coils has an impedance of 12 Ω at the feed point. Two RG-58A cables, each λ/4 long at 1.8 MHz (90 ft) can be connected in parallel to feed this antenna. An additional 10 ft length of RG-213 can make up the required 100 ft. The match is almost perfect. The disadvantage of this system is that it limits the operation to one band, but the overall efficiency is quite good.

Another use of λ/4 transformers is in matching the impedance of full-wave loop antennas to 50 Ω coax. For example, the driven element of a quad antenna or a full-wave 40-meter loop has an impedance of 100 to 150 Ω. Using a λ/4 transformer made from 62, 75, or 93 Ω coaxial cable would lower the line SWR to a level where losses are insignificant.

TWELFTH-WAVE TRANSFORMERS

The Q-section is really a special case of series-section impedance matching. (See the *ARRL Antenna Book* for an extended discussion of the technique.) A particularly handy two-section variation shown in **Figure 20.15** can be used to match two different impedances of transmission line, such as 50 Ω coax and 75 Ω hardline. Sections with special impedances are not required, only

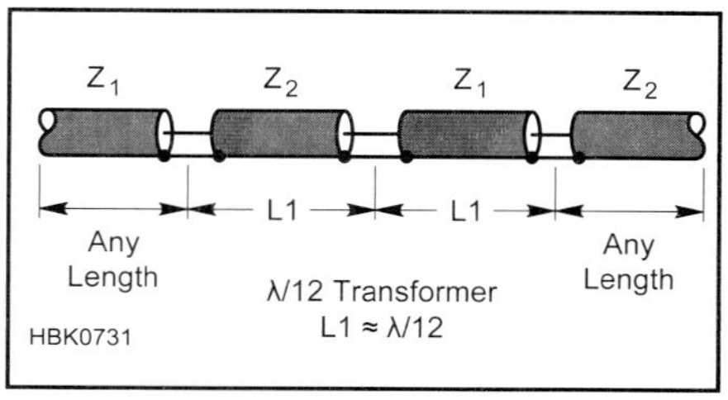

Figure 20.15 — A two-section twelfth-wave transformer can be used to match two different feed line impedances. This is a common technique for using surplus 75 Ω hardline in 50 Ω systems.

sections of the two feed lines to be matched. (Also see the sidebar "Using 75 Ω Hardline in 50 Ω Systems.")

This configuration is referred to as a *twelfth-wave transformer* because when the ratio of impedances to be matched is 1.5:1 (as is the case with 50 and 75 Ω cables) the electrical length of the two matching sections between the lines to be matched is 0.0815 λ (29.3°), quite close to λ/12 (0.0833 λ or 30°). As with all synchronous transformers, the matching function works on only one band.

MATCHING TRANSFORMERS

Another matching technique uses wideband toroidal transformers. Transformers can be made that operate over very wide frequency ranges and that match various impedances.

A very simple matching transformer consists of three windings connected in series as shown in **Figure 20.16A**. The physical arrangement of the three windings is shown in Figure 20.13B. This arrangement gives the best bandwidth. **Figure 20.17** shows in Figure 20.17 has an SWR bandwidth of 1.5 MHz to more than 30 MHz. The 51 pF capacitor is shunted to ground at the low-impedance input to cancel unwanted stray inductive reactance.

This is a good way to stack two or three triband antennas. If they have the same length feed lines and they all point the same way, their patterns add and some gain results. However, they don't even need to be on the same tower or pointed in the same direction or fed with the same length lines to have some benefit. Even dissimilar antennas can show a benefit when connected together in this fashion.

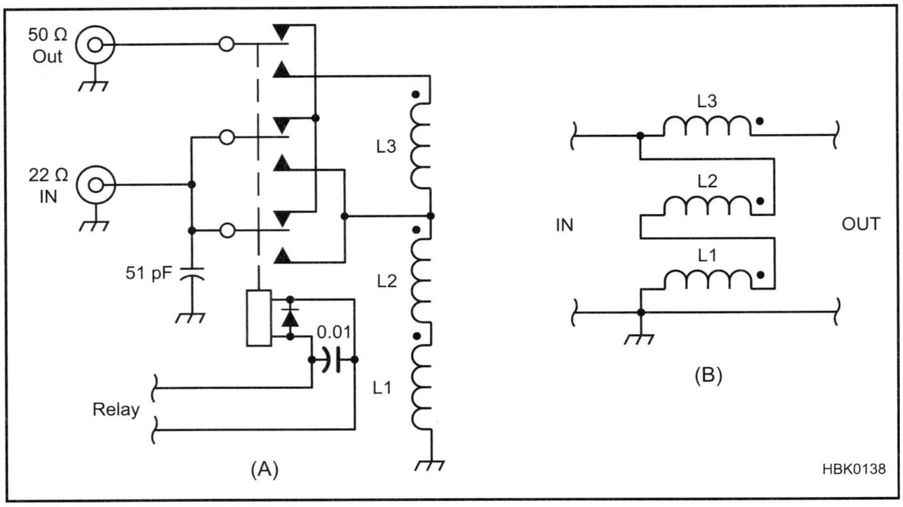

Figure 20.16 — Schematic for the impedance matching transformer described in the text. The complete schematic is shown at (A). The physical positioning of the windings is shown at (B).

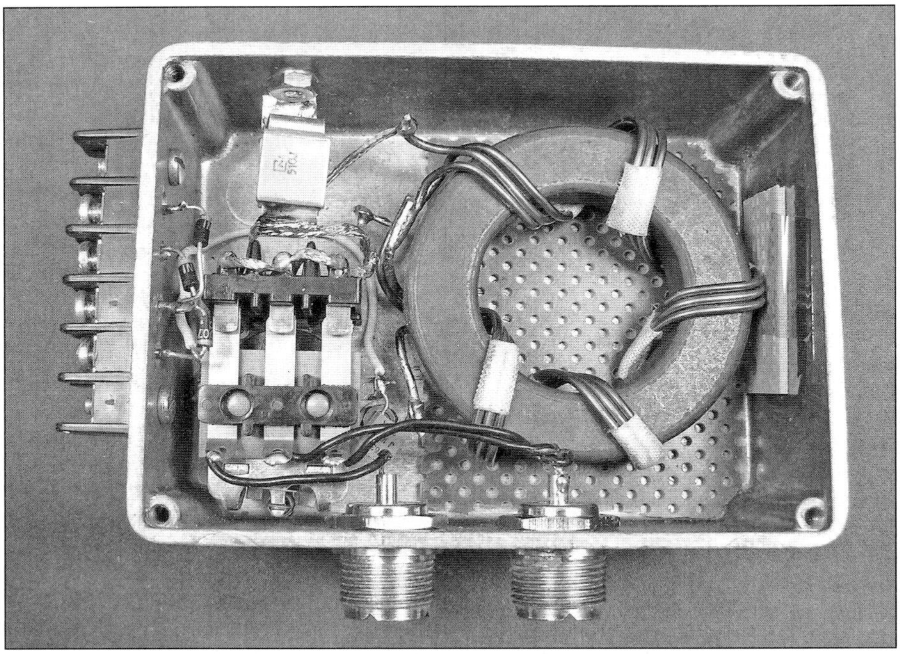

Figure 20.17 — The completed impedance matching transformer assembly for the schematic shown in Figure 20.13.

a picture of this type of transformer. An IN/OUT relay is included with the transformer. One relay pole switches the 50 Ω input port while two poles in parallel switch the 22 Ω port. Three 14-inch lengths of #14 AWG wire are taped together so they lie flat on the core. A #61 mix 2.4-inch diameter toroid core will handle full legal power.

Use magnet wire rated for high-voltage use, such as TEMCo GP/MR-200 or Superior Essex ALLEX wire, which is rated for use to 200 °C with 5.7 kV rating of the insulation. Motor repair shops may sell small quantities for single projects. Before winding on an uncoated toroid core, particularly ferrite, which is abrasive, wrap the core with one or two layers of high-quality electrical tape such as Scotch 88 to avoid damaging the wire coating. When winding the toroid, it is not necessary for the wire to be pulled tight against the core, which could cause additional stresses to the insulation.

The impedance ratio of this design is 1:2.25 or 22.2-to-50 Ω. This ratio turns out to work well for two or three 50 Ω antennas in parallel. Two in parallel give an SWR of 25/22.2 or 1.13:1. Three in parallel give an SWR of 22.2/16.7 or 1.33:1. The unit shown

20.4.4 Matching the Line to the Transmitter

So far we have been concerned mainly with the measures needed to achieve acceptable amounts of loss and a low SWR for feed lines connected to antennas. Not only is feed line loss minimized when the SWR is kept within reasonable bounds, but also the transmitter is able to perform as specified when connected to the load resistance it was designed to drive.

Most modern amateur transmitters have solid-state final amplifiers designed to work into a 50 Ω load with broadband, non-adjustable impedance matching networks. While the adjustable pi networks used in vacuum-tube amplifiers can accommodate a wide range of impedances, the final transistors in a solid-state transmitter can be damaged by too much reflected power from the antenna system and most utilize built-in protection circuitry that automatically reduces output power if the SWR rises above about 2:1.

Besides the rather limited option of using only inherently low-SWR antennas to ensure that the transmitter sees the load for which it was designed, an *impedance-matching unit* or *antenna tuner* (or "tuner" for short) can be used. The function of an antenna tuner is to transform the impedance at the input end of the transmission line, whatever that may be, to the 50 Ω value required by the transmitter for best performance.

Some solid-state transmitters incorporate automatic antenna tuners so that they too can cope with practical antennas and transmission lines. The range of impedances that can be matched by the built-in tuners is typically rather limited, however, especially at lower frequencies. Most built-in tuners specify a maximum SWR of 3:1 that can be transformed to 1:1.

Do not forget that a tuner does *not* alter the SWR on the transmission line between the tuner and the antenna; it only adjusts the impedance at the transmitter end of the feed line to the value for which the transmitter was designed. Other names for antenna tuners include *transmatch*, *impedance matcher*, *matchbox* or *antenna coupler*. Since the SWR on the transmission line between the antenna and the output of the antenna tuner is rarely 1:1, some loss in the feed line due to the mismatch is unavoidable, even though the SWR on the short length of line between the tuner and the transmitter is 1:1.

If separate feed lines are used for different bands, the tuner can be inserted in one feed line, tuned for best SWR, and left at that setting. If a particular antenna has a minimum SWR in the CW portion of a band and operation in the SSB end is desired, the tuner can be used for matching and switched out when not needed. Multiband operation generally requires retuning for each band in use.

Antenna tuners for use with balanced or open-wire feed lines include a balun or link-coupling circuit as seen in **Figure 20.18**. This allows a transmitter's unbalanced coaxial output to be connected to the balanced feed line. A fully balanced tuner has a symmetrical internal circuit with a tuner circuit for each side of the feed line and the balun at the input to the tuner where the impedance is close to 50 Ω. Most antenna tuners are unbalanced, however, with a balun located at the output of the impedance matching network, connected directly to the balanced feed line. At very high or very low impedances, the balun's power rating may be exceeded at high transmitted power levels.

Automatic antenna tuners use a microprocessor to adjust the value of the internal components. Some models sense high values of SWR and retune automatically, while others require the operator to initiate a tuning operation. Automatic tuners use a trial-and-error method to find the best settings. Most of them work by switching small capacitors in and out of the circuit with relays. When they tune, you can hear the fast clicking of the relays as the trial-and-error process proceeds. Automatic tuners are available for low- and high-power operation and generally handle the same values of impedance as their manually adjusted counterparts. One advantage of the automatic antenna tuner is that it can remember the settings to best tuning at each frequency and then automatically apply those settings as a transmitter is tuned across a band or to different bands.

THE T NETWORK

Over the years, radio amateurs have derived a number of circuits for use as tuners. The most common form of antenna tuner is some variation of a T network, as shown in **Figure 20.19A**. Note that the choke or current balun can be used at the input or output of the tuner to match balanced feed lines.

The T network can be visualized as being two L networks back to front, where the common element has been conceptually broken down into two inductors in parallel (see Figure 20.19B). The L network connected to the load transforms the output impedance Ra ± j Xa into its parallel equivalent by means of the series output capacitor C2. The first L network then transforms the parallel equivalent back into the series equivalent and resonates the reactance with the input series capacitor C1.

Note that the equivalent parallel resistance Rp across the shunt inductor can be a very large value for highly reactive loads, meaning that the voltage developed at this point can be very high. For example, assume that the load impedance at 3.8 MHz presented to the antenna tuner is Za = 20 – j 1,000. If C2 is 300 pF, then the equivalent parallel resistance across L1 is 66,326

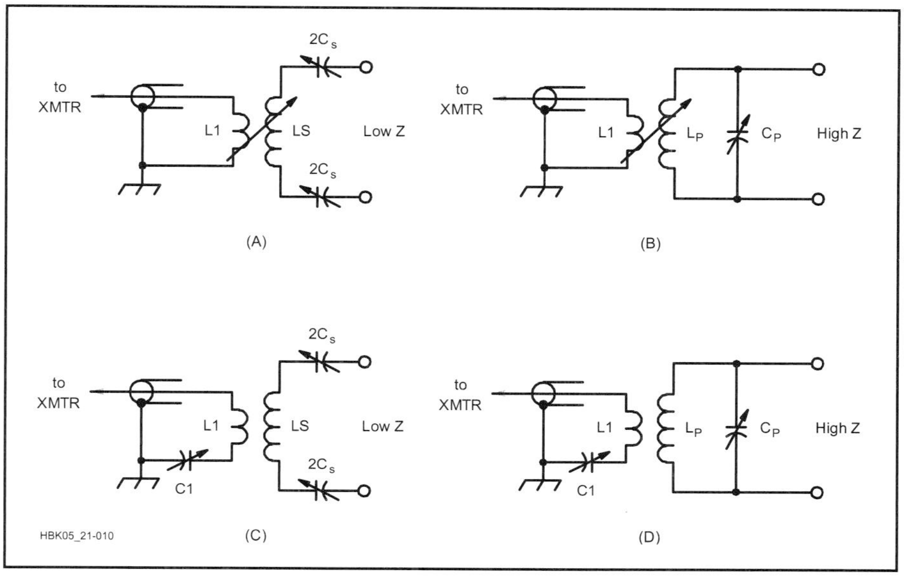

Figure 20.18 — Simple antenna tuners for coupling a transmitter to a balanced line presenting a load different from the transmitter's design load impedance, usually 50 W. (A) and (B), respectively, are series and parallel tuned circuits using variable inductive coupling between coils. (C) and (D) are similar but use fixed inductive coupling and a variable series capacitor, C1. A series tuned circuit works well with a low-impedance load; the parallel circuit is better with high impedance loads (several hundred ohms or more).

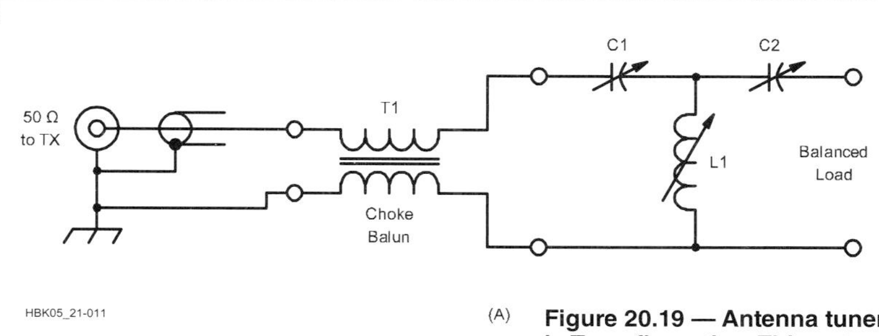

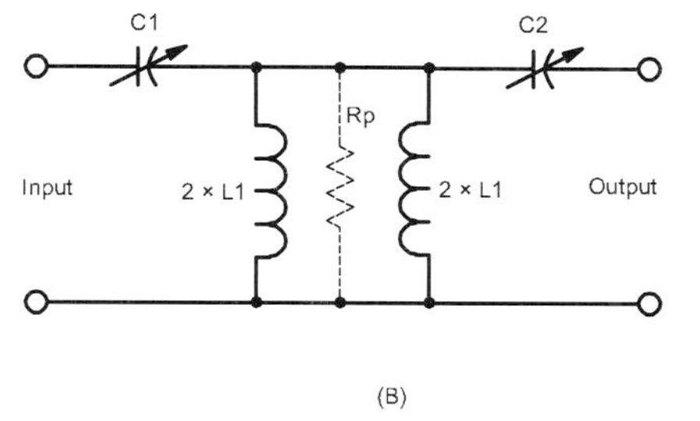

Figure 20.19 — Antenna tuner network in T configuration. This network has become popular because it has the capability of matching a wide range of impedances. At (A), the balun transformer at the input of the antenna tuner preserves balance when feeding a balanced transmission line. The balun can be placed at the input or output of the tuner. At (B), the T configuration is shown as two L networks back to back. (in the L network version, the two ½-value inductors replacing L1 are assumed to be adjustable with identical values).

W. If 1,500 W appears across this parallel resistance, a peak voltage of 14,106 V is produced, a very substantial level indeed. Highly reactive loads can produce very high voltages across components in a tuner.

The ARRL computer program *TLW* calculates and shows graphically the antenna-tuner values for operator-selected antenna impedances transformed through lengths of various types of practical transmission lines. (See the supplemental article "Using *TLW* to Design Impedance Matching Networks" with the online content.) *The ARRL Antenna Book* contains detailed information on tuner design and construction.

ANTENNA TUNER LOCATION

The tuner is usually located near the transmitter in order to adjust it for different bands or antennas. If a tuner is in use for one particular band and does not need to be adjusted once set up for minimum SWR, it can be placed in a weatherproof enclosure near the antenna. Some automatic tuners are designed to be installed at the antenna, for example. For some situations, placing the tuner at the base of a tower can be particularly effective and eliminates having to climb the tower to perform maintenance on the tuner.

It is useful to consider the performance of the entire antenna system when deciding where to install the antenna tuner and what types of feed line to use in order to minimize system losses. Here is an example, using the program TLW. Let's once again assume a flat-top antenna 50 ft high and 100 ft long and not resonant on any amateur band. As extreme examples, we use 3.8 and 28.4 MHz with 200 ft of transmission line. There are many ways to configure this system, but three examples are shown in **Figure 20.20**.

Example 1 in Figure 20.20A shows a 200 ft run of RG-213 going to a 1:1 balun that feeds the antenna. A tuner in the shack reduces the VSWR for proper matching in the transmitter. Example 2 shows a similar arrangement using 300 Ω transmitting twin lead. Example 3 shows a 50 ft run of 300 Ω line dropping straight down to a tuner near the ground and 150 ft of RG-213 going to the shack. **Table 20.6** summarizes the losses and the tuner values required.

Some interesting conclusions can be drawn. First, direct feeding this antenna with coax through a balun is very lossy — a poor solution. If the antenna is λ/2 long — a resonant half-wavelength dipole — direct coax feed would be a good method. In the second example, direct feed with 300 Ω low-loss line does not always give the lowest loss. The combination method in Example 3 provides the best solution.

There are other considerations as well. Supporting a balun at the antenna feed point adds stress to the wires or requires an additional support. Example 3 has some additional advantages. It feeds the antenna in a symmetrical arrangement which is best to reduce noise pickup on the shield of the feed line. The shorter feed line does not weigh down the antenna as much. The coax back to the shack can be buried or laid on the ground and it is perfectly matched. Burial of the cable also prevents any currents from being induced on the coax shield. Once in the shack, the tuner is adjusted for minimum SWR per the manufacturer's instructions.

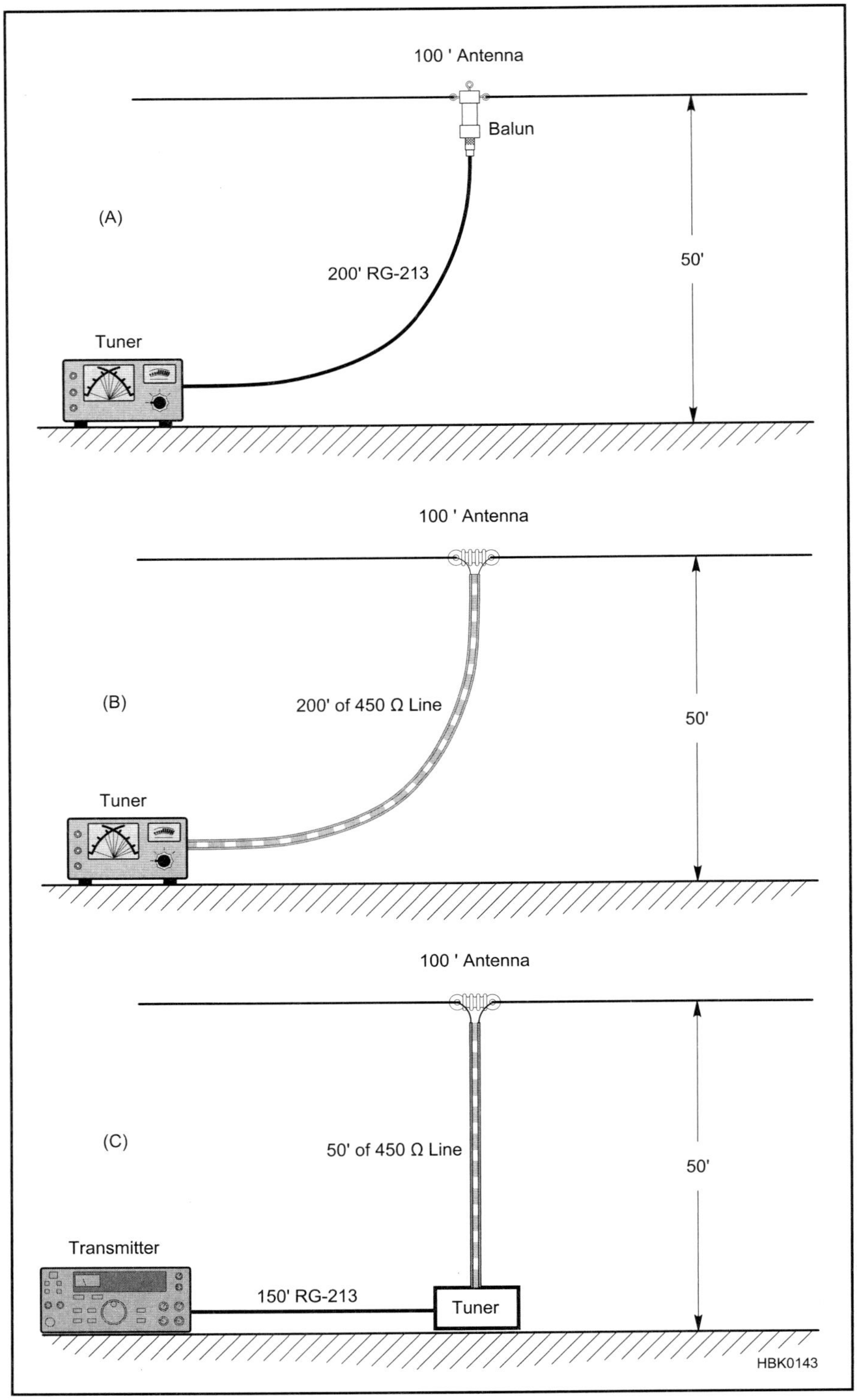

Figure 20.20 — Variations of an antenna system with different losses. The examples are discussed in the text.

20.4.5 Manually Adjusting Antenna Tuners

The process of adjusting an antenna tuner is described here and results in minimum SWR to the transmitter and also minimizes power losses in the tuner circuitry. If you have a commercial tuner and have the user's manual, the manufacturer likely provides a method of adjustment that you should follow, including initial settings.

If you do not have a user's manual, first open the tuner and determine the circuit for the tuner. The most common circuit for commercial tuners is the high-pass T network shown in Figure 20.12F. To adjust this type of tuner:

1) Set the series capacitors to maximum value. Verify that the capacitor's plates are fully meshed. (This may not correspond to the highest number on the control scale.)

2) Set the inductor to maximum value. This corresponds to placing a switch tap or roller inductor contact so that it is electrically closest to circuit ground.

3) If you have an antenna analyzer, connect it to the TRANSMITTER connector of the tuner. Otherwise, connect the transceiver and tune it to the desired frequency, but do not transmit.

In the following step, it is important to verify that you hear a peak in received noise before transmitting significant power through the tuner. Tuners can sometimes be adjusted to present a low SWR to the transmitter while coupling little energy to the output. Transmitting into a tuner in this configuration can damage the tuner's components.

4) Adjust the inductor throughout its range, watching the antenna analyzer for a dip in the SWR or listen for a peak in the received noise. Return the inductor to the setting for lowest SWR or highest received noise.

4a) If no SWR minimum or noise peak is detected, reduce the value of the capacitor closest to the transmitter in steps of about 20% and repeat.

4b) If still no SWR minimum or noise peak is detected, return the input capacitor to maximum value and reduce the output capacitor value in steps of about 20%.

4c) If still no SWR minimum or noise peak is detected, return the output capacitor to maximum value and reduce both input and output capacitors in 20% steps.

5) Once a combination of settings is found with a definite SWR minimum or noise peak:

5a) If you are using an antenna analyzer, make small adjustments to find the combination of settings that produce minimum SWR with the maximum value of input and output capacitance.

Table 20.6
Tuner Settings and Performance

Example (Figure 20.17)	*Frequency (MHz)*	*Tuner Type*	*L (µH)*	*C (pF)*	*Total Loss (dB)*
A	3.8	Rev L	1.46	2308	8.53
	28.4	Rev L	0.13	180.9	12.3
B	3.8	L	14.7	46	2.74
	28.4	L	0.36	15.6	3.52
C	3.8	L	11.37	332	1.81
	28.4	L	0.54	94.0	2.95

5b) If you do not have an antenna analyzer, set the transmitter output power to about 10 W, ensure that you won't cause interference, identify with your call sign, and transmit a steady carrier by making the same adjustments as in step 5a.

5c) For certain impedances, the tuner may not be able to reduce the SWR to an acceptable value. In this case, try adding feed line at the output of the tuner from $\lambda/8$ to $\lambda/2$ electrical wavelengths long. This does not change the feed line SWR, but it may transform the impedance to a value more suitable for the tuner components.

In general, for any type of tuner, begin with the maximum reactance to ground (maximum inductance or minimum capacitance) and the minimum series reactance between the source and load (minimum inductance or maximum capacitance). The configuration that produces the minimum SWR with maximum reactance to ground and minimum series reactance generally has the highest efficiency and broadest tuning bandwidth.

To reduce on-the-air tune-up time, record the settings of the tuner for each antenna and band of operation. If the tuner requires readjustment across the band, record the settings of the tuner at several frequencies across the band. Print out the results and keep it near the tuner. This allows you to adjust the tuner quickly with only a short transmission to check or fine tune the settings. This also serves as a diagnostic, since changes in the setting indicate a change in the antenna system.

20.4.6 Myths About SWR

This is a good point to stop and mention that there are some enduring and quite misleading myths concerning SWR.

• Despite some claims to the contrary, a high SWR does not by itself cause RFI, TVI or telephone interference. While it is true that an antenna located close to such devices can cause overload and interference, the SWR on the feed line to that antenna has nothing to do with it, providing, of course, that no components of the antenna system are arcing. The antenna is merely doing its job, which is to radiate. The transmission line is doing its job, which is to convey power from the transmitter to the radiator.

• A second myth, often stated in the same breath as the first one above, is that a high SWR causes excessive radiation from a transmission line. SWR has nothing to do with excessive radiation from a line. Imbalances in feed lines cause radiation, but such imbalances are not related to SWR. An asymmetric arrangement of a transmission line and antenna can result in current being induced on the transmission line, on the shield of coax or as an imbalance of currents in an open-wire line. Unbalanced current radiates just as if it was on an antenna. A choke balun is used on coaxial feed lines to reduce these currents as described in the section on baluns later in this chapter.

• A third, and perhaps even more prevalent myth, is that you can't "get out" if the SWR on your transmission line is higher than 1.5:1, or 2:1 or some other such arbitrary figure. On the HF bands, if you use reasonable lengths of good coaxial cable (or even better, open-wire line), the truth is that you need not be overly concerned if the SWR at the load is kept below about 6:1. This sounds pretty radical to some amateurs who have heard horror story after horror story about SWR. The fact is that you can enjoy a very effective station using antennas with feed lines having high values of SWR on them. For example, a 450 Ω open-wire line connected to the multiband dipole in the previous sections would have a 19:1 SWR on it at 3.8 MHz. Yet time and again this antenna has proven to be a great performer at many installations.

A word of caution is needed, however. Even if an antenna feed system with a high SWR acts as an effective radiator, the effect of SWR is to cause power to be reflected back to the transmitter. The final amplifiers of a solid state transmitter can be damaged if too much power is reflected back into them, and many transmitters significantly decrease their transmit power level automatically when high SWR is detected, in order to prevent damage to the finals. A tuner is needed to prevent this from happening.

20.5 Baluns and Transmission Line Transformers

A center-fed dipole or a loop in free space is *balanced*, meaning electrically and physically symmetrical with respect to the feed point. In practice, very few such antennas are balanced due to the effects of ground, nearby conductive objects and surfaces, and coupling to the feed line, whether coax or open-wire line. The last can result in *common-mode* current flowing on the feed line, which affects the antenna system impedance and can radiate in unwanted directions or pick up noise. In order to minimize these effects, the feed line can be detuned or *decoupled* from the antenna.

Many amateurs use center-fed dipoles or Yagis, fed with unbalanced coaxial line. Some method should be used for connecting the line to the antenna without upsetting the symmetry of the antenna itself. This requires a circuit that isolates the balanced load from the unbalanced line, while still providing efficient power transfer. Devices for doing this are called *baluns* (a contraction for "balanced to unbalanced"). A balanced antenna fed with balanced line, such as two-wire ladder line, maintains its inherent balance, so long as external causes of unbalance are avoided. However, even they require some sort of balun at the transmitter, since modern transmitters have unbalanced (coaxial) outputs.

If a balanced antenna is fed at the center by a coaxial feed line without a balun, as indicated in **Figure 20.21A**, the inherent symmetry and balance is upset because one side of the $\lambda/2$ radiator is connected to the shield while the other is connected to the inner conductor. On the side connected to the shield, current can be diverted from flowing into the antenna, and instead can flow away from the antenna on the outside of the coaxial shield. The field thus set up cannot be canceled by the field from the inner conductor because the fields inside the cable cannot escape through the shielding of the outer conductor. Hence currents flowing on the outside of the line are responsible for some radiation from the line, just as if they were flowing on an antenna.

This is a good point at which to say that striving for perfect balance in a line and antenna system is not always mandatory. For example, if a nonresonant, center-fed dipole is fed with open-wire line and a tuner for multiband operation, a very desirable radiation pattern for general-purpose communication is omnidirectional. A certain amount of feed line radiation might help fill in otherwise undesirable nulls in the azimuthal pattern of the antenna itself. Furthermore, the radiation pattern of a coaxial-fed dipole that is only a few tenths of a wavelength off the ground (50 feet high on the 80-meter band, for example) is not very directional anyway, because of its severe interaction with the ground.

Purists may cry out in dismay, but there are many thousands of coaxial-fed dipoles in daily use worldwide that perform very effectively without the benefit of a balun. See **Figure 20.22A** for a worst-case comparison between a dipole with and without a balun at its feed point. This is with a one λ feed line slanted downward 45° under one side of the antenna. *Common-mode currents* are conducted and induced onto the outside of the shield of the feed line, which in turn radiates. The amount of pattern distortion is not particularly severe for a dipole. It is debatable whether the bother and expense of installing a balun for such an antenna is worthwhile.

Some form of balun should be used to preserve the pattern of an antenna that is purposely designed to be highly directional, such as a Yagi or a quad. Figure 20.22B shows the distortion that can result from common-mode currents conducted and radiated back onto the feed line for a 5-element Yagi. This antenna has purposely been designed for an excellent pattern but the common-mode currents seriously distort the rearward pattern and reduce the forward gain as well. A balun is highly desirable in this case.

Choke or current baluns force equal and opposite currents to flow in the load (the antenna) by creating a high common-mode impedance to currents that are equal in both conductors or that flow on the outside of coaxial cable shields, such as those induced by the antenna's radiated field. The result of using a current balun is that currents coupled back onto the transmission line from the antenna are effectively reduced, or "choked off," even if the antenna is not perfectly balanced. Choke baluns are particularly useful for feeding asymmetrical antennas with unbalanced coax line. The common-mode impedance of the choke balun varies with frequency, but the line's differential-mode impedance is unaffected.

Reducing common-mode current on a feed line also reduces:

- Radiation from the feed line that can distort an antenna's radiation pattern
- Radiation from the feed line that can cause RFI to nearby devices
- RF current in the shack and on power-line wiring
- Coupling of noise currents on the feed line to receivers and receiving antennas
- Coupling between different antennas via their feed lines

Baluns, Chokes, and Transformers

The term "balun" applies to any device that transfers differential-mode signals between a balanced (*bal-*) system and an unbalanced (*un-*) system while maintaining symmetrical energy distribution at the terminals of the balanced system. The term only applies to the function of energy transfer, not to how the device is constructed. It doesn't matter whether the balanced-unbalanced transition is made through transmission line structures, flux-coupled transformers, or simply by blocking unbalanced current flow. A common-mode *choke balun*, for example, performs the balun function by putting impedance in the path of common-mode currents and is therefore a balun.

A *current balun* forces symmetrical current at the balanced terminals. This is of particular importance in feeding antennas, since antenna currents determine the antenna's radiation pattern. A *voltage balun* forces symmetrical voltages at the balanced terminals. Voltage baluns are less effective in causing equal currents at their balanced terminals, such as at an antenna's feed point.

An *impedance transformer* may or may not perform the balun function. Impedance transformation (changing the ratio of voltage and current) is not required of a balun nor is it prohibited. There are balanced-to-balanced impedance transformers (transformers with isolated primary and secondary windings, for example) just as there are unbalanced-to-unbalanced impedance transformers (autotransformer and transmission-line designs). A *transmission-line transformer* is a device that performs the function of power transfer (with or without impedance transformation) by utilizing the characteristics of transmission lines.

Multiple devices are often combined in a single package called a "balun." For example, a "4:1 current balun" is a 1:1 current balun in series with a 4:1 impedance transformer or voltage balun. Other names for baluns are common, such as "line isolator" for a choke balun. Baluns are often referred to by their construction — "bead balun," "coiled-coax balun," "sleeve balun," and so forth. What is important is to separate the function (power transfer between balanced and unbalanced systems) from the construction.

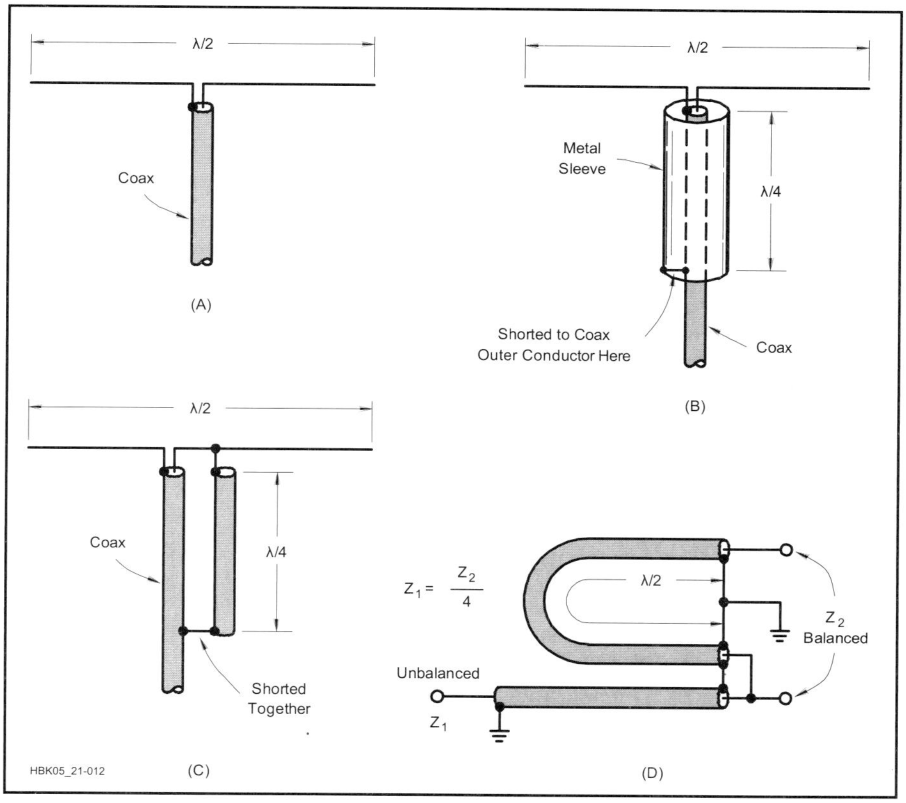

Figure 20.21 — Quarter-wavelength baluns. Radiator with coaxial feed (A) and methods of preventing unbalanced currents from flowing on the outside of the transmission line ((B) and (C)). The ½ λ phasing section shown at (D) is used for coupling to an unbalanced circuit when a 4:1 impedance ratio is desired or can be accepted.

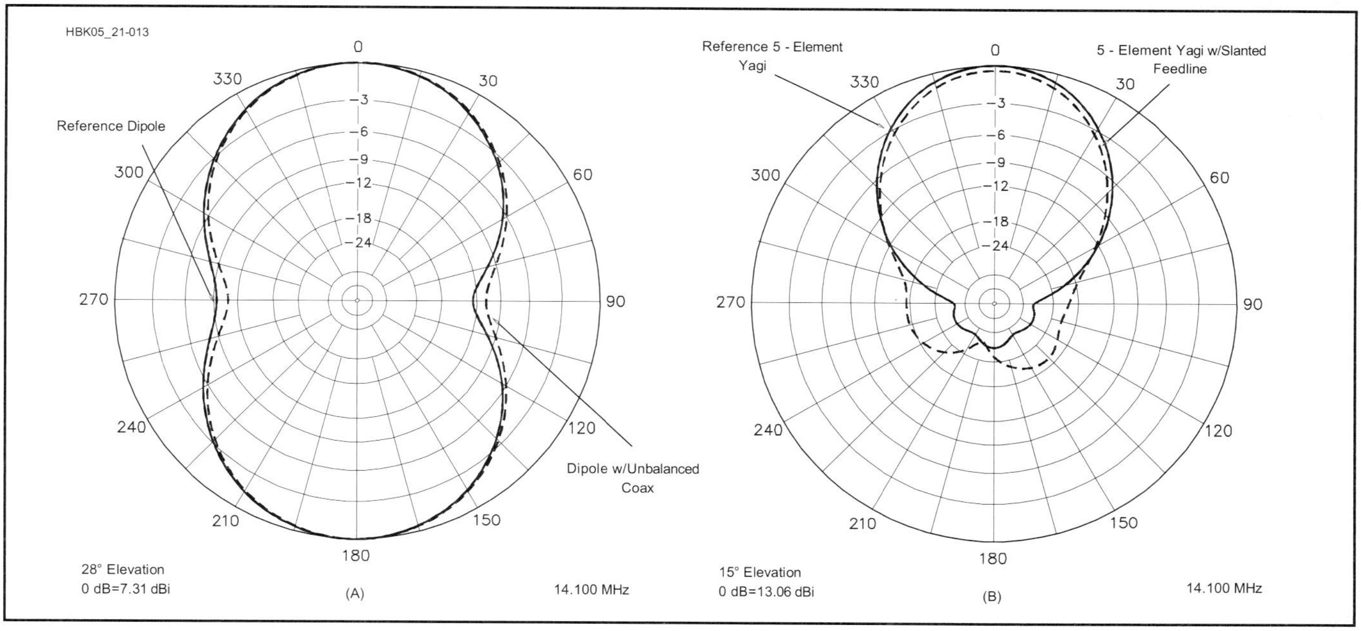

Figure 20.22 — At (A), computer-generated azimuthal responses for two λ/2 dipoles placed 0.71 λ high over typical ground. The solid line is for a dipole with no feed line. The dashed line is for an antenna with its feed line slanted 45° down to ground. Current induced on the outer braid of the 1 λ long coax by its asymmetry with respect to the antenna causes the pattern distortion. At (B), azimuthal response for two 5 element 20 meter Yagis placed 0.71 λ over average ground. Again, the solid line is for a Yagi without a feed line and the dashed line is for an antenna with a 45° slanted, 1 λ long feed line. The distortion in the radiated pattern is now clearly more serious than for a simple dipole. A balun is needed at the feed point, and most likely point, preferably λ/4 from the feed point, to suppress the common-mode currents and restore the pattern.

A single choke balun at the antenna feed point may not be sufficient to reduce common-mode current everywhere on a long feed line. If common-mode current on the line far from the antenna feed point is a problem, additional choke baluns can be placed at approximately λ/4 intervals along the line. This breaks up the line electrically into segments too short to act as effective antennas. The chokes in this case function similarly to insulators used to divide tower guy wires into non-resonant lengths.

DETERMINING BALUN POLARITY

Many baluns, impedance transformers, line isolators, and other similar items are manufactured as sealed units and markings for output polarity with respect to the input connector are not often clear. For designs in which one or more continuous coaxial cables are connected between the input and output terminals, a resistance measurement suffices to determine polarity. In flux-coupled designs there is no continuity between the input and output terminals at dc so a resistance measurement cannot be used. Similarly, for autotransformer designs there may be a low resistance across the input or output terminals. In these cases, it is necessary to test the units at RF.

The first method is to test the unit using a dual-trace oscilloscope. Analog scopes should be in "Chop" mode so that both traces show display waveforms synchronized in time. Digital scopes must also sample both waveforms at the same time. Do not use "Alt" or an alternating waveform display. Drive the input to the unit with a signal generator's output in the unit's specified operating frequency range. Connect one scope channel to the unit's input and the other channel to the output. If the waveforms are displayed in-phase, the connections to the scope have the same polarity.

The following procedure requires only a signal generator and RF voltmeter (see the **Test Equipment and Measurements** chapter) and is used to check two identical units. If the units are substantially different, the test may not be conclusive or reliable. The procedure assumes a 50 Ω system impedance. If the units operate at an impedance very different from 50 Ω, such as for 300, 450, or 600 Ω open-wire line, use a 6 to 10 dB attenuator in the generator output to isolate the generator from the mismatch.

Connect the signal generator output to the input of both units using a T connector and identical lengths of feed line. Connect one output terminal of unit "A" to an output terminal of unit "B" and measure the RF voltage across the combined balun outputs. Swap the output terminal connections on one of the units and measure the RF voltage again. One arrangement of connections should show a substantially higher output voltage — this is the arrangement with the same polarity for both units.

20.5.1 Quarter-Wave Baluns

Figure 20.21B shows a balun arrangement known as a *bazooka*, which uses a sleeve over the transmission line. The sleeve, together with the outside portion of the outer coax conductor, forms a shorted λ/4 section of transmission line. The impedance looking into the open end of such a section is very high, so the end of the outer conductor of the coaxial line is effectively isolated from the part of the line below the sleeve. The length is an electrical λ/4, and because of the velocity factor may be physically shorter if the insulation between the sleeve and the line is not air. The bazooka has no effect on antenna impedance at the frequency where the λ/4 sleeve is resonant. However, the sleeve adds inductive shunt reactance at frequencies lower, and capacitive shunt reactance at frequencies higher, than the λ/4 resonant frequency. The bazooka is mostly used at VHF, where its physical size is not excessive.

Another method that gives an equivalent effect is shown at Figure 20.21C. Since the voltages at the antenna terminals are equal and opposite (with reference to ground), equal and opposite currents flow on the surfaces of the line and second conductor. Beyond the shorting point, in the direction of the transmitter, these currents combine to cancel out each other. The balancing section acts like an open circuit to the antenna, since it is a λ/4 parallel-conductor line shorted at the far end, and thus has no effect on normal antenna operation. This is not essential to the line-balancing function of the device, however, and baluns of this type are sometimes made shorter than λ/4 to provide a shunt inductive reactance required in certain matching systems (such as the hairpin match described in the **Antennas** chapter).

Figure 20.21D shows a balun in which equal and opposite voltages, balanced to ground, are taken from the inner conductors of the main transmission line and a λ/2 phasing section. Since the voltages at the balanced end are in series while the voltages at the unbalanced end are in parallel, there is a 4:1 step-down in impedance from the balanced to the unbalanced side. This arrangement is useful for coupling between a 300 Ω balanced line and a 75 Ω unbalanced coaxial line.

Figure 20.23 shows a variation on the quarter-wave balun called the Quarter/Three-Quarter-Wave, or Q3Q, balun. It is a 1:1 decoupling balun made from two pieces of coaxial cable. One leg is λ/4 long and the other is 3λ/4 long. The two cables and the feed line are joined together with a T connector. At the antenna, the shields of the cables are connected together and the center conductors connected to the terminals of the antenna feed point. The balun has very little loss and is reported to have a bandwidth of more than 10%.

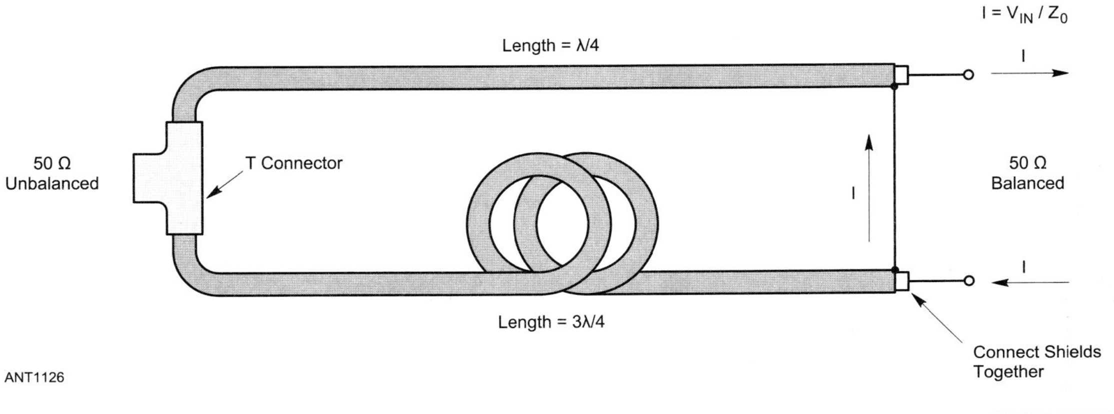

Figure 20.23 — The Quarter/Three-Quarter Wavelength (Q3Q) balun uses the current-forcing function of feed lines odd numbers of λ/4 long and the λ/2 delay of the longer line to cause equal and opposite currents to flow in the antenna terminals.

The balun works because of the current-forcing function of a transmission line is an odd number of λ/4 long. The current at the output of such a transmission line is V_{IN}/Z_0 regardless of the load impedance (less line losses — see the references for *QST* articles about this balun), similar to the behavior of a current source. Because both lines are fed with the same voltage, being connected in parallel, the output currents are also equal.

The current out of the 3λ/4 line is delayed by 180° from the current out of the λ/4 line and so is out of phase. The result is that equal and opposite currents are forced into the terminals of the load.

20.5.2 Transmission Line Transformers

The basic transmission line transformer, from which other transformers are derived, is the 1:1 *choke balun* or *current balun*, shown in **Figure 20.24A**. To construct this type of balun, a length of coaxial cable or a pair of closely spaced, parallel wires forming a transmission line are wrapped around a ferrite rod, a toroid or inserted through a number of beads. (The coiled feed line choke balun is discussed in the next section.) For the HF bands, use type 75 or type 31 material. Type 43 is used on the VHF bands. The Z_0 of the line should equal the load resistance, R.

Because of the ferrite, a high impedance exists between points A and C with a virtually identical impedance between B and D. This is true for parallel wire lines and it is also true for coax. The ferrite affects the A to C impedance of the coax inner conductor and the B to D impedance of the outer braid equally.

The conductors (two wires or coax braid and center-wire) are tightly coupled by electromagnetic fields and therefore constitute a good conventional transformer with a turns ratio of 1:1. The voltage from A to C is equal to and in phase with that from B to D. These are called the *common-mode* (*CM*) *voltages*.

A CM current is one that has the same value and direction in both wires (or on the shield and center conductor). Because of the ferrite, the CM current encounters a high impedance that acts to reduce (choke) the current. The normal *differential-mode* (*DM*) signal does not encounter this CM impedance because the electromagnetic fields due to equal and opposite currents in the two conductors cancel each other at the ferrite, so the magnetic flux in the ferrite is virtually zero. (See the section on Transmitting Ferrite Choke Baluns.)

The main idea of the transmission line transformer is that although the CM impedance may be very large, the DM signal is virtually unopposed, especially if the line length is a small fraction of a wavelength. But it is very important to keep in mind that the CM voltage across the ferrite winding, which is due to this current, is efficiently coupled to the center wire by conventional transformer action, as mentioned before. This equality of CM voltages, and also CM impedances, reduces the *conversion* of a CM signal to an *undesired* DM signal that can interfere with the *desired* DM signal in both transmitters and receivers.

The CM current, multiplied by the CM impedance due to the ferrite, produces a CM voltage. The CM impedance has L and C reactance and also R. So L, C, and R cause a broad parallel self-resonance at some frequency. The R component also produces some dissipation (heat) in the ferrite, which is an excellent way to dispose of a small amount of unwanted CM power.

Because of the high CM impedance, the two output wires of the balun in Figure 20.24A have a high impedance with respect to the generator and are therefore "isolated" from it. This feature is useful because now

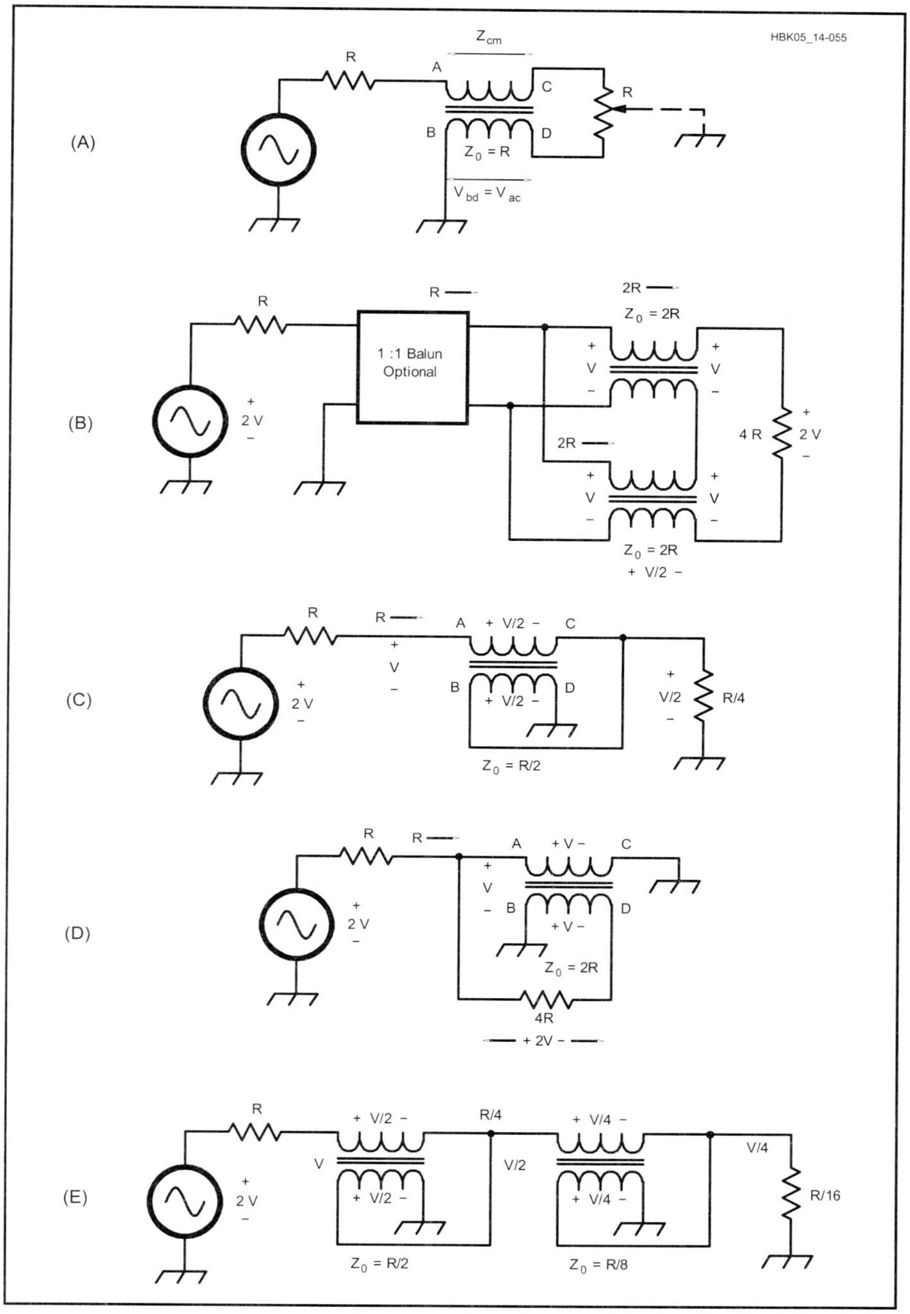

Figure 20.24 — (A) Basic current or choke balun. (B) Guanella 1:4 transformer. (C) Ruthroff 4:1 unbalanced transformer. (D) Ruthroff 1:4 balanced transformer. (E) Ruthroff 16:1 unbalanced transformer.

any point of R at the output can be grounded. In a well-designed balun circuit almost all the current in one conductor returns to the generator through the other conductor, despite this ground connection. Note also that the ground connection introduces some CM voltage across the balun cores and this has to be taken into account. The CM voltage is a maximum if point C is grounded. If point D is grounded and if all "ground" connections are at the same potential, which they often are not, the CM voltage is zero and the balun may no longer be needed. In a coaxial balun the return current flows on the inside surface of the braid.

We now look briefly at a transmission line transformer that is based on the choke balun. Figure 20.24B shows two identical choke baluns whose inputs are in parallel and whose outputs are in series. The output voltage amplitude of each balun is identical to the common input, so the two outputs add in-phase (equal time delay) to produce twice the input voltage. It is the high CM impedance that makes this voltage addition possible. If the power remains constant, the load current must be one-half the generator current, and the load resistor is 2V/0.5I = 4V/I = 4R.

THE GUANELLA TRANSFORMER

If the CM voltage in each balun is V/2, there is some flux in the cores and the right side floats. This is called a Guanella transformer. If Z_0 of the lines equals 2R and the load is a pure resistance of 4R then the input resistance R is independent of line length. If the lines are exactly λ/4, then $Z_{IN} = (2R)^2 / Z_L$, an impedance inverter, where Z_{IN} and Z_L are complex values. The quality of balance can often be improved by inserting a 1:1 balun (Figure 20.24B) at the left end so that both input leads of the 1:4 transformer are floating and a ground is at the far left side, as shown. The Guanella transformer can also be operated from a grounded right end to a floating left end. The 1:1 balun at the left then allows a grounded far left end.

THE RUTHROFF TRANSFORMER

Figure 20.24C illustrates the *Ruthroff* transformer, in which the input voltage V is divided into two equal in-phase voltages AC and BD, which are tightly coupled, so the output is V/2. Since power is constant, $I_{OUT} = 2I_{IN}$ and the load is R/4. There is a CM voltage V/2 between A and C and between B and D, so in normal operation the core is not free of magnetic flux. The input and output both return to ground so it can also be operated from right to left for a 1:4 impedance step-up.

The Ruthroff transformer is often used as an amplifier interstage transformer, for example between 200 Ω and 50 Ω. To maintain low attenuation the line length should be much less than λ/4 at the highest frequency of operation, and its Z_0 should be R/2. A balanced version is shown in Figure 20.24D, where the CM voltage is V, instead of V/2, and transmission is from left-to-right only. Because of the greater flux in the cores, just like a conventional transformer, this is not a preferred approach, although it could be used with air wound coils (for example in antenna tuner circuits) to couple 75 Ω unbalanced to 300 Ω balanced. The tuner circuit could then be used to transform 75 Ω to 50 Ω.

APPLICATIONS OF TRANSMISSION-LINE TRANSFORMERS

There are many transformer schemes that use the basic ideas of Figure 20.24. Several of them, with their toroid winding instructions, are shown in **Figure 20.25**. Two of the most commonly used devices are the 1:1 current balun and 4:1 impedance transformer wound on toroid cores, as shown in **Figure 20.26**.

For a comprehensive treatment of other transformer schemes, see *Transmission Line Transformers* and *Building and Using Baluns and Ununs* by Jerry Sevick W2FMI (SK). For applications in solid-state RF power amplifiers, see Sabin and Schoenike, *HF Radio Systems and Circuits*, Chapter 12.

20.5.3 Coiled-Coax Choke Baluns

The simplest construction method for a 1:1 choke balun made from coaxial feed line is simply to wind a portion of the cable into a coil (see **Figure 20.27**), creating an inductor from the shield's outer surface. This type of choke balun is simple, cheap, and reduces CM current. Currents on the shield are inhibited by the coil's impedance, while currents on the center conductor are unaffected.

The impedance created by this type of choke is entirely inductive reactance. If the CM impedance of the feed line is capacitive at the point where this type of choke is added, the two types of reactance partially cancel and *reduce* CM impedance at that point, *increasing* CM current. A resistive choke as described in the following sections is preferable, although a coiled coax choke may be more convenient, particularly in a temporary installation.

A scramble-wound flat coil (like a coil of rope) shows a broad resonance that easily covers three octaves, making it reasonably effective over the entire HF range. If particular problems are encountered on a single band, a coil that is resonant on that band may be added. The choke baluns described in **Table 20.7** were constructed to have a high impedance at the indicated frequencies as measured with an impedance meter. This construction technique is only valid for coaxial feed lines, since open-wire or twin-lead line would have coupling between adjacent turns.

The inductor formed by the coaxial cable's shield is self-resonant due to the distributed capacitance between the turns of the coil. The self-resonant frequency can be found by using a dip meter. To measure this, leave the ends of the choke open, couple the coil to the dip meter, and tune for a dip. This is the parallel resonant frequency and the impedance is very high.

The distributed capacitance of a flat-coil choke balun can be reduced (or at least controlled) by winding the cable as a single-layer solenoid around a section of plastic pipe, an empty bottle or other suitable cylinder. **Figure 20.28** shows how to make this type of choke balun. A coil diameter of about 5 inches is reasonable. This type of construction reduces the stray capacitance between the ends of the coil.

For both types of coiled-coaxial chokes, use cable with solid insulation, not foamed, to minimize migration of the center conductor through the insulation toward the shield. The diameter of the coil should be at least 10 times the cable diameter to avoid mechanically stressing the cable. Additionally, depending on the type of coaxial cable used, be careful not to exceed the minimum bend radius of the cable, which was discussed earlier in this chapter.

20.5.4 Transmitting Choke Baluns for HF

Transmitting chokes differ from other common-mode chokes because they must be designed to work well when the transmission line also carries RF as a differential signal. Excellent CM chokes having full power handling capability can be formed by winding multiple turns of coax through a sufficiently large ferrite core, or multiple cores. (Chokes made by winding coaxial cable on ferrite cores are referred to as "wound-coax chokes" to distinguish them from the air-core coiled-coax chokes of the preceding section.) Closely spaced paired-wire transmission lines give good performance as well and are included with the coaxial lines. This topic is discussed in more detail in the *ARRL Antenna Book* chapter on **Transmission Line System Techniques**.

Ferrite mixes designed for EMI suppression produce impedances in the MF, HF, and VHF range that are primarily resistive. For typical amateur chokes, these materials exhibit broad peaks in impedance that can be used across several amateur bands:

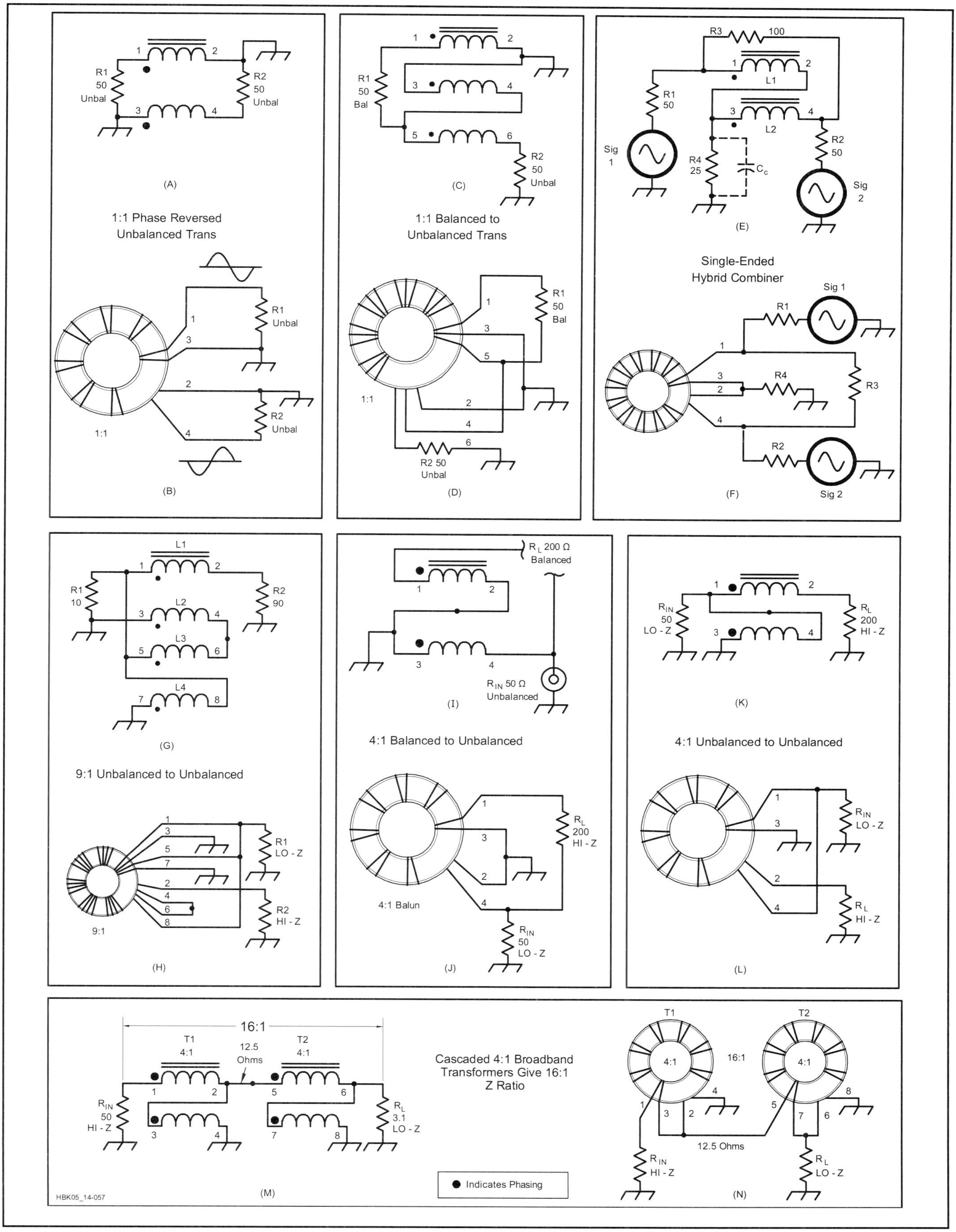

Figure 20.25 — Assembly instructions for some transmission-line transformers. See text for ferrite material type.

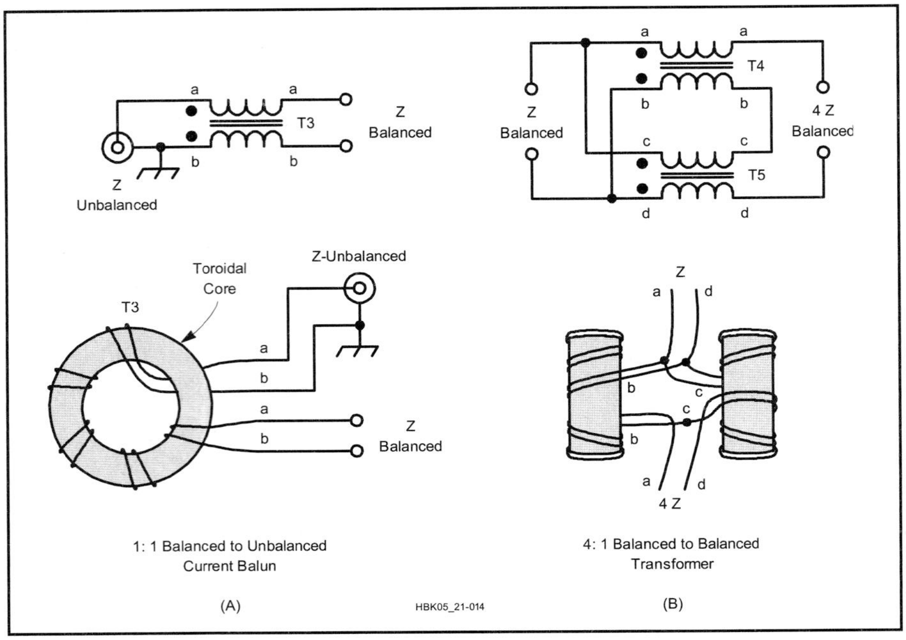

Figure 20.26 — Broadband baluns. (A) 1:1 current balun and (B) Guanella 4:1 impedance transformer wound on two cores, which are separated. Use 12 bifilar turns of #14 AWG enameled wire, wound on 2.4 inch OD cores for (A) and (B). Distribute bifilar turns evenly around core. See text for ferrite material type.

Table 20.7
Coiled-Coax Choke Baluns

Wind the indicated length of coaxial feed line into a coil (like a coil of rope) and secure with electrical tape. (Diameter 6 – 8 inches.)

The balun is most effective when the coil is near the antenna.

Lengths and diameter are not critical.

Single Band (Very Effective)

Freq (MHz)	*RG-213, RG-8*	*RG-58*
3.5	22 ft, 8 turns	20 ft, 6 – 8 turns
7	22 ft, 10 turns	15 ft, 6 turns
10	12 ft, 10 turns	10 ft, 7 turns
14	10 ft, 4 turns	8 ft, 8 turns
21	8 ft, 6 – 8 turns	6 ft, 8 turns
28	6 ft, 6 – 8 turns	4 ft, 6 – 8 turns

Multiple Band

Freq (MHz)	*RG-8, 58, 59, 8X, 213*
3.5-30	10 ft, 7 turns
3.5-10	18 ft, 9 – 10 turns
1.8-3.5	40 ft, 20 turns

- Type 31: MF, HF, and low VHF
- Type 43: High HF through VHF
- Type 61: Low UHF

As described in the **RF Techniques** chapter, a ferrite mix has a complex permeability that changes with frequency. A mix used for EMI suppression in one range may be suited for inductive applications in a different range.

In general, any combination of chokes can be used in series to provide the desired choking impedance over the desired bandwidth. Their combined choking impedance in any band is the sum of their impedances. Chokes for different frequency ranges can be placed in series to create a choke with a higher impedance across a wider frequency range, as well.

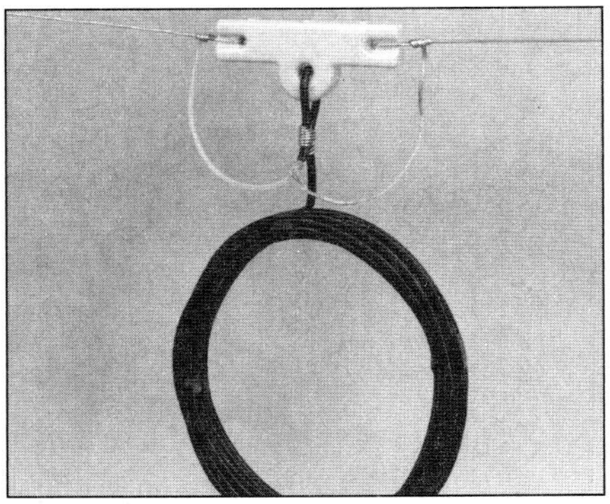

Figure 20.27 — RF choke formed by coiling the feed line at the point of connection to the antenna. The inductance of the choke isolates the antenna from the outer surface of the feed line.

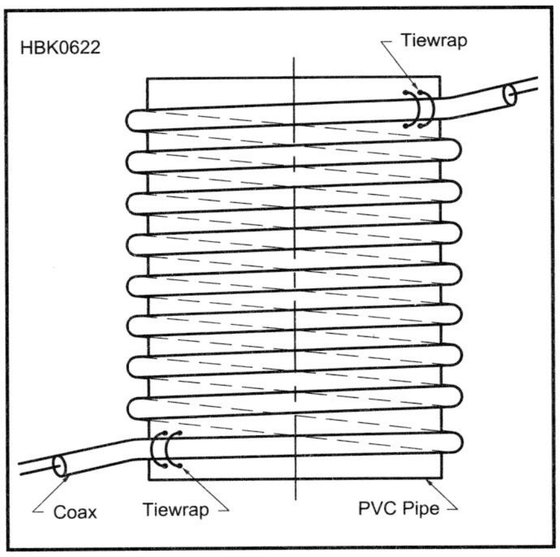

Figure 20.28 — Winding a coaxial choke balun as a single-layer solenoid typically increases impedance and self-resonant frequency compared to a flat-coil choke.

DESIGN CRITERIA

Traditionally, the rule-of-thumb for a transmission line choke has been for a choking impedance of at least 10 times Z_0 of the line, such as 500 Ω for 50 Ω coax. If the load was well-balanced and the feed point impedance reasonably close to Z_0, that would block most of the CM current without too much power dissipation. This sufficed to prevent radiation pattern distortion from feed line re-radiation, as well.

Difficulties resulted if the antenna system was *not* well-balanced, for some reason, or if the feed point impedance was much higher than Z_0. Either the choke would overheat when high power was used or there would be significant CM current on the line or both. Since these situations are fairly common, a resistive choking impedance of 5,000 Ω is recommended, such as at the feed point of a reasonably well-balanced and well-matched antenna at power levels below about 600 W.

Full power, high duty cycle, 1,500 W operation, such as for contesting and some types of digital mode operation, or use with a very unbalanced antenna, can result in high power dissipation in the choke. The best way to reduce power dissipation from CM current is to increase choking impedance. Doubling choke impedance divides the CM current by 2 and the power dissipated in the choke's series resistance by 4. Higher impedance can be obtained by winding more turns on a single core, by winding the choke on a stack of two or more cores, or placing two or more chokes in series. (Also see Dean Straw, N6BV's June 2015 *QST* article, "Don't Blow Up Your Balun," which is included in the online information for this book.)

Since the chokes are subjected to only common-mode voltage, the only effect of high SWR on power handling of coax-wound ferrite chokes is to increase the peaks of differential current and voltage along the line established by the mismatch in the antenna system. If SWR is very high, the extra mismatch loss may become an issue. Whatever feed line is used to construct the choke must be rated to handle the peak voltage and current.

For receiving applications, a choke impedance of 500 Ω to 1,000 Ω is sufficient to prevent pattern distortion, ordinary cases of RFI, and noise coupling from other sources. Chuck Counselman, W1HIS, correctly observes that radiation and noise coupling from the feed line should be viewed as a form of pattern distortion that fills in the nulls of a directional antenna, reducing its ability to reject noise and interference. (See Jim Brown, K9YC's, article on receiving chokes at **k9yc.com/RXChokesTransformers.pdf**.)

Chokes used to break up a feed line into segments too short to interact with another antenna should have a choking impedance on the order of 1,000 Ω to prevent interaction with simple antennas. A value closer to 5,000 Ω may be needed if the effects of CM current on the feed line are filling the null of directional antenna.

PRACTICAL WOUND-COAX CHOKES

Jim Brown, K9YC, and Glen Brown, W6GJB, have built, tested, and measured numerous chokes constructed with RG-8 size coax (e.g., RG-8/RG-11/RG-213); RG-400 (Teflon jacket, stranded silver-plated copper center conductor, two silver-plated copper shields, TFE dielectric); #12 and #10 enameled copper wire pairs; THHN #12 and #10 pairs; a #12 Teflon-insulated wire pair (silver-plated stranded copper, 0.109-inch OD); and a pair formed by the black and white conductors removed from #10 and #12 Romex (NM) cable. The transmission characteristics were also measured at MF and HF. The data presented here is a summary of that information and the reader is encouraged to download their papers from **k9yc.com/publish.htm**. (See the Reference entries for Brown.) The supplemental article "Measuring Ferrite Choke Impedance" by Jim Brown, K9YC is included with the online content.

Figures 20.29, **20.30**, and **20.31** are graphs of impedance magnitude for various numbers of turns, types of line, and types of core. **Table 20.8** summarizes designs for the 160 through 6-meter ham bands and several practical transmitting choke designs that are "tuned" or optimized for ranges of frequencies. The tables include designs meeting the 5,000 Ω minimum impedance requirement and a higher-impedance design if available. The number of turns is limited on the smaller 2.4-inch OD cores.

Chokes wound with higher Z_0 line (pairs of #12 THHN, NM, Teflon) work quite well at the feed point of dipoles, but may not at the feed point of a complex array. (See K9YC's "Transmitting Choke Cookbook" discussion of 75 Ω chokes for transmitting arrays.) Chokes wound with the #12 Teflon wire pair were found to have the lowest loss and the least dissipation for each band. The wire is expensive and best purchased from surplus vendors or in a quantity group purchase. Remember that for paired-wire lines, Z_0 varies with insulation thickness and the dielectric properties of the insulation. See the *ARRL Antenna Book, 25th edition* or the online K9YC paper for a table of Z_0 for lines made with wire pairs.

The other recommended choices, especially for antennas with feed point Z_0 near 50 Ω, is RG-400, followed by a paired-wire line made from the white and black conductors removed from Romex (NM) cable.

Enameled copper pairs were found to have much greater loss than other paired lines. This is because of the proximity effect that forces current to be concentrated in the side of the conductors closest to each other, raising resistive losses, similar to skin effect. It's also possible for the enamel to be scraped by the ferrite core during winding, shorting to the core at multiple points and significantly degrading choke performance. For both reasons, using enameled wire pairs for ferrite-core chokes is not recommended.

PRACTICAL CHOKES — CONSTRUCTION NOTES

Start with the information in Building Wound-Coax Chokes at the beginning of this section.

While the turn diameter on the ferrite toroid cores is smaller than the minimum specified bend radius for RG-400 coax, failures caused by center conductor migration have not been commonly reported. The consensus among experienced builders and manufacturers is that solid TFE-insulated cable works as intended as long as it is not

Winding Coaxial Cables

Using polyethylene-dielectric (PE) coax for winding common-mode choke baluns should be done with caution, if at all. Even solid PE coax such as RG-213 may experience center conductor migration through the dielectric, leading to eventual short circuits, if wound or bent too tightly. Foam-dielectric coax such as RG-8X types should never be wound into a tight coil. Coiled-coax chokes as described on the previous page can be wound with PE-dielectric cable if the radius is much larger than the manufacturer's specified minimum fixed installation bend radius (usually ½ the repeated minimum radius). A good guideline for winding PE-dielectric coax for any coil is a 3-inch minimum radius (6-inch diameter). Keep all such coils away from heat sources and protect them from direct sunlight, if practical. For coiled-coax chokes, wind the cable loosely and support it with small rope or twine so that the coax does not support all of the weight of the feed line.

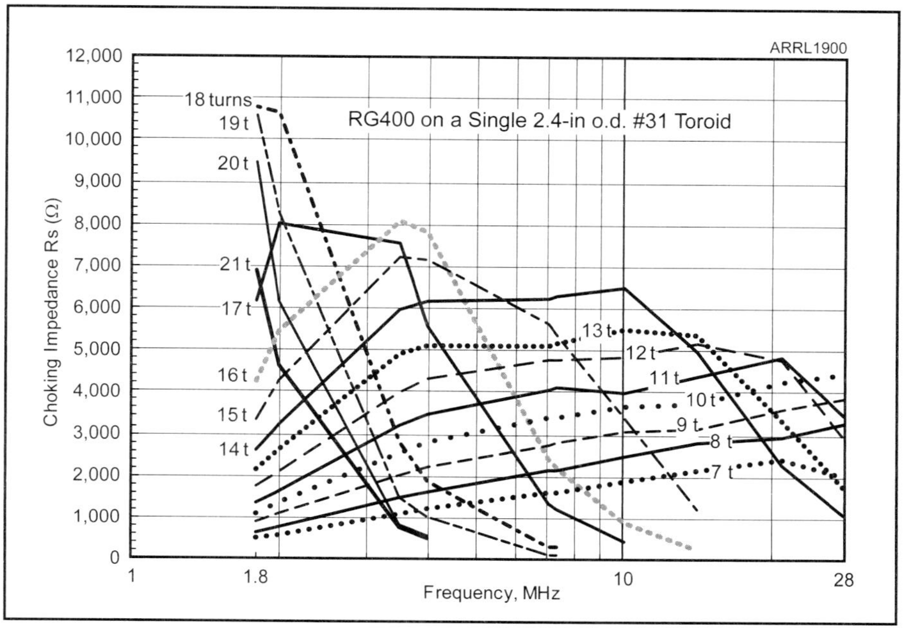

Figure 20.29 — Impedance versus frequency for currently recommended HF coax-wound ferrite transmitting chokes of RG-400 wound on a single 2.4-inch toroid core of #31 material.

operated at its full power rating and repeatedly flexed.

Starting the Winding: Wind a cable tie around the cross section of the toroid where you want to start the winding and pull it not quite tight. Feed the cable through the toroid from below, and use another cable tie to secure it to the first one, leaving enough free cable to connect the choke when it is complete. Leave enough cable tie for final tightening later. The choke in **Figure 20.32A** starts at 3 o'clock and is wound counterclockwise around the core.

Wind in Sequence: Take care that turns are wound in order around the core — out of sequence turns can cancel. Turns can be continued on a second layer when the first layer is filled by overlaying the starting turns of the winding. In Figure 20.32B, the winding starts at the upper left, completely fills the first layer around the core, and continues with five more turns overlaying the start of the winding.

Turn Spacing: Measured data are for windings tight to the core, with adjacent windings touching on the inside of the core.

Paired Lines: Take care that pairs are not twisted as they are wound. Twisting can reduce choke impedance. Using different colors for the two conductors makes it easier to see twisting, and also to count turns. Keep the wires parallel and flat against the core. Solid conductors are preferred over stranded because turns tend to stay in place. Stranded wire is much less disciplined. (The choke in the figure has short leads for measurement purposes.)

Maintain polarity between the two ends of the choke. Make sure that the same conductor of a parallel pair is connected to the coax shield at both ends of the choke. This is especially important with arrays and can be an issue with lightning protection for a choke added to the line not at the feed point. If the polarity is reversed, the choke still works, but the array doesn't work as designed and static buildup on a coax shield may not be discharged.

Pairing the wire: Loss, VF, and Z_0 data are for the paired conductors touching, held in place every 3-6 inches with Scotch 33

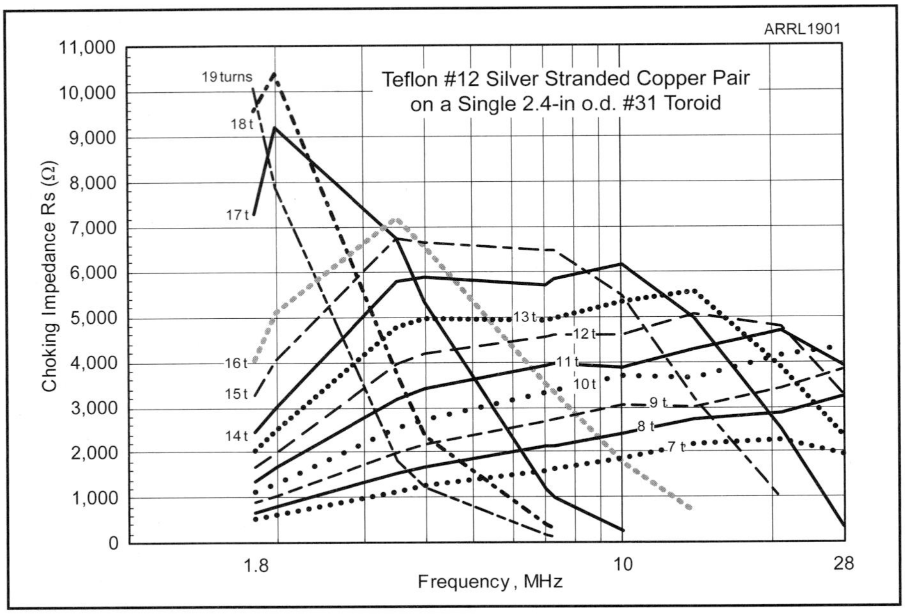

Figure 20.30 — Impedance versus frequency for currently recommended HF coax-wound ferrite transmitting chokes Teflon-insulated #12 AWG wire wound on a single 2.4-inch toroid core of #31 material.

Table 20.8

Transmitting Choke Designs for TFE Coax and Wire-Pair Lines on 2.4-inch OD Type #31 Toroid

(5 kΩ min impedance design)

Freq Band(s) (MHz)	*RG-400 Turns*	*TFE #12 Pair Turns*	*NM/THHN #12 Pair Turns*
1.8	17	17	16
3.5	13	14	13
7	13	13	13
10	12	13	13 – 14
14	12	12	11
21	11 – 12 (4.8 kΩ)	11 – 12 (4.7 kΩ)	11
28	10 (4.4 kΩ)	10 (4.3 kΩ)	10 – 11 (4.2 kΩ)
1.8 – 3.5	17	17	16
3.5 – 10		15	14
3.5 – 14	13	13	13
7 – 21	13	12	
High impedance design, if available, given as "Turns (kΩ)"			
1.8	18 (10)	18 (9.5)	18 (9.5)
3.5	16 (8)	15 – 16 (6.5)	14 (6)
7	14 (6.2)	15 (6.5)	14 (6)
10	14 (6.5)	14 (6)	13 – 14 (5.5)
14	13 (5.4)	13 (5.5)	12 – 13 (5)
3.5 – 14	14 (6, 6, 6, 6)	14 (5.8, 5.8, 5.8, 6, 5)	

Notes:
Chokes for 1.8, 3.5, and 7 MHz should have closely spaced turns.
Chokes for 14 – 28 MHz should have widely spaced turns.

Transmitting Choke Designs for TFE Coax and Wire-Pair Lines on 4-inch OD Type #31 Toroid

(5 kΩ min impedance)

Freq Band(s) (MHz)	*RG-400 Turns*	*TFE #12 Pair Turns*	*NM/THHN #12 Pair Turns*
1.8	16	15	15
3.5	13	13	20
7.0	12	15	12 – 14
10	12	13 – 14	
14	12		
1.8 – 3.5	16	21	20
1.8 – 7	16	15	
1.8 – 10	16		
3.5 – 7	19	15	13
3.5 – 10	14	13	
High impedance design, if available, given as "Turns (kΩ)"			
1.8	23 (17)	22 – 23 (15)	21 – 23 (12.5)
3.5	18 – 20 (11)	16 – 18 (7.5)	15 – 16 (6.7)
7	14 (7.5)	13 – 14 (5.7)	12 – 14 (5)
1.8 – 3.5	21 (13,10)	18 (9.5,8)	17 (8.5,6.5)
1.8 – 7	17 (7, 9.5, 6)	15 (5.5, 7.2, 5)	
1.8 – 10	16 (5.5, 8.5, 7.5, 5)		
3.5 – 7	15 (8.5, 7.5)	14 (6.5, 4.8)	14 (6.5, 5)
3.5 – 10	16 (8.5,7.5,5)	13 (5.8,5.8,5)	

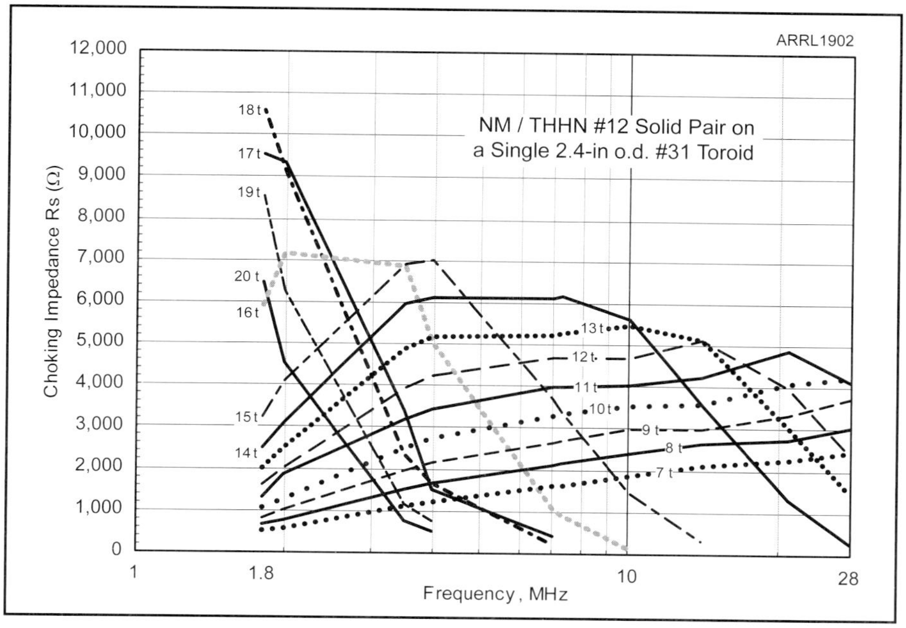

Figure 20.31 — Impedance versus frequency for currently recommended HF coax-wound ferrite transmitting chokes THHN wire wound on a single 2.4-inch toroid core of #31 material.

or 35 (thinner than Scotch 88, it can help squeeze an extra turn on 2.4-inch chokes for 160 meters). Wider spacing increases Z_0 and decrease attenuation, especially with enameled pairs (because proximity effect is reduced).

Solid PE-insulated Coax: These legacy designs have been superseded by the RG-400 and paired-wire designs. These chokes are heavier, more expensive, and have greater loss (because they use more cores and more coax). These designs are repeatable only if turns pass through the core(s) sequentially, and if they have the same radius and spacing. Space turns evenly around the toroid to minimize inter-turn capacitance. See the online paper for some suggestions for constructing these chokes.

Supporting the Choke: Ferrite-core chokes can be heavy, even if wound with the lightest line on a single core. This can lead to mechanical failure from wind or other flexing of the antenna and feed line. The referenced paper contains several photos of suggested construction techniques. Figure 20.33C shows a center insulator assembly of GPO3 fiberglass supporting two chokes in series (see the K9YC paper for series performance tables), an SO-239 receptacle for convenient feed line attachment, and sturdy attachment points for the dipole legs and a support rope or cable. The photo was taken before a waterproofing coating was applied to seal the electrical connections and provide UV resistance. Silicone adhesive was used to waterproof the SO-239. Lexan or some other UV-resistant plastic also works for the body of the assembly.

USING FERRITE BEADS

The early "current baluns" developed by Walt Maxwell, W2DU, formed by stringing multiple beads in series on a length of coax to obtain the desired choking impedance, are really common-mode chokes. Maxwell's designs utilized 50 very small beads of type 73 material as shown in **Figure 20.33**. Product data sheets show that a single type 73 bead has a very low-Q resonance around 20 MHz, and has a predominantly resistive impedance of 10 Ω to 20 Ω on all HF ham bands. Stringing 50 beads in series simply multiplies the impedance of one bead by 50, so the W2DU "current balun" has a choking impedance of 500 Ω to 1,000 Ω, and, because it is strongly resistive, any resonance with the feed line is minimal.

This is a fairly good design for moderate power levels, but suitable beads are too small to fit over most types of coax. A specialty coaxial cable such as RG-303 must be used for high-power applications. Even with high-power coax, the choking impedance is often insufficient to limit current to a low enough value to prevent overheating. Equally important, the lower choking impedance is much less effective at rejecting noise and preventing the filling of nulls in a radiation pattern.

Newer "bead balun" designs use type 31 and 43 beads, which are resonant around 150 MHz, are inductive below resonance, and have only a few tens of ohms of strongly inductive impedance on the HF bands. Even with 20 of the type 31 or 43 beads in the string, the choke is still resonant around 150

(A)

(B)

Figure 20.32 — At (A) is a wound-coax ferrite transmitting choke showing RG-400 winding technique. (B) shows how a wire-pair winding is constructed. The assembly shown in (C) is an example of mounting and supporting the wound-coax chokes.

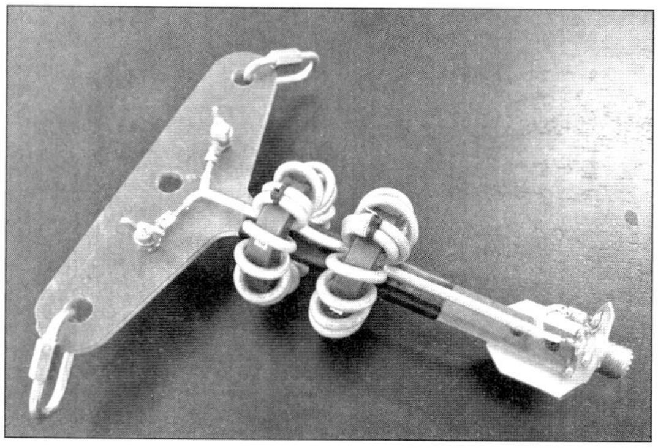

(C)

MHz, is much less effective than a wound coaxial ferrite choke, and is still inductive on the HF bands, so it is ineffective at frequencies where it resonates with the line.

Adding ferrite cores to a coiled-coax balun is a way to increase their effectiveness. The resistive component of the ferrite impedance damps the resonance of the coil and increases its useful bandwidth. The combinations of ferrite and coil baluns shown in **Table 20.9** demonstrate this very effectively. Eight feet of RG-8X in a 5-turn coil is a great balun for 21 MHz, but it is not particularly effective on other bands. If one type 43 core (Fair-Rite 2643167851) is inserted in the same coil of coax, the balun can be used from 3.5 to 21 MHz. If two of these cores are spaced a few inches apart on the coil as in **Figure 20.34**, the balun is more effective from 1.8 to 7 MHz and usable to 21 MHz. If type 31 material was used (the Fair-Rite 2631101902 is a similar core), low-frequency performance would be even better. The 20-turn, multiple-band, 1.8-3.5 MHz coiled-coax balun in Table 20.7 weighs 1 pound, 7 ounces. The single ferrite core combination balun weighs 6.5 ounces and the two-core version weighs 9.5 ounces.

20.5.5 Chokes for Receiving Applications

Common-mode current on feed line shields can cause pattern distortion, ordinary cases of RFI, and noise coupling from other sources. To reduce these effects, a choke impedance of 500 – 1,000 Ω is sufficient.

Chuck Counselman, W1HIS, correctly observes that radiation and noise coupling from the feed line should be viewed as a form of pattern distortion that fills in the nulls of a directional antenna, reducing its ability to reject noise and interference. This is particularly important for receiving antennas that are designed to have deep nulls.

Designing feed line chokes and isolation transformers for noise reduction and receiving antennas is discussed in detail by W1HIS in his referenced online article. Jim Brown, K9YC, addresses the issue in a National Contest Journal article (see References and the online information for this chapter) and in his online paper **k9yc.com/RXChokesTransformers.pdf**.

20.5.6 Transmitting Choke Baluns for VHF/UHF

This section extends the previous sections to the VHF and low UHF amateur bands. It is excerpted from the paper "Transmitting Chokes for VHF and UHF" by Jim Brown, K9YC, available at **k9yc.com/ChokesVHF.pdf**.

The primary function of these chokes is to suppress common-mode current on the feed line that in receive mode can couple to the antenna, filling in nulls in the pattern from off-axis signals and noise. For noise reduction on the HF bands, 5 kΩ of resistive impedance at the working frequency is a good starting point, with twice that value to provide greater power handling. Noise levels at VHF and UHF are usually lower than at HF, so lower values of choking impedance may be sufficient.

CHOKES FOR 6 METERS

RG8-size coax: Two 6-inch diameter turns through one 1-inch ID type #31 clamp-on (Fair-Rite 0431177081) yields about 1 kΩ of resistive impedance. Use multiple chokes in series as in **Figure 20.35A** to achieve the desired choking impedance.

RG58, RG400-size coax: Two turns through two 0.75-inch ID type #31 clamp-ons (Fair-Rite 0431173551) yields about 1.2 kΩ resistive impedance. Use multiple chokes in series to achieve the desired choking impedance.

Remember that each pass of the cable through a ferrite core counts as one turn, so each loop of cable in Figure 20.35 counts as *two* turns.

Chokes should be placed along the feed line starting as close as possible to the antenna's feed point with a few inches between cores. Coax and cores should be spaced from the boom so that capacitance to the boom does not add stray capacitance and lower the parallel resonance of the ferrite core. An insulating spacer about ½" thick should be sufficient. In Figure 20.35B, the cores are lashed under the antenna's boom (elements had not yet been attached), with the spacer in place and turns hanging below the boom. The spacer is made of nested sections of 1.5" or 2" UV-resistant PVC conduit.

The coax in the Figure 20.35 is a PTFE-insulated cable comparable to LMR-400. The cores are clamped tightly closed to minimize any air gap where their two halves

Table 20.9
Combination Ferrite and Coaxial Coil

	---------Measured Impedance---------		
Freq (MHz)	*7 ft, 4 turns of RG-8X*	*1 Core (Type 43)*	*2 Cores (Type 43)*
1.8	—	—	520 Ω
3.5	—	660	1.4 kΩ
7	—	1.6 kΩ	3.2 kΩ
14	560 Ω	1.1 kΩ	1.4 kΩ
21	42 kΩ	500 Ω	670 Ω
28	470 Ω	—	—

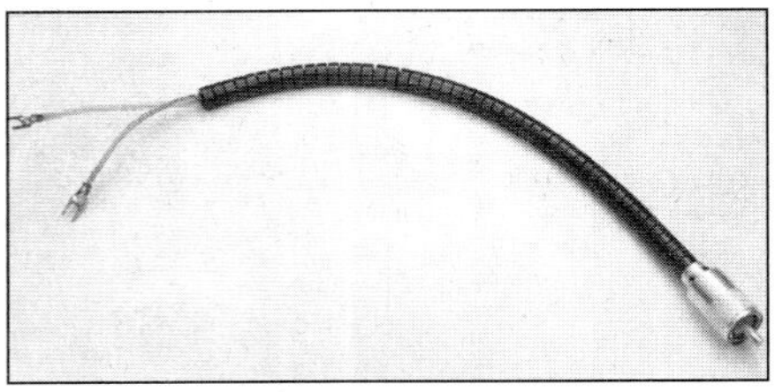

Figure 20.33 — W2DU bead balun consisting of 50 FB-73-2401 ferrite beads over a length of RG-303 coax. See text for details.

Figure 20.34 — Choke balun that includes both a coiled cable and ferrite beads at each end of the cable.

meet and the turns are also held in place on both sides of each core. Cable ties are protected with UV-resistant Scotch 88 tape.

CHOKES FOR 144 MHZ AND 220 MHZ

RG58, RG400-size coax: Apply enough Fair-Rite #31 clamp-ons to achieve the desired choking impedance. Each Fair-Rite 0431164951 core provides about 300 Ω of resistive impedance on 144 MHz and a bit less on 220 MHz.

RG213-size coax: Apply enough Fair-Rite #31 clamp-ons to achieve the desired choking impedance. Each Fair-Rite 0431164181 core provides about 360 Ω resistive impedance and a bit less on 220 MHz.

CHOKES FOR 430 MHZ

Apply enough Fair-Rite #61 clamp-ons to achieve the desired choking impedance. Each Fair-Rite 0431164951 core provides about 280 Ω resistive impedance.

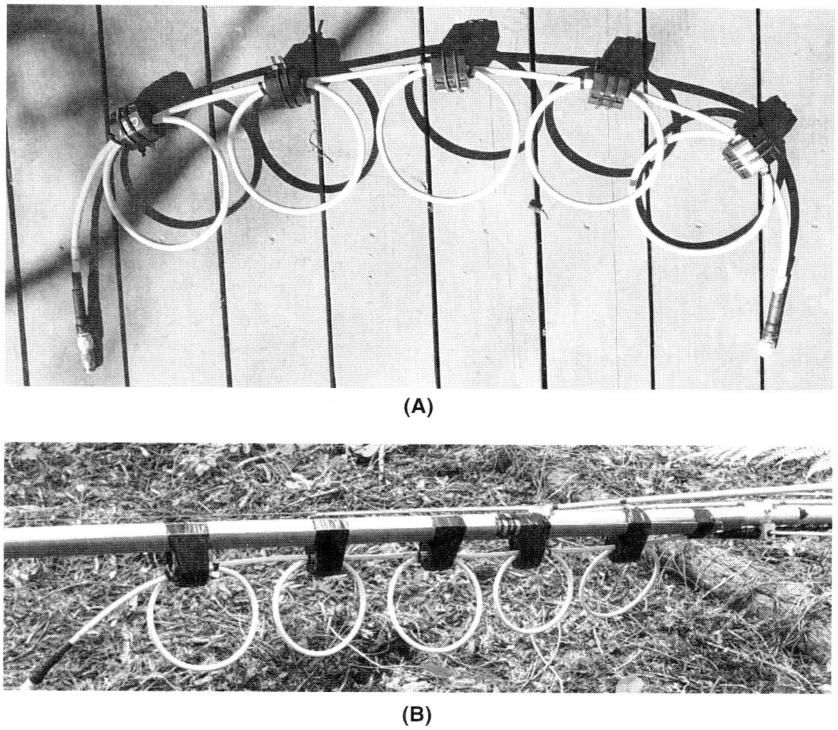

Figure 20.35 — (A) The 6-meter chokes are shown prior to installation and (B) attached to the antenna boom. The elements are not yet attached in the second photo. (Photos courtesy of Jim Brown, K9YC)

20.6 PC Transmission Lines

PC board material can be used to create a transmission line. There are several variations in which the PC trace forms one of the conductors and ground plane layers form the other. These are summarized in **Figure 20.36**, where e_r is the dielectric constant of the PC board material. (FR4 is the most common material at and above VHF.)

Microstrip (Figure 20.36A) is the most common of the PC transmission lines, consisting of an isolated trace above a ground plane.

Stripline (Figure 20.36B) is also common in multilayer boards with the PC trace embedded in the PC board material and centered between two ground plane layers.

Offset stripline (not shown) is a variation of stripline in which the PC trace is not centered between the ground plane layers.

Coplanar waveguide (Figure 20.36C) is feasible at microwave frequencies.

In microstrip and stripline the RF energy is mostly (but not completely) confined to the region between the large surface of the PC trace and the ground plane. Current is spread across the surface of the PC trace at a depth determined by the skin effect (see the **RF Techniques** chapter).

In contrast, the RF energy in coplanar waveguide is contained between the edges of the PC trace and the edges of the adjacent ground plane. The middle surfaces of the PC trace carry little, if any, current. This increases resistive losses because the current is concentrated in a smaller region but the waves travel mostly in air and so have lower losses. This becomes an important tradeoff at microwave frequencies.

Since most designs work with 50 Ω impedances, combinations of common copper

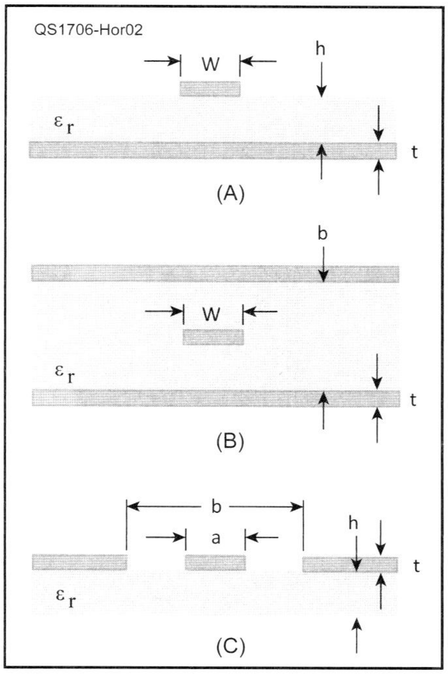

Figure 20.36 — Types of PC transmission line: microstrip (A), stripline (B), and coplanar waveguide (C). Dimensions shown are used by online calculators to determine the line's characteristic impedance. The PCB material's relative permittivity is ε_r.

foil thicknesses, trace widths, and board layer thicknesses have been calculated to produce 50 Ω. Several are shown in **Table 20.10**. For the interested reader, see Wadell's book in the Reference section of this chapter. The free program *AppCAD* (**www.hp.woodshot.com**) can handle many PC transmission line design calculations, along with S parameters and balun calculations.

With high-volume commercial and consumer electronics operating at microwave frequencies, connectors for PC transmission lines have become widely available. See the **Component Data and References** chapter's table of connector styles for possible candidates. These small connectors cannot handle a lot of power but are adequate for receiving and low-power transmitting applications. Adapters and adapter cables are available to convert these connectors to the more common SMA, UHF, BNC, N, and other styles used by amateurs.

Table 20.10
50-Ω Transmission Line Dimensions

Type of Line	*Dielectric (ε_r)*	*Layer Thickness in mils (mm)*	*Center Conductor in mils (mm)*	*Gap Gap*	*Characteristic Impedance (Ω)*
Microstrip	Prepreg (3.8)	6 (0.152)	11.5 (0.292)	n/a	50.3
	Prepreg (3.8)	10 (0.254)	20 (0.508)	n/a	50.0
Stripline	FR4 (4.5)	12 (0.305)	3.7 (0.094)	n/a	50.0
Coplanar WG	Prepreg (3.8)	6 (0.152)	14 (0.35)	20 (0.50)	49.7

Information from Maxim Integrated, Tutorial 5100:
www.maximintegrated.com/en/design/technical-documents/tutorials/5/5100.html

20.7 Waveguides

Above 2 GHz, coaxial cable is a losing proposition for communications operation. Fortunately, at this frequency the wavelength is short enough to allow practical, efficient energy transfer by an entirely different means. A *waveguide* is a conducting tube through which energy is transmitted in the form of electromagnetic waves. The tube is not considered as carrying a current in the same sense that the wires of a two-conductor line do, but rather as a *boundary* that confines the waves in the enclosed space. Skin effect prevents any electromagnetic effects from being evident outside the guide. The energy is injected at one end, either through capacitive or inductive coupling or by radiation, and is removed from the other end in a like manner. Waveguides merely confine the energy of the fields, which are propagated through them to the receiving ends by means of reflections from their inner walls.

20.7.1 Development of Waveguides

Suppose an open-wire line is used to carry UHF or microwave energy from a generator to a load. Imagine the transmission line is supported with quarter-wave stubs, shorted at the far end. The open end of such a stub presents an infinite impedance to the transmission line, provided that the shorted stub is non-reactive. Thus, the stub acts as an insulating support. Since the stubs act as an open-circuit, an infinite number of them may be connected in parallel without affecting the open-wire line.

Because the shorting link has finite length it also has some inductance. This inductance can be minimized by making the RF current flow on the surface of a plate rather than through a thin wire. If the plate is large enough, it prevents the magnetic field from encircling the RF current.

The transmission line may be supported from the top as well as the bottom and when infinitely many supports are added, they form the walls of a waveguide at its *cutoff frequency*. **Figure 20.37** illustrates how a rectangular waveguide is developed from a two-wire parallel transmission line.

20.7.2 Waveguide Operation

As a signal propagates along a waveguide, the metal walls contain the electric and magnetic fields. We'll concentrate on the electric field here, because the dominant mode of propagation for waveguide is called TE or *Transverse Electric*. **Figure 20.38** shows how the electric and magnetic fields are oriented for the TE_{10} mode. (See the discussion on waveguide modes below.) The electric field intensity in the rectangular waveguide, which is oriented parallel to the shorter side walls. The electric field is strongest in the center and must be zero at the side walls, since the walls are short circuits. This is the reason for the E-plane and H-plane terminology in Figure 20.38D.

The field strength distribution is half of a sine wave at the operating frequency, and propagates down the waveguide as if it were bouncing off the side walls. For this to work, the width of the waveguide must be at least one-half the wavelength of the propagating signal. When the width is less than half wavelength, the field cannot be zero at

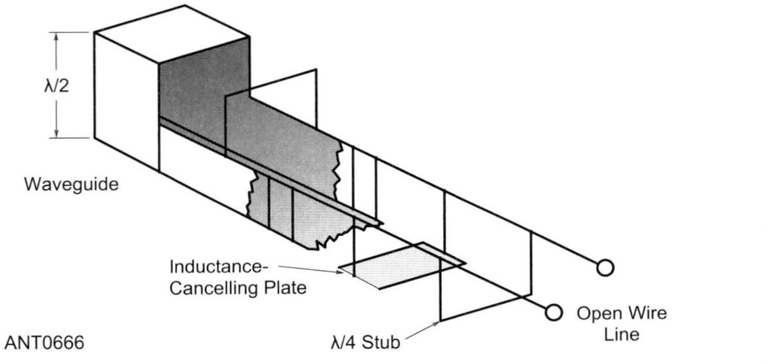

Figure 20.37 — At its cutoff frequency a rectangular waveguide can be thought of as a parallel two-conductor transmission line supported from top and bottom by an infinite number of quarter wavelength stubs.

both walls, so lower frequencies with longer wavelengths do not propagate in the waveguide.

20.7.3 Waveguide Dimensions

CUTOFF AND UPPER FREQUENCIES

The minimum frequency of operation for a waveguide is that at which the waveguide width is a half-wavelength. This is called the waveguide's *cutoff frequency*, f_c.

$$f_c = c/\lambda_c$$

where c = the speed of light in free space, 2.9979×10^8 meters per second.

A wavelength in the waveguide, λ_g, is longer than a wavelength in free space, λ_0. This implies a velocity faster than the speed of light. But only the *phase velocity* exceeds the speed of light — energy cannot travel faster. The waveguide wavelength varies with frequency as a function of the cutoff frequency:

$$\lambda_g = \frac{\lambda_0}{\sqrt{1-\left(\frac{\lambda_0}{\lambda_c}\right)^2}}$$

Near the cutoff frequency, λ_g is much longer than the free-space wavelength λ_0, becoming closer to λ_0 as frequency increases.

Waveguide operating frequencies are usually well above the cutoff frequency. Near (and below) the cutoff frequency, losses increase and the guide wavelength changes rapidly with frequency causing dispersion of the transmitted waveform. This is why a waveguide makes an excellent high-pass filter.

The height of a rectangular waveguide, between top and bottom walls, determines both the upper frequency limit and the characteristic impedance. The upper limit is the frequency at which the waveguide height is $\lambda/2$. Above this frequency, the electric field may change orientation and other modes may propagate.

WAVEGUIDE DIMENSIONS

In a rectangular guide the critical dimension is X in Figure 20.38C. This dimension must be more than $\lambda/2$ at the lowest frequency to be transmitted. In practice, the Y dimension usually is made about equal to X/2 to avoid the possibility of operation in other than the dominant mode.

Cross-sectional shapes other than a rectangle can be used, the most important being the circular pipe. Many of the same considerations apply as in the rectangular case.

Wavelength dimensions for rectangular and circular guides are given in **Table 20.11**, where X is the width of a rectangular guide and r is the radius of a circular guide. All figures apply to the dominant mode.

CHARACTERISTIC IMPEDANCE

The characteristic impedance is usually much higher than 50 ohms and is calculated for TE modes using this formula:

$$Z_0 = 377\left(\frac{\lambda_g}{\lambda_0}\right)\left(\frac{2b}{a}\right)$$

where *a* and *b* are the large and small dimensions, respectively, of the rectangular

RF Safety for Waveguides, Horns, and Dishes

Never look into the open end of a waveguide when power is applied, or stand directly in front of a dish while transmitting. Tests and adjustments in these areas should be done while receiving or at extremely low levels of transmitter power (less than 0.1 watt). Eyes are especially sensitive to the heat generated by absorbed RF energy. The FCC regulatory exposure limits are 10 mW/cm2 averaged over a 6-minute period for radio amateurs and 1 mW/cm2 averaged over a 30-minute period for anyone else. See the **Safe Practices** chapter for more information.

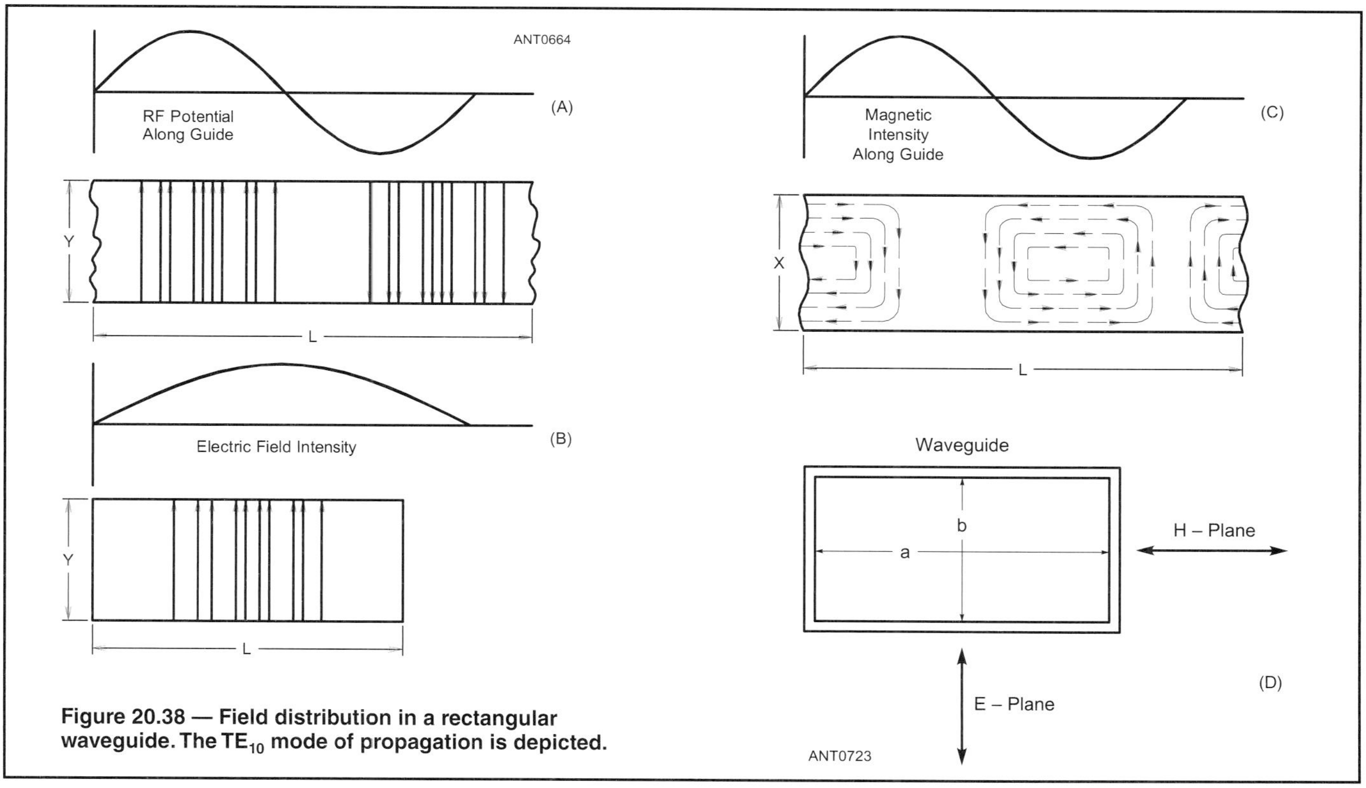

Figure 20.38 — Field distribution in a rectangular waveguide. The TE_{10} mode of propagation is depicted.

waveguide. 377 Ω is the approximate impedance of free space.

For example, at 10 GHz, WR90 waveguide is often used. The width is 0.9 inches, or 22.86 mm, and the height is 0.4 inches, or 10.16 mm. The cutoff frequency is 6.56 GHz, but the recommended operating frequency range is 8 to 12.4 GHz. At 10.368 GHz, the free space wavelength is λ_0 = 28.915 mm, the guide wavelength is λ_g = 37.33 mm and the characteristic impedance is Z_0 = 433 ohms. Several common waveguides are pictured in cross-section in **Figure 20.39**.

For a circular waveguide with radius, r, the cutoff wavelength is $\lambda_c = 3.412$ r and the characteristic impedance is

$$Z_0 = 377\left(\frac{\lambda_g}{\lambda_0}\right)$$

20.7.4 Waveguide Modes

The operating mode described above is for the dominant mode, TE_{10} (or TE_{11} in circular waveguide). This is the lowest frequency mode at which a given waveguide operates, and is the preferred mode for waveguide transmission.

If there is no upper limit to the frequency to be transmitted, there are an infinite number of ways in which the fields can arrange themselves in a waveguide. Each field configuration is a mode. All modes may be separated into two general groups: TM (*Transverse Magnetic*), which has the magnetic field entirely crosswise to the direction of propagation, but has a component of electric field in the propagation direction, and TE (*Transverse Electric*), which has the electric field entirely crosswise to the direction of propagation, but has a component of magnetic field in the direction of propagation. TM waves are sometimes called E-waves in older references and TE waves are sometimes called H-waves. The TM and TE designations are preferred, however. The particular mode of transmission is identified by the group letters followed by subscript numbers; for example TE_{11}, TM_{11}, and so on. The number of possible modes increases with frequency for a given size of guide.

Higher-order modes are useful in certain applications. For instance, in multi-mode feed horns, where the additional modes can shape the radiation pattern, special high-power waveguides, and in certain cavity filters.

WAVEGUIDE COUPLING

Energy may be introduced into or extracted from a waveguide or resonator by means of either the electric or magnetic field. One type of adapter, shown in **Figure 20.40**, is a probe like a monopole antenna in the center of a wide wall of the waveguide. The probe in Figure 20.40A is simply a short extension of the inner conductor of the coaxial line, oriented so that it is parallel to the electric lines of force. If the probe can be very thin and have no capacitance or inductance, it should be λ/4 long and spaced λ/4 from the closed end of the waveguide (a short-circuit). Actual dimensions compensate for the actual probe inductance and capacitance.

The loop shown in Figure 20.40B is arranged so that it encloses some of the magnetic lines of force. The point at which maximum coupling is obtained depends upon the mode of propagation in the guide or cavity. Coupling is maximum when the coupling device is in the most intense field.

Coupling can be varied by turning the probe or loop through a 90° angle. When the probe is perpendicular to the electric lines, the coupling is minimized. Similarly, when the plane of the loop is parallel to the magnetic lines, the coupling is minimized.

Table 20.11
Waveguide Operating Dimensions in Wavelengths

	Rectangular	*Circular*
Cutoff wavelength	2X	3.41r
Longest wavelength transmitted with little attenuation	1.6X	3.2r
Shortest wavelength before next mode becomes possible	1.1X	2.8r

Figure 20.39 — Typical waveguide sizes from WR229 on the left to WR42 on the right.

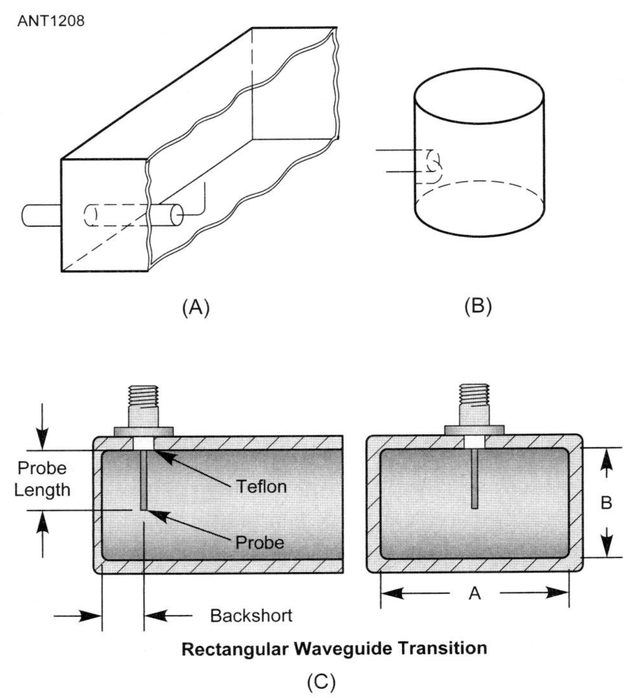

Figure 20.40 — Parts (A) and (B) shows methods of coupling coaxial line to waveguide and resonators. At (C) is a sketch of rectangular waveguide to coax transition, showing dimensions.

20.7.5 Waveguide Termination

When a waveguide is not terminated in its characteristic impedance, there is an elevated SWR on the line, just like any other transmission line. A typical termination is a horn antenna, which flares out from the end of the waveguide to match the waveguide's characteristic impedance to the impedance of free space, 377 Ω.

Waveguide connections are made by bolting their flanges together firmly. Several flanges are shown in **Figure 20.41**. There are two types of flanges, *flat flanges* and *choke flanges*, which have a groove around the waveguide. The groove acts like a shorted stub that presents a high impedance to the RF energy to prevent leakage. Choke flanges should be mated with flat flanges, but two flat flanges may be mated together.

Waveguides of two different sizes may be mated, but there should be a transition between them since different waveguide sizes have different Z_0. Just as in coaxial lines, impedance discontinuities such as at a transition cause reflections.

20.7.6 Practical Waveguides

Standard waveguide sizes dating to World War II are still in use today. The standard designator is WR*xx*, where *xx* is the wide dimension in hundredths of an inch; for instance, WR90, often used at 10 GHz, has a wide dimension of 90 hundredths of an inch, or 0.9 inches. There are no standards for circular waveguides, so common copper plumbing is often used — 3/4-inch tubing works well at 10 GHz. Practical dimensions for standard waveguides at amateur microwave calling frequencies are shown in **Table 20.12**.

Table 20.12 includes dimensions for the waveguide sizes likely to be encountered in microwave work; while a waveguide would work very well at lower frequencies, it would be very large — one meter wide for the 2-meter band! The recommended frequency range for each type is also shown in the table. For narrowband amateur work, we can stretch the frequency range a bit to take advantage of available waveguide — some of the waveguide sizes are usable on more than one band. Many smaller size guides are used at mm-wave frequencies.

Losses in waveguides are very low, much less than coaxial transmission line, but not negligible at the higher frequencies. For example, WR90 loss is approximately 10 dB per 100 feet at 10 GHz. Power handling is not a problem; even the smallest waveguide is rated at far more than the amateur power limit.

Where the waveguide has to change direction, special fittings called *bends* are used, shown in **Figure 20.42**. These are constructed so that impedance discontinuities from the change in direction are minimized.

Waveguide transmission lines used outdoors suffer from internal water condensation, like any closed metal container with temperature variations. Even a short length used in a portable station can collect condensation. Commercial installations often pressurize the waveguide with dry air. Otherwise, a drain hole should be provided at the bottom of a run.

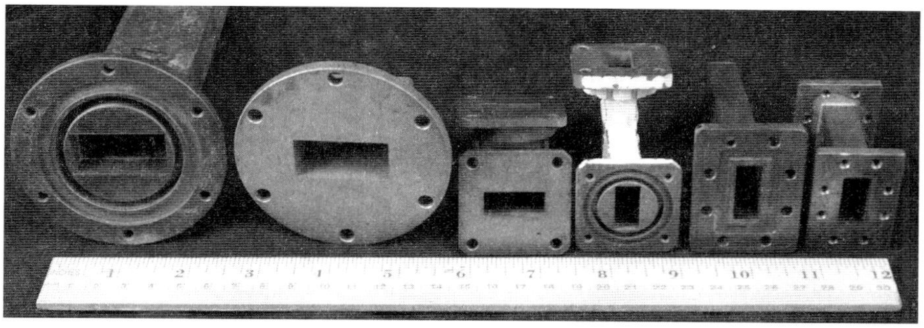

Figure 20.41 — Several typical waveguide flanges used for joining sections of waveguide. The deep grooves in choke flanges place a high-impedance in the path of any RF leakage out of the flange.

Table 20.11
Waveguide Dimensions and Coax Transitions

(Cross-section dimensions are between inner walls)

Waveguide	*Dimensions (mm)*	*Freq Range (GHz)*	*Freq (GHz)*	*Probe Diam (mm)*	*Probe Len (mm)*	*Backshort (mm)*	*Bandwidth*
WR42	10.668 × 4.318	18 – 26.5	24.192	1.27	2.413	2.489	>17%
WR75	19.05 × 9.525	10.0 – 15	10.368	1.27	5.49	5.26	14%
WR90	22.86 × 10.16	8.2 – 12.4	10.368	1.27	5.89	5.46	7%
WR112	28.24 × 12.62	7.05 – 10	10.368	1.27	6.5	6.6	15%
WR112			5.76	1.27	8.8	9.8	7%
WR137	35.85 × 15.80	5.85 – 8.2	5.76	1.27	10.5	8.5	10%
WR159	40.39 × 20.19	4.9 – 7.05	5.76	1.27	11.17	10.0	11%
WR187	47.55 × 22.15	3.95 – 5.85	5.76	2.36	11.3	11.0	16%
WR187			3.456	2.36	14.5	18.0	5%
WR229	58.17 × 29.08	3.3 – 4.9	3.456	1.27	18.2	15.0	8%
WR229			3.456	2.36	17.4	15.06	11%
WR229			3.456	3.175	17	15.6	11%
WR229			3.456	4.76	16.2	16.2	14%
WR229			3.456	6.35	15.5	16.75	17%
WR284	72.14 × 34.04	2.6 – 3.95	3.456	6.35	17.5	17.8	27%
WR284			2.304	6.35	20	28	9%
WR340	86.36 × 43.18	2.2 – 3.2	2.304	6.35	25	23	11%

Figure 20.42 — Special sections called bends make 90° and 45° angles to avoid introducing impedance discontinuities where the waveguide has to change direction.

20.8 References and Bibliography

Barter, A., G8ATD, *International Microwave Handbook*, 2nd Edition (RSGB, 2008).

Brown, J., K9YC, "A Ham's Guide to RFI, Ferrites, Baluns, and Audio Interfacing," **k9yc.com/RFI-Ham.pdf**.

Brown, J., K9YC, "Optimizing the Placement of Stubs for Harmonic Suppression," *National Contest Journal*, July/Aug. 2015, pp. 8 – 11.

Brown, J., K9YC, "A New Choke Cookbook for the 160-10M Bands," Dec. 2018, **k9yc.com/publish.htm**.

Brown, J., "Chokes and Isolation Transformers for Receiving Antennas," *National Contest Journal*, Mar./Apr. 2019, pp. 7 – 10.

Brown, J., K9YC, "Transmitting Chokes for VHF and UHF," 2020, **k9yc.com/ChokesVHF.pdf**.

Counselman, C., W1HIS, "Common Mode Chokes," **www.yccc.org/common-mode-chokes-by-w1his/**.

Cutsogeorge, G., W2VJN, "Optimizing the Performance of Harmonic Attenuation Stubs," *National Contest Journal*, Jan./Feb. 2015, pp. 3 – 4.

Cutsogeorge, G., W2VJN, *Managing Interstation Interference,* Second Edition (International Radio, 2009). **www.vibroplex.com/techdocs/INRAD/MII_W2VJN.pdf**.

Johnson, H., and Graham, M., *High Speed Digital Design* (Prentice Hall, 1993).

Lau, Z., W1VT, "Why Do Baluns Burn Up?," *QEX*, Jan./Feb. 2004, pp. 55 – 58.

Lewallan, R., W7EL, "Baluns: What They Do and How They Do It," 1985, **www.eznec.com/Amateur/Articles/Baluns.pdf**.

Reisert, J., W1JR, "VHF/UHF World," *Ham Radio*, Oct. 1987, pp. 27 – 38.

Regier, F., "Series-Section Transmission Line Impedance Matching," *QST,* July 1978, pp. 14 – 16.

Sabin and Schoenike, *HF Radio Systems and Circuits*, (SciTech Publishing, 1998).

Sevick, J., W2FMI, *Transmission Line Transformers*, 4th Edition (Noble Publishing, 2001).

Sevick, J., W2FMI, *Building and Using Baluns and Ununs* (CQ Communications, 2003).

Silver, H., NØAX, Ed., *The ARRL Antenna Book*, 25th Edition (ARRL, 2023). Chapters 23 through 28 include material on transmission lines and related topics.

Smith, P., *Electronic Applications of the Smith Chart* (Noble Publishing, 1995).

Stensby, J., N5DF, "Coax Loss Calculated Directly in Terms of Impedance Measurements" *QEX*, Jan./Feb. 2020, pp. 32 – 33.

Straw, D., N6BV, "Don't Blow Up Your Balun," *QST*, June 2015, pp. 30 – 36.

Wade, P., W1GHZ, "Microwavelengths," *QST* column.

Wadell, B., *Transmission Line Design Handbook*, (Artech House, 1991).

Witt, F., "Baluns in the Real (and Complex) World," *The ARRL Antenna Compendium Vol 5* (ARRL, 1996).

Zavrel, R., W7SX, "Build Your Own Open-Wire Line," *QST*, Mar. 2020, pp. 30 – 33.

The ARRL UHF/Microwave Experimenter's Manual (ARRL, 2000). Chapters 5 and 6 address transmission lines and impedance matching.

ONLINE RESOURCES

A collection of articles about waveguide transitions and filters — **w1ghz.org/10g/QEX_articles.htm**

A collection of Smith chart references — **http://sss-mag.com/smith.html**

Ferrite and powdered iron cores are available from **www.fair-rite.com**, **www.amidoncorp.com** and **www.cwsbytemark.com**

Microwave oriented downloads — **www.microwaves101.com/download-area**

Table of transmission line properties — **hf-antenna.com/trans/**

Times Microwave catalog — **www.timesmicrowave.com**

Transmission line loss factors — **www.microwaves101.com/encyclopedias/transmission-line-loss**

Transmission line matching with the Smith chart — **ieee.li/pdf/essay/smith_chart_fundamentals.pdf**

Transmission line transformer theory — **www.bytemark.com/products/tlttheory.htm**

Waveguide tutorials — **www.microwaves101.com/encyclopedias/waveguide-primer**

www.rfcafe.com/references/electrical/waveguide.htm

www.feynmanlectures.caltech.edu/II_24.html

Contents

Chapter 21
Antennas

In the world of radio, the antenna is "where the rubber meets the road!" With antennas so fundamental to communication, it is important that the amateur have a basic understanding of their function. That understanding enables effective selection and application of basic designs to whatever communications task is at hand. In addition, the amateur is then equipped to engage in one of the most active areas of amateur experimentation, antenna design. The goal of this chapter is to define and illustrate the fundamentals of antennas and provide a selection of basic designs: simple verticals and dipoles, quads and Yagi beams, and other antennas, plus a section on rotator use and selection. The reader will find additional in-depth coverage of these and other topics in the *ARRL Antenna Book* and other references provided. This chapter was originally written by Chuck Hutchinson, K8CH, and has been updated by Ward Silver, NØAX. Alan Applegate, KØBG, maintains the material on mobile antennas. The section on Radio Direction Finding Antennas was written by Joe Moell, KØOV.

Chapter 21 — Online Content

Articles

- A Simple Direction-Finding Receiver for 80 Meters by Dale Hunt, WB6BYU
- A Simple Broadband Dipole for 80 Meters by Frank Witt AI1H
- A Wideband 80-Meter Dipole by Rudy Severns N6LF
- Broad-Band 80-Meter Antenna by Allen Harbach WA4DRU
- Design of a Two-band Loaded Dipole Antenna by David Birnbaum, K2LYV
- Direction Finding Techniques by Joe Moell, KØOV
- Half-Wave and Multiband Dipole Azimuth and Elevation Patterns
- Small Gap-Resonated Loop Antennas by Kai Siwiak KE4PT and Richard Quick W4RQ
- The 3/8-Wavelength Vertical — A Hidden Gem by Joe Reisert, W1JR
- The Fan Dipole as a Wideband and Multiband Antenna Element by Joel Hallas W1ZR
- Weatherizing Outdoor Inductors and Traps by Dick Sander, K5QY
- Workshop Chronicles: Alloy Designations by Don Daso, K4ZA

HF Projects

- A Compact Multiband Dipole by Zack Lau W1VT
- A No Compromise Off-Center Fed Dipole for Four Bands by Rick Littlefield, K1BQT
- An Off-Center End-Fed Dipole for Portable Operation on 40 to 6 Meters by Kai Siwiak, KE4PT
- Extended Double-Zepp for 17 Meters
- Five-Band Two-Element Quad by Al Doig, W6NBH, and William Stein, KC6T
- Having a Field Day with the Moxon Rectangle, by L.B. Cebik, W4RNL
- The J78 Antenna: An Eight-band Off-Center-Fed HF Dipole by Brian Machesney, K1LI/J75Y
- The Trimox — A Moxon Tribander for a Holiday DXpedition by Brian Machesney, K1LI
- The W4SSY Spudgun by Byron Black, W4SSY
- Top-Loaded Low-Band Antenna Construction Details by Dick Stroud, W9SR
- Triband Dipole for 30, 17, and 12 Meters by Zack Lau W1VT
- Two-Band Loop for 30 and 40 Meters by James Brenner, NT4B
- Two W8NX Multiband, Coax-Trap Dipoles
- Wire Quad for 40 Meters by Dean Straw, N6BV

VHF and UHF Projects

- 6-Meter Halo Antenna for DXing by Jerry Clement, VE6AB
- A New Spin on the Big Wheel by L.B. Cebik, W4RNL, and Bob Cerreto, WA1FXT
- A Simple Fixed Antenna for VHF/UHF Satellite Work, by L.B. Cebik, W4RNL
- A Small 70-cm Yagi by Zack Lau, W1VT
- A True Plumber's Delight for 2 Meters — An All-Copper J-Pole by Michael Hood, KD8JB
- Cheap Antennas for the AMSAT LEOs by Kent Britain, WA5VJB
- Dual-Band Antenna for 146/446 MHz by Wayde Bartholomew, K3MF
- Medium-Gain 2 Meter Yagi by L.B. Cebik, W4RNL
- Quick and Cheap Omni Antenna for 1296 MHz by Paul Wade, W1GHZ

21.1 Antenna Basics

This section covers a range of topics that are fundamental to understanding how antennas work and defines several key terms. (A glossary is included in this book's online content.) While the discussion in this section uses the dipole as the primary example, the concepts apply to all antennas.

21.1.1 Directivity and Gain

All antennas, even the simplest types, exhibit directive effects in that the intensity of radiation is not the same in all directions from the antenna. This property of radiating more strongly in some directions than in others is called the *directivity* of the antenna. Directivity is the same for receiving as transmitting.

The directive pattern of an antenna at a given frequency is determined by its size and shape and on its position and orientation relative to the Earth and any other reflecting or absorbing surfaces.

The more an antenna's directivity is enhanced in a particular direction, the greater the *gain* of the antenna. This is a result of the radiated energy being concentrated in some directions at the expense of others. Similarly, gain describes the ability of the antenna to receive signals preferentially from certain directions. Gain does not create additional power beyond that delivered by the feed line — it only focuses that energy.

Gain is usually expressed in decibels, and is always stated with reference to a *standard* antenna — usually a dipole or an *isotropic radiator*. An isotropic radiator is a theoretical antenna that would, if placed in the center of an imaginary sphere, evenly illuminate that sphere with radiation. The isotropic radiator is an unambiguous standard, and for that reason is frequently used as the comparison for gain measurements.

When the reference for gain is the isotropic radiator in *free space*, gain is expressed in dBi. When the standard is a dipole also located in free space, gain is expressed in dBd. Because the dipole has some gain (2.15 dB with respect to the isotropic antenna as explained in the section on dipoles) in its favored direction, its gain can be expressed as 2.15 dBi. Gain in dBi can be converted to dBd by subtracting 2.15 dB and from dBd to dBi by adding 2.15 dB.

Gain also takes losses in the antenna or surrounding environment into account. For example, if a practical dipole antenna's wire element dissipated 0.5 dB of the transmitter power as heat, that specific dipole's gain with respect to an isotropic antenna would be 2.15 – 0.5 = 1.65 dBi.

21.1.2 Antenna Polarization

An electromagnetic wave has two components: an electric field and a magnetic field at right angles to each other. For most antennas, the field of primary interest is the electric, or *E-field*. The magnetic field is called the *H-field*. (The abbreviations E- and H- come from Maxwell's equations that describe electromagnetic waves.) By convention, the orientation of the E-field is the reference for determining the electromagnetic wave's *polarization*. The E-field of an electromagnetic wave can be oriented in any direction, so orientation with respect to the Earth's surface is the usual frame of reference. The wave's polarization can be vertical, horizontal, or some intermediate angle. If the E- and H-field orientations rotate as the wave travels, that is elliptical or circular polarization.

Antennas are also considered to have polarization determined by the orientation of the E-field of the electromagnetic field radiated by the antenna. Because the E-field of the radiated wave is parallel to the direction of current flow in the antenna's elements, the polarization of the wave and the orientation of the antenna elements is usually the same. For example, the E-field radiated by an antenna with linear elements is parallel to those elements, so that the polarization of the radiated wave is the same as the orientation of the elements. (This is somewhat over-simplified, and additional considerations apply for elements that are not linear.) Thus a radiator that is parallel to the Earth radiates a horizontally polarized wave, while a vertical antenna radiates a vertically polarized wave. If a wire antenna is slanted, it radiates waves with an E-field that has both vertical and horizontal components.

Antennas function symmetrically — a received signal will create the strongest antenna current when the antenna's elements are parallel to the E-field of the incoming wave, just as the radiated wave's E-field will be strongest parallel to current in the antenna's radiating elements. This also means that for the strongest received signal, the antenna elements should have the same polarization as that of the incoming wave. Misalignment of the receiving antenna's elements with the passing wave's E-field reduces the amount of signal received. This is called *cross-polarization*. When the polarizations of antenna and wave are at right angles, very little antenna current is created by the incoming signal.

For best results in line-of-sight communications, antennas at both ends of the circuit should have the same polarization. However, it is not essential for both stations to use the same antenna polarity for ionospheric propagation or sky wave (see the **Propagation** chapter). This is because the radiated wave is bent and rotated considerably during its travel through the ionosphere. At the far end of the communications path the wave may be horizontal, vertical, or somewhere in between at any given instant. For that reason, the main consideration for a good DX antenna is a low angle of radiation rather than the polarization.

Most HF-band antennas are either vertically or horizontally polarized. Although circular polarization is possible, just as it is at VHF and UHF, it is seldom used at HF. While most amateur antenna installations use the Earth's surface as their frame of reference, in cases such as satellite communication or EME the terms "vertical" and "horizontal" have no meaning with respect to polarization.

21.1.3 Current and Voltage Distribution

Using the dipole as an example, when power is fed to an antenna, the current and voltage vary along its length. The current is minimum at the ends, regardless of the antenna's length. The current does not actually reach zero at the current minima, because of capacitance at the antenna ends. Insulators, loops at the antenna ends, and support wires all contribute to this capacitance, which is also called the *end effect*. The opposite is true of the RF voltage. That is, there is a voltage maximum at each end.

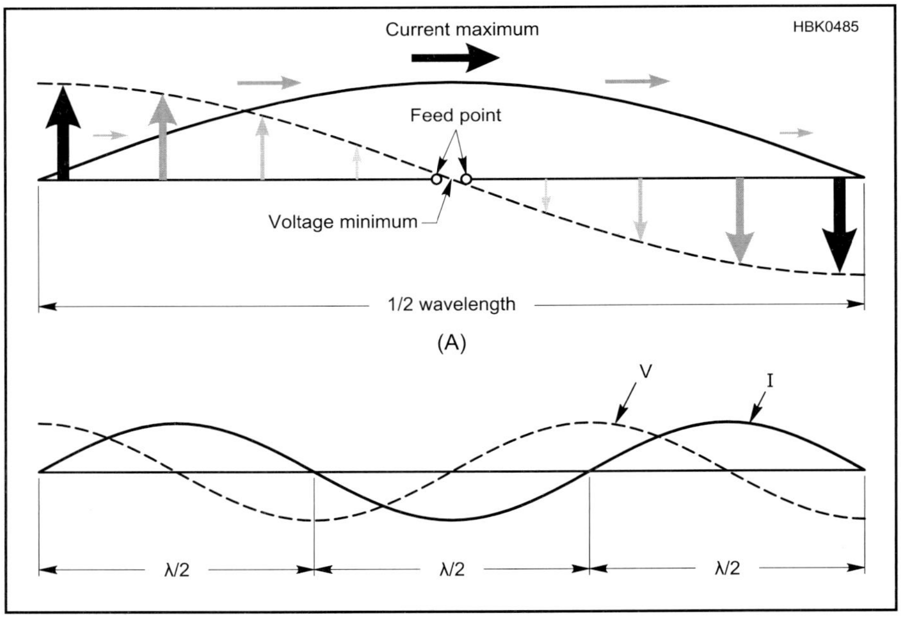

Figure 21.1 — The magnitude of current and voltage along a half-wave dipole (A) and for an antenna made from a series of half-wave dipoles (B). Both antennas are shown operating at a resonant frequency. Non-resonant antennas will exhibit a different pattern of voltage and current.

In the case of a half-wave dipole at its resonant frequency there is a current maximum at the center and a voltage minimum at the center, as illustrated in **Figure 21.1**. The graphs of current and voltage along the wire represent the magnitude of the current and voltage waveforms similarly to the envelope of a modulated signal. The magnitude of the voltage and current distributions are 90° out of phase, although the actual current and voltage waveforms are in phase, resulting in a purely resistive feed point impedance.

The pattern of alternating current and voltage maxima a quarter-wavelength apart repeats every half-wavelength along a resonant linear antenna, as shown in Figure 21.1B. The phase of the current and voltage are inverted in each successive half-wavelength section. (If the antenna is non-resonant, there will still be a current minimum at an open end but the voltage and current patterns on the antenna will be different than shown here.)

Power is dissipated as heat or as signals by the resistance of the antenna, which consists of both the RF resistance of the wire (ohmic loss resistance) and the *radiation resistance*. The radiation resistance is the equivalent resistance that would dissipate the power the antenna radiates, with a current flowing in it equal to the antenna current at a current maximum. Radiation resistance represents the work done by creating current in the antenna that transfers the energy from the signal source to the radiated electromagnetic wave. The loss resistance of a half-wave antenna is ordinarily small, compared with the radiation resistance, and can usually be neglected for practical purposes except in electrically small antennas, such as mobile HF antennas.

21.1.4 Impedance

The *impedance* at a given point in the antenna is determined by the ratio of the voltage to the current at that point. For example, if there were 100 V and 1.4 A of RF current at a specified point in an antenna and if they were in phase, the impedance would be approximately 71 Ω. The antenna's *feed point impedance* is the impedance at the point where the feed line is attached. If the feed point location changes, so does the feed point impedance.

Antenna impedance may be either resistive or complex (that is, containing resistance and reactance). The impedance of a *resonant* antenna is purely resistive anywhere on the antenna, no matter what value that impedance may be. For example, the impedance of a resonant half-wave dipole may be low at the center of the antenna and high at the ends, but it is purely resistive in all cases, even though its magnitude changes.

The feed point impedance is important in determining the appropriate method of matching the impedance of the antenna and the transmission line. The effects of mismatched antenna and feed line impedances are described in detail in the **Transmission Lines** chapter of this book. Some mistakenly believe that a mismatch, however small, is a serious matter. This is not true. The significance of a perfect match becomes more pronounced only at VHF and higher, where feed line losses are a major factor. Minor mismatches at HF are rarely significant.

21.1.5 Impedance and Height Above Ground

The feed point impedance of an antenna varies with height above ground because of the effects of energy reflected from and absorbed by the ground. For example, a ½ λ (or half-wave) center-fed dipole will have a feed point impedance of approximately 75 Ω in free space far from ground, but **Figure 21.2** shows that only at certain electrical heights above ground will the feed point impedance be 75 Ω. The feed point impedance will vary from very low when the antenna is close to the ground to a maximum of nearly 100 Ω at 0.34 λ above ground, varying between ±5 Ω as the antenna is raised farther. The 75 Ω feed point impedance is most likely to be realized in a practical installation when the horizontal dipole is approximately ½, ¾ or 1 wavelength above ground. This is why few amateur λ/2 dipoles exhibit a center-fed feed point impedance of 75 Ω, even though they may be resonant.

Figure 21.2 compares the effects of perfect ground and typical soil at low antenna heights. The effect of height on the radiation resistance of a horizontal half-wave antenna is not drastic so long as the height of the antenna is greater than 0.2 λ. Below this height, while decreasing rapidly to zero over perfectly conducting ground, the resistance decreases less rapidly with height over actual lossy ground. At lower heights the resistance stops decreasing at around 0.15 λ, and thereafter increases as height decreases further. The reason for the increasing resistance is that more and more energy from the antenna is absorbed by the ground as the height drops below ¼ λ, seen as an increase in feed point impedance.

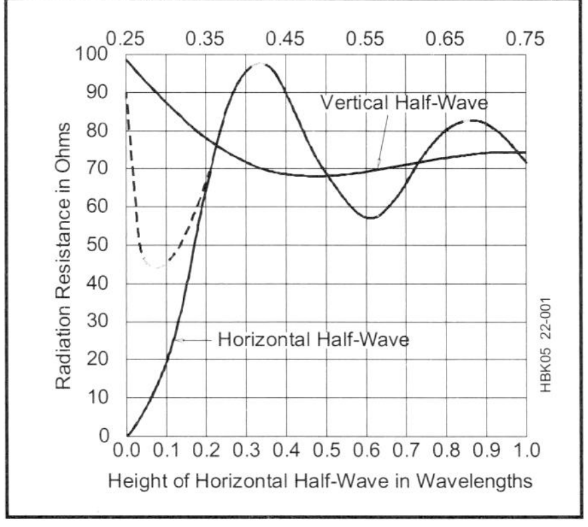

Figure 21.2 — Curves showing the radiation resistance of vertical and horizontal half-wavelength dipoles at various heights above ground. The broken-line portion of the curve for a horizontal dipole shows the resistance over *average* real ground, the solid line for perfectly conducting ground.

21.1.6 Antenna Bandwidth

The *bandwidth* of an antenna refers generally to the range of frequencies over which the antenna exhibits a specified level of performance. The bandwidth can be specified in units of frequency (MHz or kHz) or as a percentage of the antenna's design frequency.

Popular amateur usage of the term antenna bandwidth most often refers to the 2:1 *SWR bandwidth*, such as, "The 2:1 SWR bandwidth is 3.5 to 3.8 MHz" or "The antenna has a 10% SWR bandwidth" or "On 20 meters, the antenna has an SWR bandwidth of 200 kHz." Other specific bandwidth terms are also used, such as the *gain bandwidth* (the bandwidth over which gain is greater than a specified level) and the *front-to-back ratio bandwidth* (the bandwidth over which front-to-back ratio is greater than a specified level).

As operating frequency is lowered, an equivalent bandwidth in percentage becomes narrower in terms of frequency range in kHz or MHz. For example, a 5% bandwidth at 21 MHz is 1.05 MHz (more than wide enough to cover the whole band) but at 3.75 MHz only 187.5 kHz! Because of the wide percentage bandwidth of the lower frequency bands (160 meters is 10.5% wide, 80 meters is 3.4% wide), it is difficult to design an antenna with a bandwidth sufficient to include the whole band.

It is important to recognize that SWR bandwidth does not always relate directly to gain bandwidth. Depending on the amount of feed line loss, an 80 meter dipole with a relatively narrow 2:1 SWR bandwidth can still radiate a good signal at each end of the band, provided that an antenna tuner is used to allow the transmitter to load properly and feed line loss is not excessive. Broadbanding techniques, such as fanning the far ends of a dipole to simulate a conical type of dipole, can help broaden the SWR bandwidth.

21.1.7 Effects of Conductor Diameter

The impedance and resonant frequency of an antenna also depend on the diameter of the conductors that make up its elements in relation to the wavelength. As the diameter of a conductor increases, its capacitance per unit length increases and inductance per unit length decreases. This has the net effect of lowering

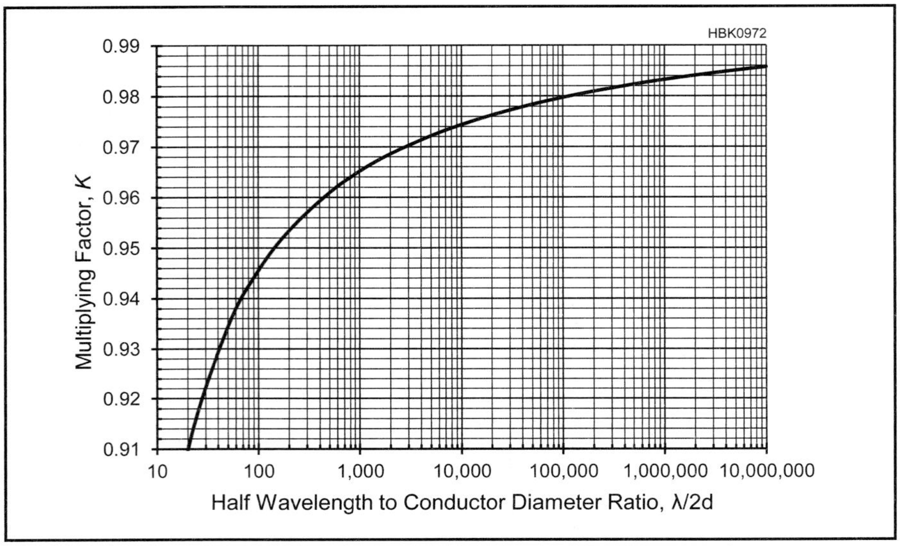

Figure 21.3 — Effect of antenna diameter on length for half-wavelength resonance in free-space, shown as a multiplying factor, *K*. The thicker the conductor relative to the wavelength, the shorter the physical length of the antenna at resonance. For antennas over ground, additional factors affect the antenna's electrical length.

the frequency at which the antenna element is resonant, as illustrated in **Figure 21.3**. The larger the conductor diameter in terms of wavelength, the smaller its *length-to-diameter ratio (l/d)* and the lower the frequency at which a specific length of that conductor is ½ wavelength long electrically, in free space.

K FACTOR

Based on the standard value for the speed of light, the wavelength λ of an electromagnetic wave in free-space at frequency f is given by:

$$\lambda(\text{in feet}) = \frac{983.571}{f(\text{in MHz})} \quad (1A)$$

and for one-half wavelength:

$$\frac{\lambda}{2}(\text{in feet}) = \frac{491.786}{\text{f(in MHz)}} \quad (1B)$$

A dipole with a length of exactly one-half of a free-space wavelength λ/2 has an impedance of $73.079 + j42.515\ \Omega$, essentially independent of wire diameter. The dipole must be shortened to achieve resonance where reactance is equal to zero. As the diameter of a conductor increases, its capacitance per unit length increases and inductance per unit length decreases. This increases the ratio of stored electric field energy to magnetic field energy, with the result of lowering the frequency at which the antenna element is resonant. Stated another way: For a given frequency, the larger a conductor's diameter, the lower the frequency at which a dipole made of that conductor is resonant.

A dipole's resonant length is given by the product of the free-space half-wavelength and a constant,

$$L_{Resonant}(\text{in feet}) = K \times \frac{491.786}{f(\text{in MHz})} \quad \text{or}$$

$$L_{Resonant}(\text{in inches}) = K \times \frac{5901.43}{f(\text{in MHz})} \quad (2)$$

where *K* is a constant between zero and one, $0 < K < 1$, that depends on the dipole's "fatness" or "thickness." There are several ways to express dipole thickness. Antenna builders prefer to specify dipole thickness in terms of the half-wavelength-to-diameter ratio (λ/2)/*d* because a builder generally knows *d* and wants to calculate L. The graph in Figure 21.3 satisfies this need and allows dipole resonant length to be determined for a given physical thickness and frequency. Most half-wavelength dipoles at HF typically have (λ/2)/d ratios in the range of 2500 to 25,000 with values of *K* from 0.97 to 0.98.

The graph shown in Figure 21.3 and Eq 2A below were determined by Steve Stearns, K6OIK, who evaluated and compared theoretical and numerical methods for calculating dipole and monopole impedance. (See the Bibliography entries for Stearns, Tai and Long, Elliott, and Schelkunoff). If better accuracy than the graph is needed, the formula in Eq 2A should be used. (A more complete discussion of the *K* factor and related topics is provided in the *ARRL Antenna Book's* 24th edition.)

$$K = 1 - \frac{0.225706}{\ln\left(\frac{\lambda/2}{d}\right) - 0.429451} \quad (2A)$$

Example 1: A half-wavelength dipole for 7.2 MHz has an uncorrected length of 491.786 / 7.2 = 68.3 feet. If it is made from #12 AWG wire (0.081 inch diameter), it has a (λ/2)/*d* ratio of:

$$\frac{491.786}{7.2}(\text{ft}) \times \frac{12\text{ in/ft}}{0.081\text{ in}} = 10{,}119$$

From Figure 21.3 or Eq 2A, a (λ/2)/*d* ratio of 10,119 gives *K* = 0.974. Thus, by Eq 2, the resonant length of the half-wavelength dipole is 0.974 × 68.3 = 66 feet 7 inches.

It should be understood that *K* is not a velocity factor because it is unrelated to waves or the speed of wave travel. Rather, *K* arises because if a dipole is exactly one-half wavelength long, the stored energies in the electric and magnetic fields are not exactly equal. A dipole must be shortened to obtain equality and resonance.

21.1.8 Radiation Patterns

Radiation patterns are graphic representations of an antenna's directivity. Two examples are given in **Figures 21.4** and **21.5**. Shown in polar coordinates (see the **Radio Mathematics** online supplement for information about polar coordinates), the angular scale shows direction, and the scale from the center of the plot to the outer ring, calibrated in dB, shows the relative strength of the antenna's radiated signal (gain) at each

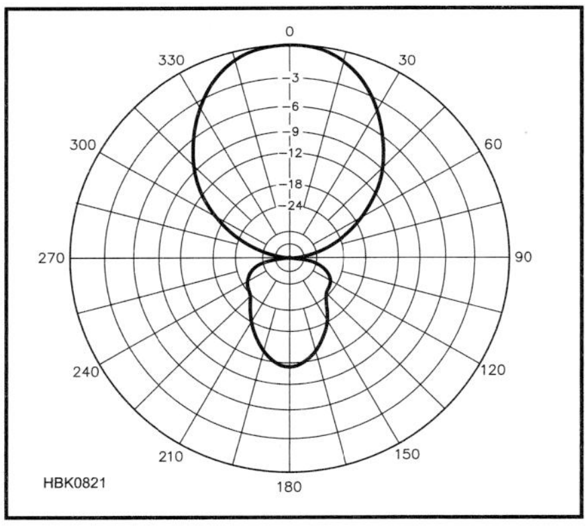

Figure 21.4 — Azimuthal pattern of a typical three-element Yagi beam antenna in free space. The Yagi's boom is along the 0° to 180° axis.

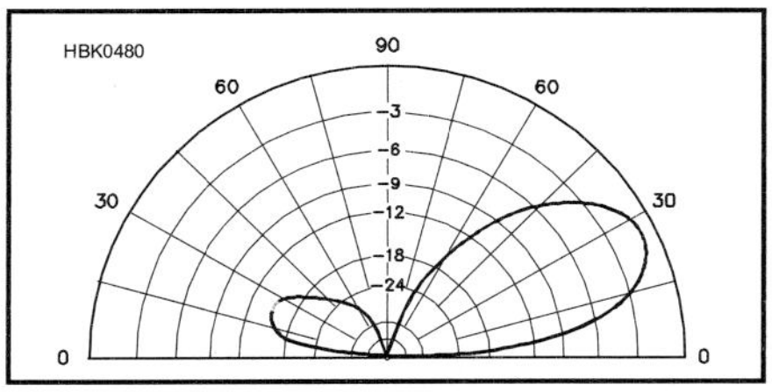

Figure 21.5 — Elevation pattern of a three-element Yagi beam antenna placed ½ λ above perfect ground.

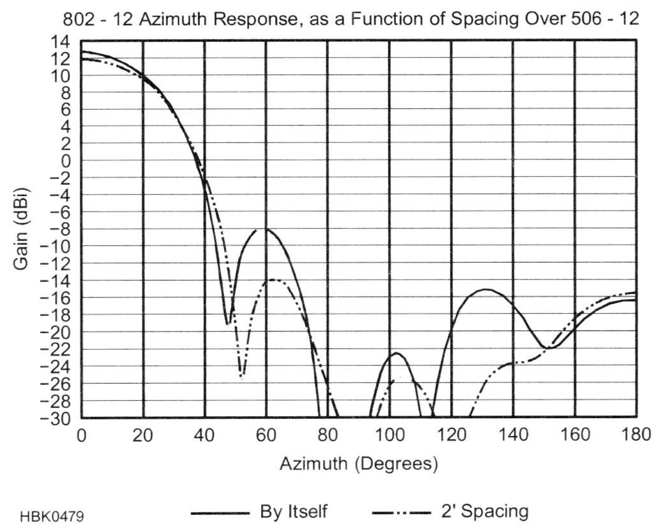

Figure 21.6 — Rectangular azimuthal pattern of an 8 element 2 meter Yagi beam antenna by itself and with another identical antenna stacked two feet above it. This example shows how a rectangular plot allows easier comparison of antenna patterns away from the main lobe.

angle. A line is plotted showing the antenna's relative gain (transmitting and receiving) at each angle. The antenna is located at the exact center of the plot with its orientation specified separately.

The pattern is composed of *nulls* (angles at which a gain minimum occurs) and *lobes* (a range of angles in which a gain maximum occurs). The *main lobe* is the lobe with the highest amplitude unless noted otherwise and unless several plots are being compared the peak amplitude of the main lobe is placed at the outer ring as a 0 dB reference point. The peak of the main lobe can be located at any angle. All other lobes are *side lobes* which can be at any angle, including to the rear of the antenna.

Figure 21.4 is an *azimuthal* or *azimuth pattern* that shows the antenna's gain in all horizontal directions (azimuths) around the antenna. As with a map, 0° is at the top and bearing angle increases clockwise. (This is different from polar plots generated for mathematical functions in which 0° is at the right and angle increases counterclockwise.)

Figure 21.5 is an *elevation pattern* that shows the antenna's gain at all vertical angles. In this case, the horizon at 0° is located to both sides of the antenna and the zenith (directly overhead) at 90°. The plot shown in Figure 21.5 assumes a ground plane (drawn from 0° to 0°), but in free-space the plot would include the missing semicircle with –90° at the bottom. Without the ground reference, the term "elevation" has little meaning, however.

You'll also encounter E-plane and H-plane radiation patterns. These show the antenna's radiation pattern in the plane parallel to the E-field or H-field of the antenna. It's important to remember that the E-plane and H-plane do not have a fixed relationship to the Earth's surface. For example, the E-plane pattern from a horizontal dipole is an azimuthal pattern, but if the same dipole is oriented vertically, the E-plane pattern becomes an elevation pattern.

Antenna radiation patterns can also be plotted on rectangular coordinates with gain on the vertical axis in dB and angle on the horizontal axis as shown in **Figure 21.6**. This is particularly useful when several antennas are being compared. Multiple patterns in polar coordinates can be difficult to read, particularly close to the center of the plot.

The amplitude scale of antenna patterns is almost always in dB. The scale rings can be calibrated in several ways. The most common is for the outer ring to represent the peak amplitude of the antenna's strongest lobe as 0 dB. All other points on the pattern represent *relative gain* to the peak gain. The antenna's *absolute gain* with respect to an isotropic (dBi) antenna or dipole (dBd) is printed as a label somewhere near the pattern. If several antenna radiation patterns are shown on the same plot for comparison, the pattern with the largest gain value is usually assigned the role of 0 dB reference.

The gain amplitude scale is usually divided in one of two ways. The ARRL standard scale has rings at 0, –3, –6, –12, –18, and –24 dB. This makes it easy to see where the gain has fallen to one-half of the reference or peak value (–3 dB), one-quarter (–6 dB), one-sixteenth (–12 dB), and so on. Another popular division of the amplitude scale is 0, –10, –20, –30, and –40 dB with intermediate rings or tick marks to show the –2, –4, –6, and –8 dB levels. You will encounter a number of variations on these basic scales.

RADIATION PATTERN MEASUREMENTS

Given the basic radiation pattern and scales, it becomes easy to define several useful measurements or metrics by which antennas are compared, using their azimuthal patterns. Next to gain, the most commonly used metric for directional antennas is the *front-to-back ratio* (*F/B*) or just "front-to-back." This is the difference in dB between the antenna's gain in the specified "forward" direction and in the opposite or "back" direction. The front-to-back ratio of the antenna in Figure 21.4 is about 11 dB. *Front-to-side ratio* is the difference between the antenna's "forward" gain and gain at right angles to the forward direction. This assumes the radiation pattern is symmetric, and is of most use to antennas such as Yagis and quads that have elements arranged in parallel planes. The front-to-side ratio of the antenna in Figure 21.4 is more than 30 dB. Because the antenna's rearward pattern can have large amplitude variations, the *front-to-rear ratio* is sometimes used. Front-to-rear uses the average of rearward gain over a specified angle, usually the 180° semicircle opposite the direction of the antenna's maximum gain, instead of a single gain figure at precisely 180° from the forward direction.

The antenna's *beamwidth* is the angle over which the antenna's main lobe gain is within 3 dB of the peak gain. Stated another way, the beamwidth is the angle between the directions at which the antenna's gain is –3 dB. In Figure 21.4, the antenna's main lobe beamwidth is about 54°, since the pattern crosses the –3 dB gain scale approximately 27° to either side of the peak direction. Antenna patterns with comparatively small beamwidths are referred to as "sharp" or "narrow."

An antenna with an azimuthal pattern that shows equal gain in all directions is called *omnidirectional*. This is not the same as an isotropic antenna that has equal gain in all directions, both vertical both horizontal.

21.1.9 Elevation Angle

For long-distance HF communication, the (vertical) *elevation angle* of maximum radiation, or *radiation angle*, is of considerable importance. You will want to erect your antenna

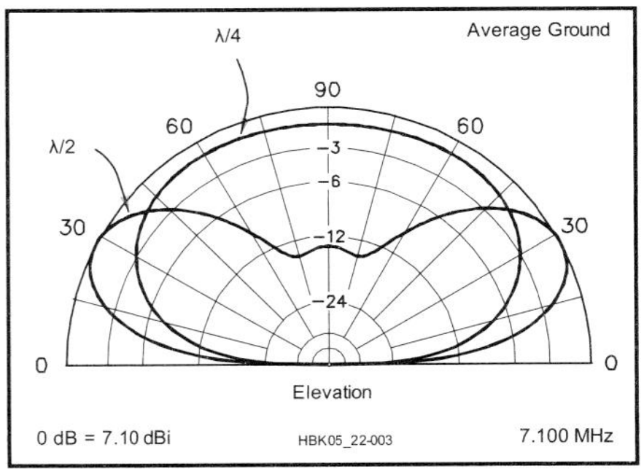

Figure 21.7 — Elevation patterns for two 40 meter dipoles over average ground (conductivity of 5 mS/m and dielectric constant of 13) at ¼ λ (33 foot) and ½ λ (66 foot) heights. The higher dipole has a peak gain of 7.1 dBi at an elevation angle of about 26°, while the lower dipole has more response at high elevation angles.

so that its strongest radiation occurs at vertical angles resulting in the best performance at the distances over which you want to communicate. In general, the greater the height of a horizontally polarized antenna, the stronger its gain will be at lower vertical angles. **Figure 21.7** shows this effect at work in horizontal dipole antennas. (See the **Propagation of Radio Signals** chapter and the *ARRL Antenna Book* for more information about how to determine the best elevation angles for communication.)

Since low radiation angles usually are most effective for long-distance communications, this generally means that horizontal antennas should be high — higher is usually better. (The optimum angle for intercontinental contacts on the HF bands is generally 15° or lower.) Experience shows that satisfactory results can be attained on the bands above 14 MHz with antenna heights between 40 and 70 feet.

Higher vertical angles can be useful for medium to short-range communications. For example, elevation angles between 20° and 65° are useful on the 40 and 80 meter bands over the roughly 550 mile path between Cleveland and Boston. Even higher angles may be useful on shorter paths when using these lower HF frequencies. A 75 meter dipole between 30 and 70 feet high works well for ranges out to several hundred miles.

For even shorter-range communications centered on your location, such as for emergency communications and regional nets, a very low antenna is used, generating its strongest radiation straight up. This is referred to as *Near-Vertical Incidence Skywave* (*NVIS*) communication. The antenna should be around 0.2 λ above ground, and the frequency used should be below the ionosphere's critical frequency so that the signal is completely reflected back toward the ground over a wide area.

Azimuthal patterns must also specify at what elevation angle the antenna gain is measured or calculated. While an azimuthal pattern may be in the plane of the antenna (an elevation angle of 0°), for antennas located above ground the gain will vary strongly with elevation angle.

21.1.10 Imperfect Ground

Ground (soil) conducts, but is far from being a perfect conductor. This influences the radiation pattern of the antennas that we use. The effect is most pronounced at high vertical angles (the ones most important for short-range communications and least important for long-distance communications) for horizontally polarized antennas. The consequences for vertical antennas are greatest at low angles, and are quite dramatic as can be clearly seen in **Figure 21.8**, where the elevation pattern for a 40 meter vertical half-wave dipole located over average ground is compared to one located over saltwater. At 10° elevation, the saltwater antenna has about 7 dB more gain than its landlocked counterpart.

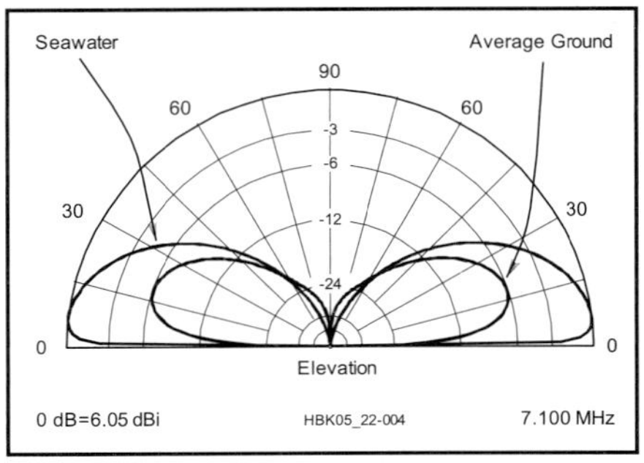

Figure 21.8 — Elevation patterns for a vertical dipole over sea water compared to average ground. In each case the center of the dipole is just over ¼ λ high. The low-angle response is greatly degraded over average ground compared to sea water, which is virtually a perfect ground.

An HF vertical antenna may work very well for a ham living in an area with rich soil. Ground of this type has very good conductivity. By contrast, a ham living where the soil is rocky or in a desert area may not be satisfied with the performance of a vertical HF antenna over such poorly conducting ground. Regardless of ground conductivity, the use of a ground screen of radial wires for HF verticals on land is necessary for good results.

When evaluating or comparing antennas, it is also important to include the effects of ground on antenna gain. Depending on height above ground and the qualities of the ground, reflections can increase apparent antenna gain by up to 6 dB. Because the actual installation of the antenna is unlikely to duplicate the environment in which the gain with reflections is claimed or measured, it is preferable to rely on free-space gain measurements or specifications that are independent of reflecting surfaces.

Antenna Systems for Small Stations

When getting started on HF, one of the most common questions is, "What is the best antenna?" The answer always begins with "It depends!" There are so many possibilities, but the choices frequently come down to a vertical dipole or ground plane versus horizontal dipoles. Many things affect the choice: ground conductivity, terrain, available supports, and esthetics have the most impact. Jim Brown, K9YC, addresses many of these concerns with data and measurements in his paper "Planning Antenna Systems for the Little Gun Station." Limits on the space available for "antenna farms" also present challenges to the station builder, as discussed by Jim in "Wire Antennas for Limited Space." Both papers are available online at **k9yc.com/publish.htm**.

21.1.11 Antenna Modeling

Computer modeling of antennas has made it a lot easier to evaluate antenna performance and design antennas. Most modeling software available to hams is based on *NEC* (Numerical Electromagnetics Code). *NEC* is a general-purpose program, capable of modeling almost any antenna type from simple dipoles to complex Yagis and more. By itself, *NEC* is difficult for non-professionals to use, so software has been developed that provides a more user-friendly interface. The best-known versions are *EZNEC* (**eznec.com**) and *4nec2* (**www.qsl.net/4nec2**). Both are available at no cost and run on any recent version of Windows.

NEC uses a *Method of Moments* algorithm. The basic principle is simple: An antenna is broken down into a set of straight-line wire *segments*. The fields resulting from the current in each segment and from the mutual interaction between segments are vector-summed in the far field to create azimuth and elevation-plane patterns. Feed point impedance and other parameters are also calculated. The antenna can be evaluated in free space or above ground with a wide variety of electrical characteristics.

There are a number of tutorials on using *EZNEC* and *4nec2*. Some are listed on the software's main web page. An internet search will turn up videos on how to use and apply these programs. In addition, there are many designs for antennas already created for these programs. If you are just getting started, practicing with one of the simple designs is an excellent way to learn the program and modeling process. The subject of antenna modeling is also discussed in some detail in the *ARRL Antenna Book*.

21.2 Dipoles and the Half-Wave Antenna

A fundamental form of antenna is a wire whose length is half the transmitting wavelength. It is the unit from which many more complex forms of antennas are constructed and is known as a *dipole antenna*. (The name di- meaning *two* and -pole meaning *electrical charge* or *polarity* comes from the antenna having two distinct regions of electrical polarity as shown in Figure 21.1.) A dipole is resonant when the current and voltage in the antenna are exactly in phase. Even though the magnitudes of current and voltage along the antenna vary as shown in Figure 21.1, the actual RF current and voltage are in phase, and the feed point impedance is entirely resistive at resonance.

The actual length of a resonant ½ λ antenna will not be exactly equal to the half wavelength of the radio wave of that frequency in free space, but depends on the thickness of the conductor in relation to the wavelength as shown in Figure 21.3. An additional shortening effect occurs with wire antennas supported by insulators at the ends because of current flow through the capacitance at the wire ends due to the end effect. Interaction with the ground and any nearby conductors also affects the resonant length of the physical antenna.

The following formula is sufficiently accurate for dipoles below 10 MHz at heights of ⅛ to ¼ λ and made of common wire sizes. To calculate the length of a half-wave antenna in feet,

$$\text{Length (ft)} = \frac{492 \times 0.95}{f\,(\text{MHz})} = \frac{468}{f\,(\text{MHz})} \quad (3)$$

Example: A half-wave antenna for 7150 kHz (7.15 MHz) is 468/7.15 = 65.5 feet, or 65 feet 6 inches. This length is from end insulator to end insulator, including the loops of wire through the insulator and the length of wires that connect to the feed line. The loops and extra connections will add a small amount of electrical length to the antenna, so be prepared to make small adjustments in antenna length before making a final installation.

For antennas at higher frequencies and/or higher above ground, the best approach is to build the initial dipole using a numerator value of 485 to 490. Make temporary connections at the end insulators. Install the antenna and measure its SWR or impedance. Calculate the percentage difference between the desired and measured resonant frequencies. Adjust the antenna's length by the same amount. For example, for a high dipole at 14.150 MHz, start with an antenna that is 490 / 14.150 = 34.63 feet long. If the antenna's measured frequency of minimum SWR is 14.6 MHz, the antenna is 100 × 14.6 / 14.150 = 3.2 % too long and should be shortened by that amount, removing an equal amount of wire from each side of the antenna.

Above 30 MHz use the following formulas, particularly for antennas constructed from rod or tubing. K is taken from Figure 21.3.

$$\text{Length (ft)} = \frac{492 \times K}{f\,(\text{MHz})} \quad (4A)$$

$$\text{Length (in)} = \frac{5904 \times K}{f\,(\text{MHz})} \quad (4B)$$

Example: Find the length of a half-wave antenna at 50.1 MHz, if the antenna is made of ½ inch-diameter tubing. At 50.1 MHz, a half wavelength in space is

$$\frac{492}{50.1} = 9.82 \text{ ft}$$

The ratio of half wavelength to conductor diameter (changing wavelength to inches) is

$$\frac{(9.82 \text{ ft} \times 12 \text{ in./ft})}{0.5 \text{ in.}} = 235.7$$

From Figure 21.3, K = 0.945 for this ratio. The length of the antenna from equation 3 is

$$\frac{492 \times 0.945}{50.1} = 9.28 \text{ ft}$$

or 9 feet 3-⅜ inches. The answer is obtained directly in inches by substitution in equation 4B

$$\frac{5904 \times 0.945}{50.1} = 111.4 \text{ in}$$

Regardless of the formula used to calculate the length of the half-wave antenna, the effects of ground and conductive objects within a wavelength or so of the antenna usually make it necessary to adjust the installed length in order to obtain the lowest SWR at the desired frequency. Use of antenna modeling software may provide a more accurate initial length than a single formula.

The value of the SWR indicates the quality of the match between the impedance of the antenna and the feed line. If the lowest SWR obtainable is too high, an impedance-matching network may be used, as described in the **Transmission Lines** chapter. (High SWR may cause modern transmitters with solid-state power amplifiers to reduce power output as a protective measure for the output transistors.)

21.2.1 Radiation Characteristics

The radiation pattern of a dipole antenna in free space is strongest at right angles to the wire as shown in **Figure 21.9**. In an actual installation, the figure-Eight pattern is less directive due to reflections from ground and other conducting surfaces. As the dipole is raised to ½ λ

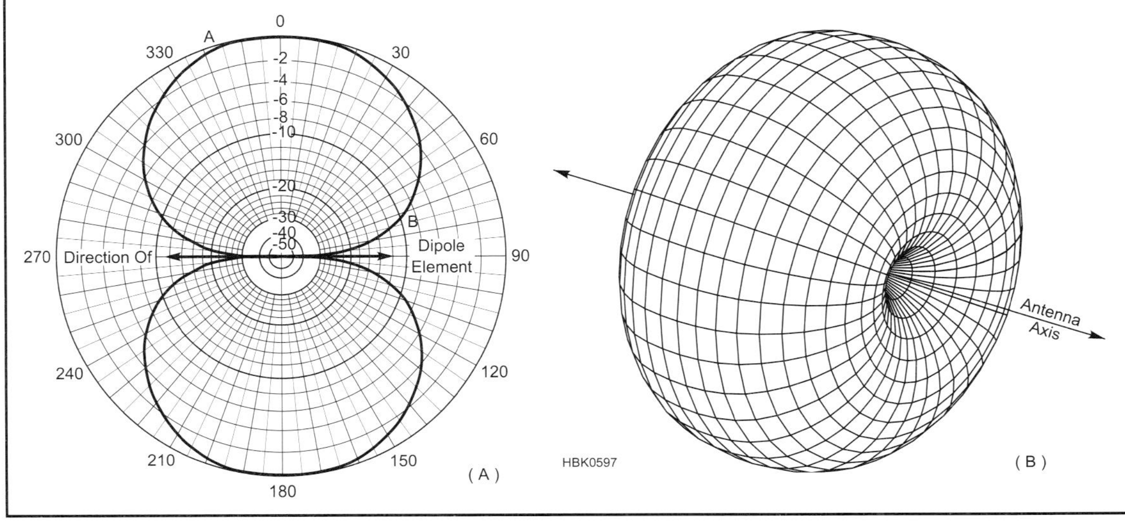

Figure 21.9 — Response of a dipole antenna in free space in the plane of the antenna with the antenna oriented along the 90° to 270° axis (A). The full three-dimensional pattern of the dipole is shown at (B). The pattern at A is a cross-section of the three-dimensional pattern taken perpendicularly to the axis of the antenna.

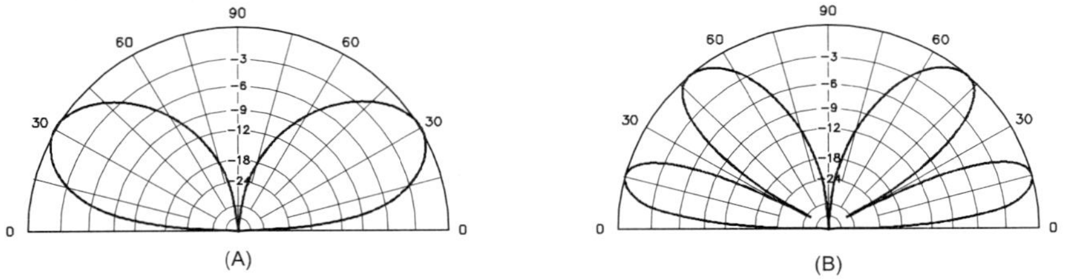

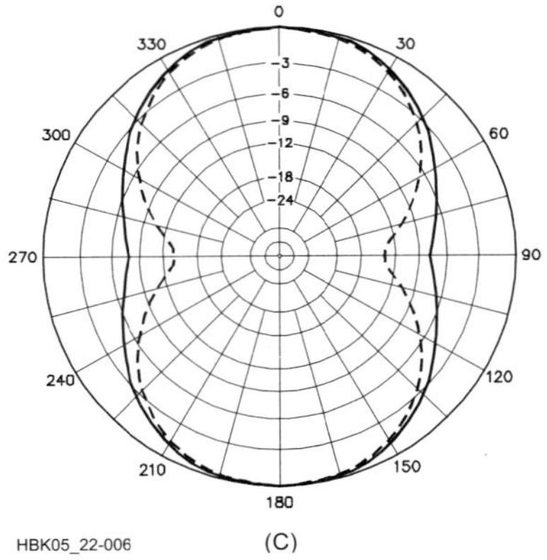

Figure 21.10 — At A, the elevation response pattern of a dipole antenna placed ½ λ above perfectly conducting ground. At B, the pattern for the same antenna when raised to 1 λ. For both A and B, the conductor is coming out of the paper at a right angle. C shows the azimuth patterns of the dipole for the two heights at the most-favored elevation angle, the solid-line plot for the ½ λ height at an elevation angle of 30°, and the broken-line plot for the 1 λ height at an elevation angle of 15°. The conductor in C lies along the 90° to 270° axis.

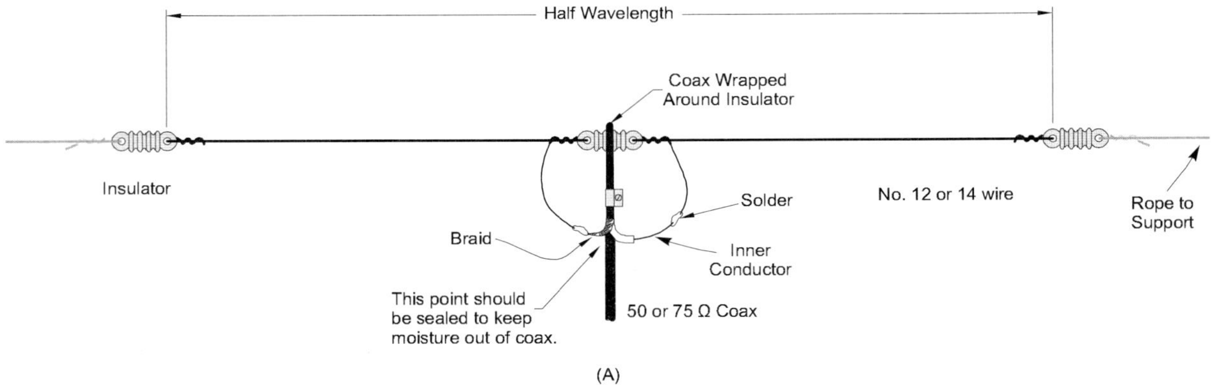

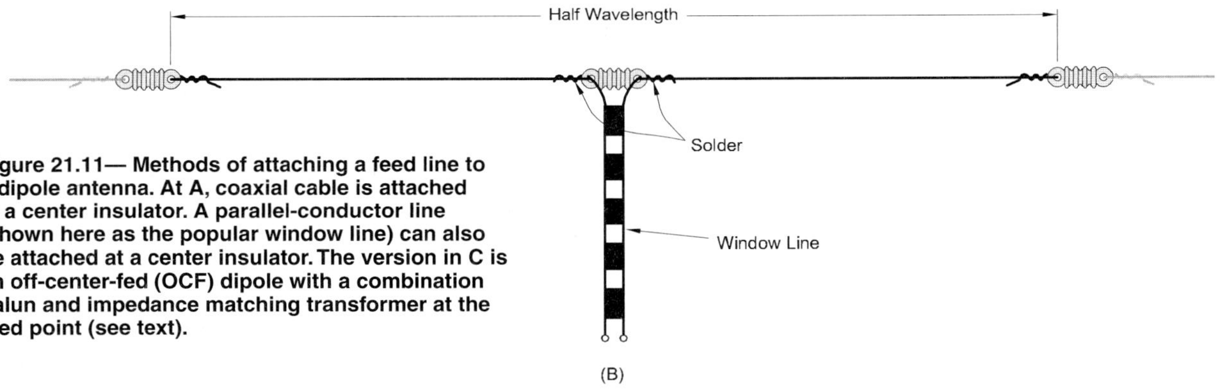

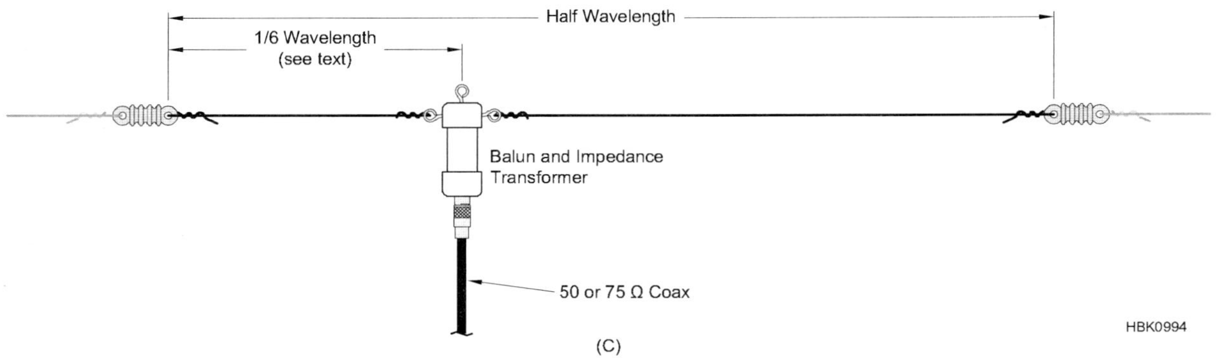

Figure 21.11— Methods of attaching a feed line to a dipole antenna. At A, coaxial cable is attached at a center insulator. A parallel-conductor line (shown here as the popular window line) can also be attached at a center insulator. The version in C is an off-center-fed (OCF) dipole with a combination balun and impedance matching transformer at the feed point (see text).

or greater above ground, nulls off the ends of the dipole become more pronounced. Sloping the antenna and coupling to the feed line tend to distort the pattern somewhat.

As a horizontal antenna is brought closer to ground, the elevation pattern peaks at a higher elevation angle as shown in Figure 21.7. **Figure 21.10** illustrates what happens to the directional pattern as antenna height changes. Figure 21.10C shows that there is significant radiation off the ends of a low horizontal dipole. For the ½ λ height (solid line), the radiation off the ends is only 7.6 dB lower than that in the broadside direction.

Figure 21.10 also shows that for short-range communication that depends on high vertical angles of radiation (NVIS communications), a dipole can be too high. For these applications, the dipole should be installed at about 0.2 λ so that the antenna radiates strongly at high vertical angles and with little horizontal directivity.

21.2.2 Feed Methods

A feed line is attached directly to the dipole, generally at the center, where an insulator separates the antenna's conductor into two sections or "legs." This is the antenna's *feed point*. One conductor of the feed line is attached to each leg. **Figures 21.11A** and 21.11B show how the two types of feed lines are attached. There are numerous variations, of course

The low feed point impedance of the half-wave dipole at its resonant frequency, f_0, and odd harmonics results in a low SWR when fed with coaxial cable. The feed point impedance and resulting SWR with coaxial cable or "coax" will be high at even harmonics and other frequencies.

When fed with ladder line and a wide-range impedance-matching unit, a center-fed antenna can be used on nearly any frequency, including non-resonant frequencies. (An example of such an antenna system is presented as a project farther along in this section.) Another variation, shown in Figure 21.11B, is the *Zepp* (or *end-fed half wave, EFHW*), named for its original application as an antenna deployed from Zeppelin airships. The feed point impedance of a "Zepp" is quite high, requiring impedance matching to deliver power effectively.

The EFHW has become popular for portable, low-power operation since it can be temporarily installed with only one support and a short coax feed line from an impedance-matching transformer at the feed point. The EFHW can be used as a variation of the *random-wire antenna*, where it is attached directly to the output of the transmitter. Care should be taken due to exposure concerns.

A dipole can be fed anywhere along its length, although the impedance of the antenna will vary as discussed earlier. One common variation is the *off-center-fed* (*OCF*) dipole (Figure 21.11C), where the feed point is offset from center by some amount and an impedance transformer matches the resulting moderately high impedance to that of coaxial cable. The impedance transformer is usually paired with a choke balun to decouple the feed line, as described in the next section. The OCF can also be fed with open-wire line. OCF dipoles are further discussed later in this chapter.

21.2.3 Feed Point Baluns

Open-wire transmission lines and center-fed dipole antennas are *balanced*, that is, each conductor or section has the same impedance to the system reference, usually Earth ground. This is different from *unbalanced* coaxial cable, in which the shield is generally connected to an Earth ground at some point, generally at the transmitter. To use balanced open-wire transmission lines with unbalanced equipment — most amateur equipment is unbalanced — a *balun* is required to make the transition between the balanced and unbalanced parts of the antenna system. "Balun" is an abbreviation of "balanced-to-unbalanced," the function of the device — it allows power to be transferred between the balanced and unbalanced portions of an antenna system in either direction. The most common application of baluns is to connect an unbalanced feed line to a balanced antenna. (Baluns and the related *ununs* are discussed in the **Transmission Lines** chapter.) Because dipoles are balanced, a balun is often used at the feed point when a dipole is fed with coax.

Due to the skin effect discussed in the **RF Techniques** chapter, the inside and outside of the coaxial cable shield act as separate conductors at RF. This "third conductor" of a coaxial cable unbalances the symmetry of the dipole antenna when the coax is connected directly to the dipole, as shown in Figure 21.11A. As a result, RF current can flow on the outside of the cable shield to the enclosures of station equipment connected to the cable. This combination of the dipole and feed line shield is likely to have a different resonant frequency and feed point impedance than what was intended.

In order to *decouple* the coax shield from the antenna, a *choke* or *current balun* is recommended as shown in **Figure 21.12**. The choke balun creates a high impedance on the outside of the coax shield while leaving currents inside the coax unaffected. See the **Transmission Lines** chapter for choke balun designs.

Shield currents can also impair the function of instruments connected to the line (such as SWR meters and SWR-protection circuits in the transmitter). The shield current also produces some feed line radiation, which changes the antenna radiation pattern and allows objects near the cable to affect the antenna-system performance.

The consequences may be negligible: A slight skewing of the antenna pattern usually goes unnoticed. Or, they may be significant: False SWR readings may cause the transmitter to reduce power unnecessarily; radiating coax near a TV feed line may cause strong local interference from overload. Therefore, it is better to eliminate feed line radiation whenever possible, and a choke balun should be used at any transition between balanced and unbalanced systems. Even so, balanced or unbalanced systems without a balun often operate with no apparent problems. For temporary or emergency stations, do not let the lack of a balun deter you from operating.

21.2.4 Building Dipoles and Other Wire Antennas

The purpose of this section is to offer information on the actual physical construction of wire antennas. Because the dipole, in one of its configurations, is probably the most common amateur wire antenna, it is used in the following examples. The techniques described here, however, enhance the reliability and safety of all wire antennas.

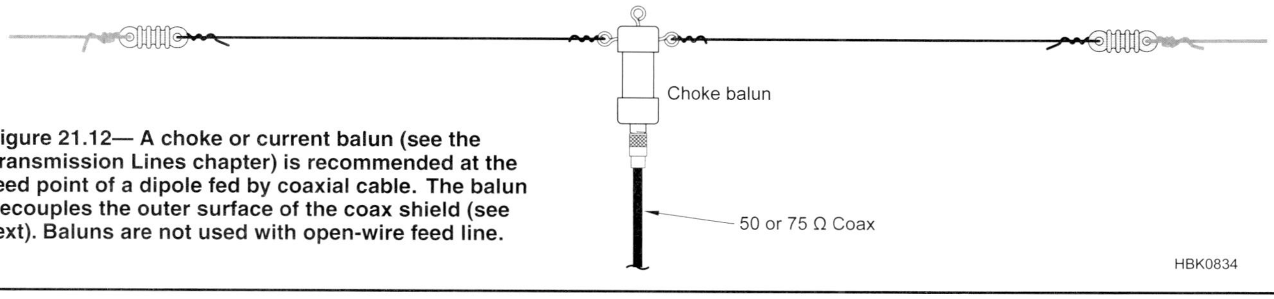

Figure 21.12— A choke or current balun (see the Transmission Lines chapter) is recommended at the feed point of a dipole fed by coaxial cable. The balun decouples the outer surface of the coax shield (see text). Baluns are not used with open-wire feed line.

WIRE

Choosing the right type of wire for the project at hand is the key to a successful antenna — the kind that works well and stays up through a winter ice storm or a gusty spring wind storm. What gauge of wire to use is the first question to settle; the answer depends on strength, ease of handling, cost, availability, and visibility. Generally, antennas that are expected to support their own weight plus the weight of the feed line should be made from #12 AWG or larger wire. (The National Electrical Code [NEC] specifies a minimum size of #10 AWG for external antennas.) Horizontal dipoles, Zepps, some long wires, and the like fall into this category. Antennas supported in the center, such as inverted-V dipoles and delta loops, may be made from lighter material such as #14 AWG wire.

The type of wire to be used is the next important decision. The wire specifications table at the end of this chapter shows popular wire styles and sizes. The strongest wire suitable for antenna service is *copper-clad steel*, also known as *copperweld*. The copper coating is necessary for RF service because steel is a relatively poor conductor. Practically all of the RF current is confined to the copper coating because of skin effect. Copper-clad steel is outstanding for permanent installations, but it can be difficult to work with because of the stiffness of the steel core. Kinks, abrasion, or repeated flexing can crack the copper cladding, leading to rusting of the inner steel core. Damage to the cladding can result in increased resistance and loss. Rusting of the inner core will eventually weaken the antenna. Regularly inspecting the antenna visually or checking SWR for changes can help spot problems.

Solid-copper wire, either hard-drawn or soft-drawn, is another popular material. Easier to handle than copper-clad steel, solid copper is available in a wide range of sizes. It is generally more expensive however, because it is all copper. Soft-drawn stretches under tension, so periodic pruning of the antenna may be necessary in some cases. Enamel-coated *magnet-wire* is a suitable choice for experimental antennas because it is easy to manage, and the coating protects the wire from the weather. Although it stretches under tension, the wire may be pre-stretched before final installation and adjustment. A local electric motor rebuilder might be a good source for magnet wire.

Hook-up wire, speaker wire, or even ac lamp cord are suitable for temporary installations. Almost any copper wire may be used, as long as it is strong enough for the demands of the installation.

Aluminum wire can be used for antennas, but is not as strong as copper or steel for the same diameter, and soldering it to feed lines requires special techniques. Galvanized and steel wire, such as that used for electric fences, is inexpensive, but it is a much poorer conductor at RF than copper and should be avoided.

Kinking, which severely weakens wire, is a potential problem when handling any solid conductor. When uncoiling solid wire of any type — copper, steel, or aluminum — take care to unroll the wire or untangle it without pulling on a kink to straighten it. A kink is actually a very sharp twist in the wire and the wire will break at such a twist when flexed, such as from vibration in the wind.

Solid wire also tends to fail at connection or attachment points at which part of the wire is rigidly clamped. The repeated flexing from wind and other vibrations eventually causes metal fatigue and the wire breaks. Stranded wire is preferred for antennas that will be subjected to a lot of vibration and flexing. If stranded wire is not suitable, use a heavier gauge of solid wire to compensate.

Insulated vs Bare Wire

Losses are the same (in the HF region at least) whether the antenna wire is insulated or bare. If insulated wire is used, a 3 to 5% shortening from the length calculated for a bare wire is required to obtain resonance at the desired frequency. This is caused by the increased distributed capacitance resulting from the higher dielectric constant of the plastic insulating material compared to air. The actual length for resonance must be determined experimentally by pruning and measuring because the dielectric constant of the insulating material varies from wire to wire. Wires that might come into contact with humans or animals should be insulated to reduce the chance of shock or burns.

INSULATORS

Wire antennas must be insulated at the ends and usually at the feed point. Commercially available insulators are made from ceramic, glass, or plastic. Insulators are available from many amateur radio dealers. Local hardware and farm stores are other possible sources of insulators for electric fences. Ceramic or glass insulators will usually outlast the wire, so they are highly recommended for a safe, reliable, permanent installation. **Figure 21.13C** shows the proper way to attach a wire to a strain or "egg" insulator. Figures 21.13A and 21.13B show examples of center insulators. Figure 21.13A is popularly known as a "dog bone" insulator and is available in glass, ceramic, and plastic. The ridges prevent water or dust from forming a conductive path across the insulator. Figure 21.13B shows a Budwig HQ-1 insulator with a built-in SO-239 and wires for attaching to the antenna. Acceptable homemade insulators may be made from a variety of material including (but not limited to) polycarbonate sheet or rod, PVC tubing or plumbing fixtures, or fiberglass rod. Temporary or emergency antennas can even make use of stiff plastic from a discarded container or dry wood. **Figures 21.14A** and 21.14B show some homemade insulators. If you make your own insulators, be sure the material is sturdy enough to withstand the mechanical stress and is UV-resistant to be suitable for prolonged outdoor use.

ATTACHING FEED LINES

Most wire antennas require an insulator at the feed point. Although there are many ways to connect the feed line, there are a few things to keep in mind. If you feed your antenna with coaxial cable, you have two choices. You can install an SO-239 connector on the center insulator and use a PL-259 on the end of your coax, or you can separate the center conductor from the braid, creating a pigtail connection,

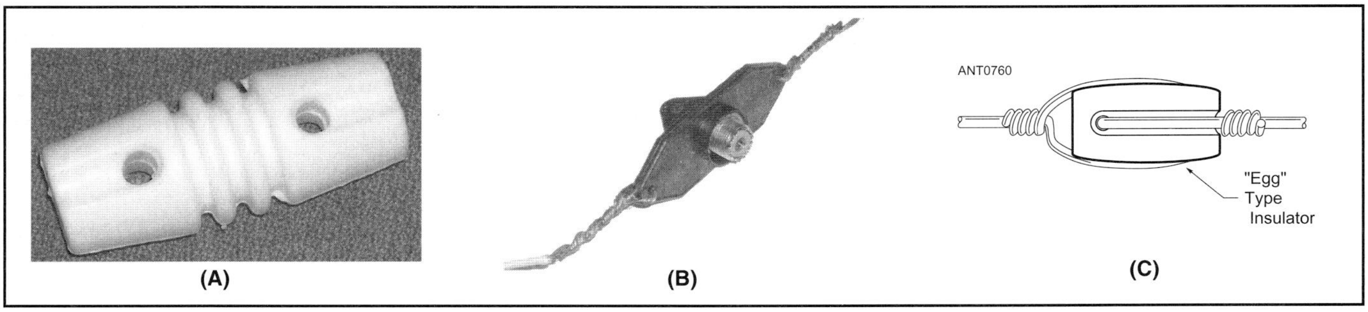

Figure 21.13 — (A) and (B) show examples of commercial center insulators: a dog-bone insulator and a Budwig HQ-1 insulator. (C) The correct way to attach a wire to a strain or "egg" insulator.

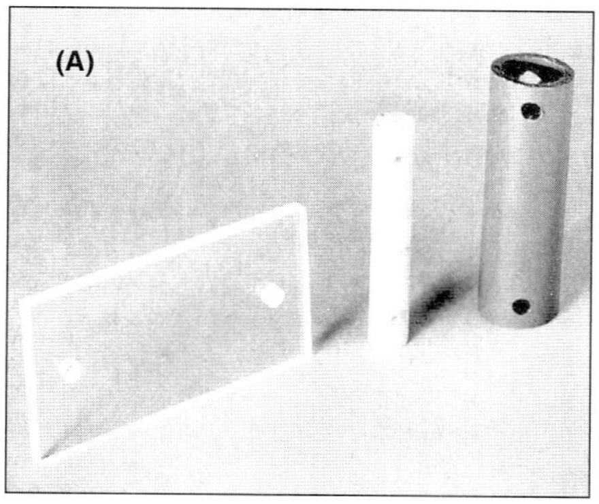

Figure 21.14 — (A) Some ideas for homemade antenna insulators. (B) Some homemade dipole center insulators. The one in the center includes a built-in SO-239 connector. Others are designed for direct connection to the feed line. Be sure to use UV-resistant plastic for insulators.

and attach the feed line directly to the antenna wire. Although it costs less to connect directly to the antenna, the use of connectors offers several advantages. (See the **Transmission Lines** chapter for more information about selecting and installing feed lines.)

Coaxial cable braid acts as a wick to soak up water. If you do not adequately seal the antenna end of the feed line, water will find its way into the braid. Water in the feed line will contaminate the braid, causing high losses and rendering the coax useless long before its normal lifetime is up. It is not uncommon for water to drip from the end of the coax inside the station after a year or so of service if the antenna connection is not properly waterproofed. Use of a PL-259/SO-239 combination (or other connector of your choice) makes the task of waterproofing connections much easier. Another advantage to using the PL-259/SO-239 combination is that feed line replacement is much easier, should that become necessary or desirable.

Whether you use coaxial cable, ladder line, or twin lead to feed your antenna, an often-overlooked consideration is the mechanical strength of the connection. Wire antennas and feed lines tend to move a lot in the wind, and unless the feed line is attached securely, the connection will weaken with time. The resulting failure can range from a frustrating intermittent electrical connection to a complete separation of feed line and antenna. Figure 21.14B illustrates several different ways of attaching the feed line to the antenna. An idea for supporting ladder line is shown in **Figure 21.15**.

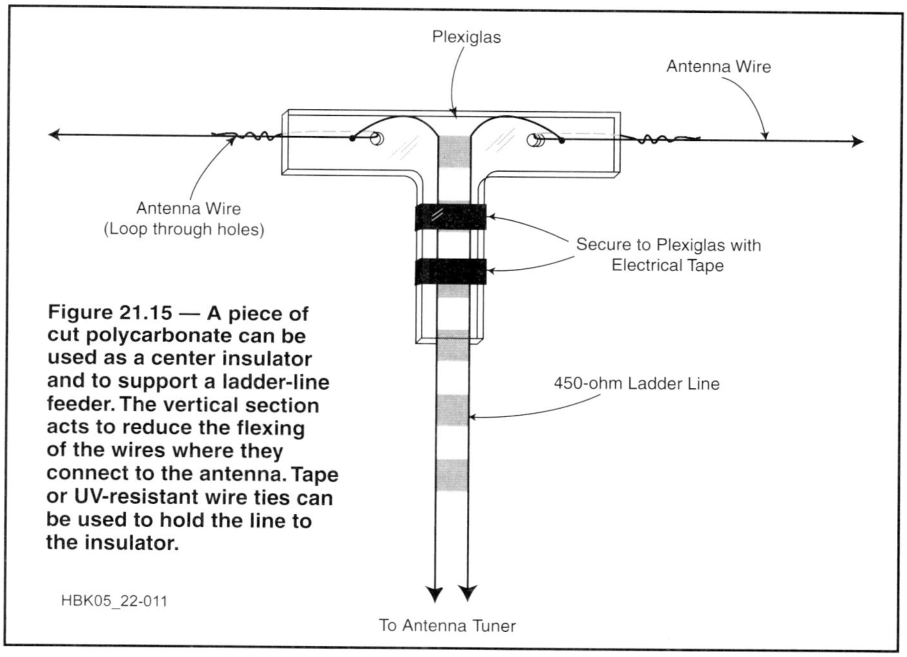

Figure 21.15 — A piece of cut polycarbonate can be used as a center insulator and to support a ladder-line feeder. The vertical section acts to reduce the flexing of the wires where they connect to the antenna. Tape or UV-resistant wire ties can be used to hold the line to the insulator.

PUTTING IT TOGETHER

If made from the right materials and installed in the clear, the dipole should give years of maintenance-free service. As you build your antenna, keep in mind that if you get it right the first time, you won't have to do it again for a long time.

Figure 21.11 shows details of antenna construction. Although a dipole is used for the examples, the techniques illustrated here apply to any type of wire antenna. **Table 21.1** shows dipole lengths for the amateur HF bands. These lengths do not include the extra wire required to attach the wire to the insulator as shown in Figure 21.11. Determine the extra amount of wire required by experimenting with the insulator you intend to use. Add twice this amount of wire to the leg lengths in Table 21.1, one extra length for each insulator. (Four such attachments are required for each dipole — two on each leg.)

Most dipoles require a little pruning to reach the desired resonant frequency due to the effects of ground and nearby conducting objects and surfaces. (See the beginning of this section for an explanation.) Table 21.1 includes lengths based on the classic 468/f formula and on 490/f intended to be a starting length. Record the constructed length with all insulators attached. (The constructed length is measured between the ends of the loops at each end of the wire and includes the length of any feed line-to-antenna connections.) Next, raise the dipole to the working height and find the frequency at which minimum SWR occurs. Multiply the frequency of the SWR minimum by the antenna length and divide the result by the desired f_0. The result is the finished length; trim both ends equally to reach that length and you're done. For example, if you want the SWR minimum to occur at 14.1 MHz and the first attempt with a constructed length of 33.8 feet results in an SWR minimum at 13.9 MHz, the final length for the antenna is

$$33.8 \times (13.9/14.1) = 33.3 \text{ ft}$$

In determining how well your antenna will work over the long term, how well you put the

Table 21.1

Dipole Dimensions for Amateur Bands

Table A – Dipole Lengths based on 468/f

Freq	*------Overall Length ------*			*------- Leg Length -------*		
(MHz)	*ft*	*in*	*m*	*ft*	*in*	*m*
1.82	257	2	78.4	128	7	39.2
3.6	130	0	39.7	65	0	19.8
5.378	7	2	43.6	43	7	21.8
7.1	65	10	20.1	32	11	10.05
10.1	46	4	14.13	23	2	7.06
14.1	33	2	10.12	16	7	5.06
18.1	25	10	7.89	12	11	3.94
21.1	22	2	6.76	11	1	3.38
24.9	18	9½	5.73	9	4¾	2.87
28.4	16	6	5.03	8	3	2.51

Table B – Dipole Lengths based on 490/f

Freq	*------Overall Length ------*			*------- Leg Length -------*		
(MHz)	*ft*	*in*	*m*	*ft*	*in*	*m*
1.82	269	3	82.1	134	7	41.1
3.6	136	1	41.5	66	1	20.8
5.37	91	3	27.8	45	7	13.9
7.1	69	0	21.0	34	6	10.5
10.1	48	6	14.8	24	3	7.4
14.1	34	9	10.6	17	5	5.3
18.1	27	1	8.26	13	6½	4.13
21.1	23	2½	7.08	11	7¼	3.54
24.9	19	8	6.00	9	10	3.00
28.4	17	3	5.26	8	7½	2.63

pieces together is second only to the ultimate strength of the materials used. Even the smallest details, such as how you connect the wire to the insulators, contribute significantly to antenna longevity. Where wires are soldered, use only enough heat to make a good connection. Excessive heat, such as from a propane torch, can anneal the wire, weakening it at the solder point.

By using plenty of wire at the end insulators and wrapping it tightly, you will decrease the possibility of the wire pulling loose in the wind. There is no need to solder the wire once it is wrapped. There is no electrical connection here, only mechanical.

Similarly, the feed line connection at the center insulator should be made to the antenna wires after they have been secured to the insulator. This way, you will be assured of a good electrical connection between the antenna and feed line without compromising the mechanical strength. Do a good job of soldering the antenna and feed line connections. Use a heavy iron or a soldering gun (200 W is a good size to use for this job), and be sure to clean the materials thoroughly before starting the job. If possible, solder the connections at a workbench, where the best possible joints may be made. Poorly soldered or unsoldered connections will become headaches as the wire oxidizes and the electrical integrity degrades with time. Besides degrading your antenna performance, poorly made joints can even be a cause of TVI because of rectification. Spray the connections with a UV-resistant acrylic coating or use a brush-on insulation coating for waterproofing.

NON-SOLDERED ATTACHMENTS

There are other options than soldering for securing wires to insulators and attaching feed lines, particularly parallel conductor lines. You can use crimping or clamping connectors as well as waterproof terminals. As with soldered connections, some form of mechanical strain relief will be necessary to prevent repeated flexing from breaking the wires at the connection.

Split-bolt connectors used for grounding conductors can be used to hold large wires together. (Search for "split-bolt" at **www.grainger.com**.) They are available in bronze and copper alloys (for copper-copper connections). The smallest standard sizes such as Burndy KS90 accept up to #8 AWG wires and can be used to clamp combinations of smaller wires down to #16 AWG. These connectors are not waterproof, so coat all of the conductors with a contact protective compound intended for outdoor use. Sleeve-style crimp connectors can also be used with the same caution about using an outdoor-listed protective compound.

For lighter duty and temporary connections, Anderson Powerpole and quick-disconnect crimp terminals are good choices. You must solder or crimp the wires into the terminals. These non-locking connections will pull apart under tension, so some strain relief is necessary. Both types of connectors can be waterproofed with a heat-shrink or other protective sleeve.

ATTACHING ANTENNA SUPPORTS

If towers, masts, or buildings are available, attaching the antenna support ropes or cables is fairly straightforward. Pulleys or guides such as stainless steel or aluminum carabiners or large screw eyes rated for the expected load should be used. Counterweights or springs can be used to allow the antenna to move in the wind.

Trees are a time-honored and very common solution to holding the antenna up in the air. They present some difficulties in attaching the support rope or cable, however. The most important consideration is safety. Any time a tree support is being considered, take care to ensure that there are no power lines at any voltage in or near the tree. The lines are not always easily visible. If a weight is propelled over or through the tree, it may fly well beyond the tree and cause the trailing line to contact power lines that way. In any case, don't become a statistic! Check the entire area around the tree for hazards.

If you can climb the tree and attach the support hardware, the main problem involves providing enough tension and flexibility to allow the supports to move with the tree during wind and storms. Climbing or hiring a tree climber can pay benefits over the long term with tree-supported antennas, since a pulley and halyard can be securely attached to the tree. (More information on tree-mounting antennas can be found in the *ARRL Antenna Book*.)

If climbing is not an option, such as for a temporary or portable station or for trees not suitable for climbing, getting a line into the tree is a challenge. Hams have devised many different methods of getting a line over a branch, from the "Armstrong method" (throwing a weight such as an arborist's weight bag or a tennis ball throwing toy), to various fishing techniques, and even archery. Two very popular current methods are the slingshot and the compressed-air "spud gun." Both launch a small projectile attached to lightweight monofilament fishing line through or over a tree. (Hint — paint the projectile a bright color so it's easy to find in brush or long grass.) That line is used to pull a length of twine or string over the branch and then the rope or cable that forms the final support. (The *QST* article "The W4SSY Spudgun" is available in this book's online content and several vendors offer commercial versions of the compressed-air or slingshot launcher.)

Another option is the use of an RPV (remotely piloted vehicle), better known as a "drone." The lightest line is attached to the drone, which is then piloted over the tree and back to the ground where the next line is attached, as with the launcher or slingshot. Drones can rarely be

Wide-Band Dipoles for 80 and 75 Meters

Amateur radio's widest HF band (about 13%), the 80 – 75-meter band can vex the station builder limited to a single wire antenna. Several approaches have been shown to work without extraordinary tricks. These range from increasing the conductor size (cage dipoles), to parallel elements (fan dipole and coupled resonators), and transmission line transformers. Several articles are included in this book's online information to provide options that may be suitable for your needs and circumstances:

"Broad-Band 80-Meter Antenna," by Allen Harbach WA4DRU (*QST*, Dec. 1980)

"A Simple Broadband Dipole for 80 Meters," by Frank Witt AI1H (*QST*, Sep. 1993)

"A Wideband 80-Meter Dipole," by Rudy Severns N6LF (*QST*, July 1995)

"The Fan Dipole as a Wideband and Multiband Antenna Element," by Joel Hallas W1ZR (*QST*, May 2005)

The ARRL Antenna Book chapter on Single-Band MF and HF Antennas goes into detail about several techniques with example designs. Finally, the article "The Story of the Broadband Dipole," by Dave Leeson W6NL (*QEX*, Nov 2018) includes a list of references for the truly interested reader to learn about solutions to one of ham radio's longest-running challenges.

flown *through* a tree, limiting line placement to the uppermost branches. This results in the highest possible support point but the branches are the smallest on the tree.

If you use a pulley, select a size that prevents the rope from slipping between the sheave and housing as the antenna and ropes move in the wind. So that the antenna stays up after installation, keep it away from tree branches and other objects that might rub or fall on the antenna. If the supports for the antenna, such as trees, move in the wind, leave enough slack in the antenna that it is not pulled overly tight in normal winds. Other options are to use pulleys and counterweights to allow the antenna supports to flex without pulling on the antenna. (This and other installation topics are covered in the *ARRL Antenna Book*.)

For long spans or heavy wire antennas, it may be necessary to know the expected amount of sag across the span. There are online cable sag calculators, such as at **www.easycalculation.com/analytical/cable-sag-error.php**. You will need to know the weight per unit length of the wire (and rope, if significant), the span to be covered, and the weight of feed line attached to the antenna. You can experiment with the tension on the support ropes to get the desired sag or enter the desired tension and calculate the resulting sag. The tables Copper Wire Specifications and Antenna Wire Strength at the end of this chapter will be useful.

ANTENNA SUPPORT ROPE

A good rope for holding up wire antennas with spans up to 150 or 200 feet is ¼-inch nylon rope, UV-resistant 5⁄32- or 3⁄16-inch Dacron rope is also popular. After an installation with any new rope, it will be necessary to repeatedly take up the slack created by stretching. This process will continue over a period of several weeks, at which time most of the stretching will have taken place. Even a year after installation, however, some slack may still arise from stretching.

Make certain that the rope ends will not unravel. Most supply stores will cut the length with a hot knife; that will do the best job of sealing the ends. You can do it at home by simply melting the ends with a match or cigarette lighter. An alternative is to tightly wrap a few layers of electrical tape or heat shrink tubing around the ends. Be sure to tape the ends of all your ropes to protect them.

21.2.5 Bent Dipoles

Dipole antennas need not be installed in a straight horizontal line. They are generally tolerant of bending, sloping or drooping. Bent dipoles may be used where antenna space is at a premium. **Figure 21.16** shows a couple of possibilities; there are many more. Bending distorts the radiation pattern somewhat and may affect the impedance as well, but compromises may be acceptable when the situation demands them. When an antenna bends back on itself (as in Figure 21.16B) some of the signal is canceled; avoid this if possible. Remember that the ends of the dipole are high-voltage points. Keep them away from conductors, flammable materials, and where they can be accidentally touched.

21.2.6 Inverted-V Dipole

An *inverted-V* dipole shown in Figure 21.16C is supported at the center with a single support, such as a tree or mast. While *V* describes the shape of this antenna, it should not be confused with long-wire horizontal-V antennas, which are highly directive.

The inverted-V's radiation pattern and feed point impedance depend on the *apex angle* between the legs: As the apex angle decreases, so does feed point impedance, and the radiation pattern becomes less directive. At apex angles below 90°, the antenna efficiency begins to decrease as well.

The proximity of ground to the antenna ends will lower the resonant frequency of the antenna so that a dipole may have to be shortened in the inverted-V configuration. Losses in the ground increase when the antenna ends are close to the ground. Keeping the ends eight feet or higher above ground reduces ground loss and also prevents accidental contact with the antenna.

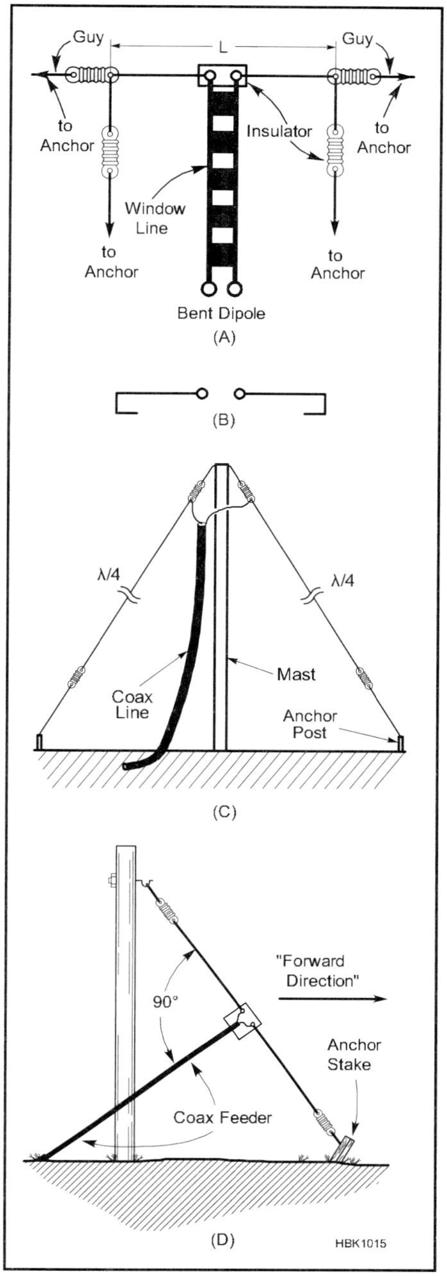

Figure 21.16 — When limited space is available for a dipole antenna, the ends can be bent downward as shown at A, or back on the radiator as shown at B. The inverted-V at C can be erected with the ends bent parallel with the ground when the available supporting structure is not high enough. At D is an example of a sloping ½ λ dipole, or *full sloper*.

Remember that antenna current produces the radiated signal, and current is maximum at the dipole center. Therefore, performance is best when the central area of the antenna is high and clear of nearby objects.

21.2.7 Sloping Dipole

A sloping dipole is shown in Figure 21.16D. This antenna is often used to favor one direction (the *forward direction* in the figure). With a non-conducting support and poor ground, signals off the back are somewhat weaker than those off the front. With a non-conducting mast and good ground, the response is approximately omnidirectional.

A conductive support such as a tower can act as a parasitic element. (So does the coax shield, unless it is routed at 90° from the antenna or decoupled with a choke balun.) The parasitic effects vary with ground quality, support height, and other conductors on the support (such as a beam at the top or other wire antennas). With such variables, performance is very difficult to predict, although modeling may give some insight into how the antenna will perform.

Losses increase as the antenna ends approach the support or the ground since the antenna ends are high-voltage points, the same cautions about their height apply as for the inverted-V antenna. To minimize feed line coupling and radiation, route the coax away from the feed point at 90° from the antenna as far as possible or use a choke balun.

21.2.8 End-Fed Half-Wave (EFHW)

Figure 21.17 shows the EFHW antenna, a half-wavelength dipole fed at one end. The open end is supported by a tree, mast, or other convenient attachment point and the feed point can be close to the ground or elevated. The EFHW is popular for portable and temporary operation in the sloping configuration shown. It can also be installed horizontally with the feed point elevated. Both ends of the antenna will be high voltage points and should be kept clear of tree branches and leaves, and away from where they can be touched accidentally.

The EFHW can be used on any band for which it is an integer number of half-wavelengths long. It may also present an acceptable impedance on non-resonant bands. For example, an EFHW for 80 meters will also be useable on 40, 20, 15, and 10 meters, as well as 30, 17, and 12 meters with a somewhat higher SWR.

The feed point's high impedance must be transformed to a lower value so that coaxial feed line can be used. A 49:1 impedance transformer (7:1 turns ratio) as shown in **Figure 21.18** is a popular choice, although 9:1 and 16:1 ratios may work acceptably. Placing the feed point a few feet from one end will also lower the impedance somewhat.

The EFHW requires a counterpoise to stabilize the feed point impedance. The outer surface of the coaxial feed line often serves as the counterpoise, so do not use a feed point choke balun without a counterpoise wire. If it is necessary to reduce common-mode RF current on the feed line, a choke balun or "line isolator" can be added 10 feet or more from the feed point, or where the feed line reaches ground level if the feed point is elevated. In the common configuration with the feed point near the ground, a few feet of wire connected to the feed line shield will supplement the feed line shield in stabilizing the feed point impedance. If installed with the feed point elevated, the feed line shield serves as a counterpoise and a separate wire is not used. A ground rod is not necessary.

The EFHW antenna is explained in more detail in a presentation from Steve Dick, K1RF, at **gnarc.org/wp-content/uploads/The-End-Fed-Half-Wave-Antenna.pdf** and on the website of Steve Yates, AA5TB, at **www.aa5tb.com/efha.html**.

21.2.9 Inductively Loaded Dipoles

Inductive loading increases the electrical length of a conductor without increasing its physical length. Therefore, we can build physically short dipole antennas by placing inductors in the antenna. These are called *loaded dipoles*, and *The ARRL Antenna Book* includes information on how to design them. There are some trade-offs involved: Inductively loaded antennas are less efficient and have narrower bandwidths than full-size antennas. Generally they should not be shortened more than 50%. David Birnbaum, K2LYV, explains the process of designing loaded dipoles that work on two bands in an article contained in this book's online information.

21.2.10 Folded Dipoles

Figure 21.19 shows a *folded dipole* constructed from open-wire transmission line. The dipole is made from a ½ λ section of open-wire line with the two conductors connected together at each end of the antenna. The top

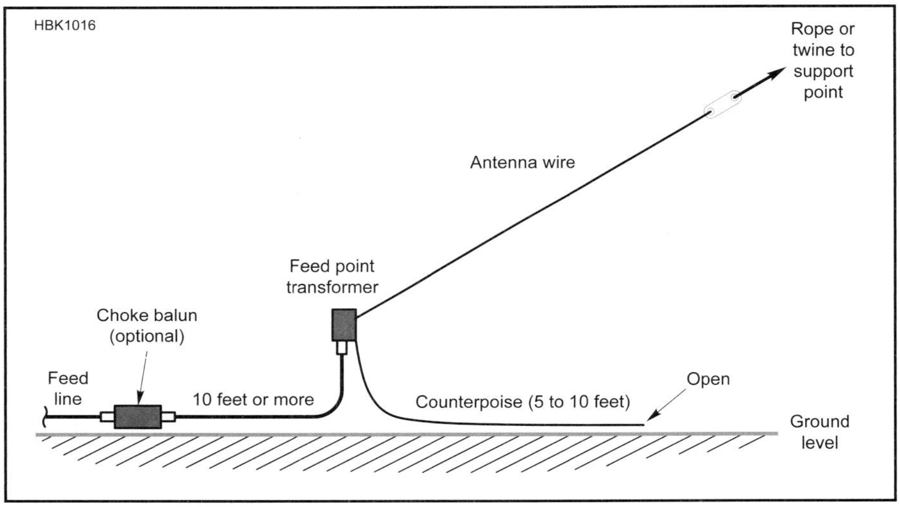

Figure 21.17 — The end-fed half-wave (EFHW) antenna. For permanent use, it can be installed horizontally with both ends well above ground. For portable or temporary use, it is common to keep the feed point near ground level and support the other end with a multi-section fiberglass mast or with a rope over a tree branch.

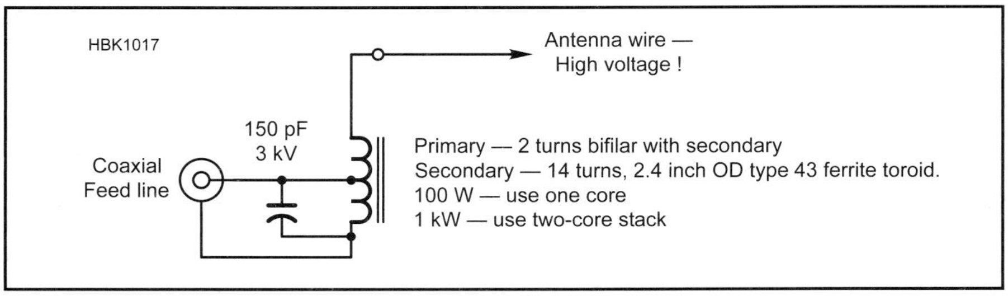

Figure 21.18 — A common 49:1 impedance transformer design. The transformer consists of a 2-turn primary and 14-turn secondary. The primary and first two secondary turns are bifilar. The 150 pF capacitor compensates for transformer inductance above 20 MHz. A low-loss 3 kV ceramic capacitor is recommended.

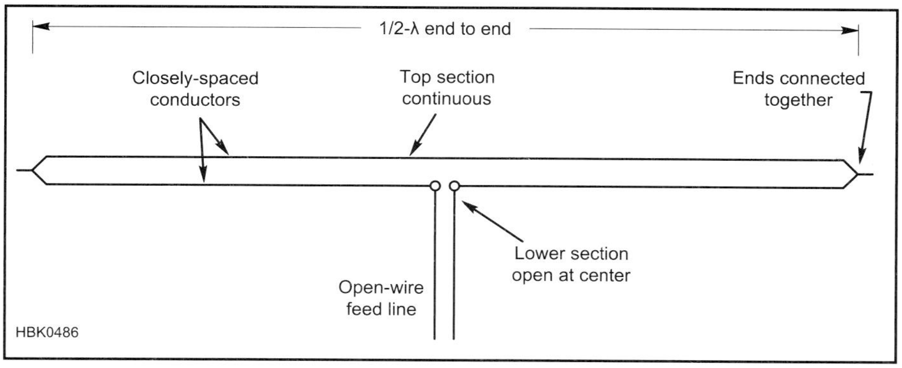

Figure 21.19 — The folded dipole is constructed from open-wire transmission line with the ends connected together. The close proximity of the two conductors and the resulting coupling act as an impedance transformer to raise the feed point impedance over that of a single-wire dipole by the square of the number of conductors used.

conductor of the open-wire length is continuous from end to end. The lower conductor, however, is cut in the middle and the feed line, attached at that point. Open-wire transmission line is then used to connect the transmitter.

A folded dipole has exactly the same gain and radiation pattern as a single-wire dipole. However, because of the mutual coupling between the upper and lower conductors that divides antenna current equally between the conductors, the feed point impedance of a single-wire dipole is multiplied by the square of the number of conductors in the antenna. In this case, there are two conductors in the antenna, so the feed point impedance is $2^2 = 4$ times that of a single-wire dipole. (A three-wire folded dipole would have a nine times higher feed point impedance and so forth.)

A common use of the folded dipole is to raise the feed point impedance of the antenna to present a better impedance match to a high impedance feed line. For example, if a very long feed line to a dipole is required, open-wire feed line would be used. By raising the dipole's feed point impedance, the SWR on the open-wire line is reduced from that of a single-wire dipole fed with open-wire feed line. Impedance matching to the usual 50 Ω can then be done at the transmitter.

A variation of the folded dipole called the *twin-folded terminated dipole* (*TFTD*) adds a resistor in the top conductor. Values of 300 to 600 Ω are used. The function of the resistor is to act as a *swamping* load, reducing the higher feed point impedances over a wide frequency range. A TFTD ½ λ long at 80 meters can be constructed to cover the entire 2 to 30 MHz range with SWR of 3:1 or less. The resistor dissipates some of the transmitter power (more than 50% at some frequencies!), but the improvement in SWR allows a coaxial feed line to be used without an impedance-matching unit. The increased convenience and installation outweigh the reduction in radiated signal. TFTD antennas are popular for emergency communications and where only a single HF antenna can be installed and high performance is not required, such as for regional coverage via NVIS as described below.

21.2.11 Parallel (Fan) Dipoles

Parallel dipoles (also called *fan dipoles*) as shown in **Figure 21.20** are a simple and convenient answer. Center-fed dipoles have low feed point impedance on the fundamental and odd harmonics, and high impedances at other frequencies. This lets us construct simple multiband systems in which the antenna for the desired band is automatically active. Consider a 50 Ω resistor connected in parallel with a 5 kΩ resistor. A generator connected across the two resistors will see 49.5 Ω, and 99% of the current will flow through the 50 Ω resistor. When resonant and non-resonant antennas are connected in parallel, the same result occurs: The non-resonant antenna has a high impedance, so little current flows in it and it has little effect on the total feed point impedance. Thus, we can connect several dipoles together at the feed point, and power naturally flows to the resonant antenna.

There are some limits, however. Wires in close proximity tend to be strongly coupled. In parallel dipoles, this coupling means that the resonant length of the shorter dipoles lengthens by a few percent. Shorter antennas don't affect longer ones much, so adjust for resonance in order from longest to shortest. Coupling also reduces the bandwidth of shorter dipoles, so an impedance-matching unit may be needed to achieve an acceptable SWR across all bands covered. These effects can be reduced by spreading the ends of the dipoles apart. (This is where the name "fan dipole" comes from.)

Also, the power-distribution mechanism requires that only one of the parallel dipoles is near resonance on any amateur band. Separate dipoles for 80 and 30 meters should not be connected in parallel because the higher frequency band is near an odd harmonic of the lower frequency band and center-fed dipoles have low impedance near odd harmonics. (The 40 and 15 meter bands have a similar relationship.) This means that you must either accept the performance of the low-band antenna operating on a harmonic or erect a separate antenna for those odd-harmonic bands. For example, four parallel-connected dipoles cut for 80, 40, 20 and 10 meters (fed by a single impedance-matching unit and coaxial cable) work reasonably on all HF bands from 80 through 10 meters.

The fan dipole technique can also be used to construct a wide-band dipole for 80 and 75 meters. Begin with a full-size 80-meter dipole and tune to a preferred frequency near 3.500 MHz, leaving several feet of wire for adjustment at the end insulators by wrapping it back around the main antenna wire. Add a 75-meter

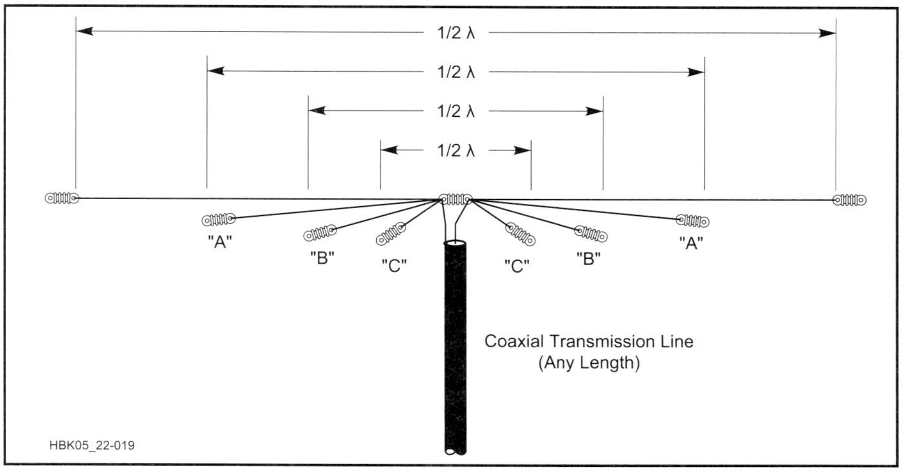

Figure 21.20 — Multiband antenna using paralleled dipoles, all connected to a common 50 or 75 Ω coax line. The ½ λ dimensions are chosen for the desired operating frequency on each band — the fundamental, A, B, and C. Be prepared to adjust the length of the various elements — both longer and shorter — because of interaction among them. See text.

dipole for 3.800 MHz, again with extra wire for adjustment, and raise the dipole into the final position. Shorten each dipole in small steps of around one foot until minimum SWR is obtained at the desired frequencies. The adjustments will interact, so avoid cutting off the excess wire until the final lengths are reached.

21.2.12 Trap Dipoles

Trap dipoles (also called "trapped dipoles") provide multiband operation from a coax-fed single-wire dipole. **Figure 21.21** shows a two-band trap antenna. A trap consists of parallel inductance and capacitance with a resonant frequency on the higher of the two bands of operation. The high impedance of the trap at its resonant frequency effectively disconnects the wire beyond the trap. Thus, on the higher of the two bands of operation at which traps are resonant, only the portion of the antenna between the traps is active.

Above resonance, the trap presents a capacitive reactance. Below resonance, the trap is inductive. On the lower of the two bands of operation, then, the inductive reactance of the trap acts as a loading coil to create a shortened or loaded dipole with the wire beyond the trap.

Traps may be constructed from coiled sections of coax (see the W8NX article in the online information) or from discrete inductors and capacitors. (Traps are also available commercially.) Choose capacitors (C1 in the figure) that are rated for high current and voltage. Mica transmitting capacitors are good. Ceramic "doorknob" transmitting capacitors will work, but be sure to use NP0 temperature coefficient parts to minimize changes in capacitance with temperature. Use large wire for the inductors to reduce loss. Any reactance (X_L and X_C) above 100 Ω at the resonant frequency will work, but bandwidth increases with reactance (up to several thousand ohms). Check trap resonance before installation. This can be done with a grid-dip meter and a receiver, or with an SWR analyzer or impedance bridge.

To construct a trap antenna, build a dipole for the higher band of operation and connect the pre-tuned traps to its ends. It is fairly complicated to calculate the additional wire needed for each band, so just add enough wire to achieve ½-wave resonance on the lower band of operation, pruning it as necessary. Because the inductance in each trap reduces the physical length needed for resonance, the finished antenna will be shorter than a simple ½ λ dipole on the lower band.

21.2.13 Dipoles for NVIS

The use of very low dipole antennas that radiate at high elevation angles for regional communication has become popular in emergency communications ("emcomm") systems. This works at frequencies (3 to 10 MHz) below the ionosphere's critical frequency — the highest frequency for which a signal traveling vertically will be reflected. (See the **Propagation** chapter.) The most common band for NVIS communication is 75 meters because the critical frequency is almost always above 4 MHz. 60 meters is growing in popularity, and 40 meters is often useful for NVIS communication through the day.

No special construction techniques are required to use a dipole for NVIS communication. Just build a regular λ/2 dipole and install it at a height of around 0.2–0.3 λ. At these low heights, the effects of ground will lower the dipole's resonant frequency and feed point impedance. Expect to shorten the antenna to achieve resonance. As height increases to 0.5 λ, a high-angle null develops in the antenna pattern, reducing NVIS effectiveness considerably. See the References entry for Witvliet for a recent paper on the subject.

Below 0.2 λ, feed point impedance will drop rapidly (see Figure 21.2) and ground losses may become significant. If feed point impedance (at resonance) drops to 40 Ω or lower, that is an indication of increasing ground loss. Ground screens or reflector wires are not needed due to the efficiency of NVIS propagation. A simple inverted-V dipole at a modest height over normal conductive soil works well, and the dB or two of improvement is not worth the additional complexity of screens or reflectors.

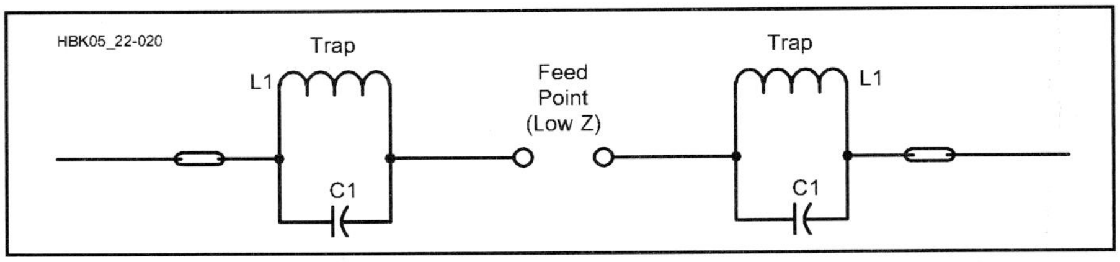

Figure 21.21 — Example of a trap dipole antenna. L1 and C1 can be tuned to the desired frequency by means of a grid-dip meter or SWR analyzer *before* they are installed in the antenna.

Turn a Horizontal Antenna Vertical

An option for adding at least one more band to a flat-top or inverted-V dipole is to turn it into a flat-top T vertical antenna. (See the T and Inverted-L Antennas section.) To do this, disconnect the feed line at the ground level, short the feed line conductors together, and connect them to the transmitter as a single wire. The remaining connection to the transmitter should be connected to a ground rod (as shown), counterpoise, or system of ground radials. (The antenna system's safety ground connection is still required.) A coaxial feed line is shown but the same technique works just as well for open-wire feed lines. An antenna tuner is required for either type of feed line.

This technique often allows a dipole to be used effectively at frequencies below those at which its horizontal section is resonant. For example, a 40 meter dipole can be used this way on 160 and 80 meters.

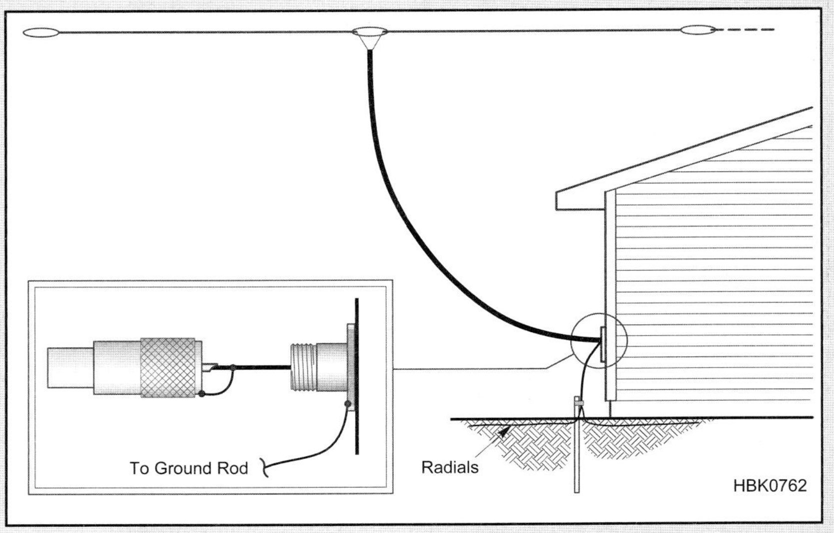

Figure 21.A – A dipole can be fed as a flat-top T vertical antenna by reconfiguring the feed line connections and exciting the antenna against ground.

21.2.14 Off-Center-Fed (OCF) Dipoles

The usual practice is to feed a $\lambda/2$ dipole in the center where the feed point impedance is low and makes a suitable match to coaxial cables. The dipole will accept energy from a feed point anywhere along its length, however, assuming that the source is matched to the higher impedance that is presented away from the center point.

The *off-center-fed* dipole takes advantage of placing the feed point in a location along the dipole at which the impedance is similar on more than one band, generally in the neighborhood of 150-300 Ω. A suitable impedance matching device such as an impedance transformer is then used to reduce the feed point impedance to something closer to 50 Ω. Note that the feed point impedance of the antenna varies with height above ground and so will SWR.

Figure 21.22 shows two off-center-fed or *OCF* dipoles. Because it is similar in appearance, this antenna is often mistakenly called a "Windom," or sometimes a "coax-fed Windom." The two antennas are not the same, since the Windom is driven against an Earth ground, while the OCF dipole is fed like a regular dipole — just not at its center. The extreme case of an OCF is the end-fed Zepp where the feed point is moved all the way to the end of the antenna.

The OCF dipole, fed ⅓ of its length from one end, may be used on its fundamental and even harmonics. Its free-space antenna-terminal impedance at 3.5, 7, and 14 MHz is on the order of 150 to 200 Ω. At the 6th harmonic, 21 MHz, the antenna is three wavelengths long and fed at a voltage maximum instead of a current maximum. The feed point impedance at this frequency is high, a few thousand ohms, so the antenna is unsuitable for use on this band. OCF dipole designs for various HF bands by NT4B, K1BQT, K1LI, and KE4PT are included in the online material for this chapter.

IMPEDANCE TRANSFORMER AND BALUN REQUIREMENTS

All OCF dipole antennas require an impedance transformer at the feed point. A 4:1 or 6:1 ratio typically provides good results. A choke balun on the feed line is also required to reduce coupling between the antenna and feed line, which can affect feed point impedance. Height above ground also affects feed point impedance — be prepared to adjust the dipole length to achieve an acceptable match on the bands to be used.

Because the OCF dipole is not fed at the center of the radiator, the feed line is not placed symmetrically with respect to the antenna's radiated field. As a result, common-mode current will flow on the feed line, usually coaxial cable. How much current flows depends on the impedance of the coax shield's outer surface, which, in turn, depends on the orientation of the cable, how long it is, height above ground, and so forth. (Some of the common-mode current results from the slightly unequal impedances presented by the OCF legs, but most of the shield current results from the feed line's asymmetric location in the antenna's field.)

Regardless of the cause, common-mode current flow on the feed line is generally viewed as undesirable and a current or choke balun is used to increase the impedance of coax shield's outer surface. Radiation from the feed line may not be a problem in your installation and may even improve the antenna's radiation pattern by filling in nulls. In that case, no choke balun is required. (Feed line chokes are discussed in the chapter **Transmission Lines**.)

Project: Multiband Center-Fed Dipole

An 80 meter dipole fed with ladder line is a versatile antenna. If you add a wide-range matching network, you have a low-cost antenna system that works well across the entire HF spectrum, even 6 meters. Countless hams have used one of these in single-antenna stations and for Field Day operations.

Figure 21.23 shows a typical installation for such an antenna. The inverted-V configuration is shown, lowering the total antenna length to 130 feet from the 135 feet used if the entire antenna is horizontal. Either configuration will work well. You can use a balanced impedance-matching unit with a balun between it and the transmitter. Many amateurs are successful in using unbalanced impedance-matching units with a balun at either the output or the input of the tuner. Don't be afraid to experiment!

This configuration is popular with other lengths for the antenna:

- 105 feet — 80 through 10 meters
- 88 feet — 80 through 10 meters
- The next lower band may also be covered if the impedance-matching unit has sufficient range, although the adjustment will be fairly sharp. Six meter coverage is possible but de-

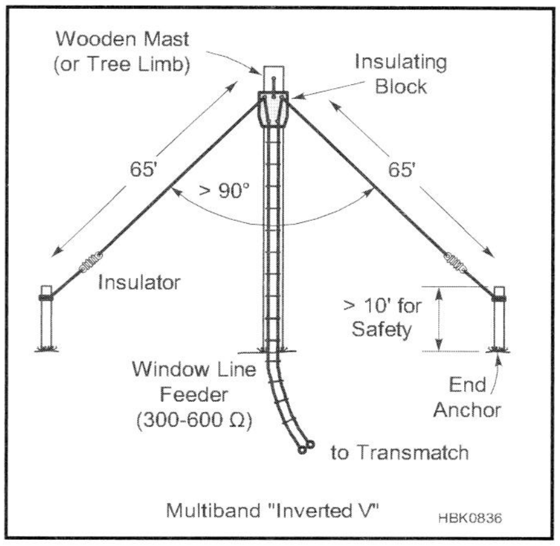

Figure 21.23 — Details for an inverted-V fed with open-wire line for multiband HF operation. The included angle between the two legs should be greater than 90° for best performance.

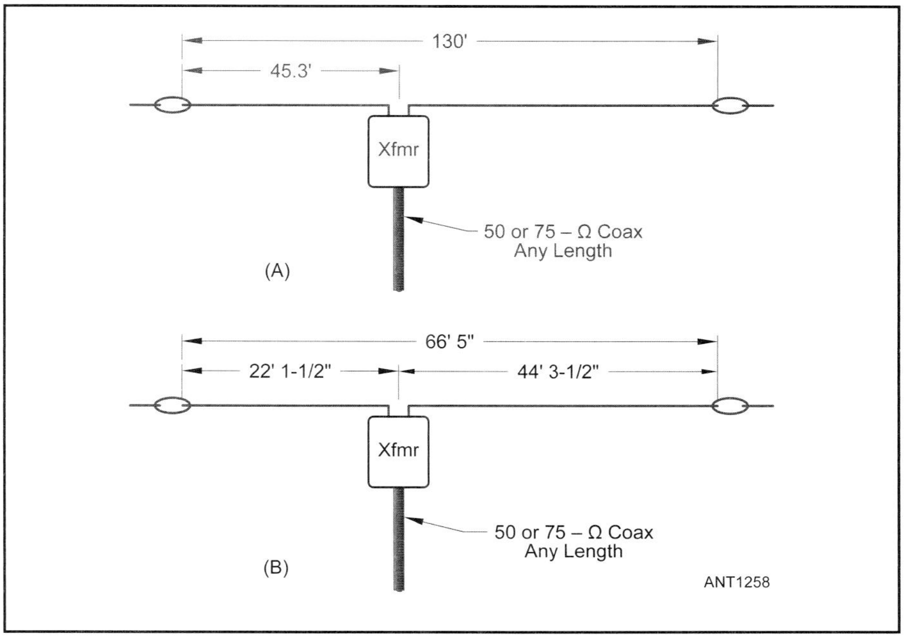

Figure 21.22 – Off-center-fed (OCF) dipoles for use on several bands. Length A is generally $\lambda/2$ on the lowest frequency of operation. Length B is usually approximately ⅓ of A. The impedance transformer is typically a 4:1 or 6:1 design. A feed line choke at the impedance transformer will provide feed line isolation, if desired. Feed point impedance and SWR will be affected by the antenna's height above ground.

pends on the station layout, length of feed line, and impedance-matching unit abilities. Again, don't be afraid to experiment!

For best results place the antenna as high as you can, and keep the antenna and ladder line clear of metal and other conductive objects. Despite significant SWR on some bands, the open-wire feed line keeps system losses low, as described in the **Transmission Lines** chapter.

ARRL staff analyzed a 135 foot dipole at 50 feet above typical ground and compared that to an inverted-V with the center at 50 feet, and the ends at 10 feet. The results show that on the 80 meter band, it won't make much difference which configuration you choose. The inverted-V exhibits additional losses because of its proximity to ground. (Radiation patterns are available in the online information.)

Installed horizontally, or as an inverted-V, the 135 foot center-fed dipole is a simple antenna that works well from 3.5 to 30 MHz (and on 1.8 MHz if the impedance-matching unit has sufficient range). The feed line impedance at the transmitter end may be out of range for the impedance-matching unit. In such a case, add ⅛ wavelength of feed line at that frequency. This may change the impedance to a value the impedance-matching unit can handle.

Project: 40 to 15 Meter Dual-Band Dipole

As mentioned earlier, dipoles have harmonic resonances at odd multiples of their fundamental resonances. Because 21 MHz is the third harmonic of 7 MHz, 7 MHz dipoles are harmonically resonant in the popular ham band at 21 MHz. This is attractive because it allows you to install a 40 meter dipole, feed it with coax, and use it without an antenna tuner on both 40 and 15 meters.

But there's a catch: The third harmonic resonance is actually higher than three times the fundamental resonant frequency. This is because there is no end effect in the center portion of the antenna where there are no insulators.

An easy fix for this, as shown in **Figure 21.24**, is to add capacitive loading to the antenna about ¼ λ wavelength (at 21.2 MHz) away from the feed point in both halves of the dipole. Known as *capacitance hats*, the simple loading wires shown lower the antenna's resonant frequency on 15 meters without substantially affecting resonance on 40 meters. This scheme can also be used to build a dipole that can be used on 80 and 30 meters and on 75 and 10 meters. (A project for a 75 and 10 meter dipole is included with the online content.)

Measure, cut, and adjust the dipole to resonance at the desired 40 meter frequency. Then, cut two 2-foot-long pieces of stiff wire (such as #12 or #14 AWG house wire) and solder the ends of each one together to form two loops. Twist the loops in the middle to form figure-8s, and strip and solder the wires where they cross. Install these capacitance hats on the dipole by stripping the antenna wire (if necessary) and soldering the hats to the dipole about a third of the way out from the feed point (placement isn't critical) on each wire. To resonate the antenna on 15 meters, use an antenna analyzer to adjust the loop shapes until the SWR is acceptable in the desired segment of the 15 meter band. Conversely, you can move the hats back and forth along the antenna until the desired SWR is achieved and then solder the hats to the antenna.

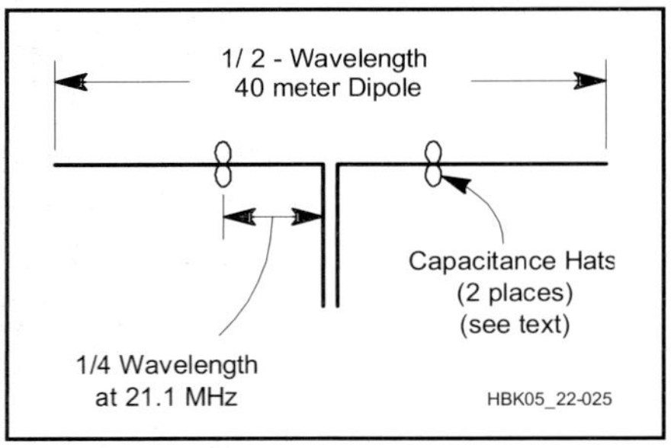

Figure 21.24 — Figure-8-shaped capacitance hats made and placed as described in the text, can make a 40 meter dipole resonate anywhere in the 15 meter band.

Project: W4RNL Rotatable Dipole Inverted-U Antenna

This simple rotatable dipole was designed and built by L.B. Cebik, W4RNL (SK), for use during the ARRL Field Day. For this and other portable operations we look for three antenna characteristics: simplicity, small size, and light weight. Today, a number of lightweight collapsible masts are available. When properly guyed, some will support antennas in the 5 to 10 pound range. Most are suitable for 10 meter tubular dipoles and allow the user to hand-rotate the antenna. Extend the range of the antenna to cover 20 through 10 meters, and you put these 20 to 30 foot masts to even better use. The inverted-U meets this need. **Figure 21.25** shows the basic kit for the antenna. Complete construction details and more information about antenna performance are available with the online content.

A dipole's highest current occurs within the first half of the distance from the feed point to the outer tips. Therefore, very little performance is lost if the outer end sections are bent. The W4RNL inverted-U starts with a 10 meter tubular dipole. You add wire extensions for 12, 15, 17 or 20 meters to cover those bands.

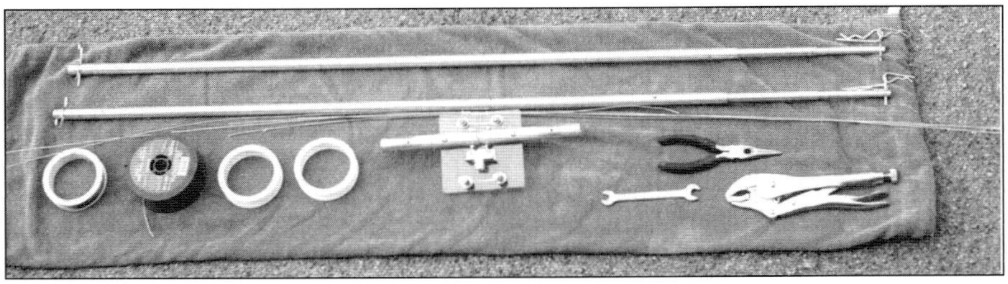

Figure 21.25 — The entire inverted-U antenna parts collection in semi-nested form, with its carrying bag. The tools stored with the antenna include a wrench to tighten the U-bolts for the mast-to-plate mount and a pair of pliers to help remove end wires from the tubing.

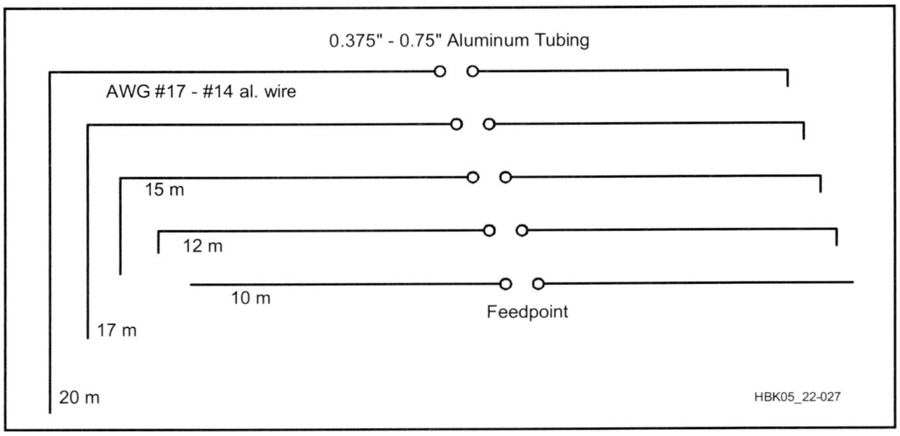

Figure 21.26 — The general outline of the inverted-U field dipole for 20 through 10 meters. Note that the vertical end extension wires apply to both ends of the main 10 meter dipole, which is constant for all bands.

You only need enough space to erect a 10 meter rotatable dipole. The extensions hang down. **Figure 21.26** shows the relative proportions of the antenna on all bands from 10 to 20 meters. The 20 meter extensions are the length of half the 10 meter dipole.

Not much signal strength is lost by drooping up to half the overall element length straight down. What is lost in bidirectional gain shows up in decreased side nulls. There is an undetectable decrease in gain between the 10 meter and 15 meter versions. The 20 meter version shows a little over ½ dB gain decrease and a signal increase off the antenna ends.

The feed point impedance of the inverted-U remains well within acceptable limits for virtually all equipment, even at 20 feet above ground. Also, the SWR curves are very broad, so it's not as critical to find exact dimensions, even for special field conditions.

Changing bands is a simple matter. Remove the extensions for the band you are using and install the ends for the new band. An SWR check and possibly one more adjustment of the end lengths will put you back on the air.

With a dipole having drooping ends, safety is very important. At any power level, the ends of a dipole have high RF voltages, and we must keep them out of contact with human body parts. Do not use the antenna unless the wire ends for 20 meters are higher than any person can touch when the antenna is in use. Even with QRP power levels, the RF voltage on the wire ends can be dangerous. With the antenna at 20 feet at its center, the ends should be at least 10 feet above ground.

Project: Two W8NX Multiband, Coax-Trap Dipoles

Two different antennas are described here. The first covers the traditional 80, 40, 20, 15 and 10 meter bands, and the second covers 80, 40, 17 and 12 meters. (The complete article describing these antenna projects is available as part of this book's online information.) Each uses the same type of W8NX trap — connected for different modes of operation — and a pair of short capacitive stubs to enhance coverage. The W8NX coaxial-cable traps have two different modes: a high- and a low-impedance mode. The inner-conductor windings and shield windings of the traps are connected in series for both modes. However, either the low- or high-impedance point can be used as the trap's output terminal. For low-impedance trap operation, only the center conductor turns of the trap windings are used. For high-impedance operation, all turns are used in the conventional manner for a trap. The short stubs on each antenna are strategically sized and located to permit more flexibility in adjusting the resonant frequencies of the antenna.

80, 40, 20, 15 AND 10 M DIPOLE

Figure 21.27 shows the configuration of the 80, 40, 20, 15 and 10 meter antenna. The radiating elements are made of #14 AWG stranded copper wire. The element lengths are the wire span lengths in feet. These lengths do not include the lengths of the pigtails at the balun, traps, and insulators. The 32.3-foot-long inner 40 meter segments are measured from the eyelet of the input balun to the tension-relief hole in the trap coil form. The 4.9 foot segment length is measured from the tension-relief hole in the trap to the 6 foot stub. The 16.1 foot outer-segment span is measured from the stub to the eyelet of the end insulator.

80, 40, 17 AND 12 METER DIPOLE

Figure 21.28 shows the configuration of the 80, 40, 17 and 12 meter antenna. Notice that the capacitive stubs are attached immediately outboard after the traps and are 6.5 feet long, ½ foot longer than those used in the other antenna. The traps are the same as those of the other antenna, but are connected for the high-impedance parallel-resonant output mode. Since only four bands are covered by this antenna, it is easier to fine tune it to precisely the desired frequency on all bands.

Project: Triband Dipole for 30, 17, and 12 Meters

(This antenna was originally described in the article "A Triband Dipole for 30, 17, and 12 Meters" by Zack Lau, W1VT, in the Mar/Apr 2015 issue of *QEX*. A similar design by W1VT for 10, 20, and 40 meters described in the article "A Compact Multiband Dipole" from the March 2016 issue of *QST*. Both articles are available as PDF files with the online content.)

The triband dipole in **Figure 21.29** is a 58 foot doublet made out of #14 THHN solid house wire fed with a 35.5 foot matching section of 600 Ω ladder line having a velocity factor of 0.91. It has modest gain over a dipole on all three of the design bands, and SWR at the balun input is approximately 1.5:1 on 17 meters and 2.4:1 on 12 meters and 30 meters. These values of SWR are all easily matched to 1:1 with the auto-tuner built into many transceivers, or with an external impedance matching unit.

If the antenna will be installed at a significantly different height, the length of the doublet and the open-wire feed line may have to be adjusted to obtain resonances in all three bands. The length of the 50 Ω cable feed line isn't critical.

The #14 THHN house wire has two layers of insulation: 15 mils of PVC and another 4 mils of nylon. While the nylon typically flakes off in less than a year, the antenna was modeled in *EZNEC* using an insulation thickness of 19 mils and a dielectric constant of 3.5. Changing the insulation thickness to 15 mils doesn't appreciably change the resonance points.

The somewhat low velocity factor of the open-wire line assumes it is constructed using the same insulated wire as the doublet. A spacing of 3 inches is recommended for constructing the 600 Ω open-wire line. If other wire types are used to construct the open-wire line, an accurate approximation for the impedance of air-insulated open-wire line is:

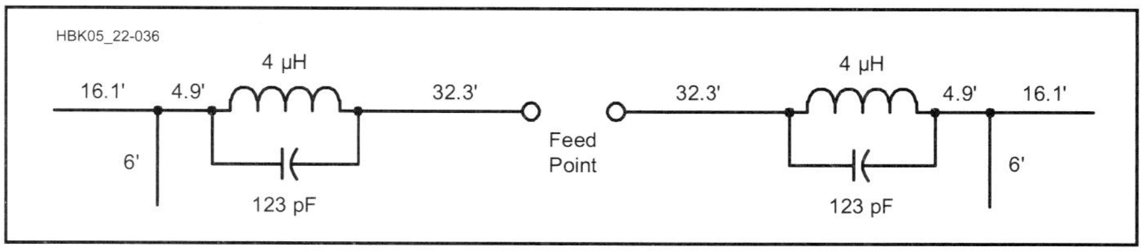

Figure 21.27 — A W8NX multiband dipole for 80, 40, 20, 15 and 10 meters. The values shown (123 pF and 4 µH) for the coaxial-cable traps are for parallel resonance at 7.15 MHz. The low-impedance output of each trap is used for this antenna.

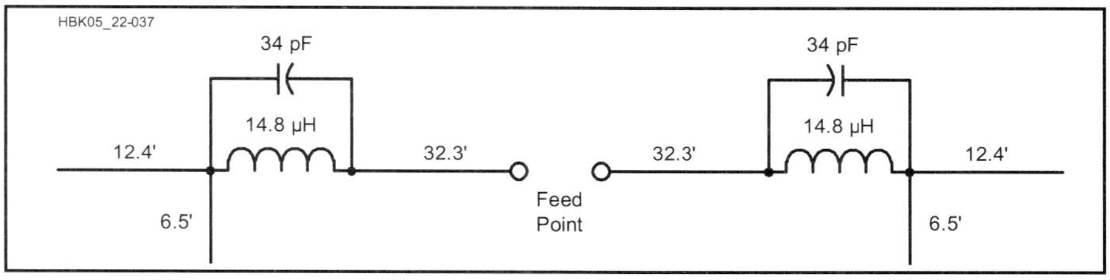

Figure 21.28 — A W8NX multiband dipole for 80, 40, 17 and 12 meters. For this antenna, the high-impedance output is used on each trap. The resonant frequency of the traps is 7.15 MHz.

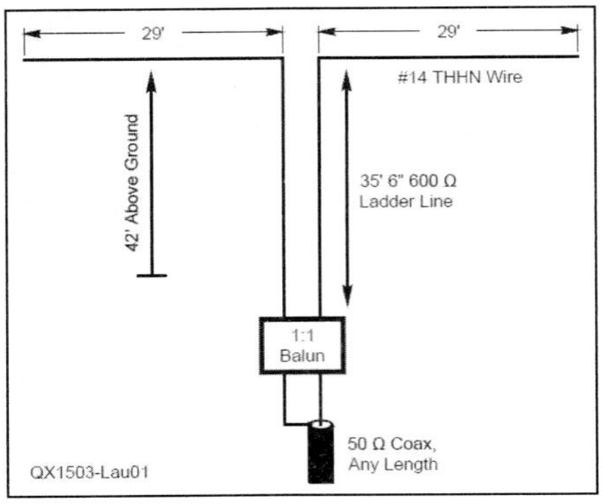

Figure 21.29 — Basic configuration of the triband dipole, consisting of a 58-foot doublet fed with 35.5 feet of 600 Ω open-wire feed line. SWR at the balun input is approximately 1.5:1 on 17 meters and 2.4:1 on 12 meters and 30 meters.

$$Z_0 = 276 \log_{10} (2S/d) \qquad (5)$$

where S is the center-to-center distance between the conductors and d is the diameter of the conductors in the same units as S. (Z_0 for open-wire line is discussed in detail in the "Fundamentals" section of the **Transmission Lines** chapter.)

A 1:1 current or choke balun is required between the open wire feed line and the coax to prevent the outside of the coax shield from becoming a radiating antenna element and affecting feed point impedance. (See the section "Baluns and Transmission Line Transformers" in the **Transmission Lines** chapter.) The author recommends 11 turns of RG-58A/U on an FT-140-43 core as working well from 10 to 30 MHz. Steve Hunt, G3TXQ published an excellent balun design using 11 turns of RG-58 on a stacked pair of FT-240-52 toroids that provides a higher choking impedance, which may be necessary if the antenna is to be used on other bands than 30, 17, or 12 meters where the open-wire line length does not result in a low impedance.

Assuming a height of 42 feet, on 17 meters the antenna has a clean, bidirectional pattern, with maximum gain of 8.9 dBi broadside to the wires, just like a dipole. (All antenna patterns are included in the full article.) On 30 meters, the antenna also has gain broadside to the wires of 6.9 dBi. On 12 meters the antenna has an azimuthal radiation pattern with four main lobes having 8.2 dBi gain, 50° off broadside. The broadside lobes are 3.5 dB weaker than the main lobes.

21.3 Vertical (Ground-Plane) Antennas

One of the more popular amateur HF antennas is the *vertical*. It usually refers to a single radiating element erected vertically over a *ground plane* of radial wires. A typical vertical is ¼ λ long electrically and is constructed of wire or tubing. The vertical antenna is more accurately named the ground plane because the conductive surface (the ground plane) creates a path for return currents to the feed point, effectively creating the "missing half" of a ½ λ antenna. Another name for this type of antenna is the *monopole* (sometimes *unipole*).

The ground plane can be a solid, conducting surface, such as a vehicle body at VHF/UHF mobile antenna. At HF, this is impractical and systems of *ground radials* are used: wires laid out on the ground radially from the base of the antenna. One conductor of the feed line is attached to the vertical radiating element of the antenna, and the remaining conductor is attached to the ground plane. The radial wires form a *ground screen* that gives the return currents a low-resistance path instead of flowing through lossy soil.

Single vertical antennas are omnidirectional radiators. This can be beneficial or detrimental, depending on the situation. On transmission there are no nulls in any direction, unlike most horizontal antennas. However, QRM on receive can't be nulled out from the directions that are not of interest unless multiple verticals are used in an array.

Ground-plane antennas need not be mounted vertically. A ground-plane antenna can operate in any orientation as long as the ground plane is perpendicular to the radiating element. Other considerations, such as minimizing cross-polarization between stations, may require a specific mounting orientation. In addition, due to the size of HF antennas, mounting them vertically is usually the most practical solution.

A vertical antenna can be mounted at the Earth's surface, in which case it is a *ground-mounted vertical*. The ground plane is then constructed on the surface of the ground. A vertical antenna and the associated ground plane can also be installed above the ground. This often reduces ground losses, but it is more difficult to install the necessary number of radials. *Ground-independent* verticals are often mounted well above the ground because their operation does not rely on a ground plane.

21.3.1 Ground Systems

When compared to horizontal antennas, verticals also suffer more acutely from two main types of losses — *ground return losses* for currents in the near field, and *far-field ground losses*. Ground losses in the near field can be minimized by using many ground radials. This is covered in the sidebar, "Optimum Ground Systems for Vertical Antennas."

Far-field losses are highly dependent on the conductivity and dielectric constant of the soil around the antenna, extending out as far as 100 wavelengths from its base. There is very little that someone can do to change the character of the ground that far away — other than moving to a small island surrounded by saltwater! Far-field losses greatly affect low-angle radiation, causing the radiation patterns of practical vertical antennas to fall far short of theoretical patterns over *perfect ground*, often seen in classical texts.

Figure 21.30 shows the elevation pattern response for two different 40 meter quarter-wave verticals. One is placed over a theoretical infinitely large, infinitely conducting ground. The second is placed over an extensive radial system over average soil, having a conductivity of 5 mS/m and a dielectric constant of 13. This sort of soil is typical of heavy clay found in pastoral regions of the US mid-Atlantic states. At a 10° elevation angle, the real antenna losses are almost 6 dB compared to the theoretical antenna; at 20° the difference is about 3 dB. See *The ARRL Antenna Book* chapter on the effects of ground for further details.

While real verticals over real ground are not a magic method to achieve low-angle radiation, cost versus performance and ease of installation are incentives that inspire many

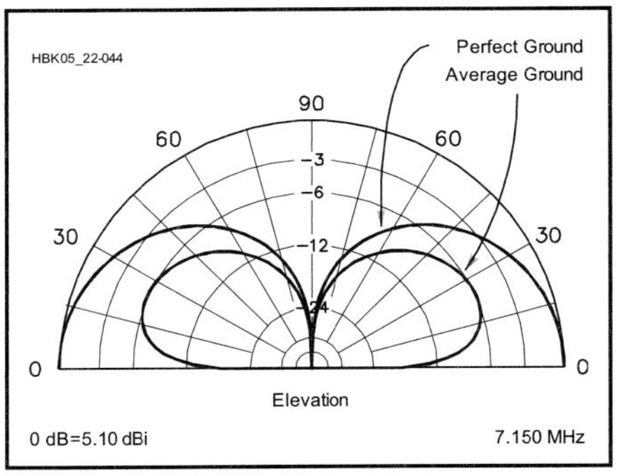

Figure 21.30 — Elevation patterns for two quarter-wave vertical antennas over different ground. One vertical is placed over perfect ground, and the other is placed over average ground. The far-field response at low elevation angles is greatly affected by the quality of the ground — as far as 100 λ away from the vertical antenna.

antenna builders. For use on the lower frequency amateur bands — notably 160 and 80 meters — it is not always practical to erect a full-size vertical. At 1.8 MHz, a full-sized quarter-wave vertical is 130 feet high. In such instances it is often necessary to accept a shorter radiating element and use some form of loading.

Figure 21.31 provides curves for the physical height of verticals in wavelength versus radiation resistance and reactance. Although the plots are based on perfectly conducting ground, they show general trends for installations where many radials have been laid out to make a ground screen. As the radiator is made shorter, the radiation resistance decreases — with 6 Ω being typical for a 0.1 λ high antenna. The lower the radiation resistance, the more the antenna efficiency depends on ground conductivity and the effectiveness of the ground screen. Also, the bandwidth decreases markedly as the length is reduced toward the left of the scale in Figure 21.31. It can be difficult to develop suitable matching networks when radiation resistance is very low.

Generally a large number of shorter radials results in a better ground system than a few longer ones. For example, eight ¼ λ radials are preferred over four ¼λ radials. Optimum radial lengths are described in the sidebar.

The conductor size of the radials is not especially significant. Wire gauges from #4 to #20 AWG have been used successfully by amateurs. Copper wire is preferred, but where soil is low in acid (or alkali), aluminum wire can be used. The wires may be bare or insulated, and they can be laid on the surface or buried a few inches below ground. Insulated wires will have greater longevity by virtue of reduced corrosion and dissolution from soil chemicals.

When property dimensions do not allow a classic installation of equally spaced radial wires, they can be placed on the ground as space permits. They may run away from the antenna in only one or two compass directions. They may be bent to fit on your property. Hardware cloth and chicken wire are also quite effective, although the galvanizing must be of high quality to prevent rapid rusting.

A single ground rod, or group of them bonded together, is seldom as effective as a collection of random-length radial wires. (A ground rod at the vertical's base should be used for lightning protection.)

All radial wires should be connected together at the base of the vertical antenna. The electrical bond needs to be of low resistance. Best results will be obtained when the wires are soldered together at the junction point. When a grounded vertical is used, the ground wires should be affixed securely to the base of the driven element.

Ground return losses are lower when vertical antennas and their radials are elevated above ground, a point that is well known by those using ground plane antennas on their

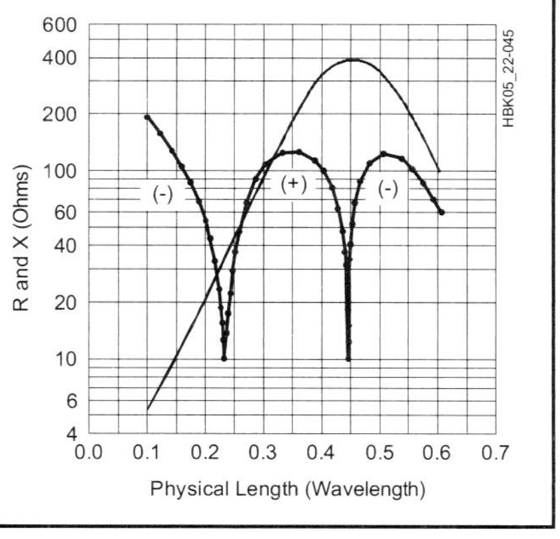

Figure 21.31 — Radiation resistance (solid curve) and reactance (dotted curve) of vertical antennas as a function of their electrical height.

Optimum Ground Systems for Vertical Antennas

A frequent question brought up by old-timers and newcomers alike is: "So, how many ground radials do I *really* need for my vertical antenna?" Most hams have heard the old standby tales about radials, such as "if a few are good, more must be better" or "lots of short radials are better than a few long ones."

John Stanley, K4ERO, eloquently summarized a study he did of the professional literature on this subject in his article "Optimum Ground Systems for Vertical Antennas" in December 1976 *QST*. His approach was to present the data in a sort of "cost-benefit" style in **Table 21.A**, reproduced here. John somewhat wryly created a new figure of merit — the total amount of wire needed for various radial configurations. This is expressed in terms of wavelengths of total radial wire.

Table 21.A

Optimum Ground-System Configurations

Configuration Designation	*A*	*B*	*C*	*D*	*E*	*F*
Number of radials	*16*	*24*	*36*	*60*	*90*	*120*
Length of each radial in wavelengths	0.1	0.125	0.15	0.2	0.25	0.4
Spacing of radials in degrees	22.5	15	10	6	4	3
Total length of radial wire installed, in wavelengths	1.6	3	5.4	12	22.5	48
Power loss in dB at low angles with a quarter-wave radiating element	3	2	1.5	1	0.5	0*
Feed point impedance in ohms with a quarter-wave radiating element	52	46	43	40	37	35

Note: Configuration designations are indicated only for text reference.
*Reference. The loss of this configuration is negligible compared to a perfectly conducting ground.

The results almost jumping out of this table are:

• If you can only install 16 radials (Case A), they needn't be very long — 0.1 λ is sufficient. You'll use 1.6 λ of radial wire in total, which is about 450 feet at 3.5 MHz.

• If you have the luxury of laying down 120 radials (Case F), they should be 0.4 λ long, and you'll gain about 3 dB over the 16 radial case. You'll also use 48 λ of total wire — For 80 meters, that would be about 13,500 feet!

• If you can't put out 120 radials but can install 36 radials that are 0.15 λ long (Case C), you'll lose only 1.5 dB compared to the optimal Case F. You'll also use 5.4 λ of total wire, or 1,500 feet at 3.5 MHz.

• A 50 Ω SWR of 1:1 isn't necessarily a good thing — the worst-case ground system in Case A has the lowest SWR.

Table 21.A represents the case for "average" quality soil, and it is valid for radial wires either laid on the ground or buried several inches in the ground. Note that such ground-mounted radials are detuned because of their proximity to that ground and hence don't have to be the classical quarter-wavelength that they need to be in "free space."

In his article John also made the point that ground-radial losses would only be significant on transmit, since the atmospheric noise on the amateur bands below 30 MHz is attenuated by ground losses, just as actual signals would be. This limits the ultimate signal-to-noise ratio in receiving.

So, there you have the tradeoffs — the loss in transmitted signal compared to the cost (and effort) needed to install more radial wires. You take your pick.

roofs. Even on 160 or 80 meters, effective vertical antenna systems can be made with as few as four ¼ λ long radials elevated 10 to 20 feet off the ground.

21.3.2 Full-Size Vertical Antennas

When it is practical to erect a full-size ¼ λ vertical antenna, the forms shown in **Figure 21.32** are worthy of consideration. The example at A is the well-known *vertical ground plane*. The ground system consists of four above-ground radial wires. The length of the driven element and ¼ λ radials is derived from the standard equation

$$L(ft) = 234 / f(MHz) \quad (6)$$

With four equidistant radial wires drooped at approximately 30° (Figure 21.32A), the feed point impedance is roughly 50 Ω. When the radials are at right angles to the radiator (Figure 21.32B), the impedance approaches 36 Ω.

Besides minimizing ground return losses, another major advantage in this type of vertical antenna over a ground-mounted type is that the system can be elevated well above nearby conductive objects (power lines, trees, buildings and so on). When drooping radials are used, they can also serve as guy wires for the mast that supports the antenna. The coax shield braid is connected to the radials, and the center conductor to the driven element.

The *Marconi* vertical antenna shown in Figure 21.32C is the classic form taken by a ground-mounted vertical. It can be grounded at the base and *shunt fed*, or it can be isolated from ground, as shown, and *series fed*. As always, this vertical antenna depends on an effective ground system for efficient performance. If a perfect ground were located below the antenna, the feed impedance would be near 36 Ω. In a practical case, owing to imperfect ground, the impedance is more likely to be in the vicinity of 50 Ω.

Vertical antennas can be longer than ¼ λ, too; ⅜ λ, ½ λ, and ⅝ λ verticals can all be used with good results, although none will present a 50 Ω feed point impedance at the base. Non-resonant lengths have become popular for the same reasons as non-resonant horizontal antennas; when fed with low-loss feed line and a wide-range impedance matching unit, the antenna can be used on multiple bands. Various matching networks, described in the **Transmission Lines** chapter, can be employed. Antenna lengths above ½ λ are not recommended because the radiation pattern begins to break up into more than one lobe, developing a null at the horizon at 1 λ.

A gamma-match feed system for a grounded ¼ λ vertical is presented in Figure 21.32D. (The gamma match is also discussed in the **Transmission Lines** chapter.) Some rules of thumb for arriving at workable gamma-arm and capacitor dimensions are to make the rod length 0.04 to 0.05 λ, its diameter ⅓ to ½ that of the driven element, and the center-to-center spacing between the gamma arm and the driven element roughly 0.007 λ. The capacitance of C1 at a 50 Ω matched condition will be about 7 pF per meter of wavelength. The absolute value of C1 will depend on whether the vertical is resonant and on the precise value of the radiation resistance. For best results, make the radiator approximately 3% shorter than the resonant length.

Amateur antenna towers lend themselves to use as shunt-fed verticals, even though an HF-band beam antenna is usually mounted on the tower. The overall system should be close to resonance at the desired operating frequency if a gamma feed is used. The HF-band beam will contribute somewhat to *top loading* of the tower. The natural resonance of such a system can be checked by dropping a #12 or #14 AWG wire from the top of the tower (connecting it to the tower top) to form a folded monopole (Figure 21.32E). A four- or five-turn link can be inserted between the lower end of the drop wire and the ground system. A dip meter is then inserted in the link or an antenna analyzer coupled to the link to determine the resonant frequency. Shunt feeding towers is discussed extensively in the *ARRL Antenna Book* and in ON4UN's *Low Band DXing*.

If the tower is equipped with guy wires, they should be broken up with strain insulators to prevent unwanted loading of the vertical. In such cases where the tower and beam antennas are not able to provide ¼ λ resonance, portions of the top guy wires can be used as top-loading capacitance. Experiment with the guy-wire lengths (using the dip-meter technique) while determining the proper dimensions. Modeling can save a lot of time in this part of the design.

A folded-monopole is depicted in Figure 21.32E. This system has the advantage of increased feed point impedance. Furthermore, an impedance-matching unit can be connected between the bottom of the drop wire and the ground system to permit operation on more than one band. For example, if the tower is resonant on 80 meters, it can be used as shown on 160 and 40 meters with reasonable results, even though it is not electrically long enough on 160 to act as a full-size antenna. The drop wire need not be a specific distance from the tower, but you might try spacings between 12 and 30 inches.

The method of feed shown at Figure 21.32F is commonly referred to as *slant-wire feed*. The guy wires and the tower combine to provide quarter-wave resonance. A matching network is placed between the lower end of one guy wire and ground and adjusted for an

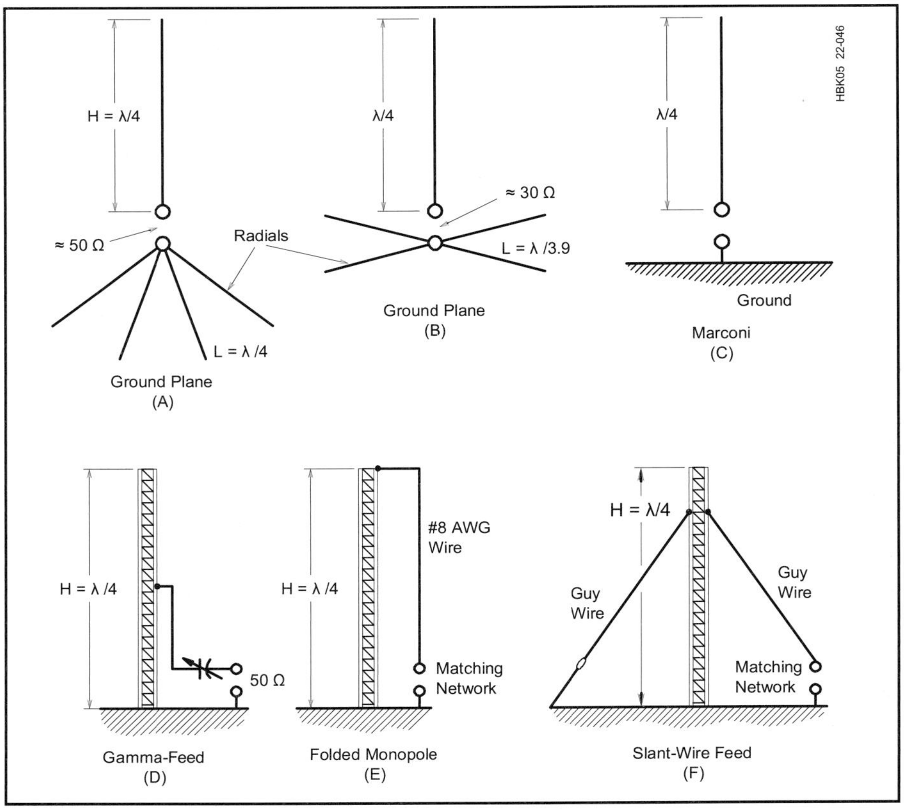

Figure 21.32 — Various types of one-quarter wavelength vertical antennas.

SWR of 1:1. It does not matter at which level on the tower the guy wires are connected, assuming that the impedance-matching unit is capable of effecting a match to 50 Ω.

3/8-Wavelength Verticals

The ⅜-wavelength vertical antenna is often an overlooked design. It has several advantages over the common quarter-wave vertical and just adds 50% to the height. Here are some advantages:

• It has a low takeoff angle of radiation at 23° compared to 26° for the ¼-wavelength antenna.

• It will work well even ground-mounted because its maximum radiation point is ⅛ wavelength above the ground. This is above the typical clutter present at ground level.

• It's feed point impedance is easy to match for a wide bandwidth with low SWR.

• Finally, it has a much higher radiation impedance. Therefore, only four ¼-λ radials are required for good performance, although more radials will reduce ground losses.

For 20 meters, a vertical height of 26 feet with 17.5-foot radials has a series impedance of approximately 200 Ω resistive with a series inductive reactance of 300 to 700 Ω. Therefore, a 4:1 transformer will match the resistive component, and a series capacitor tunes out the inductive reactance. Typically, the required series capacitance is approximately 40 to 50 pF at 14 MHz and is not critical. The design is easy to scale to other bands. See the article by Joe Reisert, W1JR, in the online material for the details of construction and matching.

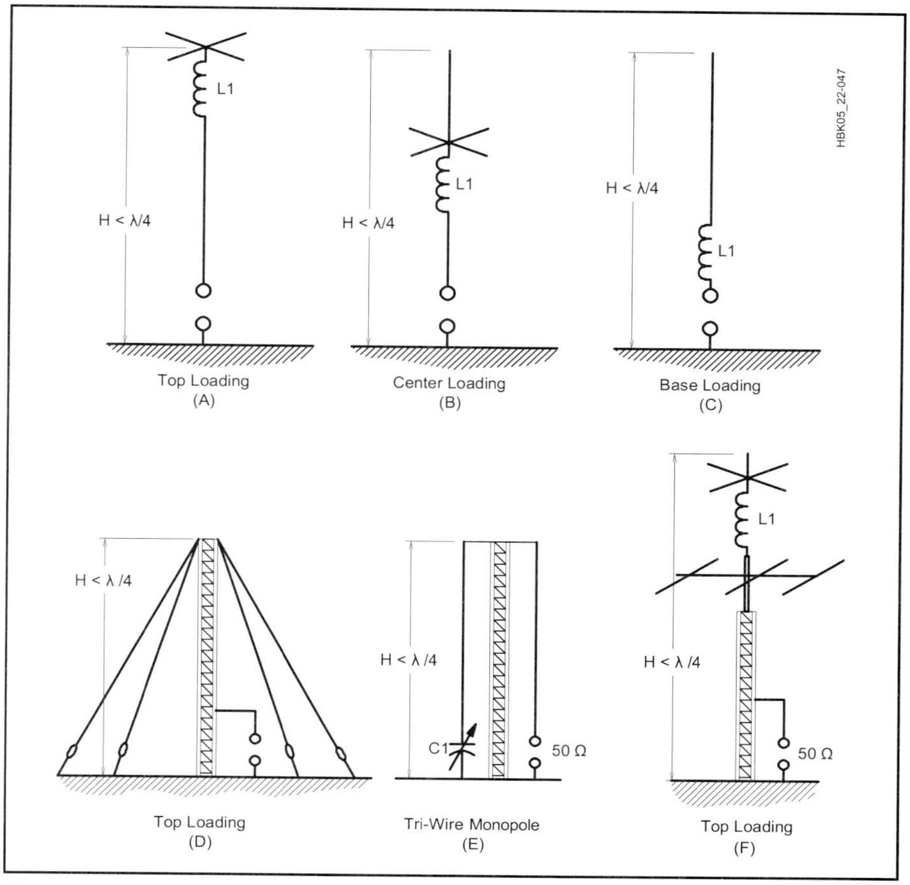

Figure 21.33 — Vertical antennas that are less than one-quarter wavelength in height.

21.3.3 Physically Short Verticals

A group of short vertical radiators is presented in **Figure 21.33**. Illustrations A and B are for top and center loading. A capacitance hat is shown in each example. The hat should be as large as practical to increase the radiation resistance of the antenna and improve the bandwidth. The wire in the loading coil is chosen for the largest gauge consistent with ease of winding and coil-form size. The larger wire diameters will reduce the resistive (I^2R) losses in the system. The coil-form material should have a medium or high dielectric constant. Phenolic or fiberglass tubing is entirely adequate.

A base-loaded vertical is shown at C of Figure 21.33. The primary limitation is that the high current portion of the vertical exists in the coil rather than the driven element. With center loading, the portion of the antenna below the coil carries high current, and in the top-loaded version the entire vertical element carries high current. Since the high-current part of the antenna is responsible for most of the radiating, base loading is the least effective of the three methods. The radiation resistance of the coil-loaded antennas shown is usually less than 16 Ω.

A method for using guy wires to top load a short vertical is illustrated in Figure 21.33D. This system works well with gamma feed. The loading wires are trimmed to provide an electrical quarter wavelength for the overall system. This method of loading will result in a higher radiation resistance and greater bandwidth than the systems shown at A through C. If an HF or VHF array is at the top of the tower, it will simply contribute to the top loading.

A three-wire monopole is shown in Figure 21.33E. Two #8 AWG drop wires are connected to the top of the tower and brought to ground level. The wires can be spaced at any convenient distance from the tower — normally 12 to 30 inches from the sides. C1 is adjusted for best SWR. This type of vertical has a fairly narrow bandwidth, but because C1 can be motor driven and controlled from the operating position, frequency changes can be accomplished easily. This technique will not be suitable for matching to 50 Ω line unless the tower is less than an electrical quarter wavelength high.

A different method for top loading is shown in Figure 21.33F. Barry Boothe, W9UCW, described this method in the December 1974 *QST*. An extension is used at the top of the tower to create an electrical quarter-wavelength vertical. L1 is a loading coil with sufficient inductance to provide antenna resonance. This type of antenna lends itself to operation on 160 meters.

Project: Top-Loaded Low-Band Antenna

The short, top-loaded vertical antenna described here by Dick Stroud, W9SR, is of interest to hams with limited space or who are portable operators. It has been used on 40, 80 and 160 meters. The antenna uses a single 10 foot TV mast section for the vertical radiator, along with a capacitance hat, loading coil, and short top mast. The overall height is less than 15 feet, as seen in **Figure 21.34**. The capacitance hat and loading coil assembly can be used with longer vertical radiators with changes to the coil inductance. A drawing showing complete dimensions and construction details is available with the online content.

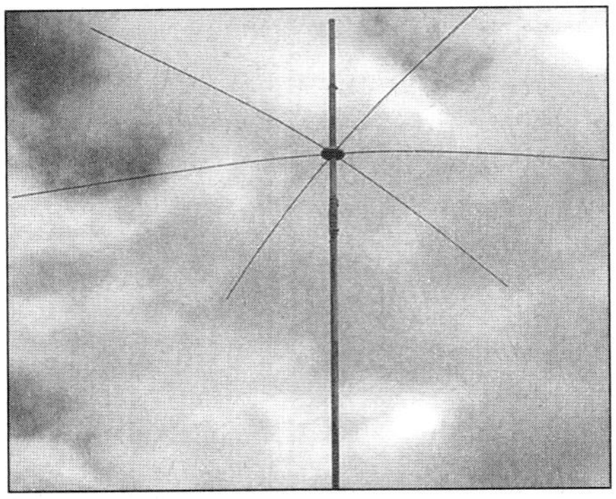

Figure 21.34 — W9SR's short vertical uses top loading and a capacitance hat with a 10 foot TV mast to make a compact antenna for 160, 80 or 40 meters.

CAPACITANCE HAT

The capacitance hat consists of a hub that mounts to the top mast above the coil and six elements made from aluminum rod. The machining, drilling, and tapping of the hub assembly can be done by nearly any machine shop if you don't have the facilities. Be sure to use stainless steel hardware throughout. (Thanks to Fred Gantzer, WØAWD, for building the original hub.)

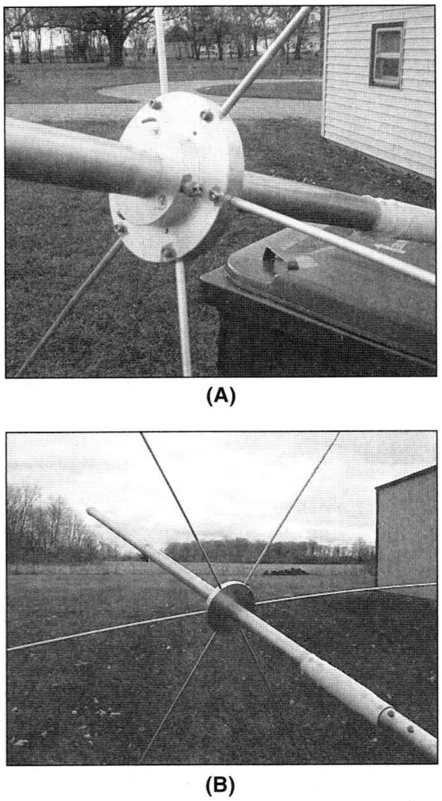

(A)

(B)

Figure 21.35 — The completed capacitance hat assembly is shown at A, and B shows a view of the upper mast assembly with the loading coil and capacitance hat.

The hub is made of two pieces of 1/2 inch thick aluminum as shown in **Figure 21.35**. The two pieces are bolted together to form the hub, which slides over the 1 1/8 inch top mast. It is held in place with three 10-32 screws. The six elements of the capacitance hat are made from 3/16 inch aluminum rod, each 4.5 feet long. These are held in place with 6-32 screws.

LOADING COIL

The coil form is made from fiberglass tubing. A 6 inch length of 1.25 inch OD fiberglass tubing is centered over a10 inch length of 1 inch OD tubing. The tubes telescope tightly, and it may be necessary to lightly sand the smaller tube for a smooth fit.

The loading coil is wound on the 1.25 inch OD fiberglass tube. After the coil is optimized, it is covered with a length of shrink sleeving for weather protection. Coil winding information in the drawing in the online content is for use with a 10 foot mast.

The bottom section of the coil assembly fits directly into the tapered upper section of the TV mast, and the exposed upper fiberglass section of the coil assembly fits into a 2.5 foot long, 1.125 inch OD, aluminum upper mast. Stainless 8-32 screws join the pieces together and also provide a connection point for the loading coil wires. If you use a painted steel mast, be sure to remove paint from the connection point, and then weather-seal it after adjustment is complete.

The capacitance hat hub slides over the upper mast and is held in place with three screws. The hub is about 6 inches above the coil, but the location can be moved to change the resonance of the antenna slightly.

The completed capacitance hat and loading coil assembly are shown in Figure 21.35. A dab of Glyptal (exterior varnish or Loctite also works) locks the screws in place once adjustments have been made, and the antenna is then ready for installation.

The coaxial feed line attaches to the bottom of the TV mast. Again, use an 8-32 screw for attachment and clean any paint from the metal. To match the antenna to a 50 Ω transmission line, a small parallel (shunt) inductance is needed at the base. The inductor is air-wound with #16 AWG wire (see **Table 21.2**).

Table 21.2
Shunt Inductor Winding Details

Frequency (MHz)	*Turns*
1.8	10 turns #16, spaced 1/8 inch
3.5	8.75 turns #16, spaced 1/8 inch
7	7.5 turns #16, spaced 1/8 inch

Note: All inductors are air wound, 1.75 inch ID. Dimensions shown are for use with 10 foot mast.

MOUNTING THE ANTENNA

A glass beverage bottle serves well as the base insulator, with the neck of the bottle fitting snugly into the TV mast. To support the base insulator, drill a hole large enough to accept the base of the bottle in the center of a 2 × 6 board about 14 inches long. Nail a piece of 1/4 inch plywood over the bottom to keep the bottle from slipping through. To keep the base support board from moving around, drill a couple of holes and secure it to the ground with stakes.

The antenna is top-heavy and will need to be guyed. A simple insulated guy ring can be made from a 2 inch PVC coupling and placed on the mast just below the loading coil. The PVC is locked to the mast with three 1/4-20 bolts. They are 1 inch long and have nuts on the inside and outside of the PVC. Three 1/4 inch holes are drilled for the guy lines, made from lengths of 3/16 inch nonabsorbent rope.

OPERATION

On-air results with a 10 foot mast have been very good, even with low power. The ground system for the early tests was nothing more than an 8 foot ground rod hammered into the soil. With this setup, and using about 90 W, many DX contacts were made over one week's time, and stateside contacts were plentiful.

There is plenty of room to experiment, however. Performance could be improved by using an extended radial system or raised and insulated radials. (Expect much better performance over average ground with a system of radials) Two, or even three, mast sections could be used with additional guys and proper loading coil inductance. If you use multiple masts, be sure to make a good electrical connection at the joints. The upper assembly is now permanently used to top load a 60 foot pole for transmitting on 160 meters.

21.3.4 Cables and Control Wires on Towers

Most vertical antennas of the type shown in Figure 21.32 and 21.33C-E consist of towers, usually with HF or VHF beam antennas at the top. The rotator control wires and the coaxial feeders to the top of the tower will not affect antenna performance adversely. In fact, they become a part of the composite antenna. To prevent unwanted RF currents from following the wires into the station, simply dress them close to the tower legs and bring them to ground level. (Running the cables inside the tower works even better.) This decouples the wires at RF. The wires should then be routed along the ground's surface (or buried underground) to the operating position. It is not necessary to use bypass capacitors or RF chokes in the rotator control leads if this is done, even when maximum legal power is employed.

21.3.5 Multiband Trap Verticals

The two-band trap vertical antenna of **Figure 21.36** operates in much the same manner as a trap dipole or trap Yagi. The notable difference is that the vertical is one-half of a dipole. The radial system (in-ground or above-ground) functions as a ground plane for the antenna and provides an equivalent for the missing half of the dipole. Once again, the more effective the ground system, the better will be the antenna performance.

Trap verticals usually are designed to work as ¼ λ radiators. The portion of the antenna below the trap is adjusted as a ¼ λ radiator at the higher proposed operating frequency. That is, a 20/15 meter trap vertical would be a resonant quarter wavelength at 15 meters from the feed point to the bottom of the trap. The trap and that portion of the antenna above the trap (plus the 15 meter section below the trap) constitute the complete antenna during 20 meter operation. But because the trap is in the circuit, the overall physical length of the vertical antenna will be slightly less than that of a single-band, full-size 20 meter vertical.

"Ground-independent" multiband vertical antennas also have traps, but are designed to be electrically longer than ¼ λ. A common electrical length is ⅜ λ, for example. The "traps" in these antennas are generally not parallel-LC circuits as described above. A variety of techniques are used with both parallel-LC traps and short resonant structures similar to stubs being used to change the antenna's electrical length at different frequencies.

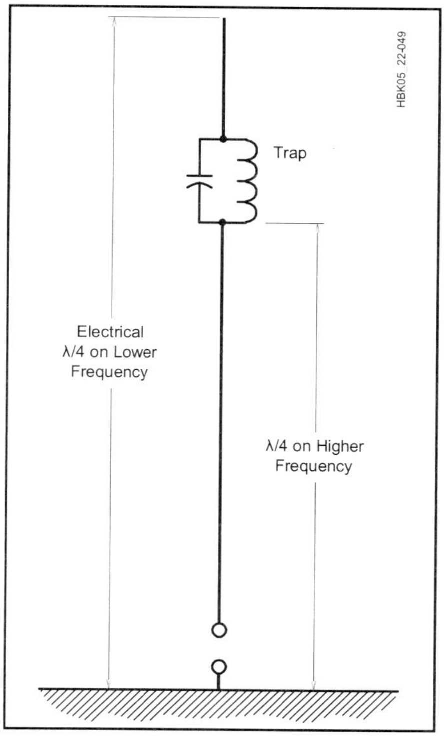

Figure 21.36 — A two-band trap vertical antenna. The trap should be resonated by itself as a parallel resonant circuit at the center of the operating range for the higher frequency band. The reactance of either the inductor or the capacitor range from 100 to 300 Ω. At the lower frequency the trap will act as a loading inductor, adding electrical length to the total antenna.

TRAPS

The trap functions as the name implies: the high impedance of the parallel resonant circuit "traps" the 15 meter energy and confines it to the part of the antenna below the trap. (See the **Electrical Fundamentals** chapter for more information on resonant LC circuits.) During 20 meter operation it allows the RF energy to reach all of the antenna. The trap in this example is tuned as a parallel resonant circuit to 21 MHz. At this frequency it electrically disconnects the top section of the vertical from the lower section because it presents a high impedance at 21 MHz, blocking 21 MHz current. Generally, the trap inductor and capacitor have a reactance of 100 to 300 Ω. Within that range it is not critical.

The trap is built and adjusted separately from the antenna. It should be resonated at the center of the portion of the band to be operated. Thus, if one's favorite part of the 15 meter band is between 21.0 and 21.1 MHz, the trap should be tuned to 21.05 MHz.

Resonance is checked by using a dip meter and detecting the signal in a calibrated receiver. An SWR analyzer can also be used. Once the trap is adjusted it can be installed in the antenna, and no further adjustment will be required. It is easy, however, to be misled after the system is assembled: Attempts to check the trap resonance in the antenna will suggest that the trap has moved much lower in frequency (approximately 5 MHz lower in a 20/15 meter vertical). This is because the trap is now part of the overall antenna, and the resultant resonance is that of the total antenna. Measure the trap's resonant frequency separately from the rest of the antenna.

Multiband operation made quite practical by using the appropriate number of traps and tubing sections. The construction and adjustment procedure is the same, regardless of the number of bands covered. The highest frequency trap is always closest to the feed end of the antenna, and the lowest frequency trap is always the farthest from the feed point. As the operating frequency is progressively lowered, more traps and more tubing sections become a functional part of the antenna.

Traps should be weatherproofed to prevent moisture from detuning them. Several coatings of high dielectric compound are effective, such as polystyrene Q Dope or insulating varnish (see the article on weatherproofing by Dick Sander, K5QY, included with the online content). Alternatively, a protective sleeve of heat-shrink tubing can be applied to the coil after completion. The coil form for the trap should be of high insulating quality and rugged enough to sustain stress during periods of wind.

21.4 T and Inverted-L Antennas

The T and inverted L are variations on the vertical antenna. **Figure 21.37** shows a flat-top T vertical. The T is basically a shortened ¼ λ vertical with the flat-top T section acting as capacitive loading to length the antenna electrically. Dimension H should be as large as possible (up to ¼ λ) for best results. The horizontal section, L, is adjusted to a length that provides resonance. Maximum radiation is polarized vertically despite the horizontal top-loading wire because current in each horizontal half creates out-of-phase radiation that cancels.

A variation of the T antenna is depicted in **Figure 21.38**. This antenna is commonly referred to as an *inverted-L* and is basically a $\frac{5}{16}$ λ vertical bent in the middle so that the top section runs parallel to the ground. As with the T antenna, the vertical section should be as long as possible. L is then added to provide an electrical $\frac{5}{16}$ λ overall.

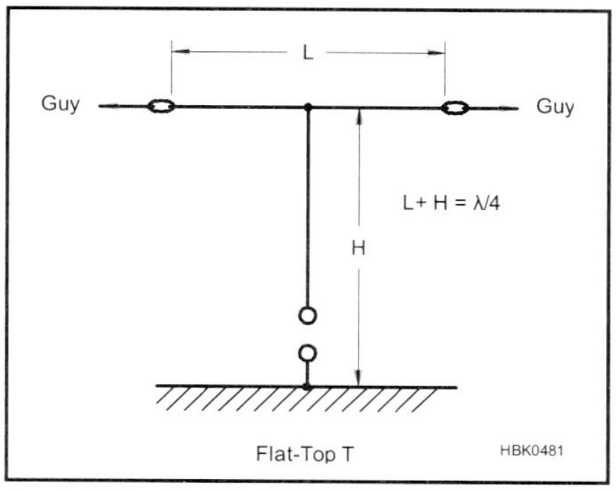

Figure 21.37 — The T antenna is basically a shortened ¼ λ vertical with the flat-top T section acting as capacitive loading to length the antenna electrically.

Because the horizontal section does carry some current, there will be some horizontally polarized radiation at high angles. This is often considered desirable because it provides local and regional coverage. The horizontal section need not be perfectly horizontal — sloping the wire at a shallow angle from horizontal does not greatly affect antenna performance. This allows the inverted-L to be constructed with a single vertical support.

A sidearm or a length of line attached to a tower can be used to support the vertical

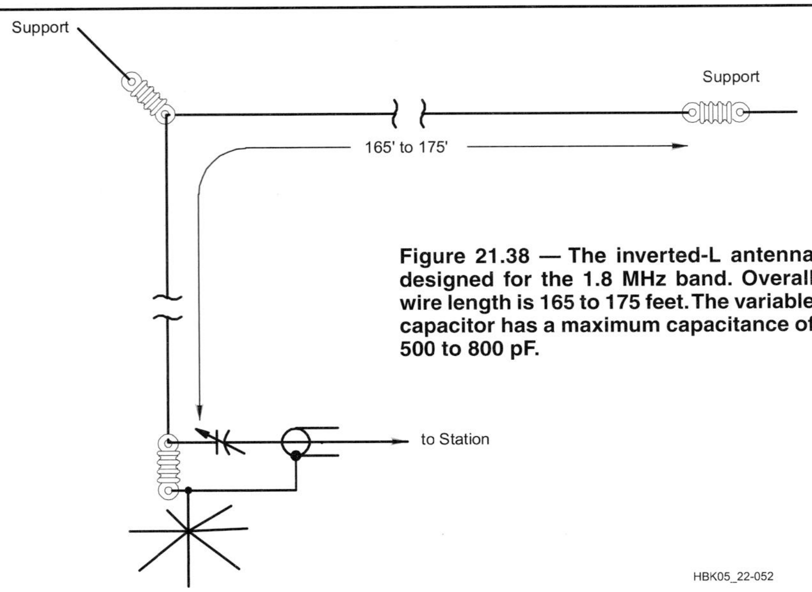

Figure 21.38 — The inverted-L antenna designed for the 1.8 MHz band. Overall wire length is 165 to 175 feet. The variable capacitor has a maximum capacitance of 500 to 800 pF.

Table 21.3
Inverted-L Dimensions for 50-Ω Feed Point Impedance at 1.82 MHz

Vertical Height (m)	*Horizontal Length (m)*	*Inductive Reactance (Ω)*	*Capacitance Required (pF)*
10	59.9	1238	71
20	39.7	517	169
30	21	235	372

Results modeled by *EZNEC 5.0* over real ground. Capacitance specified as series value to cancel inductive reactance.

section of the T or inverted-L antenna. (Keep the vertical section of the antennas as far from the tower as is practical. Certain combinations of tower height and top loading can create a resonance that interacts severely with the antennas — a 70 foot tower and a 5-element Yagi, for example.)

Both the T and inverted-L are ground-plane antennas and require a good ground system to be effective. If the T or inverted-L are used with a very good ground system, the feed-point impedance will approach 35 to 40 Ω so that the SWR approaches 1.4:1. The inverted-L or T can be constructed to be longer than needed for resonance as illustrated in Figure 21.38. This increases the feed point resistance to 50 Ω plus some inductive reactance due to the extra length. A series capacitor at the feed point then cancels the reactance, leaving a 50 Ω impedance suitable for direct connection by coaxial cable. **Table 21.3** provides values of capacitance for some combinations of vertical and horizontal height. A transmitting variable capacitor is recommended to allow the impedance match to be tuned.

21.5 Slopers and Vertical Dipoles

21.5.1 Half-Sloper Antenna

Many hams have had excellent results with *half-sloper* antennas, while others have not had such luck. Investigations by ARRL Technical Advisor John S. Belrose, VE2CV, have brought some insight to the situation through computer modeling and antenna-range tests. The following is taken from VE2CV's Technical Correspondence in Feb 1991 *QST*, pp. 39 and 40. Essentially, the half-sloper is a top-fed vertical antenna that uses the structure at the top of the tower plane (such as a grounded Yagi antenna) as a ground plane and the tower acts as a reflector (see **Figure 21.39**).

For half-slopers, the input impedance, the resonant length of the sloping wire and the antenna pattern all depend on the tower height, the angle (between the sloper and tower), the type of Yagi, and the Yagi orientation. Here are several configurations extracted from VE2CV's work:

At 160 meters — a 40 meter beam on top of a 95 foot tower with a 55° sloper apex angle. The radiation pattern varies little with Yagi type. The pattern is slightly cardioid with about 8 dB front-to-back ratio at a 25° takeoff angle (see Figure 21.39B and C). Input impedance is about 50 Ω.

At 80 meters — a 20 meter beam on top of a 50 foot tower with a 55° sloper apex angle. The radiation pattern and input impedance are similar to those of the 160 meter half-sloper.

At 40 meters — a 20 meter beam on top of a 50 foot tower with a 55° sloper apex angle. The radiation pattern and impedance depend strongly on the azimuth orientation of the Yagi. Impedance varies from 76 to 127 Ω depending on Yagi direction.

There are many configurations that will produce a reasonable feed point impedance and acceptable radiation pattern. Experimentation is usually required to find a design that works for a particular installation.

21.5.2 Half-Wave Vertical Dipole

Unlike its horizontal counterpart, which has a figure-8 pattern, the azimuthal pattern of a vertical dipole is omnidirectional. Look again at Figures 21.7 and 21.8 and note the comparison between horizontal and vertical dipole elevation patterns. These two figures illustrate the fact that performance of a horizontal dipole depends to a great extent on its height above ground. By contrast, *half-wave vertical dipole (HVD)* performance is highly dependent on ground conductivity and dielectric constant.

Experiments by K8CH show that an HVD can compete favorably with horizontal dipole or inverted-V antennas at a similar height. The HVD only requires one main support (see the caution below on feed line orientation), which makes it a good choice for portable or temporary operation.

Figure 21.40 shows the elevation patterns for the vertical dipole and for the reference dipole at a pattern peak and at a null. The large lobe in the HVD pattern at 48° is caused by the antenna being elevated 14 feet above ground. This lobe will shrink at lower heights. The patterns of both antennas depend strongly on antenna height.

One challenge to building the HVD is feed line decoupling. The feed line attached in the middle must be routed away from the antenna at right angles to prevent excessive shield current from distorting the antenna's pattern, even if a choke balun is used. This can be mechanically challenging.

Figure 21.39 — The half-sloper antenna (A). B is the vertical radiation pattern in the plane of a half sloper, with the sloper to the right. C is the azimuthal pattern of the half sloper (90° azimuth is the direction of the sloping wire). Both patterns apply to 160 and 80 meter antennas described in the text.

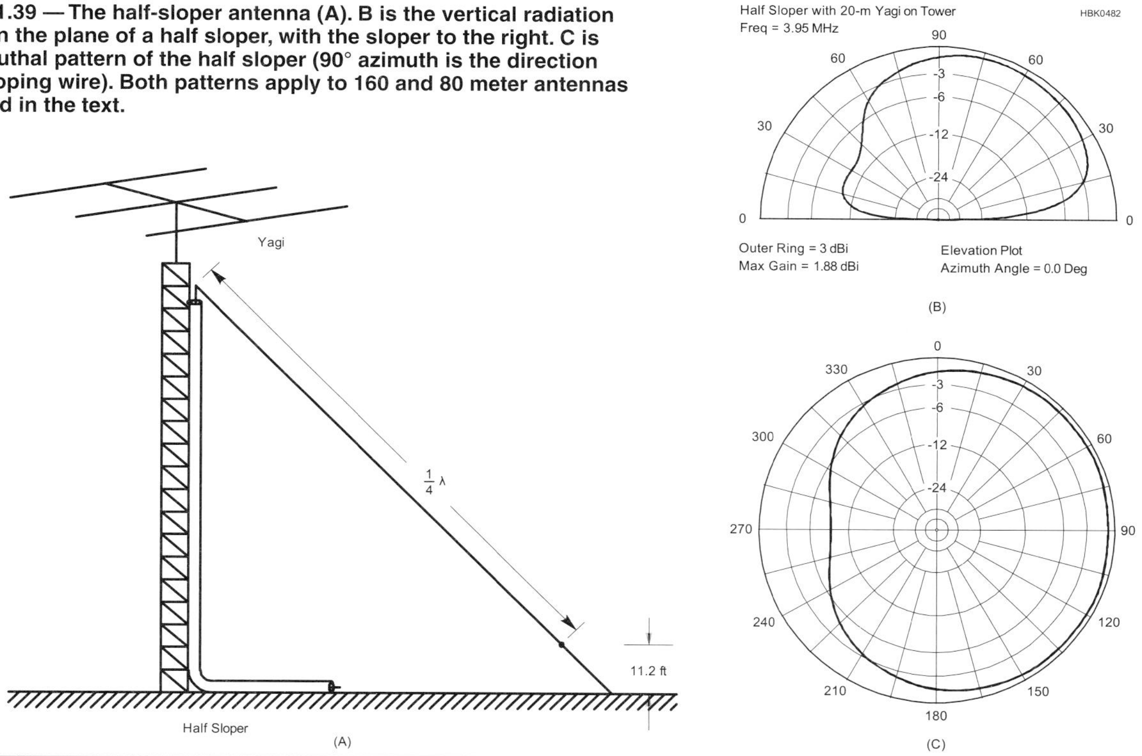

Project: Half-Wave Vertical Dipole (HVD)

Chuck Hutchinson, K8CH, describes a 15 meter vertical dipole (HVD) that he built for the ARRL book, *Simple and Fun Antennas for Hams*. The performance of this antenna, with its base at 14 feet, compares favorably with a horizontal dipole at 30 feet when making intercontinental QSOs.

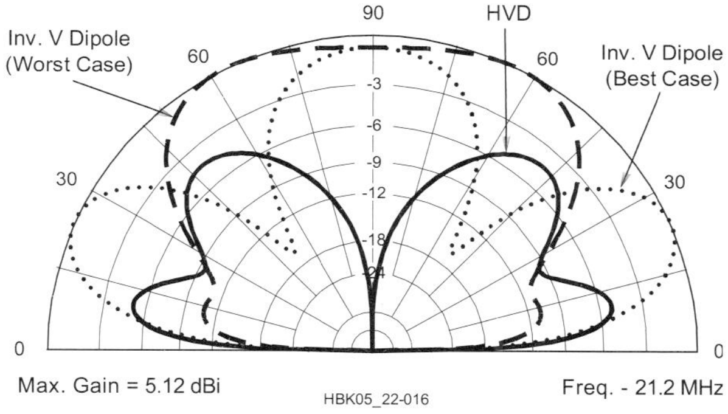

Figure 21.40 — Elevation patterns for the HVD (solid line) and the inverted-V comparison antenna in its best case (dotted line) and worst case (dashed line).

CONSTRUCTION OF A 15 METER HVD

The 15 meter HVD consists of four 6 foot lengths of 0.875 inch aluminum tube with 0.058 wall thickness. In addition there are two 1 foot lengths of 0.75 inch tubing for splices, and two 1 foot lengths of 0.75 inch fiberglass rod for insulators (see **Table 21.4** for dimensions).

Start by cutting off 1 foot from a 6 foot length of 0.875 inch tubing. Next, insert six inches of one of the 1-foot-long 0.75 inch tubes into the machine-cut end of your tubing and fasten the tubes together using sheet metal screws or pop rivets. Now, slide an end of a 6 foot length of 0.875 inch tube over the protruding end of the 0.75 inch tube and fasten them together. Repeat this procedure with the remaining 0.875 inch tubing. You should now have two 11-foot-long sections.

Because handmade cuts are not perfectly square, put those element ends at the center of the antenna. Slip these cut ends over the ends of a 1 foot length of 0.75 inch fiberglass rod. This rod serves as the center insulator. Leave about a 1 inch gap at the center. Drill aluminum and fiberglass for #8 hardware as shown in **Figure 21.41**.

Now, slip half of the remaining 1 foot length of 0.75 inch fiberglass rod into one end of the dipole. (This end will be the bottom end or base.) Drill and secure with #8 hardware (see **Figure 21.42**).

The final step is to secure the guy wires to your vertical. You can see how K8CH did that in **Figure 21.43**. Start by drilling a pilot hole and then drive a sheet metal screw into the antenna about a foot above the center. The purpose of that screw is to prevent the clamp and guys from sliding down the antenna.

The guys are continuous lengths of 3/16 inch Dacron line. (The Dacron serves a dual purpose: it supports the antenna vertically, and it acts as an insulator.) Tie secure knots into the guy ends and secure these knotted ends to the antenna with a stainless-steel worm-screw-type hose clamp. Take care to not over tighten the clamps. You don't want the clamp to slip

Table 21.4
HVD Dimensions
Length using 0.875-inch aluminum tubing

MHz	*Feet*	*Inches*
18.11	33	11
21.2	22	0
24.94	18	9
28.4	16	5

These lengths should be divided by two to determine the length of the dipole legs.

(the knots and the sheet-metal screw will help), but you especially don't want to cut your guy lines. Your antenna is ready for installation.

INSTALLATION

Installation requires two things. First, a place to sit or mount the base insulator. Second, you need anchors for the support guys.

Figure 21.41 — The center insulator of the 15 meter HVD is a 1 foot length of 0.75 inch fiberglass rod. Insulator and elements have been drilled to accept #8 hardware.

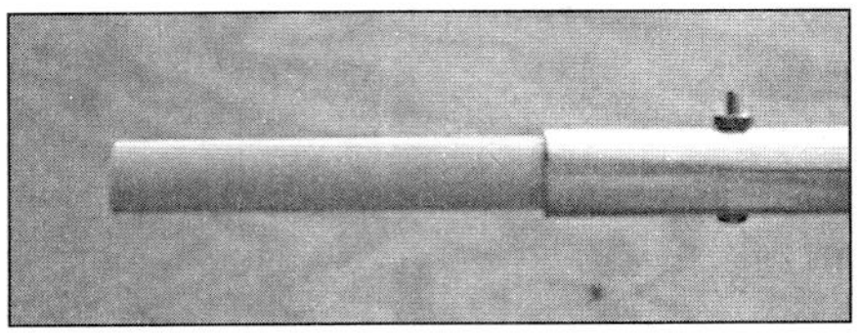

Figure 21.42 — The HVD base insulator is a 1 foot length of 0.75 inch fiberglass rod.

Figure 21.43 — Guys are made of Dacron line that is attached to the HVD by a stainless-steel worm-screw-type hose clamp. A self-tapping sheet-metal screw (not visible in the photo) prevents the clamp from sliding down the antenna.

Figure 21.44 — At K8CH, the HVD base insulator sits in this saddle-shaped wooden fixture. This was photo was taken before the fixture was painted — a necessary step to protect against the weather.

K8CH used a piece of 2 × 6 lumber to make a socket to hold the HVD base securely in place. He drilled a ¾ inch-deep hole with a ¾ inch spade bit. A couple of pieces of 2 × 2 lumber at the ends of the base form a saddle that nicely straddles the ridge at the peak of his garage roof. You can see this in **Figure 21.44**. The dimensions are not critical. Paint your base to protect it from the weather.

BALUN

This antenna needs a common-mode choke balun to ensure that RF doesn't flow on the shield of the coax. (See the **Transmission Lines** chapter for more information on choke baluns.) Unlike a horizontal dipole, don't consider it an option to omit the common-mode choke when building and installing an HVD.

You can use 8 feet of the RG-213 feed line wound into 7 turns for a balun. Secure the turns together with electrical tape so that each turn lies parallel with the next turn, forming a solenoid coil. Secure the feed line and balun to one of the guy lines with UV-resistant cable ties.

Because the feed line slants away from the antenna, you'll want to do *all* that you can to eliminate common-mode currents from the feed line. For that reason, make another balun about 11.5 feet from the first one. This balun also consists of 8 feet of the RG-213 feed line wound into 7 turns. See **Figure 21.45** for a photo of the installed antenna.

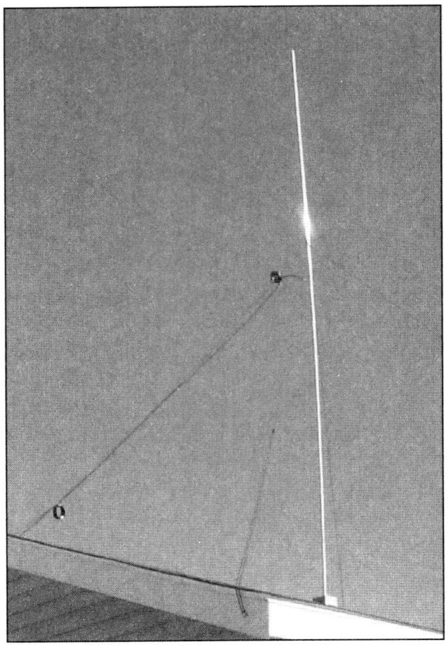

Figure 21.45 — The HVD installed at K8CH. An eye screw that is used for securing one of the guy lines is visible in the foreground. You can also see the two choke baluns that are used in the feed system (see text).

Project: Compact Vertical Dipole (CVD)

An HVD for 20 meters will be about 33 feet tall, and for 30 meters it will be around 46 feet tall. Even the 20 meter version can prove to be a mechanical challenge. The compact vertical dipole (CVD), designed by Chuck Hutchinson, K8CH, uses capacitance loading to shorten the antenna. Starting with the 15 meter HVD described in the previous project, Chuck added capacitance loading wires to lower the resonance to 30 meters. Later, he shortened the wires to move resonance to the 20 meter band. This project describes those two CVDs.

PERFORMANCE ISSUES

Shortened antennas frequently suffer reduced performance caused by the shortening. A dipole that is less than ½ λ in length is a compromise antenna. The issue becomes how much is lost in the compromise. In this case there are two areas of primary interest, radiation efficiency and SWR bandwidth.

Radiation Efficiency

Capacitance loading at the dipole ends is the most efficient method of shortening the antenna. Current distribution in the high-current center of the antenna remains virtually unchanged. Since radiation is related directly to current, this is the most desirable form of loading. Computer modeling shows that radiation from a 30 meter CVD is only 0.66 dB less than that from a full-size 30 meter HVD when both have their bases 8 feet above ground. The angle of maximum radiation shifts up a bit for the CVD. Not a bad compromise when you consider that the CVD is 22 feet long compared to the approximately 46 foot length of the HVD.

SWR and SWR Bandwidth

Shortened antennas usually have lower radiation resistance and less SWR bandwidth than the full-size versions. The amount of change in the radiation resistance is related to the amount and type of loading (shortening) being lower with shorter antennas. This can be a benefit in the case of a shortened vertical dipole. In Figure 21.2 you can see that vertical dipoles have a fairly high radiation resistance. With the dipole's lower end ⅛ λ above ground, the radiation resistance is roughly 80 Ω. In this case, a shorter antenna can have a better SWR when fed with 50 Ω coax.

SWR bandwidth tends to be wide for vertical dipoles in general. A properly designed CVD for 7 MHz or higher should give you good SWR (1.5:1 or better) across the entire band!

As you can see, in theory the CVD provides

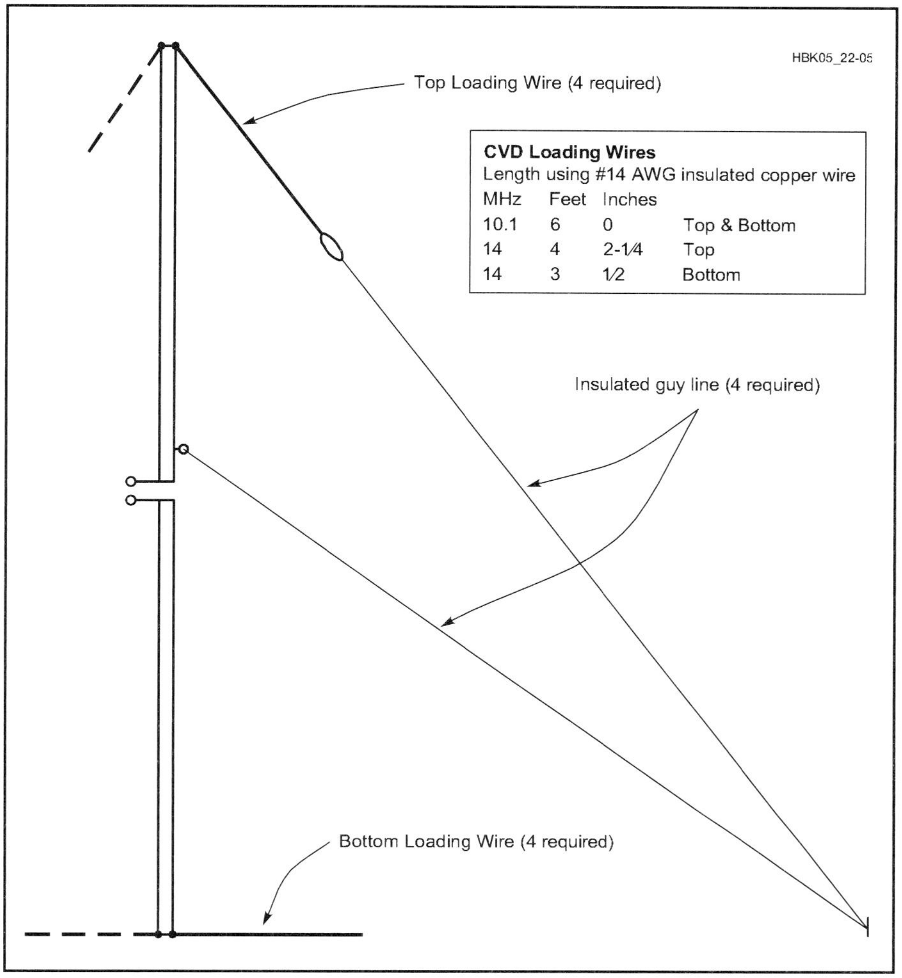

CVD Loading Wires
Length using #14 AWG insulated copper wire

MHz	Feet	Inches	
10.1	6	0	Top & Bottom
14	4	2-1⁄4	Top
14	3	1⁄2	Bottom

Figure 21.46 — The CVD consists of a vertical dipole and loading wires. Only one set of the four loading wires and only one guy line is shown in this drawing. See text for details.

Figure 21.47 — CVD loading wires can be attached using #8 hardware. Crimp and solder terminals on the wire ends to make connections easier.

excellent performance in a compact package. Experience confirms the theory.

CONSTRUCTION

To convert the K8CH 15 meter HVD to 20 or 30 meters, you'll need to add four loading wires at the top and four more at the bottom of the HVD. The lengths are given in the table in **Figure 21.46**. The upper wires droop at a 45° angle and the lower wires run horizontally. The antenna is supported by four guy lines. You can connect the wires to the vertical portion with #8 hardware. Crimp and solder terminals on the wire ends to make connections easier. The technique is illustrated in **Figure 21.47**.

The upper loading wires can be extended with insulated line and used for additional guying. The lower wires are extended with insulated line and fasten to the guy lines so that the lower wires run horizontally.

Prune the lower wires for best SWR across the band of interest. The K8CH CVD has its base at 14 feet. This antenna has an SWR of less than 1.2:1 on 30 meters and less than 1.3:1 across the entire 20 meter band.

EXPERIENCE

The 30 meter CVD was compared to a ground-mounted quarter-wave vertical and a horizontal dipole at 30 feet. In tests, the CVD was always the superior antenna.

Later, the CVD loading wires were shortened for operation on 20 meters. Once again the results were very encouraging.

Project: All-Wire 30 Meter CVD

If you have a tree or other support that will support the upper end of a CVD at 32 feet above the ground, you might want to consider an all-wire version of the 30 meter CVD. The vertical is 24 feet long, and it will have an SWR of less than 1.1:1 across the band. The four loading wires at top and bottom are each 5 feet, 2 inches long.

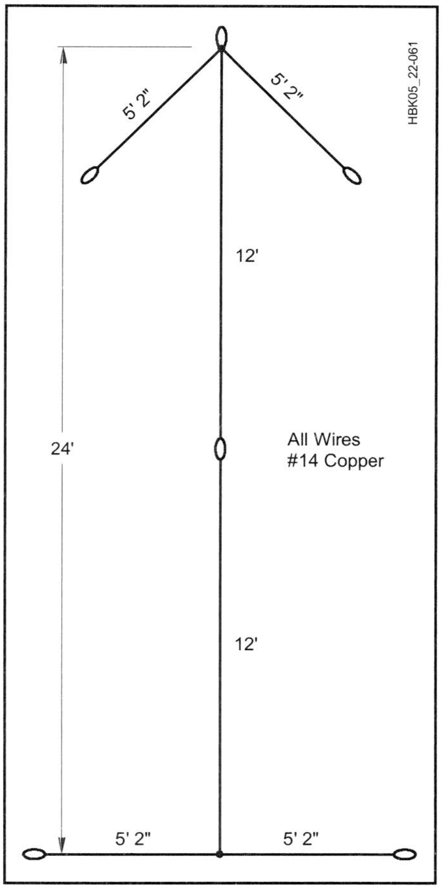

Figure 21.48 — The all-wire 30 meter CVD consists of a vertical dipole and loading wires. It can be made entirely with #14 AWG wire. Support lines have been omitted for simplicity. See text for details.

The configuration is shown in **Figure 21.48**. As with any vertical dipole, you'll need to use a balun between the feed line and the antenna.

Alternatively you can use two loading wires at the top and two at the bottom. In this case each of the loading wires is 8 feet, 7.5 inches long.

21.6 Yagi Antennas

Most antennas described earlier in this chapter have unity gain compared to a dipole, or just slightly more. For the purpose of obtaining gain and directivity it is convenient to use a Yagi-Uda *beam* antenna. The former is commonly called a *Yagi*. There are other forms of directive antennas, but the Yagi is by far the most popular used by amateurs. (For more information on phased arrays and other types of directive antennas, see the *ARRL Antenna Book*.)

Most operators prefer to erect these antennas for horizontal polarization, but they can be used as vertically polarized antennas merely by rotating the elements by 90°. In effect, the beam antenna is turned on its side for vertical polarization. The number of elements used will depend on the gain desired and the limits of the supporting structure. At HF, many amateurs obtain satisfactory results with only two elements in a beam antenna, while others have four or five elements operating on a single amateur band, called a *monoband beam*. On VHF and above, Yagis with many elements are common, particularly for simplex communication without repeaters. For fixed point-to-point communications such as repeater links, Yagis with three or four elements are more common.

Regardless of the number of elements used, the height-above-ground considerations discussed earlier for dipole antennas remain valid with respect to the angle of radiation. This is demonstrated in **Figure 21.49** at A and B where a comparison of radiation characteristics is given for a 3 element Yagi at one-half and one wavelength above average ground. It can be seen that the higher antenna (Figure 21.49B) has a main lobe that is more favorable for DX work (roughly 15°) than the lobe of the lower antenna in Figure 21.49A (approximately 30°). The pattern at B shows that some useful high-angle radiation exists also, and the higher lobe is suitable for short-skip contacts when propagation conditions dictate the need.

The azimuth pattern for the same antenna is provided in **Figure 21.50**. (This is a free-space pattern, so the pattern is taken in the plane of the antenna. Remember that azimuth patterns taken over a reflecting surface must also specify the elevation angle at which the pattern was measured or calculated.) Most of the power is concentrated in the main lobe at 0° azimuth. The lobe directly behind the main lobe at 180° is often called the *back lobe* or *rear lobe*. The front-to-back ratio (F/B) of this antenna is just less than 12 dB — the peak power difference, in decibels, between the main lobe at 0° and the rearward lobe at 180°. It is infrequent that two 3 element Yagis with different element spacing and tuning will yield the same lobe patterns. The patterns also change with frequency of operation. The pattern of Figure 21.50 is shown only for illustrative purposes.

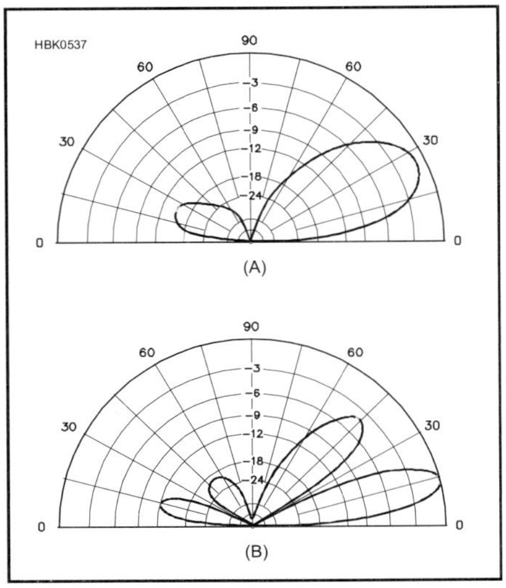

Figure 21.49 — Elevation-plane response of a 3 element Yagi placed ½ λ above perfect ground at A and the same antenna spaced 1 λ above ground at B.

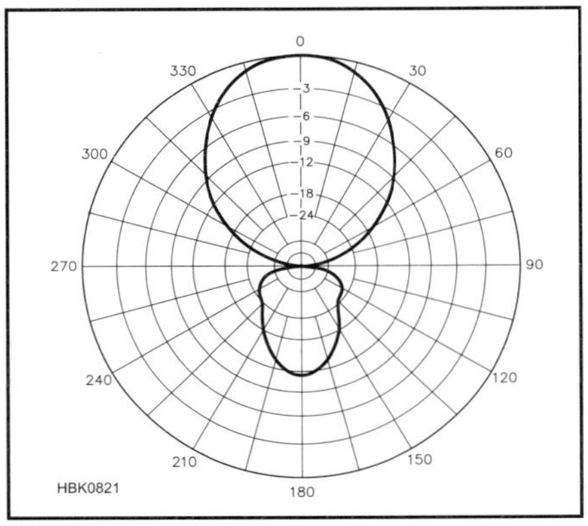

Figure 21.50 — Azimuthal pattern of a typical three-element Yagi in free space. The Yagi's boom is along the 0° to 180° axis.

21.6.1 Parasitic Excitation

In a Yagi antenna only one element (the *driven element*) is connected to the feed line. The additional elements are *coupled* to the driven element because they are so close. (Element-to-element spacing in a Yagi antenna is generally on the order of 1/10 to 1/8 wavelength.) This *mutual coupling* results in currents being induced in the non-driven elements from the radiated field of the driven element. These elements are called *parasitic elements,* and the Yagi antenna is therefore a *parasitic array*. (An antenna in which multiple elements all receive power from the transmitter is called a *driven array*.) The currents induced in the parasitic elements also result in radiated fields, just as if the current were the result of power from a feed line. This is called *re-radiation*, and it has a 180° phase shift from the current-inducing field. The combination of the field radiated by the driven element, the fields from the parasitic elements, and the physical spacing of the elements results in the fields having the proper phase relationship so as to focus the radiated energy in the desired direction and reject it in other directions.

The parasitic element is called a *director* when it reinforces radiation along a line pointing to it from the driven element, and a *reflector* in the opposite case. Whether the parasitic element is a director or reflector depends on the parasitic element tuning, which is usually adjusted by changing its length. The structure on which the elements are mounted is called the *boom* of the antenna.

21.6.2 Yagi Gain, Front-to-Back Ratio, and SWR

The gain of a Yagi antenna with parasitic elements varies with the spacing and tuning of the elements. Element tuning is a function of length, diameter, and *taper schedule* (the steps in length and diameter) if the element is constructed with telescoping tubing. For any given number of elements and the spacing between them, there is a tuning condition that will result in maximum gain. However, the maximum front-to-back ratio seldom, if ever, occurs at the same condition that gives maximum forward gain. The impedance of the driven element in a parasitic array, and thus the SWR, also varies with the tuning and spacing.

It is important to remember that all these parameters change as the operating frequency is varied. For example, if you operate both the CW and phone portions of the 20 meter band with a Yagi antenna, you probably will want an antenna with consistent performance over most of the band. Such designs typically must sacrifice a little gain in order to achieve good F/B and SWR performance across the band.

Gain and F/B performance generally improve with the number of elements. In Yagi antennas with more than three elements (a driven element and one director and reflector), the additional elements are added as directors, since little additional benefit is obtained from multiple reflectors. Wider spacing also improves gain and F/B up to a certain point, depending on a number of factors, beyond which performance begins to fall. Optimizing element spacing is a complex problem, and no single spacing satisfies all design requirements. For the lower HF bands, the size of the antenna quickly becomes impractical for truly *optimal* designs, and compromise is necessary.

OWA Designs

The Optimized Wideband Antenna (OWA) was originally developed by Jim Breakall,

WA3FET. It is a matching method that increases feed point impedance and widens the VSWR bandwidth. It can be applied to dipoles and monopoles, but it is most widely used for monoband Yagi designs. (The technique is described in more detail at **www.naic.edu/~angel/kp4ao/ham/owa.html**.)

A parasitic element placed ahead of and very close (less than 0.01 wavelength) to a driven element results in wide-band performance. That is, a low SWR relative to 50 Ω, smooth gain, and front-to-back performance over a wide HF amateur band (such as 20, 15, or 10 meters). The developers of the antenna suggest that the driver and first director perform as if they were a single element having a diameter equal to the spacing between the two elements.

The OWA is implemented in a Yagi antenna by placing the first director 0.05 λ in front of (in the forward direction) the driven element. By raising the feed point impedance is to 50 Ω, coaxial feed lines can be attached directly using a simple 1:1 choke balun. This eliminates the more complex gamma or beta matching systems used to match the low impedance of most Yagi antennas.

OWA Yagis give up a small amount of forward gain, typically less than 0.5 dB, from the lower-impedance designs. This is rarely significant in regular operation and is a good trade-off for the simplicity of direct feed and improved bandwidth.

21.6.3 Two-Element Beams

A two-element beam is useful — especially where space or other considerations prevent the use of a three-element, or larger, beam. The general practice is to tune the parasitic element as a reflector and space it about 0.15 λ from the driven element, although some successful antennas have been built with 0. l-λ spacing and director tuning.

Gain vs. element spacing for a two-element antenna is given in **Figure 21.51** for the special case where the parasitic element is resonant. It is indicative of the performance to be expected under maximum-gain tuning conditions. Changing the tuning of the driven element in a Yagi or quad will not materially affect the gain or F/R. Thus, only the spacing and the tuning of the single parasitic element have any effect on the performance of a 2 element Yagi.

In Figure 21.51, the greatest gain is in the direction A (in which the parasitic element is acting as a director) at spacings of less than 0.14 λ, and in direction B (in which the parasitic element is a reflector) at greater spacings. The front-to-back ratio is the difference in decibels between curves A and B along the boom of the antenna. The figure also shows variation in radiation resistance of the driven element.

These curves are for the special case of a self-resonant parasitic element, but are representative of how a two-element Yagi works. At most spacings the gain as a reflector can be increased by slight lengthening of the parasitic element; the gain as a director can be increased by shortening. This also improves the front-to-rear ratio.

Most two-element Yagi designs achieve a compromise F/R of about 10 dB, together with an acceptable SWR and gain across a frequency band with a percentage bandwidth less than about 4%.

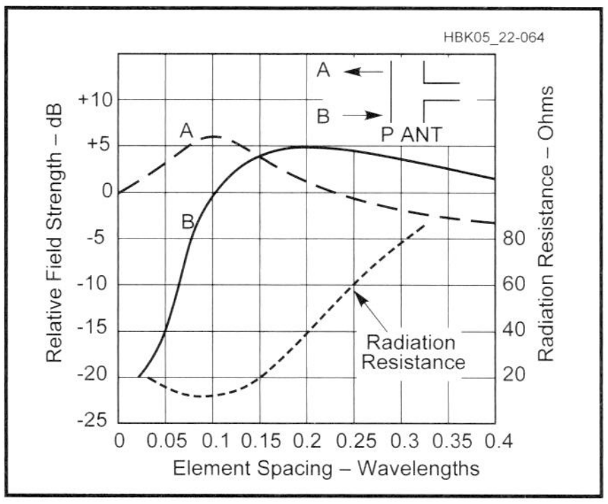

Figure 21.51 — Gain vs element spacing for a two-element Yagi, having one driven and one parasitic element. The reference point, 0 dB, is the field strength from a half-wave antenna alone.

21.6.4 Three-Element Beams

A theoretical investigation of the three-element case (director, driven element, and reflector) has indicated a maximum gain of about 9.7 dBi (7.6 dBd). A number of experimental investigations have shown that the spacing between the driven element and reflector for maximum gain is in the region of 0.15 to 0.25 λ. With 0.2 λ reflector spacing, **Figure 21.52** shows that the gain variation with director spacing is not especially critical. Also, the overall length of the array (boom length in the case of a rotatable antenna) can be anywhere between 0.35 and 0.45 λ with no appreciable difference in the maximum gain obtainable.

If maximum gain is desired, wide spacing of both elements is beneficial because adjustment of tuning or element length is less critical and the input resistance of the driven element is generally higher than with close spacing. A higher input resistance improves the efficiency of the antenna and makes a greater bandwidth possible. However, a total antenna length, director to reflector, of more than 0.3 λ at frequencies of the order of 14 MHz introduces difficulty from a construction standpoint. Lengths of 0.25 to 0.3 λ are therefore used frequently for this band, even though they are less than optimum from the viewpoint of maximum gain.

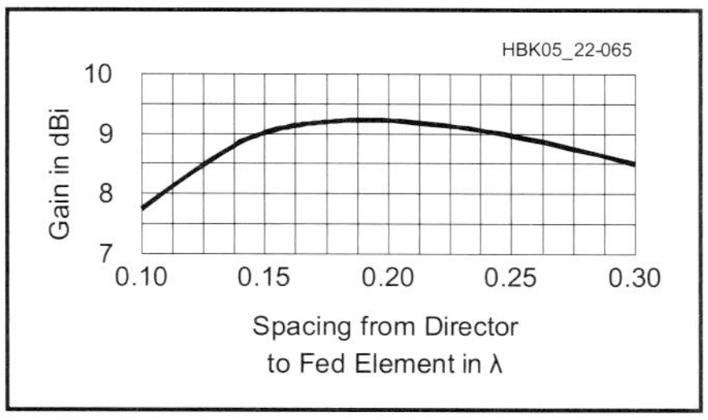

Figure 21.52 — General relationship of gain of three-element Yagi vs. director spacing, the reflector being fixed at 0.2 λ. This antenna is tuned for maximum forward gain.

In general, Yagi antenna gain drops off less rapidly when the reflector length is increased beyond the optimum value than it does for a corresponding decrease below the optimum value. The opposite is true of a director. It is therefore advisable to err, if necessary, on the long side for a reflector and on the short side for a director. This also tends to make the antenna performance less dependent on the exact frequency at which it is operated. An increase above the design frequency has the same effect as increasing the length of both parasitic elements, while a decrease in frequency has the same effect as shortening both elements. By making the director slightly short and the reflector slightly long, there will be a greater spread between the upper and lower frequencies at which the gain starts to show a rapid decrease.

21.6.5 Construction of Yagi Antennas

Most beams and verticals are made from sections of aluminum tubing. Compromise beams have been fashioned from less-expensive materials such as electrical conduit (steel) or bamboo poles wrapped with conductive tape or aluminum foil. The steel conduit is heavy, is a poor conductor, and is subject to rust. Similarly, bamboo with conducting material attached to it may deteriorate rapidly in the weather. Given the drawbacks of alternative materials, aluminum tubing (or rod for VHF and UHF Yagis) is far and away the best choice for antenna construction.

For reference, **Table 21.5** details the standard sizes of aluminum tubing, available in many metropolitan areas. Metal distributors, local and online, usually stock popular sizes of aluminum tubing. Several *QST* advertisers stock an extensive line of tubing specifically for antenna construction and cut to shippable lengths. Tubing usually comes in 12 foot lengths, although 20 foot lengths are available in some sizes. Your aluminum dealer will probably also sell aluminum plate in various thicknesses needed for boom-to-mast and boom-to-element connections.

Aluminum is rated according to its hardness. The most common material used in antenna

Table 21.5
Standard Sizes of Aluminum Tubing

6061-T6 (61S-T6) Round Aluminum Tube in 12-ft Lengths

OD (*in*)	*Wall Thickness* (*in*)	*stubs ga*	*ID* (*in*)	*Approx Weight (lb)* *per ft*	*per length*
3/16	0.035	no. 20	0.117	0.019	0.228
	0.049	no. 18	0.089	0.025	0.330
1/4	0.035	no. 20	0.180	0.027	0.324
	0.049	no. 18	0.152	0.036	0.432
	0.058	no. 17	0.134	0.041	0.492
5/16	0.035	no. 20	0.242	0.036	0.432
	0.049	no. 18	0.214	0.047	0.564
	0.058	no. 17	0.196	0.055	0.660
3/8	0.035	no. 20	0.305	0.043	0.516
	0.049	no. 18	0.277	0.060	0.720
	0.058	no. 17	0.259	0.068	0.816
	0.065	no. 16	0.245	0.074	0.888
7/16	0.035	no. 20	0.367	0.051	0.612
	0.049	no. 18	0.339	0.070	0.840
	0.065	no. 16	0.307	0.089	1.068
1/2	0.028	no. 22	0.444	0.049	0.588
	0.035	no. 20	0.430	0.059	0.708
	0.049	no. 18	0.402	0.082	0.948
	0.058	no. 17	0.384	0.095	1.040
	0.065	no. 16	0.370	0.107	1.284
5/8	0.028	no. 22	0.569	0.061	0.732
	0.035	no. 20	0.555	0.075	0.900
	0.049	no. 18	0.527	0.106	1.272
	0.058	no. 17	0.509	0.121	1.452
	0.065	no. 16	0.495	0.137	1.644
3/4	0.035	no. 20	0.680	0.091	1.092
	0.049	no. 18	0.652	0.125	1.500
	0.058	no. 17	0.634	0.148	1.776
	0.065	no. 16	0.620	0.160	1.920
	0.083	no. 14	0.584	0.204	2.448
7/8	0.035	no. 20	0.805	0.108	1.308
	0.049	no. 18	0.777	0.151	1.810
	0.058	no. 17	0.759	0.175	2.100
	0.065	no. 16	0.745	0.199	2.399
1	0.035	no. 20	0.930	0.123	1.467
	0.049	no. 18	0.902	0.170	2.040
	0.058	no. 17	0.884	0.202	2.424
	0.065	no. 16	0.870	0.220	2.640
	0.083	no. 14	0.834	0.281	3.372
1 1/8	0.035	no. 20	1.055	0.139	1.668
	0.058	no. 17	1.009	0.228	2.736
1 1/4	0.035	no. 20	1.180	0.155	1.860
	0.049	no. 18	1.152	0.210	2.520
	0.058	no. 17	1.134	0.256	3.072
	0.065	no. 16	1.120	0.284	3.408
	0.083	no. 14	1.084	0.357	4.284
1 3/8	0.035	no. 20	1.305	0.173	2.076
	0.058	no. 17	1.259	0.282	3.384
1 1/2	0.035	no. 20	1.430	0.180	2.160
	0.049	no. 18	1.402	0.260	3.120
	0.058	no. 17	1.384	0.309	3.708
	0.065	no. 16	1.370	0.344	4.128
	0.083	no. 14	1.334	0.434	5.208
	*0.125	1/8"	1.250	0.630	7.416
	*0.250	1/4"	1.000	1.150	14.823
1 5/8	0.035	no. 20	1.555	0.206	2.472
	0.058	no. 17	1.509	0.336	4.032
1 3/4	0.058	no. 17	1.634	0.363	4.356
	0.083	no. 14	1.584	0.510	6.120
1 7/8	0.508	no. 17	1.759	0.389	4.668
2	0.049	no. 18	1.902	0.350	4.200
	0.065	no. 16	1.870	0.450	5.400
	0.083	no. 14	1.834	0.590	7.080
	*0.125	1/8"	1.750	0.870	9.960
	*0.250	1/4"	1.500	1.620	19.920
2 1/4	0.049	no. 18	2.152	0.398	4.776
	0.065	no. 16	2.120	0.520	6.240
	0.083	no. 14	2.084	0.660	7.920
2 1/2	0.065	no. 16	2.370	0.587	7.044
	0.083	no. 14	2.334	0.740	8.880
	*0.125	1/8"	2.250	1.100	12.720
	*0.250	1/4"	2.000	2.080	25.440
3	0.065	no. 16	2.870	0.710	8.520
	*0.125	1/8"	2.700	1.330	15.600
	*0.250	1/4"	2.500	2.540	31.200

*These sizes are extruded; all other sizes are drawn tubes.Shown here are standard sizes of aluminum tubing that are stocked by most aluminum suppliers or distributors in the United States and Canada. Don Daso, K4ZA, explains the different aluminum alloy types in this books's online article "Workshop Chronicles: Alloy Designations."

construction is grade 6061-T6. This material is relatively strong and has good workability. In addition, it will bend without taking a *set*, an advantage in antenna applications where the pieces are constantly flexing in the wind. The softer grades (5051, 3003, and so on) will bend much more easily, while harder grades (7075 and so on) are more brittle.

Wall thickness is of primary concern when selecting tubing. It is of utmost importance that the tubing fits snugly where the element sections join. Sloppy joints will make a mechanically unstable antenna. The magic wall thickness is 0.058 inch. For example (from Table 21.5), 1 inch outside diameter (OD) tubing with a 0.058 inch wall has an inside diameter (ID) of 0.884 inch. The next smaller size of tubing, 7/8 inch, has an OD of 0.875 inch. The 0.009 inch difference provides just the right amount of clearance for a snug fit.

Figure 21.53 shows several methods of fastening antenna element sections together. The slot and hose clamp method shown at the upper left is probably the best for joints where adjustments are needed. Generally, one adjustable joint on each side of the element is sufficient to tune the antenna — usually the tips at each end of an element are made adjustable. Stainless steel hose clamps (beware — some "stainless steel" models do not have a stainless screw and will rust) are recommended for longest antenna life.

The remaining photos show possible fastening methods for joints that are not adjustable. At the upper right, machine screws and nuts hold the elements in place. At the lower left, sheet metal screws are used. At the lower right, rivets secure the tubing. If the antenna is to be assembled permanently, rivets are the best choice. Once in place, they are permanent. Properly installed, they rarely work free, regardless of vibration or wind. If aluminum rivets with aluminum mandrels are employed, they will never rust. Also, being aluminum, there is no danger of corrosion from interaction between dissimilar metals. If the antenna is to be disassembled and moved periodically, either machine or sheet metal screws will work. If machine screws are used, however, take precautions to keep the nuts from vibrating free. Use of lock washers, lock nuts, and flexible adhesive such as silicone bathtub sealant or a thread-locking compound will keep the hardware in place. For portable or temporary use, such as Field Day, rivets may be held in place with electrical tape and removed when the operation is finished.

Use of a conductive grease at the element joints is essential for long life. Left untreated, the aluminum surfaces will oxidize in the weather, resulting in a poor connection. Some trade names for this conductive grease are Penetrox, OxGard, and Noalox. The copper-bearing anti-sieze compund SS-30 is another good choice. Many electrical supply houses and hardware stores carry these products.

DRIVEN ELEMENT

The ARRL recommends *plumbers delight* construction, in which all elements are mounted directly on, and grounded to, the boom. This puts the entire array at dc ground potential, affording better lightning protection. A gamma- or T-match section can be used for matching the feed line to the array.

An alternative method is to insulate the driven element from the boom, but using a *hairpin* or *beta match*, the center point of which is electrically neutral and can be attached directly to the boom, restoring the dc ground for the driven element.

Direct feed designs in which the feed point impedance of the driven element is close to 50 Ω require no impedance matching structure. Like hairpin-matched designs, the driven element must be insulated from the boom, requiring some additional mechanical complexity.

Regardless of the design, a current or choke balun should be used (see the **Transmission Lines** chapter) to prevent the outer surface of the feed line shield from interacting with the antenna directly or picking up the radiated signal. Such interaction can degrade the antenna's radiation pattern, especially by compromising signal rejection to the side and rear.

BOOM MATERIAL

The boom size for a rotatable Yagi or quad should be selected to provide stability to the entire system. The best diameter for the boom depends on several factors, but mostly the element weight, number of elements, and overall length. Two-inch-diameter booms should not be made any longer than 24 feet unless additional support is given to reduce both vertical and horizontal bending forces. Suitable reinforcement for a long 2 inch boom can consist of a truss or a truss and lateral support, as shown in **Figure 21.54**.

A boom length of 24 feet is about the point where a 3 inch diameter begins to be very worthwhile. This dimension provides a considerable amount of improvement in overall mechanical stability as well as increased clamping surface area for element hardware. The latter is extremely important to prevent rotation of elements around the boom if heavy icing is commonplace. Pinning an element to the boom with a large bolt helps in this regard. On smaller diameter booms, however, the elements sometimes work loose and tend to elongate the pinning holes in both the element and the boom. After some time the elements shift their positions slightly (sometimes from day to day) and give a ragged appearance to the system, even though this may not harm the electrical performance.

A 3 inch diameter boom with a wall thickness of 0.065 inch is very satisfactory for antennas up to about a five-element, 20 meter array that is spaced on a 40 foot boom. A truss is recommended for any boom longer than 24 feet. One possible source for large boom material is irrigation tubing sold at farm supply houses.

PUTTING IT TOGETHER

Once you assemble the boom and elements, the next step is to fasten the elements to the boom securely and then fasten the boom to the mast or supporting structure using mounting plates, as shown in **Figure 21.55**. Be sure to leave plenty of material on either side of the U-bolt holes on the element-to-boom mounting plates. The U-bolts selected should be a snug fit for the tubing. If possible, buy muffler-clamp U-bolts that come with saddles.

The *boom-to-mast* plate shown in Figure 21.55B is similar to the *boom-to-element* plate in 21.55A. The size of the plate and number of U-bolts used will depend on the size of the antenna. Generally, antennas for the bands up through 20 meters require only two U-bolts each for the mast and boom. Longer antennas for 15 and 20 meters (35 foot booms and up) and most 40 meter beams should have four U-bolts each for the boom and mast because of the torque that the long booms and elements exert as the antennas move in the wind. When tightening the U-bolts, be careful not to crush the tubing. Once the wall begins to collapse, the connection begins to weaken. Many aluminum suppliers sell ¼ inch or ⅜ inch thick plates just right for this application. Often they will shear pieces to the correct size on request. As with tubing,

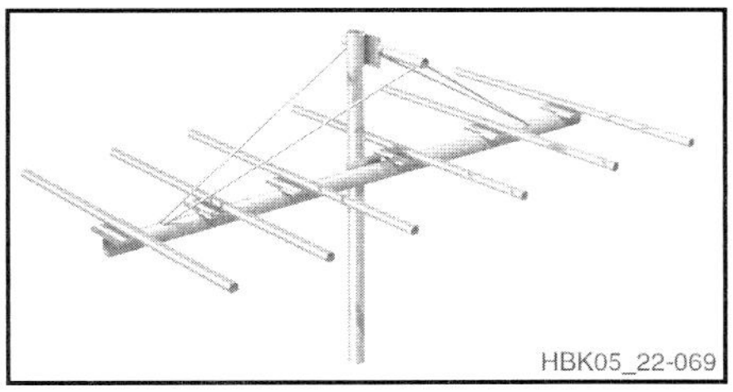

Figure 21.54 — A long boom needs both vertical and horizontal support. The crossbar mounted above the boom can support a double truss, which will help keep the antenna in position.

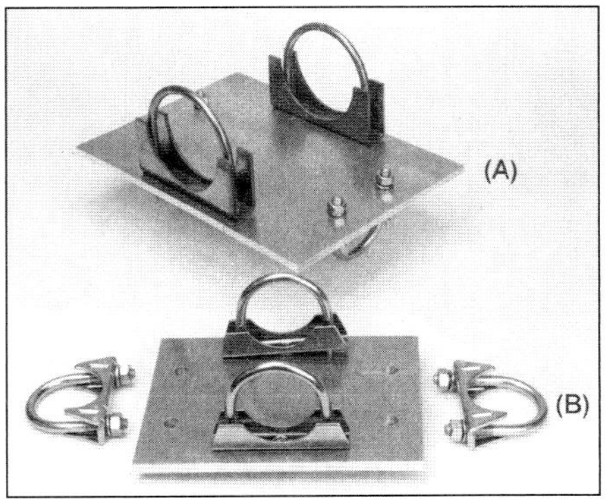

Figure 21.55 — The boom-to-element plate at A uses muffler-clamp-type U-bolts and saddles to secure the round tubing to the flat plate. The boom-to-mast plate at B is similar to the boom-to-element plate. The main difference is the size of materials used.

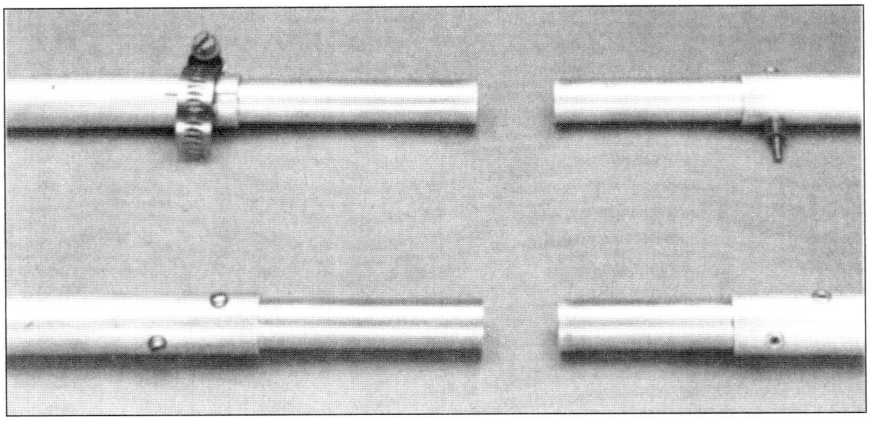

Figure 21.53 — Some methods of connecting telescoping tubing sections to build beam elements. See text for a discussion of each method.

the relatively hard 6061-T6 grade is a good choice for mounting plates.

The antenna should be put together with good-quality hardware. Stainless steel or galvanized hardware is best for long life and is available from several amateur radio dealers. Stainless steel hardware can develop surface defects called *galling* that can cause threads on nuts and bolts to seize. On hardware ¼ inch and larger, the use of an anti-seize compound is recommended. Anti-seize compound is available from auto parts stores, or a small amount of grease or anti-oxidation compound can help lubricate the threads.

Rust will attack plated steel hardware after a short time, making nuts difficult, if not impossible, to remove. If you use plated hardware, paint it with a good zinc-chromate primer and a finish coat or two.

Good-quality hardware is more expensive initially, but if you do it right the first time, you won't have to take the antenna down after a few years and replace the hardware. Also, when repairing or modifying an installation, nothing is more frustrating than fighting rusty hardware at the top of a tower.

Project: Family of Computer-Optimized HF Yagis

Yagi designers are now able to take advantage of powerful personal computers and software to optimize their designs for the parameters of gain, F/R, and SWR across frequency bands. Dean Straw, N6BV, has designed a family of Yagis for HF bands. These can be found in **Tables 21.6** to **21.10**, for the 10, 12, 15, 17 and 20 meter amateur bands, respectively.

For 12 through 20 meters, each design has been optimized for better than 20 dB F/R and an SWR of less than 2:1 across the entire amateur frequency band. For the 10 meter band, the designs were optimized for the lower 800 kHz of the band, from 28.0 to 28.8 MHz. Each Yagi element is made of telescoping 6061-T6 aluminum tubing with 0.058 inch thick walls. This type of element can be telescoped easily, using techniques shown in Figure 21.53. Measuring each element to an accuracy of ⅛ inch results in performance remarkably consistent with the computations, without any need for tweaking or fine-tuning when the Yagi is on the tower.

The dimensions shown are designed for specific telescoping aluminum elements, but the elements may be scaled to different sizes by using the information about tapering and scaling in *The ARRL Antenna Book*, although with a likelihood of deterioration in performance over the whole frequency band.

Each element is mounted above the boom with a heavy rectangular aluminum boom-to-element plate, by means of galvanized U-bolts with saddles, as shown in Figure 21.55. This method of element mounting is rugged and stable, and because the element is mounted away from the boom, the amount of element detuning due to the presence of the boom is minimal. The element dimensions given in each table already take into account any element detuning due to the mounting plate. The element mounting plate for all the 10 meter Yagis is a 0.250 inch thick flat aluminum plate, 4 inches wide by 4 inches long. For the 12 and 15 meter Yagis, a 0.375 inch thick flat aluminum plate 5 inches wide by 6 inches long is used, and for the 17 and 20 meter Yagis a 0.375 inch thick flat aluminum plate, 6 inches wide by 8 inches long is used. Where the plate is rectangular, the long dimension is in line with the element.

Each design table shows the dimensions for *one-half* of each element, mounted on one side of the boom. The other half of each element is the same, mounted on the other side of the boom. Use a tubing sleeve inside the center portion of the element so that the element is not crushed by the mounting U-bolts. Each telescoping section is inserted 3 inches into the next size of tubing. For example, in the 310-08

Table 21.6

10 Meter Optimized Yagi Designs

	Spacing Between Elements (in)	*Seg 1 Length (in)*	*Seg 2 Length (in)*	*Seg 3 Length (in)*	*Midband Gain F/R*
310-08					
Refl	0	24	18	66.750	7.2 dBi
DE	36	24	18	57.625	22.9 dB
Dir 1	54	24	18	53.125	
410-14					
Refl	0	24	18	64.875	8.4 dBi
DE	36	24	18	58.625	30.9 dB
Dir 1	36	24	18	57.000	
Dir 2	90	24	18	47.750	
510-24					
Refl	0	24	18	65.625	10.3 dBi
DE	36	24	18	58.000	25.9 dB
Dir 1	36	24	18	57.125	
Dir 2	99	24	18	55.000	
Dir 3	111	24	18	50.750	

Note: For all antennas, the tube diameters are Seg 1=0.750 inch, Seg 2=0.625 inch, Seg 3=0.500 inch.

Table 21.7

12 Meter Optimized Yagi Designs

	Spacing Between Elements (in)	*Seg 1 Length (in)*	*Seg 2 Length (in)*	*Seg 3 Length (in)*	*Midband Gain F/R*
312-10					
Refl	0	36	18	69.000	7.5 dBi
DE	40	36	18	59.125	24.8 dB
Dir 1	74	36	18	54.000	
412-15					
Refl	0	36	18	66.875	8.5 dBi
DE	46	36	18	60.625	27.8 dB
Dir 1	46	36	18	58.625	
Dir 2	82	36	18	50.875	
512-20					
Refl	0	36	18	69.750	9.5 dBi
DE	46	36	18	61.750	24.9 dB
Dir 1	46	36	18	60.500	
Dir 2	48	36	18	55.500	
Dir 3	94	36	18	54.625	

Note: For all antennas, the tube diameters are: Seg 1 = 0.750 inch,Seg 2 = 0.625 inch, Seg 3 = 0.500 inch.

Table 21.8
15 Meter Optimized Yagi Designs

	Spacing Between Elements (in)	*Seg 1 Length (in)*	*Seg 2 Length (in)*	*Seg 3 Length (in)*	*Seg 4 Length (in)*	*Midband Gain F/R*
315-12						
Refl	0	30	36	18	61.375	7.6 dBi
DE	48	30	36	18	49.625	25.5 dB
Dir 1	92	30	36	18	43.500	
415-18						
Refl	0	30	36	18	59.750	8.3 dBi
DE	56	30	36	18	50.875	31.2 dB
Dir 1	56	30	36	18	48.000	
Dir 2	98	30	36	18	36.625	
515-24						
Refl	0	30	36	18	62.000	9.4 dBi
DE	48	30	36	18	52.375	25.8 dB
Dir 1	48	30	36	18	47.875	
Dir 2	52	30	36	18	47.000	
Dir 3	134	30	36	18	41.000	

Note: For all antennas, the tube diameters (in inches) are: Seg 1 = 0.875, Seg 2 = 0.750, Seg 3 = 0.625, Seg 4 = 0.500.

Table 21.9
17 Meter Optimized Yagi Designs

	Spacing Between Elements (in)	*Seg 1 Length (in)*	*Seg 2 Length (in)*	*Seg 3 Length (in)*	*Seg 4 Length (in)*	*Seg 5 Length (in)*	*Midband Gain F/R*
317-14							
Refl	0	24	24	36	24	60.125	8.1 dBi
DE	65	24	24	36	24	52.625	24.3 dB
Dir 1	97	24	24	36	24	48.500	
417-20							
Refl	0	24	24	36	24	61.500	8.5 dBi
DE	48	24	24	36	24	54.250	27.7 dB
Dir 1	48	24	24	36	24	52.625	
Dir 2	138	24	24	36	24	40.500	

Note: For all antennas, tube diameters (inches) are: Seg 1=1.000, Seg 2=0.875, Seg 3=0.750, Seg 4=0.625, Seg 5=0.500.

Table 21.10
20 Meter Optimized Yagi Designs

	Spacing Between Elements (in)	*Seg 1 Length (in)*	*Seg 2 Length (in)*	*Seg 3 Length (in)*	*Seg 4 Length (in)*	*Seg 5 Length (in)*	*Seg 6 Length (in)*	*Midband Gain F/R*
320-16								
Refl	0	48	24	20	42	20	69.625	7.3 dBi
DE	80	48	24	20	42	20	51.250	23.4 dB
Dir 1	106	48	24	20	42	20	42.625	
420-26								
Refl	0	48	24	20	42	20	65.625	8.6 dBi
DE	72	48	24	20	42	20	53.375	23.4 dB
Dir 1	60	48	24	20	42	20	51.750	
Dir 2	174	48	24	20	42	20	38.625	

Note: For all antennas, tube diameters (inches) are: Seg 1=1.000, Seg 2=0.875, Seg 3=0.750, Seg 4=0.625, Seg 5=0.500. Seg 6=0.375.

design for 10 meters (3 elements on an 8 foot boom), the reflector tip, made out of ½ inch OD tubing, sticks out 66.75 inches from the ⅝ inch OD tubing. For each 10 meter element, the overall length of each ⅝ inch OD piece of tubing is 21 inches, before insertion into the ¾ inch piece. Since the ¾ inch OD tubing is 24 inches long on each side of the boom, the center portion of each element is actually 48 inches of uncut ¾ inch OD tubing.

The boom for all these antennas should be constructed with at least 2 inch OD tubing, with 0.065 inch wall thickness. Because each boom has three inches of extra length at each end, the reflector is actually placed three inches from one end of the boom. For the 310-08 design, the driven element is placed 36 inches ahead of the reflector, and the director is placed 54 inches ahead of the driven element. The antenna is attached to the mast with the *boom-to-mast* mounting plate shown in Figure 21.55.

Each antenna is designed with a driven element length appropriate for a gamma or T matching network, as shown in **Figure 21.56**. The variable gamma or T capacitors can be

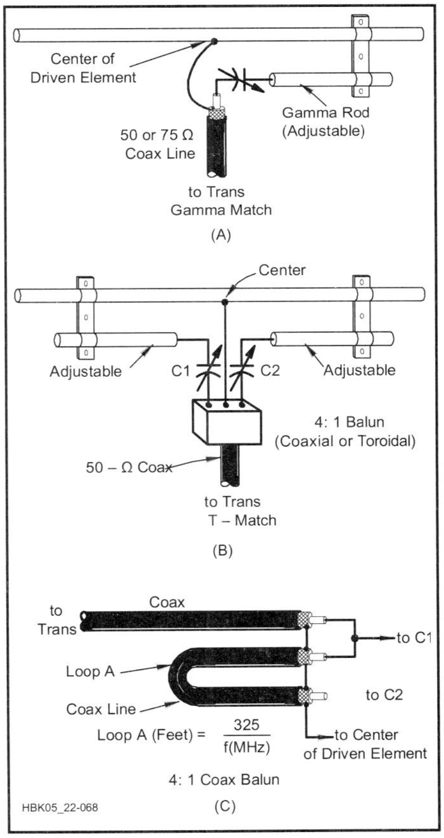

Figure 21.56 — Illustrations of gamma and T matching systems. At A, the gamma rod is adjusted along with the capacitor until the lowest SWR is obtained. A T match is shown at B. It is the same as two gamma-match rods. A 4:1 coaxial balun transformer for use with the T match is shown at C.

housed in small plastic enclosures for weatherproofing; receiving-type variable capacitors with close plate spacing can be used at powers up to a few hundred watts. Maximum capacitance required is usually 140 pF at 14 MHz and proportionally less at the higher frequencies.

The driven-element's length may require slight readjustment for best match, particularly if a different matching network is used. *Do not change either the lengths or the telescoping tubing schedule of the parasitic elements* — they have been optimized for best performance and will not be affected by tuning of the driven element.

TUNING ADJUSTMENTS

To tune the gamma match, adjust the gamma capacitor for best SWR, then adjust the position of the shorting strap or bar that connects the gamma rod to the driven element. Repeat this alternating sequence of adjustments until a satisfactory SWR is reached.

To tune the T-match, the position of the shorting straps and C1 and C2 are adjusted alternately for a best SWR. To maintain balance of the antenna, the position of the straps and capacitor settings should be the same for each side and adjusted together. A coaxial 4:1 balun transformer is shown at Figure 21.56C. A toroidal balun can be used in place of the coax model shown. The toroidal version has a broader frequency range than the coaxial one. The T match is adjusted for 200 Ω, and the balun steps this balanced value down to 50 Ω, unbalanced. Or the T match can be set for 300 Ω, and the balun used to step this down to 75 Ω unbalanced.

Dimensions for the gamma and T match rods will depend on the tubing size used and the spacing of the parasitic elements of the beam. Capacitors C1 and C2 can be 140 pF for 14 MHz beams. Somewhat less capacitance will be needed at 21 and 28 MHz.

Preliminary matching adjustments can be done on the ground. The beam should be aligned vertically so that the reflector element is closest to and a few feet off the ground, with the beam pointing upward. The matching system is then adjusted for best SWR. When the antenna is raised to its operating height, only slight touch-up of the matching network may be required.

A *choke balun* (see the **Transmission Lines** chapter) should be used to isolate the coaxial feed line shield from the antenna. Secure the feed line to the boom of the antenna between the feed point and the supporting mast.

21.7 Quad and Loop Antennas

One of the more effective DX antennas is the *quad*. It consists of two or more loops of wire, each supported by a bamboo or fiberglass cross-arm assembly. The loops are ¼ λ per side (one full wavelength overall). One loop is driven and the other serves as a parasitic element — usually a reflector. The design of the quad is similar to that of the Yagi, except that the elements are loops instead of dipoles. A two-element quad can achieve better F/R, gain, and SWR across a band, at the expense of greater mechanical complexity compared to a two-element Yagi and very nearly the same performance as a three-element Yagi. Small loops with a circumference of less than 0.3 λ are discussed in the next section.

A variation of the quad is called the *delta loop*. The electrical properties of both antennas are the same. Both antennas are shown in **Figure 21.57**. They differ mainly in their physical properties, one being of plumber's delight construction, while the other uses insulating support members. One or more directors can be added to either antenna if additional gain and directivity are desired, though most operators use the two-element arrangement.

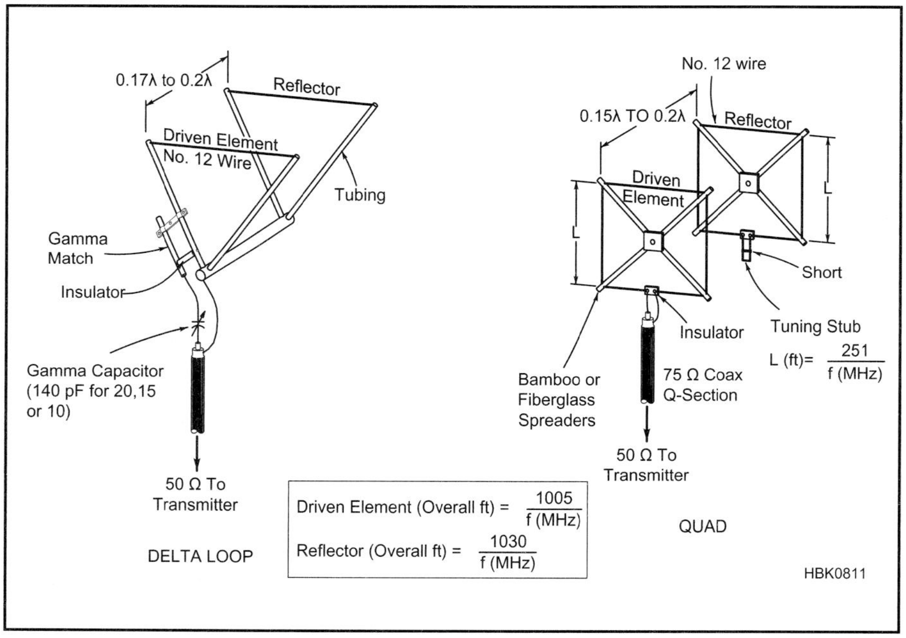

Figure 21.57 — Information on building a quad or a delta-loop antenna. The antennas are electrically similar, but the delta-loop uses plumber's delight construction. The λ/4 length of 75 Ω coax or Q-section acts as a synchronous transmission-line transformer from the approximate 100 Ω feed point impedance of the quad to the 50 Ω feed line.

It is possible to interlace quads or deltas for two or more bands, but if this is done the lengths calculated using the formulas given in Figure 21.57 may have to be changed slightly to compensate for the proximity effect of the second antenna. Using a tuning capacitor as shown in the following project allows the antenna to be adjusted for peak performance without cumbersome adjustment of wire lengths.

If multiple arrays are used, each antenna should be tuned separately for maximum forward gain, or best front-to-rear ratio, as observed on a field-strength meter. The reflector stub on the quad should be adjusted for this condition. The resonance of the antenna can be found by checking the frequency at which the lowest SWR occurs. By lengthening or shortening it, the driven element length can be adjusted for resonance in the most-used portion of the band.

A gamma match can be used at the feed point of the driven element to match the impedance to that of coaxial cable. Because the loop's feed point impedance is *higher* than that of 50 Ω coaxial cable, a *synchronous transmission line transformer* or *Q-section* (see the **Transmission Lines** chapter) with an impedance intermediate to that of the loop and the coaxial cable can be used.

***Project:* Five-Band, Two-Element HF Quad**

Two multi-band quad designs covering 20 through 10 meters are described in this project. One was constructed by William A. Stein, KC6T, from scratch, and the other was built by Al Doig, W6NBH, using modified commercial triband quad hardware. The principles of construction and adjustment are the same for both models, and the performance results are also essentially identical. One of the main advantages of this design is the ease of (relatively) independent performance adjustments for each of the five bands. These quads were described by William A. Stein, KC6T, in *QST* for April 1992. Both models use 8-foot-long, 2 inch diameter booms and conventional X-shaped spreaders (with two sides of each quad loop parallel to the ground).

Each driven element is fed separately, but running five separate feed lines to the station would be unwieldy. A remote coax switch on the boom is used to select the feed line for each element. A gamma match or quarter-wave synchronous transmission line transformer is used to match the feed point impedance of the element to 50 Ω.

These designs can also be simplified to monoband quads by using the formulas in Figure 21.57 for loop dimensions and spacing. It is recommended to the antenna builder unfamiliar with quads that a monoband quad be attempted first in order to become acquainted with the techniques of building a quad. Once comfortable with constructing and erecting the quad, success with a multiband design is much easier to achieve.

Complete construction details and more information about the performance of these designs are available with the online content.

21.7.1 Large Loop Antennas

The loop antennas described in this section are continuous loops at least one wavelength in circumference and formed into open shapes with sides that are approximately equal, such as triangles, diamonds, squares, or circles. Smaller loops used for receiving purposes are discussed in the *ARRL Antenna Book*. Loops with ratios of side lengths greater than 2 or 3:1 begin to have special characteristics beyond the scope of this chapter.

A 1 λ loop can be thought of as two ½ λ dipoles with their ends connected together and pulled apart into an open shape, as described above. The feed point of one dipole is replaced with a short circuit so that there is only one feed point on the antenna. As such, the current and voltage distribution around the loop is an extension of Figure 21.1. Three typical loop shapes and the current distributions on them are shown in **Figure 21.58**. Note that the current flow reverses at points ¼ λ to either side of the feed point. That means the current direction opposite the feed point is the same as at the feed point.

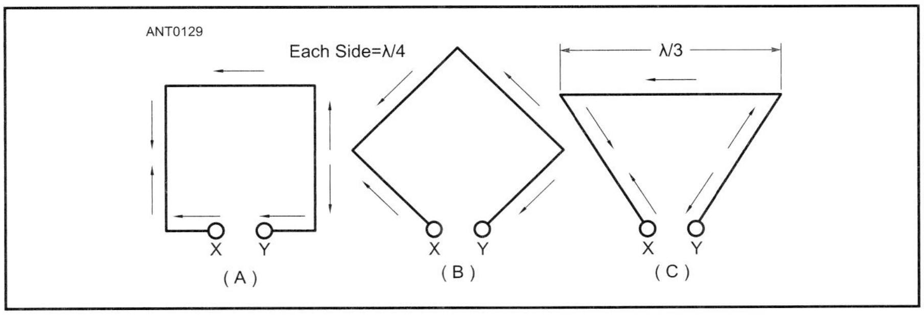

Figure 21.58 — At A and B, loops have sides ¼ λ long , and at C having sides ⅓ λ long for a total conductor length of 1 λ. The polarization depends on the orientation of the loop and on the position of the feed point (terminals X-Y) around the perimeter of the loop.

The maximum radiation strength of a 1 λ loop is perpendicular to the plane of the loop and minimum in the plane of the loop. If the loop is horizontal, the antenna radiates best straight up and straight down and poorly to the sides. The gain of a 1 λ loop in the direction of maximum radiation is approximately 1 dBd.

If the plane of the three loops shown in Figure 21.58 is vertical, the radiation is horizontally polarized because the fields radiated by the vertical components of current are symmetrical and opposing, so they cancel, leaving only the horizontally polarized fields that reinforce each other perpendicular to the loop plane. If the feed point of the antenna is moved to a vertical side or the antenna is rotated 90°, it is the horizontally polarized fields that will cancel, leaving a vertically polarized field, still maximum perpendicular to the plane of the loop. Feeding the loop at some other location, rotating the loop by some intermediate value, or constructing the loop in an asymmetrical shape will result in polarization somewhere between vertical and horizontal, but the maximum radiation will still occur perpendicular to the plane of the loop.

In contrast to straight-wire antennas, the electrical length of the circumference of a 1 λ loop is shorter than the actual length. For a loop made of bare #18 AWG wire and operating at a frequency of 14 MHz, so that the length-to-diameter ratio is very large, the loop will be close to resonance in free space when:

$$\text{Length (feet)} = 1032/\text{f (MHz)} \qquad (7)$$

The radiation resistance of a resonant 1 λ loop is approximately 120 Ω under these conditions. Since the loop dimensions are larger than those of a ½ λ dipole, the radiation efficiency is high and the SWR bandwidth of the antenna significantly larger than for the dipole.

The loop antenna is resonant on all frequencies at which it is an integral number of wavelengths in circumference; f_0, $2f_0$, $3f_0$, etc. That means an 80 meter 1 λ loop will also have a relatively low feed point impedance on 40, 30, 20, 15, 12, and 10 meters. As each side of the loop becomes longer electrically, the radiation pattern of the loop begins to develop nulls perpendicular to the plane of the loop and lobes that are closer to the plane of the loop. A horizontal diamond-shaped loop with legs more than a wavelength long is a *rhombic* antenna, which can develop significant gain along the long axis of the antenna. (The diamond-shaped rhombic is the origin of the symbol of the ARRL and many other radio organizations.)

21.7.2 Small Transmitting Loop Antennas

The small gap-resonated, circular loop antenna for HF transmitting was originally patented by John H. Dunlavy, Jr. His efficient small loop design can be tuned over wide bandwidths. The July/August 2018 issue of *QEX* features several articles on these antennas. **Figure 21.59** shows a typical "mag" loop being used for portable operation. Their small size and profile make it easy to mount them on a tripod (as seen in the figure) or other small mast. Models are available for both permanent and temporary installation.

Small loops are most popular on 7 MHz and higher frequencies, although larger models can be used on 3.5 and even 1.8 MHz. (See the article "A High-Power 160- and 80-Meter Transmitting Loop Antenna, by Steve Adler, VK5SFA.)

These loops consist of a one-turn primary (main) loop antenna having a circumference of less than ⅜-λ that is interrupted along its length (usually opposite the feed point) by a gap, with a tuning capacitor connected across the gap so the antenna can be tuned over a wide tuning range. A single-turn secondary (feed) loop, much smaller than the primary loop, is inductively coupled to the primary loop. Both loops are in the same plane. The secondary loop diameter is optimized based on the diameter of the primary loop so that variation in

Figure 21.59 — A typical small loop being used for portable operation with a QRP transceiver. Be sure to account for minimum safe distances to avoid excessive RF exposure levels.

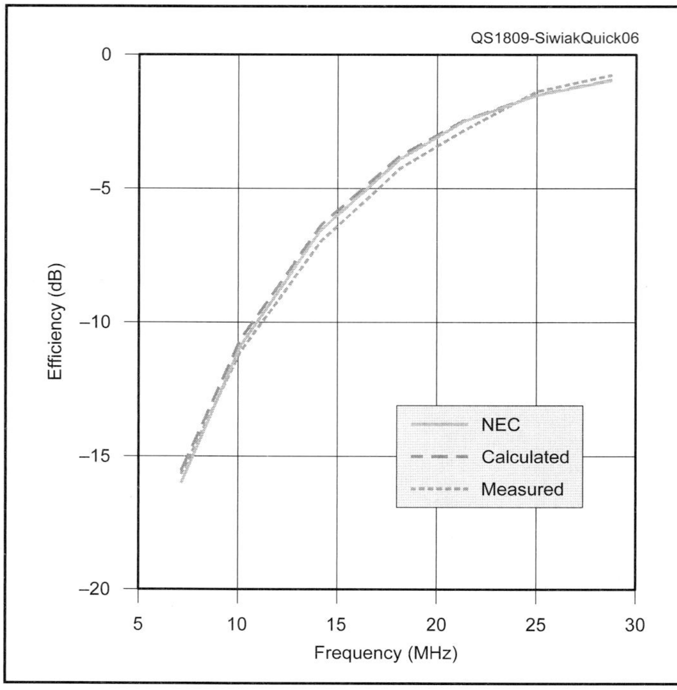

Figure 21.61 — Small loop efficiency simulated in NEC (solid), calculated from loop equations (long dashes), and measured using the Q method (short dashes). This variation represents a change of about three S units across this frequency range.

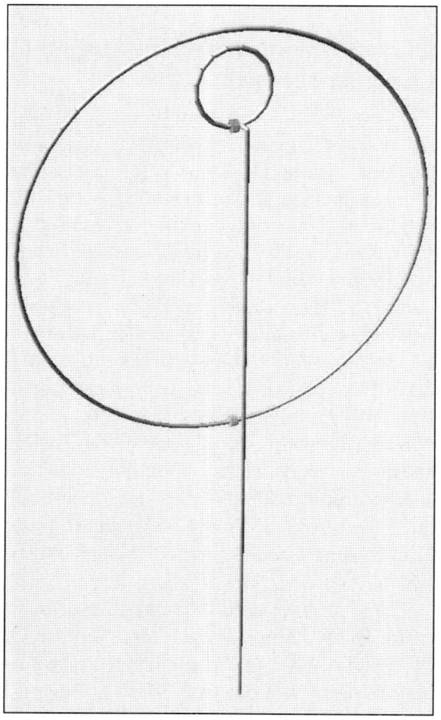

Figure 21.60 — The electrically small HF loop includes a main loop and a secondary feeding loop, both in the same plane, and a coaxial cable feed line also in the same plane, but slightly offset so that the cable does not touch the bottom of the main loop. A resonating capacitor connects across a gap at the bottom of the main loop.

feed impedance is minimized over the band of operation. A low impedance transmission line (50 Ω) connects to the terminals of the secondary loop. **Figure 21.60** shows a model of the loop. Because of the antenna's high Q, the operating bandwidth is very narrow, requiring retuning every few kHz.

The pattern of these small loops is a bidirectional "figure 8" similar to small direction-finding loops with the null broadside to the plane of the loop. The depth of the null varies from 5 – 20 dB across the HF spectrum as described in the Sep 2018 *QST* article "Small Gap-Resonated HF Loop Antennas" by Kai Siwiak, KE4PT, and Richard Quick, W4RQ. (included with this book's online material). The *ARRL Antenna Book* chapter on Loop Antennas presents more information and construction guidelines.

Because the radiation resistance of these small loops is so small (less than 1 Ω), low-loss construction techniques are very important, particularly in the adjustable tuning capacitor. Siwiak and Quick's article presents efficiencies as shown in **Figure 21.61** which seem quite low on the lower-frequency bands, if, however, an S unit represents a 6 dB change, the overall variation across the HF spectrum is only about 3 S units.

When using these small loops in the field, pay particular attention to RF exposure and minimum safe distances at different power levels. A table of distances is provided in the Portable Stations section of the chapter **Assembling a Station**. Because these antennas have such a high Q, the near field strengths can be very high, even at low power levels.

Project: Low-Band Quad and Delta Loops

(The following material is summarized from Chapter 10 of *ON4UN's Low-Band DXing, Fifth Edition.*) Dimensions for these designs assume an operating frequency of 3.75 MHz. The dimensions for the loops in this section may be scaled to frequencies in the 160, 60, 40 or 30 meter bands. The performance of the loops will vary with height above ground and ground conductivity.

SQUARE LOOP

Figure 21.62 shows the vertical-plane radiation patterns for a quad loop over very poor ground and over very good ground on the same dB scale for both horizontal and vertical polarization. Polarization of the loop depends on the location of the feed point, as shown in the figure.

The vertically polarized quad loop can be considered as two shortened, top-loaded vertical dipoles spaced ¼ λ apart. Broadside radiation from the horizontal elements of the quad is very low because the currents in the horizontal legs are approximately equal but in opposite directions in each half of the leg. The radiation angle in the broadside direction will be essentially the same as for either of the vertical members.

The resulting radiation angle will depend on the quality of the ground up to several wavelengths away from the antenna, as is the case with all vertically polarized antennas. The quality of the ground is as important as it is for any other vertical antenna, meaning

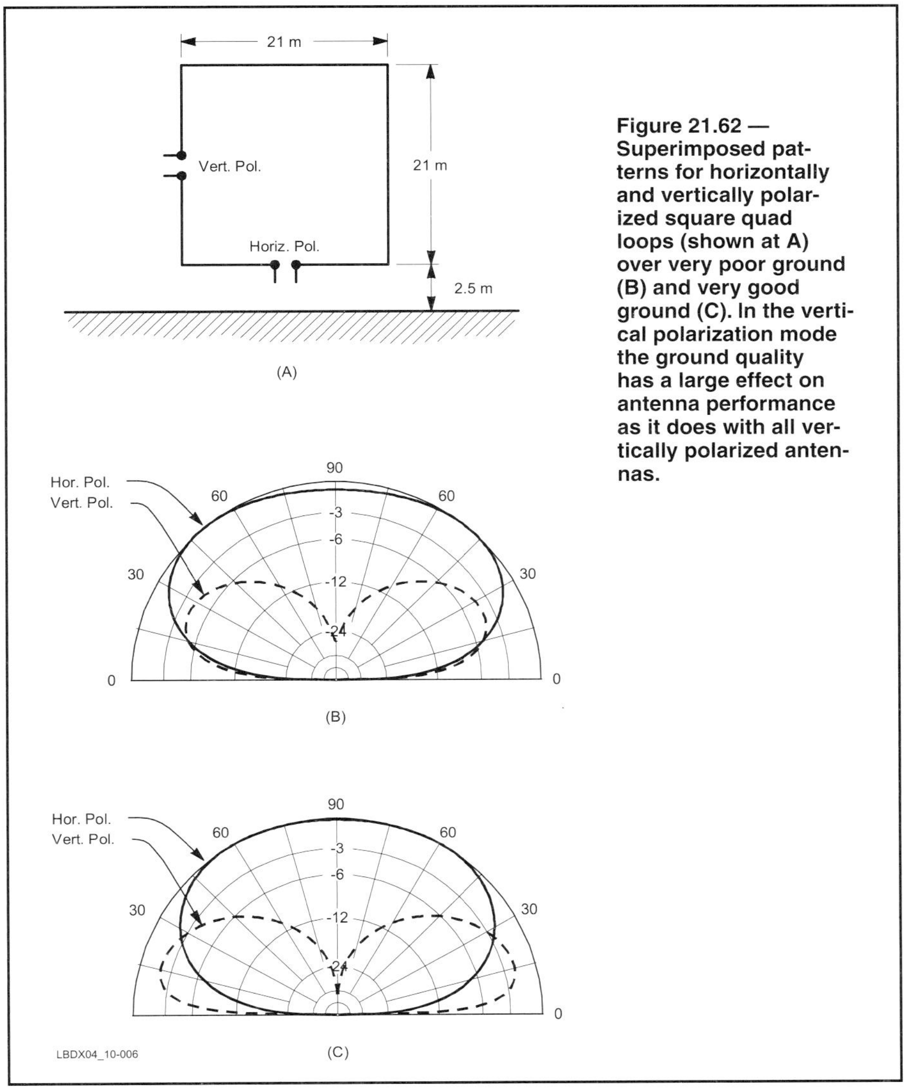

Figure 21.62 — Superimposed patterns for horizontally and vertically polarized square quad loops (shown at A) over very poor ground (B) and very good ground (C). In the vertical polarization mode the ground quality has a large effect on antenna performance as it does with all vertically polarized antennas.

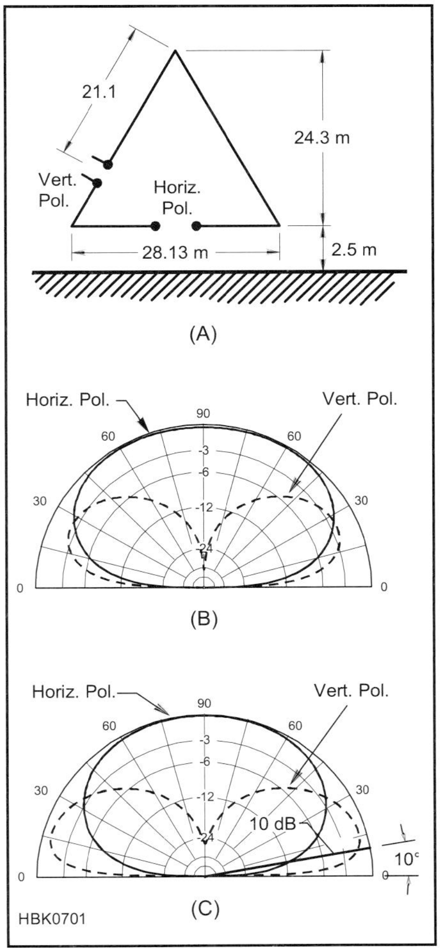

Figure 21.63 — Superimposed patterns for horizontally and vertically polarized delta quad loops (shown at A) over very poor ground (B) and very good ground (C). Over better ground, the vertically polarized loop performs much better at low radiation angles, while over both good and poor ground the vertically polarized loop gives good discrimination against high-angle local signals.

that vertically polarized loops close to the ground will not work well over poor soil. In a typical situation on 80 meters, a vertically polarized quad loop will radiate an excellent low-angle signal (lobe peak at approximately 21°) when operated over average ground. Over poorer ground, the peak elevation angle would be closer to 30°. The horizontal directivity is rather poor and amounts to approximately 3.3 dB of side rejection at any elevation angle.

A horizontally polarized quad-loop antenna can be thought of as two stacked short dipoles with a peak elevation angle dependent on the height of the loop. The low horizontally polarized quad (top at 0.3 λ) radiates most of its energy right at or near zenith angle (straight up). At low wave angles (20° to 45°) the horizontally polarized loop shows more front-to-side ratio (5 to 10 dB) than the vertically polarized rectangular loop.

With a horizontally polarized quad loop the angle of peak radiation is very dependent on the antenna height but not so much on the quality of the ground. At very low heights, the angle of peak radiation varies between 50° and 60° (but is rather constant all the way up to 90°). This is very good for NVIS and regional communication but not very good for DX. As far as gain is concerned, there is a 2.5-dB gain difference between very good and very poor ground, which is only half the difference found with the vertically polarized loop.

Comparing the gains of the horizontally and vertically polarized loops, Figure 21.62 shows that at very low antenna heights the gain is about 3 dB better for the horizontally polarized loop. But this gain exists at a high wave angle (50° to 90°), while the vertically polarized loop at very low heights radiates at 17° to 25°.

At heights from 3 to 6 meters for the bottom leg, feed point resistance for the horizontally polarized loop is approximately 100 to 120 Ω over average ground. For vertical polarization, feed point resistance varies from 200 to 170 Ω.

The quad loop feed point should be symmetrical, whether you feed the quad in the middle of the vertical or the horizontal wire. At the feed point, use a common-mode choke balun (see the **Transmission Lines** chapter) as current flowing on the outside of the coaxial feed line could upset the radiation pattern.

DELTA LOOP

Figure 21.63 shows the configuration as well as the superimposed elevation patterns for vertically and horizontally polarized low-height equilateral triangle delta loops over two different types of ground (same dB scale). The model was constructed for a frequency of 3.75 MHz. The base is 2.5 meters above ground, which puts the apex at 26.8 meters. Over good ground, the vertically polarized delta loop shows nearly 3 dB front-to-side ratio at the peak radiation angle of 22°. With average ground the gain is 1.3 dBi.

Over very poor ground, the horizontally polarized delta loop is better than the vertically polarized loop for all wave angles above 35°. Below 35° the vertically polarized loop takes over, but quite marginally. The maximum gain of the vertically and the horizontally polarized loops differs by only 2 dB, but the big difference is that for the horizontally polarized loop, the gain occurs at almost 90°, while for the vertically polarized loop it occurs at 25°. The vertically polarized antenna also gives good high-angle rejection (rejection of local signals), while the horizontally polarized loop will not.

Over very good ground, the performance at low angles is greatly improved for both polarizations. The vertically polarized loop is still better at any elevation angle under 30° than when horizontally polarized. At a 10° radiation angle the difference is as high as 10 dB. This makes the vertically polarized delta over good ground far superior for DX operating.

Most practical delta loops show a feed point impedance between 50 and 100 Ω, depending on the exact geometry and coupling to other antennas. The antenna can be fed directly with a 50 or 70 Ω coaxial cable, or via a 70 Ω quarter-wave transformer (see the **Transmission Lines** chapter) if the feed point impedance is near 100 Ω. At the feed point, use a common-mode choke balun (see the **Transmission Lines** chapter) as current flowing on the outside of the coaxial feed line could upset the radiation pattern.

THE BOTTOM-CORNER-FED DELTA LOOP

Figure 21.64 shows the layout of the delta loop being fed at one of the two bottom corners. The antenna is slightly compressed from the previous section with a slightly lower apex and longer base than the loop described in the previous section. Because of the "incorrect" location of the feed point, cancellation of radiation from the base wire is incomplete, resulting in a significant horizontally polarized radiation component. The total field has a very uniform gain coverage (within 1 dB) from 25° to 90°. This may be a disadvantage for the rejection of high-angle signals when working DX at low angles.

Due to the feed point location, the endfire radiation (radiation in line with the loop) has become asymmetrical with a side null of nearly 12 dB at the peak radiation angle of 29°. The loop actually radiates its maximum signal about 18° off the broadside direction. This feed point configuration greatly affects the pattern of the loop, so use bottom-corner-feed with care.

Project: Two-Band Loop for 30 and 40 Meters

The following antenna design is from a *QST* Hints and Kinks entry by James Brenner, NT4B, in the May 1989 issue. The version shown in **Figure 21.65** is fed at the apex of a delta loop but can be adapted to a square or quad loop shape.

The original design was derived from "The Mini X-Q Loop" in *All About Cubical Quad Antennas* by Bill Orr, W6SAI (now out of print)

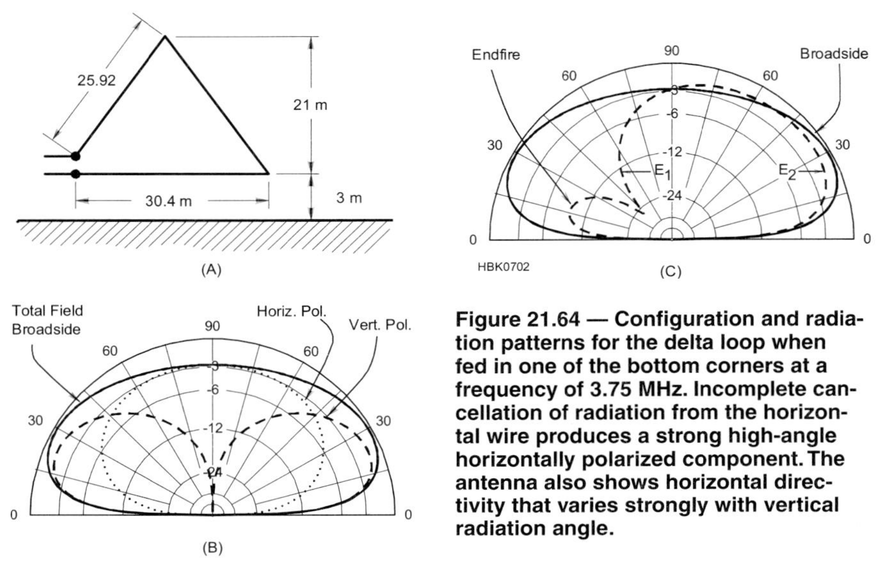

Figure 21.64 — Configuration and radiation patterns for the delta loop when fed in one of the bottom corners at a frequency of 3.75 MHz. Incomplete cancellation of radiation from the horizontal wire produces a strong high-angle horizontally polarized component. The antenna also shows horizontal directivity that varies strongly with vertical radiation angle.

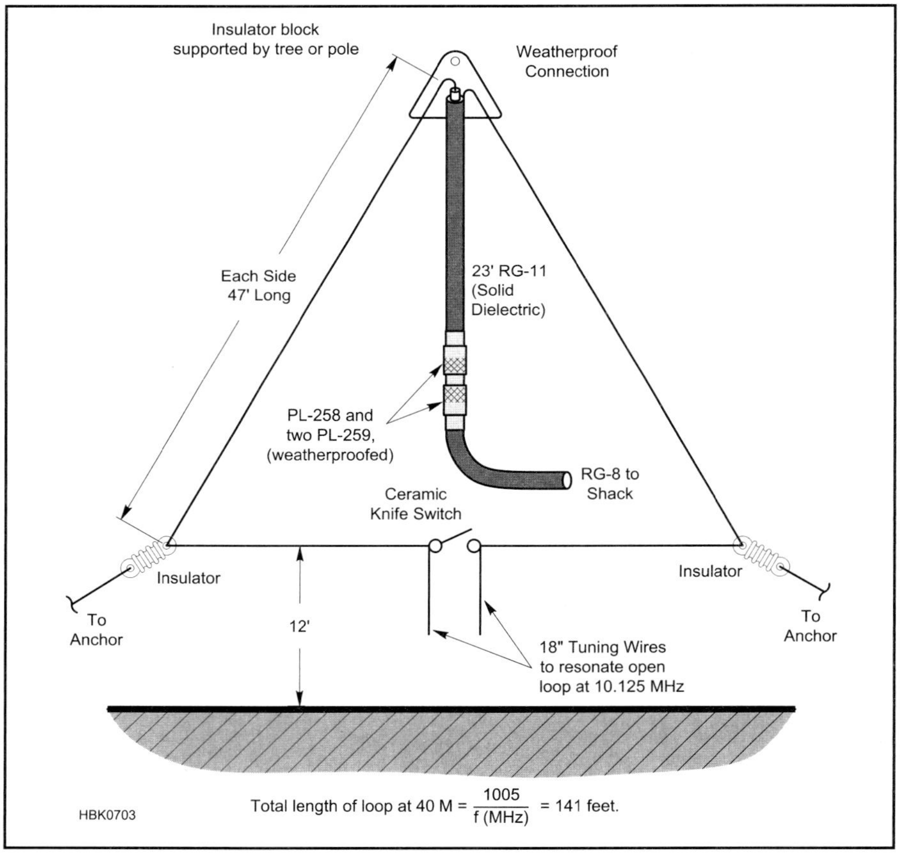

Figure 21.65— NT4B's 30 and 40 meter loop is fed at the top via a quarter-wave 40 meter matching transformer made of 75 Ω coax. Note the 18 inch tuning wires used to lower the antenna's 30 meter resonance from 10.5 to 10.1 MHz. Adjust the length of these wires to set the 30 meter resonant frequency.

which is 1½ λ in circumference, with an open circuit opposite the feed point. That antenna has approximately 1 dB of additional gain over a 1 λ loop. Since 30 and 40 meters are close to the same 1:1½ λ ratio, one loop can be converted between 1 λ on 40 meters and 1½ λ on 30 meters with a switch.

A large, ceramic SPST knife switch is installed in the center of the delta loop's bottom leg as shown in Figure 21.65. With the switch open, the loop acts as 1½ λ loop at 10.5 MHz, so 18 inch wires were added to the loop on either side of the switch to lengthen the antenna and lower the resonant frequency to 10.1 MHz. Closing the switch shorts out the wires, and the loop becomes a regular 1 λ continuous loop for 40 meters.

Note that there is fairly high voltage present at the switch when transmitting on 30 meters. If a relay is used, be sure the contact spacing is sufficient to avoid arcing or use additional pairs of contacts to increase the overall spacing.

The antenna is fed through a quarter-wave transformer (see the **Transmission Lines** chapter) of 75 Ω RG-11 coax, approximately 23 feet long. According to the author, when configured for 40 meters, the loop has a satisfactory SWR of less than 2:1 on 15 meters. In addition, the 30 meter configuration can be used successfully on 80 meters with the use of an antenna tuner.

Project: Multiband Horizontal Loop Antenna

Along with the multiband, non-resonant dipole, many amateurs operate on HF with great success using a horizontal loop antenna. All that is required are at least three supports able to hold the corners of the antenna 20 or more feet above the ground (and even that is negotiable) and enough room for a loop of wire one wavelength or more in circumference at the lowest frequency of operation. (Smaller loops can be used with an impedance-matching unit.)

Start by calculating the total length of wire using Equation 7. You'll need one insulator for each support and lengths of rope that are at least twice the height to which the insulator will be raised. You can feed the antenna at one of the corner insulators or anywhere along the wire with a separate insulator. Examples of corner insulators are shown in **Figure 21.66**. Using floating insulators allows the wire to move as the antenna flexes. One of the insulators should be of the fixed type, or the antenna can be fed at one corner with the loop wires attached to a pair of insulators sharing a common support rope. This holds the antenna feed point in place.

If the loop is only going to be used on the band for which it is resonant, coaxial cable can be used as the feed line, since SWR will be low. A choke balun at the feed point is recommended. For multiband use, open-wire feed line should be used, with an impedance-matching unit in the station. See Figure 21.15 for an example of how open-wire line can be attached at the feed point.

The impedance at the station end of the open-wire line can be matched to 50 Ω for coaxial cable with a balanced ATU or a balun and regular ATU. The impedance will vary widely, so do not expect a fixed-ratio balun (1:1, 4:1, 9:1, and so on) to provide a match over wide frequency range. The balun's function is to connect the open-wire line to unbalanced equipment or feed lines.

On its fundamental frequency, the antenna's maximum radiation will be straight up, making it most useful for regional communications at high elevation angles with the occasional DX contact. At higher frequencies, the loop will radiate more strongly at lower angles for better signal strengths at long distances. An example design is shown in **Figure 21.67**.

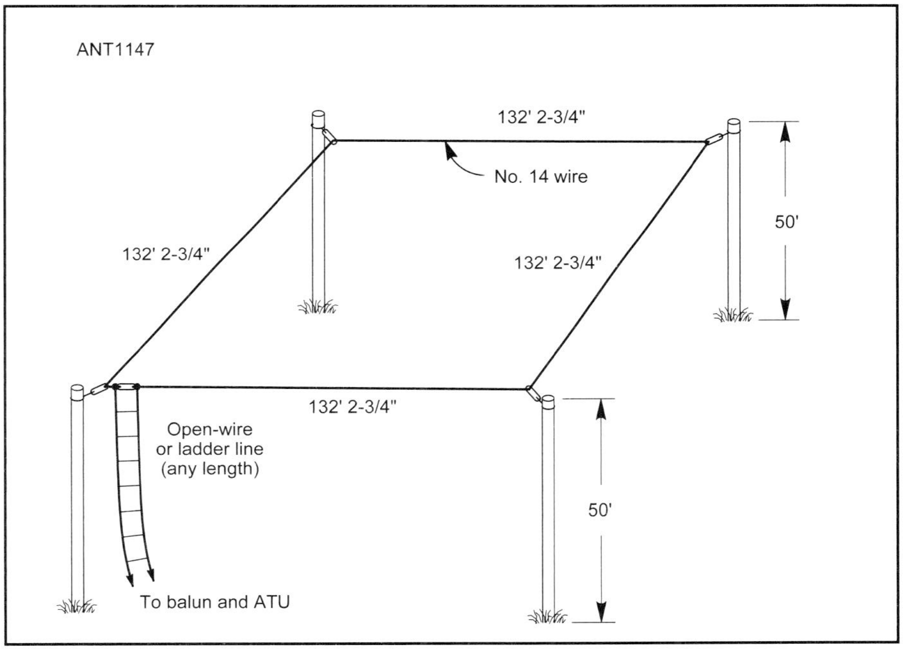

Figure 21.67 — An example of a horizontal full-wave loop designed for 1.9 MHz. The antenna can be fed with any length of open-wire or ladder line. The loop is usable on all HF bands.

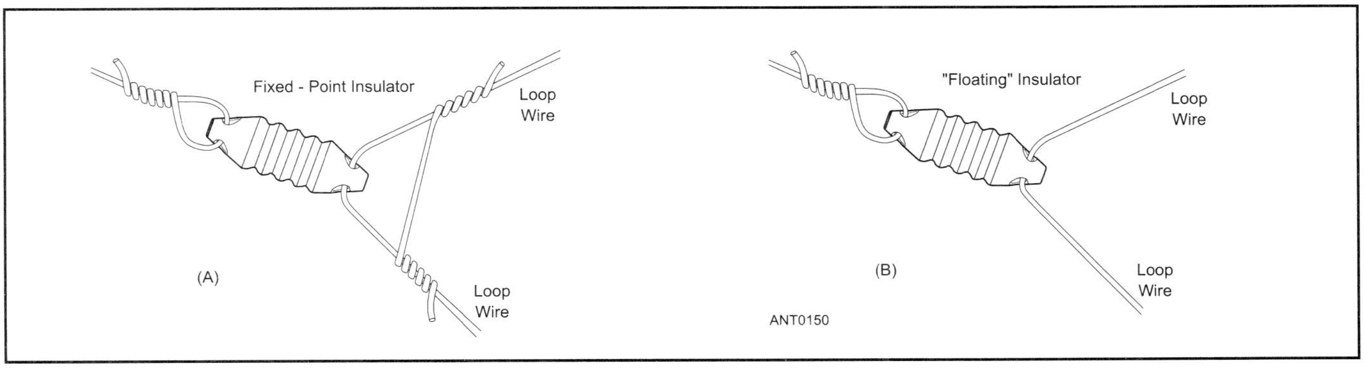

Figure 21.66 — Two methods of installing the insulators at the loop corners.

21.8 HF Mobile Antennas

HF mobile operation has been part of amateur radio since the 1930s. Mobile operation has become very popular with advances in antenna design and excellent mobile radios. Material in this section was contributed and updated by Alan Applegate, KØBG, whose website (**www.k0bg.com**) has many useful pages on HF mobile stations and operating.

Portable stations frequently use the same type of antennas as for mobile operation. Helically wound whips and base-loaded whips are the most common. The same guidelines for design and use apply to portable operation: minimize losses in the antenna's construction, configure any loading for the most current in radiating structures, and minimize ground losses with screens or radial wires.

High frequency (HF) mobile antennas come in every size and shape imaginable, from simple whips to elaborate, computer-controlled behemoths. Regardless of the type and construction, an HF mobile antenna should have a few important attributes.

- Sturdiness: It should be permanently mounted (without altering the vehicle's safety equipment) to stay upright at highway speeds with a minimum of sway.
- Mechanically stable: Sudden stops or sharp turns won't cause it to sway about, endangering others.
- Flexibly mounted: Permits bending around branches and obstacles at low speeds.
- Weatherproof: Withstands the effects of wind, rain, snow, and ice.
- Tunable: If multiband operation is desired, be tunable to different HF bands without stopping the vehicle.
- Be easily removable when required.
- Be as efficient as possible.

Of all the antenna choices available, the *whip* antenna — a self-supporting rod or wire mounted at its base — has passed the test of time as providing all of these attributes in one way or another. The following sections discuss the different types of whips, how they are attached to and interact with the vehicle, and how they are connected to the transmitter.

21.8.1 Simple Whips

The simplest of antennas is a quarter-wave whip, but it's only practical on the upper HF bands because of the required length. For example, a 10 meter quarter-wavelength antenna is about 8 feet long. It doesn't require a loading coil, so its efficiency is approximately 80%. The reason efficiency isn't 100% is because of resistive losses in the whip itself, stray capacitance losses in the mounting hardware, and ground losses, which we'll cover later. The end result is that the feed point impedance at the antenna's base is very close to 50 Ω.

As we move lower in frequency, the physical length must increase for an equivalent electrical length, but there is a limit. In most localities, the maximum height at the tip of the antenna needs to be less than 13.5 feet (4.1 meters). This generally limits whip length to 10.5 feet (3.2 meters) for an average installation on a vehicle. Creating a resonant antenna below this length on 10 or 12 meters isn't a problem. On the 15 meter and lower frequency bands, 10.5 feet is not long enough for a resonant whip antenna, and additional electrical measures are required.

WHIP RADIATION RESISTANCE

The power radiated by the antenna is equal to the radiation resistance times the square of the antenna current. The radiation resistance, R_r, of an electrically small antenna is given by:

$$R_r = 395 \times \left(\frac{h}{\lambda}\right)^2 \quad (8)$$

where

h = radiator height in meters,

λ = wavelength in meters = 300 / f in MHz.

Since radiation resistance of these electrically small antennas is a function of height, the antenna must be lengthened physically or electrically to increase it. Increasing radiation resistance improves efficiency, as shown next.

The efficiency of the antenna, η, equals the radiation resistance, R_r, divided by the resistive component of the feed point impedance, R_{fp}, which for actual antennas includes ground losses and losses in the antenna:

$$\eta = \frac{R_r}{R_{fp}} \times 100\% \quad (9)$$

(See the section on Ground Losses for more about mobile and portable "ground.") Since an

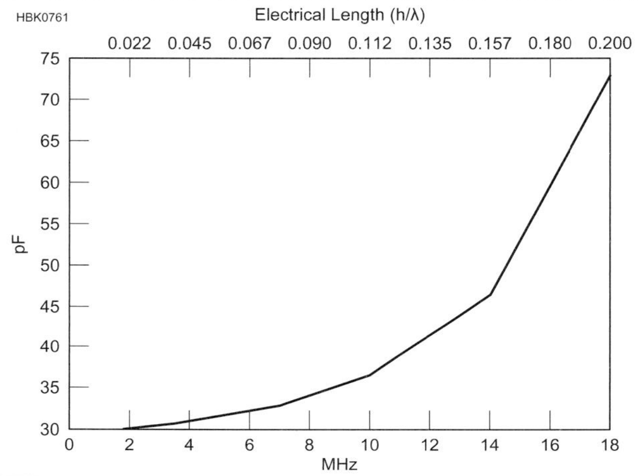

Figure 21.68 — Relationship between frequency and capacitance for a 3.2 meter vertical whip.

Table 21.11
Characteristics of an 8 foot Mobile Whip

f(MHz)	*Loading L µH*	*R_c (Q=50) Ω*	*R_c (Q=300) Ω*	*R_r Ω*	*Feed R* Ω*	*Matching L µH*
Base Loading						
1.8	345	77	13	0.1	23	3
3.8	77	37	6.1	0.35	16	1.2
7.2	20	18	3	1.35	15	0.6
10.1	9.5	12	2	2.8	12	0.4
14.2	4.5	7.7	1.3	5.7	12	0.28
18.1	3.0	5.0	1.0	10.0	14	0.28
21.25	1.25	3.4	0.5	14.8	16	0.28
24.9	0.9	2.6	—	20.0	22	0.25
29.0	—	—	—	—	36	0.23
Center Loading						
1.8	700	158	23	0.2	34	3.7
3.8	150	72	12	0.8	22	1.4
7.2	40	36	6	3.0	19	0.7
10.1	20	22	4.2	5.8	18	0.5
14.2	8.6	15	2.5	11.0	19	0.35
18.1	4.4	9.2	1.5	19.0	22	0.31
21.25	2.5	6.6	1.1	27.0	29	0.29

R_C = loading coil resistance; R_r = radiation resistance.
*Assuming loading coil Q = 300, and including estimated ground-loss resistance.

electrically short antenna has a low radiation resistance, careful attention must be paid to minimizing losses in the antenna system that can greatly reduce the antenna's effectiveness.

WHIP CAPACITANCE

As we shorten an antenna to less than ¼ λ, its radiation resistance decreases and the capacitance drops as shown in **Table 21.11**. **Figure 21.68** shows that capacitance is not very sensitive to frequency for h/λ less than 0.075.

The capacitance in pF of an electrically small antenna is given approximately by:

$$C = \frac{55.78 \times h}{[(den1) \times (den2)]} \quad (10)$$

where

(den1) = (ln (h/r) – 1),
(den2) = [1 – (f × h/75)^2],
ln = natural logarithm,
r = conductor radius in meters, and
f = frequency in MHz.

Radiation resistance rises in a nonlinear fashion, and the capacitance drops just as dramatically with an increase in the ratio h/λ. Figure 21.68 can be used for estimating antenna capacitance for other heights and shows that capacitance is not very sensitive to frequency for h/λ less than 0.075 which occurs at 8 MHz in this case.

21.8.2 Coil-Loaded Whips

To bring an electrically short whip antenna to resonance, we must add inductance in the form of a loading coil. The coil can take many forms, and it may be placed almost anywhere along the length of the radiating element. It cancels out the capacitive reactance by introducing an equal but opposite inductive reactance. Some coils are mounted at the base of the mast — a base-loaded antenna — and some are mounted near the center (center-loaded) or the top (top-loaded).

Base-loaded whips are also popular for portable operation as described in the Portable Station section of the chapter **Assembling a Station**. The loading coil is very similar in size to those used for adjustable "screwdriver" mobile antennas, resulting in fairly low loss. Adjustment of the coil is made by sliding a metal collar or moving a tap contact along the coil. These antennas are typically installed with the feed point and loading coil at ground level, along with several radial wires. Base-loading trades efficiency for convenience of installation and reduced length as described in the following sections.

As the coil is moved higher, the radiation resistance increases (a good thing), but the necessary coil reactance also increases as do resistive losses in the coil. Therefore it becomes a balancing act, which requires a thorough understanding of the parameters involved, to choose the optimal coil location.

Table 21.11 lists the characteristics of an 8 foot (2.4 meter) mobile whip in both base-loaded and center-loaded configurations. The table shows the required loading coil inductance to bring the antenna to resonance on the different bands. The matching coil inductance is placed across the feed point to bring the impedance to 50 Ω.

Note that center-loading approximately doubles both the required inductance and the coil's resistive losses. The radiation resistance also increases, but only in that part of the antenna above the loading coil. If ground losses are included in the calculations, the coil's optimal position changes, but it is typically close to the center of the antenna.

It is important to note that the total amount of stray capacitance from the mounting hardware and proximity of the whip to the vehicle may be much higher depending on where and how the antenna is mounted. The higher the stray capacitance, the less efficient the antenna will be.

Table 21.12 compares different types of HF mobile antennas against an unloaded whip antenna. The simple whip is assumed to be a full-size ¼ wavelength. The effects of mounting are not included.

Table 21.12
HF Mobile Antenna Comparison

Antenna Type	*Length*	*Frequency Coverage*	*Efficiency*	*Mounting Difficulty*	*Matching Required*
Simple Whip	< 11 ft	15 m & up	Excellent	Easy	No
Base-Loaded	9 to 10.6 ft	160 - 6 m	Fair to good	Average	Yes
Center-Loaded	9 to 10.5 ft	160 - 6 m	Good to excellent	Average	Yes
Top-Loaded	<9 ft	160 - 6 m	Fair	Average	No
Continuous Loading	< 7 feet	80 - 6 m	Poor to fair	Easy	No
Remote-tuned Small	< 7 feet	80 - 6 m	Poor to fair	Easy	No
Remote-tuned Large	9 to 10.5 ft	160 - 6 m	Excellent	Difficult	Yes

LOADING COIL Q

Antenna system Q is limited by the Q of the coil. The bandwidth between 2:1 SWR points of the system = 0.36 × f/Q. On 80 meters, the bandwidth of the 10.5 foot whip = 0.36 × 3.5/200 = 6.3 kHz. If we could double the Q of the coil, the efficiency would double and the bandwidth would be halved. The converse is also true. In the interest of efficiency, the highest possible Q should be used!

Loading coil Q is especially important on the lower HF bands where coil losses can exceed ground losses. The factors involved include wire size, wire spacing, length-to-diameter ratio and the materials used in constructing the coil. All of the factors interact with one another, making coil design a compromise — especially when wind loading and weight become major considerations.

Table 21.13
Suggested Loading Coil Dimensions

Req'd L (µH)	*Turns*	*Wire Size*	*Dia. (Inches)*	*Length (Inches)*
700	190	22	3	10
345	135	18	3	10
150	100	16	2½	10
77	75	14	2½	10
77	29	12	5	4¼
40	28	16	2½	2
40	34	12	2½	4¼
20	17	16	2½	4¼
20	22	12	2½	2¾
8.6	16	14	2	2
8.6	15	12	2½	3
4.5	10	14	2	1¼
4.5	12	12	2½	4
2.5	8	12	2	2
2.5	8	6	2⅜	4½
1.25	6	12	1¾	2
1.25	6	6	2⅜	4½

In general, the larger the coil, the higher the Q. The more mass within the field of the coil (metal end caps for example) the lower the Q. Short, fat coils are better than long, skinny ones. However, the coil's length-to-diameter ratio (L/D) for the highest Q increases as the inductance increases. It ranges from 1:1 on the upper HF bands to as much as 4:1 on the lower bands where the inductance is large. Practical mechanical considerations for coils with large reactance values above 1000 Ω (160 meter coils for example) require the length-to-diameter ratio to increase, which lowers Q. **Table 21.13** suggests loading coil dimensions that maximize Q.

Another significant factor arises from high Q. Let's assume that we deliver 100 W on 80 meters to the 7.43 Ω at the antenna terminals. The current is 3.67 A and flows through the 1375 Ω reactance of the coil, giving rise to 1375 × 3.67= 5046 V_{RMS} (7137 V_{peak}) across the coil! This is a significant voltage and may cause arcing if the coil is wet or dirty.

With only 30.6 pF of antenna capacitance, the presence of significant stray capacitance at the antenna base shunts current away from the antenna. RG-58 coax presents about

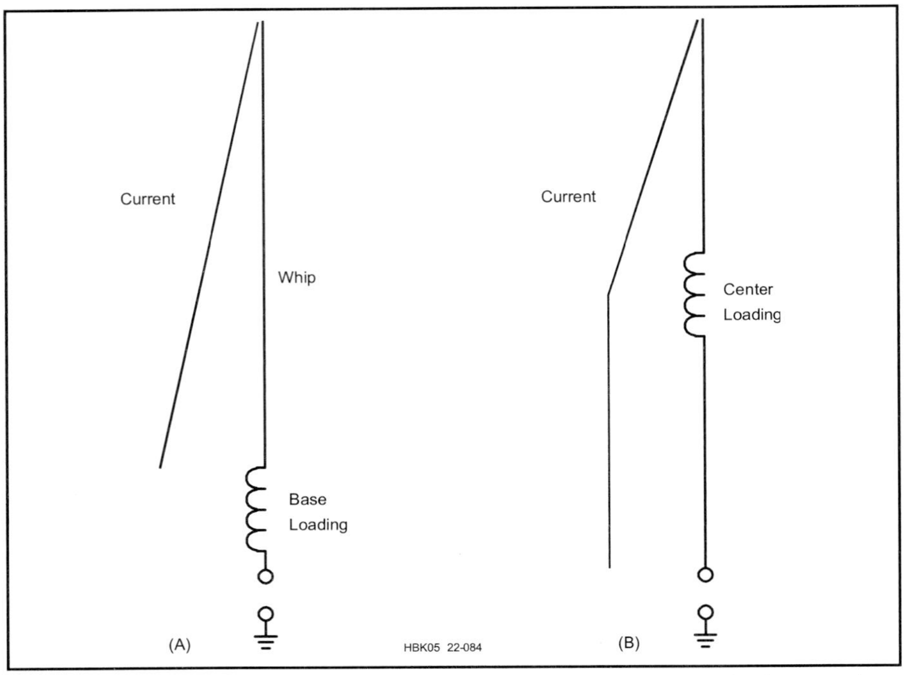

Figure 21.69 — Relative current distribution on a base-loaded antenna is shown at A and for a center-loaded antenna at B.

21 pF/foot. A 1.5 foot length of RG-58 would halve the radiation efficiency of our example antenna. For cases like the whip at 3.5 MHz, the matching network has to be right at the antenna!

21.8.3 Base vs Center vs Continuous Loading

There are a few important aspects to be kept in mind when selecting or building an HF mobile antenna. As the antenna becomes longer, less loading inductance is required and the coil Q can become higher, improving efficiency. Also, the longer the antenna, the better the mounting location has to be in order to optimize efficiency. We'll cover mounting and efficiency later. (See the section on Physically Short Verticals earlier in this chapter for more on loading vertical antennas.)

Placing the loading coil at the base results in the current distribution shown in **Figure 21.69**. If we move the coil to the center, the current curve looks like the one in Figure 21.69. The location of the optimal position between the two extremes depends on the ground losses, and to a lesser degree on loading coil Q and overall length. For example, if the ground losses were zero, the best position would be at the bottom. As the ground losses increase, the optimal position gets closer to the center. If the ground losses are high enough, the optimal position is in the top one-third of the antenna's length, but efficiency is very poor.

Center-loading increases the current in the lower half of the whip, as shown in Figure 21.69. Capacitance for the section above the coil can be calculated just as for the base-loaded antenna. This permits calculation of the loading inductance. The center-loaded antenna is often operated without any base matching, in which case the resistive component can be assumed to be 50 Ω for purposes of calculating the current rating and selecting wire size for the inductor. The reduced size of the top section results in reduced capacitance, which requires a much larger loading inductor.

Because of the high value of inductance required for center-loading, high-Q coils are very large. The large wind resistance necessitates a very sturdy mount for operation at highway speed. One manufacturer of this type of coil does not recommend its use in rain or inclement weather. The higher Q of these large coils results in a lower feed point impedance, necessitating the use of a base matching element in the form of either a tapped inductor or a shunt capacitor to match to 50 Ω. (See the following section on Mobile HF Antenna Matching.) Another manufacturer places the coil above the center and uses a small extendable whip or wand for tuning.

Antennas known by the trade name "Hamsticks," shown in **Figure 21.70**, aren't entirely continuously loaded. Instead, a small diameter enameled wire loosely wound around a fiberglass tube forms the base section of the antenna. Approximately half way up the antenna, the wire is close wound to form a lumped-element loading coil, and a metal whip or stinger is attached at the top to complete the whip. A heat-shrink sleeve covers the wound section of the antenna. The stinger's length can be adjusted to tune the antenna to the desired operating frequency.

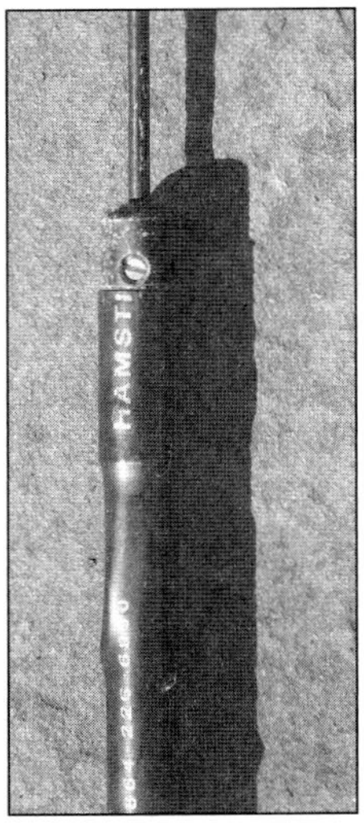

Figure 21.70 — Continuously-loaded whip antennas are short and lightweight. The base section consists of a fiberglass tube wound with wire to form the loading inductor. At the top of the base section the length of a steel whip or stinger can be adjusted to bring the antenna to resonance. [Joel Hallas, W1ZR, photo]

In recent years, these lightweight antennas have become very popular. Their input impedance is near 50 Ω in part because of their low-Q base and loading coil sections. Thus they don't require matching once the length of the top whip or stinger is adjusted. They are light in weight (about a pound), short in length (typically 6 to 8 feet), and thus easy to mount. For temporary use, easy mounting may be more important than high efficiency. This type of antenna is most effective on

Designing a Base Loading System

This design procedure was contributed by Jack Kuecken, KE2QJ. To begin, estimate the capacitance, capacitive reactance and radiation resistance as shown at the beginning of this section. Then calculate the expected loss resistance of the loading coil required to resonate the antenna. There is generally additional resistance amounting to about half of the coil loss, which must be added in. As a practical matter, it is usually not possible to achieve a coil Q in excess of 200 for such applications.

Using the radiation resistance plus 1.5 times the coil loss and the power rating desired for the antenna, one may select the wire size. For high efficiency coils, a current density of 1000 A/inch^2 is a good compromise. For the 3.67 A of the example we need a wire 0.068 inch diameter, which roughly corresponds to #14 AWG. Higher current densities can lead to a melted coil.

Design the coil with a pitch equal to twice the wire diameter and the coil diameter approximately equal to the coil length. These proportions lead to the highest Q in air core coils.

The circuit of **Figure 21.A2** will match essentially all practical HF antennas on a car or truck. The circuit actually matches the antenna to 12.5 Ω, and the transformer boosts it up to 50 Ω. Actual

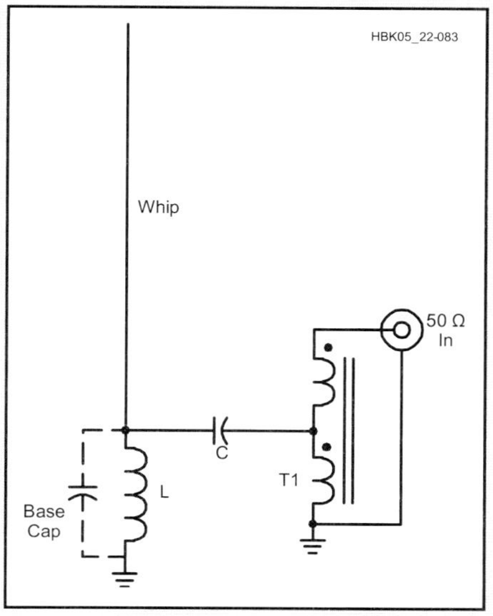

Figure 21.A2 — The base-matched mobile whip antenna

losses alter the required values of both the shunt inductor and the series capacitor. At a frequency of 3.5 MHz with an antenna impedance of 0.55 –*j*1375 Ω and a base capacitance of 2 pF results in the values shown in **Table 21.B.** Inductor and capaci-

Table 21.B

Values of L and C for the Circuit of Figure 21.A4 on 3.5 MHz

Coil Q	*L (µH)*	*C (pF)*	*System Efficiency (%)*
300	44	11.9	8.3
200	29.14	35	3.72
100	22.2	58.1	1.4

tor values are highly sensitive to coil Q. Furthermore, the inductor values are considerably below the 62.5 µH required to resonate the antenna.

This circuit has the advantage that the tuning elements are all at the base of the antenna. The whip radiator itself has minimal mass and wind resistance. In addition, the radio is protected by the fact that there is a dc ground on the radiator, so any accidental discharge or electrical contact is kept out of the cable and radio. Variable tuning elements allow the antenna to be tuned to other frequencies.

Connect the antenna, L and C. Start with less inductor than required to resonate the antenna. Tune the capacitor to minimum SWR. Increase the inductance and tune for minimum SWR. When the values of L and C are right, the SWR will be 1:1.

20 meters and higher frequency bands.

The wound base section of these antennas is a very weak radiator, acting more like a transmission line or distributed element than a linear radiating element. The long and thin coil has low Q, as well, increasing antenna loss. The close-wound section acts like a lumped element and does not radiate.

21.8.4 Top-Loaded Whips

In the interests of efficiency, electrical length matters because radiation resistance increases as the square of electrical length. Higher radiation resistance in a mobile antenna results in higher efficiency. All else being equal, a 12 foot antenna will have 4 times the radiation resistance than one 6 feet long. As pointed out above, the maximum physical height should be less than 13.5 feet (4.1 meters). As discussed in this chapter's section on Physically Short Antennas, one way to increase the electrical length but not the physical length, is top-loading. A mobile HF antenna is top-loaded by using a *capacitance hat* or "cap hat."

As their name implies, cap hats add capacitance at the top of the antenna, above any loading coil. This increases radiation resistance by

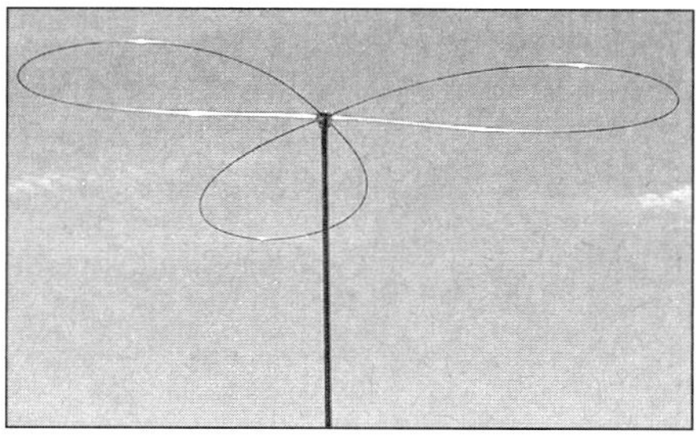

Figure 21.71 — A typical capacitance hat or "cap hat" added at the top of a mobile whip antenna. The antenna is a Scorpion (scorpionantennas.com).

as much as four times under ideal conditions, but at the expense of increased weight, wind loading and complexity. Not all antennas, especially small screwdriver types, are sturdy enough to support a large cap hat. For those that are, or if the antenna can be guyed or stiffened, cap hats offer increased efficiency and bandwidth.

The actual placement of the cap hat is important, too. If mounted too close to the loading coil, efficiency is lower than with no cap hat at all. Thus, the best mounting location is at the very top of the antenna.

The physical design of the cap hat is limited by practical concerns. The largest are approximately 3 feet in diameter and require additional bracing or guying of the antenna. Most are constructed of straight radial wires without an outer rim. The loop design shown in **Figure 21.71** is more efficient than straight wires, but tends to snag more on errant limbs. See the referenced *QEX* article by Griffith and the *QST* article by Clement for more on capacitance hats and top loading of mobile antennas.

21.8.5 Remotely Controlled HF Mobile Antennas

Remotely controlled (motorized) HF mobile antennas are commonly referred to as *screwdriver antennas*. Don Johnson, W6AAQ (SK), is credited by many as the father of the screwdriver antenna. His design was not the first motorized antenna, but he certainly popularized it. There are now over 50 commercial versions available.

They're called screwdrivers because the first examples used a stripped-down rechargeable electric screwdriver assembly to adjust the resonant frequency of the antenna. The motor turns a threaded rod in and out of a nut attached to the bottom of the coil. This in turn moves

Figure 21.72 — The screwdriver-style remotely-controlled whip antenna. A small motor in the base mast moves a coil past contacts at the top of the metal base section. The top whip section is attached to the top of the coil. As the coil moves out of the mast, more inductance is connected in series between the base section and top section. Screwdriver antennas are popular because they offer multiband coverage. [Joel Hallas, W1ZR, photo]

Determining the Radiation Efficiency of a Center Loaded Mobile Whip

We can measure the radiation efficiency by measuring ground wave field strength E (dB referenced to μV/m). For the average radio amateur, a field strength meter is not a part of the ham station gear. We can predict performance using readily available antenna modeling software (one of the many available versions of *NEC*), provided we have a measure of actual losses.

There are a number of loss parameters we do not know. We do not know the Q factor for the center loading coil (R_L), and we do not know the ground-induced loss resistance (R_g). In fact we do not know with certainty the radiation resistance (R_r), since the antenna sees an image of itself in the ground. *NEC* only gives us the sum of the various resistances.

$$R_{as} = R_r + R_C + R_g + R_L$$

R_C, the only parameter not discussed above, is the conductor loss resistance.

We need to know R_r if we are going to compute radiation efficiency, since radiation efficiency is given by:

$$\eta = \frac{R_r}{R_{as}}$$

So what do we do? We can measure R_{as} using the SWR analyzer, by adjusting the tuning so the reactance at the base of the antenna is equal to zero. We can then use *NEC* to predict the base impedance (resistive component) by changing the Q factor of the inductor so that R_{as} predicted equals R_{as} measured. We can then predict the ground field strength (dBμV/m) at say 100 m for a transmitter power of 1 kW. We reference this predicted field strength to that for an electrically small lossless vertical antenna (129.54 dBμV/m at 100 m for 1 kW transmitter power — which corresponds to the commonly quoted value of 300 mVm at 1 km). This gives us a pretty good estimate of the radiation efficiency of our mobile whip. — *Jack Belrose, VE2CV*

the coil in and out of the lower mast section. Contacts at the top of the mast slide on the outside of the coil, thus adjusting the resonance point. Position sensors may be used to keep track of the location of the coil tap. Nowadays, calling them screwdrivers is a bit of a misnomer as the electric screwdriver motors have been, replaced with much more reliable gear motors.

There are several remotely controlled HF mobile antennas that don't change length as true screwdrivers do. Both base and center loaded models are available (see **Figure 21.72**). Whether or not they're more efficient is dependent on the factors discussed in the previous section, rather than the method used to adjust the coil.

RF CHOKES

The motors and position sensors of all remotely controlled antennas operate above vehicle body potential. The amount of RF present on the leads depends on several factors, especially where and how the antenna is mounted and its overall (electrical) length. Thus the RF current coupled onto the leads must be minimized with an RF choke before entering the vehicle. An inadequate choke may result in erratic controller operation and possible interference with transceiver operation.

The choke should have an impedance of at least two orders of magnitude greater than the impedance of the circuit. In other words, at least 5 kΩ, and perhaps two or three times that in some cases (stubby antennas and poor mounting schemes are examples). Mix 31 ferrite split beads are ideal for this application, but it takes eight turns to obtain a 5 kΩ choking impedance. Depending on the wire size and insulation, you'll need to use the ½ or ¾ inch ID cores. Snap-on ferrite beads are available from most amateur radio dealers. (More information on this type of RF choke may be found in the **RF Techniques** chapter.)

The choke shown in **Figure 21.73** consists of 13 turns of #18 AWG wire, wound on a ¾ inch ID, mix 31 split bead. It has an impedance of approximately 10 kΩ at 10 MHz. When winding the chokes, try not to overlap or twist the wires, as this reduces the effectiveness.

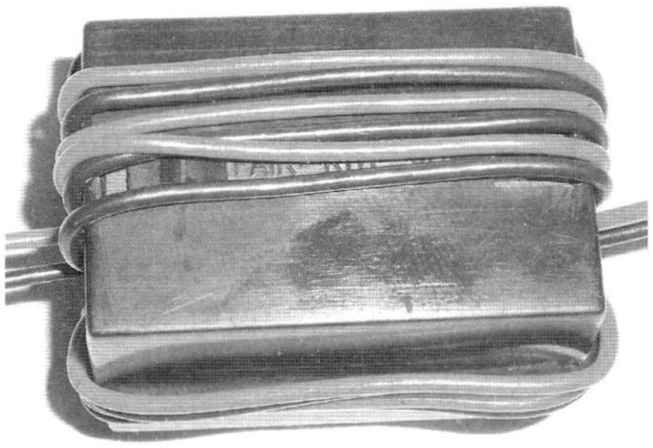

Figure 21.73 — RF choke for screwdriver antenna control and power leads.

21.8.6 Ground Losses

High frequency mobile ground loss data first appeared in a 1953 issue of *QST*, in an article written by Jack Belrose, VE3BLW (VE2CV). Belrose said that the current flowing at the base of the antenna must be returned to the base of the antenna by currents induced in the ground beneath the radiator (antenna). These currents must be collected by the car body and through the capacitance of the car body to the ground. Since the maximum dimension of car body is considerably less than a quarter wavelength on most HF bands, only a portion of these currents will be collected by the car frame itself; the rest will be collected by ground currents flowing through the capacitance of the car to the ground. Since the ground is not lossless, quite a large loss resistance (R_g) is found.

Belrose points out that the accepted ground loss figure for HF mobile applications varies between 12 Ω (for 80 meters) and 2 Ω (for 10 meters). However, these figures do not include stray capacitance from the mounting location and method. Stray capacitance has the same effect as ground losses: reduced efficiency. As a result, in the real world, ground losses can be double the accepted values, reducing an otherwise efficient antenna to mediocrity.

21.8.7 Antenna Mounting

PERMANENT OR TEMPORARY

There are many reasons to install any mobile antenna permanently on a vehicle. The decision to drill holes in sheet metal to mount antennas is hotly debated. While no-hole mounts can be used satisfactorily, it is best to look at all sides of the issue before installing any antenna.

A common concern about drilling a hole for an antenna mount is with regard to a leased vehicle. Leases don't necessarily preclude properly installed antenna mounting holes. Lease agreements are primarily concerned with body damage caused by accident or mistreatment. Properly installed NMO mounts, for example, are often acceptable. It's always prudent to ask before leasing the vehicle.

Drilled holes and waterproof mounts also help minimize common-mode current on the coaxial feed line. This helps reduce RFI to or from on-board computers and electrical devices. Aside from the hole itself, a permanent mount also minimizes damage to the finish.

Here is an important caveat to keep in mind: While the roof of a vehicle is a very good place to mount an antenna, more and more new vehicles are equipped with side curtain air bags. They typically are mounted along the edges of the headliner, including the rear seat area if there is one. The wiring to these devices is routed through any one (or more) of the roof pillars. Extra care is required when installing antennas in vehicles so equipped. If you are the least bit apprehensive about installing a roof-mounted antenna, seek professional help from your dealer or a qualified installer.

Mobile antenna mounting hardware runs the gamut from mundane to extravagant. Choosing the correct hardware is based on need as well as on personal preference. There are too many variables with respect to mounting HF mobile antennas on modern vehicles to cover in a short discussion. It is easier to explain what not to do and adapt those guidelines to your own personal circumstances. The antenna mount should:

- Be permanently mounted;
- Be strong enough to support the antenna;
- Have as much metal area under it as possible;
- Be well-grounded to the chassis or body;
- Not interfere with doors, trunks, or access panels; and
- Be removable with minimal damage to the vehicle.

Aside from maximizing performance by minimizing ground and stray capacitance losses, there are also safety reasons for using a permanent mount with HF antennas. If you need to use a temporary mount, use the multiple-magnet mounts for their superior holding strength. Any mobile HF antenna attached to a vehicle traveling at highway speed by a single-magnet mount is a tenuous situation at best.

TYPE OF MOUNT

The type of mount is dictated by several conditions. These include a decision whether or not to drill holes and the size, weight, and length of the antenna. If you want to operate HF mobile operation regularly, you are better off with a permanent mount. If you're not, a trunk lip, angle bracket, or license plate mount and a lightweight, continuously loaded antenna may meet your needs.

Ground plane losses directly affect a mobile antenna's efficiency. From this standpoint, mounts positioned high on the vehicle are preferred over a trailer hitch, bumper, or other locations that place the antenna where the vehicle body will be close to the radiating element.

The ground plane of a mobile vertical antenna begins where the coax shield connects. When the feed line shield couples to the vehicle body capacitively as in a mag-mount, or when the mass of the vehicle is far below the feed point (long extension sections attached to trailer hitch mounts), ground losses escalate dramatically. Running a ground strap to the nearest connecting point to the vehicle body does not eliminate ground losses. Remember that the ground connection is part of the antenna system, and it is the efficiency of the whole system that is important.

Figure 21.74 — A simple ball mount is sturdy but requires drilling holes in the vehicle body.

While it is difficult to mount an HF antenna without at least part of the mast being close to the body, the coil must be kept free and clear or tuning problems and reduced efficiency will result. In some cases, vans and SUVs for example, front-mounting may become necessary to avoid excessive coil-to-body interaction.

One drawback to trunk lip and similar clip mounts is the stress imposed on them as the lids and doors are open and closed. In most cases, angle brackets that attach to the inner surfaces with screws are a better choice.

Sometimes, the only solution is a custom-made bracket like the one shown later in this chapter. Here too, the circumstances dictate the requirements. Keep in mind that removing permanently installed antenna mounts such as the ball mount shown in **Figure 21.74** will always leave some body damage, but temporary ones do, too. Only the severity is in question, and that's in the eyes of the beholder.

21.8.8 Mobile HF Antenna Matching

Modern solid-state transceivers are designed for loads close to 50 Ω impedance. Depending on the design of the antenna (primarily depending on coil position and Q), overall length, and the ground losses present, the input impedance is usually closer to 25 Ω but may vary from 18 Ω to more than 50 Ω. Note that a vehicle is an inadequate ground plane for any HF mobile antenna. Typical ground loss varies from 20 Ω (160 meters) to 2 Ω (10 meters). Stray capacitance losses may further increase the apparent ground losses.

It's important to remember two important facts. First, as the coil is moved past the center of the antenna toward the top, the coil's resistive losses begin to dominate, and the input impedance gets closer to 50 Ω. Second, short stubby antennas require more inductance than longer ones, which increases resistive losses in the coil (low Q). While no matching is required in either case, efficiency suffers and may actually drop below 1% on the lower bands. Another way to look at the situation is that an antenna with no matching required implies a low efficiency.

Ground loss, coil position, coil Q, mast size, whip size, and a few other factors determine the feed point impedance, which averages about 25 Ω for a typical quality antenna and mount. This represents an input SWR of 2:1, so some form of impedance matching is required for the transceiver. There are three ways to accomplish the impedance transformation: capacitive, transmission-line transformer and inductive matching as shown in **Figure 21.75**. Each has its own unique attributes and drawbacks.

Transmission line transformers, in this case

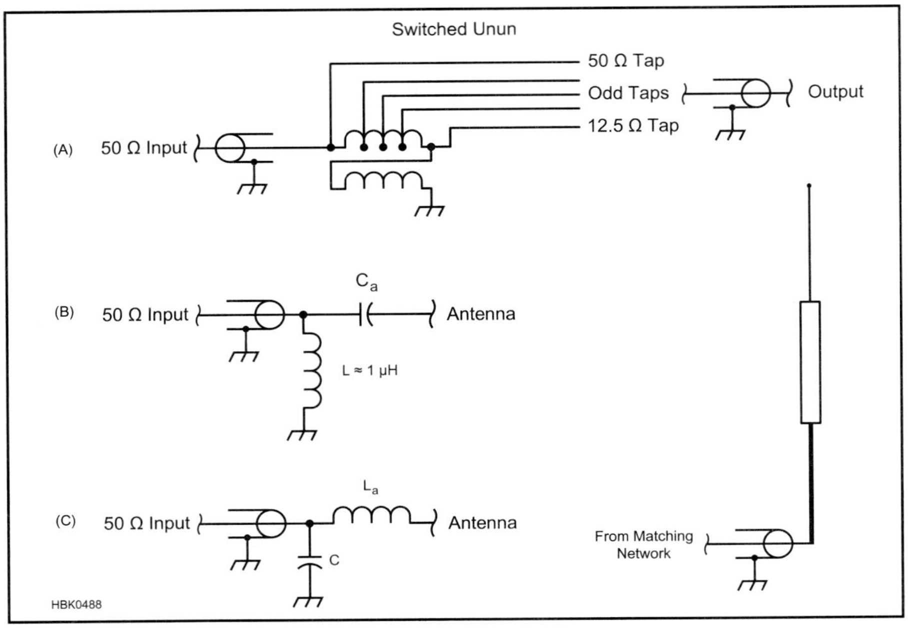

Figure 21.75 — A well-constructed and mounted HF mobile antenna will have an average input impedance around 25 Ω, requiring some matching to the feed line. The unun transmission line transformer (A) is a 4:1 configuration with taps added to match intermediate impedance values between 50 and 12.5 Ω. The high-pass L-network (B) uses some of the antenna's capacitive reactance as part of the network, and the low-pass L-network (C) uses some of the antenna's inductive reactance as part of the network. Both (A) and (B) result in an antenna at dc ground, an important safety issue.

an unun in Figure 21.75, do provide a dc ground for the antenna. They can be tapped or switched to match loads as low as a few ohms. Their broadband nature makes them ideal for HF mobile antenna matching. Since a remotely controlled HF mobile antenna's input impedance varies over a wide range, transmission line transformers are best utilized for matching monoband antennas.

Inductive matching in Figure 21.75B borrows a little capacitance from the antenna (C_a) — the antenna is adjusted to a frequency slightly higher than the operating frequency, making the input impedance capacitive. This forms a high-pass L-network, which transforms the input impedance to the 50 Ω transmission line impedance. It is ideal for use with remotely controlled antennas, as its reactance increases with frequency. By selecting the correct inductance, a compromise can be reached such that the impedance transformation will result in a low SWR from 160 through 10 meters. The approximate value is 1 μH, but may vary between 0.7 and 1.5 μH.

Adjusting the shunt coil can be done in about 10 minutes without tranmitting by using an antenna analyzer. (Full instructions for properly adjusting a shunt coil may be found at **www.k0bg.com/coil.html**.) Because no further adjustment is necessary, shunt coil matching is ideal for remotely controlled HF mobile antennas.

Capacitive matching in Figure 21.75 borrows a little inductance from the antenna (L_a) — the antenna is adjusted to a frequency slightly lower than the operating frequency, making the input impedance inductive. This forms a low-pass L-network, which transforms the input impedance to the 50 Ω transmission line impedance. While it works quite well, it has two drawbacks. First, capacitive matching presents a dc ground for the antenna, which tends to increase the static levels on receive. Second, the capacitance changes with frequency, so changing bands also requires a change in capacitance. This can be a nuisance with a remotely controlled antenna.

An important point should be made about dc grounding in addition to the static issue. If the antenna element should come in contact with a low-hanging high tension wire, or if lightning should strike it, dc grounding offers an additional level of protection for you and your transceiver.

21.8.9 Remotely Tuned Antenna Controllers

There are three basic types of remote controllers: manual, position sensing, and SWR sensing. Manual controllers consist of a DPDT center-off switch that changes the polarity of the current to the motor. Some commercial models include an interface with the radio that causes the radio to transmit a low-power carrier for tuning. Reading the SWR is left to the user. Some manual controllers incorporate a position readout to aid the operator in correctly positioning the antenna.

Position sensing controllers incorporate a magnet attached to the motor output shaft. The magnet opens and closes a reed switch. During set up, the antenna is set to one end of its range or the other. Then the resonant points are found (you have to do this yourself) and stored in multiple memory locations. As long as power is applied to the controller, a simple button push will move the antenna to a specific preset point. Some controllers use band or frequency data from a port on the radio and reset the antenna to the nearest preset based on that information.

SWR-sensing controllers either read data from the radio or from a built-in SWR bridge. Depending on the make and model, a push of the radio's tuner button (or one on the control-

Figure 21.76 — A screwdriver antenna controller made to work with the IC-7000 transceiver.

ler) causes the radio to transmit at a reduced power setting. The controller then powers the antenna's tuning motor. When the preset SWR threshold is reached, the controller stops the transmission and shuts off the motor. **Figure 21.76** shows an example of a controller made to work with a specific transceiver.

Automatic controllers are far less distracting than manual ones. Most offer a parking function that collapses the coil of a screwdriver antenna into the mast (highest frequency position, lowest overall length). If you garage your vehicle, this is a welcome feature.

21.8.10 Efficiency

Length matters! All else being equal, a 9 foot antenna will be twice as efficient as a 6 foot antenna, because radiation resistance relates directly to the square of the physical length. Further, longer antennas require less reactance to resonate, hence coil Q is higher and resistive losses lower.

Mounting methodology matters! It is the mass under the antenna, not alongside, that counts. The higher the mounting, the less capacitive coupling there will be between the antenna and the surface of the vehicle, and the lower ground losses will be.

Project: Mounts for Remotely Tuned Antennas

Remotely tuned antennas have become very popular, but they all have one thing in common: they're difficult to mount. They require both a coaxial feed line and a dc power connection, and no one makes a universal mount for them. The short, stubby ones aren't any more difficult to mount than a small whip antenna, but the "full-sized" ones (8 feet and longer) require special consideration.

These antennas are heavy (up to 18 pounds), so the mounting medium must be extra strong, and well anchored. As a result, many hams opt for a bumper or trailer hitch mount, even though the low mounting position reduces efficiency.

For some, efficiency is paramount, which dictates mounting the antenna as high as possible. Doing either low or high mounting often requires custom fabrication. The accompanying photos illustrate the two different strategies that show amateur radio ingenuity at its finest.

Figures 21.77 and **21.78** depict the mobile installation of Fokko Vos, PA3VOS. Except for a Hi-Q heavy-duty quick-disconnect, the complete mount was custom engineered by Fokko. The mount bolts to a frame extension, which in turn is bolted to the undercarriage using existing bolts. Note that the rear hatch may be opened without the antenna being removed. Had the trailer hitch been used, this would not be the case.

Figure 21.79 depicts installation on a Ford F350 based motor home owned by Hal Wilson, KE5DKM. Hal designed the bracket and had a local machine shop do the hard work. It is made of ¼ inch, high-strength aluminum, and the seams are welded. A powder-coat finish tops off the fabrication.

Shown here during installation, the bracket just fits into the right-side hood seam. The piece jutting out from the mount was to be used to further brace the mount. However, after all of the bolts were installed, additional bracing became unnecessary.

Project: Retuning a CB Whip Antenna

The most efficient HF mobile antenna is a full-size quarter-wavelength whip. Wouldn't it be nice if we could use one on every band? Alas, we cannot, but we can use one easily on 10 and 12 meters since the overall length will be less than 10 feet. If we start with a standard-length 102 inch whip and its base spring, we just need to shorten it a little for 10 meters and lengthen it a little for 12 meters. Here's how to do it.

The formula for calculating the length of a ¼ λ antenna in feet is 234/f, where f is the frequency (MHz). Since the formula is for wire antennas and the whip is larger in diameter, the resulting length will be slightly too long. This is a good thing because it is easier to remove a little length than it is to add some. This makes tuning easier.

Using the formula, we discover the needed length for 10 meters (28.5 MHz) is 98.5 inches. Thus, we need to remove 3.5 inches and account for the length of the base spring (about 6 inches), for a total of 9.5 inches to be removed. This is best accomplished by filing a notch on opposite sides of the tip of the whip and snapping it in two. Protect your eyes when you do this, as shards and splinters can fly off the broken ends. If a fiberglass whip is being modified, clip the internal wire at the top of the remaining base section. Remove the plastic cap from the discarded top section and replace it over the new top of the antenna.

Depending on the mount used, the ac-

Figure 21.77 — The PA3VOS antenna mount easily supports a large screwdriver antenna and is offset to allow the hatch to open and close without removing the antenna.

Figure 21.78 — A close-up of the PA3VOS mount.

Figure 21.79 — The KE5DKM bracket mounts under the hood of a motor home.

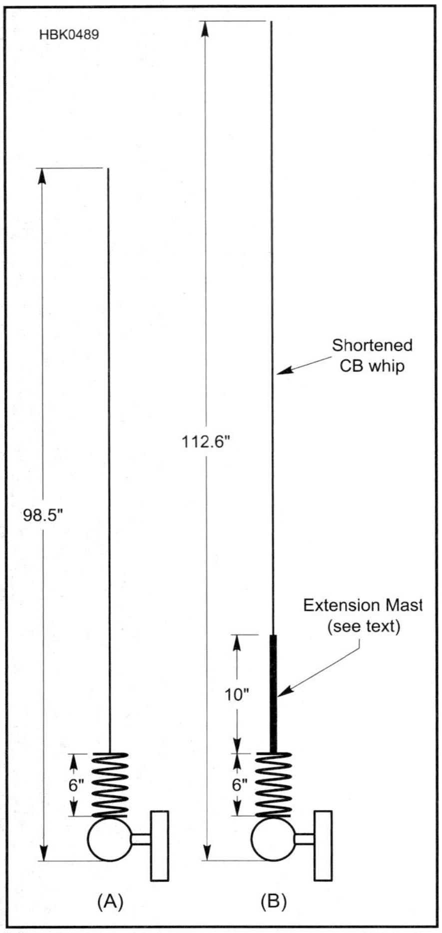

Figure 21.80 — CB whips are easily retuned for 10 meter (A) and 12 meter (B) mobile operation.

tual resonant frequency will be lower than 28.5 MHz as the mounts adds effective length. A standard CB antenna ball mount will easily support a whip. The finished antenna is shown in **Figure 21.80A.**

Once the antenna is mounted on the vehicle, simply trimming the overall length $\frac{1}{2}$ inch at a time will eventually produce a low SWR at your desired frequency.

LENGTHENING FOR 12 METER OPERATION

Lengthening a CB whip for 12 meters requires a little more work. Thankfully, there's an easy solution if you have a CB radio or truck stop near you. Wilson and other vendors sell short masts designed for the CB market. They have the requisite $\frac{3}{8} \times 24$ threads to accept a standard whip, and they come with a female-to-female coupler. The 10 inch model is ideal for our use, and costs under $10.

Using our formula 234/f, the overall length needs to be 112.6 inches for 24.93 MHz. Adding 6 inches for the spring, 10 inches for the extension mast, and 102 inches for the whip gives us 118 inches. So we need to remove 5.4 inches from the whip. Then just trim off $\frac{1}{2}$ inch at a time as above to resonate the antenna. The finished antenna is shown in Figure 21.80B.

FINISHING UP

There are two more things to consider. If a metal whip was modified, you'll need to replace the corona ball at the tip of the antenna. It helps reduce static from corona discharge. It's held on by a set screw and has little effect on tuning. Most CB shops sell them.

The other consideration is ground loss. Theoretically, a $\frac{1}{4}\lambda$ vertical will have an input impedance of 36 Ω. In a mobile installation, we have ground losses and stray capacitance losses in the mounting hardware. As a result, the real-world input impedance should be very close to 42 Ω, yielding a rather low SWR and good efficiency.

21.9 VHF/UHF Mobile Antennas

The simple $\frac{1}{4}\lambda$, $\frac{1}{2}\lambda$, and $\frac{5}{8}\lambda$ ground-plane whips are the most common VHF/UHF mobile antennas. Collinear antennas with higher gain are available. However, high gain and the lower radiation angle that goes with it isn't always a desirable attribute.

When using repeaters in urban areas, where higher angles of radiation are preferred, you're typically better off with a unity-gain antenna — a $\frac{1}{4}\lambda$ whip. It all depends on the HAAT (height above average terrain) of the repeater being used with respect to the mobile station's HAAT. In mountainous areas you're always better off with a unity gain antenna, as the repeaters are much higher in elevation than the mobile station is.

If you're working simplex and living in a suburban or rural area, a high-gain antenna might have a slight edge. However, where, and how the antenna is mounted is more important than gain. A $\frac{1}{4}\lambda$ ground plane mounted in the center of the roof will typically outperform a gain antenna mounted on the trunk lid.

Sturdiness is also an important attribute, given the unintended abuse to which mobile antennas are subjected. The simple quarter-wave ground plane has the advantage here, as it isn't much more than a springy piece of wire. If you look closely at some of the higher-gain antennas, you'll notice they have very small phasing coils, usually held together by small set screws. Hit one hard enough with a low-hanging limb and your antenna will break, whereas a simple quarter-wave whip will only bend. A little straightening and you're back on the air.

21.9.1 VHF/UHF Mobile Antenna Mounts

Without doubt, the best VHF/UHF mobile mount ever devised is the NMO (New Motorola) shown in **Figure 21.81**. When properly installed in vehicle sheet metal, the mount will not leak even when the antenna is removed for car washing. SO-239 and threaded or snap-in mounts often leak even with the antenna attached.

Glass-mounted antennas are rather lossy, especially at lower VHF frequencies (2 meters). Many new vehicles use window glass with a metallic, anti-glare (passivated) coating that interferes with capacitive coupling through the glass. These antennas also transfer mechanical abuse to the glass, risking breaking the glass the antenna is mounted on.

MOUNTING LOCATION

As mentioned above, the center of the roof is an ideal mounting location for a VHF/UHF antenna. However, many mobile operators share a reluctance to drill the proper mounting holes, fearing that doing so will depreciate the value of the vehicle. Instead, they rely on a mag-mount, which has its own set of negatives including where and how to route the coax cable into the vehicle. They tend to collect metallic road debris, primarily brake pad dust, marring and scratching the painted surface under them, often causing more damage than would the hole for an NMO mount. For these reasons alone, mag-mounts should

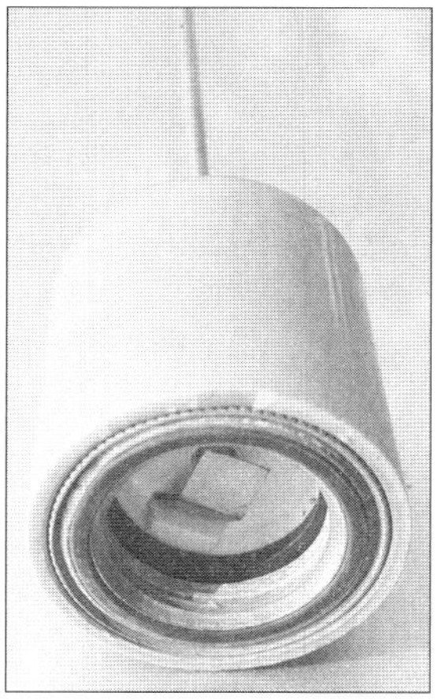

Figure 21.81 — The NMO (New Motorola) mount is widely used for VHF and UHF mobile antennas. It is available in mag-mount, trunk and lip mount and through-body mounting styles.

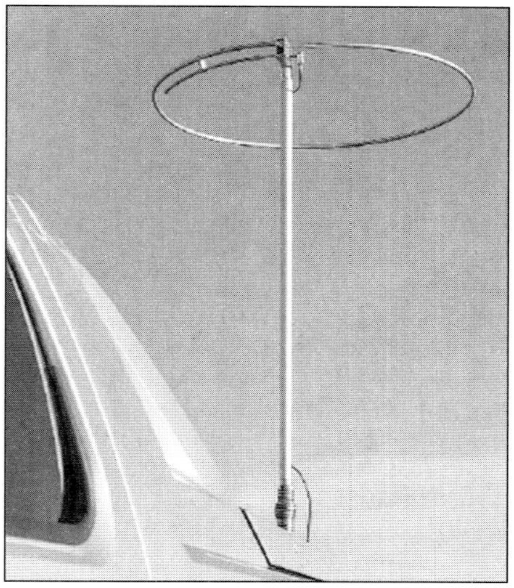

Figure 21.82 — The halo is a popular horizontally polarized mobile antenna for 6 meters through 70 cm. This version was made by Jerry Clement, VE6AB, and the construction article is available online (see text).

only be used for temporary installations, such as emergency communication. Paint scratches can be avoided during temporary operation by using a Tyvek pad under the magnet. Tyvek is a tough material used for envelopes, and other similar material will work as well.

Roof mounting has a few caveats. Modern vehicle roofs have strengthening supports to protect the occupants should the vehicle roll-over due to a crash. Do not drill through these supports!

The various side pillars supporting the roof contain wiring for lighting, side airbags, and other accessories, which can make routing of the coax difficult. If you're at all reticent about roof mounting, see a dealer or professional installer.

As an alternative, a trunk lip mount may fit the bill. However, they too have some drawbacks. They're stressed every time the trunk is opened or closed, and tend to work loose over time. Thus regular maintenance is required to assure a good electrical ground to the trunk lid. In any case, do not sand the paint down to bare metal, as this removes the zinc undercoating which in turn promotes rust. It is also important to bond across the trunk hinges to assure a good ground.

Angle brackets work well, too. They're thin enough to fit into the seam of hoods, trunk lids, and back hatches. Coax routing under or around the weather seals can be a problem with the latter two.

ADJUSTING SWR

Setting the SWR of a VHF/UHF antenna is an important installation procedure, but not one to worry about unless the SWR suddenly changes significantly. Unlike the HF bands, most VHF antennas will cover the whole band segment without the need to retune. That is to say, the SWR will be low across the FM portion of the respective bands. An in-line SWR meter is generally not required, as the vertically polarized antenna is not intended to be used in the weak-signal portion of the bands where horizontal polarization is the norm.

21.9.2 VHF/UHF Mobile Antennas for SSB and CW

Operating SSB and CW on 6 meters, 2 meters, and 70 cm offers some exciting prospects for all license classes. While FM communications on the VHF bands are often considered line-of-site, propagation *beyond* line-of-site is common, as discussed in the **Propagation of Radio Waves** chapter.

Modern mobile SSB/CW transceivers usually output 100 W PEP on 6 meters and at least 50 W PEP on 2 meters and 70 cm. Under good band conditions using horizontally polarized antennas, *beyond* line-of-sight distances can exceed 200 miles even without any sky-wave or tropospheric scatter present!

There's a catch, however. FM communications utilize vertically polarized antennas. Vertical polarization can be used for SSB and CW, but depending on the propagation path, signal strength from a vertically polarized mobile antenna can have a disadvantage of up to 20 dB compared to a horizontally polarized antenna due to cross-polarization.

Fortunately, horizontally polarized antennas are of manageable size on the VHF and UHF bands, although they are not as simple to construct as vertically polarized whips. Dipoles and small beams present too much wind resistance to withstand the normal mobile environment. The usual solution is a loop antenna.

Figure 21.82 shows a horizontally polarized 6 meter loop called a *halo* that was made by Jerry Clement, VE6AB. (Complete construction plans are available in his article "6 Meter halo Antenna for DXing" in the February 2017 issue of *QST* which is available in this book's online material.) Square versions, such as those commercially available from M^2 Antenna Systems (**www.m2inc.com**), are called *squalos*. Both have a roughly omnidirectional pattern. The "Big Wheel" design is another option, and the article "A New Spin on the Big Wheel" by L.B. Cebik, W4RNL, and Bob Cerreto, WA1FXT, from the March 2008 issue of *QST* is included with the online content. Equivalent antennas for 2 meters and 70 cm are common. Mounting loop antennas on a vehicle can be less cumbersome than mounting an HF antenna because they don't require a ground plane. A simple hitch-mounted mast will suffice, with no body holes needed!

21.10 VHF/UHF Antennas

Improving an antenna system is one of the most productive moves open to the VHF enthusiast. It can increase transmitting range, improve reception, reduce interference problems and bring other practical benefits. The work itself is by no means the least attractive part of the job. Even with high-gain antennas, experimentation is greatly simplified at VHF and UHF because an array is a workable size, and much can be learned about the nature and adjustment of antennas. No large investment in test equipment is necessary.

Whether we buy or build our antennas, we soon find that there is no one *best* design for all purposes. Selecting the antenna best suited to our needs involves much more than scanning gain figures and prices in a manufacturer's catalog. The first step should be to establish priorities for the antenna system as a whole. Once the objectives have been sorted out in a general way, we face decisions on specific design features, such as polarization, length and type of transmission line, matching methods, and mechanical design.

21.10.1 Gain

As has been discussed previously, shaping the pattern of an antenna to concentrate radiated energy, or received signal pickup, in some directions at the expense of others is the only possible way to develop gain. Radiation patterns can be controlled in various ways. One is to use two or more driven elements, fed in phase. Such arrays provide gain without markedly sharpening the frequency response, compared to that of a single element. More gain per element, but with some sacrifice in frequency coverage, is obtained by placing parasitic elements into a Yagi array.

21.10.2 Radiation Pattern

Antenna radiation can be made omnidirectional, bidirectional, practically unidirectional, or anything between these conditions. A VHF net operator may find an omnidirectional system almost a necessity, but it may be a poor choice otherwise. Noise pickup and other interference problems tend to be greater with omnidirectional antennas. Maximum gain and low radiation angle are usually prime interests of the weak-signal DX aspirant. A clean pattern with lowest possible pickup and radiation off the sides and back may be important in high-activity areas, where the noise level is high, or for challenging modes like EME (Earth-Moon-Earth).

21.10.3 Height Gain

In general, the higher a VHF antenna is installed, the better the results. If raising the antenna clears its view over nearby obstructions, it may make dramatic improvements in coverage. Within reason, greater height is almost always worth its cost, but height gain must be balanced against increased transmission line loss. Line losses can be considerable at VHF and above, and they increase with frequency. The best available line may be none too good if the run is long in terms of wavelength. Consider line losses in any antenna planning.

21.10.4 Physical Size

A given antenna design for 432 MHz, say a 5-element Yagi on a 1 λ boom, will have the same gain as one for 144 MHz, but being only one-third the size it will intercept only one-ninth as much energy in receiving. Thus, to be equal in communication effectiveness, the 432 MHz array should be at least equal in physical size to the 144 MHz one, requiring roughly three times the number of elements. With all the extra difficulties involved in going higher in frequency, it is better to be on the big side in building an antenna for the UHF bands.

21.10.5 Polarization

Whether to position the antenna elements vertically or horizontally has been a question since early VHF operation. Originally, VHF communication was mostly vertically polarized, but horizontal gained favor when directional arrays became widely used. Tests of signal strength and range with different polarizations show little evidence on which to set up a uniform polarization policy. On long paths there is no consistent advantage, either way. Shorter paths tend to yield higher signal levels with horizontal in some kinds of terrain. Man-made noise, especially ignition interference, tends to be lower with horizontal polarization. Vertically polarized antennas, however, are markedly simpler to use in omnidirectional systems and in mobile work, resulting in a standardization on vertical polarization for mobile and repeater operation on FM and for digital communications. Horizontal polarization is the standard for weak signal VHF and UHF operation. (Circular polarization is preferred for satellite work, as described below.) A loss in signal strength of 20 dB or more can be expected with cross-polarization so it is important to use antennas with the same polarization as the stations with which you expect to communicate.

21.10.6 Circular Polarization

Polarization is described as *horizontal* or *vertical*, but these terms have no meaning once the reference of the Earth's surface is lost. Many propagation factors can cause polarization change — reflection or refraction and passage through magnetic fields (Faraday rotation), for example. Polarization of VHF waves is often random, so an antenna capable of accepting any polarization is useful. Circular polarization, generated with helical antennas or with crossed elements fed 90° out of phase, will respond to any linear polarization.

The circularly polarized wave in effect threads its way through space, and it can be left- or right-hand polarized. These polarization senses are mutually exclusive, but either will respond to any plane (horizontal or vertical) polarization. A wave generated with right-hand polarization, when reflected from the moon, comes back with left-hand polarization, a fact to be borne in mind in setting up EME circuits. Stations communicating on direct paths should have the same polarization sense.

Both senses can be generated with crossed dipoles, with the aid of a switchable phasing harness. With helical arrays, both senses are provided with two antennas wound in opposite directions.

21.10.7 Transmission Lines

The most common type of transmission line at VHF through the low microwave bands is unbalanced coaxial cable. Small coax such as RG-58 or RG-59 should never be used in VHF work if the run is more than a few feet. Half-inch lines (RG-8 or RG-11) work fairly well at 50 MHz, and runs of 50 feet or less are acceptable at 144 MHz. Lines with foam rather than solid insulation have about 30% less loss. Low-loss cable is required for all but the shortest runs above 222 MHz, and *waveguide* is used on microwave frequencies. (See the **Transmission Lines** chapter for a discussion of waveguides.)

Solid aluminum-jacketed *hardline* coaxial cable with large inner conductors and foam insulation are well worth the cost. Hardline can sometimes even be obtained for free from local Cable TV operators as *end runs* — pieces at the end of a roll. The most common CATV variety is ½ inch OD 75 Ω hardline. Hardline is considered *semi-rigid* in that it can be bent, but only with a large radius to avoid kinking, repeated bending should be avoided.

Waterproof commercial connectors for hardline are fairly expensive, but enterprising amateurs have *home-brewed* low-cost connectors. If they are properly waterproofed, connectors and hardline can last almost indefinitely. See *The ARRL Antenna Book* for details on connectors and techniques for working with hardline.

Properly-built open-wire line can operate with very low loss in VHF and even UHF installations. A line made of #12 AWG wire, spaced ¾ inch or less with Teflon spreaders and running essentially straight from antenna to station, can be better than anything but the most expensive hardline at a fraction of the cost. Line loss under 2 dB per 100 feet at 432 MHz is readily obtained. This assumes the use of high-quality baluns to match into and out of the balanced line, with a short length of low-loss coax for the rotating section from the top of the tower to the antenna. Such an open-wire line could have a line loss under 1 dB at 144 MHz.

Effects of weather on transmission lines should not be ignored. A well-constructed open-wire line works well in nearly any weather, and it stands up well. TV-type twinlead is almost useless in heavy rain, wet snow, or icing conditions. The best grades of coax and hardline are impervious to weather. They can be run underground, fastened to metal towers without insulation, or bent into almost any convenient position with no adverse effects on performance. However, beware of bargain coax. Lost transmitter power can be made up to some extent by increasing power, but once lost in the transmission line a weak signal can never be recovered in the receiver.

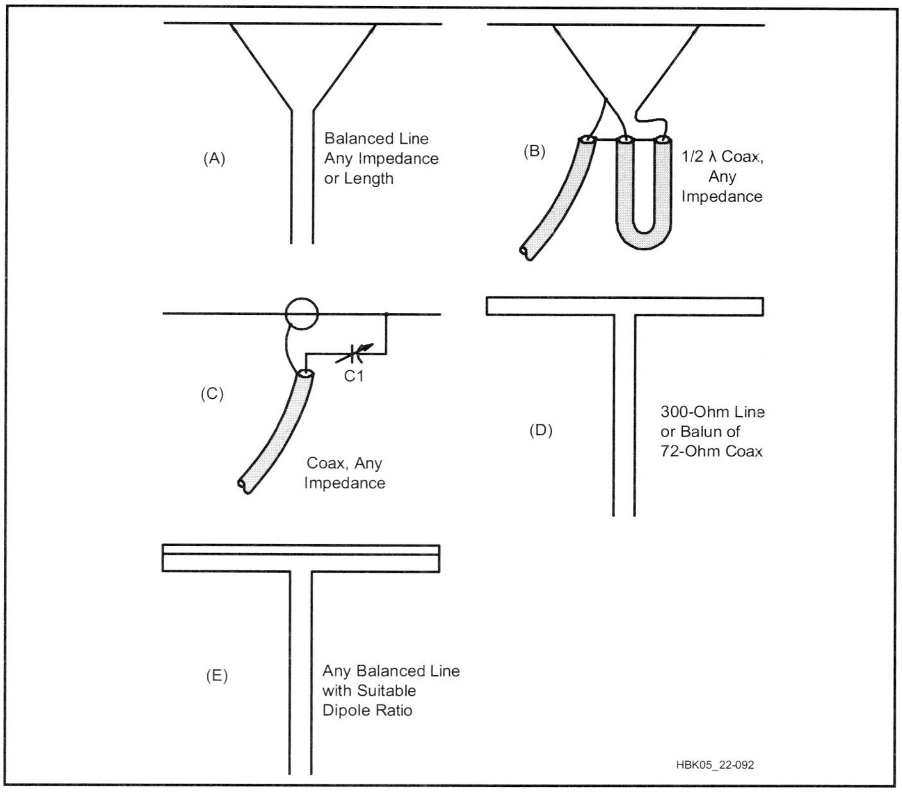

Figure 21.83 — Matching methods commonly used in VHF antennas. In the delta match, A and B, the line is fanned out to tap on the dipole at the points of best impedance match. The gamma match, C, is for direct connection of coax. C1 tunes out inductance in the arm. Folded dipole of uniform conductor size, D, steps up antenna impedance by a factor of four. Using a larger conductor in the unbroken portion of the folded dipole, E, gives higher orders of impedance transformation.

21.10.8 Impedance Matching

Theory and practice in impedance matching are discussed in detail in the **Transmission Lines** chapter, and in theory at least are the same for frequencies above 50 MHz. Practice may be similar, but physical size can be a major modifying factor in choice of methods.

DELTA MATCH

Probably the first impedance match was made when the ends of an open line were fanned out and tapped onto a half-wave antenna at the points of most efficient power transfer, as in **Figure 21.83A**. Both the side length and the points of connection either side of the center of the element must be adjusted for minimum reflected power in the line, but the impedances need not be known. The delta makes no provision for tuning out reactance, so the length of the dipole is pruned for best SWR.

Once thought to be inferior for VHF applications because of its tendency to radiate if adjusted improperly, the delta has come back to favor now that we have good methods for measuring the effects of matching. It is very handy for phasing multiple-bay arrays with low-loss open lines, and its dimensions in this use are not particularly critical.

GAMMA MATCH

The gamma match is shown in Figure 21.83C and is covered in more detail in the preceding section on HF Yagi antennas and in the **Transmission Lines** chapter. The center of a half-wave dipole being electrically neutral, the outer conductor of the coax is connected to the element at this point, which may also be the junction with a metallic or non-conductive boom. The inner conductor is connected to the element at the matching point. Inductance of the connection to the element is canceled by means of C1. Both the point of contact with the element and the setting of the capacitor are adjusted for minimum SWR using an antenna analyzer or SWR bridge.

The capacitor C1 can be a variable unit during adjustment and can then be replaced with a suitable fixed unit when the required capacitance value is found. Maximum capacitance should be about 100 pF for 50 MHz and 35 to 50 pF for 144 MHz. The capacitor and arm can be combined with the arm connecting to the driven element by means of a sliding clamp, and the inner end of the arm sliding inside a sleeve connected to the inner conductor of the coax. It can be constructed from concentric pieces of tubing and insulated by plastic sleeving or shrink tubing. RF voltage across the capacitor is low once the match is adjusted properly, so with a good dielectric insulation presents no great problem. A clean, permanent, high-conductivity bond between arm and element is important, as the RF current is high at this point.

Because it is inherently somewhat unbalanced, the gamma match can sometimes introduce pattern distortion, particularly on long-boom, highly directive Yagi arrays. The T-match, essentially two gamma matches in series creating a balanced feed system, has become popular for this reason. (See the preceding discussion on T-matches in the HF Yagi section.) A coaxial balun like that shown in Figure 21.83B is used from the balanced T-match to the unbalanced coaxial line going to the transmitter. To maintain a symmetrical pattern, the feed line should be run along the antenna boom at the centerline of the elements to the mast. A choke balun is often used to minimize currents that might be induced on the outer surface of the feed line shield.

FOLDED DIPOLE

The impedance of a half-wave dipole feed point at its center is 72 Ω. If a single conductor of uniform size is folded to make a half-wave dipole, as shown in Figure 21.83D, the impedance is stepped up four times. Such a folded dipole can thus be fed directly with 300 Ω line with no appreciable mismatch. Coaxial feed line of 70 to 75 Ω impedance may then be used with a 4:1 impedance transformer. Higher impedance step-up can be obtained if the unbroken portion is made larger in cross-section than the fed portion, as in Figure 21.83E. The folded dipole is discussed further in the *ARRL Antenna Book*.

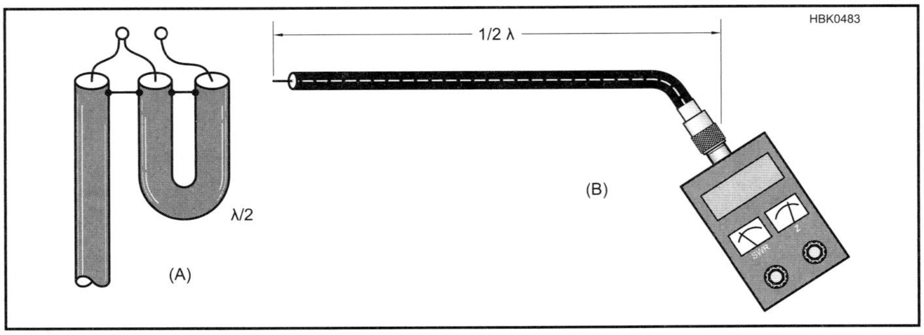

Figure 21.84 — Conversion from unbalanced coax to a balanced load can be done with a half-wave coaxial balun, A. The half-wave balun gives a 4:1 impedance step up. Electrical length of the looped section should be checked with an antenna analyzer with the far end of the line open, as in B. The lowest frequency at which the line impedance is a minimum is the frequency at which the line is ¼ λ long. Multiply that frequency by two to obtain the ½ λ frequency.

21.10.9 Baluns and Impedance Transformers

Conversion from balanced loads to unbalanced lines, or vice versa, can be performed with electrical circuits, or their equivalents made of coaxial line. A balun made from flexible coax is shown in **Figure 21.84A**. The looped portion is an electrical half-wave. This type of balun gives an impedance step-up of 4:1, 50 to 200 Ω, or 75 to 300 Ω typically. (See the **RF Techniques** and **Transmission Lines** chapters for a detailed discussion of baluns and impedance transformers.)

The physical length of the line section depends on the propagation factor of the line used, so it is best to check its resonant frequency, as shown at B. One end of the line is left open and an antenna analyzer is used to find the lowest frequency at which the impedance at the other end of the line is a minimum, the frequency at which the section of line is ¼ λ long. Multiply the frequency by two to find the frequency at which the section is ½ λ long.

Coaxial baluns giving a 1:1 impedance transfer are shown in **Figure 21.85**. The coaxial sleeve, open at the top and connected to the outer conductor of the line at the lower end (Figure 21.85A) is the preferred type. A conductor of approximately the same size as the line is used with the outer conductor to form a quarter-wave stub, in Figure 21.85B. Another piece of coax, using only the outer conductor, will serve this purpose. Both baluns are intended to present a high impedance to any RF current that might otherwise tend to flow on the outer conductor of the coax. Choke baluns made of ferrite beads of the proper material type or mix may also be used. (See the **RF Techniques** chapter for information about ferrite use at VHF and UHF.)

Project: Simple, Portable Ground-Plane Antenna

The ground-plane antenna, shown in **Figure 21.86**, uses a female chassis-mount connector to support the element and two radials. With only two radials, it is essentially two dimensional, which makes it easier to store when not in use. UHF connectors work well for 144 and 222 MHz, but you may prefer to use Type N connectors. N connectors are recommended for 440 MHz and higher frequencies. BNC connectors can be used for the shorter antennas on 915 and 1280 MHz but are not particularly sturdy.

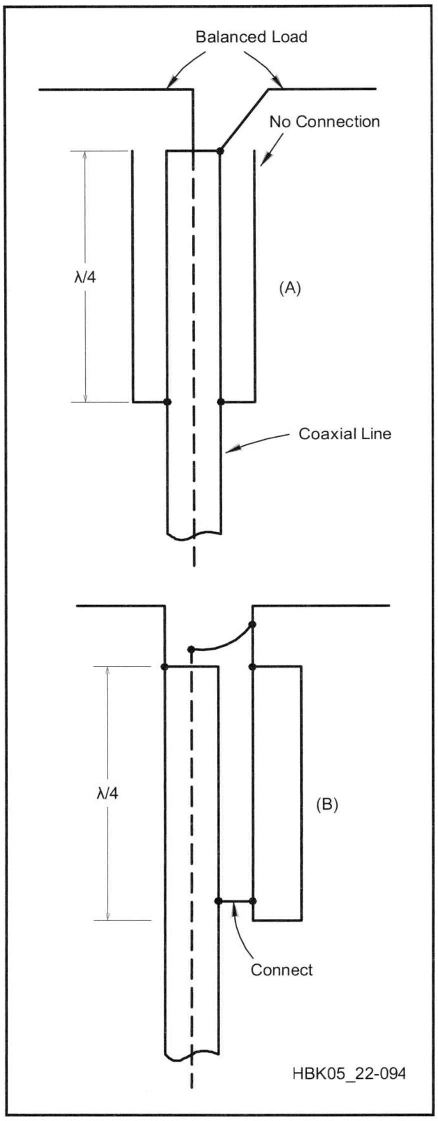

Figure 21.85 — The balun conversion function, with no impedance change, is accomplished with quarter-wave lines, open at the top and connected to the coax outer conductor at the bottom. The coaxial sleeve shown at A is preferred.

If the antenna is sheltered from weather, copper wire is sufficiently rigid for the radiating element and radials. Antennas exposed to the wind and weather can be made from brazing rod, which is available at welding supply stores. Alternatively, #12 or #14 AWG copper-clad steel wire could be used to construct this antenna.

To eliminate sharp ends, it's a good idea to bend the element and radial ends into a circle or to terminate them with a crimp terminal, as in **Figure 21.87**. The crimp terminal approach is easier with stiff wire. Crimp and then solder the terminal to the wire. Make the overall length of the element and radials the same as shown in Figure 21.86, measuring to the outer tip of the loop or terminal.

Radials may be attached directly to the mounting holes of the coaxial connector. Bend a hook at one end of each radial for insertion through the connector. Solder the radials to the connector using a large soldering iron or propane torch.

Solder the element to the center pin of the connector. If the element does not fit inside the solder cup, use a short section of brass tubing as a coupler (a slotted ⅛ inch ID tube will fit over an SO-239 or N receptacle center pin).

If necessary, prune the antenna to raise the frequency of minimum SWR. Then adjust the radial droop angle for minimum SWR — this should not affect the frequency at which the minimum SWR occurs.

One mounting method for fixed-station antennas appears in Figure 21.86. The feed line and connector are inside the mast, and a hose clamp squeezes the slotted mast end to tightly grip the plug body. Once the antenna is mounted and tested, thoroughly seal the open side of the coaxial connector with silicone sealant and weatherproof the connections with rust-preventative paint.

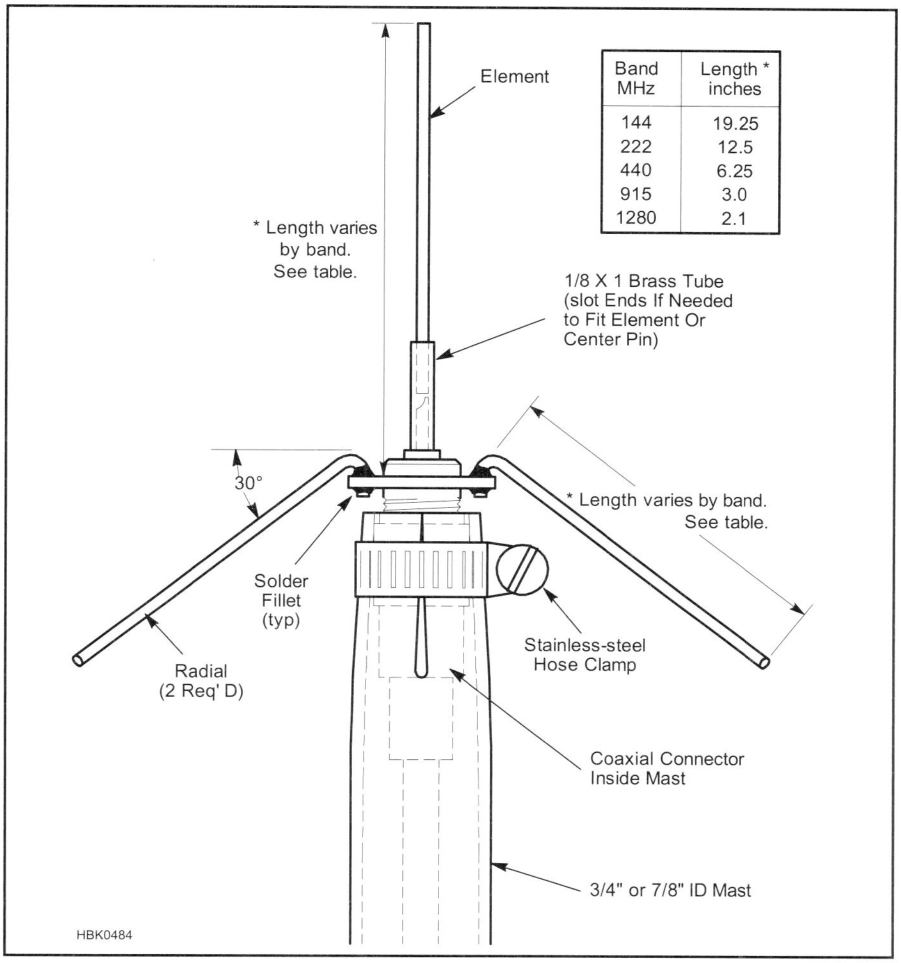

Band MHz	Length * inches
144	19.25
222	12.5
440	6.25
915	3.0
1280	2.1

Figure 21.86 — A simple ground-plane antenna for the 144, 222 or 440 MHz bands. The feed line and connector are inside the mast, and a hose clamp squeezes the slotted mast end to tightly grip the plug body. Element and radial dimensions given in the drawing are good for the entire band.

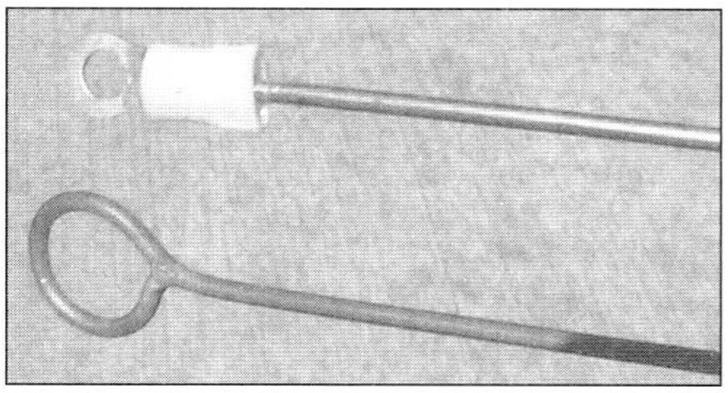

Figure 21.87 — Alternate methods for terminating element and radial tips on the simple ground-plane antenna. See text. *(Photo by K8CH)*

A related simple antenna is the crossed-dipole, which has an omnidirectional pattern and requires no ground plane. Paul Wade, W1GHZ, rescaled a design found in an IEEE journal that was originally designed for 1.7 GHz. His construction article from the *Proceedings of Microwave Update 2016* is included with the online information for this book.

Project: Coaxial Dipole for VHF or UHF

(The following antenna was originally described in July 2009 *QST* by John Portune, W6NBC, and was also reprinted in *The ARRL Antenna Compendium Volume 8*.)

Here is a homebrew coaxial dipole built from a small stainless whip, a length of threaded table-lamp tubing, and some ¾ inch copper and PVC fittings. The one shown is for 440 MHz, but it can readily be scaled for 146 or 222 MHz.

For homebrew vertical VHF antennas, coaxial dipoles often play second fiddle to J-poles. That's because the center connection to coax is often difficult to fabricate in the home workshop. Yet both antennas have the same performance. They're both full sized, half wave vertical dipoles, and the coaxial is shorter.

MAKING A COAXIAL DIPOLE

If you start with a common half wave ($\lambda/2$) stainless whip and extend it all the way down through a $\lambda/2$ long support tubing, here made from a threaded table lamp tube, the lower part of the whip becomes the center conductor of a short length of rigid coax feeding the center of the antenna. Now connection to normal coax is easily made below the antenna. To form the rigid coax section, you'll need to insulate the center conductor (lower part of the stainless whip) from the lamp tubing with some ¼ inch inside diameter (ID) polyethylene tubing. Hardware stores normally carry it. This short length of rigid coax formed in this way isn't precisely 50 Ω characteristic impedance, but the difference is totally insignificant. The drawing in **Figure 21.88** shows the details.

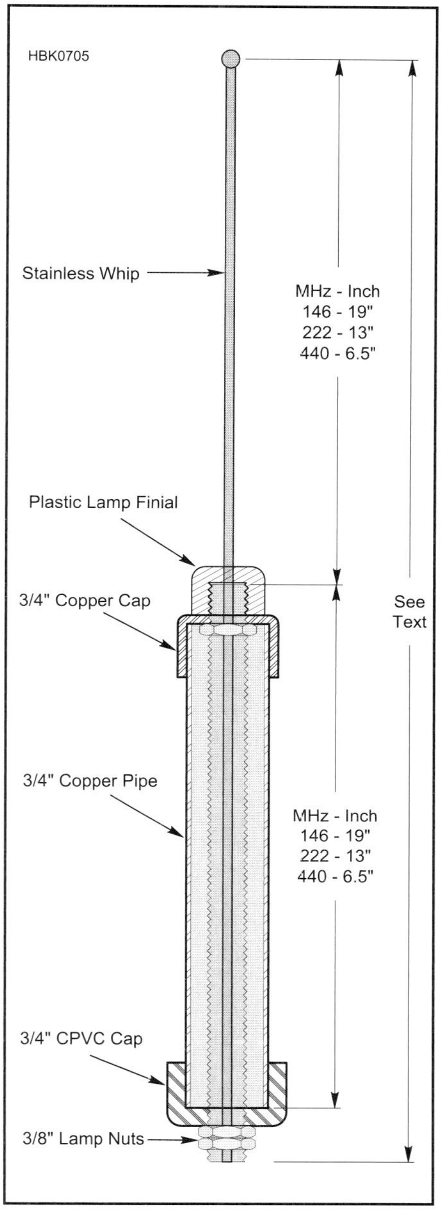

Figure 21.88 — Dimensioned drawing of coaxial dipole for three bands.

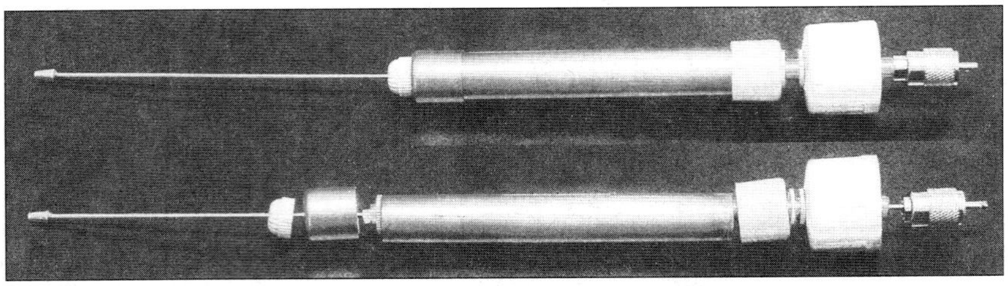

(A)

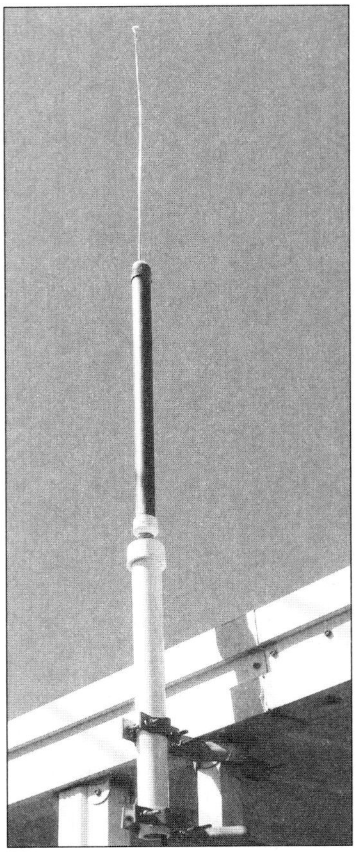

(B)

Figure 21.89 — Details of final assembly of coaxial dipole (A) and the finished product (B).

Assembly Details

The bottom half ($\lambda/4$) of the radiating dipole is a coaxial sleeve made from ¾ inch copper pipe and a pipe cap. The coax feed runs up its center to the connector at the bottom of the lamp tubing. Support and insulation of the bottom of the sleeve is provided by a ¾ inch CPVC plastic pipe cap. For those not familiar with CPVC fittings, they're made to mate with copper pipe and can handle high water temperatures. That's not true of common PVC fittings. Most hardware stores now carry CPVC. Drill a ⅜ inch hole in the center top of the copper and the CPVC caps for the lamp tubing to pass through.

The whole antenna is held together by two lamp tubing nuts and a plastic lamp finial, also readily available at hardware stores (see **Figure 21.89**). Note that a lamp tubing nut is also required inside the copper pipe cap. Drill a small hole in the middle of the lamp finial for the stainless whip. On the bottom of lamp tubing below the antenna install a 1¼ inch common PVC pipe cap and secure it with two more lamp tubing nuts. This gives you a way to easily mount the antenna on top of any convenient length of ¼ inch PVC pipe. Run the coax feed down through the PVC pipe. (Note that lamp parts are generally not intended for use outside and may rust or corrode without being painted or otherwise protected against the weather.)

Hooking it Up

A conventional PL-259 UHF type coax connector for RG-8 coax will actually screw onto the bottom of the lamp tubing. The threads are not a perfect fit, but will tighten satisfactorily. The stainless whip runs down all the way to the very tip of the PL-259 connector. Solder it in there. Before doing so, however, install all the pieces of the antenna onto the threaded lamp tubing.

Many hams may think that stainless steel won't solder. It definitely will with a hot iron and acid flux. Scrape the end of the whip and dip it in hydrochloric swimming pool acid. With a little action from the tip of the soldering iron the whip will tin perfectly well. Before soldering, however, grind two or three small side notches in the bottom end of the whip. A Dremel tool works well for this. The notches will help the solder securely lock the whip into the tip of the PL-259 connector. Neutralize any leftover acid with baking soda solution.

Perhaps surprising to some, it really isn't necessary to solder any other parts of the antenna. There is adequate mating surface at the joints for the RF to cross over efficiently. Do, however, seal all possible water access spots with common silicone sealant and/or plastic electrical tape.

MAKE IT FOR THE BAND YOU LIKE

There isn't an exact length required for the lamp tubing or the stainless whip. These merely need to provide enough space for all the pieces of the antenna to go together. The author had a 48 inch whip on hand that he used uncut for the 146 MHz coaxial dipole and a similar 17 inch uncut whip for 440 MHz. He cut the lamp tubing to an appropriate length to fit the whips. What does matter, however, is the length of the whip above the top of the lamp tubing as well as the length of the coaxial sleeve. These need to be close to a $\lambda/4$ — for 440 MHz, 6½ inches; for 222 MHz, 13 inches; and 19 inches for 146 MHz. These antennas are quite broadband and will cover the entire band in each case with these sizes. No cutting or pruning is necessary.

For ruggedness, or perhaps for stealth, you can install the whole antenna inside of 2 inch PVC water or ABS soil pipe and close the ends with end caps. The author lives in a mobile home park where antennas are not permitted, but these coaxial dipoles (in ABS pipe) look like vent pipes.

Try out one of these homebrew coaxial dipoles. You may find you prefer its smaller size, less obvious appearance, and superior weatherproofing as compared to a J-pole.

21.11 VHF/UHF Beams

Without doubt, the Yagi is king of home-station antennas these days. Today's best designs are computer optimized. For years amateurs as well as professionals designed Yagi arrays experimentally. Now we have powerful (and inexpensive) personal computers and sophisticated software for antenna modeling. These have brought us antennas with improved performance, with little or no element pruning required. A more complete discussion of Yagi design can be found earlier in this chapter and in the *ARRL Antenna Book*.

21.11.1 Stacking Yagis

Where suitable provision can be made for supporting them, two Yagis mounted one above the other and fed in-phase may be preferable to one long Yagi having the same theoretical or measured gain. The pair will require a much smaller turning space for the same gain, and their lower radiation angle can provide interesting results. On long ionospheric paths a stacked pair occasionally may show an apparent gain much greater than the 2 to 3 dB that can be measured locally as the gain from stacking.

Optimum spacing for Yagis with booms longer than 1 λ is one wavelength, but this may be too much to handle for many builders of 50 MHz antennas. Worthwhile results are possible with separations of as little as $\frac{1}{2}$ λ (10 feet), but $\frac{5}{8}$ λ (12 feet) is markedly better. At 50 MHz, the difference between 12 and 20 foot spacing may not be worth the added structural problems.

The closer spacings give lowered measured gain, but the antenna patterns are cleaner (less power in the high-angle elevation lobes) than with 1 λ spacing. Extra gain with wider spacings is usually the objective on 144 MHz and higher bands, where the structural problems are not quite as severe as on 50 MHz.

One method for feeding two 50 Ω antennas, as might be used in a stacked Yagi array, is shown in **Figure 21.90**. The transmission lines from each antenna, with a balun feeding each one (not shown in the drawing for simplicity), to the common feed point must be equal in length and an odd multiple of $\frac{1}{4}$λ. This line acts as a quarter-wave (Q-section) impedance transformer, raises the feed impedance of each antenna to 100 Ω, and forces current to be equal in each driven element. When the feed lines are connected in parallel at the coaxial tee connector, the resulting impedance is close to 50 Ω.

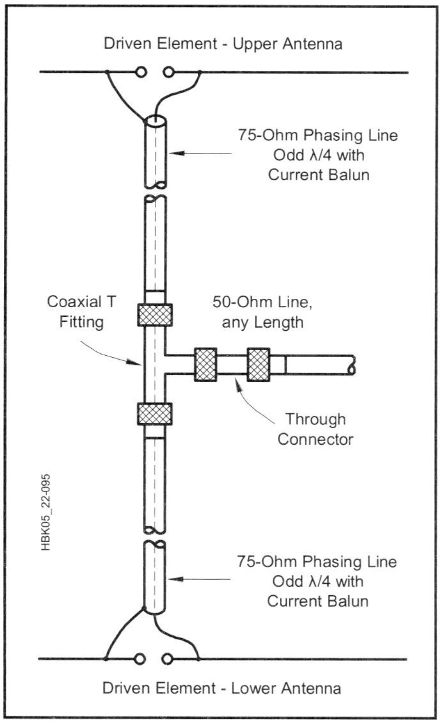

Figure 21.90 — A method for feeding a stacked Yagi array. Note that baluns at each antenna are not specifically shown. Good practice is to use choke baluns made up of ferrite beads slipped over the outside of the coax and taped to prevent movement. See the RF Techniques and Transmission Lines chapters for details.

Project: Three and Five-Element Yagis for 6 Meters

Boom length often proves to be the deciding factor when one selects a Yagi design. Dean Straw, N6BV, created the designs shown in **Table 21.14**. Straw generated the designs in the table for convenient boom lengths (6 and 12 feet). The 3 element design has about 8 dBi gain, and the 5 element version has about 10 dBi gain. Both antennas exhibit better than 22 dB front-to-rear ratio, and both cover 50 to 51 MHz with better than 1.6:1 SWR.

Element lengths and spacings are given in the table. Elements can be mounted to the boom as shown in **Figure 21.91**. Two muffler clamps hold each aluminum plate to the boom, and two U bolts fasten each element to the plate, which is 0.25 inches thick and 4.4 inches square. Stainless steel is the best choice for hardware. However, galvanized hardware can be substituted. Automotive muffler clamps do not work well in this application, because they are not galvanized and quickly rust once exposed to the weather.

The driven element is mounted to the boom on a phenolic plate of similar dimension to the other mounting plates. A 12 inch piece of Plexiglas rod is inserted into the driven element halves. The Plexiglas allows the use of a single clamp on each side of the element and also seals the center of the elements against moisture. Self-tapping screws are used for electrical connection to the driven element.

Refer to **Figure 21.92** for driven element

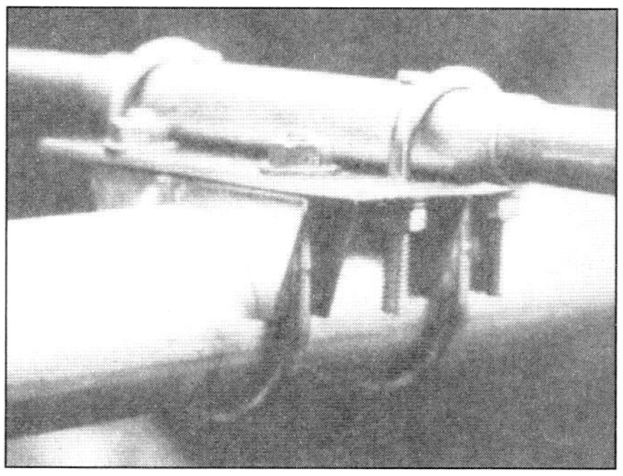

Figure 21.91 — The boom-to-element clamp. Galvanized U-bolts are used to hold the element to the plate, and 2 inch galvanized muffler clamps hold the plates to the boom.

Table 21.14
Optimized 6 Meter Yagi Designs

	Spacing From Reflector (in.)	*Seg 1 Length (in.)*	*Seg 2 Length (in.)*	*Midband Gain F/R*
306-06				
Refl	0	36	23.500	8.1 dBi
DE	24	36	16.000	28.3 dB
Dir 1	66	36	15.500	
506-12				
OD		0.750	0.625	
Refl	0	36	23.625	10.0 dBi
DE	24	36	17.125	26.8 dB
Dir 1	36	36	19.375	
Dir 2	80	36	18.250	
Dir 3	138	36	15.375	

Note: For all antennas, telescoping tube diameters (in inches) are: Seg1=0.750, Seg2=0.625.
Boom length should be 6 inches longer than the maximum Refl-Dir spacing to allow 3 inches on each end for element mounting hardware.

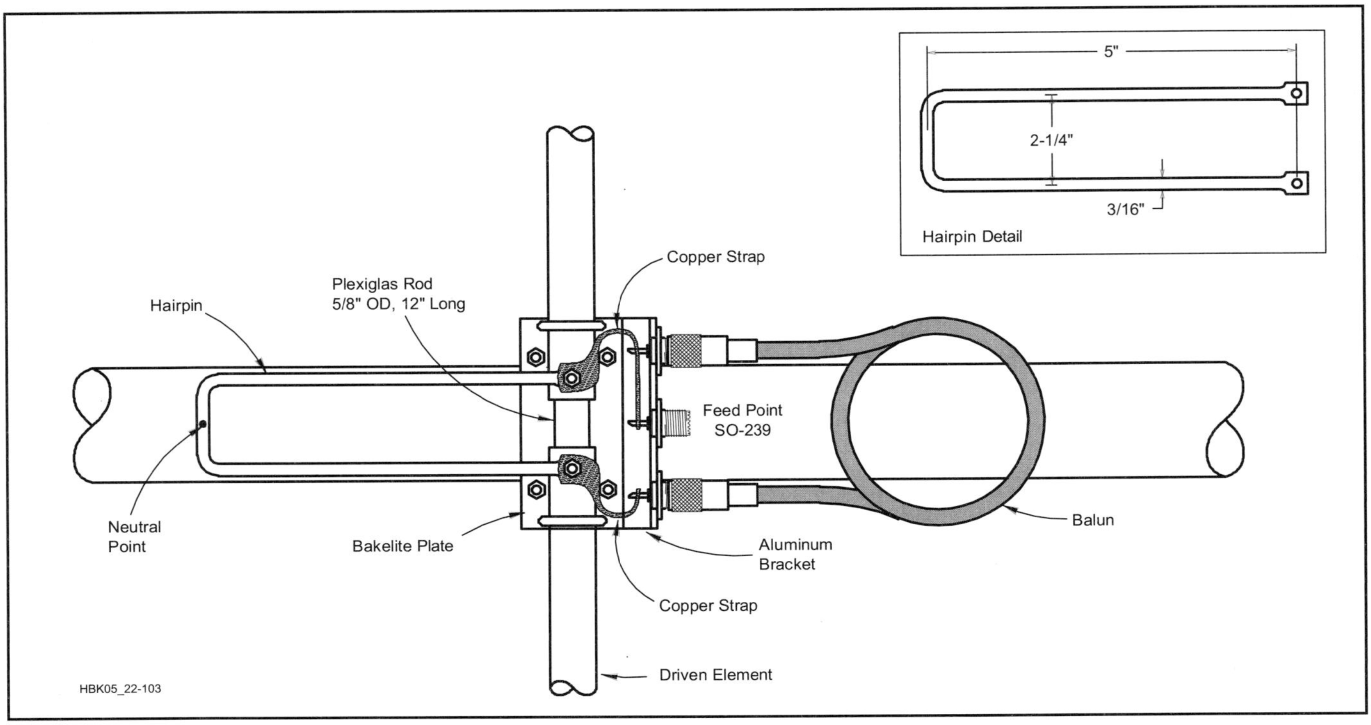

Figure 21.92 — Detailed drawing of the feed system used with the 50 MHz Yagi. Balun lengths: For cable with 0.80 velocity factor — 7 feet, 10⅜ inches. For cable with 0.66 velocity factor — 6 feet, 5¾ inches.

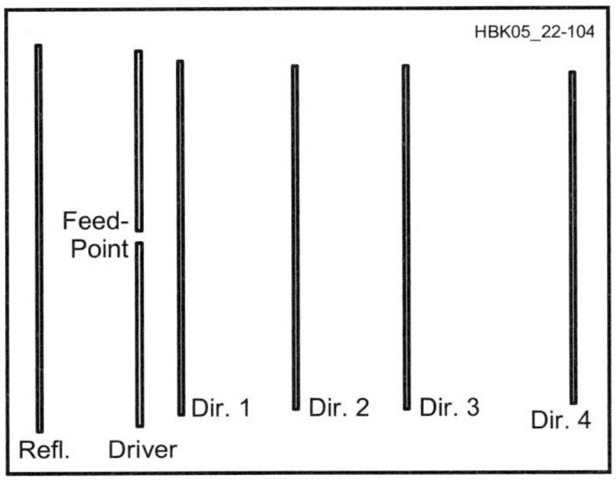

Figure 21.93 — The general outline of the 2 meter, 6 element OWA Yagi. Dimensions are given in Table 21.15.

and hairpin match details. A bracket made from a piece of aluminum is used to mount the three SO-239 connectors to the driven element plate. A 4:1 transmission-line balun connects the two element halves, transforming the 200 Ω resistance at the hairpin match to 50 Ω at the center connector. Note that the electrical length of the balun is λ/2, but the physical length will be shorter due to the velocity factor of the particular coaxial cable used. The hairpin is connected directly across the element halves. The exact center of the hairpin is electrically neutral and should be fastened to the boom. This has the advantage of placing the driven element at dc ground potential.

The hairpin match requires no adjustment as such. However, you may have to change the length of the driven element slightly to obtain the best match in your preferred portion of the band. Changing the driven-element length will not adversely affect antenna performance. *Do not adjust the lengths or spacings of the other elements — they are optimized already*. If you decide to use a gamma match, add three inches to each side of the driven element lengths given in the table for both antennas.

Project: Medium-Gain 2 Meter Yagi

This project was designed and built by L. B. Cebik, W4RNL (SK). Practical Yagis for 2 meters abound. What makes this one a bit different is the selection of materials. The elements, of course, are high-grade aluminum. However, the boom is PVC, and there are only two #6 nut-bolt sets and two #8 sheet metal screws in the entire antenna. The remaining fasteners are all hitch-pin clips. The result is a very durable six-element Yagi that you can disassemble with fair ease for transport. The complete construction details and more discussion of the antenna are included with the online content.

THE BASIC ANTENNA DESIGN

The 6 element Yagi presented here is a derivative of the *optimized wide-band antenna* (OWA) designs developed for HF use by NW3Z and WA3FET. **Figure 21.93** shows the general outline. The reflector and first director largely set the impedance. The next 2 directors contribute to setting the operating bandwidth. The final director (Dir. 4) sets the gain. This account is oversimplified, since every element plays a role in every facet of Yagi performance. However, the notes give some idea of which elements are most sensitive in adjusting the performance figures.

Designed using *NEC-4*, the antenna uses 6 elements on a 56 inch boom. **Table 21.15** gives the specific dimensions for the version described in these notes. The parasitic elements are all 3⁄16 inch aluminum rods. For ease of construction, the driver is ½ inch aluminum tubing. Do not alter the element diameters without referring to a source, such as RSGB's *The VHF/UHF DX Book*, edited by Ian White, G3SEK, (chapter 7), for information on how to recalculate element lengths.

The OWA design provides about 10.2 dBi of free-space gain with better than 20 dB front-to-back (or front-to-rear) ratio across the entire 2 meter band. Azimuth (or E-plane) patterns show solid performance across the entire band. This applies not only to forward gain but rejection from the rear.

One significant feature of the OWA design is its direct 50 Ω feed point impedance, which requires no matching network. Of course, a choke balun to suppress any currents on the feed line is desirable, and a simple ferrite bead balun (see the **Transmission Lines** chapter) works well in this application. The SWR,

Table 21.15
2 Meter OWA Yagi Dimensions

Element	*Element Length (in)*	*Spacing from Reflector (in)*	*Element Diameter (in)*
Version described in the text:			
Refl.	40.52	—	0.1875
Driver	39.70	10.13	0.5
Alt. Driver	*39.96*	10.13	0.1875
Dir. 1	37.36	14.32	0.1875
Dir. 2	36.32	25.93	0.1875
Dir. 3	36.32	37.28	0.1875
Dir. 4	34.96	54.22	0.1875
Version using ⅛-inch diameter elements throughout:			
Refl.	40.80	—	0.125
Driver	40.10	10.20	0.125
Dir. 1	37.63	14.27	0.125
Dir. 2	36.56	25.95	0.125
Dir. 3	36.56	37.39	0.125
Dir. 4	35.20	54.44	0.125

shown in **Figure 21.94**, is very flat across the band and never reaches 1.3:1. The SWR and the pattern consistency together create a very useful utility antenna for 2 meters, whether installed vertically or horizontally. The only remaining question is how to effectively build the beam in the average home shop.

The six-element OWA Yagi for 2 meters performs well. It serves as a good utility antenna with more gain and directivity than the usual three-element general-use Yagi. When vertically polarized, the added gain confirms the wisdom of using a longer boom and more elements. With a length under five feet, the antenna is still compact. The ability to disassemble the parts simplifies moving the antenna to various portable sites.

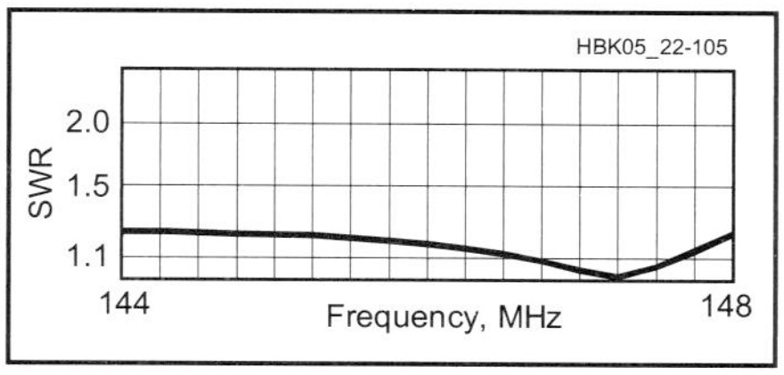

Figure 21.94 — SWR curve for the 2 meter, 6 element OWA Yagi as modeled using *NEC-4*.

Project: Utility Yagi for 432 MHz

The following design was developed by Zack Lau, W1VT (see References), and is included in the online information for this chapter. This six-element Yagi has a wide bandwidth for gain, F/B ratio, and SWR, covering almost all of the 70 cm band. The short 30-inch boom is small enough to fit in the trunk of a compact sedan, perfect for portable or emergency operation. The optimized design is shown in **Figure 21.95**, which includes the table of element lengths and placement. Changing the boom or element mounting may require adjusting the element lengths. The antenna uses a simple T-match to preserve a symmetrical radiation pattern.

Project: Cheap Yagis by WA5VJB

If you're planning to build an EME array, don't use these antennas. But if you want to put together a VHF rover station with less than $500 in the antennas, read on as Kent Britain, WA5VJB, shows you how to put together a VHF/UHF Yagi with QRO performance at a QRP price. (This material is adapted from Kent's on-line paper "Controlled Impedance 'Cheap' Antennas" at **www.wa5vjb.com/references.html**.) First, a bit of history on the design of these antennas. In 1993 at the Oklahoma City Central States VHF Society Conference, Arnie, CO2KK, spoke on the difficulties building VHF antennas in non-industrialized nations. Just run down to the store and pick up some Delrin insulators and 0.141 inch Teflon coax? Arnie's tales were the motivation to use advanced technology to come up with something simple.

The simplified feed uses the structure of the antenna itself for impedance matching. So the design started with the feed and the elements were built around it. The antennas were designed with *YagiMax*, tweaked in *NEC*, and the driven elements experimentally determined on the antenna range.

Typically a high-gain antenna is designed in the computer, then you try to come up with a driven element matching arrangement for whatever feed point impedance the computer comes up with. In this design, compromises for the feed impedance, asymmetrical feed, simple measurements, wide bandwidth, the ability to grow with the same spacing, and trade-offs for a very clean pattern cost many dB of gain. But you can build these antennas for about $5!

Construction of the antennas is straightforward. The boom is ¾ inch square, or ½ inch by ¾ inch wood. To install an element, drill a hole through the boom and insert the element. A drop of cyanoacrylate "super glue," epoxy, or silicone adhesive is used to hold the elements in place. There is no boom-to-mast plate — drill holes in the boom and use a U-bolt to attach it to the mast!

The life of the antenna is determined by what you coat it with. The author had a 902 MHz version, varnished with polyurethane, in the air for two years with little deterioration.

The parasitic elements on prototypes have been made from silicon-bronze welding rod, aluminum rod, brass hobby tubing, and #10 or #12 AWG solid copper ground wire. So that you can solder to the driven element, use the welding rod, hobby tubing, or copper wire. The driven element is folded at one end with its ends inserted through the boom.

Figure 21.96 shows the basic plan for the antenna and labels the dimensions that are given in the table for each band. All table dimensions are given in inches.

Figure 21.97 shows how the driven element is constructed for each antenna. Trim the free end of the driven element to tune it for minimum SWR at the desired frequency. **Figure 21.98** shows how to attach coaxial cable to the feed point. Sliding a quarter- wave sleeve along the coax had little effect, so there's not much RF on the outside of the coax. You may use a ferrite bead choke balun if you like, but these antennas are designed for minimum expense!

144 MHz Yagi

While others have reported good luck with 16 element long-boom wood antennas, six elements was about the maximum for most rovers. The design is peaked at 144.2 MHz, but performance is still good at 146.5 MHz. All parasitic elements are made from 3⁄16 inch aluminum rod, and the driven element is made from ⅛ inch rod. Lengths and spacings are given in **Table 21.16**.

222 MHz Yagi

This antenna is peaked at 222.1 MHz, but performance has barely changed at 223.5 MHz. You can drill the mounting holes to mount it with the elements horizontal or vertical. All parasitic elements are made from 3⁄16 inch aluminum rod, and the driven element is made from ⅛ inch rod. Lengths and spacings are given in **Table 21.17**.

432 MHz Yagi

At this band the antenna is getting very practical and easy to build. All parasitic elements

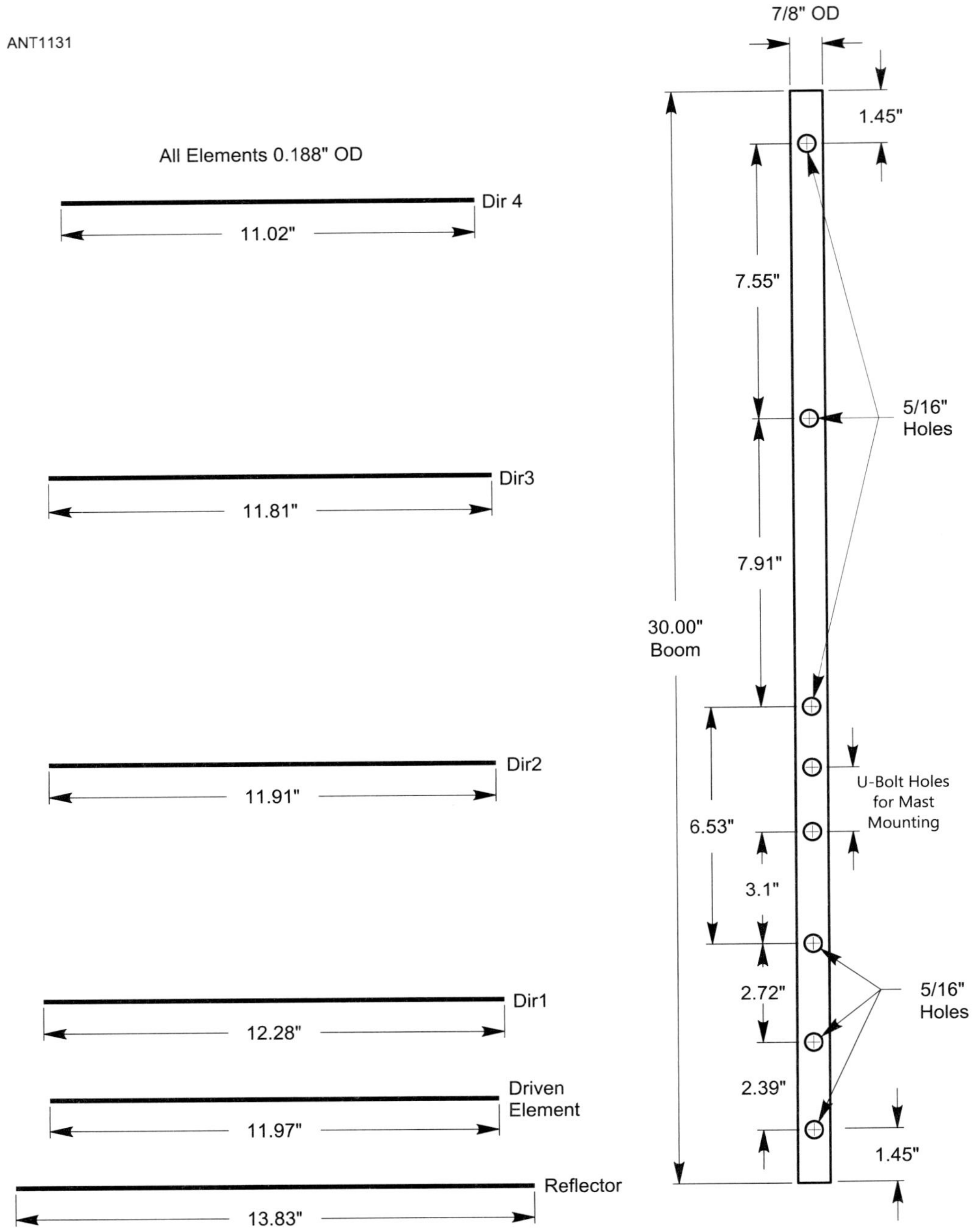

Figure 21.95 — Assembly and design information for the 70 cm Yagi boom and elements.

432-MHz Yagi Dimensions

	Spacing Between Elems. (inches)	*Spacing from Reflector (inches)*	*Full Element Length (inches)*
Reflector	0	—	13.832
Driven Ele.	2.394	2.394	11.968
Dir. 1	2.715	5.109	12.284
Dir. 2	6.528	11.637	11.908
Dir. 3	7.907	19.544	11.810
Dir. 4	7.546	27.09	11.01

...e made from ⅛ inch diameter rod, and the driven element is made from #10 AWG solid copper wire. Lengths and spacings are given in **Table 21.18**.

435 MHz Yagi for AMSAT

KA9LNV provided help and motivation for these antennas. A high front-to-back ratio (F/B) was a major design consideration of all versions. The model predicts 30 dB F/B for the 6-element and over 40 dB for the others. For gain, *NEC* predicts 11.2 dBi for the six-element, 12.6 dBi for the eight-element, 13.5 dBi for the 10 element, and 13.8 dBi for the 11 element.

Using ¾ inch square wood for the boom makes it easy to build two antennas on the same boom for cross-polarization. Offset the two antennas 6½ inches along the boom and feed them in-phase for circular polarization, or just use one for portable operations. All parasitic elements are made from ⅛ inch diameter rod, and the driven element is made from #10 AWG solid copper wire. Lengths and spacings are given in **Table 21.19**. The same element spacing is used for all four versions of the antenna.

450 MHz Yagi

For FM, this six-element Yagi is a good, cheap antenna to get a newcomer into a repeater or make a simplex-FM QSO during a contest. The author used ⅛ inch diameter aluminum ground wire in the prototype for all the elements except the driven element, which is made from #10 AWG solid copper wire. Other ⅛ inch diameter material could be used. Lengths and spacings are given in **Table 21.20**.

902 MHz Yagi

This was the first antenna the author built using the antenna to control the driven element impedance. The 2.5 foot length has proven very practical. All parasitic elements are made from ⅛ inch diameter rod, and the driven ele-

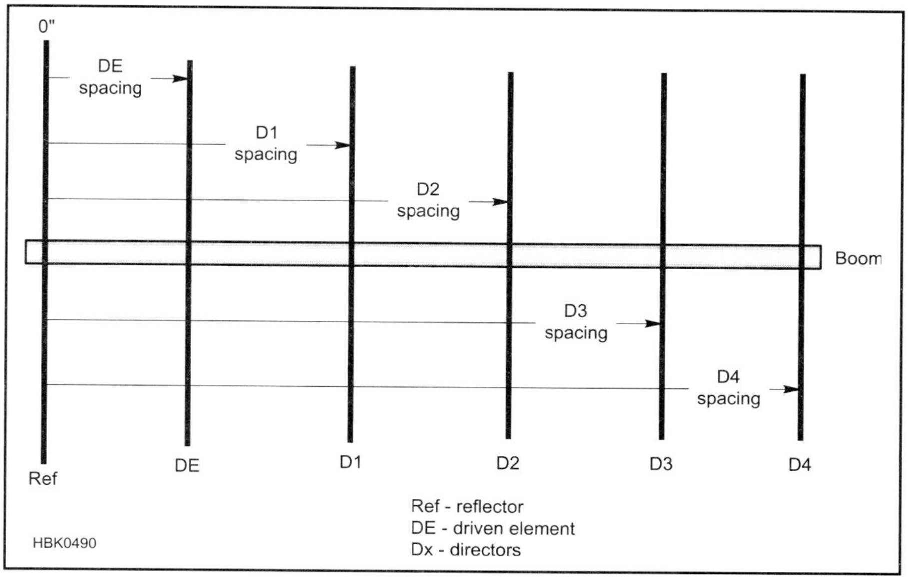

Figure 21.96 — Element spacing for the Cheap Yagis. Refer to Tables 21.16 to 21.23 for exact dimensions for the various bands.

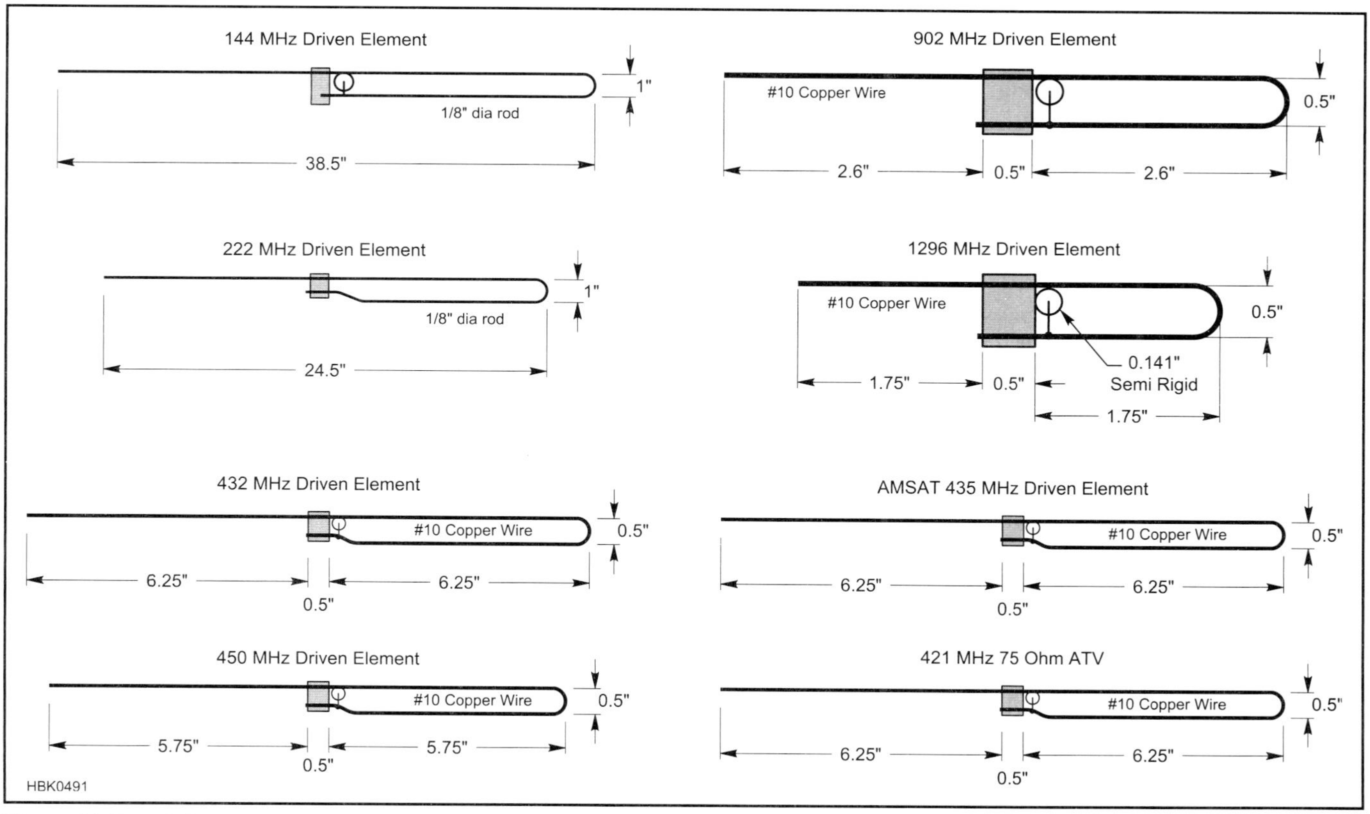

Figure 21.97 — Driven element dimensions for the Cheap Yagis. Attaching the coax shield to the center of the driven element is appropriate because that is the lowest impedance point of the element.

ment is made from #10 AWG solid copper wire. Lengths and spacings are given in **Table 21.21**.

1296 MHz Yagi

This antenna is the veteran of several "Grid-peditions" and has measured 13.5 dBi on the Central States VHF Society antenna range. Dimensions must be followed with great care. The driven element is small enough to allow 0.141 inch semi-rigid coax to be used. The prototype antennas use ⅛ inch silicon-bronze welding rod for the elements, but any ⅛ inch-diameter material can be used. The driven element is made from #10 AWG solid copper wire. Lengths and spacings are given in **Table 21.22**.

421.25 MHz 75 Ω Yagi for ATV

421 MHz vestigial sideband video is popular in North Texas for receiving the FM video input repeaters. These antennas are made for 421 MHz use, and the driven element is designed for 75 Ω. RG-59, or an F adapter to RG-6 can be directly connected to a cable-TV converter or cable-ready TV on channel 57. All parasitic elements are made from ⅛ inch diameter rod, and the driven element is made from #10 AWG solid copper wire. Lengths and spacings are given in **Table 21.23**. The same spacing is used for all versions.

Project: Fixed Moxons for Satellite Operation

The following project is based on the article "A Simple Fixed Antenna for VHF/UHF Satellite Work" by the late L.B. Cebik, W4RNL, from the August 2001 issue of *QST*. The complete article is available with the online content. This design produces a simple, reliable, fixed antenna that provides nearly hemispherical coverage for satellite operation on 145 and 435 MHz.

Many fixed-position satellite antennas for VHF and UHF have used a version of the *turnstile*. The word "turnstile" actually refers to two different ideas. One is a particular antenna: two crossed dipoles fed 90° out of phase, usually with a reflecting screen behind the dipoles. The other is the principle of obtaining omnidirectional patterns by diving almost any crossed antennas 90° out of phase. The second

Table 21.16
WA5VJB 144 MHz Yagi Dimensions

		Ref	*DE*	*D1*	*D2*	*D3*	*D4*
3-element	Length	41.0	—	37.0			
	Spacing	0	8.5	20.0			
4-element	Length	41.0	—	37.5	33.0		
	Spacing	0	8.5	19.25	40.5		
6-element	Length	40.5	—	37.5	36.5	36.5	32.75
	Spacing	0	7.5	16.5	34.0	52.0	70.0

Dimensions in inches.

Table 21.17
WA5VJB 222 MHz Yagi Dimensions

		Ref	*DE*	*D1*	*D2*	*D3*	*D4*
3-element	Length	26.0	—	23.75			
	Spacing	0	5.5	13.5			
4-element	Length	26.25	—	24.1	22.0		
	Spacing	0	5.0	11.75	23.5		
6-element	Length	26.25	—	24.1	23.5	23.5	21.0
	Spacing	0	5.0	10.75	22.0	33.75	45.5

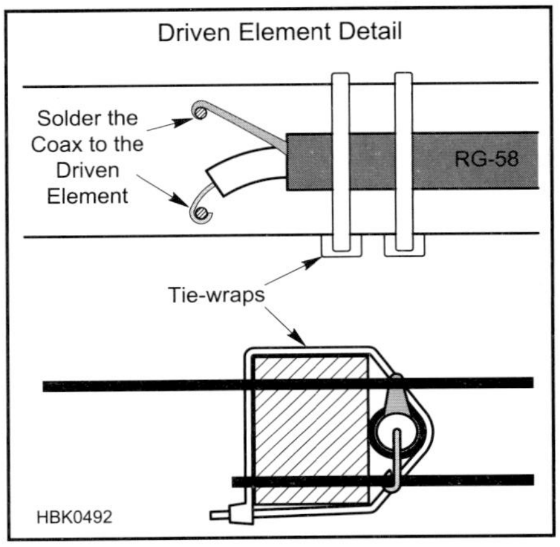

Figure 21.98 — Construction details and feed line attachment for the Cheap Yagi driven element.

Table 21.18
WA5VJB 432 MHz Yagi Dimensions

		Ref	*DE*	*D1*	*D2*	*D3*	*D4*	*D5*	*D6*	*D7*	*D8*	*D9*
6-element	Length	13.5	—	12.5	12.0	12.0	11.0					
	Spacing	0	2.5	5.5	11.25	17.5	24.0					
8-element	Length	13.5	—	12.5	12.0	12.0	12.0	12.0	11.25			
	Spacing	0	2.5	5.5	11.25	17.5	24.0	30.75	38.0			
11-element	Length	13.5	—	12.5	12.0	12.0	12.0	12.0	12.0	11.75	11.75	11.0
	Spacing	0	2.5	5.5	11.25	17.5	24.0	30.75	38.0	45.5	53.0	59.5

Dimensions in inches.

Table 21.19
WA5VJB 435 MHz Yagi Dimensions

		Ref	*DE*	*D1*	*D2*	*D3*	*D4*	*D5*	*D6*	*D7*	*D8*	*D9*
6-element	Length	13.4	—	12.4	12.0	12.0	11.0					
8-element	Length	13.4	—	12.4	12.0	12.0	12.0	12.0	11.1			
10-element	Length	13.4	—	12.4	12.0	12.0	12.0	12.0	11.75	11.75	11.1	
11-element	Length	13.4	—	12.4	12.0	12.0	12.0	12.0	11.75	11.75	11.75	11.1
	Spacing	0	2.5	5.5	11.25	17.5	24.0	30.5	37.75	45.0	52.0	59.5

Dimensions in inches.

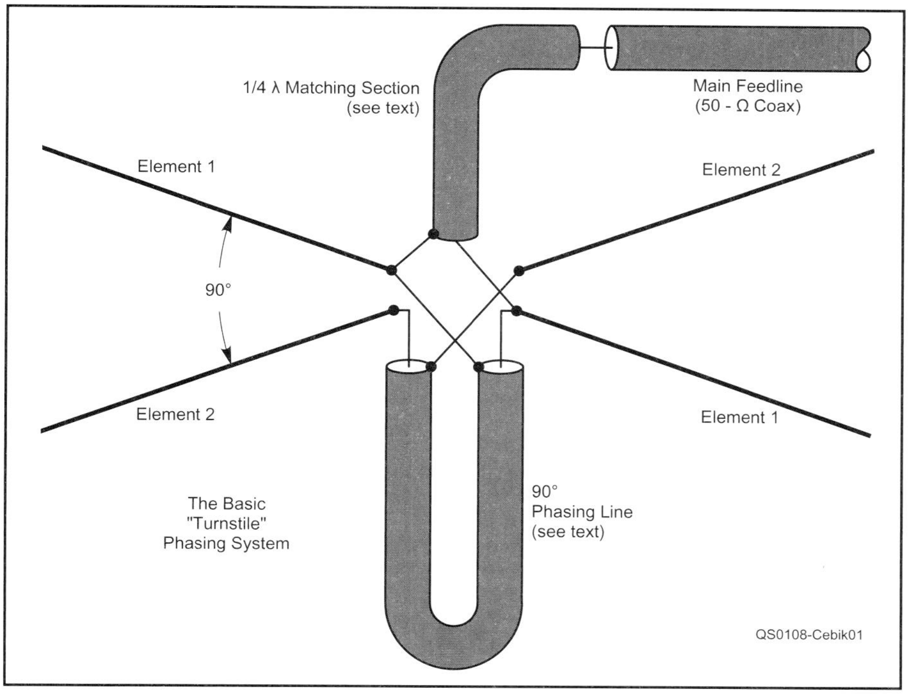

Figure 21.99 — The basic turnstile phasing (and matching) system for any antenna set requiring a 90° phase shift between driven elements in proximity.

idea opens the door to adapting many possible antenna designs to omnidirectional coverage.

Figure 21.99 shows a general method of obtaining the 90° phase shift for omnidirectional patterns. Note that the coax center conductor connects to only one of the two crossed elements. A ¼-λ section of transmission line that has the same characteristic impedance as the natural feed point impedance of the first antenna element alone connects one element to the next. The opposing ends of the two elements go to the shield at each end of the transmission line. The resulting impedance at the overall antenna feed point will be exactly half the impedance of one element alone. If dipoles are used, the feed point impedance will be approximately 35 Ω.

The dual Moxon rectangle array, shown in outline form on the right of **Figure 21.100**, offers some advantages over the traditional turnstile. (The article "Having a Field Day with the Moxon Rectangle," by L.B. Cebik describes the Moxon in detail and is included with the online content.) First, it yields a somewhat better dome-like pattern. Second, it is relatively easy to build and compact to install. The Moxon pair, with lower but smoother gain across the dome of the sky, offers the fixed-antenna user the chance to build a successful beginning satellite antenna.

Figure 21.101 shows the elevation patterns of a turnstile-and-screen and of a pair of Moxon rectangles when both are 2 λ above the ground. A 1 λ height will reduce the low angle ripples even more, if that height is feasible. The elevation patterns show the considerably smoother

Table 21.20
WA5VJB 450 MHz Yagi Dimensions

		Ref	*DE*	*D1*	*D2*	*D3*	*D4*
6-element	Length	13.0	—	12.1	11.75	11.75	10.75
	Spacing	0	2.5	5.5	11.0	18.0	28.5

Dimensions in inches.

Table 21.21
WA5VJB 902 MHz Yagi Dimensions

		Ref	*DE*	*D1*	*D2*	*D3*	*D4*	*D5*	*D6*	*D7*	*D8*
10-element	Length	6.2	—	5.6	5.5	5.5	5.4	5.3	5.2	5.1	5.1
	Spacing	0	2.4	3.9	5.8	9.0	12.4	17.4	22.4	27.6	33.0

Dimensions in inches.

Table 21.22
WA5VJB 1296 MHz Yagi Dimensions

		Ref	*DE*	*D1*	*D2*	*D3*	*D4*	*D5*	*D6*	*D7*	*D8*
10-element	Length	4.3	—	3.9	3.8	3.75	3.75	3.65	3.6	3.6	3.5
	Spacing	0	1.7	2.8	4.0	6.3	8.7	12.2	15.6	19.3	23.0

Dimensions in inches.

Table 21.23
WA5VJB 421.25 MHz 75-Ω Yagi Dimensions

		Ref	*DE*	*D1*	*D2*	*D3*	*D4*	*D5*	*D6*	*D7*	*D8*	*D9*
6-element	Length	14.0	—	12.5	12.25	12.25	11.0					
9-element	Length	14.0	—	12.5	12.25	12.25	12.0	12.0	11.25			
11-element	Length	14.0	—	12.5	12.25	12.25	12.0	12.0	12.0	11.75	11.75	11.5
	Spacing	0	3.0	6.5	12.25	17.75	24.5	30.5	36.0	43.0	50.25	57.25

Dimensions in inches.

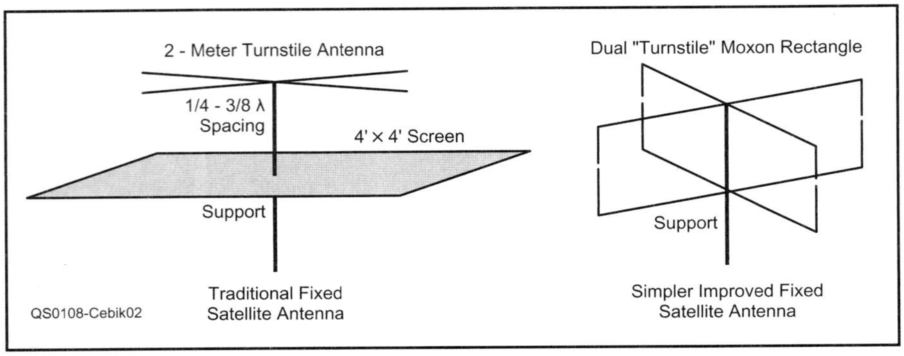

Figure 21.100 — Alternative schemes for fixed-position satellite antennas: the traditional turnstile-and-screen and a pair of "turnstiled" Moxon rectangles.

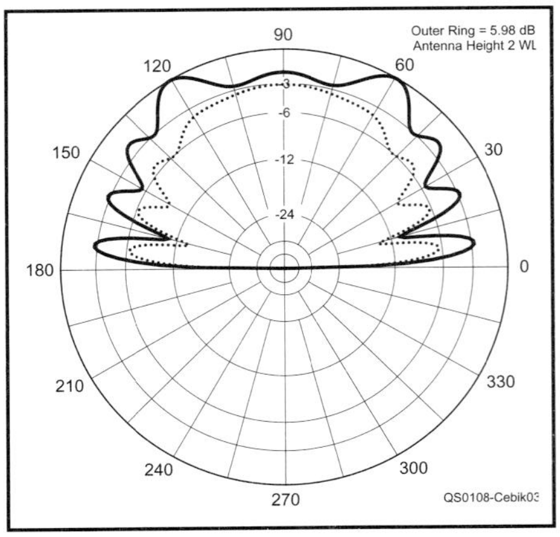

Figure 21.101 — A comparison of elevation patterns for the turnstile-and-screen system (with ⅜ λ wavelength spacing, shown by the solid black line) and a Moxon pair (dashed line), both at 2 λ height.

pattern dome of the Moxon pair over the traditional turnstile. The middle of the turnstile dome has nearly 2 dB less gain than its peaks, while the top valleys are nearly 3 dB lower than the peaks. The peaks and valleys can make the difference between successful communications and broken-up transmissions.

Without requiring a reflecting screen, which would narrow the antenna's beamwidth, the azimuthal pattern will be circular within under a 0.2-dB difference for 145.5 to 146.5 MHz, and within 0.5 dB for the entire 2 meter band. Since satellite work is concentrated in the 145.8 to 146.0 MHz region, the broadbanded antenna will prove fairly easy to build with success. A 435.6 MHz version, designed to cover the 435 to 436.2 MHz region of satellite activity, will have an even larger bandwidth.

Like the dipole-based turnstile, the Moxons are fed 90° out of phase with a ¼-λ phasing line of 50-Ω coaxial cable, as shown in Figure 21.99. Since the natural feed point impedance of a single Moxon rectangle of the design used here is 50 Ω, the pair will show a 25-Ω feed-point impedance. Paralleled ¼-λ sections of 70- to 75-Ω coaxial cable, such as RG-59, will transform the low impedance to a good match for the main 50-Ω coaxial feed line.

Figure 21.102 shows the critical dimensions for a Moxon rectangle. The lettered references are keys to the dimensions in **Table 21.24**. The design frequencies for the two satellite antenna pairs are 145.9 MHz and 435.5 MHz, the centers of the satellite activity on these two bands. The 2 meter Moxon prototype uses $^{3}/_{16}$-inch diameter rod, while the 435 MHz version uses #12 AWG wire with a nominal 0.0808-inch diameter. Going one small step up or down in element diameter will still produce a usable antenna, but major diameter changes will require that the dimensions be recalculated. (Complete construction details and drawings are available with the online content.)

The antennas can be mounted on the same mast. However, for similar patterns, they should each be the same number of wavelengths above ground. For example, if the 2 meter antenna is about two wavelengths up at about 14 feet or so, then the bottom of the 435-MHz antenna should be only about 4.5 feet above the ground. Placing the higher-frequency antenna below the 2 meter assembly will create some small irregularities in the desired dome pattern, but not serious enough to affect general operation.

There is no useful adjustment to these antennas except for making the gap between the drivers and reflectors as accurate as possible. Turnstile antennas show a very broad SWR curve. Across 2 meters, for example, the highest SWR is under 1.1:1. However, serious errors in the phasing line length can result in distortions to the desired circular pattern. There is no substitute for checking the lengths of the phasing line and the matching section several times before cutting. The correct length is from one junction to the next, including the portions of exposed cable interior.

Table 21.24
Dimensions for Moxon Rectangles for Satellite Use

Two are required for each antenna. The phase-line is 50-Ω coaxial cable, and the matching line is parallel sections of 75-Ω coaxial cable. Low power cables less than 0.15 inches in outer diameter were used in the prototypes. See Figure 21.102 for letter references. All dimensions are in inches.

Dimension	*145.9 MHz*	*435.6 MHz*
A	29.05	9.72
B	3.81	1.25
C	1.40	0.49
D	5.59	1.88
E (B + C + D)	10.80	3.62
¼ wavelength	20.22	6.77
Phasing and matching lines (0.66 velocity factor)	13.35	4.47

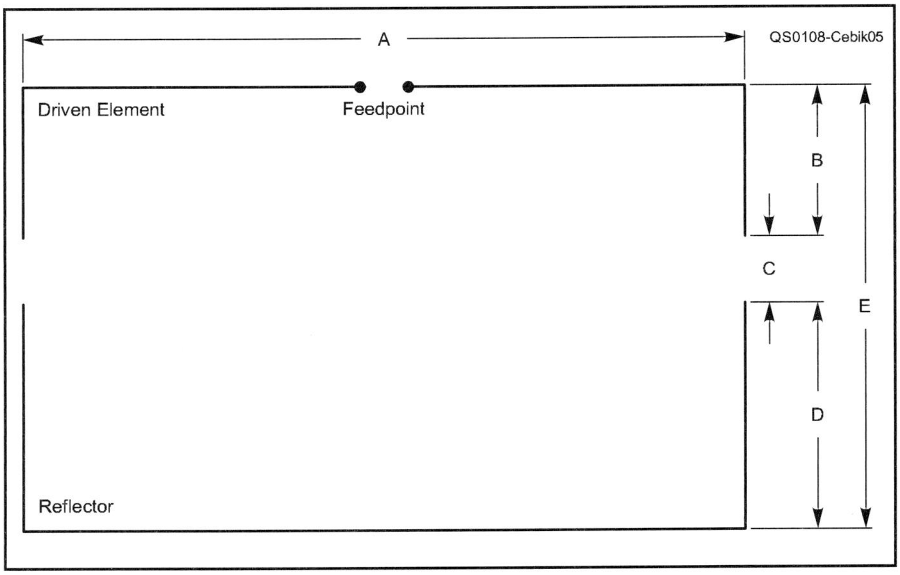

Figure 21.102 — The basic dimensions of a Moxon rectangle. Two identical rectangles are required for each "turnstiled" pair. See Table 21.24 for letter references.

21.12 Radio Direction Finding Antennas

Radio direction finding (RDF) is almost as old as radio communication. It gained prominence when the British Navy used it to track the movement of enemy ships in World War I. Since then, governments and the military have developed complex RDF systems. Fortunately, simple equipment, purchased or built at home, is quite effective in amateur radio RDF.

In European and Asian countries, direction-finding contests are foot races. The object is to be first to find four or five transmitters in a large wooded park. Young athletes have the best chance of capturing the prizes. This sport is known as *foxhunting* (after the British hill-and-dale horseback events) or *ARDF* (Amateur Radio Direction Finding). It is growing in popularity here in North America. Today, most competitive hunts worldwide are for 144 MHz FM signals, though other VHF bands are also used. Some international foxhunts include 3.5 MHz events

In North America and England, most RDF contests involve mobiles — cars, trucks, and vans, even motorcycles. It may be possible to drive all the way to the transmitter, or there may be a short hike at the end, called a *sniff*. These competitions are also called foxhunting by some, while others use *bunny hunting, T-hunting,* or the classic term *hidden transmitter hunting*.

Even without participating in RDF contests, you will find knowledge of the techniques useful. They simplify the search for a neighborhood source of power-line interference or TV cable leakage. RDF must be used to track down emergency radio beacons, which signal the location of pilots and boaters in distress. Amateur radio enthusiasts skilled in transmitter hunting are in demand by agencies such as the Civil Air Patrol and the US Coast Guard Auxiliary for search and rescue support. RDF is an important part of the evidence-gathering process in interference cases.

The most basic RDF system consists of a directional antenna and a method of detecting and measuring the level of the radio signal, such as a receiver with signal strength indicator. RDF antennas range from a simple tuned loop of wire to an acre of antenna elements with an electronic beam-forming network. Other sophisticated techniques for RDF use the Doppler effect or measure the time of arrival difference of the signal at multiple antennas.

All of these methods have been used from 2 to 500 MHz and above. However, RDF practices vary greatly between the HF and VHF/UHF portions of the spectrum. For practical reasons, high gain beams, Dopplers, and switched dual antennas find favor on VHF/UHF, while loops and phased arrays are the most popular choices on 6 meters and below. Signal propagation differences between HF and VHF also affect RDF practices. But many basic transmitter-hunting techniques, discussed later in this chapter, apply to all bands and all types of portable RDF equipment.

Several RDF projects may be found with the online content, along with a thorough article on Direction-Finding Techniques and mobile RDF system installation, including some examples of mobile RDF antenna mounting.

21.12.1 RDF Antennas for HF Bands

Below 50 MHz, gain antennas such as Yagis and quads are of limited value for RDF. The typical tribander installation yields only a general direction of the incoming signal, due to ground effects and the antenna's broad forward lobe. Long monoband beams at greater heights work better, but still cannot achieve the bearing accuracy and repeatability of simpler antennas designed specifically for RDF.

RDF LOOPS

An effective directional HF antenna can be as uncomplicated as a small loop of wire or tubing, tuned to resonance with a capacitor. When immersed in an electromagnetic field, the loop acts much the same as the secondary winding of a transformer. The voltage at the output is proportional to the amount of flux passing through it and the number of turns. If the loop is oriented such that the greatest amount of area is presented to the magnetic field, the induced voltage will be the highest. If it is rotated so that little or no area is cut by the field lines, the voltage induced in the loop is zero and a null occurs.

To achieve this transformer effect, the loop must be small compared with the signal wavelength. In a single-turn loop, the conductor should be less than 0.08 λ long. For example, a 28 MHz loop should be less than 34 inches in circumference, giving a diameter of approximately 10 inches. The loop may be smaller, but that will reduce its voltage output. Maximum output from a small loop antenna is in directions corresponding to the plane of the loop; these lobes are very broad. Sharp nulls, obtained at right angles to that plane, are more useful for RDF.

For a perfect bidirectional pattern, the loop must be balanced electrostatically with respect to ground. Otherwise, it will exhibit two modes of operation, the mode of a perfect loop and that of a non-directional vertical antenna of small dimensions. This dual-mode condition results in mild to severe inaccuracy, depending on the degree of imbalance, because the outputs of the two modes are not in phase.

The theoretical true loop pattern is illustrated in **Figure 21.103**. When properly balanced, there are two nulls exactly 180° apart. When the unwanted antenna effect is appreciable and the loop is tuned to resonance, the loop may exhibit little directivity, as shown in Figure 21.103B. By detuning the loop to shift the phasing, you may obtain a useful pattern similar to Figure 21.103C. While not symmetrical, and not necessarily at right angles to the plane of the loop, this pattern does exhibit a pair of nulls.

By careful detuning and amplitude balancing, you can approach the unidirectional pattern of Figure 21.103D. Even though there may not be a complete null in the pattern, it resolves

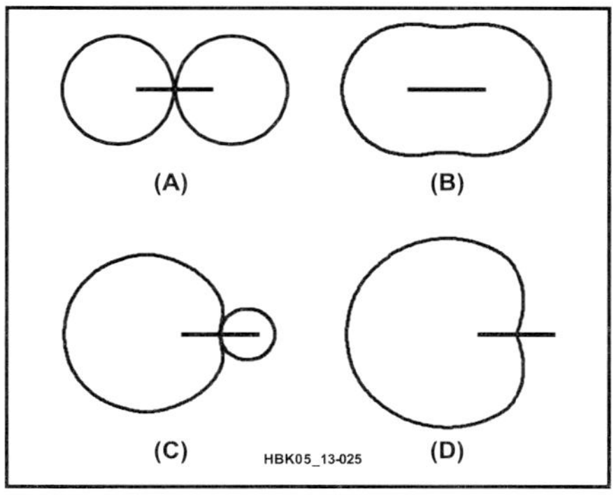

Figure 21.103 — Small loop field patterns with varying amounts of antenna effect — the undesired response of a loop acting merely as a mass of metal connected to the receiver antenna terminals. The horizontal lines show the plane of the loop turns.

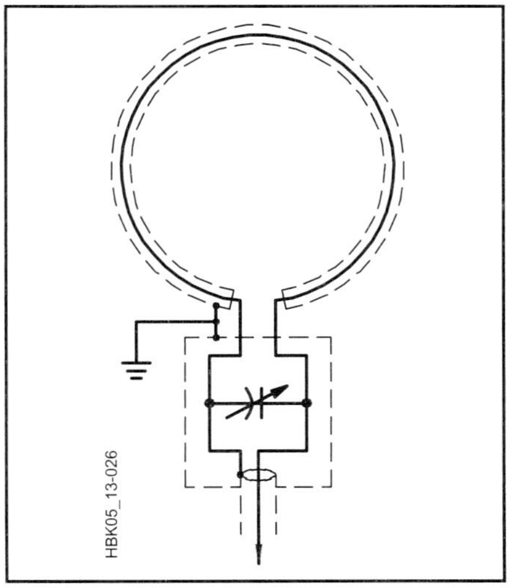

Figure 21.104 — Electrostatically shielded loop for RDF. To prevent shielding of the loop from magnetic fields, leave the shield unconnected at one end.

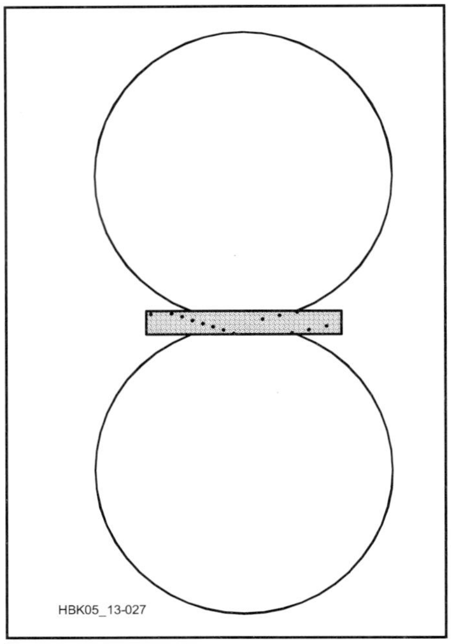

Figure 21.105 — Field pattern for a ferrite-rod antenna. The dark bar represents the rod on which the loop turns are wound.

the 180° ambiguity of Figure 21.103A. Korean War era military loop antennas, sometimes available on today's surplus market, use this controlled antenna effect principle.

An easy way to achieve good electrostatic balance is to shield the loop, as shown in **Figure 21.104**. The shield, represented by the dashed lines in the drawing, eliminates the antenna effect. The response of a well-constructed shielded loop is quite close to the ideal pattern of Figure 21.103A.

For 160 through 30 meters, single-turn loops that are small enough for portability are usually unsatisfactory for RDF work. Multi-turn loops are generally used instead. They are easier to resonate with practical capacitor values and give higher output voltages. This type of loop may also be shielded. If the total conductor length remains below 0.08 λ, the directional pattern is that of Figure 21.103A.

FERRITE ROD ANTENNAS

Another way to get higher loop output is to increase the permeability of the medium in the vicinity of the loop. By winding a coil of wire around a form made of high-permeability material such as ferrite rod, much greater flux is obtained in the coil without increasing the cross-sectional area.

Modern magnetic core materials make compact directional receiving antennas practical. Most portable AM broadcast receivers use this type of antenna, commonly called a *loopstick*. The loopstick is the most popular RDF antenna for portable/mobile work on 160 and 80 meters.

Like the shielded loop discussed earlier, the loopstick responds to the magnetic field of the incoming radio wave, and not to the electrical field. For a given size of loop, the output voltage increases with increasing flux density, which is obtained by choosing a ferrite core of high permeability and low loss at the frequency of interest. For increased output, the turns may be wound over two rods taped together. A practical loopstick antenna is described later in this chapter.

A loop on a ferrite core has maximum signal response in the plane of the turns, just like an air core loop. This means that maximum response of a loopstick is broadside to the axis of the rod, as shown in **Figure 21.105**. The loopstick may be shielded to eliminate the antenna effect; a U-shaped or C-shaped channel of aluminum or other form of "trough" is best. The shield must not be closed, and its length should equal or slightly exceed the length of the rod.

SENSE ANTENNAS

Because there are two nulls 180° apart in the directional pattern of a small loop or loopstick, there is ambiguity as to which null indicates the true direction of the target station. For example, if the line of bearing runs east and west from your position, you have no way of knowing from this single bearing whether the transmitter is east or west of you.

If bearings can be taken from two or more positions at suitable direction and distance from the transmitter, the ambiguity can be resolved and distance can be estimated by triangulation, as discussed later in this chapter. However, it is almost always desirable to be able to resolve the ambiguity immediately by having a unidirectional antenna pattern available.

You can modify a loop or loopstick antenna pattern to have a single null by adding a second antenna element. This element is called a *sense antenna*, because it senses the phase of the signal wavefront for comparison with the phase of the loop output signal. The sense element must be omnidirectional, such as a short vertical. When signals from the loop and the sense antenna are combined with 90° phase shift between the two, a heart-shaped (cardioid) pattern results, as shown in **Figure 21.106A**.

Figure 21.106B shows a circuit for adding a sense antenna to a loop or loopstick. For the best null in the composite pattern, signals from

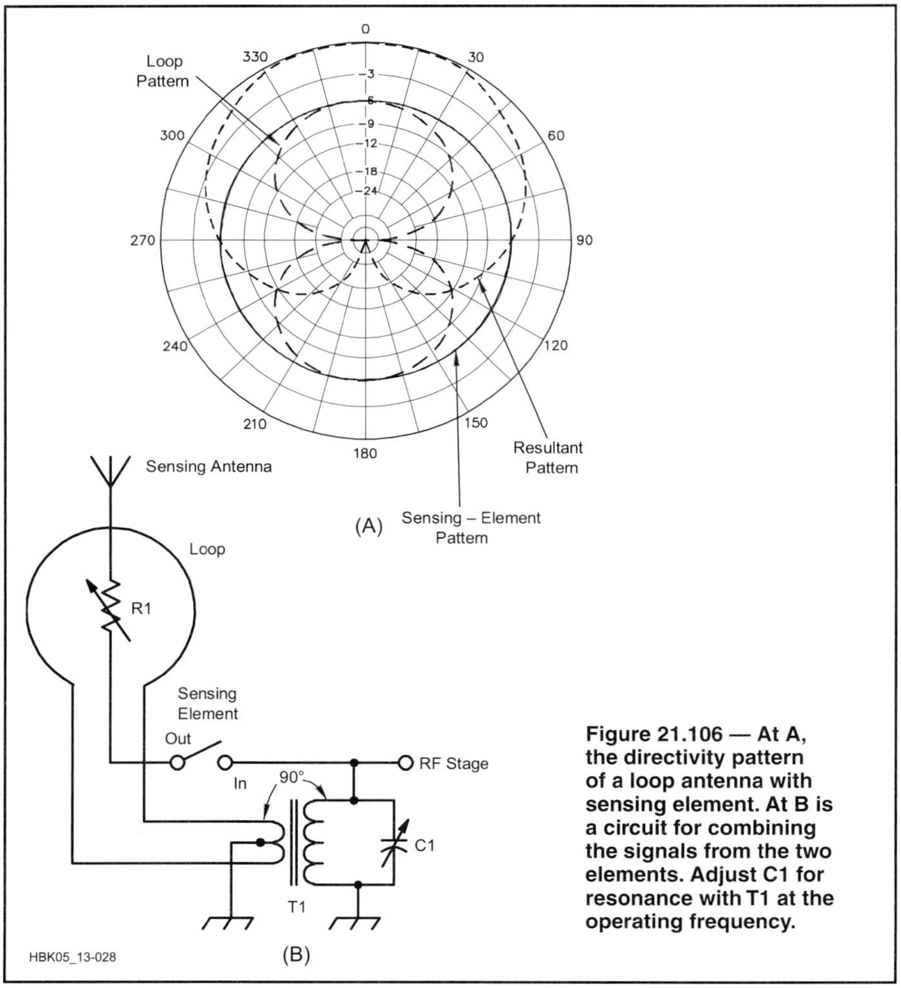

Figure 21.106 — At A, the directivity pattern of a loop antenna with sensing element. At B is a circuit for combining the signals from the two elements. Adjust C1 for resonance with T1 at the operating frequency.

the loop and sense antennas must be of equal amplitude. R1 adjusts the level of the signal from the sense antenna.

In a practical system, the cardioid pattern null is not as sharp as the bidirectional null of the loop alone. The usual procedure when transmitter hunting is to use the loop alone to obtain a precise line of bearing, then switch in the sense antenna and take another reading to resolve the ambiguity.

PHASED ARRAYS AND ADCOCK ANTENNAS

Two-element phased arrays are popular for amateur HF RDF base station installations. Many directional patterns are possible, depending on the spacing and phasing of the elements. A useful example is two ½ λ elements spaced ¼ λ apart and fed 90° out of phase. The resultant pattern is a cardioid, with a null off one end of the axis of the two antennas and a broad peak in the opposite direction. The directional frequency range of this antenna is limited to one band, because of the critical length of the phasing lines.

The best-known phased array for RDF is the Adcock, named after the man who invented it in 1919. It consists of two vertical elements fed 180° apart, mounted so the array may be rotated. Element spacing is not critical, and may be in the range from 0.1 to 0.75 λ. The two elements must be of identical lengths, but need not be self-resonant; shorter elements are commonly used. Because neither the element spacing nor length is critical in terms of wavelengths, an Adcock array may operate over more than one amateur band.

Figure 21.107 is a schematic of a typical Adcock configuration, called the H-Adcock because of its shape. Response to a vertically polarized wave is very similar to a conventional loop. The passing wave induces currents I1 and I2 into the vertical members. The output current in the transmission line is equal to their difference. Consequently, the directional pattern has two broad peaks and two sharp nulls, like the loop. The magnitude of the difference current is proportional to the spacing (d) and length (l) of the elements. You will get somewhat higher gain with larger dimensions. The Adcock of **Figure 21.108**, designed for 40 meters, has element lengths of 12 feet and spacing of 21 feet (approximately 0.15 λ).

Figure 21.109 shows the radiation pattern of the Adcock. The nulls are broadside to the axis of the array, becoming sharper with increased element spacing. When element spacing exceeds ¾ λ, however, the antenna begins to take on additional unwanted nulls off the ends of the array axis.

The Adcock is a vertically polarized antenna. The vertical elements do not respond to horizontally polarized waves, and the currents induced in the horizontal members by a horizontally polarized wave (dotted arrows in Figure 21.107) tend to balance out regardless of the orientation of the antenna.

Since the Adcock uses a balanced feed system, a coupler is required to match the unbalanced input of the receiver. T1 is an air-wound coil with a two-turn link wrapped around the middle. The combination is resonated with C1 to the operating frequency. C2 and C3 are null-clearing capacitors. Adjust them by placing a low-power signal source some distance from the antenna and exactly broadside to it. Adjust C2 and C3 until the deepest null is obtained.

While you can use a metal support for the mast and boom, wood is preferable because of its non-conducting properties. Similarly, a mast of thick-wall PVC pipe gives less distortion of the antenna pattern than a metallic mast. Place the coupler on the ground below the wiring harness junction on the boom and connect it with a short length of 300 Ω twin-lead-feed line.

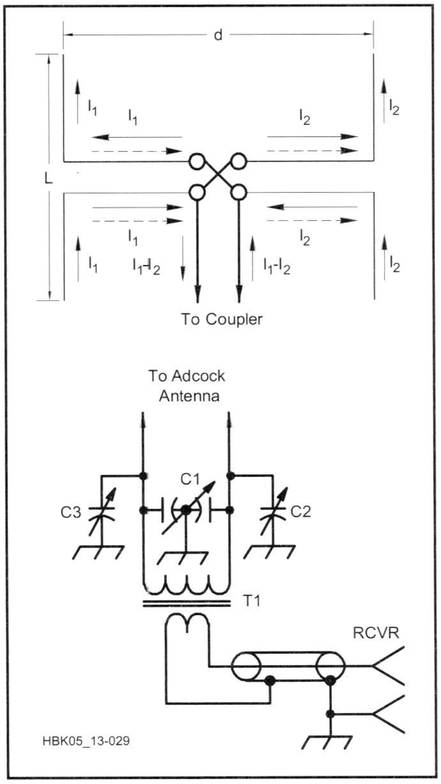

Figure 21.107 — A simple Adcock antenna and its coupler.

LOOPS VS PHASED ARRAYS

Loops are much smaller than phased arrays for the same frequency, and are thus the obvious choice for portable/mobile HF RDF. For base stations in a triangulation network, where the 180° ambiguity is not a problem, Adcocks are preferred. In general, they give sharper nulls than loops, but this is in part a function of the care used in constructing and feeding the individual antennas, as well as of the spacing of the elements. The primary construction considerations are the shielding and balancing of the feed line against unwanted signal pickup and the balancing of the antenna for a symmetrical pattern. Users report that Adcocks are somewhat less sensitive to proximity effects, probably because their larger aperture offers some space diversity.

Skywave Considerations

Until now we have considered the directional characteristics of the RDF loop only in the two-dimensional azimuthal plane. In three-dimensional space, the response of a vertically oriented small loop is doughnut-shaped. The bidirectional null (analogous to a line through the doughnut hole) is in the line of bearing in the azimuthal plane and toward the horizon in the vertical plane. Therefore, maximum null depth is achieved only on signals arriving at 0° elevation angle.

Skywave signals usually arrive at nonzero wave angles. As the elevation angle increases,

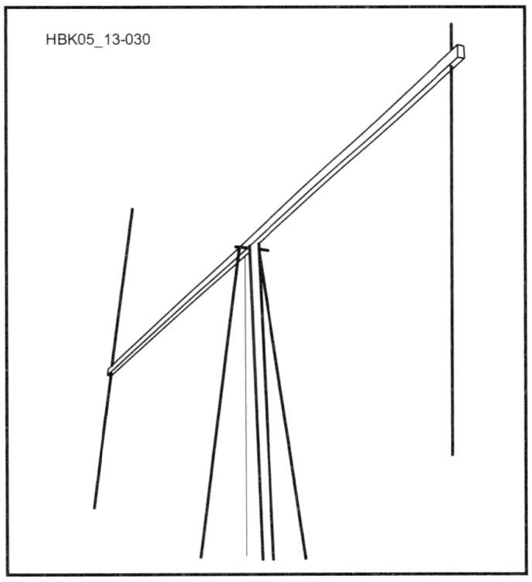

Figure 21.108 — An Adcock antenna on a wooden frame.

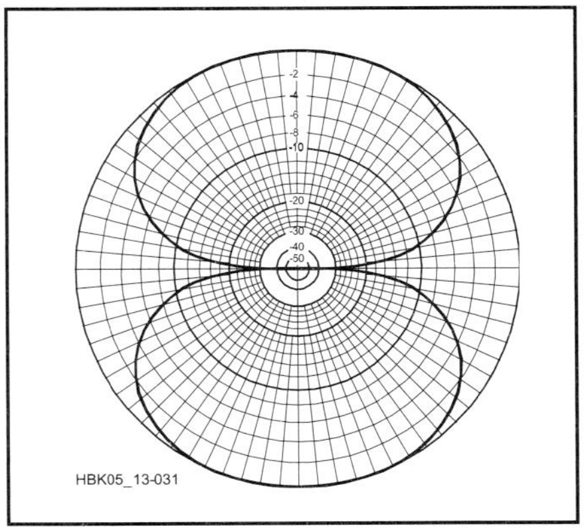

Figure 21.109 — The pattern of an Adcock array with element spacing of ½ wavelength. The elements are aligned with the vertical axis.

the null in a vertically oriented loop pattern becomes shallower. It is possible to tilt the loop to seek the null in elevation as well as azimuth. Some amateur RDF enthusiasts report success at estimating distance to the target by measurement of the elevation angle with a tilted loop and computations based on estimated height of the propagating ionospheric layer. This method seldom provides high accuracy with simple loops, however.

Most users prefer Adcocks to loops for skywave work, because the Adcock null is present at all elevation angles. Note, however, that an Adcock has a null in all directions from signals arriving from overhead. Thus for very high angles, such as under-250-mile skip on 80 and 40 meters, neither loops nor Adcocks will perform well.

ELECTRONIC ANTENNA ROTATION

State-of-the-art fixed RDF stations for government and military work use antenna arrays of stationary elements, rather than mechanically rotatable arrays. The best-known type is the *Wullenweber antenna*. It has a large number of elements arranged in a circle, usually outside of a circular reflecting screen. Depending on the installation, the circle may be anywhere from a few hundred feet to more than a quarter of a mile in diameter. Although the Wullenweber is not practical for most amateurs, some of the techniques it uses may be applied to amateur RDF.

The device, which permits rotating the antenna beam without moving the elements, has the classic name *radio goniometer*, or simply *goniometer*. Early goniometers were RF transformers with fixed coils connected to the array elements and a moving pickup coil connected to the receiver input. Both amplitude and phase of the signal coupled into the pickup winding are altered with coil rotation in a way that corresponded to actually rotating the array itself. With sufficient elements and a goniometer, accurate RDF measurements can be taken in all compass directions.

Beam Forming Networks

By properly sampling and combining signals from individual elements in a large array, an antenna beam is electronically rotated or steered. With an appropriate number and arrangement of elements in the system, it is possible to form almost any desired antenna pattern by summing the sampled signals in appropriate amplitude and phase relationships. Delay networks and/or attenuation are added in line with selected elements before summation to create these relationships.

To understand electronic beam forming, first consider just two elements, shown as A and B in **Figure 21.110**. Also shown is the wavefront of a radio signal arriving from a distant transmitter. The wavefront strikes element A first, then travels somewhat farther before it strikes element B. Thus, there is an interval between the times that the wavefront reaches elements A and B.

We can measure the differences in arrival times by delaying the signal received at element A before summing it with that from element B. If two signals are combined directly, the amplitude of the sum will be maximum when the delay for element A exactly equals the propagation delay, giving an in-phase condition at the summation point. On the other hand, if one of the signals is inverted and the two are added, the signals will combine in a 180° out-of-phase relationship when the element A delay equals the propagation delay, creating a null. Either way, once the time delay is determined by the amount of delay required for a peak or null, we can convert it to distance. Then trigonometry calculations provide the direction from which the wave is arriving.

Altering the delay in small increments steers the peak (or null) of the antenna. The system is not frequency sensitive, other than the frequency range limitations of the array elements. Lumped-constant networks are suitable for delay elements if the system is used only for receiving. Delay lines at installations used for transmitting and receiving employ rolls of coaxial cable of various lengths, chosen for the time delay they provide at all frequencies, rather than as simple phasing lines designed for a single frequency.

Combining signals from additional elements narrows the broad beamwidth of the pattern from the two elements and suppresses unwanted sidelobes. Electronically switching the delays and attenuations to the various elements causes the formed beam to rotate around the compass. The package of electronics that does this, including delay lines and electronically switched attenuators, is the beam-forming network.

21.12.2 Methods for VHF/UHF RDF

Three distinct methods of mobile RDF are commonly in use by amateurs on VHF/UHF bands: directional antennas, switched dual antennas and Dopplers. Each has advantages over the others in certain situations. Many RDF enthusiasts employ more than one method when transmitter hunting.

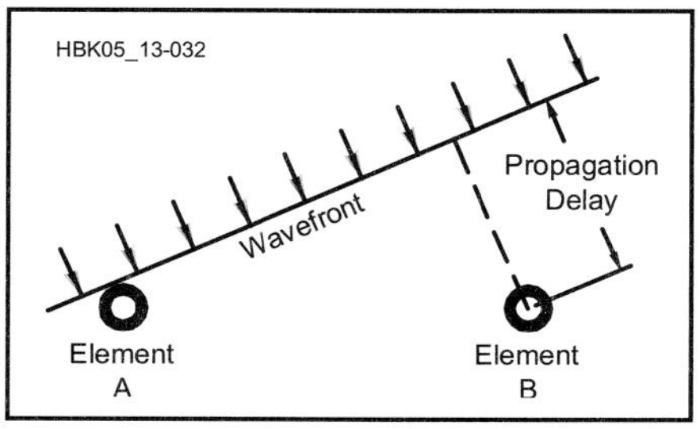

Figure 21.110 — One technique used in elec-tronic beam forming. By delaying the signal from element A by an amount equal to the propagation delay, two signals are summed precisely in phase, even though the signal is not in the broadside direction.

DIRECTIONAL ANTENNAS

Ordinary mobile transceivers and handhelds work well for foxhunting on the popular VHF bands. If you have a lightweight beam and your receiver has an easy-to-read S meter, you are nearly ready to start. All you need is an RF attenuator and some way to mount the setup in your vehicle.

Amateurs seldom use fractional wavelength loops for RDF above 60 MHz because they have bidirectional characteristics and low sensitivity, compared to other practical VHF antennas. Sense circuits for loops are difficult to implement at VHF, and signal reflections tend to fill in the nulls. Typically VHF loops are used only for close-in sniffing where their compactness and sharp nulls are assets, and low gain is of no consequence.

Phased Arrays

The small size and simplicity of two-element driven arrays make them a common choice of newcomers at VHF RDF. Antennas such as phased ground planes and ZL Specials have modest gain in one direction and a null in the opposite direction. The gain is helpful when the signal is weak, but the broad response peak makes it difficult to take a precise bearing.

As the signal gets stronger, it becomes possible to use the null for a sharper S meter indication. However, combinations of direct and reflected signals (called *multipath*) will distort the null or perhaps obscure it completely. For best results with this type of antenna, always find clear locations from which to take bearings.

Parasitic Arrays

Parasitic arrays are the most common RDF antennas used by transmitter hunters in high competition areas such as Southern California. Antennas with significant gain are a necessity due to the weak signals often encountered on weekend-long T-hunts, where the transmitter may be over 200 miles distant. Typical 144 MHz installations feature Yagis or quads of three to six elements, sometimes more. Quads are typically home-built, using data from *The ARRL Antenna Book* and *Transmitter Hunting* (see Bibliography).

Two types of mechanical construction are popular for mobile VHF quads. One model uses thin gauge wire (solid or stranded) suspended on wood dowel or fiberglass rod spreaders. It is lightweight and easy to turn rapidly by hand while the vehicle moves. Many hunters prefer to use larger gauge solid wire (such as #10 AWG) on a PVC plastic pipe frame. This quad is more rugged and has somewhat wider frequency range, at the expense of increased weight and wind resistance. It can get bent going under a branch, but it is easily reshaped and returned to service.

Yagis are a close second to quads in popularity. Commercial models work fine for VHF RDF, provided that the mast is attached at a good balance point. Lightweight and small-diameter elements are desirable for ease of turning at high speeds.

A well-designed mobile Yagi or quad installation includes a method of selecting wave polarization. Although vertical polarization is the norm for VHF-FM communications, horizontal polarization is allowed on many T-hunts. Results will be poor if a VHF RDF antenna is cross-polarized to the transmitting antenna, because multipath and scattered signals (which have indeterminate polarization) are enhanced, relative to the cross-polarized direct signal. The installation of **Figure 21.111** features a slip joint at the boom-to-mast junction, with an actuating cord to rotate the boom, changing the polarization. Mechanical stops limit the boom rotation to 90°.

Parasitic Array Performance for RDF

The directional gain of a mobile beam (typically 8 dB or more) makes it unexcelled for both weak signal competitive hunts and for locating interference such as TV cable leakage. With an appropriate receiver, you can get bearings on any signal mode, including FM, SSB, CW, TV, pulses, and noise. Because only the response peak is used, the null-fill problems and proximity effects of loops and phased arrays do not exist.

You can observe multiple directions of arrival while rotating the antenna, allowing you to make educated guesses as to which signal peaks are direct and which are from non-direct paths or scattering. Skilled operators can estimate distance to the transmitter from the rate of signal strength increase with distance traveled. The RDF beam is useful for transmitting, if necessary, but use care not to damage an attenuator in the coax line by transmitting through it.

The 3 dB beamwidth of typical mobile-mount VHF beams is on the order of 80°. This is a great improvement over 2-element driven arrays, but it is still not possible to get pinpoint bearing accuracy. You can achieve errors of less than 10° by carefully reading the S meter. In practice, this is not a major hindrance to successful mobile RDF. Mobile users are not as concerned with precise bearings as fixed station operators, because mobile readings are used primarily to give the general direction of travel to "home in" on the signal. Mobile bearings are continuously updated from new, closer locations.

Amplitude-based RDF may be very difficult when signal level varies rapidly. The transmitter hider may be changing power, or the target antenna may be moving or near a well-traveled road or airport. The resultant rapid S meter movement makes it hard to take accurate bearings with a quad. The process is slow because the antenna must be carefully rotated by hand to "eyeball average" the meter readings.

SWITCHED ANTENNA RDF UNITS

Three popular types of RDF systems are relatively insensitive to variations in signal level. Two of them use a pair of vertical dipole antennas, spaced ½ λ or less apart, and alternately switched at a rapid rate to the input of the receiver. In use, the indications of the two systems are similar, but the principles are different.

Switched Pattern Systems

The switched pattern RDF set (**Figure 21.112**) alternately creates two cardioid antenna patterns with lobes to the left and the right. The patterns are generated in much the same way as in the phased arrays described above. PIN RF diodes select the alternating patterns. The combined antenna outputs go to a receiver with AM detection. Processing after the detector output determines the phase or amplitude difference between the patterns' responses to the signal.

Switched pattern RDF sets typically have a zero center meter as an indicator. The meter swings negative when the signal is coming from the user's left, and positive when the signal source is on the right. When the plane of the antenna is exactly perpendicular to the direction of the signal source, the meter reads zero.

The sharpness of the zero crossing indication makes possible more precise bearings than those obtainable with a quad or Yagi. Under ideal conditions with a well-built unit, null direction accuracy is within 1°. Meter deflection tells the user which way to turn to zero the meter. For example, a negative (left) reading requires turning the antenna left. This solves the 180° ambiguity caused by the two zero crossings in each complete rotation of the antenna system.

Because it requires AM detection of the switched pattern signal, this RDF system finds its greatest use in the 120 MHz aircraft band, where AM is the standard mode. Commercial manufacturers make portable RDF sets with switched pattern antennas and built-in receivers for field portable use. These sets can usually be adapted to the amateur 144 MHz band. Other designs are adaptable to any VHF receiver that covers the frequency of interest and has an AM detector built in or added.

Switched pattern units work well for RDF from small aircraft, for which the two vertical antennas are mounted in fixed positions on the outside of the fuselage or simply taped inside the windshield. The left-right indication tells the pilot which way to turn the aircraft to home in. Since street vehicles generally travel only on roads, fixed mounting of the antennas on them is undesirable. Mounting vehicular switched-pattern arrays on a rotatable mast is best.

Time-of-Arrival Systems

Another kind of switched antenna RDF set uses the difference in arrival times of the signal wavefront at the two antennas. This narrow-aperture Time-Difference-of-Arrival (TDOA) technology is used for many sophisticated military RDF systems. The rudimentary TDOA implementation of **Figure 21.113** is quite effective for amateur use. The signal from transmitter 1 reaches antenna A before antenna B. Conversely, the signal from transmitter 3 reaches antenna B before antenna A. When the plane of the antenna is perpendicular to the signal source (as transmitter 2 is in

Figure 21.111 — The mobile RDF installation of WB6ADC features a thin wire quad that can be switched between vertical and horizontal polarization.

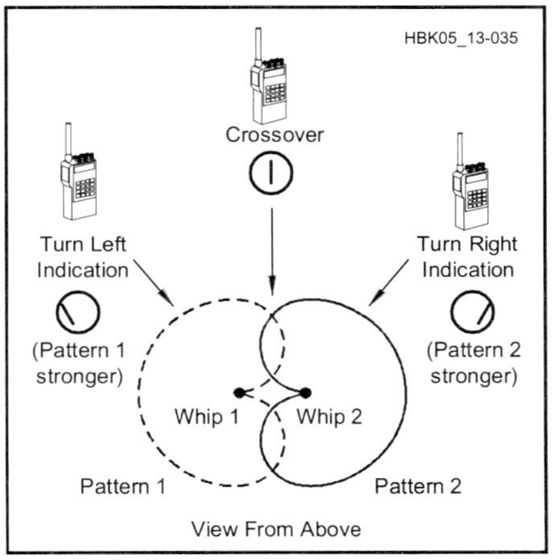

Figure 21.112 — In a switched pattern RDF set, the responses of two cardioid antenna patterns are summed to drive a zero center indicator.

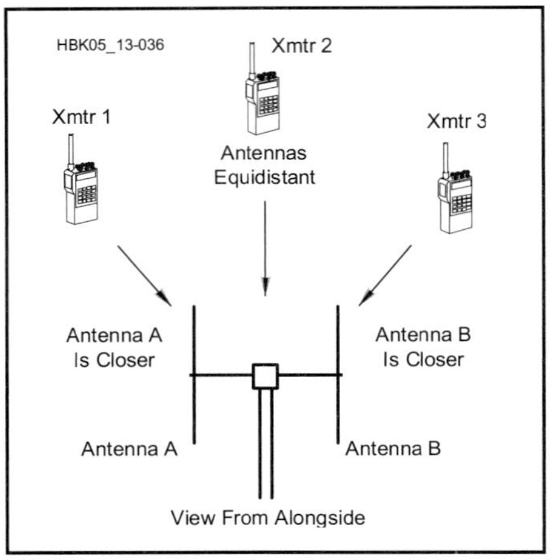

Figure 21.113 — A dual-antenna TDOA RDF system has a similar indicator to a switched pattern unit, but it obtains bearings by determining which of its antennas is closer to the transmitter.

the figure), the signal arrives at both antennas simultaneously.

If the outputs of the antennas are alternately switched at an audio rate to the receiver input, the differences in the arrival times of a continuous signal produce phase changes that are detected by an FM discriminator. The resulting short pulses sound like a tone in the receiver output. The tone disappears when the antennas are equidistant from the signal source, giving an audible null.

The polarity of the pulses at the discriminator output is a function of which antenna is closer to the source. Therefore, the pulses can be processed and used to drive a left-right zero-center meter in a manner similar to the switched pattern units described above. Left-right LED indicators may replace the meter for economy and visibility at night.

RDF operations with a TDOA dual antenna RDF are done in the same manner as with a switched antenna RDF set. The main difference is the requirement for an FM receiver in the TDOA system and an AM receiver in the switched pattern case. No RF attenuator is needed for close-in work in the TDOA case.

Popular designs for practical do-it-yourself TDOA RDF sets include the Simple Seeker (described elsewhere in this chapter) and the W9DUU design (see article by Bohrer in the Bibliography). Articles with plans for the Handy Tracker, a simple TDOA set with a delay line to resolve the dual-null ambiguity instead of LEDs or a meter, are listed in the Bibliography.

Performance Comparison

Both types of dual antenna RDFs make good on-foot "sniffing" devices and are excellent performers when there are rapid amplitude variations in the incoming signal. They are the units of choice for airborne work. Compared to Yagis and quads, they give good directional performance over a much wider frequency range. Their indications are more precise than those of beams with broad forward lobes.

Dual-antenna RDF sets frequently give inaccurate bearings in multipath situations, because they cannot resolve signals of nearly equal levels from more than one direction. Because multipath signals are a combined pattern of peaks and nulls, they appear to change in amplitude and bearing as you move the RDF antenna along the bearing path or perpendicular to it, whereas a non-multipath signal will have constant strength and bearing.

The best way to overcome this problem is to take large numbers of bearings while moving toward the transmitter. Taking bearings while in motion averages out the effects of multipath, making the direct signal more readily discernible. Some TDOA RDF sets have a slow-response mode that aids the averaging process.

Switched antenna systems generally do not perform well when the incoming signal is horizontally polarized. In such cases, the bearings may be inaccurate or unreadable. TDOA units require a carrier type signal such as FM or CW; they usually cannot yield bearings on noise or pulse signals.

Unless an additional method is employed to measure signal strength, it is easy to "overshoot" the hidden transmitter location with a TDOA set. It is not uncommon to see a TDOA foxhunter walk over the top of a concealed transmitter and walk away, following the opposite 180° null, because there is no display of signal amplitude.

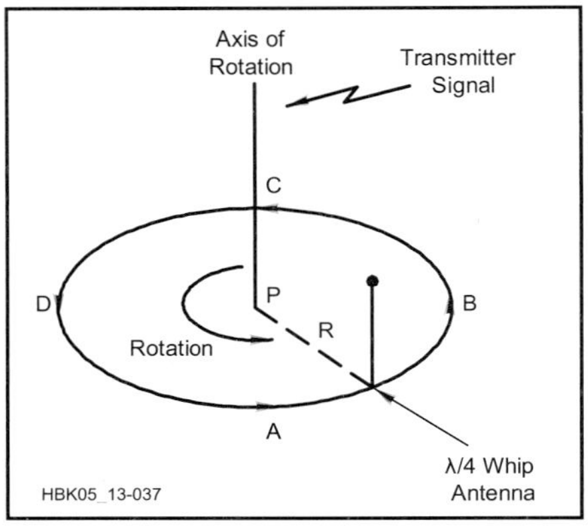

Figure 21.114 — A theoretical Doppler antenna circles around point P, continuously moving toward and away from the source at an audio rate.

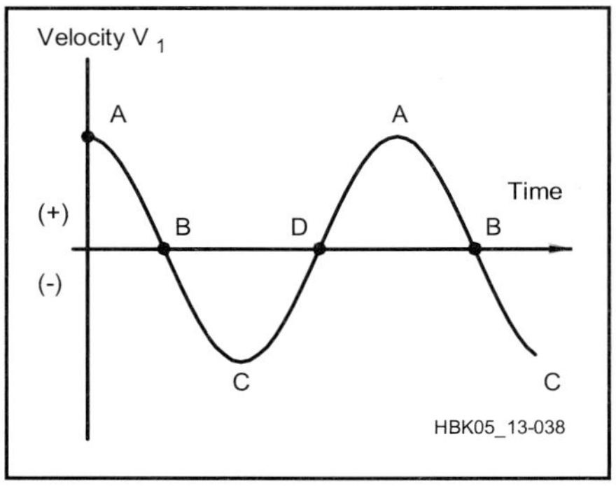

Figure 21.115 — Frequency shift versus time produced by the rotating antenna movement toward and away from the signal source.

DOPPLER RDF SETS

RDF sets using the Doppler principle are popular in many areas because of their ease of use. They have an indicator that instantaneously displays direction of the signal source relative to the vehicle heading, either on a circular ring of LEDs or a digital readout in degrees. A ring of four, eight, or more antennas picks up the signal. Quarter-wavelength monopoles on a ground plane are popular for vehicle use, but half-wavelength vertical dipoles, where practical, perform better.

Radio signals received on a rapidly moving antenna experience a frequency shift due to the Doppler effect, a phenomenon well known to anyone who has observed a moving car with its horn sounding. The horn's pitch appears higher than normal as the car approaches, and lower as the car recedes. Similarly, the received radio frequency increases as the antenna moves toward the transmitter and vice versa. An FM receiver will detect this frequency change.

Figure 21.114 shows a $\frac{1}{4}\lambda$ vertical antenna being moved on a circular track around point P, with constant angular velocity. As the antenna approaches the transmitter on its track, the received frequency is shifted higher. The highest instantaneous frequency occurs when the antenna is at point A, because tangential velocity toward the transmitter is maximum at that point. Conversely, the lowest frequency occurs when the antenna reaches point C, where velocity is maximum away from the transmitter.

Figure 21.115 shows a plot of the component of the tangential velocity that is in the direction of the transmitter as the antenna moves around the circle. Comparing Figures 21.114 and 21.115, notice that at B in Figure 21.115, the tangential velocity is crossing zero from the positive to the negative and the antenna is closest to the transmitter. The Doppler shift and resulting audio output from the receiver discriminator follow the same plot, so that a negative-slope zero-crossing detector, synchronized with the antenna rotation, senses the incoming direction of the signal.

The amount of frequency shift due to the Doppler effect is proportional to the RF frequency and the tangential antenna velocity. The velocity is a function of the radius of rotation and the angular velocity (rotation rate). The radius of rotation must be less than ¼ λ to avoid errors. To get a usable amount of FM deviation (comparable to typical voice modulation) with this radius, the antenna must rotate at approximately 30,000 RPM (500 Hz). This puts the Doppler tone in the audio range for easy processing.

Mechanically rotating a whip antenna at this rate is impractical, but a ring of whips, switched to the receiver in succession with RF PIN diodes, can simulate a rapidly rotating antenna. Doppler RDF sets must be used with receivers having FM detectors. The Dopple ScAnt and Roanoke Doppler (see Bibliography) are mobile Doppler RDF sets designed for inexpensive home construction.

Doppler Advantages and Disadvantages

Ring-antenna Doppler sets are the ultimate in simplicity of operation for mobile RDF. There are no moving parts and no manual antenna pointing. Rapid direction indications are displayed on very short signal bursts.

Many units lock in the displayed direction after the signal leaves the air. Power variations in the source signal cause no difficulties, as long as the signal remains above the RDF detection threshold. A Doppler antenna goes on top of any car quickly, with no holes to drill. Many Local Interference Committee members choose Dopplers for tracking malicious interference, because they are inconspicuous (compared to beams) and effective at tracking the strong vertically polarized signals that repeater jammers usually emit.

A Doppler does not provide superior performance in all VHF RDF situations. If the signal is too weak for detection by the Doppler unit, the hunt advantage goes to teams with beams. Doppler installations are not suitable for on-foot sniffing. The limitations of other switched antenna RDFs also apply: (1) poor results with horizontally polarized signals, (2) no indication of distance, (3) carrier type signals only, and (4) inadvisability of transmitting through the antenna.

Readout to the nearest degree is provided on some commercial Doppler units. This does not guarantee that level of accuracy, however. A well-designed four-monopole set is typically capable of ±5° accuracy on 2 meters, if the target signal is vertically polarized and there are no multipath effects.

The rapid antenna switching can introduce cross modulation products when the user is near strong off-channel RF sources. This self-generated interference can temporarily render the system unusable. While not a common problem with mobile Dopplers, it makes the Doppler a poor choice for use in remote RDF installations at fixed sites with high power VHF transmitters nearby.

21.13 Rotators

The rotator is an important component of directional antenna systems, turning the antenna to any direction with a repeatable accuracy of a few degrees. Once in position, the rotator must hold the antenna in place against the wind. The rotator must do this while supporting the weight of the mast and all of the antennas.

Although mechanical details vary, a rotator consists of a base assembly mounted to a fixed mast or tower and a rotating assembly atop it with a clamp in which the antenna support mast is held. The turning motor and gear train, brake, position indicator, and limit switches are installed in or on the base. The rotating assembly sits on a bearing race resting on the base assembly. A ring gear is the most common method of transferring the motor's rotation to the rotating housing, although worm gears are also used.

Rotators are usually installed inside a lattice tower section that uses a sleeve or thrust bearing to hold a mast and stabilize it against sideways torque. The sleeve is part of the tower's top section. A thrust bearing is mounted on a bearing plate at the top of the tower. The rotator is mounted on a rotator shelf that sits inside the tower and is sold as an accessory by the tower manufacturer. Smaller rotators usually come with a mast clamp so they can be mounted on a pipe or similar mast with the antenna directly above them. **Figure 21.116** shows typical installations.

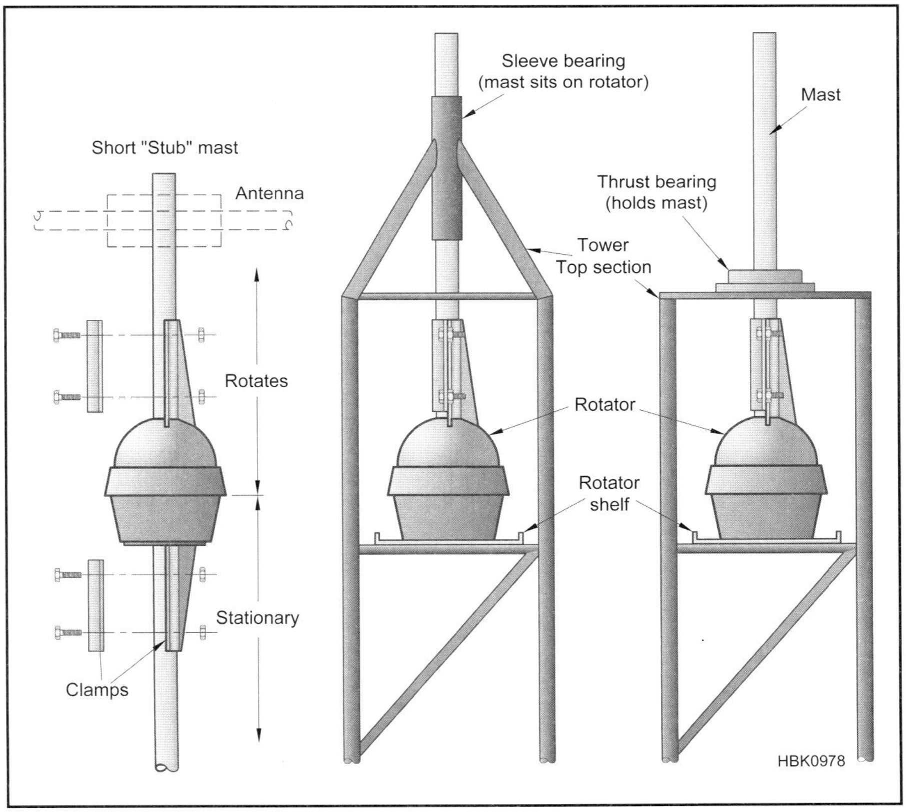

Figure 21.116 — Rotators can be mounted directly on masts or inside lattice-style towers. Rotators can also be mounted directly on the top of towers (not shown).

21.13.1 Rotator Ratings

Rotators are expected to work for many years over a very wide range of temperatures while exposed to the elements, with little or no maintenance. To achieve those performance goals, the rotator's specifications and installation requirements must be respected.

There are three primary rotator ratings: *wind load*, *braking ability*, and *turning torque*. Many rotators also specify a maximum *vertical load* in pounds or kilograms. **Table 21.25** lists manufacturers for rotators intended for fixed-station installations, and **Table 21.26** shows the primary ratings for common rotators.

Wind load is specified both with the rotator mounted inside a tower and with the rotator outside a tower mounted on a mast (see Figure 21.116). When mounted inside a tower section, the tower holds the mast in place straight above the rotator, often using a thrust bearing to hold the mast in place. This eliminates any sideways load on the rotating assembly relative to the base. Wind load ratings are given as a maximum antenna area in square feet, and antenna manufacturers specify the area of their beams for this reason. In the US, wind load calculations are based on the standard EIA/TIA-222-G:2005-08-02. In Europe, EN 1991-1-4 is the current standard, corresponding to the German standard DIN 1055-4.

Braking ability is the maximum twisting force the rotator can withstand when stopped. This force is primarily created by the wind. The rotator's braking action is provided either by a solenoid-controlled wedge or bar inside the housing, or by a worm-gear drive that does not allow backward rotation of the mast under load. Turning torque is the maximum amount of torque the rotator can produce to turn the antennas. Both braking ability and turning torque are given in inch-pounds.

Effective moment is the product of antenna weight in pounds (or kilograms) and turning radius in feet (or meters). Heavy antennas and bigger antennas are harder to turn and to hold in place against the wind, requiring a higher effective moment rating for the rotator. Hy-Gain uses effective moment as a maximum rating for its line of rotators.

It is important not to overload a rotator. If you live in a location that is prone to high winds, persistent winds, or large gusts, include a safety factor when selecting a rotator. Persistent twisting from winds can wear out a rotator's brake wedge or housing indentations that hold the brake in place. This can cause the brake to slip or jam. Rotators are not inexpensive, and a failed rotator brake can allow an antenna to "freewheel," damaging the feed line as well. A thrust bearing at the top of the tower can be used to hold the weight of a large antenna array, leaving only the turning load on the rotator.

Table 21.25
Rotator Manufacturers and Service (Fixed Station)

Manufacturers	
Alpha-Spid	**www.alfaradio.ca**
Channel Master	**www.channelmaster.com**
DX Engineering	**www.dxengineering.com**
Hy-Gain	**www.hy-gain.com**
M^2	**www.m2inc.com**
TIC	**www.ticgen.com**
Yaesu	**www.yaesu.com**
Service	
C.A.T.S.	**www.rotor-doc.com**
Norm's Rotor Service	**www.rotorservice.com**

"Rotator" or "Rotor"?

The piece of equipment installed on the tower that makes the antennas turn is a "rotator." A "rotor" is the rotary part of a motor or vehicle. For example, the blades of a helicopter form its rotor and the spinning shaft and armature of an electric motor form its rotor. The rotator includes the entire machine, both the stationary and moving parts, making the antenna system turn. Amateurs use both words, rotator and rotor, somewhat interchangeably.

Table 21.26
Common Rotator Specifications

Mfr	*Model*	*Wind Load (in tower, sq ft)*	*Wind Load (outside tower, sq ft)*	*Turning Torque (in-lb)*	*Braking Ability (in-lb)*	*Effective Moment (ft-lb)*	*Brake Type*	*Notes*
Alpha-Spid	RAK1			1400	>14,000		Worm gear	Heavy-duty
Channel Master	9521HD			100				Light-duty
DX Engineering	RT4500HD*	>35	n/a	>3300	>20,000		Gear reduction	Heavy-duty
Hy-Gain	T2X	20	10	1000	9000	3400	Wedge	Heavy-duty
	AR-40	3	1.5	350	450	300	Disc	Light-duty
	CD-45II	8.5	5	600	800	1200	Disc	Medium-duty
	HAM-IV	15	7.5	800	5000	2800	Wedge	Medium-duty
	HDR-300A	25	n/a	5000	7500		Solenoid lock	Heavy-duty
M^2	OR2800	35	n/a	3200	17000		Gear reduction	Heavy-duty
TIC	1022D			6500	6500		Gear reduction	Heavy-duty, ring mount
	1032D			7881	7530		Gear reduction	Heavy-duty, ring mount
Yaesu	G-450A	10.8	5.4	516	2604		Gear reduction	Light-duty
	G-800DXA	21.5	8	955	3472		Mech and elec	Medium-duty
	G-1000DXA	23.7	8	955	5207		Mech and elec	Medium-duty
	G-2800DXA	32.3	10.8	2170	21700		Mech and elec	Heavy-duty
Az-El Rotators								
Yaesu	G-5500	10.8	10.8	428	3468		Gear reduction	
Alpha-Spid	RAS-1	30		1400	14,000s		Worm gear	At 12 V
	RAS-2			1400	14,000s		Worm gear	
	REAL-1			1400	14,000s		Worm gear	

*Preliminary specifications; final data not available at publication time.

What is an Armstrong Rotator?

The term "armstrong" refers to anything turned or lifted manually, requiring a "strong arm." There are many examples of hams using gears or cranks to turn antennas but the most common is a rope tied to the antenna's boom and pulled from ground level. The antenna's mast turns freely in the tower or the antenna mount turns on the mast. This is a common temporary solution during antenna system repair or when operating during Field Day and portable.

21.13.2 Types of Rotators

Amateur rotators range from very light-duty models intended for TV antennas all the way to reconditioned "prop-pitch" rotators designed originally to control the pitch or angle of aircraft propeller blades. There are also models for portable and temporary use.

Light-duty antenna rotators are suitable for small VHF and UHF Yagis or log-periodic antennas. They should not be used with HF antennas, large microwave dishes, or antenna systems with a significant wind load.

Medium-duty rotators can handle a single, mid-sized HF tribander or log-periodic. A small VHF/UHF Yagi can be stacked with the HF beam. These rotators are also good choices for a stack of VHF/UHF antennas. Dish antennas should be evaluated to be sure they won't overload these rotators.

Heavy-duty rotators are able to handle the biggest amateur HF Yagis, including stacks of two or three antennas. Instead of a solenoid brake, some of these rotators use a gear train or worm-drive to provide the braking action. Be sure the tower and supporting hardware are rated to handle the antenna and mast load. These rotators are somewhat larger than the more common medium-duty models and may be difficult to install in smaller lattice-style towers.

Ring rotators, or orbital ring rotators, are a special type of rotator installed outside the tower, attached to its legs. The antenna is carried by a motorized cradle that moves around the tower on a circular, toothed track that acts as a drive gear. Ring rotators are generally used at big stations that use large, stacked HF Yagis.

It is also possible to simply rotate the entire tower with a variety of antennas mounted directly on the tower. The guy wires are attached to bearing rings, allowing the tower to turn inside them.

21.13.3 Rotator Control

TURNING CONTROL

There are two steps in controlling turning; releasing the brake, if any, and energizing a motor to turn in the desired direction.

An electrically controlled brake consists of a heavy-duty solenoid and a spring-loaded brake wedge or bar that fits into indentations inside the rotating assembly. The rotator's braking torque is determined by how securely the brake is held by the indentations or, if worm gears are used, by the resistance to the gears turning backward under load. To turn the rotator, the solenoid is energized, pulling the brake out of the indentations. Energizing the solenoid is usually the largest current draw of the rotator.

The most common rotator motor is a 2-phase ac motor with a starting capacitor. The capacitor is switched between phases to control direction of rotation. Because of gear reduction, the motor can be fairly small and does not draw much current. Unless blocked by an obstruction, rotators turn a full 360 degrees. Limit switches open at the ends of rotation, removing power from the motor at the extreme ends of travel to prevent feed line damage.

When rotation is complete, the solenoid is de-energized and the brake reengages the indentations, holding the mast in place. Over time, the indentations or brake can wear out, allowing the rotating housing to slip under heavy loads. Wear is accelerated by de-energizing the solenoid while the mast is still turning, causing the brake to impact the sides of the indentations. To reduce the impact of sudden stops, some controllers allow the mast to stop turning before de-energizing the solenoid. Typical brake delays are about 5 seconds. (Retrofit delay modules are available for the Hy-Gain family of rotators.) If your rotator does not have a brake delay, practice keeping the brake energized for a few seconds after you release the turning controls to allow the antennas to stop moving first.

POSITION INDICATION

There are two basic types of position indication — resistance and pulse counting. The most common circuit is a potentiometer ("pot") contained in the rotator's housing and turned in sync with the rotator motor. Current through the pot (typically only a few milliamps) drives an analog meter in the control unit calibrated in degrees. Rotators using pulse counters use a switch to generate pulses that the control unit counts to calculate the number of degrees from one end of travel.

Most rotator controllers in North America are configured as "North center," meaning they can turn an antenna from pointing directly south at one limit, through north at mid-travel, and all the way to south again at the opposite limit. North is the center position on the meter displaying the antenna's direction. (South center meter scales are an option for most rotators.)

To calibrate a pot-indicator rotator's direction assuming North center, first, calibrate the direction indicator. Then move the rotator to its mid-travel orientation so that it indicates north. Loosen the antenna mast clamp and rotate the mast until the antennas point directly north and re-tighten the clamp. (Using a compass or landmark is sufficient resolution for most amateur antennas.) When storing or testing a rotator, make a practice of leaving it set to mid-travel for ease of position calibration and mark or tag it for later reference. For a pulse-count rotator, the manufacturer's manual will provide the necessary directions.

ROTATOR WIRING

The ARRL appreciates being granted permission by the Hy-Gain Company to reproduce the wiring diagram of the control unit for its widely used Ham-IV and Tailtwister T2X rotators as a convenience for readers. (See **Figure 21.117.**)

Rotator Software Control

Recently-designed rotator control units have RS-232 or USB interfaces that act as COM ports to PC software. There are several different protocols, the Yaesu protocol being the most common. The Rotor-EZ and ERC interfaces can be added to most rotators that don't have a software interface. Logging software often supports several different protocols and standalone software packages and utilities are also available—internet searches for "antenna rotator software control" will find many programs.

One Control for All

Each rotator family (Ham-IV, Yaesu, M^2, and so on) comes with a custom control unit for turning and position display. There are also after-market control units that operate with any of the common rotators. The most widely used are the Green Heron controllers (**www.greenheronengineering.com**) and the EA4TX interfaces (**ea4tx.com**). Both can control most available models of rotators, allowing you to standardize in the station and customize on the tower.

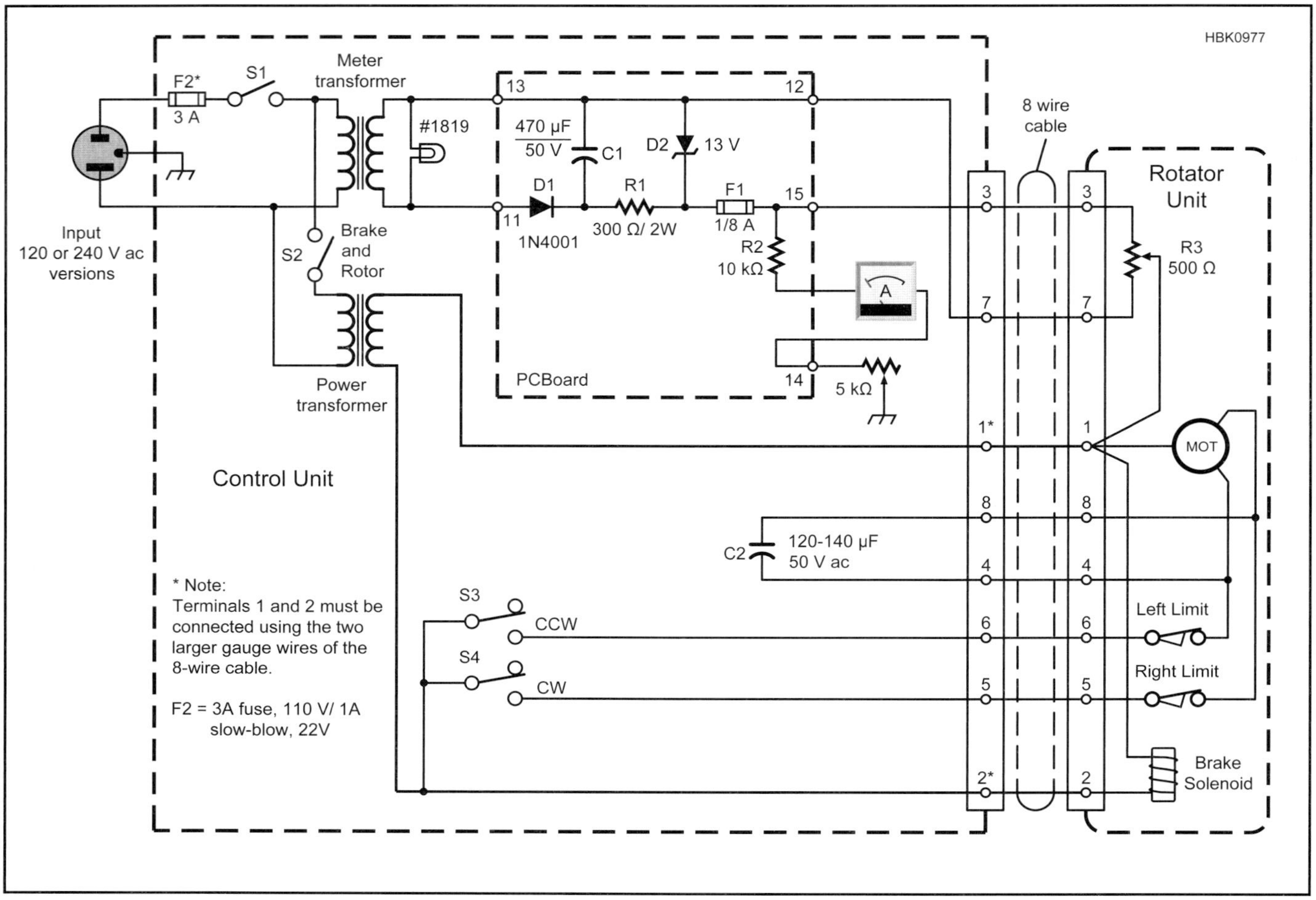

Figure 21.117 — The Hy-Gain Ham-IV/IVX control unit schematic. This control unit will also work with T2X rotators. (Circuit provided courtesy of Hy-Gain, Inc.)

The connection from the control unit to the rotator requires a multi-conductor control cable (no shield necessary). Most rotators require either 6- or 8-conductor cable. Solenoid brake circuits need heavier wire due to the higher current. If the wire used is too small, the extra resistance may cause enough voltage drop to result in erratic brake operation or slow turning. Check the manufacturer's recommendation for minimum wire size, which depends on the length of the cable. For the popular Ham-IV series, minimum recommended wire sizes are:

- Up to 125 feet: #18 (solenoid), #20 (all others)
- 125 to 200 feet: #16 (solenoid), #18 (all others)
- 200 to 300 feet: #14 (solenoid), #16 (all others)

Other rotators have similar requirements — consult the manufacturer's instructions.

At the rotator, connections can be made directly to a terminal strip under the unit or to a weatherproofed connector at the end of a short "pigtail." A nearby junction box where the pigtail and cable are attached can also be used.

Table 21.27

Ham-IV/IVX Connector and Cable Wiring

Pin numbers, colors, and resistance values from Ham-IV/IVX Instruction Manual

Pin	*Color*	*Circuit*
1	Black (heavy wire)	Solenoid and common
2	White (heavy wire)	Solenoid
3	Green	Position pot element, + end
4	Blue	Motor winding 1
5	Orange	Right limit switch
6	Yellow	Left limit switch
7	Brown	Position pot element, – end
8	Red	Motor winding 2

Resistance Checks — Read Between Terminals

1-2	Brake solenoid	0.75 Ω + cable or leads
1-8	½ Motor winding	2.5 Ω + cable or leads
1-4	½ Motor winding	2.5 Ω+ cable or leads
1-6	½ Motor + switch	2.5 Ω + cable or leads
1-5	½ Motor + switch	2.5 Ω + cable or leads
8-4	Entire motor	5 Ω + cable or leads
8-5	Right limit switch	0 Ω + cable or leads
4-6	Left limit switch	0 Ω + cable or leads
3-7	Entire pot element	500 Ω
3-1	Pot wiper to element end 1	0 to 500 Ω
7-1	Pot wiper to element end 2	0 to 500 Ω

Note: readings 3-1 and 7-1 should add to 3-7 reading

Retrofit kits to replace terminal strips are available from several vendors. The Ham-IV family suggests an 8-pin Cinch-Jones connector with wiring and color code as shown in **Table 21.27**.

For terminal strip connections, use a weatherproofing grease found at electrical and automotive stores. Run the connecting cable down and away from the rotator to guide water away from it. Use a consistent color code at the rotator and the control unit.

Lightning protection is also recommended for rotator control cables. Several vendors make 8-line lightning protectors for rotators. The protector must be well grounded and should be installed at a common entry point for all antenna system cables. See the **Safety** chapter for more about grounding and lightning protection practices.

Troubleshooting a rotator electrically can be done from the ground through resistance checks. Table 21.27 shows the nominal resistance values for the Ham-IV family of rotators. When making resistance checks, the control unit must be disconnected. Include the resistance of the cable or test leads when making resistance checks.

Antenna Material Tables

The Antenna Material Tables now appear in the online content for this chapter. The tables include:

- Copper Wire Specifications
- Antenna WIre Strength
- Guy Wires Lengths to Avoid
- Aluminum Alloy Specifications

21.15 References and Bibliography

Adler, S., "A High-Power 160- and 80-Meter Transmitting Loop Antenna," *QST*, Apr. 2019, pp. 30 – 36.

Belrose, J., VE2CV, "Short Antennas for Mobile Operation," *QST*, Sep. 1953, pp. 30 – 35.

Black, B., W4SSY, "The W4SSY Spudgun," *QST*, Mar. 2009, pp. 67 – 69.

Brown, G., "The Phase and Magnitude of Earth Currents Near Radio Transmitting Antennas," *Proc IRE*, Feb. 1935.

Brown, G., Lewis R., and Epstein, J., "Ground Systems as a Factor in Antenna Efficiency," *Proc IRE*, Jun. 1937, pp. 753 – 787.

Brown G., and Woodward, O., "Experimentally Determined Impedance Characteristics of Cylindrical Antennas," *Proc IRE*, Apr. 1945.

Buchanen, C., W3DZZ, "The Multimatch Antenna System," *QST*, Mar. 1955, p. 22.

Cebik, L., W4RNL, "A Simple Fixed Antenna for VHF/UHF Satellite Work," *QST*, Aug. 2001, pp. 38 – 42.

Christman, A., "Elevated Vertical Antenna Systems," *QST*, Aug. 1988, pp. 35 – 42.

Clement, J., VE6AB, "Gain Twist 75 Meter Mobile Monobander," *QST*, Jul. 2011, pp. 39 – 42.

Devoldere, J., *ON4UN's Low-Band DXing*, 5th Edition (ARRL, 2011).

Dome, R., "Increased Radiating Efficiency for Short Antennas," *QST*, Sep. 1934, pp. 9 – 12.

Doty, A., Frey J., and Mills, H., "Characteristics of the Counterpoise and Elevated Ground Screen," Professional Program, Session 9, Southcon '83 (IEEE), Atlanta, GA, Jan. 1983.

Doty, A., Frey, J., and Mills, H., "Efficient Ground Systems for Vertical Antennas," *QST*, Feb. 1983, pp. 20 – 25.

Doty, A., technical paper presentation, "Capacitive Bottom Loading and Other Aspects of Vertical Antennas," Technical Symposium, Radio Club of America, New York City, Nov. 20, 1987.

Doty, A., Frey, J., and Mills, H., "Vertical Antennas: New Design and Construction Data," *The ARRL Antenna Compendium, Volume 2* (ARRL, 1989), pp. 2 – 9.

Fosberg, R., "Some Notes on Ground Systems for 160 Meters," *QST*, Apr. 1965, pp. 65 – 67.

Grammer, G., "More on the Directivity of Horizontal Antennas; Harmonic Operation — Effects of Tilting," *QST*, Mar. 1937, pp. 38 – 40, 92, 94, 98.

Green, H., "Design Data for Short and Medium Length Yagi-Uda Arrays," *Trans IE Australia*, Vol EE-2, No. 1, Mar. 1966.

Griffith, A., W4ULD, "Capacitance Hats for HF Mobile Antennas," *QEX*, Jul./Aug. 1996, p. 16.

Hansen, R., "Carter Dipoles and Resonant Dipoles," Allerton Antenna Applications Symposium, pp. 282 – 287, Monticello, IL, Sep 21 – 23, 2010.

Leeson, D., "The Story of the Broadband Dipole," *QEX*, Nov./Dec. 2018, pp. 18 – 22.

McCleish, K., W7TX, "Why An Antenna Radiates," *QST*, Nov. 1992, pp. 59 – 63.

Mills, H., technical paper presentation, "Impedance Transformation Provided by Folded Monopole Antennas," Technical Symposium, Radio Club of America, New York City, Nov. 20, 1987.

Myers, B., "The W2PV Four-Element Yagi," *QST*, Oct. 1986, pp. 15 – 19.

Reisert, J., "The 3/8-Wavelength Vertical — A Hidden Gem", *QST*, Apr. 2019, pp. 44 – 47.

Richard, L., "Parallel Dipoles of 300-Ohm Ribbon," *QST*, Mar. 1957.

Richmond, J., "Monopole Antenna on Circular Disc," *IEEE Trans on Antennas and Propagation*, Vol. AP-32, No. 12, Dec 1984.

Richter, J., "Off-Center-Fed Dipole Antennas: Theory and Practice," *HF Dipole Antennas for Amateur Radio* (ARRL, 2019) pp. 105 – 123.

Roberts, B., "Input Impedance of a Folded Dipole," RCA Review, Jun 1947.

Schelkunoff, S., "Theory of Antennas of Arbitrary Size and Shape," *Proc. IRE*, Sep. 1941; corrections Nov. 1941 and Jan. 1943. Republished in *Proc. IRE*, September 1984.

Schulz, W., "Designing a Vertical Antenna," *QST*, Sep. 1978, pp. 19 – 21.

Sevick, J., "The Ground-Image Vertical Antenna," *QST*, Jul. 1971, pp. 16 – 17, 22.

Sevick, J., "The W2FMI 20-Meter Vertical Beam," *QST*, Jun. 1972, pp. 14 – 18.

Sevick, J., "The W2FMI Ground-Mounted Short Vertical," *QST*, Mar. 1973, pp. 13 – 18, 41.

Sevick, J., "A High Performance 20-, 40- and 80-Meter Vertical System," *QST*, Dec. 1973.

Sevick, J., "Short Ground-Radial Systems for Short Verticals," *QST*, Apr. 1978, pp. 30 – 33.

Siwiak, K., and Quick R., "Small Gap-resonated HF Loop Antenna Fed by a Secondary Loop," *QEX*, Jul./Aug. 2018, pp. 12 – 17.

Smith C., and Johnson, E., "Performance of Short Antennas," *Proc IRE*, Oct. 1947.

Stanley, J., "Optimum Ground Systems for Vertical Antennas," *QST*, Dec. 1976, pp. 13 – 15.

Stearns, S., K6OIK, "Antenna Modeling for Radio Amateurs," presented at ARRL Pacificon Antenna Seminar, 2017, archived at **www.fars.k6ya.org/docs**.

Stearns, S., K6OIK, "Antennas: The Story from Physics to Computational Electromagnetics," Antenna Seminar, ARRL Pacificon, San Ramon, CA, Oct. 19 – 21, 2019.

Stephens, R., "Admittance Matching the Ground-Plane Antenna to Coaxial Transmission Line," Technical Correspondence, *QST*, Apr. 1973, pp. 55 – 57.

Sumner, D., "Cushcraft 32-19 'Boomer' and 324-QK Stacking Kit," Product Review, *QST*, Nov. 1980, pp. 48 – 49.

Sykes, B., "Skeleton Slot Aerials," *RSGB Bulletin*, Jan. 1953.

Tai C., and Long, S., "Dipoles and Monopoles," Chapter 4 in *Antenna Engineering Handbook*, 3rd Edition, Johnson, R., Ed. (McGraw-Hill, 1993).

Roberts, W. Van B., "Input Impedance of a Folded Dipole," RCA Review, Jun. 1947.

Williams, E., "Radiating Characteristics of Short-Wave Loop Aerials," *Proc IRE*, Oct. 1940.

Witvliet B., et al, "Near Vertical Incidence Skywave Propagation: Elevation Angles and Optimum Antenna Height for Horizontal Dipole Antennas," *IEEE Antennas and Propagation Magazine*, Vol. 57, No. 1, Feb. 2015, pp. 129 – 146.

TEXTBOOKS ON ANTENNAS

Balanis, C., *Antenna Theory, Analysis and Design* (Harper & Row, 1982).

Bond, D., *Radio Direction Finders*, 1st ed. (New York: McGraw-Hill Book Co).

Caron, W., *Antenna Impedance Matching* (ARRL, 1989).

Cebik, L., *Antennas from the Ground Up, Volume 1 and Volume 2*, (MFJ Publishing, 2000).

Davies, K., *Ionospheric Radio Propagation* — National Bureau of Standards Monograph 80 2 (U.S. Government Printing Office, 1965).

Elliott, R., *Antenna Theory and Design* (Prentice Hall, 1981).

Harper, A., *Rhombic Antenna Design* (D. Van Nostrand Co, Inc, 1941).

Henney, K., *Principles of Radio* (John Wiley and Sons, 1938).

Hutchinson C., and Straw, R., *Simple and Fun Antennas for Hams* (ARRL, 2002).

Jasik, H., *Antenna Engineering Handbook*, 1st Edition (McGraw-Hill, 1961).

Johnson, W., *Transmission Lines and Networks*, 1st Edition (McGraw-Hill, 1950).

Johnson and Jasik, *Antenna Engineering Handbook*, 2nd Edition (McGraw-Hill, 1961).

Johnson, R., *Antenna Engineering Handbook*, 3rd Edition (McGraw-Hill, 1993).

Jordan E., and Balmain, K., *Electromagnetic Waves and Radiating Systems*, 2nd Edition (Prentice-Hall, 1968).

Keen, R., *Wireless Direction Finding*, 3rd Edition (Wireless World, 1938).

King, R., *Theory of Linear Antennas* (Harvard University Press, 1956).

King, R., Mimno, H., and Wing, A., *Transmission Lines, Antennas and Waveguides* (Dover Publications, 1965).

King, Mack and Sandler, *Arrays of Cylindrical Dipoles* (London: Cambridge Univ Press, 1968).

Knitter, M., Ed., *Loop Antennas — Design and Theory* (National Radio Club, 1983).

Knitter, M., Ed., *Beverage and Long Wire Antennas — Design and Theory* (National Radio Club, 1983).

Kraus, J., *Electromagnetics* 4th Edition (McGraw-Hill Book Co, 1991).

Kraus, J., *Antennas*, 2nd Edition (McGraw-Hill, 1988).

Laport, E., *Radio Antenna Engineering* (McGraw-Hill, 1952).

Lawson, J., *Yagi-Antenna Design*, 1st Edition (ARRL, 1986).

Lee, P., *The Amateur Radio Vertical Antenna Handbook*, 2nd Edition (Cowen Publishing Co., 1984).

Leeson, D., *Physical Design of Yagi Antennas* (ARRL, 1992).

Lowe, A., *Reflector Antennas* (IEEE Press, 1978).

Maxwell, M., *Reflections — Transmission Lines and Antennas* (ARRL, 1990).

Maxwell, M., *Reflections II — Transmission Lines and Antennas* (Worldradio Books, 2001).

Miller, G., *Modern Electronic Communication* (Prentice Hall, 1983).

Misek, V., *The Beverage Antenna Handbook* (V. A. Misek, 1977).

Moreno, T., *Microwave Transmission Design Data* (McGraw-Hill, 1948).

Moxon, L., *HF Antennas for All Locations* (Radio Society of Great Britain, 1982).

Ramo and Whinnery, *Fields and Waves in Modern Radio* (John Wiley & Sons, 1944).

Rumsey, V., *Frequency Independent Antennas* (Academic Press, 1966).

Saveskie, P., *Radio Propagation Handbook* (Tab Books, Inc, 1980).

Schelkunoff, S., *Advanced Antenna Theory* (John Wiley & Sons, 1952).

Schelkunoff S., and Friis, H., *Antennas Theory and Practice* (John Wiley & Sons, 1952).

Sevick, J., *Transmission Line Transformers* (Noble Publishing, 1996).

Skilling, H., *Electric Transmission Lines* (McGraw-Hill Book Co, 1951).

Slurzburg, M., and Osterheld, W., *Electrical Essentials of Radio* (McGraw-Hill Book Co, 1944).

Southworth, G., *Principles and Applications of Waveguide Transmission* (D. Van Nostrand Co, 1950).

Straw, R., Ed., *The ARRL Antenna Book*, 21st ed. (Newington: ARRL, 2007).

Terman, F., *Radio Engineers' Handbook*, 1st ed. (McGraw-Hill, 1943).

Terman, F., *Radio Engineering*, 3rd Edition (McGraw-Hill, 1947).

Uda S., and Mushiake, Y., *Yagi-Uda Antenna* (Sasaki Publishing Co, 1954). [Published in English — Ed.]

Viezbicke, P., "*Yagi Antenna Design*," NBS Technical Note 688 (US Dept of Commerce/National Bureau of Standards, Boulder, CO), Dec. 1976.

Welch, G., *Wave Propagation and Antennas* (D. Van Nostrand Co, 1958).

Zavrel, R., *Antenna Physics: An Introduction* (ARRL, 2016).

The GIANT Book of Amateur Radio Antennas (Tab Books, 1979).

IEEE Standard Dictionary of Electrical and Electronics Terms, 3rd Edition (IEEE, 1984).

Radio Broadcast Ground Systems, available from Smith Electronics, Inc, 8200 Snowville Rd, Cleveland, OH 44141.

Radio Communication Handbook, 5th Edition (RSGB, 1976).

RDF BIBLIOGRAPHY

Bohrer, "Foxhunt Radio Direction Finder," *73 Amateur Radio*, Jul. 1990, p. 9.

Bonaguide, "HF DF — A Technique for Volunteer Monitoring," *QST*, Mar. 1984, p. 34.

DeMaw, "Maverick Trackdown," *QST*, Jul. 1980, p. 22.

Dorbuck, "Radio Direction Finding Techniques," *QST*, Aug. 1975, p. 30.

Eenhoorn, "An Active Attenuator for Transmitter Hunting," *QST*, Nov. 1992, p. 28.

Flanagan and Calabrese, "An Automated Mobile Radio Direction Finding System," *QST*, Dec. 1993, p. 51.

Geiser, "A Simple Seeker Direction Finder," *ARRL Antenna Compendium, Volume 3*, p. 126.

Gilette, "A Fox-Hunting DF Twin'Tenna," *QST*, Oct. 1998, pp. 41 – 44.

Hunt, "A Simple Direction-Finding Receiver for 80 Meters," *QST*, Sep. 2005, pp. 36 – 42.

Johnson and Jasik, *Antenna Engineering Handbook*, Second Edition, (McGraw-Hill, 1993).

Kossor, "A Doppler Radio-Direction Finder," *QST*, Part 1, May 1999; Part 2, Jun. 1999.

McCoy, "A Linear Field-Strength Meter," *QST*, Jan. 1973, p. 18.
Moell and Curlee, Transmitter Hunting: Radio Direction Finding Simplified, Blue Ridge Summit, PA: TAB/McGraw-Hill. (This book, available from ARRL, includes plans for the Roanoke Doppler RDF unit and in-line air attenuator, plus VHF quads and other RDF antennas.)
Moell, "Transmitter Hunting — Tracking Down the Fun," *QST*, Apr. 1993, p. 48 and May 1993, p. 58.
Moell, "Build the Handy Tracker," *73 Magazine*, Sep. 1989; Nov. 1989.
O'Dell, "Simple Antenna and S-Meter Modification for 2-Meter FM Direction Finding," *QST*, Mar. 1981, p. 43.
O'Dell, "Knock-It-Down and Lock-It-Out Boxes for DF," *QST*, Apr. 1981, p. 41.
Ostapchuk, "Fox Hunting is Practical and Fun!" *QST*, Oct. 1998, pp. 68 – 69.
Rickerd, "A Cheap Way to Hunt Transmitters," *QST*, Jan. 1994, p. 65.

RDF RESOURCES

Homing In **www.homingin.com** — website by KØOV on direction finding techniques and activities Amateur Radio Direction Finding (IARU Region II)

www.ardf-r2.org/iaru/ — ARDF activities and organizations in IARU Region II Radio Direction Finding

en.wikipedia.org/wiki/Direction_finding — a general site on RDF with links to related subjects DX Zone RDF Links

www.dxzone.com/catalog/Operating_Modes/Radio_Direction_Finding — a page of links to RDF articles and websites

Contents

Chapter 22 — Online Content

Articles
- *Electric Current Abroad* — U.S. Dept of Commerce
- Field Day Tower Safety by Don Daso, K4ZA and Ward Silver, NØAX
- How to Interact with a Concerned Neighbor
- Interpreting news about RF Exposure Discoveries
- RF Safety at Field Day by Greg Lapin, N9GL
- RF Safety Standard Development
- RF Surge Suppressor Ratings for Transmissions into Reactive Loads by Gene Hinkle, K5PA
- Shop Safety by Don Daso, K4ZA
- Types of Scientific Studies

Chapter 22
Safe Practices

Previously named "Safety," this chapter focuses on safe practices to avoid common hazards associated with ac electrical power, antennas and towers, and RF exposure. The first section, updated by Jim Lux, W6RMK, Ward Silver, NØAX, and Greg Lapin, N9GL details electrical safety, grounding, bonding, batteries, and related issues in the station. The following section on antenna and tower safety was updated by Jim Idelson, K1IR. Finally, Greg Lapin, N9GL and the ARRL RF Safety Committee updated the explanations of good amateur practices, standards, and FCC regulations as they apply to RF exposure, including recent updates to station evaluation requirements.

Safety First — Always

We need to learn as much as possible about what could go wrong so we can avoid factors that might result in accidents. Amateur Radio activities are not inherently hazardous, but like many things in modern life, it pays to be informed. Stated another way, while we long to be creative and innovative, there is still the need to act responsibly. Safety begins with our attitude. Make it a habit to plan work carefully. Don't be the one to say, "I didn't think it could happen to me."

Having a good attitude about safety is not enough, however. We must be knowledgeable about common safety guidelines and follow them faithfully. Safety guidelines cannot possibly cover all situations, but if we approach each task with a measure of common sense, we should be able to work safely.

Involve your family in Amateur Radio. Having other people close by is always beneficial in the event that you need immediate assistance. Take the valuable step of showing family members how to turn off the electrical power to your equipment safely. Additionally, cardiopulmonary resuscitation (CPR) training can save lives in the event of electrical shock. Classes are offered in most communities. Take the time to plan with your family members exactly what action should be taken in the event of an emergency, such as electrical shock, equipment fire or power outage. Practice your plan!

22.1 Electrical Safety

The standard power available from commercial mains in the United States for residential service is 120/240 V ac. The "primary" voltages that feed transformers in our neighborhoods may range from 1,300 to more than 10,000 V. Generally, the responsibility for maintaining the power distribution system belongs to a utility company, electric cooperative, or city. The "ownership" of conductors usually transfers from the electric utility supplier to the homeowner where the power connects to the meter or weather head. If you are unsure of where the division of responsibility falls in your community, a call to your electrical utility will provide the answer. **Figure 22.1** shows the typical division of responsibility between the utility company and the homeowner. This section is concerned more with wiring practices in the station, as opposed to within the equipment in the station.

There are two facets to success with electrical power: safety and performance. Since we are not professionals, we need to pursue safety first and consult professionals for alternative solutions if performance is unacceptable. The ARRL's Volunteer Consulting Engineers program involves professional engineers who may be able to provide advice or direction on difficult problems.

22.1.1 Station Concerns

There never seem to be enough power outlets in your station. A good solution for small scale power distribution is a switched power strip with multiple outlets. The strip should be listed by a nationally recognized testing laboratory (NRTL) such as Underwriters Lab (UL) and should incorporate a circuit breaker. See the sidebars "What Does UL Listing Mean?" and "How Safe are Outlet Strips?" for warnings about poor quality products. It is poor practice to "daisy-chain" several power strips and may actually be a code violation. If you need more outlets than are available on a strip, have additional wall outlets installed.

Whether you add new outlets or use power strips, be sure not to overload the circuit. National and local codes set permissible branch capacities according to a rather complex process. Here's a safe rule of thumb: consider adding a new circuit

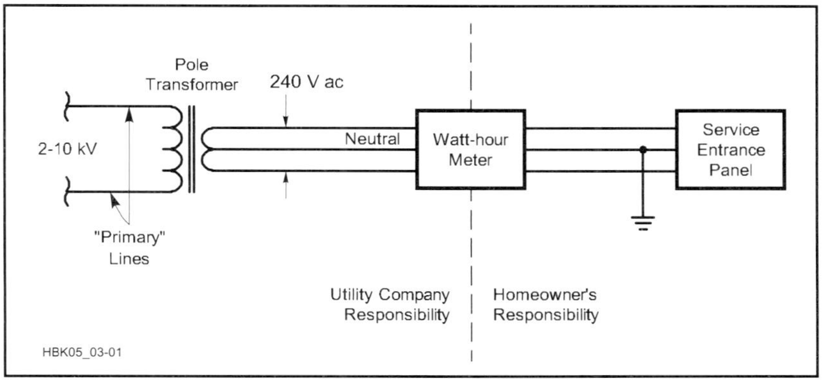

Figure 22.1 — Typical division of responsibility for maintenance of electrical power conductors and equipment. The meter is supplied by the utility company.

if the total load is more than 80% of the circuit breaker or fuse rating. (This assumes that the fuse or breaker is correct. If you have any doubts, have an electrician check it.)

22.1.2. Do-It-Yourself Wiring

Amateurs sometimes "rewire" parts of their homes to accommodate their hobby. Most local codes *do* allow for modification of wiring (by building owners), so long as the electrical codes are met. Before making changes to your wiring, it would be wise to determine what rules apply and what agency has the authority to enforce them. This is called the *authority having jurisdiction* (AHJ) and it varies from location to location. Also see the following section on the National Electrical Code.

Generally, the building owner must obtain an electrical permit before beginning changes or additions to permanent wiring. Some jobs may require drawings of planned work. Often, the permit fee pays for an inspector to review the work. Considering the risk of injury or fire if critical mistakes are left uncorrected, a permit and inspection are well worth the effort. *Don't take chances* — seek assistance from the building officials or an experienced electrician if you have *any* questions or doubts about proper wiring techniques.

Ordinary 120 V circuits are the most common source of fatal electrical accidents. Line voltage wiring must use an approved cable, be properly installed in conduit and junction boxes, within a chassis with a cover or lid, or other means described in the electrical code. But don't assume that high-current, low-voltage power sources, such as car batteries and high-current power supplies, are inherently safe. They can be just as dangerous as high-voltage sources, from melting metal, sparks, and short circuits.

Never work on electrical wiring with the conductors or system energized! Switch off the circuit breaker, remove the fuse, or disconnect the power source and take positive steps to ensure that others do not restore the power while you are working, such as using a circuit-breaker lockout. (**Figure 22.2** illustrates one way to ensure that power will be off until you want it turned on.) Check the circuit with a voltmeter to be sure that it is "dead" *each time you begin work.*

Before restoring power, check the wiring with an ohmmeter: From an installed ac outlet, there should be good continuity between the neutral conductor (white wire, "silver" screw) and the grounding conductor (green or bare wire, green screw). An ohmmeter should indicate a closed circuit between the conductors. (In the power line, high voltage world, line workers apply a shorting jumper before starting work so if the power does get reapplied, the safety jumper takes the hit.) For dc systems, test the wiring at the source. Resistance for a dc system can be quite low, a few ohms in some cases. By switching loads on and off as you measure, you should see a change in the system resistance.

With all other loads removed from the circuit (by turning off or unplugging them), an ohmmeter should indicate an *open* circuit between the hot wire and either of the other two conductors. There should be no continuity between the hot conductor (black wire, "brass" screw) and the grounding or neutral conductor.

Commercially available plug-in testers

Grounding and Bonding for the Radio Amateur

There is so much information about ac safety, lightning protection, and dealing with RF in the station that it can be difficult to understand it all. While this chapter is a helpful summary, it is still just a summary. To help hams setting up a station for the first time or trying to improve an existing station, the ARRL has published *Grounding and Bonding for the Radio Amateur*. Along with its website (**www.arrl.org/grounding-and-bonding-for-the-amateur**) the book collects information about these important practices into one reference.

How Safe Are Outlet Strips?

The switch in outlet strips is generally *not* rated for repetitive *load break* duty. Early failure and fire hazards may result from using these devices to switch heavy loads on and off. Misapplications are common (another bit of bad technique that has evolved from the use of personal computers), and manufacturers are all too willing to accommodate the market with marginal products. A lockable disconnect switch or circuit breaker is a better and safer station master switch.

Older power strips not complying with current standards can also be a safety hazard. MOVs in these older strips that are subjected to repeated transients can fail and cause a fire hazard, especially in outlet strips with plastic enclosures. Power strips made after 2009 that comply with UL standards are safe to use.

What Does UL Listing Mean?

UL is one of several *nationally recognized testing laboratories* (NRTLs), and probably the most well-known. Listing *does not* mean what most consumers expect it to mean! More often than not the listing *does not* relate to the performance of the listed product. The listing simply indicates that a sample of the device meets certain manufacturers' construction criteria. Similar devices from the same or different manufacturers may differ significantly in overall construction and performance even though all are investigated and listed against the same UL product category. There is also a difference between a listed device and a listed component.

Many local laws and regulations, as well as the National Electrical Code, require that equipment and components used in electrical installations be listed by a NRTL. Some jurisdictions (Los Angeles County) require that any electrical equipment sold to consumers be listed.

The consumer must also be aware of the fine distinctions in advertising between a device or component that is advertised as "listed" or "designed to meet" or "meets." The latter two may not actually have been tested, or if tested, may have been tested by the manufacturer, and not an independent body.

It's also important to know that in some cases UL (and other standards organizations) only publish a standardized test procedure, but don't necessarily list or test the devices. Many standards also define varying levels of compliance, so knowing that your device meets some part of the standard may not be enough to know whether it meets *your* particular needs.

Figure 22.2 — If the switch box feeding power to your shack is equipped with a lock-out hole, use it. With a lock through the hole on the box, the power cannot be accidentally turned back on. [Photo courtesy of American ED-CO]

are a convenient way to test regular three-wire receptacles but can't distinguish between the neutral and ground being reversed. Some plug-in testers draw enough current between hot and safety ground to trip a GFCI breaker.

It is wise to have an up-to-date reference guide to wiring practices. These guides may be available from home improvement stores and electrical supply houses. As of 2022, one widely available book is the *Black & Decker Complete Guide to Wiring, 7th edition*. The book is inexpensive and includes many step-by-step instructions for basic and intermediate wiring projects. It has been updated to the 2020 NEC version.

How Does the NEC Affect Me?

Exactly how does the National Electrical Code become a requirement? How is it enforced?

States, cities, and political subdivisions (jurisdictions) have the responsibility to act for the safety and welfare of the public. To address the safety and welfare of everyone, specific codes may be adopted by these jurisdictions. Because the technology for the development of general construction, mechanical, and electrical codes is beyond the scope of most jurisdictions, existing codes are adopted. These codes may be adopted by reference as they stand, or they may be amended and incorporated into local laws and ordinances. Many jurisdictions include some form of permit and accompanying inspection process within the local laws and ordinances. These codes may include, but are not limited to, the International Building Code (IBC), International Residential Code (IRC), and the National Electrical Code (NEC). For electrical issues, the National Electrical Code has been adopted by every state and most local jurisdictions within the United States. Officials within these jurisdictions will serve as "the authority having jurisdiction" (AHJ) and interpret the provisions of these Codes as they apply to specific cases.

Building codes differ from planning or zoning regulations: Building codes are directed only at safety, fire, and health issues. Zoning regulations often are aimed at preservation of property values and aesthetics.

The NEC is part of a series of reference codes published by the National Fire Protection Association, a nonprofit organization. Published codes are regularly kept up-to-date and are developed by a series of technical committees whose makeup represents a wide consensus of opinion. The NEC is updated every three years. It's important to know which version of the code your local jurisdiction uses, since it's not unusual to have the city require compliance to an older version of the code. Fortunately, the NEC is usually backward compatible: that is, if you're compliant to the 2008 code, you're probably also compliant to the 1999 code.

Do I have to update my electrical wiring as code requirements are updated or changed?

Generally, no. Codes are typically applied for new construction and for renovating existing structures. Room additions, for example, might not directly trigger upgrades in the existing service panel unless the panel was determined to be inadequate. However, the wiring of the new addition would be expected to meet current codes. Prudent homeowners, however, may want to add safety features for their own value. Many homeowners, for example, have added GFCI protection to bathroom and outdoor convenience outlets.

22.1.3 National Electrical Code (NEC)

Fortunately, much has been learned about how to harness electrical energy safely. This collective experience has been codified into the *National Electrical Code*, or *NEC*, simply known as "the code." The code details safety requirements for many kinds of electrical installations. Compliance with the NEC provides an installation that is *essentially* free from hazard, but not necessarily efficient, convenient, or adequate for good service (paraphrased from NEC Article 90-1a and b). While the NEC is national in nature and sees wide application, it is not universal.

Local building authorities set the codes for their area of jurisdiction. They often incorporate the NEC in some form, while considering local issues. For example, Washington State specifically exempts telephone, telegraph, radio and television wires and equipment from conformance to electrical codes, rules, and regulations. However, some local jurisdictions (city, county and so on) do impose a higher level of installation criteria, including some of the requirements exempted by the state.

Code interpretation is a complex subject, and untrained individuals should steer clear of the NEC itself. The NEC is not written to be understood by do-it-yourselfers, and one typically must look in several places to find *all* the requirements. (For instance, Articles 810, 250, *and* 100 all contain things applicable to typical Amateur Radio installations.) The *NEC Handbook* is a version of the code with additional drawings and discussion. It explains the requirements of the code and how to satisfy those requirements. Written for electricians, even the *NEC Handbook* may be difficult for the non-electrician to understand completely. You may wish to contact local sources of information about code compliance and acceptable practices such as local building officials or inspectors, electrical engineers, and practicing electricians. The NEC and the *NEC Handbook* are available from local libraries.

The internet has a lot of information about electrical safety, the electrical code, and wiring practices, but you need to be careful to make sure the information you are using is current and not out of date. The ARRL Volunteer Consulting Engineer (VCE) program can help you find a professional who understands the amateur radio world, as well as the regulatory environment. There are also a variety of websites with useful information (such as **www.mikeholt.com**), but you need to be aware that advice may be specific to a particular installation or jurisdiction and not applicable for yours. With that understanding, let's look at a few NEC requirements for radio installations.

HOMEBREW AND "THE CODE"

In many cases, there are now legal requirements that electrical equipment have been listed by an NRTL, such as Underwriter Laboratories. This raises an issue for hams and homebrew gear since it's unlikely you would take your latest project down to a test lab and pay them to evaluate it for safety.

International Power Standards

The power grid of the United States and Canada uses a frequency of 60 Hz and the voltage at ac power outlets is 120 V. This is also the case in other North American countries. If you travel, though, you'll encounter 220 V and 50 Hz with quite an array of plugs and sockets and color codes. If you are planning on taking amateur radio equipment with you on a vacation or DXpedition, you'll need to be prepared with the proper adapters and/or transformers to operate your equipment.

A table of international voltage and frequencies is provided with this chapter's online content, along with a figure showing the most common plug and socket configurations.

For equipment that is not permanently installed, there's not much of an issue with homebrew, as far as the code goes, because the code doesn't deal with what's inside the equipment. For a lot of low voltage equipment, the code rules are fairly easy to meet as long as the equipment is supplied by a listed power source of the appropriate type.

The problem arises with permanent installations, where the scope of the code and local regulations are ever increasing. Such things as solar panel installations, standby generators, personal computers, and home LANs all have received increased attention in local codes.

22.1.4 Station Power

Amateur radio stations generally require a 120 V ac power source, which is then converted to the proper ac or dc levels required for the station equipment. In residential systems voltages from 110 V through 125 V are treated equivalently, as are those from 220 V through 250 V. Amateurs setting up a station in a light industrial or office environment may encounter 208 V line voltage. Most power supplies operate over these ranges, but it's a good idea to measure the voltage range at your station. (The measured voltage usually varies by hour, day, season, and location.) Power supply application and use are covered in the **Power Sources** chapter.

Modern solid-state rigs often operate from dc power, provided by a suitable dc power supply, perhaps including battery backups. Sometimes, the dc power supply is part of the rig (as in a 50 V power supply for a solid-state linear). Other times, your station might have a 12 V (13.8 V) bus that supplies many devices. Just because it's low voltage doesn't mean that there aren't aspects of the system that raise safety concerns. A 15 A, 12 V power supply can start a fire as easily as a 15 A, 120 V branch circuit.

22.1.5 Connecting and Disconnecting Power

Something that is sometimes overlooked is that you need to have a way to safely disconnect all power to everything in the station. This includes not only the ac power, but also battery banks, solar panels, and uninterruptible power supplies (UPS). Most hams won't have the luxury of a dedicated room with a dedicated power feed and the "big red switch" on the wall, so you'll have several switches and cords that would need to be disconnected.

The realities of today's stations, with computers, multiple wall transformers ("wall warts"), network interfaces, and the radio equipment itself makes this tricky to do. One convenient means is a switched outlet strip, as used for computer equipment, if you have a limited number of devices. If you need more switched outlets, you can control multiple low-voltage controlled switched outlets from a common source. Or you can build or buy a portable power distribution box similar to those used on construction sites or stage sets; they are basically a portable subpanel with individual circuit breakers (or GFCIs, discussed later) for each receptacle, and fed by a suitable cord or extension cord. No matter what scheme you use, however, it's important that it be labeled so that someone else will know what to do to turn off the power.

AC LINE POWER

If your station is located in a room with electrical outlets, you're in luck. If your station is in the basement, an attic, or other area without a convenient 120 V source, you may need to have a new line run to your operating position.

Stations with full-power (1.5 kW) amplifiers should have a 240 V ac power source in addition to the 120 V supply. Some amplifiers can be powered from 120 V but will require current levels that may exceed the limits of standard house wiring or cause objectionable voltage drops on the circuit. To avoid overloading the circuit and to reduce household light dimming or blinking when the amplifier is in use, and for the best possible voltage regulation in the equipment, it is advisable to install a separate 240 V line, or a 120 V line with an appropriate current rating, if you use an amplifier.

The usual circuits feeding household outlets are rated at 15 A or 20 A., This may or may not be enough current to power your station. To determine how much current your station requires, check the VA (volt-amp) ratings for each piece of gear. (See the **Electrical Fundamentals** chapter for a discussion of VA.) Usually, the manufacturer will specify the required current at 120 V. If the power consumption is rated in watts, divide that rating by 120 V to get amperes. Modern switching power supplies draw more current as the line voltage drops, so if your line voltage is markedly lower than 120 V, you need to take that into account.

Note that the code requires you to use the "nameplate" current, even if you've measured the actual current and it's less. If the total current required is near 80% of the circuit's rating (12 A on a 15 A circuit or 16 A on a 20 A circuit), you need to install another circuit. Keep in mind that other rooms may be powered from the same branch of the electrical system, so the power consumption of any equipment connected to other outlets on the branch must be considered. If you would like to measure just how much power your equipment consumes, the inexpensive Kill-A-Watt meters by P3 International measure volts, amps, VA, and power factor.

If you decide to install a separate 120 V line or a 240 V line, consult the local requirements as discussed earlier. In some areas, a licensed electrician must perform

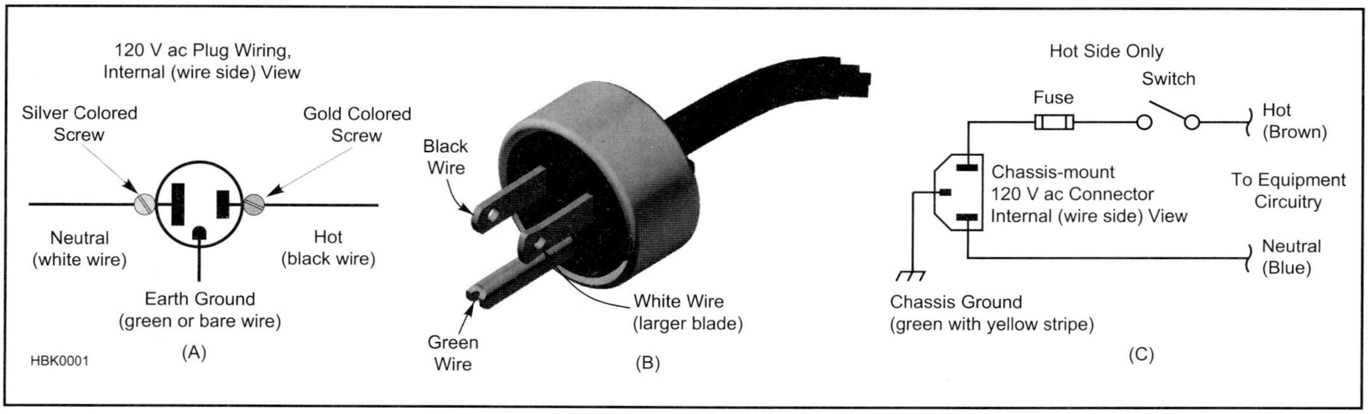

Figure 22.3 — 120 V ac plug wiring as viewed from the wire side (A) and viewed from the blade side (B). Wiring for an IEC type chassis connector is shown at C.

this work. Others may require a special building permit. Even if you are allowed to do the work yourself, it might need inspection by a licensed electrician. Go through the system and get the necessary permits and inspections! Faulty wiring can destroy your possessions and take away your loved ones. Many fire insurance policies are void if there is unapproved wiring in the structure.

If you decide to do the job yourself, work closely with local building officials. Most home-improvement centers sell books to guide do-it-yourself wiring projects. If you have any doubts about doing the work yourself, get a licensed electrician to do the installation.

THREE-WIRE 120 V POWER CORDS

Most metal-cased electrical tools and appliances are equipped with three-conductor power cords. Two of the conductors carry power to the device, while the third conductor is connected to the case, enclosure, or frame. **Figure 22.3** shows two commonly used connectors. (See the **Construction Techniques** chapter for a more comprehensive drawing of ac plugs and receptacles.)

When both plug and receptacle are properly wired, the three-contact polarized plug bonds the equipment to the system ground. If an internal short from line to case occurs, the "ground" pin carries the fault current and hopefully has a low enough impedance to trip the branch circuit breaker or blow the fuse in the device. A second reason for grounding the case is to reduce the possibility of shock for a user simultaneously connected to ground and the device. In modern practice, however, shock prevention is often done with GFCI circuit breakers as described below. These devices trip at a much lower level and are more reliable. Most commercially manufactured test equipment and ac-operated amateur equipment is supplied with three-wire cords.

It's a good idea to check for continuity from case to ground pin, particularly on used equipment, where the ground connection might have been broken or modified by the previous owner. If there is no continuity, have the equipment repaired before use.

Use such equipment only with properly installed three-wire outlets. If your house does not have such outlets, either consult a local electrician to learn about safe alternatives or have a professional review information you might obtain from online or other sources.

Equipment with plastic cases is considered "double insulated" and fed with a two-wire cord. Such equipment is safe because both conductors are insulated from the user by two layers. Nonetheless, there is still a hazard if, say, a double insulated drill was used to drill an improperly grounded case of a transmitter that was still plugged in. Remember, all insulation is subject to age, damage, and wear that may erode its initial protection.

TRANSFER SWITCHES AND GENERATORS

More hams are adding standby generators and using alternate power sources such as solar panels or wind turbines, not as standalone systems like at Field Day, but interconnected with their home electrical system. These present some potential safety problems, such as preventing the local power source from "back-feeding" the utility's system during a power failure, and the fact that a solar panel puts out power whenever there is light falling on it.

For generators, the recommended approach is to use a *transfer switch*, which is a multi-pole switch that connects a selection of the house's circuits to the generator, rather than the utility power. If you have a solar panel installation, connecting a generator is more complex and depends on the installation. Consult an electrician for this type of system. The NEC and local regulations should be consulted for transfer switch selection and connection.

The required wiring practices for permanently installed (stationary) generators are different from those for portable generators. Some issues that need to be considered are whether the neutral should be switched (many transfer switches do not switch the neutral, only the hot wire), and how the generator chassis is bonded to the building's grounding/bonding system. Most proper transfer switches are of the ON-OFF-ON configuration, with a mechanical interlock that prevents directly switching from one source to the other in a single operation.

Backfeeding your home's power panel should *never* be done unless the main breakers are in the OFF position or preferably removed. If you have an older home in which the circuit-breaker panel does not have main breakers that can disconnect the external power line, *do not* use this technique to connect your generator to the home's wiring. Your generator is likely to be damaged when power is restored, and backfeeding also endangers power utility workers. Instead, connect appliances to the generator directly with extension cords.

The most dangerous thing to do with a generator is to use a so-called "suicide cord" with a male plug at each end: one end plugged into the generator's output receptacle and the other plugged into a convenient receptacle in the home. This is frequently illegal and, regardless, should be avoided because of the inherent danger of having exposed, live contacts, and the ease of overloading the circuit being fed.

22.1.6 Ground-Fault and Arc-Fault Circuit Interrupters

GFCIs are devices that can be used with common 120 V household circuits to reduce the chance of electrocution when the path of current flow leaves the branch circuit (say, through a person's body to another branch or ground). An AFCI is similar in that it monitors current to watch for a fault condition. Instead of current imbalances, the AFCI detects patterns of current that indicate an arc, one of the leading causes of home fires. The AFCI is not supposed to trip because of "normal" arcs that occur when a switch is opened, or a plug is removed.

The NEC requires GFCI outlets in all

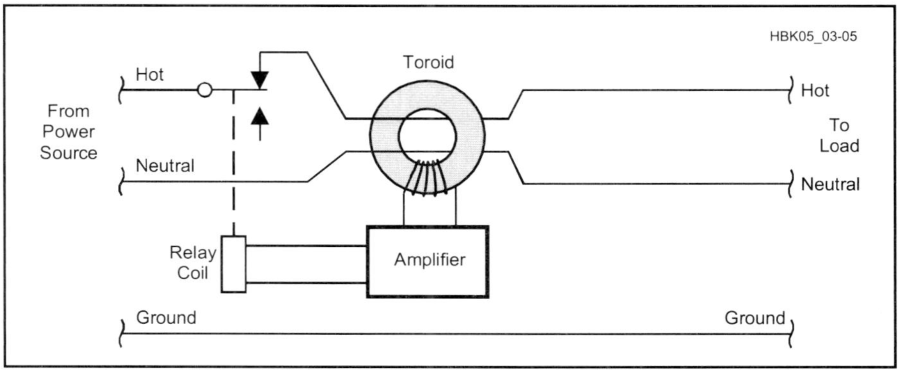

Figure 22.4 — Simplified diagram of a 120 V ac ground fault circuit interrupter (GFCI). When a stray current flows from the load (or outlet) side to ground, the current through the toroid becomes unbalanced allowing detection, amplification, and relay actuation to immediately cut off power to the load (and to the stray path!). GFCI units require a manual reset after tripping. GFCIs are required in wet locations (near kitchen sinks, in garages, in outdoor circuits and for construction work). They are available as portable units or combined with over-current circuit breakers for installation in entrance panels.

wet or potentially wet locations, such as bathrooms, kitchens, any outdoor outlet with ground-level access, garages, and unfinished basements. AFCI protection is required for all circuits that supply bedrooms. Any area with bare concrete floors or concrete masonry walls should be GFCI equipped. GFCIs are available as portable units, duplex outlets, and as individual circuit breakers. Ham radio stations in potentially wet areas (basements or outbuildings) should be GFCI equipped. **Figure 22.4** is a simplified diagram of a GFCI.

Older equipment with capacitors in the 0.01 μF to 0.1 μF range connected between line inputs and chassis as an EMI filter (or that has been modified with bypass capacitors) will often cause a GFCI to trip, because of the leakage current through the capacitor. The must-trip current is 5 mA, but many GFCIs trip at lower levels. At 60 Hz, a 0.01 μF capacitor has an impedance of about 265 kΩ, so there could be a leakage current of about 0.5 mA from the 120 V line. If you had several pieces of equipment with such capacitors, the leakage current would trip the GFCI.

Some early GFCI breakers were susceptible to RFI, but as the technology has improved, fewer and fewer such reports have been received. While it is possible to add filtering or RF suppression to the breaker wiring, a simpler and less expensive solution is to replace the GFCI breaker with a new unit less susceptible to RF. Reports have not yet been received on AFCI products. For more information on RFI and GFCI/AFCI devices, check the ARRL web page **www.arrl.org/gfci-and-afci-devices**.

22.1.7 Low-Voltage Wiring

Many ham stations use low-voltage control wiring for rotators or antenna relays. The electrical code isn't consistent in what it calls low voltage, but a guideline is "less than 50 V." Article 725 of the code contains most of the rules for low voltage/low power remote control and signaling, which is what hams are typically doing. These circuits are divided into three classes, with Class 1 being further subdivided, as shown in **Table 22.1**. There used to be code rules defining the classes in terms of power and voltage, but recent versions of the code are written so that the class of the circuit is defined by the power source, which must be listed and labeled with the class. For instance, if you have something powered by a wall transformer that is listed and labeled as Class 2, the circuit is Class 2.

A typical example of a Class 1 Power Limited circuit that you might find in your home is a 12 V garden lighting or halogen lightning system. A lot of amateur homebrew gear probably is also in this class, although because it's not made with "listed" components, it technically doesn't qualify. The other Class 1 devices apply to circuits using isolation transformers of some sort.

Class 2 is very common: doorbells, network wiring, thermostats, and so on are almost all Class 2. To be Class 2, the circuit must be powered from a listed power supply that's marked as being Class 2 with a capacity less than 100 VA. For many applications that hams encounter, this will be the familiar "wall wart" power supply. If you have a bunch of equipment that runs from dc power, and you build a dc power distribution panel with regulators to supply them from a storage battery or a big dc power supply, you're most likely not Class 2 anymore, but logically Class 1. Since your homebrew panel isn't likely to be listed, you're really not even Class 1, but rather something that isn't covered by the code.

A common example of a Class 3 circuit that is greater than 30 V is the 70 V audio distribution systems used in paging systems. Class 3 wiring must be done with appropriately rated cable.

WIRING PRACTICES

Low voltage cables must be separated from power circuits. Class 2 and 3 cannot be run alongside Class 1 low voltage cables. They can't share a cable tray or the same conduit. A more subtle point is that the 2005 code added a restriction [Article 725.56(F)] that audio cables (speakers, microphone, etc.) cannot be run in the same conduit with other Class 2 and Class 3 circuits (like network wiring).

Low voltage and remote-control wiring should be included in your transient suppression system. This involves putting appropriate protective devices where wiring enters and leaves a building, and consideration of the current paths to minimize loops that can pick up the field from transients (or RF from your antenna).

Grounding or Bonding?

You may notice the term "bonding" is replacing "grounding" in many instances. A primary safety concern is for whatever carries fault currents to be mechanically rugged and reasonably conductive. It's also important that the fault-carrying conductor be connected to a ground rod, but that's a different consideration. Bonding is the term to use when contact between pieces of equipment or between conductors is the primary concern. Grounding is the term to use when referring to an earth connection for electrical safety or lightning protection.

Table 22.1
Traditional Divisions Among the Classes of Circuits

Class	*Power*	*Notes*
Class 1		
Power Limited	<30V, <1000VA	Transformer protected per Article 450. If not transformer, other overcurrent and fault protection requirements apply
Remote Control and Signaling	<600V	No limit on VA Transformers protected as defined in Article 450
Class 2	Power supply <100VA Voltage <30V	
Class 3	Power supply <100VA Voltage <100V	

22.1.8 Grounding and Bonding

As hams we are concerned with at least four kinds of connections called "ground," even if they really aren't in the sense of connection to the Earth. These are easily confused because we call each of them "ground."

1) Electrical safety ground (equipment ground or earth connection)

2) Lightning and transient dissipation ground

3) RF voltage and current management ("RF ground")

4) Common reference potential (chassis ground or circuit common)

This section of the chapter is primarily concerned with connections for electrical or safety grounding and lightning and transient dissipation. The remaining types are covered elsewhere in chapters on circuit- and antenna-building and in station construction.

Several commercial and military standards can be used as guidebooks for grounding and bonding:

National Electrical Code (NEC) — This is the primary standard for residential and commercial electrical work in the

United States. NEC Article 250 deals with grounding and bonding. NEC Article 810 deals with antenna installation. (See the previous section on the NEC.)

MIL-HDBK-419A — Grounding, Bonding, and Shielding for Electronic Equipments and Facilities (Vol 1 and 2) — This military standard applies to communication facilities and equipment installations at any frequency. It provides many useful drawings and guidelines covering ground connections and how equipment should be bonded together. It is a public-domain document and is available from a number of sources. An internet search for "MIL-HDBK-419A" will find several available copies.

R56 Standards and Guidelines for Communications Sites — Motorola is a large vendor of communications systems, mostly for VHF/UHF/microwave applications. This standard applies primarily to equipment and facilities used at those frequencies. It is not a public domain document but may be downloaded without charge from numerous sources.

IEEE Std 1100-2005 (also known as the "Emerald Book," see the Reference listing, section 22.1.13) provides detailed information from a theoretical and practical standpoint for grounding and powering electrical equipment, including lightning protection and RF EMI/EMC concerns. It's expensive to buy but is available through libraries.

BONDING

The definition of bonding is "to connect equipment together electrically in order to minimize the potential (voltage) difference between them." A good bonding connection must have very low impedance and approximately equal voltage everywhere along it at the frequency of interest. As amateurs know, electrical length and impedance of any type of conductor, wire, strap, braid, or sheet, varies with frequency. In addition, the amount of current flowing through the conductor can create significant voltage differences along the conductor. Because of these concerns, it is important to consider the purposes of bonding when making a bonding connection.

In general, for amateurs constructing a station, even a temporary one, it is a good practice to make conductors as short as practical and as heavy as is needed to satisfy all bonding requirements. By doing so, the connection will serve its purpose for all three of the primary bonding needs in your station: ac safety, lightning protection, and RF management.

ELECTRICAL SAFETY GROUND

Power-line ground is required by building codes to ensure the safety of life and property surrounding electrical systems. The NEC requires that all grounds be bonded together; this is a very important safety feature.

The usual term one sees for the "third prong" or "green-wire ground" is the *electrical safety ground* or *equipment ground*. The purpose of the third, non-load current carrying wire is to provide a path to ensure that the overcurrent protection will trip in the event of a line-to-case short circuit in a piece of equipment. This could either be the fuse or circuit breaker back at the main panel, or the fuse inside the equipment itself.

There is a secondary purpose: Shock reduction. The electrical safety ground provides a common reference potential for all parts of the ac system. The conductive case of equipment is required to be connected to the bonded grounding system, which is also connected to earth ground at the service entrance, so someone who is connected to "earth" (for example, standing in bare feet on a conductive floor) that touches the case won't get shocked.

An effective safety ground system is necessary for every amateur station. If you have equipment at the base of a tower you generally need to provide a separate bonding conductor to connect the chassis and cases at the tower to the bonding system in the station. **Figure 22.5A** shows an overall grounding system, emphasizing the requirement to bond all ground connections together, regardless of their purpose.

Equipment-to-equipment bonding can be

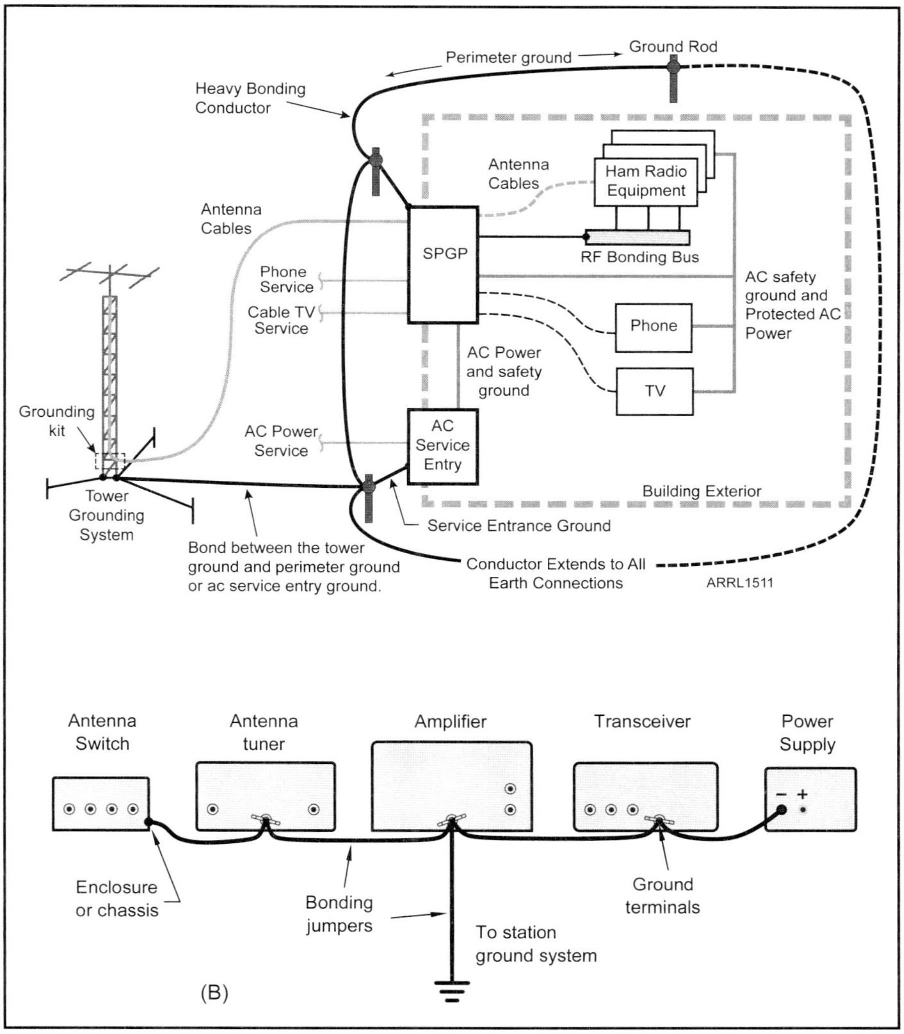

Figure 22.5 — A grounding system that includes ac safety, lightning protection, and RF management. All ground electrodes must be bonded together and to the residence's ac service entry ground rod. If a protected ac branch circuit is included, the protector should be mounted on the SPGP (single-point ground panel). Direct equipment-to-equipment bonding (B) can be done with heavy wire or strap or to an RF bonding bus. See the chapter on Assembling a Station for more information on making these connections.

done directly with wire (#6 to #14 AWG), strap (20 gauge), strips of flashing (24 gauge), or through a bonding bus or single-point ground panel (SPGP). The equipment ground or "third-wire" connection in a residence's branch circuit must be made with the same size wire as the hot and neutral conductors. Connections directly to a ground electrode, such as a ground rod, are made with heavy wire (#6 AWG minimum) or 20-gauge strap. Clamps and terminals should be rated or listed for grounding use, particularly for earth connections that are exposed to the weather or buried.

At ac power frequencies of 50 or 60 Hz, the wavelength is miles long, so the electrical length of the connection is insignificant. Reactance created by the bonding conductor (about 300 nH per foot for a straight wire) is likewise negligible. Unfortunately, an effective bonding conductor at 60 Hz may present very high impedance at RF because of the inductance, or worse yet, wind up being an excellent antenna that picks up the signals radiated by your antenna. The most important characteristics of the bonding connection is resistance and mechanical strength.

RF MANAGEMENT AND CHASSIS GROUND

"RF ground" is an obsolete term from days gone by when most operation was at low frequencies and a wire from the chassis or antenna tuner in the station to a ground rod had low RF impedance. The RF voltage difference between the chassis and "Earth ground" was small. And even if there were small potential differences, the surrounding equipment of those days was relatively insensitive to them.

Today, we have a lot of circuits that are sensitive to interfering signals at millivolt levels, such as audio signals to and from sound cards. As a result, we can no longer ignore the RF voltage differences, and shouldn't be using the equipment enclosures or shielding conductors as part of the RF circuit.

Instead, we design our stations to manage the RF picked up on cables, connecting wires, and enclosures so that it does not cause problems. The first step is to create a common reference potential, called the *ground plane* or *reference plane*. Equipment connected to the ground plane is maintained at a common potential. This minimizes RF current that would flow between pieces of equipment. (See the **RFI and EMC** chapter for more information.) The ground plane can be an actual sheet of metal or a low-inductance conductor to which all the equipment can be connected, often called an *RF bonding bus*. The ground plane is then bonded to the station ground system.

It is sometimes suggested that RF grounds should be isolated from the ac safety and lightning protection ground system. That is not correct! All grounds, including safety, RF, lightning protection, and commercial communications, must be bonded together in order to protect life and property. The electrical code requires that antenna grounds be bonded to the rest of the grounding system, although that connection can have an RF choke. Remember that the focus of the electrical code bonding requirement is safety in the event of a short to a power distribution line or other transient.

For decades, amateurs have been advised to bond all equipment cabinets to an RF ground located near the station. Given today's operating frequencies and equipment sensitivity, this practice is inadequate. Even a few meters of wire can have an impedance of hundreds of ohms (e.g., 1 µH/meter = 88 Ω/meter at 14 MHz). A better approach is to connect all chassis together in a well-organized fashion to ensure that the chassis-to-chassis connections minimize RF voltage differences as in Figure 22.5B. An RF bonding bus can be used, as well. (See the **RFI and EMC** and **Assembling a Station** chapters for more information.)

LIGHTNING DISSIPATION GROUND

Lightning dissipation ground is concerned with conducting currents to the surrounding earth. There are distinct similarities between lightning dissipation ground systems and a good ground system for a vertical antenna. Lightning strokes produce electrical energy from a few kHz to more than 10 MHz, so the length of the connection along with its resistance is important. The difference is that an antenna ground plane may handle perhaps a few tens of amps, while the lightning ground needs to handle a peak current of hundreds of kiloamperes.

A typical lightning stroke is a pulse with a rise time of 8 to 10 µs, an average duration of 30 µs, a peak current of 200 kA, and a fall time of 20 to 350 µs. Even though the average current is not that high, large conductors (usually specified as #6 AWG minimum by building codes) are used in lightning grounds for many reasons: to reduce inductance, to handle the mechanical forces from the magnetic fields, and for ruggedness to prevent inadvertent breakage. A large diameter wire, or even better, a wide flat strap, has lower inductance. The voltage along a wire is proportional to the change in current and its inductance:

$$|V| = L\frac{\Delta i}{\Delta t}$$

where

$\Delta i/\Delta t$ = rate of change in current, about 200kA/2µs for lightning, or 10^{11} A/s, and

L = the inductance of the wire.

For a straight 1-meter length of #6 AWG wire, with inductance of 1.23 µH, the voltage generated for a typical lightning strike would be about 123 kV. By comparison, changing the wire to #12 AWG, with an inductance of 1.37 µH, would generate 137 kV for the same lightning strike. Consider a connection box on a tower that contains some circuitry terminating a control cable from the station, appropriately protected internally with overvoltage protection. The lightning transient will raise the box potential (relative to the wiring coming from the station), possibly beyond the point where the transient suppression in the box can handle it. The potential will be reduced by lowering the inductance of the connection to ground, by using larger diameter wire, keeping the wire length as short as possible, and avoiding sharp bends in the wire.

The other reason for large conductors on lightning grounds is to withstand the very high mechanical forces from the high currents. This is also the reason behind the recommendation that lightning conductors be run directly, with minimal bends and large radii for bends that are needed, and certainly no loops. A wire carrying 200,000 A has a powerful magnetic field surrounding it, and if current is flowing in multiple wires that

When to Use Strap, Wire, or Braid

The standard for grounding in the communication industry is solid strap or heavy wire. Both can be used indoors or outdoors. Flat-weave, tinned grounding braid can be used if the equipment is subject to vibration or needs to be moved around.

Do not use any type of braid if it will be exposed to moisture or corrosive chemicals. Corrosion on the surface of the small wires used to make up the braid reduces its effectiveness by raising the surface resistance, where the RF currents flow. Poor contact between the individual wires can also result in noise and mixing products. Unless it is mechanically necessary, such as for a gate or door, use solid strap or wire.

It is not recommended to reuse braid removed from coaxial cable. Once removed from its protective jacket, the braid wires immediately begin to loosen and oxidize or corrode. This reduces the braid's effectiveness at RF quite a bit, making it a poor choice for long-term grounding conductors.

are close to each other, the forces pushing the wires together or apart can break the conductors or permanently deform them.

The instantaneous force between two conductors carrying 200,000 A, spaced a centimeter apart, is 80,000 newtons per meter of length (over 5,000 pounds/foot). Such forces can easily break cable strands or rip up brackets and screws. This problem is aggravated if there are loops in the wire, since the interaction of the current and its magnetic field tends to make the loop get larger, to the point where the wire will actually fail from the tension stresses.

EARTH GROUND

Earth ground usually takes one of several forms, all identified in the NEC and NFPA 780. The preferred earth ground, both as required in the NEC and verified with years of testing in the field, is a *concrete encased grounding electrode* (CEGR), also known as a *Ufer ground*, after Herb Ufer, who invented it to provide grounding for military installations in dry areas where ground rods are ineffective. The CEGR can take many forms, but the essential aspect is that a suitable conductor at least 20 feet long is encased in concrete which is buried in the ground. The conductor can be a copper wire (#8 AWG) or the reinforcing bars (rebar) in the concrete and is often the foundation footing for the building. The connection to the rebar is either with a stub of the rebar or the copper wire protruding through the concrete's top surface. There are other variations of the CEGR described in the NEC and in the electrical literature, but they are all functionally the same: a long conductor embedded in a big piece of concrete.

The electrode works because the concrete has a huge contact area with the surrounding soil, providing very low impedance and a low current density so that localized heating doesn't occur. Concrete tends to absorb water, so it is also less susceptible to problems with the soil drying out around a traditional ground rod.

The techniques required for welding rebar and making the necessary connection to the CEGR are somewhat specialized. If you are not familiar with those skills, hire a professional to do the job correctly. Similarly, unless you have documented evidence that the necessary rebar connection or embedded wire are present in a concrete footing or slab, do not attempt to use it as a CEGR.

Ground rods are a traditional approach to making a suitable ground connection and are appropriate as supplemental grounds, say at the base of a tower, or as part of an overall grounding system. The best ground rods to use are those available from an electrical supply house. The code requires that at least 8 feet of the rod be in contact with the soil, so if the rod sticks out of the ground, it must be longer than 8 feet (10 feet is standard). The rod doesn't have to be vertical and can be driven at an angle if there is a rock or hard layer, or even buried laying sideways in a suitable trench, although this is a compromise installation. Suitable rods are generally 10 feet long and made from steel with a heavy copper plating. (Stainless steel and galvanized rods are also available and may be required by soil conditions in your area. Check with a local electrician or electrical inspector.) Do not depend on shorter, thinly plated rods sold by some home electronics suppliers, as they can quickly rust and become worthless.

If multiple ground rods are installed, they must be spaced by at least half the length of each rod, or the effectiveness could be compromised. IEEE Std 142 and IEEE Std 1100 (see the Reference listing), and other references, have tables to give effective ground resistances for various configurations of multiple rods.

Once the ground rods are installed, they must be connected with either an exothermic weld (such as CadWeld) or with a listed pressure clamp. The exothermic weld is preferred because it doesn't require annual inspection like a clamp does. Some installers use brazing to attach the wiring to the ground rods. Although this is not permitted for a primary ground, it is acceptable for secondary or redundant grounds. Soft solder (tin-lead, as used in plumbing or electrical work) should not be used for grounding conductors because it gets brittle with temperature cycling and can melt if a current surge (such as from a lightning strike) heats the conductor. Soft solder is specifically prohibited for use in

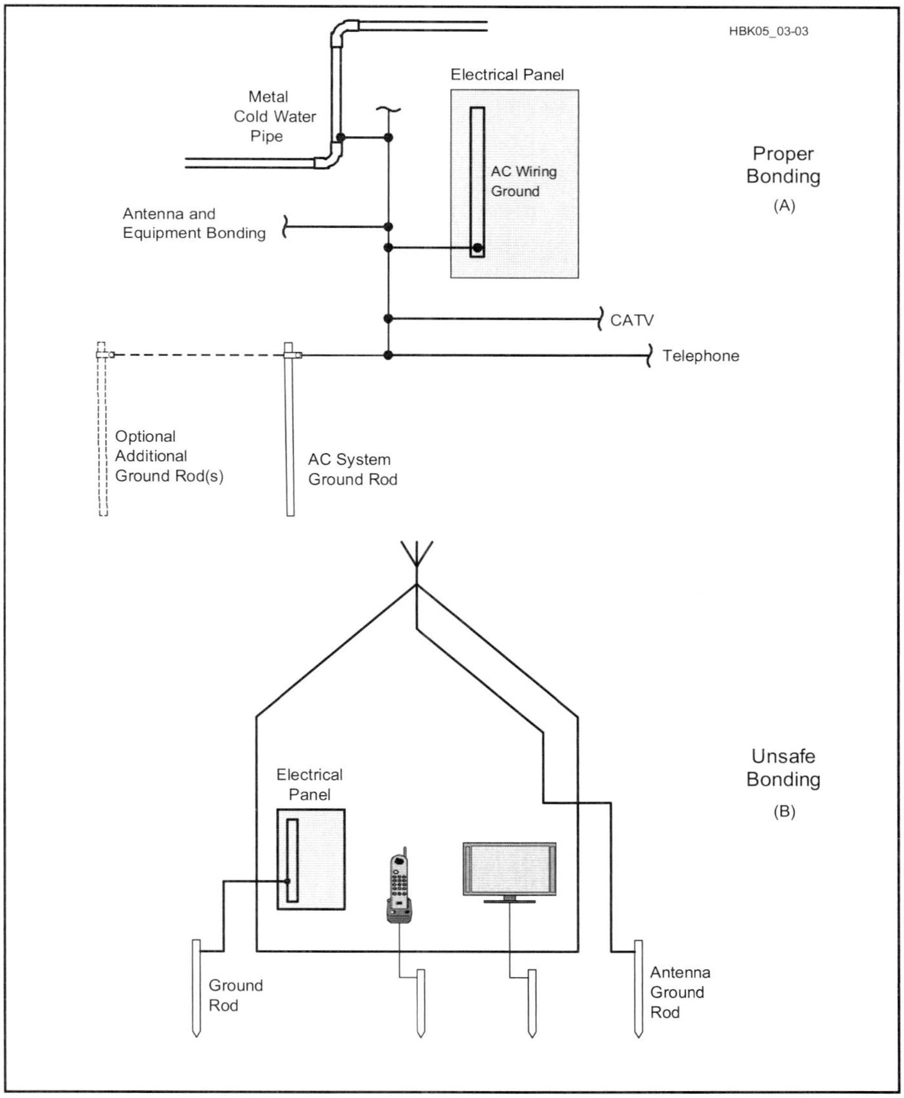

Figure 22.6 — At A, proper bonding of all grounds to electrical service panel. The installation shown at B is unsafe — the separate grounds are not bonded. This could result in a serious accident or electrical fire.

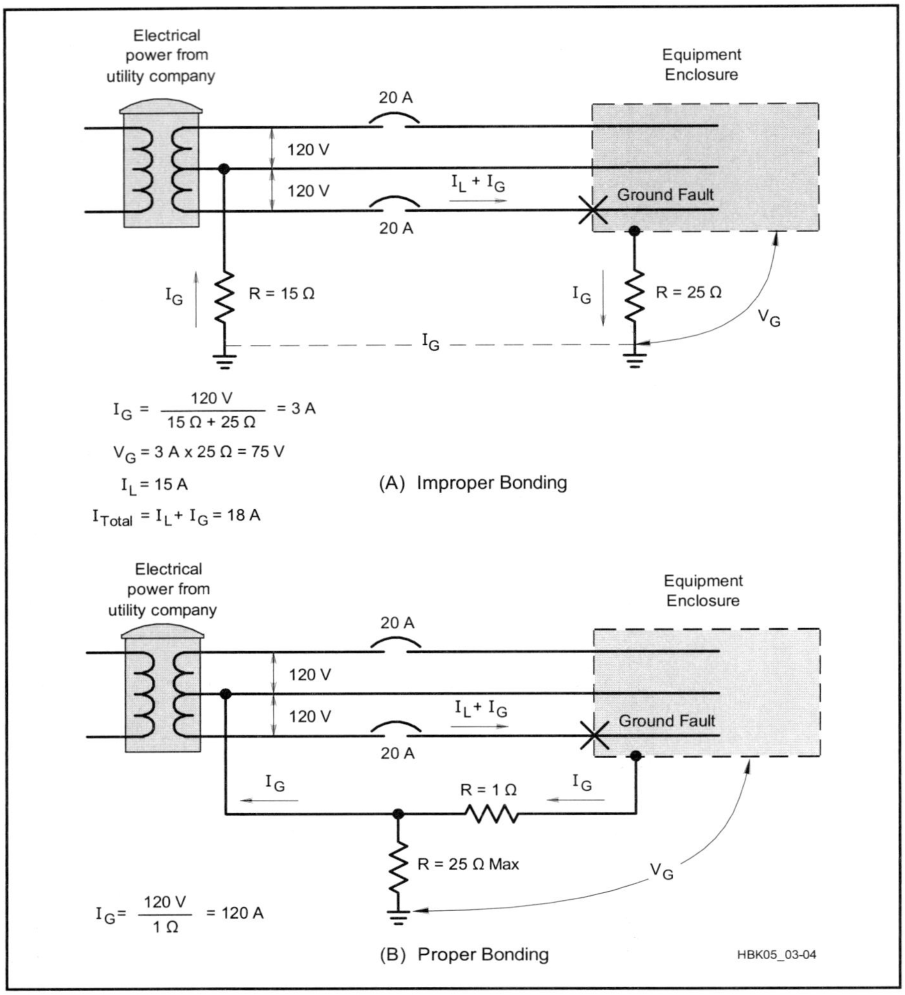

Figure 22.7 — These drawings show the importance of properly bonded ground rods. In the system shown in A, the 20 A breaker will not trip. In the system in B, the 20 A circuit breaker trips instantly. There is an equipment internal short to ground — the ground rod is properly bonded back to the power system ground. Of course, the main protection should be in a circuit ground wire in the equipment power cord itself!

Coax Shields as a Grounding Conductor

The importance of significant current-carrying capability in a grounding conductor is determined by your local circumstances. If lightning is a problem or contact with power-carrying conductors due to wind or ice is a possibility, the ability of the coax shield to carry a lot of current is important and a separate grounding conductor is a good idea. In areas where hazards are reduced, a smaller conductor may suffice. Contact a licensed electrician if you aren't sure about what is prudent for your area.

grounding systems by the building codes.

Building cold water supply systems has been used as station grounds in years past, but this is no longer recommended, or even permitted in some jurisdictions, because of increased use of plastic plumbing both inside and outside houses, and concerns about stray currents causing pipe corrosion. If you do use the cold-water line, it must be bonded to the electrical system ground, typically at the service entrance panel.

22.1.9 Ground Conductors

Building codes and wiring standards are quite specific as to the types of conductors that can be used for bonding the various parts of the system together. Grounding conductors may be made from copper, aluminum, copper-clad steel, bronze, or similar corrosion-resistant materials. Note that the sizes of the conductors required are based largely on mechanical strength considerations (to ensure that the wire isn't broken accidentally) rather than electrical resistance. Insulation is not required.

There is a "unified" grounding electrode requirement: It is necessary to bond *all* grounds to the electric service entrance ground. All utilities, antennas, and any separate grounding rods used must be bonded together. **Figure 22.6** shows correct (A) and incorrect (B) ways to bond ground rods. **Figure 22.7** demonstrates the importance of correctly bonding ground rods. (Note: The NEC requirements do not address effective RF bonding. See the **RFI and EMC** and **Assembling a Station** chapters for information about RF practices. Keep in mind that RFI is not an acceptable reason to violate the NEC.) For additional information on good grounding practices, the *NEC Handbook* and IEEE "Emerald Book" (IEEE STD 1100-2005) are good references. Both are available through libraries. Home wiring guides are also available at home improvement stores.

Additionally, the NEC covers safety inside the station. All grounding conductors inside the building must be at least 4 inches away from conductors of any lighting or signaling circuit except when they are separated from other conductors by conduit or insulator. Other code requirements include enclosing transmitters in metal cabinets that are bonded to the grounding system. Of course, conductive handles and knobs must be grounded as well.

22.1.10 Antennas

Article 810 of the NEC includes several requirements for wire antennas and feed lines that you should keep in mind when designing your antenna system. The "protective grounding conductor" (main conductor running to the ground rod) must be as large as the antenna lead-in, but not smaller than #10 AWG. The grounding conductor (used to bond equipment chassis together) must be at least #14 AWG.

The single most important thing to consider for safety is to address the potential for contact between the antenna system and power lines. As the code says, "One of the leading causes of electric shock and electrocution, according to statistical reports, is the accidental contact of radio, television, and amateur radio transmitting and receiving antennas and equipment with light or power conductors." (See Article 810.13, Fine Print Note.) The requirements in the code for wire sizes, bonding requirements, and installation practices are mostly aimed at preventing tragedy by avoiding the contact in the first place, and by mitigating the effects of a contact if it occurs.

Article 820 of the NEC applies to cable TV installations, which almost always use coaxial cable, and which require wiring practices different from Article 810 (for instance, the coax shield can serve as the grounding conductor). An electrical inspector may look to Article 820 for guidance on safe installation of coax, since there are many more satellite TV and cable TV installations than amateur radio. Ultimately, it is the inspector's call as to whether your installation is safe.

Article 830 applies to Network Powered Communication Systems, and as amateurs do things like install 802.11 wireless LAN equipment at the top of their tower, they'll have to pay attention to the requirements in this Article. The NEC requirements discussed in these sections do not address lightning protection and high-voltage transient events. See the section "Lightning/Transient Protection" later in this chapter for more information.

ANTENNA CONDUCTORS

Transmitting antennas should use hard-drawn copper wire: #14 AWG for unsupported spans less than 150 feet, and #10 AWG for longer spans. Copper-clad steel, bronze or other high-strength conductors must be #14 AWG for spans less than 150 feet and #12 AWG for longer spans. Open-wire transmission line conductors must be at least as large as those specified for antennas. Stealth antennas made with light-gauge wire are not code-compliant but may be okay as a temporary installation since the NEC does not apply in such situations.

LEAD-INS

There are several NEC requirements for antenna lead-in conductors (transmission lines are lead-in conductors). For transmitting stations, their size must be equal to or greater than that of the antenna. Lead-ins attached to buildings must be firmly mounted at least 3 inches clear of the surface of the building on nonabsorbent insulators. Lead-in conductors must enter through rigid, noncombustible, nonabsorbent insulating tubes or bushings, through an opening provided for the purpose that provides a clearance of at least two inches; or through a drilled windowpane. All lead-in conductors to transmitting equipment must be arranged so that accidental contact is difficult. As with stealth antennas, installations with feed lines smaller than RG-58 are likely not code compliant depending on how your local inspector interprets the code.

ANTENNA DISCHARGE UNITS (LIGHTNING ARRESTORS)

All antenna systems are required to have a means of draining static charges from the antenna system. A listed antenna discharge unit (lightning arrestor) must be installed on each lead-in conductor that is not protected by a permanently and effectively grounded metallic shield, unless the antenna itself is permanently and effectively grounded, such as for a shunt-fed vertical. Note that the usual transient protectors are *not* listed antenna discharge units. (The code exception for shielded lead-ins does *not* apply to coax, but to shields such as thin-wall conduit. Coaxial braid is neither "adequate" nor "effectively grounded" for lightning protection purposes.) An acceptable alternative to lightning arrestor installation is a switch (capable of withstanding many kilovolts) that connects the lead-in to ground when the transmitter is not in use. A garden-variety knife switch for household appliances is not adequately rated for this job.

ANTENNA BONDING (GROUNDING) CONDUCTORS

In general, the code requires that the conductors used to bond the antenna system to ground be at least as big as the antenna conductors, but also at least #10 AWG in size. Note that the antenna grounding conductor rules are different from those for the regular electrical safety bonding, or lightning dissipation grounds, or even for CATV or telephone system grounds.

MOTORIZED CRANK-UP ANTENNA TOWERS

If you are using a motorized crank-up tower, the code has some requirements, particularly if there is a remote control. In general, there must be a way to positively disconnect power to the motor that is within sight of the motorized device, so that someone working on it can be sure that it won't start moving unexpectedly. From a safety standpoint, you should be able to see or monitor the antenna from the remote-control point as well.

22.1.11 Lightning/Transient Protection

Nearly everyone recognizes the need to protect themselves from lightning. From miles away, the sight and sound of lightning boldly illustrates its destructive potential. Many people don't realize that destructive transients from lightning and other events can reach electronic equipment from many sources, such as outside antennas, power, telephone, and cable TV installations. Many hams don't realize that the standard protection scheme of several decades, a ground rod, and simple "lightning arrestor," is *not* adequate.

Lightning and transient high-voltage protection follows a familiar communications scenario: identify the unwanted signal, isolate it, and dissipate it. The difference here is that the unwanted signal can be over a megavolt at up to 200,000 A. What can we do?

Effective lightning protection system design is a complex topic. There are a variety of system tradeoffs which must be made, and which determine the type and amount of protection needed. An amateur station in a home is a very different proposition from an air traffic control tower, which must be available 24 hours a day, 7 days a week. Hams can easily follow some general guidelines that will protect their stations against high-voltage events that are induced by nearby lightning strikes or that arrive via utility lines. Let's talk about where to find professionals first, and then consider construction guidelines.

PROFESSIONAL HELP

Start with your local government. Find out what building codes apply in your area and have someone explain the regulations about antenna installation and safety. For more help, look in your telephone yellow pages for professional engineers, lightning protection suppliers and contractors.

Companies that sell lightning-protection products may offer considerable help to apply their products to specific installations. One such source is PolyPhaser Corporation.

Manufacturers of Lightning Protection Equipment

For current vendor contact information, use your favorite internet search tool.

- Alpha Delta Communications: Coax lightning arrestors, coax switches with surge protectors.
- The Wireman: copper wire up to #4 AWG, 2-inch flat copper strap, 8-ft copper clad ground rods and 1 × ¼-inch buss bar.
- ERICO International Corporation: CadWeld bonding system and lightning protection equipment.
- Harger Lightning & Grounding: lightning protection components.
- Industrial Communication Engineers, Ltd (ICE): Coax lightning arrestors.
- KF7P Metalwerks: Entrance panels, lightning arrestors, surge protectors, grounding and bonding hardware, and so on.
- PolyPhaser Corporation: Many lightning protection products for feed lines, towers, equipment, and so on.
- Zero Surge Inc: Power line surge protector.

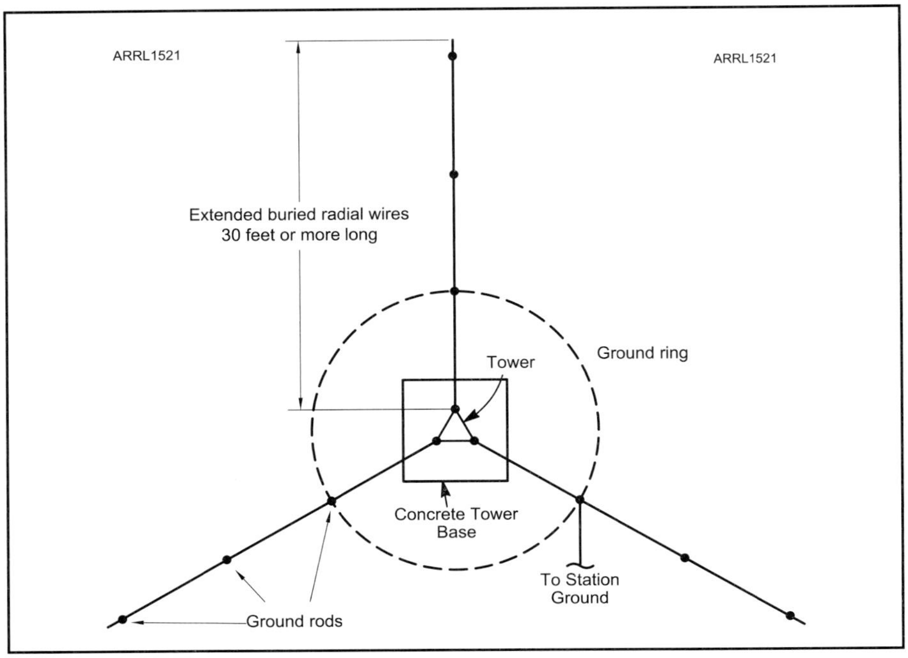

Figure 22.8 — Schematic of a properly grounded tower. A bonding conductor connects each tower leg to a ground rod and a buried (1 foot deep) bare, tinned copper ring (dashed line), which is also connected to the station ground and then to the ac safety ground. Make the ring diameter large enough that ground rods aligned with the tower legs will be approximately 6 feet apart. Buried radial wires extending beyond the ring are a common commercial practice and provide extra protection but are optional. All connectors should be compatible with the tower and conductor materials to prevent corrosion. See text for conductor sizes and details of lightning and voltage transient protection.

Look under "References" later in this chapter for a partial list of PolyPhaser's publications.

CONSTRUCTION GUIDELINES

Bonding Conductors

Copper strip (or *flashing*) comes in a number of sizes. The *minimum* recommended grounding conductor for lightning protection is 1.5 inches wide 20-gauge strap (0.032 inch (0.8 mm) thick) or #6 AWG stranded wire. The requirements for minimum size are driven by the need for mechanical durability.

Do not use braided strap outside or in wet areas because the individual strands oxidize over time, greatly reducing the effectiveness of braid as an ac conductor. Bear in mind that copper strap has about the same inductance as a wire of the same length. While strap may be easier to install and provides a lower RF loss (if it's part of a vertical antenna grounding system, for instance), it doesn't provide significant improvement over round wire for power line frequencies (the NEC's concern) or lightning (where inductance dominates the effects).

Use bare copper for buried ground wires. (There are some exceptions; seek an expert's advice if your soil is corrosive.) Exposed runs above ground that are subject to physical damage may require additional protection (a conduit) to meet code requirements. Wire size depends on the application, but never use anything smaller than #6 AWG for bonding earth electrodes. The NEC specifies conductors using wire gauge and doesn't describe the use of flashing. Local lightning-protection experts or building inspectors can recommend sizes for each application.

Tower and Antennas

Because a tower is usually the highest metal object on the property, it is the most likely strike target. Proper tower grounding is essential to lightning protection. The goal is to establish short multiple paths to the Earth so that the strike energy is divided and dissipated.

Connect each tower leg and each fan of metal guy wires to a separate ground rod. Space the rods at least 6 feet apart. Bond the leg ground rods together with #6 AWG or larger copper bonding conductor (form a ring around the tower base, see **Figure 22.8**). Connect a continuous bonding conductor between the tower ring ground and the entrance panel. Make all connections with fittings approved for grounding applications. *Do not use solder for these connections.* Solder will be destroyed in the heat of a lightning strike.

Because galvanized steel (which has a zinc coating) reacts with copper when combined with moisture, use stainless steel hardware between the galvanized metal and the copper grounding materials. Rohn and other manufacturers now offer ground clamps designed for connecting galvanized tower legs to copper ground conductors.

To prevent strike energy from entering a station via the feed line, ground the feed line *outside* the home. Ground the coax shield *to the tower* at the antenna and the base to keep the tower and line at the same potential. Several companies offer grounding blocks that make this job easy.

All grounding media at the home must be bonded together. This includes lightning-protection conductors, electrical service, telephone, antenna system grounds, and underground metal pipes. Any ground rods used for lightning protection or entrance-panel grounding should be spaced at least 6 feet from each other and the electrical service, or other utility grounds, and then bonded to the ac system ground as required by the NEC.

A Single-Point Ground Panel

The basic concept with transient protection is to make sure that all the radio and other equipment is tied together and "moves together" in the presence of a transient voltage. It's not so important that the station be at "ground" potential, but, rather, that everything is at the same potential. For fast rise-time transients such as the individual strokes that make up a lightning strike, even a short wire has enough inductance that the voltage drop along the wire is significant; so, whether you are on the ground floor or the 10th floor of a building, your station is "far" from Earth potential.

The easiest way to ensure that everything is at the same potential is to tie all the signal ground connections to a common reference.

> **Voltage Rise On Wires With Fast Transients**
>
> A rule of thumb is that a single wire has an inductance of about 1 µH per meter of length. The voltage across an inductor $V = L\ \Delta i/\Delta t$. ($\Delta i/\Delta t$ is the change in current per unit of time.) A lightning stroke has a rise time of about 1-2 µs, so the current might go from zero to 10 kA in a microsecond or two, a $\Delta i/\Delta t$ of over 1 kA/µs (10^9 A/s). An inductance as low as 1 µH would create a voltage of 1000 volts from this current transient.

In large facilities, this reference would be provided by a grid of large diameter cables under the floor, by wide copper bars, or even a solid metal floor. A more practical approach for smaller facilities such as a ham station is to have a single-point ground panel (SPGP). The SPGP may be a separate metal panel, or it can be enclosed in an electrical box. See the articles by Block in the References section for a complete discussion of the SPGP. The articles are available online through the ARRL's Radio Technology Topics website (**www.rrl.org/radio-technology-topics**) under "Safety" and in the ARRL book *Grounding and Bonding for the Radio Amateur*.

The SPGP should be mounted outside the building. The easiest way to do this is to install a large metal enclosure or a metal panel as a bulkhead and grounding block. The panel should be connected to the lightning dissipation ground with a short wide conductor (for minimum impedance), and, like all grounds, bonded to the electrical system's ground. Mount all protective devices, switches, and relay disconnects on the outside facing wall. The enclosure or panel should be installed in a way that if lightning currents cause a component to fail, the molten metal and flaming debris do not start a fire.

Every conductor that enters the structure, including antenna system control lines, telephone, and CATV cables, should have their own surge suppressor on an entrance panel. Suppressors are available from several manufacturers, including Industrial Communication Engineers (ICE) and PolyPhaser, as well as the usual electrical equipment suppliers, such as Square-D.

Fire Extinguishers in the Amateur Station

Every amateur station should have a fire extinguisher at hand. Three classes of fire extinguishers are appropriate for amateur stations:

- Class A: Combustible solid materials such as wood or paper
- Class B: Combustible liquid or gases
- Class C: Energized electrical fire

A fixed or portable station should have "ABC" class extinguishers. Mobile stations are most likely to experience class B and C fires. Liquid water extinguishers or hoses should never be used on an electrical fire because most water is a conductor of electricity and creates an electrical shock hazard. For more information on fire extinguishers, see the chapter "Assembling a Station."

We want to control the flow of the energy in a strike and eliminate any possible paths for surges to enter the building. Keep route feed lines, rotator control cables, and so on at least six feet away from other nearby grounded metal objects. Keep incoming, unprotected cables well away from cables on the protected side of any lightning arrestors, as well.

Lightning Arrestors

Feed line lightning arrestors are available for both coax cable and balanced line. Most of the balanced line arrestors use a simple spark gap arrangement, but a balanced line *impulse* suppressor is available from ICE.

DC blocking arrestors for coaxial cable have a fixed frequency range. They present a high impedance to lightning (less than 1 MHz) while offering a low impedance to RF.

DC continuity arrestors (gas tubes and spark gaps) can be used over a wider frequency range than those that block dc. Where the coax carries supply voltages to remote devices (such as a mast-mounted preamp or remote coax switch), dc-continuous arrestors *must* be used.

22.1.12 Other Hazards in the Station

UPS AND ALTERNATE ENERGY SOURCES

Many hams have alternate energy sources for their equipment, or an uninterruptible power supply (UPS), so that they can keep operating during a utility power outage. This brings up some additional safety concerns because it means that the "turning off the breaker" approach to make sure that power is disconnected might not work.

In commercial installations, fire regulations or electrical codes often require that the emergency power off (EPO) system (the big red button next to the door) also disconnect the batteries of the UPS system, or at least, disable the ac output. This is so that firefighters who may be chopping holes with conductive tools or spraying conductive water don't face the risk of electrocution. (According to NEC, Articles 645-10 and 645-11, UPSs above 750 VA installed within information technology rooms must be provided with a means to disconnect power to all UPS supply and

Electrical Shock Hazards and Effects

What happens when someone receives an electrical shock?

Electrocutions (fatal electric shocks) usually are caused by the heart ceasing to beat in its normal rhythm. This condition, called ventricular fibrillation, causes the heart muscles to quiver and stop working in a coordinated pattern, in turn preventing the heart from pumping blood.

The current flow that results in ventricular fibrillation varies between individuals but may be in the range of 100 mA to 500 mA. At higher current levels the heart may have less tendency to fibrillate but serious damage would be expected. Studies have shown 60-Hz alternating current to be more hazardous than dc currents. Emphasis is placed on application of cardiopulmonary resuscitation (CPR), as this technique can provide mechanical flow of some blood until paramedics can "restart" the heart's normal beating pattern. Defibrillators actually apply a carefully controlled waveform to "shock" the heart back into a normal heartbeat. It doesn't always work but it's the best procedure available.

What are the most important factors associated with severe shocks?

You may have heard that the current that flows through the body is the most important factor, and this is generally true. The path that currents takes through the body affects the outcome to a large degree. While simple application of Ohm's law tells us that the higher the voltage applied with a fixed resistance, the greater the current that will flow. Most electrical shocks involve skin contact. Skin, with its layer of dead cells and often fatty tissues, is a fair insulator. Nonetheless, as voltage increases the skin will reach a point where it breaks down. Then the lowered resistance of deeper tissues allows a greater current to flow. This is why electrical codes refer to the term "high voltage" as a voltage above 600 V.

How little voltage can be lethal?

This depends entirely on the resistance of the two contact points in the circuit, the internal resistance of the body, and the path the current travels through the body. Historically, reports of fatal shocks suggest that as little as 24 V *could* be fatal under extremely adverse conditions. To add some perspective, one standard used to prevent serious electrical shock in hospital operating rooms limits leakage flow from electronic instruments to only 50 µA due to the use of electrical devices and related conductors inside the patient's body.

output circuits. This disconnecting means shall also disconnect the battery from its load. The code further requires that the control for these disconnecting means shall be grouped and identified and shall be readily accessible at the principal exit doors.)

A similar problem exists with solar panel installations. Just because the breaker is turned off doesn't mean that dangerous voltages don't exist on the solar panel. As long as light is falling on them, there is voltage present. With no load, even a relatively dim light falling on part of the panels might present a shock or equipment damage hazard. Modern grid-tie solar systems with no batteries often have the panels wired in series, so several hundred volts is not unusual.

Recent revisions of the NEC have addressed many of the aspects of photovoltaic (PV) installations that present problems with disconnects, bonding, and grounding. Consulting your local authorities is always wise, and there are several organizations such as the Southwest Technology Development Institute at New Mexico State University that have prepared useful information (see the references at the end of this section). In general, PV systems at 12 or 24 V aren't covered by the NEC.

ENERGIZED CIRCUITS

Working with energized circuits can be very hazardous since, without measuring devices, we can't tell which circuits are live. The first thing we should ask ourselves when faced with troubleshooting, aligning, or other "live" procedures, is if there is a way to reduce the hazard of electrical shock? Here are some ways of doing just that.

1) If at all possible, troubleshoot with an ohmmeter. With a reliable schematic diagram, and careful consideration of how various circuit conditions may reflect resistance readings, it will often be unnecessary to do live testing.

2) Keep a fair distance from energized circuits. What is considered "good practice" in terms of distance? The NEC specifies minimum working space around electric equipment depending on the voltage level. The principle here is that a person doing live work needs adequate space so that they are not forced to be dangerously close to energized equipment.

3) If you need to measure the voltage of a circuit, install the voltmeter with the power safely off, back up, and only then energize the circuit. Remove the power before disconnecting the meter.

4) If you are building equipment that has hinged or easily removable covers that could expose someone to an energized circuit, install interlock switches that safely remove power in the event that the enclosure is opened with the power still on. Interlock switches are generally not used if tools are required to open the enclosure.

5) Never assume that a circuit is at zero potential even if the power is switched off and the power cable disconnected. Capacitors can retain a charge for a considerable period and may even partially recharge after being discharged. Bleeder resistors should be installed, but don't assume they have discharged the voltage. Instead, after power is removed and disconnected, use a "shorting stick" to ground all exposed conductors and terminals to ensure that voltage is not present. If you will be working with charged capacitors that store more than a few joules of energy, you should consider using a "discharging stick" with a high wattage, low value resistor in series to ground that limits the discharge current to around 5 to 10 A. A dead short across a large, charged capacitor can damage the capacitor because of internal thermal and magnetic stress. Avoid using screwdrivers, as this brings the holder too close to the circuit and could ruin the screwdriver's blade. For maximum protection against accidentally energizing equipment, install a shorting lead between high-voltage circuits and ground while you are working on the equipment.

6) Shorting a string of capacitors in series does not ensure that the capacitors are discharged. Consider two 400 μF capacitors in series, one charged to +300 V and the other to –300 V with the midpoint at ground. The net voltage across the series string is zero, yet each has significant (and lethal) energy stored in it. The proper practice is to discharge each capacitor in turn, putting a shorting jumper on it after discharge, and then moving to the next one.

7) If you must hold a probe to take a measurement, always keep one hand in your pocket. As mentioned in the sidebar on high-voltage hazards, the most dangerous path current can take through your body is from one hand to the other, since the flow would pass through the chest.

8) Make sure someone is in the room with you, and that they know how to remove the power safely. If they grab you with the power still on, they will be shocked as well.

9) Test equipment probes and their leads must be in very good condition and rated for the conditions they will encounter.

10) Be wary of the hazards of "float-ing" (ungrounded) test equipment.

11) Ground-fault circuit interrupters can offer additional protection for stray currents that flow through the ground on 120 V circuits. Know their limitations. For example, they cannot offer protection for the plate supply voltages in linear amplifiers.

12) Older radio equipment containing ac/dc power supplies have their own hazards. If you are working on these live, use an isolation transformer, as the chassis may be connected directly to the hot or neutral power conductor.

13) Be aware of electrolytic capacitors that might fail if used outside their intended applications.

14) Replace fuses only with those having proper ratings. The rating is not just the current, but also considers the speed with which it opens, and whether it is rated for dc or ac. DC fuses are typically rated at lower voltages than those for ac, because the current in ac circuits goes through zero once every half cycle, giving an arc time to quench. Switches and fuses rated for 120 V ac duty are typically not appropriate for high-current dc applications (such as a main battery or solar panel disconnect).

22.1.13 Battery Safety

As described above, we use batteries in our stations for backup power. However, in amateur radio activities, batteries are also prevalent in many areas outside of the station, such as portable and handheld radios. The characteristics and charging methods for many battery types are contained in the **Power Sources** chapter. There are many potential hazards related to batteries, and different battery technologies have various concerns.

LEAD-ACID STORAGE BATTERIES

The most common form of storage battery is based on lead plates that are bathed in an electrolyte of a strong solution of sulfuric acid. This type of battery is popular because it has a large energy capacity, is relatively inexpensive, and can be recharged many times.

Wet Cells

Wet cells are the type of battery most often used in automobiles. These batteries should never be tipped over or opened because the highly acidic electrolyte inside can spill out and burn anyone who touches it, in addition to the damage that strong acids cause to anything else they encounter.

When wet cell lead-acid batteries are charged, they naturally release hydrogen gas, which is very explosive when concentrated amounts are ignited. Because of the out-gassing during charging, these batteries should be located outside or have special ventilation if used indoors. The electrolyte is depleted over time due to charging and must be replenished with water. If not, the battery can be permanently damaged. Refilling water in a battery should be performed with care to avoid splashing the electrolyte. Always wear eye protection and rubber gloves to avoid injury in case acid splashes out of the battery.

Sealed Lead-Acid (SLA) Batteries

A form of lead-acid battery that does not have similar hazards is the sealed lead-acid (SLA) battery. In these batteries, the highly acidic electrolytic is bound up in a gel structure, often called "Gel-cells." Another type of SLA binds the electrolyte to an absorbent glass mat (AGM). Since they are sealed, the danger of spilling the acidic electrolyte is minimized. SLA batteries also do not suffer from out-gassing, so ventilation is not an issue. Since the by-products of charging are retained in the battery cell, they are eventually recombined with the electrolyte, so it does not require the addition of water. However, if overcharged, or if the battery temperature becomes too high, SLA batteries can build up high pressures. Even though they are designed to allow for this, overcharging, excessive temperature, or damage to the battery case could lead to the battery exploding.

DRY CELL BATTERIES

Most dry cells are alkaline or carbon-based. These batteries are not made to be recharged and can explode if that is attempted. Although rechargeable alkaline batteries have been developed, they are not widely used.

NICKEL-BASED BATTERIES

At one time, the most common, small rechargeable battery used by radio amateurs was either nickel-cadmium (NiCd) or nickel-metal-hydride (NiMH). Battery chargers for Ni-based batteries are common and safely recharge the batteries without damaging them.

LITHIUM-BASED BATTERIES

There have been many stories in the news of lithium-based batteries catching on fire and causing a lot of damage; cell phones, computer laptops, E-bikes, EV-cars, and other devices that use lithium-based batteries have all been reported as catching on fire. Though the number of fires caused by lithium-based batteries is tiny on a percentage basis (estimated to be one in ten million batteries), you should still be aware of the causes of the fires and how to avoid them. There are several reasons that lithium-based batteries catch on fire, the most common being poorly manufactured batteries, overcharging, short-circuiting, and damage to the battery case.

Lithium is a highly volatile element that can burn violently. The most common danger with lithium-based batteries is thermal runaway, which can occur when the battery is overloaded while the ambient temperature is too high. Some types of lithium-ion chemistry are safer than others, but all lithium-based batteries should be treated as if they are capable of thermal runaway. Most lithium battery fires cannot be extinguished with water; chemical fire extinguishers should be used.

It is crucial that lithium-based batteries be charged with the correct charger. They are very sensitive to charge rates, their temperature during charging, and overcharging. Many amateurs have chargers that were made for other battery types, and these should not be used with lithium-based batteries.

Another common cause of lithium-based battery fires is damage to the battery case. Again, some types of lithium chemistry are more sensitive than others to physical damage. An indication that a battery has been over-stressed is that its case starts to bulge. Batteries with obvious damage to their cases, or with cases that are bulging, should be safely discarded. Never throw a lithium-based battery into the garbage. Find a recycling center that can safely discharge the cells and then reuse the elements to make another battery.

A revealing video of stress testing several types of lithium chemistry by overcharging, short circuiting their outputs, damaging them mechanically, and igniting them with external fire sources can be viewed at **www.youtube.com/watch?v=Qzt9RZ0FQyM**.

HAZARDS COMMON TO ALL BATTERIES

Unlike many electronic power sources, which are designed to fold-back their voltages if current limits are exceeded, batteries generally have no limit on the amount of current that they can supply. If battery terminals are shorted together, huge currents can flow until something burns up and the connection is severed. To avoid such dangers, all battery connections should be fused, and battery terminals should be covered when being transported so that they are not accidentally shorted together.

All batteries contain chemicals that could be poisonous. If a battery's case is ever damaged, regardless of the type, the battery should be safely discarded.

AIRPLANE TRAVEL WITH BATTERIES

Since Lithium-based batteries have been identified as having the potential to spontaneously burst into flames, the FAA has placed a limit on the batteries that can be carried on an airplane. The rules state that no lithium-based batteries can be placed in checked luggage (if such a battery starts a fire in the luggage compartment while the plane is flying, there is no way to put the fire out). A maximum capacity of 100 Wh (approximately 8 Ah for a 12 V battery) per battery can be carried on the plane. Larger lithium-based batteries cannot be brought onto commercial airplanes.

OLD BATTERY DISPOSAL

As described in the **Power Sources** chapter, many battery types should be recycled in order to keep poisonous, heavy metals out of our landfills and water supplies. Another danger of discarding a battery is that any remaining charge that it has can start a fire. An added benefit to recycling is that many of the metals used in battery manufacture can be in short supply, and recycling reintroduces these metals to make new batteries, decreasing the need for mining.

No battery should ever be disposed of by burning it, as most have explosive properties and can harm nearby people and the environment.

22.1.14 Electrical Safety References

ARRL Technical Information Service Web page on electrical safety in the Technology area of the ARRL website.

Black & Decker Complete Guide to Wiring, 6th Edition (Cold Spring Press, 2014).

Block, R., "Lightning Protection for the Amateur Radio Station," *QST*, Part 1, Jun. 2002; Part 2, Jul. 2002; and Part 3 Aug. 2002.

Block. R. "The "Grounds" for Lightning and EMP Portection," (Polyphaser Corporation, 1993).

Federal Information Processing Standards (FIPS) publication 94: *Guideline on Electrical Power for ADP Installations*. FIPS are available from the National Technical Information Service.

IAEI: *Soares' Book on Grounding*, available from International Association of Electrical Inspectors (IAEI).

"IEEE Std 1100 - 2005 IEEE Recommended Practice for Powering and Grounding Electronic Equipment," *IEEE Std 1100-2005 (Revision of IEEE Std 1100-1999)*, pp 0_1-589, 2006. "This document presents recommended design, installation, and maintenance practices for electrical power and grounding (including both safety and noise control) and protection of electronic loads such as industrial controllers, computers, and other information technology equipment (ITE) used in commercial and industrial applications."

National Electrical Code Handbook 2017, NFPA 70, National Fire Protection Association, Quincy, MA (**www.nfpa.org**), also available through libraries.
Silver, W., Grounding and Bonding For the Radio Amateur, 2nd edition, ARRL, 2021.
Solar energy websites — **www.nmsu.edu/~tdi/PV=NEC_HTML/pv-nec/pv-nec.html** and **solarabcs.org**
Standard for the Installation of Lightning Protection Systems, NFPA 780, National Fire Protection Association, Quincy, MA (**www.nfpa.org**).
See the Protection Group's list of white papers contained in the Knowledge Base **www.protectiongroup.com**.

22.2 Antenna and Tower Safety

An antenna at height is a critical component of an effective amateur radio station. Hams have a long history of finding creative ways to get their antennas to the necessary height. We don't always think of safety first, but regardless of whether the chosen antenna support is as simple as a tree, or as sophisticated as a tall, purpose-built tower, every antenna project can present many potential risks to life and limb. Some of the hazards that may be encountered include:

- Electrical hazards
- Structural failures
- Hazards associated with lifting and lowering equipment
- Falls from substantial height
- Challenging weather
- Falling objects
- Equipment failure

The topic of antenna and tower safety is by no means an academic pursuit. In 2019 and 2020 alone there were six recorded fatalities related to amateur radio tower work, and serious injuries in other instances. The causes of these tragic incidents were all identifiable in advance, and avoidable. With careful attention and care, no ham needs to lose their life or sustain serious injury while working on an antenna system.

In the initial section, we introduce some important ways of preparation and working on antennas and their supports. This material presents a good overview of safety issues and practices associated with antennas and supporting structures. The remainder of the sections focus primarily on safety issues encountered during tower construction and maintenance, although the techniques may also apply to other types of antenna supports.

The material in this section covers a lot of topics and should be treated as an overview. If you need more information about a particular piece of equipment or technique, see the Reference section for more articles and online information. The *ARRL Antenna Book* chapter on **Building Antenna Systems and Towers** goes into much more detail about these same topics and many others. There is sufficient information in these books, articles, and online so that no one needs to be uninformed about good practices for working on and building antenna systems.

22.2.1 Planning for Safety

Safety isn't something you can just *set-and-forget*. It's not just *buy-and-be-done*. It's a way of thinking, it's pervasive, and it should be ingrained in everything you do. Done right, safety practices result in long-term benefits.

Every antenna project has a lifetime that comprises several important phases, including *Planning, Construction, Maintenance*, and *Removal*. Let's take a look at each of these phases, and the safety practices that can be incorporated into each one.

THE PLANNING PHASE

Every antenna project starts with some sort of *Planning*. Whether it's a simple wire antenna or a much bigger installation involving a tower and rotatable antennas, you'll need to make a range of important decisions before you begin to build.

Design is perhaps the most critical part of the planning phase. In antenna system design, you will consider questions about the location of the antenna system; the size, weight, and strength of the pieces and parts that will be employed; and the strength of all supporting components. Factors such as climate and weather, desired lifetime, maintenance requirements, budgets, and local regulations will all factor into your design. Getting the design right is particularly important because it sets the stage for what will happen throughout the full lifetime of the system. A good design is likely to result in a system that will be effective, reliable, and safe over the course of its entire lifetime.

Conservativism is the watchword for a safe project. Maximum safety can best be ensured by creating the path forward from a conservative perspective. Conservative goals and specifications will not only increase safety margins for the project, they will also reduce the effort and cost associated with the later phases of construction, maintenance, and removal.

THE CONSTRUCTION PHASE

Construction describes the time during which the antenna or antenna support structure is erected, usually one or more days of concentrated effort by a team of hams. Construction is one of the two phases in the life of an antenna or tower during which risk of injury is highest. Some of those involved might have experience, while others are new to the process of antenna and tower construction. Fueled by excitement and anticipation of the moment when the project will be ready to test, the build team is often motivated to get the job done quickly. With pressure to complete any job in the least possible amount of time comes the temptation to cut corners or distractions from safe procedures. The best planning is easily compromised when team members decide to skip steps, work with insufficient skills or tools, or omit critical procedures. This is a period when everyone must be vigilant about keeping safe practices front-and-center.

THE MAINTENANCE PHASE

Once the new antenna or tower has been fully installed and tested, its service life begins. Every owner/operator of an antenna system hopes for a long lifetime of trouble-free operation. But wear-and-tear is a fact of life for all mechanical and electrical devices. Some forms of degradation are predictable, such as loosening of nuts and bolts, corrosion, and wear on ropes or cables exposed to friction. These can and should be addressed with *preventative maintenance* (PM) steps. Annual checkups of hardware are a must. Replace antenna support ropes before they fail, perhaps semiannually. PM will prevent much bigger problems and help avoid the need for much bigger and riskier repair projects.

THE REMOVAL PHASE

The remaining risky phase in the life of an antenna project, particularly with respect to a tower, is removal. A variety of risks present themselves during the take-down of an existing tower. Let's consider two of the most common safety risks related to removal of a tower:

Unknown Construction Materials and Methods

It is not uncommon for hams to assist in removal of a tower or other antenna structure when a ham is moving or is otherwise no longer able to use it. After asking the first few questions, it is very common to find that the original construction details were not documented. Is there concrete in the base and anchors of a guyed tower? How much? How is the roof tower attached to the roof? What material was used for the mast? Can it be identified simply by visual inspection? When the answers to these questions are unknown, workers face significant risks. If there isn't sufficient information to make a safety determination, there is only one correct plan of action: Call in the professionals.

Old Installations

When an installation has been in place for many years, the risks increase. Corrosion is almost certain to have invaded the hardware. Frequently, the corrosion is invisible; even a trained professional may have difficulty detecting it. The opportunity to take on a removal project in exchange for the salvage value of a tower and antennas is a temptation that should be avoided. If you are not highly trained and experienced in working with tower and antenna installations, the right course of action is to call in the professionals.

22.2.2 Antenna Supports and Structures

TREES

The original antenna supports were trees: if you've got them, use them. They're free and unregulated, so it couldn't be easier. Single-trunked varieties such as fir and pine are easier to use than the multi-trunked varieties. Multi-trunked trees are not impossible to use but they require more work. For dipoles or other types of wire antennas, plan for the tree to support an end of the wire; trying to install a center-supported inverted V or similar configuration is almost impossible due to all the intervening branches.

Even when working with trees, it's important to be aware of the potential for safety hazards. There are two primary safety risks associated with using trees as antenna supports:

Proximity to Power Lines

Antennas are metallic and must be kept a substantial distance from power lines. Any wire or tubing coming in accidental contact with a power line is a lethal hazard. Even antennas protected by non-conductive sleeves must be kept from contacting power lines. When you choose trees as your antenna supports, inspect the tree carefully. It is important to avoid using trees that are in close proximity to any power lines that may be difficult to see in or near the tree. Without expert knowledge, you should assume that any wires on utility poles and any wires entering your home at a service entry point are dangerous and must be avoided.

Climbing

In general, safely climbing a tree to install an antenna requires special equipment and training in techniques that protect both the tree and the climber. Professional arborists and tree services have all the necessary experience and equipment. Professional tree climbers are often willing to help for a small fee.

There are quite a few options to support your antenna high in a tree without climbing. Rope saws, pole trimmers, ropes, pulleys, and other tools can establish a strong and secure support point for your antenna.

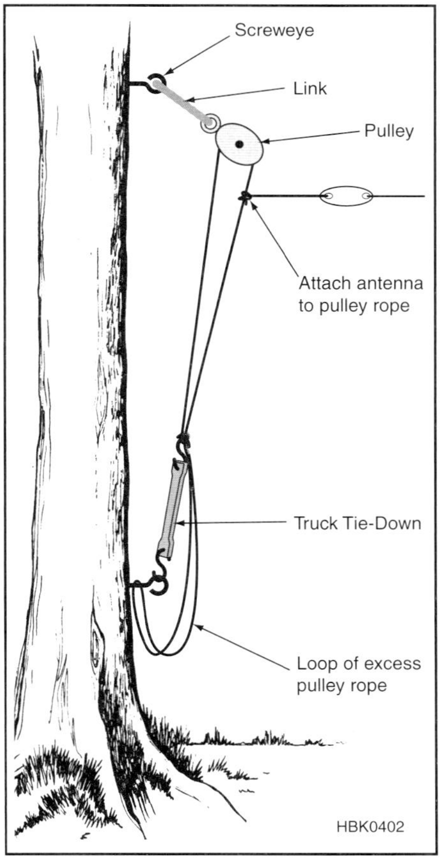

Figure 22.9 — Loop and halyard method of supporting wire antennas in trees. Should the antenna break, the continuous loop of rope allows antenna repair or replacement without climbing the tree.

If you can safely gain access to the tree at the desired height, an effective method is to install a screw-eye with a pulley into the trunk or a strong branch, then trim enough limbs to create a "window" for the antenna through the branches. Finally, attach a rope halyard to the antenna insulator. Here's a useful tip: Make the *halyard* a continuous loop as shown in **Figure 22.9**. Since it's almost always the antenna wire that breaks, a continuous halyard makes it easy to reattach the wire and insulator. With just a single halyard, if the antenna breaks, the tree will have to be climbed to reach the pulley, then reinstall and attach the line.

Another popular option to get your wires high in the trees is by using some sort of launcher. Using a bow-and-arrow is a traditional method of shooting a fishing line over a tree to pull up a bigger line. Some use a slingshot or a fishing rod with a well-chosen weight to get the initial line over a high branch. There are commercial compressed-air launchers or "spud guns" that can reach heights of more than 200 feet! One of the secrets to success in all of these techniques is to start with a light fishing line holding a weight large enough to pull the line back to the ground through the tree. Then use the light line to pull a mid-weight line over the branch, and finally pull the heavy-duty rope into place. The initial weight should be brightly colored so you can find it either on the ground or in the tree.

Somewhat slower, but a lot more controllable, a multi-rotor drone can lift light lines over the highest trees, although not through them. A light line is carried over the tree by the drone, and then the light line is used to pull a heavier line over the tree, just like when using a launcher. The process is described in a pair of *QST* articles: See the Reference section entries for articles by Kam Sirageldin, N3KS (better known as TI5W and TI7W), Michael Shandblatt, W3MAS, and Joe Warwick, KB3ZED.

MASTS AND POLES

Ground-mounted masts and poles come in a wide variety of sizes and strengths. Telescoping fiberglass poles are popular supports for wire antennas, particularly for portable and temporary use. Light-duty TV antenna push-up masts can support small beam antennas. AB-155 military surplus masts use stackable sections of fiberglass and aluminum tubing supported by guys and tripods. Another military-style mast, the AB-577, is an all-aluminum crank-up mast, often called a "rocket launcher," that can support a medium-sized HF beam.

All types of masts present challenges; don't overload them. None of the choices listed here can be climbed, so antennas or mounting gear must be attached at ground level before the mast is raised. Masts with antennas installed can bend or break when being "walked up" to vertical from a horizontal position. Stackable sections require sturdy supports to hold the mast vertical as the mast is lifted to add more sections from the bottom. The mast manufacturer will provide directions and safe loading information. These instructions must be followed to avoid damaging the mast or causing injuries. It is also wise to have a supervisor or observer monitor progress and watch out for unsafe situations as the mast is being erected.

A mast's flexibility makes the side guys especially important since they keep it from bending sideways under load. Masts require extra care during installation to be sure they are kept straight. Once a curve develops, a collapse can occur very quickly. If a team is required to manage guys as the mast is raised, a practice session should be arranged so that everyone knows the procedure and coordinating instructions are followed.

Steel push-up masts are heavy, especially when raising the final sections that are carrying the full weight of the extended mast. A slipping section can seriously injure fingers and hands. Push-up masts can be walked up, but only with light antennas attached (the top section is likely to bend if loaded when it is not vertical).

Wooden utility poles, new or used, are very sturdy but not cheap, and require special installation with a pole-setting truck, and there is also no commercial antenna mounting hardware available for them. That makes them a poor choice for most installations. Do not use existing utility poles as antenna supports if they support utility lines of any type.

TOWERS

Towers require a significant effort to design and install a safe, effective system, but are the only viable choice for large, rotatable antennas and arrays. Along with being able to support the weight and torque of larger antennas and accessory equipment, the options offered by commercial tower manufacturers allow each individual tower installation to be purpose-built to suit the needs of the station owner. A wide variety of requirements can be satisfied with selection of the right tower type and proper construction. There are many safety issues involved that require careful planning and preparation. These are discussed in the following sections.

There are three basic types of towers commonly used by amateurs to support antennas: *roof-mounted*, *self-supporting*, and *guyed*. Each is described in the section below on Design Considerations.

Using towers for portable and temporary stations has its own set of safety challenges. It is common on Field Day, for example, to erect towers of 20 to 40 feet to support a small HF beam. The June 2013 *QST* article "Field Day Towers — Doing It Right," by Daso and Silver, presents information about and guidelines for temporary tower installation.

22.2.3 Build It by the Book

Every aspiring antenna system builder/owner needs to understand and follow two types of information that constrain or guide construction and installation.

Regulations

Most antenna support structures fall under local building regulations as well as neighborhood restrictions. Many housing developments have Homeowner's Associations (HOAs) as well as Covenants, Conditions, and Restrictions (CC&Rs) that may have a direct bearing on whether a tower or similar structure can be erected at all, and, if allowed, how it must be constructed for safety and other concerns. This is a broad topic with many facets. Detailed backgrounds on these topics is provided in *Antenna Zoning for the Radio Amateur* by Fred Hopengarten, K1VR, an attorney with extensive experience in towers and zoning. You may also want to contact one of the ARRL Field Organization's Volunteer Counsels.

Even without neighborhood restrictions, a building permit is likely to be required. With the proliferation of cellular and other commercial wireless devices, and their attendant RF sites, many local governments now require that all antenna support structures meet local building codes. Again, K1VR's book is extremely helpful in sorting all this out. Building permit applications may also require Professional Engineer (PE) calculations and stamp (certification). The ARRL Field Organization's Volunteer Consulting Engineer program may be useful with the engineering side of your project.

It is important to understand that building regulations are the cumulative result of decades of experience, and that compliance with them will result in a structure that is both stronger and safer for you and your neighbors. Working closely with the local building department is often a source of useful knowledge and tips for a better installation. It also gives you permits and signoffs that can provide a level of liability protection in the event of a future incident.

Manufacturers' Guidelines

As you explore the various options in the world of commercial antenna supports, you will find that every manufacturer provides clear instructions on how their products should be installed, and specifications of the minimum performance that should be expected when installed per the manufacturer's instructions. This information is developed by design engineers and is usually signed off by a registered Professional Engineer to certify that the design is in alignment with standard building codes, such as the International Building Code (IBC). The IBC is the base building code used by most jurisdictions in the United States. (See the sidebar on the NEC in the Electrical Safety section earlier in this chapter for more information on building codes.)

Former tower climber Steve Morris, K7LXC, says it well in what is affectionately known in the amateur radio world as the K7LXC Prime Directive: "DO what the manufacturer says." Follow the corollary as well: "DON'T do what the manufacturer doesn't say to do." Professional engineers have analyzed and calculated the proper specifications and conditions for tower structures and their environment. Taking shortcuts or making different decisions will result in a less reliable installation. It is also important to note here that every installation is unique. Your analysis may reveal that even the manufacturer's recommendations are not conservative enough for your particular situation and needs.

DESIGN CONSIDERATIONS

Safety starts with design. Two critically important design parameters to consider when planning a tower installation are the maximum local *wind speed* and the proposed antenna *wind load*. Tower capacities are generally specified in square feet of antenna load. Check with your local building department to find out what the maximum wind speed is for your area. Another source is a list of maximum wind speeds for all counties in the US from the TIA-222, *Structural Standard for Antenna Supporting Structures and Antennas*. (TIA-222-I is the latest revision as of 2024.) A list of wind speeds by county based on TIA-222-G is available online at **www.wirelessestimator.com/content/standards**.

Antenna wind load specifications are provided by the antenna manufacturer. It's important to note that when installing a mast with several antennas, the total effective wind load on the tower is not a simple sum of the wind loads of the various antennas. The true load of each antenna will increase with its height above the top of the tower. A 10 sq. ft. antenna mounted 10 feet above the tower on a long mast will subject the tower to much greater forces than the same antenna mounted right at the top plate of the tower. A proper calculation of wind load should be done by a trained structural engineer. (A procedure for calculating load and mast strength is provided in the *ARRL Antenna Book*.)

Towers come in the guyed and self-supporting varieties shown in **Figure 22.10**. Each has its own safety issues. Guyed towers require a bigger footprint because the guys have to be anchored away from the tower, typically 80% of the tower height. Self-supporting towers need bigger bases to counteract the overturning moment and are more expensive than a guyed tower because there is more steel in them (the cost of a tower is largely determined by the cost of the steel).

The most popular guyed towers are the Rohn 25G and 45G. The 25G is a light-duty tower and the 45G can carry fairly large loads. The online Rohn catalog (see the References, the latest version 3 was released in 2013) has most of the information you'll need to plan an installation and is considered a "bible" of information for tower construction.

The primary risks associated with guyed towers are related to the guys themselves. Guy material, construction, termination, anchoring, and long-term maintenance must be done correctly to keep risks low. At the low end of the guyed tower category is the house-bracketed tower. In this case, the guyed tower sections are placed on a base immediately adjacent to a building, and specially designed brackets are used to support the tower. House brackets are required to be secured to the building in a structurally sound manner. Manufacturers can supply specifications. A local structural engineer can also provide custom plans designed specifically for the unique structure involved.

Fixed, self-supporting towers are made by several manufacturers and allow building a tower up to 100 feet or higher with a small footprint. Rohn, Trylon, and AN Wireless are popular vendors. In general, fixed self-supporting towers are reliable and long-lasting as long as the antenna load is kept well within the tower's specs and the tower is regularly inspected for corrosion. A self-supporting tower requires a substantial concrete base or footing.

Another type of self-supporting tower is the *crank-up*, shown in **Figure 22.11**. One large manufacturer of crank-up towers is US Towers. Using a system of cables, pulleys, and winches, crank-up towers can extend from 20 feet to over 100 feet high. These are

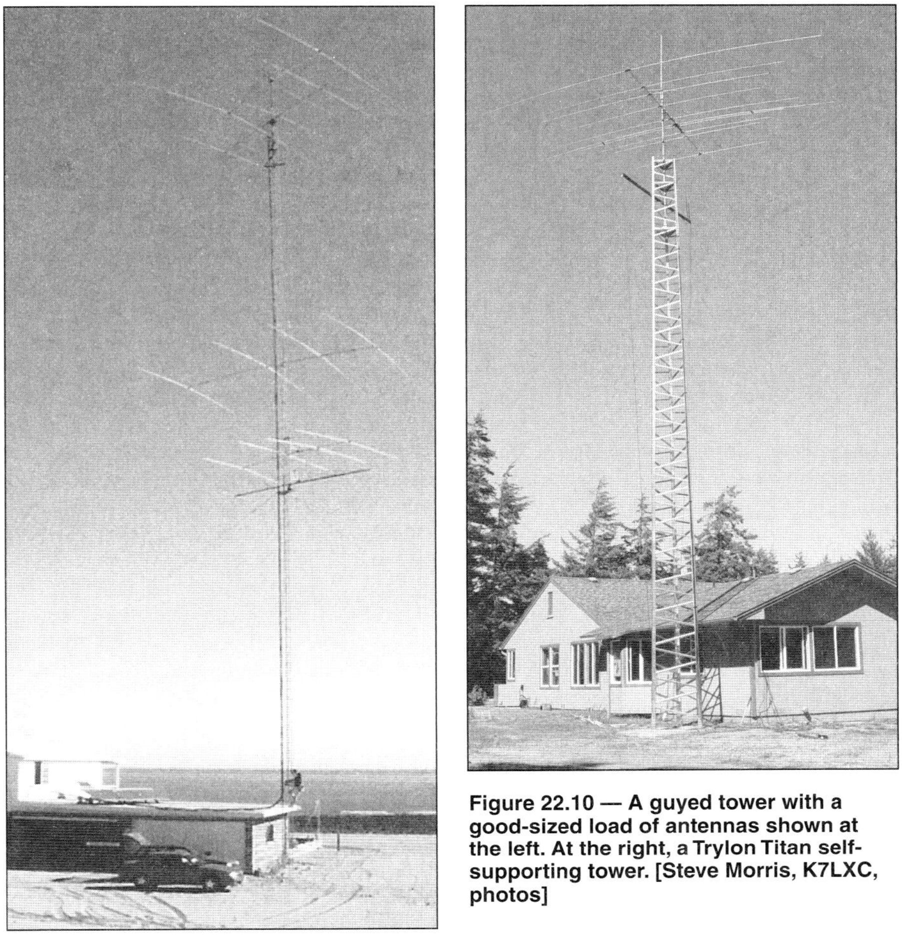

Figure 22.10 — A guyed tower with a good-sized load of antennas shown at the left. At the right, a Trylon Titan self-supporting tower. [Steve Morris, K7LXC, photos]

Figure 22.11 — The bottom of a crank-up tower is shown at left. The motor drive mechanism is on the left and a fishing net on the right catches and coils the feed lines and control cables as the tower is lowered. On the right, a fully loaded crank-up extended to its maximum height of 90 feet. [Steve Morris, K7LXC, photos]

moderately complex devices, and as such, present certain risks. Because they consist of a variety of moving parts, crank-up towers require more maintenance than guyed or other self-supporting designs. As with the fixed, self-supporting tower, a crank-up tower requires a large concrete base.

Owners and climbers of crank-up towers must observe additional safety procedures to ensure that the mechanical state of the tower will not present risks during any operations that involve climbing. The tower sections should be fully nested and blocked to ensure there is no way for any of the sections to slip within the tower unexpectedly during work.

Antennas can also be installed on a *roof-mounted tower*, seen in **Figure 22.12**. These are four-legged aluminum structures of heights from four to more than 20 feet. While they are designed to be lag-screwed directly into the roof trusses, it is often preferable to through-bolt them into a long 2×4 or 2×6 that acts a backing plate, straddling and attached to three to four roof trusses. In any case, if it is not clear how best to install the tower on the structure, have a roofing professional or engineer provide advice. Working on a roof-mounted tower also requires extra caution because you are climbing on a roof while also working on a tower. Many of these lightweight, aluminum towers cannot be climbed, so you must use a ladder to access the antenna. Again, extreme caution is required.

22.2.4 Tower Construction and Erection

THE BASE

Once all the necessary materials and the required approvals have been gathered, tower installation can begin. Let's assume you and your friends are going to install it. The first job is to construct the base. A base for a guyed tower can be hand-dug, as can the guy anchors. For a self-supporting tower, renting an excavator of some sort will make it much easier to move the several cubic yards of dirt.

In the bottom of the hole, be sure to add several inches of gravel or crushed stone that will allow water to drain from the hollow tower legs. If the base section is not going to sit on the gravel, be sure to provide a drainage path for the legs. Water that builds up in the tower legs from condensation or leakage will rust out the steel from the inside where it is not visible until significant damage has been done, creating a severe safety hazard. Similarly, standing water in the legs may also freeze and split the tubing with the same effect on safety.

Next, some sort of rebar cage will be needed for the concrete. Guyed towers only require rudimentary rebar while a self-supporting tower will need a bigger, heavier, and more elaborate cage. Consult the manufacturer's specifications for the exact materials and dimensions. As noted earlier, documentation of how the base is constructed is important for future reference when the tower is maintained or removed. It may also be necessary for a building permit and inspection along with home and liability insurance.

Typical tower concrete specs call for 3000 psi (minimum) compressive strength concrete and 28 days for a full cure. (See the manufacturer's specifications.) A local pre-mix concrete company can deliver it. Pouring the concrete is easiest if the concrete truck can back up to the hole. If that's not possible, a truck- or trailer-mounted line pump can pump it up to 400 feet at minimal expense if using a wheelbarrow is not possible or practical. Packaged concrete from the hardware store mixed manually may also be used, but the volume of concrete required will typically make this a very substantial job. Quikrete Mix #1101 is rated at 2500 psi after seven days and 4000 psi after 28 days.

TOOLS

Once the base and anchors are finished and the concrete has cured, the tower can be constructed. There are several tools that will make the job easier and safer. Since most hams do not build or climb towers on a regular basis, practice using all the tools and

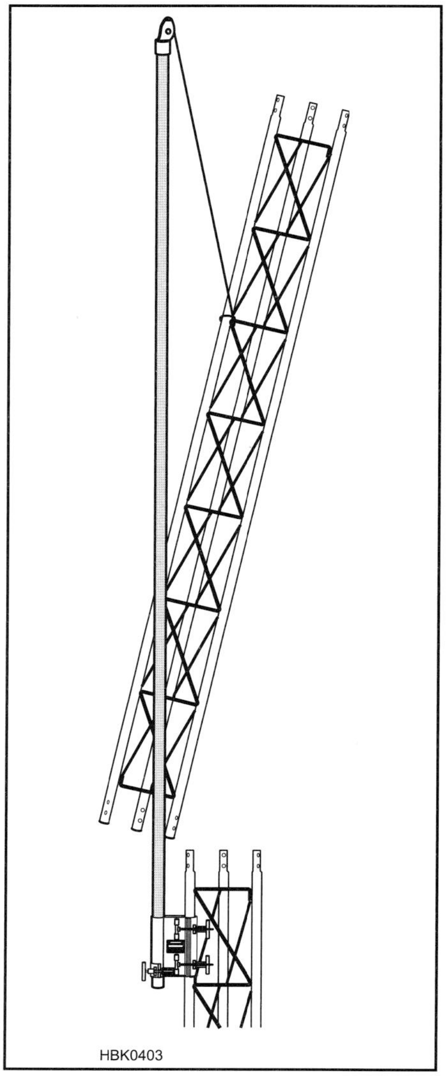

Figure 22.13 — A gin-pole consists of a leg clamp fixture, a section of aluminum mast and a pulley. It is used to lift the tower section high enough to be safely lowered into place and attached. (Based on Rohn EF2545.)

Figure 22.12 — A roof-mounted tower holding a mid-sized HF triband beam. [AA2OW photo]

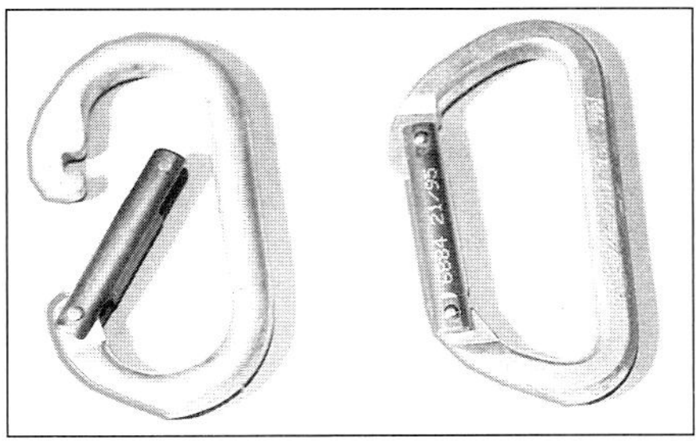

(A)

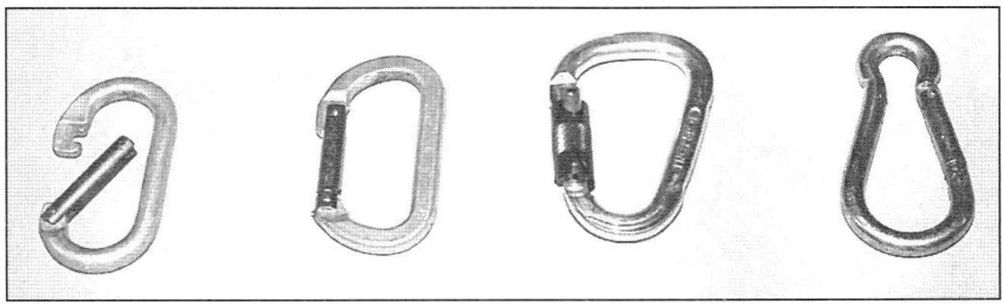

(B)

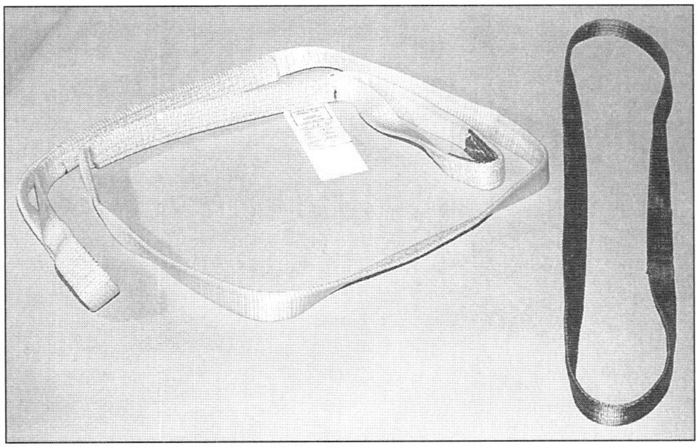

Figure 22.14 — (A) Oval, mountain climbing type carabiners are ideal for tower workloads and attachments. The gates are spring loaded — the open gate is shown for illustration. (B) An open aluminum oval carabiner; a closed oval carabiner; an aluminum locking carabiner; a steel snaplink. (C) A heavy-duty nylon sling on the left for big jobs and a lighter-duty loop sling on the right for everything else. [Steve Morris, K7LXC, photos]

accessories before starting a tower project. Know how to tie and untie basic knots. Practice lifting and lowering operations at ground level before attempting them at height.

If the tower is a guyed tower, it can be erected either with a crane (discussed in the section below on installation) or a *gin-pole*. The gin-pole, shown in **Figure 22.13**, is a pipe that attaches to the leg of the tower and has a pulley at the top for the haul rope. Use the gin-pole to pull up one section at a time (see below).

Another useful tool for rigging and hoisting is the *carabiner*. Shown in **Figure 22.14** (A and B), carabiners are oval steel or aluminum snap-links popularized by rock and mountain climbers. They have spring-loaded gates and can be used for many tower tasks. For instance, there should be one at the end of the haul rope for easy and quick attachment to rotators, parts, and tool buckets, virtually anything that needs to be raised or lowered. It can even act as a "third hand" on the tower. (Do not use light-duty carabiners marked "Not For Climbing.")

Pulleys ("blocks") are required for lifting and maneuvering loads with ropes. At a minimum, one pulley is needed for the haul rope at the top of the tower, and having two or more high-quality pulleys is usually required. Use pulleys rated for heavy loads, such as for marine rigging and rock climbing. A *snatch block* is also useful; this is a pulley with one side that swivels so it can open up to "snatch" (attach to) the rope at any point without having to thread the rope through the pulley housing over the rotating sheave. **Figure 22.15** shows two snatch-block pulleys used for tower work.

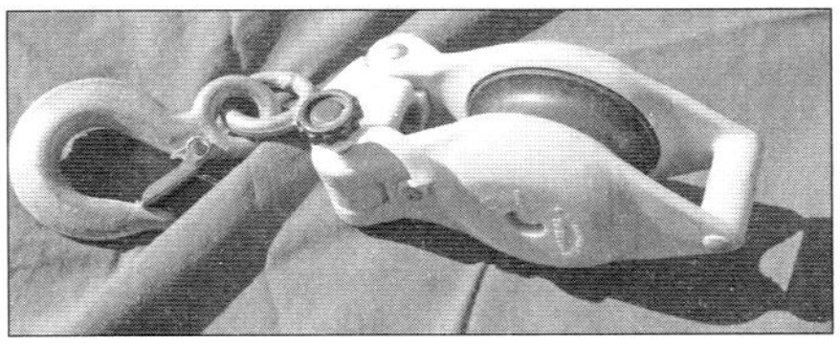

Figure 22.15 — Snatch blocks can be opened to place a rope directly on the sheave without having to thread it through the housing. A lightweight plastic block is shown at left, and two metal housing blocks at right.

Winches and come-along devices can make life a lot easier during certain operations. They provide mechanical advantage to reduce the human force required for lifting and other tasks. Winches can be manually operated or electric. Electric winches can dramatically reduce physical effort, and they can cut hours out of a big tower job, but powered winches come with increased risks. It is often very difficult to assess the tension being applied by a powered winch. Practice and extra care are required when using these powered tools.

ROPES AND SLINGS

The right number and types of ropes used in a tower project can have a dramatic impact on the time, effort, and risks of a tower job. Experienced tower teams will employ several ropes to get the job done. It is quite common to use two ropes in most projects. The first is a utility rope. This rope goes straight up and down the tower through a pulley located at the location where work is being done. It is used to lift and lower tools and other small items.

The second rope is the haul rope. This rope will typically be one-half inch in diameter, or larger, and will afford a good grip for pulling and lifting the heavy items. **Table 22.2** shows the safe working load ratings for common types of rope.

There are several choices of rope material. The best choice is a synthetic material such as nylon or Dacron. A typical twisted rope is

Table 22.2
Rope Sizes and Safe Working Load Ratings in Pounds

3-Strand Twisted Line

Diameter	*Manila*	*Nylon*	*Dacron*	*Polypropylene*
1/4	120	180	180	210
3/8	215	405	405	455
1/2	420	700	700	710
5/8	700	1140	1100	1050

Double-Braided Line

Diameter	*Nylon*	*Dacron*
1/4	420	350
3/8	960	750
1/2	1630	1400
5/8	2800	2400

fine for most applications. A synthetic rope with a braid over the twisted core is known as *braid-on-braid* or *kernmantle*. While it's more expensive than twisted ropes, the outer braid provides better abrasion resistance.

The least expensive type of rope is polypropylene. While it is reasonably durable and cheap, it's a stiff rope that doesn't take a knot as well as other types. Polypropylene is also subject to breakdown over time due to exposure to natural ultraviolet in sunlight. These factors can present real safety risks; it's best to use a higher quality rope.

Two factors that can come into play when selecting a rope are elasticity and flexibility. A more elastic rope will stretch when under load. This can be a helpful characteristic when a shock-absorbing feel is desired. It is not helpful when it is important to maintain the rope's position. Flexibility is a generally good feature, allowing the rope to be more easily maneuvered and tied.

When doing tower work, being able to tie knots is required. Of all the knots, the *bowline* is the one to know for tower work. It's a great advantage to know at least the bowline and various hitches. You can learn how to tie many useful knots at **www.animatedknots.com**.

The *loop sling* in Figure 22.14C can be wrapped around large or irregularly shaped objects such as antennas, masts, or rotators, and can be attached to ropes with carabiners. For a complex job, a professional will often climb with eight to ten slings and use every one!

Web slings are typically made of nylon webbing and come in a variety of lengths and load ratings. The smallest are a few inches long, with half-inch wide webbing, and rated for a few hundred pounds. Large slings are several feet long and rated at a few thousand pounds. They are typically wrapped around the load, such as a mast, or through a hole or section of a load like a tower section. A carabiner around the webbing is used to create a lift point for the sling. There are a variety of "hitches" used for different load shapes and orientations.

INSTALLING TOWER SECTIONS

The easiest way to erect a tower is to use a crane or boom truck, such as used for putting up signs. It's fast and safe, but more expensive than doing it in sections by hand. A crane or boom-truck from a sign company, for example, allows a tower to be constructed on the ground and lifted into place either in one piece or in sections. The crane can hold loads indefinitely while the climbers take as much time as needed to deal with the heavy sections. A crane can also lift an antenna clear of guys and other antennas, avoiding a lot of rope work by the climbers and ground crew.

The crane operator will need access to stable ground near the tower base and clearance to move the tower around. The tower is picked up using a heavy web sling attached above the tower's center of gravity so that it will hang vertically in a stable configuration. Pre-assemble and install guys on the tower so they can be secured before the crane sling is released. A crew guides the bottom of the tower onto the base with the crane holding the weight, then secures the base followed by attaching the sets of guys.

To erect a tower by sections, a gin-pole is needed (see Figure 22.13). It consists of two pieces, a clamp, or some device to attach it to the tower leg and a pole with a pulley at the top. The pole is typically longer than the work piece being hoisted, allowing it to be held above the tower top while being attached or manipulated.

With the gin-pole mounted on the tower, the haul rope runs up from the ground, through the gin-pole mast and pulley at the top of the gin-pole, and back down to the load. The haul rope has a knot (preferably a bowline) on the end for attaching things to be hauled up or down. A carabiner hooked into the bight of the knot can be attached to objects quickly so that you don't have to untie and re-tie the bowline with every use.

It's a good idea to pass the haul rope through a snatch-block at the bottom of the tower, changing the direction of the rope from vertical to horizontal. This allows the ground crew to pull horizontally. It also provides them an easier view up the tower and keeps them away from the tower base in case the climbers accidentally drop something.

GUYS

For guyed towers, it's the guys that do the work of keeping the tower where it should be under working load and the stresses of weather. Important construction parameters are *guy wire material* and *guy tension*. Do *not* use rope or any other material not rated for use as guy cable as tower guys, even during tower erection. Guyed towers for amateurs typically use either three-sixteenth-inch or quarter-inch *EHS* (extra high strength) steel guy cable.

There are also several commercially manufactured non-metallic cables designed to support towers. The most popular and well-tested in the amateur radio market is Phillystran, a lightweight cable made of Kevlar fibers. Phillystran is available with breaking strength similar to EHS cable. The advantage of Phillystran is that it is non-conducting and does not create unwanted electrical interaction with antennas on the tower. It is an excellent choice for towers supporting multiple Yagi and wire antennas and does not have to be broken up into short lengths with insulators.

EHS wire is very stiff. To cut it, use a hand-grinder with thin cutting blades or a circular saw with a metal-cutting aggregate blade. Wear safety glasses and gloves when cutting since there will be lots of sparks of burning steel being thrown off. Phillystran can be cut with a utility knife or a hot knife for cutting plastic. Because Phillystran can be melted or cut, if there is a risk of fire or vandalism, the bottom section of the guy should be EHS.

Guys should be tightened to the manufacturer's specifications. If the guys are too loose, the result will be wind-induced shock loading. Guys that are too tight exert extra compressive load on the tower legs, reducing the overall capacity and reliability of the tower. The proper tension for EHS or Phillystran guys is 10% of the material's *ultimate breaking strength*. For three-sixteenth-inch EHS the ultimate breaking strength is 4900 pounds, and for quarter-inch it's 6,000 pounds, so the respective guy tension should be 490 pounds and 600 pounds. The easiest to use, most accurate, and least expensive way to measure guy tension is by using a Loos tension gauge, developed for sailboat rigging (**www.loosnaples.com/products/tension-gauges/**).

Guy wires used to be terminated in a loop with cable clamps, but those have been largely replaced by pre-formed Big Grips, shown in **Figure 22.16**. These simply twist onto the guy wire and are very secure. They grip the guy cable by squeezing the cable as tension is applied. Be sure to use the right type of Big Grips for the thickness and material of the guy cable. It is also tempting to reuse Big Grips. A limited number of reapplications is permitted by the manufacturer before a grip should be taken out of service. Users should check manufacturer's guidelines for details.

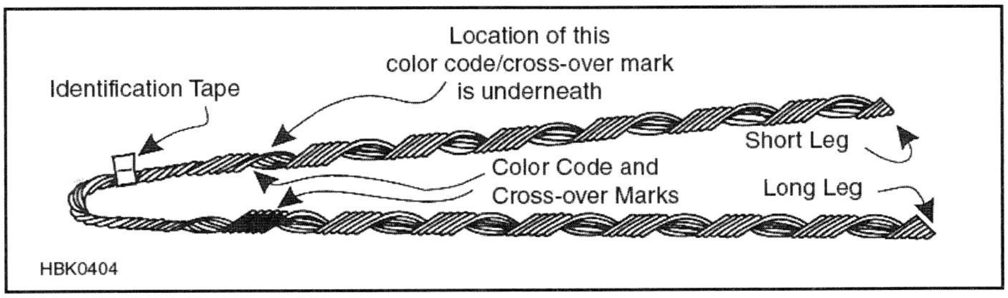

Figure 22.16 — A PreFormed Line Products Big Grip for guy wires.

22.2.5 Antenna Installation

Once the tower is up, it's time to install the antennas. VHF/UHF whips and HF wire antennas are pretty straightforward, but installing an HF Yagi is a more challenging project. With a self-supporting tower, there are no guy wires to contend with. Generally, the antenna can just be hauled up the tower face. Sometimes it is simply that easy!

In most cases, short of hiring a crane, the preferred way to get a Yagi up and down a tower is to use the *tram* method. A single tramline above or between the top guys is suspended from the tower to the ground and the load is suspended under the tramline. **Figure 22.17** illustrates the tram method of raising antennas.

The tram line is typically attached to the mast above the top of the tower. In the case of a big load, the line may exert enough force to bend the mast. If in doubt, *back-guy* the mast with another line in the opposite direction for added support.

MASTS

A mast is a pipe that sticks out of the top of the tower and connects the rotator to the antenna. For small antenna loads and moderate wind speeds, any pipe will work. But as wind speed and wind load increase, more force will be exerted on the mast. Selecting a mast that's up to the job is an extremely important design consideration. There is no tower operation more risky and potentially more expensive than recovering from a bent mast.

There are two materials used for masts: *Pipe* and *tubing*. Tubing is designed to carry structural loads and should be used for large antennas or multiple antennas above the tower. Pipe can be water pipe or conduit (EMT). Pipe is a heavy material with not much strength since its job is just to carry water or wires. Pipe is acceptable as mast material for small loads only. Another problem is that 1.5-inch pipe (pipe is measured by its inside diameter or ID)

Table 22.3
Yield Strengths of Mast Materials

Material	*Specification*	*Yield Strength (lb/in.2)*
Drawn aluminum tube	6063-T5	15,000
6063-T832	35,000	
6061-T6	35,000	
6063-T835	40,000	
2024-T3	42,000	
Aluminum pipe	6063-T6	25,000
6061-T6	35,000	
Extruded alum. Tube	7075-T6	70,000
Aluminum sheet and plate	3003-H14	17,000
5052-H32	22,000	
6061-T6	35,000	
Structural steel	A36	33,000
Carbon steel, cold drawn	1016	50,000
1022	58,000	
1027	70,000	
1041	87,000	
1144	90,000	
Alloy steel	2330 cold drawn	119,000
4130cold worked	75,000	
4340 1550 °F quench 1000 °F temper	162,000	
Stainless steel AISI	405 cold worked	70,000
AISI 440C heat-treated	275,000	

(From *Physical Design of Yagi Antennas* by David B. Leeson, W6NL)

Principles of Working Safely

The following safety tenets founded on three fundamental principles: Do it safely or not at all; There's always time to do it right; and if it's worth doing, do it better.

1. Never load or operate structures or equipment outside the design limits. Be careful with tools, ropes, pulleys, and other equipment that can cause injury or damage if they fail due to overload. Use the right stuff!

2. Always move to a safe, controlled condition and seek assistance when a situation is not understood. This is particularly important when working on towers and antennas. If something doesn't look right or isn't going according to plan, return to a safe state and figure out what to do.

3. Always operate with the safety mechanisms engaged. If a safety mechanism prevents you from doing something, either the task is unsafe, or you may not be using the right equipment.

4. Always follow safe work practices and procedures. Make a plan before you start and don't do something you know is unsafe.

5. Act to stop unsafe practices. The team's safety depends on every team member. Do not hesitate to stop work if you see it is unsafe. Don't be afraid to speak up or ask for help! Regroup and do it right.

6. Clarify and understand procedures before proceeding. This is particularly important when working with a crew. Be sure everyone understands the procedure and how to communicate.

7. Involve people with expertise and firsthand knowledge in decisions and planning. Ask for advice and guidance from experienced hams when planning a task with which you are unfamiliar.

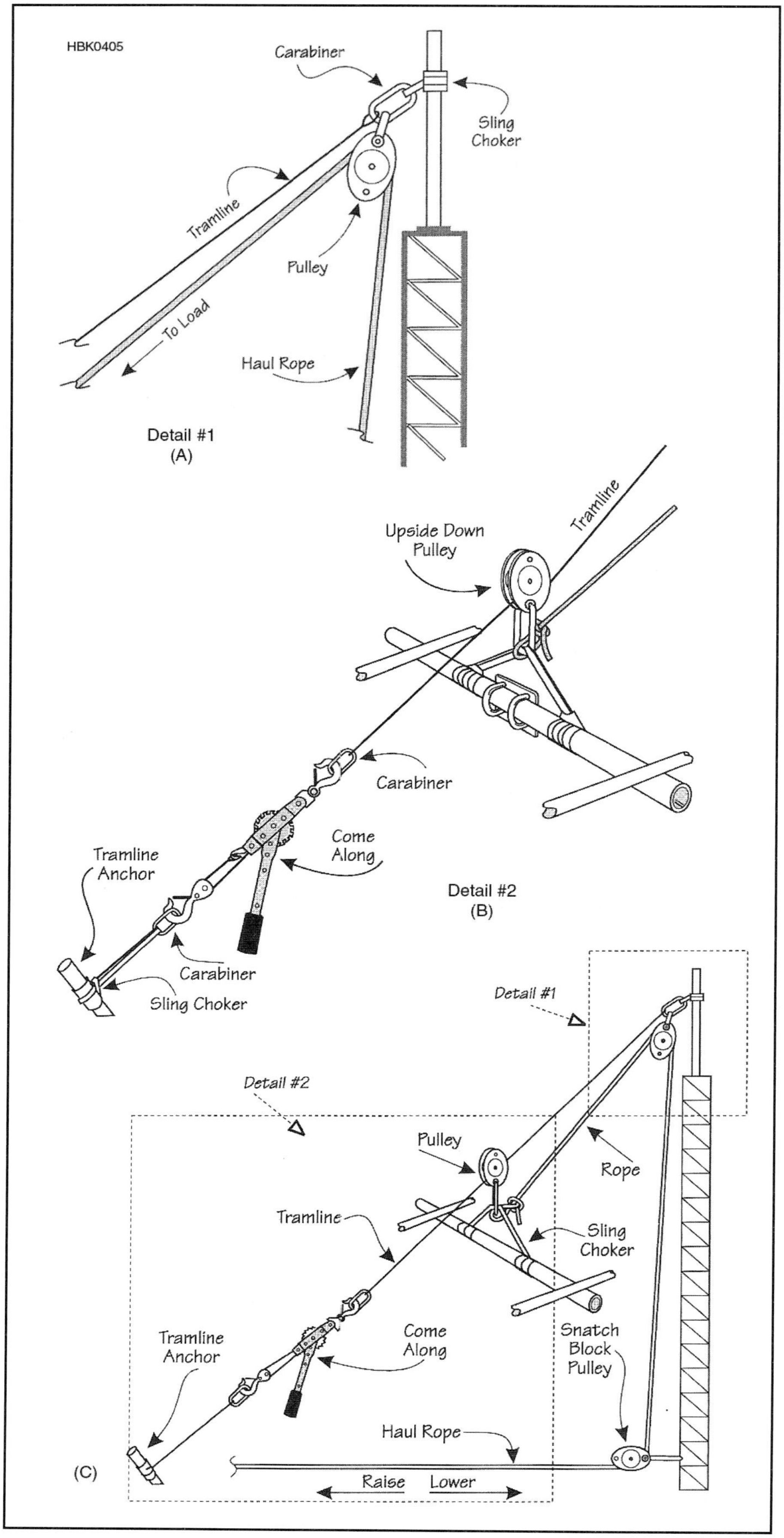

Figure 22.17 — At A, rigging the top of the tower for tramming antennas. Note the use of a sling and carabiner. (B) Rigging the anchor of the tramline. A come-along is used to tension the tramline. (C) The tram system for getting antennas up and down. Run the antenna part way up the tramline for testing before installation. It just takes a couple of minutes to run an antenna up or down once the tramline is rigged.

is only 1.9-inches OD. Since most antenna boom-to-mast hardware is designed for a 2-inch mast, the less-than-perfect fit may lead to slippage.

For any larger load, carbon-alloy steel tubing rated for high strength is highly recommended. A moderate antenna installation in an 80 MPH wind might exert 40,000 to 50,000 pounds per square inch (psi) on the mast. Pipe has a yield strength of about 35,000 psi, so you can see that pipe is not adequately rated for this type of use. Chromoly (short for *chromium-molybdenum* alloy) steel tubing is available with yield strengths from 40,000 psi up to 115,000 psi, but it is expensive. **Table 22.3** shows the ratings of several materials used as masts for amateur radio antennas.

Calculating the required mast strength can be done by using a software program such as the *Mast, Antenna and Rotator Calculator (MARC)* software. (See the References.) The software requires the local wind speed, antenna wind load, and placement on the mast as inputs. The software then calculates the mast bending moment and will recommend a suitable mast material.

22.2.6 Climbing Safety

Working at height is a potentially dangerous activity. As you think about climbing a roof, tree, or tower, it is important to frame every decision in the context of safety. Continually asking the question, “Is this the safest way?” should become second nature. In the workplace this is known as developing a *safety mindset*. Mature organizations and teams build safety considerations into every project they undertake, fully recognizing the costs in time and resources that come with keeping people and property safe. The world of amateur radio is no different. Payback on investments in safety is returned over the long term, manifested in reduced rates of injury and fatalities. Developing a reputation as a safe pursuit also helps to foster the necessary trust for us to pursue the experimental aspects of amateur radio.

In this section, we will focus on tower climbing, but many of the concepts are transferable to other forms of work aloft such as on roofs and in trees, which all have their own safety risks.

TAKE ADVANTAGE OF STANDARDS

Standards organizations provide a great deal of guidance in the area of safety equipment and technique. In the United States, OSHA, the Federal Occupational Safety and Health Administration, collects and analyzes real-world safety data. The agency then works with industry to create and publish high-level regulations for workplace safety. These rules include minimum standards for equipment and procedures. ANSI (American National Standards Institute) is an independent non-profit organization that encourages further development of detailed requirements for safety in many areas, including tower safety. The Telecommunication Industry Association (TIA) produces the most widely accepted standards package covering tower safety, currently at revision ANSI/TIA-222-I updated in 2024.

Although amateurs working on their own non-commercial projects without compensation are not bound by these rules and recommendations, you'll be much better off by following them. There are many suppliers of excellent safety equipment that meets and exceeds the industry standards. As long as you choose compliant equipment and learn safe techniques, you will be on the path to a safe and successful project.

BEFORE YOU BEGIN: PREPARATION

As you contemplate any new project, you will need to define the scope of the entire job.

- What are your goals?
- How will you define success?
- What strategies could you use to achieve the goals?

Consider a job that's diagnostic. You believe there is a problem with a coaxial cable getting hung up on some tower hardware. The goal is to observe the coax as you turn the antenna rotator through 360 degrees of rotation. The traditional approach would be to send a climber up the tower for direct, close-up observation. But new technology can offer faster, easier, and safer options. Consider observing with binoculars from the safety of the ground or an upper floor in a nearby building. Or send a drone right to the location in question and capture recorded video through the full rotation.

Another concept that can greatly reduce risks and make a job go more smoothly is to break the work down into well-defined modules. Successful completion of one module should be required before proceeding to the next. When you break the work down this way, you gain the ability to manage your time more precisely and you have a built-in series of go/no-go decision points.

When you decide that climbing is the right approach, take stock of all elements of the job.

- Skills and experience
- Parts and supplies
- Safety equipment
- Tools
- People
- Time
- Conditions (weather, physical impediments, safety status)

One of the most important aspects of safety that you must assess is whether you have the knowledge and awareness to do a job safely and efficiently. You must have the mental ability to climb and work at altitude while constantly rethinking all connections, techniques, and safety factors. Safely climbing and working on towers is 90% mental. Mental preparedness is something that must be learned. This is an occasion where there is no substitute for experience. Establishing and maintaining the right mental attitude is the hardest step for most people, but when safety gets the focus it deserves, tower work can be done safely, and accidents are relatively rare.

Tower work is the easiest when the weather is nice, and the sun is shining. For raising tower sections or antennas, a relatively windless day is preferred. Professional climbers usually do their trickiest lifts first thing in the morning when the chance of wind is the least. Don't push on in marginal conditions; you may wind up doing more harm than good. Obviously, you don't ever want to climb during a lightning storm. Don't hesitate to postpone your project and wait for the right weather.

PRE-CLIMB SAFETY INSPECTION

Before climbing any tower, perform a careful inspection of the base, anchors, and guys. A pair of binoculars allows inspection of the tower and guys above eye level. There should be no visible rust on the base section, especially where it contacts concrete. Look for splits from water freezing in the tower legs. If any of these are apparent, do not climb until further checks and any repairs are done. A small hole (eighth-inch typically) may be drilled in the bottom of a tower leg without compromising strength in order to drain any standing water. Guys should be correctly tensioned, and all guy attachments and anchors secure and free of rust.

If the tower is a crank-up model, lower it completely until the lift cable is slack. Tower sections can bind and appear to be lowered completely but slip from a climber's weight, causing serious injury. Before climbing, block all tower sections with a 2×4 or heavier material through the tower so that no downward motion is possible. Never assume a crank-up tower is fully nested.

For tilt-over towers or tower-lowering fixtures, perform a careful inspection of the winch, cable, and fixture before raising or lowering the tower. When the tower is near horizontal, the tension in the cable can be thousands of pounds. Make sure no one is near the tower as it is raised or lowered and wear protective gear when operating the winch.

100% TIE-OFF: A RULE TO LIVE BY

Since its inception in the 1970s, OSHA has focused on workplace safety. In the 1980s, OSHA introduced the *100% tie-off* standard requiring a worker's body belt to be secured by at least two lanyards. This protection factor became an essential for employees who executed tasks at height. Fortunately, the body belt soon evolved into a full harness, which further protected workers from spinal and internal organ injuries.

OSHA's fall protection standards even cover fall protection for residential construction projects. Personal fall arrest systems must bring a worker to a complete and controlled stop, restrict workers from falling more than 6 feet, and have the strength necessary to endure twice the impact energy of the worker falling 6 feet.

Why is this type of system necessary? 100% tie-off is part of your plan to "expect the unexpected." While you can be in top physical condition, equipped with the most modern tools, and trained in the latest techniques, unexpected situations still arise. When everything is going well, there is no reason to expect to suddenly "just let go" when perched on the side of the tower. But workers can and do experience sudden health emergencies. A heart attack or a fainting episode due to dehydration can render a worker unconscious. Unavoidable distractions can occur, as well. What if you were to accidentally disturb a bee or wasp nest and suddenly get stung while in your work position? Any of these situations could cause you to lose consciousness, or the attention needed to maintain your footing. In moments like these, your 100% tie-off will save your life.

Fall protection is continuing to evolve in today's workplace, as employers and regulators seek new ways to protect workers. OSHA has ongoing research on how to identify hazards and is also improving current regulations. As a ham, if you know and take to heart the concept of 100% tie-off, you will be well on your way to making your tower work as safe as possible. Done right, climbing can be enjoyable, rewarding, and safe as shown by Marty, NN1C, and Jim, K1IR, in **Figure 22.18**.

PERSONAL SAFETY EQUIPMENT

Check your safety equipment every time before you use it. Inspect it for any nicks or cuts to your belt and safety strap. Professional tower workers are required to check their safety equipment every day; follow their example.

To implement 100% tie-off, you need the right equipment. OSHA specifies the equipment that must be used. Minimum requirements for Personal Protective Equipment (PPE) are given in OSHA Standard 1910.140 - Personal fall protection systems (**osha.gov/laws-regs/regulations/standardnumber/1910/1910.140**). *DO NOT* cut corners on buying or using safety equipment; you bet your life on it every time you use it! **Figure 22.19** shows a typical set of climbing and safety gear used when working on towers.

The most important pieces of safety equipment are the *personal fall arrest system* (PFAS) you wear, and the accompanying lanyards that attach to it. **Figures 22.20A** and **B** show a climber wearing a PFAS from the front and back. The body harness portion of the PFAS has leg loops and suspenders to help spread the fall forces over more of your body and can hold you in a natural position with your arms and legs hanging below you where you're able to breathe normally. (See the warning below about lineman belts.)

Two or more lanyards are used. The first is the fall arrest lanyard (top item in Figure 22.19), which attaches to a D-ring between your shoulder blades. The other ends are attached to large snap hooks ("gorilla hooks") that clip on to the tower above the work or climbing position. The proper choice is a dual shock absorbing lanyard with bar-tacked stitches that pull apart under force to catch and decelerate you in case of a fall. A simple 6-foot rope lanyard is inexpensive but doesn't offer any shock absorption and does not provide 100% tie-off.

The second lanyard is a positioning lanyard (bottom item in Figure 22.19) that holds you in working position and attaches to the D-rings at your waist. It can be adjustable or fixed and may be made from different materials such as nylon rope, steel chain, or special synthetic materials. An adjustable positioning lanyard, as shown in the figure, will adjust to almost any situation, whereas a fixed length one is typically either too long or too short. A rope lanyard is the least expensive. **Figure 22.21A** shows a climber with a positioning lanyard around the tower and the fall arrest lanyards both attached to the tower with one above him. This is a safe working position.

Hams should note that leather climbing equipment was outlawed some years ago by OSHA, so don't use it. This includes the old-fashioned safety belt that utility linemen

Figure 22.18 — Marty, NN1C (left) and Jim, K1IR are fully equipped with the right climbing gear and tools for tower and antenna work. (Jim Idelson, K1IR, photo)

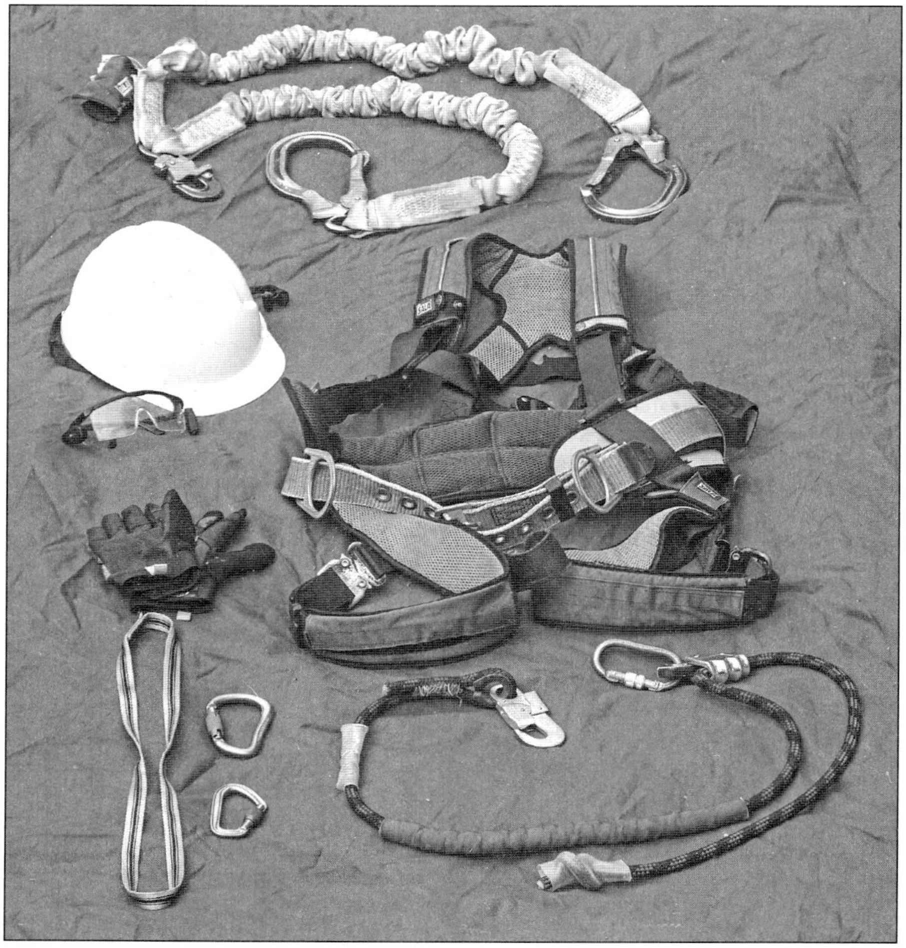

Figure 22.19 — Typical climbing equipment. From the top and left to-right, fall arrest lanyards with "gorilla hooks," hard hat and safety glasses, body harness, gloves, nylon sling, carabiners, and positioning lanyard. (Jim Idelson, K1IR, photo)

(A)

(B)

Figure 22.20 — The personal fall arrest system (PFAS) from the front (A) and back (B). Note that tools and slings are attached in the back where they will not interfere with the positioning lanyard. (Jim Idelson, K1IR, photos)

used for years, which offers no fall protection capability. If you drop down while wearing a safety belt, your body weight can cause it to rise up from your waist to your ribcage where it could immobilize your diaphragm and suffocate you. On the other hand, you can use a safety belt for positioning when it's worn in conjunction with a compliant fall arrest harness (FAH). In fact, most modern FAHs also incorporate a waist belt with positioning lanyard attachment rings for a single, convenient, comfortable climber safety package.

But having the right gear doesn't ensure it is in good working order. Take a page from the OSHA regulations: professional tower workers are required to check their safety equipment every day. Make it your habit to check your safety equipment before you use it every time. Inspect for any nicks or cuts to your harness and lanyards immediately prior to each use.

DRESS FOR SUCCESS

You'll want to equip yourself with some additional gear for a safe and compliant climb:

Boots should be leather with a steel or fiberglass shank. Diagonal bracing on Rohn 25G is only five-sixteenth-inch rod. Spending all day standing on that narrow support will take a toll on your feet. The stiff shank will support your weight and protect your feet; tennis shoes will not. Leather boots are mandatory on towers like Rohn BX that have sharp X-cross braces. Your feet are always on a slant and against a sharp angle of sheet metal. That is hard on feet.

A hard hat and safety goggles are both required and highly recommended. Make sure they are ANSI or OSHA approved and that you and your crew wear them. As you'll be looking up and down a lot, a chin strap is essential to keep the hard hat from falling or blowing off. You may not need the safety goggles until you are at the work position but have them available in your tool kit.

Tower work can be hard on hands, especially if you don't work on towers frequently. Gloves are essential. Keep several spare pairs for ground crew members who show up without them. Cotton gloves are fine for gardening, but not for tower work; they don't provide enough friction for climbing or working with a haul rope. Gloves designed for the job and the environment are the only kind to use. You will find a range of options made from leather and other synthetic materials that are up to the job. The softer the gloves, the more useful they'll be during intricate parts of the job. Framer's gloves with the tips of the thumb and first two fingers removed protect the palm and back of the hand while allowing fine fingertip control but expose fingers to cold weather. You will need warmer gloves if you are doing a job in cold weather. The bottom line is that you should maintain a selection of gloves that give you the right combination of gripping power, dexterity, and warmth to handle all the different jobs and weather you will encounter.

TOWER CLIMBING BEST PRACTICES

Just having the right gear does not ensure that it will save your life; you must use it correctly. The most important rule for using a full body harness and dual fall arrest lanyard is to follow these 100% tie-off steps as you climb. Figures 22.21B and C show the proper way to climb while maintaining 100% tie-off.

1) Attach one lanyard hook above you as you start your ascent.

2) In the case of most ham towers, attach to a tower leg, as the leg provides the strongest anchor point. A tower rung may not be strong enough to sustain the downward force of a fall as specified in the OSHA requirements.

3) As you climb, attach the second lanyard hook to a tower leg a few feet above the first. It helps to repeat "On before off" so you reinforce doing things in the proper order.

4) Disconnect the lower hook and continue to climb.

5) As you climb, alternately place the hooks such that you are always securely connected by one of the lanyards ends and try to keep your attachment point above you as much as possible. This will limit the maximum distance you can fall at any moment.

6) When you reach your work position, attach both lanyards above you for redundant fall protection while you do the job.

7) Reverse these steps as you descend.

Note that the sequence described above does not call for use of a positioning lanyard. That is because while a positioning lanyard is helpful in providing comfort while climbing or working, it does not meet

(A) (B) (C)

Figure 22.21 — Proper 100% tie-off technique (see text). Lanyard attachment at the work position is shown in A. Getting ready to climb with both fall-arrest lanyards attached is shown at B. Moving one of the fall arrest lanyards while maintaining tie-off is shown at C. (Jim Idelson, K1IR, photos)

the stringent requirements for fall arrest. So, you can, and should, use positioning lanyards to make your work easier, but don't depend on them for fall protection.

Many climbers have developed ways of thinking about movement on a tower that force safer behavior. One such model is the concept of having four points of attachment and security, two hands and two feet. When climbing, move only one point at a time. That leaves you with three points of contact and a wide margin of safety if you ever need it. This is in addition to having your fall-arrest lanyard always connected.

Another recommended technique is to always do everything the same way every time. That is, always wear your positioning lanyard on the same D-ring, and always connect it in the same way. Always look at your belt D-ring while clipping in with your safety strap to confirm that you're securely attached. A visual check helps ensure you are connected to a D-ring and not some other piece of gear. Always look!

Safety Tips

1) Don't climb with anything in your hands; attach it to your safety belt if you must climb with it or have your ground crew send it up to you in a bucket.

2) Don't put any hardware in your mouth; it can easily be swallowed.

3) Don't climb while wearing rings and/or neck chains; they can get hooked on things. Long hair should also be secured.

4) Be on the lookout for bees, wasps, hornets, and their nests. If you do run into a stinging insect, use Adolph's Meat Tenderizer on the sting. It contains the enzyme papain, which neutralizes the venom. Have a small jar in your tool kit.

5) Don't climb when tired; that's when most accidents occur.

6) Don't try to lift anything by yourself; one person on a tower has very little leverage or strength. Let the ground crew use their strength; save yours for when you really need it.

7) If a procedure doesn't work, assess the situation and re-rig, if necessary, before trying again.

A TEAM EFFORT: CLIMBERS AND GROUND CREW

Every antenna or tower project is a team effort, and the ground crew is critical to a successful outcome. Don't underestimate the impact the ground crew will have on your project. Consider these points when you put together your support team.

1) Put together a team that has the right mix of skills. If you are going to be tramming an antenna, the ground crew should know how to do it.

2) Make sure the team is committed to the job. Commitment means staying for the duration and not being distracted while the job is under way.

3) Communication is critical. The ground crew should be equipped to communicate with the tower climber(s) with appropriate gear. Many teams use handheld VHF or FRS radios for this purpose.

4) Tools are a major responsibility for the

ground crew. Climbers should only need to do the minimum amount of climbing necessary to get the job done. Ground crew need to know where to find a needed tool and have a way to get it into the hands of a climber quickly. Ground crew will also need to be fully up-to-speed on operation of specialized tools like the gin pole, come-along, and winches.

WORKING AS A TEAM

Before tower work starts, have a safety meeting with the ground crew. This "tailgate" meeting is important to get everyone on the same page, detailing the work to be done and how to do it. This is also the time and place to introduce the ground crew to gear or hardware they may be unfamiliar with, such as a capstan winch, carabiners, rock climbing gear, special tools, and so on. Explain what is going to be done and how to do it.

The climbers are frequently the most experienced members of the tower team. As such, one of the climbers is usually designated to be the team leader. This leadership role is critical for everyone to understand. The climbers are most at risk and often have a unique visual perspective on what's happening at any moment. It's very important for the team to stay focused on the job and to be actively listening for instruction from the climbers. There can be long stretches of work that only involve the climbers, but a need for ground crew support can arise in an instant and the team needs to be ready to respond immediately. Remember, don't do anything unless directed by the leader in charge. This includes handling ropes, tidying up, moving hardware, and so on.

Another important role for the ground crew is to communicate updates and any other information that can be helpful to the team on the tower. As a climber, it is very helpful to hear about anything that might affect upcoming actions or decisions. This might include notifications about changing weather, changes in the ground crew, or updates regarding supplies or parts of the project being handled on the ground. Let the tower crew know when it's time to break for lunch!

If radios are not being used to communicate, when talking to the climbers on the tower, look up and talk directly to them in a loud voice. The ambient noise level is significantly higher up on the tower because of traffic, wind, and nearby equipment. Have an agreed-upon set of hand signals for raising, lowering, holding, starting, and stopping, etc.

While leadership generally comes from the climber, it is very important for all team members to understand that responsibility for watching for safety issues belongs to everyone. Any team member who senses an impending safety risk should raise the issue right away. The team should respect any safety-related observation, pause, and evaluate the situation. The team should then decide what to do. Action should only be taken when the team has determined best right course to pursue.

Be conservative throughout the course of the job. Operate slowly; trying to go too fast can result in poor work and cutting corners on safety.

WRAPPING UP

When is it time to quit? We start all these projects with the goal of finishing up when the job is done, when the project goals have been met, and all within the planned time frame. But, despite the best of planning, complex jobs often take longer than expected. The question every antenna team will invariably face is what to do when it is becoming clear that the project can't be completed as planned. Sometimes it's a matter of running out of daylight. Or the weather is changing. Perhaps some members of the team must leave, or fatigue is setting in after a long day of work. Decisions made in these moments can be the difference between success and disaster.

The temptation is to work faster to try to squeeze the rest of the job into the allotted time, but this is a recipe for failure. Many accidents happen during these exact circumstances. It is far better to choose a path that allows for securing the site in a safe state, getting the climbing team safely back on the ground, and preparing to complete the job at the next opportunity. All this wind-down effort takes time, so the experienced crew should be anticipating the best point to suspend work, leaving sufficient time to wrap things up.

22.2.7 Antenna and Tower Safety References

ORGANIZATIONS AND STANDARDS

ANSI/ASSP

Fall Protection and Fall Restraint Code Z359 — **www.assp.org/standards/standards-topics/fall-protection-and-fall-restraint-z359**

ANSI/TIA

Structural Standards for Antenna Supporting Structures, Antennas and Small Wind Turbine Support Structures, TIA-222-I — **www.store.accuristech.com/standards/tia-tia-222-i?product_id=2592271**

OSHA (Occupational Health and Safety Administration)

Fall Protection Standards — **osha.gov/fall-protection/standards**

Personal Protective Equipment — **osha.gov/laws-regs/regulations/standardnumber/1910/1910.140**

ARRL Resources

ARRL Volunteer Counsel program — **www.arrl.org/volunteer-counsel-program**

ARRL Volunteer Consulting Engineer program — **www.arrl.org/volunteer-consulting-engineer-program**

Amateur Publications

Publications listed here may not include current safety practices but include much useful information on techniques, tools, and materials.

Brede, D., W3AS, "The Care and Feeding of an Amateur's Favorite Antenna Support — the Tree," *QST*, Sep 1989, pp. 26 – 28, 40.

Daso, D., K4ZA *Antenna Towers for Radio Amateurs* (ARRL, 2010).

Daso, D., K4ZA, "Workshop Chronicles" (column), *National Contest Journal.*

Daso, D., K4ZA, and W. Silver, NØAX, "Field Day Towers — Doing It Right," *QST*, June 2013, pp. 67 – 69.

Hopengarten, F., K1VR, *Antenna Zoning for the Radio Amateur* (ARRL, 2002).

Morris, S., K7LXC, *Up the Tower: The Complete Guide to Tower Construction* (Champion Radio Products, 2009).

M. Shanblatt and J. Warwick, "Hanging a Treetop Antenna with a Drone," *QST*, Mar. 2018, pp 67-68.

Silver, H., NØAX, ed., *The ARRL Antenna Book*, 24th Edition (ARRL, 2019).

Sirageldin, K., "Using a Drone for Antenna Installation at TI5W," *QST*, Nov. 2017, pp. 85 – 86.

ONLINE RESOURCES

Champion Radio Products (a useful resource for tower construction and safety products) — **www.championradio.com**

Knot-tying website (tie the right knot for any antenna system application) — **www.animatedknots.com**

NATE: The Communications Infrastructure Contractors Association Safety Resources — **natehome.com/safety-education/safety-resources**

Wireless Estimator (commercial tower safety and standards and incident reporting) — **www.wirelessestimator.com**

Zero Falls Alliance (repository of amateur radio tower safety statistics, root causes, equipment and techniques) — **www.zerofalls.org**

22.3 RF Safety

This section was written by Gregory Lapin, PhD, PE, N9GL, chair of the ARRL RF Safety Committee. The ARRL's RF Safety Committee reviewed this content. Additional material is available as supplemental information on the ARRL's RF Exposure web page at **www.arrl.org/rf-exposure**.

22.3.1 The Need for an RF Safety Program

We have all seen water boiling in a microwave oven. Microwave ovens generate a type of RF energy that can harm our bodies if we are exposed to it at sufficiently high levels. This is why reasonable precautions are taken so that RF energy at the frequencies and levels used by amateur radio stations are safe.

Based on over 100 years of experience in amateur radio, operators who are regularly exposed to lower levels of RF energy have similar, and sometimes less, disease than the general population (see the sidebar "Standards, Science, and the Community"). Similar encouraging health outcomes from the vast majority of people on Earth who use cellphones on a regular basis indicate that when properly controlled, exposure to RF energy does not need to be a concern. Clearly, RF can be used safely.

To be dangerous, RF must be absorbed into the body's tissue with sufficient power to cause heating for which the body cannot compensate. As licensed radio amateurs, we all need to ensure that our transmissions do not expose any people beyond the levels that have been deemed to be safe. Standards organizations have reviewed and analyzed scientific research that has been performed for over 60 years and have derived safe levels of exposure for people. Regulatory agencies, such as the FCC in the United States, have developed regulations that require all operators of RF transmitters to maintain exposure levels below specified limits.

Modern research into the biological effects of exposure to electromagnetic energy has been taking place since the 1950s. The United States Navy sponsored much of the early research and formed the first electromagnetic exposure safety committee with the help of academics and the IEEE. This later evolved into the IEEE C95 family of standards on electromagnetic safety, which includes both recommended safety levels for exposure as well as recommendations for establishing RF safety programs.

An RF safety program should be developed for each amateur radio station. The program consists of an analysis of potential exposure levels around the station. If overexposure to individuals is possible, the program determines what forms of mitigation will be necessary to prevent anyone's exposure from exceeding the regulatory thresholds. Although not required by the FCC, documentation of the RF safety program should be filed away so that it can be referred to if there is ever any question about RF exposure from that station.

Part of RF safety is recognizing that non-ham acquaintances and family members may have concerns about RF exposure. The material in this section will help you explain what hazards exist, and what steps you have taken to ensure that you are operating safely. Some people may have concerns that are difficult to address. For such cases, the ARRL provides "How to Interact with a Concerned Neighbor," which provides some guidance in responding. In addition, searching on the internet often leads to information that is taken out of context or is simply false, which can raise unwarranted concerns. The paper "Interpreting the News About RF Exposure Discoveries" may be helpful when such issues are brought to you. Both papers are available at the ARRL's RF Exposure website, **www.arrl.org/rf-exposure**.

22.3.2 Effects of RF on the Body

NONIONIZING RADIATION

When we speak of *radiation*, we are referring to the property of energy to move from one place to another or radiate. The terminology has unfortunately been associated by the public with the energy that comes from radioactive devices, such as nuclear bombs, X-ray machines, and other sources of ionizing radiation. When members of the public hear the word, *radiation*, they often think of danger.

There are two distinct types of photon energy that radiate: *Nonionizing* and *Ionizing radiation*. These phrases refer to the ability of the energy contained in a photon to push electrons out of their orbits and ionize chemical compounds. When the chemicals in biological tissue are ionized, generally the tissue no longer functions properly. The photon energy of the radiation is what determines its potential to cause ionization; photon energy is directly proportional to frequency. Thus, electromagnetic energy with frequencies above the ionization limit are far more dangerous than photons at lower frequencies. The division between nonionizing and ionizing radiation is in the ultraviolet light spectrum, with ionizing radiation having frequencies above 8×10^{14} Hz (800 THz).

The highest RF communications frequencies in use today are below 500×10^9 Hz (500 GHz), and most amateur operations occur at frequencies far below that. Thus, amateur radio makes use of nonionizing radiation that is thousands or millions of times below the frequency that would cause it to ionize tissue. This is significant, since amateur radio transmissions have insufficient energy to ionize, or alter, DNA molecules in our tissue. Alteration of DNA molecules by ionization, such as with ultraviolet light, is generally believed to have the potential for causing cancer (which is why we wear sunscreen to filter out UV light from the sun to avoid skin cancer).

TISSUE HEATING

Even though nonionizing radiation cannot alter tissue by rearranging its electrons, it can still cause damage with heat. The heat generated in tissue is proportional to the rate of RF energy absorption in the tissue; the rate of energy absorption is determined by the incident power density of the electromagnetic waves and the electrical properties of the tissue. Radiation interacts with any substance in three ways: it can reflect from its surface, it can pass through, and it can be absorbed. The portion that is absorbed is generally converted to heat. The human body core is referred to as *homeothermic*, meaning that body temperature is relatively uniform throughout the body and remains within a relatively narrow range. Your core body temperature is typically 98.6°F (37°C) and if your core temperature rises to be more than 104°F (40°C), your life could be danger.

The body has developed efficient methods to remove excess heat. Control of blood flow through tissue, sweating, and breathing all are used in the tight control of tissue temperature. As long as the combined heat load resulting from the body's metabolic rate, heat absorbed from the environment, and any heat generated by absorbed RF can be quickly and effectively removed by the body there is no danger of adverse health effects from the exposure. When the additional heat load from RF energy absorption exceeds the ability of the body to remove it, health can be compromised.

Heating vs Non-Thermal RF Safety

RF safety exposure standards were originally based solely on the generation of heat from absorbed RF energy. The initial standards were derived from a single incident exposure limit of 10 mW/cm^2 for the entire spectrum of RF frequencies and assuming the body was exposed to a uniform field. Over time, this limit was revised to account for how the body absorbs RF energy differently at different frequencies, analogous to an antenna capturing more power from an RF field at its resonant frequency and less at non-resonant frequencies.

While the effects of heating from absorption of RF energy are easily calculated, there have been several demonstrations of biological effects from exposure to electromagnetic energy. For instance, change of the operation of calcium channels in isolated cells exposed to levels of electromagnetic energy too low to cause measurable heating has been demonstrated in the laboratory; a number of other non-thermal effects have also been demonstrated.

Although reports of RF effects in isolated cells in the laboratory is of interest to the scientific understanding of the biological effects of electromagnetic energy exposure, such reports remain inconclusive as to the potential for causing adverse health effects in humans. To address this, the standards bodies have based their decisions about safe exposure levels on scientific studies that demonstrate deleterious effects on animals or on calculated body and tissue temperature increases presumed to be unsafe. Once the thresholds of potentially adverse effects have been determined, safety factors have been applied to arrive at acceptable exposure limits that include a margin of safety.

22.3.3 RF Safety Standard Development

Most of what we know about operating with safe exposure levels comes from over 60 years of scientific study of how electromagnetic energy affects biological tissue. Thousands of studies have been summarized by standards bodies that then identified levels of exposure considered to be safe. The two most recognized standards bodies are The Institute of Electrical and Electronics Engineers (IEEE), the International Committee on Electromagnetic Safety (ICES), and the International Commission on Non-Ionizing Radiation Protection (ICNIRP). The FCC has based its exposure regulations on both the IEEE C95.1-1991 standard and recommendations from a scientific group chartered by the U.S. Congress, the National Council on Radiation Protection (NCRP), in their NCRP Report #86. To read more about the history of electromagnetic standards development, see the RF Safety Standard Development section in the *Handbook's* online content.

TYPES OF SCIENTIFIC STUDIES

There are two basic types of scientific studies that are used to determine the limits of safe exposure to electromagnetic energy. Laboratory studies use either isolated cells or animals to test for effects of highly controlled exposures. Epidemiological studies focus on the incidence of adverse effects or diseases in the population to try to identify trends that may be related to some type of exposure. All scientific studies must deal with the biological variations between various individuals. Natural variations are separated from the effects of the stimulus under examination by examining large populations of subjects that have been exposed to the stimulus (study population), and those that have not (control population). If the number of subjects studied is large enough, and all variables and stimuli except the stimulus under study are identical between the study and control populations, statistical analysis will point to any effects that are associated with the stimulus in question. To read more about scientific studies, including epidemiological studies of amateur radio operators, see the Types of Scientific Studies section in the *Handbook's* online content.

Neither type of study can be conclusive as to the formation of a disease or production of some form of adverse reaction. An epidemiological study can indicate an association between a given stimulus and disease, and laboratory experimentation can shed light on the mechanisms that may cause disease, but neither type of study on its own is capable of proving a causal link between RF exposure and human disease.

All scientific studies are made public through the process of peer review. The results of a study are written as a scientific article, which briefly reviews previous work on that subject, specifies the methods that were used to obtain the reported results, presents the results, and then includes an interpretation of the results. The written report of a study is submitted to a scientific journal, generally which specializes in the topic of the current study. The journal sends copies of the report to a number of peer reviewers, who use their expertise in the subject to critique the study and, after it is acceptable to the peer experts, is then published in the journal. The process of peer review has been widely considered an acceptable gate keeper that separates good from bad science.

In recent years, there has been evidence that the peer review process is not infallible. Many scientific publications that clearly do not have the scientific basis to make the claims that they do have been published as having been peer reviewed but appear to have bypassed the process. There are several ways that this has been done. For instance, papers that are not accepted by journals in their field are sometimes resubmitted to journals in other fields where the peers are not experts in the topic of the paper.

The scientific publication process also includes the ability for other peers, who were not asked to review a paper, to comment on science that they do not agree with. The comments are published in future editions of the journal and the original authors are given the opportunity to reply. Often, the comment and reply process is not followed with the original paper that is being challenged.

As weaknesses in the peer review process have become more evident, one tool that the scientific community has come to rely on is independent replication of results. If a scientific study provides results that contradict what has been seen in the past,

The ARRL RF Safety Committee

The ARRL maintains an RF Safety Committee that is composed of scientific and medical experts in the many aspects of the study of RF safety. The RFSC serves as a resource to the ARRL Board of Directors, the ARRL Laboratory, and to the amateur community. It regularly monitors new scientific research and many of its members participate in the scientific committees that write safety standards for RF exposure. The RFSC participates in generating the RF safety questions for FCC amateur question pools and works with the FCC in developing its environmental regulations. For more information about the RFSC, see **www.arrl.org/arrl-rf-safety-committee**.

Standards, Studies, and the Community

You or your neighbors are likely to have questions about where standards for RF Exposure come from or how science evaluates the effect of RF on people. There are two papers on the **www.arrl.org/rf-exposure** web page and in this book's online information to help understand these questions: "RF Safety Standard Development" and "Types of Scientific Studies." To help you explain to your neighbors that amateur radio is safe because you operate your equipment safely, the discussion "Helping Amateurs Interact with Neighbors about Radio Transmissions" can make the conversation easier for everyone. Finally, there is so much information online with more coming out every day. "Interpreting the News About RF Exposure 'Discoveries'" will help you deal with information that might be out of context or simply untrue. These documents will be helpful as you navigate questions about RF and amateur radio transmissions.

it is important that the new results be confirmed by an independent laboratory that follows that first study's procedures. Often, independent replication is able to identify errors in the original study that led to the unique results.

22.3.4 FCC RF Exposure Requirements

The FCC is given the responsibility under the National Environmental Policy Act of 1969 to control the operations of stations that it regulates in order to prevent adverse effects on the environment. In 1996, this duty was expanded to prevent the signals from all transmitters, including those of the Amateur Radio Service, from causing excessive human exposure. Since the FCC is not a health and safety agency, it relied on accepted RF safety standards and advice from the FDA to develop its rules. The bases of the FCC safety thresholds were NCRP Report #86 and ANSI/IEEE C95.1-1992, which were discussed earlier.

SAR vs MPE

As per the safety standards, the FCC defined *maximum permissible exposure* (MPE) limits in terms of the power absorbed in tissue, the *specific absorption rate* (SAR). A fixed, safe SAR value applies across the frequency spectrum and is defined for exposure of the whole body or of a specific area (local exposure). At lower frequencies in the range of whole-body resonance, the whole body averaged SAR (in watts per kilogram of body mass) is the appropriate limit. At higher frequencies with exposures from antennas that are close to the body, localized SAR (in watts per kilogram of tissue, measured in a small amount of tissue, either a single gram or ten grams in size) becomes a more important measure of exposure.

It is very difficult to measure or model SAR. To measure SAR, one would have to place a probe directly in the tissue at many locations and monitor the rate of absorbed electromagnetic energy over time. Modeling SAR is also complicated since the varying shapes of anatomical structures and their electromagnetic properties must be modeled over the entire body. While these tasks are not impossible, they are complex enough that few amateur licensees would have the ability or means to determine if exposure from their stations exceeds the FCC limits.

In contrast to determining SAR, electromagnetic energy in the air is much easier to either measure or model. To make the exposure limits easier to follow, the standards bodies developed equivalences between electromagnetic energy incidents on the surface of the body and the whole-body SAR that would result. Even though the safe SAR limit is fixed across the spectrum, the actual SAR that results from exposure, both within the body as a whole and within different organs, depends on frequency.

The relationship between wavelength and the sizes and shapes of the body and its organs causes resonances in some frequency ranges that result in more energy being absorbed. In **Figure 22.22**, which represents the MPEs in the FCC regulations, this effect is clear. The average human body and its larger structures are most resonant to frequencies in the VHF range, and because of this, the MPEs in that frequency range are the lowest. Where the difference between wavelength and the size of body structures is greater, both at VLF, LF, and low HF frequencies, and at microwave and higher frequencies, absorption is less and the MPEs are higher.

At lower frequencies, the interaction between electromagnetic energy and the human body differs for the electric and magnetic components of the energy. Because of this, different MPEs exist for the electric (E) field (in V/m) and the magnetic (H) field (in A/m) up to 30 MHz, and both limits must be met. There is a transition region between 30 and 300 MHz where limits can be measured either with E and H fields or power density (in mW/cm^2). The MPE limits above 300 MHz are expressed only by the power density. It is common to measure or model power density at frequencies below 30 MHz, but in doing so one must be careful to obtain plane-wave equivalent power density. In the far-field, the power density at all frequencies is determined from a plane-wave but in the near-field, before the energy resolves into a plane-wave, the power density must be derived from both the E- and the H-fields at any location.

The graph in Figure 22.22 is also expressed in tabular form, as in **Tables 22.4** and **22.5**. The FCC has published their MPE limits in that form in their rules, in §1.1310(e). The lines in the graph represent the power density columns in those tables, with the note that power densities for frequencies below 30 MHz must be plane wave equivalent values. The tables also include columns for E and H fields for frequencies below 300 MHz.

LOCALIZED EXPOSURE

The MPEs in the FCC rules presume uniform whole-body exposure. This may or may not be the case for typical amateur radio stations. In some circumstances, exposure is mainly limited to a portion of the body. Handheld radios are an example of this. When you hold a handheld radio in front of your face, most of the exposure from its antenna is to your head. The FCC uses a distance of 20 cm (about 8 inches) between the antenna and any portion of a person's body to distinguish between whether whole body MPE can be used for assessing compliance with their rules, or if a localized SAR determination must be made. For any antenna that is used less than 20 cm

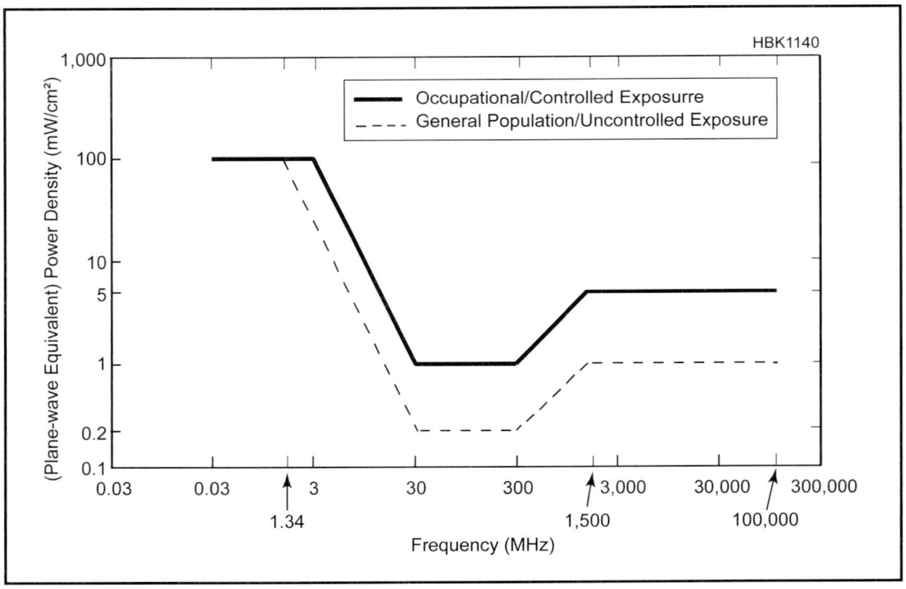

Figure 22.22 — Graph of FCC Maximum Permissible Exposure Levels, derived from IEEE C95.1-1991 "IEEE Standard for Safety Levels with Respect to Human Exposure to Radio Frequency Electromagnetic Fields, 3 kHz to 300 GHz." This graph shows only power density limits, with the stipulation that below 30 MHz power density is a plane-wave equivalent derivation from the E and H fields.

Table 22.4

(From §1.1310) Limits for Maximum Permissible Exposure (MPE)

Limits for Occupational/Controlled Exposure

Frequency Range (MHz)	*Electric Field Strength (V/m)*	*Magnetic Field Strength (A/m)*	*Power Density (mW/cm2)*	*Averaging Time (minutes)*
0.3-3.0	614	1.63	(100)*	6
3.0-30	1842/f	4.89/f	($900/f^2$)*	6
30-300	61.4	0.163	1.0	6
300-1500	—	—	f/300	6
1500-100,000	—	—	5	6

f = frequency in MHz

* = Plane-wave equivalent power density (see Notes 1 and 2 in Table 22.5).

Table 22.5

Limits for General Population/Uncontrolled Exposure

Frequency Range (MHz)	*Electric Field Strength (V/m)*	*Magnetic Field Strength (A/m)*	*Power Density (mW/cm2)*	*Averaging Time (minutes)*
0.3-1.34	614	1.63	(100)*	30
1.34-30	824/f	2.19/f	($180/f^2$)*	30
30-300	27.5	0.073	0.2	30
300-1500	—	—	f/1500	30
1500-100,000	—	—	1.0	30

f = frequency in MHz

* = Plane-wave equivalent power density (see Notes 1 and 2).

Note 1: This means the equivalent far-field strength that would have the E or H-field component calculated or measured. It does not apply well in the near field of an antenna. The equivalent far-field power density can be found in the near or far field regions from the relationships: $P_d = |E_{total}|^2 / 3770$ mW/cm^2 or from $P_d = |H_{total}|^2 \times 37.7$ mW/cm^2.

Note 2: $|E_{total}|^2 = |E_x|^2 + |E_y|^2 + |E_z|^2$, and $|H_{total}|^2 = |H_x|^2 + |H_y|^2 + |H_z|^2$

from a person, the only acceptable form of exposure determination is SAR.

Since SAR determination is beyond the capabilities of most radio amateurs, it is expected that manufacturers of handheld radios will perform that test and provide the results to the amateur, who will be able to use those data to comply with the FCC exposure requirements. This requirement became official on May 3, 2021, and any handheld radios manufactured before that time are assumed to meet the exposure requirements even without SAR testing by the manufacturer. In support of this supposition, a survey study of the FCC equipment database to see what SAR measurements have revealed for commercial handheld devices that are similar to amateur radio handhelds found that commonly used handheld transmitter power levels did not cause overexposure. This limited study showed that in commercial handheld devices at frequencies near the amateur 2m, 1.25m, 70 cm and 33 cm bands with similar power outputs and rubber duck antennas, the SAR measurements were below the FCC SAR limits. See the *QEX* article by Tell, referenced below.

Occupational vs General Population

The ANSI/IEEE C95.1-1992 standard makes a distinction between what is termed the occupational population, or *controlled environment*, and the general public, or *uncontrolled environment*. The reasoning for creating two sets of safety limits was to provide an additional margin of safety for people who were unaware of their exposure. The use of controlled and uncontrolled environments in the IEEE standard also helps emphasize that the IEEE standard does not restrict the higher set of exposure limits only to those who may be occupationally exposed. For example, the IEEE context is that the upper set of exposure limits, those for controlled environments, apply to anyone who is aware of the potential for exposure, not just to those who may be exposed because of their occupation.

This contrasts with the FCC exposure rules, wherein controlled exposures generally apply only to exposure encountered as a result of one's occupation, with the caveat that anyone considered to be in the occupational population must also be trained about RF safety.

The additional safety margins for general population/uncontrolled exposures are generally five times greater than those applicable to the controlled environment, which themselves already result in exposures that are 10-times lower than the exposure levels at which any deleterious effect had been detected in scientific studies. Thus, those individuals in uncontrolled environments are protected by limits that are 50-times lower than exposures at which deleterious effects have been demonstrated in scientific studies.

Radio amateurs, and the people living in their households, were included in the FCC rules to be considered part of the occupational population subject to the more permissive exposure limits. This gives amateurs more leeway in designing their RF safety programs, since the people living in their houses are subject to the higher MPE values. Questions on RF exposure were added to each of the amateur radio licensing exam question pools as evidence of training for licensed amateurs. Radio amateurs are expected to educate the people living in their homes about the potential for RF exposure caused by operation of their stations, and how to reduce exposure if they so desire.

TIME AVERAGING

The MPEs are based on eliminating the whole-body heating effects caused by RF exposure. Because of the thermal time constant of the body, exposures are averaged over any six-minute period. This process results in controlling the approximate average level of thermal load imposed on the body. The time averaging permits short exposure to very high-power signals followed by very low exposure, which has the same effect on the body as a continuous lower exposure, as long as the average does not exceed the MPE over the designated averaging time. For the occupational population, the averaging time is 6 minutes; for the general population it is 30 minutes.

MODULATION DUTY CYCLE

Averaging is defined in two ways for amateur radio operations. Many forms of modulation cause the peak power of the carrier to vary over time, and the average of those variations is lower than the exposure that would be caused by the unmodulated carrier. Some common averaging duty cycles for typical amateur radio modulation types are shown in **Table 22.6**.

OPERATIONAL DUTY CYCLE

During the exposure averaging times of 6 or 30 minutes, typical amateur communication switches from transmit-to-receive several times. Since exposure occurs only during the transmit portion of the conversation, that represents another duty cycle that can be combined with the modulation duty cycle to determine the overall time-averaged value of exposure. A typical voice conversation between two

Table 22.6

Duty Cycles for Common Forms of Modulation

SSB (no compression)	20%
SSB (compression)	40%
CW	40%
AM (with 100% modulation)	25%
AM (unmodulated)	100%
FM	100%
FSK/RTTY	100%
AFSK on SSB	100%
SSTV on SSB	100%
PSK31	75%
FT4	34%
FT8	42%
JT4	39%
JT9	41%
JT65	39%
QRA64	40%
WSPR	92%

radio amateurs would have each station transmitting for 50% of the time. In a contesting situation, the transmit duty cycle could either be much less, with short calls and longer times listening, or higher, with frequent CQ calls and short pauses for a reply. Each amateur should make a best estimate of the duty cycle of transmission in their form of operating.

The operational duty cycle has meaning only in situations where transmissions are significantly shorter than the averaging period. For instance, if an amateur transmits continuously for six minutes and then listens for six minutes, the duty cycle for assessing the occupational exposure, with a six-minute averaging time, is 100%, since the transmission is occurring during the full averaging time. However, for this same transmit-receive timing in calculations for exposure of the general population, with a 30-minute averaging period, the duty cycle is 60% (30 minutes = 6 minutes on, 6 minutes off, 6 minutes on, 6 minutes off, 6 minutes on = 18/30 = 0.60).

ACTUAL EXPOSURE AFTER AVERAGING

The exposure that is calculated for a person standing near an antenna can be lowered by factoring in both the modulation duty cycle and the operational duty cycle. For instance, if you calculate that when standing in a certain location near a 10-meter dipole, a person in the general public could be exposed to the equivalent of 1.0 mW/cm^2, a power density that exceeds the FCC MPE of 0.2 mW/cm^2 by a large margin, you may then apply the duty cycles of your operation. In this example, you normally operate with uncompressed SSB and converse with a friend where you each talk for one minute and listen for one minute. In this case your modulation duty cycle is 20% (factor of 0.2) and your transmit-receive duty cycle is 50% (factor of 0.5). The averaged exposure of the person in question would then be:

$$1.0 \text{ mW/cm}^2 \times 0.2 \times 0.5 = 0.1 \text{ mW/cm}^2$$

Therefore, this location meets the FCC MPE for the general population when the station is operated as described. However, if you were to switch your modulation from SSB to FM, the average exposure would change markedly. Frequency modulation has a modulation duty cycle of 100% (factor of 1.0) and, with all other operating parameters being the same, the averaged exposure of the person in question would become:

$$1.0 \text{ mW/cm}^2 \times 1.0 \text{ x} \times 0.5 = 0.5 \text{ mW/cm}^2$$

which is more than double the FCC MPE for the general population.

22.3.5 Responsibilities of the Radio Amateur

The FCC expects every amateur radio licensee to abide by its exposure rules. Even when you are not required to perform a complicated analysis, it is your responsibility to make sure that no person is ever exposed above the MPE listed in the FCC rules. Keep in mind that the MPE table refers to values of plane wave equivalent power density that are averaged over applicable times and spatially averaged over the dimensions of the body.

There are several ways to evaluate the exposure from your amateur radio station. The simplest forms of evaluation are designed to be highly conservative so that any operation passing the evaluation is highly unlikely to ever cause overexposure to any person near the antennas. However, this conservatism may cause one to think that their operating parameters need to be changed while a more exact analysis may show that no overexposure would occur. For instance, the FCC provides exemptions from more detailed analysis applying calculations that indicate what the maximum transmitted power must be for a person standing a given distance from an antenna. In a given situation, the exemption formulae may tell you that you need to reduce power from what you planned to use. If that occurs, a more detailed form of analysis might show that your power setting would not actually cause overexposure and your planned power level will be permitted.

For the purposes of exposure analysis, it is not necessary to determine the exact levels to which people might be exposed. It is only necessary to affirm that any exposure will be less than the FCC MPE thresholds. Any assessment method that provides credible confirmation without determining the exact exposure is perfectly acceptable when demonstrating compliance with the FCC human exposure rules.

Most stations will require multiple analyses to account for differences in operating modes, antennas, transmitters, and frequency bands. It is not necessary to use the same analysis methods for each set of operating parameters and, generally, the simplest method that shows compliance with the FCC MPEs is the best to use.

The FCC does not require that amateurs submit the results of their exposure assessments to the Commission. However, it is wise to document the assessments and file them so that if there is ever a question about overexposure from your station, you will be able to provide documentation to show why that is not the case.

Many radio amateurs will find that the simplest forms of exposure assessment are satisfactory for their station's operations. Descriptions of the two most common methods are found below. More complicated assessments are beyond the scope of this discussion. If they should become necessary, other documents, including some that are referenced here, can provide appropriate guidance that you will need.

As you read this chapter about the federal regulations and requirements for compliance with RF exposure rules, it may seem a bit overwhelming in terms of both regulatory and technical detail. Perhaps the most useful guidance to new amateur radio licensees is to design your station to comply with these rules in the first place and choose conservative operational practices that will help you avoid unnecessary trouble in the future. If you are a seasoned amateur licensee with a long-established station that may have never given any serious attention to RF safety in the past, now is the time to use the information in this chapter to help you review all aspects of your station that could lead to excessive RF exposures and, where appropriate, take those steps that will get your station into compliance.

PERFORMING AN EXPOSURE ASSESSMENT

Before choosing what type of exposure assessment is needed, there is certain information that must be gathered, which will be used with any assessment. These are:

1. *The amount of power that will be emitted from the antenna (ERP)*

The measure of power that is important in the analysis of exposure is the amount that is emitted from the antenna. You need to know the output power of the transmitter after factoring in the modulation duty cycle (see Table 22.6), the loss in your feed line, and the gain of your antenna. The FCC has standardized on effective radiated power (ERP) to define emitted power; ERP is the net power delivered to the antenna multiplied by the power gain of the antenna relative to a half wave dipole in free space. The simplest way to calculate ERP is to first convert transmitter output power, corrected for modulation duty cycle, to dBW (dB with respect to a watt, e.g., 100 watts = 20 dBW). Next, subtract the feed line loss, which is reported by coaxial cable manufacturers in dB/100 feet at various frequencies. Choose the closest frequency that is reported by the manufacturer and factor in the length of your coaxial cable. Finally, add the gain of the antenna in dBd (dB with respect to a half wave dipole). Some antenna manufacturers report the gains of their antennas in dBi, the gain with respect to an isotropic radiator. To convert from dBi to dBd, subtract 2.15 dB (i.e., dBd = dBi - 2.15).

If you have a power amplifier that you use occasionally, try using that high power output in the analysis. If the method you are using suggests that exposure may not comply with the MPE, you can analyze with and without the amplifier separately. If analysis with the higher transmit power clearly indicates no potential exposure problems, then it is not necessary to repeat the analysis with the barefoot power level.

2. The shortest distance there will ever be between any part of the antenna and a person

Some methods of exposure assessment distinguish between exposure to people in the occupational population and exposure to the general population. This can be important for some stations since the FCC MPE values for the occupational population are significantly less stringent. The FCC has designated licensed radio amateurs and those living in their households as members of the occupational population, so it may be necessary to perform two analyses for each antenna on each band, with the distance from the antenna to the nearest people in the occupational population used in one calculation, and the distance from the antenna to the nearest people in the general population used in the other calculation.

FCC EXEMPTIONS

The current FCC rules allow for "exemptions" from performing more detailed exposure evaluations. However, there is no exemption from complying with the FCC exposure regulations!

The first requirement for the use of FCC exemptions is that the distance between a person and the nearest point of the antenna be outside the reactive field region of the antenna, which is greater than $\lambda/2\pi$, where λ represents the wavelength of the signal being transmitted. **Table 22.7** gives the minimum distances from the antenna for which the FCC exemptions can be used for most amateur bands.

If the shortest distance between a person and any part of the antenna (R), expressed in meters, will be less than the values on Table 22.7, then a different exposure evaluation method must be used. To determine if you are exempt from further, more detailed, evaluation, you can use the expressions in **Table 22.8** for your frequency range of interest to calculate the threshold ERP that will ensure compliance with the exposure rules. The calculation result tells you the maximum power that can be emitted from the antenna (ERP) in order to maintain the exemption. Take care to match your units. The FCC exemption formulas are based on distances in meters, while you may have made your measurements in feet. The result of the exemption is ERP in watts, while you may have calculated your ERP in dBW.

Normally, FCC exemptions must be recalculated for each frequency band that will be used by a station. Some peculiarities of the equations allow you to decrease the number of calculations. For instance, in the equation for the HF bands (1.34 to 30 MHz) the frequency parameter is in the denominator. If you are using the same multiband antenna for several HF bands, by calculating the threshold ERP for the highest frequency band, as the frequency decreases for the other HF bands then the threshold ERP will only get larger. Calculation at the highest frequency used by an antenna in the HF bands gives the worst case, or lowest allowable threshold ERP. No additional calculations are necessary for lower frequencies on the same antenna as long as the exemption is valid for the highest frequency used and the minimum exposure distance remains greater than $\lambda/2\pi$ for all lower frequency bands being considered.

For VHF bands (30 to 300 MHz) the exemption formula does not vary with frequency, so you need only perform one calculation per antenna (with unchanging power levels), no matter how many VHF bands are transmitted by that antenna.

On UHF bands (300 to 1500 MHz) the exemption formula is proportional to frequency, so a calculation at the lowest frequency in that range for an antenna is the only one needed for all frequencies transmitted from that antenna. In these bands, an exposure distance greater than $\lambda/2\pi$ could still be less than 20 cm. Recall that any exposure distance less than 20 cm must be evaluated with SAR, and these exemption formulae would not apply (see the note near the bottom of Table 22.8).

Table 22.7
Minimum Exemption Distances ($\lambda/2\pi$)

Band (MHz)	*Distance*	*Band (MHz)*	*Distance*
1.8	86.1 ft	24.89	6.2 ft
3.5	44.3 ft	28.0	5.5 ft
3.75	41.4 ft	50.0	3.1 ft
7.0	22.2 ft	144	1.1 ft
10.1	15.4 ft	222	8.4 in
14.0	11.1 ft	420	7.9 in*
18.068	8.6 ft	902	7.9 in*
21.0	7.4 ft	1240	7.9 in*

** Distances less than 20 cm (7.9") must be evaluated with SAR*

Table 22.8
Maximum Exempt ERP

	Frequency (MHz)	*Maximum ERP (Watts)*
VLF	0.3 – 1.34	$1920 \times R^2$
HF	1.34 – 30	$3450 \times R^2 / f^2$
VHF	30 – 300	$3.83 \times R^2$
UHF	300 – 1500	$0.0128 \times R^2 \times f$
MW	1500 – 100,000	$19.2 \times R^2$

Note: R is distance in meters and f is frequency in MHz.
ERP is based on transmitter output power, modulation duty cycle, feedline loss, and antenna gain

Example Calculations:
On 14.1 MHz at 10 meters from the antenna, the maximum exempt ERP is $3450 \times 10^2 / 14.1^2 = 1735$ W.
On 21.2 MHz at 10 meters from the antenna, the maximum exempt ERP is $3450 \times 10^2 / 21.2^2 = 768$ W.
On 50.1 MHz at 5 meters from the antenna, the maximum exempt ERP is $3.83 \times 5^2 = 96$ W.
On 146 MHz at 0.5 meters from the antenna, the maximum exempt ERP is $3.83 \times 0.5^2 = 0.96$ W.

DISADVANTAGES OF USING EXEMPTIONS

Even though determining if you qualify for the exemption from more detailed evaluation is easy, the analysis process makes certain assumptions that tend to overestimate exposure. If the calculations with the exemption criteria yield acceptable ERP values for your station, then there is nothing else to do and your exposure analysis is complete. You can presume that operation of your station will comply with the FCC RF exposure rules.

If, however, the application of the exemption criteria indicates that you must decrease your ERP below what you planned to transmit, then the highly conservative assumptions associated with the process may be the culprit, and not the potential exposure that would result from your station's operation. When an exemption calculation specifies a maximum allowable ERP, the assumption is that the antenna gain is equal in all directions. Even though a directional antenna is often rotated through a full circle, the regions above and below the antenna are never subject to that much gain, and there can be significant attenuation in the transmitted pattern that is not accounted for with this analysis method.

The exemption process assumes that members of the general population will be exposed at the previously determined shortest access distance to the antenna and applies the more restrictive general population MPE for finding a compliant ERP. It could be that only the amateur and/or members of the household will have such access and, in this case, the less stringent occupational/controlled exposure MPEs are actually applicable.

Further, the exemption determination does not factor in the operational duty cycle, which usually decreases the actual averaged exposure.

Finally, the exemption criteria also make the generally unrealistic assumption that perfect reflection of RF fields from the ground occurs as if it was a perfectly conducting surface. This assumption, by itself, increases the potential exposure field power density by up to a factor of four. Hence, the exemption criteria represent a highly conservative approach that will in most cases overestimate potential exposure levels but can be regarded as a fail-safe approach for compliance assessment.

If the exemption process indicates that your station needs to decrease power to operate below the general population MPE, you should consider a more accurate form of analysis (see below), and confirm whether the individuals who have access to your antenna at the distance you determined will be members of the general population or just the licensed operator and members of the operator's household, for which the less restrictive occupational MPEs apply.

ONLINE CALCULATORS

Several RF field calculators for estimating potential exposure are available on the internet. Most are based on calculation in the far-field; some assume a free space environment while others permit inclusion of a ground reflection factor that increases the calculated field strength or power density. It is important to use a known online calculator. The calculations performed in the background of an online calculator may contain unwarranted assumptions, or even outright errors that are invisible to the user. The calculator must compare the results of its power density calculations to the proper exposure limits; in the United States that must be the current FCC MPEs and, if they are ever changed, the calculator must be modified to produce the correct results. The calculator that you use should explicitly state which thresholds are being used with its calculations.

Even though the calculation is based on an equation for far-field power densities, the results for simple antennas are applicable in the near-field and even in the reactive near-field, which is not valid for FCC exemptions and may still be overly conservative. Other assumptions used in the FCC exemptions are also true for most online calculators, mainly that the antenna gain can occur in all directions. However, most online calculators will distinguish between the exposure distances for the occupational population and the general population.

A good example of an online RF exposure calculator can be found on the ARRL website at **www.arrl.org/rf-exposure-calculator**. A sample calculation is shown in **Figure 22.23**. This calculator requires that you determine the feed line loss and enter the power at the antenna. However, for simplicity, you may want to first perform the calculations as if there was no feed line loss and see if it gives favorable results. If not, then you can go through the additional work of calculating the feed line loss to see if that makes an important difference for your station's exposure calculation.

This calculator factors in the modulation and operational duty cycles. It asks for antenna gain in dBi. The results are the minimum compliance distances from the antenna to people in the occupational population and to people in the general population in both feet and meters. Remember that the averaging times for the FCC MPEs are different for the two different population groups, and that the averaging times refer to either a 6- or 30-minute "sliding window" of time. Thus, the operational duty cycle must consider the ratio of the maximum transmit time

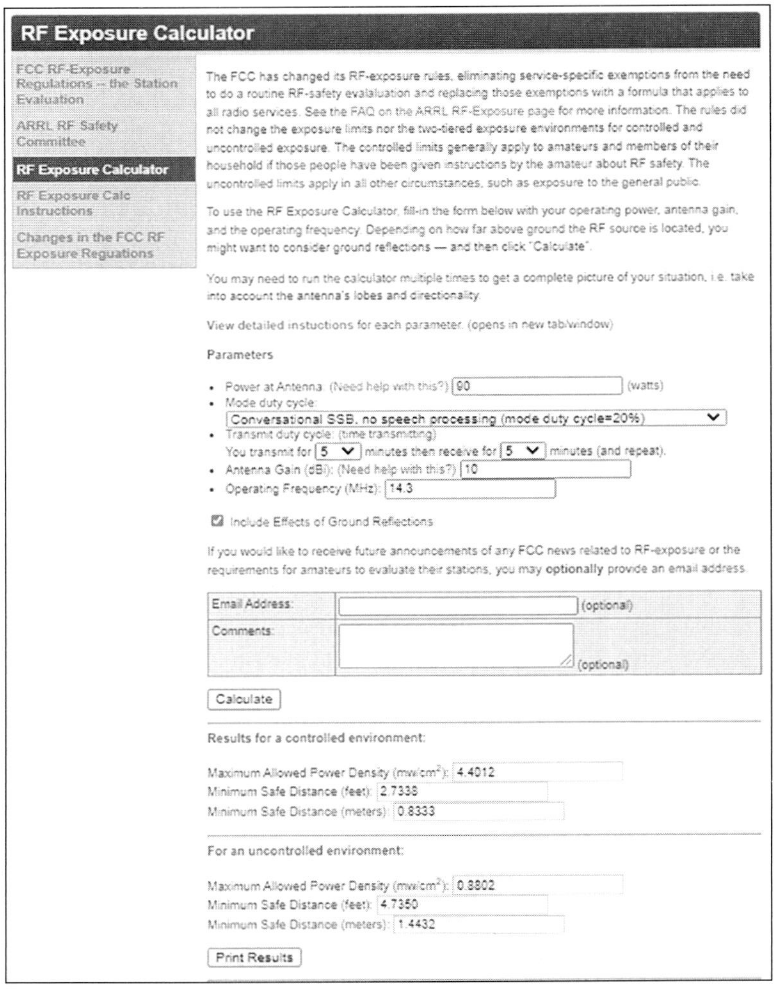

Figure 22.23 — ARRL RF Exposure Calculator.

during any 6- or 30-minute period to the averaging time period. This distinction is made automatically by the ARRL online calculator.

The ARRL online calculator also allows for applying a realistic ground reflection factor of 2.56 when calculating power density or no reflection factor at all (as in free space). This means that the calculator is intended for more realistic estimation of expected RF field power density, which is suitable for certifying compliance with the FCC MPEs.

WHEN SHOULD ONLINE CALCULATORS NOT BE USED?

Online calculators that are based on the far-field equation also give accurate exposure results as close as the surface of the antenna for some simple antenna types, such as resonant dipoles. It is important, however, to not use this form of analysis too close to more complex antennas. Complex antenna types include those with interaction between different parts of the antenna, antennas that are very close to the ground (i.e., less than $\lambda/2\pi$), or antennas that are much shorter than a half wavelength. The ARRL online calculator does not ask what type of antenna you are using and if it reports a short compliance distance (i.e., less than $\lambda/2\pi$), that result would only be accurate for a simple wire antenna. If the shortest distance between a person and the antenna is much larger than the calculated compliance distance, the result may still be valid. A good distance to use for this determination is the same one that is used by the FCC exemptions, the reactive near-field, or $\lambda/2\pi$, as shown in Table 22.7. If the distance to the nearest person is greater than the value in that table and the calculator gives a compliance distance less than that, it can still be used, even for a complex antenna.

WHAT IF THE ONLINE CALCULATOR RESULTS ARE NOT ACCEPTABLE

If the online calculator tells you that people are too close to your antennas, or if the calculator results are not valid for your particular type of antenna, as discussed above, then you still have the option of improving the accuracy of your exposure calculations.

For example, if you have a Yagi antenna mounted on a small tower, the calculator could incorrectly tell you that someone would have excess exposure while standing at the base of the tower. The calculator, as well as the FCC exemption formulae, assumes that the antenna gain is the same in all directions, which would not be true in this case.

The two options available to improve the accuracy of the exposure determination are to make calibrated measurements around the antenna while it is transmitting, or to model the antenna near-field patterns. These topics are discussed in detail in the *ARRL Antenna Book*.

22.3.6 RF Exposure Mitigation

It is not the purpose of the FCC human exposure regulations to keep you off the air. Their goal is to prevent overexposure of anyone while you are operating. If your calculations indicate that locations at which people may be present could cause exposure exceeding the FCC MPE (an overexposure), you must perform some type of mitigation to prevent that occurrence. There are many ways that you can mitigate an overexposure situation, some of which may actually improve the performance of your station.

If possible, when areas in which an overexposure could occur exist on the ground, raising the antenna could do away with the problem. For antennas that have sections mounted close to the ground, such as the ends of an inverted-V, raising those ends by even a few feet could convert a noncompliant station into one that complies with the FCC MPEs.

POSITIVE ACCESS CONTROL

The FCC requires *positive access control* (PAC) to prevent people from accessing areas in which they might be overexposed. PAC is a general term that refers to an active measure that keeps people from entering overexposure areas. PAC can be as simple as a locked door to a rooftop on which antennas are located. A common way to achieve PAC is to place effective fencing around the areas where access must be controlled.

Wherever PAC is used, the FCC requires that signs be posted to warn about the possibility of overexposure. Lettering must be large enough to be read from a distance at which overexposure cannot occur. The content of the sign is specified in the FCC rules. For areas in which the FCC general population MPE may be exceeded, the sign must contain the word "NOTICE" in white letters on a blue banner, along with a description of the hazard and your contact information. If exposure to RF fields within the controlled area have the potential to exceed the FCC occupational population MPE, then the sign must have the word "CAUTION" in black letters on a yellow banner, along with similar text as the "NOTICE" sign.

OPERATING MODIFICATIONS

Two conditions must be met to cause exposure: the station must be transmitting and there must be a person present in a potential overexposure area. PAC is a means of preventing the latter. However, control of the former can also satisfy the requirement that no person be exposed to electromagnetic energy exceeding the MPEs.

The FCC writes its regulations to cover all forms of radio transmission that they regulate. Some of these include broadcast transmitters and cellular base stations. Both differ from amateur radio in two ways: they transmit during all hours of every day of the year and, most importantly, they are usually unattended. In contrast, amateur radio transmissions are generally intermittent, and the amateur radio operator is usually present at the transmitter while it is operating.

If a person enters a potential overexposure area and the amateur radio operator then stops operating, no hazard exists, and the station remains compliant with the FCC rules on exposure. Similarly, the amateur radio operator may determine that a lower transmit power removes the hazard, and then decreases the power any time a person is seen to enter the overexposure area.

Amateur radio stations that are remotely controlled require special considerations when identifying areas in which individuals might be exposed above the FCC MPEs. For remote operations, the best approach is to design the transmitting site to be inherently compliant with the MPEs without need for real time monitoring of activity within the area.

22.3.7 RF Safety References

The preceding section has presented an introduction to the reasons for developing an RF Safety program for your station, and some basic procedures to help determine what areas around your station may be a cause for concern that a person might be exposed to your transmitted signals beyond the limits that the FCC has set. To perform more exact exposure analyses, which may be necessary if your station has marginal exposure results from the more common methods, you can look to the following references:

FCC OET Bulletin 65

The FCC has provided guidance for complying with their rules on human exposure to electromagnetic energy in their OET Bulletin 65. OET Bulletin 65 Supplement B, which was principally written by members of the ARRL RF Safety Committee, provides more specialized information that applies to the operation of an amateur radio station. Both documents are available for free download from the FCC website: **fcc.gov/general/oet-bulletins-line**.

ARRL Antenna Book

The *ARRL Antenna Book* has an RF Safety section that builds on the information provided here. Additional information is provided to help perform more exact exposure modeling with the Numerical Electromagnetic Code (NEC), with a discussion of the importance of ground reflections and spatial averaging. A discussion of the use of Pre-Assessed Configurations (PACs) to estimate the potential RF exposure more accurately around specific antenna types and installations is also included.

RF Exposure and You

The ARRL has published a book entitled *RF Exposure and You*, which delves into all aspects of RF exposure and how to comply with FCC regulations. This book is available electronically on the ARRL website. Be aware that this book was written before recent FCC exposure rule changes, and parts of it are no longer valid.

SCIENTIFIC REPORTS AND STUDIES

The two standards/reports that the FCC exposure regulations are based on:

ANSI/IEEE C95.1-1992: IEEE Standard for Safety Levels with Respect to Human Exposure to Radio Frequency Electromagnetic Fields, 3 kHz to 300 GHz. Note: The IEEE C95.1-1991 standard was adopted by the American National Standards Institute, ANSI, in 1992.

NCRP Report #86: Biological Effects and Exposure Criteria for Radiofrequency Electromagnetic Fields, 1986.

The IEEE standard is periodically updated, even though the FCC has decided that the older standard is sufficiently protective. The latest version of the standard is:

IEEE C95.1™-2019: IEEE Standard for Safety Levels with Respect to Human Exposure to Electric, Magnetic, and Electromagnetic Fields, 0 Hz to 300 GHz.

A similar report that is the basis of most European exposure regulations is:

ICNIRP, "Guidelines for limiting exposure to electromagnetic fields (100 kHz to 300 GHz)." Health Phys 118(5):483-524, 2020.

A study to show that amateur handheld radios are compliant with SAR exposure measurement based on testing of similar commercial handheld radios:

Tell, R., "Amateur portable radios (handheld transceivers): exposure considerations based on SAR." *QEX*, Jul./Aug. 2021, pp. 11 – 15.

Contents

Chapter 23 — Online Content

Articles

- 3D Printed Antennas by Paul Wade, W1GHZ
- 3D Printed Coax-to-Wire Connection Blocks by John Portune, W6NBC
- 3D Printed Fixture Simplifies Ground-Plane Antenna Construction by John Portune, W6NBC
- A No-Special-Tools SMD Desoldering Technique by Wayne Yoshida, KH6WZ
- Introduction to 3D Printing for Hams by John Portune, W6NBC
- Making Your Own Printed Circuit Boards
- Reflow Soldering for the Radio Amateur by Jim Koehler, VE5JP
- Reflow Soldering for the Radio Amateur — Revisited by Jim Koehler, VE5FP
- Repurposing Obsolete Instrument Enclosures by Scott Roleson, KC7CJ
- Soldering Surface-Mount Devices by Dino Papas, KLØS
- Surface Mount Technology — You Can Work With It (Parts 1 – 4) by Sam Ulbing, N4UAU

Chapter 23
Construction Techniques

Home construction of electronics projects and kits can be a fun part of amateur radio. A recent ARRL survey shows that more than half of active hams build electronic projects. There are many kit vendors with products ranging from simple introductory projects all the way to complete transceivers. Becoming familiar with building practices will help you repair and install commercial equipment as well.

Even experienced constructors will find valuable tips in this chapter. It discusses tools and their uses, electronic construction techniques, tells how to turn a schematic into a working circuit, and summarizes common mechanical construction practices. This chapter's construction and fabrication sections are maintained by Joe Eisenberg, KØNEB, based on material originally written and compiled by Ed Hare, W1RFI, and Jim Duffey, KK6MC. Dale Grover, KD-8KYZ, maintains the comprehensive introduction to the use of PCB design and layout software. Material on 3D printing was contributed by John Portune, W6NBC, with additional material provided by Curt Laumann, K7ZOO, and Ron Jones, K7RJ.

23.1 Electronic Shop Safety

Building, repairing, and modifying equipment in home workshops is a longstanding ham radio tradition. In fact, in the early days building your own equipment this was the only option available. While times and interests have changed, home construction of radio equipment and related accessories remains a popular and enjoyable pastime. Building your own gear need not be hazardous if you become familiar with the hazards, learn how to perform the necessary functions, and follow some basic safe practices, including the ones listed below and in the section on soldering. Let's start with our own abilities:

Consider your state of mind. Working on projects or troubleshooting (especially where high voltage is present) requires concentration. Don't work when you're tired or distracted. Be realistic about your ability to focus on the job at hand. Put another way, if we aren't highly alert, we should postpone doing hazardous work until we can focus on the hazards.

Think! Pay attention to what you are doing. No list of safety rules can cover all possibilities. Safety is always your responsibility. You must think about what you are doing, how it relates to the tools, and the specific situation at hand. When working with tools, avoid creating situations in which you can be injured or the project damaged if things don't go "just right."

Take your time. If you hurry, not only will you make more mistakes and possibly spoil the appearance of your new equipment, you also won't have time to think things through. Always plan ahead. Do not work with shop tools if you can't concentrate on what you are doing — it's not a race! Working when you are tired can also lead to problems, such as misidentification of parts or making poor or erroneous connections. Listening to the regular time notices on broadcast radio while you work will prompt you to take plenty of breaks so that you don't work when you're too tired.

Protect yourself. Use of drills, saws, grinders, and other wood- or metal-working equipment can release small fragments that could cause serious eye damage. Always wear safety glasses or goggles when doing work that might present a flying object hazard, and that includes soldering, where small bits of molten solder can be flung a surprising distance. If you use hammers, wire cutters, chisels, and other hand tools, you will also need the protection that safety eyewear offers. Dress appropriately — loose clothing (or even hair) can be caught in exposed rotating equipment such as drill presses. Keep an eye-wash kit near any dangerous chemicals or power tools that can create chips or splinters. Even a small burn or scratch on your eye can develop into a serious problem if not treated promptly.

Don't work alone. Have someone nearby who can help if you get into trouble when working with dangerous equipment, chemicals, or voltages.

Know what to do in an emergency. Despite your best efforts to be careful, accidents may still occur from time to time. Ensure that everyone in your household knows basic first aid procedures and understands how to summon help in an emergency. They should also know where to find and how to safely shut down electrical power in your shack and shop. Get medical help when necessary. Every workshop should contain a good first-aid kit. If you become injured, apply first aid and then seek medical help if you are not sure that you are okay.

What about the equipment and tools involved in shop work? Here are some basic safety considerations that apply to them, as well:

Read instructions and manuals carefully...and follow them. The manufacturers of tools are the most knowledgeable about how to use their products safely. Tap their knowledge by carefully reading all operating instructions and warnings. Avoiding injuries with power tools requires safe tool design as well as proper operation by the user. Keep the instructions in a place where you can refer to them in the future.

Respect safety features of the equipment you work on and use. Never disable any safety feature of any tool. If you do, sooner or later you, or someone else, will make the mistake the safety feature was designed to prevent.

Keep your shop or work area neat and organized. A messy shop is a dangerous shop. A knife left lying in a drawer can cut someone looking for another tool; a hammer left on top of a shelf can fall down at the worst possible moment; a sharp tool left on a chair can be a dangerous surprise for the weary constructor who sits down.

Keep your tools in good condition. Always take care of your investment. Store tools in a way to prevent damage or use by untrained persons (young children, for example). Keep the cutting edges of saws, chisels, and drill bits sharp. Protect metal surfaces from corrosion. Frequently inspect the cords and plugs of electrical equipment and make any necessary repairs. If you find that your power cord is becoming frayed, repair it right away. One solution is to buy a replacement cord with a molded connector already attached.

Make sure your shop is well ventilated. Paint, solvents, cleaners, or other chemicals can create dangerous fumes. If you feel dizzy, get into fresh air immediately, and seek medical help if you do not recover quickly.

Respect power tools. Power tools are not forgiving. A drill can go through your hand a lot easier than metal. A power saw can remove a finger with ease. Keep away from the business end of power tools. Tuck in your shirt, roll up your sleeves, and remove your tie before using any power tool. If you have long hair, tie it back so it can't become entangled in power equipment.

Poison Control

If you think you have a chemical emergency, call your local poison control center immediately. Dial 1-800-222-1222, dial 911, or use the online tools at **www.poison.org**.

23.1.1 Chemicals

Chemicals such as cleaners, adhesives, construction materials, and coolants are used every day by amateurs without ill effects. Take the opportunity to become knowledgeable of the hazards associated with these materials and treat them with respect. **Table 23.1** summarizes the uses and hazards of chemicals and other materials found in the ham shack. It includes preventive measures that can minimize risk. For advanced information, the Centers for Disease Prevention and Control maintains an extensive database at **www.cdc.gov**. Meridian Engineering maintains a collection of materials safety information at **www.meridianeng.com**. When in doubt, contact the manufacturer or distributor of the material for safety information, or use an internet search engine by entering the material name and "safety" into the search bar.

Here are a few key suggestions for safely storing, handling, and using chemicals:

Read the information that accompanies the chemical and follow the manufacturer's recommended safety practices. If you would like more information than is printed on the label, ask for a material safety data sheet (MSDS). Manufacturers of brand-name chemical products usually post MSDS's on their product websites.

Store chemicals properly, away from sunlight and sources of heat. Secure their containers to prevent spills, and so that children and untrained persons will not gain access. Always keep containers labeled so there is no confusion about the contents. It is best to use the container in which the chemical was purchased. If you transfer solvents to other containers such as wash bottles, label the new container according to its contents.

Handle chemicals carefully to avoid spills. Clean up any spills or leaks promptly but don't overexpose yourself in the process. Never dispose of chemicals in household sinks or drains. Instead, contact your local waste plant operator, transfer station, or fire department to determine the proper disposal procedures for your area. Many communities have household hazardous waste collection programs. Of course, the best solution is to only buy the amount of chemical that you will need, and use it all if possible. Always label any waste chemicals, especially if they are no longer in their original containers. Oil-filled capacitors and transformers were once commonly filled with oil containing PCBs. Never dispose of any such items that may contain PCBs in landfills. Contact your county or city recycling office, or local electric utility, for information on proper disposal.

Always use recommended personal protective equipment (such as gloves, face shield, splash goggles, and aprons). If corrosives (acids or caustics) are splashed on you *immediately* rinse with cold water for a minimum of 15 minutes to flush the skin thoroughly. If splashed in the eyes, direct a gentle stream of cold water into the eyes for at least 15 minutes. Gently lift the eyelids so trapped liquids can be flushed completely. Start flushing before removing contaminated clothing. Seek professional medical assistance. If using hazardous chemicals, it is unwise to work alone since people splashed with chemicals need the calm influence of another person.

Food and chemicals don't mix. Keep food, drinks, and cigarettes *away* from areas where chemicals are used, and don't bring chemicals to places where you eat.

Table 23.1

Properties and Hazards of Chemicals often Used in the Shack or Workshop

Generic Chemical Name	*Purpose or Use*	*Hazards*	*Ways to Minimize Risks*
Lead-tin solder	Bonding electrical components	Lead exposure (mostly from hand contact) Flux exposure (inhalation)	Always wash hands after soldering or touching solder. Use good ventilation.
Isopropyl alcohol	Flux remover	Dermatitis (skin rash) Vapor inhalation Fire hazard	Wear molded gloves suitable for solvents. Use good ventilation and avoid aerosol generation. Use good ventilation, limit use to small amounts, keep ignition sources away, dispose of rags only in tightly sealed metal cans.
Freons	Circuit cooling and general solvent	Vapor inhalation Dermatitis	Use adequate ventilation. Wear molded gloves suitable for solvents.
Phenols and methylene chloride	Enameled wire/ paint stripper	Strong skin corrosive	Avoid skin contact; wear suitable molded gloves; use adequate ventilation.
Beryllium oxide	Ceramic insulator found in some power transistors and vacuum tubes that conducts heat well	Toxic when in fine dust form and inhaled	Avoid grinding, sawing or reducing to dust form.
Beryllium metal	Lightweight metal, alloyed with copper	Same as beryllium oxide	Avoid grinding, sawing, welding or reducing to dust. Contact supplier for special procedures.
Various paints	Finishing	Exposures to solvents Exposures to sensitizers (especially urethane paint) Exposure to toxic metals (lead, cadmium, chrome, and so on) in pigments Fire hazard (especially when spray painting)	Adequate ventilation; use respirator when spraying. Adequate ventilation and use respirator. Adequate ventilation and use respirator. Adequate ventilation; control of residues; eliminate ignition sources.
Ferric chloride	Printed circuit board etchant	Skin and eye contact	Use suitable containers; wear splash goggles and molded gloves suitable for acids.
Ammonium persulphate and mercuric chloride	Printed circuit board etchants	Skin and eye contact	Use suitable containers; wear splash goggles and molded gloves suitable for acids.
Epoxy resins	General purpose cement or paint	Dermatitis and possible sensitizer	Avoid skin contact. Mix only amount needed.
Sulfuric acid	Electrolyte in lead-acid batteries	Strong corrosive when on skin or eyes. Will release hydrogen when charging (fire, explosion hazard).	Always wear splash goggles and molded plastic gloves (PVC) when handling. Keep ignition sources away from battery when charging. Use adequate ventilation.

23.2 AC and Power Connectors

The following tables show the most common connectors for ac power in the United States, along with wiring color codes and maximum current-carrying capacity. Pay particular attention to whether the diagram shows the plug or receptacle and the internal (wiring) side or external side of the connector. If you are unsure of how to connect the wiring, ask for professional help to avoid a shock or fire hazard.

Table 23.2A

Common US AC Power Receptacles

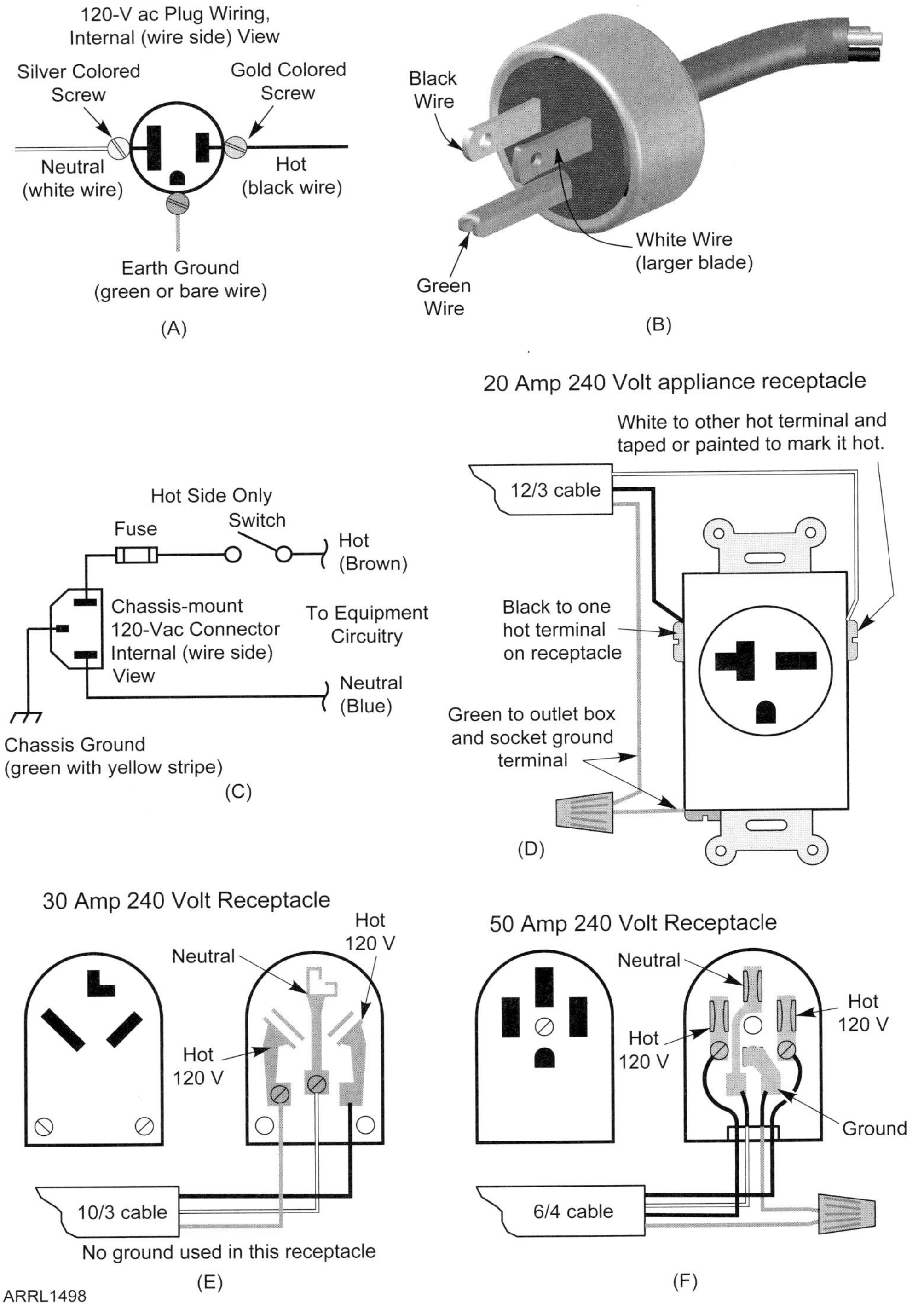

Table 23.2B

Circuit Wiring for 120 V and 240 V Connectors

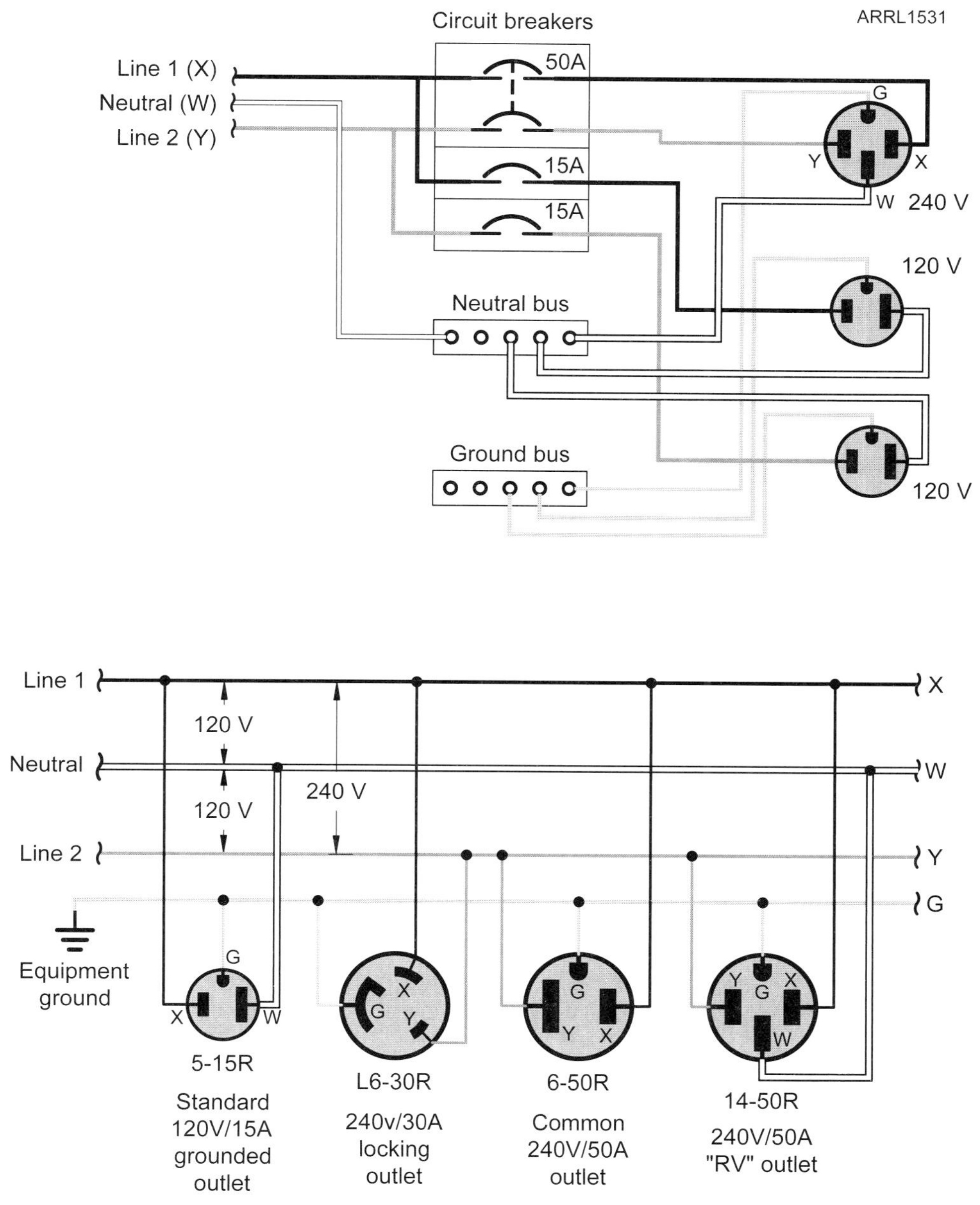

Table 23.2C

NEMA 240 V Receptacles

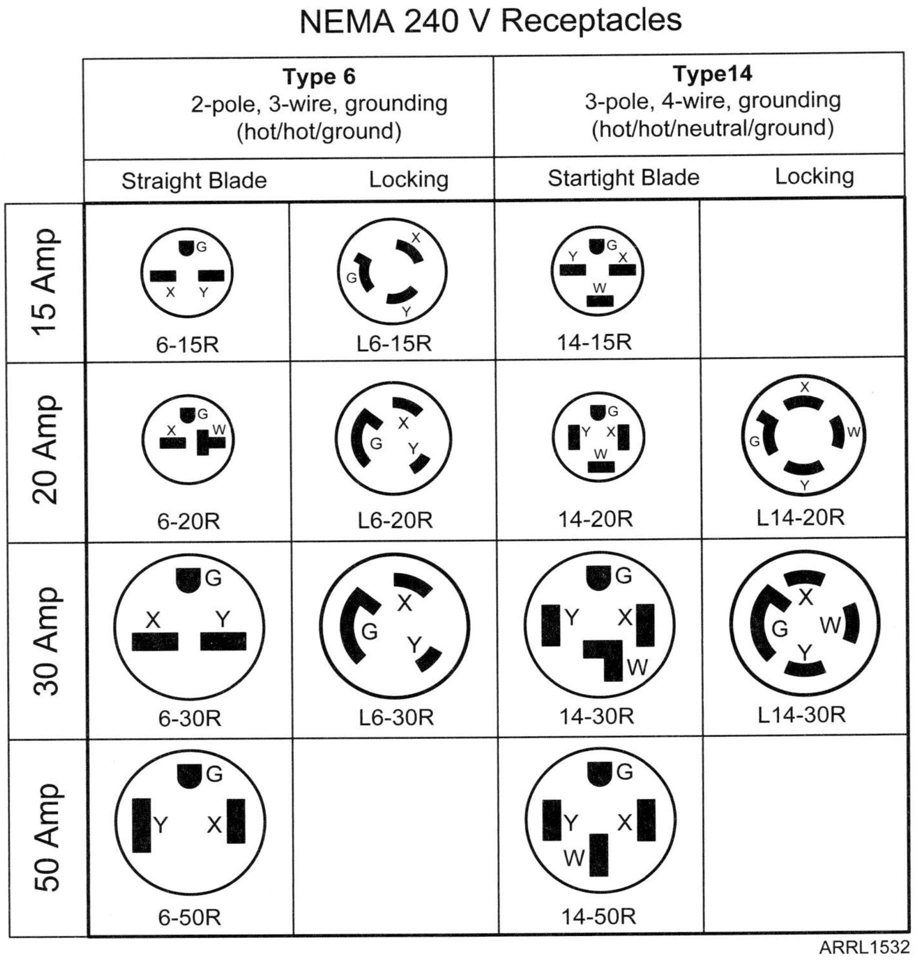

23.3 Soldering Tools and Techniques

23.3.1 Soldering Tools

A soldering tool (we'll use the general term "iron" unless discussing a specific type of tool) must be hot enough to do the job and lightweight enough for agility and comfort. An assortment of soldering irons will help you do a wide variety of soldering tasks correctly. The powerful soldering tool you need for assembling wire antennas is too large for printed-circuit board (PC board) work, for example, while a small fine-point tool works well for smaller jobs. The key is to be able to deliver the right amount of heat to do the job: Too much can damage the work piece and too little can result in poor connections.

If you buy an iron for use on circuits that contains static-sensitive components, get one that has a grounded tip and use it on a static-dissipating workbench mat. Otherwise, you risk damage from static discharge. AC leakage current can also damage components. These irons are usually specified as having a grounded tip and will have a three-prong ac power plug. It is usually not necessary to have a grounded tip on a high-power iron or soldering gun, as those are not normally used on sensitive components.

For PCB soldering, the ideal temperature is around 600 °F (316 °C) when using either 60/40 or 63/37 leaded solder and about 700 °F (371 °C) for lead-free solder. For use with solder wick when desoldering, a higher temperature setting may be necessary, depending on the size of connection being desoldered. These temperatures enable rapid melting of solder and heating of the component leads as well as the PCB pads or terminals being soldered. Using excessive heat, or applying heat for too long, can damage PCB pads, terminals, or the components themselves. A temperature that is too low will result in cold solder joints that don't conduct properly, or that fail mechanically. See the Solder and Soldering sections for more discussion.

SOLDERING IRONS

Soldering pencils and soldering stations (see **Figures 23.1A** and **B**) are by far the most common soldering tools for PCB and small component soldering. Pencil irons come in a variety of power ratings from 20 W to 40 W. The inexpensive pencil irons provided in soldering training kits will get you started but should be replaced with a higher quality iron for regular soldering of kits and projects.

Some pencil irons are available with a variable temperature control in the handle or on a cord-mounted module. Irons with accurate control of soldering temperature are a must for the best chances of successfully soldering a variety of components. Without accurate control of the temperature, the tips can overheat, corrode, wear much faster, and they may overheat or underheat your connections.

There are also small battery-powered pencil irons that are useful away from the workbench. Battery-powered irons generally aren't powerful enough for larger connections and take time to recharge. Pencil irons powered by a 12 V dc cigarette lighter plug are useful for working on mobile stations and keeping in a kit for emergencies.

A soldering station is the most preferable soldering tool for most kits and projects. Soldering stations have a base that houses the controls, power supply, and a tip cleaner, and they include a holder to safely place the iron when not in use. Some soldering stations even have a holder for a roll of solder. The temperature controls range from a simple knob to calibrated controls that can keep the iron's temperature set within 2 °C. Some soldering stations feature programmable temperature settings, allowing the user to specify different preset heat levels and idle cool-down timing. Features to look

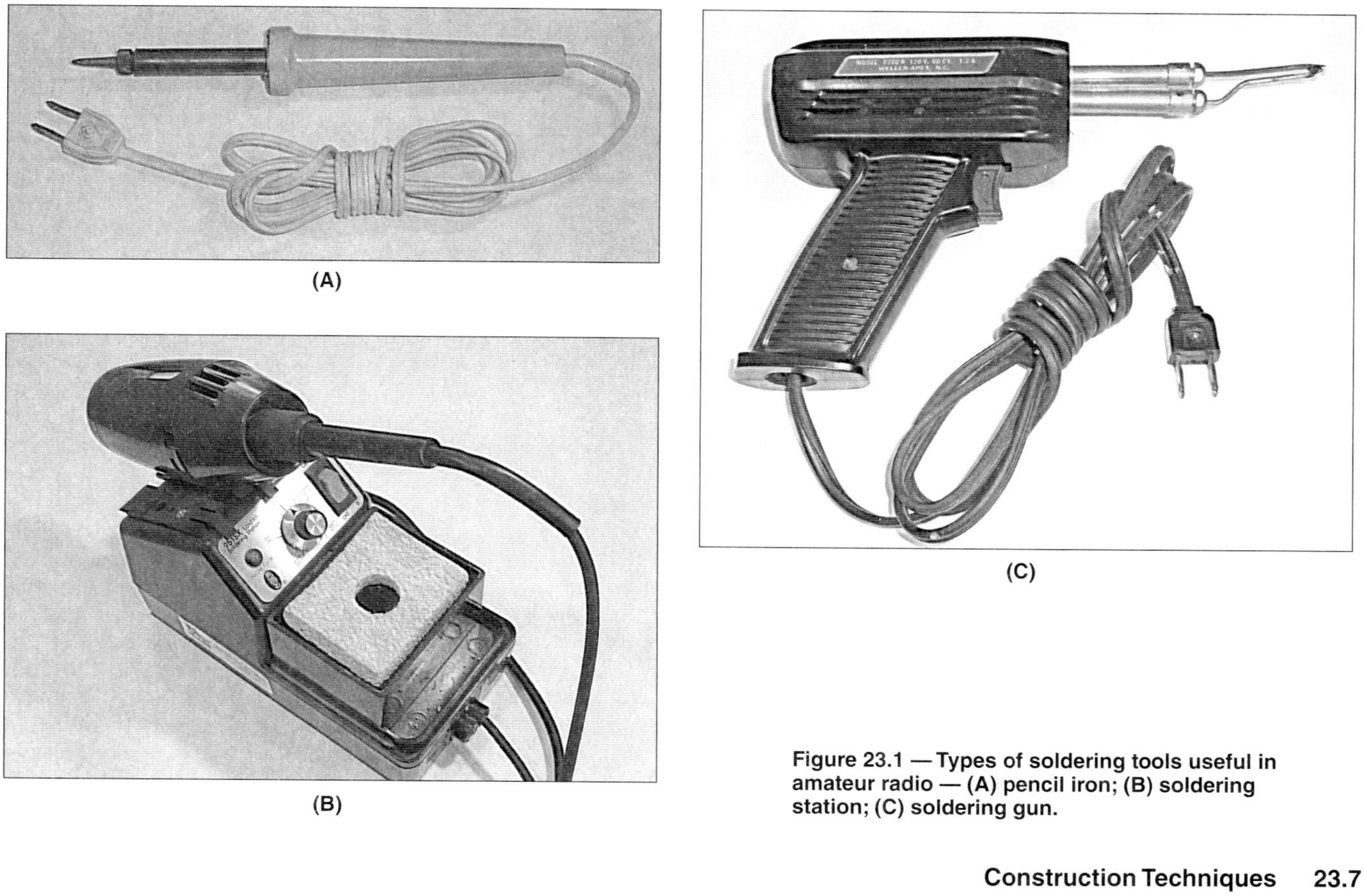

Figure 23.1 — Types of soldering tools useful in amateur radio — (A) pencil iron; (B) soldering station; (C) soldering gun.

for include temperature memories, automatic idle cool-down, and automatic quick heating after idling when the hand piece is lifted for use.

The Metcal RF-powered soldering stations heat the tip with RF for rapid heating, very accurate temperature control, and ease of use. This type of soldering station is most commonly found in commercial manufacturing and product development; used or reconditioned stations are often available. If you do a lot of soldering, one of these might be worth the investment.

A *soldering gun* (**Figure 23.1C**) is used when a smaller soldering iron cannot supply the amount of heat needed for good solder flow. Typical power ratings are 100 to 250 W. Typical jobs for a gun include antenna wires, larger coax connectors, and heavy point-to-point wiring such as for vacuum-tube equipment. Soldering guns consist of an ac transformer with a low-voltage, high-current secondary formed by the copper wire tip. The high current heats the tip. The tips attach to terminals on the front of the soldering gun. Many soldering guns have a high and low power setting controlled by how far the trigger is pulled during operation. Soldering guns usually do not provide heat when not triggered.

Larger ac-powered soldering irons are also available with power ratings of 250 W or more. These irons are most often used for soldering very heavy connections or tubing and sheet metal. This type of iron is used where thorough heating is needed to allow solder to flow fully into the connection. If you plan on soldering a lot of PL-259 coax connectors, a large iron heats the connector body quickly to make a good connection without overheating the cable dielectric. Popular manufacturers include American Beauty, Weller, Hexacon, and others.

TIPS FOR SOLDERING IRONS

Pencil irons and soldering stations are used with a variety of tip shapes and sizes. Several common types are shown in **Figure 23.2**. A *conical* tip comes to a small round point; the *chisel* tip or *mini-chisel* tip is also useful. Low-wattage pencil irons generally have replaceable tips with several sizes and geometries available. For printed circuit board work, a good rule of thumb is to use a tip with a point of the same size as the component leads to be soldered. High-power irons of 100 W and larger usually have a non-interchangeable chisel tip.

Specialized tips are available for use with surface mount technology (SMT) components. These tips can take on the shape of the component being soldered, such as a square tip for working with square IC packages, or a special tip for working with SMT resistors and capacitors. When working on SMT assemblies, a hot air gun or reflow oven is another common technique discussed in the section on Surface Mount Technology.

Figure 23.2—Typical soldering iron tips (this group is from Weller, weller-tools.com). Conical tips are shown at the upper left and chisel tips at the upper right. Larger connections require the heavier tips.

Temperature and Heat Capacity

If two different tips are both at 600 °F, why is the larger one able to solder a larger connection better than the other? Temperature is a measurement of how fast the molecules in the tip are moving, and that can be the same regardless of tip size. But in the larger tip there are more moving molecules that can transfer their energy to the connection so it has more *heat capacity*. (*Thermal mass* is a similar concept.) Think of temperature as voltage and heat capacity as power. Tip temperature tells you the maximum temperature of the connection. Heat capacity tells you how fast the connection will heat and how big a connection can be heated to the maximum temperature.

FLAME-POWERED SOLDERING

When soldering outdoors and working with larger connections, a more powerful iron is needed. In addition, running a long extension cord outdoors can be a problem. Making antennas from copper tubing requires a lot of heat to allow the solder to flow completely into and through the connection. Soldered connections that are exposed to the weather should use a silver-bearing, high-temperature solder. All of these require a large amount of heat that is best supplied by combustion instead of resistance heating.

In this case, a *butane iron* or *butane torch* is ideal for work outdoors, especially for antenna wire or coaxial connectors. Wind will cool a conventional gun or pencil, but a butane iron can become hotter if the moving air enhances combustion. Butane irons use the same fuel canisters as refillable cigarette lighters. Fueling is quick and can supply heat for a useful amount of time. Most butane irons use a flint-type striker to ignite the fuel once the valve is opened. Butane irons generally come with a variety of tips for soldering and cutting.

For really large jobs, a *propane torch* may be required. Propane torches can be found in most hardware stores and are most commonly used for household or commercial plumbing. A propane torch can deliver plenty of heat in a short amount of time, giving you enough time to flow the solder evenly with no gaps or irregularities in the connection. Be sure to follow the supplied directions carefully with a propane torch, including the correct amount of gas flow and the proper way to ignite the gas. A propane torch's open flame can overheat connectors, melt coaxial cable, and damage plastic components. (Some torches have a soldering style tip, but that limits the amount of heat they can deliver to the connection.) Use them with extreme caution and respect the amount of heat they can deliver.

The two types of gas most often used in larger torches include MAPP gas and propane. MAPP gas burns hotter than propane and is used with high-temperature solder. Propane or MAPP gas can be used for brazing or welding connections, depending on the type of metal involved. Information on brazing can be found at **www.thefabricator.com/thewelder/article/tubepipefabrication/6-steps-to-successful-brazing**.

Finally, for the very large connections used in ground systems and ground rods, *thermite welding* is used. Known as "one-shots" by electrical contractors, they use an exothermic chemical reaction to melt copper wires and ground rods into a solid, welded connection that can be buried. The most common types used by amateurs are made by ERICO ("Cadweld") and Harger ("Uni-Shot").

Maintaining Soldering Tips

Keep soldering iron and gun tips in good condition by keeping them well-tinned with solder. Tinning is performed by melting solder directly onto the tip and letting it form a coating. Do not keep the tips at full temperature for long periods when not in use. Automatic cool-down is a great feature that reduces the need to tin the tip.

After moderate usage, remove the tip and clean off any scale that may have accumulated. There are "tip tinner" products available that can also be used for this function.

A tip tinner often comes in a small can and is made up of a mixture of solder, flux, and cleaner. Simply rolling the tip in the tinner makes it look and work like new again.

If a tip becomes oxidized during use, it can be restored to its shiny state by wiping the tip with a damp sponge or rag and then re-tinning. (Some tips are not supposed to be cleaned with water — check the manufacturer's recommendations.) A gentle scraping is also useful for stubborn cases. A copper or stainless-steel coil kitchen scrubber (do not use a scrubber that contains soap) works fine for this. Swipe the iron through the scrubber to clean it. There are coiled brass tip cleaners available that perform at least as well as using a soldering sponge. A brass scrubber or sponge can be placed in a clean tuna can next to the solder station.

If a copper tip becomes pitted after repeated use, file it smooth and bright and then tin it immediately with solder. The solder prevents further oxidation of the tip. Most modern soldering iron tips are nickel or iron clad and should not be filed, as the cladding protects the tip from pitting.

23.3.2 Solder

Solders have different melting points depending on the ratio of tin to lead. Tin melts at 450 °F (232 °C), and lead at 621 °F (327 °C). Solder made from 63% tin and 37% lead melts at 361 °F (183 °C), the lowest melting point for a tin and lead mixture. This type of solder, called 63-37 (or eutectic), also provides the most rapid solid-to-liquid transition and the best stress resistance. 63-37 solder also stays in the plastic state (between solid and liquid) for the briefest time, thus making it ideal for hand soldering because it is a lot less likely to cool unevenly.

Solders made with different lead/tin ratios have a plastic state at some temperatures. If the solder is deformed while it is in the plastic state, the deformation remains when the solder hardens into the solid state. Any stress or motion applied to "plastic solder" causes a poor solder joint.

60-40 solder has the best wetting qualities. Wetting is the ability to spread rapidly, coat the surfaces to be joined, and bond materials uniformly. 60-40 solder also has a low melting point. These factors make it the most commonly used solder in electronics. However, 63-37 solder is the best suited for manual construction techniques.

Some connections that carry high current can't be made with ordinary tin-lead solder because the heat generated by the current would melt the solder. Automotive starter brushes and transmitter tank circuits are two examples. Silver-bearing solders prevent this problem by having higher melting points. High-temperature silver alloys become liquid in the 1,100 to 1,200 °F (593 to 649 °C) range, and a silver-manganese (AgMn 85-15) alloy requires almost 1,800 °F (982 °C).

Because silver dissolves easily in tin, tin-bearing solders can leach silver plating from components. This problem can be greatly reduced by partially saturating the tin in the solder with silver, or by eliminating the tin. Commercial solders are available which incorporate these features. Tin-silver (SnAg) or tin-lead-silver (SnPbAg) alloys become liquid at temperatures from 430 °F (221 °C) for 96.5-3.5 SnAg, to 588 °F (309 °C) for 1.0-97.5-1.5 (SnPbAg). A 15.0-80.0-5.0 alloy of lead-indium-silver (PbInAg) melts at 314 °F (157 °C).

LEAD-FREE SOLDER

In 2006, the European Union Restriction of Hazardous Substances directive (RoHS) went into effect. This directive prohibits manufacture and import of consumer electronics that incorporate lead, including the common tin-lead solder used in electronic assembly. California enacted a similar RoHS law. As a result of these directives, there has been a move to lead-free solders in commercial use. They can contain two or more elements that are not as hazardous as lead, including tin, copper, silver, bismuth, indium, zinc, antimony, and traces of other metals. Two lead-free solders commonly used for electronic work are tin-silver-copper (SnAgCu) alloy SAC305, and tin-copper (SnCu) alloy Sn100. SAC305 contains 96.5% tin, 3% silver, and 0.5% copper, and melts at 423 °F (217 °C). Sn100 contains 99.3% tin, and 0.6% copper, as well as traces of nickel and silver, and melts at 442 °F (228 °C). Both of these melting points are higher than the 349 °F (176 °C) melting point of 60-40 and 63-37 lead-bearing solder, but most conventional soldering stations are able to reach the melting points of the new solders easily. Tin-lead solders are still available, but the move away from them by commercial manufacturers will probably lead to the day when they will be unavailable to hams who build their own gear. Be prepared.

The new RoHS solders can be used in much the same manner as conventional solders. The resulting solder joint appears somewhat duller than a conventional solder joint, and the lead-free solders tend to wick higher than the leaded solders. Due to the higher heat, it is important that the soldering iron tip be clean, shiny, and freshly tinned so that heat is transferred to the joint as quickly as possible to avoid excess heating of the parts being soldered.

Solder and soldering equipment vendors provide numerous guides to hand soldering with lead-free solders. Weller's "Weller University" online presentation (**file.elecfans.com/web1/M00/20/C9/ooYBAFmk1wiAYUh8AA9r_GFLSAQ283.pdf**) provides great details about how soldering iron tips work with lead-free solder. More information is available from Kester (**www.kester.com**) in the Knowledge Base under the Hand Soldering link.

Most, if not all, RoHS-compliant components can be soldered with lead-tin solder. If the RoHS part has leads that are tinned or coated with an alloy to make soldering easier, it is necessary to use a hotter iron than would normally be required in lead-tin soldering. A soldering iron tip temperature of 600 °F (315 °C) or greater will be adequate for soldering RoHS parts with lead-tin solder. In contrast, a working tip temperature of 527 °F (275 °C) is generally adequate for working with conventional non-RoHS parts.

23.3.3 Solder Flux

There are several different types of solder flux and which one you use affects how you solder and how you clean the connection after you're done. Flux is often contained in the core of the solder that you use, but you can also brush liquid flux onto what you are soldering. Flux removes oxides from the metals being soldered, greatly improving the solder connection. It also helps the solder to flow onto the metal surfaces.

TYPES OF FLUX

There are four types of flux that are commonly used for soldering:

- Rosin Flux: This is the most common type of flux, which is formulated from tree sap. It leaves a residue on circuit boards that can be unsightly but does not affect the operation of the circuit. It has been reported that some rosin flux formulations can become corrosive when the circuit is used continuously at high temperatures. It can be cleaned with alcohol (denatured alcohol is not conductive, so the circuit can be powered on before it fully dries).
- Organic Flux: This type of flux is water soluble and, as such, conducts electricity. It must be thoroughly cleaned off any electronic circuit. It can be cleaned with water but must then be allowed to dry completely before turning on the power.
- "No-Clean" Flux: This is a formulation of flux that leaves a small amount of non-ionic residue that is considered to be nonconducting and noncorrosive. This is only true after the flux is heated; insufficient heating can lead to the flux absorbing moisture and creating electrical leakage and corrosion prob-

lems over time. There have, however, been reports that some no-clean flux residue is slightly conductive and can affect the operation of highly sensitive circuits. Properly applied no-clean flux does not have to be removed, but if you want to do so you will need to use a petroleum-based solvent (100% alcohol also works, though less effectively)

- Acid Flux: This is used for non-electronic soldering. DO NOT USE IT!

APPLYING FLUX

Most electronic solder contains flux in its core; the solder forms a tube with flux in its center. As the solder melts, the flux is released onto the junction being soldered and then prepares the metal surfaces to help the solder flow onto them.

A more effective method of soldering is to apply flux to the surfaces to be soldered before melting the solder. The flux sits on the surfaces helping the solder to flow onto them more thoroughly.

Flux is available in liquid and paste forms. There are several popular methods used to apply flux, one of which is a flux applicator pen, which is a convenient way to apply liquid flux. Other delivery methods for liquid flux include syringes with wide needles and squeeze bottles. Paste flux is usually painted onto an area to be soldered with a brush. Since excess flux usually needs to be cleaned up after soldering, it is best to apply it only to the solder junction and minimize the flux that goes elsewhere.

CLEANING FLUX

Whether flux is applied as liquid or paste, the flux residue is hardened and not easily removed after soldering. Although some flux types do not need cleaning, as described above, it is good practice to always clean excess flux from circuit boards. All flux types must be cleaned from the circuit board if a conformal coating is to be applied. When the soldered connections are to be visually inspected to ensure their quality, removing flux helps.

Different flux types require various solvents, but the methods of cleaning are similar. Flux solvent is available in pressurized sprays, and the force of the spray often cleans away all the flux. However, if that is insufficient, rubbing the solvent onto the flux with a brush is more effective. After the flux is fully dissolved, blotting the wet circuit board with a dry cloth helps to remove it from the board. Q-Tips are sometimes used to apply flux solvent and rub it onto the hardened residue; however, the cotton can sometimes get caught on the sharp edges of the wires and leaves fibers behind. When cleaning flux from circuits with static-sensitive electronics, care should be taken that all materials used to apply and remove flux do not produce static that can damage the circuitry.

23.3.4 Soldering

Flux removes oxide by suspending it in solution and floating it to the top. Flux is not a cleaning agent! Always clean the surfaces to be soldered before soldering. Rubbing alcohol on a cotton swab is a good cleaning aid. Cleaning an entire board before beginning to install components is easy and prepares the surface for the best soldering. Flux is not a part of a soldered connection, it merely aids the soldering process. Don't touch the board with your fingers after cleaning.

The two key factors in quality soldering are time and temperature. Rapid heating is desired so that all parts of the joint are made hot enough for the solder to remain molten as it flows over the joint surfaces. Most unsuccessful solder jobs fail because insufficient heat has been applied. To achieve rapid heating, the soldering iron tip should be hotter than the melting point of solder and large enough that the transfer of heat to the cooler joint materials occurs quickly. A tip temperature that is higher than the solder melting point (see the melting point temperatures of different types of solder above) by about 212 °F (100 °C) is right for mounting components on PC boards.

Use solder that is sized appropriately for the job. As the cross section of the solder decreases, so does the amount of heat required to melt it. Diameters from 0.025 to 0.040 inch are good for nearly all circuit wiring. The most common solder diameters suitable for amateur projects are 0.031 or 0.025 inches. Sensitive and smaller components can be damaged, or surfaces re-oxidized, if heat is applied for too long a period. Solder that is too thick can cause shorts between adjacent connections. (A heavier solder is useful for working on antennas, connectors, and light metal work such as shields.)

If you are a beginner, you may want to start with one of the "Learn to Solder" kits available from many electronics parts and kit vendors. The kits come with a printed-circuit board, a basic soldering iron, solder, and the components to complete a simple electronics project. Or you can seek out a group kit building session, such as those put on by the QRP ARCI at the annual Dayton Hamvention or by a local or regional group at a hamfest or specialty convention. Robotics and maker groups often have learn-to-solder sessions too.

You can also download the online article "The Art of Soldering," which provides an overview of guidelines on soldering (**www.arrl.org/files/file/Technology/tis/info/pdf/0102072.pdf**).

Here's how to make a good solder joint. This description assumes that solder with a flux core is used to solder a typical PC board connection, such as an IC pin.

- Prepare the joint. Clean all conductors thoroughly with fine steel wool or a plastic scrubbing pad. Clean the circuit board at the beginning of assembly and individual parts such as resistors and capacitors immediately before soldering. Some parts (such as ICs and surface mount components) cannot be cleaned easily; don't worry about this unless they're exceptionally dirty.
- Prepare the tool. It should be hot enough to melt solder applied to its tip quickly (half a second when the tip is dry, instantly when it is already wet with solder). Apply a little solder directly to the tip so that the surface is shiny. This process is called "tinning" the tool. The solder coating helps conduct heat from the tip to the joint and prevents the tip from oxidizing. However, do not allow too much wet solder to accumulate on the tip. If it does, wipe the excess solder off on a damp sponge or other tip cleaner.
- Place the tip in contact with one side of the joint. If you can place the tip on the underside of the joint, do so. With the tool below the joint, convection helps transfer heat to the joint.
- Place the solder against the joint directly opposite the soldering tool. It should melt within a second for normal PC connections, within two seconds for most other connections. If it takes longer to melt, there is not enough heat for the job at hand. If you have a variable heat soldering iron, adjust it with just enough heat so that the solder flows quickly for the size of wire and joints you are soldering. Much more heat can damage components and the board.
- Keep the tool against the joint until the solder flows freely throughout the joint. When it flows freely, solder tends to form concave joints called "fillets" between the conductors. With insufficient heat, solder will not flow freely, which will result in the formation of convex shapes, or blobs. Once the solder shape changes from convex to concave, remove the tool from the joint. If a fillet won't form, the joint may need additional cleaning. Look for that "Hershey's Kiss" shape to know if it is done correctly.
- Let the joint cool without movement at room temperature. It usually takes no

more than a few seconds. If the joint is moved before it is cool, it may take on a dull, satin, or grainy appearance that is characteristic of a "cold" solder joint. Reheat cold joints until the solder flows freely and hold them still until cool. Using 63-37 solder will help reduce the time until the solder is solid again.

- When the iron is set aside, or if it loses its shiny appearance, wipe away any dirt with a damp cloth or sponge. If it remains dull after cleaning, tin it again.
- Adafruit has published a good collection of photographs showing many common soldering problems and how to repair them. See **learn.adafruit.com/adafruit-guide-excellent-soldering/common-problems**.
- A basic tutorial on "Soldering 101" including a video demonstration is available from Sparkfun Electronics at **learn.sparkfun.com/tutorials/how-to-solder-through-hole-soldering**.

Overheating a transistor or diode while soldering can cause permanent damage, although as you get better at soldering, you'll be able to solder very quickly with little risk to the components. If the soldering iron will be applied for longer than a couple of seconds, use a small heat sink when you solder transistors, diodes, or components with plastic parts that can melt. Grip the component lead with a pair of needle-nose pliers up close to the unit so that the heat is conducted away (be careful — it is easy to damage delicate component leads). A rubber-band wrapped around the pliers' handles will hold the pliers on the wire. A small alligator clip or a flat spring type paper clip also makes a good heat sink.

Mechanical stress can damage components. Mount components so that there's no appreciable mechanical strain on the leads. Be especially careful with small glass diodes and small disc capacitors, as these components are easy to break when forming the leads.

Installing wire jumpers between connectors or modules and circuit boards can be straightforward, or it can lead to a rat's nest. For some excellent advice and guidance about this basic task, read "W8ZR's Tips for Wiring Circuit Board Jumpers" at **w8zr.net/stationpro/images/download files/Tips for Wiring Circuit Board Jumpers.pdf**. This is a common task in building equipment, so why not learn to do it well?

Soldering to hollow pins, such as those found on connectors, can be difficult, particularly if the connector has been used previously or has oxidized. Use a suitable small twist drill to clean the inside of the pin and then tin it. While the solder is still melted, clear the surplus solder from each pin with a whipping motion or by blowing through the pin from the inside of the connector. Watch out for flying hot solder. Use safety goggles and protect the work surface, along with your arms and legs! A glass ashtray or small baking dish works great for catching the loose solder. Do not perform this operation near open electronic equipment because the loose solder can easily form short circuits. If the pin surface is plated, file the plating from the pin tip, then insert the wire and solder it. After soldering, remove excess solder with a file, if necessary.

When soldering to the pins of plastic connectors or other assemblies, heat-sink the pin with needle-nose pliers at the base where it comes in contact with the plastic housing. Do not allow the pin to overheat; it will loosen and become misaligned. If you need to solder very quickly due to concerns about melting the housing, tin both the pin and wire first. Reheat the pin while inserting the wire. This melts the solder on both the pin and the wire, resulting in a solid connection without overheating the pin.

23.3.5 Desoldering

There are times when soldered components need to be removed; the parts may be bad, installed incorrectly, or you may want to remove them to use in another project.

There are several techniques for desoldering. The easiest way is to use a desoldering braid, which is simply a fine copper braid, often containing flux. It is available under a wide variety of trade names wherever soldering supplies are sold. A good rule of thumb is to choose a width and thickness of desoldering braid that matches the size of the connection being desoldered.

The soldering braid is placed against the joint to be desoldered. A hot iron is pressed onto the braid. If you have a variable temperature soldering station, you might get better results by turning up the temperature, as the combination of the wick and the connection absorbs more heat. If the solder does not melt easily, you can try to paint some flux on it first. As the solder melts, it is wicked into the braid and away from the joint. Copper is an excellent conductor of heat, and the braid can get quite hot, so watch your fingers when using braid for desoldering. After all the leads have been treated in this manner the part can be removed. (A thin film of solder may remain, but is easily broken loose through the use of needle-nose pliers.)

The part of the braid that wicked up the solder is clipped off and discarded. Do not allow the used portion of the braid to get too long, as it will absorb too much heat and not do as good of a job of removing solder. When desoldering connections that have been made by using lead-free or "RoHS" solder, it can sometimes be difficult to achieve the high temperature needed to use solder wick, even with the iron's heat turned all of the way up. A good tip, in this case, is to add a small amount of conventional leaded solder to the joint first, allowing it to mix. That lowers the melting point to a level that allows for much easier desoldering using solder wick, or other methods.

A desoldering vacuum pump can also be used. There are two types of desoldering vacuum tools, a simple rubber syringe bulb with a high temperature plastic tip and a desoldering pump. The desoldering pump is a simple manual vacuum pump consisting of a cylinder that contains a spring-loaded plunger attached to a metal rod inside a tip of high temperature plastic on the end of the pump. To desolder a joint, the plunger is pushed down and locked in place. The tip is placed against the joint to be desoldered along with a soldering iron. When the solder melts, a button on the pump releases the plunger, which pulls the rod back, creating a vacuum that sucks the molten solder through the tip. The part being desoldered can then be removed. Pushing the plunger again ejects the solder from the desoldering tip.

The desoldering bulb employs a similar concept: heat the joint to be desoldered, squeeze the bulb, place the tip on the joint, and release it to suck up the solder. Remove the part that was desoldered. If the first application of the desoldering pump doesn't suck up all the solder, reheat the joint and suck up the rest.

If the desoldering tool doesn't seem to be working, the tip may be clogged with solder. The tip can be unclogged by pushing the solder through. You may have to clear the tip several times when doing a job that requires desoldering many joints. One can purchase small desoldering irons that contain a bulb on the handle that leads to a tip adjacent to the iron tip. This desoldering iron combination is somewhat easier to handle than the separate bulb and iron, and it does an effective job. An ashtray is a good container for blowing out the excess solder from this tool before using it again.

Desoldering stations are also available. One type contains a vacuum pump in a console much like a soldering station. A vacuum line is connected to a hollow soldering iron tip. There is a valve trigger on the iron that is used to open the tip to the vacuum when the solder is melted. The solder is sucked up the line into a receptacle in the station or in the hand piece.

DESOLDERING SURFACE MOUNT DEVICES

Special tools are available for desoldering and replacing surface mount parts. The most common is "hot tweezers," which is really a special dual-tipped soldering iron that is squeezed to match the size of the component. Hot tweezers can be used to not only remove but to also replace surface mount parts. Another tool for this purpose is a hot-air tool with a small tip that focuses the hot air on a single component. This tool can be used to remove a surface mount part as well as to replace one. It heats the joint with hot air so the component can be lifted off with tweezers or needle-nose pliers.

Use surplus circuit boards to practice soldering and desoldering techniques. You can find old boards from various electronics available at low cost or free at many flea markets. Use these surplus boards to practice different soldering and desoldering techniques without having to worry about damaging a project. Computer boards are often made with lead-free solder, and are ideal for practicing desoldering methods by adding leaded solder to the connections before desoldering.

23.3.6 Soldering Safety

Soldering requires a certain degree of practice and, of course, the right tools. What potential hazards are involved?

When the solder used for electronic components is a lead-tin alloy, the first thing in most people's minds is lead, a well-known health hazard. There are two primary ways lead might enter our bodies when soldering: we could breathe lead fumes into our lungs, or we could ingest (swallow) lead or lead-contaminated food. Always wash your hands carefully after any soldering (or touching of solder connections).

Using a small fan can keep the fumes away from your eyes and reduce your exposure to solder smoke. A small computer fan, often only an inch or two wide, can be used. By reducing the voltage that feeds the fan, the speed and noise can be reduced, allowing the fumes to be blown away without creating other problems with the airflow, such as blowing light parts out of position or lowering the heat of the soldering iron. Look in old computers for these little fans, often attached to video cards or to the bus chips on the motherboard. CPU or case fans can also be used, but the voltage supplying them will need to be reduced to create the desired level of airflow. There are also commercially available specialized fans with built-in filters designed for this purpose. Look for a soldering "fume extractor."

Soldering equipment gets *hot*! Be careful. Treat a soldering iron as you would any other hot object. A soldering iron stand is helpful, preferably one that has a cage that surrounds the hot tip of the iron. Here's a helpful tip: If the soldering iron gets knocked off the bench, train yourself not to grab for it because chances are you'll end up grabbing the hot end!

When heated, the flux in solder gives off a vapor in the form of a light gray smoke-like plume. This flux vapor, which often contains aldehydes, is a strong irritant that can potentially cause serious problems to persons with respiratory sensitivity conditions such as asthma. In most cases, it is relatively easy to use a small fan, like the small computer fans described previously, to move the flux vapor away from your eyes and face. Opening a window provides additional air exchange.

Solvents are often used to remove excess flux after the parts have cooled to room temperature. Minimize skin contact with solvents by wearing gloves resistant to the solvent. If you use a solvent to remove flux, it is best to use the mildest one that does the job. Isopropyl, or rubbing alcohol, is often sufficient. You can purchase alcohol ranging from 70% to 92% concentration that works well in removing most types of fluxes at local drug stores.

Observe these precautions to protect yourself and others:

- Properly ventilate the work area. If you can smell fumes, you are breathing them.
- Wash your hands after soldering, especially before handling food.
- Minimize direct contact with flux and flux solvents. Wear disposable gloves when handling solvents.

For more information about soldering hazards and the ways to make soldering safer, see "Making Soldering Safer," by Brian P. Bergeron, MD, NU1N (Mar 1991 *QST*, pp. 28–30) and "More on Safer Soldering," by Gary E. Meyers, K9CZB (Aug 1991 *QST*, p. 42).

23.4 Surface Mount Technology (SMT)

Today, nearly all consumer electronic devices are made with surface mount technology. Hams have lagged behind in adopting this technology largely due to the misconception that it is difficult, requires extensive practice, and requires special equipment. In fact, surface mount devices can be soldered easily with commonly available equipment. There is no more practice required to become proficient enough to produce a circuit with surface mount devices than there is with soldering through-hole (leaded) components.

There are several advantages to working with SMT:

- Projects are much more compact than if through-hole components are used
- SMT parts are available that through-hole packages don't include
- Fewer and fewer through-hole parts are being produced
- Equivalent SMT components are often cheaper than through-hole parts
- SMT parts have less self-inductance, less self-capacitance, and better thermal properties

There are several techniques that can be used effectively to work with SMT components: conventional soldering iron, hot air reflow, and hot plate/hot air reflow. This section describes the soldering iron technique. Online descriptions of reflow techniques are available for the advanced builder who wants to try them. As is the case for through-hole soldering, kits are available to teach the beginner how to solder SMT components.

The following material on working with surface mount technology contains excerpts from a series of *QST* articles, "Surface Mount Technology — You Can Work with It! Part 1," by Sam Ulbing, N4UAU, published in the April 1999 issue. (The entire series of four articles is available with the online content. Additional information on SMT is available at **www.arrl.org/surface-mount-technology**.) Additional information and illustrations were contributed by George Heron, N2APB.

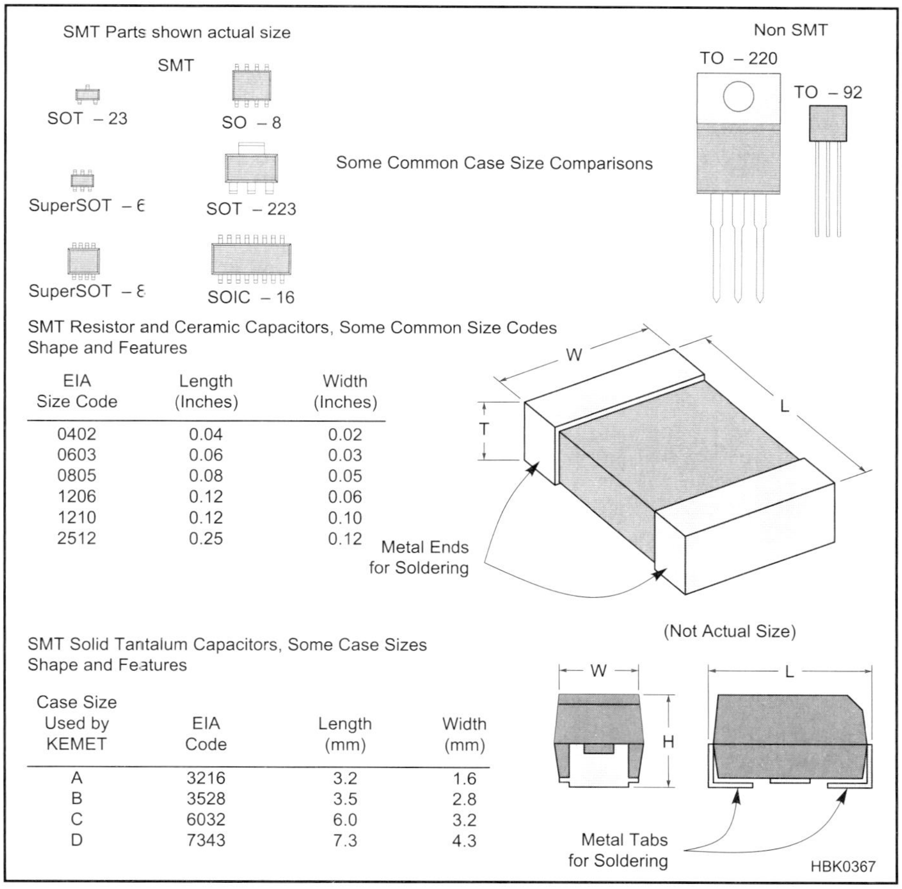

Figure 23.3 — Size comparisons of some surface-mount devices and their dimensions. See the Component Data and References chapter for more information about component packages and labeling.

23.4.1 Surface mount Parts

Figure 23.3 shows some common SMT parts. (The **Circuits and Components** chapter has more information on component packages.) Resistors and ceramic capacitors come in many different sizes, and it is important to know the part size for two reasons: Working with SMT devices by hand is easier if you use the larger parts; and it is important that the PC board pad size is larger than the part.

Discrete component packaging has shrunk to 0.12 × 0.06 inches, as shown in the "1206" capacitor in **Figure 23.4** compared to a penny. Even smaller packages are common today, requiring much less PC board area for the equivalent circuits. Integrated circuit packaging has also been miniaturized to create 10 × 5 mm SOIC packages with lead separations of 0.025 inch.

Tantalum and electrolytic capacitors are some of the larger SMT parts. Their case code, which is usually a letter, often varies from manufacturer to manufacturer because of different thicknesses. The EIA code for ceramic capacitors and resistors is a measurement of the length and width in inches, but for larger capacitors those measurements are in millimeters times 10! Keep in mind that tantalum and electrolytic capacitors are polarized; the case usually has a mark or stripe to indicate the positive end. Nearly any part that is used in through-hole technology is available in an SMT package.

If you are a beginner, it is probably best to start with the larger sizes — 1206 for resistors and ceramic capacitors, SOT-23 for transistors, and SO-8 for ICs. When you get proficient with these parts you can move to smaller packages.

Figure 23.4 — SMT components are small. Clockwise from left: MMIC RF amp, 1206 resistor, SOIC integrated circuit, 1206 capacitor, and ferrite inductor.

23.4.2 Equipment Needed

You do not need lots of expensive equipment to work with SMT devices.

- A fundamental piece of equipment for SMT work is an illuminated magnifying glass. You can use an inexpensive one with a 5-inch diameter lens, which is also convenient to use the magnifier for all soldering work, not just for SMT use. Such magnifiers are widely available from about $25. Most offer a 3X magnification and have a built-in circular light. Be sure to use a light that is fluorescent or LED-based. Incandescent lights are much hotter and make it difficult to ascertain the correct color of component markings.
- Magnifying visors provide a wearable alternative to a large magnifying glass. **Figure 23.5** shows a convenient magnifying visor.
- A low-power, temperature-controlled soldering iron is necessary. Use a soldering iron with a grounded tip, as most SMT parts are CMOS devices and are subject to possible ESD (static) failure.
- Use of thin (0.020 inch diameter) rosin-core solder is preferred because the parts are so small that regular 0.031 inch diameter solder will flood a solder pad and cause bridging. Solder paste or cream can also be used.
- A flux pen comes in handy for applying just a little flux at a needed spot.
- Good desoldering braid is necessary to remove excess solder if you get too much on a pad. 0.100-inch-wide braid works well for this.
- ESD protective devices such as wrist straps may be necessary if you live in a dry area and static is a problem.
- Tweezers help pick up parts and position them. Nonmagnetic, stainless-steel drafting dividers also work well. The nonmagnetic property of stainless steel means the chip doesn't get attracted to the dividers.
- Some hams prefer to hold components in place with a temporary adhesive such as DAP Blue-Stik while soldering rather than holding the part with tweezers.

PREPARING FOR SMT WORK

A clean work surface is of paramount importance because SMT components have a tendency to slip from pliers or tweezers and fly off even when held with the utmost care. You'll have the best chance of recovering your wayward part if your table is clear. When the inevitable happens, you'll have lots of trouble finding the tiny part if it falls onto a rug. For this reason, it's best to have your work area in a room without carpeting, as well as to protect static-sensitive parts. Using a cookie sheet with a raised lip helps to contain stray SMT parts.

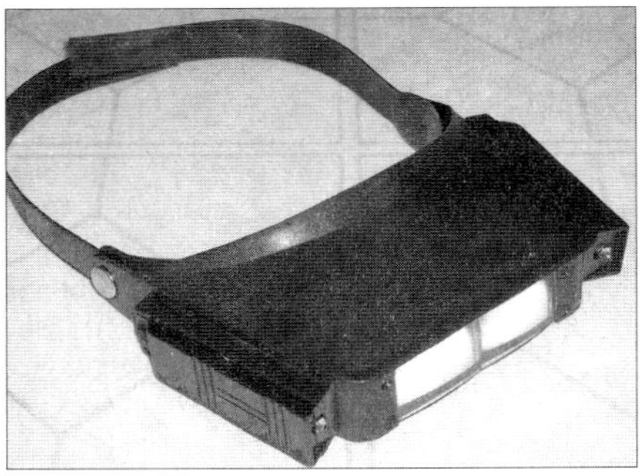

Figure 23.5 — A magnifying visor is great for close-up work on a circuit board. These headsets are often available for less than $10 at hamfests, and some even come with superbright LEDs mounted on the side to illuminate the components being soldered.

23.4.3 SMT Soldering Basics

If a project contains both SMT and conventional through-hole parts and you intend to use a heat-gun or oven for reflow soldering, always mount the SMT parts first, as through-hole components are not always designed to handle the higher reflow heat levels. Use junk PC boards to practice your soldering and desoldering techniques for SMT parts until you are comfortable beginning your own project.

SMT soldering techniques are covered in the March 2019 *QST* article "Soldering Surface-Mount Devices" by Dino Papas, KLØS. Reflow soldering is explained in the *QST* and *QEX* articles "Reflow Soldering for the Radio Amateur" (Jan 2011 *QST*) and "Reflow Soldering for the Radio Amateur — Revisited" (Mar 2019 *QEX*) by Jim Koehler, VE5FP. All three are included in the online content for this chapter.

USING A SOLDERING IRON

Let's look at how to solder a surface mount IC with a soldering iron. If the board is not already pre-tinned, use a little solder to pre-tin the PC board pads. The trick is to add just enough solder so that when you reheat it, it flows to the IC, but not so much that you wind up with a solder bridge. Putting just a little flux on the board and the IC legs makes for better solder flow, providing a smooth layer. You can tell if you have the proper soldering-iron tip temperature if the solder melts within 2 to 4 seconds.

Place the part on the board and then use dividers (or fingers) to push and prod the chip into position. Because the IC is so small and light, it tends to stick to the soldering iron and pull away from the PC board. To prevent this, use the dividers to hold the chip down while tack-soldering two IC legs at diagonally opposite corners. After each tack, check that the part is still aligned. With a dry and clean soldering iron, heat the PC board near the leg. If you do it right, when the solder melts it will flow to the IC.

The legs of the IC must lie flat on the board. The legs bend easily, so don't press down too hard. Check each connection with a continuity checker, placing one tip on the board the other on the IC leg. Check all adjacent pins to ensure there's no bridging. It is easier to correct errors early on, so perform this check often.

If you find that you did not have enough solder on the board for it to flow to the part, add a little. It's best to put a drop on the trace near the part, then heat the trace and slide the iron and melted solder toward the part. This reduces the chance of creating a bridge. Soldering resistors and capacitors is similar to soldering an IC's leads, except the resistors and capacitors don't have exposed leads. The reflow method also works well for these parts.

Attaching wires that connect to points off the board can be a bit of a challenge because even #24 AWG stranded wire is large in comparison to the SMT parts. First, make sure all the wire strands are close together, then pre-tin the wire. Next pre-tin the pad, carefully place the wire on the pad, then heat it with the soldering iron until the solder melts.

USING REFLOW TECHNIQUES

While SMT projects can be built with conventional solder and a fine-point soldering iron, if you move on to reflow techniques you will need to use solder paste. Solder paste is a grayish looking substance made of a blend of flux and solder. A small dot of solder paste is put on the board at each location a component will need to be soldered. The components are then carefully placed on the board, with the paste loosely holding the parts in position. The whole board is then heated in an oven, or the area of the board being assembled is heated with a heat gun. With sufficient heat, the solder paste melts and flows onto the pad and component contacts, then the board is cooled leaving all of the components soldered in place.

A toaster oven can also be modified to perform reflow soldering as described in the article "Reflow Soldering for the Radio Amateur" by Jim Koehler, VE5FP, in the January 2011 issue of *QST* (this article is included with the online content). There are many online tutorials for adapting and using a toaster oven, such as **www.instructables.com/id/Hack-a-Toaster-Oven-for-**

Reflow-Soldering/. SparkFun also shows how to add a temperature control to a toaster oven at **www.sparkfun.com/tutorials/60**. A toaster oven that has been used to perform reflow soldering should never be used for food. Place a label on the oven to warn anyone from placing food in it. If you plan on doing a lot of SMT assembly, learning reflow techniques is well worth the effort.

For occasional SMT use, a heat gun is a good choice. Many hobby/crafts stores sell a special heat tool that is used for melting embossing inks used in scrap booking. Look for an embossing heat tool in the scrap booking department of these stores. Do not get the heat gun too close to the board, as the airflow may move your components out of place. Hold the tool steady a couple of inches above the board until you see the solder paste turn silver and the component appears to be soldered in place, then gradually remove the heat gun. Never use a heat gun designed for paint stripping, as the airflow is way too strong for this purpose and will blow the SMT parts off the board.

Only a small amount of solder paste is required. Kester Easy Profile 256 is a good solder paste to start with. It is available at reasonable cost in a small syringe with a fine needle-point applicator. Since only a little paste is needed for each solder joint, this small amount will last through several medium sized projects. Solder paste must be kept cold or it deteriorates. If kept in a household refrigerator or freezer it has a shelf life of at least a year. Make sure that the solder paste is well wrapped and kept away from any food that is also in the refrigerator or freezer.

23.4.4 Removing SMT Components

The surface mount ICs used in commercial equipment are not easy for experimenters to replace. They have tiny pins designed for precision PC boards. Sooner or later, however, you may need to replace one. If you do, don't try to get the old IC out in one piece! This will damage the IC beyond use anyway, and is likely to damage the PC board in the process.

Although it requires a delicate touch and small tools, it is possible to change a surface mount IC at home. To remove the old one, use small, sharp wire cutters to cut the IC pins flush with the IC body. This usually leaves just enough of the old pin to grab with tweezers. Heat the soldered connection with a small iron and use the tweezers to gently pull the pin from the PC board. Use desoldering braid to remove excess solder from the pads and remove the flux with rubbing alcohol. Solder the new component in place by using the techniques discussed above.

You can also use the embossing heat gun previously described to remove SMT parts, especially ICs. Keep in mind that not only the desired component, but also some adjacent parts, may be loosened by this process, so be sure not to move the board during reheating to allow the other components to stay in place. Use long-handled tweezers to remove a component once you see the solder become silvery. Be careful not to disturb any adjacent components. Allow the board to thoroughly cool before moving it to prevent inadvertently allowing other components to shift before the solder solidifies again.

To remove individual components without a heat gun, first remove excess solder from the pads by using desoldering braid. Then the component will generally come loose from the pads if gently lifted with a hobby knife or dental tool. If the component remains attached to the pad, touch the pad with the soldering iron and lift the component off the pad. It may take one or two attempts to free the component from all pads. In extreme cases, it may be necessary to add solder to the pad to completely loosen the component. A product called ChipQuick Alloy (**www.chipquik.com**, click on ChipQuik Alloy on the menu to the left) reduces the melting point of the existing solder by mixing in a very low temperature solder. Using the flux and solder that comes with it lets you desolder an SMD easily without a hot-air tool.

23.4.5 SMT Construction Examples

DDS DAUGHTERCARD

The first project example is the DDS Daughtercard, a small module that generates precision RF signals for a variety of projects. This kit has become immensely popular in homebrew circles and is supplied with the chip components contained in color-coded packaging that makes an easy job of identifying the little parts, a nice touch by a kit supplier. Full details can be found at **njqrp.club/dds/index.html**.

Figure 23.6A shows the DDS PC board, a typical layout for SMT components. All traces are on one side, since the component leads are not "through-hole." The little square pads are the places where the 1206 package-style chips will eventually be soldered. This project demonstrates the reflow technique using a soldering iron. Figure 23.6B illustrates how to use this technique,

- Pre-solder ("tin") one of the pads on the board where the component will ultimately go by placing a small blob of solder there.
- Carefully hold the component in place with small needle-nose pliers or sharp tweezers on the tinned pad.
- Reheat the tinned pad and component to reflow the solder onto the component lead, thus temporarily holding the component in place.
- Solder the other end of the component to its pad.
- Finally, check all connections to make sure there are no bridges or shorts.

Figure 23.6C shows the completed DDS board.

K8IQY AUDIO AMP

The second project example is the K8IQY Audio Amp, a discrete component audio amplifier that is homebrew-constructed "Manhattan-style" as described later in the chapter, in which small pads are glued or soldered to the copper-clad base board wherever you need to attach component leads or wires. See Figure 23.6D. A manual for this project can be found at **k8iqy.com/testequipment/islanderamp/Islander Amp Manual v1m.pdf**

Instead of using little squares or dots of PC board material for pads, you might decide to create isolated connection points by cutting an "island" in the copper using an end mill or pad cutter. No matter how the pads are created, SMT components may be easily soldered from pad-to-pad, or from pad-to-ground plane, to build up the circuit. Figure 23.6E shows the completed board, combining SMT and leaded components.

Homebrewing with SOIC-packaged integrated circuits is a little trickier and typically requires the use of an "SOIC carrier board," such as the one shown in Figure 23.6F, onto which you solder your surface mount integrated circuit. You can then wire the carrier board onto your base board, or whatever you're using to hold your other circuit components.

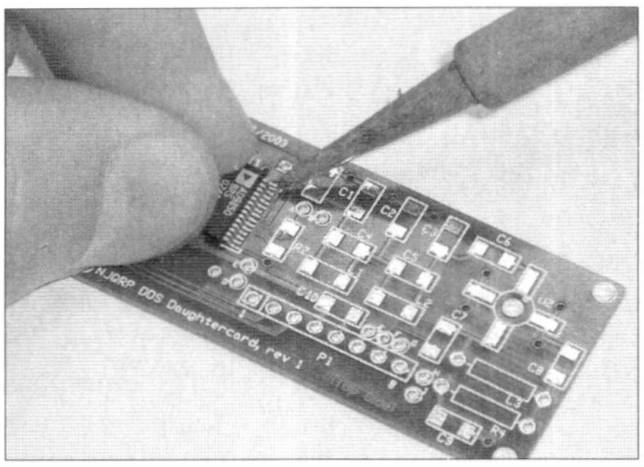

Figure 23.6A — This DDS Daughtercard has all interconnections on the top side. Connections to the ground plane on the backside of the board are made by the use of "vias," wires through the PC board. Pin 28 of the SMT IC is shown being tack-soldered to hold it to the board, keeping all other pins carefully aligned on their pads. Then the other pins are carefully soldered, starting with pin 14 (opposite pin 28). Finally, pin 28 is reheated to ensure a good connection there. If you bridge solder across adjacent pads or pins, use solder wick or a vacuum solder sucker to draw off the excess solder.

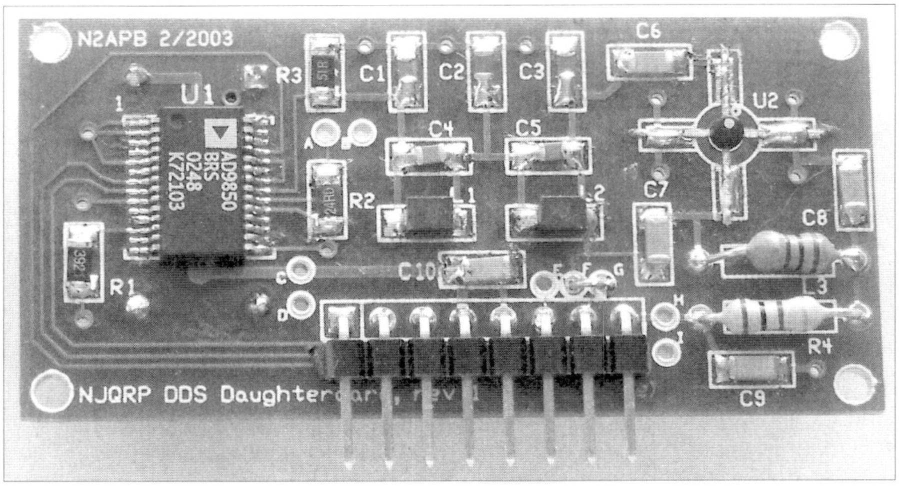

Figure 23.6C — The fully-populated DDS Daughtercard PC board contains a mix of SMT and through-hole parts, showing how both packaging technologies can be used together.

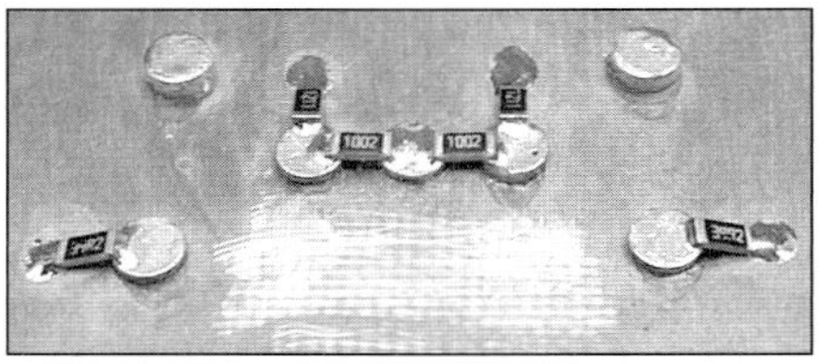

Figure 23.6D — SMT resistors soldered to base board of the Audio Amp in the beginning stages of assembly.

Figure 23.6F — Surface mount ICs can be mounted to general-purpose carrier boards, then attached as a submodule with wires to the base board of the homebrew project.

Figure 23.6B — Attaching an SMT part. It is a lot easier attaching capacitors, resistors, and other discrete components compared to multi-pin ICs. Carefully hold the component in place and properly aligned using needle-nose pliers or tweezers and then solder one end of the component. Then reheat the joint while gently pushing down on the component with the pliers or a Q-tip stick to ensure it is lying flat on the board. Finally, solder the other side of the component.

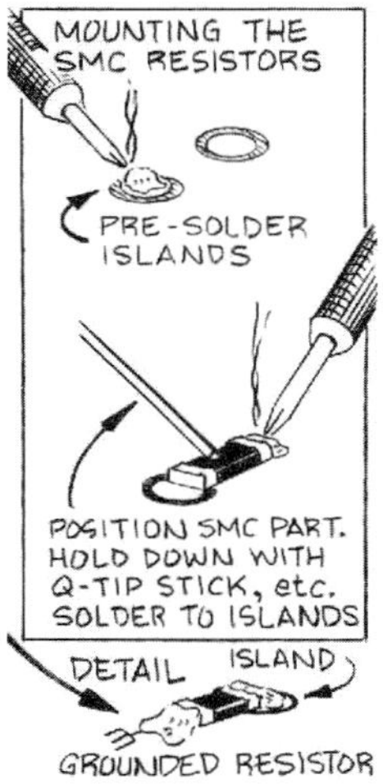

Figure 23.6E — The completed homebrew Audio Amp assembly shows simple, effective use of SMT components used together with conventional leaded components when constructed "Manhattan-style."

Common Standard Parts

When building a project or repairing equipment, it is helpful to have an assortment of standard and common parts on hand for use in modifying circuit designs and fine-tuning performance. Making repairs or completing a kit that is missing a part are also good reasons to keep an assortment of parts on hand. It is not possible to have every possible needed part, but the majority of components in any project are usually one of the common standard values.

Assortments of new parts such as resistors, capacitors, and semiconductors are available from electronic distributors such as DigiKey, Mouser, Newark, and Jameco, and from parts companies such as Velleman. These are often available with a storage container or cabinet. When taking into account the cost of the parts cabinet and the parts themselves, it is very economical to buy standard parts in this manner. If parts cabinets are not available, craft stores as well as fishing supply stores often have low-cost compartmentalized containers ideal for sorting and storing small parts. Tackle boxes are also useful for storing components and materials, particularly if purchased during end-of-season sales.

Another recommendation is to buy in quantity when ordering parts for a project. Not only can you often get a price break on the individual components but you will be building up your store of components along the way.

Many vendors also have bags of one or more values of surplus parts from electronic manufacturing. It is often less expensive to buy an entire bag of surplus parts than it is to buy even one or two of them elsewhere. Some come as truly random assortments, often called "grab bags," and others may be sold as "tapes" that were prepared for a parts-placement machine. You will have to sort out the components but the price makes it worthwhile!

You can accumulate a good selection of parts at ham radio flea markets. It is common to see parts cabinets available with entire collections of components and hardware! Grab bags and parts junk boxes are also common here and generally a very good value if you are willing to sort through the components.

The tables in this sidebar list common parts used in many projects. It is a good idea to accumulate these parts and keep them on hand. Keep in mind that you do not have to have every one of these parts, but have the list in mind as you shop for parts. (See the **Circuits and Components** chapter for more information on part types.)

Resistors (values in ohms)

10	15	18	22	27	33	39	47	56	68	82	100	120
150	180	220	270	390	470	560	680	820				
1.0k	1.2k	1.5k	1.8k	2.2k	2.7k	3.3k	3.9k	4.7k	5.6k	6.8k	8.2k	
10k	12k	15k	18k	22k	27k	33k	39k	47k	56k	68k	82k	
100k	120k	150k	180k	220k	470k	560k	680k	820k	1M	4.7M		

Potentiometers (values in ohms)

500	1k	5k	10k	100k	1M

Ceramic disc capacitors (values in pF)

5	10	18	22	27	33	39	47	56	68	82	100	120
150	220	390	470	560	680	1000	4700					

Ceramic disc capacitors (values in μF)

0.001	0.005	0.01	0.022	0.05	0.1

Electrolytic and tantalum capacitors (values in μF)

1	2.2	4.7	10	22	33	47	100	220	470	1000	2200

Diodes and rectifiers

1N34A	1N914	1N4001	1N4007	1N4148	1N5401	1N5819

Transistors

2N2222	2N3055	MJ2955	2N3904	2N3906	2N4401	2N4404	2N7000	IRF510	TIP31C	TIP32C

Voltage regulators

78L05	7805	7812	7815	LM317	723

Operational amplifiers and miscellaneous ICs

324	741	747	TL081
555 (timer)			
ULN2001 (driver array)			

23.5 Constructing Electronic Circuits

Most of the construction projects undertaken by the average amateur involve electronic circuitry. The circuit is the "heart" of most amateur equipment. It might seem obvious, but in order for you to build it, the circuit must work! Don't always assume that a "cookbook" circuit that appears in an applications note or electronics magazine is flawless. These are sometimes design examples that have not always been thoroughly debugged. Many home-construction projects are one-time deals; the author has put one together and it worked in that instance. In some other cases, component tolerances or minor layout changes might make it difficult to get a second unit working. Using a solderless breadboard can make it easier to test this type of circuit design. For RF circuits above a few MHz, a solderless breadboard is not always practical due to long lead lengths that result.

Take steps to protect the electronic and mechanical components you use in circuit construction. Some components can be damaged by rough handling. Dropping a 1/4 W resistor causes no harm, but dropping a vacuum tube or other delicate subassemblies usually causes damage.

Some components are easily damaged by heat. Some of the chemicals used to clean electronic components (such as flux removers, degreasers, or control-lubrication sprays) can damage plastic. Check them for suitability before you use them.

23.5.1 Electrostatic Discharge (ESD)

Some components, especially high-impedance components such as FETs and CMOS gates, can be damaged by electrostatic discharge (ESD). Protect these parts from static charges. Most people are familiar with the static charge that builds up when one walks across a carpet and then touches a metal object; the resultant spark can be quite lively. Walking across a carpet on a dry day can generate 35 kV! A worker sitting at a bench can generate voltages up to 6 kV, depending on conditions, such as when relative humidity is less than 20%.

You don't need this much voltage to damage a sensitive electronic component; damage can occur with as little as 30 V. The damage is not always catastrophic. A MOSFET can become noisy, or lose gain; an IC can suffer damage that causes early failure. To prevent this kind of damage, you need to take some precautions.

The energy from a spark can travel inside a piece of equipment to affect internal components. Protection of sensitive electronic components involves the prevention of static build-up together with the removal of any existing charges by dissipating any energy that does build up.

MINIMIZING STATIC BUILD-UP

Several techniques can be used to minimize static build-up. Start by removing any carpet in your work areas. You can replace it with special antistatic carpet, but this is expensive. It's less expensive to treat the carpet with antistatic spray, which is available from electronics wholesalers. Adding humidity to the air can help reduce the presence of static charges as well. Even the choice of clothing you wear can affect the amount of ESD. Polyester has a much greater ESD potential than cotton.

Many builders who have their workbench on a concrete floor use a rubber mat to minimize the risk of electric shocks from the ac line. Unfortunately, the rubber mat increases the risk of ESD. An antistatic rubber mat serves both purposes.

Many components are shipped in antistatic packaging. Leave components in their conductive packaging. Other components, notably MOSFETs, are shipped with a small metal ring that temporarily shorts all of the leads together. Leave this ring in place until the device is fully installed in the circuit.

Use antistatic bags to transport susceptible components or equipment. Keep your workbench free of objects such as paper, plastic, and other static-generating items. Use conductive containers with a dissipative surface coating for equipment storage. Storing partially assembled projects in antistatic bags is also a good idea.

These precautions help reduce the build-up of electrostatic charges. Other techniques offer a slow discharge path for the charges or keep the components and the operator handling them at the same ground potential.

DISSIPATING STATIC

One of the best techniques is to connect the operator and the devices being handled to earth ground, or to a common reference point. It is not a good idea to directly ground an operator working on electronic equipment, though, as the risk of electrocution is too great. If the operator is grounded through a high-value resistor such as 100 kΩ to 1 MΩ, ESD protection is still offered but there is less risk of being harmed.

The operator is usually grounded through a conductive wrist strap. This wrist band is equipped with a snap-on ground lead. A 1 MΩ resistor is built into the snap of the strap to protect the user should a live circuit be contacted. Build a similar resistor into any homemade ground strap.

The devices and equipment being handled should also be grounded, by working on a charge-dissipating mat that is connected to ground. The mat should be an insulator that has been impregnated with a resistive material. Suitable mats and wrist straps are available from most electronics supply houses. **Figure 23.7** shows a typical ESD protective

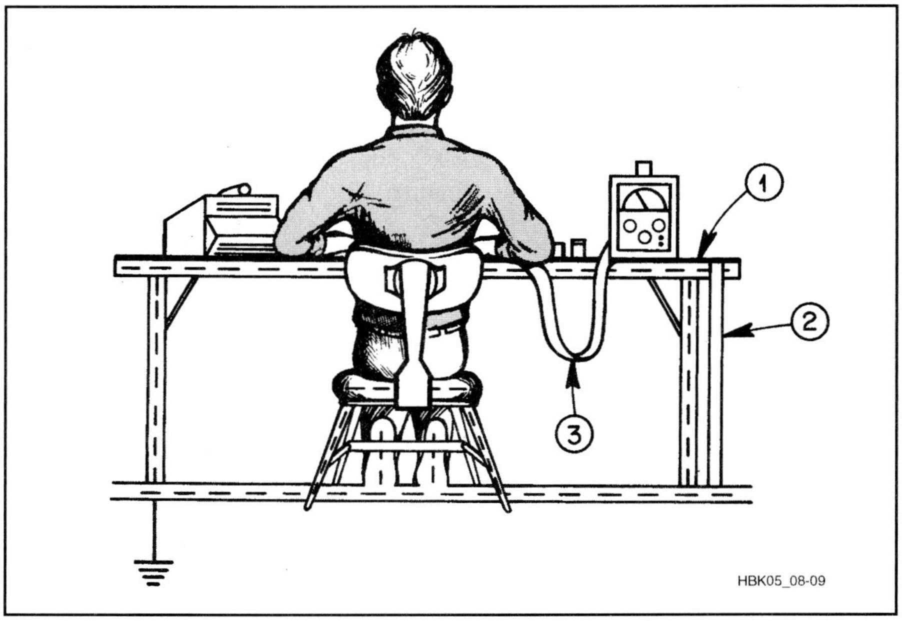

Figure 23.7 — A work station that has been set up to minimize ESD features with (1) a grounded dissipative work mat and (2) a wrist strap that (3) grounds the worker through high resistance.

safe workstation.

The work area should also be grounded, directly or through a conductive mat. Use a soldering iron with a grounded tip to solder sensitive components. Most irons that have three-wire power cords are properly grounded. When soldering static-sensitive devices, use two or three jumpers: one to ground you, one to ground the work, and one to ground the iron. If the iron does not have a ground wire in the power cord, clip a jumper from the metal part of the iron near the handle to the metal box that houses the temperature control. Another jumper connects the box to the work. Finally, a jumper goes from the box to an elastic wrist band for static grounding.

23.5.2 Sorting Parts

When building a project, especially one packaged as a kit, finding the appropriate container to sort your components can be a challenge. There are a number of things that make this task a lot easier at a low cost.

The first consideration is to be aware of the material the storage container is made of, if ESD sensitive components are being used. If the container is of metal construction, grounding helps protect components from static buildup. However, if it is made of plastic, avoid using it with ESD sensitive components.

Connecting conductive containers to ground through a high resistance, such as a 1 MΩ resistor, better protects components, since the bleed-off of static occurs more slowly (dissipative grounding). If the container is connected directly to ground (conductive grounding), the rapid movement of static charges from the component to ground can actually damage sensitive components.

Using a plastic egg carton or poking component leads into Styrofoam works quite well for sorting parts, but both methods can lead to ESD damage of sensitive components. Use this method for components such as resistors, capacitors, and inductors that are relatively immune to ESD.

Metal cupcake trays are ideal, as the tray itself can be grounded to prevent static buildup. Cupcake trays typically come in three sizes: 6, 12 and 16 cups. The best sorting technique is to place the most common or first-used components in the cups closest to the builder (usually resistors), with the least used parts (usually mounting hardware) in the distant cups.

Another great idea for parts sorting is to use a fishing tackle box with removable trays. Inexpensive tackle boxes are available at outdoor supply houses and department stores. Be sure to keep in mind the material the box is made from. The tackle box has the distinct advantage of allowing you to sort your parts into different compartments within each movable tray, and some trays have removable dividers that allow you to resize the compartments to better fit the parts for your project. Many tackle boxes have a larger open space in the top that allows you to store your partially finished boards, with the remaining components in the closeable trays below it. This arrangement is ideal to protect your project from damage and can be securely stored between work sessions.

23.5.3 Construction Order

When building a kit or DIY project, the question often arises as to what order the parts need to be mounted. In a kit, the manual often is very explicit, requiring the builder to construct the project stage by stage. Other kit manuals offer only minimal directions, leaving it up to the builder. Have the manual handy, either printed or on a nearby laptop or tablet, for reference.

In stage-by-stage construction, each stage in the project is completed in order to allow the builder to test and troubleshoot that area of the project without having a more complex problem to solve after all components are mounted. This method also allows the builder to learn the principles involved in the project and how each part of the circuit works. This is a great aid to future modification and repair.

When building a kit in stages it is often better to sort the parts by stages as well, placing the parts from each stage in their own space. That way, when each stage is completed, there should be no extra parts left over in that stage's container. Number the stages, if they are not already numbered in the manual, and place a small piece of paper with that number in each compartment, indicating the stage that those parts belong to.

When given a minimum of assembly instructions, the best approach is to mount the resistors first, then the capacitors, and then the semiconductors, followed by the more unique components. The majority of parts are then mounted early in the process, so finding the remaining part locations is a lot easier. This technique also allows the builder to double check the usage of parts. Try to mount large parts after the smaller surrounding parts to not possibly block your ability to properly mount all of them.

Inventory the parts before commencing construction. If a change is introduced after a number of kits or DIY projects have been assembled, there is a chance of errors so that the parts list or board layouts do not reflect changes in the design of the circuit. Sometimes, a number of extra parts are supplied with a kit to facilitate different options, such as the choice of bands covered. Sometimes parts are eliminated or substituted with a change of other components in the circuit. Be sure to ask the kit supplier if you are not sure why you have an empty space or surplus parts. Sometimes new parts values are added to substitute for old values already packed in the kit, making the old values surplus. Resolve questions about component placement before powering up a completed project.

23.5.4 Component Mounting

When working with a large number of components, there are a few techniques that can be helpful should troubleshooting be required. Although resistors are not polarized, it is a good idea to mount them with the color codes reading in the same direction to make it easier to spot a part that is not in its correct position. Polarity-sensitive parts, such as diodes, electrolytic capacitors, and ICs, must be placed in their specified direction. In general, mount components so their values are readable without having to remove them from the board or bend them, possibly causing damage.

Axial-lead components, such as resistors, are mounted either upright or flat. To save space on a PC board, resistors are often mounted upright with one lead bent double in a manner resembling a hairpin. The best practice is to make this bend so that the color stripes denoting the resistor value begin at the top and the precision stripe (often silver or gold) is at the bottom, making it easier to read the values once mounted. Components with alphanumeric markings, such as diodes, should be mounted with the markings visible.

Non-polarized capacitors are best placed with markings facing in the same direction, unless the markings would be blocked by another component. For polarized parts, always double-check their positioning before soldering. A commercially prepared PC board often has stripes on the diode labels, indicating which end to place the cathode stripe on the diode. A "+" sign on the board inside or next to the circle for a capacitor denotes the positive lead, which often will be longer. LEDs will have a flat spot or a notch on a lead to identify the cathode. See the **Component Data and References** chapter and manufacturers' data sheets for more information on component body styles.

When mounting ICs, using a pin straightener helps align the leads for insertion. If one is not available, use a flat surface to align them all at once. Be sure a pin does not get bent inward and that all pins go into the socket or PC board holes.

Straight-pin "header" connectors are commonly used in projects and kits. Unless you have a specialized jig for this purpose, soldering this type of connector can appear to be difficult. An easy technique is to simply solder one pin to hold the part on the board, without regard to how exactly straight the connector is. Using a finger, apply pressure on an unsoldered pin or pins, and reheat the solo pin that was previously soldered. You can feel when the connector is moved into place, straight and vertical. Be sure to only touch a pin that is not being heated as it can become very hot! (KØNEB has contributed photos that are available in the book's online material as additional guidance.)

Once the connector is in the correct position with only one pin soldered, the other pins can be soldered. This same process can also be used for soldering things like plugs and jacks to a PC board. Temporarily, or "tack," solder one pin, reheat it while adjusting its position, and then solder the other pins on the connector.

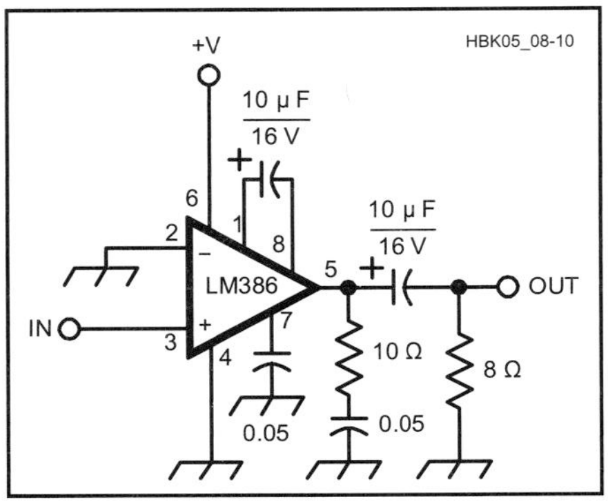

Figure 23.8 — Schematic diagram of the audio amplifier used as a design example of various construction techniques.

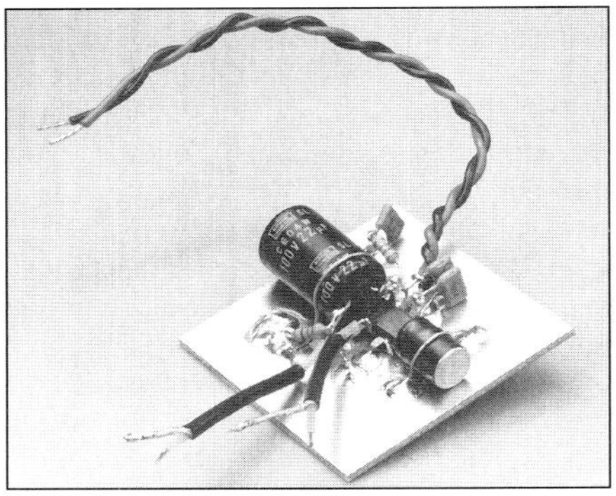

Figure 23.9 — The example audio amplifier of Figure 23.9 built using ground-plane construction.

23.5.5 Electronics Construction Techniques

Several different point-to-point wiring techniques or PC boards can be used to construct electronic circuits. Most circuit projects use a combination of techniques. The selection of techniques depends on many different factors and builder preferences.

For one-time construction, PC boards may not be necessary. It takes time to lay out, drill, and etch a PC board. Alterations are difficult to make if you change your ideas or make a mistake.

The simple audio amplifier shown in **Figure 23.8** will be built using various point-to-point or PC board techniques. This shows how the different construction methods are applied to a typical circuit. (Surface mount techniques are discussed above.)

POINT-TO-POINT TECHNIQUES

Point-to-point techniques include all circuit construction techniques that rely on tie points and wiring, or component leads, to build a circuit. This is the technique used in most home-brew construction projects. It is sometimes used in commercial construction, such as old vacuum-tube receivers and modern tube amplifiers.

Point-to-point wiring is also used to connect the "off-board" components used in a printed circuit project. It can be used to interconnect the various modules and printed circuit boards used in more complex electronic systems. Most pieces of electronic equipment have at least some point-to-point wiring.

GROUND-PLANE CONSTRUCTION

A point-to-point construction technique that uses the leads of the components as tie points for electrical connections is known as "ground-plane," "dead-bug," or "ugly" construction. "Dead-bug construction" gets its name from the appearance of an IC with its leads sticking up in the air. In most cases, this technique uses copper-clad circuit-board material as a foundation and ground plane on which a circuit is built using point-to-point wiring. (In this chapter it is called "ground-plane construction.") An example is shown in **Figure 23.9**.

Ground-plane construction is quick and simple: You build the circuit on an unetched piece of copper-clad circuit board. Wherever a component connects to ground, you solder it to the copper board. Ungrounded connections between components are made point-to-point. Once you learn how to build with a ground-plane board, you can grab a piece of circuit board and start building with little preparation.

A PC board has strict size limits as the components must fit in the space allotted. Ground-plane construction is more flexible since it allows you to use the parts on hand. The circuit can be changed easily, which is a big help when you are experimenting. The greatest virtue of ground-plane construction is that it is fast.

Ground-plane construction is something like model building, connecting parts using solder almost, but not exactly, like glue. In ground-plane construction you build the circuit directly from the schematic, so it can help you get familiar with a circuit and how it works. You can build subsections of a large circuit on small ground-plane modules and string them together into a larger design.

Circuit connections are made directly, minimizing component lead length. Short lead lengths and a low-impedance ground conductor help prevent circuit instability. There is usually less inter-component capacitive coupling than would be found between PC-board traces, so it is often better than PC-board construction for RF, high gain, or sensitive circuits.

Circuit components are used to support other circuit components. Start by mounting one component onto the ground plane, building from there. There is a two-handed technique to mount a component to the ground plane: Bend a component lead that must be connected to ground at a 90° angle and trim off the excess. Solder a 0.1 inch diameter blob of solder to the ground plane. Using one hand, hold the component in place on top of the soldered blob and heat both the component lead and the solder, which should flow nicely, soldering the component securely. Remove the iron and hold the component perfectly still until the solder cools. You can then make connections to the other lead(s) of that part.

Connections should be mechanically secure before soldering. Bend a small hook in the lead of a component, then crimp it to the next component lead(s). Do not rely only on the solder connections to provide mechanical strength. Solder is a soft metal and is likely to fail over time if force is continuously applied to it.

In most cases, a design has enough grounded components to support all of the components in the circuit. However, this is not always the case. In some circuits, high-value resistors can be used as standoff insulators. One resistor lead is soldered to the ground plane and the other lead is used as a circuit connection point. You can use 1/4 W or 1/2 W resistors with values from 1 to 10 MΩ. Such high resistances do not permit much current to flow, and in low-impedance circuits they act more like insulators than resistors. As a rule of thumb, resistors used as standoff insulators should have values at least 10 times higher than the circuit impedances that they are connected to.

Figure 23.10A shows how to use the standoff technique to wire the circuit shown at Figure 23.10C. Figure 23.10B shows how the resistor leads are bent before the standoff component is soldered to the ground plane.

Components E1 through E5 are resistors that are used as standoff insulators. They do not appear in the schematic diagram. The base circuitry at Q1 of Figure 23.10A has been spread out to reduce clutter in the drawing. In a practical circuit, all of the signal leads should be kept as short as possible. E4 should, therefore, be placed much closer to Q1 than the drawing indicates.

No standoff posts are required near R1 and R2 of **Figure 23.11**. These two resistors serve both purposes: They are the normal circuit resistances and also function as standoff posts.

"MANHATTAN" CONSTRUCTION

Another solution to building up a circuit is called "Manhattan" construction, shown previously in the Surface Mount Components section as Figure 23.6E. This method got its name from the appearance of the finished product, resembling the tall buildings in a city. Manhattan construction uses plain, unetched copper clad PC board material to make both the main board and the component connection points. The PC board material used may be single or double-sided.

After cutting the desired size and shape of the board required for the project, use the scraps left over to make the insulated contact pads. These pads can be made a number of ways, the most common being cutting the material into tiny squares about 1/4 or 3/8 inches across. Another method to create the pads is to use a heavy duty hole punch. This kind of punch often has changeable dies to create various sizes of round holes in materials such as sheet metal, and is available at many tool dealers. Once cut, the pads can be glued to the board in a pattern that accommodates the lead lengths of the parts to be connected.

Pads can be glued to the base board by using a tiny drop of instant glue, such as a cyanoacrylate "super glue." If the pads are double-sided PC board material, they can be soldered to the base board.

Component leads are soldered to the pads, and additional leads can be added to a pad by simply reheating the connection already there. The main board is used as the ground plane with all ground leads soldered to the board. This method of construction works well with RF circuits up to UHF.

WIRED TRACES — THE LAZY PC BOARD

If you already have a PC-board design but don't want to copy the entire circuit, or you don't want to make a double-sided PC board, then the easiest construction technique is to use a bare board or perfboard (see below) and hard wire the traces.

Drill the necessary holes in a piece of single-sided board, remove the copper ground plane from around the holes, and then wire up the back using component leads and bits of wire instead of etched traces (Figure 23.11).

To transfer an existing board layout, make a 1:1 photocopy and tape it to your piece of PC board. Prick through the holes with an automatic (one-handed) center punch or by firm pressure with a sharp scriber, remove the photocopy, and drill all the holes. Holes for ground leads are optional. You generally get a better RF ground by bending the component lead flat to the board and soldering it down. Remove the copper around the rest of the holes by pressing a drill bit lightly against the hole and twisting it between your fingers. A drill press can also be used, but either way, don't remove too much board material. Then wire up the circuit beneath the board. The results look very neat

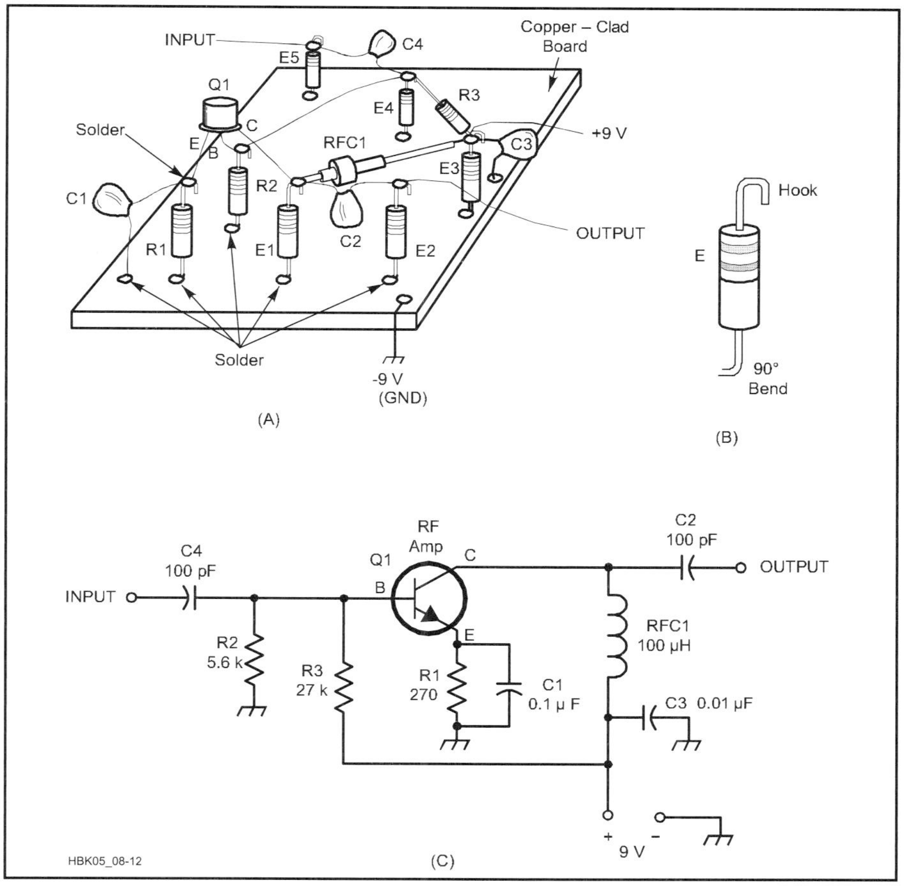

Figure 23.10 — Pictorial view of a circuit board that uses ground-plane construction is shown at (A). A close-up view of one of the standoff resistors is shown at (B). Note how the leads are bent. The schematic diagram at (C) shows the circuit displayed at (A).

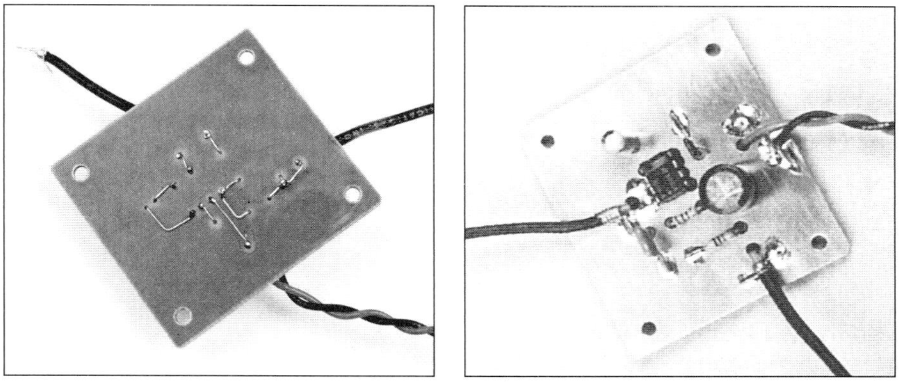

Figure 23.11 — The audio amplifier built using wired-traces construction.

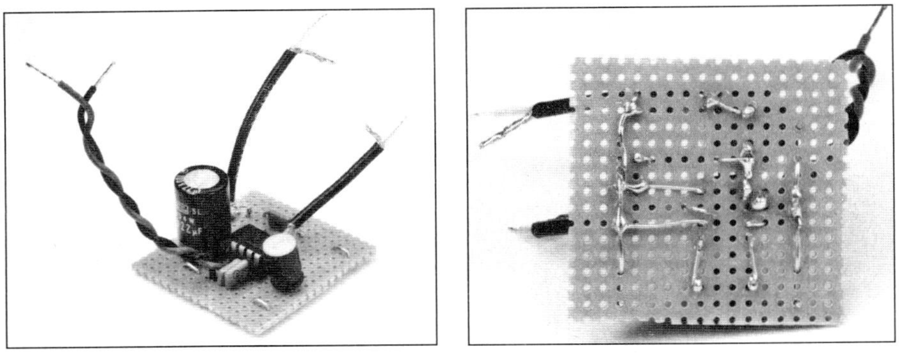

Figure 23.12 — The audio amplifier built on perforated board. Top view at left; bottom view at right.

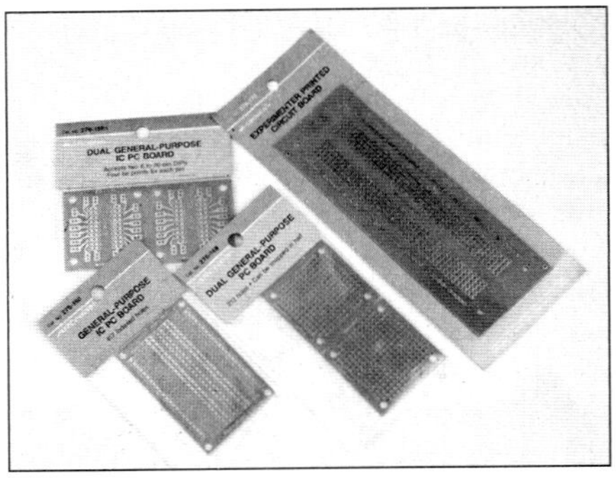

Figure 23.14 — Utility PC boards like these are available from many suppliers.

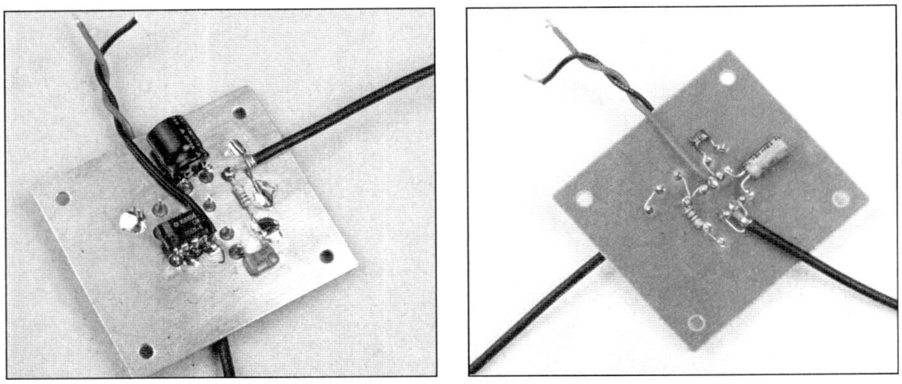

Figure 23.13 — The audio amplifier built using terminal-and-wire construction.

and tidy, from the top, at least!

Circuits that contain components originally designed for PC board mounting are good candidates for this technique. Wired traces would also be suitable for circuits involving multi-pin RF ICs, double-balanced mixers, and similar components. To bypass the pins of these components to ground, connect a miniature ceramic capacitor on the bottom of the board directly from the bypassed pin to the ground plane.

A wired-trace board is fairly sturdy, even though many of the components are only held in by their bent leads and blobs of solder. A drop of cyanoacrylate "super glue" can hold down most larger components, components with fragile leads, or any long leads or wires that can move.

PERFORATED CONSTRUCTION BOARD

A simple approach to circuit building uses a perforated phenolic or epoxy resin board known as *perfboard*. Perfboard is available with many different hole patterns. Choose the one that suits your needs. Perfboard is usually unclad, although it is sometimes made with pads that facilitate soldering. Perfboard, mounting hardware, and a variety of connection terminals for solder and solderless construction, are widely available.

Circuit construction on perfboard is easy. Start by placing the components loosely on the board and moving them around until a satisfactory layout is obtained. Most of the construction techniques described in this chapter can be applied to perfboard. The audio amplifier of Figure 23.8 is shown constructed with this technique in **Figure 23.12**.

In another technique, push terminals are inserted into the holes in a perfboard. Components can then be easily soldered to the terminals. As an alternative, drill holes into a bare or copper-clad board wherever they are needed (**Figure 23.13**). The components are usually mounted on one side of the board and wires are soldered to the bottom of the board, acting as wired PC board "traces." If a component has a reasonably rigid lead to which you can attach other components, you can use that instead of a push terminal.

If you are using a bare board to provide a ground plane, drill holes for your terminals with a high-speed PC board drill bit and drill press. Mark the position of the hole with a center punch to prevent the drill bit from skidding. The hole should provide a snug fit for the push terminal.

Mount RF components on top of the board, keeping the dc components and much of the interconnecting wiring underneath. Make dc feed-through connections with terminals having bypass capacitors on top of the board. Use small solder-in feedthrough capacitors for more critical applications.

Figure 23.15 — The audio amplifier built on a multipurpose PC breadboard. Top surface view above. Bottom surface view below.

UTILITY PC BOARDS

"Utility" PC boards are an alternative to custom-designed etched PC boards. They offer the flexibility of perfboard construction and the mechanical and electrical advantages of etched circuit connection pads. Utility PC boards can be used to build anything from simple passive filter circuits to computers.

Circuits can be built on boards on which the copper cladding has been divided into connection pads. Power supply voltages can be distributed on bus strips. Boards, like those shown in **Figure 23.14**, are commercially available.

An audio amplifier constructed on a utility PC board is shown in **Figure 23.15**. Component leads are inserted into the board and soldered to the etched pads. Wire jumpers connect the pads together to complete the circuit.

Utility boards with one or more etched plugs for use in computer-bus, interface, and general purpose applications are widely available. Connectors, mounting hardware, and other accessories are also available. Check with your parts supplier for details.

23.5.6 Solderless Prototype Board

When designing circuits, building a prototype is often necessary to test the design. Laying out and fabricating a circuit board as soon as a schematic is available is usually premature for the home builder. Similarly, soldering components together directly or with wires makes substituting values more difficult. An easier solution is to use a solderless prototype board, or "protoboard," which allows for the rapid assembly and modification of a circuit with no soldering. This section touches on important aspects of protoboards used for amateur radio circuits. A detailed guide to protoboards is available at **protosupplies.com/guide-to-solderless-breadboards**.

A protoboard makes it possible to quickly substitute parts to vary component values in a design. Having a variety of standard parts available can make experimenting or fine-tuning a design very easy and effective. Protoboards come in a variety of sizes (specified by the number of "tie points" or contacts), ranging from very small to accommodate very simple circuits up to multiple large protoboards that are placed together on a host board.

There are also a number of small circuit board adapters for mounting types of components that cannot be attached easily to a protoboard. Components like a small surface mount IC can be used with an adapter board that has pins below it to connect to a protoboard. Some small controllers, such as an Arduino Nano or a Raspberry Pi Zero, can have connecting pins soldered to the bottom of the board intended for connecting to a protoboard. There are also protoboards with controllers built-in.

PROTOBOARD LAYOUT

Protoboard contacts (see **Figure 23.16**) are spaced by 0.1 inches in rows or grids. Parallel groups of five contacts are connected together with a metal strip to form a common point of contact. These groups are insulated from each other to form common connection points. Components and jumpers are connected together by plugging them into the same group of five contacts.

Along one or both long edges of the protoboard are strips of contacts called *rails* that are connected together. The rails are often marked with a stripe. These are intended to be used for positive (+) or negative (–) dc power supply connections and the circuit common (ground) connections. If there is a stripe, the holes running down that entire length of the protoboard are connected along that stripe. Some boards also

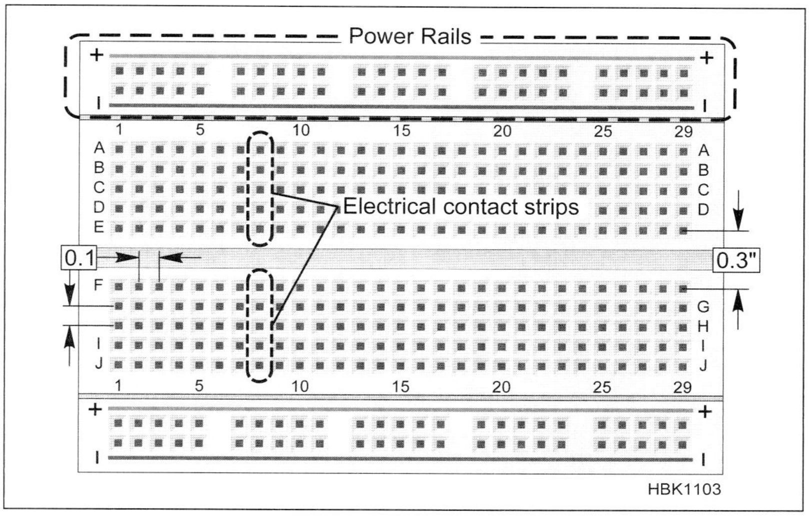

Figure 23.16 — Protoboard contacts are laid in out in groups that form common contacts and power supply rails. Contacts are laid out on a 0.1-inch grid. A 0.3-inch gap between rows of 5-contact groups accommodates standard DIP package ICs.

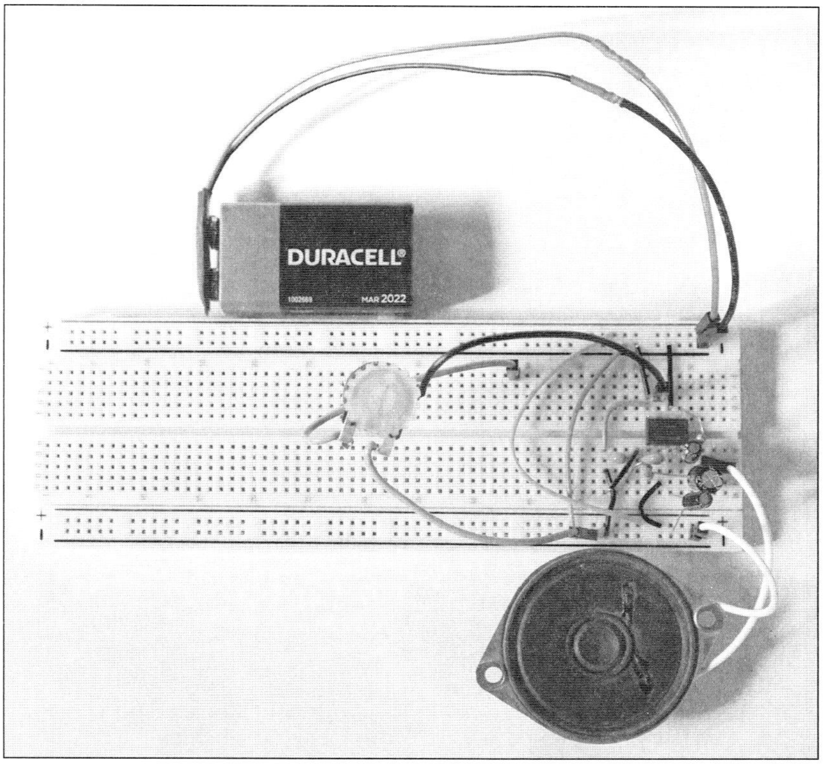

Figure 23.17 — The audio amplifier circuit of Figure 23.9 constructed on a protoboard.

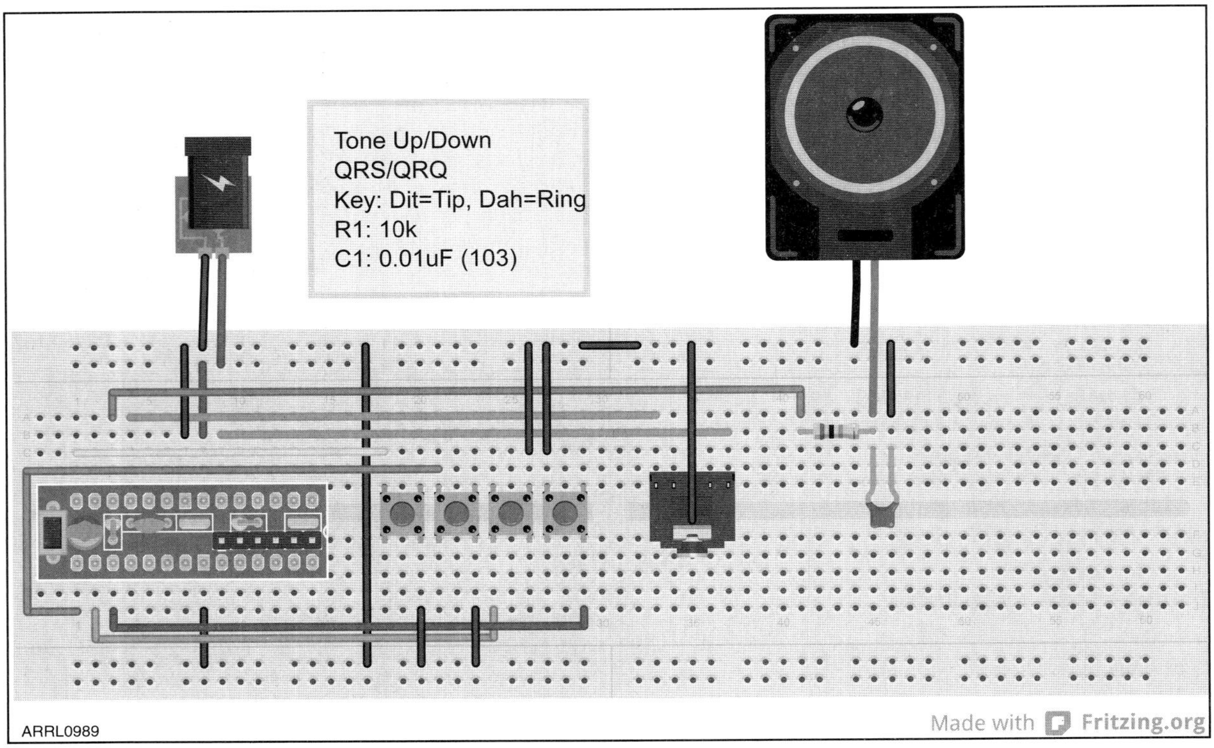

Figure 23.18 — A Fritzing diagram of a protoboard layout for a CW keyer using a small version of the Arduino (at left).

divide the long bus lines in the center of the board to create isolated power supply sections, so check with an ohmmeter to see if the rails go all the way down the side of the protoboard.

Across the centerline of the protoboard, groups of contacts are separated by 0.3 inches to fit most 8-14-16-18-, and 20-pin DIP ICs. Most through-hole style components, including discrete semiconductors, have leads spaced to fit the 0.1-inch hole spacing of the protoboard grid, either as manufactured or by bending the leads to standard width. The leads of components like resistors and capacitors can be shaped and bent to fit in the desired five-position row of contacts.

VOLTAGE AND CURRENT RATINGS

Protoboards are not intended for high-power use and should never be used with non-isolated connections to utility ac power. The maximum voltage rating for most protoboards is less than 50 V dc (35 V_{RMS} ac). Higher voltages may result in arcs between adjacent contact strips or to a metal ground plane, if there is one. Maximum current rating for the spring contacts is typically a few amperes. Higher currents will cause the contact to heat and lose its shape, resulting in poor contact in future use and can melt the plastic insulation of the protoboard. A protoboard should be connected to separate assemblies that control high-power circuits. Be sure to read the board specifications before beginning to build your circuit.

WIRING JUMPERS

A typical protoboard can accommodate solid wires from #20 to #28 AWG. It may be possible to force larger wires into the holes, but that may force the underlying spring contacts too far apart so that they make an unreliable connection in the future. Smaller wires may not be gripped tightly by the contacts, leading to intermittent connections.

Wire jumpers can be home-made by simply cutting the appropriate gauge of solid wire to the desired length and stripping the ends. Stranded wire does not work well on protoboards unless carefully tinned. A good source of jumper wires is 8-wire Ethernet cable or round 4-conductor telephone cable. Custom jumpers can be made of any length.

Commercially prepared assortments of pre-made wire jumpers of a few different lengths and colors are available. Some are simple solid wires with stripped ends. Another type is stranded wire with a solid pin attached to each end. Pre-made jumpers have the advantage of being easier to use but are also more likely to be affected by the effects of added capacitance, resistance, and inductance caused by the addition of a longer lead length. For many circuits, this is no problem at all, but these factors need to be kept in mind when analyzing the performance of the circuit.

Pre-made cables are also available in pin-to-pin, pin-to-socket, and socket-to-socket configurations. This accommodates connecting the protoboard to pins on single-board computer/CPU boards, such as an Arduino, Raspberry Pi, or other types of programmable controllers.

ASSEMBLY AND LAYOUT NOTES

Figure 23.17 shows a circuit similar to the audio amplifier in Figure 23.9 constructed on a protoboard. Simple protoboard circuits can be constructed directly from a regular schematic. Print a copy of the schematic and use a colored marker to highlight each connection as you install components and jumpers. It might also help to make a sketch of the protoboard showing each component with its schematic designator (R1, D2, C3, IC4, etc.)

If ICs are used, it is best to install them first. When installing components and jumpers, work around ICs from pin 1 to the last pin. This allows you to be sure all of the wires and components needed are connected to each pin.

Before you connect the power supply leads to the protoboard, measure resistance between the power rails and from the rails to any circuit common connection. Be sure there are no short circuits in the circuit or power supply lines before applying power to the board. It can be helpful for testing and troubleshooting to connect meter probes to

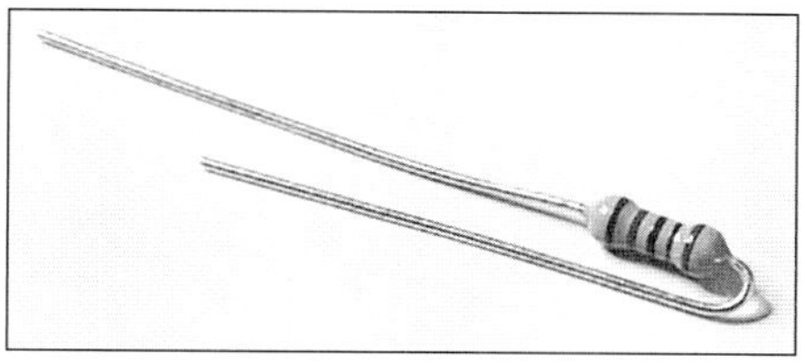

Figure 23.19— A leaded component can be mounted in a protoboard or printed-circuit board "hairpin" style by bending one lead parallel to the component.

a jumper wire for probing the circuit and making reliable contact.

It is also common for the schematic of a circuit intended to be constructed on a protoboard to be drawn as a pictorial, as in **Figure 23.18**. This is called a *Fritzing diagram,* which displays the individual wires and components as their physical packages would be laid out on the protoboard. Individual wires are drawn showing the contact groups to which they connect. The program for Fritzing is available for a small fee from **www.fritzing.org**.

Because the contacts are laid out in a grid, components such as resistors and capacitors may need to be placed at an angle between the desired contacts. This lets larger components be used without damaging the parts by bending the leads too much. It is usually okay to use any contact within the same contact group.

Parts such as resistors and diodes can also be mounted in the vertical "hairpin" style to minimize space needed (see **Figure 23.19**). This does create a small additional inductance. The alternative is to place the component flat against the board connected to an unused contact group and, using additional wire jumpers, complete the connection.

LIMITATIONS OF PROTOBOARDS

One limitation is that the layout of these boards and the interconnections are not ideal when working with RF circuits, especially at higher frequencies. The capacitance between adjacent rows of contacts can cause problems at HF and higher frequencies. Long jumpers and indirect signal routes contribute extra inductance to the circuit. Digital circuits with high clock frequencies or extremely fast signal rise and fall times can be difficult to build reliably on a protoboard.

Low-frequency audio or digital circuits work well with protoboards, although there can still be problems if the circuit is highly sensitive to stray capacitance or inductance from the long leads. Additional capacitance and inductance are also created by the arrangements of the components. Very high-impedance circuits may be sensitive to stray capacitance.

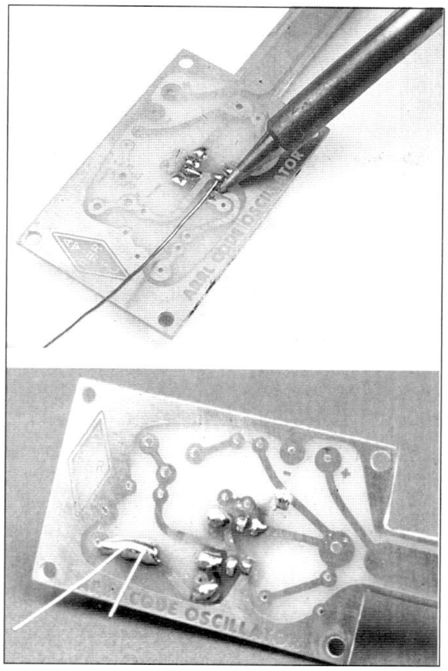

Figure 23.20 — The top photo shows how to solder a component to a PC board. Make sure that the component is flush with the board on the other side. Below is a solder bridge has formed a short circuit between PC board traces.

High-gain circuits can oscillate if coupling between input and output circuits is not controlled. Careful layout and using the minimum jumper length helps avoid these problems.

Be aware that after a lot of use, the connections inside a protoboard can become worn and won't work as well as when they were new. If you notice that the contacts have become loose or erratic, it is best to replace the entire protoboard to avoid unintended problems that can be frustrating to diagnose.

23.5.7 Printed-Circuit (PC) Boards

PC boards are everywhere, in all kinds of consumer electronics, and in most of your amateur radio equipment. They are also used in most kits and construction projects. A newcomer to electronics might think that there is some unwritten law against building equipment in any other way!

The misconception that everything needs to be built on a printed-circuit board is often a stumbling block to easy project construction. In fact, a PC board is probably the worst choice for a one-time project. In actuality, a moderately complex project (such as a QRP transmitter) can be built in much less time using other techniques, such as those described in the preceding sections. The additional design, layout, and manufacturing is usually much more work than it would take to build the project by hand.

So why does everyone use PC boards? The most important reason is that they are reproducible. They allow many units to be mass-produced with exactly the same layout, reducing the time and work of conventional wiring and minimizing the possibilities of wiring errors. If you can buy a ready-made PC board for your project, it can save a lot of construction time. This is true because someone else has done most of the work — designing the PC board layout and fixing any "bugs" caused by inter-trace capacitive coupling, ground loops, and similar problems. In most cases, if a ready-made board is not available, ground-plane construction is a lot less work than designing, debugging, and then making a PC board.

Using a PC board usually makes project construction easier by minimizing the risk of wiring errors or other construction blunders. Inexperienced constructors usually feel more confident when construction has been simplified to the assembly of components onto a PC board. One of the best ways to get started with home construction (to some the best part of amateur radio) is to start by assembling a few kits using PC boards. A list of kit manufacturers can be found on the QRP ARCI website, **www.qrparci.org,** under "Links." Then click on "QRP Kits Bits and Supplies."

PC BOARD ASSEMBLY TECHNIQUES

Cleanliness

Make sure your PC board and component leads are clean. Clean the entire PC board before assembly; clean each component before you install it. Corrosion looks dark instead of bright and shiny. Don't use sandpaper to clean your board. Use a piece of fine steel wool or a Scotchbrite cleaning pad to clean component leads or PC board before you solder them together.

Installing Components

In a construction project that uses a PC board, most of the components are installed on the board. Installing components is easy — stick the components in the right board holes, solder the leads, and cut off the extra lead length. Most construction projects have a parts-placement diagram that shows you where each component is installed.

Getting the components in the right holes is called "stuffing" the circuit board. Inserting and soldering one component at a time takes too long. Some people like to put all the components in at once and then turn the board over and solder all the leads. If you bend the leads a bit (about 20°) from the

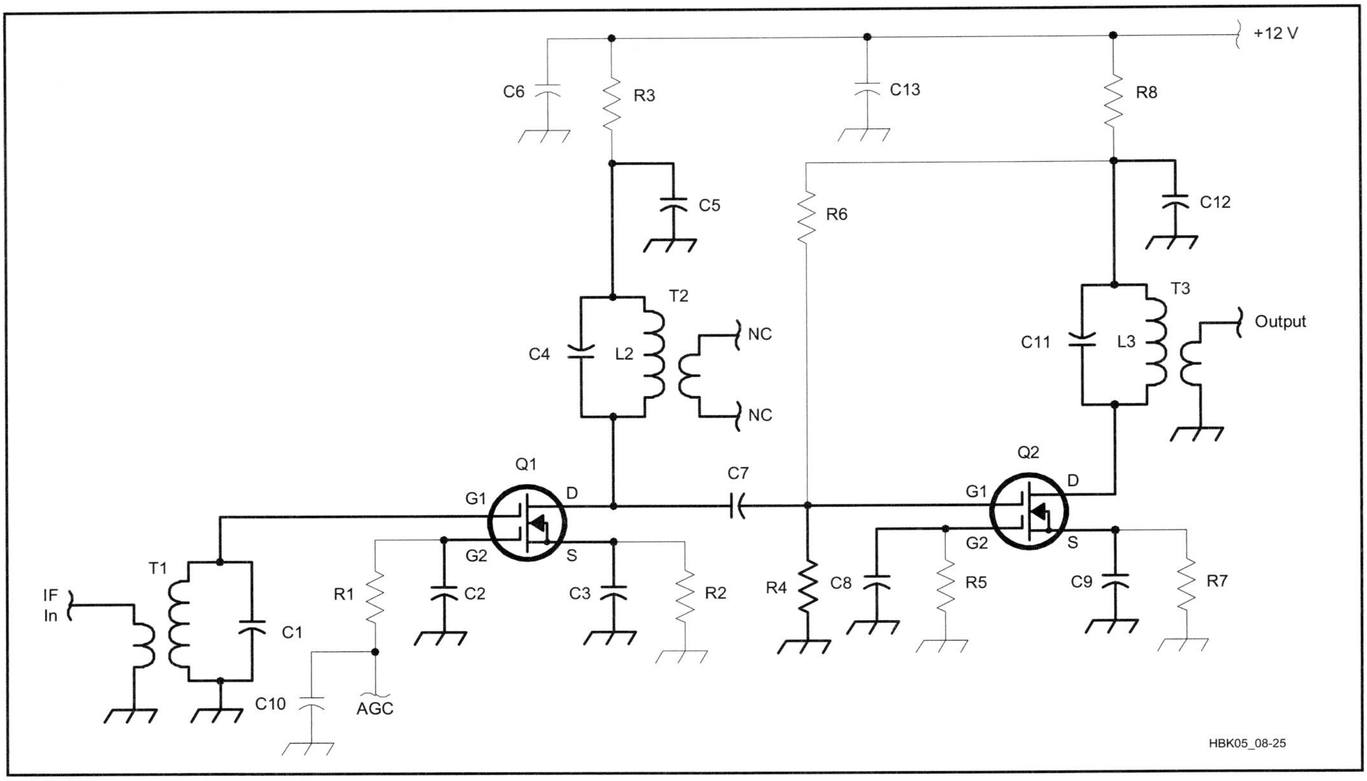

Figure 23.21 — The IF amplifier used in the design example. C1, C4 and C11 are not specified because they are internal to the IF transformers.

bottom side after you push them through the board, the components are not likely to fall out when you turn the board over.

Start with the shortest components — horizontally mounted diodes and resistors. Larger components sometimes cover smaller components, so these smaller parts must be installed first. If building a kit, follow the suggested order of mounting your parts if provided. Use adhesive tape to temporarily hold difficult components in place while you solder.

PC-Board Soldering

To solder components to a PC board, bend the leads at a slight angle, apply the soldering iron to one side of the lead, and flow the solder in from the other side of the lead (see **Figure 23.20**). Too little heat causes a bad or "cold" solder joint; too much heat can damage the PC board. Practice a bit on some spare copper stock before you tackle your first PC board project. After the connection is soldered properly, clip the lead flush with the solder.

Make sure you have the components in the right holes before you solder them. Components that have polarity, such as diodes, ICs, and some capacitors must be oriented as shown on the parts-placement diagram.

Inspect solder connections. A bad solder joint is much easier to find before the PC

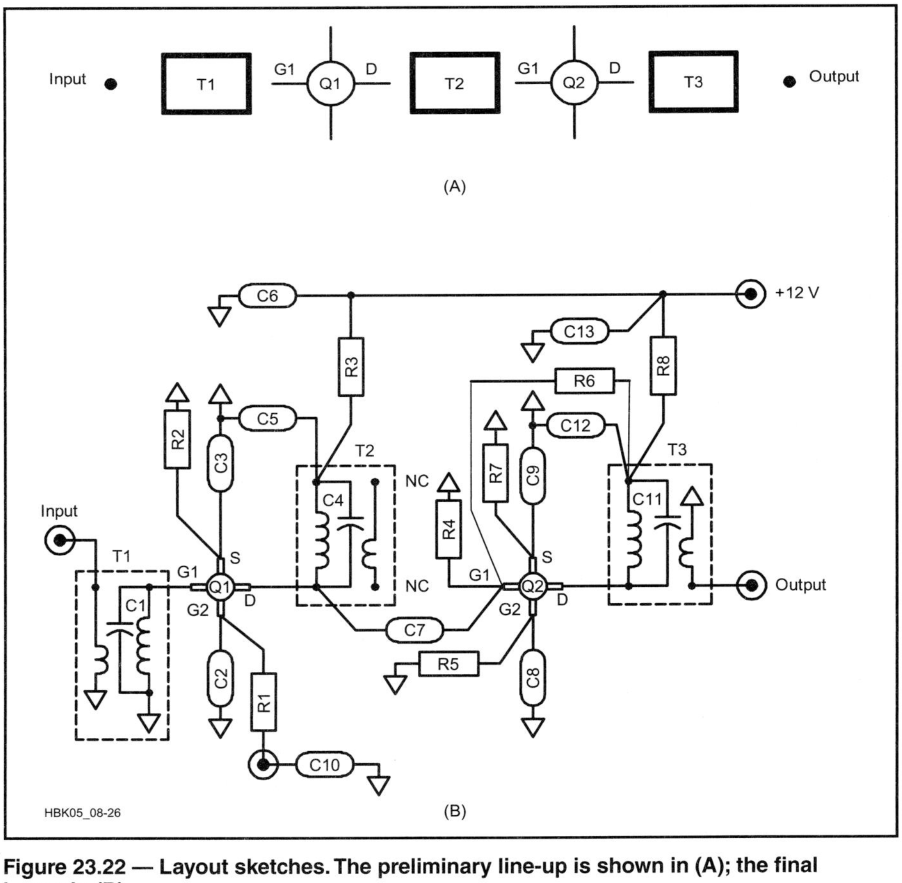

Figure 23.22 — Layout sketches. The preliminary line-up is shown in (A); the final layout in (B).

board is mounted to a chassis. Look for any damage caused to the PC board by soldering. Look for solder "bridges" between adjacent circuit-board traces. Solder bridges (Figure 23.20) occur when solder accidentally connects two or more conductors that are supposed to be isolated. It is often difficult to distinguish a solder bridge from a conductive trace on a tin-plated board. If you find a bridge, re-melt it and the adjacent trace or traces to allow the solder's surface tension to absorb it. Double check that each component is installed in the proper holes on the board and that the orientation is correct. Make sure that no component leads or transistor tabs are touching other components or PC board connections. Check the circuit voltages before installing ICs in their sockets. Ensure that the ICs are oriented properly and installed in the correct sockets.

23.5.8 From Schematic to Working Circuit

Turning a schematic into a working circuit is more than just copying the schematic with components. One thing is usually true — you can't build it the way it looks on the schematic. The schematic describes the electrical connections, but it does not describe the mechanical layout of the circuit. Many design and layout considerations that apply in the real world of practical electronics don't appear on the schematic.

HOW TO DESIGN A GOOD CIRCUIT LAYOUT

A circuit diagram is a poor guide toward a proper layout. Circuit diagrams are drawn to be readable and to describe the electrical connections. They follow drafting conventions that have very little to do with the way the circuit works. On a schematic, ground and supply voltage symbols are scattered all over the place. The first rule of RF layout is — *do not lay out RF circuits as their schematics are drawn!* How a circuit works in practice depends on the layout. Poor layout can ruin the performance of even a well-designed circuit.

The easiest way to explain good layout practices is with an example. **Figure 23.21** is the circuit diagram of a two-stage receiver IF amplifier using dual-gate MOSFETs. It is only a design example, so the values are typical. To analyze which things are most affected by the layout of this circuit, ask these questions:

- Which are the RF components, and which are only involved with low frequencies or dc?
- Which components are in the main RF signal path?
- Which components are in the ground return paths?

Use the answers to these questions to plan the layout. The RF components that are in the main RF signal path are usually the most critical. The low frequency or dc components can usually be placed anywhere. The components in the ground return path should be positioned so they are easily connected to the circuit ground. Answer the questions, apply the answers to the layout, and then follow these guidelines:

- Avoid laying out circuits to avoid feedback loops. If a stage's output is too near a previous stage's input, the output signal can feed back into the input and cause problems.
- Keep component leads as short as practical. This doesn't necessarily mean as short as possible, just consider lead length as part of your design.
- Remember that metal transistor cases conduct, and that a transistor's metal case is usually connected to one of its leads. Prevent cases from touching ground or other components, unless called for in the design.

In our design example, the RF components are shown in heavy lines, though not all of these components are in the main RF signal path. The RF signal path consists of T1/C1, Q1, T2/C4, C7, Q2, T3/C11. These need to be positioned in an almost straight line in order to avoid feedback from output to input. They form the backbone of the layout, as shown in **Figure 23.22A**.

The question about ground paths requires some further thought — what is really meant by "ground" and "ground-return paths"? Some points in the circuit need to be kept at RF ground potential. The best RF ground potential on a PC board is a copper ground plane covering one entire side. Points in the circuit that cannot be connected directly to ground for dc reasons must be bypassed ("decoupled") to ground by capacitors that provide ground-return paths for RF.

In Figure 23.22B, the components in the ground-return paths are the RF bypass capacitors C2, C3, C5, C8, C9 and C12. R4 is primarily a dc biasing component, but it is also a ground return for RF, so its location is important. The values of RF bypass capacitors are chosen to have a low reactance at the frequency in use; typical values would be 0.1 μF at LF, 0.01 μF at HF, and 0.001 μF or less at VHF. Not all capacitors are suitable for RF decoupling; the most common are disc ceramic capacitors. RF decoupling capacitors should always have short leads. Surface mount ceramic capacitors with no leads are ideal for bypassing.

Almost every RF circuit has an input, output, and a common ground connection. Many circuits also have additional ground connections, at both the input and the output side. Maintain a low-impedance path between input and output ground connections. The input ground connections for Q1 are the grounded ends of C1 and the two windings of T1. The two ends of an IF transformer winding are generally not interchangeable; one is designated as the "hot" end and the other must be connected or bypassed to RF ground. The capacitor that resonates with the adjustable coil is often mounted inside the can of the IF transformer, leaving only two component leads to be grounded, as shown in Figure 23.22B.

The RF ground for Q1 is its source lead connection via C3. Since Q1 is in a plastic package that can be mounted in any orientation, you can make the common ground either above or below the signal path in Figure 23.22B. Although the circuit diagram shows the source lead at the bottom, the practical circuit works much better with the source lead at the top, because of the connections to T2.

It's a good idea to locate the hot end of the main winding of T2 close to the drain lead of the transistor package, so the other end is toward the top of Figure 23.22B. If the source lead of Q1 is also toward the top of the layout, there is a common ground point for C3 (the source bypass capacitor) and the output bypass capacitor C5. Gate-2 of Q1 can be safely bypassed through C2 toward the bottom of the layout.

C7 couples the signal from the output of Q1 to the input of Q2. The source lead of Q2 should be bypassed toward the top of the layout, in the same way as the source lead of Q1. R4 is not critical, but it should be connected on the same side as the other components. Note how the pinout of T3 has placed the output connection as far as possible from the input. With this layout for the signal path and the critical RF components, the circuit has an excellent chance of working properly.

DC Components

The rest of the components carry dc, so their layout is less critical. Even so, try to keep everything well separated from the main RF signal path. One good choice is to put the 12 V connections along the top of the layout and the AGC connection at the bottom. The source bias resistors, R2 and R7, can be placed alongside C3 and C9. The Q2 gate-2 bias resistors, R5 and R6, are not RF components so their locations are not critical. R7 has to cross the signal path in order to reach C12, however, and the best way to avoid signal pickup would be to mount R7 on the opposite side of the copper ground plane from the signal wiring. Generally speaking, 1/8 W or 1/4 W metal-film or carbon-film resistors are best for low-level RF circuits.

It is not right to say that resistors such as R3 and R8 are not "RF" components.

They provide a high impedance to RF in the positive supply lead. Because of R8, for example, the RF signal in T2 is conducted to ground through C5 rather than ending up on the 12 V line, possibly causing unwanted RF feedback. Just to be sure, C6 bypasses R3 and C13 serves the same function for R8. Note that the Q2 gate-1 bias resistor, R6, is bypassed by C12 and connected to R8 rather than directly to the 12 V supply, to take advantage of the extra decoupling provided by R8 and C13.

If you build something, you want it to work the first time, so don't cut corners! Some commercial PC boards take liberties with layout, bypassing and decoupling. Don't assume that you can do the same. Don't try to eliminate "extra" decoupling components such as R3, C6, R8, and C13, even though they might not all be absolutely necessary. If other people's designs have left them out, put them in again. In the long run it's far easier to take a little more time and use a few extra components to build in some insurance that your circuit will work. For a one-time project, the few extra parts won't hurt your pocket too badly, and they may save untold hours in debugging time.

Let's summarize how we got from Figure 23.21 to Figure 23.22B:

- Lay out the signal path in a straight line.
- By experimenting with the placement and orientation of the components in the RF signal path, group the RF ground connections for each stage close together, without mixing up the input and output grounds.
- Place the non-RF components well clear of the signal path, freely using decoupling components for extra measure.

Practical Construction Hints

Now it's time to actually construct a project. The layout concepts discussed earlier can be applied to nearly any construction technique. Although you'll eventually learn from your own experience, the following guidelines give a good start:

- Divide the unit into modules built into separate shielded enclosures — RF, IF, VFO, for example. Modular construction improves RF stability and makes the individual modules easier to build and test. It also means that you can make major changes without rebuilding the whole unit. RF signals between the modules can usually be connected using small coaxial cable.
- Use a full copper ground plane. This is your largest single assurance of RF stability and good performance.
- Keep inputs and outputs well separated for each stage, and for the whole unit. If possible, lay out all stages in a straight line. If an RF signal path doubles back or re-crosses itself, it usually results in instability.
- Keep the stages at different frequencies well separated to minimize interstage coupling and spurious signals.
- Use interstage shields where necessary, but don't rely on them to cure a bad layout.
- Make all connections to the ground plane short and direct. Locate the common ground for each stage between the input and the output ground. Single-point grounding may work for a single stage, but it is rarely effective in a complex RF system.
- Locate frequency-determining components away from heat sources and mount them so as to maximize mechanical strength.
- Avoid unwanted coupling between tuned circuits. Use shielded inductors or toroids rather than open coils. Keep the RF high-voltage points close to the ground plane. Orient air-wound coils at right angles to minimize mutual coupling.
- Use lots of extra RF bypassing, especially on dc supply lines.
- Try to keep RF and dc wiring on opposite sides of the board, so the dc wiring is well away from RF fields.
- Compact designs are convenient, but don't overdo it! If the guidelines cited above mean that a unit needs to be bigger, make it bigger.

COMBINING TECHNIQUES

You can use a mixture of construction techniques on the same board, and in most cases you probably should. Even though you choose one style for most of the wiring, there will probably be places where other techniques would be better. If so, do whatever is best for that part of the circuit. The resulting hybrid may not be pretty (these techniques aren't called "ugly construction" for nothing), but it will work!

Mount dual-in-line package (DIP) ICs in an array of drilled holes, then connect them using wired traces as described earlier. It is okay to mount some of the components using a ground-plane method, push pins, wired traces, or perfboards. On any one board, you may use a combination of these techniques, drilling holes for some ICs or gluing others upside down, then surface mounting some of the pins, and using other techniques to connect the rest. These combination techniques are often found in a project that contains audio, RF, and digital circuitry.

A Final Check

No matter what construction technique is chosen, do a final check before applying power to the circuit! Things go wrong, and a careful inspection minimizes the risk of a project beginning and ending its life as a puff of smoke! Check wiring carefully. Make a photocopy of the schematic and mark each connection and lead on the schematic with a red X, or use a highlighter to mark the circuit when you've verified that it's connected properly.

23.5.9 Tuning and Alignment

The task of performing adjustment or alignment can be difficult unless you are using the proper tools. There are variable inductors of various sizes and types and different types of variable capacitors, as well as specialized potentiometers to adjust. Each different type of adjustable component requires its own specialized tool for making the adjustment. Failure to use the proper tool in the proper manner can result in damage to the component.

In RF circuits, the proximity of your hand or a metal object can greatly influence the apparent value of the variable component and make it difficult if not impossible to adjust. For this reason, a number of plastic or ceramic tools are available to make this task as easy as possible and at a very low cost. An example is the Velleman "Plastic Tuning Needle Set," which can be found online (**www.velleman.eu**) and is widely available for a few dollars. Sources such as Mouser, Digi-Key, Newark, and Allied sell both individual tools and assortments of tools. Specialized tools, if required, are usually available from the same vendor selling the adjustable component.

When choosing a tuning tool, use one that best fits the component. Tips that are too small can end up damaging the inside of the adjustment slot or hole, and tools that are too large can also damage small components or their adjustment mechanism. Choose a tuning tool with a tip that exactly matches the component.

VARIABLE INDUCTORS

Most variable inductors used in low-power RF circuits are a tiny coil of wire wound around a plastic form, with a threaded ferrite (or sometimes brass) slug in the center. For ferrite slug coils, inserting the slug into to coil increases the inductance and vice versa. (Brass slugs behave the opposite way but are uncommon.) Some variable inductors are inside metal cans, which act as shields to reduce coupling to nearby components. The slug either has a hexagonal hole or a slot for a tuning tool to adjust the position of the slug, like turning a screw.

Do not use small metal screwdrivers or hex keys (Allen wrenches) to adjust these coils. The metal tool will alter the inductance when inserted, making adjustment unpredictable and frustrating. A metal tool can also damage the slug by cracking it, as ferrite slugs are quite brittle, ruining the inductor. Plastic or plastic-tipped tools allow you to make the adjustment while keeping your hand far away and not damaging the slug. Most tuning tools also have a mark on them in the form of a dimple, a logo/identification, or a stripe to help you count the number of times you rotate the tool.

VARIABLE CAPACITORS

Variable capacitors not tuned with a shaft and knob usually have a screw-slot for adjustment. These include ceramic and mica trimmers, piston trimmers, and small air variables. Try using a plastic tool first, but these adjustments are sometimes too stiff for a PLASTIC tool to make. Ceramic and plastic-tipped tools are available but not common in most electronics stores. Metal screwdrivers introduce enough capacitance to alter the value of these small capacitors but may be used in a series of small adjustments.

POTENTIOMETERS

Multi-turn miniature potentiometers (a.k.a "trimpots") usually have small metal adjusting screws. Resistance values are rarely sensitive to metal tools and so a miniature jeweler's screwdriver can be used. For RF circuits, however, use a plastic or ceramic screwdriver to avoid any possible interactions with the signals.

Larger potentiometers that are found inside equipment and do not have shafts are designed to be adjusted with a screwdriver-style tool. While a metal screwdriver can be used, a plastic tool eliminates any chance that the metal screwdriver could slip and come in contact with nearby components or leads. The metal screwdriver's shaft can also make accidental contact with wiring inside equipment.

ADJUSTING HIGH VOLTAGE CIRCUITS

If high voltages are involved, such as during tube neutralization or bias voltage adjustment, safety precautions are important. If the adjustable component is located near an RF stage, the presence of your hand can influence the circuit. Use plastic or ceramic screwdrivers for both slotted and Philips-type screws. Be sure to use a non-conductive tool that is long enough to keep your hands well away from any high voltage source! There are extra long tools available for use in high voltage areas. Shops that repair vacuum tube musical instrument amplifiers may be able to help locate suitable tools — remember, safety first!

RESPECT ADJUSTMENT LIMITS

When making adjustments on any variable component, be sure not to force an adjustment beyond its range, as that can cause permanent damage. For example, turning a slug all the way into or out of an adjustable inductor can strip the tiny threads that hold the slug, making it difficult to perform future adjustments. If you feel resistance, stop, then look to see if the adjustment is at a limit. When an adjustment reaches a component's limit, that is usually an indication of a problem somewhere else in the circuit or, less frequently, that the component value has changed. Determine whether either is the case before proceeding.

In some circuits, adjustment is quite smooth and easy, while others can be quite touchy, requiring patience and a steady hand to get the circuit adjusted properly. Take your time and watch your test equipment for the desired changes. Use the dimples or marks on the tool to give a visual indication of how the adjustment is set, or to count turns of a multi-turn adjustment. Keep notes if you are making multiple adjustments or if different adjustments interact.

Finally, avoid being a "screwdriver technician" who adjusts components seemingly at random, hoping to get lucky and correct a problem. That usually results in more problems and even a completely non-functional piece of equipment requiring professional realignment or repair. Make small adjustments, and if you don't see the results you were expecting, stop and figure out why before proceeding.

23.5.10 Other Construction Techniques

WIRING

Select the wire used in connecting amateur equipment by considering the maximum current it must carry, the voltage its insulation must withstand, and its use.

To minimize leakage of RF that causes EMI, the power wiring and low-level signal wiring of all transmitters should use shielded wire or coaxial cable. Receiver and audio circuits may also require the use of shielded wire at some points for stability or the elimination of coupling to adjacent circuits. Coaxial cable is recommended for all 50 Ω circuits. It can also be used for *short* runs of high-impedance audio wiring.

When choosing wire, consider how much current it will safely carry, called *ampacity*. (See the **Antennas** chapter for wire tables.) See **www.electricaltechnology.org/2021/01/standard-wire-gauge-swg-calculator.html** for a handy online calculator and table. Stranded wire is usually preferred over solid wire because stranded wire better withstands the inevitable bending that is part of building and troubleshooting a circuit. Solid wire is more rigid than stranded wire; use it where mechanical rigidity is needed or desired.

Wire with typical plastic insulation is good for voltages up to about 500 V. Use Teflon-insulated or another high-voltage wire for higher voltages. Teflon insulation does not melt when a soldering iron is applied. This makes it particularly helpful in tight places or large wiring harnesses. Although Teflon-insulated wire is more expensive, it is often available from industrial surplus houses.

Solid wire is often used to wire RF circuits in both receivers and transmitters. Bare soft-drawn tinned wire, #22 to #12 AWG (depending on mechanical requirements) is suitable. Avoid kinks by stretching a piece 10 or 15 feet long and then cutting it into short, convenient lengths. Run RF wiring directly from point to point with a minimum of sharp bends and keep the wire well-spaced from the chassis or other grounded metal surfaces. Where the wiring must pass through a chassis wall or shield, cut a clearance hole and line it with a rubber grommet. If insulation is necessary, slip spaghetti insulation or heat-shrink tubing over the wire. For power-supply leads, bring the wire through walls or barriers via a feed-through capacitor.

In transmitters where the peak voltage does not exceed 500 V, shielded wire is satisfactory for power circuits. Shielded wire is not readily available for higher voltages — use point-to-point wiring instead. In the case of filament circuits carrying heavy current, it is necessary to use #10 or #12 AWG bare or enameled wire. Slip the bare wire through spaghetti and cover it with copper braid pulled tightly over the spaghetti. Slide the shielding back over the insulation and flow solder into the end of the braid; the braid will stay in place, making it unnecessary to cut it back or secure it in place. Clean the braid first so solder will flow into the braid with a minimum of heat.

ENAMELED WIRE

When connecting enameled wire leads, care must be taken to be sure a good connection is made. There are two methods that can be used to remove the insulation. With Thermaleze-type enamel, the heat from a soldering iron can be used to remove the insulation. You will first need to turn up the heat on your soldering iron for best results. After adding some melted solder to your soldering tip, move the drop slowly along the desired length of the wire lead. Moving it slowly gives the insulation time to melt and the solder a chance to tin the now-exposed wire. The tinned leads will then easily solder to a PC board or other mounting system.

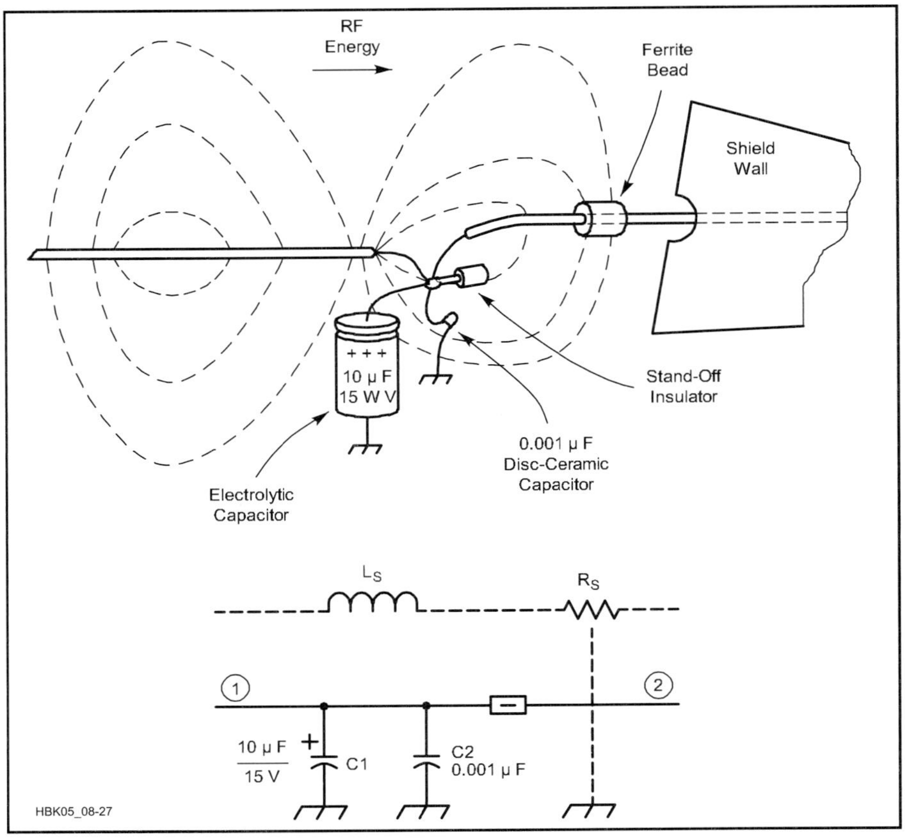

Figure 23.23 — Two capacitors in parallel afford better bypassing across a wide frequency range.

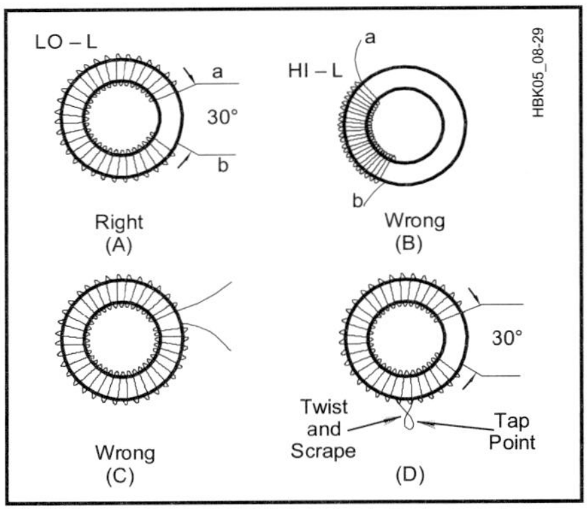

Figure 23.24 — The maximum-Q method for winding a single-layer toroid is shown at (A). A 30° gap is best. Methods at (B) and (C) have greater distributed capacitance. (D) shows how to place a tap on a toroidal coil winding.

If using other kinds of enameled wire, an emery board or small file can be used to remove the insulation. (Using a knife to scrape off the enamel usually nicks the wire, which will eventually break at that point.) Be sure you have removed it completely from the desired lead. Follow that up with your soldering iron and solder to tin the wire using a thin coating of solder. Be sure to not make the tinned surface so thick that the lead cannot fit through the PC board holes. Preparing the leads in this manner will give the best possibility for a clean and secure connection.

HIGH-VOLTAGE TECHNIQUES

High-voltage wiring and construction requires special care. Read and follow the guidelines for high-voltage construction in the **Power Sources** chapter. You must use wire with insulation rated for the voltage it is carrying. Most standard hookup wire is inadequate above 300 or 600 V. High-voltage wire is usually insulated with Teflon or special multilayer plastic. Some coaxial cable is rated at 3700 V_{RMS} (or more) *internally* between the center conductor and shield, but the outer jacket rating is usually considerably lower.

CABLE ROUTING

Power or control leads run together for more than a few inches present a better appearance when bound together in a single cable. Plastic cable ties or tubing cut into a spiral are used to restrain and group wiring. Check with your local electronic parts supplier for items that are in stock.

To give a commercial look to the wiring of any unit, route any dc leads and shielded signal leads along the edge of the chassis. If this isn't possible, the cabled leads should then run parallel to an edge of the chassis. Further, the generous use of the tie points mounted parallel to an edge of the chassis, for the support of one or both ends of a resistor or fixed capacitor, adds to the appearance of the finished unit. In a similar manner, arrange the small components so that they are parallel to the panel or sides of the chassis.

Tie Points

When power leads have several branches in the chassis, it is convenient to use fiber-insulated terminal strips as anchors for junction points. Strips of this kind are also useful as insulated supports for resistors, RF chokes, and capacitors. Hold exposed points of high-voltage wiring to a minimum; otherwise, make them inaccessible to accidental contact.

WINDING COILS

A detailed tutorial for winding coils by Robert Johns, W3JIP, titled "Homebrew Your Own Inductors!" from August 1997 *QST* can be found in the Radio Technology section of the ARRL TIS at **www.arrl.org/radio-technology-topics** under Circuit Construction. Understanding these techniques greatly simplifies coil construction.

Close-wound coils are readily wound on the specified form by anchoring one end of the length of wire (in a vise or to a doorknob) and the other end to the coil form. Straighten any kinks in the wire and then pull to keep the wire under slight tension. Wind the coil to the required number of turns while walking toward the anchor, always maintaining a slight tension on the wire.

To space-wind the coil, wind the coil simultaneously with a suitable spacing medium (heavy thread, string, or wire) in the manner described above. When the winding is complete, secure the end of the coil to the coil-form terminal and then carefully unwind the spacing material. If the coil is wound under suitable tension, the spacing material can be easily removed without disturbing the winding. Finish space-wound coils by judicious applications of RTV sealant or hot-melt glue to hold the turns in place.

The "cold" end of a coil is the end at (or close to) chassis or ground potential. Wind coupling links on the cold end of a coil to minimize capacitive coupling.

Winding Toroidal Inductors

Toroidal inductors and transformers are specified for many amateur radio projects. The advantages of these cores include compactness and a self-shielding property. **Figures 23.23** and **23.24** illustrate the proper way to wind and count turns on a toroidal core.

The task of winding a toroidal core, when more than just a few turns are required, can be greatly simplified by the use of a homemade bobbin upon which the wire is first wound. A simple yet effective bobbin can be fashioned from a wooden popsicle stick. Cut a "V" notch at each end and first wind the wire coil on the popsicle stick lengthwise through the notches. Once this is done, the wound bobbin can be easily passed through the toroid's inside diameter. While firmly grasping one of the wire ends against the toroidal core, the bobbin can be moved up, around, and through the toroidal core repeatedly until the wire has been completely transferred from the bobbin. The choice of bobbin used is somewhat dependent on the inside diameter of the toroid, the wire size, and the number of turns required.

Another method is to create a holder that allows you to grip the core, yet thread the wire easily around it. When winding toroids, be sure to seat the wire so it hugs the shape of the core, and do not allow turns to overlap unless part of a twisted group of wires as in bifilar or trifilar windings. Do not pull the wire too tight because the thin wire used in some toroids can break.

When you wind a toroid inductor, count each pass of the wire through the toroid center as a turn. (see **Figure 23.25A**).

Multiwire Windings

A bifilar winding is one that has two identical lengths of wire, which when placed on the core result in the same number of turns for each wire. The two wires are wound on the core side by side at the same time, just as if a single winding was being applied.

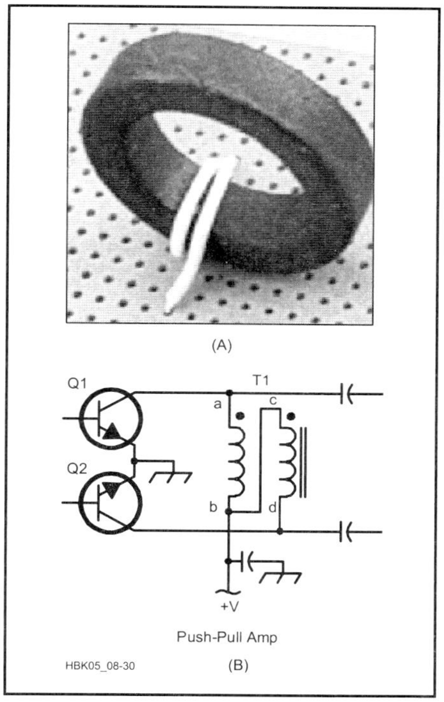

Figure 23.25 — (A) shows a toroidal core with two turns of wire (see text). Large black dots, like those at T1 in (B), indicate winding polarity (see text).

An easier and more popular method is to twist the two wires (8 to 15 turns per inch is adequate), then wind the twisted pair on the core. The wires can be twisted handily by placing one end of each in a bench vise. Tighten the remaining ends in the chuck of a small hand drill and turn the drill to twist the pair.

A trifilar winding has three wires, and a quadrifilar winding has four. The procedure for preparation and winding is otherwise the same as for a bifilar winding. Figure 23.25 shows a bifilar toroid in schematic and pictorial form. The wires had been twisted together prior to placing them on the core. It is helpful, though by no means essential, to use wires of different color for multifilar windings. It is more difficult to identify multiple windings on a core after it has been wound. Various colors of enamel insulation are available, but it is not easy for amateurs to find this wire locally or in small-quantity lots. This problem can be solved by taking lengths of wire (enameled magnet wire), cleaning the ends to remove dirt and grease, then spray painting them. Ordinary aerosol-can spray enamel works fine. Spray lacquer is not as satisfactory because it is brittle when dry and tends to flake off the wire.

You may also identify bifilar and trifilar toroid lead pairs of identical colors by using a continuity checker. It is a good idea to check all toroids with an ohmmeter or continuity tester to be sure there are no shorts between different windings and that there is a good connection on the ends of each winding. Testing the leads after mounting can be difficult due to the circuit layouts, so be sure to test all toroids before mounting.

The winding sense of a multifilar toroidal transformer is important in most circuits. Figure 23.25B illustrates this principle. The black dots (called phasing dots) at the top of the T1 windings indicate polarity. That is, points a and c are both start or finish ends of their respective windings. In this example, points a and d are of opposite polarity to provide push-pull voltage output from Q1 and Q2.

23.6 PCB CAD and Fabrication

With numerous PCB design software packages available and low quantity, low-cost PCB manufacturing services accepting orders electronically, the development of PCBs has never been easier for the amateur. As with any assembly or manufacturing process, it is important to understand the vocabulary and technology in order to achieve the desired result. Thus, this section provides a detailed description of the entire process of PCB design.

The primary goal of using software for printed circuit board (PCB) design is the production of PCB "artwork" — the graphic designs used to create the patterns of traces that establish connectivity on the PCB. Historically, PCB artwork was created by hand on clear film using black tape and special decals that were then photographically reduced. However, free and low-cost programs specifically for the PCB design process are now widely available. These programs not only allow the creation of artwork efficiently and accurately, but produce the required ancillary files for commercial production, exchange information with schematic capture software, produce Bills of Materials (parts lists), and even include such features as three-dimensional visualization of the finished board. While artwork files can be shared with other people for PCB production, the "source" files used by the CAD program can typically only be used by other people who share the same program.

The decision to produce a PCB must take into account the nature of the circuit itself (for example, high frequency, low noise, and high current circuits require additional care). Other considerations are time available; expense; available alternatives; quantity required; ability to easily share, replicate, and improve the design; and non-electrical characteristics such as thermal and mechanical properties and physical robustness. For complicated designs, the typical PCB process ensures the connectivity of the schematic is properly reflected in the PCB, which can save a lot of debugging.

23.6.1 Overview of the PCB Design Process

The PCB design process begins with establishing the list of components in the circuit, the connections between the components, the physical outline/size of the board, and any other physical, thermal, and electrical constraints or design goals. Much of the connectivity and component information is reflected in the schematic for a circuit, so in many cases the PCB layout process begins by entering the schematic in a schematic capture program which may be integrated with the PCB CAD program or standalone. (Schematic capture is not required for PCB layout but offers so many advantages that it is almost always done.) Once the schematic is entered, there may be other options possible such as simulating the circuit as described in the preceding sections. A clean, well-organized schematic that is easily modified is an asset regardless of the circuit production and construction methods.

With input from the schematic and other information, the board outline is created, mounting and other holes placed, the components positioned, and the pattern of traces created. Once the layout is complete, in many cases it is possible to run a design rules check — the equivalent of a "spell checker." Design rules include component connections and other information to check for problems related to connectivity and manufacture. This step can save a great deal of time and expense by catching errors that could be fixed by hand, but would otherwise negate some of the benefits of a PCB.

The final step in the PCB layout program is to produce the collection of up to a dozen or so different files required for PCB production. In brief, the list includes the artwork for the pattern of traces, files for producing the board outline, solder masks, silk screens, and holes.

The designer then uploads the set of files to a PCB manufacturer. As quickly as two to three days later a package will be delivered with the freshly minted boards ready for assembly! Alternatively, the designer may create the board "in house" using photomechanical, CNC (Computer Numerical Control) machining, or other processes based on the output files from the software.

23.6.2 Options for PCB Design Software

PCB software varies in features, function, and cost. The good news is that the radio amateur has excellent no-cost options that include both open-source and commercial PCB software. Needless to say, the availability and terms of PCB design software are continually in flux; check for the current terms, particularly for free software.

The first time designing and ordering a PCB can be daunting, so keep the initial job simple and pay attention to details (and read the instructions). When starting to use a specific software package, join a user's support group or forum if available. Request sample designs from other users and experiment with them to see how they are constructed and what files are required in the output data set. Once you are comfortable with the tools, you can begin on a design of your own.

FREE AND OPEN SOURCE (FOSS) SOFTWARE

This software is no-cost, and the source is available for the designer to read and improve upon. Support is generally found through online forums, and there are no artificial limitations on the PCB design. The functionality can't be taken away from the designer, but at the same time there is no guarantee the software will continue to be improved. Two popular examples, *KiCad* (**kicad.org**) and LibrePCB *(***librepcb.org***)*, have been developed over decades, and *KiCad* in particular is quite popular and powerful. *KiCad* and other FOSS software may often be available on different operating systems (i.e., Linux, Mac, and Windows); commercial software is often limited to the Windows platform.

FREE COMMERCIAL SOFTWARE

Commercial software may be made available in a free version for several reasons, including:

- the limited function of the free offering encourages upgrade to the paid version
- the free offering is licensed only for non-commercial (e.g., educational/ hobbyist) use
- output is limited to the manufacturer's PCB production
- older versions are sometimes made available as-is with limited or no support

Because there is an income stream supporting development, commercial software may have more polish or advanced features than its open-source counterparts. Letting a license (even if free) lapse, however, may mean losing access to one's designs for software that uses a subscription model. Changes in company ownership or direction have in the past resulted in changes to use policy, often in the direction of limitation. Limitations on size or layers could be annoying as one's designs increase in sophistication. Examples of free commercial software include:

Altium CircuitMaker (**www.altium.com/circuitmaker**). Few limits, but only five projects can be private; the rest are shared. Commercial use okay.

DesignSpark (**www.rs-online.com/designspark/pcb-software**). No artificial limitations; commercial use okay. From RS Components, a major electronic component vendor, providing an easy way to order parts through the software.

ExpressPCB (**www.expresspcb.com/pcb-cad-software**). Maximum of six layers. Free when used to order boards through the ExpressPCB manufacturing service, otherwise there is a cost per design.

AutoDesk Eagle (**www.autodesk.com/products/eagle).** The free version is limited in size and layers, is licensed for non-commercial use, and the user must have an active license for the free version of Fusion 360 CAD (must be affiliated with an educational institution).

PAID COMMERCIAL SOFTWARE

The sky is the limit in terms of features and cost, but many of the advanced features are likely of less value to the occasional amateur radio PCB designer than to someone who uses the software day in and day out, or is engaged in cutting-edge PCB designs.

Nearly every piece of software will be available at least as a trial download so that the potential designer can do a test drive. For the open-source software, of course, the full version is freely available. PCB design software manuals and tutorials discuss the basic operation but also special keystrokes and other shortcuts that make operations such as routing traces much more efficient.

23.6.3 Schematic Capture

The first step in PCB design is to create a schematic. It is possible to design a PCB directly from a paper schematic, but it is much easier if the schematic is entered (or "captured") in electronic form. Schematic capture software has two outputs, the visual schematic and the component and connectivity data for subsequent PCB layout. These two separate requirements can make some operations during schematic entry more complicated than what would seem at first glance necessary. Bear in mind, however, that the designer is creating not only a clear graphic representation of the circuit but of the underlying electrical connectivity.

Schematics are generally entered on a virtual page that usually corresponds to common paper sizes, such as 11×17 inches. More complicated schematics can span multiple pages using special labels or components to indicate both visual and electrical connectivity. Often, one can group logically related elements into a module that can then be referenced as a "black box" on a higher-level schematic. For complex circuits, these features are extremely useful and make the difference between a jumbled diagram that is difficult to use and an organized, compact diagram that efficiently communicates the function and operation of the circuit.

COMPONENTS

The components (resistors, capacitors, etc.) on a schematic are either selected from an existing library or created by the designer and stored in a custom library. It is also possible to find components and/or additional libraries on the internet, although each program has its own specific format.

Each schematic component includes a great deal more than shape and pin numbers. A typical component library entry includes:

Symbol — This is the graphic representation shown on the schematic. Many components may have the same symbol (e.g., the op amp symbol may be shared by many different types of op amps).

Pins — For each pin or point of electrical connection, the component model may specify the pin number, label (e.g., "V_{DD}"), pin type (inverting, noninverting), or pin functions (common).

PCB footprint — A given component may be available in a number of different packages (e.g., DIP or surface mount). Many components may have the same physical footprint (e.g., op amps, comparators, and optoisolators could all map to the same eight-pin DIP footprint). Footprints include the electrical connections (pins) as well as mechanical mounting holes and pad sizes, and the component outline. Note that in some software, the association of schematic symbol to PCB footprint is done as a separate step, so the schematic symbol is independent of the PCB footprint.

Value — Many components such as resistors and capacitors will have identical information except for a difference in value. For example, all 1/4 W resistors may be instances of the same component, differing only in value and designator.

Designator — The unique reference to the component, such as R1, C7, D3. Designators can be assigned manually, or usually there is a provision to automatically assign them sequentially once the design is done.

Source information — Part number, vendor, cost, etc. This information is for the Bill of Materials (discussed below).

Components are typically placed on the schematic by opening a library and searching for the desired component. It may be tempting for the beginner to select a component that looks "about right" when faced with a long list of components in some libraries. There are two caveats to keep in mind; first, the schematic will be easier to read if symbols are consistent (e.g., consistently using either the US or European style of resistor symbol, but not both), second, in many schematic capture programs the component is associated with a particular PCB footprint. In that case, during schematic capture one must also decide on the desired package for the component. In other cases, this mapping from schematic symbol to physical package takes place in a later step. For example, either "1/8 W Resistor, Axial" or "1 W Resistor, Upright" will result in the same neatly drawn resistor symbol on the schematic, but the PCB footprints will be dramatically different.

It is not uncommon to add new components to the library in the course of creating a schematic. Since many components are closely related to existing devices, the process often consists of selecting an existing schematic symbol, editing the shape and/or component data, creating a new label, and associating the part with an existing footprint. Adding a specific type of op amp is an example. This usually only needs to be done once, since symbols can be saved in a personal library to be reused in other designs or shared with others. It is often faster to modify a part that is close to what is desired than to "build" a new part from scratch.

Component symbols can generally be rotated and flipped when placing the component instance on the schematic.

CONNECTIONS

The schematic software has a mode for making electrical connections, usually called "nets." For example, one might click on the "draw net" symbol and then draw a line from one pin to another pin, using intermediate mouse clicks to route the line neatly with 90° turns on a uniform grid (e.g., 0.1 inch). Internally, the software must not only draw the visual line but recognize what electrical connectivity that connection represents. So, one must click (exactly) on a component pin to start or end a line. When making a connection between two lines that intersect, one must explicitly indicate a net-to-net connection, often with a special "dot" component. The connections on a schematic can often be assigned additional information, such as the desired width of the trace for this connection on the PCB or a name assigned by the designer, such as "Input Signal." For most simple designs this is not needed.

Not all connections on a schematic are drawn. To make any schematic, electronic or hand drawn, more readable, conventions are often employed such as ground or power symbols, or grouping similar connections into busses. Schematic capture software often supports these conventions. In some cases, components may be created with implicit power connections; in these cases, the connections may not even be noted on the schematic but will be exported to the PCB software.

Since it is often possible for component pins to be assigned attributes such as "Power Input," "Output," "Input," and so on, some schematic entry programs allow one to do an early design check. The program can then point out connections between incompatible pins, inputs that are missing connections, and so on.

Most schematic capture software should allow one to mark unused pins as "No Connect" or "Not Connected" (NC or N/C), which then makes it easier to see those pins that have been unintentionally left without a connection.

Text can be placed freely on the schematic to capture important notes or to label sections, connectors, etc., and there will be a text block in a corner for date, designer, version, title, and the other information that identifies the schematic. The information in the text block is sometimes maintained as data for an entire project, or it may be entered manually.

NETLISTS

Once the components are placed and connections made, the schematic may be printed and any output files for the PCB layout software produced. The connectivity and component information needed for PCB layout is captured in a *netlist* file. The flow from schematic entry to PCB may be tightly integrated, in which case the designer may switch between schematic and PCB like two views of the same design (which they are) and not have to worry about explicitly generating the netlist. The netlist can also be exported to an external circuit simulation program or be used by an integrated simulator program. (See the **Computer Modeling and Simulation** chapter for more information on circuit simulation.)

ANNOTATION AND BILL OF MATERIALS

The important features of *forward* and *backward annotation* enter at the interface between schematic entry and PCB layout. It is not uncommon during the PCB layout process to either come across some design deficiency or realize that a change to the schematic could produce a design that would be easier to lay out. Likewise, a review of the schematic partway through the PCB layout process could reveal some needed design change. In the case of changes to the PCB (perhaps changing some pins on a connector to make routing easier), backward annotation can propagate the changes "backward" to the schematic. The connectivity data will be updated; however, the designer may need to manually route the connection lines to neaten up the schematic. Likewise, changes to the schematic when the PCB is already (partially) routed are known as forward annotations. When the connectivity is updated the designer will likely need to manually route the traces. Neither forward nor backward annotation is absolutely necessary but can be very useful in keeping the schematic and PCB consistent. In their absence, the designer is strongly urged to keep the schematic and PCB up to date manually to avoid time-consuming problems later on.

Finally, the underlying data in the schematic can be used to produce a *Bill of Materials* (BOM). A BOM lists all the components of the schematic, typically ordered by reference designators, and may even be exportable for online ordering. The entry for each component usually includes a part number, cost per part, distributor, and other information used by the manufacturing process.

ASSOCIATION OF SCHEMATIC SYMBOLS AND PCB FOOTPRINTS

As was noted earlier, the selection of a schematic symbol must be associated with a particular PCB footprint. If the assignment is not automatically made, there is an intermediate step between schematic capture and PCB layout wherein PCB footprints are assigned to each symbol in the schematic. Unfortunately, the description of component footprints may be confusing. The use of highly condensed industry standards or non-uniform naming conventions often means the designer needs to browse through the component library to see the different types of components. Resistors, diodes, and capacitors seem particularly prone to a propagation of perplexing options. One solution is to open an example PCB layout and see what library element that designer used for resistors, LEDs, and so on. Here, the PCB layout software directed at hobbyists may be superior in that there are fewer options than in professional programs.

23.6.4 PCB Characteristics

PCB CONSTRUCTION

It is useful to know a little bit about PCB construction in order to make sense of the PCB design process. **Figure 23.26** shows the basic structure of a PCB and some of its design elements. The laminate material provides a stable, electrically insulating substrate with other known characteristics (thermal, dielectric, etc.). Copper is bonded to one or both sides and then areas are selectively removed (usually chemically) to leave traces and pads. The pads provide points of connection for components. Though electrical connectivity is crucial, it is important to remember that the solder and pads provide mechanical and thermal connectivity as well. Thermal properties will be critical to components that must dissipate a lot of heat, especially if no other means (such as a heat sink) are provided. Components such as switches and potentiometers may transmit a lot of force to the solder joints, which could fail with repeated cycling, and even heavy components could exert a lot of force if the device is dropped, for example. Pads may be drilled for mounting *through-hole* components or left undrilled for *surface mount* components.

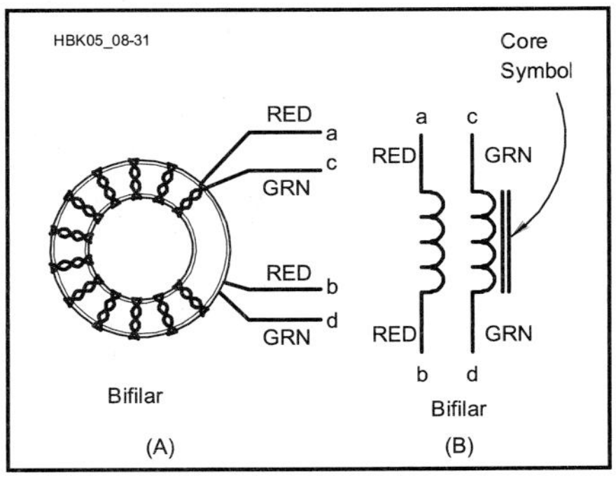

Figure 23.26 — (A) Pictorial and (B) Schematic representations of a bifilar-wound toroidal transformer.

A separate electrochemical process plates the inside surface of *plated-through holes* to provide connectivity between upper and lower pads. Plated-through holes whose sole purpose is to provide electrical connectivity between layers of a PCB are known as *vias*, shown in **Figure 23.27**. Since they do not need to accommodate a component lead, their hole and pad size can be smaller.

While two-layer boards can mount components on either side, most PCBs have a primary side called the *component side* upon which most of the components are placed, and a *solder side* dominated by soldered pins and traces. Where high density is required, surface mount, and sometimes through-hole, devices are mounted on both sides, but such designs are considerably more complex to build.

Multi-layer boards are essentially a stack of two or more two-layer boards, with an insulating layer between each board. Plated-through holes make connections possible on every layer, and the laminate material is proportionally thinner so the entire multi-layer board is roughly the same thickness as a regular two-layer board. Vias that join selected, adjacent copper layers without connecting the entire stack of layers are called "buried" or "blind" vias and are typically only needed for very dense designs. Multi-layer boards provide much more flexibility in routing signals and some other benefits such as dedicated layers for power distribution and grounding, shielding, and impedance control, but often have additional cost and fabrication time.

PCB MANUFACTURING SPECIFICATIONS

Unless the board is manufactured by the hobbyist, the PCB files are sent out to be manufactured by a *board house*. The most

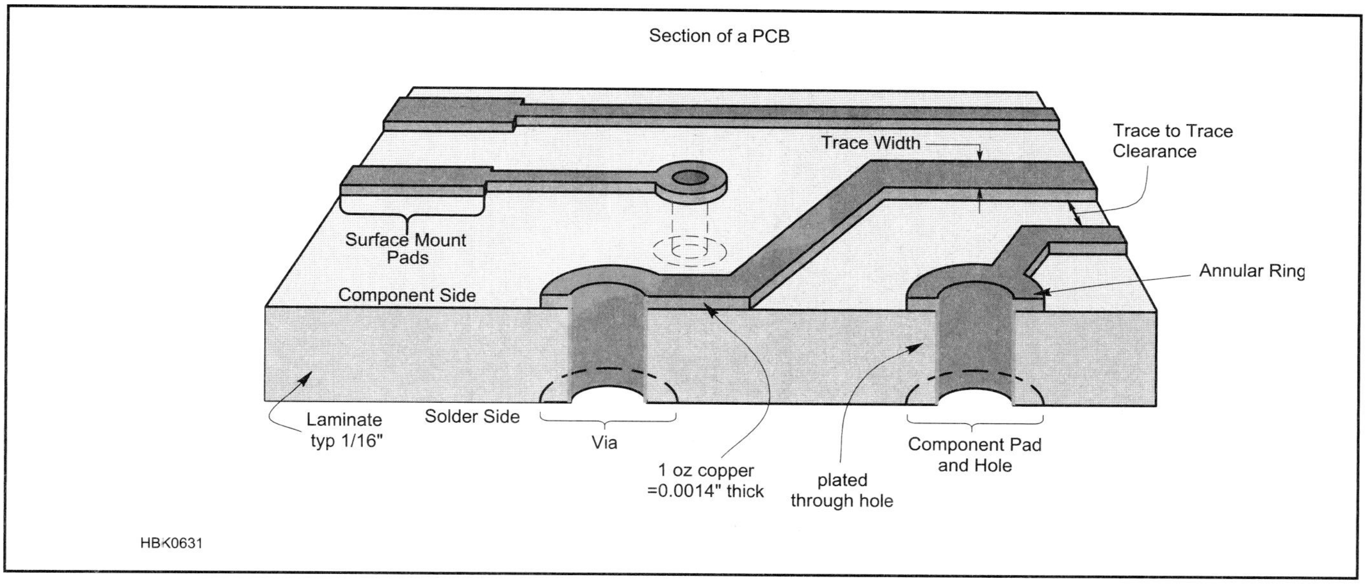

Figure 23.27 — The various elements of PCB construction and specification.

important issue for the amateur may be the pricing policies of the board house. Board size, quantity, delivery time, number of layers, number and density of holes, presence of solder masks and silk screens, minimum trace/separation width, type of board material, and thickness of copper all influence pricing. One cost-saving option of the past, a single-layer board, may not be offered with low-cost, low-volume services — two layers may be the simplest option, and it results in a more robust board. Note that most ordering specifications use English units of inches and ounces. Offshore board houses may use both English and metric units, or be metric-only. English units are used here because they are the most common encountered by hobbyists in the US.

The second issue to consider is manufacturing capabilities and ordering options. These vary with pricing and delivery times but include the following:

Board material and thickness — FR-4 is the most popular board material for low volume PCBs; it consists of flame-resistant woven fiberglass with epoxy binder. (G-10 is similar but lacks the self-extinguishing property.) Typical thickness is 0.062 inch (1/16 inch), but thinner material is often available. Flexible laminates are also available at greater cost and longer delivery time. Special board laminates for microwave use or high-temperature applications are also available.

Copper thickness — The thickness of copper on the board is expressed in ounces per square foot. Typical values are 1 to 2 oz (1 oz corresponds to 0.0014 inch of thickness). Other values may not be available inexpensively for small volumes. Inner layers on multi-layer boards may be thinner — check if this is important. Most board designs can assume 1 oz copper for double-sided boards; trace width is then varied to accommodate any high current requirements.

Layers — Two-layer boards are the most common. Because of the way multi-layer PCBs are manufactured, the number of copper layers will be multiples of two. For quick-turn board houses, usually only two- or four-layer boards are available. PCBs with more than two layers are more expensive and often take longer to manufacture.

Minimum hole size, number, and density of holes — Minimum hole size is rarely an issue, but unusual board designs with high hole density or many different hole sizes may incur additional costs. Be sure to include vias when specifying minimum hole size. Some board houses may have a specific list of drill sizes they support. Note that you can often just hand edit the drill file, which lists the different hole sizes in the design, to specify the drill sizes that a board house supports.

Minimum trace width and clearance — The smallest trace width that can be placed on a board and the smallest distance between traces are often close in value. Most board houses are comfortable with traces at least 0.010 inch in width, but 0.008 and 0.006 inch are often available, sometimes at a higher cost.

Minimum annular ring — A minimum amount of copper is required around each plated-through hole, since the PCB manufacturing process has variations. This may be expressed as the ratio of the pad size to hole size, but more commonly as the width of the ring.

Edge clearances — Holes, pads, and traces must all be at least a minimum distance from the edge of the board. That distance is usually specified by the board house.

Board outline and copies — There may be options to route the outline of the board in other shapes than a rectangle, perhaps to accommodate a specific enclosure or optimize space. If multiple copies of a board are ordered, some board houses can *panelize* a PCB, duplicating it multiple times on a single larger PCB (reducing cost per board). These copies may be cut apart at the board house, or small tabs or v-grooves may be left to connect the boards so that assembly of multiple boards can be done as a single unit, and then the individual board separated.

Tin plating — Once the traces and pads have been etched and drilled, tin plating is usually applied to the exposed copper surfaces for good soldering. Lead-free finishes are almost always available. Gold plating may be available for edge connectors.

Solder mask — A solder-resistant coating can be applied over the tin plating on both sides of the board covering everything except the component mounting pads. It prevents molten solder from bridging the gaps between pads and traces. Solder mask is offered by all but the quickest turn services. Green is the most common color, but other colors may be available with an additional cost or small delay.

Silkscreen — This is the ink layer, usually white, on top of the solder mask that lays out component shapes and designators and other symbols or text. A minimum line width may be specified, but try to avoid thin lines. All but the quickest turn services offer silk screening on one or both sides of the PCB.

23.6.5 PCB Design Elements

The schematic may not note the specific package of a part, nor the width or length of a connection. The PCB, being a physical object, is composed of specific instances of components (not just "a 10 kΩ resistor," but a "10 kΩ, 1/4 W, axial-lead resistor mounted horizontally," for example) plus traces, which are specified as connections between pins of components with a specific width and separation from other conductors. Before discussing the process of layout, we must understand the nature of components and their connections in a PCB.

COMPONENTS

A component in a PCB design is very similar to its counterpart on the schematic. **Figure 23.28** shows the PCB footprint of an opto-interrupter, including graphics and connectivity information. The footprint of a component needs to specify the connectivity on all applicable copper layers, any necessary holes including non-electrical mounting holes or slots, and any additional graphics such a silkscreen layer.

Take a common 1/4 W axial-lead resistor as an example. This footprint has two pins, each associated with a pad corresponding to the resistor's two leads. This pad appears on both the top and bottom layers of the board, but has a smaller pad associated with inner layers, should there be any. The hole's size is based on the nominal lead diameter, plus some allowance (typically 0.006 inch). The pad size is big enough to provide a manufacturable annular ring, but is usually larger so as to allow good quality soldering. The pins are labeled in a way that corresponds to the pin numbering on the schematic symbol (even if there is no polarity). A silkscreen layer is defined, usually a box within which the value or designator will appear. The silkscreen layer is particularly important to indicate orientation of parts with polarity.

More complicated parts may require additional holes that are not associated with a schematic pin (a mounting hole, for example). These are usually added to the part differently than adding a hole with a pad, such as a mounting hole without an annular ring or plating. The silkscreen layer may be used to outline the part above and beyond what is obvious from the pads, for example, a TO-220 power transistor lying on its back, or the plastic packaging around the opto-interrupter in Figure 23.28.

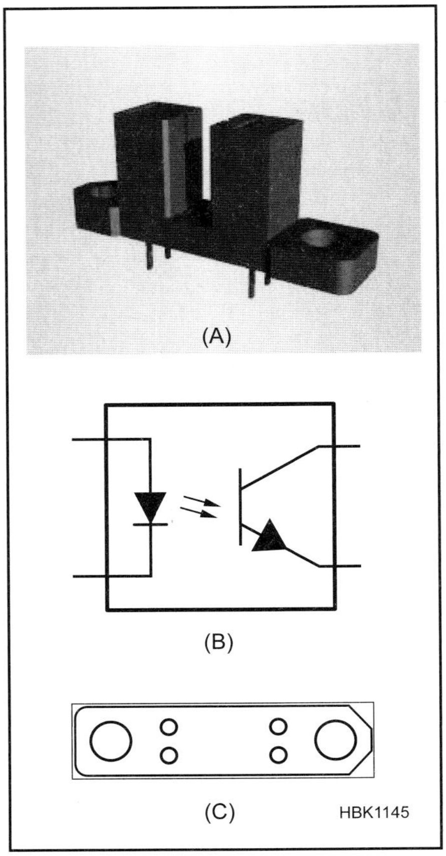

Figure 23.28 — The PCB representation of a component, such as the opto-interrupter shown here (A), combines electrical connectivity as defined in the part's schematic (B) and the part's PC Board footprint (C).

As with schematic entry, it is not uncommon to have to modify or create a new PCB footprint. Good technical drawings are often available for electronic parts; when possible, the designer should verify these dimensions against a real part with a digital caliper. It is also useful to print out the finished circuit board artwork at actual size and do a quick check against any new or unusual parts.

TRACES

Traces are the other main element of PCB construction. They are the copper pathways that connect components electrically. PCB traces are merely flat wires, subject to all the properties as an equivalent thickness of copper wire in the same circumstance (resistance, capacitance, inductance, etc.). At VHF/UHF/microwave frequencies, and for high-speed digital signals, PCB traces act as transmission lines. If a ground layer is placed on the adjacent layer above or below the PCB trace, the trace impedance can be controlled by the width of the trace and the distance between it and the ground layer. At sufficiently high current PCB traces heat up, and have even been used (or abused) as heating elements, though unintended heat can damage the trace and the PCB material.

There are few constraints on traces apart from those such as minimum width and clearance imposed by the board house. They are created by chemical etching and can take arbitrary shapes. In fact, text and symbols created on copper layers may be handy if a silkscreen is not included. Traces may be of any length, vary in width, incorporate turns or curves, and so on. Traces can be formed to generate RF inductors, can be arranged to produce a simple set of windings for a motor, or create a strain gauge. However, most traces will be a uniform width over their entire length and make neat 45° corners. On two-layer boards, for most efficient routing of signals, trace direction tends to be horizontal one side and vertical on the other side.

The same considerations when building a circuit in other methods applies to PCB design, such as current capacity (width of trace and thickness of copper), voltage (clearance to other signals), noise (shielding, guarding, proximity to other signals), impedance, ground and power supply bypassing (noise reduction), and so on.

23.6.6 PCB Layout

With a schematic and netlist ready, and all of the PCB characteristics defined, the actual layout of the PCB can begin.

BOARD SIZE AND LAYERS

The first step in PCB layout is to create the board outline to contain not only the circuit itself but also any additional features such as mounting holes. For prototype or one-off designs, the board is often best made a bit larger to allow more space between components for ease in testing and debugging. (Some low cost or freeware commercial PCB software imposes limits on board size and number of components.) The board outline may be provided in a default size that the designer can modify, or the designer may specify any desired outline.

While the software may not require deciding at the start how many layers the PCB uses, this is a decision the designer should make as early as possible, since the jump from two to four or more layers has a big impact on routing the traces, control of trace impedance, and cost.

For your initial design, start with a two-layer board for a simple circuit that you have already built and tested. This reduces the number of decisions you have to make and remove some of the unknowns from the design process.

COMPONENT PLACEMENT

Good component placement is much more than half the battle of PCB layout. Poor placement will require complicated routing of traces, make assembly difficult, and even introduce noise or other degradations, while good placement can lead to clean, easily routed, easy-to-assemble designs.

The first elements placed should be mounting holes or other fixed location features. These are often placed using a special option selected from a palette of tools in the software rather than as parts from a library. Holes sufficient for a #4 or #6 screw are often fine; be sure to leave room around them for the heads of the screws and nut driver or standoff below. These will often be "non-plated-through holes" (NPTH) with no annular pad (though the board house may plate all the holes in a board, regardless). In some designs, the mounting screws are used to connect the PCB ground plane to chassis ground, the mounting holes for those boards must be plated through with a pad on the top that is the size of the screw head.

(A brief aside on plated-through holes: The normal flow of PCB manufacture has holes drilled early in the process and all the insides plated. If for some reason one needs a hole to specifically not be plated through, this requires a separate drilling step, and perhaps extra cost or time in manufacturing. Although a mechanical element, such as a mounting screw, may not need a plated-through hole, it is often more economical to lump all the holes together and make all of them plated-through. Hence the option in some PCB software to combine PTH and NPTH holes into the same file.)

Depending on the software and whether schematic capture was performed, the board outline may already contain the footprints of all the circuit components (sometimes stacked in a heap in one corner of the board) and the netlist will already be loaded. In this case, components may be placed by clicking and dragging the components to the desired location on the board. Otherwise, one may need to manually import the netlist.

Most PCB programs have a "rat's nest" option that draws a straight line for each netlist connection of a component, and this is a great aid in placement because the connections between components are apparent as the components are moved around. (See **Figure 23.29**) However, connections are shown to the nearest pin sharing that electrical connection; thus, components such as decoupling capacitors (which are often meant to be near a specific component) will show rats nest connections to the nearest power and ground pins and not the pins the designer may have intended. In this case, the component must be placed by referring to the schematic.

The PCB layout software may offer *auto-placement*, where the components are initially arranged automatically. The beginner should certainly feel free to experiment and see how well this tool performs, but is cautioned that manual tweaking will likely be needed.

PCBs need not be arranged to precisely mimic the schematic, but it is appropriate to place components in a logical flow, when possible, so as to minimize the length of traces in the signal path. Sensitive components may need to be isolated or shielded from other components, and grounding and decoupling attended to, just as one would do with a point-to-point soldered version.

During placement, the designer will find that different orientations of components simplify routing (for example, minimizing the number of traces that have to cross over each other or reducing the trace lengths). Components are generally rotated in increments of 90°, although free rotation may be an option (for either functional or artistic intent). However, there is a good reason to maintain the same orientation on similar devices. If pin 1 on all IC packages and the positive sides of all components with polarity point in the same direction there is much less chance that any of them will be pointing the wrong way when the board is assembled.

Placement and orientation of components can also affect the ease of assembly. Allow plenty of space around sockets so ICs can be more easily inserted and removed. Components with a mechanical interface, such as potentiometers and switches, should be positioned to allow access for adjustment. Any components such as connectors, switches, or indicators (e.g., LEDs) should be positioned carefully, especially if they are to protrude through a panel. Often this will involve having the component overhang the edge of the PCB.

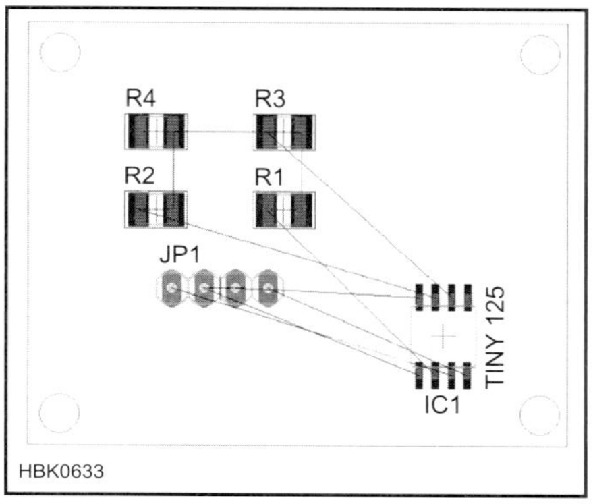

Figure 23.29 — The rat's nest view during PCB layout shows the direct connections between component pins. This helps the designer with component placement and orientation for the most convenient routing.

Components ideally should include a silkscreen outline that shows the size of the whole component — for example, a transistor in a TO-220 package mounted flat against the PCB should have an outline that shows the mounting hole and the extent of the mounting tab. The designer should also consider the clearance required by any additional hardware for mounting a component, such as nuts and bolts or heat sinks — including clearance to allow assembly tools to be easily used. (It is a good policy to avoid running traces under mounting hardware, since the only protection the trace has from shorting out is the solder mask, which could become abraded.)

Take care to minimize the mechanical stress on the PCB, since this can result in cracked traces, separated pads, or other problems. Utilize mounting holes or tabs, when possible, for components such as connectors, switches, and pots. Use two-layer boards with plated-through holes even if the design can be single-layer. Component leads soldered to plated-through holes produce much stronger mechanical connections than single-layer boards in which the soldered pad is held only by the bond between copper and laminate and is easily lifted if too much heat or stress is imposed.

When prototyping a new design, add a few unconnected pads on the circuit board for extra components (e.g., a 16 pin DIP, 0.4-inch spaced pads for resistors and other discrete components). Include test points and ground connections. These can be simply pads to which cut off leads can be soldered to provide convenient test points for ground clips or to monitor signals.

Wires or cables can be directly soldered to the PCB, but this is inconvenient when swapping out boards, and is not very robust since stress is concentrated at the solder joint, leading to fatigue breaks. Connectors are preferred whenever possible and often provide strain relief for the wire or cable. However, if a wire is directly soldered to the PCB, the designer should consider adding a (non-plated) hole nearby just large enough to pass the wire, including insulation. The wire can be passed from the solder side through the non-plated hole, then soldered into the regular plated-through hole. This provides good strain relief, which can be augmented with a dollop of glue if desired.

ROUTING TRACES

After placing components, mounting holes, and other fixed location features that limit component or trace placement, traces can be *routed* in order to complete all the connections between pins without producing short circuits.

Most PCB design programs allow components and traces to be placed on a regular grid, similarly to drawing programs. There may be two grids — a visible coarse pitch grid and a "snap" fine pitch grid, to which components and other objects will be aligned when placed. It is good practice to use a 0.1- or 0.050-inch grid for component placement and to route traces on a 0.025-inch grid. While the "snap to grid" feature can usually be turned off to allow fine adjustment of placement, a board routed on a grid is likely to look cleaner and be easier to route. Snapping traces to a grid also help the you to meet the minimum trace clearance distance required by the board house.

The trace starts at a component pin and wends its way to any other pins to which it should be connected. Traces should start and end at the center of pads, not at the edge of a pad, so that the connection is properly recorded in the program's database. If a netlist has been loaded, most PCB software displays a rat's nest line showing a direct connection between pads. Once the route is completed, the rat's nest line for that connection disappears. The rat's nest line is rarely the desired path for the trace, since is a straight shot and often not the best placement. For example, when routing power traces, the designer should use good design sense rather than blindly constructing a Byzantine route linking pins together in random order. For this reason, routing the power and ground early is a good practice.

High speed, high frequency, and low noise circuits require additional care in routing. In general, traces connecting digital circuits such as microprocessors and memories, should not cross or be in close proximity to traces carrying analog or RF signals. (Refer to earlier sections of this chapter, the **RF Techniques** chapter, and the references listed at the end of this section.)

Manual routing is a core skill of PCB design, whether or not auto-routing is used. The process is generally made as simple as possible in the software, since routing will take up most of the PCB design time. A trace is routed on the currently selected copper layer. For a single-sided board, there is only one layer for routing; for a two-layer board the component side and solder side can have traces; and for multi-layer boards additional inner layers can have traces. Often a single keystroke can change the active copper layer (sometimes automatically inserting a via if a route is in progress). The trace is drawn in straight segments and ends at the destination pin. When routing, 90° corners are normally avoided, rather use a pair of 45° angles. **Figure 23.30** shows some sample traces.

It is good practice on a double-sided board to have one side of the board laid out with mostly horizontal traces, and the other side laid out with mostly vertical traces. A trace that needs to travel primarily vertically can do so on the side with vertical traces and use a via to move to the other side to complete the horizontal part of the route.

It is easiest during testing and debugging if most traces are routed on the bottom (solder) side of the board. Traces on the component side often run under components, making them hard to access or follow. It is often much clearer to connect adjacent IC or connector pins by routing a trace that leaves one pad, moves away from the IC or connector, then heads back in to the adjacent pad to connect. This makes it clear the connection between adjacent pads on the assembled board is not an accidental solder bridge.

It may be the case that no number of vias or wending paths can complete a route (particularly for single-sided boards). The remaining tool for the PCB designer is a jumper, a wire added as a component during assembly just for the purpose of making a connection between two points on the board. Jumpers are required more often for single-sided boards; when the "jump" is rather small, uninsulated wire can be used. Jumpers are usually straight lines, and can be horizontal or vertical. Professional production PCBs may use machine-insertable zero-ohm resistors as jumpers. (The radio amateur may as well — all the major distributors carry zero-ohm resistors in a range of packages.) Jumpers on double-sided boards are usually not viewed very favorably, but this is an aesthetic and efficiency issue, not a functional one.

Multi-layer boards (i.e., four or more) clearly offer additional routing options, but again having some dominant routing direction (vertical or horizontal) on each layer is recommended, since mixing directions tends to cause routing problems. However, it is not uncommon to devote one or two inner layers to power and ground, rather than merely be additional layers for routing signals. This allows power and ground to be routed with minimal resistance and exposes the traces carrying interesting signals on the component and solder sides, where they are available to be probed or modified. It is very difficult to modify traces on inner layers.

Before routing too many traces, it is helpful to run the *Design Rule Check (DRC)* on the board. (See the section on Design Rule Checking below.) Applied early and often, DRC can identify areas of concern when they are easiest to correct. For example, a given trace width may provide insufficient clearance when passing between two IC pads. Waiting until the end of the initial layout process to run a DRC can lead to time-consuming rework if problems aren't easily corrected.

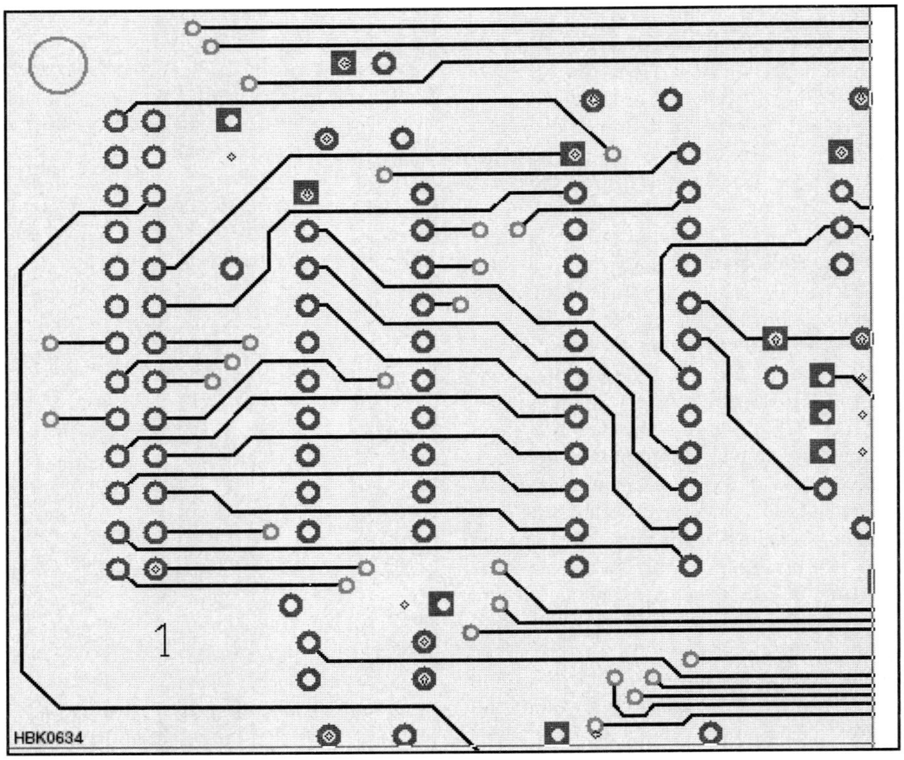

Figure 23.30 — This example shows traces on the side of the PCB for horizontal routing. Traces are routed between pins of ICs. The smaller pads are for vias to a different layer of the PCB.

Some PCB design packages offer *auto-router* capability in which the software uses the component and connectivity data of the netlist and attempts to route the traces automatically. There are some circumstances when they save time, but view these tools with some caution. Auto-routers are good at solving the routing puzzle for a given board, but merely connecting all the pins correctly does not necessarily produce a good PCB design. Traces carrying critical signals may take "noisy" routes; components that should have short, low resistance connections to each other may be assigned lengthy paths, and so on. More sophisticated auto-routers can be provided with extensive lists of "hints" to minimize these problems. For the beginner, the time spent conveying this design information to the auto-router is likely better spent manually routing the traces.

If an auto-router is used, at a minimum, critical connections should be first routed manually. These include sensitive signals, connections whose length should be minimized, and often power and ground (for both RFI and trace width reasons). Better still is to develop a sense of what a good layout looks like (which will come with practice and analyzing well designed boards) and learning at what stage the auto-router can be "turned loose" to finish the routing puzzle.

TRACE WIDTH AND SPACING

All traces have some width — the width may be the default width, the last width selected, or a width provided from data in the netlist. It may be tempting to route all but the power traces using the smallest trace width available from the board house (0.008 inch or smaller), since this allows the highest density of traces and simplifies routing. A better design practice is to use wider traces to avoid hard-to-detect trace cracking and improve board reliability. The more common 0.012-inch wide traces can be run in parallel on a 0.025-inch grid and can pass between many pads with 0.1-inch center spacing. Even wider traces will make the board easier to produce "in house," though the exact process used (CNC routing, chemical etching, etc.) will limit the resolution. Note that it is possible to "neck down" traces where they pass between IC or connector pads. That is, the regular, thick trace is run up close to the narrow gap between the pads, passes between the pads with a narrow width, then expands back to the original width. There is little reason to use traces wider than 0.030 inch or so for most signals (see **Table 23.3**), but power and ground trace widths should be appropriate for the current.

All traces have resistance, and this resistance is a function of the cross section of the trace (width times thickness) and the length. This resistance will convert electrical power to heat. If the heat exceeds a relatively high threshold, the trace becomes a fairly expensive and difficult-to-replace fuse. The trace width should be selected such that for the worst case expected current, heat rise is limited to some threshold, often 10 °C. In practice, power traces (especially grounds) are often made as wide as practical to reduce resistance, and they greatly exceed the width required by heat rise limits alone.

Table 23.3 summarizes maximum currents for external (component and solder side) and internal traces for some common trace widths. Internal traces (on inner layers of multi-layer boards) can carry only about half the current of external traces for the same width, since the internal layers do not dissipate heat to the ambient air like external traces can. (Note that trace widths are also sometimes expressed in "mils." 1 mil = 0.001 inch; it is shorthand for "milli-inch," not millimeter!)

There is no upper bound on the effective trace width. It is common to have large areas of the board left as solid copper. These *copper fill* areas can serve as grounds, heat sinks, or may just simplify board production (especially homemade boards). It is not a good idea to place a component hole in the middle of a copper fill — the copper is a very efficient heat sink when soldering. Instead, a "wagon wheel" pattern known as a thermal relief is placed (sometimes automatically) around the solder pad, providing good electrical connectivity but reducing the heat sinking. Often, copper fill areas can be specified using a polygon, and the fill will automatically flow around pads and traces in that area.

In practice, most boards will have only two or perhaps three different trace widths: narrow widths for signals, and a thicker width for power (usually with a healthy margin).

One final note on trace width. Vias are typically one size (i.e., small), but multiple vias can be used to create low resistance connections between layers. Spacing the vias so their pads do not touch works well; the pads are then shorted on both top and bottom layers with a trace.

Voltage also figures into the routing equation, but instead of trace width, higher voltages should be met with an increased clearance between the trace and other copper. The IPC-2221 standard calls for a clearance of 0.024 inch for traces carrying 31-150 V (peak) and 0.050 inch for traces carrying 151-300 V (peak); these are external traces with no coating. With the appropriate polymer solder mask coating, the clearances become 0.006 inch for 31-100 V and 0.016 inch for 101-300 V. Internal traces also have reduced clearance requirements. Fully addressing the safety (and regulatory) issues around high voltage wiring is outside the scope of this brief review, however, and the reader is urged to consult UL or IPC standards.

SILKSCREEN AND SOLDER MASK

The silkscreen (or "silk") layer contains the text and graphics that will be silkscreened on the top (and bottom) of the board, shown in **Figure 23.31**. Component footprints generally have elements on the silkscreen layer that automatically appear, such as designators and values, but other elements must be created and placed manually. Common silkscreen elements include circuit name, date, version, designer (and call sign), company name, power requirements (voltage, current, and fusing), labels for connections (e.g., "Mic input"), warnings and cautions, labels for adjustments, and switches. A solid white rectangle on the

Table 23.3
Maximum Current for 10 °C Rise, 1 oz/ft^2 Copper

Based on IPC-2221 standards (not an official IPC table)

Trace Width (inches)	*Max. Current (External Trace) (A)*	*Max. Current (Internal Trace) (A)*	*Resistance (ohms/inch)*
0.004	0.46	0.23	0.13
0.008	0.75	0.38	0.063
0.012	1.0	0.51	0.042
0.020	1.5	0.73	0.025
0.040	2.4	1.2	0.013
0.050	2.8	1.4	0.010
0.100	4.7	2.4	0.0051
0.200	7.8	3.9	0.0025
0.400	13	6.4	0.0013

IPC-2221 Generic Standard on Printed Circuit Design, Institute for Interconnecting and Packaging Electronic Circuits, **www.ipc.org**

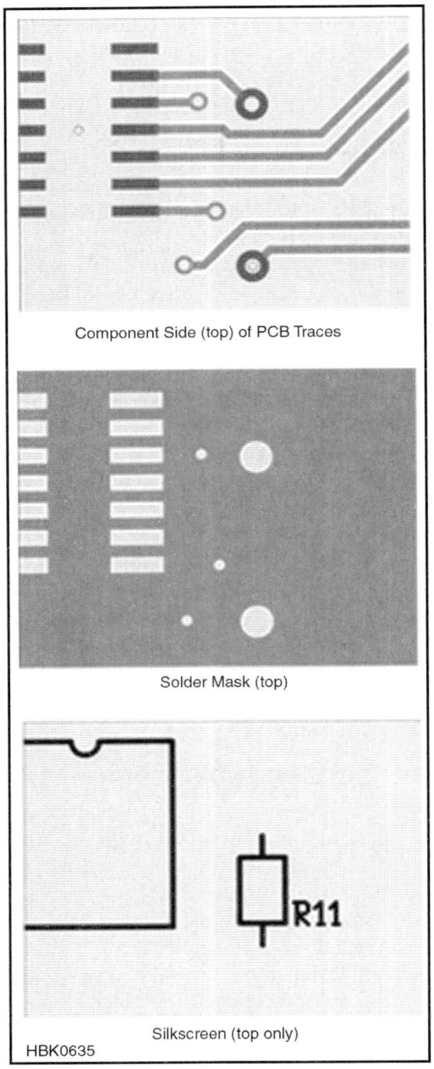

Figure 23.31 — The relationship between the layout's top copper layer with traces and pads, the solder mask that covers the copper (a separate solder mask is required for the top and bottom layers of the PCB), and the silkscreen information that shows component outlines and designators.

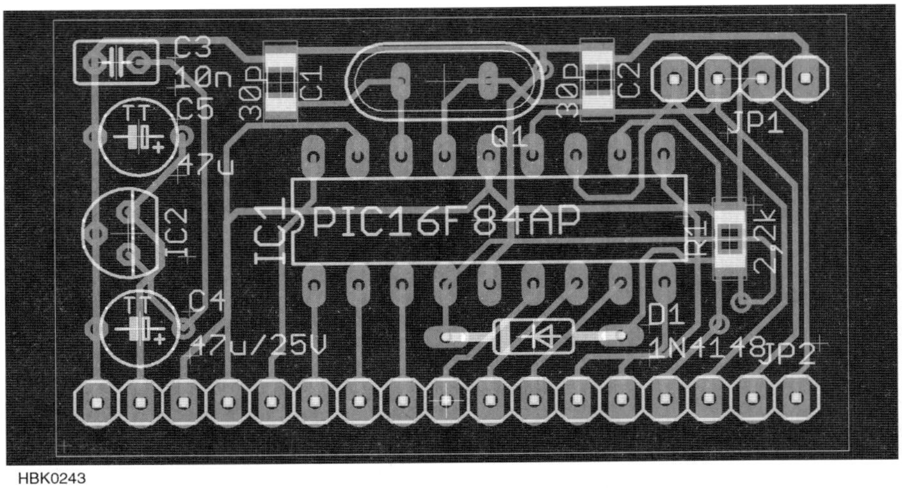

Figure 23.32 — A completed microprocessor board as it is seen in a typical PCB layout editor (Eagle). Solder mask layers are omitted for visibility. Traces that appear to cross each other are on different sides of the board and are in different colors in the layout software. The silkscreen layer is shown in white.

silkscreen layer can provide a good space to write a serial number or test information.

The board house will specify the minimum width for silkscreen lines, including the width of text. Text and graphics can be placed anywhere on the solder mask, but not on solder pads and holes.

In the past, some quick-turn board houses omit the silkscreen for prototype boards. And if the PCB is made in-house, it is unlikely to be silkscreened. As noted earlier, many of the text elements above formed in copper and placed on the external layers. Component outlines are not possible since the resulting copper would short out traces, but component polarization can be noted with symbols such as a hand-made "+" made from two short traces, or a "1" from a single short trace. Some component footprints already follow a practice of marking pin 1 or positive polarity with a square pad, while other pads are round or oval.

The solder mask is a polymer coating that is screened onto the board before the silkscreen graphics. As shown in Figure **23.32**, it covers the entire surface of the board except for pads and vias. Solder masking prevents solder bridges between pads and from pads to traces during assembly, and is particularly important for production processes that use wave soldering or reflow soldering. There is one solder mask layer for the top layer and another for the bottom layer. Internal layers do not need a solder mask. For prototype boards without solder masking, care must be taken to keep solder from creating unwanted bridges or short circuits.

During the PCB layout process, solder mask layers are generally not shown because they do not affect connectivity. Figure 23.32 shows a typical PCB as it appears when the PCB layout process is complete.

DESIGN RULE CHECK

If a netlist has been provided from the schematic capture program, a Design Rule Check (DRC) can be made of the board's layout. The PCB software applies a list of rules to the PCB, verifying that all the connections in the netlist are made, that there is sufficient clearance between all the traces, and so on. These rules can be modified based on the specific board house requirements. As stated above, it is useful to run the DRC even before all the traces have been routed. This can identify clearance or other issues that might require substantial re-routing or a different approach.

If the designer has waited until all the routing is done before running the DRC, the list of violations can be daunting. However, it is often the case that many, if not all, of the violations represent issues that may prevent the board from operating as desired. All DRC violations should be rectified before fabrication. In some cases, the board house may reject designs that violate design rules.

23.6.7 Preparation for Fabrication

LAYOUT REVIEW

Once the board has passed DRC, the electrical connectivity and basic requirements for manufacture have likely been satisfied. However, the design may benefit from an additional review pass. Turn off all the layers but one copper layer and examine the traces. Often simplifications in routing will be apparent without the distractions of the other layers. For example, a trace can be moved to avoid going between two closely spaced pins. Densely-spaced traces could be spaced farther apart. There may be opportunities to reduce vias by routing traces primarily on one layer even if that now means both vertical and horizontal travel. Repeat the exercise for all the copper layers.

Review the mechanical aspects of the board as well, including the proximity of traces to hardware. If your prototype PCB does not have a solder mask, traces that run underneath components such as crystals in a conductive case, or too close to mounting hardware, can form a short circuit. Such traces should be re-routed, or insulators must be provided.

GENERATING OUTPUT FILES

Once the PCB design is complete, the complete set of design description files can be generated for producing the PCB:

Copper layers — One file is generated per copper layer. These are known as *Gerber files* and are text files of commands that were originally intended to drive a *photoplotter*. Gerber was the primary manufacturer of photoplotters, machines that moved a light source of variable width (apertures) from one location to another to draw patterns on photographic film. While photoplotters have been replaced by digital technology, the format used by Gerber has been standardized as RS-274X and is universally used, except by PCB software tied to a specific manufacturer. RS-274X is related to the RS-274 ("G-Code") used by machinists to program CNC machinery but is an *additive* description (essentially saying "put copper here"), rather than describing the movements of a tool to remove material. A program is thus required to translate between Gerber and G-Code if a CNC machine is used to make a PCB by mechanically removing copper. Another format for PCB files, ODB++, is sometimes also supported by software and board houses, but Gerber is still the most common.

Drill file — The Drill File contains the coordinates and drill sizes for all the holes, plated or not. Also called the *NC* or *Excellon* file, some board houses may require a specific format for the coordinates, but these are usually available as options in the PCB program. There is only one drill file for a PCB, since the holes are drilled from one side. (Exotic options such as buried vias require more information.) Like RS-274X apertures, the drill file generally contains a drill table. (If non-plated-through-holes (NPTH) are required, these need a separate NC file.)

Silkscreen — The Silkscreen is also in RS-274 format. Some board houses can provide silkscreen on both sides of the board, which requires two files.

Solder mask — The solder mask file is used by the board house to create the solder mask. One file per side is required.

A Gerber preview program such as *Gerbv* (**gerbv.github.io**, open source, *Linux*, Mac OS X), *GerbView* (found in KiCad), or *GC-Prevue* (**www.graphicode.com**, *Windows*) can be used to review the trace layout Gerber files. This is a good test — the board house will make the boards from the Gerber files, not the PCB design file. Gerber previewers can import the copper layers, silkscreen, and drill files to verify they correspond and make sense.

Any of the layers can usually be printed out within the PCB program (and/or Gerber preview program) for reference and further inspection.

In addition to files for PCB production, PCB layout programs can also generate assembly diagrams, and in some cases can provide 3D views of what the assembled board will look like. These can be useful for documentation as well as verification of mechanical issues such as height clearance.

Sending the files to the PCB manufacturer or board house and ordering PCBs is explained on the manufacturing website, or a customer service representative can walk you through the process. Some firms may also offer a design review service for first-time customers or on a fee basis.

Finally, many PCB layout programs can generate a "pick & place" or "component placement list" file, which is used by automatic pick and place machines to populate a circuit board with surface mount components. While this would obviously make sense at large scale, some boards with a large number of components could be candidates for board assembly, a service offered by many board houses. Automated assembly may require some additional PCB features for alignment.

23.6.8 Checklist

SCHEMATIC CAPTURE

- Verify all non-connected inputs can be left disconnected (versus, for example, needing to tie some inputs to GND or VCC). Unused pins should be marked as no-connect (NC).
- In some programs, it may be useful to include mounting hardware in the schematic so they can be imported into the layout program. If mounting holes are to be grounded, make these connections in the schematic.
- Verify that all necessary decoupling capacitors are placed.
- Especially for prototypes, add plenty of test points to signals of interest, including ground points, for test equipment.
- LEDs (with appropriate current limiting resistors) are cheap, take up little space, and provide visual indication of slow (and sometimes fast!) signals and power status.
- Again, for prototypes add extra components (resistors, capacitors, various IC packages) to the schematic, that are not connected to anything else. These will allow adding RC filters, op-amps, etc., should the need arise. Bring unused pins of microcontrollers to nearby pads to allow cleaner modifications if needed later. Be aware that such unused pins produce errors on the DRC, so you should not ignore the errors from the real circuitry.
- Check that connections to potentiometers are such that clockwise adjustment produces the correct effect.

PRINTED CIRCUIT BOARD LAYOUT

- Run a DRC, if available, using design rules from the board house will produce the board.
- Verify decoupling caps are located near the components they are associated with.
- Are all the traces routed? In some cases, programs will have a status line showing how many traces remain unrouted. Some rats nest lines may be difficult to see, especially short ones between adjacent pins.
- Verify mounting holes are located properly. (0.125-inch diameter works well for 4-40 hardware.) Allow clearance around holes for screw heads or standoffs to traces and other components.
- Check side clearance for board guides or slotted enclosure.
- Verify clearance to components and traces around connectors, modules, or parts with heat sinks. Large heat sinks may require separate mounting holes.
- Adjust silk-screened component references (e.g., "R8") so they will be visible after the component is installed, and are clearly associated with the correct footprint. Component values (if used) should also be properly associated, but in tight quarters could be covered by the component.
- Verify that connectors, power components (with thermal requirements), and other components are located properly. For example, a right-angle connector may need to stick out past the edge of the circuit board for proper mounting in a panel.
- Check that the board outline is located on the proper layer (e.g., "edge cuts"), as specified by the board house.
- Verify that ground and power traces are thick enough to carry the required current without voltage drops or generating excessive heat. Use multiple vias for less resistance (both electrical and thermal) than for a single via.
- Check for sufficient clearance for high voltage traces.
- Make sure the orientation of components is noted for connectors, ICs, diodes, electrolytic capacitors, and other components with polarity or specific orientation. These notations should be visible after the component is installed.
- Avoid traces (e.g., between adjacent

pins) that could resemble solder bridges.

- On the silk screen layer, add descriptions to connectors (e.g., "serial port," "test 1").
- Add tick marks on the silk screen every 5 or 10 pins for high-pin count components to aid in debugging.
- Add identification on the front (and/or back) silk screen, including
  - Name/function of board
  - Date
  - Board version
  - Company/designer
  - Call sign
  - Website or email
  - Copyright (as needed)
  - Power requirements
  - (e.g., "+12VDC 1.5A")
  - A solid rectangular block on the front silk screen allows for a hand-written serial number

Several PCB layout checklists can be found online, including the comprehensive one at **aqdi.com/wordpress/PCBCheck-listLive.htm**.

PCB CAD REFERENCES

Analog Devices, *High Speed Design Techniques*, Analog Devices, 1996.

Johnson, Howard, and Graham, Martin, *High Speed Digital Design: A Handbook of Black Magic*, Prentice Hall, 1993.

Ott, Henry, *Electromagnetic Compatibility Engineering*, Wiley Press, 2009.

Pease, Robert A, *Troubleshooting Analog Circuits*, Butterworth-Heinemann, 1991.

Silver, W. NØAX, "Hands-On Radio Experiments 107-110: PCB Layout," *QST*, Dec. 2011 through Mar. 2010.

23.7 Microwave Construction

Paul Wade, W1GHZ, updated this short tutorial to construction practices suitable for microwave frequency operation, originally written by ARRL Laboratory Engineer Zack Lau, W1VT. For more information on microwave equipment, read Paul's series of columns, "Microwavelengths," in *QST*.

Microwave construction is not only fun, but within the capabilities of most amateurs. To get on the air requires some degree of construction, since you can't buy a ready-to-go box. At a minimum, a few modules must be connected together with coax and power cables. At the other extreme, the whole station may be home brewed from scratch, or from a kit.

The growth of mobile phones and wireless networking has caused a proliferation of microwave integrated circuits offering high performance at low cost. Some of these are also useful for ham applications, so that microwave construction has evolved from traditional waveguide "plumbing" to printed circuitry requiring surface mount assembly of tiny components. Waveguide techniques are still valuable for some components, such as filters and antennas, so proficiency in both types of construction is valuable.

A problem we all face when building new microwave equipment is finding a nearby station to try it out. One solution is to convince a buddy to build a similar system, so that you will both have someone to work with. If you have complementary skills, say one with metal and the other with surface mount soldering, you can help each other. A larger group or club effort can be even better. Someone with experience and test equipment can be an Elmer for the group.

In addition to the material here and in magazine columns, several excellent references for amateurs interested in working on microwave frequencies include:

- *Microwave Know How*, by Andy Barter, G8ATD (RSGB)
- *International Microwave Handbook*, by Andy Barter, G8ATD (RSGB)
- *The ARRL UHF/Microwave Experimenter's Manual* (out of print but available used)
- *The ARRL UHF/Microwave Projects* CD (out of print but available used)

23.7.1 Lead Lengths

Microwave construction is becoming more popular, but at those frequencies the size of physical component leads and PC-board traces cannot be neglected. Microwave construction techniques either minimize these stray values or make them part of the circuit design.

The basic consideration in microwave circuitry is short lead lengths, particularly for ground returns. Current always flows in a complete loop with a return path through the "ground" (see **Figure 23.33**). The loop must be much smaller than a wavelength for a circuit to work properly. As the frequency gets higher, dimensions must get smaller.

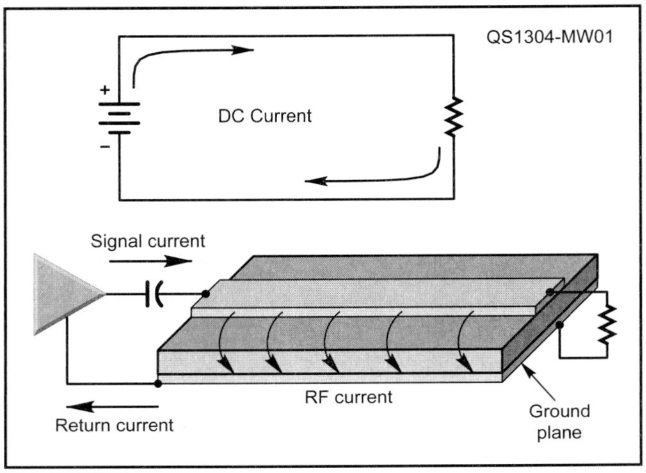

Figure 23.33— Current must always flow in a complete loop. This is particularly important at microwave frequencies where wavelengths are very short.

One area that requires particular care to ensure good ground return continuity is the transition from coax cable outside a cabinet to a PC board inside the cabinet.

At microwave frequencies, the mechanical aspects and physical size of circuits become very much a part of the design. A few millimeters of conductor has significant reactance at these frequencies. This even affects VHF and HF designs in which the traces and conductors resonate on microwave frequencies. If a high-performance FET has lots of gain in this region, a VHF preamplifier might also function as a 10 GHz oscillator if the circuit's stray reactances were just right (or wrong!). You can prevent this by using shields between the input and output, or by adding microwave absorptive material to the lid of the shielded module. (www.laird.com/products/absorbers)

23.7.2 Metalworking

Waveguide construction requires some basic metalworking skills, which are also useful for assembly and packaging of microwave systems. Some minimal tools would be a hacksaw, drills, files, and layout tools. While a hand drill can suffice, an inexpensive drill press will make work more precise and safer. In metals, DeWalt Pilot Point drill bits make clean, accurate holes. Files and nibbling tools can make rectangular or odd-shaped holes.

Solder has a poor reputation for microwave losses, but, of course, it is essential. It does have higher resistance than copper or aluminum, but that only matters in locations of high current, for instance, the ground end of a high-Q 1/4 wave resonator. Horn antennas and most waveguide structures have low Q so a clean solder joint, without excess blobs, has no effect. Traditional tin-lead solder works well; silver solder with 1% silver has no performance advantage.

For things like waveguide and antennas that are too large for a soldering iron, a hot air gun does an excellent job. **Figure 23.34** shows the backshort of a waveguide transition as it is being soldered with a hot air gun. A small amount of paste flux (Kester SP-44) was applied to the waveguide end, and a ring of solder placed along the joint. Then heat is applied to the whole area until the solder melts and flows into the joint.

A torch can also be used for soldering, but tends to cause more oxidation of the metal. However, when a high-temperature brazed joint is desired, a torch is required. A MAPP gas torch and Silvaloy® 15 (or Harris Stay-Silv® 15) silver brazing material, which requires no flux, are suitable for small and medium-sized components.

Many hams seem to think that silver plating is desirable, but it rarely makes a difference. The only application in which silver plating has been shown to make a significant difference is in UHF cavities with sliding contacts. In most other places, copper losses are low enough so that little improvement is possible, and aluminum is fine for most uses.

Figure 23.34 — Larger assemblies may be soldered with a heat gun. A torch is better for high-temperature brazing.

23.7.3 Circuit Construction

Modular construction is a useful technique for microwave circuits. Often, circuits are tested by connecting their inputs and output to known 50 Ω sources and loads. Modules are typically kept small to prevent the chassis and PC board from acting as a waveguide, which provides a feedback path from the output to the input of a circuit, resulting in instability.

PRINTED-CIRCUIT BOARDS AT MICROWAVES

A microwave printed circuit board (PCB) is typically double-sided, with transmission-line circuitry printed on one side and a ground plane on the far side. The ground return path is best provided by plated-thru holes (PTH) connecting the two sides with minimum length. Surface mount components — integrated circuits, chip capacitors, and chip resistors — are soldered directly to printed pads and transmission lines, and to ground pads with embedded PTH. Many of the microwave integrated circuits come in really tiny packages, with lead pitch as small as 25 mils. See the section on Surface Mount Technology earlier in this chapter for more information on this type of construction.

When using glass-epoxy PC board at microwave frequencies, the crucial board parameter is the thickness of the dielectric. It can vary quite a bit, in excess of 10%. This is not surprising: digital and lower-frequency analog circuits work just fine if the board is a little thinner or thicker than usual. Some of the board types used in microwave-circuit construction are a generic Teflon PC board, Duroid 5870 and 5880.

TRANSMISSION LINES AND CONNECTORS

From a mechanical accuracy point of view, the most tolerant type of construction is based on waveguide. Tuning is usually accomplished via one or more screws threaded into the waveguide. It becomes unwieldy to use waveguide on the amateur bands below 10 GHz because the dimensions get too large.

Proper connectors are a necessary expense at microwave frequencies. At 10 GHz, the use of the proper connectors is essential for repeatable performance. Do not connect microwave circuits with coax and pigtails. It might be made to work, but it probably can't be duplicated. SMA connectors are common because they are small and work well. SMA jacks are sometimes soldered in place, although mounting with 2-56 hardware is more common.

At 24 GHz and above, even waveguides become small and difficult to work with. At these frequencies, most readily available coax connectors work unreliably, so these higher bands are really a challenge. Special SMA connectors are available for use up to 24 GHz.

DEVICE SUBSTITUTION

It is important to copy microwave circuits exactly, unless you really know what you are doing. "Improvements," such as better shielding or grounding, can sometimes cause poor performance. It may be tempting to substitute components, particularly with the active devices, but the results are often disappointing. While substituting different grades of the same wafer, such as the ATF13135 and the ATF13335, would seem work just as well, these are really the same transistor with different performance measurements. While two transistors may have exactly the same gain and noise figure at the desired operating frequency, often the impedances needed to maintain stability at other frequencies can be different. Thus, the "substitute" may oscillate, while the proper transistor would work just fine.

You can often substitute MMICs (monolithic microwave integrated circuits) for one another because they are designed to be stable and operate with the same input and output impedances (50 Ω).

The size of components used at microwaves can be critical — in some cases, a chip resistor 80 mils across is not a good substitute for one 60 mils across. Hopefully, the author of a construction project tells you which dimensions are critical, but you can't always count on this since the author may not know. It's not unusual for a person to spend years building just one prototype, so it's not surprising that the author might not have experience trying possible substitutions.

23.7.4 Capacitors for Microwave Construction

Ordinary ceramic chip capacitors cost a few cents, while microwave chip capacitors are more than a dollar each. All of them have parasitic resistance and inductance, but the microwave versions use lower-loss materials that make a difference at higher microwave frequencies. Ordinary ceramic capacitors work fine in non-critical applications like blocking and bypass capacitors, even up to 10 GHz. In critical areas like low-noise or power amplifiers, microwave capacitors are preferred.

For applications requiring high-Q for low loss, like the printed comb-line filter in **Figure 23.35**, two ordinary capacitors in parallel as shown have lower loss than one expensive microwave capacitor. By paralleling, the parasitic resistances are also paralleled, cutting the resistance in half. An additional advantage is that combinations can be chosen to yield non-standard capacitance values. We might apply this trick to really demanding applications, like high-power solid-state amplifiers, by paralleling microwave capacitors to reduce losses.

Figure 23.35 — Printed comb-line filter uses two chip capacitors in parallel to reduce loss.

23.7.5 Tuning and "No-Tune" Designs

Microwave construction does not always require tight tolerances and precision construction. A fair amount of error can often be tolerated if you are willing to tune your circuits, as you do at lower frequencies. This usually requires the use of variable components that can be expensive and tricky to adjust.

Proper design and construction techniques using high precision can result in a "no-tune" microwave design. To build one of these no-tune projects, all you need do is buy the parts and install them on the board. The circuit tuning has been precisely controlled by the board and component dimensions, so the project should work.

One tuning technique you can use with a microwave design, if you have the suitable test equipment, is to use bits of copper foil or EMI shielding tape as "stubs" to tune circuits. Solder these small bits of conductor into place at various points in the circuit to make reactances that can actually tune a circuit. After their position has been determined as part of the design, tuning is accomplished by removing or adding small amounts of conductor, or slightly changing the placement of the tuning stub. The size of the foil needed depends on your ability to determine changes in circuit performance, as well as the frequency of operation and the circuit board parameters. A precision setup that lets you see tiny changes allows you to use very small pieces of foil to get the best tuning possible.

23.8 Tools and Their Use

All electronic construction makes use of tools, from mechanical tools for chassis fabrication to the soldering tools used for circuit assembly. A good understanding of tools and their uses will enable you to perform most construction tasks.

While sophisticated and expensive tools often work better or more quickly than simple hand tools, with proper use simple hand tools can turn out a fine piece of equipment. **Table 23.4** lists tools indispensable for construction of electronic equipment. These tools can be used to perform nearly any construction task. Add tools to your collection from time to time, as finances permit.

23.8.1 Sources of Tools

Electronic-supply houses, mail-order/web stores, and most hardware stores carry the tools required to build or service amateur radio equipment. Bargains are available at ham flea markets or local neighborhood sales, but beware! Some flea-market bargains are really shoddy and won't work very well or last very long. Some used tools are offered for sale because the owner is not happy with their performance.

There is no substitute for quality! A high-quality tool, while a bit more expensive, will last a lifetime. Poor quality tools don't last long and often do a poor job even when brand new. You don't need to buy machinist-grade tools, but stay away from cheap tools; they are not the bargains they might appear to be.

CARE OF TOOLS

The proper care of tools is more than a matter of pride. Tools that have not been cared for properly will not last long or work well. Dull or broken tools can be safety hazards. Tools that are in good condition do the work for you; tools that are misused or dull are difficult to use.

Store tools in a dry place. Tools do not fit in with most living room decors, so they are often relegated to the basement or garage. Unfortunately, many basements or garages are not good places to store tools; dampness and dust are not good for tools. If your tools are stored in a damp place, use a dehumidifier. Sometimes you can minimize rust by keeping your tools lightly oiled, but this is a second-best solution. If you oil your tools, they may not rust, but you will end up covered in oil every time you use them. Wax or silicone spray is a better alternative.

Store tools neatly. A messy toolbox in which tools are strewn about haphazardly can be more than an inconvenience. You may waste a lot of time looking for the right tool, and sharp edges can be dulled or nicked by tools banging into each other in the bottom of the box. As the old adage says, a place for every tool, and every tool in its place. If you must search the workbench, garage, attic, and car to find the right screwdriver, you'll spend more time looking for tools than building projects.

SHARPENING

Many cutting tools can be sharpened. Send a tool that has been seriously dulled to a professional sharpening service. These services can sharpen saw blades, some files, drill bits, and most cutting blades. Touch up the edge of cutting tools with a whetstone to extend the time between sharpening.

Sharpen drill bits frequently to minimize the amount of material that must be

Table 23.4
Recommended Tools and Materials

Simple Hand Tools

Screwdrivers

Slotted, 3-in, 1/8-in blade
Slotted, 8-in, 1/8-in blade
Slotted, 3-in, 3/16-in blade
Slotted, stubby, 1/4-in blade
Slotted, 4-in, 1/4-in blade
Slotted, 6-in, 5/16-in blade
Phillips, 2½-in, #0 (pocket clip)
Phillips, 3-in, #1
Phillips, stubby, #2
Phillips, 4-in, #2
Long-shank screwdriver with holding clip on blade
Jeweler's set
Right-angle, slotted and Phillips

Pliers, Sockets and Wrenches

Long-nose pliers, 6- and 4-in
Diagonal cutters, 6- and 4-in
Channel-lock pliers, 6-in
Slip-joint pliers
Locking pliers (Vise Grip or equivalent)
Socket nut-driver set, 1/16- to 1/2-in
Set of socket wrenches for hex nuts
Allen (hex) wrench set
Wrench set
Adjustable wrenches, 6- and 10-in
Tweezers, regular and reverse-action
Retrieval tool/parts holder, flexible claw
Retrieval tool, magnetic

Cutting and Grinding Tools

File set consisting of flat, round, half-round, and triangular. Large and miniature types recommended
Burnishing tool
Wire strippers
Wire crimper
Hemostat, straight
Scissors
Tin shears, 10-in
Hacksaw and blades
Hand nibbling tool (for chassis-hole cutting)
Scratch awl or scriber (for marking metal)
Heavy-duty jackknife
Knife blade set (X-ACTO or equivalent)
Machine-screw taps, #4-40 through #10-32 thread
Socket punches, 1/2 in, 5/8 in, 3/4 in, 1 1/8 in, 1 1/4 in, and 1 1/2 in
Tapered reamer, T-handle, 1/2-in maximum width
Deburring tool

Miscellaneous Hand Tools

Combination square, 12-in, for layout work
Hammer, ball-peen, 12-oz head
Hammer, tack
Bench vise, 4-in jaws or larger
Center punch
Plastic alignment tools
Mirror, inspection
Flashlight, penlight and standard
Magnifying glass
Ruler or tape measure
Dental pick
Calipers
Brush, wire
Brush, soft
Small paintbrush
IC-puller tool

Hand-Powered Tools

Hand drill, 1/4-in chuck or larger
High-speed drill bits, #60 through 3/8-in diameter

Power Tools

Motor-driven emery wheel for grinding
Electric drill, hand-held
Drill press
Miniature electric motor tool (Dremel or equivalent) and accessory drill press

Soldering Tools and Supplies

Soldering station, adjustable temp, assorted tips
Soldering iron, 100 W or higher, 5/8-in tip
Soldering gun, 200 W or higher
Solder, 60/40, rosin core
Desoldering tool
Desoldering wick, 1/8-in and 1/4-in width
Liquid flux, pen or bottle
Isopropyl alcohol for flux removal

Safety

Safety glasses
Hearing protector, earphones or earplugs
Fire extinguisher
First-aid kit

Useful Materials

Medium-weight machine oil
Contact cleaner, liquid or spray can
RTV sealant or equivalent
Electrical tape, vinyl plastic
Sandpaper, assorted
Emery cloth
Steel wool, assorted
Cleaning pad, Scotchbrite or equivalent
Cleaners and degreasers
Contact lubricant
Sheet aluminum, solid and perforated, 16- or 18-gauge, for brackets and shielding.
Aluminum angle stock, 1/2 × 1/2-in and 1/4-in diameter round brass or aluminum rod (for shaft extensions)
Machine screws: Round-head and flat head, with nuts and lockwashers to fit. Most useful sizes: 4-40, 6-32 and 8-32, in lengths from 1/4-in to 1 1/2 in (Nickel-plated steel is satisfactory except in strong RF fields, where brass should be used.)
Bakelite, Lucite, polystyrene, and copper-clad PC-board scraps.
Soldering lugs, panel bearings, rubber grommets, terminal-lug wiring strips, varnished-cambric insulating tubing, heat-shrinkable tubing
Shielded and unshielded wire
Tinned bare wire, #22, #14 and #12
Enameled wire, #20 through #30

removed each time. Frequent sharpening also makes it easier to maintain the critical surface angles required for best cutting with least wear. Most inexpensive drill-bit sharpeners available for shop use do a poor job, either from the poor quality of the sharpening tool or the inexperience of the operator. Also, drills should be sharpened at different angles for different applications. Commercial sharpening services do a much better job.

INTENDED PURPOSE

Don't use tools for anything other than their intended purpose! If you use a pair of wire cutters to cut sheet metal, pliers as a vise, or a screwdriver as a pry bar, you ruin a good tool and sometimes the work piece as well. Although an experienced constructor can improvise with tools, most take pride in not abusing them. Having a wide variety of good tools at your disposal minimizes the problem of using the wrong tool for the job.

23.8.2 Tool Descriptions and Uses

Specific applications for tools are discussed throughout this chapter. Hand tools are used for so many different applications that they are discussed first, followed by some tips for proper use of power tools.

SCREWDRIVERS AND NUT DRIVERS

For construction or repair, you need to have an assortment of screwdrivers. Each blade size is designed to fit a specific range of screw head sizes. Using the wrong size blade usually damages the blade, the screw head, or both. You may also need stubby lengths to fit into tight spaces. Right-angle screwdrivers are inexpensive and can get into tight spaces that can't otherwise be reached.

Electric screwdrivers are relatively inexpensive and very useful, particularly for repetitive tasks. If you have a lot of screws to fasten, they can save a lot of time and effort. They come with a wide assortment of screwdriver and nut driver bits. An electric drill can also function as an electric screwdriver, although it may be heavy and overpowered for many applications.

Keep screwdriver blades in good condition. If a blade becomes broken or worn out, replace the screwdriver. A screwdriver only costs a few dollars; do not use one that is not in perfect condition. Save old screwdrivers to use as pry bars and levers, but use only good ones on screws. Filing a worn blade seldom gives good results.

Nut drivers, the complement to screwdrivers, are often much easier to use than wrenches, particularly for nuts smaller than 3/8 inch. They are also less damaging to the nut than any type of pliers, with a better grip on the nut. Nut drivers also minimize the chances of damage to front panels when tightening the nuts on control shafts. A set of interchangeable nut drivers with a shared handle is a very handy addition to the toolbox.

PLIERS AND LOCKING-GRIP PLIERS

Pliers and locking-grip pliers are used to hold or bend things. They are not wrenches! Using pliers to remove a nut or bolt usually damages the nut or the pliers. To remove a nut, use a wrench or nut driver. There is one exception to this rule of thumb: To remove a nut that is stripped too badly for a wrench, use a pair of pliers, locking-grip pliers, or a diagonal cutter to bite into the nut and start it turning. Reserve an old tool or one dedicated to just this purpose, which can damage the tool.

Pliers are not intended for heavy-duty applications. Use a metal brake to bend heavy metal; use a vise to hold a heavy component. If the pliers' jaws or teeth become worn, replace the tool.

There are many different kinds of fine pliers, usually called "needle-nose" pliers or something similar, that are particularly useful in electronics work. These are intended for light jobs, such as bending or holding wires or small work pieces. Two or three of these tools with different sizes of jaws will suffice for most jobs.

WIRE CUTTERS AND STRIPPERS

Wire cutters are primarily used to cut wires or component leads. The choice of blade style depends on the application. Diagonal blades or "dikes" are most often used to cut wire. Some delicate components can be damaged by cutting their leads with dikes because of the abrupt shock of the cut. Scissors or shears designed to cut wire should be used instead.

Specialized wire cutters are available to trim wires leads on circuit boards. These cutters are often called "flush cutters." Their cutting end is *not* designed to cut thicker wires. Use them *only* to clip smaller gauge wires, such as that on components used in circuits.

Wire strippers are available in manual and automatic styles. The manual strippers have a series of holes designed to remove insulation from a specific gauge of wire. Using the holes that are too big or too small will create nicks in the wire, which usually leads to the wire breaking at the nick. Automatic strippers grab and hold the wire for a consistent strip — some even judge the wire thickness automatically. If you strip a lot of wires, an automatic stripper may be worth the extra expense.

Wire strippers are handy, but with a little practice you can usually strip wires using a diagonal cutter or a knife. This is not the only use for a knife, so keep an assortment handy. Do not use wire cutters or strippers on anything other than wire! If you use a cutter to trim a protruding screw head or cut a hardened-steel spring, you will usually damage the blades.

Table 23.5
Numbered Drill Sizes

No.	*Diameter (Mils)*	*Will Clear Screw*	*Drilled for Tapping from Steel or Brass*
1	228.0	12-24	—
2	221.0	—	—
3	213.0	—	14-24
4	209.0	12-20	—
5	205.0	—	—
6	204.0	—	—
7	201.0	—	—
8	199.0	—	—
9	196.0	—	—
10	193.5	—	—
11	**191.0**	**10-24** **10-32**	—
12	189.0	—	—
13	185.0	—	—
14	182.0	—	—
15	180.0	—	—
16	177.0	—	12-24
17	173.0	—	—
18	169.5	—	—
19	166.0	8-32	12-20
20	161.0	—	—
21	**159.0**	—	**10-32**
22	157.0	—	—
23	154.0	—	—
24	152.0	—	—
25	**149.5**	—	**10-24**
26	147.0	—	—
27	144.0	—	—
28	**140.0**	**6-32**	—
29	**136.0**	—	**8-32**
30	128.5	—	—
31	120.0	—	—
32	116.0	—	—
33	**113.0**	**4-40**	—
34	111.0	—	—
35	110.0	—	—
36	**106.5**	—	**6-32**
37	104.0	—	—
38	101.5	—	—
39	**099.5**	**3-48**	—
40	098.0	—	—
41	096.0	—	—
42	093.5	—	—
43	**089.0**	—	**4-40**
44	**086.0**	**2-56**	—
45	082.0	—	—
46	081.0	—	—
47	078.5	—	3-48
48	076.0	—	—
49	073.0	—	—
50	**070.0**	—	**2-56**
51	067.0	—	—
52	063.5	—	—
53	059.5	—	—
54	055.0	—	—

FILES

Files are used for a wide range of tasks. In addition to enlarging holes and slots, they are used to remove burrs, shape metal, wood, or plastic, and clean some surfaces in preparation for soldering. Files are especially prone to damage from rust and moisture. Keep them in a dry place. The cutting edge of the blades can also become clogged with the material you are removing. Use file brushes (also called file cards) to keep files clean. Most files cannot be sharpened easily, so when the teeth become worn the file must be replaced.

DRILL BITS

Drill bits are made from carbon steel, high-speed steel, or carbide. Carbon steel is more common and is usually supplied unless a specific request is made for high-speed bits. Carbon steel drill bits cost less than high-speed or carbide types; they are sufficient for most equipment construction work. Carbide drill bits last much longer under heavy use. One disadvantage of carbide bits is that they are brittle and break easily, especially if you are using a handheld power drill. When drilling abrasive material such as fiberglass, the carbide bits last much longer than the steel bits.

Twist drills are available in several sizes, as listed in **Table 23.5**. Those listed in bold type are the most commonly used in construction of amateur equipment. You may not use all of the drills in a standard set, but it is nice to have a complete set on hand. You should also buy several spares of the more common sizes. Although Table 23.6 lists drills down to #54, the series extends to number #80. While the smaller sizes cannot usually be found in hardware stores or home improvement stores, they are commonly available through industrial tool suppliers and various sources on the internet.

A "step drill" consists of multiple drill diameters stacked as one bit, looking somewhat like a Christmas tree. A step drill also has the advantage of being able to drill a number of different standard size holes without having to change the bit. These bits are useful for drilling metal case material used in radio projects, since the fluted edge of the bit between sizes removes most burrs from the hole you are drilling. Use them carefully and slowly to take advantage of this feature.

SPECIALIZED TOOLS

Most constructors know how to use common tools, such as screwdrivers, wrenches, and hammers. Although specialized tools usually do a job that can be done with other tools, once the specialty tool is used you will wonder how you ever did the job without it! Let's discuss other tools that are not so common.

A hand nibbling tool is shown in **Figure 23.36A**. Use this tool to remove small "nibbles" of metal. It is easy to use; position the tool where you want to remove metal and squeeze the handle. The tool takes a small bite out of the metal. When you use a nibbler, be careful that you don't remove too much metal, clip the edge of a component mounted to the sheet metal, or grab a wire that is routed near the edge of a chassis. It is easy to remove metal but nearly impossible to put it back. Do it right the first time!

Deburring Tool

A deburring tool (see Figure 23.36B) is just the thing to remove the sharp edges left on a hole after drilling or punching operations. Position the tool over the hole and rotate it around the edge of the hole to remove burrs or rough edges. As an alternative, select a drill bit that is somewhat larger than the hole, position it over the hole, and spin it lightly to remove the burr. Be sure to deburr both sides of the hole.

Socket or Chassis Punches

Greenlee is the most widely known of the socket punch manufacturers. Most socket punches are round, but they do come in other shapes. To use one, drill a pilot hole large enough to clear the bolt that runs through the punch. Then, mount the punch as shown in Figure 23.36C, with the cutter on one side of the sheet metal and the socket on the other. Tighten the nut with a wrench until the cutter cuts all the way through the sheet metal. These punches are often sold in sets at a significant discount to the same punches purchased separately. Hand punches, that operate by squeezing, will also cut small holes by hand in light gauge sheet metal, printed circuit boards, and plastic.

Crimping Tools

The use of crimped connectors is common in the electronics industry. In many commercial and aerospace applications, soldered joints are no longer used. Hams have been reluctant to adapt crimped connections, largely due to mistrust of contacts that are not soldered, the use of cheap crimp connectors on consumer electronics, and the high cost of quality crimping tools or

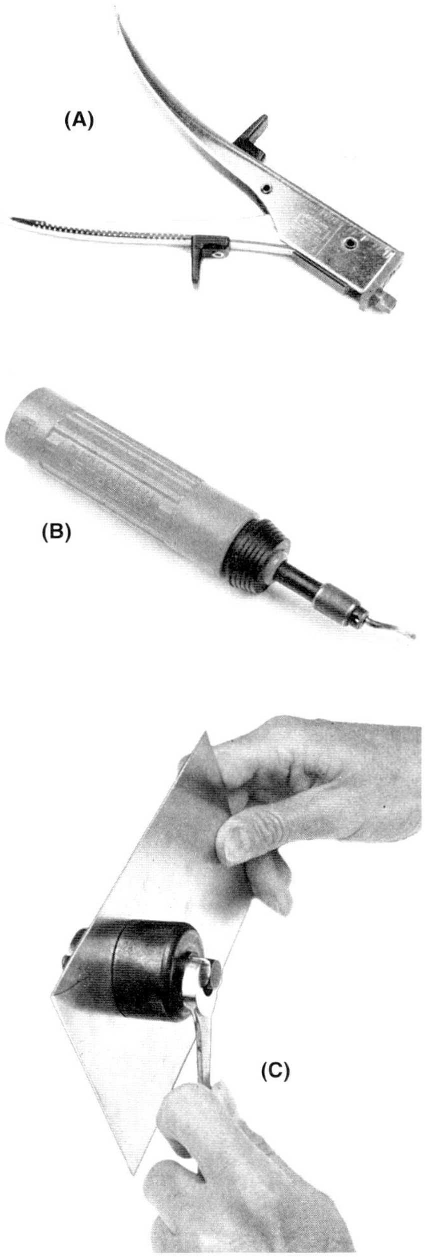

Figure 23.36 — At (A), a nibbling tool is used to remove small sections of sheet metal. At (B), a deburring tool is used to remove the burrs left after drilling a hole. At (C), a socket punch is used to punch a clean, round hole in sheet metal.

"crimpers." If high quality connectors and tools are used, the crimped connector will be as reliable a connection as a soldered one. The crimped connection is easier to make than a soldered one in most cases.

Crimped coaxial connectors are the most common crimped connector. MIL-spec, or equivalent, crimp connectors are available for the UHF, BNC, F, and N-series connectors. Power connectors, such as the Anderson PowerPoles and Molex connectors, are probably the second most commonly used crimped connections.

When purchasing a crimper, look for a ratcheting model with dies that are intended for the connectors you will be using. The common pliers-type crimper designed for household electrical terminals will have trouble crimping power connectors and is unsuitable for coaxial connectors. A good ratcheting crimper can be obtained for $50 to $100 with the necessary interchangeable dies. Large ratcheting crimpers suitable for the larger coaxial connectors can cost several hundred dollars. A good crimper and set of dies is an excellent investment for a club or group of like-minded hams.

Compression tools are now the most common method of installing Type F connectors. (See the **Transmission Lines** chapter.) BNC, TNC, and SMA connectors can now be installed using compression connectors and tools, as well. Quad-shielded RG-6 is the most common cable used for this purpose, and a compression tool makes the job very easy. A compression crimp tool works differently as it presses an internal sleeve into the body of the connector to make contact with the shield. Most compression tools are lower in cost than conventional ratcheting tools, and are available at most home improvement stores. You will need to purchase compression connectors designed to work with the tool and cable type you selected. Installation instructions are usually provided on the tool's package and with packages of connectors.

Useful Shop Materials

Small stocks of various materials are used when constructing electronics equipment. Most of these are available from hardware or radio supply stores. A representative list is shown at the end of Table 23.5.

Small parts such as machine screws, nuts, washers, and soldering lugs can be economically purchased in large quantities (it doesn't pay to buy more than a lifetime supply). For items you don't use often, many radio supply stores or hardware stores sell small quantities and assortments. Stainless steel hardware can be kept on hand for outdoor use.

Tuning and Alignment Tools

It's helpful to have an assortment of special tools for adjusting variable capacitors, inductors, and potentiometers. See the section Tuning and Alignment earlier in this chapter.

23.9 Mechanical Fabrication

Most projects end up in some sort of an enclosure, and most hams choose to purchase a ready-made chassis for small projects, but some projects require a custom enclosure. Even a ready-made chassis may require a fabricated sheet-metal shield or bracket, so it's good to learn something about sheet-metal and metal-fabrication techniques.

Most often, you can buy a suitable enclosure. These are sold by most electronics distributors. Select an enclosure that has plenty of room. A removable cover or front panel can make any future troubleshooting or modifications easy. A project enclosure should be strong enough to hold all of the components without bending or sagging; it should also be strong enough to stand up to expected use and abuse.

23.9.1 Cutting and Bending Sheet Metal

Enclosures, mounting brackets, and shields are usually made of sheet metal. Most sheet metal is sold in large sheets, 4 to 8 feet or larger. It must be cut to the size needed.

Most sheet metal is thin enough to cut with metal shears or a hacksaw. A jigsaw or band saw makes the task easier. If you use any kind of saw, select a blade that has teeth fine enough so that at least two teeth are in contact with the metal at all times.

If a metal sheet is too large to cut conveniently with a hacksaw, it can be scored and broken. Make scratches as deep as possible along the line of the cut on both sides of the sheet. Then, clamp it in a vise and work it back and forth until the sheet breaks at the line. Do not bend it too far before the break begins to weaken, or the edge of the sheet might bend. A pair of flat bars, slightly longer than the sheet being bent, make it easier to hold a sheet firmly in a vise. Use "C" clamps to keep the bars from spreading at the ends.

Smooth rough edges with a file or by sanding with a large piece of emery cloth or sandpaper wrapped around a flat block.

23.9.2 Finishing Aluminum

Give aluminum chassis, panels, and parts a sheen finish by treating them in a caustic bath. (See the information on chemical safety at the beginning of this chapter.) Use a plastic container to hold the solution and wear both safety goggles and protective clothing while treating aluminum. Ordinary household lye can be dissolved in water to make a bath solution. Follow the directions on the container. A strong solution will do the job more rapidly.

Stir the solution with a non-metal utensil until the lye crystals are completely dissolved. If the lye solution gets on your skin, wash with plenty of water. If you get any in your eyes, immediately rinse with plenty of clean, room-temperature water and seek medical help. It can also damage your clothing, so wear something old. Prepare sufficient solution to cover the piece completely. When the aluminum is immersed, a very pronounced bubbling takes place. Provide ventilation to disperse the escaping hydrogen gas. A half hour to two hours in the bath is sufficient, depending on the strength of the solution and the desired surface characteristics.

23.9.3 Chassis Working

With a few essential tools and proper procedure, building radio gear on a metal chassis is a relatively simple matter. Aluminum is better than steel, not only because it is a superior shielding material, but also because it is much easier to work and provides good chassis contact when used with secure fasteners.

Spend sufficient time planning a project to save trouble and energy later. The actual construction is much simpler when all details are worked out beforehand. Here we discuss a large chassis-and-cabinet project, such as a high-power amplifier. The techniques are applicable to small projects as well.

Cover the top of the chassis with a piece of wrapping paper or graph paper. Fold the edges down over the sides of the chassis and fasten them with adhesive tape. Place the front panel against the chassis front and draw a line there to indicate the chassis top edge.

Assemble the parts to be mounted on the chassis top and move them about to find a satisfactory arrangement. Consider that some will be mounted underneath the chassis and ensure that the two groups of components won't interfere with each other.

Place controls with shafts that extend through the cabinet first, and arrange them so that the knobs will form the desired pattern on the panel. Position the shafts perpendicular to the front chassis edge. Locate any partition shields and panel brackets next, then sockets and any other parts. Mark the mounting-hole centers of each part accurately on the paper. Watch out for capacitors with off-center shafts that do not line up with the mounting holes. Do not forget to mark the centers of socket holes and holes for wiring leads. Make the large center hole for a socket *before* the small mounting holes. Then use the socket itself as a template to mark the centers of the mounting holes. With all chassis holes marked, center-punch and drill each hole.

Next, mount on the chassis the capacitors and any other parts with shafts extending to the panel. Fasten the front panel to the chassis temporarily. Use a machinist's square to extend the line (vertical axis) of any control shaft to the chassis front and mark the location on the front panel at the chassis line. If the layout is complex, label each mark with an identifier. Also mark the back of the front panel with the locations of any holes in the chassis front that must go through the front panel. Remove the front panel.

MAKING ENCLOSURES WITH PC BOARD MATERIAL

Much tedious sheet-metal work can be eliminated by fabricating chassis and enclosures from copper-clad printed circuit board material. While it is manufactured in large sheets for industrial use, some hobby

Repurposing Obsolete Equipment Cabinets

One of the best deals at a hamfest or flea market might not be anything having to do with radio — yet! Obsolete industrial and commercial equipment may not be useable in the ham station, but their metal enclosures are top-quality. Purchasing one of these cabinets new would cost hundreds of dollars, but the surplus equipment is often sold very cheaply. Reusing the cabinets is easy, and stripping the electronics often results in a large collection of hardware and electronic components that can be re-used in ham equipment. The overall process is illustrated in a January/February 2017 *QEX* article by Scott Roleson, KC7CJ, included with this book's online material. Keep an eye out for these bargains everywhere, even at garage sales!

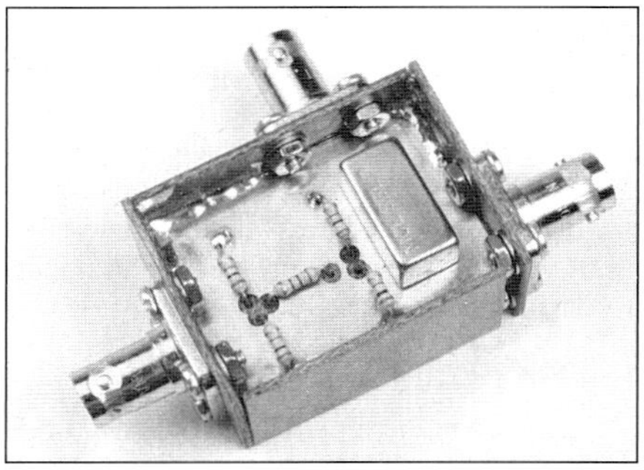

Figure 23.37 — An enclosure made entirely from PC-board stock.

electronics stores and surplus outlets market usable scraps at reasonable prices. PC board stock cuts easily with a small hacksaw. The nonmetallic base material isn't malleable, so it can't be bent. Corners are easily formed by holding two pieces at right angles and soldering the seam. This technique makes excellent RF-tight enclosures. If mechanical rigidity is required of a large copper-clad surface, solder stiffening ribs at right angles to the sheet.

Figure 23.37 shows the use of PC board stock to make a project enclosure. This enclosure was made by cutting the pieces to size, then soldering them together. Start by laying the bottom piece on a workbench, then putting one of the sides in place at a right angle. Tack-solder the second piece in two or three places, then start at one end and run a bead of solder down the entire seam. Use plenty of solder and plenty of heat. Continue with the rest of the pieces until all but the top cover is in place.

In most cases, it is better to drill all needed holes in advance. It can sometimes be difficult to drill holes after the enclosure is soldered together.

You can use this technique to build enclosures, subassemblies, or shields. This technique is easy with practice; hone your skills on a few scrap pieces of PC board stock.

23.9.4 Drilling Techniques

Before drilling holes in metal with a hand drill, indent the hole centers with a center punch. This prevents the drill bit from "walking" away from the center when starting the hole. Predrill holes greater than 1/2 inch in diameter with a smaller bit that is large enough to contain the flat spot at the larger bit's tip. When the metal being drilled is thinner than the depth of the drill-bit tip, back up the metal with a wood block to smooth the drilling process.

The chuck on the common hand drill is limited to 3/8 inch bits. Some bits are much larger, with a 3/8 inch shank. If necessary, enlarge holes with a reamer or round file. For very large or odd-shaped holes, drill a series of closely spaced small holes just inside of the desired opening. Cut the metal remaining between the holes with a cold chisel and file or grind the hole to its finished shape. A nibbling tool also works well for such holes.

Use socket-hole punches to make socket holes and other large holes in an aluminum chassis. Drill a guide hole for the punch center bolt, assemble the punch with the bolt through the guide hole, and tighten the bolt to cut the desired hole. Oil the threads of the bolt occasionally.

Cut large circular holes in steel panels or chassis with an adjustable circle cutter ("fly cutter") in a drill press at low speed. Occasionally apply machine oil to the cutting groove to speed the job. Test the cutter's diameter setting by cutting a block of wood or scrap material first.

Remove burrs or rough edges that result from drilling or cutting with a burr-remover, round or half-round file, a sharp knife, or a chisel. Keep an old chisel sharpened and available for this purpose.

23.9.5 Construction Notes

If a control shaft must be extended or insulated, a flexible shaft coupling with adequate insulation should be used. Satisfactory support for the shaft extension, as well as electrical contact for safety, can be provided by means of a metal panel bushing made for the purpose. These can be obtained singly for use with existing shafts, or they can be bought with a captive extension shaft included. In either case the panel bushing gives a solid feel to the control. The use of fiber washers between ceramic insulation and metal brackets, screws, or nuts will prevent the ceramic parts from breaking.

PAINTING

Painting is an art, but, like most arts, successful techniques are based on skills that can be learned. The surfaces to be painted must be clean to ensure that the paint will adhere properly. In most cases, you can wash the item to be painted with soap, water, and a mild scrub brush, then rinse thoroughly. When it is dry, it is ready for painting. Avoid touching it with your bare hands after it has been cleaned. Your skin oils will interfere with paint adhesion. Wear rubber or clean cotton gloves.

Sheet metal can be prepared for painting by abrading the surface with medium-grade sandpaper, making certain the strokes are applied in the same direction (not circular or random). This process will create tiny grooves on the otherwise smooth surface. As a result, paint or lacquer will adhere well. On aluminum, one or two coats of zinc chromate primer applied before the finish paint will ensure good adhesion.

Keep work areas clean and the air free of dust. Any loose dirt or dust particles will probably find their way onto a freshly painted project. Even water-based paints produce some fumes, so properly ventilate work areas.

Select paint suitable to the task. Some paints are best for metal, others for wood, and so on. Some dry quickly, with no fumes; others dry slowly and need to be thoroughly ventilated. You may want to select rust-preventative paint for metal surfaces that might be subjected to high moisture or salts.

Most metal surfaces are painted with some sort of spray, either from a spray gun or from spray cans of paint. Either way, follow the manufacturer's instructions for a high-quality job.

PANEL LAYOUT AND LABELING

There are many ways to layout and label a panel. Some builders don't label any controls or jacks, relying on memory to figure out what does what. Others use a marking pen to label controls and inputs. Decals and dry transfers have long been a staple of home brewing. Label makers that print on clear or colored tape are used by many.

With modern computers and available software, it is not hard to lay out professional looking panels. One can use a standard drawing program for the layout. The grids available on these drawing programs are sufficient to make sure that everything is lined up squarely. If the panel label is laid out before the panel is drilled for controls, a copy of the label can be used as a drill template.

Computer-aided design (CAD) programs can also be used to lay out and label panels, although they can have a steep learning curve and may be overkill for many applications.

Surplus meters often find their way into projects. Unfortunately, the meter faces usually do not have an appropriate scale for the project at hand. Relabeling meters has long been a mainstay to make home brew gear look professional. With the advent of computers this job has been made very easy. A software package, *MeterBasic*, by Jim Tonne, W4ENE, available with the online content, is very easy to use and results in professional looking meters that indicate exactly what you want them to indicate.

23.10 3D Printing

This section is extracted from the in-depth article "Introduction to 3D Printing for Hams" by John Portune, W6NBC, in the online material for this book. Detailed instructions on how to do 3D printing are not the intent of this section. 3D printing is too broad a topic. The object of this section is to define basic terms and concepts, present the fundamentals of 3D printing, and discuss common materials used for ham applications. Several articles with examples of 3D printed items for amateur radio use are included in the online information, and a list of references is included at the end of this section.

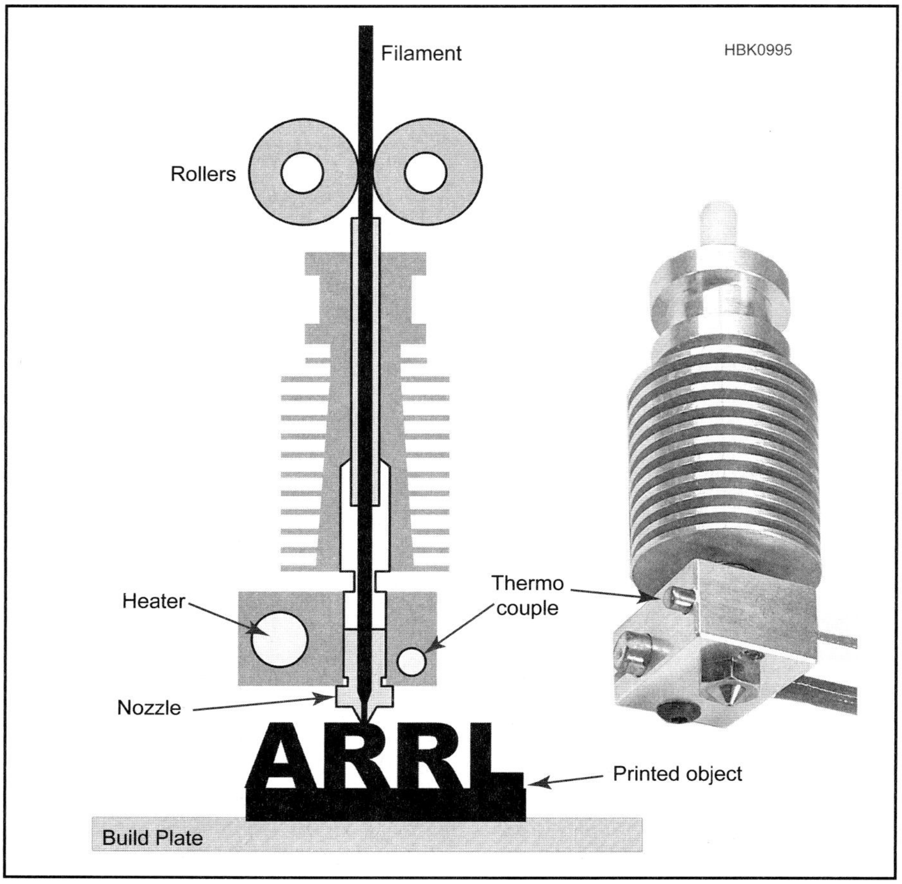

Figure 23.38 — A typical FDM extruder.

23.10.1 What is 3D Printing?

Fabricating one-piece physical objects has traditionally been a *subtractive* process. That is, the builder starts with a solid volume of material and gradually removes or subtracts material until it is in the final form. An *additive* process, however, builds the object by depositing or *printing* a sequence of layers of material, usually as 2D layers called *slices*, each adhering to the previous layer, until the entire 3D object has been constructed or *printed.*

For the home hobbyist, the most widely used method is *Fused Deposition Modeling (FDM)*, also known as *Fused Filament Fabrication (FFF)*. It is affordable, easy to learn, and offers high print quality. An FDM 3D printer creates an object by depositing fine strings of plastic in layers. The strings are created by heating the plastic until it is soft enough to be forced out of or *extruded* from a nozzle. The softened plastic immediately sticks (*fuses*) to the previously deposited layers below.

23.10.2 Understanding the FDM 3D Printer

THE EXTRUDER

Figure 23.38 shows a typical home 3D printer extruder. Plastic filament (commonly 1.75 mm in diameter) is pushed into the top (cold end) of the extruder by knurled rollers at a controlled rate. At the bottom (hot end), a temperature-controlled heating block softens the filament to make it sticky. Then, under pressure from more incoming filament, the soft plastic extrudes through a tiny hole in the extruder's nozzle to fuse to the 3D object forming on the build plate below.

A typical extruder nozzle diameter is 0.4 mm, slightly less than 1/64 inch. Nozzle diameter and the material from which the

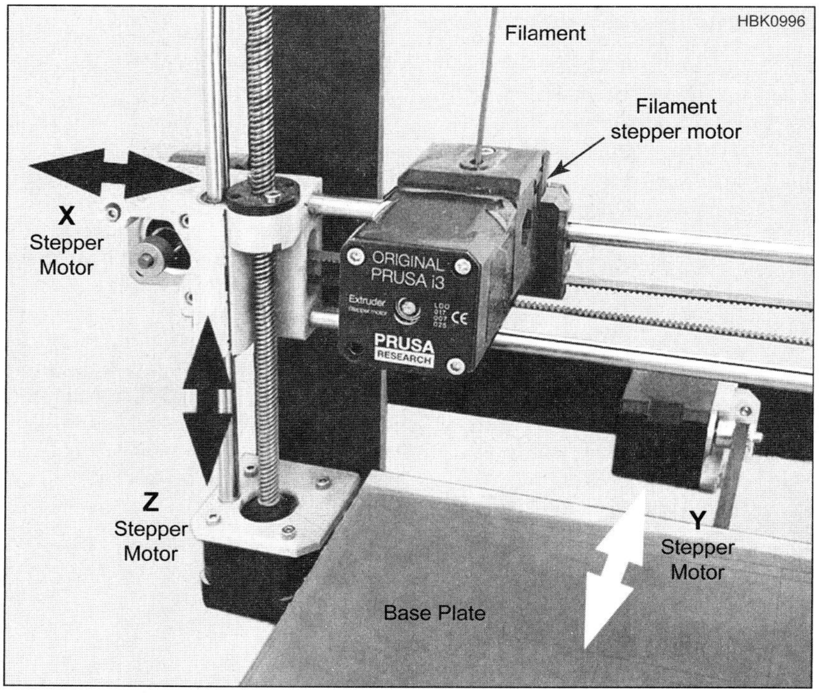

Figure 23.39 — Simplified pictorial of the X, Y, and Z stepper motors and extruder (filament) stepper motor of the popular Prusa i3 MK3S 3D printer.

nozzle is made can be changed to suit the type of plastic and the detail of the print. Fineness of the extruded string is also affected by the speed with which the bead is extruded.

THE BUILD PLATE

Most *build plates* have an aluminum base, but the extruded plastic is not deposited directly on the base. Various surfaces are added, such as a flat glass plate, a self-adhesive plastic film, or one or two types of tape. Further, an adhesive is commonly applied to make sure the first layer stays in place. A well-adhered first layer is a critical factor in obtaining a high-quality print and in avoiding *failed,* or *misprints*, in which the object is distorted, extruded plastic fails to fuse properly, and so on.

The build plate on many FDM printers is heated to facilitate adhesion of the first layer and to reduce warping of the object during the print. It also increases the variety of different plastics that can be used and reduces the number of failed prints.

COMPUTER NUMERIC CONTROL (CNC)

The motions of the 3D printer's extruder and values of other parameters, such as temperature and extrusion rates, are controlled by a microcomputer from a programmed sequence of positions and values in numeric form. This physical/electronic architecture is called *Computer Numeric Control* (*CNC*).

To be able to print in three dimensions (X, Y, and Z), the printer must be able to rapidly move the extruder from side to side (X), front to back (Y), and up and down (Z). The extruder typically moves only in the X and Z dimensions; the build plate moves in the Y direction. (See **Figure 23.39**.)

A 3D printer's X, Y, and Z motion is generally produced by *stepper motors*. Stepper motors, or just "steppers," are a specialized type of motor designed to move in tiny precise increments. Most 3D printers use three steppers to provide the X, Y, and Z motions. A fourth stepper feeds the plastic filament into the extruder at a controlled rate. Notched or toothed belts, pulleys, and threaded lead screws transfer the controlled rotations of the stepper motors to the build plate, the extruder, and the rollers feeding filament into the extruder.

G-CODE

The sequence of CNC motions and a number of other machine parameter values are encoded as text instructions, called *G-code*. The G-code file is loaded into the 3D printer's memory for each print. It is the complete set of instructions to the printer for printing a finished object. G-code for most home 3D printers today is a customized subset of industry-standard G-code for CNC machines and is controlled by the RS-274 standard. The complete set of industry standard G-code commands can be found at **en.wikipedia.org/wiki/G-code.**

23.10.3 3D Object Design Software

In common practice, the G-code file is created in two steps by separate software programs. The first is 3D design software, which creates a computer model of an object to be printed. The result is saved as an *STL* file or an *OBJ* file.

The STL file format (*Standard Triangle* or *Tessellation Language*) is the most common format for 3D printing models. STL is the native file format of CAD software created by 3D Systems. It uses the X, Y, Z coordinates of a series of interconnected triangles (tessellations) to represent the surfaces of the model. All modern CAD software programs will export an STL file. The OBJ format (short for "object") supports multicolor printing and is used for printing more complex objects. You can find out more about STL and OBJ formats at **3dinsider.com/stl-vs-obj**. Also see **all3dp.com/1/obj-file-format-3d-printing-cad/** for more in-depth discussion and a section on resources for obtaining the files. A list of websites that offer free STL files is included in the online information.

The second program is a called a *slicer* because it converts the 3D object model into the layers or slices needed for 3D printing. Unlike the generic 3D object design software, the slicer is specific to each model of printer. The slicer converts the STL file, or other native 3D design output file, into the G-code required by the individual printer. Being printer specific, the slicer's output G-code is not suitable for sharing between different brands or models of printers.

3D DESIGN SOFTWARE — TINKERCAD

Easy-to-use, free hobbyist-level 3D object design software, suitable for most ham projects, is readily available on the internet. As an illustration of a suitable program for ham 3D printing newcomers, we will examine the *Tinkercad* package (**see Figure 23.40**). It is perhaps the most widely used 3D object design program for beginners. There are many tutorials for it on the internet. It was developed by AutoDesk and is cloud-based. To obtain a copy, create a free account at **www.tinkercad.com**. Finished projects may be saved on the *Tinkercad* cloud or a local computer for later use or modification, either as a project file or an STL file. 3D printed ham project objects are typically combinations of geometric shapes called *primitives* — rectangular solids, cylinders, spheres, cones, pyramids, and so on — that are combined to make complex single objects. One can accurately size, scale, rotate, and align objects in *Tinkercad* in units of millimeters or inches. All of the primitive objects can be subtracted from other objects, for example, to make a hole. The workspace is also easily zoomed and rotated for easy viewing.

SLICER SOFTWARE

The second step in creating the G-code file for the 3D printer is performed by the slicer software. It converts the X, Y, Z coordinates of the tessellated triangles in an STL file, called a "mesh," into the G-code instructions that the printer needs for each print. Configuration settings for the 3D printer can also be made in the slicer program and included in the G-code file. It is highly advisable at first to stay with the de-

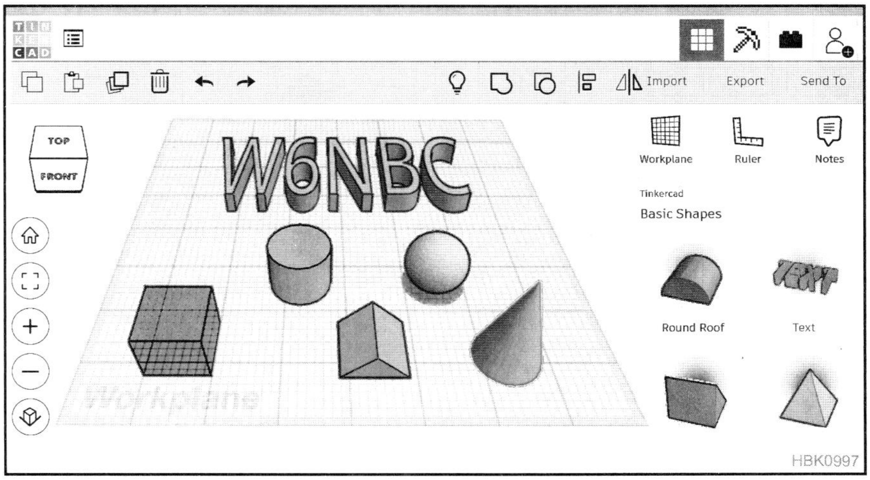

Figure 23.40 — Workspace of *Tinkercad* showing "primitive" geometric shapes and text used to create a 3D printer object and an STL file.

fault settings of the slicer. After printing a test object, the presets of the slicer can be changed to improve the final print quality or to correct errors.

Configuration settings for the slicer software itself can also be changed to suit a particular material or object. Many free test object STL files are available on the web to help make adjustments. (See the directory of STL file websites in the online article.) A simple test object such as a cube is recommended for initial tests. For more advanced testing and calibration, the best-known test object is a small tugboat called Benchy, available from **www.3dbenchy.com**. Other test objects are available online for specific needs, such as minimizing stringing, which is a common cause of poor prints and printer crashes.

NORMAL SLICER ADJUSTMENTS

There are a number of required inputs to the slicer for every object. The working screen of an example slicer program is shown in **Figure 23.41**. From the top of the selection area at the right of the window:

- PRINT SETTINGS: The user may choose nozzle diameters and print quality modes. This sets the size of the plastic bead to be extruded and the thickness of the layers or slices.
- FILAMENT (Plastic Type): Slicers contain a list of plastic filaments with the correct temperature settings for the extruder, the heated build plate, and a number of other plastic and printer specific settings.
- SUPPORTS: The slicer can be instructed to add *supports* that can be removed after the print is finished. Printers can *bridge* over small open horizontal spaces, typically on the order of 1/4 inch (6 mm).
- IN-FILL DENSITY: Solid shells are printed on the surfaces of the object, but a honeycomb-like grid work is printed internally. The density of the honeycomb is set for each object from 0% (hollow) to 100% (solid).
- BRIM: For objects that will have only a small surface touching the build plate, set the slicer to generate a *brim* or *skirt*. It is several rows of plastic bead printed around the periphery of, and in light contact with, the first layer of the object.
- COLOR AND FILAMENT CHANGE: The slicer for a single-extruder printer can be set to halt at a particular slice for the filament to be changed. Afterward, the print seamlessly resumes.

At the top of the workspace, on the left side of the window, are controls to add, delete, copy and paste, and to arrange multiple objects on the build plate. At the left edge of the workspace are controls to change the position and rotation of the object, set the object on the build plate, and make limited modification to the object.

23.10.4 Choosing a Printer

The most worthwhile guide to choosing a printer is the internet. YouTube is an exceptionally valuable resource, both for learning 3D printing and for evaluating printers, accessories, and materials. Spend time searching on 3D printing in general, on manufacturer's websites, and in 3D printing magazines and blogs. Ham friends and radio club members who already own 3D printers are also an excellent resource. Magazines often publish lists of the top performing printers at that time. Also look at the comments and reviews from readers. (Wikipedia publishes a list of "notable" printer manufacturers at **en.wikipedia.org/wiki/List_of_3D_printer_manufacturers**.)

If there are a large number of videos on the printer and design software you are considering, and if the video makers find these reliable and easy to use, they are probably acceptable choices. The same is true for 3D design software and for the reputation of printer manufacturers.

Generally, medium-range domestic printers, currently in the $500 to $1000 range, are the best choice for most ham buyers. At the low end of the price range, there are a rapidly growing number of 3D printers, some in the low-$100 range. Some are excellent, and new models appear frequently. Some companies, including medium-range manufacturers, offer lower-cost and junior-model printers.

It is important to consider the largest object that can be printed. In 3D printing this is called the *print volume*. Low-priced printers generally only print small objects. Other important features, often not found on low-priced models, are a heated build plate and automatic build plate leveling. Most experienced 3D printers consider these two features to be necessities.

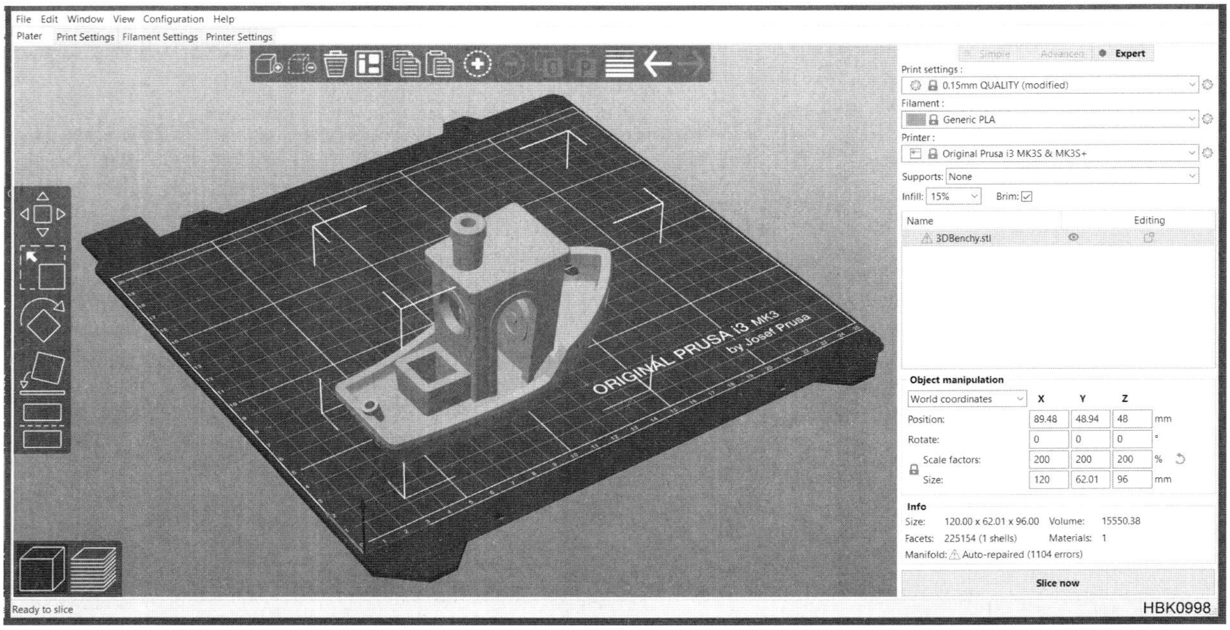

Figure 23.41 — Workspace of the slicer program for a Prusa i3 MK3S printer with a single "Benchy" test object loaded.

Table 23.6

Printing Properties of 3D Filament Plastics

	Printability	*$ / kg*	*Strength*	*Rigidity*	*Durability*	*Max. Temp.*
PLA	9	10-40	65	7.5	4	52C/126F
PETG	9	20-60	53	5	8	73C/163F
ABS	8	10-40	40	5	8	98C/208F
Nylon	8	25-65	62	5	10	95C/203F
Carbon fiber filled	8	30-80	46	10	3	52C/126F
Polypropylene	8	60-120	32	4	9	100C/212F
Wood filled	8	25-50	46	8	3	52C/126F
ASA	7	35-40	55	5	10	95C/203F
Metal filled	7	50-120	25	10	4	52C/126F
Flexible	6	30-70	35	1	9	65C/149F
HIPS	6	25-35	32	10	7	100C/212F
Polycarbonate	6	40-75	32	6	10	120C/248F
PVA	5	40-110	78	3	7	75C/167F

portant to consider both the physical and the electrical properties in selecting a 3D printing filament for a ham radio project.

Table 23.8 is organized from lowest to highest dissipation factor. One or two plastics not commonly used for 3D printing are included to illustrate the general use of dissipation factor in selecting plastics for electronic applications.

23.10.5 Plastic Filament for 3D Printing

Essential to the hobbyist entering 3D printing is a familiarity with the commonly available plastic filaments. The best choice for each project will depend on the nature of the project. **Table 23.6** is a quick guide, sorted by ease of printability. The two filaments that stand out for printability are PLA and PETG. This accounts for their popularity, and they are good choices for beginners.

Websites such as **www.simplify3d.com/support/materials-guide/properties-table** and **all3dp.com/1/3d-printer-filament-types-3d-printing-3d-filament** also provide a lot of information about the properties of different plastics for 3D printing.

RESISTANCE TO ULTRAVIOLET LIGHT

Of particular concern to hams is the plastic's resistance to the ultraviolet light (UV) in sunlight. Many ham 3D printed objects will be used outdoors and experience long-term exposure to sunlight. Without exception, all sources agree that the best plastic for UV resistance is ASA (Acrylonitrile Styrene Acrylate). Acrylics are virtually impervious to UV. ASA is somewhat more difficult to print with and is more expensive, but for ham projects involving long-term exposure to sunlight, it is the best choice.

Table 23.7 gives a general picture of the UV properties of the common 3D printing materials. Another abbreviated summary for other common 3D plastics is available at **3dprinterly.com/best-filament-for-outdoor-use-is-pla-suitable**. PLA, ABS and PETG are discussed in more detail in the online article.

RF POWER HANDLING CAPACITY OF PLASTICS

Another important factor in choosing a 3D printing plastic for ham projects is how well a plastic will work in a high RF environment. The relevant property is called *dissipation factor* (*DF*) or *loss tangent* (*tan* δ). It is a measure of how much the materials dissipate RF as heat. The higher the DF, the lower an RF field strength the plastic will tolerate before melting. Materials with a high DF are referred to as "lossy." It is important to consider both the physical and the electrical properties in selecting a 3D printing filament for a ham radio project.

23.10.6 The Build Plate

BUILD PLATE SURFACES

Most printers have a base aluminum build plate. Printing is not done directly on the base build plate. An additional surface is added to the base. Common surface materials include borosilicate glass plates, polyetherimide (PEI) sheets, and Kapton tape, all of which are available online. Blue painter's tape, a common hardware store item, is also used.

A removable flexible steel sheet is another common added print surface. A self-adhesive PEI plastic sheet is then typically applied to the steel. A removable flexible sheet makes the removal of printed object easy: a small flex and the printed parts pop off. Tightly adhered objects can be difficult to remove from a fixed build plate, increasing the risk of damage to the build plate.

BUILD PLATE ADHESIVES

In most cases it will also be necessary to apply an adhesive to the build plate so that the first layer of plastic sticks firmly. Problems in the first layer are the cause of a high percentage of failed prints. Poor adhesion can also allow objects to warp, especially with some plastics, such as ABS. A table of common adhesives, and the plastics they are used with, is included in the online article.

The two most-used adhesives are glue stick (PVA) and hair spray (PVP). Both are water soluble. Before applying any

Table 23.7

UV Resistance of 3D Printing Plastics

Material	*UV Resistance*
ABS	Fair
Acrylic	Good
ASA	Excellent
PETG	Very Good
PLA	Poor
Polycarbonate	Fair
Polyethylene	Poor* – Fair
Polypropylene	Poor* – Fair
Polystyrene	Poor – Fair
PTFE	Very Good
PVC (plasticized)	Fair

*Basic UV resistance can be improved with additives. Part of this table is from "Using Plastics for Dielectrics," Robert Zavrel, W7SX, *QEX*, January/February 2021

Table 23.8

Dissipation Factor of Common Plastics

Plastic	*Dissipation Factor (tan δ)*	
	1 kHz	*1 MHz*
PTFE (polytetrafluoroethylene) Teflon	<0.0001	<0.0001
LDPE (low density polyethylene)	0.0003	0.0003
PP (polypropylene)	0.0003	0.0003
PLA (polylactic acid)	0.0003	0.0004
HDPE (high density polyethylene)	0.0005	0.0005
PS (polystyrene) Styron	0.0005	0.0005
PVC (polyvinylchloride)	0.013	0.006
PC (polycarbonate) Lexan	0.0015	0.01
PI (polyimide)	0.0025	0.01
PET (polyethyleneterephthalate) Mylar	0.005	0.016
PMMA (polymethylmethacrylate) Plexiglas, Acrylic [ASA]	0.04	0.03
Nylon6 (polycaprolactam)	0.016	0.036

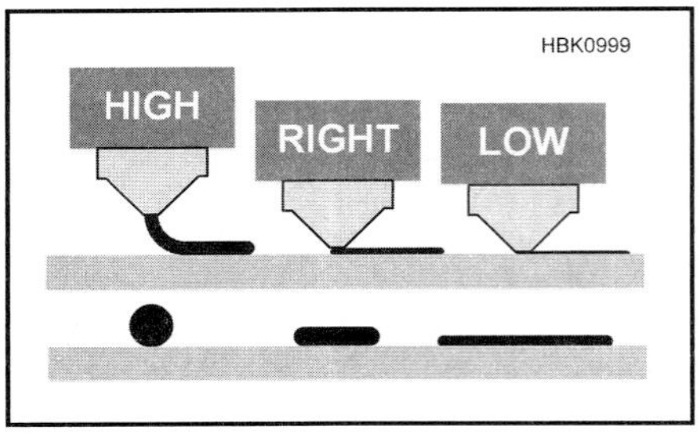

Figure 23.42 — "Squishing" of the extruding plastic bead: (left) No squish; (center) correct amount of squish; (right) too much squish (not-to-scale).

adhesive, clean the surface with isopropyl alcohol (most common) or hot water. Occasionally, also clean the surface with acetone. Due to ABS plastic's tendency to shrink as it cools, a special adhesive is highly recommended, variously called ABS slurry or ABS juice. (See the full article for more on this.)

EXTRUDER HEIGHT ADJUSTMENT

The instruction manual for every printer includes the steps to adjust the working height of the extruder during a print. However, before it can be set, the mechanical geometry of the printer must have been carefully aligned. This is called the *X, Y, Z calibration*. Follow the manufacturer's calibration instructions meticulously. This cannot be overemphasized.

The next adjustment that must be correct before extruder height can be set is *bed leveling*. It is only a manual adjustment on lower-cost printers; it is automatic on others. Many experienced 3D hobbyists consider automatic bed leveling essential, and it is highly recommended.

To visualize why extruder height adjustment is critical, see **Figure 23.42**. To cause the soft plastic to positively adhere to the build plate and to subsequently fuse to a previously extruded layer of plastic, the tip of the extruder's nozzle must flatten the bead, called *squish*. The gap between nozzle tip and the bed or object must be roughly half the height of the diameter of the extruding plastic bead as illustrated in the figure. Recognizing a properly squished bead is an essential 3D printing skill and requires close attention. Become very familiar with first layer calibration of extruder height.

23.10.7 A Typical Print Sequence

The following basic steps must be performed for every print. This is a reference for the newcomer and highlights some tips for all experienced 3D printers.

1. Load the object's G-code file.
2. Let the printer go through its start-up sequence.
3. Clean the build plate surface.
4. Check for any particles or scraps under the added surface, if any.
5. Clean the extruder tip.
6. Select the G-code file to print.
7. Check the *purge strip*.
8. Check the skirt or brim, if any.
9. Confirm the printer proceeds to the second layer.
10. Remove the printed part.
11. Clean up the printed part.

23.10.8 3D Printing References

"3D Printing," Wikipedia, **en.wikipedia.org/wiki/3D_printing**.

"Introduction to 3D Printing," Ham Radio Workbench (podcast, list of additional online resources) **www.hamradioworkbench.com/podcast/introduction-to-3d-printing**, 2016.

"Introduction to 3D Printing," Instructables (**www.instructables.com/3D-Printing-1**), 2012.

Pingree, J. WB2TVB, "Get Started with 3D Printing," *QEX*, July/Aug. 2019, pp. 16–19.

Wade, P., W1GHZ, "3D-Printed Antennas," Microwavelengths, *QST*, Jan. 2019, pp. 67–68.

Contents

Chapter 24 — Online Content

Articles

- A Look at Gasoline Powered Inverter Generators by Bob Allison, WB1GCM
- Field Day Towers — Doing it Right by Don Daso, K4ZA and Ward Silver, NØAX

Chapter 24

Assembling a Station

Although many hams might not build a major project, such as a transmitter, receiver, or amplifier, they do have to assemble the various components into a working station. There are many benefits to be derived from assembling a safe, comfortable, easy-to-operate collection of radio gear, whether the station is at home, in the car, or in a field. The section on Portable Stations has been brought up to date by Matt Heere, N3NWV, addressing the operating style's growing popularity. Alan Applegate, KØBG, has expanded the section on Mobile Stations to cover electric vehicles. Configuring and operating a station by remote control is now common, and Ken Norris, KK9N, keeps the Remote Stations section up to date. Brian Daly, WB7OML, describes what needs to be done to make a station ready to handle emergency situations. Topics such as station location, grounding and bonding, power sources, station layout, and cable routing are also covered.

Aids for Hams with Disabilities

A station used by an amateur with physical disabilities or sensory impairments may require adapted equipment or particular layout considerations. The station may be highly customized to meet the operator's needs or just require a bit of "tweaking." Each situation must be approached individually based on the operator's particular needs. However, many types of situations have already been encountered and worked through by others, potentially eliminating the need to start from scratch.

An excellent resource is the Courage Kenny Handiham Program. Part of the Courage Kenny Rehabilitation Institute, this program provides a number of services to hams (and aspiring hams) with disabilities. You'll find a wealth of useful information on their comprehensive website, **www.handiham.org**.

24.1 Fixed Stations

Regardless of the type of installation you are attempting, good planning greatly increases your chances of success. Take the time to think the project all the way through, consider alternatives, and make rough measurements and sketches during your planning and through the process. You save headaches and time by avoiding "shortcuts." What might seem to save time now may come back to haunt you with extra work when you could be enjoying your station.

One of the first considerations should be to determine what type of operating you intend to do. While you do not want to strictly limit your options later, you need to consider what you want to do, how much you want to spend, and the amount of room you have to work with. For example, there is a big difference between a casual operating position and a "big gun" contest station.

24.1.1 Selecting a Location

Selecting the right location for your station is the first, and perhaps the most important, step in assembling a safe, comfortable, and convenient station. The exact location depends on the type of home you have and how much space you can devote to your station. Fortunate amateurs have a spare room to devote to housing the station; some may even have a separate building for their exclusive use. Most must make do with a spot in the cellar or attic, or a corner of the living room is pressed into service.

Examine the possibilities from several angles. A station should be comfortable; odds are good that you'll be spending a lot of time there over the years. Some unfinished basements are damp and drafty — not an ideal environment for several hours of leisurely hamming. Attics have their drawbacks, too; they can be stifling during warmer months. If possible, locate your station away from the heavy traffic areas of your home. Operation of your station should not interfere with family life. A night of chasing DX on 80 meters may be exciting to you, but the other members of your household may not share your enthusiasm.

Figure 24.1 — Scott Redd, KØDQ, operated this station at WW1WW in pursuit of top scores in DX contests. Notice that the equipment on the operating desk is laid out logically and comfortably for long periods "in the chair." [Woody Beckford, WW1WW, photo]

Keep in mind that you must connect your station to the outside world. The location you choose should be convenient to a good power source and an adequate ground. If you use a computer, you may need access to the internet. There should be a fairly direct route to the outside for running antenna feed lines, rotator control cables, and the like.

Although most homes do not have an "ideal" space meeting all requirements, the right location for you will be obvious after you scout around. The amateurs

Figure 24.2 — Top RTTY contest operator Don Hill, AA5AU operates with low power from his effective station. RTTY operation emphasizes use of the computer mouse so Don's desk has lots of room for his "mouse hand." [Shay Hill, photo]

Figure 24.3 — Spreading out horizontally, John Sluymer, VE3EJ, has arranged his effective contest and DXing station to keep all of the controls at the same level for easy adjustment. [John Sluymer, VE3EJ, photo]

whose stations are depicted in **Figures 24.1** through **24.3** all have found the right spot for them. Weigh the trade-offs and decide which features you can do without and which are necessary for your style of operation. If possible, pick an area large enough for future expansion.

24.1.2 Station Grounding and Bonding

As discussed in the **Safe Practices** chapter, the station's ground system is very important and serves three purposes: ac safety, lightning protection, and RF management. With some attention to detail and planning, your station can be designed to satisfy all three needs from the start. This section is limited to providing general guidelines for grounding and bonding. Additional information is available from the ARRL's website in the Technology section under the Radio Technology Topics page's group of Safety articles and resources. The ARRL's *Grounding and Bonding for the Radio Amateur* collects a great deal of information into a single book for convenient reference with guidelines for station builders.

AC SAFETY

The **Safe Practices** chapter covers the basic elements of how ac power is configured in residences, including the requirements for grounding and bonding of equipment and earth connections. Make sure your station wiring complies with all ac wiring requirements.

Before building a station or modifying an existing station, start by making sure your existing ac wiring is in good shape. If you are new to ac wiring or uncomfortable around it, get a professional to do the job, or have an experienced person show you how to do it. When treated with respect and following simple safety rules, working on ac wiring is safe. This is a good time to read the safety section of any wiring handbook or guide, even if you *think* you know what you're doing.

All branch circuits must have a ground conductor that runs back to the service panel. The ground conductor should be at least as large as the current-carrying conductors so it can handle the same current overloads without causing a fire hazard. The branch circuit ground conductor is almost always one wire in a multi-conductor non-metallic (NM) cable, usually referred to by its trade name, Romex. The wire is usually bare copper inside the cable sheath, but it can also have insulation that is green or green with yellow stripes. Make sure the ac service entry ground rod or other earth connection is present, that all wiring to it is sufficiently heavy, and that all connecting hardware is in good condition.

Along with the residence's ac service entry ground rod, additional ground rods are often installed for panels where feed lines enter a building, for antennas and towers, and a direct connection to the station's bonding bus. All of these external earth connections must be bonded together using heavy wire or strap.

Although it has been long recommended in amateur radio articles and books, do not use a metal cold water pipe as a ground electrode. Those recommendations were made in an era when copper and galvanized pipe were used for all water service plumbing. Plastic pipe is now the standard, and mixed systems of copper and plastic pipe are common. Your local building codes specify how the pipes inside your home should or shouldn't be connected together for electrical grounding purposes.

A bonding bus is a good way for connecting equipment together and also provides a good way to make ac safety grounding connections. Installed behind or under your equipment, the bus can be any heavy metal conductor — copper and aluminum are both available and relatively inexpensive; strap, pipe, or even heavy wire will do. Connect the bus to the ac safety ground conductor at a power outlet. Only one connection is needed. Then connect the enclosure of each piece of equipment to the bus using a piece of #14 AWG wire, strap, or flat-weave grounding braid.

The typical amateur station includes a lot of accessory equipment that is itself unpowered but which connects to equipment that is connected to the ac line. Your goal in station building should be to ensure that any exposed metal does not present a shock hazard. This requires you to provide a safety ground connection for any unpowered equipment with a metal enclosure, such as an SWR bridge, antenna tuner, or antenna switch. Most larger pieces of equipment have a ground terminal for you to use. If not, add a screw to the enclosure or use a mounting screw for the connection.

LIGHTNING PROTECTION

The **Safe Practices** chapter covers the basic elements of lightning protection. From reading that, you can learn more about what lightning is, and what protection against its effect consists of. More detailed information about lightning and protection plans is available in the 2002 series of *QST* articles by Ron Block, NR2B (then KB2UYT) available at **www.arrl.org/lightning-protection**. Once you have a plan, you can implement it using standard grounding and bonding techniques.

Figure 24.4 shows an overview of how

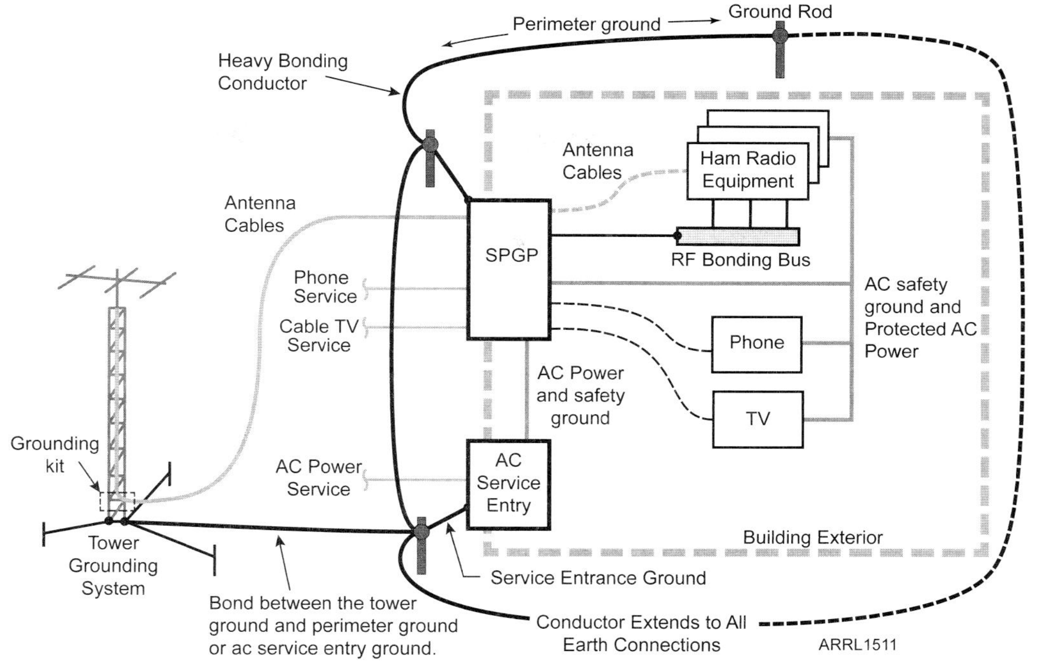

Figure 24.4 — Overview of a station grounding system than incorporates ac safety, lightning protection, and RF management. A single-point ground-panel (SPGP) is used to provide a common bonding connection to equalize voltage during a lightning transient. All ground connections must be bonded together.

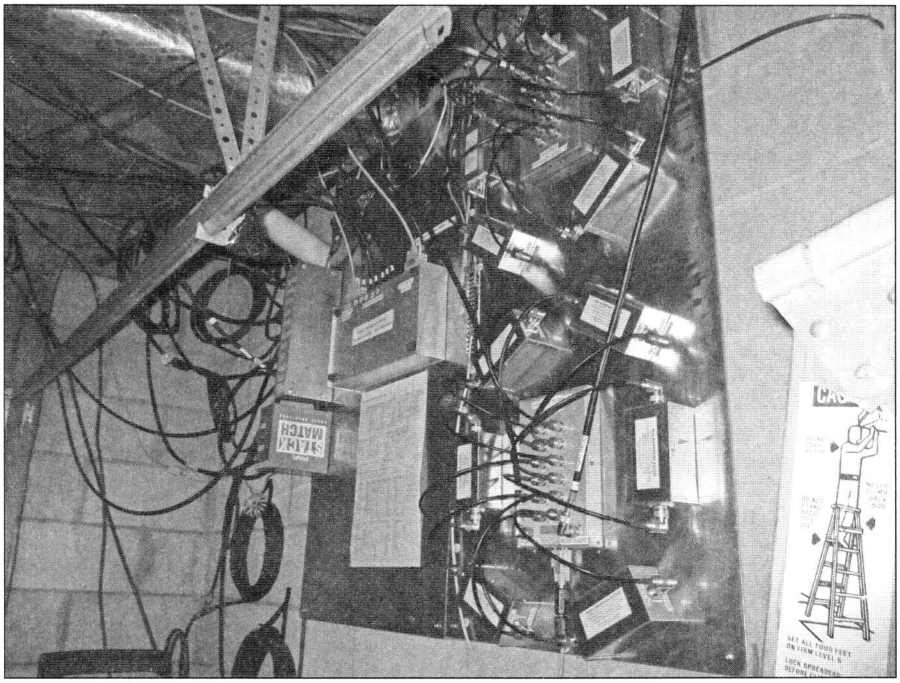

Figure 24.5 — A single-point ground-panel (SPGP) for an active HF station. Mounted on an exterior wall under the station, all filters, switches, and protectors are attached to the aluminum sheet. A heavy ground conductor connects the panel to a ground rod outside. [Kirk Pickering, K4RO, photo]

station grounding and bonding works together to provide lightning protection. All earth connections, such as ground rods and buried conductors, must be connected together. Inside the station, all equipment is bonded together either directly or through a bonding bus.

Devise a lightning protection plan as instructed in the articles by Block. Remember that every conductor entering or leaving your protected zone must have some means of protection against lightning or lightning-induced transients. **Figure 24.5** shows a single-point ground-panel (SPGP) installed on an exterior wall below the station. **Figure 24.6** shows an example of how to mount lightning protectors on an SPGP. Your circumstances are unique and the exact solution is up to you to design and build.

Don't depend on power strips with simple MOV surge protectors to protect your equipment. You can protect one or two branch circuits with a professional/industrial-quality ac surge protector (see **Figure 24.7**) mounted on your ground panel, and connect your ac power distribution system to that. Whole house surge protectors do not protect your station equipment from lightning since they are only connected to the residence's incoming ac service. You still need to create an effective ground system in your station.

Make sure all of your equipment is bonded together so that voltage is equalized between pieces of equipment. Bonding can be done directly between pieces of equipment if you do not have too many individual radios and accessories. An alternative is to use the bonding bus approach, as illustrated by **Figure 24.8**. Bonding also serves to provide ac safety if it includes a connection to the ac safety ground.

RF MANAGEMENT

It is typical for a station to begin with one or two radios, a power supply, and some feed lines to antennas outside. Eventually, more accessories are added, along with a

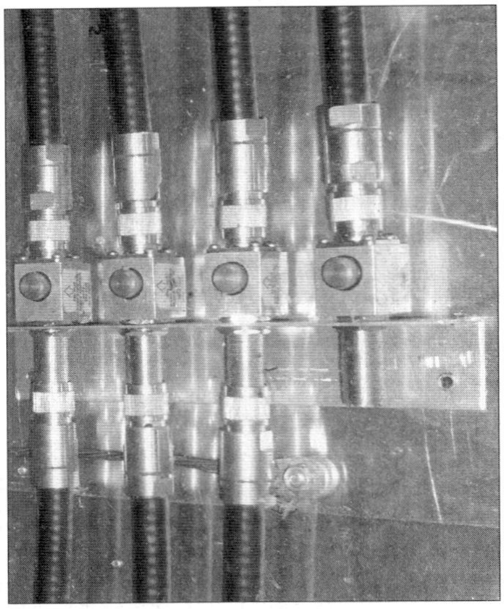

Figure 24.6 — Lightning protectors mounted on a SPGP using an aluminum angle bracket. Some protectors can be mounted directly to panels or as a through-hole using the threaded bulkhead-style connectors. [Ward Silver, NØAX, photo]

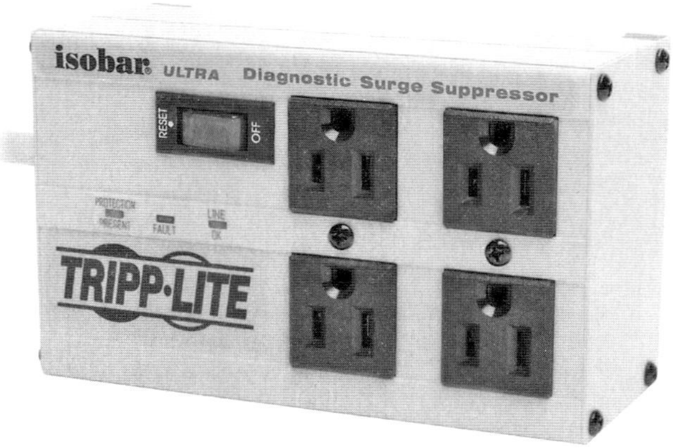

Figure 24.7 — An in-line ac power line surge protector suitable for lightning protection. It may be connected to a grounded panel via one of the enclosure screws.

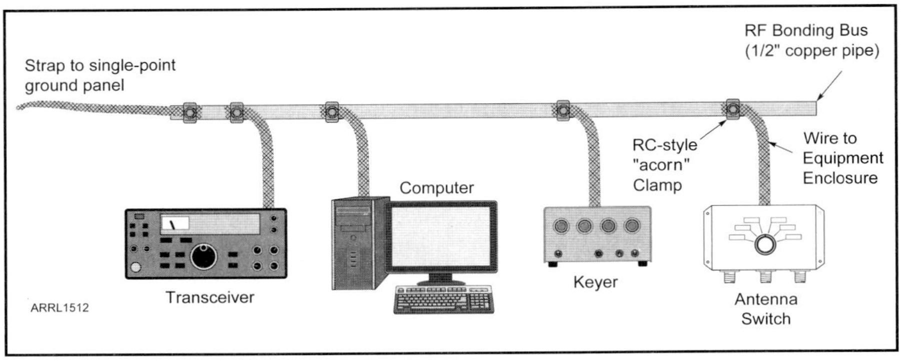

Figure 24.8 — A typical bonding bus for a station with equipment on a table or desk. Connections between the equipment and the bonding bus should be made with heavy wire (#14 AWG is suitable), 20 ga solid strap, or flat-weave grounding braid.

computer, possibly an amplifier, and so on. This eventually leads to "RF in the shack," or RFI to various pieces of equipment. The problem is caused by the numerous interconnections between pieces of equipment. These pick up transmitted signals as common-mode RF current that can affect amateur equipment, just as it can cause RFI to consumer equipment. (See the **RF and EMC** chapter for more information.)

Along with picking up the transmitted signal, your station's wiring also picks up RF from noise sources. There is a surprising amount of noise being radiated by household appliances, consumer electronics, lighting, computers and networking equipment, and power lines. Many opportunities exist for noise on the outside of shields or on unshielded wires to get in to the receiver input where the effect is to raise the station's noise floor, sometimes significantly. Good bonding and careful attention to proper shield connections minimizes those opportunities. For more information about minimizing receive noise, see Jim Brown, K9YC's, online paper, "Build Contest Scores by Killing Receive Noise" at **k9yc.com/KillingReceiveNoise.pdf**.

By starting your station with a plan to manage the RF that is present, you can reduce or eliminate the effects. You probably cannot shield your station against RF being picked up by the various cables and enclosures. Instead, use bonding to minimize RF voltage between pieces of equipment so that RF current doesn't flow between them. Bonding can be done by direct connection

Figure 24.9 — A simple RF ground plane made of aluminum roof flashing with a copper pipe bonding bus attached at the rear of the table. Equipment is placed on the flashing and attached to the pipe with short jumpers. The ground plane and bus are attached to the station ground system with a heavy wire. [Ward Silver, NØAX, photo]

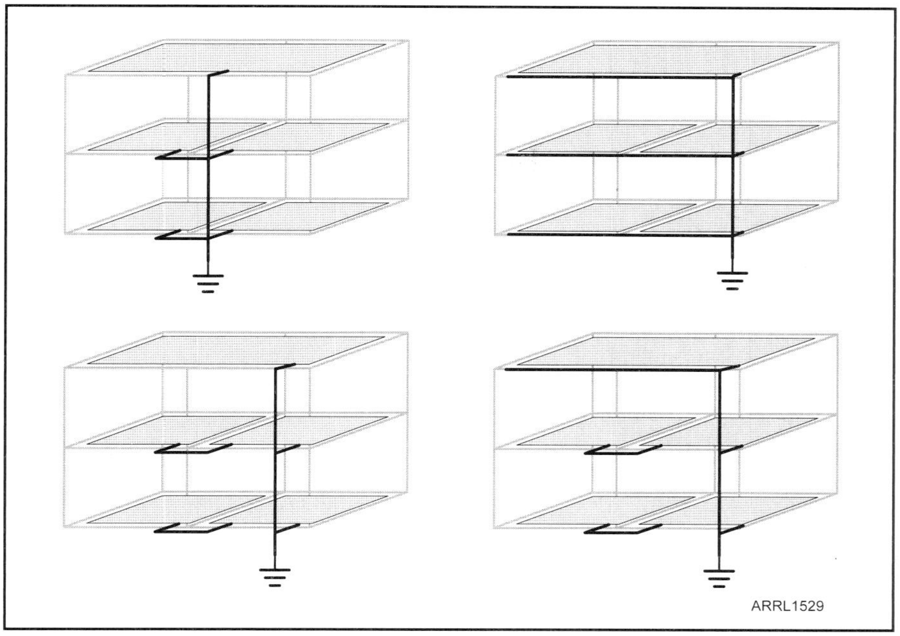

Figure 24.10 — For equipment on shelves, create a ground plane on each shelf. Bond each ground plane segment together (heavy lines) and route cables between the segments along the bonding strap or wire. Several possible examples are shown.

or by using a bonding bus.

As an added bonus, good bonding between pieces of equipment also minimizes the small audio buzz and hum voltages caused by leakage current and stray magnetic fields. Many stations use low-level audio signals for digital operation. Bonding helps keep unwanted hum and buzz from contaminating these signal paths. Bonding for RF management can satisfy the requirements for ac safety and lightning protection, too.

One technique to help equalize RF voltage around your station is the use of an RF *ground plane* or *reference plane*. **Figure 24.9** shows a simple ground plane made from a sheet of aluminum roof flashing attached to a tabletop. A bonding bus of copper pipe is clamped to the flashing at the back of the table. Sheet metal screws in the pipe are provided for attaching bonding jumpers to each piece of equipment. The ground plane and bonding bus are then connected to the station ground system. A metal desk or table can serve as the RF ground plane, as well. (The RF ground plane is different than a set of ground radials that provide a return path for RF to the feed point of a vertical antenna.) RF voltage differences are kept low by the ground plane. Extra cable length can be coiled up and placed directly on the ground plane to help keep them from picking up common-mode RF.

Portable and temporary stations pose a challenge for RF management. Station configuration usually has to work around various constraints and physical obstacles. Installing a comprehensive ground and bonding system is usually not possible. Under these circumstances, a roll of strap or flexible braid may serve as an effective bonding bus. Screen or even aluminum foil works as a ground plane. Have plenty of short jumpers and clip leads to connect the equipment enclosures to whatever common connection you can manage.

Many stations have shelving to hold equipment above a desk or table. **Figure 24.10** shows several possible techniques for adding metal sheet or strap to the shelves and connecting it together. You can use metal sheet, screen, or even surplus PC board material as the ground plane. The important thing is to get some metal under the various pieces of equipment to help the bonding connections work at RF.

24.1.3 Station Power

Power supply design and use is covered in the **Power Sources** chapter, and safety issues of station wiring are covered in the **Safe Practices** chapter. A single 20 A, 120 V circuit provides sufficient power for all of the radio and computer equipment in almost any single-operator amateur station, including most 500 W power amplifiers. A single 20 A, 240 V circuit provides sufficient power for two legal limit amplifiers that do not transmit simultaneously. A generous number of 120 V outlets should be provided. The 120 V and 240 V outlets should have their equipment grounds bonded together. This simple power arrangement has the major advantage of minimizing problems with audio hum and buzz caused by power system leakage current. Be sure to install good quality 20 A outlets and 20 A breakers.

Figure 24.11 shows how you can build your own heavy-duty, high-quality power distribution box. Use metal "back boxes" with high-quality outlets and heavy #12 or #14 AWG wiring. Multiple boxes can be mechanically attached together to support as many outlets as needed. If more than one multi-outlet box is used on the same circuit, use rigid EMT or flexible metal BX conduit between them with all of the metalwork solidly assembled and connected to the ground conductor. A heavy switch can be included to turn the entire package on and off. A GFCI outlet can be included in the box if the branch circuit is not already protected by a GFCI.

Wherever commercial power strips are used in your station, use industrial-quality products with a metal enclosure, heavy wiring, good quality outlets, and a built-in switch and circuit breaker (typically rated at 15 A). Avoid inexpensive plastic power strips that may cost much less but are of far lower quality.

Using ground fault circuit interrupter (GFCI) circuit breakers or outlets to supply

Figure 24.11 — A heavy-duty homemade power distribution box. Several gangable "back boxes" are assembled into a larger box that can hold several duplex outlets. A light switch or GFCI-protected outlet can be included for power control and shock protection. [Jim Brown, K9YC, photo]

power is not required by the NEC. However, given the many opportunities for stray current in an amateur station and antenna system, GFCI protection is not a bad idea. You can install GFCI breakers in the ac service panel or install GFCI outlets at the station.

24.1.4 Station Layout

Station layout is largely a matter of personal taste and needs. It depends mostly on the amount of space available, the equipment involved, and the types of operating to be done. With these factors in mind, some basic design considerations apply to all stations.

THE OPERATING TABLE

The operating table may be an office or computer desk, a kitchen table, or a custom-made bench. What you use depends on space, materials at hand, and cost. The two most important considerations are height and size of the top. Most commercial desks are about 29 inches above the floor. Computer tables are usually a couple inches lower for a more comfortable keyboard and mouse placement. This is a comfortable height for most adults. Heights much lower or higher than this may cause an awkward operating position.

The dimensions of the top are an important consideration. A deep (36 inches or more) top allows plenty of room for equipment interconnections along the back, equipment about midway and room for a keyer, writing and a keyboard and mouse toward the front. The length of the top depends on the amount of equipment being used. An office or computer desk makes a good operating table. These are often about 36 inches deep and 60 inches wide. Drawers can be used for storage of logbooks, headphones, writing materials, and so on. Desks specifically designed for computer use often have built-in shelves that can be used for equipment stacking. Desks of this type are available ready-to-assemble at most discount and home improvement stores. The low price and adaptable design of these desks make them an attractive option for an operating position. An example is shown in **Figure 24.12**.

If possible, arrange your operating desk or table so that it is easy to access the rear of the equipment. You may be able to provide a walkway behind the equipment, or the desk or table might be movable. Either way, you will appreciate being able to work behind the equipment without having to get under a table or work in uncomfortable positions. It also helps avoid wiring errors caused by not having a clear view of the equipment. Equipment can also be left in place while connections are made.

Figure 24.12 — Mike Adams, N1EN makes the most of his desktop to operate on the HF and VHF+ bands. His laptop and tablet computers are an alternative to the larger desktop systems. He uses a full-size keyboard with the laptop. [Mike Adams, N1EN, photo]

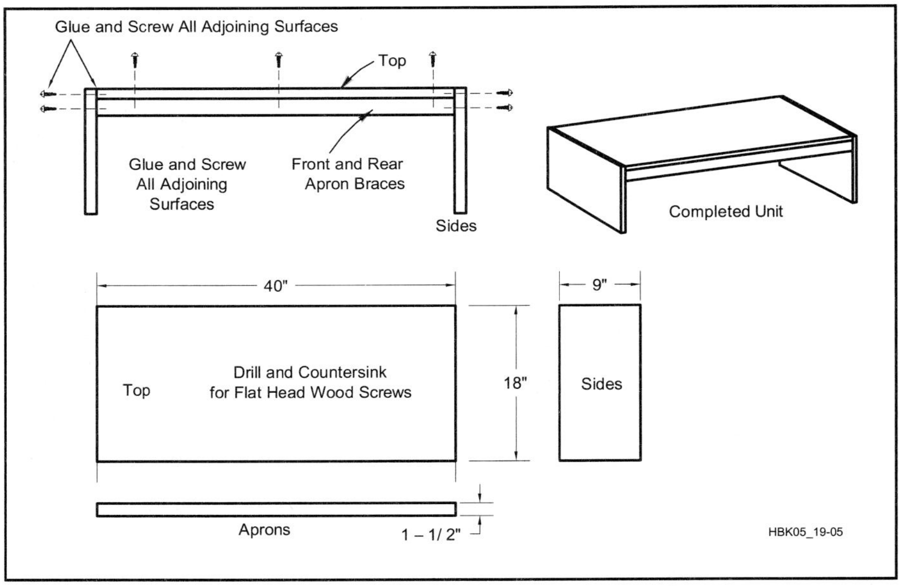

Figure 24.13 — A simple but strong equipment shelf can be built from readily available materials. Use ¾-inch plywood along with glue and screws for the joints for adequate strength.

ARRANGING THE EQUIPMENT

No matter how large your operating table, some vertical stacking of equipment may be necessary to allow you to reach everything from your chair. Stacking pieces of equipment directly on top of one another is not a good idea because most amateur equipment needs airflow around it for cooling. A shelf, like that shown in **Figure 24.13**, can improve equipment layout in many situations. Dimensions of the shelf can be adjusted to fit the size of your operating table.

When you have acquired the operating table and shelving for your station, the next task is arranging the equipment in a convenient, orderly manner. The first step is to provide power outlets and a good ground as described earlier. Be generous in estimating the number of power outlets for your installation; radio equipment has a habit of multiplying with time, so plan for the future at the outset. Try to obtain high-quality power distribution strips, or install some outlets with enough space between them to accommodate multiple plug-in "wall-wart" power supplies.

Every station is different and each operator has different preferences, so giving "cookbook" instructions for station layout is unrealistic. Even so, there are some gen-

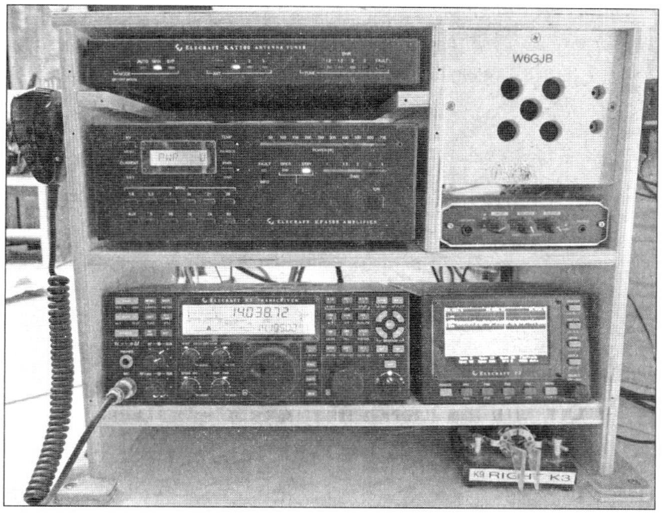

Figure 24.14 — This portable 500 W HF station was constructed by W6GJB and K9YC for portable operation during the California QSO Party. [Jim Brown, K9YC, photo]

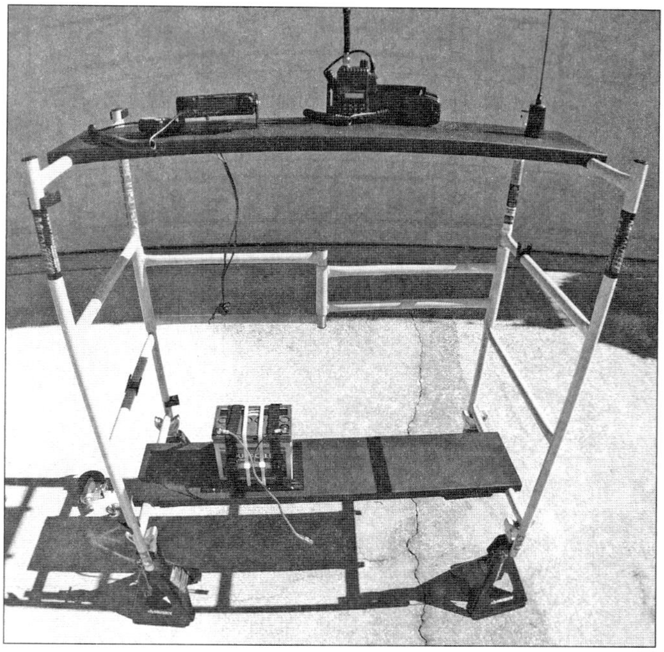

Figure 24.15 — K1CE adapted this portable metal scaffold as a portable operating desk. The top shelf also serves as a ground plane for the mag-mount VHF/UHF antennas. [Rick Palm, K1CE, photo]

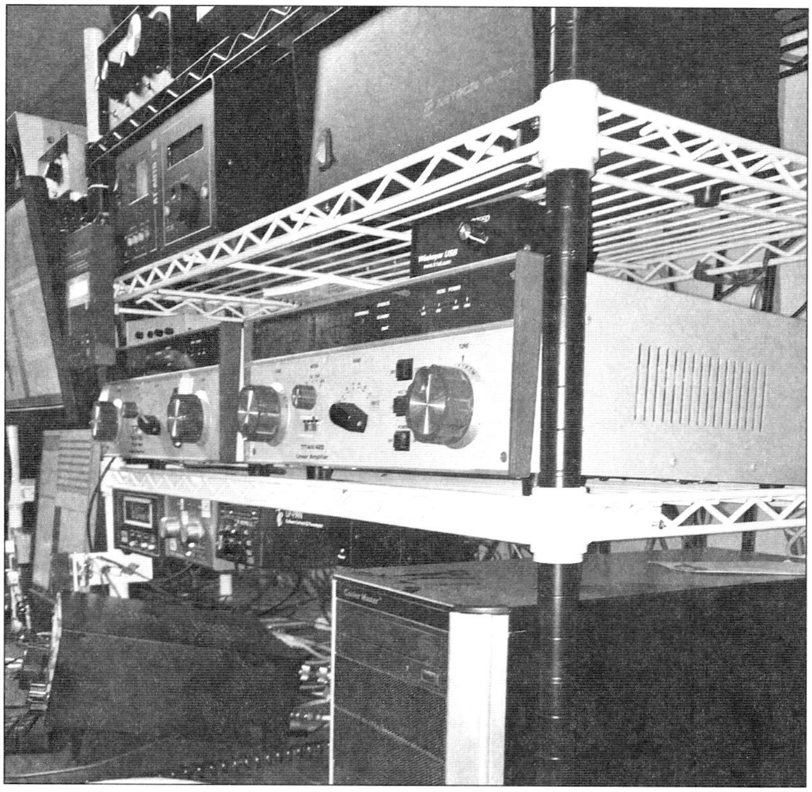

Figure 24.16 — Taking advantage of a good opportunity, NØNI adapted this set of metal wire shelving to hold his equipment. The wire shelves also act as an RF ground plane and provide ventilation. [Bob Lee, WØGXA, photo]

eral principles that make a station layout effective. Start by looking at the photographs of stations in this chapter, in *QST*, and online. Visit stations of other operators to get ideas and try different arrangements. Think about what equipment you use most while operating and make it easy to reach and use. Consider making a temporary arrangement to try out different configurations before you settle on a final layout. Make it easy to update your layout as your habits, tastes, and body change. Configure the station to suit your interests, and keep thinking of ways to refine the layout.

Portable stations have the additional requirement of being lightweight and sturdy, often part of a "go-kit" for emergency communication or other public service. **Figures 24.14** and **24.15** show two different solutions for very different sets of circumstances, but both are well-considered and appropriate to the need.

You may find yourself taking advantage of non-traditional materials for your station. **Figure 24.16** shows equipment supported by open metal wire shelving used as commercial or industrial shelving. Not only is the wire shelving strong, it also serves as an RF ground plane and is open to airflow that provides ventilation to the equipment.

ERGONOMICS

Ergonomics is a term that loosely means

"fitting the work to the person." When tools and equipment are designed around what people can accommodate, the results are much more satisfactory. Considerable attention has been focused on ergonomics in recent years because we have come to realize that long periods of time spent in unnatural positions can lead to repetitive motion injuries. Much of this attention has been focused on people whose job tasks have required them to operate computers and other office equipment. While most amateur radio operators do not devote as much time to their hobby as they might to a full-time job, it does make sense to consider comfort and flexibility when choosing furniture and arranging it in the station or workshop. Adjustable height chairs are available with air cylinders to serve as shock absorbers. Footrests come in handy if the chair is so high that your feet cannot support your lower leg weight. The heights of tables and keyboards are often not adjustable.

Place computer screens so that they are at the same level as your eyes to limit strain on your neck, and to minimize reflected light coming from windows. It is always wise to build into your sessions time to walk around in order to stimulate blood circulation and exercise your muscles and joints.

Selection of hand tools is another area where there are choices that affect how comfortable you are while working in your station; look for screwdrivers with pliable grips. Be sure to take into account how heavy things are before picking them up — your back will thank you for this.

24.1.5 Interconnecting Your Equipment

Once you have your equipment arranged you then have to interconnect it all. No matter how simple the station, you need to make antenna, power, microphone, and keyer connections. Equipment such as amplifiers, computers, TNCs, and so on add complexity. By keeping your equipment interconnections well organized and of high quality, you avoid problems later on.

Ready-made cables are often available, but in many cases, you have to make your own cables. An advantage of making your own cables is that you can customize the length. This allows more flexibility in arranging your equipment and avoids extra cable length that can pick up RF. Many manufacturers supply connectors with their equipment along with pinout information in the manual. This allows you to make the necessary cables in the lengths you need for your particular installation.

Always use high quality wire, cables, and connectors. Take your time and make good mechanical and electrical connections on your cable ends. Sloppy cables are a source of trouble, and the problems they cause are often intermittent and difficult to track down. You can bet that they crop up right in the middle of a contest, or during a rare DX QSO! Even worse, a poor-quality connection could cause RFI or even create a fire hazard. A cable with a poor mechanical connection could come loose and short a power supply to ground or apply a voltage where it should not be. Wire and cables should have good quality insulation that is rated to withstand high enough voltages to prevent shock hazards.

Interconnections should be neatly bundled and labeled. Wire ties, masking tape, or paper labels with string work well. See **Figure 24.17**. Whatever method you use, proper labeling makes disconnecting and reconnecting equipment much easier. **Figure 24.18** illustrates the number of potential interconnections in a modern, full-featured transceiver.

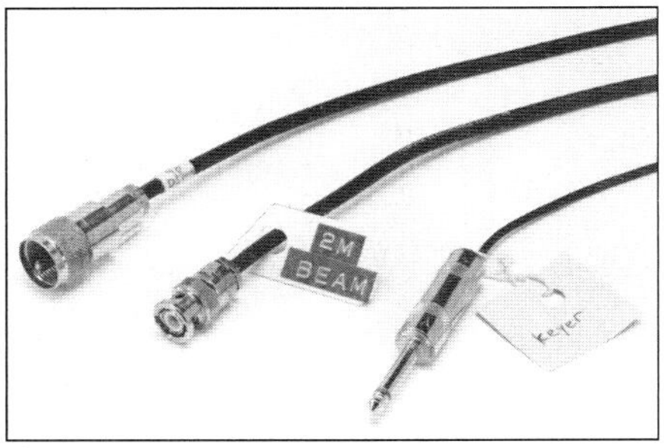

Figure 24.17 — Labels on the cables make it much easier to rearrange things in the station. Labeling ideas include masking tape, cardboard labels attached with string and labels attached to fasteners found on plastic bags (such as bread bags).

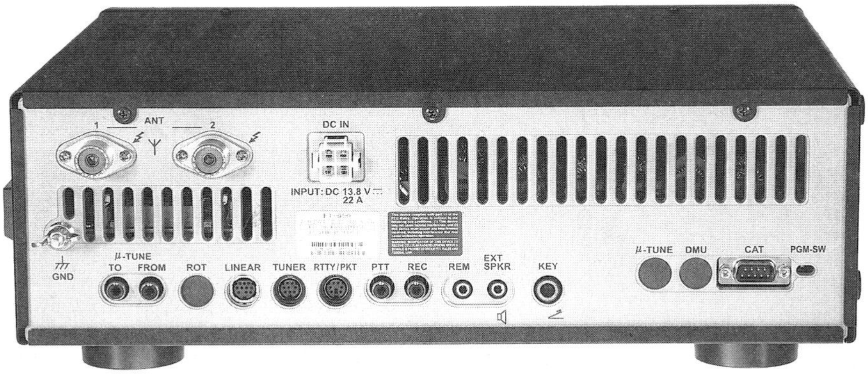

Figure 24.18 — The back of this Yaesu FT-950 transceiver shows some of the many types of connectors encountered in the amateur station. Note that the variety is found on a single piece of equipment.

Fire Extinguishers in the Amateur Station

Three classes of fire extinguishers are appropriate for amateur stations:

- Class A: Combustible solid materials such as wood or paper
- Class B: Combustible liquid or gases
- Class C: Energized electrical fire

A fixed or portable station should have "ABC" class extinguishers. Mobile stations are most likely to experience class B and C fires.

Most ABC extinguishers contain a solid powder that does not conduct electricity and is safe for electrical equipment. The powder does leave a residue, however. A "clean agent" extinguisher uses a gas, such as Halon, to displace oxygen. The extinguisher contains a liquid form of the agent which changes to a gas on being released. Less common, CO_2 extinguishers use carbon dioxide gas to smother the fire and do not conduct electricity. Gas extinguishers leave no residue.

Liquid water extinguishers or hoses should never be used on an electrical fire because most water conducts electricity and creates an electrical shock hazard. Water mist fire extinguishers disperse de-ionized water in a fog that cools the fire and displaces oxygen. Since the mist is de-ionized, it is safe to use on electrical (Class C) fires.

For use with portable generators, such as at Field Day, keep a fire extinguisher near the generator in case of a fuel spill or other fire. Place the extinguisher far enough away from items that are likely to catch fire so that it can be accessed while a fire is blazing.

If the fire is caused by an electrical overload or short circuit, remove power before using the fire extinguisher, if possible. This takes away the ignition source and helps keep the fire out once the flames have been extinguished.

Check fire extinguishers regularly to be sure that they are always fully charged. The Fire Equipment Manufacturer's Association publishes useful documents on selection and use of fire extinguishers at **femalifesafety.org/fire-equipment/portable-fire-extinguishers**.

WIRE AND CABLE

The type of wire or cable to use depends on the job at hand. The wire must be of sufficient size to carry the necessary current. Use the tables in the **Construction Techniques** chapter to find this information. Never use underrated wire; it is a fire hazard. Be sure to check the insulation, too. For high-voltage applications, the insulation must be rated above the intended voltage. A good rule of thumb is to use a rating at least twice the voltage that can be encountered.

Use good quality coaxial cable of sufficient size for connecting transmitters, transceivers, antenna switches, antenna tuners, and so on. RG-58 might be fine for a short patch between your transceiver and SWR bridge, but is too small to use between your legal-limit amplifier and antenna tuner. For more information, see the **Transmission Lines** chapter.

Hookup wire may be stranded or solid. Generally, stranded is a better choice since it is less prone to break under repeated flexing. Many applications require shielded wire to reduce the chances of RF getting into the equipment. RG-174 is a good choice for control, audio, and some low-power RF applications. Shielded microphone or computer cable can be used where more conductors are necessary.

For RF grounding connections, #12 – #16 AWG solid or stranded wire, or solid strap, are preferred. For indoor connections not exposed to the weather, flat-weave tinned braid strap is acceptable. Do not use braid salvaged from coaxial cable for RF grounding connections.

CONNECTORS

There are quite a few connector styles, but common terms apply to all of them. Pins are contacts that extend out of the connector body, and connectors in which pins make the electrical contact are called *male* connectors. Sockets are hollow, recessed contacts, and connectors with sockets are called *female*. Connectors designed to attach to each other are called *mating connectors*. Connectors with specially shaped bodies or inserts that require a complementary shape on a mating connector are called *keyed connectors*. Keyed connectors ensure that the connectors can only go together one way, reducing the possibility of damage from incorrect mating.

Plugs are connectors installed on the end of cables and *jacks* are installed on equipment. *Adapters* make connections between two different styles of connector, such as between two different families of RF connectors. Other adapters join connectors of the same family, such as double-male, double-female, and gender changers. *Splitters* divide a signal between two connectors.

While the number of different types of connectors is mind-boggling, many manufacturers of amateur equipment use a few standard types. If you are involved in any group activities such as public service or emergency-preparedness work, check to see what kinds of connectors others in the group use and standardize connectors wherever possible. Assume connectors are not waterproof, unless they are clearly marked for outdoor use (and assembled correctly).

Power Connectors

Amateur Radio equipment uses a variety of power connectors. Some examples are shown in **Figure 24.19**. Most low power amateur equipment uses coaxial power connectors. These are the same type found on consumer electronic equipment that is supplied by a wall transformer power supply. Transceivers and other equipment that requires high current in excess of a few amperes often use Molex connectors (**www.molex.com** — enter "MLX" in the search window) with a plastic body housing pins and sockets crimped on to the end of wires.

An emerging standard, particularly among ARES (Amateur Radio Emergency Service) and other emergency communications groups, is the use of Anderson Powerpole connectors (**www.andersonpower.com**). These connectors are "sexless," meaning that any two connectors of the same series can be mated — there are no male or female connectors. By standardizing on a single connector style, equipment can be shared and replaced easily in the field. The standard orientation for pairs of these connectors is shown in Figure 24.19. Using this orientation increases the compatibility of your wiring with that of other hams.

Molex and Powerpole connectors use crimp terminals (both male and female) installed on the end of wires. A special crimping tool is used to attach the wire to the terminal and the terminal is then inserted into the body of the connector. Making a solid connection requires the use of an appropriate tool — do not use pliers or some other tool to make a crimp connection.

Some equipment uses terminal strips for direct connection to wires or crimp terminals, often with screws. Other equipment uses spring-loaded terminals or binding posts to connect to bare wire ends. **Figure 24.20** shows some common crimp terminals that are installed on the ends of wires using special tools.

Audio and Control Connectors

Consumer audio equipment and amateur radio equipment share many of the same connectors for the same uses. Phone plugs and jacks are used for mono and stereo audio circuits. These connectors, shown in **Figure 24.21**, come in 1⁄4 inch, 1⁄8 inch (miniature), and subminiature varieties. The contact at the end of the plug is called the tip and the connector at the base of the plug is the sleeve. If there is a third contact between the tip and sleeve, it is the ring (these are

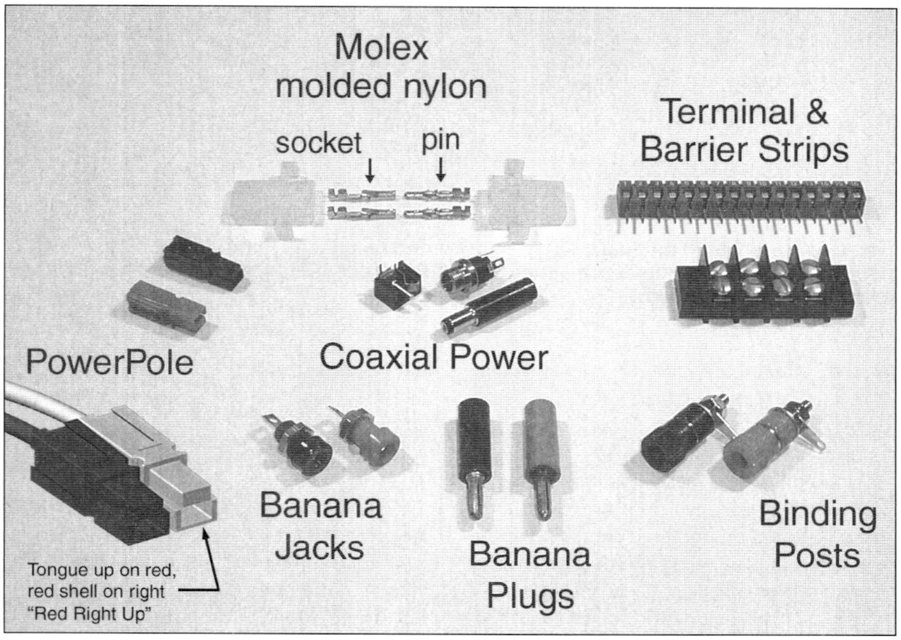

Figure 24.19 — These are the most common connectors used on amateur equipment to make power connections. The proper orientation for paired Powerpole connectors is with the red connector on the right and its tongue on top — "red-right-up." [Courtesy of Wiley Publishing, Ham Radio for Dummies, or Two-Way Radios and Scanners for Dummies]

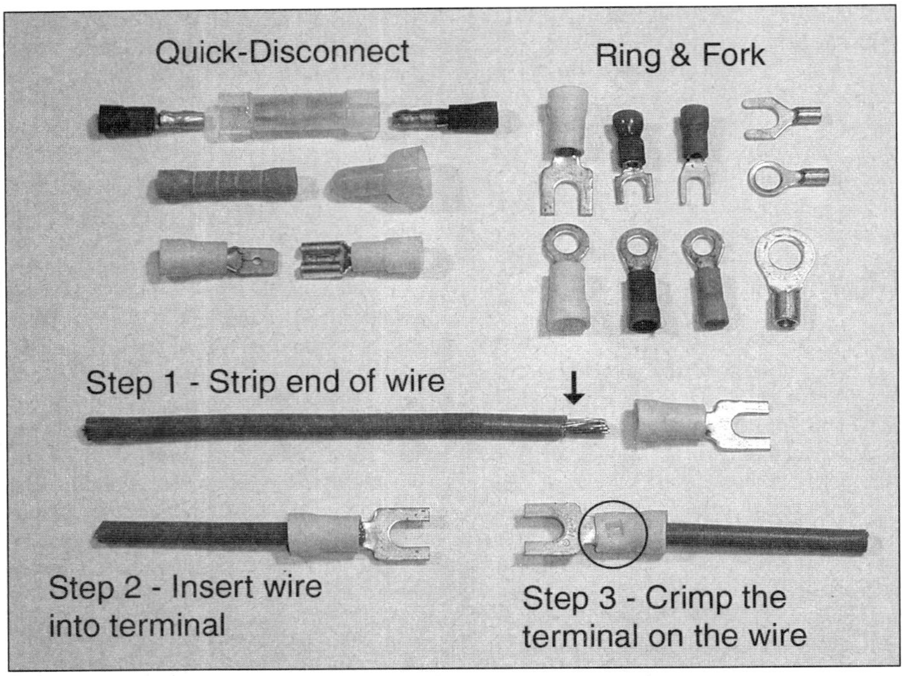

Figure 24.20 — Power connectors often use terminals that are crimped onto the end of wires with special crimping tools. [Courtesy of Wiley Publishing, Ham Radio for Dummies, or Two-Way Radios and Scanners for Dummies]

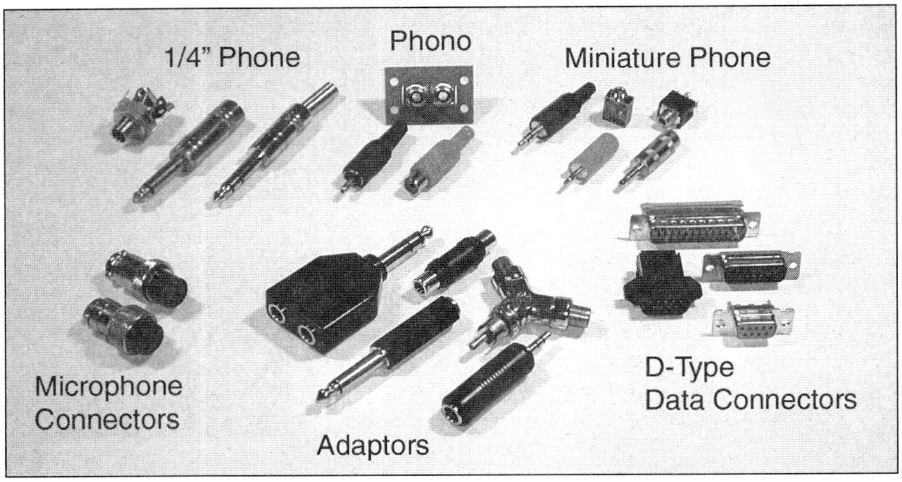

Figure 24.21 — Audio and data signals are carried by a variety of different connectors. Individual cable conductors are either crimped or soldered to the connector contacts. [Ward Silver, NØAX, photo]

"stereo" phone connectors). Stereo phone connectors are often called TRS (Tip-Ring-Sleeve) connectors by audio equipment manufacturers.

Phono plugs and jacks (sometimes called RCA connectors since they were first used on RCA brand equipment) are used for audio, video and low-level RF signals. They are also widely used for control signals.

The most common microphone connector on mobile and base station equipment is an 8-pin round connector. On older transceivers you may see 4-pin round connectors used for microphones. RJ-45 modular connectors (see the section on telephone connectors below) are often used in mobile and smaller radios.

RF Connectors

Feed lines used for radio signals require special connectors for use at RF frequencies. The connectors must have the same characteristic impedance as the feed line they are attached to. Inexpensive audio and control connectors cannot meet that requirement, nor can they handle the high-power levels often encountered in RF equipment. Occasionally, phono connectors are used for HF receiving and low-power transmitting equipment.

The most common connector for RF in amateur equipment by far is the UHF family shown in **Figure 24.22**. (The UHF designator has nothing to do with frequency.) A PL-259 is the plug that goes on the end of feed lines, and the SO-239 is the jack mounted on equipment. A "barrel" (PL-258) is a double-female adapter that allows two feed lines to be connected together. UHF connectors are rated for use up to 200 MHz and can handle legal-limit transmitter power at HF.

UHF connectors have several drawbacks, including lack of weatherproofing, poor performance above the 2-meter band, and limited power handling at higher frequencies. The Type-N series of RF connectors addresses all of those needs. Type-N connectors are somewhat more expensive than UHF connectors, but they require less soldering and perform better in outdoor use since they are moisture resistant. Type-N connectors can be used to 10 GHz.

For low-power uses, BNC connectors are often used. BNC connectors are the standard for laboratory equipment, and they are often used for dc and audio connections. BNC connectors are common on handheld radios for antenna connections. (However, some of the newest handheld transceivers use small, screw-on SMA type connectors for their antennas.)

The type of connector used for a specific job depends on the size of the cable, the frequency of operation, and the power levels involved. More information on RF connectors may be found in the **Transmission Lines** chapter.

Data Connectors

Digital data is exchanged between computers and pieces of radio equipment more than ever before in the amateur station. The connector styles follow those found on computer equipment.

D-type connectors are used for RS-232 (COM ports) and parallel (LPT port) interfaces. A typical D-type connector has a model number of "DB" followed by the number of connections and a "P" or "S" depending on whether the connector uses pins or sockets. For example, the DB-9P is used for PC COM1 serial ports.

USB connectors are the most popular in amateur equipment as the computer industry has eliminated the bulkier and slower RS-232 interface. A number of manufacturers make USB-to-serial converters that allow devices with RS-232 interfaces to be used with computers that only have USB interfaces.

Null modem, or *crossover*, adapters or cables are designed to fit inline with a cable connection, with both sexes of the same connector type. The internal connections

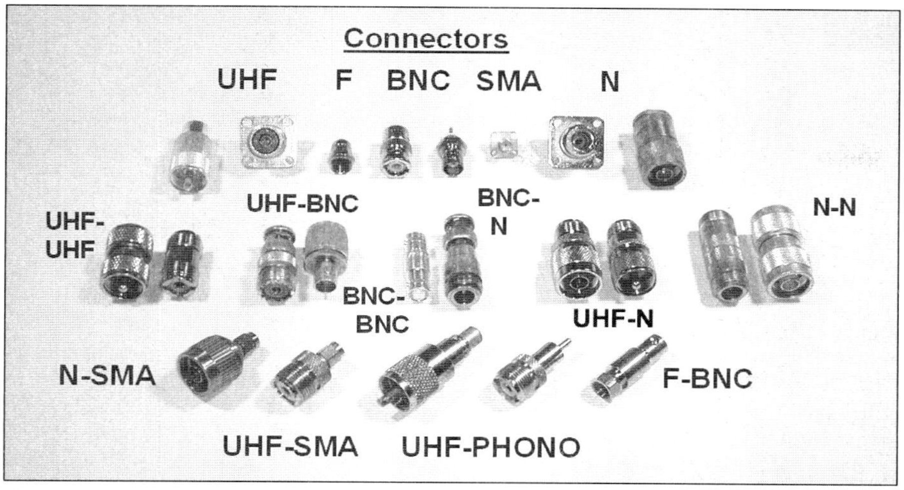

Figure 24.22 — Each type of RF connector is specially made to carry RF signals and preserve the shielding of coaxial cable. Adapters are available to connect one style of connector to another. [Ward Silver, NØAX, photo]

between signal pins are swapped between ends so that inputs and outputs are swapped. This allows interfaces to be connected together directly without any intermediary equipment, such as an Ethernet switch or an RS-232 modem. Several practical data interface projects are shown in the **Station Accessories** and **Digital Communications** chapters in this book's online content.

Telephone and Computer Network Connectors

Modular connectors are used for telephone and computer network connections. Connector part numbers begin with "RJ." The connectors are crimped on to multi-conductor cables with special tools. The RJ11 connector is used for single- and double-line telephone system connection with 4 or 6 contacts. The RJ10 is a 4-contact connector for telephone handset connections. Ethernet computer network connections are made using RJ45 connectors with 8 contacts.

24.1.6 Documenting Your Station

An often neglected, but very important, part of putting together your station is properly documenting your work. Ideally, you should diagram your entire station from the ac power lines to the antenna on paper and keep the information in a special notebook with sections for the various facets of your installation. Having the station well documented is an invaluable aid when tracking down a problem or planning a modification. Rather than having to search your memory for information on what you did a long time ago, you'll have the facts on hand.

In addition to recording the interconnections and hardware around your station, you should also keep track of the performance of your equipment. Each time you install a new antenna, measure the SWR at different points in the band and make a table or plot a curve. Later, if you suspect a problem, you'll be able to look in your records and compare your SWR with the original performance.

In your station, you can measure the power output from your transmitter(s) and amplifier(s) on each band. These measurements can be helpful if you ever suspect you have a problem. If you have access to a signal generator, you can measure receiver performance for future reference.

Coax Connectors — Not as Simple as They Appear

by Hal Kennedy, N4GG

"You get what you pay for" was never more true than when it comes to common UHF connectors, including PL-259s, SO-239s, adapters, and related parts. Every hamfest seems to have at least one vendor selling "mystery" UHF connectors, sometimes for as little as a dollar each. What are you getting when you buy the cheapest PL-259? It's pretty much a guess. For the difference of a dollar or two, "mystery" UHF connectors are a very poor investment.

PL-259s have four parts: the outer sleeve called the "knurled nut," the connector body, the insulator/dielectric and the center pin. All four components can be compromised to the point of making a bargain connector useless.

Problems frequently encountered:

• Finish: Bargain connectors sometimes have a finish you can't solder to! They may have a chrome-like appearance, but the plating may not take solder well and has to be filed down for a good connection.

• Threading: The internal threads at the rear of the body are there to accept a UG-style insert that narrows the connector barrel to accept smaller diameter coax such as RG-8X or RG-58. The threads may be metric! UG inserts also sometimes appear in the US market with metric threads. Either way, the insert does not screw into the body.

• Dielectric: Good connectors use quality phenolic or Teflon insulation between the center pin and the body. Bargain connectors might use anything, including materials such as a thermoplastic material, which can melt when the center pin is soldered.

• Center pin diameter: This is one of the most common and insidious problems in mystery PL-259s. The center pin outer diameter (OD) is almost always slightly smaller than it should be and it's hard to notice. The center pin connection between a PL-259 and an SO-239 or barrel connector depends on the male side pin OD being correct and the matching fingers on the female side being the correct diameter and made of the proper spring material.

• Center socket spring tension: If the SO-239 socket metal relaxes over time or temperature, an intermittent connection can be created that can be very hard to track down.

• Mating indentations: The indentations on the end of the SO-239 that mate to a PL-259 (the annulus flange) may only have four indentations to match up with the short prongs on the body of the male connector. A quality SO-239 or barrel connector has indentations all the way around. If the PL-259 and SO-239 don't seat completely, an intermittent connection is likely to develop.

• T and right-angle (elbow) UHF adapters: The center conductor has to make a right-angle turn inside the shell. In poor-quality adapters the right-angle connection is done with a spring contact — these do not hold up. Quality T and right-angle adapters are reliable because the internal conductors are tapped and threaded — the conductors are screwed together within the body at the right angle junction.

How can we tell the good connectors? If the price is too good to be true — well, it is. PL-259s with good silver plating have a dull appearance. Good connectors have a part number and manufacturer's name stamped into them. You can look up the connector's specifications if it's marked. An example is the connectors made by Amphenol — all of which have parts numbers such as 83-1SP (PL-259) or 83-1R (SO-239) stamped into or onto the connector body.

24.2 Mobile Installations

Solid-state electronics and miniaturization have allowed mobile operators to equip their vehicles with stations rivaling base stations. Indeed, it is possible to operate from 160 meters through 70 cm with one compact transceiver. Adding versatility, most transceivers designed for mobile use are set up so that the main body of the radio can be safely tucked under a seat, with the operating "head" conveniently placed for ease of use as shown in **Figure 24.23**.

Common power levels reach 100-150 W on HF, and 50-75 W on VHF. With proper antenna selection and placement (see the **Antennas** chapter), mobile stations can work the world, just like their base station counterparts. A major difference between them is that you're trying to drive at the same time as you are operating, and safe operating requires attention to the details.

For some of us living in antenna-restricted areas, mobile operating may offer the best solution for getting on the air. For others it is an enjoyable alternative to home-station operation, with the advantage of being able to change your location to activate new grid squares, counties, or other desired locations. Mobile operation also has the advantage of moving away from noisy RF environments to operate from locations that have lower noise floors. No matter which category you're in, you can enjoy success if you plan your installation with safety and convenience in mind.

There is a considerable amount of information about mobile operation, both HF and VHF+, online. Many contributions to this section have come from the website of Alan Applegate, KØBG (**www.k0bg.com**.) Specialty groups, such as RVers and off-road groups, have additional perspectives. Your vehicle dealer also has service bulletins regarding the installation of radio equipment. Although the bulletins mainly apply to VHF equipment, they are a valuable source of information about convenient wiring routes and access points for power and other wiring.

24.2.1 Installation

Installing amateur radio equipment in modern vehicles can be quite challenging, yet rewarding, if basic safety rules are followed. First, all gear must be securely attached to the vehicle; unsecured cup holder mounts, mounts wedged between cushions, elastic cords, hook-and-loop tape, magnets, or any other temporary mounting scheme *must be avoided*! Remember, if it isn't bolted down, it can become a missile in the event of a crash. The radio mounting location must avoid SRS (airbag) deployment zones — virtually eliminating the top of the dash in most modern vehicles — as well as vehicle controls (see the sidebar "Air Bags and Mobile Installations).

There are many no-holes-needed mounts available. Some mounts are even designed for a specific transceiver make and model. **Table 24.1** lists some suppliers. Two other points to keep in mind when choosing a mounting location are convenience and lack of distraction. Microphone and power cabling should be placed out of the way and be properly secured. The transceiver's controls should be convenient to use and view. See **Figure 24.24** for examples.

Mounting radios inside unvented center consoles and overhead bins should also be avoided. Modern mobile transceivers designed for remote mounting allow the main body to be located under a seat, in the trunk, or in another out-of-the-way place (**Figure 24.25**), but be sure there is plenty of ventilation.

If you drive off-road or on rough roads, you may wish to consider using shock mounts, also known as "Lord Vibration Mounts" for their original manufacturer. For more information, see the product information from the Astrotex Company at **www.astrotex.com/lord-vibration**.

Figure 24.23 — In this mobile installation, the transceiver control head is mounted in the center console, next to a box with switches for adjusting the antenna. The main body of the radio can be located under the seat or in the trunk.

24.2.2 Coaxial Cable

Cable lengths in mobile installations seldom exceed 15 feet, so coax losses are not a major factor except on the 70 cm and higher frequency bands. Good quality RG-58A or RG-8X size coax is more than adequate for HF and VHF. While there is nothing wrong with using RG-8 size coax (0.405 inch diameter), it is stiffer and has a larger bending radius, making it harder to work with in most mobile applications.

There are some caveats when selecting coax. Avoid solid center conductors such as in standard RG-58. It has a propensity to kink, is susceptible to failure from vibration, and can be difficult to solder properly. Both RG-58A and RG-8X use foam dielectric, and care is needed when soldering PL-259 connectors, specially when reducers are being used. The **Transmission Lines** chapter illustrates the correct installation procedure.

24.2.3 Wiring

Proper wiring is an essential part of any mobile installation. Consider the following points when selecting materials and planning cable routing:

- Wire needs to be correctly sized and fused. Stranded wire is preferred.
- All cables need to be protected from abrasion, heat, and chemicals.
- Wiring needs to be shortened and/or bundled with appropriate wire ties to avoid interaction with passengers and mechanical devices. **Figure 24.26** shows a typical vehicle wiring tray.

Power cables should be connected directly to the battery following manufacturers' recommendations, with the requisite positive and negative lead fuses located close to the battery. **Figure 24.27** shows a typical fuse block. Accessory (cigarette lighter) sockets and power taps should not be used except for very low current loads (<5 A), and then only with care. It pays to

Table 24.1
Mobile Mount Sources

Gamber Johnson	**www.gamberjohnson.com**
Havis-Shields	**www.havis.com**
Jotto Desk	**www.jottodesk.com**
PanaVise Products	**www.panavise.com**
RAM Mounting Systems	**www.ram-mount.com**

(A)

(B)

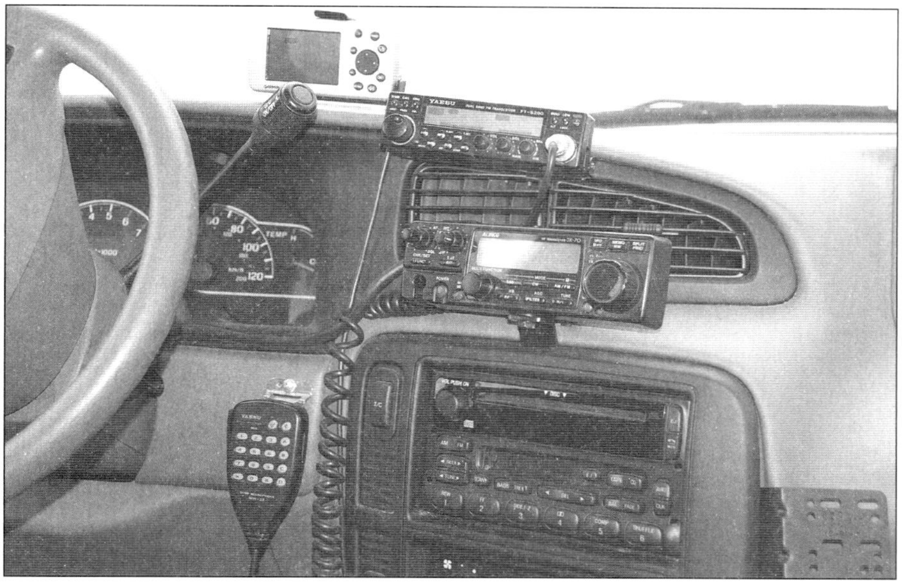

Figure 24.24 — At (A), the transceiver control head is attached to one of many available mounts designed for this purpose. Mounts are typically highly adjustable, allowing the control head or radio to be positioned close to the operator. An antenna controller is mounted below the microphone. At (B), HF and VHF transceiver control heads and the microphone are all mounted to the dashboard, within easy reach.

Figure 24.25 — The main body of the radio may be mounted in the vehicle's trunk or other out-of-the-way spot. Allow for plenty of ventilation.

remember that a vehicle fire is both costly and dangerous!

The fuses supplied with most mobile transceivers are ATC style. Most automotive fuses are ATO. The ATC fuse element is completely sealed in plastic and the ATO is not. Since the power cable fuse holders are not waterproof, only an ATC fuse should be used if the fuse holder is exposed to the weather or located anywhere in the engine compartment. If an ATO fuse is used, and water gets into the fuse, the fuse element corrodes and eventually fails.

Marine Grounding and Equipment Installation

Marine vessels, either fresh water or saltwater, pose challenges to radio and electrical systems. Luckily, there are many online and print references explaining how to do things right, whether the vessel is power or sail, wooden, fiberglass, or metal-hulled. Each type requires certain practices be followed to avoid corrosion and maintain good electrical connections at dc and RF. Your best sources of information can be found through marine outfitters and communication shops that install marine HF and VHF radios.

Proper wiring also minimizes voltage drops and helps prevent ground loops. Modern solid-state transceivers operate effectively down to 12.0 V dc (engine off). When the voltage drops below 11.6 V under load, some transceivers reduce power, shut down, or operate incorrectly. The vehicle chassis should not be used for ground returns; paint or other insulation can isolate different chassis sections and using a chassis return can create a ground loop.

Running cables through the engine firewall can be easy in some vehicles and nearly impossible in others. Using factory wiring grommets should be avoided unless they're not being used. In some cases, the only alternative is to drill your own hole. If you have any questions or concerns, have your local mobile sound shop or two-way radio dealer install the wiring for you.

A good guide to installation of radio equipment in an automobile may be found at **www.manualslib.com/manual/851938/Ford-Mobile-Radio.html**. Power for ancillary equipment (wattmeters, remotely tuned antennas, and so on) should follow the same wiring rules. The use of a multiple outlet power distribution panel such as a RigRunner (**www.westmountainradio.com/rigrunner.php**) is also recommended. They're convenient and offer a second level of protection.

WIRE SIZE

The **Construction Techniques** chapter lists the current-handling capabilities of various gauges of wire and cable. The correct wiring size is one that provides a low voltage drop (less than 0.5 V under full load). Don't use wire at its maximum current carrying capacity.

Here's the formula for calculating the cable assembly voltage drop (V_d):

$$V_d = [(R_W \times 2\,\ell \times 0.001) + 2\,k] \times I$$

where

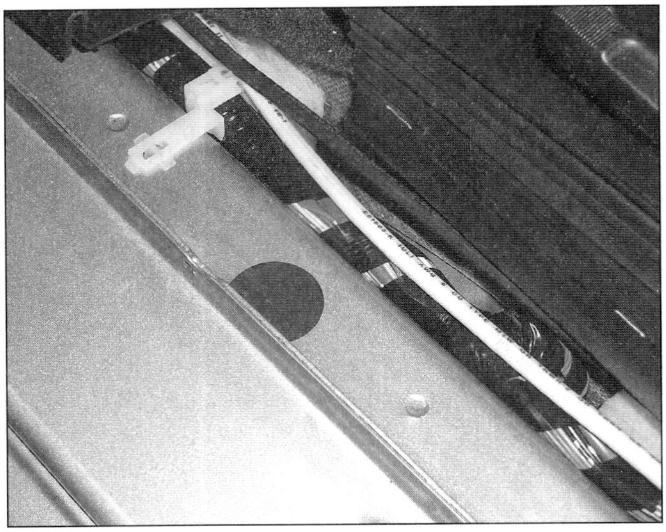

Figure 24.26 — Most vehicles have wiring troughs hidden behind interior body panels.

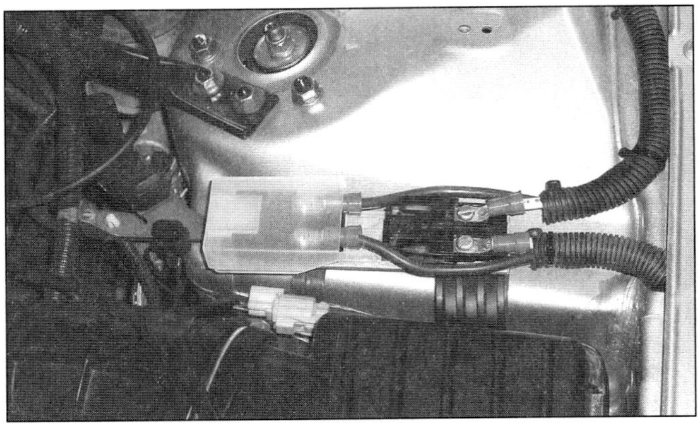

Figure 24.27 — Wiring attached to a fuse block.

R_W = resistance value (Ω per 1000 feet) from the **Construction Techniques** chapter.

ℓ = overall length of the cable assembly including connectors, in feet.

k = nominal resistive value for one fuse and its holder. Note: Most power cables have two fuses. If yours doesn't, use 1 k in the formula. If you don't know the fuse and holder resistance, use a conservative value of 0.002 Ω.

I = peak current draw in amperes for an SSB transceiver, or steady state for an FM radio.

For example, the peak current draw for a 100 W transceiver is about 22 A, and a typical power cable length is 10 feet. Using the resistive values for 1000 feet of #10 AWG wire (0.9987 Ω), and a conservative value for the fuse resistances (0.002 Ω each), the calculated drop is 0.527 V.

It's important to reiterate that the wire size should be selected for minimum voltage drop, not maximum power handling capability. The voltage drop is often referred to as "I-squared-R loss" — the current in amperes, squared, times the resistance — and should be held to a minimum whenever possible. In cabling, excessive I^2R losses can cause the wire to overheat with predictable results.

The insulation material of wire used in mobile installations should have a temperature rating of at least 90 °C, and preferably 105 °C. It should be protected with split-loom covering whenever possible, especially under-hood wiring.

Selecting the correct size fuse is also important. The average current draw for any given fuse should not exceed 60% of its rating. Thus, the correct fuse rating for a 22 A load is 30 A. That same 30 A fuse can han-

Battery Connections

For many years, connecting mobile station power leads directly to the battery (or to the battery positive and ground tie-point) has been the standard recommendation. In vehicles equipped with EIS (Engine Idle Shutoff), however, additional sensing modules in the vehicle electrical system may require alternate connections.

With EIS, as soon as the vehicle stops for a short time, the engine shuts off and the battery voltage drops. To support the additional starting cycles, starters and batteries are more robust, but so are the sophistication of the electrical devices supporting them. The most important device is the ELD (Electronic Load Detector), typically located within the main fuse panel.

ELDs have been in use for many years to measure the current drawn by the accessories (air conditioning, lights, and so on), which allow the engine CPU to more accurately adjust the air/fuel mixture. However, on vehicles equipped with EIS, the ELD is located in battery's negative lead or its connector. (See your vehicle's service manual for the exact location of the ELD.) The ELD is used for *coulomb-counting* to estimate the battery's State of Charge (SoC). Measuring current during starting provides an estimate of the battery's Reserve Capacity (RC). This ensures the battery has enough reserves to restart the engine when the engine has to start again. (See the **Power Sources** chapter for more information on vehicle batteries.)

During engine shutdown, most EIS systems use a dc-to-dc converter to assure that the accessories have a constant voltage source for electric motors that power brakes, air conditioning, transmission servo pressure, fuel pumps, and engine cooling systems. A second trunk-mounted battery may be used, or even super-capacitors (low-voltage capacitors with many farads of capacitance). The converter and accessory operation is under the control of the Battery Monitoring System (BMS) and amateur transceiver wiring must avoid circumventing its operation.

Connecting the transceiver directly to the battery would bypass the BMS which is not recommended, nor is connecting the radio to the dc-to-dc converter, to the trunk-mounted accessory-power battery, or to existing vehicle wiring. In vehicles equipped with a BMS and/or ELD the correct method of connecting the radio is to connect the positive lead directly to the battery, and the negative lead to the battery's chassis grounding point. The negative lead fuse should not be removed! This avoids damage caused by a loose or broken battery connection, which could cause a Load Dump Transient (LDT) to occur. (See the **RFI and EMC** chapter.) Should an LDT occur, the fuse might blow depending on the location of the battery lead failure. Without the fuse, damage to the transceiver could be the result.

If there is any doubt, check with your dealer about how the Battery Monitoring System (BMS) is connected and the recommended connection points for your radio's power leads. Be sure the connections you make are to points adequately rated for the load your radio presents. More information on batteries, alternators, and the newer vehicle power systems is available online at **www.k0bg.com**.
— *Alan Applegate, KØBG*

dle a 40 A load for about 120 seconds, and a 100 A load for about 2 seconds. Therefore, it pays to be conservative when selecting the carrying capacities of both wire and fuses. **Figure 24.28** shows the characteristics of several sizes of automotive fuses.

24.2.4 Amplifiers

Mobile HF amplifiers have been around for many years, and with the advent of high-power solid-state devices, they are also very common. However, running high power in a mobile environment requires careful planning. Considerations include, but are not limited to:

- alternator current ratings and battery capacity
- wiring (in addition to safe current ratings, excessive voltage drop can create distortion of the output signal)
- antenna and feed line power ratings
- placement and secure mounting in the vehicle
- wiring and placing of remote controls
- safe distances from the antenna to maintain exposure below FCC limits

See **www.k0bg.com/amplifiers.html** and the **Safe Practices** chapter for more information on these topics.

Before purchasing an amplifier, take a close look at your antenna installation and make sure it is operating efficiently. Using an amplifier with a poor antenna installation is counterproductive. Here's a rule of thumb applicable to any type of antenna: If the unmatched input SWR is less than 1.7:1 on 17 meters, or any lower frequency band, then it isn't mounted correctly, or you need a better antenna. Whatever antenna you use, it must be capable of handling the amplifier power level. More information on HF mobile antennas and installation techniques may be found in the **Antennas** chapter.

Mobile amplifiers for VHF/UHF operation are not as popular as they once were because most mobile transceivers have adequate output power (about 50 W). Boosting this to 150 – 300 W or more should be done with caution. Mobile VHF/UHF antennas for high power (>100 W) are rare, so check antenna ratings carefully. Those that are available need to be permanently mounted, preferably on the roof to avoid inadvertent contact and to minimize excessive exposure inside the car.

As with any high-power mobile installation, pay careful attention to RF safety, both for occupants in the car and people nearby. Be aware of safe exposure distances and the locations of people around your car while operating, and pause transmissions if people are too close. More information on RF exposure can be found in the **Safe Practices** chapter.

24.2.5 Interference Issues

In a mobile installation, radio frequency interference falls under two basic categories: *egress* (interference from the vehicle to your amateur station) and *ingress* (from your amateur gear to the vehicle). Most hams are familiar with ignition interference, since it is the most common form of egress. RF interference to an auto sound system is a common form of ingress.

Both types of interference have unique solutions. One solution common to both

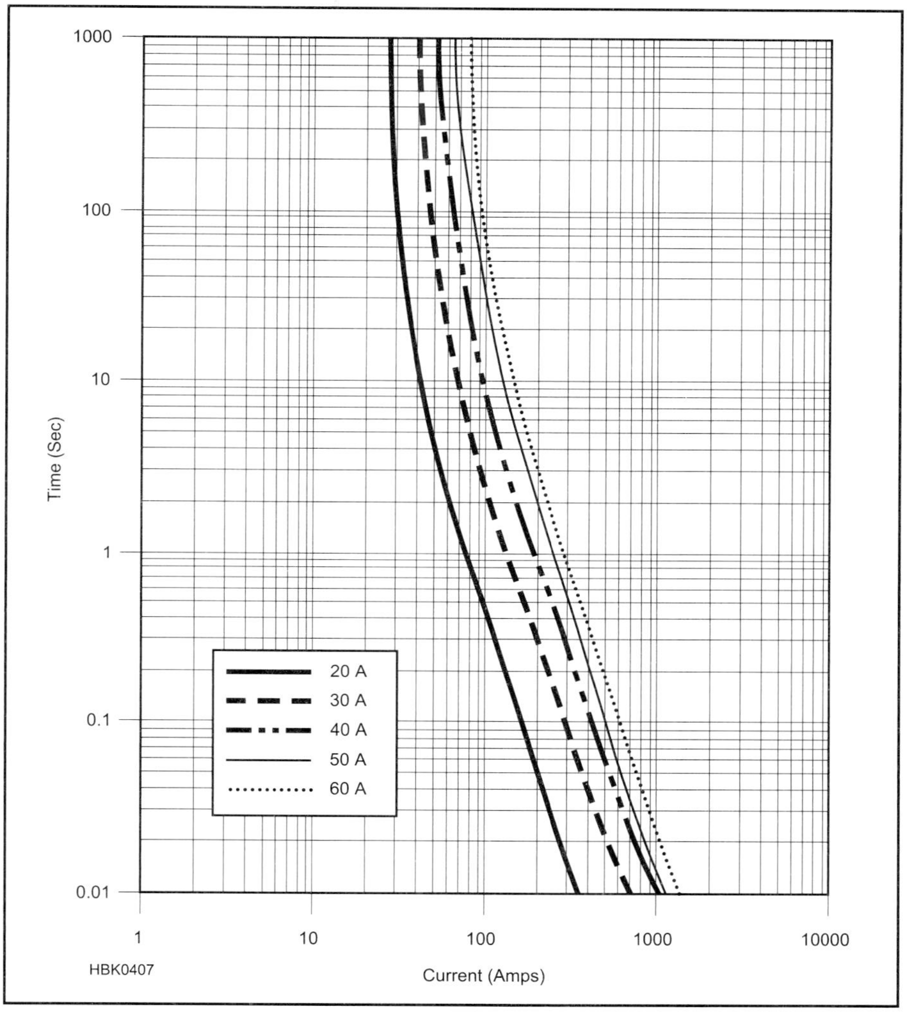

Figure 24.28 — Chart of opening delay versus current for five common sizes of Maxi fuses, the plastic-body high-current fuses common in vehicles. [Based on a chart from Littelfuse Corp]

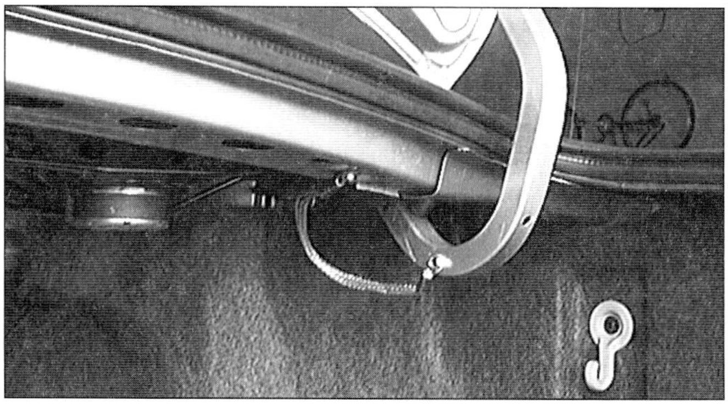

Figure 24.29 — Bonding vehicles parts — in this case, the trunk lid to the main body — can help reduce interference.

types is *chassis bonding*. Chassis bonding refers to connecting accessory equipment or assemblies to the frame or chassis of the vehicle. For example, the exhaust system is isolated from the structure of the vehicle and can act like an antenna for the RFI generated by the ignition system. It should be bonded to the chassis in at least three places. **Figure 24.29** shows an example of bonding. More information about these techniques is available at **www.k0bg.com/bonding.html**.

Other RFI egress problems are related to fuel pumps, air conditioning, engine cooling fans, ABS sensors, data distribution systems, and control system computers. These are best cured at the source by liberal use of snap-on ferrite cores on the wiring harnesses of the offending devices. Snap-on cores come is a variety of sizes and formulations, called mixes. The best all-around ferrite core material for mobile RFI issues is mix 31. Suitable cores are available from many distributors. Unspecified ferrite cores typically offer little HF attenuation and should be avoided. See the **RF Techniques** and **RFI and EMC** chapters for more information on ferrite cores.

Alternator whine can be another form of RFI egress. It is typically caused by an incorrectly mounted antenna that results in a ground loop, rather than a defective alternator diode or inadequate dc power filtering, which has been the traditional solution. Attempting to solve alternator whine with a dc filter can mask the problem and increases I^2R losses. Additional information on proper antenna mounting is in the **Antennas** chapter.

RFI ingress to the various on-board electronic devices is less common. The major causes are unchecked RF flowing on the control wires and common mode currents flowing on the coax cable of remotely tuned HF antennas. Again, these problems point out the need to properly mount mobile antennas.

For more information on RFI issues, see the **RFI and EMC** chapter, *The ARRL RFI Handbook*, and the ARRL Technical Information Service (**www.arrl.org/tis**).

24.2.6 Electric Vehicles

As of 2023, electric vehicles comprised about 16% of all new vehicles sold in the United States, and this percentage is expected to increase over the next decade. What is

Air Bags and Mobile Installation

Since 1998 all passenger vehicles are required to have Supplemental Restraint Systems (SRS), better known as *airbags*. Side airbags and airbags for rear seat passengers have become commonplace. When used in conjunction with seat belts, they've become a great life saving device, but they do have a drawback — they literally explode when they deploy!

Airbags deploy within 200 ms, expanding at about 200 mph, driven by gas from a controlled explosion. **Figure 24.A** drawing shows a typical vehicle with several air bags deployed in the passenger compartment. Any radio gear within range of an airbag can be ripped free with great force and flung about the interior. This should eliminate from consideration any dash-top mounting scheme, including windshield suction cup (mobile phone) mounts, so often employed.

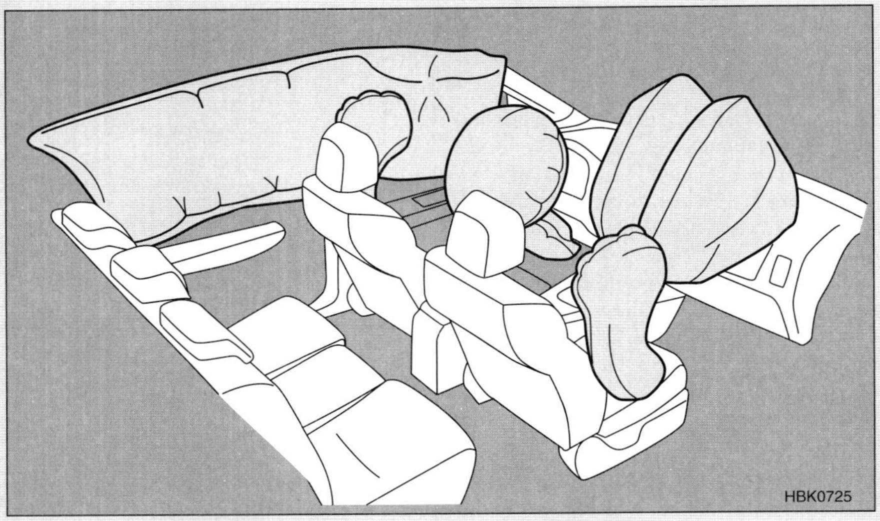

Figure 24.A — Airbag deployment zones in a modern vehicle.

Figure 24.B shows the passenger compartment of a vehicle with airbags deployed after a minor collision that caused less than $300 damage to the bumper. Note the loose piece of dashboard on top of the deflated air bag and the broken windshield. These are typical effects of a deploying airbag, whether from the top or center of the dash. Knowing how airbags deploy, avoid mounting radio gear anywhere near them.

It is always a difficult task finding a suitable mounting location for a transceiver and/or control head that is out of airbag range yet easily seen and operated. One workaround is a gooseneck mount (see **Table 24.1**, for a list of suppliers). These attach via a seat bolt (no hole needed). They're a good alternative as long as they're placed away from the passenger airbag deployment area (the whole right side of the dashboard). The dealer for your make and model may have additional guidelines for mounting radios and control heads in the car.

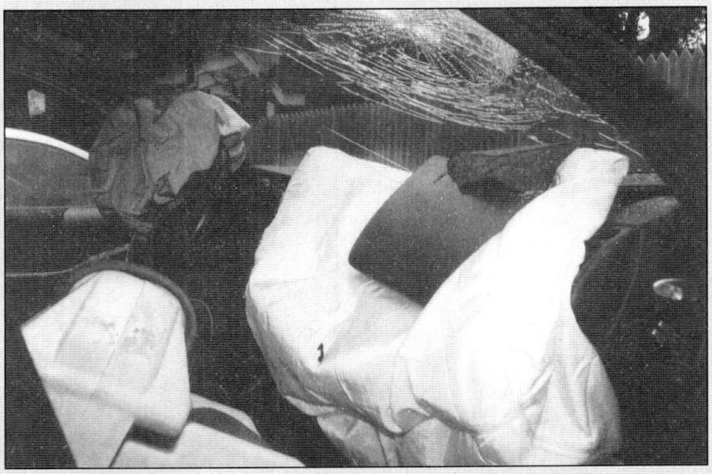

Figure 24.B — Airbag deployment following a minor collision caused significant disruption inside the passenger compartment.

considered an "electric vehicle" includes a number of hybrid variations that use electric motors for a significant amount of their drive power. Here are some examples:

Hybrid Electric Vehicles (HEVs) incorporate a traditional internal combustion engine (ICE), gasoline or diesel, with electric motor and battery (≈5 to 35 kW) subsystems, all integrated into the drive train. (Note that for electric vehicles, battery capacity is represented as the ability to deliver power in kW, instead of total energy storage in kW-hr.) The electric motor provides extra power when needed, as well as slow speed, battery-only operation. The batteries are recharged by the ICE, and by using regenerative braking.

Plug-in Hybrid Electric Vehicles (PHEV) are similar to HEVs, except their batteries may also be charged from the electrical grid. PHEVs often have larger capacity batteries (≈10-70 kW) to allow them to move on battery power alone.

All-Electric Vehicles (EV) run solely on battery power. Their batteries are much larger in capacity (up to 760 kW) than HEVs or PHEVs. Although these vehicles employ regenerative braking like their counterparts, the electrical grid is their primary energy source.

The control systems for electric vehicles are complex, using computer-controlled, solid-state electronics to convert battery dc to the ac that drives the motor(s). (See **Figure 24.30** for an example.) As a result, all types of electric vehicles produce prodigious amounts of RFI, well into VHF. Manufacturers do their best to properly filter these switching products, however, the level of RFI may make amateur mobile operation difficult in certain models.

Electric vehicles are classified by the FCC as *incidental radiators*, falling under Part 15 rules. The rules have specific exemptions, such as Part 15.103(a), which reads: "A digital device utilized exclusively in any transportation vehicle including motor vehicles and aircraft." In other words, vehicles are exempt from Part 15 limits, whether receiving or causing RFI.

Electric vehicle warranties can be a major hurdle to installing third-party amateur radio equipment, as well. Consult the warranty conditions and the vehicle dealer service department about installing radio equipment without voiding the vehicle warranty. The combination of warranty restrictions and RFI can make operating from an electric vehicle a challenge, but it is not impossible.

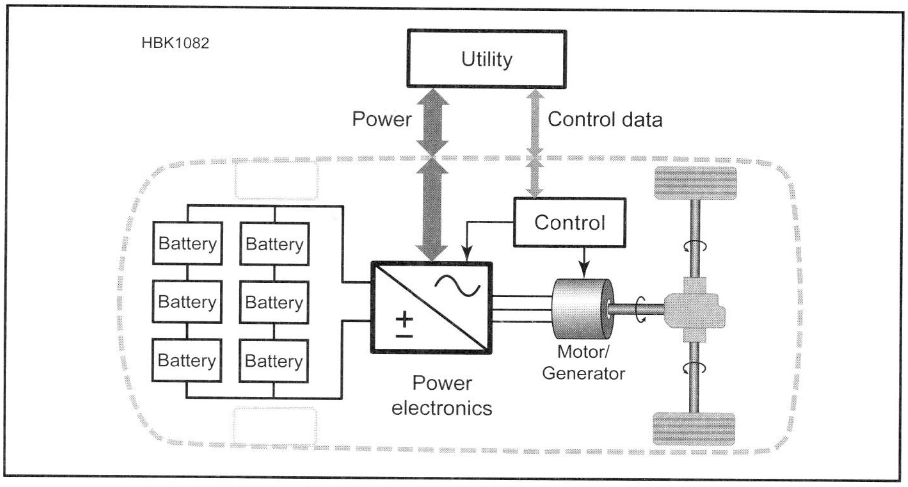

Figure 24.30 — Power system architecture for a typical electric vehicle.

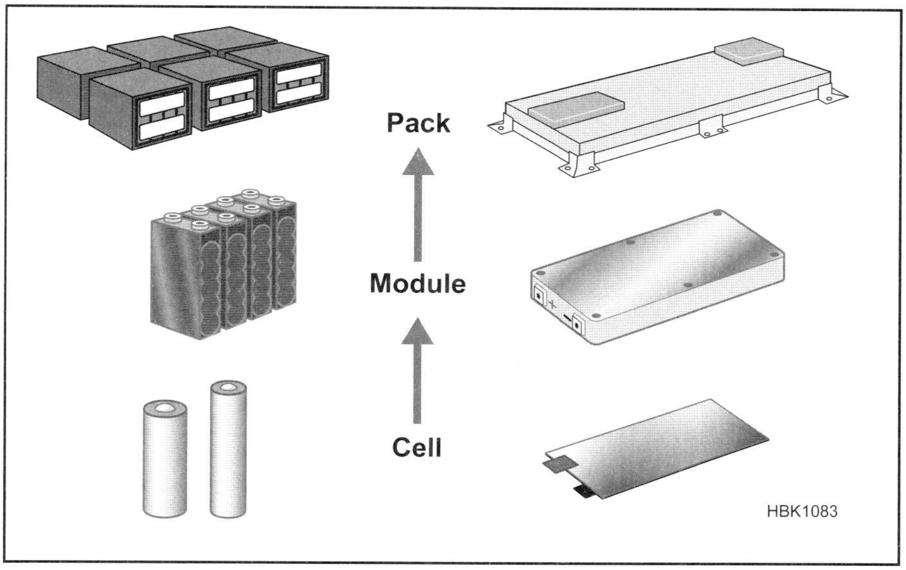

Figure 24.31 – Battery packs are composed of individual cells that are assembled into modules. The modules are then combined to form the overall battery pack that can realize overall voltages as high as 650 volts and currents as high as 2,000 amps.

Hazards from High Voltage

The *propulsion batteries* in most electric vehicles are lithium-ion based (Li-ion), but other types are used. Individual cells are connected in a series/parallel arrangement, so that current and voltage from the overall *battery pack* meet specific performance requirements. A microprocessor-controlled charging circuit ensures an equal charge for each cell, and each cell bank.

Battery pack voltages range from 150 volts to as high as 450 volts. Current capability may be as high as 2,000 amps. Once converted to ac, the peak voltage driving the motors may be as high as 650 volts. Needless to say, these are lethal voltages (see **Figure 24.31**).

Keeping occupants safely isolated from these voltages is a design requisite. However, installing amateur radio equipment in an electric vehicle may place the installer within harm's way of these lethal voltages. Professional help may be required to prevent injury or death.

Power in Electric Vehicles

Some models use a small 12-volt battery to power accessories (lights, radio, etc.) while the vehicle is stopped and the engine isn't running. Others may use inverters to convert the propulsion battery's high voltage to a lower voltage (not necessarily 12 V), to power these accessories. Check to see if your vehicle has a 12-volt system that can power the expected radio equipment. If not, you have to provide a 12-volt system that is independent of the vehicle's power system, and you are required to charge it externally.

Distracted Driving — Don't Do It!

Driver distraction has replaced drunk driving as the number one cause of vehicular crashes and deaths. This has prompted The National Highway Traffic Safety Administration (NHTSA) to address the problem with a nationwide ad campaign (**www.nhtsa.gov/risky-driving/distracted-driving**). Although texting is the number one cause, it isn't the only driver-distraction factor.

Manipulating any vehicular control, especially touch-screen operations, can — and does — cause driver distraction. Automobile manufacturers have partially addressed the issue by equipping vehicles with voice control, but you must still push a button to activate voice control. Even software-driven, self-driving, automatous vehicles, require input from you, the driver.

Please, don't allow amateur radio to become the cause of a distracted crash. When driving conditions warrant your attention, hang up the microphone.

Using what are now called *Accessory Sockets* (previously known as cigar or cigarette lighter sockets) to power amateur gear has many caveats. These are used for low-power equipment from dash cams to radar detectors and even computers. However, they are not designed to power amateur radio gear, nor should they be used to do so. Regardless of the circuit's fuse size used to feed them power, the electrical harness' wire size is too small (≈ #14 or #16 AWG) for continuous loads larger than 4 amps. Heavier loads may not blow the fuse, but it can overheat the power wiring bundle, possibly causing a fire or other electrical damage. Built-in USB charging ports may only supply 2 A or less at 5 V, which is not enough to power mobile-type radios.

Using these sources to power amateur radio gear may also void the vehicle's warranty. Checking with your dealer is the first line of defense in these cases. Even if allowed under warranty provisions, keep in mind that rated current is limited to a few amps, and thus the choice of transceivers is also limited.

One exception to these limits is Ford's electric F150 Hybrid pickup truck and the all-electric (EV) F150 Lightning model truck. Both have provisions for providing 120/240 V ac electric power even when in motion. Other manufacturers are likely to follow suit.

Antennas in Electric Vehicles

As noted elsewhere in this chapter, it is important to minimize common-mode RF current on feed lines. This is necessary in electric vehicles because of the stronger noise signals that may be picked up by the feed line and conducted to the radio equipment. In addition, radiated RF from the common-mode current may also cause RFI to the operation of the vehicle.

Any permanent mobile antenna installation should be directly attached to the vehicle body. This typically requires drilling holes in sheet metal and using chokes to block common-mode current. Except for temporary operation, mag-mount antennas should be avoided, especially on HF, because the feed line shield carries much more common-mode current than for a directly attached antenna in the same location.

Even though the antenna is close to the source of RF noise from the vehicle, you may be able to find mounting locations that pick up less noise. When testing antenna placement, be sure to decouple the feed line by using ferrite chokes. Finding a lower-noise location may also result in less RF pickup by the vehicle wiring.

RFI from Electric Vehicles

The electronics that convert dc battery power to multi-phase ac motor power use switching waveforms with very fast rise and fall times. This generates large amounts of high-frequency noise that covers the entire MF-HF spectrum, and even well into VHF. Switching noise is the primary source of RFI to amateur radio equipment in electric vehicles.

Electric vehicles also use a lot of the same types of electric motors, solenoids, switches, and control electronics that are found in non-electric vehicles. The same types of noise are radiated from these devices and can interfere with any amateur radio equipment within feet, or even inches, of the noise source. One difference with electric vehicles is that adding chokes or other RFI remedies may not be permitted under the vehicle's warranty due to the more complex power and control electronics systems. If you add components to the vehicle wiring to reduce RFI, do it in small steps, testing vehicle and noise performance along the way.

RFI to Electric Vehicles

While all modern vehicles have RF-sensitive electronics on board, electric vehicles have electronics that control every facet of vehicle operation, usually communicated over digital networks such as the CAN (Controller Area Network). These circuits are well-filtered, which reduces the chance for stray RF to negatively affect them. However, the vehicle wiring and electronics are very close to the antennas that are radiating a strong signal. As more vehicles adopt partial or complete automated driving, RFI can present serious safety challenges.

We need to be aware that on-board electronics may be susceptible to the high levels of RF generated by amateur transmitters, regardless of the frequency of operation. Even when allowed, proper antenna installation is very important in keeping common-mode RF current pickup by cables and wiring harnesses to a minimum.

If the level of radiated RF is high enough, the digital network signals may be corrupted. This might cause annoying behavior of the car, such as the windshield wipers operating or the entertainment system squealing. However, it can also make other circuitry fail to operate correctly, which can lead to unexpected consequences, particularly in safety equipment and systems.

As amateur radio and vehicles use more digital protocols and technology, susceptibility to RFI in electric vehicles should be reduced. In the interim, amateurs have to remain cautious and be prepared to be flexible in their approach to mobile operating.

24.2.7 Mobile Operating

The most important consideration while operating mobile-in-motion is safety! Driver distraction is a familiar cause of vehicle crashes. While amateur radio use is less distracting than mobile phones or texting, there are times when driving requires all of our attention. When bad weather, excessive traffic, or a construction zone require extra care, play it safe — hang up the microphone and turn off the radio!

In addition to properly installing gear, a few operating hints can make your journey less distracting. One of those is familiarization with your transceiver's menu functions and its microphone keys (if so equipped). Even then, complicated programming or adjustments are not something to undertake while driving.

Logging mobile contacts has always been difficult. Inexpensive compact digital voice recorders have made that function easier. Units with up to 24 hours of recording time are available for less than $50.

For maximum intelligibility at the other end, avoid excessive speech processing and too much microphone gain. Don't shout into the microphone! It's human nature to increase your speaking level when excited or when the background noise level increases. In the closed cabin of a vehicle, your brain interprets the reflected sound from your own voice as an increase in background level. Add in a little traffic noise and by the end of your transmission you're in full shout mode! One solution is to use a headset and the transceiver's built-in monitor function. Doing so gives you direct feedback (not a time-delayed echo), and your brain won't get confused. Note that headset use is not legal in some jurisdictions and never legal if both ears are covered.

Overcoming vehicle ambient noise levels often requires the use of an external speaker and all too often it is an afterthought. Selecting a speaker that is too small accentuates high frequency noise, which makes reception tiresome. For speakers that are too large, mounting becomes a safety issue.

For best results, use at least a 4-inch speaker. Rather than mounting it out in the open, mount it under the driver's seat so that it's out of the way. This attenuates the high frequency noise, and enhances the mid-range response, which increases intelligibility.

Many vehicle audio systems now support audio input from external sources including analog (headphone audio), USB, and Bluetooth. You may be able to route your mobile rig's audio through the vehicle audio system using one of these methods. The audio sounds far better than through a small speaker!

Most modern transceivers contain some form of DSP (digital signal processing) noise reduction as covered in the **DSP and SDR Fundamentals** chapter. Some are audio-based and some are IF-based, with the latter being preferred. Both types do a decent job of reducing high-frequency hiss, static spikes, and even ignition hash.

24.3 Portable Stations

Many amateurs experience the joys of portable operation once each year in the annual emergency exercise known as ARRL Field Day. All year long, amateurs are operating with portable stations from parks, mountain tops, islands, and more. As lightweight equipment becomes more effective, expect more of this type of activity.

A good portable setup is simple. Although you may bring lots of gear to Field Day and set it up the day before, during a real emergency speed is of the essence. The less equipment to set up, the faster it will be operational. Another benefit of simplicity is that the easier it is for you to collect the equipment and head for the operating site, the more likely you will be to do it!

If you expect to frequently operate a portable station, check out some of the online communities that have developed to support popular "on the air" programs like Parks on the Air (POTA; **parksontheair.com**) and Summits On the Air (SOTA; **www.sota.org.uk**). QRP clubs and societies like QRP ARCI (**www.qrparci.org**) publish many helpful projects, equipment reviews, and even sponsor operating events and conventions. There are many helpful videos available online and social media programs, as well.

We'll begin with power sources for field operating. Generators and batteries are by far the most popular. Solar panels are increasing in efficiency and small wind turbines are sometimes used — even human power can be used, such as from a bicycle-powered generator — but nearly always to charge a battery, which won't be covered here. The books *Emergency Power for Radio Communications* by Mike Bryce, WB8VGE, and *Energy Choices for the Radio Amateur* by the late Bob Bruninga, WB4APR, are good resources for information about off-grid power sources.

24.3.1 AC Generators

Essentially, a generator is a motor that's operating "backward." When you apply electricity to a motor, it turns the motor's shaft (allowing it to do useful work). Take the same motor, physically rotate its shaft, and it generates electricity across the same terminals used to supply power when used as a motor. Turn the shaft faster and the voltage and frequency increase. Turn it slower and they decrease. To some degree, all motors are generators and all generators are motors. The differences are in the details and in the optimization for specific functions. Consumer-class ac generators are classified as *home backup* (for powering an entire home during power outages), *portable* (basic gas-powered generators for tools and appliances), *inverter* (generators coupled to solid-state inverters), and *portable power stations* (batteries packaged with an inverter, discussed in the battery section below).

PORTABLE GENERATORS

A "motor" that is optimized for generating electricity is an alternator, just like the one in your car. The most basic generators use a small gas engine directly coupled to an ac alternator, so the output voltage and frequency depend on the engine speed. If the engine is running too fast or too slow, the voltage and frequency of the output will be off. When everything is running at or near the correct speed, the voltage and frequency of the output is a close approximation of the power supplied by the ac mains — a 120 V ac sine wave with a frequency of 60 Hz. These are referred to as *constant-speed generators*. Most consumer models use two-pole armatures that run at 3,600 RPM to produce a 60 Hz sine wave.

A standard generator *must* turn at a specific speed in order to maintain output regulation, so when more power is drawn from the generator, the engine must supply more torque to overcome the increased physical/magnetic resistance in the generator's core. The generator *can't* simply spin faster to supply the extra power.

Most generators have engines that use mechanical or vacuum "governors" to keep the generator shaft turning at the correct speed. If the shaft slows down because of increasing generator demand, the governor "hits the gas" and draws energy stored in a heavy rotating flywheel, for example, to bring (or keep) the shaft speed up to par. The opposite happens if the generator is spinning too fast.

In addition to mechanical and vacuum speed regulating systems, generators that are a step up in sophistication additionally have electronic automatic voltage regulation (AVR) systems that use special windings in the generator core (and a microprocessor or circuit to monitor and control them) to help keep things steady near 120 V and 60 Hz. AVR systems can respond to short term load changes much more quickly than mechanical or vacuum governors alone. They are mostly used in higher quality 5-15 kW "home backup systems" and in many recreational vehicles.

INVERTER GENERATORS

Inverter generators produce high-voltage, multiphase ac that is rectified to dc, similar to an automobile alternator. The dc power is then converted back to very clean and consistent ac power by a microprocessor-controlled solid-state power inverter. Unlike the constant-speed generators, inverters can run

Things to Consider for Operating Portable Stations

Safety

• Being the cause of a medical emergency will not ingratiate you or any future amateur operators with the public or the facility administration. Think ahead!

• RF Safety: Ensure that your personal RF exposure and the RF exposure of anyone who happens by is within published safety limits. Have some idea of safe distances before setting up and flag areas of caution.

• Physical Safety: Radial wires, coaxial cable, power cords, and guy lines are all trip hazards. Antennas and masts are often top heavy. Generators must be well-ventilated and fuel supplies secured. Have a basic first-aid kit with you.

• Electrical Safety: High RF voltages exist in many antenna systems and batteries usually contain significant electrical energy. Use high-quality power distribution cables and boxes and keep them dry!

Permission and Permits

• Whether public or private, almost all property is covered by some set of rules. You need to ensure that access to the location is authorized during the hours you expect to be there and that the operation of an amateur radio station is allowed.

• Have a copy of any necessary permits. The use of trees for supporting antennas or driving stakes in the ground may not be allowed.

• RF emissions are restricted around certain research and medical facilities.

• Be 100% certain that your operation is allowed.

Physical Presence

Being seen and not heard is great advice for operating in a shared public space. Avoid needlessly annoying other visitors. Loud generators, noisy static-CW-garbled voices, and shouting into a microphone are all bad ideas around non-hams.

• Taking up an entire pavilion or most of an open area with antennas will not impress other park patrons and staff, whether you have a permit to operate or not.

Practice Before you Go

• Set up your station outside at home to be sure you have all the parts and pieces and connections. Whatever you need to complete the station should be close at hand.

• Operate the station and make a few contacts to ensure that you really do have everything. You'll find all sorts of things to learn or adjust or fix before doing it "for real." This is the time to solve any RFI problems or intermittent problems.

• Assess what spares, supplies, and tools you needed or are likely to need. Unless you are going with a bare minimum setup; a roll of electrical tape, a multi-tool of some kind, and a spare length of coax are essentials, for example.

• As you take the station apart afterwards, make a checklist of all equipment. Note the individual pieces of gear and also the containers into which stuff has been packed. Include the non-radio equipment in your list: chair, bug spray, cell phone power bank, water and sunscreen, hat and gloves, etc. Include everything you were using when you were outside at home.

• After the actual operation, supplement your checklist with things you noticed on-site. What worked well and what didn't, things forgotten or needed, and improvements for next time are all part of what public service teams call a "hot wash." The time for this is right away while your memory is still fresh.

Preparing Yourself

• Make a plan for food and have plenty of water. If you have any health issues, plan to compensate for them and mitigate any risks.

• Dress appropriately for the climate and be prepared for changes in the weather.

• Get experience in operating with the modes you plan to use. In the field is not the time to learn new tactics.

• If participating in an "on the air" program (POTA, SOTA, IOTA, etc.) or competition, be sure you know the rules and any requirements for logging contacts.

• Be sure to tell someone your plans, including where you are going and when you'll be home!

Interacting With the Public

Assume visitors want to know what you are doing. Bring a printout of some kind to explain what ham radio is and where they can go for more information. Someone curious enough to stop and ask you a question is a good prospect for amateur radio. A minute or two to be friendly and courteous won't cost you many contacts, and who knows what you will gain!

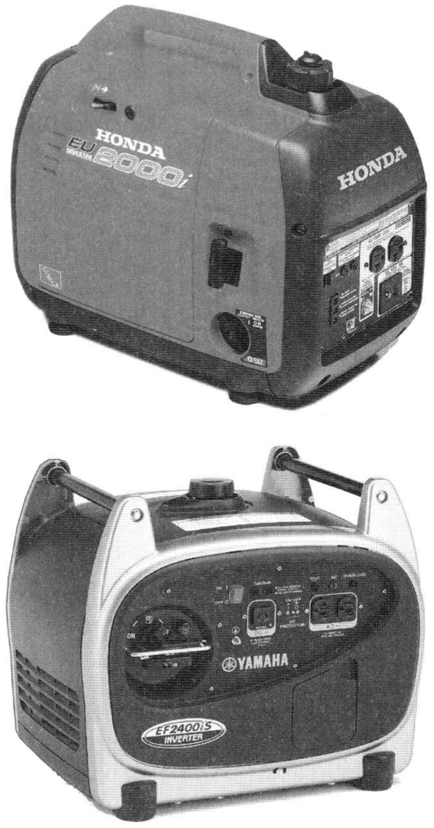

Figure 24.32 — Modern inverter generators from Honda and Yamaha. See Table 24.2 for a partial list of specifications.

at idle while still providing power, increasing speed to meet additional demand. This improves economy and reduces emissions. The most common models are available with capacities up to 2000 W output, and some can be paralleled with special cables for higher capacity. The June 2012 *QST* Product Review "A Look at Gasoline Powered Inverter Generators" compares several popular models available at that time and can be found in this book's online content.

Basic portable generators are intended to power lights, saws, drills, ac motors, electric frying pans, and other devices that are not dependent on clean sine-wave power. If you want the highest margin of protection when powering computers, transceivers, and other sensitive electronics, an *inverter generator* is the best way to go. Some popular examples are shown in **Figure 24.32**, and their key specifications are shown in **Table 24.2**. Available in outputs ranging from 1 to 5 kW, these generators use one or more of the mechanical regulation systems mentioned previously, but their ultimate benefit comes from the use of a built-in ac-dc-ac inverter system that produces clean 60 Hz sine waves at 120 V ac, with a 1% to 2% tolerance, even under varying load conditions.

Table 24.2

Specifications of the Inverter Generators Shown in Figure 24.32

Make and Model	*Output (W) (Surge/Cont)*	*Run Time (h) Full / 25% Load*	*Noise Range (dBA@21 feet)*	*Engine Type*	*Weight (Pounds)*	*Notes**
Honda EU2000i	2000 / 1600	4 /15	53-59	100 cc, OHC	46	a,b,c,d
Yamaha EF2400iS	2400 / 2000	N/A / 8.6	53-58	171 cc, OHV	70	a,b,c

*a — has 12 V dc output; b — has "smart throttle" for better fuel economy;
c — has low oil alert/shutdown; d — replaced by EU2200i with similar specifications

Instead of using two windings in the generator core, an inverter generator uses 24 or more windings to produce a high frequency ac waveform of up to 20 kHz. A solid-state inverter module converts the high frequency ac to smooth dc, which is in turn converted to clean, tightly regulated 120 V ac power. Most inverter generators are compact, lightweight, and quiet. Low-noise operation is important if you plan on operating at a site where other people might be present.

GENERATOR CONSIDERATIONS

In addition to capacity and output regulation, factors such as engine type, noise level, fuel options, fuel capacity, run time, size, weight, cost, or connector type, may factor into your decision. Consider additional uses for your new generator beyond Field Day or other portable operation.

Power Rating

Your generator must be able to safely power all the devices that are attached to it. Simply add up the power requirements of *all* the devices, add a reasonable safety margin (25 to 30%), and choose a suitably powerful generator that meets your other requirements.

Some devices, especially electric appliances, take a lot more power to start up than they do to keep running. A motor that takes 100 W to run may require a surge of 200 W or more for a brief period when started. Many items don't require extra start-up power but be sure to plan accordingly.

Always plan to have more capacity than you require or, conversely, plan to use less gear than you have capacity for. Running close to maximum power is bad for your generator *and* your gear. Some generators are somewhat overrated, probably for marketing purposes. Give yourself a margin of safety and don't rely on built-in circuit breakers to save your gear during overloads. When operating at or beyond capacity, a generator's frequency and voltage can vary widely before the current breaker trips.

Size and Weight

Size and weight vary according to power output. Low power units are lightweight and physically small. Watt-for-watt, however, most modern units are smaller and lighter than their predecessors. Models suitable for ham radio typically weigh between 25 and 125 pounds.

Engines and Fuel

Low-end portable generators are typically powered by low-tolerance, side-valve engines of the type usually found in discount store lawnmowers. They're noisy, need frequent servicing, and often fail relatively quickly. Better models have overhead valve (OHV) or overhead cam (OHC) engines, pressure lubrication, low oil shutdown, cast iron cylinder sleeves, oil filters, electronic ignition systems, and even fuel injection. These features may be overkill for occasional use but are desirable for more consistent power needs.

Generators used for emergency power are frequently left sitting idle for long periods of time. Running the generator up to 30 minutes every so often — six months is frequently suggested — is a good way to keep it ready. After you use the generator for an emergency, or to keep it in good condition, run it until the gas tank and carburetor are completely dry. Another option is to just close the fuel shut-off valve and run the generator until the carburetor is dry. This prevents the build-up of films known as "varnish" in the carburetor.

If you store the generator with fuel in the tank, use a fuel stabilizer additive to prevent varnish build up and to absorb any condensation that may occur. Ethanol-free gasoline, if available, holds up better for long storage periods. Generator shops, power sports dealers, power equipment dealers, and RV websites are good sources of recommendations for additives and sources of ethanol-free gas. Gasoline stored in a container should also be regularly rotated out by using it in a vehicle and refilling the container, or an additive for storing gasoline should be added.

Run Time

Smaller generators usually have smaller gas tanks, but that doesn't necessarily mean they need more frequent refueling. Some small generators are significantly more efficient than their larger counterparts and may run for half a day while powering small loads. Some generators offer separate gas tanks as an accessory.

As with output power, run times for many units are somewhat exaggerated and are usually specified for 50% loads. If you're running closer to maximum capacity, your run times may be seriously degraded. The opposite is also true. Typical generators run from three to nine hours on a full tank of gas at a 50% load. Make a practice run with a typical load (electric lights are good test loads) to see what the actual gas consumption is and use that information to plan for your actual portable operation.

Audible Noise

Portable generators for tools, pumps, and appliances are almost always loud. Audible noise levels for many models are stated on the box, but try to test them yourself or talk to someone who owns the model you're interested in before buying. Environmental conditions, distance to the generator, and the unit's physical orientation can affect perceived noise levels. Inverter generators are much quieter.

Generators housed in special sound dampened compartments in large boats and RVs can be much quieter than typical "outside" models. However, they are expensive and heavy, use more fuel than compact models, and most don't have regulation specs comparable to inverter models. If you are concerned about audible noise levels, consider building a small plywood enclosure with at least one open side to direct sound away from people. Be sure the enclosure has adequate ventilation.

Output Regulation

For hams, voltage and frequency regulation are very important since most ham equipment is designed to be operated directly or from a power supply using ac line power. AVR units with electronic output regulation (at a minimum) and inverter generators are highly desirable and should be used exclusively, if only for peace of mind.

Unloaded portable generators can put out as much as 160 V ac at 64 Hz. As loads increase, frequency and voltage decrease. Un-

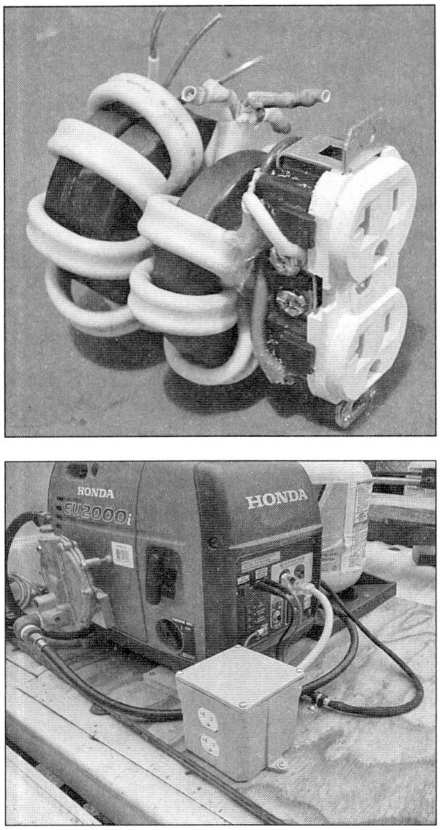

Figure 24.33 — A heavy-duty common-mode RF choke wound on Type 31 ferrite cores for ac generators. K9YC and W6GJB use this choke during portable operating. See the text for construction details.

der full load, output values may fall as low as 105 V at 56 Hz. Normal operating conditions are somewhere in between.

Some have tried inserting uninterruptible power supplies (UPSs) between the generator and their sensitive gear. These devices are often used to maintain steady, clean ac power for computers and telecommunication equipment. As the mains voltage moves up and down, the UPS's regulator system adjusts the output voltage accordingly. The unit's internal batteries provide power to the loads if the ac mains (or your generator) go down.

In practice, however, most UPSs can't handle the variation in frequency and voltage of a generator powered system. When fed by a portable generator, most UPSs constantly switch in and out of battery power mode — or won't switch back to what is perceived to be ac power. When the UPS battery goes flat, the unit shuts off. Not all UPSs and generators interact like this, but an inverter generator is a better solution.

RF Noise

Some generators create RF noise in the HF bands as common-mode current on connected power cords. The choke in **Figure 24.33** was designed by Jim Brown, K9YC, and built by Glen Brown, W6GJB, to suppress the noise. The assembly consists of a pair of chokes in series, each consisting of 8 turns of #14 AWG Romex-type cable wound on a type #31, 2.4-inch OD ferrite toroid. One choke is wound on two cores for 80- and 40-meter coverage. The second choke is wound on one core for 10 MHz and above. (Figure 24.33 upper photo) The chokes are attached directly to a duplex outlet installed in an electrical enclosure (Figure 24.33 lower photo). A heavy-duty extension cord connects the choke to the generator.

DC Output

Some generators have 12 V dc outputs for charging batteries. These range from 2 A trickle chargers to 100 A quick-charge outputs. Typical outputs run about 10 to 15 A. As with the ac outputs, be sure to test the dc outputs for voltage stability (under load, if possible) and ripple. Large batteries can tolerate some ripple in the charging circuit, but your radio might not tolerate it at all! Be sure your battery can be charged with the type of charging output from the generator. Lead acid batteries are very tolerant, but lithium-chemistry and nickel-metal-hydride batteries require special chargers.

Miscellaneous

Other considerations include outlets (120 V, 240 V, and dc output), circuit breakers (standard or ground fault interrupter type), fuel level gauges, favorite brands, warranties, starters (pull or electric), wheels, handles, and whatever else you require for your expected operating conditions.

SETUP, SAFETY, AND TESTING

Before starting the engine, read the user manual. Carefully follow the instructions regarding engine oil, throttle, and choke settings (if any). Be sure you understand how the unit operates and how to use the receptacles, circuit breakers, and connectors. If there is a fuel tank shutoff valve, be sure it is open before attempting to start the engine.

Make sure the area is clean, dry, and unobstructed. Generators should *always* be set up *outdoors*. Do not operate gas powered engines in closed spaces, inside vans, in covered pickup beds, etc. If rain is a possibility, set up an appropriate canopy or other *outdoor protective structure*. Operating generators and electrical devices in the rain or snow can be dangerous. Keep the generator and any attached cords dry.

Exhaust systems can get hot enough to ignite certain materials. Keep the unit several feet away from buildings and keep the gas can (and other flammable items) at a safe distance. Don't touch hot engines or mufflers.

When refueling, shut down the generator and let things cool off for a few minutes. Don't smoke, and don't spill gasoline onto hot engine parts. A flash fire or explosion may result. Keep a small fire extinguisher nearby. If you refuel at night, use a light source that isn't powered by the generator and can't ignite the gasoline. Generators with separate fuel tanks can be refueled with the generator running if the tank is located away from the generator.

Testing

Before starting (or restarting) the engine, disconnect or turn off all electrical loads. Starting the unit while loads are connected may not damage the generator, but transients or voltage over-shoot may damage solid-state devices. The engine is much harder to start under load, as well. After the engine has warmed up and stabilized, test the output voltage (and frequency), if possible, *before* connecting loads.

Because unloaded values may differ from loaded values, be sure to test your generator under load (high wattage quartz lights or electric heaters are good high-power loads). Notice that when you turn on a hefty load, a portable generator "hunts" a bit as the engine stabilizes. An inverter generator should always output regulated ac. Measure ac voltage and frequency again to see what the power conditions is like under load. See your unit's user manual or contact the manufacturer if adjustments are required.

Safety Grounds and Field Operation

The National Electric Code (NEC) addresses safety grounding for these type of generators in section 250.34. It describes "portable" as a generator that is easily carried from one location to another by a person. "Mobile" applies to generators that are capable of being moved on wheels or rollers, and includes generators mounted in vehicles.

A ground rod or other direct earth connection is not required for portable generators as long as the generator has receptacles mounted on the generator panel and the receptacles have equipment grounding terminals (i.e., the third ground pin of an ac receptacle) that are bonded to the frame of the generator. Equipment must be connected to the generator through a suitable cord and plug, such as a standard extension cord. Any exposed metal surface of the equipment must be connected through the ground wire of the power cord to the receptacle ground terminal. If the generator is mounted in a vehicle the same rules apply as long as the equipment supplied by the generator is mounted on the vehicle and the frame of the generator is bonded to the frame of the vehicle.

Ground rods may be used if desired. If the generator is more than a short distance from the station, or if more than one separate station is powered by the same generator, a ground rod at the generator and at each station may be prudent. Bond all ground rods together, don't depend on ground conductivity.

Regardless of the grounding method you choose, a few electrical safety rules remain the same. Your extension cords *must* have intact, waterproof insulation, three "prongs" and three wires, and must be sized appropriately for the loads and cable runs. Use #14 to #16 AWG, three wire extension cords for low wattage (under 500 watts) runs of 100 feet or less. For high wattage loads, use heavier #12 AWG, three-wire cords designed for air compressors, air conditioners, or RV service feeds. Try to position extension cords so they won't become a trip hazard (particularly at night) or be run over by vehicles. Secure long cords to stakes so if someone does trip, the cord does not pull equipment off an operating position or disconnect from the generator. Never run electrical cords through standing water.

During operations, try to let all operators know when the generator will be shut down for refueling so radio and computer gear can be turned off in an orderly manner. Keep the loads disconnected at the generator or powered down until the generator has been refueled and restarted.

24.3.2 Battery Power Sources

If a generator is not available to supply ac voltage for portable operation, batteries are often used for portable or Field Day operation. (See the **Power Sources** chapter for complete information on batteries, including charging, and the **Safe Practices** chapter for safe handling of batteries.) Alternative energy sources such as solar panels and wind generators can be used to charge batteries, as well.

Battery technology has advanced considerably in recent years in support of alternative energy and electric vehicle applications. *Lithium-Iron-Phosphate* (LiFePO4 or LFP) and *lithium-polymer* (LiPo) batteries in particular have become widely used for portable operation, including ultra-light backpacking. (LiPo refers to the method of battery construction using polymer separators between the anode and cathode and can use any lithium chemistry.) The *sealed lead-acid* (*SLA*) battery is rapidly being displaced for portable stations that are not in a vehicle or marine vessel. **Table 24.3** is a comparison of the lead-acid and LiFePO4 battery types with data excerpted from the more complete table in the Batteries section of the **Power Sources** chapter. See that chapter's sections on battery selection for a more complete comparison of battery chemistries in mobile and portable stations.

Table 24.3 Lead-Acid and LiFePO4 Comparison

Specification	*Lead-Acid*	*LiFePO4*
Specific energy (Wh/kg)	30-50	90-120
Cycle life (1,2)	200-300	1000-3000
Fast-charge time	8-16 hr	1 hr or less
Overcharge tolerance	High	Low
Trickle charge	Yes	No
Safety requirements	Thermally stable	Protection circuit mandatory (3)
Toxicity	Very high	Low

1 — Based on depth of discharge (DoD). Shallow DoD improves cycle life.
2 — Based on current (Feb 2022) manufacturer information
3 — Usually built-in to battery packs

Battery technology continues to evolve. While the LFP chemistry is currently the most popular choice for portable operating, other variations and completely different chemistries and physical form factors are likely to appear in the coming years.

VOLTAGE STABILITY

Voltage stability is a significant concern for battery-power operation. Most solid-state transceivers are designed to operate at a nominal 13.8 V from a lead-acid battery being charged by a vehicle's engine. As battery voltage drops during use, transmit signal quality begins to drop, sometimes drastically, and output power drops off as well. Further, most transceivers simply shut themselves off once their input voltage drops to approximately 11.6 V. Even for a partially charged battery, voltage drops in power cables and connectors when transmitting can cause the radio to shut off intermittently or operate improperly.

Several manufacturers and distributors sell "battery boosters" that provide a regulated output of 13.8 V as battery voltage varies. Booster efficiency varies with the input voltage and current draw but is typically around 80%, a significant loss of capacity.

Another option is a dc-to-dc converter that runs from a nominal input of 24 or 48 V, providing a regulated output of 13.8 V. In this case, two (or more) 12 V batteries are wired in series or in series-parallel to extend operating time. Converter efficiency is typically greater than 88%. Both ground-isolated and non-isolated models are available. The latter units use a common bus for the negative connection between power source batteries and output connections.

LEAD-ACID BATTERIES

Most lead-acid batteries used today are "maintenance-free" types in which the liquid electrolyte (sulfuric acid) is sealed in the battery as a liquid or gel. Batteries used in vehicle systems are called *starter-lights-ignition* (SLI), or just *starter* batteries, and are designed to supply momentary power surges with continuous charging while the engine is running. *Deep-cycle* batteries are designed to supply power continuously. *Sealed lead-acid* (SLA) batteries are completely sealed with a gelled electrolyte and can be operated or stored in any orientation. *Absorbent Glass Mat* (AGM) batteries are spill-proof with the electrolyte stored in mats of glass fibers. For portable operation, any of these types can be used if they can be transported to the operating site. Of these types, SLA is the one most likely to be used for portable stations.

As the battery's state of charge (SoC) is reduced, so is the voltage, especially under load. Battery voltage should be monitored to assure long battery charge-cycle life. For example, a nominal 12 V SLA battery is considered 100% discharged when the voltage under load reaches 10.5 V. Below this level the charge-cycle life is reduced. Even a deep-cycle lead-acid battery can be severely damaged by repeated discharge below 50% of its rated capacity.

Multiple SLA batteries wired in parallel may be used to extend operating time. The SLA battery terminal voltage when not being charged is still around 12 to 12.2 V, well below the radio's design voltage.

An advantage of using lead-acid batteries of any type (except for very small batteries) is that they can be charged directly from a vehicle charging system without a special charger. If a vehicle is available, a pair of medium-capacity batteries can be rotated, with one battery powering the radio while the other charges. Charging requires the vehicle engine to be running on occasion throughout the operation.

LFP AND LIPO BATTERIES

LFP is the most common type of lithium-chemistry battery used for 100 W portable stations. LiPo battery packs using LFP or similar chemistries are much smaller than SLA batteries and used for lightweight and ultra-lightweight QRP operation. A lithium-chemistry battery is half the size and a quarter (or less) the weight of a comparable capacity lead-acid battery. For a more complete discussion of LFP and LiPo batteries, see the Bienno Power web page at **www.bioennopower.com/pages/faq-lifepo4-batteries**. LFP in the remainder of this section refers to both LFP and LiPo battery types.

The LFP battery is very well suited for portable use, providing a nearly constant 12.8 V output over its discharge cycle. When fully charged, LFP batteries output around 13.5 V with no load. As they reach depletion, voltage drops to about 10 V. However, for 95% of the total range the voltage stays between 12.6 and 12.9 V. This is within 10% of the 13.8 V design voltage of most amateur radio gear. If an LFP battery is rated for 25 A-hr, it supplies 25 usable ampere-hours when new.

The LFP is tolerant of deep discharge with less damage. Typically, an LFP battery supports 10 times the number of cycles available from an SLA battery of equivalent capacity. In other words, a single LFP battery would replace 10 lead-acid equivalents over its useful life. Total cost of ownership for LFP batteries is therefore much lower, even though the initial purchase price is higher.

Two issues with LFP batteries require consideration:

• They require an appropriate charger, and a vehicle charging system is NOT an appropriate charger. Recharging an LFP battery in the field is more complicated than charging a lead-acid battery. Chargers cost more for lithium chemistries than for lead-acid.

• The usable capacity of an LFP battery drops considerably as the battery gets colder, and charging when it's cold can easily damage the battery. LFP batteries can be discharged when their temperatures are as low as 0 °F (–20 °C) but they should not be charged if their temperatures are 32 °F (0 °C).

PORTABLE POWER STATIONS

These integrated power sources combine a large battery, an inverter, and electronics to allow charging from ac, solar panels, and other sources. Having the entire system in one package makes them very convenient. Since they do not have engines, they are completely silent and are safe to use indoors. See **www.cnet.com/home/energy-and-utilities/best-portable-power-stations** for more information about these devices.

These units are rated in watt-hours (W-hr) with peak power up to around 1500 W. To compare with battery-only capacity, you'll have to convert W-hr to amp-hour (A-hr) by dividing by 12 V (12 V is the standard conversion from A-hr to W-hr even when batteries supply slightly different voltages). To account for an estimated 10% loss and other inefficiencies in the electronics, multiply the resulting A-hr value by 0.9. For example, a 500 W-hr station would be roughly equivalent to a 500 / 12 = 41.7 A-hr × 0.9 = 37.5 A-hr battery-only capacity. Remember that there are additional losses when the ac power is converted back to dc for powering a radio.

Output ac waveforms are clean, and there are usually several accessory dc outputs to run 12 V electronics and USB-charging ports. The ac outlets are typically limited in the amount of power that can be drawn from them, so check the specifications to be sure they can supply enough power for your electronics. To let you know the state of charge for the internal battery, a display of voltage and charge remaining is included. Complete packages that include solar panels are available.

BATTERY CAPACITY REQUIREMENTS

Watts (W) is the unit of power and the watt-hour (W-hr) is the unit of energy and battery capacity. Most batteries are rated in ampere-hours (A-hr). This is the battery's ability to supply a certain level of current at a nominal output voltage. The battery's energy storage capacity is the product of its A-hr capacity and the nominal voltage. (A more detailed procedure for evaluating the energy needs of a complete portable station is supplied in the section "Choosing a Battery for Portable Operation" in the **Power Sources** chapter.)

If capacity is not specified for a battery, multiply the battery's A-hr rating by its nominal voltage. For example, an 8 A-hr, 1.5-V D-cell battery has an 8 × 1.5 = 12 W-hr capacity. A typical 3 A-hr, 12.8-V LFP battery would have a capacity of 3 × 12.8 = 38.4 W-hr.

The following method produces a rough estimate of required capacity for operating with a 100 W HF transceiver:

• For SSB operation, assume that energy consumption is 100% of the value of peak output power in watt-hours. For example, if peak output power is 50 W, assume 50 W-hr per hour of operation.

• For other modes, assume 200% of peak output power. For example, if peak output power is 35 W for CW or FT8, assume 35 × 2 = 70 W-hr per hour of operation.

Multiply the energy consumption by duty cycle (the percentage of time transmitting). For casual operation this is generally assumed to be 0.25 or 25% of total operating time. For POTA/SOTA, Field Day, or contesting, assume 0.50 or 50%. Then multiply by the number of hours of operation. This is the total required capacity.

For example, a 5-hour SSB POTA operation at 100 W PEP requires (100 W-hr per hour) × 50% duty cycle × 5 hours = 250 W-hr capacity. If using an LFP battery that can be nearly fully depleted, 250 W-hr / 12.8 V = 19.5 A-hr, and a 20 A-hr battery, is sufficient. An SLA battery has a lower nominal voltage (12 V) and can only be discharged 50% without damage, so a rating of 20 A-hr × (12.8 V / 12 V) × (1 / 50%) = 42 A-hr is required.

24.3.3 Antennas for Portable Stations

An effective antenna system is essential to all types of operation but is likely the most important part of a portable station. It's important to strive for the best antenna system possible, because operations in the field are often restricted to low power by power source and equipment considerations. An effective antenna makes the difference between success and frustration.

Effective antennas are more challenging for portable stations than for permanent stations. A portable antenna must be light, compact, and easy to assemble. Portable station antennas need to go up and come down quickly, often times in just minutes. If you only have four hours to spend at a park, for example, and the antenna takes two hours to setup and an hour to tear down, three-quarters of the available time is lost. It is also important to remember that the antenna must be capable of being erected at a variety of sites, not all of which offer ready-made supports.

This section covers supports for the antenna, review of a few basic types of popular full-size and compact antennas, and antennas for VHF/UHF operation that are popular for mountain-top operating (for instance, the SOTA program encourages activity on these bands). More information about the design and construction of these antennas is available in the **Antennas** chapter, and there is an entire chapter on portable antennas in the *ARRL Antenna Book*.

ANTENNA SUPPORTS

The simplest support is no support, or more accurately, an antenna that supports itself. Vertical antennas that use steel or telescoping radiating elements often need no external support if they are attached to a substantial base. Mobile antennas, attached to a vehicle, are the best-known example of this type of antenna, and they represent a very reasonable option for some types of portable use.

Another relatively simple option is to use a support that already exists at the location. Usually this means trees, although creative amateurs have pressed observation towers, telephone or light poles (never power poles!), and even buildings into service. Make 100% certain that you have permission to use the desired structure. Even with trees, you need to be certain that park or other green space regulations allow you to use them as part of an antenna system.

The complexity of getting an antenna onto or held up by one of these supports varies, but the tools used by arborists are the best and safest bet for use in a public space. Arborist rope and "throw bags" are effective, do not damage delicate trees, and work quickly and easily. Tennis ball launchers (the manual ones for throwing a ball a long way) can also be quite effective at getting a light line over tree branches. Avoid using slingshots or compressed-air launchers. These devices are often not allowed in parks or public spaces and are considered weapons under a number of ordinances. Their use, other than on private property, is not advised, and could lead to a visit from public safety officers.

The telescoping fiberglass mast (see **Figure 24.34**) is a popular choice to provide a portable support solution. Multiple heights and weights are available from nearly 20 meters tall down to systems that are intended more for fishing than holding up antennas. Consider the operating requirements for height, weight capacity, and cost to select the correct mast. Remember that carbon fiber, despite its strength to weight benefits, is conductive and can affect antennas in unpredictable ways.

A fiberglass mast cannot hold up even a small HF triband beam made of aluminum tubing. Wire antennas or a lightweight rotary dipole, a 2-element, 6-meter beam or loop, or a 2 meter / 70 cm beam is about the limit for the heavy-duty versions of these masts. To support bigger antennas, a traditional push-up mast such as the 30-foot Rohn H30 is required. These thin-wall steel masts are sturdy, but they are much heavier than fiberglass.

(A) (B)

Figure 24.34 — Telescoping fiberglass poles can be used to support a variety of wire antennas or small VHF/UHF Yagis. The one shown in (A) is 40 feet long, yet collapses to 8 feet for storage. At (B), a typical twist-lock mechanism secures the sliding sections in place.

Figure 24.35 — The K5ND/r push-up mast (a painter's pole) on a drive-on base with a stabilizing brace attached to the roof rack.

Do not attempt to "walk-up" an extended mast with a significant load attached to the top. A fiberglass mast section may snap, and a steel push-up mast can fold over, ruining them. The collapsing support can also be dangerous. With the mast vertical, extend one section at a time with the load attached. If the sections slip from the weight of the load, a small hose clamp at the bottom of the section can act as a stop.

Securing a mast is important as it can be tall enough to cover a lot of ground if it falls, and damage to the mast itself is almost certain in an unplanned crash. If a vehicle is available, popular methods of supporting these masts involve "flagpole" holders that either mount on a trailer-hitch receiver or consist of a solid plate base onto which you drive one wheel of the vehicle. **Figure 24.35** shows the VHF/UHF rover antenna setup of K5ND with a heavy-duty fiberglass mast in a drive-on mount. A brace attached to the vehicle roof rack keeps the mast vertical.

Figures 24.36 show a drive-on base built by W7NS for small towers. A 3-foot section of Rohn 25 tower is welded to a pair of large hinges, which in turn are welded to a steel plate measuring approximately 18 × 30 inches. One of the rear wheels of a vehicle is driven onto the plate. Additional tower sections are then attached with the base horizontal and antennas added to the tower. The loaded tower is then walked up and secured to the truck or guyed. Never climb a temporary tower.

Without a vehicle, a tall mast requires either a solid attachment to an existing structure or a set of guy lines. Lashing or tying fiberglass masts to picnic tables, water spigots, or signposts can avoid the need for guying. Remember that a guy line is only as strong as the anchor to which it is attached and, like using trees, permission is usually required to drive things like tent stakes into

Field Day Tower Safety

In the haste to get a Field Day station on the air, safety can be overlooked when raising a tower or mast. The *QST* article "Field Day Towers — Doing It Right" by K4ZA and NØAX addresses many common situations and illustrates ways to put up small towers safely in temporary installations. The article is included in the *Handbook's* online information.

public ground. Whenever guying an antenna or mast in a public place, be sure to tie strips of caution tape to the guys to help people avoid walking into them.

A base-loaded whip antenna is shown mounted on its own miniature tripod in **Figure 24.37A**. Camera tripods are another excellent option for compact antennas as shown in Figures 24.37B and C. Antennas make the assembly top-heavy, and wind can easily blow them over. Either secure the legs as for the heavy-duty tripod described above or add guy wires to the antenna.

WIRE ANTENNAS

Wire antennas are very popular for portable operation because they are simple and inexpensive while being fairly effective radiators. Quarter-wave vertical antennas and half-wave dipoles of all sorts find their way into the field so often because they do well in addressing the basics of portable stations:

- **Footprint** — How much room is required for this antenna?
- **Deployment effort** — How long does it typically take to deploy and take down this antenna?
- **RF Performance** — How efficient is the antenna and what does its radiation pattern look like?

(A) (B)

Figure 24.36 — The portable tower mounting system by W7NS at (A). A fairly substantial antenna system with rotor is mounted on a Rohn 25 tower, which is held up in (B) by a metal base plate with a truck driven onto it to weigh it down. The tower can be stabilized by attaching it to the roof of the truck or by guying it. Never attempt to climb a temporary tower installation.

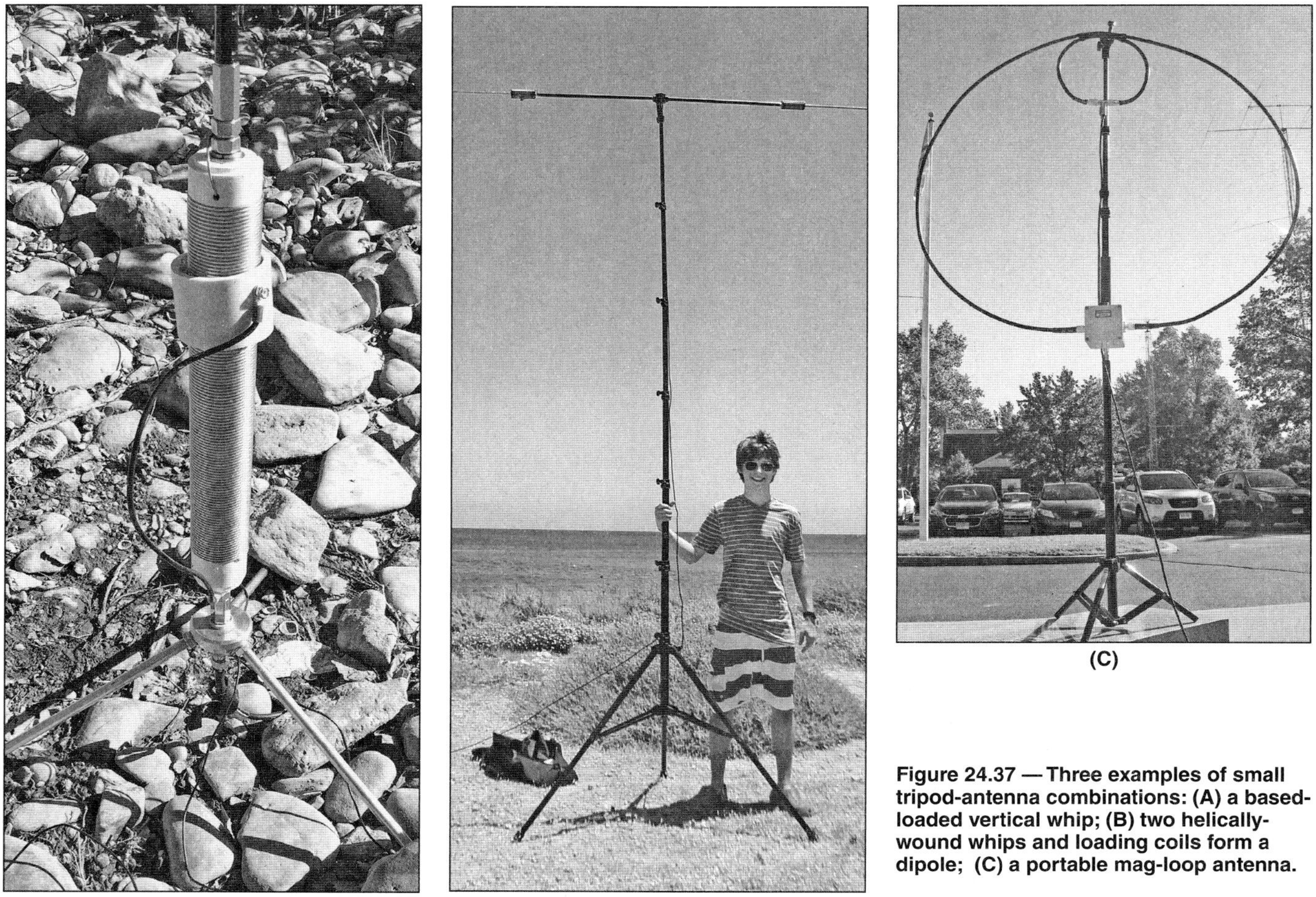

(A) (B) (C)

Figure 24.37 — Three examples of small tripod-antenna combinations: (A) a based-loaded vertical whip; (B) two helically-wound whips and loading coils form a dipole; (C) a portable mag-loop antenna.

• **Cost** — What investment is required to either build or buy this antenna?

Wire antennas for portable use can be made using smaller wire than is typically used in permanent installations. Larger wire offers better physical strength, but even #30 AWG magnet wire can easily handle 100 W. The lighter carry weight is well worth the effort to handle a portable antenna with a little care, and the low cost of small wire makes replacements fairly painless.

Quarter-Wave Vertical

At HF operating frequencies, the quarter wave vertical is one of the few antenna designs that does not require elevation for good performance. Only the center radiating element (or elements, if you build a multiband version) needs vertical support. Supports that are too short to be of any use for horizontal dipoles can work well with a quarter wave vertical.

Quarter wave verticals also operate across a relatively wide bandwidth and typically do not require impedance matching. This represents a significant size and weight savings for the entire station, as well as simplified operation. This is especially true if higher power operation is to occur.

The quarter wave ground-plane vertical requires a ground plane of radial wires that increases its footprint. If you get a total of two wavelengths of wire down on the ground in eight different radial elements, you've accomplished 90% of what can be done. The length of any one radial is almost irrelevant. All else being equal, more numerous and shorter radials work just fine and are easier to deal with than are longer ones. Put your radials out in a circle if you can, but if that is not possible, just lay them out where you can. Remember to use caution tape to help people avoid walking over the radials. Rolls of chicken wire or hardware cloth at the base of the antenna is also very effective.

Of all the portable antennas discussed here, ground mounted verticals in public spaces are the most likely to require positive access control in order to comply with FCC regulations (see **Safe Practices** chapter). Temporary fencing and signs to inform passers-by that standing closer to the antenna would expose them to radio frequencies that exceed the FCC's safe limits are needed.

Center-Fed Half-Wave Dipole

Dipoles are very popular and present three potential advantages over the quarter wave vertical. First, less wire is needed to construct a dipole when compared to a radial field. This means far less wire to buy, carry, and deploy. Second, if deployed at a suitable height, a dipole can exhibit gain over a vertical. This can facilitate operation with smaller radios powered by smaller batteries. Third, if the dipole can be mounted high enough in the air, it does not cause excessive exposure to people on the ground, and positive access control methods are not necessary.

To achieve these benefits, a dipole should be mostly horizontal and at least one quarter wave above ground. That requires two trees spaced far enough apart and two sets of arborist ropes and weights, or at least two, if not three, portable masts to support the center and ends of the dipole.

The dipole can be installed with a single support as an inverted-V rather than a horizontal. The support still needs to be as tall as possible for best performance. An inverted-V is a nearly omni-directional antenna. The support has to hold the weight of the full dipole, as well as the feed line and, in some cases, a balun. Depending on the heights of the ends of the inverted-V, positive access control may be required.

The End-Fed Half-Wave (EFHW)

If the dipole is fed at one end, the entire antenna system becomes far simpler to deploy. Half-wave wires are also resonant on all harmonics, not just the odd ones, making the EFHW a natural multiband option. The EFHW is likely the single most popular of all the simple wire antennas for portable use.

For portable operation, the feed point impedance transformer is usually mounted a few feet above ground and the antenna held up in the air at the far end. Remember that the feed point is a point of high RF voltage. Ensure that there is no one in the vicinity of the feed point when transmitting.

The radiating wire can be deployed in nearly any layout. If high and horizontal it performs on par with a traditional dipole. If deployed vertically, it compares well to a quarter wave vertical and does so without a radial field. Typically, some "zigging and zagging" is necessary to get the wire in the air. Sloped runs, inverted Ls, inverted-Vs, and "that's where the branches are" paths are all acceptable, albeit with sometimes unpredictable radiation patterns.

The EFHW does not require a radial field, but a return path for the feed line current must be provided. The feed point transformer has an output for the radiating element and a ground connection. This ground could be strapped to an actual earth ground or ground radials. Most commonly, portable operators simply rely on the coax feed line shield to act as the return path portion of the antenna system.

End-Fed "Random" Wire

Similar to the EFHW, the "Random Wire" antenna is a popular end-fed wire for portable use. It is considered a "random" wire because it is not intended to be resonant on any frequency where it is to be used. The wire is installed wherever the operator can place it and connected directly to the output of an antenna tuning unit or transformer. Because the length is unknown in advance, predicting its performance is difficult. The feed point impedance of the antenna varies over a wide range, so be prepared to adjust the tuning unit. As described in the *ARRL Antenna Book*, this type of antenna could present an overexposure situation to the radio operator. You should perform a careful analysis of the exposure patterns near the end of the antenna before using it.

Compact Antennas

Compact antennas need less in the way of support and are easy to deploy. Large antennas, strung through the trees, are simply not an option in many portable scenarios. On the other hand, a small tripod placed directly on the ground holding a whip antenna is much less likely to draw attention.

For portable use, the advantages of having a physically smaller antenna are obvious. It reduces the station footprint and, while they may not ultimately be any lighter, smaller components are easier to pack. This makes them easier to transport in smaller cars, backpacks, or however the operator is getting to the location. These physically compact antennas are a compromise that trade performance for size and, depending on one's perspective, operational convenience as well.

Commercially produced antennas are generally more expensive than home-built versions of the same design. This is true for any antenna, but it is more pronounced for the compact portable antenna.

Loading (see the **Antennas** chapter) uses some combination of inductance and capacitance to make a radiating element electrically resonant at a frequency lower than physical dimensions of the element would indicate. The best-known example of this technique is a base-loaded vertical antenna, where a loading coil is placed at or near the bottom of the vertical radiator. This design allows for the vertical radiator to be shorter, in many cases considerably shorter, than a full quarter wave.

Coil loading is also possible with horizontal antennas. Dipoles with "traps" to allow multi-band operation are shorter than their monoband versions at the same frequencies because the inductance in the traps electrically elongates the radiating element below the trap's resonant frequency. (See the **Antennas** chapter for more about traps.) Capacitance "hats" can also produce the same electrical elongation effect, and due to size constraints are the more common choice for the 160-meter band.

A very popular arrangement is to combine an adjustable loading coil with an adjustable, telescoping whip for the radiating element. (see Figure 24.37A) The coil for this arrangement is typically adjusted by means of a collar or collars that can be moved along the coil and secured with a thumbscrew at any point, thereby selecting just the right amount of inductance for the target operating frequency. The telescopic whip allows the radiating element to be adjusted so that the least amount of loading coil is needed. For higher HF bands such as 6 meters and 10 meters, the whip does not need to be fully extended, and the coil is essentially bypassed. Moving lower in frequency, the whip needs to be longer. At some point there is no more whip length available and turns of the coil are introduced to compensate for a radiator that is shorter than resonant. The base-loaded antenna still requires ground radials, however.

Another popular option is to combine a pair of helically-wound loaded mobile whips, making a shortened dipole. A special fixture with a SO-239 connector attaches one whip to the center conductor of the feed line and the other whip to the shield. You can see this kind of antenna in Figure 24.37B.

"Mag" (magnetic) loops are another option for portable operation. Typically, 6 feet in diameter or smaller, they can be mounted on a short pipe support or tripod as shown in Figure 24.37C. The loops are very high-Q antennas resulting in a very narrow bandwidth that requires re-tuning every few kHz. Loops can be homemade, but be very careful to avoid lossy construction because of the extremely low radiation resistance of the antenna. Another caution is to be sure to keep the antenna away from people due to the high field strengths near the antenna. **Table 24.4** shows the minimum distance needed from this type of antenna in order to be in compliance with RF exposure limits at different power levels and frequencies.

VHF/UHF Antennas

Portable operation often means that the operator is in a remote location, which mistakenly leads many to believe that VHF/UHF+ are not useful bands. From higher elevations, line-of-sight is a long distance and operation at the shorter wavelengths can be great fun. High gain antennas at these frequencies are much smaller than their HF counterparts, so taking them portable is practical, and their benefits are more noticeable than in local repeater work.

The simplest VHF antenna is no dedicated antenna! Many HF antennas, especially wire antennas, behave as "long wires" once the frequency is greater than 100 MHz. A 40-meter dipole is 10 wavelengths of wire on 2 meters! Give the HF antenna you are already using a try, it might surprise you!

If the goal is an omni-directional antenna, then variations on the classic "J-pole" antenna are very popular. A J-pole is nothing more than a vertical EFHW, and has the same gain and radiation pattern as a vertical dipole. However, unlike most end-fed antennas, the J-pole has a quarter wave transmission line matching section built into the antenna itself.

A Yagi for 2 meters can be handheld and, if constructed so that it can be disassembled for transport, can be packed up in a relatively small space. Just 3 elements in a 2-meter Yagi net the operator 6 dBi of gain, which is the same as more than doubling the transmitter power, while also providing enough directionality to reduce the effect of noise sources to the sides. When operating on top of a summit laden with commercial communication towers, the directionality of a Yagi can be far more beneficial than its gain. Such Yagi's are usually mono-band designs.

TIPS FOR PORTABLE ANTENNAS

Remember that in a public space you are also an ambassador for the hobby. Appropriate equipment goes a long way to making a positive impression with the public and with any managers of the space you're using. Since the antenna is likely the most visible part of the station, as a rule, a smaller footprint is a better choice. Consider placing caution tape around the antenna or support to keep people from walking into it. In public locations especially, be sure to perform an exposure analysis for all antennas in operation and, if warranted, place barriers and signage to keep people away from areas in which their exposure would exceed FCC MPE limits. For more information on these requirements, see the **Safe Practices** chapter.

Portable antennas need to be easily transportable. This generally means they need to be relatively lightweight and pack down to a manageable size. Traveling in a full-size pickup truck is different from traveling by motorcycle, so each operator needs to define their own requirements for size and weight. Backpacking an antenna to a summit poses different challenges than driving to a parking lot!

Sometimes, operating portable simplifies the antenna. Since these are temporary antennas, there is no need for them to survive for several seasons and through extended bad weather. Durability for a portable antenna is more about being able to survive repeated deployments than it is surviving the time while it's up. Taking a portable antenna down for a repair is easy if the need arises.

The acceptable footprint for a portable antenna varies by operator and by location. Operating in a vast expanse of uninhabited space may allow antennas much larger than would be practical at home. An operator working from a popular tourist attraction has far fewer plausible options.

A club Field Day-style operation may be able to manage large antennas and even use portable and temporary towers. For personal portable setups, big or heavy antennas should be passed over in favor of smaller arrays. The few decibels of gain a multi-element, 20-meter beam may have over a wire antenna is not as significant as the mechanical considerations. Stick with arrays of reasonable size that are easily assembled and supported within your abilities.

Wire antennas should be cut to size and tuned prior to their use in the field. Be careful when coiling these antennas for transport or you may end up with a tangled mess when you need an antenna in a hurry.

Beam antennas should be assembled and tested before taking them afield. Break the beam into as few pieces as necessary for transportation and mark each joint for speed when reassembling. Hex nuts can be replaced with wing nuts to reduce the number of tools necessary.

For coaxial feed lines, use RG-58 for the low bands and RG-8X for higher-band antennas. Although these cables exhibit higher loss than standard RG-8, they are far more compact and weigh much less for a given length. Avoid the extra loss of ultra-light cable like RG-174 unless the feed line is very short or saving weight is a high priority, such as when hiking and summiting. Some ultra-light QRP radios can drive 300-ohm feed line, as well. The feed line should be

Table 24.4

RF Exposure and Compliance References for Small Transmitting Loops

Power (W)	*Band*	*Controlled/Uncontrolled Distance (m)*
5	40 – 10	1.2 / 1.7
10	40 – 10	1.5 / 2.1
150	40 – 10	2.4 / 4.2
1,500	80 – 10	4.9 / 6.6

Source: Kai Siwiak, KE4PT, "RF Exposure Compliance Distances for Transmitting Loops, and Transmitting Loop Current," *QST* "Technical Correspondence" May 2014, page 64.

Less Is More — Ultralight Portable Operation

Nowhere are weight and efficiency more important than on the top of a mountain and Summit On The Air (SOTA: **www.sota.org.uk**) operations illustrate that very well. There are a number of QRP 1 to 5-W "pocket" transceivers, generally operating on 40 to 20 meters, although 5- and 6-band versions are available. Larger radios such as the Elecraft KX-3 have more features, including top-class receivers. VHF/UHF handhelds are also highly portable and have the batteries (and chargers) built-in. You can find out a lot about this style of operating from KE6MP in his presentation at **www.ke6mt.us/wp-content/uploads/2018/12/KT5X-WS0TA-SOTA-UltraLyte.pdf**.

Steve Galchutt, WGØAT, and Fred Maas, KT5X, both with years of experience and hundreds of peaks activated, contribute the following valuable advice about ultralight stations. (See their **QRZ.com** web pages for more photos and links to ultralight operating resources.)

• A support mast or pole must be light and fold up quite short so it doesn't hang up on low branches while hiking. (Models are available that weigh only a few ounces, extend to nearly 20 feet, and collapse to under 20 inches.) A "tip-top" for a fishing pole can be glued on top, providing a ferrule to guide the antenna wire and strengthening the small tip section which is also telescoped and glued into the second section for extra strength.

• Antenna height is not as important on a mountain top. We have experienced good success using an antenna only a few feet off the ground when high winds require such. Lean lightweight support masts about 20 degrees in the direction of the antenna wire and orient the antenna wire down wind.

• Antenna wire needs to be *light* for lightweight supports but still strong enough to be installed and stay up. #26 or even #28 wire can easily handle 10 watts. The best is Teflon-insulated stranded wire as the insulation strengthens the wire, insulates it from branches, the wire itself is very flexible, and silver-plated strands present a low resistance.

• The EFHW installed in an inverted-L configuration offers advantages. The low end of the vertical section has almost no current, so ground losses are minimal. The last few feet of the wire can even be used as the feed line, connected directly to the radio without coax. At QRP power levels the body and radio act as a sufficient counterpoise.

Figure 24.D — It helps to have someone willing to carry the gear and Steve Galchutt, WGØAT, enlists his pygmy goat, Peanut, who enjoys the trail as much as anyone and sometimes gives operating tips as shown here. (Photo courtesy Steve Galchutt, WGØAT)

Figure 24.C — Steve Schlang, K7TX, skied to this summit and operates a KX 2 radio with an EFHW antenna in an inverted-V configuration. (Photo courtesy Fred Maas, KT5X)

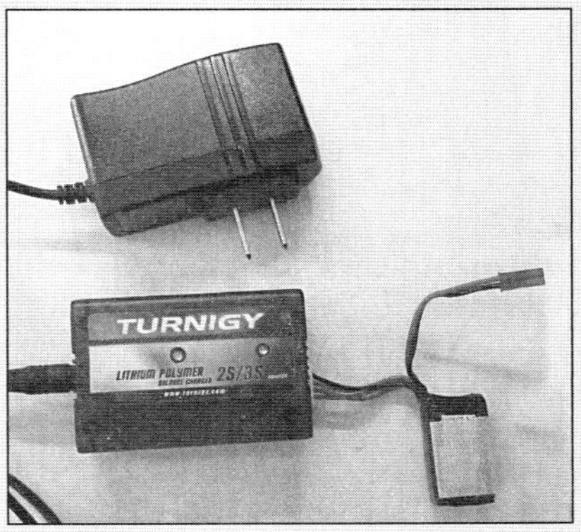

Figure 24.E — Lithium-ion batteries used to power small drones are available that only weigh a few ounces but store enough energy (up to 200 mA-hr) to power a QRP radio for an hour. (Photo courtesy Fred Maas KT5X)

attached to the antenna with a connector to increase the speed of assembly.

24.3.4 Computers for Portable Stations

The portable station often includes a computer of some kind. Smartphones, tablets, laptops, and modules such as the Raspberry Pi present a wide range of options for automating field operation.

The first decision is whether a computer is needed at all. In the case of strictly CW or phone operation, a computer is optional. Paper logs work fine for logging CW and phone contacts, and the simplicity of paper lets you leave the extra electronics (and weight) at home. All the popular portable operating programs, along with most contests, require electronic logs, so you would need to transcribe the paper logs after returning home if you decide against them.

Many operators choose to log directly with a computer in the field, which increases accuracy. For instance, the computer automatically dates and time stamps contacts. Logging software can also include call sign lookup functions, which helps to verify that a call sign is valid and correct. (Some contests do not allow online call sign lookup during the contest, though.)

Ask first: Does the computer need to be connected to the radio? Control of the radio requires using its CAT (computer aided transceiver) protocol via an RS-232 COM port, USB interface, or CI-V single-wire serial interface for Icom radios. This collects frequency and mode data, and possibly other items like power level. Computer interfaces can key the radio for CW or FSK. Software can play back and record audio messages, as can some radios, with a computer selecting messages under keyboard control. Tablet and smartphone devices generally do not support these external connections, leaving the operator to manually enter data into the logging program and operate radio functions. This is changing as the newest radios begin to support wireless connectivity like Bluetooth, but, as a rule, functionality is traded for portability when selecting a tablet or phone over a traditional computer.

Digital communication modes, such as FT8 or RTTY, make having a computer a requirement. (Some radios have built-in RTTY or PSK31/63 capabilities but generally do not make the information available to logging software.) AFSK modes also require an audio and PTT connection between the computer and radio, either as individual audio connections or via a USB audio codec. Software such as the *WSJT-X* suite that supports FT8, WSPR, and MSK144 requires a regular laptop or one of the more powerful Raspberry Pi modules (that need their own keyboard and monitor).

Having a computer in the vicinity of a radio introduces the potential for RFI ingress due to being close to antennas and compromised station bonding. Simply moving the station and antenna farther apart is often enough to solve computer RFI problems. Plan for good bonding practices, as described in the earlier sections of this chapter. Decrease common-mode RF current on antenna feed lines and control cables with ferrite chokes and by minimizing cable length. In some cases, isolating connections between the radio and the computer with transformers or opto-electronics may be required. Support for Bluetooth connectivity in the newest radios provides isolation, as well. (See the **RFI and EMC** chapter for more discussion about dealing with RFI.)

RFI egress problems from the computing equipment are common, too. Switchmode power supplies can easily create RFI. The same solutions listed above can help. Additionally, it is often helpful to power auxiliary devices from the same battery as the radio using linear voltage regulators that do not generate RF noise. Unshielded computing devices can generate RFI directly from their circuit-boards and may need to have shielding added, or be operated at a distance from the antenna.

As connectivity options continue to improve for radios, so too will the capabilities of connected computers and associated software. Options continue to expand and the trend is clearly towards more portable stations including a computer of some kind. Nevertheless, "bring along a pad of paper and a pencil or two, just in case" remains solid advice for the portable operator.

24.4 Preparing Your Station for Emergencies

24.4.1 Day-to-Day vs. Emergency Communications

Emergency Communications is primarily focused on providing communication support during emergencies, disasters, or situations where normal communication infrastructure is unavailable or overloaded. The emphasis is on relaying critical information quickly and accurately to aid in disaster response and recovery. This differs from Day-to-Day Communications, which typically involves casual conversations, technical discussions, contests, and other non-emergency activities for personal enjoyment and skill development.

Emergency Communications emphasizes the use of reliable, portable, and often battery-operated equipment. Operators must be prepared for quick deployment and operation in adverse conditions. Day-to-Day Communications operators may use a wide variety of equipment, including fixed stations with more complex setups that are not necessarily portable or suited for emergency use.

Many of the procedures specified previously for assembling fixed, mobile, and portable stations can be applied to emergency communications, but the impetus of the design must have a different perspective. Emergency communication requires the highest reliability during the worst of conditions. Communicators typically know exactly who they must communicate with, and what is required to make that connection. Communications capabilities must be sustained for the duration of the emergency, which requires considerations about maintaining power over the long term and appropriate antennas for communications links under potentially unfavorable conditions. Advance preparations to maintain such communications are paramount.

24.4.2 Designing an Emergency Communications Station

Designing an amateur radio emergency communications station involves several key considerations to ensure effective and reliable communication during emergencies. The design of the station should begin with research on anticipated emergency and disaster scenarios in the local area. Information on what to expect in the local area may be found from federal, state, and local emergency management agencies. Knowing these scenarios and the impacts to existing communications infrastructure aids in the design and equipment needed for the emergency communications station. It is also important to understand the local communication plans and procedures, including predefined emergency communication frequencies, modes, and channels for various scenarios (local, regional, and national). As the station is designed, coordinate with local emergency services, government agencies, ARES teams, and other amateur radio operators to understand their communication needs and protocols to be used in an emergency. In doing this, the station can integrate smoothly into the wider emergency communications network. By considering these factors, an amateur radio emergency station can be set up to provide reliable and efficient communication support in critical situations.

Equipment Choices

Setting up an amateur radio emergency communications station involves selecting equipment that is reliable, versatile, and suitable for various emergency scenarios. By proper selection of equipment, an amateur radio emergency communications station can be effectively set up to provide crucial communication support during emergencies and disasters. Considerations include: Will the station be set up in a fixed location, such as an Emergency Operations Center (EOC) or hospital? Should the equipment be easily transportable and quick to set up, such as in a shelter or field location? Can the station be moved, if necessary, especially if the initial location becomes unsuitable? What is the budget for equipment purchase?

The first and most obvious equipment choice is the transceiver. This is where it is important to understand the emergency communications plan, frequencies, and modes in the local area.

Many local emergency communication plans rely on the use of local VHF or UHF repeaters (assuming they are operational), or as back-up, local VHF/UHF simplex mode. Thus, a dual-band VHF/UHF transceiver is a good choice for almost any emergency communications scenario. In addition, the Technician class license privileges allow for VHF/UHF communications, so emergency communications are open to all members of the amateur community. The question is, should the choice for VHF/UHF be a handheld transceiver (HT), a mobile, or a base unit? When setting up an amateur radio emergency communications station, choosing between a handheld transceiver (HT), a mobile transceiver, or a base station involves several trade-offs. Each type of radio has its strengths and weaknesses, and the choice depends on the specific requirements of the emergency communication scenario.

An HT is highly portable and can be easily carried, making it ideal for field operations and on-the-go communications. It is simple to operate, with few setup requirements. An HT can operate independently of commercial power, making it especially useful when the power infrastructure is not working, or when used in remote areas where there is no access to power. The trade-off is that an HT typically is low power (usually 5 watts or less), which restricts its range, especially in urban or remote environments, and has limited battery life, especially under continuous use. It is also not a good choice for high duty-cycle use, such as extended net control operation. An HT is the best choice for field operations where mobility is essential, for short-range communications within a local area or using local repeaters, and for the initial response phase of an emergency when portability and rapid deployment are critical. While HTs have limitations, they can be highly effective for emergency communications when properly equipped and used in conjunction with other communication tools and infrastructure. Their portability and ease of use make them indispensable in many emergency communication scenarios.

A mobile VHF/UHF transceiver can be used in vehicles, as a temporary base station, or as a portable station with a power source. Mobiles typically offer higher power output (25-50 watts), providing better range and signal strength when needed, and are generally built to withstand harsher environments and mobile use. Mobiles are less portable than HTs and require a vehicle or fixed setup with a 12 V power source, such as a battery or an external power supply. A mobile VHF/UHF transceiver is a good choice for vehicle-based operations, extended range communications where being mobile is needed, temporary field stations where a higher power output is necessary, and situations where the portability of an HT is less critical but some mobility is still required. A roof or tower-mounted antenna expands the utility and signal strength of a mobile while still preserving the need for potential portability.

A VHF/UHF base station is the best choice for permanent or semi-permanent emergency communication centers, especially when long-range communications and coordination across wide areas is required. A base station should be a consideration at command posts when operations requiring continuous, high-volume communication capabilities, including net control stations, are needed. Base station transceivers offer the highest power output (100 watts or more), providing maximum range and signal clarity, and typically include advanced features such as multi-band capabilities, digital modes, and robust filtering. They are designed for continuous operation with stable power from mains electricity, generators, or large battery systems. Since they are usually used at fixed locations, base station transceivers are typically connected to large, high-gain antennas mounted on towers or rooftops, significantly improving performance. They are typically not portable, and fixed installation limits use to a single location. They are also the highest cost among the three options.

In summary, for VHF/UHF emergency communications, HTs are best for their portability, ease of use, and immediate deployment, but they are limited by range and power. Mobile transceivers offer a good balance of power, versatility, and portability, making them suitable for vehicle-based or temporary field operations. Base stations transceivers provide the highest power and best performance, ideal for fixed locations requiring robust, long-range communications. In an emergency communications setup, having a combination of these can provide comprehensive coverage and flexibility, allowing operators to choose the most appropriate equipment for each specific situation.

Longer Distance Communications

While VHF/UHF communications find widespread use in local emergency communications plans, regional emergency communications require either the use of linked VHF/UHF repeater systems for broader regional coverage, or the use of HF communications. HF radio is used for amateur radio emergency communications in several key scenarios where its unique characteristics provide critical advantages. HF radio is essential for communicating over long distances, which is crucial when coordinating with emergency services, government agencies, or other amateur radio operators outside the local area, and can be used to relay information from one region to another, ensuring that remote areas stay connected to a central command post or other affected regions. HF is useful in areas with significant geographic obstacles (mountains, forests) where VHF/UHF signals might be obstructed. In remote areas with sparse population and infrastructure, HF is often the only reliable means of communication. HF radio also provides a reliable backup when local communication infrastructures (cell towers, internet) are down due to natural disasters or other emergencies. HF NVIS (Near Vertical Incidence Skywave) is an excellent choice for regional communication (200-500 miles) and is good in mountainous terrain. During hurricanes, earthquakes, floods, or wildfires, HF radio can be used to coordinate relief efforts, request aid, and disseminate information to affected areas. The National Traffic System (NTS) uses HF to pass formal written messages (radiograms) across the country, ensuring that information can flow even when other systems are compromised.

Setting up an effective amateur HF emergency communications station requires careful selection of equipment that ensures reliability, versatility, and ease of use in various emergency scenarios. It is important to establish and use standardized emergency frequencies, and to know the regional and national HF frequencies designated for emergency communications. Coordinate with local emergency services and other agencies to ensure compatibility and interoperability with their communication systems and integration with existing emergency communication networks, such as ARES, RACES (Radio Amateur Civil Emergency Service), SHAred RESources (SHARES), High Frequency (HF) Radio program, and Military Auxiliary Radio System (MARS). An understanding of HF characteristics, such as signal propagation variation with time of day, solar activity, and atmospheric conditions, as well as familiarization with tools like propagation charts and online resources (e.g., VOACAP), is useful to predict and understand propagation. Recognize that ionospheric conditions affect HF communication, and solar flares and geomagnetic storms can disrupt HF signals. By considering factors such as frequency selection, propagation, antenna setup, power sources, and operational planning, amateur radio operators can ensure they are well-prepared to provide effective communication support during emergencies. Regular training, drills, and coordination with other emergency services are crucial to maintaining a high state of readiness.

HF equipment choices include mobile or base stations. When deciding between a mobile HF radio setup and a base HF radio setup for amateur radio emergency communications, there are several key factors to consider, similar to the choices for VHF/UHF mobile and base stations. Both have their advantages and drawbacks depending on the specific needs and circumstances of the emergency communication scenario.

Mobile HF stations are designed to be compact and can be easily transported and deployed in various locations, such as vehicles or temporary field stations. Mobile HF radios can be used in both fixed positions and while mobile, offering versatility in dynamic emergency situations. Typically, mobile HF radios have lower power output compared to base stations, often less than 100 watts, which could limit communication range and signal strength. Mobile antennas are generally less efficient than base station antennas and, while they can be mounted on vehicles or portable masts, they usually do not perform as well as larger, permanent fixed station antennas.

HF base stations are best at locations that serve as central communication hubs, such as an EOC or command post, coordinating with multiple remote sites. HF base transceivers often support higher power outputs, up to 200 watts or more, providing stronger signals and better long-range communication capabilities. Base setups can use larger, more efficient antennas such as full-size dipoles, verticals, beams, or NVIS setups, which enhance signal strength and coverage. HF base stations are fixed installations and cannot be easily moved, limiting their use to specific locations. They are usually dependent on commercial power, though they can be backed up with generators or battery systems, which add to the complexity and cost.

Choosing between a mobile and a base HF radio setup for emergency communications depends on the specific needs of the operation. Mobile setups offer flexibility and rapid deployment, making them invaluable in initial response and dynamic field operations. Base setups, on the other hand, provide robust, high-performance communication capabilities essential for centralized, long-term emergency management. Ideally, a combination of both can provide a comprehensive solution, leveraging the strengths of each to ensure reliable and effective emergency communication.

Digital Modes

Digital modes play a crucial role in amateur radio emergency communications due to their efficiency, reliability, and ability to transmit data over various distances under challenging conditions. They offer various options to suit different scenarios and conditions, such as Health and Welfare Traffic, Resource Coordination, Situation Reports, and Position Tracking.Modes like PSK31, FT8, and Winlink are particularly valued for their efficiency, reliability, and ability to operate under adverse conditions. APRS can be used to monitor the locations of field teams and assets in real time. Incorporating these communications modes into emergency communications plans helps to ensure robust, flexible, and effective communication during emergencies.

Digital modes are used on HF, VHF, or UHF depending on the mode and range. A digital mode interface (e.g., *SignaLink USB*, *RigExpert*) is needed to connect the radio to a laptop or desktop computer running the necessary software. Software is chosen depending on the mode, and includes: *Fldigi*, for a wide range of digital modes and NBEMS integration; *WSJT-X*, for FT8 and related modes; *Winlink Express*, for sending and receiving emails over radio; and *JS8Call* for extended FT8-style messaging.

Portable Repeaters

Portable repeaters are invaluable assets in emergency communications at VHF/UHF, offering extended range, enhanced reliability, and flexibility. By incorporating portable repeaters into emergency preparedness plans, organizations can ensure robust and effective communication capabilities in a wide range of scenarios. When deploying a portable repeater, ensure you have a reliable and field-proven portable repeater system. Use high-gain antennas and portable masts to maximize the repeater's coverage area. Have multiple power options available, including rechargeable batteries, solar panels, and generators. Deploy repeaters in elevated locations or areas with clear lines of sight to maximize coverage. Equip vehicles with mobile repeater setups to provide on-the-go communication capabilities. Coordinate frequencies with local authorities and other operators to avoid interference and ensure effective use of available spectrum and use preassigned and well-known emergency frequencies for quick setup and deployment.

Power Sources

In emergency situations, reliable communication can be crucial. The effectiveness of your communication equipment greatly depends on having a dependable power source. The Station Power section above discusses the proper ways to make power available to equipment in your station. However, in an emergency situation you should be prepared for the ac mains to malfunction, requiring that you use alternate power sources. The Portable Stations section above describes several forms of alternate power and how to use them effectively and safely.

Here are excellent power sources suited for emergency communications, each with its own set of advantages: Rechargeable Batteries (Lead-Acid, SLA, NiMH, Li-ion, LiFePO4) or Alkaline Batteries; Solar Panels; Hand-Crank Generators; Power Banks; Gasoline, Diesel, or Propane Generators; and Wind Turbines. Vehicles can serve as mobile power generators, allowing you to charge devices using the car's battery and alternator.

Lead-acid batteries store electricity through a chemical reaction between lead, lead dioxide, and sulfuric acid. They come in various sizes and capacities, making them versatile for different applications. There are mainly two types - flooded (or wet) and sealed (SLA or VRLA). For emergency communications, SLA batteries are preferred due to their maintenance-free nature and safety. SLA batteries for emergency communication are 12 volts, with capacities varying from a few Ah (Ampere-hours) to over 100Ah, depending on the requirement.

The choice between NiMH, Li-ion, and LiFePO4 batteries for emergency amateur communications depends on specific needs such as portability, safety, longevity, and budget. By understanding the advantages and maintenance requirements of each battery type, amateurs can select the optimal power solution to ensure reliable communication in emergencies, enhancing their preparedness and response capabilities. For handheld devices, or when mobility is critical, Li-ion and NiMH batteries are preferred due to their lighter weight and good energy density. For stationary setups, or where safety is a major concern, LiFePO4 batteries are superior due to their thermal stability and long lifespan. NiMH batteries may be more cost-effective for users with budget constraints, offering a good balance between price and performance. Consider the operating environment. LiFePO4 batteries are well-suited for extreme conditions, while Li-ion and NiMH batteries are better for general use in mild to moderate climates.

In an emergency, when AC mains power fails, amateur radio operators need a reliable method to switch to alternate power sources to maintain communication capabilities. While professional settings like EOCs may have sophisticated automatic switchover circuitry, there are more accessible and cost-effective alternatives suitable for a ham shack. These include:

Manual Transfer Switches - A manual transfer switch is a simple and cost-effective solution. It requires you to physically switch from the mains to an alternate power source (like a generator, battery system, or solar power). This method is reliable and gives you control over when and how to switch power sources. The main disadvantage is the need for your presence to switch power sources during an outage.

Automatic Charge Controllers with Transfer Switches for Battery Systems - For those using battery banks as an alternate power source, some charge controllers come with built-in automatic transfer switches. These devices can automatically switch your radio equipment to battery power when the mains fail, then switch back once mains power is restored, also ensuring your batteries are charged when mains power is available.

UPS (Uninterruptible Power Supplies) - typically used for computers and electronic equipment, can also be effective for amateur radio setups. A UPS can provide immediate power during a switchover with no interruption, which is essential for critical communications. Although UPS units may not offer a long-term power solution on their own (depending on their size and the power consumption of your equipment), they can be used in tandem with generators or larger battery systems to provide an immediate but temporary power bridge until a longer-term source is online.

Solar Power Systems with Backup Batteries - A solar power system with a sufficient battery backup can offer an excellent alternative power source. Modern solar charge controllers often include features to switch loads between power sources automatically. While the initial setup cost can be higher, solar systems provide a renewable energy source that can maintain battery charge levels during extended power outages, reducing the need to manually switch power sources.

Generator with a Manual Interlock Kit - For those with access to a portable generator, installing a manual interlock kit in your electrical panel allows you to safely switch between mains power and generator power. This method requires some upfront cost and should be installed by a professional electrician to ensure it meets local electrical codes. While not as automatic as some options, it provides a more integrated solution for those who might use a generator to power more than just their ham shack.

Assess the power consumption of your equipment to determine the capacity needed from your alternate power source. This helps to select the right size UPS, battery bank, or generator. By considering these alternatives and implementing the one that best fits your needs and budget, you can ensure your amateur radio operations remain functional during mains power failures, enhancing your ability to communicate in emergencies. A combination of power sources tailored to the specific requirements of your emergency communication equipment, the duration of potential emergencies, and the environment you're operating in provides the best chance of maintaining communication during critical times. Power the most critical equipment first, such as communication devices and medical equipment. Turn off devices when not in use and consider energy-efficient alternatives. Keep generators and rechargeable batteries in good working order through regular testing and maintenance.

Safety and Grounding

Safety considerations for emergency communications are paramount to protect both the operators and the public, ensuring the integrity and reliability of communication during critical situations. Safety considerations should include equipment, operational, electrical, antenna, and environmental considerations such as weather conditions and physical hazards. By adhering to these safety considerations, emergency communication operations can be conducted more effectively and safely, enhancing the overall response to emergencies and protecting both responders and the public.

Grounding is a critical safety and operational consideration for emergency communication systems, especially for radio equipment and antennas. Proper grounding provides protection against electrical surges and lightning strikes, helps to reduce radio frequency interference (RFI), and improves the overall performance of the communication system. Grounding methods and considerations include electrical grounding such as using a grounding rod or stake; single-point grounding to prevent ground loops and potential differences between equipment; RF grounding (counterpoise) to improve antenna performance; surge protection by incorporating surge protectors and lightning arrestors; and bonding all metal parts and equipment enclosures to the ground system to ensure everything has the same electrical potential.

Generator Safety

Generators, whether portable or stationary, pose safety risks if not used correctly. By following safety tips and best practices, you can ensure that your use of a generator for emergency communications is safe, efficient, and effective, keeping you connected when it matters most.

Here are essential safety tips and best practices to consider:

Before Use:

Read the Manual - Before operating a generator, thoroughly read the manufacturer's manual to understand its operation, fuel type, maintenance requirements, and safety features.

Proper Placement - Always operate the generator outdoors and away from windows, doors, and vents to prevent carbon monoxide (CO) poisoning. CO is a colorless, odorless gas that can be deadly.

Dry Operation Area - Ensure the generator is on a dry, stable surface. Never use it in rainy or wet conditions without adequate protection. Consider using a generator tent or cover for operation during inclement weather, ensuring it does not trap harmful gases.

Install Carbon Monoxide Detectors - Equip your home and operation area with battery-operated or battery-backup CO detectors and test them regularly.

During Use:

Cord Safety - Use heavy-duty, outdoor-rated extension cords free of cuts and tears. Ensure the cord's wattage rating is compatible with the generator output and the demands of connected devices.

Electrical Safety - To prevent electrical shock, make sure your hands are dry before touching the generator. Use a Ground Fault Circuit Interrupter (GFCI) to protect against electrical shock hazards.

Avoid Overloading - Do not exceed the rated capacity of your generator. Prioritize your power needs, focusing on essential communication equipment and safety devices.

Fuel Safety - Store fuel in approved containers, away from living areas, and only refuel the generator when it is off and cool. Gasoline and other generator fuels should be stored properly to prevent fire risks.

After Use:

Cool Down - Allow the generator to cool down before storing it. This minimizes the risk of burns and fires.

Maintenance - Regularly check and maintain your generator according to the manufacturer's instructions. This includes checking oil levels, cleaning air filters, and ensuring the spark plugs are in good condition.

Storage - When not in use, store the generator in a dry, ventilated area. If storing for a prolonged period, follow the manufacturer's guidance on fuel treatment and parts lubrication to ensure it remains in good working order.

Additional Generator Safety Considerations:

Understanding Power Needs - Know the power requirements of your emergency communication equipment to select a generator that meets these needs without overloading.

Noise and Vibration - Use noise reduction techniques and vibration isolators if necessary, especially in residential areas or if operating for long periods.

Community Guidelines - Be mindful of local noise ordinances and community guidelines regarding generator use, especially during nighttime hours.

Professional Installation - For standby generators, consider professional installation to ensure they are correctly integrated into your home or facility's electrical system, including the installation of a transfer switch to avoid back feeding, which can be dangerous to utility workers and neighbors.

Handling Gasoline Safely

Handling gasoline safely is crucial to prevent accidents, fires, and health hazards. Gasoline is highly flammable, and its vapors can be explosive under certain conditions. Only use containers specifically designed and approved for gasoline. These are typically made from high-density polyethylene (HDPE) and are colored red for easy identification. Ensure the container's cap is tight-fitting to prevent leaks and vapor release. Always handle gasoline in well-ventilated areas to avoid inhaling fumes. Outdoor is best. Keep gasoline away from sparks, open flames, or any source of ignition. This includes not smoking near gasoline.

Store gasoline in a cool, well-ventilated area outside of your living quarters. Avoid storing it in a basement or near any potential sources of ignition. Store only a small quantity of gasoline, enough for immediate needs. Large quantities pose a greater risk. When transporting gasoline, ensure containers are secured and upright to prevent tipping and spilling. Keep your vehicle's interior well-ventilated during transport, and remove the gasoline from the vehicle as soon as possible.

Never fuel an engine while it's running or hot. Allow it to cool down first to minimize the risk of vapors igniting. When transferring gasoline, pour slowly to minimize spillage and the generation of static electricity. When pouring gasoline into a container, place it on the ground to avoid static electricity buildup that could spark and ignite gasoline vapors.

Keep a class B fire extinguisher nearby when handling or storing gasoline, and educate yourself on how to use it, along with basic first aid in case of gasoline exposure. If gasoline contacts your skin, wash it off immediately with soap and water. If it splashes into your eyes, rinse them with water for at least 15 minutes and then seek medical attention.

Have materials on hand (like sand or commercial absorbents) to manage small spills, and know how to dispose of them properly.

Never dispose of gasoline down drains, in toilets, or on the ground. It can contaminate water sources and soil. Take unwanted gasoline to a hazardous waste disposal facility; some communities have specific collection days for hazardous materials.

Read the material safety data sheet (MSDS) for gasoline to be aware of its properties, hazards, and safety precautions. Be aware of local laws and regulations regarding the storage and disposal of gasoline.

By following these safety tips you will be able to handle gasoline safely, protecting yourself, others, and the environment from the potential hazards associated with improper handling and storage. Always prioritize safety and exercise caution when dealing with flammable substances like gasoline.

Antennas for Emergency Communications

A good antenna choice can make a significant difference in the ability to send and receive critical information. Some excellent antenna options tailored for different scenarios and needs in emergency communications include Portable Antennas, such as a Roll-up J-pole Antenna or Magnetic Loop Antennas; Mobile Antennas, such as Vertical Whip Antennas, Mag-mount antennas, or Screwdriver Antennas; Base Station Antennas, such as Dipole Antennas or Vertical Antennas; and Specialized Antennas including NVIS Antennas and Yagi-Uda Antennas. **Figure 24.38** shows how at mobile HF screwdriver antenna can be modified to operate in NVIS mode.

An NVIS antenna is a highly effective antenna choice in emergency communications for several reasons, particularly in scenarios where conventional communication infrastructure is compromised or unavailable. In many scenarios, HF emergency communications need Short to Medium Range Communication, and NVIS antennas are designed to reliably provide that, typically within a radius of 0 - 300 miles (0 - 500 kilometers). This range covers an area that is often too far for reliable VHF/UHF line-of-sight communications but too close for traditional HF skywave (skip) communications to be effective. NVIS fills

Figure 24.38: Mobile screwdriver antenna mounted vertically on the back of this car can be made to operate in NVIS mode by flipping it over and raising the back door that the antenna is mounted on, causing the antenna to sit horizontally. (courtesy Dan Huber KX4DAN)

the "skip zone" gap, where direct waves fall short and refracted waves have not yet returned to Earth. This makes it indispensable for regional emergency communications, enabling contact within a disaster-affected area or between adjacent regions. NVIS also allows signals to cover rough terrain, valleys, and urban areas where line-of-sight communications may be obstructed. NVIS antennas can be relatively simple to construct and deploy, often consisting of dipole antennas placed at low heights above the ground (usually less than a quarter-wavelength). This simplicity allows for quick setup in emergency situations using readily available materials.

Preparing for natural disasters, like hurricanes or tornadoes, is crucial for maintaining emergency communication operations. These events can indeed damage or completely destroy base station antennas, which are vital for effective communication.

Ensure that your antenna installations are as sturdy as possible. Use high-quality mounts and consider guy wires as additional support for freestanding antennas. Some antennas are designed to withstand high winds and harsh weather conditions better than others. Do the research and invest in antennas known for their durability. While placing antennas higher can increase their risk of wind damage, it also often provides better performance. Balance the need for elevation with the risk based on your local weather patterns and consider using lower-profile or more aerodynamic antennas that can handle the wind better.

Keep a variety of portable antennas on hand (long wire, dipoles, NVIS, etc.) that can be quickly deployed in the event your main antenna is damaged. Portable antennas can be surprisingly effective and can be set up relatively quickly. Keep DIY antenna kits or ready-to-assemble antenna parts that can be quickly put together after a storm passes. This might include simple dipole or vertical antennas that require minimal setup. Consider equipping a vehicle with a mobile unit that can serve as a temporary base station. Vehicles can be moved to a safe location during a storm and then used for communication in the immediate aftermath. If possible, arrange for access to an alternate site that might be less affected by the storm. This could be a site at a higher elevation or a more structurally secure building.

Preparation and flexibility are key to maintaining emergency communication operations through severe weather events or other natural disasters. By reinforcing your existing setup, having quick-deploy and portable options ready, and coordinating with the broader amateur radio and emergency response community, you can significantly increase your resilience against the destructive power of hurricanes, tornadoes, and other natural disasters.

24.5 Remote Stations

The following section was updated by Ken Norris, KK9N. Although this section focuses on "remoting" HF equipment, the same considerations apply to most remotely operated stations, with the exception of repeater installations.

24.5.1 Introduction to Remote Stations

Remote stations (stations operated by remote control) have been a part of amateur radio for decades, but usually in the form of VHF/UHF repeaters. In the past, a few remote stations operating on HF have been developed in impressive locations with significantly more land, equipment, and expense than for a repeater installation. Prior to the expansion of internet technologies, some of these remote pioneers utilized VHF/UHF links, dialup, or commercial microwave equipment for connectivity between the remote and home stations. There has been a significant increase in the number of remote stations over the last 10 years, and the trend is accelerating rapidly.

Since this section was originally written, there have been several new developments and refinements of the technology available to access your home or remote station. As latency across the internet has been reduced, operating in fast-paced contests over very long links — even between continents — is now commonplace.

Reasons for operating remotely are varied. You may want to monitor your home station while you're away, avoid homeowner restrictions on outdoor antennas, or fight interference issues from electrical power lines, plasma TVs, network routers, and the like. With today's technology, your home station or a remote HF station may be as simple as a 100 W HF radio and all-band dipole/vertical, or as complex as a fully automated contest station, with control and monitoring available to the operator via the internet. There is a form of remote station to meet almost every taste and budget.

This overview of remote station operation looks at different methods of remote operation and how they are each unique in equipment and technologies. There is no cookbook approach to creating a remote station. This discussion is intended to introduce topics to be considered in the circumstances encountered by each station builder.

Two important definitions used throughout this chapter are the naming of the locations of equipment. The location for your radio, host computer, amplifier, and antennas are your *station site*, whether this is at your house or at a distant site. The other end of the connection is the *remote client* location. The remote client site consists of the computer in front of you with the remote computer running client software and possibly a remote radio head (front panel) that is also connected to the station site.

LICENSING AND REMOTE OPERATION

Before discussing the technical details, it must be emphasized that operating a remote station also carries with it an extra responsibility to be properly licensed and to identify your station correctly. Because the transmissions are made from the location of the station site, you must follow the regulations that apply at that location. This is particularly important when the station is outside the jurisdiction of the regulatory authority that granted your license, such as a US ham operating a station in South America or a European ham operating a station in the US. You must be properly licensed to transmit from the station site location! This may require a separate license or reciprocal operating permission from the country where the station site is located. FCC rules require the control operator of a remotely controlled station in the US to have a US license, regardless of where the operator is located.

Don't assume that because you have a US license you automatically have a corresponding license to operate a remote station outside the US, though. For instance, European CEPT agreements generally do not apply to operators who are not physically present in the country of the transmitter, and you may be required to be physically present at the transmitter, as well. There may also be contest or award rules that apply to remote stations — be sure to know and follow all rules that apply.

24.5.2 Types of Remote Operation

REMOTE RECEIVERS

There are several different options for remote operation today. The most basic are the many remote receivers available online that are maintained by other amateurs and communication hobbyists; these can be used by anyone with a computer, web browser, and the internet. Websdr and GlobalTuner are two examples of remote monitoring sites, but there are many more. (See this section's Resources list for URLs of these and other websites.) Such websites give you the opportunity to be able to monitor radio activities in other countries, or to hear your own signal from another location.

A great source for listening to DX stations is on the **DXHeat.com** cluster page. If you right-click on the DX station you want to listen to, you get a popup that shows a pair of headphones, chart page, and link to the DX's QRZ page. If you select the headphones, it opens a dedicated receiver on the wide-band WebSDR website. The wide-band SDR radio is tuned to the DX station, so you can hear it. The caveat is that this wide-band receiver is in the Netherlands; so if you like to hear what the European stations are hearing, this is a great web page to use.

Amateurs are using the Raspberry Pi microcomputers, USB SDR receivers, and simple-wide-band antennas to build remote radio servers. The technologies available on these microcomputers allow remote access through a WiFi or cellular modem, connectivity by VPN (Virtual Private Network) or VNC (Virtual Network Computing) software, and RDP (Remote Desktop Protocol), or use the operating system's built-in web publisher to produce the screen of the radio software. The best part is that all the components can be powered by a solar cell on top of a mountain, or even at a friend's farm.

REMOTING YOUR HOME STATION

With the availability of high-speed internet through cell phones, hotel, or office WiFi, operating your amateur radio station from the living room, back yard, at work, on business travel, while vacationing, or just visiting the family, has become an enjoyable part of the hobby. Your remote connection, referred to as "remoting," can be as simple as a computer, tablet, or cell phone monitoring the audio from your radio for the frequency you left it on, or as complex as full automated remote radio control so you can catch that last needed DX country to fulfill your DXCC award. Some amateur radio operators have removed their HF equipment from their vehicles and installed remote cellular connections at their home stations.

DEDICATED REMOTE STATION

There are many amateur radio operators who are restricted by homeowners' association requirements, or living in a temporary location, apartment, or retirement home that restricts the installation of outdoor antennas. Other operators live in areas where noise from electrical utilities and electronic devices render an HF receiver almost useless. For these reasons, setting up a station away from the restrictions and noise can be very appealing.

Some contesters have built a remote station on top of a mountain, or in another country, in order to be more competitive while still operating from the comfort of their home. Setting up a remote station in another country does not add any more equipment other than a station a few miles out of town. As previously mentioned, it does however add different rules and regulations to your operating that must be followed.

Maintenance is also more difficult with a remote station. If you want to find out why the amplifier is alarming when operating on 40 meters, a 6-hour drive or flight to a remote site is no trivial matter. Technology is available for remote monitoring of the remote site. As long as you have internet connectivity to the remote site, there are ways of troubleshooting your problems.

24.5.3 Remote Networking Basics

Today, the internet has become the preferred source for connectivity of remote stations. Landline telephones can still be used for remote DTMF control of equipment, or as a source for transferring the audio to and from your remote radio.

With the availability of high-speed internet at your home, office, or on the road from WiFi and mobile phones, reliable connectivity to a remote station has become easier. This ease of connectivity comes with the need to understand more about networking and how computers and other TCP devices communicate with each other. This section has been included to expose you to terminologies and technology that aid in getting your internet equipment connected and operational. (See this section's Glossary in the online content for definitions of terms, abbreviations, and acronyms.)

ROUTER AND PORT FORWARDING

When you turn on your computer at home to browse the internet, it does not connect directly. As seen in **Figure 24.39**, the computer first connects to a gateway most commonly known as a *router*. The router then allows the computer to connect to it and waits to see what the computer wants to do with the connection. When you open your internet browser, the router then assigns a numeric *port* within the router for the com-

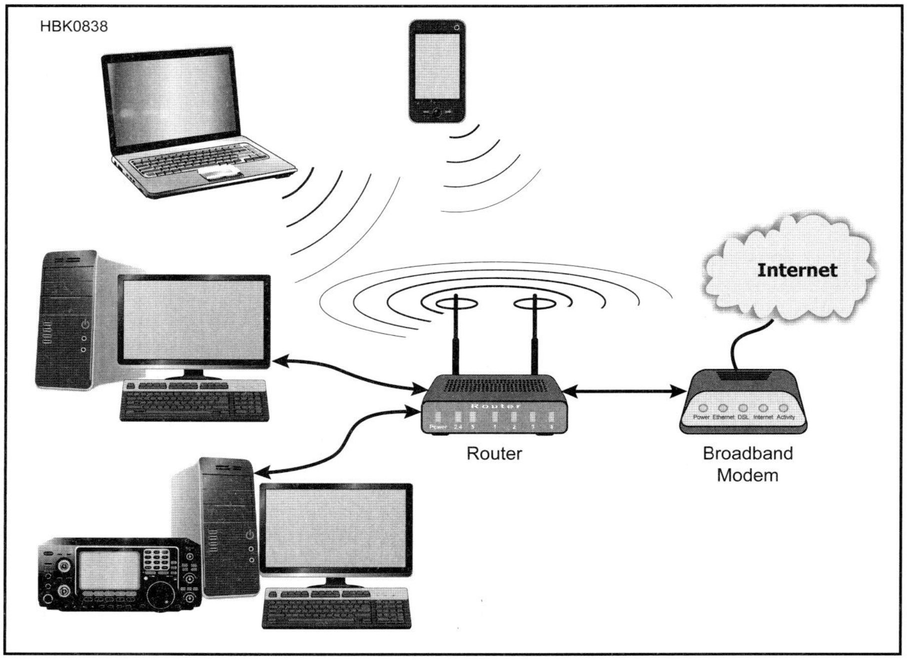

Figure 24.39 — Basic network structure and components.

puter's IP address to route data to the IP address your Internet Service Provider (ISP) has assigned to the router.

This is basically how computers in your private home or work network communicate with the World Wide Web, public computer servers, and devices throughout the world. The router is the guardian that controls what information comes in to and goes out of your local network.

Port forwarding is an option in most routers that enables you to create a permanent translation table that maps an incoming protocol port from your ISP to a specific IP address on your local network. It is a transparent process, meaning internet clients cannot see that port forwarding is being performed. This process enables you to remotely connect to TCP/IP-enabled equipment and TCP servers on a computer on your local network that are otherwise hidden from the internet by your router.

STATIC OR DDNS IP ADDRESSING

When you sign up with an internet service, the ISP provides you with a modem to connect their network with the local network in your house. Normally when the modem is activated, it requests an IP from the ISP's *Dynamic Host Configuration Protocol* (DHCP) servers. Even though you can find out what the IP address is, you have no control over what is assigned and how often it changes. Because it can be changed, it is called a *dynamic IP address*.

For the best connection to the internet to provide access to your station remotely, a *static IP address* from your ISP is the preferred option. With a static IP address, you are able to connect to your remote station without the IP address changing periodically. ISPs are currently assigning more static IP addresses due to better management of active IP addresses and larger companies moving to the new IPv6 routing that has more addresses available. This means the old static IPv4 IP addresses they were using would also have to be changed.

If you are unable to get a static IP address, there are avenues for finding out what your assigned dynamic IP address is. WhatisMyIPAddress, MyIPAddress, and other websites can tell you the current dynamic IP address you have been assigned from the ISP.

A better way to manage the ISP dynamic IP address is to use a service known as *Dynamic Domain Name System* (DDNS). This service maintains an external connection to your computer through the modem/router and *firewall*, providing a named address for you to connect to your equipment at home. This requires a small piece of software to run on your computer at home and an external DDNS service provider. Most DDNS service providers charge a fee to maintain and support this connection.

INTERNET BANDWIDTH

How much internet bandwidth do you need? How you set up your station, what devices and protocols you use, and what means you use for connecting to the internet all determine if you get a reliable connection or one that freezes or delays when you are trying to copy the call of a DX station.

If you are using a mobile connection for your roaming client station, you need *high-speed* or *high-bandwidth* service, the same as required for a streaming music service. 3G cellular service is the minimum service you can use, with a maximum 144 kbits/s transfer rate. Mobile network services such as 4G, LTE, and 5G vary in most areas, but give you network speeds and reliability closer to, if not more than, what most ISP networks can provide.

Speed is advertised by download speed (the bit rate from the ISP to your computer). *Upload speed* (from your computer to the ISP) may be only a third of the download speed but is important for getting your audio and control to the transmitter. Using internet sites such as Speakeasy.net and Fast.com can give you a good indication of the download and upload speeds on your computer.

It is important to realize that sharing an internet connection can also reduce bandwidth. If others in your home are online at the same time, their activity consumes some of the bandwidth available to your network devices. A dedicated line or service upgrade to support remote station operation may be required unless you can account for all the concurrent household bandwidth requirements and determine that your service is capable of handling it all. A router with *quality of service* (QoS) prioritizing capabilities might solve, or at least minimize, multi-user problems on a shared internet connection.

Rural ISPs usually consist of a wireless network over large areas. Network speeds can vary depending on the time of day or day of the week. This can be important, especially when you are trying to operate in a contest over a weekend when everyone else is at home streaming movies. If you plan to use a webcam for monitoring and security purposes, or stream a panadapter/waterfall type display, you must account for that bandwidth in your overall connectivity needs as well.

Hotels are known to vary network speeds depending on how many guests are in the hotel that night. You can check with the hotel staff to see if there is a secondary WiFi system that can give you more bandwidth.

INTERNET LATENCY

You need *low delay* (latency) to operate remotely, the same as any online gamer. For casual operating, you may be able to tolerate a latency of 250 ms or 300 ms. Serious contesters and DXers most likely find this much delay frustrating, especially on CW. If you are a high-speed CW contest operator, very low latency is required to be competitive. Below 200 ms is adequate, below 100 ms is better. Otherwise, tuning "lags" and you experience poor timing in the pileups. Voice and digital operating are more tolerant of higher latencies.

You can easily evaluate basic internet latency yourself using your computer. In a *Windows 7* (PC) environment, under START | ALL PROGRAMS | ACCESSORIES find COMMAND PROMPT, or in a *Windows 8* and newer systems, select START and type CMD. From within this DOS-like window you can run the *ping*, *tracert* and or *pathping* commands to access a particular IP address at the station site and average the response times to measure the latency. By adding the -t command to the ping command (for example, "ping 192.168.0.1 –t"), you get continuous ping results until you use the CTRL-C keys to stop the ping process. This is critical to know in advance or to troubleshoot connectivity issues once you are operational.

For the purposes of conducting internet system audio latency tests, N6RK's method illustrated in **Figure 24.40** is very effective. By using a receiver at the remote site and one at the control point, the recorded latency can easily be determined. **Figure 24.41** shows a latency of 40 ms for the return audio over K6VVA's private control link using a RemoteRig audio interface that requires a network speed of about 180 kbps. The audio is crystal clear, and at 40 ms latency, little difference from having the radio control head connected directly to the radio at home is noticed at the remote site.

24.5.4 Connecting to Your Remote Station

There are many ways to connect, control, and monitor a remote station. The equipment, communications medium to connect them, and how you use your remote station is only limited by time, money, and resources available. The following methods are to be used as guidance as you plan and build your remote amateur radio station.

BASIC MONITORING

The basic remote monitoring station shown in **Figure 24.42** can consist of very few components. If you just want to set your HF radio on a specific frequency, let's say a 40-meter net, and monitor it from work

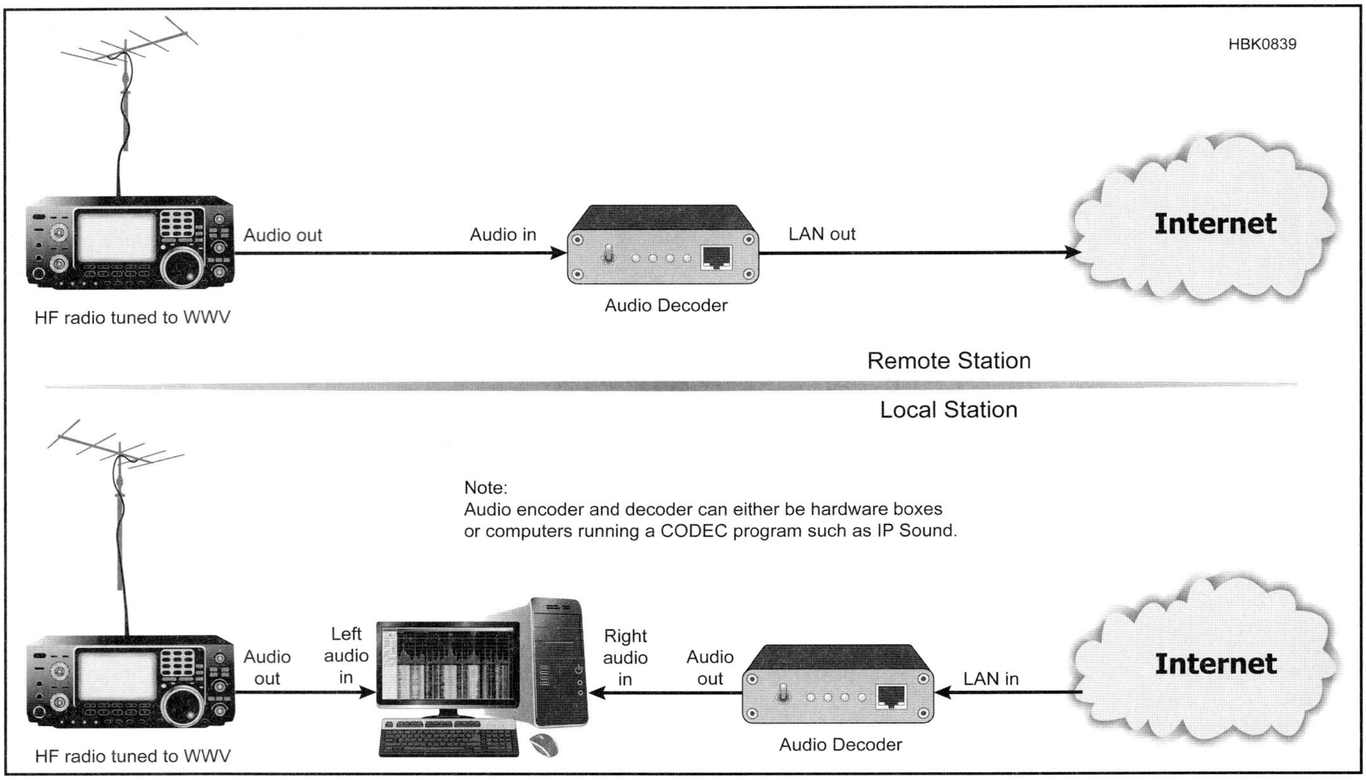

Figure 24.40 — N6RK's method for testing latency.

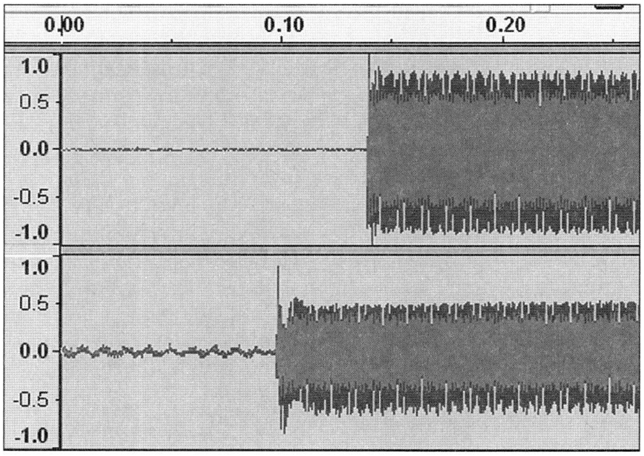

Figure 24.41 — K6VVA's latency test results.

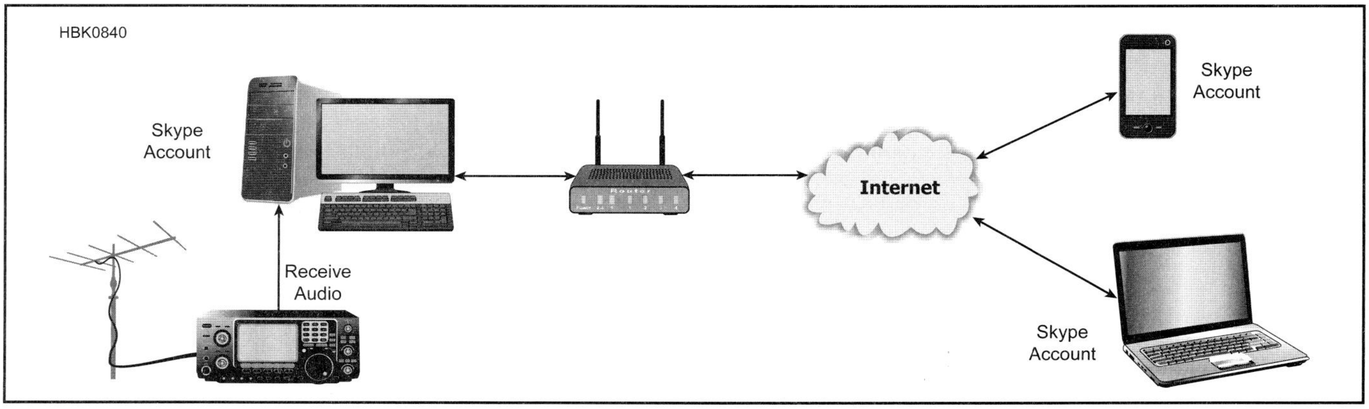

Figure 24.42 — Basic remote monitoring.

during your lunch break, then all you need is an audio cable connected from the headphone/speaker jack on your radio to the microphone jack on your computer. On your computer you need audio software to provide a connection to your mobile phone or work on a computer. Using the *Auto-answer* function on the host computer gives you unmanned audio connect. Simply connect the computer/handheld device to the audio software account on your home computer and listen in on the net.

Many remote station operators have found *Skype* to be a workable solution for remote audio routing as well as *IP-Sound*. Another way to completely avoid internet latency and bandwidth issues for audio is to use a POTS (Plain Old Telephone System) connection with a dedicated analog phone line that might also be shared with a DSL data service for control.

One of the newest technologies available to give you basic remote monitoring is the ability to use a Raspberry Pi as a TCP server for an SDR receiver. There are a few different versions of RTL-SDR TCP servers available that make it possible for programs like Airspy's SDR# to remotely connect to a Raspberry Pi/SDR receiver through the internet.

VNC CONNECTIONS

If you enjoy listening to your HF radio and would like to be able to tune remotely through the bands for different conversations, start with an audio cable from your radio's headphone/speaker jack to the microphone input on the computer. Your radio needs *computer aided transceiver* (CAT) control capabilities. Radios with CAT control have a serial, USB, or Ethernet port available for the computer to be able to connect and send control data to the radio. Icom radios have a CI-V control port that requires a COM-to-CI-V or USB-to-CI-V interface.

You then need radio control software installed on the computer. Several radio manufacturers have radio control software available. There are also companies that have software to control just about any radio with CAT control, such as *Ham Radio Deluxe* and *TRX-Manager*.

Once you have installed the radio control software and connected the computer to the radio, you need remote access to the host computer. A simple scenario, shown in **Figure 24.43**, uses *Virtual Network Computing* (VNC) software, such as *Chrome Remote Desktop*, *TightVNC*, *TeamViewer*, or *LogMeIn*, that allows users to take remote control of the host computer running the radio software. VNC software allows full control of the host computer, with some software including the sound interface and others requiring additional software for monitoring the audio. Audio server software like *Skype* and *IP-Sound* can also be used. Be aware that some VNC software packages charge extra for access to the remote sound. This software takes a lot of bandwidth to pass the screens and control between the connected computers or handheld devices.

All the VNC software options mentioned here can be accessed with an iPhone, iPad, Android phone, or tablet. Once you connect to your station with the VNC software, you can change frequency and bands using the radio control software on the host computer.

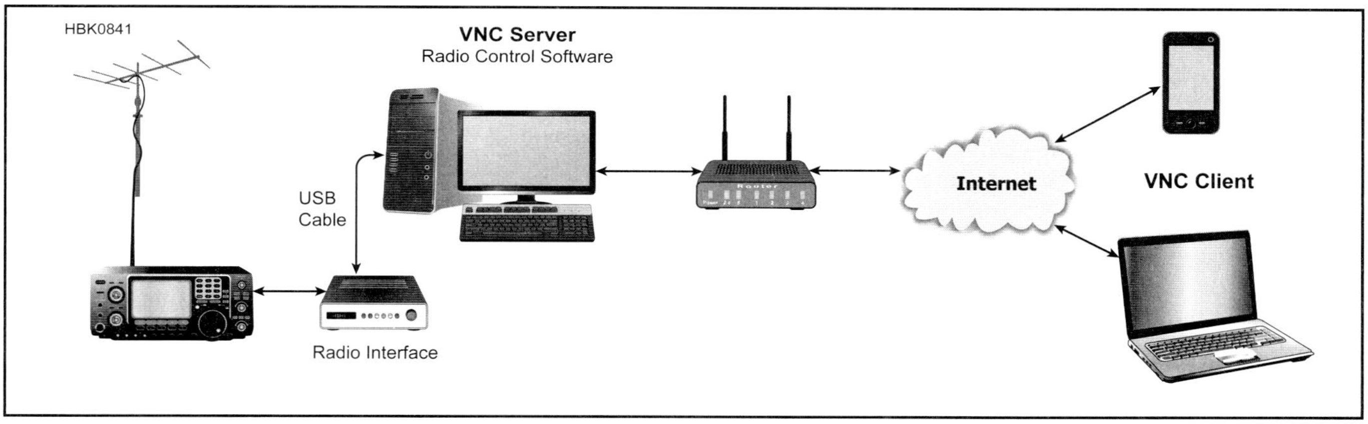

Figure 24.43 — VNC software connection.

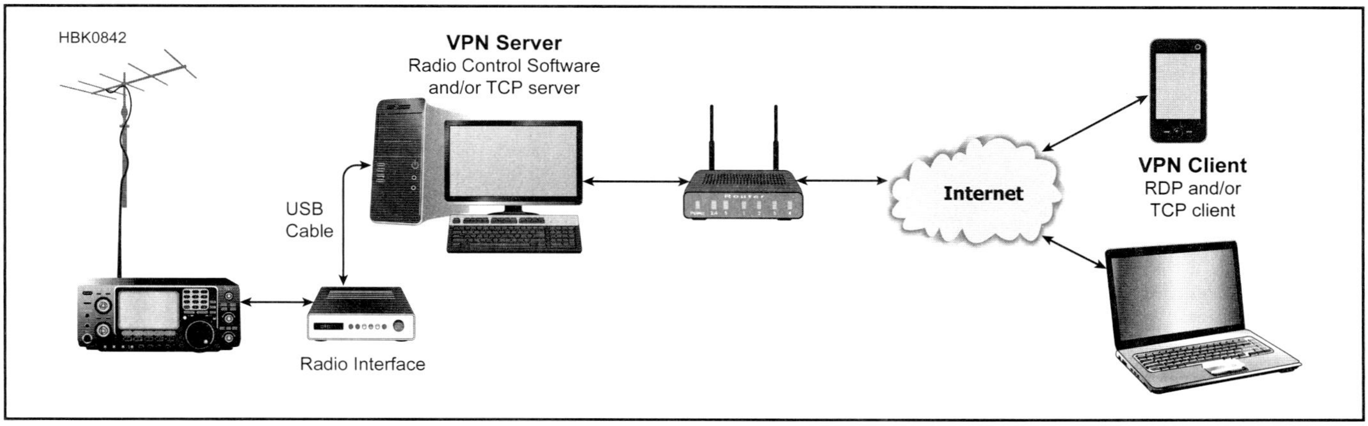

Figure 24.44 — VPN software connection.

VPN CONNECTIONS

A virtual private network (VPN) is like a tunnel through public internet access to connect to a private network, such as your station network. There are several VPN software packages and hardware options that enable you to create the internet connection for transporting data between the computer in front of you and the computer/devices at your home or remote station. A VPN uses encryption and other security mechanisms to ensure that only authorized users can access the network and that the data cannot be intercepted.

The VPN server at the station end can be either on the host computer or in the network router. It depends if you need connectivity to only the host computer or the whole station network. Microsoft has included VPN servers and clients in recent versions of their operating systems and some network routers include VPN servers. Other VPN software packages include *OpenVPN* and LogMeIn *Hamachi*.

Once you establish a VPN connection to the host computer or router from your client computer, as shown in **Figure 24.44**, you are able to connect your TCP client software to the radio control TCP server software and/or accessories at your station. Another option is connecting to the radio control software on the host computer with a *Remote Desktop Protocol* (RDP) connection, to view your radio control software on the screen and use *Skype*, *IP-Sound*, or other audio streaming client/server software for the audio connection. This gives you a secure connection to your station.

Using a VPN connection can be combined with a remote radio head. This gives you secure access to the equipment connected to your radio and allows you to operate the radio from the remote radio head.

TCP SERVER/CLIENT SOFTWARE

TCP server/client software comprises two programs that provide a connection over the internet between your host computer and the client computer you have in front of you. Each piece of equipment can have its own TCP server/client software installed at both ends.

The TCP server software is installed on a computer that is connected to your radio, tuner, rotator, or other station equipment, where it communicates with the station equipment by serial, Ethernet, or USB connections. Most of the TCP server software allows you to control the equipment locally on this computer when you are at the station, as well as through a remote internet connection.

The TCP client software is installed on the computer connects to the TCP server software through your internet connection. This provides you access to the station equipment with a dedicated TCP network connection to each device you connected to a TCP server computer.

There are advantages to using dedicated TCP server/client connections to your station. They use very small data packets, multiple TCP servers can be run in parallel for each device, and they can be connected only when you really need to see the data from the remote piece of equipment.

There are several radio manufacturers that have radio control software available utilizing TCP server/client technology. There are also software companies that provide software to control just about any radio with a CAT interface. Examples of this software are *Ham Radio Deluxe* (shown in **Figure 24.45**) and *TRX-Manager*.

REMOTE RADIO HEADS

Operating remotely from a computer or tablet can be fun and exciting, but it is not the same as turning a VFO or tweaking a bandpass filter knob to dig out the weak DX station. Manufacturers have developed internet-based interfaces to remotely connect a detachable radio head to the base section of the radio, hundreds of miles away.

One company to advance this idea is Microbit. Their RemoteRig standalone interfaces use separate *Session Initiation Protocol* (SIP), VoIP, and control channels to connect the two parts of the radio over Ethernet. Passing of the audio and control signals, along with CW paddle signals, is transparent to the remote radio head. Elecraft makes the K3/0 remote head for the K3 transceiver that is identical to the front of the radio, giving you full functionality of your K3 station wherever you have internet access. The W4AAW multi-multi contest station uses the K3/RemoteRig combination at one of the operating positions for off-site operators during contests.

Remote radio interfaces work with several other radios. Remoting the Icom IC-706 and Kenwood TS-480 transceivers is very popular among amateurs. There are groups that have installed the base part of the Icom IC-706 in different locations and have the ability to connect to any one of them from the radio head. They have the capability to change the profile in the RemoteRig at the radio head for it to connect to the different remote site base units.

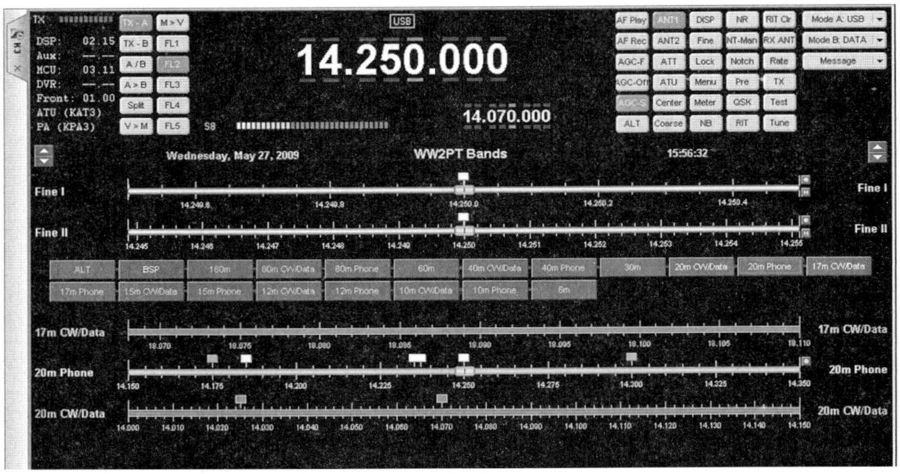

Figure 24.45 — Ham Radio Deluxe.

Figure 24.46 — Wireless FlexRadio Maestro.

The FlexRadio Maestro (as seen in **Figure 24.46**) gives you an all-in-one Plug and Play control console for the Flex-6000 series radios. This unit has a built-in local area network (LAN) connection, as well as an integrated Wi-Fi client module for local wireless network connectivity. The LCD touch screen allows for pan, zoom, monitoring, and menu operation. The back of the unit provides connections for CW paddle, PTT, microphone, headphone, and other accessory connections to give you full access of the remote radio. An optional battery can make this remote unit a fully wireless option.

HAM RADIO ONLINE

If you are unable to invest in the equipment to operate remotely and really want to experience the rush of operating a contest or DX from a station with large antennas and lots of power, there are options available online.

One such service is RemoteHams. They offer the opportunity to use their RCForb client software on a PC computer or Android device to remotely connect to and operate stations that are a part of their community from around the world.

RemoteHamRadio currently offers you a chance to operate up to 25 dedicated world-class amateur radio stations across North America using an Elecraft K3/0, FlexRadio Maestro, or from your home computer/smart device using the RHR WebDX online or a custom desktop application. (See **Figure 24.47**.)

Using these systems, you have access to a community of stations through the internet, offering an experience not available through more modest stations.

24.5.5 Station Automation

Automation is key to having a fully functional remote station. The ability to control your rotator, antennas, and amplifier from your computer has been around for some time. Connecting a local host computer to the equipment and remotely monitoring the host computer works, but now several manufacturers have added remote TCP server/client software to control your station from off-site. The following section describes software and hardware to enable automation and remote control of your station.

SWITCHING ANTENNAS

You can simplify your remote station by using one antenna for all the bands. Using a multiband dipole or vertical antenna eliminates the need for antenna switching. Using the SteppIR BigIR vertical antenna with its controller, for example, gives you a versatile antenna that follows the frequency from your radio.

To automate your antenna system, one of the first pieces of hardware you need is a band decoder. A band decoder detects what band your radio is on and passes this information to the antenna switching network. The band decoders can detect the band either by a binary coded decimal (BCD) output from the radio, through a CI-V interface on an Icom radio, or by reading the frequency data from the radio. Band decoding can be accomplished with standalone units, such as the Array Solutions BandMaster or Elecraft KRC2, or a decoder built into an antenna switching controller, as with the Ameritron RCS-12, EA4TX RemoteBox, or Hamation integrated controllers.

An alternative to using a band decoder takes advantage of station control or logging software with the capability to send band data to antenna switching units via Ethernet or serial connections. The Green Heron Engineering antenna switches can

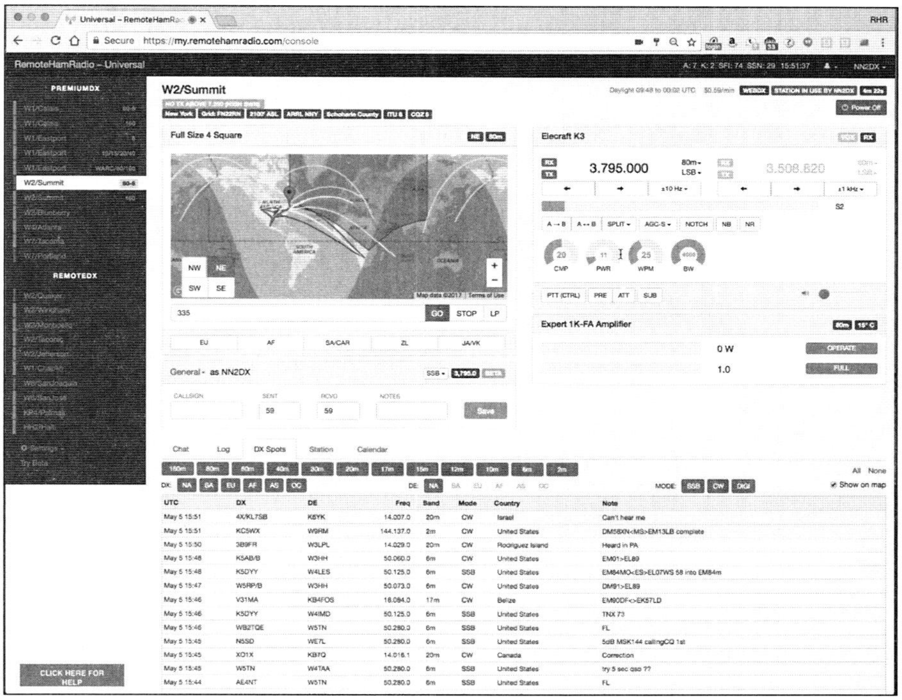

Figure 24.47 — RemoteHamRadio WebDX.

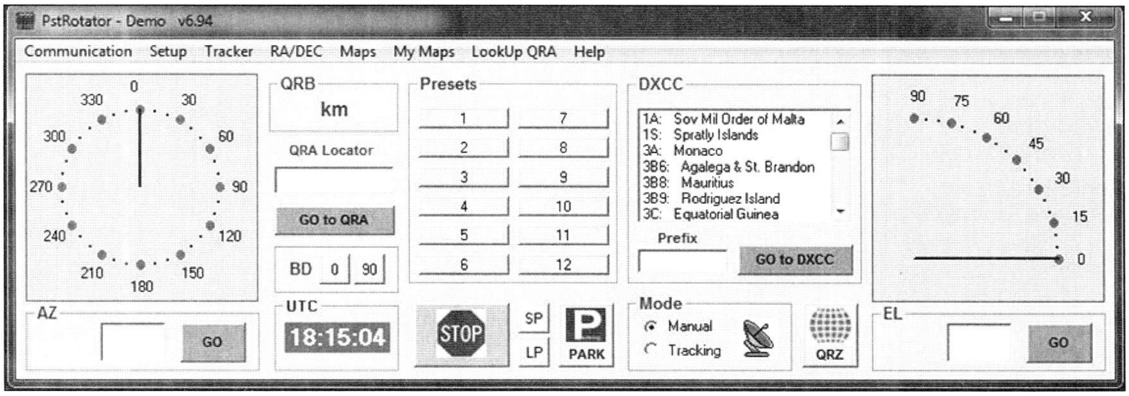

Figure 24.48 — PSTRotator software.

be controlled by local software, as well as through TCP server/client software. Snaptekk offers a 4-port antenna switch with WiFi and a TCP server built in.

Additional advantages of Hamation antenna switches and EA4TX RemoteBox is the availability of USB connections to the host computer and TCP server/client software. With this software you can reconfigure your automatic antenna selection remotely or manually control what antenna you want to use.

The Antenna Genius from 4O3A Signature is the next generation of automatic antenna switches that eliminates the need for a band decoder or computer. The firmware in this line of antenna switches support connection and monitoring of band/antenna data through BCD, PIN to port, or advanced API software. The API software allows deep integration into supported amateur radios such as the Flex-6000 series.

ROTATORS

Automating your rotator system usually starts with installing an interface in the rotator control box to give your computer control of the rotator. Electronic Rotator Control (ERC) interface kits from Vibroplex or EA4TX give you the capability of connecting to your computer via USB or serial connections.

Most logging and contesting software, either directly or indirectly, gives you control of your rotator as you enter a call sign into the call field of your logging software. *PSTRotator* control software (see **Figure 24.48**) can interface with several logging software packages. It also has built-in TCP server/client software to tie it into your logging software on your remote client computer for automated rotator control from an offsite operating location.

RF AMPLIFIERS

Several modern amplifiers have the capability of full automation by detecting and following the frequency of the radio's output signal. Add this to the automatic antenna switching and tuners found in some of today's high-end amplifiers, and automation of your station becomes much easier.

Most modern RF amplifiers can read the CAT data straight from the transceiver, set the band on the amp, a preset memory location on the tuner, and the appropriate Transmit antenna. Combine this with a remote head radio connection and you have no need for a VPN/RDP or VNC connection to your station for manual control, providing you with a clean, low bandwidth connection for your remote radio head.

A good example of remote monitoring and control is the Elecraft KPA500 amplifier and KAT500 tuner. Both have Aux connections to the radio for frequency tracking and serial connections for the host computer for control and monitoring. This allows custom TCP server software to be run on a host computer that can be mirrored on a remote client computer, an example of which is seen in **Figure 24.49**. For amplifiers with serial or USB connections and no TCP server software, they can also be connected to your host computer. A remote VPN/ RDP session or VNC connection gives you remote access to configure and monitor the amplifier.

Several of today's RF amplifiers are being produced with an Ethernet port incorporated into their design. This Ethernet port provides a built-in TCP server so any computer with the manufacturers TCP client software can connect directly to the amplifier for control and monitoring of the unit.

STATION MONITORING

It's important to know a station is operating correctly when you are operating remotely. There are a few manufacturers of power and SWR meters that include serial ports and software to monitor the meter on the host computer. The TelePost LP-100A power meter includes an RS-232 serial port and LP-500/700 models have USB for connection to the host computer. This software can be monitored via a VPN or VNC connection. TCP server/client software from Wizkers will allows remote access to the LP-100A through your host computer.

Several other RF power meters have built-in Ethernet connections and TCP server/client software for direct connection to the meter from a remote site without going through a host computer. The SMΩRF meter from MicroHam, shown in **Figure 24.50**, is one such example that gives the operator a full range of information about how the station is operating and evaluates the antenna system in the event of a problem.

POWER AND RELAY CONTROL

There are many ways to remotely turn devices on and off in your station. If you want to turn your radio, amplifier, and computer on and off, or switch between antennas, there is equipment available to do this.

Using your phone line to control devices is as simple as using the DTMF keypad on a phone. Several manufactures produce equipment to connect to the phone line with onboard relays that are turned on and off by DTMF tones. You can buy units with 1 to 8 relays from Velleman and Viking Electronics.

Using the internet to control relays is also possible. There are Digital Loggers and Belken Ethernet-enabled power strips that can be turned on and off by an app on your Android or iPhone. **Figure 24.51** shows how G4IRN uses a pair of Wi-Fi units to turn his amplifier and 12 V power supply on and off. This works great for turning on a power supply or switching between two antennas. For a lower cost option, Velleman makes a relay kit that connects to your host computer by USB and provides software that runs on the computer with an app to control the relays from your phone.

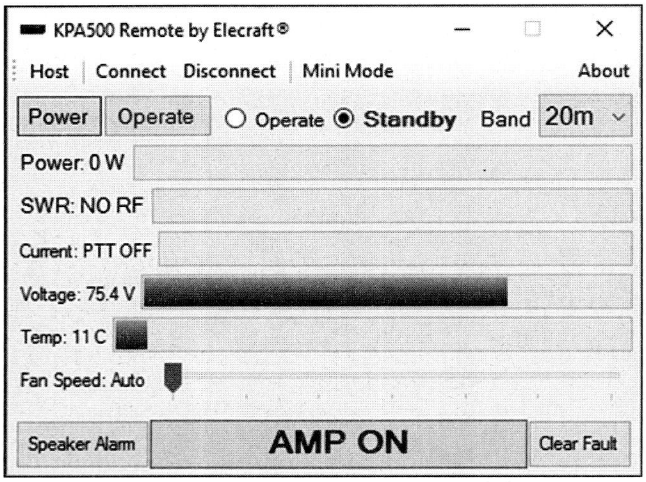

Figure 24.49 — Elecraft KPA500 TCP server/client.

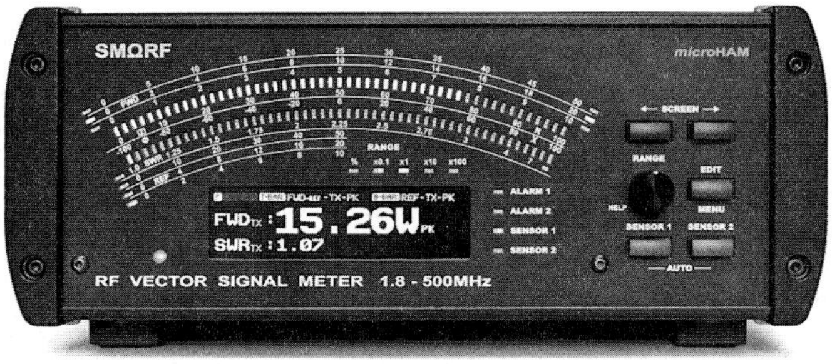

Figure 24.50 — MicroHAM SMΩRF power meter.

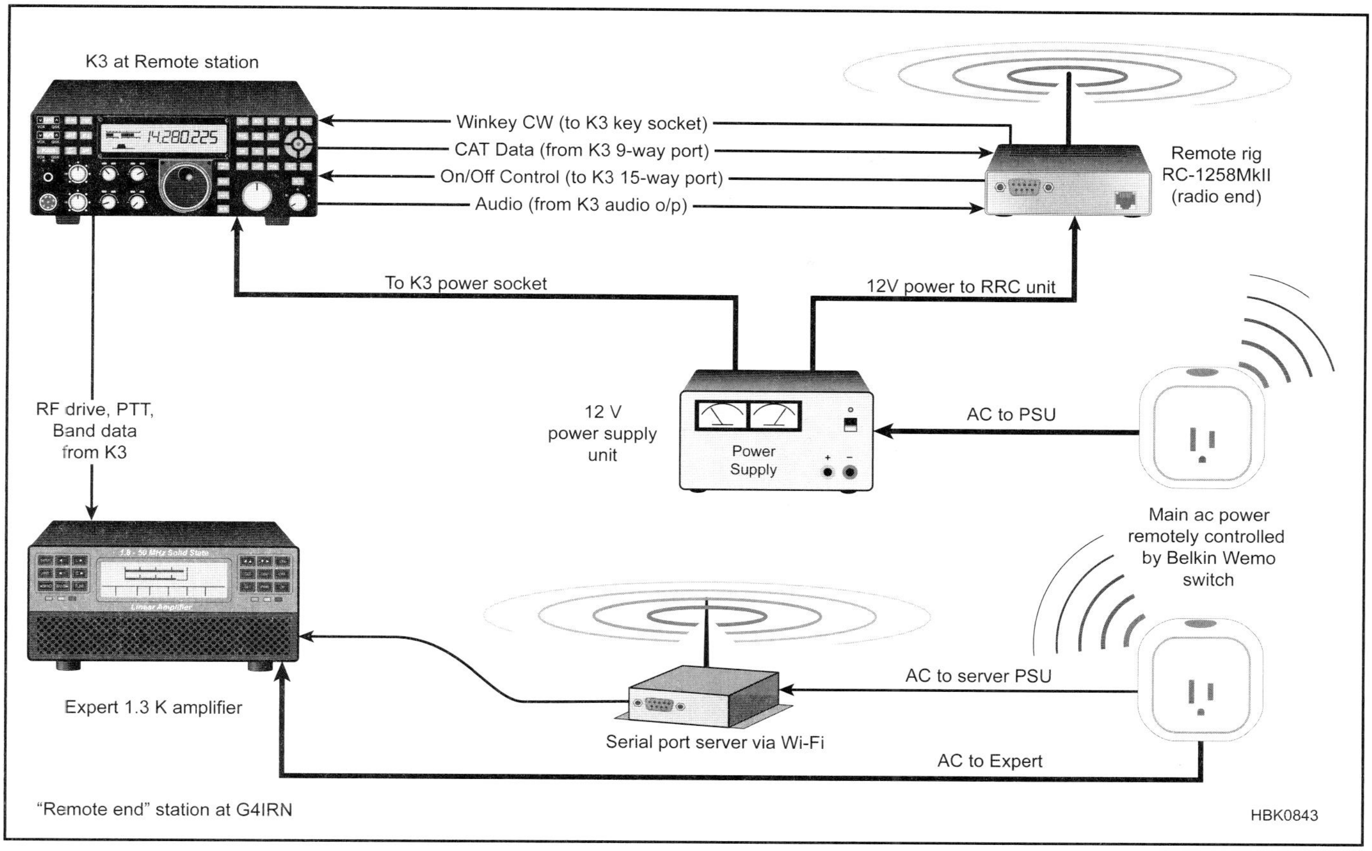

Figure 24.51 — Belken WiFi switches.

RADIO INTERFACES

Connecting the radio to the computer can be as simple as running serial cables and audio cables from the radio to the computer, but serial connections can be difficult to configure, and the audio connections can allow noise to get into the receive and transmit audio. It is recommended that an interface between the radio and the computer be installed. Several manufacturers make interfaces for most radio models that have CAT control in them, such as the ones offered from MicroHam, Tigertronics, and West Mountain Radio. Using interfaces simplifies connections and configurations. They also provide higher quality analog-to-digital audio converters than what most computers come with.

Several newer radios have the interface built in. A USB cable is all that's required for audio and CAT between the radio and the computer.

SERIAL PORT SERVERS

Serial port servers, such as the Moxa *Nport* or *Digi Connect* series, are interfaces that connect a RS-232/422/485 device to an Ethernet port. These port servers can act as a gateway between two protocols or pass encapsulated serial traffic between a serial device and a remote computer over Ethernet networks. Unless the software on the client computer is designed to communicate with the serial device over Ethernet, redirect software such as the *Digi RealPort* is required to manage the encapsulated serial traffic. Serial port servers are used when you do not want to use a computer at the station to connect the serial port on a piece of remote station equipment.

RASPBERRY PI

The Raspberry Pi is a tiny and affordable computer that has taken on a powerful role in the amateur radio world. This small computer is a series of small single-board computers developed in the United Kingdom by the Raspberry Pi Foundation in association with Broadcom. Taking advantage of the Linux operating system, this computer utilizes the Node-Red programming environment for communicating between a wide range of devices through RS-232, USB, or TCP/IP using direct communication or APIs. This data can be displayed and controlled with user defined interface screens as part of the Node-Red Dashboard as in **Figure 24.52**.

Almost any piece of amateur radio equipment that has an external communications port can be part of the Node-Red program. The following is a general list of modern devices that can currently be monitored and/or controlled by Node-Red:

- •Amateur radio transceivers
- •Tuners
- •Amplifiers
- •SteppIR Antennas
- •Rotors (Using PstRotator software)
- •Antenna Switches
- •Preamps
- •Watt meters
- •USB Relays for external control
- •Ethernet based power switches
- •Weather Stations

The Node-Red Dashboard is a browser-based GUI interface that can be viewed at the operator's home station or as a window into the remote station through the internet. This Dashboard can turn your station on or off, show you how much power your radio is putting out, the VSWR of your antenna, which antenna is connected, what direction the antenna is pointed, current weather conditions, DX spots, and a host of other information. The Dashboards can display basic information from a contester who wants to know the antenna switched, when the VSWR goes high to the casual operator monitoring everything at the station and getting an audio alert when the radio tunes outside the allowable frequencies. It is as simple as adding a Port Forwarding rule to

Figure 24.52 — Node-Red Dashboard.

Figure 24.53 — Stream Deck.

your router and a Static IP (or DDNS service if no Static IP is available) to get remote access to the Node-Red Dashboard.

There is a large group of amateurs that have been building the individual "flows" for different pieces of equipment. If you have equipment that you would like to monitor or control through one piece of software, Node-Red is the best answer today.

If you are interested in learning more about Raspberry Pi and Node-Red, the first place to start would be to read the September 2021 *QST* article, "Node-Red for Ham Radio," by David De Coons, WO2X, and Michael Walker, VA3MW. This article explains Node-Red and its uses in amateur radio. Another great source is to join the online user group. The group's home page is at **groups.io/g/nodered-hamradio**. The amateurs that monitor the online groups can answer questions and direct you to the sources for your situation. The user group also maintains a database of custom "flows" for different pieces of equipment and functions helpful to the amateur radio operator, written by amateur radio operators.

YouTube is also full of helpful content on using Node-Red in amateur radio. Kyle Krieg, AAØZ, Dave De Coons, WO2X, and Michael Walker, VA3MW, have all produced wonderful videos on installation and programming Node-Red applications for amateur radio.

ETHERNET IN THE STATION

Several of today's amateur radio manufacturers have embraced Ethernet communications in the equipment they produce. This allows for TCP or UDP communications between the radio, amplifier, tuner, antenna switch, filter controller, and other station accessories through a local Ethernet network. They have also produced custom API software for computers to interact with their equipment. This eliminates the need for the serial connections and radio interfaces previously used, reducing cabling and configuration headaches.

Using Ethernet communications can bring its own set of problems. Without proper shielding, RF can get onto the CATx Ethernet cable and interfere with communications between devices. Using shielded CATx cable (grounded only at one end) or properly installing toroid coils helps to prevent RF from getting into your equipment. (See the **RFI and EMC** chapter.)

Another problem that can occur with Ethernet networks is excessive network traffic. Sharing your radio equipment with other devices in the home can cause interference and delays in communications with your radio equipment. One way to resolve this issue is to use an isolated Ethernet network dedicated to your radio equipment, and a separate network for internet connectivity for your main computer and the remainder of the house. Adding a USB-to-Ethernet adapter to your main computer can link the second network to the internet and keep the two networks isolated, reducing traffic on the radio network.

If you still want access to your radio from outside your house, and also isolate your radio network from other parts of the house, a good router can be installed and programmed for one WAN to your ISP and two separate LAN (House LAN and Station LAN) networks, each with different IP schemes. This gives both local LAN's access to the internet keeps both networks from interfering with each other.

REMOTE CW

One problem CW operators run into while operating remotely is their dots and dashes not getting to the radio and transmitted as intended. Internet delays and breakup of CW data distort the CW data causing undecodable transmissions. The most efficient way to send CW to a remote transmitter is by using *Winkeyer* software. Most logging software support redirecting CW characters to the local *Winkeyer* client application that sends it to the remote radio or interface to be converted from characters to dots and dashes for the transmitter to send.

Another problem CW operator's run into is getting the transmitter sidetone back to the operator's location while the remote radio is generating the sidetone. One answer is to add a computer at the radio to input the sidetone audio and stream it back to the operator. FlexRadio users can use the Maestro to send the CW data to the remote radio and provide a local sidetone.

A third option is to use a local K1EL Winkeyer, like the WKUSB, and setting up K1EL *Com Port Scanner* software from K1EL Systems. This software creates virtual Winkeyer channels, one for the remote radio and a second one to the local K1EL Winkeyer. This gives you a local sidetone for the CW characters you are sending to the remote radio.

STREAM DECK

The Elgato Stream Deck is another power tool for local and remote operation of the radio station. The desktop device comes in several different models ranging from 6 to 32 buttons. Each button can be programmed for an operation that lets you open apps on your computer, adjust audio, mute your mic, turn on lights, and so much more locally. A good resource for amateur radio use of the Stream Deck is Github *Hamdeck* and *Ham Radio Workbench*.

If you are a FlexRadio 6000 Series user, the Stream Deck can be programmed to monitor and control the remote radio (see **Figure 24.53**). See the FlexRadio Peripherals and Station Integration for details on setting up the Stream Deck. Other manufacturers will be supported soon as operators learn how to configure their radios and publish the plug-ins for the Stream Deck.

24.5.6 Remote Site Requirements

PURCHASE VS RENT OR LEASE

Although having direct control over every aspect of what you want to do is most desirable, you may want to consider a rental or lease with a cooperative landowner. In either case, having a detailed agreement or contract prepared by a competent attorney is a must. Be sure to include every contingency you can think of, especially what happens if the landowner decides to sell. Leasing or renting with an option to purchase allows you to fully evaluate the site before making a long-term commitment to it.

RELIABLE ACCESS BY ROAD

If your remote site is not on developed property, can you get there safely, rain or shine? What happens if any problems develop that require immediate attention at the remote site? Pavement, or a good all-weather, well-maintained gravel road should be available. Having more than one means of getting in or out should emergencies arise while you are at the site is also an important consideration. These concerns must be addressed before agreeing to purchase, rent, or lease a remote site.

GEOLOGICAL AND OTHER HAZARDS

Evaluate the geological factors involved. In a hilltop or mountaintop site, is there evidence of soil movement that could affect access or damage your equipment shelter or tower installations? Does the property sit near an earthquake fault? Is there a significant risk of fire during fire season? Is the site exposed to high wind or storms? In flat-land sites, is there a risk of flooding or drainage problems? Remember, you won't be there to rescue your equipment! In consideration of both your initial construction activities and ongoing maintenance tasks, are there any wildlife hazards from animals or insects? What about allergies or reactions to plants such as poison ivy or poison oak? Diligently investigate all aspects of a potential remote site *before* you end up with bad surprises down the line. For a reasonable fee, there are companies from which you can secure a Geological and Property Hazards report to aid in your assessment.

RAW SITE PREPARATION

If you are fortunate enough to locate suitable property, be sure to evaluate the site preparation needs. How much time and money will be necessary to develop the remote site to meet your specific needs? Will there be brush and vegetation to dispense with? Will trees need to be cut down? Will bulldozers, excavators, and trenchers need access to the site? Concrete trucks? Service vehicles? Diligent evaluation of all costs involved will reduce unnecessary surprises.

SITES WITH EXISTING FACILITIES

Finding a site with a self-supporting tower or two in place that can accommodate your needs, along with a building large enough for your equipment and operational needs, can be like finding a pot of remote gold. However, unless you are experienced with towers and construction, hire professionals to evaluate everything. If you are considering co-locating at a remote site with existing antennas and equipment in use, consider hiring a professional to help you evaluate the likelihood of problems. Birdies and intermodulation could end up being a source of potential conflict and stress if problems can't be resolved.

TERRAIN CONSIDERATIONS

Evaluate the terrain on the bands you wish to operate. For HF, Dean Straw, N6BV, has developed a very useful program called *HFTA — High Frequency Terrain Assessment*. (See the *ARRL Antenna Book*.) Use *HFTA* to assist you in placement of your antennas. Don't expect excellent ground/soil conductivity if your remote site is on top of a rocky mountaintop.

If you plan to use vertical antennas, make

Figure 24.54 — Solar panels can be used to charge a bank of batteries to supply power. Effective off-grid power requires an accurate and conservative estimate of power needs and available energy sources. [Rick Hilding, K6VVA, photo]

sure there is sufficient room on the property to place the radials, whether on a mountaintop or flat area. Placing verticals near water, especially saltwater, is the most desirable way to achieve the low takeoff angles for long-haul DX. However, exposure to salt air or spray creates significant maintenance issues.

NOISE LEVELS

Since one of the reasons for building a dedicated remote station away from your home is to lower noise levels, you'll want to carefully evaluate existing and potential noise sources. Start by visiting the site with a receiver in several different types of weather and listening on all the bands you intend to use with a full-size dipole antenna. If noise levels are low, this is no guarantee that problems won't crop up in different seasons or from hardware failures. You should avoid nearby substations or high-voltage and power distribution lines. If you are sharing a site with other users, coordinate with them to test intermodulation and adjacent channel interference before you go into full operation.

ELECTRICAL POWER

Having electrical power already available at your remote site can certainly save a lot of money. The most ideal situation is to have underground power to the remote site. Be sure to evaluate the full costs of installing ac power if it is not already present at the site or on the property. Truly remote sites may require wind, solar, or generator power systems like that shown in **Figure 24.54**.

If your remote site requires solar/wind/battery power, it is critical that you properly assess sunlight path and duration, wind patterns, and provide sufficient battery storage to meet your intended operating needs during lengthy periods of inclement weather. The article "Designing Solar Power System for FM Translators" is a valuable resource for planning of remote HF station off-grid power. (See the Resources section below.) You must thoroughly account for the power requirements of each and every piece of equipment.

There are various estimates for the duty cycle of transceivers. A conservative duty cycle of 50% for CW/SSB is recommended for power capacity evaluation. (RTTY/digital/AM/FM duty cycles are likely to be higher.) Internet switches, routers, rotators, computers, and anything electrical also consumes power. A supplementary diesel or propane remote-start generator to a basic off-grid system may be an appropriate option, particularly if you plan on using an amplifier. If your remote site is in a rugged terrain rural area, make sure a diesel or propane service provider vehicle can get to the location *before* you make final plans! It may be possible to use smaller propane tanks you can haul in a truck, but any manifold-type system necessitates insuring that the proper BTU requirements can be met based upon the propane generator manufacturer's ratings. Installing a solar and wind remote battery monitor system is advisable.

When you have made an accurate assessment of your complete system needs, add a generous additional contingency factor. It is also advisable to get quotes on your proposed system from at least two different "green power" equipment suppliers, and to make sure they fully understand your requirements and intended use *before* you purchase anything.

INTERNET ACCESS

If you are planning to operate your remote station via the internet, be sure that broadband service such as cellular, rural wireless, DSL or cable TV can reach your site by asking the local service provider to do a site evaluation. Satellite internet downlink speed may be fast, but uplink speed is often quite a bit slower. Be sure the minimum link speed is adequate in both directions.

If your control point is close enough to the remote site, another option is to implement your own private wireless "bridge" system. Even if multiple "hops" are involved, with the right equipment you can potentially have better results than using even broadband internet. Hamnet technologies (see the **Digital Modes and Protocols** chapter) may be an option to connect a remote station that's not too far away.

INSURANCE AND SECURITY

You've heard the expression, "Stuff Happens." Unless you have blanket coverage in an existing homeowner's insurance policy that includes your remote site equipment, look into the ARRL "All-Risk" Ham Radio Equipment Insurance Plan. If you are hiring others to do work at your remote site, be sure that liability coverage issues are addressed. The same applies to any guests or visitors, such as fellow hams. Have a competent attorney draw up a "Liability Release" agreement with strong "Hold Harmless" language.

If your remote site is yours alone and not a co-habitation arrangement with other services, be sure and evaluate all security needs (a necessity even if co-located). You might need one or more security gates, webcam surveillance, and auto-remote notification of security breaches.

PERSONAL SAFETY

Regardless of your age, it is prudent to have someone with you at the site. You could severely cut yourself or fall and break a leg. If there are wildlife hazards in the area, take appropriate steps to protect yourself.

Carry a first aid kit and have one readily available at your remote site. Keep some emergency supplies such as various sealed food items in a varmint-proof container and a case of bottled water in the event you find yourself stranded.

If your remote site is within the coverage area of a local repeater, take along a handheld radio (with batteries charged) as part of your SOP (Standard Operating Procedure). This is especially important for rural and remote operating sites.

24.5.7 Station in Your Hand

Operating or just monitoring band conditions from your mobile phone or tablet is almost as easy as from your computer. With today's smartphone technologies, Skype, VPN, RDP, and VNC connections to your station are quite possible.

There are several apps available, or being developed, for connections directly to your radio or radio interface. *RRC-Nano* is an app for Android devices from RemoteRig that makes it possible to remote your station. It uses the RemoteRig interface at the station site to provide remote access to the radio through your phone.

Available for the Flex-6000 series users with IOS devices is *SmartSDR* app for the iPhone or iPad. A small fee on the Apple Store gets you a third party IOS application (see **Figure 24.55**), that allows you full access to your remote FlexRadio Systems 6000-series radios.

24.5.8 Future of Remote Operation

With the increasing interest and equipment available for remote operation today, manufacturers of amateur radio products are moving forward with upgrades to existing and new equipment to take advantage of the internet and expanding broadband coverage. Future radios will be easier to interface to other equipment in the station using APIs and will provide access from anywhere in the world. Several manufactures are developing new APIs and software for TCP Server/Client connections to their equipment.

With all the talent and resources in the amateur radio world, the Raspberry Pi and Stream Deck continue to advance into more powerful station tools.

24.5.9 Special Events and Demonstrations

Remote stations come in very handy dur-

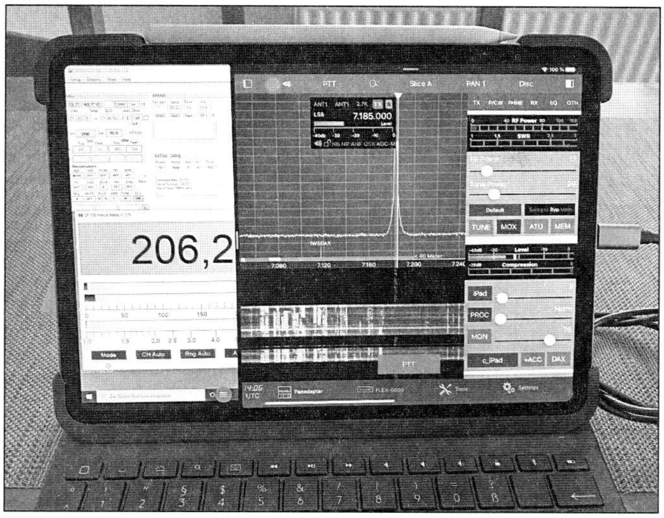

Figure 24.55 — Flex IOS app

ing special events. If you are putting on a demonstration of amateur radio, setting up an effective antenna may not always be viable, and you may not be able to communicate very well. Local noise levels or RFI problems may present significant challenges.

Setting up a remote link to your home or dedicated remote station gives you a definite advantage and make the experience more enjoyable for the audience. For example, a remote client station was used at a Boy Scout Jamboree-On-The-Air event. Young scouts were given the chance to communicate with other scouts around the world with the help of a capable station without the usual compromises associated with portable stations. This makes for a more effective demonstration of what amateur radio can do.

24.5.10 Remote Station Resources

For other great resources on how to set up a remote station, see the following videos, articles, or books:

Aaker, M., K6UFO, "Remote Operating for Amateur Radio: Ten Things to Know" see **www.k6ufo.com/attachments/Remote_Op_Ten_Things.pdf**

Aaker, M., K6UFO, "Remote Access — Six Ways to Implement" see **www.k6ufo.com/attachments/K6UFO_Visalia_2016.pdf**

Aaker, M., K6UFO, "Remote RTTY Contesting" **www.k6ufo.com/attachments/K6UFO_Dayton_RTTY.pdf**

ARRL Website Radio and Technology Topics, Link and Remote Control, **www.arrl.org/link-remote-control**

Craft, B., Evans, J., and Cook, M., *Raspberry Pi Projects for Dummies* (Wiley, 2015).

De Coons, D., WO2X, Basics of Node Red for Ham Radio **www.youtube.com/watch?v=nqJPPHG4aBw**

Ford, S., WB8IMY, *Remote Operating for Amateur Radio* (ARRL, 2010).

Hilding, R., K6VVA, "How Not to Build a Remote Station — Part 1," *National Contest Journal*, Jan./Feb. 2010, pp. 8 – 10.

Hilding, R., K6VVA, "How Not to Build a Remote Station — Part 2," *National Contest Journal*, Mar./Apr. 2010, pp. 19 – 23.

James, E., KA8JMW and Pendley, M., K5ATM, "Raspberry Pi: A Low Cost Platform for Amateur Radio Projects," **www.nm5hd.com/documents/PRESENTATIONS/RaspberryPi.pdf**

Krieg, K., AA0Z, - Getting started on a Remote HF station: **www.youtube.com/watch?v=0UJs5C1Yqzo**

Krieg, K., AAØZ, - Intro to Node Red for Ham Radio **www.youtube.com/watch?v=1WsEBCixuGc&t=2099s**

Sepmeier, Bill, "Designing Solar Power System for FM Translators" *Radio Guide*, Sep. 2008, pp. 6, 8.

Walker, M., VA3MW, Setting up my First HF Remote Station: **www.youtube.com/watch?v=C7d5C7nc5R0**

Yerger, Alfred T., WA2EHI, "Remote Control Your Rig via the Internet," see **www.arrl.org/files/file/Technology/LinkRemoteControl/RemoteControl.pdf**

Web URLs for Companies and Websites in this Section

4O3A Signature — **4o3a.com**
ACOM — **acom-bg.com**
Airspy — **airspy.com**
Ameritron — **ameritron.com**
Array Solutions — **arraysolutions.com**
Belken — **belkin.com/us**
Chrome Remote Desktop — **remotedesktop.google.com/**
Digi — **digi.com**
Digital Loggers — **digital-loggers.com**
EA4TX — **ea4tx.com/en**
Elecraft — **elecraft.com**
FlexRadio — **flexradio.com**
Hamdeck — **github.com/ftl/hamdeck**
Ham Radio Deluxe — **ham-radio-deluxe.com**
Ham Radio Workbench — **hamradioworkbench.com**
Hamation — **hamation.com**
HAMNET — **broadband-hamnet.org**
Icom — **icomamerica.com/en**
IP Sound — **iw5edi.com/software/ip-sound**
Kenwood — **kenwood.com/usa/com/amateur**
LogMeIn — **secure.logmein.com/home/en**
LogMeIn Hamachi — **vpn.net**
MicroHAM — **microham-usa.com**
Moxa — **moxa.com**
MyIPAddress — **myipaddress.com/show-my-ip-address**
OpenVPN — **openvpn.net**
PSTRotator — **qsl.net/yo3dmu/index_Page346.htm**
Raspberry Pi — **raspberrypi.org**
RealVNC — **realvnc.com/index.html**
RemoteHamRadio — **remotehamradio.com**
RemoteHams — **remotehams.com**
Skype — **skype.com**
Snaptekk — **snaptekk.com**
SPE Expert — **linear-amplifier.com**
SteppIR — **steppir.com**
Teamviewer — **teamviewer.com/en**
TelePost (N8LP) — **telepostinc.com/n8lp.html**
Tigertronics — **tigertronics.com**
TightVNC — **tightvnc.com**
TRX Manager — **trx-manager.com/demoe.htm**
Velleman — **velleman.eu**
Vibroplex — **vibroplex.com**
Viking Electronics — **VikingElectronics.com**
Websdr — **Websdr.org**
West Mountain Radio — **westmountainradio.com**
WhatisMyIPAddress — **whatismyipaddress.com**
Wizkers — **wizkers.io**

Contents

Chapter 25 — Online Content

Supplemental Files

- Antenna Analyzer Pet Tricks by Paul Wade, W1GHZ
- Apparatus for RF Measurements by Bruce Pontius, NØADL, and Kai Siwiak, KE4PT
- ARRL Lab Test Procedures Manual
- Build a Return Loss Bridge by James Ford, N6JF
- E- and H-Field Probes by Ward Silver, NØAX
- Low Frequency Adapter for your Vector Network Analyzer (VNA) by Jacques Audet, VE2AZX
- Noise Instrumentation and Measurement by Paul Wade W1GHZ
- RF Field Strength Meter by John Noakes, VE7NI
- RF Sampler Construction details by Thomas Thompson, WØIVJ
- RF Step Attenuator by Denton Bramwell, K7OWJ
- Test and Measurement Bibliography
- Test and Measurement Further Reading
- Testing and Calculating Intermodulation Distortion in Receivers by Dr. Ulrich Rohde, N1UL
- Two-Tone Oscillator — PCB artwork and layout graphics by ARRL Lab
- Receiver Testing and Performance by Rob Sherwood, NCØB (Separate folder)
- Receiver Noise Floor and Band Noise
- Reciprocal Mixing Test Procedure
- Sherwood Lab Setup for Dynamic Range Measurements
- Terms Explained for the Sherwood Table of Receiver Performance
- Voltage-Power Conversion Table

Project Files

- Compensated RF Voltmeter articles by Sidney Cooper, K2QHE
- Gate Dip Oscillator articles and PCB artwork by Alan Bloom, N1AL
- Logic Probe — supporting photos and graphics by Alan Bloom, N1AL
- Logic Probe and Gate-Dip Oscillator (GDO)
- RF Power Meter — supporting files by William Kaune, W7IEQ
- Tandem Match articles by John Grebenkemper, KI6WX
- Transistor Tester PCB artwork and layout graphics by Alan Bloom, N1AL

Chapter 25

Test Equipment and Measurement

Every radio amateur should undertake to master basic electronic and test instruments: the multimeter, oscilloscope, signal generator, RF power meter, SWR, and impedance analyzers. Even though much of this material is not relevant to the amateur who is not interested in radio design and intends to operate commercial equipment, this chapter contains topics that every radio amateur should be familiar with. The reason for this is the special relationship that the Amateur Radio Service has with the Federal Communications Commission. The FCC considers amateur radio to be a valuable service because amateurs, in part, contribute to the state of the art of radio communications. Amateurs have the right to design or modify their equipment without special permission from the FCC. This is the only service regulated by the FCC that has this privilege. As such, the radio equipment used by amateurs does not require FCC certification, unlike every other transmitter used in the United States. With these special rights, however, also include additional responsibilities. Radio amateurs must be assured that their operation on the air does not violate any of the FCC signal purity or exposure requirements. The FCC trusts the amateur radio community to self-regulate and not cause interference to any other services. All amateurs should be well versed in the operation of test equipment, as described in this chapter, and in the control of human exposure to their transmissions, as described in the **Safe Practices** chapter.

This chapter is the reader's guide to test instruments and the physical parameters they measure. The chapter goes beyond basic instrumentation to advanced instruments and measurements that are often encountered by amateurs. The material does not attempt to be complete in covering all available instruments and measurements. The goal of the chapter is to instruct and educate, present useful projects, and encourage the reader to understand more about these important facets of the radio art. A revolution in test equipment has taken place since the advent of the personal computer. While the acquisition of many types of test equipment was once out of the realm of many radio amateurs, less expensive versions of test equipment that operate with a personal computer are now widely available.

The chapter has two major sections: Instruments and Measurements. Reorganized from the original chapter written by Alan Bloom, N1AL, the first section covers the different types of instruments used in amateur stations. Material on using antenna and vector network analyzers (VNA) was updated by Jim Brown, K9YC. (Basic ac, dc, and RF measurement units of measurements are covered in the chapters on **Electronic Fundamentals**, **Radio Fundamentals**, and **Circuits and Components**.) The second section describes the types of measurements performed on RF electronics and in amateur radio stations. Transmitter and receiver performance tests are provided, including the standard tests performed by the ARRL Lab for *QST* Product Review and other evaluations. Transceiver tests have been updated based on material provided by Adam Farson, AB4OJ/VA7OJ. Finally, a selection of test equipment construction projects is presented.

25.1 Measurement Fundamentals

25.1.1 Measurement Standards

The measurement process involves evaluating the characteristic being tested using a *standard*, which is a rule for determining the proper numbers to assign to the measurement. In the past, instruments were calibrated to standards represented by physical objects. For example, the reference standard for length was a platinum bar, exactly 1.0 meters long, stored in an environmentally controlled vault in Paris, France. In the latter part of the 20th century, most measurements were redefined based on fundamental physical constants. In 1983, the meter was defined as the distance traveled by light in free space in 1/299,792,458 second.

While most common everyday measurements in the United States still use the old Imperial system (feet, inches, pounds, gallons, and so on), electronic measurements are based on the international system of units, called *SI* after the French name *Système International d'Unités*, which is the modern, revised version of the metric system. The SI defines seven base units, which include length (meter, m), time (second, s), mass (kilogram, kg), temperature (kelvin, K), amount of substance (mole, mol), current (ampere, A), and light intensity (candela, cd). All other units are derived from those seven. For example, the unit of electric charge, the coulomb, is proportional to s × A and the volt has the dimensions $m^2 \times kg \times s^{-3} \times A^{-1}$.

In the United States, measurement standards are managed by the *National Institute of Standards and Technology* (NIST), a non-regulatory agency of the federal government that, until 1988, was called the National Bureau of Standards (NBS: **www.nist.gov**). (Notes may be found in the References section.) NIST's services include calibration of *transfer standards*. The transfer standards in turn are used to calibrate *working standards*, which are used by companies to calibrate and measure their products. Such products are said to be *NIST traceable* if the rules and procedures specified by NIST have been followed. Most low-cost instruments used by hobbyists are another level down in accuracy, being calibrated by instruments that themselves may or may not be NIST traceable.

25.1.2 Accuracy, Precision, and Uncertainty

Accuracy and precision are different measurement specifications that are often confused. *Accuracy* is the maximum expected error in the measurement. *Precision* is the ability of a device to make consecutive measurements that are close to one another.

Figure 25.1A illustrates the difference by placing hits on a target. Higher accuracy means being able to hit a specific location on any single attempt. Higher precision means being able to hit the same location on multiple attempts.

Resolution is the smallest distinguishable difference in a measured value. In Figure 25.1B, the target hits are all in the same location, but the increased number of rings allows for higher resolution in stating their location.

Expressing these characteristics in terms of measurements, assume an 8-digit frequency counter can measure the frequency of a 100 MHz signal to a resolution of 1 Hz, which is 0.01 ppm (parts per million). However, accuracy is determined by the time-base oscillator used as a reference, which typically would be significantly less accurate than 0.01 ppm in a low-cost instrument. The instrument's precision would be determined by its ability to produce the same result if the same signal was measured multiple times.

Similarly, many digital voltmeters can display measurements to more digits of resolution than their accuracy permits. The extra resolution can be useful when comparing two or more values that differ only slightly because closely spaced measurements tend to have nearly the same error. The difference can then be measured more accurately than the individual values.

An instrument's accuracy can be specified in absolute or relative terms, or sometimes in both. An example of an absolute specification is an RF power meter with an accuracy of "5% of full scale." If the full-scale reading is 100 W, then the accuracy is plus or minus 5 W at all power levels and a reading of 10 W would represent an actual power ranging from 5 W to 15 W.

An example of a relative accuracy specification is an analog voltmeter with an accuracy of 3%. That means a voltage of 1.000 V can be measured to an accuracy of plus or minus 0.030 V. A voltage of 10.000 V can be measured to an accuracy of plus or minus 0.300 V. An example of a combined absolute and relative accuracy specification is a frequency counter with an accuracy of "1 ppm plus one count." When measuring a 100 kHz signal with 1 Hz resolution, one count is 10 ppm of the 100 kHz signal. The total accuracy is 1 + 10 = 11 ppm, or ±1.1 Hz.

25.1.3 Measurement Error and Uncertainty

An important point is that measurement error is not a mistake but rather a natural result of the imperfections inherent in any measurement. The sources of error can be sorted into several general classes. *Systematic error* is repeatable; it is always the same when the measurement is taken in the same way. An example is the inaccuracy of the voltage reference in a digital voltmeter that causes all measurements to be off by the same percentage in the same direction. *Random error* is caused by noise and results in a different measured value each time it is measured. Receiver sensitivity measurements involve measuring the signal-to-noise ratio of the audio output which varies due to the random fluctuations of the noise level.

Dynamic error results when the value being measured changes with time. The peak envelope power (PEP) of a single sideband transmitter varies with the modulation that is present at the particular time the measurement is taken. *Instrument insertion error* (also called *loading error*) is an often-overlooked factor. For example, a voltmeter must draw at least a little current from the circuit under test in order to perform the measurement, which can affect circuit operation. When performing high-frequency measurements, the capacitance of the measurement probe often can be significant.

Measurement uncertainty must be considered when the test system uses multiple instruments. Each instrument has its own accuracy limitations. Different instruments may have different precision, as well. As a result, even if the parameter being measured has exactly the same value, the measurements are likely to vary. Even successive measurements with the same instrument in the same circumstances vary. Measurement uncertainty is a statistical expression of how the measured values fall within a range of accuracy.

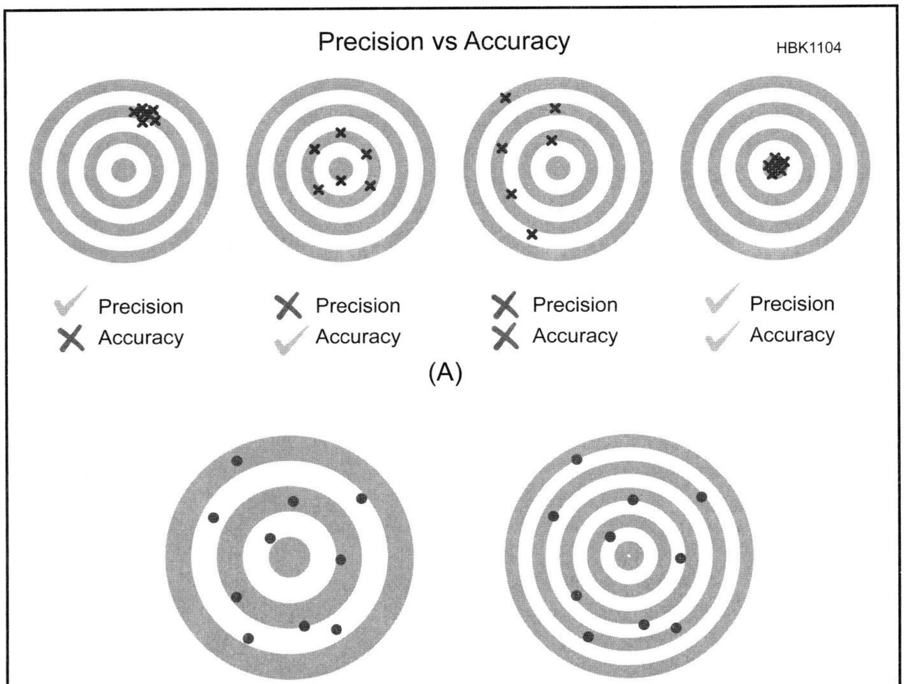

Figure 25.1 — An illustration of precision versus accuracy (A) and of resolution (B).

25.2 Basic Test Meters

Before beginning, be aware that both multi-function test instruments and standalone "panel" meters are both referred to as "meters." To make things even more confusing, multi-function instruments are often called "voltmeters," as well! Avoid confusion by considering the context of the material.

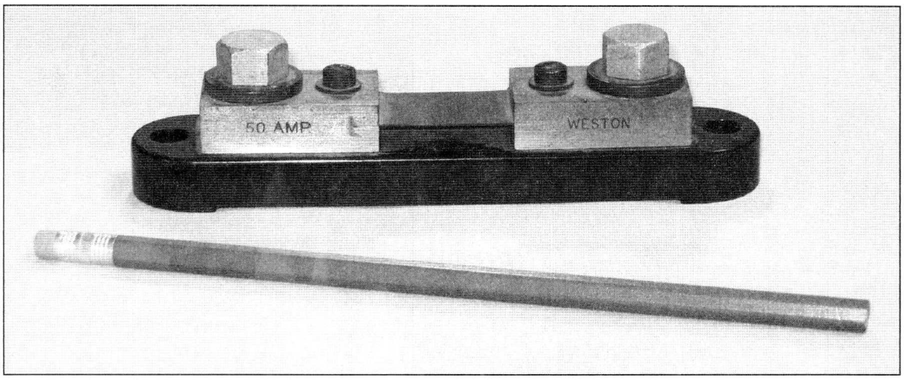

Figure 25.2 — This 50-A, 50-mV current shunt has a resistance of 0.05 / 50 = 0.001 Ω. The two large terminals are for connecting to the circuit under test and the two small terminals are for connecting to a voltmeter.

25.2.1 Voltmeters

Voltage is always measured "with respect to" some other point. Voltage is a measure of the *difference* in potential between two points, not a property of a single point. Most circuits have a *reference point*, and voltages are specified or measured with respect to that point. Normally the reference point is understood to be the chassis or circuit common.

The common connection for a circuit does not have to be at 0 V potential, or "ground." In some circuits, use of the word "ground" is a misnomer and should be more correctly described as the "local reference voltage" or "common point." For protection from electrocution, the equipment chassis must always be connected to Earth ground, as described in the **Safe Practices** chapter. If the local reference voltage is not at the same potential as Earth ground, it cannot be connected to the equipment chassis, and the reference electrode on the voltmeter should not make circuit measurements with respect to the chassis.

25.2.2 Ammeters

An *ammeter* is an instrument for measuring current. An ammeter may be calibrated by measuring the attractive force between two electromagnets carrying the current to be tested, but in practice it is easier to place a known resistor in series, measure the voltage drop across the resistor, and calculate the current using Ohm's law. Resistance and voltage can both be calibrated accurately, so that method gives good accuracy. However, the current in the circuit may be changed by the introduction of more resistance. Therefore, to be accurate, the ammeter's resistor must have much lower resistance than that of the circuit being measured.

Digital ammeters use an analog-to-digital converter to measure the voltage drop across a low-value resistor and scale the result to display as a value of current. They are generally more accurate than analog meters and are more rugged due to the lack of delicate moving parts.

To measure a current, the meter must be inserted in series with the circuit. In a series-connected circuit, all components carry the same current, so it doesn't matter which component is disconnected to allow inserting the ammeter. Select the one that is most convenient, or the one that is at a low-voltage point if you're measuring a high-voltage circuit. For measuring ac power circuits without disconnecting them, clamp-on current probes are used.

Most inexpensive meters do not have a high enough current range to measure the 20 A or so that is drawn from a 12 V power supply by a typical 100 W transceiver. The solution is to use an external *meter shunt*, which is a low value resistor placed in series with the current. See **Figure 25.2**. The multimeter reads the voltage, E, across the shunt, and then the current is calculated from Ohm's law, $I = E/R_S$, where R_S is the resistance of the shunt. Resistors designed for this service may have four leads rather than two to allow a true four-wire measurement. (This technique is discussed later in the chapter.)

25.2.3 Ohmmeters

An *ohmmeter* is a meter designed for measuring resistance. Various circuits can be used, but most are variations of the simplified schematics in **Figure 25.3**. In the circuit at A, the battery is in series with the resistor under test. If the resistance is zero (the test leads are shorted) then the battery is in parallel with the voltmeter, and it reads a maximum value. If the resistance is infinite (test leads not connected) the meter reads zero. If the resistor equals R, the internal reference resistance, the meter reads half-scale.

In the circuit at B, the resistor under test is in parallel with the voltmeter. The meter indication is reversed from the series-connected circuit, that is, the meter reads zero for zero resistance, full-scale for infinite resistance, and mid-scale when the resistor equals R.

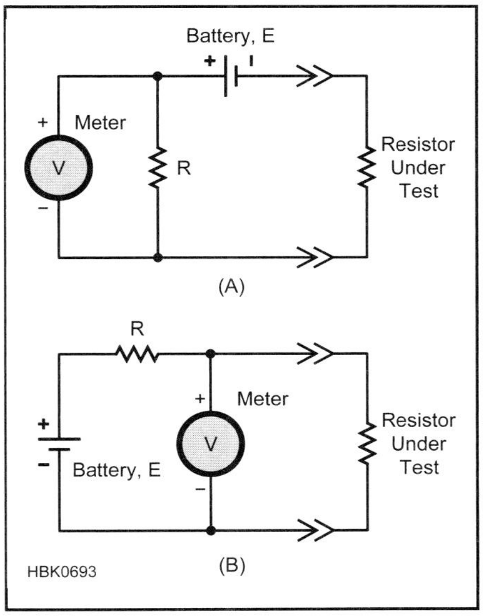

Figure 25.3 — Two ohmmeter circuits. At A, the meter reads full scale with a zero-ohm resistor and reads zero with no resistor connected. The circuit at B is the opposite; the meter reads full scale with no resistor connected (infinite resistance) and reads zero with zero resistance.

In both circuits, an adjustment is normally provided to set the meter to full scale with the test leads shorted or open, as appropriate. In addition, a switch selects different values of the internal resistance R for measuring high or low-valued resistances.

A common mistake made with ohmmeters is that they can only be used to make measurements when circuits are not powered. If an ohmmeter is placed across an in-circuit resistor that also has power applied, the ohmmeter may read the wrong value, and it could be damaged, as well.

25.2.4 Multimeters

A *multimeter* is probably the single most useful test equipment for an electronics experimenter. Besides measuring dc voltage and current, most also measure resistance and low-frequency ac voltage as well. Another common feature is a tone that sounds whenever the test leads are shorted, which is useful as a quick continuity tester. With most modern multimeters, the voltage they use for the resistance measurement is too low to forward-bias the junction of a silicon diode or transistor, which is a desirable characteristic for measuring in-circuit resistance. Many multimeters also have a special "diode" mode with a higher test voltage so that diodes and transistors can be tested by measuring the forward voltage drop across them. It is increasingly common for even low-cost digital multimeters to include functions such as frequency and capacitance measurement.

ANALOG INSTRUMENTS

An analog volt/ohm/current meter (referred to as a "VOM" here) contains no electronic circuitry other than a switch and resistors to set the scale, a battery for the resistance-measuring circuit, and perhaps a diode to convert ac voltage to dc. Despite the name, most can also measure current as well as voltage and resistance.

A disadvantage of the analog instrument is that, when measuring voltage, the current to operate the meter must be drawn from the circuit under test. A figure of merit for a VOM is its *ohms-per-volt* (Ω/V) rating, which is just the reciprocal of the full-scale current of the meter movement. For example, if the VOM uses a meter that reads 50 µA full-scale then it has $1 / 50 \times 10^{-6} = 20{,}000$ Ω/V. On the 1 V scale the meter has a resistance of 20 kΩ, and on the 10 V scale it is 200 kΩ.

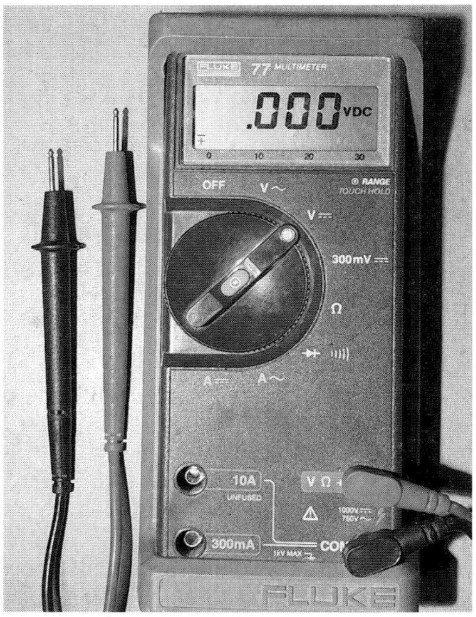

Figure 25.4 — A modern digital multimeter typically has a liquid crystal display readout.

Depending on the circuit being tested, drawing 50 µA may be enough to disrupt the measurement or the operation of the circuit. To solve that problem, some analog meters include a built-in amplifier with high input impedance. In the days of vacuum tubes, such meters were called *vacuum tube voltmeters* (VTVM). The modern equivalent is called an *electronic voltmeter* and generally uses an amplifier with field-effect transistors at the input.

Some older VOMs include a high voltage battery that is used on the highest resistance ranges. When testing solid-state devices, that voltage can be high enough to cause semiconductors to conduct, giving erroneous readings of resistance. In some cases, the applied measurement voltage can be high enough to damage the circuit being measured. If there is any doubt, use an instrument with a diode test function because that meter uses only a low voltage to measure resistance.

DIGITAL MULTIMETER (DMM)

A *digital multimeter* (DMM) has a digital readout, usually an LCD display. A microprocessor controls the measurement process and the display (see **Figure 25.4**). To convert the analog voltage or current being measured to a digital number requires an analog-to-digital

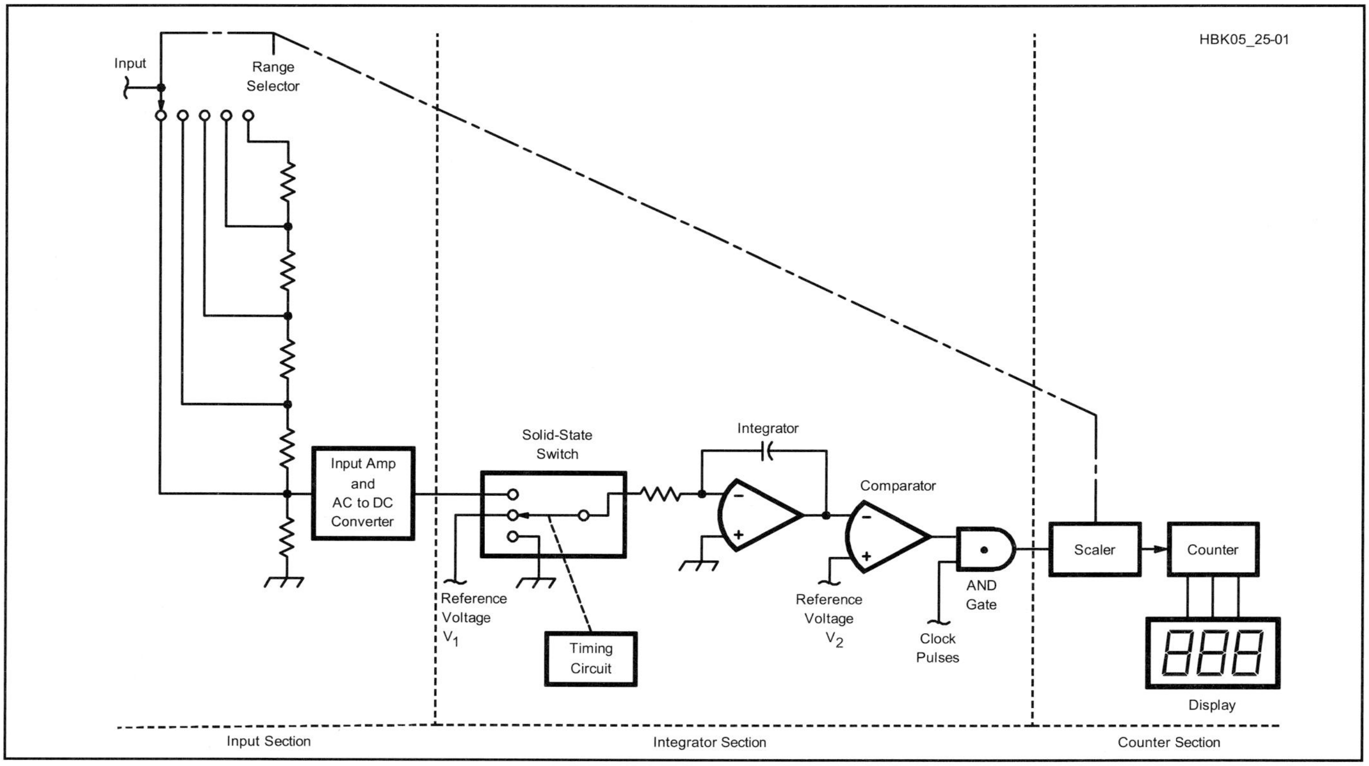

Figure 25.5— A typical digital voltmeter consists of three parts: an input section for scaling, an integrator to convert voltage to a pulse whose width varies with voltage, and a counter to measure the width of the pulse and display the measured voltage.

converter (ADC) controlled by a microprocessor. Most use a dual slope type of ADC, which trades off a relatively slow measuring speed for excellent accuracy and low cost. (Analog-to-digital conversion is discussed in the **DSP and SDR Fundamentals** chapter.)

A DMM is constructed as shown in **Figure 25.5**. The input section is the same as in an analog electronic voltmeter. A range selector switch scales the input signal appropriately for amplification by an input preamplifier, which also includes a rectifier circuit that is used when in ac mode to convert the ac signal into dc, suitable for conversion by the ADC.

An additional feature that becomes possible with DMMs is *autoranging*. The selector switch is used only to choose between voltage, current, resistance, and any other available functions. The scale or range is selected automatically based on the amplitude of the signal or value of resistance being measured, which is a nice convenience. If the signal is fluctuating such that it causes frequent range changes, there is usually a way to turn off autoranging. On DMMs that allow specific ranges to be selected in place of autoranging, selecting a higher range still allows most lower values to be measured, only with less precision.

DMMs that feature serial data interfaces can also act as a *data loggers*, taking and storing measurements for use by a PC or under the control of a PC. (See Reference for Hageman) This is a very useful feature for experimenters or for troubleshooting.

DMMs generally have greater precision than analog meters, often showing results to four or more significant figures. This is a detriment, though, for measuring signals that vary over time or for trying to adjust for a peak or minimum value. The rapidly changing numerical display can be difficult to interpret as the signal changes. Some DMMs provide a bar plot beneath the numerical display to simulate an analog meter, improving readability when the numerical display is changing too quickly.

Multimeter Specifications and Features

Nearly all multimeters measure dc voltage, current, and resistance, and most also measure low-frequency ac voltage. Other common features on digital meters include autoranging and automatic turn-off to save the battery. More capable meters include features such as data hold, peak voltage, true RMS voltage, 4-wire resistance, capacitor tester, inductor tester, diode and transistor tester, logic tester, frequency counter, computer data logging, and a graphical display.

One important feature that is often not specified is over-voltage and over-current protection. On analog meters, there may be internal back-to-back diodes across the meter movement to protect it from over-voltage. Most digital meters include autoranging, which should prevent damage from any voltage below the specified maximum. The current-measuring input on both types is normally protected with a fuse.

The next most critical specification is the available measurement ranges of the voltage, current, resistance, and any other functions provided. Analog meters tend to space the voltage and current ranges by a factor of three. For example, the full-scale voltage readings might be 3 V, 10 V, 30 V, 100 V, 300 V or perhaps 5 V, 15 V, 50 V, 150 V, 500 V. The latter is a particularly good choice since two voltages that are commonly measured, 12 V dc and 120 V ac, are near the top of a range where accuracy is highest. The more range selections provided, the greater the span of values that can be measured. With autoranging instruments, the measurement range may not be obvious without consulting the manual.

The input impedance is important in minimizing the effect of the multimeter on the circuit under test. Impedance should be high for voltage measurements and low for current measurements. For VOMs, the figure of merit for voltage measurements is the ohms-per-volt rating. Multiply the full-scale voltage range by ohms-per-volt to get the input resistance. Both digital and analog electronic voltmeters usually have the same input impedance on all voltage ranges, typically between 1 and 11 MΩ. The input resistance for current measurements is often specified by the burden voltage, which is the voltage drop across the test leads with a full-scale signal. Typical values vary widely, from a few millivolts to more than a volt. The voltage drop can often be reduced by switching to a higher current range, at the expense of measurement resolution.

Measurement accuracy can be an important specification for some applications. An inexpensive analog meter may have voltage and current accuracy of 3% or so. The best analog meters have accuracy specifications in the range of 1%. At that level, accuracy may be limited by *parallax*, which causes the meter reading to appear to change as you change the angle of view. To mitigate that, some high-end analog meters have a *mirror scale*. The reflection of the pointer in the mirror has a parallax error equal and opposite to the unreflected pointer so that the correct reading is halfway between the two. If your meter does not have a mirror scale, determine

Meter Safety

The following paragraphs are not a comprehensive treatment of safety, but provide some examples of important safety practices.

When measuring high voltage, special precautions must be taken. As little as 35 V should be considered dangerous because it can produce lethal current in the human body under some conditions. Grasp the test probes by the insulated handles, being careful to keep fingers away from the metal probe tips.

A maximum voltage rating is printed near the probes (See **Figure 25.A1**). Heed it! That is the maximum voltage the probes, connectors, and body of the meter can withstand. A higher voltage may cause a *flashover* from the wiring to anything touching the case, such as your hand. Electrocution caused by using meters beyond their voltage rating can happen. In the photo, note that the probe connectors are recessed for extra insulation — use the right probes for the full rating.

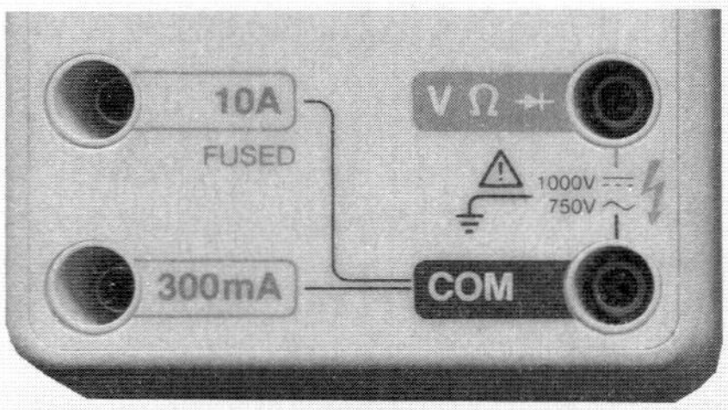

Figure 25.A1 — Voltage ratings are shown next to a voltmeter's probe jacks (1000 Vdc and 750 Vac). Exceeding these voltages can result in a flashover, presenting a severe electrocution hazard.

Special high-voltage probes, such as the Fluke 80K-40 or B&K HV44-A, are available for measuring the multi-kV voltages sometimes found in tube RF amplifiers. When testing high-voltage equipment, don't touch the meter or oscilloscope when the equipment is energized and the probes are attached.

As probes, fixtures, and connectors age, their insulation can become brittle and crack. That compromises the insulation, and you can even come in direct contact with the internal wires. Before beginning to work on a piece of equipment using high voltage, carefully inspect your test probes and other equipment. Make sure the insulation is clean, flexible, and not scuffed or cracked so it can protect you as it should. Replace anything that doesn't look safe.

High current can be dangerous as well. Make sure cables and connectors can carry the current without significant heating. If a probe accidentally shorts a power supply, sparks can fly, damaging the equipment and endangering the operator. Be careful of metal jewelry such as rings and bracelets. If connected across a high-current circuit you could get a nasty burn. Most meters have a fuse to protect the instrument from an over-current condition in current-measuring mode. If the multimeter turns on but always reads zero, consult the manual on how to replace the fuse.

the angle between your eyes and the meter, at which a zero reading appears and then try to make all other meter readings with the same angle between your eyes and the meter.

You can't necessarily tell the accuracy of a digital meter from the number of digits in the display. Usually, the accuracy is limited by the analog circuitry. However, the number of digits defines the resolution, which can be important when comparing two nearly equal readings or when evaluating a peak or minimum value. A typical DMM might have a specified accuracy of 0.1% to 1% for dc measurements and perhaps 1% to 3% for ac voltage. (See the section on ac measurements.) Many inexpensive DMMs do not have published specifications and may not be very accurate.

Many bench-type multimeters and some handheld units can be connected to a computer. That allows the computer to control the instrument to take automated readings and store the results in a computer file. Some older test equipment may have a GPIB (general purpose interface bus or HPIB on Hewlett-Packard equipment) interface, also known as IEEE-488. GPIB-to-USB converters are available to allow connection to a PC. Modern instruments typically have a USB or RS-232 interface.

25.2.5 Panel Meters

Standalone Analog Meters

The most common type of standalone analog meter is the *D'Arsonval galvanometer,* in which the pointer is attached to a rotating electromagnet mounted between the poles of a fixed permanent magnet. The modern form of this meter was invented by Edward Weston and uses two spiral springs to provide the restoring force for the pointer, providing good scale linearity and accuracy. Most commonly available meter movements of this type have a full-scale deflection between about 50 μA and several mA. The D'Arsonval meter movement can be used to measure ac and dc voltage, current, and resistance.

Analog panel meters are quite expensive to buy new, so many experimenters keep an eye open for flea-market bargains. You often can find old "boat anchor" equipment with good salvageable panel meters selling for less than the value of the meters.

The scale markings on surplus meters often represent what the meter was measuring in the equipment rather than the actual current flowing through the meter itself. Sometimes the full-scale current of the meter movement is shown in small text at the bottom of the scale. However, that may not be the same as the current measured at the meter terminals because some meters include an internal shunt.

The only sure way to know the full-scale current and resistance of a surplus meter is to measure it. The resistance can be measured with the ohmmeter function of a multimeter but be careful that the meter does not exceed the meter's current rating, which can be very low for some meters. (You may want to measure the current from the meter before using it to test a panel meter.) Do not use a diode test function on anything except diodes and transistors since the test current of that function may be high enough to damage many sensitive meters.

If the multimeter is an autoranging type you have no way to control test current unless you can turn autoranging off. With a non-autoranging ohmmeter, start the measurement at the highest resistance scale and then reduce the scale one step at a time until a valid reading is obtained, while keeping an eye on the meter under test to be sure it is not over-ranged.

A safer way to measure both the full-scale current and resistance of a panel meter is to place a high-value resistor in series with it and connect the combination to a dc power supply, perhaps a battery. A 1.5 V battery in series with a 100 kΩ resistor (15 μA of current) is a good starting point. Keep trying smaller and smaller resistance values until a good reading is obtained on the meter. Assuming the scale that is marked on the meter's scale face is linear, the full-scale current is

$$I_{FS} = I_{TEST} \frac{D_{FS}}{D_{TEST}}$$

where I_{FS} is the full-scale meter current, D_{FS} is the scale's full-scale marking, D_{TEST} is the needle indication measured with the test current, and I_{TEST} is the test current, which is equal to the voltage across the resistor divided by the resistance. The meter's resistance is the voltage across the meter divided by I_{TEST}.

Digital Panel Meters (DPM)

Digital panel meters (DPMs) are preassembled modules that are almost as easy to use as analog meters. The displays are generally of the liquid crystal type, with or without a back light, and typically have 3 to 4-1/2 digits. A "1/2" digit is one that can display only a 1 or a blank. Displays with a half-digit usually have a full-scale input voltage of either 2 V or 200 mV minus one count, so that the full-scale voltage is 199.9 mV, for example. Most have programmable decimal points after each digit and some have indicators to indicate the units, such as μ, m, V, A, and so forth. DPMs have a high input impedance so there is minimal loading on the circuit under test.

The required power supply voltage varies by model. Some require a floating supply, so if the power supply and the voltage being measured need a common ground connection, be sure the meter is capable of that.

Accuracy is typically 0.1% or better. The total accuracy is usually limited by the external circuitry that drives the meter, such as the amplifier, current shunt, or attenuator that is required to get the signal within the input voltage range of the DPM.

25.2.6 Calibration of Meters

The simplest way to calibrate an inexpensive meter is to compare with a more accurate meter. For example, when constructing a home-built power supply, the analog panel meter may be calibrated with a digital voltmeter (DVM) because most DVMs have better accuracy than an analog meter. Perhaps the most practical voltage reference for the home workshop is an integrated circuit voltage reference. Special circuit techniques in the IC are used to generate a very stable, low-temperature-coefficient reference based on the band-gap voltage of silicon, approximately 1.25 V, on the chip. Inexpensive devices with specified accuracy of 0.1% and better are available from companies such as Analog Devices, Linear Technology, Maxim and National Semiconductor.

Most D'Arsonval meters have an adjustment screw located near the pointer's pivot point that may be accessed from the front of the meter. It should be adjusted so that the meter reads zero with no signal applied.

Probe Adapters for Multimeters

Multimeters come with test probes intended for precise contact with terminals, components, wires, and so forth. They work well if the item to be probed is easily exposed or otherwise available to the probe. Measuring signals on connector pins, however, is often a challenge. Inserting a probe into the miniature sockets on many connectors is often not possible and if the connector has exposed pins, trying to ensure the probe does not slip to or between adjacent pins is nearly impossible.

The solution is to build an adapter as shown in the following three examples.

Figure 25.A2 provides a convenient way to hold probes steady in a spring-loaded, push-button, two-wire speaker connector connected to a pair of Powerpole connectors. Keep the colors of the wires, buttons, and connectors consistent to prevent confusion. An enclosure such as an inexpensive plastic box protects the exposed terminals.

Figure 25.A3 shows how to adapt an automotive-style fuse to make a current-measuring adapter for your multimeter. First, remove the back of the fuseholder and the fuse element and pry the fuseholder open. Solder leads to the exposed terminals then glue or snap the fuseholder back together. Be sure to include the external in-line fuseholder — available in auto parts stores — so that the circuit is protected and you don't blow a multimeter fuse. (Note — multimeter fuses are rated at the full voltage limits of the meter for your safety. Do not replace them with low-voltage fuses.)

Figure 25.A4 is a typical adapter for a multi-pin connector using a terminal strip. Take care to arrange the terminals in order of pin number and label them so you don't have to guess when using the adapter. Make an adapter for the common connectors in your station and you'll never regret it!

These are just three types of adapters — you will no doubt think of many more that will help you with your particular needs. Remember to protect yourself against exposed voltages and short-circuits when constructing and using the adapters. (Thanks to W4QO and KG4VHV and the *QRP Quarterly* for the suggestions.)

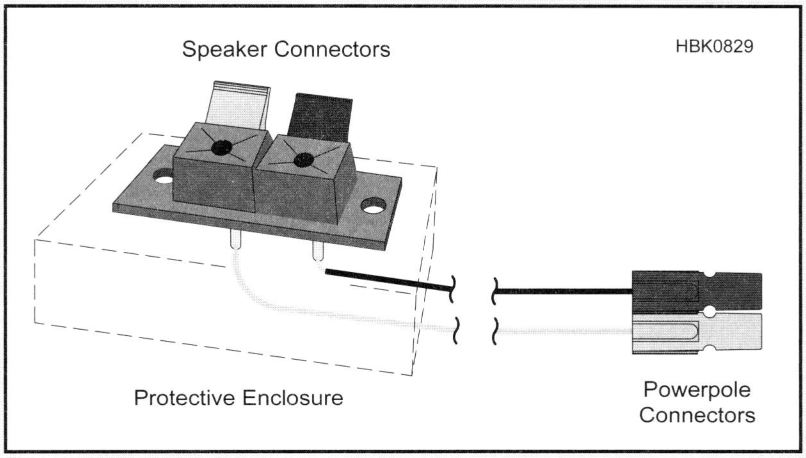

Figure 25.A2 — A convenient way to hold probes steady

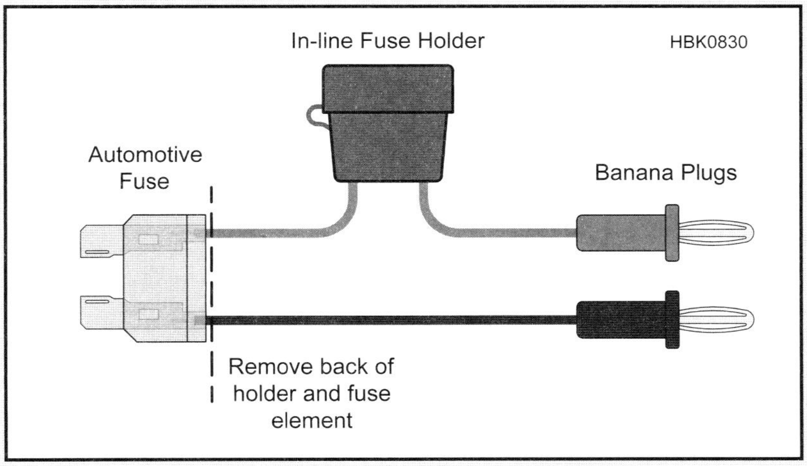

Figure 25.A3 — Adapting an automotive-style fuse to make a current-measuring adapter.

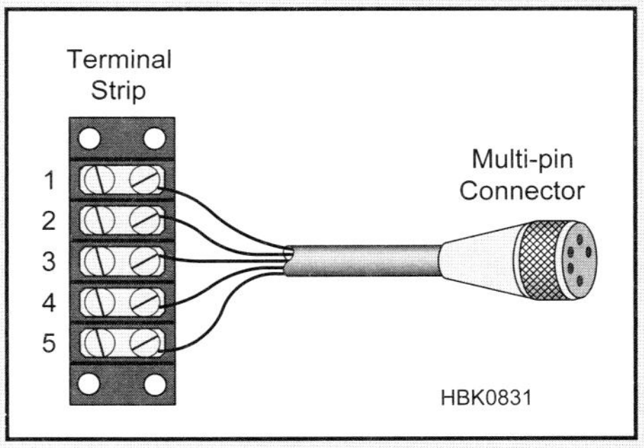

Figure 25.A4 — A typical adapter for a multi-pin connector.

25.3 Frequency Counters

The basic instrument for measuring frequency is the *frequency counter*. A block diagram of a very basic design is shown in **Figure 25.6**. Three digits are shown, but typically there are more. The signal to be measured is routed through three cascaded decade counters. For 1 Hz frequency resolution, the counters count for 1 second. For 10 Hz resolution, they count for 1/10 second, and so on. The count time is determined by a high-stability crystal oscillator, whose frequency is divided down to 1 Hz, 1/10 Hz, or whatever resolution is desired. At the end of each count time, control circuitry displays the final count and resets the counters for the next count period.

With this scheme, the displays are updated once per second when 1 second resolution is chosen, 10 times per second with 10 Hz resolution, and so on. If the frequency is between two adjacent displayed values, the least-significant digit flickers back and forth between the two values on successive counts. Sometimes the designer chooses not to show the least-significant digit for that reason, even though that slows down the display update rate by a factor of 10 for any given display resolution.

MEASURING PERIOD

Some frequency counters include the ability to measure time as well. In the block diagram, the connections to the first divide-by-10 stage input and the control circuit input are swapped. In that way the signal being measured controls the count time, and the reference oscillator provides the signal being counted. If the divided-down reference oscillator has a frequency of 1 kHz, for example, then the period can be measured to a resolution of 1 ms. This same technique can be used to measure low-frequency signals as well. For example, when measuring the frequency of a subaudible tone encoder, you need at least 0.1 Hz measurement resolution. Normally, that would require a 10-second count time, which can be inconvenient. Some counters can measure the period, calculate the reciprocal, and display the resulting frequency. Since the count time is only one cycle of the measured signal, the display updates in real time.

PRESCALERS

Many frequency counters include a *prescaler*, a frequency divider between the input and the main part of the circuitry, to allow operation at higher frequencies. Usually, the prescaler has a 50 Ω input. For low frequencies, there is a separate high-impedance input, typically 1 MΩ, that bypasses the prescaler. A switch selects between the two inputs.

INPUT IMPEDANCE AND SENSITIVITY

It is important to realize that the so-called high-impedance input only has a high impedance at low frequencies. For example, if the stray input capacitance is 30 pF, then the impedance is only 177 Ω at 30 MHz. If you try to measure the frequency of an oscillator by connecting the frequency counter's input directly to the circuitry, it may alter the oscillator tuning enough to invalidate the measurement. If possible, connect the counter to the output of a buffer amplifier or at some other point in the circuitry that won't be adversely affected. If that isn't possible, another trick is to use a pickup coil placed near the oscillator. The coil could be a few turns of insulated wire soldered between the center conductor and shield of a coaxial cable that connects to the frequency counter input. Hold the coil just close enough to get a stable reading. Even this may detune the oscillator slightly.

When connecting a frequency counter to a circuit, observe the maximum voltage and power ratings, both for dc and ac. An oscilloscope probe with a 10:1 attenuation ratio connected to the high-impedance input is a good method to reduce both the signal level as well as the capacitive loading.

Frequency counters tend to have sensitive inputs. Only a small fraction of a volt is typically enough for valid readings. It is quite practical to measure the frequency of a nearby transmitter off the air with a small whip antenna. (Don't place it too close to the

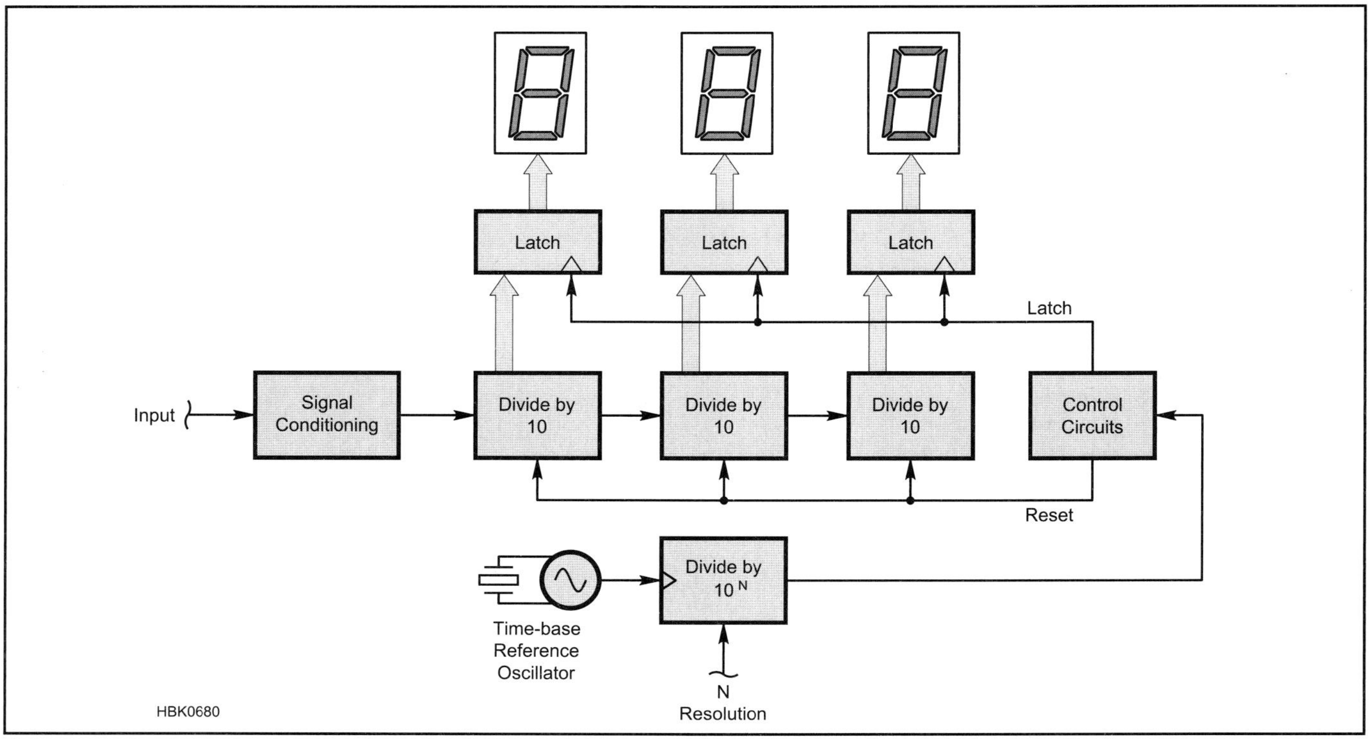

Figure 25.6 — A simplified block diagram of a frequency counter. The display update rate and the resolution are controlled by the divide ratio of the divider at the output of the time-base oscillator.

transmit antenna or it may pick up enough energy to damage the counter.) The transmitter should not be modulated while measuring its frequency. SSB transmitters should be measured in CW mode.

ACCURACY AND TIME BASE

The principal figure of merit for a frequency counter is its frequency accuracy, which is primarily determined by the reference oscillator, also known as the *time base*. The time base accuracy is affected by the temperature, power supply voltage, crystal aging, and quality of calibration. For a temperature-compensated crystal oscillator (TCXO), the total accuracy is typically a few parts per million (ppm). If it is 10 ppm, for example, then the error at 144 MHz could be as high as 1.44 kHz.

Normally temperature is the factor with the greatest effect on the short-term stability. For best accuracy, calibrate the reference oscillator at the same temperature at which measurements are taken.

CHOOSING A COUNTER

Other important specifications are the number of digits in the display, frequency resolution, display update time, frequency range, input sensitivity, input impedance and, for portable units, power supply voltage and current. In choosing a frequency counter you need to decide if you want a desktop model or a portable handheld unit. Additional features to consider include the size and visibility of the display, the ability to measure period, a data hold feature, an external time base input, adjustable trigger level and polarity, input attenuator, switchable low-pass filter, and frequency ratio measurement.

Units with a wide range of prices, feature sets and performance levels are available both on-line and from local electronics distributors. Older used and surplus frequency counters tend to be less of a bargain than other types of test equipment because advances in solid-state electronics have made modern instruments inexpensive, lightweight, and packed with features and performance.

25.4 Signal Generators

25.4.1 Function Generators

A *function generator*, also known as a *waveform generator*, is a type of oscillator that can generate several waveforms. In addition to sine waves, most can also generate square waves, triangle waves, and sawtooth ramp waveforms. A pulse output with variable duty factor is another common feature. Many models can also linearly sweep the frequency. Frequency coverage is typically from below audio frequencies up to a few MHz. Function generators are generally less accurate than the signal generators discussed in the next section.

The heart of most analog function generators is a triangle wave oscillator. To create the sine wave, a diode shaping circuit rounds off the top and bottom of the triangles, resulting in a reasonably accurate sine wave with distortion on the order of a percent or two. Circuitry that determines when the triangle wave is rising or falling is used to generate the square wave as well. The rise/fall duty factor of the triangle wave can be varied, which also varies the duty factor of the square wave. If the triangle wave duty factor is set near 100% or 0%, it becomes either a rising or falling sawtooth wave.

The easiest way to build your own analog function generator is to use a waveform generator IC that includes most of what you need in one package. The most common ICs that have been used to build function generators are MAX038, XR2206 and ICL8038. These parts are not being manufactured any more but can often be found by searching for excess stock on the internet or at hamfests. The AD9833 direct digital synthesis (DDS) IC (see **Figure 25.7**) is also designed to be used as a low frequency function generator. (See the Reference for Richardson. DDS is discussed in the **Oscillators and Synthesizers** chapter.)

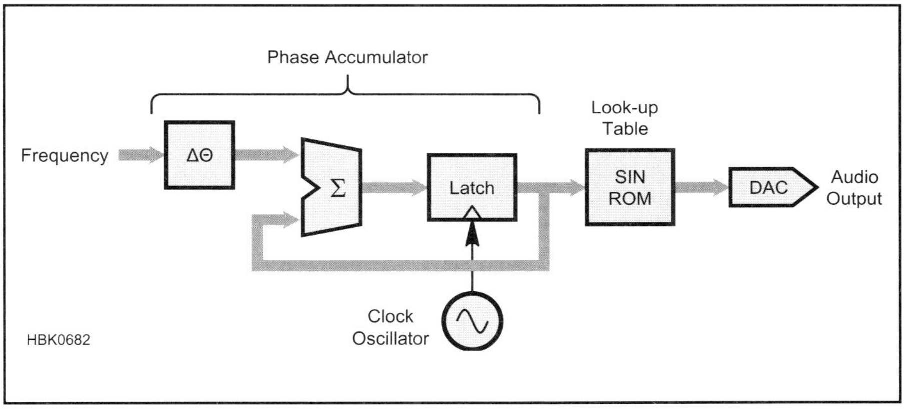

Figure 25.7 — A direct digital synthesizer (DDS) sine wave generator. Not shown is a low-pass filter at the output that attenuates unwanted spurious frequencies above one-half the clock frequency.

An *arbitrary waveform generator* (AWG or ARB) has capabilities that are a superset of a waveform generator. The sinusoidal waveform ROM look-up table in Figure 25.7 is replaced by RAM so that data which creates waveforms other than a sine wave can be used. Usually, a large amount of memory is included so that long non-repetitive waveforms can be generated in addition to periodic signals. ARBs generally include a microprocessor and provide means to generate complicated sequences that may combine repeating and nonrepeating segments.

CHOOSING A FUNCTION GENERATOR

For many applications, the basic analog sine/square wave generator based on a Wien bridge oscillator works fine. The sine wave is adequate for testing the gain, frequency response, and maximum signal level of an audio-frequency circuit. The square wave is useful for checking the transient response of circuits.

The additional wave shapes offered by a function generator are useful for more esoteric applications. Sawtooth and triangle waves can be used to sweep a voltage through a range to test the response of a circuit to various voltages. The pulse output is useful for testing digital circuitry, especially if the instrument has provision for adjusting the on and off voltage levels. Frequency sweep capability is very handy for testing the frequency response of an audio circuit, especially if a sweep ramp or sweep trigger output is provided for synchronizing an oscilloscope. Some instruments provide an input to control the frequency by means of an external voltage.

When selecting a function generator, the first specification to look at is the frequency range. Many units cover low radio frequencies as well as audio. The frequency accuracy may be important for many applications. Digitally synthesized models use quartz crystal clock oscillators that are much more accurate than analog models. Some analog models do have built-in frequency counters, but the accuracy may still be limited by frequency drift.

The amplitude range and accuracy should also be considered. An amplitude of several volts is useful for testing power devices and the ability to accurately set the amplitude to a few millivolts may be needed for driving the microphone input of a transmitter. The output impedance is usually either 600 Ω or 50 Ω.

Function generators are available for a wide range of prices, from inexpensive hand-held units to sophisticated bench instruments costing thousands of dollars. Surplus tube-type models can be good values, starting with the venerable Hewlett-Packard 200A, manufactured until the early 1970s. More-recent models can also be found on the used-equipment market made by HP (and its successor companies Agilent and Keysight Technologies), Tektronix, B&K, Keithley, Wavetek, Leader, Fluke, and others.

An even less-expensive solution is to use a computer sound card as a function generator. A variety of free function generator software can be found on the internet. Since the sound card output is ac-coupled, it is not possible to adjust the dc offset voltage as it is with most function generators. Also, the frequency range, the level range and accuracy, and the output drive capabilities are not as good as you would expect from a special-purpose instrument. However, quite sophisticated waveforms may be generated with the right software for very little cost.

25.4.2 RF Signal Generators

An *RF signal generator* is a test oscillator that generates a sine wave signal that has an accurately calibrated frequency and amplitude over a wide radio-frequency range. Usually, AM and FM modulation capability is provided, and sometimes other modulation types as well. Some instruments also have built-in sweep capability. If not, narrowband sweep can be obtained by feeding a sawtooth waveform into the FM input.

Another feature that is very useful when testing the receiver portion of a transceiver is *reverse power protection* (RPP). It protects the sensitive output circuits in the event that the transceiver accidentally goes into transmit mode while connected to the signal generator. RPP circuits typically have a protection range of 1 to as much as 50 watts. Few RPP circuits protect a signal generator from a 100-watt signal. Be completely aware of all the ways that a transceiver can be put into transmit, and remove all microphones, keying, and PTT lines before connecting it to any test instrument other than a watt meter or SWR meter. If possible, also turn the transmit power to be as low as possible.

Before the days of modern digital electronics, all signal generators used free-running oscillators. There is a wide variation in frequency stability and accuracy between the best and the worst. Nearly any modern synthesized signal generator has good enough accuracy for most amateur purposes. However, that accuracy sometimes comes at the price of phase noise, a type of wideband noise caused by short-term fluctuations in phase that usually drops off gradually from the carrier frequency. Phase noise is discussed in the **Oscillators and Synthesizers** chapter.

Some signal generators, even professional laboratory-grade types, may not have enough output attenuation to be used to accurately measure the minimum discernible signal (MDS) on a sensitive communications receiver. A 10 dB noise figure in a 500 Hz bandwidth implies an MDS of –137 dBm, which is well below the minimum amplitude level on many signal generators. The problem is easily resolved by adding an external 20 dB fixed or step attenuator. Just remember to subtract any attenuation used from all amplitude readings.

Many signal generators include some way to connect them to a computer so they can be controlled for automated testing. That is useful for testing a receiver at many frequencies across a band to be able to plot sensitivity or dynamic range as a function of frequency, for example. Older instruments may have a GPIB interface, while modern ones are more likely to have RS232, USB, or Ethernet.

Probably the most common use for signal generators is receiver testing. That is covered in detail in a later section. They are also useful for general-purpose test signals. For example, when developing a new receiver design, you could test the RF, IF, and audio stages before the VFO is completed by using a signal generator as the local oscillator.

Critical specifications include the frequency range, frequency accuracy and stability, amplitude range, amplitude accuracy, and the spectral purity, including phase noise, harmonics, and non-harmonic spurious emissions. Modulation accuracy, and the capabilities of the internal modulation generator, if present, may be important as well, depending on the application.

Many inexpensive signal generators made for the hobbyist and consumer electronics service industry are not suitable for measuring receiver sensitivity because when the output attenuator is at maximum, there can be more signal leaking out the cabinet and radiating from the power cord than is coming out the coax connector. Such instruments also tend to have poor frequency stability, and the tuning dial tunes so fast that they are difficult to set to a precise frequency. They are useful for simple troubleshooting but are not capable of making accurate measurements.

Better quality signal generators intended for servicing land-mobile and other communications equipment sometimes become available on the surplus market. They don't have the precision specifications of lab instruments, but are generally rugged, reliable, and easy to use.

Laboratory-grade signal generators tend to be too expensive for the home hobbyist if purchased new, but older models such as the HP8640B or its military version (AN/USM323) are widely available and have excellent specifications. Of course, they may be less reliable than new instruments, but in many cases the older technology is repairable without special equipment. Service manuals are sometimes available from the manufacturers on their websites.

25.5 Inductance and Capacitance Testers

25.5.1 LCR Measurement Using a DMM

Some multimeters have built-in capability to measure capacitance. They apply a dc pulse and measure the resulting RC time constant as the capacitor under test charges. This type of measurement has an accuracy of about 1%. Measurements are generally made without a calibration option to compensate for the effect of probes and stray capacitance. Inductance is more difficult for the low-power electronics of a DMM to measure, and most models do not offer inductance measurement. A guide to measuring capacitance with a DMM is available at **fluke.com/en-us/learn/blog/digital-multimeters/how-to-measure-capacitance**.

25.5.2 LCR Meters and Bridges

There is a wide variety of instruments to measure component values. These range from low-cost handheld instruments for hobbyist use to benchtop units that support manufacturing. The accuracy of even simple handheld meters is suitable for amateur radio applications. **Figure 25.8** shows the popular Peak Atlas LCR 45 model that can measure 0–2 MΩ, 0–10,000 µF, and 0-10 H with accuracies of 1 to 1.5%.

Both handheld and benchtop LCR meters measure inductance and capacitance by using a phase-shift method. An ac signal is applied to the component under test through a resistor internal to the meter. The resulting phase shift between voltage and current waveforms is measured as a time interval that is converted to a value of phase. Since the internal resistor value is known, the L or C value can be calculated. There are several values of resistance available so that a wide range of inductance can be measured. The manual explains the measurement process, which includes "calibrating out" the effects of leads and test fixtures.

Benchtop LCR meters generally offer features such as programmable frequencies, measurement accuracy to 0.01%, computer control, and data collect for automated applications. Advanced features such as dc bias voltage, dc bias current, and sweep capability are common. LCR meters in this category are used for ac calibration of inductance, capacitance, and resistance standards, dielectric constant measurements with a variety of dielectric cells, and production testing of components and sensors.

LCR BRIDGES

The traditional way to measure inductance (L), capacitance (C), or resistance (R) is with an LCR bridge consisting of a Wheatstone bridge driven by a sine wave voltage and with an ac voltmeter for the null detector. In **Figure 25.9**, the box labeled Z is the reference component. It must be the same type as the device under test, an inductor, capacitor, or resistor. The bridge is nulled when the ratio of the variable resistor R_V to R is the same as the ratio of the impedance of the device under test to that of Z. The value of R_V is proportional to the resistance or inductive reactance and inversely proportional to the capacitive reactance. To make it proportional to C, simply swap the positions of R_V and R in the circuit. In all cases, when $R_V = R$, the null is achieved when the value of the device under test equals the value of Z.

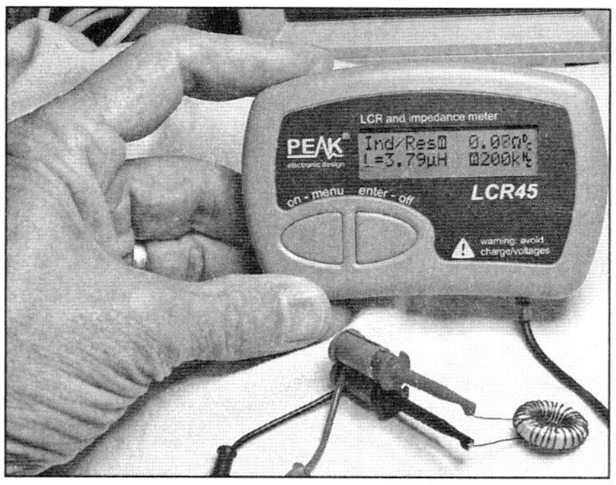

Figure 25.8 — Peak Atlas LCR45 LCR and Impedance Meter.

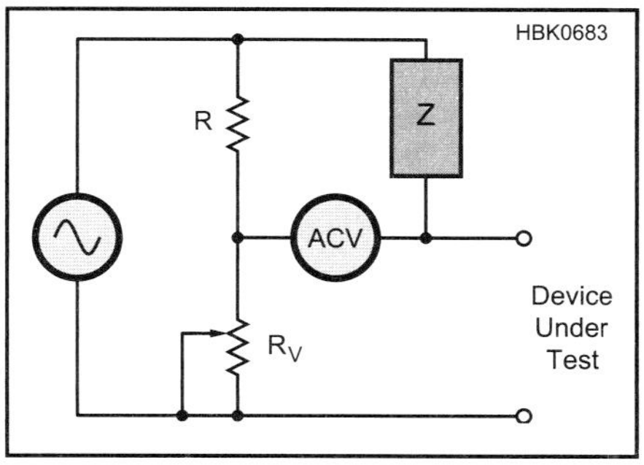

Figure 25.9 — An LCR bridge is a type of Wheatstone bridge that is used to measure inductors (L), capacitors (C) and resistors (R). The box labeled "Z" is the reference component to which the device under test is compared.

MEASURING LCR WITH A DIP METER

A dip meter (described in the RF and Microwave Test Accessories section) may be used for measuring either L or C as long as a component of the opposite type whose value is accurately known is available. The technique is to make a tuned circuit by connecting the inductor and capacitor in parallel and then measure the resonant frequency with a dip meter. The inductance or capacitance can be determined from the dip frequency, f, using the following formulas:

$$L = \frac{1}{(2\pi f)^2 C} \quad \text{or} \quad C = \frac{1}{(2\pi f)^2 L}$$

25.5.3 Component Test Frequency

Electrical components need to be tested over the frequency range at which the component is used. In particular, the measured inductance of coils using high permeability ferrite cores often varies significantly with frequency. An instrument with a wide frequency range and multiple programmable frequencies provides this capability. Common measurement frequencies are dc, 50/60 Hz, 120 Hz, 1 kHz, 100 kHz, and 1 MHz. LCR meters with programmable frequencies provide the most flexibility in matching frequency of measurement to the frequency at which the component is used. In R&D applications, variable measurement frequency allows determination of useful frequency range or resonance. Most LCR meters today use an AC test signal over a frequency range of 10 Hz to 2 MHz.

25.5.4 Equivalent Series Inductance (ESL) and Resistance (ESR)

The use of capacitors in switchmode supplies makes it important to measure their equivalent series resistance (ESR) and equivalent series inductance (ESL). ESR and ESL cause loss and affect the switching circuit's ability to regulate properly. An ESR meter measures a capacitor's ESR by using short pulses or ac signals. Some ESR meters can be used with the capacitor in-circuit, although they should not be used with the capacitor charged or energized. ESL is typically measured with an inductance meter as described above.

25.6 Oscilloscopes

An oscilloscope ("scope" for short) is an instrument that displays voltage versus time on a screen, similar to the waveforms seen in electronics textbooks. Scopes are broken down into two major classifications: analog and digital. This does not refer to the signals they measure, but rather to the methods used inside the instrument to process signals for display.

While scopes being sold today are digital, much of the terminology used to describe their operation was defined for analog scopes. Describing settings on analog scopes also explains the functions of controls with the same name on digital scopes, even though they may work differently.

It is also important to note that inexpensive analog oscilloscopes may not have better than 5% accuracy. Lab-quality analog scopes, while they might be fairly old, are still available and can be calibrated to be close to their original specifications. Such scopes are available economically at most hamfest flea markets. Regardless of price and age, analog scopes are still useful, with many models suitable for HF and even VHF use.

The most important feature of an oscilloscope is that it presents an image of what is going on in a circuit, which is very useful for troubleshooting waveforms and other time-varying phenomena. It can show modulation levels, relative gain between stages, clipping distortion, intermittent oscillations, and other useful information.

There are two basic functions of any oscilloscope, whether analog or digital. The first is signal capture, which is when the voltage of the signal to be measured is brought into the oscilloscope and processed. The second function is display, which is when the processed signal is shown on a screen in a desired fashion that indicates the variations of the voltage over time. One major difference for the user of the oscilloscope is the retention of the data that is being displayed. In an analog oscilloscope, the phosphors on the CRT screen have a persistence that keeps the screen glowing as the trace moves across it, typically a few milliseconds. Since the screen persistence is fairly short, in order to see a steady waveform, it must repeat over and over to keep the phosphors on the CRT screen glowing. For signals that do not repeat, very short-lived waveforms may not be visible to the eye.

In a digital oscilloscope, a memory stores the values of the voltages that are captured over time and then that memory is displayed on a screen. The waveform can be continually displayed, even after the incoming signal is no longer being captured.

The sweep circuit in the analog oscilloscope controls the horizontal position of the display. A digital oscilloscope has two sweep circuits, one that selects the memory locations that signal samples are placed in as they are captured, and another that selects the memory locations that signal samples are displayed from. In the simplest digital oscilloscope operation, these two sweep circuits are synchronized so that they sweep the memory at the same speed. However, the digital oscilloscope has powerful functionality that allows the captured signal to be frozen on the screen or magnified to examine portions of it. **Figure 25.10** shows a simplified diagram of the stages of an analog oscilloscope and **Figure 25.11** shows a similar diagram for a digital oscilloscope.

SIGNAL INPUT

All oscilloscopes need to capture the signal to be viewed. The input stages are common for both analog and digital oscilloscopes. Most oscilloscopes contain circuitry for two input

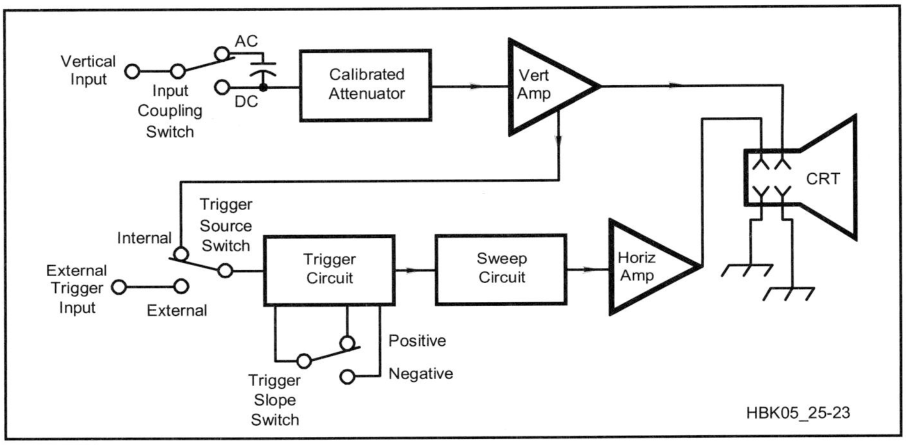

Figure 25.10 — Typical block diagram of a simple analog oscilloscope. The vertical input circuits can be duplicated to support multiple channels.

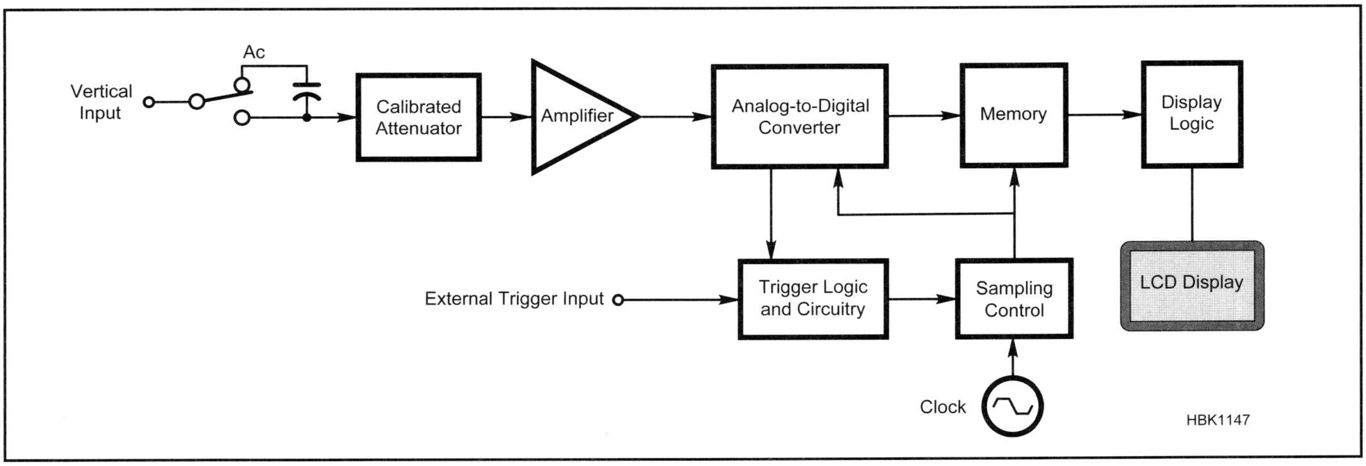

Figure 25.11 — Typical block diagram of a simple digital oscilloscope. The vertical input circuits can be duplicated to support multiple channels.

channels. There is often a need to show how a signal changes through a circuit and two traces can be positioned one above the other to see how they differ.

Each input signal is connected through an input COUPLING switch. The switch allows selection of either the ac component of an ac/dc signal or the complete signal. If you wanted to measure, for example, the RF swing at the collector of an output stage including the dc level, you would use the dc-coupling mode. In the ac-coupled mode, dc is blocked from reaching the vertical amplifier chain so that you can measure a small ac signal superimposed on a much larger dc level. For example, you might want to measure a 25 mV 120-Hz ripple on a 13 V dc power supply. Note that you should not use ac coupling at frequencies below the low-frequency cutoff of the instrument in that mode, typically around 30 Hz, because the value of the blocking capacitor represents a high series impedance to very low-frequency signals.

After the coupling switch, the signal is connected to a calibrated attenuator. This is used to reduce the signal to a level within the range of the scope's vertical amplifier. The vertical amplifier boosts the signal, resulting in a level that can drive the CRT in an analog oscilloscope, or that matches the range of the ADC in a digital oscilloscope. In an analog oscilloscope, a bias voltage is also added to position the waveform vertically on the screen. The result is that the vertical displacement on the screen represents input voltage.

In a digital oscilloscope, the ADC is designed for a number of bits, which is a measure of the number of digital levels shown along the vertical axis. More bits give you better resolution and accuracy of measurement. An 8-bit vertical resolution means each vertical screen has 256 discrete values. Similarly, 10-bit resolution yields 1024 discrete values.

TRIGGERING CIRCUITRY

In an analog oscilloscope, a small sample of the signal from the vertical amplifier is sent to the trigger circuitry. Conversely, in a digital oscilloscope, a microprocessor compares the voltages in the memory to determine the appropriate sample on which to trigger.

The trigger circuit controls the horizontal sweep, and in a digital oscilloscope starts data capture into the memory. It looks at the trigger source (internal or external) to find out if it is positive- or negative-going and to see if the signal has passed a particular level. **Figure 25.12A** shows a typical signal and the dotted line on the figure represents the trigger level. It is important to note that once a trigger circuit is "fired" it cannot fire again until the sweep has moved all the way across the screen from left to right. There normally is a TRIGGER LEVEL control to move the trigger level up and down until a stable display is seen. Triggering circuitry generally is activated when a signal crosses the selected triggering level with a specified direction (rising or falling). Some scopes have an AUTOMATIC position that chooses a level to lock the display in place without manual adjustment.

Figure 25.12B shows what happens when the level has not been properly selected. Because there are two points during a single cycle of the waveform that meet the triggering requirements, the trigger circuit tends to jump from one trigger point to another. This makes the waveform jitter from left to right. Adjustment of the TRIGGER LEVEL control fixes that problem.

Single sweep is useful for looking at a non-repetitive signal. The single sweep can be armed, that is, reset and ready to start the next sweep, by pushing a button or sometimes with an external signal. Trigger delay allows horizontal centering of the display at a different place from the trigger point. Trigger hold-off inhibits the trigger for a selectable time after the sweep to prevent unwanted multiple triggers. However, be aware that the CRT screen in an analog oscilloscope does not glow for very long (persistence) and the single sweep waveform is only visible for a second or two. In contrast, one of the advantages of a digital oscilloscope is that the single sweep is captured into the memory and can be displayed on the screen indefinitely.

It is also possible to trigger the sweep system from an external source (such as the system clock in a digital system). This is done via an external input jack with the trigger select switch in the EXTERNAL position.

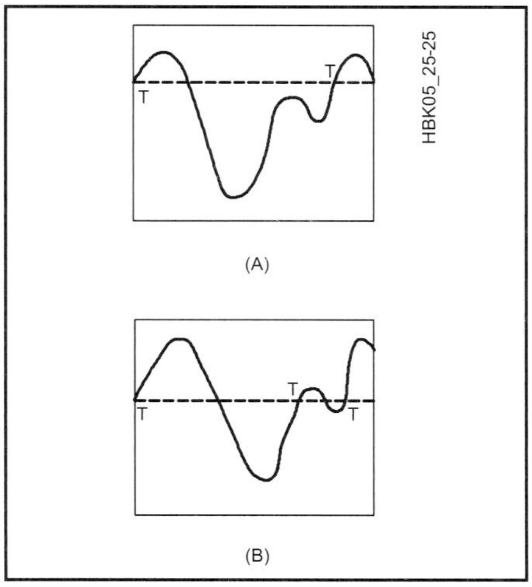

Figure 25.12 — In order to produce a stable display the selection of the triggering parameters is very important. In these examples, the trigger level is indicated with a dashed line, and the trigger direction is rising. Selecting the triggering parameters in (A) produces a stable display, but the trigger shown at (B) produces a display that "jitters" from side to side.

Multiple Trace Functions

The best dual-trace analog oscilloscopes use a CRT with two electron beams, but these are rare. It is possible to trick the eye into seeing two traces simultaneously using a single-beam CRT. The "channel switching circuit" determines whether we display channel A, channel B, or both simultaneously.

The appearance of dual traces is created by the analog scope using one of two methods, referred to as chopped mode and alternate mode. In the chopped mode a small portion of the channel A waveform is written to the CRT, then a corresponding portion of the channel B waveform is written to the CRT. These alternate until both waveforms are completely written across the CRT. Switching from one channel to the other is so fast that each trace looks continuous. The chopped mode is essential for single-shot signals (signals that do not repeat periodically). It is most useful at slow sweep speeds. However, as the sweep speed approaches the frequency of the chopping rate, the switching from channel to channel becomes visible, making each trace into a dotted line.

In the alternate mode, the complete channel A waveform is written across the CRT followed immediately by the complete channel B waveform. This happens so quickly that it appears that the waveforms are displayed at the same time. This mode of operation is not very useful at very slow sweep speeds since the two traces no longer appear simultaneous. It also does not work for single-shot events.

Most dual-trace oscilloscopes also have a feature called "X-Y" mode. This feature allows one channel to drive the horizontal amplifier of the scope (called the X channel) while the other channel (called Y in this mode of operation) drives the vertical amplifier. Some single-trace oscilloscopes support this mode as well. X-Y operation allows the scope to display Lissajous patterns for frequency and phase comparison, and to use specialized test adapters, such as curve tracers or spectrum analyzer front ends. Because of frequency limitations of most scope horizontal amplifiers, the X channel is usually less than a 10 MHz bandwidth.

A digital oscilloscope has similar chopped and alternate modes for capturing multiple channels. Instead of determining the appearance of the channels on the display, however, these modes choose which input channel is digitized and captured into the memory. Once in memory, they can be displayed simultaneously. The considerations for which mode to use are similar to those for the analog scope. The vertical position control for each channel is read by the microprocessor, which it uses to place the traces in relative locations on the screen.

WAVEFORM DISPLAY

Figure 25.10 shows how an analog scope places its waveforms on a cathode-ray tube (CRT) display. An electron beam inside the CRT strikes the phosphorescent screen causing a glowing spot. The exact location of the spot is a result of the voltage applied to the vertical and horizontal deflection plates. To trace how a signal travels through the oscilloscope circuitry, start by assuming that the trigger select switch is in the INTERNAL position.

Early digital oscilloscopes displayed their captured voltages from memory onto a CRT display, though the positions selected by the deflection's plates were controlled by a microprocessor rather than the incoming voltages themselves. Later, digital oscilloscopes saved money and size by replacing the CRT with a flat LED or LCD display.

The trigger circuit feeds a start pulse to the sweep generator when the input signal reaches a certain level (level triggering) or exhibits a positive- or negative-going edge (edge triggering). In an analog oscilloscope, the sweep generator gives a precisely timed voltage sawtooth. The rising edge of the ramp signal feeds the horizontal amplifier that, in turn, drives the CRT. This causes the scope trace to sweep from left to right, with the zero-voltage point representing the left side of the screen and the maximum voltage representing the right side of the screen. The result is that the horizontal position of the beam on the CRT represents time. At the end of the ramp, the sharp edge of the sawtooth quickly moves the beam back to the left side of the screen.

In a digital oscilloscope, the sweep starts a clock that causes the incoming waveform to be sampled at a rate based on the selected sweep speed, after which the samples are digitized and put into memory. A second sweep timer reads the digitized signal samples from the memory and positions them across the display screen.

25.6.1 Digital Oscilloscope Considerations

Signal processing in a digital oscilloscope is performed with microprocessor-based digital circuitry. (See **Figure 25.13**) The microprocessor determines which pixels to light up to draw the traces on the screen. This results in dramatically improved accuracy for both amplitude and time measurements as well as enabling sophisticated features that would be difficult or impossible in an analog scope. For example, a trace can be displayed with infinite persistence and stored as a data file if desired.

You can also use your personal computer as an oscilloscope with an external signal digitizer module (a high-speed, multi-channel, analog-to-digital converter with enough high-speed memory to store the desired data in real time) that transfers the data to the PC via a high-speed USB or Firewire interface. This reduces the cost by making use of the cabinet, power supply, processor, and display of the PC.

Most standalone digital scopes can transfer configuration and trace data to a memory card or thumb drive. Many models can download their data to a PC for storage and analysis. Many high-end scopes now incorporate analysis functions, such as the Fast Fourier Transform (FFT). This allows spectrum analysis or other advanced mathematical techniques to be applied to the displayed waveform.

It is important to understand some of the limitations resulting from sampling the signal rather than taking a continuous, analog measurement. When you try to reconstruct a signal from individual discrete samples, you must take samples at least twice as fast as the highest frequency signal being displayed. If you digitize a 100 MHz sine wave, your minimum sampling rate should be 200 million samples a second (megasamples/second). It is actually more accurate to take samples even more often, usually at a rate of at least five times higher than the highest frequency component of the input signal. (See the **DSP and SDR Fundamentals** chapter for more information on sampled signals.)

If the sample rate is not high enough, very fast signal changes between sampling points do not appear on the display. For example, **Figure 25.14** shows one signal measured using both analog and digital scopes. The large spikes seen in the analog-scope display are not visible on the digital scope. The sampling frequency of the digital scope is not fast enough to store the higher frequency components of the waveform. If you take samples at a rate less than twice the input frequency, called the Nyquist frequency, the reconstructed signal's apparent frequency is wrong; this is referred to as aliasing. In Figure 25.14 you can see that there is about one sample taken per cycle of the signal's ringing frequency. This does not meet the sampling rate rule and the result is that the scope displays an incomplete waveform or one with a different apparent frequency.

A simple check for aliasing is to use the highest practical sweep speed (shortest time

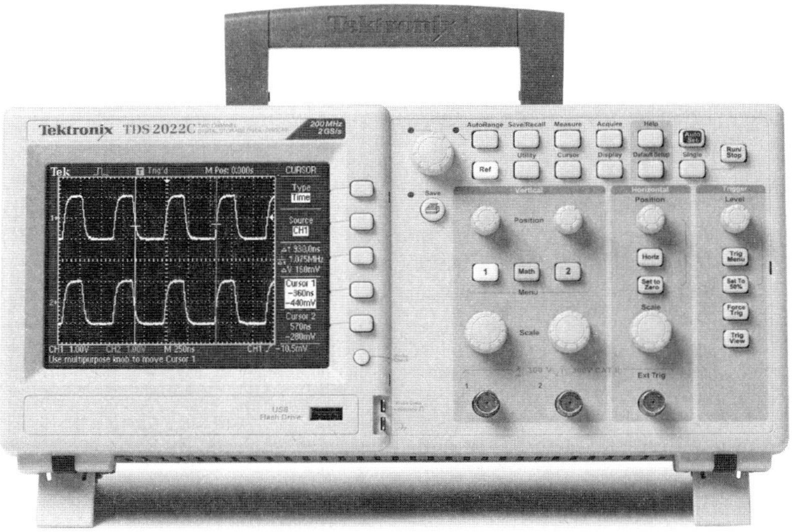

Figure 25.13 — The Tektronix TDS 2000C series is a typical example of digital oscilloscopes using an LCD display.

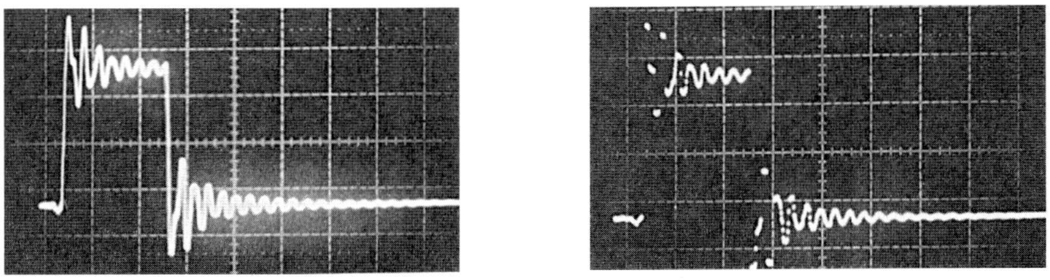

(A) (B)

Figure 25.14 — Comparison of an analog scope waveform (A) and that produced by a digital oscilloscope with a low sampling rate (B). Notice that the digital samples in B are not continuous, which may leave the actual shape of the waveform in doubt for the fastest signal changes the scope displays.

per division) and then to change to other sweep speeds to verify that the apparent frequency doesn't change. Some modern oscilloscopes use a special technique to increase the effective sample rate for repetitive signals. The phase of the sample clock is adjusted slightly on each successive sweep, so that the new samples occur in between the previous ones. After several sweeps the missing data in the spaces between the original samples are filled in, producing a continuous trace. This only works with periodic signals that trigger at exactly the same point on each sweep.

25.6.2 Using an Oscilloscope

An oscilloscope can measure a signal's shape, amplitude, frequency, and whether it is dc, ac, or a mixture of both. For example, in **Figure 25.15** it is clear from the shape that the signal is a sine wave. Assuming that the center horizontal line or axis represents zero volts, the signal has no dc component (assuming the vertical input has been set to DC) since there is as much above the axis as below it. If the vertical gain has been set to 1 V per division, then the peak value is 2 V and the peak-to-peak value is 4 V.

The horizontal travel of the trace is calibrated in units of time. If the sweep speed is known and we count the number of divisions (vertical bars) between peaks of the waveform (or any similar well-defined points that repeat once per cycle) we can find the period of one cycle. The frequency is the reciprocal of the period. In Figure 25.15, for example, the distance between the peaks is 8 divisions. If the sweep speed is 10 µs/division then the period is 80 µs. That means that the frequency of the waveform is 1/80 µs, or 12,500 Hz. The accuracy of the measured frequency depends on the accuracy of the scope's ramp generator, typically a few percent for an analog instrument. This accuracy cannot compete with even the least-expensive frequency counter, but the scope can still be used to determine whether a circuit is functioning properly or to ensure that the counter is not displaying the frequency of a harmonic or other unintended signal. Most digital oscilloscopes have software tools built in to make measurements of voltage levels and frequencies, displaying the results numerically.

Care should be taken when trying to measure AC line voltages. In some oscilloscopes, the ground reference for each probe is connected to the chassis ground, which can be connected to the neutral of the AC line (you can test this with an ohmmeter to see if this is true for your scope). If the same AC line that is being measured is used to power the oscilloscope, a dangerous short circuit between the hot lead and the neutral of the AC circuit may occur. This can cause a fire hazard, as well as potentially destroying circuitry in the oscilloscope. If AC line voltages must be measured with an oscilloscope, it is safest to use a battery powered device to make the measurements, or to power the scope from the AC line through an isolation transformer. Of course, you should avoid touching high voltage circuitry when you measure it. It is safest to remove power from a high voltage device that you are measuring when you connect probes to the circuit. Once your hands are away from the high voltage area, you can turn on the power and see the high voltage signals on the oscilloscope screen (See the **Safe Practices** chapter for more information about AC line safety).

LIMITATIONS OF OSCILLOSCOPES

Oscilloscopes have fundamental limits, primarily in frequency of operation and range of input voltages. For most purposes the voltage range can be expanded by using appropriate probes. The frequency response (also called the bandwidth) of a scope is usually the most important limiting factor. For example, a 100 MHz 1 V sine wave fed into an oscilloscope with a 100 MHz 3 dB bandwidth reads approximately 0.7 V on the display. The same instrument at frequencies below 30 MHz should be accurate to about 5%.

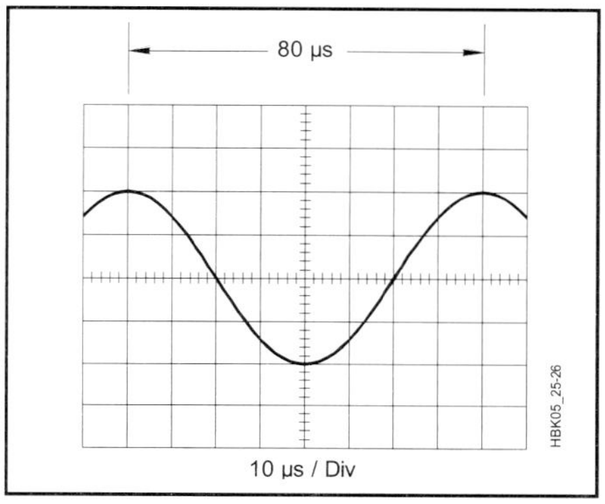

Figure 25.15 — An oscilloscope can measure frequency as well as amplitude. Here the waveform shown has a period of 80 microseconds (8 divisions × 10 µs per division) and therefore a frequency of 1/80 µs or 12.5 kHz.

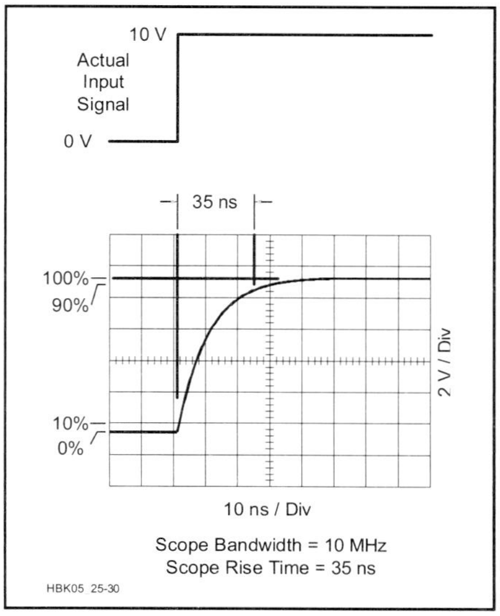

Figure 25.16 — The bandwidth of the oscilloscope vertical channel limits the rise time of the signals displayed on the scope.

A parameter called rise time is related to a scope's bandwidth. This term describes a scope's ability to accurately display voltages that rise very quickly. For example, a very sharp and square waveform may appear to take some time in order to reach a specified fraction of the input voltage level. The rise time is usually defined as the time required for the display to show a change from the 10% to 90% points of the input waveform, as shown in **Figure 25.16**.

For an analog scope, assuming the frequency response is primarily limited by a 1- or 2-pole filter roll off in the amplifier circuitry, the mathematical definition of rise time is given by:

$$t_r = \frac{0.35}{BW}$$

where t_r = rise time in µs and BW = bandwidth in MHz of the amplifier.

For digital scopes, substitute 0.45 for 0.35 in the equation because digital scopes have a much steeper roll-off above the specified cut-off frequency. The steeper roll-off is used to reduce aliasing by eliminating any signal above the Nyquist frequency as discussed in the previous sections.

Be wary of using a scope to judge harmonic content or distortion of a sine wave. A waveform that "looks good" to the eye may have significant distortion or high-frequency components unsuitable for on-the-air signals. A spectrum analyzer (described below) should be used for determining the spectral content of signals.

25.6.3 Oscilloscope Probes

Oscilloscopes are usually connected to a circuit under test with a short length of shielded cable and a probe. At low frequencies, a piece of small-diameter coax cable and some sort of insulated test probe might do. However, at higher frequencies the capacitive reactance of the cable would be much less than the one-megohm input impedance of the oscilloscope. In addition, each scope has a certain built-in capacitance at its input terminals (usually between 5 and 35 pF). The total capacitance causes problems when probing an RF circuit with relatively high impedance.

Most new oscilloscopes come with specially designed scope probes, one for each vertical channel. They can also be purchased separately. The most common type is a *×10 probe* (called a "times ten" probe), which forms a 10:1 voltage divider using the built-in resistance of the probe and the input resistance of the scope. When using a ×10 probe, all voltage readings must be multiplied by 10. For example, if the scope is on the 1 V/division range and a ×10 probe is in use, the signals would be displayed on the scope face at 10 V/division. Some scopes can sense whether a ×10 probe is in use and automatically adjust the scale of the scope's display. **Figure 25.17** shows a photo of a scope probe that can be switched between ×1 and ×10 operation, as well as having an adjustable compensating capacitor, as discussed below. (A good overview of oscilloscope probes, how they are designed and intended to be used, along with links to descriptions of different types of probes is published by Tektronix at **tek.com/en/blog/what-is-an-oscilloscope-probe**.)

Unfortunately, a resistor alone in series with the scope input seriously degrades the scope's rise-time performance and bandwidth because of the low-pass filter formed by the series resistance, along with the parallel capacitance of the cable and scope input. This may be corrected by using a compensating capacitor in parallel with the series resistor. If the capacitor value is chosen so that the R-C time constant is the same as the R-C network formed by the input resistance and capacitance of the scope, as shown in **Figure 25.18**, then the probe and scope should have a flat response curve throughout the whole bandwidth of the scope. A ×10 probe reduces the voltage by a factor of 10, increases the input resistance, and reduces the capacitance, all of which reduces loading on the circuit under test.

To account for manufacturing tolerances in the scope and probe, the compensating capacitor is made variable. Most scopes include a "calibrator" output that produces a fast-rise-time square wave (see **Figure 25.19**) for the purpose of adjusting the compensating capacitor in a probe. **Figure 25.20** shows possible responses when the probe is connected to the oscilloscope's calibrator jack. An improperly adjusted compensating capacitor can greatly affect the frequency response of the scope and create artifacts in signal displays that are not actually present in the original signal.

If a probe cable is too short, do not attempt to extend the length of the cable by adding a piece of common coaxial cable. The compensating capacitor in the probe is chosen to compensate for the provided length of cable. It usually does not have enough range to compensate for extra lengths.

The shortest ground lead possible should be used from the probe to the circuit ground. Long ground leads act as inductors at high frequencies where they create ringing and other undesirable artifacts in the displayed signal.

For the best high frequency performance, the scope probe can be eliminated entirely, and the oscilloscope input converted to a 50 Ω impedance. Some scopes have a switch to choose between a high impedance and a 50 Ω input. For others, you can purchase a 50 Ω through-line termination, which is just a connector with a male BNC on one end, a female BNC on the other, and an internal 50 Ω termination resistor in parallel. Plug the male connector to the scope's high impedance vertical input and connect the 50 Ω cable from the device under test to the female connector.

Some situations may require the use of a scope to measure a current waveform, rather than a voltage. Specialized current probes are available that make this task possible, with some capable of measuring both dc and ac.

25.6.4 Choosing an Oscilloscope

FEATURES

When choosing an oscilloscope, the first decision is analog versus digital. Used and surplus instruments are often analog but digital scopes are now widely available.

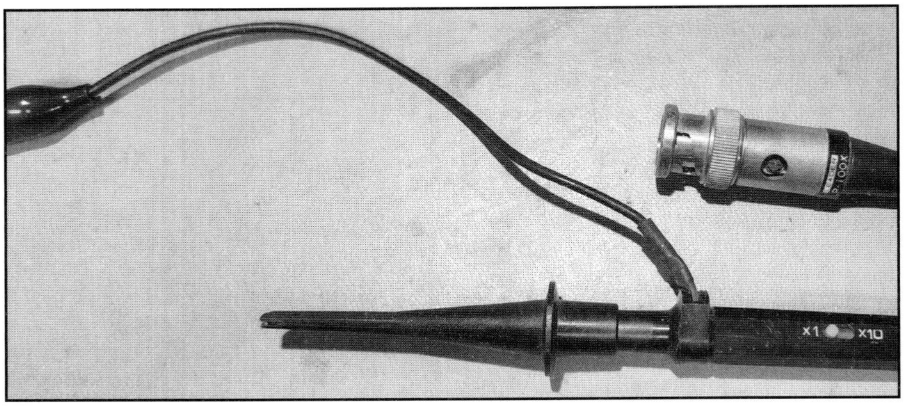

Figure 25.17 — Oscilloscope probe showing the switch to select between ×1 and ×10 modes. Also a compensating capacitor adjustment can be seen just below the BNC connector. This capacitor is adjusted to make a square wave as square as possible (see Figure 25.20).

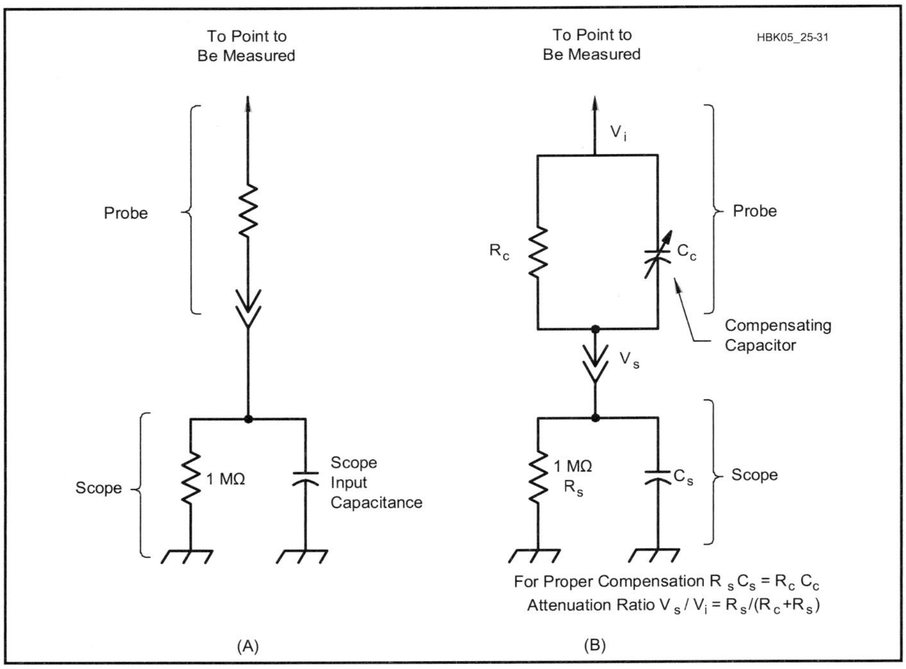

Figure 25.18 — Uncompensated probes such as the one at (A) are sufficient for low-frequency and slow-rise-time measurements. However, for accurate display of fast rise times with high-frequency components the compensated probe at (B) must be used. The variable capacitor is adjusted for proper compensation (see text for details).

The next big decision is how many input channels you need. There are many situations in which having a second channel is extremely helpful and even more than two channels is often useful, especially when troubleshooting digital circuits. With two or more channels you typically also get X-Y mode.

Trace storage is essential for viewing very slow signals or single-shot transients. This is very easy to implement in a digital scope and nearly all of them have the feature. Ideally, one or more stored traces can be displayed simultaneously with the current trace for easy comparison.

Digital scopes may draw a horizontal line on the screen so you can see exactly where the trigger level is with respect to the signal, which is very handy. Automatic adjustment of trigger level is a common feature as well. Some scopes offer a noise-rejection feature that adds hysteresis to reduce false triggering. (See the Comparators section of the **Circuits and Components** chapter for more information about hysteresis.)

Some high-speed scopes include a switchable input low-pass filter to reduce wideband noise when measuring lower frequency signals. A feature sometimes found on multiple channel scopes is the ability to add or subtract two or more channels.

Digital oscilloscopes often include various data analysis features. Averaging reduces noise and peak mode makes transient peaks visible. A common feature is the ability to calculate and display the peak, peak-to-peak, and RMS values of a signal. For repetitive signals, the period and frequency can be calculated as well. Some scopes include amplitude and time markers that allow accurate measurements of specific points on a waveform. Sometimes the instrument state can be stored to one of several internal memories, which can be very handy on complicated instruments that take a long time to set up.

Computer connectivity allows storing traces for later reference, and you can produce nice graphical screen shots for a magazine article or notebook entry about your latest creation. The proper software allows automated testing as well. Older instruments often have a GPIB (general purpose interface bus) interface, also known as IEEE-488 or HPIB on Hewlett-Packard equipment. GPIB-to-USB adapters are available from several companies. More modern scopes usually come with an RS-232 or USB interface, or a provision to store screen shots and data to a USB thumb drive.

SPECIFICATIONS

Bandwidth is the main "money spec" for an oscilloscope. The higher the frequency range, the more expensive it tends to be. Rise time is also sometimes specified. For digital scopes, the sample rate is just as important. Theoretically it must be at least twice the bandwidth, but practically it should be much higher than that to avoid aliasing. For repetitive waveforms, some scopes can do tricks with the sample phase to eliminate aliasing with a lower sample rate.

For many amateur radio applications, a 100 MHz oscilloscope is adequate to measure carriers for most HF radios. However, be aware that for some oscilloscopes rated with 100 MHz bandwidth, signals with that frequency may be 3 dB down, resulting in amplitudes that are half of what is expected. Not all oscilloscope ratings have this characteristic, and many high-end oscilloscopes show the correct calibrated voltage up to the bandwidth rating.

The range of input signals is normally specified as the maximum and minimum volts per division. Amplitude accuracy is usually specified as a percent of the reading, typically 5-10% for inexpensive analog scopes and 1-2% for high-quality digital ones. Another important specification is the maximum input signal that can be accepted without damage. It is typically presented as a maximum dc plus peak ac voltage, that is, the maximum peak voltage of the complete signal. This is increased when using a ×10 probe, subject to the probe specifications.

For digital scopes, the resolution of the input ADC is typically specified as a number of bits. Sometimes the lowest-voltage input ranges are obtained simply by using only the low end of the ADC range, which results in the displayed signal having a stepped response rather than a smooth curve.

The sweep speed is generally specified in seconds (or milliseconds or microseconds) per division. The accuracy is typically a few percent for analog scopes and much better than that for digital scopes, often limited just by the screen resolution. The triggering system is a major factor that determines how useful a scope is in actual operation, but it can be hard to tell how well it works by studying the specifications. The trigger sensitivity is the main parameter to look for. It is specified in fractions of a division for internal trigger and in mV for external trigger.

Figure 25.19 — Most oscilloscopes provide a signal for calibrating the probe's compensating capacitor.

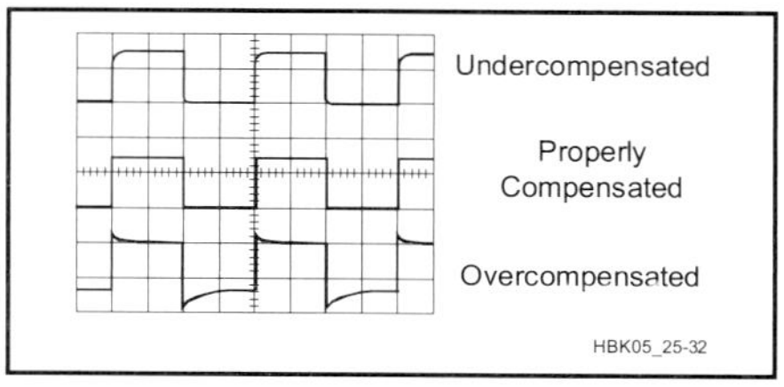

Figure 25.20 — Displays of a square-wave input illustrating undercompensated, properly compensated and overcompensated probes.

BUYING A USED SCOPE

Many hams end up buying a used scope due to price, and used high-end analog scopes are often quite affordable. Digital scopes are now showing up on the surplus market at reasonable prices. Early digital models had somewhat cumbersome user interfaces, so look for one with an analog-like feel with separate buttons or knobs for most of the important functions. On an older scope, be sure the CRT trace is bright and there are no burned-in spots or lines on the CRT from extended use.

If you intend to service an old scope yourself, be aware all scopes that use CRT contain lethal voltages as high 15 kV. Treat an oscilloscope with the same care you would use with a tube-type, high-power amplifier. The CRT should be handled carefully because if dropped it can crack and implode, resulting in pieces of glass and other materials being sprayed everywhere in the immediate vicinity. You should always wear a full-face safety shield and other appropriate safety equipment to protect yourself when handling this material.

Another concern when servicing an older scope is the availability of parts. The CRTs in older units may no longer be available. Many scopes made since about 1985 used special ICs, LCDs, and microprocessors. Some of these may not be available or may be prohibitive in cost. You should buy a used scope from a reputable vendor. Better yet, try it out before you buy it. Make sure that you also get the operator's manual.

Older tube-type models are generally quite serviceable, often needing nothing more than a new tube or two. The massive lab-grade instruments from days of yore made by Tektronix and Hewlett-Packard can still give good service with a little care. They are so large and heavy that a special scope cart was often used to house them and allow easy movement from lab bench to lab bench.

25.7 Spectrum Analyzers

25.7.1 Time and Frequency Domain

A spectrum analyzer is similar in appearance to an oscilloscope. Both present a graphical view of an electrical signal. The oscilloscope is used to observe electrical signals in the time domain (amplitude versus time), which gives little information about the frequencies that make up complex signals. That information is obtained by viewing electrical signals in the frequency domain (amplitude versus frequency). One instrument that can display the frequency domain is the spectrum analyzer.

To better understand the concepts of time and frequency domain, see **Figure 25.21**. The three-dimensional coordinates in Figure 25.21A show time as the horizontal line sloping toward the bottom right, frequency as the depth dimension going back into the page, and amplitude as the vertical axis. The two discrete frequencies shown are harmonically related, so we'll refer to them as f_1 and $2f_1$.

In the representation of time domain in Figure 25.21B, all frequency components of a signal are summed together. In fact, if the two discrete frequencies shown were applied to the input of an oscilloscope, we would see the solid line (which corresponds to the sum of f_1 and $2f_1$) on the display.

In the frequency domain, complex signals (signals composed of more than one frequency) are separated into their individual frequency components. A spectrum analyzer measures and displays the power level at each discrete frequency; this display is shown at Figure 25.21C.

The frequency domain contains information not apparent in the time domain and therefore the spectrum analyzer offers advantages over the oscilloscope for certain measurements, such as harmonic content or distortion, as mentioned previously. For measurements that are best made in the time domain, the oscilloscope is the tool of choice.

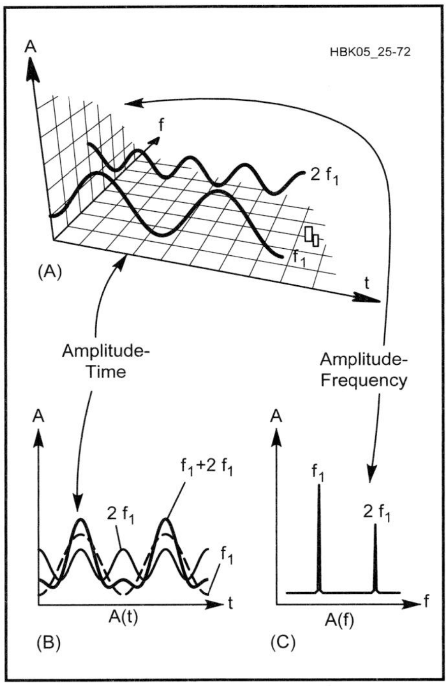

Figure 25.21 — A complex signal in the time and frequency domains. (A) is a three-dimensional display of amplitude, time and frequency. The solid trace in (B) is what is seen on an oscilloscope display of time vs amplitude. (C) is spectrum analyzer display of the frequency domain and shows frequency vs amplitude.

25.7.2 Spectrum Analyzer Functions

As with oscilloscopes, analog spectrum analyzers have been replaced by digital models that use DSP/SDR techniques to perform similar signal processing and display functions. Just as for oscilloscopes, most of the operation and display terminology used today was defined for analog instruments. We describe analog spectrum analyzer functions in order to explain the controls and functions used on digital analyzers today.

ANALOG SPECTRUM ANALYZERS

The most common analog spectrum analyzer is basically an electronically tuned superheterodyne receiver. The receiver is tuned by means of a ramp voltage. That performs two functions: First, it sweeps the frequency of the analyzer local oscillator; second, it deflects a beam across the horizontal axis of a CRT display, as shown in **Figure 25.22**. The vertical axis deflection of the CRT beam is determined by the strength of the received signal. In this way, the CRT displays frequency on the horizontal axis and signal strength on the vertical axis.

Most analog spectrum analyzers use an up-converting technique in which a wide-band input is converted to an IF higher than the highest input frequency. Up-conversion is used so that a fixed-tuned input filter can remove any image signals and only the first local oscillator needs to be tuned to tune the analyzer.

As with most up-converting communications receivers, it is not easy to achieve the desired ultimate selectivity at the first IF because of the high frequency. For this reason, multiple conversions are used to generate an IF low enough so that the desired selectivity is practical. See the **Receiving** chapter for more detail on multiple conversion heterodying.

For the spectrum analyzer to be most useful, it should display signals of widely different levels. As an example, consider two signals that differ by 60 dB, which is a thousand to one difference in voltage or a million to one in power. That means that if power were displayed, one signal would be one million times larger than the other. In the case of voltage one signal would be a thousand times larger. In either case it would be difficult to

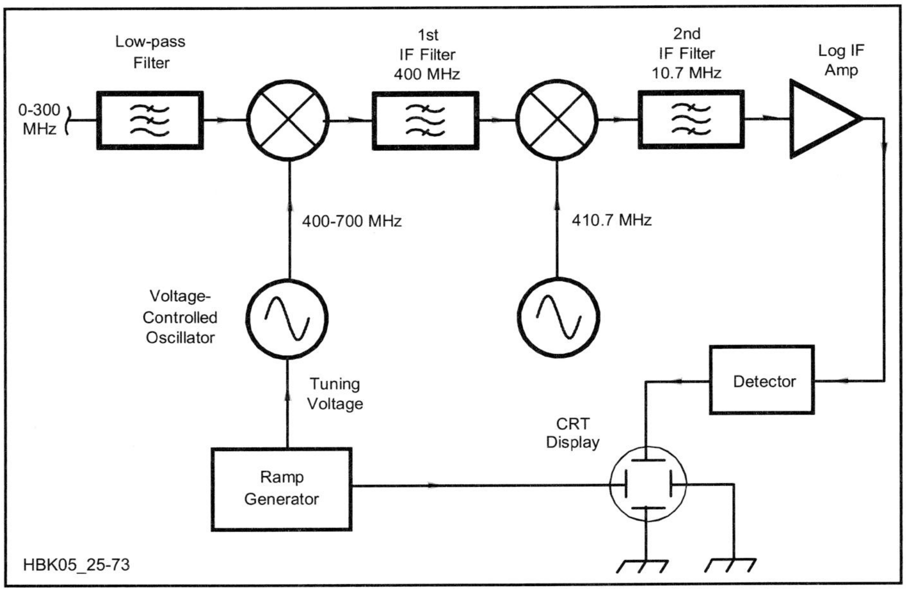

Figure 25.22 — A block diagram of a typical superheterodyne spectrum analyzer. Input frequencies of up to 300 MHz are up-converted by the local oscillator and mixer to a fixed frequency of 400 MHz.

display both signals on a linear display, like an oscilloscope. The solution to this problem is to use a logarithmic display that shows the relative signal levels in decibels. Using this technique, a 1,000:1 ratio of voltage reduces to a 60 dB difference.

The conversion of the signal to a logarithm is usually performed in the IF amplifier or detector, resulting in an output voltage proportional to the logarithm of the input RF level. This output voltage is then used to drive the CRT display.

SDR Receivers as Spectrum Analyzers

SDR receivers can also be used as spectrum analyzers, although they do not have the same specifications as a spectrum analyzer designed for use as a test instrument. This application of the SDR is commonly called a panadapter.Even inexpensive SDRs can make absolute signal level measurements although the difference in levels between signals is probably more accurate. They can measure frequency and frequency response, as well. Portable SDRs can serve as portable spectrum analyzers when searching for interference sources— see the **RFI and EMC** chapter for more information about this application.

25.7.3 Spectrum Analyzer Performance

The performance parameters of a spectrum analyzer are specified in terms similar to those used for radio receivers, in spite of the fact that there are many differences between a receiver and a spectrum analyzer.

The sensitivity of a receiver is often specified as the minimum discernible signal (MDS), which means the smallest signal that can be heard. In the case of the spectrum analyzer, it is not the smallest signal that can be heard, but the smallest signal that can be seen. The dynamic range of the spectrum analyzer determines the largest and smallest signals that can be simultaneously viewed on the analyzer. As with a receiver, one factor that affects dynamic range is second- and third-order intermodulation distortion (IMD). IMD dynamic range is the maximum difference in signal level between the minimum detectable signal and the level of two signals of equal strength that generate an IMD product equal to the minimum detectable signal. (See the **Receiving** chapter for more information on IMD.)

Although the communications receiver is an excellent example to introduce the spectrum analyzer, there are several differences such as the previously explained lack of a demodulator. Unlike the communications receiver, the spectrum analyzer is not a sensitive radio receiver. To preserve a wide dynamic range, the first and second mixers in the spectrum analyzer are passive. Therefore, referring to Figure 25.22, the noise figure of the analyzer is no better than the losses of the input low-pass filter plus the first mixer, the first IF filter, the second mixer, and the loss of the second IF filter. This often results in a combined noise figure of more than 20 dB. With that kind of noise figure the spectrum analyzer is obviously not a communications receiver for extracting very weak signals from the noise, but a measuring instrument for the analysis of frequency spectrum. For some applications, it may be necessary to add an external wide-band, low-noise preamplifier to improve the noise figure.

The selectivity of the analyzer is called the *resolution bandwidth* (RBW). This term refers to the minimum frequency separation of two signals of equal level so that the signals are separated by a drop in amplitude of 3 dB between them. The IF filters used in a spectrum analyzer differ from a communications receiver in that they have very gentle skirts and rounded passbands, rather than the flat passband and very steep skirts of an IF filter in a high-quality communications receiver. The rounded passband is necessary because the signals pass through the filter passband as the spectrum analyzer sweeps the desired frequency range. If the signals suddenly pop into the passband (as they do if the filter has steep skirts), the filter tends to ring. A filter with gentle skirts has less ringing. Another effect, which occurs even with rounded-passband filters, is that the signal amplitudes are reduced at fast sweep rates, which distorts the display and requires that the analyzer not sweep frequency too quickly. When adjusting the resolution bandwidth or the width of the frequency span (the range of frequencies being measured), the scan rate may need to be reduced so that the signal amplitude is not affected by fast sweeping.

The signal produced by the detector is known as the video signal. Most spectrum analyzers include a low-pass video filter to reduce the displayed noise level. The video bandwidth (VBW) is the bandwidth of this filter. Like the RBW, the VBW must also be taken into consideration when setting the sweep speed.

25.7.4 Choosing a Spectrum Analyzer

The frequency range is perhaps the most important specification when choosing a spectrum analyzer. Of course, it must cover the amateur bands you wish to test, but it must also cover harmonics of those frequencies if you wish to test transmitter harmonics. Many higher-frequency spectrum analyzers only cover down to 10 kHz, so they are not useful for audio spectrum analysis (spectrum analyzers for audio use are also available). Some instruments use a harmonic sampler on the higher microwave bands. You may need to use an external filter to remove any unwanted signals at harmonics or sub-harmonics of the signal you wish to examine.

Frequency stability is vital for narrowband measurements such as looking at modulation spectra. Some older instruments include a frequency lock scheme that stabilizes the display after the analyzer is tuned to the desired signal.

The range of resolution bandwidths (RBW) available determines what kinds of measurements are possible. To measure two-tone intermodulation distortion of an SSB signal, the minimum RBW must be narrow enough to resolve the two tones, that is, no more than about 100 Hz. If you wish to view the demodulated time-domain video from a fast-scan television signal, the RBW must be wide enough to include the entire signal, perhaps 5 or 6 MHz.

Input sensitivity is normally not very important. An external preamplifier can always be added if needed for a particular measurement. However, dynamic range is a key specification, as previously discussed. Also, the screen display range is important. For example, if there are 8 divisions on the display and the maximum dB-per-division setting is 10 dB/div, then the maximum display range is 80 dB. Many spectrum analyzers can display signals in linear as well as logarithmic (dB) mode, which can be useful to look at modulation.

For accurate power measurements, the amplitude accuracy is important. For doing relative measurements, which are important for measuring spurious signals relative to the carrier, logarithmic linearity is the key specification.

There are several features that expand the spectrum analyzer's utility. *Markers* are pips that can be moved with a knob back and forth across the signals on the display. By placing a marker at the top of a signal you can read the frequency and amplitude directly. *Peak search* is a feature that automatically places a marker at the peak of the strongest signal. *Marker delta* measures the difference in frequency and amplitude of two markers.

Signal amplitudes that vary over time, such as with modulation or intermittent operation, benefit from a feature in many spectrum analyzers that performs peak capture. Rather than showing the signal amplitude as it varies in real time, only the highest value of the signal is displayed. This facilitates measurement of peak-envelope-power (PEP) and viewing intermittent signals.

Zero span means setting the sweep width to zero hertz to view the change in signal versus time with the selected resolution bandwidth.

This is useful for looking at modulation. Many spectrum analyzers include a sweep triggering feature for this purpose, similar to the triggering circuit in an oscilloscope. Some include a speaker so you can hear what the modulating signal sounds like.

Modern spectrum analyzers have many of the features found in digital oscilloscopes, such as trace storage and retrieval. Most include computer connectivity, both to store measured data and to control the instrument with the proper software.

In addition to traditional swept-frequency spectrum analyzers, some modern units use the *Fast Fourier Transform* (FFT) either in place of, or in addition to, the swept-frequency architecture. The FFT allows for much faster screen updates when using narrow spans. Some RF and microwave instruments have a *digital IF* based on the FFT. For narrow spans, the local oscillator is held at a constant frequency and the "sweep" is performed mathematically by the FFT.

Spectrum analyzers can falsely indicate high amplitude frequencies existing near dc, particularly those that are FFT-based. The high signal level can be an artifact of the way the spectrum is determined, and incorrectly indicates that very low frequency signals are present when they are not.

There are many used spectrum analyzers on the surplus market, at all different vintages and price points. As usual, try before you buy if possible, and obtain the operator's and service manuals if available. Another alternative is an external digitizing pod connected to the PC via a serial data link. As with PC-based oscilloscopes, they can represent excellent value because they save the cost of the display, cabinet, power supply, and microprocessor provided by the computer.

The least-expensive alternative is an audio spectrum analyzer that uses a PC's sound card. Free spectrum analysis software is available on the internet. The main limitation is the frequency coverage, which is typically about 40 – 45% of the sample rate of the sound card. The sound card that comes standard in a PC typically has a 48 kHz sample rate, so that the frequency response is limited to about 20 kHz. That is plenty for checking out the audio circuits in a communications receiver or transmitter.

25.8 Using a Spectrum Analyzer

Spectrum analyzers are used in situations where the signals to be analyzed are complex, for very low-level signals, or when the frequency of the signals to be analyzed is very high. Although high-performance oscilloscopes are capable of operation into the UHF region, moderately priced spectrum analyzers can be used well into the gigahertz region.

Unlike the oscilloscope, which is a wide-bandwidth instrument, the spectrum analyzer measures the waveform using a narrow bandwidth; thus, it can reduce the noise power displayed.

Also, unlike the oscilloscope, most spectrum analyzers are highly sensitive to the characteristics of the input signal and can therefore be easily damaged. **Figure 25.23** shows the input connectors of a spectrum analyzer and an oscilloscope. In 25.23A, a warning box on the spectrum analyzer shows the maximum limits of the signal that can enter the input connector. Unlike the oscilloscope in 25.23B, which can usually tolerate hundreds of volts of input signal, this spectrum analyzer, like many, cannot tolerate any dc voltage on the input signal. Likewise, most oscilloscopes can tolerate hundreds of voltages of swing from the highest to lowest peaks, yet this spectrum analyzer, like many, can only tolerate a swing of 1 W, which in a 50 Ω termination is equivalent to a little more than 7 volts peak-to-peak. RF signals that are measured by the spectrum analyzer should be ac-coupled (to ensure the 0 VDC maximum) and should be fed through sufficient attenuation (to ensure the 30 dBm maximum). For instance, if you want to measure the conducted output of your 100 W transmitter (50 dBm), you should feed the signal from the transmitter connector through at least a 30 dB attenuator, which is also capable of dissipating 100 watts of heat (20 dB attenuation is the minimum acceptable attenuation to bring a 100 watt signal down to 1 watt, but some extra attenuation is a good idea in case your transmitter is putting out more than 100 watts, and to help ensure that the spectrum analyzer does not generate any non-linear spurious products that are not actually part of the input signal). The measured results on the spectrum analyzer screen can then have the attenuation value added back to them to represent true wattage of the main signal and its spurious emissions.

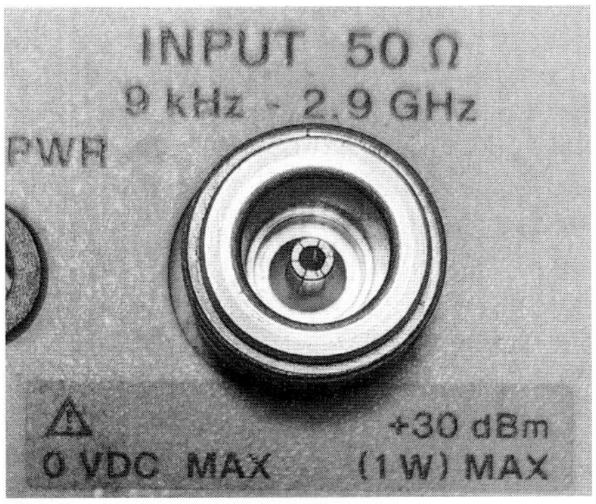

(A)

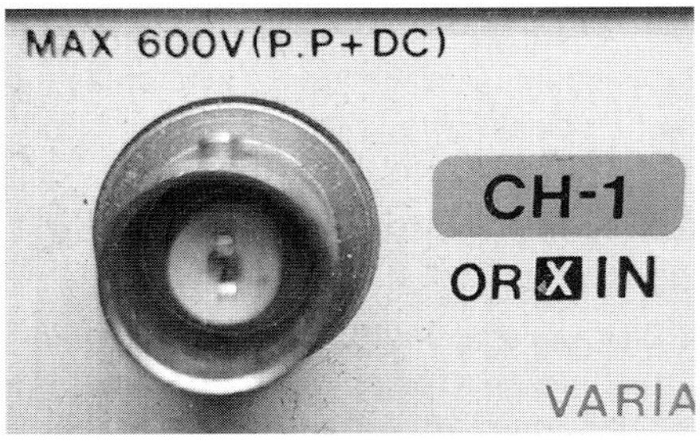

(B)

Figure 25.23 — Input connectors of (A) the typical spectrum analyzer and (B) the typical oscilloscope. The spectrum analyzer is easily damaged by large amplitude signals, typically a 1 watt limit, and cannot handle any dc offset voltage. In contrast, the oscilloscope can measure high voltage signals (600 volts, sum of amplitude and dc offset). Be careful when probing a signal on an oscilloscope and then moving the probe to a spectrum analyzer to see frequency content.

25.8.1 Harmonic and Spurious Signal Measurements

Probably the most common amateur radio application of a spectrum analyzer is the measurement of the harmonic content and other spurious signals in the output of a radio transmitter. **Figure 25.24** shows two ways to connect the transmitter and spectrum analyzer. The method shown at Figure 25.24A should not be used for wide-band measurements since most line sampling devices do not exhibit a constant amplitude output over a broad frequency range. Using a line sampler is fine for narrowband measurements, however.

The method shown at Figure 25.24B is used in the ARRL Lab. The attenuator must be capable of dissipating the transmitter power. It must also have sufficient attenuation to protect the spectrum analyzer input.

When using a spectrum analyzer, it is very important that the proper amount of attenuation be applied before a measurement is made.

One needs to consider both the maximum-power input rating of the spectrum analyzer and its linearity specifications. At a 0 dBm input level, for example, some spectrum analyzers may create harmonics internally that are not actually present at its input.

It is a good practice to start with maximum attenuation and view the entire spectrum of a signal before the attenuator is adjusted. The signal being viewed could appear to be at a safe level, but another spectral component, which is not visible, could be above the damage limit. It is also very important to limit the input power to the analyzer when pulse power is being measured. The average power may be small enough so the input attenuator is not damaged, but the peak pulse power, which may not be readily visible on the analyzer display, can destroy a mixer in microseconds.

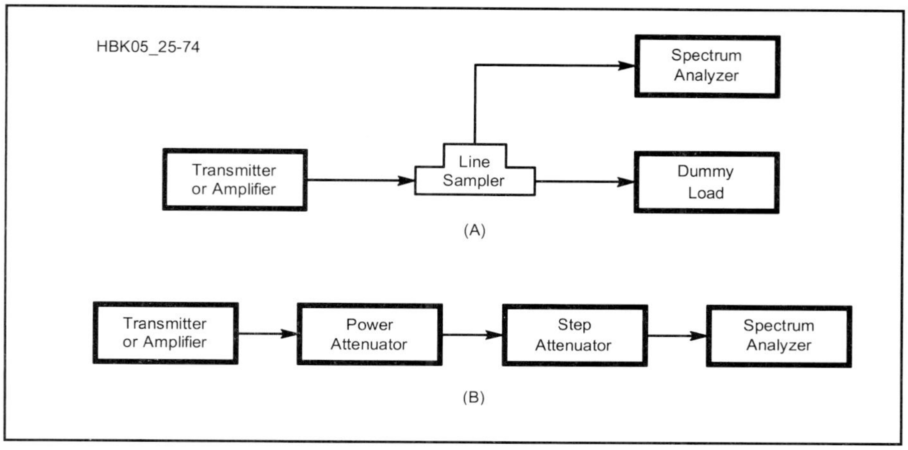

Figure 25.24 — Alternate bench setups for viewing the output of a high-power transmitter or oscillator on a spectrum analyzer. (A) uses a line sampler to pick off a small amount of the transmitter or amplifier power. In (B), most of the transmitter power is dissipated in the power attenuator.

SPURIOUS RESPONSES IN SPECTRUM ANALYZERS

It is necessary to ensure that the spectrum analyzer does not generate additional spurious signals that are then attributed to the system under test. Some of the spurious signals that can be generated by a spectrum analyzer are harmonics and IMD. It is good practice to check for the generation of spurious signals within the spectrum analyzer. When an input signal causes the spectrum analyzer to generate a spurious signal, adding attenuation at the analyzer input causes the internally generated spurious signals to decrease by an amount greater than the added attenuation. If attenuation added ahead of the analyzer causes all of the visible signals to decrease by the same amount, this indicates a spurious-free display.

If it is desired to measure the harmonic levels of a transmitter at a level below the spurious level of the analyzer itself, a notch filter can be inserted between the attenuator and the spectrum analyzer as shown in **Figure 25.25**. This reduces the level of the fundamental signal and prevents that signal from generating harmonics within the analyzer, while still allowing the harmonics from the transmitter to pass through to the analyzer without attenuation. Use caution with this technique; detuning the notch filter or inadvertently changing the transmitter frequency allows potentially high levels of power to enter the analyzer. In addition, use care when choosing filters; some filters (such as cavity filters) respond not only to the fundamental but notch out odd harmonics as well.

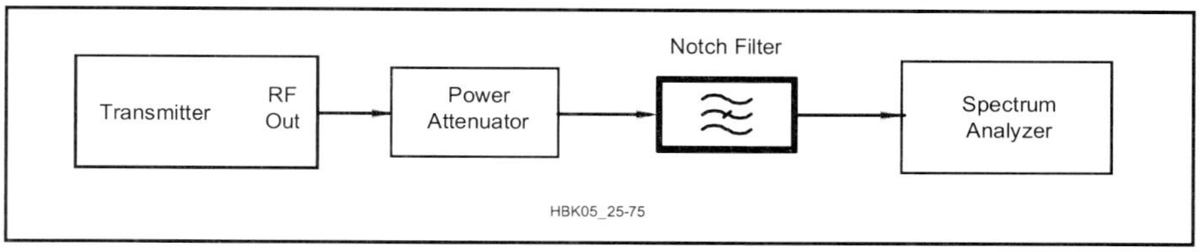

Figure 25.25 — A notch filter is another way to reduce the level of a transmitter's fundamental signal so that the fundamental does not generate harmonics within the analyzer. However, in order to know the amplitude relationship between the fundamental and the transmitter's actual harmonics and spurs, the attenuation of the fundamental in the notch filter must be known.

SIGNAL AND INTERFERENCE TRACING

The input impedance for most RF spectrum analyzers is 50 Ω, however not all circuits have convenient 50 Ω connections that can be accessed for testing purposes. Using a probe such as the one shown in **Figure 25.26** allows the analyzer to be used as a troubleshooting

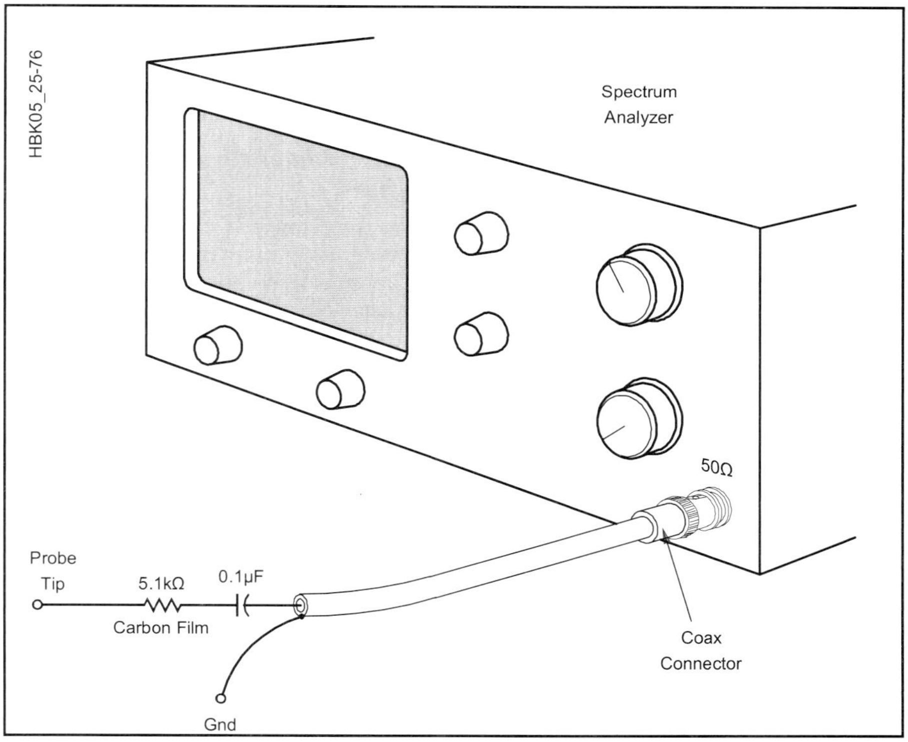

Figure 25.26 — A schematic representation of a voltage probe designed for use with a spectrum analyzer. Keep the probe tip (resistor and capacitor) and ground leads as short as possible.

tool. The probe can be used to track down signals within a transmitter or receiver, much like an oscilloscope is used. The probe shown offers a 100:1 voltage reduction and loads the circuit with about 5000 Ω. The resistor creates a severe mismatch in the coaxial cable so use the shortest cable practical to keep the low-pass RC-filter created by the resistor and cable's capacitance from degrading accuracy at the highest frequency being measured. Measurements using this setup should be treated as relative or qualitative and not as calibrated levels.

A different type of probe is shown in **Figure 25.27**. This inductive pickup coil (sometimes called a "sniffer") is very handy for troubleshooting. The coil is used to couple signals from the radiated magnetic field of a circuit into the analyzer. A short length of miniature coax is wound into a pickup loop and soldered to a larger piece of coax that connects to the spectrum analyzer. The dimensions of the loop are not critical, but smaller loop dimensions enable the loop to more precisely locate the source of radiated RF. Coax is used for the loop to provide shielding from the electric field component (capacitive coupling). Connecting the coax shield on only one end provides a complete electrostatic shield without introducing a shorted turn.

The sniffer allows the spectrum analyzer to sense RF energy without contacting the circuit being analyzed. If the loop is brought near an oscillator coil, the oscillator can be tuned without directly contacting (and thus disturbing) the circuit. The oscillator can then be checked for reliable startup and the generation of spurious sidebands. With the coil brought near the tuned circuits of amplifiers or frequency multipliers, those stages can be tuned using a similar technique.

Even though the sniffer does not contact the circuit being evaluated, it does extract some energy from the circuit. For this reason, the loop should be placed as far from the tuned circuit as is practical. If the loop is placed too far from the circuit, the signal may be too weak or the pickup loop picks up energy from other parts of the circuit and does not give an accurate indication of the circuit under test.

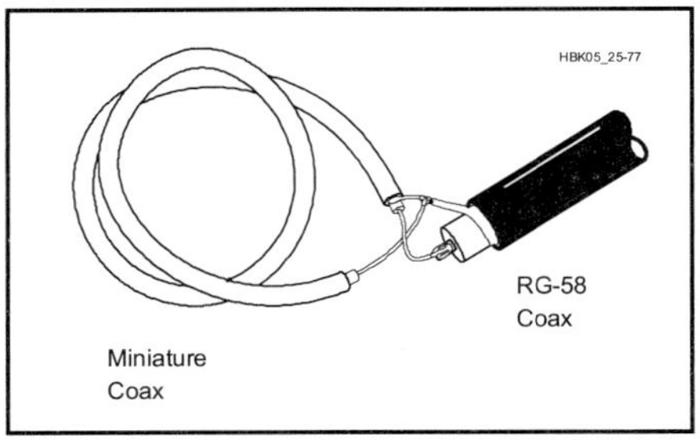

Figure 25.27 — A "sniffer" probe consisting of an inductive pick-up. It has the advantage of not directly contacting the circuit under test.

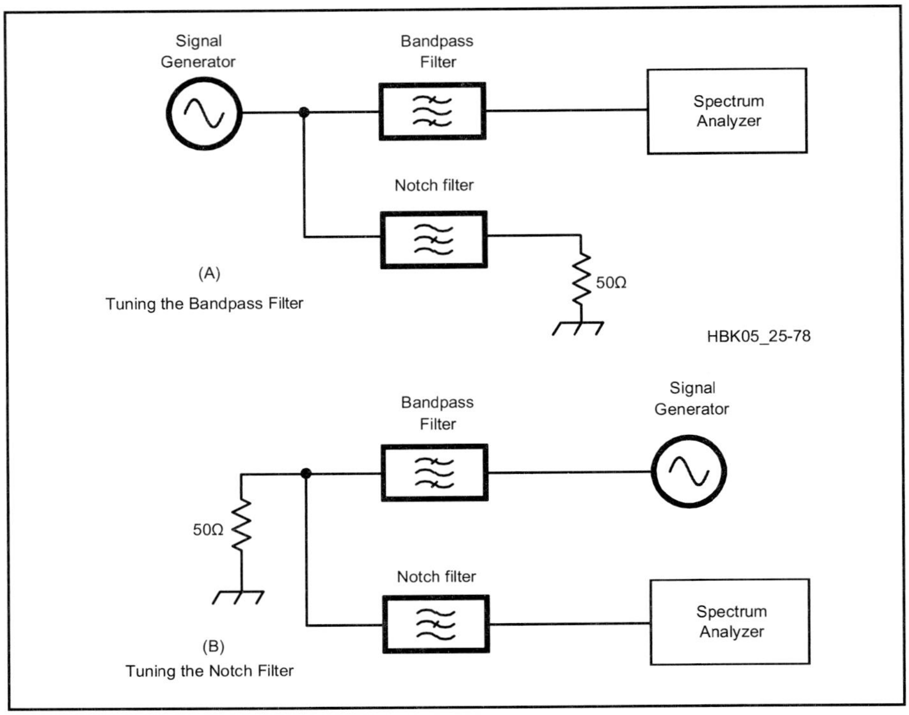

Figure 25.28 — Block diagram of a spectrum analyzer and signal generator being used to tune the (A) bandpass and (B) notch filters of a duplexer. All ports of the duplexer must be properly terminated and good quality coax with intact shielding used to reduce leakage.

The sniffer is very handy to locate sources of RF leakage. By probing the shields and cabinets of RF generating equipment (such as transmitters) egress and ingress points of RF energy can be identified by increased indications on the analyzer display.

25.8.2 Measuring Low-Level Signals

One very powerful characteristic of the spectrum analyzer is the instrument's capability to measure low-level signals. This characteristic is advantageous when very high levels of attenuation are measured. **Figure 25.28** shows the setup for tuning the notch and passband of a VHF duplexer. The spectrum analyzer, being capable of viewing signals well into the low microvolt region, is capable of measuring the insertion loss of the notch cavity more than 100 dB below the signal generator output. Making a measurement of this sort requires care in the interconnection of the equipment and a well-designed spectrum analyzer and signal generator. RF energy leaking from the signal generator cabinet, line cord or even the coax itself, can get into the spectrum analyzer through similar paths and corrupt the measurement. This leakage can make the measurement look either better or worse than the actual attenuation, depending on the phase relationship of the leaked signal.

25.8.3 Frequency Response Measurement

Some spectrum analyzers can be used in conjunction with a suitable signal generator to measure the frequency response of a circuit. Generators connected in this way create a *tracking generator* because the output frequency tracks the spectrum analyzer input frequency. The tracking generator makes it possible to make swept frequency measurements of the attenuation characteristics of circuits, even when the attenuation involved is large. A tracking generator also simplifies many other measurements and adjustments that require involve a signal generator tuned to the same frequency as the spectrum analyzer. Medium-cost spectrum analyzers are available that include a tracking generator integrated with the analyzer for a modest extra cost.

Figure 25.29 shows the connection of a tracking generator to a circuit under test. In order for the tracking generator to create an output frequency exactly equal to the input frequency of the spectrum analyzer, the internal local oscillator frequencies of the spectrum analyzer must be known. This is the reason for the interconnections between the tracking generator and the spectrum analyzer. The test setup shown measures the gain or loss of the circuit under test.

Only the magnitude of the gain or loss is

available; in some cases, the phase angle between the input and output would also be an important and necessary parameter. That is the function of a vector network analyzer, covered in a following section.

You can also take advantage of the flat spectrum of a noise source to measure the frequency response of RF devices. It requires either a spectrum analyzer or a receiver that can be tuned manually across the frequency range of interest. With sufficient averaging, the output noise level at each frequency accurately reflects the response of the device at that frequency. For example, it is quite easy to measure the total response of a receiver from antenna to audio output using an RF noise source and free software running on a PC that turns the computer sound card into an audio spectrum analyzer. With the noise source connected to the receiver antenna and the audio output connected to the sound card input, the total response, including all filters and audio stages, is shown by the average noise level in the spectrum analyzer window.

25.8.4 Service Monitors

A *service monitor* is a "one-box tester" for

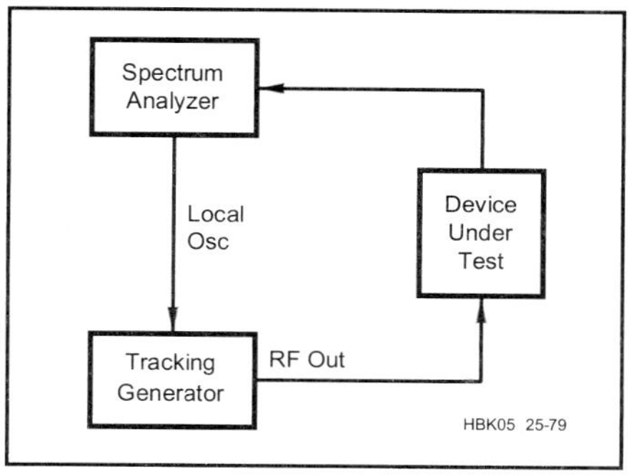

Figure 25.29 — A signal generator (shown in the figure as the "Tracking Generator") locked to the local oscillator of a spectrum analyzer can be used to determine filter response over a range of frequencies.

transceivers. It includes a signal generator for testing the receiver and a spectrum analyzer for testing the transmitter, using the same RF connector so that only one connection to the transceiver's antenna jack is required. Other common features are an RF wattmeter and dummy load, a frequency counter, an FM deviation meter, audio tone generators to connect to the microphone, and an audio voltmeter and distortion analyzer/SINAD meter to connect to the speaker output. Some units contain additional features such as DTMF (touch-tone) and CTCSS (sub-audible tone) generators, an audio frequency counter and adjacent-channel power (ACP) measurement capability.

Older service monitors found on the surplus market were designed for testing analog two-way radios and repeaters. Many are portable for easy transportation to a mountaintop repeater site. More recent units may be more oriented to testing cellular telephone base stations. Modern instruments cover the latest digital modes, with bit error rate (BER) testers for the receiver and various modulation quality tests for the transmitter.

While all the functions of a service monitor are available in separate instruments, having everything integrated in one box is more convenient and allows faster testing, which is something a commercial enterprise is willing to pay extra for. A brand-new service monitor is not inexpensive, but older used units made by such companies as Singer-Gertsch, Cushman and IFR that are suitable for testing analog radios can be found for reasonable prices.

25.9 Impedance, Antenna, and Network Analyzers

Another type of analyzer measures parameters of components, antennas, and circuits over a frequency range. These analyzers are often used to measure impedance and are sometimes referred to as impedance analyzers. Analyzers optimized for use with antenna systems are called antenna analyzers. More capable analyzers are used to measure gain, filter response, isolation, and other properties of circuits and systems that are referred to by the general name of networks. These analyzers are called network analyzers.

Over the last decade, a new generation of relatively low-cost analyzers has emerged, both as free-standing handheld instruments, and as instruments that interface to a computer via a USB port and require software running on the computer to complete the measurement system. Some of the free-standing instruments can operate under computer control and with the computer software greatly expanding their capabilities.

These instruments make measurements of the *scattering parameters* (*S-parameters*) of a single-port or two-port network. They allow any circuit or component to be treated as a "black box" with the S-parameters describing its behavior in terms of amplitude and phase.

The scattering parameters for a two-port network, are:

S_{11}, the reflection coefficient of the input port;

S_{21}, the voltage gain from input to output, also known as Forward Transmission or Forward Gain;

S_{12}, the voltage gain from output to input, also known as Reverse Transmission;

S_{22}, the reflection coefficient of the output port.

Scattering parameters are explained in the **RF Techniques** chapter's Two-Port Networks section.

25.9.1 SWR Meters

The analog SWR meter is a descendant of the *QST* project by Lew McCoy, W1ICP, that introduced the "Monimatch," to amateurs in 1956, originally created by Walter Bruene, W5OLY. (**collinsradio.org/wp-content/uploads/2015/05/Understanding-the-Bruene-Coupler-Transmission-Line-Bold.pdf**) Directional RF wattmeters, also used to determine SWR, are covered in this chapter's section on Measuring RF Power.

Even though they tend to be used interchangeably, Standing Wave Ratio (SWR) is a concept that is based on the various impedances found between the transmitter and the antenna. What we measure using a directional RF wattmeter is in reality Voltage Standing Wave Ratio (VSWR). Since both terms indicate the same behavior of the antenna feed system, the term that is used is not important. See the **Transmission Lines** chapter for a more detailed description of SWR.

Figure 25.30 shows the schematic of a typical unit. SWR can be computed from the power measured in the forward (P_F) and reflected (P_R) directions, as shown by the following equation:

$$SWR = \frac{1+\sqrt{P_R / P_F}}{1-\sqrt{P_R}}$$

If voltages representing the two powers are provided, it is straightforward to create a circuit that computes SWR. This is the reason for the CAL (calibration) control on the meter.

Protecting Antenna Analyzers

Portable antenna analyzers are sensitive instruments with low-power components at their input. They can be damaged by static electricity if care is not exercised. Antennas can collect a significant static charge from rain or wind. Be sure to momentarily ground the transmission line before connecting it to an analyzer to reduce the risk of damage to the sensitive components on the input.

Static buildup can be minimized by providing a dc path between center conductor and the grounded shield for the antenna being measured, offering some protection both for the analyzer and for station equipment. The dc path can be a shunt inductor (typically an RF choke) or a high-value resistor (10 kΩ or more). If the resistor is to be a permanent part of the antenna, it must have a power rating suitable for the transmitter power.

Antennas with non-insulated elements mounted directly on a grounded boom or support are already protected against static. Insulated elements need protection against static. Measure resistance between the feed line center conductor and shield to be sure of whether there is a dc path to drain static.

Using the analyzer when another transmitter is active can also cause damage if enough signal is picked up by the antenna under test. Some analyzers include an isolation relay that protects the input when a measurement is not in progress.

The meter is set to FWD (indicating the voltage representing forward power), power is applied, and the CAL control adjusted so that the forward power indication is the full-scale value of 1. By assuming $P_F = 1$ and P_R is a value between 0 and 1, the equation for SWR simplifies to:

$$SWR = \frac{1+\sqrt{P_R}}{1-\sqrt{P_R}}$$

The voltage representing reflected power is always some fraction of the full-scale voltage, so the meter scale can then be calibrated to read SWR directly, instead of voltage.

THE BRUENE DIRECTIONAL COUPLER

Inductive and capacitive coupling are used to create a *directional coupler* that can provide the independent measurement voltages. The coupling provides samples of forward and reflected voltage and current from the undisturbed center conductor of the coaxial feed line. Figure 25.30 shows how voltages representing reflected power are obtained. As described by W5OLY, "A pickup wire placed parallel to the inner conductor samples the line current by inductive coupling. The voltage e_i induced in the pickup is determined by spacing, length, line current, and frequency. The mechanical dimensions determine the mutual inductance, M. The induced voltage due to line current is:

$$e_i = -j2\pi fM$$

where f is frequency in Hz and *j* represents a phase shift of 90°. This shows that the higher the frequency, the larger the induced voltage.

"The sample of voltage is picked up by capacitive coupling (C_{CPLG}) from the inner conductor to the pickup wire. A current due to this capacitance flows through R and develops a voltage across it; this voltage also increases with frequency because the reactance of the coupling capacitance goes *down* with frequency. That is:

$$e_v = E{\cdot}R/X_C{\cdot}R = E{\cdot}R / -j(1/2\pi{\cdot}f{\cdot}C) = j2\pi{\cdot}f{\cdot}E{\cdot}R{\cdot}C$$

when X_C is much larger than R, D, and C it forms an RF detector that sums e_i and e_v, creating a single voltage proportional to the power in the line.

Since the line current and voltage contain components of both forward and reflected power, the single resulting output voltage also contains components of both. The current and voltage components of reflected power are 180° out-of-phase, compared to those of forward power. So e_i is added to e_v, producing a voltage proportional only to reflected power in the figure.

The different polarities of e_i are obtained by reversing the current sensing pickup. This is accomplished by two identical pickup circuits in the meter; e_v is the same in both circuits, but the e_i pickup direction is reversed from one to the other. One circuit produces a voltage proportional only to forward power and the other proportional only to reflected power. Display or forward or reflected power is controlled by the switch that selects which voltage is applied to the meter.

Continuing with a note from W5OLY, "Since the current and voltage pickups both increase with frequency, their ratio will stay the same. The variation in pickup just means that the sensitivity goes down at lower fre-

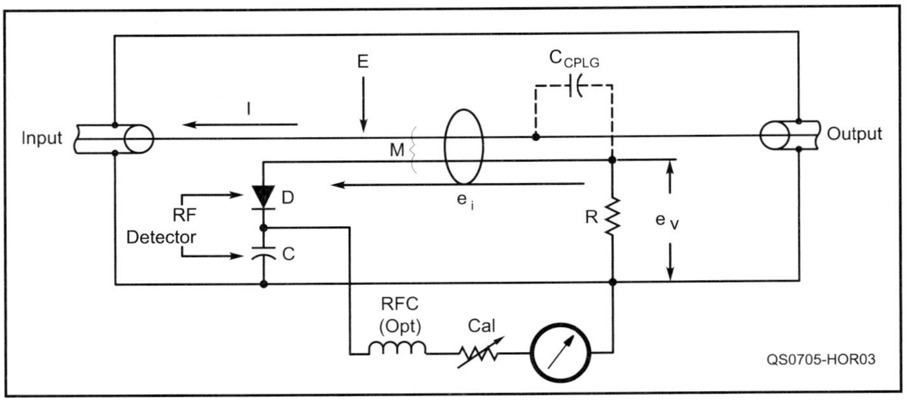

Figure 25.30 — One-half of a Monimatch directional coupler circuit that senses the reflected voltage and current components. The samples of induced voltage e_i and line voltage e_v sum in the RF detector formed by D and C, producing a voltage that drives the meter. The Cal control is adjusted so that a full-scale reading is obtained. The meter is then switched to display the output of an identical circuit oriented in the opposite direction to picks up the forward components. The meter is then calibrated according to the SWR equation in the text.

Determining Reactance Type

Some antenna analyzers can show the positive or negative sign of the reactance. Less expensive units may only show a reactance magnitude. To determine whether reactance is inductive or capacitive on a unit that doesn't indicate the sign, make a slight increase in frequency. If the reactance increases, it is inductive, and if reactance decreases, it is capacitive. This is a helpful trick when adjusting antennas with a portable analyzer that doesn't indicate reactance sign.

quencies." That's why the CAL adjustment is necessary not only at different power levels, but at different frequencies. The meter scale is calibrated according to the equation for SWR when the calibration adjustment places the voltage representing P_F at full-scale.

For best performance from a Monimatch type of directional coupler, the value of R in one of the detector circuits should be adjusted so that the two circuits produce the same value of e_i for any given current. In addition, balanced detector diodes would also provide better performance at low power levels where the detected voltage level is small enough for variations in the diode forward voltage to introduce significant errors. However, for low-cost equipment, using fixed components is generally "good enough."

For SWR meters that have been damaged, repair is very simple. The manufacturer calibrated the meter assuming very similar component values. Replace both detector diodes with the same type of diode, preferably from the same batch of components. The diodes can be matched by using a multimeter with a diode test function that displays the diode's forward voltage. If one of the original diodes is undamaged, choose a pair of diodes with a similar forward drop.

25.9.2 Time-Domain Reflectometry

Time domain reflectometry (TDR) shows what happens to a short, abrupt pulse as it travels through a transmission line. The pulse is reflected by any changes in impedance, such as an open or short (complete reflection) or a change in the line's characteristic impedance (partial reflection). The resulting series of pulses and reflections is displayed as a sequence in time, thus the name of the technique.

In an ideal transmission line terminated by its characteristic impedance, Z_0, the pulse travels to the far end and is dissipated in the termination, so the trace is a perfectly flat line. But at any point along the line where the impedance changes (called a *discontinuity*) some of the pulse's energy is reflected back toward the line's input. The reflected component of the pulse creates an artifact (visually, a "bump") on the otherwise straight line.

The sequence of pulses and their reflections is the *impulse response* of the line. An *impulse* is basically a very short pulse that begins and ends before the system can respond and stabilize. A mathematical impulse is an infinitely narrow made up of all frequencies from zero to infinity.

While it is not possible to create an ideal (perfect) impulse, a very fast rising edge of a longer pulse is a good enough approximation to measure the line's impulse response and the same information can be measured. The longer pulse is called a *step function*. The ideal step function is an infinitely fast change from one level to another, after which it remains at that level. The response of the line to the longer pulse is called the *step response*. Like the impulse, the infinitely fast change in level also contains all frequencies.

A *time domain reflectometer* is the instrument that generates the pulse and displays the results. A TDR displays amplitude (voltage) on the vertical axis and time on the horizontal axis. The position of each artifact along the TDR trace corresponds to the distance from the transmission line input to the discontinuity that produced it. Large discontinuities occur when a line is open-circuited or short-circuited, or when it is connected to an antenna. (Most antennas are matched to the line at their operating frequency, but at other frequencies they are not. Since the pulse contains many frequencies, an antenna is a large discontinuity.) Small discontinuities occur at splices or when a line is damaged.

The delay between the input pulse's rising edge and the artifact is the round-trip time in the line from the TDR to the discontinuity. If the line's *velocity of propagation* or *velocity factor* (VF) is known, the physical distance from the input to the discontinuity can be calculated. The shape of each reflection can sometimes provide a clue as to the nature of the discontinuity.

In this sense, the TDR is very much like a radar display in which a pulse is transmitted (shown at the center on a radar display) and any echoes from discontinuities in the air (i.e. targets) reflect some of the pulse back toward the transmitter. The farther away the target, the longer it takes for the pulse to travel to the target and back to the receiver. A radar screen shows echoes from all directions. The TDR only shows echoes from one direction, along the line. The larger the echo, the larger the target or discontinuity.

25.9.3 Network Analyzer Basics

Connections to the analyzer are made at a *port* which is a pair of terminals that serve as an input or output of a network. The port terminals can be anything from alligator clips to a coaxial connector. A network can have more than one port. A dummy load is a *single-port network*, for example, and an amplifier, filter, or transmission line are examples of a *two-port network* each with an input and an output. See the **RF Techniques** chapter section on Two-Port Networks for more information about networks and ports.

A *scalar analyzer* measures only the magnitude of a parameter (a scalar is a value expressed as a single number). Both oscilloscopes and spectrum analyzers are scalar instruments. Scalar parameters include SWR, attenuation, and power. A *vector analyzer* measures the magnitude of two parameters (typically voltages) and the phase difference between them. That combination creates a vector with a magnitude and angle. (See the **Radio Mathematics** online supplement for more information about vectors, scalars, and phasors.) The measurements are repeated over a range of frequencies.

25.9.4 Antenna Analyzers

Because measurements of antenna systems, including feed lines, are so common, a class of analyzers has been designed with features specifically for that task. The popular name for these instruments is *antenna analyzers*, whether they use analog or digital circuitry.

In many products available today, the measurements and graphs are displayed on a built-in screen about the size of those on mobile phones. The real-time graphing functions are especially helpful when making tuning adjustments in the field. Data is stored internally and can be downloaded to an external PC for processing and storage. (See the sidebar on Touchstone data files later in this section.)

Antenna analyzers suitable for amateur use are available from a number of manufacturers. Search for "antenna analyzers" on the internet to find them. When shopping for an antenna analyzer, pay careful attention to the capabilities and limitations. Several basic designs are available with tradeoffs between performance and cost.

Be aware that some inexpensive or older units measure only the SWR or impedance magnitude while others measure both the resistive and reactive parts of the impedance. Some of these units may give the magnitude of the reactance but not the sign, requiring the operator to change frequency a small amount and watch the change in impedance magnitude to determine the sign and thus the type of the impedance, inductive or capacitive.

These analyzers have many more uses than impedance and SWR measurements. Paul Wade, W1GHZ, has contributed the paper "Antenna Analyzer Pet Tricks" in this book's online information. It includes how-to guides for a number of useful antenna system measurements.

Some users have reported difficulties in obtaining accurate impedance measurements on low-band antennas when there is a nearby AM broadcast station. This is due to the wideband detector responding to the incoming signal from the AM station. Some manufacturers offer external high-pass broadcast-reject filters to allow the analyzers to be used in the presence of these strong signals. The filter can affect measurements near the broadcast band, particularly in the 3.5 MHz and lower-frequency bands. Check with the manufacturer about the limitations of using filters with the analyzer.

SCALAR ANTENNA ANALYZERS

The simplest way to measure the frequency response of a coax-fed antenna to show its resonant frequency or frequency of minimum SWR is to use a standing wave ratio (SWR) meter. By measuring SWR, the meter shows how well-matched the load or the antenna's feed point impedance is to the characteristic impedance of the feed line. The frequency of the measurement is controlled by changing the frequency of the transmitter. Because SWR is a single numeric value, the combination of a transmitter and SWR meter creates a simple scalar antenna analyzer.

SWR meters are usually calibrated for 50 Ω since most transmitters are designed to drive a 50 Ω load. (Some can be switched to operate in a 75 Ω system.) An SWR meter by itself, however, cannot provide any information about the values of resistance and reactance that make up the impedance.

The first self-contained antenna analyzers replaced the transmitter with a manually tuned low-power signal source and added a detector and analog meter in a single piece of portable equipment. These simple scalar analyzers display SWR or impedance magnitude, but not separate values of resistance and reactance. While they have been made obsolete by the more advanced equipment available today, they are still quite suitable for simple adjustments and troubleshooting antennas.

Today's scalar antenna analyzers add more signal processing capabilities to measure both resistance and reactance, along with SWR. Some units have a manually tuned signal source while others make swept frequency measurements under the control of a microprocessor. **Figure 25.31** shows the block diagram of a typical scalar antenna analyzer.

Figure 25.32 shows a relatively simple circuit that uses diodes for detecting voltages corresponding to voltage and current at the external load. This inexpensive circuit is useful but at low signal levels the diodes introduce some non-linearity and temperature drift. This type of diode detector responds to signals over a wide frequency range and there may be stray pickup from nearby broadcasting stations that makes the measurement results inaccurate. For a more detailed analysis of the antenna system, an instrument with a narrow-band detector such as for a radio receiver gives much better performance than the broadband diode detector.

The rectified voltages can be digitized by a microprocessor and the results displayed numerically. The signal source is typically a varactor-tuned LC oscillator in the analog units or a direct digital synthesizer (DDS) in more sophisticated models. The DDS signal source is very stable since it is controlled by a crystal oscillator and it can be set to the desired frequency quickly with a keyboard entry. The DDS version costs somewhat more but compared to the varactor-tuned oscillator, it has significant performance and operating conveniences. (See the **Oscillators and Synthesizers** chapter for information on DDS signal sources.)

VECTOR ANTENNA ANALYZERS

Today, the most common antenna analyzer is a *vector impedance analyzer* (VIA) that measures both the magnitude and the phase angle of impedance at a single-port across a range of frequencies. The impedance is then displayed numerically (R + *j*X), as a phasor (|Z|∠Θ), or some computed value, such as SWR. (Magnitude is denoted by vertical bars, | |, and the angle by ∠.) The measurements can also be displayed on a graph and many instruments can transfer the measurements to a computer for further processing and storage. The VIA can measure the impedance of a single-port network such as an antenna, the input port of an antenna tuner, or a two-terminal component like a resistor or capacitor.

An older analog instrument called a *vector impedance meter* made a single-frequency measurement of complex impedance and displayed resistance and reactance on a pair of analog meters. The HP4800 series of instruments was a typical example.

The impedance data can be used to calculate several parameters for the antenna system. Using the specified value for the system reference impedance, which can be any value (it doesn't have to be 50 Ω), the reflection coefficient (ρ or rho) can be calculated as:

$$\rho = (Z_L - Z_0) / (Z_L + Z_0)$$

where Z_L is the measured impedance of the load and Z_0 is the specified impedance of the transmission line, which can be any value. Z_L is a complex number; therefore, ρ is, in

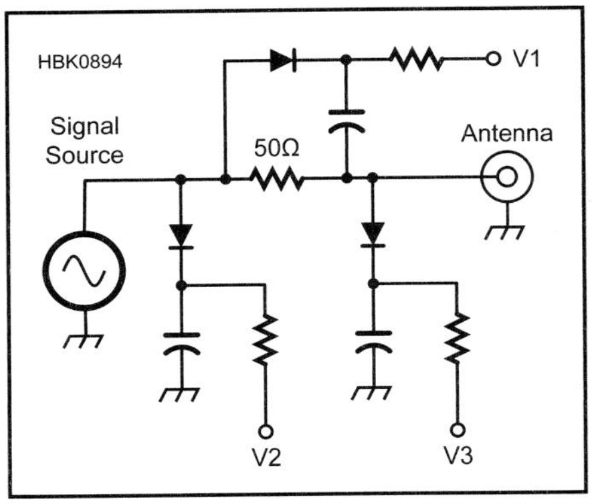

Figure 25.32 — A relatively simple diode detector circuit for measuring impedance that gives adequate results for many applications.

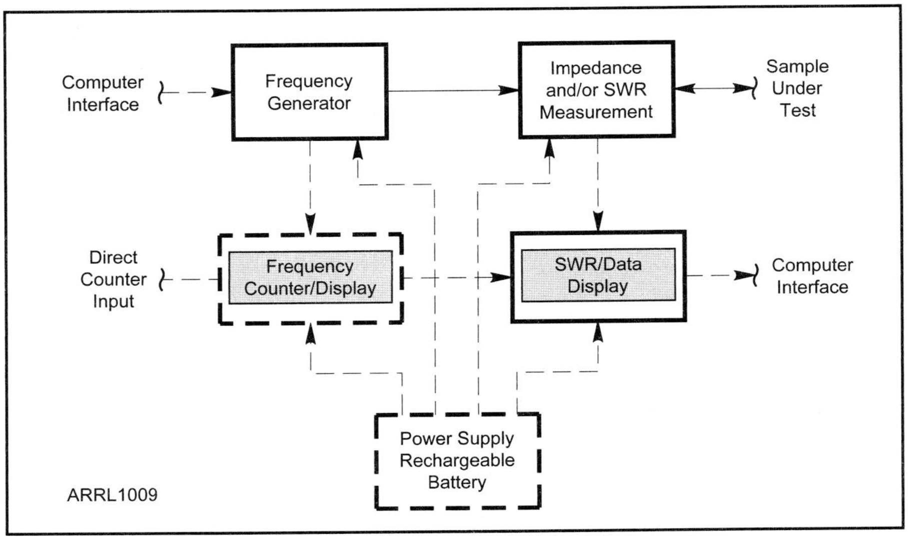

Figure 25.31 — Block diagram showing the elements of a scalar antenna analyzer.

general, a complex number with a magnitude between zero and one. Rho is approximately equal to zero when the line is matched to the antenna because there is no reflection, all the transmitter power is absorbed by the antenna. When the antenna is poorly matched to the line, ρ is larger and it approaches 1 when the mismatch is large.

$$SWR = (1 + |\rho|) / (1 - |\rho|)$$

Note that SWR only depends on the magnitude of ρ denoted by the vertical bars as $|\rho|$ so it is not a complex number.

25.9.5 Network Analyzers

A *scalar network analyzer* (SNA) is an instrument that can measure, as a function of frequency, the magnitude of the gain or loss between the two ports of a two-port device as well as the magnitude of the return loss (equivalent to the SWR) of each port. (Because this is a scalar analyzer, only magnitude is measured.) In **Figure 25.33A**, the signal labeled "Ref" is proportional to the sweep oscillator output and the one labeled "A" is proportional to the signal that is reflected back from the device under test due to an imperfect 50 Ω match. The ratio of A to Ref can be used to calculate the return loss and SWR of the device. Signal "B" is proportional to the signal that travels between the two ports of the device. The ratio of B to Ref gives the device gain. To find the gain in the opposite direction and the return loss of the other port, simply turn the device around and swap the two ports. Some network analyzers come with a *test set* or include RF switches to do that automatically.

The box in the figure labeled "Signal processing and display" includes circuitry to take the logarithm of the signals so they can be displayed in dB. In modern instruments it normally includes a microprocessor, which does much of the signal processing as well as controlling the sweep oscillator and other circuitry.

A *vector network analyzer* (VNA) is an instrument that can measure both the magnitude and phase of the gain and return loss of a two-port network. The block diagram in Figure 25.33B shows an example of a typical lab instrument VNA. The VNA's signal source has a 50 Ω impedance that drives the network port and detector circuits.

The generator can drive P1 as the input and measure P2 as the output or the generator output switch can reverse the signal flow. The first configuration (P1 as input) allows the VNA to measure S_{11} and S_{21}. The signal flow is reversed to measure S_{12} and S_{22}. The splitters also supply the signal generator output to the detector circuits.

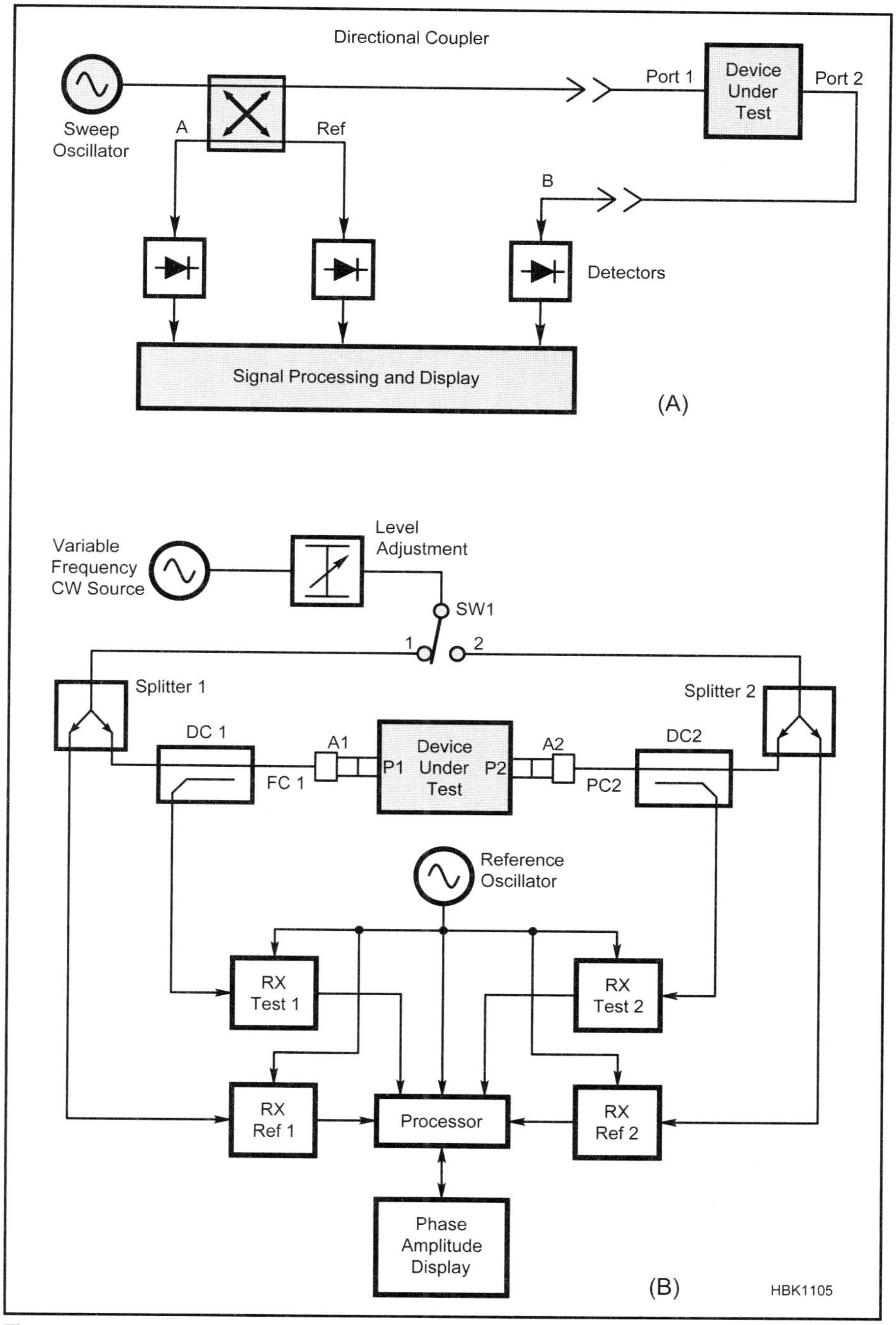

Figure 25.33 — Block diagram showing the elements of a typical scalar (A) and vector (B) antenna analyzers. Block diagram (B) was created by Chris Angove from original research and study (From English Wikipedia, CC BY-SA 3.0, commons.wikimedia.org/w/index.php?curid1703095)

Touchstone Data Files

The most useful analyzers can store sets of measurements in a file for review later and transfer to a PC for further processing and display. The standard format for swept impedance data (frequency and impedance) is called *Touchstone* (**en.wikipedia.org/wiki/Touchstone_file**). Files in the Touchstone format can be read and processed by a wide variety of software, including as an input to design or simulation software.

Touchstone files are plain text with a line for each data point. A single-line header defines the sub-format of the data. The sub-format most widely supported by hardware and software for impedance data is s1p in which the parameter is complex S11, and each line consists of Frequency, Real S11, and Imaginary S11, with spaces as the separator.

The detectors mix the reference signals Ref1 and Ref2 with each directional coupler output and the input signal. The mixer/detector outputs and the reference signals are all fed to a processor circuit that measures the magnitude and phase of each signal. The measured data are then converted to S parameters as explained previously. From that data, other parameters, such as impedance, can be calculated.

Usually, the VNA or its host software provides the capability to plot the S parameters on a Smith chart for easy evaluation. S_{11} and S_{21} are often available in a magnitude versus frequency graph, as well.

A VNA is similar to a VIA, but it has circuitry to measure two ports and all S parameters, instead of just S11. It can also compute complex impedance, including the sign of the reactance, to tell whether the response is inductive or capacitive. The VNA can be configured to do single port swept frequency measurements, to obtain the same results as the VIA.

Many VNAs have a low-frequency limit in the range of 100 kHz to 1 MHz. This limit is imposed by the variable frequency oscillator of the analyzer. The range of the VNA can be extended lower to the audio frequency range by using an external adapter that shifts the output RF signal from the VNA, applies it to the circuit under test, and then re-shifts the circuit's output signal into the VNA's RF range. A low frequency adapter project is described later in the projects section at the end of this chapter.

DUAL-DDS VNA DESIGN

Figure 25.34 shows an alternative VNA design that uses two DDS signal sources with one applied to the antenna system and the other used as a reference. Typically, the clock oscillator is crystal-controlled for high accuracy. One DDS is programmed to output a signal at the actual test frequency and the other DDS is programmed to a slightly higher frequency in the 1 to 10 kHz range, shown in the figure as a 2 kHz offset. See the reference entries for articles by Bob Clunn, W5BIG, Thomas Baier, DG8SAQ, and Michael Knitter, DG5MK, giving a detailed design description of this type of analyzer.

The signals are then mixed to produce a low frequency output that can easily digitized by an inexpensive ADC. The measurements are then processed mathematically in a microprocessor or PC to yield full information about the complex impedance being measured. This type of analyzer measures both the magnitude and phase of a test impedance over a wide range of frequencies.

One signal corresponds to the voltage applied to the test port (V1) and the other signal voltage is derived from the current flowing in the test port (V2). The ratio of these signals (V1 / V2) is the impedance at the port and the difference of their phases is the phase angle of the impedance:

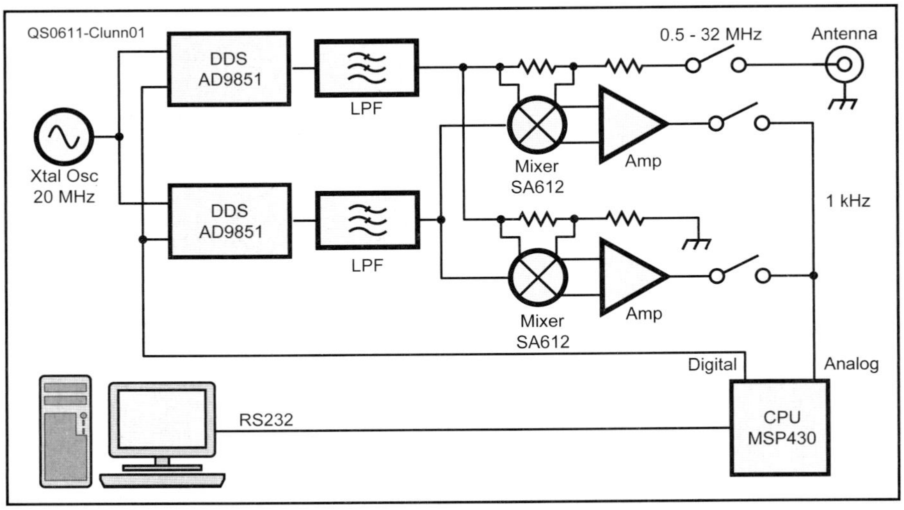

Figure 25.34 — Block diagram of a dual DDS-based antenna analyzer. By mixing two signal sources that are very close in frequency, with one of the sources applied to the load (antenna), low-frequency signals are generated. The signals contain the necessary information to measure impedance, magnitude, and phase.

Used Network Analyzers

Used lab-quality network analyzers, both scalar and vector, are readily available on the surplus market. Some require an external sweep oscillator while others include an internal signal source. The test sets may be internal or external as well. These instruments are generally well designed and offer excellent quality measurements, but tends to be typically bulky and difficult to repair. The high quality of analyzers available at low prices today makes the older units much less desirable unless needed for a laboratory setup or to replace a specific instrument.

Choosing A Computer Controlled Analyzer

The most capable and useful analyzers available at moderate cost are controlled by a computer. They convert the measurement to baseband audio, which is fed to the computer via an integrated USB interface. Software on the computer controls the analyzer, performs computations to turn the audio into a measured result, and plots the results. These are things to look for:

- It should be a precision instrument, solidly built and with a high-quality bridge, so that it can be reliably calibrated.
- A frequency range of 100 kHz to 600 MHz is a good starting point; 10 kHz to 1.5 GHz is even better.
- The dynamic range of the instrument, including the USB interface, places an upper limit on the range of impedance values (or gain/loss response) that can be displayed in a single measurement. A 12-bit interface (about 88 dB dynamic range) is sufficient for most amateur measurements with a 16-bit interface (about 90 dB dynamic range) desirable for precision work.
- It should be able to make and display measurements at both linear and logarithmic scales; linear sweeps are best for narrow sweeps (a ham band); log sweeps are important for wide sweeps (showing a filter's out-of-band rejection or a choke's impedance and over a wide range).
- An analyzer that can be powered from the USB port simplifies setup both in the field and in the station.

The analyzer's software should include:

- The capability to store and recall calibration data for multiple measurement fixtures or configurations.
- Support the use of correction factors to account for various test fixtures and calibration standards results in greater accuracy for these special setups.
- Perform Time Delay Reflectometry (TDR) from a wide-frequency sweep of a transmission line's S11 (Return Loss). The TDR function should find the line's electrical length and any discontinuities (splices or faults). Cursors should show position in feet or meters with multiple windowing functions for the most useful view of the response.

Magnitude (Z) = Magnitude (V1) / Magnitude (V2)

Phase (Z) = Phase (V1) – Phase (V2)

The frequency is swept in small steps across the specified range and measurements of V1 and V2 are made at each frequency.

25.9.6 Calibration and Measurement Plane

The accuracy of a VNA depends on the performance of the directional couplers, detectors, and other circuitry. The coupler directivity is a key contributor to measurement errors. For example, if the directivity is 20 dB, then even a perfect 50 Ω load shows 20 dB of return loss, equivalent to a 1.22:1 SWR. Fortunately, it is possible to eliminate most errors by measuring several terminations of known impedance and calculating a table of correction factors that are applied to the raw measurements. The mathematics involved is quite complex, but it is all handled by the instrument software and is invisible to the user. A thorough discussion of VNA error-correction techniques can be found in the Keysight application note, "Applying Error Correction to Vector Network Analyzer Measurements" (**www.keysight.com/us/en/assets/7018-06761/application-notes/5965-7709.pdf**)

The point at which the measurements are to be performed is called the *measurement plane* or *reference plane*. That point can be at the measurement connector of a VNA, at a test fixture connected to the analyzer, or at the ends of cables. (The end of the cable can even be a long distance from the analyzer at the top of a tower!) Calibration consists of taking measurements of known impedances. The analyzer makes the measurements across the desired frequency range and then applies mathematical corrections at each frequency so that the net result is the exact known values. This set of measurements is saved as a *calibration file* that can be stored and retrieved whenever that measurement plane is used in the future. Most analyzers can save multiple configuration files for different measurements. This process is very quick and easy and the final results of any measurement are much more accurate than from a simple analyzer that does not have a calibration procedure.

One of the sources of error removed by the calibration is any error due to the length of the coaxial cables that connect to the device under test. For that reason, the measurement plane should always be at the end of the cables that are to be used for the test.

Calibration includes the effects of length for any coax between the analyzer and the measurement plane; substituting a different piece of coax requires another calibration. Adding a connector adapter to connect an antenna to the plane that has been calibrated introduces an error that is small enough not to matter for most antenna measurements, but can be significant on the higher HF bands, and much more so at VHF and UHF.

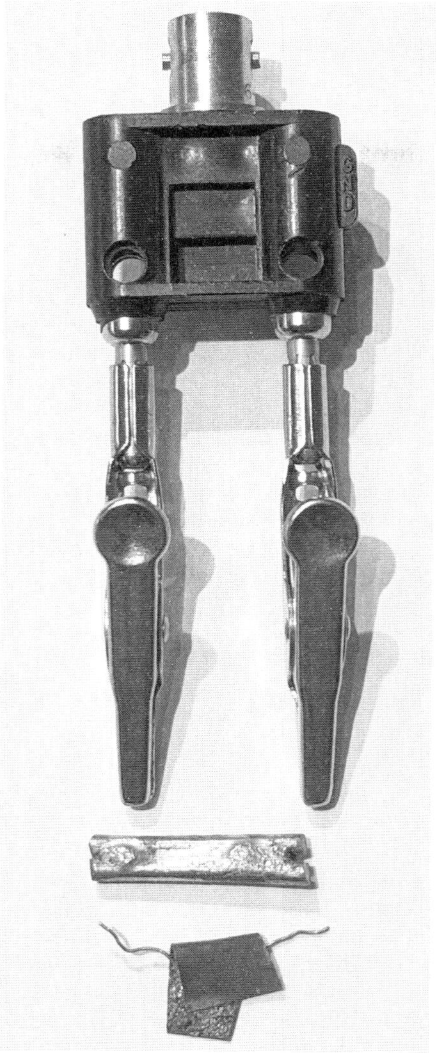

Figure 25.35 — A simple vector impedance analyzer (VIA) test fixture for measuring component impedance into the VHF range. The fixture is a BNC-to-banana plug adapter with alligator clips on the plugs. Below the clips are the short-circuit calibration load and a non-inductive 50 Ω resistor to perform the SOL calibration.

Calibration for S_{21} and S_{12} adds additional measurements with the analyzer output and inputs connected together with cables of the same length (preferably the same cables) for connecting the VNA output to the Device Under Test (DUT) and the output of the DUT to the VNA input, with the two cables mated through a double female adapter barrel adapter.

For antenna measurements, the measurement plane might have to move when one or more connector adapters are attached to the same cable. Since the analyzer is then connected at a different spot on the line and has different geometry, it needs its own calibration.

SOLT CALIBRATION

The most popular technique is called a Short-Open-Load-Through (SOLT) calibration because it uses four standards, a short-circuit (0 Ω), an open-circuit (∞ Ω), a 50 Ω load, and a through (direct) connection between the two ports. If the analyzer is a single-port VIA, there is no through connection and only the SOL calibration is performed. A set of standards with those three impedances over a very wide frequency range are provided with the analyzer or available as a calibration test kit.

Figure 25.35 shows a simple component test fixture consisting of a BNC-to-banana plug adapter with alligator clips slipped on to the plugs. Under the clips is shown the Short test load (#10 bare copper wire folded back on itself and coated with solder to minimize both resistance and inductance) and a large-format 1% 49.9-Ω chip resistor with short leads soldered to it of the minimum length that allows them to attach to the clips. When calibrated, this fixture can make reliable measurements well into the VHF range. Calibration accuracy is then verified to the highest frequency of interest by measuring high value chip resistors (5–10 kΩ), looking primarily for the effects of parallel capacitance or series inductance. For this test fixture, the measurement plane is at the tips of the alligator clips. The simple fixture like that in the figure is typically used at the end of a convenient length of coaxial cable (RG-400 is a good choice) for S_{11} measurement of chokes and other components.

Most antenna analyzers have a single-end-

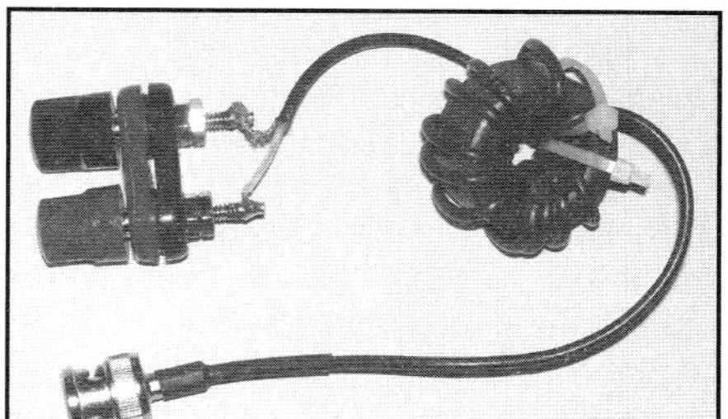

Figure 25.36 — A common-mode choke near the test fixture should be used when making measurements through a length of coaxial cable or measuring two-wire transmission line impedances. This choke is made of 12 turns of RG-193 miniature coax through a 2.4-inch OD, Type 31 core (Fair-Rite 2631803802) and is effective in the upper HF range. The best choke should present at least 5,000 Ω of resistive impedance at the measurement frequency. See the choke section of the Transmission Lines chapter for more information about this type of choke.

ed (unbalanced) output with a coax connector. For making measurements on a two-wire transmission line, such as window line or ladder line, a common-mode choke can be added between the analyzer's RF connector and the input to the transmission line as in **Figure 25.36**. The calibration loads are attached to the output side of this balun.

This calibration procedure can be extended to allow measuring the actual driving point impedance of the antenna. The transmission line is disconnected at the antenna and the calibration loads are attached at the far end instead of the antenna. This shifts the measurement plane to the antenna itself which is handy when designing matching networks.

25.9.7 Analyzer Measurements

Firmware built into vector analyzers, or software running on an associated computer, can display impedances computed from S_{11} in many forms. Some of the more common are: 1) Smith Chart; 2) Series Resistance (R_S) and Reactance (X_S); 3) Parallel Resistance (R_P) and Reactance (X_P); the magnitude of the Impedance (Z_{MAG} or $|Z|$) 4) Inductance (L) and parallel capacitance (C_P); 5) Capacitance (C) and series resistance (R_S). Depending on the firmware or software, the frequency sweep can be made and data presented linearly or logarithmically. (See the section Measuring Components With an Analyzer later in this chapter.)

Using built-in firmware or associated computer software, most modern analyzers can perform TDR to determine the length of a transmission line, to look for splices or faults along the line, even find the impedance of each of multiple sections connected in series.

Most modern vector analyzers and/or associated computer software can export a measurement for use by design and analysis software like *SimSmith* (**ae6ty.com/smith_charts.html**) and *ZPlots* (**ac6la.com/zplots1.html**) to study feed line loss and design matching networks. (See the sidebar on Touchstone format) These two freeware packages can transform an impedance measurement made anywhere along a feed line to the impedance at the antenna feed point if the length and parameters of the feed line are known. *ZPlots* also offers some design capability for common amateur applications. *SimSmith* allows the design of far more complex two-port networks and transmission line systems.

25.10 Testing Digital Modulation

As digital modulation modes become more and more important in amateur radio, it is increasingly important to have ways of testing the performance. There are dozens of different digital formats in use, from traditional radio teleprinting (RTTY) using frequency-shift keying (FSK) to the latest systems that employ sophisticated error detection and correction along with various modulation types that pack multiple bits into each symbol. Despite the wide differences in modulation and coding, nearly all have in common a relatively narrow bandwidth suitable for use with a voice transceiver using SSB or FM modulation.

If you are having trouble with reception or transmission of digital signals using a PC sound card, one straightforward troubleshooting technique is to install the software on two computers and see if you can transmit data from one computer to the other by connecting the sound card output of one to the input of the other and vice versa. If you do not get perfect reception, that indicates a problem with the computer software or hardware.

The next step is to transmit into a dummy load and receive the signal with a separate receiver located nearby so it picks up the stray radiation from the dummy load. A piece of wire plugged into the receiver antenna connector can be moved around to adjust the signal level. Many software programs for receiving digital signals include a spectrum display, which can indicate faults in the transmitted signal such as distortion and *skew*, the amplitude imbalance among the tones of a multi-tone modulation signal. To see what the signal is supposed to look like, connect the two computers directly, as previously described. Then when you examine the RF signal transmitted into the dummy load, any additional bandwidth due to distortion or skew in the spectrum shape should be apparent.

If the demodulation software does not include a spectrum display, there are separate programs available that can display the spectrum of the signal at the sound-card input, as discussed in the Spectrum Analyzer section of this chapter. An RF spectrum analyzer measuring the RF output signal directly gives an even better idea of modulation quality because it is not affected by the filters and other circuitry of the receiver. A receiver panadapter, as described earlier, is a less-expensive substitute.

Comprehensive testing of a digital communications system is quite complicated because of all the variables involved. The *bit error ratio* (BER) is the number of single-bit errors divided by the number of bits sent in a certain time interval. It requires special test equipment to measure because the individual bits are typically decoded deep inside the demodulation software where they are difficult to access. The *packet error ratio* (PER) is easier to measure. In a packetized data system, it is the number of incorrect packets divided by the number of packets sent. It can be measured either before or after error correction. In a non-packetized system like PSK31 the character error ratio is a useful figure of merit. BER is affected by the signal-to-noise ratio (SNR), interference, distortion, synchronization errors and multipath fading. PER is further affected by the effectiveness of the coding and error correction of the particular digital mode used.

It is interesting to measure BER or PER as a function of the SNR. For some digital systems with lots of error correction the errors are nearly zero down to a certain signal level and then degrade very sharply below that. However, in real-world operation the SNR is almost never constant. The signal is constantly changing, both in amplitude and phase, as propagation changes due to movement of the ionosphere (on HF) or of the vehicle (VHF and above), as explained in the **Propagation of Radio Signals** chapter. Measuring actual on-the-air performance is not a good way to compare systems because propagation varies so much at different times. For a repeatable test, you need a *channel simulator,* which is a device that intentionally degrades a test signal in a precise way as to simulate an over-the-air radio channel. Moe Wheatley, AE4JY offers *PathSim*, a free software HF channel simulator. (**moetronix.com/ae4jy**) A hardware HF channel simulator has been described by Johann B Forrer, KC7WW. (See the Reference for Forrer.)

25.11 Software-Based Instruments

Most amateurs these days own a personal computer with a powerful microprocessor, lot of memory and mass data storage, a large color display, and a sound card with stereo high-bandwidth audio input/output. It doesn't take a great deal of imagination to realize that these resources can be harnessed to make low-cost measuring instruments of various types.

Several manufacturers offer hybrid RF instruments that combine a computer with a separate signal interface module to create oscilloscopes, spectrum analyzers, and other types of instruments. The connection between the module and the computer is usually a USB 2.0 or 3.0 interface.

Audio-frequency instruments can be implemented directly using the computer sound card, which typically has a frequency response from perhaps 50 Hz up to about 20 kHz. While computer sound cards can be quite useful for measurement, the low-cost sound cards built into computers may not have good noise performance. Laptops may not offer a line-level input, only the more sensitive microphone input which is more susceptible to overload. Sound card outputs may distort at levels near their rated output. It is good practice to keep their output level at least 6 dB below rated output to minimize distortion on signal peaks. Both input and output signal levels should also be set so that the lowest amplitude components are at least 10 dB above the noise floor. However, many modern laptops do not offer any interface connectors that you can use to process signals. Rather, they use wireless techniques, such as WiFi or Bluetooth, to interface with those signals. Access to old laptops that are no longer powerful enough for much of today's software could give you all the interface connectors that you need, enough calculating power to perform your measurement calculations, and a lower cost.

Inexpensive standalone sound cards with a USB interface can offer better quality than built-in motherboard sound card interfaces. These work well in applications such as computer-to-radio interfaces for digital modes. Because these products are made for a mass market, models tend to be updated every year or two, but the quality brands such as Numark and Tascam tend to offer stable designs.

Free software is available on the internet for instruments such as audio function generators, DTMF and CTCSS tone generators, DTMF and CTCSS decoders, two-tone generators for SSB transmitter testing, distortion/SINAD analyzers, oscilloscopes, and audio spectrum analyzers. In addition to the frequency-response limitations of a typical sound card, another issue is that the device can be damaged by applying excessive voltage to the inputs or outputs. It is wise to add external buffer amplifiers that include overvoltage protection.

Radio-frequency test equipment can also use the sound card inputs by means of some type of frequency converter, consisting of a local oscillator and mixer. If the mixer is a quadrature type, the two outputs may be fed to the stereo sound card inputs so that software can treat the left and right channels as the in-phase and quadrature signals. A common application is a narrowband RF spectrum analyzer. The RF bandwidth is typically limited to twice the sound card's audio bandwidth.

One problem with using a sound card is that the low-level signals may be susceptible to picking up low frequency hum and buzz, noise, and radio-frequency interference (RFI). If the computer and the device under test are grounded separately to the ac power system, hum, buzz, and other noise can be generated from currents flowing in the ground connection between the two. Similarly, the necessary loops formed by the shields of multiple cables can pick up 50 or 60 Hz magnetic fields, generating a hum voltage. It is helpful to bond the sound card and PC to whatever device is being tested by using a short, low-resistance connection. It is also possible to use isolation transformers or differential amplifiers to isolate the grounds and thus break the ground loop. Good quality cables and attention to proper shielding and bonding help prevent hum and noise pickup.

In general, it is always good practice to bond together the chassis or shielding enclosure of every piece of equipment in a system with short, heavy copper wires (#14 AWG or larger). Failure to do so often results in hum, buzz, and RFI being introduced into the signals and data. For the same reasons, all interconnected equipment should be powered from the same ac outlet or from outlets that share the same 'green wire' (that is, they plug into the same multiple wall outlet box). If the equipment is powered from different outlets, the green wires of those outlet boxes should be bonded together. Good quality cables and attention to proper shielding help prevent hum, buzz, noise, and RFI. (See the discussions of bonding in the **Safe Practices** and **RFI and EMC** chapters, as well as the ARRL book *Grounding and Bonding for the Radio Amateur*.)

25.12 RF and Microwave Test Accessories

This section assumes that all coaxial devices are designed for 50 Ω characteristic impedance. While that is by far the most common value for amateur radio systems, most of the information applies as well for other impedances, such as 75 Ω which is common in the video and television industry.

25.12.1 Attenuators and Terminations

Most amateurs are familiar with the concept of a *dummy load*, which is nothing more than a high-power 50 Ω resistor with a coaxial connector. The term comes from the fact that it is often used as a dummy antenna to provide a low-SWR load for transmitter testing without radiating a signal. The resistor must have low stray reactance throughout the desired frequency range. Common wire-wound power resistors are not suitable for a dummy load because they have too much inductance. The most effective resistors are those made from carbon or silicon. Commercial loads, such as the power attenuator in **Figure 25.37**, use special resistors encased in a finned enclosure for dissipating the heat generated by the power absorbed from the transmitter. One technique used in some amateur dummy loads is to enhance the power dissipation capability of less-expensive, lower-power resistors by submerging them in a bath of oil.

Low-power dummy loads, normally called *terminations*, are used whenever a device being tested needs to have one of its coaxial connectors terminated in a load that has low SWR over a wide range of frequencies. (This is often specified as *return loss* (RL), which is similar to SWR as described in the **RF Techniques** chapter.) A *feed-through termination*, such as the one in **Figure 25.38**, is one that has two coaxial connectors with a straight-through connection between them as well as a 50 Ω load resistor connected between the center conductors and the shield. It is used to provide a 50 Ω load when using a measuring device with a high-impedance input, such as an oscilloscope. To avoid standing waves on the coaxial transmission line connected to the device under test, it is important to place the feed-through termination directly at the oscilloscope's input connector so that the transmission line is properly terminated.

Another technique to maintain a low SWR is to use a T connector at the oscilloscope input. The device under test is connected to one side of the T and some other device that provides a good 50 Ω load is connected to the other side. The high-impedance oscilloscope input does not excessively disturb the 50 Ω system. This is a good way to "tap into" a signal traveling between two devices on a coaxial line while maintaining the connection between the devices.

A *fixed attenuator* is useful both for reducing a signal level as well as for improving the 50 Ω match. For example, a 10 dB attenuator guarantees a minimum of 20 dB of return loss (1.22:1 SWR) even if the load SWR is infinite. A 20 dB attenuator makes a high-quality 50 Ω termination, even with nothing connected to its output. Attenuators are rated for the amount of power that they can dissipate and it is important to make sure that the attenuator you are using can handle the power being transmitted through it. It is not necessary for all attenuators in a system to have the same power rating. After the first, high power, attenuator decreases the RF power, attenuators connected after it only need to handle reduced power. For example, if 30 dB of attenuation is needed from a 100 W transmitter, a 100 W 20 dB attenuator decreases the power to 1 W and a 10 dB 1W attenuator can be put in series to realize the full 30 dB attenuation.

A *step attenuator* is used when you need to adjust the signal level in fixed steps. The term "10 dB step attenuator" means that the step size is 10 dB. The maximum attenuation would typically be perhaps 70 to 120 dB. An attenuator with 1 dB steps and 10 dB of maximum attenuation would be called a "1 dB step attenuator" or a "0 to 10 dB step attenuator." Two attenuators, one with 10 dB steps and one with 1 dB steps, can be connected in series to obtain 1 dB resolution over a very wide range of attenuation. At microwave frequencies, continuously variable attenuators are available for waveguide transmission lines. They can sometimes be obtained used at reasonable prices.

25.12.2 Connectors and Adapters

Performing accurate and repeatable tests and measurements demands good quality in every aspect of the test process. This includes coaxial cables and connectors. Test cables should use very high-quality coaxial

RF Measurement Test Set

Many low-level RF measurements require similar test setups so why not combine them into one package? This also makes it easier to perform the tests and get consistent results. N0ADL and KE4PT show an example in the *QEX* article "Apparatus for RF Measurements" that you can download from the online information.

Figure 25.37 — A 100-watt power attenuator rated for dc-3000 MHz. The transmitter is connected to the male type-N connector. The female connector is for the attenuated connection.

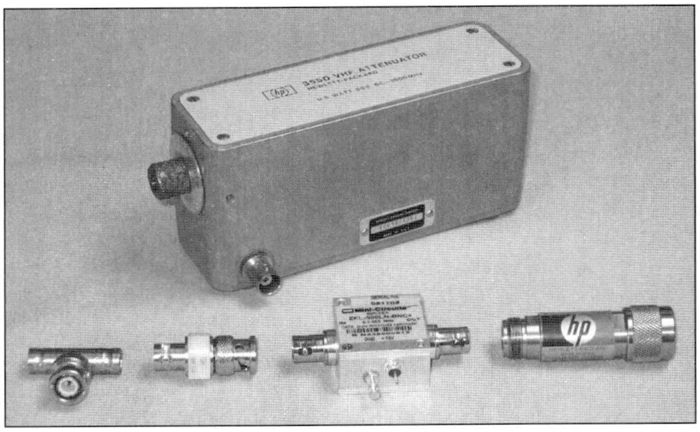

Figure 25.38 — Various RF test accessories. At the rear is a 10 dB step attenuator with a 0-120 dB range. In the foreground from left to right are a BNC "T" connector, a 50 Ω feed-through termination, a 0.1-500 MHz low-noise amplifier, and a 10 dB 2W fixed.

cable, such as RG-400 or the equivalent, particularly if measurements are taken above the HF range. To make really accurate measurements of signal levels, the loss of the coaxial cables used in the measurement needs to be accounted for, as well.

The correct coaxial connector should be used in each situation. For instance, larger cables, such as RG-213, cannot be used with small connectors, such as BNC or SMA, and require larger connectors such as UHF (PL-259) or N. Like coaxial cable, connectors have characteristic impedances. For instance, the BNC connector that is made to be used with RG-58 cable has a 50Ω characteristic impedance and the BNC connector for RG-59 cable has a 75Ω characteristic impedance. Those two connectors look very much alike and can be physically interchanged for each other, but using the wrong one increases the SWR for each incorrect connector in the feedline path.

Various connectors have maximum frequencies at which they can be used. The PL-259 (incorrectly named the "UHF connector") should not be used above 300 MHz. For larger diameter cables that must be used above that frequency, the N connector should be used, which has a range up to 18 GHz. The BNC connector is the most common connector used on medium diameter coaxial cable and its maximum frequency is 4 GHz. The SMA connector is used with small diameter cables and has a frequency range up to 24 GHz. The F connector is commonly found in cable television systems and is not usually used by radio amateurs because it is a 75Ω connector. Other connectors are less common and tend to be used at even higher frequencies.

Coaxial connectors and adapters can be a source of problems if poorly made or worn. It is worth the extra expense to obtain a few high-quality connectors and adapters for making measurements. The manufacturer should be clearly identified on the body of the connector. Silver-plating is desirable and may be required for measurements at extremely low signal levels on in the UHF and microwave range.

Coaxial cables tend to be made with male connectors on both ends. To lengthen a feedline path, it is sometimes necessary to use a double-female connector, often called a barrel, in between. Like other connectors, barrels are made with characteristic impedances so you should make sure that the barrel you are using has the same impedance as the rest of your feed system.

Cables, connectors, and adapters can be tested by using an antenna or network analyzer. With the device being tested connected to the analyzer, move the connector around and bend cables, paying particular attention to where the connector is attached to the cable. Watch the analyzer display for any intermittent changes in behavior that indicate a possible poor or loose connection. Check the response of adapters at the highest frequencies for which they are expected to be used.

25.12.3 Noise Sources

Most of the time, noise is something to be avoided in electronic circuits. It can be quite useful for testing, however, precisely because the calibrated noise that you create on purpose has the same properties as the unwanted circuit noise that you are trying to minimize. Thus, calibrated sources of noise are used as test instruments for various measurements. (See the **RF Techniques** chapter for a discussion of noise and its associated terminology.) The noise source can be combined with a bridge circuit, creating a *noise bridge,* which is described in this chapter's section on RF Impedance Measurement. The wideband noise from a noise source can also be used as a test signal to measure frequency response of a filter or amplifier with a spectrum analyzer or tunable receiver.

25.12.4 Dip Meters

This instrument was originally a grid-dip oscillator (GDO), so-called because the indicating "dip" was in the grid current of its vacuum-tube oscillator. Most dip meters today are solid-state but the principle is the same: If you hold the external oscillator coil near a tuned circuit and adjust the dip meter tuning dial to the frequency of the tuned circuit, there is a dip in the meter reading as the resonant circuits interact with each other. The coil extends from the end of the instrument and a set of plug-in coils is provided to cover the unit's frequency range.

Most dip meters can also serve as absorption wavemeters by turning off the oscillator and looking for a peak instead of a dip in the meter reading. Sometimes frequencies can be detected in this way that would be difficult to read on a frequency counter because of the presence of harmonics. Further, some dip meters have a connection for headphones. The operator can usually hear signals that do not register on the meter.

A dip meter may be coupled to a circuit either inductively or capacitively. Inductive coupling results from the magnetic field generated by current flow. Therefore, inductive coupling should be used when a coil or a conductor with relatively high current is convenient. Maximum inductive coupling results when the axis of the pick-up coil is placed perpendicular to the current path and the coil is adjacent to the wire. To avoid detuning the circuit being tested, always use the minimum coupling that yields a noticeable indication.

High-impedance circuits have high voltage and low current. Use capacitive coupling when a point of relatively high voltage is convenient. An example might be the output of a 12 V powered RF amplifier. (For safety's sake, *do not* attempt dip-meter measurements on true high-voltage equipment such as vacuum-tube amplifiers or switching power supplies while they are energized.) Capacitive coupling is strongest when the end of the pick-up coil is near a point of high impedance. In either case, the circuit under test is affected by the presence of the dip meter.

To measure resonance, use the following procedure. First, bring the dip meter gradually closer to the circuit while slowly varying the dip meter frequency. When a current dip occurs, hold the meter steady and tune for minimum current. Once the dip is found, move the meter away from the circuit and confirm that the dip comes from the circuit under test (the depth of the dip should decrease with distance from the circuit until the dip is no longer noticeable). Finally, move the meter back toward the circuit until the dip just reappears. Retune the meter for minimum current and read the dip-meter frequency from the dial or with a calibrated receiver or frequency counter.

The current dip of a good measurement is smooth and symmetrical above and below the minimum. An asymmetrical dip indicates that the dip meter oscillator frequency is being significantly influenced by the test circuit, degrading the accuracy of the measurement. Increase the distance between the dip meter and the circuit until a shallow symmetrical dip is obtained.

Along with resonant frequency measurements, the dip meter can also serve as a crude signal generator, capacitor and inductor meter, and antenna and transmission line tester, and other uses. If you are purchasing a dip meter, look for one that is mechanically and electrically stable. On used units, the socket where the coils plug in is a common cause of intermittent operation. The coils should be in good condition. A headphone connection is helpful. Battery operated models are convenient for antenna measurements. Solid state dip meters are widely available. Be sure the plug-in coils are included. A project to build your own dip meter is included in the online material.

25.12.5 Other Test Accessories

Accessories available for waveguides include coax-to-waveguide adapters, detectors, directional couplers, isolators, absorption wavemeters, mixers and terminations of various kinds. Each size of waveguide covers about a 1.5:1 frequency range. Be aware that over the years there have been several systems for assigning letters to the various microwave frequency bands. For example, "X" band in the Keysight Technologies (formerly part of Hewlett-Packard) catalog is 8.2 to 12.4

GHz, but the old US Navy definition was 6.2 to 10.9 GHz, and the ITU assignment for X-band radar is 8.5 to 10.68 GHz. The IEEE standard definition is 8.0 to 12.0 GHz. When buying surplus waveguide accessories be sure they cover the frequency range you need and have flanges compatible with your equipment. See the WR-series of waveguide sizes in the **Transmission Lines** chapter.

Adapters such as those shown in **Figure 25.39** are useful for making connections among various pieces of test equipment.

Figure 25.39 — A sampling of coaxial connector adapters. From the left are: BNC male to type-N male, BNC female to UHF male, BNC female to SMA male, type-N male to SMA female, and type-N female to SMC male. Many other combinations are available.

A *coaxial detector* is just what the name implies, a diode detector in a coaxial package, usually with a 50 Ω RF load impedance. Silicon Schottky diodes are usually used to obtain good sensitivity, although gallium arsenide devices are sometimes employed for the high microwave and millimeter-wave frequencies. Below a certain signal level, typically about –15 dBm, the detected output voltage is proportional to the square of the input voltage, that is, it is proportional to the input power. Sensitivity is typically specified in mV/mW, assuming a high-impedance load for the detected signal.

Wideband amplifiers can often be useful in test systems. They are available with various connector types in packages sized appropriately for the power level. A low-noise amplifier is useful as a preamplifier for a spectrum analyzer, for example. A higher-power amplifier can be used at the output of a signal generator either for receiver dynamic range testing or as an input signal for testing an RF power amplifier.

Directional couplers and bridges are useful network analyzer accessories, as previously described, but can also be used to acquire a small amount of signal for other test purposes. For example, a 30 dB coupler at the output of a 100 W transmitter produces a 100 mW signal that can be fed to a coaxial detector or a spectrum analyzer, while passing the rest of the 100 W signal through to an antenna or dummy load.

A *power divider*, also known as a *splitter*, is a device that divides an input signal equally among its output ports. A *combiner* is basically a power divider hooked up backward; the signals from the input ports are combined at the output port. Some power dividers consist only of a network of resistors. The insertion loss is typically at least 6 dB for 2-port splitters, meaning that a quarter of the power comes out each port and half the power is absorbed in the resistors.

A *hybrid combiner*, such as the *Wilkinson combiner* shown in **Figure 25.40**, includes a transformer as part of the network. It has the advantage that, so long as there is a low SWR match on the output port, there is excellent isolation between the two input ports. That is very useful when combining the signals from two signal generators for high-level dynamic range testing of a receiver. The port-to-port isolation prevents the two signals from combining in the output amplifiers of the signal generators, where they could generate distortion products greater than the ones generated by the receiver that you are trying to measure. The insertion loss is nominally 3 dB.

The traditional Wilkinson combiner has a 2:1 ratio of input to output impedance. The variation in Figure 25.40B uses a transformer with a turns ratio (2:1 impedance ratio) to obtain 50 Ω on all ports.

The Wilkinson combiner can also be used as a power divider. The transmission line version in Figure 25.40C replaces the transformer of Figure 25.40A with a pair of quarter wavelength transmission line transformers (see the **Transmission Lines** chapter) to divide the input signal between the pair of output ports. All three ports must have the same impedance, Z_0. A resistor between the output ports with a value of $2Z_0$ reduces loss and increases isolation. The quarter wavelength sections must have a characteristic impedance of

$$Z_0 \times \frac{\sqrt{2}}{2} = 0.707 Z_0$$

Since most amateur equipment and loads have an input of 50 Ω, using 75 Ω coax (RG-59 or RG-11) results in acceptable performance. If isolation between the output ports is not required and a 35 Ω transmission line can be constructed, the divider in 25.40D can be used. Two pieces of 70 Ω cable in parallel can also be used. RG-11 (75 Ω) can also be used if the resulting slight mismatch is acceptable.

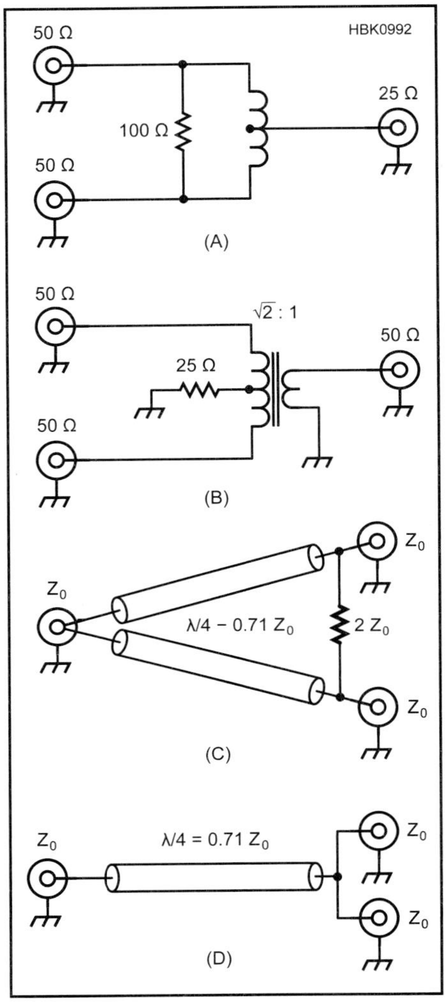

Figure 25.40 — Four forms of the Wilkinson power combiner/splitter. (A) and (B) show a lumped-element implementation. The version at (C) uses a pair of λ/4 transmission lines acting as synchronous transformers. If all three ports are terminated in Z0, the output ports are isolated from each other. If isolation between the output ports is not required, the divider at (D) can be used.

25.13 Making Basic Measurements

25.13.1 Four-wire Measurements

It can be difficult to measure low-value resistors accurately because of the resistance of the meter leads, typically one or two tenths of an ohm. With the series-connected ohmmeter circuit described in the previous section, the calibration procedure compensates for that and some digital meters may have a means to compensate as well. However, accurate low-resistance measurement requires the *four-wire* technique, as illustrated in **Figure 25.41**. Two wires are connected to each end of the resistor, one to carry the test current and one to read the voltage with a high-input impedance voltmeter. In that way the resistance of the wires does not affect the measurement.

Most inexpensive meters do not have a high-enough current range to measure the 20 A or so that is drawn from a 12 V power supply by a typical 100 W transceiver. The solution is to use an external *meter shunt*, which is a low value resistor placed in series with the current. (As discussed earlier, see Figure 25.2). The voltmeter reads the voltage, E, across the shunt, and then the current is calculated from Ohm's law, $I = E/R_S$, where R_S is the resistance of the shunt. Resistors designed for this service may have four leads rather than two to allow a true four-wire measurement as shown in Figure 25.41.

25.13.2 Wheatstone Bridge

A *Wheatstone bridge* is a method of measuring resistance that does not depend on the accuracy of the meter. Each arm of the bridge ($R_1 - R_S$ and $R_2 - R_X$) forms a voltage divider. In **Figure 25.42**, the meter is a zero-center type so that it can read both positive and negative voltages between the center connections of the two voltage dividers. When variable resistor R_S in Figure 25.42B is adjusted for a zero reading on the meter, then the two arms of the bridge have the same ratio,

$$\frac{R_1}{R_s} = \frac{R_2}{R_x}$$

The resistor under test, R_X, can be calculated from

$$R_x = R_s \frac{R_2}{R_1}$$

The Wheatstone bridge is rarely used for measuring resistance since it is not as convenient as an ohmmeter, but the bridge concept is important for several other types of measurement circuits.

25.12.3 Using Panel Meters

Whether surplus or new, it is rare that a panel meter measures exactly what you need for a particular application. Usually, you must change the current or voltage sensitivity.

To increase the full-scale current, place a *current shunt* in parallel with the meter. This is simply a resistor whose value is

$$R_{SHUNT} = R_M \frac{I_M}{I_{FS} - I_M}$$

where R_M is the meter resistance, I_M is the meter full-scale current and I_{FS} is the desired full-scale current reading. The shunt resistance is very small for high-current shunts, such that the resistance of the wires or circuit traces can cause a significant error. To reduce that error, connect the meter directly to the leads of the shunt, with no wires or circuit traces in common with the high-current path.

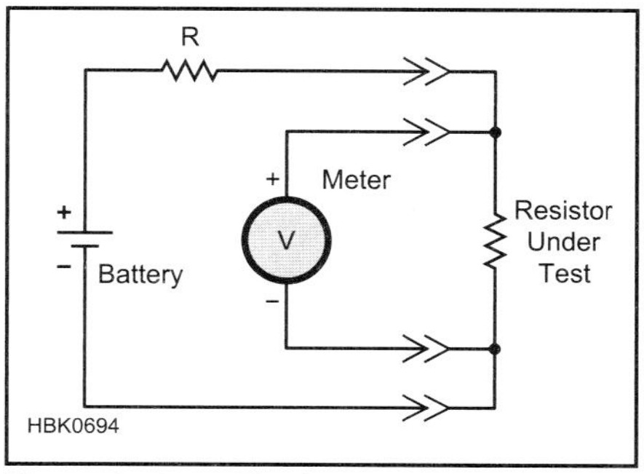

Figure 25.41 — The four-wire technique for measuring low-value resistors. By connecting the current source and the meter separately to the leads of the device under test, the error due to lead resistance is reduced.

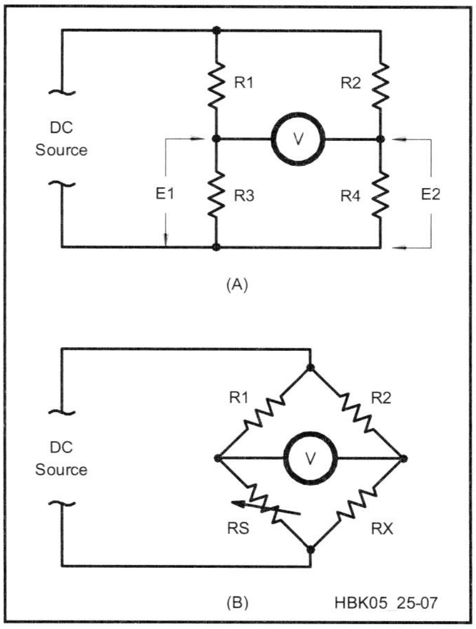

Figure 25.42 — A Wheatstone bridge circuit. A bridge circuit is actually a pair of voltage dividers (A). (B) shows how bridges are normally drawn.

You can make a low-value shunt by wrapping a length of copper wire around a resistor or other component used as a form. Unfortunately, however, the resistance of copper has a poor temperature coefficient, typically around 0.4 percent per degree C. That means the meter reading can change more than 10% between a warm and a cold day. As the wire self-heats from the high current flowing through, the meter reading can easily be in error by 20% or more. Commercial shunts are made from a metal with a low temperature coefficient such as nichrome. Copper wire shunts should only be used where accuracy is not important. (Wire tables including resistance in ohms per foot (Ω/ft) are available in the **Antennas** chapter.)

If the panel meter is to be used to measure voltage, a *voltage multiplier* resistor is inserted in series with the meter. The value is

$$R_{MULT} = \frac{V_{FS}}{I_M} - R_M$$

where V_{FS} is the desired full-scale voltage, I_M is the meter full-scale current and R_M is the meter resistance. If the meter has an internal current shunt, it should normally be removed to maximize the value of the multiplier resistor. For high-voltage applications, be aware that in addition to a power rating a resistor also has a working voltage specification, perhaps 200 to 250 V or so for a typical 1/4 W, through-hole resistor. Applying voltages higher than the rating, even if the rated power dissipation is not exceeded, can result in arcing across the body of the resistor. If you need to measure a voltage higher than the voltage rating, use several resistors in series. For example, to measure a 2000 V power supply, ten 1/4 W resistors in series, each with a value of one-tenth the desired resistance, would be suitable.

If you intend to use the same panel meter for several different purposes in your project, be sure to use a *break-before-make* switch to make the selection. That protects the meter by making sure it is never connected to two circuits at the same time, even for an instant.

Analog D'Arsonval meter movements are easily damaged if subjected to excessive current. A standard technique to protect them is to wire back-to-back diodes in parallel with the meter. Silicon junction diodes have the property that they act like an open circuit for low voltages and start to conduct when the forward voltage reaches about 0.4 V. Most meters can withstand up to two times the full-scale current without damage. So choose resistor R in **Figure 25.43** such that the voltage across the diodes is about 0.2 V when the meter current is at full scale, that is

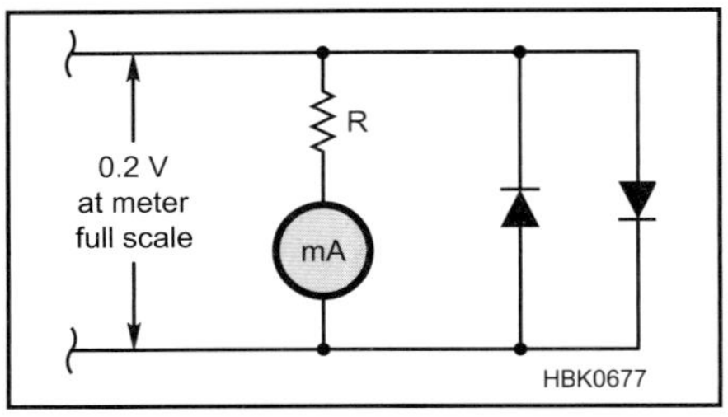

Figure 25.43 — Back-to-back silicon diodes protect the meter by limiting the maximum voltage. See the text for a discussion of how to select the value of R.

$$R = \frac{0.2V}{I_M} - R_M$$

where I_M is the full-scale meter current and R_M is the meter resistance. If that equation results in a negative value for R, replace each diode with two diodes in series and recalculate using 0.4 V instead of 0.2 V in the equation.

25.13.4 Using Multimeters for AC Measurements

RMS MEASUREMENTS

(See the **Radio Fundamentals** chapter for definitions of values used to measure ac waveforms.) Most multimeters can indicate the RMS value of ac voltage and current, but many do not measure the RMS value directly. Instead, they measure the rectified average or peak voltage and then apply a correction factor so that the display reads RMS, assuming a sine wave. Unfortunately, that means that the RMS values are not accurate if the ac signal being measured does not have a sinusoidal waveform.

Some meters full-wave rectify the ac signal and then measure the average of that, internally correcting for the difference between the average of a rectified sine wave

$$V_{AVG} = \frac{2}{\pi} V_{PK} = 0.637 \times V_{PK}$$

and the RMS value

$$V_{RMS} = \frac{1}{\sqrt{2}} V_{PK} \approx 0.707 \times 0.707\ V_{PK}$$

so that the reading is in RMS. Very inexpensive analog meters may only use a half-wave rectifier which causes RMS readings for asymmetric waveforms to vary with the orientation of the test connections.

Additional considerations may apply to RMS readings. For example, the accuracy of the RMS reading for most meters varies with frequency of the applied signal. Check the specifications of the multimeter for the frequency range over which it may be used to measure RMS values.

The only way to accurately measure RMS values of non-sinusoidal signals is with a meter that has *true RMS* capability. Such a meter uses circuitry or software to compute the RMS value of the signal. Note that the measurement bandwidth of the meter must include the significant harmonics of the signal as well as the fundamental in order to give accurate RMS readings.

An example of a measurement that requires a true RMS voltmeter is receiver sensitivity. For that, you need to measure signal and noise levels at the receiver audio output. Standard multimeters using a rectifier and averaging circuit give inaccurate results when measuring noise because noise and sine waves have different peak-to-RMS ratios. Another advantage of true RMS meters is that they tend to have better scale linearity, even for sinusoidal signals. A diode detector is nonlinear, especially at the low end of the scale.

LIMITATIONS TO ACCURACY

Frequency response is another limitation when making ac measurements with a multimeter. Most are specified from below 50 or 60 Hz, to cover power-line frequencies, up to a few hundred Hz. Many receiver measurements use a 1 kHz test frequency, so a meter specified up to at least that frequency is especially useful.

For all of these and other reasons, the ac accuracy is usually significantly worse than the dc accuracy. Generally, an oscilloscope makes more accurate ac measurements than a multimeter, especially at frequencies near the upper limit of the multimeter's capabilities. Modern digital oscilloscopes often have built-in capability to indicate peak, average and true RMS voltage.

One final issue with multimeters is the ac impedance of the probes. While the dc input resistance of a modern electronic multimeter is typically over 1 MΩ, the capacitive reactance can be a significant factor at radio frequencies. Even if all you care about is the dc voltage, if ac signals are present, reactance of the probe can affect the circuit's operation.

25.13.5 Frequency Calibration

The best test equipment is of limited use if it is not well-calibrated. The traditional frequency calibration method is to zero-beat a crystal oscillator (or its harmonic) with a radio station of known frequency, preferably a standard frequency station such as WWV or WWVH. **Table 25.1** contains the locations and frequencies of some of those stations. A receiver is tuned to one of those frequencies and the oscillator is loosely coupled to the antenna. It may be necessary to use frequency multiplication or division to obtain a common frequency.

The frequency difference between the two causes a *beat note*, a rapid variation in the strength of the tone received in the speaker that slows down as the frequencies are brought close together. Maximum beat-note modulation occurs when the off-the-air and oscillator signals are approximately equal in amplitude. When very close to the station's frequency, watch the variation of the receiver's S-meter instead of relying on the receiver's audio response at very low frequencies. The popular program *WSJT-X* has a frequency calibration tool that uses the sound card of a computer connected to the audio output of a receiver.

While the transmitted frequencies from WWV and WWVH are highly accurate, better than 1 part in 10^{11}, after propagation via the ionosphere the received accuracy is significantly degraded by Doppler shift, typically to a few parts in 10^7. Also, due to fading of the received signal, it can be difficult to zero-beat the oscillator to better than about 1 Hz accuracy. Best results generally occur on the highest frequency that provides good reception.

Table 25.1
Standard Frequency Stations
(Note: In recent years, frequent changes in these schedules have been common.)

Call Sign	*Location*	*Frequency (MHz)*
BPM	China	2.5, 5, 10, 15
BSF	Taiwan	5, 15
CHU	Ottawa, Canada	3.330, 7.850, 14.670
dcF	Germany	0.0775
HLA	South Korea	5.000
JJY	Japan	0.04, 0.06
MSF	Great Britain	0.06
RID	Irkutsk	5.004, 10.004, 15.004
RWM	Moscow	4.996, 9.996, 14.996
TDF	France	0.162
WWV	USA	2.5, 5, 10, 15, 20
WWVB	USA	0.06
WWVH	USA (Hawaii)	2.5, 5, 10, 15
ZSC	South Africa	4.291, 8.461, 12.724 (part time)

VLF time standards and surplus rubidium standards can be used for frequency references. (See the References for Miller and Nash) The Global Positioning System (GPS) satellites offer further possibilities for very precise frequency calibration. Various companies sell *disciplined oscillator* units that correct the frequency using the cesium-clock-based signals from the GPS satellites. These can sometimes be found on the surplus market. (See the Reference for Jones) Amateur-level kits are also available or you can build one from scratch. (See the Reference for Sherra)

25.14 RF Measurements

RF measurements are a special case of ac measurements. While everything in the previous section about measuring low-frequency ac applies, there are additional factors to take into consideration for high frequency measurements, such as parasitic values and operational bandwidth. This section concentrates on equipment and techniques especially suited to measuring at radio frequencies. The **RF Techniques** chapter has additional information on circuits at high frequencies. A table for converting between voltage and power in 50 Ω systems is provided in the supplemental information for this chapter.

25.14.1 Measuring RF Voltage and Current

An *RF probe* rectifies a radio-frequency signal so that its amplitude can be measured with a dc instrument such as a multimeter. An RF probe can also be used to detect AM or SSB signals so the modulation can be measured on an ac multimeter. The circuit is typically quite simple, consisting of a diode, a resistor and a couple capacitors, as in the example of **Figure 25.44**. The resistor value is chosen so that the rectified output voltage is approximately equal to the RMS value of a sine wave input signal. The 390 kΩ value shown assumes the meter has a 1 MΩ input resistance. The diode can be a high-speed switching diode, such as a 1N914 or 1N4148, or for greater sensitivity, it can be a germanium diode, such as 1N34 or 1N277.

The detector is located as close to the measuring point as possible to minimize stray inductance and capacitance. The leads to the dc voltmeter can be longer. The RF probe can be housed in any convenient enclosure that fits easily in the hand and the circuitry can be constructed on a scrap of perforated phenolic board, as long as the leads are kept short. A much more elaborate version with a coaxial input and an integrated compensated voltmeter is described with the downloadable supplemental content.

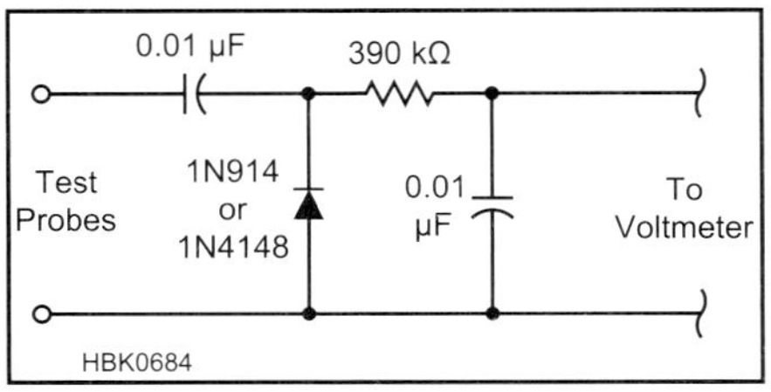

Figure 25.44 — A basic RF probe, used to convert an RF signal into a dc voltage that can be measured by a voltmeter.

Measuring RF current is a little more difficult. When measuring the current on an antenna or feed line, it may not be practical to break the connection to insert an ammeter. Even when that can be done, the meter itself often upsets the measurement, either because it must be plugged into ac power or simply because of the instrument's size.

A time-honored technique is to wire a small incandescent lamp in series with the antenna or feed line to be measured. Although you can get a rough idea of the amount of RF current by comparing the brightness to the brightness with a known dc current, this method is only useful for relative measurements where the absolute value does not need to be known. The impedance of the bulb is also in series with the antenna or feed line, affecting the measurement to some degree.

Another alternative is to construct an RF ammeter using a current transformer that clamps over the wire. (See the Reference for Lau) That method has the advantage that the wire does not need to be disconnected. It is important that the leads to the meter not couple to the wire or cable being measured.

Highly accurate commercial RF ammeters made to measure the base current of AM broadcast station antennas can sometimes be found on the surplus market. If you can find one, it would give much better accuracy than a homebrew device.

25.14.2 Measuring RF Power

Transmitter power is normally given in units of watts. Lower-power RF signals found in receiver and transmitter circuits may be specified in units of milliwatts (mW) or microwatts (mW) but it is perhaps more common to see units of *dBm*, decibels with respect to one milliwatt. For example, it is much easier to express the power of an S1 signal as –121 dBm than 0.00000000076 µW. The formula is dBm = 10 log (1000 P), where P is the power in watts. One milliwatt (P = 0.001 W) is 0 dBm, 10 mW is +10 dBm, 0.01 mW is –20 dBm and so on. **Table 25.2** lists common dBm and power equivalents.

Table 25.2
Power and dBm Equivalents

dBm	*Power*	*dBm*	*Power*
–60 dBm	1 nW	3 dBm	2 mW
–30 dBm	1 µW	6 dBm	4 mW
–20 dBm	10 µW	10 dBm	10 mW
–10 dBm	100 µW	20 dBm	100 mW
–6 dBm	¼ mW	30 dBm	1 W
–3 dBm	½ mW	60 dBm	1 kW
0 dBm	1 mW	61.7 dBm	1.5 kW

Measuring RF power can be a little confusing because there are several ways to do it. We have already covered the difference between peak and RMS voltage and current. RF power is always based on RMS values. For example, if the RF voltage into a 50 W dummy load is 70.7 V RMS (100 V at the peak of the RF sine waves), then the power is

$$P = E^2 / R = 70.72 / 50 = 100 \text{ W}$$

(See **eznec.com/Amateur/RMS_Power.pdf** for a more extensive explanation of power and RMS.)

Most RF systems use 50 Ω as the impedance at various signal interfaces and for the characteristic impedance of most coaxial cables. Since this is so common, it is useful to be able to convert directly between voltage measurements and power, assuming a 50 Ω impedance. A table of voltage-power conversions for 50 Ω systems is provided in this chapter's supplemental information.

Peak envelope power (PEP) has nothing to do with the difference between the peak and average voltage of a sine wave. It is a measure of the power of an RF signal at the modulation peak, averaged over one RF cycle at the crest of the modulation envelope, as defined in the FCC Rules §97.3(b)(6). For a CW signal, the PEP is simply the power when the key is closed, as read on any wattmeter. However, for an SSB signal, the power is constantly changing as you speak. An average-reading wattmeter displays a value much lower than the PEP.

Using an oscilloscope can be a highly accurate method to measure the RMS of a sine wave, within its bandwidth limitations, if the load impedance is known accurately. Don't forget that the oscilloscope shows peak-to-peak rather than RMS voltage, so you must divide the maximum reading by $2\sqrt{2}$, or 2.83. Some wattmeters do have PEP-reading capability. Their circuitry must have very fast response to the detected RF signal to give an accurate reading of the peaks.

A *directional wattmeter* is a device that measures power flowing in each direction on a transmission line. The manner in which RF signals propagate on transmission lines is covered in the **Transmission Lines** chapter. Many amateurs keep a wattmeter permanently connected at their station to monitor the condition of their transmitter and antennas. See the **Station Accessories** online material for a further discussion of directional wattmeters.

A *bolometer* is a device for measuring transmitter power by measuring the heat dissipated in a resistive load. A thermistor or other device measures the temperature. The device is calibrated with a dc voltage, since dc voltage and current can be measured very precisely. With careful construction and calibration, very high accuracy can be obtained. However, the response time of the measurement is very slow, so a bolometer is normally used to calibrate another wattmeter rather than being used directly for measurements. Bolometers are made commercially, however it is possible to homebrew one using a plastic picnic cooler. (See the Reference for Steinbaugh)

Commercial laboratory power meters are generally intended for measuring power levels in the milliwatt or microwatt range. There are two basic types, based on either diode or thermocouple detectors. Below a certain power level, the dc output from a diode detector is directly proportional to power, that is, the square of the RF voltage. With a suitable dc amplifier, the detector output can drive a meter with a linear scale to read power directly.

A form of bolometer, based on thermocouples, feeds the RF signal into a resistor that heats up in proportion to the power level. The temperature is measured with a thermocouple, which consists of a pair of junctions of dissimilar metals. A voltage is generated based on the temperature difference of the two junctions, is proportional to the RF power. The thermocouple method gives high accuracy and wide bandwidth, but the measurement time can be up to several seconds at low power levels, rather than being nearly instantaneous as with a diode detector. Older-model surplus diode and thermocouple-type meters can often be found for reasonable prices, but the sensors are fragile and easily damaged. Obtaining replacements for the desired frequency range and power level can cost more than the power meter itself.

Analog Devices makes a series of integrated circuits that can detect RF signals and output a dc voltage proportional to the logarithm of the power level. That makes it easy to construct an RF power meter that reads directly in dBm. For example, the AD8307 covers dc to 500 MHz with 1 dB accuracy over an 88 dB (nearly 1 billion-to-one) power range.

25.14.3 Measuring Components with an Analyzer

Antenna and network analyzers can be used to measure components at RF across a wide frequency range. For example, a swept-frequency test can show if an inductor is resonant within the frequency range that it is intended for. The **RF Techniques** chapter's section on Effects of Parasitic Characteristics has more information about how components are likely to behave at high frequencies.

Figure 25.45 shows a graph generated from swept-frequency impedance data collected by an antenna analyzer of the type discussed above in Figure 25.34. The graph shows the impedance of an air-core inductor with a nominal inductance of 7.4 μH at low frequencies. (The traces are labeled with the measurement they represent.) The self-resonant frequency is 45.4 MHz. This resonance occurs because of the coil's inter-turn capacitance.

Far below the resonant frequency the inductance has a positive reactance and the coil presents its expected value, 7.4 μH. As the test frequency approaches the self-resonant frequency, the parasitic (stray) capacitance that causes self-resonance causes the total reactance to increase, so the inductance appears to be larger. At resonance, the impedance is a resistance of high value. Above the self-resonant frequency, the component increasingly looks like a capacitor, as indicated by the negative phase angle.

The analyzer is also very handy for determining the material of a toroid core. Cores of different ferrite or powdered iron mixes cannot be distinguished by their physical appearance, but you can identify the material used by measuring their characteristics over a wide frequency range. A procedure for determining a ferrite component's mix is presented in the **RF Techniques** chapter section on Determining Ferrite Mix.

Capacitors are usually closer to the ideal component than inductors, but they do have some inductance in their leads and electrodes. This is called *equivalent series inductance* (ESL). Eventually they become self-resonant at some high frequency. Above the self-resonant frequency, a capacitor acts like an inductor. This self-resonance should be checked for capacitors that are used in the VHF/UHF range.

Depending on the dielectric material used, capacitors also have some loss. This loss is commonly described as *equivalent series resistance* (ESR) which appears in series with its capacitive reactance. The ESR in most capacitors increases with frequency and can be critical in power handling circuits.

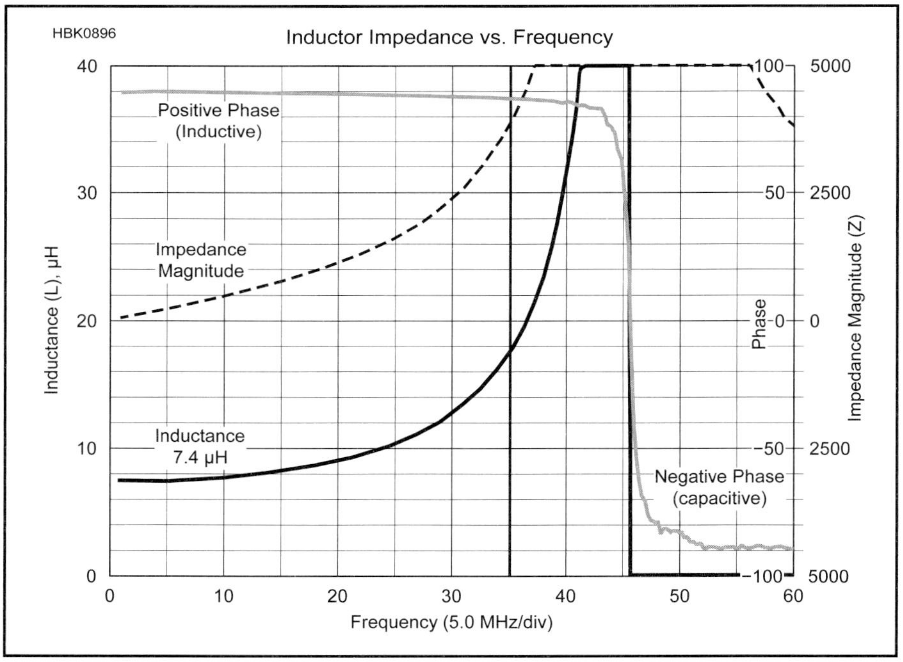

Figure 25.45 — Example of an antenna analyzer (AIM-4170 and companion software) being used to measure an air-core inductor's behavior. The inductor has a nominal value of 7.4 μH and a self-resonant frequency of 45.4 MHz.

Resistors have an effective capacitance in parallel with them as well as inductance in their leads so they are not ideal over a wide frequency range. Physically large power resistors used for dummy loads have larger parasitic components. (Thin-film power resistors in TO-220 packages are available with significantly lower reactance.) Tubular metal and carbon film resistors are often trimmed with a laser to create a spiral track in the deposited film, creating inductance. If the resistor is to be used in an RF circuit, it is prudent to verify its effective frequency range.

25.14.4 Measuring RF Impedance

The most convenient method of measuring RF impedance today is to use a vector impedance analyzer (VIA) or vector network analyzer (VNA). See the preceding sections about those instruments for more information. Regardless of how the measurement is made, at RF there are many factors that can affect an impedance measurement, particularly in excess of a few hundred ohms. Here are some guidelines to minimize those effects:

• Stray capacitance: Even a few pF of stray capacitance can affect an impedance measurement at higher radio frequencies. Keep nonessential cables and surfaces away from the device or circuit being measured or use a test fixture (see below) to control the capacitance.

• Lead inductance: Unless a conductor is part of what is being measured, keep leads as short as practical.

• Skin effect: Minimize the extra resistance created by skin effect. Use silver-plated connectors at VHF and higher frequencies. Use large diameter or width conductors.

• Control test setup: If multiple tests are going to be made, consider making a test jig or fixture so that the layout of the cables and connections and stray capacitance is as consistent as possible between tests.

• Calibration: Make sure your instruments and test accessories are calibrated. Perform a calibration on network analyzers at the intended measurement plane.

• Reference Devices: Before making a measurement on the circuit or system under evaluation, measure a similar device for which the impedance is already known. A significant difference can alert you that there may be flaws or unexpected characteristics in your setup.

• Manufacturer Resources: Particularly for sophisticated analyzers and lab-quality instruments, the user's manual may show good practices for that type of measurement. There may be application or "app" notes available that go into detail.

USING AN IMPEDANCE BRIDGE

Most traditional impedance-measuring devices are based on the Wheatstone bridge, as previously described in the section on LCR bridges. The difference is that an impedance bridge does not require that the device under test be a pure inductance, capacitance or resistance but may be a complex impedance that includes both resistance and reactance. That means that two adjustments are required to balance the bridge, rather than just one as with an LCR bridge. Both the resistance and reactance must match.

In addition, since an impedance bridge is typically designed to operate at higher frequencies, circuit layout and the connection to the device under test are more important. Usually, a coaxial connector is used to connect to the device and the impedance measurements are referenced to 50 Ω.

A *return-loss bridge* (RLB) is an RF bridge with fixed, usually 50 Ω, resistors in each leg except the one connected to the device under test. *Return loss* is the ratio of the reflected signal to the signal incident on a component, usually expressed in dB, which is always a positive number for a passive device. (See the Reference for Bird and the **RF Techniques** chapter section on Two-Port Networks for a discussion of return loss.)

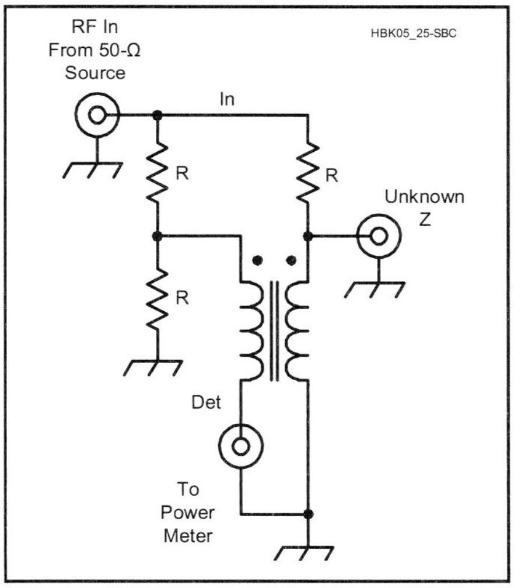

Figure 25.46 — A relatively simple diode detector circuit for measuring impedance that gives adequate results for many applications.

The schematic of a simple return-loss bridge is shown in **Figure 25.46**. The circuit can also be used as a *hybrid combiner* (see the RF and Microwave Test Accessories section of this chapter), where the port labeled unknown is the common port and the other two are isolated from each other. For good results at high frequencies, it should be built in a small box with short leads to the coax connectors.

Apply the output of a signal generator or other signal source to the rf in port of the RLB. The power level should be appropriate for driving the device connected to the unknown port, after accounting for the 6 dB loss of the bridge. Connect the bridge power meter port to a power meter or other power-measuring device through a step attenuator and leave the unknown port of the bridge open circuited. Set the step attenuator for a relatively high level of attenuation and note the power meter indication.

Now connect the unknown impedance to the bridge. The power meter reading should decrease. Adjust the step attenuator to produce the same reading obtained when the unknown port was open circuited. The difference between the two settings of the attenuator is the return loss, measured in dB.

USING A NOISE BRIDGE

The noise bridge includes an adjustable bridge circuit similar to that in Figure 25.46. A wide-band noise generator is connected as the source and a conventional receiver is attached to the Power Meter port as a detector. Tune the receiver to the desired frequency and adjust the resistance and reactance controls for minimum noise in the receiver. If the receiver has a panadapter display, the null frequency can be seen on the screen, speeding the adjustment. Noise bridges are rarely used today in favor of the more convenient antenna analyzers. See the References entry for "The Noise Bridge" by Althouse for a more complete description of how a noise bridge works and how it is used.

RF Measurement Test Set

Many low-level RF measurements require similar test setups so why not combine them into one package? This also makes it easier to perform the tests and get consistent results. NOADL and KE4PT show an example in the *QEX* article "Apparatus for RF Measurements" that you can download from the online information.

25.15 Antenna System Measurements

This section discusses measurements of antenna systems. The antenna system typically includes the antenna along with the transmission line and any other accessories such as switches, filters, matching networks, antenna tuners, and so forth. The instruments and measurements in this section assist with construction and installation, performance assessment, and troubleshooting of the entire antenna system. Related information is available in the **Transmission Lines, Antennas,** and **Assembling a Station** chapters.

25.15.1 Field Strength Measurements

To determine how well an antenna is actually radiating, a way to detect the signal level at some distance from the antenna is needed. For measurements in the vicinity of the antenna, a *field-strength meter,* generally with a built-in antenna, picks up the radiated signal off the air and indicates the level on a meter or display. Professional field-strength meters use a carefully calibrated antenna and circuitry that can read out the actual radiated signal strength in volts per meter or watts per square meter. Most amateur field-strength meters such as the one shown in **Figure 25.47** are not calibrated but give a relative indication only. They are useful for tuning an antenna, antenna tuner, or transmitter for maximum signal as well as for comparing different antennas. A field strength meter is a simple one-evening construction project. **Figure 25.48** shows the schematic of a simple field strength meter. (This article by VE7NI is also included in the book's online material.)

If you have a *reference antenna* (such as a half-wave dipole) that has a known gain, then the gain of a second antenna can be calculated by alternately transmitting with each antenna and measuring the difference in signal level at a receiver whose antenna is located far enough away to be outside the *near field*, typically up to several wavelengths from the antenna under test. At first glance, it seems like this should be an easy measurement to make but in practice there are a number of devilish details that can ruin the measurement accuracy.

The biggest issue is reflections. If there is any significant reflector of RF signals between the transmitting and receiving antennas, the signal comparison may not be accurate if the test and reference antenna patterns are significantly different. Even if there are no wires, bodies of water, fences or other conducting objects in the vicinity, the ground reflection can result in an apparent additional gain of up to 6 dB or conversely a loss of many dB if the receive antenna happens to be in a null. At microwave frequencies it is sometimes possible to use a high-gain, narrow-beamwidth receive antenna and mount the antenna under test high enough so that the ground reflection is outside the receive antenna's beam width. (See the Reference for Wade). At HF frequencies that is rarely possible.

Commercial antenna companies use elaborate *antenna test ranges* that employ various techniques to assure measurement accuracy. Absent a proper test range, the best solution is probably to measure the reference antenna and antenna under test at various heights above ground to get an idea of how much ground reflections are affecting the measurement. [The subject of antenna measurements is addressed by Paul Wade, W1GHZ in his "Microwavelengths" *QST* columns for October 2012 (covering the antenna range) and January 2013 (discussing measurements and equipment).]

The most common way to compare antennas is to ask for signal reports on the air. Unfortunately, propagation is so variable on most amateur bands (fading of more than 20 dB in a few seconds is common) that a signal report gives only a very rough idea of how well your antenna is working, unless it can be compared with another antenna (perhaps at another local amateur's station) at the same time.

An improved method of assessing antenna performance from distant stations is to use one of the automated receiving systems; the *Reverse Beacon Network* or RBN (**reverse-beacon.net** for CW signals), *PSKReporter* (**pskreporter.info** for PSK signals) and *WSPRNet* (**wsprnet.org** for WSPR signals). These systems consist of a worldwide network of independent receivers that decode signals and report the call sign and signal strength to a central server. The popular FT8 digital mode can also be used, saving decoded transmissions from other stations with objective signal strengths for analysis.

To use the systems for antenna system testing, the server can be queried or multiple decoding cycles can be used for comparative signal strengths after switching between antennas or making adjustments. Recognizing that propagation can change substantially over the short-term, to obtain reliable results, collect a large number of reports from multiple receiving stations over a period of time.

25.15.2 Measuring Antenna Feeds

Between the transmitter output and ra-

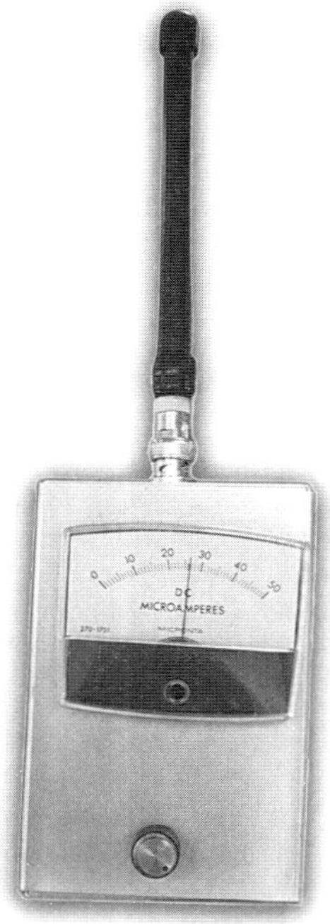

Figure 25.47 — The VE7NI field-strength meter.

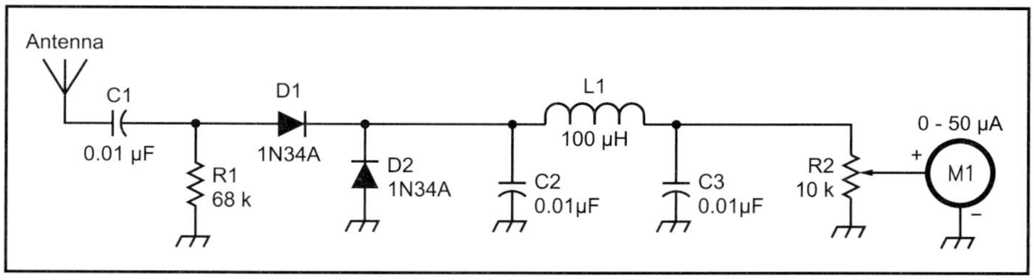

Figure 25.48 — Schematic diagram of the VE7NI field-strength meter.
C1-C3 — 0.01 μF capacitors.
D1, D2 — 1N34A diodes.
L1 — 100 μH inductor.
M1 — Analog meter, 50 μA.
R2 — Sensitivity control potentiometer, 10 kΩ.
Antenna — BNC female chassis mount socket. Antenna selection should match the frequency band for VHF and UHF. A random length of wire might work best for close field measurements on HF to 40 meters. Metal box enclosure is mandatory.

diation of the signal from the antenna, the RF signal must be efficiently transferred to a feedline, travel over the feedline with as little loss as possible, transfer to the antenna and then be radiated. Matching impedance at all of the transfer locations is key to the maximum transfer of power to be transmitted. When an impedance mismatch occurs, some of the signal is reflected back toward the transmitter, resulting in standing waves along the feedline and less power being radiated into the air. Impedance mismatches can occur if the feedline is damaged (sharply bent, crushed, partially cut, or has leaked water), if the connectors are not properly installed, or if the antenna impedance does not match the transmitter and feedline impedance.

SWR TO INDICATE FEEDLINE FAULTS

Several devices that have been described above allow one to detect faults in antenna systems. The most common, and easiest to use, is the SWR meter. SWR increases for each impedance mismatch and can indicate problems in the antenna system that should be fixed. Most modern amateur transmitters have the built-in capability to measure SWR, making this the most convenient measurement at an amateur station.

SWR INACCURACIES

SWR is not always the most reliable measurement, and its weaknesses should be taken into account when using it. An SWR measurement at the transmitter does not account for normal feedline losses. Since SWR is based on a ratio of forward power (P_F) to reflected power (P_R), P_F is measured without any losses when the SWR meter is located at the transmitter. However, if the fault causing a reflection is at the far end of a long length of coaxial cable, the signal has to travel to the fault with attenuation due to the feedline, and then, after reflection, must travel the same distance back to the SWR meter, doubling the attenuation of the reflected signal when it is measured. This can greatly underestimate the actual SWR of the feed system. An example of this is in the **Transmission Lines** chapter.

If multiple impedance mismatches exist in a feed system, the SWR is unlikely to indicate the magnitude of the problem. The reflected power from a distant mismatch also reflects again to the forward direction on the way back toward the transmitter when it encounters another mismatch that is nearer to the transmitter. Thus, a complex system of standing waves is created, and it may result in fixing one mismatch in order to discover others.

Impedance mismatch at the antenna can be variable, based on frequency. For most properly designed antennas, the antenna feed point impedance at the resonant frequency should be close to the transmitter and feedline impedance, approaching the ideal SWR of 1:1. Usually, as the frequency moves away from the resonant point, the antenna feed point impedance changes and the SWR increases. This variation can make it difficult to distinguish feed system faults from the normal variation of SWR due to frequency-based changes in antenna impedance. It is important to locate the resonant frequency of the antenna and then to determine the minimum SWR for that antenna. If the minimum is too high, this may indicate other problems in the feed system.

How high is too high, for the SWR? That can be based on the antenna design and not under your control. However, keep in mind that an SWR measurement of 3:1 indicates that 25% of the transmitted power is being reflected back to the transmitter, so for a 100 W transmitter, at most 75 W is being radiated, which does not account for attenuation losses in the coaxial cable that would make the radiated power even lower. See the **Transmission Lines** chapter for a discussion of total losses in feedlines.

TDR to locate feedline faults

Although the SWR meter can alert one to faults in the antenna feed system, it may be difficult to locate the causes of the faults, or even how many faults exist, and then to correct them. For instance, a feedline system may have many connectors, e.g., between the transmitter and the feedline, on either side of an antenna switch, on either side of a filter, on either side of an antenna tuner, on either side of the lightning arrester, on either side of splices needed due to the length of the feedline, and at the antenna. If one or more of the connectors are not properly installed, the SWR meter indicates that problems exist but does not indicate which connectors are responsible, or even how many of them need attention. TDR can indicate the exact location of the faulty connections, simplifying the task of fixing the problems.

DIRECT METHOD TDR

There are two common TDR implementations, with variations of both. In the "direct method," which is the oldest and simplest, the line is driven by a pulse. This can be a single pulse, or it may be a train of pulses like a square wave.

The pulse and all of the reflections are displayed on an oscilloscope trace that is triggered by the pulse's rising edge. The rise time of the pulse must be much shorter than the round-trip time for the impulse to travel to and from the discontinuity. **Figure 25.49** shows an example of a direct method TDR with the pulse generator and cable attached to the oscilloscope which is displaying the pulse.

Although a digital scope can capture a single pulse and its reflections, making the pulse repetitive means that multiple responses can be averaged to improve the signal-to-noise ratio. The signal-to-noise ratio of a system excited by a single pulse can be rather limited. Repeated pulses also sustain the trace so the operator to see it if the scope is an analog model. The repetition rate of the pulse must be much slower than it takes for the impulse to make a complete round trip time through the line, however. This ensures the line's response dies out completely before the next pulse excites the line again. See the reference article by King about this type of TDR and the *ARRL Antenna Book* also includes material on TDR.

Modern antenna analyzers can use the direct method, as well. For example, the AIM

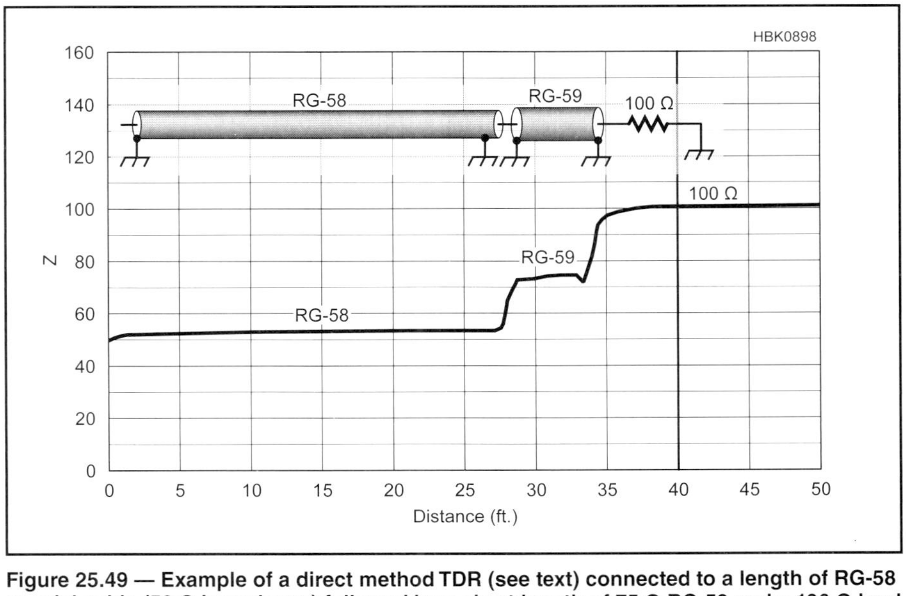

Figure 25.49 — Example of a direct method TDR (see text) connected to a length of RG-58 coaxial cable (50 Ω impedance) followed by a short length of 75 Ω RG-59 and a 100 Ω load. The discontinuities are clearly shown as the reflections add to the input pulse.

family of vector impedance analyzers (**arraysolutions.com**) excite the line with a step function waveform having a very fast rise time. This implementation provides a display of the impedance at every point on the line. It can, for example, show the relative impedance of cables having different Z_0, as well as the position of discontinuities. Figure 25.49 shows a TDR display from an AIM analyzer connected to a length of RG-58 (50 Ω cable), a short length of RG-59 (75 Ω), and a 100 Ω load. The discontinuities at the cable and load transitions are clearly shown. Software converts the raw data from the pulse amplitudes into impedance and the time is converted into distance along the line.

The TDR function can be used to determine if the line has been degraded, for example, by water leaking into the coax or if the line has been shorted or cut somewhere between the transmitter and the antenna. Damage or defects can be located within a few inches and this reduces the effort required to repair the line. Defective connectors can also be indicated by short glitches in the trace corresponding to the location of the connectors.

TRANSFORM METHOD TDR

The other common implementation could be described as the "transform method." Instead of determining the impulse response of the cable with a pulse, the excitation is a sine wave swept over a range of frequencies and the analyzer captures the *frequency response*. An inverse Fourier Transform is performed on that frequency response, producing the time-domain response (see the chapter on **DSP and SDR Fundamentals**).

Some antenna analyzers and vector network analyzers use this method. One example is the handheld SARK-110 vector impedance antenna analyzer (**sark110.com**) The sine wave exciting the cable need not be a continuous sweep. Rather, it can be stepped over a wide range of frequencies and the frequency response is computed from those data. Sweep range, spacing between data points, and the settling time at each data point are set by the user.

An inverse Fourier Transform produces spurious artifacts that must be removed by applying a mathematical windowing function to the transformed data. Several mathematically different windowing functions are commonly used, and the window that provides the most useful display depends on the shape of the impulse response.

The frequency content of the excitation strongly influences the degree of detail that the measurement can reveal. When the excitation is an impulse, a very fast rise time reveals greater detail. When the excitation is a swept sine wave, a wider frequency range reveals the greatest detail. Currently available analyzers can sweep from 1 kHz to more than 1 GHz.

For TDR studies using the sweep method, a sweep from 5 to 500 MHz (or from 500 MHz to 1 GHz) clearly shows detail that would be missed with a sweep to only 100 MHz, while a 5 MHz to 1 GHz sweep may provide too much detail (or show discontinuities that don't matter below 50 MHz). Beginning the sweep in the HF range avoids smearing of the data due to the variation of VF with frequency.

Some analyzers also have a feature called *Distance to Fault*. This measures the length of a transmission line to any point(s) along the line where the impedance differs from its characteristic impedance (Z_0). The "different impedance" can be an open or short circuit, the point at which an antenna is connected, or it can be some small change in the impedance such as that which occurs at a connector, or is the result of damage to the cable. With the best TDR systems, the line can be accurately measured with or without an antenna connected.

If the line is disconnected from the antenna, the distance to fault is the total length of the line. If the line has been damaged somewhere this measurement gives you an idea where the damage is, which is very handy when the line is buried or otherwise requires special access. TDR can also locate impedance discontinuities or other changes in the line's characteristic impedance.

EXAMPLES OF THE TRANSFORM METHOD

Figure 25.50 is the impulse response of the feed line for a 30 meter half-wave dipole at a height of 100 feet, computed from a sweep

HBK0899

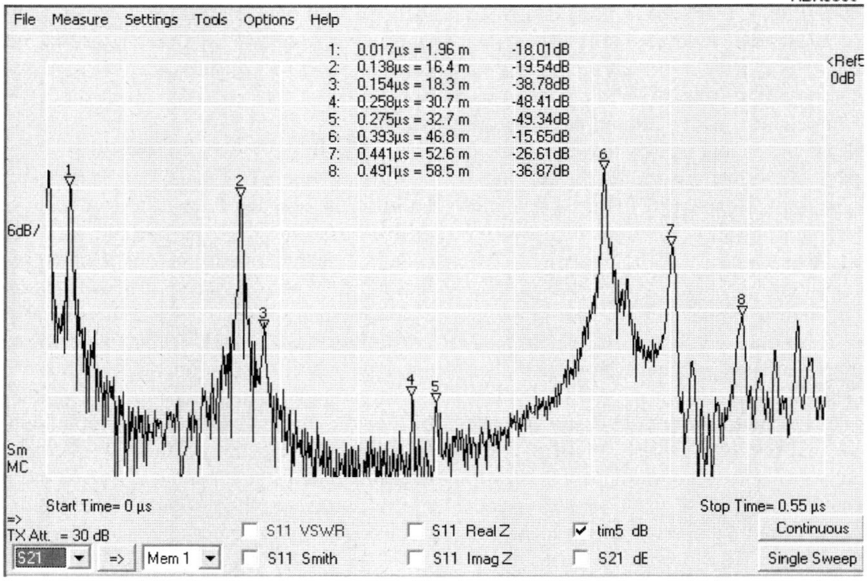

Figure 25.50 — The impulse response of a feed line attached to a 30 meter dipole. The system is swept from 50 to 500 MHz. See text for an explanation of the markers.

Transforming Analyzer Data

Like the trace on a simple oscilloscope, the TDR plot of the impulse response contains all frequencies (or the range of frequencies if it is transformed from a sweep). The scope and the TDR plot are "frequency blind" — that is, they display information only about the time response, and no information about the frequency response. An impulse or sweep measurement can, theoretically, be manipulated mathematically to compute the impedance at every point in the line over the same range of frequencies. The precision of that computation and whether it is practical, depends on how the data is gathered (sweep rate, sweep range, spacing of data points) and the software tools available. Frequency sweep ranges chosen for TDR may be inappropriate for examination of other line properties. Free software such as *SimSmith* (**www.ae6ty.com/smith_charts.html**) and *ZPlots* (**ac6la.com/zplots.html**) can accept swept measurements made at discrete frequencies to compute and plot the impedance at any point on a line if the characteristics of the line are known. Data is interchanged between a measurement device and software programs (and between one software program and another) by means of a plain text file. These files, defined by the Touchstone format (see previous sidebar) can take several forms that are defined by the first line(s), called a "header." The filename extension indicates the type of measurements:.s1p files describe single port measurements, like impedance or a time response and .s2p files describe two-port measurements such as the S21 (gain) transfer function produced by a vector network analyzer.

over the range of 50 to 500 MHz. Marker 1 shows the effect of lightning protectors at the station entry bulkhead. Marker 2 is a coax splice (two PL-259s and a PL-258 double-receptacle). Markers 3, 4, and 5 are coax defects. Marker 6 is the antenna feed point. Marker 7 is the end of the antenna (displayed distances are for the feed line, so for the antenna are divided by 0.795) the peak at Marker 8 is unexplained, but most TDR sweeps show multiple reflections after the antenna.

Figure 25.51 is the step response (similar to that from the direct method shown in Figure 25.49) computed from the same sweep as that used for Figure 25.50. The data revealed (and a visual inspection confirmed) that the cable inside the station and from the bulkhead to the coax splice is 50 Ω, but that the cable from there to the antenna is 75 Ω. A list of discontinuities is provided at the top of the graph with time delays and computed distances.

TDR measures time; to convert that measurement to physical distance, we must provide the velocity factor. The 75 Ω cable is Belden 8213, this sample of which has a measured VF of 0.795 at VHF. The 50 Ω cable has a measured VF of 0.8425 at VHF. In the setup screen for this TDR measurement, VF was set at 0.795, so distance measurements are correct for the 75 Ω cable, but wrong for the 50 Ω cable. Computed results could be made correct for the 50 Ω cable by changing VF in that setup screen, or by leaving VF at 0.795 and applying a correction factor of (0.8425/0.795) to the dimensions of the 50 Ω cable.

HBK0900

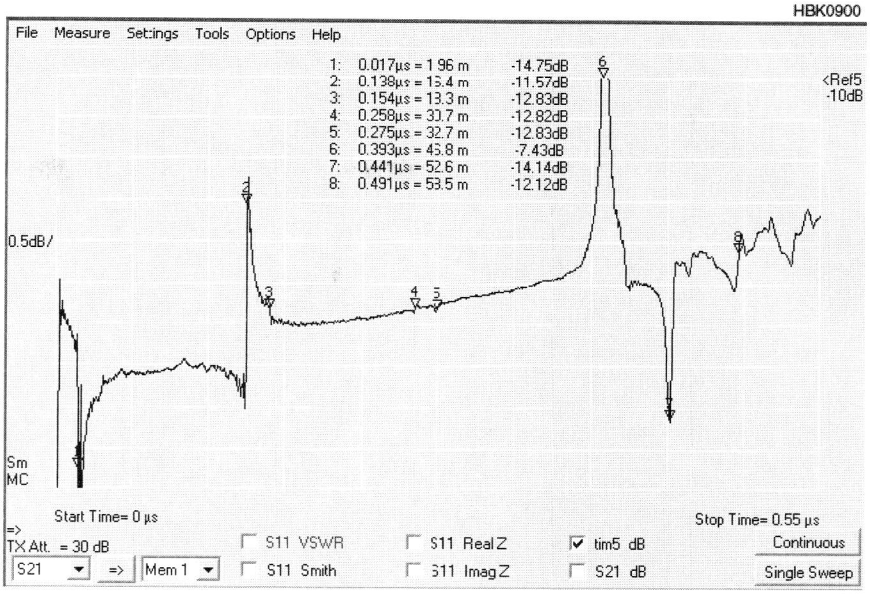

Figure 25.51 — The step response similar to the direct method for the same antenna system in Figure 25.10. See text for an explanation of the graph.

ARRL Lab Test Procedures Manual

All of the test procedures used by the ARRL Lab in evaluating equipment for *QST* Product Reviews are published in the ARRL Lab's *Test Procedures Manual*. This manual is published online for public use at **arrl.org/product-review**. While the procedures refer to specific pieces of test equipment, they are easily adapted or used as guidelines for developing your own tests.

Clean Signal Initiative

Recognizing concerns about potential interference from new amateur transmitting technology and modes, the ARRL created the Clean Signal Initiative in 2021. The goal is to provide guidance and training materials to avoid transmitting poor-quality signals that create interference to other amateurs. The initiative engages ARRL Lab staff and volunteer technical experts to establish formalized technical standards for transmitted signal purity of amateur radio equipment, certify new equipment to these standards, and work with manufacturers to ensure compliance with these standards for the good of amateur radio worldwide. It also requires that the ARRL address the educational aspect of transmitted signal purity by developing materials to train amateurs in the proper operation of amateur radio transmitting equipment. The ARRL Lab is leading a CSI Working Group comprised of knowledgeable amateurs and industry professionals to develop these principles into an industry standard that will help improve amateur radio technology and on-the-air practices.

25.16 Receiver Measurements

25.16.1 Noise Figure

A noise source can be used to measure the noise figure of an SSB or CW receiver. The *excess noise ratio (ENR)* is the ratio of the noise added by the noise source to the thermal noise level, normally expressed in dB. An explanation of noise figure is given in the **RF Techniques** chapter. The basic idea is to connect the noise source to the receiver antenna input and then measure the difference in the noise level at the receiver speaker terminals when the noise source is turned on and off. When it is off, the measured noise contains only thermal noise and the noise contributed by the receiver. When it is on, the noise includes the excess noise from the noise source. The noise figure can then be calculated from the ratio of those two values,

$$NF_M = ENR - 10\log\left(\frac{E_{ON}^{\ 2}}{E_{OFF}^{\ 2}} - 1\right)$$

where NF_M is the measured noise figure in dB, ENR is the excess noise ratio in dB, E_{ON} is the RMS output noise voltage with the noise source on, and E_{OFF} is the RMS output noise voltage with the noise source off.

Rather than using the equation, another technique is to include a step attenuator with 1 dB steps at the noise source output. See **Figure 25.52**. Adjust the attenuator so that the noise increases by 3 dB (1.414 times the voltage) when the noise source is turned on. The noise figure is then simply the ENR of the noise source minus the attenuation.

The nice thing about measuring noise with noise is that the accuracy of the voltage reading is not important so long as the ratio of the two readings is accurate. A non-RMS ac voltmeter is not accurate when measuring noise, but so long as the inaccuracy is the same for both voltage levels the ratio is correct. When using a non-RMS instrument, follow the step attenuator technique described in the previous paragraph along with a 3 dB attenuator (1/1.414 voltage divider) in front of the ac voltmeter. First take a voltmeter reading with the noise source off and the 3 dB attenuator bypassed. Then turn on the 3 dB attenuator and the noise source and adjust the step attenuator for the same reading. The noise figure is the ENR of the noise source minus the attenuation.

There are some potential errors to watch out for. The receiver AGC may need to be turned off for this test so that the gain does not change when the noise source is turned on or off. In addition, the audio gain and perhaps the IF gain must be adjusted so that there is no clipping of the noise peaks under any condition. With the noise source turned on, the RMS value of the noise at the speaker output should be adjusted to no more than 1/5 the maximum output level to prevent the peaks from clipping.

NOISE FIGURE OF RF AMPLIFIERS

In addition to receivers, noise sources can also be used to measure the noise figure of other devices such as amplifiers. The technique is basically the same, except that the output signal to be measured is at a radio frequency instead of audio. Another difference is that the noise figure of the measuring instrument can affect the results. That is almost never a concern when measuring receivers since the gain from the antenna to the speaker output is normally so high as to swamp out the effect. The corrected noise figure is

$$NF = 10\log\left[10^{NF_M/10} - 10^{(NF_{NMI}-G)/10}\right]$$

where NF_M is the measured noise figure in dB, NF_{NMI} is the noise figure of the noise-measuring instrument in dB, and G is the gain of the device under test in dB. For that equation to be valid, the bandwidth of the measuring instrument must be no greater than the bandwidth of the device under test.

NOISE FIGURE METERS

Commercial noise figure meters combine everything you need to perform noise figure measurements of amplifiers and other devices into one box. While the noise source itself is usually a separate module, the switched power is provided by the noise figure meter. Calculations are performed internally and the noise figure reads out directly on a meter or display. The noise source is repeatedly switched on and off to produce a continuously updated noise figure reading, which is handy for adjusting or tuning the device under test. A low-level 5 to 7 dB ENR source such as the HP 346A is best for testing high-performance preamplifiers. A higher level 15 dB ENR source is also common since it can be used for self-calibration of the meter.

Older units typically operate at a small number of fixed intermediate frequencies and it is up to the user to heterodyne the signal to the IF. Many modern units operate over a wide range of frequencies and some can sweep to provide a plot of noise figure versus frequency. Rohde & Schwarz (1MA178) and Keysight (5952-8255) both publish good application notes on noise figure measurement that are available online and are listed in this chapter's References section.

25.16.2 Receiver Sensitivity (MDS)

Several methods are used to determine receiver sensitivity. The modulation mode often determines the best choice. One of the most common sensitivity measurements is *minimum discernible signal* (MDS) or *noise floor*, which is suitable for CW and SSB receivers. The minimum discernible signal is defined as that which produces the same audio-output power as the internally generated receiver noise. Hence, the term "noise floor." (Many people can hear a signal below the MDS due to the additional processing power of the auditory system.)

To measure MDS, use a signal generator tuned to the same frequency as the receiver. Be certain that the receiver is peaked on the generator signal. In **Figure 25.53** a step attenuator is included at the receiver input since most signal generators cannot accurately generate the low signal levels required for MDS measurements. An audio-frequency ac voltmeter is connected to the receiver's speaker terminals. If no speaker is connected, then a resistor of the same resistance as the speaker impedance should be substituted. Set the receiver to CW mode with a bandwidth of 500 Hz, or the nearest to that bandwidth that is available. Since the noise power is directly proportional to the bandwidth, always use identical filter bandwidths when comparing readings. Turn off the AGC, if possible. With the generator output turned off, you should hear nothing but noise in the speaker. Note the voltmeter reading at the receiver audio output. Next, turn on the generator and increase the output level until the voltmeter shows a 3 dB increase (1.414 times the voltage). The signal input at this point is the minimum discernible

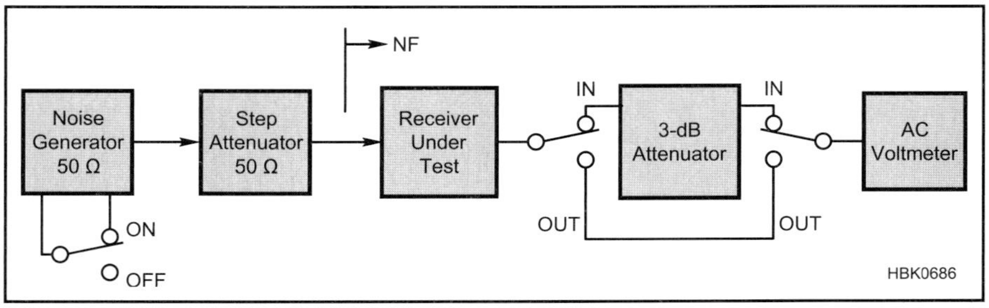

Figure 25.52 — Setup for measuring receiver noise figure.

signal, which can be expressed in μV or dBm.

Typical Test Conditions:

Filter width: SSB 2.4 kHz, CW 500 Hz (sharp passband shape, if available)

Attenuation OFF; Preamplifier OFF

Noise Blanker OFF; Noise Reduction OFF, Notch filter OFF

RF Gain MAX; AGC Medium time constant

In the hypothetical example of Figure 25.53, the signal generator was adjusted to –133 dBm to cause the 3 dB increase in audio output power and the step attenuator is set to 4 dB. MDS is calculated with this equation:

MDS = –133 dBm – 4 dB = –137 dBm

where the MDS is the minimum discernible signal and 4 dB is the loss through the attenuator.

Note that the voltmeter really should be a true-RMS type (e.g. an HP-3400B, Ballantine 323, or Rohde & Schwarz URE3 are often used) to accurately measure the RMS voltage of the noise. A typical average-reading multimeter is calibrated to indicate the RMS voltage of a sine wave but reads low on Gaussian noise. The error is small for the MDS test, however. To correct for the error, adjust the signal generator for a 3.2 dB increase (1.445 times the voltage) instead of 3.0 dB.

Another issue to watch out for is that the peak noise voltage is much greater than the RMS, typically by a factor of 4 or 5. The receiver volume should be adjusted so that the RMS output voltage is no more than about one-fifth the clipping level of the audio amplifier.

For AM modulation, receiver sensitivity is expressed as the RF signal level that results in a 10 dB signal-plus-noise to noise (S+N/N) ratio at the audio output. The AM signal is 30% modulated with a 1000 Hz tone. Otherwise, the test setup is the same as for the MDS test. The signal generator output level is adjusted until there is a 10 dB (3.16 voltage ratio) increase in audio output voltage when the modulation is switched from off to on. Again, there is an error if a typical average-reading multimeter is used instead of a true-RMS voltmeter. For the AM sensitivity measurement, the error is 0.8 dB. Adjust the signal generator output to give a 10.8 dB (3.47 voltage ratio) increase in signal when the modulation is turned on. If an audio distortion meter is used in place of the RMS voltmeter, you can leave the modulation turned on and adjust the signal generator output level for 31.6% distortion.

For FM modulation, receiver sensitivity is expressed as the RF signal level that results in 12 dB SINAD. SINAD stands for "signal plus noise and distortion" and is calculated from

$$\mathrm{SINAD} = 10\log\left[\frac{\text{signal} + \text{noise} + \text{distortion}}{\text{noise} + \text{distortion}}\right]\text{dB}$$

where signal, noise and distortion are all entered in units of power (watts or milliwatts). It is very similar to the signal-plus-noise to noise ratio used for AM testing except that any distortion in the audio signal is added to the noise measurement. It means, however, that the signal, the noise and the distortion must all be measured at the same time, without turning off the modulation. For that, a distortion analyzer (e.g. HP-339A or Rohde & Schwarz UPV) is needed as shown in **Figure 25.54**. A distortion analyzer includes a switchable band-reject filter to null out the 1 kHz tone for the noise-and-distortion measurement. The filter is bypassed for the signal-plus-noise-and-distortion measurement. For the test, the signal generator is set for FM modulation with a 1 kHz tone and the desired FM deviation, normally 3 kHz for VHF repeater operation. Adjust the signal generator output level until the distortion analyzer indicates 25% distortion, which is equivalent to 12 dB SINAD. Don't forget to subtract the attenuation of the step attenuator from the signal generator output level, as was done for the MDS test.

Noise figure is a measure of receiver sensitivity that, unlike the other methods presented so far, is independent of the receiver bandwidth and operating mode. It can be calculated from the MDS so long as the noise bandwidth of the receiver filtering is known. Noise bandwidth is the bandwidth of a hypothetical perfect filter with a rectangular spectrum shape that would produce the same total noise power as the receiver filter. Accurate measurement of a filter's noise bandwidth requires integrating the spectral response using a swept signal source, but the filter's 3 dB bandwidth can be used as a reasonable approximation. The formula is

$$\mathrm{NF} = \mathrm{MDS} - (10\log(\mathrm{BW}) - 174)\ \mathrm{dB}$$

where MDS is the minimum discernible signal in dBm and BW is the noise bandwidth in Hz. Assuming the noise bandwidth is 500 Hz, we get NF = MDS + 147 dB. For example, if the MDS is –137 dBm then NF = –137 + 147 = 10 dB.

25.16.3 Receiver Dynamic Range

Dynamic range is a measure of a receiver's ability to receive weak signals without being overloaded by strong signals. It is easy to design a receiver with good sensitivity to weak signals. It is also easy to design a receiver that is not overloaded by strong signals. It is much more difficult to design a receiver that can do both at the same time.

It is important to note that analog superheterodyne receivers and digital SDR receivers respond to strong signals differently. While both may produce distortion products, the signal levels at which the products appear and how the levels change are different in the two types of receivers. Hybrid receivers in which analog circuits are used to convert signals to an IF where they are digitized and processed using DSP techniques, sometimes referred to as "DSP-IF," respond differently than either pure analog or fully digital receivers. Be cautious in comparing IMD measurements between the different receiver architectures. See the **Receiving** chapter's section on Receiver Dynamic Range for more information.

BLOCKING DYNAMIC RANGE (BDR)

Blocking dynamic range (BDR) or *blocking gain compressions* is the difference between the noise floor and the signal level at which *blocking* occurs, that is, the level that causes a 1 dB reduction in gain for nearby weaker signals. The noise floor is just the MDS and can be measured using the technique described in the previous section.

It is useful to measure BDR for receivers using both analog superheterodyne and digital SDR architectures, including hybrid analog/digital designs and even direct-sampling receivers. Any analog circuitry in the signal path, including signal conditioning circuits in an ADC or DAC are susceptible to overload. Making the same measurement on all receivers gives a better picture of comparative performance.

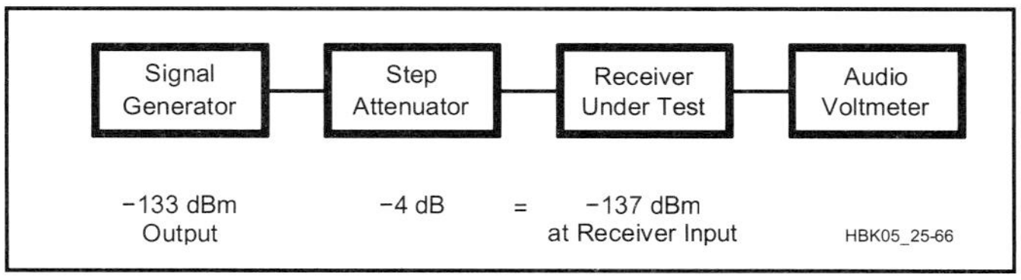

Figure 25.53 — A general test setup for measuring receiver MDS, or noise floor. Signal levels shown are for an example discussed in the text.

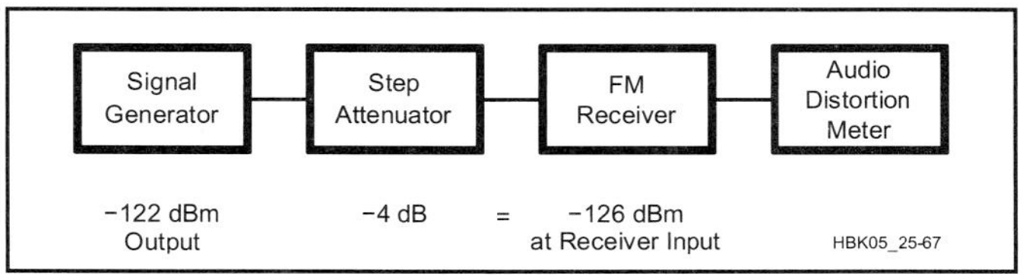

Figure 25.54 — FM SINAD test setup.

The blocking level is measured using a test setup similar to the one used for measuring MDS except that two signal generators are connected to the input through a hybrid combiner, as shown in **Figure 25.55**. The receiver AGC should be turned off for this test. The mode is set to CW and the bandwidth to 500 Hz or the closest available. Two signal generators are used. One generates the weak signal that the receiver is tuned to. The ARRL standard specifies –110 dBm at the receiver input, which requires –97 dBm at the input to the hybrid combiner, assuming it has 3 dB loss. The other signal generator generates the strong interfering signal on a nearby frequency. Standard frequency separations are plus and minus 20, 5 and 2 kHz. The level of the strong signal is increased until the level of the weaker signal measured at the receiver audio output decreases by 1 dB.

Referring to Figure 25.55, the blocking level is the level from the signal generator minus the loss of the hybrid combiner and attenuator,

BL = –7 – 3 – 10 = –20 dBm

The blocking dynamic range is given by

BDR = BL – MDS = –20 – (–137) = 117 dB

assuming an MDS of –137 dBm as in the previous examples.

One complication is that it may be difficult to measure the amplitude of the audio tone because of the presence of noise caused by the phase noise of the signal generator and the receiver's local oscillator, especially at the 2 and 5 kHz frequency spacings. The solution is to use an audio-frequency spectrum analyzer to measure the change in tone amplitude. The absolute accuracy of the spectrum analyzer is not important so long as it can accurately show a 1 dB change in signal level. An instrument based on a computer sound card and free spectrum analysis software should be adequate.

In a direct-sampling SDR receiver, no blocking occurs until the ADC is driven into saturation (clipping). This assumes correct design of the active stages preceding the ADC. (Circuitry ahead of the ADC, such as an RF amplifier, can exhibit gain compression, for example.) Thus, for SDRs, the clipping level is measured and stated in dBm, rather than blocking dynamic range (BDR) in dB.

RECIPROCAL MIXING DYNAMIC RANGE (RMDR)

Reciprocal mixing is the name for the mixing of a nearby interfering signal with the phase noise of the receiver's local oscillator or phase noise sidebands of the digitizing ADC's clock signal. The noise signals mix with strong signals close in frequency to the desired signal, producing unwanted noise in the audio output or detection channel. This degrades receiver sensitivity.

Although the ARRL BDR test eliminates this effect from the measurement, in actual on-the-air operation the phase noise is often the factor that limits the effective dynamic range. To address this issue, the ARRL test suite includes a separate measurement for reciprocal mixing. The test setup is the same as for MDS except that the signal generator in Figure 25.53 is replaced with a low-phase-noise crystal oscillator (or OCXO) with an output power level of +15 dBm. The OCXO operates at a fixed offset from the desired signal. The oscillator signal level is increased until the receiver noise floor increases by 3 dB as measured at the audio output.

Since the output of the OCXO is usually fixed, a variable attenuator is used to reduce the output level in controlled steps. The step attenuator block should include both a unit with 10 dB steps to be able to adjust the signal level over a wide range and a 1 dB step attenuator for fine adjustment. The receiver is tuned to 20, 5 or 2 kHz above or below the oscillator frequency. The output noise level is first measured with the oscillator turned off, then the oscillator is turned on and the step attenuator gradually reduced until the noise increases by 3 dB. *Reciprocal mixing dynamic range* is expressed as:

RMDR (in dB) = (OCXO – A) – MDS

where MDS is the noise floor, OCXO is the signal level of the crystal oscillator in dBm (+15 dBm in the example figure), and A is the total attenuation in dB.

The same test can also measure phase noise at the particular receiver offset from the oscillator.

Phase noise (in dBc/Hz) = (RMDR + 10 log BW)

Where BW is the receiver's filter bandwidth. If BW = 500 Hz, 10 log BW = 27)

Typical Test Conditions:

Filter width: CW 500 Hz (sharp passband shape, if available)

Attenuation OFF; Preamplifier OFF

Noise Blanker OFF; Noise Reduction OFF, Notch filter OFF

RF Gain MAX; AGC Medium time constant

In general, to assess the impact of strong off-channel noise on weak signals, one should look at the worst-case of blocking and reciprocal-mixing dynamic range.

INTERMODULATION DISTORTION DYNAMIC RANGE (DR2 and DR3)

Intermodulation distortion (IMD) means the creation of unwanted signals at new frequencies because of two or more strong interfering signals modulating each other. This happens due to non-linearities of circuits,

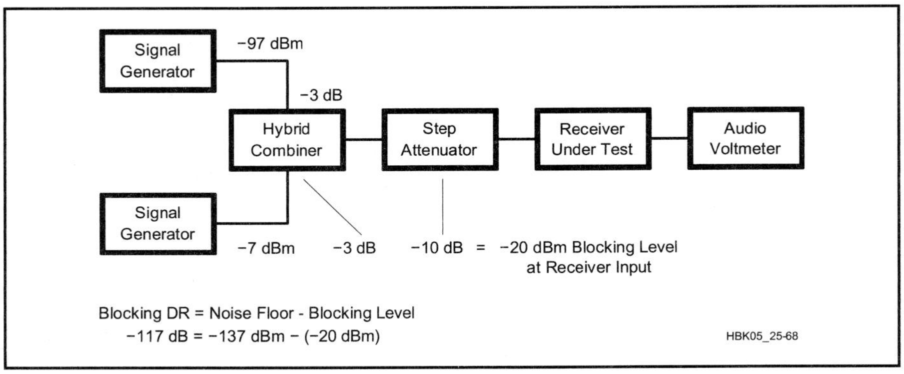

Figure 25.55 — Receiver blocking dynamic range is measured with this equipment and arrangement. Signal levels shown are for the example discussed in the text.

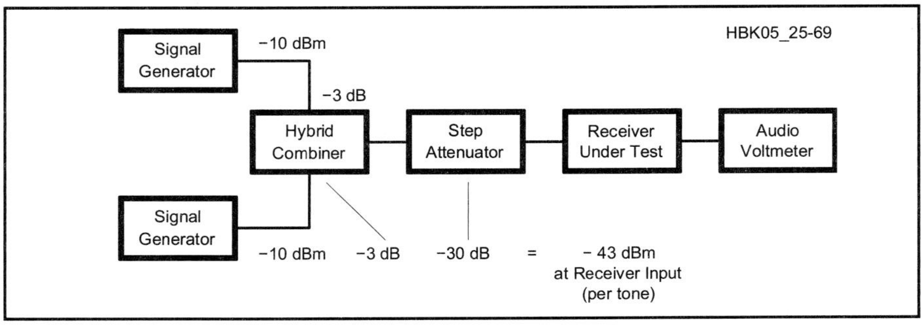

Figure 25.56 — The test setup for receiver intermodulation distortion dynamic range. Signal levels shown are for the example discussed in the text.

active and passive, in the path of the strong signals.

If there are two interfering signals at frequencies f1 and f2, then *second-order IMD* products occur at f1 + f2. If f1 and f2 are close together, then the second-order products occur near the second harmonics. *Third-order IMD* products occur at 2f1 – f2 and 2f2 – f1. If f1 and f2 are close together, then the third-order products occur close by. For example, if f1 and f2 differ by 10 kHz, then the third-order IMD products are 10 kHz above the higher frequency and 10 kHz below the lower.

The two-tone IMD test setup shown in **Figure 25.56** is the same for second and third-order IMD. For receivers with high dynamic range, to obtain sufficient output power and isolation of the two signal generators, it may be necessary to follow each one with a wideband power amplifier, not shown. The receiver under test is set to receive CW with the same bandwidth as for the MDS test and is tuned to the frequency of the distortion product to be measured. The two signal generators are always set to the same output amplitude level, which is increased until IMD products equal in amplitude to the noise floor appear, resulting in a 3 dB increase in the audio voltmeter reading. The *IMD dynamic range* is then the difference in dB between the level of the interfering signals and the MDS.

For the second-order IMD measurement, the standard ARRL test sets the two signal generators to 6.000 MHz and 8.020 MHz and the receiver is tuned to 14.020 MHz. For the third-order IMD measurements, the two signal generators are set to frequencies 20, 5 and 2 kHz apart, separated from the receiver frequency by plus and minus 20, 5 and 2 kHz so that the lower or upper IMD product falls within the receiver passband.

For example, tune the receiver to 14.020 MHz and the two signal generators to 13.980 MHz and 14.000 MHz. With the signal generators turned off, measure the noise level with the audio voltmeter. Turn on the signal generators and increase their amplitudes until the voltmeter shows a 3 dB increase. If the signal generator amplitudes are –10 dBm, the loss in the hybrid combiner is 3 dB, and the step attenuator is set to 30 dB attenuation, then the third-order IMD level is

$$IMD = -10 - 3 - 30 = -43 \text{ dBm}$$

If the MDS is –137 dBm, then the third-order IMD dynamic range is

$$IMD\_DR = IMD - MDS = -43 - (-137) = 94 \text{ dB}$$

As with the blocking dynamic range test, the phase noise of the signal generators and the receiver LO may obscure the IMD product being measured. Again, the solution is to use an audio spectrum analyzer to measure the tone amplitude. Calibrate the amplitude by temporarily tuning one of the signal generators to the receiver frequency and setting the amplitude level so that the signal level at the receiver is the MDS. Note the level on the spectrum analyzer. Then return the signal generator to the interfering frequency and adjust the signal generators' amplitudes until the IMD product is at the same level as the MDS signal. (This is the *subtractive method* discussed in ITU-R Recommendation SM.1837-1, Section 2.)

If the subtractive method is used for DR3 measurement, RMDR must be measured in the course of the same test suite to allow the test engineer to determine whether the DUT is IMD-limited or phase-noise-limited.

The ARRL test bench actually uses a third signal generator and a second hybrid com-

Dynamic Range Metrics: Superheterodyne vs. SDR Receivers

Reciprocal Mixing (RM) is the noise generated in a superheterodyne receiver when noise from the local oscillator (LO) mixes with strong, adjacent signals. All oscillators have noise sidebands, and some types and designs have more than others. The noise sidebands mix with the strong adjacent off-channel signal, creating noise products in the mixer's output. This noise can degrade the sensitivity of the receiver and is most notable when a strong signal is just outside the IF passband. Reciprocal mixing is worse with a single strong signal 2 kHz away from the tuned frequency than it is 5 or 20 kHz away.

It's interesting to compare to the two-tone third-order IMD dynamic range (3IMD DR) to the reciprocal mixing dynamic range (RMDR). In the case of 3IMD, two strong adjacent signals add up to cause an unwanted effect, but in the case of RM, a *single* strong adjacent station 5 or 2 kHz away from the desired signal has more of an impact on the ability to hear a desired weak signal. For most superheterodyne receivers, RMDR is the worst dynamic range figure at 2 kHz spacing. The major source of phase noise in a direct-sampling SDR receiver is the ADC clock subsystem. Thus, it is essential to ensure that the clock source and distribution chain have as low phase noise as practicable.

It has been observed in the ARRL Laboratory that many direct-sampling SDR receivers exhibit little, if any reciprocal mixing up to the point of ADC overload. These receivers have no front-end mixer and local oscillator, thus, no sideband noise of an oscillator to mix with adjacent signals.

We perform the reciprocal mixing test at 14.025 MHz, using a very low-noise Wenzel test oscillator with a measured output of +14 dBm to generate the off-channel, strong signal. A test oscillator must have the lowest sideband noise as possible — considerably lower than the reciprocal mixing being measured. An excellent receiver exhibits 100 dB of RMDR at 2 kHz spacing.

Blocking — an apparent reduction in receiver gain due to the presence of a strong signal — was problematic in years past. With the AGC off, the desired signal, especially if weak, would disappear from the receiver audio if a strong, off-channel signal was present. The ability to reject adjacent signals has improved greatly in recent years, so much so that blocking dynamic range (BDR) greatly exceeds the other dynamic ranges. During Lab tests, reciprocal mixing in a receiver can raise the audio output noise enough to drown out the desired signal! Some of our laboratory consultants have expressed that the BDR figure is a bit useless because of this behavior. Agreed! Still, some still wish for us to report blocking, since we have the use of signal analyzers set to narrow audio bandwidths (2 or 5 Hz).

Many SDR receivers measured for *QST* Product Review exhibit little or no blocking effects up to the point of ADC overload. This behavior, along with an SDR's reciprocal mixing behavior is most desirable. Typical BDR of modern receivers at 2 kHz spacing is 110 dB or greater.

Two-tone Third-Order Intermodulation Distortion (3IMD) happens when two strong signals (2 and 4 kHz away from the tuned frequency, for example) appear at the antenna jack, with a resulting unwanted "phantom" signal heard at the tuned frequency. With a superheterodyne receiver, IMD is generated at the input stages to the first IF, with the undesired mixing products passing through the rest of the receiver stages. In an SDR receiver, the ADC clock quality and the cumulative signal level input to the ADC determines the level of IMD at the speaker (see other sidebar).

Due to design improvements of both superheterodyne and SDR receivers, typical 3IMD DR at 2 kHz spacing is 90 dB or greater. Due to changes in technology, most modern receivers do not exhibit a 3:1 relationship between the IMD signal level and the IMD input level. The ratio can be significantly greater or less than 3:1. Since the third order intercept point (IP3) figure is calculated based on the assumption of a 3:1 ratio, this figure is meaningless with today's receivers. Emphasis must be placed on all three dynamic ranges. The lowest dynamic range at 2 kHz of the three dynamic ranges we publish is the dynamic range of the receiver. The benchmark for excellent performance is 100 dB.

biner at the receiver input to generate the calibration signal so that it and the IMD product can be seen on the spectrum analyzer at the same time for a more-accurate comparison.

An alternative method for measuring IMD is presented in the supplemental paper "Testing and Calculating Intermodulation Distortion in Receivers," by Ulrich Rohde, N1UL. Avoiding potential errors of measuring noise power and audio response characteristics of the receiver, Rohde's method measures IMD product levels at a repeatable S meter reading. An adjustable attenuator is used to control the IMD product levels, with the difference in attenuator readings being much less error-prone than noise power measurements.

THIRD-ORDER INTERCEPT POINT (IP3)

The third-order intercept (IP3) is generally not a valid concept for direct-sampling SDR receivers that do not use an analog front end. ADCs usually do not exhibit the 3 dB per dB relationship between signal level and third-order products, at least over major portions of their operating range. Comparing IP3 measurements of an SDR and a conventional analog receiver is likely to give misleading results.

In most analog components such as mixers and amplifiers, the second-order products increase in amplitude by 2 dB for each 1 dB increase in the interfering signals and third-order products increase 3 dB per dB. If the output signal levels are plotted versus the input levels on a log-log chart (that is, in units of dB), the desired signal and the undesired IMD products theoretically trace out straight lines as shown for third-order products in **Figure 25.57**. Although the IMD products increase more rapidly than the desired signal, the lines never actually cross because blocking occurs before that level, however, the point where the extensions of those two lines cross is called the *third-order intercept point* (IP3).

Although the third-order intercept is an artificial point, it is a useful measure of the strong-signal-handling capability of a receiver. It can be calculated from the equation

$$IP3 = MDS + 1.5 \times IMD\_DR \text{ dBm}$$

where IP3 is the third-order intercept point, MDS is the minimum discernible signal in dBm and IMD_DR is the third-order IMD dynamic range in dB. Using the numbers from the previous example, IP3 = –137 + 1.5 × 94 = +4 dBm.

Note that IP3 (3rd-order intercept) is not a valid test metric for a direct-sampling SDR, as IMD in an ADC follows a quasi-1st-order rather than a 3rd-order law. The transfer and IMD curves diverge and never intersect. In a conventional receiver, IP3 is the convergence point of the transfer and IMD curves.

The second-order intercept may be calculated in an analogous way.

$$IP2 = MDS + 2 \times IMD\_DR$$

where IP2 is the second-order intercept and IMD_DR is the second-order IMD dynamic range in this case.

An alternate method of measuring third-order IMD is to use S5 (–97 dBm) instead of the MDS as the reference level to which the IMD products are adjusted. That results in a higher IMD level but a smaller value of IMD dynamic range (the difference between the IMD level and the reference). It may be a more accurate method of determining the third-order intercept because the signal levels do not have to be measured in the presence of noise that is at the same level as the signal.

Figure 25.58 shows the relationship between the various dynamic range values. The base line represents different power levels, with very small signals at the left and large signals at the right. The numbers listed are from the previous examples for a typical receiver. The thermal noise and the noise floor are referenced to a 500 Hz bandwidth. The third-order IMD dynamic range is less than the blocking dynamic range, which means that signals as low as –43 dBm may cause IMD interference while signals must exceed –20 dBm to cause blocking. However, the intermodulation distortion may actually cause fewer problems since the IMD products only appear at certain discrete frequencies. A signal that exceeds the blocking level can cause interference across the entire band.

Third-order IMD dynamic range may also be measured on an FM receiver. The test setup is the same as Figure 25.56 except that the audio voltmeter is replaced with an audio distortion meter, as in Figure 25.54. The frequencies involved are the same as in that example. For this test, one of the signal generators (the one tuned to the frequency farthest from the receiver under test) is FM-modulated with a 1000 Hz tone at 3 kHz deviation. That causes the IMD products also to be FM-modulated, which can be measured with the distortion meter. The signal generator amplitudes are increased until the distortion product produces 12 dB SINAD (25% distortion) on the meter. The FM third-order IMD dynamic range is calculated using the same equation as for SSB and CW except that the SINAD sensitivity is substituted for the MDS. For example, if the 12 dB SINAD sensitivity is –120 dBm, the signal generator outputs are –10 dBm, the combiner loss is 3 dB, and the step attenuator is set to 30 dB, then

$$FM\_IMD\_DR = -10 - 3 - 30 - (-120) = 77 \text{ dB}$$

where FM_IMD_DR is the third-order FM IMD dynamic range.

INTERFERENCE-FREE SIGNAL STRENGTH (IFSS)

Since a direct-sampling SDR does not experience IMD in the same way as an analog superheterodyne, a different test is required to measure the spurious products generated by strong signals. Using the same test configuration as for single-frequency DR3 tests described above, the IFSS test creates a chart of absolute IMD product power level vs. test signal input power level, measured at 1 dB intervals over the range from ADC clip level to the noise floor; with the ITU-R P.372 site band-noise levels at the test frequency as da-

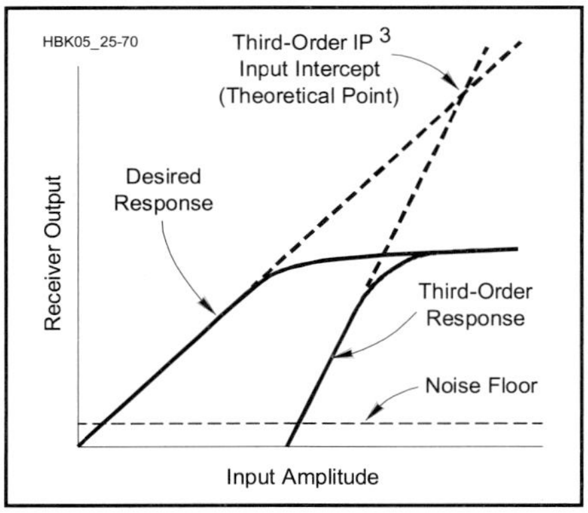

Figure 25.57 — The third-order intercept point can be determined by extending the lines representing the interfering signal level and the third-order intermodulation products on a plot of the signal levels in dB.

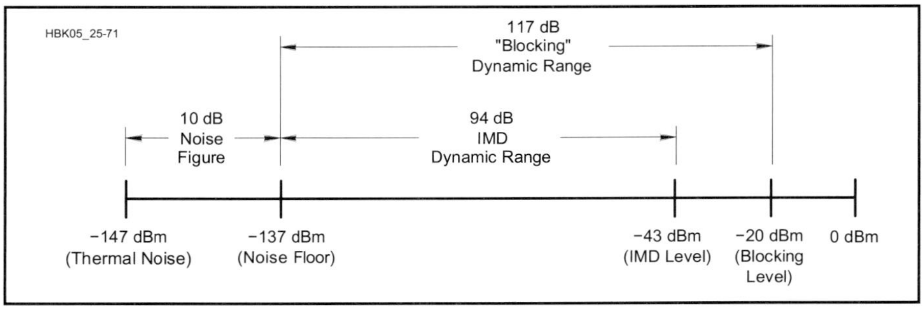

Figure 25.58 — Performance plot of the example receiver discussed in the text.

tum lines. The same instrumentation is used as for the "classical" single-point DR3 test.

ADC CLIPPING

Also known as *overflow*, *overload*, and *saturation*, *ADC overload* occurs when the signal level into an ADC exceeds the maximum encoded level for the ADC. At and beyond this point, the receiver exhibits spurious and non-linear responses. Even if the overload condition is removed, there may be persistent effects in DSP functions such as filtering or time-sensitive functions such as AGC or noise reduction. Many receivers provide an indicator that lights when the overload condition is present.

In the ADC Overload test, the receiver is offset +25 kHz above the test signal frequency and the input level required to light the overflow indication or screen icon is noted. The overload indication should only occur when a strong out-of-channel signal is present. In-channel signals stimulate AGC action to reduce signal levels at the ADC input.

Typical Test Conditions:

Receiver tuned to 14.1 MHz, Test signal frequency 14.125 MHz

Selectivity filter width: CW 500 Hz (sharp passband shape, if available)

Attenuation OFF; Preamplifier OFF

Noise Blanker OFF; Noise Reduction OFF, Notch filter OFF

RF Gain MAX; AGC Medium time constant;

NOISE POWER RATIO (NPR)

Noise Power Ratio is a very effective test of overall receiver dynamic performance on a band crowded with extremely strong signals. In other words, exactly the conditions in which a receiver's strong-signal performance is most likely to be tested on the air.

The NPR test is performed by applying a band of white noise (called *noise loading*) to the receiver with a narrow notch filter. (A bandpass filter is used to limit the bandwidth of the applied white noise.) The receiver is configured to have a bandwidth narrower than the notch, then tuned to the center frequency of the notch. The output of the receiver is called *idle channel noise (ICN)*. The test is performed with and without the notch. When testing an SDR receiver, increase the noise level until the onset of clipping is reached, then reduce the noise level until no clipping indication occurs for at least 10 seconds (typically 1dB below clip level).

The theory behind the NPR test is that the noise outside the notch causes reciprocal mixing noise and multiple IMD products, which then appear in the receiver output (the idle channel) and raise the ICN. The test requires the receiver IF or DSP filter bandwidth to be no wider than the bottom of the notch; the filter bandwidth must not be wide enough

SDR Behavior with Two-Tone Third Order Intermodulation Dynamic Range: The Third Signal

The ARRL Laboratory determined that our normal two-tone third order IMD dynamic range (3IMD DR) tests did not always apply to the way software defined radio (SDR) receivers work in the real world.* With an SDR receiver, all of the demodulation is done by software. An SDR converts a block of analog RF spectrum to a digital stream of data, then uses software to select and demodulate a desired signal from that data. The digitization of an analog signal results in a sequence of small steps in signal level. These small steps are a type of nonlinearity that forms intermodulation (IMD) products, similar to those observed in analog receivers. (See reference entry for Allison.)

For analog receivers, once intermodulation occurs, it gets strong quickly as the off-channel signals used for testing are increased. With an SDR, however, because the small steps are the same for all levels of the signal being digitized, intermodulation at low levels does not vary significantly with the level of test signals. For the steady-state sine waves seen from signal generators in a laboratory environment, the steps are the same for each cycle of RF. Any intermodulation that is created adds coherently in the receiver output.

In some SDR designs, random noise is added to the RF input to randomize the digitization process, preventing intermodulation products from adding coherently. This technique is called *dithering*. When a receiver is connected to the antenna, band noise present at the input to the ADC can, in most cases, serve to randomize the digitization. The net effect is the same — low-level intermodulation seen in a laboratory test environment is not present when that receiver is connected to an antenna and used to receive on-the-air signals. This effect was verified by measurements in the Lab by adding a receive antenna (acting as our dithering source) to our two-tone IMD test fixture; the level of band noise varied with local conditions and propagation, thus, the amount of dithering present also varied, causing IMD products to vary as well.

With such observations, we had to find out if the IMD products varied as a function of noise or as a function of level. We substituted a random noise generator as our dithering source. The result: low-level IMD products were masked by the raised noise level with little change in the level of the IMD products. Varying the level of the noise had no effect, except the unwanted effect of a raised noise floor. Sherwood & Farson have confirmed via lab testing that noise added to the 2-tone test signal at the DUT RF input did not reduce IMD product amplitude, but merely appeared as noise in the DUT audio output. For noise dithering to be effective, ADC on-chip dithering is required in which added digitized noise is digitally subtracted from the ADC output bitstream. (See reference entry for Kester.)

Since noise or its level had no effect on the level of the IMD products, the only other dithering source received on an antenna must be discrete signals! We soon discovered that by using a strong, off frequency, single signal as our dithering source (the third signal), the IMD product level could be reduced. By varying the level of the third signal (generator), the IMD product level also varied. Thus, when connected to an antenna, the levels of all the signals at the input to the ADC vary considerably with propagation and mode, and thus the dynamic range varies with propagation and mode. This significant discovery determined that adding noise (dithering) does not necessarily improve an SDR's 3IMD DR, but that a single, strong, constant, off-frequency signal does. By experimenting, we found that by varying both the third signal's level and frequency, a "sweet spot" could easily be found that maximizes the 3IMD DR. A third signal at a level of –43 dBm, 200 kHz away from the tuned frequency is a good example of a "sweet spot."

This single additional signal, at a specific level and frequency, hardly represents actual on-air conditions. In reality, the 3IMD DR can be a moving target, varying by 30 dB or more depending on conditions! With SDRs that exhibit improved 3IMD DR, *QST* Product Reviews report "up to XX dB," which represents the absolute best-case dynamic range. A user of such devices can expect much lower dynamic range under average band conditions, with an effort made to mention the expected dynamic range during "quiet" conditions.

It is interesting to note that by design, some manufacturers of SDRs now employ a "third signal" to improve 3IMD DR; the third signal is internally generated and tracks with the tuned signal, at a specific level, which minimizes IMD products and maximizes 3IMD DR.

Please keep in mind that when an antenna is attached to a receiver, atmospheric, solar and man-made noise mask much of the low-level IMD products that are measured in a laboratory environment. That is, the dynamic range can be higher under laboratory "quiet" conditions, where a receiver's noise floor is measurable.

* Allision, Bob, WB1GCM, "Testing the SDR-IQ in the ARRL Lab," QST, Feb. 2010 p. 52.

to allow noise outside the notch to spill over into the receiver's passband.

NPR for a given noise bandwidth is the reciprocal of the ratio of the noise power in the notched band to the power in an equal bandwidth adjacent to the notch. For a given noise bandwidth:

$$NPR = P_{TOT} - BWR - MDS$$

where

P_{TOT} is the total noise power in dBm in the noise bandwidth (*noise loading level*)

$BWR = 10 \log (B_{RF} / B_{IF})$

B_{RF} = RF bandwidth or noise bandwidth in Hz of the filter applied to band-limit the noise

B_{IF} = Receiver selectivity filter bandwidth in Hz (IF for superheterodyne, DSP filter for DR)

This can also be expressed as:

$$NPR = D_N + 10 \log B_{IF} - MDS$$

Where D_N is the noise density in dBm/Hz = $P_{TOT} - 10 \log B_{RF}$

The band-limited filter should be wider than the receiver's front-end bandwidth to ensure that the test subjects all stages of the receiver to noise loading, including any front-end filter or preselector. That ensures any effects of the noise in the front end of the receiver are taken into account in the measurement.

In a superheterodyne receiver, the effect of the high noise power outside the notch most strongly affects the first and second mixer. The noise also mixes with the noise sidebands of the LO to cause reciprocal mixing. The noise components also mix with each other and any internally generated noise or spurious products to produce a very large number of IMD products. Some of both the reciprocal mixing noise and IMD products fall into the receiver's passband, degrading ICN.

See the paper on Noise Power Ratio (NPR) Testing of HF Receivers by Farson in the References for a detailed analysis of NPR testing.

Typical Test Conditions:
Receiver tuned to notch filter center frequency, ±1.5 kHz
Selectivity filter width: SSB 2.4 kHz
Attenuation OFF; Preamplifier OFF
Noise Blanker OFF; Noise Reduction OFF, Notch filter OFF
RF Gain MAX; AGC Slow time constant;

25.16.4 Standard ARRL Receiver Measurements

LATENCY

Latency is the time interval between arrival of an impulsive event at the receiver RF input and the corresponding output event at the receiver audio output. Latency is measured at various detection bandwidth settings. The RF test signal is generated by a pulse-modulated or "bursty" RF signal source. A dual-channel oscilloscope is used to measure the time interval.

Various aspects of receiver design exert influence on latency. Among these are DSP processing speed and group delay across selectivity filters. Since the DSP processing speed is fixed, latency is measured for various filter configurations of bandwidth and shape factor.

To measure latency, a pulse or pulse-modulated signal generator feeds repetitive pulses via a hybrid splitter to the receiver antenna input and to Channel 1 of a dual-trace oscilloscope. Channel 2 is connected to the receiver AF output. The scope is triggered from the generator's trigger output. The time interval between the pulse displayed on Channels 1 and 2 is the receiver's latency as in **Figure 25.59**. This is *not* the same as latency of a receiver operated by remote control where the audio channel includes network delays.

Typical Test Conditions:
Receiver tuned to 3.6 MHz
Attenuation OFF; Preamplifier OFF
Noise Blanker OFF; Noise Reduction OFF, Notch filter OFF
RF Gain MAX; AGC Fast time constant
Pulse generator (if used) output 16 V_{P-P} with 60 dB attenuation between pulse generator output and receiver input, 60 ns duration, 0.2 s period
Signal generator (if used):
RF output –70 dBm
Pulse duration 200 µ
Pulse period 200 µ

FIRST IF AND IMAGE REJECTION

The first *IF rejection* and *image rejection* tests are performed on superheterodyne receivers. These tests measure the signal levels at the intermediate frequency and the image frequency that produces an audio output signal equivalent to the MDS, or noise floor. The test setup is the same as for the MDS test, as shown in Figure 25.53. The receiver is set to CW mode and 500 Hz bandwidth. The signal generator is tuned to the receiver first intermediate frequency or to the image frequency, which is the receiver frequency plus or minus two times the IF. The signal generator output level is gradually increased until there is a 3 dB increase in the audio voltmeter reading. As with the MDS test, if a multimeter is used instead of a true-RMS voltmeter, the signal generator should be adjusted for a 3.2 dB increase. The IF or image rejection is just the signal generator level at the receiver input less the MDS. For example, if the MDS is –137 dBm, the signal generator level is –40 dBm, and the attenuator is set for 10 dB, then the IF or image rejection is IR = –40 – 10 – (–137) = 87 dB.

FM ADJACENT-CHANNEL REJECTION

FM adjacent-channel rejection is a measure of an FM receiver's ability to detect a weak signal in the presence of a strong interfering FM-modulated signal on an adjacent frequency channel. The test setup is the same as for the two-tone IMD dynamic range test shown in Figure 25.56 except that a distortion meter is substituted for the audio voltmeter, as was done for the FM SINAD test illustrated in Figure 25.54. The standard channel spacing is 20 kHz. The weak signal is modulated with a

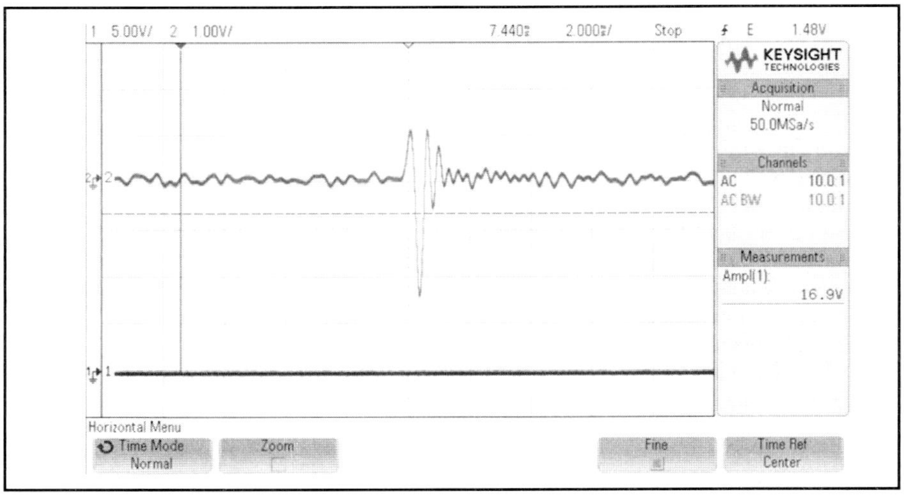

Figure 25.59 — Receiver latency is measured by inputting a very narrow pulse to the receiver (vertical line on the bottom trace with the marker on the top) and measuring the time until the response is observed in the receiver output (top trace). (Image courtesy of Adam Farson, AB4OJ/VA7OJ)

1000 Hz tone and the interfering signal with a 400 Hz tone, both with 3 kHz deviation. The receiver under test is set for FM modulation and is tuned to the frequency of the weak signal. The weak signal is adjusted for 12 dB SINAD (25% distortion) on the distortion meter and then the interfering signal level is increased until the SINAD drops to 6 dB (50% distortion). The adjacent channel rejection is the difference between the power of the interfering signal and the 12 dB SINAD sensitivity, which is just the difference in level between the two signal generators.

SQUELCH SENSITIVITY

Another test that applies especially to FM receivers is the *squelch sensitivity test.* As usual, the signal generator is set for FM modulation with a 1 kHz tone and 3 kHz deviation. If squelch is available for SSB, it can be tested in that mode as well. With the signal generator off, the squelch control on the receiver is adjusted just to the point where the noise is squelched. Then the signal generator is turned on and the level increased until the signal is heard.

AUDIO OUTPUT POWER

Audio power output is tested with the same setup as the FM SINAD test illustrated in Figure 25.54. The receiver under test is set for SSB mode with the widest bandwidth available. A load resistor of the specified resistance, usually 8 Ω, is connected to the speaker output in place of the speaker. The signal generator level is set for an S9 level and the receiver is tuned for a 1 kHz output tone frequency. The receiver volume control is increased until the specified distortion level, usually 10%, is indicated on the distortion meter. The output power is given by the equation

$$P = V_{RMS}^2 / R$$

where P is the power in watts, V_{RMS} is the RMS output voltage, and R is the load resistance.

The audio and IF frequency response measures the audio frequencies at which receiver audio output drops by 6 dB from the peak. It includes the total response of the entire receiver, from the antenna connector to the speaker output. The test setup is the same as for the audio power output test except that some method must be included to measure the output audio frequency, such as a digital oscilloscope or frequency counter. The receiver is set for SSB mode at the bandwidth to be tested and the AGC is turned off. First tune the signal generator for a peak audio output signal and record the level of the audio signal. Then tune the signal generator downward until the signal drops 6 dB (1/2 the voltage) and record the audio frequency. Then tune the signal generator upward and record the high-end frequency at which the signal drops by 6 dB. The 6 dB bandwidth is the difference between the two frequencies.

MISCELLANEOUS TESTS

There are several other miscellaneous ARRL receiver tests that are commonly reported in product reviews. The S meter test measures the signal level required to produce an S9 indication on the meter. The notch filter test uses a setup similar to the IMD dynamic range test in Figure 25.56 with an audio spectrum analyzer at the output. One signal is notched and the other is used as a level reference; the notch depth is the amplitude difference between the two tones. The DSP noise reduction test uses a similar setup except that one signal generator is replaced with a wideband noise generator. The signal generator is adjusted for S9 and the noise source for a 3 dB increase in the audio voltmeter so that the noise and signal are at the same level. The DSP noise reduction is then turned on and the reduction in noise level recorded.

25.16.5 Other Common Receiver Measurements

There are a few other useful receiver tests that are not covered in the standard ARRL test suite. For example, in addition to the IF and image response, it is possible for a receiver to have spurious responses at other frequencies as well. Testing for that can be a time-consuming process since it involves tuning the signal generator through a wide frequency range with the receiver tuned to each of a number of representative frequencies, typically at least one on each band. Start with the signal generator at maximum output power. When you find a response, reduce the level until the received signal level is at the noise floor (MDS). The spurious response amplitude is the difference between the signal generator level and the MDS.

Another time-consuming test is for internally generated spurious signals, sometimes called *birdies* because they sometimes sound like a bird chirping as you tune through the signal. Birdies are typically caused by harmonics of the local oscillator(s) and BFO and their IMD products. For this test, connect a 50 Ω termination to the antenna connector and tune the receiver through its entire frequency range, writing down the frequency and S meter reading for each spurious signal found. You must tune very slowly because many birdies pass through the IF passband much faster than regular signals, sometimes in the opposite direction.

The ARRL S meter test only measures the response at the S9 level. If the S meter is accurate throughout its range, it can be used to measure signal levels off the air. The standard definition is that, on the HF bands, S9 should correspond to –73 dBm (50 µV into 50 Ω) and each S unit corresponds to 6 dB, or a doubling of RF voltage. So S8 is –79 dBm or 25 µV, S7 is –85 dBm or 12.5 µV, and so on. The S unit calibration of most commercial equipment varies considerably from 6 dB per S unit, but the S meters accuracy can easily be measured with a calibrated signal generator.

The automatic gain control circuitry has a major effect on the operation of a receiver. The static AGC performance can be measured on a CW or SSB receiver with a signal generator and an audio voltmeter connected to the speaker output. You can plot a graph of audio output in dBV (decibels with respect to one volt) versus RF input level in dBm to see how good a job the receiver does in keeping the output level constant. Some receivers have menu settings to set the slope of the curve and the threshold, which is the small signal level at which the AGC circuitry starts to reduce the gain. The dynamic response is also important, although it is difficult to measure. A short attack time is important to reduce transients on sudden strong signals, but if it is too short then in-channel intermodulation distortion between strong signals can make signals sound "mushy."

The ARRL test suite includes a measurement of the 6 dB bandwidth at the audio output, but the sound is also affected by slope and ripple in the passband response. Using the same test setup as for the 6 dB measurement, you can measure the output voltage at a number of equally spaced frequencies and plot the results on a graph. Output power is measured at the 10% distortion level, but it also is useful to measure the distortion at a volume level closer to what is used in actual practice. An output level of 1 V_{RMS} is commonly used for that test.

Frequency accuracy and stability are important performance criteria for a receiver. If a signal generator of sufficient frequency accuracy is not available, a standard time signal such as from radio station WWV can be used. The traditional method is to put the receiver in SSB mode and *zero-beat* the carrier, which means to tune the receiver until the audio tone (the beat note) is at zero Hz. Since most receivers' audio frequency response only extends down to a couple of hundred Hertz, it is difficult to get good accuracy using that method. If a frequency counter or other means of measuring the audio output frequency is available, you can tune the receiver to obtain a 1000 Hz tone and add (for LSB) or subtract (for USB) 1000 Hz from the receiver's indicated frequency. The frequency drift from cold turn-on can be measured by plotting the audio output frequency versus time.

25.17 Transmitter Measurements

25.17.1 RF Power Output

For a transmitter, the RF power output is probably the first measurement that comes to mind. The test setup is straightforward; connect the transmitter output to the input of an RF wattmeter and connect the wattmeter output to a suitable dummy load. For CW, AM, and FM modes, simply key the transmitter and measure the power on the wattmeter. For AM and FM, the modulation should be turned off. For SSB, a two-tone audio oscillator should be connected to the microphone input to take the place of the voice signal. For the SSB test, the wattmeter must be a peak-reading type to measure the PEP power. PEP can also be measured with an oscilloscope previously calibrated against a wattmeter in CW mode.

25.17.2 Transmitter Spectral Purity

Ideally a transmitter confines its emissions to a narrow frequency range around the desired signal. Unwanted emissions can be divided into two categories, those that fall close to the desired signal and those that extend far away.

In the latter category are included harmonics and other discrete spurious frequencies. The measurement is done with the transmitter in CW mode using the test setup shown in **Figure 25.60**. Tune up the transmitter as specified in the manual, set it for the desired power level, and adjust the frequency from one end of the band to the other while observing the spectrum analyzer. This should be done on each band. It may be necessary to retune the transmitter occasionally as the frequency is adjusted. The spurious-signal and harmonic suppression is the difference in dB between the carrier and the maximum spurious signal. It is important that the power level into the spectrum analyzer be maintained at a low enough level that spurious signals are not generated in the spectrum analyzer itself. To test for that, try changing the step attenuator setting; the desired carrier and all spurious signals should change by the same amount. If not, increase the attenuation until they do.

Representative spectrum analyzer plots are shown in **Figure 25.61**. The horizontal (frequency) scale is 5 MHz per division. The desired carrier frequency at 7 MHz appears 1.4 divisions from the left of the plot. Although not shown, a large apparent signal is often seen at the extreme left edge. This signal at zero Hz is caused by leakage of the spectrum analyzer local oscillator frequency and should be ignored.

In addition to the discrete spurious signals measured by the previous test, a transmitter may also generate broadband noise. It can be due both to the phase noise of the local oscillator as well as AM noise from all the devices in the amplifier chain. Generally, the phase noise predominates, at least for frequencies close to the carrier. The composite noise test measures the total noise from both sources as well as any close-in spurious frequencies that don't show up in the wideband spurious signal and harmonic suppression test. Measuring phase noise requires high-performance test equipment and special measuring techniques as described in the Reference for Pontius. The ARRL Lab uses a special low-noise oscillator and a Hewlett-Packard (now Keysight Technologies) phase noise test set under computer control to perform this sophisticated measurement, which takes about 15 minutes to perform once it is set up. The result is a plot of the noise spectrum such as the one in **Figure 25.62**.

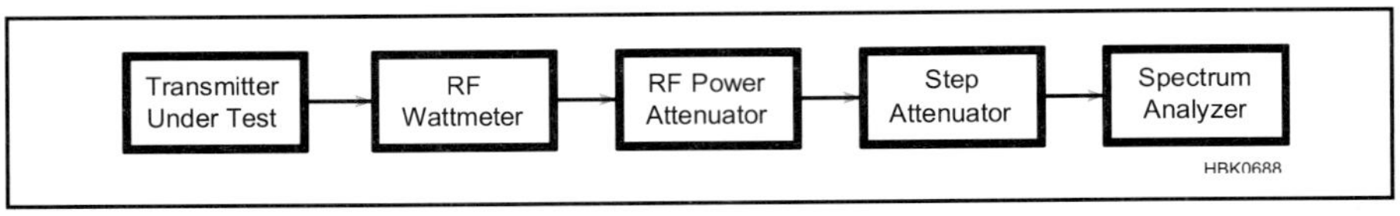

Figure 25.60 — The transmitter spectral purity test uses a spectrum analyzer to display the spurious frequencies.

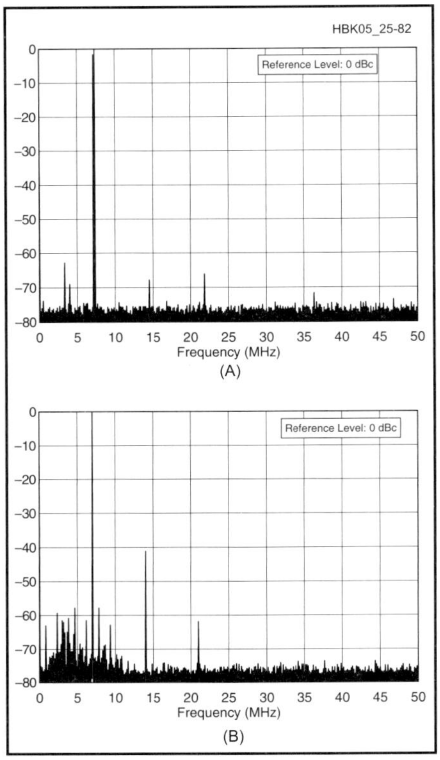

Figure 25.61 — Comparison of the spurious signal levels of two 100-watt transmitters, as shown on the spectrum analyzer display. The display on the top shows about 63 dB worst-case spurious signal suppression while the one on the bottom has a second harmonic suppressed approximately 42 dB. For transmitters below 30 MHz installed after January 1, 2003, the worst-case spurious emission must be at least 43 dB below the carrier power.

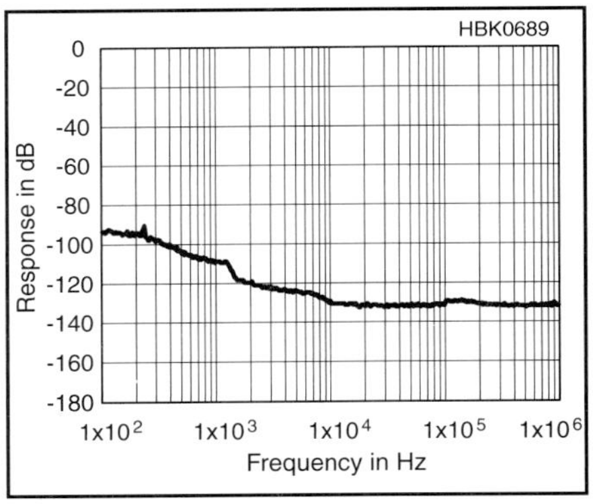

Figure 25.62 — Spectral display of an amateur transmitter output during composite-noise testing in the ARRL Lab. Power output was 200 W on the 14 MHz band. The carrier, off the left edge of the plot, is not shown. This plot shows composite transmitted noise 100 Hz to 1 MHz from the carrier on a logarithmic scale. The vertical scale is in dB with respect to the carrier.

The most important unwanted emissions close to the carrier frequency are caused by distortion in the transmitter amplifier stages. In an SSB transmitter, this distortion creates a signal that is wider than the bandwidth of the original modulation and causes interference to other stations on nearby frequencies. For this test, the same test setup is used as for the harmonics and spurious frequencies test shown in Figure 25.60 except that a two-tone audio generator is connected to the transmitter microphone input to simulate a voice signal, which contains many frequency components. The two-tone frequencies must be non-harmonically related to prevent tone harmonics from being confused with IMD products. The ARRL Lab uses 700 Hz and 1900 Hz for these tests. Because many transmitters' modulation frequency response is not flat, the relative amplitudes of the two audio-frequency tones must be adjusted to obtain equal-amplitude RF tones on the spectrum analyzer. The test should be done for both lower and upper-sideband modes.

Two-tone IMD products are measured with respect to the transmitter peak-envelope power (PEP) which is 6 dB greater than the amplitude of either of the two tones. By adjusting the spectrum analyzer reference level to place the two tones 6 dB below the zero dB reference line, as shown in **Figure 25.63**, the IMD distortion may be read out directly. For the signal shown, the third-order products are at –30 dB from PEP.

Carrier and unwanted sideband suppression may be measured with the same setup. In this case the 700 Hz tone of the two-tone generator is turned off and only the 1900 Hz tone is used. The single tone is set to the 0 dB reference line of the spectrum analyzer. For USB, the suppressed carrier shows as a small pip 1900 Hz below the desired signal and the unwanted sideband is 3800 Hz below. For LSB the unwanted signal frequencies are above the desired signal.

25.17.3 Transmitted Composite Noise

As discussed in the **Transmitting** chapter, transmitters output noise along with the desired signal. *Composite noise* is a combination of phase noise and amplitude noise. (Both are discussed in the **Oscillators and Synthesizers** chapter.) The transmitted noise is a source of interference to stations operating on the same band, particularly those operating on adjacent channels. See the discussion on RMDR in the Receiver Measurements section. Wideband noise across multiple bands is also encountered.

Transmitted noise is measured using a spectrum analyzer with very narrow filter that is swept across the frequency range of the measurement and the result converted to a dBc/Hz value. If a signal source analyzer is available (such as the Rohde & Schwarz FSUP), that is a much more convenient option, since the conversion is performed automatically. The analyzer captures phase noise as a plot of phase noise in dBc/Hz (dB below carrier in a 1 Hz bandwidth) vs. offset. A typical wideband measurement is shown in **Figure 25.64**.

If amplitude noise is to be measured, a diode detector is used. An alternative is a signal-source analyzer with phase and amplitude noise measurement capability such as the FSUP analyzer referenced in the previous paragraph. The test is made across the same spectrum as composite noise. To compute composite noise, convert both levels to absolute power/Hz, add them together, and convert back to dBc/Hz.

The transmitter output is connected to the spectrum analyzer directly through a high-power attenuator so the transmitter can be operated at full power output. An alternative is to use a dummy load and a directional coupler. Next, a swept-frequency measurement is made across the frequency range of interest. 10 Hz – 500 kHz or 10 Hz – 1 MHz offsets from the carrier frequency are typical.

Typical Test Configuration:
Frequency – each band of interest
Modulation – RTTY or another 100% duty cycle FSK/AFSK mode
Power output — full power

The plots of phase or composite noise are useful for comparing different radios and can also be used to calculate the amount of interference that may be received from a nearby transmitter. An approximation is given by

$$A_{QRM} = NL + 10 \times \log(BW)$$

where

A_{QRM} is the interfering signal level in dBc
NL is the noise level on the receiver frequency in dBc (from the noise plot)
BW = receiver IF bandwidth in Hz (for SDR receivers, use the receive filter bandwidth)

For instance, to find receiver interference level 20 kHz from the transmitter frequency, read the noise level (NL) on the noise plot at 20 kHz from the carrier. If NL is -90 dBc at 20 kHz from the carrier and the receiver is using a 2.5 kHz SSB filter, the interference level is –56 dBc. If the transmitted signal is 20 dB over S9 and each S unit represents 6 dB, the interference is (S9+20) – 56 = S9-36 dB = S3. In other words, the interference is as strong as an S3 signal.

As of April 2022, only phase noise is measured during ARRL Lab testing for *QST* Product Reviews. The specific measurements for transmitted AM noise are being developed.

25.17.4 Latency

As in the receiver test, latency is the time interval between arrival of an event at the transmitter audio or keying input and the corresponding output event at the transmitter's RF output. Latency is measured at various transmit occupied bandwidth settings.

The audio test signal is applied from a pulse or function generator to a line input, if available, and a speech input, if not. A dual-channel

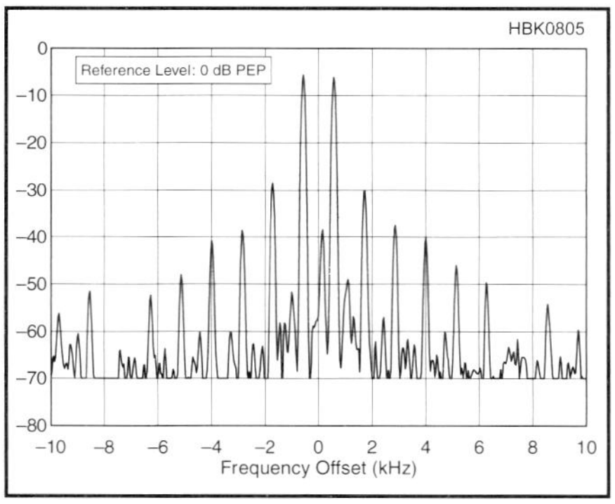

Figure 25.63 — An SSB transmitter two-tone test as seen on a spectrum analyzer. Each horizontal division represents 2 kHz and each vertical division is 10 dB. The third-order products are 30 dB below the PEP (top line), the fifth-order products are down 38 dB and seventh-order products are down 40 dB.

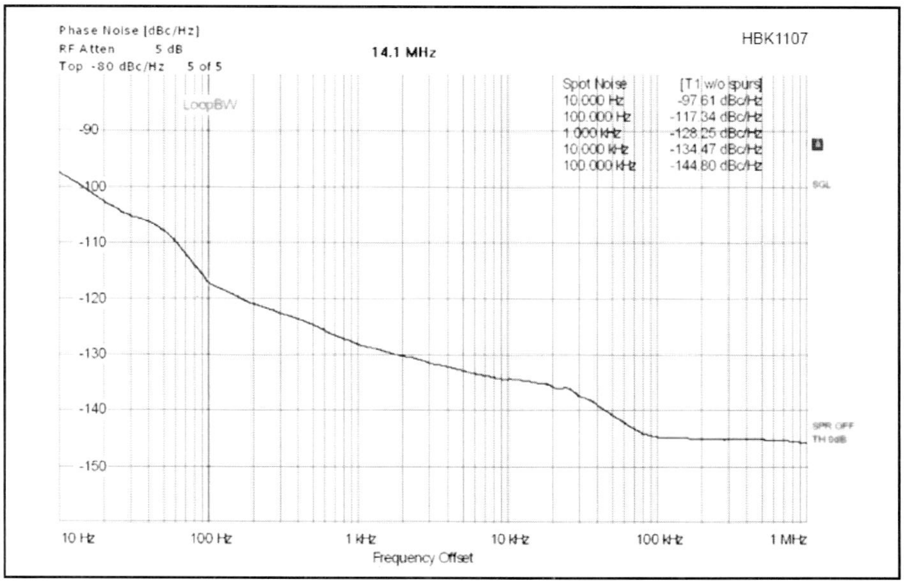

Figure 25.64 — The spectrum of transmitted phase noise from a modern transceiver operating on 20 meters. Noise is measured up to 1 MHz from the carrier signal. (Image courtesy of Adam Farson, AB4OJ/VA7OJ)

oscilloscope is used to measure the time interval. The transmitter's output is connector to the oscilloscope through a high-power attenuator or via a directional coupler and dummy load. Latency should be measured at all available transmit bandwidth settings. (See the Receiver Measurements section for more information.) **Figure 25.65** shows a typical latency test result.

25.17.5 CW Keying Tests

KEYING SIDEBANDS

Keying sidebands that include both noise and components resulting from imperfections in the keying waveform can spread out for several kHz around the carrier frequency. Usually referred to as *key clicks*, these sidebands can be a source of strong interference to adjacent stations.

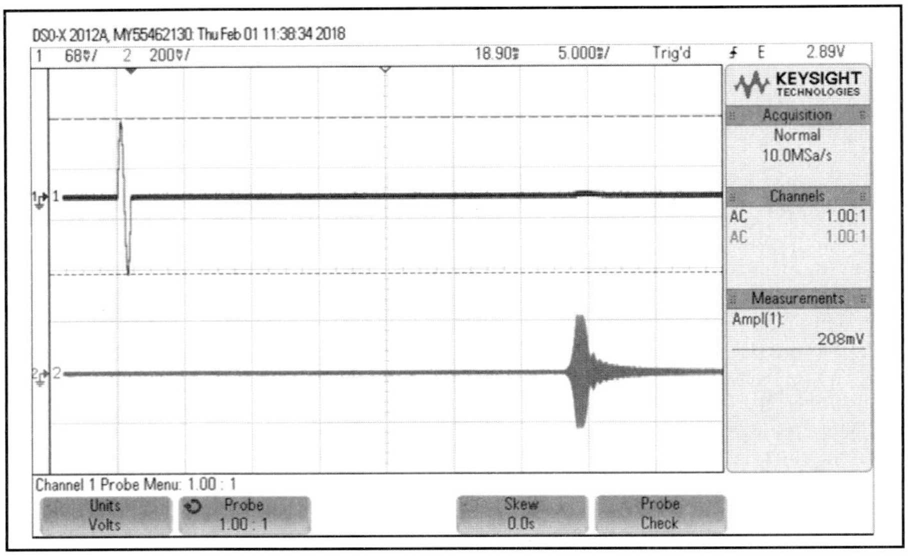

Figure 25.65 — Transmit latency is measured by inputting a very narrow pulse to the transmitter audio input (pulse waveform on the top trace with the marker on the top) and measuring the time until the response is observed in the transmitter's RF output (bottom trace). (Image courtesy of Adam Farson, AB4OJ/VA7OJ)

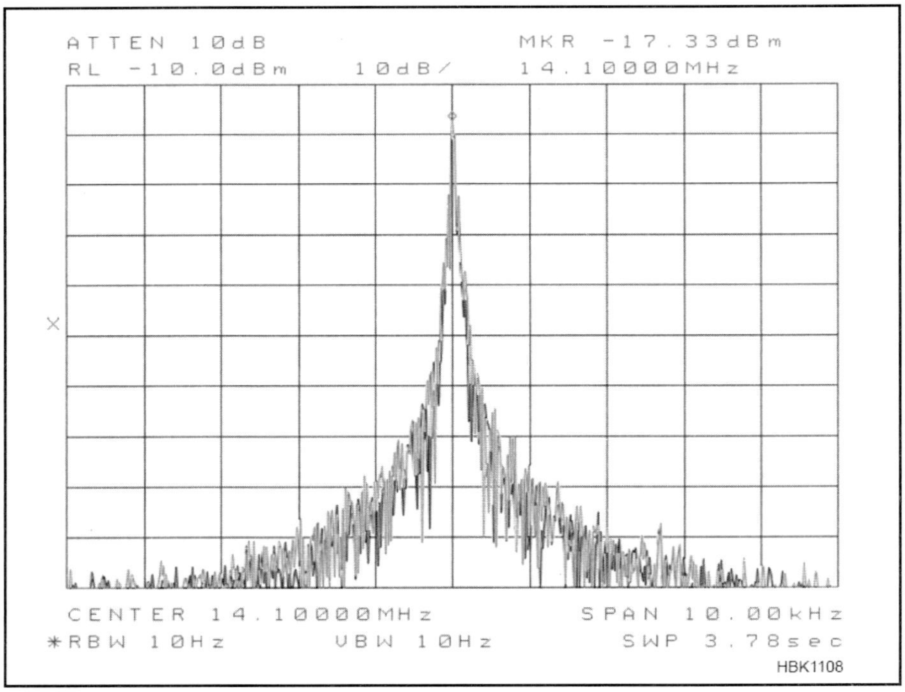

Figure 25.66 — The spectrum of a 48 WPM CW signal from a modern transceiver operating on 20 meter. The spectrum is measured up to 5 kHz above and below the carrier signal. (Image courtesy of Adam Farson, AB4OJ/VA7OJ.)

Noise can be keyed with the CW signal as well as be present whenever the transmitter is enabled, whether it is keyed or not. For more information about bandwidth of CW signals, see the section Transmitting CW in the **Transmitting** chapter.

The spectrum of the keyed signal is measured with the same setup as for Transmitted Phase Noise. Note that the waveform rise/fall times (configurable on virtually all modern transceivers) are important parameters for the test. A typical CW keying sideband measurement is shown in **Figure 25.66**. Tests may be run with different combinations of rise and fall times, if desired.

The keying speed is set to a high speed so that the sidebands can be fully captured by the analyzer's slow sweep across the band, usually ±5 kHz from the transmitted signal using a very narrow RBW (resolution bandwidth) filter. The keying waveform is generally a continuous stream of "dits" with a 50% duty cycle.

Typical Test Configuration
Frequency — each band of interest
Modulation — CW dits (50% duty cycle) at a maximum 50 WPM
Power output — full power

Note that as keying speed begins to exceed 45 WPM, the keying waveform's necessary sidebands begin to affect the test results. For this reason, keying speeds in excess of 50 WPM are not recommended.

KEYING WAVEFORM AND DELAY

A typical setup for measuring CW keying waveform and time delay is shown in **Figure 25.67**. A keying test generator is used to key the transmitter at a controlled rate. The generator can be set to any reasonable speed, but ARRL tests are usually conducted at 20 ms on and 20 ms off (25 Hz, 50% duty cycle), which corresponds to a series of dits at 60 WPM. **Figure 25.68** shows a typical display. The first two dits at the beginning of the transmission are displayed in order to show any transients or truncations that occur when the transceiver transitions from receive to the transmit state. The rise and fall times of the RF output pulse are measured between the 10% and 90% points on the leading and trailing edges, respectively. The delay times are measured between the 50% points of the keying and RF output waveforms. Look at the **Transmitting** and **Transceiver Design Topics** chapters for further discussion of CW keying issues.

MAXIMUM FULL BREAK-IN KEYING SPEED

Maximum QSK (CW full-break-in) keying speed is a measure of the maximum speed at which the receiver can still be heard between

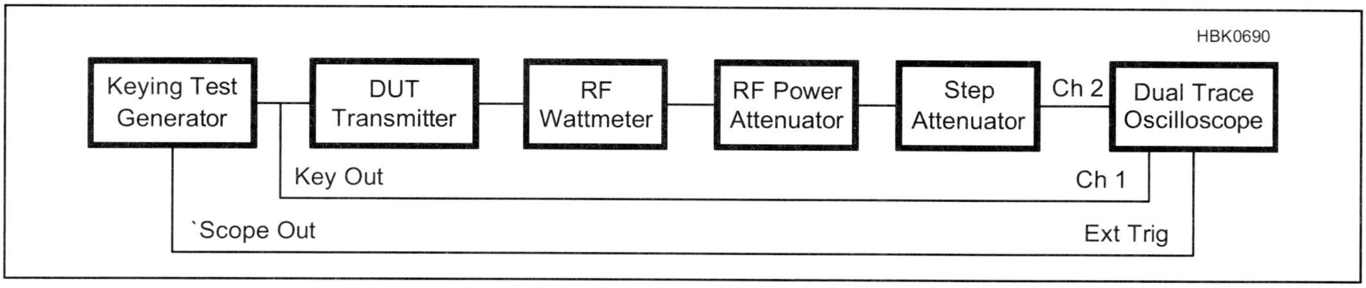

Figure 25.67 — CW keying waveform test setup.

code elements in full-QSK CW mode. This is also a good indication of *receiver recovery time* which can be important when using fast ARQ digital modes that require messages to be acknowledged with a minimum of delay.

The test is performed by injecting a CW test signal offset from the carrier by 800 Hz into the transceiver's RF port via the coupling port of a directional coupler. This produces a continuous 800 Hz tone in the transceiver's audio output. A series of "dits" (50% duty cycle) with the standard 600 Hz offset at low power (< 10W) is transmitted into the coupler's input port. The coupler's output port is terminated in 50 Ω. The keying speed is increased until the 800 Hz tone can still just be heard in the receiver. This is the maximum usable QSK speed.

25.17.6 Transmit-Receive Turnaround Time

The transmit-receive turnaround time is the time it takes for a transceiver to switch from the 50% rise time of the key line to 50% rise of audio output. Turnaround time is an important consideration in some digital modes with required turnaround times of less than 50 ms in some cases. The test setup is shown in **Figure 25.69**. This test requires extreme care to prevent excessive transmitter power from reaching the signal generator and exceeding its specifications. The step attenuator is preset to maximum and gradually decreased until the receiver's S meter reads S9. Receiver AGC is usually on and set for the fastest response for this test but experimentation with AGC and signal input level can reveal surprising variations. As for the PTT-to-RF output test, the transmitter is tuned to full power output with a single 700 Hz tone. The keying rate must be considerably slower than the turnaround time; rates of 200 ms on/200 ms off or faster, have been used with success in *QST* Product Review tests at the ARRL Lab.

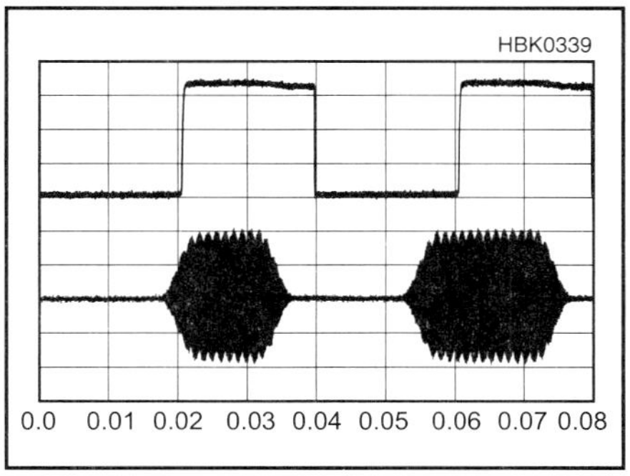

Figure 25.68 — Typical CW keying waveform for a modern amateur radio transceiver during testing in the ARRL Lab. This plot shows the first two dits in full break-in (QSK) mode using external keying. Equivalent keying speed is 60 WPM. The upper trace is the actual key closure; the lower trace is the RF envelope. (Note that the first key closure starts at the left edge of the figure.) Horizontal divisions are 10 ms. The transceiver was being operated at 100 W output on the 14 MHz band.

25.17.7 PTT-to-RF Output

For voice modes, a PTT-to-RF output test is similar to CW keying tests. It measures rise and fall times, as well as the on- and off-delay times just as in the CW test. See **Figure 25.70** for the test setup. For SSB the transmitter is modulated with a single 700-Hz tone. For FM the transmitter is unmodulated. The keying generator is set to a speed that allows plenty of time for the transceiver to recover between dits. The ON or OFF delay times are measured from the 50% point of the falling or rising edge of the key out line to the 50% point of the RF waveform.

25.17.8 Other Common Transmitter Measurements

AM MODULATION PERCENTAGE

The peak envelope power (PEP) of a 100%-modulated AM signal is four times the average power. For that reason, the specified AM power level of most SSB transmitters is generally about 25% of the PEP SSB power rating.

The modulation percentage can most easily be measured with a wide-band oscilloscope connected to the 50 Ω dummy load. It is OK to exceed the specified bandwidth somewhat as long as a clean signal is displayed since this is a relative measurement only. With 100% modulation, the negative modulation peaks just reach zero signal level and the positive

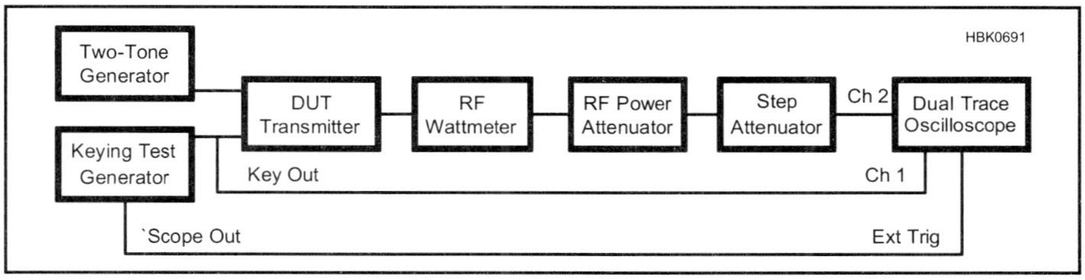

Figure 25.69 — Transmit-receive turnaround time test setup.

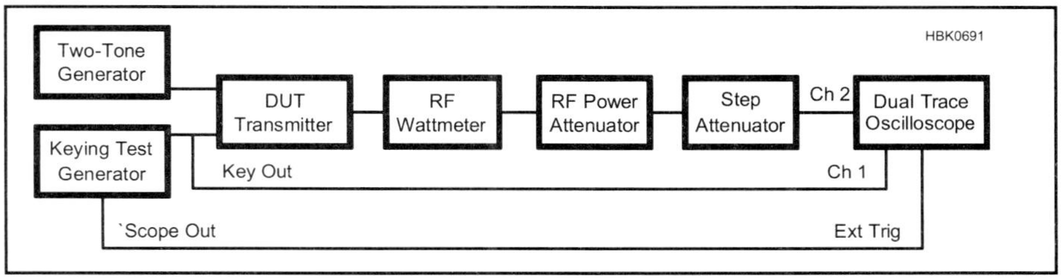

Figure 25.70 — PTT-to-RF-output test setup for voice-mode transmitters.

peaks are twice the amplitude of the unmodulated carrier. The exact value can be calculated with the equation,

$$M = 100\frac{max - min}{max + min}$$

where M is the modulation percentage, max is the signal amplitude at the peaks and min is the amplitude at the troughs. An alternative is to use an RF spectrum analyzer. With 100% sine wave modulation, the two sidebands are each –6 dB with respect to the carrier. In this case, $M = 100 \times 10^{(S+6)/20}$, where S is the sideband level with respect to the carrier (a negative number) and M is the modulation percentage.

FM DEVIATION

The equivalent measurement for an FM transmitter is the *deviation*, which is the amount the RF frequency deviates from the center carrier frequency. It is possible to purchase an instrument that measures FM deviation directly; the function is generally included in two-way radio test sets. Another way is to use slope detection with a standard analog spectrum analyzer. Start by choosing a resolution bandwidth (RBW) on the spectrum analyzer such that as you tune away from the unmodulated carrier the signal level changes approximately linearly (the same number of kHz per dB) over at least a 10 kHz range. An RBW of 10 kHz is usually about right. Record the kHz per dB sensitivity value. Then adjust the frequency for the middle of the linear range and set the spectrum analyzer for zero span. With modulation applied to the transmitter, the deviation is equal to one-half the peak-to-peak dB variation of the signal on the screen, times the kHz per dB value determined previously.

The Bessel Null method of measuring FM deviation uses the fact that the carrier disappears for a modulation index of 2.405, as explained in the **Modulation** chapter. To set the deviation to 3 kHz, for example, apply sine wave modulation at a frequency of 3.0/2.405 = 1.25 kHz and adjust the modulation level to null the carrier.

25.18 Construction Projects

25.18.1 Fixed-Frequency Audio Oscillator

An audio signal generator should provide a reasonably pure sine wave. The best oscillator circuits for this use are RC-coupled, the amplifiers operating as close to class A as possible. Variable frequencies covering the entire audio range are needed for determining frequency response of audio amplifiers.

A circuit of a simple RC oscillator that is useful for general testing is given in **Figure 25.71**. This Twin-T arrangement gives a waveform that is satisfactory for most purposes.

The oscillator can be operated at any frequency in the audio range by varying the component values. R1, R2, and C1 form a low-pass network, while C2, C3, and R3 form a high-pass network. As the phase shifts are opposite, there is only one frequency at which the total phase shift from collector to base is 180 degrees. Oscillation occurs at this frequency. When C1 is about twice the capacitance of C2 or C3 (C2 = C3) the best operation results. (Note that in the example given for 750 Hz, C1 is specified at 0.05 uF while C2 and C3 are specified at 0.02 uF. This is slightly more than double, which also works and it more practical since there is not a standard capacitor value of 0.04 uF).

R3 should have a resistance about one tenth that of R1 or R2 (R1 = R2). Output is taken across C1, where the harmonic distortion is least. Use a relatively high impedance load, 100 kΩ or more. The collector voltage, V_{CC}, can be any easily obtained low dc voltage, such as 5V or 12V. Most small-signal audio frequency transistors, such as 2N2222 or 2N3904, can be used for Q1. PNP types, such as 2N2907 or 2N3906, are satisfactory if the supply polarity is more negative than the ground reference point. R4, the collector load resistor may be changed a little to adjust the oscillator for best output waveform.

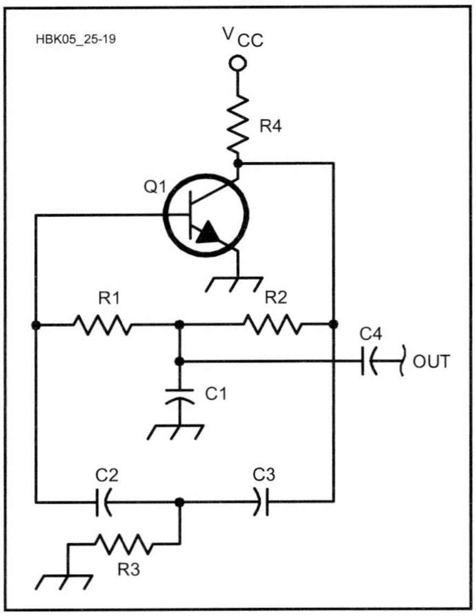

Figure 25.71 — Values for the twin-T audio oscillator circuit range from 18 kΩ for R1-R2 and 0.05 µF for C1 (750 Hz) to 15 kΩ and 0.02 µF for 1800 Hz. For the same frequency range, R3 and C2-C3 vary from 1,800 Ω and 0.02 µF to 1,500 Ω and 0.01 µF. R4 is 3,300 Ω and C4, the output coupling capacitor, can be 0.05 µF for high-impedance loads.

25.18.2 Wide-Range Audio Oscillator

A wide-range audio oscillator that provides a moderate output level can be built from a single 741 operational amplifier (see **Figure 25.72**). Power is supplied by two 9 V batteries from which the circuit draws 4 mA. The frequency range is selectable from about 7 Hz to around 70 kHz. Distortion is approximately 1%. The output level under a light load (10 kΩ) is 4 to 5 V. This can be increased by using higher battery voltages, up to a maximum of plus and minus 18 V, with a corresponding adjustment of R_F.

Pin connections shown are for the eight-pin DIP package. Variable resistor R_F is trimmed for an output level of about 5% below clipping as seen on an oscilloscope. This should be done for the temperature at which the oscillator normally operates, as the lamp is sensitive to ambient temperature. This unit was originally described by Shultz in November 1974 *QST*; it was later modified by Neben as reported in June 1983 *QST*.

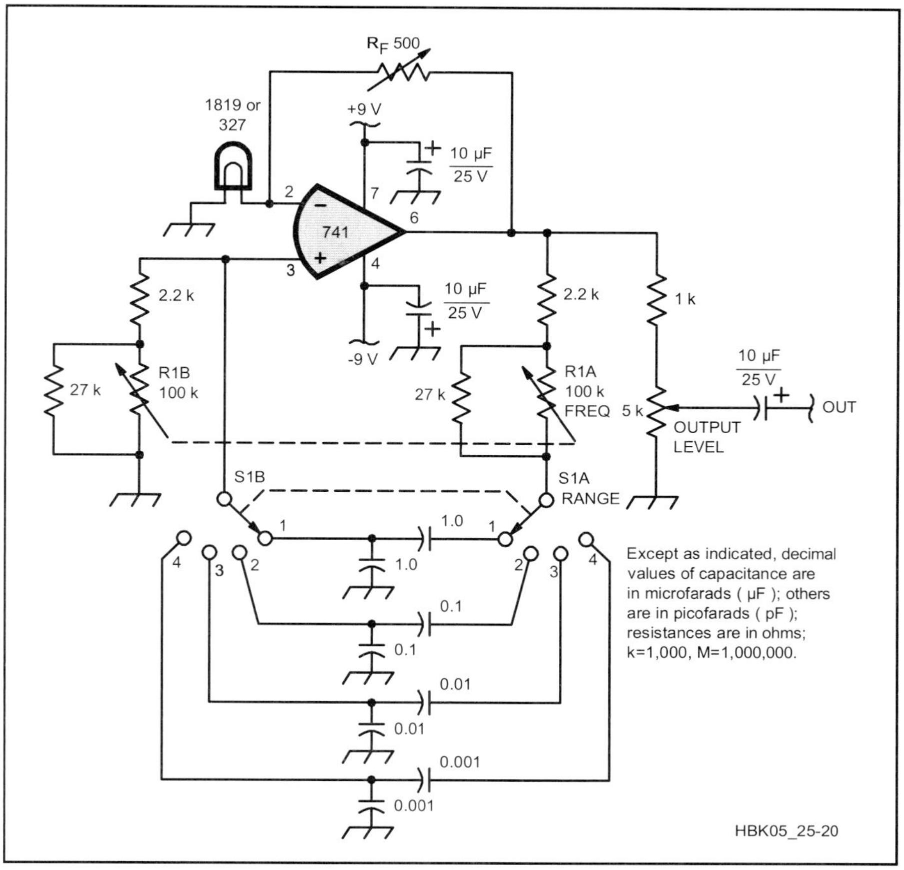

Figure 25.72 — An audio oscillator based on a single IC. The 741 op-amp is shown but most op-amps work in this application. The frequency range is set by switch S1.

25.18.3 Two-Tone Audio Generator

This generator is used in the ARRL Laboratory to test SSB transmitters for ARRL Product Reviews and makes a very convenient signal source for testing the linearity of a single-sideband transmitter. To be suitable for transmitter evaluation, a generator of this type must produce two non-harmonically related tones of equal amplitude. The level of harmonic and intermodulation distortion must be sufficiently low so as not to confuse the measurement. The frequencies used in this generator are 700 and 1900 Hz, both well inside the normal audio passband of an SSB transmitter. Spectral analysis and practical application with many different transmitters has shown this generator to meet all of the requirements mentioned above. While designed specifically for transmitter testing it is also useful any time a fixed-frequency, low-level audio tone is needed.

CIRCUIT DETAILS

Each of the two tones is generated by a separate Wien bridge oscillator, U1B and U2B. (See **Figure 25.73**.) The oscillators are followed by RC active low-pass filters, U1A and U2A. Because the filters require nonstandard capacitor values, provisions have been made on the circuit board for placing two capacitors in parallel in those cases where standard values cannot be used. (The circuit board artwork for layout and part placement is available as graphics with the online content.) The oscillator and filter capacitors should be polystyrene or Mylar film types if available.

The two tones are combined by op amp U3A, a summing amplifier. This amplifier has a variable resistor, R4, in its feedback loop which serves as the output LEVEL control. While R4 varies the amplitude of both tones together, R3, the BALANCE control, allows the level of tone A to be changed without affecting the level of tone B. This is necessary because some transmitters do not have equal audio response at both frequencies. Multi-turn pots are recommended for both R3 and R4 so that fine adjustments can be made. Following the summing amplifier is a step attenuator; S3 controls the output level in 10 dB steps. The use of two output level controls, R4 and S3, allows the output to cover a wide range and still be easy to set to a specific level.

The remaining op amp, U3B is connected as a voltage follower and serves to buffer the output while providing a high-impedance load for the step attenuator. Either high or low output impedance can be selected by S4. The values shown are suitable for most transmitters using either high- or low-impedance microphones.

CONSTRUCTION AND ADJUSTMENT

Component layout and wiring are not critical, and any type of construction can be used with good results. Because the generator is normally be near a transmitter, it should be enclosed in some type of metal case for shielding. Battery power was chosen to reduce the possibility of RF entering the unit through the ac line. With careful shielding and filtering, the builder should be able to use an ac power supply in place of the batteries.

The only adjustment required before use is the setting of the oscillator feedback trimmers, R1 and R2. These should be set so that the output of each oscillator, measured at pin 7 of U1 and U2, is about 0.5 volts RMS. A DVM or oscilloscope can be used for this measurement. If neither of these is available, the feedback should be adjusted to the minimum level that allows the oscillators to start reliably and stabilize quickly. When the oscillators are first turned on, they take a few seconds before they have stable output amplitude. This is caused by the lamps, DS1 and DS2, used in the oscillator feedback circuit. This is normal and should cause no difficulty. The connection to the transmitter should be through a shielded cable.

25.18.4 RF Current Meter

The following project was designed by Tom Rauch, W8JI (**w8ji.com/building_a_current_meter.htm**). The circuit of **Figure 25.74** is based on a current transformer (T1) consisting of a T157-2 powdered-iron toroid core with a 20-turn winding. The meter is used with the current-carrying wire or antenna inserted through the middle of the core as a one-turn primary.

When 1 A is flowing in the single-turn primary, the secondary current is 50 mA = primary current divided by the turns ratio

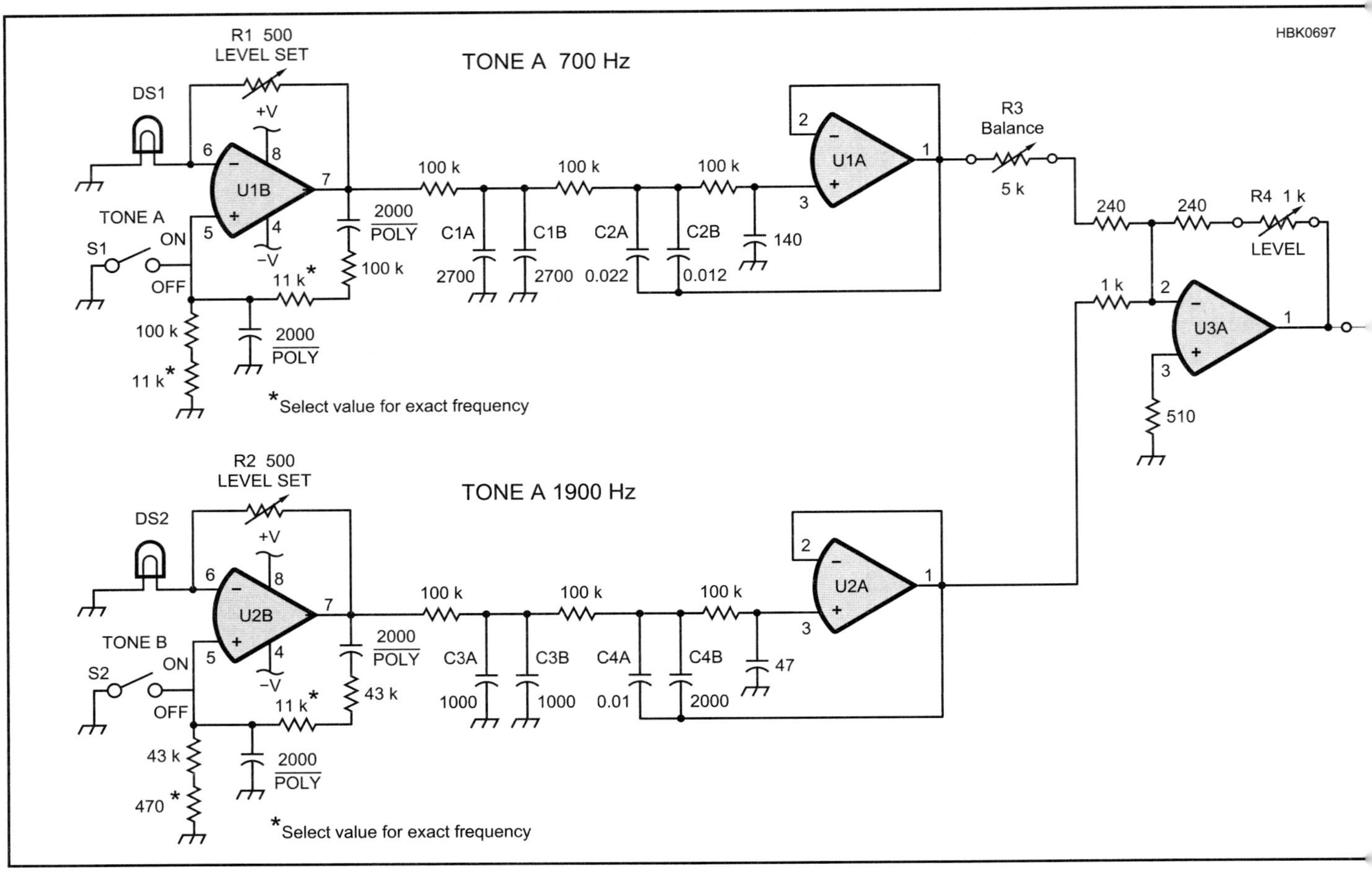

of 20:1. R1 across the transformer flattens the frequency response and limits the output voltage. The RF voltage is then detected and filtered by the D1 (a low-threshold Schottky diode for minimum voltage drop) and C1. The adjustable sum of R2 and R3 allow for full-scale (FS) calibration of the 100 μA meter. C2 provides additional filtering. The toroid core and all circuitry are glued to the back of the meter case with only R2 exposed, a screwdriver-adjustable calibration pot.

It is important to minimize stray capacitance by using a meter with all-plastic construction except for the electrical parts. The meter in **Figure 25.75** has an all-plastic case including the meter scale. The meter movement and all metallic areas are small. The lack of large metallic components minimizes stray capacitance from the proximity of the meter. Low stray capacitance ensures the instrument has the least possible effect on the circuit being tested.

A value of 100 Ω for R1 gave the flattest response from 1.8 to 30 MHz. With 50 mA of secondary current, the voltage across R1 is 0.05 × 100 = 5 V_{RMS}. The peak voltage is then 1.414 × 5 = 7.1 V. At full current, power dissipation in R1 = 50 mA × 5 V_{RMS} = 0.25 W so a 1/2-W or larger resistor should be used.

The meter used here was a 10,000 Ω/V model so for full-scale deflection from a primary current of 1 A producing a secondary voltage of ~7 V, the sum of R2 and R3 must be set to 7 × 10,000 = 70 kΩ. The low-current meter combined with high detected voltage improves detector linearity.

Calibration of the meter can be performed by using a calibrated power meter and a test fixture consisting of two RF connectors with a short piece of wire between them and through the transformer core. With 50 W applied to a 50 Ω load, the wire carries 1 A of current. Full-scale accuracy is not required in comparison measurements, since the meter references against itself, but linearity within a few percent is important.

This transformer-based meter is much more reliable and linear than thermocouple RF ammeters and perturbs systems much less. Stray capacitance added to the system being tested is very small because of the proximity of the meter and the compact wiring area. Compared to actually connecting a meter with its associated lead lengths and capacitance in line with the load, the advantages of a transformer-coupled meter become apparent.

25.18.5 RF Ammeters

When it comes to getting your own RF ammeter, there's good news and bad news as related by John Stanley, K4ERO. First, the bad news. New RF ammeters are expensive, and even surplus pricing can vary widely between \$10 and \$100 in today's market. AM radio stations are the main users of new units. The FCC defines the output power of AM stations based on the RF current in the antenna, so new RF ammeters are made mainly for that market. They are quite accurate, and their prices reflect that!

The good news is that used RF ammeters are often available. For example, Fair Radio Sales in Lima, Ohio has been a consistent source of RF ammeters. Ham flea markets are also worth trying. Some grubbing around in your nearest surplus store or some older ham's junk box may provide just the RF ammeter you need. Be sure you are really buying an RF ammeter as meters labeled "RF Amps" may just be regular current meters intended for use with an external RF current sensing unit.

RF AMMETER SUBSTITUTES

Don't despair if you can't find a used RF ammeter. It's possible to construct your own. Both hot-wire and thermocouple units can be homemade. Pilot lamps in series with antenna wires, or coupled to them in various ways, can

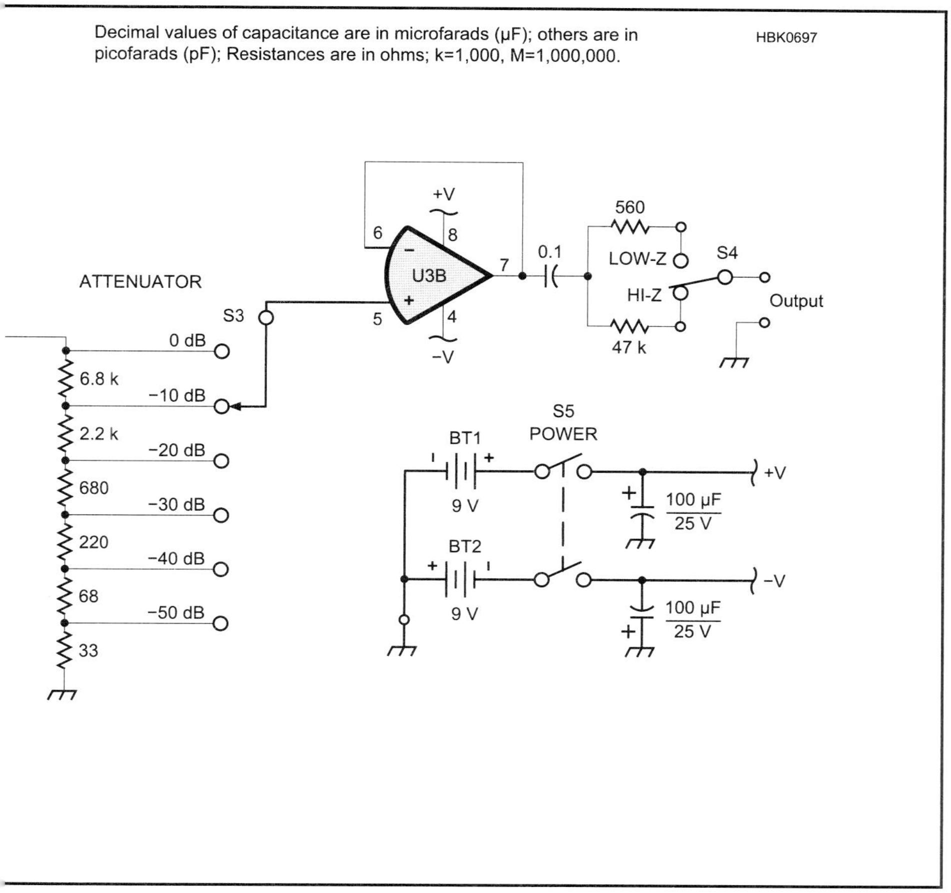

Figure 25.73 — Two-tone audio generator schematic.

BT1, BT2 — 9 V alkaline.
C1A,B — Total capacitance of 0.0054 µF, ±5%.
C2A,B — Total capacitance of 0.034 µF, ±5%.
C3A,B — Total capacitance of 0.002 µF, ±5%.
C4A,B — Total capacitance of 0.012 µF, ±5%.
DS1, DS2 — 12 V, 25 mA lamp.
R1, R2 — 500 Ω, 10-turn trim potentiometer.
R3 — 500 Ω, panel mount potentiometer.
R4 — 1 kΩ, panel mount potentiometer.
S1, S2 — SPST toggle switch.
S3 — Single pole, 6-position rotary switch.
S4 — SPDT toggle switch.
S5 — DPDT toggle switch.
U1, U2, U3 — Dual JFET op amp, type LF353N or TL082.

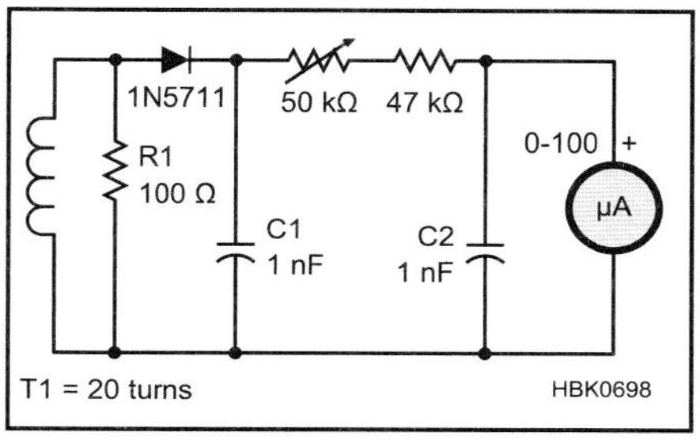

Figure 25.74 — The schematic of the RF current probe. See text for component information.

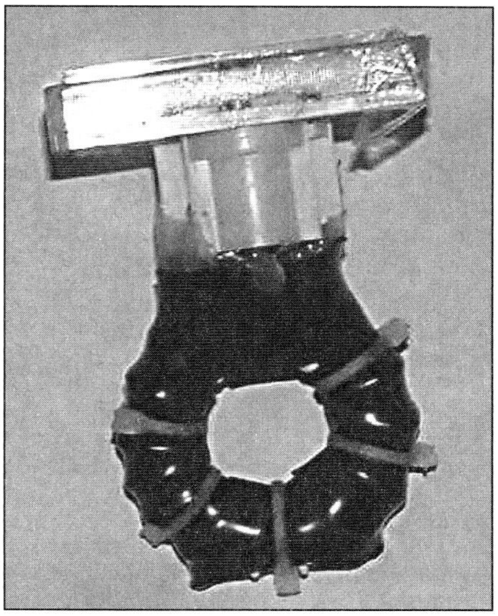

Figure 25.75 — Assembly of the RF current probe. Use an all-plastic meter and mount the circuits and toroid directly on the back of the meter case.

indicate antenna current (Sutter, F., "What, No Meters?" *QST*, Oct. 1938, p. 49) or even forward and reflected power (Wright, C., "The Twin-Lamp," *QST*, Oct. 1947, pp. 22 – 23, 110 and 112).

Another approach is to use a small low-voltage lamp as the heat/light element and use a photo detector driving a meter as an indicator. (Your eyes and judgment can serve as the indicating part of the instrument.) A feed line balance checker could be as simple as a couple of lamps with the right current rating and the lowest voltage rating available. You should be able to tell fairly well by eye which bulb is brighter or if they are about equal. You can calibrate a lamp-based RF ammeter with 60-Hz or dc power.

As another alternative, you can build an RF ammeter that uses a dc meter to indicate rectified RF from a current transformer that you clamp over a transmission line wire (Lau, Z., "A Relative RF Ammeter for Open-Wire Lines," *QST*, Oct. 1988, pp. 15 – 17).

25.18.6 RF Step Attenuator

A good RF step attenuator is one of the key pieces of equipment that belongs on your workbench. The attenuator in this project offers good performance yet can be built with a few basic tools. The attenuator is designed for use in 50 Ω systems, provides a total attenuation of 71 dB in 1 dB steps, offers respectable accuracy and insertion loss through 225 MHz and can be used at 450 MHz as shown in **Table 25.3**. This material was originally published as "An RF Step Attenuator" by Denton Bramwell, K7OWJ, in the June 1995 *QST*.

The attenuator consists of 10 resistive pi-attenuator sections such as the one in **Figure 25.76**. Each section consists of a DPDT slide switch and three 1/4 W, 1% tolerance metal-film resistors. The complete unit contains single 1, 2, 3. and 5 dB sections, and six 10 dB sections. **Table 25.4** lists the resistor values required for each section.

The enclosure is made of brass sheet stock, readily available at hardware and hobby stores. By selecting the right stock, you can avoid having to bend the metal and need only perform a minimum of cutting.

CONSTRUCTION

The enclosure can be built using only a nibbling tool, drill press, metal shears, and a soldering gun or heavy soldering iron. (Use a regular soldering iron on the switches and resistors.) One method of cutting the small pieces of rectangular tubing to length is to use a drill press equipped with a small abrasive

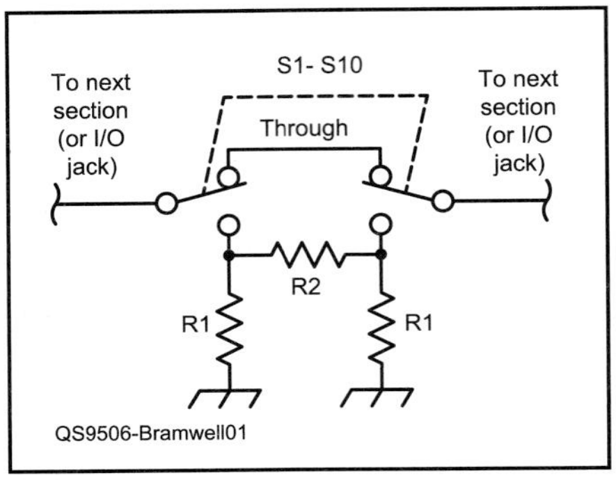

Figure 25.76 — Schematic of one section of the attenuator. All resistors are ¼-W, 1%-tolerance metal-film units. See Table 25.4 for the resistor values required for each attenuator section. There are six 10 dB sections and one each of 1, 2, 3, and 5 dB.

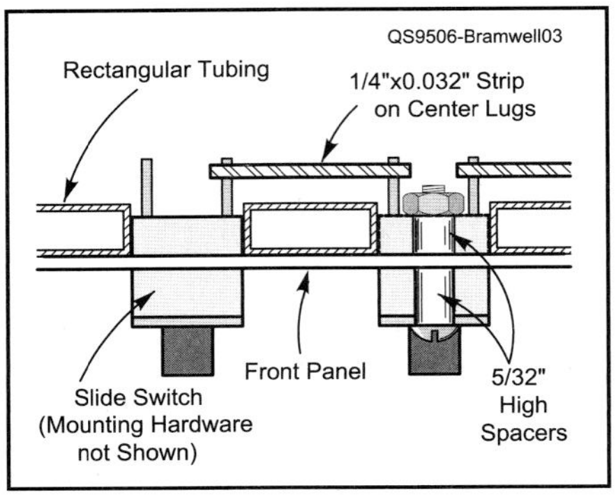

Figure 25.77 — Key to obtaining acceptable insertion loss in the "through" position is to make the whole device look as much as possible like 50-Ω coax. The rectangular tubing and the ¼ × 0.032-inch brass strip between the switch sections form a 50-Ω stripline.

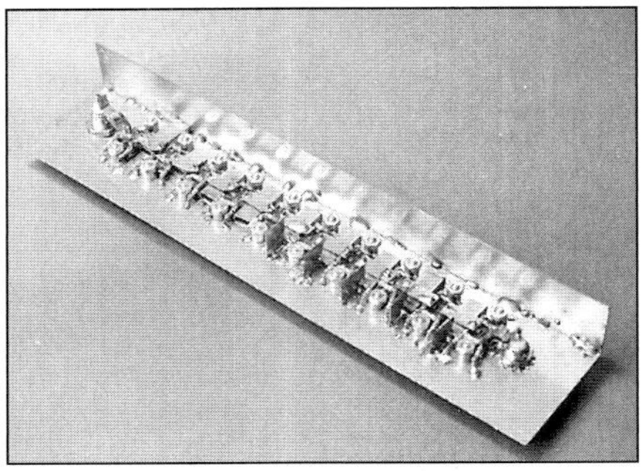

Figure 25.78 — Solder one of the 1-inch-wide chassis side pieces in place to make the assembly more rigid during construction. Solder the side piece to the edge of the top plate that faces the "through" side of the switches; this makes the rest of the assembly easier.

cutoff wheel.

Brass is easy to work and solder. For the enclosure, you'll need two precut 2 × 12 × 0.025-inch sheets and two 1 × 12 × 0.025-inch sheets. The 2-inch-wide stock is used for the front and back panels; the 1-inch-wide stock is used for the ends and sides. For the internal wiring, you need a piece of 5/32 × 5/16-inch rectangular tubing, a 1/4 × 0.032-inch strip, and a few small pieces of 0.005-inch-thick stock to provide inter-stage shields and form the 50 Ω transmission lines that run from the BNC connectors to the switches at the ends of the step attenuator.

For the front panel, nibble or shear a piece of 2-inch-wide brass to a length of about 9½ inches. Space the switches from each other so that a piece of the rectangular brass tubing lies flat and snugly between them (see **Figure 25.77**). Drill holes for the #4-40 mounting screws and nibble or punch rectangular holes for the bodies of the slide switches.

Before mounting any parts, solder in place one of the 1-inch-wide chassis side pieces to make the assembly more rigid. Solder the side piece to the edge of the top plate that faces the "through" side of the switches; this makes later assembly easier (see **Figure 25.78**). Although the BNC input and output connectors are shown mounted on the top (front) panel, better lead dress and high-frequency performance may result from mounting the connectors at the ends of the enclosure.

DPDT slide switches designed for sub-panel mounting often have mounting holes tapped for #4-40 screws. Enlarge the holes to allow a #4-40 screw to slide through. Before mounting the switches, make the "through" switch connection (see Figure 25.76) by bending the two lugs at one end of each switch toward each other and soldering the lugs together or solder a small strip of brass between the lugs and clip off the lug ends. Mount the switches above the front panel, using 5/32-inch-high by 7/32-inch-OD spacers. Use the same size spacer on the inside. On the inside, the spacer creates a small post that helps reduce capacitive coupling from one side of the attenuator to the other. The spacers position the switch so that the 50 Ω stripline can be formed later.

The trick to getting acceptable insertion loss in the "through" position is to make the attenuator look as much as possible like 50 Ω coax. That's where the rectangular tubing and the 1/4 × 0.032-inch brass strip come into the picture (see Figure 25.77); they form a 50 Ω stripline. (See the **Transmission Lines** chapter for information on stripline.)

Cut pieces of the rectangular tubing about 3/4-inch long, and sweat solder them to the front panel between each of the slide switches. Next, cut lengths of the 1/4-inch strip long enough to conveniently reach from switch to switch, then cut one more piece. Drill 1/16-inch holes near both ends of all but one of the 1/4-inch strips. The undrilled piece is used as a temporary spacer, so make sure it is flat and deburred.

Lay the temporary spacer on top of the rectangular tubing between the first two switches, then drop one of the drilled 1/4-inch pieces over it, with the center switch lugs through the 1/16-inch holes. Before soldering, check the strip to make sure there's sufficient clearance between the 1/4-inch strip and the switch lugs; trim the corners if necessary. Use a screwdriver blade to hold the strip flat and solder the lugs to the strip. Remove the temporary spacer. Repeat this procedure for all switch sections. This creates a 50 Ω stripline running the length of the attenuator.

Next, solder in place the three 1% tolerance resistors of each section, keeping the leads as short as possible. Use a generous blob of solder on ground leads to make the lead less inductive. Install a 1/2-inch-square

Table 25.4
Closest 1%-Tolerance Resistor Values

Attenuation (dB)	*R1 (Ω)*	*R2 (Ω)*
1.00	866.00	5.60
2.00	436.00	11.50
3.00	294.00	17.40
5.00	178.00	30.10
10.00	94.30	71.50

Table 25.3
Step Attenuator Performance at 148, 225, and 450 MHz

Measurements made in the ARRL Laboratory.

Attenuator set for Maximum attenuation (71 dB)		**Attenuator set for minimum attenuation (0 dB)**	
Frequency (MHz)	*Attenuation (dB)*	*Frequency (MHz)*	*Attenuation (dB)*
148	72.33	148	0.4
225	73.17	225	0.4
450	75.83	450	0.84

Note: Laboratory-specified measurement tolerance of ±1 dB.

brass shield between each 10 dB section to ensure that signals don't couple around the sections at higher frequencies.

Use parallel 1/4-inch strips of 0.005-inch-thick brass spaced 0.033 inch apart to form 50 Ω feed lines from the BNC connectors to the switch contacts at each end of the stripline as shown in Figure 25.78. (Use the undrilled piece of 0.032-inch-thick brass to insure the proper line spacing.) The attenuator with all switches and shields in place is shown ready for final mechanical assembly in **Figure 25.79**.

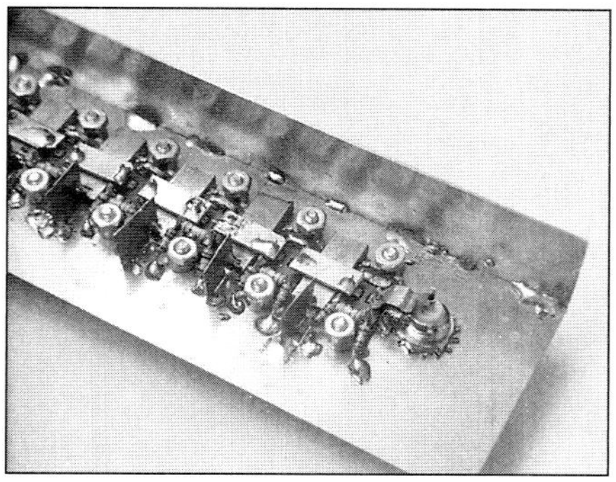

Figure 25.79 — The attenuator before final mechanical assembly. The ¼-inch strips are spaced 0.033 inch apart to form a 50 Ω connection from the BNC connector to the stripline. There are ½-inch square shields between 10-dB sections. The square shields have a notch in one corner to accommodate the end of the rectangular tubing.

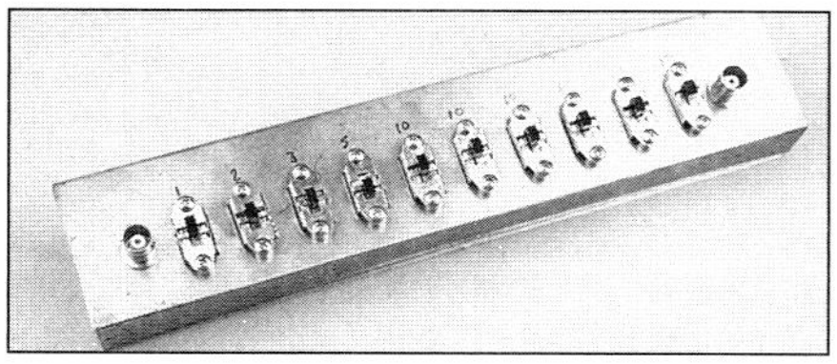

Figure 25.80 — The completed step attenuator in the enclosure of brass sheet. The BNC connectors may be mounted on the front panel at the end of the switches

Finally, solder in place the remaining enclosure side, cut and solder the end pieces, and solder brass #4-40 nuts to the inside walls of the case to hold the rear (or bottom) panel. Drill and attach the rear panel and round off the sharp corners to prevent scratching or cutting anyone or anything. Add stick-on feet and labels and your step attenuator of **Figure 25.80** is ready for use.

Remember that the unit is built with 1/4-W resistors, so it can't dissipate a lot of power. Remember, too, that for the attenuation to be accurate, the input to the attenuator must be a 50 Ω source and the output must be terminated in a 50 Ω load.

25.18.7 High-Power RF Samplers

If one wants to measure characteristics of a transmitter or high-powered amplifier, some means of reducing the power of the device to 10 or 20 dBm must be used. The most straightforward way to do this is to use a 30- or 40 dB attenuator capable of handling the high power. A 30 dB attenuator reduces a 100 W transmitter to 20 dBm. A 40 dB attenuator reduces a 1 kW amplifier to 20 dBm. If further attenuation is needed, a simple precision attenuator may be used after the signal has been reduced to the 20 dBm level.

The problem with high-powered attenuators is that they are expensive to buy or build since the front end of the attenuator must handle the output power of the transmitter or amplifier. If one already has a dummy load, an RF sampler may be used to produce a replica of the signal at a reduced power level. The sampler described here was originally presented in *QST* Technical Correspondence for May 2011 by Tom Thompson, WØIVJ.

A transformer sampler passes a single conductor (usually the insulated center conductor from a piece of coaxial cable) from the transmitter or amplifier to the dummy load through a toroidal inductor forming a transformer with a single turn primary. The secondary of the transformer is connected to a resistor network and then to the test equipment as shown in **Figure 25.81**. Assume that the source, whether a transmitter or amplifier, is a pure voltage source in series with a 50 Ω resistor. This most likely is not exactly the case but is sufficient for analysis.

If a current, I, flows into the dummy load, then a current, I / N flows in the secondary of the transformer, where N is the number of turns on the secondary. Figure 25.81 also shows the equivalent circuit, substituting a current source for the transformer. The attenuation is 40 dB and 15 turns for the secondary of the transformer. If R_{SHUNT} = 15 Ω, and R_{SERIES} = 35 Ω, then the voltage across a 50 Ω load resistor, R_{SAMPLE}, is 1/100 of the voltage across the dummy load, which is 40 dB of attenuation.

Reflecting this resistor combination back through the transformer yields 0.06 Ω in series with the 50 Ω dummy load impedance. This is an insignificant change. Furthermore, reflecting 100 Ω from the primary to the secondary places 22.5 kΩ in parallel with R_{SHUNT}, which does not significantly affect its value. The test equipment sees a 50 Ω load looking back into the sampler. Even at low frequencies, where the reactance of the secondary winding is lower than 15 Ω, the impedance looking back into the sample port remains close to 50 Ω.

The samplers described here use an FT37-61 ferrite core followed by two resistors as described above. The through-line SWR is good up to 200 MHz, the SWR is fair looking into the sampled port, and the useful bandwidth extends from 0.5 MHz to about 100 MHz. If you are interested in an accurate representation of

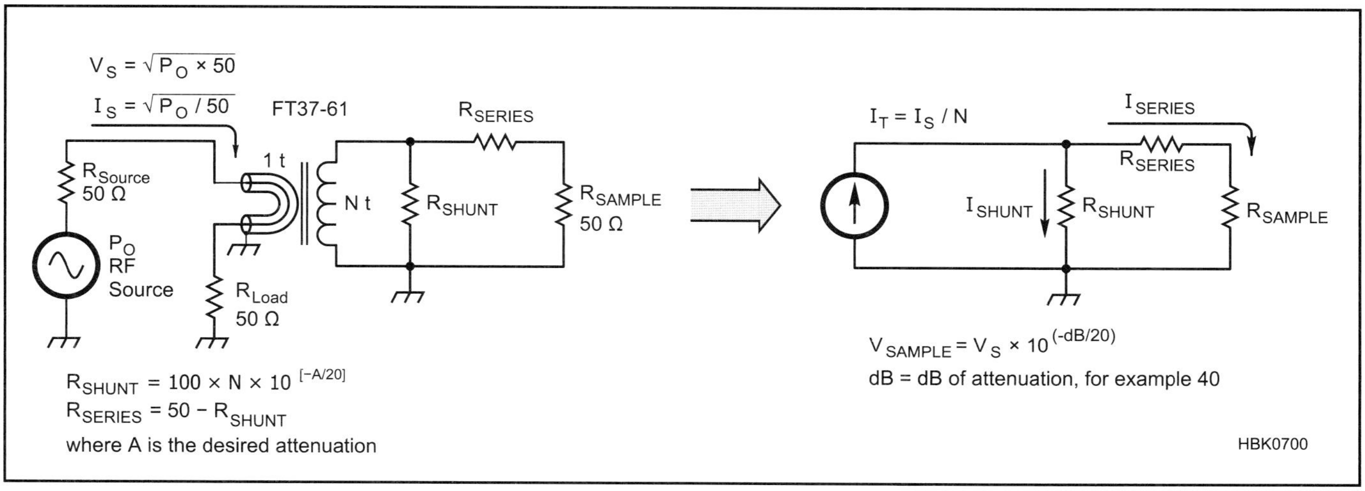

Figure 25.81 — RF sampler circuit diagram and equivalent circuit showing calculations.

Figure 25.82 — RF sampler using box construction.

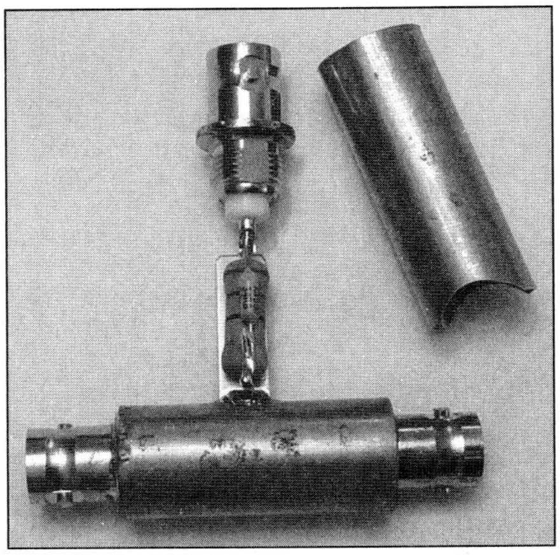

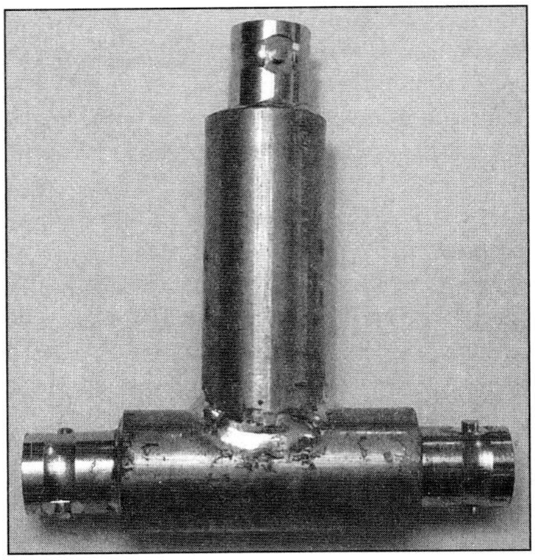

Figure 25.83 — RF sampler using tube construction.

the third harmonic of your HF transmitter or amplifier, it is important for the sampler to give accurate attenuation into the VHF range.

Figure 25.82 shows a photo of a sampler built into a 1.3 × 1.3 × 1 inch (inside dimensions) box constructed from single-sided circuit board material. The through-line connection is made with a short piece of UT-141 semi-rigid coax with the shield grounded only on one side to provide electrostatic shielding between the toroid and the center conductor of the coax. (Do not ground both ends of the shield or a shorted turn is created.) R_{SHUNT} is hidden under the toroid, and R_{SERIES} is shown connected to the sample port. This construction technique looks like a short piece of 200 Ω transmission line in the through-line which affects the SWR at higher frequencies. This can be corrected by compensating with two 3 pF capacitors connected to the through-line input and output connectors as shown in the photo. The through-line SWR was reduced from 1.43:1 to 1.09:1 at 180 MHz by adding the capacitors. This compensation, however, causes the attenuation to differ at high frequencies depending on the direction of the through-line connection. A sampler constructed using the box technique is useable from below 1 MHz through 30 MHz.

Figure 25.83 shows a different approach using 9/16-inch diameter, 0.014 inch wall thickness, hobby brass tubing. This lowers the impedance of the through-line so that no compensation is needed. The through-line SWR for the tube sampler is 1.08:1 at 180 MHz which is as good as the box sampler and the sensitivity to through-line direction is reduced. Although the high frequency attenuation is not as good as the box sampler, the construction technique provides a more consistent result. A sampler constructed using the tube technique should be useable through 200 MHz.

CONSTRUCTION OF THE TUBE SAMPLER

Both samplers use 15 turns of #28 AWG wire on an FT37-61 core, which just fits over the UT-141 semi-rigid coax. R_{SHUNT} is a 15 Ω, 2 W, noninductive metal-oxide resistor and R_{SERIES} is a 34.8 Ω, ¼ W, 1% noninductive metal-film resistor. The power dissipation of the resistors and the flux handling capability of the ferrite core are adequate for sampling a 1,500 W source. For those uncomfortable using BNC connectors at high power, an SO-239 version may be constructed using an FT50A-61 core and larger diameter tubing. Construction details are included with the online content.

25.18.8 RF Oscillators for Circuit Alignment

Receiver testing and alignment can make use of inexpensive RF signal generators which are available as complete units and in kit form. Any source of signal that is weak enough to avoid overloading the receiver can serve for alignment work and troubleshooting.

A crystal oscillator is often a satisfactory signal source for amplifier testing and receiver repair or alignment. Several example circuits can be found in the **Oscillators and Synthesizers** chapter. The output frequencies of crystal oscillators, while not adjustable, are quite precise and very stable. The Elecraft XG2 (**elecraft.com**) and NorCal S9 (**norcalqrp.org**) are good examples of a simple fixed-frequency signal source kits. The harmonics of the output signals are on known frequencies and can also be used as low-level signal sources. The fundamental signals have known output amplitudes for calibrating S meters and other gain stages.

Variable frequency oscillators can be used as signal generators and there are several kits or assembled units available based on a direct digital synthesis (DDS) integrated circuit. (See the **DSP and Software Radio Design** chapter.) The Elecraft XG3 is a programmable signal source that operates from 1.5 to 200 MHz with four programmable output levels between –107 and 0 dBm.

For receiver performance testing, precise frequency control, signal purity, noise, and low-level signal leakage become very important. A lab-quality instrument is required to make these measurements. Commercial and military-surplus units such as the HP608-series are big and stable, and they may be inexpensive. Recently, the HP8640-series of signal generators have become widely available at very attractive prices. When buying a used or inexpensive signal generator, look for these attributes: output level is calibrated, the output doesn't "ring" too badly when tapped, and doesn't drift too badly when warmed up.

25.18.9 Hybrid Combiners for Signal Generators

Many receiver performance measurements require two signal generators to be attached to a receiver simultaneously. This, in turn, requires a combiner that isolates the two signal generators (to keep one generator from being frequency or phase modulated by the other). Commercially made hybrid combiners are available from Mini-Circuits Labs (**minicircuits.com**).

Alternatively, a hybrid combiner is not difficult to construct. The combiners described here (see **Figure 25.84**) provide 40 to 50 dB of isolation between ports, assuming the common port is terminated in a 50-ohm load. Attenuation in the desired signal paths (each input to output) is 6 dB. Loads with low return loss typical of receiver inputs reduce isolation.

The combiners are constructed in small boxes made from double-sided circuit-board material as shown in Figure 25.84A. Each piece is soldered to the next one along the entire length of the seam. This makes a good RF-tight enclosure. BNC coaxial fittings are used on the units shown. However, any type of coaxial connector can be used. Leads must be kept as short as possible and precision non-inductive resistors (or matched units from the junk box) should be used. The circuit diagram for the combiners is shown in Figure 25.84B

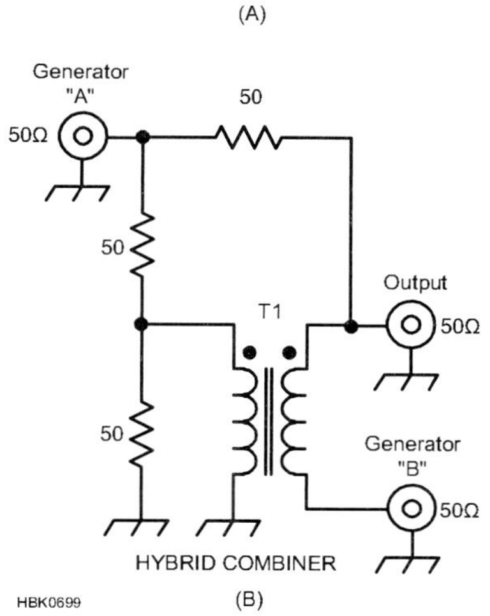

Figure 25.84 —The hybrid combiner on the left of A is designed to cover the 1 to 50-MHz range; the one on the right 50 to 500 MHz. B shows the circuit diagram of the hybrid combiner. Transformer T1 is wound with 10 bifilar turns of #30 AWG enameled wire. For the 1 to 50-MHz model, T1 is an FT-23-77 ferrite core. For the 50 to 500-MHz model, use an FT-23-67 ferrite core. Keep all leads as short as possible when constructing these units.

The combiner may also be constructed and used as a *return loss bridge* as described in the *QST* article by Jim Ford, N6JF, "Build a Return Loss Bridge," from September 1997 and included with the online content. Return loss is discussed in the **RF Techniques** chapter.

25.19 References

ARTICLES

Allison, B., WB1GCM, "Testing the SDR-IQ in the ARRL Lab," *QST*, Feb., 2010, p. 52.

Baier, T., DG8SAQ, "A Simple S-Parameter Test Set for the VNWA2 Vector Network Analyzer," *QEX*, May/June 2009, pp. 29 – 32.

Baier, T., DG8SAQ, "A Small, Simple USB-Powered Vector Network Analyzer Covering 1 kHz to 1.3 GHz," *QEX*, Jan./Feb. 2009, pp. 32 – 36. **sdrits.net/VNWA/VNWA_Description.html**.

Bird, T., "Definition and Misuse of Return Loss," *QEX*, Sep./Oct. 2010, pp. 38 – 39.

Bold, G., ZL1AN, "The Bruene Directional Coupler and Transmission Lines," Version 1.3 (2009). **collinsradio.org/wp-content/uploads/2015/05/Understanding-the-Bruene-Coupler-Transmission-Line-Bold.pdf**.

Bradley, M., K6TAF, "What Can You Do With a Dip Meter?," *QST*, May 2002, pp. 65 – 68.

Chunn, R., W5BIG, "An Antenna Impedance Meter for the High Frequency Bands," *QST*, Nov. 2006, p. 28.

Forrer, J., KC7WW, "A Low-Cost HF Channel Simulator for Digital Systems," *QEX*, May/June 2000, pp. 13 – 22.

"Fundamentals of RF and Microwave Noise Figure Measurements," Application Note 57-1, **keysight.com/us/en/assets/7018-06808/application-notes/5952-8255.pdf**. The older Hewlett-Packard application note AN57 may be more useful for legacy noise figure meters such as the HP340 series.

Hageman, S., "Build a Data Acquisition System for your Computer" (Tech Notes), *QEX*, Jan./Feb. 2001, pp. 52 – 57.

Kester, W., "The Good, the Bad, and the Ugly Aspects of ADC Input Noise — Is No Noise Good Noise?" Analog Devices Tutorial MT-004.

Knitter, M., DG5MK, "The DG5MK IV Meter — An Accurate Antenna Analyzer," *QEX*, May/June 2017, pp. 3 – 14.

Lau, Z., "A Relative RF Ammeter for Open-Wire Lines," *QST*, Oct. 1988, pp. 15 – 17.

Miller, B., KE6F, "Atomic Frequency Reference for Your Shack," *QEX*, Sep./Oct. 2009, pp. 35 – 44.

Nash, Bob, KF6CDO, "An Event per Unit Time Measurement System for Rubidium Frequency Standards," *QEX*, Nov./Dec. 2010, pp. 3 – 17.

Noakes, J., VE7NI, "The 'No Fibbin' RF Field Strength Meter, *QST*, Aug. 2002, pp. 28 – 29.

Pontius, Bruce E., NØADL, "Measure-ment of Signal-Source Phase Noise with Low-Cost Equipment," *QEX*, May/June 1998.

Richardson, Gary, AA7VM, "A Low Cost DDS Function Generator," *QST*, Nov. 2005, pp. 40 – 42.

Shera, B., W5OJM, "A GPS-Based Frequency Standard," *QST*, July 1998.

Silver, W., NØAX, Grounding and Bonding for the Radio Amateur, 2nd Edition (ARRL, 2021).

Steinbaugh, G., AF8L, "An Inexpensive Laboratory-Quality RF Wattmeter," *QEX*, May/June 2010, pp. 26 – 32.

"Vector Network Analyzer Error Correction — Calibration Techniques," Keysight Technologies, **keysight.com/us/en/assets/7018-06761/application-notes/5965-7709.pdf**.

Wade, P., W1GHZ, "Microwave System Test," Microwavelengths, *QST*, Aug. 2010, pp. 96 – 97. At microwave frequencies, the antenna and radio are sometimes tested as a system.

WEBSITES

moetronix.com/ae4jy/pathsim.htm. PathSim, free HF channel simulation software. Note that this page is out of date and no longer supported.

nist.gov. National Institute of Standards and Technology.

prologix.biz. Prologix is one manufacturer of low-cost GPIB-to-USB converters.

Contents

Chapter 26 — Online Content

Articles

- Amplifier Care and Maintenance, by Ward Silver, NØAX
- Building a Modern Signal Tracer, by Curt Terwilliger, W6XJ
- Diode and Transistor Test Circuits, by Ed Hare, W1RFI
- Hands On Radio: Power Supply Analysis, by Ward Silver, NØAX
- Troubleshooting Radios by, Mel Eiselman, NC4L

PC Board Templates

- AF/RF signal injector template
- Crystal controlled signal source template

Chapter 26

Troubleshooting and Maintenance

The robust and self-reliant ethic of amateur radio is nowhere stronger than in the amateur's ability to maintain, troubleshoot, and repair electronic equipment. Amateurs work with not just radios, but all sorts of equipment from computers and software to antennas and transmission lines. This flexibility and resilience are keys to fulfilling the FCC Part 97.1 Basis and Purpose of amateur radio.

The sections on troubleshooting approaches, tools and techniques build on earlier material written by Ed Hare, W1RFI. They will help you approach troubleshooting in an organized and effective manner, appropriate to your level of technical experience and tools at hand. This material shows how to get started and ask the right questions — often the most important part of troubleshooting.

Additional sections on troubleshooting power supplies, amplifiers, radios and antenna systems and tuners (contributed or updated by Matt Kastigar, WØMJ; Tom Schiller, N6BT; Ted Thrift, VK2ARA; and Ross Pittard, VK3CE) tackle the most common troubleshooting needs. Restoring and maintaining vintage equipment is a popular part of ham radio, and so there are some sections by John Fitzsimmons, W3JN, and Pat Bunsold, WA6MHZ, on the special needs of this equipment.

This chapter is organized in three groups of sections to be consulted as required for any particular troubleshooting need. You will not need to read it from end-to-end in order to troubleshoot successfully. The first group of sections covers test equipment details, pertinent information about components, and safety practices. The second group presents general guidelines and techniques for effective troubleshooting. The third group presents specific advice and information on equipment that is commonly repaired by amateurs.

TROUBLESHOOTING — ART OR SCIENCE?

Although some say troubleshooting is as much art as it is science, the repair of electronic gear is not magic. It is more like detective work. Knowledge of advanced math or electronics theory is not required. However, you must have, or develop, a good grasp of basic electronics and simple measurements, guided by the ability to read a schematic diagram and to visualize signal flow through the circuit. As with most skills, these abilities will develop with practice.

Not everyone is an electronics wizard; your gear may end up at the repair shop in spite of your best efforts. The theory you learned for the FCC examinations and the information in this *Handbook* can help you decide if you can fix it yourself. Even if the problem appears to be complex, most problems have simple causes. Why not give troubleshooting a try to the best of your abilities? Maybe you can avoid the effort and expense of shipping the radio to the manufacturer. It is gratifying to save time and money, but the experience and confidence you gain by fixing it yourself may prove even more valuable.

SAFETY FIRST! — SWITCH TO SAFETY

Always! Death is permanent. A review of safety must be the first thing discussed in a troubleshooting chapter. Some of the voltages found in amateur equipment can be fatal! Only 50 mA flowing through the body is painful; 100 to 500 mA is usually fatal. Under certain conditions, as little as 24 V can kill. RF exposure in a high-power amplifier can create severe burns very quickly. Batteries can deliver huge amounts of power that can melt tools and wires or create an explosion when short-circuited. Charging lead-acid cells can create a buildup of explosive hydrogen gas.

Make sure you are 100% familiar with all safety rules and the dangerous conditions that might exist in the equipment you are servicing. A list of safety rules can be found in **Table 26.1**. You should also read the **Safety** chapter of this *Handbook* — all of it — before you begin to work on equipment.

Remember, if the equipment is not working properly, dangerous conditions may exist where you don't expect them. Treat every component as potentially "live." Some older equipment uses "ac/dc" circuitry. In this circuit, one side of the chassis is connected directly to the ac line, a condition unexpected by today's amateurs, who are accustomed to modern safety standards and practices. This is an electric shock waiting to happen.

The maximum voltage rating of voltmeters and oscilloscopes is not often noted by the hobbyist, but it is crucial to safety when working on voltages higher than the household ac line voltage. Test equipment designed to measure voltage always has a maximum safe voltage rating between the circuit being measured and the equipment user — you! This is particularly

Table 26.1
Safety Rules

1. Keep one hand in your pocket when working on live circuits or checking to see that capacitors are discharged.
2. Include a conveniently located ground-fault current interrupter (GFCI) circuit breaker in the workbench wiring.
3. Use only grounded plugs and receptacles.
4. Use a GFCI protected circuit when working outdoors, on a concrete or dirt floor, in wet areas, or near fixtures or appliances connected to water lines, or within six feet of any exposed grounded building feature.
5. Use a fused, power limiting isolation transformer when working on ac/dc devices.
6. Switch off the power, disconnect equipment from the power source, ground the output of the internal dc power supply, and discharge capacitors when making circuit changes.
7. Do not subject electrolytic capacitors to excessive voltage, ac voltage or reverse voltage.
8. Test leads should be well insulated and without cracks, fraying, or exposed conductors.
9. Do not work alone!
10. Wear safety glasses for protection against sparks and metal or solder fragments.
11. Be careful with tools that may cause short circuits.
12. Replace fuses only with those having proper ratings.
13. Never use test equipment to measure voltages above its maximum rating.

important in handheld equipment in which there is no metal enclosure connected to an ac safety ground. Excessive voltage can result in a flashover to the user from the internal electronics, probes, or test leads, resulting in electric shock. Know and respect this rating.

If you are using an external high voltage probe, make sure it is in good condition with no cracks in the body. The test lead insulation should be in good condition — flexible and with no cracks or wire exposed. If practical, do not make measurements while holding the probe or meter. Attach the probe with the voltage discharged and then turn the power on. Turn power off and discharge the voltage before touching the probe again. Treat high voltage equipment with care and respect!

Soldering Safety

Remember that soldering tools and melted solder can be hot and dangerous! Wear protective goggles and clothing when soldering. A full course in first aid is beyond the scope of this chapter, but if you burn your skin, run the burn immediately under cold water and seek first aid or medical attention. Always seek medical attention if you burn your eyes; even a small burn can develop into serious trouble.

UNDERSTANDING THE BASICS

To fix electronic equipment, you need to understand the system and circuits you are troubleshooting. A working knowledge of electronic theory, circuitry and components is an important part of the process. When you are troubleshooting, you are looking for specific conditions that cause the symptoms you are experiencing. Knowing how circuits are supposed to work will help you to notice things that are out of place or that indicate a problem.

To be an effective troubleshooter, review and understand the following topics discussed elsewhere in this book:

• Ohms law and basic resistor circuits (**Electrical Fundamentals**)

• Basic transistor and diode characteristics (**Circuits and Components**)

• Fundamental digital logic and logic signals (**Digital Basics** supplement)

• Voltage and current measurements (**Test Equipment and Measurements**)

• SWR and RF power measurement (**Transmission Lines**)

You would be surprised at how many problems — even problems that appear complicated — turn out to have a simple root cause found by understanding the fundamentals and methods of one of these categories.

GETTING HELP

Other hams may be able to help you with your troubleshooting and repair problems, either with a manual or technical help. Check with your local club or repeater group. You may get lucky and find a troubleshooting wizard. (On the other hand, you may get some advice that is downright dangerous, so be selective.) Most clubs have one or two troubleshooting gurus who can provide guidance and advice, if not some on-the-workbench help.

There is a wealth of information available online, too. Many of the popular brands of equipment and even specific models have their own online communities or user's groups. The archives of these groups — almost universally free to join — contain much valuable troubleshooting, modification and operating information. If the problem doesn't appear to have been described, you can ask the group.

The Technology area of the ARRL's website also has an extensive section on Servicing Equipment (**www.arrl.org/servicing-equipment**). That page features articles and other resources, including links to schematic databases.

Your fellow hams in the ARRL Field organization may also help. Technical Coordinators (TC) and Technical Specialists (TS) are volunteers who are willing to help hams with technical questions. For the name and address of a local TC or TS, contact your Section Manager (listed in the front of any recent issue of *QST*).

Using Search Engines for Troubleshooting

The power of Internet search engines can save huge amounts of time when troubleshooting equipment. The key is in knowing how to construct the right list of words for them to find. Precision is your friend — be exact and use words others are likely to use if they had the same problem. Use the primary model number without suffixes to avoid being too specific. For example, when troubleshooting the well-known PLL potting compound problem exhibited by Kenwood TS-440 transceivers, entering the search string "TS-440 display dots" immediately finds many web pages dealing with the problem, while simply entering "Kenwood transceiver blank display" returns dozen of unrelated links.

Start with a very specific description of the problem and gradually use less exact terms if you don't find what you want. Learn how to use the "Advanced Search" functions of the search engine, too.

26.1 Test Equipment

Many of the steps involved in effiicient troubleshooting require the use of test equipment. We cannot see electricity directly, but we can measure its characteristics and effects. Our test equipment becomes our electrical senses.

The **Test Equipment and Measurements** chapter is where you can find out more about various common types of equipment, how to operate it, and even how to build some of your own. There are many articles in *QST* and in books and websites that explain test equipment and offer build-it-yourself projects, too. Surplus equipment of excellent quality is widely available at a fraction of its new cost.

You need not purchase or build every type of test equipment. Specialty equipment such as spectrum analyzers or UHF frequency counters can often be borrowed from a club member or friend — maybe one of those troubleshooting gurus mentioned earlier. If you own the basic instruments and know how to use them, you'll be able to do quite a bit of troubleshooting before you need the special instruments.

26.1.1 Senses

Although they are not test equipment in the classic sense, your own senses will tell you as much about the equipment you are trying to fix as the most-expensive spectrum analyzer. We each have some of these natural test instruments.

Eyes — Use them constantly. Look for evidence of heat and arcing, burned components, broken connections or wires, poor solder joints or other obvious visual problems.

Ears — Severe audio distortion can be detected by ear. The snaps and pops of arcing or the sizzling of a burning component may help you track down circuit faults. An experienced troubleshooter can diagnose some circuit problems by the sound they make. For example, a bad audio-output IC sounds slightly different from a defective speaker.

Nose — Your nose can tell you a lot. With experience, the smells of ozone, an overheating transformer, and a burned resistor or PC board trace each become unique and distinctive. Many troubleshooting sessions begin with "something smells hot!"

Finger — After using a voltmeter to ensure no hazardous voltages are present, you can use a fingertip to determine low heat levels—never do this in a high-voltage circuit. Use a temperature probe if using a finger is unsafe. Small-signal transistors can be fairly warm, but being very hot indicates a circuit problem. Warm or hot capacitors are always suspect. High-power devices and resistors can be quite hot during normal operation.

Brain — More troubleshooting problems have been solved with a multimeter and a brain than with the most expensive spectrum analyzer. You must use your brain to analyze data collected by other instruments.

26.1.2 Internal Equipment

Some test equipment is included in the equipment you repair. Nearly all receivers include a speaker. An S meter is usually connected ahead of the audio chain. If the S meter shows signals, that indicates that the RF and IF circuitry is probably functioning. Transmitters often have a power supply voltage and current meter, along with power output, SWR, ALC and speech compression readings that give valuable clues about what is happening inside the equipment.

The equipment also has visual indicator lights that provide additional information such as transmit status, high SWR, low voltage, squelch status, and so forth. These readings or indicators are often specifically referenced by the troubleshooting sections of manuals to help sort out problems.

Microprocessor-controlled equipment often provides error indications, either through a display or by indicator lights. In addition, faults detected by the control software are sometimes communicated through patterns of beeps or flashing of LEDs. Each sequence has a specific meaning that is described in the operating or service manual.

26.1.3 Bench Equipment

The following is a list of the most common and useful test instruments for troubleshooting. Some items serve several purposes and may substitute for others on the list. The theory and operation of most of this equipment is discussed in detail in the **Test Equipment and Measurements** chapter. Notes about the equipment's use for troubleshooting are listed here.

Multimeters —The most often used piece of test equipment, the digital multimeter or DMM, can often test capacitors of most values in addition to voltage, current and resistance. Most can test diodes and transistors on a go/no-go basis, while some can measure gain. Some can even measure frequency or use an external probe to measure temperature.

Some DMMs are affected by RF, so most technicians keep an old-style analog moving-needle VOM (volt-ohm-meter) on hand for use in strong RF fields. Some technicians prefer the moving needle for peaking or nulling adjustments.

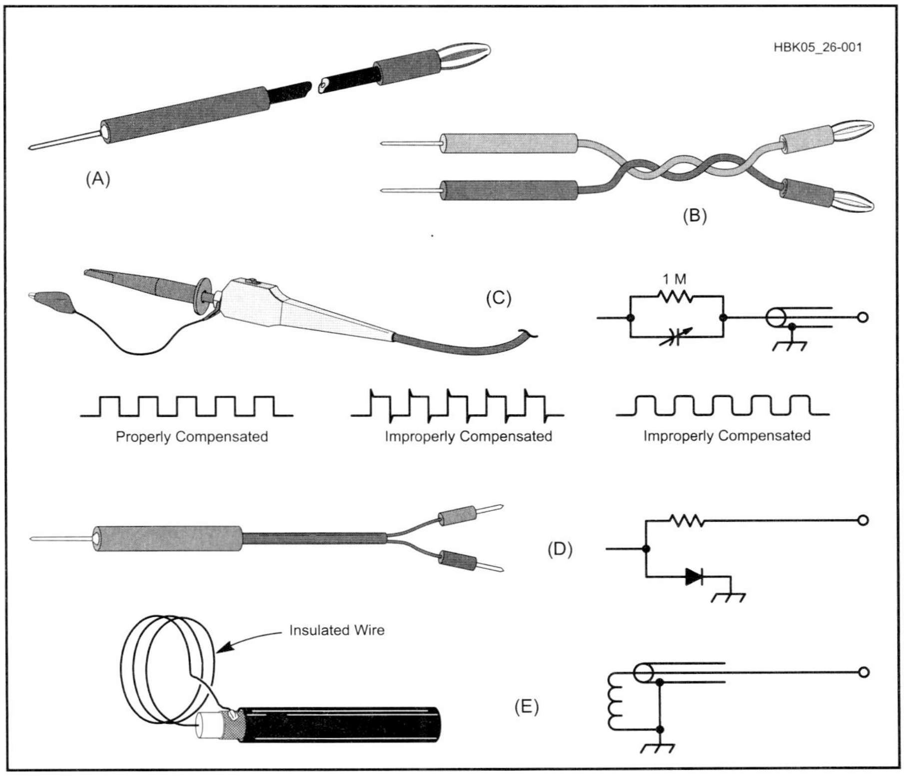

Figure 26.1 — An array of test probes for use with various test instruments.

Test or clip leads — Keep an assortment of these wires with insulated alligator clips. Commercially made leads have a high failure rate because they use small wire that is not soldered to the clips, just crimped. You can slip off the clip jackets and solder the wire yourself for better reliability. Making a set of heavier-gauge leads is a good idea for currents above several hundred milliamps.

Individual wire leads (**Figure 26.1A**) are good for dc measurements, but they can pick up unwanted RF energy. This problem is reduced somewhat if the leads are twisted together (Figure 26.1B). Coaxial cable test leads can avoid RF pickup but also place a small capacitance across the circuit being measured. The added capacitance may affect performance.

Test probes — The most common probe is the low-capacitance (×10) oscilloscope probe shown in Figure 26.1C. This probe isolates the oscilloscope from the circuit under test, preventing the scope's input and test-probe capacitance from affecting the circuit and changing the reading. A network in the probe serves as a 10:1 divider and compensates for frequency distortion in the cable and test instrument.

Demodulator probes (see the **Test Equipment and Measurements** chapter and the schematic shown in Figure 26.1D) are used to demodulate or detect RF signals, converting modulated RF signals to audio that can be heard in a signal tracer or seen on a low-bandwidth scope.

You can make a probe for inductive coupling as shown in Figure 26.1E. Connect a two or three-turn loop across the center conductor and shield before sealing the end. The inductive pickup is useful for coupling to high-current points and can also be used as a sniffer probe to pick up RF signals without contacting a circuit directly.

Other common types of probes are the non-contact clamp-on probes shown in **Figure 26.2** that use magnetic fields to measure current. A high-voltage probe for use with DMMs or VOMs is shown in **Figure 26.3** and is discussed more in this chapter's section on power supply troubleshooting.

Thermocouple and active temperature sensor probes are also commonly available. These display temperature directly on the meter in °F or °C.

RF power and SWR meters — Simple meters indicate relative power SWR and are fine for adjusting matching networks and monitoring transmission line conditions for problems. However, if you want to make accurate measurements, a calibrated directional RF wattmeter with the proper sensing elements for the frequencies of signals being measured is required.

Dummy load — Do not put a signal on the air while repairing equipment. Defective equipment can generate signals that interfere with other hams or other radio services. A dummy load also provides a known, matched load (usually 50 Ω) for use during adjustments and test measurements. See the **Transmitting** chapter.

Dip meter — As described in the **Test Equipment and Measurements** chapter, dip meters are used to adjust and troubleshoot resonant circuits. Many can perform as an absorption frequency meter, as well. Dip meters can be used as low-power signal sources but are not very stable.

New dip meters are fairly rare. When purchasing a dip meter, look for one that is mechanically and electrically stable. All of the coils should be present and in good condition. A headphone connection is helpful. Battery operated models are easier to use for antenna measurements. Dip meters are not nearly as common as they once were.

Oscilloscope — The oscilloscope, or scope, is the second most often used piece of test equipment, although a lot of repairs can be accomplished without one. The trace of a scope can give us a lot of information about a signal at a glance. For example, when signals from the input and output of a stage are displayed on a dual-trace scope, stage linearity and phase shift can be checked (see **Figure 26.4**).

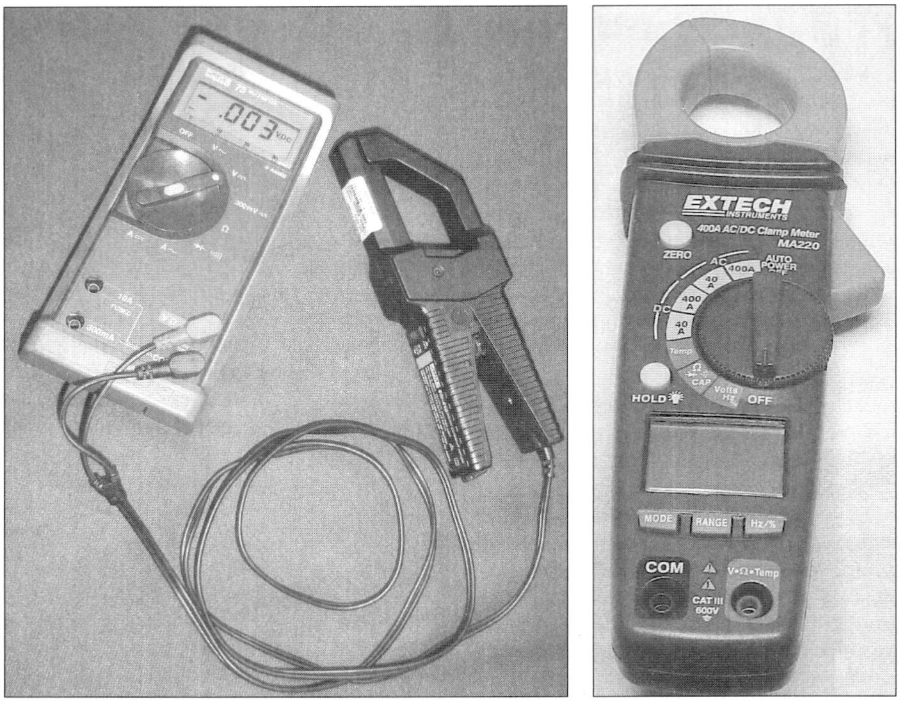

Figure 26.2 — A clamp-on meter probe is used with a digital multimeter for measuring ac current (left). Meters are also available integrated with the clamp-on probe (right).

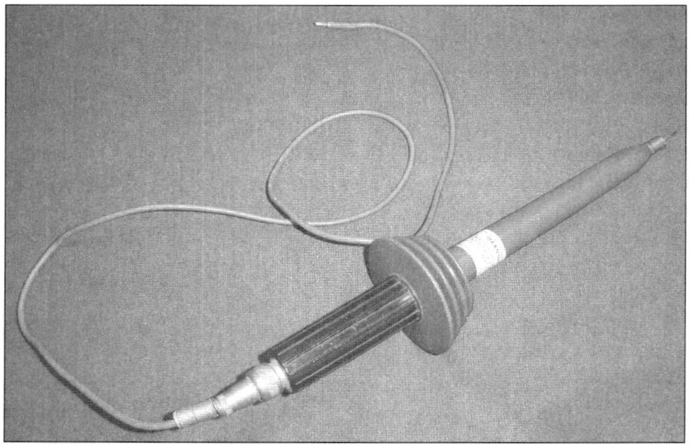
Figure 26.3 — A probe used for measuring high-voltage with a standard multimeter.

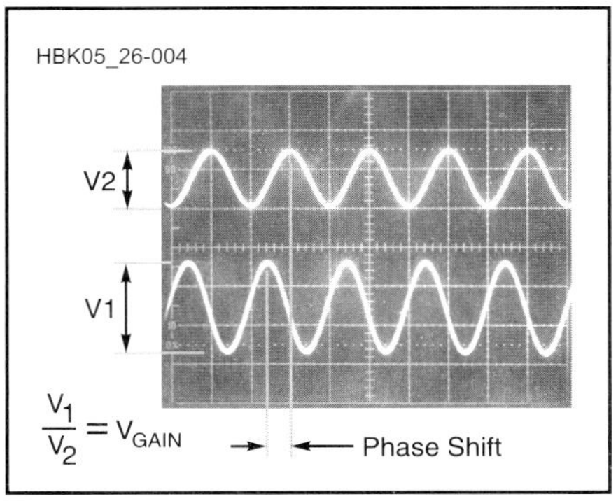

Figure 26.4 — A dual-trace oscilloscope display of amplifier input and output waveforms.

An oscilloscope will show gross distortions of audio and RF waveforms, but it cannot be used to verify that a transmitter meets FCC regulations for harmonics and spurious emissions. Harmonics that are down only 20 dB from the fundamental would be illegal in most cases, but they would not change the oscilloscope waveform enough to be seen.

When buying a scope, get the highest bandwidth you can afford. Old Hewlett-Packard or Tektronix instruments are usually quite good for amateur use.

Signal generator — Although signal generators have many uses, in troubleshooting they are most often used for signal injection (more about this later) and alignment of vintage equipment.

When buying a generator, look for one that can generate a sine wave signal. A good signal generator is double or triple shielded against leakage. Fixed-frequency audio should be available for modulation of the RF signal and for injection into audio stages. The most versatile generators can generate amplitude and frequency modulated signals. Used Hewlett-Packard (Agilent) and Tektronix units are typically available for reasonable prices but may not be repairable if they fail due to unavailable parts.

Good generators have stable frequency controls with no backlash. They also have multiposition switches to control signal level. A switch marked in dBm is a good indication that you have located a high-quality test instrument. The output jack should be a coaxial connector (usually a BNC or N), not the kind used for microphone connections.

In lieu of a fully tunable generator, you can build some simple equipment that generates a signal. For example, Elecraft makes the XG3 kit — a programmable signal source (**www.elecraft.com**) that generates 160 through 2 meter signals with 4 calibrated output levels. It's very useful for receiver calibration, sensitivity tests and signal tracing.

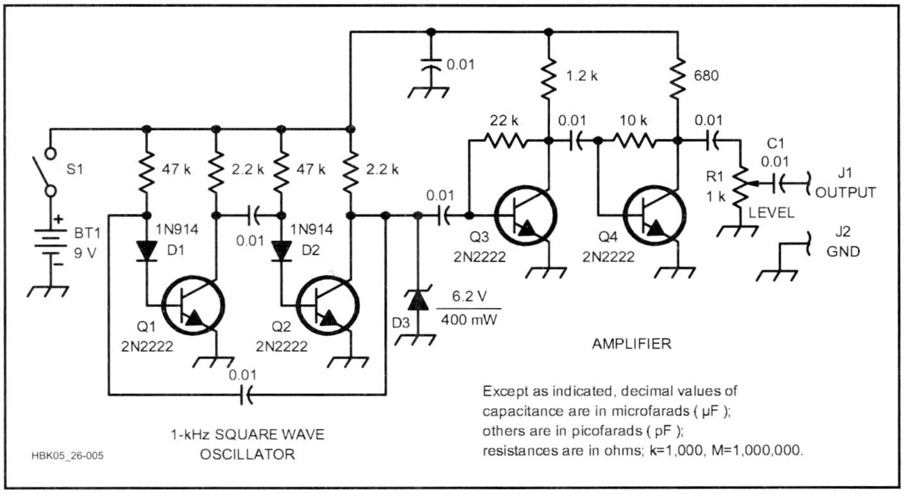

Figure 26.5 — Schematic of the AF/RF signal injector. All resistors are ¼ W, 5% carbon units, and all capacitors are disc ceramic. A full-size etching pattern and parts-placement diagram can be found in the downloadable supplemental content.

BT1 — 9 V battery.
D1, D2 — Silicon switching diode, 1N914 or equiv.
D3 — 6.2 V, 400 mW Zener diode.
J1, J2 — Banana jack.
Q1-Q4 — General-purpose silicon NPN transistors, 2N2222 or similar.
R1 — 1 kΩ panel-mount control.
S1 — SPST toggle switch.

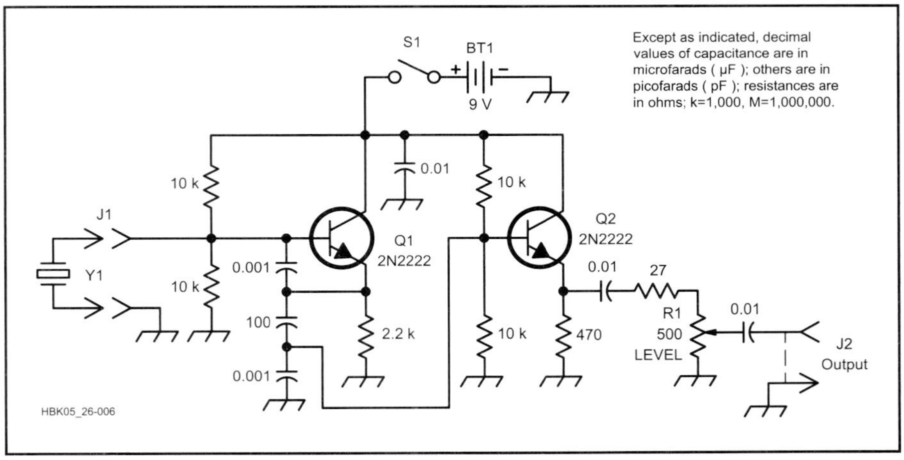

Figure 26.6 — Schematic of the crystal-controlled signal source. All resistors are ¼ W, 5% carbon units, and all capacitors are disc ceramic. A full-size etching pattern and parts-placement diagram can be found in the downloadable supplemental content.

BT1 — 9 V transistor radio battery.
J1 — Crystal socket to match the crystal type used.
J2 — RCA phono jack or equivalent.
Q1, Q2 — General-purpose silicon NPN transistors, 2N2222 or similar.
R1 — 500 Ω panel-mount control.
S1 — SPST toggle switch.
Y1 — 1 to 15-MHz crystal.

Even simpler, you can homebrew the AF/RF signal-injector schematic as shown in **Figure 26.5**. If frequency accuracy is needed, the crystal-controlled signal source of **Figure 26.6** can be used. The AF/RF circuit provides usable harmonics up to 30 MHz, while the crystal controlled oscillator will function with crystals from 1 to 15 MHz. These two projects are not meant to replace standard signal generators for alignment and precision testing, but they are adequate for generating signals that can be used for general troubleshooting. (See the section on Signal Tracing and Signal Injection.)

Signal tracer — Signals can be traced with a voltmeter and an RF probe, a dip meter with headphones or an oscilloscope, but signal tracers combine these functions especially for signal tracing through a receiver or other RF signal processing circuit. Articles describing the use of signal tracers, including a project you can build yourself, are provided with the downloadable supplemental content.

A general-coverage receiver can be also used to trace RF or IF signals, if the receiver covers the necessary frequency range. Most receivers, however, have a low-impedance input that severely loads the test circuit. To minimize loading, use a capacitive probe or loop pickup as in Figure 26.1. When the probe is held near the circuit, signals will be picked up and carried to the receiver. It may also pick up stray RF, so make sure you are listening to the correct signal by switching the circuit under test on and off while listening.

Transistor tester — Most transistor failures appear as either an open or shorted junction. Opens and shorts can be found easily with an ohmmeter or the diode junction checker of a standard DMM; a special tester is not required.

Transistor testers measure device current while the device is conducting or while an ac signal is applied at the control terminal. Transistor gain characteristics vary widely even between units with the same device number. Testers can be used to measure the gain of a transistor. DMM testers measure only transistor dc alpha and beta. Testers that apply an ac signal show the ac alpha or beta. Better testers also test for leakage.

In addition to telling you whether a transistor is good or bad, a transistor tester can help you decide if a particular transistor has sufficient gain for use as a replacement. It may also help when matched transistors are required. The final test is the repaired circuit.

Frequency counter — Most inexpensive frequency counters display frequency with 1 Hz resolution or better up to around low VHF frequencies. Some may include a prescaler that divides higher frequencies by 10 to extend the counter's range. Good quality used counters are widely available.

Power supplies — A well-equipped test bench should include a means of varying the ac-line voltage, a variable-voltage regulated dc supply and an isolation transformer.

AC-line voltage varies slightly with load. An autotransformer with a movable tap (also known by the trade name Variac) lets you boost or reduce the line voltage slightly. This is helpful to test circuit functions with supply-voltage variations.

An isolation transformer is required to work safely on vintage equipment, which often ties one side of the ac line to the chassis. An isolation transformer is also required when working on any equipment or circuits that operate directly connected to the line. Note that your test equipment will also have to be powered through the isolation transformer in such cases!

A good multi-voltage supply will help with nearly any analog or digital troubleshooting project. Many electronics distributors stock bench power supplies. A variable-voltage dc supply may be used to power various small items under repair or provide a variable bias supply for testing active devices. Construction details for a laboratory power supply appear in the **Power Sources** chapter.

Heat and cold sources — Many circuit problems are sensitive to temperature. A piece of equipment may work well when first turned on (cold) but fail as it warms up. In this case, a cold source will help you find the intermittent connection. When you cool the bad component, the circuit will suddenly start working again (or stop working). Cooling sprays are available from most parts suppliers.

A heat source helps locate components that fail only when hot. A small incandescent lamp can be mounted in a large piece of sleeve insulation to produce localized heat for test purposes. The tip of a soldering iron set to low heat can also be used.

A heat source is usually used in conjunction with a cold source. If you have a circuit that stops working when it warms up, heat the circuit until it fails, then cool the components one by one. When the circuit starts working again, the last component sprayed was the bad one.

Stethoscope — A stethoscope (with the pickup removed — see **Figure 26.7**) or a long piece of sleeve insulation can be used to listen for arcing or sizzling in a circuit. Remove any metal parts at the end of the pickup tube before use for troubleshooting live equipment.

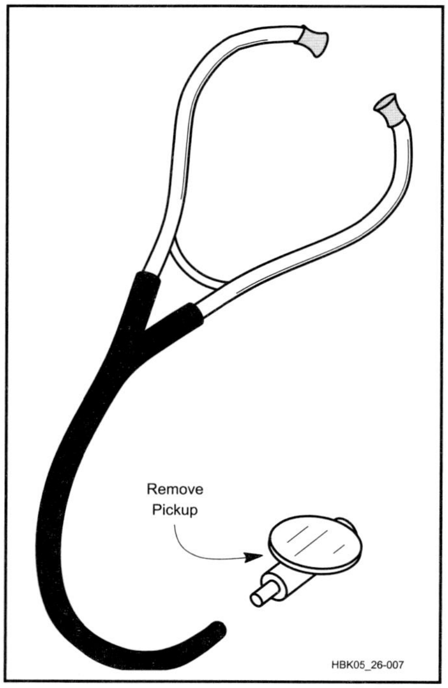

Figure 26.7 — A stethoscope, with the pickup and all metal hardware removed from the listening tube, is used to listen for arcing in crowded circuits.

The Shack Notebook

A shack notebook is an excellent way to keep track of test results, wiring, assembly notes, and so forth. If you haven't already started one, now is a good time. All it takes is an inexpensive composition book or spiral-bound notebook. The books filled with graph paper are especially good for drawing and making graphs.

The goal is to have one place where information is collected about how equipment was built, performs, or operates. The notebook is invaluable when trying to determine if performance has changed over time, or what color code was used for a control cable, for example.

Before beginning a test session or when adding a new piece of equipment to your shack, get out the shack notebook first and have it available as you work. For new equipment, record serial numbers, when installed, whether it was modified or specially configured to work in your station, etc. For a new antenna or feed line, it's a good idea to make a few SWR measurements so you can refer to them later if something seems wrong in the antenna system. Be sure to include the date of any entries as well.

You can also make good use of that digital camera, even the one in your mobile phone. Take pictures of equipment, inside and out, to document how it is assembled or configured. This can be very helpful when you have to maintain the equipment later.

26.2 Components

Once you locate a defective part, it is time to select a replacement. This is not always an easy task. Each electronic component has a function. This section acquaints you with the functions, failure modes and test procedures of resistors, capacitors, inductors and other components. Test the components implicated by symptoms and stage-level testing. In most cases, a particular faulty component will be located by these tests. If a faulty component is not indicated, check the circuit adjustments. As a last resort, use a "shotgun" approach — replace all parts in the problem area with components that are known to be good.

26.2.1 Check the Circuit

Before you install a replacement component of any type, you should be sure that another circuit defect didn't cause the failure. Check the circuit voltages carefully before installing any new component, especially on each trace or connection to the bad component. The old part may have died as a result of a lethal voltage. Measure twice — repair once! (With apologies to the old carpenter...) Of course, circuit performance is the final test of any substitution.

26.2.2 Fuses

Most of the time, when a fuse fails, it is for a reason — usually a short circuit in the load. A fuse that has failed because of a short circuit usually shows the evidence of high current: a blackened interior with little blobs of fuse element everywhere. Fuses can also fail by fracturing the element at either end. This kind of failure is not visible by looking at the fuse. Check even fuses thought to be good with an ohmmeter. You may save hours of troubleshooting.

For safety reasons, always use *exact* replacement fuses. Check the current and voltage ratings. The fuse timing (fast, normal or slow blow) must be the same as the original. Never attempt to force a fuse that is not the right size into a fuse holder. The substitution of a bar, wire or penny for a fuse invites a smoke party.

26.2.3 Wires and Cables

Wires seldom fail unless abused. Short circuits can be caused by physical damage to insulation or by conductive contamination. Damaged insulation is usually apparent during a close visual inspection of the conductor or connector. Look carefully where conductors come close to corners or sharp objects. Repair worn insulation by replacing the wire or securing an insulating sleeve (spaghetti) or heat-shrink tubing over the worn area.

When wires fail, the failure is usually caused by stress and flexing. Nearly everyone has broken a wire by bending it back and forth, and broken wires are usually easy to detect. Look for sharp bends or bulges in the insulation.

When replacing conductors, use the same material and size, if possible. Substitute only wire of greater cross-sectional area (smaller gauge number) or material of greater conductivity. Insulated wire should be rated at the same, or higher, temperature and voltage as the wire it replaces.

Cables used for audio, control signals, and feed lines sometimes fail from excessive flexing, being crimped or bent too abruptly, or getting pulled out of connectors. As with replacing wires, use the same cable type or one with higher ratings.

26.2.4 Connectors

Connection faults are one of the most common failures in electronic equipment. This can range from something as simple as the ac-line cord coming out of the wall, to a connector having been inserted into the wrong socket, to a defective IC socket. Connectors that are plugged and unplugged frequently can wear out, becoming intermittent or noisy. Inspect male connectors for bent pins, particularly miniature connectors with very small pins. Check connectors carefully when troubleshooting.

Connector failure can be hard to detect. Most connectors maintain contact as a result of spring tension that forces two conductors together. As the parts age, they become brittle and lose tension. Any connection may deteriorate because of nonconductive corrosion at the contacts. Solder helps prevent this problem, but even soldered joints suffer from corrosion when exposed to weather.

Signs of excess heat are sometimes seen near poor connections in circuits that carry moderate current. The increase in dissipated power at the poor connection heats the contacts, and this leads to more resistance and soon the connection fails. Check for short and open circuits with an ohmmeter or continuity tester. Clean those connections that fail as a result of contamination.

Occasionally, corroded connectors may be repaired by cleaning, but replacement of the conductor/connector is usually required, especially for battery holders supplying moderate currents. Solder all connections that may be subject to harsh environments and protect them with acrylic enamel, RTV compound or a similar coating. An anti-corrosion compound or grease is a good idea for connections located outside. See the entry on Weatherproofing RF Connectors in the Antenna and Tower Safety section of the **Safety** chapter.

Choose replacement connectors with consideration of voltage and current ratings. Use connectors with symmetrical pin arrangements only where correct insertion will not result in a safety hazard or circuit damage.

26.2.5 Resistors

Resistors usually fail by becoming an open circuit. More rarely they change value. Both failures are usually caused by excess heat. Such heat may come from external sources or from power dissipated within the resistor. Sufficient heat burns the resistor until it becomes an open circuit.

Resistors can also fracture and become an open circuit as a result of physical shock. Contamination on or around a high-value resistor (100 kΩ or more) can cause a change in value by providing a leakage path for current around a resistor. This contamination can occur on the resistor body, mounts or printed-circuit board. Resistors that have changed value should be replaced. Leakage is cured by cleaning the resistor body and surrounding area.

In addition to the problems of fixed-value resistors, potentiometers and rheostats can develop noise problems, especially in dc circuits. Dirt often causes intermittent contact between the wiper and resistive element. To cure the problem, spray electronic contact cleaner into the control, through holes in the case, and rotate the shaft a few times.

The resistive element in wire-wound potentiometers eventually wears and breaks from the sliding action of the wiper. In this case, the control needs to be replaced.

Replacement resistors should be of the same value, tolerance, type and power rating as the original. The value should stay within tolerance. Replacement resistors may be of a different type than the original, if the characteristics of the replacement are consistent with circuit requirements. (See the **Electrical Fundamentals** chapter for more information on resistor types.)

Substitute resistors can usually have a greater power rating than the original, except in high-power emitter circuits where the resistor also acts as a fuse, or in cases where the larger size presents a problem.

Variable resistors should be replaced with the same kind (carbon or wire wound) and taper (linear, log, reverse log and so on) as the original. Keep the same, or better, tolerance and pay attention to the power rating.

In all cases, mount high-temperature resistors away from heat-sensitive components. Keep carbon composition and film resistors away from heat sources. This will extend their life and ensure minimum resistance variations.

26.2.6 Capacitors

Capacitors usually fail by shorting, opening or becoming electrically (or physically) leaky. They rarely change value. Capacitor failure is usually caused by excess current, voltage, temperature or aging of the dielectric or materials making up the capacitor. Leakage can be external to the capacitor (contamination on the capacitor body or circuit) or internal to the capacitor.

TESTS

If you do not have a multimeter with a capacitor test function or a component tester, the easiest way to test capacitors is out of circuit with an ohmmeter. In this test, the resistance of the meter forms a timing circuit with the capacitor to be checked. Capacitors from 0.01 µF to a few hundred µF can be tested with common ohmmeters. Set the meter to its highest range and connect the test leads across the discharged capacitor. When the leads are connected, current begins to flow. Current is high when the capacitor is discharged, but drops as the capacitor voltage builds up. This shows on the meter as a low resistance that builds, over time, toward infinity.

The speed of the resistance build-up corresponds to capacitance. Small capacitance values approach infinite resistance almost instantly. A 0.01 µF capacitor checked with a meter having an 11 MΩ input impedance would increase from zero to a two-thirds scale reading in 0.11 second, while a 1 µF unit would require 11 seconds to reach the same reading. If the tested capacitor does not reach infinity within five times the period taken to reach the two-thirds point, it has excess leakage. If the meter reads infinite resistance immediately, the capacitor is open. (Aluminum electrolytics normally exhibit high-leakage readings.)

Capacitance can also be measured for approximate value with a dip meter by constructing a parallel-resonant circuit using an inductor of known value. The formula for resonance is discussed in the **Electrical Fundamentals** chapter of this book.

It is good practice to keep a collection of known components that have been measured on accurate L or C meters. Alternatively, a standard value can be obtained by ordering 1 or 2% components from an electronics supplier. A 10% tolerance component can be used as a standard; however, the results will be known only to within 10%. The accuracy of tests made with any of these alternatives depends on the accuracy of the standard value component. Further information on this technique appears in Bartlett's article, "Calculating Component Values," in Nov 1978 *QST*.

Older capacitors can also be checked and the dielectric reformed, if necessary, with a capacitor checker of similar vintage. See this chapter's section on Restoration and Repair of Vintage Radios for more information about this technique.

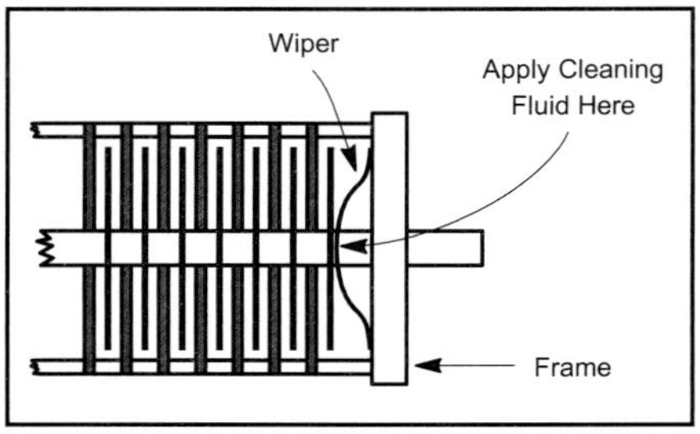

Figure 26.8 — Partial view of an air-dielectric variable capacitor. If the capacitor is noisy or erratic in operation, apply electronic cleaning fluid where the wiper contacts the rotor plates.

CLEANING

The only variety of common capacitor that can be repaired is the air-dielectric variable capacitor. Electrical connection to the moving plates is made through a spring-wiper arrangement (see **Figure 26.8**). Dirt normally builds on the contact area, and it needs occasional cleaning. Before cleaning the wiper/contact, use gentle air pressure and a soft brush to remove all dust and dirt from the capacitor plates. Apply some electronic contact cleaning fluid. Do not lubricate the contact point. Rotate the shaft quickly several times to work in the fluid and establish contact. Use the cleaning fluid sparingly and keep it off the plates except at the contact point.

REPLACEMENTS

Replacement capacitors should match the original in value, tolerance, dielectric, working voltage and temperature coefficient. Use only ac-rated capacitors for line service (capacitors connected directly to the ac line) to prevent fire hazards. If exact replacements are not available, substitutes may vary from the original part in the following respects: bypass capacitors may vary from one to three times the capacitance of the original, coupling capacitors may vary from one-half to twice the value of the original, capacitance values in tuned circuits (especially filters) must be exact (even then, any replacement will probably require circuit realignment).

If the same kind of capacitor is not available, use one with better dielectric characteristics. Do not substitute polarized capacitors for non-polarized parts. Capacitors with a higher working voltage may be used, although the capacitance of an electrolytic capacitor used significantly below its working voltage will usually increase with time.

The characteristics of each type of capacitor are discussed in the **Electrical Fundamentals** and **RF Techniques** chapters. Consider these characteristics if you're not using an exact replacement capacitor.

Batteries and Tools

When working on equipment powered by a battery with a capacity of more than a few ampere-hours (Ah), take special care to avoid short circuits and always have a fuse or circuit-breaker in-line to the battery. If possible, fuse both the positive and negative connections. This is particularly important for vehicle batteries and mobile equipment. A short circuit can do thousands of dollars of damage to a vehicle's electrical power system in a matter of seconds! When working on mobile equipment, disconnect the battery's positive terminal or at least disconnect the circuit that powers the equipment by removing the fuse.

A tool that accidentally short circuits the battery terminals can cause an instantaneous current flow of thousands of amps, often destroying both the battery and the tool and creating a significant fire and burn hazard.

26.2.7 Inductors and Transformers

The most common inductor or transformer failure is a broken conductor. More rarely, a short circuit can occur across one or more turns of a coil. In an inductor, this changes the value. In a transformer, the turns ratio and resultant output voltage changes. In high-power circuits, excessive inductor current can generate enough heat to melt plastics used as coil forms.

Inductors may be checked for open circuit failure with an ohmmeter. In a good inductor, dc resistance rarely exceeds a few ohms. Shorted turns and other changes in inductance show only during alignment or inductance measurement.

The procedure for measurement of inductance with a dip meter is the same as that given for capacitance measurement, except that a capacitor of known value is used in the resonant circuit.

Replacement inductors must have the same inductance as the original, but that is only the first requirement. They must also carry the same current, withstand the same voltage and present nearly the same Q as the original part. Given the original as a pattern, the amateur can duplicate these qualities for many inductors. Note that inductors with ferrite or iron-powder cores are frequency sensitive, so the replacement must have the same core material.

If the coil is of simple construction, with the form and core undamaged, carefully count and write down the number of turns and their placement on the form. Also note how the coil leads are arranged and connected to the circuit. Then determine the wire size and insulation used. Wire diameter, insulation and turn spacing are critical to the current and voltage ratings of an inductor. (There is little hope of matching coil characteristics unless the wire is duplicated exactly in the new part.) Next, remove the old winding — be careful not to damage the form — and apply a new winding in its place. Be sure to dress all coil leads and connections in exactly the same manner as the original. Apply Q dope (a solution of polystyrene plastic) or a thin coating of plastic-based glue to hold the finished winding in place.

Follow the same procedure in cases where the form or core is damaged, except that a suitable replacement form or core (same dimensions and permeability) must be found.

Ready-made inductors may be used as replacements if the characteristics of the original and the replacement are known and compatible. Unfortunately, many inductors are poorly marked. If so, some comparisons, measurements and circuit analysis are usually necessary.

When selecting a replacement inductor, you can usually eliminate parts that bear no physical resemblance to the original part. This may seem odd, but the Q of an inductor depends on its physical dimensions and the permeability of the core material. Inductors of the same value, but of vastly different size or shape, will likely have a great difference in Q. The Q of the new inductor can be checked by installing it in the circuit, aligning the stage and performing the manufacturer's passband tests. Although this practice is all right in a pinch, it does not yield an accurate Q measurement. Methods to measure Q appear in the **Test Equipment and Measurements** chapter.

Once the replacement inductor is found, install it in the circuit. Duplicate the placement, orientation and wiring of the original. Ground-lead length and arrangement should not be changed. Isolation and magnetic shielding can be improved by replacing solenoid inductors with toroids. If you do, however, it is likely that many circuit adjustments will be needed to compensate for reduced coupling and mutual inductance. Alignment is usually required whenever a tuned-circuit component is replaced.

A transformer consists of two inductors that are magnetically coupled. Transformers are used to change voltage and current levels (this changes impedance also). Failure usually occurs as an open circuit or short circuit of one or more windings. Insulation failures can also occur that result in short circuits between windings or between windings and the core or case. It is also common for the insulated wire leads to develop cracks or abrasion where they come out of the case. This can be repaired easily by replacing the lead, or if the conducting wire strands are not burned or broken, by sliding an insulation sleeve over the wire to protect the insulation from further wear.

Amateur testing of power transformers is mostly limited to ohmmeter tests for open circuits and voltmeter checks of secondary voltage. Make sure that the power-line voltage is correct, then check the secondary voltage against that specified. There should be less than 10% difference between open-circuit and full-load secondary voltage. A test setup and procedure for evaluating power transformers is also provided in the **Power Sources** chapter.

Replacement transformers must match the original in voltage, volt-ampere (VA), duty cycle and operating-frequency ratings. They must also be compatible in size. (All transformer windings should be insulated for the full power supply voltage.)

When disconnecting a transformer for testing or repair, be sure to carefully record the color and connection for each of the transformer leads. In power transformers it is common for leads to be mostly one color, but with a contrasting stripe or other pattern that can be overlooked. If in doubt, use tape or paper labels to note the connection for each lead. Recording transformer color codes and connections is a good use of the shack notebook.

26.2.8 Relays

Although relays have been replaced by semiconductor switching in low-power circuits, they are still used extensively in high-power amateur radio equipment for applications such as amplifier TR switching and in antenna systems or ac power control. Relay action may become sluggish. AC relays can buzz (with adjustment becoming impossible). A binding armature or weak springs can cause intermittent switching. Excessive use or hot switching ruins contacts and shortens relay life.

You can test relays with a voltmeter by measuring voltage across contacts (power on, in-circuit) or with an ohmmeter (out of circuit). Look for erratic readings across the contacts, open or short circuits at contacts, or an open circuit at the coil. A visual inspection with a magnifying glass should show no oxidation or corrosion. Limited pitting is usually OK.

Most failures of simple relays can be repaired by a thorough cleaning. Clean the contacts and mechanical parts with a residue-free cleaner. Keep it away from the coil and plastic parts that may be damaged. Dry the contacts with lint-free paper, such as a business card, then burnish them with a smooth steel blade. Do not use a file to clean contacts because it will damage the contact surface.

Replacement relays should match or exceed the original specifications for voltage, current, switching time and stray impedance (impedance is significant in RF circuits only). Many relays used in transceivers are specially made for the manufacturer. Substitutes may not be available from any other source.

Before replacing a multi-contact relay, make a drawing of the relay, its position, and the leads and their routings through the surrounding parts. This drawing allows you to complete the installation properly, even if you are distracted in the middle of the operation. (This is a good use of the shack notebook!)

26.2.9 Semiconductors

Testing diodes and transistors with the ohmmeter function of an analog VOM used to be the normal method. Today's inexpensive multimeters nearly always provide a forward and reverse junction voltage drop test function. This almost eliminates the need for resistance-based tests for functional troubleshooting with the attendant variability and dependence on meter circuits. However, it is occasionally useful to perform threshold and voltage-current testing to match components or troubleshoot a specialized circuit. In support of those tests, the short article "Diode and Transistor Test Circuits" containing test circuits and procedures for measuring leakage, gain, Zener point voltage and so forth is included with the downloadable supplemental content. This section will focus on simple go/no-go testing.

DIODES

The primary function of diodes is to pass current in one direction only. They can be easily tested with an ohmmeter, and most multimeters have a diode junction test function built-in as well.

Signal or switching diodes — The most common diode in electronics equipment, signal diodes, are used to convert ac to dc, to detect RF signals or to take the place of relays to switch ac or dc signals within a circuit. Signal diodes usually fail open, although shorted diodes are not rare.

Power rectifiers — Most equipment contains a power supply, so power rectifier diodes are the second-most common diodes in electronic circuitry. They usually fail shorted, blowing the power-supply fuse.

Other diodes — Zener diodes are made with a predictable reverse-breakdown voltage and are used as voltage regulators. Varactor diodes are specially made for use as voltage controlled variable capacitors. (Any semiconductor diode may be used as a voltage-variable capacitance, but the value will not be as predictable as that of a varactor.) A diac is a special-purpose diode that passes only pulses of current in each direction.

Diode testing — There are several basic tests for most diodes. First, is it a diode? Does it conduct in one direction and block current flow in the other? A simple resistance measurement is suitable for this test in most cases.

Diodes should be tested out of circuit. Disconnect one lead of the diode from the circuit, then perform the test. Diodes can also be tested by measuring the voltage drop across the diode junction while the diode is conducting.

A functioning diode will show high resistance in one direction and low resistance in the other. A DMM with a diode-test function is the best instrument to use. If using an analog meter, make sure to use more than 0.7 V and less than 1.5 V to measure resistance. Use the highest resistance scale of the meter that gives a reading of less than full-scale. Check a properly functioning diode to determine the meter polarity if there is any question. Compare the forward and reverse resistance readings for a properly functioning diode to those of the diode being tested to determine whether the diode is good.

Diode junction forward voltage drops are measured by a multimeter's diode test function. Silicon junctions usually show about 0.6 V at typical test current levels, while germanium is typically 0.2 V. Junction voltage drop increases with current flow.

Multimeters measure the junction resistances at low voltage and are not useful for testing Zener diodes. A good Zener diode will not conduct in the reverse direction at voltages below its rating. See the article included with the downloadable supplemental content, and mentioned at the beginning of this section, for procedures and a circuit to determine Zener diode performance.

Replacement diodes — When a diode fails, check associated components as well. Replacement rectifier diodes should have the same current and peak inverse voltage (PIV) as the original. Series diode combinations are often used in high-voltage rectifiers. (The resistor and capacitor networks used to distribute the voltage equally among the diodes are no longer required for new rectifiers but should be retained for older parts. See the **Power Sources** chapter for more information.)

Switching diodes may be replaced with diodes that have equal or greater current ratings and a PIV greater than twice the peak-to-peak voltage encountered in the circuit. Switching time requirements are not critical except in RF, logic and some keying circuits. Logic circuits may require exact replacements to assure compatible switching speeds and load characteristics. RF switching diodes used near resonant circuits must have exact replacements, as the diode resistance and capacitance will affect the tuned circuit.

Voltage and capacitance characteristics must be considered when replacing varactor diodes. Once again, exact replacements are best. Zener diodes should be replaced with parts having the same Zener voltage and equal or higher power rating, and equal or lower tolerance. Check the associated current-limiting resistor when replacing a Zener diode.

BIPOLAR TRANSISTORS

Transistor failures occur as an open junction, a shorted junction, excess leakage or a change in amplification performance. Most transistor failure is catastrophic. A transistor that has no leakage and amplifies at dc or audio frequencies will usually perform well over its design range. For this reason, transistor tests need not be performed at the planned operating frequency. Tests are made at dc or a low frequency (usually 1000 Hz). The circuit under repair is the best test of a potential replacement part. Swapping in a replacement transistor in a failed circuit will often result in a cure.

A simple and reliable test of bipolar transistors can be performed with the transistor in a circuit and the power on. It requires a test lead, a 10 kΩ resistor and a voltmeter. Connect the voltmeter across the emitter/collector leads and read the voltage. Then use the test lead to connect the base and emitter (**Figure 26.9A**). Under these conditions, conduction of a good transistor will be cut off and the meter should show nearly the entire supply voltage across the emitter/collector leads. Next, remove the clip lead and connect the 10 kΩ resistor from the base to the collector. This should bias the transistor into conduction, and the emitter/ collector voltage should drop (Figure 26.9B). (This test indicates transistor response to changes in bias voltage.)

Transistors can be tested (out of circuit) with an ohmmeter in the same manner as diodes, or a multimeter with a transistor test function can be used. Before using the ohmmeter-transistor circuit, look up the device characteristics before testing and consider possible consequences of testing the transistor in this way. Limit junction current to 1 to 5 mA for small-signal transistors. Transistor destruction or inaccurate measurements may result from careless testing.

The reverse-to-forward resistance ratio for good transistors may vary from 30:1 to better than 1,000 1. Germanium transistors — still occasionally encountered — sometimes show high leakage when tested with an ohmmeter. Bipolar transistor leakage may be specified from the collector to the base, emitter to base or emitter to collector (with the junction reverse biased in all cases). The specification may be identified as I_{cbo}, I_{bo}, collector cutoff current or collector leakage for the base-collector junction, I_{ebo}, and so on for other junctions. Leakage current increases with junction temperature. (See the **Circuits and Components** chapter for definitions of these and other transistor parameters.)

While these simple test circuits will identify most transistor problems, RF devices should be tested at RF. Most component manufacturers include a test-circuit schematic on the data sheet. The test circuit is usually an RF amplifier that operates near the high end of the device frequency range. If testing at RF is not possible, substitution of a properly functioning device is required.

Semiconductor failure is sometimes the result of environmental conditions. Open junctions, excess leakage (except with germanium transistors) and changes in amplification performance result from overload or excessive current.

Cross-Reference Replacement Semiconductors

Semiconductors from older equipment, even ICs, may be available as a cross-reference generic replacement. The primary source for these devices is NTE Electronics (**www.nteinc.com**). Enter the part number of the device you are trying to replace in the "Cross-Reference" window. NTE also supplies cross-referenced replacements for ECG part numbers that were the original generic replacement parts.

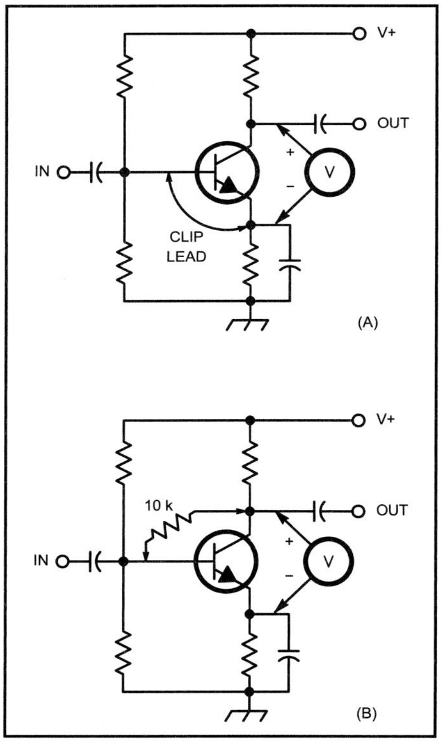

Figure 26.9 — An in-circuit semiconductor test with a clip lead, resistor and voltmeter. The meter should read V+ at (A). During test (B) the meter should show a decrease in voltage, ranging from a slight variation down to a few millivolts. It will typically cut the voltage to about half of its initial value.

Shorted junctions (low resistance in both directions) are usually caused by voltage spikes. Electrostatic discharge (ESD) or transients from lightning can destroy a semiconductor in microseconds.

Transistors rarely fail without an external cause. Check the surrounding parts for the cause of the transistor's demise, and correct the problem before installing a replacement.

JFETs

Junction FETs can be tested with a multimeter's diode junction test function in much the same way as bipolar transistors (see text and **Figure 26.10**). Reverse leakage should be several megohms or more. Forward resistance should be 500 to 1000 Ω if measured with an analog meter.

MOSFETs

Small-signal MOS (metal-oxide semiconductor) layers are extremely fragile. Normal body static is enough to damage them. Even gate protected (a diode is placed across the MOS layer to clamp voltage) MOSFETs may be destroyed by a few volts of static electricity. MOSFETs used for power circuits and RF amplifiers are much more resistant to damage. The manufacturer's sheet will specify any special static-protection measures that are required. (See the **Construction Techniques** chapter for more information about managing static at the workbench.)

When testing small MOSFETs make sure the power is off, capacitors, discharged and the leads are shorted together before installing or removing it from a circuit. Use a voltmeter to be sure the chassis is near ground potential, then touch the chassis before and during MOSFET installation and removal. This assures that there is no difference of potential between your body, the chassis and the MOSFET leads. Ground the soldering-iron tip with a clip lead when soldering MOS devices. The FET source should be the first lead connected to and the last disconnected from a circuit. The insulating layers in MOSFETs prevent testing with an ohmmeter. Substitution is the only practical means for amateur testing of MOSFETs.

FET CONSIDERATIONS

Replacement FETs should be of the same type as the original part: JFET or MOSFET, P-channel or N-channel, enhancement or depletion. Consider the breakdown voltage required by the circuit. The breakdown voltage should be at least two to four times the power-supply and signal voltages in amplifiers. Allow for transients of 10 times the line voltage in power supplies. Breakdown voltages are usually specified as $V_{(BR)GSS}$ or $V_{(BR)GDO}$.

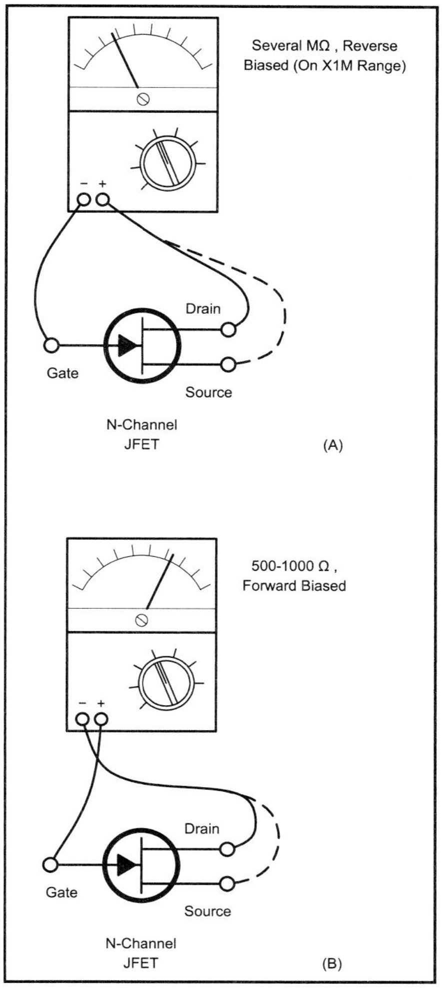

Figure 26.10 — Ohmmeter tests of a JFET. The junction is reverse biased at A and forward biased at B. (Analog meters are shown for convenience of illustration.)

The gate-voltage specification gives the gate voltage required to cut off or initiate channel current (depending on the mode of operation). Gate voltages are usually listed as $V_{GS(OFF)}$, V_p (pinch off), V_{TH} (threshold) or $I_{D(ON)}$ or I_{TH}.

Dual-gate MOSFET characteristics are more complicated because of the interaction of the two gates. Cutoff voltage, breakdown voltage and gate leakage are the important traits of each gate.

INTEGRATED CIRCUITS

The basics of integrated circuits are covered in earlier chapters of this book. Amateurs seldom have the sophisticated equipment required to test ICs. Even a multi-channel oscilloscope can view only their simplest functions. We must be content to check every other possible cause, and only then assume that the problem lies with an IC. Experienced troubleshooters will tell you that — most of the time anyway — if a defective circuit uses an IC, it is the IC that is bad.

Linear ICs — There are two major classes of ICs: linear and digital. Linear ICs are best replaced with identical units. Original equipment manufacturers are the best source of a replacement; they are the only source with a reason to stockpile obsolete or custom-made items. If substitution of an IC is unavoidable, first try the cross-reference website of NTE mentioned in the sidebar. You can also look in manufacturers' websites and compare pinouts and other specifications.

Digital ICs — It is usually not a good idea to substitute digital devices. While it may be okay to substitute an AB74LS00YZ from one manufacturer with a CD74LS00WX from a different manufacturer, you will usually not be able to replace an LS (low-power Schottky) device with an S (Schottky), C (CMOS), or any of a number of other families. The different families all have different speed, current-consumption, input and output characteristics. You would have to analyze the circuit to determine if you could substitute one type for another.

SEMICONDUCTOR SUBSTITUTION

In all cases try to obtain exact replacement semiconductors. Specifications vary slightly from one manufacturer to the next. Cross-reference equivalents such as NTE (**www.nteinc.com**) are useful but not guaranteed to be an exact replacement. Before using an equivalent, check the specifications against those for the original part. When choosing a replacement, consider:

- Is it a PNP or an NPN?
- What are the operating frequency and input/output capacitance?
- How much power can it dissipate (often less than Vmax × Imax)?
- Will it fit the original socket or pad layout?
- Are there unusual circuit demands (low noise and so on)?
- What is the frequency of operation?

Remember that cross-reference equivalents are not guaranteed to work in every application. In cases where an absolutely exact replacement is required for an obsolete part, Rochester Electronics (**www.rocelec.com**) or 4 Star Electronics (**www.4starelectronics.com**) maintain extensive stocks, although the cost is likely to be rather high.

There may be cases where two dissimilar devices have the same part number, so it pays to compare the listed replacement specifications with the intended use. If the book says to use a diode in place of an RF transistor, it isn't going to work! Derate power specifications, as recommended by the manufacturer, for high-temperature operation.

26.2.10 Tubes

The most common tube failures in amateur service are caused by cathode depletion and gas contamination. Whenever a tube is operated, the coating on the cathode loses some of its ability to produce electrons. It is time to replace the tube when electron production (cathode current, I_c) falls to 50 to 60% of that exhibited by a new tube.

Gas contamination in a tube can often be identified easily because there may be a greenish or whitish-purple glow between the elements during operation. (A faint deep-purple glow is normal in most tubes.) The gas reduces tube resistance and leads to runaway plate current evidenced by a red glow from the anode, interelectrode arcing or a blown power supply fuse. Less common tube failures include an open filament, broken envelope and inter-electrode shorts.

The best test of a tube is to substitute a new one. Another alternative is a tube tester; these are sometimes available at hamfests or through antique radio or audiophile groups. You can also do some limited tests with an ohmmeter. Tube tests should be made out of circuit so circuit resistance does not confuse the results.

Use an ohmmeter to check for an open filament (remove the tube from the circuit first). A broken envelope is visually obvious, although a cracked envelope may appear as a gassy tube. Interelectrode shorts are evident during voltage checks on the operating stage. Any two elements that show the same voltage are probably shorted. (Remember that some inter-electrode shorts, such as the cathode-suppressor grid, are normal.)

Generally, a tube may be replaced with another that has the same type number. Compare the data sheets of similar tubes to assess their compatibility. Consider the base configuration and pinout, inter-electrode capacitances (a small variation is okay except for tubes in oscillator service), dissipated power ratings of the plate and screen grid and current limitations (both peak and average). For example, the 6146A may usually be replaced with a 6146B (heavy duty), but not vice versa.

In some cases, minor type-number differences signify differences in filament voltages, or even base styles, so check all specifications before making a replacement. (Even tubes of the same model number, prefix and suffix vary slightly, in some respects, from one supplier to the next.)

26.3 Getting Started

INSTINCTIVE OR SYSTEMATIC

A systematic approach to troubleshooting uses a defined process to analyze and isolate the problem. An instinctive approach relies on troubleshooting experience to guide you in selecting which circuits to test and which tests to perform.

When instinct is based on experience, searching by instinct may be the fastest procedure. If your instinct is correct, repair time and effort may be reduced substantially. As experience and confidence grow, the merits of the instinctive approach grow with them. However, inexperienced technicians who choose this approach are at the mercy of chance.

A systematic approach is a disciplined procedure that allows us to tackle problems in unfamiliar equipment with a reasonable hope of success. The systematic approach is usually chosen by beginning troubleshooters.

26.3.1 The Systematic Approach

Armed with a collection of test equipment, you might be tempted to immediately dig in and start looking for the problem. While it is sometimes obvious what piece of equipment or subassembly inside equipment is at fault, the many connections that make up nearly all ham stations today make it far more effective to begin troubleshooting by looking at the problem from the system perspective. By *system*, we mean more than one piece of equipment or subassemblies connected together — nothing fancier than that.

Amateur stations are full of systems: a digital mode system is made up of a radio, power supply, and PC. An antenna system consists of the antenna tuner, feed line, and antenna. Inside a radio there is a system made up of the power circuits, receiver, transmitter, control panel, and transmit-receive switching circuits. Connections between parts of the system need not be cables; wireless data links can also be part of a system.

In general, it's best to approach any problem — even the supposedly obvious problems — from the perspective of the system it affects. The first step is to determine the system with the smallest number of components that exhibit the problem. Then you can start looking for the problem in one of those components or the connection between them. We'll start with an example to illustrate the process.

Problem — After a year of trouble-free operation, when you transmit with more than 50 W of output power using PSK31, other stations now report lots of distortion products around your signal on the waterfall display, even though the ALC is not active at all and no software level settings have been changed. This could be RF interference, a problem with the digital interface between the PC and the radio, settings in the PC, settings in the radio, a bad connection...there are lots of possibilities.

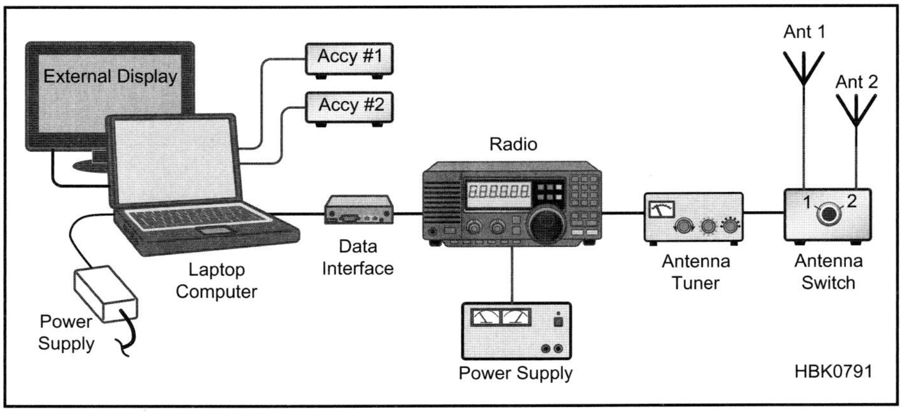

Figure 26.11 — The complete system of radio and power supply, data interface, laptop PC with a couple of accessories (Accy) and power supply, antenna tuner, antenna switch, antenna 1 and 2, and interconnecting cables.

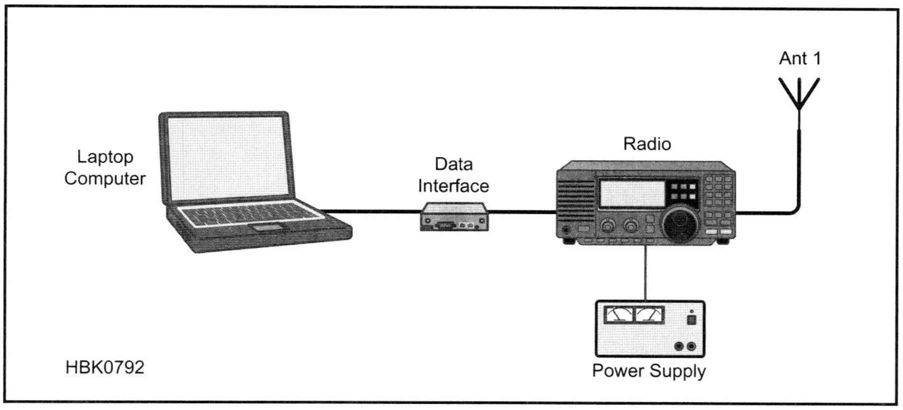

Figure 26.12 — The minimum system of radio and power supply, digital interface, laptop on battery power, antenna 1, and interconnecting cables.

Read the Manual!

Your equipment may be working as designed. Many electronic "problems" are caused by a switch or control in the wrong position, a misconfigured menu item, or a unit that is being asked to do something it was not designed to do. Before you open up your equipment for major surgery, make sure you are using it correctly. Most user's manuals have a procedure for setting up the equipment initially and for performing partial and full resets of microprocessor-controlled gear.

Get a pad of paper and sketch out the system that is exhibiting the problem, such as in **Figure 26.11**. Write down what the symptom(s) is (are) and the conditions under which they occur. Figure 26.11 shows that you have a system that has a lot of parts involved, including the interconnecting cables. Let's simplify that system! For this type of problem, it would be very helpful to have a friend available to listen on the air as you troubleshoot. To find the minimum system, start by removing any accessories that aren't being used from the radio and PC and trying to reproduce the problem, writing down each step and noting whether the problem changed. Not only will you simplify the problem but in the process you will have an excellent opportunity to inspect and check connections and configurations — perhaps even discovering the problem!

Let's say that the distortion persists with the mouse and external display disconnected and the laptop running from its internal battery. Then you were able to remove the antenna tuner and antenna switch, manually attaching just one antenna to the radio at a time, and found that the problem only occurred when using antenna 1. Removing any other part of the system makes it non-functional, so you have found the minimum system with the problem as shown in **Figure 26.12**.

Now it's time to start making small changes in the minimum system to observe the effect, again taking notes as you go. Changing the microphone gain a little bit up and down has hardly any effect. Changing the PC sound card output audio level a little bit also has hardly any effect. Reducing RF output power a little bit has a *big* effect: as power is reduced, so are the distortion products. Below 35 W, the problem was completely gone! This is an important clue, dutifully recorded on the pad of paper. Replacing antenna 1 with a dummy load had a similar effect — the problem disappeared according to your friend, who lives close enough to hear the weak signal leaking out of your greatly reduced antenna system. Putting antenna 1 back on the radio causes the problem to come back above 35 W, too.

The behavior of the problem points strongly to RF being picked up by a cable in your minimum system and interfering with the low-level digital audio. It's time for a close inspection of each remaining cable. Removing the radio's microphone connector shell and checking the wires inside didn't turn up anything — all of the connections were intact and only touching what they were supposed to touch. Looking at the cable between the digital interface and the sound card, though, you see that the shield of the cable at the PC end is badly frayed and barely making contact at all! In a few minutes, you've trimmed and re-soldered the cable, plugging it back into the PC. Testing even at full power output shows no distortion products on either antenna.

You fixed the problem by finding the minimum system and then inspecting one thing at a time until the root cause was found and eliminated.

What if the root cause didn't appear when you checked the cables? Then you would have to continue to test the system at the interfaces where the system components are connected together. You could measure signal levels with a sensitive voltmeter. You could add ferrite beads or cores on cables that might be picking up RF. You could check to see if the radio's power supply output is stable and clean as the transmitter power increases.

Another strategy would be to substitute known good components, one at a time, and see if the problem changes or disappears: you might swap in a different data interface, you could change the cables; or try a different power supply or radio or laptop.

Most systematic testing combines direct testing or inspection and a process of substitution. The goal is to either find the root cause or find the system component in which the root cause is hiding. Once you have identified system component as the culprit, treat the component as a system itself and start the process all over again. Eventually, you'll find the problem.

Here are some guidelines to systematic troubleshooting in the ham shack:

1. Take the time to define and understand the system exhibiting the problem.
2. Remove system components one at a time.
3. Verify that the problem still exists. If not, restore that component.
4. Continue to remove components and test until the minimum system exhibiting the problem has been reached.
5. Inspect what is happening at each interface between system components by testing or substitution. In the example, you inspected the cables connecting the radio and other system components.
6. If a problem is identified, correct it and re-test.
7. Otherwise, continue to test or substitute each system component until the root cause has been identified or isolated to one component.
8. Treat that component as a new system and return to step 1.

Systematically reducing the number of components and testing each interface between them is almost always the fastest way to determine where the problem's root cause really lies. You may be a lucky guesser and sometimes a little plume of smoke is a dead giveaway, but in the long run the system approach will save you time and money.

26.3.2 Assessing the Symptoms

An important part of the troubleshooting process is a careful definition of just what the symptoms of the problem really are. It is important to note exactly how the problem manifests itself and the conditions under which the problem occurs. Avoid vague descriptions such as "broken" and "not working." Train yourself to use precise descriptions such as "relay fails to activate" or "no speaker output."

Ask yourself these questions:

1. What functions of the equipment do not work as they should; what does not work at all?

2. What kind of performance can you realistically expect?

3. Has the trouble occurred in the past? (Keep a record of troubles and maintenance in the owner's manual, shack notebook or log book.)

Write down the answers to the questions. The information will help with your work, and it may help service personnel if their advice or professional service is required.

Question your assumptions and verify what you think you know. Are you *sure* that power supply output is OK under all conditions? Did you actually confirm that there is continuity at every position of the antenna switch? If there is any doubt, make a confirming measurement or inspection. Countless hours have been wasted because of unjustified assumptions!

Intermittent problems are generally harder to track down, so try to note the conditions under which the problem occurs — those are often important clues.

Troubleshooting is a good reason to do regular maintenance. Not only will you have fewer problems, but when something is just a little off or out of place you are much likelier to notice it. Learn to listen to the little voice in your head noting something out of the ordinary. Don't discount the wild cards that occasionally cause problems.

Occasionally step away from the workbench and relax. Take a walk. Your mind will continue to think about the problem, and you may surprise yourself when you sit back down to work!

If you can bounce ideas off a friend, try explaining the problem and letting the friend ask you questions. It's common for someone else with a different perspective to ask questions you haven't thought of.

Newly Constructed Equipment

What if you built a piece of equipment and it doesn't work? In most repair work, the troubleshooter is aided by the knowledge that the circuit once worked so that it is only necessary to find and replace the faulty part(s). This is not the case with newly constructed equipment.

Repair of equipment with no working history is a special, and difficult, case. You may be dealing with a defective component, construction error or even a faulty design. Carefully checking for these defects can save you hours. This is a good reason to test homebrew equipment at each possible step and on a section-by-section basis, if possible. In that way, you'll know more about what does work if the completed project has a problem.

26.3.3 External Inspection

Inspection is the easiest part of troubleshooting to do, and careful, detailed inspection will often find the problem or a clue that leads to it. Make sure you have some paper to keep notes on as you go so when something occurs to you it can be recorded.

Try the easy things first. If you are able to solve the problem by replacing a fuse or reconnecting a loose cable, you might be able to avoid a lot of effort. Many experienced technicians have spent hours troubleshooting a piece of equipment only to learn the hard way that the on/off switch was set to OFF or the squelch control was set too high, or that they were not using the equipment properly.

Next, make sure that equipment is plugged in, that the ac outlet does indeed have power, that the equipment is switched on and that all of the fuses are good. If the equipment uses batteries or an external power supply, make sure they supply the right voltage under load.

Check that all wires, cables and accessories are working and plugged into the right connectors or jacks. In a system of components it is often difficult to be sure which component or subsystem is bad. Your transmitter may not work on SSB because the transmitter is bad, but it could also be a bad microphone.

Connector faults or misconnections are common. Consider them prime suspects in your troubleshooting detective work. Do a thorough inspection of the connections. Is the antenna connected? How about the speaker, fuses and TR switch? Are transistors and ICs firmly seated in their sockets? Are all interconnection cables sound and securely connected? Are any pins bent or is a connector inserted improperly? Many of these problems are obvious to the eye, so look around carefully.

While you are performing your inspection, don't forget to use all of your senses. Do you smell anything burnt or overheated? Is something leaking electrolyte or oil? Perhaps a component or connector looks overheated and discolored. Is mounting hardware secure?

Once you're done with your inspection, retest the equipment to be sure the problem is still there or note if it has changed in some way.

26.4 Inside the Equipment

At this point, you've determined that a specific piece of equipment has a problem. A visual inspection of all the operating controls and connections hasn't turned up anything, but the problem is still there. It is time to really dig in, take it apart, and fix it!

26.4.1 Documentation

In order to test any piece of equipment, you'll probably need at least a user's manual. If at all possible, locate a schematic diagram and service manual. It is possible to troubleshoot without a service manual, but a schematic is almost indispensable.

The original equipment manufacturer is the best source of a manual or schematic. However, many old manufacturers have gone out of business. Several sources of equipment manuals can be located by a web search or from one of the *QST* vendors that sell equipment needs. In addition K4XL's Boat Anchor Manual Archive (**www2.faculty.sbc.edu/kgrimm/boatanchor**) has hundreds of freely downloadable electronic copies of equipment manuals. If there is a user's group or email list associated with your equipment, a request to the group may turn up a manual and maybe even troubleshooting assistance.

If all else fails, you can sometimes reverse engineer a simple circuit by tracing wiring paths and identifying components to draw your own schematic. By downloading datasheets for the active devices used in the circuit, the pin-out diagrams and applications notes will sometimes be enough to help you understand and troubleshoot the circuit.

THE BLOCK DIAGRAM

An important part of the documentation, the block diagram is a road map. It shows the signal paths for each circuit function. These paths may run together and cross occasionally or not at all. Those blocks that are not in the paths of faulty functions can be eliminated as suspects. Sometimes the symptoms point to a single block and no further search is necessary. In cases where more than one block is suspect, several approaches may be used. Each requires testing a block or stage.

26.4.2 Disassembly

This seemingly simple step can trap the unwary technician. Most experienced service technicians can tell you the tale of the equipment they took apart and were unable to easily put back together. Don't let it happen to you.

Take photos and lots of notes about the way you take it apart. Take notes about each component you remove. Take a photo or make a sketch of complicated mechanical assemblies before disassembly and then record how you disassembled them. It is particularly important to record the position of shields and ground straps.

Write down the order in which you do things, color codes, part placements, cable routings, hardware notes, and anything else you think you might need to reassemble the equipment weeks from now when the back-ordered part comes in.

Put all of the screws and mounting hardware in one place. A plastic jar with a lid works well; if you drop it the plastic is not apt to break, and the lid will keep all the parts from flying around the work area (you will never find them all). It may pay to have a separate labeled container for each subsystem. Paper envelopes and muffin pans also work well.

26.4.3 Internal Inspection

Many service problems are visible if you look for them carefully. Many a technician has spent hours tracking down a failure, only to find a bad solder joint or burned component that would have been spotted in careful inspection of the printed circuit board. Internal inspections are just as important as external inspections.

It is time consuming, but you really need to look at every connector, every wire, every solder joint and every component. A low power magnifying glass or head-mounted magnifier enables you to quickly scan the equipment to look for problems. A connector may have loosened, resulting in an open circuit. You may spot broken wires or see a bad solder joint. Flexing the printed circuit board or tugging on components a bit while looking at their solder joints will often locate a defective solder job. Look for scorched components.

Make sure all of the screws securing the printed circuit board are tight and making good electrical contact. Check for loose screws on chassis-mounted connectors. (Do not tighten any electrical or mechanical adjusting screws or tuning controls, however!) See if you can find evidence of previous repair jobs; these may not have been done properly. Make sure that ICs are firmly seated in sockets if they are used. Look for pins folded underneath the IC rather than inserted into the socket. If you are troubleshooting a newly constructed circuit, make sure each part is of the correct value or type number and is installed correctly.

POWER SUPPLIES

If your careful inspection doesn't reveal anything, it is time to apply power to the unit under test and continue the process. Observe all safety precautions while troubleshooting equipment. There are voltages inside some equipment that can kill you. If you are not qualified to work safely with the voltages and conditions inside of the equipment, do not proceed. See Table 26.1 and the **Safety** chapter.

You may be able to save quite a bit of time if you test the power supply right away. If the power supply is not working at all or not working properly, no other circuit in the equipment can be expected to work properly either. Once the power supply has either been determined to be OK or repaired, you can proceed to other parts of the equipment. Power supply diagnosis is discussed in detail later in this chapter.

With power applied to the equipment, listen for arcs and look and smell for smoke or overheated components. If no problems are apparent, you can move on to testing the various parts of the circuit. For tube equipment, you may want to begin with ac power applied at a reduced voltage as described in the sections on repair of vintage equipment.

26.4.4 Signal Tracing and Signal Injection

There are two common systematic approaches to troubleshooting radio equipment at the block level. The first is signal tracing; the second is signal injection. The two techniques are very similar. Differences in test equipment and the circuit under test determine which method is best in a given situation. They can often be combined.

Both of these approaches are used on equipment that is designed to operate on a signal (RF, audio or data) in a sequence of steps. A

Knowing When to Quit

It is common for experienced repair techs to be given "basket cases" — equipment that the original troubleshooter disassembled but then couldn't reassemble for whatever reason or couldn't find the problem. One of the most important decisions in the troubleshooting process is knowing when to quit. That is, realizing that you are about to go beyond your skills or understanding of the equipment.

If you proceed past this point, the chances of a successful outcome go down pretty quickly. It's far better to ask for help, work with a more knowledgeable friend, or carefully re-assemble the equipment and take it to a repair tech. Don't let your equipment wind up as a box full of partially connected pieces and mounting hardware under the table at the hamfest with a sign that says, "Couldn't fix — make offer!"

Block-Level Testing for DSP and SDR

More and more functions of today's radio equipment are implemented by microprocessor-based digital signal processing, up to and including full SDR with direct digitization of RF signals very close to the antenna input. A look at the PC boards of such a radio show a few large, many-leaded ICs surrounded by control and interface components, power supply circuits, filtering, and transmitter power amplifiers. The individual stages of the classic superheterodyne architecture are nowhere to be seen. How do you troubleshoot such a radio?

Start with the same basic approach as for an older radio — begin at the input or output and work your way towards the "other end" of the radio. In a DSP-based radio, however, you'll rapidly encounter the point at which the signal "goes digital" and disappears into the microprocessor or an analog/digital converter. Jump to the point where the signal returns to analog form on the "other side" of the microprocessor (or PC in the case of most SDR equipment) and resume testing. In this way you can simply treat the microprocessor as one very large stage in the radio. If the problem turns out to be in the surrounding circuitry or in the interface between a PC and the RF circuits, you can troubleshoot it as you would any other piece of equipment.

If the problem turns out to be in the microprocessor or analog/digital converter, in all probability you will have to get a replacement board from the manufacturer. It is possible to replace the converter or processor (if you can obtain the pre-programmed part), but most manufacturers treat the entire circuit board as the lowest-level replaceable component. This makes repair more expensive (or impossible if no replacements are available), but the positive tradeoff is that the microprocessors rarely fail by themselves, resulting in fewer repairs being required in the first place.

transceiver based on the superheterodyne architecture is probably the best example of this type of equipment in the ham shack. Audio equipment is also built this way. More information about the signal injector and signal sources appears in "Some Basics of Equipment Servicing," from February 1982 *QST* (Feedback, May 1982).

Newer equipment that incorporates DSP and specialized or proprietary ICs is much less amenable to stage-by-stage testing techniques. (See the sidebar "Block-Level Testing for DSP and SDR) Nevertheless, the techniques of signal tracing and signal injection are still useful where the signal path is accessible to the troubleshooter.

SIGNAL TRACING

In signal tracing, start at the beginning of a circuit or system and follow the signal through to the end. When you find the signal at the input to a specific stage, but not at the output, you have located the defective stage. You can then measure voltages and perform other tests on that stage to locate the specific failure. This is much faster than testing every component in the unit to determine which is bad.

It is sometimes possible to use over-the-air signals in signal tracing, in a receiver for example. However, if a good signal generator is available, it is best to use it as the signal source. A modulated signal source is best.

Signal tracing is suitable for most types of troubleshooting of receivers and analog amplifiers. Signal tracing is the best way to check transmitters because all of the necessary signals are present in the transmitter by design. Most signal generators cannot supply the wide range of signal levels required to test a transmitter.

Equipment

A voltmeter with an RF probe is the most common instrument used for signal tracing. Low-level signals cannot be measured accurately with this instrument. Signals that do not exceed the junction drop of the diode in the probe will not register at all, but the presence, or absence, of larger signals can be observed.

A dedicated signal tracer can also be used. It is essentially an audio amplifier. (See the downloadable supplemental content for a project to build your own signal tracer.) An experienced technician can usually judge the level and distortion of the signal by ear. You cannot use a dedicated signal tracer to follow a signal that is not amplitude modulated (single sideband is a form of AM). A signal tracer is not suitable for tracing CW signals, FM signals or oscillators. To trace these, you will have to use a voltmeter and RF probe or an oscilloscope.

An oscilloscope is the most versatile signal tracer. It offers high input impedance, variable sensitivity, and a constant display of the traced waveform. If the oscilloscope has sufficient bandwidth, RF signals can be observed directly. Alternatively, a demodulator probe can be used to show demodulated RF signals on a low-bandwidth oscilloscope. Dual-trace scopes can simultaneously display the waveforms, including their phase relationship, present at the input and output of a circuit.

Procedure

First, make sure that the circuit under test and test instruments are isolated from the ac line by internal transformers, an isolation transformer, or operate from battery power. Set the signal source to an appropriate level and frequency for the unit you are testing. For a receiver, a signal of about 100 μV should be plenty. For other circuits, use the schematic, an analysis of circuit function and your own good judgment to set the signal level.

In signal tracing, start at the beginning and work toward the end of the signal path. Switch on power to the test circuit and connect the signal-source output to the test-circuit input. Place the tracer instrument at the circuit input and ensure that the test signal is present. Observe the characteristics of the signal if you are using a scope (see **Figure 26.13**). Compare the detected signal to the source signal during tracing.

Move the test instrument to the output of the next stage and observe the signal. The signal level should increase in amplifier stages and may decrease slightly in other stages. The signal will not be present at the output of a dead stage.

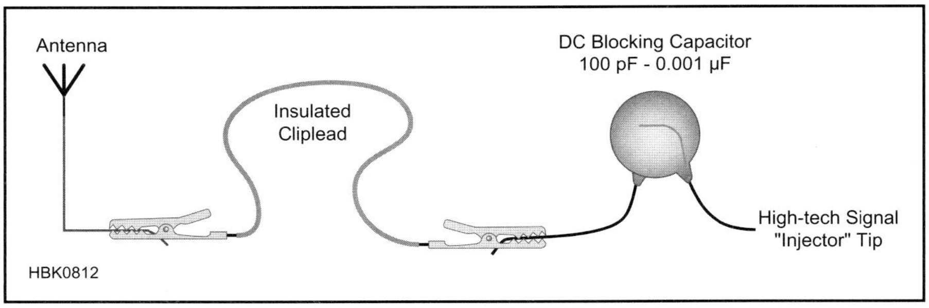

Figure 26.14 — A simple signal injector that uses an antenna to pick up signals to be applied to the circuit under test. [Courtesy Elecraft, www.elecraft.com]

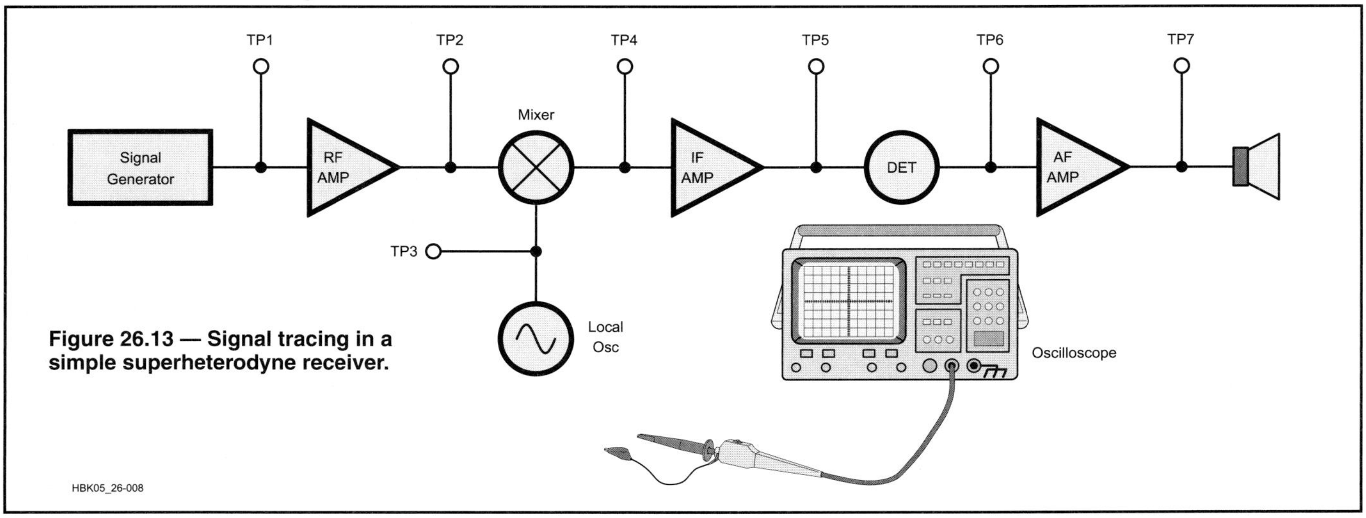

Figure 26.13 — Signal tracing in a simple superheterodyne receiver.

Low-impedance test points may not provide sufficient signal to drive a high-impedance signal tracer, so tracer sensitivity is important. Also, in some circuits the output level appears low where there is an impedance change from input to output of a stage. For example, in a properly-working common-collector (emitter follower) circuit, the input (high-impedance) and output (low-impedance) signals are in phase and have roughly equal voltages. The voltages at TP1 and TP2 are approximately equal and in phase.

There are two signals — the test signal and the local oscillator signal — present in a mixer stage. Loss of either one will result in no output from the mixer stage. Switch the signal source on and off repeatedly to make sure that the test instrument reading varies (it need not disappear) with source switching.

SIGNAL INJECTION

Signal injection is a good choice for receiver troubleshooting because the receiver already has a detector as part of the design. If the detector is working or a suitable detector is devised (see the signal tracing section), a test signal can be injected at different points in the equipment until the faulty stage is discovered.

Equipment

Most of the time, your signal injector will be a signal generator. There are other injectors available, some of which are square-wave audio oscillators rich in RF harmonics (see Figure 26.5). These simple injectors do have their limits because you can't vary their output level or determine their frequency. They are still useful, though, because most circuit failures are caused by a stage that is completely dead.

Consider the signal level at the test point when choosing an instrument. The signal source used for injection must be able to supply appropriate frequencies and levels for each stage to be tested. For example, a typical superheterodyne receiver requires AF, IF and RF signals that vary from 6 V at AF to 200 µV at RF. Each conversion stage used in a receiver requires a different IF from the signal source. When testing the signal path of an AM radio, such as a broadcast receiver, you'll need a modulated RF signal before the detector stage. Use an unmodulated RF signal to simulate the local oscillator.

The simple test circuit of **Figure 26.14** can be used as a quick-and-dirty signal injector for RF stages in a receiver. The antenna can be anything that will receive a signal at the stage's frequency of operation.

Procedure

If an external detector is required, set it to the proper level and connect it to the test circuit. Set the signal source for AF and inject a signal directly into the signal detector to test operation of the injector and detector. Move the signal source to the input of the preceding stage and observe the signal. Continue moving the signal source to the inputs of successive stages.

When you inject the signal source to the input of the defective stage, there will be no output. Prevent stage overload by reducing the level of the injected signal as testing progresses through the circuit. Use suitable frequencies for each tested stage.

Make a rough check of stage gain by injecting a signal at the input and output of an amplifier stage. You can then compare how much louder the signal is when injected at the input. This test may mislead you if there is a radical difference in impedance from stage input to output. Understand the circuit operation before testing.

Mixer stages present a special problem because they have two inputs rather than one. A lack of output signal from a mixer can be caused by either a faulty mixer or a faulty local oscillator (LO). Check oscillator operation with an oscilloscope or by listening on another receiver. If none of these instruments is available, inject the frequency of the LO at the LO output. If a dead oscillator is the only problem, this should restore operation.

If the oscillator is operating but off frequency, a multitude of spurious responses will appear. A simple signal injector that produces many frequencies simultaneously is not suitable for this test. Use a well-shielded signal generator set to an appropriate level at the LO frequency.

> **Divide and Conquer**
>
> If the equipment has a single primary signal path, the block-by-block search may be sped up considerably by testing between successively smaller groups of circuit blocks. Each test thus exercises some fraction of the remaining circuit.
>
> This "divide and conquer" tactic cannot be used in equipment that splits the signal path between the input and the output. Test readings taken inside feedback loops are misleading unless you understand the circuit and the waveform to be expected at each point in the test circuit. It is best to consider all stages within a feedback loop as a single block during the block search.
>
> Divide-and-conquer is a good tactic for those inclined to take the instinctive approach to troubleshooting. As you gain more experience, you'll find yourself able to quickly isolate problems this way. You can then test each block in more detail.

26.4.5 Microprocessor-Controlled Equipment

The majority of today's amateur equipment and accessories have at least one microprocessor and sometimes several. While reliability of this equipment is greatly improved over the older analog designs, troubleshooting microprocessor-based circuitry takes a different approach. While a tutorial on microprocessor troubleshooting is well outside the scope of this *Handbook*, the following basic guidelines will help determine whether the problem is inside the microprocessor (or its firmware) or in the supporting circuitry. In addition, the downloadable **Digital Basics** chapter contains lots of information about the operation of digital circuits.

1) Start by obtaining the microprocessor datasheet and identifying all of the power and control pins if they are not identified on the equipment schematic. Determine which state the control pins should be in for the device to run.

2) Test all power and control pins for the proper state (voltage). Verify that the microprocessor clock signal is active. If not, determine the reason and repair before proceeding.

3) If there is an address and data bus for external memory and input-output (I/O) devices (less common in newer equipment), use a logic probe or scope to verify that they are all active (changing state). If not, there is a program or logic fault.

4) Determine which pins are digital inputs or outputs of the microprocessor and verify that a valid digital logic level exists at the pin. If not, check the external circuit to which the pin is attached.

5) Determine which pins are analog inputs or outputs of the microprocessor. Analyze the external circuit to determine what constitutes a proper voltage into or out of the processor. If the voltage is not valid, check the external circuit to which the pin is attached. If an external voltage reference is used, verify that it is working.

6) External circuits can often be checked by disconnecting them from the microprocessor and either driving them with a temporary voltage source or measuring the signal they are attempting to send to the microprocessor.

7) If all control and power signals and the external circuitry checks out OK, it is likely that a microprocessor or firmware fault has occurred.

If you determine that the microprocessor is faulty, you will have to contact the manufacturer in most cases, since firmware is most often contained within the processor, which must be programmed before it is installed. Older equipment in which the program and data memory are external to the microprocessor can be very difficult to repair due to obsolescence of the parts themselves and the requirement to program EPROM or PROM devices.

26.5 Testing at the Circuit Level

Once you have followed all of the troubleshooting procedures and have isolated your problem to a single defective stage or circuit, a few simple measurements and tests will usually pinpoint one or more specific components that need adjustment or replacement.

First, check the parts in the circuit against the schematic diagram to be sure that they are reasonably close to the design values, especially in a newly built circuit. Even in a commercial piece of equipment, someone may have incorrectly changed them during attempted repairs. A wrong-value part is quite likely in new construction, such as a home-brew or kit project.

26.5.1 Voltage Levels

Check the circuit voltages. If the voltage levels are printed on the schematic, this is easy. If not, analyze the circuit and make some calculations to see what the circuit voltages should be. Remember, however, that the printed or calculated voltages are nominal; measured voltages may vary from the calculations.

When making measurements, remember the following points:

- Make measurements at device leads, not at circuit-board traces or socket lugs.
- Use small test probes to prevent accidental shorts.
- Never connect or disconnect power to solid-state circuits with the switch on.
- Remember that voltmeters, particularly older analog meters, may load down a high-impedance circuit and change the voltage, as will ×1 and low-impedance scope probes.

Voltages may give you a clue to what is wrong with the circuit. If not, check the active device. If you can check the active device in the circuit, do so. If not, remove it and test it, or substitute a known good device. After connections, most circuit failures are caused directly or indirectly by a bad active device. The experienced troubleshooter usually tests or substitutes these first. Analyze the other components and determine the best way to test each as described earlier.

There are two voltage levels in most analog circuits (V+ and ground, for example). Most component failures (opens and shorts) will shift dc voltages near one of these levels. Typical failures that show up as incorrect dc voltages include open coupling transformers; shorted capacitors; open, shorted or overheated resistors; and open or shorted semiconductors.

Digital logic circuits require that signals be within specific voltage ranges to be treated as a valid logic-low or logic-high value.

> **Controlling Key Clicks**
>
> Key clicks are a special type of amplitude distortion caused by fast rising and falling edges of the output waveform or from abrupt disturbances in an otherwise smooth waveform. See the **Transmitting** chapter for more information about controlling key clicks. Most modern radios offer user configuration options to adjust keying rise and fall times. Older radios may require modification of a timing circuit that controls rise and fall time. Be a good neighbor to other operators and eliminate key clicks on your signal!

26.5.2 Noise

A slight hiss is normal in all electronic circuits. This noise is produced whenever current flows through a conductor that is warmer than absolute zero. Noise is compounded and amplified by succeeding stages. Repair is necessary only when noise threatens to obscure normally clear signals.

Semiconductors can produce hiss in two ways. The first is normal — an even white noise that is much quieter than the desired signal. Faulty devices frequently produce excessive noise. The noise from a faulty device is often erratic, with pops and crashes that are sometimes louder than the desired signal. In an analog circuit, the end result of noise is usually sound. In a control or digital circuit, noise causes erratic operation: unexpected switching and so on.

Noise problems usually increase with temperature, so localized heat may help you find the source. Noise from any component may be sensitive to mechanical vibration. Tapping various components with an insulated screwdriver may quickly isolate a bad part. Noise can also be traced with an oscilloscope or signal tracer.

Nearly any component or connection can be a source of noise. Defective components are the most common cause of crackling noises. Defective connections are a common cause of loud, popping noises.

Check connections at cables, sockets and switches. Look for dirty variable capacitor wipers and potentiometers. An arcing mica trimmer capacitor can create static crashes in received or transmitted audio. Test it by installing a series 0.01 µF capacitor. If the noise disappears, replace the trimmer.

Potentiometers are particularly prone to noise problems when used in dc circuits. Clean them with a small amount of spray contact cleaner and rotate the shaft several times.

Rotary switches may be tested by jumpering the contacts with a clip lead. Loose contacts may sometimes be repaired, either by cleaning, carefully rebending the switch contacts or gluing loose switch parts to the switch deck. Operate variable components through their range while observing the noise level at the circuit output.

26.5.3 Oscillations

Oscillations occur whenever there is sufficient positive feedback in a circuit that has gain. (This can even include digital devices.) Oscillation may occur at any frequency from a low-frequency "putt-putt" (often called *motorboating*) well up into the RF region.

Unwanted oscillations are usually the result of changes in the active device (increased junction or interelectrode capacitance), failure of an oscillation suppressing component (open decoupling or bypass capacitors or neutralizing components) or new feedback paths (improper lead dress or dirt on the chassis or components). It can also be caused by improper design, especially in home-brew circuits. A shift in bias or drive levels may aggravate oscillation problems.

Oscillations that occur in audio stages do not change as the radio is tuned because the operating frequency — and therefore the component impedances — do not change. RF and IF oscillations, however, usually vary in amplitude as operating frequency is changed.

Oscillation stops when the positive feedback is removed. Locating and replacing a defective (or missing) bypass capacitor may effect an improvement. The defective oscillating stage can be found most reliably with an oscilloscope.

26.5.4 Amplitude Distortion

Amplitude distortion is the product of nonlinear operation. The resultant waveform contains not only the input signal but new signals at other frequencies as well. All of the frequencies combine to produce the distorted waveform. Distortion in a transmitter gives rise to splatter, harmonics and interference.

Figure 26.15 shows some typical cases of distortion. Clipping (also called flat-topping) is the consequence of excessive drive, a change in bias, or insufficient supply voltage to the circuit. The corners on the waveform show that harmonics are present. (A square wave contains the fundamental and all odd harmonics.) If this were a transmitter circuit, these odd harmonics would be heard well away from the operating frequency, possibly outside of amateur bands.

Harmonic distortion produces radiation at frequencies far removed from the fundamental; it is a major cause of electromagnetic interference (EMI). Harmonics are generated in nearly every amplifier. When they occur in a transmitter, they are usually caused by insufficient transmitter filtering (either by design,

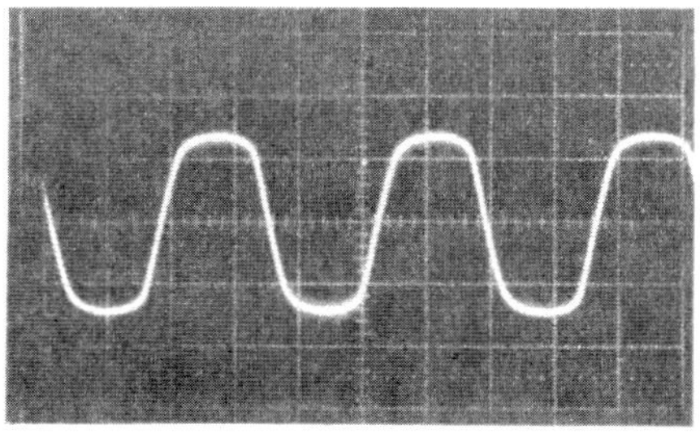
(A)

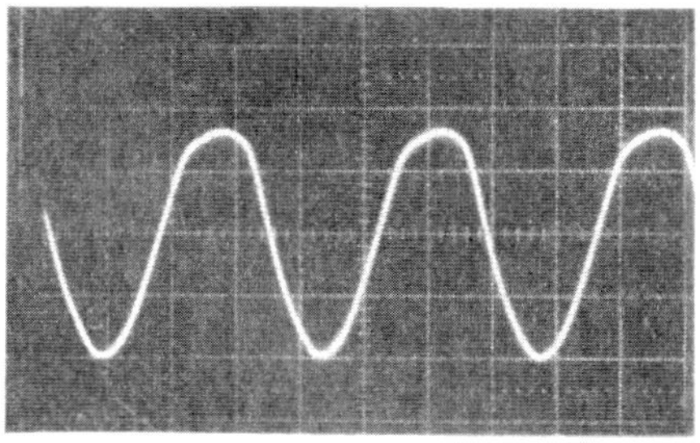
(B)

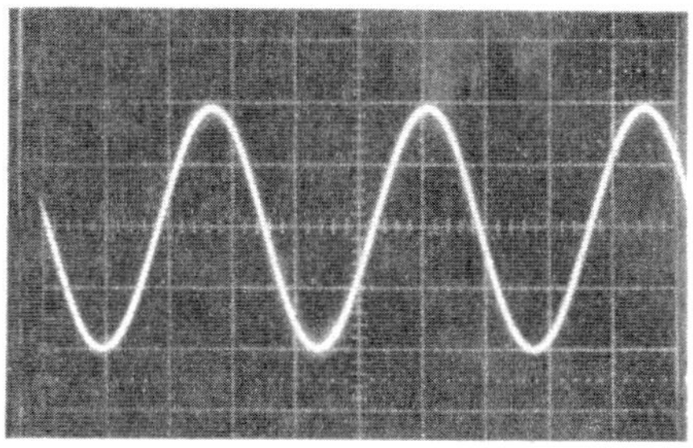
(C)

Figure 26.15 — Examples of distorted waveforms. The result of clipping is shown in A. Nonlinear amplification is shown in B. A pure sine wave is shown in C for comparison.

or because of filter component failure).

Anything that changes the proper bias of an amplifier can cause distortion. This includes failures in the bias components, leaky transistors or vacuum tubes with interelectrode shorts. In a receiver, these conditions may mimic AGC trouble. Improper bias of an analog circuit often results from a resistor that changed value or a leaky or shorted capacitor. RF feedback can also produce distortion by disturbing bias levels. Distortion is also caused by circuit imbalance in Class AB or B amplifiers.

Oscillations in an IF amplifier may produce distortion. They cause constant, full AGC action, or generate spurious signals that mix with the desired signal. IF oscillations are usually evident on the S meter, which will show a strong signal even with the antenna disconnected.

26.5.5 Frequency Response

Every circuit, even a broadband circuit, has a desired frequency response. Audio amplifiers used in amateur SSB circuits, for example, typically are designed for signals between 300 and 3000 Hz, more or less. A tuned IF amplifier may have a bandwidth of 50 to 100 kHz around the stage's center frequency. Any change in the circuit's frequency response can alter its effect on the signals on which it operates.

Frequency response changes are almost always a consequence of a capacitor or inductor changing value. The easiest way to check frequency response is to either inject a signal into the circuit and measure the output at several frequencies or use a spectrum analyzer or a signal generator's sweep function. In LC networks, it is relatively simple to lift one lead of the component and use a component checker to determine the value.

26.5.6 Distortion Measurement

A distortion meter is used to measure distortion of AF signals. A spectrum analyzer is the best piece of test gear to measure distortion of RF signals. If a distortion meter is not available, an estimation of AF distortion can sometimes be made with a function generator and an oscilloscope. Inject a square wave signal into the circuit with a fundamental frequency roughly in the middle of the expected frequency response.

Compare the input square wave to the output signal with an oscilloscope. **Figure 26.16** shows several effects on the square wave that are related to frequency response. These provide clues to what components or devices may be causing the problem. Severe distortion indicates some other problem besides frequency response changes.

26.5.7 Alignment

Alignment — the tuning or calibration of frequency sensitive circuits — is rarely the cause of an electronics problem with receiving or transmitting equipment, particularly modern equipment. Alignment does not shift suddenly and should be a last resort for treating sensitivity or frequency response problems. Do not attempt to adjust alignment without the proper equipment and alignment procedures. The process often requires steps to be performed exactly and in a specific order. Equipment misaligned in this way usually must be professionally repaired as a consequence.

26.5.8 Contamination

Contamination is another common service problem. Soda or coffee spilled into a piece

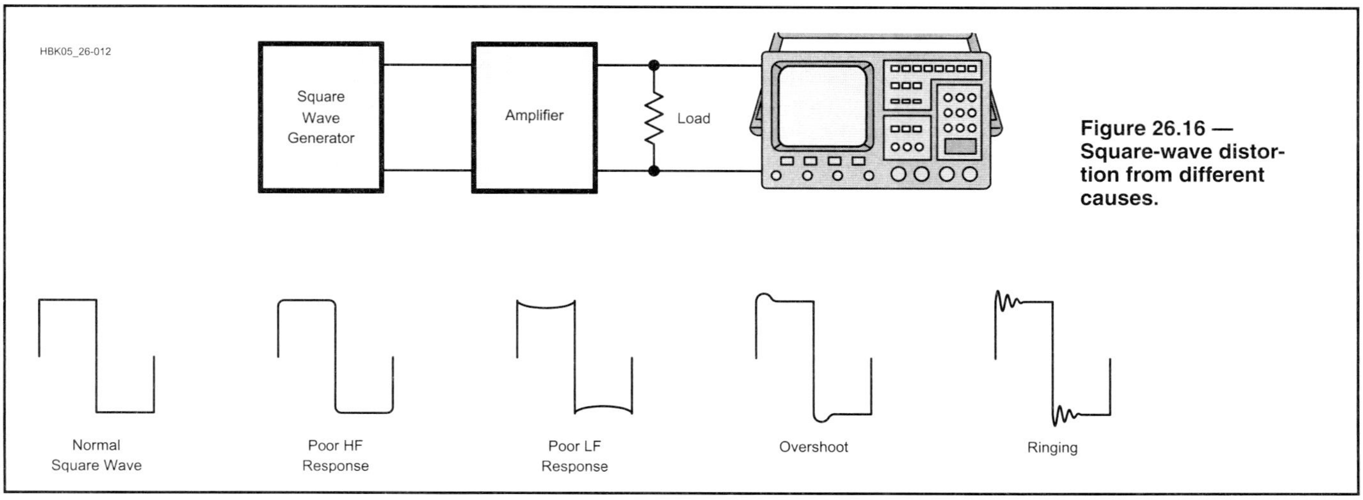

Figure 26.16 — Square-wave distortion from different causes.

of electronics is an extreme example (but one that does actually happen).

Conductive contaminants range from water to metal filings. Most can be removed by a thorough cleaning. Any of the residue-free cleaners can be used, but remember that the cleaner may also be conductive. Do not apply power to the circuit until the area is completely dry.

Keep cleaners away from variable-capacitor plates, transformers, and parts that may be harmed by the chemical. The most common conductive contaminant is solder, either from a printed circuit board solder bridge or a loose piece of solder deciding to surface at the most inconvenient time.

High-voltage circuits attract significant amounts of dust, described in the section on high-power amplifier maintenance later in this chapter. Cooling fans and ventilation holes also allow dust to accumulate, creating conductive paths between components. If not removed, the conductive paths gradually become lower in resistance until they begin to affect equipment performance. Vacuuming or blowing out the equipment is usually sufficient to clear away dust, although a carbon track may require a more thorough cleaning.

26.5.9 Solder Bridges

In a typical PC-board solder bridge, the solder that is used to solder one component has formed a short circuit to another PC-board trace or component. Unfortunately, they are common in both new construction and repair work. Look carefully for them after you have completed any soldering, especially on a PC-board. It is even possible that a solder bridge may exist in equipment you have owned for a long time, unnoticed until it suddenly became a short circuit.

Related items are loose solder blobs, loose hardware or small pieces of component leads that can show up in the most awkward and troublesome places.

26.5.10 Arcing

Arcing is a serious sign of trouble. It may also be a real fire hazard. Arc sites are usually easy to find because an arc that generates visible light or noticeable sound also pits and discolors conductors.

Arcing is caused by component failure, dampness, dirt or lead dress. If the dampness is temporary, dry the area thoroughly and resume operation. Dirt may be cleaned from the chassis with a residue-free cleaner. Arrange leads so high-voltage conductors are isolated. Keep them away from sharp corners and screw points.

Arcing occurs in capacitors when the working voltage is exceeded. Air-dielectric variable capacitors can sustain occasional arcs without damage, but arcing indicates operation beyond circuit limits. Antenna tuners working beyond their ability may suffer from arcing. A failure or high SWR in an antenna circuit may also cause transmitter arcing. Prolonged, repeated, or high-power arcing can cause pits or deposits on capacitor plates that reduce capacitor voltage rating and lead to more arcing.

26.5.11 Digital Circuitry

Although every aspect of digital circuit operation may be resolved to a simple 1 or 0, or tristate (open circuit), the symptoms of failure are far more complicated. The most common problems are false triggers or counts, digital inputs that do not respond to valid logic signals, and digital outputs stuck at ground or the supply voltage.

In most working digital circuitry the signals are constantly changing between low and high states, often at RF rates. Low-frequency or dc meters should not be used to check digital signal lines. A logic probe is often helpful in determining signal state and whether it is active or not. An oscilloscope or logic analyzer is usually needed to troubleshoot digital circuitry beyond simple go/no-go testing.

If you want to use an oscilloscope to give an accurate representation of digital signals, the scope bandwidth must be at least several times the highest clock frequency in the circuit in order to reproduce the fast rise and fall times of digital signals. Lower-bandwidth scopes can be useful in determining whether a signal is present or active, but they often miss short glitch signals (very fast transients) that are often associated with digital circuit malfunction.

LOGIC LEVELS

Begin by checking for the correct voltages at the pins of each IC. The correct logic voltages are specified in the device's datasheet, which will also identify the power pins (V_{cc} and ground). The voltages on the other pins should be a logic high, a logic low, or tristate (more on this later).

Most digital circuit failures are caused by a failed logic IC. IC failures are almost always catastrophic. It is unlikely that an AND gate will suddenly start functioning like an OR gate. It is more likely that the gate will have a signal at its input, and no signal at the output. In a failed device, the output pin will have a steady voltage. In some cases, the voltage is steady because one of the input signals is missing. Look carefully at what is going into a digital IC to determine what should be coming out. Manufacturers' datasheets describe the proper functioning of most digital devices.

TRISTATE DEVICES

Many digital devices are designed with a third logic state, commonly called tristate. In this state, the output of the device acts as an open circuit so that several device outputs can be connected to a common bus. The outputs that are active at any given time are selected by software or hardware control signals. A computer's data and address busses are good examples of this. If any device output connected to the bus fails by becoming locked or stuck in a 0 or 1 logic state, the entire bus becomes nonfunctional. Tristate devices can also be locked on by a failure of the signal that controls the tristate output status.

SIMPLE GATE TESTS

Most discrete logic ICs (collections of individual gates or other logic functions) are easily tested by in-circuit inspection of the input and output signals. The device's truth table or other behavior description specifies what the proper input and output signals should be. Testing of more complicated ICs requires the use of a logic analyzer, multi-trace scope or a dedicated IC tester. If a simple logic IC is found to be questionable, it is usually easiest to simply substitute a new device for it.

CLOCK SIGNALS

In clocked circuits, check to see if the clock signal is active. If the signal is found at the clock chip, trace it to each of the other ICs to be sure that the clock system is intact. Clock frequencies are rarely wrong but clock signals derived from a master clock can be missing or erratic if the circuitry that creates them is defective.

RF INTERFERENCE

If digital circuitry interferes with other nearby equipment, it may be radiating spurious signals. These signals can interfere with your amateur radio operation or other services. Computer networking and microprocessor-controlled consumer equipment generates a significant amount of noise due to RF being radiated from cables and unshielded equipment.

Digital circuitry can also be subject to interference from strong RF fields. Erratic operation or a complete lock-up is often the result. Begin by removing the suspect equipment from RF fields. If the symptoms stop when there is no RF energy present, apply common-mode chokes as described in the **RF Interference** chapter..

The ARRL RFI Book has a chapter on computer and digital interference to and from digital devices and circuits. The subject is also covered in the **RF Interference** chapter of this book.

26.5.12 Replacing Parts

If you have located a defective component within a stage, you need to replace it. When replacing socket mounted components, be sure to align the replacement part correctly. Make sure that the pins of the device are properly inserted into the socket. See the **Construction Techniques** chapter for guidance on working with SMT components.

Some special tools can make it easier to remove soldered parts. A chisel-shaped soldering tip helps pry leads from printed-circuit boards or terminals. A desoldering iron or bulb forms a suction to remove excess solder, making it easier to remove the component. Spring-loaded desoldering pumps are more convenient than bulbs. Desoldering wick draws solder away from a joint when pressed against the joint with a hot soldering iron.

Removing soldered ICs is a lot simpler if you simply clip its leads next to the IC body using fine-point wire cutters, although it does destroy the IC. Then melt the solder and lift out the pin with tweezers or the wire cutter. Use the desoldering pumps or wick to remove the solder. This works well for both through-hole and SMT components.

26.6 After the Repairs

Once you have completed your troubleshooting and repairs, it is time to put the equipment back together. Take a little extra time to make sure you have done everything correctly.

26.6.1 All Units

Give the entire unit a complete visual inspection. Look for any loose ends left over from your troubleshooting procedures — you may have left a few components temporarily soldered in place, forgotten to reattach a wire or cable, or overlooked some other repair error. Look for cold solder joints and signs of damage incurred during the repair. Double-check the position, leads and polarity of components that were removed or replaced.

Make sure that all ICs and connectors are properly oriented and inserted in their sockets. Test fuse continuity with an ohmmeter and verify that the current rating matches the circuit specification.

Look at the position of all of the wires and components. Make sure that wires and cables will be clear of hot components, screw points and other sharp edges. Make certain that the wires and components will not be in the way and pinched or crimped when covers are installed and the unit is put back together.

Separate the leads that carry dc, RF, input and output as much as possible. Plug-in circuit boards should be firmly seated with screws tightened and lock washers installed if so specified. Shields and ground straps should be installed just as they were on the original.

26.6.2 Transmitter Checkout

Since the signal produced by an HF transmitter can be heard the world over, a thorough check is necessary after any service has been performed. Do not exceed the transmitter duty cycle while testing. Limit transmissions to 10 to 20 seconds unless otherwise specified by the owner's manual.

1. Set all controls as specified in the operation manual, or at midscale.
2. Connect a dummy load and a power meter to the transmitter output.
3. Set the drive or carrier control for low output.
4. Switch the power on.
5. Transmit and quickly set the final-amplifier bias to specifications if necessary
6. For vacuum tube final amplifiers, slowly tune the output network through resonance. The current dip should be smooth and repeatable. It should occur simultaneously with the maximum power output. Any sudden jumps or wiggles of the current meter indicate that the amplifier is unstable. Adjust the neutralization circuit (according to the manufacturer's instructions) if one is present, or check for oscillation. An amplifier usually requires neutralization whenever active devices, components or lead dress (that affect the output/input capacitance) are changed.
7. Check to see that the output power is consistent with the amplifier class used in the PA (efficiency should be about 25% for Class A, 50 to 60% for Class AB or B, and 70 to 75% for Class C). Problems are indicated by the efficiency being significantly low.
8. Repeat steps 4 through 6 for each band of operation from lowest to highest frequency.
9. Check the carrier balance (in SSB transmitters only) and adjust for minimum power output with maximum RF drive and no microphone gain.
10. Adjust the VOX controls.
11. Measure the passband and distortion levels if equipment (wideband scope or spectrum analyzer) is available.

26.6.3 Other Repaired Circuits

After the preliminary checks, set the circuit controls per the manufacturer's specifications (or to midrange if specifications are not available) and switch the power on. Watch and smell for smoke, and listen for odd sounds such as arcing or hum. Operate the circuit for a few minutes, consistent with allowable duty cycle. Verify that all operating controls function properly.

Check for intermittent connections by subjecting the circuit to heat, cold and slight flexure. Also, tap or jiggle the chassis lightly with an alignment tool or other insulator.

If the equipment is meant for mobile or portable service, operate it through an appropriate temperature range. Many mobile radios do not work on cold mornings, or on hot afternoons, because a temperature-dependent intermittent was not found during repairs.

26.6.4 Close It Up

After you are convinced that you have repaired the circuit properly, put it all back together. If you followed the advice in this book, you have all the screws and assorted doodads in a secure container. Look at the notes you took while taking it apart; put it back together in the reverse order. Don't forget to reconnect all internal connections, such as ac power, speaker or antenna leads.

Once the case is closed, and all appears well, don't neglect the final, important step — make sure it still works. Many an experienced technician has forgotten this important step, only to discover that some minor error, such as a forgotten antenna cable, has left the equipment nonfunctional.

26.7 Professional Repairs

Repairs that deal with very complex and temperamental circuits, or that require sophisticated test equipment, should be passed on to a professional. Factory authorized service personnel have a lot of experience. What seems like a servicing nightmare to you is old hat to them. There is no one better qualified to service your equipment than factory personnel.

If the manufacturer is no longer in business, check with your local dealer or look through amateur radio magazines and websites. You can usually find one or more companies or repair services that handle all makes and models. Your local club or a user's group may also be able to make a recommendation.

If you are going to ship your equipment somewhere for repair, notify the repair center first. Get authorization for shipping and an identification name or number for the package.

26.7.1 Packing Equipment

You can always blame shipping damage on the shipper, but it is a lot easier for all concerned if you package your equipment properly for shipping in the first place.

• Take photos of the equipment before packing it to document its condition before shipping. Additional photos during the packing steps might also be useful to show that you took the proper care in packing the equipment.

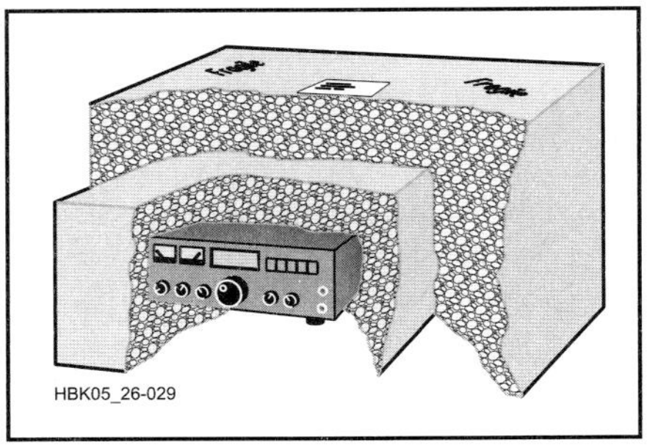

Figure 26.17 — Ship equipment packed securely in a box within a box.

• Firmly secure all heavy components, either by tying them down or blocking them off with shipping foam.

• Large transformers, such as for RF power amplifiers, should probably be removed and shipped separately.

• Large vacuum tubes should be wrapped in packing material or shipped separately.

• Make sure that all circuit boards and parts are firmly attached.

If you have the original shipping container, including all of the packing material, you should use that if the repair facility approves doing so. Otherwise use a box within a box for shipping (see **Figure 26.17**). Place the equipment and some packing material inside a box and seal it with tape. Place that box inside another that is at least six inches larger in each dimension.

Don't forget to enclose a statement of the trouble, a short history of operation and any test results that may help the service technician. Include a good description of the things you have tried. Be honest! At current repair rates you want to tell the technician everything to help ensure an efficient repair. Place the necessary correspondence, statement of symptoms, and your contact information in a mailing envelope and place it just inside the top covers of the outer box, or tape it to the top of the inner box.

Fill any remaining gaps with packing material and seal, address and mark the outer box. Choose a good freight carrier and insure the package. If available, get a tracking number for the package so you can tell when it was delivered.

Even if you tried to fix it yourself but ended up sending it back to the factory, you can feel good about your experience. You learned a lot by trying, and you have sent it back knowing that it really did require the services of a pro. Each time you troubleshoot and repair a piece of electronic circuitry, you learn something new. The downside is that you may develop a reputation as a real electronics whiz. You may find yourself spending a lot of time at club meetings offering advice, or getting invited over to a lot of shacks for a late-evening pizza snack. There are worse fates.

26.8 Typical Symptoms and Faults

26.8.1 Power Supplies

Many equipment failures are caused by power supply trouble. Fortunately, most power supply problems are easy to find and repair. This section focuses on the common linear power supply. Some notes are also made about switchmode supplies. Both types are discussed in detail in the **Power Sources** chapter, including projects with typical schematics.

LINEAR POWER SUPPLIES

The block diagram for a linear power supply is shown in **Figure 26.18.** First, use a voltmeter to measure output. Complete loss of output voltage is usually caused by an open circuit. (A short circuit draws excessive current that opens the fuse, thus becoming an open circuit.) If output voltage appears normal, apply a small load (1/10th supply capacity or smaller) and test output voltage again. If the small load causes voltage to drop, there is generally a problem in the regulator circuitry, often the pass transistors.

If the ac input circuit fuse is blown, that is usually caused by a shorted diode in the filter block, a failure of the output protection circuitry, or a short circuit in the device being powered by the supply. More rarely, one of the filter capacitors can short. If the fuse has opened, turn off the power, replace the fuse, and measure the load-circuit's dc resistance. The measured resistance should be consistent with the power-supply ratings. A short or open load circuit indicates a problem.

If the measured resistance is too low, troubleshoot the load circuit. (Nominal circuit resistances are included in some equipment manuals.) If the load circuit resistance is normal, suspect a defective regulator IC or a problem in the rest of the unit.

IC regulators can oscillate, sometimes causing failure. The small-value capacitors on the input, output or adjustment pins of the regulator prevent oscillations. Check or replace these capacitors whenever a regulator has failed.

AC ripple (120 Hz buzz) is usually caused by low-value filter capacitors in the power

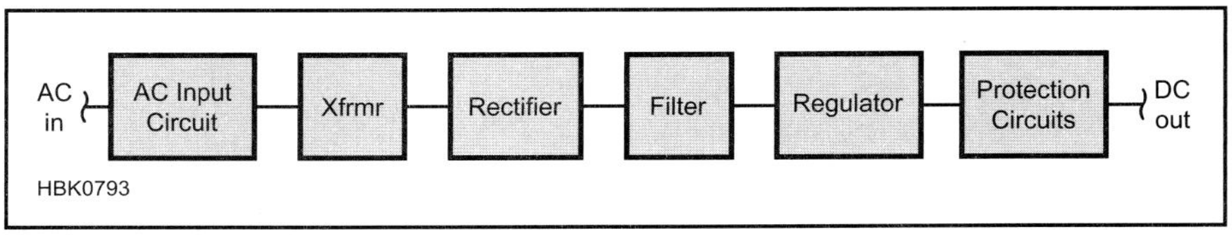

Figure 26.18 — Block diagram of a typical linear power supply.

supply. Ripple can also become excessive due to overload or regulation problems. Look for a defective filter capacitor (usually open or low-value), defective regulator, or shorted filter choke if chokes are used (not common in modern equipment). In older equipment, the defective filter capacitor will often have visible leaking electrolyte: look for corrosion residue at the capacitor leads. In new construction projects make sure RF energy is not getting into the power supply.

Here's an easy filter capacitor test: Temporarily connect a replacement capacitor (about the same value and working voltage) across the suspect capacitor. If the hum goes away, replace the bad component permanently.

Once the faulty component is found, inspect the surrounding circuit and consider what may have caused the problem. Sometimes one bad component can cause another to fail. For example, a shorted filter capacitor increases current flow and burns out a rectifier diode. While the defective diode is easy to find, the capacitor may show no visible damage.

If none of these initial checks find the problem, here are the usual systematic steps to locate the part of the supply with a problem:

- Check the input ac through switches and fuses to the transformer primary.
- Verify that the transformer secondary outputs ac of the right voltage. Disconnect the output leads, if necessary.
- Check the rectifiers for shorted diodes. Disconnect the rectifier output and test for output with a resistor load.
- Reconnect the filter and disconnect the regulator. Verify that the right dc voltage is present at the filter output.
- Reconnect the regulator and in the regulator IC or circuit test every pin or signal, especially enable/disable/soft-start and the voltage reference.
- Disconnect any output protective circuitry and verify that the pass transistors are working with a resistor load.
- Reconnect and test the output protective circuitry.

SWITCHMODE POWER SUPPLIES

Switchmode or switching power supplies are quite different from conventional supplies. In a switcher, the regulator circuit is based on a switching transistor and energy storage inductor instead of a pass transistor to change power from one dc level to another that can be higher or lower than the input voltage. Switching frequencies range from 20-120 kHz for most supplies and up to the MHz range for miniature dc-dc converter modules.

Switchmode supplies operating from the ac line have similar input circuits to linear supplies. The transformer that supplies isolation is then located in the low-voltage, high frequency section.

Apply the same input and output block-level tests as for linear supplies. The regulator circuitry in a swichmode supply is more complex than for a linear supply, but it usually is implemented in a single regulator IC. Failure of the regulator IC, transistor switch, or feedback path usually results in a completely dead supply. While active device failure is still the number one suspect, it pays to carefully test all components in the regulator subsystem.

HIGH-VOLTAGE POWER SUPPLIES

Obviously, testing HV supplies requires extreme caution. See the safety discussion at the beginning of this chapter, the **Safety** chapter of this book, and the discussion of HV supplies in the **Power Sources** chapter. If you do not feel comfortable working on HV supplies, then don't. Ask for help or hire a professional repair service to do the job.

Most HV supplies used in amateur equipment are linear supplies with the same general structure as low voltage supplies. A typical supply is presented as a project in the **Power Sources** chapter and the same basic steps can be applied — just with a lot more caution.

Components in a string, such as rectifiers or filter capacitors, should all be tested if any are determined to have failed. This is particularly true for capacitors, which can fail in sequence if one capacitor in the string shorts. Voltage equalizing resistors are not required for rectifier diodes available today, such as the 1N5408. Consider replacing older rectifier strings with new rectifiers as a preventive maintenance step.

Interlocks rarely fail, but verify that they are functioning properly before assuming they are in good working order.

26.8.2 Amplifier Circuits

Amplifiers are the most common circuits in electronics. The output of an ideal amplifier would match the input signal in every respect except magnitude: no distortion or noise would be added. Real amplifiers always add noise and distortion. Typical discrete and op-amp amplifier circuits are described in the **Analog Basics** chapter.

AMPLIFIER GAIN

Amplifier failure usually results in a loss of gain or excessive distortion at the amplifier output. In either case, check external connections first. Is there power to the stage? Has the fuse opened? Check the speaker and leads in audio output stages, the microphone and push-to-talk (PTT) line in transmitter audio sections. Excess voltage, excess current or thermal runaway can cause sudden failure of semiconductors. The failure may appear as either a short circuit or open circuit of one or more PN junctions.

Thermal runaway occurs most often in bipolar transistor circuits. If degenerative feedback (the emitter resistor reduces base-emitter voltage as conduction increases) is insufficient, thermal runaway will allow excessive current flow and device failure. Check transistors by substitution, if possible. If not, voltage checks as described below usually turn up the problem.

Faulty coupling components can reduce amplifier output. Look for component failures that would increase series impedance, or decrease shunt impedance, in the coupling network. Coupling faults can be located by signal tracing or parts substitution. Other passive component defects reduce amplifier output by shifting bias or causing active-device failure. These failures are evident when the dc operating voltages are measured.

If an amplifier is used inside a feedback loop, faults in the feedback loop can force a transistor into cutoff or saturation, or force an op amp's output to either power supply rail. In a receiver, the AGC subsystem is such a feedback loop. Open the AGC line to the device and substitute a variable voltage for the AGC signal. If amplifier action varies with voltage, suspect the AGC-circuit components; otherwise, suspect the amplifier.

In an operating amplifier, check carefully for oscillations or noise. Oscillations are most likely to start with maximum gain and the amplifier input shorted. Any noise that is due to 60 Hz sources can be heard, or seen with an oscilloscope triggered by the ac line.

Unwanted amplifier RF oscillations should be cured with changes of lead dress or circuit components. Separate input leads from output leads; use coaxial cable to carry RF between stages; neutralize inter-element or junction capacitance. Ferrite beads on the control element of the active device often stop unwanted oscillations.

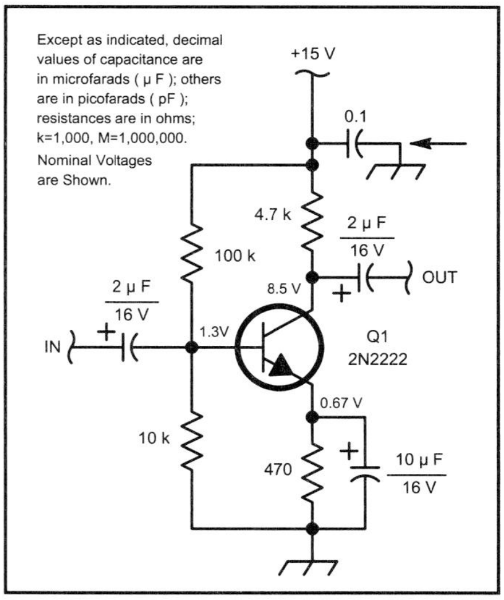

Figure 26.19 — The decoupling capacitor in this circuit is designated with an arrow.

Low-frequency oscillations (motorboating) indicate poor stage isolation or inadequate power supply filtering. Try a better lead-dress arrangement and/or check the capacitance of the decoupling network (see **Figure 26.19**). Use larger capacitors at the power supply leads; increase the number of capacitors or use separate decoupling capacitors at each stage. Coupling capacitors that are too low in value can also cause poor low-frequency response. Poor response to high frequencies is usually caused by circuit design.

COMMON-EMITTER AMPLIFIER

The common-emitter circuit (or common-source using an FET) is the most widely used configuration. It can be used as an amplifier or as a switch. Both are analyzed here as an example of how to troubleshoot transistor amplifier circuits. Other types of circuit can be analyzed similarly

Figure 26.20 is a schematic of a common-emitter transistor amplifier. The emitter, base and collector leads are labeled e, b and c, respectively. Important dc voltages are measured at the emitter (V_e), base (V_b) and collector (V_c) leads. V+ is the supply voltage.

First, analyze the voltages and signal levels in this circuit. The junction drop is the potential measured across a semiconductor junction that is conducting. It is typically 0.6 V for silicon and 0.2 V for germanium transistors.

This is a Class-A linear circuit. In Class-A circuits, the transistor is always conducting some current. R1 and R2 form a voltage divider that supplies dc bias (V_b) for the transistor. Normally, V_e is equal to V_b less the emitter-base junction drop. R4 provides degenerative dc bias, while C3 provides a low-impedance path for the signal. From this information, normal operating voltages can be estimated.

The bias and voltages will be set up so that the transistor collector voltage, V_c, is somewhere between V+ and ground potential. A good rule of thumb is that V_c should be about one-half of V+, although this can vary quite a bit, depending on component tolerances. The emitter voltage is usually a small percentage of V_c, say about 10%.

Any circuit failure that changes collector current, I_c, (ranging from a shorted transistor to a failure in the bias circuit) changes V_c and V_e as well. An increase of I_c lowers V_c and raises V_e. If the transistor shorts from collector to emitter, V_c drops to about 1.2 V, as determined by the voltage divider formed by R3 and R4.

You would see nearly the same effect if the transistor were biased into saturation by collector-to-base leakage, a reduction in R1's value or an increase in R2's value. All of these circuit failures have the same effect. In some cases, a short in C1 or C2 could cause the same symptoms.

To properly diagnose the specific cause of

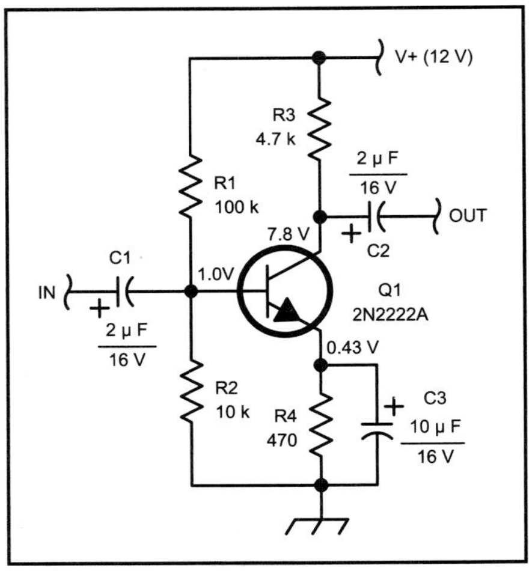

Figure 26.20 — A typical common-emitter audio amplifier.

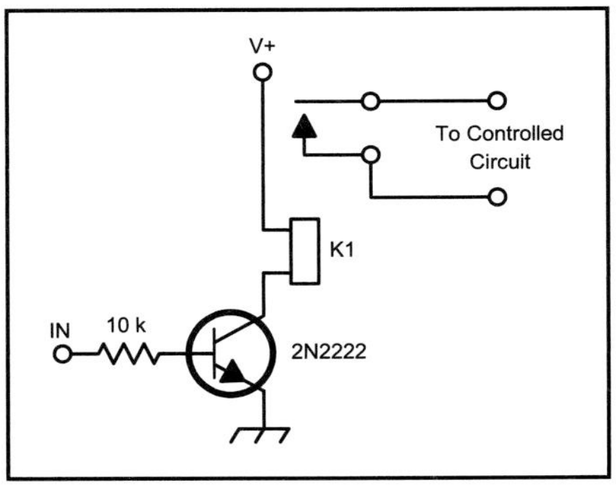

Figure 26.21 — A typical common-emitter switch or driver.

low V_c, consider and test all of these parts. It is even more complex; an increase in R3's value would also decrease V_c. There would be one valuable clue, however: if R3 increased in value, I_c would not increase; V_e would also be low.

Anything that decreases I_c increases V_c. If the transistor failed open, R1 increased in value, R2 were shorted to ground or R4 opened, then V_c would be high.

COMMON-EMITTER SWITCH

A common-emitter transistor switching circuit is shown in **Figure 26.21**. This circuit functions differently from the circuit shown in Figure 26.20. A linear amplifier is designed so that the output signal is a faithful reproduction of the input signal. Its input and output may have any value from V+ to ground.

The switching circuit of Figure 26.21, however, is similar to a digital circuit. The active device is either on or off, 1 or 0, just like digital logic. Its input signal level should either be 0 V or positive enough to switch the transistor on fully (saturate). Its output state should be either full off (with no current flowing through the relay), or full on (with the relay energized). A voltmeter placed on the collector will show either approximately +12 V or 0 V, depending on the input.

Understanding this difference in operation is crucial to troubleshooting the two circuits. If V_c were +12 V in the circuit in Figure 26.20, it would indicate a circuit failure. A V_c of +12 V in the switching circuit is normal when V_b is 0 V. (If V_b measured 0.8 V or higher, V_c should be low and the relay energized.)

DC-COUPLED AMPLIFIERS

In dc-coupled amplifiers, the transistors are directly connected together without coupling capacitors. They comprise a unique troubleshooting case. Most often, when one device fails, it destroys one or more other semiconductors in the circuit. If you don't find all of the bad parts, the remaining defective parts can cause the installed replacements to fail immediately. To reliably troubleshoot a dc coupled circuit, you must test every semiconductor in the circuit and replace them all at once.

26.8.3 Oscillators

In many circuits, a failure of the oscillator will result in complete circuit failure. A transmitter will not transmit, and a superheterodyne receiver will not receive if you have an internal oscillator failure. (These symptoms do not always mean oscillator failure, however.)

Whenever there is weakening or complete loss of signal from a radio, check oscillator operation and frequency. There are several methods:

- Use a receiver with a coaxial probe to listen for the oscillator signal.
- A dip meter can be used to check oscillators by tuning to within ±15 kHz of the oscillator, coupling it to the circuit, and listening for a beat note in the dip-meter headphones.
- Look at the oscillator waveform on a scope. The operating frequency can't be determined with great accuracy, but you can see if the oscillator is working at all. Use a low capacitance (10×) probe for oscillator observations.

Many modern oscillators are phase-locked loops (PLLs). Read the **Oscillators and Synthesizers** chapter of this book in order to learn how PLLs operate.

To test for a failed LC oscillator, use a dip meter in the active mode. Set the dip meter to the oscillator frequency and couple it to the oscillator output circuit. If the oscillator is dead, the dip-meter signal will take its place and temporarily restore some semblance of normal operation.

STABILITY

Drift is caused by variations in the oscillator. Poor voltage regulation and heat are the most common culprits. Check regulation with a voltmeter (use one that is not affected by RF). Voltage regulators are usually part of the oscillator circuit. Check them by substitution.

Chirp is a form of rapid drift that is usually caused by excessive oscillator loading or poor power-supply regulation. The most common cause of chirp is poor design. If chirp appears suddenly in a working circuit, look for component or design defects in the oscillator or its buffer amplifiers. (For example, a shorted coupling capacitor increases loading drastically.) Also check for new feedback paths from changes in wiring or component placement (feedback defeats buffer action).

Frequency instability may also result from defects in feedback components. Too much feedback may produce spurious signals, while too little makes oscillator start-up unreliable.

Sudden frequency changes are frequently the result of physical variations. Loose components or connections are probable causes. Check for arcing or dirt on printed circuit boards, trimmers and variable capacitors, loose switch contacts, bad solder joints or loose connectors.

FREQUENCY ACCURACY

In manually tuned LC oscillators, tracking at the high-frequency end of the range is controlled by trimmer capacitors. A trimmer is a variable capacitor connected in parallel with the main tuning capacitor (see **Figure 26.22**). The trimmer represents a higher percentage of the total capacitance at the high end of the tuning range. It has relatively little effect on tuning characteristics at the low-frequency end of the range.

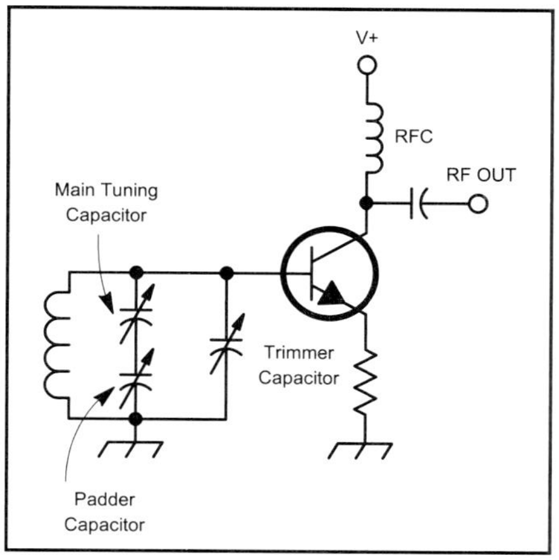

Figure 26.22 — A partial schematic of a simple oscillator showing the locations of the trimmer and padder capacitors.

Low-end range is adjusted by a series trimmer capacitor called a *padder*. A padder is a variable capacitor that is connected in series with the main tuning capacitor. Padder capacitance has a greater effect at the low-frequency end of the range. The padder capacitor is often eliminated to save money. In that case, the low-frequency range is set by adjusting the main tuning coil.

26.8.4 Transmit Amplifier Modules

Most VHF/UHF mobile radios and many small HF radios use commercial amplifier modules as the final amplifier instead of discrete transistors. These modules are quite reliable and can withstand various stresses such as disconnected antennas. However, replacement units are rarely available more than a few years after a particular radio model goes out of production. You may be able to find a damaged radio of the same model to scavenge for parts, but once the modules fail, you are usually out of luck. User's groups are often good sources of information about common failure modes of certain radios and possibly even sources of replacement parts.

The usual failure mode of an amplifier module is caused by thermal cycling that eventually leads to an internal connection developing a crack. The module becomes intermittent and then eventually fails completely. If you can open the module, you can sometimes identify and repair such a problem by soldering over the crack.

When reinstalling or replacing an amplifier module, be very careful to attach it to the heat sink as it was at the factory. Do not use excessive amounts of thermal compound or grease and make sure the mounting screws are secure. If you can find a datasheet for the module, check to see if there are recommendations for mounting screw torque and any other installation procedures. (See the **RF Power Amplifiers** chapter for guidelines on mounting power transistors.)

26.9 Radio Troubleshooting Hints

Tables 26.2, **26.3**, **26.4** and **26.5** list some common problems and possible cures for older radios that are likely to have developed problems with some later-model equivalents. These tables are not all-inclusive. They are a collection of hints and shortcuts that may save you some troubleshooting time. If you don't find your problem listed, continue with systematic troubleshooting.

Remember that many problems are caused by improper setting of a switch, a control, or a configuration menu item. Before beginning a troubleshooting session, be sure you've checked the operating manual for proper control settings and checked through the manual's troubleshooting guide. If possible, obtain a service manual with its detailed procedures, measurements, and schematics.

26.9.1 Receivers

A receiver can be diagnosed using any of the methods described earlier, but if there is not even a faint sound from the speaker, signal injection is not a good technique. If you lack troubleshooting experience, avoid following instinctive hunches. Begin with power supply tests and proceed to signal tracing.

SELECTIVITY

Failure of control or switching circuits that determine the signal path and filters can cause selectivity problems. In older equipment, tuned transformers or the components used in filter circuits may develop a shorted turn, capacitors can fail and alignment is required occasionally. Such defects are accompanied by a loss of sensitivity. Except in cases of catastrophic failure (where either the filter passes all signals, or none), it is difficult to spot a loss of selectivity. Bandwidth and insertion-loss measurements are necessary to judge filter performance.

SENSITIVITY

A gradual loss of sensitivity results from gradual degradation of an active device or long-term changes in component values. Sudden partial sensitivity changes are usually the result of a component failure, usually in the RF or IF stages. Excessive signal levels or

Table 26.2
Symptoms and Their Causes for All Electronic Equipment

Symptom	*Cause*
Power Supplies	
No output voltage	Open circuit (usually a fuse, pass transistor, or transformer winding)
Hum or ripple	Faulty regulator, capacitor or rectifier, low-frequency oscillation
Amplifiers	
Low gain	Transistor, coupling capacitors, emitter-bypass capacitor, AGC component
Noise	Transistors, coupling capacitors, resistors
Oscillations	Dirt on variable capacitor or chassis, shorted op-amp input
Oscillations, untuned (oscillations do not change with frequency)	Audio stages
Oscillations, tuned	RF, IF and mixer stages
Static-like crashes	Arcing trimmer capacitors, poor connections
Static in FM receiver	Faulty limiter stage, open capacitor in ratio detector, weak RF stage, weak incoming signal
Intermittent noise	All components and connections, band-switch contacts, potentiometers (especially in dc circuits), trimmer capacitors, poor antenna connections
Distortion (constant)	Oscillation, overload, faulty AGC, leaky transistor, open lead in tab-mount transistor, dirty potentiometer, leaky coupling capacitor, open bypass capacitors, imbalance in tuned FM detector, IF oscillations, RF feedback (cables)
Distortion (strong signals only)	Open AGC loop
Frequency change	Physical or electrical variations, dirty or faulty variable capacitor, broken switch, loose compartment parts, poor voltage regulation, oscillator tuning (trouble when switching bands)
No Signals	
All bands	Dead VFO or LO, PLL won't lock
One band only	Defective crystal, oscillator out of tune, band switch
No function control	Faulty switch or control, poor connection to front panel subassembly

Table 26.3
Receiver Problems

Symptom	*Cause*
Low sensitivity	Semiconductor degradation, circuit contamination, poor antenna connection
Signals and calibrator heard weakly	
(low S-meter readings)	RF chain
(strong S-meter readings)	AF chain, detector
No signals or calibrator heard, only hissing	RF oscillators
Distortion	
On strong signals only	AGC fault
AGC fault	Active device cut off or saturated
Difficult tuning	AGC fault
Inability to receive	Detector fault
AM weak and distorted	Poor detector, power, or ground connection
CW/SSB unintelligible	BFO off frequency or dead
FM distorted	Open detector diode

transients can damage input RF switching, amplifier, or mixing circuits. Complete and sudden loss of sensitivity is caused by an open circuit anywhere in the signal path or by a dead oscillator.

AGC

AGC failure usually causes distortion that affects only strong signals. All stages operate at maximum gain when the AGC influence is removed. An S meter can help diagnose AGC failure because it is operated by the AGC loop. If the S meter does not move at all or remains at full scale, the AGC system has a problem.

In DSP radios, the AGC function is often controlled by software which you cannot troubleshoot, but inputs to the software such as signal level detectors may be causing a problem instead.

In analog receivers, an open bypass capacitor in the AGC amplifier causes feedback through the loop. This often results in a receiver squeal (oscillation). Changes in the loop time constant affect tuning. If stations consistently blast, or are too weak for a brief time when first tuned in, the time constant is too fast. An excessively slow time constant makes tuning difficult, and stations fade after tuning. If the AGC is functioning but the timing seems wrong, check the large-value capacitors found in the AGC circuit — they usually set the AGC time constants. If the AGC is not functioning, check the AGC-detector circuit. There is often an AGC voltage that is used to control several stages. A failure in any one stage could affect the entire loop.

DETECTOR PROBLEMS

Detector trouble usually appears as complete loss or distortion of the received signal. AM, SSB and CW signals may be weak and unintelligible. FM signals will sound distorted. Look for an open circuit in the detector circuit. If tests of the detector parts indicate no trouble, look for a poor connection in the detector's power supply or ground connections. A BFO that is dead or off frequency prevents SSB and CW reception. In modern rigs, the BFO frequency is usually derived from the main VFO system.

26.9.2 Transmitters

Many potential transmitter faults are discussed in several different places in this chapter. There are, however, a few techniques used to ensure stable operation of RF amplifiers in transmitters that are not covered elsewhere.

RF final amplifiers often use parasitic chokes to prevent instability. Older parasitic chokes usually consist of a 51- to 100-Ω non-inductive resistor with a coil wound around the body and connected to the leads. It is used to prevent VHF and UHF oscillations in a vacuum-tube amplifier. The suppressor

Table 26.4
Transmitter Problems

Symptom	*Cause*
Key clicks	Keying filter, distortion in stages after keying, ALC overshoot or instability
Modulation Problems	
Loss of modulation	Broken cable (microphone, PTT, power), open circuit in audio chain, defective modulator
Distortion on transmit	Defective microphone, RF feedback, modulator imbalance, bypass capacitor, improper bias, excessive drive
Arcing	Dampness, dirt, improper lead dress
Low output	Incorrect control settings, improper carrier shift (CW signal outside of passband), audio oscillator failure, transistor or tube failure, SWR protection circuit
Antenna Problems	
Poor SWR	Damaged antenna element, matching network, feed line, balun failure (see below), resonant conductor near antenna, poor connection at antenna
Balun failure	Excessive SWR, weather or cold-flow damage in coil choke, broken wire or connection
RFI	Arcing or poor connections anywhere in antenna system or nearby conductors

Table 26.5
Transceiver Problems

Symptom	*Cause*
Inoperative S meter	Faulty TR switching or relay
PA noise in receiver	Faulty TR switching or relay
Excessive current on receive	Faulty TR switching or relay
Arcing in PA	Faulty TR switching or relay
Reduced signal strength on transmit and receive	IF failure
Poor VOX operation	VOX amplifiers
Poor VOX timing	Adjustment, component failure in VOX timing circuits or amplifiers
VOX consistently tripped by receiver audio	AntiVOX circuits or adjustment

is placed in the plate lead, close to the plate connection.

In recent years, problems with this style of suppressor have been discovered. See the **RF Power Amplifiers** chapter for information about suppressing parasitics. If parasitic suppressors are present in your transmitter, continue to use them as the exact layout and lead dress of the RF amplifier circuitry may require them to avoid oscillation. If they are not present, do not add them. When working on RF power amplifiers, take care to keep leads and components arranged just as they were when they left the factory.

Parasitic chokes often fail from excessive current flow. In these cases, the resistor is charred. Occasionally, physical shock or corrosion produces an open circuit in the coil. Test for continuity with an ohmmeter.

Transistor amplifiers are protected against parasitic oscillations by low-value resistors or ferrite beads in the base or collector leads. Resistors are used only at low power levels (about 0.5 W), and both methods work best when applied to the base lead. Negative feedback is used to prevent oscillations at lower frequencies. An open component in the feedback loop may cause low-frequency oscillation, especially in broadband amplifiers.

KEYING

The simplest form of modulation is on/off keying. Although it may seem that there cannot be much trouble with such an elementary form of modulation, two very important transmitter faults are the result of keying problems.

Key clicks are produced by fast rise and times of the keying waveform (see the previous sidebar, "Controlling Key Clicks"). Most transmitters include components in the keying circuitry to prevent clicks. When clicks are experienced, check the keying filter components first, then the succeeding stages. An improperly biased power amplifier, or a Class C amplifier that is not keyed, may produce key clicks even though the keying waveform earlier in the circuit is correct. Clicks caused by a linear amplifier may be a sign of low-frequency parasitic oscillations. If they occur in an amplifier, suspect insufficient power-supply decoupling. Check the power-supply filter capacitors and all bypass capacitors.

The other modulation problem associated with on/off keying is called backwave. Backwave is a condition in which the signal is heard, at a reduced level, even when the key is up. This occurs when the oscillator signal feeds through a keyed amplifier. This usually indicates a design flaw, although in some cases a component failure or improper keyed-stage neutralization may be to blame.

LOW OUTPUT POWER

Check the owner's manual to see if the condition is normal for some modes or bands, or if there is a menu item to set RF output power. Check the control settings. Solid-state transmitters require so little effort from the operator that control settings are seldom noticed. The CARRIER (or DRIVE) control may have been bumped. Remember to adjust tuned vacuum tube amplifiers after a significant change in operating frequency (usually 50 to 100 kHz). Most modern transmitters are also designed to reduce power if there is high (say 2:1) SWR. Check these obvious external problems before you tear apart your rig.

Power transistors may fail if the SWR protection circuit malfunctions. Such failures occur at the weak link in the amplifier chain: it is possible for the drivers to fail without damaging the finals. An open circuit in the reflected side of the sensing circuit leaves the transistors unprotected; a short shuts them down.

Low power output in a transmitter may also spring from a misadjusted carrier oscillator or a defective SWR protection circuit. If the carrier oscillator is set to a frequency well outside the transmitter passband, there may be no measurable output. Output power will increase steadily as the frequency is moved into the passband.

26.9.3 Transceivers

SWITCHING

Elaborate switching schemes are used in transceivers for signal control. Many transceiver malfunctions can be attributed to relay or switching problems. Suspect the switching controls when:

• The S meter is inoperative, but the unit otherwise functions. (This could also be a bad S meter or a consequence of a configuration menu item.)

• There is arcing in the tank circuit. (This could also be caused by a fault in the antenna system.)

• There is excessive broadband PA noise in the receiver.

Since transceiver circuits are shared, stage defects frequently affect both the transmit and receive modes, although the symptoms may change with mode. Oscillator problems usually affect both transmit and receive modes, but different oscillators, or frequencies, may be used for different emissions. Check the block diagram.

VOX

Voice operated transmit (VOX) controls are another potential trouble area. If there is difficulty in switching to transmit in the VOX mode, check the VOX-SENSITIVITY and ANTI-VOX control settings. Next, see if the PTT and manual (MOX) transmitter controls work. If the PTT and MOX controls function, examine the VOX control circuits. Test the switches, control lines and control voltage if the transmitter does not respond to other TR controls.

VOX SENSITIVITY and ANTI-VOX settings should also be checked if the transmitter switches on in response to received audio. Suspect the ANTI-VOX circuitry next. Unacceptable VOX timing results from a poor VOX-delay adjustment, or a bad resistor or capacitor in the timing circuit or VOX amplifiers.

26.9.4 Amplifiers

While this section focuses on vacuum-tube amplifiers using high-voltage (HV) supplies, it also applies to solid-state amplifiers that operate at lower voltages and generally have fewer points of failure. Amplifiers are simple, reliable pieces of equipment that respond well to basic care, regular maintenance, and common sense. A well-maintained amplifier will provide reliable service and maximum tube lifetime. (The complete version of a *QST* article on amplifier repair is included with the downloadable supplemental content.)

The key to finding the trouble with your amplifier (or any piece of sophisticated equipment) is to be careful and methodical, and to avoid jumping to false conclusions or making random tests. The manufacturer's customer service department will likely be helpful if you are considerate and have taken careful notes detailing the trouble symptoms and any differences from normal operation. There may be helpful guidelines on the manufacturer's web pages or from other internet resources. Sometimes there is more than one problem — they work together to act like one very strange puzzle. Just remember that most problems can be isolated by careful, step-by-step tests.

SAFETY FIRST

It is important to review good safety practices. (See the **Safe Practices** and **Power Sources** chapters for additional safety information.) Tube amplifiers use power supply voltages well in excess of 1 kV, and the RF output can be hundreds of volts, as well. Almost every voltage in a vacuum-tube amplifier can be lethal! Take care of yourself and use caution!

Power Control — Know and control the state of both ac line voltage and dc power supplies. Physically disconnect line cords and other power cables when you are not working on live equipment. Use a lockout on circuit breakers. Double-check visually and with a meter to be absolutely sure power has been removed.

Interlocks — Unless specifically instructed by the manufacturer's procedures to do so, never bypass an interlock. This is rarely required except in troubleshooting and should only be done when absolutely necessary. Interlocks are there to protect you.

The One-Hand Rule — Keep one hand in your pocket while making any measurements on live equipment. The hand in your pocket removes a path for current to flow through you. It's also a good idea to wear shoes with insulating soles and work on dry surfaces. Current can be lethal even at levels of a few mA — don't tempt the laws of physics.

Test Equipment Rating — Be sure your test equipment is adequately rated for the voltages and power levels encountered in amplifiers! This is particularly important in handheld equipment in which there is no metal enclosure connected to an ac safety ground. Excessive voltage can result in a *flashover* to the user from the internal electronics, probes, or test leads, resulting in electric shock. Know and respect this rating.

If you are using an external high voltage probe, make sure it is in good condition with no cracks in the body. The test lead insulation should be in good condition — flexible and with no cracks or wire exposed. If practical, do not make measurements while holding the probe or meter. Attach the probe with the voltage discharged and then turn the power on. Turn power off and discharge the voltage before touching the probe again. Treat high voltage with care and respect!

Patience — Repairing an amplifier isn't a race. Take your time. Don't work on equipment when you're tired or frustrated. Wait several minutes after turning the amplifier off to open the cabinet — capacitors can take several minutes to discharge through their bleeder resistors. On most amplifiers, the meter switch has a HV position — wait until it is at zero before opening the cabinet. Some amplifiers have a safety interlock that shorts the high voltage to ground — if the capacitor bank has not discharged, this can be quite spectacular — and destroy power supply components.

A Grounding Stick — Make the simple safety accessory shown in the High Voltage section of the **Power Sources** chapter and use it whenever you work on equipment in which hazardous voltages have been present. The ground wire should be heavy duty (#12 AWG or larger) due to the high peak currents (hundreds of amperes) present when discharging a capacitor or tripping a circuit breaker. When equipment is opened, touch the tip of the stick to every exposed component and connection that you might come in contact with. Assume nothing — accidental shorts and component failures can put voltage in places it shouldn't be.

The Buddy System and CPR — Use the buddy system when working around any equipment that has the potential for causing serious injury. The buddy needn't be a ham, just anyone who will be nearby in case of trouble. Your buddy should know how to remove power and administer basic first aid or CPR.

CLEANLINESS

The first rule in taking good care of an amplifier is cleanliness. Amplifiers need not be kept sparkling new, but their worst enemy is heat. Excess heat accelerates component aging and increases stress during operation.

Outside the amplifier, prevent dust and obstructions from blocking the paths by which heat is removed. This means keeping all ventilation holes free of dust, pet hair and insects. Fan intakes are particularly susceptible to inhaling all sorts of debris. Use a vacuum cleaner to clean the amplifier and surrounding areas. Keep liquids well away from the amplifier.

Keep papers or magazines off the amplifier — even if the cover is solid metal. Paper acts as an insulator and keeps heat from being radiated through the cover. Amplifier heat sinks must have free air circulation to be effective. There should be at least a couple of inches of free space surrounding an amplifier on its sides and top. If the manufacturer recommends a certain clearance, mounting orientation or air flow, follow those recommendations.

Inside the amplifier, HV circuits attract dust

Figure 26.23 — A small paintbrush and a vacuum cleaner crevice attachment make dust removal easy.

that slows heat dissipation and will eventually build up to where it arcs or carbonizes. Use the vacuum cleaner to remove any dust or dirt. If you find insects (or worse) inside the amp, try to determine how they got in and plug that hole. Window screening works fine to allow airflow while keeping out insects. While you're cleaning the inside, perform a visual inspection as described in the next section.

Vacuuming works best with an attachment commonly known as a crevice cleaner. **Figure 26.23** shows a crevice cleaning attachment being used with a small paintbrush to dislodge and remove dust. The brush will root dust out of tight places and off components without damaging them or pulling on connecting wires. Don't use the vacuum cleaner brush attachment; they're designed for floors, not electronics. Some vacuums also have a blower mechanism, but these rarely have enough punch to clean as thoroughly as a brush. Blowing dust just pushes the dust around and into other equipment.

If you can't get a brush or attachment close enough, a spray can of compressed air will usually dislodge dust and dirt so you can vacuum it up. If you use a rag or towel to wipe down panels or large components, be sure not to leave threads or lint behind. Never use a solvent or spray cleaner to wash down components or flush out crevices unless the manufacturer advises doing so — it might leave behind a residue or damage the component.

Cigarette smoke causes its own set of unique problems as tar and nicotine accumulate along with dust and dirt. Removal is difficult at best; commercial solvents such as *Krud Kutter* can be thinned 4-to-1 and will remove the gummy deposits. In extreme cases, disassemble the amplifier (remove tube(s), power transformer and meters) and wash it in a sink with soap and water. Be careful to protect or remove meters and other components that can be damaged by water. Thoroughly dry the amplifier in warm air. Re-lubricate variable capacitors, variable inductors, and any rotary controls.

INSPECTION

Remove any internal covers or access panels and...stop! Get out the grounding stick (see the **Power Sources** chapter), clip its ground lead securely to the chassis, and touch every exposed connection. Now, using a strong light and possibly a magnifier, look over the components and connections. Use all of your senses to analyze the interior — smell, look, listen.

Amplifiers have far fewer components than transceivers, so look at every component and insulator. Look for cracks, signs of arcing, carbon traces (thin black lines), discoloration, loose connections, melting of plastic, and anything else that doesn't look right. This is a great time to be sure that mounting and grounding screws are tight. Does anything smell burnt? Learn the smells of overheated components. Make a note of what you find, repair or replace — even if no action is required. If the amplifier was recently powered up, an infrared thermometer can spot components that were running hot. Or you may hear a crackling or sizzling noise from heat —not a good sign! Try to isolate the source of the noise.

ELECTRICAL COMPONENTS

Let's start with the power supply. There are three basic parts to amplifier power supplies — the ac transformer and line devices, the rectifier/filter, and the metering/regulation circuitry. (See the **Power Sources** chapter for more information.) Transformers need little maintenance except to be kept cool and be mounted securely, but inspect for overheating, discoloration, or seepage from insulation or tar. Line components such as switches, circuit breakers and fuses, if mechanically sound and adequately rated, are usually electrically okay as well. Check fuses and switches with a multimeter.

Rectifiers and HV filter capacitors require occasional cleaning. Look for discoloration around components mounted on a printed circuit board (PCB) and make sure that all wire connections are secure. HV capacitors are generally electrolytic or oil and should show no signs of leakage, swelling or outgassing around terminals.

Components that perform metering and regulation of voltage and current can be affected by heat or heavy dust. If there has been a failure of some other component in the amplifier — such as a tube — these circuits can be stressed severely. Resistors may survive substantial temporary overloads but may show signs of overload such as discoloration or swelling — and change value. Verify the correct value with a multimeter against the color code; if it is not readable or reads erratically, replace the component.

Amplifiers contain two types of relays — control and RF. Control relays switch ac and dc voltages and do not handle input or output RF energy. The usual problem encountered with control relays is oxidation or pitting of their contacts. A burnishing tool can be used to clean relay contacts. In a pinch a strip of ordinary paper can be pulled between contacts gently held closed. Avoid the temptation to over-clean silver-plated relay and switch contacts. It is easy to remove contact plating with excessive polishing, and while silver-plated relay and switch contacts may appear to be dark in color, oxidized silver (black) is still a good conductor. Once the silver's gone, it's gone; contact erosion will then be pervasive.

If visual inspection shows heavy pitting or discoloration or resistance measurements show the relay to have intermittent contact quality, it should be replaced.

If the resting current is too high or intermittent, check power relays for good contact from the bias supply. On some older amplifiers, the relay coil is used as a resistor to create a voltage drop that cuts off the tube. If the relay coil covering is discolored, measure the resistance of the coil and possibly replace it.

RF relays are used to perform transmit/receive (TR) switching and routing of RF signals through or around the amplifier circuitry. Amplifiers designed for full break-in operation will usually use a high-speed vacuum TR relay. Vacuum relays are sealed and cannot be cleaned or maintained. When you replace RF relays, use a direct replacement part or one rated for RF service with the same characteristics as the original.

If the SWR measured between the transceiver and the amplifier suddenly increases, is erratic, or QSK (break-in) stays in the transmit state or is just open in receive, check the RF relay contacts.

Cables and connectors are subjected to heavy heat and electrical loads in amplifiers. Plastics may become brittle and connections may oxidize. Cables should remain flexible and not be crimped or pinched if clamped or tied down. Gently wiggle cables while watching the connections at each end for looseness or bending. Connectors can be unplugged and reseated once or twice to clear oxide on contact surfaces.

Carefully inspect any connector that seems loose. Be especially careful with connectors and cables in amplifiers with power supplies in separate enclosures from the RF deck. Those interconnects are susceptible to both mechanical and electrical stress, and you don't want an energized HV cable loose on the operating desk. Check the electrical integrity of those cables and make sure they are tightly fastened.

As with relays, switches found in amplifiers either perform control functions or route RF signals. Adequately rated control switches, if mechanically sound, are usually okay. Band switches are the most common RF switch — usually a rotary phenolic or ceramic type. A close visual inspection should show no pitting or oxidation on the wiper (the part of the switch that rotates between contacts) or the individual contacts. Arcing or overheating will quickly destroy rotary switches. **Figure 26.24** shows a heavy-duty band switch that has suffered severe damage from arcing.

Slight oxidation is acceptable on silver-plated switches. Phosphor-bronze contacts can sometimes be cleaned with a light scrub from a pencil eraser, but plating can be easily removed, so use caution with this method and be sure to remove any eraser crumbs.

Figure 26.24 — The band switch section on the left clearly shows the signs of destructive arcing.

Wafer Switch Repair

Because they are often custom parts, replacements for individual wafers and contacts for rotary switches can be difficult or impossible to find. You may be able to find switches with the same size contacts, however. These can serve as replacements for contacts on the damaged switch. Save pieces of the damaged switch, since they can be used for parts.

Start by disassembling the damaged switch. If it is on a shaft, there are usually two threaded rods holding it together. Save all pieces. Note the order in which flat and lock washers are used. Taking a close-up photo or making a careful drawing is a good idea. Take care to save all the parts. Place the damaged wafer on your work surface with the hollow-side of the rivet facing up. Use a drill bit just a bit larger than the rivet. Drill-out the rivet at low speed, using just enough pressure to cut off the lip of the rivet – do not drill into the contact itself. Remove the burned contacts first to get a "feel" for how to do it without damaging the contact.

Now disassemble the switch to be used for part replacement and once again, save all the parts for future use. Once the wafers are in-hand, inspect them and pick the best ones to be used as replacements. Remove the replacement contacts in the same way as the damaged contacts were removed.

Now replace the damaged contacts with the replacements using #2-56 screws and nuts (smaller switches may require even smaller hardware), being careful not to over tighten the screws and crack the ceramic or phenolic wafer; "just tight" is good enough. Use a dab of LockTite Red (or paint or fingernail polish) to secure. **Figure 26.A1** shows a wafer that has been repaired using this technique.

Figure 26.A1 — Five contacts starting at the 4:30 position on the wafer have been replaced using contacts removed from another switch and #2-56 hardware. [Photo courtesy of Matt Kastigar, WØMJ]

Rotary switch contacts cannot be replaced easily, although individual wafer sections may be replaced if an exact matching part can be obtained (see sidebar). De-oxidation chemicals should be applied with a pen-type applicator and not sprayed to avoid absorption by porous phenolics or plastics that can create resistive paths.

When replacing capacitors and resistors, be sure to use adequately rated parts. Voltage and power-handling ratings are particularly important, especially for components handling high RF currents. An RF tank capacitor replacement should be checked carefully for adequate RF voltage and current ratings, not just dc. HV resistors are generally long and thin to prevent arcing across their surfaces. Even if a smaller (and cheaper) resistor has an equivalent power rating, resist the temptation to substitute it. In a pinch, a series string of resistors of the appropriate combined value can be used to replace one HV unit. Don't use carbon resistors for metering circuits, use metal or carbon film types. The carbon composition types are too unstable. Electrolytic capacitors also have a temperature rating, usually 85 or 105 °C; use the higher rating if available.

If you are repairing or maintaining an old amplifier and manufacturer-specific parts are no longer available, the ham community has many sources for RF and HV components. Fair Radio Sales (**www.fairradio. com**) and Surplus Sales of Nebraska (**www. surplus-sales.com**) are familiar names. Hamfests and websites often have amplifier components for sale. (See the **RF Power Amplifiers** chapter's sidebar on using surplus or used parts for amplifiers.) You might consider buying a non-working amplifier of the same model for parts. MFJ (**www.mfjenterprises.com**) sells some parts that are used in the Ameritron brand of amplifiers — you may be able to substitute.

TUBES

Good maintenance starts with proper operation of the amplifier. Follow the manufacturer's instructions for input drive levels, duty cycles, tuning and output power level. Frequently check all metered voltages and zcurrent to be sure that the tubes are being operated properly and giving you maximum lifetime. The manufacturer spent time and effort to develop the manual — read it! If you do not have one, do a web search (or call the manufacturer) and get one. It is time well spent.

The internal mechanical structures of tubes generally do not deal well with mechanical shock and vibration, so treat them gently. The manufacturer may also specify how the amplifier is to be mounted, so read the operating manual. Tubes generate a lot of heat, so it's important that whatever cooling mechanism, employed is kept at peak efficiency. Airways should be clean, including between the fins on metal tubes. All seals and chimneys should fit securely and be kept clean. Wipe the envelope of glass tubes clean after handling them — fingerprints should be removed to prevent baking them into the surface. On metal tubes that use finger-stock contacts, be sure the contacts are clean and make good contact all the way around the tube. Partial contact or dirty finger stock can cause asymmetric current and heating inside the tube, resulting in warping of internal grids and possibly causing harmonics or parasitics.

Plate cap connections and VHF parasitic suppressors should be secure and show no signs of heating. Overheated parasitic suppressors may indicate that the neutralization circuit is not adjusted properly. Inspect socket contacts and the tube pins to be sure all connections are secure, particularly high-current filament connections. Removing and inserting the tubes once or twice will clean the socket contacts.

While the tubes are removed, check the pins for melted solder — this is common on over-heated 3-500Z tubes. If needed, resolder with silver solder, and be careful not to use too much solder, which might expand the diameter of the pin.

Adjustments to the neutralizing network, which suppresses VHF oscillations by negative feedback from the plate to grid circuit, are rarely required except when you are replacing a tube or after you do major rewiring or repair of the RF components. The manufacturer will provide instructions on making these adjustments. If symptoms of VHF oscillations occur without changing a tube, then perhaps the tube characteristics or associated components have changed. Parasitic oscillations in high-power amplifiers can be strong enough to cause arcing damage. Perform a visual inspection prior to readjusting the neutralizing circuit.

Metering circuits rarely fail on their own, but they play a key part in maintenance. By keeping a record of normal voltages and currents, you will have a valuable set of clues when things go wrong. Record tuning settings, drive levels, and tube voltages and currents on each band and with every antenna. When settings change, you can refer to the notebook instead of relying on memory.

MECHANICAL

The most common faults appear with the moving parts: connectors, relays and switches. Thermal cycling and heat-related stresses can result in mechanical connections loosening over time or material failures. Switch shafts, shaft couplings (especially if they are plastic) and panel bearings all need to be checked for tightness and proper alignment. All mounting hardware needs to be tight, particularly if it supplies a grounding path. Examine all panel-mounted components, particularly RF connectors, and be sure they're attached securely. BNC and UHF connectors mounted with a single nut in a round panel hole are notorious for loosening with repeated connect/disconnect cycles. Rubber and plastic parts are particularly stressed by heat. If there are any belts, gears or pulleys, make sure they're clean and that dust and lint are kept out of their lubricant. Loose or slipping belts should be replaced. Check O-rings, grommets and sleeves to be sure they are not brittle or cracked. If insulation sleeves or sheets are used, check to be sure they are covering what they're supposed to. Never discard them or replace them with improperly sized or rated materials.

Enclosures and internal shields should all be fastened securely with every required screw in place. Watch out for loosely overlapping metal covers. If a sheet metal screw has stripped out, either drill a new hole or replace the screw with a larger size, taking care to maintain adequate clearance around and behind the new screw. Or, if space permits, a "speed" or clip nut can sometimes be used. These are available from auto parts or home improvement stores, usually in the "specialty hardware" section.

Tip the amplifier from side to side while listening for loose hardware or metal fragments, all of which should be removed.

Clean the front and back panels to protect the finish. If the amplifier cabinet is missing a foot or an internal shock mount, replace it. A clean unit with a complete cabinet will have a significantly higher resale value, so it's in your interest to keep the equipment looking good.

SHIPPING

When you are traveling with an amplifier or shipping it, some care in packing will prevent damage. Improper packing can also result in difficulty in collecting on an insurance claim, should damage occur. The original shipping cartons are a good method of protecting the amplifier for storage and sale, but they were not made to hold up to frequent shipping. If you travel frequently, it is best to get a sturdy shipping case made for electronic equipment. Pelican (**www.pelican-shipping-cases.com**) and Anvil (**www.anvilsite.com**) make excellent shipping cases suitable for carrying amplifiers and radio equipment.

Some amplifiers require the power transformer to be removed before shipping. Check your owner's manual or contact the manufacturer to find out. Failure to remove it before shipping can cause major structural damage to the amplifier's chassis and case.

Tubes should also be removed from their sockets for shipment. It may not be necessary to ship them separately if they can be packed in the amplifier's enclosure with adequate plastic foam packing material. If the manufacturer of the tube or amplifier recommends separate shipment, however, do it!

CLEANING AND MAINTENANCE PLAN

For amateur use, there is little need for maintenance more frequently than once per year. Consider the maintenance requirements of the amplifier and what its manufacturer recommends. Review the amplifier's manuals and make up a checklist of what major steps and tools are required.

TROUBLESHOOTING

A benefit of regular maintenance will be familiarity with your amplifier should you ever need to repair it. Knowing what it looks (and smells) like inside will give you a head start on effecting a quick repair.

The following discussion is intended to illustrate the general flow of a troubleshooting effort, not be a step-by-step guide. Before starting on your own amplifier, review the amplifier manual's "Theory of Operation" section and familiarize yourself with the schematic. If there is a troubleshooting procedure in the manual, follow it. **Figure 26.25** shows a general-purpose troubleshooting flow chart. Do not swap in a properly functioning tube or tubes until you are sure that a tube is actually defective. Installing a good tube in an amplifier with circuit problems can damage a good tube.

Many "amplifier is dead" problems turn out to be simply a lack of ac power. Before even opening the cabinet of an unresponsive amplifier, be sure that ac is really present at the wall socket and that the fuse or circuit breaker is really closed. If ac power is present at the wall socket, trace through any internal fuses, interlocks and relays all the way through to the transformer's primary terminals. If the amplifier operates from 240-V circuits, be sure you check both hot wires. (See the **Safety** chapter for more information about ac wiring practices.)

Hard failures in a high-voltage power supply are rarely subtle, so it's usually clear if there is a problem. When you repair a power supply, take the opportunity to check all related components. If all defective components are not replaced, the failures may be repeated when the circuit is re-energized.

Rectifiers may fail open or shorted — test them using a DMM diode checker. An open rectifier will result in a drop in the high-voltage output of 50% or more but will probably not overheat or destroy itself. A shorted rectifier failure is usually more dramatic and may cause additional rectifiers or filter capacitors to fail. If one rectifier in a string has failed, it may be a good idea to replace the entire string as the remaining rectifiers have been subjected to a higher-than-normal voltage.

High-voltage filter capacitors usually fail shorted, although they will occasionally lose capacitance and show a rise in ESR (equivalent series resistance). Check the rectifiers and any metering components — they may have been damaged by the current surge caused from a short circuit.

Power transformer failures are usually due to arcing in the windings, insulation failures, or overheating. High-voltage transformers can be disassembled and rewound by a custom transformer manufacturer.

Along with the plate supply, tetrode screen supplies also occasionally fail. The usual cause is the regulation circuit that drops the voltage from the plate level. Operating without a screen supply can be damaging to a tube, so be sure to check the tube carefully after repairs.

If the power supply checks out okay and the tube's filaments are not lit, check the tube socket and the pins on the tube itself — overheating can cause solder to flow, resulting in intermittent failure. If the tube's filament is lit, check the resting or bias current. If it is excessive or very low, check all bias voltages and dc current paths to the tube, such as the plate choke, screen supply (for tetrodes) and grid or cathode circuits.

If you do not find power supply and dc problems, check the RF components or RF deck. Check the input SWR to the amplifier. If it has changed, then you likely have a problem in the input circuitry (overheated coil, shorted capacitor or bad switch contact) or one or more tubes have failed. Perform a

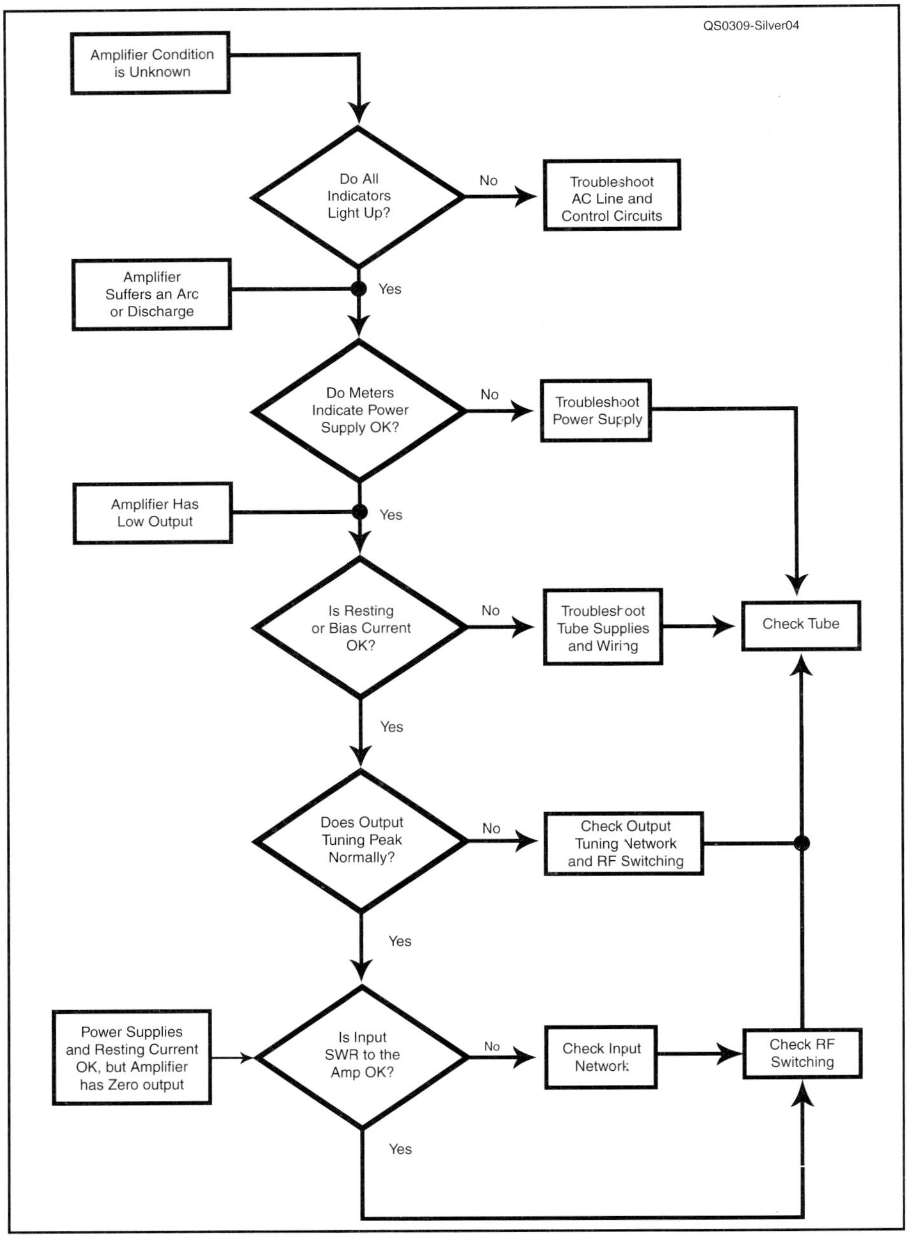

Figure 26.25 — This general-purpose flow chart will help identify amplifier problems. For solid-state units, substitute "Check Output Transistors" for "Check Tube."

visual check of the input circuitry and the band switch, followed by an ohmmeter check of all input components. Use an SWR analyzer or a dip oscillator to see if the input tuning has changed, indicating a possible bad tuning network component in the input circuits. If input SWR is normal and applying drive does not result in any change in plate current, you may have a defective tube, tube socket or connection between the input circuits and the tube.

Check the TR control circuits and relay. If plate current changes, but not as much as normal, try adjusting the output tuning circuitry. If this has little or no effect, the tube may be defective or a connection between the tube and output circuitry may have opened. If retuning has an effect, but at different settings than usual, the tube may be defective or there may be a problem in the tuning circuitry. A visual inspection and an ohmmeter check are in order.

SOLID STATE AMPLIFIERS

Almost all diagnostic techniques used for tube amplifiers are also applicable to solid state amplifiers. Input and output circuits are very similar. Power supplies are usually lower voltage (but high current).

Power supplies typically can be checked with a multimeter for voltage, but current will require reading the voltage drop across a low-value, high-wattage resistor (this can cause problems from the voltage drop) or using a clamp-on current probe. Bias adjustments typically are made with a potentiometer on a voltage regulator circuit (you'll need the schematic or user's manual), and resting current needs to be checked and verified against the manufacturer's specifications.

As with tubes, heat is the enemy. Keep all ventilation openings clear and clean. Check fans for proper operating and speed. Remove any dust or dirt buildup, especially on heat sinks and transistor packages.

Bipolar transistors can be checked with a multimeter (see the **Component Data and References** chapter), as can MOSFETS, which are now more common in RF amplifiers. Replacements will require de-soldering the old part, cleaning up the PC board, and installation of the new part. If the transistors are mounted against a heat sink, use a thin, bubble-free film of heat sink compound between the transistor and heat sink. Make sure all screws are in place and tight. If shoulder washers are used to insulate screws, be sure they are used when the amplifier is reassembled.

POWERING UP OLD AMPLIFIERS

When a piece of equipment has been idle for a long period of time, proper care must be taken to "revive" it. DO NOT just "plug it in" and "power it up" — the results will most likely be smoke and sparks! This is especially true for amplifiers with high-voltage power supplies. (See Powering Up the Equipment in the section Repair and Restoration of Vintage Equipment.)

First, examine the device for physical damage such as bent sheet metal or dents. Clean metal work and panels is usually a good sign. Assuming it is ac-powered, look at the line cord for cuts, scuffs, or cracks, then bend it between your fingers and listen for "cracking." If it is dried out or brittle, replace it before proceeding. If the line cord is okay, tie a piece of wire through the plug's pins so that it cannot be plugged in to a receptacle. Make sure the plug is the right one for the input voltage (120 or 240 V). Check the fuses (if present) — a good indication of condition when last used.

Open the cabinet, remove the tube(s) and store in a safe place. Look closely for any irregularities such as dark PC board areas; discolored resistors; breached, swollen, or leaky capacitors; disconnected wires; loose components or screws; or cracked components or hardware. Begin with these and replace as required. Verify that all of the power supply connections are in order, including that power transformer connections are configured for the proper voltage.

Typically, old electrolytic capacitors, even if otherwise good, have dried out and need attention. They can be tested with a multimeter for shorts (or open), or with an ESR meter if available. These are best replaced if the amplifier is really old or if they don't charge up properly in the power-up test (see below and this chapter's section on repairing vintage equipment).

Use a multimeter to test for shorts to ground in the power supply section. With the power switch in the ON position (still not attached to ac power), measure resistance across the pins of the ac plug. It should read 4 to 5 Ω if it is connected to the primary of a transformer and open when the power switch is OFF. If you measure anything other than an open circuit, check for decoupling capacitors on the ac line that might be leaky. Check the secondary resistance of the transformer — it will be higher in resistance. Then work your way to the rectifier and filter capacitors and any chokes.

Check the resistance across the input and output RF connectors. They should be open or nearly open depending on the input or output network type. (Some amplifiers have an RF choke across the output connector that will present a low resistance.) Check the resistance of the HV plate connections of the tubes to ground — it should be open (the plate capacitors may be observed charging up).

If all looks good and it is possible to do so, disconnect the rectifier board or assembly and check the transformer output windings. Then check any interlock components that might have to be closed or opened (depending on function).

Before starting power-up tests, *make sure* your multimeter is rated for the voltage. If not, don't use it as the meter could be destroyed, or an arc through the meter or its leads (called *flashover*) can present a severe shock hazard.

Using a variable transformer (such as a Variac), bring up input power slowly to about 25% of rated input voltage. The secondary voltage of the transformer should be 25% of the recommended output. (Make high voltage measurements either hands-free or with one hand not touching anything.) Because full output voltage will exceed the maximum rating of most multimeters, disconnect the multimeter unless you are using a high-voltage probe. Increase the voltage slowly to full input while watching the transformer — it should not get hot or make noise. Use an infrared thermometer to avoid direct contact but verify the transformer does not get hot without a load.

Next, power down and reconnect the rectifier / filter board or assemblies. Using the variable transformer, start at 25% input voltage again. If the amplifier has a meter with an HV or B+ position, look for about 25% of the normal voltage reading. Dial lights might just barely light. Leave the power supply to "cook" for a few hours, monitoring it closely. Initially, there will be a high current surge that will slowly drop after the electrolytic filter capacitors start to reform their dielectric layer. When the current drops, bump the voltage up 10 to 20 V, wait a half hour, then turn it up again, monitoring the power supply with each increase. Continue until you have reached full input voltage.

When full power supply high voltage has been achieved, the electrolytic caps should be reformed and the dial lights should be at full brightness. If at any time there is smoke, crackling noise or sparks, disconnect at once, discharge the capacitors, and investigate the issue(s).

Once the power supply is up and running, disconnect the plate B+ lead from the output of the rectifier / filter. In order to check filament voltage, solder temporary wires to the filament socket pins and route them out of the amplifier cabinet. Insert the tube(s), close the cabinet (checking the interlock safety switches), and power up. Quickly verify the filament voltage — it should be within 5% of the specified voltage; if not, power off and investigate. Check the fan(s) for proper operation and air flow.

Tubes get gassy and may arc internally when first powered up after a long period of storage; leave the B+ disconnected and let the tubes "cook" for a few hours. If the tube has a "getter," this may activate it to remove gas. Not all power tubes have a getter — check the data sheet for your tube.

Using lots of caution and a high-voltage

probe, check that there is no dc voltage at the output RF connector — if the plate coupling capacitor is shorted, there may be full B+ on the amplifier output!

Power down, wait for the HV / B+ to read zero, and open the cabinet. Use a grounding stick to *be sure* the B+ is at zero, then reconnect it (see the **Power Sources** chapter for safety tips on working with high-voltage power supplies). Put the plate caps on, reattach the cover, and power it up. Listen and watch (if you can see the tubes) for any issues. If you notice anything unusual, stop at once and investigate (after discharging the high voltage, of course). If all is well, monitor the amp and let it cook for a few hours.

After a thorough test, power the amp off, let it sit overnight, then power it up the next day. Check the idling current for the recommended value once again.

Connect the amp to a dummy load that can handle full power output. Put the amp in transmit mode. The idling current should be the recommended value. Switch to standby. Connect a transmitter set for low output (about 20 W) and drive the amplifier. At low drive, the output should be proportional to the input times the gain. If all goes well, apply more drive until full output is achieved. At this point, if the tube is bad, it may fail to reach full power or it may arc. Listen and carefully verify all operations.

26.10 Antenna Systems

This section is an abbreviated version of the Antenna System Troubleshooting chapter of the *ARRL Antenna Book* which was added to the 22nd edition. Because of the enormous variety of antenna systems, general guidelines must be presented, but the successful troubleshooting process usually follows a systematic approach just as for any other radio system.

26.10.1 Basic Antenna Systems

Start with an inventory of the antenna system. Any of these can be the cause of your problem: supports, insulators, elements, feed point, balun (if any), feed line, grounding or transient protectors, impedance matching and switching equipment, RF jumper cables at any point. As with any troubleshooting process, be alert for mistaken or loose connections, loose or disconnected power and control cables, wires touching each other that shouldn't be, and so forth. Reduce the antenna system to the simplest system with the problem and it will likely look something like the system in **Figure 26.26**.

It is particularly important to remember that your station ground is often part of the antenna system. The length of the connection between the equipment ground bus and the ground rod is usually several feet at minimum, which can be an appreciable fraction of a wavelength on the higher HF bands. This can greatly affect tuning if there is common-mode current on the feed line or if a random-wire or end-fed type of antenna is being used. If touching equipment enclosures or the ground wire affects SWR or impedance readings, that will affect your antenna measurements as well.

DUMMY-LOAD TESTING

Begin by replacing components of your antenna system with a dummy load, starting at the output of the radio using a known-good jumper cable. Verify that the radio works properly into a 50 Ω load using a known-good directional wattmeter. Then move the dummy load to the output side of any antenna tuning or switching equipment, one component at a time, until you have replaced the antenna

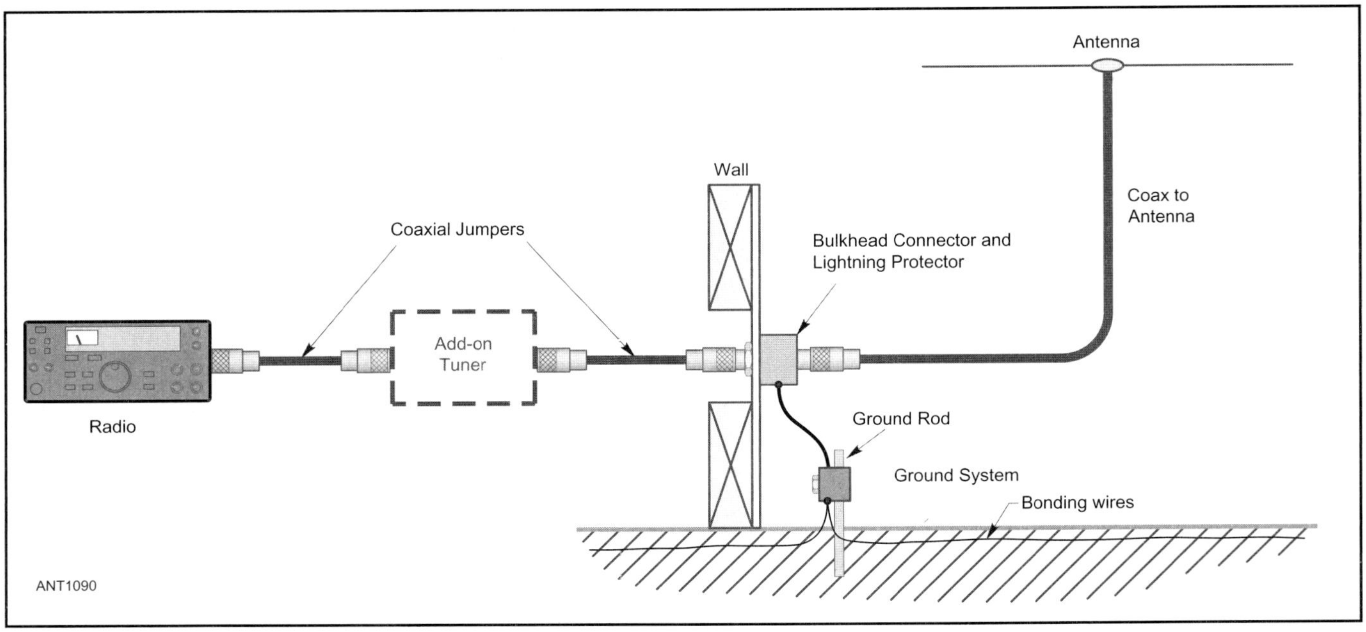

Figure 26.26 — A typical simple antenna system. An add-on antenna tuner is likely to be used if one is not built in to the radio. More complex antenna systems have all of the same components plus some switching equipment.

feed line with the dummy load. If everything checks OK to this point, the problem is in the feed line or antenna. Don't forget to verify that the problem with the antenna is still present after each dummy load test. If the problem was a loose or intermittent connection, it is likely that swapping the dummy load in and out changed or eliminated the problem.

ANTENNA VISUAL INSPECTION

Now it is time to perform a visual inspection of the feed line and antenna itself. Start with the feed line connector at the last point where the dummy load was swapped in. Disassemble the connector and inspect it for damage from water or corrosion. If either are present, replace the connector and check the condition of the cable before proceeding. Note that if water can get into the cable at the antenna's elevated feed point, it is not unknown for it to flow downward through the braid both by gravity and by capillary action all the way to the shack! If the cable braid is wet at both ends, the cable must be replaced.

If you have a wire antenna, lower it and make a visual inspection of all the pieces:

- Cut away the waterproofing around the coax termination and inspect for water and/or corrosion.
- On the insulators at each end, there should be no possibility of contact between the antenna wire and the supporting wire/ropes.
- If there are any splices in the wire elements, they should be well crimped or soldered.
- At the center insulator, there should be no possibility of contact between the element wires.
- At the balun or coax connection the element connections should be soldered or firmly connected.

If you have a Yagi or vertical antenna, similar steps are required. Carefully check any feed point matching assembly, such as a gamma match, hairpin, or stub and make sure connections are clean and tight.

ANTENNA TEST

Assuming any mechanical problems have been rectified, proceed to retest the antenna and feed line. Replace the antenna with a dummy load and check the feed line loss with a wattmeter or antenna analyzer at the antenna end of the feed line. If the feed line checks out OK, the problem must be in the antenna. Reattach the feed line to the antenna and verify that the problem remains. Note that for wire antennas lowered to near ground level, the resonant point will change — this is to be expected.

If the problem is still present, repeat the visual inspection at a closer level of detail. Check all dimensions and connections. Double-check any telescoping sections of tubing, transmission line stubs, in-line coax connectors clamped connections between wires and between wires and tubing. If possible, give joints and connections a good shake while watching for intermittent readings on the wattmeter or antenna analyzer. If you cannot see inside a component or assembly, perform resistance tests for continuity. Remember to identify your signals, since you are testing on the air at this point in the process.

The next step is to reinstall or raise the antenna at least ¼-wavelength off the ground and verify that the problem remains. If you have repaired the antenna, perform a re-check at this point to be sure everything is in good working order before returning the antenna to full height. Once you have re-installed the antenna, including full weatherproofing of any coaxial cable terminations or connectors, record in the shack notebook your measurements of the antenna, along with what the problem was discovered to be and how you repaired it.

AM Broadcast Interference to Antenna Analyzers

Living or testing within a couple of miles of an AM broadcast station can create a lot of problems for the sensitive RF detectors in portable antenna analyzers. This type of RFI usually appears as values of SWR and impedance that don't change with frequency or that change in unexpected ways or an upscale meter reading that varies with the station modulation. The analyzer SWR reading will not agree with SWR measured by using a directional wattmeter and more than a few watts of power. The solution is sometimes to use a broadcast-rejection filter (available from analyzer manufacturers) although this tends to color measurements a bit and typically can't be used for measurements on 160 meters because it is so close to the AM broadcast band. In cases where the station is nearby or on 160 meters, directional wattmeters or analyzers with narrow-band tuned inputs must be used.

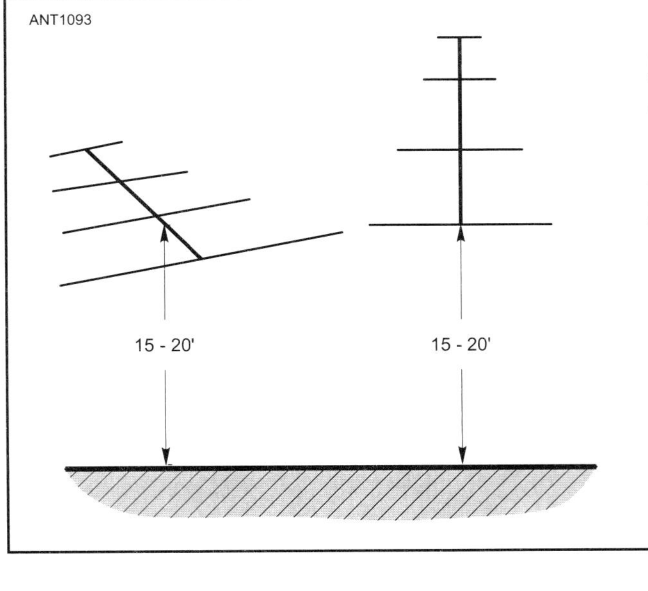

Figure 26.27 — When testing a Yagi or quad antenna, make sure it is at least 15 to 20 feet above ground. If oriented vertically, the reflector should be the closest to ground. Performance will still change as the antenna is raised.

26.10.2 General Antenna Systems

Think of the following topics as a kind of toolbox for troubleshooting. Many of them assume you are testing some type of Yagi or other beam antenna, but the general guidelines apply to all types of antennas

It is important to remember this simple rule for adjustments and troubleshooting: Do the simplest and easiest adjustment or correction *first*, and only *one* at a time.

When making on-air comparisons, select signals that are at the margin and not pushing your receiver well over S9, where it can be difficult to measure differences of a few dB. Terrain has a lot to do with performance as well. If you are comparing with a large station, keep in mind that its location was probably selected carefully and the antennas were placed exactly where they should be for optimum performance on the property.

TEST MEASUREMENTS

A) Test the antenna at a minimum height of 15 to 20 feet (see **Figure 26.27**). This will move the antenna far enough away from the ground (which adds capacitance to the antenna) and enable meaningful measurements. Use sawhorses *only* for construction purposes.

- 15 to 20 feet above ground does not mean 5 feet above a 10 to 15 foot high roof, it means above ground with nothing in between;

• Antenna resonant frequency will shift upward as it is raised;

• Feed point impedance will change with a change in height, which applies to both horizontal and vertical antennas;

• Some antennas are more sensitive to proximity to ground than others;

• Some antennas are more sensitive to nearby conductive objects (such as other antennas) than others.

B) Aiming the antenna upward with the reflector on the ground might coincide with some measurements on rare occasions, but there are no guarantees with this method. The reflector is literally touching a large capacitor (earth) and the driver element is very close, too. Raise the antenna at least 15 to 20 feet off the ground.

C) When using a hand-held SWR analyzer you are looking for the dip in SWR, not where the impedance or resistance meter indicates 50 Ω. (Dip = frequency of lowest SWR value, or lowest swing on the meter.) On the MFJ-259/269 series, the left-hand meter (SWR) is the one you want to watch, not the right-hand meter (impedance).

D) Does the SWR and frequency of lowest dip change when the coax length is changed? If so, the balun might be faulty, as in not isolating the load from the coax feed line.

Additionally, with an added length of coax and its associated small (hopefully small) amount of loss :

• The value of SWR is expected to be lower with the additional coax and,

• The width of the SWR curve is expected to be wider with the additional coax, when measured at the transmitter end of the coax.

E) Be sure you are watching for the right dip, as some antennas can have a secondary resonance (another dip). It is quite possible to see a Yagi reflector's resonant frequency, or some other dip caused by interaction with adjacent antennas.

MECHANICAL

A. Are the dimensions correct? Production units should match the documentation (within reason). When using tubing elements, measure each *exposed* element section during assembly and the element *half-length* (the total length of each half of the element) after assembly. Measuring the entire length is sometimes tricky, depending on the center attachment to the boom on Yagis, as the element can bow, or the tape might not lay flat along the tubing sections. Self-designed units might have a taper error.

B. Making the average taper diameter larger will make the equivalent electrical element longer. This makes the antenna act as if the physical element is too long.

C. Making the average taper diameter smaller will make the equivalent electrical element shorter. This makes the antenna act as if the physical element is too short.

D. If the element is a mono-taper (tubing element is the same for the entire length), larger diameter elements will be physically shorter than smaller diameter mono-taper elements to give the same electrical performance at the same frequency.

E. The type of mounting of the element to the boom affects the element length, whether it is attached directly to the boom, or insulated from the boom. Incorrect mounting/mounting plate allocation will upset the antenna tuning:

• A mounting plate 4 × 8 inches has an equivalent diameter of approximately 2.5 inches and 4 inches in length for each element half.

• A mounting plate that is 3 × 6 inches has an equivalent diameter of about 1.8 inches and a length of 3 inches for each element half.

• The mounting plate equivalent will be the first section in a model of the element half.

F. In a Yagi, if the elements are designed to be touching, are the elements touching the boom in the correct locations?

G. In a Yagi, if the elements are designed to be insulated, are the elements insulated from the boom in the correct locations?

H. The center of hairpin matching devices (i.e. on a Yagi) can be grounded to the boom.

I. The boom is neutral, but it is still a conductor! The center of a dipole element is also neutral and can be touched while tuning without affecting the reading. With a hairpin match, the center of the hairpin can also be touched while tuning, and touching the whole hairpin might not affect the readings much at all.

PROXIMITY

A. What else is nearby (roof, wires, guy lines, gutters)? If it can conduct at all, it can and probably will couple to the antenna!

B. Does the SWR change when the antenna is rotated? If so, this indicates interaction. Note that in some combinations of antennas, there can be destructive interaction even if the SWR does not change. Computer models can be useful here.

C. What is within ¼ wavelength of the antenna? Imagine a sphere (like a big ball) with the antenna in question at the center of the sphere, with the following as a radius, depending on frequency. Think in three dimensions like a sphere — up and down and front and rear. Any resonant conductor (antenna or not) with the following radii will couple to and probably affect the antenna you are testing or installing.

160 meters = 140 feet
80 meters = 70 feet
40 meters = 35 feet
20 meters = 18 feet
15 meters = 12 feet
10 meters = 9 feet

D. Interaction occurs whether or not you are transmitting on the adjacent antennas. When receiving, it simply is not as apparent as when transmitting.

E. Wire antennas under a Yagi can easily affect it. This includes inverted V dipoles for the low bands and multi-band dipoles. The wire antennas are typically for lower frequency band (s) and will not be affected by the Yagi(s), as the Yagis are for the higher bands.

FEED SYSTEM

The feed system includes:

• the feed line,

• switching mechanisms,

• pigtails from the feed point on the antenna to the main feed line or switch and,

• all feed lines inside the radio room.

The feed system is the *entire connection* between the radio and the feed point of the antenna.

A. Is the feed line (coax) known to be good? (Start with the easiest first.) Check the dc resistance across the cable with the far end open and shorted. Is there water in the coax? This can give strange readings, even frequency-dependent ones. If there is any question, swap the feed line for a known good one and re-test.

B. Are the connectors installed properly? Has a connector been stressed (pulled)? Is the rotation loop done properly to not stress the coax? Is it an old existing loop or a new one? Usually it's alright if new. Type N connectors (especially the older type) are prone to having the center conductor pull out due to the weight of the coax pulling down on the connector. Connectors are easy to do, using the right technique.

C. Is there a barrel connector (a PL-258 dual-SO-239 adapter) in the feed line anywhere? Has a new or different barrel been inserted? These are a common failure point, even with new barrels. The failures range from micro-bridges across the face of the barrel shorting out the center and shield, to resistance between the two ends of the barrel. Have the new barrels been tested in a known feed system? Always test them before installing. Use only quality RF adapters, as these are common system failure points.

D. Is the coax intact and not frayed such that the shield can come into contact with anything? This can cause intermittent problems as the coax shield touches the tower, such as on rotation loops and coax on telescoping towers.

E. Is the tuner OFF on the radio? This is often overlooked when adding a new antenna.

F. Are there any new devices in the line? It might be a good idea to remove everything but the essential items when troubleshooting.

G. Is there a remote antenna switch? Swap to another port.

H. Is there a low pass filter in the line? The filter can be defective, especially on 10 meters, causing strange SWR readings.

26.10.3 Antenna Tuners

Antenna tuners are usually well designed and built with adequately rated parts. Most will last a lifetime, but even QRP power levels can result in arcing if the tuner is trying to match a very high impedance or is accidentally disconnected from an antenna. *Hot switching* can cause damage when contacts in the tuner are moved while transmitting. Lightning transients and excessive power beyond the tuner's ratings cause damage. Frequently adjusted components and contacts can wear out over time or just get dirty. Any of these can cause problems with normal operation.

For more information on the different types and circuits used in tuners, see the section on Matching Impedances in Antenna Systems in the chapter on **Transmission Lines**. Remember that "tuner" is the most common term, but ATU, transmatch, antenna coupler, impedance matching unit, matchbox, and other terms also refer to the same piece of equipment.

If you think your tuner is misbehaving, begin the process of troubleshooting according to the previous section on General Antenna System Troubleshooting. Make sure the problem really is in the tuner! Once you're sure the tuner has a problem, give it a good checkout to get an idea of what you're looking for. Keep notes as you test, since these will provide clues to what the problem may be and also point to tests you should perform after repairs to be sure you've fixed the problem.

Whether the tuner is internal to your transceiver or an external unit, start with a close and detailed visual inspection. Are the connectors clean? Do any connectors look like there has been an arc, or are they discolored from heat or corrosion? If there are multiple antenna connectors, does the unit perform differently when the antenna is connected to different connectors? Are there any loose parts inside that rattle when the unit is tipped or shaken? Smell the unit, too — if anything smells burned, you may have a damaged component inside. These are also excellent inspections to make before buying a used tuner.

Operate the tuner into a dummy load at low and high power. Does it behave the same way at both power levels? Does it operate erratically? Does wiggling any of the feed lines to or from the tuner cause SWR to change at the transmitter output? Operate the tuner into a known-good antenna and feed line, then repeat the tests.

T network tuners have a reputation of being able to "tune up into themselves," meaning they can match even an open circuit because they have such a wide tuning range. This results in very high voltages and currents inside the tuner, creating all sorts of problems.

Once you've put the tuner through its paces, you'll have a better idea of what problems you're facing. It is also worth considering that there may be more than one problem. When you find something wrong and fix it, repeat your tests to see if the problem is still there or has changed.

AUTOMATIC ANTENNA TUNERS

Most of the same components in a manual tuner are present in an automatic tuner along with sensing, selection, and indicating circuits. The most important new moving parts are relays. Automatic tuners are a bit more difficult to diagnose, but by breaking them down into functional parts troubleshooting is straightforward. (This section also applies to the internal antenna tuners common in HF transceivers.)

Typically, automatic tuners have fixed-value capacitors and inductors in an L network. A phase sensor and voltage detectors are read by a microprocessor, which switches in L and C based on an algorithm to arrive at the best SWR match. The inductors and capacitors are switched in or out by relays. In "zero-power" tuners (tuners that do not draw much if any current once tuning is complete) latching relays are used. These remain open or closed once power is removed.

There are three usual symptoms for automatic tuner problems. The tuner may not be able to achieve a match at all and display some kind of error condition. The tuner can "hunt" without stopping for a match according to its control program — you'll hear the relays clicking or buzzing without stopping. Or the tuner will suddenly start re-tuning at higher power levels during transmission because a component or relay is failing.

If the tuner cannot find a match or keeps "hunting" to find a match, the issue is typically a bad relay contact. The controller "thinks" it has the correct L or C value switched-in, but the contact is not being made. A relay driver line can also be open. Also check for a bad interconnect cable between the external antenna connector and the main board.

The most common issue is relay failure. Relays have a coil that is energized to move an *armature* holding the contacts. The usual relay acts as a DPDT switch with a set of NC (*normally closed*) and NO (*normally open*) contacts. To test the relay, first measure the coil resistance. Then energize each relay one at a time and measure the resistance across the contacts with and without power. A coil failure (usually an open circuit) requires relay replacement.

If the coil is good and you can get to the relay contacts, a strip of bond paper and cleaning solution can sometimes clean the contacts of corrosion or oxide. Carefully remove the cover using a hobby knife along the bottom edge, not poking too far into the relay — you do not want to cut the coil winding. In miniature relays in particular, be careful to avoid bending the armature while cleaning. Put some cleaning solution on the paper strip, slide the strip between one set of contacts, hold them together, and slide the strip back and forth. If there is a dark deposit on the paper, keep sliding it back and forth until no additional deposit is seen. Repeat with the other set of contacts. If the contacts are severely pitted or there is large buildup of material on the contact from arcing, the relay should be replaced. Filing the contacts should be avoided because it removes any plating from the contact surfaces.

If the relays are good, the controller may not be actuating them. Check the printed-circuit board traces for burned or open traces. Then work your way back to the driver ICs or transistors and test them; microcontrollers generally cannot drive relays directly. Typically, a transistor is used to energize the relay coils. There should also be a suppressor diode across each coil, which can be tested with a VOM. Replace components as necessary.

MAINTENANCE AND OPERATION

Think of your antenna tuner as the output stage of an amplifier — it contains many of the same type of components, operates at the same power levels, and needs the same care to operate reliably.

In large manual tuners, keep air-variable capacitors, exposed inductors, and switches clean. Dirt and dust can form paths for arcing and cause heat buildup. Remove dust with a gentle brush and vacuum. If the tuner has been in a smoking environment, smoke deposits can accumulate. Clean with a brush first, then apply cleaning solution using a brush and thoroughly dry before operation. Lightly re-oil any bearings in air-variable capacitors after using cleaning solution. Do not put oil on a roller inductor shaft or contact. Check for arcing and broken or heated connections and wires; repair or replace as needed. Use a high-quality cleaning solution on switch contacts, but not to excess, and remove any residue. Check connectors and the cables in and out of the tuner.

Check grounding and bonding. The cabinet of an (external) tuner should be bonded to the transceiver, and to any RF bonding bus or plane in the station. If common-mode RF current is upsetting SWR measurements, use ferrite beads or a toroid to block it.

26.11 Repair and Restoration of Vintage Equipment

When purchasing a classic receiver or transmitter, unless you absolutely know otherwise, assume the radio will need work. Often you can get a top-of-the-line radio needing a bit of inexpensive repair or cleanup. Don't worry — these radios were designed to be repaired by their owners — and curiously, except for cosmetic parts such as cabinets and knobs, parts are much easier to find for 60-year-old radios than a 20-year-old imported transceiver!

Chances are the radio has gone for years without use. Even if it has been recently used, don't completely trust components that might be 60 or more years old. Don't start by plugging in your new acquisition! To do so might damage a hard-to-replace power transformer, or cause a fire.

Instead, if the radio didn't come with its owner's manual, get one. Several *QST* vendors sell old manuals in good condition. K4XL's Boat Anchor Manual Archive (**www2.faculty.sbc.edu/kgrimm/boatanchor**) is probably *the* best free resource for these manuals. Armed with the manual, remove the radio from its cabinet. You very likely will find evidence of unsightly repairs, modifications, or even dangling wires. While modifications aren't necessarily bad, they can certainly add some drama to any necessary subsequent troubleshooting. It's up to you to reverse or remove them.

Another option for working with vintage equipment is to refer to editions of the *ARRL Handbook* published around the time that the equipment was in common use. The circuit design and construction practices described in the *Handbook* are likely to be representative of those in the radio and may provide guidance for troubleshooting, repair, and adjustment. Similarly, the troubleshooting sections and chapters in previous editions provide valuable guidance for working with equipment of the same or earlier vintages.

26.11.1 Component Replacement

Correct any obvious problems such as dangling components. Replace the line cord with a three-wire, grounded plug, if not a transformerless "ac/dc" type as discussed below. If the radio is one with a live chassis, you should operate it from an isolation transformer for safety. If you don't have an isolation transformer, use a voltmeter to determine the orientation of the ac plug that places the chassis at ground potential. Avoid touching the chassis and do not use knobs with set screws that contact metal control shafts. It's also a good idea to add a fuse, if the radio doesn't originally have one. Are we ready to give it the smoke test? Not so fast!

CAPACITOR RATINGS

Obviously, aged components deteriorate, and capacitors are particularly prone to developing leakage or short-circuits with age. There are as many opinions on capacitor replacement as there are radio collectors, but *at the very least* you should replace the electrolytic filter capacitors. Here's why: they *will* short circuit sometime, and when they do, they'll probably take the rectifier tube and the power transformer with them. Modern high voltage electrolytic capacitors are reliable and much smaller than their classic counterparts. Old paper-wax and black plastic tubular capacitors should also be replaced. Again, a short circuit in one of them could take out other components. Modern film capacitors of the appropriate voltage are great replacements. Opinions vary as to whether all should be replaced, but replacements are cheap and you have the radio apart now, so why not? If keeping the original components is important, follow the procedure for using a variable transformer to reform electrolytics as described below.

You can mount the new capacitors under the chassis by mounting a new terminal strip (do *not* just wire them to the old capacitor terminals), you can re-stuff an old electrolytic capacitor's can with new capacitors, or you can buy a new can from places such as **www.hayseedhamfest.com** or Antique Electronics Supply (**www.tubesandmore.com**). In any event, follow the manufacturer's schematic — don't assume that the – (minus) end of the capacitor goes to ground, as in some radios the ground path is through a resistor so as to develop bias for the audio output stage or RF gain circuit. Observe the polarity or you'll soon be cleaning up a stinky mess!

TESTING OLD CAPACITORS

All capacitors have a voltage rating written on the side of the cap, unless it is a small disc. Surplus stores often have bins full of capacitors of unmarked voltage rating. Don't assume they are a high enough voltage to use — check them with a capacitor checker. There are many models out there by Knight Kit, Lafayette, Sencore, and Eico, but the best were the Heathkit IT-11 or IT-28. They are basically the same model but different colors. Both use a 6E5 Magic Eye tube to indicate the status of the capacitor. A selectable voltage from 3 to 600 V dc can be applied to check for leakage and operation. These are good for small disc or paper caps and large electrolytics.

Take the unknown voltage cap and place it in the test terminals. Advance the voltage control from minimum until the eye tube shows it breaking down. You now know what voltage it is good to.

If the capacitor tests good through the 600 V dc range, it probably is a 1000 V dc capacitor. It is best, though, to know for sure the rating of the cap. In tube equipment, most capacitors should be 500 or 1000 V rated. Mouser (**www.mouser.com**) and Digikey (**www.digikey.com**) do still sell caps for those voltage ranges, but they have become very expensive. You could also find new old stock (NOS) capacitors of the correct voltage rating at surplus stores.

REFORMING ELECTROLYTIC CAPACITORS

The best idea is to replace old electrolytic capacitors with a new unit. They are available cheaply in the voltage ranges required and

Using a Tube Tester

Vacuum-tube testers are scarce but can be located through antique or vintage radio associations, audiophile groups, and sellers of tubes.

Most simple tube testers measure the cathode emission of a vacuum tube. Each grid is shorted to the plate through a switch, and the current is observed while the tube operates as a diode. By opening the switches from each grid to the plate (one at a time), we can check for opens and shorts. If the plate current does not drop slightly as a switch is opened, the element connected to that switch is either open or shorted to another element. (We cannot tell an open from a short with this test.) The emission tester does not necessarily indicate the ability of a tube to amplify.

Other tube testers measure tube gain (transconductance). Some transconductance testers read plate current with a fixed bias network. Others use an ac signal to drive the tube while measuring plate current.

Most tube testers also check inter-element leakage. Contamination inside the tube envelope may result in current leakage and shorts between elements. The paths can have high resistance, and may be caused by gas or deposits inside the tube. Tube testers use a moderate voltage to check for leakage. Leakage can also be checked with an ohmmeter using the ×1M range, depending on the actual spacing of tube elements.

more compact and reliable than the original electrolytic caps. If necessary, however, old electrolytics can often be revived by reforming the dielectric using a capacitor checker.

Disconnect the wires attached to the capacitor under test and connect it to the capacitor checker. Start at the lowest voltage rating and let it charge up the capacitor. You will know when it is charged by viewing the eye tube: if it is wide open, the cap is charged; if it is closed, the capacitor is either shorted or still charging. Advance gradually to the next voltage rating and wait until the eye fully opens—take plenty of time for the capacitor to stabilize. Continue on through the voltage ranges; each time it will take longer for the eye to open. The dielectric is being reformed. Finally, when you reach the voltage range of the capacitor and the eye is fully open, the cap has been fully revived and is ready for use.

The same process can be performed with a variable autotransformer (Variac) by advancing it a few volts at a time over several hours, but that is a coarse and unreliable process. A diode must be placed in series with the transformer to convert the ac voltage to dc. Monitor the voltage across the capacitor with a meter. If it suddenly jumps to zero, the capacitor has shorted and is now useless. Usually the capacitor can be revived successfully and will work just fine.

RESISTORS

Over time, carbon resistors in older radios can change value significantly, which can affect circuit operation. Disconnect one end of the questionable resistor and measure it with an accurate ohmmeter. If it is out of tolerance, replace it. Most resistors in the tube era were ½ W or greater. Most circuits today use ¼ W or smaller resistors, which will not dissipate the power tubes produce.

Carbon composition resistors are becoming rarer, but there are still ample quantities of NOS in surplus stores. Be careful about power rating. Use metal oxide resistors if needed. Remember that wirewound resistors are very inductive and not good for RF circuits. They are excellent for power supply circuits and are usually found in the 1 to 25 W range.

REPLACING DIODES

Many old tube radios use rectifier tubes. It is not a good idea to replace these with solid-state rectifiers, as a shorted diode can take out the transformer. Selenium rectifiers, however, are good candidates to replace with a silicon diode. The 1N400x series diode are usually fine for use and very cheap. For higher voltage supplies, be sure to use 1N4007 or higher rated diodes. This may result in higher ouput voltage from the supply. Add a series dropping resistor if necessary to reduce voltage.

TRANSFORMER REPLACMENT

The best bet is to find another radio of the same variety from which you can harvest the transformer. This is especially common in transmitters like the Heathkit DX-35 and DX-40, whose transformers frequently failed. Replacement transformers are generally no longer available. Some companies will rewind transformers, but that is usually prohibitively expensive. Output voltages are quite critical in the design of tube radios, so it is not a good idea to replace a 400 V ac transformer with a 600 V ac unit. It may be best to find a donor radio for a replacement transformer.

WIRE REPLACEMENT

Power cords should be replaced at the first sign of hardening and cracking. It is often a good idea to replace the two-wire power cords with three-wire cords, but this *cannot* be done on ac/dc transformerless radios in which one side of the ac line is connected directly to the chassis! Those must retain the two-wire cords. As noted previously, operating these radios with an isolation transformer is the safest option. If replacing the cord with a two-wire version, the neutral (larger blade) must be connected to the chassis.

Pre-WWII vintage radios often used a cotton covered power cord. To keep the radio as authentic as possible, cotton covered power cords and matching plugs can be found at suppliers such as Antique Radio Supply.

Many early radios had a two-pin power plug with fuses in the plug (the radio has no internal fusing). These are made by Elmeco and are still available (check eBay). Standard 3AG type fuses go in the power plug. Usually a 1 or 2 A fast-blow fuse will work fine except for a higher powered transmitter.

For using PVC-covered wire with terminal strips, solid wire is easier to attach before soldering than stranded wire, although stranded is very usable. You can also use Teflon covered wire that doesn't burn when the soldering iron hits the insulation.

TUBE REPLACEMENT

The sidebar "Using a Tube Tester" explains what a tube tester does. Watch swap meets and garage sales for tube substitution books. Many tubes are interchangeable or similar in purpose. Be sure to document any tube substitutions you make on a vintage radio. Remember that a new tube in the circuit may require re-peaking of the tuned circuits associated with it.

If a tube has a loose tube cap on top, you can easily repair it. Unsolder the cap and make sure that ample wire is still coming from the tube glass envelope. The tube cap should have a tiny hole in the center of the top which the wire will pass through. Mix a small amount of JB Weld epoxy (**www.jbweld.com**) and glue the tube cap back in place. Make sure the wire is sticking out of the hole. After the epoxy has hardened, solder the wire back onto the tube top. Don't let a loose tube cap break it off.

One might be anxious to wipe off the tube and clean it up. Be careful, as you might wipe off the tube number, and then you won't have any idea what the tube is. Many tubes have been lost because they have become unidentifiable. If you do decide to clean up the tube, make sure you steer well clear of the tube number.

Tube sockets and tube pins easily become oxidized, which result in radios not working or being intermittent. It is a good idea to pull each tube and spray the socket with DeoxIT or tuner cleaner. Re-insert the tube and wiggle it around in the socket to rub away any oxidation remaining.

REMOVING AND REPLACING COMPONENTS

Replacing capacitors and/or other components isn't difficult, unless they are buried under other components. The Hallicrafters SX-28 and SX-42 receivers are examples of receivers that have extremely difficult-to-reach components. There are different schools of thought on the proper component replacement method. You can use solder wick and/or a desoldering tool to remove the solder from a terminal, unwrap the wires, and install the new component by wrapping the lead around the terminal and soldering it securely. The proponents of this method point out that this is the preferred military and commercial method. I find it often will needlessly damage other components such as tube sockets and create solder droplets inside of the radio.

Back in the day, radio repairmen clipped out a component leaving a short stub of wire, made little coils in the new lead, then soldered the coiled lead to the old stub. This is a much faster, easier and neater method. Refer to books and websites on repairing vintage equipment for other useful tips and tricks.

26.11.2 Powering Up the Equipment

Before applying any power, use a VOM and measure the resistance from the B+ line to ground. Filter capacitors will cause an initial low resistance reading that increases as the capacitors charge. If the resistance stays low or does not increase beyond tens of kΩ, find the short circuit before you proceed.

DETERMINING THE RIGHT INPUT AC VOLTAGE

Vintage radios were often designed for use with lower ac mains voltages than are the standard today. If they are powered from 120 V ac, the resulting higher voltages can burn out filaments and overload the plate and screen

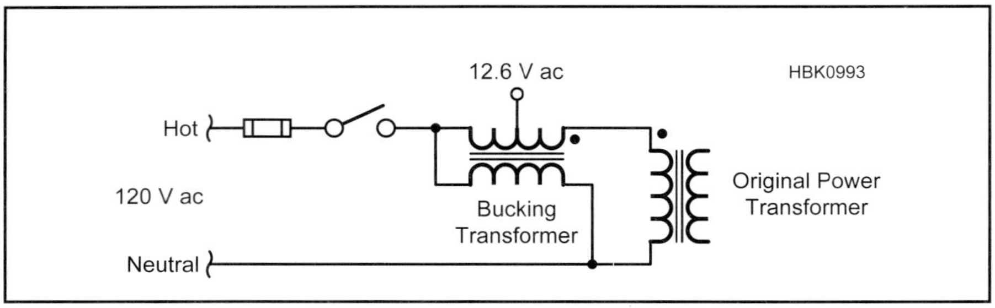

Figure 26.28 — The bucking transformer secondary is connected out of phase with the radios power transformer's primary. This reduces voltage output from the power transformer and avoids over-voltage stress on the radio's tubes and other components.

circuits. Here's how to find out what voltage the radio was designed for:

• Remove all tubes and determine the proper filament voltage for one of them.

• Connect a voltmeter to the filament pins of the tube for which filament voltage is known.

• Using a Variac to power the radio, slowly increase voltage until the filament voltage matches the specified voltage for the tube.

• Measure the input voltage to the radio. That is the ac voltage from which the radio should be powered. If the input voltage is 115 V ac or lower, a bucking transformer can be used to reduce the input voltage.

Wiring for a bucking transformer is shown in **Figure 26.28**. The bucking transformer's primary is connected to the regular input ac voltage. Its secondary is connected in series and out of phase with the radio's power transformer primary. That reduces the input voltage to the power transformer primary so the output voltages are closer to what the components are rated for. Since phasing of a transformer secondary may not be obvious, hook the transformers up and measure output voltage from the radio's power transformer filament winding. If it has increased, reverse the connections for the bucking transformer secondary.

A small filament transformer with a secondary rating of 3 to 5 A will do for most radios. (Large radios may require transformers with higher current ratings.) To power a radio designed for 115 V ac from 120 V ac, use a 5 V or 6.3 V filament transformer. If the radio needs 110 V ac, use a 12.6 V filament transformer.

INITIAL POWER-UP CHECKS

With the right input ac voltage determined, it's time to gradually power up the radio. DO NOT plug it in and turn it on "to see if it works!" Applying full voltage to a radio that has not been used for a long time can destroy it! It's best to use a variable transformer such as a Variac and ramp up the voltage slowly, or use a "dim bulb tester" (a 100 W light-bulb wired in series with one leg of the ac power). Turn on the radio, and watch for any sparking, flashing, or a red glow from the plates of the rectifier tube, or smoke. If any of these occur, immediately remove power and correct the problem. Observe that the tube filaments should light (although you won't see the glow from metal tubes, you should be able to feel them warm up). Again, any tubes that fail to light should be replaced before you continue.

Now hook up a speaker and antenna, and test the radio. With any luck you'll be greeted by a perfectly performing radio. Seldom, however, is that the case. You may encounter any number of problems at this point. Dirty bandswitches and other controls manifest themselves by intermittently cutting out; they can be cleaned by DeoxIT contact cleaner applied with a cotton swab (don't spray the switch directly!). Scratchy volume or RF gain controls can be cleaned with some DeoxIT; in some cases you might need to remove the control and uncrimp the cover to reveal the carbon element inside.

If a receiver is totally dead at this point but the filaments and dial lights are lit, double-check to see that the Receive-Standby switch is in the receive position, and any battery plug or standby switch jumpers (as described in the manual) are in their correct place.

Although comprehensive troubleshooting is covered elsewhere in this chapter, the next step is comparing voltages with those stated in the user manual. If the manual doesn't have a voltage table denoting the expected voltage at each tube pin, expect 200-350 V at the tube plate terminals, a few volts at the cathode (unless it's directly grounded), 70-200 V at the screen, and slightly negative voltage at the grid. If you're faced with this situation and a newcomer to troubleshooting vintage gear, help can be found at **amfone.net**, **www.antiqueradios.com**, or other forums that cater to boat-anchors and/or vintage radio repair and restoration.

26.11.3 Alignment

Over the years hams have been cautioned that alignment is usually the *last* thing that should be attempted to repair a radio. In general this is true — but it's also a certainty that a 50-year-old radio *will* need alignment in order for it to perform at its best. In any case, replace the capacitors and any other faulty components before you attempt alignment — it'll never be right if it still has bad parts! You'll need a good signal generator and a volt-ohm-meter or oscilloscope. Follow the manufacturer's instructions, and with care you'll be rewarded with a radio that performs as good as it did when it was new.

RECEIVER ALIGNMENT

One last caution — alignment should not be attempted by the novice technician or if you do not have the proper equipment or experience. That said, alignment may be justified under the following conditions:

• The set is very old and has not been adjusted in many years.

• The circuit has been subject to abusive treatment or environment.

• There is obvious misalignment from a previous repair.

• Tuned-circuit components or crystals have been replaced.

• An inexperienced technician attempted alignment without proper equipment.

• There is a malfunction, but all other circuit conditions are normal. (Faulty transformers can be located because they will not tune.)

Even if one of the above conditions is met, do not attempt alignment unless you have the proper equipment. Receiver alignment should progress from the detector to the antenna terminals. When working on an FM receiver, align the detector first, then the IF and limiter stages and finally the RF amplifier and local oscillator stages. For an AM receiver, align the IF stages first, then the RF amplifier and oscillator stages.

Both AM and FM receivers can be aligned in much the same manner. Always follow the manufacturer's recommended alignment procedure. If one is not available, follow these guidelines:

1. Set the receiver RF gain to maximum, BFO control to zero or center (if applicable to your receiver), and tune to the high end of the receiver passband.

2. Disable the AGC.

3. Set the signal source to the center of the IF passband, with no modulation and minimum signal level.

4. Connect the signal source to the input of the IF section.

5. Connect a voltmeter to the IF output.

6. Adjust the signal-source level for a slight indication on the voltmeter.

7. Peak each IF transformer in order, from the meter to the signal source. The adjustments interact; repeat steps 6 and 7 until adjustment brings no noticeable improvement.

8. Remove the signal source from the IF section input, reduce the level to minimum,

set the frequency to that shown on the receiver dial, and connect the source to the antenna terminals. If necessary, tune around for the signal — if the local oscillator is not tracking, it may be off.

9. Adjust the signal level to give a slight reading on the voltmeter.

10. Adjust the trimmer capacitor of the RF amplifier for a peak reading of the test signal. (Verify that you are reading the correct signal by switching the source on and off.)

11. Reset the signal source and the receiver tuning for the low end of the passband.

12. Adjust the local-oscillator padder for peak reading.

13. Steps 8 through 11 interact, so repeat them until the results are as good as you can get them.

26.11.4 Using Vintage Receivers

Connect a speaker, preferably the same impedance as the output impedance of the receiver. Some receivers have a 600 Ω and 3.2- or 4 Ω output. An 8 Ω speaker is fine — connect it to the low impedance output. Do not operate the receiver without a speaker, however, as the audio output transformer could be damaged by high voltage transients with no load. Alternatively, plug in a pair of headphones, keeping in mind that old receivers usually have high impedance headphone outputs and new headphones are usually low impedance. They'll work fine, but the volume may be considerably lower with the newer headphones.

If you're going to use the receiver in conjunction with a transmitter, you need to be able to mute the receiver while you're transmitting — otherwise, you'll end up with copious feedback from the receiver. Most receivers have mute terminals — some mute with a closed switch, others mute on open. Figure out which method your receiver and transmitter use. You'll need a relay if the receiver mute arrangement doesn't match that of the transmitter.

Some receivers — such as the older Hammarlund Super Pros and pre-WWII Hallicrafters models — use the mute terminals to open the B+ when putting the receiver in standby mode. This is extremely dangerous with 300 V or so on exposed terminals! An easy modification will save you from an almost certain shock. Open up the receiver and remove the wires from the standby terminals. Solder them together and insulate the connection with electrical tape or a wire nut. Better, solder them to an unused, ungrounded terminal if there's one handy. Next, examine the RF gain control and notice that one terminal is probably grounded. Cut this wire and solder a 47 kΩ resistor between ground and the RF gain control terminal. Connect a pair of wires from the terminals of the mute connection across the 47 kΩ resistor just installed. Now, with the mute terminals open the RF gain is all the way down so the receiver is essentially muted. Short the terminals to receive. The voltage here is low and not dangerous.

Next, connect an antenna and antenna relay in the same manner and tune the bands. You'll find that the best fidelity from the receiver occurs at its maximum bandwidth. The crystal filter, if fitted, can help notch out heterodynes, as can tuning the receiver slightly higher or lower. The bandspread control can be used to fine tune. Now, just enjoy using your classic, vintage equipment!

26.11.5 Plastic Restoration

Sometimes an old radio has a meter lens or dial face that is badly damaged. If it is cracked, there isn't much you can do but find a replacement. If it is just scratched, you have a good chance at fully restoring it.

First, remove the meter lens from the meter movement if possible. Most just snap on. To avoid damaging the very delicate needle and meter movement, place it in a protected area and be sure metal filings cannot get to the movement's magnet.

Make sure the scratch hasn't gone all the way though the plastic to become a crack, although even a deep scratch can be buffed out. You will need various grades of wet/dry sandpaper. Obtain sheets of 320, 400 and 600 grit (600 is the finest). One sheet will last a long time. Cut a piece about ½ inch by 2 inches and fold it in half. Start with 320 and gently sand in *one* direction only, over the scratch. This is very tedious and will take a while before you sand away enough plastic to get through the scratch.

Once the scratch is removed, it is time to start reconditioning the plastic. Again sanding in one direction, use the 400 and finally the 600 to completely remove all traces of the earlier sanding scratches. The 600 should leave almost a powdered effect, but the plastic will still be hazy and opaque. Sand a little more with the 600 just to be sure all traces of any sanding scratches are gone completely.

To remove the haze you will need a polishing compound called Novus #2 (**www.novuspolish.com** or hobby stores) and another compound called Novus #1, which is a plastic shiner and static remover. This will be important to remove the static on the meter cases when you finish. Static causes the needle to react strangely and erratically.

Start with just a drop of NOVUS #2 and a soft cloth such as the disposable shop towels found at auto parts stores for polishing. Cut a 2 inch by 3 inch piece and start polishing. Continue for a long time and bit by bit the haze will disappear — you will be left with a perfectly clear lens.

FRONT PANEL RESTORATION

The most important part of a radio restoration, cosmetically, is the front panel. The case should also look good, but the front panel is the highlight of the radio and needs the most attention. Usually, paint on a radio from the '50s and '60s is well oxidized and there may be some fine scratches as well.

Scratches can be touched up by using enamel model car paint. Buy a bottle of gloss white, gloss black, and the color closest to the panel you have. Into a small paper cup put a drop or two of the stock color. Using the white or the black, stirring in a small drop at a time, lighten or darken the color until it most closely matches the panel. Use a model paint brush with the finest tip you can find to fill just the scratch and not get it on the rest of the paint. Remove excess with a Q-tip. Let it dry completely.

Remove the panel completely, if possible, or at least remove all the knobs. You will work on them individually later. If the panel was originally a gloss panel, you are in luck. If it is a wrinkle finish or flat, this technique might not work for you — those will be addressed later.

Use the Novus #2 compound to remove a few microns of paint — just the oxidized layer. Place a drop on the panel, and with a soft cloth or shop towel, start working the Novus into the paint. You just want to remove the oxidation, so don't rub too hard. Be careful when working on paint that is a second layer above a base paint. It can be very thin and removed with the Novus. After a small amount of gentle polishing, get another cloth or towel and wipe off the panel, which should appear just as it did when it came from the factory.

If your panel is a wrinkle finish, you cannot use the Novus. Use a gentle soapy type cleaner and careful brush the ridges and peaks of the finish to remove layers of dirt and smoke. Sometimes cleaners like 409 and Simple Green will work very well, but be cautious that it doesn't take off the lettering. Be very careful around the lettering. One way to work it in is with a plumber's acid brush with the bristles cut very short. The bristles will get down into the crevices of the wrinkle finish and clean it. Once the panel is clean, it can be shined up a bit with some WD-40. This also works very well on wrinkle finish cabinets. Just lightly brush it on and remove it with a rag. This can collect dust but gives a nice wet finish to the wrinkle finish paints.

KNOBS

Most knobs are plastic or have metal plates around the bottom. Use the Novus #2 to shine up the flat parts of the knob. Knobs with flutes on the sides are very tedious to clean, but look beautiful once restored. Take a fine pick and drag it down each flute to scrape out the accumulated dirt. Once cleaned, the Novus can be used to shine up the flutes as well. On knobs with metalized bases, the bases are also shined up with the Novus. Make sure you don't polish off any markings on the trim bases. Sometimes the knobs will have white or red lines in the tops. Those can be filled with model paint to restore them to full beauty.

HOLES IN FRONT PANELS

It may be most disconcerting to find someone has drilled a hole in the panel for one reason or another. Extra holes greatly devalue the rig because to properly restore it, you have to find a replacement panel. Panels with wrong holes can be salvaged with quite nice results by using JB Weld epoxy metal filler (**www.jbweld.com**).

First remove the panel. If the hole is a considerable distance from the other knobs, repairs can be done on the radio but it is not advised. You will need a special tape called Kapton tape. This is a high temperature polymide film tape that is widely available but not cheap. A little tape lasts a long time.

You will need a piece of tape slightly larger than the hole. Other types of tape may work but won't give as smooth a finish. If the panel is wrinkle finished, you may want a rough finish tape such as masking tape. The key is a tight fit to the front of the panel. Place a piece of tape across the hole on the front of the panel and seal it securely all around the hole.

Lay the panel on a flat surface tape-down so the tape will be held flat and not bulge at all. Mix a suitable amount of the JB Weld epoxy and flow it into the hole from the rear of the panel. Stir it while it is still very wet and pliable to make sure no air bubbles are in the hole. Let it cure securely overnight so the hole will be filled and rock hard.

Flip the panel over and remove the tape. If you have secured the tape well enough, none of the epoxy will have been drawn out onto the front surface. There should be a flat filled area exactly level with the rest of the panel.

Now you can touch up the repaired hole with an exact match of spray paint. Mask off the rest of the panel so you don't get any on the lettering. You can also use the model paint method, described for scratch repair. Once painted, the offending hole should be virtually invisible.

BROKEN PLASTIC

Sometimes a microphone or other item will have a chunk of the plastic broken out of it. It is most unsightly and usually is the reason for discarding the item. But, using our repair technology, it can be saved and fully restored as in this example of repairing a damaged microphone case. This procedure will work well on broken Bakelite cases too.

Once again, we will use JB Weld Epoxy and Kapton tape. Remove the microphone elements and switch from the case along with the cord and anything else not plastic. Clean the area around the broken part well. With the Kapton tape, make a backing area on the inside of the case to form a backing for the epoxy. The tape will hold it well and not deform.

Mix some JB Weld and pour it into the space around the break. Fill *higher* than the surrounding plastic. This will be difficult as the fill area will not be level. It may take a couple of fills to build up the area past the level of the surrounding plastic. Once cured and very hard (wait at least 24 hours), you now can begin to file the epoxy. File it down until it is nearly level with the plastic and then switch to sandpaper. Carefully sand the epoxy so it is exactly flush with the original plastic, and make sure the shape is correct. You can always add more JB Weld if too much is filed off.

Once sanded flush, finish the sanding with fine sandpaper (400 and 600 grade). The JB Weld will shine just like the original plastic, but will be the wrong color. Spray paint the entire case with a color close to the original. Once painted, the repair will be virtually undetectable!

26.12 References

Bartlett, J., "Calculating Component Values," *QST*, Nov. 1978, pp. 21 – 28.

Bradshaw, N., "Some Basics of Equipment Servicing — Part 4," *QST*, Mar. 1982, pp. 40 – 44.

Carr, J., *How to Troubleshoot and Repair Amateur Radio Equipment* (TAB Books, 1980).

Collins, G., "Some Basics of Equipment Servicing — Part 2," *QST*, Jan. 1982, pp. 38 – 41.

Collins, G., "Some Basics of Equipment Servicing — Part 3," *QST*, Feb. 1982, pp. 40 – 44, Feedback May 1982, p. 43.

DeMaw, D., "Some Basics of Equipment Servicing — Part 1," *QST*, Dec. 1981, pp. 11 – 14.

DeMaw, D., "Understanding Coils and Measuring their Inductance," *QST*, Oct. 1983, pp. 23 – 26.

Eiselman, M., "Troubleshooting Radios," *QST*, May 2009, pp. 30 – 33.

Goodman, R., *Practical Troubleshooting with the Modern Oscilloscope* (TAB Books, 1979).

Haas, A., *Oscilloscope Techniques* (Gernsback Library, 1958).

Lenk, J., *Handbook of Electronic Test Procedures* (Prentice-Hall, 1982).

Lenk, J., *Encyclopedia of Circuits and Troubleshooting Guide, Vol 1-4* (McGraw-Hill, 1993 – 1997).

Loveday, G., and Seidman, A., *Troubleshooting Solid-State Circuits* (Wiley, 1981).

Smith, C., *Test Equipment for the Radio Amateur*, 4th edition (RSGB, 2011).

Contents

Chapter 27 — Online Content

Articles

- Building Contest Scores by Killing Receive Noise — Parts 1 and 2, by Jim Brown, K9YC
- Can Home Solar Power and Ham Radio Coexist?, by Tony Brock-Fisher, K1KP
- Tracking RFI with an SDR One Source at a Time, by Alan Higbie, KØAV
- TV Channel, Amateur Band, and Harmonic Chart
- What to Do if You Have an Electronic Interference Problem — *CEA Handbook*

Projects

- A Home-made Ultrasonic Power Line Arc Detector and Project Update, by Jim Hanson, W1TRC
- A Simple TRF Receiver for Tracking RFI, by Rick Littlefield, K1BQT
- Active Attenuator for VHF-FM by Fao Eenhoorn, PAØZR (article and template)
- Handheld Loop Antenna for RFI Location, by Gary Johnson, NA6O
- RF Sniffer Construction Notes, by Mark Kupferschmid, AC9PR
- Simple Seeker by Dave Geiser, W5IXM
- Tape Measure Beam for Power Line Hunting, by Jim Hanson, W1TRC

Chapter 27

RFI and EMC

This chapter begins with an overview of RF Interference (RFI)-the regulations that manage it, important definitions, and key technical elements. These introductory sections help us understand what interference is along with our responsibilities and privileges as licensed amateurs.

The second part of this chapter is about evaluating and troubleshooting the problems associated with interference. It includes a discussion on identifying the causes and types of interference, RFI-related sources, and specific examples of common systems that are affected by RFI.

A new third section introduces topics associated with electromagnetic compatibility (EMC) that extend beyond RFI. The new material in this edition focuses on transients and protective devices.

The material in this chapter may provide enough information for you to deal with your RFI problem, but if not, the ARRL website offers extensive resources on RF interference at **www.arrl.org/radio-frequency-interference-rfi**. Many topics covered in this chapter are covered in more detail in the *ARRL RFI Book* from a practical amateur perspective.

For this edition, a new section showing how to use a portable SDR to locate sources of RFI was provided by Alan Higbie, KØAV. Alan Applegate, KØBG, also updated the material on RFI in electric and hybrid-electric vehicles.

What was once primarily a conversation about "interference" has expanded to include power systems; incidental, intentional, and unintentional radiators; bonding and grounding; managing transients; and many other related topics and phenomena. These are all grouped together under the general title of *electromagnetic compatibility (EMC)*. The scope of EMC includes all the ways in which electronic devices interact with each other and their environment.

The general term for interference caused by electrical signals or fields is *electromagnetic interference* or *EMI*. You will encounter this term in professional literature and standards. The most common term for EMI involving amateur radio is *radio frequency interference (RFI)* and when a television or video display is involved, *television interference (TVI)*. RFI is the term used most commonly by amateurs today, and that's what we'll use in this chapter.

A few words will be used to apply in a general sense, so they are defined here:

- Neighbor: a person or business either affected by RFI from your station or generating RFI that affects your station.
- Signal: any kind of RF energy, whether it is radiated as an electromagnetic wave or conducted as voltage and current. It doesn't matter if the energy is created intentionally or not.
- Noise: an undesired signal, regardless of what it may or may not sound like as audio and whether or not it is wide- or narrow-band.
- Interference: disruption or degradation of communication by an unwanted signal.

Throughout this chapter you'll also find references to "Ott," meaning the book *Electromagnetic Compatibility Engineering* by EMC consultant Henry Ott, WA2IRQ. EMC topics are treated in far greater depth in Ott's book than is possible in this *Handbook*. Readers interested in the theory of EMC, analysis of EMC mechanisms, test methodology, and EMC standards should be able to borrow a copy through a library.

How To Use this Chapter

RFI is a very broad topic, and there are too many variations of it to provide a "cookbook" approach that can be followed for even the most-common causes. Instead, this chapter provides enough background for you to understand the various ways in which RFI occurs and the tools for dealing with it, no matter what the exact type of RFI it may be. There are special sections for the widespread RFI from power lines and for the unique topics of RFI in mobile stations.

If you are dealing with a case of RFI and are already familiar with the general nature and tools of managing RFI, the sections on specific types of RFI or devices will be a good place to start. If you are just getting started and need background, read the first few sections so you will understand the terms and concepts before beginning your RFI detective work.

RFI is a very interesting part of amateur radio because it touches so many different aspects of the service — you can get a sense of how "connected" RFI is from the many sidebars in the chapters. It affects both the technical and operating aspects as well as all types of electronics fundamentals. Plus, it really develops and exercises your troubleshooting and analysis skills!

27.1 FCC Rules and Regulations

Before presenting the technical elements of RFI, here is an overview of the FCC rules and regulations governing interference. (See the sidebar Regulations and Interference Complaints for references to more detailed information.) This section summarizes the three sets of rules that apply to RFI affecting amateur stations.

The Amateur service is regulated by FCC Part 97. Part 97 rules are available at **www.arrl.org/part-97-amateur-radio.** To be legal, the amateur station's signal must meet all Part 97 technical requirements, such as for spectral purity and power output.

In the United States most unlicensed electrical and electronic devices are regulated by Part 15 of the FCC's rules. These are referred to as "Part 15 devices." Most RFI issues reported to the ARRL involve a Part 15 device. Some consumer equipment, such as certain medical and lighting devices, is covered under FCC Part 18, which pertains to ISM (Industrial, Scientific and Medical) devices. Although some wireless devices do operate in the ISM bands, all such devices that communicate information are not regulated by Part 18, but by other regulations such as Part 15.

As a result, it isn't surprising that most interference complaints involve multiple parts of the FCC rules. (The FCC's jurisdiction does have limits, though — ending below 9 kHz.) It is also important to note that each of the three parts (15, 18 and 97) specifies different requirements with respect to interference, including absolute emissions limits and spectral purity requirements. The FCC does not specify *any* RFI immunity requirements, although the Food and Drug Administration (FDA) does specify immunity levels for some medical devices. Non-radio consumer devices such as wired telephones, audio systems, and wired alarm systems therefore receive no FCC protection from a legally licensed transmitter, including an amateur transmitter operating legally according to Part 97.

PROTECTIONS FROM INTERFERENCE

Amateur radio, being a licensed service, is protected from interference to its signals from unlicensed devices, from spurious emissions from licensed and unlicensed transmitters, and from any transmission by a transmitter that is secondary to the Amateur service. Amateur radio is *not* offered protection from the fundamental signal from transmitters licensed to a service that is primary to amateur radio, such as radar signals operated by the military on the 420 – 450 MHz band. For example, consider RFI from an amateur transmitter's spurious emissions, such as harmonics, that meet the requirements of Part 97 but are still strong enough to be received by nearby FM receivers. The FM receiver itself is not protected from interference from the fundamental signal of nearby licensed transmitters, such as amateur radio, under the FCC rules. However, within its service area the licensed FM broadcast station's signal is protected from harmful interference caused by spurious emissions from other licensed transmitters. In this case, the amateur transmitter's interfering spurious emission would have to be eliminated or reduced to a level at which harmful interference has been eliminated.

27.1.1 FCC Part 97 Rule Summary

While most interference to consumer devices may be caused by a problem associated with the consumer device as opposed to the signal source, all amateurs must still comply with Part 97 rules. Regardless of who is at fault, strict conformance to FCC requirements, coupled with a neat and orderly station appearance, will go far toward creating a good and positive impression in the event of an FCC field investigation. Make sure your station and signal exhibit good engineering and operating practices.

The bandwidth of a signal is defined by §97.3(a)(8), while the paragraphs of §97.307 define the technical standards amateur transmissions must meet. Paragraph (c) defines the rules for interference caused by spurious emissions. As illustrated in **Figure 27.1**, modulation sidebands outside the necessary bandwidth are considered *out-of-band* emissions, while harmonics and parasitic emissions are considered *spurious emissions*. Paragraphs (d) and (e) specify absolute limits on spurious emissions, illustrated in **Figure 27.2**. Spurious emissions must not exceed these levels, whether or not the emissions are causing interference. Even if you meet the limits for spurious emissions, if they are causing interference, it's your responsibility to clean them up.

Strict observance of these rules can not only help minimize interference to the amateur service, but other radio services and consumer devices as well.

Regulations and Interference Complaints

Over the years, this book's RFI chapter has grown from the traditional treatment of TVI (interfering with an analog television receiver) and power line noise (still as big a problem as ever) to cover a wide range of interference and related issues. The subject of managing interference and the associated rules and regulations outgrew many technical sections! The information about regulations and handling interference complaints has been collected and published separately on the ARRL web page as a pair of stand-alone documents. They are available to everyone as a public resource at **arrl.org/rfi**.

FCC Rules and Regulations for Interference includes the FCC regulations and definitions related to interference both to and from your station.

Managing an RFI Complaint discusses how to prepare for and respond to RFI complaints. The various steps of getting your own station in order and obtaining technical assistance are presented along with basic "diplomacy" skills of assessing the problem and suggesting your responses.

These two papers are excellent resources when you are confronted with an interference problem to a neighbor's equipment and are wondering "What do I do now?" The information in the document is based on the experiences of ARRL Lab staff in assisting amateurs with RFI problems.

The material on RFI in this chapter of the *ARRL Handbook* focuses on the technical and practical aspects of interference: what causes it, how can you tell what the source is, techniques for preventing and eliminating it, and good practices that mitigate against it. You can then use these techniques and practices to deal with RFI issues.

When Is a Signal Noise?

In this chapter, the term "signal" can mean any kind of RF energy in a circuit, on a wire, or inside a cable. A signal can be conducted on or in a cable, and it can be radiated. An *interfering signal* disrupts or degrades the reception of a desired signal or disrupts the operation of some electronic device. The term "noise" doesn't necessarily mean buzzing power-line pulses or hissing, cracking static. In a very general sense, "*Noise is any electrical signal present in a circuit other than the desired signal*" (page 3 of Ott). So, noise is any signal you *don't* want, whether it sounds like static or not.

Noise could start out as an intentionally transmitted *signal*, be picked up by a cable shield, and become noise when it causes a device to misbehave. It doesn't matter if the noise is random electrical energy or an intentionally created signal. Your transmitted signal becomes noise when it gets into your home entertainment audio!

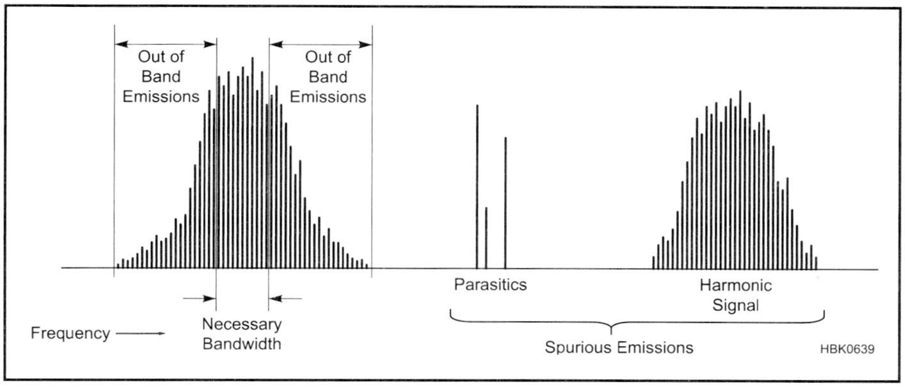

Figure 27.1 — An illustration of out-of-band versus spurious emissions. Some of the modulation sidebands are outside the necessary bandwidth. These are considered out-of-band emissions, not spurious emissions. The harmonic and parasitic emissions shown here are considered spurious emissions; these must be reduced to comply with §97.307.

27.1.2 FCC Part 15 Rule Summary

In the United States, most unlicensed devices are regulated by Part 15 of the FCC's rules. While understanding these rules doesn't necessarily solve an RFI problem, they do provide some important insight and background on interference to and from a Part 15 device.

There are literally thousands of Part 15 devices with the potential to be at the heart of an RFI problem. A Part 15 device can be almost anything not already covered in another Part of the FCC rules. In fact, many Part 15 devices may not normally even be associated with electronics, RF, or in some cases electricity. While televisions, radios, telephones, and even computers obviously constitute a Part 15 device, the rules extend to anything that is capable of generating RF, including electric motors and consumer devices such as baby monitors, wireless microphones and intercoms, RF remote controls, garage door openers, etc. With so many Part 15 consumer devices capable of generating and responding to RF, it isn't surprising therefore that most reported RFI problems involving amateur radio also involve a Part 15 device.

PART 15 SUMMARY

FCC's Part 15 rules pertain to unlicensed devices and cover a lot of territory. Although reading and understanding Part 15 can appear rather formidable — especially at first glance — the rules pertaining to RFI can be roughly summarized as follows:

• Part 15 devices operate under an unconditional requirement to not cause harmful interference to a licensed radio service, such as amateur radio. Even if the manufacturer meets the Part 15 limits for radiated and conducted emissions, if harmful interference occurs, the operator of the Part 15 device is responsible for eliminating the interference.

• Part 15 devices receive no protection from interference from a licensed radio service. There are no FCC rules or limits with regard to Part 15 device RFI immunity.

When is the operator of a licensed transmitter responsible for interference to a Part 15 device?

• The rules hold the transmitter operator responsible only if interference is caused by spurious emissions such as a harmonic that exceeds the Part 97 limits. An example would be a illegal harmonic from an amateur's transmitter interfering with a baby monitor. In this case, the transmitter is generating interfering RF energy beyond its permitted levels. A cure must be installed at the transmitter.

• The transmitter operator is not responsible when a Part 15 device is improperly responding to a legal and intentional output of the transmitter. An example of this case would be interference to a weather radio by the strong-but-legal signal from a nearby amateur transmitter. In this case, the Part 15 device is at fault and the cure must be installed there. It is important to note that this situation is typical of most

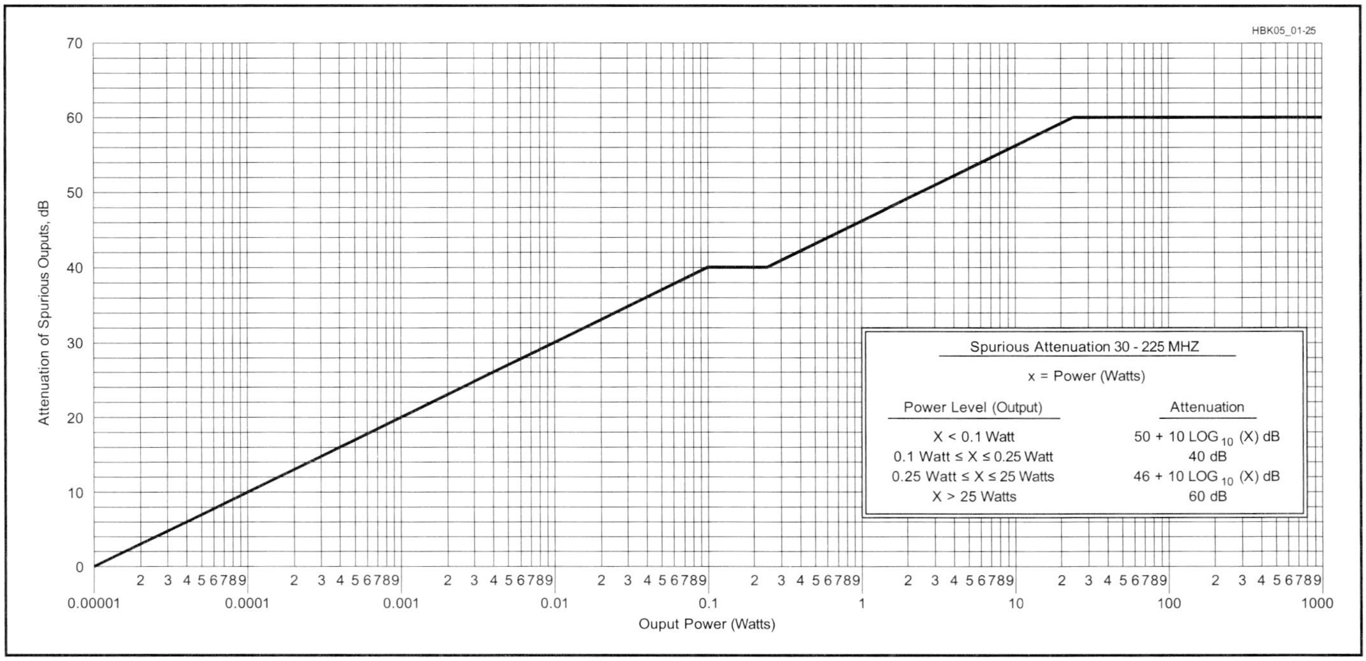

Figure 27.2 — Required attenuation of spurious outputs, 30-225 MHz. Below 30 MHz, spurious emissions must be suppressed by 43 dB for amateur transmitters installed after January 1, 2003.

interference to Part 15 devices.

Even though the causes and cures for these situations are different, the common element for all three situations is the need for personal diplomacy in resolving the problem.

TYPES OF PART 15 DEVICES

Part 15 describes three different types of devices that typically might be associated with an RFI problem. A fourth type of device, called a *carrier current device*, uses power lines and wiring for communications purposes. As we'll see, the rules are different for each type.

Intentional Emitters — Intentionally generate RF energy and radiate it. Examples include garage door openers, cordless phones, and baby monitors.

Unintentional Emitters — Intentionally generate RF energy internally, but do not intentionally radiate it. Examples include computers and network equipment, superheterodyne receivers, switchmode power supplies, and TV receivers.

Incidental Emitters — Generate RF energy only as an incidental part of their normal operation. Examples include power lines, arcing electric fence, arcing switch contacts, dc motors, and mechanical light switches.

Carrier Current Devices — Intentionally generate RF and conduct it on power lines and/or house wiring for communications purposes. Examples include Powerline and X.10 networks, Access or In-House Broadband-Over-Power-Line (BPL), campus radio-broadcast systems, and other power-line communications devices.

27.1.3 FCC Part 18 Rule Summary

Some consumer devices are regulated by Part 18 of the FCC Rules, which pertains to the Industrial, Scientific and Medical (ISM) bands. These devices convert RF energy directly into some other form of energy, such as heat, light, or ultrasonic sound energy. Some common household Part 18 devices therefore include microwave ovens, electronic fluorescent light ballasts, CFLs, and ultrasonic jewelry cleaners. (Note that LED bulbs are covered under Part 15 because of the process by which they generate light.)

Recently, indoor grow light ballasts and related equipment have been causing a lot of RFI. Those sold to consumers are required meet lower Part 18 rules than the higher emissions limits that apply to Part 18 devices sold to industrial and commercial customers. There have been reports of industrial Part 18 devices being sold at home-improvement and other consumer outlets. The ARRL is working with the FCC and manufacturers to resolve this issue.

Consumer Part 18 devices are generators of RF — but not for communications purposes — and can cause interference in some cases. However, there are no rules that protect them from interference. The purpose of Part 18 is to permit those devices to operate and to establish rules prohibiting interference.

From the standpoint of an RFI problem, although Part 18 devices can have unlimited emissions with various bands designated for ISM devices, the Part 18 rules aren't much different from Part 15. As with a Part 15 device, a Part 18 device is required to meet specified emissions limits outside of the ISM bands. Furthermore, it must not cause harmful interference to a licensed radio service operating in other than an ISM band. (The 902 MHz and 2.4 GHz bands are ISM bands, so if interference were to occur from an ISM device, the licensed user would not be protected in these bands). This gets a bit confusing, though, because some Part 15 devices also operate in the "ISM" bands, but as such, they are still Part 15 devices, so licensed services are protected.

27.2 Elements of RFI

This section defines basic terms and concepts associated with RFI that amateurs have to deal with. From these definitions, the chapter will proceed to address the various types of RFI and methods for dealing with them.

27.2.1 Source-Path-Victim

All cases of RFI involve a *source* of radio frequency energy, a device that responds to the electromagnetic energy (*victim*), and a transmission *path* that allows energy to flow from the source to the victim. Sources include radio transmitters, receiver local oscillators, computing devices, electrical noise, lightning, and other natural sources. Note that receiving unwanted electromagnetic energy does not necessarily cause the victim equipment to function improperly.

A device is said to be *immune* to a specific source if it functions properly in the presence of electromagnetic energy from the source. In fact, designing devices for various levels of immunity is one aspect of electromagnetic compatibility engineering. Only when the victim experiences a *disturbance* in its function as a consequence of the received electromagnetic energy does RFI exist. In this case, the victim device is *susceptible* to RFI from that source.

There are several ways that RFI can travel from the source to the victim: *radiation*, *conduction*, *inductive coupling*, and *capacitive coupling*. *Radiated RFI* propagates by electromagnetic radiation from the source through space to the victim. *Conducted RFI* travels over a physical conducting path between the source and the victim, such as wires, enclosures, ground planes, and so forth. Inductive coupling occurs when two circuits are magnetically coupled. Capacitive coupling occurs when two circuits are coupled electrically through capacitance. Typical RFI problems you are likely to encounter often include multiple paths, such as conduction and radiation. (See the section Shields and Filters, also Ott, sections 2.1 – 2.3.)

Many instances of RFI are a combination of radiated and conducted RFI. Conducted RFI exits the source as current on one or more conductors connected to the source. The conductors act as transmitting antennas for the common-mode current. A conductor connected to the victim then picks up the radiated noise as common-mode current which is conducted to the victim.

27.2.2 Differential-Mode vs Common-Mode Signals

The path from source to victim almost always includes some conducting portion, such as wires or cables. RF energy can be conducted directly from source to victim, be conducted onto a wire or cable that acts as an antenna where it is radiated, or be picked up by a conductor connected to the victim that acts like an antenna. When the noise is traveling along the conducted portion of the path, it is important to understand the differences between *differential-mode* and *common-mode* signals (see **Figure 27.3**).

Differential-mode signals usually have two easily identified conductors. In a two-wire transmission line, for example, the signal leaves the generator on one wire and returns on the other. When the two conductors are in close proximity, they form a transmission line and the two signals have opposite polarities as shown in Figure 27.3A. Most desired signals, such as the TV signal inside a coaxial cable or an Ethernet signal carried on CAT5 network cable, are conducted as differential-mode signals.

A common-mode circuit consists of two or more wires in a multi-wire cable acting as if they were a single path as in Figure 27.3B. Common-mode circuits also exist when the outside surface of a cable's shield acts as a conductor as in Figures 27.3C and 27.3D. (See the chapter on **Transmission Lines** for a discussion about isolation between the shield's inner and outer surfaces for RF signals.) The return path for a common-mode signal often involves earth ground. Common-mode currents are the net result of currents for which there is not an equal-and-opposite current in the same conductors or group of conductors

Figure 27.3D illustrates the case for a signal and its return path both enclosed in a separate shield. This is a common arrangement for digital data or control signals in a multiconductor cable with a dedicated signal return or signal ground wire. The cable shield should be kept separate from all data and data return connections.

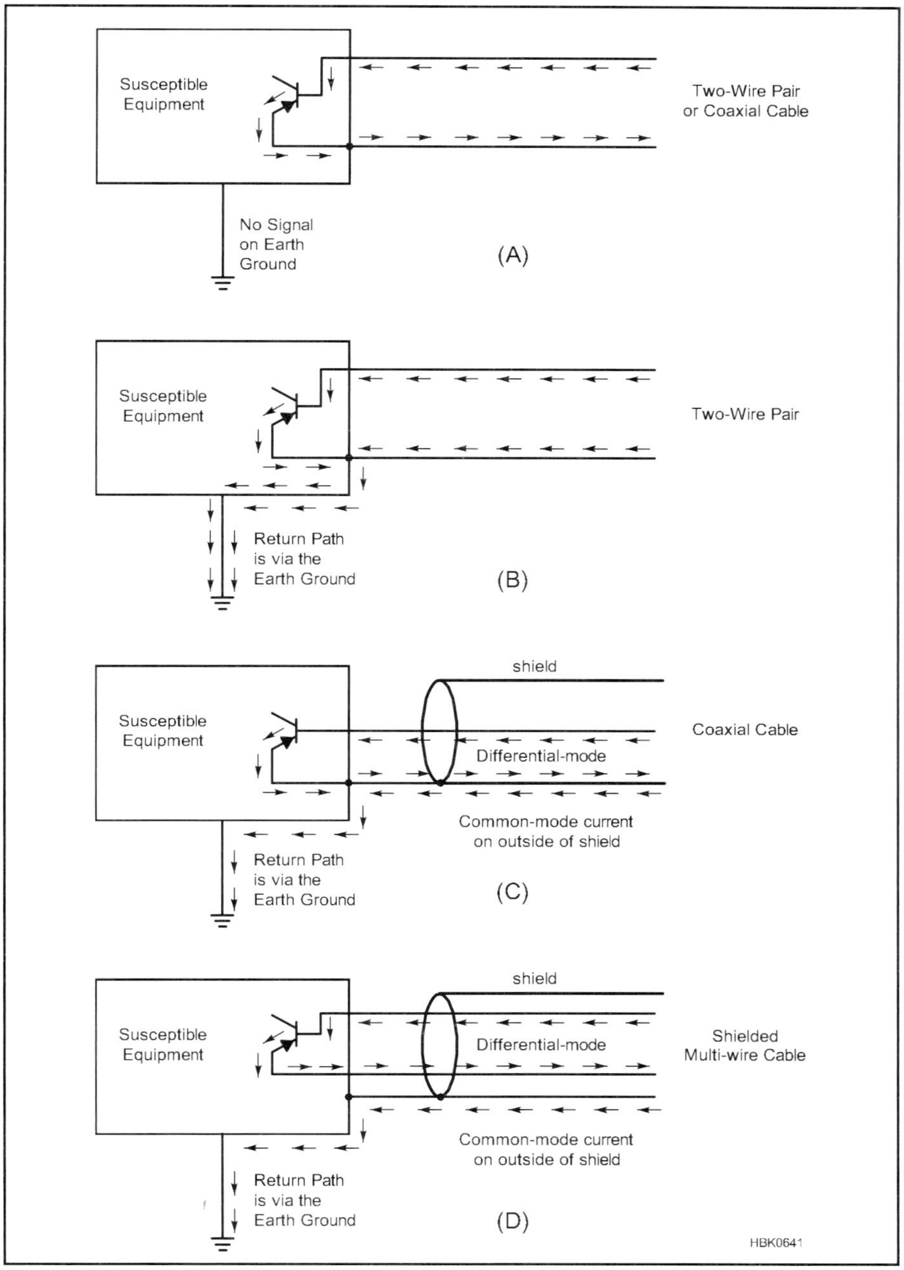

Figure 27.3 — Typical configurations of common-mode and differential-mode current. The drawing in A shows the currents of a differential-mode signal, while B shows a common-mode signal with currents flowing equally on all of the source wires. In C, a common-mode signal flows on the outside of a coaxial cable shield with a differential-mode signal inside the cable. In D, the differential-mode signal, such as a data signal with a dedicated return circuit, flows on the internal wires, while a common-mode signal flows on the outside of the cable shield.

External noise generated by electronic equipment and RFI caused to electronic equipment is often associated with common-mode current on a cable shield that is improperly connected. For example, consider what happens when the cable shield in Figures 27.3C and 27.3D is not connected to the enclosure but enters the device before connecting to a circuit's common or ground connection. In this case, any noise generated by the circuit has a path out of the enclosure and onto the surface of the cable shield where it is then radiated. Similarly, any noise or signals picked up by the cable shield are conducted into the equipment, where it is connected to the circuit and can disrupt normal operation or interfere with the desired signals. This is discussed more in this chapter's section Elements of RFI Control.

COMMON-MODE AND POWER WIRING

There is an important difference between the preceding definition and what ac power companies consider common-mode. Corcom is a major manufacturer of power line RFI filters and related components. (See **www.te.com/usa-en/products/emi-filters/power-line-filters.html**.) Common-mode is defined in their Product Guide's appendix on "Understanding Insertion Loss," as "signals present on both sides of the line (hot and neutral) referenced to ground." The RF definition of common-mode in this book includes signals on the ground conductor, as well as hot and neutral.

Because the ground connection through a power-line RFI filter is typically a direct connection between input and output, a typical ac-line filter may be ineffective against common-mode RF flowing on the ground conductor. To block RF on the ground conductor of a three-wire ac power cord or cable, the usual remedy is to wind the ac power cord on a ferrite core. This places an impedance in all three conductors, not just hot and neutral.

27.2.3 Differential- and Common-Mode Paths

As shown in **Figure 27.4**, RFI's path from source to victim almost always includes some conducting portion, such as wires or cables. In addition, the interfering signal may be radiating from conductors and picked up by an antenna or by cables unintentionally acting as antennas. Four common paths include:

- Conducted directly from source to victim,
- Conducted on a wire or cable acting as an antenna where it is radiated, then received by the victim's antenna,
- Radiated then picked up by a conductor connected to the victim,
- Inductive and capacitive coupling when the source and victim are very close together.

Many RFI problems are caused by con-

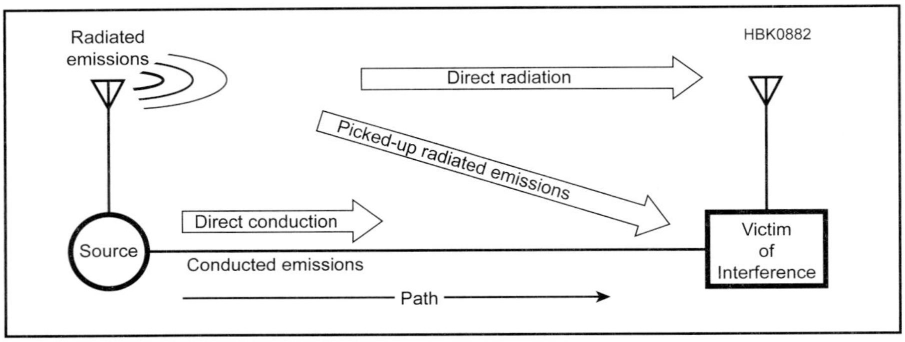

Figure 27.4 — A block diagram showing every RFI problem involves three elements — a source, a path, and a victim. Fixing the problem requires removing one or more of them.

ducted emissions which leave the source as common-mode signals on wiring connected to the source. The interfering signal may travel directly from the source to the victim, or it might be radiated by the wiring and picked up by wiring connected to the victim. The same design or production problems that enable conducted emissions can also work in reverse, allowing the signal that has been picked up to be conducted into the victim where it causes the interference.

The path also includes the interfering signal's return path back to the source, which is different for differential- and common-mode signals. A differential-mode signal's return path is almost always in the same cable. The return path for common-mode signals often involves earth ground, or even the chassis of equipment if it is large enough to form part of an antenna at the frequency of the RFI.

27.2.4 Grounding and Bonding

An electrical ground is not a huge sink that somehow swallows noise and unwanted signals. Ground is a *circuit* concept, whether the circuit is small, like a radio receiver, or large, like the propagation path between a transmitter and AM/FM receiver. Ground refers to a shared reference voltage between circuits. Bonding refers to connections between pieces of equipment.

GROUNDING

While grounding is not a cure-all for RFI problems, ground is an important safety component of any electronics installation. It is part of the lightning protection system in your station and a critical safety component of your house wiring. Any changes made to a grounding system must not compromise these important safety considerations. Refer to the **Safe Practices** chapter for important information about safety grounding. The ARRL book *Grounding and Bonding for the Radio Amateur* goes into detail about ground systems for electrical safety, lightning protection, and managing RF voltages and currents.

Many amateur stations have several connections referred to as "grounds:" the required safety ground that is part of the ac wiring system, an earth connection for lightning protection, and shared connections between equipment, such as the negative terminal of a dc power supply. These connections can interact with each other in ways that are difficult to predict. Rearranging the station ground connections may cure some RFI problems in the station by changing the RF current distribution so that the affected equipment is at a low-impedance point and away from RF "hot spots." It is better to address the problem by implementing proper bonding within the station.

BONDING

Bonding refers to a connection intended to minimize potential (voltage) differences. The purpose of bonding in the amateur station is to minimize the potential difference between equipment and all elements of the ground system — ground rods, entry wiring for electrical power, telephone or data systems, cable or satellite TV systems, and amateur antennas. This minimizes voltage differences in the event of lightning surges. It also minimizes hum and buzz and reduces RFI resulting from voltage differences between pieces of equipment.

Creating a low-impedance connection between your station's equipment is easy to do and will help reduce voltage differences (and current flow) between pieces of equipment. Bonding is discussed in this book's **Assembling a Station** and **Safe Practices** chapters.

Bonding also reduces voltage differences between the ends of cable shields that are connected to different pieces of equipment. This voltage difference is effectively in series with the cable shield and can be added to the desired signals carried by the cable or cause common-mode RF to flow on the shield. The voltage difference can be at ac power, audio, or RF so bonding helps reduce RFI across a wide frequency range.

The conductors used for bonding should be heavy enough to have low inductance and resistance. The standard for commercial and military facilities is solid copper strap, but heavy stranded or solid wire (bare or insulated)

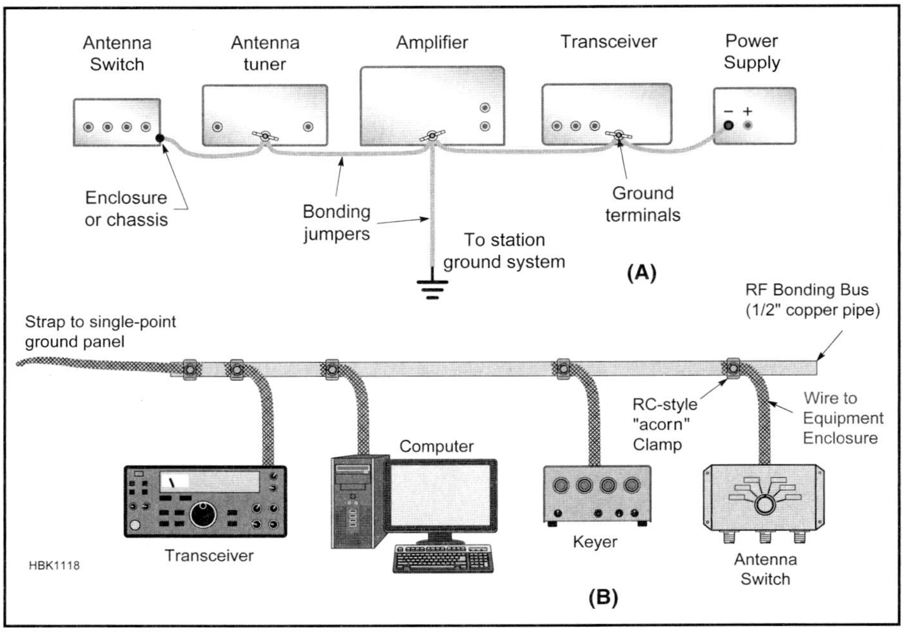

Figure 27.5 — Direct equipment-to-equipment bonding (A). Use heavy wire or strap connected to a single ground terminal or enclosure screw on each piece of equipment, including computers and related gear. The same approach can be used in home entertainment systems or with computers and accessories. Part B shows the common use of a bonding bus. This allows equipment to be added or removed without reworking other connections. The bonding bus works better with a ground reference plane.

Bonding and Chokes and RFI

Grounding and bonding in your station is also important for lightning safety and to help manage RF currents and voltages picked up by the wiring and cables. Bonding equipment in your station (see the **Assembling a Station** chapter) helps minimize common-mode RF current in the station that can cause improper operation. Make sure cable shields are connected properly and that RF current picked up from your transmitted signal by audio and power wiring is minimized. Learn how to apply ferrite RF chokes in sensitive spots. These are useful techniques in many different situations to eliminate interference.

RFI and End-Feeding in the Station

An end-fed antenna brought directly into the station is often adjusted to produce a current-maximum (low-impedance point) at the feed point. The feed point is often the output of an antenna tuner, so the feed point is actually *in* the station. If a resonant counterpoise is not attached to the tuner, the resulting current will couple to and flow on whatever conductors happen to be connected to or close to the enclosures of the transmitter and tuner, including ground system and signal connections. Common-mode current on the outside of an antenna feed line can act in the same way. To address this situation, make sure equipment is bonded together properly, as well as the ground system, and provide the necessary counterpoise for the end-fed wire or a common-mode choke for the antenna feed line (see the **Transmission Lines** chapter).

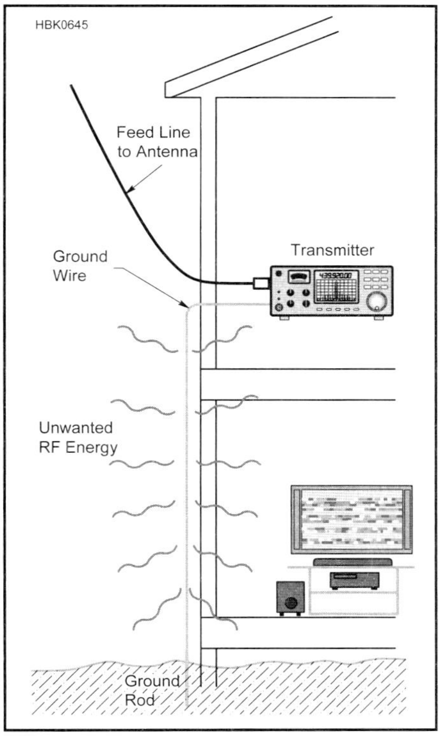

Figure 27.6 — An earth ground connection can radiate signals that might cause RFI to nearby equipment. This can happen if the ground connection is part of an antenna system, or if it is connected to a coaxial feed line carrying RF current on the outside of the shield.

works well. Flat woven braid will also work if kept dry and jacketed braid is available for use outdoors or in vehicles. Braid removed from coaxial cable is not recommended — leave the braid in the cable jacket and use it as a large wire.

As shown in **Figure 27.5**, equipment enclosures can be connected either directly together or by using a bonding bus or ground reference plane as shown in the chapter on **Assembling a Station**. Amateur radio and home entertainment equipment usually has a ground screw or terminal that can be used for bonding. Chassis connections for many computers can be found at the shell of video or data connectors.

LENGTH OF GROUND CONNECTIONS

The required ground connection for lightning protection between the station equipment and an outside ground rod is at least several feet long in most practical installations. (See the **Safe Practices** and **Assembling a Station** chapters for safety and lightning protection ground requirements.)

In general, however, a long connection to earth should be considered as part of an RFI problem since it will act as part of the antenna system. For example, should a long-wire HF antenna terminate in the station, a ground connection of *any* length is a necessary and useful part of that antenna and will radiate RF.

At VHF a ground wire can be several wavelengths long — a very effective antenna for any harmonics that could cause RFI! For example, in **Figure 27.6**, signals radiated from the required safety ground wire could very easily create an interference problem in the downstairs electronic equipment. Noise from the electronic equipment can also be picked up by the ground wire and carried back to the station as interference.

GROUND LOOPS, HUM, AND BUZZ

A *ground loop* is created by a continuous conductive path around a series of equipment enclosures. While this does create an opportunity for lightning damage and RF pickup, the ground loop itself is rarely a cause of problems at RF. The potential for problems created by loops can be minimized by proper bonding of equipment and by minimizing the loop area formed by cables and wiring.

Ground loops are usually associated with audio hum caused by coupling to power-frequency magnetic fields from transformers or motors and sometimes from coupling to high-current power wiring. The hum appears as a nearly pure sine wave at the frequency of the ac power system, 50 or 60 Hz. To avoid low-frequency ground-loop issues, use short cables that are the minimum length required to connect the equipment and bundle them together to minimize the area of any enclosed loop. Moving the cables away from the source of the magnetic field or reorienting the source or cables can reduce hum. Bonding equipment together to short-circuit a loop is sometimes effective.

Audio "buzz" is caused primarily by currents at harmonics of the ac power frequency. The current results from leakage in ac-powered equipment with rectifiers or switching circuits that conduct during parts of the ac power waveform. Buzz is addressed by bonding and by ensuring the ac power ground connections have a minimum of these voltages on them.

27.3 Tools for RFI Control

Building from the basic definitions and mechanisms of RFI in the previous section, we now turn our attention to the techniques for managing and controlling it. This section will become your "tool kit" for dealing with RFI.

Differential-Mode and Common-Mode Combinations

Most RFI problems involve a combination of differential-mode and common-mode signals or noise. The right type of filter is required for each type of signal. In many cases, a noisy device creates both differential-mode and common-mode currents due to various imbalances in the device and the wiring it is connected to. A susceptible device experiencing interference from a nearby transmitter can be the result of both common-mode and differential-mode signals being picked up on its wiring. Imbalance in ac or signal wiring usually results in both types of signals being created and coupled into and out of devices. For this reason, in many cases both common-mode and differential-mode filtering may be required, even on the same cable.

27.3.1 Breaking the Path

Even if you can't eliminate the source, by preventing the RFI-causing signals or noise from reaching the victim, the problem can be solved. As described above, the path may be conducting or radiated at different points along the way. Be aware there may be more than one path! Use the tools and techniques described in the following sections to "break the path."

Breaking the path between source and victim is often an attractive option, especially if either is a consumer electronics device. The path will involve one or more of the possibilities — radiation, conduction, or coupling. Determining the path can require analysis and experimentation in some cases. Obviously, you must know what the path is before you can break it. While the path may be readily apparent in some cases, more complex situations may not be so clear. Multiple attempts at finding a solution may be required, but identifying the path is crucial to successfully managing the problem.

When the interfering signal is traveling along the conducting portion of its path, it is important to understand whether it is as a differential- or common-mode signal. This determines what techniques will be effective against the signal *at that point in the path.*

Differential-mode techniques (a high-pass filter, low-pass filter, or a bypass capacitor across the ac power line, for example) do not attenuate common-mode signals. As another example, a differential-mode filter does not usually filter the ground lead in a three-wire ac cord or cable (see **Figure 27.7**), so any common-mode signals or noise present on the ground conductor of the ac wiring would be coupled directly through the filter. Similarly, a common-mode choke will not affect differential-mode signals. In practice, conducted emissions problems are more likely to be the result of common-mode than differential-mode signals.

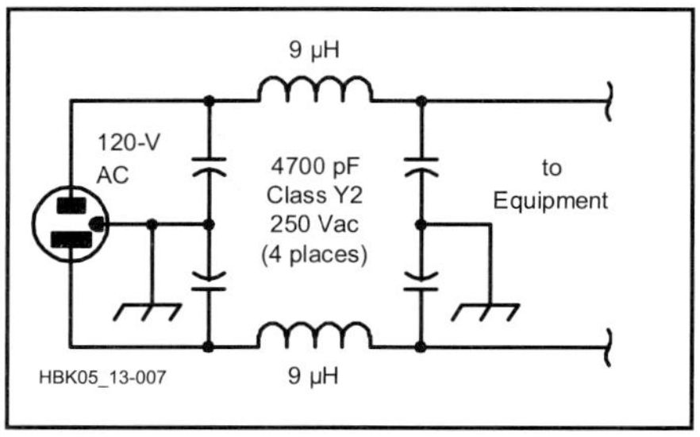

Figure 27.7 — A "brute-force" ac-line filter. Note there is no filtering on the chassis ground connections. See the text's discussion of differential- and common-mode paths.

27.3.2 Shielding

Shielding can be used to control radiated emissions — that is, signals radiated by wiring inside the device — or to prevent radiated signals from being picked up by signal leads in cables or inside a piece of equipment (direct detection). Shielding can also be used to reduce inductive or capacitive coupling, usually by acting as an intervening conductor between the source and victim.

Shields are used to set boundaries for radiated energy and to contain electric and magnetic fields. At RF, the small skin depth allows thin shields to be effective at these frequencies. (see the **RF Techniques** chapter). Thin conductive films, copper braid, and sheet metal are the most common shield materials for the electric field (capacitive coupling), and for electromagnetic fields (radio waves).

Thicker shielding material is needed for magnetic fields (inductive coupling) at low frequencies to minimize the voltage caused by induced current. At audio frequencies and below, the higher skin depth of common shield materials is large enough (at 60 Hz, the skin depth for aluminum is 0.43 inches) that special high-permeability materials such as steel or mu-metal (a nickel-iron alloy that exhibits high magnetic permeability) are required.

Maximum shield effectiveness usually requires solid sheet metal that completely encloses the source or susceptible circuitry or equipment. While electrically small holes generally do not affect shield effectiveness (fine-mesh screening makes good shielding at VHF and below, for example), seams can act as a slot antenna if they are a significant fraction of a wavelength long. In addition, mating surfaces between different parts of a shield must be conductive. To ensure conductivity, file or sand off paint or other nonconductive coatings on mating surfaces.

The effectiveness of a shield is determined by its ability to reflect or absorb the undesired energy. Reflection occurs at a shield's surface. In this case, the shield's effectiveness is independent of its thickness. Reflection is typically the dominant means of shielding for radio waves and capacitive coupling, but is ineffective against magnetic coupling. Most RFI shielding works, therefore, by reflection. Any good conductor will serve in this case, even thin plating.

Magnetic material is required when attempting to break a low-frequency inductive coupling path by shielding. A thick layer of high permeability material is ideal in this case. Although the near field of low frequency magnetic fields can extend for long distances, magnetic fields generally have their greatest effect at relatively short range. Simply increasing the distance between the source and victim may help avoid the expense and difficulty of implementing a shield.

Adding shielding may not be practical in many situations, especially for many consumer products that have non-conductive enclosures. Adding a shield to a cable can minimize capacitive coupling and RF pickup, but it has no effect on magnetic coupling. Replacing parallel-conductor cables (such as zip cord speaker wire) with twisted-pair is quite effective against magnetic coupling and also reduces electromagnetic coupling.

Additional material on shielding may be found in chapter 2 of Ott (see the References).

27.3.3 Filters

Filters can be very effective in dealing with RFI by blocking or suppressing the unwanted signal. Fortunately, filters are simple, economical, and easy to try. In many cases, filters can also be applied externally to the victim device, avoiding the need to open an enclosure. Solving an RFI problem without having to alter a device's internal wiring is almost always preferable to the alternative. (See the **Analog and Digital Filtering** chapter for more information on filters of all types.)

Most stand-alone filters are differential-mode filters for signals in a cable or connector. They may or may not affect common-mode signals depending on filter design. Be sure you know which type you want to use and how the filter will perform on the type of signal you are trying to reject. A common-mode cable filter, such as a choke, is designed not to affect differential-mode signals and noise in the cable, for example. (See the section below on Common-Mode Chokes.)

Filters vary in attenuation characteristics, frequency characteristics, and power-handling capabilities. The names given to various filters are based on these characteristics and their intended application. (More information on filters may be found in the **Analog and Digital Filtering** chapter.) Unless otherwise noted, the filter types in this section are differential-mode filters.

It's relatively simple to build a differential-mode filter that passes desired signals and blocks unwanted signals with a high series impedance. An alternative filter technique presents a low shunt or parallel impedance to the signal, effectively "shorting it out." Some filters, such as the popular pi-circuit ac line filters, combine both techniques.

Low-pass filters pass frequencies below some cutoff frequency, while attenuating frequencies above that cutoff frequency. A typical low-pass filter curve is shown in **Figure 27.8**. A schematic is shown in Figure 27.13B. Filters capable of handling 100 W of RF or more can be difficult to construct properly, so many hams choose to buy them. Many *QST* advertisers stock low-pass filters.

High-pass filters pass frequencies above some cutoff frequency while attenuating frequencies below that cutoff frequency. A typical high-pass filter curve is shown in **Figure 27.9**. Figure 27.9B shows a schematic of a typical high-pass filter. For receiving applications, designs for building your own filters are abundant, and components are fairly inexpensive. Commercial products are widely available.

Band-stop or ***band-reject filters*** reject a narrow range of frequencies. These are also referred to as "notch filters" or "traps" and are often used to reject strong signals from nearby FM, TV, or commercial VHF/UHF transmitters that operate on one channel. The filter can remove an entire band (for example, 88-108 MHz for FM broadcast) or a single channel (such as the common shore station frequency of 160.650 MHz, Marine VHF channel 1). Amateurs use band-reject filters to reduce the strength of nearby AM broadcast signals when operating on the HF bands.

AC-line filters, sometimes called "brute-force" filters, are used to remove RF energy from ac power lines. A typical schematic is shown in Figure 27.7. Note that no filtering is performed on the ground or common connection. (See the Elements of RFI section for the definition for common-mode by utilities and ac power systems.)

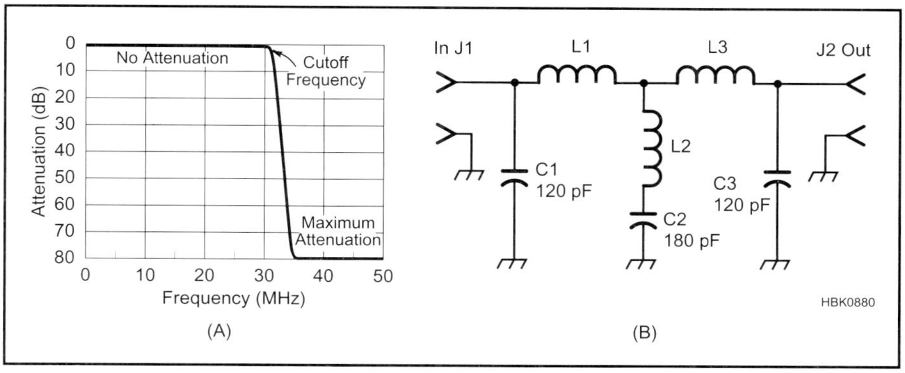

Figure 27.8 — (A) An example of a low-pass filter's response curve. (B) A low-pass filter schematic for amateur transmitting use. Complete construction information appears in the Transmitters chapter of *The ARRL RFI Book*. A high-performance 1.8-54 MHz filter project can be found in the Analog and Digital Filtering chapter of this *Handbook*.

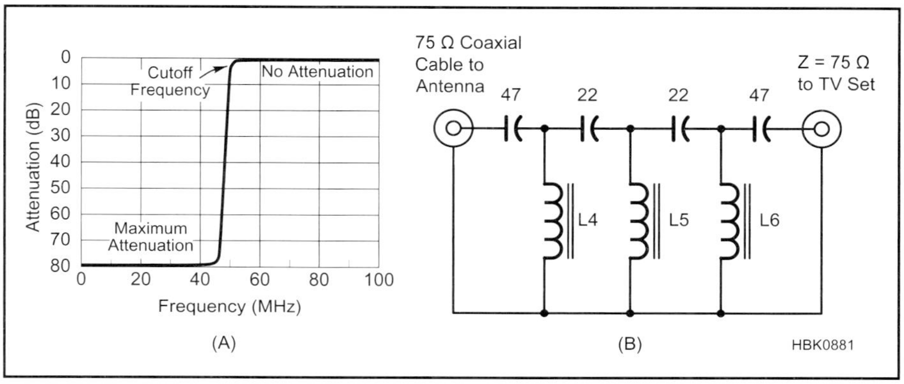

Figure 27.9 — (A) An example of a high-pass filter's response curve. (B) A differential-mode high-pass filter for 75-Ω coaxial cable. It rejects HF signals picked up by a TV antenna or that leak into a cable-TV system. All capacitors are high-stability, low-loss, NP0 ceramic disc components. Values are in pF. The inductors are all #24 AWG enameled wire on T-44-0 toroid cores. L4 and L6 are each 12 turns (0.157 µH) and L5 is 11 turns (0.135 µH).

AC-line filtering products are widely available from various sources. In addition to amateur radio equipment distributors, filters can be sourced direct from manufacturers such as Corcom **(www.corcom.com)** and Mor-gan Systems (**www.surgestop.com**). Some of these are stand-alone filters with ac plugs and sockets, such as the Morgan Systems M-471 filter shown in **Figure 27.10A**. Others, such as the Corcom model no. 10VS1 (Figure 27.10B), would need to be wired in

Figure 27.10 — Two AC line filters. A Morgan Systems M-471 (A) is a plug-in filter. The Corcom 10VS1 (B) is wired in place.

(A)

(B)

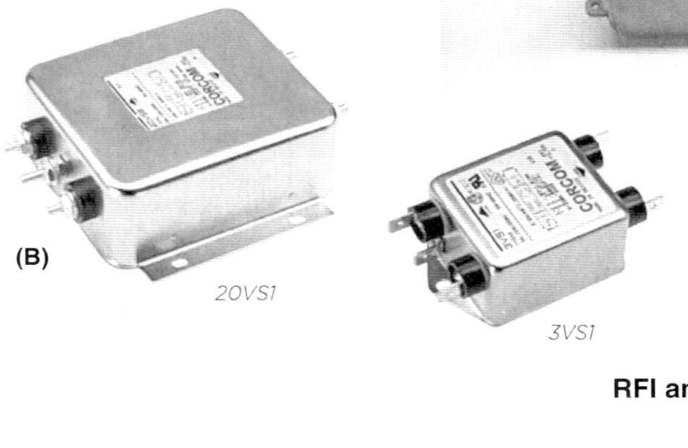

Warning: Bypassing Speaker Leads

Older amateur literature might suggest connecting a 0.01-µF capacitor across an amplifier's speaker output terminals to cure RFI from common-mode current picked up by speaker cables. Don't do this! Doing so can cause some modern solid-state amplifiers to break into a destructive, full-power, sometimes ultrasonic oscillation if they are connected to a highly capacitive load. Use common-mode chokes and twisted-pair speaker cables instead.

place by the installer. SMT EMI filters are now available for brush DC motors as well, which are increasingly common in consumer appliances. See the *Microwaves & RF* article by Jeff Elliott in the References section for more information.

AC-line filters come in a wide variety of sizes, current ratings, and attenuation. It is wise to look for an ac-line filter with both common- and differential-mode filtering (the Morgan Systems M-471 is a good example of such a filter). In general, a filter must be physically larger to handle higher currents at lower frequencies. The Corcom 1VB1, for example, is a compact filter small enough to fit in the junction box for many low voltage lighting fixtures, provides good common-mode attenuation at MF and HF (only on the hot and neutral lines), and its 1 A at 250 V ac rating is sufficient for many LV lighting fixtures. In general, you will get more performance from a filter that is physically small if you choose the filter with the lowest current rating sufficient for your application. (Section 13.3 of Ott covers ac-line filters.)

The best location for an ac-line filter is at the enclosure of the equipment causing the RFI. Filter modules can be bonded to the enclosure through the case of the filter (best) or via a bonding jumper to the enclosure. Bond the enclosure of the filter to the enclosure of the equipment by the shortest possible path. Many metal enclosures are painted, which defeats the bonding of either a filter case, terminal, or solder lug. Remove paint from wherever a bonding connection is made.

Some commercial filters are built with an integral IEC power socket, and can replace a standard IEC connector if there is sufficient space behind the panel. (IEC is the International Electrotechnical Commission, an international standards organization that has created specifications for power plugs and sockets. See the **Safe Practices** chapter for a drawing of an IEC connector.) The case of such a filter is bonded to the enclosure, and interconnecting leads are shielded by the enclosure, optimizing its performance.

AC-line filters connected externally to equipment through a cord require extra attention. Any wiring between a filter and the equipment being filtered acts as an antenna and forms an inductive loop that degrades the performance of the filter. All such wiring should be as short as possible, and should be twisted. As described above, they do not apply filtering to the ground conductor, which remains active for RFI. Wind several turns of the wiring on a ferrite core to block common-mode RF current on the ground conductor.

AC-line capacitors — A capacitor between line and neutral or between line and ground at the noise source or at victim equipment can solve some RFI problems. ("Ground" in this sense is not "earth," it is the power system equipment safety ground — the green wire — at the equipment.) Power lines, cords, and cables are often subjected to short-duration transients of very high voltage (4 kV). Ordinary capacitors are likely to fail when subjected to these voltages, and the failure could cause a fire. Only Type X1, X2, Y1 and Y2 capacitors should be used on power wiring. AC-rated capacitors can safely handle being placed across an ac line along with the typical voltage surges that occur from time to time. Type X1 and X2 capacitors are rated for use between line and neutral, and are available in values between 0.1 µF and 1 µF. Type X2 capacitors are tested to withstand 2.5 kV, type X1 capacitors to 4 kV. Type Y1 and Y2 capacitors are rated for use between line and ground; Y1 capacitors are impulse tested to 8 kV, Type Y2 to 5 kV. Note that 4700 pF is the largest value permitted to be used between line and ground — larger values can result in excessive leakage currents.

Bypass capacitors provide a low-impedance path for RF signals away from an affected lead or cable. A bypass capacitor is usually placed between a signal or power lead and the equipment chassis or enclosure. If the bypass capacitor is attached to a shielded cable, the shield should also be connected to the enclosure. Bypass capacitors for HF signals are usually 0.01 µF, while VHF bypass capacitors are usually 0.001 µF. Leads of bypass capacitors should be kept short, particularly when dealing with VHF or UHF signals.

27.3.4 Cable Shield Connections

The outer surface of cable shields are potential antennas for RFI, either radiating or receiving unwanted signals regardless of the shield's quality. Improperly connected shields on audio, RF, and data cables are a common source of radiated and conducted emissions. They provide a path into equipment for common-mode RF that has been picked up on a cable shield as common-mode current. As a result, an improperly connected shield is a likely cause of RFI if the victim device is unable to reject the unwanted common-mode signals. This is why it is important to connect cable

Filter Both Modes at Once

If an amateur decides on filtering the ac line supplying equipment belonging to a neighbor, it is often best to resolve that problem as quickly and efficiently as possible. Experimenting to find out which type of filter is needed often leads to frustration on the part of the neighbor, who may quickly lose confidence in the ham's ability to resolve the problem. For this reason, it is usually advisable to install both common-mode and differential-mode filtering on a neighbor's device as shown in **Figure 27.A1**, resolving the interference that is present and making future interference less likely.

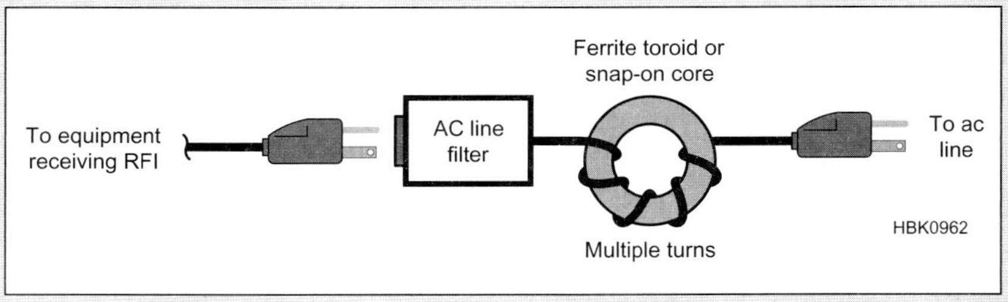

Figure 27.A1 — An ac line filter and a ferrite RF choke combined in the ac line to the affected equipment block both differential- and common-mode current. The line filter can be a stand-alone model with plugs and receptacles or a module in a grounded metal enclosure. The ferrite mix should be Type #31 (for HF) or Type #43 (for middle HF through VHF). Either a toroid or clamp-on core can be used if multiple turns can be wound on the core to create sufficient choking impedance.

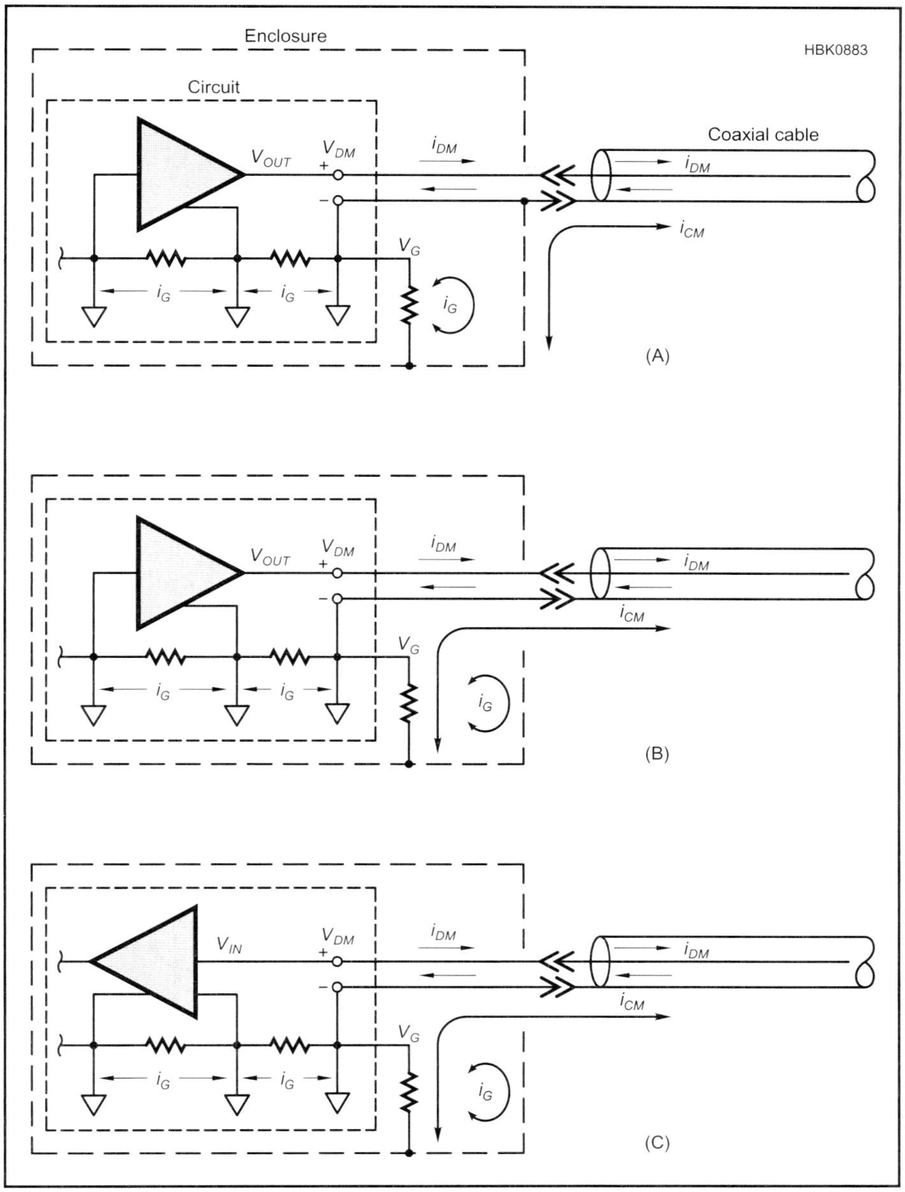

Figure 27.11 — Properly (A) and improperly (B, C) connected shields showing undesired paths for common-mode current into and out of equipment. Improperly connected shields create an ingress and egress path for RF common-mode current. See the text for discussion.

shields in such a way that common-mode currents flowing on the shields are not allowed to enter the victim device or leave the source device.

Figure 27.11 illustrates the basic reasoning as to how cable shields should be connected. This is a very simplified overview to illustrate the basic concepts. Figure 27.11A shows a metal enclosure and an internal circuit, usually on a PC board. In the circuit itself, there is a common connection that could be a ground plane or point-to-point wiring or traces. Either way, there is some small impedance between any two points in the circuit's ground system. This is shown by the resistor symbols. Current flowing in the ground system is shown as i_G. The usual practice is to connect the ground system to the enclosure (chassis ground) through a wire or hardware. That, too, has some impedance, shown as a resistor.

The actual voltage applied to the cable connection, V_{DM}, is not equal to V_{OUT}. The ground currents flowing through the ground system impedances create small voltages. So does the ground current flowing through the connection between the circuit ground system and the enclosure. All of these voltages combine to create V_G, which then combines with the amplifier output voltage, V_{OUT}, to create V_{DM}.

Outside the enclosure, a coaxial cable is terminated in an RF connector, such as a PL-259, BNC, Type N, or even a shielded, metal phono plug. The differential-mode signal currents inside the cable (shown as i_{DM}) flow on the cable's center conductor and the *inside* of the cable shield. If a signal is picked up by the cable, the resulting common-mode current (shown as i_{CM}) flows on the *outside* of the cable shield. Due to the skin effect, the inside and outside of the cable shield are independent conductors for RF currents.

On the inside of the mating connector (such as an SO-239) a cable or wire pair connect the i_{DM} currents to the circuit. In Figure 27.11A, the center conductor in the cable is connected to some kind of amplifier output (V_{OUT}) and the enclosure-connected part of the connector to the circuit's common or ground terminal. If the connector and shield are attached properly, there is no path between the circuit's differential-mode common terminal and the outside, common-mode surface of the cable shield. Similarly, whatever current i_{DM} is flowing inside the cable, it cannot flow to the outside of the cable where it would become common-mode current and radiate an unwanted signal — a radiated emission.

Figure 27.11B shows what happens when the shield connection goes through the enclosure and is connected directly to the inner circuit ground system. The common-mode current path is now open to i_G that flows in the connection from the circuit ground system to the enclosure. This allows noise and signals from the circuit to escape the enclosure as a radiated emission. The path is also open for i_{CM} to flow in the enclosure connection and add to the voltage that makes up V_G. Depending on the ground system impedance, some of i_{CM} can add to i_G, as well. This allows external noise and signals to get into the circuit's ground system and disrupt its operation.

Figure 27.11C illustrates the situation when the differential-mode signal is an input to the circuit. Noise and signals on the circuit ground system can still escape the enclosure. In addition, the common-mode current i_{CM} combines with i_G, which modifies V_G and V_{IN}. This can easily cause a lot of problems if the circuit is a receiver front end or other sensitive function.

Figure 27.3D showed a similar situation with two wires inside a shield. This is common for data or control connections where there might be several data lines and a dedicated signal return or signal ground circuit that is separate from the overall shield. Treat the signal return circuit as an independent circuit and do not connect it to the enclosure. Most shielded data cables have separate signal ground and shield connections. Only the shield should connect to the enclosure.

Not connecting the shield at all breaks the path for the desired differential-mode current on the inside of the shield. This forces it to find another return path, often through ground systems or power wiring. The gap in the shield creates an antenna similar to a magnetic loop that not only radiates the differential-mode signal but allows the undesired common-mode current to combine with the differential-mode signals. The whole system acts as if it were unshielded.

The need to connect the shields of cables to a shielding enclosure is clear whether they are coaxial cables, shielded twisted-pair, or multiconductor data cables. If the shield connection penetrates the enclosure and is connected directly to the circuit's ground system, both radiated emissions and incoming RFI are the likely result. In addition, the unintended current path can affect the impedance of the connection to the cable as discussed in the material on common-mode chokes later in this section and in the **Transmission Lines** chapter. It is possible to devise alternate connections, such as for transceiver microphone cables, but RF filtering or chokes and careful circuit ground system design and construction are required.

27.3.5 Common-Mode Chokes

Common-mode chokes on ferrite cores are the most effective answer to RFI from a common-mode signal. Differential-mode filters are *not* effective against common-mode signals. (AC-line filters usually only perform differential-mode filtering as described in the preceding section.) Common-mode chokes work differently, but equally well, with coaxial cable and paired conductors. (Additional material on common-mode chokes is found in sections 3.5 and 3.6 of Ott.) Common-mode chokes can often solve RFI problems, especially at HF where common-mode current is more likely to be the culprit.

The most common form of common-mode choke is multiple turns of cable wound on a ferrite toroid core as shown in **Figure 27.12**. The following discussion applies to chokes wound on rods as well as toroids, but avoid rod cores since they may couple to nearby circuits at HF. At HF, toroid cores are recommended — beads and clamp-on ferrite cores are usually adequate.

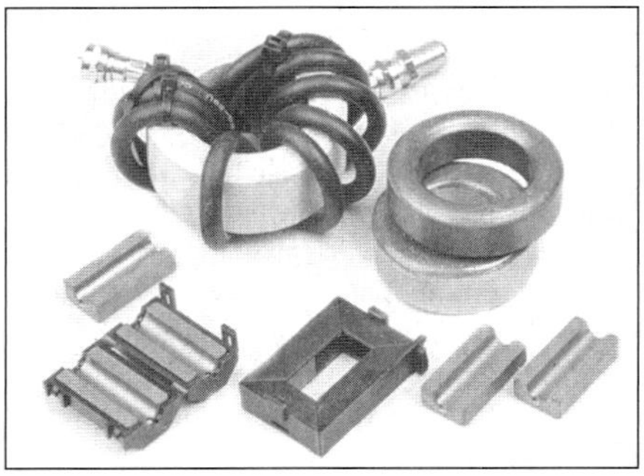

Figure 27.12 — A common-mode RF choke wound on a toroid core is shown at top left. Several styles of ferrite cores for common-mode chokes are also shown.

Common-mode chokes using ferrite cores are discussed at length in the **RF Techniques** and **Transmission Lines** chapters. They block common-mode RF current by adding a large value of resistive impedance in series with the common-mode circuit. The choke actually behaves as a parallel-self-resonant circuit that includes the winding inductance and stray capacitance along with the resistance of the core material at that frequency. (The electrical characteristics of ferrite at RF are discussed in the **RF Techniques** chapter.)

Most of the time, a common-mode signal on a coaxial cable or a shielded, multi-wire cable is a current flowing on the outside of the cable's shield. By wrapping the cable around a magnetic core the current creates a flux in the core, creating a high impedance in series with the outside of the shield. (The impedance required for an effective choke depends on the circumstances and can range from a few hundred to several thousand ohms.) The impedance then blocks or reduces the common-mode current. Because equal-and-opposite fields are coupled to the core from each of the differential-mode currents, the common-mode choke has no effect on differential-mode signals inside the cable.

When the cable consists of two-wire, unshielded conductors such as zip cord or twisted-pair, the equal-and-opposite differential currents each create a magnetic flux in the

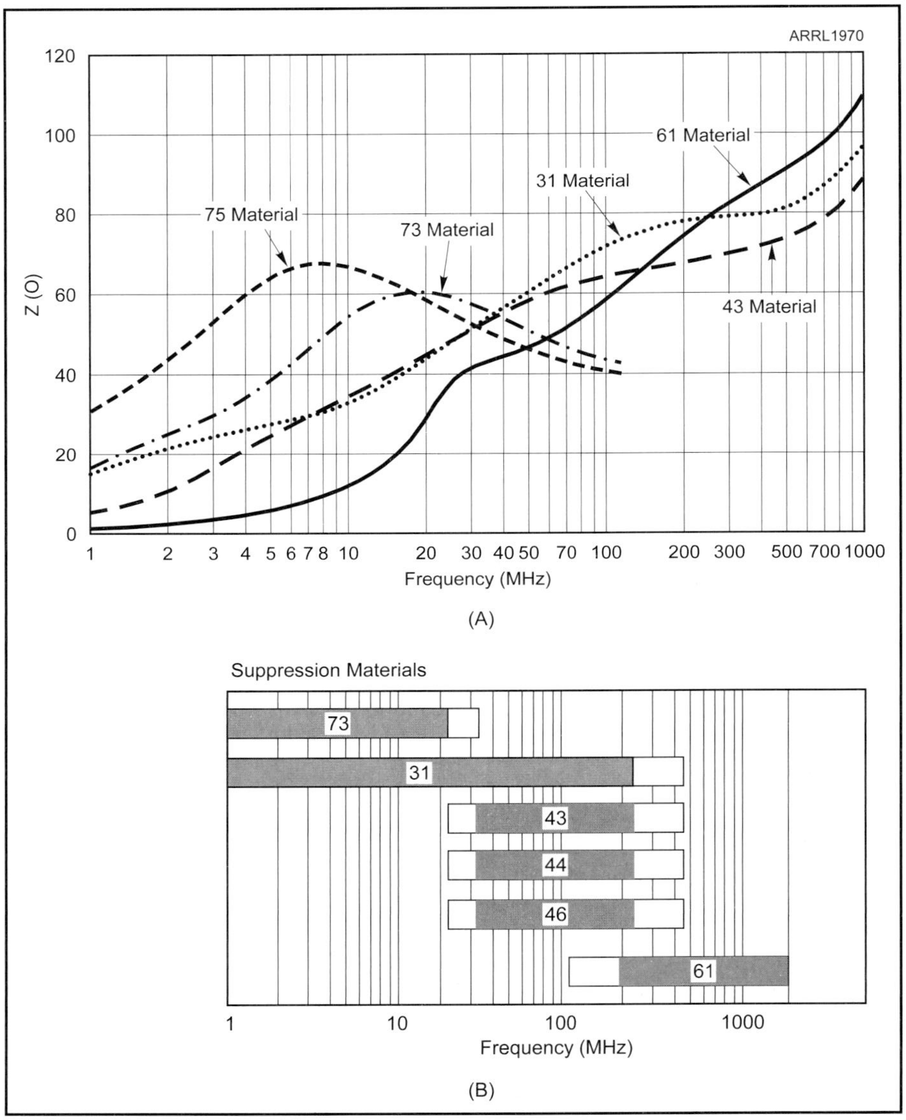

Figure 27.13 — Impedance vs. frequency plots (A) for a single wire passed through one ferrite bead, 3.50 mm diameter x 6.00 mm long (Fair-Rite size 301). Not all mixes are available is all types of cores and beads. The chart at (B) shows frequency ranges for various ferrite mixes used for EMI suppression. Type #73 and 75 material cover approximately the same range. [Charts provided courtesy of the Fair-Rite Products Corporation].

core. The equal-and-opposite fluxes cancel each other, and the differential-mode signal experiences zero net effect. To common-mode signals, however, the choke appears as a high impedance in series with the signal: the higher the impedance, the lower the common-mode current.

It is important to note that common-mode currents on a transmission line, coaxial or parallel-conductor, will result in radiation of a signal from the feed line. (See this section's sidebar for an explanation of balanced and unbalanced transmission lines.) The radiated signal can then cause RFI in nearby circuits. This is most common when using coaxial cable as a feed line to a balanced load, such as the dipole in the sidebar. Reducing common-mode current with a common-mode choke at the antenna feed point and where the feed line enters the residence can help reduce RFI caused by signals radiated from the feed line's shield.

Common-mode noise is also picked up by feed lines. If it is allowed to flow into the station, it can also be converted to differential-mode signals at connectors and by coupling to other cables and wiring in the station. Chokes can block this current and eliminate this source of received noise and unwanted signals, which can contribute to the noise level considerably.

FERRITE MATERIALS FOR RFI CONTROL

The self-resonance of a conductor passing once through most ferrite cores (considered to be one turn) used for suppression is in the range of 150 MHz, and this is where a core simply clamped around a cable will be effective. To obtain good suppression in the range of 1 – 50 MHz, we must wind multiple turns through the ferrite core to lower the resonant frequency. Common-mode chokes are typically wound on toroid cores with a 1-inch or greater inside diameter or a split-core clamp-on core. (Ferrite materials are discussed extensively in the **RF Techniques** chapter.)

Figure 27.13 is a combined graph of the impedance of single turns through a ferrite bead of different mixes. It gives an idea of the range of available ferrites. By using the right material and the right number of turns through the core, choke impedance can be optimized for resistance and for the frequency range desired. Resistive impedance is desired to avoid interacting with the conductor's reactance at the frequency of use. The right material to use is specified for "suppression" over the frequency range required. (From low to high frequencies, Fair-Rite mixes #73 (beads only), #75, #31, #43, and #61 are of the most use to amateurs.) This means the core's impedance is primarily resistive in that range. Chokes wound on these cores have a very low Q (because they are dissipating energy instead

Warning: Surplus Ferrite Cores

Don't use a core to make a common-mode choke if you don't know what type of material it is made of. Such cores may not be effective in the frequency range you are working with. This may lead to the erroneous conclusion that a common-mode choke doesn't work when a core with the correct material would have done the job. The **RF Techniques** chapter includes a test procedure for determining the type of ferrite. It is often better to simply purchase known cores.

Feed Line Radiation — The Difference Between Balanced and Unbalanced Transmission Lines

The physical differences between balanced and unbalanced feed lines are obvious. Balanced lines are parallel-type transmission lines, such as ladder line or twin lead. The two conductors that make up a balanced line run side-by-side, separated by an insulating material (plastic, air, whatever). Unbalanced lines, on the other hand, are coaxial-type feed lines. One of the conductors (the shield) completely surrounds the other (the center).

In an ideal world, both types of transmission lines would deliver RF power to the load (typically your antenna) without radiating any energy along the way. It is important to understand, however, that both types of transmission lines require a balanced condition in order to accomplish this feat. That is, the currents in each conductor must be equal in magnitude, but opposite in polarity.

The classic definition of a balanced transmission line tells us that both conductors must be symmetrical (same length and separation distance) relative to a common reference point, usually ground. It's fairly easy to imagine the equal and opposite currents flowing through this type of feeder. When such a condition occurs, the fields generated by the currents cancel each other — hence, no radiation. An imbalance occurs when one of the conductors carries more current than the other. This additional "imbalance current" causes the feed line to radiate.

Things are a bit different when we consider a coaxial cable. Instead of its being a symmetrical line, one of its conductors (usually the shield) is grounded. In addition, the currents flowing in the coax are confined to the outside portion of the center conductor and the inside portion of the shield.

When a balanced load, such as a resonant dipole antenna, is connected to unbalanced coax, the outside of the shield can act as an electrical third conductor (see **Figure 27.A2**). This phantom third conductor can provide an alternate path for the imbalance current to flow. Whether the small amount of stray radiation that occurs is important or not is subject to debate. In fact, one of the purposes of a balun (a contraction of balanced to unbalanced) is to reduce or eliminate imbalance current flowing on the outside of the shield. See the **Transmission Lines** chapter of this book for more information on baluns.

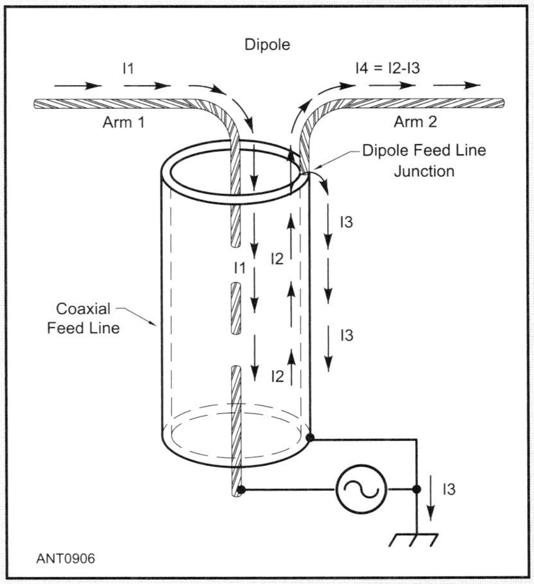

Fig 27.A2 — Various current paths are present at the feed point of a balanced dipole fed with unbalanced coaxial cable. The diameter of the coax is exaggerated to show the currents clearly.

of storing it) so the choke's bandwidth can be as much as an octave. Chokes can be combined to work over different frequency ranges by placing chokes optimized for different ranges in series.

Type #31 material is a good all-purpose material for HF and low-VHF applications, especially at low HF frequencies. Type #43 is widely used for HF through VHF and UHF. See the **RF Techniques** chapter for more information on ferrite materials and characteristics. This discussion only touches on the basic characteristics of ferrite-core common-mode chokes as they apply to dealing with RFI.

27.4 Types of RFI

There are several basic types of RFI involving amateur radio. The appropriate remedies for each are different, and the correct one must be used for each mechanism. It is important to note that RFI may be caused by more than one type of RFI, particularly from one of the strong-signal causes.

The first three occur when a strong signal causes RFI. This can occur even for a transmitted signal that is in complete compliance with all rules for signal quality. In order of likelihood:

1) Fundamental Overload — Disruption or degradation of a receiver's function in the presence of a transmitted fundamental signal (the intended signal from the transmitter). The unwanted signal is received normally along with desired signals, but the transmitted signal is simply too strong for the victim receiver's circuitry to reject — a true overload situation.

2) Common-mode breakthrough — The interfering signal is picked up as common-mode current on an external cable which is then conducted into the victim device's circuitry, where it disrupts normal operation. This is also referred to as *common-mode ingress* and RF "getting into" a piece of equipment.

3) Direct pickup or *direct detection* — The interfering signal is picked up directly by the victim device's internal wiring without any external cabling or wiring required.

The following types of RFI occur, again in order of likelihood, when the receiver functions normally but reception of a desired signal is interfered with by RF energy received along with the desired signal.

1) Spurious Emissions — Signals from the transmitter as defined in the previous section on Part 97 definitions and Figure 27.1. For example, harmonics of a transmitted signal can be received with and interfere with a desired signal even though they are not strong enough to cause fundamental overload.

2) External Noise Sources — RF energy transmitted incidentally or unintentionally by a device that is not a licensed transmitter. For example, impulse noise from motors or power lines.

3) Intermodulation — Intermodulation distortion (IMD) products generated by signals mixing together internally or externally to the receiver.

As an RFI troubleshooter, start by determining which of these is involved in your interference problem. Once you know the type of RFI, identifying the cause and selecting the most appropriate cure for the problem becomes much easier.

27.4.1 Interference from Strong Signals

FUNDAMENTAL OVERLOAD

Most cases of interference caused by an amateur transmission are due to fundamental overload. Properly designed radio receivers of any sort should be able to select the desired signal while rejecting all others. Unfortunately, because of design limitations or deficiencies such as inadequate shields or filters, some radio receivers are unable to reject strong out-of-band signals. Even amateur receivers can exhibit fundamental overload. There does not need to be anything wrong with the interfering signal — it's just too strong for the receiver to reject.

A signal causing fundamental overload is received along with the desired signals in the normal signal path, such as an FM or TV antenna, usually as a differential-mode signal such as inside coaxial cable. A good solution might be to put a filter in the path of the incoming signals to block the strong amateur signal while passing the desired signals, which are presumably in some other band. An attenuator at the receiver's signal input path can also help as long as the desired signals can still be received properly.

Significant improvement can often result from just moving the victim equipment's antenna and the signal source farther away from each other. The effect of an interfering signal is directly related to its strength, diminishing with the square of the distance from the source. If the distance from the source doubles, the strength of the RF signal decreases to one-fourth of its power density at the original distance from the source.

Reducing the level of the offending strong signal often returns the receiver to normal operation and eliminates the undesirable effects. This characteristic can often be used to help identify an RFI problem as fundamental overload. If gradually reducing the strength of the signal source causes the interference to abruptly disappear, that is a signature of fundamental overload.

COMMON-MODE BREAKTHROUGH

A strong signal can enter equipment in several different ways. Most commonly, it is conducted into the equipment after being picked up by connecting wires and cables. Possible RFI conductors include antennas and feed lines, interconnecting signal and control cables, power cords, and ground wires. TV antennas and feed lines, telephone or speaker wiring, and ac power cords are the most common points of pickup and subsequent entry.

Once inside the equipment, the signal is conducted to electronic circuits that are probably not designed to reject unwanted RF signals. The unwanted signal might then be detected or transformed by the circuit to ac or dc signals that disrupt its normal functions. Since the cause of the problem is an unintended RF signal path from outside the equipment, the usual remedy is to block the path with an RF choke or filter, depending on the type of connection. The common-mode RF signal must be kept out of the affected equipment to cure the problem.

DIRECT DETECTION

Direct detection occurs when wiring or circuitry that is internal to the equipment being affected picks up RF energy directly. Direct detection RFI can be very difficult to eliminate since it does not involve an external signal path that can be blocked or filtered. Examples of equipment that experience direct detection are battery-powered consumer devices in unshielded (plastic or wood) or poorly shielded metal enclosures. This permits the RF signal to be picked up by the wiring in the device. Eliminating the interference would require modifying the device, adding shielding, or perhaps simply moving the device farther from the transmitted signal source.

27.4.2 Interference from Spurious Signals

All transmitters generate some (hopefully few) RF signals that are outside their intended transmission bandwidth — these are the out-of-band emissions and spurious emissions as shown in Figure 27.1. Out-of-band signals result from distortion in the modulation process or consist of broadband noise generated by the transmitter's oscillators that is radiated along with the intended signal. Harmonics, the most common spurious emissions, are signals at integer multiples of the operating (or fundamental) frequency.

Transmitters may also produce broadband noise and/or parasitic oscillations as spurious emissions. (Parasitic oscillations are discussed in the **RF Power Amplifiers** chapter.) Overdriving an amplifier often creates spurious emissions. Amplifiers not meeting FCC certification performance standards but sold illegally are frequent sources of spurious emissions.

Regardless of how the unwanted signals are created, if they cause interference, FCC regulations require the operator of the transmitter to correct the problem. The usual cure is to adjust or repair the transmitter or use filters at the transmitter output to block the spurious emissions from being radiated by the antenna.

27.4.3 Interference from External Noise Sources

Most cases of interference to amateur stations reported to the FCC are eventually determined to involve some sort of external noise source, rather than signals from a radio transmitter. Noise in this sense means an RF signal that is not essential to the generating device's operation.

The most common external noise sources are electrical, primarily power lines. Motors, switching equipment, and street lighting can also generate electrical noise. The section Power Line Noise addresses this type of interference.

Noise radiated by switchmode systems such as power supplies, digital equipment, and variable-speed drives is increasingly common. Other common sources of external noise are grow lights being used in various forms of indoor gardening and some solar power installations. External noise can also come from unlicensed Part 15 RF sources such as computers and networking equipment, video games, appliances, and other types of consumer electronics.

Regardless of the source, if you determine that the problem is caused by external noise, elimination of the noise must take place at the source. As an alternative, several manufacturers also make noise canceling devices that can help in some circumstances.

27.4.4 Interference from Intermodulation Distortion

Intermodulation distortion (IMD) is caused by two signals combining in such a way as to create intermodulation products — signals at various combinations of the two original frequencies. The two original signals may be perfectly legal, but the resulting intermodulation distortion products could occur on frequencies used by other services and cause interference in the same way as a spurious signal from a transmitter. The process of combination can occur outside or inside the receiver.

External IMD is sometimes called the "rusty bolt effect" because it can occur in any nonlinear connection that is located very near transmitting sources or very near victim receivers. This is often caused by contact between dissimilar metals, with corroded metals being the worst offenders.

The resulting IMD product signals include the modulation of both signals. For example, intermodulation from two SSB or FM voice signals produces somewhat distorted signals with the modulating signals of both stations. Since the two signals are not synchronized, the intermodulation products come and go unpredictably, existing only when both of the external signals are present.

INTERFERENCE FROM EXTERNAL IMD

IMD can be generated externally by signals mixing together in non-linear junctions or connections. IMD products generated externally cannot be filtered out at the receiver and must be eliminated where the original signals are being combined.

IMD is a common problem at sites that host multiple high-power transmitters such as AM broadcast stations or commercial and public safety communications sites, which are often hosts to amateur repeaters. In some cases, two or more signals can combine in the output stages of one of the transmitters (also known as *cross-modulation*) to create the IMD products. The solution is to increase isolation between the affected transmitter(s).

INTERFERENCE FROM INTERNAL IMD

As discussed in the chapter on **Receiving**, IMD can also be generated *inside* a receiver by strong signals. This is a situation similar to fundamental overload in which the signals are so strong that the receiver circuits can't process them as independent signals. The solution is generally to reduce signal levels into the receiver by using an attenuator or filter so that it is no longer acting non-linearly.

27.5 RFI Troubleshooting Guidelines

Troubleshooting an RFI problem is a multistep process, and all steps are equally important. Each step in troubleshooting an RFI problem will involve asking and answering several questions: Is the problem caused by harmonics, fundamental overload, common-mode breakthrough, conducted emissions, radiated emissions, or a combination of all of these factors? Should it be fixed with a lowpass filter, high-pass filter, common-mode chokes, or ac-line filter? How about shielding, isolation transformers, a different ground or antenna configuration? By the time you finish with these questions, the possibilities could number in the millions. You probably will not see your exact problem and cure listed in this book or any other. You must not only diagnose the problem but find a cure as well!

Now that you have learned some RFI fundamentals, you can work on specific technical solutions. A systematic approach will identify the problem and suggest a cure. Most RFI problems can be solved in this way by the application of standard techniques. The following sections suggest specific approaches for different types of common interference problems. This advice is based on the experience of the ARRL RFI Desk, but is not guaranteed to provide a solution to your particular problem. Armed with your RFI knowledge, a kit of filters and tools, your local TC, and a determination to solve the problem, it is time to begin. You should also get a copy of the *ARRL RFI Book*. It's comprehensive and picks up where this chapter leaves off.

RFI Email Reflector

A good source of information and help is the RFI email reflector maintained on the **contesting.com** website **lists.contesting.com/mailman/listinfo/RFI**. Simply subscribing in digest form provides a daily dose of RFI discussion. Before asking your question, however, be sure to check the reflector's searchable archives to see if your topic has been covered recently.

27.5.1 Basic Troubleshooting Principles

Amateur radio teaches a lot of skills, and a very valuable skill is thinking in terms of the whole communications system, from the power supply all the way through the antenna, to the other station, and back again. RFI is just one type of system-level troubleshooting and as you will discover even involves non-amateur equipment! The troubleshooting skills you learn from finding and fixing RFI can also be applied to other problems in your station, so here are some general principles to learn:

Know the Symptoms — Before beginning any kind of troubleshooting, be sure you know what it is you're trying to fix! Write down what the problem is and list all of the symptoms you can.

Be Sure It's Not Improper Configuration — Make sure the features of affected equipment are configured properly and its controls and switches are set correctly. Many a "dead" radio simply has the squelch turned all way up so that no signals are strong enough to be heard.

Make a Plan — After you have the symptoms defined and are sure the affected equipment is not just misconfigured, sketch out a few steps from making tests to trying solutions. Be sure to think about how you will know the problem is really fixed!

Take Controlled Steps — Each step in the process should be intended to discover or measure something specific. Don't just "shotgun" a problem or try things on a whim hoping that you'll get lucky.

Change One Thing at a Time — Try to change as few things as possible in each step. If you have to change several things, have a plan for figuring out the effect of each change.

Keep Notes — Have a notebook or open a memo file to record what you're doing and what you learn. Clearly labeled photos or sketches are also helpful. This is the place to capture "What if..." ideas and things you notice.

Don't Jump to Conclusions — Along with taking controlled steps, avoid the temptation to abandon your plan and decide what the cause of the problem is without taking the intermediate steps to be sure.

Test Your Conclusions — At the end, after you're sure you've identified and corrected the problem, make tests to verify the problem has, indeed, been solved.

Report What You've Learned — Don't assume you'll remember what you found out. Capture it in writing or in a voice memo or a video/photo. If you made changes to your station, record them in your station notebook. Share the results with others if the solution has value beyond your station!

27.5.2 RFI Troubleshooting Steps

Here are the general steps in RFI troubleshooting. Remember that "noise" refers to whatever type of signal is causing the problem.

1. Determine the type(s) of RFI problem you have.
2. Determine the type of noise source(s) and the victim(s) that are involved.
3. Diagnose the problem by locating the noise source and the means by which it creates the noise.
4. Identify the path by which the noise or signals reach the victim device.
5. Eliminate the source or break the path from source to victim.

AT THE BEGINNING

At Your Station — Make sure that your own station and consumer equipment are operating correctly. This eliminates them as possible sources of interference, and when this is done, you won't need to diagnose or troubleshoot your station. Also, any cures successful at your house may work at your neighbor's as well. If you do have problems in your own home, continue through the troubleshooting steps and specific cures and take care of your own problem first.

Is It Really RFI? — Before trying to solve a suspected case of RFI, verify that the symptoms actually result from external causes. A variety of equipment malfunctions or external noise can look like interference. (Remember that we are using "RFI" to apply to any kind of RF-related EMI, including TVI.) For example, a failing power supply can turn something on and off erratically. A misconfigured TV can behave in unexpected ways. Just mistuning a receiver can create distorted audio. There are lots of ways to cause problems besides RFI, so start with an open mind about what the problem actually is.

RFI Survival Kit — **Table 27.1** is a list of material most often needed to troubleshoot and solve most RFI problems. Having all of these materials in one container, such as a small tackle or craft box, makes the troubleshooting process go a lot smoother.

ASSESS THE SITUATION

Look Around — Aside from the brain, the eyes are a troubleshooter's best tool. Installation defects contribute to many RFI problems. Look for loose connections, shield breaks in a cable-TV installation, corroded battery contacts, damaged cables, or anything that looks like it might cause a problem. Fix these first.

Look for Unintended Antennas — Look for wiring connected to the victim equipment that might be long enough to be resonant on one or more amateur bands. If so, a common-mode choke may be an easy cure. Ideally, you'll generally want to place the choke as close to the victim device as practical. If this placement proves too difficult or additional suppression is required, chokes placed at the middle of the wiring may help break up resonances. These are just a few of the possible deficiencies in a home installation.

Is It a Transmitted Signal from Your Station? — "Your" RFI problem might be caused by another ham or a radio transmitter of another radio service, such as a local CB or police transmitter. If it appears that your station is involved, operate your station on each band, mode, and power level that you use. Note all conditions that produce interference. If no transmissions produce the problem, your station *may* not be the cause. (Although some contributing factor may have been missing in the test.) Have your neighbor take notes about when and how the interference appears: what time of day, what station, what other appliances were in use, what was the weather? You should do the same whenever you operate. If you can readily reproduce the problem with your station, you can start to troubleshoot the problem.

Table 27.1
RFI Survival Kit

Quantity	*Item*
(2)	75-Ω high-pass filter
(2)	Commercially available clamp-on ferrite cores: #31 and #43 material, 0.3" ID
(12)	Assorted ferrite cores: #31 and #43 material, FT-140 and FT-240 size
(3)	Telephone RFI filters
(2)	Brute-force ac line filters
(6)	0.01-µF ceramic capacitors
(6)	0.001-µF ceramic capacitors

Miscellaneous:
- Hand tools, assorted screwdrivers, wire cutters, pliers
- Hookup wire
- Electrical tape
- Soldering iron and solder (use with caution!)
- Assorted lengths of 75-Ω coaxial cable with connectors
- Spare F connectors, male, and crimping tool
- F-connector female-female "barrel"
- Clip leads
- Notebook and pencil
- Portable multimeter

WHILE TROUBLESHOOTING

Take One Away — Can you remove the source or victim entirely? The best cure for an RFI problem is often removing the source of the noise. If the source is something broken, for example, the usual solution is to repair it. Power-line noise and an arcing electric fence usually fall into this category. If a switchmode power supply is radiating noise, replace it with a linear supply. Victim devices can sometimes be replaced with a more robust piece of equipment as well, such as replacing an old 49 MHz cordless phone with a current model that uses spread-spectrum multi-GHz signals instead.

Simplify the Problem — Don't tackle a complex system all at once, such as a home entertainment system with a TV receiver, audio system, DVD player, video game console, satellite TV receiver, and remote audio to other rooms. You could spend the rest of your life running in circles and never find the true cause of the problem!

There's a better way. Start with the minimum configuration for the system — just the TV receiver, for example. Remove every cable (label them as you go!) from the TV except power and a video source such as an antenna or cable system connection. If you still have the problem, try cures until the problem is solved, then start adding cables and equipment back one at a time, fixing the problems as you go along. If you are lucky, you will solve all of the problems in one pass. If not, at least you can point to one piece of equipment as the source of the problem. (While rebuilding the system, watch carefully for loose or intermittent connections and connectors and damaged cables you might have missed during the first inspection.)

Multiple Causes, Multiple Cures — Many RFI problems have multiple causes. These are usually the ones that give new RFI troubleshooters the most difficulty. For example, consider a TVI problem caused by the combination of harmonics from the transmitter due to an arc in the transmitting antenna, an overloaded TV preamplifier, fundamental overload in the TV tuner, common-mode RF on the ac-power connections, and more common-mode RF picked up on the shield of the TV's coaxial feed line. You would never find a cure for this multiple-cause problem by trying only one cure at a time.

In this case, the solution requires that all of the cures be present at the same time. When troubleshooting, if you try a cure, leave it in place even if it doesn't solve the problem. When you add a cure that finally solves the problem entirely, start removing the "temporary" attempts one at a time. If the interference returns, you know that there were multiple causes.

Take Notes — In the process of troubleshooting an RFI problem, it's easy to lose track of what remedies were applied, to what equipment, and in what order. Configurations of equipment can change rapidly when you're experimenting. To minimize the chances of going around in circles or getting confused, take lots of notes as you proceed. Sketches and drawings can be very useful. Take pictures with your camera or smartphone. When you do find the cause of a problem and a cure for it, be sure to write all that down so you can refer to it in the future.

Common Mistake — Using Ferrite Improperly

Amateurs often read about "ferrite cores" and hear only the word "ferrite." They may read that they need to use 10 turns of wire on an FT-240-43 ferrite core and instead choose to use one clamp-on ferrite bead. This will not create enough impedance to form an effective choke. Grabbing a junk-box core and wrapping wire on it may work, and that's okay, but if doesn't, don't discount using a common-mode choke because using one with the right material and construction may work very well. Read the sections on Common-Mode Chokes and follow the directions!

27.5.3 Simple Troubleshooting with Chokes and Filters

Filters and chokes are the number one weapons of choice for many RFI problems, whether the device is the source or the victim. They are relatively inexpensive, easy to install, and do not require permanently modifying the device. The following suggestions will solve many RFI problems.

Common-mode choke — Making a common-mode choke is simple as described in the earlier section, "Common-Mode Chokes." Select the type of core and ferrite material for the frequency range of the interference. (Type #31 is a good HF/low-VHF material, Type #43 from 5 MHz through VHF) Wrap several turns of the cable or wire pairs around the toroid. Six to 8 turns is a good start at 10-30 MHz and 10 to 15 turns from 1.8 to 7 MHz. (Ten to 15 turns is probably the practical limit for most cables.) Ferrite clamp-on split cores and beads that slide over the cable or wire are not as effective as toroid-core chokes at HF but are the right solution at VHF and higher frequencies. For a clamp-on core, the cable doesn't even need to be disconnected from its end. Use Type #31 or Type #43 material at VHF, Type #61 at UHF. At 50 MHz, use two turns through Type #31 or #43 cores.

"Brute-Force" ac-line filters — RF signals often enter and exit a device via an ac power connection. "Brute-force" ac-line filters (see the preceding section on Filters) are simple and easy to install. Most ac filters only provide differential-mode suppression as described in the text. It is essential to use a filter rated to handle the device's required current.

General installation guidelines for using chokes and filters:

1. If you have a brute-force ac-line filter, put one on the device or power cord. If RFI persists, add a common-mode choke to the power cord at the device.

2. Simplify the problem by removing cables one at a time until you no longer detect RFI. Start with cables longer than 1/10th-wavelength at the highest frequency of concern. If the equipment can't operate without a particular cable, add common-mode chokes at the affected or source device.

3. Add a common-mode choke to the last cable removed and verify its effect on the RFI. Some cables may require several chokes in difficult cases.

4. Begin reconnecting cables one at a time. If RFI reappears, add a common-mode choke to that cable. Repeat for each cable.

5. Once the RFI goes away, remove the common-mode chokes you added one at a time. If the RFI does not return, you do not need to reinstall the choke. If the RFI returns after removing a choke, reinstall it.

If you are troubleshooting RFI at a neighbor's house, it may be prudent to leave the extra chokes installed. Some neighbors may want to have you come in and leave quickly, and if the neighbor starts to lose confidence in your skills, the neighbor may ask you to leave and fix the "problem at your station."

27.6 Identifying the Type of RFI Source

This section assumes that you can use a receiver of some sort to observe the presence and characteristics of the RFI signal. That might mean an AM or FM receiver that coverts the signal to audio you can hear and evaluate. Another type of receiver is an SDR with waterfall or spectrum displays that can show you what the signal "looks like" in the frequency domain. If you do not have a receiver but the RFI is causing a particular symptom in the victim device, you can use the device to detect the presence of the interfering signal or noise. The same advice about using repetitive patterns and timing apply, however.

It is useful to recognize an offending noise source as one of several broad categories at the early stages of any RFI investigation. Since locating and resolution techniques can vary somewhat for each type of noise, the process of locating and resolving RFI problems begins with identifying the *general* type of RFI source. It is not useful, however, to go farther and attempt to identify a particular type of equipment or device at the beginning of the search. The purpose of this section is to help you determine the general category of the noise being received and choose the right set of tools and methods to locate its source.

It is often impossible to identify the exact type of device generating the RFI from the sound of the interference. Because there are many potential sources of RFI, it is often more important to obtain and interpret clues from the general noise characteristics and the patterns in which it appears.

A source that exhibits a repeatable pattern during the course of a day or week, for example, suggests something associated with human activity or time of day. A sound that varies with or is affected by weather suggests an outdoor source. Noise that occurs in a regular and repeating pattern of peaks and nulls as you tune across the spectrum, every 50 kHz for example, is often associated with a switchmode power supply or similar pulsed-current devices. A source that exhibits fading or other sky wave characteristics suggests something that is not local. A good ear and careful attention to detail will often turn up some important clues. A detailed RFI log can often help, especially if maintained over time.

27.6.1 RFI From Transmitted Signals

We start with transmitters not because most interference comes from transmitters, but because your station transmitter is under your direct control. In addition, it is relatively simple to determine if your station is the cause of the RFI. If it is, the location of the source is known right away! If it is another antenna, they

Common Mistake — What Is It?

When someone has an unusual RFI problem, the first question he or she almost always asks is: What is it? That's an interesting question, and you may need to ask it during the troubleshooting process, but it is not the first question you should ask. Even if someone were to say that it's a Model XYZ Panashibi switching power supply, what would that tell you? You would still have to go into the world and find it. It could be useful, though, to know whether you are searching for a switching power supply, DSL or cable leakage, or a plasma TV. But be general, not specific, because you don't want to be misled. (An extended version of this sidebar is available on this book's online material.)

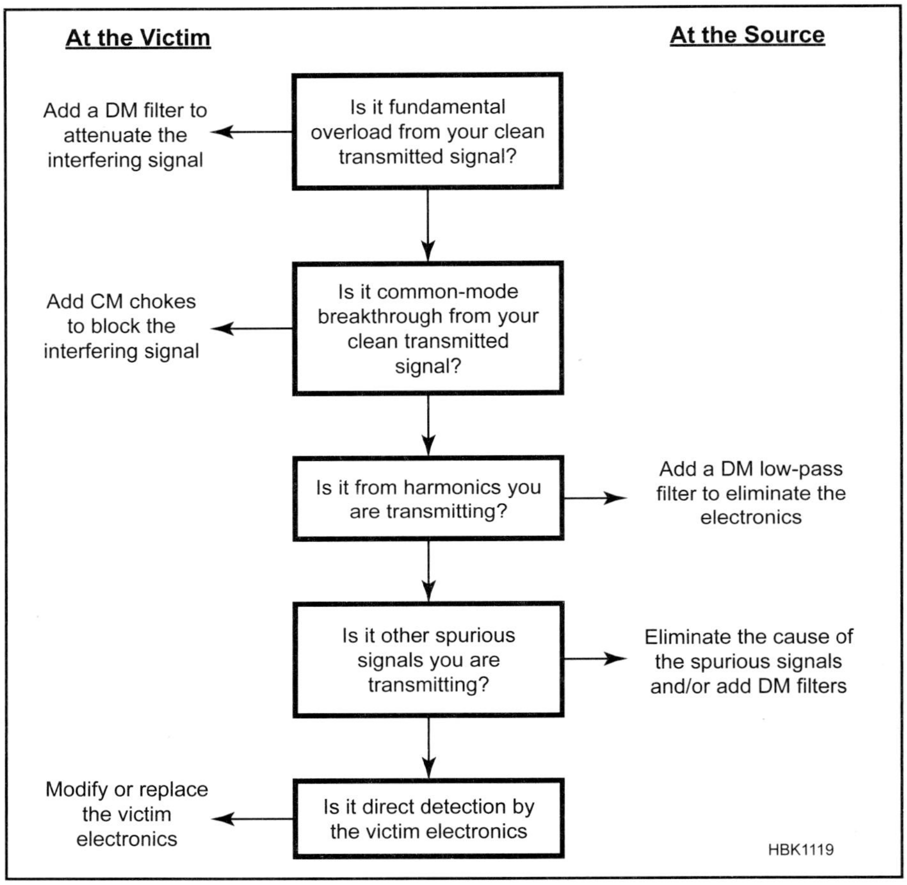

Figure 27.14 — A general process for dealing with RFI from your station's transmitted signal. Remember that RFI can result from both differential-mode and common-mode signals. You may need to perform several of these steps to eliminate the RFI.

Keeping an RFI Log

The importance of maintaining a good and accurate RFI log cannot be overstated. Be sure to record time and weather conditions. Correlating the presence of the noise with periods of human activity and weather often provide very important clues when trying to identify power-line noise. It can also be helpful in identifying noise that is being propagated to your station via sky wave. A log showing the history of the noise can also be of great value should professional services or FCC involvement become necessary at some point.

are generally fairly easy to locate and identify. **Figure 27.14** provides a general guide for dealing with RFI caused by your station's transmitted signal.

Many of the troubleshooting steps in other parts of this chapter assume that your transmitter is "clean" (free of unwanted RF output). This point has been made before, but it is simply good practice to be sure your station is transmitting a signal that meets all FCC regulations *at a minimum*. That includes harmonics, wideband noise, distortion and IMD products, and parasitic oscillations. Be sure there is no arcing to generate noise, either across an insulator or between an antenna and some nearby conductor, even a tree limb. Overloaded or damaged connectors or components can also generate spurious signals, including connectors or antennas that have been damaged by lightning or from failed weatherproofing.

Start by looking for patterns in the interference. Problems that occur only on harmonics of a fundamental signal usually indicate the transmitter is the source of the interference, although harmonics can be generated as non-linear 2nd-order IMD or in the front end of an overloaded receiver. Harmonics can also be generated in nearby semiconductors, such as an unpowered VHF receiver left connected to an antenna, rectifiers in a rotator control box, or a corroded connection in a tower guy wire. Harmonics can also originate in the front-end components of the TV or radio experiencing interference.

If HF transmitter spurious signals or "spurs" at VHF are causing interference, a low-pass filter at the transmitter output will usually cure the problem. If an amplifier is used, be sure it is adjusted properly and not mistuned. If a filter at the transmitter is insufficient, either the amplifier or transmitter may need repair or something else may be creating the harmonics.

Transmitting filters are generally designed for 50 Ω input and output impedances, so install any filters on the input side of an antenna tuner, if one is used. Install a low-pass filter as your first step in any interference problem that involves another radio service at VHF or higher frequencies.

Interference from non-harmonic spurious emissions is extremely rare in commercial radios. Any such problem indicates a malfunction that should be repaired.

27.6.2 Noise from Part 15 Devices

The most common RFI problem reported to the ARRL comes from an unknown and unidentified source. Part 15 devices and other consumer equipment noise sources are ubiquitous. Although the absolute signal level from an individual noise source may be small, their increasing numbers makes this type of noise a serious problem in many suburban and urban areas. The following paragraphs describe several common types of electronic noise sources.

Electronic devices containing oscillators, microprocessors, or digital circuitry produce RF signals as a byproduct of their operation. The RF noise they produce may be radiated from internal wiring as a result of poor shielding. The noise may also be conducted to external, unshielded, or improperly shielded wiring as a common-mode signal where it radiates noise. Noise from these devices is usually narrowband that changes characteristics (frequency, modulation, on-off pattern) as the device is used in different ways. HF and lower frequency problems are typically caused by conducted emissions, although they may travel to the victim as radiated signals. As we'll see, the cure in these cases usually involves common-mode chokes and filters. At VHF and higher frequencies, the probable cause is radiated emissions from the source device. In these cases, shielding is often the solution of choice.

Another major class of noise source is equipment or systems that control or switch large currents. Among them are variable-speed motors in products as diverse as washing machines, elevators, and heating and cooling systems. Charging regulators and control circuitry for battery and solar power systems are a prolific source of RF noise. So are switchmode power supplies for computers and low-voltage lighting. This type of noise is only present when the equipment is operating.

Switchmode or "switching" supplies, solar controllers, and inverters often produce noise signals every N kHz, with N typically being from 5 to 50 or more kHz, the frequency at which current is switched. This is different from noise produced by spark or arc sources that is uniform across a wide bandwidth. This pattern is often an important clue in distinguishing switching noise from power-line or electrical noise. Switching supplies tend to be somewhat unstable in frequency, so you may see the noise signals drift a bit across the band, especially when the supply is first turned on. They may also change a bit in frequency with load changes or with ac voltage dips or surges.

Wired computer networks radiate noise directly from their unshielded circuitry and from network and power supply cables. The noise takes two forms — broadband noise and modulated carriers at multiple frequencies within the amateur bands. As an example, Ethernet network interfaces often radiate signals heard on a receiver in CW mode. 10.120, 14.030, 21.052, and 28.016 MHz have been reported as frequencies of RFI from Ethernet networks. Each network interface uses its own clock, so if you have neighbors with networks you'll hear a cluster of carriers around these frequencies, ±500 Hz or so. All of these signals from digital equipment will be fairly stable with frequency, not drifting across the band. They may be broadband, though, to the point where they may just be a fairly constant hiss across the entire band.

In cable TV systems video signals are converted to RF across a wide spectrum and distributed by coaxial cable into the home. Some cable channels overlap with amateur bands, but the signals should be confined within the cable system. No system is perfect, and it is common for a defective coax connection to allow leakage to and from the cable. When this happens, a receiver outside the cable will hear RF from the cable and the TV receiver may experience interference from local transmissions. Interference to and from cable TV signals is discussed in detail later in this chapter.

Broadband and Narrowband Noise

Noise can be characterized as broadband or narrowband — another important clue. *Broadband noise* is defined as noise having a bandwidth much greater than the affected receiver's operating bandwidth and reasonable uniformity across a wide frequency range. Noise from arcs and sparks, such as power-line noise, tend to be broadband. *Narrowband noise* is defined as noise having a bandwidth less than the affected receiver's bandwidth. Narrowband noise is present on specific, discrete frequencies or groups of frequencies, with or without additional modulation. In other words, if you listened to the noise on an SSB receiver, tuning would cause its sound to vary, just like a regular signal. Narrowband noise often sounds like an unmodulated carrier with a frequency that may drift or suddenly change. Microprocessor clock harmonics, network equipment leakage, oscillators, and transmitter harmonics are all examples of narrowband noise.

27.6.3 Identifying Power-Line and Electrical Noise

POWER-LINE NOISE (PLN)

Next to external noise from an unknown source, the most frequent cause of an RFI problem reported to the ARRL involving a known source is power-line noise. (For more information on power-line noise, see this chapter's section on Power-Line Noise and *The ARRL RFI Book*.) Virtually all power-line noise originating from utility equipment is caused by spark or arcing across some hardware connected to or near a power line. A breakdown and ionization of air occurs and current flows

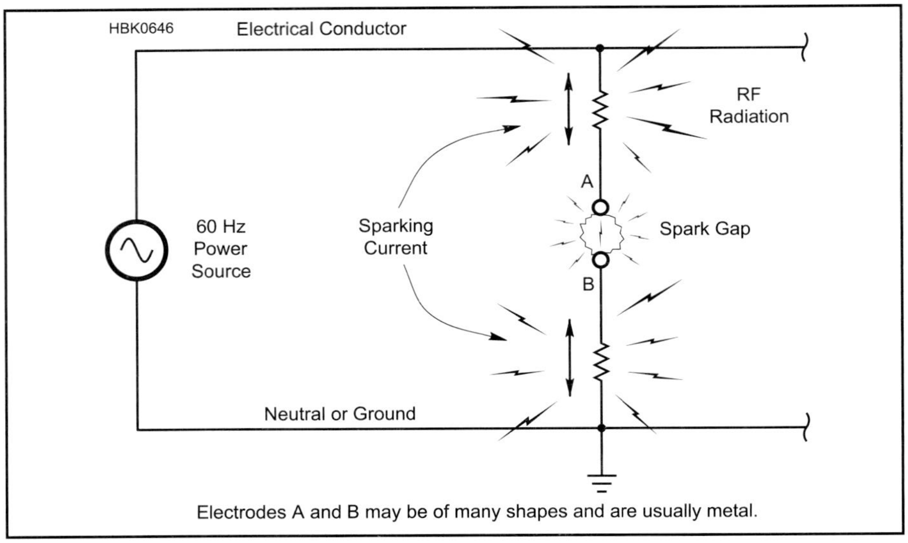

Figure 27.15 — The gap noise circuit on a power line — simplified. (From Loftness, *AC Power Interference Handbook*)

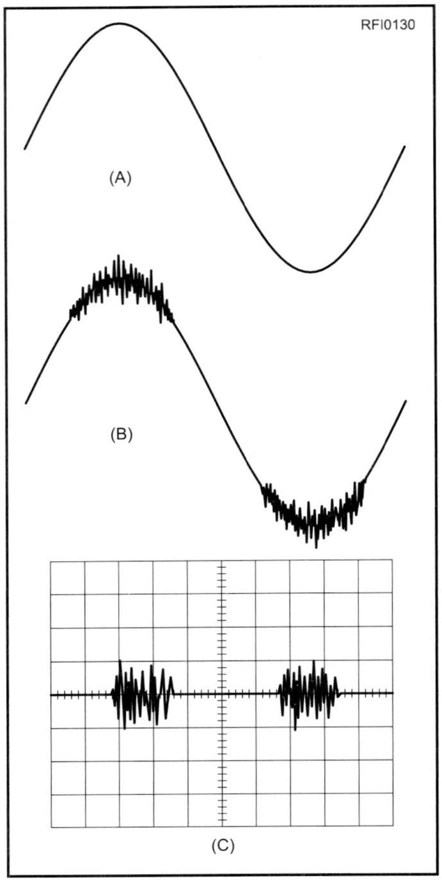

Figure 27.16 — The 60-Hz signal found on quiet power lines is almost a pure sine wave, as shown in A. If the line, or a device connected to it, is noisy, this will often add visible noise to the power-line signal, as shown in B. This noise is usually strongest at the positive and negative peaks of the sine wave where line voltage is highest. If the radiated noise is observed on an oscilloscope, the noise will be present during the peaks, as shown in C.

across a gap between two conductors, creating RF noise as shown in **Figure 27.15**. Such noise is often referred to as "gap noise" in the utility power industry. The gap may be caused by broken, improperly installed, or loose hardware. Typical culprits include insufficient and inadequate hardware spacing such as a gap between a ground wire and a staple. Contrary to common misconception, corona discharge is almost never a source of power-line noise.

While there may not be one single conclusive test for power-line noise, there are a number of important telltale signs. On an AM or SSB receiver, the characteristic raspy buzz or frying sound, sometimes changing in intensity as the arc or spark sputters a bit, is often the first and most obvious clue.

Power-line noise is typically a broadband type of interference, relatively constant across a wide spectrum. Since it is broadband noise, you simply can't change frequency to eliminate it. Power-line noise is usually, but not always, stronger on lower frequencies. It occurs continuously across each band and up through the spectrum. It can cause interference from below the AM broadcast band through UHF, gradually tapering off as frequency increases. If the noise is not continuous across all of an amateur band, exhibits a pattern of peaks and nulls at different frequencies, or repeats at some frequency interval, you probably do not have power-line noise.

The frequency at which power-line noise diminishes can also provide an important clue as to its proximity. The closer the source, the higher the frequency at which it can be received. If it affects VHF and UHF, the source is relatively close by. If it drops off just above or within the AM broadcast band, it may be located some distance away — up to several miles.

Power-line noise is often affected by weather if the source is outdoors. It frequently changes during rain or humid conditions, for example, either increasing or decreasing in response to moisture. Wind may also create fluctuations or interruptions as a result of line and hardware movement. Temperature effects can also result from thermal expansion and contraction.

Another good test for power-line noise requires an oscilloscope. Remember that power-line noise occurs in bursts most frequently at a rate of 120 bursts per second and sometimes at 60 bursts per second. Observe the suspect noise from your radio's audio output. (Note: The record output jack works best if available). Use the AM mode with wide filter settings and tune to a frequency without a station so the noise can be heard clearly. Use the LINE setting of the oscilloscope's trigger subsystem to synchronize the sweep to the line. Power-line noise bursts will remain stable on the display and should repeat every 8.33 ms (a 120-Hz repetition rate) or less commonly, 16.67 ms (60 Hz) if the gap is only arcing once per cycle. (This assumes the North American power-line frequency of 60 Hz.) See **Figure 27.16** for an explanation. If a noise does not exhibit either of these characteristics, it is probably not power-line noise.

ELECTRICAL NOISE

Electrical noise sounds like power-line noise, but is generally only present in short bursts or during periods when the generating equipment or machinery is in use. Noise that varies with the time of day, such as daytime-only or weekends-only, usually indicates some electrical device or appliance being used on a regular basis and not power-line noise. Unless it is associated with a climate control or HVAC system, an indoor RFI source of electrical noise

Common Mistake — The Source Must Be Illegal Because It's So Loud

The FCC limits for consumer electronics are not sufficient to protect against interference to nearby receivers. In a typical suburban neighborhood, the interference can be loud even from a neighboring residence. Proving they meet the applicable limits, these devices can be legally marketed and sold. If and when interference occurs, the burden then falls on the operator of the device to correct the problem.

Conclusion: Don't automatically assume that the device illegally exceeds the FCC limits just because it is causing harmful interference, even if it seems excessive.

is less likely to be affected by weather than power-line noise.

ELECTRIC FENCES

A special type of electrical noise that is easy to identify is the "pop...pop...pop" of an electric fence. High voltage is applied to the fence about once a second by a charging unit. Arcs will occur at corroded connections in the fence, such as at a gate hook or splice. If brush or weeds touch the fence, the high-voltage will cause an arc at those points until the vegetation burns away (the arc will return when the vegetation re-grows). Each arc results in a short burst of broadband noise, received as a "tick" or "pop" in an HF receiver.

27.6.4 Identifying Intermodulation Distortion (IMD)

IMD GENERATED OUTSIDE A RECEIVER

Mixing of signals can occur in any non-linear junction where the original signals are both strong enough to cause current to flow in the junction. This is a particular problem at multi-transmitter sites, such as broadcast facilities and industrial or commercial communications sites. Non-linear junctions can be formed by loose mechanical contacts in metal hardware, corroded metal junctions, and by semiconductor junctions connected to wires that act as antennas for the strong signals. Non-linear junctions also detect or demodulate RF signals to varying degrees, creating interfering audio or dc signals in some cases.

An intermodulation product generated externally to a receiver often appears as an intermittent transmission, similar to a spurious emission. It is common for strong signals from commercial paging or dispatch transmitters sharing a common antenna installation to combine and generate short bursts of voice or data signals. AM broadcast transmissions can combine to produce AM signals with the modulation from both stations audible.

Signals near a broadcast or other communications facility can be strong enough that metal fencing or roofing or a poor cable or wire connection is sufficient to act as a mixer. Curing RFI caused by externally generated IMD requires identifying the signals that are being mixed and then the equipment or structure that generates the unwanted products. Finding where the mixing is occurring can be difficult but is generally easy to fix once identified.

IMD can also be created in a broadcast FM or TV receiver or preamplifier. The solution is to add suitable filters or reduce overall signal levels and gain so that the strong interfering signal or signals can be processed linearly.

IMD GENERATED INSIDE A RECEIVER

Since the products are generated internally to the receiver, the strong signals must be filtered out or attenuated before they can enter the receiver circuits in which IMD products are generated. Like external intermodulation products, those generated by a receiver acting non-linearly appear as combinations of two or more strong external signals. Intermodulation generated within a receiver can often be detected simply by activating the receiver's incoming signal attenuator. If attenuating the incoming signals causes the intermodulation product to by reduced in strength by a greater amount than that of the applied attenuation, intermodulation in the receiver is a strong possibility.

Superheterodyne receiver IMD generally increases and decreases with signal strength. Direct-sampling SDR receivers, however, do not exhibit IMD in the same way as superhets. The SDR will be essentially IMD-free until the limits of the input ADC are reached, and then many distortion products are generated. Hybrid receivers with analog front-ends and DSP functions performed at an IF frequency have IMD performance somewhere between all-analog and direct-sampling SDR.

The following simple "attenuator test" can be used to identify an IMD product, even in cases where it appears similarly in multiple receivers. This procedure applies mainly to superheterodyne receivers with analog circuitry in the sensitive front-end stages.

• If your receiver does not have one, install

Common Mistake — Radiated vs. Conducted Emissions

The FCC does not impose radiated emissions limits below 30 MHz. There are only conducted emissions below 30 MHz. Specifically, the FCC requirements are for the RF conducted on the ac power connection, a power cord for most consumer devices. The limits in the case are expressed in terms of a voltage from both the phase (or "hot") conductor and neutral conductor to the ground conductor.

Note: Typical consumer devices are too small to be an effective antenna at HF. While the device may generate the RF, it doesn't radiate it. The actual radiation takes place from the wires and cables connected to the source device. The FCC rules in this regard only address the connection to ac power.

Above 30 MHz, the situation is reversed. There are no limits for conducted emissions — only for radiated emissions. The limits in this case are expressed as a field strength at a specified distance from the device.

Unbypassed Diode Junctions and RFI

Any conductor in an amateur station more than a small fraction of a wavelength long will pick up considerable RF from the strong transmitted signal. If this conductor is then connected to a diode (or bipolar transistor) junction, a non-linear current will flow in the junction. This, in turn, creates a multitude of harmonics. The harmonics will then be radiated by the same conductor where they are received by nearby stations, including the station that is transmitting.

It is very common to see rectifiers or low-power signal diodes used to protect circuits against reversed power polarity or lightning-induced transients. In addition, rotator control boxes and other station accessories contain rectifiers, signal diodes, and LED indicators. All of these can create harmonics.

In a single-transceiver station, these harmonics may not be apparent since the operator is typically not receiving and transmitting at the same time. If a second receiver is in use, however, these harmonics can be a real problem! Multioperator stations, including Field Day setups, are frequently bedeviled by harmonics, and diode-generated harmonics are an important source of the problem, along with non-linear operation of transmitters.

Luckily, the fix is simple and inexpensive in most cases. Bypass each diode with a small-value capacitor (0.001 to 0.01 μF) — just solder it across the diode leads. Disc or monolithic ceramic capacitors of an adequate voltage rating are sufficient. The trick is to find all of the diodes! An alternative is to add a capacitor from each conductor to the equipment enclosure, which is then connected to the station ground reference plane or bonding bus. Adding ferrite chokes to reduce RF current on the conductors is also good practice.

One place where you should *not* install bypass capacitors is across shunt protection diodes on RF feed lines or other conductors. The capacitor will also shunt the RF! Diodes should not be used as protection devices at RF because of their non-linearity, but they are sometimes used at receiver inputs.

Product Review Testing and RFI

The ARRL Laboratory mostly considers RFI as an outside source interfering with the desired operation of radio reception. Power line noise, power supplies, motor controls, and light dimmers are all *outside* a radio receiver and antenna system and can be a major annoyance that distracts from the pleasure of operating. Have you ever considered RFI generated *inside* your own receiver or by another legally operating amateur radio station?

Harmonics

RFI is just that: radio frequency interference. For instance, a harmonic from another radio amateur's transmitter could be interfering with the desired signal you're trying to receive. One might think in this day and age harmonics are minimal and do not cause interference, but you should reconsider.

The maximum spurious output of a modern amateur transmitter must be 43 dB lower than the output on the carrier frequency at frequencies below 30 MHz. While that figure may be "good enough" for an FCC standard, it's not good enough to eliminate the possibility of causing interference to other radio amateurs or possibly other services. Here at the ARRL Laboratory, the measurement of harmonic emission level is a measurement we make of RFI generated *outside* your receiver.

A radio amateur contacted the Lab, concerned about a report that he was causing interference to operators on the 80-meter CW band while operating at full legal limit power during a 160 meter CW contest. Knowing FCC rule part 97.307, he made the effort to measure his 80-meter harmonic. Easily meeting FCC standards at 50 dB below carrier output on 160, he wondered how his transmitter could cause interference.

Here's a breakdown of power output and signal reduction:

0 dB down from 1500 W = 1500 W
10 dB down from 1500 W = 150 W
20 dB down from 1500 W = 15 W
30 dB down from 1500 W = 1.5 W
40 dB down from 1500 W = 150 mW
43 dB down from 1500 W = 75.2 mW
(this is the FCC legal limit)
50 dB down from 1500 W = 15 mW

While 15 mW may seem too low of a power to cause interference, it wasn't in this case; the interfering signal was reported to be S9. QRP enthusiasts know that at 15 mW, signals can propagate well with the right conditions. Using a single band, resonant antenna will reduce interference caused by harmonics located on other bands, but today many stations employ antennas resonant on more than one ham band. (In this case, the use of a bandpass filter designed for 160 meters would significantly attenuate the harmonic on 80 meters, eliminating the interference.)

RFI Generated in the Receiver

What about RFI generated *inside* the receiver you're operating? You're tuning across the 15-meter band when, all of a sudden, you hear what seems to be an AM broadcast station. Is it a jammer? It is definitely interference — radio frequency interference — caused by two strong shortwave stations! In this particular case, a Midwest radio amateur experienced interference on 15 meters and figured out what was happening. One station was transmitting near 6 MHz and another transmitting above 15 MHz. Signals from these two strong stations combined to create a second order IMD (intermodulation distortion) product at the first IF stage, and this unwanted signal was passed along to subsequent stages and to the speaker. The RFI in this case was caused by insufficient receiver performance (second order IMD dynamic range) where the frequencies of the two stations added up to exactly the frequency that the operator was tuned to. Third-order IMD products from strong in-band signals are another form of RFI created within a radio receiver.

Nearby stations transmitting at or near the IF frequency will cause interference not because the transmitter is at fault, but because of a receiver's insufficient IF rejection. The same interference will be heard if a nearby transmitter is operating at an image frequency.

Power Supplies

RFI can also be created from another part of a radio system, such as an external power supply. In addition to transmitter and receiver testing, the Lab also measures the conducted emission levels of power supplies. This is an indication of the amount of RF at given frequencies conducted onto power lines from a power supply as described in this chapter.

Through the ARRL Lab's published Product Review test results in *QST* magazine, readers can compare the above figures of modern HF transceivers when considering the purchase of a new or used transceiver. Our published data tables spawn friendly competition between radio manufacturers, who in turn strive to perfect their circuit designs. The result is a better product for the manufacturer and a better product for you, the radio amateur.

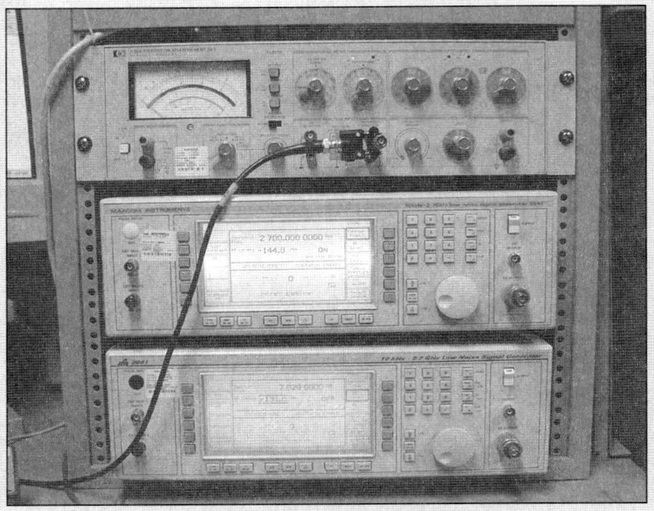

The Lab uses these signal generators to test receivers for internally generated intermodulation distortion, as well as other key performance parameters.

The ARRL Lab maintains a complete set of up-to-date equipment as well as an RF-tight screen room for Product Review testing.

Common Mistake — It's Everywhere, Joe Even Hears It Across Town!

RFI from a typical otherwise legal consumer device usually only propagates for a few hundred feet or less. However, the ARRL often receives reports of non-power line noise RFI covering an extraordinarily wide area. In this case multiple hams hear a noise at opposite ends of a town, across a state wide area, or in some cases, even a multi-state region.

The fact of the matter is that RFI is a very common problem. And if you look hard enough, noise from a consumer device can probably be heard in most suburban neighborhoods these days. Many of them sound alike, so much so that they can easily be mistaken for each other.

Conclusion: Don't rely on RFI reported by hams in other locations as being the same RFI at your location. Before concluding that the RFI covers a widespread area, always make sure that an RFI source is not local and nearby.

Note: While power-line noise at 80 or 40 meters can propagate for miles, it won't propagate much more than that. Most power-line noise sources can sound alike, so beware. Power-line noise will not propagate across state or multi-state regions. In this case, the ARRL recommends the use of signature analysis to determine which source or sources are contributing to an RFI problem before making unnecessary repairs. See the section on Power-Line Noise for more information on signature analysis.

Prospecting for RFI

By using several types of radios at the same time, you can characterize a wide area. For example, RFI professional KB4T uses the following radios when driving: the vehicle AM/FM radio tuned to 1710 kHz AM, an IC-7000 mobile HF transceiver tuned to 15 MHz AM (WWV), and an IC-2820 VHF receiver tuned to 121.500 MHz in AM mode. All three are operated un-squelched with the volume turned down so only a slight hiss is heard. A noise source will be very apparent on one or more of the radios when approached or passed.

Hams aren't the only ones with RF noise problems. Two extensive studies of RF noise mitigation were also conducted by the US Navy; the authors of these studies were all radio amateurs. "The Signal-to-Noise Enhancement Program" and "The Mitigation of Radio Noise and Interference from On-Site Sources at Radio Receiving Sites" are both available from the ARRL. Scroll to the bottom of **www.arrl.org/radio-frequency-interference-rfi** to the section "Naval Postgraduate School RFI Handbooks".

a step attenuator at its RF input. If you use an attenuator internal to your receiver, it must attenuate the RF at the receiver's input.

- Tune the receiver to the suspected intermod product with the step attenuator set to 0 dB. Note the signal level.
- Add a known amount of attenuation to the signal. Typically, 10 or 20 dB makes a good starting point for this test.
- If the suspect signal drops by more than the amount of added attenuation, the suspect signal is an IMD product. For example, if you add 10 dB of attenuation and the signal drops by 30 dB, you have identified an IMD product. An S-unit of change usually represents around 6 dB.
- You can also compare the reduction in signal level between the suspect IMD product and a known genuine signal with and without the added attenuation. Use a known genuine signal that is about the same strength as the suspect signal with no added attenuation. If the suspect signal drops by the same as the added attenuation, it is not an IMD product.

Intermodulation distortion can be cured in a number of ways. The goal is to reduce the strength of the signals causing IMD so that the receiver circuits can process them linearly and without distortion.

If IMD is occurring in your receiver, adding additional filters to remove the strong unwanted signals causing the IMD while passing the desired signal are generally the best approach, since they do not compromise receiver sensitivity. (These are generally referred to as *roofing filters* and are options on many superheterodyne transceivers.) Turning off preamplifiers, adding or increasing the receiver's attenuation, and reducing its RF gain will reduce the signal strength in your receiver. Antenna tuners and external band-pass filters can also act as a filter to reduce IMD from out-of-band signals. Directional beams and antennas with a narrower bandwidth can also help, depending on the circumstances of your particular problem.

27.7 Locating Sources of RFI

Once you have determined the *general* nature of the noise, it's time to find out where it's coming from. "Pinpoint the source precisely and you will know what it is." Good advice! There is no need to know exactly what the source is until it is found or narrowed down to a few possibilities. So, the proper first question is, "How do I locate the source?"

Along with the material in this chapter, the Radio Direction Finding chapter of *The ARRL RFI Book* describes the many methods available to locate sources. There are numerous websites with information about direction finding at all frequencies. This information helps educate the amateur that locating the source is the first step rather than speculating about what it might be.

Locating an offending device or noise source might sometimes seem like trying to find a needle in a haystack. With a little patience and know-how, it is often possible to find the source of a problem in relatively short order. RF detective work is often required, and some cases require a little more perseverance than others. In any case, armed with some background and technique, it is often easier to find an offending source than the first-time RFI investigator might expect. Once the source is pinpointed and its exact nature is known, proper treatment can be applied to eliminate the RFI.

27.7.1 Preparing to Search

Consider the following questions:

1) First and foremost, does the searcher have an open mind? This kind of search is not a visual search until the structure containing the source is located. Even then, equipment must still be relied upon to pinpoint the source. An open mind means a willingness to trust the equipment and not waste time guessing or

deviating from the search path to pursue something seen or guessed at.

2) Is the searcher equipped to produce useful results? A battery-operated AM-FM-SW-VHF receiver with a useful signal strength indicator and a directional antenna suitable for a walking search that receives fairly well at the highest relevant frequency is essential. See the Power-Line Noise section for examples of what works for that type of source. Handheld SDR receivers are now available that can act as a portable spectrum analyzer as well.

3) Does the searcher know how to direction-find? The type of power lines or potential sources in the area is irrelevant at this point. Finding the structure that contains the source is the only objective. Only by trusting your equipment and using it correctly to lead you to the source will the job get done with a minimum of wasted time and effort.

4) Has the searcher done enough listening and record keeping to have a good idea when the source is active, the best (highest?) frequency at which to listen to hear and track the source? Information includes:

- Signal strength
- Highest frequency at which each source can be heard
- A simple description of each source's character
- Any other descriptive data
- Brief notes about weather conditions

Recording the noise over a long period and noting the time whenever it occurs will provide a lot of information about the source.

5) Can it be decided what tools (receiver, antenna, attenuator) are best to locate the source? Even if the right tools are in hand, mastery of them is essential. Make sure you know how to interpret the measurements or work with an experienced user to learn how.

NOISE SOURCE SIGNATURES

Be wary of assuming too much about a source based on what the interference "sounds like" on a receiver or even "looks like" on an oscilloscope or spectrum analyzer. At the beginning of the search, "keep it simple" by not making too many assumptions about the nature of the source. Once the location of a source has been determined, you can begin to narrow the search based on the source's "fingerprint" or signature.

Keep an open mind, be deliberate in your search, and don't act on assumptions about what a source is or isn't until you have exhausted the search techniques. Much time has been wasted in looking for sources of RFI because of assuming a location or type of source at the beginning.

MULTIPLE NOISE SOURCES

Consider that the source may be more than a single piece of equipment! You might have power-line noise combined with switchmode power supply noise, for example. Trying to find the source based on the combination of noise signatures will be frustrating at best. Most suburban and urban locations have multiple noises active at any time. Survey the area to see if different sources are stronger in different areas.

Take careful notes about the nature and behavior of the source, then let your location-finding equipment lead you to each source in turn. It is often most productive to work on the strongest noise source first. Be prepared to discover there are multiple levels of noise sources — after dealing with the top-level source, you may find that additional (and equally objectionable) sources remain. Be patient!

VERIFY NOISE IS EXTERNAL TO THE RADIO

Whenever an unknown source of interference becomes an issue, you can begin the process of identifying the source by verifying that the problem is external to your radio. This advice might not sound very useful, but it can save you a lot of wasted time. Start by removing the antenna connection. If the noise disappears, the source is external to your radio and you are ready to begin hunting for the noise source.

27.7.2 Locating Noise Sources Inside Your Home

Professional RFI investigators and the experiences of the ARRL RFI desk confirm that most RFI sources are ultimately found to be in the complainant's home. Furthermore, locating an in-house source of RFI is so simple that it makes sense to start an investigation by simply turning off your home's main circuit breaker while listening to the noise with a battery-powered portable radio. Dave Cole, NK7Z, has developed a helpful and informative flowchart for investigating RFI in your home. See **Figure 27.17** and Dave's website at **www.NK7Z.net** for further information on his experiences investigating RFI.

Remember that battery-powered equipment may also be a noise source — also turn off or disable battery-powered consumer devices. If you have a UPS (uninterruptable power supply), it will continue to operate with ac power removed. It may be necessary to actually disconnect the UPS battery to shut it completely down.

Common Mistake — It's Not Me, I've Already Eliminated My House as the Source

Consumer devices are by far one of the most common sources of RFI. And once found, a surprisingly high percentage of them are actually in the complainant's home. For this reason, we recommend that hams affected by noise open the main breaker to their residence while listening with a battery-powered radio. (See the text's caution about making sure battery-powered devices are off.) If the noise goes away, you can further isolate the source by opening the individual circuit breakers.

While all this sounds simple enough, many hams actually skip this step. Unplugging or turning off suspect sources is not the same as opening the main breaker. Another mistake is to conclude that the source could not be in your home because you haven't bought or changed anything. Even though quiet when new, devices that fail or otherwise break down can become a source of RFI.

Conclusion: When it comes to consumer devices, do not assume that a noise source is not in your home until you open the main breaker. This is a simple and easy test to perform. Many hams have spent months — even years — searching for a noise that they could have easily found with a simple breaker test. Why skip it?

Common Mistake — Nothing in My Home Could Cause Noise Like That

One of the most common sources of RFI is the ubiquitous switchmode power supply. They are found in almost everything these days. Devices such as "wall warts," many light bulbs, computers, battery chargers, televisions, and various other appliances all may and probably do include switchers capable of causing RFI. In addition, many other electronic devices have clocks and other oscillators that can cause RFI.

Despite this reality, it is surprising how many hams are convinced that the source of an RFI problem could not possibly be located in their home. They are often proven wrong.

Conclusion: Do not automatically assume that nothing in your home could cause an RFI problem. Almost every electronic device in a home can be suspect. Furthermore, a typical home these days can contain hundreds of potential RFI sources.

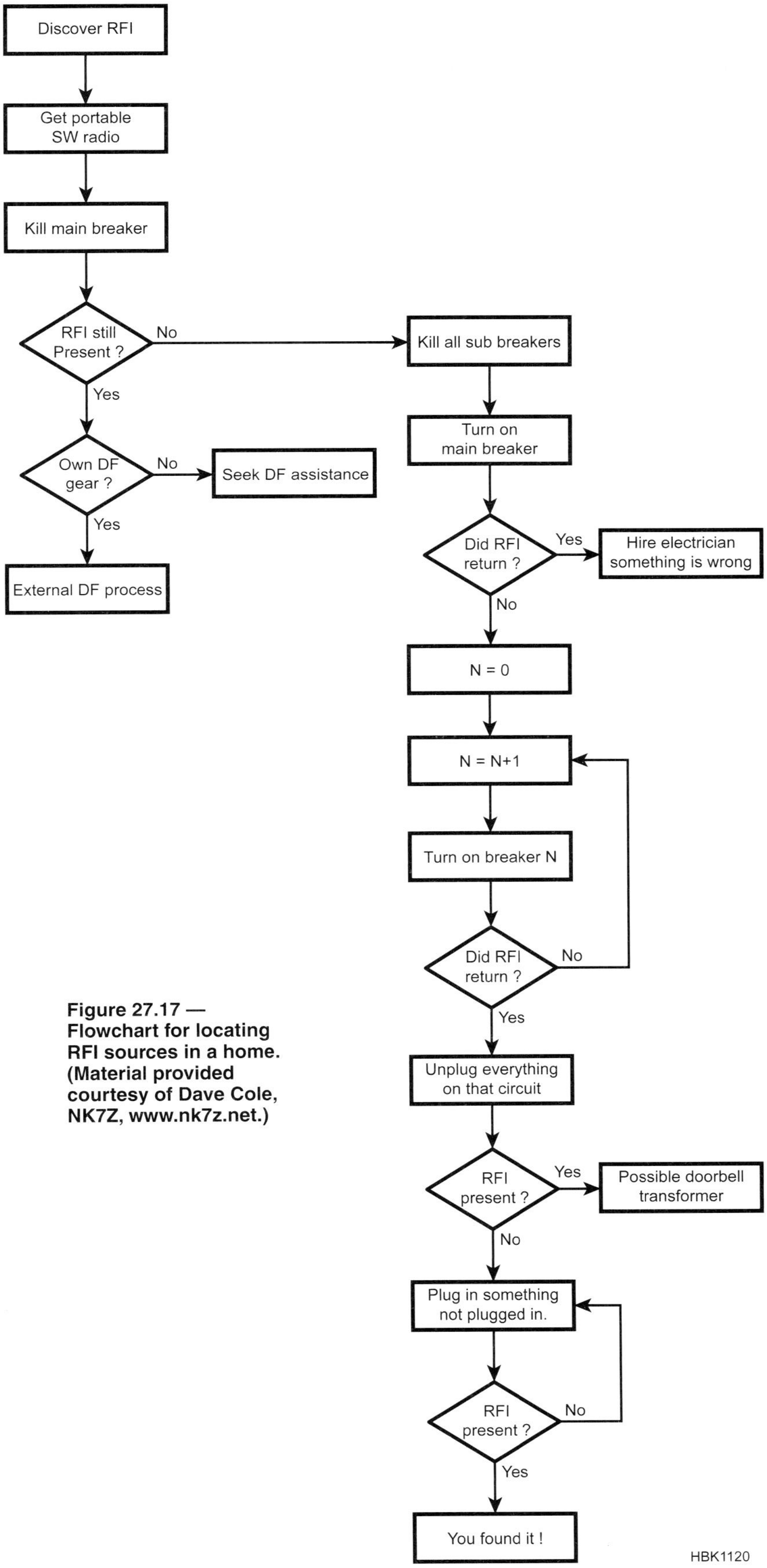

Figure 27.17 — Flowchart for locating RFI sources in a home. (Material provided courtesy of Dave Cole, NK7Z, www.nk7z.net.)

Common Mistake — I Don't Have the Equipment

Most consumer devices actually meet the FCC emission limits, so the RFI problem is usually limited to about 300 feet or less. They are also often in the complainant's residence, and if not, frequently in a home connected to the same power transformer secondary system as the complainant.

With this in mind, a simple portable (battery powered) shortwave radio capable of hearing the noise is all you typical need to find it. Opening the main breaker of your home will locate many of the problems. And if it's not in your home, it's likely in an adjacent or nearby residence. By collapsing or removing the antenna, adding a step attenuator, or reducing the RF gain of the locating receiver, you can often reduce the area in which you search for the source. Reducing the signal level to something at or below the receiver's AGC threshold (or turning AGC off) can often help.

An approach that often works is to find the loudest power pole. Next, look to see what houses are connected to that pole. And finally, if possible, hold the radio a short distance from the power meters at each house. Always be sure to minimize signal levels but do not make any adjustments when making comparisons between the different meters. If access to the power meters is not possible, you can use lawn lights or something similar. The house with the highest noise level will be the one containing the source.

Please note that spectrum analyzers are not needed. In fact, in many cases a spectrum analyzer can add to confusion, especially if there are other sources at other frequencies.

Conclusion: Finding noisy consumer devices is often easier than most people think. Furthermore, it does not require specialized equipment. A cheap battery powered shortwave radio will often do the job.

If the noise goes away, you know the source is in your residence. After resetting the breaker, you can further isolate the source by turning off individual breakers one at a time. Once you know the source circuit, you can then unplug devices on that circuit to find it.

CAUTION: *Do not attempt to remove cartridge fuses or operate exposed or open-type disconnects if it is possible to make physical contact with exposed electrical circuits.*

27.7.3 Locating Noise Sources Outside Your Home

It is often possible to locate a noise source outside your home with a minimum of equipment and effort. The following process will help you determine whether the noise is radiated directly to your equipment or conducted, most frequently by power lines or wiring which then radiate the noise signal themselves.

Because of Part 15's absolute emissions limits, most noise sources that are compliant with Part 15 are within a few hundred feet of the complainant's antenna. They are also often on the same power transformer secondary system as the complainant if the noise is conducted. This typically reduces the number of possible residences to relatively few.

If the noise source is not compliant with Part 15 limits, it may be blocks, or in some rare cases, even thousands of feet from your station. The ballasts of grow lights, for example, have been known to exceed the FCC limits by a considerable margin and have been heard for over ½ mile. Some Part 15 devices, battery chargers for electric scooters and wheelchairs, for example, are well known for exceeding Part 15 absolute emissions limits on conducted noise.

Electrical noise sources in a home, such as an arcing thermostat or a noisy washing machine controller, can also be tracked down in the same way as noise from consumer electronic devices. Electrical noise from an incidental emitter, such as a power line, can propagate much farther than noise from an otherwise legal, unintentional emitter.

The following procedure can be used to trace a noise source to a private home, town house, apartment, or condominium. The number of homes that could be host to a source generating noise could make searching house by house impractical. In such cases, use noise tracking techniques to narrow the search to a more reasonable area.

1) Verify that the noise is active before attempting to locate it. Don't forget this all-important first step. You cannot find the source when it's not present.

2) If possible, use a beam to record bearings to the noise before leaving your residence. Walk or drive through the neighborhood with particular emphasis in the direction of the noise, if known. Try to determine the rough geographic area over which the noise can be heard. If the geographic area over which you can hear the noise is confined to a radius of several hundred feet or less, or it diminishes quickly as you leave your neighborhood, this confirms you are most likely dealing with a Part 15 consumer device.

3) Since the noise will be strongest at an electrical device connected to the residence containing the source, you want to measure the noise at a device common to the exterior of all the potential homes. Suitable devices include electric meters, main service breakers (whether outside or in a utility room), front porch lights, electric lamp posts, outside air conditioner units, or doorbell buttons. Whatever radiator you choose, it should be accessible at each home. The device you select to test as the noise radiator will be referred to in these instructions as the "radiator." Using the same type of device as a test point at each home helps obtain consistent results.

4) You are now ready to compare the relative signal strengths at the radiator on each of the potential source residences. Use a detector suitable to receive the noise, typically a battery-powered receiver. Preferably, the receiver should have a variable RF gain control. An external step attenuator will also work if the antenna is external to the radio. If the antenna can be removed, a probe can also be made from a small piece of wire or paper clip to reduce the receiver's sensitivity. Start by holding the detector about two inches from the radiator at the residence where the noise source may be located. Turn the detector's RF gain control down to a point where you can just barely hear the noise. Alternately, increase the attenuation if using an external step attenuator. Record the RF gain or attenuator setting for each test. Because any wiring connected to a portable receiver becomes part of the receiver's antenna, including headphones, external battery, even the operator's hands and body, keep the configuration the same as you evaluate each potential source.

5) Proceed to the next residence. Again, hold the detector approximately two inches from the radiator. (The detector should be placed at the same location at each residence, as much as is practical.) Since you had previously set the detector to just barely hear the noise at the residence having the interference problem, you can move on to the next residence if you do not hear the noise. Remember, in order for your detector to hear the noise at the next house, the noise level will have to be the same or higher than the previous location. If you need to increase the detector's sensitivity to be able to hear the noise, you are moving away from the noise source.

6) When you reach the next residence, if the level is lower or not heard, you're moving further from the source. Continue your search to residences in other directions or across the street. If the level is higher, then you're headed in the right direction. Be sure to turn the gain control down to the point of just barely hearing the noise as its strength increases.

7) Continue on to the next house, repeating the previous steps as necessary. The residence with the source will be the one with the strongest noise at the radiator.

Depending on the circumstances of a particular situation, it may be possible to first isolate the power pole to which the source residence is connected. Walk or drive along the power lines in the affected area while listening to the noise with a battery-powered radio. Continue to decrease the receiver's RF gain as the noise gets louder, thus reducing the area over which you can hear it. Finally, isolate the loudest pole by reducing the RF gain to a point at which you can hear it at only one pole. Once the pole has been isolated, look to see which houses are connected to its transformer. Typically, this will reduce the number of potential residences to a very small number.

Common Mistake — It Must Be The XYZ Company — They Were Just Here

Many hams might notice activity in their neighborhood by a cable, power, or telephone company. Other variations include an installation of something electrical at an adjacent or nearby residence. Typical examples might include a swimming pool pump, broadband, satellite TV, or an alarm system. So far, so good...

The problem, however, will become apparent a few days later when the hams turn on the radio. Interference now plagues the amateur bands! The ham may now erroneously conclude that the interference is a result of the activity observed in the neighborhood several days prior. In some cases, the ham is so convinced of the source that they fail to take even the most rudimentary steps to confirm their suspicions. This often leads to significant delays in resolving the problem.

The reality in a typical suburban neighborhood is that there could easily be over a thousand devices in range of an amateur's antenna if those sources are all at the legal limit. This RFI environment is also constantly in transition as family members and nearby neighbors buy or turn on electronic devices and other appliances.

Conclusion: Do not automatically assume that you know what the source of an interference problem is based on activity that you've seen in the neighborhood. Always verify the source to the best extent possible.

Be aware that not all power poles have a grounding conductor (a wire running down the side of the pole into the ground). This wire will carry currents from the entire section of the power system for that pole. A receiver held close to that wire will hear more noise than at poles without a ground wire. This may create a "false positive" with respect to homes fed from that pole — the actual noise source may be in any home connected to that section of the power system. Follow up by checking the power drop to the home and, if possible, at the meter before identifying that residence as the source of the noise. Noise on power lines can travel a long way, and it is easy to be fooled by where it is heard.

CAUTION: *Always observe good safety practices! Only qualified people familiar with the hazards of working around energized electrical equipment should inspect power-line or other energized circuitry.*

When attempting to isolate the pole, it is often best to use the highest frequency at which you can hear the noise. Noise can exhibit peaks and nulls along a power line that are a function of its wavelength. Longer wavelengths can therefore make it difficult to pinpoint a particular point along a line. Furthermore, longer wavelength signals typically propagate further along power lines. You can often reduce your search area by simply increasing the frequency at which you look for the noise.

In some cases, tuning upward in frequency can also be used to attenuate noise. This can be especially helpful in cases where your receiver does not have an RF gain control. As mentioned previously, switchmode power supplies typically generate noise that exhibits a regular and repeating pattern of peaks and nulls across the spectrum. While a typical interval might be every 50 kHz or so, the noise will often start to diminish at the highest frequencies. The peaks in some cases might drift over time, but tuning to the highest frequency at which you can hear the noise will often attenuate it enough to help locate it. If the peak drifts, be sure to keep your receiver set on the peak as you attempt to locate the source.

Under FCC rules, the involved utility is responsible for finding and correcting harmful interference that is being generated by its own equipment. In cases where a utility customer is using an appliance or device that generates noise, the operator of the device is responsible for fixing it — even if the noise is conducted and radiated by the power company's power lines.

APPROACHING A NEIGHBOR

Once you identify the source residence and approach your neighbor, the importance of personal diplomacy simply cannot be overstated. The first contact regarding an RFI problem between a ham and a neighbor is often the most important; it is the start of all future relations between the parties. The way you react and behave when you first discuss the problem can set the tone for everything that follows. It is important, therefore, to use a diplomatic path from the very start. A successful outcome can depend upon it!

A self-help guide for the consumer published jointly by the ARRL and the Consumer Electronics Association (CEA) often proves helpful when discussing an interference problem to a neighbor's equipment with the neighbor. Entitled *What To Do if You Have an Electronic Interference Problem*, it may be printed and distributed freely. It is available on the ARRL website at **www.arrl.org/information-for-the-neighbors-of-hams** and also in the online information accompanying this book. Be sure to download and print a copy for your neighbor before you approach him or her.

With the noise active and with a copy of the pamphlet handy, approach your neighbor with a radio in hand, preferably an ordinary AM broadcast or short-wave receiver. Let them hear it, but not so loud that it will be offensive. Tell them this is the problem you are experiencing and you believe the source may be in their home. Don't suggest what you think the cause is. If you're wrong, it often makes matters worse. Give them the pamphlet and tell them it will only take a minute to determine whether the source is in their home. Most neighbors will agree to help find the source, and if they agree to turn off circuit breakers, it can be found very quickly. Start with the main breaker to verify you have the correct residence, then the individual breakers to find the circuit. The procedure then becomes the same as described for your own residence.

27.7.4 Radio Direction Finding

Radio direction finding (RDF) can be a highly effective method to locate an RFI source, although it requires more specialized equipment than other methods. RDF techniques can be used to find both broadband and single-signal, narrowband noise sources. Professional interference investigators almost always use radio direction finding techniques to locate power-line noise sources. See the **Antennas** chapter for more information on direction finding antennas. The following recommendations are most effective in locating noise from a broadband source. For narrowband, single-signal sources, regular RDF techniques are the most effective way to determine a source's location.

A good place to start, whenever possible, is at the affected station. Use an AM receiver, preferably one with a wide IF bandwidth. An RF gain control is particularly helpful, but an outboard step attenuator can be a good substitute. If there is a directional beam capable of receiving the noise, use it on highest frequency band at which the noise can be heard using the antenna. If you can hear the noise at VHF or UHF, you'll typically want to use those frequencies for RDF.

Select a frequency at which no other stations or signals are present and the antenna can discern a directional peak in the noise. Rotate the beam as required to get a bearing on the noise, keeping the RF gain at a minimum. Repeat with a complete 360° sweep using the minimum RF gain possible to hear the noise in its loudest direction. Try to decrease the RF gain to a point at which the noise clearly comes from one and only one direction. You can simultaneously increase the AF gain as desired to hear the noise.

Distant sources, including power-line noise, are generally easier to radio direction find at HF than nearby sources. Whenever possible, it's almost always better to use VHF or UHF when in close proximity to a source. Tracking a source to a specific residence by RDF at HF is sometimes possible. Such factors as balance and geometry of a home's internal wiring, open switch circuits, and distance may cause the residence to appear somewhat as a point source.

If the search is being conducted while mobile or portable, VHF and UHF are typically the easiest and most practical antennas. Small handheld Yagi antennas for 2 meters and 440 MHz are readily available and can serve double duty when operating portable. Many handheld receivers can be configured to receive AM on the VHF bands. Be sure to check your manual for this feature. VHF Aircraft band or "Air band" receivers are also a popular choice since they receive AM signals.

Using RDF to locate an HF noise source while in motion presents significant challenges. Conducted emissions are typical from a consumer device or appliance. In this case, the emissions can be conducted outside the residence and on to the power line. The noise can then propagate along neighborhood distribution lines, which in turn act as an antenna. The noise can often exhibit confusing peaks and nulls along the line, and if in the vicinity of a power line radiating it, RDF can be extremely difficult, if not impossible. Depending on the circumstances, you could literally be surrounded by the near field of an antenna! You would generally want to stay away from power lines and other potential radiators when searching at HF.

Antennas for HF RDF while walking typically include small loops and ferrite rod antennas. In some cases, a portable AM broadcast radio with a ferrite rod antenna can be used for direction finding. An HF dipole made from a pair of whip antennas may be able to get an approximate bearing toward the noise. Mount the dipole about 12 feet above ground (remembering to watch out for overhead conductors!)

and rotate to null out the noise. For all three types of antennas, there will be two nulls in opposite directions. Note the direction of the null. Repeat this procedure from another location and then triangulate to determine the bearing to the noise.

27.7.5 Using SDR Receivers

A portable SDR receiver with supporting software and a few accessories can be a powerful tool for locating RFI sources. Recent developments have greatly simplified the setup needed to effectively and inexpensively visualize and track down RFI sources. All you need to get started is a portable computer, an SDR receiver, SDR software, and a direction-finding antenna (see **Figure 27.18**). An SDR and display software provide the ability to see the characteristics of the RFI signal. This helps you focus on locating one RFI source at a time. This allows faster locating of the source. Once located, the source can usually be eliminated.

TYPES OF RECEIVERS

There are many SDR receivers to choose from. Be sure your receiver and accompanying SDR display and control software has the features and specifications suitable for tracking RFI. It may be best to select your SDR software first, and then let that decision drive your choice of SDR receiver.

The receiver should have frequency coverage from 100 kHz (where switchmode power supplies often operate) up through at least the UHF range (for tracking power line sources close-in). The accompanying software should allow you to see up to 5 to 10 MHz of spectrum at once. Observing this wide frequency range allows you to correlate which of the various pulses, spurs, and spikes (both in and out of the amateur bands) may be from the same source. It will help if your SDR is powered by a USB port on the portable computer so that a separate power source is not required.

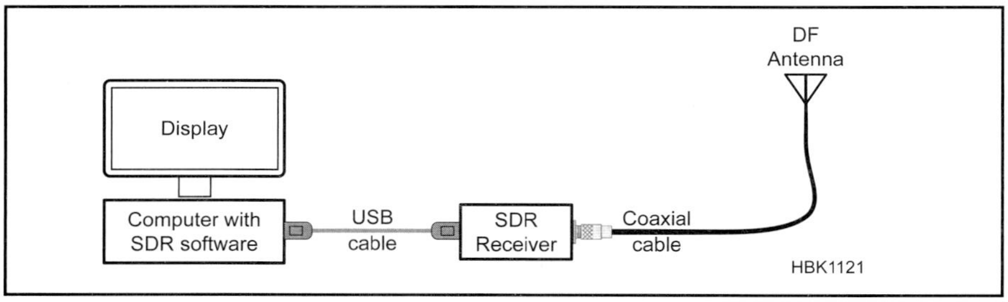

Figure 27.18 — Block diagram of PC+SDR+DF Antenna.

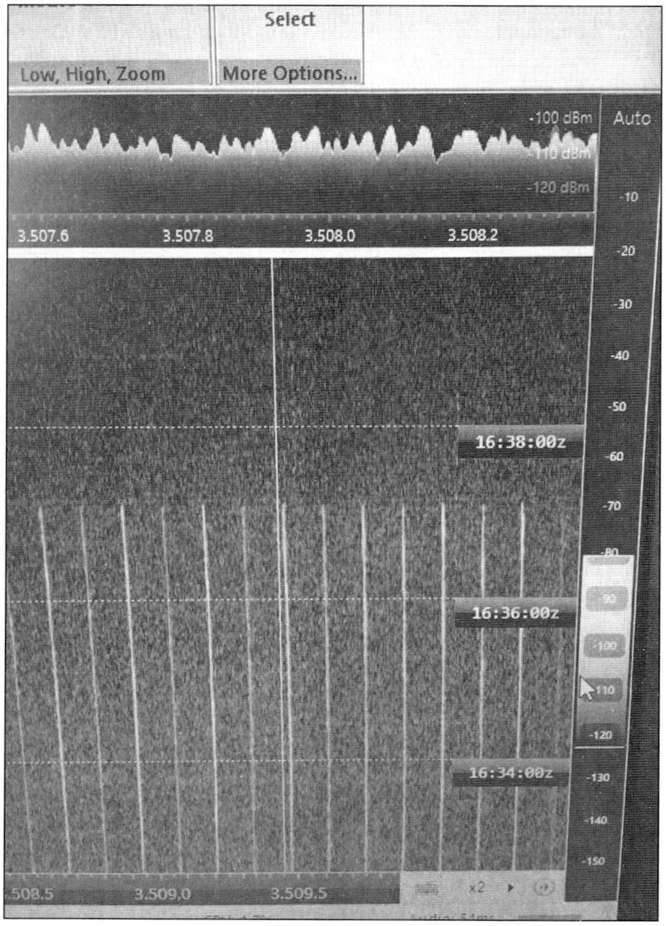

Figure 27.19 — Photo showing 2 kHz span with spikes from refrigerator's microprocessor, cured with a choke made from two snap-on Type #31 cores on ac power cord.

SDR SOFTWARE

Most SDR receivers require support from a user interface software application. Your choice of software will depend on your computer's operating system; whether the software supports the SDR receiver; and the software's features. The following SDR software features are useful for tracking RFI:

- Both spectrum and waterfall display
- Adjustable visible bandwidth with the ability to zoom in to a very narrow span of 1 or 2 kHz (helpful for identifying microprocessors) (see **Figure 27.19**)
- Ability to zoom out to see a wide span of 2 to 8 MHz (to match patterns of RFI from sources outside the amateur bands to what you are experiencing inside the bands) (see **Figure 27.20**)
- CW, SSB, AM and FM modes
- Ability to pre-set particular bands of interest
- Adjustable settings for IF and RF gain (useful for reducing receiver overload and artifacts, as well as allowing you to calibrate the software's signal strength readings)
- Ability to obtain screenshots of the spectrum and waterfall display (useful for documenting the signature of RFI source)
- Has speaker audio output

DISPLAY COMPUTER

For tracking down RFI in the field you will want a portable computer to display the signals received by the SDR. Your setup is most versatile when the SDR software is installed on a lightweight, battery-operated computer, such as a tablet or laptop with a bright screen. You will need to be able to read the screen in sunlight. The computer's system resources should be adequate to allow the SDR software to run smoothly without freezing or hesitation. The broader the spectrum being visualized, the more resources that will be required.

In addition to the spectrum and waterfall images on the computer display, it can also be

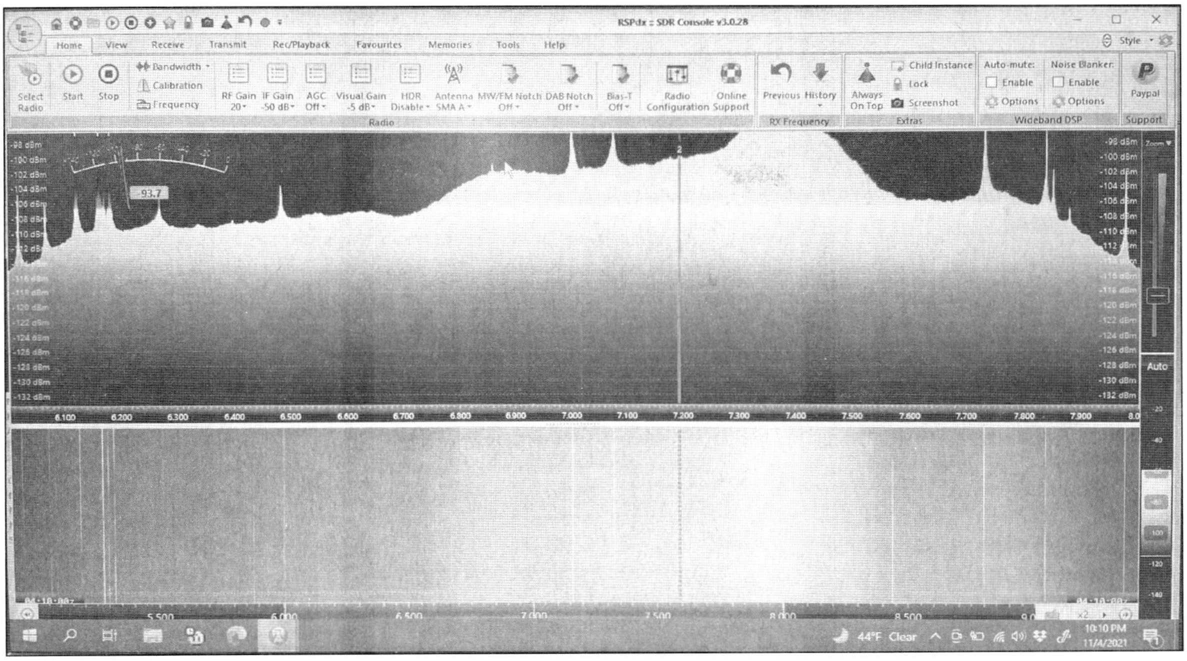

Figure 27.20 — Showing source of 160-meter RFI originating from a switchmode power supply source outside the band.

useful to have substantial audio output available either from the computer or through an accessory speaker. This is helpful when you are attempting to demonstrate the sound of the RFI. For example, you might be standing on the ground below a suspected power line source, while a utility lineman is up above moving hardware components to determine which component is the RFI source. With adequate volume, the lineman can directly hear how the movement affects the RFI.

ANTENNAS AND PROBES

Having a variety of different antennas available can be useful at various stages of the RFI hunt. The appropriateness of a particular antenna will depend upon the frequency of the RFI and your distance from the source. These will be the tools that will help you find out where the source is located.

You should start with your usual station antennas connected to your regular transceiver. Begin here and observe the behavior of the RFI. Try to correlate it to time of day, weather conditions, human activities, modulation characteristics, and bands where it is most prominent. Avoid becoming distracted by jumping to early conclusions. If your station has a directional antenna, that may be useful in determining the general direction of the source. Connect your station antennas to your SDR set-up and capture an image of its "signature" for future reference. Saving these snapshots of the RFI can be useful for future reference in the case of intermittent or recurring noise sources.

Most station antennas are tuned for particular bands. Having an antenna with a broadband response can be very useful for analyzing a wide range of the spectrum with the SDR. As mentioned above, this may help correlate your interference with primary RFI sources that are outside of the amateur bands. There are broadband dipole designs but these are not easily rotatable.

A very good broadband antenna choice is the portable flag antenna designed in 2021 by Don Kirk, WD8DSB. It was made specifically for hunting RFI and has impressive directional characteristics from 1 to 30 MHz. It is portable and has a sharp unidirectional sharp null. Due to its small size, it requires a preamplifier. Construction details are found in the March 2021 *QST* article by Don Kirk, WD8DSB, "Portable Flag Antenna for MF/HF Radio Direction Finding." DX Engineering (**www.dxengineering.com**) also offers a complete kit, part number DXE-NOISELOOP, based on this design. They also offer a broadband 30 dB preamplifier and attenuator, part number DXE-NL-PRE-ATT-1, designed for use with the portable flag.

Another direction-finding (DF) antenna option is a magnetic loop. These can be inexpensive, easy to construct, and relatively portable. Like the portable flag, these require preamplifiers. They are very narrow-banded and bi-directional, thus requiring triangulation to locate a source.

As you get closer to the RFI source, you may be able to detect it in the VHF/UHF range. In that case, a Yagi antenna is very useful. The Yagi is unidirectional and requires no preamplifier; however, they are relatively narrowband antennas. They are particularly useful for pinpointing close-in power line sources. You may even be able to identify an individual component generating the RFI.

Another useful antenna for RFI sleuthing is a *sniffer probe*. This can be handy as you get very close to a suspected source such as a computer, power supply, or other device. The sniffer probe is very small (typically a short wire or small loop with a maximum dimension of about an inch). This allows you to get very close to the RFI source, possibly identifying a specific wire or component as the source. Sniffer probes as shown in **Figure 27.21** are inexpensive and easy to construct.

A sniffer probe can also be used to probe individual circuit breakers on a structure's electrical service panel. This can help identify

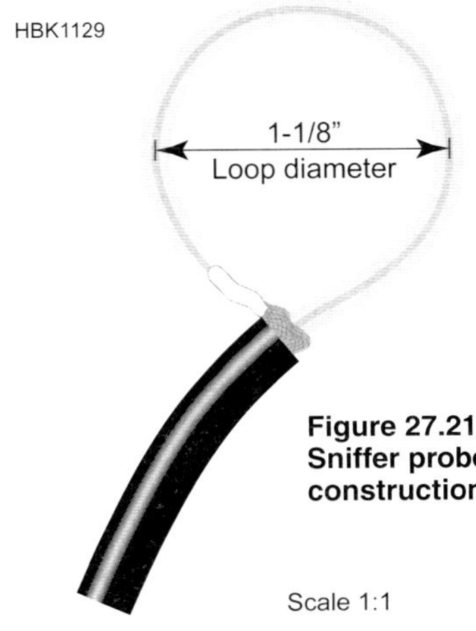

Figure 27.21 — Sniffer probe construction.

Common Mistake — Rotating the DF Antenna Too Quickly

SDR software works by performing sophisticated digital signal processing which introduces significant latency into the signal chain. A delay of perhaps 1 second occurs between when you point the antenna at an RFI source and the time signals from it appear on the SDR software's display or in the audio. Accurate direction-finding requires slow rotation of the antenna. This is not a time for wildly swinging the antenna back and forth!

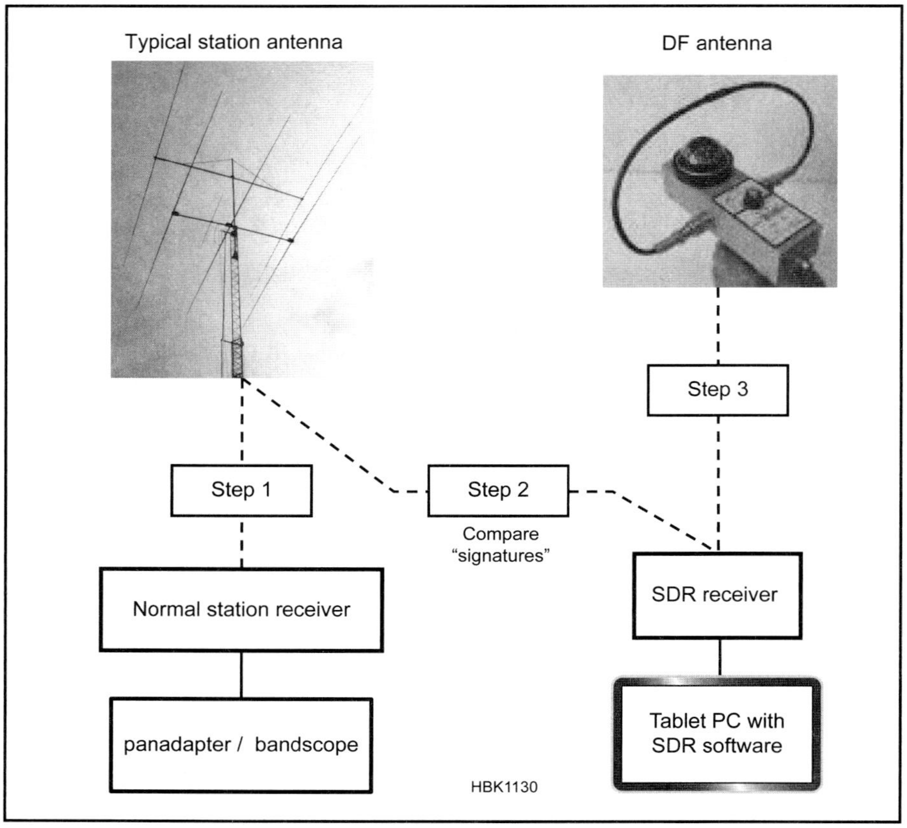

Figure 27.22 — Confirming signature between station transceiver + antenna versus SDR + DF antenna.

which branch circuit is connected to the RFI source. See WD8DSB's video of the configuration and process on YouTube at **www.youtube.com/user/wd8dsb**.

Several accessories will help round out your RFI detective kit. A camera tripod can hold your DF antenna steady as you make careful measurements. This will also assist in accurately repeating and confirming your direction measurements.

A step attenuator may be necessary to reduce signal strength from the RFI source as you move in closer. Strong RFI signals can easily overload a receiver and produce misleading artifacts. Place the step attenuator in the feed line between the DF antenna and the SDR's antenna input, then switch in the desired attenuation. If signals suddenly appear or disappear when attenuation is changed, the receiver may be overloading, creating spurious responses mistaken for RFI. The signals displayed should change only by the amount of attenuation added or removed.

Depending on the type of DF antenna you use, a preamplifier may be necessary. The flag loop and magnetic loop antennas require them to boost signal strength to a useable level. Broadband preamplifiers are more versatile because they allow for looking at larger segment of spectrum. Preamplifiers can make the receiver more susceptible to overload, so use them only when necessary.

An AM broadcast band hardware filter can help reduce potential artifacts caused by overload from nearby AM transmitters. The filter should be placed between the antenna and SDR input. Software filters will not protect the input of the SDR from overloading but can be useful in cleaning up the SDR display and removing strong signals not related to the RFI signals.

PRACTICING WITH YOUR EQUIPMENT

Once you have assembled your SDR, software, and DF antennas, you should, as with any other gear, practice using it before going into the field. This is especially true of the software. It is much easier to configure and experiment with the SDR software while sitting relaxed inside, not trying to do so the first time your equipment is out in the field. It will take a while to get comfortable with the system, so give yourself plenty of time. Then go outside to practice by locating a known RF source such as a signal generator, broadcast station, or friend's amateur station.

CHARACTERIZE THE RFI

Once you have RFI affecting your receiver's noise floor, begin by carefully studying its characteristics. This might include getting an image of what it looks like on your regular transceiver's spectrum scope or recording its audio characteristics. The results will become part of the signature you use to track the RFI source. Your goal will be to ensure you are

Common Mistake — Failing to Recognize SDR Artifacts

Beware of false signatures caused by artifacts. An artifact is something appearing on the SDR spectrum or waterfall display that is generated by the SDR, the computer, or a power supply close by. Some sources of artifacts include an overload of the SDR input and RFI generated by your computer or its power supply.

Take these steps to test for and prevent artifacts in the RFI hunting setup. Be sure the SDR receiver is not overloaded by strong signals such as local broadcast stations. You might use a broadcast band filter. You could also test by inserting an attenuator between the antenna and the SDR. Artifacts will usually not be reduced — but real RFI is.

A common source of artifacts is a computer's power supply. To test and eliminate these you can remove the power supply and run the computer on a battery.

A laptop's touchpad may generate close-in RFI. Test for this by touching the touchpad to see if the potential artifact is affected. This source may usually be eliminated by using a mouse instead.

Many computer artifacts can be eliminated by moving the antenna away from the computer.

And, finally, consider using a second receiver to confirm whether a "signal" is real or an artifact.

tracking the same RFI as is raising the noise floor in your station. There are many potential sources, but you should stay focused on whatever impedes your receiving ability. After you eliminate the top layer of RFI getting to your receiver, you can then move on the next RFI source.

Start by focusing on specific RFI while using your normal station receiver and antenna. Then change from that receiver (but with the same station antenna) to the SDR. Be sure to confirm that you are seeing the same RFI source on the SDR. Next, connect the SDR to your DF antenna. Again, confirm that you are seeing the same RFI source (see **Figure 27.22**). Now, you are ready to locate the same RF source that is on your radio.

Common Mistake — Blocking Directional Pattern of the DF Antenna

Having the direction-finding antenna blocked by objects may make its pattern unreliable. This will confound the results of your direction finding. Be sure to place the DF antenna in the clear. Keep your body and building structures out of the way. Great distances are required at lower frequencies.

LOOKING FOR THE RFI SOURCE

Before you start looking for an RFI source outside your own home, be sure to conduct a "power-off" test. Doing this will allow you to dramatically narrow the scope of your RFI hunt. This test consists of completely shutting off the power to your house, while listening on your regular receiver (operating on batteries) and using your normal antenna. The power is switched off at the main electrical switch panel. Be sure that no devices in the house are left powered on. Check for devices that may have standby batteries, UPS devices, chargers, battery tenders, and even vehicles parked at your house that may have active electronics.

If the RFI signature disappears during this test, you can be assured that the RFI source is located within your house. Once the power is restored, and after all devices have come out of standby mode, you can use a sniffer probe to test each of the circuits on your breaker panel. This allows you to sniff for a branch circuit with high RFI and to narrow your search to that circuit.

Additionally, a sniffer probe may also allow you to locate the offending source(s) inside your station. These sources can be switchmode power supplies, computer monitors, USB ports, etc. The source may also be isolated by sequentially powering-off individual devices.

To locate sources outside your house, obtain the RFI source's signature while using your regular station antennas, then match that image to what you can see with the SDR receiver and one of the portable antennas. Next, take the SDR, computer, and portable DF antenna into the field, e.g., near your regular antenna.

Mount the DF antenna on a rigid support, such as a tripod, then slowly rotate it to determine from which direction the RFI is coming. To maximize directional characteristics of the DF antenna, be sure to place the antenna away from other objects and up and away from your own body. A tripod helps assure repeatability of the measurements (either nulling or peaking, depending on the type of antenna being used).

When a suspected RFI source is further away, you will need to expand your search area. Initially, this may require taking your SDR and DF equipment in a vehicle. Have a partner drive while you operate the equipment.

Continually confirm that you are tracking the same RFI signature. Obtain an initial DF bearing and head in that direction while watching for changes in signal strength. This should eventually lead you closer and closer to the source. To avoid being confounded by possible re-radiation, attenuation, and reflections, you should obtain multiple bearings from various locations. Then triangulate on a map to find the approximate area where the bearings intersect.

When looking for SDR over long periods, an example of using an SDRPlay for "site surveys" is described by Dave Cole, NK7Z on the web page "Using a SDR as a RFI site survey tool" at **www.nk7z.net/sdr-for-rfi-survey-p1**. In that example, a PC stores data over an extended period across spectrum extending from audio through 2 GHz. Other resources are available, including RFI snapshots, at **www.nk7z.net**.

POWER-LINE NOISE

These same techniques may also be used for locating power line sources. However, be aware that power lines present special challenges. Power lines consist of many discrete hardware components such as lightning arrestors, tie wires, ground wires, and staples, among other items. Any of these may be an RFI source.

RFI sources on power lines are also connected to long conductors, which often carry and re-radiate RFI over great distances. This may confound the process of pinpointing the source(s). Often, one stretch of power lines can contain multiple RFI sources that may be active at different times. Keep an open mind and remain alert for the possibility that you have more than one source. Use the SDR image to help differentiate multiple sources.

The SDR software can "zoom in" to show a very narrow bandwidth, and many have a DSP component that may allow demodulating the audio. This may allow you to see whether the signal has the characteristics of a 120 Hz signature. Additional clues may be found by correlating RFI intensity with weather patterns (humidity, temperature, wind). As you close in on a power line RFI source, it likely will be stronger at VHF/UHF where a highly directional beam for those frequencies becomes very useful.

27.8 Television Interference (TVI)

Beginning in 2009, over-the-air TV broadcasting in the United States began a transition from the NTSC analog AM format mostly on VHF and low-UHF channels to a digital DTV format exclusively on UHF channels. (See the **Modulation** and **Image Communications** chapters for more information about digital TV signals.) Due to the interference-rejection characteristics of the DTV system and the greater frequency separation between TV channels and most amateur operation, the scourge of TVI from amateur transmissions has been greatly reduced.

Digital TV has better immunity to interference than the obsolete analog system, but for both formats clear reception requires a strong signal at the TV antenna-input connector so the receiver must be in what is known as a *strong-signal area*. Outside this area, signal strength eventually falls below a threshold at which picture quality degrades rapidly. It is in the weak-signal areas where TV receivers are most likely to experience TVI, from amateur or other noise signals.

Along with over-the-air reception, some TV receivers can also be connected directly to cable TV systems through the antenna input (for "cable-ready" receivers). They can also act as video displays for cable TV converters and satellite receivers that translate the cable or satellite signals to RF signals for the TV receiver to display. If the converter experiences interference, it will likely be observed on the TV equipment.

The TV receiver is often central to a home entertainment system and may have audio, video, and control cables connected to it from various media players or gaming devices, creating additional opportunities for interference. Because of the different types of TV equipment and the many ways in which the equipment can be configured and connected, the general term "TV" will be used in this section.

models. Switchmode power supplies in current TV equipment, however, can still be a source of RFI to amateurs.

Note that most TV broadcasts are now in the UHF spectrum, regardless of what channel number they use in their identification. You can find the actual channel used by a TV station at **www.fcc.gov/media/engineering/dtvmaps**, which provides coverage maps based on Zip code.

The manufacturer of the TV or video equipment can sometimes help with an interference problem. The Consumer Technology Association (CTA) can also help you contact equip-

27.8.1 Causes of TVI

TVI to the reception and display of an RF television signal caused by problems receiving the over-the-air signal include:

• Spurious signals within the TV channel coming from your transmitter or station.

• The TV may be overloaded by your transmitter's fundamental signal.

• Signals within the TV channel from some source other than your station, such as electrical noise, an overloaded mast-mounted TV preamplifier, or a transmitter in another service.

• One or more connecting cables are loose or defective. Be sure cables are of good-quality and that connectors are installed properly, especially older crimp-type F connectors.

The *TVI Troubleshooting Flowchart* in **Figure 27.23** is a good resource for troubleshooting interference from a strong signal, although it is somewhat outdated and does not address TVI via non-RF or power connections. If you are following this diagram, treat "CATV" as including satellite or other over-the-air receiving equipment that produces an RF signal connected to the TV's antenna or similar input. If you get to the box with the label beginning "Possible Direct Pickup Problem," refer to the section below on Non-Radio Devices for further troubleshooting directions.

Interference to basic functions or features or disruption of a video signal from another piece of equipment is usually caused by common-mode breakthrough or direct detection of:

• Strong signals picked up on video or other cables connected to the TV as common-mode current

• Interference to an RF signal received by a device connected to the TV

• Common-mode breakthrough or direct detection of a strong signal by an unshielded TV or device connected to the TV.

And you should always check to see if the TV might be defective or misconfigured, making it look like there is an interference problem.

Certain types of television receivers and video monitors are reported to cause broadband RFI *to* amateur signals — large-screen plasma display models were at one time reported to be the most frequent offender — and this may be difficult to cure due to the nature of the display technology. Fortunately, less-expensive, more power-efficient, and RF-quieter LCD and OLED technology has displaced plasma technology for current

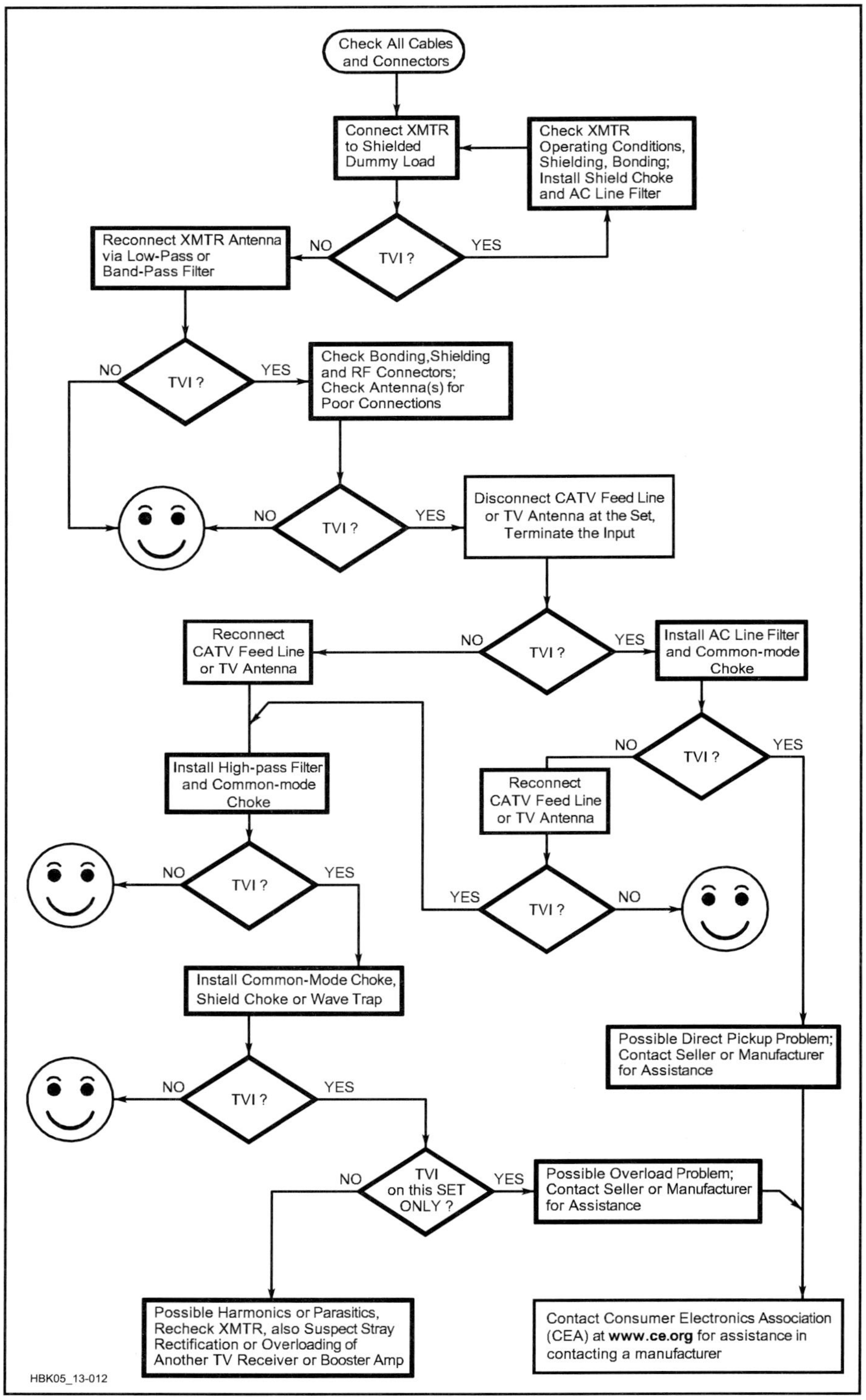

Figure 27.23 — TVI troubleshooting flowchart for over-the-air or cable TV interference.

ment manufacturers. Contact them directly for assistance in locating help at **www.cta.tech**.

COMMON SOURCES OF TVI

HF transmitters — A nearby HF transmitter is most likely to cause fundamental overload. This is usually indicated by interference to all channels, or at least all VHF channels. To cure fundamental overload from an HF transmitter to an antenna-connected TV, install a high-pass filter directly at the TV set's antenna input. (Do not use a high-pass filter on a cable TV input because the HF range is used for data and other system signals.)

A strong HF signal can also result in a strong common-mode signal on the TV's feed line. A common-mode choke will block signals on the outside of the feed line shield, leaving the desired signals inside the feed line unaffected. **Figure 27.24** shows how a common-mode choke is constructed for a coaxial feed line. The same choke can be applied to audio, control, and power cables as well, to prevent RFI caused by common-mode ingress via these non-RF paths. (See the section Common-Mode Chokes earlier in this chapter.)

These filters and chokes can probably cure most cases of TVI! **Figure 27.25** shows a "bulletproof" installation for both over-the-air and cable TV receivers. If one of these methods doesn't cure the problem, the problem may be common-mode breakthrough on a non-feed line connection. It may also be direct pickup in which a signal is received by the TV set's circuitry without any conducting path being required. In that case, don't try to fix it yourself — it is a problem for the TV manufacturer.

High-pass filters *should not* be used in a cable TV feed line (Figure 27.21A) with two-way cable devices such as cable modems, set-top boxes, and newer two-way or cable-ready CableCARD-equipped TVs. The high-pass filter may prevent the device from communicating via the cable network's upstream signal path.

VHF Transmitters — Most TV tuners are not very selective, and a strong VHF or UHF signal, including those from nearby FM and TV transmitters, can overload the tuner easily, particularly when receiving VHF or UHF broadcasts over the air and not via a cable or satellite system. In this case, a notch filter (also known as stop-band or band-reject filters) at the TV can help by attenuating the fundamental signal that overloads the TV tuner. Channel Plus (**www.solidsignal.com**), PAR Electronics (**www.parelectronics.com**), and Scannermaster (**www.scannermaster.com**) sell notch filters.

If the VHF transmitter is generating a harmonic or other spurious emission causing RFI, a transmission line stub filter may be a good solution. The stub can be designed to remove a signal at the transmitter. If the transmitter's fundamental signal is overloading the receiver, a notch filter stub can also be applied at the receiver. See the **Transmission Lines** chapter for more information about these filters.

A common-mode choke may also be necessary if the TV is responding to the fundamental signal present as common-mode RF on the TV's feed line or other connecting cable.

TV Preamplifiers — Preamplifiers are only needed in weak-signal areas and they often cause trouble, particularly when used unnecessarily in strong-signal areas. They are subject to the same overload problems as TVs, and when located on the antenna mast it can be difficult to install the appropriate cures. You may need to install a high-pass or notch filter at the input of the preamplifier, as well as a common-mode choke on the input, output, and power-supply wiring (if separate) to effect a complete cure. All filters, connections, and chokes must be weatherproofed. A common-mode choke will reduce RF current on the feed line's shield.

For a common-mode feed line choke, use two 1-inch-long Type #43 clamp-on ferrite

Figure 27.24 — To eliminate HF and VHF signals on the outside of a coaxial cable, use a 1- to 2-inch OD toroid core and wind as many turns of the cable on the core as practical.

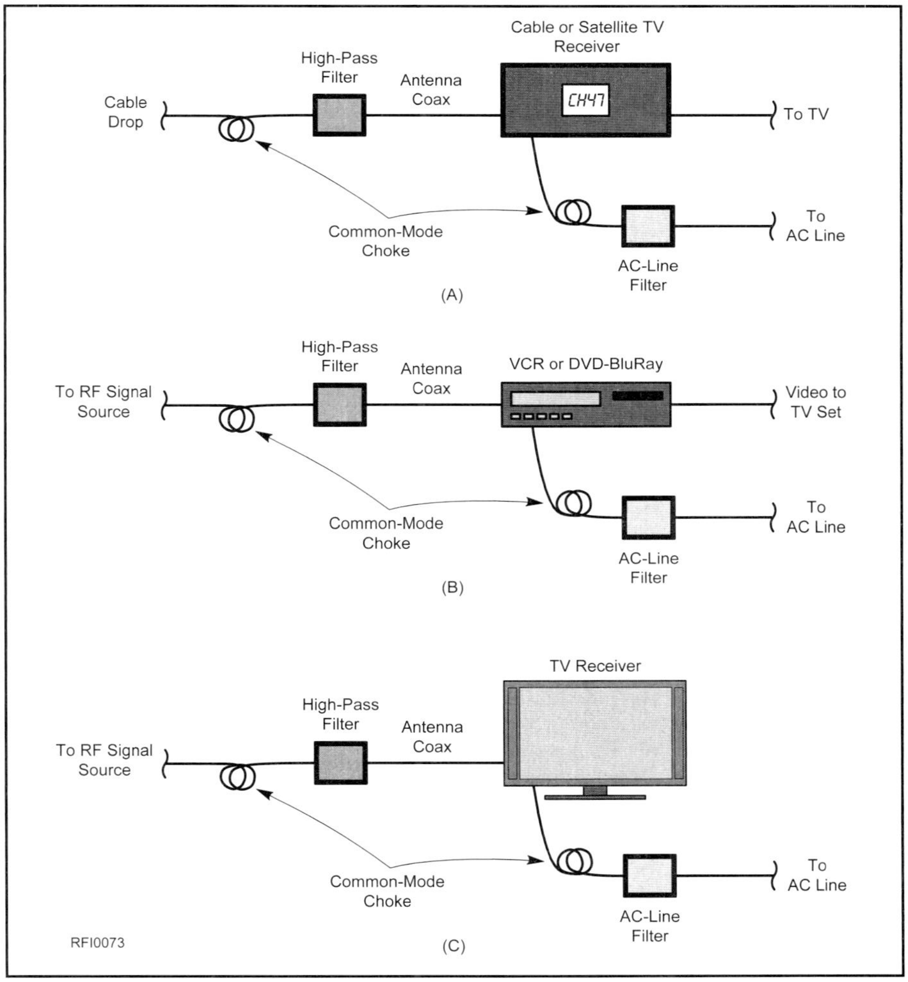

Figure 27.25 — Installing common-mode chokes and high-pass filters will cure most fundamental overload and common-mode breakthrough interference from HF sources. Apply common-mode chokes in the ac power cord before adding ac-line filters. This technique does not address direct pickup or spurious emission problems.

cores if VHF signals are causing the interference and Type #61 material for UHF. HF choke design is discussed in the section on Common-Mode Chokes.

Spurious Emissions — You are responsible for spurious emissions produced by your station. If your station is generating any interfering spurious signals, the problem must be cured there. Start by analyzing which TV channels are affected. A TV Channel Chart showing the relationship of the amateur allocations and their harmonics to over-the-air and cable channels is provided in this book's online material. Each channel is 6 MHz wide. If the interference is only on channels that are multiples of your transmitting frequency, you probably have interference caused by harmonics of your transmitted signal.

Harmonics from commercial transceivers, however, are quite rare. As mentioned earlier, significant harmonics from such equipment probably indicates a failure that needs to be repaired. Harmonics can also be generated by a poor connection somewhere in the transmission line or by arcing from the antenna. Both of these typically cause erratic or intermittently high SWR.

It is not certain that these harmonics are coming from your station, however. Harmonics can be generated by overloaded preamplifiers or tuner input circuits. Harmonics can also be generated like IMD products by non-linear junctions near your station transmitter or very near the TV antenna. (See the section on Interference from Intermodulation Distortion.) If your transmitter and station do not cause interference on a TV set in your own home, then you must look elsewhere for the source of the harmonics.

An inexpensive SDR receiver is a good way to see if spurious emissions of any sort — not necessarily just harmonics — are present. See the section on Using SDR Receivers earlier in this chapter for information on how to use these tools.

Electrical Noise — Digital TV signals are fairly resistant to electrical noise, but in extreme cases can cause the picture to freeze or fail to be displayed as discussed in the following section on Digital TV Receivers.

On an AM receiver (including SSB or CW receivers), electrical noise usually sounds like a buzz, sometimes changing in intensity as the arc or spark sputters a bit. If you have a problem with electrical noise, refer to the section on Electrical Noise.

DIGITAL TV (DTV) RECEIVERS

Nearly all over-the-air TV broadcasters in the US, with the exception of low-power TV stations and translators, are using the DTV (digital TV) format. Digital TV signals can operate with much lower signal-to-noise ratios, but are still susceptible to interference.

Interference to digital TV signals from amateur signals — narrowband interference, for instance, a CW carrier — to a 6 MHz-wide digital TV signal generally has two effects. If the interfering signal is strong enough, it will cause degraded modulation error ratio (MER) and degraded bit error rate (BER) in the digital video signal. If the amplitude of the interference is sufficient, the digital receiver's forward error correction (FEC) circuitry will be unable to fix the broken bits, and the digital video signal will "crash." (See the **Image Communications** chapter for more information on coding and error correction in digital TV signals.)

TV viewers watching any of the multiple video streams that may be contained within the digital video signal won't see any problems in the picture (or hear anything wrong in the sound), until the so-called "crash point" is reached. At that point, the picture will begin to show intermittent "tiling" (the picture breaking up into small squares) or blocking (freezing) in the image. As the amplitude of the interfering signal increases perhaps another 0.5 dB to 1 dB, the crash point or "digital cliff" is reached, and the picture and sound are gone! As you can see, there is a tiny window between receiving a perfect picture and receiving no picture. The same effect is produced by signal fading and may be difficult to distinguish from RFI.

Interference to the digital signal does not make its presence known through visual or audible artifacts such as streaks, lines, tearing in the picture, or garbled audio. This means that a viewer experiencing interference may not be able to identify its source, but troubleshooting interference may also become more difficult. Nevertheless, the more robust digital modulation is often less susceptible to interference from narrowband amateur signals. A clue to the source of the interference is that interference caused by an amateur signal will occur in sync with the amateur's transmissions, while other types of interference will have no correlation.

ANALOG TV RECEIVERS

Even though over-the-air TV broadcasting largely switched to a digital format in 2009, many analog TV receivers are still in use for cable TV, satellite TV, with converter boxes for digital broadcast signals, and for displaying video from DVDs and other video sources. Older VCR and DVD players may also include an analog TV tuner to receive analog TV signals.

Interference to video displays and monitors that do not receive RF signals from an antenna or RF modulator should be assumed to be common-mode breakthrough or direct pickup. The same applies to interference to a TV displaying video signals (not through the antenna input). Interference that is present only on the audio is probably a case of common-mode RFI. (See the Non-Radio Devices section of this chapter.)

27.8.2 Cable TV

Cable TV (CATV) has generally benefited amateur radio with respect to TVI. The cable system delivers a strong, consistent signal to the TV receiver, reducing susceptibility to interference from amateur signals. It is also a shielded system so an external signal shouldn't be able to cause interference. Most cable companies are responsible about keeping signal leakage (*egress*) and *ingress* — the opposite

> ***Common Mistake — It's the Cable System***
>
> The cable company is often incorrectly blamed for causing interference that is actually generated by consumer devices. Remember that all the individual circuit grounds in your home are bonded at the service entrance panel. The National Electrical Code also requires the cable TV ground to be bonded to this same ground system.
>
> When you now consider that most RFI problems, especially at HF, involve conducted emissions, it's easy to see what can happen. Conducted emissions from a consumer device can propagate along ground conductors and wind up on the shield of the cable TV coax. The cable (and possibly other components of the cable TV system) can then radiate the noise.
>
> Ron Hranac, NØIVN, also adds the following from an industry point of view:
>
> "A key point that needs to be emphasized related to this topic is that in many cases wideband noise from consumer devices that is being coupled to and radiated from the cable TV network is often assumed to be leaking cable modem digital signals. The cable company channelizes its digital signals (downstream channel bandwidths are 6 MHz, and upstream channel bandwidths are typically either 3.2 MHz or 6.4 MHz). The cable company generally does not use upstream frequencies below about 15 MHz for cable modem data transmission, although there might be a narrowband data carrier from set-top boxes in the roughly 8 MHz to 12 MHz range. Cable modem upstream signals are generally found in the 20 MHz to 42 MHz range, sometimes as low as 15 MHz or so. The cable company does not transmit signals in the diplex filter cutoff region of about 42 MHz to 50 MHz, nor does it transmit signals below about the previously mentioned 8 MHz or so."
>
> Conclusion: Don't automatically assume that the cable TV system is the cause of an RFI problem just because it is radiating the noise.

Table 27.2

Amateur Radio Bands Relative to Cable TV Downstream Channels

Amateur Band	*Over-The-Air Frequency Range*	*Cable Channel*	*Cable Frequency Range*
6 meters	50 – 54 MHz	Below Ch. 2	50 – 54 MHz, sometimes used for narrowband telemetry carriers
2 meters	144 – 148 MHz	Ch. 18	144 – 150 MHz
1.25 meters	222 – 225 MHz	Ch. 24	222 – 228 MHz
70 cm	420 – 450 MHz	Ch. 57	420 – 426 MHz
		Ch. 58	426 – 432 MHz
		Ch. 59	432 – 438 MHz
		Ch. 60	438 – 444 MHz
		Ch. 61	444 – 450 MHz
33 cm	902 – 928 MHz	Ch. 142	900 – 906 MHz
		Ch. 143	906 – 912 MHz
		Ch. 144	912 – 918 MHz
		Ch. 145	918 – 924 MHz
		Ch. 146	924 – 930 MHz

of leakage — under control, but problems do happen. Cable companies are not responsible for direct pickup or common-mode interference problems but are responsible for leakage, ingress, and any noise radiated by common-mode currents from their equipment.

Cable companies are able to take advantage of something known as frequency reuse. That is, all radio frequencies higher than 5 MHz are used to transmit TV signals. The latter is possible because the cables and components used to transport signals to and from paying subscribers comprise what is known as a closed network. In other words, a cable company can use frequencies inside of its cables that may be used for entirely different purposes in the over-the-air environment. As long as the shielding integrity of the cable network is maintained, the cable company's signals won't interfere with over-the-air services, and vice-versa.

The reality is that the shielding integrity of a cable network *is* sometimes compromised, perhaps because of a loose or damaged connector, a cracked cable shield, rodent damage, poorly shielded customer premises equipment (CPE) such as cable-ready TVs and various media players, and problems that may happen when someone tries to steal cable service! §76.605(a)(12) of the FCC Rules defines the maximum allowable signal leakage (*egress*) field strength at specified measurement distances, and §76.613 covers harmful interference. FCC Rules also mandate that cable operators "...shall provide for a program of regular monitoring for signal leakage by substantially covering the plant every three months," and leaks greater than 20 microvolts per meter (µV/m) at a 10 ft. measurement distance repaired in a reasonable period of time. As well, an annual "snapshot" of leakage performance must be characterized via a flyover measurement of the cable system, or a ground-based measurement of 75% of the network.

CABLE TV FREQUENCY USAGE

A typical modern North American cable network is designed to use frequencies in the 5 to 1002 MHz spectrum. Signals that travel from the cable company to the subscriber occupy frequencies from just above 50 MHz to as high as 1002 MHz range (this is the downstream or forward spectrum), and signals that travel from the subscriber to the cable company are carried in the 5 to as high as 42 MHz range, known as the upstream or return spectrum. The downstream is divided into 6 MHz-wide channel slots, which carry 64- or 256-QAM digitally modulated signals used for digital video, high-speed data, and telephone services. (Analog cable service has been phased out.) Upstream signals from cable modems and two-way set-top boxes are generally carried on specific frequencies chosen by the cable company. **Table 27.2** summarizes cable downstream channel allocations that overlap amateur radio bands. The complete North American channel plan is controlled by EIA standard 542-B. A summary of the channel structure for North America is maintained at **en.wikipedia.org/wiki/North_American_television_frequencies**.

COMMON MECHANISMS FOR LEAKAGE AND INGRESS

As noted previously, cable TV leakage and ingress occur when the shielding integrity of the cable network is compromised. A large cable system that serves a major metropolitan area has literally millions of connectors, tens of thousands of miles of coaxial cable, thousands of amplifiers, hundreds of thousands of passives (splitters, directional couplers, and similar devices), and uncountable customer premises equipment connected to the cable network! Any of these may be a source of leakage and ingress.

DIGITAL SIGNAL LEAKAGE

The digitally modulated signals carried in a cable TV network use 64-QAM or 256-QAM, the latter more common. If a QAM signal were to leak from a cable TV network, it is possible for interference to an over-the-air service to occur, but very unlikely to be identified as from a digital TV signal. The reason for this is that a QAM signal is noise-like, and sounds like normal background noise or hiss on a typical amateur receiver. The QAM signal's digital channel power — its average power over the entire occupied bandwidth — is typically 6 to 10 dB lower than what an analog TV signal's visual carrier peak envelope power (PEP) would be on the same channel. As well, a QAM signal occupies most of the 6 MHz channel slot, and there are no carriers *per se* within that channel bandwidth. Note that over-the-air 8-VSB digital TV broadcast signals transmit a pilot carrier near the lower end of the digital "haystack," but the QAM format used by cable operators has no comparable pilot carrier.

What makes the likelihood of interference occurring (or not occurring) has in large part to do with the behavior of a receiver in the presence of broadband noise. While each downstream cable TV QAM signal occupies close to 6 MHz of RF bandwidth, the IF bandwidth of a typical amateur FM receiver might be approximately 20 kHz. Thus, the noise power in the receiver will be reduced by $10 \log_{10}(6{,}000{,}000/20{,}000) = 24.77$ dB because of the receiver's much narrower IF bandwidth compared to the QAM signal's occupied bandwidth. In addition, there is the 6 to 10 dB reduction of the digital signal's average signal PEP.

Field tests during 2009 confirmed this behavior, finding that a leaking QAM signal would not budge the S meter of a Yaesu FT-736R at low to moderate field strength leaks, even when the receiver's antenna — a resonant half-wave dipole — was located just 10 feet from a calibrated leak. In contrast, a CW carrier that produced a 20 µV/m leak resulted in an S meter reading of S9 +15 dB, definitely harmful interference! When the CW carrier was replaced by a QAM signal whose digital channel power was equal to the CW carrier's PEP and which produced the same leakage field strength (the latter integrated over the full 6 MHz channel bandwidth), the S-meter read <S1. When the leakage field strength was increased to 100 µV/m, the CW carrier pegged the S meter at S9 +60 dB, while the QAM signal was S3 in FM mode and between S1 and S2 in USB mode. It wasn't until the leaking QAM signal's field strength reached several hundred µV/m that the "noise" (and it literally sounded like typical white noise) could be construed to be harmful interference.

One of the most common signs of possible

Table 27.3

VHF Midband Cable Channels

Channel Number	*Standard Video Carrier (STD) (MHz)*	*Harmonically Related Video Carrier (HRC) (MHz)*	*Incrementally Related Video Carrier (IRC) (MHz)*	*Audio Carrier (MHz)*
98	109.25	108.0054	109.25	113.75
99	115.25	114.0057	115.25	119.75
14	121.25	120.006	121.25	125.75
15	127.25	126.0063	127.25	131.75
16	133.25	132.0066	133.25	137.75
17	139.25	138.0069	139.25	143.75
18	145.25	144.0072	145.25	149.75
19	151.25	150.0075	151.25	155.75
20	157.25	156.0078	157.25	161.75
21	163.25	162.0081	163.25	167.75
22	169.25	168.0084	169.25	173.75

leakage is interference to the 2-meter amateur band, especially in the vicinity of standard (STD) cable channel 18's visual carrier on 145.25 MHz. If you suspect cable leakage, listen for the telltale broadband noise from the digital video signal over the 144–150 MHz range, and check other STD, incrementally related carrier (IRC), and harmonically related carrier (HRC) visual carrier frequencies on nearby channels listed in **Table 27.3** using a wide range receiver or scanner. (Leakage of an analog TV signal on cable channel 18 sounds like buzzing at the carrier frequencies in the table on or near 145.25 MHz. Also listen for TV channel sound on the FM aural carriers 4.5 MHz above the visual carriers.)

LOCATING LEAKAGE SOURCES

When a cable company technician troubleshoots signal leakage, the process is similar to amateur radio fox hunting. The technician uses radio direction finding techniques that may include equipment such as handheld dipole or Yagi antennas, Doppler antenna arrays on vehicles, near-field probes, and commercially manufactured signal leakage detectors. Many leakage detectors incorporate what is known as "tagging" technology to differentiate a leaking cable signal from an over-the-air signal or electrical noise that may exist on or near the same frequency. Most leakage detection is done on a dedicated cable channel in the 108 – 138 MHz frequency range.

ELIMINATING LEAKAGE

A large percentage of leakage and ingress problems are not the result of a single shielding defect, although this does happen. For example, a squirrel might chew a hardline feeder cable, or a radial crack might develop in the shield as a consequence of environmental or mechanical damage. Most often, leakage and ingress are caused by several small shielding defects in an area: loose or corroded hardline connectors and splices, old copper braid subscriber drop cabling, improperly installed F connectors, subscriber-installed substandard "do-it-yourself" components, and the previously mentioned poorly shielded cable-ready TVs and other *customer premises equipment (CPE)*.

Other leakage and ingress problems can be caused by improper shield connections at the cable TV set-top box. The return data signal in the low HF region (3.7 – 5.5 MHz) can be radiated in this way, as well. Common-mode chokes at the equipment with the poor shield connection can block the RF current.

After the cable technician locates the source(s) of the leakage, it is necessary to repair or replace the culprit components or cabling. In the case of poorly shielded TVs or DVD players, the cable technician cannot repair those devices, only recommend that they be fixed by a qualified service shop. Often the installation of a set-top box will take care of a cable-ready CPE problem because the subscriber drop cabling is no longer connected directly to the offending device.

It is important to note that interference from leakage that is received over the air cannot be eliminated at the receiver. It is an "in-band" signal just like the desired signal and can't be filtered out or suppressed with chokes. It must be eliminated at the source of the leakage.

Similarly, RFI from cable ingress — where a clean, transmitted signal gets into the cable system signals through similar defects to those that cause leakage — must also be eliminated at the point at which the transmitted signal enters the cable system.

In both cases, a little RFI detective work may be necessary. Refer to the various RFI troubleshooting sections of this chapter; radio direction finding techniques may come in handy, as well. Once the source of leakage or point of ingress is determined, like powerline noise, it becomes the job of the cable company to repair.

VERIFYING AN RFI SOURCE TO BE LEAKAGE

Spurious signals, birdies, harmonics, intermodulation, electrical noise, and even interference from Part 15 devices are sometimes mistaken for cable signal leakage. One of the most common is emissions from Part 15 devices that become coupled to the cable TV coax shield in some way. Non-leakage noise or spurious signals may radiate from the cable TV lines or an amplifier location, but only because the outer surface of the cable shield is carrying the coupled interference as a common-mode current.

A common non-leakage interference that may radiate from a cable network is broadband electrical interference or other noise in the MF and lower end of the HF spectrum. A common misconception is that since cable companies carry digital signals on frequencies that overlap portions of the over-the-air spectrum below 30 MHz, any "noise" that radiates from the cable plant must be leaking digital signals. This type of interference is almost always powerline electrical interference or other noise that is coupled to the cable network's shield as a common-mode signal.

Leakage of downstream digital signals sounds like broadband noise as described above, over a range of frequencies given in the channels of Table 27.3. Upstream digital signals from cable modems, which have channel bandwidths of 1.6, 3.2, or 6.4 MHz, are typically transmitted in the roughly 20 to 40 MHz range and are bursty in nature rather than continuous like downstream digital signals. Set-top box upstream telemetry carriers are narrowband frequency shift keying (FSK) or quadrature phase shift keying (QPSK) carriers usually in the approximately 8 to 11 MHz range.

If normal leakage troubleshooting techniques do not clearly identify the source of the interference, sometimes the cable company may temporarily shut off its network in the affected neighborhood. If the interference remains after the cable network is turned off, it is not leakage, and the cable company is not responsible for that type of interference. If the interference disappears when the cable network is turned off, then it most likely is leakage or something related to the cable network. Turning off even a small portion of the cable network is a last resort and may not be practical because of the service disruption to subscribers. It may be easier for the cable company to temporarily shut off a suspect cable channel briefly. Here, too, if the interference remains after the channel is turned off, the interference is not leakage.

HOW TO REPORT LEAKAGE

If you suspect cable signal leakage is causing interference to your amateur station, *never attempt your own repairs to any part of the cable network, even the cabling in your own home*! Document what you have observed. For instance, note the frequency or frequencies involved, the nature of the interference, any changes to the interference with time of day, how long it has been occurring, and so forth. If you have fox-hunting skills and equipment, you might note the probable source(s) of the interference, or at least the direction from which it appears to be originating.

Next, contact the cable company. You will most likely reach the cable company's customer service department, but ask to speak with the local cable system's Plant Manager (may also be called Chief Engineer, Director of Engineering, Chief Tech, VP of Engineering, or similar), and explain to them that you are experiencing what you believe to be signal leakage-related interference. If you cannot reach this individual, ask that a service ticket be created, and a technician familiar with leakage and ingress issues be dispatched. Share the information you have gathered about the interference. And as with all RFI issues, remember diplomacy!

In the vast majority of cases when cable leakage interference to amateur radio occurs, it can be resolved by working with local cable system personnel. Every now and then for whatever reason, the affected ham is unable to get the interference resolved locally. Contact the ARRL for help in these cases.

27.9 Consumer Electronics RFI

27.9.1 RFI to DVD and Other Media Players

A DVD or similar video player often contains a television tuner. Older models may have an analog TV channel output. (Newer models typically have an HDMI or similar digital video interface which is less susceptible to RFI.) It is also connected to an antenna or cable system and the ac line, so it is subject to all of the interference problems of a TV. Many media players can also playback to computer video monitors.

Start by proving that the media player is the susceptible device. Temporarily disconnect the device from the television or video monitor. If there is no interference to the TV, then the media player is the most likely culprit. Cables between the media player and TV can also be the means by which RF is getting into the monitor, so a cable dummy (see the sidebar Dummy Detectives below) may be a useful way to determine if that is the case and which cable or cable(s) are the problem.

Next, find out how the interfering signal is getting into the media player. Temporarily disconnect the antenna or cable feed line from the media player. If the interference goes away, then the antenna feed line is involved. In this case, you can probably fix the problem with a common-mode choke or high-pass filter.

Figure 27.25 shows a bulletproof video player installation. If you have tried all of the cures shown and still have a problem, the player is probably subject to direct pickup. In this case, you can replace it or contact the manufacturer through the CTA.

Older analog-type VCRs are quite susceptible to RFI from HF signals. The video baseband signal extends from 30 Hz to 3.5 MHz, with color information centered around 3.5 MHz and the FM sound subcarrier at 4.5 MHz. The entire video baseband is frequency modulated onto the tape at frequencies up to 10 MHz. Direct pickup of strong signals by VCRs is a common problem and may not be easily solved, short of replacing the VCR with a better-shielded model or a modern DVD player.

27.9.2 RFI to Non-Radio Devices

Interference to non-radio devices is not the fault of the transmitter. (A portion of the *FCC Interference Handbook,* 1990 Edition, is shown in **Figure 27.26**. Although the FCC no longer offers this *Handbook*, an electronic version is available in this book's online information, from the ARRL at **www.arrl.org/fcc-and-rfi-matters** or search the ARRL website for "cib interference handbook".) In essence, the FCC views non-radio devices that pick up nearby radio signals as improperly functioning; contact the manufacturer and return the equipment. The FCC does not require that non-radio devices include RFI protection, and they don't offer legal protection to users of these devices that are susceptible to interference.

TELEPHONES

Landline or "wired" telephones are much less common than they used to be, but they may still present possible interference prob-

PART II

INTERFERENCE TO OTHER EQUIPMENT

CHAPTER 6

TELEPHONES, ELECTRONIC ORGANS, AM/FM RADIOS, STEREO AND HI-FI EQUIPMENT

Telephones, stereos, computers, electronic organs and home intercom devices can receive interference from nearby radio transmitters. When this happens, the device improperly functions as a radio receiver. Proper shielding or filtering can eliminate such interference. The device receiving interference should be modified in your home while it is being affected by interference. This will enable the service technician to determine where the interfering signal is entering your device.

The device's response will vary according to the interference source. If, for example, your equipment is picking up the signal of a nearby two-way radio transmitter, you likely will hear the radio operator's voice. Electrical interference can cause sizzling, popping or humming sounds.

Figure 27.26 — Part of page 18 from the FCC *Interference Handbook* (1990 edition) explains the facts and places responsibility for interference to non-radio equipment.

lems from amateur radio. As more people switch over to mobile telephone service instead of landline, this problem is gradually diminishing. Nevertheless, landline and cordless phones will continue to be with us for many years. Most cases of telephone interference to these phones can be cured by correcting any installation defects and installing telephone RFI filters where needed. The remainder of this section assumes the telephone is connected to landline service known as POTS (Plain Old Telephone Service) and in many cases to telephone *premises wiring* throughout a home to connect several phones to one telephone line.

Telephones can improperly function as radio receivers. Semiconductor devices inside many telephones act like diodes. When such a telephone is connected to the telephone wiring (a large antenna), an AM radio receiver can be formed. When a nearby transmitter goes on the air, these telephones can be affected.

Troubleshooting techniques were discussed earlier in the chapter. The suggestion to simplify the problem applies especially to telephone interference. Disconnect all telephones except one, right at the service entrance if possible, and start troubleshooting the problem there.

If any single device or bad connection in the phone system detects RF and puts the detected signal back onto the phone line as audio, that audio cannot be removed with filters. Once the RF has been detected and turned into audio, it cannot be filtered out because the interference is at the same frequency as the desired audio signal. To effect a cure, you must locate the detection point and correct the problem there.

Defective telephone company lightning arrestors can act like diodes, rectifying any nearby RF energy. Telephone-line amplifiers or other electronic equipment may also be at fault. Do not attempt to diagnose or repair any telephone company wiring or devices on the "telco" side of your service box or that were installed by the phone company. Request a service call from your phone company.

Inspect the telephone system installation. Years of exposure in damp basements, walls or crawl spaces may have caused deterioration. Be suspicious of anything that is corroded or discolored. In many cases, homeowners have installed their own telephone wiring, often using substandard wiring. If you find sections of telephone wiring made from nonstandard cable, replace it with standard twisted-pair telephone or CAT5 cable. If you do use telephone cable, be sure it is high-quality twisted-pair to minimize differential-mode pickup of RF signals.

Next, evaluate each of the telephone instruments. If you find a susceptible telephone, install a telephone RFI filter on that telephone, such as those sold by K-Y Filters (**www.ky-filters.com**) or use DSL filters that keep the DSL data signals out of the telephones. If the home uses a DSL broadband data service, be sure that the filters do not affect DSL performance by testing online data rates with and without a filter installed at the telephone instrument. The DSL gateway may have a filtered output for the landline interface that does not include the data signals.

If you determine that you have interference only when you operate on one particular ham band, the telephone wiring system either has an "RF hot spot" at that point when excited on that band, or some cable in the system could be resonant and thus especially responsive on that band. Install common-mode chokes on the wiring to add a high impedance in series with the "antenna." A telephone RFI filter may also be needed. (See the section on DSL Equipment for filtering suggestions.)

Telephone Accessories — Answering machines and fax machines (two more telephone-related instruments that are slowly disappearing) are also prone to interference problems. All of the troubleshooting techniques and cures that apply to telephones also apply to these telephone devices. In addition, many of these devices connect to the ac mains. Try a common-mode choke and/or ac-line

Maintaining Balance with Twisted-Pair

Speaker cables made of parallel-conductor zip cord and open-wire RF feed lines are good examples of what we think of as "balanced" wiring. If the cables are perfectly balanced, common-mode signals on the cable will not affect the differential-mode signals in the cable and vice versa. This is because the balance maintains equal-and-opposite currents in each conductor. Similarly, if a radiated RF signal encounters balanced wiring, it will cause the same amount of common-mode current to flow on each conductor. Unfortunately, it's hard to maintain that exact state of balance.

If the impedance connected to each of the wires isn't the same (of either an input or output circuit), common-mode RF current will result in different voltages in those impedances. This converts some of the common-mode signal into a differential-mode signal which combines with the intended signal in the cable. RFI from this converted signal is much harder to get rid of!

Another way for cables to be imbalanced is for each of the wires to be exposed differently to the radiated RF signal. (This is also true for picking up low-frequency magnetic fields from power transformers and ac wiring, for example.) Zip-cord speaker cables and dc power cords are particularly prone to this problem. They look balanced, but there is usually a little bit of difference in the way the wires are exposed to the RF and the resulting RF current. That little bit of difference creates a differential-mode RF signal that can cause RFI.

To avoid converting common-mode signals to differential-mode, don't use unshielded cables for low-level RF and audio signals. Make sure the shields are connected properly. For unbalanced zip cord cables, replace them with *twisted-pair* cables. RF is still picked up by both conductors, but the twist ensures that both conductors pick up the *same* amount of RF.

Twisted-pair cables are easy to make: Clamp the wires in a variable-speed drill, holding the other end fixed with a vise or clamp. Slowly run the drill until the *pitch* of the twist (the number of twists per length) is one turn for every inch or two. Regular zip cord may not want to "hold the twist" so you can make your own cables using #14 or #16 AWG wire for speakers (handles most power levels, high power systems can use #12 AWG) and whatever gauge you need for a power cable. Use tape or heat-shrink tubing at the end to keep the wires together.

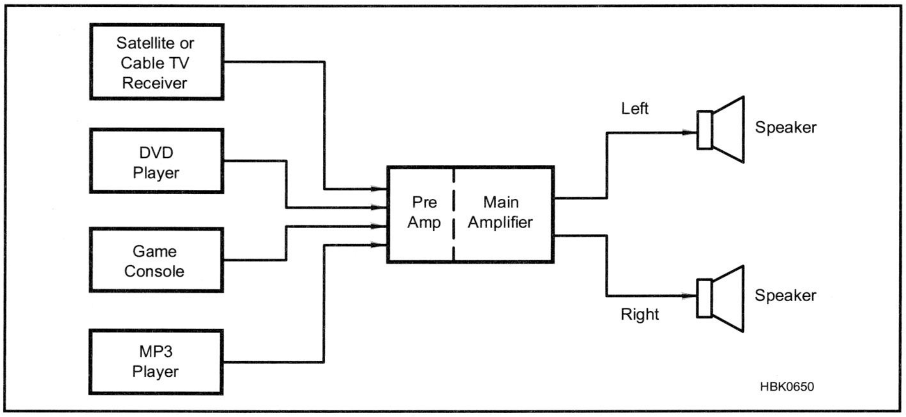

Figure 27.27 — A typical home-entertainment system.

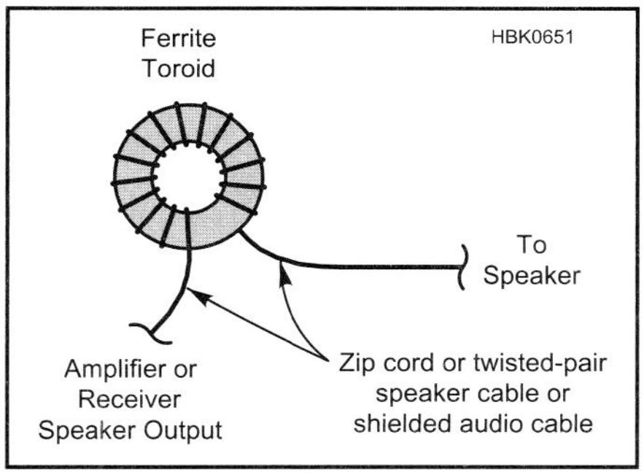

Figure 27.28 — Making a speaker-lead common-mode choke. Use ferrite material appropriate for the frequency of the RF interference.

Dummy Detectives

One particularly useful troubleshooting tool for determining where RFI is entering a victim device is called a "dummy" — a short test cable that can be inserted in a signal path. The dummy's cable shield is connected on each end of the cable but the center conductor is not connected. The dummy will allow common-mode current to flow on the shield, but will not pass a differential-mode signal. This technique was devised by Bill Whitlocak of Jensen Transformers.

Dummies are quite useful in identifying where RFI is being introduced in a signal path. If the interference goes away when the dummy is inserted, the RFI is being introduced upstream of the dummy. If the interference is common-mode, it will still be present when the dummy is inserted upstream of the victim equipment that is detecting it because the dummy's shield passes common-mode noise to the victim.

An F connector dummy is easily made with two F plugs installed on a short length of coax. A double-receptacle F "barrel" adaptor can be attached to one of the plugs to create an extension cable. At one or both of the plugs, the center conductor is snipped flush with the shield so that it does not make contact with the mating connector.

Dummies made with RCA and 1/8-in connectors (mono and stereo) are also useful for the same purpose in home entertainment systems and other consumer electronics. Use high-quality connectors, such as those made by Switchcraft and Neutrik, and use good shielded cable that fits the connector, such as RG-174 subminiature coax. Cut the center conductor of the coaxial cable so that it does not make contact with the connector pins.

filter on the power cord (which may be an ac cord set, a small transformer, or a power supply).

Cordless Telephones — A cordless telephone is an unlicensed *radio* device that is manufactured and used under Part 15 of the FCC regulations. The FCC does not intend Part 15 devices to be protected from interference. Older models that operate at 49 MHz usually have receivers with very wide front-end filters, which make them very susceptible to interference. These should be replaced with new models operating at 900 MHz and higher frequencies, which are much less susceptible to interference. A likely path for interference to cordless phones is as common-mode current on the base unit's connecting cables, which can be blocked by common-mode chokes. In addition, a telephone filter on the base unit and an ac line filter may help. The best source of help is the manufacturer, but they may point out that the Part 15 device is not protected from interference.

AUDIO EQUIPMENT

Consumer and commercial audio equipment such as stereos, home entertainment systems, intercoms, and public-address systems can also pick up and detect strong signals from nearby transmitters. The FCC considers these non-radio devices and does not protect them from licensed radio transmitters that may interfere with their operation. The RFI can be caused by one of several things: pickup on speaker leads or interconnecting cables, pickup by the ac mains wiring, or direct pickup. If the interference involves wiring connected to the affected device, common-mode chokes are the most likely solution.

Use the standard troubleshooting techniques discussed earlier in this chapter to isolate problems. In a multi-component home entertainment system (as in **Figure 27.27**), for example, you must determine what combination of components is involved with the problem. First, disconnect all auxiliary components to

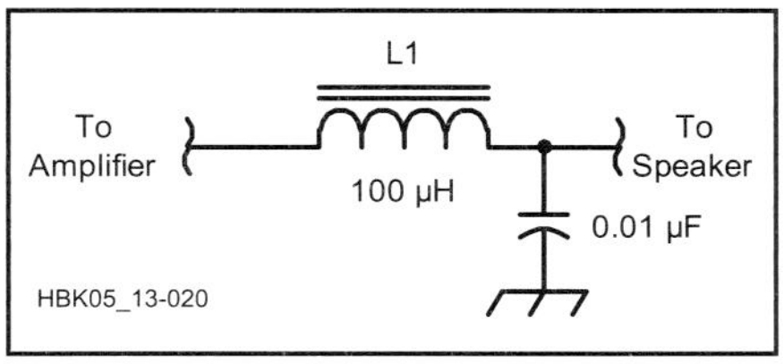

Figure 27.29 — A low-pass LC filter.

determine if there is a problem with the main receiver/amplifier. Long speaker or interconnect cables are prime suspects.

Stereos and Home Entertainment Systems — If the problem remains with the main amplifier isolated, determine if the interference level is affected by the volume control. If so, the interference is getting into the circuit *before* the volume control, usually through accessory wiring. If the volume control has no effect on the level of the interfering sound, the interference is getting in *after* the control, usually through speaker wires.

Speaker wires are often effective antennas on HF and sometimes into VHF and above. The speaker terminals are often connected directly to the output amplifier transistors. Modern amplifier designs use a negative feedback loop to improve fidelity. This loop can conduct the detected RF signal back to the high-gain stages of the amplifier. The combination of all of these factors often makes the speaker cables the dominant receiving antenna for RFI.

There is a simple test that will help determine if the interfering signal is being coupled into the amplifier by the speaker wires. Temporarily disconnect the speaker wires at the amplifier (not at the speakers) and plug in a test set of headphones with short leads. If there is no interference with the headphones, filtering the speaker wires will likely cure the problem.

Start by applying common-mode chokes. **Figure 27.28** shows how to wrap speaker wires around a large (1.4-inch O.D. or larger) ferrite core to cure speaker-wire RFI. Type #31 material is preferred at HF. (See the section on Common-Mode Chokes in this chapter.)

In some cases, the speaker wires may be picking up RF as a differential-mode signal. To reduce differential-mode pickup, replace the zip cord speaker wire with twisted-pair wire. (#14 or #16 AWG will work for most systems with higher-power systems requiring #12 AWG.) See the sidebar Maintaining Balance with Twisted-Pair to learn how to make twisted-pair cables.

Powered Speakers — A powered speaker is one that has its own built-in power amplifier. Powered subwoofers are common in home entertainment systems, and small powered speakers are often used with computer and gaming systems. If a speaker runs on batteries and/or an external power supply, or is plugged into main power, it is a powered loudspeaker. Powered loudspeakers are notoriously susceptible to common-mode interference from internally misconnected cable shields and poor shielding. Apply suitable common-mode chokes to all wiring, including power wiring. If the RFI persists, try an RF filter at the input to the speaker, such as the LC low-pass filter in **Figure 27.29**. Unshielded speakers may not be curable, however.

Intercoms and Security Systems — RFI to these systems is nearly always caused by common-mode current on interconnect wiring. Common-mode chokes are the most likely cure, but you may also need to contact the manufacturer to see if they have any additional, specific information. Twisted-pair wiring (CAT5 network cable contains four such pairs) should be used, including for audio output wiring. Wiring can often be complex, so any work on these systems should be done by a qualified sound contractor.

Public-Address Systems — Common-mode current is also the culprit here. Powered speakers are increasingly used and can be treated as described above. Work to remove interference should be done by the installing contractor and may require coordination between the amateur and contractor to characterize the interference and provide test assistance while the work is being done.

27.9.3 RFI From Computers and Other Unlicensed RF Sources

Computers and microprocessor-based devices such as video games or media players can be sources or victims of interference. These devices contain oscillators that can and do radiate RF energy. In addition, the internal functions of a computer generate different frequencies, based on the various data signals. All of these signals are digital — with fast rise and fall times that are rich in harmonics.

Computing devices are covered under Part 15 of the FCC regulations as unintentional emitters. As for any other unintentional emitter, the FCC has set absolute radiation limits for these devices. As previously discussed in this chapter, FCC regulations state that the operator or owner of Part 15 devices must take whatever steps are necessary to reduce or eliminate any interference they cause to a licensed radio service. This means that if your neighbor's video game interferes with your radio, the neighbor is responsible for correcting the problem. (Of course, your neighbor may appreciate your help in locating a solution!)

The FCC has set up two tiers of limits for computing devices. Class A is for computers used in a commercial environment. FCC Class B requirements are more stringent — for computers used in residential environments. If you buy a computer or peripheral, be sure that it is Class B certified or it will probably generate interference to your amateur station or home-electronics equipment.

If you find that a computer system is generating interfering signals, start by simplifying the problem. Temporarily remove power from as many peripheral devices as possible and disconnect their cables from the back of the computer. (It is necessary to physically remove the power cable from the device, since many devices remain in a low-power state when turned off from the front panel or by a software command.) If possible, use just the computer, keyboard, and monitor. This test may identify

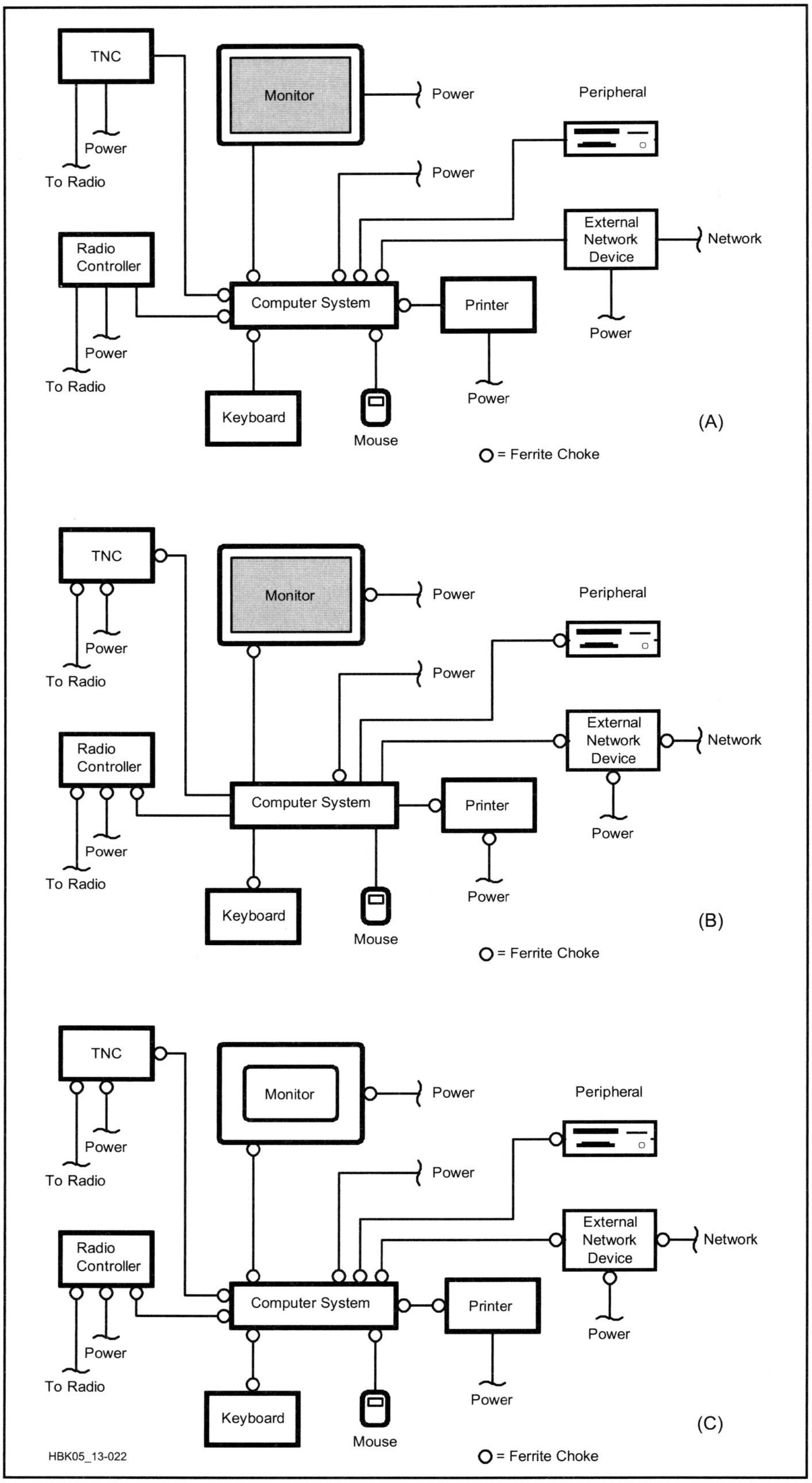

Figure 27.30 — Where to locate ferrite chokes in a computer system. At A, the computer is noisy, but the peripherals are quiet. At B, the computer is quiet, but external devices are noisy. At C, both the computer and externals are noisy.

specific peripherals as the source of the interference.

It can be difficult to determine whether peripheral connecting cables are shielded. If possible, use shielded cables for all peripherals. Replace any unshielded cables with shielded ones; this often significantly reduces RF noise from computer systems. The second line of defense is the common-mode choke. The choke should be installed as close to the computer and/or peripheral device as practical. **Figure 27.30** shows the location of common-mode chokes in a complete computer system where both the computer and peripherals are noisy. USB-power devices can also create noise from internal power supplies and are discussed in K9YC's *National Contest Journal* paper "Killing Receive Noise – Part 1 and 2," included in the online information for this chapter.

A multi-turn common-mode choke wound for the HF bands is often ineffective at VHF. When VHF antennas are located close to these devices (or if a more distant beam is pointed at them), it may also be necessary to add one or two 2-turn chokes to suppress noise on 6 meters and multiple, single-turn, clamp-on cores for 2-meter noise. The cores for the higher frequencies should always be placed closest to the noise source, and when two pieces of digital equipment are connected, each should be considered a potential source, so VHF cores may be needed on both ends.

Switchmode power supplies in computers are often sources of interference. A common-mode choke and/or ac-line filter may cure this problem. In extreme cases of computer RFI you may need to improve the shielding of the computer. (Refer to the *ARRL RFI Book* for more information about this.) Don't forget that some peripherals (such as DSL gateways) are connected to the phone line, so you may need to treat them like telephones.

WALL TRANSFORMER SWITCHING SUPPLIES

While small, low-current linear power supplies known as "wall warts" have been widely used for many years, it is now becoming common for these devices to contain switchmode or "switcher" supplies. Because they must be manufactured very inexpensively, these supplies often have little or no RF filtering at either the ac input or dc output, frequently creating significant RFI to nearby amateurs.

The least expensive course of action may be to simply replace the switchmode supply with a linear model. If the system in which the supply is used is still under warranty, the distributor or manufacturer may be able to replace it. Otherwise, a third-party linear replacement may be available with the same voltage rating and current output equal to or higher than the original supply. Adapters may be available to convert output connector styles where necessary or new connectors may be installed. Older linear-style supplies can be re-used if you are willing to splice them in to replace the switchmode model.

If replacing the supply with a linear model is not an option, you will have to apply RF filters to the supply. These supplies are rarely serviceable, so filters much be installed externally. Noise is usually radiated from the output cable, so winding the cable onto a ferrite core creates a common-mode RF choke as described in this chapter's Elements of RFI Control section. Since the wall-wart style supply plugs directly into a wall-mounted receptacle, a short ac extension cord or power strip cord can be made into a common-mode choke as well.

27.9.4 RFI to Ground-Fault (GFCI) and Arc-Fault (AFCI) Circuit Interrupters

GFCIs are occasionally reported to "trip" (open the circuit) when a strong RF signal, such as an amateur's HF transmission, is present. GFCI circuit breakers operate by sensing unbalanced currents in the hot and neutral conductors of an ac circuit. In the absence of RF interference, such an imbalance indicates the presence of a fault somewhere in the circuit, creating a shock hazard. The breaker then trips (opens) to remove the shock hazard.

An Arc Fault Circuit Interrupter (AFCI) circuit breaker is similar in that it monitors current to watch for a fault condition. Instead of current imbalances, the AFCI detects patterns of current that indicate an arc — one of the leading causes of home fires. The AFCI is not supposed to trip because of "normal" arcs that occur when a switch is opened or a plug is removed.

Under current codes, GFCI protection is required for all basement outlets, outdoor outlets, and for outlets in kitchens and bathrooms. AFCI protection is now required for all circuits that supply bedrooms and other areas of a home as well. Code requirements can vary, so be sure to check with your local building inspector for those that apply in a specific case.

RF interference to GFCI breakers is caused by RF current or voltage upsetting normal operation of the imbalance detection circuit, resulting in the false detection of a fault. Similarly, RF current or voltage could upset the arc detection circuitry of an AFCI breaker. Some early GFCI breakers were susceptible to RFI, but as the technology has improved, fewer and fewer such reports have been received. While it is possible to add filtering or RF suppression to the breaker wiring, a simpler and less expensive solution is to replace the GFCI breaker with a new unit less susceptible to RF.

The ARRL Lab has received favorable reports on the following GFCI products:

- Leviton (**www.leviton.com**) GFCI outlets which are available in both 15 and 20 A versions for 120 V ac circuits as well as cord sets and user-attachable plugs and receptacles.
- Bryant (**www.bryant-electric.com**) ground fault receptacles, which feature published 0.5 V immunity from 150 kHz to 230 MHz.
- Cooper (**www.cooperindustries.com**) GFCI products that are labeled "UL 943 compliant" on the package.

A web page on the ARRL website is maintained on GFCI/AFCI technology (**www.arrl.org/gfci-and-afci-devices**). Manufacturers seem to have largely fixed RFI issues with these devices.

27.9.5 RFI from Solar Power Installations

Solar energy is becoming increasingly popular as we transition from our dependence on fossil fuels to renewable energy sources. Residential solar installations are on the rise, especially in regions most affected by climate change. To understand how solar technology can impact an amateur radio station, it's important to first understand how energy is created from photovoltaic (PV) modules.

PV materials and devices convert sunlight into electrical energy. A single PV device is known as a *cell* or *solar cell*. (See the **Circuits and Components** chapter for more detailed information.) An individual PV cell is usually small, typically producing about 1 or 2 W of power. These cells are made of different semiconductor materials and are often less than the thickness of four human hairs. In order to withstand the outdoors for many years, cells are sandwiched between protective materials in a combination of glass and/or plastics.

To create useful amounts of power from PV cells, they are connected together to form larger units known as *modules* or *panels*. Modules can be used individually, or several can be connected to form arrays. One or more arrays is then connected to the electrical grid as part of a complete PV system. Because of this modular structure, PV systems can be built to meet almost any electric power need, small or large.

PV modules and arrays are just one part of a PV system. Systems also include mounting structures that point panels toward the Sun, along with the components that take the dc produced by modules and convert it to the ac electricity used to power all of the appliances in a home (the above excerpted from **www.energy.gov** — more on this technology can be found on the United States Department of Energy website at **www.energy.gov/eere/solar/solar-energy-technologies-office**).

Of relevance to amateur radio operators is

the fact that residential solar is on the rise. According to the Solar Energy Industries Association (SEIA), in 2021 residential solar installations exceeded 1 GWdc and more than 130,000 systems in a single quarter for the first time. One out of every 600 US homeowners is now installing solar each quarter (see the SEIA Solar Market Insight Report at **www.seia.org/research-resources/solar-market-insight-report-2021-q4**).

This rise in residential solar has also brought to light an increase in RFI cases reported to the ARRL due to the various design technologies used to convert and condition the electricity produced by the panels. It should be noted that the PV cells themselves do not create RFI — they are basically large diodes — it is the power conversion process which changes the solar energy from dc to ac that creates RFI. **Figure 27.31** summarizes the different types of solar energy systems. More detail can be found at **news.energysage.com/string-inverters-power-optimizers-microinverters-compared**.

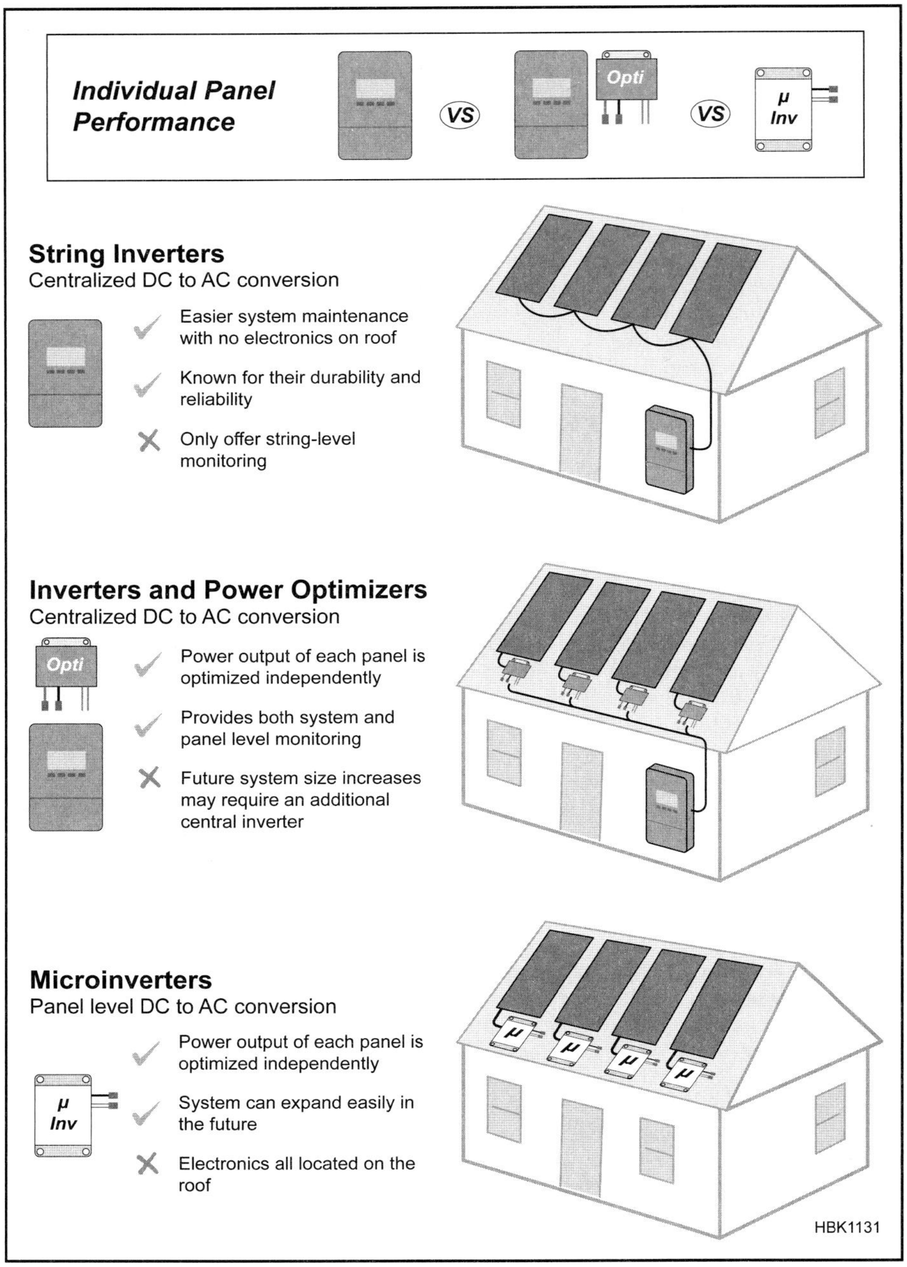

Figure 27.31 — Three common types of solar energy systems in common use today. Each type is discussed in the text.

CAUSES OF SOLAR SYSTEM RFI

RFI has been experienced by hams who have installed systems on their own homes, and by hams whose neighbors have installed systems in their homes. ARRL Lab staff have noted there are at least three basic mechanisms by which residential PV arrays can generate RFI. Some or all may be present in a given array, depending on the manufacturer's system architecture. Figure 27.31 illustrates the three basic types of arrays and power conversion that are in common use today — *string inverters*, *power optimizers*, and *microinverters*.

1. RFI from inverters. These devices are responsible for switching the high voltage dc from the array to 60 Hz phase-synchronous ac, meaning ac power in-phase with the utility ac waveform. One or more inverters may be incorporated in a given array, depending on its size. Typically, RFI from these devices is radiated by the dc wiring to the PV panels with an 18 kHz to 60 kHz fundamental switching frequency. Harmonics can extend well into the HF bands and lower VHF bands.

2. RFI from power optimizers. Power optimizers might be installed at every PV panel, or there may be a single power optimizer for several PV panels, depending on the component manufacturer and system size. Under full sunlight, power optimizers can have fundamental operating frequencies ranging from 39 kHz to 200 kHz. As with inverters, there may be harmonics extending through HF and into the lower VHF bands.

It should also be noted that when power optimizers are in the "off" or non-power generating mode, the PV array is disconnected from the ac wiring but may still generate RFI, as there is still some part of the device powered by the sunlight. RFI may be reduced significantly but it can still be noticeable.

3. Data collection and system control devices. These devices may be stand-alone or integral to other components in the PV system. They also have the least potential to cause widespread RFI, and typically only the operator of the PV array might be subjected to RFI from these components. Typical communications channels observed thus far include 60 kHz to 74 kHz and 1.8 MHz.

Vendors generally start by isolating and testing to ensure the RFI issue an amateur might be experiencing is actually being caused by the vendor's system. This can be as simple as shutting down the solar installation briefly to ensure the RFI goes away at the amateur's station. Once the vendor performs the diagnostic tests and identifies the RFI as coming from their system, solutions employed by vendors might include component changes or enhancements, addition of ferrite chokes, or modifying wiring to include twisted-pair wiring harnesses. Vendors of these systems, from local installers to component manufacturers such as SolarEdge, Enphase, and Generac, have all been cooperative in working with hams and the ARRL to work through RFI issues related to solar installations. For an example, see the PDF on a SolarEdge retrofit at the home of Paul Cianciolo, former ARRL RFI Engineer, included in online materials.

Table 27.4
Part 15B Conducted Emissions Limits for Unintentional Radiators

Frequency of emission (MHz)	*Conducted limit (dBμV)*	
	Quasi-peak	*Average*
0.15 – 0.5	66 to 56*	56 to 46*
0.5 – 5	56	46
5 – 30	60	50

* Decreases with the logarithm of the frequency

27.9.6 RFI From Grow Lights and Other Low-Voltage Lighting

There are myriad lighting types that may be in operation near amateur radio stations. While incandescent bulbs have a long history of not causing interference to amateurs, global concerns about climate change have led to a move towards renewable energy and energy efficiency, along with the need for more efficient lighting devices. Of particular relevance today are LED lighting and grow lights.

Two sets of rules apply for LED bulbs in residential environment. LED bulbs involving electronics are typically subject to Part 15 rules (described at the beginning of this chapter). If the lighting device is operated at greater than 9 kHz, it is classified as an "unintentional radiator," with limits specified in **Table 27.4**. If the device operates at less than 9 kHz, it is classified as an "incidental radiator" and has no specified emissions limits, but still has a requirement that it not cause "harmful interference," as defined by Part 15. Incandescent bulbs are an example of incidental radiators.

For the lighting equipment, when designed to be connected to the public utility ac power line, the radio frequency voltage that is conducted back onto the ac power line on any frequency or frequencies shall not exceed the limits in Table 27.4. These limits are based on the measurement of the radio frequency voltage between each power line and ground at the power terminal using a 50 μH/50 Ω line impedance stabilization network (LISN).

The ARRL Laboratory has conducted testing on numerous lighting devices in recent years, and has concluded that at least some of these devices fail to comply with FCC standards (see **www.arrl.org/lighting-devices**). A discussion of these tests, and a broader discussion about lighting devices and RFI can be found in an article on the ARRL website at **www.arrl.org/files/file/RFI/Light_Bulbs.pdf**.

GROW LIGHTS

Large high-power grow lights in residential areas are also on the rise. These lights are typically rated at over 1,000 W (commercial lighting devices can be 5,000 W or more) and employ electronics that can act as ballasts or perform some other vital function. Grow lights (as with other lighting devices) can also incorporate switching power supplies, another potential source of RFI to amateur stations. (See **Figure 27.32A** for an example of a grow light and a pair of electronic ballasts.)

Unfortunately, these lights have proven to be a notorious source of RFI to amateur stations. Based on the distances over which the noise can propagate and tests conducted in the ARRL Lab, these devices can exceed the FCC limits by a considerable margin. Although these tests were limited to just a few samples, one light was measured to be more than 30 dB over the FCC limits.

Interference from grow lights has been shown to be problematic at distances of well over 1,000 feet from the offending device. To put this into perspective, this is over three times the distance one might expect from a compliant Part 15 or 18 device.

While grow lights may be used in other growing operations, the most recent increases in reports of RFI from grow lights can be attributed to states that have legalized or decriminalized marijuana. California and Colorado are two such states. Other uses of these lights can include growing indoor vegetables and household, ornamental, and exotic plants. It should be stressed that the mere presence of a grow light is not proof that illegal activity may be taking place in someone's home.

(A)

Figure 27.32 — A typical single LED grow light (A) and two types of electronic ballasts (B and C) that connect the lights to ac power. (Photos are courtesy Tom Thompson, WØIVJ)

(B)

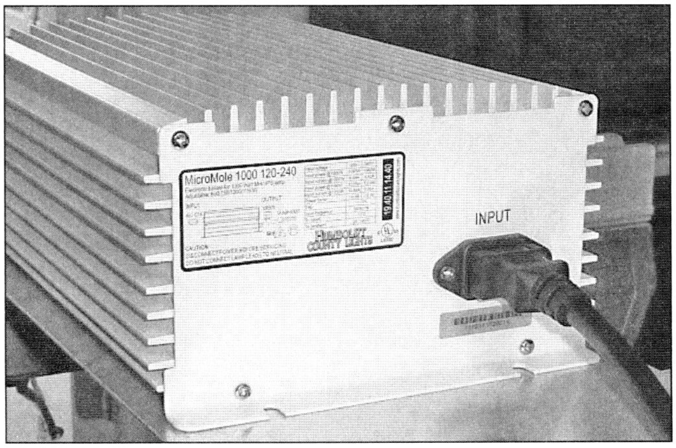

(C)

IS IT RFI FROM A GROW LIGHT?

Grow light RFI tends to exhibit several unique and identifiable characteristics. Some clues as to whether or not RFI may be coming from grow lights include the following:

• Grow lights are typically controlled by a timer, cycling on and off at precise times every 12 hours or so. While this time can vary depending upon where plants are located in relation to a particular growing cycle, there should be at least a short-term, distinct pattern to the noise being generated from grow lights.

• Grow lights typically produce a broadband noise, particularly across the 40-meter band. The noise propagates over much greater distances than one might expect from devices that meet the applicable FCC limits. Once you have established a turn-on time, listen to the noise as the lamp turns on. It will generally be noisier when cold, with the RFI more focused on part of the band. This generally lasts about 5 minutes until the lamp is warm.

• The polarization is predominately vertical even though the lamp line may be horizontal.

• In the case of marijuana, the grow time is typically 6 to 8 weeks. The lights may be shut off for a few days while the plants are harvested and new ones made ready.

JURISDICTIONAL CONSIDERATIONS

Under federal law, only the FCC can create and enforce rules regarding harmful radio interference. Furthermore, as described previously in this section, both Part 15 and Part 18 of the FCC's rules unconditionally prohibit interference to a licensed radio service caused by lighting devices. This includes interference to an amateur caused by grow lights.

While the FCC may not have jurisdiction in the enforcement of drug laws, an interference complaint to the FCC may result in an investigation. Grow lights can also produce a distinctive noise signature that may result in an unwelcome calling card, depending on the type of plants involved. Although marijuana may not be illegal under some state laws, it is still illegal under federal law. Should a field investigation uncover a grow light operation involving marijuana, we simply don't know what inter-agency agreements may exist between the FCC and other law enforcement bodies. Many such growers might wish to cease operation of the offending device rather than risk a federal investigation of any kind.

APPROACHING A NEIGHBOR

As discussed elsewhere in this chapter, we typically recommend that hams discuss any RFI problems with their neighbor as a good first step. Obviously, even under the best of circumstances, the importance of diplomacy cannot be overemphasized during these discussions. It's also important never to suggest what you think the cause might be. If you're wrong, it often makes matters worse, regardless of what you suspect the source might be. Simply let the neighbor know that a source of radio frequency interference appears to be coming from their home. It can also be helpful to demonstrate the problem with a portable radio, especially one that is or looks like an AM broadcast receiver. Be sure not to turn up the radio so loud that it becomes offensive.

Obviously, depending on circumstances, approaching your neighbor in a case involving a grow light can be a potentially dangerous situation. Exercise extreme caution and good judgment in this case. For example, you may not want to directly approach a neighbor if you suspect a grow light is the source of your RFI problem. Remember that the presence of a grow light does not guarantee that marijuana is being grown. Every case is obviously different, and no two neighbors are exactly alike, but there is simply no substitute for good judgment in any situation like this.

TECHNICAL SOLUTIONS TO GROW LIGHT RFI

Technical solutions typically involve the addition of filters and chokes to the cables connected to the ballast. In the case of a grow light, this typically requires that filters be added to both the ac and the lamp side of the ballast.

See the *Grow Light Electronic Ballasts Page* by Larry Benko, WØQE (**www.w0qe.com/RF_Interference/grow_light_electronic_ballasts.html**) for details on filters being used by Larry and Tom Thompson, WØIVJ, to help in these cases. **Figure 27.33A** shows a homemade conducted RFI filter constructed by WØIVJ. The *Grow Light Ballast RFI Filter* web page, also by Tom Thompson, includes some additional filter info (**tomthompson.com/radio/GrowLight/GrowLightBallastFilter.html**).

Nanolux also sells an RF Filter Set (see Figure 27.33B, **nanoluxtech.com/rf-filter-set**) designed for single-ended lamp ballasts. Note the Nanolux web site states the following with regard to RFI from electronic ballasts:

There are two types of RFI created by electronic ballasts. Both are regulated by the FCC, but each has its own distinctive characteristics. First, there's emitted RFI, which can disrupt electronic devices within your home that run on an RF signal. Such devices include some cable boxes, internet, Bluetooth devices, and even some hydroponic controller units currently on the market.

The emitted RF filter attaches to the lamp cord, while the second type of RFI, conducted RFI, can be controlled by the conducted RF filter that connects to the power cord. The conducted RF filter protects your neighbors' and public safety communications. If conducted RFI reaches the copper wiring of your home, it will turn the house into one giant antenna. This antenna of RFI can extend for miles. It will disrupt all things operating on a radio frequency around 30 kHz. This includes your neighbor's cable, internet, and other RF devices, plus many emergency communications that run on this frequency.

Regardless of which cure you attempt, be sure to choose a filter or other components that are properly rated for the current and voltages involved.

Although common-mode chokes are another popular and effective solution in cases involving conducted emissions, the size of grow light cables and associated connectors may make them impractical without additional

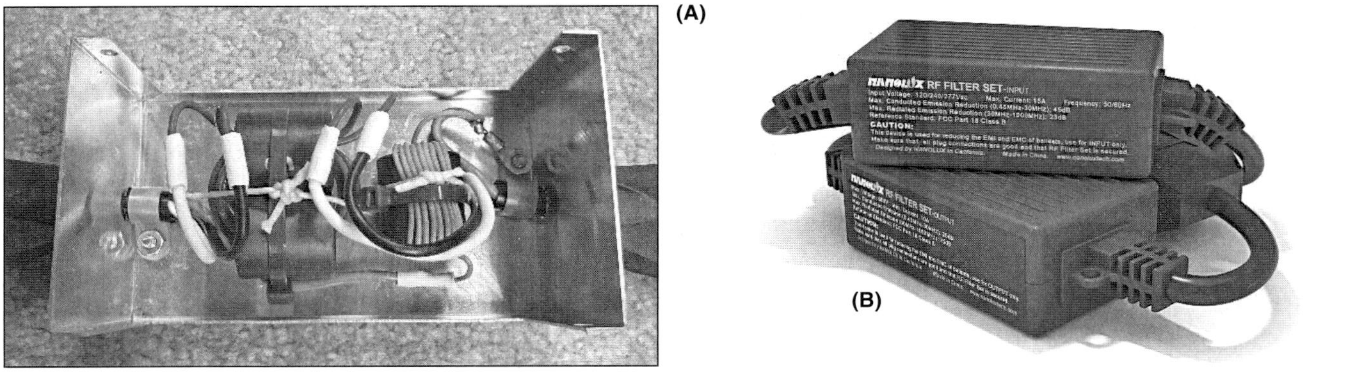

Figure 27.33 — Common-mode filter for electronic lighting ballast, designed and built by Tom Thompson, WØIVJ.

consideration. For example, you may need to temporally remove a connector or add an extension to the cable for the choke.

Keep in mind that split ferrite beads and clamp-on cores are not as effective as multi-turn toroids made of Type #31 ferrite at HF. You'll need to pass the cable end through the center hole multiple times. The most widely available and largest such toroid is 2.4 in diameter, supporting approximately eight turns of cable. The size of the cable (and connector), therefore, may prohibit or limit the number of turns that are possible with grow lights. A 4-inch diameter toroid is also available. (See the section on Common-Mode Chokes earlier in this chapter for information on ferrite materials and how to make them.) Be sure to keep the toroid as close to the RFI source as practical. Additional chokes can be placed on the cable in series.

RF Interference and the FCC

by Riley Hollingsworth, K4ZDH
(retired Special Counsel to the FCC Enforcement Bureau)

Since 1999, the FCC has worked with the ARRL in a cooperative agreement whereby the staff at the ARRL Lab takes the first cut at resolving RFI. The lab works with the complainant to make sure that the noise is narrowed down to the most probable source. The success rate with this program has been very high, and in many cases — perhaps most — the ARRL and the complainant solve the problem without any FCC involvement.

The lab can help you with the proper testing you need to do in your shack and with the documentation you need in the event the matter is referred to the FCC. I was always amazed at the number of situations in which noise that at first seemed to be power line related, was in fact (these are real examples) a nearby electric fence, a battery charger for a golf cart, an Ethernet adapter, a paper shredder, or a circuit board in a brand-new clothes washer in a room adjacent to the radio shack.

Don't assume anything. Test every possibility and document your testing. Not only will you learn a lot about the devices in your house and what causes noise and what doesn't, but good documentation will make your case stronger and easier to work. Follow to the letter the ARRL articles and website tutorials on tracking down noise in and around your shack. You can even hear noise samples on the website.

The documentation requirement is especially important if it is power line related and you have to start dealing with the power company. Take notes of every call you make, who you talked to, and when. This helps not only the power company but also the FCC in the unfortunate event that FCC action is required. In many cases, power company staff has a lot of experience running down such noise, and they take pride in locating it. In other cases, the power company has been bought, sold, or merged and does not have staff with a lot of experience in these matters. Sometimes its staff has no experience.

The cost of the equipment required to track down power line noise is less than that of two employees and a bucket truck for a day. Often the source of power line noise is a simple piece of equipment that is loose or about to fail, so finding the source helps the power company maintain its system. Keep in mind, though, that some areas are just not suitable for amateur radio. If you live next to an old substation, or a conglomeration of old poles and transformers, or an industrial area, your situation is tenuous. Whether it's our roads or power grids, lots of the infrastructure in this country is just plain old and out-of-date.

You must test diligently and document thoroughly. Never go to the FCC with a situation when you have not already worked with the ARRL and the power company.

27.9.7 RFI From Wireless Power Transfer

Wireless Power Transfer (WPT) is already widely deployed in low-power wireless chargers used to charge cell phones and similar devices. These devices use either capacitive coupling (metal plates or sleeves) or inductive coupling (coils) as the path to transfer power. There haven't been any reported cases of interference from low-power WPT chargers so far.

However, the emerging technology of WPT for much higher-powered charging of electric vehicles (WPT-EV) has the potential to cause significant interference problems. These systems are based on inductive power transfer (IPT), which uses a pair of coils close together to transfer power. The systems are controlled so that the power frequency is at or near the resonant frequency of the coupled coils, increasing efficiency. Typical resonant frequencies are from 20-100 kHz today. More about WPT-EV technologies is available at **www.intechopen.com/chapters/75267**. The current SAE standard for WPT is SAE J2954: Wireless Power Transfer for Light-Duty Plug-In/Electric Vehicles and Alignment Methodology.

The largest concern for amateurs is that harmonics and noise present in a high-power system could fall into the amateur spectrum. One potential mechanism for this to happen is in the coupling between the charging power source primary coil and the vehicle's receiving secondary coil, which can form an imperfect coupling system. For example, the coils could be too far apart or misaligned, allowing energy to leak from the charging system. The vehicle may be in motion (charging coils embedded in roadways), and the charging system will have to be controlled so that power is only transferred when alignment is optimized.

The amount of stray energy is particularly important, since the estimated power transfer for a typical residential WPT-EV charger system is expected to be in range of 10-15 kW. It would not take a lot of leakage at those power levels for the spurious emissions from such a system to create interference. It should be stressed that WPT-EV is still in the experimental and prototype stages. No RFI from these systems has yet been reported. The ARRL continues to monitor WPT development through industry organizations and committees. As more information becomes available, it will be published on the ARRL's website and in publications such as this one.

27.10 Power-Line Noise

This chapter's section "Identifying the Type of RFI Source" describes power-line noise, its causes, and methods to identify it. (For more information about power-line interference, see the Reference entry for Loftness.) Many more details, including how to file a complaint, are available on the ARRL's Power-line Noise FAQ web page at **www.arrl.org/power-line-noise-faq**.

Power-line noise is a unique problem in several respects. First and foremost, the offending source is never under your direct control. You can't just simply "turn it off" or unplug the offending device. Nor will the source be under the direct control of a neighbor or someone you are likely to know. In the case of power-line noise, the source is usually operated by a company, municipality, or in some cases a cooperative. Furthermore, shutting down a power line is obviously not a practical option.

Another unique aspect of power-line noise is that it almost always involves a defect of some sort. The cure for power-line noise is to fix the defect. This is almost always a utility implemented repair, and one over which you do not have any direct control.

FCC rules specify that the operator of a device causing interference is responsible for fixing it. Whenever encountering a power-line noise problem, you will be dealing with a utility and won't have the option of applying a relatively simple technical solution to facilitate a cure, as you would if the device were located in your home. Utilities have a mixed record when it comes to dealing with power-line noise complaints. In some cases, a utility will have a budget, well-trained personnel, and equipment to quickly locate and address the problem. In other cases, however, the utility is simply unable to effectively deal with power-line noise complaints or even denies their equipment can cause RFI.

What does this mean for an amateur with a power-line noise complaint? Utilities can be of any size from large corporations to local cooperatives or city-owned systems. Regardless of the category in which your utility may fall, it must follow Part 15 of the FCC rules. Dealing with a company, co-op or municipality, however, as opposed to a device in your home or a nearby neighbor that you know personally, can present its own set of unique challenges. Multiple parties and individuals are often involved, including an RFI investigator, a line crew, and associated management. In some cases, the utility may never have received an RFI complaint before yours.

27.10.1 Before Filing a Complaint

Obviously, before filing a complaint with your local utility, it is important to verify that the problem is power-line noise and is not caused by a problem with electrical equipment in your home. Other sources, such as lighting devices and motors, can mimic power-line noise, especially to an untrained ear. Don't overlook these important steps. Attempting to engage your utility in the resolution of an RFI problem can not only waste time but can be embarrassing if the source is right in your own home!

Utilities are not responsible for noise generated by customer-operated devices — *even if the noise is being radiated by the power lines.* They are responsible for fixing only that noise which is being generated by their equipment.

27.10.2 Filing a Complaint

Once you have verified that the problem is power-line noise (see this chapter's section on Identifying Power-line and Electrical Noise) and is not coming from a source in your home or a nearby residence, contact your utility's customer service department. In addition to your local phone book, customer service phone numbers are included on most power company websites.

It is important to maintain a log during this part of the process. Be sure to record any "help ticket" numbers that may be assigned to your complaint as well as names, dates, and a brief description of each conversation you have with electric company personnel. If you identify specific equipment or power poles as a possible noise source, record the address and any identifying numbers on it.

Hopefully, your complaint will be addressed in a timely and professional manner. Once a noise source has been identified, it is up to the utility to repair it within a reasonable period. You and the utility may not agree on what constitutes a reasonable period, but attempt to be patient. If no action is taken after repeated requests, reporting the complaint to the ARRL and requesting assistance may be in order. (Before contacting the ARRL review The Cooperative Agreement, a section of this chapter.)

It is also important to cooperate with utility personnel and treat them with respect. Hostile and inappropriate behavior is almost always counter-productive in these situations. Remember, you want utility and other related personnel to help you — not avoid you. Even if the utility personnel working on your case seem unqualified, hostile behavior has historically never been a particularly good motivator in these situations. In fact, most protracted power-line noise cases reported to the ARRL began with an altercation in the early stages of the resolution process. In no case did it help or expedite correction of the problem.

27.10.3 Techniques for Locating Power-Line Noise Sources

Radio direction finding (RDF) techniques typically offer the best and most efficient approach to locating most power-line noise sources. It is the primary method of choice used by professionals. While RDF is usually the most effective method, it also requires some specialized equipment, such as a hand-held beam antenna. Although specialized professional equipment is available for RDF, hams can also use readily available amateur and homebrew equipment successfully. The online information for this *Handbook* includes some power-line noise locating equipment projects you can build. This includes a simple and easy to build RDFing antenna, an ultrasonic dish, and attenuators.

Although it is the utility's responsibility to

Common Mistake — It's Coming from That Pole!

RFI problems from consumer devices, especially below 30 MHz, frequently involve conducted emissions. The RF in this case is generated by a consumer device in a home. It is then conducted by the house wiring, where it is radiated. In some cases, the RF can also be conducted to the service entrance of the home and out to a utility pole. In this case, the noise will peak at the pole and appear to be the source of the interference. It is not. The real source of the problem is in a residence connected to the pole.

Always verify that the actual source is not in a home connected to that pole. Frequently, the source residence will be connected to the power transformer secondary system on that pole. Also keep in mind that the hardware on a pole can make for a better "antenna." This means that noise tends to peak at poles in general. Noise also tends to peak at about every half wavelength as you traverse a power line. Be sure to use higher frequencies (i.e., shorter wavelengths) whenever possible.

Conclusion: Don't automatically assume that the source of an RFI problem is on a pole where the noise peaks. This is a common mistake made by many beginners!

locate a source of noise emanating from its equipment, many companies simply do not possess the necessary expertise or equipment to do so. As a practical matter, many hams have assisted their utility in locating noise sources. In some cases, this can help expedite a speedy resolution.

There is a significant caveat to this approach, however. Should you mislead the power company into making unnecessary repairs, they will become frustrated. This expense and time will be added to their repair list. Do not make a guess or suggestions if you don't know what is causing the noise. While some power companies might know less about the locating process than the affected ham, indiscriminate replacement of hardware almost always makes the problem worse. Nonetheless, depending on your level of expertise and the specifics of your situation, you may be able to facilitate a speedy resolution by locating the RFI source for the utility.

27.10.4 Amateur Power-Line Noise Locating Equipment

Additional information on the equipment, antennas, and techniques used to locate power line noise is also discussed in the earlier section Radio Direction Finding. Before discussing how to locate power-line noise sources, here are a few additional equipment guidelines.

Receiver — You'll need a battery-operated portable radio capable of receiving VHF or UHF in the AM mode. Ideally, it should also be capable of receiving HF frequencies, especially if the interference is a problem at HF and not VHF. Some amateurs also use the aircraft band from 108 to 137 MHz. The lower frequencies of this band can sometimes enable an RFI investigator to hear the noise at greater distances than on 2 meters or 70 cm. An RF gain control is essential, but an outboard step attenuator can be used as a substitute. A good S-meter is also required.

Attenuator — Even if your receiver has an RF gain control, an additional outboard step attenuator can often be helpful. It can not only minimize the area of a noise search but also provide added range for the RF gain control. As with other RFI sources, you'll need to add more and more attenuation as you approach the source.

VHF/UHF Antennas — You'll need a hand-held directional beam antenna. A popular professional noise-locating antenna is an eight-element Yagi tuned for 400 MHz. Since power-line noise is a broadband phenomenon, the exact frequency is not important. Either a 2 meter or 70 cm Yagi is capable of locating a power-line noise source on a specific power pole.

Although professional grade antennas can cost several hundred dollars, some hams can build their own for a lot less. See this book's online information for the article "Adapting a Three-Element Tape Measure Beam for Power-line Noise Hunting," by Jim Hanson, W1TRC. This low cost and easy to build antenna for locating power-line noise can be adapted for a variety of frequencies and receivers. Commercial 2 meter and 70 cm antennas for portable use are also suitable if a handle is added, such as a short length of PVC pipe.

Before using an antenna for power-line noise locating, determine its peak response frequency. Start by aiming the antenna at a known power-line noise source. Tune across its range and just beyond. Using minimum RF gain control, find its peak response. Label the antenna with this frequency using a piece of tape or marking pen. When using this antenna for noise locating, tune the receiver to this peak response frequency.

If you don't have a VHF or UHF receiver that can receive AM signals, see the online content for the article "A Simple TRF Receiver for Tracking RFI," by Rick Littlefield, K1BQT. It describes the combination of a simple 136 MHz beam and receiver for portable RFI tracking.

HF Antennas — Depending on the circumstances of a particular case, a mobile HF whip such as a 7 or 14 MHz model can be helpful. Magnet-mount models are acceptable for temporary use. An RFI investigator can typically get within VHF range by observing the relative strength of the noise from different locations. Driving in a circle centered on the affected station will typically indicate the general direction in which the noise is strongest. As with beam antennas, determine the peak response frequency for best results.

Ultrasonic Pinpointer — Although an ultrasonic pinpointer is not necessary to locate the pole or structure containing the source, some hams prefer to go one more step by finding the offending noise source on that structure. Guidelines for the use of an ultrasonic device are described later in this section.

Professional-grade ultrasonic locators are often beyond the budget of the average ham. Home brewing options, however, can make a practical ultrasonic locator affordable in most situations — and make a great weekend project too. See the online information accompanying this book for "A Home-made Ultrasonic Power Line Arc Detector" by Jim Hanson, W1TRC.

Oscilloscope — A battery-powered portable oscilloscope is only required for signature analysis. See the next section, Signature or Fingerprint Method, for details.

Thermal/Infrared Detectors and Corona Cameras — This equipment is not recommended for the sole purpose of locating power-line noise sources. It is rare that an RFI source is even detectable using infrared techniques. Although these are not useful tools for locating noise sources, many utilities still use them for such purposes with minimal or no results. Not surprisingly, ARRL experience has shown that these utilities are typically unable to resolve interference complaints in a timely fashion.

27.10.5 Signature or Fingerprint Method

Each sparking interference source exhibits a unique pattern. By comparing the characteristics between the patterns taken at the affected station with those observed in the field, it becomes possible to conclusively identify the offending source or sources from the many that one might encounter. It therefore isn't surprising that a pattern's unique characteristic is often called its "fingerprint" or "signature." See **Figure 27.34** for an example.

This is a very powerful technique and a real money saver for the utility. Even though there may be several different noise sources in the field, this method helps identify only those sources that are actually causing the interference problem. The utility need only correct the problem(s) matching the pattern of noise affecting your equipment.

You as a ham can use the signature method by observing the noise from your radio's audio output with an oscilloscope. Record the pattern by drawing it on a notepad or taking a photograph of the screen. Take the sketch or photograph with you as you hunt for the source and compare it to signatures you might observe in the field.

Professional interference-locating receivers, such as the Radar Engineers Model 243 shown in Figure 27.34, have a built-in time-domain display and waveform memory similar to a digital oscilloscope. This is the preferred method used by professional interference investigators. These receivers provide the ability to switch between the patterns saved at the affected station and those from sources located in the field.

Once armed with the noise fingerprint taken at the affected station, you are ready to begin the hunt. If you have a directional beam, use it to obtain a bearing to the noise. If multiple sources are involved, you'll need to record the bearings to each one. Knowing how high in frequency a particular noise can be heard also provides a clue to its proximity. If the noise can be heard at 440 MHz, for example, the source will typically be within walking distance. If it diminishes beginning 75 or 40 meters, it can be up to several miles away.

Since each noise source will exhibit unique characteristics, you can now match this noise "signature" with one from the many sources you may encounter in the investigation. Compare such characteristics as the duration of each noise burst, pulse shape, and number of pulses.

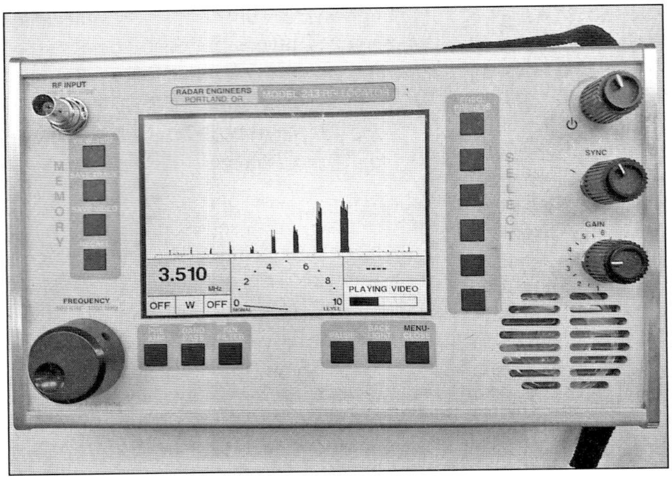

(A)

(B)

Figure 27.34 — The spectrum of an unknown noise source is shown on the display of a Model 243 RFI locating receiver (A) along with a typical noise receiving antenna (B). During the RFI investigation, noise signatures not matching this pattern can be ignored. Once the matching signature originally shown in A is found, the offending noise source has been located.

Be wary of assuming that all devices have similar signatures when arcing. The fingerprint of a particular piece of hardware arcing may be quite different even for the same device on another pole.

If you have a non-portable oscilloscope, you may still be able to perform signature matching by using an audio recorder. Make a high-quality recording of the noise source at your station and at each suspected noise source in the field, using the same receiver if possible. Replay the sounds for signature analysis.

27.10.6 Locating the Source's Power Pole or Structure

Start your search in front of the affected station. If you can hear the noise at VHF or UHF, begin with a handheld beam suitable for these frequencies. As discussed previously, the longer wavelengths associated with the AM broadcast band and even HF can create misleading "hotspots" along a line when searching for a noise source, as shown in **Figure 27.35**.

As a general rule, only use lower frequencies when you are too far away from the source to hear it at VHF or UHF. Generally, work with the highest frequency at which the noise can be heard. As you approach the source, keep increasing the frequency to VHF or UHF, depending on your available antennas. Typically, 2 meters and 70 cm are both suitable for isolating a source down to the pole level.

If you do not have an initial bearing to the noise and are unable to hear it with your portable or mobile equipment, start traveling in a circular pattern around the affected station, block-by-block, street-by-street, until you find the noise pattern matching the one recorded at the affected station.

Once in range of the noise at VHF or UHF, start using a handheld beam. You're well on your way to locating the structure containing the source. In many cases, you can now continue your search on foot. *Again, maintain minimum RF gain to just barely hear the noise over a minimum area.* This is important step is crucial for success. If the RF gain is too high, it will be difficult to obtain accurate bearings with the beam.

Power-line noise will often be neither vertically nor horizontally polarized but somewhere in between. Be sure to rotate the beam's polarization for maximum noise response. Maintain this same polarization when comparing poles and other hardware.

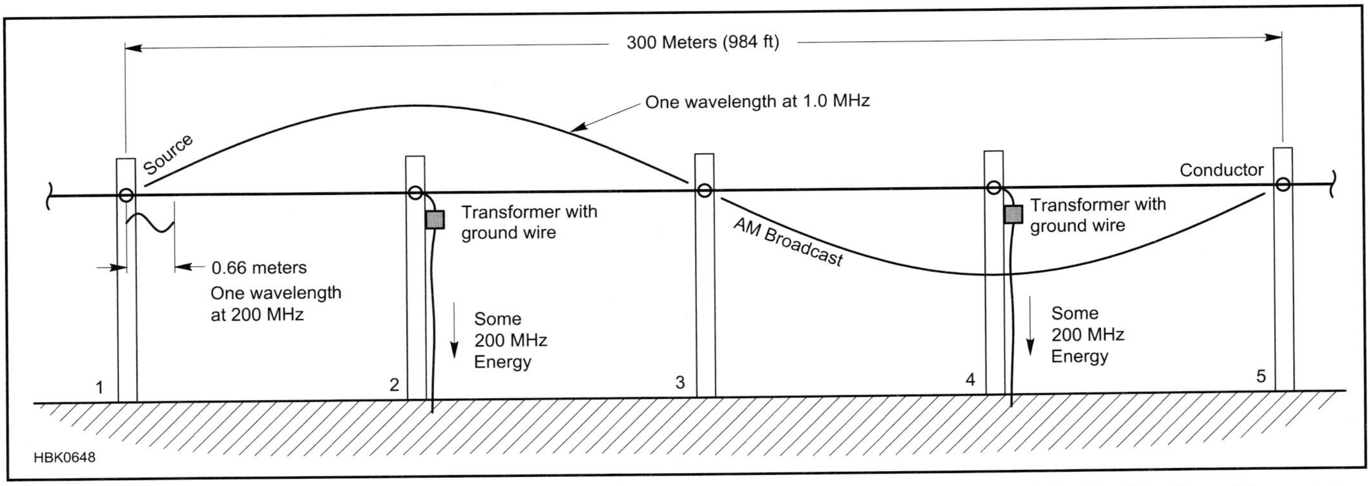

Figure 27.35 — Listening for noise signals on a power distribution line at 1 MHz vs 200 MHz can result in identification of the wrong power pole as the noise source. (From Loftness, *AC Power Interference Handbook*)

27.10.7 Pinpointing the Source on a Pole or Structure

Once the source pole has been identified, the next step becomes pinpointing the offending hardware on that pole. A pair of binoculars on a dark night may reveal visible signs of arcing, and in some cases you may be able to see other evidence of the problem from the ground. These cases are rare. More than likely, a better approach will be required. Professional and utility interference investigators typically have two types of specialized equipment for this purpose:

• An ultrasonic dish or pinpointer. The RFI investigator, even if not a lineman, can pinpoint sources on the structure down to a component level from the ground using this instrument. See **Figure 27.36**.

• An investigator can instruct the lineman on the use of a hot stick-mounted device used to find the source. This method is restricted to only qualified utility personnel, typically from a bucket truck. See **Figure 27.37**.

Both methods are similar but hams only have one option — the ultrasonic pinpointer.

CAUTION — *Hot sticks and hot stick mounted devices are not for hams! Do not use them. Proper and safe use of a hot stick requires specialized training. In most localities, it is generally unlawful for anyone unqualified by a utility to come within 10 feet of an energized line or hardware. This includes hot sticks.*

Remember that it may be interesting to you to determine what piece of equipment is generating noise, but it is the utility's job to positively complete the identification and then work on the equipment. Once you have identified the noise as coming from a particular location, be prepared to step back and let the professionals do their job.

ULTRASONIC PINPOINTER TIPS

An ultrasonic dish is the tool of choice for pinpointing the source of an arc from the ground. While no hot stick is required, an unobstructed, direct line-of-sight path is required between the arc and the dish. This is not a suitable tool for locating the structure containing the source. It is only useful for pinpointing a source once its pole or structure has been determined.

Caveat: Corona discharge, while typically not a source of RF power-line noise, can and often is a significant source of ultrasonic sound. It can often be difficult to distinguish between the sound created by an arc and corona discharge. This can lead to mistakes when trying to pinpoint the source of an RFI problem with an ultrasonic device.

The key to success, just as with locating the structure, is using gain control effectively. Use minimum gain after initially detecting the noise. If the source appears to be at more than one location on the structure, reduce the gain. In part, this will eliminate any weaker noise signals from hardware not causing the problem.

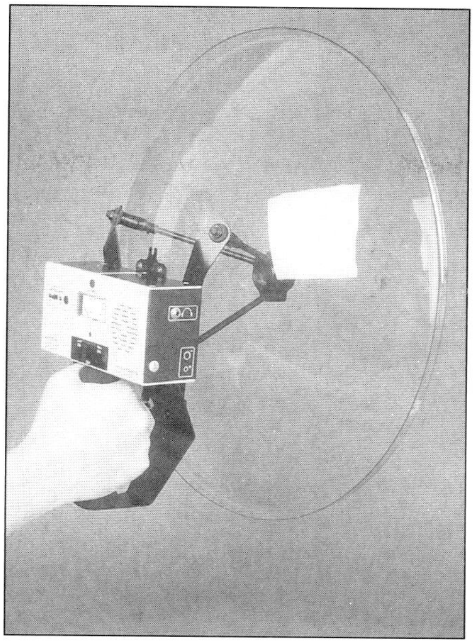

Figure 27.36 — The clear plastic parabolic dish is an "ear" connected to an ultrasonic detector that lets utility personnel listen for the sound of arcs.

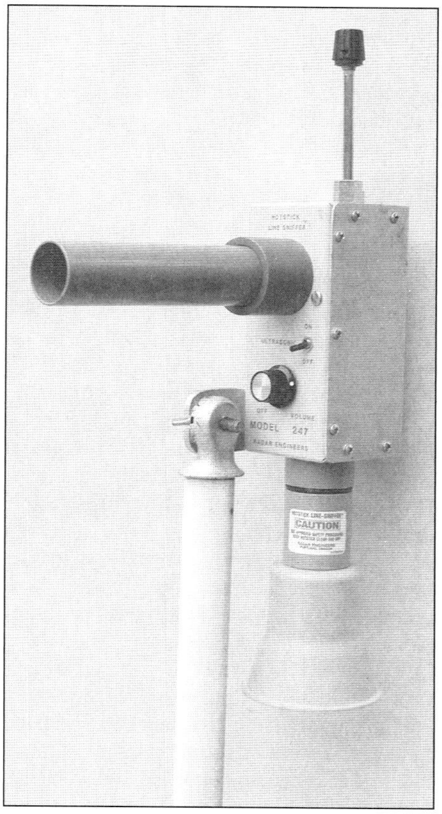

Figure 27.37 — The Radar Engineers Model 247 Hotstick Line Sniffer is an RF and ultrasonic locator. It is used by utility workers to pinpoint the exact piece of hardware causing a noise problem.

27.10.8 Common Causes of Power-Line Noise

The following are some of the more common power-line noise sources. They're listed in order from most common to least common. Note that some of the most common sources are not connected to a primary conductor. This in part is due to the care most utilities take to ensure sufficient primary conductor clearance from surrounding hardware. Note, too, that power transformers do not appear on this list:

- Loose staples on ground conductor
- Loose pole-top pin
- Ground conductor touching nearby hardware
- Corroded slack span insulators
- Guy wire touching neutral
- Loose hardware
- Bare tie wire used with insulated conductor
- Insulated tie wire on bare conductor
- Loose cross-arm braces
- Lightning arrestors

27.10.9 The Cooperative Agreement

While some cases of power-line noise are resolved in a timely fashion, the reality is that many cases can linger for an extended period of time. Many utilities simply do not have the expertise, equipment, or motivation to properly address a power-line noise complaint. There are often no quick solutions. Patience can often be at a premium in these situations. Fortunately, the ARRL has a Cooperative Agreement with the FCC that can help. While the program is not a quick or easy solution, it does offer an opportunity and step-by-step course of action for relief. It emphasizes and provides for voluntary cooperation without FCC intervention.

Under the terms of the Cooperative Agreement, the ARRL provides technical help and information to utilities in order to help them resolve power-line noise complaints. It must be emphasized that the ARRL's role in this process is strictly a technical one — it is not in the enforcement business. In order to participate, complainants are required to treat utility personnel with respect, refrain from hostile behavior, and reasonably cooperate with any reasonable utility request. This includes making his or her station available for purposes of observing and recording noise signatures. The intent of the Cooperative Agreement is to solve as many cases as possible before they go to the FCC. In this way, the FCC's limited resources can be allocated where they are needed the most — enforcement.

As the first step in the process, the ARRL

sends the utility a letter advising of pertinent Part 15 rules and offering assistance. The FCC then requires a 60-day waiting period before the next step. If by the end of 60 days the utility has failed to demonstrate a good faith effort to correct the problem, the FCC then issues an advisory letter. This letter allows the utility another 60-day window to correct the problem.

A second FCC advisory letter, if necessary, is the next step. Typically, this letter provides another 20- or 30-day window for the utility to respond. If the problem still persists, a field investigation would follow. At the discretion of the field investigator, he or she may issue an FCC Citation or Notice of Apparent Liability (NAL). In the case of an NAL, a forfeiture or fine can result.

It is important to emphasize that the ARRL Cooperative Agreement Program does not offer a quick fix. There are several built-in waiting periods and a number of requirements that a ham must follow precisely. It does however provide a step-by-step and systematic course of action under the auspices of the FCC in cases where a utility does not comply with Part 15.

27.11 Automotive RFI

Automobiles have evolved from a limited number of primitive electrical components to high technology, multi-computer systems-on-wheels. Every new technology deployed can potentially interfere with amateur equipment.

Successful mobile operation depends on a multitude of factors such as choosing the right vehicle, following installation guidelines, troubleshooting, and deploying the appropriate RFI fixes as needed.

A number of these factors will be covered in this section, as well as newly emerging electrical and hybrid-electric vehicles,which pose unique challenges to amateur equipment installations and operation.

27.11.1 Before Purchasing a Vehicle

When shopping for a new vehicle intended for a mobile amateur installation, begin with research. A wealth of information is available on the internet, and specifically at **www.arrl.org/automotive**, where the ARRL has compiled years of data from automotive manufacturers and other hams. Email reflectors and websites may provide information from hams willing to share their experiences concerning mobile communications in their own vehicle, which may be the very make and model you were considering.

Armed with research, your next stop is your dealer. The manufacturer of each vehicle is the expert on how that vehicle will perform. The dealer should have good communication with the manufacturer and should be able to answer your questions. Ask about service bulletins and installation guidelines. You can also ask your dealer about fleet models of their vehicles. Some manufacturers offer special modifications for vehicles intended for sale to police, taxicabs and other users who will be installing radios (usually operating at VHF and UHF) in their cars.

When shopping for a vehicle, it is useful to take along some portable (preferably battery operated) receivers or scanners and have a friend tune through your intended operating frequencies while you drive the vehicle. This will help identify any radiated noise issues associated with that model vehicle, which can be more difficult to resolve than conducted noise. If you intend to make a permanent transceiver installation, give some consideration to how you will mount the transceiver and route the power and/or antenna cables. While looking for ways to route the wiring, keep in mind that in some newer cars the battery is located in the trunk or under the rear seat, which may make power wire routing easier.

Test the car before you buy it. A dealer expects you to take the car for a test ride; a cooperative dealer may let you test it for radio operation too. A fair amount of checking can easily be performed without digging too deeply into the car. Check the vehicle for noise with a portable receiver on VHF, where your handheld transceiver will do the job nicely. On HF, you can usually locate noise with a portable short-wave receiver, or operate your HF transceiver with a portable antenna and cigarette-lighter plug. With the engine running, tune across the bands of interest. You may hear some noise or a few birdies, but if the birdies don't fall on your favorite frequency, this is an encouraging sign! Check with the vehicle completely off, with the key in the ignition, and with the vehicle running — electronic subsystems operate in different ways with the vehicle running or not running.

To test the vehicle for susceptibility to your transmitted signal, you must transmit. It is important to note that without a full and complete installation, you will not be able to fully assess the effects of full-power transmissions on a vehicle. Any testing done with temporary equipment installations cannot be considered an absolute guarantee because an installed transmitter connected directly to the vehicle's power source may cause the vehicle to act differently.

To perform transmit tests, bring your radio and a separate battery (if permitted by dealer) so you can transmit at full power while in motion without having to run cables to the vehicle battery. Use a magnet-mount antenna (several *QST* advertisers sell mounts suitable for HF) for temporary testing. (Use the magnet-mount carefully; it is possible to scratch paint if any particles of dirt get on the bottom of the magnets.) Transmit on each band you will use to see if the RF has any effect on the vehicle. Lack of response to your transmissions is a good sign, but does not mean the vehicle is immune to RF, as a permanent installation will result in different (likely stronger) field strengths and distributions in and around the vehicle and a permanent antenna more effectively coupled to the vehicle.

On both transmit and receive, you may want to experiment with the placement of the antenna. Antenna placement plays an important role in operation, and you may be able to find an optimal location for the antenna that predicts good performance with a permanent installation.

27.11.2 Transceiver Installation Guidelines

While most amateurs are familiar with the process of installing a transceiver, there are preferred practices that will help minimize potential problems. These include support from the automotive dealer, typical "best practices" installations, and consideration of special situations.

The first step is to ensure that your installation complies with both the vehicle manufacturer's and radio transceiver manufacturer's installation guidelines. Links to domestic automotive manufacturer installation guidelines are found at **www.arrl.org/automotive**. Automotive manufacturers that import vehicles for sale here do not publish installation guidelines because their vehicles are not typically used in police, fire, and taxicab applications within the US.

The installation guidelines of different manufacturers vary as to how to install a radio transceiver's power leads. Most manufacturers recommend that the positive and negative leads from the radio be run directly to the battery. This minimizes the potential for the interaction between the radio's negative lead currents and vehicle electronics. If the manufacturer recommends that both wires be connected to the battery, they will also require that both wires be fused. This is necessary because, in the unlikely event that the connection between the battery and the engine block were to fail, excessive current could be drawn on the radio's negative lead when the vehicle starter is engaged.

Some vehicles provide a "ground block" near the battery for a negative cable to be connected. On these vehicles, run the negative power lead, unfused, to the "ground block." When this technique is recommended by the manufacturer, the interaction between the power return currents and vehicle electronics has been evaluated by the manufacturer. In all cases, the most important rule to remember is this: If you want the manufacturer to support your installation, do it exactly the way the installation guidelines tell you to do it!

If no installation guidelines are available for your vehicle, the practices outlined below will improve compatibility between in-vehicle transceivers and vehicle electronics:

1) Transceivers

• Transceivers should be securely mounted in a location that does not interfere with vehicle operator controls and visibility, and provides transceiver ventilation.

• Ensure all equipment and accessories are removed from the deployment path of the air-bag and safety harness systems.

2) Power Leads

• The power leads should be twisted together from the back of the rig all the way to the battery. This minimizes the area formed by the power leads, reducing susceptibility to transients and RFI.

• Do not use the vehicle chassis as a power return.

• The power leads should be routed along the body structure, away from vehicle wiring harnesses and electronics.

• Any wires connected to the battery should be fused at the battery using fuses appropriate for the required current.

• Use pass-through grommets when routing wiring between passenger and engine compartments.

• Route and secure all under-hood wiring away from mechanical hazards.

3) Coaxial Feed Lines

• The coaxial feed line should have at least 95% braid coverage. The cable shield should be connected to every coaxial connector for the entire circumference (no "pigtails").

• Keep antenna feed lines as short as practical and avoid routing the cables parallel to vehicle wiring.

4) Antennas

• Antenna(s) should be mounted as far from the engine and the vehicle electronics as practical. Typical locations would be the rear deck lid or roof. Metal tape can be used to provide an antenna ground plane on non-metallic body panels.

• Care should be used in mounting antennas with magnetic bases, since magnets may affect the accuracy or operation of the compass in vehicles, if equipped.

• Since the small magnet surface results in low coupling to the vehicle at HF, it is likely that the feed line shield will carry substantial RF currents. A large (2-inch OD or larger toroid) common-mode choke at the antenna will help reduce this current, but will also reduce any radiation produced by that current.

• Adjust the antenna for a low SWR.

27.11.3 Diagnosing Automotive RFI

Most VHF/UHF radio installations should result in no problems to either the vehicle systems or the transceiver, while HF installations are more likely to experience problems. In those situations where issues do occur, the vast majority are interference to the receiver from vehicle on-board sources of energy that are creating emissions within the frequency bands used by the receiver. Interference to one of the on-board electronic systems can be trivial, or it can cause major problems with an engine control system.

The dealer should be the first point of contact when a problem surfaces, because the dealer should have access to information and factory help that may solve your problem. The manufacturer may have already found a fix for your problem and may be able to save your mechanic a lot of time (saving you money in the process). If the process works properly, the dealer/customer-service network can be helpful. In the event the dealer is unable to solve your problem, the next section includes general troubleshooting techniques you can perform independently.

GENERAL TROUBLESHOOTING TECHNIQUES

An important aspect is to use the source-path-victim model presented earlier in this chapter. The path from the source to the receiver may be via radiation or conduction. If the path is radiation, the electric field strength (in V/m) received is reduced as a function of the distance from the source to the receiver. In most cases, susceptible vehicle electronics is in the near-field region of the radiating source, where the electric and magnetic fields can behave in complex ways. In general, however, the strength of radiated signals falls off with distance.

The best part of all this is that with a general-coverage receiver or spectrum analyzer, a fuse puller, and a shop manual, the vehicle component needing attention may be identified using a few basic techniques. The only equipment needed could be as simple as:

• A mobile rig, scanner, or handheld transceiver, or

• Any other receiver with good stability and an accurate readout, and

• An oscilloscope for viewing interference waveforms

BROADBAND NOISE

Automotive broadband noise sources include:

• Electric motors such as those that operate fans, windows, sunroof, AM/FM antenna deployment, fuel pumps, etc.

• Ignition spark

If you suspect electric motor noise is the cause of the problem, obtain a portable AM or SSB receiver to check for this condition. Switch on the receiver and then activate the electric motors one at a time. When a noisy motor is switched on, the background noise increases. It may be necessary to rotate the radio, since portable AM radios use a directional ferrite rod antenna.

To check whether fuel pumps, cooling fans, and other vehicle-controlled motors are the source of noise, pull the appropriate fuse and see whether the noise disappears.

A note concerning fuel pumps: virtually every vehicle made since the 1980s has an electric fuel pump, powered by long wires. It may be located inside the fuel tank. Don't overlook this motor as a source of interference just because it may not be visible. Electric fuel-pump noise often exhibits a characteristic time pattern. When the vehicle ignition switch is first turned on, without engaging the starter, the fuel pump will run for a few seconds and then shut off when the fuel system is pressurized. At idle, the noise will generally follow the pattern of being present for a few seconds before stopping, although in some vehicles the fuel pump will run almost continuously if the engine is running.

NARROWBAND NOISE

Automotive narrowband noise sources include:

• Microprocessor based engine control systems
• Instrument panel
• RADAR obstacle detection
• Remote keyless entry
• Key fob recognition systems
• Tire pressure monitoring systems
• Global positioning systems
• Pulse width modulating motor speed controls
• Fuel injectors
• Specialized electric traction systems found in newer hybrid/electric vehicles

Start by moving the antenna to different

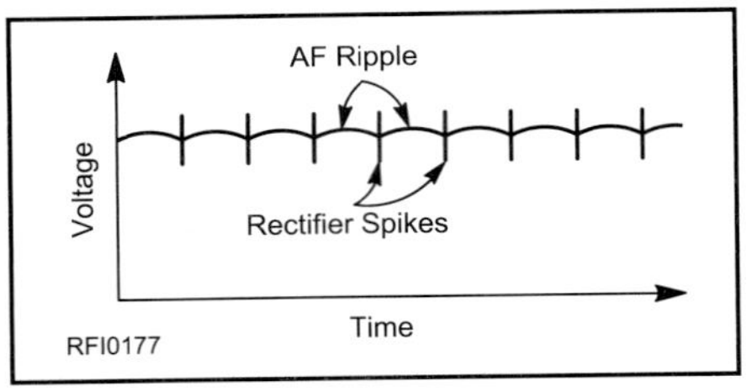

Figure 27.38 — Alternator whine consists of full-wave rectified ac, along with pulses from rectifier switching, superimposed on the vehicle's dc power voltage.

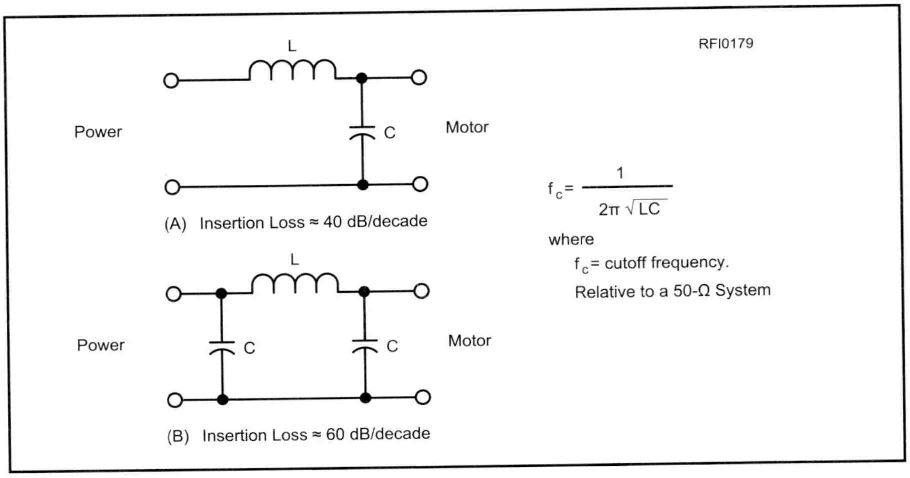

Figure 27.39 — Filters for reducing noise from dc motors.

locations. Antenna placement is often key to resolving narrowband RFI problems. However, if antenna location is not the solution, consider pulling fuses. Tune in and stabilize the noise, then find the vehicle fuse panels and pull one fuse at a time until the noise disappears. If more than one module is fed by one fuse, locate each module and unplug it separately. Some modules may have a "keep-alive" memory that is not disabled by pulling the fuse. These modules may need to be unplugged to determine whether they are the noise source. Consult the shop manual for fuse location, module location, and any information concerning special procedures for disconnecting power.

A listening test may verify alternator noise, but if an oscilloscope is available, monitor the power line feeding the affected radio. Alternator whine appears as full-wave rectified ac ripple and rectifier switching transients superimposed on the power system's dc power voltage (see **Figure 27.38**).

Alternators rely on the low impedance of the battery for filtering. Check the wiring from the alternator output to the battery for corroded contacts and loose connectors when alternator noise is a problem.

Receivers may allow conducted harness noise to enter the RF, IF, or audio sections (usually through the power leads), and interfere with desired signals. Check whether the interference is still present with the receiver powered from a battery or power supply instead of from the vehicle. If the interference is no longer present when the receiver is operating from a battery or external supply, the interference is conducted via the radio power lead. Power line filters installed at the radio may resolve this problem.

27.11.4 Eliminating Automotive RFI

The next section includes various techniques to resolve the more common RFI problems. As a caveat, when performing your own RFI work, in or out of warranty, you assume the same risks as you do when you perform any other type of automotive repair. Most state laws (and common sense) say that those who work on cars should be qualified to do so. In most cases, this means that work should be done either by a licensed dealer or automotive repair facility.

CONDUCTED INTERFERENCE

To reduce common-mode current, impedance can be inserted in series with the wiring in the form of common-mode chokes. (See this chapter's section on Common-Mode Chokes.) Wire bundles may also be wound around large toroids for the same effect.

Mechanical considerations are important in mobile installations. A motor vehicle is subject to a lot of vibration. If a choke is installed on a wire, this vibration may cause the choke to flex the wire, which may ultimately fail. It is critical that any additional shielding and/or chokes placed on wiring have been installed by qualified personnel who have considered these factors. These must be properly secured, and sometimes cable extenders are required to implement this fix.

RFI TO ON-BOARD CONTROL SYSTEMS

RFI to a vehicle's on-board control and electronic modules should be treated with common-mode chokes at the connection to the module. Some success has been reported by using braid or metal foil to cover a wire bundle as a shield and connecting the shield to the vehicle chassis near the affected module. Vehicle electronic units should not be modified except by trained service personnel according to the manufacturer's recommendations. The manufacturer may also have specific information available in the form of service bulletins.

FILTERS FOR DC MOTORS

If the motor is a conventional brush- or commutator-type dc motor, the following cures shown in **Figure 27.39** are those generally used. As always, the mechanic should consult with the vehicle manufacturer. To diagnose motor noise, obtain an AM or SSB receiver to check the frequency or band of interest. Switch on the receiver, and then activate the electric motors one at a time. When a noisy motor is switched on, the background noise increases as well.

The pulses of current drawn by a brush-commutator motor generate broadband RFI that is similar to ignition noise. However, the receiver audio sounds more like bacon frying rather than popping. With an oscilloscope displaying receiver audio, the noise appears as a series of pulses with random space between the pulses. Such broadband noise generally has a more pronounced effect on AM than on FM receivers. Unfortunately, the pulses may affect FM receivers by increasing the "background noise level" and will reduce perceived receiver sensitivity because of the degraded signal-to-noise ratio.

ALTERNATOR AND GENERATOR NOISE

As mentioned previously, brush-type motors employ sliding contacts that can generate noise. The resulting spark is primarily responsible for the "hash" noise associated with these devices. Hash noise appears as overlapping pulses on an oscilloscope connected to the receiver audio output. An alternator also has brushes, but they do not interrupt current. They ride on slip rings and supply a modest current, typically 4 A to the field winding. Hence, the hash noise produced by alternators is relatively minimal.

Generators use a relay regulator to control field current and thus output voltage. The voltage regulator's continuous sparking creates broadband noise pulses that do not overlap in time. They are rarely found in modern automobiles.

Alternator or generator noise may be conducted through the vehicle wiring to the power input of mobile receivers and transmitters and may then be heard in the audio output. If alternator or generator noise is suspected and an oscilloscope is not available, temporarily remove the alternator belt as a test. (This may not be possible in vehicles with a serpentine belt.)

IGNITION NOISE

Ignition noise is created by fast-rise-time pulses of coil current discharging across air gaps (distributor and spark plug). The theoretical models (zero rise time) of such pulses are called impulse functions in the time domain. When viewed in the frequency domain, the yield is a constant spectral energy level starting nearly at 0 Hz and theoretically extending up in frequency to infinity. In practice, real ignition pulses have a finite rise time, so the spectral-energy envelope decreases above some frequency.

It turns out that noise generated by ignition sparks and fuel injector activation manifests as a regular, periodic "ticking" in the receiver audio output, which varies with engine RPM. If an oscilloscope were connected to the audio output, a series of distinct, separate pulses would appear. At higher speeds it sounds somewhat musical, like alternator whine, but with a harsher note (more harmonic content).

A distinguishing feature of ignition noise is that it increases in amplitude under acceleration. This results from the increase in the required firing voltage with higher cylinder pressure. (Noise at higher frequencies may also be reproduced better by the audio circuits.) Since ignition noise is usually radiated noise, it should disappear when the antenna element is disconnected from the antenna mount. The radiation may be from the secondary parts of the system, or it may couple from the secondary to the primary of the coil and be conducted for some distance along the primary wiring to the ignition system, then radiated from the primary wiring.

Two main methods are employed to suppress this noise — one involves adding an inductance, and the other involves adding a resistance — both in the secondary (high voltage) wiring. This is shown in **Figure 27.40**. The addition of these elements does not have a measurable effect on the engine operation, because the time constants involved in the combustion process are much longer than those associated with the suppression components. (Note that modifying your vehicle's ignition system may be considered as tampering with your vehicle's emission control system and may affect your warranty coverage — work with your dealer or limit your efforts to changing spark plug wires or possibly shielding them.)

The resistance method suppresses RFI by dissipating energy that would have been radiated and/or conducted. Even though the amount of energy dissipated is small, it is still enough to cause interference to sensitive amateur installations. The other method uses inductance and even though the energy is not dissipated, suppression occurs because the inductor will store the pulse energy for a short time. It then releases it into the ignition burn event, which is a low impedance path, reducing the RFI.

For traditional "Kettering" inductive discharge ignition systems, a value of about 5 kΩ impedance (either real and/or reactive) in the spark plug circuit provides effective suppression and, with this value, there is no detectable engine operation degradation. (Capacitor discharge systems, in comparison, are required to have very low impedance on the order of tens of ohms in order to not reduce spark energy, so they are not tolerant of series impedance). Most spark plug resistances are designed to operate with several kV across the plug gap, so a low-voltage ohmmeter may not give proper resistance measurement results.

The term "resistor wire" is somewhat misleading. High-voltage ignition wires usually contain both resistance and inductance. The resistance is usually built into suppressor spark plugs and wires, while there is some inductance and resistance in wires, rotors, and connectors. The elements can be either distributed or lumped, depending on the brand, and each technique has its own merit. A side benefit of resistance in the spark plug is reduced electrode wear.

COIL-ON-PLUG IGNITION NOISE

Many newer spark-ignition systems incorporate a "coil on plug" (COP) or "coil near plug" (CNP) approach. There are advantages to this from an engine operation standpoint, and this approach may actually reduce some of the traditional sources of ignition system RFI. This is because of the very short secondary wires that are employed (or perhaps there are no wires — the coil is directly attached to the spark plug). This reduces the likelihood of coupling from the secondary circuit to other wires or vehicle/engine conductive structures.

There will always be some amount of energy from the spark event that will be conducted along the lowest impedance path. It may mean that the energy that would have been in the secondary circuit will be coupled back on the primary wiring harness attached to the coils. This means that the problem may go from a radiated to a conducted phenomenon.

The fix for this in some cases may actually be easier or harder than one might think. Two

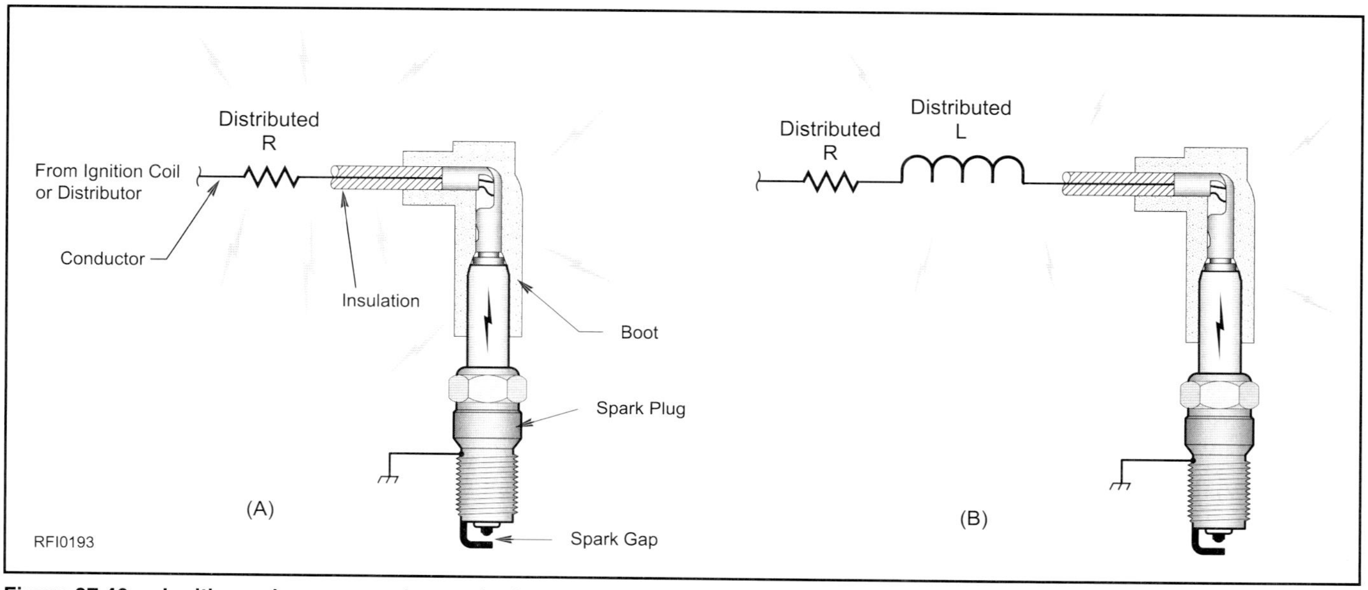

Figure 27.40 — Ignition noise suppression methods.

approaches that have been used with success are ferrite cores and bypass capacitors.

Ferrite cores are recommended as the first choice, since they require no electrical modification to the vehicle. Ferrite clamp-on split cores are added to the 12-V primary harness attached to the coils. Depending on the frequency of the noise and selection of the ferrite material, there can be significant improvement (as much as 10 dB). Key to optimizing the amount of suppression is to determine where the noise "peaks" and selecting the correct ferrite material for that frequency range (see this chapter's section Using Ferrite for RFI Suppression).

The second method is to add a bypass capacitor between the primary wire of the 12-V coil and ground in the harness near the coil assemblies (there may be two, three, or four coils). This must be done carefully because it could affect the functionality of the ignition system and — perhaps most importantly — may void the vehicle warranty. This "bypass" capacitor performs the same function that bypass capacitors in any other application perform — separating the noise from the intended signal/power.

27.11.5 Electric and Hybrid-Electric Vehicles

Electric vehicles (EV) and hybrid-electric vehicles (HEV) are quickly becoming a practical means of transportation. EV/HEVs are advanced vehicles that pose unique challenges for amateur equipment. While EV/HEVs provide improved emissions and fuel economy, they utilize switched high voltage and high current to control propulsion. The switching techniques used generate RFI within much of our frequency bands — a cause for concern, particularly for HF operation.

This section is designed to enlighten vehicle owners to the challenges and to make suggestions when installing mobile amateur equipment in an EV/HEV.

EV AND HEV ARCHITECTURE

Most EVs and HEVs have similar electrical traction system (ETS) architectures consisting of a high voltage battery supplying energy to an inverter, which controls an electric motor within a transmission connected to the drive wheels. The main difference between the two is that an HEV includes an internal combustion engine to aid in propulsion, and while a pure EV is strictly electrically powered.

The heart of the ETS is a device called an inverter. It simply converts dc voltage from the high voltage battery (typical voltage range from 42 to 350 V dc) to an ac waveform supplying the electric motor. This dc-to-ac conversion is performed by a matrix of six transistor switches. The switches chop the dc voltage into systematically varying pulses called pulse-width-modulation (PWM) to form an adjustable frequency and RMS voltage suitable to power an electric motor.

In most cases, the ac voltage from the inverter is a three-phase waveform similar to industrial applications because three-phase motors can be smaller, more efficient, and provide greater torque than single-phase motors.

IMPORTANT — Bright orange cables connect the battery pack to the inverter and the inverter to the drive motor, transferring voltage and current to and from the inverter. Because of the non-sinusoidal waveforms being transferred, these cables are shielded and terminated at each end. Under no condition should these cables be disconnected or modified, because the high voltage system depends on a delicate balance of sensors and safety mechanisms. Possible malfunction and damage to the ETS may occur if modified.

RFI FROM EV AND HEV

The inverter uses PWM to convert dc battery voltage to an ac waveform. The phase-to-phase terminal voltage appears in **Figure 27.41** as rectangular blocks with positive and negative amplitude equal to the battery voltage. For example, a 300 V dc battery pack will provide 600 V peak-to-peak at the motor terminals. In Figure 27.41, the same terminal voltage signal is sent through a low pass filter to show how PWM forms a sinusoidal waveform. Each pulse is essentially a square wave. Harmonics from these pulses fall within most of our amateur HF bands and well into VHF, affecting radio performance. This switching noise is the primary source of RFI to amateur radio equipment in electric vehicles. Because EV/HEV systems are evolving rapidly, check the ARRL's Automotive RFI web page (**www.arrl.org/automotive**) for more information.

EV AND HEV RFI REMEDIES

Troubleshooting techniques described earlier apply in diagnosing RFI from EV/HEV systems. Limited RFI remedies are available associated with components within the ETS. Work with your dealer when you suspect the ETS as the RFI source. Do not attempt to modify or repair your ETS; the dealer's service center is most qualified to inspect and repair your EV/HEV electrical traction system.

Electric vehicles also use a lot of the same type of electric motors, solenoids, switches, and control electronics as are found in non-electric vehicles. The same types of noise will be radiated from these devices, and any amateur radio equipment is essentially within feet or even inches of the noise source. One difference with electric vehicles is that adding chokes or other RFI remedies may not be permitted under the vehicle's warranty due to the more complex power and control electronics systems. If you do add components to the vehicle wiring to reduce RFI, do it in small steps, testing vehicle and noise performance along the way.

During installation, mobile equipment power cables and antenna coaxial cables should be routed as far as possible from the bright orange cables. Common-mode chokes can decrease noise on 12-V dc power cables. Additionally, antenna placement plays a critical role in mobile equipment performance. Areas such as the top of a roof or trunk sometimes provide additional shielding.

RFI to EV and HEV

While all modern vehicles have RF-sensitive electronics on board, electric vehicles have electronics that control every facet of vehicle operation, usually controlled over digital networks such as the CAN (Controller Area Network). These circuits are well filtered, which reduces the chance for stray RF to negatively affect them. However, the vehicle wiring and electronics are very close to the antennas that are radiating a strong signal. As more vehicles adopt partial or complete automated driving, RFI can present serious safety challenges!

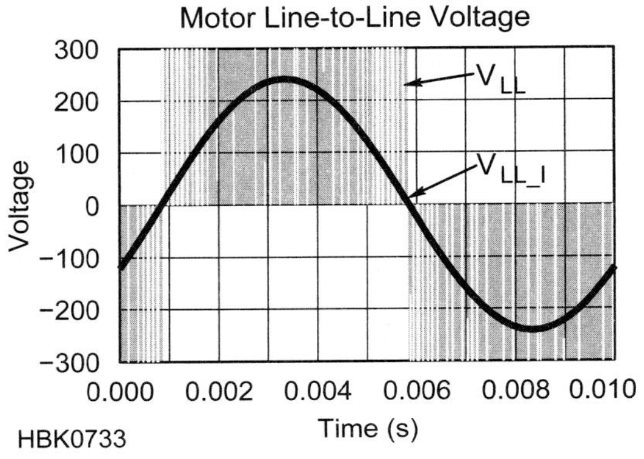

Figure 27.41 — Phase-to-phase motor terminal voltage.

Be aware that on-board electronics may be susceptible to the high levels of RF generated by amateur transmitters, regardless of the frequency of operation. Even when allowed, proper antenna installation is very important in keeping common-mode RF current pickup by cables and wiring harnesses to a minimum. Disrupting network data signals or affecting control circuit operation can have serious safety consequences.

As amateur radio and vehicles use more digital protocols and technology, it should reduce susceptibility to RFI in electric vehicles. In the interim, amateurs will have to remain cautious and be prepared to be flexible in their approach to mobile operating. Amateur operators will still have to contend with RFI, but the future may bring improvements as the digital systems become more tolerant for both the radios and vehicles.

27.11.6 Automotive RFI Summary

Most radio installations should result in no problems to either the vehicle systems or any issues with the transceiver. However, manufacturer, make, and models differ, thus introducing challenges during amateur equipment installations.

Begin by researching your vehicle of interest and visiting the dealer. Insist on transmitting and receiving your favorite frequencies as you test drive. Request information pertaining to the manufacturer's transceiver installation guidelines. If manufacturer information is not available, follow the guidelines described earlier.

After installation, RFI problems may appear. Report your problem to the dealer, because they have access to manufacturer service bulletins that may describe a repair solution. Additional troubleshooting and remedies are also described previously to assist in successful communication.

Limited RFI remedies are available associated to components within the ETS. Work with your dealer when you suspect the ETS is the RFI source. Do not attempt to modify or repair your ETS; the dealer's service center is most qualified to inspect and repair your EV/HEV electrical traction system.

Lastly, the latest version of the *ARRL RFI Book* contains additional information on RFI in automobiles. More details are given about noise sources, troubleshooting techniques, a troubleshooting flow chart, additional filtering techniques, and information on EV/HEVs.

27.12 EMC Topics

"Electromagnetic compatibility (EMC) is the ability of an electronic system to (1) function properly in its intended electromagnetic environment, and (2) not be a source of pollution to that electromagnetic environment." (Ott, page 4) This chapter's previous sections deal exclusively with the RF interference aspects of EMC — whether our signals cause RFI and whether our stations are susceptible to RFI from amateur and non-amateur sources. In this section other topics associated with EMC are discussed.

27.12.1 Standard Transient Types

A transient, by definition, "Refers to momentary over-voltages or voltage reductions in an electric power system," (CRC Electrical Engineering Dictionary) Just because a transient is momentary doesn't mean it can't cause harm. This section describes several transients that amateurs are likely to encounter and devices used to protect circuits against them.

AUTOMOTIVE TRANSIENTS

DC voltage in vehicles that use a lead-acid starter battery as the energy source varies between 10.5 V (a discharged battery) to more than 15 V during heavy charging. The typical 13.8 V ±15% input voltage specification corresponds to a fresh battery during normal charging. Numerous ac and transient signals are superimposed on that dc supply voltage – for example, the rapid current switching in the vehicle's alternator creates sharp (short-duration) transients on the power bus. If not adequately filtered by the radio, the result is a high-pitched *alternator whine* added to both receive and transmit signals that follows engine speed.

SAE (Society of Automotive Engineers) standard J1113, "Immunity to Conducted Transients on Power Leads," describes transients encountered on a vehicle's power bus such as these common occurrences:

- **Load Dump** — occurs when a loose battery connection opens up during charging and the alternator's energy is "dumped" on the power bus with no battery to hold down the voltage.
- **Alternator Field Decay** — occurs every time the vehicle is turned off and the alternator's stored energy has to be dissipated via the power bus.
- **Inductive Load Switching** — the kickback voltage from an inductive load (like an electric window motor) being turned off.
- **Mutual Coupling** — transient energy that is coupled between conductors in a wiring harness.

Table 27.5 summarizes the electrical characteristics of these transients, and **Figure 27.42** shows what they look like in time. Vehicle manufacturers build in some transient protection so that every electronic device in the vehicle does not have to protect itself against all of these transients. As explained in the Littelfuse application note *Suppression of Transients in an Automotive Environment* (available at **littelfuse.com**), vehicle electronics are already protected by a *central suppressor* in the vehicle, usually located as close to the master control computer module as possible. There are usually suppressors in other modules around the vehicle as well.

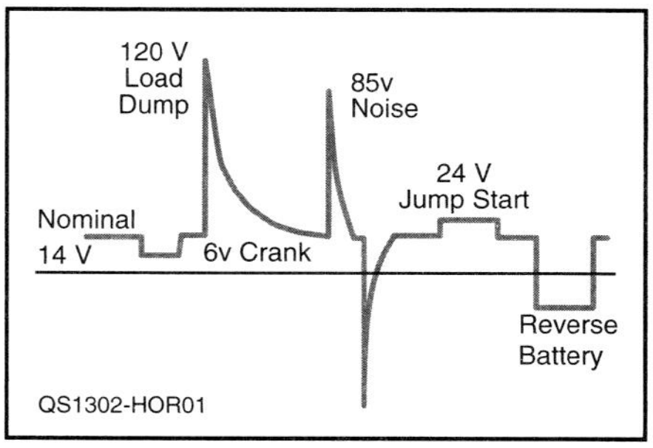

Figure 27.42 — Typical vehicle power system voltage levels for a 12-V electrical system. (Data courtesy of Littlefuse Corp.)

Table 27.5
Typical Vehicle Transients (12V System)

Type	*Voltage*	*Energy (Joules)*	*Duration*	*Occurs*
Load Dump	<125 V	>10 J	200 – 400 msec	Infrequently
Field Decay	–100 to +40V	<1 J	200 msec	At turn-off
Inductive	–300 to +80 V	<1 J	<320 µsec	Often
Mutual Coupling	<200 V	<1 J	1 msec	Often

ELECTROSTATIC DISCHARGE (ESD)

Another source of transients, especially in areas with dry air, is *electrostatic discharge* or *ESD*. A sudden discharge of static electricity by a spark, such as from walking across a carpet then touching a grounded surface, is a typical example of ESD. In fact, the standard ESD test generator uses a finger-shaped probe.

There can be enough energy in an ESD to destroy semiconductors or scramble the operation of a circuit. A typical ESD transient lasts for less than 50 µsec but can generate voltages up to 15 kV! ESD transients can appear on power and signal wiring, connectors, controls and switches, displays – anything a finger can touch, even metal enclosures. The standard ESD transient waveform is shown in **Figure 27.43**. Note the fast rise time and long decay.

Protection of circuits against ESD is discussed in a pair of application notes from Littlefuse (**www.littlefuse.com**): *ESD Protection Design Guide* and *Tips for Enhancing ESD Protection*.

27.12.2 Transient Protective Devices

There are several ways to protect electronics against transients – block them, route them away from the circuitry, and dissipate their energy as heat. The goal is to limit the remaining voltage to levels the electronics can handle. Several different types of protective components have been developed to handle transients. Available from Littlefuse, *Transient Suppression Devices and Principles*, application note AN9768 is an excellent overview of transient protection and the different types of devices that are used. A widely available and inexpensive text is *Transient Protection of Electronic Circuits*, published by Dover Press.

Vendors of protective components include Bourns (**bourns.com**), Littlefuse, TDK (**product.tdk.com**), and Vishay (**vishay.com**), who all offer selection guides and datasheets. MOV distributors such as Digi-Key (**digikey.com**) and Mouser Electronics (**mouser.com**) also provide online comparison features as well as product lines from a wide variety of vendors.

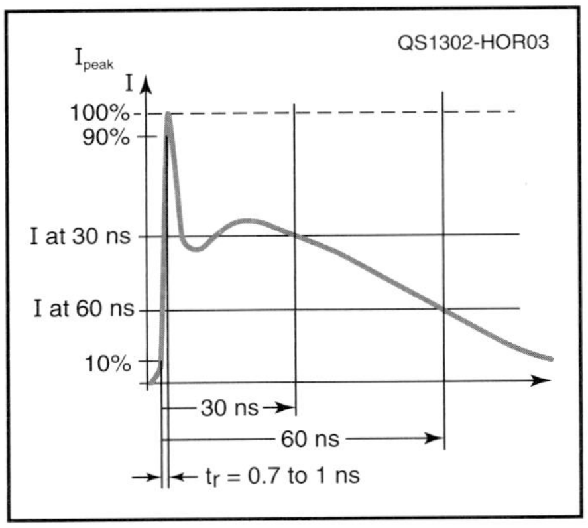

Figure 27.43 — ESD Transient waveform. (Data from standard IED 61000-4-2.)

SERIES AND SHUNT DEVICES

There are two types of transient or surge protection devices, *series-mode* and *shunt-mode,* which refer to how they are connected with respect to the equipment or circuit being protected. A series-mode protector blocks the transient from traveling to the protected equipment along the signal path by changing from a low to a high impedance state. This might be done with inductance or a switch that opens very rapidly. A series-mode protector must be able to withstand the full signal level to or from the equipment without dissipating too much energy or significantly altering the load characteristics.

A shunt-mode protector is installed between the protected conductor and a local ground system, such as a chassis connection. The protector acts as an open circuit until a threshold voltage is exceeded, then it changes to a low-impedance state to dissipate the transient energy as heat like a resistor, or direct it to ground as a switch. A shunt-mode protector "shares" the transient voltage with the protected equipment until the protector's threshold voltage is exceeded. The protected device is exposed to the transient energy until the shunt device switches to the low-impedance state.

Most protective devices used by amateurs are shunt-mode, such as lightning arrestors and spark gaps. An example of a series-mode device is a sub-miniature incandescent light bulb in the feed line to a receiver to protect it against strong signals from a nearby transmitter. Be wary of using non-linear shunt-mode devices such as diodes on any conductor that may carry significant RF voltages or currents. The resulting harmonics may cause significant RFI — see the sidebar Unbypassed Diode Junctions and RFI in the earlier section on Interference from IMD.

METAL OXIDE VARISTOR (MOV)

An MOV consists of partially conductive powder pressed into a disc or cylinder so that it is non-conductive up to its *clamping voltage*. At voltages (of either polarity) higher than the clamping voltage, its resistance drops, limiting the voltage by dissipating energy as heat. MOVs are generally connected between the circuit being protected and ground so that the lower resistance causes the MOV to absorb the transient's energy and keep voltage at a safe level. After repeated transients, MOVs generally fail in a low-resistance state.

As shown in **Figure 27.44**, the varistor consists of metal oxide grains pressed together and heated (sintered) into a disc that has electrodes and wire leads attached. An epoxy coating completes the component, which looks almost identical to a disc capacitor. The size and composition of the grains determines how the MOV behaves electrically as shown in the figure.

At voltages below the MOVs breakdown voltage, the MOV acts like an open circuit. As a transient exceeds its breakdown voltage, the MOV begins to conduct and acts as a low-value resistor. (The voltage-current graph for a fixed resistor is also shown for reference.) This allows the MOV to dissipate the transient's energy as heat. The MOV is non-polarized and acts the same with either positive or negative voltage across it.

MOVs are available with a wide range of breakdown voltages and energy ratings. Their specified operating voltage is well below the breakdown voltage, so they won't begin to conduct under normal circumstances. Their energy rating also stipulates how often the rated energy can be safely dissipated. Repeated transients, cause the MOV breakdown voltage to decrease and eventually the normal operating voltage can cause the MOV to fail at a low resistance.

The most common use for MOVs by amateurs is in ac power surge protectors and in protecting data and control lines. They are not used in high RF power environments. An MOV subjected to a very large transient may fail open-circuited so that it no longer provides any protection. Surge suppressors containing MOVs often have an indicator to show when the MOVs are no longer providing protection. Thermal fuses are included to prevent fires if the MOV fails at low-resistance.

TRANSIENT VOLTAGE SUPPRESSOR (TVS) DIODES

Essentially a heavy-duty Zener diode designed for transient suppression, the transient voltage suppressor or TVS diode (also known by the trade name Transzorb®) acts as a voltage clamp but does not dissipate the transient energy. TVS diodes offer more precise clamping action than MOVs. Unless overloaded, TVS diodes can handle repeated transients without changing their characteristics.

Transzorbs are available in both unidirectional (like a regular Zener diode) and bidirectional (two back-to-back Zener diodes in series). In the reverse direction, the Transzorb acts as an open circuit until its breakdown voltage is exceeded when it begins to conduct. In the forward direction it is a regular semiconductor diode.

The Transzorb acts faster than an MOV or GDT, both in clamping and recovery, but does not dissipate the energy of a transient. It can clamp repeated transients without the breakdown voltage changing as long as its power dissipation rating is not exceeded. Transzorbs are generally used in low-power signal applications, where more control over clamping

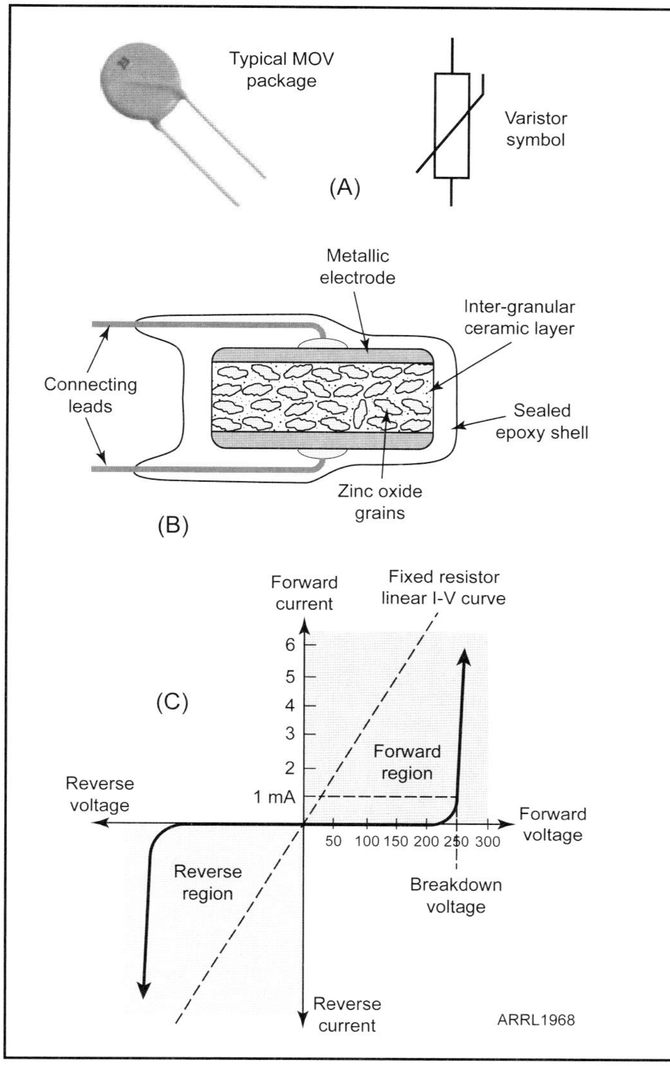

Figure 27.44 — The schematic symbol and a typical metal-oxide varistor (MOV) component (A). MOVs are made from a sintered disc of metal oxides between metal electrodes (B). The current-voltage graph in C shows the change to a low-resistance state above a threshold voltage of either polarity.

> **Lightning and EMP**
>
> While there are some similarities between the lightning and *electromagnetic pulse* (EMP) events, lightning is comparatively slow. EMP events create large voltage transients or gradients that act 10 times faster (or more) than lightning. The damage from both can be similar but protection techniques are different. You can read about EMP in the *QST* articles by Dennis Bodson, W4PWF, which are online in the ARRL's *QST* archives.

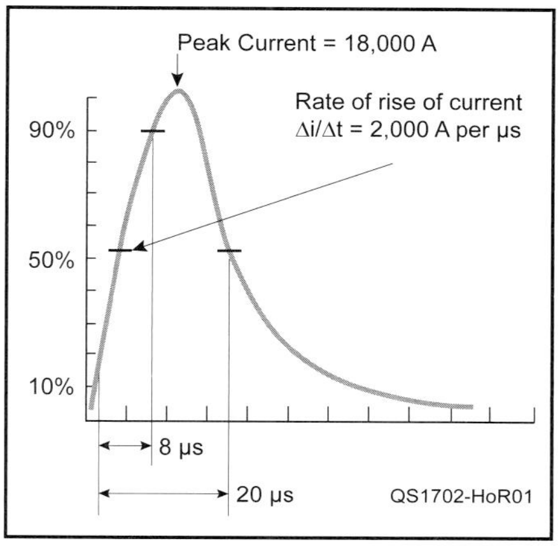

Figure 27.45 — The IEEE 8/20 model waveform for a typical lightning pulse.

action is required, and repeated transients are expected.

DIODE CLAMPING AND RC FILTERING

Most effective for ESD and fast transients, clamping diodes route energy away from the protected circuit into the power supply, where it is absorbed by the filter components. This limits the circuit voltage to one forward voltage drop beyond the power supply voltage.

RC low-pass filtering is useful for both RFI and ESD. Inserted in series with an input circuit, the RC filter consists of a series resistor followed by a capacitor to ground. The filter attenuates RF voltages above its cutoff frequency and smooths out transient voltages, dissipating some of the energy as heat.

RC filters are particularly useful in high-impedance inputs for which the additional series impedance of the filter does not cause significant errors. However, the shunt capacitance across a high-impedance input may affect high-frequency performance considerably. Be sure to consider the effects of any protection circuit on the protected circuit's behavior.

27.12.3 Lightning

Lightning and protection against it are also discussed in the chapters **Assembling a Station** and **Safe Practices**. The ARRL publication *Grounding and Bonding for the Radio Amateur, 2nd Edition* offers an extensive discussion on lightning. You can also find practical information at **www.arrl.org/lightning-protection**.

A typical lightning strike is composed of three to four impulses per strike. Peak current for the first pulse averages around 18 kA (98% of the strikes fall between 3 kA to 140 kA at their peak). For the second and subsequent impulses, the current will be about half the initial peak. The typical interval between impulses is approximately 50 msec. **Figure 27.45** shows a typical impulse, referred to as the *IEEE 8/20 model waveform*. Remember, this is an *average* and fully half of lightning strikes have more energy than this waveform!

Voltages created by this current pulse can be enormous and depend on the resistance (R) and inductance (L) through which the current flows. According to Faraday's Law, the faster that current changes (Δi / Δt, where the symbol Δ means "the change in") through an inductance, the higher the voltage that is created. Higher current through a resistance also means higher voltage, per Ohm's Law. The combination is stated as:

$$V = I \times R + L \times \Delta i/\Delta t$$

While the majority of lightning's energy is pulsed dc, there is a substantial amount of RF created by the fast rise time of the pulses. A typical lightning strike rise time is 1.8 μsec. That translates into a radiated RF signal peaking at 139 kHz. Rise times can vary from a very fast 0.25 μsec to a very slow 12 μsec, yielding an RF range from 1 MHz down to 20 kHz. When lightning "attaches" to the *air terminal* (where the leader channel reaches the grounded object), the rise time for current can be as short as 10 nsec. The result is that lightning's energy extends upward in frequency to 10 MHz and higher.

An indirect or nearby strike can generate fast-rising transients with energies equal to or larger than the automotive inductive transient and with voltages nearly as high as an ESD pulse. Protection for ac power circuits is usually provided by *surge suppressors,* which consist of multiple power outlets in a single enclosure (aka "power strips) protected by MOVs. Protection ratings are specified in joules (J) or watt-seconds. Surge suppressors used in the home or amateur station should meet the UL 1449 safety standard. Older and non-rated units may fail gradually in ways that can create a fire hazard and should be discarded.

27.12.4 Gas Discharge Tubes (GDT)

Several examples of *gas discharge tubes* (GDT) used by amateurs are seen in **Figure 27.46**. The photograph of a GDT shows the shape and placement of the metal electrodes. They are typically made of tungsten or some other tough metal that can withstand repeated short arcs. The schematic symbol includes a small dot next to the electrodes. This indicates the presence of gas. A proprietary combination of gases is used, along with electrode spacing, shape, and material to control the *breakdown* or *sparkover voltage*. This is the voltage at which the GDT "fires" (an arc forms between the electrodes), limiting the voltage in the feed line. GDTs are available with breakdown voltages of less than 100 V to several kV. The GDT is connected across the protected signal or feed line so it acts as a shunt-mode device.

After the GDT fires, the voltage across the arc is limited to about 10 – 15 V, depending on the model and the current level of the arc. Once the transient energy is discharged, the arc is extinguished or *quenched* and the GDT returns to an open state. As long as the maximum current is not exceeded or the electrode surfaces are not damaged, the GDT will continue to provide the rated protection.

The GDT is a type of *crowbar* protection. From the Littlefuse GDT catalog, "A crowbar device limits the energy delivered to the protected circuit by abruptly changing from a high impedance state to a low impedance state in response to an elevated voltage level. Having been subjected to a sufficient voltage level the crowbar begins to conduct. While conducting, the voltage across the crowbar remains quite low (so) the majority of the transient's power is dissipated in the circuit's resistive elements and not in the protected circuit or the crowbar itself. This allows the crowbar to be able to withstand and protect loads from higher voltage and/or higher current levels for a greater duration of time than clamping devices." A typical GDT limits the voltage across the electrodes to 15 V or so, which is well within the safe voltage range for nearly all radio equipment.

There are several types of packages as shown in the figure. The GDTs used in commercial lightning protectors for coaxial cables are removable cartridges available from the protector vendors and manufacturers. Amateur applications are considered "low to medium voltage" for rotator or antenna switch control lines and "high voltage" for feed lines. Because GDTs rarely fail shorted, they are recommended over MOVs for protecting rotator control lines.

GDTs are used in *lightning arrestors* (See the chapters **Assembling a Station** and **Safe Practices**) that are inserted in feed lines and attached to a protective ground system. The GDT is usually a cartridge-style component that can be inspected and replaced, if necessary. Because the GDT is voltage-sensitive, it is important to specify an arrestor as being rated for low-power (typically 150 W) or full-power operation.

Axial wire lead GDTs are also available for PCB and terminal-strip use. Testing a GDT requires a special *hi-pot* (short for "high-potential") tester that limits current through the device being tested for insulation breakdown. Motor rewinding shops often have this type of equipment and can test a GDT, but it is usually more cost-effective to replace the GDT if you think it has been damaged.

APPLYING GDTS

As you can see in Figure 27.46, there are many different styles of GDT packages. The *leaded* and *cartridge* packages are the most common in amateur stations. GDTs are rated by breakdown voltage and ability and energy handling ability. The Littlefuse CG-series is suitable for most amateur applications.

At RF and especially in circuits that must withstand transmitter output voltages, a cartridge style GDT is used. **Figure 27.47** shows a typical lightning protector with UHF connectors for attaching to feed lines. The GDT is connected between the shield and center conductor of the feed line, with the protector body also mounted on a grounded metal mounting bracket. Several *QST* advertisers offer these protectors for either low power or for full legal limit operation.

Because SWR affects maximum voltage in the feed line, the lightning arrestor's power rating must specify the maximum SWR for that rating. For example, the Alpha-Delta TT-series specifies power rating with a maximum SWR of 3:1. From the Alpha-Delta datasheet:

200 W models: 200 W RF at a VSWR of 3:1 generates a voltage of 173.2 volts. The gas tube in this model is rated at 350 volts.

2 kW models: 2 kW RF at a VSWR of 3:1 generates a voltage of 547.7 volts. The gas tube in this model is rated at 1,000 volts.

Don't confuse the SWR rating for power with the maximum SWR that results from inserting the arrestor in the line — that specification will be much lower.

For protecting dc or low-frequency control and telephone lines, the leaded package is the most common. These components can be connected to screw terminals or barrier strips. For rotator control lines, measure the RMS voltage from the control box and multiply by 1.414 to get peak voltage to ground. (2.8 for peak-to-peak) Add another 10-20% to provide margin and determine the required breakdown voltage.

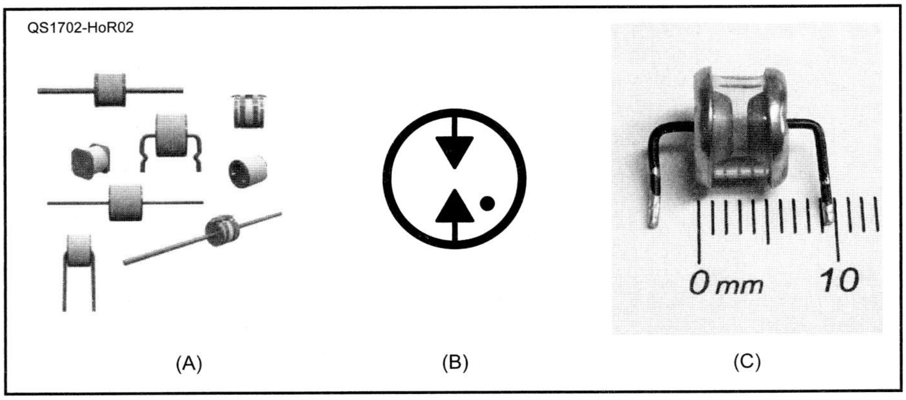

Figure 27.46 — Typical gas discharge tube components (A), the GDT schematic symbol (B), and a close-up photograph showing the electrode shape (C). (Photo provided by Ulfbastel via Wikimedia Commons)

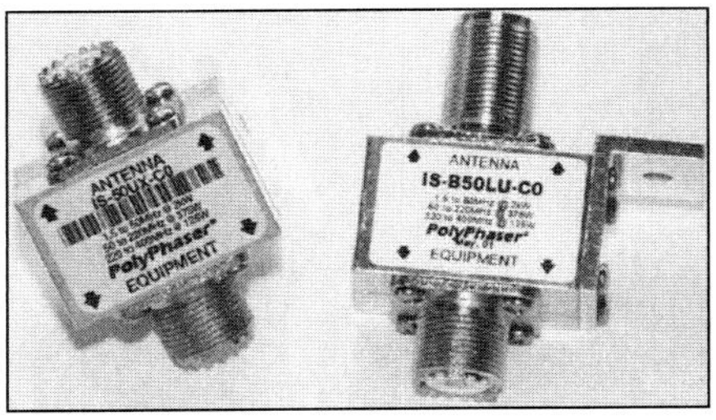

Figure 27.47 — A pair of coaxial lightning arrestors (aka lightning protectors) manufactured by Polyphaser — the IS-50UX and IS-B50LU.

Remember that telephone line ringing voltage can be as high as 150 V_{PK-PK}.

When using GDT-based protectors you must provide a good lightning protection ground system for full protection, including a lightning protection plan and bonding for all protected equipment. (See the section on Grounding and Bonding in the chapter on **Assembling a Station**.)

You should also be aware that there are two basic versions of commercial lightning protectors: one that passes dc current and one that does not. If you plan on using a remote device such as a preamp or antennas switch that requires feed line power, be sure not to purchase the dc blocking version. A call to the vendor will help you make the right choice.

27.12.5 Spark Gaps

A similar type of crowbar device is a *spark gap* or *air gap protector*. Again, from the Littlefuse catalog, "An air gap protector consists of two conductive surfaces with spacing between them that will permit an arc when a specified potential is placed across the surfaces. The air gap is not a sealed device and therefore it must operate at atmospheric pressure and under the effects of the environment." A spark gap can also provide protection by limiting maximum voltages on towers or antennas or feed lines.

Spark gaps can be easily made from common materials and ARRL publications have many examples, including the use of automotive spark plugs for open-wire feed lines. Two examples are shown in **Figure 27.48**. Spark gaps are nothing more than metal surfaces, one connected to a ground system, that are spaced just far enough apart that normal transmit signal voltages do not cause an arc between them.

With dry air having a breakdown voltage of about 3 kV/mm, a 3 mm separation will arc over at about 9 kV. Once the arc starts, the voltage across it is very low until the current is discharged and the arc is extinguished. Be aware that the breakdown voltage may vary significantly based on the environment at the time of use, such as during rain or fog.

The spark gap in Figure 27.48A, made by W8JI, consists of a simple piece of copper pipe that is connected to a tower ground system. The pipe is positioned close to the base of the non-grounded tower section. The separation is about 1/8th inch (approximately 3 mm).

An alternate method in Figure 27.48B was constructed by NØAX and uses #2 AWG tinned copper wire. One piece of wire is connected to a listed ground clamp on the tower leg above and the other below a fiberglass insulator section. The wire position and spacing can be adjusted by loosening the clamp screws or just bending the wire.

Another option shown in Figure 27.48C is to mount a conventional spark plug or GDT across a stand-off insulator to a grounded bracket and attach the protected conductor to its top terminal with a short flexible lead. The spark plug gap is easily adjusted to the desired width using a feeler gauge. Do not use resistor-type spark plugs.

Debris or water in the spark gap will reduce its breakdown voltage. Arrange the conductors so that water flows away from the gap and does not pool or gather in the gap. Keep the gap from becoming an attractive nesting site for insects. Regular inspection and cleaning of the gap is a good idea as well.

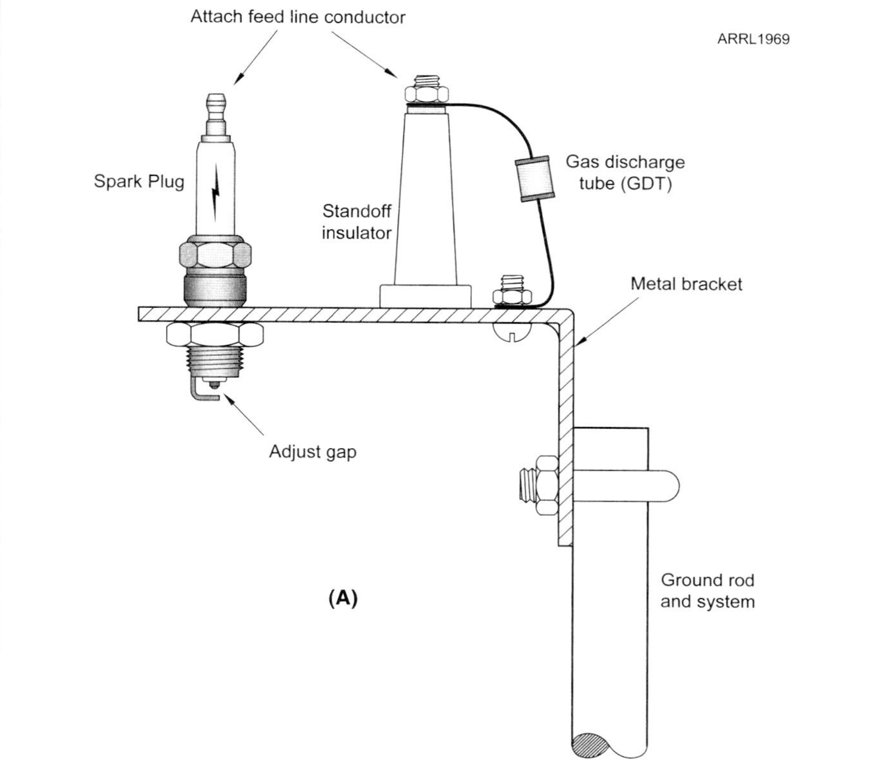

(B)

(C)

Figure 27.48 — Pipe spark gap constructed by Tom Rauch, W8JI (A) and a wire-and-clamp spark gap constructed by the author (B). Spark plugs (C) can be mounted on a grounded bracket and adjusted to the desired breakdown voltage. [Photo B courtesy Tom Rauch, W8JI]

27.13 References and Bibliography

AES48-2005: AES standard on interconnections — Grounding and EMC practices — Shields of connectors in audio equipment containing active circuitry.

"Identifying Power-Line Noise," ARRL, **www.youtube.com/watch?v=mD4GHcKMepU&feature=youtu.be**.

Radio Frequency Interference web page, **www.arrl.org/radio-frequency-interference-rfi**.

Brown, J., K9YC, "Power, Grounding, Bonding, and Audio," 2014, **k9yc.com/publish.htm**.

Brown, J., K9YC, "Building Contest Scores by Killing Receive Noise, Parts 1 and 2," National Contest Journal, May/June and July/Aug 2016, **k9yc.com/KillingReceiveNoise.pdf**.

Brown, J., K9YC, "RFI, Ferrites, and Common Mode Chokes For Hams," 2010, **k9yc.com/publish.htm**.

Brown, J., K9YC, "RF Interference in Audio Systems," 2008 Fall Convention of the Audio Engineering Society, **k9yc.com/AES-RFI-SF08.pdf**.

Brown, J., K9YC, "New Understandings of the Use of Ferrites in the Prevention and Suppression of RF Interference to Audio Systems," 119th AES Convention in New York, October 2005, **k9yc.com/AESPaperFerritesASGWeb.pdf**.

Counselman, C., W1HIS, "Common Mode Chokes," **www.yccc.org/Articles/W1HIS/CommonModeChokesW1HIS2006Apr06.pdf**.

Elliott, J., "Advanced EMI Filters Cut Costs on Brush DC Motors," *Microwaves & RF*, Mar. 2020, pp. 36 – 39.

Foerster, M., WØIH, "Hunting Down RF Noises," *QST*, Feb. 2015, pp. 45 – 46, **www.arrl.org/files/file/RFI/FOESTER.PDF**.

Gruber, M.,Ed., *The ARRL RFI Book*, 3rd Edition (ARRL, 2010).

Higbie, A., KØAV, "Tracking RFI with an SDR One Source at a Time," *National Contest Journal*, Mar./Apr. 2021, pp. 3 – 7.

Loftness, M., *AC Power Interference*, 3rd Edition (Percival, 2001).

Muncy, N., "Noise Susceptibility in Analog and Digital Signal Processing Systems," presented at the 97th AES Convention of the Audio Engineering Society in San Francisco, CA, Nov. 1994.

Ott, H., *Electromagnetic Compatibility Engineering*, (Wiley, 2009).

Paul, C., *Introduction to Electromagnetic Compatibility*, (Wiley, 1992).

Thompson, T., WØIVJ, "Locating RF Interference at HF," *QST*, Nov. 2014, p. 33. **www.arrl.org/files/file/RFI/Thompson%20Noise.pdf**.

Underwriter Labs Standard UL943, "Ground-Fault Circuit-Interrupters."

Vinson, W., et al. *The Mitigation of Radio Noise and Interference from on-site Sources at Radio Receiving Sites*, Naval Postgraduate School, 2009.

Vinson, W., et al., *The Mitigation of Radio Noise from External Sources at Radio Receiving Sites*, 5th Edition, Wilbur R. Vinson, et al., Naval Postgraduate School, 2007.

HAM RADIO OUTLET®

WWW.HAMRADIO.COM

*Free Shipping and Fast Delivery!

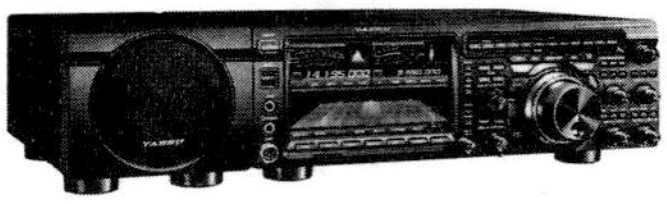

FTDX101MP | *200W HF/50MHz Transceiver*

• Hybrid SDR Configuration • Unparalleled 70 dB Max. Attenuation VC-Tune • New Generation Scope Display 3DSS • ABI (Active Band Indicator) & MPVD (Multi-Purpose VFO Outer Dial) • PC Remote Control Software to Expand the Operating Range • Includes External Power With Matching Front Speaker

FTDX10 | *HF/50MHz 100 W SDR Transceiver*

• Narrow Band and Direct Sampling SDR • Down Conversion, 9MHz IF Roofing Filters Produce Excellent Shape Factor • 5" Full-Color Touch Panel w/3D Spectrum Stream • High Speed Auto Antenna Tuner • Microphone Amplifier w/3-Stage Parametric Equalizer • Remote Operation w/optional LAN Unit (SCU-LAN10)

FT-991A | *HF/VHF/UHF All ModeTransceiver*

Real-time Spectrum Scope with Automatic Scope Control • Multi-color waterfall display • State of the art 32-bit Digital Signal Processing System • 3kHz Roofing Filter for enhanced performance • 3.5 Inch Full Color TFT USB Capable • Internal Automatic Antenna Tuner • High Accuracy TCXO

FTDX101D | *HF + 6M Transceiver*

• Narrow Band SDR & Direct Sampling SDR • Crystal Roofing Filters Phenomenal Multi-Signal Receiving Characteristics • Unparalleled - 70dB Maximum Attenuation VC-Tune • 15 Separate (HAM 10 + GEN 5) Powerful Band Pass Filters • New Generation Scope Displays 3-Dimensional Spectrum Stream

FT-710 Aess | *HF/50MHz 100W SDR Transceiver*

• Unmatched SDR Receiving Performance • Band Pass Filters Dedicated for the Amateur Bands • High Res 4.3-inch TFT Color Touch Display • AESS: Acoustic Enhanced Speaker System with SP-40 For High-Fidelity Audio • Built-in High Speed Auto Antenna Tuner

FT-891 | *HF+50 MHz All Mode Mobile Transceiver*

Stable 100 Watt Output • 32-Bit IF DSP • Large Dot Matrix LCD Display with Quick Spectrum Scope • USB Port Allows Connection to a PC with a Single Cable • CAT Control, PTT/RTTY Control

FTM-300DR | *C4FM/FM 144/430MHz Dual Band*

• 50W Output Power • Real Dual Band Operation • Full Color TFT Display • Band Scope • Built-in Bluetooth • WiRES-X Portable Digital Node/Fixed Node with HRI-200

FT-2980R | *Heavy-Duty 80W 2M FM Transceiver*

• 80 watts of RF power • Large 6 digit backlit LCD display for excellent visibility • 200 memory channels for serious users

FTM-200DR | *C4FM/FM 144/430MHz Dual Band*

• 1200/9600bps APRS® Data Communications • 2" High-Res Full-Color TFT Display • High-Speed Band Scope • Advanced C4FM Digital Mode • Voice Recording Function for TX/RX

FTM-500DR | *C4FM/FM 144/430MHz Dual Band Xcvr*

• Front Firing Acoustically Enhanced Speaker System • True Dual Band Operation, C4FM/C4FM Digital D-D Dual Receive • 2.4" High-Resolution Full-Color Touch Panel Display • Built-in High Precision GPS Receiver • Wireless Operation Capability with Optional Bluetooth® Headset

FT-70DR *C4FM/FM 144/430MHz Xcvr*

• System Fusion Compatible • Large Front Speaker delivers 700 mW of Loud Audio Output • Automatic Mode Select detects C4FM or Fm Analog and Switches Accordingly • Huge 1,105 Channel Memory Capacity • External DC Jack for DC Supply and Battery Charging

FT-5DR *C4FM/FM 144/430 MHz Dual Band*

• High-Res Full-Color Touch Screen TFT LCD Display • Easy Hands-Free Operation w/Built-In Bluetooth© Unit • Built-In High Precision GPS Antenna • 1200/9600bps APRS Data Communications • Supports Simultaneous C4FM Digital • Micro SD Card Slot

FT-65R | *144/430 MHz Transceiver*

Compact Commercial Grade Rugged Design • Large Front Speaker Delivers 1W of Powerful Clear Audio • 5 Watts of Reliable RF Power Within a compact Body • 3.5-Hour Rapid Charger Included • Large White LED Flashlight, Alarm and Quick Home Channel Access

FTM-6000R | *50W VHF/UHF Mobile Transceiver*

• All New User Operating Interface-E20-III (Easy to Operate-III) • Robust Speaker Delivers 3W of Clear, Crisp Receive Audio • Detachable Front Panel Can Be Mounted in Multiple Positions • Supports Optional Bluetooth® Wireless Operation Using the SSM-BT10 or a Commercially Available Bluetooth® Headset

5 Ways to Shop!

- RETAIL LOCATIONS – Store hours 10:00AM - 5:30PM - Closed Sunday
- PHONE – Toll-free phone hours 9:30AM - 5:30PM
- ONLINE – WWW.HAMRADIO.COM
- FAX – All store locations
- MAIL – All store locations

YAESU
The radio

ANAHEIM, CA (800) 854-6046	PORTLAND, OR (800) 765-4267	PHOENIX, AZ (800) 559-7388	MILWAUKEE, WI (800) 558-0411	WOODBRIDGE, VA (800) 444-4799	WINTER SPRINGS, FL (800) 327-1917
SACRAMENTO, CA (877) 892-1745	DENVER, CO (800) 444-9476	PLANO, TX (877) 455-8750	NEW CASTLE, DE (800) 644-4476	SALEM, NH (800) 444-0047	ATLANTA, GA (800) 444-7927

*On most orders over $100 in the continental US. (Rural locations excluded.) Contact HRO for promotion details. Toll-free including Hawaii, Alaska and Canada. All HRO 800-lines can assist you. If the first line you call is busy, you may call another. Prices, specifications and descriptions subject to change without notice.

ADVANCED SPECIALTIES INC.
New York's Communications Store
• OPEK • COMET • PROCOMM • MFJ •
• UNIDEN • ANLI • YAESU • ASTATIC •
YAESU
The radio
Authorized
FT-2980R
80W 2M Mobile
FTM-891
100-W HF & 6M All Mode
Yaesu
FT-5DR
Dual Band
5W Digital
Transceiver
VX-6R
Tri Band
Submersible
HT
FTM-300DR
Digital Fusion Dualband
Uniden Bearcat
SDS100
Police Scanner
Orders/Quotes: (516) 998-0370
Closed Sunday & Monday – 880 West Beech Street · Long Beach, NY 11561
www.advancedspecialties.net
AMATEUR RADIO – SCANNERS – BOOKS – ANTENNAS – MOUNTS
FILTERS – ACCESSORIES AND MORE!
MasterCard
VISA
DISCOVER

Your radio installation.
Revolutionized.
We did all the hard work, so you don't have to.
Icom? Kenwood? Yaesu? Install with SwapMyRigs and you're
always plug-and-play ready, regardless of brand.
Swap rigs anytime. Go anywhere. And never pull cables again.
SMR
www.swapmyrigs.com
The world's only patented, brand-universal, single-cable installation solution.

Index of Advertisers

Index

Editor's Note: Except for commonly used phrases and abbreviations, topics are indexed by their noun names. Many topics are also cross-indexed.

The letters ff after a page number indicate coverage of the indexed topic on succeeding pages.

Separate Project and Author indexes follow the main index.

B

D

E

F

G

H

I

N

O

P

Q

R

S

T

U

V

W

Project Index — Online indicates design and construction information is available only in the online content.

Topic	Title	Page	Author
Antenna, VHF/UHF	Coaxial Dipole for VHF or UHF	21.55	W6NBC
	6 Meter Halo	Online	VE6AB
	Big Wheel for 2 Meters	Online	W4RNL
	Quick and Cheap Omni Antenna for 1296 MHz	Online	W1GHZ
	A Medium-Gain 2 Meter Yagi	21.58	W4RNL
	Utility Yagi for 432 MHz	21.59	W1VT
	Cheap Yagis by WA5VJB	21.59	WA5VJB
	Fixed Moxons for Satellite Operation	21.62	W4RNL
	Homebrew Coaxial Dipole for VHF or UHF	21.54	W6NBC
	Medium Gain 2 Meter Yagi	21.58	W4RNL
	Simple, Portable Ground-Plane Antenna	21.54	W1VT
	Three and Five-Element Yagis for 6 Meters	21.57	N6BV
	Dual-band antenna for 146/446 MHz	Online	K3MF
	All-copper, 2 meter J-pole antenna	Online	KD8JB
	Three- and Five-element Yagis for 6 Meters	21.57	N6BV
	3D Printed Antennas	Online	W1GHZ
Components and Interfaces	Uses for Thermistors	Online	WØIYH
	Simple Serial Interface	Online	NØXAS
	Trio of Transceiver/Computer Interfaces	Online	AA8DX
	USB Interfaces for Your Ham Gear	Online	NØXAS
	Audio Interface Unit for Field Day and Contesting	Online	KØIZ
Construction Techniques	Making PC Boards With Printed Artwork	Online	K7QO
	Repurposing Obsolete Instrument Enclosures	Online	KC7CJ
CW Keyer	TiCK-4 — A Tiny CMOS Keyer	Online	N2JGU, WB8YGG
	Universal Keying Adapter	Online	NØXAS
Filter, Audio	Audio Intelligibility Enhancer	Online	N4GG
	Audio Waveshaping Filter for CW Reception	Online	W4ENE
Filter, RF	Band-pass Filter for 145 MHz	10.44	RSGB
	Broadcast-Band Reject Filters	10.41	ARRL
	6-Meter Filter with Harmonic Suppression	Online	W1GHZ
	Combline Filters for 50-432 MHz	10.41	W1GHZ
	Altoids Tin Filters	Online	W1GHZ
	Field Day Stub Assembly	20.13	W2VJN
	High-Performance, Low-Cost 1.8 to 54 MHz Low-Pass Filter	10.43	K8CU
	High-Power Harmonic Filters	Online	W2VJN
	Optimized Harmonic Transmitting Filters	10.40	W4ENE
	Wave Trap for Broadcast Stations	10.42	ARRL
	Half-Lattice Single-Crystal Filter	Online	WØIYH
	Manually-Tuned Preselector	Online	W5OZF
	Software-Controlled Preselector	Online	MØWWA
	Crystal Parameter Measurements	Online	K7QO
	HF Yagi Triplexer	Online	K6KV
	Commercial Triplexer Design	Online	W2VJN
	High-Power Harmonic Filters	Online	W2VJN
Oscillator	Crystal Test Oscillators	Online	W6HPH
	JFET Hartley VFO	9.11	W7ZOI, KA7EXM, W7EL
	Low-noise Differential Oscillator	9.10	K7HFD
	Modified Vackar VFO	9.9	N1UL
	VHF/UHF grounded-base oscillator	Online	N1UL
Power Source	12-V, 15-A Power Supply	7.47	WA1TWX
	13.8-V, 5-A Regulated Power Supply	7.51	G4YNM
	Adjustable Resistive Load	7.52	G4YNM
	Adjustable Tracking Power Supply	7.58	KCØZNG
	Automatic Sealed-Lead-Acid Battery Charger	Online	AA4PB
	Deluxe High Voltage Supply	Online	W8ZR
	Four-Output Switching Bench Supply	Online	K3PTO
	High-Voltage Power Supply	7.54	W8ZR
	Inverting DC-DC Converter	7.52	Stewart, Jim
	Overvoltage Crowbar Circuit	7.60	ARRL
	Overvoltage Protection for AC Generators	7.59	ARRL
	Power from USB Sources	7.48	NØGSG
	Reverse-Polarity Protection Circuits	7.55	WAØITP
	Sealed Lead-Acid Battery Float Charger and Switch	7.56	K9JEB

Topic	Title	Page	Author
Receiver	10 GHz Preamplifier	Online	W1VT
	430 MHz Preamplifier	Online	ARRL
	Configuring an RTL-SDR for Amateur Radio	Online	HA7ILM
	2 Meter Down-converter	Online	N1UL
	Dual-Band LNA for 2 Meters and 70 Centimeters	Online	AMSAT
	Micro R2 SSB or CW Receiver	Online	KK7B
	Rock-Bending Receiver for 7 MHz	Online	WI5W
	Binaural I-Q receiver	Online	KK7B
	High-Performance 45 MHz IF	Online	G3SBI
RF Interference	RF Sniffer and Construction Notes	Online	ARRL, AC9PR
	Active Attenuator	Online	PAØZR
	Simple Seeker RFI Receiver	Online	W5IXM
	Ultrasonic Power Line Arc Detector	Online	W1TRC
	TRF Receiver for Tracking RFI	Online	K1BQT
	Handheld Loop Antenna for RFI Location	Online	NA6O
	Beam for Power Line RFI Hunting	Online	W1TRC
Test Equipment	Bipolar Transistor Tester	Online	N1AL
	Fixed-Frequency Audio Oscillator	25.54	ARRL
	Gate-Dip Oscillator	Online	N1AL
	High-Power RF Samplers	25.58	WØIVJ
	Hybrid Combiners for Signal Generators	25.60	N6JF
	Inductance Tester	Online	WA1PIO
	Logic Probe	Online	N1AL
	RF Current Meter	28.55	W8JI
	RF Ammeters	28.55	K4ERO
	RF Oscillators for Circuit Alignment	25.59	ARRL
	RF Power Meter	Online	W7IEQ
	RF Step Attenuator	25.56	K7OWJ
	RF Voltmeter	Online	K2QHE, GM4THU
	Two-Tone Audio Generator	25.54	ARRL
	Wide-Range Audio Oscillator	25.54	Schultz, W9QB
	Test probe adaptors	25.7	W4QO, KG4VHV, QRP ARCI
	Low-frequency VNA Adaptor	Online	VE2AZX
	Return Loss Bridge	Online	N6JF
	Tandem Match	Online	KI6WX
	RF Field Strength Meter	Online	VE7NI
	A West Coast Lightwave Transceiver	Online	VE7CA, VE7SL
	TAK-40 SSB/CW Transceiver	Online	WA2EUJ
Transmitter	VHF/UHF Beacon Transmitters	Online	WA3TTS
	MicroT2 Single-Band SSB Transmitter	Online	KK7B
	MkII Universal QRP Transmitter	Online	W7ZOI
	Tuna Tin 2	Online	W1FB
	Near-Space Tracker	16.12	KD4STH
	Pebble Crusher 7 MHz QRP CW Transmitter	Online	G3RJV
	VHF Signal Sources (50 and 144 MHz)	Online	KK7B
Transverter	A Microwave Transverter Controller	Online	OH2GAQ

Author Index — Online indicates material available with the downloadable supplemental content.

Author	Call	Topic	Section Ref	Page Ref
Gionettni, Riccardo	IØFDH	Automatic tracking filter for DDS generator	Chapter 9	Online
Gordon-Smith, Dave	G3UUR	Crystal filters	10.6	10.23ff
Grant, Doug	K1DG	Space Communications	Operating Supplement	Online
		DSP and SDR fundamentals	Chapter 8	8.1ff
		SDR transmitters	Chapter 13	13.1ff
		SDR receivers	Chapter 12	12.1ff
		Digital modulation	Chapter 11	11.1ff
		Digital filters	10.5	10.16ff
Grover, Dale	KD8KYZ	PC-board CAD	23.6	23.32ff
Grumm, Linley	K7HFD	Low-noise oscillator	9.2.4	9.9
Halstead, Roger	K8RI	Surplus amplifier parts	Chapter 17	Online
		Amplifier tuning	17.4.3	17.14
Hare, Ed	W1RFI	RF Interference	Chapter 27	27.1ff
		Troubleshooting and Maintenance	Chapter 26	26.1ff
		Construction techniques	Chapter 23	23.1ff
Hass, Frank	KB4T	Locating power line RFI sources	27.10	27.46ff
Hayward, Wes	W7ZOI	RF semiconductor design	Chapter 5	5.1ff
		Segment-tuned VCO	9.7.6	9.58
		Wide-range VXO	9.4.5	9.23
		Oscillator temperature compensation	9.3.2	9.16
Heere, Matt	N3NWV	Portable stations	24.3	24.19ff
		Battery selection	7.13.10	7.45
Hicks, Steve	N5AC	SDR user interface and control interface	Chapter 14	14.1ff
		SDR functions and noise	Chapter 13	13.1ff
		SDR functions, effects of noise, and noise reduction	12.1.4	12.8
Hilding, Rick	K6VVA	Remote stations	24.5.3	24.37
Hollingsworth, Riley	K4ZDH	RF Interference and the FCC	27.9	27.45
Holmes, Walter	K5WH	Digital voice	15.4.5	15.17
Honnaker, Scott	N7SS	Digital modes	Chapter 15	15.1ff
Hranac, Ron	NØIVN	Cable television RFI	27.8.2	27.34
Humbertson, Ken	WØKAH	FLDIGI	15.1.4	15.5
Hutchinson, Chuck	K8CH	Antennas	Chapter 21	21.1ff
Idelson, Jim	K1IR	Antenna and tower safety	22.2	22.15ff
Jones, Bill	K8CU	Low-Pass Filter	10.11.4	10.43
Karlquist, Rick	N6RK	Mixers and demodulators	12.4	12.25ff
		Audio latency measurement system	24.5.3	24.37
Karn, Phil	KA9Q	Convolution digital filters	11.6.3	11.17
Kastigar, Matt	N9ES	Amplifier and antenna tuner repair	26.9.4	26.28
Kay, Leonard	K1NU	RF Techniques	Chapter 5	5.1ff
Keuken, Jack	KE2QJ	Base-loading system for whip antennas	21.8.3	21.44
Klitzing, James	W6PQL	Pallet RF Amplifiers	17.10.7	17.37ff
Langner, John	WB2OSZ	Packet radio and APRS	15.5.9	15.25
Lapin, Gregory	N9GL	RF safety	22.3	22.30ff
Larkin, Bob	W7PUA	RF semiconductor design	Chapter 5	5.1ff
Loveall, Pete	AE5PL	D-STAR	15.5.12	15.31ff
Luetzelschwab, Carl	K9LA	Propagation	Chapter 19	19.1ff
Lux, Jim	W6RMK	Electrical safety	22.1	22.1ff
Mack, Ray	W5IFS	SDR topics	Chapter 8	8.1ff
		SDR transmitting functions	Chapter 13	13.1ff
		SDR receiving functions	Chapter 12	12.1ff
		Digital filters	10.5	10.16ff
Martin, Mike	K3RFI	Power-line RF noise	27.10	27.46ff
McCune, Earl	WA6SUH	DDS and analog synthesizers	9.6	9.28ff
Miller, Ethan	K8GU	Propagation	Chapter 19	19.1ff
		HF transmitting filter	10.10.1	10.36
Moell, Joe	KØOV	Radio direction-finding techniques	Chapter 27	Online
		Radio direction-finding antennas	21.12	21.65ff
Morris, Steve	K7LXC	Tower and antenna safety	22.2	22.16ff
Mueller, Wayne	W1QC	Slow-scan television	Operating Supplement	Online
Mullett, Chuck	KR6R	CAD for power supplies	7.11.7	7.27
Newkirk, David	W9BRD	Mixers and demodulators	12.4	12.25ff
Norris, Ken	KK9N	Remote stations	24.5	24.35ff
Ott, Henry	WA2IRQ	Electromagnetic compatibility	Chapter 27	27.1ff
Overbeck, Wayne	N6NB	Transverters	14.5	14.19
Pearce, Gary	KN4AQ	Repeaters	Chapter 18	18.1ff
Petrich, John	W7FU	GNU radio and SDR design tools	14.4.1	14.12
Poddar, Ajay	AC2KG	Low-noise VHF oscillators	9.2.3	9.7ff
Portune, John	W6NBC	3D printing	23.10	23.50ff
		Cavity Filters	10.9.1	10.30ff
Pratt, Warren	NRØV	RF Amplifier linearity	17.11	17.42ff
Rauch, Tom	W8JI	Plate chokes and high-power RF inductors	17.5.3	17.17
		RF Chokes	5.3.4	5.8
Rautio, Jim	AJ3K	Electromagnetic Simulation	6.4	6.19ff

Author	Call	Topic	Section Ref	Page Ref
Rohde, Ulrich	N1UL	Crystal oscillator design	Chapter 9	Online
		Mathematical stability problems in modern non-linear simulation programs	Chapter 6	Online
		Mixer performance capability	Chapter 12	Online
		Oscillator phase noise	Chapter 9	Online
		FM and AM noise signals of colpitts oscillators	Chapter 9	Online
		The Dangers of Simple Usage of Microwave Software	Chapter 6	Online
		Using Simulation at RF	Chapter 6	Online
		Low noise VHF/UHF oscillators	9.2.3	9.7ff
		Simulation at RF	6.3	6.14ff
		Testing and calculating intermodulation distortion	12.1.4	12.7
		Modified Vackar oscillator	9.2.4	9.9
		Crystal oscillators	9.4.4	9.20
Rowe, David	VK5DGR	Digital voice - CODEC2 and FreeDV	15.4.5	15.17
Russel, Rich	AC0UB	Radio Astronomy Receivers	12.1.1	12.4ff
Sabin, Bill	WØIYH	Transmission-line transformers	20.5.2	20.25ff
		Diplexer Filter	10.10.4	10.38
Schumacher, Hermann	DF2DR	Propagation	Chapter 19	19.1ff
Severns, Rudy	N6LF	Switchmode power conversion	7.11	7.21ff
Sherwood, Rob	NCØB	Receiver testing	24.5.2	24.36ff
		RF Amplifier linearity	17.11	17.42ff
		Receiver testing	12.1.4	12.11
Sickles, Cory	WA3UVV	System Fusion repeaters	18.6	18.20
		System Fusion	15.5.14	15.33
Silver, Ward	NØAX	Circuits and Components	Chapter 4	4.1ff
		Radio Fundamentals	Chapter 3	3.1ff
		Station grounding, bonding, and power	24.1.2	24.2ff
		Antennas	Chapter 21	21.1ff
		Electrical Fundamentals	Chapter 2	2.1ff
		Filters	10.1-10.3	10.1ff
		Amplifier maintenance	26.9.4	26.28
Siwiak, Kai	KE4PT	Discussion of Q	3.7	3.23
Stanley, John	K4ERO	Vacuum-tube amplifiers	17.1	17.1ff
		Optimum ground systems (antennas)	21.3.1	21.20
Stearns, Steve	K6OIK	Effect of (antenna) conductor diameter	21.1.7	21.3
Steffka, Mark	WW8MS	Automotive RFI	27.11	27.50ff
Stockton, David	GM4ZNX	Oscillators	9.1	9.1ff
		Poles and zeros	3.9.3	3.33
Stoddard, Patrick	WD9EWK	Satellite Communications	Operating Supplement	Online
Straw, Dean	N6BV	Antenna modeling and design	Chapter 21	21.1ff
Stroh, Steve	N8GNJ	VARA and ARDOP	15.5.7, 15.5.8	15.22ff
Stuart, Ken	W3VVN	Power sources	Chapter 7	7.1ff
Summers, Hans	GØUPL	PIN diode RF switching	13.11.7	13.42
		Software controlled power switch	7.15.6	7.56
Tayloe, Dan	N7VE	Active filter design process	10.4	Online
		Tayloe mixer	12.4.5	12.37
Taylor, Joe	K1JT	Noise	5.8	5.34ff
		WSJT-X modes	15.4.1	15.14
		Earth-Moon-Earth (EME) Communications		
Thompson, Chris	AC2CZ	FOX-1 telemetry	Chapter 16	Online
Thompson, Tom	WØIVJ	Adaptive predistortion	17.12	17.43ff
		RF Amplifier linearity	17.11	17.42ff
Tonne, Jim	W4ENE	Audio processing circuits	13.8.7	13.31ff
		Filters and filter design software	Chapter 10	10.1ff
Verhage, Paul	KD4STH	Amateur Radio Data Platforms	Chapter 16	16.1ff
Wade, Paul	W1GHZ	Noise	5.8	5.30ff
		Waveguide	20.7	20.34ff
		Antenna analyzer - alternative uses	25.8.2	25.20
		Microwave construction techniques	23.7	23.42
		Transverters	14.5	14.19
		Combline filters	10.11.2	10.41
Wetherhold, Ed	W3NQN	HF transmitting filters	10.10	10.36ff
Whitten, Mel	KØPFX	Digital voice	15.4.5	15.17
Wilson, Mark	K1RO	Introduction to Amateur Radio	Chapter 1	1.1ff
Winer, Ethan		Microphone conditioning circuits	13.8.7	13.31
Woodrick, Ed	WA4IYH	D-STAR	18.4	18.12ff
		D-STAR	15.5.12	15.31
Yoshida, Wayne	KH6WZ	Surface-mount device desoldering	Chapter 23	Online
Youngblood, Gerald	K5SDR	Tayloe mixer	12.4.5	12.37

Notes

Notes

Notes

Notes